The Oxford Encyclopedic English Dictionary

The Oxford Encyclopedic English Dictionary

Third Edition

Edited by

Judy Pearsall and Bill Trumble

New York

OXFORD UNIVERSITY PRESS

1996

Oxford University Press, Walton Street, Oxford OX2 6DP
Oxford New York
Athens Auckland Bangkok Bombay
Calcutta Cape Town Dar es Salaam Delhi
Florence Hong Kong Istanbul Karachi
Kuala Lumpur Madras Madrid Melbourne
Mexico City Nairobi Paris Singapore
Taipei Tokyo Toronto
and associated companies in
Berlin Ibadan

Oxford is a trade mark of Oxford University Press

© Oxford University Press 1991, 1995, 1996
First Edition 1991, Second Edition 1995, Third Edition 1996

Published by Oxford University Press, Inc., 200 Madison Avenue,
New York, New York 10016

British Library Cataloguing in Publication Data
Data available

Library of Congress Cataloging in Publication Data
Data available
ISBN 0-19-860057-7

Printed on acid-free paper in the USA

Contents

Project Team

Chief Editor, Current English Dictionaries
Patrick Hanks

Editors
Judy Pearsall, Bill Trumble

Senior Assistant Editors
David Shirt, Catherine Soanes, Angus Stevenson

Assistant Editors
Julia Elliott, Patricia Moore, Rachel Unsworth

Project Coordinators
Rena Jones, Magda Seaton

Pronunciation Consultant
Judith Scott

Proof Checking
Trish Stableford

Editorial Assistants
Sara Hawker, Louise Jones, Sandy Vaughan

Keyboarders
Steven Siddle (co-ordinator), Diane Diffley,
Pam Marjara, Bobby Novakovic

Preface

This book is a fully integrated reference book, a "dictionary in depth", which combines within a single volume the breadth of information one would expect to find in a dictionary of current English together with greater depth of information than is normal in a dictionary. The dictionary text is based on the *Concise Oxford Dictionary* (8th edition), substantially updated and revised to take account of changes and developments in the language, including the addition of around 3,000 new words and senses. Coverage of scientific and technical vocabulary in particular has been greatly extended and approximately 1,000 new words and senses have been added from these fields.

An important feature of this book is the explanatory coverage of specific subjects, concepts, movements, events, etc., which are normally given only very brief treatment (if any) in conventional dictionaries, but here are presented as concise and self-contained articles following the dictionary entry, be it on Marxism, black holes, existentialism, computers, dinosaurs, post-modernism, Buddhism, American football, or the Gaelic language.

In addition there are around 4,000 biographical entries, just under 5,000 place-name entries, and more than 4,000 other proper names. The biographies represent a wide-ranging selection of figures of international importance, both historical and contemporary. As well as giving the important facts about a person's life and work, the entries provide information which puts in a wider context the life and achievements of that person and his or her overall importance within a particular field or historical period. The entries for place-names are based on David Munro's *Oxford Dictionary of the World*. They not only identify and locate the place in question and give population figures; they also provide information about the character of the place and its historical, economic, or political importance. Longer articles are provided for countries and major regions and cities. The other proper-name entries cover a wide range of topics, and include, for example, entries on political parties, religious organizations, historical events, treaties and Acts of Parliament, celestial bodies, and mythological and fictional characters.

The aim of the editors has been to provide not merely a dictionary, but a handy single-volume reference work that will serve as an index to the words, names, and concepts that are of importance in the English-speaking world. The style of entries is designed to be not only factually accurate and informative but also interesting and readable: with this aim in mind abbreviations and special signs and symbols have been kept to a minimum.

Additional useful information presented in tabular or diagrammatic form is provided in the Appendices. Great care has been taken to ensure that the contents of the Appendices are integrated with the main text of the dictionary, so that the two sources of information can be used to complement one another. In particular there is a comprehensive Chronology of World Events from prehistoric times to the present day; in most cases more detailed information on the events, people, etc. listed there can be found in their own entries in the dictionary.

The editors would like to thank all those who helped in the preparation of this book; those not listed as part of the project team who deserve special mention include: Sara Tulloch and Jeremy Marshall, for advisory reading and useful comments and suggestions on all aspects of the text; Michael Proffitt and Georgia Hole, for research on and drafting of new word entries; Ann Lackie and Hazel Wright, for critical reading of the whole text; Della Thompson, for help in revision and updating of the dictionary text; and Andrew Hodgson, for assistance in incorporation of changes and revisions.

J.M.P.   W.R.T.

Guide to the Use of
This Dictionary

This dictionary is designed to be as straightforward and self-explanatory as possible and the use of special conventions has been kept to a minimum. The following notes will enable the reader to understand more fully the principles involved in assembling the information.

1. HEADWORD

The headword is printed in large bold roman type, or in large bold italic type if the word is originally a foreign word and not naturalized in English.

> **granary** /ˈɡrænərɪ/ n. (pl. **-ies**) **1** a storehouse for threshed grain. **2** a region producing, and esp. exporting, much corn. **3** propr. a type of brown bread or flour containing whole grains of wheat (*granary bread*). [L *granarium* f. *granum* grain]
>
> **comme il faut** /ˌkɒm iːl ˈfəʊ/ adj. & adv. ● predic.adj. (esp. of behaviour, etiquette, etc.) proper, correct. ● adv. properly, correctly. [F, = as is necessary]

Order of headwords

● Strict alphabetical order is followed even when the headword consists of more than one word or of a hyphenated word (see also *Form of headwords* below).

● Capitalized headwords come before lower-case headwords of the same spelling.

> **Gable** /ˈɡeɪb(ə)l/, (William) Clark (1901–60), American actor. He became famous through his numerous roles in Hollywood films of the 1930s; they include *It Happened One Night* (1934), for which he won an Oscar, and *Gone with the Wind* (1939), in which he starred as Rhett Butler.
>
> **gable** /ˈɡeɪb(ə)l/ n. **1 a** the triangular upper part of a wall at the end of a ridged roof. **b** (in full **gable-end**) a gable-topped wall. **2** a gable-shaped canopy over a window or door. □ **gabled** adj. (also in comb.). [ME *gable* f. ON *gafl*]

● Numerals in headwords take precedence over letters in ordering.

> **Charles XII** /tʃɑːlz/ (also **Karl XII** /kɑːl/) (1682–1718), king of Sweden 1697–1718.
>
> **Charles Martel** /mɑːˈtel/ (c.688–741), Frankish ruler of the eastern part of the Frankish kingdom from 715 and the whole kingdom from 719.

● Headwords consisting of more than one word where the main headword has been reversed (as in **Versailles, Treaty of**, **Andrew, St**, and **Edward, Lake**; see *Form of headwords* below) are listed after headwords with numerals.

> **Charles XII** /tʃɑːlz/ (also **Karl XII** /kɑːl/) (1682–1718), king of Sweden 1697–1718.
>
> **Charles, Prince** /tʃɑːlz/, Charles Philip Arthur George, Prince of Wales (b.1948), heir apparent to Elizabeth II.
>
> **Charles Martel** /mɑːˈtel/ (c.688–741), Frankish ruler of the eastern part of the Frankish kingdom from 715 and the whole kingdom from 719.

Form of headwords

● In biographical entries the surname of the person forms the main headword, except where the person is better known or only known by a Christian name, title, sobriquet, etc. (e.g. **Anne of Cleves**, **Henry IV**).

● Main headwords consisting of more than one word are entered in the alphabetical place deemed to be most helpful to the user, which may not always correspond to the alphabetical place of the first word. Thus the main entry for **Treaty of Versailles** is entered under 'V' not 'T' and the headword becomes **Versailles, Treaty of**, while **St Andrew** and **Lake Edward** are entered as **Andrew, St** and **Edward, Lake** (see also *Order of headwords* above).

Variant spellings

Any variant spellings or forms are given at the main headword, in bold type in brackets before the definition. They are also given their own headword entry and cross-reference when these are more than three entries away from the main headword. The main headword represents the preferred form in British usage.

> **phoney** /ˈfəʊnɪ/ adj. & n. (also **phony**) colloq. ● adj. (**phonier**, **phoniest**) **1** sham; counterfeit. **2** fictitious; fraudulent. ● n. (pl. **-eys** or **-ies**) a phoney person or thing. □ **phonily** adv. **phoniness** n. [20th c.: orig. unkn.]
>
> **phony** var. of PHONEY.
>
> **charleston** /ˈtʃɑːlstən/ n. & v. (also **Charleston**) ● n. a lively American dance of the 1920s with side-kicks from the knee. ● v.intr. dance the charleston. [CHARLESTON 2]

● If the variant spelling or form applies only to one sense or part of an entry it will appear at that sense or part of the entry.

> **alpine** /ˈælpaɪn/ adj. & n. ● adj. **1 a** of or relating to high mountains. **b** growing or found on high mountains. **2** (**Alpine**) of or relating to the Alps. **3** (**Alpine**) (of skiing) involving fast downhill work. ● n. a plant native to mountain districts, esp. one suited to rock gardens. [L *Alpinus*: see ALP]
>
> **Bohemian** /bəʊˈhiːmɪən/ n. & adj. ● n. **1** a native of Bohemia, a Czech. **2** (also **bohemian**) a socially unconventional person, esp. an artist or writer. ● adj. **1** of, relating to, or characteristic of Bohemia or its people. **2** socially unconventional. □ **bohemianism** n. (in sense 2). [BOHEMIA + -AN: sense 2 f. F *bohémien* gypsy]

● Variant spellings used in the US (and often in other places where US rather than British spelling conventions are followed) are introduced by the restrictive label *US*.

> **catalogue** /ˈkætəˌlɒɡ/ n. & v. (US **catalog**) ● n. **1** a complete list of items (e.g. articles for sale, books held by a library), usu. in alphabetical or other systematic order and often with a description of each. **2** an extensive list (*a catalogue of crimes*). **3** US a university course-list etc. ● v.tr. (**catalogues**, **catalogued**, **cataloguing**; US **catalogs**, **cataloged**, **cataloging**) **1** make a catalogue of. **2** enter in a catalogue. □ **cataloguer** n. (US **cataloger**). [F f. LL *catalogus* f. Gk. *katalogos* f. *katalegō* enrol (as CATA-, *legō* choose)]

• Variant spellings or forms of place-names are introduced by a language label where appropriate.

> **Geneva** /dʒɪ'niːvə/ (French **Genève** /ʒənɛv/) a city in SW Switzerland, on Lake Geneva; pop. (1990) 167,200.

• Pronunciations of variants are given (at the main headword) where they differ significantly from the pronunciation of the headword. Where the variant form or spelling has a restrictive language label the extra pronunciation given represents the standard pronunciation in that language. (See also section 5 below.)

> **Archangel** /'ɑːk,eɪndʒəl/ (Russian **Arkhangelsk** /ar'xangjiljsk/) a port of NW Russia, on the White Sea; pop. (1990) 419,000.

Alternative names

Alternative names or forms are given at the main headword. They are also given their own headword entry and cross-reference when these are more than three entries away from the main headword. Pronunciation, where appropriate, will be given at the cross-reference entry, not at the main headword. The main headword represents the usual preferred form in British usage. The placing of alternative names depends to a large extent on the context, but the general guidelines are as follows:

• Alternative names in place-name entries are given in italics in brackets before the definition, preceded either by 'also called', or by 'called in Italian' etc. where the alternative name represents the usual form in the given language. Alternative names needing more explanation may also be introduced in the text.

> **Black Forest** (called in German *Schwarzwald*) a hilly wooded region of SW Germany, lying to the east of the Rhine valley.
> **Schwarzwald** /'ʃvartsvalt/ the German name for the BLACK FOREST.
> **Galilee, Sea of** (also called *Lake Tiberias*) a lake in northern Israel. The River Jordan flows through it from north to south.
> **Tiberias, Lake** /taɪ'bɪərɪəs/ an alternative name for the Sea of Galilee (see GALILEE, SEA OF).

• Alternative names in scientific, especially zoological and botanical, entries are given in italics in brackets, usually at the end of the definition.

> **apatosaurus** /ə,pætə'sɔːrəs/ n. a huge plant-eating dinosaur of the genus *Apatosaurus* (formerly *Brontosaurus*), of the Jurassic and Cretaceous periods, with a long whiplike tail and trunklike legs. Also called *brontosaurus*. [mod.L f. Gk *apatē* deceit + *sauros* lizard]

• Alternative names in biographical entries are given in brackets in roman before the definition, introduced by 'known as'.

> **Frederick II** /'fredrɪk/ (known as Frederick the Great) (1712–86), king of Prussia 1740–86.

Abbreviations

Commonly used abbreviations are included at the main headword entry for the full form, in bold, preceded by 'abbr.', with a cross-reference from the abbreviation to the main entry.

> **African National Congress** (abbr. **ANC**) a South African political party and black nationalist organization, founded in 1912 with the aim of securing racial equality and black representation in Parliament.
> **ANC** see AFRICAN NATIONAL CONGRESS.

Homographs

Words that are different in origin but spelt the same way (called homographs) are treated as separate headwords, distinguished by superscript numbers immediately following the headword. Each homograph is given a separate pronunciation.

> **chase**[1] /tʃeɪs/ v. & n. ● v. **1** tr. pursue in order to catch. **2** tr. (foll. by *from, out of, to*, etc.) drive. **3** intr. **a** (foll. by *after*) hurry in pursuit of (a person). **b** (foll. by *round* etc.) colloq. act or move about hurriedly. **4** tr. (usu. foll. by *up*) colloq. pursue (overdue work, payment, etc. or the person responsible for it). **5** tr. colloq. **a** try to attain. **b** court persistently and openly. ● n. **1** pursuit. **2** unenclosed hunting-land. **3** (prec. by *the*) hunting, esp. as a sport. **4** an animal etc. that is pursued. **5** = STEEPLECHASE. □ **go and chase oneself** (usu. in *imper.*) colloq. depart. [ME f. OF *chace chacier*, ult. f. L *capere* take]
> **chase**[2] /tʃeɪs/ v.tr. emboss or engrave (metal). [app. f. earlier *enchase* f. F *enchasser* enshrine, set (as EN-[1], CASE[2])]
> **chase**[3] /tʃeɪs/ n. *Printing* a metal frame holding composed type. [F *châsse* f. L *capsa* CASE[2]]
> **chase**[4] /tʃeɪs/ n. **1** the part of a gun enclosing the bore. **2** a trench or groove cut to receive a pipe etc. [F *chas* enclosed space f. Prov. *ca(u)s* f. med.L *capsum* thorax]

• Biographical entries dealing with different people of the same name are treated as separate homographs and are alphabetically ordered by forename (see also section 2 below).

> **Burke**[1] /bɜːk/, Edmund (1729–97), British man of letters and Whig politician.
> **Burke**[2] /bɜːk/, John (1787–1848), Irish genealogical and heraldic writer.

• Geographical entries dealing with places of the same name are treated within the same entry.

> **Cambridge** /'keɪmbrɪdʒ/ **1** a city in eastern England, the county town of Cambridgeshire; pop. (1991) 101,000. Cambridge University is located there. **2** a city in eastern Massachusetts, forming part of the conurbation of Boston; pop. (1990) 95,800. Harvard University and the Massachusetts Institute of Technology are located there.

2. BIOGRAPHICAL ENTRIES

Biographical entries follow place-name entries of the same spelling.

Names

• Forenames which are not widely known or used are given in brackets and are discounted for purposes of alphabetical ordering. A similar practice is used to indicate that a person's initials are better known than his or her full forenames.

> **Bader** /'bɑːdə(r)/, Sir Douglas (Robert Steuart) (1910–82), British airman.
> **Auden** /'ɔːd(ə)n/, W(ystan) H(ugh) (1907–73), British-born poet.

• Nicknames and shortened forms of forenames are given in brackets after other forenames.

> **Weissmuller** /'waɪs,mʊlə(r)/, John Peter ('Johnny') (1904–84), American swimmer and actor.

• Nicknames, bynames, and alternative names which substitute for a person's full name are given in brackets after forenames, preceded by the words 'known as'. Such names are given their own headword entry and cross-reference where they are more than three entries away from the main headword.

> **Peter I** /'piːtə(r)/ (known as Peter the Great) (1672–1725), tsar of Russia 1682–1725.
> **Bismarck**[2] /'bɪzmɑːk/, Otto (Eduard Leopold) von, Prince of Bismarck, Duke of Lauenburg (known as 'the Iron Chancellor') (1815–98), German statesman.
> **Iron Chancellor** the nickname of Bismarck after he used the phrase 'blood and iron' in a speech in 1862, referring to war as an instrument of foreign policy.

- Full names, names in other languages, etc. are given in brackets following the forenames as appropriate.

- The term 'born' is used when a person has changed his or her name or is normally known by a title or name other than the one he or she was born with. The introductions 'pseudonym of' (for writers) and 'title of' (for political leaders, rulers, etc.), as well as a number of other self-explanatory introductions, are also used.

> **Allen** /ˈælən/, Woody (born Allen Stewart Konigsberg) (b.1935), American film director, writer, and actor.

Dates

- Dates given for literary works are those of first publication. Dates for dramatic works, musical compositions, etc. are also of first publication, unless the date of first performance is more appropriate; where this is the case it is indicated in the text.

- The abbreviation *fl.* (= *floruit,* flourished) and imprecise periods such as '2nd century BC' are occasionally used in place of birth and death dates, where the latter are not known.

Nationality

- Where a person has changed his or her nationality both the original and later nationality are indicated.

- Nationalities relating to former names of countries are used where appropriate to the period in question, e.g. Ian Smith is described as 'Rhodesian', Mikhail Gorbachev as 'Soviet'.

- British (rather than English, Scottish, etc.) is used to describe politicians and other public officials of the UK and also to describe people whose nationality was not originally British but who took on British citizenship or for those born abroad to parents of British nationality. In most other cases, the appropriate nationality within the UK (English, Welsh, Scottish, or Northern Irish) is used.

Monarchs

Monarchs and other hereditary rulers of the same country with the same name are grouped together within the same main entry, having subentries of their own:

> **Edmund** /ˈedmənd/ the name of two kings of England:
> **Edmund I** (921–46), reigned 939–46. Soon after Edmund succeeded Athelstan, a Norse army took control of York and its dependent territories. From 941 Edmund set about recovering these northern territories, but after his death Northumbria fell again under Norse control.
> **Edmund II** (known as Edmund Ironside) (*c.*980–1016), son of Ethelred the Unready, reigned 1016. Edmund led the resistance to Canute's forces in 1015 and on his father's death was proclaimed king. After some initial success he was defeated at Ashingdon in Essex (1016) and was forced to divide the kingdom with Canute, retaining only Wessex. On Edmund's death Canute became king of all England.

Foreign titles of works

Titles of works in a foreign language are normally translated into English, and the original publication date is given. Exceptions to this rule include cases in which the title is better known or only known in the original language (such as *À la recherche du temps perdu* or *Les Enfants du paradis*).

3. PLACE-NAME ENTRIES

Place-names come before biographical names of the same spelling.

Statistics

- Population figures are given for all countries and US states, for towns and cities, and for some islands and other regions. The most recent census figures available are given; where no census figure is available, the most reliable up-to-date estimate is given.

- Other significant statistics, such as length of rivers, height of mountains, are also given.

Names

- For Chinese names the Pinyin transliteration is given as the main headword; if other transliterations or anglicized forms are well attested they are given as variant forms (e.g. the main headword is **Beijing** but the form **Peking** is also given).

- For other foreign names the form most familiar to speakers of English is given as the main headword, with variant and alternative names if necessary.

- Where a place-name has changed, the former name (or names) is given as appropriate.

Languages

In entries for countries all languages (i.e. not just official languages) spoken by a significant proportion (roughly 20 per cent) of people in that country are listed.

4. OTHER ENCYCLOPEDIC MATERIAL: GENERAL NOTES

This dictionary includes in-depth entries containing up-to-date information about a wide variety of topics, including politics, history, religion, medicine, technology, zoology, astronomy, archaeology, sport, the arts, philosophy, language, etc.

- Entries will often include references to people, places, etc. which have separate entries of their own. They will also include from time to time references to names which do not have their own separate entries (usually because they are not important enough in their own right); in these cases extra explanatory information will be given contextually, as, for example, a person's birth (and death) date, or a place's location.

- Where a normal word definition is followed by '(*See note below.*)', this means more is to be found on this particular topic in a separate paragraph following the end of the normal dictionary entry:

> **fossil** /ˈfɒs(ə)l/ *n. & adj.* ● *n.* **1** the remains or impression of a prehistoric animal or plant, usu. petrified while embedded in rock, amber, etc. (often *attrib.*: *fossil bones*; *fossil shells*). (*See note below.*) **2** *colloq.* an antiquated or unchanging person or thing. **3** a word that has become obsolete except in set phrases or forms, e.g. *hue* in *hue and cry*. ● *adj.* **1** of or like a fossil. **2** antiquated; out of date. □ **fossil fuel** a natural fuel such as coal or gas formed in the geological past from the remains of living organisms. **fossil ivory** see IVORY. □ **fossilize** *v.tr. & intr.* (also **-ise**). **fossilization** /ˌfɒsɪlaɪˈzeɪʃ(ə)n/ *n.* **fossiliferous** /ˌfɒsɪˈlɪfərəs/ *adj.* [F *fossile* f. L *fossilis* f. *fodere* foss- dig]
>
> ▪ Fossils are usually the hard parts of organisms, such as bones, teeth, shells, or wood, preserved in petrified form or as moulds or casts in rock. *Trace fossils* represent burrows, footprints, etc. Rarely, the fossils of such insubstantial organisms as jellyfish are preserved in fine-grained rocks such as shales, or the entire bodies of animals may be preserved frozen or in amber or tar. Fossils may be used to date accurately a particular rock stratum, and they have provided much of the information upon which the present subdivisions of the geological time-scale are based.

- Proper names which represent the capitalized form of a normal word (often with *the*, as **the Fall**, **the Constitution**) are often given their own separate entry, especially if the information given is more than just a short definition.

> **Constitution, the** (in full **Constitution of the USA**) the basic written set of principles and precedents of federal government in the US. The Constitution, consisting of seven main articles, was drawn up in 1787, largely based on the earlier Articles of Confederation (1781–7); after ratification by the states of the Union, it came into operation in 1789. Since 1789 twenty-six amendments have been added, including, in particular, the Bill of Rights (1791), which contains the first ten amendments. The underlying principle and aim of the Constitution is the preservation of individual citizens' and states' rights while ensuring an effective federal government; the division into legislative (Congress), judicial (Supreme Court), and executive (President) arms of government was designed to limit overall federal powers.

5. PRONUNCIATION

Guidance on the pronunciation of a headword will be found in most cases immediately after the headword, enclosed in oblique strokes / /, and is based on the standard pronunciation associated especially with southern England (sometimes referred to as 'received pronunciation').

- A pronunciation is not given for headwords consisting of more than one word where the separate words are listed elsewhere in the dictionary with their own pronunciation. For example, entries such as **Second World War**, **Anglo-Irish Agreement**, and **Amnesty International** are not given separate pronunciations.

- In some cases, more than one pronunciation is given: that given first is generally the preferred pronunciation; if the variant is preceded by the label *disp.* (= disputed), this indicates that some people disapprove of this variant or consider it incorrect.

- Pronunciations for variant forms are given at the main entry; those for alternative names are given at the alternative name (cross-reference) entry.

- Guidance on the pronunciation of words printed in bold type within an entry (for example, in the derivatives list) is limited to cases in which the pronunciation is substantially different from that of the headword. In derivative sections, the pronunciation for subsequent derivatives relates to the first preceding derivative having a pronunciation; if this is inappropriate, a separate pronunciation will be given.

Foreign pronunciations

In some entries, especially place-name entries, an alternative or variant, non-English, form of the headword is given; where this is the case, a foreign pronunciation is also given.

- In all other cases the first pronunciation given is always the standard pronunciation in English, even if the entry relates to a foreign name or place-name.

- A second pronunciation reflecting that used in the original language may be given where this is substantially different from the English pronunciation.

The International Phonetic Alphabet

Pronunciations are given using the International Phonetic Alphabet (IPA). The symbols used, with their values, are as follows:

Consonants

b, *d*, *f*, *h*, *k*, *l*, *m*, *n*, *p*, *r*, *s*, *t*, *v*, *w*, and *z* have their usual English values. Other symbols are used as follows:

g	(*get*)	ŋ	(ri*ng*)	ʃ	(*she*)
tʃ	(*chip*)	θ	(*thin*)	ʒ	(deci*sion*)
dʒ	(*jar*)	ð	(*this*)	j	(*yes*)
x	(lo*ch*)				

Vowels

short vowels		*long vowels*		*diphthongs*	
æ	(*cat*)	ɑː	(*arm*)	eɪ	(*day*)
e	(*bed*)	iː	(*see*)	aɪ	(*my*)
ə	(*ago*)	ɔː	(*saw*)	ɔɪ	(*boy*)
ɪ	(*sit*)	ɜː	(*her*)	əʊ	(*no*)
ɒ	(*hot*)	uː	(*too*)	aʊ	(*how*)
ʌ	(*run*)			ɪə	(*near*)
ʊ	(*put*)			eə	(*hair*)
				ʊə	(*poor*)
				aɪə	(*fire*)
				aʊə	(*sour*)

- (ə) signifies the indeterminate sound as in gard*e*n, carn*a*l, and altruis*m*.

- (r) at the end of a word indicates an r that is sounded when a word beginning with a vowel follows, as in *clutter up* and *an acre of land*.

The main or primary stress of a word is shown by ' preceding the relevant syllable; any secondary stress in words of three or more syllables is shown by ˌ preceding the relevant syllable.

- In addition the following IPA symbols are used in the representation of foreign pronunciations:

Consonants

ç	(Dan*z*ig)
β	(Se*v*illa)
ʎ	(Pug*li*a, Ma*ll*orca)
ɣ	(I*g*uazú)
ɲ	(Gasco*gn*e, D*n*epr)

Vowels

short vowels		*long vowels*		*nasalized vowels*	
a	(S*a*rre, L*á*risa, Arkh*a*ngelsk)	aː	(B*a*sel, Den H*aa*g)		
ɑ	(*A*ntakya, *A*ntwerpen)			ã	(Orl*éan*s)
ɛ	(Gen*è*ve, Br*e*slau, V*e*neto)	eː	(Bod*e*nsee)	ɛ̃	(Louv*ain*)
i	(P*i*cardie, Bras*i*l, Napol*i*)				
o	(T*o*rino, Du*e*ro)	oː	(R*o*ma, Brass*ó*, *Oo*stende)		
u	(P*o*rto, B*u*cureşti)				
œ	(Sol*eu*re, G*ö*teborg)			œ̃	(D*un*kerque)
ɔ	(C*o*rse, Mik*o*nos)			ɔ̃	(L*y*on)
ø	(K*ø*benhavn)	øː	(Helsing*ø*r)		
				õ	(Brag*an*ça)
y	(J*y*lland, N*ü*rnberg)	yː	(M*üh*lhausen)		
ɥ	(S*ui*sse)				

6. PART OF SPEECH

● A part of speech identifier, such as *n.* (= noun), *v.* (= verb), *adj.* (= adjective), is given for all main head-words except place-name entries, biographical entries, and other proper-name entries except for languages and peoples (as **Boer War, Rosetta Stone, World Health Organization, Constitution, the**).

● Part-of-speech identification is not intended to be exhaustive. It is part of normal usage, for example, for some adjectives to be used as nouns and vice versa, even where entries do not explicitly mention this.

● All derivatives are given a part of speech identifier. Compounds and phrases listed under a main headword are given a part of speech when necessary to aid clarity.

● When a headword has more than one part of speech, a list is given at the beginning of the entry, and the treatment of successive parts of speech (in the same order as the list) is introduced by a bullet marker in each case:

lack /læk/ *n. & v.* ● *n.* (usu. foll. by *of*) an absence, want, or deficiency (*a lack of talent; felt the lack of warmth*). ● *v.tr.* be without or deficient in (*lacks courage*). □ **for lack of** owing to the absence of (*went hungry for lack of money*). **lack for** lack. [ME *lac, lacen*, corresp. to MDu., MLG *lak* deficiency, MDu. *laken* to lack]

● The following standard parts of speech are used (the abbreviation used is given in brackets):

adjective (*adj.*)
adverb (*adv.*)
conjunction (*conj.*)
interjection (*int.*)
noun or noun phrase (*n.*)
preposition (*prep.*)
pronoun (*pron.*)
verb (*v.*)

● The following standard identifiers are used in place of a part of speech:

abbreviation (*abbr.*)
combining form (*comb.form*)
contraction (*contr.*)
prefix
suffix
symbol (*symb.*)

The following additional explanations should also be noted (see also section 10 below):

● *attrib.* (= attributive) is used to describe an adjective placed in front of the word it modifies; *attrib.adj.* is used as a part of speech identifier for adjectives which can normally only be used attributively, as **breakneck** in *breakneck speed* or **consulting** in *consulting physician*.

● *attrib.* is also used to describe a noun which is placed before and used to modify another noun but where its function is not fully adjectival (e.g. **model** in *a model student; the student is very model* is not acceptable usage).

● *predic.* (= predicative) is used to describe an adjective placed in the predicate of a sentence (usually after a verb); *predic.adj.* is used as a part of speech identifier for adjectives which can normally only be used predicatively, as **aware** in *she's not aware of it*, or **asleep** in *he's asleep.*

● *absol.* (= absolute) is used to refer to uses of transitive verbs where an object is implied but not stated, as in *smoking kills* and *let me explain.*

● 'in *comb.*' (= in combination) refers to uses of words

(especially adjectives) as an element joined by a hyphen with another word, as **crested**, in *red-crested, great-crested*, and so on.

7. INFLECTION

Inflection of words (i.e. plurals, past tenses, etc.) is given after the part of speech concerned:

safari /sə'fɑːrɪ/ *n.* (*pl.* **safaris**) **1** a hunting or scientific expedition, esp. in East Africa (*go on safari*). **2** a sightseeing trip to see African animals in their natural habitat.

sag /sæg/ *v. & n.* ● *v.intr.* (**sagged, sagging**) **1** sink or subside under weight or pressure, esp. unevenly. **2** have a downward bulge or curve in the middle.

● The forms given are normally those in use in British English. Variant American forms are identified by the label *US.*

● Pronunciation of inflected forms is given when this differs significantly from the pronunciation of the head-word. The designation 'pronunc. same' denotes that the pronunciation, despite a change of form, is the same as that of the headword.

● The inflection of nouns, verbs, adjectives, and adverbs is given when it is irregular (as described below) or when, though regular, it causes difficulty (as with forms such as **budgeted, coos**, and **taxis**).

Plurals of nouns

Nouns that form their plural regularly by adding -*s* (or -*es* when they end in -*s*, -*x*, -*z*, -*sh*, or soft -*ch*) receive no comment. Other plural forms are given, notably:

● nouns ending in -*i* or -*o*.

● nouns ending in -*y*.

● nouns ending in Latinate forms such as -*a* and -*um*.

● nouns with more than one plural form, e.g. **fish** and **aquarium**.

● nouns with plurals involving a change in the stem, e.g. **foot, feet**.

● nouns with a plural form identical to the singular form, e.g. **sheep**.

● nouns ending in -*ful*, e.g. **handful**.

Forms of verbs

The following forms are regarded as regular:

● third person singular present forms adding -*s* to the stem (or -*es* to stems ending in -*s*, -*x*, -*z*, -*sh*, or soft -*ch*).

● past tenses and past participles adding -*ed* to the stem, dropping a final silent *e* (e.g. **changed, danced**).

● present participles adding -*ing* to the stem, dropping a final silent *e* (e.g. **changing, dancing**).

Other forms are given, notably:

● doubling of a final consonant, e.g. **bat, batted, batting**.

● strong and irregular forms involving a change in the stem, e.g. **come, came, come**, and go, **went, gone**.

Comparative and superlative of adjectives and adverbs

● Words of one syllable adding -*er* or -*est*, those ending in silent *e* dropping the *e* (e.g. **braver, bravest**), are regarded as regular. Most one-syllable words have these forms, but participial adjectives (e.g. **pleased**) do not.

- Those that double a final consonant (e.g. **hot, hotter, hottest**) are given, as are two-syllable words that have comparative and superlative forms in *-er* and *-est* (of which very many are forms ending in *-y*, e.g. **happy, happier, happiest**), and their negative forms (e.g. **unhappier, unhappiest**).

- It should be noted that specification of these forms indicates only that they are used; it is usually also possible to form comparatives with *more* and superlatives with *most* (as in *more happy, most unhappy*), which is the standard way of proceeding with adjectives and adverbs that cannot be inflected.

Adjectives in *-able* formed from transitive verbs

These are given as derivatives when there is sufficient evidence of their currency; in general they are formed as follows:

- Verbs drop silent final *-e* except after *c* and *g* (e.g. **achievable** but **exchangeable**).

- Verbs of more than one syllable ending in *-y* (preceded by a consonant or *qu*) change *y* to *i* (e.g. **enviable, fanciable**).

- A final consonant is often doubled as in normal inflection (e.g. **conferrable, regrettable**).

8. DEFINITION

Definitions are listed in a numbered sequence in order of comparative familiarity and importance, with the most current and important senses first. They are subdivided into lettered senses (**a**, **b**, etc.) when these are closely related or call for collective treatment.

> **pan**[1] /pæn/ *n. & v.* ● *n.* **1 a** a vessel of metal, earthenware, or plastic, usu. broad and shallow, used for cooking and other domestic purposes. **b** the contents of this. **2** a panlike vessel in which substances are heated etc. **3** any similar shallow container such as the bowl of a pair of scales or that used for washing gravel etc. to separate gold. **4** *Brit.* the bowl of a lavatory. **5** part of the lock that held the priming in old guns. **6** a hollow in the ground (*salt-pan*). **7** a hard substratum of soil (*hardpan*). **8** *US sl.* the face. **9 a** a metal drum in a steel band. **b** steel-band music and the associated culture.

9. ILLUSTRATIVE EXAMPLES

Many examples of words in use are given to support, and in some cases supplement, the definitions. These appear in italics in brackets. They are meant to amplify meaning and (especially when following a grammatical point) illustrate how the word is used in context, as in the following sense of **alert**:

> **alert** /ə'lɜːt/ …. ● *v.tr.* (often foll. by *to*) make alert; warn (*were alerted to the danger*).

10. GRAMMATICAL INFORMATION

Definitions are often accompanied by explanations in brackets of how the word or phrase in question is used in context. Often, the comment refers to words that usually follow (foll. by) or precede (prec. by) the word being explained. For example, at **sack**[1]:

> **sack**[1] /sæk/ *n. & v.* ● *n.* **1 a** a large strong bag, usu. made of hessian, paper, or plastic, for storing or conveying goods. **b** (usu. foll. by *of*) this with its contents (*a sack of potatoes*). **c** a quantity contained in a sack. **2** (prec. by *the*) *colloq.* dismissal from employment. **3** (prec. by *the*) *N. Amer. sl.* bed. **4 a** a woman's short loose dress with a sacklike appearance. **b** *archaic* or *hist.* a woman's loose gown, or a silk train attached to the shoulders of this. **5** a man's or woman's loose-hanging coat not shaped to the back.

sense 1b usually appears as *a sack of* (something), as the example further shows; and senses 2 and 3 always appear as *the sack*.

- With verbs, the fact that a sense is transitive or intransitive can affect the construction. In the examples given below, **prevail** is intransitive (and the construction is *prevail on a person*) and **urge** is transitive (and the construction is *urge a person on*).

> **prevail** /prɪ'veɪl/ *v.intr.* **1** (often foll. by *against, over*) be victorious or gain mastery. **2** be the more usual or predominant. **3** exist or occur in general use or experience; be current. **4** (foll. by *on, upon*) persuade.
>
> **urge** /ɜːdʒ/ *v. & n.* ● *v.tr.* **1** (often foll. by *on*) drive forcibly; impel; hasten (*urged them on; urged the horses forward*). **2** (often foll. by *to* + infin. or *that* + clause) encourage or entreat earnestly or persistently (*urged them to go; urged them to action; urged that they should go*).

- The formula (foll. by *to* + infin.) means that the word is followed by a normal infinitive with *to*, as in *want to leave* and *eager to learn*.

- The formula (foll. by *that* + clause) indicates the routine addition of a clause with *that*, as in *said that it was late*. (For the omission of *that*, as in *said it was late*, see the usage note in the entry for **that**.)

- 'pres. part.' and 'verbal noun' denote verbal forms in *-ing* that function as adjectives and nouns respectively, as in *set him laughing* and *tired of asking*.

11. USAGE

If the use of a word is restricted in any way, this is indicated by any of various labels printed in italics, as follows:

Geographical labels

- *Brit.* indicates that the use is found chiefly in British English (and often also in Australian and New Zealand English and in other parts of the Commonwealth) but not in American English.

- *US* indicates that the use is found chiefly in American English (and also in Australian and New Zealand English) but not in British English except as a conscious Americanism.

- *N. Amer.* indicates that the use is found chiefly in the US and Canada (and also in Australian and New Zealand English) but not in British English.

- *Ind.* indicates that the use is found in the subcontinent comprising India, Pakistan, and Bangladesh.

- Other geographical designations (e.g. *Austral., NZ, Sc., S.Afr.*) restrict uses to the areas named.

- These usage labels should be distinguished from comments of the type '(in the UK)' or '(in the US)' preceding definitions, which denote that the thing defined is associated with the country named.

Register labels

Levels of usage, or *registers*, are indicated as follows:

- *formal* indicates uses that are normally restricted to formal (esp. written) English, e.g. **commence**.

- *colloq.* (= colloquial) indicates a use that is normally restricted to informal (esp. spoken) English.

- *sl.* (= slang) indicates a use of the most informal kind, unsuited to written English and often restricted to a particular social group.

- *coarse sl.* indicates a slang use which is normally taboo.

- *dial.* (= dialect) indicates a word that is restricted to non-standard or dialect use.

- *archaic* indicates a word that is generally considered old-fashioned and is often restricted to special contexts such as legal or religious use, or is used for special effect.

- *literary* indicates a word or use that is found chiefly in literature.

- *poet.* (= poetic) indicates uses confined to poetry or other contexts with similar connotations.

- *joc.* (= jocular) indicates uses that are intended to be humorous or playful.

- *derog.* (= derogatory) denotes uses that are intentionally disparaging.

- *offens.* (= offensive) denotes uses that cause offence, whether intentionally or not.

- *disp.* (= disputed) indicates a use that is disputed (and often regarded as erroneous or controversial). Often this is enough to alert the user to a danger or difficulty; when further explanation is needed a usage note (see note below) is used as well or instead.

- *hist.* (= historical) denotes a word or use that is confined to historical reference, normally because the thing referred to no longer exists.

- *propr.* (= proprietary) denotes a term that has the status of a trade mark (see Note on proprietary status, p. xx).

Subject labels

The many subject labels, e.g. *Law, Med., Math., Naut.,* show that a word or sense is current only in a particular field of activity.

Usage notes

These are added to give extra information not central to the definition, and to explain points of grammar and usage. They are introduced by the symbol ¶. The purpose of these notes is not to prescribe usage but to alert the user to a difficulty or controversy attached to a particular use.

12. PHRASES AND IDIOMS

These are listed (together with compounds) in alphabetical order after the treatment of the main senses, introduced by the symbol □. The words *a, the, one,* and *person* do not count for purposes of alphabetical order:

□ **on the safe side** with a margin of security against risks. **safe bet** a bet that is certain to succeed. **safe-breaker** (or **-blower** or **-cracker**) a person who breaks open and robs safes. **safe conduct 1** a privilege of immunity from arrest or harm, esp. on a particular occasion. **2** a document securing this. **safe deposit** a building containing strongrooms and safes let separately. **safe haven 1** a place of refuge or security. **2** a protected zone in a country designated for members of a religious or ethnic minority. **safe house** a place of refuge or rendezvous for spies, criminals, etc. **safe keeping** preservation in a safe place. **safe light** *Photog.* a filtered light for use in a darkroom. **safe period** the time during and near the menstrual period when conception is least likely. **safe seat** a seat in Parliament etc. that is usually won with a large margin by a particular party. **safe sex** sexual activity in which precautions are taken to reduce the risk of spreading sexually transmitted diseases, esp. Aids.

They are normally defined under the first important word in the phrase, except when a later word is more clearly the key word or is the common word in a phrase with variants (in which case a cross-reference often appears at the entry for the first word):

make do 1 manage with the limited or inadequate means available. **2** (foll. by *with*) manage with (something) as an inferior substitute. **make an example of** punish as a warning to others. **make a fool of** see FOOL[1]. **make for 1** tend to result in (happiness etc.). **2** proceed towards (a place). **3** assault; attack. **4** confirm (an opinion). **make friends** (often foll. by *with*) become friendly. **make fun of** see FUN.

13. COMPOUNDS

Compound terms forming one word (such as **bathroom, newspaper**) are listed as main entries.

Compound terms consisting of two or more separate words or two or more words joined by a hyphen are generally given under the headword of the first word. Exceptions to this, in which compounds are given as main entries, include:

- where the compound is a proper name (and is therefore capitalized), e.g. **East River, Baltic Exchange, Lord Privy Seal, Armistice Day.**

- where the first element of the compound is a place-name or person's name (e.g. **Faraday's constant, Adonis blue, London pride**).

- where additional encyclopedic or etymological information is introduced (e.g. **air raid, general strike, burning bush**).

- where the compound has multiple senses and parts of speech.

14. DERIVATIVES

Words formed by adding a suffix to another word are in many cases listed at the end of the entry for the main word, introduced in a separate section by the symbol □. In this position they are not given a definition since they can be understood from the sense of the main word and that given at the suffix concerned, as in the entry for **citizen**.

□ **citizenhood** *n.* **citizenry** *n.* **citizenship** *n.*

When further definition is called for they are given main entries in their own right (e.g. **changeable**).

15. ETYMOLOGY

A brief account of the etymology, or origin, of words is given in square brackets at the end of entries. (If there is a separate paragraph of information, the etymology comes before this, as, for example, at the entry for **fossil**.) Etymologies are not given for compound words of obvious formation (such as **bathroom** and **jellyfish**), for routinely formed derivatives (such as **changeable, muddy,** and **seller**), or for words consisting of clearly identified elements already explained (such as **Anglo-Irish, overrun,** and many words in *in-, re-, un-,* etc.). Etymologies are also not always given for every word of a set sharing the same basic origin (**proprietary** and **proprietor**, for example). Noteworthy features, such as an origin in Old English, are however always given.

Etymologies are not regularly given for proper names such as place-names or personal names, though they are given for names of languages and peoples. A place-name, biographical entry, or other proper-name entry will occasionally include etymological information where this is thought to be particularly relevant or interesting, though in this case the information is often presented contextually rather than in square brackets (see, for example, **America, Threadneedle Street, Methodist, Heliopolis**).

The following notes act as a guide to the etymological information in this dictionary. More detailed information about word origins can be found in the *New Shorter Oxford English Dictionary* (ed. Lesley Brown, 1993).

● The immediate source language is given first. Forms in other languages are not given if they are exactly or nearly the same as the English form given in the headword.

● Words of Germanic origin are described as 'f. Gmc' or 'f. WG' (West Germanic) as appropriate; unrecorded or postulated forms are not normally given.

● OE (Old English) is used for words that are known to have been used before AD 1150, and ME (Middle English) for words traceable to the period 1150–1500 (no distinction being made between early and late Middle English).

● Words of Romance origin are referred to their immediate source, usually F (French) or OF (Old French before 1400), and then to earlier sources when known.

● AF (Anglo-French) denotes the variety of French current in England in the Middle Ages after the Norman Conquest.

● Rmc (Romanic) denotes the vernacular descendants of Latin that are the source of French, Spanish, Italian, etc. Romanic forms are almost always of the 'unrecorded' or 'postulated' kind, and are not specified except to clarify a significant change of form. Often the formula 'ult. f. L' etc. (ultimately from Latin etc.) is used to indicate that the route from Latin is via Romanic forms.

● L (Latin) denotes classical Latin up to about AD 200; OL (Old Latin) Latin before about 75 BC; LL (Late Latin) Latin of about 200–600; med.L (medieval Latin) Latin of about 600–1500; mod.L (modern Latin) Latin in use (mainly for technical and scientific purposes) since about 1500.

● Similar divisions for 'late', 'medieval', and 'modern' are made for Greek.

● Many English words have corresponding forms in both French and Latin, and it cannot always be established which was the immediate source. In such cases the formula 'F or L' is used (e.g. at **section**: 'F *section* or L *sectio*'); in these cases the Latin form is the source of the French word and (either directly or indirectly) of the English word.

● When the origin of a word cannot be reliably established, the forms 'orig. unkn.' (= origin unknown) and 'orig. uncert.' (= origin uncertain) are used. In these cases the century of the first recorded occurrence of the word in English is given.

● An equals sign (=) precedes words in other languages that are parallel formations from a common source (cognates) rather than sources of the English word.

16. PREFIXES, SUFFIXES, AND COMBINING FORMS

A large selection of these is given in the main body of the text; prefixes are given in the form **ex-**, **re-**, etc., and suffixes in the form **-ion**, **-ness**, etc. These entries should be consulted to explain the many routinely formed derivatives given at the end of entries (see above, section 14).

● Combining forms (e.g. **bio-**, **-graphy**) are semantically significant elements that can be attached to words or elements as explained in the usage note at the entry for **combine**.

● The pronunciation given for a prefix, suffix, or combining form is an approximate one for purposes of articulating and (in some cases) identifying the headword; pronunciation and stress may change considerably when they form part of a word.

17. CROSS-REFERENCES

These are introduced by any of a number of reference types, as follows:

● '=' denotes that the meaning of the item at which the cross-reference occurs is the same as that of the item referred to.

● 'see' indicates that information will be found at the point referred to, and is widely used for encyclopedic matter and in the idiom sections of entries to deal with items that can be located at any of a number of words included in the idiom (see also above, section 12).

● 'see also' indicates that further information can be found at the point referred to.

● 'cf.' denotes an item related or relevant to the one being consulted, and the reference often completes or clarifies the exact meaning of the item being treated.

● 'opp.' refers to a word or sense that is opposite to the one being treated, and again often completes or clarifies the sense.

● References of the kind '*pl.* of' (= plural of), '*past* of' (= past tense of), etc., are given at entries for inflections and other related forms.

● 'var. of' (= variant of) refers to a word that is the main headword form of one or more variant forms.

Cross-references preceded by any of these reference types appear in small capitals if the reference is to a main headword, and in italics if the reference is to a compound or phrase within an entry.

● References in italics to compounds and defined phrases are to the entry for the first word unless another is specified.

List of Abbreviations and Symbols

Some abbreviations (especially of language-names) occur only in etymologies. Others may appear in italics. Abbreviations in general use (such as etc., i.e., and those for books of the Bible) are explained in the dictionary itself.

abbr.	abbreviation	compar.	comparative	Heb.	Hebrew
ablat.	ablative	compl.	complement	Hind.	Hindustani
absol.	absolute(ly)	conj.	conjunction	hist.	with historical
accus.	accusative	contr.	contraction		reference
adj.	adjective	corresp.	corresponding		
adv.	adverb	corrupt.	corruption	Icel.	Icelandic
Aeron.	Aeronautics	Crystallog.	Crystallography	imit.	imitative
AF	Anglo-French			imper.	imperative
Afr.	Africa, African	Da.	Danish	incept.	inceptive
Afrik.	Afrikaans	demonstr.	demonstrative	Ind.	Indian (of the subcontinent comprising India, Pakistan, and Bangladesh)
AL	Anglo-Latin	derog.	derogatory		
alt.	alteration	dial.	dialect		
Amer.	America, American	dimin.	diminutive		
Anat.	Anatomy	disp.	disputed (use)	infin.	infinitive
Anglo-Ind.	Anglo-Indian	Du.	Dutch	infl.	influenced
Anthropol.	Anthropology			int.	interjection
Antiq.	Antiquity	Eccl.	Ecclesiastical	interrog.	interrogative
app.	apparently	Ecol.	Ecology	intr.	intransitive
Arab.	Arabic	Econ.	Economics	Ir.	Irish
Aram.	Aramaic	Egypt.	Egyptian	iron.	ironical(ly)
Archaeol.	Archaeology	Electr.	Electricity	irreg.	irregular(ly)
Archit.	Architecture	ellipt.	elliptical(ly)	It.	Italian
assim.	assimilated	emphat.	emphatic(ally)		
assoc.	associated, association	Engin.	Engineering	Jap.	Japan, Japanese
Assyr.	Assyrian	Engl.	English	joc.	jocular(ly)
Astrol.	Astrology	erron.	erroneous(ly)		
Astron.	Astronomy	esp.	especial(ly)	L	Latin
attrib.	attributive(ly)	est.	estimated	LG	Low German
attrib.adj.	attributive adjective	euphem.	euphemism	lit.	literally
augment.	augmentative	exclam.	exclamation	LL	Late Latin
Austral.	Australia, Australian				
aux.	auxiliary	F	French	M	Middle (with languages)
		f.	from	masc.	masculine
back-form.	back-formation	f. as	formed as	Math.	Mathematics
Bibl.	Biblical	fem.	feminine	MDu.	Middle Dutch
Biochem.	Biochemistry	*fl.*	*floruit* (flourished)	ME	Middle English
Biol.	Biology	Flem.	Flemish	Mech.	Mechanics
Bot.	Botany	foll.	followed, following	Med.	Medicine
Brit.	British, in British use	Fr.	French	med.	medieval
Bulg.	Bulgarian	Frank.	Frankish	med.L	medieval Latin
		frequent.	frequentative	Meteorol.	Meteorology
c.	century			Mex.	Mexican
c.	*circa*	G	German	MHG	Middle High German
Can.	Canada, Canadian	Gael.	Gaelic	Mil.	Military
Celt.	Celtic	Geog.	Geography	Mineral.	Mineralogy
Chem.	Chemistry	Geol.	Geology	mistransl.	mistranslation (of)
chem.	chemical	Geom.	Geometry	MLG	Middle Low German
Chin.	Chinese	Ger.	German	mod.	modern
Cinematog.	Cinematography	Gk	Greek	mod.L	modern Latin
collect.	collective(ly)	Gmc	Germanic	MSw.	Middle Swedish
colloq.	colloquial(ly)	Goth.	Gothic	Mus.	Music
comb.	combination; combining	Gram.	Grammar	Mythol.	Mythology

xix

List of Abbreviations and Symbols

n.	noun	Pharm.	Pharmacology	S. Afr.	South Africa, South African
N. Amer.	North America, North American	Philol.	Philology		
		Philos.	Philosophy	S. Amer.	South American
Naut.	Nautical	Phonet.	Phonetics	Sc.	Scottish
neg.	negative	Photog.	Photography	Scand.	Scandinavian
N. Engl.	North of England	phr.	phrase	sing.	singular
neut.	neuter	Physiol.	Physiology	Sinh.	Sinhalese
north.	northern	pl.	plural	Skr.	Sanskrit
Norw.	Norwegian	poet.	poetical	sl.	slang
n.pl.	noun plural	Pol.	Polish	Slav.	Slavonic
NZ	New Zealand	Polit.	Politics	Sp.	Spanish
		pop.	population	Stock Exch.	Stock Exchange
O	Old (with languages)	pop.L	popular Latin, informal spoken Latin	superl.	superlative
obj.	objective			Sw.	Swedish
obs.	obsolete	Port.	Portuguese	symb.	symbol
OCelt.	Old Celtic	poss.	possessive		
OE	Old English	prec.	preceded	Theatr.	Theatrical
OF	Old French	predic.	predicative	Theol.	Theology
offens.	offensive	prep.	preposition	tr.	transitive
OFris.	Old Frisian	pres.	present	transl.	translation (of)
OHG	Old High German	prob.	probably; probable	Turk.	Turkish
OIcel.	Old Icelandic	pron.	pronoun		
OIr.	Old Irish	pronunc.	pronunciation	ult.	ultimate (ly)
OIt.	Old Italian	propr.	proprietary term	uncert.	uncertain
OL	Old Latin	Prov.	Provençal	unexpl.	unexplained
OLG	Old Low German	Psychol.	Psychology	unkn.	unknown
ON	Old Norse			US	American, in American use
ONF	Old Northern French	RC Ch.	Roman Catholic Church		
OPers.	Old Persian	redupl.	reduplicated, reduplication	usu.	usually
opp.	opposite (of); (as) opposed (to)				
		ref.	reference	v.	verb
orig.	origin; original(ly)	refl.	reflexive	var.	variant(s)
OS	Old Saxon	rel.	related; relative	v.aux.	auxiliary verb
OSlav.	Old Slavonic	rel.adj.	relative adjective	v.intr.	intransitive verb
OSp.	Old Spanish	Relig.	Religion	v.refl.	reflexive verb
OSw.	Old Swedish	rel.pron.	relative pronoun	v.tr.	transitive verb
		repr.	representing		
Parl.	Parliament	Rhet.	Rhetoric	WG	West Germanic
part.	participle	rhet.	rhetorical		
perh.	perhaps	Rmc	Romanic	Zool.	Zoology
Pers.	Persian	Rom.	Roman		
pers.	person(al)	Russ.	Russian		

Symbols used in the dictionary

¶ introduces notes on usage (see **Guide to the Use of This Dictionary: Usage Notes**).

□ introduces defined compounds, phrases, and idioms or undefined derivatives formed by adding a suffix to the main word.

• introduces a new part of speech.

■ introduces a paragraph of extra information relating to the headword.

Note on proprietary status

This dictionary includes some words which have or are asserted to have proprietary status as trademarks or otherwise. Their inclusion does not imply that they have acquired for legal purposes a non-proprietary or general significance nor any other judgement concerning their legal status. In cases where the editorial staff have some evidence that a word has proprietary status this is indicated in the entry for that word but no judgement concerning the legal status of such words is made or implied thereby.

The Oxford Encyclopedic English Dictionary

Aa

A¹ /eɪ/ n. (also **a**) (pl. **As** or **A's**) **1** the first letter of the alphabet. **2** *Mus.* the sixth note of the diatonic scale of C major. **3** the first hypothetical person or example. **4** the highest class or category (of roads, academic marks, etc.). **5** (usu. **a**) *Algebra* the first known quantity. **6** a human blood type of the ABO system. □ **A1** /eɪ ˈwʌn/ n. *Naut.* a first-class vessel in Lloyd's Register of Shipping. ● *adj.* **1** *Naut.* (of a ship) first-class. **2** *colloq.* excellent, first-rate. **A1, A2**, etc. the standard European paper sizes, each half the previous one, e.g. A4 = 297 × 210 mm, A5 = 210 × 148 mm. **from A to B** from one place to another (*a means of getting from A to B*). **from A to Z** over the entire range, completely.

A² /eɪ/ *abbr.* (also **A.**) **1** *Cards* ace. **2** = A LEVEL. **3** answer. **4** atomic (*A-bomb*).

A³ *symb.* **1** *Physics* ampere(s). **2** *Brit.* (of films) classified as suitable for an adult audience but not necessarily for children. ¶ Now replaced by *PG*.

Å *abbr.* angstrom(s).

a¹ /ə, eɪ/ *adj.* (also **an** before a vowel) (called the indefinite article) **1** (as an unemphatic substitute) one, some, any. **2** one like (*a Judas*). **3** one single (*not a thing in sight*). **4** the same (*all of a size*). **5** in, to, or for each (*twice a year; £20 a man; seven a side*). [weakening of OE *ān* one; sense 5 orig. = A²]

a² /ə/ *prep.* (usu. as *prefix*) **1** to, towards (*ashore; aside*). **2** (with verb in pres. part. or infin.) in the process of; in a specified state (*a-hunting; a-wandering; abuzz; aflutter*). **3** on (*afire; afoot*). **4** in (*nowadays*). [weakening of OE prep. *an, on* (see ON)]

a³ *abbr.* (also **a.**) **1** arrives. **2** before. [sense 2 from Latin *ante*]

a⁴ *symb.* **1** atto-. **2** acceleration. **3** are(s) (measure of area).

a-¹ /eɪ, æ/ *prefix* not, without (*amoral; agnostic; apetalous*). [Gk *a-*, or L f. Gk, or F f. L f. Gk]

a-² /ə/ *prefix* implying motion onward or away, adding intensity to verbs of motion (*arise; awake*). [OE *a-*, orig. *ar-*]

a-³ /ə/ *prefix* to, at, or into a state (*adroit; agree; amass; avenge*). [ME *a-* (= OF prefix *a-*), (f. F) f. L *ad-* to, at]

a-⁴ /ə/ *prefix* **1** from, away (*abridge*). **2** of (*akin; anew*). **3** out, utterly (*abash; affray*). **4** in, on, engaged in, etc. (see A²). [sense 1 f. ME *a-*, OF *a-*, f. L *ab*; sense 2 f. ME *a-* f. OE *of* prep.; sense 3 f. ME, AF *a-* = OF *e-, es-* f. L *ex*]

a-⁵ /ə, æ/ *prefix* assim. form of AD- before *sc, sp, st.*

-a¹ /ə/ *suffix* forming nouns from Greek, Latin, and Romanic feminine singular, esp.: **1** ancient or Latinized modern names of animals and plants (*amoeba; campanula*). **2** oxides (*alumina*). **3** geographical names (*Africa*). **4** ancient or Latinized modern feminine names (*Lydia; Hilda*).

-a² /ə/ *suffix* forming plural nouns from Greek and Latin neuter plural, esp. names (often from modern Latin) of zoological groups (*phenomena; Carnivora*).

-a³ /ə/ *suffix colloq.* **1** of (*kinda; coupla*). **2** have (*mighta; coulda*). **3** to (*oughta*).

AA *abbr.* **1** (in the UK) Automobile Association. **2** see ALCOHOLICS ANONYMOUS. **3** *Mil.* anti-aircraft. **4** *Brit. hist.* (of films) classified as suitable for persons of over 14 years.

aa /ˈɑːɑː/ n. *Geol.* lava forming a mass of rough, jagged, scoriaceous blocks (cf. PAHOEHOE). [Hawaiian *'a-'a*]

AAA *abbr.* **1** (in the UK) Amateur Athletic Association. **2** American Automobile Association. **3** Australian Automobile Association. **4** anti-aircraft artillery.

Aachen /ˈɑːxən, ˈɑːkən/ (called in French *Aix-la-Chapelle*) an industrial city and spa in western Germany, in North Rhine-Westphalia, close to the Belgian and Dutch borders; pop. (1991) 244,440. German emperors were crowned in Aachen from the time of Charlemagne (who was born and buried there) until 1531.

Aalborg /ˈɔːlbɔːɡ/ (also **Ålborg**) an industrial city and port in north Jutland, Denmark; pop. (1991) 155,000.

Aalto /ˈɑːltəʊ/, (Hugo) Alvar (Henrik) (1898–1976), Finnish architect and designer. His early work was in the neoclassical style, but in the late 1920s he adopted the 'international style', in such buildings as the Viipuri (Vyborg, now in Russia) Library (1927–35) and the Paimio convalescent home (1929–33). In Finland he often used mixed materials such as brick, copper, and timber in his designs to blend with the landscape. Outside Finland his best-known work is the Baker House hall of residence at Massachusetts Institute of Technology, where he taught 1945–9.

A. & M. *abbr.* (Hymns) Ancient and Modern.

A. & R. *abbr.* **1** artists and recording. **2** artists and repertoire.

aardvark /ˈɑːdvɑːk/ n. a nocturnal mammal of southern Africa, *Orycteropus afer*, with a tubular snout and a long extendible tongue, that feeds on termites. Also called *ant-bear* or *earth-hog*. [Afrik. f. *aarde* earth + *vark* pig]

aardwolf /ˈɑːdwʊlf/ n. (pl. **aardwolves**) a southern African mammal, *Proteles cristatus*, of the hyena family. It has grey fur with black stripes, and feeds largely on insects. [Afrik. f. *aarde* earth + *wolf* wolf]

Aarhus /ˈɔːhuːs/ (also **Århus**) a city on the coast of east Jutland, Denmark; pop. (1990) 261,440.

Aaron /ˈeərən/ (in the Bible) brother of Moses and traditional founder of the Jewish priesthood (see Exod. 28:1). The first anointed high priest, he acted, according to the biblical account, as a spokesman for his brother after the latter's commissioning to lead the Hebrew people out of Egypt. He was persuaded by the people to make an image of God in the form of a golden calf, thereby earning Moses' displeasure.

Aaron's beard n. (see Ps. 133:2) a popular name for several plants, especially a St John's wort, *Hypericum calycinum*.

Aaron's rod n. (see Numbers 17:8) a popular name for several tall plants with flowering stems, especially a mullein, *Verbascum thapsus*.

A'asia *abbr.* Australasia.

aasvogel /ˈɑːsˌfəʊɡ(ə)l/ n. *S. Afr.* a vulture. [Afrik. f. *aas* carrion + *vogel* bird]

AAU *abbr.* (in the US) Amateur Athletic Union.

AB¹ /eɪˈbiː/ n. a human blood type of the ABO system.

AB² *abbr.* **1** able rating or seaman. **2** *US* Bachelor of Arts. [sense 1 f. *able-bodied*; sense 2 f. L *Artium Baccalaureus*]

Ab¹ /æb/ n. (also **Av** /æv/) (in the Jewish calendar) the eleventh month of the civil and fifth of the religious year, usually coinciding with parts of July and August. [Heb. *'āb*]

Ab² *abbr.* antibody.

ab- /əb, æb/ *prefix* off, away, from (*abduct; abnormal; abuse*). [F or L]

aba /ˈæbə/ n. (also **abba, abaya** /ə'beɪjə, ə'baɪjə/) a sleeveless garment worn by Arabs. [Arab. *'abā*]

abaca /ˈæbəkə/ n. **1** Manila hemp. **2** the Philippine tree, *Musa textilis*, yielding this. [Sp. *abacá*]

aback /ə'bæk/ adv. **1** *archaic* backwards, behind. **2** *Naut.* (of a sail) pressed against the mast by a head wind. □ **take aback 1** surprise, disconcert (*your request took me aback; I was greatly taken aback by the news*). **2** (as

taken aback *adj.*) (of a ship) with the sails pressed against the mast by a head wind. [OE *on bæc* (as A², BACK)]

abacus /ˈæbəkəs/ *n.* (*pl.* **abacuses**) **1** an oblong frame with rows of wires or grooves along which beads are slid, used for calculating. (*See note below.*) **2** *Archit.* the flat slab on top of a capital, supporting the architrave. [L f. Gk *abax abakos* slab, drawing-board, f. Heb. *'ăḇāḳ* dust (from use of board sprinkled with sand or dust for drawing geometrical diagrams)]

▪ The provenance and date of origin of the abacus are uncertain. The ancient Egyptians, Greeks, and Romans used a counting-board with vertical columns; the frame may possibly be a development of this. It was in use in Europe during the Middle Ages until the adoption of arabic numerals, and is still used in some parts of the world, especially the Far East.

Abadan /ˌæbəˈdɑːn/ a major port and oil-refining centre on an island of the same name on the Shatt al-Arab waterway in western Iran; pop. (1986) 308,000.

Abaddon /əˈbæd(ə)n/ *n.* **1** hell. **2** the Devil (Rev. 9:11). [Heb., = destruction]

abaft /əˈbɑːft/ *adv. & prep. Naut.* ● *adv.* in the stern half of a ship. ● *prep.* nearer the stern than; aft of. [A² + *-baft* f. OE *beæftan* f. *be* BY + *æftan* behind]

Abakan /ˌæbəˈkɑːn/ an industrial city in south central Russia, capital of the republic of Khakassia; pop. (1989) 154,000. It was known as Ustabaskanskoe until 1931.

abalone /ˌæbəˈləʊnɪ/ *n.* an edible gastropod mollusc of the genus *Haliotis*, with a shallow ear-shaped shell having respiratory holes, and lined with mother-of-pearl, e.g. the ormer. [Amer. Sp. *abulón*]

abandon /əˈbændən/ *v. & n.* ● *v.tr.* **1** give up completely or before completion (*abandoned hope; abandoned the game*). **2 a** forsake or desert (a person or a post of responsibility). **b** leave or desert (a motor vehicle, ship, building, etc.). **3 a** give up to another's control or mercy. **b** *refl.* yield oneself completely to a passion or impulse. ● *n.* lack of inhibition or restraint; reckless freedom of manner. □ **abandoner** *n.* **abandonment** *n.* [ME f. OF *abandoner* f. *à bandon* under control ult. f. LL *bannus, -um* BAN]

abandoned /əˈbændənd/ *adj.* **1 a** (of a person or animal) deserted, forsaken (*an abandoned child*). **b** (of a building, vehicle, etc.) left empty or unused (*an abandoned cottage; an abandoned ship*). **2** (of a person or behaviour) unrestrained, profligate.

abase /əˈbeɪs/ *v.tr. & refl.* humiliate or degrade (another person or oneself). □ **abasement** *n.* [ME f. OF *abaissier* (as A-³, *baissier* to lower ult. f. LL *bassus* short of stature): infl. by BASE²]

abash /əˈbæʃ/ *v.tr.* (usu. as **abashed** *adj.*) embarrass, disconcert. □ **abashment** *n.* [ME f. OF *esbaïr* (*es-* = A-⁴, *baïr* astound or *baer* yawn)]

abate /əˈbeɪt/ *v.* **1** *tr. & intr.* make or become less strong, severe, intense, etc. **2** *tr. Law* **a** quash (a writ or action). **b** put an end to (a nuisance). □ **abatement** *n.* [ME f. OF *abatre* f. Rmc (as A-³, L *batt(u)ere* beat)]

abatis /ˈæbətɪs/ *n.* (also **abattis**) (*pl.* same or **abatises**, **abattises**) *hist.* a defence made of felled trees with the boughs pointing outwards. [F f. *abatre* fell: see ABATE]

abattoir /ˈæbəˌtwɑː(r)/ *n.* a slaughterhouse. [F (as ABATIS, -ORY¹)]

abaxial /æbˈæksɪəl/ *adj. Bot.* facing away from the stem of a plant, esp. of the lower surface of a leaf (cf. ADAXIAL). [AB- + AXIAL]

abaya (also **abba**) var. of ABA.

Abba /ˈæbə/ a Swedish pop group which first became popular in the 1970s with catchy, well-crafted songs such as 'Waterloo' (1974) and 'Knowing Me Knowing You' (1977). The group was originally made up of two couples, the initials of whose Christian names formed the word *Abba*.

abbacy /ˈæbəsɪ/ *n.* (*pl.* **-ies**) the office, jurisdiction, or period of office of an abbot or abbess. [ME f. eccl.L *abbacia* f. *abbat-* ABBOT]

Abbas /əˈbæs/, Ferhat (1899–1989), Algerian nationalist leader. After attempting to cooperate with the French in setting up an Algerian state, he became disenchanted and in 1956 joined the revolutionary Front de Libération Nationale (FLN). He was elected President of the provisional government of the Algerian republic (based in Tunisia) in 1958 and President of the constituent assembly of independent Algeria in 1962. In 1963 he was detained for opposing the FLN's proposed constitution, but was released the following year.

Abbasid /əˈbæsɪd/ *n. & adj.* ● *n.* a member of a dynasty of caliphs ruling in Baghdad 750–1258. The dynasty claimed descent from Abbas (566–652), uncle of Muhammad. ● *adj.* of or relating to this dynasty.

abbatial /əˈbeɪʃ(ə)l/ *adj.* of an abbey, abbot, or abbess. [F *abbatial* or med.L *abbatialis* (as ABBOT)]

Abbe /ˈæbɪ/, Ernst (1840–1905), German physicist. He worked with Carl Zeiss from 1866, and in 1868 invented the apochromatic lens. He also designed several optical instruments, including a light condenser for use in microscopes and a refractometer. He became sole owner of the Zeiss company in 1888.

abbé /ˈæbeɪ/ *n.* (in France) an abbot; a man entitled to wear ecclesiastical dress. [F f. eccl.L *abbas abbatis* ABBOT]

abbess /ˈæbɪs/ *n.* a woman who is the head of certain communities of nuns. [ME f. OF *abbesse* f. eccl.L *abbatissa* (as ABBOT)]

Abbevillian /æbˈvɪlɪən/ *adj. & n. Archaeol.* of, relating to, or denoting the first palaeolithic culture in Europe, from which the earliest European hand-axes have been found, dated to *c.*500,000 BP. (See also ACHEULEAN.) [F *Abbevillien* f. *Abbeville*, location of the type-site on the River Somme in northern France]

abbey /ˈæbɪ/ *n.* (*pl.* **-eys**) **1** the building(s) occupied by a community of monks or nuns. **2** the community itself. **3** a church or house that was once an abbey. [ME f. OF *abbeie* etc. f. med.L *abbatia* ABBACY]

Abbey Theatre a theatre in Abbey Street, Dublin, first opened in 1904, staging the work of Irish dramatists such as J. M. Synge and Sean O'Casey. W. B. Yeats was associated with its foundation, and in 1925 it became the first state-subsidized theatre in the English-speaking world.

abbot /ˈæbət/ *n.* a man who is the head of an abbey of monks. □ **abbotship** *n.* [OE *abbod* f. eccl.L *abbas -atis* f. Gk *abbas* father f. Aram. *'abba*]

abbreviate /əˈbriːvɪˌeɪt/ *v.tr.* shorten, esp. represent (a word etc.) by a part of it. [ME f. LL *abbreviare* shorten f. *brevis* short: cf. ABRIDGE]

abbreviation /əˌbriːvɪˈeɪʃ(ə)n/ *n.* **1** an abbreviated form, esp. a shortened form of a word or phrase. **2** the process or result of abbreviating.

ABC¹ /ˌeɪbiːˈsiː/ *n.* **1** the alphabet. **2** the rudiments of any subject. **3** an alphabetical guide.

ABC² *abbr.* **1** Australian Broadcasting Corporation. **2** American Broadcasting Company.

ABC Islands an acronym for the Dutch islands of Aruba, Bonaire, and Curaçao, which lie in the Caribbean Sea near the northern coast of Venezuela.

abdicate /ˈæbdɪˌkeɪt/ *v.tr.* **1** (usu. *absol.*) give up or renounce (the throne). **2** renounce (a responsibility, duty, etc.). □ **abdicator** *n.* **abdication** /ˌæbdɪˈkeɪʃ(ə)n/ *n.* [L *abdicare abdicat-* (as AB-, *dicare* declare)]

abdomen /ˈæbdəmən/ *n.* **1** the part of the body containing the stomach, bowels, reproductive organs, etc. **2** *Zool.* the posterior part of the body of an insect, crustacean, spider, etc. □ **abdominal** /æbˈdɒmɪn(ə)l/ *adj.* **abdominally** *adv.* [L]

abduct /əbˈdʌkt/ *v.tr.* **1** carry off or kidnap (a person) illegally by force or deception. **2** *Anat.* (esp. of a muscle) move (a limb etc.) away from the midline of the body. □ **abductor** *n.* **abduction** /-ˈdʌkʃ(ə)n/ *n.* [L *abducere abduct-* (as AB-, *ducere* draw)]

Abduh /ˈæbduː/, Muhammad (1849–1905), Egyptian Islamic scholar, jurist, and liberal reformer. As Grand Mufti of Egypt from 1899, he introduced reforms in Islamic law and education and is best known for his *Treatise on the Oneness of God*.

Abdul Hamid II /ˌæbdʊl ˈhæmɪd/ (known as 'the Great Assassin') (1842–1918), the last sultan of Turkey 1876–1909. An autocratic ruler, he suspended Parliament and the constitution and is remembered for the brutal massacres of Christian Armenians in 1894–6. In 1909 he was deposed after the revolt (in 1908) of the Young Turks.

Abdullah /æbˈdʊlə/, Sheikh Muhammad (known as 'the Lion of Kashmir') (1905–82), Kashmiri Muslim leader. In the 1930s he actively opposed the rule of the Hindu maharajah of Kashmir. After accepting Indian sovereignty (1947), he eventually won for Kashmir a form of autonomy within India, although he was imprisoned for much of the time between 1953 and 1968 on suspicion of seeking its full independence.

Abdullah ibn Hussein /æbˌdʊlə ˌɪb(ə)n hʊˈseɪn/ (1882–1951), king of Jordan 1946–51. He served as emir of Transjordan 1921–46, and was Jordan's first king when it became independent in 1946. He was assassinated in 1951.

Abdul Rahman /ˌæbdʊl ˈrɑːmən/, Tunku (1903–90), Malayan statesman, Prime Minister of Malaya 1957–63 and of Malaysia 1963–70.

A skilled negotiator, he secured Malayan independence from Britain (1957) and was one of the architects of the federation of Malaysia (1963).

abeam /əˈbiːm/ adv. **1** on a line at right angles to a ship's or an aircraft's length. **2** (foll. by *of*) opposite the middle of (a ship etc.). [A² + BEAM]

abed /əˈbed/ adv. archaic in bed. [OE (as A², BED)]

Abel¹ /ˈeɪb(ə)l/ (in the Bible) the second son of Adam and Eve, murdered by his brother Cain.

Abel² /ˈɑːb(ə)l/, Niels Henrik (1802–29), Norwegian mathematician. He proved that equations of the fifth degree cannot be solved by conventional algebraic methods, and made advances in the fields of power series and elliptic functions. He died in poverty of tuberculosis at the age of 26.

Abelard /ˈæbɪˌlɑːd/, Peter (1079–1142), French scholar, theologian, and philosopher. His independence of mind, particularly his application of philosophical principles to theological questions, impressed his contemporaries but brought him into frequent conflict with the authorities and led to his being twice condemned for heresy. He lectured in Paris until his academic career was cut short in 1118 by his tragic love affair with his pupil Héloïse, niece of Fulbert, a canon of Notre-Dame. Abelard was castrated at Fulbert's instigation; he entered a monastery, and made Héloïse become a nun. Abelard continued his controversial teaching, applying reason to questions of faith, notably to the doctrine of the Trinity. In the early 1130s he and Héloïse put together a collection of their love letters and other correspondence, which was published in 1616. Abelard and Héloïse are buried together in Paris.

abele /əˈbiːl, ˈeɪb(ə)l/ n. the white poplar, *Populus alba*. [Du. *abeel* f. OF *abel*, *aubel* ult. f. L *albus* white]

abelia /əˈbiːlɪə/ n. a shrub of the genus *Abelia*, esp. *A. grandiflora*. [Clarke *Abel*, Engl. botanist (1780–1826)]

abelian /əˈbiːlɪən/ adj. Math. (of a group) in which the operation is commutative. [ABEL²]

Abeokuta /ˌæbɪˈəʊkuːtə/ a city in SW Nigeria, capital of the state of Ogun; pop. (1983) 308,800.

Aberdeen¹ /ˌæbəˈdiːn/ a city and seaport in NE Scotland, the administrative capital of Grampian region; pop. (1991) 201,100. In modern times Aberdeen has been named the 'Granite City' after its granite buildings and quarrying industry. It is a centre of the off-shore North Sea oil industry.

Aberdeen² /ˌæbəˈdiːn/, 4th Earl of (title of George Hamilton Gordon) (1784–1860), British Conservative statesman, Prime Minister 1852–5. He reluctantly involved his country in the Crimean War (1853–6) and was subsequently blamed for its mismanagement and obliged to resign.

Aberdeen Angus n. a Scottish breed of hornless black beef cattle. [ABERDEEN¹, ANGUS]

Aberdeenshire /ˌæbəˈdiːnʃɪə(r)/ a former county of NE Scotland. It became a part of Grampian region in 1975.

Aberdonian /ˌæbəˈdəʊnɪən/ adj. & n. ● adj. of Aberdeen. ● n. a native or inhabitant of Aberdeen. [med.L *Aberdonia*]

Aberfan /ˌæbəˈvæn/ a village in South Wales where, in 1966, a slag-heap collapsed, overwhelming houses and a school and killing 28 adults and 116 children.

aberrant /əˈberənt/ adj. **1** esp. Biol. diverging from the normal type. **2** departing from an accepted standard. □ **aberrance** n. **aberrancy** n. [L *aberrare aberrant-* (as AB-, *errare* stray)]

aberration /ˌæbəˈreɪʃ(ə)n/ n. **1** a departure from what is normal or accepted or regarded as right. **2** a moral or mental lapse. **3** Biol. deviation from a normal type. **4** Optics the failure of rays to converge at one focus because of a defect in a lens or mirror. **5** Astron. the apparent displacement of a celestial body from its true position caused by the observer's relative motion. (*See note below*.) □ **chromatic aberration** see CHROMATIC. □ **aberrational** adj. [L *aberratio* (as ABERRANT)]

■ The aberration of a planet, star, or other body can have a number of components: *diurnal* and *annual aberration* are both due to the motion of the earth; *planetary aberration* is due to the motion of the body itself, during the time taken for its light to reach the earth.

abet /əˈbet/ v.tr. (**abetted**, **abetting**) (usu. in **aid and abet**) encourage or assist (an offender or offence). □ **abetment** n. [ME f. OF *abeter* f. *à* to + *beter* BAIT]

abetter /əˈbetə(r)/ n. (also **abettor**) a person who abets.

abeyance /əˈbeɪəns/ n. (usu. prec. by *in, into*) a state of temporary disuse or suspension. □ **abeyant** adj. [AF *abeiance* f. OF *abeer* f. *à* to + *beer* f. med.L *batare* gape]

ABH abbr. actual bodily harm.

abhor /əbˈhɔː(r)/ v.tr. (**abhorred**, **abhorring**) detest; regard with disgust and hatred. □ **abhorrer** n. [ME f. F *abhorrer* or f. L *abhorrere* (as AB-, *horrere* shudder)]

abhorrence /əbˈhɒrəns/ n. **1** disgust; detestation. **2** a detested thing.

abhorrent /əbˈhɒrənt/ adj. **1** (often foll. by *to*) (of conduct etc.) inspiring disgust, repugnant; hateful, detestable. **2** (foll. by *to*) not in accordance with; strongly conflicting with (*abhorrent to the spirit of the law*). **3** (foll. by *from*) archaic inconsistent with.

abide /əˈbaɪd/ v. (past and past part. **abided** or rarely **abode** /əˈbəʊd/) **1** tr. (usu. in neg. or interrog.) tolerate, endure (*can't abide him*). **2** intr. (foll. by *by*) **a** act in accordance with (*abide by the rules*). **b** remain faithful to (a promise). **3** intr. archaic **a** remain, continue. **b** dwell. **4** tr. archaic sustain, endure. □ **abidance** n. [OE *ābīdan* (as A-², *bidan* BIDE]

abiding /əˈbaɪdɪŋ/ adj. enduring, permanent (*an abiding sense of loss*). □ **abidingly** adv.

Abidjan /ˌæbiːˈdʒɑːn/ the chief port of the Ivory Coast and capital (1935–83); pop. (est. 1982) 1,850,000.

abigail /ˈæbɪˌgeɪl/ n. a lady's maid. [character in Beaumont and Fletcher's *Scornful Lady*; cf. 1 Sam. 25]

ability /əˈbɪlɪtɪ/ n. (pl. **-ies**) **1** (often foll. by *to* + infin.) capacity or power (*has the ability to write songs*). **2** cleverness, talent; mental power (*a person of great ability*; *has many abilities*). [ME f. OF *ablete* f. L *habilitas -tatis* f. *habilis* able]

-ability /əˈbɪlɪtɪ/ suffix forming nouns of quality from, or corresponding to, adjectives in *-able* (*capability*; *vulnerability*). [F *-abilité* or L *-abilitas*: cf. -ITY]

ab initio /æb ɪˈnɪʃɪəʊ/ adv. from the beginning. [L]

abiogenesis /ˌeɪbaɪəʊˈdʒenɪsɪs/ n. the formation of organic matter without the action of living organisms; esp. the supposed spontaneous generation of living organisms. □ **abiogenic** adj. [A-¹ + Gk *bios* life + GENESIS]

Abiola /ˌæbɪˈəʊlə/, Moshood (Kashimawo Olawale) (b.1937), Nigerian politician. The leader of the Social Democratic Party, he declared himself President-elect in June 1993 after election results showed him to be on the way to a comfortable victory. The election was annulled by the ruling military regime; a year later Abiola again declared himself President, but was placed under house arrest.

abject /ˈæbdʒekt/ adj. **1** miserable, wretched. **2** degraded, self-abasing, humble. **3** despicable. □ **abjectly** adv. **abjectness** n. [ME f. L *abjectus* past part. of *abicere* (as AB-, *jacere* throw)]

abjection /əbˈdʒekʃ(ə)n/ n. a state of misery or degradation. [ME f. OF *abjection* or L *abjectio* (as ABJECT)]

abjure /əbˈdʒʊə(r)/ v.tr. **1** renounce on oath (an opinion, cause, claim, etc.). **2** swear perpetual absence from (one's country etc.). □ **abjuration** /ˌæbdʒʊˈreɪʃ(ə)n/ n. [L *abjurare* (as AB-, *jurare* swear)]

Abkhazia /æbˈkɑːzɪə/ an autonomous territory in NW Georgia, south of the Caucasus mountains on the Black Sea; pop. (1990) 537,500; capital, Sokhumi. It corresponds to the ancient region of Colchis. In 1992 Abkhazia unilaterally declared itself independent, sparking armed conflict with Georgia, and the following year drove Georgian forces from its territory. □ **Abkhazian** adj. & n.

ablation /æbˈleɪʃ(ə)n/ n. **1** the surgical removal of body tissue. **2** Geol. the wasting or erosion of a glacier, iceberg, or rock by melting or the action of water. **3** the loss of surface material from a spacecraft, meteorite, etc. through evaporation or melting caused by friction with the atmosphere. □ **ablate** /-ˈleɪt/ v.tr. & intr. [F *ablation* or LL *ablatio* f. *ablat-* (as AB-, *lat-* past part. stem of *ferre* carry)]

ablative /ˈæblətɪv/ n. & adj. Gram. ● n. the case (esp. in Latin) of nouns and pronouns (and words in grammatical agreement with them) indicating an agent, instrument, or location. ● adj. of or in the ablative. □ **ablative absolute** an absolute construction in Latin with a noun and participle or adjective in the ablative case (see ABSOLUTE 5). [ME f. OF *ablatif -ive* or L *ablativus* (as ABLATION)]

ablaut /ˈæblaʊt/ n. Linguistics a change of vowel in related words or forms. In Indo-European languages this arises from differences of accent and stress in the parent language, e.g. in *sing, sang, sung*. [G]

ablaze /əˈbleɪz/ predic.adj. & adv. **1** on fire (*set it ablaze*; *the house was ablaze*). **2** (often foll. by *with*) glittering, glowing, radiant. **3** (often foll. by *with*) greatly excited.

able /'eɪb(ə)l/ adj. (**abler, ablest**) **1** (often foll. by to + infin.; used esp. to provide an infinitive for can: to be able) having the capacity or power (was not able to come). **2** having great ability; clever, skilful. □ **able-bodied** fit, healthy. **able-bodied rating** (or **seaman**) Naut. one able to perform all duties. [ME f. OF hable, able f. L habilis handy f. habēre to hold]

-able /əb(ə)l/ suffix forming adjectives meaning: **1** that may or must be (eatable; forgivable; payable). **2** that can be made the subject of (dutiable; objectionable). **3** that is relevant to or in accordance with (fashionable; seasonable). **4** (with active sense, in earlier word-formations) that may (comfortable; suitable). [F -able or L -abilis forming verbal adjectives f. verbs of first conjugation]

abled /'eɪb(ə)ld/ adj. having a full range of physical and mental abilities; able-bodied. □ **differently abled** euphem. disabled.

ableism /'eɪbə,lɪz(ə)m/ n. discrimination in favour of the able-bodied.

abloom /ə'blu:m/ predic.adj. blooming; in flower.

ablush /ə'blʌʃ/ predic.adj. blushing.

ablution /ə'blu:ʃ(ə)n/ n. (usu. in pl.) **1** the ceremonial washing of parts of the body or sacred vessels etc. **2** colloq. the ordinary washing of the body. **3** a building containing washing-places etc. in a camp, ship, etc. □ **ablutionary** adj. [ME f. OF ablution or L ablutio (as AB-, lutio f. luere lutwash)]

ably /'eɪblɪ/ adv. capably, cleverly, competently.

-ably /əblɪ/ suffix forming adverbs corresponding to adjectives in -able.

ABM abbr. anti-ballistic missile.

abnegate /'æbnɪ,geɪt/ v.tr. **1** give up or deny oneself (a pleasure etc.). **2** renounce or reject (a right or belief). □ **abnegator** n. [L abnegare abnegat- (as AB-, negare deny)]

abnegation /,æbnɪ'geɪʃ(ə)n/ n. **1** denial; the rejection or renunciation of a doctrine. **2** = SELF-ABNEGATION. [OF abnegation or LL abnegatio (as ABNEGATE)]

abnormal /æb'nɔ:m(ə)l/ adj. **1** deviating from what is normal or usual; exceptional. **2** relating to or dealing with what is abnormal (abnormal psychology). □ **abnormally** adv. [earlier and F anormal, anomal f. Gk anōmalos ANOMALOUS, assoc. with L abnormis: see ABNORMITY]

abnormality /,æbnɔ:'mælɪtɪ/ n. (pl. **-ies**) **1 a** an abnormal quality, occurrence, etc. **b** the state of being abnormal. **2** a physical irregularity.

abnormity /æb'nɔ:mɪtɪ/ n. (pl. **-ies**) **1** an abnormality or irregularity. **2** a monstrosity. [L abnormis (as AB-, normis f. norma rule)]

Abo /'æbəʊ/ n. & adj. (also **abo**) Austral. sl. offens. ● n. (pl. **Abos**) an Aborigine. ● adj. Aboriginal. [abbr.]

Åbo /'ɔ:bu:/ the Swedish name for TURKU.

aboard /ə'bɔ:d/ adv. & prep. **1** on or into (a ship, aircraft, train, etc.). **2** alongside. □ **all aboard!** a call that warns of the imminent departure of a ship, train, etc. [ME f. A² + BOARD & F à bord]

abode¹ /ə'bəʊd/ n. **1** habitual residence; a house or home. **2** archaic a stay or sojourn. [verbal noun of ABIDE: cf. ride, rode, road]

abode² past of ABIDE.

abolish /ə'bɒlɪʃ/ v.tr. put an end to the existence or practice of (esp. a custom or institution). □ **abolishable** adj. **abolisher** n. **abolishment** n. [ME f. F abolir f. L abolere destroy]

abolition /,æbə'lɪʃ(ə)n/ n. **1** the act or process of abolishing or being abolished. **2** an instance of this. [F abolition or L abolitio (as ABOLISH)]

abolitionist /,æbə'lɪʃənɪst/ n. a person who favours the abolition of a practice or institution, esp. of capital punishment or (formerly) of slavery. □ **abolitionism** n.

abomasum /,æbə'meɪsəm/ n. (pl. **abomasa** /-sə/) the fourth stomach of a ruminant. [mod.L f. AB- + OMASUM]

A-bomb /'eɪbɒm/ n. = ATOM BOMB. [A (for ATOMIC) + BOMB]

Abomey /ə'bəʊmeɪ, ,æbə'meɪ/ a town in southern Benin, capital of the former kingdom of Dahomey; pop. (1982) 54,400.

abominable /ə'bɒmɪnəb(ə)l/ adj. **1** detestable; loathsome; morally reprehensible. **2** colloq. very bad or unpleasant (abominable weather). □ **abominable snowman** an unidentified manlike or bearlike animal said to exist in the Himalayas; a yeti. □ **abominably** adv. [ME f. OF f. L abominabilis f. abominari deprecate (as AB-, ominari f. OMEN)]

abominate /ə'bɒmɪ,neɪt/ v.tr. detest, loathe. □ **abominator** n. [L abominari (as ABOMINABLE)]

abomination /ə,bɒmɪ'neɪʃ(ə)n/ n. **1** (often foll. by to) an object of disgust. **2** an odious or degrading habit or act. **3** loathing. [ME f. OF (as ABOMINATE)]

aboral /æb'ɔ:rəl/ adj. Zool. furthest from or opposite the mouth. [AB- + ORAL]

aboriginal /,æbə'rɪdʒɪn(ə)l/ adj. & n. ● adj. **1** (of races and natural phenomena) inhabiting or existing in a land from the earliest times or from before the arrival of colonists. **2** (**Aboriginal**) of the Australian Aboriginals or their languages. ● n. **1** an aboriginal inhabitant. **2** (**Aboriginal**) an aboriginal inhabitant of Australia. (See note below.) ¶ With reference to indigenous Australians, some people prefer Aborigine and others Aboriginal as the noun. The adjective in this sense is always Aboriginal. **3** (**Aboriginal**) any of the Australian Aboriginal languages. □ **Aboriginality** /,æbə,rɪdʒɪ'nælɪtɪ/ n. [as ABORIGINE + -AL¹]

■ The Australian Aboriginals arrived in Australia between 25,000 and 40,000 years ago. Before the arrival of Europeans in the 18th century, they were scattered over the whole continent, including Tasmania, and lived as hunter-gatherers; their numbers fell dramatically following European occupation, especially as a result of European diseases such as smallpox. After the 1930s, numbers began to rise, and in 1991 their population was roughly 265,000. Their languages, estimated at several hundred in number and now mostly extinct, are related to each other but not, apparently, to any other language family. Throughout the 19th and early 20th centuries, Aboriginals were regarded as inferior and were persecuted; in more recent years, they have become politically more active and their situation with regard to ownership of land and legal representation has improved.

aborigine /,æbə'rɪdʒɪnɪ/ n. (usu. in pl.) **1** an aboriginal inhabitant. **2** (usu. **Aborigine**) = ABORIGINAL 2. ¶ In the plural, Aboriginals is preferred to Aborigines. **3** an aboriginal plant or animal. [back-form. pl. aborigines f. L, prob. f. phr. ab origine from the beginning]

abort /ə'bɔ:t/ v. & n. ● v. **1** intr. **a** (of a woman) undergo abortion; miscarry. **b** (of a foetus) suffer abortion. **2** tr. **a** effect the abortion of (a foetus). **b** effect abortion in (a mother). **3 a** tr. cause to end fruitlessly or prematurely; stop in the early stages. **b** intr. end unsuccessfully or prematurely. **4 a** tr. abandon or terminate (a space flight or other technical project) before its completion, usu. because of a fault. **b** intr. terminate or fail to complete such an undertaking. **5** Biol. **a** intr. (of an organ) remain undeveloped; shrink away. **b** tr. cause to do this. ● n. **1** a prematurely terminated space flight or other undertaking. **2** the termination of such an undertaking. [L aboriri miscarry (as AB-, oriri ort- be born)]

abortifacient /ə,bɔ:tɪ'feɪʃ(ə)nt/ adj. & n. Med. ● adj. causing abortion of a foetus. ● n. a drug or other agent that does this.

abortion /ə'bɔ:ʃ(ə)n/ n. **1** the expulsion of a foetus (naturally or esp. by medical induction) from the womb before it is able to survive independently. (See note below.) **2** a stunted or deformed creature or thing. **3** the failure of a project or an action. **4** Biol. the arrest of the development of an organ. [L abortio (as ABORT)]

■ The practice of inducing abortion to end unwanted pregnancy was widespread from ancient times, and in general was not criminalized until the 19th century. Abortion began to be legalized in many countries in the 1950s and 1960s (in the Soviet Union from 1920), but with strict controls; in the UK it is governed by the 1967 Abortion Act and permitted under certain circumstances up to the twenty-fourth week of pregnancy, although some foetuses delivered at an earlier stage may be enabled to survive. In some parts of the world, for example the Republic of Ireland, it remains illegal. Abortion is an issue which arouses great controversy between defenders of the mother's right of choice and those (particularly Roman Catholics and Christian fundamentalists) holding that human life is sacred and begins at the moment of conception.

abortionist /ə'bɔ:ʃənɪst/ n. **1** a person who carries out abortions, esp. illegally. **2** a person who favours the legalization of abortion.

abortive /ə'bɔ:tɪv/ adj. **1** fruitless, unsuccessful, unfinished. **2** resulting in abortion. **3** Biol. (of an organ etc.) rudimentary; arrested in development. □ **abortively** adv. [ME f. OF abortif -ive f. L abortivus (as ABORT)]

ABO system /,eɪbi:'əʊ/ n. a system of four groups (A, AB, B, and O) into which human blood is classified, based on the presence or absence of certain inherited antigens.

Aboukir Bay, Battle of /,æbu:'kɪə(r)/ (also **Abukir Bay**) (also called Battle of the Nile) a naval battle in 1798 off Aboukir Bay on the Mediterranean coast of Egypt at the mouth of the Nile, in which the British under Nelson defeated the French fleet.

aboulia /ə'bu:lɪə/ n. (also **abulia**) the loss of will-power as a mental disorder. □ **aboulic** adj. [Gk a- not + boulē will]

abound /ə'baʊnd/ v.intr. **1** be plentiful. **2** (foll. by in, with) be rich; teem or be infested. [ME f. OF *abunder* etc. f. L *abundare* overflow (as AB-, *undare* f. *unda* wave)]

about /ə'baʊt/ *prep. & adv.* ● *prep.* **1 a** on the subject of; in connection with (*a book about birds; what are you talking about?; argued about money*). **b** relating to (*something funny about this*). **c** in relation to (*symmetry about a plane*). **d** so as to affect (*what are you going to do about it?*). **2** at a time near to (*come about four*). **3 a** in, round, surrounding (*wandered about the town; a scarf about her neck*). **b** all round from a centre (*look about you*). **4** here and there in; at points throughout (*toys lying about the house*). **5** at a point or points near to (*fighting going on about us*). **6** carried with (*have no money about me*). **7** occupied with (*what are you about?*). ● *adv.* **1 a** approximately (*costs about a pound; is about right*). **b** *colloq.* used to indicate understatement (*just about had enough; it's about time they came*). **2** here and there; at points nearby (*a lot of flu about; I've seen him about recently*). **3** all round; in every direction (*look about; wandered about; scattered all about*). **4** on the move; in action (*out and about*). **5** in partial rotation or alteration from a given position (*the wrong way about*). **6** in rotation or succession (*turn and turn about*). **7** *Naut.* on or to the opposite tack (*go about; put about*). □ **be about to** be on the point of (doing something) (*was about to laugh*). [OE *onbūtan* (on = A², *būtan* BUT¹)]

about-face /əbaʊt'feɪs/ *n., v., & int.* ● *n. & v.intr.* = ABOUT-TURN *n. & v.* ● *int.* = ABOUT-TURN *int.*

about-turn /əbaʊt'tɜ:n/ *n., v., & int.* ● *n.* **1** a turn made so as to face the opposite direction. **2** a change of opinion or policy etc. ● *v.intr.* make an about-turn. ● *int.* (**about turn**) *Mil.* a command to make an about-turn. [orig. as *int.*]

above /ə'bʌv/ *prep., adv., adj., & n.* ● *prep.* **1** over; on the top of; higher (vertically, up a slope or stream etc.) than; over the surface of (*head above water; above the din*). **2** more than (*above twenty people; above average*). **3** higher in rank, position, importance, etc., than (*thwarted by those above him*). **4 a** too great or good for (*above one's station; is not above cheating at cards*). **b** beyond the reach of; not affected by (*above my understanding; above suspicion*). **5** *archaic* to an earlier time than (*not traced above the third century*). ● *adv.* **1** at or to a higher point; overhead (*the floor above; the clouds above*). **2 a** upstairs (*lives above*). **b** upstream. **3** (of a text reference) further back on a page or in a book (*as noted above*). **4** on the upper side (*looks similar above and below*). **5** in addition (*over and above*). **6** *rhet.* in heaven (*Lord above!*). ● *adj.* mentioned earlier; preceding (*the above argument*). ● *n.* (prec. by *the*) what is mentioned above (*the above shows*). □ **above all** most of all, more than anything else. **above-board** *adj. & adv.* without concealment; fair or fairly; open or openly. **above ground** alive. **above one's head** see HEAD. **above oneself** conceited, arrogant. [A² + OE *bufan* f. *be* = BY + *ufan* above]

ab ovo /æb 'əʊvəʊ/ *adv.* from the very beginning. [L, = from the egg]

Abp. *abbr.* Archbishop.

abracadabra /ˌæbrəkə'dæbrə/ *int. & n.* ● *int.* a supposedly magic word used by conjurors in performing a trick. ● *n.* **1** a spell or charm. **2** jargon or gibberish. [a mystical word engraved and used as a charm: L f. Gk]

abrade /ə'breɪd/ *v.tr.* scrape or wear away (skin, rock, etc.) by rubbing. □ **abrader** *n.* [L f. *radere ras-* scrape]

Abraham /'eɪbrə,hæm/ the Hebrew patriarch from whom all Jews trace their descent (Gen. 11:27–25:10). In Gen. 22, he is ordered by God to sacrifice his son Isaac as a test of faith, a command later revoked.

Abraham, Plains of see PLAINS OF ABRAHAM.

Abrahams /'eɪbrə,hæmz/, Harold (Maurice) (1899–1978), English athlete. In 1924 he became the first Englishman to win the 100 metres in the Olympic Games. His story was the subject of the film *Chariots of Fire* (1981).

abrasion /ə'breɪʒ(ə)n/ *n.* **1** the scraping or wearing away (of skin, rock, etc.). **2** a damaged area resulting from this. [L *abrasio* (as ABRADE)]

abrasive /ə'breɪsɪv/ *adj. & n.* ● *adj.* **1 a** tending to rub or graze. **b** capable of polishing by rubbing or grinding. **2** harsh or hurtful in manner. ● *n.* an abrasive substance. [as ABRADE + -IVE]

abreact /ˌæbrɪ'ækt/ *v.tr. Psychol.* release (an emotion) by abreaction. [back-form. f. ABREACTION]

abreaction /ˌæbrɪ'ækʃ(ə)n/ *n. Psychol.* the free expression and consequent release of a previously repressed emotion. □ **abreactive** /-'æktɪv/ *adj.* [AB- + REACTION after G *Abreagierung*]

abreast /ə'brest/ *adv.* **1** side by side and facing the same way. **2 a** (often foll. by *with*) up to date. **b** (foll. by *of*) well-informed (*abreast of all the changes*). [ME f. A² + BREAST]

abridge /ə'brɪdʒ/ *v.tr.* **1** shorten (a book, film, etc.) by using fewer words

or making deletions. **2** curtail (liberty). □ **abridgeable** *adj.* **abridger** *n.* [ME f. OF *abreg(i)er* f. LL *abbreviare* ABBREVIATE]

abridgement /ə'brɪdʒmənt/ *n.* (also **abridgment**) **1 a** a shortened version, esp. of a book; an abstract. **b** the process of producing this. **2** a curtailment (of rights). [F *abrégement* (as ABRIDGE)]

abroad /ə'brɔːd/ *adv.* **1** in or to a foreign country or countries. **2** over a wide area; in different directions; everywhere (*scatter abroad*). **3** at large; freely moving about; in circulation (*there is a rumour abroad*). **4** *archaic* in or into the open; out of doors. **5** *archaic* wide of the mark; erring. □ **from abroad** from another country. [ME f. A² + BROAD]

abrogate /'æbrə,geɪt/ *v.tr.* repeal, annul, or abolish (a law or custom). □ **abrogator** *n.* **abrogation** /ˌæbrə'geɪʃ(ə)n/ *n.* [L *abrogare* (as AB-, *rogare* propose a law)]

abrupt /ə'brʌpt/ *adj.* **1** sudden and unexpected; hasty (*his abrupt departure*). **2** (of speech, manner, etc.) uneven; lacking continuity; curt. **3** steep, precipitous. **4** *Bot.* truncated. □ **abruptly** *adv.* **abruptness** *n.* [L *abruptus* past part. of *abrumpere* (as AB-, *rumpere* break)]

Abruzzi /ə'brʊtsɪ/ a mountainous region of east central Italy; capital, Aquila.

ABS *abbr.* anti-lock brake (or braking) system.

abs- /əbs, æbs/ *prefix* = AB-. [var. of L *ab-* used before *c*, *q*, *t*]

abscess /'æbsɪs/ *n.* a swollen area accumulating pus within bodily tissue. □ **abscessed** *adj.* [L *abscessus* a going away (as AB-, *cedere cess-* go)]

abscisic acid /æb'saɪzɪk/ *n. Biochem.* a plant hormone which promotes leaf detachment and bud dormancy and inhibits germination. [L *abscis-* past part. stem of *abscindere* (as AB-, *scindere* to cut)]

abscissa /əb'sɪsə/ *n.* (*pl.* **abscissae** /-siː/ or **abscissas**) *Math.* (in a system of coordinates) the distance from a point to the vertical or y-axis, measured parallel to the horizontal or x-axis; the Cartesian x-coordinate of a point (cf. ORDINATE). [mod.L *abscissa (linea)* fem. past part. of *abscindere absciss-* (as AB-, *scindere* cut)]

abscission /əb'sɪʒ(ə)n/ *n.* **1** the act or an instance of cutting off. **2** *Bot.* the natural detachment of leaves, branches, flowers, etc. [L *abscissio* (as ABSCISSA)]

abscond /əb'skɒnd/ *v.intr.* depart hurriedly and furtively, esp. unlawfully or to avoid arrest. □ **absconder** *n.* [L *abscondere* (as AB-, *condere* stow)]

abseil /'æbseɪl, 'æbzaɪl/ *v. & n. Mountaineering* ● *v.intr.* descend a steep rock-face by using a doubled rope coiled round the body and fixed at a higher point. ● *n.* a descent made by abseiling. [G *abseilen* f. *ab* down + *Seil* rope]

absence /'æbs(ə)ns/ *n.* **1** the state of being away from a place or person. **2** the time or duration of being away. **3** (foll. by *of*) the non-existence or lack of. □ **absence of mind** inattentiveness. [ME f. OF f. L *absentia* (as ABSENT)]

absent *adj. & v.* ● *adj.* /'æbs(ə)nt/ **1** not present. **b** (foll. by *from*) not present at or in. **2** not existing. **3** inattentive to the matter in hand. ● *v.refl.* /əb'sent/ **1** stay away. **2** withdraw. □ **absently** /'æbs(ə)ntlɪ/ *adv.* (in sense 3 of *adj.*). [ME ult. f. L *absent-* pres. part. of *abesse* be absent]

absentee /ˌæbsən'tiː/ *n.* a person not present, esp. a person who is absent from work or school. □ **absentee landlord** a landlord who lets a property while living elsewhere.

absenteeism /ˌæbsən'tiːɪz(ə)m/ *n.* the practice of absenting oneself from work or school etc., esp. frequently or illicitly.

absent-minded /ˌæbs(ə)nt'maɪndɪd/ *adj.* habitually forgetful or inattentive; with one's mind on other things. □ **absent-mindedly** *adv.* **absent-mindedness** *n.*

absinth /'æbsɪnθ/ *n.* **1** a shrubby composite plant, *Artemisia absinthium*, or its essence. See also WORMWOOD. **2** (usu. **absinthe**) a green aniseed-flavoured potent liqueur turning milky when water is added, originally made with wormwood. [F *absinthe* f. L *absinthium* f. Gk *apsinthion* wormwood]

absit omen /ˌæbsɪt 'əʊmen/ *int.* may what is threatened not become fact. [L, = may this (evil) omen be absent]

absolute /'æbsə,luːt, -,ljuːt/ *adj. & n.* ● *adj.* **1** complete, utter, perfect (*an absolute fool; absolute bliss*). **2** unconditional, unlimited (*absolute authority*). **3** despotic; ruling arbitrarily or with unrestricted power (*an absolute monarch*). **4** (of a standard or other concept) universally valid; not admitting exceptions; not relative or comparative. **5** *Gram.* **a** (of a construction) syntactically independent of the rest of the sentence, as in *dinner being over, we left the table; let us toss for it, loser to pay.* **b** (of an adjective or transitive verb) used or usable without an expressed noun or object (e.g. *the deaf, guns kill*). **6** (of a legal decree etc.) final. ● *n. Philos.* **1** a value, standard, etc., which is objective and universally valid, not subjective or relative. **2** (prec. by *the*) **a** *Philos.* that which can exist without being related to anything else. **b** (in Christian theology) ultimate reality; God. □ **absolute alcohol** *Chem.* ethanol which is at least 99 per cent pure. **absolute magnitude** see MAGNITUDE. **absolute majority 1** a majority over all others combined. **2** more than half. **absolute pitch** *Mus.* **1** the ability to recognize the pitch of a note or produce any given note. **2** a fixed standard of pitch defined by the rate of vibration. **absolute temperature** temperature measured from absolute zero. **absolute zero** the lowest temperature theoretically possible (0 Kelvins, −273.15°C), at which the particle motion constituting heat is minimal. □ **absoluteness** *n.* [ME f. L *absolutus* past part.: see ABSOLVE]

absolutely /ˈæbsəˌluːtlɪ, -ˌljuːtlɪ/ *adv.* **1** completely, utterly, perfectly (*absolutely marvellous*; *he absolutely denies it*). **2** independently; in an absolute sense (*God exists absolutely*). **3** (foll. by *neg.*) (no or none) at all (*absolutely no chance of winning*; *absolutely nowhere*). **4** *colloq.* in actual fact; positively (*it absolutely exploded*). **5** *Gram.* in an absolute way, esp. (of a verb) without a stated object. **6** *colloq.* (used in reply) quite so; yes.

absolution /ˌæbsəˈluːʃ(ə)n, -ˈljuːʃ(ə)n/ *n.* **1** a formal release from guilt, obligation, or punishment. **2** an ecclesiastical declaration of forgiveness of sins. **3** a remission of penance. **4** forgiveness. [ME f. OF f. L *absolutio -onis* (as ABSOLVE)]

absolutism /ˈæbsəluːˌtɪz(ə)m, -ljuːˌtɪz(ə)m/ *n.* the acceptance of or belief in absolute principles in political, philosophical, ethical, or theological matters. □ **absolutist** *n. & adj.*

absolve /əbˈzɒlv/ *v.tr.* **1** (often foll. by *from, of*) **a** set or pronounce free from blame or obligation etc. **b** acquit; pronounce not guilty. **2** pardon or give absolution for (a sin etc.). [L *absolvere* (as AB-, *solvere solut-* loosen)]

absorb /əbˈsɔːb, -ˈzɔːb/ *v.tr.* **1** include or incorporate as part of itself or oneself (*the country successfully absorbed its immigrants*). **2** take in; suck up (liquid, heat, knowledge, etc.) (*she quickly absorbed all she was taught*). **3** reduce the effect or intensity of; deal easily with (an impact, sound, difficulty, etc.). **4** consume (income, time, resources, etc.) (*his debts absorbed half his income*). **5** engross the attention of (*television absorbs them completely*). □ **absorber** *n.* **absorbable** *adj.* **absorbability** /-ˌsɔːbəˈbɪlɪtɪ, -ˌzɔːbə-/ *n.* [ME f. F *absorber* or L *absorbere absorpt-* (as AB-, *sorbere* suck in)]

absorbance /əbˈsɔːb(ə)ns, -ˈzɔːb(ə)ns/ *n. Physics* a measure of the capacity of a substance to absorb light, equal to the logarithm of the reciprocal of the transmittance. [ABSORB + -ANCE]

absorbed /əbˈsɔːbd, -ˈzɔːbd/ *adj.* intensely engaged or interested (*he was absorbed in his work*). □ **absorbedly** /-bɪdlɪ/ *adv.*

absorbent /əbˈsɔːb(ə)nt, -ˈzɔːb(ə)nt/ *adj. & n.* ● *adj.* having a tendency to absorb (esp. liquids). ● *n.* an absorbent substance or thing. □ **absorbency** *n.* [L *absorbent-* f. *absorbere* ABSORB]

absorbing /əbˈsɔːbɪŋ, -ˈzɔːbɪŋ/ *adj.* engrossing; intensely interesting. □ **absorbingly** *adv.*

absorption /əbˈsɔːpʃ(ə)n, -ˈzɔːpʃ(ə)n/ *n.* **1** the process or action of absorbing or being absorbed. **2** disappearance through incorporation into something else. **3** mental engrossment. □ **absorption spectrum** a spectrum of electromagnetic radiation transmitted through a substance, with dark lines or bands showing absorption at specific wavelengths. □ **absorptive** /-ˈsɔːptɪv, -ˈzɔːp-/ *adj.* [L *absorptio* (as ABSORB)]

abstain /əbˈsteɪn/ *v.intr.* **1 a** (usu. foll. by *from*) restrain oneself; refrain from indulging in (*abstained from cakes and sweets*). **b** refrain from drinking alcohol. **2** formally decline to use one's vote. □ **abstainer** *n.* [ME f. AF *astener* f. OF *abstenir* f. L *abstinere abstent-* (as AB-, *tenere* hold)]

abstemious /æbˈstiːmɪəs/ *adj.* (of a person, habit, etc.) moderate, not self-indulgent, esp. in eating and drinking. □ **abstemiously** *adv.* **abstemiousness** *n.* [L *abstemius* (as AB-, *temetum* strong drink)]

abstention /əbˈstenʃ(ə)n/ *n.* the act or an instance of abstaining, esp. from voting. [F *abstention* or LL *abstentio -onis* (as ABSTAIN)]

abstinence /ˈæbstɪnəns/ *n.* **1** the act of abstaining, esp. from food or alcohol. **2** the habit of abstaining from pleasure, food, etc. [ME f. OF f. L *abstinentia* (as ABSTINENT)]

abstinent /ˈæbstɪnənt/ *adj.* practising abstinence. □ **abstinently** *adv.* [ME f. OF f. L (as ABSTAIN)]

abstract *adj., v., & n.* ● *adj.* /ˈæbstrækt/ **1 a** to do with or existing in thought rather than matter, or in theory rather than practice; not tangible or concrete (*abstract questions rarely concerned him*). **b** (of a word,

esp. a noun) denoting a quality or condition or intangible thing rather than a concrete object. **2** (of art) not representational; not aiming to depict a representation of external reality. (See also ABSTRACT ART.) ● *v.* /əbˈstrækt/ **1** *tr.* (often foll. by *from*) take out of; extract; remove. **2 a** *tr.* summarize (an article, book, etc.). **b** *intr.* do this as an occupation. *refl.* (often foll. by *from*) disengage (a person's attention etc.); distract. **4** *tr.* (foll. by *from*) consider abstractly or separately from something else. **5** *tr. euphem.* steal. ● *n.* /ˈæbstrækt/ **1** a summary or statement of the contents of a book etc. **2** an abstract work of art. **3** an abstraction or abstract term. □ **in the abstract** in theory rather than in practice. □ **abstractly** /ˈæbstræktlɪ/ *adv.* **abstractor** /əbˈstræktə(r)/ *n.* (in sense 2 of *v.*). [ME f. OF *abstract* or L *abstractus* past part. of *abstrahere* (as AB-, *trahere* draw)]

abstract art *n.* in its broadest sense, any art that does not attempt to represent external, recognizable reality, and which thereby tends to concentrate on shape, colour, and form. Specifically, the term relates to a 20th-century phenomenon (the first abstract works of art are generally considered to date from *c*.1910–14, in which the traditional Western view of art imitating nature was abandoned, and which is manifested in a great variety of movements and styles. From these, two general types can be distinguished: hard-edged and geometric (akin to the linear and classical) and flowing and organic (akin to the painterly and romantic).

abstracted /əbˈstræktɪd/ *adj.* inattentive to the matter in hand; preoccupied. □ **abstractedly** *adv.*

abstract expressionism *n.* a movement in American painting associated with New York in the 1940s and 1950s, whose subscribers laid particular emphasis on the creative spontaneous act and the role of the unconscious and unfettered emotion in the process of painting. Strongly influenced by surrealism, and most often associated with the spontaneous gestural style of action painting and figures such as Pollock and de Kooning, the term also refers to the vast 'all-over' canvases of colour-field painting associated with Mark Rothko, Barnett Newman, and Clyfford Still (1904–80).

abstraction /əbˈstrækʃ(ə)n/ *n.* **1** the act or an instance of abstracting or taking away. **2 a** an abstract or visionary idea. **b** the formation of abstract ideas. **3 a** abstract qualities (esp. in art). **b** an abstract work of art. **4** absent-mindedness. [F *abstraction* or L *abstractio* (as ABSTRACT)]

abstractionism /əbˈstrækʃəˌnɪz(ə)m/ *n.* **1** the principles and practice of abstract art. **2** the pursuit or cult of abstract ideas. □ **abstractionist** *n.*

abstruse /əbˈstruːs/ *adj.* hard to understand; obscure; profound. □ **abstrusely** *adv.* **abstruseness** *n.* [F *abstruse* or L *abstrusus* (as AB-, *trusus* past part. of *trudere* push)]

absurd /əbˈsɜːd/ *adj. & n.* ● *adj.* **1** (of an idea, suggestion, etc.) wildly unreasonable, illogical, or inappropriate. **2** (of a person) unreasonable or ridiculous in manner. **3** (of a thing) ludicrous, incongruous (*an absurd hat*; *the situation was becoming absurd*). ● *n.* (**the absurd**) that which is absurd, esp. human existence in a purposeless chaotic universe. □ **absurdly** *adv.* **absurdness** *n.* [F *absurde* or L *absurdus* (as AB-, *surdus* deaf, dull)]

absurdism /əbˈsɜːdɪz(ə)m/ *n.* the belief that humans exist in a purposeless chaotic universe in which attempts to impose order are frustrated, particularly as articulated by the Theatre of the Absurd. □ **absurdist** *n. & adj.*

absurdity /əbˈsɜːdɪtɪ/ *n.* (pl. **-ies**) **1** wild inappropriateness or incongruity. **2** extreme unreasonableness. **3** an absurd statement or act. [F *absurdité* or LL *absurditas* (as ABSURD)]

ABTA /ˈæbtə/ *abbr.* Association of British Travel Agents.

Abu Dhabi /ˌæbuː ˈdɑːbɪ/ **1** the largest of the seven member states of the United Arab Emirates, lying between Oman and the Gulf coast; pop. (1985) 670,125. The former sheikhdom joined the federation of the United Arab Emirates in 1971. Much of the land is desert or salt flats, and the economy is largely based on oil, discovered in the 1960s. **2** the capital of Abu Dhabi; pop. (est. 1980) 242,975. It is also the federal capital of the United Arab Emirates.

Abuja /əˈbuːdʒə/ a newly built city in central Nigeria, designated in 1982 to replace Lagos as the national capital; pop. (1991) 378,670.

Abukir Bay see ABOUKIR BAY, BATTLE OF.

abulia var. of ABOULIA.

Abu Musa /ˌæbuː ˈmuːsə/ a small island in the Persian Gulf. Formerly held by the emirate of Sharjah, it has been occupied by Iran since 1971.

abundance /əˈbʌndəns/ *n.* **1** a very great quantity, usu. considered to be more than enough. **2** wealth, affluence. **3** wealth of emotion (*abundance of heart*). **4** a call in solo whist undertaking to make nine tricks. [ME f. OF *abundance* f. L *abundantia* (as ABUNDANT)]

abundant /ə'bʌndənt/ *adj.* **1** existing or available in large quantities; plentiful. **2** (foll. by *in*) having an abundance of (*a country abundant in fruit*). □ **abundantly** *adv.* [ME f. L (as ABOUND)]

abuse *v. & n.* ● *v.tr.* /ə'bjuːz/ **1 a** use to bad effect or for a bad purpose; misuse (*abused his position of power*). **b** take (a drug) for a purpose other than a therapeutic one; be addicted to (a substance). **2** insult verbally. **3** maltreat; assault (esp. sexually). ● *n.* /ə'bjuːs/ **1 a** incorrect or improper use (*the abuse of power*). **b** an instance of this. **2** insulting language (*a torrent of abuse*). **3** unjust or corrupt practice. **4** maltreatment or (esp. sexual) assault of a person (*child abuse*). □ **abused** /ə'bjuːzd/ *adj.* **abuser** /ə'bjuːzə(r)/ *n.* [ME f. OF *abus* (n.), *abuser* (v.) f. L *abusus*, *abuti* (as AB-, *uti us-* USE)]

Abu Simbel /ˌæbuː 'sɪmb(ə)l/ the site of two huge rock-cut temples in southern Egypt, built during the reign of Ramses II in the 13th century BC, and commemorating him and his first wife Nefertari. In the 1960s, the building of the High Dam at Aswan endangered the site, and a massive archaeological salvage operation was undertaken, in which engineers carved up the monument into numbered sections and carried it up the hillside to be rebuilt.

abusive /ə'bjuːsɪv/ *adj.* **1** using or containing insulting language. **2** (of language) insulting. **3** involving or given to physical abuse. □ **abusively** *adv.* **abusiveness** *n.*

abut /ə'bʌt/ *v.* (**abutted, abutting**) **1** *intr.* (foll. by *on*) (of estates, countries, etc.) adjoin (another). **2** *intr.* (foll. by *on, against*) (of part of a building) touch or lean upon (another) with a projecting end or point (*the shed abutted on the side of the house*). **3** *tr.* abut on. [OF *abouter* (BUTT[1]) and AL *abuttare* f. OF *but* end]

abutment /ə'bʌtmənt/ *n.* **1** the lateral supporting structure of a bridge, arch, etc. **2** the point of junction between such a support and the thing supported.

abutter /ə'bʌtə(r)/ *n.* *Law* the owner of an adjoining property.

abuzz /ə'bʌz/ *adv. & adj.* in a state of excitement or hurried activity. [BUZZ]

abysmal /ə'bɪzm(ə)l/ *adj.* **1** *colloq.* extremely bad (*abysmal weather*; *the standard is abysmal*). **2** profound, utter (*abysmal ignorance*). □ **abysmally** *adv.* [archaic or poet. *abysm* = ABYSS, f. OF *abi(s)me* f. med.L *abysmus*]

abyss /ə'bɪs/ *n.* **1** a deep or seemingly bottomless chasm. **2 a** an immeasurable depth (*abyss of despair*). **b** a catastrophic situation as contemplated or feared (*his loss brought him a step nearer the abyss*). **3** (prec. by *the*) primal chaos, hell. [ME f. LL *abyssus* f. Gk *abussos* bottomless (as A-[1], *bussos* depth)]

abyssal /ə'bɪs(ə)l/ *adj.* **1** at or of the ocean depths or floor. **2** *Geol.* at or from a great depth in the earth's crust; plutonic.

Abyssinia /ˌæbɪ'sɪnɪə/ a former name for ETHIOPIA.

AC *abbr.* **1** (also **ac**) alternating current. **2** *Brit.* aircraftman. **3** before Christ. **4** Companion of the Order of Australia. **5** appellation *contrôlée*. [sense 3 f. L *ante Christum*]

Ac *symb.* *Chem.* the element actinium.

a/c *abbr.* account. [*account current*: see ACCOUNT *n.* 2, 3]

ac- /ək/ *prefix* assim. form of AD- before *c, k, q.*

-ac /æk/ *suffix* forming adjectives which are often also (or only) used as nouns (*cardiac*; *maniac*) (see also -ACAL). [F *-aque* or L *-acus* or Gk *-akos* adj. suffix]

acacia /ə'keɪʃə/ *n.* **1** a leguminous tree of the genus *Acacia*, often thorny and with yellow or white flowers; esp. *A. senegal*, which yields gum arabic. **2** (also **false acacia**) the locust tree, *Robinia pseudoacacia*, grown for ornament. [L f. Gk *akakia*]

academe /ˈækəˌdiːm/ *n.* **1 a** the world of learning. **b** universities collectively. **2** *literary* a college or university. □ **grove** (or **groves**) **of Academe** a university environment. [Gk *Akadēmos* (see ACADEMY): used by Shakespeare (*Love's Labour's Lost* I. i. 13) and Milton (*Paradise Regained* iv. 244)]

academia /ˌækə'diːmɪə/ *n.* the academic world; scholastic life. [mod.L: see ACADEMY]

academic /ˌækə'demɪk/ *adj. & n.* ● *adj.* **1 a** scholarly; to do with learning. **b** of or relating to a scholarly institution (*academic dress*). **2** abstract; theoretical; not of practical relevance. **3** *Art* conventional, over-formal. **4 a** of or concerning Plato's philosophy. **b** sceptical. ● *n.* a teacher or scholar in a university or institute of higher education. □ **academic year** a period of nearly a year reckoned from the time of the main student intake, usu. from the beginning of the autumn

term to the end of the summer term. □ **academically** *adv.* [F *académique* or L *academicus* (as ACADEMY)]

academical /ˌækə'demɪk(ə)l/ *adj. & n.* ● *adj.* belonging to a college or university. ● *n.* (in *pl.*) university costume.

academician /əˌkædə'mɪʃ(ə)n/ *n.* a member of an Academy, esp. of the Royal Academy of Arts, the Académie française, or the Russian Academy of Sciences. [F *académicien* (as ACADEMIC)]

academicism /ˌækə'demɪˌsɪz(ə)m/ *n.* (also **academism** /ə'kædəˌmɪz(ə)m/) academic principles or their application in art.

Académie française /əˌkædəmɪ frɒn'seɪz/ a French literary academy with a constant membership of forty, founded by Cardinal Richelieu in 1635. Its functions include the compilation and periodic revision of a definitive dictionary of the French language, the first edition of which appeared in 1694. Its tendency is to defend traditional literary and linguistic rules and to discourage innovation. [F, = French Academy]

Academy /ə'kædəmɪ/ the park and gymnasium on the outskirts of ancient Athens, named after the hero Academus, where Plato founded a school. The name is applied by extension to the philosophical system of Plato and to the philosophical scepticism of the school in the 3rd and 2nd centuries BC.

academy /ə'kædəmɪ/ *n.* (*pl.* **-ies**) **1 a** a place of study or training in a special field (*military academy*; *academy of dance*). **b** *hist.* a place of study. **2** (usu. **Academy**) a society or institution of distinguished scholars, artists, scientists, etc. (*Royal Academy*). **3** *Sc.* a secondary school. [F *académie* or L *academia* f. Gk *akadēmeia* f. *Akadēmos*; cf. ACADEMY]

Academy award *n.* (also called *Oscar*) any of the awards of the Academy of Motion Picture Arts and Sciences (Hollywood, US) given annually since 1928 for achievement in the film industry.

Acadia /ə'keɪdɪə/ a former French colony established in 1604 in the territory now forming Nova Scotia in Canada, and later occupying the whole of the region between the St Lawrence River and the Atlantic Ocean, lying in what is now SE Quebec, east Maine, New Brunswick, Nova Scotia, and Prince Edward Island. Acadia was contested by France and Britain until it was eventually ceded to Britain in 1763. French Acadians were deported to other parts of North America, especially Louisiana, where the name has survived as Cajun. Some French settlers eventually returned to Nova Scotia, New Brunswick, and Prince Edward Island.

Acadian /ə'keɪdɪən/ *n. & adj.* ● *n.* a native or inhabitant of Acadia, esp. a French-speaking descendant of the early French settlers in Acadia (living esp. in the Maritime Provinces of Canada or Louisiana). ● *adj.* of, relating to, or originating from Acadia.

-acal /ək(ə)l/ *suffix* forming adjectives, often used to distinguish them from nouns in *-ac* (*heliacal*; *maniacal*).

acanthocephalan /əˌkænθəʊ'sefələn, -'kefələn/ *n. & adj.* (also **acanthocephalid** /-lɪd/) *Zool.* ● *n.* a parasitic worm of the phylum Acanthocephala, with a thornlike proboscis. Also called *thorny-headed worm*. ● *adj.* of or relating to this phylum. [mod.L f. Gk *akanthos* (see ACANTHUS) + *kephalē* head]

acanthus /ə'kænθəs/ *n.* **1** a herbaceous plant or shrub of the genus *Acanthus*, with spiny leaves. **2** *Archit.* a conventionalized representation of an acanthus leaf, used esp. as a decoration for Corinthian column capitals. [L f. Gk *akanthos* f. *akantha* thorn perh. f. *akē* sharp point]

a cappella /ˌɑː kə'pelə/ *adj. & adv.* (also **alla cappella** /ˌælə/) *Mus.* (of choral music) unaccompanied. [It., = in church style]

Acapulco /ˌækə'pʊlkəʊ/ (in full **Acapulco de Juárez** /deɪ 'hwɑːrez/) a port and resort in southern Mexico, on the Pacific coast; pop. (1990) 592,290.

acaricide /ə'kærɪˌsaɪd/ *n.* a preparation for destroying mites.

acarid /'ækərɪd/ *n.* a small arachnid of the order Acarina, which includes mites and ticks. □ **acarology** /ˌækə'rɒlədʒɪ/ *n.* [mod.L *Acarus* f. Gk *akari* mite + -ID[3]]

ACAS /'eɪkæs/ *abbr.* (in the UK) Advisory, Conciliation, and Arbitration Service, a body set up in 1975 to provide facilities for arbitration, conciliation, and mediation in industrial disputes.

Accadian var. of AKKADIAN.

accede /æk'siːd/ *v.intr.* (often foll. by *to*) **1** take office, esp. become monarch. **2** assent or agree (*acceded to the proposal*). **3** (foll. by *to*) formally subscribe to a treaty or other agreement. [ME f. L *accedere* (as AC-, *cedere cess-* go)]

accelerando /əkˌselə'rændəʊ, əˌtʃel-/ *adv., adj., & n. Mus.* ● *adv. & adj.* with a gradual increase of speed. ● *n.* (*pl.* **accelerandos** or **accelerandi** /-dɪ/) a passage performed accelerando. [It.]

accelerate /ək'selə,reɪt/ *v.* **1** *intr.* **a** (of a moving body, esp. a vehicle) move or begin to move more quickly; increase speed. **b** (of a process) happen or reach completion more quickly. **2** *tr.* **a** cause to increase speed. **b** cause (a process) to happen more quickly. [L *accelerare* (as AC-, *celerare* f. *celer* swift)]

acceleration /ək,selə'reɪʃ(ə)n/ *n.* **1** the process or act of accelerating or being accelerated. **2** an instance of this. **3** (of a vehicle etc.) the capacity to gain speed (*the car has good acceleration*). **4** *Physics* the rate of change of velocity per unit of time. [F *accélération* or L *acceleratio* (as ACCELERATE)]

accelerative /ək'selərətɪv/ *adj.* tending to increase speed; quickening.

accelerator /ək'selə,reɪtə(r)/ *n.* **1** a device for increasing speed, esp. the pedal that controls the speed of a vehicle's engine. **2** *Physics* = PARTICLE ACCELERATOR. **3** *Chem.* a substance that speeds up a chemical reaction.

accelerometer /ək,selə'rɒmɪtə(r)/ *n.* an instrument for measuring the acceleration experienced by a moving body. [ACCELERATE + -METER]

accent *n. & v.* ● *n.* /'æks(ə)nt, -sent/ **1** a particular mode of pronunciation, esp. one associated with a particular region or group (*Liverpool accent; German accent; upper-class accent*). **2** prominence given to a syllable by stress or pitch. **3** a mark on a letter or word to indicate pitch, stress, or the quality of a vowel. **4** a distinctive feature or emphasis (*an accent on comfort*). **5** *Mus.* emphasis on a particular note or chord. ● *v.tr.* /ək'sent/ **1** pronounce with an accent; emphasize (a word or syllable). **2** write or print accents on (words etc.). **3** accentuate. **4** *Mus.* play (a note etc.) with an accent. □ **accentual** /ək'sentʃʊəl, -tjʊəl/ *adj.* [L *accentus* (as AC-, *cantus* song) repr. Gk *prosōidia* (PROSODY), or through F *accent, accenter*]

accentor /ək'sentə(r)/ *n.* a songbird of the genus *Prunella*, e.g. the dunnock. [med.L *accentor* f. L ad to + *cantor* singer]

accentuate /ək'sentʃʊ,eɪt, -'sentjʊ,eɪt/ *v.tr.* emphasize; make prominent. □ **accentuation** /-,sentʃʊ'eɪʃ(ə)n, -,sentjʊ-/ *n.* [med.L *accentuare accentuat-* (as ACCENT)]

accept /ək'sept/ *v.tr.* **1** (also *absol.*) consent to receive (a thing offered). **2** (also *absol.*) give an affirmative answer to (an offer or proposal). **3** regard favourably; treat as welcome (*her mother-in-law never accepted her*). **4 a** believe, receive (an opinion, explanation, etc.) as adequate, valid, or correct. **b** be prepared to subscribe to (a belief, philosophy, etc.). **5** receive as suitable (*the hotel accepts traveller's cheques; the machine only accepts tokens*). **6 a** tolerate; submit to (*accepted the umpire's decision*). **b** (often foll. by *that* + clause) be willing to believe (*we accept that you meant well*). **7** undertake (an office or responsibility). **8** agree to meet (a draft or bill of exchange). □ **accepter** *n.* [ME f. OF *accepter* or L *acceptare* f. *accipere* (as AC-, *capere* take)]

acceptable /ək'septəb(ə)l/ *adj.* **1 a** worthy of being accepted. **b** pleasing, welcome. **2** adequate, satisfactory. **3** tolerable (*an acceptable risk*). □ **acceptably** *adv.* **acceptableness** *n.* **acceptability** /-,septə'bɪlɪtɪ/ *n.* [ME f. OF f. LL *acceptabilis* (as ACCEPT)]

acceptance /ək'septəns/ *n.* **1** willingness to receive (a gift, payment, duty, etc.). **2** an affirmative answer to an invitation or proposal. **3** (often foll. by *of*) a willingness to accept (conditions, a circumstance, etc.). **4 a** approval, belief (*found wide acceptance*). **b** willingness or ability to tolerate. **5 a** agreement to meet a bill of exchange. **b** a bill so accepted. [F f. *accepter* (as ACCEPT)]

acceptant /ək'septənt/ *adj.* (foll. by *of*) willingly accepting. [F (as ACCEPTANCE)]

acceptation /,æksep'teɪʃ(ə)n/ *n.* a particular sense, or the generally recognized meaning, of a word or phrase. [ME f. OF f. med.L *acceptatio* (as ACCEPT)]

acceptor /ək'septə(r)/ *n.* **1** *Commerce* a person who accepts a bill. **2** *Physics* an atom or molecule able to receive an extra electron, esp. an impurity in a semiconductor. **3** *Chem.* a molecule or ion etc. to which electrons are donated in the formation of a bond. **4** *Electr.* a circuit tuned to resonate at a given frequency.

access /'ækses/ *n. & v.* ● *n.* **1** a way of approaching or reaching or entering (*a building with rear access*). **2 a** (often foll. by *to*) the right or opportunity to reach or use or visit; admittance (*has access to secret files; was granted access to the prisoner*). **b** the condition of being readily approached; accessibility. **3** (often foll. by *of*) an attack or outburst (*an access of anger*). **4** (*attrib.*) *Brit.* (of broadcasting) undertaken by minority or special-interest groups by arrangement with broadcasting companies (*access television*). ● *v.tr.* **1** *Computing* gain access to (data, a file, etc.).

2 accession. □ **access road 1** a road giving access to a place, site, or building. **2** a slip-road. **access time** *Computing* the time taken to retrieve data from storage. [ME f. OF *acces* or L *accessus* f. *accedere* (as AC-, *cedere cess-* go)]

accessary var. of ACCESSORY.

accessible /ək'sesɪb(ə)l/ *adj.* (often foll. by *to*) **1** that can readily be reached, entered, used, or understood. **2** (of a person) readily available (esp. to subordinates). □ **accessibly** *adv.* **accessibility** /-,sesɪ'bɪlɪtɪ/ *n.* [F *accessible* or LL *accessibilis* (as ACCEDE)]

accession /ək'seʃ(ə)n/ *n. & v.* ● *n.* **1** entering upon an office (esp. a throne) or a condition. **2** (often foll. by *to*) a thing added (e.g. a book to a library); increase, addition. **3** *Law* the incorporation of one item of property in another. **4** assent; the formal acceptance of a treaty etc. ● *v.tr.* record the addition of (a new item) to a library or museum. [F *accession* or L *accessio -onis* (as ACCEDE)]

accessorize /ək'sesə,raɪz/ *v.tr.* provide (an outfit etc.) with accessories.

accessory /ək'sesərɪ/ *n. & adj.* (also **accessary**) ● *n.* (*pl.* **-ies**) **1** an additional or extra thing. **2** (usu. in *pl.*) **a** a small attachment or fitting. **b** a small item of (esp. a woman's) dress (e.g. shoes, gloves, handbag). **3** (often foll. by *to*) a person who helps in or knows the details of an (esp. illegal) act, without taking part in it. ● *adj.* additional; contributing or aiding in a minor way; dispensable. □ **accessory before** (or **after**) **the fact** a person who incites (or assists) another to commit a crime. □ **accessorial** /,ækse'sɔːrɪəl/ *adj.* [med.L *accessorius* (as ACCEDE)]

acciaccatura /ə,tʃækə'tʊərə/ *n. Mus.* a grace-note performed as quickly as possible before an essential note of a melody. [It.]

accidence /'æksɪd(ə)ns/ *n.* the part of grammar that deals with variable parts or inflections of words. [med.L sense of L *accidentia* (transl. Gk *parepomena*) neut. pl. of *accidens* (as ACCIDENT)]

accident /'æksɪd(ə)nt/ *n.* **1** an event that is without apparent cause, or is unexpected (*their early arrival was just an accident*). **2** an unfortunate event, esp. one causing physical harm or damage, brought about unintentionally. **3** occurrence of things by chance; the working of fortune (*accident accounts for much in life*). **4** *colloq.* an occurrence of involuntary urination or defecation. **5** an irregularity in structure. □ **accident-prone** (of a person) subject to frequent accidents. **by accident** unintentionally. [ME f. OF f. LL *accidens* f. L *accidere* (as AC-, *cadere* fall)]

accidental /,æksɪ'dent(ə)l/ *adj. & n.* ● *adj.* **1** happening by chance, unintentionally, or unexpectedly. **2** not essential to a conception; subsidiary. ● *n.* **1** *Mus.* a sign indicating a momentary departure from the key signature by raising or lowering a note. **2** something not essential to a conception. **3** = VAGRANT *n.* 2. □ **accidentally** *adv.* [ME f. LL *accidentalis* (as ACCIDENT)]

accidie /'æksɪ,di:/ *n.* **1** laziness, sloth, apathy. **2** black despair. [ME f. AF *accidie* f. OF *accide* f. med.L *accidia* alteration of LL ACEDIA]

accipiter /æk'sɪpɪtə(r)/ *n.* a short-winged, long-legged hawk of the genus *Accipiter*, e.g. a sparrowhawk, a goshawk. [L, = hawk, bird of prey]

acclaim /ə'kleɪm/ *v. & n.* ● *v.tr.* **1** welcome or applaud enthusiastically; praise publicly. **2** (foll. by complement) hail as (*acclaimed him king; was acclaimed the winner*). ● *n.* **1** applause; welcome; public praise. **2** a shout of acclaim. [ME f. L *acclamare* (as AC-, *clamare* shout: spelling assim. to *claim*)]

acclamation /,æklə'meɪʃ(ə)n/ *n.* **1** loud and eager assent to a proposal. **2** (usu. in *pl.*) shouting in a person's honour. **3** the act or process of acclaiming. □ **by acclamation** *N. Amer. Polit.* (elected) unanimously and without ballot. [L *acclamatio* (as ACCLAIM)]

acclimate /'æklɪ,meɪt, ə'klaɪ-/ *v.tr.* **1** esp. *US* acclimatize. **2** *Biol.* adapt physiologically to environmental stress. [F *acclimater* f. *à* to + *climat* CLIMATE]

acclimation /,æklɪ'meɪʃ(ə)n, ,æklaɪ-/ *n.* **1** esp. *US* acclimatization. **2** *Biol.* physiological adaptation to environmental stress. [irreg. f. ACCLIMATE]

acclimatize /ə'klaɪmə,taɪz/ *v.* (also **-ise**) **1** *tr.* accustom to a new climate or to new conditions. **2** *intr.* become acclimatized. □ **acclimatization** /ə,klaɪmətaɪ'zeɪʃ(ə)n/ *n.* [F *acclimater*: see ACCLIMATE]

acclivity /ə'klɪvɪtɪ/ *n.* (*pl.* **-ies**) an upward slope. □ **acclivitous** *adj.* [L *acclivitas* f. *acclivis* (as AC-, *clivis* f. *clivus* slope)]

accolade /'ækə,leɪd/ *n.* **1** the awarding of praise; an acknowledgement of merit. **2** a touch made with a sword at the bestowing of a knighthood. [F f. Prov. *acolada* (as AC-, L *collum* neck)]

accommodate /əˈkɒməˌdeɪt/ *v.tr.* **1** provide lodging or room for (*the flat accommodates three people*). **2** adapt, harmonize, reconcile (*must accommodate ourselves to new surroundings; cannot accommodate your needs to mine*). **3 a** do a service or favour to; oblige (a person). **b** (foll. by *with*) supply (a person) with. [L *accommodare* (as AC-, *commodus* fitting)]

accommodating /əˈkɒməˌdeɪtɪŋ/ *adj.* obliging, compliant. □ **accommodatingly** *adv.*

accommodation /əˌkɒməˈdeɪʃ(ə)n/ *n.* **1** (in *sing.* or *US* in *pl.*) room for receiving people, esp. a place to live or lodgings. **2 a** an adjustment or adaptation to suit a special or different purpose. **b** a convenient arrangement; a settlement or compromise. **3** the act or process of accommodating or being accommodated. **4** the automatic adjustment of the focus of the eye by flattening or thickening the lens. □ **accommodation address** an address used on letters to a person who is unable or unwilling to give a permanent address. **accommodation bill** a bill to raise money on credit. **accommodation ladder** a ladder up the side of a ship from a small boat. **accommodation road** a road for access to a place not on a public road. [F *accommodation* or L *accommodatio -onis* (as ACCOMMODATE)]

accompaniment /əˈkʌmpənɪmənt/ *n.* **1** *Mus.* an instrumental or orchestral part supporting or partnering a solo instrument, voice, or group. **2** an accompanying thing; an appendage. [F *accompagnement* (as ACCOMPANY)]

accompanist /əˈkʌmpənɪst/ *n.* a person who provides a musical accompaniment.

accompany /əˈkʌmpənɪ/ *v.tr.* (**-ies, -ied**) **1** go with; escort, attend. **2** (usu. in *passive*; foll. by *with*, *by*) **a** be done or found with; supplement (*speech accompanied with gestures*). **b** have as a result (*pills accompanied by side effects*). **3** *Mus.* support or partner with accompaniment. [ME f. F *accompagner* f. *à* to + OF *compaing* COMPANION¹: assim. to COMPANY]

accomplice /əˈkʌmplɪs, əˈkɒm-/ *n.* a partner or helper, esp. in a crime or wrongdoing. [ME and F *complice* (prob. by assoc. with ACCOMPANY), f. LL *complex complicis* confederate (cf. COMPLICATE)]

accomplish /əˈkʌmplɪʃ, əˈkɒm-/ *v.tr.* perform; complete; succeed in doing. □ **accomplishable** *adj.* [ME f. OF *acomplir* f. L *complere* COMPLETE]

accomplished /əˈkʌmplɪʃt, əˈkɒm-/ *adj.* clever, skilled; well trained or educated.

accomplishment /əˈkʌmplɪʃmənt, əˈkɒm-/ *n.* **1** the fulfilment or completion (of a task etc.). **2** an acquired skill, esp. a social one. **3** a thing done or achieved.

accord /əˈkɔːd/ *v.* & *n.* ● *v.* **1** *intr.* (often foll. by *with*) (esp. of a thing) be in harmony; be consistent. **2** *tr.* **a** grant (permission, a request, etc.). **b** give (a welcome etc.). ● *n.* **1** agreement, consent. **2** harmony or harmonious correspondence in pitch, tone, colour, etc. **3** a formal treaty or agreement. □ **in accord** of one mind, united; in harmony. **of one's own accord** on one's own initiative; voluntarily. **with one accord** unanimously; in a united way. [ME f. OF *acord, acorder* f. L *cor cordis* heart]

accordance /əˈkɔːd(ə)ns/ *n.* harmony, agreement. □ **in accordance with** in a manner corresponding to (*we acted in accordance with your wishes*). [ME f. OF *acordance* (as ACCORD)]

accordant /əˈkɔːd(ə)nt/ *adj.* (often foll. by *with*) in tune; agreeing. [ME f. OF *acordant* (as ACCORD)]

according /əˈkɔːdɪŋ/ *adv.* **1** (foll. by *to*) **a** as stated by or in (*according to my sister; according to their statement*). **b** in a manner corresponding to; in proportion to (*he lives according to his principles*). **2** (foll. by *as* + clause) in a manner or to a degree that varies as (*he pays according as he is able*).

accordingly /əˈkɔːdɪŋlɪ/ *adv.* **1** as suggested or required by the (stated) circumstances (*silence is vital so please act accordingly*). **2** consequently, therefore (*accordingly, he left the room*).

accordion /əˈkɔːdɪən/ *n.* **1** a portable musical instrument with reeds blown by bellows and played by means of keys and buttons. **2** (*attrib.*) folding like the bellows of an accordion (*accordion pleat; accordion wall*). □ **accordionist** *n.* [G *Akkordion* from It. *accordare* to tune]

accost /əˈkɒst/ *v.tr.* **1** approach and address (a person), esp. boldly. **2** (of a prostitute) solicit. [F *accoster* f. It. *accostare* ult. f. L *costa* rib: see COAST]

accouchement /əˈkuːʃmɒn/ *n.* **1** childbirth. **2** the period of childbirth. [F f. *accoucher* act as midwife]

accoucheur /ˌækuːˈʃɜː(r)/ *n.* a male midwife. [F (as ACCOUCHEMENT)]

account /əˈkaʊnt/ *n.* & *v.* ● *n.* **1** a narration or description (*gave a long account of the ordeal*). **2 a** an arrangement or facility at a bank or building society etc. for commercial or financial transactions, esp. for depositing and withdrawing money (*opened an account*). **b** the assets credited by such an arrangement (*has a large account; paid the money into her account*). **c** an arrangement at a shop for buying goods on credit (*has an account at the newsagent's*). **3 a** (often in *pl.*) a record or statement of money, goods, or services received or expended, with the balance (*firms must keep detailed accounts*). **b** (in *pl.*) the practice of accounting or reckoning (*is good at accounts*). **4** a statement of the administration of money in trust (*demand an account*). **5** the period during which transactions take place on a stock exchange; the period from one account day to the next. **6** counting, reckoning. **7** estimation, importance; consideration. ● *v.tr.* (foll. by *to be* or complement) consider, regard as (*account it a misfortune; account him wise; account him to be guilty*). ¶ Use with *as* (*we accounted him as wise*) is considered incorrect. □ **account day** a day of periodic settlement of stock exchange accounts. **account executive** a business executive, esp. in advertising, who manages a client's account. **account for 1** serve as or provide an explanation or reason for (*that accounts for their misbehaviour*). **2 a** give a reckoning of or answer for (money etc. entrusted). **b** answer for (one's conduct). **3** succeed in killing, destroying, disposing of, or defeating. **4** supply or make up a specified amount or proportion of (*rent accounts for 50% of expenditure*). **account rendered** a bill which has been sent but is not yet paid. **accounts payable** money owed by a company. **accounts receivable** money owed to a company. **by all accounts** in everyone's opinion. **call to account** require an explanation from (a person). **give a good** (or **bad**) **account of oneself** make a favourable (or unfavourable) impression; be successful (or unsuccessful). **keep account of** keep a record of; follow closely. **leave out of account** fail or decline to consider. **money of account** denominations of money used in reckoning, but not current as coins. **of no account** unimportant. **of some account** important. **on account 1** (of goods) to be paid for later. **2** (of money) in part payment. **on account of** because of. **on no account** under no circumstances; certainly not. **on one's own account** for one's own purposes; at one's own risk. **on a person's account** for a person's benefit (*don't do it on my account*). **settle** (or **square**) **accounts with 1** receive or pay money etc. owed to. **2** have revenge on. **take account of** (or **take into account**) consider along with other factors (*took their age into account*). **turn to account** (or **good account**) turn to one's advantage. [ME f. OF *acont, aconter* (as AC-, *conter* COUNT¹)]

accountable /əˈkaʊntəb(ə)l/ *adj.* **1** responsible; required to account for (one's conduct) (*accountable for one's actions*). **2** explicable, understandable. □ **accountably** *adv.* **accountableness** *n.* **accountability** /əˌkaʊntəˈbɪlɪtɪ/ *n.*

accountancy /əˈkaʊntənsɪ/ *n.* the profession or duties of an accountant.

accountant /əˈkaʊntənt/ *n.* a professional keeper or inspector of accounts. [legal F f. pres. part. of OF *aconter* ACCOUNT]

accounting /əˈkaʊntɪŋ/ *n.* **1** the process of or skill in keeping and verifying accounts. **2** in senses of ACCOUNT *v.*

accoutre /əˈkuːtə(r)/ *v.tr.* (*US* **accouter**) (usu. as **accoutred** *adj.*) attire, equip, esp. with a special costume. [F *accoutrer* f. OF *acoustrer* (as A-³, *cousture* sewing: cf. SUTURE)]

accoutrement /əˈkuːtrəmənt, -təmənt/ (*US* **accouterment** /-təmənt/) (usu. in *pl.*) **1** equipment, trappings. **2** *Mil.* a soldier's outfit other than weapons and garments. [F (as ACCOUTRE)]

Accra /əˈkrɑː/ the capital of Ghana, a port on the Gulf of Guinea; pop. (1984) 867,460 (pop. of Greater Accra 1,431,100).

accredit /əˈkredɪt/ *v.tr.* (**accredited, accrediting**) **1** (foll. by *to*) attribute (a saying etc.) to (a person). **2** (foll. by *with*) credit (a person) with (a saying etc.). **3** (usu. foll. by *to* or *at*) send (an ambassador etc.) with credentials; recommend by documents as an envoy (*was accredited to the sovereign*). **4** gain belief or influence for or make credible (an adviser, a statement, etc.). □ **accreditation** /əˌkredɪˈteɪʃ(ə)n/ *n.* [F *accréditer* (as AC-, *crédit* CREDIT)]

accredited /əˈkredɪtɪd/ *adj.* **1** (of a person or organization) officially recognized. **2** (of a belief) generally accepted; orthodox. **3** (esp. of cattle or milk) having guaranteed quality.

accrete /əˈkriːt/ *v.* **1** *intr.* grow together or into one. **2** *intr.* (often foll. by *to*) form round or on, as round a nucleus. **3** *tr.* attract (such additions). [L *accrescere* (as AC-, *crescere cret-* grow]

accretion /əˈkriːʃ(ə)n/ *n.* **1 a** the addition of external matter or things. **b** growth by this means. **c** a thing formed by such growth. **2** an extraneous addition. **3** the growth or cohesion of separate things into one. **4** growth by organic enlargement. **5** *Law* **a** = ACCESSION *n.* 3. **b** the increase of a legacy etc. by the share of a failing co-legatee. □ **accretive** /əˈkriːtɪv/ *adj.* [L *accretio* (as ACCRETE)]

accrue /əˈkruː/ *v.intr.* (**accrues, accrued, accruing**) (often foll. by *to*) come as a natural increase or advantage, esp. financial. □ **accrual** *n.* **accrued** *adj.* **accruement** *n.* [ME f. AF *acru(e)*, past part. of *acreistre* increase f. L *accrescere* ACCRETE]

acculturate /əˈkʌltʃəˌreɪt/ *v.* **1** *intr.* adapt to or adopt a different culture. **2** *tr.* cause to do this. □ **acculturative** /-rətɪv/ *adj.* **acculturation** /əˌkʌltʃəˈreɪʃ(ə)n/ *n.*

accumulate /əˈkjuːmjʊˌleɪt/ *v.* **1** *tr.* **a** acquire an increasing number or quantity of; heap up. **b** produce or acquire (a resulting whole) in this way. **2** *intr.* grow numerous or considerable; form an increasing mass or quantity. [L *accumulare* (as AC-, *cumulus* heap)]

accumulation /əˌkjuːmjʊˈleɪʃ(ə)n/ *n.* **1** the act or process of accumulating or being accumulated. **2** an accumulated mass. **3** the growth of capital by continued interest. [L *accumulatio* (as ACCUMULATE)]

accumulative /əˈkjuːmjʊlətɪv/ *adj.* **1** arising from accumulation; cumulative (*accumulative evidence*). **2** arranged so as to accumulate. **3** acquisitive; given to hoarding. □ **accumulatively** *adv.*

accumulator /əˈkjuːmjʊˌleɪtə(r)/ *n.* **1** *Brit.* a rechargeable electric cell. **2** a bet placed on a sequence of events, the winnings and stake from each being placed on the next. **3** a register in a computer used to contain the results of an operation. **4** a person who accumulates things.

accuracy /ˈækjʊrəsɪ/ *n.* exactness or precision, esp. arising from careful effort.

accurate /ˈækjʊrət/ *adj.* **1** careful, precise; lacking errors. **2** conforming exactly with a qualitative standard, physical or quantitative target, etc. □ **accurately** *adv.* **accurateness** *n.* [L *accuratus* done carefully, past part. of *accurare* (as AC-, *cura* care)]

accursed /əˈkɜːsɪd, əˈkɜːst/ *adj.* (*archaic* **accurst** /əˈkɜːst/) **1** lying under a curse; ill-fated. **2** *colloq.* detestable, annoying. [past part. of *accurse*, f. A-² + CURSE]

accusal /əˈkjuːz(ə)l/ *n.* accusation.

accusation /ˌækjʊˈzeɪʃ(ə)n/ *n.* **1** the act or process of accusing or being accused. **2** a statement charging a person with an offence or crime. [ME f. OF f. L *accusatio -onis* (as ACCUSE)]

accusative /əˈkjuːzətɪv/ *n. & adj. Gram.* ● *n.* **1** a case of nouns, pronouns, and adjectives that identifies the object of an action or the goal of motion. **2** a noun in this case. ● *adj.* **1** of or in this case. **2** denoting a language in which the object of a verb is typically the recipient of an action. Cf. ERGATIVE. [ME f. OF *accusatif -ive* or L (*casus*) *accusativus*, transl. Gk (*ptōsis*) *aitiatikē*]

accusatorial /əˌkjuːzəˈtɔːrɪəl/ *adj. Law* (of proceedings) involving accusation by a prosecutor and a verdict reached by an impartial judge or jury (opp. INQUISITORIAL 3). [L *accusatorius* (as ACCUSE)]

accusatory /əˈkjuːzətərɪ/ *adj.* (of language, manner, etc.) of or implying accusation.

accuse /əˈkjuːz/ *v.tr.* **1** (foll. by *of*) charge (a person etc.) with a fault or crime; indict (*accused them of murder; was accused of stealing a car*). **2** lay the blame on. □ **the accused** the person or persons charged with a crime. □ **accuser** *n.* **accusingly** *adv.* [ME *acuse* f. OF *ac(c)user* f. L *accusare* (as AC-, CAUSE)]

accustom /əˈkʌstəm/ *v.tr. & refl.* (foll. by *to*) make (a person or thing or oneself) used to (*the army accustomed him to discipline*). [ME f. OF *acostumer* (as AD-, *custume* CUSTOM)]

accustomed /əˈkʌstəmd/ *adj.* **1** (usu. foll. by *to*) used to (*accustomed to hard work*). **2** customary, usual.

AC/DC *abbr.* **1** alternating current/direct current. **2** *sl.* bisexual.

ace /eɪs/ *n. & adj.* ● *n.* **1 a** a playing card, domino, etc., with a single spot and generally having the value 'one' or in card-games the highest value in each suit. **b** a single spot on a playing card etc. **2 a** a person who excels in some activity. **b** *Aeron.* a pilot who has shot down many enemy aircraft. **3 a** *Tennis* a service that is too good for the opponent to return. **b** a point scored in this way. ● *adj. sl.* excellent. □ **ace up one's sleeve** (*N. Amer.* **in the hole**) something effective kept in reserve. **play one's ace** use one's best resource. **within an ace of** on the verge of. [ME f. OF f. L *as* unity, AS²]

-acea /ˈeɪʃə/ *suffix* forming the plural names of orders and classes of animals (*Crustacea*) (cf. -ACEAN). [neut. pl. of L adj. suffix *-aceus* of the nature of]

-aceae /ˈeɪsɪˌiː/ *suffix* forming the plural names of families of plants (*Rosaceae*). [fem. pl. of L adj. suffix *-aceus* of the nature of]

-acean /ˈeɪʃ(ə)n/ *suffix* **1** forming adjectives, = -ACEOUS. **2** forming nouns as the sing. of names in *-acea* (*crustacean*). [L *-aceus*: see -ACEA]

acedia /əˈsiːdɪə/ *n.* = ACCIDIE. [LL *acedia* f. Gk *akēdia* listlessness]

Aceldama /əˈkeldəmə/ *n.* (in the New Testament) a field near ancient Jerusalem purchased for a cemetery with the blood-money received by Judas (Matt. 27:8, Acts 1:19). [Aram., = field of blood]

acellular /eɪˈseljʊlə(r)/ *adj. Biol.* **1** not divided into cells. **2** (esp. of protozoa) consisting of one cell; unicellular.

-aceous /ˈeɪʃəs/ *suffix* forming adjectives, esp. from nouns in *-acea*, *-aceae* (*herbaceous; rosaceous*). [L *-aceus*: see -ACEA]

acephalous /əˈsefələs, əˈkef-/ *adj.* **1** headless. **2** having no governing head or chief. **3** *Zool.* having no part of the body specially organized as a head. **4** *Prosody* lacking a syllable or syllables in the first foot. [med.L *acephalus* f. Gk *akephalos* headless (as A-¹, *kephalē* head)]

acer /ˈeɪsə(r)/ *n.* a tree or shrub of the genus *Acer*, which includes the maples and the European sycamore. [L, = maple]

acerb /əˈsɜːb/ *adj.* = ACERBIC.

acerbic /əˈsɜːbɪk/ *adj.* **1** astringently sour; harsh-tasting. **2** bitter in speech, manner, or temper. □ **acerbically** *adv.* **acerbity** *n.* (pl. **-ies**). [L *acerbus* sour-tasting]

acetabulum /ˌæsɪˈtæbjʊləm/ *n.* (pl. **acetabula** /-lə/) **1** *Anat.* the socket of the hip-bone, with which the femur articulates. **2** *Zool.* a cup-shaped sucker of a tapeworm, cuttlefish, or other animal. [ME f. L, = vinegar cup f. *acetum* vinegar + *-abulum* dimin. of *-abrum* holder]

acetal /ˈæsɪˌtæl/ *n. Chem.* a compound formed by the condensation of two alcohol molecules with an aldehyde molecule; specifically, an odoriferous liquid (chem. formula: $CH_3CH(OC_2H_5)_2$) formed by reaction of acetaldehyde and ethanol. [as ACETIC + -AL²]

acetaldehyde /ˌæsɪˈtældɪˌhaɪd/ *n.* a colourless volatile liquid aldehyde (chem. formula: CH_3CHO). Also called *ethanal*. [ACETIC + ALDEHYDE]

acetaminophen /əˌsetəˈmɪnəˌfen, ˌæsɪt-/ *n. N. Amer.* = PARACETAMOL. [*para-acetylaminophenol* (cf. PARACETAMOL)]

acetate /ˈæsɪˌteɪt/ *n.* **1** a salt or ester of acetic acid, esp. the cellulose ester used to make textiles, gramophone records, etc. Also called *ethanoate*. **2** a fabric made from cellulose acetate. □ **acetate fibre** (or **silk**) fibre (or silk) made artificially from cellulose acetate. [ACETIC + -ATE¹]

acetic /əˈsiːtɪk/ *adj.* of or like vinegar or acetic acid. □ **acetic acid** the clear liquid acid that gives vinegar its characteristic taste (chem. formula: CH_3COOH); also called *ethanoic acid*. [F *acétique* f. L *acetum* vinegar]

aceto- /ˈæsɪtəʊ/ *comb. form Chem.* acetic, acetyl.

acetone /ˈæsɪˌtəʊn/ *n.* a colourless volatile liquid ketone (chem. formula: CH_3COCH_3), used as an organic solvent. Also called *propanone*. [ACETO- + -ONE]

acetous /ˈæsɪtəs/ *adj.* **1** having the qualities of vinegar. **2** producing vinegar. **3** sour. [LL *acetosus* sour (as ACETIC)]

acetyl /ˈæsɪˌtaɪl, -tɪl/ *n. Chem.* the monovalent radical of acetic acid (chem. formula: CH_3CO-). □ **acetyl silk** = acetate silk. [ACETIC + -YL]

acetylcholine /ˌæsɪtaɪlˈkəʊliːn, ˌæsɪtɪl-/ *n. Biochem.* a compound which is the one of the main neurotransmitters in the body. [ACETYL + CHOLINE]

acetylene /əˈsetɪˌliːn/ *n.* a colourless hydrocarbon gas (chem. formula: C_2H_2), which burns with a bright flame, used in welding and formerly in lighting. Also called *ethyne*. [ACETIC + -YL + -ENE]

acetylide /əˈsetɪˌlaɪd/ *n. Chem.* a saltlike compound of acetylene and a metal.

acetylsalicylic acid /ˌæsɪtaɪlˌsælɪˈsɪlɪk, ˌæsɪtɪl-/ *n.* = ASPIRIN. [ACETYL + SALICYLIC ACID]

Achaea /əˈkiːə/ a region of ancient Greece on the north coast of the Peloponnese.

Achaean /əˈkiːən/ *adj. & n.* ● *adj.* **1** of or relating to Achaea. **2** *literary* (esp. in Homeric contexts) Greek. ● *n.* **1** an inhabitant of Achaea. (See note below.) **2** (usu. in *pl.*) *literary* a Greek. [L *Achaeus* f. Gk *Akhaios*]

▪ The Achaeans were among the earliest Greek-speaking inhabitants of Greece, being established there well before the 12th century BC. Some scholars identify them with the Mycenaeans of the 14th–13th centuries BC, who were the Greek protagonists in the Trojan War. The Greeks in the *Iliad* are regularly referred to as Achaeans, but it may be that only the military leaders were Achaeans. In the Peloponnese and elsewhere, the Achaeans are thought to have supplanted in the 11th or 12th century BC by the Dorians.

Achaemenid /əˈkiːmənɪd/ *n. & adj.* (also **Achaemenian** /ˌækɪˈmiːnɪən/) ● *n.* a member of the dynasty that ruled in Persia 553–330 BC, from Cyrus the Great to Darius III, and ended with the defeat of Darius III by Alexander the Great. The dynasty was named after Achaemenes, ancestor of Cyrus. ● *adj.* of or relating to this dynasty.

acharnement /əˈʃɑːnmɒn/ *n.* **1** bloodthirsty fury; ferocity. **2** gusto. [F]

Achates /əˈkeɪtiːz/ *Gk & Rom. Mythol.* a companion of Aeneas. His fidelity to his friend was so exemplary as to become proverbial, hence the term *fidus Achates*.

ache /eɪk/ *n. & v.* ● *n.* **1** a continuous or prolonged dull pain. **2** mental distress. ● *v.intr.* **1** suffer from or be the source of an ache (*I ached all over; my left leg ached*). **2** (foll. by *to* + infin.) desire greatly (*we ached to be at home again*). □ **achingly** *adv.* [ME f. OE *æce, acan*. ¶ Dr Johnson is mainly responsible for the modern spelling, as he erroneously derived *ache* and the earlier form *ake* from Gk *akhos* (= pain, distress) and declared that the latter was 'more grammatically written *ache*']

Achebe /əˈtʃeɪbɪ/, Chinua (born Albert Chinualumgu) (b.1930), Nigerian novelist, poet, short-story writer, and essayist. His novels, all written in English, show traditional African society in confrontation with European customs and values. They include *Things Fall Apart* (1958) and *A Man of the People* (1966). In 1989 Achebe won the Nobel Prize for literature.

achene /əˈkiːn/ *n. Bot.* a small dry one-seeded fruit that does not open to liberate the seed (e.g. a strawberry pip). [mod.L *achaenium* (as A-¹, Gk *khainō* gape)]

Achernar /ˈeɪkəˌnɑː(r)/ *Astron.* the ninth brightest star in the sky, and the brightest in the constellation of Eridanus. It marks the southern limit of Eridanus, and is only visible to observers in the southern hemisphere. [Arab., = end of the river (of Eridanus)]

Acheson /ˈeɪtʃɪs(ə)n/, Dean Gooderham (1893–1971), American statesman. In 1949 he became Secretary of State to President Truman. He urged international control of nuclear power, was instrumental in the formation of NATO, and implemented the Marshall Plan and the Truman Doctrine.

Acheulean /əˈʃuːlɪən/ *adj. & n.* (also **Acheulian**) *Archaeol.* of, relating to, or denoting the later stages of the lower palaeolithic tradition of hand-axe industries in Europe, differentiated from the preceding Abbevillian by the use of implements made of wood, antler, or bone as well as stone. In the context of Africa, the entire lower palaeolithic hand-axe sequence is referred to as Acheulean, with the African lower Acheulean representing the Abbevillian of Europe; remains occur at Olduvai Gorge in northern Tanzania and at Kalambo Falls near Lake Tanganyika. The industries as a whole are dated to between *c*.1.5 million and 150,000 years ago. [F *acheuléen* f. St-*Acheul*, the type-site near Amiens in northern France]

achieve /əˈtʃiːv/ *v.tr.* **1 a** reach or attain by effort (*achieved victory*). **b** acquire, gain, earn (*achieved notoriety*). **2** accomplish or carry out (a feat or task). **3** *absol.* be successful; attain a desired level of performance. □ **achievable** *adj.* **achiever** *n.* [ME f. OF *achever* f. *a chief* to a head]

achievement /əˈtʃiːvmənt/ *n.* **1** something achieved; an instance of achieving. **2** the act or process of achieving. **3** *Psychol.* performance in a standardized test. **4** *Heraldry* **a** an escutcheon with adjuncts, or bearing, esp. in memory of a distinguished feat. **b** = HATCHMENT.

achillea /ˌækɪˈliːə/ *n.* a composite plant of the genus *Achillea*, usu. aromatic and with flower-heads of small white or yellow flowers, e.g. yarrow. [L f. Gk *Akhilleios* a plant supposed to have been used medicinally by Achilles]

Achilles /əˈkɪliːz/ *Gk Mythol.* a hero of the Trojan War, son of Peleus and Thetis. During his infancy his mother plunged him in the Styx, thus making his body invulnerable except for the heel by which she held him. The account of the Iliad tells how Achilles withdrew from fighting in the Trojan War after a bitter quarrel with Agamemnon. He later re-entered the battle after his friend Patroclus, clad in Achilles' armour, was killed by Hector. In revenge, Achilles killed Hector and dragged his body behind the wheels of his chariot round the walls of Troy. He was wounded in the heel during the Trojan War by an arrow shot by Paris and died of this wound. □ **Achilles' heel** a weak or vulnerable point. **Achilles' tendon** a tendon attaching the calf muscles to the heel.

achondroplasia /əˌkɒndrəˈpleɪzɪə/ *n. Med.* a hereditary condition in which the growth of long bones by ossification of cartilage is retarded, resulting in very short limbs and often a small face. □ **achondroplasic** *adj.* **achondroplastic** /-ˈplæstɪk/ *adj.* [f. A-¹ + Gk *khondros* cartilage + *plasis* moulding + -IA¹]

achromat /ˈækrəʊˌmæt/ *n.* a lens made achromatic by correction.

achromatic /ˌækrəʊˈmætɪk/ *adj. Optics* **1** that transmits light without separating it into constituent colours (*achromatic lens*). **2** without colour (*achromatic fringe*). □ **achromatically** *adv.* **achromaticity** /əˌkrəʊməˈtɪsɪti/ *n.* **achromatism** /əˈkrəʊməˌtɪz(ə)m/ *n.* [F *achromatique* f. Gk *akhromatos* (as A-¹, CHROMATIC)]

achy /ˈeɪki/ *adj.* full of or suffering from aches.

acid /ˈæsɪd/ *n. & adj.* ● *n.* **1** *Chem.* any of a class of substances (often sour and corrosive liquids) that contain hydrogen atoms replaceable by metals, liberate hydrogen ions in water, turn litmus red, and have a pH of less than 7. (*See note below.*) **2** (in general use) any sour substance. **3** *sl.* the drug LSD.● *adj.* **1** sharp-tasting, sour. **2** biting, sharp (*an acid wit*). **3** *Chem.* having the essential properties of an acid. **4** *Geol.* (of rocks) containing much silica. **5** (of a colour) intense, bright. □ **acid drop** *Brit.* a kind of sweet with a sharp taste. **acid-head** *sl.* a habitual user of the drug LSD. **acid house** a style of house music featuring a mesmeric synthesized sound and distinctive gurgling bass. **acid radical** *Chem.* an organic radical formed by the removal of hydroxyl from an acid. **acid test 1** a severe or conclusive test. **2** a test in which acid is used to test for gold etc. **put the acid on** *Austral. sl.* seek to extract a loan or favour etc. from. □ **acidly** *adv.* **acidness** *n.* **acidic** /əˈsɪdɪk/ *adj.* **acidimetry** /ˌæsɪˈdɪmɪtri/ *n.* [F *acide* or L *acidus* f. *acere* be sour]

■ The term *acid* has been defined in various ways. It was originally applied to sour, corrosive substances like sulphuric acid, whose properties were attributed in the 17th century to their consisting of sharp particles. The commonest modern definition of an acid is (as given above) a substance which releases hydrogen ions when dissolved in water (to form the ion H_3O^+) and is neutralized by bases to form salts. In chemistry acids are sometimes more widely defined as compounds which are donors of hydrogen ions or protons, or acceptors of electron pairs. Most organic acids contain the carboxyl group −COOH. Some of the most important industrial compounds are acids, such as sulphuric and nitric acids.

acidify /əˈsɪdɪˌfaɪ/ *v.tr. & intr.* (**-ies, -ied**) make or become acid. □ **acidification** /əˌsɪdɪfɪˈkeɪʃ(ə)n/ *n.*

acidity /əˈsɪdɪti/ *n.* (*pl.* **-ies**) an acid quality or state, esp. an excessively acid condition of the stomach.

acidophil /ˈæsɪdəʊˌfɪl, əˈsɪdəfɪl/ *n.* (also **acidophile** /-ˌfaɪl/) *Physiol.* a white blood cell that is acidophilic.

acidophilic /ˌæsɪdəʊˈfɪlɪk, əˌsɪdəfɪl-/ *adj.* **1** *Biol.* (of a cell or its contents) readily stained with acid dyes. **2** esp. *Bot.* growing best in acid conditions.

acidosis /ˌæsɪˈdəʊsɪs/ *n.* an over-acid condition of the body fluids or tissues. □ **acidotic** /-ˈdɒtɪk/ *adj.*

acid rain *n.* rainfall made acidic by atmospheric pollution. The phenomenon first came to prominence in the 1970s with reports of widespread deaths of trees and freshwater life in northern forests and lakes. The chief cause is the industrial burning of fossil fuels, especially coal burned in power stations. The waste gases contain sulphur and nitrogen oxides which combine with atmospheric water to form acids. The contaminants are often carried long distances and over national boundaries before causing environmental damage at ground level. International action to reduce the problem has taken the form of agreements to limit oxide emissions; practical action has focused on removing potentially harmful elements from fuel before burning and on removing the oxides by scrubbing the exhaust gases.

acidulate /əˈsɪdjʊˌleɪt/ *v.tr.* make somewhat acid. □ **acidulation** /əˌsɪdjʊˈleɪʃ(ə)n/ *n.* [L *acidulus* dimin. of *acidus* sour]

acidulous /əˈsɪdjʊləs/ *adj.* somewhat acid.

acinus /ˈæsɪnəs/ *n.* (*pl.* **acini** /-ˌnaɪ/) **1** any of the small elements that make up a compound fruit of the blackberry, raspberry, etc. **2** the seed of a grape or berry. **3** *Anat.* a small saclike cavity, esp. one in a gland surrounded by secretory cells. [L, = berry, kernel]

-acious /ˈeɪʃəs/ *suffix* forming adjectives meaning 'inclined to, full of' (*vivacious; pugnacious; voracious; capacious*). [L -*ax* -*acis*, added chiefly to verbal stems to form adjectives + -OUS]

-acity /ˈæsɪti/ *suffix* forming nouns of quality or state corresponding to adjectives in -*acious*. [F -*acité* or L -*acitas* -*tatis*]

ack-ack /ˈækæk/ *adj. & n. colloq.* ● *adj.* anti-aircraft. ● *n.* an anti-aircraft gun etc. [formerly signallers' name for the letters *AA*]

ackee /ˈæki:/ *n.* (also **akee**) **1** a tropical tree, *Blighia sapida*. **2** its fruit, edible when cooked. [Kru *ákee*]

ack emma /æk ˈemə/ *adv. & n. Brit. colloq.* = A.M. [formerly signallers' name for the letters *AM*]

acknowledge /əkˈnɒlɪdʒ/ *v.tr.* **1 a** recognize; accept; admit the truth of (*acknowledged the failure of the plan*). **b** (often foll. by *to be* + complement) recognize as (*acknowledged it to be a great success*). **c** (often foll. by *that* + clause or *to* + infin.) admit that something is so (*acknowledged that he was wrong; acknowledged him to be wrong*). **2** confirm the receipt of (*acknowledged her letter*). **3 a** show that one has noticed (*acknowledged my arrival with a grunt*). **b** express appreciation of (a service etc.). **4** own; recognize the validity of (*the acknowledged king*). □ **acknowledgeable** *adj.* [obs. verb *knowledge* after obs. *acknow* (as A-⁴, KNOW), or f. obs. noun *acknowledge*]

acknowledgement /əkˈnɒlɪdʒmənt/ *n.* (also **acknowledgment**) **1** the act or an instance of acknowledging. **2 a** a thing given or done in return for a service etc. **b** a letter confirming receipt of something. **3** (usu. in *pl.*) an author's statement of indebtedness to others.

aclinic line /əˈklɪnɪk/ *n.* = *magnetic equator*. [Gk *aklinēs* (as A-¹, *klinō* bend)]

acme /ˈækmɪ/ *n.* the highest point or period (of achievement, success, etc.); the peak of perfection (*displayed the acme of good taste*). [Gk, = highest point]

acne /ˈækni/ *n.* a skin condition, usu. of the face, characterized by red pimples. □ **acned** *adj.* [mod.L f. erron. Gk *aknas* for *akmas* accus. pl. of *akmē* facial eruption: cf. ACME]

acolyte /ˈækəˌlaɪt/ *n.* **1** a person assisting a priest in a service or procession. **2** an assistant; a beginner. [ME f. OF *acolyt* or eccl.L *acolytus* f. Gk *akolouthos* follower]

Aconcagua /ˌækɒnˈkɑːgwə/ an extinct volcano in the Andes, on the border between Chile and Argentina, rising to 6,960 m (22,834 ft). It is the highest mountain in the western hemisphere.

aconite /ˈækəˌnaɪt/ *n.* **1 a** a poisonous ranunculaceous plant of the genus *Aconitum*, esp. monkshood or wolfsbane. **b** the drug obtained from this. Also called *aconitine*. **2** (in full **winter aconite**) a small plant of the related genus *Eranthis*, with yellow flowers early in the year. [F *aconit* or L *aconitum* f. Gk *akoniton*]

aconitine /əˈkɒnɪˌtiːn/ *n. Chem.* a poisonous alkaloid obtained from the aconite.

acorn /ˈeɪkɔːn/ *n.* the fruit of the oak, a smooth nut in a rough cuplike base. □ **acorn barnacle** a multivalve marine cirriped, *Balanus balanoides*, living on rocks. **acorn worm** a wormlike marine animal of the phylum Hemichordata, having a proboscis and gill slits, and inhabiting seashores. [OE *æcern*, rel. to *æcer* ACRE, later assoc. with OAK and CORN¹]

acotyledon /əˌkɒtɪˈliːd(ə)n/ *n.* a plant with no distinct seed-leaves. □ **acotyledonous** *adj.* [mod.L *acotyledones* pl. (as A-¹, COTYLEDON)]

acoustic /əˈkuːstɪk/ *adj. & n.* ● *adj.* **1** relating to sound or the sense of hearing. **2** (of a musical instrument, gramophone, or recording) not having electrical amplification (*acoustic guitar*). **3** (of building materials) used for soundproofing or modifying sound. **4** *Mil.* (of a mine) that can be exploded by sound waves transmitted under water. ● *n.* **1** (usu. in *pl.*) the properties or qualities (esp. of a room or hall etc.) in transmitting sound (*good acoustics; a poor acoustic*). **2** (in *pl.*; usu. treated as *sing.*) the science of sound (*acoustics is not widely taught*). □ **acoustic coupler** *Computing* a modem which converts digital signals into audible signals and vice versa, so that the former can be transmitted and received over telephone lines. □ **acoustical** *adj.* **acoustically** *adv.* [Gk *akoustikos* f. *akouō* hear]

acoustician /ˌækuːˈstɪʃ(ə)n/ *n.* an expert in acoustics.

acquaint /əˈkweɪnt/ *v.tr. & refl.* (usu. foll. by *with*) make (a person or oneself) aware of or familiar with (*acquaint me with the facts*). □ **be acquainted with** have personal knowledge of (a person or thing); have made the acquaintance of (a person). [ME f. OF *acointier* f. LL *accognitare* (as AC-, *cognoscere cognit-* come to know)]

acquaintance /əˈkweɪntəns/ *n.* **1** (usu. foll. by *with*) slight knowledge (of a person or thing). **2** the fact or process of being acquainted (*our acquaintance lasted a year*). **3** a person one knows slightly. □ **acquaintance rape** rape of a woman by someone known to her. **make a person's acquaintance** first meet or introduce oneself to another person; come to know. □ **acquaintanceship** *n.* [ME f. OF *acointance* (as ACQUAINT)]

acquiesce /ˌækwɪˈes/ *v.intr.* **1** agree, esp. tacitly. **2** raise no objection. **3** (foll. by *in*) accept (an arrangement etc.). □ **acquiescence** *n.* **acquiescent** *adj.* [L *acquiescere* (as AC-, *quiescere* rest)]

acquire /əˈkwaɪə(r)/ *v.tr.* **1** gain by and for oneself; obtain. **2** come into possession of (*acquired fame; acquired much property*). □ **acquired characteristic** *Biol.* a characteristic caused by the environment, not inherited. **acquired immune deficiency syndrome** *Med.* see AIDS. **acquired taste 1** a liking gained by experience. **2** the object of such a liking. □ **acquirable** *adj.* [ME f. OF *aquerre* ult. f. L *acquirere* (as AC-, *quaerere* seek)]

acquirement /əˈkwaɪəmənt/ *n.* **1** something acquired, esp. a mental attainment. **2** the act or an instance of acquiring.

acquisition /ˌækwɪˈzɪʃ(ə)n/ *n.* **1** something acquired, esp. if regarded as useful. **2** the act or an instance of acquiring. [L *acquisitio* (as ACQUIRE)]

acquisitive /əˈkwɪzɪtɪv/ *adj.* keen to acquire things; avaricious; materialistic. □ **acquisitively** *adv.* **acquisitiveness** *n.* [F *acquisitive* or LL *acquisitivus* (as ACQUIRE)]

acquit /əˈkwɪt/ *v.* (**acquitted, acquitting**) **1** *tr.* (often foll. by *of*) declare (a person) not guilty (*were acquitted of the offence*). **2** *refl.* **a** conduct oneself or perform in a specified way (*we acquitted ourselves well*). **b** (foll. by *of*) discharge (a duty or responsibility). [ME f. OF *aquiter* f. med.L *acquitare* pay a debt (as AC-, QUIT)]

acquittal /əˈkwɪt(ə)l/ *n.* **1** the process of freeing or being freed from a charge, esp. by a judgement of not guilty. **2** performance of a duty.

acquittance /əˈkwɪt(ə)ns/ *n.* **1** payment of or release from a debt. **2** a written receipt attesting settlement of a debt. [ME f. OF *aquitance* (as ACQUIT)]

Acre 1 /ˈeɪkə(r)/ (also **Akko** /ˈækəʊ/) an industrial seaport of Israel; pop. (1982) 39,100. Captured by the Crusaders in the 12th century, Acre was the last Christian stronghold in the Holy Land until taken by the Ottoman Turks in 1517. **2** /ˈɑːkrə/ a state of western Brazil, on the border with Peru; capital, Rio Branco.

acre /ˈeɪkə(r)/ *n.* **1** a measure of land, 4,840 sq. yards, 0.405 hectare. **2** (in *pl.*) a large area or amount. □ **acred** *adj.* (esp. in *comb.*). [OE *æcer* f. Gmc]

acreage /ˈeɪkərɪdʒ/ *n.* **1** a number of acres. **2** an extent of land.

acrid /ˈækrɪd/ *adj.* **1** bitterly pungent; irritating; corrosive. **2** bitter in temper or manner. □ **acridly** *adv.* **acridity** /əˈkrɪdɪtɪ/ *n.* [irreg. f. L *acer acris* keen + -ID¹, prob. after *acid*]

acridine /ˈækrɪˌdiːn/ *n.* a solid colourless organic compound used in the manufacture of dyes and drugs. [ACRID + -INE⁴]

acriflavine /ˌækrɪˈfleɪvɪn, -viːn/ *n.* a reddish powder used as an antiseptic. [irreg. f. ACRIDINE + FLAVINE]

acrimonious /ˌækrɪˈməʊnɪəs/ *adj.* bitter in manner or temper. □ **acrimoniously** *adv.* [F *acrimonieux, -euse* f. med.L *acrimoniosus* f. L *acrimonia* ACRIMONY]

acrimony /ˈækrɪmənɪ/ *n.* (*pl.* **-ies**) bitterness of temper or manner; ill feeling. [F *acrimonie* or L *acrimonia* pungency (as ACRID)]

acrobat /ˈækrəˌbæt/ *n.* **1** a performer of spectacular gymnastic feats. **2** a person noted for constant change of mind, allegiance, etc. □ **acrobatic** /ˌækrəˈbætɪk/ *adj.* **acrobatically** *adv.* [F *acrobate* f. Gk *akrobatēs* f. *akron* summit + *bainō* walk]

acrobatics /ˌækrəˈbætɪks/ *n.pl.* **1** acrobatic feats. **2** (as *sing.*) the art of performing these. **3** a skill requiring ingenuity (*mental acrobatics*).

acromegaly /ˌækrəˈmegəlɪ/ *n. Med.* abnormal growth of the hands, feet, and face, caused by over-production of growth hormone in the pituitary gland. □ **acromegalic** /-mɪˈgælɪk/ *adj.* [F *acromégalie* f. Gk *akron* extremity + *megas* megal- great]

acronym /ˈækrənɪm/ *n.* a word, usu. pronounced as such, formed from the initial letters of other words (e.g. *Ernie, laser, Nato*) (cf. INITIALISM). [Gk *akron* end + -onum- = *onoma* name]

acropetal /əˈkrɒpɪt(ə)l/ *adj. Bot.* developing from below upwards. □ **acropetally** *adv.* [Gk *akron* tip + L *petere* seek]

acrophobia /ˌækrəˈfəʊbɪə/ *n. Psychol.* an abnormal dread of heights. □ **acrophobic** *adj.* [Gk *akron* peak + -PHOBIA]

Acropolis /əˈkrɒpəlɪs/ the ancient citadel at Athens, containing the Parthenon, Erechtheum, and other notable buildings, mostly dating from the 5th century BC.

acropolis /əˈkrɒpəlɪs/ *n.* a citadel or upper fortified part of an ancient city. [Gk *akropolis* f. *akron* summit + *polis* city]

across /əˈkrɒs/ *prep. & adv.* ● *prep.* **1** to or on the other side of (*walked across the road; lives across the river*). **2** from one side to another side of (*the cover stretched across the opening*; *a bridge across the river*). **3** at or forming an angle (esp. a right angle) with (*deep cuts across his legs*). ● *adv.* **1** to or on the other side (*ran across*; *shall soon be across*). **2** from one side to another (*a blanket stretched across*). **3** forming a cross (*with cuts across*). **4** (of a crossword clue or answer) read horizontally (*cannot do nine across*). □ **across the board** general; generally; applying to all. [ME f. OF *a croix, en croix*, later regarded as f. A² + CROSS]

acrostic /əˈkrɒstɪk/ *n.* **1** a poem or other composition in which certain letters in each line form a word or words. **2** a word-puzzle constructed in this way. □ **double acrostic** one using the first and last letters of each line. **single acrostic** one using the first letter only. **triple acrostic** one using the first, middle, and last letters. [F *acrostiche* or Gk *akrostikhis* f. *akron* end + *stikhos* row, line of verse, assim. to -IC]

Acrux /ˈeɪkrʌks/ *Astron.* the twelfth brightest star in the sky, and the brightest in the constellation of Crux, only visible to observers in the southern hemisphere. [*a* alpha + *Crux*]

acrylic /əˈkrɪlɪk/ *adj. & n.* ● *adj.* **1** of or derived from acrylic acid, esp. made from synthetic polymers of acrylic acid or acrylates. **2** (of a picture) in acrylic paint. ● *n.* **1** an acrylic textile fibre. **2** (also **acrylics**) acrylic paint. □ **acrylic acid** a pungent liquid organic acid (chem. formula: CH₂CHCOOH). **acrylic paint** paint based on acrylic resin as a medium. □ **acrylate** /ˈækrɪˌleɪt/ *n.* [*acrolein* f. L *acer acris* pungent + *olere* to smell + -YL + -IC]

ACT *abbr.* Australian Capital Territory.

act /ækt/ *n. & v.* ● *n.* **1** something done; a deed; an action. **2** the process of doing something (*caught in the act*). **3 a** a piece of entertainment, usu. one of a series in a programme. **b** the performer(s) of this. **4** a pretence; behaviour intended to deceive or impress (*it was all an act*). **5** a main division of a play or opera. **6 a** a written ordinance of a parliament or other legislative body. **b** a document attesting a legal transaction. **7** (often in *pl.*) the recorded decisions or proceedings of a committee, an academic body, etc. **8** (**Acts**) (in full **Acts of the Apostles**) a New Testament book immediately following the Gospels and relating the history of the early Church. ● *v.* **1** *intr.* behave (*see how they act under stress*). **2** *intr.* perform actions or functions; operate effectively; take action (*act as referee*; *the brakes failed to act*; *we must act quickly*). **3** *intr.* (also foll. by *on*) exert energy or influence (*the medicine soon began to act*; *alcohol acts on the brain*). **4** *intr.* **a** perform a part in a play, film, etc. **b** pretend. **5** *tr.* **a** perform the part of (*acted Othello*; *acts the fool*). **b** perform (a play etc.). **c** portray (an incident) by actions. **d** feign (*we acted indifference*). □ **act for** be the (esp. legal) representative of. **act of God** the operation of uncontrollable natural forces; an instance of this. **act of grace** a privilege or concession that cannot be claimed as a right. **act on** (or **upon**) perform or carry out; put into operation (*acted on my advice*). **act out 1** translate (ideas etc.) into action. **2** *Psychol.* represent (one's subconscious desires etc.) in action. **3** perform (a drama). **act up** *colloq.* misbehave; give trouble (*my car is acting up again*). **get one's act together** *sl.* become properly organized; make preparations for an undertaking etc. **get into** (or **in on**) **the act** *sl.* become a participant (esp. for profit). **put on an act** *colloq.* carry out a pretence. □ **actable** *adj.* (in sense 5 of *v.*). **actability** /ˌæktəˈbɪlɪtɪ/ *n.* (in sense 5 of *v.*). [ME ult. f. L *agere act-* do]

Actaeon /ækˈtiːən, ˈæktɪən/ *Gk Mythol.* a hunter who, because he accidentally saw Artemis bathing, was changed into a stag and killed by his own hounds.

ACTH *abbr. Med.* adrenocorticotrophic hormone.

actin /ˈæktɪn/ *n. Biochem.* a protein which, with myosin, forms the contractile filaments of muscle fibres. [f. Gk *aktis -inos* ray + -IN]

acting /ˈæktɪŋ/ *n. & attrib.adj.* ● *n.* **1** the art or occupation of performing parts in plays, films, etc. **2** in senses of ACT *v.* ● *attrib.adj.* serving temporarily or on behalf of another or others (*acting manager*; *Acting Captain*).

actinia /ækˈtɪnɪə/ *n.* (*pl.* **actiniae** /-nɪˌiː/) a sea anemone, esp. of the genus *Actinia*. [mod.L f. Gk *aktis -inos* ray]

actinide /ˈæktɪˌnaɪd/ *n.* (also **actinoid** /-ˌnɔɪd/) *Chem.* any of the series of fifteen metallic elements from actinium to lawrencium in the periodic table. They are all radioactive; the heavier members are very unstable and do not occur naturally. Chemically they are more diverse than the lanthanides and behave more like transition metals. [ACTINIUM + -IDE as in *lanthanide*]

actinism /ˈæktɪˌnɪz(ə)m/ *n.* the property of short-wave radiation that produces chemical changes, as in photography. □ **actinic** /ækˈtɪnɪk/ *adj.* [Gk *aktis -inos* ray]

actinium /ækˈtɪnɪəm/ *n.* a radioactive metallic chemical element (atomic number 89; symbol **Ac**). Actinium was discovered in the

mineral pitchblende by the French scientist André Debierne in 1899. The first member of the actinide series, it is rare in nature, occurring as an impurity in uranium ores. [f. as ACTINISM]

actinoid var. of ACTINIDE.

actinometer /ˌæktɪˈnɒmɪtə(r)/ *n.* an instrument for measuring the intensity of radiation, esp. ultraviolet radiation. [Gk *aktis -inos* ray + -METER]

actinomorphic /ˌæktɪnəˈmɔːfɪk/ *adj. Biol.* radially symmetrical. [as ACTINOMETER + Gk *morphē* form]

actinomycete /ˌæktɪnəʊˈmaɪsiːt/ *n.* a usu. non-motile filamentous anaerobic bacterium of the order Actinomycetales. [as ACTINOMORPHIC + -*mycetes* f. Gk *mukēs -ētos* fungus]

action /ˈækʃ(ə)n/ *n. & v.* ● *n.* **1** the fact or process of doing or acting (*demanded action*; *put ideas into action*). **2** forcefulness or energy as a characteristic (*a woman of action*). **3** the exertion of energy or influence (*the action of acid on metal*). **4** something done; a deed or act (*not aware of his own actions*). **5 a** a series of events represented in a story, play, etc. **b** *sl.* exciting activity (*arrived late and missed the action*; *want some action*). **6 a** armed conflict; fighting (*killed in action*). **b** an occurrence of this, esp. a minor military engagement. **7 a** the way in which a machine, instrument, etc. works (*explain the action of an air pump*). **b** the mechanism that makes a machine, instrument, etc. (e.g. a musical instrument, a gun, etc.) work. **c** the mode or style of movement of an animal or human (usu. described in some way) (*a runner with good action*). **8** a legal process; a lawsuit (*bring an action*). **9** a word of command to begin acting a scene or some other activity, esp. as used by a film director. □ **action committee** (or **group** etc.) a body formed to take active steps, esp. in politics. **action-packed** full of action or excitement. **action point** a proposal for action, esp. arising from a discussion etc. **action replay** a playback of part of a television broadcast, esp. a sporting event, often in slow motion. **action stations** positions taken up by troops etc. ready for battle. **go into action** start work. **out of action** not working. **take action** begin to act (esp. energetically in protest). [ME f. OF f. L *actio -onis* (as ACT)]

actionable /ˈækʃənəb(ə)l/ *adj.* giving cause for legal action.

Action Directe /ˌæksjɔ̃ diˈrekt/ a group of extreme left-wing French terrorists. [F, = direct action]

action painting *n.* a technique and style of abstract painting in which paint is randomly splashed, thrown, or poured on the canvas. The term was first used in 1952 and was made famous by Jackson Pollock. It formed part of the more general movement of abstract expressionism, and the terms are sometimes taken as synonymous.

Actium, Battle of /ˈæktɪəm/ a naval battle which took place in 31 BC off the promontory of Actium in western Greece, in the course of which Octavian defeated Mark Antony. This cleared the way for Octavian to become sole ruler of Rome.

activate /ˈæktɪˌveɪt/ *v.tr.* **1** make active; bring into action. **2** *Chem.* cause reaction in; excite (a substance, molecules, etc.). **3** *Physics* make radioactive. □ **activated carbon** carbon, esp. charcoal, treated to increase its adsorptive power. **activated sludge** aerated sewage containing aerobic bacteria and protozoa that are used to break it down. □ **activator** *n.* **activation** /ˌæktɪˈveɪʃ(ə)n/ *n.*

active /ˈæktɪv/ *adj. & n.* ● *adj.* **1 a** consisting in or marked by action; energetic; diligent (*leads an active life*; *an active helper*). **b** able to move about or accomplish practical tasks (*infirmity made him less active*). **2** working, operative (*an active volcano*). **3** originating action; not merely passive or inert (*active support*; *active ingredients*). **4** radioactive. **5** *Gram.* designating the voice that attributes the action of a verb to the person or thing from which it logically proceeds (e.g. of the verbs in *guns kill*; *we saw him*). ● *n. Gram.* the active form or voice of a verb. □ **active carbon** = *activated carbon* (see ACTIVATE). **active citizen** a person who plays an active role in the community through crime prevention etc. **active list** *Mil.* a list of officers available for service. **active service** actual participation in warfare as a member of the armed forces. **active transport** *Biol.* the movement of ions or molecules across a cell membrane into a region of higher concentration, assisted by enzymes and requiring energy. □ **actively** *adv.* **activeness** *n.* [ME f. OF *actif -ive* or L *activus* (as ACT *v.*)]

activism /ˈæktɪˌvɪz(ə)m/ *n.* a policy of vigorous action in a cause, esp. in politics. □ **activist** *n.*

activity /ækˈtɪvɪtɪ/ *n.* (*pl.* **-ies**) **1 a** the condition of being active or moving about. **b** the exertion of energy; vigorous action. **2** (often in *pl.*) a particular occupation or pursuit (*outdoor activities*). **3** = RADIOACTIVITY. [F *activité* or LL *activitas* (as ACTIVE)]

Act of Settlement, Act of Uniformity, etc. see SETTLEMENT, ACT OF; UNIFORMITY, ACT OF, etc.

actor /ˈæktə(r)/ n. **1 a** a person who acts a part in a play etc. **b** a person whose profession is performing such parts. **2** a person who pretends to be something he or she is not. [L, = doer, actor (as ACT, -OR[1])]

Actors' Studio an acting workshop in New York City, founded in 1947 by Elia Kazan and others. Under the directorship of Lee Strasberg from 1948 to 1982, the workshop became a leading centre of method acting.

actress /ˈæktrɪs/ n. a female actor.

actual /ˈæktʃʊəl, ˈæktjʊəl/ adj. (usu. attrib.) **1** existing in fact; real (often as distinct from ideal). **2** existing now; current. ¶ Redundant use, as in tell me the actual facts, is disp., but common. □ **actual bodily harm** Law any hurt or injury inflicted intentionally on a person (although less serious in degree than grievous bodily harm). □ **actualize** v.tr. (also **-ise**). **actualization** /ˌæktʃʊəlaɪˈzeɪʃ(ə)n, ˌæktjʊ-/ n. [ME f. OF actuel f. LL actualis f. agere ACT]

actuality /ˌæktʃʊˈælɪtɪ, ˌæktjʊ-/ n. (pl. **-ies**) **1** reality; what is the case. **2** (in pl.) existing conditions. [ME f. OF actualité entity or med.L actualitas (as ACTUAL)]

actually /ˈæktʃʊəlɪ, ˈæktjʊ-/ adv. **1** as a fact, really (I asked for ten, but actually got nine). **2** as a matter of fact, even (strange as it may seem) (he actually refused!). **3** at present; for the time being.

actuary /ˈæktʃʊərɪ, ˈæktjʊ-/ n. (pl. **-ies**) an expert in statistics and probability theory, esp. a person who calculates insurance risks and premiums. □ **actuarial** /ˌæktʃʊˈeərɪəl, ˌæktjʊ-/ adj. **actuarially** adv. [L actuarius bookkeeper f. actus past part. of agere ACT]

actuate /ˈæktʃʊˌeɪt, ˈæktjʊ-/ v.tr. **1** communicate motion to (a machine etc.). **2** cause the operation of (an electrical device etc.). **3** cause (a person) to act. □ **actuator** n. **actuation** /ˌæktʃʊˈeɪʃ(ə)n, ˌæktjʊ-/ n. [med.L actuare f. actus: see ACTUAL]

acuity /əˈkjuːɪtɪ/ n. sharpness, acuteness (of a needle, senses, understanding). [F acuité or med.L acuitas f. acuere sharpen: see ACUTE]

aculeate /əˈkjuːlɪət/ adj. & n. ● adj. **1** Zool. having a sting. **2** pointed, incisive. ● n. Zool. a hymenopterous insect of the section Aculeata, which includes bees, wasps, and ants. [L aculeatus f. aculeus sting, dimin. of acus needle]

acumen /ˈækjʊmən, əˈkjuːmən/ n. keen insight or discernment, penetration. [L acumen -minis anything sharp f. acuere sharpen: see ACUTE]

acuminate /əˈkjuːmɪnət/ adj. Biol. tapering to a point. [L acuminatus pointed (as ACUMEN)]

acupressure /ˈækjʊˌpreʃə(r)/ n. a form of therapy in which symptoms are relieved by applying pressure with the thumbs or fingers to specific points on the body. [alt. of ACUPUNCTURE]

acupuncture /ˈækjʊˌpʌŋktʃə(r)/ n. a method of treating various medical conditions by pricking the skin or tissues with needles. Acupuncture has been practised in China for more than 4,500 years and is now becoming increasingly popular in the West. Success is claimed in the alleviation of pain and the treatment of many physical, mental, and emotional conditions, although how the method is effective is not known. Acupuncturists believe that many such conditions arise through an imbalance of energy in the body and that treatment may restore this balance; recently some scientists have suggested that the acupuncturist's needles may stimulate the body's endorphins or naturally occurring painkillers. □ **acupuncturist** n. [L acu with a needle + PUNCTURE]

acushla /əˈkʊʃlə/ n. Ir. darling. [Ir. a chuisle O pulse (of my heart)!]

acutance /əˈkjuːt(ə)ns/ n. sharpness of a photographic or printed image; a measure of this. [ACUTE + -ANCE]

acute /əˈkjuːt/ adj. & n. ● adj. (**acuter, acutest**) **1 a** (of senses etc.) keen, penetrating. **b** (of pain) intense, severe; sharp or stabbing rather than dull, aching, or throbbing. **2** shrewd, perceptive (an acute critic). **3** (of a disease) coming sharply to a crisis; severe, not chronic. **4** (of a difficulty or controversy) critical, serious. **5 a** (of an angle) less than 90°. **b** sharp, pointed. **6** (of a sound) high, shrill. ● n. = acute accent. □ **acute accent** a mark (ˊ) placed over letters in some languages to show quality, vowel length, pronunciation (e.g. maté), etc. **acute rheumatism** Med. = rheumatic fever. □ **acutely** adv. **acuteness** n. [L acutus past part. of acuere sharpen f. acus needle]

ACW abbr. Brit. (preceding a name) Aircraftwoman.

-acy /əsɪ/ suffix forming nouns of state or quality (accuracy; piracy; supremacy), or an instance of it (conspiracy; fallacy) (see also -CRACY). [a

branch of the suffix -CY from or after F -acie or L -acia or -atia or Gk -ateia]

acyl /ˈæsaɪl, ˈæsɪl/ n. Chem. the monovalent radical of a carboxylic acid. [G (as ACID, -YL)]

AD abbr. (of a date) of the Christian era. (See DIONYSIUS EXIGUUS.) ¶ Strictly, AD should precede a date (e.g. AD 410), but uses such as the tenth century AD are well established. [Anno Domini 'in the year of the Lord']

ad /æd/ n. colloq. an advertisement. [abbr.]

ad- /əd, æd/ prefix (also **a-** before sc, sp, st, **ac-** before c, k, q, **af-** before f, **ag-** before g, **al-** before l, **an-** before n, **ap-** before p, **ar-** before r, **as-** before s, **at-** before t) **1** with the sense of motion or direction to, reduction or change into, addition, adherence, increase, or intensification. **2** formed by assimilation of other prefixes (accurse; admiral; advance; affray). [(sense 1) (through OF a-) f. L ad to: (sense 2) a- repr. various prefixes other than ad-]

-ad[1] /əd, æd/ suffix forming nouns: **1** in collective numerals (myriad; triad). **2** in fem. patronymics (Dryad). **3** in names of poems and similar compositions (Iliad; Dunciad; Jeremiad). [Gk -as -ada]

-ad[2] /əd/ suffix forming nouns (ballad; salad) (cf. -ADE[1]). [F -ade]

Ada /ˈeɪdə/ n. Computing a high-level programming language. [Ada Lovelace (LOVELACE[1])]

adage /ˈædɪdʒ/ n. a traditional maxim, a proverb. [F f. L adagium (as AD-, root of aio say)]

adagio /əˈdɑːdʒɪˌəʊ/ adv., adj., & n. Mus. ● adv. & adj. in slow time. ● n. (pl. **-os**) an adagio movement or passage. [It.]

Adam[1] /ˈædəm/ (in the biblical and Koranic traditions) the first man. (See note below.) □ **Adam's ale** water. **Adam's apple** the projection at the front of the neck formed by the thyroid cartilage of the larynx. **not know a person from Adam** be unable to recognize the person in question. [Heb. 'ādām man]

▪ In the Bible, the Book of Genesis describes how Adam was formed from the dust of the ground and God's breath; Eve, the first woman, was created from one of Adam's ribs as his companion. They lived together in the Garden of Eden until the serpent tempted Eve to eat an apple from the forbidden tree; she persuaded Adam to do the same. As a result of this original sin of disobedience, they were both expelled from the garden.

Adam[2] /ˈædəm/, Robert (1728–92), Scottish architect. He undertook a grand tour of Europe, starting in 1754, mainly in France and Italy, which gave him knowledge of ancient buildings and modern neoclassical theory. Robert and his brother James (1730–94) set up in business in London in 1758 and in the course of the next forty years initiated the 'Adam Revolution', introducing a lighter, more decorative style than the Palladianism favoured by the British architecture of the previous half-century. Typical examples of their work include Home House in Portland Square, London, and Register House in Edinburgh.

adamant /ˈædəmənt/ adj. & n. ● adj. stubbornly resolute; resistant to persuasion. ● n. archaic diamond or other hard substance. □ **adamance** n. **adamantly** adv. **adamantine** /ˌædəˈmæntaɪn/ adj. [OF adamaunt f. L adamas adamant- untameable f. Gk (as A-[1], damaō to tame)]

Adams[1] /ˈædəmz/, Ansel (Easton) (1902–84), American photographer. In the 1930s he developed the use of large-format cameras and small apertures to produce sharp images with maximum depth of field, very different from the fashionable soft-focus work of the time. Many of Adams's collections, such as My Camera in the National Parks (1950) and This is the American Earth (1960), depict the American wilderness and reflect his interest in conservation.

Adams[2] /ˈædəmz/, John (1735–1826), American Federalist statesman, 2nd President of the US 1797–1801. He was a key figure in the drafting of the Declaration of Independence (1776), and was minister to Britain 1785–8.

Adams[3] /ˈædəmz/, John Couch (1819–92), English astronomer. In 1843 he postulated the existence of an eighth planet from perturbations in the orbit of Uranus. Similar calculations performed almost simultaneously by Le Verrier resulted in the discovery of Neptune three years later. Adams carried out numerous advanced mathematical calculations, including improved figures for the moon's parallax and motion.

Adams[4] /ˈædəmz/, John Quincy (1767–1848), American Republican statesman, 6th President of the US 1825–9. He was the eldest son of President John Adams. As Secretary of State under President Monroe, he helped to shape the Monroe doctrine.

Adam's Bridge a line of shoals lying between NW Sri Lanka and the SE coast of Tamil Nadu in India, separating the Palk Strait from the Gulf of Mannar.

Adam's Peak a mountain in south central Sri Lanka, rising to 2,243 m (7,360 ft). The mountain is regarded as sacred by Buddhists, Hindus, and Muslims.

Adana /'ædənə/ a town in southern Turkey, capital of a province of the same name; pop. (1990) 916,150.

adapt /ə'dæpt/ v. **1** tr. **a** (foll. by to) fit, adjust (one thing to another). **b** (foll. by to, for) make suitable for a purpose. **c** alter or modify (esp. a text). **d** arrange for broadcasting etc. **2** intr. & refl. (usu. foll. by to) become adjusted to new conditions. [F adapter f. L adaptare (as AD-, aptare f. aptus fit)]

adaptable /ə'dæptəb(ə)l/ adj. **1** able to adapt oneself to new conditions. **2** that can be adapted. □ **adaptably** adv. **adaptability** /ə,dæptə'bɪlɪtɪ/ n.

adaptation /,ædæp'teɪʃ(ə)n/ n. **1** the act or process of adapting or being adapted. **2** a thing that has been adapted. **3** Biol. the process by which an organism or species becomes suited to its environment. [F f. LL adaptatio -onis (as ADAPT)]

adaptive /ə'dæptɪv/ n. characterized by adaptation. □ **adaptive radiation** Biol. the diversification of a group of organisms into forms filling different ecological niches.

adaptor /ə'dæptə(r)/ n. (also **adapter**) **1** a device for making equipment compatible. **2** a device for connecting several electrical plugs to one socket. **3** a person who adapts.

Adar /æ'dɑ:(r)/ n. (in the Jewish calendar) **1** the sixth month of the civil and twelfth of the religious year, usually coinciding with parts of February and March (known in leap years as Second Adar). **2** (in leap years) an intercalary month (also called First Adar) preceding Second Adar. [Heb. 'ādār]

adaxial /æd'æksɪəl/ adj. Bot. facing toward the stem of a plant, esp. of the upper side of a leaf (cf. ABAXIAL). [AD- + AXIAL]

ADC abbr. **1** aide-de-camp. **2** analogue-digital converter.

add /æd/ v.tr. **1** join (one thing to another) as an increase or supplement (add your efforts to mine; add insult to injury). **2** put together (two or more numbers) to find a number denoting their combined value. **3** say in addition (added that I was wrong; 'What's more, I don't like it,' he added). □ **add in** include. **add-on** something added to an existing object or quantity (also attrib.: several add-on features are available). **add to** increase; be a further item among (this adds to our difficulties). **add up 1** find the total of. **2** (foll. by to) amount to; constitute (adds up to a disaster). **3** colloq. make sense; be understandable. □ **added** adj. [ME f. L addere (as AD-, dare put)]

Addams /'ædəmz/, Jane (1860–1935), American social reformer and feminist. Inspired by Toynbee Hall in London, she founded Hull House, a social settlement, in Chicago in 1889; this became a national model for the combat of urban poverty and the treatment of young offenders. She was prominent in the suffrage movement, serving as president of the National American Woman Suffrage Association (1911–14); also a pacifist, she presided over the first Women's Peace Congress at The Hague in 1915. She shared the Nobel Peace Prize with Nicholas Butler (1862–1947) in 1931.

addax /'ædæks/ n. a large desert antelope, Addax nasomaculatus, of northern Africa, with twisted horns. [L f. an African word]

addendum /ə'dendəm/ n. (pl. **addenda** /-də/) **1** a thing (usu. something omitted) to be added, esp. (in pl.) as additional matter at the end of a book. **2** an appendix; an addition. [L, gerundive of addere ADD]

adder /'ædə(r)/ n. a small venomous snake, esp. the common viper, Vipera berus, the only poisonous snake in Britain. □ **adder's tongue 1** a fern of the genus Ophioglossum, having a narrow spikelike sporing body. **2** a liliaceous plant of the esp. North American genus Erythronium, with mottled leaves. [OE nædre: n lost in ME by wrong division of a naddre: cf. APRON, AUGER, UMPIRE]

addict v. & n. ● v.tr. & refl. /ə'dɪkt/ (usu. foll. by to) devote or apply habitually or compulsively; make addicted. ● n. /'ædɪkt/ **1** a person addicted to a habit, esp. one dependent on a (specified) drug (drug addict; heroin addict). **2** colloq. an enthusiastic devotee of a sport or pastime (film addict). [L addicere assign (as AD-, dicere dict- say)]

addicted /ə'dɪktɪd/ adj. (often foll. by to) **1** dependent on as a habit; unable to do without a thing (addicted to heroin). **2** devoted (addicted to football).

addiction /ə'dɪkʃ(ə)n/ n. the fact or process of being addicted, esp. the condition of taking a drug habitually and being unable to give it up without incurring adverse effects. [L addictio: see ADDICT]

addictive /ə'dɪktɪv/ adj. (of a drug, habit, etc.) causing addiction or dependence.

Addington /'ædɪŋtən/, Henry, 1st Viscount Sidmouth (1757–1844), British Tory statesman, Prime Minister 1801–4. As Home Secretary (1812–21), he introduced harsh legislation to suppress the Luddites and other protest groups.

Addis Ababa /,ædɪs 'æbəbə/ (also **Adis Abeba**) the capital of Ethiopia, situated at an altitude of about 2,440 m (8,000 ft); pop. (est. 1990) 1,700,000. The headquarters of the Organization of African Unity and the United Nations Economic Commission for Africa are located there.

Addison[1] /'ædɪs(ə)n/, Joseph (1672–1719), English poet, dramatist, essayist, and Whig politician. In 1711 he founded the Spectator with Sir Richard Steele. His tragedy Cato (1713) was an immediate success. In the history of English literature he is remarkable for his simple unornamented prose style, which marked the end of the more mannered and florid writing of the 17th century.

Addison[2] /'ædɪs(ə)n/, Thomas (1793–1860), English physician. He described the disease now named after him, ascribing it correctly to defective functioning of the adrenal glands. Distinguished for his zeal in the investigation of disease, Addison had a great reputation as a clinical teacher, and Guy's Hospital in London attained fame as a school of medicine during the time of his connection with it.

Addison's disease n. a disease characterized by progressive anaemia, great weakness, and bronze discoloration of the skin. [ADDISON[2]]

addition /ə'dɪʃ(ə)n/ n. **1** the act or process of adding or being added. **2** a person or thing added (a useful addition to the team). □ **in addition** moreover, furthermore, as well. **in addition to** as well as, as something added to. [ME f. OF addition or f. L additio (as ADD)]

additional /ə'dɪʃən(ə)l/ adj. added, extra, supplementary. □ **additionally** adv.

additive /'ædɪtɪv/ n. & adj. ● n. a thing added, esp. a substance added to another so as to give it specific qualities (food additive). ● adj. **1** characterized by addition (additive process). **2** to be added. [LL additivus (as ADD)]

addle /'æd(ə)l/ v. & adj. ● v. **1** tr. muddle, confuse. **2** intr. (of an egg) become addled. ● adj. **1** muddled, unsound (addle-brained; addle-head). **2** empty, vain. **3** (of an egg) addled. [OE adela filth, used as adj., then as verb]

addled /'æd(ə)ld/ adj. **1** (of an egg) rotten, producing no chick. **2** muddled. [ADDLE adj., assim. to past part. form]

Addled Parliament the Parliament of James I of England (James VI of Scotland), summoned in 1614, so known because it refused to accede to the king's financial requests, did not succeed in its attempts to curb his existing powers of taxation, and was dissolved without having passed any legislation.

Addo National Park /æ'dəʊ/ a national park in the province of Eastern Cape, South Africa, north of Port Elizabeth. It was established in 1931.

address /ə'dres/ n. & v. ● n. **1 a** the place where a person lives or an organization is situated. **b** particulars of this, esp. for postal purposes. **c** Computing the location of an item of stored information. **2** a discourse delivered to an audience. **3** skill, dexterity, readiness. **4** (in pl.) a courteous approach, courtship (pay one's addresses to). **5** archaic manner in conversation. ● v.tr. **1** write directions for delivery (esp. the name and address of the intended recipient) on (an envelope, packet, etc.). **2** direct in speech or writing (remarks, a protest, etc.). **3** speak or write to, esp. formally (addressed the audience; asked me how to address a duke). **4** direct one's attention to. **5** Golf take aim at or prepare to hit (the ball). □ **address oneself to 1** speak or write to. **2** attend to. □ **addresser** n. [ME f. OF adresser ult. f. L (as AD-, directus DIRECT): (n.) perh. f. F adresse]

addressee /,ædre'si:/ n. the person to whom (esp. a letter) is addressed.

Addressograph /ə'dresəʊ,grɑːf/ n. propr. a machine for printing addresses on envelopes.

adduce /ə'djuːs/ v.tr. cite as an instance or as proof or evidence. □ **adducible** adj. [L adducere adduct- (as AD-, ducere lead)]

adduct /ə'dʌkt/ v.tr. Anat. (esp. of a muscle) move (a limb etc.) towards the midline of the body. □ **adductor** n. **adduction** /ə'dʌkʃ(ə)n/ n.

-ade[1] /eɪd/ suffix forming nouns: **1** an action done (blockade; tirade). **2** the body concerned in an action or process (*cavalcade*). **3** the product or result of a material or action (*arcade*; *lemonade*; *masquerade*). [from or after F *-ade* f. Prov., Sp., or Port. *-ada* or It. *-ata* f. L *-ata* fem. sing. past part. of verbs in *-are*]

-ade² /eɪd/ *suffix* forming nouns (*decade*) (cf. -AD¹). [F *-ade* f. Gk *-as -ada*]

-ade³ /eɪd/ *suffix* forming nouns: **1** = -ADE¹ (*brocade*). **2** a person concerned (*renegade*). [Sp. or Port. *-ado*, masc. form of *-ada*: see -ADE¹]

Adelaide /ˈædəˌleɪd/ a city in Australia, the capital and chief port of the state of South Australia; pop. (1990) 1,049,870. Adelaide was named after the wife of William IV.

Adélie Land /æˈdeɪlɪ/ (also **Adélie Coast**) a section of the Antarctic continent south of the 60th parallel, between Wilkes Land and King George V Land. It was discovered in 1840 by the French naval explorer J.-S.-C. Dumont d'Urville, who named it after his wife.

Aden /ˈeɪd(ə)n/ a port in Yemen at the mouth of the Red Sea; pop. (1987) 417,370. Aden was formerly under British rule, first as part of British India (from 1839), then from 1935 as a Crown Colony. It was capital of the former South Yemen from 1967 until 1990.

Aden, Gulf of a part of the eastern Arabian Sea lying between the south coast of Yemen and the Horn of Africa.

Adenauer /ˈædɪˌnaʊə(r)/, Konrad (1876–1967), German statesman, first Chancellor of the Federal Republic of Germany 1949–63. He co-founded the Christian Democratic Union in 1945. As Chancellor, he is remembered for the political and economic transformation of his country. He secured the friendship of the US and was an advocate of strengthening political and economic ties with Western countries through NATO and the European Community.

adenine /ˈædɪˌniːn/ *n.* a purine found in all living tissue as a component base of DNA or RNA. [G *Adenin* formed as ADENOIDS: see -INE⁴]

adenoids /ˈædɪˌnɔɪdz/ *n.pl. Med.* a mass of enlarged lymphatic tissue between the back of the nose and the throat, often hindering speaking and breathing in the young. □ **adenoidal** /ˌædɪˈnɔɪd(ə)l/ *adj.* [Gk *adēn -enos* gland + -OID]

adenoma /ˌædɪˈnəʊmə/ *n.* (*pl.* **adenomas** or **adenomata** /-mətə/) *Med.* a glandlike benign tumour of epithelial tissue, which may in some cases become malignant. [f. Gk *adēn* gland]

adenosine /əˈdenəˌsiːn/ *n. Biochem.* a nucleoside of adenine and ribose present in all living tissue in combined form (see ADP, AMP). □ **adenosine triphosphate** (abbr. **ATP**) a nucleotide whose breakdown in living cells to the diphosphate provides energy for physiological processes. [ADENINE + RIBOSE]

adept /ˈædept, əˈdept/ *adj. & n.* ● *adj.* (foll. by *at, in*) thoroughly proficient. ● *n.* usu. /ˈædept/ a skilled performer; an expert. □ **adeptly** *adv.* **adeptness** *n.* [L *adeptus* past part. of *adipisci* attain]

adequate /ˈædɪkwət/ *adj.* **1** sufficient, satisfactory. **2** (foll. by *to*) proportionate. **3** barely sufficient. □ **adequacy** *n.* **adequately** *adv.* [L *adaequatus* past part. of *adaequare* make equal (as AD-, *aequus* equal)]

à deux /æ ˈdɜː/ *adv. & adj.* **1** for two. **2** between two. [F]

ad fin. /æd ˈfɪn/ *abbr.* at or near the end. [L *ad finem*]

adhere /ədˈhɪə(r)/ *v.intr.* **1** (usu. foll. by *to*) (of a substance) stick fast to a surface, another substance, etc. **2** (foll. by *to*) behave according to; follow in detail (*adhered to our plan*). **3** (foll. by *to*) give support or allegiance to. [F *adhérer* or L *adhaerere* (as AD-, *haerere haes-* stick)]

adherent /ədˈhɪərənt/ *n. & adj.* ● *n.* **1** a supporter of a party, person, etc. **2** a devotee of an activity. ● *adj.* **1** (foll. by *to*) faithfully observing a rule etc. **2** (often foll. by *to*) (of a substance) sticking fast. □ **adherence** *n.* [F *adhérent* (as ADHERE)]

adhesion /ədˈhiːʒ(ə)n/ *n.* **1** the act or process of adhering. **2** the capacity of a substance to stick fast. **3** *Med.* abnormal union of surfaces due to inflammation or injury. **4** the maintenance of contact between the wheels of a vehicle and the road. **5** the giving of support or allegiance. ¶ More common in physical senses (e.g. *the glue has good adhesion*), with *adherence* used in abstract senses (e.g. *adherence to principles*). [F *adhésion* or L *adhaesio* (as ADHERE)]

adhesive /ədˈhiːsɪv/ *adj. & n.* ● *adj.* sticky, enabling surfaces or substances to adhere to one another. ● *n.* an adhesive substance, esp. one used to stick other substances together. □ **adhesively** *adv.* **adhesiveness** *n.* [F *adhésif -ive* (as ADHERE)]

adhibit /ədˈhɪbɪt/ *v.tr.* (**adhibited, adhibiting**) **1** affix. **2** apply or administer (a remedy). □ **adhibition** /ˌædhɪˈbɪʃ(ə)n/ *n.* [L *adhibere adhibit-* (as AD-, *habere* have)]

ad hoc /æd ˈhɒk/ *adv. & adj.* for a particular (usu. exclusive) purpose (*an ad hoc appointment*). [L, = to this]

ad hominem /æd ˈhɒmɪˌnem/ *adv. & adj.* **1** relating to or associated with a particular person. **2** (of an argument) appealing to the emotions and not to reason. [L, = to the person]

adiabatic /ˌeɪdaɪəˈbætɪk/ *adj. & n. Physics* ● *adj.* **1** impassable to heat. **2** occurring without heat entering or leaving the system. ● *n.* a curve or formula for adiabatic phenomena. □ **adiabatically** *adv.* [Gk *adiabatos* impassable (as A-¹, *diabainō* pass)]

adiantum /ˌædɪˈæntəm/ *n.* **1** a fern of the genus *Adiantum*; maidenhair fern. **2** (in general use) a spleenwort. [L f. Gk *adianton* maidenhair (as A-¹, *diantos* wettable)]

adieu /əˈdjuː/ *int. & n.* ● *int.* goodbye. ● *n.* (*pl.* **adieus** or **adieux** /əˈdjuːz/) a goodbye. [ME f. OF f. *à* to + *Dieu* God]

Adi Granth /ˌɑːdɪ ˈɡrʌnθ/ the principal sacred scripture of Sikhism, also called the *Granth Sahib* or 'Revered Book'. The original compilation was made under the direction of Arjan Dev (1563–1606), the fifth Sikh guru; it contains hymns and religious poetry as well as the teachings of the first five gurus. Successive gurus added to the text: the tenth and last guru, Gobind Singh (1666–1708), declared that henceforward there would be no more gurus, the Adi Granth taking their place. [Hindi (= first book), f. Skr.]

ad infinitum /æd ˌɪnfɪˈnaɪtəm/ *adv.* without limit; for ever. [L]

ad interim /æd ˈɪntərɪm/ *adv. & adj.* for the meantime. [L]

adios /ˌædɪˈɒs/ *int.* goodbye. [Sp. *adiós* f. *a* to + *Dios* God]

adipocere /ˈædɪpəˌsɪə(r)/ *n.* a greyish fatty or soapy substance generated in dead bodies subjected to moisture. [F *adipocire* f. L *adeps adipis* fat + F *cire* wax f. L *cera*]

adipose /ˈædɪˌpəʊz, -ˌpəʊs/ *adj.* of or characterized by fat; fatty. □ **adipose tissue** fatty connective tissue in animals. □ **adiposity** /ˌædɪˈpɒsɪtɪ/ *n.* [mod.L *adiposus* f. *adeps adipis* fat]

Adirondack Mountains /ˌædɪˈrɒndæk/ (also **Adirondacks**) a range of mountains in New York State, source of the Hudson and Mohawk rivers.

Adis Abeba see ADDIS ABABA.

adit /ˈædɪt/ *n.* **1** a horizontal entrance or passage in a mine. **2** a means of approach. [L *aditus* (as AD-, *itus* f. *ire it-* go)]

Adivasi /ˌɑːdɪˈvɑːsɪ/ *n.* (*pl.* **Adivasis**) a member of the aboriginal tribal peoples of India. [Hindi *ādivāsī* original inhabitant]

Adj. *abbr.* (preceding a name) Adjutant.

adjacent /əˈdʒeɪs(ə)nt/ *adj.* (often foll. by *to*) lying near or adjoining. □ **adjacency** *n.* [ME f. L *adjacere* (as AD-, *jacere* lie)]

adjective /ˈædʒɪktɪv/ *n. & adj.* ● *n.* a word or phrase naming an attribute, added to or grammatically related to a noun to modify it or describe it. ● *adj.* additional; not standing by itself; dependent. □ **adjectival** /ˌædʒɪkˈtaɪv(ə)l/ *adj.* **adjectivally** *adv.* [ME f. OF *adjectif -ive* ult. f. L *adjicere adject-* (as AD-, *jacere* throw)]

adjoin /əˈdʒɔɪn/ *v.tr.* **1** (often as **adjoining** *adj.*) be next to and joined with. **2** *archaic* = ADD 1. [ME f. OF *ajoindre, ajoign-* f. L *adjungere adjunct-* (as AD-, *jungere* join)]

adjourn /əˈdʒɜːn/ *v.* **1** *tr.* **a** put off; postpone. **b** break off (a meeting, discussion, etc.) with the intention of resuming later. **2** *intr.* of persons at a meeting: **a** break off proceedings and disperse. **b** (foll. by *to*) transfer the meeting to another place. [ME f. OF *ajorner* (as AD-, *jorn* day ult. f. L *diurnus* DIURNAL): cf. JOURNAL, JOURNEY]

adjournment /əˈdʒɜːnmənt/ *n.* adjourning or being adjourned. □ **adjournment debate** *Brit.* a debate in the House of Commons on the motion that the House be adjourned, used as an opportunity for raising various matters.

adjudge /əˈdʒʌdʒ/ *v.tr.* **1** adjudicate (a matter). **2** (often foll. by *that* + clause, or *to* + infin.) pronounce judicially. **3** (foll. by *to*) award judicially. **4** *archaic* condemn. □ **adjudgement** *n.* (also **adjudgment**). [ME f. OF *ajuger* f. L *adjudicare*: see ADJUDICATE]

adjudicate /əˈdʒuːdɪˌkeɪt/ *v.* **1** *intr.* act as judge in a competition, court, tribunal, etc. **2** *tr.* **a** decide judicially (a claim etc.). **b** (foll. by *to be* + complement) pronounce (*was adjudicated to be bankrupt*). □ **adjudicator** *n.* **adjudicative** /-kətɪv/ *adj.* **adjudication** /əˌdʒuːdɪˈkeɪʃ(ə)n/ *n.* [L *adjudicare* (as AD-, *judicare* f. *judex -icis* judge)]

adjunct /ˈædʒʌŋkt/ *n.* **1** (foll. by *to, of*) a subordinate or incidental thing. **2** an assistant; a subordinate person, esp. one with temporary appointment only. **3** *Gram.* a word or phrase used to amplify or modify the meaning of another word or words in a sentence. □ **adjunctive** /əˈdʒʌŋktɪv/ *adj.* [L *adjunctus*: see ADJOIN]

adjure /ə'dʒʊə(r)/ v.tr. (usu. foll. by to + infin.) charge or request (a person) solemnly or earnestly, esp. under oath. □ **adjuratory** /-rətərɪ/ adj. **adjuration** /ˌædʒʊ'reɪʃ(ə)n/ n. [ME f. L adjurare (as AD-, jurare swear) in LL sense 'put a person to an oath']

adjust /ə'dʒʌst/ v. **1** tr. **a** arrange; put in the correct order or position. **b** regulate, esp. by a small amount. **2** tr. (usu. foll. by to) make suitable. **3** tr. harmonize (discrepancies). **4** tr. assess (loss or damages). **5** intr. (usu. foll. by to) make oneself suited to; become familiar with (adjust to one's surroundings). □ **adjustable** adj. **adjuster** n. **adjustment** n. **adjustability** /əˌdʒʌstə'bɪlɪtɪ/ n. [F adjuster f. OF ajoster ult. f. L juxta near]

adjutant /'ædʒʊt(ə)nt/ n. **1 a** Mil. an officer who assists superior officers by communicating orders, conducting correspondence, etc. **b** an assistant. **2** (in full **adjutant bird**) a large black and white southern Asian stork of the genus Leptoptilos. □ **adjutancy** n. [L adjutare frequent. of adjuvare: see ADJUVANT]

Adjutant-General /ˌædʒʊt(ə)nt'dʒenrəl/ n. a high-ranking army administrative officer.

adjuvant /'ædʒʊv(ə)nt/ adj. & n. ● adj. **1** helpful, auxiliary. **2** Med. (of therapy) applied after the initial treatment of cancer, esp. to suppress secondary tumour formation. ● n. **1** an adjuvant person or thing. **2** Med. a substance which enhances the body's immune response to an antigen. [F adjuvant or L adjuvare (as AD-, juvare jut- help)]

Adler /'ædlə(r)/, Alfred (1870–1937), Austrian psychologist and psychiatrist. At first a disciple of Sigmund Freud, he came to disagree with Freud's idea that mental illness was caused by sexual conflicts in infancy, arguing that society and culture were equally, if not more, significant factors. In 1907 he introduced the concept of the inferiority complex, asserting that the key to understanding both personal and mass problems was the sense of inferiority and the individual's striving to compensate for this. In 1911 he and his followers formed their own school to develop the ideas of individual psychology, and in 1921 he founded the first child guidance clinic, in Vienna. □ **Adlerian** /æd'lɪərɪən/ adj.

ad lib /æd 'lɪb/ v, adj., adv., & n. ● v.intr. (**ad libbed, ad libbing**) speak or perform without formal preparation; improvise. ● adj. improvised. ● adv. as one pleases; to any desired extent. ● n. something spoken or played extempore. [abbr. of AD LIBITUM]

ad libitum /æd 'lɪbɪtəm/ adv. = AD LIB adv. [L, = according to pleasure]

ad litem /æd 'laɪtəm/ adj. (of a guardian etc.) appointed for a lawsuit. [L]

Adm. abbr. (preceding a name) Admiral.

adman /'ædmæn/ n. (pl. **admen**) colloq. a person who produces advertisements commercially.

admass /'ædmæs/ n. esp. Brit. the section of the community that is regarded as readily influenced by advertising and mass communication.

admeasure /əd'meʒə(r)/ v.tr. archaic apportion; assign in due shares. □ **admeasurement** n. [ME f. OF amesurer f. med.L admensurare (as AD-, MEASURE)]

admin /'ædmɪn/ n. colloq. administration. [abbr.]

adminicle /əd'mɪnɪk(ə)l/ n. **1** a thing that helps. **2** (in Scottish law) collateral evidence of the contents of a missing document. □ **adminicular** /ˌædmɪ'nɪkjʊlə(r)/ adj. [L adminiculum prop]

administer /əd'mɪnɪstə(r)/ v. **1** tr. attend to the running of (business affairs etc.); manage. **2** tr. **a** be responsible for the implementation of (the law, justice, punishment, etc.). **b** Eccl. give out, or perform the rites of (a sacrament). **c** (usu. foll. by to) direct the taking of (an oath). **3** tr. **a** provide, apply (a remedy). **b** give, deliver (a rebuke). **4** intr. act as administrator. □ **administrable** adj. [ME f. OF aministrer f. L administrare (as AD-, MINISTER)]

administrate /əd'mɪnɪˌstreɪt/ v.tr. & intr. administer (esp. business affairs); act as an administrator. [L administrare (as ADMINISTER)]

administration /ədˌmɪnɪ'streɪʃ(ə)n/ n. **1** management of a business. **2** the management of public affairs; government. **3** the government in power; the ministry. **4** US a President's period of office. **5** Law the management of another person's estate. **6** (foll. by of) **a** the administering of justice, an oath, etc. **b** application of remedies. [ME f. OF administration or L administratio (as ADMINISTRATE)]

administrative /əd'mɪnɪstrətɪv/ adj. concerning or relating to the management of affairs. □ **administratively** adv. [F administratif -ive or L administrativus (as ADMINISTRATION)]

administrator /əd'mɪnɪˌstreɪtə(r)/ n. **1** a person who administers a business or public affairs. **2** a person capable of organizing (is no administrator). **3** Law a person appointed to manage the estate of a person who has died intestate. **4** a person who performs official duties in some sphere, e.g. in religion or justice. □ **administratorship** n. **administratrix** n. [L (as ADMINISTER)]

admirable /'ædmərəb(ə)l/ adj. **1** deserving admiration. **2** excellent. □ **admirably** adv. [F f. L admirabilis (as ADMIRE)]

admiral /'ædmərəl/ n. **1 a** the commander-in-chief of a country's navy. **b** a naval officer of high rank, the commander of a fleet or squadron. **c** (**Admiral**) an admiral of the second grade. **2** a boldly patterned butterfly of the family Nymphalidae. (See red admiral, white admiral.) □ **admiralship** n. [ME f. OF a(d)mira(i)l etc. f. med.L a(d)miralis etc., f. Arab. 'amīr commander (cf. AMIR), assoc. with ADMIRABLE]

Admiral of the Fleet n. an admiral of the first grade.

Admiral's Cup a yacht-racing competition held every two years since 1957 between international teams of three yachts. The competition consists of inshore stages in the Solent and offshore races across the English Channel and around the Fastnet rock.

Admiralty /'ædmərəltɪ/ n. (pl. **-ies**) **1** (hist. except in titles) (in the UK) the department administering the Royal Navy. **2** (**admiralty**) Law trial and decision of maritime questions and offences. [ME f. OF admiral(i)té (as ADMIRAL)]

Admiralty Board hist. a committee of the Ministry of Defence superintending the Royal Navy.

Admiralty Islands an island group of Papua New Guinea, in the western Pacific. In 1884 the islands became a German protectorate, but after 1920 they were administered as an Australian mandate.

admiration /ˌædmɪ'reɪʃ(ə)n/ n. **1** pleased contemplation. **2** respect, warm approval. **3** an object of this (was the admiration of the whole town). [F admiration or L admiratio (as ADMIRE)]

admire /əd'maɪə(r)/ v.tr. **1** regard with approval, respect, or satisfaction. **2** express one's admiration of. [F admirer or L admirari (as AD-, mirari wonder at)]

admirer /əd'maɪərə(r)/ n. **1** a woman's suitor. **2** a person who admires, esp. a devotee of an able or famous person.

admiring /əd'maɪərɪŋ/ adj. showing or feeling admiration (an admiring follower; admiring glances). □ **admiringly** adv.

admissible /əd'mɪsɪb(ə)l/ adj. **1** (of an idea or plan) worth accepting or considering. **2** Law allowable as evidence. **3** (foll. by to) capable of being admitted. □ **admissibility** /-ˌmɪsɪ'bɪlɪtɪ/ n. [F admissible or med.L admissibilis (as ADMIT)]

admission /əd'mɪʃ(ə)n/ n. **1** an acknowledgement (admission of error; admission that he was wrong). **2 a** the process or right of entering or being admitted. **b** a charge for this (admission is £5). **3** a person admitted to a hospital. ¶ Has more general application in senses of admit than admittance. [ME f. L admissio (as ADMIT)]

admit /əd'mɪt/ v. (**admitted, admitting**) **1** tr. **a** (often foll. by to be, or that + clause) acknowledge; recognize as true. **b** accept as valid or true. **2** intr. **a** (foll. by to) acknowledge responsibility for (a deed, fault, etc.). **b** (foll. by of) allow for something to exist, have influence, etc. **3** tr. **a** allow (a person) entrance or access. **b** allow (a person) to be a member of (a class, group, etc.) or to share in (a privilege etc.). **c** (of a hospital etc.) bring in (a person) for residential treatment. **4** tr. (of an enclosed space) have room for; accommodate. **5** intr. (foll. by of) allow as possible. □ **admittable** adj. [ME f. L admittere admiss- (as AD-, mittere send)]

admittance /əd'mɪt(ə)ns/ n. **1** the right or process of admitting or being admitted, usu. to a place (no admittance except on business). **2** Electr. a measure of the ability of a circuit to conduct alternating current; the reciprocal of impedance. ¶ A more formal and technical word than admission.

admittedly /əd'mɪtɪdlɪ/ adv. as an acknowledged fact (admittedly there are problems).

admix /æd'mɪks/ v. **1** tr. & intr. (foll. by with) mingle. **2** tr. add as an ingredient.

admixture /æd'mɪkstʃə(r)/ n. **1** a thing added, esp. a minor ingredient. **2** the act of adding this. [L admixtus past part. of admiscere (as AD-, miscere mix)]

admonish /əd'mɒnɪʃ/ v.tr. **1** reprove. **2** (foll. by to + infin., or that + clause) urge. **3** give advice to. **4** (foll. by of) warn. □ **admonishment** n. **admonitory** adj. **admonition** /ˌædmə'nɪʃ(ə)n/ n. [ME f. OF amonester ult. f. L admonere (as AD-, monere monit- warn)]

ad nauseam /æd 'nɔːzɪˌæm, 'nɔːsɪ-/ adv. to an excessive or disgusting degree. [L, = to sickness]

adnominal /ˌæd'nɒmɪn(ə)l/ adj. Gram. attached to a noun. [L adnomen -minis (added name)]

ado /ə'duː/ n. fuss, busy activity; trouble, difficulty. □ **without more ado** immediately. [orig. in much ado = much to do, f. north. ME at do (= to do) f. ON at AT as sign of infin. + DO¹]

-ado /'ɑːdəʊ/ suffix forming nouns (desperado) (cf. -ADE³). [Sp. or Port. -ado f. L -atus past part. of verbs in -are]

adobe /ə'dəʊbɪ, ə'dəʊb/ n. **1** an unburnt sun-dried brick. **2** the clay used for making such bricks. [Sp. f. Arab.]

adolescent /ˌædə'les(ə)nt/ adj. & n. ● adj. between childhood and adulthood. ● n. an adolescent person. □ **adolescence** n. [ME f. OF f. L adolescere grow up]

Adonis /ə'dəʊnɪs/ Gk Mythol. a beautiful youth loved by both Aphrodite and Persephone. He was killed by a boar, but Zeus decreed that he should spend the winter of each year in the underworld with Persephone and the summer months with Aphrodite. He is often identified with the god Tammuz and his cult involved the celebration of the seasonal death and rebirth of crops. [L f. Gk f. Phoenician adōn lord]

Adonis blue n. a small bright blue European butterfly, Lysandra bellargus.

adopt /ə'dɒpt/ v.tr. **1** take (a person) into a relationship, esp. another's child as one's own. **2** choose to follow (a course of action etc.). **3** take over (an idea etc.) from another person. **4** choose as a candidate for office. **5** Brit. (of a local authority) accept responsibility for the maintenance of (a road etc.). **6** accept; formally approve (a report, accounts, etc.). □ **adoption** /ə'dɒpʃ(ə)n/ n. [F adopter or L adoptare (as AD-, optare choose)]

adoptive /ə'dɒptɪv/ adj. as a result of adoption (adoptive son; adoptive father). □ **adoptively** adv. [ME f. OF adoptif -ive f. L adoptivus (as ADOPT)]

adorable /ə'dɔːrəb(ə)l/ adj. **1** deserving adoration. **2** colloq. delightful, charming. □ **adorably** adv. [F f. L adorabilis (as ADORE)]

adore /ə'dɔː(r)/ v.tr. **1** regard with honour and deep affection. **2 a** worship as divine. **b** RC Ch. offer reverence to (the Host etc.). **3** colloq. like very much. □ **adoring** adj. **adoringly** adv. **adoration** /ˌædə'reɪʃ(ə)n/ n. [ME f. OF aourer f. L adorare worship (as AD-, orare speak, pray)]

adorer /ə'dɔːrə(r)/ n. **1** a worshipper. **2** an ardent admirer.

adorn /ə'dɔːn/ v.tr. **1** add beauty or lustre to; be an ornament to. **2** furnish with ornaments; decorate. □ **adornment** n. [ME f. OF ao(u)rner f. L adornare (as AD-, ornare furnish, deck)]

Adorno /ə'dɔːnəʊ/, Theodor Wiesengrund (born Theodor Wiesengrund) (1903–69), German philosopher, sociologist, and musicologist. Director of the Frankfurt Institute for Social Research (see FRANKFURT SCHOOL) 1958–69, he is known for such works as Dialectic of Enlightenment (1947, written with Max Horkheimer) and Negative Dialectics (1966). In these works Adorno develops his concept of reason as a key factor in social control, concluding that philosophical authoritarianism is inevitably oppressive and all theories should be systematically and consciously rejected.

ADP abbr. **1** adenosine diphosphate. **2** automatic data processing.

ad personam /ˌæd pə'səʊnæm/ adv. & adj. ● adv. to the person. ● adj. personal. [L]

Adrar des Iforas /æˌdrɑː deɪz ɪ'fɔːrɑː/ a massif region in the central Sahara, on the border between Mali and Algeria.

ad rem /æd 'rem/ adv. & adj. to the point; to the purpose. [L, = to the matter]

adrenal /ə'driːn(ə)l/ adj. & n. ● adj. **1** at or near the kidneys. **2** of the adrenal glands. ● n. (in full **adrenal gland**) either of two ductless glands above the kidneys, secreting adrenalin. [AD- + RENAL]

adrenalin /ə'drenəlɪn/ n. (also **adrenaline**) Biochem. **1** a hormone secreted by the adrenal glands, affecting circulation and muscular action, and causing excitement and stimulation. Also called epinephrine. **2** the same substance obtained from animals or by synthesis, used as a stimulant.

adrenocorticotrophic hormone /əˌdriːnəˌkɔːtɪkəʊ'trəʊfɪk, -'trɒfɪk/ n. (also **adrenocorticotropic** /-'trəʊpɪk, -'trɒpɪk/) (abbr. **ACTH**) Biochem. a hormone secreted by the pituitary gland and stimulating the adrenal glands. [ADRENAL + CORTEX + -TROPHIC, -TROPIC]

adrenocorticotrophin /əˌdriːnəˌkɔːtɪkəʊ'trəʊfɪn/ n. Biochem. = ADRENOCORTICOTROPHIC HORMONE.

Adrian IV /'eɪdrɪən/ (born Nicholas Breakspear) (c.1100–59), pope 1154–9. The only Englishman to have held this office, he assisted Henry II of England to gain control of Ireland and opposed Frederick I's (Barbarossa's) claims to power.

Adriatic /ˌeɪdrɪ'ætɪk/ adj. & n. ● adj. of or relating to the Adriatic Sea or its region. ● n. this region.

Adriatic, Marriage of the see MARRIAGE OF THE ADRIATIC.

Adriatic Sea an arm of the Mediterranean Sea between the Balkans and the Italian peninsula.

adrift /ə'drɪft/ adv. & predic.adj. **1** drifting. **2** at the mercy of circumstances. **3** colloq. **a** unfastened. **b** out of touch. **c** absent without leave. **d** (often foll. by of) failing to reach a target. **e** out of order. **f** ill-informed. [A² + DRIFT]

adroit /ə'drɔɪt/ adj. dexterous, skilful. □ **adroitly** adv. **adroitness** n. [F f. à droit according to right]

adsorb /əd'sɔːb, -'zɔːb/ v.tr. (usu. of a solid) hold (molecules of a gas or liquid or solute) to its surface, causing a thin film to form. □ **adsorbable** adj. **adsorbent** adj. & n. **adsorption** /-'sɔːpʃ(ə)n, -'zɔːp-/ n. (also **adsorbtion**). **adsorptive** /-'sɔːptɪv, -'zɔːp-/ adj. [AD-, after ABSORB]

adsorbate /æd'sɔːbeɪt, -'zɔːbeɪt/ n. a substance adsorbed.

adsuki var. of ADZUKI.

ADT abbr. Atlantic Daylight Time, one hour ahead of Atlantic Standard Time.

adulate /'ædjʊˌleɪt/ v.tr. flatter obsequiously. □ **adulator** n. **adulatory** adj. **adulation** /ˌædjʊ'leɪʃ(ə)n/ n. [L adulari adulat- fawn on]

Adullamite /ə'dʌləˌmaɪt/ n. a member of a dissident political group, a term applied to a group of Liberal MPs who seceded from their party and brought about defeat of the Reform Bill in the House of Commons in 1866. [f. cave of Adullam (1 Sam. 22:1–2), where all who were distressed or discontented came to join David]

adult /'ædʌlt, ə'dʌlt/ adj. & n. ● adj. **1** mature, grown-up. **2 a** of or for adults (adult education). **b** euphem. sexually explicit; indecent (adult films). ● n. **1** an adult person. **2** Law a person who has reached the age of majority. □ **adulthood** n. **adultly** adv. [L adultus past part. of adolescere grow up: cf. ADOLESCENT]

adulterant /ə'dʌltərənt/ adj. & n. ● adj. used in adulterating. ● n. an adulterant substance.

adulterate v. & adj. ● v.tr. /ə'dʌltəˌreɪt/ debase (esp. foods) by adding other or inferior substances. ● adj. /ə'dʌltərət/ spurious, debased, counterfeit. □ **adulterator** /-ˌreɪtə(r)/ n. **adulteration** /əˌdʌltə'reɪʃ(ə)n/ n. [L adulterare adulterat- corrupt]

adulterer /ə'dʌltərə(r)/ n. (fem. **adulteress** /-ərɪs/) a person who commits adultery. [obs. adulter (v.) f. OF avoutrer f. L adulterare: see ADULTERATE]

adulterine /ə'dʌltəˌraɪn/ adj. **1** illegal, unlicensed. **2** spurious. **3** born of adultery. [L adulterinus f. adulter: see ADULTERY]

adulterous /ə'dʌltərəs/ adj. of or involved in adultery. □ **adulterously** adv. [ME f. adulter: see ADULTERER]

adultery /ə'dʌltərɪ/ n. voluntary sexual intercourse between a married person and a person (married or not) other than his or her spouse. [ME f. OF avoutrie etc. f. avoutre adulterer f. L adulter, assim. to L adulterium]

adumbrate /'ædʌmˌbreɪt/ v.tr. **1** indicate faintly. **2** represent in outline. **3** foreshadow, typify. **4** overshadow. □ **adumbration** /ˌædʌm'breɪʃ(ə)n/ n. **adumbrative** /ə'dʌmbrətɪv/ adj. [L adumbrare (as AD-, umbrare f. umbra shade)]

ad valorem /ˌæd və'lɔːrem/ adv. & adj. (of taxes) in proportion to the estimated value of the goods concerned. [L, = according to the value]

advance /əd'vɑːns/ v., n., & adj. ● v. **1** tr. & intr. move or put forward. **2** intr. make progress. **3** tr. **a** pay (money) before it is due. **b** lend (money). **4** tr. give active support to; promote (a person, cause, or plan). **5** tr. put forward (a claim or suggestion). **6** tr. cause (an event) to occur at an earlier date (advanced the meeting three hours). **7** tr. raise (a price). **8** intr. rise (in price). **9** tr. (as **advanced** adj.) **a** far on in progress (the work is well advanced). **b** ahead of the times (advanced ideas). ● n. **1** an act of going forward. **2** progress. **3** a payment made before the due time. **4** a loan. **5** (esp. in pl.; often foll. by to) an amorous or friendly approach. **6** a rise in price. ● attrib.adj. done or supplied beforehand (advance warning; advance copy). □ **advanced level** (in the UK except Scotland) a GCE examination of a standard higher than ordinary level and GCSE. **advanced supplementary level** (in the UK except Scotland) a GCE examination with a smaller syllabus than A levels. **advance guard** a body of soldiers preceding the main body of an army. **advance man** esp. US a person who goes ahead to make arrangements for the visit of a dignitary etc. **advance on** approach threateningly. **in advance**

ahead in place or time. □ **advancer** *n.* [ME f. OF *avancer* f. LL *abante* in front f. L *ab* away + *ante* before: (n.) partly through F *avance*]

Advance Australia Fair an Australian patriotic song composed *c.*1878 by P. D. McCormick, a Scot, under the pen-name 'Amicus'.

advancement /əd'va:nsmənt/ *n.* the promotion of a person, cause, or plan. [ME f. F *avancement* f. *avancer* (as ADVANCE)]

advantage /əd'va:ntɪdʒ/ *n.* & *v.* ● *n.* **1** a beneficial feature; a favourable circumstance. **2** benefit, profit (*is not to your advantage*). **3** (often foll. by *over*) a better position; superiority in a particular respect. **4** *Tennis* the next point won after deuce. ● *v.tr.* **1** be beneficial or favourable to. **2** further, promote. □ **have the advantage of** be in a better position in some respect than. **take advantage of 1** make good use of (a favourable circumstance). **2** exploit or outwit (a person), esp. unfairly. **3** *euphem.* seduce. **turn to advantage** benefit from. (*was seen to advantage*). **to advantage** in a way which exhibits the merits. □ **advantageous** /ˌædvən'teɪdʒəs/ *adj.* **advantageously** *adv.* [ME f. OF *avantage*, *avantager* f. *avant* in front f. LL *abante*: see ADVANCE]

advection /əd'vekʃ(ə)n/ *n.* *Meteorol.* transfer of heat by the horizontal flow of air. □ **advective** /-'vektɪv/ *adj.* [L *advectio* f. *advehere* (as AD-, *vehere vect-* carry)]

Advent /'ædvent/ *n.* **1** the season before Christmas, including the four preceding Sundays. **2** the coming or second coming of Christ. **3** (**advent**) the arrival of esp. an important person or thing. □ **Advent calendar** *Brit.* a calendar for Advent, usu. made of card with flaps to open each day to reveal a picture or scene. **Advent Sunday** the first Sunday in Advent. [OE f. OF *advent*, *auvent* f. L *adventus* arrival f. *advenire* (as AD-, *venire vent-* come)]

Adventist /'ædvəntɪst/ *n.* a member of any of various Christian sects that believe in the imminent second coming of Christ. The modern movement originated in the US among followers of the millenarian preacher William Miller (1782–1849), who originally prophesied that Christ would return in 1843 or 1844. Some of Miller's followers formed the Seventh-Day Adventists (see SEVENTH-DAY ADVENTIST). □ **Adventism** *n.*

adventitious /ˌædven'tɪʃəs/ *adj.* **1** accidental, casual. **2** added from outside. **3** *Biol.* formed accidentally or in an unusual anatomical position. **4** *Law* (of property) coming from a stranger or by collateral succession rather than directly. □ **adventitiously** *adv.* [L *adventicius* (as ADVENT)]

adventure /əd'ventʃə(r)/ *n.* & *v.* ● *n.* **1** an unusual and exciting experience. **2** a daring enterprise; a hazardous activity. **3** enterprise (*the spirit of adventure*). **4** a commercial speculation. ● *v.intr.* **1** (often foll. by *into*, *upon*) dare to go or come. **2** (foll. by *on*, *upon*) dare to undertake. **3** incur risk; engage in adventure. □ **adventure playground** a playground where children are provided with functional materials for climbing on, building with, etc. □ **adventuresome** *adj.* [ME f. OF *aventure*, *aventurer* f. L *adventurus* about to happen (as ADVENT)]

adventurer /əd'ventʃərə(r)/ *n.* (*fem.* **adventuress** /-ərɪs/) **1** a person who seeks adventure, esp. for personal gain or enjoyment. **2** a financial speculator. [F *aventurier* (as ADVENTURE)]

adventurism /əd'ventʃəˌrɪz(ə)m/ *n.* a tendency to take risks, esp. in foreign policy. □ **adventurist** *n.*

adventurous /əd'ventʃərəs/ *adj.* **1** rash, venturesome; enterprising. **2** characterized by adventures. □ **adventurously** *adv.* **adventurousness** *n.* [ME f. OF *aventuros* (as ADVENTURE)]

adverb /'ædvɜ:b/ *n.* a word or phrase that modifies or qualifies another word (esp. an adjective, verb, or other adverb) or a word-group, expressing a relation of place, time, circumstance, manner, cause, degree, etc. (e.g. *gently*, *quite*, *then*, *there*). □ **adverbial** /æd'vɜ:bɪəl/ *adj.* [F *adverbe* or L *adverbium* (as AD-, VERB)]

adversarial /ˌædvə'seərɪəl/ *adj.* **1** involving conflict or opposition. **2** opposed, hostile. [ADVERSARY + -IAL]

adversary /'ædvəsəri/ *n.* & *adj.* ● *n.* (*pl.* **-ies**) **1** an enemy. **2** an opponent in a sport or game; an antagonist. ● *adj.* opposed, antagonistic. [ME f. OF *adversarie* f. L *adversarius* f. *adversus*: see ADVERSE]

adversative /əd'vɜ:sətɪv/ *adj.* (of words etc.) expressing opposition or antithesis. [F *adversatif -ive* or LL *adversativus* f. *adversari* oppose f. *adversus*: see ADVERSE]

adverse /'ædvɜ:s/ *adj.* (often foll. by *to*) **1** contrary, hostile. **2** hurtful, injurious. □ **adversely** *adv.* **adverseness** *n.* [ME f. OF *advers* f. L *adversus* past part. of *advertere* (as AD-, *vertere vers-* turn)]

adversity /əd'vɜ:sɪti/ *n.* (*pl.* **-ies**) **1** the condition of adverse fortune. **2** a misfortune. [ME f. OF *adversité* f. L *adversitas -tatis* (as ADVERSE)]

advert[1] /'ædvɜ:t/ *n.* *Brit.* an advertisement. [abbr.]

advert[2] /əd'vɜ:t/ *v.intr.* (foll. by *to*) *literary* refer in speaking or writing. [ME f. OF *a(d)vertir* f. L *advertere*: see ADVERSE]

advertise /'ædvəˌtaɪz/ *v.* **1** *tr.* (also *absol.*) draw attention to or describe favourably (goods, services, or vacant positions) in a public medium in order to sell, promote sales, or seek employees. **2** *tr.* make generally or publicly known. **3** *intr.* (foll. by *for*) seek by public notice, esp. in a newspaper. **4** *tr.* (usu. foll. by *of*, or *that* + clause) notify. □ **advertiser** *n.* [ME f. OF *a(d)vertir* (stem *a(d)vertiss-*): see ADVERT[2]]

advertisement /əd'vɜ:tɪsmənt, -tɪzmənt/ *n.* **1 a** a public notice or announcement, esp. one advertising goods or services in newspapers, on posters, or in broadcasts. **b** (usu. foll. by *for*) *colloq.* a person or thing regarded as a means of conveying the merits or demerits of something (*he's a good advertisement for a healthy lifestyle*). **2** the act or process of advertising. **3** *archaic* a notice to readers in a book etc. [earlier *avert-* f. F *avertissement* (as ADVERTISE)]

advertising /'ædvəˌtaɪzɪŋ/ *n.* the practice of influencing people through public media in order to promote sales of products and services or promote political or other messages. By the mid-17th century weekly news-sheets in London carried announcements and advertisements, but advertising did not become an established industry until the 19th century, when the growth of popular newspapers (especially in the US) enabled advertisements to reach a wider audience, and led to the setting up of advertising agencies. Advertising has grown enormously in variety and sophistication in recent years, largely as a result of the growth in the media in general; television and radio commercials, posters and billboards, and direct mail are some of its different forms. It is being used increasingly by non-commercial organizations such as governments, trade unions, and charities.

Advertising Standards Authority (abbr. **ASA**) (in the UK) an independent regulatory body set up in 1962 to monitor standards within advertising and to ensure that advertisements comply with the requirement that they be legal, decent, honest, and truthful.

advertorial /ˌædvə'tɔ:rɪəl/ *n.* an advertisement in the style of editorial comment. [blend of ADVERTISEMENT and EDITORIAL]

advice /əd'vaɪs/ *n.* **1** words given or offered as an opinion or recommendation about future action or behaviour. **2** information given; news. **3** formal notice of a transaction. **4** (in *pl.*) communications from a distance. □ **take advice 1** obtain advice, esp. from an expert. **2** act according to advice given. [ME f. OF *avis* f. L *ad* to + *visum* past part. of *videre* see]

advisable /əd'vaɪzəb(ə)l/ *adj.* **1** (of a course of action etc.) to be recommended. **2** expedient. □ **advisably** *adv.* **advisability** /-ˌvaɪzə'bɪlɪti/ *n.*

advise /əd'vaɪz/ *v.* **1** *tr.* (also *absol.*) give advice to. **2** *tr.* recommend; offer as advice (*they advise caution*; *advised me to rest*). **3** *tr.* (usu. foll. by *of*, or *that* + clause) inform, notify. **4** *intr.* (foll. by *with*) *US* consult. [ME f. OF *aviser* f. L *ad* to + *visare* frequent. of *videre* see]

advised /əd'vaɪzd/ *adj.* **1** judicious (*well-advised*). **2** deliberate, considered. □ **advisedly** /-zɪdli/ *adv.*

adviser /əd'vaɪzə(r)/ *n.* (also *disp.* **advisor**) a person who advises, esp. one appointed to do so and regularly consulted. ¶ The disputed form *advisor* is prob. influenced by the adj. *advisory*.

advisory /əd'vaɪzəri/ *adj.* & *n.* ● *adj.* **1** giving advice; constituted to give advice (*an advisory body*). **2** consisting in giving advice. ● *n.* (*pl.* **-ies**) *N. Amer.* an advisory statement, esp. a bulletin about bad weather.

advocaat /'ædvəˌka:, -ˌka:t/ *n.* a liqueur of eggs, sugar, and brandy. [Du., = ADVOCATE (being orig. an advocate's drink)]

advocacy /'ædvəkəsi/ *n.* **1** (usu. foll. by *of*) verbal support or argument for a cause, policy, etc. **2** the function of an advocate. [ME f. OF *a(d)vocacie* f. med.L *advocatia* (as ADVOCATE)]

advocate *n.* & *v.* ● *n.* /'ædvəkət/ **1** (foll. by *of*) a person who supports or speaks in favour. **2** a person who pleads for another. **3 a** a professional pleader in a court of justice. **b** *Sc.* a barrister. ● *v.tr.* /'ædvəˌkeɪt/ **1** recommend or support by argument (a cause, policy, etc.). **2** plead for, defend. □ **advocateship** /-kətˌʃɪp/ *n.* [ME f. OF *avocat* f. L *advocatus* past part. of *advocare* (as AD-, *vocare* call)]

advowson /əd'vaʊz(ə)n/ *n.* *Brit.* *Eccl.* (in ecclesiastical law) the right of recommending a member of the clergy for a vacant benefice, or of making the appointment. [ME f. AF *a(d)voweson* f. OF *avoeson* f. L *advocatio -onis* (as ADVOCATE)]

advt. *abbr.* advertisement.

Adygea /ˌɑːdɪˌgeɪə, ˌɑːdɪ'gjeɪə/ (also **Adygei Autonomous Republic** /ˈɑːdɪˌgeɪ, ˌɑːdɪ'gjeɪ/) an autonomous republic in the NW

Caucasus in SW Russia, with a largely Muslim population; pop. (1989) 432,000; capital, Maikop.

adytum /ˈædɪtəm/ n. (pl. **adyta** /-tə/) the innermost part of an ancient temple. [L f. Gk *aduton* neut. of *adutos* impenetrable (as A-¹, *duō* enter)]

adze /ædz/ n. & v. (US **adz**) ● n. a tool for cutting away the surface of wood, like an axe with an arched blade at right angles to the handle. ● v.tr. dress or cut with an adze. [OE *adesa*]

adzuki /ədˈzuːkɪ/ n. (also **adsuki, azuki** /əˈzuːkɪ/) **1** an annual leguminous plant, *Vigna angularis*, native to China and Japan. **2** the small round dark red edible bean of this plant. [Jap. *azuki*]

-ae /iː/ suffix forming plural nouns, used in names of animal and plant families, tribes, etc. (*Felidae; Rosaceae*) and instead of *-as* in the plural of many non-naturalized or unfamiliar nouns in *-a* derived from Latin or Greek (*larvae; actiniae*). [pl. *-ae* of L nouns in *-a* or pl. *-ai* of some Gk nouns]

aedile /ˈiːdaɪl/ n. either of a pair of Roman magistrates who administered public works, maintenance of roads, public games, the corn-supply, etc. □ **aedileship** n. [L *aedilis* concerned with buildings f. *aedes* building]

AEEU abbr. (in the UK) Amalgamated Engineering and Electrical Union (formed from a merger of the AEU and EETPU).

Aegean /ɪˈdʒiːən/ adj. & n. ● adj. of or relating to the Aegean Sea or its region. ● n. this region.

Aegean Islands a group of islands in the Aegean Sea, forming a region of Greece. The principal islands of the group are Chios, Samos, Lesbos, the Cyclades, and the Dodecanese.

Aegean Sea a part of the Mediterranean Sea lying between Greece and Turkey, bounded to the south by Crete and Rhodes and linked to the Black Sea by the Dardanelles, the Sea of Marmara, and the Bosporus. It is scattered with numerous islands which are now part of Greece.

aegis /ˈiːdʒɪs/ n. a protection; an impregnable defence. □ **under the aegis of** under the auspices of. [L f. Gk *aigis* mythical shield of Zeus or Athene]

Aegisthus /iːˈɡɪsθəs/ Gk Mythol. the son of Thyestes and lover of Agamemnon's wife Clytemnestra.

aegrotat /ˈiːɡrəʊˌtæt/ n. Brit. **1** a certificate that a university student is too ill to attend an examination. **2** an examination pass awarded in such circumstances. [L, = he is sick f. *aeger* sick]

Aelfric /ˈælfrɪk/ (c.955–c.1020) Anglo-Saxon monk, writer, and grammarian. His chief works are the *Catholic Homilies* (990–2) and the *Lives of the Saints* (993–6), both written in Old English. He also wrote a Latin grammar, which earned him the name 'Grammaticus'.

-aemia /ˈiːmɪə/ comb. form (also **-haemia** /ˈhiːmɪə/, US **-emia, -hemia**) forming nouns denoting that a substance is (esp. excessively) present in the blood (*bacteriaemia; pyaemia*). [mod.L f. Gk *-aimia* f. *haima* blood]

Aeneas /ɪˈniːəs/ Gk & Rom. Mythol. a Trojan leader, son of Anchises and Aphrodite, and legendary ancestor of the Romans. When Troy fell to the Greeks he escaped and, after long wandering, reached the River Tiber. The story of his voyage is recounted in Virgil's *Aeneid*.

Aeneid /ɪˈniːɪd, ˈiːnɪɪd/ an epic poem in twelve books by Virgil, which relates the wanderings of the Trojan hero Aeneas after the fall of Troy, his love affair with the Carthaginian queen Dido, his visit to his dead father Anchises in the underworld, his arrival in Italy, and his eventual victory over the hostile Italian peoples led by Turnus. Written in iambic hexameters, it is modelled on the epics of Homer. Virgil's wish that the poem (unfinished at his death) be burned was not respected.

aeolian /iːˈəʊlɪən/ adj. (US **eolian**) wind-borne. □ **aeolian harp** a stringed instrument or toy that produces musical sounds when the wind passes through it. [L *Aeolius* f. AEOLUS]

Aeolian Islands the ancient name for the LIPARI ISLANDS.

Aeolian mode n. Mus. the mode represented by the natural diatonic scale A–A. [L *Aeolius* f. *Aeolis* in Asia Minor f. Gk *Aiolis*]

Aeolus /ˈiːələs/ Gk Mythol. the guardian (later thought of as a god) of the winds. [Gk *Aiolos* f. *aiolos* swift, changeable]

aeon /ˈiːɒn/ n. (also **eon**) **1** a very long or indefinite period. **2** an age of the universe. **3** Astron. a thousand million years. **4** an eternity. **5** Philos. (in Neoplatonism, Platonism, and Gnosticism) a power existing from eternity, an emanation or phase of the supreme deity. [eccl.L f. Gk *aiōn* age]

aepyornis /ˌiːpɪˈɔːnɪs/ n. a gigantic flightless bird of the genus *Aepyornis*, known from remains found in Madagascar and exterminated probably less than a thousand years ago. Also called *elephant-bird*. [L f. Gk *aipus* high, *ornis* bird]

aerate /ˈeəreɪt/ v.tr. **1** charge (a liquid) with a gas, esp. carbon dioxide, e.g. to produce effervescence. **2** expose to the mechanical or chemical action of the air. □ **aerator** n. **aeration** /eəˈreɪʃ(ə)n/ n. [L *aer* AIR + -ATE³, after F *aérer*]

aerenchyma /eəˈreŋkɪmə/ n. Bot. a soft plant tissue containing air spaces, found esp. in many aquatic plants. [Gk *aēr* air + *egkhuma* infusion]

aerial /ˈeərɪəl/ n. & adj. ● n. a metal rod, wire, or other structure by which signals are transmitted or received as part of a radio or television transmission or receiving system. ● adj. **1** by or from or involving aircraft (*aerial navigation; aerial photography*). **2 a** existing, moving, or happening in the air. **b** of or in the atmosphere, atmospheric. **3 a** thin as air, ethereal. **b** immaterial, imaginary. **c** of air, gaseous. □ **aerially** adv. **aeriality** /ˌeərɪˈælɪtɪ/ n. [L *aerius* f. Gk *aerios* f. *aēr* air]

aerialist /ˈeərɪəlɪst/ n. a high-wire or trapeze artist.

aerie var. of EYRIE.

aeriform /ˈeərɪˌfɔːm/ adj. **1** of the form of air; gaseous. **2** unsubstantial, unreal. [L *aer* AIR + -FORM]

aero- /ˈeərəʊ/ comb. form **1** air. **2** aircraft. [Gk *aero-* f. *aēr* air]

aerobatics /ˌeərəˈbætɪks/ n.pl. **1** feats of expert and usu. spectacular flying and manoeuvring of aircraft. **2** (as sing.) a performance of these. [AERO- + ACROBATICS]

aerobe /ˈeərəʊb/ n. a micro-organism usu. growing in the presence of oxygen, or needing oxygen for growth. [F *aérobie* (as AERO-, Gk *bios* life)]

aerobic /eəˈrəʊbɪk/ adj. **1** of or relating to aerobics. **2** Biol. relating to or requiring free oxygen.

aerobics /eəˈrəʊbɪks/ n.pl. vigorous exercises often performed to music, designed to increase the body's oxygen intake.

aerobiology /ˌeərəʊbaɪˈɒlədʒɪ/ n. the study of airborne micro-organisms, pollen, spores, etc., esp. as agents of infection.

aerodrome /ˈeərəˌdrəʊm/ n. Brit. a small airport or airfield. ¶ Now largely replaced by *airfield* and *airport*.

aerodynamics /ˌeərəʊdaɪˈnæmɪks/ n.pl. (usu. treated as sing.) the study of the interaction between the air and solid bodies moving through it. □ **aerodynamic** adj. **aerodynamically** adv. **aerodynamicist** /-ˈnæmɪsɪst/ n.

aero-engine /ˈeərəʊˌendʒɪn/ n. an engine used to power an aircraft.

aerofoil /ˈeərəˌfɔɪl/ n. Brit. a structure with curved surfaces (e.g. a wing, fin, or tailplane) designed to give lift in flight.

aerogramme /ˈeərəˌɡræm/ n. (also **aerogram**) an air letter in the form of a single sheet that is folded and sealed.

aerolite /ˈeərəˌlaɪt/ n. a stony meteorite.

aerology /eəˈrɒlədʒɪ/ n. the study of the atmosphere, esp. away from ground level. □ **aerological** /ˌeərəˈlɒdʒɪk(ə)l/ adj.

aeronautics /ˌeərəʊˈnɔːtɪks/ n.pl. (usu. treated as sing.) the science or practice of motion or travel in the air. □ **aeronautic** adj. **aeronautical** adj. [mod.L *aeronautica* (as AERO-, NAUTICAL)]

aeronomy /eəˈrɒnəmɪ/ n. the science of the upper atmosphere.

aeroplane /ˈeərəˌpleɪn/ n. esp. Brit. a powered heavier-than-air flying vehicle with fixed wings. The first true aeroplane to achieve controlled sustained flight was that designed, built, and flown by the Wright brothers in 1903. Aeroplanes were used in the First World War, and since then development has been stimulated largely by military requirements. Early aeroplanes were largely piston-engined externally braced biplanes, which began to be superseded in the 1920s and 1930s by internally braced monoplanes of increasing speed and strength. The first jet fighters were operational before the end of the Second World War. In the postwar period civil aviation expanded massively with the introduction of large long-range airliners powered by propellers or, increasingly, jets. Supersonic flight, first achieved in the US in 1947, is confined to military aviation with the exception of the Anglo-French Concorde airliner. [F *aéroplane* (as AERO-, PLANE¹)]

aerosol /ˈeərəˌsɒl/ n. **1 a** a substance packed under pressure, usually with a propellant gas, and able to be released as a fine spray. (See note below.) **b** a container holding such a substance. **2** a system of colloidal particles dispersed in a gas (e.g. fog or smoke). [AERO- + SOL¹, orig. in sense 2]

▪ Aerosols as a means of applying paints, insecticides, polishes, cosmetics, etc. were first developed commercially in the US in the early 1940s. The propellant gas should ideally be inert, non-toxic, and non-flammable; for this reason chlorofluorocarbons were often used, but they have been banned in many countries after having been shown to damage the earth's ozone layer. Hydrocarbon gases have also been used as propellants; some newer aerosols use carbon dioxide generated in the canister.

aerospace /'eərəʊ,speɪs/ n. **1** the earth's atmosphere and outer space. **2** the technology of aviation in this region.

aerotrain /'eərəʊ,treɪn/ n. a train that is supported on an air-cushion and guided by a track. [F *aérotrain* (as AERO-, TRAIN)]

aeruginous /ɪə'ruːdʒɪnəs/ adj. of the nature or colour of verdigris. [L *aeruginosus* f. *aerugo -inis* verdigris f. *aes aeris* bronze]

Aeschines /'iːskɪ,niːz/ (*c*.390–*c*.314 BC), Athenian orator and statesman. He opposed Demosthenes' efforts to unite the Greek city-states against Macedon, with which he attempted to make peace. Aeschines was tried for treason in 343 but acquitted, and left Athens for Rhodes in 330 after failing to defeat Demosthenes.

Aeschylus /'iːskɪləs/ (*c*.525–*c*.456 BC), Greek dramatist. The earliest writer of Greek tragic drama whose works survive, he is best known for his trilogy the *Oresteia* (458 BC), consisting of *Agamemnon*, *Choephoroe*, and *Eumenides*. These tell the story of Agamemnon's murder at the hands of his wife Clytemnestra and the vengeance of their son Orestes. Aeschylus is distinguished by the scale and grandeur of his conceptions. He departed from tradition by giving more weight to dialogue than to choral song and in adding a second actor to the existing one plus chorus.

Aesculapian /,iːskjʊ'leɪpɪən/ adj. of or relating to medicine or physicians. [L *Aesculapius* f. Gk *Asklēpios* god of medicine]

Aesir /'iːsɪə(r)/ *Scand. Mythol.* the Norse gods collectively, including Odin, Thor, Balder, Freya, and Tyr.

Aesop /'iːsɒp/ (6th century BC), Greek storyteller. The moral animal fables associated with him were probably collected from many sources, and initially communicated orally; they were later popularized by the Roman poet Phaedrus (1st century BC), who translated some of them into Latin. Aesop is said to have lived as a slave on the island of Samos.

aesthete /'iːsθiːt/ n. (US **esthete**) a person who has or professes to have a special appreciation of beauty. [Gk *aisthētēs* one who perceives, or f. AESTHETIC]

aesthetic /iːs'θetɪk/ adj. & n. (US **esthetic**) ● adj. **1** concerned with beauty or the appreciation of beauty. **2** having such appreciation; sensitive to beauty. **3** in accordance with the principles of good taste. ● n. **1** (in *pl.*) the philosophy of the beautiful, esp. in art. **2** a set of principles of good taste and the appreciation of beauty. □ **aesthetically** adv. **aestheticism** /-'θetɪ,sɪz(ə)m/ n. [Gk *aisthētikos* f. *aisthanomai* perceive]

Aesthetic Movement n. a literary and artistic movement which flourished in England in the 1880s, devoted to 'art for art's sake', and rejecting the notion that art should have a social or moral purpose. Its chief exponents included Wilde, Beerbohm, and Beardsley, and others associated with the journal the *Yellow Book*. They were heavily influenced by the Pre-Raphaelites and Walter Pater. They were notorious for their dedication to the ideal of beauty, which was often carried to extravagant lengths. Although ridiculed for their affectation of speech and manner and eccentricity of dress, they nevertheless helped to focus on important formal concerns in art.

aestival /'iːstɪv(ə)l, e'staɪv(ə)l, iː'staɪv(ə)l/ adj. (US **estival**) *formal* belonging to or appearing in summer. [ME f. OF *estival* f. L *aestivalis* f. *aestivus* f. *aestus* heat]

aestivate /'iːstɪ,veɪt, 'iːs-/ *v.intr.* (US **estivate**) **1** *Zool.* spend the summer or dry season in a state of torpor. **2** *formal* pass the summer. [L *aestivare aestivat-*]

aestivation /,iːstɪ'veɪʃ(ə)n, ,iːs-/ n. (US **estivation**) **1** *Bot.* the arrangement of petals in a flower-bud before it opens (cf. VERNATION). **2** *Zool.* spending the summer or dry season in a state of torpor.

aet. abbr. (also **aetat.**) *aetatis*.

aetatis /iː'tɑːtɪs, aɪ'tɑː-/ adj. of or at the age of.

aether var. of ETHER 2, 3.

aetiology /,iːtɪ'ɒlədʒi/ n. (US **etiology**) **1** the assignment of a cause or reason. **2** the philosophy of causation. **3** *Med.* the science of the causes of disease. □ **aetiologic** /,iːtɪə'lɒdʒɪk/ adj. **aetiological** adj. **aetiologically** adv. [LL *aetiologia* f. Gk *aitiologia* f. *aitia* cause]

AEU abbr. *hist.* (in the UK) Amalgamated Engineering Union.

AF abbr. audio frequency.

af- /əf/ prefix assim. form of AD- before *f*.

Afar /'ɑːfɑː(r)/ n. & adj. (also called *Danakil*) ● n. (*pl.* same or **Afars**) **1** a member of a widely spread Hamitic people of the Republic of Djibouti and NE Ethiopia. **2** the Cushitic language of this people. ● adj. of or relating to the Afar. [Afar *qafar*]

afar /ə'fɑː(r)/ adv. at or to a distance. □ **from afar** from a distance. [ME f. A², A⁻⁴ + FAR]

Afars and Issas, French Territory of the /'ɑːfɑːz, 'iːsɑːz/ the former name (1946–77) for the Republic of DJIBOUTI.

AFC abbr. **1** (in the UK) Air Force Cross. **2** (in the UK) Association Football Club.

AFDCS abbr. (in the UK) Association of First Division Civil Servants (cf. FDA).

affable /'æfəb(ə)l/ adj. friendly, good-natured. □ **affably** adv. **affability** /,æfə'bɪlɪti/ n. [F f. L *affabilis* f. *affari* (as AD-, *fari* speak)]

affair /ə'feə(r)/ n. **1** a concern; a business; a matter to be attended to (*that is my affair*). **2 a** a celebrated or notorious happening or sequence of events. **b** *colloq.* a noteworthy thing or event (*was a puzzling affair*). **3** = *love affair*. **4** (in *pl.*) **a** ordinary pursuits of life. **b** business dealings. **c** public matters (*current affairs*). [ME f. AF *afere* f. OF *afaire* f. *à faire* to do: cf. ADO]

affaire /ə'feə(r)/ n. (also **affaire de cœur** /də 'kɜː(r)/) a love affair. [F]

affairé /ə'feəreɪ/ adj. busy; involved. [F]

affect¹ /ə'fekt/ *v.tr.* **1 a** produce an effect on. **b** (of a disease etc.) attack (*his liver is affected*). **2** move; touch the feelings of (*affected me deeply*). ¶ Often confused with *effect*, which as a verb means 'bring about; accomplish'. □ **affecting** adj. **affectingly** adv. [F *affecter* or L *afficere affect-* influence (as AD-, *facere* do)]

affect² /ə'fekt/ *v.tr.* **1** pretend to have or feel (*affected indifference*). **2** (foll. by *to* + infin.) pretend. **3** assume the character or manner of; pose as (*affect the freethinker*). **4** make a show of liking or using (*she affects fancy hats*). [F *affecter* or L *affectare* aim at, frequent. of *afficere* (as AFFECT¹)]

affect³ /'æfekt/ n. *Psychol.* a feeling, emotion, or desire, esp. as leading to action. [G *Affekt* f. L *affectus* disposition f. *afficere* (as AFFECT¹)]

affectation /,æfek'teɪʃ(ə)n/ n. **1** an assumed or contrived manner of behaviour, esp. in order to impress. **2** (foll. by *of*) a studied display. **3** pretence. [F *affectation* or L *affectatio* (as AFFECT²)]

affected /ə'fektɪd/ adj. **1** in senses of AFFECT¹, AFFECT². **2** artificially assumed or displayed; pretended (*an affected air of innocence*). **3** (of a person) full of affectation; artificial. **4** (prec. by adv.; often foll. by *towards*) *archaic* disposed, inclined. □ **affectedly** adv. **affectedness** n.

affection /ə'fekʃ(ə)n/ n. **1** (often foll. by *for*, *towards*) goodwill; fond or kindly feeling. **2** a disease; a diseased condition. **3** a mental state; an emotion. **4** a mental disposition. **5** the act or process of affecting or being affected. □ **affectional** adj. (in sense 3). [ME f. OF f. L *affectio -onis* (as AFFECT¹)]

affectionate /ə'fekʃənət/ adj. loving, fond; showing love or tenderness. □ **affectionately** adv. [F *affectionné* or med.L *affectionatus* (as AFFECTION)]

affective /ə'fektɪv/ adj. **1** concerning the affections; emotional. **2** *Psychol.* relating to affects. □ **affectivity** /,æfek'tɪvɪti/ n. [F *affectif -ive* f. LL *affectivus* (as AFFECT¹)]

affenpinscher /'æfən,pɪnʃə(r)/ n. a breed of toy dog resembling the griffon, with a profuse wiry coat. [G f. *Affe* monkey + *Pinscher* terrier]

afferent /'æfərənt/ adj. *Physiol.* conducting inwards or towards (*afferent nerves*; *afferent vessels*) (opp. EFFERENT). [L *afferre* (as AD-, *ferre* bring)]

affiance /ə'faɪəns/ *v.tr.* (usu. in *passive*) *literary* promise solemnly to give (a person) in marriage. [ME f. OF *afiancer* f. med.L *affidare* (as AD-, *fidus* trusty)]

affidavit /,æfɪ'deɪvɪt/ n. a written statement confirmed by oath, for use as evidence in court. [med.L, = has stated on oath, f. *affidare*: see AFFIANCE]

affiliate v. & n. ● v. /ə'fɪlɪ,eɪt/ **1** *tr.* (usu. in *passive*; foll. by *to*, *with*) attach or connect (a person or society) with a larger organization. **2** *tr.* (of an institution) adopt (persons as members, societies as branches). **3** *intr.* **a** (foll. by *to*) associate oneself with a society. **b** (foll. by *with*) associate oneself with a political party. ● n. /ə'fɪlɪ,eɪt, -ɪət/ an affiliated person or organization. □ **affiliative** /-ɪ,eɪtɪv/ adj. [med.L *affiliare* adopt (as AD-, *filius* son]

affiliation /ə,fɪlɪ'eɪʃ(ə)n/ n. the act or process of affiliating or being affiliated. □ **affiliation order** *Brit.* a legal order that the man judged to be the father of an illegitimate child must help to support it. [F f. med.L *affiliatio* f. *affiliare*: see AFFILIATE]

affined /ə'faɪnd/ adj. related, connected. [*affine* (adj.) f. L *affinis* related: see AFFINITY]

affinity /əˈfɪnɪtɪ/ n. (pl. **-ies**) **1** (often foll. by *between, to, for*) a spontaneous or natural liking for or attraction to a person or thing. **2** relationship, esp. by marriage. **3** resemblance in structure between animals, plants, or languages. **4** a similarity of characteristics suggesting a relationship. **5** *Chem.* the tendency of certain substances to combine with others, often with high specificity. □ **affinity (credit) card 1** a credit card available to members of an affinity group and entitling them to a range of discounts etc. **2** a credit card issued to supporters of a particular charity and earning a small donation from the issuing bank on each use. **affinity group** a group of people linked by a common interest or purpose. [ME f. OF *afinité* f. L *affinitas -tatis* f. *affinis* related, lit. 'bordering on' (as AD- + *finis* border)]

affirm /əˈfɜːm/ v. **1** tr. assert strongly; state as a fact. **2** intr. **a** *Law* make an affirmation. **b** make a formal declaration. **3** tr. *Law* confirm, ratify (a judgement). □ **affirmatory** adj. **affirmer** n. [ME f. OF *afermer* f. L *affirmare* (as AD-, *firmus* strong)]

affirmation /ˌæfəˈmeɪʃ(ə)n/ n. **1** the act or process of affirming or being affirmed. **2** *Law* a solemn declaration by a person who conscientiously declines to take an oath. [F *affirmation* or L *affirmatio* (as AFFIRM)]

affirmative /əˈfɜːmətɪv/ adj. & n. ● adj. **1** affirming; asserting that a thing is so. **2** (of a vote) expressing approval. ● n. **1** an affirmative statement, reply, or word. **2** (prec. by *the*) a positive or affirming position. □ **affirmative action** esp. *N. Amer.* action favouring those who tend to suffer from discrimination; positive discrimination, esp. in recruitment to jobs. **in the affirmative** with affirmative effect; so as to accept or agree to a proposal; yes (*the answer was in the affirmative*). □ **affirmatively** adv. [ME f. OF *affirmatif -ive* f. LL *affirmativus* (as AFFIRM)]

affix v. & n. ● v.tr. /əˈfɪks/ **1** (usu. foll. by *to, on*) attach, fasten. **2** add in writing (a signature or postscript). **3** impress (a seal or stamp). ● n. /ˈæfɪks/ **1** an appendage; an addition. **2** *Gram.* an addition or element placed at the beginning (*prefix*) or end (*suffix*) of a root, stem, or word, or in the body of a word (*infix*), to modify its meaning. □ **affixture** /əˈfɪkstʃə(r)/ n. [F *affixer, affixe* or med.L *affixare* frequent. of L *affigere* (as AD-, *figere fix-* fix)]

afflatus /əˈfleɪtəs/ n. a divine creative impulse; inspiration. [L f. *afflare* (as AD-, *flare flat-* to blow)]

afflict /əˈflɪkt/ v.tr. inflict bodily or mental suffering on. □ **afflicted with** suffering from. □ **afflictive** adj. [ME f. L *afflictare*, or *afflict-* past part. stem of *affligere* (as AD-, *fligere flict-* dash)]

affliction /əˈflɪkʃ(ə)n/ n. **1** physical or mental distress, esp. pain or illness. **2** a cause of this. [ME f. OF f. L *afflictio -onis* (as AFFLICT)]

affluence /ˈæflʊəns/ n. an abundant supply of money, commodities, etc.; wealth. [ME f. F f. L *affluentia* f. *affluere*: see AFFLUENT]

affluent /ˈæflʊənt/ adj. & n. ● adj. **1** wealthy, rich. **2** abundant. **3** flowing freely or copiously. ● n. a tributary stream. □ **affluently** adv. [ME f. OF f. L *affluere* (as AD-, *fluere flux-* flow)]

afflux /ˈæflʌks/ n. a flow towards a point; an influx. [med.L *affluxus* f. L *affluere*: see AFFLUENT]

afford /əˈfɔːd/ v.tr. **1** (prec. by *can* or *be able to*; often foll. by *to* + infin.) **a** have enough money, means, time, etc., for; be able to spare (*can afford £50; could not afford a holiday; can we afford to buy a new television?*). **b** be in a position to do something (esp. without risk of adverse consequences) (*can't afford to let him think so*). **2** yield a supply of. **3** provide (*affords a view of the sea*). □ **affordable** adj. **affordability** /əˌfɔːdəˈbɪlɪtɪ/ n. [ME f. OE *geforthian* promote (as Y-, FORTH), assim. to words in AF-]

afforest /əˈfɒrɪst/ v.tr. **1** convert into forest. **2** plant with trees. □ **afforestation** /əˌfɒrɪˈsteɪʃ(ə)n/ n. [med.L *afforestare* (as AD-, *foresta* FOREST)]

affranchise /əˈfræntʃaɪz/ v.tr. release from servitude or an obligation. [OF *afranchir* (as ENFRANCHISE, with prefix A-³)]

affray /əˈfreɪ/ n. a breach of the peace by fighting or rioting in public. [ME f. AF *afrayer* (v.) f. OF *esfreer* f. Rmc]

affricate /ˈæfrɪkət/ n. *Phonet.* a combination of a plosive with an immediately following fricative or spirant of corresponding position, e.g. *ch* as in *chair*. [L *affricare* (as AD-, *fricare* rub)]

affront /əˈfrʌnt/ n. & v. ● n. an open insult (*feel it an affront; offer an affront to*). ● v.tr. **1** insult openly. **2** offend the modesty or self-respect of. **3** face, confront. [ME f. OF *afronter* slap in the face, insult, ult. f. L *frons frontis* face]

Afghan /ˈæfɡæn/ n. & adj. ● n. **1 a** a native or national of Afghanistan. **b** a person of Afghan descent. **2** the official language of Afghanistan (also called *Pashto*). **3** (**afghan**) a knitted and sewn woollen blanket or shawl. **4** (in full **Afghan coat**) a kind of sheepskin coat with the skin outside and usu. with a shaggy border. ● adj. of or relating to Afghanistan or its people or language. [Pashto *afghānī*]

Afghan hound n. a breed of tall hunting dog with long silky hair.

afghani /æfˈɡɑːnɪ/ n. (pl. **afghanis**) the chief monetary unit of Afghanistan, equal to 100 puls. [Pashto]

Afghanistan /æfˈɡænɪˌstæn/ a mountainous landlocked republic in central Asia; pop. (est. 1991) 16,600,000; official languages, Pashto and Dari (the local form of Persian); capital, Kabul. Part of the Indian Mogul empire, Afghanistan became independent in the mid-18th century, and in the 19th and early 20th centuries was a focal point for conflicting Russian and British interests on the North-West Frontier, the British fighting three wars against the Afghans between 1839 and 1919. A constitutional monarchy since 1930, Afghanistan became politically unstable in the 1970s and was invaded by the Soviet Union in Dec. 1979. The period of Soviet occupation was marked by continual warfare against Afghan guerrillas, and about two to three million people left the country as refugees. Soviet forces withdrew in 1988–9; the regime they had set up collapsed in 1992, leaving the country in turmoil with various groups struggling for power.

aficionado /əˌfɪsjəˈnɑːdəʊ/ n. (pl. **-os**) a devotee of a sport or pastime (orig. of bullfighting). [Sp.]

afield /əˈfiːld/ adv. **1** away from home; to or at a distance (esp. *far afield*). **2** in the field. [OE (as A², FIELD)]

afire /əˈfaɪə(r)/ adv. & predic.adj. **1** on fire. **2** intensely roused or excited.

aflame /əˈfleɪm/ adv. & predic.adj. **1** in flames. **2** = AFIRE 2.

aflatoxin /ˌæfləˈtɒksɪn/ n. *Chem.* any of a class of related toxic compounds produced by the mould *Aspergillus flavus*, sometimes present in stored grain and peanuts and causing liver damage and cancer. [*Aspergillus* + *flavus* + TOXIN]

AFL–CIO abbr. American Federation of Labor and Congress of Industrial Organizations (see AMERICAN FEDERATION OF LABOR).

afloat /əˈfləʊt/ adv. & predic.adj. **1** floating in water or air. **2** at sea; on board ship. **3** out of debt or difficulty. **4** in general circulation; current. **5** full of or covered with a liquid. **6** in full swing. [OE (as A², FLOAT)]

AFM abbr. (in the UK) Air Force Medal.

afoot /əˈfʊt/ adv. & predic.adj. **1** in operation; progressing. **2** astir; on the move.

afore /əˈfɔː(r)/ prep., conj., & adv. archaic or dial. before; previously; in front (of). [OE *onforan* (as A², FORE)]

afore- /əˈfɔː(r)/ comb. form before, previously (*aforementioned; aforesaid*).

aforethought /əˈfɔːθɔːt/ adj. premeditated (following a noun: *malice aforethought*).

a fortiori /ˌeɪ fɔːtɪˈɔːraɪ/ adv. & adj. with a yet stronger reason (than a conclusion already accepted); more conclusively. [L]

afoul /əˈfaʊl/ adv. esp. *N. Amer.* foul. □ **run afoul of** run foul of.

afraid /əˈfreɪd/ predic.adj. **1** (often foll. by *of*, or *that* or *lest* + clause) alarmed, frightened. **2** (foll. by *to* + infin.) unwilling or reluctant for fear of the consequences (*was afraid to go in*). □ **be afraid** (foll. by *that* + clause) *colloq.* admit or declare with (real or politely simulated) regret (*I'm afraid there's none left*). [ME, past part. of obs. *affray* (v.) f. AF *afrayer* f. OF *esfreer*]

AFRC abbr. hist. (in the UK) Agricultural and Food Research Council.

afreet /ˈæfriːt/ n. (also **afrit**) a demon in Arabian mythology. [Arab. *ʿifrīt*]

afresh /əˈfreʃ/ adv. anew; with a fresh beginning. [A-² + FRESH]

Africa /ˈæfrɪkə/ the second largest continent, a southward projection of the Old World land mass, divided roughly in two by the equator and surrounded by sea except where the Isthmus of Suez joins it to Asia. The human race may well have originated in Africa; fossil hominid remains dating from about 14,000,000 years BP have been found in East Africa. Egypt in the north-east was one of the world's earliest centres of civilization, and the Mediterranean coast has been subject to European influence since classical times, but much of the continent remained unknown to the outside world until voyages of discovery between the 15th and 17th centuries. The interior was explored and partitioned by European nations in the second half of the 19th century. In the three decades following the Second World War, almost all the former colonies became established as independent states.

African /ˈæfrɪkən/ n. & adj. ● n. **1** a native of Africa (esp. a dark-skinned person). **2** a person of African descent. ● adj. of or relating to Africa. □ **African American** a black American. **African-American** of or relating to black Americans. **African elephant** the elephant, *Loxodonta africana*, of Africa, which is larger than the Indian elephant and has much larger ears. **African violet** a saintpaulia, *Saintpaulia ionantha*, native to tropical central and East Africa, with heart-shaped velvety leaves and blue, purple, pink, or white flowers that resemble violets. [L *Africanus*]

Africana /ˌæfrɪˈkɑːnə/ *n.pl.* things connected with Africa.

Africander /ˌæfrɪˈkændə(r)/ *n.* (also **Afrikander**) one of a South African breed of sheep or longhorn cattle. [Afrik. *Afrikaander* alt. of Du. *Afrikaner* after *Hollander* etc.]

Africanize /ˈæfrɪkəˌnaɪz/ *v.tr.* (also **-ise**) **1** make African in character. **2** place under the control of African blacks. **3** (usu. as **Africanized** *adj.*) hybridize (honey bees) with a stock of African origin to give an unusually aggressive strain.

African National Congress (abbr. **ANC**) a South African political party and black nationalist organization, founded in 1912 with the aim of securing racial equality and black representation in Parliament. After the ANC was banned by the South African government in 1960, its members turned to sabotage and guerrilla warfare, operating both in and outside South Africa, with the headquarters in Zambia. In 1990 the ban on the ANC was lifted, and Nelson Mandela, the ANC's most prominent leader, was released from prison. During 1991–3 progress towards reform was hampered by fighting between factions of the ANC's mainly Xhosa supporters and the rival (largely Zulu) Inkatha organization and by serious clashes with the security forces. Following the country's first democratic elections in April 1994 the ANC gained 62.6 per cent of the vote (252 seats in the National Assembly) and Mandela became the new President of South Africa.

Afrikaans /ˌæfrɪˈkɑːns/ *n.* one of the official languages of the Republic of South Africa. It is a development of 17th-century Dutch brought to South Africa by settlers from the Netherlands, and is spoken by about 6 million people. [Du., = African]

Afrika Korps /ˈæfrɪkə ˌkɔː(r)/ a German army force sent to North Africa in 1941 under the command of General Rommel. Early success was ended by defeat at El Alamein in 1942, and the force was driven out of North Africa in 1943.

Afrikander var. of AFRICANDER.

Afrikaner /ˌæfrɪˈkɑːnə(r)/ *n.* **1** an Afrikaans-speaking white person in South Africa, esp. one descended from the Dutch and Huguenot settlers of the 17th century (see BOER). **2** a southern African gladiolus. [Afrik., formed as AFRICANDER]

afrit var. of AFREET.

Afro /ˈæfrəʊ/ *adj. & n.* ● *adj.* (of a hairstyle) long and bushy, as naturally grown by some blacks. ● *n.* (pl. **-os**) an Afro hairstyle. [AFRO-, or abbr. of AFRICAN]

Afro- /ˈæfrəʊ/ *comb. form* African (*Afro-Asian*). [L *Afer Afr-* African]

Afro-American /ˌæfrəʊəˈmerɪkən/ *adj. & n.* = African American.

Afro-Asiatic /ˌæfrəʊˌeɪʃɪˈætɪk, -ˌeɪzɪ-/ *adj.* (also called *Hamito-Semitic*) denoting or relating to a family of languages spoken in the Middle East and northern Africa. They can be divided into five groups: Semitic, Egyptian, Berber, Cushitic, and Chadic. (See also SEMITIC.)

Afro-Caribbean /ˌæfrəʊˌkærɪˈbɪən, -kəˈrɪbɪən/ *n. & adj.* ● *n.* a person of African descent living in or coming from the Caribbean. ● *adj.* of or relating to Afro-Caribbeans.

Afrocentric /ˌæfrəʊˈsentrɪk/ *adj.* centring on African or Afro-American culture; regarding African or black culture as pre-eminent. [AFRO- + -CENTRIC]

afrormosia /ˌæfrɔːˈməʊzɪə/ *n.* **1** an African tree, *Pericopsis elata*, yielding a hard wood resembling teak and used for furniture. **2** this wood. [mod.L *Afrormosia* former genus name, f. AFRO- + *Ormosia* genus of trees]

aft /ɑːft/ *adv. Naut. & Aeron.* at or towards the stern or tail. [prob. f. ME *baft*: see ABAFT]

after /ˈɑːftə(r)/ *prep., conj., adv., & adj.* ● *prep.* **1 a** following in time; later than (*after six months; after midnight; day after day*). **b** *N. Amer.* in specifying time (*a quarter after eight*). **2** (with causal force) in view of (something that happened shortly before) (*after your behaviour tonight what do you expect?*). **3** (with concessive force) in spite of (*after all my efforts I'm no better off*). **4** behind (*shut the door after you*). **5** in pursuit or quest of (*run after them; inquire after him; hanker after it; is after a job*). **6** about, concerning (*asked after her; asked after her health*). **7** in allusion to (named him William after the prince). **8** in imitation of (a person, word, etc.) (*a painting after Rubens; 'aesthete' is formed after 'athlete'*). **9** next in importance to (*the best book on the subject after mine*). **10** according to (*after a fashion*). ● *conj.* in or at a time later than that when (*left after they arrived*). ● *adv.* **1** later in time (*soon after; a week after*). **2** behind in

place (*followed on after; look before and after*). ● *adj.* **1** later, following (*in after years*). **2** *Naut.* nearer the stern (*after cabins; after mast; after-peak*). □ **after all 1** in spite of all that has happened or has been said etc. (*after all, what does it matter?*). **2** in spite of one's exertions, expectations, etc. (*they tried for an hour and failed after all; so you have come after all!*).

after-care 1 care of a patient after a stay in hospital or of a person on release from prison. **2** support or advice offered to a customer following the purchase of a product. **after-damp** choking gas left after an explosion of firedamp in a mine. **after-effect** an effect that follows after an interval or after the primary action of something. **after-image** an image retained by a sense-organ, esp. the eye, and producing a sensation after the cessation of the stimulus. **after one's own heart** see HEART. **after-taste** a taste remaining or recurring after eating or drinking. **after you** a formula used in offering precedence. [OE *æfter* f. Gmc]

afterbirth /ˈɑːftəˌbɜːθ/ *n.* the placenta and foetal membranes discharged from the womb after the birth of offspring.

afterburner /ˈɑːftəˌbɜːnə(r)/ *n.* an auxiliary burner in a jet engine to increase thrust.

afterglow /ˈɑːftəˌɡləʊ/ *n.* a light or radiance remaining after its source has disappeared or been removed.

afterlife /ˈɑːftəˌlaɪf/ *n.* **1** *Relig.* life after death. **2** life at a later time.

aftermarket /ˈɑːftəˌmɑːkɪt/ *n.* **1** a market in spare parts and components. **2** *Stock Exch.* a market in shares after their original issue.

aftermath /ˈɑːftəˌmæθ, -ˌmɑːθ/ *n.* **1** consequences; after-effects (*the aftermath of war*). **2** esp. *dial.* new grass growing after mowing or after a harvest. [AFTER *adj.* + *math* mowing f. OE *mæth* f. Gmc]

aftermost /ˈɑːftəˌməʊst/ *adj.* **1** last. **2** *Naut.* furthest aft. [AFTER *adj.* + -MOST]

afternoon /ˌɑːftəˈnuːn/ *n. & int.* ● *n.* **1** the time from noon or lunchtime to evening (*this afternoon; during the afternoon; afternoon tea*). **2** this time spent in a particular way (*had a lazy afternoon*). **3** a time compared with this, esp. the later part of something (*the afternoon of life*). ● *int.* good afternoon (see GOOD *adj.* 14).

afterpains /ˈɑːftəˌpeɪnz/ *n.pl.* pains caused by contraction of the womb after childbirth.

afters /ˈɑːftəz/ *n.pl. Brit. colloq.* the course following the main course of a meal; dessert.

aftershave /ˈɑːftəˌʃeɪv/ *n.* an astringent lotion for use after shaving.

aftershock /ˈɑːftəˌʃɒk/ *n.* **1** a lesser shock following the main shock of an earthquake. **2** an after-effect.

afterthought /ˈɑːftəˌθɔːt/ *n.* an item or thing that is thought of or added later.

afterwards /ˈɑːftəwədz/ *adv.* (*US* **afterward**) later, subsequently. [OE *æfterwearde* (as AFTER, -WARD)]

afterword /ˈɑːftəˌwɜːd/ *n.* concluding remarks in a book, esp. by a person other than its author.

Ag[1] *symb. Chem.* the element silver. [L *argentum*]

Ag[2] *abbr.* antigen.

ag- /əɡ/ *prefix* assim. form of AD- before g.

Aga /ˈɑːɡə/ *n. propr.* a type of heavy heat-retaining cooking stove or range burning solid fuel or powered by gas, oil, or electricity and intended for continuous heating. [Sw. f. *Svenska Aktiebolaget Gasackumulator* (Swedish Gas Accumulator Company), the original manufacturer]

aga /ˈɑːɡə/ *n.* (in Muslim countries, esp. under the Ottoman Empire) a commander, a chief. [Turk. *ağa* master]

Agadir /ˌæɡəˈdɪə(r)/ a seaport and resort on the Atlantic coast of Morocco; pop. (1982) 110,500.

again /əˈɡeɪn, əˈɡen/ *adv.* **1** another time; once more. **2** as in a previous position or condition (*back again; home again; quite well again*). **3** in addition (*as much again; half as many again*). **4** further, besides (*again, what about the children?*). **5** on the other hand (*I might, and again I might not*). □ **again and again** repeatedly. [orig. a northern form of ME *ayen* etc., f. OE *ongēan, ongægn*, etc., f. Gmc]

against /əˈɡeɪnst, əˈɡenst/ *prep.* **1** in opposition to (*fight against the invaders; am against hanging; arson is against the law*). **2** into collision or in contact with (*ran against a rock; lean against the wall; up against a problem*). **3** to the disadvantage of (*his age is against him*). **4** in contrast to (*against a dark background; 99 as against 102 yesterday*). **5** in anticipation of or preparation for (*against his coming; against a rainy day; protected against the cold; warned against pickpockets*). **6** as a compensating factor to (*income against expenditure*). **7** in return for (*issued against payment of* *the fee).* □ **against the clock** see CLOCK[1] *n.* 3. **against the grain** see GRAIN. **against time** see TIME. [ME *ayenes* etc. f. *ayen* AGAIN + *-t* as in *amongst*: see AMONG]

Aga Khan /ˌɑːgə ˈkɑːn/ the title of the imam or leader of the Nizari sect of Ismaili Muslims. The first Aga Khan was given his title in 1818 by the shah of Persia, subsequently moving with the majority of the Nizaris to the Indian subcontinent. The present (4th) Aga Khan (Karim Al-Hussain Shah, b.1937 in Geneva) inherited the title from his grandfather in 1957. The title of Aga Khan carries with it responsibility for various services and welfare provisions for members of the Nizari community. [f. Turk. *aǧa* master, *khān* ruler]

agal /əˈgɑːl/ *n.* a fillet or band worn by Bedouin Arabs to keep the keffiyeh (head-dress) in place. [Arab. *'iḳāl* bond, rope]

agama /ˈægəmə/ *n.* an Old World lizard of the genus *Agama* or a related genus, resembling an iguana. [Carib]

Agamemnon /ˌægəˈmemnən/ *Gk Mythol.* king of Mycenae and brother of Menelaus. In the Homeric poems he is Commander-in-Chief of the Greek expedition against Troy. On his return home from Troy he was murdered by his wife Clytemnestra and her lover Aegisthus; his murder was avenged by his son Orestes and daughter Electra.

agamic /əˈgæmɪk/ *adj. Bot.* characterized by asexual reproduction. [Gk *agamos* unmarried + -IC]

agamospermy /ˈægəməˌspɜːmɪ/ *n. Bot.* asexual reproduction by division of an unfertilized ovule. [Gk *gamos* unmarried + *sperma* seed]

agapanthus /ˌægəˈpænθəs/ *n.* a lily-like plant of the southern African genus *Agapanthus*, with blue or white flowers. [mod.L f. Gk *agapē* love + *anthos* flower]

agape[1] /əˈgeɪp/ *adv. & predic.adj.* gaping, open-mouthed, esp. with wonder or expectation.

agape[2] /ˈægəˌpeɪ/ *n.* **1** a Christian feast in token of fellowship, esp. one held by early Christians in commemoration of the Last Supper. **2** (in Christian theology) Christian love, esp. as distinct from erotic love. [Gk, = brotherly love]

agar /ˈeɪgɑː(r)/ *n.* (also **agar-agar** /ˌeɪgɑːrˈeɪgɑː(r)/) a gelatinous substance obtained from various kinds of red seaweed used esp. to make soups and to form biological culture media. [Malay]

agaric /ˈægərɪk/ *n.* a gill-bearing fungus of the order Agaricales, which includes mushrooms and many toadstools. [L *agaricum* f. Gk *agarikon*]

Agartala /ˈʌgətəˌlɑː/ a city in the far north-east of India, capital of the state of Tripura, situated near the border with Bangladesh; pop. (1991) 157,640.

Agassiz /ˈægəsɪ/, Jean Louis Rodolphe (1807–73), Swiss-born zoologist, geologist, and palaeontologist. In 1837 Agassiz was the first to propose that much of Europe had once been in the grip of an ice age. He lived in America from 1846 onwards and became an influential teacher and writer on many aspects of natural history. He was an opponent of Darwin's theory of evolution, holding that organisms were immutable and independent of each other.

agate /ˈægət/ *n.* **1** a hard usu. banded chalcedony. **2** a coloured toy marble resembling this. [F *agate, -the,* f. L *achates* f. Gk *akhatēs*]

agave /əˈgeɪvɪ/ *n.* a plant of the genus *Agave*, with a rosette of narrow spiny leaves and a tall flowering stem. A prominent example is the American aloe or century plant, *A. americana*, a large tropical plant with a flowering stem that can be 12 m (40 ft) high, produced only once in 10–70 years, and yielding sap from which tequila is distilled. Some species are a source of fibre (especially sisal) or alcoholic drink (pulque and mescal); a few are ornamental. [L f. Gk *Agauē*, proper name in myth f. *agauos* illustrious]

agaze /əˈgeɪz/ *adv. poet.* gazing.

age /eɪdʒ/ *n. & v.* ● *n.* **1 a** the length of time that a person or thing has existed or is likely to exist. **b** a particular point in or part of one's life, often as a qualification (*old age; voting age*). **2 a** (often in *pl.*) *colloq.* a long time (*took an age to answer; have been waiting for ages*). **b** a distinct period of the past (*golden age; Bronze age; Middle Ages*). **c** *Geol.* a period of time. **d** a generation. **3** the latter part of life; old age (*the peevishness of age*). ● *v.* (*pres. part.* **ageing, aging**) **1** *intr.* show signs of advancing age (*has aged a lot recently*). **2** *intr.* grow old. **3** *intr.* mature. **4** *tr.* cause or allow to age. □ **age-long** lasting for a very long time. **age of consent** see CONSENT. **age of discretion** see DISCRETION. **age-old** having existed for a very long time. **come of age** reach adult status (esp. in law at 18, formerly 21). **of age** old enough, of adult status. **over age** (usu. hyphenated when *attrib.*) **1** old enough. **2** too old. **under age** (usu. hyphenated when *attrib.*) not old enough, esp. not yet of adult status. [ME f. OF ult. f. L *aetas -atis* age]

-age /ɪdʒ/ *suffix* forming nouns denoting: **1** an action (*breakage; spillage*). **2** a condition or function (*bondage; a peerage*). **3** an aggregate or number of (*coverage; the peerage; acreage*). **4** fees payable for; the cost of using (*postage*). **5** the product of an action (*dosage; wreckage*). **6** a place; an abode (*anchorage; orphanage; parsonage*). [OF ult. f. L *-aticum* neut. of adj. suffix *-aticus* -ATIC]

aged *adj.* **1** /eɪdʒd/ **a** of the age of (*aged ten*). **b** that has been subjected to ageing. **c** (of a horse) over 6 years old. **2** /ˈeɪdʒɪd/ having lived long; old.

ageing /ˈeɪdʒɪŋ/ *n.* (also **aging**) **1** the action or process of growing or causing to grow old. **2** a change of properties occurring in some metals after heat treatment or cold working.

ageism /ˈeɪdʒɪz(ə)m/ *n.* prejudice or discrimination on the grounds of age. □ **ageist** *adj. & n.*

ageless /ˈeɪdʒlɪs/ *adj.* **1** never growing or appearing old or outmoded. **2** eternal, timeless.

agency /ˈeɪdʒənsɪ/ *n.* (*pl.* **-ies**) **1 a** the business or establishment of an agent (*employment agency*). **b** the function of an agent. **2 a** active operation; action (*free agency*). **b** intervening action; means (*fertilized by the agency of insects*). **c** action personified (*an invisible agency*). **3** a specialized department of the United Nations. [med.L *agentia* f. L *agere* do]

agenda /əˈdʒendə/ *n.* **1** (*pl.* **agendas**) **a** a list of items of business to be considered at a meeting. **b** a series of things to be done. **2** (as *pl.*) **a** items to be considered. **b** things to be done. ¶ Now very common as a countable noun in sense 1 (cf. DATA, MEDIA[1]). [L, neut. pl. of gerundive of *agere* do]

agent /ˈeɪdʒənt/ *n.* **1 a** a person who acts for another in business, politics, etc. (*estate agent; insurance agent*). **b** a spy. **2 a** a person or thing that exerts power or produces an effect. **b** the cause of a natural force or effect on matter (*oxidizing agent*). **c** such a force or effect. □ **agent-general** a representative of an Australian state or Canadian province, usu. in London. **agent noun** a noun denoting an agent or agency (e.g. *lawyer, accelerator*). [L *agent-* part. stem of *agere* do]

agent provocateur /ˌæʒɒn prɒˌvɒkəˈtɜː(r)/ *n.* (*pl.* **agents provocateurs** *pronunc.* same) a person employed to detect suspected offenders by tempting them to overt self-incriminating action. [F, = provocative agent]

agglomerate *v., n., & adj.* ● *v.tr. & intr.* /əˈglɒməˌreɪt/ **1** collect into a mass. **2** accumulate in a disorderly way. ● *n.* /əˈglɒmərət/ **1** a mass or collection of things. **2** *Geol.* a mass of large volcanic fragments bonded under heat (cf. CONGLOMERATE). ● *adj.* /əˈglɒmərət/ collected into a mass. □ **agglomerative** /-rətɪv/ *adj.* **agglomeration** /əˌglɒməˈreɪʃ(ə)n/ *n.* [L *agglomerare* (as AD-, *glomerare* f. *glomus -meris* ball)]

agglutinate /əˈgluːtɪˌneɪt/ *v.* **1** *tr.* unite as with glue. **2** *tr. & intr. Biol.* cause or undergo rapid clumping (of bacteria, erythrocytes, etc.). **3** *tr. Linguistics* (of language) combine (grammatical elements) to express compound ideas. □ **agglutination** /əˌgluːtɪˈneɪʃ(ə)n/ *n.* [L *agglutinare* (as AD-, *glutinare* f. *gluten -tinis* glue)]

agglutinative /əˈgluːtɪnətɪv/ *adj. Linguistics* denoting a language in which grammatical elements are combined with little or no change of form to express compound ideas.

agglutinin /əˈgluːtɪnɪn/ *n. Biol.* an antibody, lectin, or other substance causing agglutination. [AGGLUTINATE + -IN]

aggrandize /əˈgrændaɪz/ *v.tr.* (also **-ise**) **1** increase the power, rank, or wealth of (a person or state). **2** cause to appear greater than is the case. □ **aggrandizer** *n.* **aggrandizement** /-dɪzmənt/ *n.* [F *agrandir* (stem *agrandiss-*), prob. f. It. *aggrandire* f. L *grandis* large: assim. to verbs in -IZE]

aggravate /ˈægrəˌveɪt/ *v.tr.* **1** increase the gravity of (an illness, offence, etc.). **2** *disp.* annoy, exasperate (a person). □ **aggravation** /ˌægrəˈveɪʃ(ə)n/ *n.* [L *aggravare aggravat-* make heavy f. *gravis* heavy]

aggregate *n., adj., & v.* ● *n.* /ˈægrɪgət/ **1** a collection of, or the total of, disparate elements. **2** pieces of crushed stone, gravel, etc. used in making concrete. **3 a** *Geol.* a mass of minerals formed into solid rock. **b** a mass of particles. ● *adj.* /ˈægrɪgət/ **1** (of disparate elements) collected into one mass. **2** constituted by the collection of many units into one body. **3** *Bot.* **a** (of fruit) formed from several carpels derived from the same flower (e.g. raspberry). **b** (of a group of species) comprising several very similar species that were formerly regarded as a single species. ● *v.* /ˈægrɪˌgeɪt/ **1** *tr. & intr.* collect together; combine into one mass. **2** *tr. colloq.* amount to (a specified total). **3** *tr.* unite (*was aggregated to the group*). □ **in the aggregate** as a whole.

□ **aggregative** /ˈægrɪˌɡeɪtɪv/ *adj.* **aggregation** /ˌægrɪˈɡeɪʃ(ə)n/ *n.* [L *aggregare aggregat-* herd together (as AD-, *grex gregis* flock)]

aggression /əˈɡreʃ(ə)n/ *n.* **1** the act or practice of attacking without provocation, esp. beginning a quarrel or war. **2** an unprovoked attack. **3** self-assertiveness; forcefulness. **4** *Psychol.* hostile or destructive tendency or behaviour. [F *agression* or L *aggressio* attack f. *aggredi aggress-* (as AD-, *gradi* walk)]

aggressive /əˈɡresɪv/ *adj.* **1** of a person: **a** given to aggression; openly hostile. **b** forceful; self-assertive. **2** (of an act) offensive, hostile. **3** of aggression. □ **aggressively** *adv.* **aggressiveness** *n.*

aggressor /əˈɡresə(r)/ *n.* a person who attacks without provocation. [L (as AGGRESSION)]

aggrieved /əˈɡriːvd/ *adj.* having a grievance. □ **aggrievedly** /-vɪdlɪ/ *adv.* [ME, past part. of *aggrieve* f. OF *agrever* make heavier (as AD-, GRIEVE¹)]

aggro /ˈæɡrəʊ/ *n. Brit. sl.* **1** aggressive troublemaking. **2** trouble, difficulty. [abbr. of AGGRAVATION (see AGGRAVATE) or AGGRESSION]

aghast /əˈɡɑːst/ *predic.adj.* (often foll. by *at*) filled with dismay or consternation. [ME, past part. of obs. *agast, gast* frighten: see GHASTLY]

Aghios Nikolaos /ˌæɡɪɒs ˌnɪkəˈlaɪɒs/ (Greek **Áyios Nikólaos** /ˌajiɒs niˈkɒlaɒs/) a fishing port and holiday resort on the north coast of Crete, east of Heraklion; pop. (1981) 8,100. [Gk, = St Nicholas]

agile /ˈædʒaɪl/ *adj.* quick-moving, nimble, active. □ **agilely** *adv.* **agility** /əˈdʒɪlɪtɪ/ *n.* [F f. L *agilis* f. *agere* do]

agin /əˈɡɪn/ *prep. colloq.* or *dial.* against. [corrupt. of AGAINST or synonymous *again* obs. prep.]

Agincourt, Battle of /ˈædʒɪnˌkɔː(r)/, -ˌkɔːt, French aʒɛ̃kur/ a battle in northern France in 1415, in which the English army under Henry V defeated a large French army. The victory, largely thanks to the successful use of the longbow on the English side, allowed Henry to occupy Normandy and consolidate his claim to the French throne.

aging var. of AGEING.

agio /ˈædʒɪˌəʊ/ *n.* (*pl.* **agios**) **1** the percentage charged on the exchange of one currency, or one form of money, into another more valuable. **2** the excess value of one currency over another. **3** money-exchange business. [It. *aggio*]

agist /əˈdʒɪst/ *v.tr.* take in and feed (livestock) for payment. □ **agistment** *n.* [OF *agister* f. *giste* lodging]

agitate /ˈædʒɪˌteɪt/ *v.* **1** *tr.* (often as **agitated** *adj.*) disturb or excite (a person or feelings). **2** *intr.* (often foll. by *for, against*) stir up interest or concern, esp. publicly (*agitated for tax reform*). **3** *tr.* shake or move, esp. briskly. □ **agitatedly** *adv.* [L *agitare agitat-* frequent. of *agere* drive]

agitation /ˌædʒɪˈteɪʃ(ə)n/ *n.* **1** the act or process of agitating or being agitated. **2** mental anxiety or concern. [F *agitation* or L *agitatio* (as AGITATE)]

agitato /ˌædʒɪˈtɑːtəʊ/ *adv. & adj. Mus.* in an agitated manner. [It.]

agitator /ˈædʒɪˌteɪtə(r)/ *n.* **1** a person who agitates, esp. publicly for a cause. **2** an apparatus for shaking or mixing liquid etc. [L (as AGITATE)]

agitprop /ˈædʒɪtˌprɒp, ˈæɡɪt-/ *n. hist.* the dissemination of Communist political propaganda, esp. in plays, films, books, etc. [Russ. (as AGITATION, PROPAGANDA)]

aglet /ˈæɡlɪt/ *n.* **1** a metal tag attached to each end of a shoelace etc. **2** = AIGUILLETTE. [ME f. F *aiguillette* small needle, ult. f. L *acus* needle]

agley /əˈɡleɪ, əˈɡliː/ *adv. Sc.* askew, awry. [A² + Sc. *gley* squint]

aglow /əˈɡləʊ/ *adv. & predic.adj.* glowing; in a glow (of warmth, excitement, etc.).

AGM *abbr.* annual general meeting.

agma /ˈæɡmə/ *n. Phonet.* **1** the speech sound of *ng* as in *thing*, represented by the symbol /ŋ/. **2** this symbol. [Gk, lit. 'fragment']

agnail /ˈæɡneɪl/ *n.* **1** a piece of torn skin at the root of a fingernail. **2** the soreness resulting from this. [OE *angnægl* f. *nægl* NAIL *n.* 2a: cf. HANGNAIL]

agnate /ˈæɡneɪt/ *adj. & n.* ● *adj.* **1** descended esp. by male line from the same male ancestor (cf. COGNATE). **2** descended from the same forefather; of the same clan or nation. **3** of the same nature; akin. ● *n.* a person who is descended esp. by male line from the same male ancestor. □ **agnatic** /æɡˈnætɪk/ *adj.* **agnation** /-ˈneɪʃ(ə)n/ *n.* [L *agnatus* f. *ad* to + *gnasci* be born f. stem *gen-* beget]

Agnes, St¹ /ˈæɡnɪs/ (died *c.*304), Roman martyr. Said to have been a Christian virgin who refused to marry, she was martyred during the reign of Diocletian. She is the patron saint of virgins and her emblem is a lamb (L *agnus*). Feast day, 21 Jan.

Agnes, St² /ˈæɡnɪs/ (*c.*1211–82), patron saint of Bohemia. She was canonized in 1989. Feast day, 2 Mar.

Agnesi /æˈnjeɪʒi/, Maria Gaetana (1718–99), Italian mathematician and philosopher. She is regarded as the first female mathematician of the Western world, though she worked on a variety of scientific subjects. Her major work, which appeared in two volumes in 1748, was a comprehensive treatment of algebra and analysis, of which perhaps the most important part was concerned with differential calculus.

Agni /ˈæɡnɪ/ the Vedic god of fire, the priest of the gods and the god of the priests. As mediator between gods and men, he takes offerings to the gods in the smoke of sacrifice and returns to the earth as lightning. [Skr., = fire, cognate with L *ignis*]

agnolotti /ˌænjəˈlɒtɪ/ *n.pl.* small pasta shapes containing meat stuffing; a dish of these, served in broth or with a sauce. [It.]

agnosia /æɡˈnəʊsɪə/ *n. Med.* the loss of the ability to interpret sensations. [mod.L f. Gk *agnōsia* ignorance]

agnostic /æɡˈnɒstɪk/ *n. & adj.* ● *n.* **1** a person who believes that nothing is known, or can be known, of the existence or nature of God or of anything beyond material phenomena. **2** a person who is uncertain or noncommittal about a certain thing. ● *adj.* of or relating to agnostics or agnosticism. □ **agnosticism** /-ˈnɒstɪˌsɪz(ə)m/ *n.* [A⁻¹ + GNOSTIC]

Agnus Dei /ˌæɡnʊs ˈdeiiː/ *n.* **1** a figure of a lamb bearing a cross or flag, as an emblem of Christ. **2** the part of the Roman Catholic mass beginning with the words 'Lamb of God'. [L, = lamb of God]

ago /əˈɡəʊ/ *adv.* earlier, before the present (*ten years ago; long ago*). ¶ Note the construction *it is ten years ago that* (not *since*) *I saw them.* [ME (*ago, agone*), past part. of obs. *ago* (v.) (as A⁻², GO¹)]

agog /əˈɡɒɡ/ *adv. & adj.* ● *adv.* eagerly, expectantly. ● *predic.adj.* eager, expectant. [F *en gogues* f. *en* in + pl. of *gogue* fun]

agogic /əˈɡɒdʒɪk/ *adj. & n. Mus.* ● *adj.* designating or relating to an accent effected by lengthening the time value of a note. ● *n.* (in *pl.*, usu. treated as *sing.*) the use of agogic accents; the deliberate modification of time values in a musical performance. [G *agogisch* f. Gk *agogos* leading]

à gogo /ə ˈɡəʊɡəʊ/ *adv.* in abundance (*whisky à gogo*). [F]

agonic /əˈɡɒnɪk/ *adj.* having or forming no angle. □ **agonic line** a line passing through the two poles, along which a magnetic needle points directly north or south. [Gk *agōnios* without angle (as A⁻¹, *gōnia* angle)]

agonist /ˈæɡənɪst/ *n.* **1** *Physiol.* a muscle whose contraction moves a part directly. **2** *Biochem.* a substance which initiates a physiological response when combined with a receptor. **3** = PROTAGONIST. [Gk *agōnistēs* contestant f. *agōn* contest]

agonistic /ˌæɡəˈnɪstɪk/ *adj.* **1** polemical, combative. **2** *Zool.* (of animal behaviour) associated with conflict. **3** *Biochem.* of or relating to an agonist; acting as an agonist. □ **agonistically** *adv.* [LL *agonisticus* f. Gk *agōnistikos* f. *agōnistēs* (see AGONIST)]

agonize /ˈæɡəˌnaɪz/ *v.* (also **-ise**) **1** *intr.* (often foll. by *over*) undergo (esp. mental) anguish; suffer agony. **2** *tr.* (often as **agonizing** *adj.*) cause agony or mental anguish to. **3** *tr.* (as **agonized** *adj.*) expressing agony (*an agonized look*). **4** *intr.* struggle, contend. □ **agonizingly** *adv.* [F *agoniser* or LL *agonizare* f. Gk *agōnizomai* contend f. *agōn* contest]

agony /ˈæɡənɪ/ *n.* (*pl.* **-ies**) **1** extreme mental or physical suffering. **2** a severe struggle. □ **agony aunt** *colloq.* a person (esp. a woman) who answers letters in an agony column. **agony column** *colloq.* **1** a column in a newspaper or magazine offering personal advice to readers who write in. **2** = *personal column.* **agony uncle** *colloq.* the male equivalent of an agony aunt. [ME f. OF *agonie* or LL f. Gk *agōnia* f. *agōn* contest]

agoraphobe /ˈæɡərəˌfəʊb/ *n.* a person who suffers from agoraphobia.

agoraphobia /ˌæɡərəˈfəʊbɪə/ *n. Psychol.* an abnormal fear of open spaces or public places. □ **agoraphobic** *adj. & n.* [mod.L f. Gk *agora* place of assembly, market-place + -PHOBIA]

agouti /əˈɡuːtɪ/ *n.* (also **aguti**) (*pl.* **agoutis**) **1** a long-legged rodent of the genus *Dasyprocta*, of Central and South America, related to the guinea pig. **2** a fur in which each hair has alternate dark and light bands, giving a grizzled appearance. **b** an animal with this. [F *agouti* or Sp. *aguti* f. Tupi *aguti*]

AGR *abbr.* advanced gas-cooled (nuclear) reactor.

Agra /ˈɑːɡrə/ a city on the River Jumna in Uttar Pradesh state, northern India; pop. (1991) 899,000. Agra was founded by the Mogul emperor Akbar in 1566 and was the capital of the Mogul empire until 1658. It is the site of the Taj Mahal.

agrarian /əˈɡreərɪən/ *adj. & n.* ● *adj.* **1** of or relating to the land or its cultivation. **2** relating to landed property. ● *n.* a person who advocates a redistribution of landed property. [L *agrarius* f. *ager agri* field]

Agrarian Revolution the transformation of British agriculture during the 18th century, characterized by the acceleration of enclosure of common land and the consequent decline of the open-field system, as well as by the introduction of technological innovations such as the seed drill and the scientific rotation of crops. (See also ENCLOSURE.)

agree /əˈgriː/ *v.* (**agrees, agreed, agreeing**) **1** *intr.* hold a similar opinion (*I agree with you about that; they agreed that it would rain*). **2** *intr.* (often foll. by *to*, or *to* + infin.) consent (*agreed to the arrangement; agreed to go*). **3** *intr.* (often foll. by *with*) **a** become or be in harmony. **b** suit; be good for (*caviar didn't agree with him*). **c** *Gram.* have the same number, gender, case, or person as. **4** *tr.* reach agreement about (*agreed a price*). **5** *tr.* consent to or approve of (terms, a proposal, etc.). **6** *tr.* bring (things, esp. accounts) into harmony. **7** *intr.* (foll. by *on*) decide by mutual consent (*agreed on a compromise*). □ **agree to differ** leave a difference of opinion unresolved. **be agreed** have reached the same opinion. [ME f. OF *agreer* ult. f. L *gratus* pleasing]

agreeable /əˈgriːəb(ə)l/ *adj.* **1** (often foll. by *to*) pleasing. **2** (often foll. by *to*) (of a person) willing to agree (*was agreeable to going*). **3** (foll. by *to*) conformable. □ **agreeably** *adv.* **agreeableness** *n.* **agreeability** /əˌgriːəˈbɪlɪtɪ/ *n.* [ME f. OF *agreable* f. *agreer* AGREE]

agreement /əˈgriːmənt/ *n.* **1** the act of agreeing; the holding of the same opinion. **2** mutual understanding. **3** an arrangement between parties as to a course of action etc. **4** *Gram.* having the same number, gender, case, or person. **5** mutual conformity of things; harmony. [ME f. OF (as AGREE)]

agribusiness /ˈægrɪˌbɪznɪs/ *n.* **1** agriculture conducted on strictly commercial principles, esp. using advanced technology. **2** an organization engaged in this. **3** the group of industries dealing with the produce of, and services to, farming. [AGRICULTURE + BUSINESS]

Agricola /əˈgrɪkələ/, Gnaeus Julius (AD 40–93), Roman general and governor of Britain 78–84. As governor he completed the subjugation of Wales, advanced into Scotland, and defeated the Caledonian Highland tribes at the battle of Mons Graupius.

agriculture /ˈægrɪˌkʌltʃə(r)/ *n.* the science or practice of cultivating the soil and rearing animals. (*See note below.*) □ **agricultural** /ˌægrɪˈkʌltʃərəl/ *adj.* **agriculturalist** *n.* **agriculturally** *adv.* **agriculturist** *n.* [F *agriculture* or L *agricultura* f. *ager agri* field + *cultura* CULTURE]

▪ The beginnings of agriculture date from the neolithic period, approximately 9,000 years ago, and are associated with the establishment of settled communities, as in the Tigris–Euphrates valley and beside the Nile. The type and development of farming have varied in different parts of the world and methods used in one place have been (and still are) sometimes centuries ahead of those of another. In Britain, strip farming associated with the feudal system was replaced in the 18th century by enclosed fields (see AGRARIAN REVOLUTION); labour-saving machinery and improved methods of transport made increased production possible. In developed countries, modern-day intensive farming (increased mechanization, the use of pesticides and fertilizers, as well as scientific developments in animal breeding and rearing), though helping to increase production and efficiency, has not proved entirely beneficial; recent government legislation and other interventionist strategies in Europe and elsewhere reflect environmental concern with problems such as soil erosion, disruption of natural habitat, and pollution, along with an awareness of ethical issues related to practices such as factory farming.

agrimony /ˈægrɪmənɪ/ *n.* (*pl.* **-ies**) a perennial rosaceous plant of the genus *Agrimonia*, esp. *A. eupatoria*, which has small yellow flowers. [ME f. OF *aigremoine* f. L *agrimonia* alt. of *argemonia* f. Gk *argemōnē* poppy]

Agrippa /əˈgrɪpə/, Marcus Vipsanius (63–12 BC), Roman general. Augustus' adviser and son-in-law, he played an important part in the naval victories over Mark Antony, and held commands in western and eastern provinces of the empire.

agro- /ˈægrəʊ/ *comb. form* agricultural (*agro-climatic; agro-ecological*). [Gk *agros* field]

agrochemical /ˌægrəʊˈkemɪk(ə)l/ *n.* a chemical used in agriculture.

agronomy /əˈgrɒnəmɪ/ *n.* the science of soil management and crop production. □ **agronomist** *n.* **agronomic** /ˌægrəˈnɒmɪk/ *adj.* **agronomical** *adj.* **agronomically** *adv.* [F *agronomie* f. *agronome* agriculturist f. Gk *agros* field + *-nomos* f. *nemō* arrange]

Agro Pontino /ˌæːgrəʊ pɒnˈtiːnɔ/ the Italian name for the PONTINE MARSHES.

aground /əˈgraʊnd/ *predic.adj. & adv.* (of a ship) on or on to the bottom of shallow water (*be aground; run aground*). [ME f. A² + GROUND¹]

Aguascalientes /ˌægwəskæˈljenteɪz/ **1** a state of central Mexico. **2** its capital, a health resort noted for its hot springs; pop. (1990) 506,380.

ague /ˈeɪgjuː/ *n.* archaic or hist. **1** a malarial fever, with cold, hot, and sweating stages. **2** a shivering fit. [ME f. OF f. med.L *acuta* (*febris*) acute (fever)]

Agulhas, Cape /əˈgʌləs/ the most southerly point of the continent of Africa, in the province of Western Cape, South Africa.

Agulhas Current /əˈgʌləs/ an ocean current flowing southward along the east coast of Africa.

aguti var. of AGOUTI.

AH *abbr.* in the year of the Hegira (AD 622); (of a date) of the Muslim era. [L *anno Hegirae*]

ah /ɑː/ *int.* expressing surprise, pleasure, sudden realization, resignation, etc. ¶ The sense depends much on intonation. [ME f. OF *a(h)*]

aha /ɑːˈhɑː, əˈhɑː/ *int.* expressing surprise, triumph, mockery, irony, etc. ¶ The sense depends much on intonation. [ME f. AH + HA¹]

Ahaggar Mountains see HOGGAR MOUNTAINS.

ahead /əˈhed/ *adv.* **1** further forward in space or time. **2** in the lead; further advanced (*ahead on points*). **3** in the line of one's forward motion (*roadworks ahead*). **4** straight forwards. □ **ahead of** **1** further forward or advanced than. **2** in the line of the forward motion of. **3** prior to. [orig. *Naut.*, f. A² + HEAD]

ahem /əˈhəm, əˈhem/ (not usu. clearly articulated) *int.* used to attract attention, gain time, or express disapproval. [lengthened form of HEM²]

ahimsa /əˈhɪmsɑː/ *n.* (in the Hindu, Buddhist, and Jainist tradition) respect for all living things and avoidance of violence towards others both in thought and deed. [Skr. f. *a* non- + *himsa* violence]

Ahmadabad /ˈɑːmədəˌbæd/ (also **Ahmedabad**) an industrial city in the state of Gujarat in western India; pop. (1991) 2,873,000.

-aholic /əˈhɒlɪk/ *comb. form* (also **-oholic**) forming nouns meaning 'a person addicted to something', as *workaholic, chocoholic,* etc. [f. ALCOHOLIC]

ahoy /əˈhɔɪ/ *int. Naut.* a call used in hailing. [AH + HOY¹]

Ahriman /ˈɑːrɪmən/ the evil spirit in the dualistic doctrine of Zoroastrianism. [Pers., f. Avestan *angramainya* dark or destructive spirit]

à huis clos /ˌæ hwiː ˈkləʊ/ *adv.* in private. [F, = with closed doors]

Ahura Mazda /əˌhʊərə ˈmæzdə/ the creator god of ancient Iran and, in particular, of Zoroastrianism. Later called Ormazd, he is the force for good and the opponent of Ahriman. [Avestan, = the living wise one]

Ahvaz /ɑːˈvɑːz/ (also **Ahwaz** /ɑːˈwɑːz/) a town in western Iran; pop. (1991) 725,000.

Ahvenanmaa /ˈɑːhvɛnɑmˌmɑː/ the Finnish name for the ÅLAND ISLANDS.

AI *abbr.* **1** artificial insemination. **2** artificial intelligence.

ai /ˈɑːɪ/ *n.* (*pl.* **ais**) the three-toed sloth of South America, of the genus *Bradypus*. [Tupi *ai*, repr. its cry]

AID *abbr.* artificial insemination by donor.

aid /eɪd/ *n. & v.* ● *n.* **1** help. **2** financial or material help, esp. given by one country to another. **3** a material source of help (*teaching aid*). **4** a person or thing that helps. **5** (as second element in *comb.*) denoting an organisation or event that raises money for charity (*Band Aid*). **6** *hist.* a grant of subsidy or tax to a king. ● *v.tr.* **1** (often foll. by *to* + infin.) help. **2** promote or encourage (*sleep will aid recovery*). □ **aid and abet** see ABET. **in aid of** in support of. **what's this** (or **all this**) **in aid of?** *colloq.* what is the purpose of this? [ME f. OF *aïde, aïdier,* ult. f. L *adjuvare* (as AD-, *juvare* jut- help)]

Aidan, St /ˈeɪd(ə)n/ (d. AD 651), Irish missionary. While a monk in the monastery at Iona, he was assigned the mission of Christianizing Northumbria by the Northumbrian king Oswald (*c.*604–41). Aidan founded a church and monastery at Lindisfarne in 635 and became its first bishop; he also established a school for training missionaries of the Celtic Church. He later founded further churches and monasteries in Northumbria.

aide /eɪd/ *n.* **1** an aide-de-camp. **2** *N. Amer.* an assistant. [abbr.]

aide-de-camp /ˌeɪddəˈkɒm/ n. (pl. **aides-de-camp** pronunc. same) an officer acting as a confidential assistant to a senior officer. [F]

aide-mémoire /ˌeɪdmemˈwɑː(r)/ n. (pl. same or **aides-mémoire** pronunc. same) **1 a** an aid to the memory. **b** a book or document meant to aid the memory. **2** (in diplomacy) a memorandum. [F f. aider to help + mémoire memory]

Aids /eɪdz/ n. (also **AIDS**) acquired immune deficiency syndrome, a disease in which there is severe loss of cellular immunity leaving the sufferer susceptible to certain opportunistic infections and malignancies. The cause is a virus (called the human immunodeficiency virus or HIV) transmitted in blood and in sexual fluids, and though the incubation period may be long the fully developed disease is invariably fatal. Aids was first identified in the early 1980s but certainly existed somewhat earlier; it now affects millions of people. In the developed world the disease first spread among homosexuals, intravenous drug users, and recipients of infected blood transfusions, before reaching the wider population. This has tended to overshadow a greater epidemic in parts of Africa, where transmission is mainly through heterosexual contact. □ **Aids-related complex** a set of symptoms including diseased lymph nodes, fever, weight loss, and malaise that seems to precede the full development of Aids. [acronym]

aigrette /ˈeɪɡret, eɪˈɡret/ n. **1** an egret. **2** its white plume. **3** a tuft of feathers or hair. **4** a spray of gems or similar ornament. [F]

Aigues-Mortes /eɪɡˈmɔːt/ a town in SE France, in the Rhône delta; pop. (1982) 4,106. Its name is derived from the Latin aquae mortuae (= dead waters), referring to the saline marshland which surrounds it. The site was chosen by Louis IX as the embarkation port for his two Crusades (1248 and 1270). The harbour has long been silted up, and the sea is now about 3 miles away.

aiguille /eɪˈɡwiːl/ n. a sharp peak of rock, esp. in the Alps. [F: see AGLET]

aiguillette /ˌeɪɡwɪˈlet/ n. a tagged point hanging from the shoulder on the breast of some uniforms. [F: see AGLET]

AIH abbr. artificial insemination by husband.

aikido /aɪˈkiːdəʊ/ n. a Japanese form of self-defence and martial art, using locks, holds, throws, and the opponent's own movements. [Jap. f. ai together, unify + ki spirit + dō way]

ail /eɪl/ v. **1** tr. (only in 3rd person interrog. or indefinite constructions) archaic trouble or afflict in mind or body (what ails him?). **2** intr. (usu. be ailing) be ill. [OE egli(a)n f. egle troublesome]

ailanthus /eɪˈlænθəs/ n. a tall deciduous tree of the genus Ailanthus; esp. the ornamental A. altissima, native to China and Australasia. [mod.L ailantus f. Amboinese aylanto]

aileron /ˈeɪləˌrɒn/ n. a hinged surface in the trailing edge of an aeroplane wing, used to control lateral balance. [F, dimin. of aile wing f. L ala]

ailing /ˈeɪlɪŋ/ adj. **1** ill, esp. chronically. **2** in poor condition.

ailment /ˈeɪlmənt/ n. an illness, esp. a minor one.

aim /eɪm/ v. & n. ● v. **1** intr. (foll. by at + verbal noun, or to + infin.) intend or try (aim at winning; aim to win). **2** tr. (usu. foll. by at) direct or point (a weapon, remark, etc.). **3** intr. take aim. **4** intr. (foll. by at, for) seek to attain or achieve. ● n. **1** a purpose, a design, an object aimed at. **2** the directing of a weapon, missile, etc., at an object. □ **take aim** direct a weapon etc. at an object. [ME f. OF ult. f. L aestimare reckon]

aimless /ˈeɪmlɪs/ adj. without aim or purpose. □ **aimlessly** adv. **aimlessness** n.

ain't /eɪnt/ contr. colloq. **1** am not; are not; is not (you ain't doing it right; she ain't nice). **2** has not; have not (we ain't seen him). ¶ Usually regarded as an uneducated use, and unacceptable in spoken and written English, except to represent dialect speech. [contr. of are not]

Aintab /ˈaɪntɑːb/ the former name (until 1921) for GAZIANTEP.

Aintree /ˈeɪntriː/ a suburb of Liverpool, in Merseyside, site of a racecourse over which the Grand National has been run since 1839.

Ainu /ˈaɪnuː/ n. & adj. ● n. (pl. same or **Ainus**) **1** a member of an aboriginal people of the Japanese archipelago. (See note below.) **2** the language of this people, regarded by some as being related to the Altaic languages. It is no longer used on an everyday basis. ● adj. of or relating to the Ainu or their language. [Ainu, = man]

■ The Ainu are physically distinct (with light skin colour and round eyes) from the majority population of Japan. Archaeological evidence suggests that the Ainu were resident in the area as early as 5000 BC, thereby predating the great Mongoloid expansion. During the 19th century the Ainu were largely displaced from their traditional lands

in the north of Hokkaido island, and in the 20th century intermarriage and cultural assimilation with the Japanese have resulted in a decline in numbers to about 50,000 and a shift from their traditional lifestyle as hunter-gatherers.

air /eə(r)/ n. & v. ● n. **1** an invisible gaseous substance surrounding the earth, a mixture mainly of oxygen and nitrogen. **2 a** the earth's atmosphere. **b** the free or unconfined space in the atmosphere (birds of the air; in the open air). **c** the atmosphere as a place where aircraft operate. **3 a** a distinctive impression or characteristic (an air of absurdity). **b** one's manner or bearing, esp. a confident one (with a triumphant air; does things with an air). **c** (esp. in pl.) an affected manner; pretentiousness (gave himself airs). **4** Mus. a tune or melody; a melodious composition. **5** a breeze or light wind (cf. light air (see LIGHT²)). ● v.tr. **1** warm (washed laundry) to remove damp, esp. at a fire or in a heated cupboard. **2** expose (a room etc.) to the open air; ventilate. **3** express publicly (an opinion, grievance, etc.). **4** parade; show ostentatiously (esp. qualities). **5** refl. go out in the fresh air. □ **air bag** a safety device that fills with air on impact to protect the occupants of a vehicle in a collision. **air-bed** an inflatable mattress. **air bladder** a bladder or sac filled with air in fish or some plants (cf. swim-bladder). **air brake 1** a brake worked by air pressure. **2** a movable flap or other device on an aircraft to reduce its speed. **air-brick** a brick perforated with small holes for ventilation. **air-bridge** a portable bridge or walkway put against an aircraft door. **air-conditioned** (of a room, building, etc.) equipped with air-conditioning. **air-conditioner** an air-conditioning apparatus. **air-conditioning 1** a system for regulating the humidity, ventilation, and temperature in a building. **2** the apparatus for this. **air-cooled** cooled by means of a current of air. **air corridor** = CORRIDOR 4. **air-cushion 1** an inflatable cushion. **2** the layer of air supporting a hovercraft or similar vehicle. **air force** a branch of the armed forces concerned with fighting or defence in the air. **air hostess** Brit. a stewardess in a passenger aircraft. **air lane** a path or course regularly used by aircraft (cf. LANE 4). **air letter** a sheet of light paper forming a letter for sending by airmail. **air line** a pipe supplying air, esp. to a diver. **air mass** a very large body of air with roughly uniform temperature and humidity. **air plant** a plant growing naturally without soil. **air pocket 1** a cavity containing air. **2** a region of low pressure causing an aircraft to drop suddenly. **air power** the ability to defend and attack by means of aircraft, missiles, etc. **air pump** a device for pumping air into or out of an enclosed space. **air rifle** a rifle using compressed air to propel pellets. **air sac** an extension of the lungs in birds or of the tracheae in insects. **airs and graces** affected elegance designed to attract or impress. **air-sea rescue** rescue from the sea by aircraft. **air speed** the speed of an aircraft relative to the air through which it is moving. **air terminal 1** an airline office in a town to which passengers report and which serves as a base for transport to and from an airport. **2** = TERMINAL n. **3. air time** time allotted for a broadcast. **air-to-air** from one aircraft to another in flight. **air traffic controller** an airport official who controls air traffic by giving radio instructions to pilots concerning route, altitude, take-off, and landing. **air waves** colloq. radio waves used in broadcasting. **by air** by aircraft; in an aircraft. **in the air** (of opinions, feelings, etc.) prevalent; gaining currency. **on** (or **off**) **the air** in (or not in) the process of broadcasting. **take the air** go out of doors. **tread** (or **walk**) **on air** feel elated. **up in the air** (of projects etc.) uncertain, not decided. [ME f. F and L f. Gk aēr]

airbase /ˈeəbeɪs/ n. a base for the operation of military aircraft.

airborne /ˈeəbɔːn/ adj. **1** transported by air. **2** (of aircraft) in the air after taking off.

airbrush /ˈeəbrʌʃ/ n. & v. ● n. an artist's device for spraying paint by means of compressed air. ● v.tr. paint or paint over with an airbrush, esp. in order to cover up imperfections or otherwise create a more favourable impression.

Airbus /ˈeəbʌs/ n. propr. an aircraft designed to carry a large number of passengers economically, esp. over relatively short routes.

Air Chief Marshal n. an RAF officer of high rank, below Marshal of the RAF and above Air Marshal.

air commodore n. an RAF officer next above group captain.

aircraft /ˈeəkrɑːft/ n. (pl. same) a machine capable of flight, esp. an aeroplane or helicopter. □ **aircraft-carrier** a warship that carries and serves as a base for aeroplanes.

aircraftman /ˈeəˌkrɑːftmən/ n. (pl. **-men**) the lowest rank in the RAF.

aircraftwoman /ˈeəkrɑːftˌwʊmən/ n. (pl. **-women**) the lowest rank in the WRAF.

aircrew /'eəkru:/ n. **1** the crew manning an aircraft. **2** (pl. **aircrew**) a member of such a crew.

Airedale /'eədeɪl/ n. a breed of large terrier with a short dense coat. [*Airedale* in W. Yorkshire]

airer /'eərə(r)/ n. a frame or stand for airing or drying clothes etc.

airfield /'eəfi:ld/ n. an area of land where aircraft take off and land, are maintained, etc.

airflow /'eəfləʊ/ n. the flow of air, esp. that encountered by a moving aircraft or vehicle.

airfoil /'eəfɔɪl/ n. US = AEROFOIL. [AIR + FOIL²]

airframe /'eəfreɪm/ n. the body of an aircraft as distinct from its engine(s).

airfreight /'eəfreɪt/ n. & v. ● n. cargo carried by an aircraft. ● v.tr. transport by air.

airglow /'eəgləʊ/ n. radiation from the upper atmosphere, detectable at night.

airgun /'eəgʌn/ n. a gun using compressed air to propel pellets.

airhead /'eəhed/ n. **1** Mil. a forward base for aircraft in enemy territory. **2** sl. a silly or foolish person.

airing /'eərɪŋ/ n. **1** exposure to fresh air, esp. for exercise or an excursion. **2** exposure (of laundry etc.) to warm air. **3** public expression of an opinion etc. (*the idea will get an airing at tomorrow's meeting*).

airless /'eəlɪs/ adj. **1** stuffy; not ventilated. **2** without wind or breeze; still. □ **airlessness** n.

airlift /'eəlɪft/ n. & v. ● n. the transport of troops and supplies by air, esp. in a blockade or other emergency. ● v.tr. transport in this way.

airline /'eəlaɪn/ n. an organization providing a regular public service of air transport on one or more routes.

airliner /'eə,laɪnə(r)/ n. a large passenger aircraft.

airlock /'eəlɒk/ n. **1** a stoppage of the flow in a pump or pipe, caused by an air bubble. **2** a compartment with controlled pressure and parallel sets of doors, to permit movement between areas at different pressures.

airmail /'eəmeɪl/ n. & v. ● n. **1** a system of transporting mail by air. **2** mail carried by this. ● v.tr. send by airmail.

airman /'eəmən/ n. (pl. **-men**) **1** a pilot or member of the crew of an aircraft, esp. in an air force. **2** a member of the RAF below commissioned rank.

Air Marshal n. an RAF officer of high rank, below Air Chief Marshal and above Air Vice-Marshal.

Air Miles n.pl. propr. points (equivalent to miles of free air travel) accumulated by buyers of airline tickets and of some other products and services and redeemable against the cost of air travel with a particular airline.

airmiss /'eəmɪs/ n. a circumstance in which two or more aircraft in flight on different routes are less than a prescribed distance apart.

airmobile /eə'məʊbaɪl/ adj. (of troops) that can be moved about by air.

air officer n. any RAF officer above the rank of group captain.

airplane /'eəpleɪn/ n. N. Amer. = AEROPLANE.

airplay /'eəpleɪ/ n. broadcasting (of recorded music).

airport /'eəpɔ:t/ n. a complex of runways and buildings for the take-off, landing, and maintenance of civil aircraft, with facilities for passengers.

air raid n. an attack by aircraft with bombs etc. The first air raid took place when the Austrian army sent up hot-air balloons, with bombs but no pilots, to drift over Venice in the campaign of 1849. The first successful manned bombing flights were by Zeppelin raiders over London in the First World War.

airscrew /'eəskru:/ n. Brit. an aircraft propeller.

airship /'eəʃɪp/ n. a power-driven steerable aircraft that uses gases which are lighter than air to keep it buoyant. There are three main types of airship: *rigid*, having a rigid framework for maintaining the shape of its hull; *non-rigid* (also called a *blimp*), in which internal pressure maintains the shape; and *semi-rigid*, with a rigid keel running the length of the ship. The first successful airship was made in France in 1884; by the turn of the century Germany had taken the lead with the designs of Count von Zeppelin, whose machines were used in the First World War for bombing raids on England, as well as for passenger services, including transatlantic flights. A series of disasters with hydrogen-filled airships in the 1930s, including the British *R101* at Beauvais, France (1930), and the German *Hindenburg* at Lakehurst, New Jersey (1937), led to a decline in their use; in recent years airships have been made with the less dangerous non-flammable helium.

airshow /'eəʃəʊ/ n. a public display of aircraft or aerobatics.

airsick /'eəsɪk/ adj. affected with nausea due to travel in an aircraft. □ **airsickness** n.

airside /'eəsaɪd/ n., adj., & adv. (designating, to, or towards) the sections of an airport, to which only passengers and airport personnel have admittance.

airspace /'eəspeɪs/ n. the air available to aircraft to fly in, esp. the part subject to the jurisdiction of a particular country.

airstrip /'eəstrɪp/ n. a strip of ground suitable for the take-off and landing of aircraft.

airtight /'eətaɪt/ adj. not allowing air to pass through.

Air Vice-Marshal n. an RAF officer of high rank, just below Air Marshal.

airway /'eəweɪ/ n. **1 a** a recognized route followed by aircraft. **b** (often in pl.) = AIRLINE. **2** a ventilating passage in a mine. **3** (often in pl.) Med. the passage(s) through which air passes into the lungs.

airwoman /'eə,wʊmən/ n. (pl. **-women**) **1** a woman pilot or member of the crew of an aircraft, esp. in an air force. **2** a member of the WRAF below commissioned rank.

airworthy /'eə,wɜ:ðɪ/ adj. (of an aircraft) fit to fly.

Airy /'eərɪ/, Sir George Biddell (1801–92), English astronomer and geophysicist. He investigated the diffraction pattern of a point source of light and devised cylindrical lenses to correct for astigmatism. In geophysics, he proposed the concept of isostasy to account for the gravitational anomalies associated with mountain masses, and gave an improved estimate of the earth's density. Airy was Astronomer Royal for forty-six years.

airy /'eərɪ/ adj. (**airier**, **airiest**) **1** well-ventilated, breezy. **2** flippant, superficial. **3 a** light as air. **b** graceful, delicate. **4** insubstantial, ethereal, immaterial. □ **airy-fairy** colloq. unrealistic, impractical, foolishly idealistic. □ **airily** adv. **airiness** n.

aisle /aɪl/ n. **1** part of a church, esp. one parallel to and divided by pillars from the nave, choir, or transept. **2** a passage between rows of pews, seats, etc. □ **aisled** adj. [ME *ele*, *ile* f. OF *ele* f. L *ala* wing: confused with *island* and F *aile* wing]

ait /eɪt/ n. (also **eyot**) Brit. a small island, esp. in a river. [OE *iggath* etc. f. *īeg* ISLAND + dimin. suffix]

aitch /eɪtʃ/ n. the name of the letter H. □ **drop one's aitches** fail to pronounce the initial h in words. [OF *ache*]

aitchbone /'eɪtʃbəʊn/ n. **1** the buttock or rump bone of cattle. **2** a cut of beef lying over this. [ME *nage-*, *nache-bone* buttock, ult. f. L *natis*, *-es* buttock(s): for loss of *n* cf. ADDER, APRON]

Aitken /'eɪtkɪn/, William Maxwell, see BEAVERBROOK.

Aix-en-Provence /,eɪksɒnprɒ'vɒns/ a city in Provence in southern France; pop. (1990) 126,850. It was the home of the painter Paul Cézanne.

Aix-la-Chapelle /ɛkslaʃapɛl/ the French name for AACHEN.

Aizawl /'aɪdʒəl/ a city in the far north-east of India, capital of the state of Mizoram; pop. (1991) 154,000.

Ajaccio /ə'dʒæksɪ,əʊ/ a port on the west coast of Corsica; pop. (1990) 59,320. The capital of Corsica until 1975, it is now the capital of the southern department of Corse-du-Sud. Napoleon was born there in 1769.

Ajanta Caves /ə'dʒʌntə/ a series of caves in the state of Maharashtra, south central India, containing Buddhist frescos and sculptures dating from the 1st century BC to the 7th century AD.

ajar¹ /ə'dʒɑ:(r)/ adv. & predic.adj. (of a door) slightly open. [A² + obs. *char* f. OE *cerr* a turn]

ajar² /ə'dʒɑ:(r)/ adv. out of harmony. [A² + JAR²]

Ajax /'eɪdʒæks/ Gk Mythol. **1** a Greek hero of the Trojan war, son of Telamon, king of Salamis. He was proverbial for his size and strength. **2** a Greek hero, son of Oileus, king of Locris, who was killed after a shipwreck on his homeward journey after the fall of Troy.

Ajman /æd'ʒmɑ:n/ **1** the smallest of the seven emirates of the United Arab Emirates; pop. (1985) 64,320. **2** its capital city.

Ajmer /ʌdʒ'mɪə(r)/ a city in NW India, in Rajasthan; pop. (1991) 402,000.

AK abbr. US Alaska (in official postal use).

a.k.a. abbr. also known as.

Akbar /ˈækbɑː(r)/, Jalaludin Muhammad (known as Akbar the Great) (1542–1605), Mogul emperor of India 1556–1605. Akbar expanded the Mogul empire to incorporate northern India, and established administrative efficiency and a coherent commercial system. He was the first ruler of India to promote religious and racial toleration. Akbar abolished slavery, prohibited the practice of suttee, legitimized the remarriage of widows, and banned polygamy except in cases of infertility.

akee var. of ACKEE.

akela /ɑːˈkeɪlə/ n. the adult leader of a group of Cub Scouts. [name of the leader of a wolf-pack in Kipling's *The Jungle Book*]

Akhenaten /ˌækəˈnɑːt(ə)n/ (also **Akhenaton**, **Ikhnaton** /ɪkˈnɑːt(ə)n/) (14th century BC), Egyptian pharaoh of the 18th dynasty, reigned 1379–1362 BC. The husband of Nefertiti, he came to the throne as Amenhotep IV, and after six years introduced the monotheistic solar cult of Aten, the sun disc, with the king as sole intermediary, changing his name to Akhenaten. The capital of Egypt was moved from Thebes to his newly built city of Akhetaten (now Tell el-Amarna). He was succeeded by his son-in-law, Tutankhamen, who abandoned the new religion early in his reign.

Akhetaten /ˌækəˈtɑːt(ə)n/ an ancient Egyptian capital at Tell el-Amarna (see AMARNA, TELL EL-).

Akhmatova /ækˈmɑːtəvə/, Anna (pseudonym of Anna Andreevna Gorenko) (1889–1966), Russian poet. A member of the Acmeist group of poets with Osip Mandelstam, Akhmatova favoured concrete detail, direct expression, and precision of language as a reaction against the mysticism of contemporary symbolist poetry. The personal and Christian tone of *Anno Domini* (1922), however, contributed to her official disfavour, which was to last for over thirty years. Her works include *Poem without a Hero* (1940–62) and *Requiem* (1940).

akimbo /əˈkɪmbəʊ/ adv. (of the arms) with hands on the hips and elbows turned outwards. [ME *in kenebowe*, prob. f. ON]

akin /əˈkɪn/ predic.adj. **1** related by blood. **2** (often foll. by *to*) of similar or kindred character. [A-⁴ + KIN]

Akkad /ˈækæd/ the capital city which gave its name to an ancient kingdom, traditionally founded by Sargon in north central Mesopotamia (modern Iraq). Its site is lost.

Akkadian /əˈkeɪdɪən/ adj. & n. (also **Accadian**) hist. ● adj. of or relating to Akkad or its people or language. ● n. **1** the Akkadian language. (See note below.) **2** an inhabitant of Akkad.

▪ The Akkadian language, now extinct but known from cuneiform inscriptions, is the oldest Semitic language for which there is evidence. It was used in Mesopotamia from about 3000 BC; two dialects, Assyrian and Babylonian, were widely spoken in the Middle East for the next 2,000 years, and the Babylonian form in particular functioned as a lingua franca until replaced by Aramaic around the 6th century BC. Akkadian survived as a written language until Hellenistic times.

Akko see ACRE 1.

Ak-Mechet /ˌækməˈtʃet/ a former name for SIMFEROPOL.

Akron /ˈækrən/ a city in NE Ohio; pop. (1990) 223,000. It is a centre of the rubber industry.

Aksai Chin /ˌæksaɪ ˈtʃɪn/ a region of the Himalayas occupied by China since 1950, but claimed by India as part of Kashmir.

Aksum /ˈɑːksəm/ (also **Axum**) a town in the province of Tigray in northern Ethiopia. It was a religious centre and the capital of a powerful kingdom between the 1st and 6th centuries AD. □ **Aksumite** adj. & n.

akvavit var. of AQUAVIT.

AL abbr. US Alabama (in official postal use).

Al symb. Chem. the element aluminium.

al- /æl, əl/ prefix assim. form of AD- before -l.

-al¹ /əl/ suffix **1** forming adjectives meaning 'relating to, of the kind of': **a** from Latin or Greek words (*central*; *regimental*; *colossal*; *tropical*) (cf. -IAL, -ICAL). **b** from English nouns (*tidal*). **2** forming nouns, esp. of verbal action (*animal*; *rival*; *arrival*; *proposal*; *trial*). [sense 1 f. F *-el* or L *-alis* adj. suffix rel. to *-aris* (-AR¹); sense 2 f. F *-aille* or f. (or after) L *-alis* etc. used as noun]

-al² /æl, əl/ suffix **1** Chem. forming the names of compounds which are aldehydes or occasionally other derivatives of alcohols (*acetal*, *chloral*,

etc.). **2** used with little or no chemical significance to form the names of drugs etc. (*barbital*, *veronal*, etc.). [f. ALCOHOL, ALDEHYDE]

Ala. abbr. Alabama.

à la /ˈɑː læ/ prep. after the manner of (*à la russe*). [F, f. À LA MODE]

Alabama /ˌæləˈbæmə/ a state in the south-eastern US, on the Gulf of Mexico; pop. (1990) 4,040,600; capital, Montgomery. Visited by Spanish explorers in the mid-16th century and later settled by the French, it passed to Britain in 1763 and to the US in 1783, becoming the 22nd state in 1819. □ **Alabaman** adj. & n.

alabaster /ˈæləˌbɑːstə(r), -ˌbæstə(r)/ n. & adj. ● n. a translucent usu. white form of gypsum, often carved into ornaments. ● adj. **1** of alabaster. **2** like alabaster in whiteness or smoothness. □ **alabastrine** /ˌæləˈbɑːstrɪn, -ˈbæstrɪn, -straɪn/ adj. [ME f. OF *alabastre* f. L *alabaster*, *-trum*, f. Gk *alabast(r)os*]

à la carte /ˌæ ˈkɑːt/ adv. & adj. ordered as separately priced item(s) from a menu, not as part of a set meal (cf. TABLE D'HÔTE). [F]

alack /əˈlæk/ int. (also **alack-a-day** /əˈlækəˌdeɪ/) archaic an expression of regret or surprise. [prob. f. AH + LACK]

alacrity /əˈlækrɪtɪ/ n. briskness or cheerful readiness. [L *alacritas* f. *alacer* brisk]

Aladdin /əˈlædɪn/ the hero of a story in the *Arabian Nights*. Now a popular pantomime choice, the story tells of Aladdin's adventures after he finds (in a cave) an old lamp which, when rubbed, brings a genie who obeys the will of the owner. □ **Aladdin's cave** a treasure-house of jewels or other valuables. **Aladdin's lamp** a talisman enabling its holder to gratify any wish.

Alagoas /ˌæləˈɡəʊəs/ a state in eastern Brazil, on the Atlantic coast; capital, Maceió.

Alain-Fournier /ˌælæ̃ˈfʊənɪˌeɪ/ (pseudonym of Henri-Alban Fournier) (1886–1914), French novelist. A literary columnist, he completed only one novel before he was killed in action in the First World War. The book, *Le Grand Meaulnes* (1913), is a lyrical semi-autobiographical narrative set in the countryside of Alain-Fournier's adolescence. His early poems and short stories were published as *Miracles* in 1924.

Alamein see EL ALAMEIN, BATTLE OF.

Alamo, the /ˈæləˌməʊ/ a mission in San Antonio, Texas, site of a siege in 1836 by Mexican forces during the Texan struggle for independence from Mexico. It was defended by about 180 volunteers (including Davy Crockett), all of whom were killed.

à la mode /ˌæ læ ˈməʊd/ adv. & adj. **1** in fashion; fashionable. **2 a** (of beef) braised in wine. **b** N. Amer. served with ice-cream. [F, = in the fashion]

Åland Islands /ˈɔːlənd/ (called in Finnish *Ahvenanmaa*) a group of islands in the Gulf of Bothnia, forming an autonomous region of Finland; capital, Mariehamn (known in Finnish as Maarianhamina). The group includes more than 6,500 islands and rocky islets, of which only 80 are inhabited. Swedish is the main language. In 1809 the islands became part of Russia, but in 1921 they were assigned to Finland by the League of Nations.

alanine /ˈæləˌniːn/ n. Biochem. a hydrophobic amino acid occurring in proteins. [G *Alanin*]

Alar /ˈeɪlɑː(r)/ n. propr. a growth retardant sprayed on fruit and vegetables to enhance the quality of the crop.

alar /ˈeɪlə(r)/ adj. **1** relating to wings. **2** winglike or wing-shaped. **3** axillary. [L *alaris* f. *ala* wing]

Alarcón /ˌælɑːˈkɒn/, Pedro Antonio de (1833–91), Spanish novelist and short-story writer. His best-known work, the humorous short story *The Three-Cornered Hat* (1874), was the source for Manuel de Falla's ballet and Hugo Wolf's opera, both of the same name.

Alarcón y Mendoza see RUIZ DE ALARCÓN Y MENDOZA.

Alaric /ˈælərɪk/ (c.370–410), king of the Visigoths 395–410. Alaric invaded Greece (395–6) and then Italy (400–3), but was checked on each occasion by the Roman general Stilicho (c.365–408). He invaded Italy again in 408 and in 410 captured Rome.

alarm /əˈlɑːm/ n. & v. ● n. **1** a warning of danger etc. (*gave the alarm*). **2 a** warning sound or device (*the burglar alarm was set off accidentally*). **b** = *alarm clock*. **3** frightened expectation of danger or difficulty (*were filled with alarm*). ● v.tr. **1** frighten or disturb. **2** arouse to a sense of danger. □ **alarm clock** a clock with a device that can be made to sound at the time set in advance. [ME f. OF *alarme* f. It. *allarme* f. *all'arme!* to arms]

alarming /əˈlɑːmɪŋ/ adj. disturbing, frightening. □ **alarmingly** adv.

alarmist /əˈlɑːmɪst/ *n.* & *adj.* ● *n.* a person given to spreading needless alarm. ● *adj.* creating needless alarm. □ **alarmism** *n.*

alarum /əˈlɑːrəm/ *n. archaic* = ALARM. □ **alarums and excursions** *joc.* confused noise and bustle.

Alas. *abbr.* Alaska.

alas /əˈlæs, əˈlɑːs/ *int.* an expression of grief, pity, or concern. [ME f. OF *a las(se)* f. *a* ah + *las(se)* f. L *lassus* weary]

Alaska /əˈlæskə/ the largest state of the US, in the extreme north-west of North America, with coasts on the Arctic Ocean, Bering Sea, and North Pacific; pop. (1990) 550,000; capital, Juneau. About one-third of Alaska lies within the Arctic Circle. Russian explorers led by Vitus Bering reached Alaska in 1741, and it was further explored by Captain James Cook, George Vancouver, and others in the late 18th century. The territory was purchased from Russia in 1867 and became the 49th state of the US in 1959. □ **Alaskan** *adj.* & *n.*

Alaska, Gulf of a part of the NE Pacific between the Alaska Peninsula and the Alexander Archipelago.

Alaska Peninsula a peninsula on the south coast of Alaska. It extends south-westwards into the NE Pacific, between the Bering Sea and the Gulf of Alaska, and is continued in the Aleutian Islands.

alate /ˈeɪleɪt/ *adj.* having wings or winglike appendages. [L *alatus* f. *ala* wing]

alb /ælb/ *n.* a white vestment reaching to the feet, worn by clergy and servers in some Christian churches. [OE *albe* f. eccl.L *alba* fem. of L *albus* white]

Albacete /ˌælbəˈseɪtɪ/ a city in a province of the same name in SE Spain; pop. (1991) 134,600.

albacore /ˈælbəˌkɔː(r)/ *n.* a long-finned tuna, *Thunnus alalunga*. Also called *germon*. [Port. *albacor*, *-cora*, f. Arab. *al* the + *bakr* young camel or *bakūr* premature, precocious]

Alba Iulia /ˌælbə ˈjuːlɪə/ a city in west central Romania, to the north of the Transylvanian Alps; pop. (1989) 72,330. Founded by the Romans in the 2nd century AD, it was the capital of Transylvania.

Alban, St /ˈɔːlbən/ (3rd century), the first British Christian martyr. A pagan of Verulamium (now St Albans, Herts.), he was converted and baptized by a fugitive priest whom he sheltered. When soldiers searched his house, he put on the priest's cloak and was arrested and condemned to death. Feast day, 22 June.

Albania /ælˈbeɪnɪə/ a republic in SE Europe, bordering on the Adriatic Sea; pop. (est. 1991) 3,300,000; official language, Albanian; capital, Tirana. Much of the land is mountainous and forested. Although Albania was nominally part of the Byzantine and later the Ottoman Empire, its mountain tribes always remained fiercely independent and central rule was never completely effective. It gained independence as a result of the Balkan Wars in 1912, becoming a rather unstable monarchy under King Zog I in 1928. Invaded by Italy in 1939, it became a Stalinist regime under Enver Hoxha after the Second World War, remaining extremely isolationist in policy and outlook. The Communists lost power to the Democratic Party in elections in 1992.

Albanian /ælˈbeɪnɪən/ *n.* & *adj.* ● *n.* **1 a** a native or national of Albania. **b** a person of Albanian descent. **2** the official language of Albania. (*See* note below.) ● *adj.* of or relating to Albania or its people or language.

■ Albanian constitutes a separate branch of the Indo-European language group, spoken by some 5 million people in Albania, the former Yugoslavia, Greece, and elsewhere. There are two distinct dialects, Tosk in the north and Gheg in the south.

Albany /ˈɔːlbənɪ/ the state capital of New York, on the Hudson River; pop. (1990) 101,080.

albata /ælˈbɑːtə/ *n.* German silver; an alloy of nickel, copper, and zinc. [L *albata* whitened f. *albus* white]

albatross /ˈælbəˌtrɒs/ *n.* **1 a** a large long-winged seabird of the family Diomedeidae, related to petrels and inhabiting the Pacific and Antarctic Oceans. **b** an encumbrance (in allusion to Coleridge's 'The Rime of the Ancient Mariner'); a source of frustration or guilt. **2** *Brit.* *Golf* a score of three strokes under par at any hole. [alt. (after L *albus* white) of 17th-c. *alcatras*, applied to various seabirds, f. Sp. and Port. *alcatraz*, var. of Port. *alcatruz* f. Arab. *alḳādūs* the pitcher]

albedo /ælˈbiːdəʊ/ *n.* (pl. **-os**) the proportion of light or radiation reflected by a surface, esp. of a planet or moon. [eccl.L, = whiteness, f. L *albus* white]

Albee /ˈɔːlbiː, ˈæl-/, Edward Franklin (b.1928), American dramatist. He was initially associated with the Theatre of the Absurd; *Who's Afraid of Virginia Woolf* (1962) marked a more naturalistic departure and showed

an interest in closely observed human relationships. Later plays, such as *Tiny Alice* (1965) and *A Delicate Balance* (1966), demonstrate his increasing preoccupation with abstract issues.

albeit /ɔːlˈbiːɪt/ *conj. literary* though (*he tried, albeit without success*).

Albena /ælˈbemə/ a resort town in Bulgaria, on the coast of the Black Sea.

Albers /ˈælbəz/, Josef (1888–1976), German-born American artist, designer, and teacher. He was a teacher at the Bauhaus until 1933, and his work was strongly influenced by constructivism. His designs are characterized by intellectual calculation and the use of simple geometric shapes, as in his furniture design (including the first laminated chair for mass production) and pictures made with pieces of coloured glass. He experimented with colour juxtapositions in his *Homage to the Square* series (begun in 1950). As a teacher at Black Mountain College in North Carolina (1934–50), he was instrumental in disseminating Bauhaus principles throughout the US.

albert /ˈælbət/ *n.* a watch-chain with a bar at one end for attaching to a buttonhole. [ALBERT, PRINCE]

Albert, Lake /ˈælbət/ (also called *Lake Mobutu Sese Seko*) a lake in the Rift Valley of east central Africa, on the border between Zaire and Uganda. It is linked to Lake Edward by the Semliki river and to the White Nile by the Albert Nile. The lake was named after the Prince Consort by the English explorer Samuel Baker, who in 1864 was the first European to sight it. In 1973 it was renamed in Zaire Lake Mobutu Sese Seko, in honour of that country's President.

Albert, Prince /ˈælbət/ (1819–61), consort to Queen Victoria. First cousin of the queen and prince of Saxe-Coburg-Gotha, he revitalized the British court in the first twenty years of his wife's reign. He was one of the driving forces behind the Great Exhibition of 1851; its profits allowed the construction of the Royal Albert Hall (1871) and of museum buildings in South Kensington. In 1861, just before his premature death from typhoid fever, his moderating influence was crucial in keeping Britain out of the American Civil War.

Alberta /ælˈbɜːtə/ a prairie province in western Canada, bounded on the south by the US and on the west by the Rocky Mountains; capital, Edmonton; pop. (1991) 2,522,300.

Alberti /ælˈbeatɪ/, Leon Battista (1404–72), Italian architect, humanist, painter, and art critic. He wrote the first account of the theory of perspective in the Renaissance in his *Della Pittura* (1435). He is also credited with reawakening interest in Roman architecture, after the publication of his *De Re Aedificatoria* (1485). His own architecture, classical in style, includes the façade of the church of Santa Maria Novella in Florence and San Francesco in Rimini.

Albert Nile the upper part of the Nile flowing through NW Uganda between Lake Albert and the Ugandan–Sudanese border. (See NILE[1].)

Albertus Magnus, St /ælˌbɜːtəs ˈmægnəs/ (known as 'Doctor Universalis') (*c.*1200–80), Dominican theologian, philosopher, and scientist. A teacher of St Thomas Aquinas, he was a pioneer in the study of Aristotle and contributed significantly to the comparison of Christian theology and pagan philosophy. His particular interest in the physical sciences, including alchemy, earned him a reputation for magical powers. Feast day, 15 Nov.

albescent /ælˈbes(ə)nt/ *adj.* growing or shading into white. [L *albescere* *albescent-* f. *albus* white]

Albi /ˈælbɪ/ a town in southern France; pop. (1990) 48,700. It was the birthplace, in the 12th century, of the Albigensian movement.

Albigenses /ˌælbɪˈdʒensiːz/ the adherents of a heretical Christian sect in southern France in the 12th–13th centuries, identified with the Cathars. Their teaching was a form of Manichaean dualism (see MANICHAEISM), with an extremely strict moral and social code, including the condemnation of both marriage and procreation. The heresy spread rapidly until ruthlessly crushed by the elder Simon de Montfort's crusade (1209–31) and by an Inquisition. □ **Albigensian** *adj.* & *n.* [L f. ALBI]

albino /ælˈbiːnəʊ/ *n.* (pl. **-os**) **1** a person or animal having a congenital absence of pigment in the skin and hair (which are white) and the eyes (which are usu. pink). **2** a plant lacking normal colouring. □ **albinism** /ˈælbɪˌnɪz(ə)m/ *n.* **albinotic** /ˌælbɪˈnɒtɪk/ *adj.* [Sp. & Port. (orig. of white Negroes) f. *albo* L f. *albus* white + *-ino* = -INE[1]]

Albinoni /ˌælbɪˈnəʊnɪ/, Tomaso (1671-1751), Italian composer. He wrote more than fifty operas, but is now best known for his melodic instrumental music. Several of his many concertos have seen a revival, but the *Adagio in G* with which he is popularly associated was in fact composed by the Italian musicologist Remo Giazotto (b.1910), based on a manuscript fragment by Albinoni.

Albinus see ALCUIN.

Albion /'ælbɪən/ n. (also **perfidious Albion**) Britain or England. [OE f. L f. Celt. *Albio* (unrecorded): F *la perfide Albion* with ref. to alleged treachery to other nations]

albite /'ælbaɪt/ n. *Mineral.* a feldspar, usu. white, rich in sodium. [L *albus* white + -ITE¹]

Ålborg see AALBORG.

album /'ælbəm/ n. **1** a blank book for the insertion of photographs, stamps, etc. **2** a long-playing gramophone record; a compact disc or tape comprising several pieces of music. [L, = a blank tablet, neut. of *albus* white]

albumen /'ælbjʊmɪn/ n. **1** egg-white. **2** *Bot.* the substance found between the skin and germ of many seeds, usu. the edible part; = ENDOSPERM. [L *albumen -minis* white of egg f. *albus* white]

albumin /'ælbjʊmɪn/ n. any of a class of water-soluble proteins found in egg-white, milk, blood, etc. □ **albuminous** /æl'bjuːmɪnəs/ adj. [F *albumine* f. L *albumin-*: see ALBUMEN]

albuminoid /æl'bjuːmɪˌnɔɪd/ n. = SCLEROPROTEIN.

albuminuria /ˌælbjuːmɪ'njʊərɪə/ n. *Med.* the presence of albumin in the urine, usu. a symptom of kidney disease.

Albuquerque¹ /'ælbəˌkɜːkɪ/ the largest city in the state of New Mexico; pop. (1990) 384,740.

Albuquerque² /'ælbəˌkɜːkɪ/, Alfonso de (known as Albuquerque the Great) (1453–1515), Portuguese colonial statesman. He first travelled east in 1502, and, after being appointed viceroy of the Portuguese Indies four years later, conquered Goa and made it the capital of the Portuguese empire in the east. Albuquerque made further conquests in Ceylon, Malacca, Ormuz, the Sunda Islands, and the Malabar Coast, but was relieved of office as a result of a court intrigue at home and died on the passage back to Portugal.

alburnum /æl'bɜːnəm/ n. = SAPWOOD. [L f. *albus* white]

Alcaeus /æl'siːəs/ (c.620–c.580 BC), Greek lyric poet. His most important contribution was a new form of lyric metre (see ALCAIC); he also wrote political odes, drinking-songs, and love-songs. His works were an important model for the Roman poet Horace as well as for French and English verse of the Renaissance.

alcahest var. of ALKAHEST.

alcaic /æl'keɪɪk/ adj. & n. ● adj. of the verse metre invented by Alcaeus, occurring in four-line stanzas. ● n. (in pl.) alcaic verses. [LL *alcaicus* f. Gk *alkaikos* f. *Alkaios* ALCAEUS]

Alcalá de Henares /ˌælkæˌlɑː deɪ e'nɑːres/ a city in central Spain, on the River Henares, 25 km (15 miles) north-east of Madrid; pop. (1991) 162,780.

alcalde /æl'kældɪ/ n. a magistrate or mayor in a Spanish, Portuguese, or Latin American town. [Sp. f. Arab. *al-ḳāḍī* the judge: see CADI]

Alcatraz /ˌælkə'træz/ a rocky island in San Francisco Bay, California, named after the pelicans (Spanish *alcatraz*) which once lived there. It was, between 1934 and 1963, the site of a top-security Federal prison.

Alcestis /æl'sestɪs/ *Gk Mythol.* wife of Admetus, king of Pherae in Thessaly, whose life she saved by consenting to die on his behalf. She was brought back from Hades by Hercules.

alchemy /'ælkəmɪ/ n. **1** the medieval forerunner of chemistry, esp. as concerned with turning base metals into gold or silver. (*See note below.*) **2** miraculous transformation or the means of achieving it. □ **alchemist** n. **alchemize** v.tr. (also **-ise**). **alchemic** /æl'kemɪk/ adj. **alchemical** adj. **alchemically** adv. [ME f. OF *alkemie, alkamie* f. med.L *alchimia, -emia,* f. Arab. *alkīmiyā'* f. *al* the + *kīmiyā'* f. Gk *khēmia, -meia* art of transmuting metals]

■ Alchemy was based on the possible transmutation of all matter, and was far wider in scope than the attempt to turn base metals into gold. It attracted such medieval scholars as Roger Bacon and St Albertus Magnus, and was patronized by princes. Paracelsus (16th century) applied it to the search for a chemical therapy for disease; his followers developed chemical medicines and sought a universal elixir which they dreamed would prolong life and restore youth. The science of chemistry separated slowly from alchemy; many leading scientists, including Newton, retained a belief or interest in transmutation. The rise of mechanical philosophy in the 17th century gradually undermined alchemy; it became an aspect of the occult, its mystical aspects surviving in esoteric movements such as that of the Rosicrucians.

alcheringa /ˌæltʃə'rɪŋgə/ n. = dream-time. [Aboriginal, = dream-time]

Alcibiades /ˌælsɪ'baɪəˌdiːz/ (c.450–404 BC), Athenian general and statesman. Educated in the household of Pericles, he became the pupil and friend of Socrates. In the Peloponnesian War he sponsored the unsuccessful Athenian expedition against Sicily, but fled to Sparta after being recalled for trial on a charge of sacrilege. He later held commands for Athens against Sparta and Persia, before his enemies finally forced him from Athens and had him murdered in Phrygia.

Alcock /'ɔːlkɒk/, Sir John William (1892–1919), English aviator. Together with Sir Arthur Whitten Brown, he made the first non-stop transatlantic flight (16 hours 27 minutes) on 14–15 June 1919, from Newfoundland to Clifden, Ireland, in a converted Vickers Vimy bomber.

alcohol /'ælkəˌhɒl/ n. **1** a colourless volatile inflammable liquid (chem. formula: C_2H_5OH) forming the intoxicating element in wine, beer, spirits, etc., and also used as a solvent, as fuel, etc. Also called *ethanol, ethyl alcohol.* **2** drink containing this (*had always enjoyed food and alcohol*). **3** *Chem.* any of a large class of organic compounds that contain one or more hydroxyl groups attached to carbon atoms. □ **alcohol-free 1** not containing alcohol. **2** where, or during which, alcoholic drinks are not consumed. [F or med.L f. Arab. *al-kuḥl* f. *al* the + *kuḥl* KOHL]

alcoholic /ˌælkə'hɒlɪk/ adj. & n. ● adj. of, relating to, containing, or caused by alcohol. ● n. a person suffering from alcoholism.

Alcoholics Anonymous (abbr. **AA**) a self-help organization founded in the US in 1935 for people fighting alcoholism. The organization, based on a system of local groups, has now spread around the world.

alcoholism /'ælkəhɒˌlɪz(ə)m/ n. **1** an addiction to the consumption of alcoholic liquor. **2** the diseased condition resulting from this. [mod.L *alcoholismus* (as ALCOHOL)]

alcoholometer /ˌælkəhɒ'lɒmɪtə(r)/ n. an instrument for measuring alcoholic concentration. □ **alcoholometry** n.

Alcott /'ɔːlkɒt/, Louisa May (1832–88), American novelist. From an early age she published sketches and stories to support her family, including *Hospital Sketches* (1863), which recounted her experiences as a nurse in the Civil War. Her most popular novel was *Little Women* (1868–9), a largely autobiographical work about a New England family, written for adolescent girls. She wrote a number of sequels to this as well as novels for adults. Alcott was involved in diverse reform movements, including women's suffrage.

alcove /'ælkəʊv/ n. a recess, esp. in the wall of a room or of a garden. [F f. Sp. *alcoba* f. Arab. *al-ḳubba* f. *al* the + *ḳubba* vault]

Alcuin /'ælkwɪn/ (c.735–804) (also **Albinus** /æl'biːnəs/), English scholar, theologian, and adviser to Charlemagne. While attached to the court of the Frankish leader, he improved the palace library and transformed the court into a cultural centre for Charlemagne's empire (a period referred to as the Carolingian Renaissance). On becoming abbot of Tours in 796, he established an important library and school there. He developed the type of script known as Carolingian minuscule, which influenced the roman type now used in printed books.

Aldabra /æl'dæbrə/ a coral island group in the Indian Ocean, north-west of Madagascar. Formerly part of the British Indian Ocean Territory, it became an outlying dependency of the Seychelles in 1976. Since then it has been administered as a nature reserve.

Aldebaran /æl'debərən/ *Astron.* the brightest star in the constellation of Taurus, forming the 'eye of the bull'. It is a binary star of which the main component is a red giant. [Arab., = the follower (of the Pleiades)]

Aldeburgh /'ɔːldbrə/ a town on the coast of Suffolk, England; pop. (1981) 3,000. It was the home of Benjamin Britten; the Aldeburgh Festival, established by the composer in 1948, is held there annually.

aldehyde /'ældɪˌhaɪd/ n. *Chem.* any of a class of compounds formed by the oxidation of alcohols (and containing the group –CHO). □ **aldehydic** /ˌældɪ'hɪdɪk/ adj. [abbr. of mod.L *alcohol dehydrogenatum* alcohol deprived of hydrogen]

al dente /æl 'dentɪ/ adj. (of pasta etc.) cooked so as to be still firm when bitten. [It., lit. 'to the tooth']

alder /'ɔːldə(r)/ n. a catkin-bearing tree of the genus *Alnus*, related to the birch; esp. *A. glutinosa*, which is common in damp ground. □ **alder buckthorn** a shrub, *Frangula alnus*, related to the buckthorn. **alder fly** a neuropterous insect of the genus *Sialis*, found near streams. [OE *alor, aler,* rel. to L *alnus*, with euphonic d]

alderman /'ɔːldəmən/ n. (pl. **-men**) **1** esp. *hist.* a co-opted member of an English county or borough council, next in dignity to the Mayor.

2 *N. Amer. & Austral.* the elected governor of a city. □ **aldermanship** *n.*
aldermanic /ˌɔːldəˈmænɪk/ *adj.* [OE *aldor* patriarch f. *ald* old + MAN]

Aldermaston /ˈɔːldəˌmɑːstən/ a village in Berkshire, England, site of the Atomic Weapons Research Establishment. From 1958 to 1963 a march between Aldermaston and London took place each year in protest against the development and production of nuclear weapons.

Alderney[1] /ˈɔːldənɪ/ an island in the English Channel, to the north-east of Guernsey; pop. (1986) 2,130. It is the third largest of the Channel Islands.

Alderney[2] /ˈɔːldənɪ/ *n.* a breed of dairy cattle originally from Alderney or the Channel Islands in general.

Aldershot /ˈɔːldəˌʃɒt/ a town in southern England, in Hampshire; pop. (1981) 54,358. It is the site of a military training centre.

Aldine /ˈældaɪn/ *adj.* of or relating to the Italian printer Aldus Manutius, or to the books printed by him, or to certain styles of display types. The device characteristic of Aldine books is a figure of a dolphin on an anchor. [L *Aldinus*, f. *Aldus*]

Aldis lamp /ˈɔːldɪs/ *n. propr.* a hand-lamp for signalling in Morse code. [A. C. W. *Aldis*, Brit. inventor (1878–1953)]

Aldiss /ˈɔːldɪs/, Brian W(ilson) (b.1925), English novelist and critic. Known primarily for his works of science fiction such as *Frankenstein Unbound* (1973) and *Moreau's Other Island* (1980), he has done much to promote the cause of science fiction as a literary genre, including writing a history of the subject, *Billion Year Spree* (1973).

aldrin /ˈældrɪn/ *n. Chem.* a white crystalline chlorinated hydrocarbon formerly used as an insecticide. [K. *Alder*, Ger. chemist (1902–58) + -IN]

Aldus Manutius /ˌɔːldəs məˈnjuːʃɪəs/ (Latinized name of Teobaldo Manucci; also known as Aldo Manuzio) (1450–1515), Italian scholar, printer, and publisher. He is remembered for the fine first editions of many Greek and Latin classics which were printed at his press in Venice. Aldus Manutius pioneered the widespread use of italic type, which he used as a text type in many of his printed books.

ale /eɪl/ *n.* beer (now usu. as a trade word). [OE *alu*, = ON *öl*]

aleatoric /ˌeɪlɪəˈtɒrɪk/ *adj.* **1** depending on the throw of a die or on chance. **2** *Mus. & Art* involving random choice by a performer or artist. [L *aleatorius aleator* dice-player f. *alea* die]

aleatory /ˈeɪlɪətərɪ/ *adj.* = ALEATORIC. [as ALEATORIC]

aleatory music *n.* music of which the composition or performance is randomly determined, for example by the use of a computer, dice, or randomly tuned radio receivers. The method was pioneered by John Cage.

alec /ˈælɪk/ *n.* (also **aleck**) *Austral. sl.* a stupid person. [shortening of SMART ALEC]

Alecto /əˈlektəʊ/ (also **Allecto**) *Gk Mythol.* one of the Furies. [Gk, = the implacable one]

alee /əˈliː/ *adv. & predic.adj.* **1** on the lee or sheltered side of a ship. **2** to leeward. [ME, f. A[2] + LEE]

alehouse /ˈeɪlhaʊs/ *n. hist.* a tavern.

Alekhine /ˈælɪˌkiːn, ˈæljəkɪn/, Alexander (born Aleksandr Aleksandrovich Alyokhin) (1892–1946), Russian-born French chess player. Notorious for his multiple strategies of attack, he was world champion 1927–35 and from 1937 until his death.

Aleksandropol /ˌælɪkˈsɑːndrəpɒl/ (also **Alexandropol** /ˌælɪɡˈzɑːndrəpɒl/) a former name (1840–1924) for GYUMRI.

Aleksandrovsk /ˌælɪkˈsɑːndrəfsk/ the former name (until 1921) for ZAPORIZHZHYA.

Alembert /ˈæləmˌbeə(r)/, Jean le Rond d' (1717–83), French mathematician, physicist, and philosopher. His most famous work was the *Traité de dynamique* (1743), in which he developed his own laws of motion. From 1746 to 1758 he was Diderot's chief collaborator on the *Encyclopédie*, a seminal text of the Age of Enlightenment.

alembic /əˈlembɪk/ *n.* **1** *hist.* an apparatus formerly used in distilling. **2** a means of refining or extracting. [ME f. OF f. med.L *alembicus* f. Arab. *al-'anbīk* f. *al* the + *'anbīk* still f. Gk *ambix*, *-ikos* cup, cap of a still]

Alentejo /ˌælənˈteɪʒuː/ a region and former province of east central Portugal. Its name is derived from an Arabic phrase meaning 'beyond the Tagus'.

aleph /ˈɑːlef/ *n.* the first letter of the Hebrew alphabet. [Heb. *'āleþ*, lit. 'ox']

Aleppo /æˈlepəʊ/ (Arabic **Halab** /hɑːˈlɑːb/) a city in northern Syria; pop. (1990) 1,355,000. This ancient city was formerly an important commercial centre on the trade route between the Mediterranean and the countries of the East.

alert /əˈlɜːt/ *adj., n., & v.* ● *adj.* **1** watchful or vigilant; ready to take action. **2** nimble (esp. of mental faculties); attentive. ● *n.* **1** a warning call or alarm. **2 a** a warning of an air raid. **b** the duration of this. ● *v.tr.* (often foll. by *to*) make alert; warn (*were alerted to the danger*). □ **on the alert** on the lookout against danger or attack. □ **alertly** *adv.* **alertness** *n.* [F *alerte* f. It. *all' erta* to the watch-tower]

-ales /ˈeɪliːz/ *suffix* forming the plural names of orders of plants (*Rosales*). [pl. of L adj. suffix *-alis*: see -AL[1]]

Aletschhorn /ˈɑːletʃˌhɔːn/ a mountain in Switzerland, in the Bernese Alps, rising to 4,195 m (13,763 ft). The Aletsch glaciers are amongst the largest in Europe.

aleurone /əˈljʊərəʊn/ *n.* (also **aleuron** /-rən/) *Biochem.* protein stored as granules in the seeds of plants etc. [Gk *aleuron* flour]

Aleut /əˈljuːt/ *n.* **1** a native of the Aleutian Islands. **2** the language of the Aleuts, related to Eskimo (Inupiaq and Yupik). □ **Aleutian** /əˈljuːʃ(ə)n/ *adj. & n.* [18th c.: orig. unkn.]

Aleutian Islands (also **Aleutians**) a chain of volcanic islands in US possession, extending south-west from the Alaska Peninsula.

A level /eɪ/ *n. Brit.* = advanced level (see ADVANCE).

alewife /ˈeɪlwaɪf/ *n.* (*pl.* **alewives**) *US* a fish of the NW Atlantic, *Alosa pseudoharengus*, related to the herring. [corrupt. of 17th-c. *aloofe*: orig. uncert.]

Alexander[1] /ˌælɪɡˈzɑːndə(r)/ (known as Alexander the Great) (356–323 BC), son of Philip II, king of Macedon 336–323. He was a pupil of Aristotle. After his succession he invaded Persia, liberating the Greek cities in Asia Minor, and then defeating the Persians in Egypt, Syria, and Mesopotamia. While in Egypt he founded Alexandria, his first and best-known city. He went on to extend his conquests eastwards, taking Bactria and the Punjab. He died of a fever at Babylon, and his empire quickly fell apart after his death. Regarded as a god in his lifetime, he became a model for many subsequent imperialist conquerors of antiquity, and the subject of many fantastic legends.

Alexander[2] /ˌælɪɡˈzɑːndə(r)/ the name of three kings of Scotland:
Alexander I (c.1077–1124), son of Malcolm III, reigned 1107–24. During his reign he paid homage to King Henry I of England as his overlord. Nevertheless, he maintained the independence of the Scottish Church from the English Church.
Alexander II (1198–1249), son of William I of Scotland, reigned 1214–49. He generally restored friendly relations with England, negotiating a treaty with King Henry III. He extended royal authority within his own realm, particularly in the western Highlands and the south-west.
Alexander III (1241–86), son of Alexander II, reigned 1249–86. He built up a strong, united, and prosperous kingdom, and succeeded in annexing the Hebrides and the Isle of Man in 1266. His death in a riding accident left Scotland without a male heir to the throne, and plunged the country into three decades of dynastic upheaval and struggle against English domination.

Alexander[3] /ˌælɪɡˈzɑːndə(r)/ the name of three tsars of Russia:
Alexander I (1777–1825), reigned 1801–25. Alexander came to the throne after the murder of his tyrannical father Paul, and pursued reforming policies in education, science, administration, and the system of serfdom. The first half of his reign was dominated by the military struggle with Napoleon, which culminated in the French emperor's unsuccessful invasion of Russia in 1812. In the latter part of his reign Alexander became disillusioned by the failure of the Holy Alliance and less interested in reforms at home.
Alexander II (known as Alexander the Liberator) (1818–81), son of Nicholas I, reigned 1855–81. He introduced a programme of modernization and political and economic reforms, including limited emancipation of the serfs (1861), nationwide construction of railways and schools, and the creation of local government in the provinces (1864). Alexander's harsh suppression of peasant revolts and revolutionary movements led, however, to the hardening of radical opposition in the form of terrorist groups, such as the People's Will, a group of which succeeded in assassinating the tsar with a bomb in Mar. 1881.
Alexander III (1845–94), son of Alexander II, reigned 1881–94. Ascending the throne after his father's assassination, Alexander III fell back on repressive conservatism, heavily centralizing the imperial administration, weakening local government, abolishing autonomous peasant administrations, and persecuting Jews. Although his reign witnessed considerable economic development, his social policies

produced a dangerous situation in Russia, which he bequeathed to his doomed son and successor Nicholas II.

Alexander[1] /ˌælɪɡˈzɑːndə(r)/, Harold (Rupert Leofric George), 1st Earl Alexander of Tunis (1891–1969), British Field Marshal and Conservative statesman. In the Second World War he supervised the evacuation from Dunkirk, the withdrawal from Burma, and the victorious campaigns in North Africa (1943), Sicily, and Italy (1943–5). After the war he became Governor-General of Canada (1946–52) and British Minister of Defence (1952–4).

Alexander Archipelago a group of about 1,100 US islands off the coast of SE Alaska, the remnants of a submerged mountain system.

Alexander Nevsky /ˈnjefskɪ/ (also **Nevski**) (canonized as St Alexander Nevsky) (c.1220–63), Russian national hero. Born Aleksandr Yaroslavich, he was elected prince of Novgorod in 1236, ruling it until 1263. He was called 'Nevsky' after the River Neva, on whose banks he defeated the Swedes in 1240. Feast day, 30 Aug. or 23 Nov.

alexanders /ˌælɪɡˈzɑːndəz/ n. an umbelliferous plant, *Smyrnium olusatrum*, formerly used in salads but superseded by celery. [OE f. med.L *alexandrum*]

Alexander technique n. a system of body awareness designed to promote well-being by ensuring minimum effort in maintaining postures and carrying out movements. [F. Matthias *Alexander*, Australian-born physiotherapist (1869–1955)]

Alexandretta /ˌælɪɡzɑːnˈdretə/ the former name for ISKENDERUN.

Alexandria /ˌælɪɡˈzɑːndrɪə/ the chief port of Egypt; pop. (est. 1986) 2,893,000. Founded in 332 BC by Alexander the Great, after whom it is named, it became a major centre of Hellenistic culture. It is renowned for the Mouseion library, founded by Alexander's successor Ptolemy I, which held a vast store of Greek scholarship in the form of papyrus and vellum scrolls and was burned down by Arab invaders (c.641). A nearby island was the site of the Pharos lighthouse.

Alexandrian /ˌælɪɡˈzɑːndrɪən/ adj. **1** of or characteristic of Alexandria in Egypt. **2 a** belonging to or akin to the schools of literature and philosophy of Alexandria. (See also HELLENISTIC.) **b** (of a writer) derivative or imitative; fond of recondite learning.

alexandrine /ˌælɪɡˈzændraɪn/ adj. & n. ● adj. (of a line of verse) having six iambic feet. ● n. an alexandrine line. [F *alexandrin* f. *Alexandre* Alexander (the Great), the subject of an Old French poem in this metre]

alexandrite /ˌælɪɡˈzɑːndraɪt/ n. Mineral. a green variety of chrysoberyl. [Tsar *Alexander* I of Russia + -ITE[1]]

Alexandropol see ALEKSANDROPOL.

alexia /əˈleksɪə/ n. Med. the inability to see words or to read, caused by a defect of the brain (cf. DYSLEXIA). [A-[1] + Gk *lexis* speech f. *legein* to speak, confused with L *legere* to read]

alfalfa /ælˈfælfə/ n. a leguminous plant, *Medicago sativa*, with clover-like leaves and flowers, grown for fodder and for use as a salad vegetable. Also called *lucerne*. [Sp. f. Arab. *al-faṣfaṣa*, a green fodder]

Al Fatah see FATAH, AL.

Alfonso XIII /ælˈfɒnsəʊ/ (1886–1941), king of Spain 1886–1931. Alfonso ruled under the regency of his mother until 1902, during which time Spain lost her colonial possessions in the Philippines and Cuba to the US. In 1923 he supported Miguel Primo de Rivera's assumption of dictatorial powers, but by 1931 Alfonso had agreed to elections. When these indicated the Spanish electorate's clear preference for a republic, the king was forced to abdicate.

Alfred /ˈælfrɪd/ (known as Alfred the Great) (849–99), king of Wessex 871–99. Alfred's military resistance saved SW England from Viking occupation. He negotiated the treaty giving the Danelaw to the Norsemen (886). A great reformer, he reorganized his land-based garrisons, founded the English navy, issued a new code of laws, introduced administrative and financial changes, revived learning, and promoted the use of English for literature and education.

alfresco /ælˈfreskəʊ/ adv. & adj. in the open air (*we lunched alfresco; an alfresco lunch*). [It. *al fresco* in the fresh (air)]

Al Fujayrah see FUJAIRAH.

Alfvén /ˈælven/, Hannes Olof Gösta (1908–95), Swedish theoretical physicist. He worked mainly in plasma physics, and pioneered the study of magnetohydrodynamics. His contributions have proved important for studies of plasmas in stars and in controlled thermonuclear fusion, though he opposed the development of nuclear reactors and weapons. Alfvén shared the Nobel Prize for physics in 1970.

alga /ˈælɡə/ n. (pl. **algae** /ˈældʒiː, ˈælɡiː/) (usu. in pl.) a primitive non-flowering photosynthetic plant of a large assemblage that includes mainly aquatic forms such as seaweeds and many plankton. Most algae are microscopic and single-celled, and there are also multicellular forms with no vascular tissue or absorbent root system. Algae are now separated into a number of divisions or phyla, partly on the basis of their pigmentation; many (in some schemes, all) of the algae are placed in the kingdom Protista along with protozoans. The primitive blue-green algae are related more to bacteria than to other algae; traces of them are reported in rocks several thousand million years old. The brown algae include the largest and most complex seaweeds, and the green algae are thought to be the ancestors of most land plants. Planktonic algae of various kinds provide the food which supports nearly all marine life. □ **algal** adj. **algoid** adj. [L]

Algarve, the /ælˈɡɑːv/ the southernmost province of Portugal, on the Atlantic coast; capital, Faro. The Algarve is noted as a holiday resort area. [Arab. *al* the + *gharb* west]

algebra /ˈældʒɪbrə/ n. **1** a branch of mathematics that uses letters and other general symbols to represent numbers and quantities in formulae and equations. (*See note below.*) **2** a system of this based on given axioms (*linear algebra; the algebra of logic*). □ **algebraic** /ˌældʒɪˈbreɪk/ adj. **algebraical** adj. **algebraically** adv. **algebraist** n. [It. & Sp. and med.L, f. Arab. *al-jabr* f. *al* the + *jabr* reunion of broken parts f. *jabara* reunite]

▪ Algebra is the generalization of arithmetic to variable or abstract quantities represented by symbols. One of the earliest works on algebra was the *Arithmetica* by Diophantus of Alexandria (fl. *c*.250 AD). Modern or abstract algebra deals with systems (such as fields, rings, vector spaces, and groups) that consist of elements (typically numbers, vectors, or geometrical transformations) and operations that may be performed with them. Algebra strictly does not include analysis (see ANALYSIS 4).

Algeciras /ˌældʒɪˈsɪərəs/ a ferry port and resort in southern Spain; pop. (1991) 101,365. It lies on the Strait of Gibraltar.

Algeria /ælˈdʒɪərɪə/ a republic on the Mediterranean coast of North Africa; pop. (est. 1991) 25,800,000; official language, Arabic; capital, Algiers. The country extends from the heart of the Sahara Desert to a narrow fertile coastal plain separated from the desert by the Atlas Mountains. Algeria came under nominal Turkish rule in the 16th century, but the native peoples retained much independence, dominating the Barbary Coast until the region was colonized by France in the mid-19th century. Heavily settled by French immigrants, Algeria was for a time closely integrated with metropolitan France, but the refusal of the European settlers to grant equal rights to the native population led to increasing political instability and civil war in the 1950s, and in 1962 the country achieved independence. Many of the French settlers left, damaging the previously prosperous economy, and from 1962 to 1989 the country was a one-party state. A brief period of multi-party democracy was ended by a military takeover in 1992 after the fundamentalist Islamic Salvation Front had won the first round of the national elections. □ **Algerian** adj. & n.

-algia /ˈældʒə/ comb. form Med. denoting pain in a part specified by the first element (*neuralgia*). □ **-algic** comb. form forming adjectives. [Gk f. *algos* pain]

algicide /ˈældʒɪsaɪd/ n. a preparation for destroying algae.

algid /ˈældʒɪd/ adj. Med. cold, chilly. □ **algidity** /ælˈdʒɪdɪtɪ/ n. [L *algidus* f. *algere* be cold]

Algiers /ælˈdʒɪəz/ the capital of Algeria and one of the leading Mediterranean ports of North Africa; pop. (1989) 1,722,000.

alginate /ˈældʒɪneɪt/ n. Chem. a salt or ester of alginic acid. The sodium salt is widely used as a thickener in foods, cosmetics, etc.

alginic acid /ælˈdʒɪnɪk/ n. Chem. an insoluble carbohydrate found (chiefly as salts) in many brown seaweeds. [ALGA + -IN + -IC]

Algol[1] /ˈælɡɒl/ Astron. a variable star in the constellation of Perseus, regarded as the prototype of eclipsing binary stars. [Arab., = the destruction]

Algol[2] /ˈælɡɒl/ n. a high-level computer programming language. [ALGORITHMIC (see ALGORITHM) + LANGUAGE]

algolagnia /ˌælɡəˈlæɡnɪə/ n. sexual pleasure obtained from inflicting pain on oneself or others; masochism or sadism. □ **algolagnic** adj. & n. [G *Algolagnie* f. Gk *algos* pain + *lagneia* lust]

algology /ælˈɡɒlədʒɪ/ n. the study of algae. □ **algologist** n. **algological** /ˌælɡəˈlɒdʒɪk(ə)l/ adj.

Algonquian /ælˈɡɒŋkwɪən/ adj. & n. (also **Algonkian** /-kɪən/) ● adj. of or relating to a large scattered group of North American Indian peoples, or their languages. (*See note below.*) ● *n.* **1** a member of any of these peoples. **2** the group of languages spoken by these peoples. [f. ALGONQUIN + -IAN]

▪ The Algonquian peoples were pushed northward and westward by European colonial expansion in the 18th and 19th centuries. Algonquian forms one of the largest groups of American Indian languages and includes Ojibwa, Cree, Blackfoot, Cheyenne, Fox, Menomini, and Delaware, which are spoken mainly in the north Middle West of the US, Montana, and south central Canada. Many words in English have been adopted from these languages, e.g. *moccasin*, *moose*, *pow-wow*, *squaw*, and *toboggan*.

Algonquin /æl'gɒŋkwɪn/ *n.* & *adj.* ● *n.* **1 a** a member of an American Indian people inhabiting the districts of Ottawa and Quebec. **b** the Algonquian language of this people. **2 a** = ALGONQUIAN *n.* 1. **b** = ALGONQUIAN *n.* 2. ● *adj.* **1** of or relating to the Algonquin people. **2** = ALGONQUIAN *adj.* ¶ The use of *Algonquin* to refer to the Algonquian peoples or their languages is widespread but strictly incorrect. [F, said to be f. an Algonquian word = 'at the place of spearing fish and eels']

algorithm /'ælgə,rɪðəm/ *n.* **1** *Math.* a process or set of rules used for calculation or problem-solving, esp. with a computer. **2** *hist.* the Arabic or decimal notation of numbers. □ **algorithmic** /,ælgə'rɪðmɪk/ *adj.* [*algorism* ME ult. f. Pers. *al-Kuwārizmī* 9th-c. mathematician: *algorithm* infl. by Gk *arithmos* number (cf. F *algorithme*)]

alguacil /'ælgwə,sɪl/ *n.* (also **alguazil** /-,zɪl/) **1** a mounted official at a bullfight. **2** a constable or an officer of justice in Spain or Spanish-speaking countries. [Sp. f. Arab. *al-wazīr* f. al the + *wazīr*: see VIZIER]

Alhambra /æl'hæmbrə/ a fortified Moorish palace, the last stronghold of the Muslim kings of Granada, built between 1248 and 1354 near Granada in Spain. It is an outstanding piece of Moorish architecture with its marble courts and fountains, delicate columns and archways, and wall decorations of carved and painted stucco. [Arab., = the red castle]

Al-Hudayda See HODEIDA.

Ali[1], Muhammad, see MUHAMMAD ALI[1].

Ali[2], Muhammad, see MUHAMMAD ALI[2].

alias /'eɪlɪəs/ *adv.* & *n.* ● *adv.* also named or known as. ● *n.* a false or assumed name. [L, = at another time, otherwise]

Ali Baba /,ælɪ 'bɑːbə/ the hero of a story supposed to be from the *Arabian Nights*, who discovered the magic formula ('Open Sesame!') which opened a cave in which forty robbers kept the treasure they had accumulated.

alibi /'ælɪ,baɪ/ *n.* & *v.* ● *n.* (*pl.* **alibis**) **1** a claim, or the evidence supporting it, that when an alleged act took place one was elsewhere. **2** *disp.* an excuse of any kind; a pretext or justification. ● *v.* (**alibis**, **alibied**, **alibiing**) *colloq.* **1** *tr.* provide an alibi or offer an excuse for (a person). **2** *intr.* provide an alibi. [L, = elsewhere]

Alicante /,ælɪ'kæntɪ/ a seaport on the Mediterranean coast of SE Spain, the capital of a province of the same name; pop. (1991) 270,950.

Alice /'ælɪs/ the heroine of Lewis Carroll's *Alice's Adventures in Wonderland* and *Through the Looking Glass*.

Alice Springs a railway terminus and supply centre serving the outback of Northern Territory, Australia; pop. (1991) 20,450. The site, discovered in 1871, was named after the wife of the South Australian Superintendent of Telegraphs.

alicyclic /,ælɪ'saɪklɪk/ *adj.* *Chem.* of, denoting, or relating to organic compounds combining a cyclic structure with aliphatic properties, e.g. cyclohexane. [G *alicyclisch* (as ALIPHATIC, CYCLIC)]

alidade /'ælɪ,deɪd/ *n.* *Surveying* & *Astron.* a sighting device or pointer for determining directions or measuring angles. [F f. med.L f. Arab. *al-'iḍāda* the revolving radius f. *'aḍud* upper arm]

alien /'eɪlɪən/ *adj.* & *n.* ● *adj.* **1 a** (often foll. by *to*) unfamiliar; not in accordance or harmony; unfriendly, hostile; unacceptable or repugnant (*army discipline was alien to him; struck an alien note*). **b** (often foll. by *to* or *from*) different or separated. **2** foreign; from a foreign country (*help from alien powers*). **3** of or relating to beings supposedly from other worlds; extraterrestrial. **4** (of a plant or animal species) introduced from elsewhere and naturalized in a region. ● *n.* **1 a** a foreigner, esp. a person who is not a naturalized citizen of the country where he or she is living. **2** a being supposedly from another world. **3** an alien plant or animal species. □ **alienness** /-ənnɪs/ *n.* [ME f. OF f. L *alienus* belonging to another (*alius*)]

alienable /'eɪlɪənəb(ə)l/ *adj.* *Law* able to be transferred to new ownership. □ **alienability** /,eɪlɪənə'bɪlɪtɪ/ *n.*

alienage /'eɪlɪənɪdʒ/ *n.* the state or condition of being an alien.

alienate /'eɪlɪə,neɪt/ *v.tr.* **1 a** cause (a person) to become unfriendly or hostile. **b** (often foll. by *from*) cause (a person) to feel isolated or estranged from (friends, society, etc.). **2** transfer ownership of (property) to another person etc. □ **alienator** *n.* [ME f. L *alienare alienat-* (as ALIEN)]

alienation /,eɪlɪə'neɪʃ(ə)n/ *n.* **1 a** the act of alienating. **b** the result of alienating; a state of estrangement in feeling or affection. **2** (**alienation effect**) *Theatr.* a dramatic effect whereby an audience remains objective, not identifying with the characters or action of a play.

alienist /'eɪlɪənɪst/ *n.* *US* a psychiatrist, esp. a legal adviser on psychiatric problems. [F *aliéniste* (as ALIEN)]

aliform /'eɪlɪ,fɔːm/ *adj.* wing-shaped. [mod.L *aliformis* f. L *ala* wing: see -FORM]

Aligarh /,ɑːlɪ'gɑː(r)/ a city in northern India, in Uttar Pradesh; pop. (1991) 480,000. The city comprises the ancient fort of Aligarh and the former city of Koil.

Alighieri /,ælɪ'gjeɑːrɪ/, Dante, see DANTE.

alight[1] /ə'laɪt/ *v.intr.* **1 a** (often foll. by *from*) descend from a vehicle. **b** dismount from a horse. **2** descend and settle; come to earth from the air. **3** (foll. by *on*) find by chance; notice. [OE *ālīhtan* (as A-[2], *līhtan* LIGHT[2] *v.*)]

alight[2] /ə'laɪt/ *adv.* & *predic.adj.* **1** on fire; burning (*they set the old shed alight; is the fire still alight?*). **2** lighted up; excited (*eyes alight with expectation*). [ME, prob. f. phr. *on a light* (= lighted) *fire*]

align /ə'laɪn/ *v.tr.* **1** put in a straight line or bring into line (*three books were neatly aligned on the shelf*). **2** (usu. foll. by *with*) esp. *Polit.* bring (oneself etc.) into agreement or alliance with (a cause, policy, political party, etc.). □ **alignment** *n.* [F *aligner* f. phr. *à ligne* into line: see LINE[1]]

alike /ə'laɪk/ *adj.* & *adv.* ● *adj.* (usu. *predic.*) similar, like one another; indistinguishable. ● *adv.* in a similar way or manner (*all were treated alike*). [ME f. OE *gelīc* and ON *glíkr* (LIKE[1])]

aliment /'ælɪmənt/ *n.* *formal* **1** food. **2** support or mental sustenance. □ **alimental** /,ælɪ'ment(ə)l/ *adj.* [ME f. F *aliment* or L *alimentum* f. *alere* nourish]

alimentary /,ælɪ'mentərɪ/ *adj.* of, relating to, or providing nourishment or sustenance. □ **alimentary canal** *Anat.* the whole passage through the body, from mouth to anus, by which food is received, digested, etc. [L *alimentarius* (as ALIMENT)]

alimentation /,ælɪmen'teɪʃ(ə)n/ *n.* **1** nourishment; feeding. **2** maintenance, support; supplying with the necessities of life. [F *alimentation* or med.L *alimentatio* f. *alimentare* (as ALIMENT)]

alimony /'ælɪmənɪ/ *n.* money payable by a man to his wife or former wife or by a woman to her husband or former husband after they are separated or divorced. ¶ In UK use replaced by *maintenance*. [L *alimonia* nutriment f. *alere* nourish]

A-line /'eɪlaɪn/ *adj.* (of a garment) having a narrow waist or shoulders and a flared skirt.

aliphatic /,ælɪ'fætɪk/ *adj.* *Chem.* of, denoting, or relating to organic compounds in which carbon atoms form open chains, not aromatic rings. [Gk *aleiphar -atos* fat]

aliquot /'ælɪ,kwɒt/ *adj.* & *n.* ● *adj.* *Math.* (of a part or portion) contained by the whole an integral or whole number of times (*4 is an aliquot part of 12*). ● *n.* **1** *Math.* an aliquot part; an integral factor. **2** any known fraction of a whole; a sample. [F *aliquote* f. L *aliquot* some, so many]

alive /ə'laɪv/ *adj.* (usu. *predic.*) **1** (of a person, animal, plant, etc.) living, not dead. **2** (of a thing) existing; continuing; in operation or action (*kept his interest alive*). **b** under discussion; provoking interest (*the topic is still very much alive today*). **3** (of a person or animal) lively, active. **4** = LIVE[2] 4d. **5** (foll. by *to*) aware of; alert or responsive to. **6** (foll. by *with*) **a** swarming or teeming with. **b** full of. □ **alive and kicking** *colloq.* very active; lively. **alive and well** still alive or active (esp. despite contrary assumptions or rumours). □ **aliveness** *n.* [OE *on life* (as A[2], LIFE)]

alizarin /ə'lɪzərɪn/ *n.* **1** the red colouring matter of madder root, used in dyeing. **2** (*attrib.*) (of a dye) derived from or similar to this pigment. [F *alizarine* f. *alizari* madder f. Arab. *al-'iṣāra* pressed juice f. *'aṣara* to press fruit]

Al Jizah see GIZA.

alkahest /'ælkə,hest/ *n.* (also **alcahest**) the universal solvent sought by alchemists. [sham Arab., prob. invented by Paracelsus]

alkali /'ælkə,laɪ/ *n.* (*pl.* **alkalis**) *Chem.* any of a class of substances (often forming caustic and corrosive solutions) that neutralize acids, turn litmus blue, and have a pH of more than 7. (See note below.) □ **alkalimeter** /ˌælkəˈlɪmɪtə(r)/ n. **alkalimetry** n. [ME f. med.L, f. Arab. al-ḳalī calcined ashes f. ḳala fry]

▪ Since the 17th century alkalis such as quicklime and caustic soda have been thought of as being in some sense opposed to acids. Often loosely synonymous with base, the term alkali usually denotes bases which are soluble in water, liberating hydroxide ions (see BASE¹).

alkali metal n. Chem. any of the elements lithium, sodium, potassium, rubidium, caesium, and francium, occupying Group IA (1) of the periodic table. They are very reactive, electropositive, monovalent metals forming strongly alkaline hydroxides.

alkaline /ˈælkəˌlaɪn/ adj. of, relating to, or having the nature of an alkali; rich in alkali. □ **alkalinity** /ˌælkəˈlɪnɪtɪ/ n.

alkaline earth n. (in full **alkaline earth metal**) Chem. any of the elements beryllium, magnesium, calcium, strontium, barium, and radium, occupying Group IIA (2) of the periodic table. They are reactive, electropositive, divalent metals; they form basic oxides which react with water to form comparatively insoluble hydroxides.

alkaloid /ˈælkəˌlɔɪd/ n. Chem. any of a series of nitrogenous organic compounds of plant origin, many of which are used as drugs, e.g. morphine, quinine. [G (as ALKALI)]

alkalosis /ˌælkəˈləʊsɪs/ n. Med. an excessive alkaline condition of the body fluids or tissues.

alkane /ˈælkeɪn/ n. Chem. any of a series of saturated aliphatic hydrocarbons having the general formula C_nH_{2n+2}, including methane, ethane, and propane. [ALKYL + -ANE²]

alkanet /ˈælkəˌnet/ n. **1 a** a plant of the genus Alkanna, of the borage family; esp. A. tinctoria, which yields a red dye from its roots. **b** the dye itself. **2** a related plant, esp. Anchusa officinalis, with usu. violet flowers, and (in full **green alkanet**) Pentaglottis sempervirens, with blue flowers. [ME f. Sp. alcaneta dimin. of alcana f. Arab. al-ḥinnā' the henna shrub]

alkene /ˈælkiːn/ n. Chem. any of a series of unsaturated aliphatic hydrocarbons containing a double bond and having the general formula C_nH_{2n}, including ethylene and propene. [ALKYL + -ENE]

alky /ˈælkɪ/ n. (also **alkie**) (pl. **-ies**) sl. an alcoholic.

alkyd /ˈælkɪd/ n. (usu. attrib.) Chem. any of a group of synthetic resins derived from various alcohols and acids. [ALKYL + ACID]

alkyl /ˈælkaɪl, -kɪl/ n. (attrib.) Chem. a radical derived from an alkane by the removal of a hydrogen atom. [G Alkohol ALCOHOL + -YL]

alkylate /ˈælkɪˌleɪt/ v.tr. Chem. introduce an alkyl radical into (a compound).

alkyne /ˈælkaɪn/ n. Chem. any of a series of unsaturated aliphatic hydrocarbons containing a triple bond and having the general formula C_nH_{2n-2}, including acetylene. [ALKYL + -YNE]

all /ɔːl/ adj., n., & adv. ● adj. **1 a** the whole amount, quantity, or extent of (waited all day; all his life; we all know why; take it all). **b** (with pl.) the entire number of (all the others left; all ten men; the children are all boys; film stars all). **2** any whatever (beyond all doubt). **3** greatest possible (with all speed). ● n. **1 a** all the persons or things concerned (all were present; all were thrown away). **b** everything (all is lost; that is all). **2** (foll. by of) **a** the whole of (take all of it). **b** every one of (all of us). **c** colloq. as much as (all of six feet tall). **d** colloq. affected by; in a state of (all of a dither). **3** one's whole strength or resources (prec. by my, your, etc.). **4** (in games) on both sides (two goals all). ¶ Widely used with of in sense 2a, b, esp. when followed by a pronoun or by a noun implying a number of persons or things, as in all of the children are here. However, use with mass nouns (as in all of the bread) is often avoided. ● adv. **1 a** entirely, quite (dressed all in black; all round the room; the all-important thing). **b** as an intensifier (a book all about ships; stop all this grumbling). **2** colloq. very (went all shy). **3** (foll. by the + compar.) **a** by so much; to that extent (if they go, all the better). **b** in the full degree to be expected (that makes it all the worse). □ **all along** all the time (he was joking all along). **all-American 1** representing the whole of (or only) America or the US. **2** truly American (all-American boy). **all and sundry** everyone. **all-around** US = all-round. **all but** very nearly (it was all but impossible; he was all but drowned). **all-clear** a signal that danger or difficulty is over. **all for** colloq. strongly in favour of. **all-important** crucial; vitally important. **all in** colloq. exhausted. **all-in** (attrib.) inclusive of all. **all in all** everything considered. **all-in wrestling** wrestling with few or no restrictions. **all manner of** see MANNER. **all of a sudden** see SUDDEN. **all one** (**or the same**) (usu. foll. by to) a matter of indifference (it's all one to me). **all out** involving all one's strength or resources; at full speed (also with hyphen) attrib.: an all-out effort. **all over 1** completely finished. **2** in or on all parts of (esp. the body) (went hot and cold all over; mud all over the carpet). **3** colloq.

typically (that is you all over). **4** sl. effusively attentive to (a person). **all-party** involving all (esp. political) parties. **all-purpose** suitable for many uses. **all right** predic.adj. satisfactory; safe and sound; in good condition; adequate. ● adv. satisfactorily, as desired (it worked out all right). ● int. **1** an interjection expressing consent or assent to a proposal or order. **2** as an intensifier (that's the one all right). **all-right** attrib.adj. colloq. fine, acceptable (an all-right guy). **all round 1** in all respects (a good performance all round). **2** for each person (he bought drinks all round). **all-round** (attrib.) (of a person) versatile. **all-rounder** Brit. a versatile person. **all the same** nevertheless, in spite of this (he was innocent but was punished all the same). **all set** colloq. ready to start. **all there** colloq. **1** mentally alert. **2** in possession of all one's faculties. **all-time** (of a record etc.) hitherto unsurpassed. **all the time** see TIME. **all together** all at once; all in one place or in a group (they came all together) (cf. ALTOGETHER). **all told** in all. **all-up weight** the total weight of an aircraft with passengers, cargo, etc., when airborne. **all very well** colloq. an expression used to reject or to imply scepticism about a favourable or consoling remark. **all the way** the whole distance; completely. **at all** (with neg. or interrog.) in any way; to any extent (did not swim at all; did you like it at all?). **be all up with** see UP. **in all** in total number; altogether (there were 10 people in all). **on all fours** see FOUR. **until** (or till) **all hours** colloq. until very late. [OE all, eall, prob. f. Gmc]

alla breve /ˌælə ˈbreɪveɪ/ n. Mus. a time signature indicating 2 or 4 minim beats in a bar. [It., = at the breve (see BREVE 1)]

alla cappella var. of A CAPPELLA.

Allah /ˈælə/ n. the name of God among Arabs and Muslims. [Arab. 'allāh contr. of al-'ilāh f. al the + 'ilāh god]

Allahabad /ˈæləhəˌbæd/ a city in the state of Uttar Pradesh, north central India; pop. (1991) 806,000. Situated at the confluence of the sacred Jumna and Ganges rivers, it is a place of Hindu pilgrimage.

allantois /əˈlæntəʊɪs/ n. (pl. **allantoides** /ˌælənˈtəʊɪˌdiːz/) Zool. the foetal membrane lying beneath the chorion in reptiles, birds, or mammals. □ **allantoic** /ˌælənˈtəʊɪk/ adj. [mod.L f. Gk allantoeidēs sausage-shaped]

allay /əˈleɪ/ v.tr. **1** diminish (fear, suspicion, etc.). **2** relieve or alleviate (pain, hunger, etc.). [OE alecgan (as A-², LAY¹)]

All Blacks colloq. the New Zealand international Rugby Union football team.

allegation /ˌælɪˈɡeɪʃ(ə)n/ n. **1** an assertion, esp. an unproved one. **2** the act or an instance of alleging. [ME f. F allégation or L allegatio f. allegare allege]

allege /əˈledʒ/ v.tr. **1** (often foll. by that + clause, or to + infin.) declare to be the case, esp. without proof. **2** advance as an argument or excuse. □ **alleged** adj. [ME f. AF alegier, OF esligier clear at law; confused in sense with L allegare: see ALLEGATION]

allegedly /əˈledʒɪdlɪ/ adv. as is alleged or said to be the case.

Allegheny Mountains /ˌæləˈɡeɪnɪ/ (also **Alleghenies**) a mountain range of the Appalachian system in the eastern US.

allegiance /əˈliːdʒəns/ n. **1** loyalty (to a person or cause etc.). **2** the duty of a subject to his or her sovereign or government. [ME f. AF f. OF ligeance (as LIEGE): perh. assoc. with ALLIANCE]

allegorical /ˌælɪˈɡɒrɪk(ə)l/ adj. (also **allegoric**) consisting of or relating to allegory; by means of allegory. □ **allegorically** adv.

allegorize /ˈælɪɡəˌraɪz/ v.tr. (also **-ise**) treat as or by means of an allegory. □ **allegorization** /ˌælɪɡəraɪˈzeɪʃ(ə)n/ n.

allegory /ˈælɪɡərɪ/ n. (pl. **-ies**) **1** a story, play, poem, picture, etc., in which the meaning or message is represented symbolically. **2** the use of such symbols. **3** a symbol. □ **allegorist** n. [ME f. OF allegorie f. L allegoria f. Gk allēgoria f. allos other + -agoria speaking]

allegretto /ˌælɪˈɡretəʊ/ adv., adj., & n. Mus. ● adv. & adj. in a fairly brisk tempo. ● n. (pl. **-os**) an allegretto passage or movement. [It., dimin. of ALLEGRO]

allegro /əˈleɪɡrəʊ, əˈleɡ-/ adv., adj., & n. Mus. ● adv. & adj. in a brisk tempo. ● n. (pl. **-os**) an allegro passage or movement. [It., = lively, gay]

allele /ˈæliːl/ n. (also **allel** /ˈælel/) one of the (usu. two) alternative forms of a gene, found at the same place on a chromosome. □ **allelic** /əˈliːlɪk/ adj. [G Allel, abbr. of ALLELOMORPH]

allelomorph /əˈliːləˌmɔːf/ n. = ALLELE. □ **allelomorphic** /əˌliːləˈmɔːfɪk/ adj. [Gk allēl- one another + morphē form]

alleluia /ˌælɪˈluːjə/ int. & n. (also **alleluya**, **hallelujah** /ˌhæl-/) ● int. God be praised. ● n. **1** praise to God. **2** a song of praise to God. **3** RC Ch.

the part of the mass including this. [ME f. eccl.L f. (Septuagint) Gk *allēlouia* f. Heb. *hall'lûyāh* praise ye the Lord]

allemande /ˈælmɑːnd/ n. **1 a** the name of several German dances. **b** the music for any of these, esp. as a movement of a suite. **2** a figure in a country dance. [F, = German (dance)]

Allen /ˈælən/, Woody (born Allen Stewart Konigsberg) (b.1935), American film director, writer, and actor. Starring in most of his own films, he is best known for his comedy films such as *Play it Again, Sam* (1972) and *Annie Hall* (1977), for which he won three Oscars. His other films include the Oscar-winning *Hannah and her Sisters* (1986).

Allenby /ˈælənbɪ/, Edmund Henry Hynman, 1st Viscount (1861–1936), British soldier. A veteran of the Boer War, during the First World War he commanded the First Cavalry Division and later the Third Army on the Western Front. In 1917 he was sent to the Middle East to lead the Egyptian Expeditionary Force. Having captured Jerusalem in Dec. 1917, he went on to defeat the Turkish forces in Palestine in 1918. He was promoted to Field Marshal and later served as High Commissioner in Egypt (1919–25).

Allende /æˈjendɪ/, Salvador (1908–73), Chilean statesman, President 1970–3. Co-founder of the Chilean Socialist Party (1933) and the first avowed Marxist to become president after a free election, Allende used his term to introduce socialist measures, including the nationalization of several industries, the redistribution of land, and the opening of diplomatic relations with Communist countries. Soon faced with a severe economic crisis, industrial unrest, and the withdrawal of foreign investment, he was overthrown and killed in 1973 in a military coup led by General Pinochet.

Allen key n. propr. a spanner designed to fit into and turn an Allen screw.

Allen screw n. propr. a screw with a hexagonal socket in the head. [*Allen* Manufacturing Co., Hartford, Connecticut, Amer. manufacturer]

Allenstein see OLSZTYN.

allergen /ˈælədʒən/ n. a substance that causes an allergic reaction. □ **allergenic** /ˌæləˈdʒenɪk/ adj. [ALLERGY + -GEN]

allergic /əˈlɜːdʒɪk/ adj. **1** (foll. by *to*) **a** having an allergy to. **b** colloq. having a strong dislike for (a person or thing). **2** caused by or relating to an allergy.

allergy /ˈælədʒɪ/ n. (pl. **-ies**) **1** Med. a damaging immune response of the body to a substance (esp. a particular food, pollen, fur, or dust) to which it has become hypersensitive. **2** colloq. an antipathy. □ **allergist** n. [G *Allergie*, after *Energie* ENERGY, f. Gk *allos* other]

alleviate /əˈliːvɪˌeɪt/ v.tr. lessen or make less severe (pain, suffering, etc.). □ **alleviator** n. **alleviative** /-vɪətɪv/ adj. **alleviatory** /-vɪətərɪ/ adj. **alleviation** /əˌliːvɪˈeɪʃ(ə)n/ n. [LL *alleviare* lighten f. L *levare* (as AD-, *levare* raise)]

alley[1] /ˈælɪ/ n. (pl. **-eys**) **1** (also **alley-way**) **a** a narrow street. **b** a narrow passageway, esp. between or behind buildings. **2** a path or walk in a park or garden. **3** an enclosure for skittles, bowling, etc. **4** Tennis either of the two side strips of a doubles court. □ **alley cat** a stray town cat, often mangy or half wild. **up** (or **right up**) **one's alley** (or **street**) see STREET. [ME f. OF *alee* walking, passage f. *aler* go f. L *ambulare* walk]

alley[2] var. of ALLY[2].

All Fools' Day 1 April.

All Hallows (also **All Hallows' Day**) All Saints' Day, 1 Nov.

alliaceous /ˌælɪˈeɪʃəs/ adj. **1** Bot. of or relating to the genus *Allium*, which includes the onions and related plants. **2** tasting or smelling like onion or garlic. [mod.L *alliaceus* f. L *allium* garlic]

alliance /əˈlaɪəns/ n. **1 a** union or agreement to cooperate, esp. of states by treaty or families by marriage. **b** the parties involved. **2** (**Alliance**) a political party formed by the allying of separate parties. **3** a relationship resulting from an affinity in nature or qualities etc. (*the old alliance between logic and metaphysics*). **4** Bot. a group of allied families. [ME f. OF *aliance* (as ALLY[1])]

allied /ˈælaɪd/ adj. **1 a** united or associated in an alliance. **b** (**Allied**) of or relating to the Allies. **2** connected or related (*studied medicine and allied subjects*).

Allier /alje/ a river of central France which rises in the Cévennes and flows 410 km (258 miles) north-west to meet the Loire.

Allies, the a group of nations taking military action together against an aggressor, in particular the nations (including Britain, France, Russia, and later Italy and the US) opposing the Central Powers in the First World War, the nations (including Britain, the Soviet Union, and

the US) opposing the Axis Powers in the Second World War, and those joined in opposition against Iraq in the Gulf War of 1991.

alligator /ˈælɪˌɡeɪtə(r)/ n. **1 a** a large crocodilian of the genus *Alligator* and family Alligatoridae, native to America and China. It has upper teeth that lie outside the lower teeth, and a head broader and shorter than that of the crocodile. **b** (in general use) any large crocodilian. **2 a** the skin of such an animal or material resembling it. **b** (in pl.) shoes of this. □ **alligator clip** a clip with teeth for gripping. **alligator pear** N. Amer. an avocado. **alligator tortoise** a large freshwater snapping turtle, native to the region of the Gulf of Mexico. [Sp. *el lagarto* the lizard f. L *lacerta*]

alliterate /əˈlɪtəˌreɪt/ v. **1** intr. **a** contain alliteration. **b** use alliteration in speech or writing. **2** tr. **a** construct (a phrase etc.) with alliteration. **b** speak or pronounce with alliteration. □ **alliterative** /-rətɪv/ adj. [back-form. f. ALLITERATION]

alliteration /əˌlɪtəˈreɪʃ(ə)n/ n. the occurrence of the same letter or sound at the beginning of adjacent or closely connected words (e.g. *cool, calm, and collected*). [mod.L *alliteratio* (as AD-, *littera* letter)]

allium /ˈælɪəm/ n. a liliaceous plant of the genus *Allium*, usu. bulbous and strong-smelling, e.g. onion and garlic. [L, = garlic]

allo- /ˈæləʊ, əˈlɒ/ comb. form other (*allophone*; *allogamy*). [Gk *allos* other]

allocate /ˈæləˌkeɪt/ v.tr. (usu. foll. by *to*) assign or devote to (a purpose, person, or place). □ **allocator** n. **allocable** /-kəb(ə)l/ adj. **allocation** /ˌæləˈkeɪʃ(ə)n/ n. [med.L *allocare* f. *locus* place]

allochthonous /əˈlɒkθənəs/ adj. Geol. (of a deposit) formed at a distance from its present position (cf. AUTOCHTHONOUS 2b). [ALLO- + Gk *khthōn, -onos* earth]

allocution /ˌæləˈkjuːʃ(ə)n/ n. formal or hortatory speech or manner of address. [L *allocutio* f. *alloqui allocut-* speak to]

allogamy /əˈlɒɡəmɪ/ n. Bot. cross-fertilization in plants. [ALLO- + Gk *-gamia* f. *gamos* marriage]

allograft /ˈæləˌɡrɑːft/ n. Med. & Biol. a tissue graft from a genetically dissimilar donor of the same species as the recipient (cf. XENOGRAFT). Also called *homograft, homotransplant*. [ALLO- + GRAFT[1]]

allomorph /ˈæləˌmɔːf/ n. Linguistics any of two or more alternative forms of a morpheme. □ **allomorphic** /ˌæləˈmɔːfɪk/ adj. [ALLO- + MORPHEME]

allopath /ˈæləˌpæθ/ n. a person who practises allopathy. [F *allopathe* back-form. f. *allopathie* = ALLOPATHY]

allopathy /əˈlɒpəθɪ/ n. the treatment of disease by conventional means, i.e. with drugs having opposite effects to the symptoms (cf. HOMEOPATHY). □ **allopathist** n. **allopathic** /ˌæləˈpæθɪk/ adj. [G *Allopathie* (as ALLO-, -PATHY)]

allopatric /ˌæləˈpætrɪk/ adj. Biol. occurring in separate geographical areas (cf. SYMPATRIC). [f. ALLO- + Gk *patra* fatherland]

allophone /ˈæləˌfəʊn/ n. Linguistics any of the variant sounds forming a single phoneme. □ **allophonic** /ˌæləˈfɒnɪk/ adj. [ALLO- + PHONEME]

allosaurus /ˌæləˈsɔːrəs/ n. (also **allosaur** /ˈæləˌsɔː(r)/) a large bipedal carnivorous dinosaur of the genus *Allosaurus*, of the late Jurassic period. [mod.L f. ALLO- + Gk *sauros* lizard]

allot /əˈlɒt/ v.tr. (**allotted, allotting**) **1** give or apportion to (a person) as a share or task; distribute officially to (*they allotted us each a pair of boots; the men were allotted duties*). **2** (foll. by *to*) give or distribute officially (*a sum was allotted to each charity*). [OF *aloter* f. *a* to + LOT]

allotment /əˈlɒtmənt/ n. **1** a small piece of land rented (usu. from a local authority) for cultivation. **2** a share allotted. **3** the action of allotting.

allotrope /ˈæləˌtrəʊp/ n. Chem. each of two or more different physical forms in which an element can exist. Graphite, charcoal, and diamond are all allotropes of carbon. [back-form. f. ALLOTROPY]

allotropy /əˈlɒtrəpɪ/ n. Chem. the existence of two or more different physical forms of a chemical element. □ **allotropic** /ˌæləˈtrəʊpɪk, -ˈtrɒpɪk/ adj. **allotropical** /-ˈtrɒpɪk(ə)l/ adj. [Gk *allotropos* of another form f. *allos* different + *tropos* manner f. *trepō* to turn]

allottee /əlɒˈtiː/ n. a person to whom something is allotted.

allow /əˈlaʊ/ v. **1** tr. permit (a practice, a person to do something, a thing to happen or be done, etc.) (*smoking is not allowed; we allowed them to speak*). **2** tr. give or provide; permit (a person) to have (a limited quantity or sum) (*we were allowed £500 a year*). **3** tr. provide or set aside for a purpose; add or deduct in consideration of something (*allow 10% for inflation*). **4** tr. admit, agree, concede (*he allowed that it was so; 'You know best,' he allowed*). **b** dial. state; be of the opinion. **5** refl. permit oneself, indulge oneself in (conduct) (*allowed herself to be persuaded; allowed myself a few angry words*). **6** intr. (foll. by *of*) admit of. **7** intr. (foll. by *for*) take into consideration or account; make addition or deduction corresponding to (*allowing for wastage*). □ **allowable** *adj.* **allowably** *adv.* [ME, orig. = 'praise', f. OF *alouer* f. L *allaudare* to praise, and med.L *allocare* to place]

allowance /ə'laʊəns/ *n. & v.* ● *n.* **1** an amount or sum allowed to a person, esp. regularly for a stated purpose. **2** an amount allowed in reckoning. **3** a deduction or discount (*an allowance on your old cooker*). **4** (foll. by *of*) tolerance of. ● *v.tr.* **1** make an allowance to (a person). **2** supply in limited quantities. □ **make allowances** (often foll. by *for*) **1** take into consideration (mitigating circumstances) (*made allowances for his demented state*). **2** look with tolerance upon, make excuses for (a person, bad behaviour, etc.). [ME f. OF *alouance* (as ALLOW)]

allowedly /ə'laʊɪdlɪ/ *adv.* as is generally allowed or acknowledged.

alloy /'ælɔɪ, ə'lɔɪ/ *n. & v.* ● *n.* **1** a mixture of two or more chemical elements, at least one of which is a metal, e.g. brass (a mixture of copper and zinc). **2** an inferior metal mixed esp. with gold or silver. ● *v.tr.* **1** mix (metals). **2 a** debase (a pure substance) by admixture. **b** reduce the quality or spoil the character of a thing by adding something else. **3** moderate. [F *aloi* (n.), *aloyer* (v.) f. OF *aloier, aleier* combine f. L *alligare* bind]

All Saints' Day 1 November.

allspice /'ɔːlspaɪs/ *n.* **1** the aromatic spice obtained from the ground berry of the pimento plant, *Pimenta dioica*. **2** the berry of this plant. **3** an aromatic shrub, esp. of the genus *Calycanthus*.

all-star /'ɔːlstɑː(r)/ *n. & adj.* ● *n.* **1** *Sport* a player chosen as among the finest in his or her league. **2** a superstar. ● *adj.* relating to or consisting of all-stars (*an all-star cast*).

Allston /'ɔːlstən/, Washington (1779–1843), American landscape painter. He was the first major artist of the American romantic movement; his early works (e.g. *The Deluge*, 1804, and his vast unfinished canvas, *Belshazzar's Feast*, 1817–43) exhibit a taste for the monumental, apocalyptic, and melodramatic, in the same vein as the English painters J. M. W. Turner and John Martin. More influential in America, however, were his later visionary and dreamlike paintings such as *Moonlit Landscape* (1819).

allude /ə'luːd, ə'ljuːd/ *v.intr.* (foll. by *to*) **1** refer, esp. indirectly, covertly, or briefly to. **2** *disp.* mention. [L *alludere* (as AD-, *ludere lus-* play)]

allure /ə'ljʊə(r)/ *v. & n.* ● *v.tr.* attract, charm, or fascinate. ● *n.* attractiveness, personal charm, fascination. □ **allurement** *n.* **alluring** *adj.* **alluringly** *adv.* [ME f. OF *alurer* attract (as AD-, *luere* LURE[1])]

allusion /ə'luːʒ(ə)n, ə'ljuː-/ *n.* (often foll. by *to*) a reference, esp. a covert, passing, or indirect one. ¶ Often confused with *illusion*. [F *allusion* or LL *allusio* (as ALLUDE)]

allusive /ə'luːsɪv, ə'ljuː-/ *adj.* **1** (often foll. by *to*) containing an allusion. **2** containing many allusions. □ **allusively** *adv.* **allusiveness** *n.*

alluvial /ə'luːvɪəl/ *adj. & n.* ● *adj.* of or relating to alluvium. ● *n.* alluvium, esp. containing a precious metal.

alluvion /ə'luːvɪən/ *n. Law* the formation of new land by the movement of the sea or of a river. [F f. L *alluvio -onis* f. *luere* wash]

alluvium /ə'luːvɪəm/ *n.* (pl. **alluvia** /-vɪə/ or **alluviums**) a deposit of usu. fine fertile soil left during a time of flood, esp. in a river valley or delta. [L neut. of *alluvius* adj. f. *luere* wash]

ally[1] /'ælaɪ/ *n. & v.* ● *n.* (pl. **-ies**) **1** a state formally cooperating or united with another for a special purpose, esp. by a treaty. **2** a person or organization that cooperates with or helps another. ● *v.tr.* (also /ə'laɪ/) (**-ies, -ied**) (often foll. by *with*) combine or unite in alliance. [ME f. OF *al(e)ier* f. L *alligare* bind: cf. ALLOY]

ally[2] /'ælɪ/ *n.* (also **alley**) (pl. **-ies** or **-eys**) a playing-marble made of marble, alabaster, or glass. [perh. dimin. of ALABASTER]

-ally /əlɪ/ *suffix* forming adverbs from adjectives in *-al* (cf. -AL[1], -LY[2], -ICALLY).

allyl /'ælaɪl, 'ælɪl/ *n.* (attrib.) *Chem.* the unsaturated monovalent radical CH₂=CH-CH₂-. [L *allium* garlic + -YL]

Alma-Ata see ALMATY.

almacantar var. of ALMUCANTAR.

Al Madinah see MEDINA.

Almagest /'ælmə,dʒɛst/ *n.* **1** the title of an Arabic version of Ptolemy's astronomical treatise. **2** (in the Middle Ages; also **almagest**) any celebrated textbook on astrology and alchemy. [f. Arab. *al* the, Gk *megistē* (*suntaxis*) the greatest (system)]

Alma Mater /ˌælmə 'mɑːtə(r), 'meɪtə(r)/ *n.* the university, school, or college one attends or attended. [L, = bounteous mother]

almanac /'ɔːlmə,næk, 'ɒl-/ *n.* (also **almanack**) **1** an annual table, or book of tables, containing a calendar of months and days and usu. astronomical data and other information. **2** a usu. annual directory or handbook containing statistical and other information of either general or specialist interest. [ME f. med.L *almanac*(h) f. Gk *almenikhiaka*]

Almanach de Gotha /ˌɔːlmə,næk də 'ɡəʊtə, ,ɒl-/ an annual publication (in French) giving information about European royalty, nobility, and diplomats, published in Gotha 1763–1944. A smaller-scale version was revived in Paris in 1968.

almandine /'ælmən,diːn, -,daɪn/ *n.* a kind of garnet with a violet tint. [F, alt. of obs. *alabandine* f. med.L *alabandina* f. *Alabanda*, ancient city in Asia Minor]

Alma-Tadema /ˌælmə'tædɪmə/, Sir Lawrence (1836–1912), Dutch-born British painter. Influenced by a trip to Naples and Pompeii in 1863, he turned to lush genre scenes set in the ancient world, which earned him many imitators. His major paintings include *Pyrrhic Dance* (1869) and *Roses of Heliogabalus* (1888).

Almaty /'ælmɑːtɪ/ (also **Alma-Ata** /ˌælmə ə'tɑː/) the capital of the central Asian republic of Kazakhstan; pop. (1991) 1,515,300. It was known as Verny until 1921.

Almería /ˌælmə'rɪə/ a town in a province of the same name in Andalusia, southern Spain; pop. (1991) 157,760.

almighty /ɔːl'maɪtɪ/ *adj. & adv.* ● *adj.* **1** having complete power; omnipotent. **2** (**the Almighty**) God. **3** *sl.* very great (*an almighty crash*). ● *adv. sl.* extremely; very much. [OE *ælmihtig* (as ALL, MIGHTY)]

Almirante Brown /ˌælmɪˈrɑːntɪ/ a city in eastern Argentina, forming part of the conurbation of Buenos Aires; pop. (1991) 449,100. It was first settled in 1873 by residents of Buenos Aires fleeing from an epidemic of yellow fever.

Almohad /'ælmə,hæd/ *n.* (also **Almohade** /-,heɪd/) a member of a group of Muslim Berber peoples that conquered the Spanish and North African empire of the Almoravids in the 12th century, taking the capital Marrakesh in 1147. They were driven out of Spain in 1212 but held on to Marrakesh until 1269.

almond /'ɑːmənd/ *n.* **1** the oval kernel of the stone-fruit from the tree *Prunus dulcis*, of which there are sweet and bitter varieties. **2** the tree itself, allied to the peach and plum and frequently grown for its ornamental blossom. □ **almond eyes** narrow almond-shaped eyes. **almond oil** the oil expressed from the almond seed (esp. the bitter variety), used for toilet preparations, flavouring, and medicinal purposes. **almond paste** = MARZIPAN. [ME f. OF *alemande* etc. f. med.L *amandula* f. L *amygdala* f. Gk *amugdalē*]

almoner /'ɑːmənə(r)/ *n.* **1** *Brit.* a social worker attached to a hospital and seeing to the after-care of patients. ¶ Now usu. called *medical social worker*. **2** *hist.* an official distributor of alms. [ME f. AF *aumoner*, OF *aumonier*, ult. f. med.L *eleēmosynarius* (as ALMS)]

Almoravid /æl'mɒrə,vɪd/ *n.* (also **Almoravide** /-,vaɪd/) a member of a federation of Muslim Berber peoples that established an empire in Morocco and Algeria in the second half of the 11th century, founding Marrakesh as their capital, and went on to take much of Spain from 1086. The Almoravids were in turn driven out by the Almohads, losing Marrakesh in 1147.

almost /'ɔːlməʊst/ *adv.* all but; very nearly. [OE *ælmǣst* for the most part (as ALL, MOST)]

alms /ɑːmz/ *n.pl. hist.* charitable donations of money or food to the poor. [OE *ælmysse, -messe*, f. Gmc ult. f. Gk *eleēmosunē* compassionateness f. *eleēmōn* (adj.) f. *eleos* compassion]

almshouse /'ɑːmzhaʊs/ *n. hist.* a house founded by charity for the poor.

almucantar /ˌælmə'kæntə(r)/ *n.* (also **almacantar**) *Astron.* a line of constant altitude above the horizon. [ME f. med.L *almucantarath* or F *almucantara* etc., f. Arab. *almuḳanṭarāt* sundial f. *ḳanṭara* arch]

aloe /'æləʊ/ *n.* **1** a liliaceous plant of the genus *Aloe*, usu. having toothed fleshy leaves. **2** (in pl.) (in full **bitter aloes**) a strong laxative obtained from the bitter juice of various species of aloe. **3** (in full **American aloe**) an agave native to Central America. (See also AGAVE.) [OE *al(e)we* f. L *aloē* f. Gk]

aloe vera /'vɪərə/ *n.* **1** a Caribbean aloe, *Aloe vera*, yielding a gelatinous substance used esp. in cosmetics as an emollient. **2** this substance. [mod.L *vera* true, real]

aloft /ə'lɒft/ *predic.adj. & adv.* **1** high up; overhead. **2** upwards. [ME f. ON *á lopt(i)* f. *á* in, on, to + *lopt* air: cf. LIFT, LOFT]

alogical /eɪ'lɒdʒɪk(ə)l/ *adj.* **1** not logical. **2** opposed to logic.

aloha /ə'ləʊə/ *int. & n.* a Hawaiian expression of love or affection, used as a greeting and at parting. □ **aloha shirt** a loose brightly coloured Hawaiian shirt. [Hawaiian]

alone /ə'ləʊn/ *predic.adj. & adv.* **1 a** without others present (*they wanted to be alone; the tree stood alone*). **b** without others' help (*succeeded alone*). **c** lonely and wretched (*felt alone*). **2** (often foll. by *in*) standing by oneself in an opinion, quality, etc. (*was alone in thinking this*). **3** only, exclusively (*you alone can help me*). □ **go it alone** act by oneself without assistance. □ **aloneness** *n.* [ME f. ALL + ONE]

along /ə'lɒŋ/ *prep. & adv.* ● *prep.* **1** from one end to the other end of (*a handkerchief with lace along the edge*). **2** on or through any part of the length of (*was walking along the road*). **3** beside or through the length of (*shelves stood along the wall*). ● *adv.* **1** onward; into a more advanced state (*come along; getting along nicely*). **2** at or to a particular place; arriving (*I'll be along soon*). **3** in company with a person, esp. oneself (*bring a book along*). **4** beside or through part or the whole length of a thing. □ **along with** in addition to; together with. [OE *andlang* f. WG, rel. to LONG¹]

alongshore /əlɒŋˈʃɔː(r)/ *adv.* along or by the shore.

alongside /əlɒŋˈsaɪd/ *adv. & prep.* ● *adv.* at or to the side (of a ship, pier, etc.). ● *prep.* close to the side of; next to. □ **alongside of** side by side with; together or simultaneously with.

aloof /ə'luːf/ *adj. & adv.* ● *adj.* distant, unsympathetic. ● *adv.* away, apart (*he kept aloof from his colleagues*). □ **aloofly** *adv.* **aloofness** *n.* [orig. Naut., f. A² + LUFF]

alopecia /ˌæləˈpiːʃə/ *n. Med.* the (complete or partial) absence of hair from areas of the body where it normally grows; baldness. [L f. Gk *alōpekia* fox-mange f. *alōpēx* fox]

Alor Setar /ˌɑːlɔː səˈtɑː(r)/ the capital of the state of Kedah, in Malaysia, near the west coast of the central Malay Peninsula; pop. (1980) 71,682.

aloud /ə'laʊd/ *adv.* **1** audibly; not silently or in a whisper. **2** *archaic* loudly. [A² + LOUD]

alow /ə'ləʊ/ *adv. & predic.adj. Naut.* in or into the lower part of a ship. [A² + LOW¹]

alp /ælp/ *n.* **1** a high mountain. (See also ALPS, THE.) **2** (in Switzerland) an area of pasture on a mountainside. [sing. of *Alps*]

alpaca /æl'pækə/ *n.* **1** a domesticated South American mammal, *Lama pacos*, related to the llama, kept for its long fine woolly hair. **2** wool from this animal. **3** fabric made from this wool, with or without other fibres. [Sp. f. Aymara or Quechua]

alpargata /ˌælpɑːˈɡɑːtə/ *n.* a light canvas shoe with a plaited fibre sole; an espadrille. [Sp.]

alpenhorn /'ælpənˌhɔːn/ *n.* a long wooden horn used by Alpine cattle-herders to call their animals. [G, = Alp-horn]

alpenstock /'ælpənˌstɒk/ *n.* a long iron-tipped staff used in hillwalking. [G, = Alp-stick]

alpha /'ælfə/ *n.* **1** the first letter of the Greek alphabet (*A*, *α*). **2** a first-class mark given for a piece of work or in an examination. **3** (**Alpha**) *Astron.* the first (usu. brightest) star in a constellation (foll. by Latin genitive: *Alpha Leonis*). □ **alpha and omega** the beginning and the end; the most important features. **alpha decay** *Physics* a form of radioactive decay in which an alpha particle is emitted. **alpha particle** *Physics* a positively charged particle consisting of two protons and two neutrons (a helium nucleus). **alpha rhythm** (or **waves**) *Physiol.* the normal electrical activity of the brain when conscious, with a frequency of approx. 8 to 13 hertz. **alpha test** *n.* an initial test of machinery, software, etc., carried out in-house by the developer. ● *v.tr.* subject (a product) to an alpha test. [ME f. L f. Gk]

alphabet /'ælfəˌbet/ *n.* **1** the set of letters used in writing a language (*the Russian alphabet*). (See note below.) **2** a set of symbols or signs representing letters. [LL *alphabetum* f. Gk *alpha*, *bēta*, the first two letters of the alphabet]

▪ Alphabetic systems use symbols to represent the sounds in a language (unlike pictographs or Chinese characters) and this makes them the most versatile and economical writing systems available. The origin of the alphabet goes back to the second millennium BC, to the system known as Phoenician, which had twenty-two signs. The systems of modern Hebrew and Arabic came from Phoenician via Canaanite and Aramaic and are virtually without vowels; the Greek alphabet, which emerged in 1000–900 BC, developed two branches, Cyrillic (which became the script of Russian, Ukrainian, etc.) and Etruscan (from which derives the Roman alphabet). The International Phonetic Alphabet, developed in the late 19th century,

attempts to apply strictly the principle of one symbol for one sound. (See also WRITING.)

alphabetical /ˌælfəˈbetɪk(ə)l/ *adj.* (also **alphabetic**) **1** of or relating to an alphabet. **2** in the order of the letters of the alphabet. □ **alphabetically** *adv.*

alphabetize /'ælfəbəˌtaɪz/ *v.tr.* (also **-ise**) arrange (words, names, etc.) in alphabetical order. □ **alphabetization** /ˌælfəbətaɪˈzeɪʃ(ə)n/ *n.*

Alpha Centauri /ˌælfə sen'tɔːraɪ/ *Astron.* the third brightest star in the sky, in the constellation of Centaurus, visible only to observers in the southern hemisphere. It is the nearest bright star to the solar system (distance 4.34 light-years), and is a visual binary. See also PROXIMA CENTAURI.

alphanumeric /ˌælfənjuːˈmerɪk/ *adj.* (also **alphameric** /ˌælfəˈmerɪk/, **alphanumerical**) containing both alphabetical and numerical symbols. [*alphabetic* (see ALPHABETICAL) + NUMERICAL]

alpine /'ælpaɪn/ *adj. & n.* ● *adj.* **1 a** of or relating to high mountains. **b** growing or found on high mountains. **2** (**Alpine**) of or relating to the Alps. **3** (**Alpine**) (of skiing) involving fast downhill work. ● *n.* a plant native to mountain districts, esp. one suited to rock gardens. [L *Alpinus*: see ALP]

Alpinist /'ælpɪnɪst/ *n.* (also **alpinist**) a climber of high mountains, esp. in the Alps. [F *alpiniste* (as ALPINE; see -IST)]

Alps, the a mountain system in Europe extending in a curve from the coast of SE France through NW Italy, Switzerland, Liechtenstein, and southern Germany into Austria. Its highest peak, Mont Blanc, rises to a height of 4,807 m (15,771 ft).

Al Qahira see CAIRO.

already /ɔːl'redɪ/ *adv.* **1** before the time in question (*I knew that already*). **2** as early or as soon as this (*already at the age of 6*). **3** *N. Amer.* (at the end of a sentence) as an intensive (*give it to me already*). [ALL + READY: sense 3 transl. Yiddish *shoyn*]

alright /ɔːl'raɪt/ *adj., adv., & int. disp.* = all right.

Alsace /æl'sæs, French alzas/ a region of NE France, on the borders with Germany and Switzerland. Alsace was annexed by Prussia, along with part of Lorraine (forming Alsace-Lorraine), after the Franco-Prussian War of 1870–1, and restored to France after the First World War.

Alsatian /æl'seɪʃ(ə)n/ *adj. & n.* ● *adj.* of or relating to Alsace. ● *n.* **1** a breed of wolfhound often used for guard or police dogs (also called *German shepherd dog*). **2** a native of Alsace. [*Alsatia* (= Alsace) + -AN]

alsike /'ælsɪk/ *n.* a clover, *Trifolium hybridum*, grown for fodder. [*Alsike* in Sweden]

also /'ɔːlsəʊ/ *adv.* in addition; likewise; besides. □ **also-ran 1** a horse or dog etc. not among the winners in a race. **2** an undistinguished person. [OE *alswā* (as ALL, SO¹)]

Alta. *abbr.* Alberta.

Altai /'æltaɪ/ (also **Altay**) a krai (administrative territory) of Russia in SW Siberia, on the border with Kazakhstan; pop. (1990) 2,835,000; capital, Barnaul.

Altaic /æl'teɪɪk/ *adj. & n.* ● *adj.* **1** of or relating to the Altai Mountains in central Asia. **2** of or designating a family of languages which includes Turkish, Mongolian, and Tungus. They are distinguished by agglutination and vowel harmony. ● *n.* the Altaic family of languages.

Altai Mountains a mountain system of central Asia extending about 1,600 km (1,000 miles) eastwards from Kazakhstan into western Mongolia and northern China.

Altair /'ælteə(r)/ *Astron.* the brightest star in the constellation Aquila. [Arab., = flying eagle]

Altamira /ˌæltəˈmɪərə/ **1** the site of a cave with palaeolithic rock paintings, south of Santander in northern Spain, discovered in 1879. The paintings are realistic depictions of deer, wild boar, and especially bison; they are dated to the upper Magdalenian period. **2** a town in NE Brazil, which in 1989 attracted world attention as the venue of a major protest against the devastation of the Amazonian rainforest.

altar /'ɔːltə(r), 'ɒl-/ *n.* **1** a table or flat-topped block, often of stone, on which to make offerings or sacrifices to a deity. **2** a Communion table. □ **altar boy** a boy who serves as a priest's assistant in a service. **lead to the altar** marry (a woman). [OE *altar -er*, Gmc adoption of LL *altar*, *altarium* f. L *altaria* (pl.) burnt offerings, altar, prob. rel. to *adolere* burn in sacrifice]

altarpiece /'ɔːltəˌpiːs, 'ɒl-/ *n.* a piece of art, esp. a painting, set above or behind an altar.

Altay see ALTAI.

altazimuth /ˌæl'tæzɪməθ/ n. (usu. attrib.) an instrument for measuring the altitude and azimuth of celestial bodies, or horizontal and vertical angles in surveying. □ **altazimuth mount** (or **mounting**) a mount for a telescope etc. that rotates about the horizontal and vertical axes, used for most terrestrial and some astronomical telescopes (cf. equatorial mount). [ALTITUDE + AZIMUTH]

Altdorfer /'æltdɔːfə(r)/, Albrecht (c.1485–1538), German painter and engraver. Inspired by travels along the Danube and in the Austrian Alps in 1511, he emerged as one of the first European landscape painters of modern history and principal artist of the Danube School. His romantic treatment of landscape and the emotional harmony of landscape and human action (as in Saint George in the Forest) epitomize the methods and sentiments of the school.

alter /'ɔːltə(r), 'ɒl-/ v. **1** tr. & intr. make or become different; change. **2** tr. US & Austral. castrate or spay. □ **alterable** adj. **alteration** /ˌɔːltə'reɪʃ(ə)n, ˌɒl-/ n. [ME f. OF alterer f. LL alterare f. L alter other]

alterative /'ɔːltərətɪv, 'ɒl-/ adj. tending to produce alteration. [ME f. med.L alterativus (as ALTER)]

altercate /'ɔːltə,keɪt, 'ɒl-/ v.intr. (often foll. by with) dispute hotly; wrangle. □ **altercation** /ˌɔːltə'keɪʃ(ə)n, ˌɒl-/ n. [L altercari altercat-]

alter ego /ˌæltə 'iːɡəʊ, ˌɔːltə, 'eɡəʊ/ n. (pl. **alter egos**) **1** a person's secondary or alternative personality. **2** an intimate and trusted friend. [L, = other self]

alternate v., adj., & n. ● v. /'ɔːltə,neɪt, 'ɒl-/ **1** intr. (often foll. by with) (of two things) succeed each other by turns (rain and sunshine alternated; elation alternated with depression). **2** intr. (foll. by between) change repeatedly (between two conditions) (the patient alternated between hot and cold fevers). **3** tr. (often foll. by with) cause (two things) to succeed each other by turns (the band alternated fast and slow tunes; we alternated criticism with reassurance). ● adj. /ɔː'ltɜːnət, ɒl-/ **1** (with noun in pl.) every other (comes on alternate days). **2** (of things of two kinds) each following and succeeded by one of the other kind (alternate joy and misery). **3** (of a sequence etc.) consisting of alternate things. **4** Bot. (of leaves etc.) placed alternately on the two sides of the stem. **5** esp. N. Amer. = ALTERNATIVE adj. ● n. /ɔː'ltɜːnət, ɒl-/ esp. N. Amer. something or someone that is an alternative; a deputy or substitute. □ **alternate angles** two angles, not adjoining one another, that are formed on opposite sides of a line that intersects two other lines. **alternating current** (abbr. **AC, ac**) an electric current that reverses its direction at regular intervals. □ **alternately** /ɔː'ltɜːnətlɪ, ɒl-/ adv. **alternation** /ˌɔːltə'neɪʃ(ə)n, ˌɒl-/ n. [L alternatus past part. of alternare do things by turns f. alternus every other f. alter other]

alternation of generations n. Biol. a pattern of reproduction occurring in the life cycles of many lower plants (e.g. ferns) and some invertebrates (especially coelenterates), involving a regular alternation between two distinct forms, differently produced and often very different from each other. Generation is alternately sexual and asexual (as in ferns) or dioecious and parthenogenetic (as in some jellyfish).

alternative /ɔː'ltɜːnətɪv, ɒl-/ adj. & n. ● adj. **1** (of one or more things) available or usable instead of another (an alternative route). ¶ Use with reference to more than two options (e.g. many alternative methods) is common, and acceptable. **2** (of two things) mutually exclusive. **3** of or relating to practices that offer a substitute for the conventional ones (alternative theatre). ● n. **1** any of two or more possibilities. **2** the freedom or opportunity to choose between two or more things (I had no alternative but to go). □ **alternative birth** (or **birthing**) any method of childbirth in which the delivery occurs at home or in a homely environment, often without the use of painkilling drugs or obstetric instruments. **alternative comedy** a style of comedy which seeks to reject established (esp. racist or sexist) stereotypes. **alternative energy** energy fuelled in ways that do not use up the earth's natural resources or otherwise harm the environment. **alternative fuel** a fuel other than petrol for motor vehicles. **alternative medicine** any of a range of medical therapies not regarded as orthodox by the medical profession, e.g. chiropractic, faith-healing, herbalism, homeopathy, and reflexology (also called complementary medicine). **the alternative society** a group of people dissociating themselves from conventional society and its values. □ **alternatively** adv. [F alternatif -ive or med.L alternativus (as ALTERNATE)]

Alternative Service Book a book containing the public liturgy of the Church of England published in 1980 for use as the alternative to the Book of Common Prayer.

alternator /'ɔːltə,neɪtə(r), 'ɒl-/ n. a dynamo that generates an alternating current.

althorn /'ælthɔːn/ n. Mus. an instrument of the saxhorn family, esp. the alto or tenor saxhorn in E flat. [G f. alt high f. L altus + HORN]

although /ɔːl'ðəʊ/ conj. = THOUGH conj. [ME f. ALL + THOUGH]

Althusser /'æltʊ,seə(r)/, Louis (1918–90), French philosopher. His work in reinterpreting traditional Marxism in the light of structuralist theories had a significant influence on literary and cultural theory from the 1970s. He sought to reassert an anti-humanist approach to Marxism and develop it into a structural analysis of society. His most important works include For Marx (1965) and Reading Capital (1970). Found guilty of the murder of his wife, he spent his last years in a mental asylum.

altimeter /'æltɪ,miːtə(r)/ n. an instrument for showing height above sea or ground level, esp. one fitted to an aircraft. [L altus high + -METER]

altiplano /ˌæltɪ'plɑːnəʊ/ n. (pl. **-os**) (usu. prec. by the) a high-altitude plateau or plain in South or Central America, especially that in the Andes in western Bolivia and southern Peru. [Amer. Sp., f. L altus high + planum plain]

altitude /'æltɪtjuːd/ n. **1** the height of an object in relation to a given point, esp. sea level or the horizon. **2** Geom. the length of the perpendicular from a vertex to the opposite side of a figure. **3** a high or exalted position (a social altitude). □ **altitude sickness** a sickness experienced at high altitudes. □ **altitudinal** /ˌæltɪ'tjuːdɪn(ə)l/ adj. [ME f. L altitudo f. altus high]

alto /'æltəʊ/ n. (pl. **-os**) **1** = CONTRALTO. **2 a** the highest adult male singing voice, above tenor. **b** a singer with this voice. **c** a part written for it. **3 a** (attrib.) denoting the member of a family of instruments pitched second or third-highest. **b** an alto instrument, esp. an alto saxophone. □ **alto clef** a clef placing middle C on the middle line of the staff, used chiefly for viola music. [It. alto (canto) high (singing)]

altocumulus /ˌæltəʊ'kjuːmjʊləs/ n. (often attrib.) Meteorol. a cloud type consisting of rounded masses with a level base formed at medium altitude. [f. L altus high + CUMULUS]

altogether /ˌɔːltə'ɡeðə(r)/ adv. **1** totally, completely (you are altogether wrong). **2** on the whole (altogether it had been a good day). **3** in total (there are six bedrooms altogether). ¶ Note that all together is used to mean 'all in one place' or 'all at once', as in there are six bedrooms all together; they came in all together. □ **in the altogether** colloq. naked. [ME f. ALL + TOGETHER]

alto-relievo /ˌæltəʊrɪ'liːvəʊ, -'ljeɪvəʊ/ n. (also **alto-rilievo**) (pl. **-os**) Sculpture **1** = high relief (see RELIEF n. 6a). **2** a sculpture or carving in high relief. [It. alto high + RELIEVO]

altostratus /ˌæltəʊ'strɑːtəs, -'streɪtəs/ n. (often attrib.) Meteorol. a cloud type forming a continuous uniform layer at medium altitude. [f. L altus high + STRATUS]

altricial /æl'trɪʃ(ə)l/ adj. & n. Zool. ● adj. **1** (of a young bird or animal) requiring care and feeding by the parents after hatching or birth (cf. PRECOCIAL). **2** having such young. ● n. an altricial bird. [L altrix altricis (fem.) nourisher f. altor f. alere nourish]

altruism /'æltruː,ɪz(ə)m/ n. **1** regard for others as a principle of action. **2** unselfishness; concern for other people. □ **altruist** n. **altruistic** /ˌæltruː'ɪstɪk/ adj. **altruistically** adv. [F altruisme f. It. altrui somebody else (infl. by L alter other)]

alum /'æləm/ n. Chem. **1** a double sulphate of aluminium and potassium. **2** any of a group of crystalline salts which are double sulphates of a monovalent metal (or group) and a trivalent metal. [ME f. OF f. L alumen aluminis]

alumina /ə'luːmɪnə/ n. the compound aluminium oxide (chem. formula: Al_2O_3), occurring naturally as corundum and emery. [L alumen alum, after soda etc.]

aluminium /ˌæljʊ'mɪnɪəm/ n. (N. Amer. **aluminum** /ə'luːmɪnəm/) a silvery-grey metallic chemical element (atomic number 13; symbol **Al**). (See note below.) □ **aluminium bronze** an alloy of copper and aluminium. [aluminium, alt. (after sodium etc.) f. aluminum, earlier aluminium f. ALUM]

▪ Aluminium metal was first prepared by H. C. Oersted in 1825, alumina having been recognized as the oxide of an unknown metal by A. Lavoisier in 1782. Aluminium is the most abundant metal in the earth's crust, and is a major constituent of clays and feldspars; the main commercial ore is bauxite. The metal is resistant to corrosion because of a surface layer of oxide, and its lightness and strength (especially when alloyed) have led to widespread use in domestic utensils, engineering parts, and aircraft construction.

aluminize /ə'luːmɪ,naɪz/ v.tr. (also **-ise**) coat with aluminium. □ **aluminization** /əˌluːmɪnaɪ'zeɪʃ(ə)n/ n.

aluminosilicate /əˌluːmɪnəʊˈsɪlɪˌkeɪt/ n. Chem. & Mineral. a silicate containing aluminium, esp. a mineral of this kind, e.g. a feldspar, a clay mineral.

alumnus /əˈlʌmnəs/ n. (pl. **alumni** /-naɪ/; fem. **alumna** /-nə/, pl. **alumnae** /-niː/) a former pupil or student. [L, = nursling, pupil f. alere nourish]

Al Uqsur see LUXOR.

Alvarez /ælˈvɑːrez/, Luis Walter (1911–88), American physicist. In particle physics, he discovered the phenomenon whereby an atomic nucleus can capture an orbiting electron, and made (with F. Bloch) the first measurement of the neutron's magnetic moment. He also developed the bubble chamber for detecting charged particles, for which he received the Nobel Prize for physics in 1968. In 1980 Alvarez and his son Walter, a geologist, discovered iridium in sediment from the Cretaceous–Tertiary boundary and proposed that this resulted from a catastrophic meteorite impact (see CRETACEOUS).

alveolar /ælˈvɪələ(r), ˌælvɪˈəʊlə(r)/ adj. **1** esp. Anat. of an alveolus. **2** Phonet. (of a consonant) pronounced with the tip of the tongue in contact with the ridge of the upper teeth, e.g. n, s, t. [ALVEOLUS + -AR[1]]

alveolus /ælˈvɪələs, ˌælvɪˈəʊləs/ n. (pl. **alveoli** /-laɪ/) **1** a small cavity, pit, or hollow. **2** Anat. any of the many tiny air sacs of the lungs which allow for rapid gaseous exchange. **3** Anat. the bony socket for the root of a tooth. **4** the cell of a honeycomb. □ **alveolate** /ælˈvɪəlɪt, -ˌleɪt/ adj. [L dimin. of alveus cavity]

always /ˈɔːlweɪz/ adv. **1** at all times; on all occasions (they are always late). **2** whatever the circumstances (I can always sleep on the floor). **3** repeatedly; often (they are always complaining). **4** for ever, for all time (I am with you always). [ME, prob. distributive genitive f. ALL + WAY + -s[3]]

alyssum /ˈælɪsəm/ n. **1** a cruciferous plant of the genus Alyssum, widely cultivated and usu. having yellow or white flowers. **2** (in full **sweet alyssum**) a small cruciferous Mediterranean plant, Lobularia maritima, with fragrant white flowers. [L f. Gk alusson]

Alzheimer's disease /ˈælts.haɪməz/ n. a serious disorder of the brain manifesting itself in premature senility. [A. Alzheimer, Ger. neurologist (1864–1915)]

AM abbr. **1** amplitude modulation. **2** US Master of Arts. **3** Member of the Order of Australia. [(sense 2) L artium Magister]

A. M. abbr. (Hymns) Ancient and Modern.

Am symb. Chem. the element americium.

am 1st person sing. present of BE.

a.m. abbr. before noon. [L ante meridiem]

amadavat var. of AVADAVAT.

amadou /ˈæməˌduː/ n. a spongy and combustible tinder prepared from dry fungi. [F f. mod.Prov., lit. = 'lover' (because quickly kindled) f. L (as AMATEUR)]

amah /ˈɑːmə/ n. (in the Far East and India) a nursemaid or maid. [Port. ama nurse]

Amal /əˈmɑːl/ a Lebanese Shiite Muslim organization founded in 1975 and having political and paramilitary wings. [Arab. 'amal hope]

Amalfi /əˈmælfɪ/ a port and resort on the west coast of Italy, on the Gulf of Salerno; pop. (1990) 5,900.

amalgam /əˈmælgəm/ n. **1** a mixture or blend. **2** an alloy of mercury with one or more other metals, used esp. in dentistry. [ME f. F amalgame or med.L amalgama f. Gk malagma an emollient]

amalgamate /əˈmælgəˌmeɪt/ v. **1** tr. & intr. combine or unite to form one structure, organization, etc. **2** intr. (of metals) form an alloy with mercury. □ **amalgamation** /əˌmælgəˈmeɪʃ(ə)n/ n. [med.L amalgamare amalgamat- (as AMALGAM)]

Amalthea /æˈmælθɪə/ **1** Gk Mythol. a goat which suckled the infant Zeus. **2** Astron. satellite V of Jupiter, the third closest to the planet, discovered in 1892. It is red in colour and heavily cratered, 262 km long and 146 km across.

amanuensis /əˌmænjuːˈensɪs/ n. (pl. **amanuenses** /-siːz/) **1** a person who writes from dictation or copies manuscripts. **2** a literary assistant. [L f. (servus) a manu secretary + -ensis belonging to]

Amapá /ˌæməˈpɑː/ a state of northern Brazil, on the Atlantic coast, lying between the Amazon delta and the border with French Guiana; capital, Macapá. Situated on the equator, it is a region of dense rainforest.

amaranth /ˈæməˌrænθ/ n. **1** a plant of the genus Amaranthus, usu. having small green, red, or purple tinted flowers, e.g. prince's feather. **2** an imaginary flower that never fades. **3** a purple colour. □ **amaranthine** /ˌæməˈrænθaɪn/ adj. [F amarante or mod.L amaranthus f. L f. Gk amarantos everlasting f. a- not + marainō wither, alt. after polyanthus etc.]

amaretto /ˌæməˈretəʊ/ n. (pl. **amaretti** /-tɪ/) **1** an Italian almond-flavoured liqueur. **2** (in pl.) Italian almond-flavoured biscuits. [It., dim. of amaro bitter (with ref. to bitter almonds)]

Amarna, Tell el- /əˈmɑːnə, ˌtel el/ the site of the ruins of the ancient Egyptian capital Akhetaten, on the east bank of the Nile. The city was built by Akhenaten in c.1375 BC, when he established the new worship of the sun disc Aten. It was abandoned four years after his death, when the court returned to the former capital, Thebes. A series of cuneiform tablets known as the Amarna Letters was discovered on the site in 1887, providing valuable insight into Near Eastern diplomacy of the 14th century BC.

amaryllis /ˌæməˈrɪlɪs/ n. **1** a bulbous lily-like plant, Amaryllis belladonna, native to southern Africa, with white or rose-pink flowers. Also called belladonna lily. **2** a related plant, esp. one of the genus Hippeastrum. [L f. Gk Amarullis, name of a country girl]

amass /əˈmæs/ v.tr. **1** gather or heap together. **2** accumulate (esp. riches). □ **amasser** n. **amassment** n. [F amasser or med.L amassare ult. f. L massa MASS[1]]

Amaterasu /əˌmɑːtəˈrɑːsuː/ the principal deity of the Japanese Shinto religion, the sun-goddess and ancestor of Jimmu, founder of the imperial dynasty.

amateur /ˈæmətə(r)/ n. & adj. • n. **1 a** a person who engages in a pursuit (e.g. an art or sport) as a pastime rather than a profession. **b** derog. a person who does something unskilfully, in the manner of an amateur rather than a professional. **2** (foll. by of) a person who is fond of (a thing). • adj. for or done by amateurs; amateurish, unskilful (amateur athletics; did an amateur job). □ **amateurism** n. [F f. It. amatore f. L amator -oris lover f. amare love]

amateurish /ˈæmətərɪʃ/ adj. characteristic of an amateur, esp. inept or badly organized. □ **amateurishly** adv. **amateurishness** n.

Amati /əˈmɑːtɪ/ a family of Italian violin-makers. The three generations, all based in Cremona, included Andrea (c.1520–c.1580), his sons Antonio (1550–1638) and Girolamo (1551–1635), and, most notably, the latter's son Nicolò (1596–1684). From Nicolò's workshop came the violin-makers Antonio Stradivari and Andrea Guarneri (c.1626–98), uncle of Giuseppe Guarneri 'del Gesù'. The Amatis developed the basic proportions of the violin, viola, and cello. They refined the body outlines, sound-holes, purfling, and scroll.

amatory /ˈæmətərɪ/ adj. of or relating to sexual love or desire. [L amatorius f. amare love]

amaurosis /ˌæməˈrəʊsɪs/ n. Med. partial or total loss of sight, from disease of the optic nerve, retina, spinal cord, or brain. □ **amaurotic** /-ˈrɒtɪk/ adj. [mod.L f. Gk f. amauroō darken f. amauros dim]

amaze /əˈmeɪz/ v.tr. (often foll. by at, or that + clause, or to + infin.) surprise greatly; overwhelm with wonder (am amazed at your indifference; was amazed to find them alive). □ **amazement** n. **amazing** adj. **amazingly** adv. **amazingness** n. [ME f. OE āmasod past part. of āmasian, of uncert. orig.]

Amazon[1] /ˈæməz(ə)n/ a river in South America, flowing over 6,683 km (4,150 miles) through Peru, Colombia, and Brazil into the Atlantic Ocean. It drains two-fifths of the continent and in terms of water-flow it is the largest river in the world. It bore various names after its discovery in 1500 and was finally named after a legendary tribe of female warriors believed to live on its banks. □ **Amazonian** /ˌæməˈzəʊnɪən/ adj.

Amazon[2] /ˈæməz(ə)n/ n. **1** a member of a legendary race of female warriors in Scythia and elsewhere. (See note below.) **2** (**amazon**) a very tall, strong, or athletic woman. □ **amazon ant** an ant of the genus Polyergus, which captures pupae of other ant species to raise as slaves. □ **Amazonian** /ˌæməˈzəʊnɪən/ adj. [ME f. L f. Gk: see below]

▪ The Amazons were alleged to exist somewhere on the borders of the known world. Their name was explained by the Greeks as meaning 'without a breast', in connection with a story that they destroyed the right breast so as not to interfere with the use of the bow, but this is probably the popular etymology of an unknown word. They caught the Greek imagination and appear in many legends. Amazons appear as allies of the Trojans in the Trojan War, and their queen, Penthesilea, was killed by Achilles. One of the labours of Hercules was to obtain the girdle of Hippolyta, queen of the Amazons. According to Athenian legend, Attica once suffered an invasion of Amazons, which Theseus repelled.

Amazonas /ˌæməˈzəʊnəs/ a state of NW Brazil; capital, Manaus. It is traversed by the Amazon and its numerous tributaries.

Amazonia /ˌæməˈzəʊnɪə/ **1** the area around the River Amazon in South America, principally in Brazil, but also extending into Peru, Colombia, and Bolivia. Crossed by the equator in the north, this region comprises approximately one-third of the world's remaining tropical rainforest. **2** a national park protecting 10,000 sq. km (3,850 sq. miles) of tropical rainforest in the state of Pará, northern Brazil.

ambassador /æmˈbæsədə(r)/ *n.* **1** an accredited diplomat sent by a state on a mission to, or as its permanent representative in, a foreign country. **2** a representative or promoter of a specified thing (*an ambassador of peace*). □ **ambassador-at-large** US an ambassador with special duties, not appointed to a particular country. □ **ambassadorship** *n.* **ambassadorial** /ˌæmbæsəˈdɔːrɪəl/ *adj.* [ME f. F *ambassadeur* f. It. *ambasciator*, ult. f. L *ambactus* servant]

ambassadress /æmˈbæsədrɪs/ *n.* **1** a female ambassador. **2** an ambassador's wife.

ambatch /ˈæmbætʃ/ *n.* an African tree, *Aeschynomene elaphroxylon*, with very light spongy wood. [Ethiopic]

Ambato /æmˈbɑːtəʊ/ a market town in the Andes of central Ecuador; pop. (1990) 229,190.

amber /ˈæmbə(r)/ *n. & adj.* ● *n.* **1 a** fossilized resin from extinct (esp. coniferous) trees, usually translucent and yellow. (*See note below.*) **b** a honey-yellow colour. **2** a yellow cautionary signal light, esp. a yellow traffic light showing between green for 'go' and red for 'stop'. ● *adj.* made of or coloured like amber. [ME f. OF *ambre* f. Arab. *ʿanbar* ambergris, amber]

■ Amber is found chiefly along the southern shores of the Baltic Sea. It burns with an agreeable odour, and when rubbed becomes charged with static electricity (the word *electric* is derived from the Greek word for amber). Amber has been used for ornaments since the mesolithic period and seems to have been the basic commodity of regular prehistoric trade routes from the Baltic to the Adriatic Sea. Amber sometimes contains the bodies of insects and other creatures which were trapped in the resin before it hardened.

ambergris /ˈæmbəgrɪs, -ˌgriːs/ *n.* a strong-smelling waxlike secretion of the intestine of the sperm whale, found floating in tropical seas and used in perfume manufacture. [ME f. OF *ambre gris* grey AMBER]

amberjack /ˈæmbəˌdʒæk/ *n.* US a large brightly coloured marine fish of the genus *Seriola*, found in tropical and subtropical Atlantic waters.

ambiance var. of AMBIENCE.

ambidextrous /ˌæmbɪˈdekstrəs/ *adj.* (also **ambidexterous** /-ˈdekstərəs/) **1** able to use the right and left hands equally well. **2** working skilfully in more than one medium. □ **ambidextrously** *adv.* **ambidextrousness** *n.* **ambidexterity** /ˌæmbɪdekˈsterɪtɪ/ *n.* [LL *ambidexter* f. ambi- on both sides + *dexter* right-handed]

ambience /ˈæmbɪəns/ *n.* (also **ambiance**) **1** the surroundings or atmosphere of a place. **2** background noise or depth added to a musical recording to give the impression that it was recorded live. [AMBIENT + -ENCE or F *ambiance*]

ambient /ˈæmbɪənt/ *adj.* surrounding; of the surroundings (*ambient temperature*). □ **ambient music** background music used to create atmosphere. [F *ambiant* or L *ambiens -entis* pres. part. of *ambire* go round]

ambiguity /ˌæmbɪˈɡjuːɪtɪ/ *n.* (*pl.* **-ies**) **1 a** double meaning, either deliberate or caused by inexactness of expression (*tried to avoid ambiguity in setting out the rules*). **b** an example of this; an expression able to be interpreted in more than one way (e.g. *dogs must be carried on the escalator*). [ME f. OF *ambiguité* or L *ambiguitas* (as AMBIGUOUS)]

ambiguous /æmˈbɪɡjʊəs/ *adj.* **1** having an obscure or double meaning. **2** difficult to classify. □ **ambiguously** *adv.* **ambiguousness** *n.* [L *ambiguus* doubtful f. *ambigere* f. ambi- both ways + *agere* drive]

ambisonics /ˌæmbɪˈsɒnɪks/ *n.pl.* a system of high-fidelity sound reproduction designed to reproduce the directional and acoustic properties of the sound source using two or more channels. [L *ambi-* on both sides + SONIC]

ambit /ˈæmbɪt/ *n.* **1** the scope, extent, or bounds of something. **2** precincts or environs. [ME f. L *ambitus* circuit f. *ambire*: see AMBIENT]

ambition /æmˈbɪʃ(ə)n/ *n.* **1** (often foll. by *to* + infin.) a determination to achieve success or distinction, usu. in a chosen field. **2** the object of this determination. **3** aggressive self-seeking or self-centredness. [ME f. OF f. L *ambitio -onis* f. *ambire ambit-* canvass for votes: see AMBIENT]

ambitious /æmˈbɪʃəs/ *adj.* **1 a** full of ambition. **b** showing ambition (*an ambitious attempt*). **2** (foll. by *of*, or *to* + infin.) strongly determined.

□ **ambitiously** *adv.* **ambitiousness** *n.* [ME f. OF *ambitieux* f. L *ambitiosus* (as AMBITION)]

ambivalence /æmˈbɪvələns/ *n.* (also **ambivalency** /-sɪ/) the coexistence in a person's mind of opposing feelings, esp. love and hate, in a single context. □ **ambivalent** *adj.* **ambivalently** *adv.* [G *Ambivalenz* f. L *ambo* both, after *equivalence*, *-ency*]

ambivert /ˈæmbɪˌvɜːt/ *n.* Psychol. a person who fluctuates between being an introvert and an extrovert. □ **ambiversion** /ˌæmbɪˈvɜːʃ(ə)n/ *n.* [L *ambi-* on both sides + *-vert* f. L *vertere* to turn, after EXTROVERT, INTROVERT]

amble /ˈæmb(ə)l/ *v. & n.* ● *v.intr.* **1** move at an easy pace, in a way suggesting an ambling horse. **2** (of a horse etc.) move by lifting the two feet on one side together. **3** ride an ambling horse; ride at an easy pace. ● *n.* an easy pace; the gait of an ambling horse. [ME f. OF *ambler* f. L *ambulare* walk]

amblyopia /ˌæmblɪˈəʊpɪə/ *n.* Med. dimness of vision without obvious defect or change in the eye. □ **amblyopic** /-ˈɒpɪk/ *adj.* [Gk f. *ambluōpos* (adj.) f. *amblus* dull + *ōps, ōpos* eye]

ambo /ˈæmbəʊ/ *n.* (*pl.* **-os** or **ambones** /æmˈbəʊniːz/) a stand for reading the lesson in an early Christian church etc. [med.L f. Gk *ambōn* rim (in med.Gk = pulpit)]

Amboinese /ˌæmbɔɪˈniːz/ *n. & adj.* ● *n.* (*pl.* same) **1** a native or inhabitant of the island of Ambon in the Molucca Islands, Indonesia. **2** the Malayo-Polynesian language of the Amboinese. ● *adj.* of or relating to the Amboinese or their language.

Ambon /æmˈbɒn/ (also **Amboina** /-ˈbɔɪnə/) **1** a mountainous island in eastern Indonesia, one of the Molucca Islands. **2** a port on this island, the capital of the Molucca Islands; pop. (1980) 79,636.

amboyna /æmˈbɔɪnə/ *n.* the decorative wood of the SE Asian tree *Pterocarpus indicus*. [*Amboina*: see AMBON]

Ambrose, St /ˈæmbrəʊz/ (c.339–97), Doctor of the Church. He was a Roman governor at Milan and a converted Christian, though not yet baptized, when he was elected bishop of Milan (374) and became a champion of orthodoxy. He was partly responsible for the conversion of St Augustine of Hippo. His knowledge of Greek enabled him to introduce much Eastern theology and liturgical practice into the West; Ambrosian (antiphonal) plainsong is associated with his name, and the Athanasian Creed has been attributed to him. Feast day, 7 Dec.

ambrosia /æmˈbrəʊzɪə/ *n.* **1** Gk & Rom. Mythol. the food of the gods; the elixir of life. **2** anything very pleasing to taste or smell. **3** the food of certain bees and beetles. □ **ambrosia beetle** a small beetle of the family Scolytidae or Platypodidae, whose larvae are pinhole borers. □ **ambrosial** *adj.* **ambrosian** *adj.* [L f. Gk, = elixir of life f. *ambrotos* immortal]

ambry var. of AUMBRY.

ambulance /ˈæmbjʊləns/ *n.* **1** a vehicle specially equipped for conveying the sick or injured to and from hospital, esp. in emergencies. **2** a mobile hospital following an army. [F (as AMBULANT)]

ambulant /ˈæmbjʊlənt/ *adj.* Med. **1** (of a patient) able to walk about; not confined to bed. **2** (of treatment) not confining a patient to bed. [L *ambulare ambulant-* walk]

ambulatory /ˈæmbjʊlətərɪ/ *adj. & n.* ● *adj.* **1** = AMBULANT. **2** of or adapted for walking. **3 a** movable. **b** not permanent. ● *n.* (*pl.* **-ies**) a place for walking, esp. an aisle or cloister in a church or monastery. [L *ambulatorius* f. *ambulare* walk]

ambuscade /ˌæmbəˈskeɪd/ *n. & v.* ● *n.* an ambush. ● *v.* **1** *tr.* attack by means of an ambush. **2** *intr.* lie in ambush. **3** *tr.* conceal in an ambush. [F *embuscade* f. It. *imboscata* or Sp. *emboscada* f. L *imboscare*: see AMBUSH, -ADE¹]

ambush /ˈæmbʊʃ/ *n. & v.* ● *n.* **1** a surprise attack by persons (e.g. troops) in a concealed position. **2 a** the concealment of troops etc. to make such an attack. **b** the place where they are concealed. **c** the troops etc. concealed. ● *v.tr.* **1** attack by means of an ambush. **2** lie in wait for. [ME f. OF *embusche, embuschier*, f. a Rmc form = 'put in a wood': rel. to BUSH¹]

ameba US var. of AMOEBA.

amebic US var. of AMOEBIC.

ameer var. of AMIR.

ameliorate /əˈmiːlɪəˌreɪt/ *v.tr. & intr. formal* make or become better; improve. □ **ameliorator** *n.* **ameliorative** /-rətɪv/ *adj.* **amelioration** /əˌmiːlɪəˈreɪʃ(ə)n/ *n.* [alt. of MELIORATE after F *améliorer*]

amen /ɑːˈmen, eɪˈmen/ *int. & n.* ● *int.* **1** uttered at the end of a prayer or hymn etc., meaning 'so be it'. **2** (foll. by *to*) expressing agreement or assent (*amen to that*). ● *n.* an utterance of 'amen'. [ME f. eccl.L f. Gk f. Heb. 'āmēn certainly]

amenable /əˈmiːnəb(ə)l/ *adj.* **1** responsive, tractable. **2** (often foll. by *to*) (of a person) responsible to law. **3** (foll. by *to*) (of a thing) subject or liable. □ **amenably** *adv.* **amenableness** *n.* **amenability** /əˌmiːnəˈbɪlɪtɪ/ *n.* [AF (Law) f. *amener* bring to f. *a-* to + *mener* bring f. LL *minare* drive animals f. L *minari* threaten]

amend /əˈmend/ *v.tr.* **1** make minor improvements in (a text or a written proposal). **2** correct an error or errors in (a document). **3** make better; improve. ¶ Often confused with *emend*, a more technical word used in the context of textual correction. □ **amendable** *adj.* **amender** *n.* [ME f. OF *amender* ult. f. L *emendare* EMEND]

amende honorable /æˌmɒnd ˌɒnɒˈrɑːblə/ *n.* (pl. **amendes honorables** pronunc. same) a public or open apology, often with some form of reparation. The term applied originally to a punishment known in France from the 12th century to the Revolution, involving a public and humiliating acknowledgement of crime, imposed in cases of public scandal and frequently required before execution. [F, = honourable reparation]

amendment /əˈmendmənt/ *n.* **1** a minor improvement in a document (esp. a legal or statutory one). **2** an article added to the US Constitution. [AMEND + -MENT]

amends /əˈmendz/ *n.* □ **make amends** (often foll. by *for*) compensate or make up (for). [ME f. OF *amendes* penalties, fine, pl. of *amende* reparation f. *amender* AMEND]

Amenhotep /ˌɑːmenˈhəʊtep/ (Greek name Amenophis) the name of four Egyptian pharaohs of the 18th dynasty:

Amenhotep I (16th century BC), son of Ahmose I (founder of the 18th dynasty), reigned 1546–1526. He fought wars in Nubia and raided Libya.

Amenhotep II (15th century BC), son of Hatshepsut and Tuthmosis III, reigned 1450–1425. Brought up as a warrior, he fought successful campaigns in Syria and the Middle East; he completed some of the buildings begun by his father.

Amenhotep III (15th–14th century BC), son of Tuthmosis IV, reigned 1417–1379. After early military campaigns, his reign was generally peaceful and prosperous; he embarked on an extensive building programme centred on his capital, Thebes, including the colossi of Memnon and the Luxor temple.

Amenhotep IV see AKHENATEN.

amenity /əˈmiːnɪtɪ, əˈmen-/ *n.* (pl. **-ies**) **1** (usu. in *pl.*) a pleasant or useful feature. **2** pleasantness (of a place, person, etc.). □ **amenity-bed** *Brit.* a bed available in a hospital to give more privacy for a small payment. [ME f. OF *amenité* or L *amoenitas* f. *amoenus* pleasant]

amenorrhoea /eɪˌmenəˈrɪə/ *n.* (US **amenorrhea**) *Med.* an abnormal absence of menstruation. [A-¹ + MENO- + Gk -*rrhoia* f. *rheō* flow]

ament /əˈment/ *n.* (also **amentum** /-təm/) (pl. **aments** or **amenta** /-tə/) *Bot.* a catkin. [L, = thong]

amentia /əˈmenʃə/ *n.* *Med.* the condition of having a mental handicap. [L f. *amens* *ament-* mad (as A-¹, *mens* mind)]

Amerasian /ˌæməˈreɪʃ(ə)n, -ˈreɪʒ(ə)n/ *adj. & n.* ● *adj.* of mixed American and Asian parentage. ● *n.* an Amerasian person, esp. a child fathered by an American serviceman in Asia. [portmanteau word]

amerce /əˈmɜːs/ *v.tr.* **1** *Law* punish by fine. **2** punish arbitrarily. □ **amercement** *n.* **amerciable** /-sɪəb(ə)l/ *adj.* [ME *amercy* f. AF *amercier* f. *a* at + *merci* MERCY]

America /əˈmerɪkə/ **1** (also **the Americas**) a land mass of the New World or western hemisphere, consisting of the continents of North and South America, joined by the Isthmus of Panama. (*See note below.*) **2** the United States of America.

▪ The date of the first human occupation of the Americas is uncertain; ancestors of the native peoples are believed to have migrated over a land bridge from Siberia between 20,000 and 35,000 years ago, although it may have been earlier. The earliest reliably dated cultural artefacts date from about 12,000 years BP. The NE coastline of North America was visited by Norse seamen in the 8th or 9th century, but, for the modern world, the continent was discovered by Christopher Columbus, who reached the West Indies in 1492 and the South American mainland in 1498. The name of America dates from the early 16th century and is believed to derive from the Latin form (*Americus*) of the name of Amerigo Vespucci, who sailed along the west coast of South America in 1501.

American /əˈmerɪkən/ *adj. & n.* ● *adj.* **1** of, relating to, or characteristic of the United States or its inhabitants. **2** (usu. in *comb.*) of or relating to the continents of America (*Latin American*). ● *n.* **1** a native or national of the United States. **2** (usu. in *comb.*) a native or inhabitant of the continents of America (*North Americans*). **3** the English language as it is used in the United States. □ **American dream** the traditional ideals of the American people, such as equality, democracy, and material prosperity.

Americana /əˌmerɪˈkɑːnə/ *n.pl.* things connected with America, esp. with the United States.

American Civil War the war between the northern US states (usually known as the Union) and the Confederate states of the South, 1861–5. The war was fought over the issues of slavery and states' rights, and precipitated by the election of Abraham Lincoln, an opponent of slavery, as President. The pro-slavery southern states seceded from the Federal Union and maintained a military resistance to the superior industrial strength of the North for four years. Although its armies won some impressive victories in the first two years of the war, the Confederacy failed to gain foreign recognition and was gradually overwhelmed by superior military might and naval blockade. By the time the main Confederate army surrendered in Apr. 1865, most of the South, including its capital, Richmond, Virginia, had already fallen to Union soldiers.

American English *n.* the English language as spoken and written in the US. The term *North American English* is used to refer to the language as used in Canada as well as the US. Specifically American uses of words and meanings fall into a number of categories: adoptions from languages with which the early settlers came in contact (e.g. *moccasin*, *pecan*, *raccoon* from American Indian languages; *prairie*, *shanty* from French; *corral*, *lasso*, *ranch* from Spanish; *boss*, *sleigh* from Dutch), changes in meaning (e.g. *corn*, British = cereal plants, American = maize), survivals of 17th–18th-century English (e.g. *guess* = suppose, *fall* = autumn, *gotten* = got), different words for the same referent (*elevator* = lift, *sidewalk* = pavement). In addition, there are some notable differences in grammar and construction (e.g. the American *to teach school*; *I just ate*; *a quarter of ten*; *she wrote me last week*). There are also differences in spelling (e.g. *color* = colour) and pronunciation. Many American words have come into such general use outside the US (e.g. *blizzard*, *bogus*, *crank*, *deadline*) that few people realize their transatlantic origin.

American Federation of Labor a federation of North American trade unions, organized largely by craft rather than industry, founded in 1886 and merged in 1955 with the Congress of Industrial Organizations to form the American Federation of Labor and Congress of Industrial Organizations (AFL–CIO).

American football *n.* a form of football originating and chiefly played in the US between two teams of eleven players with an oval ball and an H-shaped goal, on a field marked out as a gridiron. Scoring is by points. The modern game developed from 1876 onwards in the colleges and the Rugby Union code was adopted as the basis of the rules. Emphasis is placed on strategy and tactics in both attack and defence, and the game is a contact sport which requires the wearing of helmets and other protective clothing.

American Independence, War of the war of 1775–83, in which the American colonists won independence from British rule. In the US and Canada it is called the *American Revolution*. It was triggered by resentment at the economic policies of Britain, particularly the right of Parliament to tax the colonies, and by the exclusion of the colonists from participation in political decisions affecting their interests. Following disturbances such as the Boston Tea Party of 1773, fighting broke out at Lexington and Concord in 1775; a year later, in July 1776, the Declaration of Independence was signed. The British were unable to provide sufficient manpower or coordinated leadership to crush the Americans, who gained the support of France and Spain after defeating the British at Saratoga; French sea-power eventually played a crucial role in the decisive surrender of a British army at Yorktown in 1781. The war was ended by the Peace of Paris in 1783 and George Washington became the first President of the US in 1789.

American Indian *n. & adj.* ● *n.* a member of a group of indigenous peoples of North and South America and the Caribbean. (See also NATIVE AMERICAN.) ● *adj.* of or relating to these peoples. ¶ The term *Native American* is now often preferred.

Americanism /əˈmerɪkəˌnɪz(ə)m/ *n.* **1 a** a word, sense, or phrase peculiar to or originating from the United States. **b** a thing or feature characteristic of or peculiar to the United States. **2** attachment to or sympathy for the United States.

Americanize /əˈmerɪkəˌnaɪz/ *v.* (also **-ise**) **1** *tr.* **a** make American in character. **b** naturalize as an American. **2** intr. become American in character. □ **Americanization** /əˌmerɪkənaɪˈzeɪʃ(ə)n/ n.

American Legion (in the US) an association of ex-servicemen formed in 1919.

American organ n. a type of reed organ resembling the harmonium but in which air is sucked (not blown) through reeds.

American Revolution see AMERICAN INDEPENDENCE, WAR OF.

American Samoa an unincorporated overseas territory of the US comprising a group of islands in the southern Pacific Ocean, to the east of Western Samoa and south of the Kiribati group; pop. (1990) 46,770; capital, Fagatogo. In 1899 the US acquired rights to the islands by agreement with Germany and Britain, and in April 1900 the two main islands were ceded to the US by their chiefs. Further islands were handed over in succeeding years.

American Sign Language n. (also called *Ameslan*) a form of sign language developed in America for the use of the deaf, consisting of over 4,000 signs.

American Standard Version an English translation of the Bible published in the US in 1901, based on the Revised Version with the incorporation of material produced by American scholars.

America's Cup an international yachting race held every three to four years, named after the yacht *America*, which won it in 1851; the cup was originally presented by the Royal Yacht Squadron for a race round the Isle of Wight. The *America*'s owners gave the trophy to the New York Yacht Club as a perpetual international challenge trophy, and it remained in the club's possession for 132 years until the successful challenge by an Australian crew in Sept. 1983. The Americans won back the trophy in 1987, and held it until 1995, when New Zealand were successful.

americium /ˌæməˈrɪsɪəm, -ˈrɪʃɪəm/ n. an artificial radioactive metallic chemical element (atomic number 95; symbol **Am**). A member of the actinide series, americium was first obtained by Glenn Seaborg and his colleagues in 1945 by bombarding plutonium with neutrons. It has been used in industrial measuring equipment as a source of gamma rays. [*America* (where first made)]

Amerind /ˈæmərɪnd/ adj. & n. (also **Amerindian** /ˌæməˈrɪndɪən/) = AMERICAN INDIAN. □ **Amerindic** /ˌæməˈrɪndɪk/ adj. [portmanteau word from AMERICAN + INDIAN]

Ameslan /ˈæmɪˌslæn/ n. an alternative name for AMERICAN SIGN LANGUAGE.

amethyst /ˈæmɪθɪst/ n. a precious stone of a violet or purple variety of quartz. □ **amethystine** /ˌæmɪˈθɪstiːn/ adj. [ME f. OF *ametiste* f. L *amethystus* f. Gk *amethustos* not drunken, the stone being supposed to prevent intoxication]

Amex[1] /ˈæmeks/ n. US colloq. the American Stock Exchange in New York, the second largest US securities exchange (the largest being the New York Stock Exchange). [*American + exchange*]

Amex[2] /ˈæmeks/ n. propr. American Express, an international organization providing personal charge-card facilities. [*American Express Company*]

Amharic /æmˈhærɪk/ n. the principal (and official) language of Ethiopia, spoken by about 9 million people in and to the north of Addis Ababa. It belongs to the Semitic language group and is directly descended from Ge'ez. [*Amhara*, Ethiopian province + -IC]

amiable /ˈeɪmɪəb(ə)l/ adj. friendly and pleasant in temperament; likeable. □ **amiably** adv. **amiableness** n. **amiability** /ˌeɪmɪəˈbɪlɪtɪ/ n. [ME f. OF f. LL *amicabilis* amicable: confused with F *aimable* lovable]

amianthus /ˌæmɪˈænθəs/ n. (also **amiantus** /-ˈæntəs/) a fine silky-fibred variety of asbestos. [L f. Gk *amiantos* undefiled f. *a-* not + *miainō* defile, i.e. purified by fire, being incombustible: for *-h-* cf. AMARANTH]

amicable /ˈæmɪkəb(ə)l/ adj. showing or done in a friendly spirit (*an amicable meeting*). □ **amicably** adv. **amicableness** n. **amicability** /ˌæmɪkəˈbɪlɪtɪ/ n. [LL *amicabilis* f. *amicus* friend]

amice[1] /ˈæmɪs/ n. a white linen cloth worn on the neck and shoulders by a priest celebrating the Eucharist. [ME f. med.L *amicia*, *-sia* (earlier *amit* f. OF), f. L *amictus* outer garment]

amice[2] /ˈæmɪs/ n. a cap, hood, or cape worn by members of certain religious orders. [ME f. OF *aumusse* f. med.L *almucia* etc., of unkn. orig.]

amicus curiae /æˌmiːkʊs ˈkjʊərɪˌiː/ n. (pl. **amici curiae** /æˌmiːkaɪ/) Law an impartial adviser in a court of law. [mod.L, = friend of the court]

amid /əˈmɪd/ prep. (also **amidst** /əˈmɪdst/) **1** in the middle of. **2** in the course of. [ME *amidde(s)* f. OE *on* ON + MID]

amide /ˈeɪmaɪd, ˈæm-/ n. Chem. a compound formed from ammonia by replacement of one (or sometimes more than one) hydrogen atom by a metal or an acyl radical. [AMMONIA + -IDE]

amidships /əˈmɪdʃɪps/ adv. (US **amidship**) in or into the middle of a ship. [MIDSHIP after AMID]

amidst var. of AMID.

Amiens /ˈæmɪˌen/ a town in northern France; pop. (1990) 136,230.

amigo /æˈmiːɡəʊ/ n. (pl. **-os**) esp. US colloq. (often as a form of address) a friend or comrade, esp. in Spanish-speaking areas. [Sp.]

Amin /æˈmiːn/, Idi (full name Idi Amin Dada) (b.1925), Ugandan soldier and head of state 1971–9. Having risen through the ranks of the army to become its commander, in 1971 he overthrew President Obote and seized power. His rule was characterized by the advancing of narrow tribal interests, expulsion of non-Africans (most notably the Ugandan Asians), and the murder of thousands of his political opponents. With Tanzanian assistance he was overthrown in 1979.

Amindivi Islands /ˌæmɪnˈdiːvɪ/ the northernmost group of islands in the Indian territory of Lakshadweep in the Indian Ocean.

amine /ˈeɪmiːn, ˈæm-/ n. Chem. a compound formed from ammonia by replacement of one or more hydrogen atoms by an organic radical or radicals. [AMMONIA + -INE[4]]

amino /əˈmiːnəʊ/ n. (attrib.) Chem. of, relating to, or containing the monovalent group –NH₂. [AMINE]

amino acid n. Biochem. a simple organic compound containing both a carboxyl (COOH) and an amino (NH_2) group. Amino acids occur naturally in plant and animal tissues and form the basic constituents of proteins. There are about twenty common amino acids, of which the simplest is glycine (H_2NCH_2COOH); some of them can be synthesized in the human body but others must be supplied in the diet.

amir /əˈmɪə(r)/ n. (also **ameer**) the title of some Arab rulers. □ **amirate** /əˈmɪərət/ n. [Arab. *'amīr* commander f. *amara* command: cf. EMIR]

Amirante Islands /ˈæmɪˌrænt/ a group of coral islands in the Indian Ocean, forming part of the Seychelles.

Amis[1] /ˈeɪmɪs/, Sir Kingsley (1922–95), English novelist and poet. He achieved popular success with his first novel *Lucky Jim* (1954), a satire on middle-class and academic aspirations set in a provincial university and seen through the eyes of its anti-Establishment hero. Amis's later novels show a more sombre tendency, exemplified in works such as *The Folks that Live on the Hill* (1990). He won the Booker Prize for his novel *The Old Devils* (1986). He also published four volumes of poetry.

Amis[2] /ˈeɪmɪs/, Martin (Louis) (b.1949), English novelist. Son of Kingsley Amis, he published his first novel, *The Rachel Papers*, in 1973. His other novels include *Money* (1984) and *Time's Arrow* (1991). Chiefly set against a background of contemporary urban life, his works are notable for their black humour and inventive use of language, especially slang and dialect.

Amish /ˈɑːmɪʃ, ˈeɪm-/ n. & adj. ● n.pl. (with *the*) a strict North American Mennonite sect. (*See note below.*) ● adj. of or relating to this sect. [G *Amisch* f. Jakob *Amman* (see below)]

▪ The Amish sect was founded by a Swiss Mennonite preacher, Jakob Amman (or Amen) (c.1645–c.1730), in the 1690s, who sought to enforce stricter codes of conduct and worship (including excommunication and ostracism for transgressors). His teachings divided the Mennonites, attracting followers in several European countries. Amish migration to North America began c.1720 and major settlements grew up in Pennsylvania, Ohio, and elsewhere.

amiss /əˈmɪs/ predic.adj. & adv. ● predic.adj. wrong; out of order; faulty (*knew something was amiss*). ● adv. wrong; wrongly; inappropriately (*everything went amiss*). □ **take amiss** be offended by (*took my words amiss*). [ME prob. f. ON á *mis* so as to miss f. à *on* + *mis* rel. to MISS[1]]

amitosis /ˌæmɪˈtəʊsɪs/ n. Biol. a form of nuclear division that does not involve mitosis. [A-[1] + MITOSIS]

amitriptyline /ˌæmɪˈtrɪptɪˌliːn/ n. Pharm. an antidepressant drug that has a mild tranquillizing action. [AMINE + TRI- + *heptyl* (see HEPTANE) + -INE[4]]

amity /ˈæmɪtɪ/ n. friendship; friendly relations. [ME f. OF *amitié* ult. f. L *amicus* friend]

Amman /əˈmɑːn/ the capital of Jordan; pop. (est. 1986) 1,160,000.

ammeter /ˈæmɪtə(r)/ n. an instrument for measuring electric current in amperes. [AMPERE + -METER]

ammo /ˈæməʊ/ n. colloq. ammunition. [abbr.]

Ammon var. of AMUN.

ammonia /əˈməʊnɪə/ n. *Chem.* **1** a colourless gas (chem. formula: NH₃) with a characteristic pungent smell. **2** (in full **ammonia water**) (in general use) a strongly alkaline solution of ammonia gas in water. [mod.L f. SAL AMMONIAC]

ammoniacal /ˌæməˈnaɪək(ə)l/ adj. of, relating to, or containing ammonia or sal ammoniac. [ME *ammoniac* f. OF (*arm-*, *amm-*) f. L f. Gk *ammōniakos* of Ammon (cf. SAL AMMONIAC) + -AL¹]

ammoniated /əˈməʊnɪˌeɪtɪd/ adj. combined or treated with ammonia.

ammonite /ˈæməˌnaɪt/ n. an extinct cephalopod mollusc of the mainly Mesozoic order Ammonoidea, having a flat coiled spiral shell commonly found as a fossil. [mod.L *ammonites*, after med.L *cornu Ammonis*, = L *Ammonis cornu*, horn of (Jupiter) Ammon (cf. AMUN)]

ammonium /əˈməʊnɪəm/ n. *Chem.* the ion NH₄⁺, present in solutions and salts of ammonia. [AMMONIA + -IUM]

ammunition /ˌæmjʊˈnɪʃ(ə)n/ n. **1** a supply of projectiles (esp. bullets, shells, and grenades). **2** points used or usable to advantage in an argument. [obs. F *amunition*, corrupt. of (*la*) *munition* (the) MUNITION]

amnesia /æmˈniːzɪə/ n. a partial or total loss of memory. □ **amnesic** adj. & n. **amnesiac** /-zɪˌæk/ n. [mod.L f. Gk, = forgetfulness]

amnesty /ˈæmnɪstɪ/ n. & v. ● n. (pl. **-ies**) a general pardon, esp. for political offences. ● v.tr. (**-ies**, **-ied**) grant an amnesty to. [F *amnestie* or L f. Gk *amnēstia* oblivion]

Amnesty International an international human rights organization, founded in London in 1961. Politically unaligned, it actively seeks the release of political prisoners and condemns torture and the death penalty. It was awarded the Nobel Peace Prize in 1977.

amnio /ˈæmnɪəʊ/ n. (pl. **-os**) *colloq.* amniocentesis. [abbr.]

amniocentesis /ˌæmnɪəʊsenˈtiːsɪs/ n. (pl. **amniocenteses** /-siːz/) *Med.* the sampling of amniotic fluid by insertion of a hollow needle into the uterus, performed in order to determine the condition of a foetus. [AMNION + Gk *kentēsis* pricking f. *kentō* to prick]

amnion /ˈæmnɪən/ n. (pl. **amnia** /-nɪə/) *Zool.* & *Physiol.* the innermost membrane that encloses the embryo of a reptile, bird, or mammal. □ **amniotic** /ˌæmnɪˈɒtɪk/ adj. [Gk, = caul (dimin. of *amnos* lamb)]

amniote /ˈæmnɪˌəʊt/ n. & adj. *Zool.* ● n. an animal whose embryo develops in an amnion. ● adj. of or relating to such animals. [back-form. f. AMNIOTIC]

amoeba /əˈmiːbə/ n. (*US* **ameba**) (pl. **amoebas** or **amoebae** /-biː/) a usu. aquatic protozoan of the genus *Amoeba*, which constantly changes its shape, esp. *A. proteus*. □ **amoeboid** adj. [mod.L f. Gk *amoibē* change]

amoebic /æˈmiːbɪk/ adj. (*US* **amebic**) of or relating to amoebae. □ **amoebic dysentery** dysentery caused by infection of the gut with certain amoebae.

amok /əˈmɒk/ adv. (also **amuck** /əˈmʌk/) □ **run amok** run about wildly in an uncontrollable violent rage. [Malay *amok* rushing in a frenzy]

among /əˈmʌŋ/ prep. (also **amongst** /əˈmʌŋst/) **1** surrounded by; in the company of (*lived among the trees*; *be among friends*). **2** in the number of (*among us were those who disagreed*). **3** an example of; in the class or category of (*is among the richest men alive*). **4 a** between; within the limits of (collectively or distributively); shared by (*had £5 among us*; *divide it among you*). **b** by the joint action or from the joint resources of (*among us we can manage it*). **5** with one another; by the reciprocal action of (*was decided among the participants*; *talked among themselves*). **6** as distinguished from; pre-eminent in the category of (*she is one among many*). [OE *ongemang* f. *on* ON + *gemang* assemblage (cf. MINGLE): *-st* = adverbial genitive *-s* + *-t* as in AGAINST]

amontillado /əˌmɒntɪˈlɑːdəʊ/ n. (pl. **-os**) a medium dry sherry. [Sp. f. *Montilla* in Spain + *-ado* = -ATE²]

amoral /eɪˈmɒrəl/ adj. **1** not concerned with or outside the scope of morality (cf. IMMORAL). **2** having no moral principles. □ **amoralism** n. **amoralist** n. **amorality** /ˌeɪmɒˈrælɪtɪ/ n.

amoretto /ˌæmɒːˈretəʊ/ n. (pl. **amoretti** /-tɪ/) a Cupid. [It., dimin. of *amore* love f. L (as AMOUR)]

amorist /ˈæmərɪst/ n. a person who professes or writes of (esp. sexual) love. [L *amor* or F *amour* + -IST]

Amorite /ˈæməˌraɪt/ n. & adj. ● n. a member of an ancient people whose semi-nomadic culture flourished in Mesopotamia, Palestine, and Syria in the 3rd millennium BC and who founded a number of states and dynasties, including Mari on the Euphrates and the first dynasty of Babylon. ● adj. of or relating to this people. [Heb. *'emōrī* f. Akkadian (Sumerian *martu* west)]

amoroso¹ /ˌæməˈrəʊsəʊ/ adv. & adj. *Mus.* in a loving or tender manner. [It.]

amoroso² /ˌæməˈrəʊsəʊ/ n. (pl. **-os**) a full rich type of sherry. [Sp., = amorous]

amorous /ˈæmərəs/ adj. **1** showing, feeling, or inclined to sexual love. **2** of or relating to sexual love. □ **amorously** adv. **amorousness** n. [ME f. OF f. med.L *amorosus* f. L *amor* love]

amorphous /əˈmɔːfəs/ adj. **1** shapeless. **2** vague, ill-organized. **3** *Mineral.* & *Chem.* non-crystalline; having neither definite form nor structure. □ **amorphously** adv. **amorphousness** n. [med.L *amorphus* f. Gk *amorphos* shapeless f. *a-* not + *morphē* form]

amortize /əˈmɔːtaɪz/ v.tr. (also **-ise**) *Commerce* **1** gradually extinguish (a debt) by money regularly put aside. **2** gradually write off the initial cost of (assets). **3** transfer (land) to a corporation in mortmain. □ **amortization** /əˌmɔːtaɪˈzeɪʃ(ə)n/ n. [ME f. OF *amortir* (stem *amortiss-*) ult. f. L *ad* to + *mors mort-* death]

Amos /ˈeɪmɒs/ **1** a Hebrew minor prophet (*c.*760 BC), a shepherd of Tekoa, near Jerusalem. **2** a book of the Bible containing his prophecies.

amount /əˈmaʊnt/ n. & v. ● n. **1** a quantity, esp. the total of a thing or things in number, size, value, extent, etc. (*a large amount of money*; *came to a considerable amount*). **2** the full effect or significance. ● v.intr. (foll. by *to*) **1** be equivalent to in number, size, significance, etc. (*amounted to £100*; *amounted to a disaster*). **2** (of a person) develop into, become (*might one day amount to something*). □ **any amount of** a great deal of. **no amount of** not even the greatest possible amount of. [ME f. OF *amunter* f. *amont* upward, uphill, f. L *ad montem*]

amour /əˈmʊə(r)/ n. a love affair, esp. a secret one. [F, = love, f. L *amor amoris*]

amour propre /ˌæmʊə ˈprɒprə/ n. self-respect. [F]

Amoy /əˈmɔɪ/ an alternative name for XIAMEN.

AMP abbr. *Biochem.* adenosine monophosphate.

amp¹ /æmp/ n. *Electr.* an ampere. [abbr.]

amp² /æmp/ n. *colloq.* an amplifier. [abbr.]

ampelopsis /ˌæmpɪˈlɒpsɪs/ n. (pl. same) a vine of the genus *Ampelopsis* or *Parthenocissus*, e.g. Virginia creeper. [mod.L f. Gk *ampelos* vine + *opsis* appearance]

amperage /ˈæmpərɪdʒ/ n. *Electr.* the strength of an electric current in amperes.

Ampère /ˈæmpeə(r), ɑ̃per/, André-Marie (1775–1836), French physicist, mathematician, and philosopher. He was a child prodigy who became one of the founders of electromagnetism and electrodynamics, and is best known for his analysis of the relationship between magnetic force and electric current. Ampère also developed a precursor of the galvanometer.

ampere /ˈæmpeə(r)/ n. *Electr.* a unit of electric current (symbol: **A**), equal to a flow of one coulomb per second. The ampere is the SI base unit of electric current, and is that constant current which, if maintained in two straight parallel conductors of infinite length, of negligible circular cross-section, and placed 1 metre apart in a vacuum, would produce between these conductors a force of 2×10^{-7} newton per metre. [AMPÈRE]

ampersand /ˈæmpəˌsænd/ n. the sign & (= *and*, Latin *et*). [corrupt. of *and per se and* ('&' by itself is 'and')]

amphetamine /æmˈfetəmɪn, -ˌmiːn/ n. a synthetic drug used esp. as a stimulant. [abbr. of chem. name alpha-methyl *phenethylamine*]

amphi- /ˈæmfɪ/ comb. form **1** both. **2** of both kinds. **3** on both sides. **4** around. [Gk]

amphibian /æmˈfɪbɪən/ adj. & n. ● adj. **1** living both on land and in water. **2** *Zool.* of or relating to the class Amphibia. **3** (of a vehicle) able to operate on land and water. ● n. **1** *Zool.* a vertebrate of the class Amphibia. (See note below.) **2** (in general use) a creature living both on land and in water. **3** an amphibian vehicle. [mod.L *amphibium* f. Gk *amphibion* f. AMPHI- + *bios* life]

▪ The amphibians typically have an aquatic larval stage with gills (e.g. a tadpole) and an air-breathing four-legged adult stage. There are about 2,400 living species, divided into two main groups: tailless forms (frogs and toads) and those with tails (e.g. newts and salamanders). Amphibians, which evolved from fish in the Devonian period, were the first vertebrates to live on land, preceding the reptiles. Their eggs lack shells and require water or damp conditions to survive, and so, unlike reptiles, amphibians cannot be completely independent of a moist environment.

amphibious /æmˈfɪbɪəs/ adj. **1** living both on land and in water. **2** of or relating to or suited for both land and water. **3** *Mil.* **a** (of a military operation) involving forces landed from the sea. **b** (of forces) trained for such operations. **4** having a twofold nature; occupying two positions. □ **amphibiously** *adv.*

amphibole /ˈæmfɪˌbəʊl/ *n. Mineral.* any of a class of rock-forming silicate and aluminosilicate minerals with fibrous or columnar crystals. [F f. L *amphibolus* ambiguous]

amphibology /ˌæmfɪˈbɒlədʒɪ/ *n.* (*pl.* **-ies**) **1** a quibble. **2** an ambiguous wording. [ME f. OF *amphibologie* f. LL *amphibologia* for L f. Gk *amphibolia* ambiguity]

amphimixis /ˌæmfɪˈmɪksɪs/ *n. Biol.* true sexual reproduction with the fusion of gametes from two individuals (cf. APOMIXIS). □ **amphimictic** /-ˈmɪktɪk/ *adj.* [AMPHI- + Gk *mixis* mingling]

amphioxus /ˌæmfɪˈɒksəs/ *n.* (*pl.* **amphioxi** /-saɪ/ or **amphioxuses**) *Zool.* a lancelet of the genus *Branchiostoma* (formerly *Amphioxus*). [mod.L, formed as AMPHI- + Gk *oxus* sharp]

amphipathic /ˌæmfɪˈpæθɪk/ *adj. Chem.* (of a molecule) having both hydrophilic and hydrophobic parts; consisting of such molecules. [AMPHI- + Gk *pathikos* (as PATHOS)]

amphipod /ˈæmfɪˌpɒd/ *n. Zool.* a crustacean of the largely marine order Amphipoda, having a laterally compressed abdomen with two kinds of limb, e.g. the freshwater shrimp (*Gammarus pulex*). [AMPHI- + Gk *pous podos* foot]

amphiprostyle /æmˈfɪprəˌstaɪl/ *n. & adj. Archit.* ● *n.* a classical building with a portico at each end. ● *adj.* of or in this style. [L *amphiprostylus* f. Gk *amphiprostulos* (as AMPHI-, *prostulos* PROSTYLE)]

amphisbaena /ˌæmfɪsˈbiːnə/ *n.* (*pl.* **amphisbaenae** /-niː/ or **amphisbaenas**) **1** *Mythol. & poet.* a fabulous serpent with a head at each end. **2** *Zool.* a burrowing wormlike lizard of the genus *Amphisbaena*, having no apparent division of head from body (making both ends look similar). [ME f. L f. Gk *amphisbaina* f. *amphis* both ways + *bainō* go]

amphitheatre /ˈæmfɪˌθɪətə(r)/ *n.* (*US* **amphitheater**) **1** a round, usu. unroofed building with tiers of seats surrounding a central space. **2** a semicircular gallery in a theatre. **3** a large circular hollow. **4** the scene of a contest. [L *amphitheatrum* f. Gk *amphitheatron* (as AMPHI-, THEATRE)]

Amphitrite /ˌæmfɪˈtraɪtɪ/ *Gk Mythol.* a sea-goddess, wife of Poseidon and mother of Triton.

amphora /ˈæmfərə/ *n.* (*pl.* **amphorae** /-ˌriː/ or **amphoras**) a Greek or Roman receptacle with two handles and a narrow neck. [L f. Gk *amphoreus*]

amphoteric /ˌæmfəˈterɪk/ *adj. Chem.* able to react as a base and an acid. [Gk *amphoteros* compar. of *amphō* both]

ampicillin /ˌæmpɪˈsɪlɪn/ *n. Pharm.* a semi-synthetic penicillin used esp. in treating infections of the urinary and respiratory tracts. [*amino* + *penicillin*]

ample /ˈæmp(ə)l/ *adj.* (**ampler**, **amplest**) **1 a** plentiful, abundant, extensive. **b** *euphem.* (esp. of a person) large, stout. **2** enough or more than enough. □ **ampleness** *n.* **amply** *adv.* [F f. L *amplus*]

amplifier /ˈæmplɪˌfaɪə(r)/ *n.* an electronic device for increasing the strength of electrical signals, esp. for conversion into sound in radio and musical equipment.

amplify /ˈæmplɪˌfaɪ/ *v.* (**-ies**, **-ied**) **1** *tr.* increase the volume or strength of (sound, electrical signals, etc.). **2** *tr.* enlarge upon or add detail to (a story etc.). **3** *intr.* expand what is said or written. □ **amplification** /ˌæmplɪfɪˈkeɪʃ(ə)n/ *n.* [ME f. OF *amplifier* f. L *amplificare* (as AMPLE, -FY)]

amplitude /ˈæmplɪˌtjuːd/ *n.* **1 a** *Physics* the maximum extent of a vibration or oscillation from the position of equilibrium. **b** *Electr.* the maximum departure of the value of an alternating current or wave from the average value. **2 a** spaciousness, breadth; wide range. **b** abundance. □ **amplitude modulation** (abbr. **AM**) *Electronics* the modulation of a radio etc. wave by variation of its amplitude, esp. as a means of carrying an audio signal. [F *amplitude* or L *amplitudo* (as AMPLE, -TUDE)]

ampoule /ˈæmpuːl/ *n.* a small capsule in which measured quantities of liquids or solids, esp. for injecting, are sealed ready for use. [F f. L AMPULLA]

ampster /ˈæmstə(r)/ *n.* (also **amster**) *Austral. sl.* the accomplice of a sideshow operator who acts as a purchaser in an attempt to persuade others to follow his example. [f. *Amsterdam*, rhyming sl. for RAM *n.* 6]

ampulla /æmˈpʊlə/ *n.* (*pl.* **ampullae** /-liː/) **1 a** a Roman globular flask with two handles. **b** a receptacle for sacred uses. **2** *Anat.* the dilated end of a vessel or duct. [L]

amputate /ˈæmpjʊˌteɪt/ *v.tr.* cut off by surgical operation (a part of the body, esp. a limb), usu. because of injury or disease. □ **amputator** *n.*

amputation /ˌæmpjʊˈteɪʃ(ə)n/ *n.* [L *amputare* f. *amb-* about + *putare* prune]

amputee /ˌæmpjʊˈtiː/ *n.* a person who has lost a limb etc. by amputation.

Amritsar /æmˈrɪtsə(r)/ a city in the state of Punjab in NW India; pop. (1991) 709,000. Founded in 1577 by Ram Das (1534–81), fourth guru of the Sikhs, it became the centre of the Sikh faith and the site of its holiest temple, the Golden Temple. It was the scene of a riot in 1919, in the course of which 400 people were killed by British troops. A Sikh leader was killed there in 1984 during fighting between Sikh militants and the Indian army.

amster *Austral. sl.* var. of AMPSTER.

Amsterdam /ˌæmstəˈdæm/ the capital and largest city of the Netherlands; pop. (1991) 702,440. Built on some ninety islands separated by canals, at the southern tip of the IJsselmeer, Amsterdam has been an important port since the 14th century and a financial centre from the early 17th. It is especially known for its diamond industry, and is a leading cultural centre, housing important art collections, e.g. at the Rijksmuseum. Although Amsterdam is the capital, the country's seat of government and administrative centre is at The Hague.

amtrac /ˈæmtræk/ *n.* (also **amtrak**) *US* an amphibious tracked vehicle used for landing assault troops on a shore. [*amphibious* + *tractor*]

Amtrak /ˈæmtræk/ (also **Amtrack**) *propr.* a government-controlled Federal railway service in the US, operated by the National Railroad Passenger Corporation. [*American tra(c)k*]

amu *abbr. Physics* atomic mass unit.

amuck var. of AMOK.

Amu Darya /ˌɑːmuː ˈdɑːrɪə/ a river of central Asia, rising in the Pamirs and flowing 2,400 km (1,500 miles) into the Aral Sea. In classical times it was known as the Oxus.

amulet /ˈæmjʊlɪt/ *n.* **1** an ornament or small piece of jewellery worn as a charm against evil. **2** something which is thought to give such protection. [L *amuletum*, of unkn. orig.]

Amun /ˈæmən/ (also **Ammon**) *Mythol.* a supreme god of the ancient Egyptians. His worship spread to Greece, where he was identified with Zeus, and to Rome, where he was known as Jupiter Ammon. As a national god of Egypt he was associated in a triad with Mut and Khonsu.

Amundsen /ˈɑːmʊnds(ə)n/, Roald (1872–1928), Norwegian explorer. Amundsen made his name as a polar explorer when he became the first to navigate the North-west Passage in the small sailing vessel *Gjöa* (1903–6), during which expedition he also travelled over the ice by sledge and located the site of the magnetic North Pole. In 1911 he beat the British explorer Robert F. Scott in the race to be the first to reach the South Pole. In the 1920s Amundsen devoted himself to aerial exploration of the polar regions, eventually disappearing on a search for the missing Italian airship expedition led by Umberto Nobile (1885–1978).

Amur /əˈmʊə(r)/ (called in Chinese *Heilong*) a river of NE Asia, forming for the greater part of its length the boundary between Russia and China. Its length, including its northern headstream, the Shilka, is about 4,350 km (2,737 miles).

amuse /əˈmjuːz/ *v.* **1** *tr.* cause (a person) to laugh or smile. **2** *tr. & refl.* (often foll. by *with*, *by*) interest or occupy; keep (a person) entertained. □ **amusing** *adj.* **amusingly** *adv.* [ME f. OF *amuser* cause to muse (see MUSE²) f. causal *a* to + *muser* stare]

amusement /əˈmjuːzmənt/ *n.* **1** something that amuses, esp. a pleasant diversion, game, or pastime. **2 a** the state of being amused. **b** the act of amusing. **3** a mechanical device (e.g. a roundabout) for entertainment at a fairground etc. □ **amusement arcade** *Brit.* an indoor area for entertainment with automatic game-machines. [F f. *amuser*: see AMUSE, -MENT]

amygdaloid /əˈmɪgdəˌlɔɪd/ *adj.* shaped like an almond. □ **amygdaloid nucleus** *Anat.* a roughly almond-shaped mass of grey matter deep inside each cerebral hemisphere, associated with the sense of smell. [L *amygdala* f. Gk *amugdalē* almond]

amyl /ˈeɪmaɪl, ˈæmɪl/ *n.* (usu. *attrib.*) *Chem.* the monovalent group $-C_5H_{11}$, derived from pentane. Also called *pentyl.* [L *amylum* starch]

amylase /ˈæmɪˌleɪz/ *n. Biochem.* an enzyme that converts starch and glycogen into simple sugars. [AMYL + -ASE]

amyloid /ˈæmɪˌlɔɪd/ *n. & adj. Med.* ● *n.* a glycoprotein deposited in connective tissue in certain diseases. ● *adj.* of or involving this substance. [AMYL + -OID]

amylopsin /ˌæmɪˈlɒpsɪn/ *n. Biochem.* a digestive enzyme, secreted by the pancreas, that converts starch into maltose. [AMYL after *pepsin*]

Amytal /ˈæmɪˌtæl/ *n. propr.* a barbiturate drug used as a sedative and a hypnotic. [AMYL + euphonic *-t-* + -AL[2]]

an /æn, ən/ *adj.* the form of the indefinite article (see A[1]) used before words beginning with a vowel sound (*an egg; an hour; an MP*). ¶ Now less often used before aspirated words beginning with *h* and stressed on a syllable other than the first (so *a hotel*, not *an hotel*).

an-[1] /ən, æn/ *prefix* not, without (*anarchy*) (cf. A-[1]). [Gk *an-*]

an-[2] /ən, æn/ assim. form of AD- before *n*.

-an /ən/ *suffix* (also **-ean, -ian**)/ən, ɪən/ forming adjectives and nouns, esp. from names of places, systems, zoological classes or orders, and founders (*Mexican; Anglican; crustacean; European; Lutheran; Georgian; theologian*). [ult. f. L adj. endings *-(i)anus, -aeus*: cf. Gk *-aios, -eios*]

ana /ˈɑːnə/ *n.* **1** (as *pl.*) anecdotes or literary gossip about a person. **2** (as *sing.*) a collection of a person's memorable sayings. [= -ANA]

ana- /ˈænə/ *prefix* (usu. **an-** before a vowel) **1** up (*anadromous*). **2** back (*anamnesis*). **3** again (*anabaptism*). [Gk *ana* up]

-ana /ˈɑːnə/ *suffix* forming plural nouns meaning 'things associated with' (*Victoriana; Americana*). [neut. pl. of L adj. ending *-anus*]

Anabaptism /ˌænəˈbæptɪz(ə)m/ *n.* the doctrine of the Anabaptists. [eccl.L *anabaptismus* f. Gk *anabaptismos* (as ANA-, BAPTISM)]

Anabaptist /ˌænəˈbæptɪst/ *n.* a member of a sect of radical Protestant reformers who believed that baptism should only be administered to believing adults. The sect emerged during the 1520s and 1530s, following the ideas of reformers such as Zwingli. Anabaptists not only rejected infant baptism, but also advocated complete separation of Church and state; many of them were persecuted and executed. Many of their beliefs are today carried on by the Mennonites, but they also have links with Baptist sects.

anabas /ˈænəˌbæs/ *n.* = *climbing perch* (see CLIMB). [mod.L f. Gk past part. of *anabainō* walk up]

anabasis /əˈnæbəsɪs/ *n.* (*pl.* **anabases** /-ˌsiːz/) **1** the march of Cyrus the Younger into Asia in 401 BC as narrated by Xenophon in his work *Anabasis*. **2** a military up-country march. [Gk, = ascent f. *anabainō* (as ANA-, *bainō* go)]

anabatic /ˌænəˈbætɪk/ *adj. Meteorol.* (of a wind) caused by local upward motion of warm air (cf. KATABATIC). [Gk *anabatikos* ascending (as ANABASIS)]

anabiosis /ˌænəbaɪˈəʊsɪs/ *n. Biol.* **1** a temporary state of suspended animation or greatly reduced metabolism. **2** the ability to revive from this. □ **anabiotic** /-ˈɒtɪk/ *adj.* [med.L f. Gk *anabiōsis* f. *anabioō* return to life]

anabolic /ˌænəˈbɒlɪk/ *adj. Biochem.* of or relating to anabolism. □ **anabolic steroid** a synthetic steroid hormone used to increase muscle size.

anabolism /əˈnæbəˌlɪz(ə)m/ *n. Biochem.* the synthesis of complex molecules in living organisms from simpler ones together with the storage of energy; constructive metabolism (opp. CATABOLISM). [Gk *anabolē* ascent (as ANA-, *ballō* throw)]

anabranch /ˈænəˌbrɑːntʃ/ *n.* esp. *Austral.* a stream that leaves a river and re-enters it lower down. [ANASTOMOSE + BRANCH]

anachronic /ˌænəˈkrɒnɪk/ *adj.* **1** out of date. **2** involving anachronism. [ANACHRONISM after *synchronic* etc.]

anachronism /əˈnækrəˌnɪz(ə)m/ *n.* **1 a** the attribution of a custom, event, etc., to a period to which it does not belong. **b** a thing attributed in this way. **2 a** anything out of harmony with its period. **b** an old-fashioned or out-of-date person or thing. □ **anachronistic** /əˌnækrəˈnɪstɪk/ *adj.* **anachronistically** *adv.* [F *anachronisme* or Gk *anakhronismos* (as ANA-, *khronos* time)]

anacoluthon /ˌænəkəˈluːθɒn/ *n.* (*pl.* **anacolutha** /-θə/) a sentence or construction which lacks grammatical sequence (e.g. *while in the garden the door banged shut*). □ **anacoluthic** *adj.* [LL f. Gk *anakolouthon* (as AN-[1], *akolouthos* following)]

anaconda /ˌænəˈkɒndə/ *n.* a large South American boa of the genus *Eunectes*, esp. the semi-aquatic *E. murinus*, which is one of the largest snakes known. [alt. of *anacondaia* f. Sinh. *henakandayā* whip-snake f. *hena* lightning + *kanda* stem: orig. of a snake in Sri Lanka]

Anacreon /əˈnækrɪən/ (c.570–478 BC), Greek lyric poet. The surviving fragments of his work include iambic invectives and elegiac epitaphs,

but he is most famous for his poetry written in celebration of love and wine.

anacreontic /əˌnækrɪˈɒntɪk/ *n. & adj.* ● *n.* a poem written after the manner of Anacreon. ● *adj.* **1** after the manner of Anacreon. **2** convivial and amatory in tone. [LL *anacreonticus* f. Gk *Anakreōn*]

anacrusis /ˌænəˈkruːsɪs/ *n.* (*pl.* **anacruses** /-siːz/) **1** (in poetry) an unstressed syllable at the beginning of a verse. **2** *Mus.* an unstressed note or notes before the first bar-line. [Gk *anakrousis* (as ANA-, *krousis* f. *krouō* strike)]

anadromous /əˈnædrəməs/ *adj.* (of a fish, e.g. the salmon) that swims up a river from the sea to spawn (opp. CATADROMOUS). [Gk *anadromos* (as ANA-, *dromos* running)]

anaemia /əˈniːmɪə/ *n.* (*US* **anemia**) *Med.* a deficiency in the blood, usu. of red cells or their haemoglobin, resulting in pallor and weariness. □ **pernicious anaemia** defective formation of red blood cells through a lack of vitamin B_{12} or folic acid. [mod.L f. Gk *anaimia* (as AN-[1], -AEMIA)]

anaemic /əˈniːmɪk/ *adj.* (*US* **anemic**) **1** *Med.* relating to or suffering from anaemia. **2** pale; lacking in vitality.

anaerobe /ˈænəˌrəʊb, əˈneərəʊb/ *n. Biol.* an organism that grows without air, or requires oxygen-free conditions to live. □ **anaerobic** /ˌæneəˈrəʊbɪk/ *adj.* [F *anaérobie* formed as AN-[1] + AEROBE]

anaesthesia /ˌænɪsˈθiːzɪə/ *n.* (*US* **anesthesia**) the absence of sensation, esp. artificially induced insensitivity to pain. (*See note below.*) □ **anaesthesiology** /-ˌθiːzɪˈɒlədʒɪ/ *n.* [mod.L f. Gk *anaisthēsia* (as AN-[1], *aisthēsis* sensation)]

▪ Attempts at anaesthesia were made from ancient times by stupefying patients with wine or narcotic juices, but it was not until the 1840s that unconsciousness was produced by inhalation of anaesthetic gases in dentistry and surgery. In 1842 an American surgeon, C. W. Long (1815–78), removed a tumour from the neck of a patient who had inhaled ether. Nitrous oxide and chloroform were also used, but all these substances had disadvantages and other anaesthetic agents were sought. Today new materials and methods are constantly being introduced. Anaesthetics may be given as gases, by inhalation; as drugs in solution, by injection or enema; or (for localized effect) sprayed on the skin. Anaesthesia may also be produced by other means, including acupuncture and hypnosis. The introduction of anaesthesia removed the necessity for the surgeon to concentrate on speed, in order to end the patient's ordeal as soon as possible, and widened the scope of surgery enormously.

anaesthetic /ˌænɪsˈθetɪk/ *adj. & n.* (*US* **anesthetic**) ● *n.* a substance that produces anaesthesia. ● *adj.* producing anaesthesia. □ **general anaesthetic** an anaesthetic that affects the whole body, usu. with loss of consciousness. **local anaesthetic** an anaesthetic that affects a restricted area of the body. [Gk *anaisthētos* insensible (as ANAESTHESIA)]

anaesthetist /əˈniːsθɪtɪst/ *n.* a specialist in the administration of anaesthetics.

anaesthetize /əˈniːsθəˌtaɪz/ *v.tr.* (also **-ise**, *US* **anesthetize**) **1** administer an anaesthetic to. **2** deprive of physical or mental sensation. □ **anaesthetization** /əˌniːsθətaɪˈzeɪʃ(ə)n/ *n.*

anaglyph /ˈænəglɪf/ *n.* **1** *Photog.* a composite stereoscopic photograph printed in superimposed complementary colours. **2** an embossed object cut in low relief. □ **anaglyphic** /ˌænəˈglɪfɪk/ *adj.* [Gk *anagluphē* (as ANA-, *gluphē* f. *gluphō* carve)]

Anaglypta /ˌænəˈglɪptə/ *n. propr.* a type of thick embossed wallpaper, usu. for painting over. [L *anaglypta* work in low relief: cf. ANAGLYPH]

anagram /ˈænəˌgræm/ *n.* a word or phrase formed by transposing the letters of another word or phrase. □ **anagrammatic** /ˌænəgrəˈmætɪk/ *adj.* **anagrammatical** *adj.* **anagrammatically** *adv.* **anagrammatize** /ˌænəˈgræməˌtaɪz/ *v.tr.* (also **-ise**). [F *anagramme* or mod.L *anagramma* f. Gk ANA- + *gramma -atos* letter: cf. -GRAM]

Anaheim /ˈænəˌhaɪm/ a city in California, on the SE side of the Los Angeles conurbation; pop. (1990) 266,400. It is the site of Disneyland, an amusement park opened in 1955.

anal /ˈeɪn(ə)l/ *adj.* relating to or situated near the anus. □ **anal-retentive** *Psychol.* (of a person) excessively orderly and fussy (interpreted as the result of conflict over toilet-training in infancy). □ **anally** *adv.* [mod.L *analis* (as ANUS)]

analects /ˈænəˌlekts/ *n.pl.* (also **analecta** /ˌænəˈlektə/) a collection of short literary extracts. [L f. Gk *analekta* things gathered f. *analegō* pick up]

analeptic /ˌænəˈleptɪk/ *adj. & n.* ● *adj.* (of a drug etc.) restorative. ● *n.* a restorative medicine or drug. [Gk *analēptikos* f. *analambanō* take back]

analgesia /ˌænæl'dʒiːzɪə, -'dʒiːsɪə/ *n.* the absence or relief of pain. [mod.L f. Gk, = painlessness]

analgesic /ˌænæl'dʒiːzɪk, -'dʒiːsɪk/ *adj. & n.* ● *adj.* relieving pain. ● *n.* an analgesic drug.

analog *US var. of* ANALOGUE.

analogize /ə'næləˌdʒaɪz/ *v.* (also **-ise**) **1** *tr.* represent or explain by analogy. **2** *intr.* use analogy.

analogous /ə'næləgəs/ *adj.* (usu. foll. by *to*) **1** partially similar or parallel; showing analogy. **2** *Biol.* (of organs etc.) performing a similar function but having a different evolutionary origin (cf. HOMOLOGOUS 2). □ **analogously** *adv.* [L *analogus* f. Gk *analogos* proportionate]

analogue /'ænəˌlɒg/ *n. & adj.* (*US* **analog**) ● *n.* **1** an analogous or parallel thing. **2** *Chem.* a compound with a molecular structure closely similar to that of another. **3** a synthetic food product resembling a natural food in taste and texture. ● *adj.* (usu. **analog**) designating, relating to, operating with, or created using, signals or information represented by a continuously variable quantity such as spatial position, voltage, etc. (cf. DIGITAL). [F f. Gk *analogon* neut. adj.: see ANALOGOUS]

analogy /ə'nælədʒɪ/ *n.* (*pl.* **-ies**) **1** (usu. foll. by *to, with, between*) correspondence or partial similarity. **2** *Logic* a process of arguing from similarity in known respects to similarity in other respects. **3** *Philol.* the imitation of existing words in forming inflections or constructions of others, without the existence of corresponding intermediate stages. **4** *Biol.* the resemblance of function between organs essentially different. **5** = ANALOGUE *n.* 1. □ **analogical** /ˌænə'lɒdʒɪk(ə)l/ *adj.* **analogically** *adv.* [F *analogie* or L *analogia* proportion f. Gk (as ANALOGOUS)]

analysand /ə'nælɪˌsænd/ *n.* a person undergoing psychoanalysis.

analyse /'ænəˌlaɪz/ *v.tr.* (*US* **analyze**) **1** examine in detail the constitution or structure of. **2 a** *Chem.* ascertain the constituents of (a sample of a mixture or compound). **b** take apart; break (something) down into its constituent parts. **3** find or show the essence or structure of (a book, music, etc.). **4** *Gram.* resolve (a sentence) into its grammatical elements. **5** psychoanalyse. □ **analysable** *adj.* **analyser** *n.* [obs. *analyse* (n.) or F *analyser* f. *analyse* (n.) f. med.L ANALYSIS]

analysis /ə'nælɪsɪs/ *n.* (*pl.* **analyses** /-ˌsiːz/) **1 a** a detailed examination of the elements or structure of a substance etc. **b** a statement of the result of this. **2 a** *Chem.* the determination of the constituent parts of a mixture or compound. **b** the act or process of breaking something down into its constituent parts. **3** psychoanalysis. **4** *Math.* the part of mathematics concerned with the theory of functions and the use of limits, continuity, and the operations of calculus (cf. ALGEBRA). **5** *Cricket* a statement of the performance of a bowler, usu. giving the numbers of overs and maiden overs bowled, runs conceded, and wickets taken. □ **in the final** (or **last** or **ultimate**) **analysis** after all due consideration; in the end. [med.L f. Gk *analusis* (as ANA-, *luō* set free)]

analyst /'ænəlɪst/ *n.* **1** a person skilled in (esp. chemical) analysis. **2** a psychoanalyst. [F *analyste*]

analytic /ˌænə'lɪtɪk/ *adj.* **1** of or relating to analysis. **2** *Philol.* analytical. **3** *Logic* (of a statement etc.) such that its denial is self-contradictory; true by definition (as ANALYSIS)]

analytical /ˌænə'lɪtɪk(ə)l/ *adj.* **1** using analytic methods. **2** *Philol.* using separate words instead of inflections (cf. SYNTHETIC *adj.* 4). □ **analytical cubism** see CUBISM. **analytical geometry** geometry using coordinates. □ **analytically** *adv.*

analyze *US var. of* ANALYSE.

anamnesis /ˌænəm'niːsɪs/ *n.* (*pl.* **anamneses** /-siːz/) **1** recollection (esp. of a supposed previous existence). **2** a patient's account of his or her medical history. **3** *Eccl.* the part of the anaphora recalling the Passion, Resurrection, and Ascension of Christ. [Gk, = remembrance]

anamorphosis /ˌænə'mɔːfəsɪs/ *n.* a distorted projection or drawing which appears normal when viewed from a particular point or in a suitable mirror. □ **anamorphic** *adj.* [Gk *anamorphōsis* transformation]

anandrous /ə'nændrəs/ *adj. Bot.* having no stamens. [Gk *anandros* without males f. *an-* not + *anēr andros* male]

Anangu /'ɑːnɑːŋˌuː/ *n.* (*pl.* same) *Austral.* an Aborigine, esp. one from central Australia. [Aboriginal language, = person]

Ananias /ˌænə'naɪəs/ *n.* (in the New Testament) **1** the husband of Sapphira, struck dead because he lied (Acts 5). **2** the Jewish high priest before whom St Paul was brought (Acts 23).

anapaest /'ænəˌpiːst/ *n.* (*US* **anapest** /-ˌpest/) *Prosody* a foot consisting of two short or unstressed syllables followed by one long or stressed syllable. □ **anapaestic** /ˌænə'piːstɪk/ *adj.* [L *anapaestus* f. Gk *anapaistos* reversed (because the reverse of a dactyl)]

anaphase /'ænəˌfeɪz/ *n. Biol.* the stage of meiotic or mitotic cell division when the chromosomes move away from one another to opposite poles of the spindle. [ANA- + PHASE]

anaphora /ə'næfərə/ *n.* **1** *Rhet.* the repetition of a word or phrase at the beginning of successive clauses. **2** *Gram.* the use of a word referring to or replacing a word used earlier in a sentence, to avoid repetition (e.g. *do* in *I like it and so do they*). **3** *Eccl.* the part of the Eucharist which contains the consecration, anamnesis, and Communion. □ **anaphoric** /ˌænə'fɒrɪk/ *adj.* [L f. Gk, = repetition (as ANA-, *pherō* to bear)]

anaphrodisiac /ænˌæfrə'dɪzɪˌæk/ *adj. & n.* ● *adj.* tending to reduce sexual desire. ● *n.* an anaphrodisiac drug.

anaphylaxis /ˌænəfɪ'læksɪs/ *n. Med.* an extreme and often life-threatening reaction to an antigen, e.g. to a bee sting, owing to hypersensitivity following an earlier dose. □ **anaphylactic** /-'læktɪk/ *adj.* [F *anaphylaxie* (as ANA- + Gk *phulaxis* guarding)]

anaptyxis /ˌænəp'tɪksɪs/ *n. Phonet.* the insertion of a vowel between two consonants to aid pronunciation (as in *went thataway*). □ **anaptyctic** /-'tɪptɪk/ *adj.* [mod.L f. Gk *anaptuxis* (as ANA-, *ptussō* fold)]

anarchism /'ænəˌkɪz(ə)m/ *n.* the doctrine that all government should be abolished. Contrary to popular belief, political anarchism does not imply a state of violent disorder; anarchists believe that after the abolition of the state an anarchist society could be organized on a voluntary, cooperative basis without recourse to force or compulsion. Anarchist ideas, detectable in the work of Percy Bysshe Shelley and William Godwin, were developed from the mid-19th century by Bakunin, Kropotkin, and Proudhon and played a significant part in radicalizing the labour movement in Europe and the US. Anarchist groups fighting on the Republican side in the Spanish Civil War gained control of many areas in its early stages. More recently anarchist activity has been largely restricted to involvement with other organizations in causes such as the peace movement and animal rights. [F *anarchisme* (as ANARCHY)]

anarchist /'ænəkɪst/ *n. & adj.* ● *n.* an advocate of anarchism or of political disorder. ● *adj.* relating to anarchism or anarchists. □ **anarchistic** /ˌænə'kɪstɪk/ *adj.* [F *anarchiste* (as ANARCHY)]

anarchy /'ænəkɪ/ *n.* **1** disorder, esp. political or social. **2** lack of government in a society. □ **anarchic** /ə'nɑːkɪk/ *adj.* **anarchical** *adj.* **anarchically** *adv.* [med.L f. Gk *anarkhia* (as AN-[1], *arkhē* rule)]

Anasazi /ˌænə'sɑːzɪ/ *n.* an ancient American Indian culture and people of the south-western US, which flourished between *c.*200 BC and AD 1500. The earliest phase of the culture, when the people lived in pit dwellings, is known as the Basket Maker period. In its later phases the people lived in pueblos and cliff dwellings; the present-day Pueblo culture developed from it. [Navajo, = ancient one (also = 'enemy ancestor')]

anastigmat /ə'næstɪgˌmæt/ *n.* a lens or lens-system made free from astigmatism by correction. [G f. *anastigmatisch* ANASTIGMATIC]

anastigmatic /ˌænəstɪg'mætɪk/ *adj.* (of a lens) free from astigmatism. [AN-[1] + ASTIGMATIC]

anastomose /ə'næstəˌməʊz/ *v.intr.* link by anastomosis. [F *anastomoser* (as ANASTOMOSIS)]

anastomosis /əˌnæstə'məʊsɪs/ *n.* (*pl.* **anastomoses** /-siːz/) a cross-connection of arteries, branches, rivers, etc. [mod.L f. Gk f. *anastomoō* furnish with a mouth (as ANA-, *stoma* mouth)]

anastrophe /ə'næstrəfɪ/ *n. Rhet.* the inversion of the usual order of words or clauses. [F *anastrophē* turning back (as ANA-, *strephō* to turn)]

anathema /ə'næθəmə/ *n.* (*pl.* **anathemas**) **1** a detested thing or person (*is anathema to me*). **2 a** a curse of the Church, excommunicating a person or denouncing a doctrine. **b** a cursed thing or person. **c** a strong curse. [eccl.L, = excommunicated person, excommunication, f. Gk *anathema* thing devoted, (later) accursed thing, f. *anatithēmi* set up]

anathematize /ə'næθəməˌtaɪz/ *v.tr. & intr.* (also **-ise**) curse. [F *anathématiser* f. L *anathematizare* f. Gk *anathematizo* (as ANATHEMA)]

Anatolia /ˌænə'təʊlɪə/ the western peninsula of Asia that now forms the greater part of Turkey, bounded by the Black Sea, the Aegean, and the Mediterranean. Most of it consists of a high, largely mountainous, plateau. [Gk *anatolē* east]

Anatolian /ˌænə'təʊlɪən/ *n. & adj.* ● *n.* **1** a native or inhabitant of Anatolia. **2** an ancient group of languages, a branch of the Indo-European language family including Hittite. ● *adj.* of or relating to Anatolia, its inhabitants, or Anatolian.

anatomical /ˌænəˈtɒmɪk(ə)l/ adj. **1** of or relating to anatomy. **2** of or relating to bodily structure. □ **anatomically** adv. [F anatomique or LL anatomicus (as ANATOMY)]

anatomist /əˈnætəmɪst/ n. a person skilled in anatomy. [F anatomiste or med.L anatomista (as ANATOMIZE)]

anatomize /əˈnætəˌmaɪz/ v.tr. (also **-ise**) **1** examine in detail. **2** dissect. [F anatomiser or med.L anatomizare f. anatomia (as ANATOMY)]

anatomy /əˈnætəmɪ/ n. (pl. **-ies**) **1** the science of the bodily structure of animals and plants. **2** this structure. **3** colloq. a human body. **4** analysis. **5** the dissection of the human body, animals, or plants. [F anatomie or LL anatomia f. Gk (as ANA-, -TOMY)]

anatta (also **anatto**) var. of ANNATTO.

Anaxagoras /ˌænækˈsæɡərəs/ (c.500–c.428 BC), Greek philosopher. He taught in Athens, where his pupils included Pericles. He believed that all matter was infinitely divisible and initially held together in a motionless, uniform mixture until put into a system of circulation directed by Spirit or Intelligence, which created the sky and earth, from which the sun, moon, and stars were formed. His concept of an independent moving cause prepared the way for a fully teleological view of nature.

Anaximander /æˌnæksɪˈmændə(r)/ (c.610–c.545 BC), Greek scientist, who lived at Miletus. He is reputed to have drawn the earliest map of the inhabited world, to have introduced the sundial into Greece, and to have taught that life began in water and that man originated from fish. He believed that all phenomena result from vortical motion in the primordial substance, and that the earth is cylindrical and poised in space.

Anaximenes /ˌænækˈsɪmɪˌniːz/ (c.546 BC), Greek philosopher and scientist, who lived at Miletus. Anaximenes believed the earth to be flat and shallow, and his view of astronomy was a retrograde step from that of Anaximander. He also proposed a theory of condensation and rarefaction, as a means of change from the basic form of matter to the diversity of natural substances.

ANC see AFRICAN NATIONAL CONGRESS.

-ance /əns/ suffix forming nouns expressing: **1** a quality or state or an instance of one (arrogance; protuberance; relevance; resemblance). **2** an action (assistance; furtherance; penance). [from or after F -ance f. L -antia, -entia (cf. -ENCE) f. pres. part. stem -ant-, -ent-]

ancestor /ˈænsestə(r)/ n. (fem. **ancestress** /-trɪs/) **1** any (esp. remote) person from whom one is descended. **2** an early type of animal or plant from which others have evolved. **3** an early prototype or forerunner (ancestor of the computer). [ME f. OF ancestre f. L antecessor -oris f. antecedere (as ANTE-, cedere cess- go)]

ancestral /ænˈsestrəl/ adj. belonging to or inherited from one's ancestors. [F ancestrel (as ANCESTOR)]

ancestry /ˈænsestrɪ/ n. (pl. **-ies**) **1** one's (esp. remote) family descent. **2** one's ancestors collectively. [ME, alt. of OF ancesserie (as ANCESTOR)]

Anchises /ænˈkaɪsiːz/ Gk & Rom. Mythol. the father of the Trojan hero Aeneas.

anchor /ˈæŋkə(r)/ n. & v. ● n. **1** a heavy metal weight used to moor a ship to the sea-bottom or a hot-air balloon to the ground. **2** a thing affording stability. **3** a source of confidence. **4** = ANCHORPERSON. ● v. **1** tr. secure (a ship or hot-air balloon) by means of an anchor. **2** tr. fix firmly. **3** intr. cast anchor. **4** intr. be moored by means of an anchor. □ **anchor escapement** a form of escapement in clocks and watches in which the teeth of the crown- or balance-wheel act on the pallets by recoil. **anchor-plate** a heavy piece of timber or metal, e.g. as support for suspension-bridge cables. **at anchor** moored by means of an anchor. **cast** (or **come to**) **anchor** let down the anchor. **weigh anchor** take up the anchor. [OE ancor f. L anchora f. Gk agkura]

Anchorage /ˈæŋkərɪdʒ/ the largest city in Alaska, a seaport on an inlet of the Pacific Ocean; pop. (1990) 226,340.

anchorage /ˈæŋkərɪdʒ/ n. **1** a place where a ship may be anchored. **2** the act of anchoring or lying at anchor. **3** anything dependable.

anchorite /ˈæŋkəˌraɪt/ n. (also **anchoret** /-rɪt/) (fem. **anchoress** /-rɪs/) **1** a hermit; a religious recluse. **2** a person of secluded habits. □ **anchoretic** /ˌæŋkəˈretɪk/ adj. **anchoritic** /-ˈrɪtɪk/ adj. [ME f. med.L anc(h)orita, eccl.L anchoreta f. eccl.Gk anakhōrētēs f. anakhōreō retire]

anchorman /ˈæŋkəmən/ n. (pl. **-men**; fem. **anchorwoman**, pl. **-women**) **1** a person who coordinates activities, esp. as a compère in a broadcast. **2** a person who plays a crucial part, esp. at the back of a tug-of-war team or as the runner of the last stage in a relay race.

anchorperson /ˈæŋkəˌpɜːs(ə)n/ n. (pl. **-people**, **-persons**) an anchorman or anchorwoman.

anchoveta /ˌæntʃəˈvetə/ n. a small Pacific anchovy caught for use as bait or to make fish-meal. [Sp., dimin. of anchova: cf. ANCHOVY]

anchovy /ˈæntʃəvɪ, ænˈtʃəʊvɪ/ n. (pl. **-ies**) a small silvery fish of the herring family; esp. the mainly Mediterranean Engraulis encrasicholus, which is usu. preserved in salt and oil and has a strong taste. □ **anchovy pear** an edible West Indian fruit like a mango, borne by the tree Grias cauliflora. **anchovy toast** toast spread with paste made from anchovies. [Sp. & Port. ancho(v)a, of uncert. orig.]

anchusa /æŋˈkjuːzə, ænˈtʃuːzə/ n. a plant of the genus Anchusa, of the borage family, e.g. bugloss, alkanet. [L f. Gk agkhousa]

anchylose var. of ANKYLOSE.

anchylosis var. of ANKYLOSIS.

ancien régime /ɒ̃sjẽ reɪˈʒiːm/ n. (pl. **anciens régimes** pronunc. same) **1** the political and social system in France before the Revolution of 1789. **2** any superseded regime. [F, = old rule]

ancient[1] /ˈeɪnʃ(ə)nt/ adj. & n. ● adj. **1** of long ago. **2** having lived or existed long. ● n. archaic an old man. □ **ancient history 1** the history of the ancient civilizations of the Mediterranean area and the Near East before the fall of the Western Roman Empire in AD 476. **2** something already long familiar. **ancient lights** a window that a neighbour may not deprive of light by erecting a building. **ancient monument** Brit. an old building etc. preserved usu. under government control. **the ancients** the people of ancient times, esp. the Greeks and Romans. **ancient world** the region around the Mediterranean and the Near East before the fall of the Roman Empire in AD 476. □ **ancientness** n. [ME f. AF auncien f. OF ancien, ult. f. L ante before]

ancient[2] /ˈeɪnʃ(ə)nt/ n. archaic = ENSIGN. [corrupt. of form ensyne etc. by assoc. with ancien = ANCIENT[1]]

anciently /ˈeɪnʃ(ə)ntlɪ/ adv. long ago.

ancillary /ænˈsɪlərɪ/ adj. & n. ● adj. **1** (of a person, activity, or service) providing essential support to a central service or industry, esp. the medical service. **2** (often foll. by to) subordinate, subservient. ● n. (pl. **-ies**) **1** an ancillary worker. **2** something which is ancillary; an auxiliary or accessory. [L ancillaris f. ancilla maidservant]

ancon /ˈæŋkən/ n. (pl. **ancones** /æŋˈkəʊniːz/) Archit. **1** a console, usu. of two volutes, supporting or appearing to support a cornice. **2** each of a pair of projections on either side of a block of stone etc. for lifting or repositioning. [L f. Gk agkōn elbow]

Ancona /æŋˈkəʊnə/ a port on the Adriatic coast of central Italy, capital of Marche region; pop. (1990) 103,270. It was founded by Greeks from Syracuse about 390 BC. [Gk agkōn elbow]

-ancy /ənsɪ/ suffix forming nouns denoting a quality (constancy; relevancy) or state (expectancy; infancy) (cf. -ANCE). [from or after L -antia: cf. -ENCY]

Ancyra /ænˈsaɪrə/ the ancient Roman name for ANKARA.

and /ænd, ənd/ conj. **1** a connecting words, clauses, or sentences, that are to be taken jointly (cakes and buns; white and brown bread; buy and sell; two hundred and forty). **b** implying progression (better and better). **c** implying causation (do that and I'll hit you; she hit him and he cried). **d** implying great duration (he cried and cried). **e** implying a great number (miles and miles). **f** implying addition (two and two are four). **g** implying variety (there are books and books). **h** implying succession (walking two and two). **2** colloq. to (try and open it). **3** in relation to (Britain and the EC). □ **and/or** either or both of two stated possibilities (usually restricted to legal and commercial use). [OE and]

-and /ænd/ suffix forming nouns meaning 'a person or thing to be treated in a specified way' (analysand; ordinand). [L gerundive ending -andus]

Andalusia /ˌændəˈluːsɪə/ (Spanish **Andalucía** /ˌandaluˈθiːa/) the southernmost region of Spain, bordering on the Atlantic and the Mediterranean; capital, Seville. The region, under Moorish rule from 711 to 1492, was probably named by the Moors after the Vandals, who had settled there in the 5th century. □ **Andalusian** adj. & n.

Andaman and Nicobar Islands /ˈændəmən, ˈnɪkəˌbɑː(r)/ two groups of islands in the Bay of Bengal, constituting a Union Territory in India; pop. (1991) 279,110; capital, Port Blair.

andante /ænˈdæntɪ/ adv., adj., & n. Mus. ● adv. & adj. in a moderately slow tempo. ● n. an andante passage or movement. [It., part. of andare go]

andantino /ˌændænˈtiːnəʊ/ adv., adj., & n. Mus. ● adv. & adj. rather quicker (orig. slower) than andante. ● n. (pl. **-os**) an andantino passage or movement. [It., dimin. of ANDANTE]

Andersen /ˈændəs(ə)n/, Hans Christian (1805–75), Danish author. The son of a poor shoemaker, he published several volumes of poetry and was acknowledged in Scandinavia as a novelist and travel writer before publishing the first of his fairy tales. These appeared from 1835 and include 'The Snow Queen', 'The Ugly Duckling', and 'The Little Match Girl'. Although deeply rooted in Danish folklore, the stories were also shaped by Andersen's own psychological alienation. His clear, simple prose set a new style of writing for both children and adults.

Anderson[1] /'ændəs(ə)n/, Carl David (1905–91), American physicist. In 1932, while using a cloud chamber to investigate cosmic rays, he accidentally discovered the positron – the first antiparticle known. He shared the Nobel Prize for physics for this in 1936, and also discovered the muon later the same year.

Anderson[2] /'ændəs(ə)n/, Elizabeth Garrett (1836–1917), English physician. Debarred from entry to medical courses because of her sex, she studied privately and in 1865 obtained a licence to practise from the Society of Apothecaries. In 1866 she opened a dispensary for women and children in London, which later became a hospital; it was renamed the Elizabeth Garrett Anderson Hospital in 1918. In 1870 she received the degree of MD from Paris University. Her influence was considerable in securing the admission of women to professional medical bodies.

Anderson[3] /'ændəs(ə)n/, Marian (1902–93), American operatic contralto. Despite early recognition of her singing talents, racial discrimination meant that she was initially unable to give concerts in her native land. She gained international success from several European tours between 1925 and 1935, but it was not until after her New York début in 1936 that her American career flourished. In 1955 she became the first black singer to perform at the New York Metropolitan Opera.

Anderson[4] /'ændəs(ə)n/, Philip Warren (b.1923), American physicist. He made contributions to the study of solid-state physics, and research on molecular interactions has been facilitated by his work on the spectroscopy of gases. He also investigated magnetism and superconductivity, and his work is of fundamental importance for modern solid-state electronics. He shared the Nobel Prize for physics in 1977.

Andes /'ændiːz/ a major mountain system running the length of the Pacific coast of South America. It extends over some 8,000 km (5,000 miles), with a continuous height of more than 3,000 m (10,000 ft). Its highest peak is Aconcagua, which rises to a height of 6,960 m (22,834 ft). □ **Andean** /æn'diːən, 'ændiən/ adj.

andesite /'ændɪˌzaɪt/ n. Geol. a fine-grained brown or greyish intermediate volcanic rock. [ANDES + -ITE[1]]

Andhra Pradesh /ˌɑːndrə prə'deʃ/ a state in SE India, on the Bay of Bengal; capital, Hyderabad.

andiron /'ændˌaɪən/ n. a metal stand (usu. one of a pair) for supporting burning wood in a fireplace; a firedog. [ME f. OF andier, of unkn. orig.: assim. to IRON]

Andorra /æn'dɔːrə/ a small autonomous principality in the southern Pyrenees, between France and Spain; pop. (est. 1991) 50,000; official languages, Catalan and French; capital, Andorra la Vella. Its independence dates from the late 8th century, when Charlemagne is said to have granted the Andorrans self-government for their help in defeating the Moors. Under a treaty of 1278 the sovereignty of Andorra is shared between the President of France and the Spanish bishop of Urgel. In 1993 the adoption of a formal constitution which legalized political parties was followed by the election of an executive head of government. □ **Andorran** adj. & n.

Andre /'ɑːndreɪ/, Carl (b.1935), American minimalist sculptor. His most famous works are installations created by stacking ready-made units such as bricks, cement blocks, or ceramic magnets according to a mathematically imposed modular system and without adhesives or joints. Andre's *Equivalent VIII* (1966), which consists of 120 bricks arranged two deep in a rectangle, was the subject of protests in Britain at the alleged waste of public money when it was purchased by the Tate Gallery.

Andrew, Prince /'ændruː/, Andrew Albert Christian Edward, Duke of York (b.1960), second son of Elizabeth II. Educated at Gordonstoun School, Scotland and the Royal Naval College, Dartmouth, he gained a commission in the Royal Marines and in 1982 served as a helicopter pilot in the Falklands War. He married Sarah Ferguson (b.1959) in 1986; the couple formally separated in 1993. They have two children, Princess Beatrice Elizabeth Mary (b.1988) and Princess Eugenie Victoria Helena (b.1990).

Andrew, St /'ændruː/, an Apostle, the brother of St Peter. An apocryphal work dating probably from the 3rd century describes his death by crucifixion; the X-shaped cross became associated with his name during the Middle Ages. Since c.750 he has been regarded as the patron saint of Scotland; he is also a patron saint of Russia. Feast day, 30 Nov.

Andrews[1] /'ændruːz/, Julie (born Julia Elizabeth Wells) (b.1935), English actress and singer. Early success in the British and American theatre, including her creation of the role of Eliza Doolittle in *My Fair Lady* on Broadway (1956), was followed by her film début in *Mary Poppins* (1964), for which she won an Oscar. Her wholesome appeal and talents as a singer were further displayed in *The Sound of Music* (1965). In her later films, such as *10* (1979), she moved away from her typecast prim image to play more diverse roles.

Andrews[2] /'ændruːz/, Thomas (1813–85), Irish physical chemist. He is best known for his work on the continuity of the gaseous and liquid states. His discovery of the critical temperature of carbon dioxide suggested that any gas could be liquefied if the temperature was sufficiently low, and he also showed that ozone is an allotrope of oxygen.

Andrić /'ændrɪtʃ/, Ivo (1892–1975), Yugoslav novelist, essayist, and short-story writer. A diplomat turned writer, he wrote his best-known novels while living under voluntary house arrest in German-occupied Belgrade. Set in his native Bosnia, these include *The Bridge on the Drina* (1945), whose narrative symbolically bridges past and present, as well as Western and Eastern cultures, and *Bosnian Chronicle* (1945). He was awarded the Nobel Prize for literature in 1961.

Androcles /'ændrəˌkliːz/ a runaway slave in a story by Aulus Gellius (2nd century AD), who extracted a thorn from the paw of a lion, which later recognized him and refrained from attacking him when he faced it in the arena.

androecium /æn'driːsɪəm/ n. (pl. **androecia** /-sɪə/) Bot. the stamens taken collectively. [mod.L f. Gk *andro-* male + *oikion* house]

androgen /'ændrədʒən/ n. any of a group of male sex hormones, esp. testosterone, involved in causing the development and maintenance of certain male sexual characteristics. □ **androgenic** /ˌændrə'dʒenɪk/ adj. [Gk *andro-* male + -GEN]

androgyne /'ændrəˌdʒaɪn/ adj. & n. ● adj. = ANDROGYNOUS. ● n. an androgynous person. [OF *androgyne* or L *androgynus* f. Gk *androgunos* (*anēr andros* male, *gunē* woman)]

androgynous /æn'drɒdʒɪnəs/ adj. **1** having a partly male, partly female appearance; of ambiguous gender. **2** Bot. with stamens and pistils in the same flower or inflorescence. **3** Biol. = HERMAPHRODITE. [as ANDROGYNE]

androgyny /æn'drɒdʒɪnɪ/ n. **1** androgynous character. **2** Biol. hermaphroditism.

android /'ændrɔɪd/ n. a robot with a human appearance. [Gk *andro-* male, man + -OID]

Andromache /æn'drɒməkɪ/ Gk Mythol. wife of Hector. She became the slave of Neoptolemus (son of Achilles) after the fall of Troy.

Andromeda /æn'drɒmɪdə/ **1** Gk Mythol. a daughter of Cepheus, king of the Ethiopians. Her mother Cassiopeia boasted that she herself (or her daughter) was more beautiful than the nereids, whereupon Poseidon in vengeance sent a sea monster to ravage the country. To abate his wrath Andromeda was fastened to a rock and exposed to the monster, from which she was rescued by Perseus. **2** Astron. a large northern constellation with few bright stars. It is chiefly notable for the Andromeda Galaxy (or Nebula), a conspicuous spiral galaxy probably twice as massive as our own and located 2 million light-years away.

Andropov[1] /æn'drɒpɒf/ a former name (1984–9) for RYBINSK.

Andropov[2] /æn'drɒpɒf/, Yuri (Vladimirovich) (1914–84), Soviet statesman, General Secretary of the Communist Party of the USSR 1982–4 and President 1983–4. Born in Russia, he served as ambassador to Hungary 1954–7, playing a significant role in the crushing of that country's uprising in 1956. He was appointed chairman of the KGB in 1967; its suppression of dissidents enhanced Andropov's standing within the Communist Party, and he gained the presidency on Brezhnev's death. While in office, he initiated the reform process carried through by Mikhail Gorbachev, his chosen successor.

-androus /'ændrəs/ comb. form Bot. forming adjectives meaning 'having specified male organs or stamens' (*monandrous*). [mod.L f. Gk *-andros* f. *anēr andros* male + -OUS]

-ane[1] /eɪn/ suffix var. of -AN, used esp. in words that have a parallel form in *-an* (*germane*, *humane*, *urbane*); also in *mundane*.

-ane[2] /eɪn/ *suffix Chem.* forming names of saturated hydrocarbons (*methane*; *propane*). [after -*ene*, -*ine*, etc.]

anecdotage /ˈænɪkˌdəʊtɪdʒ/ *n.* **1** *joc.* garrulous old age. **2** anecdotes. [ANECDOTE + -AGE: sense 1 after DOTAGE]

anecdote /ˈænɪkˌdəʊt/ *n.* a short account (or painting etc.) of an entertaining or interesting incident. □ **anecdotist** *n.* **anecdotal** /ˌænɪkˈdəʊt(ə)l/ *adj.* **anecdotalist** *n.* **anecdotic** /-ˈdɒtɪk/ *adj.* [F *anecdote* or mod.L f. Gk *anekdota* things unpublished (as AN-[1], *ekdotos* f. *ekdidōmi* publish)]

anechoic /ˌænɪˈkəʊɪk/ *adj.* free from echo.

anele /əˈniːl/ *v.tr. archaic* anoint, esp. in extreme unction. [ME f. OE *anon* + *elien* f. OE *ele* f. L *oleum* oil]

anemia *US* var. of ANAEMIA.

anemic *US* var. of ANAEMIC.

anemograph /əˈneməˌgrɑːf/ *n.* an instrument for recording on paper the direction and force of the wind. □ **anemographic** /əˌneməˈgræfɪk/ *adj.* [Gk *anemos* wind + -GRAPH]

anemometer /ˌænɪˈmɒmɪtə(r)/ *n.* an instrument for measuring or indicating the force of the wind. [Gk *anemos* wind + -METER]

anemometry /ˌænɪˈmɒmɪtrɪ/ *n.* the measurement of the force of the wind. □ **anemometric** /-məˈmetrɪk/ *adj.* [Gk *anemos* wind + -METRY]

anemone /əˈneməni/ *n.* **1** a ranunculaceous plant of the genus *Anemone*, often with flowers of various vivid colours; e.g. the wood anemone. **2** = sea anemone. [L f. Gk *anemōnē* wind-flower f. *anemos* wind]

anemophilous /ˌænɪˈmɒfɪləs/ *adj. Bot.* wind-pollinated. [Gk *anemos* wind + -*philous* (as -PHILIA)]

anent /əˈnent/ *prep. archaic* or *Sc.* or *N. Amer.* concerning. [OE *on efen* on a level with]

-aneous /ˈeɪnɪəs/ *suffix* forming adjectives (*cutaneous*; *miscellaneous*). [L -*aneus* + -OUS]

aneroid /ˈænəˌrɔɪd/ *adj.* & *n.* ● *adj.* (of a barometer) that measures air-pressure by its action on the elastic lid of an evacuated box. ● *n.* an aneroid barometer. [F *anéroïde* f. Gk *a-* not + *nēros* water]

anesthesia etc. *US* var. of ANAESTHESIA etc.

aneurin /əˈnjʊərɪn, ˈænjʊrɪn/ *n.* = THIAMINE. [anti + polyneuritis + vitam*in*]

aneurysm /ˈænjʊˌrɪz(ə)m/ *n.* (also **aneurism**) *Med.* an excessive localized enlargement of an artery. □ **aneurysmal** /ˌænjʊˈrɪzm(ə)l/ *adj.* (also **aneurismal**). [Gk *aneurusma* f. *aneurunō* widen out f. *eurus* wide]

anew /əˈnjuː/ *adv.* **1** again. **2** in a different way. [ME, f. A-[4] + NEW]

anfractuosity /ænˌfræktʃʊˈɒsɪtɪ, ˌænˌfræktjʊ-/ *n.* **1** circuitousness. **2** intricacy. [F *anfractuosité* f. LL *anfractuosus* f. L *anfractus* a bending]

anfractuous /ænˈfræktʃʊəs, ˌænˈfræktjʊ-/ *adj.* winding, sinuous; roundabout, circuitous. [f. late L *anfractuosus*, f. L *anfractus* a bending]

angary /ˈæŋgərɪ/ *n. Law* the right of a belligerent (subject to compensation for loss) to seize or destroy neutral property under military necessity. [F *angarie* ult. f. Gk *aggareia* f. *aggaros* courier]

angel /ˈeɪndʒəl/ *n.* **1 a** an attendant or messenger of God. (See note below.) **b** a conventional representation of this in human form with wings. **c** an attendant spirit (*evil angel*; *guardian angel*). **d** a member of the lowest order of the ninefold celestial hierarchy (see ORDER *n.* 19). **2 a** a very virtuous person. **b** an obliging person (*be an angel and answer the door*). **3** a messenger or bringer of something (*angel of death*; *angel of mercy*). **4** an old English coin bearing the figure of the archangel Michael piercing the dragon. **5** *sl.* a financial backer of an enterprise, esp. in the theatre. **6** an unexplained radar echo. □ **angel cake** a very light sponge cake. **angel dust** *sl.* the hallucinogenic drug phencyclidine hydrochloride. **angel-shark** = MONKFISH 2. **angels-on-horseback** a savoury of oysters wrapped in slices of bacon. [ME f. OF *angele* f. eccl.L *angelus* f. Gk *aggelos* messenger]

▪ In Judaism, Christianity, Zoroastrianism, and Islam, angels are thought of as spiritual messengers or mediators between the sacred and profane realms. Judaic tradition developed a hierarchy of angels which included archangels, seraphim, and cherubim, and subsequent traditions in Christianity were largely based on this (see ORDER *n.* 19). The Church in the Middle Ages was involved in extensive philosophical speculation and controversy over the substance, form, and nature of angels.

Angeleno /ˌændʒəˈliːnəʊ/ *n.* (also **Los Angeleno** /lɒs/) a native or inhabitant of Los Angeles. [Amer. Sp.]

Angel Falls a waterfall in the Guiana Highlands of SE Venezuela. It is the highest waterfall in the world, with an uninterrupted descent of 978 m (3,210 ft). The falls were discovered in 1935 by the American aviator and prospector James Angel (*c.*1899–1956).

angelfish /ˈeɪndʒəlˌfɪʃ/ *n.* a fish with winglike or elongated fins; esp. *Pterophyllum scalare*, a South American cichlid popular in aquariums.

angelic /ænˈdʒelɪk/ *adj.* **1** like or relating to angels. **2** having characteristics attributed to angels, esp. sublime beauty or innocence. □ **angelical** *adj.* **angelically** *adv.* [ME f. F *angélique* or LL *angelicus* f. Gk *aggelikos* (as ANGEL)]

angelica /ænˈdʒelɪkə/ *n.* **1** an aromatic umbelliferous plant, *Angelica archangelica*, used in cooking and medicine. **2** its candied stalks. [med.L (*herba*) *angelica* angelic herb]

Angelic Doctor the nickname of St Thomas Aquinas.

Angelico /ænˈdʒelɪˌkəʊ/, Fra (born Guido di Pietro, monastic name Fra Giovanni da Fiesole) (*c.*1400–55), Italian painter. He was a Dominican friar and his work was intended chiefly for contemplation and instruction. His simple and direct, mature style shows an awareness and understanding of contemporary developments in Renaissance painting, especially in his mastery of new ideas such as perspective. His most celebrated works are the frescos in the convent of San Marco, Florence (*c.*1438–47), and the *Scenes from the Lives of SS Stephen and Lawrence* (1447–9) in the private chapel of Pope Nicholas V in the Vatican.

Angelou /ˈændʒəˌluː/, Maya (b.1928), American novelist and poet. After working variously as a waitress, actress, teacher, and night-club singer, she became involved in the black civil-rights movement in the 1950s and 1960s. She received critical acclaim as a writer with the first volume of her autobiography, *I Know Why the Caged Bird Sings* (1970), which recounts her harrowing childhood experiences. More volumes of autobiography followed, as well as several volumes of poetry, including *Oh Pray My Wings Are Gonna Fit Me Well* (1975).

angelus /ˈændʒɪləs/ *n.* **1** a Roman Catholic devotion commemorating the Incarnation, said at morning, noon, and sunset. **2** a bell rung to announce this. [opening words *Angelus domini* (L, = the angel of the Lord)]

anger /ˈæŋgə(r)/ *n.* & *v.* ● *n.* extreme or passionate displeasure. ● *v.tr.* make angry; enrage. [ME f. ON *angr* grief, *angra* vex]

Angers /ˈɒnʒeɪ/ a town in western France, capital of the former province of Anjou; pop. (1990) 146,160.

Angevin /ˈændʒɪvɪn/ *n.* & *adj.* ● *n.* **1** a native or inhabitant of Anjou. **2** a Plantagenet, esp. any of the English kings from Henry II to John. ● *adj.* **1** of Anjou. **2** of the Plantagenets. [F]

angina /ænˈdʒaɪnə/ *n. Med.* **1** (in full **angina pectoris** /ˈpektərɪs/) pain in the chest brought on by exertion, owing to an inadequate blood supply to the heart. **2** an attack of intense constricting pain, esp. in the throat, often causing suffocation. [L, = spasm of the chest f. *angina* quinsy f. Gk *agkhonē* strangling]

angiogram /ˈændʒɪəˌgræm/ *n. Med.* a radiograph of blood and lymph vessels, made by introducing a substance opaque to X-rays. □ **angiography** /ˌændʒɪˈɒgrəfɪ/ *n.* [f. Gk *aggeion* vessel + -GRAM]

angioma /ˌændʒɪˈəʊmə/ *n.* (*pl.* **angiomas** or **angiomata** /-mətə/) *Med.* a tumour produced by the dilatation or new formation of blood vessels. [f. Gk *aggeion* vessel]

angioplasty /ˈændʒɪəʊˌplæstɪ, -ˌplɑːstɪ/ *n. Med.* the surgical repair of a damaged blood vessel. [Gk *aggeion* vessel, *plastía* f. *plastos* formed, moulded]

angiosperm /ˈændʒɪəˌspɜːm/ *n. Bot.* a plant of the subdivision Angiospermae, producing flowers and reproducing by seeds enclosed within a carpel. The angiosperms include herbaceous plants, herbs, shrubs, grasses, and most trees (cf. GYMNOSPERM). □ **angiospermous** /ˌændʒɪəˈspɜːməs/ *adj.* [mod.L f. Gk *aggeion* vessel + *sperma* seed]

Angkor /ˈæŋkɔː(r)/ the capital of the ancient kingdom of Khmer in NW Cambodia. It is noted for its temples, especially the Angkor Wat (early 12th century), decorated with relief sculptures. The site was overgrown with jungle when it was rediscovered in 1860.

Angle /ˈæŋg(ə)l/ *n.* a member of a Germanic people, originally inhabitants of what is now Schleswig-Holstein, who came to England in the 5th century AD, founding kingdoms in Mercia, Northumbria, and East Anglia and giving their name to England and the English. □ **Anglian** /ˈæŋglɪən/ *adj.* [L *Anglus* f. Gmc (OE *Engle*: cf. ENGLISH) f. *Angul* a district of Schleswig (now in N. Germany) (as ANGLE[2])]

angle[1] /ˈæŋg(ə)l/ *n.* & *v.* ● *n.* **1 a** the space between two meeting lines or surfaces. **b** the inclination of two lines or surfaces to each other. **2 a** a corner. **b** a sharp projection. **3 a** the direction from which a photograph etc. is taken. **b** the aspect from which a matter is considered. ● *v.* **1** *tr. & intr.* move or place obliquely; point in a particular direction. **2** *tr.* present (information) from a particular point of view (*was angled in favour of the victim*). □ **angle brackets** brackets in the form < > (see BRACKET *n.* 3). **angle-iron** a piece of iron or steel with an L-shaped cross-section, used to strengthen a framework. **angle of attack** the angle between the chord of an aerofoil etc. and the direction of the surrounding undisturbed flow of air, water, etc. **angle of incidence** see INCIDENCE. **angle of repose** the steepest angle at which a sloping surface formed of loose material is stable. [ME f. OF *angle* or f. L *angulus*]

angle² /ˈæŋg(ə)l/ *v. & n.* ● *v.intr.* **1** (often foll. by *for*) fish with hook and line. (See ANGLING.) **2** (foll. by *for*) seek an objective by devious or calculated means (*angled for a pay rise*). ● *n. archaic* a fish-hook. [OE *angul*]

angled /ˈæŋg(ə)ld/ *adj.* **1** placed at an angle to something else. **2** presented to suit a particular point of view. **3** having an angle.

Anglepoise /ˈæŋg(ə)l,pɔɪz/ *n.* (often *attrib.*) *propr.* a type of desk lamp with a sprung and jointed adjustable arm.

angler /ˈæŋglə(r)/ *n.* **1** a person who fishes with a hook and line. **2** = *angler fish.* □ **angler fish** a large-mouthed fish that preys upon smaller fish, attracting them by filaments arising from the dorsal fin (cf. FROGFISH, MONKFISH).

Anglesey /ˈæŋg(ə)lsɪ/ an island of NW Wales, separated from the mainland by the Menai Strait.

Anglican /ˈæŋglɪkən/ *adj. & n.* ● *adj.* of or relating to the Church of England or any Church in communion with it. ● *n.* a member of an Anglican Church. □ **Anglicanism** *n.* [med.L *Anglicanus* (Magna Carta) f. *Anglicus* (Bede) f. *Anglus* ANGLE]

Anglican Communion a group of Christian Churches derived from or related to the Church of England, including the Episcopal Church in the US and other national, provincial, and independent Churches. The body's senior bishop is the Archbishop of Canterbury, and its forum for discussion the Lambeth Conference.

anglice /ˈæŋglɪsɪ/ *adv.* in English. [med.L]

Anglicism /ˈæŋglɪ,sɪz(ə)m/ *n.* **1** a peculiarly English word or custom. **2** Englishness. **3** preference for what is English. [L *Anglicus* (see ANGLICAN) + -ISM]

anglicize /ˈæŋglɪ,saɪz/ *v.tr.* (also **-ise**) make English in form or character.

angling /ˈæŋglɪŋ/ *n.* fishing with hook, line, and usually rod, especially as a sport. The practice of angling is of great antiquity; an Egyptian angling scene dates from *c.*2000 BC. The first treatise in English on angling is that by Dame Juliana Barnes (1496), and perhaps the most famous is Sir Izaak Walton's *The Compleat Angler* (1653). Today it is the most popular participant sport in many countries, including Britain. Marine and coarse freshwater fish are caught with a baited hook, whereas salmon and trout are fished for with an artificial fly, usually under strictly controlled conditions.

Anglist /ˈæŋglɪst/ *n.* a student of or scholar in English language or literature. □ **Anglistics** /æŋˈglɪstɪks/ *n.* [G f. L *Anglus* English]

Anglo /ˈæŋgləʊ/ *n.* (*pl.* **-os**) **1** *N. Amer.* a person of British or northern-European origin. **2** *N. Amer.* an English-speaking Canadian. **3** *Brit.* a Scottish, Welsh, or Irish sports player who plays for an English club. [abbr. of ANGLO-SAXON]

Anglo- /ˈæŋgləʊ/ *comb. form* **1** English (*Anglo-Catholic*). **2** of English origin (*an Anglo-American*). **3** English or British and (*an Anglo-American agreement*). [f. mod.L f. L *Anglus* English]

Anglo-Catholic /ˌæŋgləʊˈkæθəlɪk, -ˈkæθlɪk/ *adj. & n.* ● *adj.* of or relating to Anglo-Catholicism. ● *n.* a member of an Anglo-Catholic Church.

Anglo-Catholicism /ˌæŋgləʊkəˈθɒlɪ,sɪz(ə)m/ *n.* a tradition within the Anglican Church which is close to Catholicism in its doctrine and worship and is broadly identified with High Church Anglicanism. As a movement, Anglo-Catholicism grew out of the Oxford Movement of the 1830s and 1840s, emphasizing its unbroken connection with the early Christian Church.

Anglocentric /ˌæŋgləʊˈsentrɪk/ *adj.* centred on or considered in terms of England.

Anglo-French /ˌæŋgləʊˈfrentʃ/ *adj. & n.* ● *adj.* **1** of or relating to both England (or Britain) and France. **2** of or relating to Anglo-French. ● *n.* (also called *Anglo-Norman*) the variety of Norman French used in England after the Norman Conquest, which remained the language of

the English nobility for several centuries and is still preserved in some legal phraseology.

Anglo-Indian /ˌæŋgləʊˈɪndɪən/ *adj. & n.* ● *adj.* **1** of or relating to England and India. **2 a** of British descent or birth but living or having lived long in India. **b** of mixed British and Indian parentage. **3** (of a word) adopted into English from an Indian language. ● *n.* an Anglo-Indian person.

Anglo-Irish /ˌæŋgləʊˈaɪərɪʃ/ *adj. & n.* ● *adj.* **1** of English descent but born or resident in Ireland. **2** of mixed English and Irish parentage. **3** of or belonging to both Britain and the Republic of Ireland. ● *n.* **1** (prec. by *the;* treated as *pl.*) Anglo-Irish people. **2** the English language as used in Ireland.

Anglo-Irish Agreement an agreement made between Britain and the Republic of Ireland in 1985, admitting the Republic to discussions on Northern Irish affairs and providing for greater cooperation between the security forces in border areas. The agreement has been strongly opposed by Unionists in the North.

Anglo-Irish Treaty an agreement signed in 1921 by representatives of the British government and the provisional Irish Republican government, by which Ireland was partitioned and the Irish Free State created.

Anglo-Latin /ˌæŋgləʊˈlætɪn/ *adj. & n.* ● *adj.* of Latin as used in medieval England. ● *n.* this form of Latin.

Anglomania /ˌæŋgləʊˈmeɪnɪə/ *n.* excessive admiration of English customs.

Anglo-Norman /ˌæŋgləʊˈnɔːmən/ *adj. & n.* ● *adj.* **1** English and Norman. **2** of the Normans in England after the Norman Conquest. **3** = ANGLO-FRENCH *adj.* ● *n.* = ANGLO-FRENCH *n.*

Anglophile /ˈæŋgləʊ,faɪl/ *n. & adj.* (also **Anglophil** /-fɪl/) ● *n.* a person who is fond of or greatly admires England or the English. ● *adj.* being or characteristic of an Anglophile.

Anglophobe /ˈæŋgləʊ,fəʊb/ *n. & adj.* ● *n.* a person who greatly hates or fears England or the English. ● *adj.* being or characteristic of an Anglophobe.

Anglophobia /ˌæŋgləʊˈfəʊbɪə/ *n.* intense hatred or fear of England or the English.

anglophone /ˈæŋgləʊ,fəʊn/ *adj. & n.* ● *adj.* English-speaking. ● *n.* an English-speaking person. [ANGLO-, after FRANCOPHONE]

Anglo-Saxon /ˌæŋgləʊˈsæks(ə)n/ *adj. & n.* ● *adj.* **1** of the English Saxons (as distinct from the Old Saxons of the continent, and from the Angles) before the Norman Conquest. **2** of the Old English people as a whole before the Norman Conquest. **3** of English descent. ● *n.* **1** an Anglo-Saxon person. **2** the Old English language. **3 a** *colloq.* plain (esp. crude) English. **b** *US* the modern English language. [mod.L *Anglo-Saxones*, med.L *Angli Saxones* after OE *Angulseaxe, -an*]

Angola /æŋˈgəʊlə/ a republic on the west coast of southern Africa; pop. (est. 1991) 10,301,000; languages, Portuguese (official), Bantu languages; capital, Luanda. Before colonization, the territory was divided between several Bantu-speaking kingdoms. Angola was a Portuguese possession from the end of the 16th century until it achieved independence in 1975 after a bitter anti-colonial war. Independence was followed by years of civil war, chiefly between the ruling Marxist MPLA and the UNITA movement, while South African raids took place periodically against Namibian resistance forces operating from Angola. □ **Angolan** *adj. & n.*

Angora /æŋˈgɔːrə/ the former name (until 1930) for ANKARA.

angora /æŋˈgɔːrə/ *n.* **1** (often *attrib.*) a long-haired variety of cat, goat, or rabbit. **2** a fabric made from the hair of the angora goat or rabbit. □ **angora wool** a mixture of sheep's wool and angora rabbit hair. [ANGORA]

Angostura /ˌæŋgəˈstjʊərə/ the former name for CIUDAD BOLÍVAR.

angostura /ˌæŋgəˈstjʊərə/ *n.* (in full **angostura bark**) an aromatic bitter bark used as a flavouring, and formerly used as a tonic and to reduce fever. [ANGOSTURA]

Angostura Bitters *n. propr.* a kind of tonic first made in Angostura.

angry /ˈæŋgrɪ/ *adj.* (**angrier, angriest**) **1** feeling or showing anger; extremely displeased or resentful. **2** (of a wound, sore, etc.) inflamed, painful. **3** suggesting or seeming to show anger (*an angry sky*). □ **angrily** *adv.* [ME, f. ANGER + -Y¹]

Angry Young Men a group of British writers of the early 1950s, notably John Osborne, Kingsley Amis, and John Wain (1925–94). Their work was marked by irreverence towards the Establishment, disgust at the survival of class distinctions and privilege, and contempt for what they saw as the narrowness and drabness of postwar life. It is epitomized by Osborne's play *Look Back in Anger* (1956), and its central character, Jimmy Porter.

angst /æŋst/ *n.* **1** anxiety. **2** a feeling of guilt or remorse. □ **angst-ridden** *adj.* [G *Angst* dread]

Ångström /ˈæŋstrəm/, Anders Jonas (1814–1874), Swedish physicist. He wrote on terrestrial magnetism and the conduction of heat, but his most important work was in spectroscopy. He proposed a relationship between the emission and absorption spectra of chemical elements, discovered hydrogen in the sun's atmosphere, and published an atlas of the solar spectrum. He measured optical wavelengths in the unit later named in his honour.

angstrom /ˈæŋstrəm/ *n.* (also **ångström**, **angstrom unit**) a unit of length equal to one hundred-millionth of a centimetre, 10^{-10} metre (symbol: Å). [ÅNGSTRÖM]

Anguilla /æŋˈgwɪlə/ the most northerly of the Leeward Islands in the West Indies; pop. (est. 1989) 7,020; languages, English (official), English Creole; capital, The Valley. Formerly a British colony, and briefly united with St Kitts and Nevis (1967), the island is now a self-governing dependency of the UK. □ **Anguillan** *adj. & n.*

anguine /ˈæŋgwɪn/ *adj.* of or resembling a snake. [L *anguinus* f. *anguis* snake]

anguish /ˈæŋgwɪʃ/ *n. & v.* ● *n.* severe misery or mental suffering. ● *v.tr.* (often as **anguished** *adj.*) cause to suffer physical or mental pain. [ME f. OF *anguisse* choking f. L *angustia* tightness f. *angustus* narrow]

anguished /ˈæŋgwɪʃt/ *adj.* suffering or expressing anguish. [past part. of *anguish* (v.) f. OF *anguissier* f. eccl.L *angustiare* to distress, formed as ANGUISH]

angular /ˈæŋgjʊlə(r)/ *adj.* **1 a** having angles or sharp corners. **b** (of a person) having sharp features; lean and bony. **c** awkward in manner. **2** forming an angle. **3** measured by angle (*angular distance*). □ **angular momentum** the quantity of rotation of a body, the product of its moment of inertia and angular velocity. **angular velocity** the rate of change of angular position of a rotating body. □ **angularly** *adv.*

angularity /ˌæŋgjʊˈlærɪtɪ/ *n.* [L *angularis* f. *angulus* ANGLE[1]]

Angus /ˈæŋgəs/ a former county of NE Scotland, known from the 16th century until 1928 as Forfarshire. It became an administrative district of Tayside region in 1975.

angwantibo /æŋˈgwæntɪˌbəʊ/ *n.* (*pl.* **-os**) a small rare primate, *Arctocebus calabarensis*, of western central Africa, related to the potto. [Efik]

anhedral /ænˈhiːdrəl/ *n. & adj. Aeron.* ● *n.* the angle between an aircraft wing and the horizontal where the wing has a downward inclination. ● *adj.* of or having an anhedral. [AN-[1] + *-hedral* (see -HEDRON)]

anhinga /ænˈhɪŋgə/ *n.* US a darter, esp. *Anhinga anhinga* of America. Also called *snake bird*. [Port. f. Tupi *áyinga*]

Anhui /ænˈhweɪ/ (also **Anhwei**) a province in eastern China; capital, Hefei.

anhydride /ænˈhaɪdraɪd/ *n. Chem.* a compound obtained by removing the elements of water from another compound, esp. from an acid. [as ANHYDROUS + -IDE]

anhydrite /ænˈhaɪdraɪt/ *n.* a naturally occurring usu. rock-forming anhydrous mineral form of calcium sulphate. [as ANHYDROUS + -ITE[1] 2]

anhydrous /ænˈhaɪdrəs/ *adj. Chem.* without water, esp. water of crystallization. [Gk *anudros* (as AN-[1], *hudōr* water)]

ani /ˈɑːnɪ/ *n.* a glossy black large-billed bird of the genus *Crotophaga*, of the cuckoo family, of Central and South America. [Sp. *aní* f. Tupi *anú*]

aniline /ˈænɪˌliːn, -lɪn, -ˌlaɪn/ *n.* a colourless oily liquid (chem. formula: $C_6H_5NH_2$), used in the manufacture of dyes, drugs, and plastics. □ **aniline dye** a synthetic dye, esp. any of those made from aniline. [G *Anilin* f. *Anil* indigo (from which it was orig. obtained), ult. f. Arab. *an-nīl*]

anima /ˈænɪmə/ *n. Psychol.* **1** the inner personality (opp. PERSONA 1). **2** Jung's term for the feminine part of a man's personality (opp. ANIMUS 4). [L, = mind, soul]

animadvert /ˌænɪmædˈvɜːt/ *v.intr.* (foll. by *on*) criticize, censure (*conduct, a fault, etc.*). □ **animadversion** /-ˈvɜːʃ(ə)n/ *n.* [L *animadvertere* f. *animus* mind + *advertere* (as AD-, *vertere* *vers-* turn)]

animal /ˈænɪm(ə)l/ *n. & adj.* ● *n.* **1** a living organism which feeds on organic matter, usu. one with specialized sense-organs and a nervous system, and able to respond rapidly to stimuli. (*See note below.*) **2 a** such an organism other than a human. **b** a quadruped. **3** a brutish or uncivilized person. **4** *colloq.* a person or thing of any kind (*there is no such animal*). ● *adj.* **1** characteristic of animals. **2** of animals as distinct from vegetables (*animal charcoal*). **3** characteristic of the physical needs of animals; carnal, sensual. □ **animal husbandry** the science of breeding and caring for farm animals. **animal liberation** (a movement supporting) the liberation of animals from exploitation by humans. **animal magnetism 1** *hist.* mesmerism. **2** sexual attraction. **animal rights** (a movement upholding) the natural right of animals to live free from human exploitation. **animal spirits** natural exuberance. [L f. *animale* neut. of *animalis* having breath f. *anima* breath]

▪ Animals are generally distinguished from plants by being unable to synthesize organic molecules from inorganic ones, so that they have to feed on plants or on other animals. The animal kingdom (Animalia) was formerly held to include the unicellular protozoans, but these are now usually assigned to the kingdom Protista (see PROTIST). The great majority of animals are invertebrates, of which there are some thirty phyla; the vertebrates constitute but a single subphylum.

animalcule /ˌænɪˈmælkjuːl/ *n. archaic* a microscopic animal. □ **animalcular** *adj.* [mod.L *animalculum* (as ANIMAL, -CULE)]

animalism /ˈænɪməˌlɪz(ə)m/ *n.* **1** the nature and activity of animals. **2** the belief that humans are not superior to other animals. **3** concern with physical matters; sensuality.

animality /ˌænɪˈmælɪtɪ/ *n.* **1** the animal world. **2** the nature or behaviour of animals. [F *animalité* f. *animal* (adj.)]

animalize /ˈænɪməˌlaɪz/ *v.tr.* (also **-ise**) **1** make (a person) bestial; sensualize. **2** convert to animal substance. □ **animalization** /ˌænɪməlaɪˈzeɪʃ(ə)n/ *n.*

animate *adj. & v.* ● *adj.* /ˈænɪmət/ **1** having life. **2** lively. ● *v.tr.* /ˈænɪˌmeɪt/ **1** enliven, make lively. **2** give life to. **3** inspire, actuate, encourage. **4** (esp. as **animated** *adj.*) *Cinematog.* give (a film, cartoon figure, etc.) the appearance of movement using animation techniques. □ **animator** /-ˌmeɪtə(r)/ *n.* (esp. in sense 4 of v.). [L *animatus* past part. of *animare* give life to f. *anima* life, soul]

animated /ˈænɪˌmeɪtɪd/ *adj.* **1** lively, vigorous. **2** having life. □ **animatedly** *adv.*

animation /ˌænɪˈmeɪʃ(ə)n/ *n.* **1** vivacity, ardour. **2** the state of being alive (*archaic exc. in suspended animation*). **3** *Cinematog.* the technique of filming successive drawings or positions of models to create an illusion of movement when the film is shown as a sequence.

animatron /ˈænɪməˌtrɒn/ *n.* = AUDIO-ANIMATRON.

animatronics /ˌænɪməˈtrɒnɪks/ *n.pl.* (usu. treated as *sing.*) = AUDIO-ANIMATRONICS. □ **animatronic** *n. & adj.*

animé /ˈænɪˌmeɪ/ *n.* a resin, esp. a West Indian resin used in making varnish. [F, of uncert. orig.]

animism /ˈænɪˌmɪz(ə)m/ *n.* **1** the attribution of a living soul to plants, inanimate objects, and natural phenomena. **2** the belief in a supernatural power that organizes and animates the material universe. □ **animist** *n.* **animistic** /ˌænɪˈmɪstɪk/ *adj.* [L *anima* life, soul + -ISM]

animosity /ˌænɪˈmɒsɪtɪ/ *n.* (*pl.* **-ies**) a spirit or feeling of strong hostility. [ME f. OF *animosité* or LL *animositas* f. *animosus* spirited, formed as ANIMUS]

animus /ˈænɪməs/ *n.* **1** a display of animosity. **2** ill feeling. **3** a motivating spirit or feeling. **4** *Psychol.* Jung's term for the masculine part of a woman's personality (opp. ANIMA). [L, = spirit, mind]

anion /ˈænˌaɪən/ *n. Chem.* a negatively charged ion; an ion that is attracted to the anode in electrolysis (opp. CATION). [ANA- + ION]

anionic /ˌænaɪˈɒnɪk/ *adj. Chem.* **1** of an anion or anions. **2** having an active anion.

anise /ˈænɪs/ *n.* an umbelliferous plant, *Pimpinella anisum*, having aromatic seeds (see ANISEED). [ME f. OF *anis* f. L f. Gk *anison* anise, dill]

aniseed /ˈænɪˌsiːd/ *n.* the seed of the anise, used to flavour liqueurs and sweets. [ME f. ANISE + SEED]

anisette /ˌænɪˈzet/ *n.* a liqueur flavoured with aniseed. [F, dimin. of *anis* ANISE]

anisotropic /ˌænaɪsəˈtrɒpɪk/ *adj.* having physical properties that are different in different directions, e.g. the strength of wood along the grain differing from that across the grain (opp. ISOTROPIC). □ **anisotropically** *adv.* **anisotropy** /ˌænaɪˈsɒtrəpɪ/ *n.* [AN-[1] + ISOTROPIC]

Anjou /ˈɒnʒuː/ a former province of western France, on the Loire. It was an English possession from 1154, when it was inherited by Henry

II as count of Anjou, until 1204, when it was lost to France by King John.

Ankara /'æŋkərə / the capital of Turkey since 1923; pop. (1990) 2,559,470. Prominent in Roman times as Ancyra, the capital of Galatia, it later declined in importance until chosen by Kemal Atatürk in 1923 as his seat of government. It was known until 1930 as Angora.

ankh /æŋk/ n. a device consisting of a looped bar with a shorter crossbar, used in ancient Egypt as a symbol of life. [Egypt., = life, soul]

ankle /'æŋk(ə)l/ n. & v. ● n. **1** the joint connecting the foot with the leg. **2** the part of the leg between this and the calf. ● v.intr. sl. walk. □ **ankle-biter** Austral. colloq. a child. **ankle-bone** a bone forming the ankle. **ankle sock** a short sock just covering the ankle. [ME f. ON ankul- (unrecorded) f. Gmc: rel. to ANGLE[1]]

anklet /'æŋklɪt/ n. an ornament or fetter worn round the ankle. [ANKLE + -LET, after BRACELET]

ankylosaur /'æŋkɪlə,sɔ:(r)/ n. a short-legged heavily armoured dinosaur of the suborder Ankylosauria, of the Jurassic and Cretaceous periods. [mod.L f. ANKYLOSIS + Gk sauros lizard]

ankylose /'æŋkɪ,ləuz, -,ləus/ v.tr. & intr. (also **anchylose**) Med. (of bones or a joint) stiffen or unite by ankylosis. [back-form. f. ANKYLOSIS after anastomose etc.]

ankylosis /,æŋkɪ'ləusɪs/ n. (also **anchylosis**) Med. **1** the abnormal stiffening and immobility of a joint by fusion of the bones. **2** such fusion. □ **ankylotic**/-'lɒtɪk/ adj. [mod.L f. Gk agkulōsis f. agkuloō crook]

anna /'ænə/ n. a former monetary unit of India and Pakistan, one-sixteenth of a rupee. [Hind. ānā]

Annaba /'ænəbə/ (formerly called Bône) a port of NE Algeria; pop. (est. 1989) 348,000. The modern town is adjacent to the site of Hippo Regius, a prominent city in Roman Africa and the home and bishopric of St Augustine of Hippo from 396 to 430.

An Najaf see NAJAF.

annal /'æn(ə)l/ n. **1** the annals of one year. **2** a record of one item in a chronicle. [back-form. f. ANNALS]

annalist /'ænəlɪst/ n. a writer of annals. □ **annalistic** /,ænə'lɪstɪk/ adj. **annalistically** adv.

annals /'æn(ə)lz/ n.pl. **1** a narrative of events year by year. **2** historical records. [F annales or L annales (libri) yearly (books) f. annus year]

Annapolis /ə'næpəlɪs/ the state capital of Maryland, on Chesapeake Bay; pop. (1990) 33,190. It is the home of the US Naval Academy.

Annapurna /,ænə'pɜ:nə/ a ridge of the Himalayas, in north central Nepal. Its highest peak rises to 8,078 m (26,503 ft).

annates /'æneɪts/ n.pl. RC Ch. the first year's revenue of a see or benefice, paid to the pope. [F annate f. med.L annata year's proceeds f. annus year]

annatto /ə'nætəu/ n. (also **anatta** /-tə/, **anatto**) an orange-red dye from the pulp of a tropical fruit, used for colouring foods. [Carib name of the fruit-tree]

Anne /æn/ (1665–1714), queen of England and Scotland (known as Great Britain from 1707) and Ireland, 1702–14. The last of the Stuart monarchs, daughter of the Catholic James II (but herself a Protestant), she succeeded her brother-in-law William III to the throne, there presiding over the Act of Union, which completed the unification of Scotland and England. None of her five children born alive survived childhood, and by the Act of Settlement (1701) the throne passed to the House of Hanover on her death.

Anne, Princess, Anne Elizabeth Alice Louise, the Princess Royal (b.1950), daughter of Elizabeth II. A skilled horsewoman, she rode for Great Britain in the 1976 Olympics. She has also been involved in the work of charitable organizations, notably in her capacity as president of Save the Children Fund since 1971. She was married to Captain Mark Philips (b.1948) 1973–92. In Dec. 1992 she married Commander Timothy Laurence. She has two children, Peter Mark Andrew Philips (b.1977) and Zara Anne Elizabeth Philips (b.1981).

Anne, St, traditionally the mother of the Virgin Mary, first mentioned by name in the apocryphal gospel of James (2nd century). The extreme veneration of St Anne in the late Middle Ages was attacked by Luther and other reformers. She is the patron saint of Brittany and the province of Quebec in Canada. Feast day, 26 July.

anneal /ə'ni:l/ v. & n. ● v. **1** tr. heat (metal or glass) and allow it to cool slowly, esp. to toughen it. **2** tr. toughen. **3** Biochem. **a** tr. recombine (DNA) in the double-stranded form. **b** intr. undergo this process. ● n. treatment by annealing. □ **annealer** n. [OE onǣlan f. on + ǣlan burn, bake f. āl fire]

Anne Boleyn see BOLEYN.

annelid /'ænəlɪd/ n. Zool. a segmented worm of the phylum Annelida, which includes earthworms, lugworms, etc. □ **annelidan** /ə'nelɪd(ə)n/ adj. & n. [F annélide or mod.L annelida (pl.) f. F annelés ringed animals f. OF anel ring f. L anellus dimin. of anulus ring]

Anne of Cleves /kli:vz/ (1515–57), fourth wife of Henry VIII. Henry's marriage to her (1540) was the product of his minister Thomas Cromwell's attempt to forge a dynastic alliance with one of the Protestant German states. Henry, initially deceived by a flattering portrait of Anne painted by Holbein, took an instant dislike to his new wife and dissolved the marriage after six months.

annex /æ'neks/ v.tr. **1 a** add as a subordinate part. **b** (often foll. by to) append to a book etc. **2** incorporate (territory of another) into one's own. **3** add as a condition or consequence. **4** colloq. take without right. □ **annexation** /,ænek'seɪʃ(ə)n/ n. [ME f. OF annexer f. L annectere (as AN-[2], nectere nex- bind)]

annexe /'æneks/ n. (also **annex**) **1** a separate or added building, esp. for extra accommodation. **2** an addition to a document. [F annexe f. L annexum past part. of annectere bind: see ANNEX]

Annigoni /,ænɪ'gəuni/, Pietro (1910–88), Italian painter. One of the few 20th-century artists to practise the techniques of the Old Masters, he painted mainly in tempera, and his religious paintings include altarpieces and frescos. However, he is most famous for his portraits of Queen Elizabeth II (1955, 1970), President Kennedy (1961), and other prominent figures.

annihilate /ə'naɪə,leɪt/ v.tr. **1** completely destroy. **2** defeat utterly; make insignificant or powerless. □ **annihilator** n. [LL annihilare (as AN-[2], nihil nothing)]

annihilation /ə,naɪə'leɪʃ(ə)n/ n. **1** the act or process of annihilating. **2** Physics the conversion of a particle and an antiparticle into radiation. [F annihilation or LL annihilatio (as ANNIHILATE)]

anniversary /,ænɪ'vɜ:səri/ n. (pl. **-ies**) **1** the date on which an event took place in a previous year. **2** the celebration of this. [ME f. L anniversarius f. annus year + versus turned]

Annobón /,ænə,bɒn/ an island of Equatorial Guinea, in the Gulf of Guinea. It was known as Pagalu between 1973 and 1979.

Anno Domini /,ænəu 'dɒmɪ,naɪ/ adv. & n. ● adv. in the year of our Lord, in the year of the Christian era. ● n. colloq. advancing age (suffering from Anno Domini). [L, = in the year of the Lord]

annotate /'ænə,teɪt/ v.tr. add explanatory notes to (a book, document, etc.). □ **annotatable** adj. **annotative** adj. **annotator** n. **annotation** /,ænə'teɪʃ(ə)n/ n. [L annotare (as AD-, nota mark)]

announce /ə'nauns/ v.tr. **1** (often foll. by that) make publicly known. **2** make known the arrival or imminence of (a guest, dinner, etc.). **3** make known (without words) to the senses or the mind; be a sign of. [ME f. OF annoncer f. L annuntiare (as AD-, nuntius messenger)]

announcement /ə'naunsmənt/ n. **1** the action of announcing; something announced. **2** an official communication or statement. **3** (esp. US) an advertisement or other piece of promotional material.

announcer /ə'naunsə(r)/ n. a person who announces, esp. introducing programmes in broadcasting.

annoy /ə'nɔɪ/ v.tr. **1** cause slight anger or mental distress to. **2** (in passive) be somewhat angry (am annoyed with you; was annoyed at my remarks). **3** molest; harass repeatedly. □ **annoyer** n. **annoying** adj. [ME f. OF anuier, anui, anoi, etc., ult. f. L in odio hateful]

annoyance /ə'nɔɪəns/ n. **1** the action of annoying or the state of being annoyed; irritation, vexation. **2** something that annoys, a nuisance.

annual /'ænjuəl/ adj. & n. ● adj. **1** reckoned by the year. **2** occurring every year. **3** living or lasting for one year. ● n. **1** a book etc. published once a year; a yearbook. **2** a plant that lives only for a year or less (cf. PERENNIAL n.). □ **annual general meeting** a yearly meeting of members or shareholders, esp. for holding elections and reporting on the year's events. **annual ring** a ring in the cross-section of a plant, esp. a tree, produced by one year's growth. □ **annually** adv. [ME f. OF annuel f. LL annualis f. L annalis f. annus year]

annualized /'ænjuə,laɪzd/ adj. (of rates of interest, inflation, etc.) calculated on an annual basis, as a projection from figures obtained for a shorter period.

annuitant /ə'nju:ɪt(ə)nt/ n. a person who holds or receives an annuity. [ANNUITY + -ANT, by assim. to accountant etc.]

annuity /ə'nju:ɪti/ n. (pl. **-ies**) **1** a yearly grant or allowance. **2** an investment of money entitling the investor to a series of equal annual sums over a stated period. **3** a sum payable in respect of a particular year. [ME f. F *annuité* f. med.L *annuitas -tatis* f. L *annuus* yearly (as ANNUAL)]

annul /əˈnʌl/ v.tr. (**annulled, annulling**) **1** declare (a marriage etc.) invalid. **2** cancel, abolish. □ **annulment** n. [ME f. OF *anuller* f, LL *annullare* (as AD-, *nullus* none)]

annular /ˈænjʊlə(r)/ adj. ring-shaped; forming a ring. □ **annular eclipse** an eclipse of the sun in which the edge of the sun remains visible as a bright ring around the moon. □ **annularly** adv. [F *annulaire* or L *annularis* f. *an(n)ulus* ring]

annulate /ˈænjʊlət/ adj. having rings; marked with or formed of rings. □ **annulation** /ˌænjʊˈleɪʃ(ə)n/ n. [L *annulatus* (as ANNULUS)]

annulet /ˈænjʊlɪt/ n. **1** Archit. a small fillet or band encircling a column. **2** a small ring. [L *annulus* ring + -ET[1]]

annuli /ˈænjʊləs/ n. (pl. **annuli** /-laɪ/) esp. Math. & Biol. a ring. [L *an(n)ulus*]

annunciate /əˈnʌnsɪˌeɪt/ v.tr. **1** proclaim. **2** indicate as coming or ready. [LL *annunciare* f. L *annuntiare annuntiat-* announce]

annunciation /əˌnʌnsɪˈeɪʃ(ə)n/ n. **1** (**Annunciation**) **a** the announcing of the Incarnation, made by the angel Gabriel to Mary, related in Luke 1:26–38. **b** the festival commemorating this (Lady Day) on 25 Mar. **2 a** the act or process of announcing. **b** an announcement. [ME f. OF *annonciation* f. LL *annuntiatio -onis* (as ANNUNCIATE)]

annunciator /əˈnʌnsɪˌeɪtə(r)/ n. **1** a device giving an audible or visible indication of which of several electrical circuits has been activated, of the position of a train, etc. **2** an announcer. [LL *annuntiator* (as ANNUNCIATE)]

annus mirabilis /ˌænəs mɪˈrɑːbɪlɪs/ n. (pl. ***anni mirabiles*** /ˌænaɪ mɪˈrɑːbɪˌliːz/) a remarkable or auspicious year. [mod.L, = wonderful year]

anoa /əˈnəʊə/ n. a small wild buffalo of the genus *Bubalus*, native to Sulawesi. [name in Sulawesi]

anode /ˈænəʊd/ n. (opp. CATHODE) Electr. **1** a positively charged electrode. **2** the terminal by which electric current enters a device. □ **anodal** /əˈnəʊd(ə)l/ adj. **anodic** /əˈnɒdɪk/ adj. [Gk *anodos* way up f. *ana* up + *hodos* way]

anodize /ˈænəˌdaɪz/ v.tr. (also **-ise**) coat (a metal, esp. aluminium) with a protective oxide layer by electrolysis. □ **anodizer** n. [ANODE + -IZE]

anodyne /ˈænəˌdaɪn/ adj. & n. ● adj. **1** able to relieve pain. **2** mentally soothing. ● n. an anodyne drug or medicine. [L *anodynus* f. Gk *anōdunos* painless (as AN-[1], *odunē* pain)]

anoesis /ˌænəʊˈiːsɪs/ n. Psychol. consciousness with sensation but without thought. □ **anoetic** /-ˈetɪk/ adj. [A-[1] + Gk *noēsis* understanding]

anoint /əˈnɔɪnt/ v.tr. **1** apply oil or ointment to, esp. as a religious ceremony (e.g. at baptism, or the consecration of a priest or king, or in ministering to the sick). **2** (usu. foll. by *with*) smear, rub. □ **anointer** n. [ME f. AF *anoint* (adj.) f. OF *enoint* past part. of *enoindre* f. L *inungere* (as IN-[2], *ungere unct-* smear with oil)]

Anointing of the Sick n. esp. RC Ch. the sacramental anointing of the ill or infirm with blessed oil; unction.

anole /əˈnəʊlɪ/ n. a lizard of the American genus *Anolis*, esp. the green *A. carolinensis*. See also CHAMELEON 1b. [Carib]

anomalistic /əˌnɒməˈlɪstɪk/ adj. Astron. of the anomaly of a planet etc. (ANOMALY 3). □ **anomalistic month** a month measured between successive perigees of the moon. **anomalistic year** a year measured between successive perihelia of the earth.

anomalous /əˈnɒmələs/ adj. having an irregular or deviant feature; abnormal. □ **anomalously** adv. **anomalousness** n. [LL *anomalus* f. Gk *anōmalos* (as AN-[1], *homalos* even)]

anomalure /əˈnɒməˌljʊə(r)/ n. a squirrel-like rodent of the family Anomaluridae, having a tail with rough overlapping scales on the underside. [mod.L *anomalurus* f. Gk *anōmalos* ANOMALOUS + *oura* tail]

anomaly /əˈnɒməlɪ/ n. (pl. **-ies**) **1** an anomalous circumstance or thing; an irregularity. **2** irregularity of motion, behaviour, etc. **3** Astron. the angular distance of a planet or satellite from its last perihelion or perigee. [L f. Gk *anōmalia* f. *anōmalos* ANOMALOUS]

anomie /ˈænəmɪ/ n. (also **anomy**) lack of the usual social or ethical standards in an individual or group. □ **anomic** /əˈnɒmɪk/ adj. [Gk *anomia* f. *anomos* lawless: *-ie* f. F]

anon /əˈnɒn/ adv. archaic or literary soon, shortly (*will say more of this anon*). [OE *on ān* into one, *on āne* in one]

anon. /əˈnɒn/ abbr. anonymous; an anonymous author.

anonym /ˈænənɪm/ n. **1** an anonymous person or publication. **2** a pseudonym. [F *anonyme* f. Gk *anōnumos*: see ANONYMOUS]

anonymous /əˈnɒnɪməs/ adj. **1** of unknown name. **2** of unknown or undeclared source or authorship. **3** without character; featureless, impersonal. □ **anonymously** adv. **anonymity** /ˌænəˈnɪmɪtɪ/ n. [LL *anonymus* f. Gk *anōnumos* nameless (as AN-[1], *onoma* name)]

anopheles /əˈnɒfɪˌliːz/ n. a mosquito of the genus *Anopheles*, which includes many species that are carriers of the malarial parasite. [mod.L f. Gk *anōphelēs* unprofitable]

anorak /ˈænəˌræk/ n. **1** a waterproof jacket of cloth or plastic, usu. with a hood, of a kind orig. used in polar regions. **2** sl. a boring, studious, or socially inept young person with unfashionable and solitary interests (caricatured as typically wearing an anorak). [Greenland Eskimo *anoraq*]

anorectic var. of ANOREXIC.

anorexia /ˌænəˈreksɪə/ n. **1** a lack or loss of appetite for food. **2** (in full **anorexia nervosa** /nɜːˈvəʊsə/) a psychological illness, esp. in young women, characterized by an obsessive desire to lose weight by refusing to eat. [LL f. Gk f. AN-[1] + *orexis* appetite]

anorexic /ˌænəˈreksɪk/ adj. & n. (also **anorectic** /-ˈrektɪk/) ● adj. **1** involving, producing, or characterized by a lack of appetite, esp. in anorexia nervosa. **2** colloq. extremely thin. ● n. **1** an anorexic agent. **2** a person with anorexia. [F *anoréxique*; *anorectic* f. Gk *anorektos* without appetite (as ANOREXIA)]

anorthosite /æˈnɔːθəˌsaɪt/ n. Geol. a granular igneous rock composed largely of plagioclase. [F *anorthose* plagioclase + -ITE[1]]

anosmia /æˈnɒzmɪə/ n. Med. the loss of the sense of smell. □ **anosmic** adj. [LL f. Gk f. an- not + *osmē* smell]

another /əˈnʌðə(r)/ adj. & pron. ● adj. **1** an additional; one more (*have another cake; after another six months*). **2** a person like or comparable to (*another Callas*). **3** a different (*quite another matter*). **4** some or any other (*will not do another man's work*). ● pron. **1** an additional one (*have another*). **2** a different one (*take this book away and bring me another*). **3** some or any other one (*I love another*). **4** Brit. an unnamed additional party to a legal action (*X versus Y and another*). **5** (also **A. N. Other** /ˌeɪ en ˈʌðə(r)/) a player unnamed or not yet selected. □ **another place** Brit. the other House of Parliament (used in the Commons to refer to the Lords, and vice versa). **such another** another of the same sort. [ME f. AN + OTHER]

anothery /əˈnʌðərɪ/ n. (also **anotherie**) Austral. colloq. another one (*I'll have anothery*).

Anouilh /ˈɒnwiː, French anuj/, Jean (1910–87), French dramatist. His first success was *Traveller without Luggage* (1937), and he soon achieved widespread popularity. His works include romantic comedies such as *Ring Round the Moon* (1947), fantasies (*Thieves' Carnival*, 1932), historical dramas (*Beckett or The Honour of God*, 1959), and *Antigone* (1944), a reworking of the Greek myth with undertones of the contemporary situation in Nazi-occupied Paris.

anova /ˈænəˌvɑː/ n. Statistics analysis of variance, a statistical method in which the variation in a set of observations is divided into distinct components. [acronym]

anovulant /æˈnɒvjʊlənt/ n. & adj. Pharm. ● n. a drug preventing ovulation. ● adj. preventing ovulation. [AN-[1] + *ovulation* (see OVULATE) + -ANT]

anoxia /əˈnɒksɪə/ n. Med. an absence or deficiency of oxygen reaching the tissues; severe hypoxia. □ **anoxic** adj. [f. AN-[1] + OXYGEN + -IA[1]]

Anschluss /ˈænʃlʊs/ the annexation of Austria by Germany in 1938, after Hitler had forced the resignation of the Austrian Chancellor by demanding that he admit Nazis into his Cabinet. The new Chancellor, a pro-Nazi, invited German troops to enter the country on the pretext of restoring law and order. Any form of union between Austria and Germany had been expressly forbidden under the terms of the Treaty of Versailles (1919); the ban was reiterated when the Allied Powers recognized the second Austrian republic in 1946. [G f. *anschliessen* join]

Anselm, St /ˈænselm/ (c.1033–1109), Italian-born philosopher and theologian, Archbishop of Canterbury 1093–1109. A distinguished theologian and reformer who worked to free the Church from secular control, he preferred to defend the faith by intellectual reasoning rather than by basing arguments on scriptural and other written authorities. The most famous of his writings is a mystical study on the Atonement (*Cur Deus Homo?*). Feast day, 21 Apr.

anserine /ˈænsəˌraɪn/ adj. **1** of or like a goose. **2** silly. [L *anserinus* f. *anser* goose]

Anshan /ænˈʃæn/ a city in Liaoning, China; pop. (1990) 1,370,000.

Anshan is situated close to major iron-ore deposits and China's largest iron and steel complex is nearby.

answer /'ɑːnsə(r)/ n. & v. ● n. **1** something said or done to deal with or in reaction to a question, statement, or circumstance. **2** the solution to a problem. ● v. **1** tr. make an answer to (*answer me; answer my question*). **2** intr. (often foll. by *to*) make an answer. **3** tr. respond to the summons or signal of (*answer the door; answer the telephone*). **4** tr. be satisfactory for (a purpose or need). **5** intr. **a** (foll. by *for, to*) be responsible (*you will answer to me for your conduct*). **b** (foll. by *for*) vouch (for a person, conduct, etc.). **6** intr. (foll. by *to*) correspond, esp. to a description. **7** intr. be satisfactory or successful. □ **answer back** answer a rebuke etc. impudently. **answering machine** a tape recorder which supplies a recorded answer to a telephone call. **answering service** a business that receives and answers telephone calls for its clients. **answer to the name of** be called. [OE *andswaru, andswarian* f. Gmc, = swear against (charge)]

answerable /'ɑːnsərəb(ə)l/ adj. **1** (usu. foll. by *to, for*) responsible (*answerable to them for any accident*). **2** that can be answered. □ **answerability** /,ɑːnsərə'bɪlɪtɪ/ n.

answerphone /'ɑːnsə,fəʊn/ n. a telephone answering machine.

ant /ænt/ n. & v. ● n. a small hymenopterous insect of the family Formicidae, usu. living in complex social colonies, wingless (except for adults at the time of mating), and proverbial for industry. ● v. intr. & refl. (usu. as **anting** n.) (of a bird) place or rub ants on the feathers to repel parasites. □ **ant-bear** = AARDVARK. **ant** (or **ant's**) **eggs** pupae of ants. **ant-lion** a dragonfly-like neuropterous insect, the predatory larva of which forms a conical pit to trap ants etc. **have ants in one's pants** colloq. be fidgety, be restless. **white ant** = TERMITE. [OE *æmet(t)e, ēmete* (see EMMET) f. WG]

ant- /ænt/ assim. form of ANTI- before a vowel or *h* (*Antarctic*).

-ant /ənt/ suffix **1** forming adjectives denoting attribution of an action (*pendant; repentant*) or state (*arrogant; expectant*). **2** forming nouns denoting an agent (*assistant; celebrant; deodorant*). [F -*ant* or L -*ant-, -ent-*, pres. part. stem of verbs: cf. -ENT]

antacid /ænt'æsɪd/ n. & adj. ● n. a substance that prevents or corrects acidity esp. in the stomach. ● adj. having these properties.

Antaeus /æn'tiːəs/ Gk Mythol. a giant, son of Poseidon and Earth, living in Libya. He compelled all comers to wrestle with him and overcame and killed them until he was defeated by Hercules.

antagonism /æn'tægə,nɪz(ə)m/ n. active opposition or hostility. [F *antagonisme* (as ANTAGONIST)]

antagonist /æn'tægənɪst/ n. **1** an opponent or adversary. **2** Biol. a substance or organ that partially or completely opposes the action of another. □ **antagonistic** /-,tægə'nɪstɪk/ adj. **antagonistically** adv. [F *antagoniste* or LL *antagonista* f. Gk *antagōnistēs* (as ANTAGONIZE)]

antagonize /æn'tægə,naɪz/ v. tr. (also **-ise**) **1** evoke hostility or opposition or enmity in. **2** (of one force etc.) counteract or tend to neutralize (another). □ **antagonization** /-,tægənaɪ'zeɪʃ(ə)n/ n. [Gk *antagōnizomai* (as ANTI-, *agōnizomai* f. *agōn* contest)]

Antakya see ANTIOCH.

Antall /'æntæl/, Jozsef (1933–93), Hungarian statesman, Prime Minister 1990–3. In 1990 he became leader of the Hungarian Democratic Forum and was elected Premier in the country's first free elections for over forty years.

Antalya /æn'tæljə/ a port in southern Turkey; pop. (1990) 378,200.

Antananarivo /,æntə,nænə'riːvəʊ/ the capital of Madagascar, situated in the central plateau; pop. (1990) 802,390. Until 1975 the city was known as Tananarive.

Antarctic /ænt'ɑːktɪk/ adj. & n. (prec. by *the*) of, relating to, or denoting the south polar region or Antarctica. [ME f. OF *antartique* or L *antarcticus* f. Gk *antarktikos* (as ANTI-, *arktikos* ARCTIC)]

Antarctica /ænt'ɑːktɪkə/ a continent round the South Pole, situated mainly within the Antarctic Circle. Antarctica is almost entirely covered by ice sheets; it holds 90 per cent of the world's ice, representing 70 per cent of the world's fresh water. The coastline was explored in the 19th century; in 1911 Roald Amundsen was the first to reach the Pole, and Robert F. Scott reached it a month later. Although there is no permanent human population, Norway, Australia, France, New Zealand, and the UK claim sectors of the continent (Argentina and Chile claim parts of the British sector); its exploitation is governed by an international treaty of 1959, renewed in 1991. (See also POLE².)

Antarctic Circle the parallel of latitude 66° 33′ south of the equator. It marks the southernmost point at which the sun is visible on the southern winter solstice and the northernmost point at which the midnight sun can be seen on the southern summer solstice.

Antarctic Ocean (also called *Southern Ocean*) the sea surrounding Antarctica, consisting of parts of the South Atlantic, the South Pacific, and the southern Indian Ocean.

Antarctic Peninsula a mountainous peninsula of Antarctica between the Bellingshausen Sea and the Weddell Sea, extending northwards towards Cape Horn and the Falkland Islands.

Antares /æn'teəriːz/ Astron. the brightest star in the constellation of Scorpius, a binary star of which the main component is a red supergiant. [Gk, = simulating Mars (in colour)]

ante /'æntɪ/ n. & v. ● n. **1** a stake put up by a player in poker etc. before receiving cards. **2** (esp. in phr. **up the ante**) an amount to be paid in advance; a stake. ● v. tr. (**antes, anted**) **1** put up as an ante. **2** US **a** bet, stake. **b** (foll. by *up*) pay. [L = before]

ante- /'æntɪ/ prefix forming nouns and adjectives meaning 'before, preceding' (*ante-room; antenatal; ante-post*). [L *ante* (prep. & adv.), = before]

anteater /'ænt,iːtə(r)/ n. a mammal that feeds on ants and termites; esp. one of the edentate family Myrmecophagidae, with a long snout and long sticky tongue.

ante-bellum /,æntɪ'beləm/ adj. occurring or existing before a particular war, esp. the US Civil War. [L f. *ante* before + *bellum* war]

antecedent /,æntɪ'siːd(ə)nt/ n. & adj. ● n. **1** a preceding thing or circumstance. **2** Gram. a word, phrase, clause, or sentence to which another word (esp. a relative pronoun, usu. following) refers. **3** (in pl.) past history, esp. of a person. **4** Logic the statement contained in the 'if' clause of a conditional proposition. ● adj. **1** (often foll. by *to*) previous. **2** presumptive, a priori. □ **antecedence** n. **antecedently** adv. [ME f. F *antécédent* or L *antecedere* (as ANTE-, *cedere* go)]

antechamber /'æntɪ,tʃeɪmbə(r)/ n. a small room leading to a main one. [earlier *anti-,* f. F *antichambre* f. It. *anticamera* (as ANTE-, CHAMBER)]

antechapel /'æntɪ,tʃæp(ə)l/ n. the outer part at the west end of a college chapel.

antedate v. & n. ● v. tr. /,æntɪ'deɪt/ **1** exist or occur at a date earlier than. **2** assign an earlier date to (a document, event, etc.), esp. one earlier than its actual date. ● n. /'æntɪ,deɪt/ a date earlier than the actual one.

antediluvian /,æntɪdɪ'luːvɪən, -'ljuːvɪən/ adj. **1** of or belonging to the time before the biblical Flood. **2** colloq. very old or out of date. [ANTE- + L *diluvium* DELUGE + -AN]

antelope /'æntɪ,ləʊp/ n. (pl. same or **antelopes**) **1** a deerlike ruminant of the family Bovidae. Antelopes, which are mainly found in Africa, are typically slender, graceful, and swift-moving, with smooth hair and upward-pointing horns, e.g. gazelles, gnus, kudus, and impala. **2** leather made from the skin of any of these. [ME f. OF *antelop* or f. med.L *ant(h)alopus* f. late Gk *antholops*, of unkn. orig.]

antenatal /,æntɪ'neɪt(ə)l/ adj. **1** existing or occurring before birth. **2** relating to the period of pregnancy.

antenna /æn'tenə/ n. (pl. **antennae** /-niː/) **1** Zool. each of a pair of mobile appendages on the heads of insects, crustaceans, etc., sensitive to touch and taste; a feeler. **2** (pl. **antennas**) = AERIAL n. □ **antennal** adj. (in sense 1). **antennary** adj. (in sense 1). [L, = sail-yard]

antenuptial /,æntɪ'nʌpʃ(ə)l/ adj. existing or occurring before marriage. □ **antenuptial contract** S. Afr. a contract between two persons intending to marry each other, setting out the terms and conditions of their marriage. [LL *antenuptialis* (as ANTE-, NUPTIAL)]

antependium /,æntɪ'pendɪəm/ n. (pl. **antependia** /-dɪə/) a veil or hanging for the front of an altar. [med.L (as ANTE-, *pendere* hang)]

antepenult /,æntɪpɪ'nʌlt/ n. the last syllable but two in a word. [abbr. of LL *antepaenultimus* (as ANTE-, *paenultimus* PENULT)]

antepenultimate /,æntɪpɪ'nʌltɪmət/ adj. & n. ● adj. last but two. ● n. anything that is last but two.

ante-post /,æntɪ'pəʊst/ adj. Brit. (of betting) done at odds determined at the time of betting, in advance of the event concerned. [ANTE- + POST¹]

anterior /æn'tɪərɪə(r)/ adj. **1** nearer the front. **2** (often foll. by *to*) earlier, prior. □ **anteriorly** adv. **anteriority** /-,tɪərɪ'ɒrɪtɪ/ n. [F *antérieur* or L *anterior* f. *ante* before]

ante-room /'æntɪ,ruːm, -,rʊm/ n. **1** a small room leading to a main one. **2** Mil. a sitting-room in an officers' mess.

antheap /'ænthiːp/ n. = ANTHILL.

anthelion /ænt'hiːlɪən, æn'θiːl-/ n. (pl. **anthelia** /-lɪə/) a luminous halo projected on a cloud or fog-bank opposite to the sun. [Gk, neut. of *anthēlios* opposite to the sun (as ANTI-, *hēlios* sun)]

anthelmintic /ˌænθelˈmɪntɪk/ (also **anthelminthic** /-ˈmɪnθɪk/) n. & adj. ● n. a drug or agent used to destroy parasitic, esp. intestinal, worms, e.g. tapeworms, roundworms, and flukes. ● adj. having the power to eliminate or destroy parasitic worms. [ANTI- + Gk *helmins helminthos* worm]

anthem /ˈænθəm/ n. **1** an elaborate choral composition usu. based on a passage of scripture for church use. **2 a** a solemn hymn of praise etc., esp. = *national anthem*. **b** a grandiose or rousing song identified with a person, group, etc. (*rock anthem*). **3** a composition sung antiphonally. [OE *antefn, antifne* f. LL *antiphona* ANTIPHON]

anthemion /ænˈθiːmɪən/ n. (pl. **anthemia** /-mɪə/) a flower-like ornament used in art. [Gk, = flower]

Anthemius /ænˈθiːmɪəs/ (known as Anthemius of Tralles) (6th century AD), Greek mathematician, engineer, and artist. His experiments included study of the effects of compressed steam, and he had a high reputation for both these and his artistic pursuits. In 532 he was chosen by Justinian to design St Sophia in Constantinople.

anther /ˈænθə(r)/ n. Bot. the apical portion of a stamen, containing pollen. □ **antheral** adj. [F *anthère* or mod.L *anthera*, in L 'medicine extracted from flowers' f. Gk *anthēra* flowery, fem. adj. f. *anthos* flower]

antheridium /ˌænθəˈrɪdɪəm/ n. (pl. **antheridia** /-dɪə/) Bot. the male sex organ of algae, mosses, ferns, etc. [mod.L f. *anthera* (as ANTHER) + Gk *-idion* dimin. suffix]

anthill /ˈænthɪl/ n. **1** a moundlike nest built by ants or termites. **2** a community teeming with people.

anthologize /ænˈθɒləˌdʒaɪz/ v.tr. & intr. (also **-ise**) compile or include in an anthology.

anthology /ænˈθɒlədʒɪ/ n. (pl. **-ies**) a published collection of passages from literature (esp. poems), songs, reproductions of paintings, etc. □ **anthologist** n. [F *anthologie* or med.L f. Gk *anthologia* f. *anthos* flower + *-logia* collection f. *legō* gather]

Anthony, St (also **Antony**) (c.251–356), Egyptian hermit, the founder of monasticism. At the age of 20 he gave away his possessions and went to live as a hermit in the Egyptian desert, attracting a colony of followers. These he organized into a community which became the first to live under a monastic rule. Shortly before his death he used his influence in association with Athanasius against the Arians. During the Middle Ages the belief arose that praying to St Anthony would effect a cure for ergotism. Feast day, 17 Jan.

Anthony of Padua, St (also **Antony**) (1195–1231), Portuguese Franciscan friar. His charismatic preaching in the south of France and Italy made many converts. His devotion to the poor is commemorated by the alms known as St Anthony's bread; he is invoked to find lost articles. Feast day, 13 June.

anthozoan /ˌænθəˈzəʊən/ n. & adj. Zool. ● n. a sessile marine coelenterate of the class Anthozoa, which includes sea anemones and corals. ● adj. of or relating to this class. [mod.L *Anthozoa* f. Gk *anthos* flower + *zōia* animals]

anthracene /ˈænθrəˌsiːn/ n. Chem. a colourless crystalline aromatic hydrocarbon (chem. formula: $C_{14}H_{10}$) obtained by the distillation of crude oils and used in the manufacture of chemicals. [Gk *anthrax -akos* coal + -ENE]

anthracite /ˈænθrəˌsaɪt/ n. coal of a hard variety consisting of relatively pure carbon and burning with little flame and smoke. □ **anthracitic** /ˌænθrəˈsɪtɪk/ adj. [Gk *anthrakitis* a kind of coal (as ANTHRACENE)]

anthrax /ˈænθræks/ n. a fatal disease of sheep and cattle caused by the bacterium *Bacillus anthracis*. It is transmissible to humans, usu. affecting the skin or lungs. [LL f. Gk, = carbuncle]

anthropic principle /ænˈθrɒpɪk/ n. the cosmological principle that theories of the origin of the universe are constrained by the necessity to allow individual human existence.

anthropo- /ˈænθrəpəʊ/ comb. form human, humankind. [Gk *anthrōpos* human being]

anthropocentric /ˌænθrəpəʊˈsentrɪk/ adj. regarding humankind as the centre of existence. □ **anthropocentrically** adv. **anthropocentrism** n.

anthropogenesis /ˌænθrəpəʊˈdʒenɪsɪs/ n. = ANTHROPOGENY.

anthropogenic /ˌænθrəpəʊˈdʒenɪk/ adj. **1** relating to anthropogeny. **2** Ecol. originating in human activity.

anthropogeny /ˌænθrəˈpɒdʒɪnɪ/ n. the study of the origin of man.

anthropoid /ˈænθrəˌpɔɪd/ adj. & n. ● adj. **1** resembling a human being in form, esp. designating a tailless ape. **2** Zool. of or relating to the primate suborder Anthropoidea, which comprises the monkeys, apes, and humans. (See PRIMATE.) **3** colloq. (of a person) apelike. ● n. a creature that is human in form only, esp. an anthropoid ape. [Gk *anthrōpoeidēs* (as ANTHROPO-, -OID)]

anthropology /ˌænθrəˈpɒlədʒɪ/ n. **1** the study of humankind, esp. of its societies and customs. (See note below.) **2** the study of the structure and evolution of humans as animals. □ **anthropologist** n. **anthropological** /-pəˈlɒdʒɪk(ə)l/ adj.

■ Interest in the variety of human communities, physical types, and cultures is at least as old as written records, but it was not until the advances of Saint-Simon and Comte in the early 19th century that the foundation for a 'science of humanity' was laid. Modern physical anthropology originated with the evolutionary theories of Darwin, which stimulated European interest in the 'primitive' peoples of the world, who were seen as providing a living laboratory to test theories of physical and cultural evolution. In the second half of the 19th century scholars undertook ambitious comparative research into diverse topics including kinship systems, law, magic and religion, and culture, but their study was still based in the library rather than in the field. At the turn of the present century Émile Durkheim pioneered a shift from evolutionary concerns to the study of the ways in which societies maintain themselves. In the early 20th century, anthropologists such as Bronisław Malinowski in Britain and Franz Boas in the US developed an increasing emphasis on ethnographic fieldwork studies, laying the basis of modern social and cultural anthropology, and in particular utilizing the technique of 'participant observation' (where the anthropologist spends a period living with the community) that has since become the single most important feature distinguishing anthropology from sociology. From the 1950s the work of structuralists such as Claude Lévi-Strauss provided a theoretical framework for the analysis of beliefs and myths in cultural systems, according to which elements or concepts (such as *edible/inedible*) are arranged in binary opposition.

anthropometry /ˌænθrəˈpɒmɪtrɪ/ n. the scientific study of the measurements of the human body. □ **anthropometric** /-pəˈmetrɪk/ adj.

anthropomorphic /ˌænθrəpəˈmɔːfɪk/ adj. of or characterized by anthropomorphism. □ **anthropomorphically** adv. [as ANTHROPOMORPHOUS + -IC]

anthropomorphism /ˌænθrəpəˈmɔːfɪz(ə)m/ n. the attribution of a human form or personality to a god, animal, or thing. □ **anthropomorphize** v.tr.

anthropomorphous /ˌænθrəpəˈmɔːfəs/ adj. human in form. [Gk *anthrōpomorphos* (as ANTHROPO-, *morphē* form)]

anthropophagy /ˌænθrəˈpɒfədʒɪ/ n. the eating of human flesh; cannibalism. □ **anthropophagous** /-ˈpɒfəgəs/ adj. [Gk *anthrōpophagia* (as ANTHROPO-, *phagō* eat)]

anthroposophy /ˌænθrəˈpɒsəfɪ/ n. a movement inaugurated by Rudolf Steiner to develop the faculty of cognition and the realization of spiritual reality. [ANTHROPO- + Gk *sophia* wisdom f. *sophos* wise]

anti /ˈæntɪ/ prep. & n. ● prep. (also absol.) opposed to (*is anti everything*; *seems to be rather anti*). ● n. (pl. **antis**) a person opposed to a particular policy etc. [ANTI-]

anti- /ˈæntɪ/ prefix (also **ant-** /ænt/ before a vowel or h) forming nouns and adjectives meaning: **1** opposed to; against (*antivivisectionism*). **2** preventing (*antiscorbutic*). **3** the opposite of (*anticlimax*). **4** rival (*antipope*). **5** unlike the conventional form (*anti-hero*; *anti-novel*). **6** Physics the antiparticle of a specified particle (*antineutrino*; *antiproton*). [from or after Gk *anti-* against]

anti-abortion /ˌæntɪəˈbɔːʃ(ə)n/ adj. opposing abortion. □ **anti-abortionist** n.

anti-aircraft /ˌæntɪˈeəkrɑːft/ adj. (of a gun, missile, etc.) used to attack enemy aircraft.

anti-apartheid /ˌæntɪəˈpɑːteɪt/ adj. opposed to the policy or system of apartheid.

antiar /ˈæntɪˌɑː(r)/ n. = UPAS 1a, 2. [Javanese *antjar*]

antibacterial /ˌæntɪbækˈtɪərɪəl/ adj. active against bacteria.

Antibes /ɒnˈtiːb/ a fishing port and resort in SE France; pop. (1990) 70,690.

antibiosis /ˌæntɪbaɪˈəʊsɪs/ n. Biol. an antagonistic association between two organisms (esp. micro-organisms), in which one is adversely affected (cf. SYMBIOSIS). [F *antibiose* (as ANTI-, SYMBIOSIS)]

antibiotic /ˌæntɪbaɪˈɒtɪk/ n. & adj. Pharm. ● n. a substance that inhibits the growth of or destroys micro-organisms. The first antibiotics, of which penicillin was the earliest and most important, were themselves produced by other micro-organisms, but they are now often prepared synthetically. ● adj. functioning as an antibiotic. [F antibiotique (as ANTI-, Gk biōtikos fit for life f. bios life)]

antibody /ˈæntɪˌbɒdɪ/ n. (pl. -ies) a blood protein produced in response to and counteracting an antigen. Vertebrates develop immunity to infectious disease by the production of antibodies, which function by combining chemically with substances which the body recognizes as alien. This action leads to the destruction and elimination of bacteria, viruses, and foreign substances in the blood. Antibodies are formed a few days after infection and are specific in their action: immunity to diphtheria, for example, gives no protection against tetanus. All antibodies are immunoglobins, with the same basic structure of four polypeptide chains. [transl. G Antikörper (as ANTI-, Körper body)]

antic /ˈæntɪk/ n. & adj. ● n. **1** (usu. in pl.) absurd or foolish behaviour. **2** an absurd or silly action. ● adj. archaic grotesque, bizarre. [It. antico ANTIQUE, used as = grotesque]

anticathode /ˌæntɪˈkæθəʊd/ n. the target (or anode) of an X-ray tube on which the electrons from the cathode impinge and from which X-rays are emitted.

Antichrist /ˈæntɪˌkraɪst/ n. **1** an arch-enemy of Christ. **2** a postulated personal opponent of Christ expected by the early Church to appear before the end of the world. [ME f. OF antecrist f. eccl.L antichristus f. Gk antikhristos (as ANTI-, Khristos CHRIST)]

antichristian /ˌæntɪˈkrɪstɪən, -tʃən/ adj. **1** opposed to Christianity. **2** concerning the Antichrist.

anticipate /ænˈtɪsɪˌpeɪt/ v.tr. **1** deal with or use before the proper time. **2** disp. expect, foresee; regard as probable (did not anticipate any difficulty). **3** forestall (a person or thing). **4** look forward to. □ **anticipative** adj. **anticipator** n. **anticipatory** adj. [L anticipare f. anti- for ANTE- + -cipare f. capere take]

anticipation /ænˌtɪsɪˈpeɪʃ(ə)n/ n. **1** the act or process of anticipating. **2** Mus. the introduction beforehand of part of a chord which is about to follow. [F anticipation or L anticipatio (as ANTICIPATE)]

anticlerical /ˌæntɪˈklerɪk(ə)l/ adj. & n. ● adj. opposed to the influence of the clergy, esp. in politics. ● n. an anticlerical person. □ **anticlericalism** n.

anticlimax /ˌæntɪˈklaɪmæks/ n. a trivial conclusion to something significant or impressive, esp. where a climax was expected. □ **anticlimactic** /-klaɪˈmæktɪk/ adj. **anticlimactically** adv.

anticline /ˈæntɪˌklaɪn/ n. Geol. a ridge or fold of stratified rock in which the strata slope down from the crest (opp. SYNCLINE). □ **anticlinal** /ˌæntɪˈklaɪn(ə)l/ adj. [ANTI- + Gk klinō lean, after INCLINE]

anticlockwise /ˌæntɪˈklɒkwaɪz/ adv. & adj. ● adv. in a curve opposite in direction to the movement of the hands of a clock. ● adj. moving anticlockwise.

anticoagulant /ˌæntɪkəʊˈægjʊlənt/ n. & adj. ● n. a drug or agent that retards or inhibits coagulation, esp. of the blood. ● adj. retarding or inhibiting coagulation.

anticodon /ˌæntɪˈkəʊdɒn/ n. Biochem. a sequence of three nucleotides forming a unit of genetic code in a transfer RNA molecule, corresponding to a complementary codon in messenger RNA.

anticonstitutional /ˌæntɪˌkɒnstɪˈtjuːʃən(ə)l/ adj. violating a political constitution.

anticonvulsant /ˌæntɪkənˈvʌls(ə)nt/ n. & adj. ● n. a drug or agent that prevents or reduces the severity of convulsions, esp. epileptic fits. ● adj. preventing or reducing convulsions.

Anti-Corn-Law League /ˌæntɪˈkɔːnlɔː/ a pressure group formed in Britain in 1838 to campaign for the repeal of the Corn Laws, under the leadership of Richard Cobden and John Bright.

anticyclone /ˈæntɪˌsaɪkləun/ n. Meteorol. a weather system with high barometric pressure at its centre, around which air circulates in a slow clockwise (northern hemisphere) or anticlockwise (southern hemisphere) direction. □ **anticyclonic** /-saɪˈklɒnɪk/ adj.

antidepressant /ˌæntɪdɪˈpres(ə)nt/ n. & adj. ● n. a drug or agent that alleviates depression. ● adj. alleviating depression.

antidiuretic hormone /ˌæntɪˌdaɪjʊəˈretɪk/ n. = VASOPRESSIN. [ANTI- + DIURETIC]

antidote /ˈæntɪˌdəʊt/ n. **1** a medicine etc. taken or given to counteract poison. **2** (usu. foll. by to or for) anything that counteracts something unpleasant or evil (an antidote to stress). □ **antidotal** /ˌæntɪˈdəʊt(ə)l/ adj.

[F antidote or L antidotum f. Gk antidoton neut. of antidotos given against (as ANTI- + stem of didonai give)]

antifreeze /ˈæntɪˌfriːz/ n. a substance (usu. ethylene glycol) added to water to lower its freezing-point, esp. in the radiator of a motor vehicle.

anti-g /ˌæntɪˈdʒiː/ adj. (of clothing for an astronaut etc.) designed to counteract the effects of high acceleration. [ANTI- + g symb. for acceleration due to gravity]

antigen /ˈæntɪdʒən/ n. a foreign substance (e.g. a toxin) which induces the body to react with an immune response, esp. by producing antibodies. □ **antigenic** /ˌæntɪˈdʒenɪk/ adj. [G (as ANTIBODY, -GEN)]

Antigone /ænˈtɪgənɪ/ Gk Mythol. daughter of Oedipus and Jocasta, the subject of a tragedy by Sophocles. Strife between her brothers Eteocles and Polynices resulted in the latter's death in an assault on the city of Thebes. In defiance of an order by Creon, king of Thebes, Antigone buried Polynices' body. Creon ordered that she should be buried alive, but she took her own life before the sentence could be carried out, and Creon's son Haemon, who was betrothed to her, killed himself over her body.

anti-gravity /ˌæntɪˈgrævɪtɪ/ n. & adj. ● n. Physics a hypothetical force opposing gravity. ● adj. = ANTI-G.

Antigua /ænˈtiːgwə/ (also **Antigua Guatemala** /ˌgwɑːtəˈmɑːlə/) a town in the central highlands of Guatemala; pop. (1988) 26,630. Founded in 1543, it was the capital of Spanish Guatemala until devastated by an earthquake in 1773.

Antigua and Barbuda /ænˈtiːgə, bɑːˈbuːdə/ a country consisting of two islands in the Leeward Islands in the West Indies; pop. (est. 1991) 80,000; languages, English (official), Creole; capital, St John's (on Antigua). Discovered in 1493 by Columbus and settled by the English in 1632, Antigua became a British colony with Barbuda as its dependency; the islands gained independence within the Commonwealth in 1981. □ **Antiguan** adj. & n.

anti-hero /ˈæntɪˌhɪərəʊ/ n. (pl. -oes) a central character in a story or drama who noticeably lacks conventional heroic attributes.

antihistamine /ˌæntɪˈhɪstəmiːn, -ˌmiːn/ n. Med. a substance that counteracts the effects of histamine, used esp. in the treatment of allergies.

antiknock /ˈæntɪˌnɒk/ n. a substance added to motor fuel to prevent premature combustion.

Anti-Lebanon Mountains /ˌæntɪˈlebənən/ a range of mountains running north to south along the border between Lebanon and Syria, east of the Lebanon range.

Antilles /ænˈtɪliːz/ a group of islands, forming the greater part of the West Indies. The Greater Antilles, extending roughly east to west, comprise Cuba, Jamaica, Hispaniola (Haiti and the Dominican Republic), and Puerto Rico; the Lesser Antilles, to the south-east, include the Virgin Islands, Leeward Islands, Windward Islands, and various small islands to the north of Venezuela. See also NETHERLANDS ANTILLES.

anti-lock /ˈæntɪˌlɒk/ adj. (of brakes) set up so as to prevent locking and skidding when applied suddenly.

antilog /ˈæntɪˌlɒg/ n. colloq. = ANTILOGARITHM. [abbr.]

antilogarithm /ˌæntɪˈlɒgəˌrɪðəm/ n. the number to which a logarithm belongs (100 is the common antilogarithm of 2).

antilogy /ænˈtɪlədʒɪ/ n. (pl. -ies) a contradiction in terms. [F antilogie f. Gk antilogia (as ANTI-, -LOGY)]

antimacassar /ˌæntɪməˈkæsə(r)/ n. a covering put over furniture, esp. over the back of a chair, as a protection from grease in the hair or as an ornament. [ANTI- + MACASSAR OIL]

antimatter /ˈæntɪˌmætə(r)/ n. Physics matter consisting solely of the antiparticles of the electrons, protons, and neutrons making up normal matter. Matter composed of antiparticles is theoretically as stable as ordinary matter, but could only exist in isolation from it. It has been speculated that antimatter galaxies or clusters of galaxies could exist. (See also ANTIPARTICLE.)

antimetabolite /ˌæntɪmɪˈtæbəˌlaɪt/ n. Pharm. a drug that interferes with the normal metabolic processes within cells, usu. by combining with enzymes.

antimony /ˈæntɪmənɪ/ n. a brittle silvery-white semi-metallic chemical element (atomic number 51; symbol **Sb**). (See note below.) □ **antimonial** /ˌæntɪˈməʊnɪəl/ adj. **antimonious** adj. **antimonic** /-ˈmɒnɪk/ adj. [ME f. med.L antimonium (11th c.), of unkn. orig.]

▪ Antimony was known from ancient times; the naturally occurring black sulphide was used as a cosmetic (see KOHL). The element is used in alloys, usually with lead, such as pewter, type-metal, and Britannia metal. The symbol Sb is from the Latin *stibium*.

antinode /'æntɪˌnəʊd/ *n. Physics* the position of maximum displacement in a standing wave system.

antinomian /ˌæntɪ'nəʊmɪən/ *adj. & n.* ● *adj.* of or relating to antinomianism. ● *n.* (**Antinomian**) *hist.* a member of an antinomian sect.

antinomianism /ˌæntɪ'nəʊmɪəˌnɪz(ə)m/ *n.* the doctrine that Christians are released by grace from the obligation of observing the moral law. This view was attributed to St Paul by his opponents and was held by many of the Gnostic sects; the teaching was revived at the Reformation as following from the Lutheran doctrine of justification by faith, and was also associated with the Anabaptists and Ranters. [med.L *Antinomi*, name of a sect in Germany (1535) alleged to hold this view (as ANTI-, Gk *nomos* law)]

antinomy /æn'tɪnəmɪ/ *n.* (*pl.* **-ies**) **1** a contradiction between two beliefs or conclusions that are in themselves reasonable; a paradox. **2** a conflict between two laws or authorities. [L *antinomia* f. Gk (as ANTI-, *nomos* law)]

antinovel /'æntɪˌnɒv(ə)l/ *n.* a novel in which the conventions of the form are studiously avoided.

anti-nuclear /ˌæntɪ'njuːklɪə(r)/ *adj.* opposed to the development of nuclear weapons or nuclear power.

Antioch /'æntɪˌɒk/ **1** (Turkish **Antakya** /ɑn'tɑkjɑ/) a city in southern Turkey, near the Syrian border; pop. (1990) 123,871. Antioch was the ancient capital of Syria under the Seleucid kings, who founded it *c.*300 BC. **2** a city in ancient Phrygia.

Antiochus /æn'taɪəkəs/ the name of eight Seleucid kings, notably:
Antiochus III (known as Antiochus the Great) (*c.*242–187 BC), reigned 223–187 BC. He restored and expanded the Seleucid empire, regaining the vassal kingdoms of Parthia and Bactria and conquering Armenia, Syria, and Palestine. When he invaded Europe he came into conflict with the Romans, who defeated him on land and sea and severely limited his power.
Antiochus IV Epiphanes (*c.*215–163 BC), son of Antiochus III, reigned 175–163 BC. His firm control of Judaea and his attempt to Hellenize the Jews resulted in the revival of Jewish nationalism and the Maccabean revolt.

antioxidant /ˌæntɪ'ɒksɪd(ə)nt/ *n.* an agent that inhibits oxidation, esp. used to reduce deterioration of products stored in air.

antiparallel /ˌæntɪ'pærəlel/ *adj.* parallel but oppositely directed.

antiparticle /'æntɪˌpɑːtɪk(ə)l/ *n. Physics* a subatomic particle having the same mass as a given particle but opposite electric or magnetic properties, esp. a particle of antimatter. Every kind of subatomic particle has a corresponding antiparticle, e.g. the positron has the same mass as the electron but an equal and opposite charge. When particle and antiparticle collide they annihilate each other, producing other particles and an enormous release of energy. (See also ANTIMATTER.)

antipasto /ˌæntɪ'pɑːstəʊ, -'pæstəʊ/ *n.* (*pl.* **-os** or **antipasti** /-tɪ/) an hors-d'œuvre, esp. in an Italian meal. [It.]

antipathetic /ˌæntɪpə'θetɪk/ *adj.* (usu. foll. by *to*) having a strong aversion or natural opposition. □ **antipathetical** *adj.* **antipathetically** *adv.* [as ANTIPATHY after PATHETIC]

antipathic /ˌæntɪ'pæθɪk/ *adj.* of a contrary nature or character.

antipathy /æn'tɪpəθɪ/ *n.* (*pl.* **-ies**) (often foll. by *to, for, between*) a strong or deep-seated aversion or dislike. [F *antipathie* or L *antipathia* f. Gk *antipatheia* f. *antipathēs* opposed in feeling (as ANTI-, *pathos -eos* feeling)]

anti-personnel /ˌæntɪˌpɜːsə'nel/ *adj.* (of a bomb, mine, etc.) designed to kill or injure people rather than to damage buildings or equipment.

antiperspirant /ˌæntɪ'pɜːspɪrənt/ *n. & adj.* ● *n.* a substance applied to the skin to prevent or reduce perspiration. ● *adj.* that acts as an antiperspirant.

antiphon /'æntɪf(ə)n/ *n.* **1** a hymn or psalm, the parts of which are sung or recited alternately by two groups. **2** a versicle or phrase from this. **3** a sentence sung or recited before or after a psalm or canticle. **4** a response. [eccl.L *antiphona* f. Gk (as ANTI-, *phōnē* sound)]

antiphonal /æn'tɪfən(ə)l/ *adj. & n.* ● *adj.* **1** sung or recited alternately by two groups. **2** responsive, answering. ● *n.* a collection of antiphons. □ **antiphonally** *adv.*

antiphonary /æn'tɪfənərɪ/ *n.* (*pl.* **-ies**) a book of antiphons. [eccl.L *antiphonarium* (as ANTIPHON)]

antiphony /æn'tɪfənɪ/ *n.* (*pl.* **-ies**) **1** antiphonal singing or chanting. **2** a response or echo.

antipode /'æntɪˌpəʊd/ *n.* (usu. foll. by *of, to*) the exact opposite. [back-form. f. ANTIPODES]

antipodes /æn'tɪpəˌdiːz/ *n.pl.* **1 a** (also **Antipodes**) a place diametrically opposite to another, esp. Australasia as the region on the opposite side of the earth to Europe. **b** places diametrically opposite to each other. **2** (usu. foll. by *of, to*) the exact opposite. □ **antipodal** *adj.*
antipodean /æn,tɪpə'diːən/ *adj. & n.* [F or LL f. Gk *antipodes* having the feet opposite (as ANTI-, *pous podos* foot)]

antipope /'æntɪˌpəʊp/ *n.* a person set up as pope in opposition to one (held by others to be) canonically chosen. [F *antipape* f. med.L *antipapa*, assim. to POPE¹]

antiproton /ˌæntɪ'prəʊtɒn/ *n. Physics* the negatively charged antiparticle of a proton.

antipruritic /ˌæntɪprʊə'rɪtɪk/ *adj. & n.* ● *adj.* relieving itching. ● *n.* an antipruritic agent.

antipyretic /ˌæntɪpaɪ'retɪk/ *adj. & n.* ● *adj.* preventing or reducing fever. ● *n.* an antipyretic drug or agent.

antiquarian /ˌæntɪ'kweərɪən/ *adj. & n.* ● *adj.* **1** of or dealing in antiques or rare books. **2** of the study of antiquities. ● *n.* an antiquary. □ **antiquarianism** *n.* [see ANTIQUARY]

antiquary /'æntɪkwərɪ/ *n.* (*pl.* **-ies**) a student or collector of antiques or antiquities. [L *antiquarius* f. *antiquus* ancient]

antiquated /'æntɪˌkweɪtɪd/ *adj.* old-fashioned; out of date. [eccl.L *antiquare antiquat-* make old]

antique /æn'tiːk/ *n., adj., & v.* ● *n.* an object of considerable age, esp. an item of furniture or the decorative arts having a high value. ● *adj.* **1** of or existing from an early date. **2** old-fashioned, archaic. **3** of ancient times. ● *v.tr.* (**antiques, antiqued, antiquing**) give an antique appearance to (furniture etc.) by artificial means. [F *antique* or L *antiquus, anticus* former, ancient f. *ante* before]

antiquity /æn'tɪkwɪtɪ/ *n.* (*pl.* **-ies**) **1** ancient times, esp. the period before the Middle Ages. **2** great age (*a city of great antiquity*). **3** (usu. in *pl.*) physical remains or relics from ancient times, esp. buildings and works of art. **4** (in *pl.*) customs, events, etc., of ancient times. **5** the people of ancient times regarded collectively. [ME f. OF *antiquité* f. L *antiquitas -tatis* f. *antiquus*: see ANTIQUE]

anti-racism /ˌæntɪ'reɪsɪz(ə)m/ *n.* the policy or practice of opposing racism and promoting racial tolerance. □ **anti-racist** *n. & adj.*

antirrhinum /ˌæntɪ'raɪnəm/ *n.* a plant of the genus *Antirrhinum*, of the figwort family, with tubular two-lipped flowers; esp. the snapdragon, *A. majus*. [L f. Gk *antirrhinon* f. *anti* counterfeiting + *rhis rhinos* nose (from the resemblance of the flower to an animal's snout)]

antiscorbutic /ˌæntɪskɔː'bjuːtɪk/ *adj. & n.* ● *adj.* preventing or curing scurvy. ● *n.* an antiscorbutic agent or drug.

anti-Semitism /ˌæntɪ'semɪˌtɪz(ə)m/ *n.* hostility to or prejudice against Jews. The Jews were dispersed throughout the Mediterranean region from the 8th century BC, especially after the destruction of Jerusalem in 70 AD, and maintained a separate and highly distinct identity, which often made them the subject of intolerance and harassment. Throughout medieval Europe the persecution of Jews was a common response to economic or social crisis, reinforced by religious prejudice and jealousy of the commercial activities of the Jewish community. Christians were forbidden to practise usury, but Jews were permitted to lend money at interest to Christians, and having taken on this role they became associated in the popular mind with extortion. In the late 19th century anti-Semitism crystallized as an explicit doctrine, based on theories of racial determinism, and was subsequently made a major tenet of Nazism, culminating in the Holocaust. Reaction to anti-Semitism gave rise to the Zionist movement and led to the establishment of the state of Israel, which itself created new antagonisms between Jews and Arabs. (See also NAZI, ZIONISM.) □ **anti-Semite** /-'siːmaɪt, -'semaɪt/ *n.* **anti-Semitic** /-sɪ'mɪtɪk/ *adj.*

antisepsis /ˌæntɪ'sepsɪs/ *n.* the process of using antiseptics to eliminate undesirable micro-organisms such as bacteria, viruses, and fungi that cause disease. (See also ANTISEPTIC.) [ANTI- + SEPSIS]

antiseptic /ˌæntɪ'septɪk/ *adj. & n.* ● *adj.* **1** counteracting sepsis esp. by preventing the growth of disease-causing micro-organisms. (*See note below*.) **2** sterile or free from contamination. **3** lacking character. ● *n.* an antiseptic agent. □ **antiseptically** *adv.*

▪ In 1847 the Hungarian physician Ignaz Semmelweis ordered the use of a disinfectant solution by students attending women in childbirth, but without understanding the underlying cause of infection and without influencing his senior colleagues. In 1865 the antiseptic system was introduced into surgery by Joseph Lister, who worked on Pasteur's discoveries. Until their time the mortality rate from operations was very high because of infection of the exposed tissues, and septicaemia was a major hazard. Lister's methods brought about a revolution. However, antiseptics strong enough to counter infection often also damaged body tissue, and there soon arose aseptic methods, the keeping of wounds free from all contact with harmful micro-organisms.

antiserum /ˈæntɪˌsɪərəm/ n. (pl. **antisera** /-rə/) a blood serum containing antibodies against specific antigens, injected to treat or protect against specific diseases.

antisocial /ˌæntɪˈsəʊʃ(ə)l/ adj. **1** opposed or contrary to normal social instincts or practices. **2** not sociable. **3** opposed or harmful to the existing social order.

antistatic /ˌæntɪˈstætɪk/ adj. that counteracts the effects of static electricity.

antistrophe /ænˈtɪstrəfɪ/ n. the second section of an ancient Greek choral ode or of one division of it (see STROPHE). [LL f. Gk *antistrophē* f. *antistrephō* turn against]

anti-tank /ˌæntɪˈtæŋk/ attrib.adj. used against tanks.

antitetanus /ˌæntɪˈtetənəs/ adj. effective against tetanus.

antithesis /ænˈtɪθɪsɪs/ n. (pl. **antitheses** /-ˌsiːz/) **1** (foll. by *of*, *to*) the direct opposite. **2** (usu. foll. by *of*, *between*) contrast or opposition between two things. **3** a contrast of ideas expressed by parallelism of strongly contrasted words. [LL f. Gk *antitithēmi* set against (as ANTI-, *tithēmi* place)]

antithetical /ˌæntɪˈθetɪk(ə)l/ adj. (also **antithetic**) **1** contrasted, opposite. **2** connected with, containing, or using antithesis. □ **antithetically** adv. [Gk *antithetikos* (as ANTITHESIS)]

antitoxin /ˌæntɪˈtɒksɪn/ n. an antibody that counteracts a toxin. □ **antitoxic** adj.

antitrades /ˌæntɪˈtreɪdz/ n.pl. winds that blow in the opposite direction to (and usu. above) a trade wind.

antitrust laws /ˌæntɪˈtrʌst/ n.pl. legislation, principally in the US, which aims to promote fair competition in business, esp. by controlling trusts and other monopolies (see TRUST n. 8c). Such laws were passed in the US from 1820 onwards.

antitype /ˈæntɪˌtaɪp/ n. **1** that which is represented by a type or symbol. **2** a person or thing of the opposite type. □ **antitypical** /ˌæntɪˈtɪpɪk(ə)l/ adj. [Gk *antitupos* corresponding as an impression to the die (as ANTI-, *tupos* stamp)]

antivenene /ˌæntɪˈveniːn/ n. (also **antivenin** /-ˈvenɪn/) an antiserum containing antibodies against specific poisons in the venom of snakes, spiders, scorpions, etc. [ANTI- + L *venenum* poison + -ENE, -IN]

antiviral /ˌæntɪˈvaɪərəl/ adj. effective against viruses.

antivivisectionism /ˌæntɪˌvɪvɪˈsekʃəˌnɪz(ə)m/ n. opposition to vivisection. □ **antivivisectionist** n.

antler /ˈæntlə(r)/ n. **1** each of the branched horns of a stag or other (usu. male) deer. **2** a branch of this. □ **antlered** adj. [ME f. AF, var. of OF *antoillier*, of unkn. orig.]

Antofagasta /ˌæntəʊfəˈɡæstə/ a port in northern Chile, capital of a region of the same name; pop. (1991) 218,750.

Antonine /ˈæntəˌnaɪn/ adj. & n. ● adj. of or relating to the Roman emperors Antoninus Pius and Marcus Aurelius or their rule. ● n. either of the Antonine emperors.

Antonine Wall a defensive fortification about 59 km (37 miles) long, built (AD c.140) across the narrowest part of southern Scotland between the Firth of Forth and the Firth of Clyde, in the time of Antoninus Pius. It was intended to mark the frontier of the Roman province of Britain, but the Romans were unable to consolidate their position. In c.181 the wall was breached and the northern tribes forced a retreat from the Forth–Clyde frontier, eventually to that established earlier at Hadrian's Wall.

Antoninus Pius /ˌæntəˌnaɪnəs ˈpaɪəs/ (86–161), Roman emperor 138–61. The adopted son and successor of Hadrian, he was the first of the Antonines. His reign was generally peaceful and he ruled in harmony with the Senate, pursuing a policy of moderation and liberality. Although no great conqueror, he extended the empire; the frontier of Britain was temporarily advanced to the Antonine Wall.

antonomasia /ˌæntənəˈmeɪzɪə/ n. **1** the substitution of an epithet or title etc. for a proper name (e.g. *the Maid of Orleans* for Joan of Arc, *his Grace* for an archbishop). **2** the use of a proper name to express a general idea (e.g. *a Scrooge* for a miser). [L f. Gk f. *antonomazō* name instead (as ANTI-, + *onoma* name)]

Antony /ˈæntənɪ/, Mark (Latin name Marcus Antonius) (c.83–30 BC), Roman general and triumvir. A supporter of Julius Caesar, in 43 he was appointed one of the triumvirate with Octavian and Lepidus after Caesar's murder the previous year. Following the battle of Philippi he took charge of the Eastern Empire, where he established his association with Cleopatra. Quarrels with Octavian led finally to his defeat at the sea battle of Actium in NW Greece in 31 and to his suicide the following year.

Antony, St see ANTHONY, ST.

antonym /ˈæntənɪm/ n. a word opposite in meaning to another in the same language (e.g. *bad* and *good*) (opp. SYNONYM). □ **antonymous** /ænˈtɒnɪməs/ adj. **antonymy** /ænˈtɒnɪmɪ/ n. [F *antonyme* (as ANTI-, SYNONYM)]

Antony of Padua, St see ANTHONY OF PADUA, ST.

Antrim /ˈæntrɪm/ **1** one of the Six Counties of Northern Ireland, formerly an administrative area. **2** a town in this county, on the NE shore of Lough Neagh; pop. (1981) 22,340.

antrum /ˈæntrəm/ n. (pl. **antra** /-trə/) Anat. **1** a natural chamber or cavity in the body, esp. in a bone. **2** the part of the stomach just inside the pylorus. □ **antral** adj. [L f. Gk *antron* cave]

antsy /ˈæntsɪ/ adj. US colloq. irritated, impatient; fidgety, restless. [ants, pl. of ANT + -Y¹]

Antung /ænˈtʊŋ/ the former name for DANDONG.

Antwerp /ˈæntwɜːp/ (French **Anvers** /ɑ̃vɛr/, Flemish **Antwerpen** /ˈɑntwɛrpə/) **1** a province of northern Belgium. **2** its capital city, a port on the Scheldt; pop. (1991) 467,520. A former seat of the counts of Flanders, Antwerp had by the 16th century become a leading European commercial and financial centre.

Anubis /əˈnjuːbɪs/ Egyptian Mythol. the god of mummification, protector of tombs, often represented as having a jackal's head.

Anuradhapura /əˌnʊˈrɑːdəˌpʊərə/ a city in north central Sri Lanka, capital of a district of the same name; pop. (1981) 36,000. It is the ancient capital of Sri Lanka and a centre of Buddhist pilgrimage. The Sinhalese ruler Mahinda (c.270–c.204 BC) was converted to Buddhism there. A sacred bo tree, brought there as a sapling from the tree in Bodhgaya in India over 2,000 years ago, is alleged to be the oldest living tree in the world.

anuran /əˈnjʊərən/ n. & adj. Zool. ● n. a tailless amphibian of the order Anura, which includes the frogs and toads. ● adj. of or relating to this order. [mod.L *Anura* (AN-¹ + Gk *oura* tail)]

anus /ˈeɪnəs/ n. Anat. the excretory opening at the end of the alimentary canal. [L]

Anvers see ANTWERP.

anvil /ˈænvɪl/ n. **1** a block (usu. of iron) with a flat top, concave sides, and often a pointed end, on which metals are worked in forging. **2** Anat. = INCUS. [OE *anfilte* etc.]

anxiety /æŋˈzaɪətɪ/ n. (pl. **-ies**) **1** the state of being anxious. **2** concern about an imminent danger, difficulty, etc. **3** (foll. by *for*, or *to* + infin.) anxious desire. **4** a thing that causes anxiety (*my greatest anxiety is that I shall fall ill*). **5** Psychol. a nervous disorder characterized by a state of excessive uneasiness. [F *anxiété* or L *anxietas -tatis* (as ANXIOUS)]

anxiolytic /ˌæŋzɪəˈlɪtɪk/ n. & adj. Pharm. ● n. a drug reducing anxiety. ● adj. reducing anxiety. [f. ANXIETY + -LYTIC]

anxious /ˈæŋkʃəs/ adj. **1** troubled; uneasy in the mind. **2** causing or marked by anxiety (*an anxious moment*). **3** (foll. by *for*, or *to* + infin.) earnestly or uneasily wanting or trying (*anxious to please*; *anxious for you to succeed*). □ **anxiously** adv. **anxiousness** n. [L *anxius* f. *angere* choke]

any /ˈenɪ/ adj., pron., & adv. ● adj. **1** (with interrog., neg., or conditional expressed or implied) **a** one, no matter which, of several (*cannot find any answer*). **b** some, no matter how much or many or of what sort (*if any books arrive*; *have you any sugar?*). **2** a minimal amount of (*hardly any difference*). **3** whichever is chosen (*any fool knows that*). **4 a** an appreciable or significant (*did not stay for any length of time*). **b** a very large (*has any amount of money*). ● pron. **1** any one (*did not know any of them*). **2** any number (*are any of them yours?*). **3** any amount (*is there any left?*). ● adv. (usu. with neg. or interrog.) at all, in some degree (*is that any good?*; *do not make it any larger*; *without being any the wiser*). □ **any more** to any further extent; any longer (*don't like you any more*). **any time** colloq. at any time. **any time** (or **day** or **minute** etc.) **now** colloq. at any time in the near future. **not having any** colloq. unwilling to participate, not interested or persuaded. [OE *ænig* f. Gmc (as ONE, -Y¹)]

Anyaoku /ˌænjəˈəʊkuː/, Eleazar Chukwuemeka (b.1933), Nigerian diplomat. After holding posts in the Commonwealth Secretariat in the 1970s and 1980s, he was appointed Commonwealth Secretary-General in 1989, the first African to hold this position.

anybody /ˈenɪˌbɒdɪ/ n. & pron. **1 a** a person, no matter who. **b** a person of any kind. **c** whatever person is chosen. **2** a person of importance (*are you anybody?*). □ **anybody's** (of a contest) evenly balanced (*it was anybody's game*). **anybody's guess** see GUESS.

anyhow /ˈenɪˌhaʊ/ adv. **1** anyway. **2** in a disorderly manner or state (*does his work anyhow; things are all anyhow*).

anymore /ˌenɪˈmɔː(r)/ adv. esp. US = any more.

anyone /ˈenɪˌwʌn/ pron. anybody. ¶ Written as two words to imply a numerical sense, as in *any one of us can do it*.

anyplace /ˈenɪˌpleɪs/ adv. US anywhere.

anything /ˈenɪˌθɪŋ/ pron. **1** a thing, no matter which. **2** a thing of any kind. **3** whatever thing is chosen. □ **anything but** not at all (*was anything but honest*). **like anything** colloq. with great vigour, intensity, etc.

anytime /ˈenɪˌtaɪm/ adv. esp. US colloq. = any time.

anyway /ˈenɪˌweɪ/ adv. **1** in any way or manner. **2** at any rate; in any case. **3** to resume (*anyway, as I was saying*).

anyways /ˈenɪˌweɪz/ adv. US colloq. or dial. = ANYWAY.

anywhere /ˈenɪˌweə(r)/ adv. & pron. ● adv. in or to any place. ● pron. any place (*anywhere will do*).

anywise /ˈenɪˌwaɪz/ adv. archaic in any manner. [OE *on ǣnige wīsan* in any wise]

Anzac /ˈænzæk/ n. **1** a soldier in the Australian and New Zealand Army Corps (1914–18). **2** a person, esp. a member of the armed services, from Australia or New Zealand. [acronym]

Anzac Day 25 April, commemorating the Anzac landing at Gallipoli in 1915.

Anzus /ˈænzəs/ (also **ANZUS**) a treaty (formerly called the Pacific Security Treaty) signed in 1951 by Australia, New Zealand, and the US, designed to protect those countries in the Pacific area from armed attack. [acronym]

AO abbr. Officer of the Order of Australia.

AOB abbr. any other business.

AOC abbr. appellation d'origine contrôlée.

A-OK abbr. N. Amer. colloq. excellent; in good order. [all systems OK]

AOR abbr. adult-oriented rock (music).

Aorangi /aʊˈræŋɪ/ the Maori name for Mount Cook (see COOK, MOUNT).

aorist /ˈeɔrɪst/ n. & adj. Gram. ● n. an unqualified past tense of a verb (esp. in Greek), without reference to duration or completion. ● adj. of or designating this tense. □ **aoristic** /eəˈrɪstɪk/ adj. [Gk *aoristos* indefinite f. a- not + *horizō* define, limit]

aorta /eɪˈɔːtə/ n. (pl. **aortas**) the main artery of the body supplying oxygenated blood to the circulatory system. In humans it arches over the heart from the left ventricle and runs down in front of the backbone. □ **aortic** adj. [Gk *aortē* f. *a(e)irō* raise]

Aosta /ɑːˈɒstə/ a city in NW Italy, capital of Valle d'Aosta region; pop. (1990) 36,095.

Aotearoa /aʊˌteɪəˈrəʊə/ the Maori name for NEW ZEALAND. It means 'land of the long white cloud'.

à outrance /æ ˈuːtrɒns/ adv. **1** to the death. **2** to the bitter end. [F, = to the utmost]

Aouzou Strip /aʊˈzuː/ a narrow corridor of disputed desert land in northern Chad, stretching the full length of the border between Chad and Libya. In 1994 Libya agreed to withdraw its troops from the area.

AP see ASSOCIATED PRESS.

ap-¹ /æp/ prefix assim. form of AD- before *p*.

ap-² /æp/ prefix assim. form of APO- before a vowel or *h*.

apace /əˈpeɪs/ adv. literary swiftly, quickly. [OF *à pas* at (a considerable) pace]

Apache /əˈpætʃɪ/ n. & adj. ● n. **1** (pl. same or **Apaches**) a member of an American Indian people who migrated from Canada to the south-western US over 1,000 years ago. (*See note below.*) **2** the language of this people. ● adj. of or relating to this people or their language. [Mex. Sp.]

▪ The Apache initially suffered less cultural disruption than many other American Indian groups from European colonial expansion, owing to the remoteness and arid climate of the region they inhabited. When contact finally came the Apache put up fierce resistance and were, under the leadership of Geronimo, the last

American Indian people to be conquered. Since 1887 they have been confined to reservations and unable to continue their traditional hunting and gathering.

apache /əˈpæʃ/ n. a violent street ruffian, orig. in Paris.

apanage /ˈæpənɪdʒ/ n. (also **appanage**) **1** provision for the maintenance of the younger children of kings etc. **2** a perquisite. **3** a natural accompaniment or attribute. [F ult. f. med.L *appanare* endow]

apart /əˈpɑːt/ adv. **1** separately; not together (*stand apart from the crowd*). **2** into pieces (*came apart in my hands*). **3 a** to or on one side. **b** out of consideration (placed after noun: *joking apart*). **4** to or at a distance. □ **apart from 1** excepting; not considering. **2** in addition to (*apart from roses we grow irises*). [ME f. OF f. *à* to + *part* side]

apartheid /əˈpɑːteɪt/ n. a policy or system of segregation or discrimination on grounds of race, especially the policy followed by South Africa in respect of Europeans and non-Europeans until 1991. Adopted by the successful Afrikaner National Party as a slogan in the 1948 election, apartheid extended and institutionalized existing racial segregation, guaranteeing the dominance of the white minority. The extent of the system was seen in the legal and social provisions of so-called petty apartheid which regulated, for example, the use of transport, park benches, and beaches. In the early 1960s domestic and international opposition to apartheid began to intensify, but despite rioting and terrorism at home and isolation abroad, the white regime maintained the apartheid system with only minor relaxation until Feb. 1991. [Afrik. (as APART, -HOOD)]

apartment /əˈpɑːtmənt/ n. **1** (in pl.) a suite of rooms, usu. furnished and rented. **2** a single room in a house. **3** N. Amer. a flat. □ **apartment house** N. Amer. a block of flats. [F *appartement* f. It. *appartamento* f. *appartare* to separate f. *a parte* apart]

apathetic /ˌæpəˈθetɪk/ adj. having or showing no emotion or interest. □ **apathetically** adv. [APATHY, after PATHETIC]

apathy /ˈæpəθɪ/ n. (often foll. by towards) lack of interest or feeling; indifference. [F *apathie* f. L *apathia* f. Gk *apatheia* f. *apathēs* without feeling f. a- not + *pathos* suffering]

apatite /ˈæpəˌtaɪt/ n. a naturally occurring crystalline mineral consisting of calcium phosphate and fluoride, used in the manufacture of fertilizers. [G *Apatit* f. Gk *apatē* deceit (from its deceptive forms)]

apatosaurus /əˌpætəˈsɔːrəs/ n. a huge plant-eating dinosaur of the genus *Apatosaurus* (formerly *Brontosaurus*), of the Jurassic and Cretaceous periods, with a long whiplike tail and trunklike legs. Also called *brontosaurus*. [mod.L f. Gk *apatē* deceit + *sauros* lizard]

ape /eɪp/ n. & v. ● n. **1** a primate of the families Pongidae or Hylobatidae, characterized by the absence of a tail, e.g. the gorilla, chimpanzee, orang-utan, and gibbon. **2** (in general use) a monkey. **3 a** an imitator. **b** an apelike person. ● v.tr. imitate, mimic. □ **go ape** sl. become crazy. **the naked ape** present-day humans regarded as a species. [OE *apa* f. Gmc]

Apeldoorn /ˈæp(ə)lˌdɔːn/ a town in the east central Netherlands; pop. (1991) 148,200. Since 1685 it has been the site of the summer residence of the Dutch royal family.

Apelles /əˈpeliːz/ (4th century BC), Greek painter. The court painter to Alexander the Great, Apelles is now only known from written sources, but was highly acclaimed throughout the ancient world. Among his recorded pictures was a depiction of Aphrodite rising from the sea and a work entitled *Calumny*; both of these were emulated by Botticelli in the 15th century.

apeman /ˈeɪpmæn/ n. (pl. **-men**) an apelike primate held to be a forerunner of present-day humans.

Apennines /ˈæpɪˌnaɪnz/ a mountain range running 1,400 km (880 miles) down the length of Italy, from the north-west to the southern tip of the peninsula.

aperçu /ˌæpeəˈsjuː/ n. **1** a summary or survey. **2** an insight. [F, past part. of *apercevoir* perceive]

aperient /əˈpɪərɪənt/ adj. & n. Med. ● adj. laxative. ● n. a laxative medicine. [L *aperire* aperient- to open]

aperiodic /ˌeɪpɪərɪˈɒdɪk/ adj. **1** not periodic; irregular. **2** Physics (of a potentially oscillating or vibrating system, e.g. an instrument with a pointer) that is adequately damped to prevent oscillation or vibration. **3** Physics (of an oscillation or vibration) without a regular period. □ **aperiodicity** /-rɪəˈdɪsɪtɪ/ n.

aperitif /əˌperɪˈtiːf/ n. an alcoholic drink taken before a meal to stimulate the appetite. [F *apéritif* f. med.L *aperitivus* f. L *aperire* to open]

aperture /ˈæpəˌtjʊə(r)/ n. **1** an opening; a gap. **2** a space through which light passes in an optical or photographic instrument, esp. the variable space by which light can enter a camera. [L *apertura* (as APERITIF)]

apery /ˈeɪpərɪ/ *n.* pretentious or silly mimicry.

apetalous /eɪˈpetələs/ *adj. Bot.* (of flowers) having no petals. [mod.L *apetalus* f. Gk *apetalos* leafless f. *a-* not + *petalon* leaf]

APEX /ˈeɪpeks/ *abbr.* Association of Professional, Executive, Clerical, and Computer Staff.

Apex /ˈeɪpeks/ *n.* (also **APEX**) (often *attrib.*) a system of reduced fares for scheduled airline flights when paid for before a certain period in advance of departure. [Advance Purchase Excursion]

apex /ˈeɪpeks/ *n.* (*pl.* **apexes** or **apices** /ˈeɪpɪˌsiːz, ˈæp-/) **1** the highest point. **2** a climax; a high point of achievement etc. **3** the vertex of a triangle or cone. **4** a tip or pointed end. [L, = peak, tip]

apfelstrudel /ˈæpfəlˌstruːd(ə)l/ *n.* a confection of flaky pastry filled with spiced apple. [G f. *Apfel* apple + STRUDEL]

aphaeresis /əˈfɪərɪsɪs/ *n.* the omission of a letter or syllable at the beginning of a word as a morphological development (e.g. in the derivation of *adder*). [LL f. Gk *aphairesis* (as APO-, *haireō* take)]

aphasia /əˈfeɪzɪə/ *n. Med.* the loss of ability to understand or express speech, owing to brain damage. □ **aphasic** *adj. & n.* [Gk f. *aphatos* speechless f. *a-* not + *pha-* speak]

aphelion /æpˈhiːlɪən/ *n.* (*pl.* **aphelia** /-lɪə/) *Astron.* the point in a body's orbit where it is furthest from the sun (opp. PERIHELION). [Graecized f. mod.L *aphelium* f. Gk *aph' hēliou* from the sun]

aphesis /ˈæfɪsɪs/ *n.* the gradual loss of an unstressed vowel at the beginning of a word (e.g. of *e* from *esquire* to form *squire*). □ **aphetic** /əˈfetɪk/ *adj.* **aphetically** *adv.* [Gk, = letting go (as APO-, *hiēmi* send)]

aphid /ˈeɪfɪd/ *n.* a small homopterous insect of the family Aphididae, which comprises the greenfly, blackfly, and related plant-lice, feeding by sucking sap from the leaves, stems, or roots of plants. [back-form. f. *aphides*: see APHIS]

aphis /ˈeɪfɪs/ *n.* (*pl.* **aphides** /ˈeɪfɪˌdiːz/) an aphid, esp. of the genus *Aphis* which includes the greenfly. [mod.L, perh. a misreading of Gk *koris* bug]

aphonia /əˈfəʊnɪə/ *n.* (also **aphony** /ˈæfənɪ/) *Med.* the loss or absence of the voice through a disease of the larynx or mouth. [mod.L *aphonia* f. Gk f. *aphōnos* voiceless f. *a-* not + *phōnē* voice]

aphorism /ˈæfəˌrɪz(ə)m/ *n.* **1** a short pithy maxim. **2** a brief statement of a principle. □ **aphorist** *n.* **aphorize** *v.intr.* (also **-ise**). **aphoristic** /ˌæfəˈrɪstɪk/ *adj.* **aphoristically** *adv.* [F *aphorisme* or LL f. Gk *aphorismos* definition f. *aphorizō* (as APO-, *horos* boundary)]

aphrodisiac /ˌæfrəˈdɪzɪˌæk/ *adj. & n.* ● *adj.* that arouses sexual desire. ● *n.* an aphrodisiac drug. [Gk *aphrodisiakos* f. *aphrodisios* f. *Aphroditē* Gk goddess of love]

Aphrodisias /ˌæfrəˈdɪsɪəs/ an ancient city of western Asia Minor, site of a temple dedicated to Aphrodite. Now in ruins, it is situated 80 km (50 miles) west of Aydin, in modern Turkey.

Aphrodite /ˌæfrəˈdaɪtɪ/ *Gk Mythol.* the goddess of beauty, fertility, and sexual love, identified by the Romans with Venus. She is variously described as the daughter of Zeus and Dione, or as being born of the sea-foam. Her cult was of eastern origin, hence her identification with Astarte and Ishtar. The statue of Aphrodite (now lost) by Praxiteles was the first important female nude in sculpture; another statue, the Venus de Milo, is probably the most famous antique sculpture. [perh. f. Gk *aphros* foam]

aphyllous /əˈfɪləs/ *adj. Bot.* (of plants) having no leaves. [mod.L f. Gk *aphullos* f. *a-* not + *phullon* leaf]

Apia /ˈæpɪə/ the capital of Western Samoa; pop. (1986) 32,200. It was the home of Robert Louis Stevenson from 1888 until his death in 1894.

apian /ˈeɪpɪən/ *adj.* of or relating to bees. [L *apianus* f. *apis* bee]

apiary /ˈeɪpɪərɪ/ *n.* (*pl.* **-ies**) a place where bees are kept. □ **apiarist** *n.* [L *apiarium* f. *apis* bee]

apical /ˈeɪpɪk(ə)l, ˈæp-/ *adj.* of, at, or forming an apex. [L *apex apicis*: see APEX]

apices *pl.* of APEX.

apiculture /ˈeɪpɪˌkʌltʃə(r)/ *n.* bee-keeping. □ **apicultural** /ˌeɪpɪˈkʌltʃərəl/ *adj.* **apiculturist** *n.* [L *apis* bee, after AGRICULTURE]

apiece /əˈpiːs/ *adv.* for each one; severally (*had five pounds apiece*). [A[2] + PIECE]

Apis /ˈɑːpɪs, ˈæp-/ *Egyptian Mythol.* a god depicted as a bull, symbolizing fertility and strength in war. Apis was worshipped especially at Memphis, where he was recognized as a manifestation of Ptah, then of Ra, and later of Osiris. A live bull, carefully chosen, was considered to be his incarnation and kept in an enclosure. When it died it was mummified and ceremonially interred, and a young black bull with suitable markings was installed in its place.

apish /ˈeɪpɪʃ/ *adj.* **1** of or like an ape. **2** silly; affected. □ **apishly** *adv.* **apishness** *n.*

aplanat /ˈæpləˌnæt/ *n.* a reflecting or refracting surface made aplanatic by correction. [G]

aplanatic /ˌæpləˈnætɪk/ *adj.* (of a reflecting or refracting surface) free from spherical aberration. [Gk *aplanētos* free from error f. *a-* not + *planaō* wander]

aplasia /əˈpleɪzɪə/ *n. Med.* total or partial failure of development of an organ or tissue. □ **aplastic** /əˈplæstɪk/ *adj.* [Gk f. *a-* not + *plasis* formation]

aplenty /əˈplentɪ/ *adv.* in plenty.

aplomb /əˈplɒm/ *n.* assurance; self-confidence. [F, = perpendicularity, f. *à plomb* according to a plummet]

apnoea /æpˈnɪə, ˈæpnɪə/ *n.* (*US* **apnea**) *Med.* a temporary cessation of breathing. [mod.L f. Gk *apnoia* f. *apnous* breathless]

apo- /ˈæpə/ *prefix* **1** away from (*apogee*). **2** separate (*apocarpous*). [Gk *apo* from, away, un-, quite]

Apoc. *abbr.* **1** Apocalypse (New Testament). **2** Apocrypha.

apocalypse /əˈpɒkəlɪps/ *n.* **1** (**the Apocalypse**) = REVELATION 3. **2** a revelation, esp. of the end of the world. **3** a grand or violent event resembling those described in the Apocalypse. [ME f. OF ult. f. Gk *apokalupsis* f. *apokaluptō* uncover, reveal]

apocalyptic /ˌæpɒkəˈlɪptɪk/ *adj.* **1** of or resembling the Apocalypse. **2** revelatory; prophetic. □ **apocalyptically** *adv.* [Gk *apokaluptikos* (as APOCALYPSE)]

apocarpous /ˌæpəˈkɑːpəs/ *adj. Bot.* (of ovaries) having distinct carpels not joined together (opp. SYNCARPOUS). [APO- + Gk *karpos* fruit]

apochromat /ˈæpəkrəˌmæt/ *n.* a lens or lens-system that reduces spherical and chromatic aberrations. □ **apochromatic** /ˌæpəkrəˈmætɪk/ *adj.* [APO- + CHROMATIC]

apocope /əˈpɒkəpɪ/ *n.* the omission of a letter or letters at the end of a word as a morphological development (e.g. in the derivation of *curio*). [LL f. Gk *apokopē* (as APO-, *koptō* cut)]

Apocr. *abbr.* Apocrypha.

apocrine /ˈæpəˌkraɪn, -krɪn/ *adj. Biol.* (of a multicellular gland, e.g. the mammary gland) releasing part of its cells with its secretion. [APO- + Gk *krinō* to separate]

Apocrypha /əˈpɒkrɪfə/ biblical books and writings not forming part of the accepted canon. The Old Testament Apocrypha, dating from *c*.300 BC–AD *c*.100, were received by the early Church as part of the Greek version of the Hebrew Bible (the Septuagint). They are included in the Vulgate and are generally accepted by the Roman Catholic Church, while Protestants tend to dispute their authority. The New Testament Apocrypha include acts, gospels, and epistles relating to biblical figures, including the Apostles. [ME f. eccl.L *apocrypha* (*scripta*) hidden writings f. *apokruptō* hide away]

apocryphal /əˈpɒkrɪf(ə)l/ *adj.* **1** of doubtful authenticity. **2** invented; mythical (*an apocryphal story*). **3** of or belonging to the Apocrypha.

apodictic /ˌæpəˈdɪktɪk/ *adj.* (also **apodeictic** /-ˈdaɪktɪk/) **1** clearly established. **2** of clear demonstration. [L *apodicticus* f. Gk *apodeiktikos* (as APO-, *deiknumi* show)]

apodosis /əˈpɒdəsɪs/ *n.* (*pl.* **apodoses** /-ˌsiːz/) the main (consequent) clause of a conditional sentence (e.g. *I would agree* in *if you asked me I would agree*). [LL f. Gk f. *apodidōmi* give back (as APO-, *didōmi* give)]

apogee /ˈæpəˌdʒiː/ *n.* **1** the point in a body's orbit where it is furthest from the earth (opp. PERIGEE). **2** the most distant or highest point. □ **apogean** /ˌæpəˈdʒiːən/ *adj.* [F *apogée* or mod.L *apogaeum* f. Gk *apogeion* away from earth (as APO-, *gē* earth)]

apolitical /ˌeɪpəˈlɪtɪk(ə)l/ *adj.* not interested in or concerned with politics.

Apollinaris /əˌpɒlɪˈneərɪs/ (*c*.310–*c*.390), bishop of Laodicea in Asia Minor. He upheld the heretical doctrine, condemned at the Council of Constantinople (381), which asserted that Christ had a human body and soul but no human spirit, this being replaced by the divine Logos. □ **Apollinarian** *adj. & n.*

Apollo /əˈpɒləʊ/ **1** *Gk Mythol.* a god, son of Zeus and Leto and brother of Artemis; in art he is presented as the ideal type of manly beauty, and in later poetry he is associated with the sun. He is also associated especially with music, poetic inspiration, archery, prophecy, medicine, and pastoral life; the sanctuary at Delphi was dedicated to him. **2** the American space programme for landing men on the moon, proposed by President John F. Kennedy in 1961. Apollo 8 was the first mission to orbit the moon (Dec. 1968), and the first landing took place on 20 July 1969, on the Apollo 11 mission; a further five landings took place up to 11 Dec. 1972.

Apollonian /ˌæpə'ləʊnɪən/ adj. **1** of or relating to Apollo. **2** orderly, rational, self-disciplined. [L *Apollonius* f. Gk *Apollōnios*]

Apollonius[1] /ˌæpə'ləʊnɪəs/ (known as Apollonius of Perga) (c.260–190 BC), Greek mathematician. In his principal surviving work, *Conics*, he examined and redefined the various conic sections, and was the first to use the terms *ellipse*, *parabola*, and *hyperbola* for these classes of curve. He also dealt with other aspects of higher geometry, and from his astronomical studies he probably originated the concept of epicycles to account for the retrograde motion of the outer planets.

Apollonius[2] /ˌæpə'ləʊnɪəs/ (known as Apollonius of Rhodes) (3rd century BC), Greek poet. The librarian at Alexandria, he was the author of many works on grammar. He is chiefly known for his epic poem *Argonautica*; written in Homeric style and dealing with the expedition of the Argonauts, it was the first such poem to place love (Medea's love for Jason) in the foreground of the action.

Apollyon /ə'pɒljən/ in Christian thought, the Devil, the 'angel of the bottomless pit' in Rev. 9:11. [L (Vulgate) f. Gk *apolluōn* pres. part. of *apollumi* (as APO-, *ollumi* destroy)]

apologetic /əˌpɒlə'dʒetɪk/ adj. & n. ● adj. **1** regretfully acknowledging or excusing an offence or failure. **2** diffident. **3** of reasoned defence or vindication. ● n. (usu. in pl.) a reasoned defence, esp. of Christianity. □ **apologetically** adv. [F *apologétique* f. LL *apologeticus* f. Gk *apologētikos* f. *apologeomai* speak in defence]

apologia /ˌæpə'ləʊdʒɪə/ n. a formal defence of one's opinions or conduct. [L: see APOLOGY]

apologist /ə'pɒlədʒɪst/ n. a person who defends something by argument. [F *apologiste* f. Gk *apologizomai* render account f. *apologos* account]

apologize /ə'pɒlə.dʒaɪz/ v.intr. (also **-ise**) **1** (often foll. by *for*) make an apology for an offence or failure; express regret. **2** (foll. by *for*) seek to explain or justify. [Gk *apologizomai*: see APOLOGIST]

apologue /'æpə.lɒg/ n. a moral fable. [F *apologue* or L *apologus* f. Gk *apologos* story (as APO-, *logos* discourse)]

apology /ə'pɒlədʒɪ/ n. (pl. **-ies**) **1** a regretful acknowledgement of an offence or failure. **2** an assurance that no offence was intended. **3** an explanation or defence. **4** (foll. by *for*) a poor or scanty specimen of (*this apology for a letter*). [F *apologie* or LL *apologia* f. Gk (as APOLOGETIC)]

apolune /'æpə.luːn/ n. the point in a body's lunar orbit where it is furthest from the moon's centre (opp. PERILUNE). [APO- + L *luna* moon, after *apogee*]

apomixis /ˌæpə'mɪksɪs/ n. (pl. **apomixes** /-siːz/) Biol. asexual reproduction, esp. agamospermy (cf. AMPHIMIXIS). □ **apomictic** /-'mɪktɪk/ adj. [APO- + Gk *mixis* mingling]

apophthegm /'æpə.θem/ n. (US **apothegm**) a terse saying or maxim, an aphorism. □ **apophthegmatic** /ˌæpəθeg'mætɪk/ adj. [F *apophthegme* or mod.L *apothegma* f. Gk *apophthegma* -matos f. *apophtheggomai* speak out]

apoplectic /ˌæpə'plektɪk/ adj. **1** of, causing, suffering, or liable to apoplexy. **2** colloq. enraged. □ **apoplectically** adv. [F *apoplectique* or LL *apoplecticus* f. Gk *apoplēktikos* f. *apoplēssō* strike completely (as APO-, *plēssō* strike)]

apoplexy /'æpə.pleksɪ/ n. **1** Med. = STROKE n. 2. **2** colloq. a rush of extreme emotion, esp. anger. [ME f. OF *apoplexie* f. LL *apoplexia* f. Gk *apoplēxia* (as APOPLECTIC)]

apoptosis /ˌæpɒp'təʊsɪs/ n. Biol. the controlled destruction of cells as part of an organism's normal growth, development, etc. □ **apoptotic** /-'tɒtɪk/ adj. [Gk *apoptōsis* falling off, f. as APO- + PTOSIS]

aporia /ə'pɔːrɪə/ n. **1** Rhet. an expression of doubt. **2** a doubtful matter; a perplexing difficulty. [LL f. Gk, f. *aporos* impassable]

aposematic /ˌæpəsɪ'mætɪk/ adj. Zool. (of coloration, markings, etc.) serving to warn or repel predators. [APO- + Gk *sēma sēmatos* sign]

apostasy /ə'pɒstəsɪ/ n. (pl. **-ies**) **1** renunciation of a belief or faith, esp. religious. **2** abandonment of principles or of a party. **3** an instance of apostasy. [ME f. eccl.L f. NT Gk *apostasia* defection (as APO-, *stat-* stand)]

apostate /ə'pɒsteɪt/ n. & adj. ● n. a person who renounces a former

belief, adherence, etc. ● adj. engaged in apostasy. □ **apostatical** /ˌæpə'stætɪk(ə)l/ adj. [ME f. OF *apostate* or eccl.L *apostata* f. Gk *apostatēs* deserter (as APOSTASY)]

apostatize /ə'pɒstə.taɪz/ v.intr. (also **-ise**) renounce a former belief, adherence, etc. [med.L *apostatizare* f. *apostata*: see APOSTATE]

a posteriori /ˌeɪ pɒˌsterɪ'ɔːraɪ/ adj. & adv. ● adj. (of reasoning) inductive, empirical; proceeding from effects to causes. ● adv. inductively, empirically; from effects to causes (opp. A PRIORI). [L, = from what comes after]

apostle /ə'pɒs(ə)l/ n. **1 a** (**Apostle**) in the Gospels, the twelve chief disciples of Christ. (*See note below.*) **b** the first successful Christian missionary in a country or to a people. **2** a leader or outstanding figure, esp. of a reform movement (*apostle of temperance*). **3** a messenger or representative. □ **apostleship** n. [OE *apostol* f. eccl.L *apostolus* f. Gk *apostolos* messenger (as APO-, *stellō* send forth)]

▪ The twelve Apostles were Peter, Andrew, James, John, Philip, Bartholomew, Thomas, Matthew, James (the Less), Judas (or Thaddaeus), Simon, and Judas Iscariot. After the suicide of Judas Iscariot his place was taken by Matthias; the term Apostle was applied also to Paul and Barnabas.

apostle-bird /ə'pɒs(ə)lˌbɜːd/ n. Austral. a bird forming flocks of about a dozen; esp. a grey bird, *Struthidea cinerea*, of the family Grallinidae.

Apostles' Creed a statement of Christian belief used in the Western Church, dating (with minor variations in form) from the 4th century, at which time the story that it was a joint composition by the twelve Apostles was widely accepted.

Apostle spoon n. a spoon with a figure of an Apostle or saint on the handle.

apostolate /ə'pɒstələt/ n. **1** the position or authority of an apostle. **2** leadership in reform. [eccl.L *apostolatus* (as APOSTLE)]

apostolic /ˌæpə'stɒlɪk/ adj. **1** of or relating to the Apostles. **2** of the pope regarded as the successor of St Peter. [F *apostolique* or eccl.L *apostolicus* f. Gk *apostolikos* (as APOSTLE)]

Apostolic Fathers n.pl. the Christian leaders immediately succeeding the Apostles.

apostolic succession n. (in Christian thought) the uninterrupted transmission of spiritual authority from the Apostles through successive popes and other bishops. The continuity has been disputed; the necessity of it is taught by the Roman Catholic Church but denied by most Protestants.

apostrophe[1] /ə'pɒstrəfɪ/ n. a punctuation mark used to indicate: **1** the omission of letters or numbers (e.g. *can't; he's; 1 Jan. '92*). **2** the possessive case (e.g. *Harry's book; boys' coats*). [F *apostrophe* or LL *apostrophus* f. Gk *apostrophos* accent of elision f. *apostrephō* turn away (as APO-, *strephō* turn)]

apostrophe[2] /ə'pɒstrəfɪ/ n. an exclamatory passage in a speech or poem, addressed to a person (often dead or absent) or thing (often personified). □ **apostrophize** v.tr. & intr. (also **-ise**). [L f. Gk, lit. 'turning away' (as APOSTROPHE[1])]

apothecary /ə'pɒθəkərɪ/ n. (pl. **-ies**) archaic a chemist licensed to dispense medicines and drugs. □ **apothecaries' measure** (or **weight**) Brit. units of weight and liquid volume formerly used in pharmacy. ¶ 12 ounces = one pound; 20 fluid ounces = one pint. [ME f. OF *apotecaire* f. LL *apothecarius* f. L *apotheca* f. Gk *apothēkē* storehouse]

apothegm US var. of APOPHTHEGM.

apothem /'æpə.θem/ n. Geom. a line from the centre of a regular polygon at right angles to any of its sides. [Gk *apotithēmi* put aside (as APO-, *tithēmi* place)]

apotheosis /əˌpɒθɪ'əʊsɪs/ n. (pl. **apotheoses** /-siːz/) **1** elevation to divine status; deification. **2** a glorification of a thing; a sublime example (*apotheosis of the dance*). **3** a deified ideal. [eccl.L f. Gk *apotheoō* make a god of (as APO-, *theos* god)]

apotheosize /ə'pɒθɪə.saɪz/ v.tr. (also **-ise**) **1** make divine; deify. **2** idealize, glorify.

apotropaic /ˌæpətrə'peɪɪk/ adj. supposedly having the power to avert an evil influence or bad luck. [Gk *apotropaios* (as APO-, *trepō* turn)]

appal /ə'pɔːl/ v.tr. (US **appall**) (**appalled, appalling**) **1** greatly dismay or horrify. **2** (as **appalling** adj.) colloq. shocking, unpleasant; bad. □ **appallingly** adv. [ME f. OF *apalir* grow pale]

Appalachian Mountains /ˌæpə'leɪʃ(ə)n/ (also **Appalachians**) a mountain system of eastern North America, stretching from Quebec and Maine in the north to Georgia and Alabama in the south. Its highest peak is Mount Mitchell in North Carolina, which rises to

2,037 m (6,684 ft). Although not particularly high, the Appalachians served as an effective barrier for some 200 years to westward expansion by early European settlers.

Appalachian Trail a 3,200-km (about 2,000-mile) footpath through the Appalachian Mountains from Mount Katahdin in Maine to Springer Mountain in Georgia.

Appaloosa /ˌæpəˈluːsə/ n. a North American breed of horse with dark spots on a light background. [*Opelousa* in Louisiana, or *Palouse*, a river in Idaho]

appanage var. of APANAGE.

apparat /ˈæpəˌrɑːt/ n. the administrative system of a Communist party, esp. in a Communist country. [Russ. f. G, = apparatus]

apparatchik /ˌæpəˈrɑːtʃɪk/ n. (pl. **apparatchiks** or **apparatchiki** /-kɪ/) **1 a** a member of a Communist apparat. **b** a Communist agent or spy. **2 a** a member of a political party in any country who executes policy; a zealous functionary. **b** an official of a public or private organization. [Russ.: see APPARAT]

apparatus /ˌæpəˈreɪtəs/ n. **1** the equipment needed for a particular purpose or function, esp. scientific or technical. **2** a political or other complex organization. **3** *Anat.* the organs used to perform a particular process. **4** (in full **apparatus criticus**) a collection of variants and annotations accompanying a printed text and usu. appearing below it. [L f. *apparare apparat-* make ready for]

apparel /əˈpærəl/ n. & v. **1** *formal* clothing, dress. **2** embroidered ornamentation on some ecclesiastical vestments. ● *v.tr.* (**apparelled**, **apparelling**; US **appareled**, **appareling**) *archaic* clothe. [ME *aparailen* (v.) f. OF *apareillier* f. Rmc *appariculare* (unrecorded) make equal or fit, ult. f. L *par* equal]

apparent /əˈpærənt/ adj. **1** readily visible or perceivable. **2** seeming. □ **apparent horizon** see HORIZON 1b. **apparent magnitude** see MAGNITUDE. **apparent time** solar time. □ **apparently** adv. [ME f. OF *aparant* f. L (as APPEAR)]

apparition /ˌæpəˈrɪʃ(ə)n/ n. a sudden or dramatic appearance, esp. of a ghost or phantom; a visible ghost. [ME f. F *apparition* or f. L *apparitio* attendance (as APPEAR)]

appeal /əˈpiːl/ v. & n. ● v. **1** *intr.* make an earnest or formal request; plead (*appealed for calm*; *appealed to us not to leave*). **2** *intr.* (usu. foll. by *to*) be attractive or of interest; be pleasing. **3** *intr.* (foll. by *to*) resort to or cite for support. **4** *Law* **a** *intr.* (often foll. by *to*) apply (to a higher court) for a reconsideration of the decision of a lower court. **b** *tr.* refer to a higher court to review (a case). **c** *intr.* (foll. by *against*) apply to a higher court to reconsider (a verdict or sentence). **5** *intr. Cricket* call on the umpire for a decision on whether a batsman is out. ● n. **1** the act or an instance of appealing. **2** a formal or urgent request for public support, esp. financial, for a cause. **3** *Law* the referral of a case to a higher court. **4** attractiveness; appealing quality (*sex appeal*). □ **appealer** n. [ME f. OF *apel*, *apeler* f. L *appellare* to address]

appealable /əˈpiːləb(ə)l/ adj. *Law* (of a case) that can be referred to a higher court for review.

appealing /əˈpiːlɪŋ/ adj. attractive, likeable. □ **appealingly** adv.

appear /əˈpɪə(r)/ v.intr. **1** become or be visible. **2** be evident (*a new problem then appeared*). **3** seem; have the appearance of being (*appeared unwell*; *you appear to be right*). **4** present oneself publicly or formally, esp. on stage or as the accused or counsel in a lawcourt. **5** be published or issued (*it appeared in the papers*; *a new edition will appear*). [ME f. OF *apareir* f. L *apparere apparit-* come in sight]

appearance /əˈpɪərəns/ n. **1** the act or an instance of appearing. **2** an outward form as perceived (whether correctly or not), esp. visually (*smarten up one's appearance*; *gives the appearance of trying hard*). **3** a semblance. □ **keep up appearances** maintain an impression or pretence of virtue, affluence, etc. **make** (or **put in**) **an appearance** be present, esp. briefly. **to all appearances** as far as can be seen; apparently. [ME f. OF *aparance, -ence* f. LL *apparentia* (as APPEAR, -ANCE)]

appease /əˈpiːz/ v.tr. **1** make calm or quiet, esp. conciliate (a potential aggressor) by making concessions. **2** satisfy (an appetite, scruples). □ **appeasement** n. **appeaser** n. [ME f. AF *apeser*, OF *apaisier* f. *à* to + *pais* PEACE]

Appel /ˈɑːp(ə)l/, Karel (b.1921), Dutch painter, sculptor, and graphic artist. An exponent of abstract expressionism, he is best known for his paintings, executed in impasto and bright colours and characteristically depicting swirling abstract images suggestive of human and animal forms or fantasy figures. He has also produced polychrome aluminium sculptures.

appellant /əˈpelənt/ n. *Law* a person who appeals to a higher court. [ME f. F (as APPEAL, -ANT)]

appellate /əˈpelət/ adj. *Law* (esp. of a court) concerned with or dealing with appeals. [L *appellatus* (as APPEAL, -ATE[2])]

appellation /ˌæpəˈleɪʃ(ə)n/ n. *formal* a name or title; nomenclature. [ME f. OF f. L *appellatio -onis* (as APPEAL, -ATION)]

appellation contrôlée /ˌæpəˌlæsjɒn kɒnˈtrəʊleɪ/ n. (also **appellation d'origine contrôlée** /ˌdɒrɪˈʒiːn/) a mark that may be awarded to a French wine or foodstuff, guaranteeing the origin and quality of the product; a wine or food carrying this mark. [F, = controlled name]

appellative /əˈpelətɪv/ adj. **1** naming. **2** *Gram.* (of a noun) that designates a class; common. [LL *appellativus* (as APPEAL, -ATIVE)]

append /əˈpend/ v.tr. (usu. foll. by *to*) attach, affix, add, esp. to a written document etc. [L *appendere* hang]

appendage /əˈpendɪdʒ/ n. **1** something attached; an addition. **2** *Zool.* a leg or other projecting part of an arthropod.

appendant /əˈpendənt/ adj. & n. ● adj. (usu. foll. by *to*) attached in a subordinate capacity. ● n. an appendant person or thing. [OF *apendant* f. *apendre* formed as APPEND, -ANT]

appendectomy /ˌæpenˈdektəmɪ/ n. (also **appendicectomy** /əˌpendɪˈsektəmɪ/) (pl. **-ies**) the surgical removal of the appendix. [APPENDIX + -ECTOMY]

appendicitis /əˌpendɪˈsaɪtɪs/ n. inflammation of the appendix. [APPENDIX + -ITIS]

appendix /əˈpendɪks/ n. (pl. **appendices** /-dɪˌsiːz/; **appendixes**) **1** (in full **vermiform appendix**) *Anat.* a small outgrowth of tissue forming a tube-shaped sac attached to the lower end of the caecum. **2** subsidiary matter at the end of a book or document. [L *appendix -icis* f. *appendere* APPEND]

apperceive /ˌæpəˈsiːv/ v.tr. **1** be conscious of perceiving. **2** *Psychol.* compare (a perception) to previously held ideas so as to extract meaning from it. □ **apperception** /-ˈsepʃ(ə)n/ n. **apperceptive** /-ˈseptɪv/ adj. [ME (in obs. sense 'observe') f. OF *aperceveir* ult. f. L *percipere* PERCEIVE]

appertain /ˌæpəˈteɪn/ v.intr. (foll. by *to*) **1** relate. **2** belong as a possession or right. **3** be appropriate. [ME f. OF *apertenir* f. LL *appertinere* f. *pertinere* PERTAIN]

appetence /ˈæpɪt(ə)ns/ n. (also **appetency** /-sɪ/) (pl. **-ces** or **-cies**) (foll. by *for*) longing or desire. [F *appétence* or L *appetentia* f. *appetere* seek after]

appetite /ˈæpɪˌtaɪt/ n. **1** a natural desire to satisfy bodily needs, esp. for food or sexual activity. **2** (usu. foll. by *for*) an inclination or desire. □ **appetitive** /əˈpetɪtɪv/ adj. [ME f. OF *apetit* f. L *appetitus* f. *appetere* seek after]

appetizer /ˈæpɪˌtaɪzə(r)/ n. (also **-iser**) a small amount, esp. of food or drink, to stimulate an appetite. [appetize (back-form. f. APPETIZING)]

appetizing /ˈæpɪˌtaɪzɪŋ/ adj. (also **-ising**) stimulating an appetite, esp. for food. □ **appetizingly** adv. [F *appétissant* irreg. f. *appétit*, formed as APPETITE]

Appian Way /ˈæpɪən/ (Latin **Via Appia** /ˌviːə ˈæpɪə/) the principal road southward from Rome in classical times, named after the censor Appius Claudius Caecus, who in 312 BC built the section to Capua; it was later extended to Brindisi on the SE coast of Italy.

applaud /əˈplɔːd/ v. **1** *intr.* express strong approval or praise, esp. by clapping. **2** *tr.* express approval of (a person or action) verbally or by clapping. [L *applaudere applaus-* clap hands]

applause /əˈplɔːz/ n. **1** an expression of approbation, esp. from an audience etc. by clapping. **2** emphatic approval. [med.L *applausus* (as APPLAUD)]

apple /ˈæp(ə)l/ n. **1** the fruit of a tree of the genus *Malus*, rounded in form and with a crisp flesh. **2** (in full **apple tree**) the tree bearing this. □ **apple of one's eye** a cherished person or thing. **apple-pie bed** a bed made (as a joke) with the sheets folded short, so that the legs cannot be accommodated. **apple-pie order** perfect order; extreme neatness. **she's apples** *Austral. sl.* everything is fine. **upset the apple-cart** spoil careful plans. [OE *æppel* f. Gmc]

applejack /ˈæp(ə)lˌdʒæk/ n. *N. Amer.* a spirit distilled from fermented apple juice. [APPLE + JACK[1]]

Appleton /ˈæp(ə)lt(ə)n/, Sir Edward Victor (1892–1965), English physicist. His investigation of the Heaviside or E layer of the atmosphere led him to the discovery of a higher region of ionized gases (the Appleton layer, now resolved into two layers F1 and F2), from which short-wave radio waves are reflected back to earth. This work, for which he was awarded the Nobel Prize for physics in 1947, was important for long-range radio transmission and radar.

appliance /ə'plaɪəns/ n. a device or piece of equipment used for a specific task. [APPLY + -ANCE]

applicable /'æplɪkəb(ə)l, ə'plɪkə-/ adj. (often foll. by to) **1** that may be applied. **2** having reference; appropriate. □ **applicably** adv. **applicability** /ˌæplɪkə'bɪlɪtɪ, əˌplɪkə-/ n. [OF applicable or med.L applicabilis (as APPLY, -ABLE)]

applicant /'æplɪkənt/ n. a person who applies for something, esp. a job. [APPLICATION + -ANT]

application /ˌæplɪ'keɪʃ(ə)n/ n. **1** the act or an instance of applying, esp. medicinal ointment to the skin. **2** a formal request, usu. in writing, for employment, membership, etc. **3 a** relevance. **b** the use to which something can or should be put. **4** sustained or concentrated effort; diligence. □ **application program** Computing a program designed and written to fulfil a particular purpose of the user. [ME f. F f. L applicatio -onis (as APPLY, -ATION)]

applicator /'æplɪˌkeɪtə(r)/ n. a device for applying a substance to a surface, esp. the skin, or for inserting something into a cavity, esp. into the body. [APPLICATION + -OR¹]

applied /ə'plaɪd/ adj. (of a subject of study) put to practical use as opposed to being theoretical (cf. PURE adj. 10). □ **applied mathematics** see MATHEMATICS.

appliqué /æ'pliːkeɪ/ n., adj., & v. ● n. ornamental work in which fabric is cut out and attached, usu. sewn, to the surface of another fabric to form pictures or patterns. ● adj. executed in appliqué. ● v.tr. (**appliqués, appliquéd, appliquéing**) decorate with appliqué; make using appliqué technique. [F, past part. of appliquer apply f. L applicare: see APPLY]

apply /ə'plaɪ/ v. (**-ies, -ied**) **1** intr. (often foll. by for, to, or to + infin.) make a formal request for something to be done, given, etc. (apply for a job; apply for help to the governors; applied to be sent overseas). **2** intr. (often foll. by to) have relevance (does not apply in this case). **3** tr. **a** make use of as relevant or suitable; employ (apply the rules). **b** operate (apply the handbrake). **4** tr. (often foll. by to) **a** put or spread on (applied the ointment to the cut). **b** administer (applied the remedy; applied common sense to the problem). **5** refl. (often foll. by to) devote oneself (applied myself to the task). □ **applier** n. [ME f. OF aplier f. L applicare fold, fasten to]

appoggiatura /əˌpɒdʒə'tʊərə/ n. Mus. a grace note performed before an essential note of a melody and normally taking half its time value. [It.]

appoint /ə'pɔɪnt/ v.tr. **1** assign a post or office to (appoint him governor; appoint him to govern; appointed to the post). **2** (often foll. by for) fix, decide on (a time, place, etc.) (Wednesday was appointed for the meeting; 8.30 was the appointed time). **3** prescribe; ordain (Holy Writ appointed by the Church). **4** Law **a** (also absol.) declare the destination of (property etc.). **b** declare (a person) as having an interest in property etc. (Jones was appointed in the will). **5** (as **appointed** adj.) equipped, furnished (a badly appointed hotel). □ **appointer** n. **appointive** adj. **appointee** /ˌæpɔɪn'tiː/ n. [ME f. OF apointer f. à point to a point]

appointment /ə'pɔɪntmənt/ n. **1** an arrangement to meet at a specific time and place. **2 a** a post or office available for applicants, or recently filled (took up the appointment on Monday). **b** a person appointed. **c** the act or an instance of appointing, esp. to a post. **3** (usu. in pl.) **a** furniture, fittings. **b** equipment. [ME f. OF apointement (as APPOINT, -MENT)]

apport /ə'pɔːt/ n. **1** the production of material objects by supposedly occult means at a seance. **2** an object so produced. [ME (in obs. senses), f. OF aport f. aporter f. à to + porter bring]

apportion /ə'pɔːʃ(ə)n/ v.tr. (often foll. by to) share out; assign as a share. □ **apportionable** adj. **apportionment** n. [F apportionner or f. med.L apportionare (as AD-, PORTION)]

apposite /'æpəzɪt/ adj. (often foll. by to) **1** apt; well chosen. **2** well expressed. □ **appositely** adv. **appositeness** n. [L appositus past part. of apponere (as AD-, ponere put)]

apposition /ˌæpə'zɪʃ(ə)n/ n. **1** placing side by side; juxtaposition. **2** Gram. the placing of a word next to another, esp. the addition of one noun to another, in order to qualify or explain the first (e.g. William the Conqueror; my friend Sue). □ **appositional** adj. [ME f. F apposition or f. LL appositio (as APPOSITE, -ITION)]

appraisal /ə'preɪz(ə)l/ n. **1** the act or an instance of appraising. **2** the evaluation of an employee's performance over a particular period, often in the form of a formal interview.

appraise /ə'preɪz/ v.tr. **1 a** estimate the value or quality of (appraised

her skills). **b** evaluate (an employee) formally in terms of professional performance, progress, etc. **2** (esp. of an official valuer) set a price on; value. □ **appraisable** adj. **appraiser** n. **appraisive** adj. **appraisee** /əpreɪ'ziː/ n. [APPRIZE by assim. to PRAISE]

appreciable /ə'priːʃəb(ə)l/ adj. large enough to be noticed; significant; considerable (appreciable progress has been made). □ **appreciably** adv. [F f. apprécier (as APPRECIATE)]

appreciate /ə'priːʃɪˌeɪt, ə'priːsɪ-/ v. **1** tr. **a** esteem highly; value. **b** be grateful for (we appreciate your sympathy). **c** be sensitive to (appreciate the nuances). **2** tr. (often foll. by that + clause) understand; recognize (I appreciate that I may be wrong). **3 a** intr. (of property etc.) rise in value. **b** tr. raise in value. □ **appreciator** n. **appreciative** /-ʃɪətɪv/ adj. **appreciatively** adv. **appreciativeness** n. **appreciatory** /-ʃɪətərɪ/ adj. [LL appretiare appraise (as AD-, pretium price)]

appreciation /əˌpriːʃɪ'eɪʃ(ə)n, əˌpriːsɪ-/ n. **1** favourable or grateful recognition. **2** an estimation or judgement; sensitive understanding of or reaction to (a quick appreciation of the problem). **3** an increase in value. **4** a (usu. favourable) review of a book, film, etc. [F f. LL appretiatio -onis (as APPRECIATE, -ATION)]

apprehend /ˌæprɪ'hend/ v.tr. **1** understand, perceive (apprehend your meaning). **2** seize, arrest (apprehended the criminal). **3** anticipate with uneasiness or fear (apprehending the results). [F appréhender or L apprehendere (as AD-, prehendere prehens- lay hold of)]

apprehensible /ˌæprɪ'hensɪb(ə)l/ adj. capable of being apprehended by the senses or the intellect (an apprehensible theory; an apprehensible change in her expression). □ **apprehensibility** /-ˌhensɪ'bɪlɪtɪ/ n. [LL apprehensibilis (as APPREHEND, -IBLE)]

apprehension /ˌæprɪ'henʃ(ə)n/ n. **1** a sense of uneasiness or foreboding. **2** understanding, grasp. **3** arrest, capture (apprehension of the suspect). **4** an idea; a conception. [F appréhension or LL apprehensio (as APPREHEND, -ION)]

apprehensive /ˌæprɪ'hensɪv/ adj. **1** (often foll. by of, for) uneasily fearful; anxious. **2** relating to perception by the senses or the intellect. **3** archaic perceptive; intelligent. □ **apprehensively** adv. **apprehensiveness** n. [F appréhensif or med.L apprehensivus (as APPREHEND, -IVE)]

apprentice /ə'prentɪs/ n. & v. ● n. **1** a person who is learning a trade by being employed in it for an agreed period at low wages. **2** a beginner; a novice. ● v.tr. (usu. foll. by to) engage or bind as an apprentice (was apprenticed to a builder). □ **apprenticeship** n. [ME f. OF aprentis f. aprendre learn (as APPREHEND), after words in -tis, -tif, f. L -tivus: see -IVE]

apprise /ə'praɪz/ v.tr. inform. □ **be apprised of** be aware of. [F appris -ise past part. of apprendre learn, teach (as APPREHEND)]

apprize /ə'praɪz/ v.tr. archaic **1** esteem highly. **2** appraise. [ME f. OF aprisier f. à to + pris PRICE]

appro /'æprəʊ/ n. Brit. colloq. □ **on appro** = on approval (see APPROVAL). [abbr. of approval or approbation]

approach /ə'prəʊtʃ/ v. & n. ● v. **1** tr. come near or nearer to (a place or time). **2** intr. come near or nearer in space or time (the hour approaches). **3** tr. make a tentative proposal or suggestion to (approached me about a loan). **4** tr. **a** be similar in character, quality, etc., to (doesn't approach her for artistic skill). **b** approximate to (a population approaching 5 million). **5** tr. attempt to influence or bribe. **6** tr. set about, tackle (a task etc.). **7** intr. Golf play an approach shot. **8** intr. Aeron. prepare to land. **9** tr. archaic bring near. ● n. **1** an act or means of approaching (made an approach; an approach lined with trees). **2** an approximation (an approach to an apology). **3** a way of dealing with a person or thing (needs a new approach). **4** (usu. in pl.) a sexual advance. **5** Golf a stroke from the fairway to the green. **6** Aeron. the final part of a flight before landing. **7** Bridge a bidding method with a gradual advance to a final contract. □ **approach road** Brit. a road by which traffic enters a motorway. [ME f. OF aproch(i)er f. eccl.L appropiare draw near (as AD-, propius compar. of prope near)]

approachable /ə'prəʊtʃəb(ə)l/ adj. **1** friendly; easy to talk to. **2** able to be approached. □ **approachability** /əˌprəʊtʃə'bɪlɪtɪ/ n.

approbate /'æprəˌbeɪt/ v.tr. N. Amer. approve formally; sanction. [ME f. L approbare (as AD-, probare test f. probus good)]

approbation /ˌæprə'beɪʃ(ə)n/ n. approval, consent. □ **approbative** /'æprəˌbeɪtɪv/ adj. **approbatory** /-ˌbeɪtərɪ/ adj. [ME f. OF f. L approbatio -onis (as APPROBATE, -ATION)]

appropriate adj. & v. ● /ə'prəʊprɪət/ (often foll. by to, for) **1** suitable or proper. **2** formal belonging or particular. ● v.tr. /ə'prəʊprɪˌeɪt/ **1** take possession of, esp. without authority. **2** devote (money etc.) to special purposes. □ **appropriately** /-ətlɪ/ adv. **appropriateness** /-ətnɪs/ n.

appropriator /-ˌeɪtə(r)/ n. **appropriation** /əˌprəʊprɪˈeɪʃ(ə)n/ n. [LL *appropriatus* past part. of *appropriare* (as AD-, *proprius* own)]

approval /əˈpruːv(ə)l/ n. **1** the act of approving. **2** an instance of this; consent; a favourable opinion (*with your approval; looked at him with approval*). □ **on approval** (of goods supplied) to be returned if not satisfactory.

approve /əˈpruːv/ v. **1** tr. confirm; sanction (*approved his application*). **2** intr. give or have a favourable opinion. **3** tr. commend (*approved the new hat*). **4** tr. (usu. refl.) archaic demonstrate oneself to be (*approved himself a coward*). □ **approved school** hist. a residential place of training for young offenders. **approve of 1** pronounce or consider good or satisfactory; commend. **2** agree to. □ **approver** n. **approvingly** adv. [ME f. OF *aprover* f. L (as APPROBATE)]

approx. abbr. **1** approximate. **2** approximately.

approximate adj. & v. ● adj. /əˈprɒksɪmət/ **1** fairly correct or accurate; near to the actual (*the approximate time of arrival; an approximate guess*). **2** near or next (*your approximate neighbour*). ● v.tr. & intr. /əˈprɒksɪˌmeɪt/ (often foll. by *to*) bring or come near (esp. in quality, number, etc.), but not exactly (*approximates to the truth; approximates the amount required*). □ **approximately** /-mətlɪ/ adv. **approximation** /əˌprɒksɪˈmeɪʃ(ə)n/ n. [LL *approximatus* past part. of *approximare* (as AD-, *proximus* very near)]

appurtenance /əˈpɜːtɪnəns/ n. (usu. in pl.) a belonging; an appendage; an accessory. [ME f. AF *apurtenaunce*, OF *apertenance* (as APPERTAIN, -ANCE)]

appurtenant /əˈpɜːtɪnənt/ adj. (often foll. by *to*) belonging or appertaining; pertinent. [ME f. OF *apartenant* pres. part. (as APPERTAIN)]

APR abbr. annual or annualized percentage rate (esp. of interest on loans or credit).

Apr. abbr. April.

après-ski /ˌæpreɪˈskiː/ n. & adj. ● n. the evening, esp. its social activities, following a day's skiing. ● attrib.adj. (of clothes, drinks, etc.) appropriate to social activities following skiing. [F]

apricot /ˈeɪprɪˌkɒt/ n. & adj. ● n. **1 a** a juicy soft fruit, smaller than a peach, of an orange-yellow colour. **b** the tree, *Prunus armeniaca*, bearing it. **2** the ripe fruit's orange-yellow colour. ● adj. orange-yellow (*apricot dress*). [Port. *albricoque* or Sp. *albaricoque* f. Arab. *al* the + *barḳuḳ* f. late Gk *praikokion* f. L *praecoquum* var. of *praecox* early-ripe: *apri-* after L *apricus* ripe, *-cot* by assim. to F *abricot*]

April /ˈeɪprɪl/ n. the fourth month of the year. The Romans considered the month *Aprilis* to be sacred to Venus, goddess of love, and its name may be taken from that of her Greek equivalent Aphrodite. □ **April Fool** a person successfully tricked on 1 Apr. **April Fool's** (or **Fools'**) **Day** 1 Apr. The custom of playing tricks on this day has been observed in many countries for hundreds of years, but its origin is unknown. [ME f. L *Aprilis*]

a priori /ˌeɪ praɪˈɔːraɪ/ adj. & adv. ● adj. **1** (of reasoning) deductive; proceeding from causes to effects (opp. A POSTERIORI). **2** (of concepts, knowledge, etc.) logically independent of experience; not derived from experience (opp. EMPIRICAL). **3** not submitted to critical investigation (*an a priori conjecture*). ● adv. **1** in an a priori manner. **2** as far as one knows; presumptively. □ **apriorism** /ˌeɪpraɪˈɔːrɪz(ə)m, eɪˈpraɪəˌrɪz(ə)m/ n. [L, = from what is before]

apron /ˈeɪprən/ n. **1 a** a garment covering and protecting the front of a person's clothes, either from chest or waist level, and tied at the back. **b** official clothing of this kind (*bishop's apron*). **c** anything resembling an apron in shape or function. **2** Theatr. the part of a stage in front of the curtain. **3** the hard-surfaced area on an airfield used for manoeuvring or loading aircraft. **4** an endless conveyor belt. □ **tied to a person's apron-strings** dominated by or dependent on that person (usu. a woman). □ **aproned** adj. **apronful** n. (pl. **-fuls**). [ME *naperon* etc. f. OF dimin. of *nape* table-cloth f. L *mappa*: for loss of n cf. ADDER]

apropos /ˈæprəˌpəʊ/ adj., adv., & prep. ● adj. to the point or purpose; appropriate (*his comment was apropos*). ● adv. **1** appropriately (*spoke apropos*). **2** (absol.) by the way; incidentally (*apropos, she's not going*). ● prep. in respect of; concerning. [F *à propos* f. *à* to + *propos* PURPOSE]

apse /æps/ n. **1** a large semicircular or polygonal recess, arched or with a domed roof, esp. at the eastern end of a church. **2** = APSIS. □ **apsidal** /ˈæpsɪd(ə)l/ adj. [L APSIS]

apsis /ˈæpsɪs/ n. (pl. **apsides** /-sɪˌdiːz/) either of two points on the orbit of a planet or satellite that are nearest to or furthest from the body round which it moves. □ **apsidal** /ˈæpsɪd(ə)l/ adj. [L f. Gk (h)*apsis*, -*idos* arch, vault]

APT abbr. (in the UK) Advanced Passenger Train.

apt /æpt/ adj. **1** appropriate, suitable. **2** (foll. by *to* + infin.) having a tendency (*apt to lose his temper*). **3** clever; quick to learn (*an apt pupil; apt at the work*). □ **aptly** adv. **aptness** n. [ME f. L *aptus* fitted, past part. of *apere* fasten]

apterous /ˈæptərəs/ adj. **1** Zool. (of insects) without wings. **2** Bot. (of seeds or fruits) having no winglike expansions. [Gk *apteros* f. *a-* not + *pteron* wing]

aptitude /ˈæptɪˌtjuːd/ n. **1** a natural propensity or talent (*shows an aptitude for drawing*). **2** ability or fitness, esp. to acquire a particular skill. [F f. LL *aptitudo -inis* (as APT, -TUDE)]

Apuleius /ˌæpjʊˈliːəs/ (born AD *c*.123), Roman writer, born in Africa. Renowned as an orator, he wrote a variety of rhetorical and philosophical works, but is best known as the author of the *Metamorphoses* (*The Golden Ass*), a picaresque novel which recounts the adventures of a man who is transformed into an ass. Apuleius' writings are characterized by an exuberant and bizarre use of language.

Apulia /əˈpjuːlɪə/ (Italian **Puglia** /ˈpuʎʎa/) a region of SE Italy, extending into the 'heel' of the peninsula; capital, Bari.

Aqaba /ˈækəbə/ Jordan's only port, at the head of the Gulf of Aqaba; pop. (est. 1983) 40,000.

Aqaba, Gulf of a part of the Red Sea extending northwards between the Sinai and Arabian peninsulas.

aqua /ˈækwə/ n. the colour aquamarine. [abbr.]

aquaculture /ˈækwəˌkʌltʃə(r)/ n. the cultivation or rearing of aquatic plants or animals. [L *aqua* water + CULTURE, after *agriculture*]

aqua fortis /ˌækwə ˈfɔːtɪs/ n. Chem. nitric acid. [L, = strong water]

aqualung /ˈækwəˌlʌŋ/ n. & v. (US propr. **Aqua-Lung**) ● n. a portable breathing apparatus for divers, consisting of cylinders of compressed air strapped on the back, feeding air automatically through a mask or mouthpiece. ● v.intr. use an aqualung. [L *aqua* water + LUNG]

aquamarine /ˌækwəməˈriːn/ n. & adj. ● n. **1** a light bluish-green beryl. **2** its colour. ● adj. bluish-green [L *aqua marina* sea water]

aquanaut /ˈækwəˌnɔːt/ n. an underwater swimmer or explorer. [L *aqua* water + Gk *nautēs* sailor]

aquaplane /ˈækwəˌpleɪn/ n. & v. ● n. a board for riding on water, pulled by a speedboat. ● v.intr. **1** ride on an aquaplane. **2** (of a vehicle) glide uncontrollably on the wet surface of a road. [L *aqua* water + PLANE[1]]

aqua regia /ˌækwə ˈriːdʒɪə/ n. Chem. a mixture of concentrated nitric and hydrochloric acids, a highly corrosive liquid attacking many substances unaffected by other reagents. [L, = royal water]

aquarelle /ˌækwəˈrel/ n. a painting in thin, usu. transparent watercolours. [F f. It. *acquarella* watercolour, dimin. of *acqua* f. L *aqua* water]

aquarist /ˈækwərɪst/ n. a person who keeps an aquarium.

aquarium /əˈkweərɪəm/ n. (pl. **aquariums** or **aquaria** /-rɪə/) an artificial environment designed for keeping live aquatic plants and animals for study or exhibition, esp. a tank of water with transparent sides. [neut. of L *aquarius* of water (*aqua*) after *vivarium*]

Aquarius /əˈkweərɪəs/ n. **1** Astron. a large constellation (the Water-carrier or Water-bearer), said to represent a man pouring water from a jar. It contains no bright stars but has several good examples of planetary nebulae. **2** Astrol. **a** the eleventh sign of the zodiac, which the sun enters about 21 Jan. **b** a person born when the sun is in this sign. □ **Aquarian** adj. & n. [ME f. L (as AQUARIUM)]

Aquarius, Age of Astrol. an age which the world has just entered or is about to enter, believed by some to signal a period of peace and harmony. (See also NEW AGE; ARIES, FIRST POINT OF.)

aquatic /əˈkwætɪk, əˈkwɒt-/ adj. & n. ● adj. **1** growing or living in or near water. **2** (of a sport) played in or on water. ● n. **1** an aquatic plant or animal. **2** (in pl.) aquatic sports. [ME f. F *aquatique* or L *aquaticus* f. *aqua* water]

aquatint /ˈækwəˌtɪnt/ n. **1** a print resembling a watercolour, produced from a copper plate etched with nitric acid. **2** the process of producing this. (See note below.) [F *aquatinte* f. It. *acqua tinta* coloured water]

▪ The use of aquatint as a landscape medium was pioneered by the topographical artist Paul Sandby in the 1770s, while its combination with line etching reached supreme heights in Goya's graphic work. Although it was somewhat eclipsed by new graphic techniques in the 19th century, there has been a revival of interest in this medium in recent years.

aquavit /ˈækwəvɪt, -ˌviːt/ (also **akvavit** /ˈækvə-/) n. an alcoholic spirit made from potatoes etc. [Scand.]

aqua vitae /ˌækwə ˈviːtaɪ/ n. a strong alcoholic spirit, esp. brandy. [L = water of life]

aqueduct /ˈækwɪˌdʌkt/ n. **1** an artificial channel for conveying water, esp. in the form of a bridge supported by tall columns across a valley. **2** Physiol. a small canal, esp. in the head of mammals. [L aquae ductus conduit f. aqua water + ducere duct- to lead]

aqueous /ˈeɪkwɪəs/ adj. **1** of, containing, or like water. **2** Geol. produced by water (aqueous rocks). □ **aqueous humour** Anat. the clear fluid in the eye between the lens and the cornea. [med.L aqueus f. L aqua water]

aquifer /ˈækwɪfə(r)/ n. Geol. a layer of rock or soil able to hold or transmit much water. [L aqui- f. aqua water + -fer bearing f. ferre bear]

Aquila[1] /ˈækwɪlə/ Astron. a small northern constellation (the Eagle), said to represent the eagle that carried Ganymede to Olympus. It contains the bright star Altair, and some rich star fields of the Milky Way. [L]

Aquila[2] /ˈækwɪlə/ (Italian **L'Aquila** /ˈlaːkwɪla/) a city in east central Italy, capital of Abruzzi region; pop. (1990) 67,820.

aquilegia /ˌækwɪˈliːdʒə/ n. a ranunculaceous plant of the genus Aquilegia, with (often blue-coloured) flowers having backward pointing spurs. See also COLUMBINE. [mod. use of a med.L word: orig. unkn.]

aquiline /ˈækwɪˌlaɪn/ adj. **1** of or like an eagle. **2** (of a nose) curved like an eagle's beak. [L aquilinus f. aquila eagle]

Aquinas, St Thomas /əˈkwaɪnəs/ (known as 'the Angelic Doctor') (1225–74), Italian philosopher, theologian, and Dominican friar. Regarded as the greatest figure of scholasticism, he also devised the official Roman Catholic tenets as declared by Pope Leo XIII. His works include many commentaries on Aristotle as well as the Summa Contra Gentiles (intended as a manual for those disputing with Spanish Muslims and Jews). His principal achievement was to make the work of Aristotle acceptable in Christian western Europe; his own metaphysics, his account of the human mind, and his moral philosophy were a development of Aristotle's, and in his famous arguments for the existence of God ('the Five Ways') he was indebted to Aristotle and to Arabic philosophers. Feast day, 28 Jan.

Aquitaine[1] /ˌækwɪˈteɪn/, French akitɛn/ a region and former province of SW France, on the Bay of Biscay, centred on Bordeaux. A province of the Roman Empire and a medieval duchy, it became, by the marriage of Eleanor of Aquitaine to Henry II, an English possession and remained so until 1453.

Aquitaine[2], Eleanor of, see ELEANOR OF AQUITAINE.

AR abbr. **1** US Arkansas (in official postal use). **2** Autonomous Republic.

Ar symb. Chem. the element argon.

ar- /ə(r)/ prefix assim. form of AD- before r.

-ar[1] /ə(r)/ suffix **1** forming adjectives (angular; linear; nuclear; titular). **2** forming nouns (scholar). [OF -aire or -ier or L -aris]

-ar[2] /ə(r)/ suffix forming nouns (pillar). [F -er or L -ar, -are, neut. of -aris]

-ar[3] /ə(r)/ suffix forming nouns (bursar; exemplar; mortar; vicar). [OF -aire or -ier or L -arius, -arium]

-ar[4] /ə(r)/ suffix assim. form of -ER[1], -OR[1] (liar; pedlar).

ARA abbr. Associate of the Royal Academy.

Arab /ˈærəb/ n. & adj. ● n. **1** a member of a Semitic people inhabiting originally the Arabian peninsula and neighbouring countries, now also other parts of the Middle East and North Africa. **2** an Arabian horse, esp. prized for pure breeding and swiftness. ● adj. of or relating to Arabia or the Arabs. [F Arabe f. L Arabs Arabis f. Gk Araps -abos f. Arab. 'arab]

arabesque /ˌærəˈbesk/ n. **1** Ballet a posture with one leg extended horizontally backwards, torso extended forwards, and arms outstretched. **2** a design of intertwined leaves, scrolls, etc. **3** Mus. a florid melodic section or composition. [F f. It. arabesco f. arabo Arab]

Arabia /əˈreɪbɪə/ (also **Arabian peninsula**) a peninsula of SW Asia, largely desert, lying between the Red Sea and the Persian Gulf and bounded on the north by Jordan and Iraq. The original homeland of the Arabs and the historic centre of Islam, modern Arabia comprises the states of Saudi Arabia, Yemen, Oman, Bahrain, Kuwait, Qatar, and the United Arab Emirates.

Arabian /əˈreɪbɪən/ adj. & n. ● adj. of or relating to Arabia (esp. with geographical reference). ● n. a native of Arabia. ¶ Now less common than Arab in this sense. □ **Arabian camel** a domesticated camel, Camelus dromedarius, native to the deserts of North Africa and the Near East, with one hump (cf. DROMEDARY). [ME f. OF arabi prob. f. Arab. 'arabī, or f. L Arabus, Arabius f. Gk Arabios]

Arabian Desert (also called the Eastern Desert) a desert in eastern Egypt, between the Nile and the Red Sea.

Arabian Gulf an alternative name for the PERSIAN GULF.

Arabian Nights a collection of stories and romances written in Arabic, also called Arabian Nights' Entertainments or The Thousand and One Nights. The framework (in which the king of Samarkand has killed all his wives until he marries Scheherazade, who saves her life by entertaining him with her stories) is of Persian origin, though the stories themselves also derive from Indian and other sources. After their translation into French and other European languages in the early 18th century, the tales were very influential in European culture; they include the stories of Aladdin and Sinbad the Sailor.

Arabian peninsula see ARABIA.

Arabian Sea the north-western part of the Indian Ocean, between Arabia and India.

Arabic /ˈærəbɪk/ n. the Semitic language of the Arabs, now spoken in much of North Africa and the Middle East. Arabic is related to Hebrew and was originally confined to the Arabian peninsula, but during the Islamic conquests of the 7th century it was carried northwards to Iraq and Syria and westwards to North Africa and Spain; it is now the official language of a number of countries in the Middle East and North Africa, being the native language of some 120 million people. There are many dialects of spoken Arabic but one common written language, which is based on 'classical' Arabic, the written form used in the Middle Ages when Arabic was the universal language of the Near and Middle East. It is the language of the Koran and of medieval Arabian literature, and was the chief medium of scientific and philosophical thought for some centuries, bequeathing to us words such as alcohol, elixir, and zenith. Arabic is written from right to left in a traditional script of uncertain origin, which is used also for a number of other languages. [ME f. OF arabic f. L arabicus f. Gk arabikos]

arabica /əˈræbɪkə/ n. (also **arabica coffee**) **1** a coffee plant of the most widely grown species Coffea arabica. **2** beans or coffee from this (cf. ROBUSTA). [mod.L f. L arabicus]

Arabic numeral n. any of the numerals 0, 1, 2, 3, 4, 5, 6, 7, 8, and 9. Arabic numerals reached western Europe (where they replaced Roman numerals) through Arabia by about AD 1200 but probably originated in India. Different symbols are used in modern Arabic. (See also ROMAN NUMERAL.)

arabis /ˈærəbɪs/ n. a cruciferous plant of the genus Arabis, low-growing with toothed leaves and usu. white flowers. Also called rock cress or wall cress. [med.L f. Gk, = Arabian]

Arabist /ˈærəbɪst/ n. a student of Arabic civilization, language, etc.

arable /ˈærəb(ə)l/ adj. & n. ● adj. **1** (of land) ploughed, or suitable for ploughing and crop production. **2** (of crops) that can be grown on arable land. ● n. arable land or crops. [F arable or L arabilis f. arare to plough]

Arab League see LEAGUE OF ARAB STATES.

Araby /ˈærəbɪ/ n. poet. Arabia. [OF Arabie f. L Arabia f. Gk]

Aracajú /ˌærəkəˈʒuː/ a port in eastern Brazil, on the Atlantic coast, capital of the state of Sergipe; pop. (1990) 404,828.

Arachne /əˈræknɪ/ Gk Mythol. a woman of Colophon in Lydia, a skilful weaver who challenged Athene to a contest. Athene destroyed Arachne's work and Arachne tried to hang herself, but Athene changed her into a spider. [Gk arakhnē spider]

arachnid /əˈræknɪd/ n. Zool. an arthropod of the class Arachnida, having four pairs of legs, simple eyes, and usually pincers or fangs, e.g. scorpions, spiders, mites, and ticks. □ **arachnidan** adj. & n. [F arachnide or mod.L arachnida f. Gk arakhnē spider]

arachnoid /əˈræknɔɪd/ n. & adj. ● n. (in full **arachnoid membrane**) Anat. one of the three membranes (see MENINX) that surround the brain and spinal cord of vertebrates. ● adj. Bot. covered with long cobweb-like hairs. [mod.L arachnoides f. Gk arakhnoeidēs like a cobweb f. arakhnē: see ARACHNID]

arachnophobia /əˌræknəˈfəʊbɪə/ n. an abnormal fear of spiders. □ **arachnophobe** /əˈræknəˌfəʊb/ n. [f. Gk arakhnē spider + -PHOBIA]

Arafat /ˈærəˌfæt/, Yasser (b.1929), Palestinian leader, chairman of the Palestine Liberation Organization from 1968. In 1956 he co-founded Al Fatah, the Arab group which came to dominate the PLO from 1967. In 1974 he became the first representative of a non-governmental organization to address the United Nations General Assembly. Despite challenges to his authority within the PLO, he has remained its leader. After the signing of a PLO–Israeli peace accord providing for limited Palestinian autonomy in the West Bank and the Gaza Strip, in July

1994 Arafat became leader of the new Palestine National Authority. The same year he shared the Nobel Peace Prize with Yitzhak Rabin and Shimon Peres. Arafat won a landslide victory in the first Palestinian presidential election (1996).

Arafura Sea /ˌærəˈfʊərə/ a sea lying between northern Australia, the islands of east Indonesia, and New Guinea.

Aragon[1] /ˈærəgən/ an autonomous region of NE Spain, bounded on the north by the Pyrenees and on the east by Catalonia and Valencia; capital, Saragossa. It was formerly an independent kingdom, which was conquered in the 5th century by the Visigoths and then in the 8th century by the Moors. In 1137 it was united with Catalonia and in 1479 with Castile.

Aragon[2], Catherine of, see CATHERINE OF ARAGON.

arak var. of ARRACK.

Araldite /ˈærəlˌdaɪt/ n. propr. an epoxy resin used as a strong heatproof cement to mend china, plastic, etc. [20th c.: orig. uncert.]

Aral Sea /ˈærəl/ an inland sea in central Asia, on the border between Kazakhstan and Uzbekistan. The diversion for irrigation of the water flowing into the Aral Sea has led to its area being reduced to two-thirds of its original size between 1960 and 1990 with serious consequences for the environment.

Aramaean /ˌærəˈmiːən/ n. & adj. ● n. a member of a group of ancient Aramaic-speaking peoples inhabiting Aram (modern Syria) and part of Babylonia in the 11th–8th centuries BC. ● adj. of or relating to Aram or the Aramaeans. [f. as ARAMAIC + -AN]

Aramaic /ˌærəˈmeɪɪk/ n. a Semitic language of ancient Syria, which was used as the lingua franca in the Near East from the 6th century BC and gradually replaced Hebrew as the language of the Jews in those parts (Jesus and his disciples are believed to have spoken a dialect of it). Aramaic was supplanted by Arabic in the 7th century AD. A modern form of Aramaic is still spoken in parts of Syria and Turkey; one of its most important descendants is Syriac. [L *Aramaeus* f. Gk *Aramaios* of Aram (see ARAMAEAN)]

Aran /ˈærən/ adj. of a type of knitwear typically featuring raised cable-stitch patterns and large diamond designs traditionally used in the Aran Islands.

Aranda /əˈrændə/ n. & adj. (also **Arunta** /əˈrʌntə/) ● n. (pl. same or **Arandas**) **1** a member of an Aboriginal people of central Australia. **2** the language spoken by this people. ● adj. of or relating to the Aranda or their language. [Aboriginal]

Aran Islands a group of three islands, Inishmore, Inishmaan, and Inisheer, off the west coast of the Republic of Ireland.

Aranyaka /ˌærəˈnjækə/ n. each of a set of Hindu sacred treatises based on the Brahmanas, composed in Sanskrit c.700 BC. Intended only for initiates, the Aranyakas contain mystical and philosophical material and explications of esoteric rites. [Skr., lit. 'book of the forest']

arapaima /ˌærəˈpaɪmə/ n. a very large edible freshwater fish, *Arapaima gigas*, native to South America. [Tupi]

Ararat, Mount /ˈærəˌræt/ a pair of volcanic peaks in eastern Turkey, near the borders with Armenia and Iran. The higher peak, which rises to 5,165 m (16,946 ft), is the traditional site of the resting-place of Noah's ark after the Flood (Gen. 8:4).

arational /eɪˈræʃ(ə)n(ə)l/ adj. that does not purport to be rational.

Araucanian /ˌærɔːˈkeɪnɪən/ n. & adj. ● n. **1** a member of a native people of central Chile and adjacent parts of Argentina. **2** the language spoken by this people. ● adj. of or relating to this people or their language. [f. *Araucania* region of Chile]

araucaria /ˌærɔːˈkeərɪə/ n. an evergreen conifer of the genus *Araucaria*, found in the southern hemisphere, e.g. the monkey-puzzle tree. [mod.L f. *Arauco*, name of a province in Chile]

Arawak /ˈærəˌwæk/ n. & adj. (pl. same or **Arawaks**) ● n. **1** a member of a group of native peoples of the Greater Antilles and northern and western South America. The Arawak were forced out of the Antilles by the more warlike Caribs shortly before Spanish expansion in the Caribbean. **2** the languages of these peoples. ● adj. of or relating to these peoples or their languages. □ **Arawakan** /ˌærəˈwækən/ adj. & n.

arb /ɑːb/ n. colloq. = ARBITRAGEUR.

arbalest /ˈɑːbəˌlest/ n. (also **arblast** /ˈɑːblɑːst/) hist. a crossbow with a mechanism for drawing the string. [OE *arblast* f. OF *arbaleste* f. LL *arcubalista* f. *arcus* bow + BALLISTA]

arbiter /ˈɑːbɪtə(r)/ n. (fem. **arbitress** /-trɪs/) **1 a** an arbitrator in a dispute. **b** a judge; an authority (*arbiter of taste*). **2** (often foll. by *of*) a person who has entire control of something. □ **arbiter elegantiarum** (or **elegantiae**) /ˌelɪˌgæntɪˈɑːrəm, ˌelɪˈgæntɪˌiː/ a judge of artistic taste and etiquette. [L]

arbitrage /ˈɑːbɪˌtrɑːʒ, -trɪdʒ/ n. the buying and selling of stocks or bills of exchange to take advantage of varying prices in different markets. [F f. *arbitrer* (as ARBITRATE)]

arbitrageur /ˌɑːbɪtrɑːˈʒɜː(r)/ n. (also **arbitrager** /ˈɑːbɪtrɪdʒə(r)/) a person who engages in arbitrage. [F]

arbitral /ˈɑːbɪtrəl/ adj. concerning arbitration. [F *arbitral* or LL *arbitralis*: see ARBITER]

arbitrament /ɑːˈbɪtrəmənt/ n. **1** the deciding of a dispute by an arbiter. **2** an authoritative decision made by an arbiter. [ME f. OF *arbitrement* f. med.L *arbitramentum* (as ARBITER, -MENT)]

arbitrary /ˈɑːbɪt(rə)rɪ/ adj. **1** based on or derived from uninformed opinion or random choice; capricious. **2** despotic. □ **arbitrarily** /ˈɑːbɪtrərɪlɪ, ˌɑːbɪˈtreərɪlɪ/ adv. **arbitrariness** n. [L *arbitrarius* or F *arbitraire* (as ARBITER, -ARY[1])]

arbitrate /ˈɑːbɪˌtreɪt/ v.tr. & intr. decide by arbitration. [L *arbitrari* judge]

arbitration /ˌɑːbɪˈtreɪʃ(ə)n/ n. the settlement of a dispute by an arbitrator. [ME f. OF f. L *arbitratio -onis* (as ARBITER, -ATION)]

arbitrator /ˈɑːbɪˌtreɪtə(r)/ n. a person appointed to settle a dispute; an arbiter. □ **arbitratorship** n. [ME f. LL (as ARBITRATION, -OR[1])]

arbitress see ARBITER.

arblast var. of ARBALEST.

arbor[1] /ˈɑːbə(r)/ n. **1** an axle or spindle on which something revolves. **2** a device holding a tool in a lathe etc. [F *arbre* tree, axis, f. L *arbor*: refashioned on L]

arbor[2] US var. of ARBOUR.

arboraceous /ˌɑːbəˈreɪʃəs/ adj. **1** treelike. **2** wooded. [L *arbor* tree + -ACEOUS]

Arbor Day a day set apart by law for the public planting of trees, originally in Nebraska in 1872. It is now observed throughout the US, usually in late April or early May, and the practice has been adopted in Canada, Australia, and New Zealand. [L *arbor* tree]

arboreal /ɑːˈbɔːrɪəl/ adj. of, living in, or connected with trees. [L *arboreus* f. *arbor* tree]

arboreous /ɑːˈbɔːrɪəs/ adj. **1** wooded. **2** arboreal.

arborescent /ˌɑːbəˈres(ə)nt/ adj. treelike in growth or general appearance. □ **arborescence** n. [L *arborescere* grow into a tree (*arbor*)]

arboretum /ˌɑːbəˈriːtəm/ n. (pl. **arboretums** or **arboreta** /-tə/) a botanical garden devoted to trees. [L f. *arbor* tree]

arboriculture /ˈɑːbərɪˌkʌltʃə(r)/ n. the cultivation of trees and shrubs. □ **arboricultural** /ˌɑːbərɪˈkʌltʃərəl/ adj. **arboriculturist** n. [L *arbor -oris* tree, after *agriculture*]

arborization /ˌɑːbəraɪˈzeɪʃ(ə)n/ n. (also **-isation**) a treelike arrangement, esp. in anatomy.

arbor vitae /ˌɑːbə ˈviːtaɪ, ˈvaɪtiː/ n. = THUJA. [L, = tree of life]

arbour /ˈɑːbə(r)/ n. (US **arbor**) a shady garden alcove with the sides and roof formed by trees or climbing plants; a bower. □ **arboured** adj. [ME f. AF *erber* f. OF *erbier* f. *erbe* herb f. L *herba*: phonetic change to *ar-* assisted by assoc. with L *arbor* tree]

arbovirus /ˈɑːbəˌvaɪərəs/ n. Med. any of a group of pathogenic viruses transmitted by mosquitoes, ticks, etc., e.g. that causing yellow fever. [arthropod-borne + VIRUS]

Arbus /ˈɑːbəs/, Diane (1923–71), American photographer. Her early career was spent in the world of traditional fashion photography, but she is best known for her sometimes disturbing images of people on the streets of New York and other US cities. She began to take these in 1958, often showing the poor or depicting unusual individuals such as transvestites or midgets.

Arbuthnot /ɑːˈbʌθnət/, John (1667–1735), Scottish physician and writer. He was the physician to Queen Anne and is known as the author of medical works as well as for his satirical writings. A friend of Swift and acquainted with Pope and John Gay, he was the principal author of a satirical work entitled *Memoirs of Martinus Scriblerus* (c.1714). His *History of John Bull* (1712), a collection of pamphlets advocating the termination of the war with France, was the origin of John Bull, the personification of the typical Englishman.

arbutus /ɑːˈbjuːtəs/ n. an evergreen ericaceous tree or shrub of the genus *Arbutus*, having white or pink clusters of flowers and strawberry-like fruits. See also *strawberry tree*. □ **trailing arbutus** US the mayflower, *Epigaea repens*. [L]

ARC abbr. **1** hist. (in the UK) Agricultural Research Council. **2** Aids-related complex.

arc /ɑːk/ n. & v. ● n. **1** part of the circumference of a circle or any other curve. **2** *Electr.* a luminous discharge between two electrodes. ● *v.intr.* (**arced** /ɑːkt/; **arcing** /'ɑːkɪŋ/) form an arc. □ **arc lamp** (or **light**) a light source using an electric arc. **arc welding** a method of using an electric arc to melt metals to be welded. [ME f. OF f. L *arcus* bow, curve]

arcade /ɑː'keɪd/ *n.* **1** a passage with an arched roof. **2** any covered walk, esp. with shops along one or both sides. **3** *Archit.* a series of arches supporting or set along a wall. □ **arcaded** *adj.* [F f. Prov. *arcada* or It. *arcata* f. Rmc: rel. to ARCH¹]

Arcadia /ɑː'keɪdɪə/ a mountainous district in the Peloponnese of southern Greece. In poetic fantasy it represents a pastoral paradise, the home of song-loving shepherds, and in Greek mythology it is the home of Pan.

Arcadian /ɑː'keɪdɪən/ *n. & adj.* ● *n.* an idealized peasant or country dweller, esp. in poetry. ● *adj.* simple and poetically rural. □ **Arcadianism** *n.* [L *Arcadius* f. Gk *Arkadia* ARCADIA]

Arcady /'ɑːkədɪ/ *n. poet.* an ideal rustic paradise. [Gk *Arkadia* ARCADIA]

arcane /ɑː'keɪn/ *adj.* mysterious, secret; understood by few. □ **arcanely** *adv.* [F *arcane* or L *arcanus* f. *arcere* shut up f. *arca* chest]

arcanum /ɑː'keɪnəm/ *n.* (*pl.* **arcana** /-nə/) (usu. in *pl.*) a mystery; a profound secret. [L neut. of *arcanus*: see ARCANE]

Arc de Triomphe /ˌɑːk də 'triːɒmf/ a ceremonial arch standing at the top of the Champs Élysées in Paris, commissioned by Napoleon to commemorate his victories in 1805–6. Inspired by the design of the Arch of Constantine in Rome, it was completed in 1836. The Unknown Soldier was buried under the centre of the arch on Armistice Day 1920.

arch¹ /ɑːtʃ/ *n. & v.* ● *n.* **1 a** a curved structure as an opening or a support for a bridge, roof, floor, etc. **b** an arch used in building as an ornament. **2** any arch-shaped curve, e.g. as on the inner side of the foot, the eyebrows, etc. ● *v.* **1** *tr.* provide with or form into an arch. **2** *tr.* span like an arch. **3** *intr.* form an arch. [ME f. OF *arche* ult. f. L *arcus* arc]

arch² /ɑːtʃ/ *adj.* self-consciously or affectedly playful or teasing. □ **archly** *adv.* **archness** *n.* [ARCH-, orig. in *arch rogue* etc.]

arch- /ɑːtʃ/ *comb. form* **1** chief, superior (*archbishop; archdiocese; archduke*). **2** pre-eminent of its kind (esp. in unfavourable senses) (*arch-enemy*). [OE *arce-* or OF *arche-*, ult. f. Gk *arkhos* chief]

Archaean /ɑː'kiːən/ *adj. & n.* (US **Archean**) *Geol.* of, relating to, or denoting the earlier part of the Precambrian era, characterized by the absence of life (cf. PROTEROZOIC). [Gk *arkhaios* ancient (as ARCHAIC)]

archaeology /ˌɑːkɪ'ɒlədʒɪ/ *n.* (US **archeology**) the study of human history and prehistory through the excavation of sites and the analysis of physical remains. Archaeology, as a systematic discipline rather than a romantic interest, is not much more than a hundred years old. The major methodological advances began in the second half of the 19th century, and the subject has developed at an increasing pace ever since. Excavation to recover evidence is the main activity in archaeology, and whereas before the 19th century the quest was for objects it is now for information of all kinds. Stratigraphy yields evidence of sequence: in a simple undisturbed series of layers the oldest must be at the bottom and the most recent at the top, although the sequence is rarely so simple, and is likely to show phases of rebuilding, periods of abandonment, etc. Typology (the study of changes in forms of pottery, tools, etc.) can link finds from one site with those of another. Specialists in many fields have essential contributions to make: palaeontology, botany, and geology can all contribute to the dating and interpretation of evidence; in addition the archaeologist can now call on scientific aids (e.g. aerial photography and radiocarbon dating). □ **archaeologist** *n.* **archaeologize** *v.intr.* (also **-ise**). **archaeologic** /ˌɑːkɪə'lɒdʒɪk/ *adj.* **archaeological** *adj.* [mod.L *archaeologia* f. Gk *arkhaiologia* ancient history (as ARCHAEAN, -LOGY)]

archaeomagnetism /ˌɑːkɪə'mægnɪˌtɪz(ə)m/ *n.* the remanent magnetism of magnetic materials in clay and rocks which have been heated above a certain temperature. On cooling, the orientation and intensity of their magnetism becomes fixed, and is determined by the direction and intensity of the earth's magnetic field at that time. Archaeomagnetism can be used to study past changes in the earth's magnetism and, hence, as a method of dating some archaeological samples. [Gk *arkhaios* ancient + MAGNETISM]

archaeopteryx /ˌɑːkɪ'ɒptərɪks/ *n.* the oldest known fossil bird, *Archaeopteryx lithographica*, dating from the late Jurassic period. It had feathers and forelimbs modified as wings, but it also retained teeth, a long tail containing vertebrae, and other features of the small dinosaurs from which it evolved. [mod.L f. Gk *arkhaios* ancient + *pterux* wing]

archaic /ɑː'keɪɪk/ *adj.* **1 a** antiquated. **b** (of a word etc.) no longer in ordinary use, though retained for special purposes. **2** primitive. **3** of an early period of art or culture, esp. the 7th–6th centuries BC in Greece. □ **archaically** *adv.* [F *archaïque* f. Gk *arkhaïkos* f. *arkhē* beginning]

archaism /'ɑːkeɪˌɪz(ə)m/ *n.* **1** the retention or imitation of the old or obsolete, esp. in language or art. **2** an archaic word or expression. □ **archaistic** /ˌɑːkeɪ'ɪstɪk/ *adj.* [mod.L f. Gk *arkhaïsmos* f. *arkhaïzō* (as ARCHAIZE, -ISM)]

archaize /'ɑːkeɪˌaɪz/ *v.* (also **-ise**) **1** *intr.* imitate archaic style, expressions, etc. **2** *tr.* make (a work of art, literature, etc.) archaic in style etc. [Gk *arkhaïzō* be old-fashioned f. *arkhaios* ancient]

Archangel /'ɑːkˌeɪndʒəl/ (Russian **Arkhangelsk** /ar'xangjiljsk/) a port of NW Russia, on the White Sea; pop. (1990) 419,000. It is named after the monastery of the Archangel Michael, situated there. From December to April the port is usually icebound.

archangel /'ɑːkˌeɪndʒəl/ *n.* **1** an angel of the highest rank. **2** a member of the eighth order of the nine ranks of heavenly beings (see ORDER *n.* 19). □ **archangelic** /ˌɑːkæn'dʒelɪk/ *adj.* [OE f. AF *archangele* f. eccl.L *archangelus* f. eccl.Gk *arkhaggelos* (as ARCH-, ANGEL)]

archbishop /ɑːtʃ'bɪʃəp/ *n.* the chief bishop of a province. [OE (as ARCH-, BISHOP)]

archbishopric /ɑːtʃ'bɪʃəprɪk/ *n.* the office or diocese of an archbishop. [OE (as ARCH-, BISHOPRIC)]

archdeacon /ɑːtʃ'diːkən/ *n.* **1** an Anglican cleric ranking below a bishop. **2** a member of the clergy of similar rank in other Churches. □ **archdeaconry** *n.* (*pl.* **-ies**). **archdeaconship** *n.* [OE *arce-*, *ercediacon*, f. eccl.L *archidiaconus* f. eccl.Gk *arkhidiakonos* (as ARCH-, DEACON)]

archdiocese /ɑːtʃ'daɪəsɪs/ *n.* the diocese of an archbishop. □ **archdiocesan** /ˌɑːtʃdaɪ'ɒsɪs(ə)n/ *adj.*

archduke /ɑːtʃ'djuːk/ *n.* (*fem.* **archduchess** /-'dʌtʃɪs/) *hist.* the chief duke (esp. as the title of a son of the emperor of Austria). □ **archducal** /-'djuːk(ə)l/ *adj.* **archduchy** /-'dʌtʃɪ/ *n.* (*pl.* **-ies**). [OF *archeduc* f. med.L *archidux -ducis* (as ARCH-, DUKE)]

Archean US var. of ARCHAEAN.

archegonium /ˌɑːkɪ'gəʊnɪəm/ *n.* (*pl.* **archegonia** /-nɪə/) *Bot.* the female sex organ in mosses, ferns, conifers, etc. [L, dimin. of Gk *arkhegonos* f. *arkhe-* chief + *gonos* seed]

arch-enemy /ɑːtʃ'enəmɪ/ *n.* (*pl.* **-ies**) **1** a chief enemy. **2** the Devil.

archeology US var. of ARCHAEOLOGY.

Archer /'ɑːtʃə(r)/, Jeffrey (Howard), Baron Archer of Weston-super-Mare (b.1940), British writer and Conservative politician. Conservative MP for Louth from 1969 to 1974, he resigned his seat after being declared bankrupt. To pay his debts, he embarked on a career as a novelist, *Not a Penny More, Not a Penny Less* (1975) becoming the first of many best sellers. He was deputy chairman of the Conservative Party 1985–6, resigning after a libel case, and was created a life peer in 1992.

archer /'ɑːtʃə(r)/ *n.* **1** a person who shoots with a bow and arrows. **2** (**the Archer**) the zodiacal sign or constellation Sagittarius. □ **archer-fish** a SE Asian fish that catches flying insects by shooting water at them from its mouth. [AF f. OF *archier* ult. f. L *arcus* bow]

archery /'ɑːtʃərɪ/ *n.* shooting with a bow and arrows, esp. as a sport. Bow and arrows were used in prehistoric times by hunters, and for recreation as well as in war by the ancient peoples of Egypt, India, and China. The sport of shooting developed as a pastime after the decline (16th century) of the bow as a weapon. Modern archery is practised chiefly in the form of shooting at a target. (See also SHOOTING.) [OF *archerie* f. *archier* (as ARCHER, -ERY)]

archetype /'ɑːkɪˌtaɪp/ *n.* **1 a** an original model; a prototype. **b** a typical specimen. **2** (in Jungian psychology) a primitive mental image inherited from man's earliest ancestors, and supposed to be present in the collective unconscious. **3** a recurrent symbol or motif in literature, art, etc. □ **archetypal** /ˌɑːkɪ'taɪp(ə)l/ *adj.* **archetypical** /-'tɪpɪk(ə)l/ *adj.* [L *archetypum* f. Gk *arkhetupon* (as ARCH-, *tupos* stamp)]

archidiaconal /ˌɑːkɪdaɪ'ækən(ə)l/ *adj.* of or relating to an archdeacon. □ **archidiaconate** /-nət, -ˌneɪt/ *n.* [med.L *archidiaconalis* (as ARCH-, DIACONAL)]

archiepiscopal /ˌɑːkɪɪ'pɪskəp(ə)l/ *adj.* of or relating to an archbishop. □ **archiepiscopate** /-pət, -ˌpeɪt/ *n.* [eccl.L *archiepiscopus* f. Gk *arkhiepiskopos* archbishop]

archil var. of ORCHIL.

Archilochus /ɑː'kɪləkəs/ (8th or 7th century BC), Greek poet. Acclaimed in his day as equal in stature to Homer and Pindar, he wrote satirical verse and fables and is credited with the invention of iambic metre.

archimandrite /ˌɑːkɪˈmændraɪt/ *n.* **1** the superior of a large monastery or group of monasteries in the Orthodox Church. **2** an honorary title given to a monastic priest. [F *archimandrite* or eccl.L *archimandrita* f. eccl. Gk *arkhimandrites* (as ARCH-, *mandra* monastery)]

Archimedean /ˌɑːkɪˈmiːdɪən/ *adj.* of or associated with Archimedes. □ **Archimedean screw** a device invented by Archimedes for raising water, using an inclined helical screw within a cylinder, rotated by a handle. Its principle is also used in many other devices (e.g. a mincing machine).

Archimedes /ˌɑːkɪˈmiːdiːz/ (*c*.287–212 BC), Greek mathematician and inventor, of Syracuse. He is known as the inventor of the Archimedean screw and other devices, for his boast 'give me a place to stand on and I will move the earth', and for his discovery of Archimedes' principle (legend has it that he made this discovery while taking a bath, and ran through the streets shouting 'Eureka!'). Among his mathematical discoveries are the ratio of the radius of a circle to its circumference, and formulas for the surface area and volume of a sphere and of a cylinder. He devised weapons for use against the Roman fleet during the siege of Syracuse, but was killed during the attack.

Archimedes' principle *n.* the law that a body totally or partially immersed in a fluid is subject to an upward force equal in magnitude to the weight of fluid it displaces.

archipelago /ˌɑːkɪˈpeləˌɡəʊ/ *n.* (*pl.* **-os** or **-oes**) **1** a group of islands. **2** a sea with many islands. [It. *arcipelago* f. Gk *arkhi-* chief + *pelagos* sea (orig. = the Aegean Sea)]

Archipenko /ˌɑːkɪˈpjeŋkəʊ/, Aleksandr (Porfirevich) (1887–1964), Russian-born American sculptor. He adapted cubist techniques to sculpture and attempted to unite form and colour in a mixed medium. He introduced the idiom of 'negative form' into modern sculpture in works such as *Walking Woman* (1912), opening parts of it with holes and concavities to create a contrast of solid and void. From *c*.1946 he experimented with 'light' sculpture, making structures of plastic lit from within.

Archipiélago de Colón /ˌarkipiˌelaʋo ðe koˈlon/ the official Spanish name for the GALAPAGOS ISLANDS.

architect /ˈɑːkɪˌtekt/ *n.* **1** a designer who prepares plans for buildings, ships, etc., and supervises their construction. **2** (foll. by *of*) a person who brings about a specified thing (*the architect of his own fortune*). [F *architecte* f. It. *architetto*, or L *architectus* f. Gk *arkhitektōn* (as ARCH-, *tektōn* builder)]

architectonic /ˌɑːkɪtekˈtɒnɪk/ *adj. & n.* ● *adj.* **1** of or relating to architecture or architects. **2** of or relating to the systematization of knowledge. ● *n.* (in *pl.*; usu. treated as *sing.*) **1** the scientific study of architecture. **2** the study of the systematization of knowledge. [L *architectonicus* f. Gk *arkhitektonikos* (as ARCHITECT)]

architecture /ˈɑːkɪˌtektʃə(r)/ *n.* **1** the art or science of designing and constructing buildings. **2** the style of a building as regards design and construction. **3** buildings or other structures collectively. **4** *Computing* the conceptual structure and logical organization of a computer or computer-based system. □ **architectural** /ˌɑːkɪˈtektʃərəl/ *adj.* **architecturally** *adv.* [F *architecture* or L *architectura* f. *architectus* (as ARCHITECT)]

architrave /ˈɑːkɪˌtreɪv/ *n.* **1** (in classical architecture) a main beam resting across the tops of columns. **2** the moulded frame around a doorway or window. **3** a moulding round the exterior of an arch. [F f. It. *trave* f. L *trabs trabis* beam)]

archive /ˈɑːkaɪv/ *n. & v.* ● *n.* (usu. in *pl.*) **1** a collection of esp. public or corporate documents or records. **2** the place where these are kept. ● *v.tr.* **1** place or store in an archive. **2** *Computing* transfer (data) to a less frequently used file, e.g. from disc to tape. □ **archival** /ɑːˈkaɪv(ə)l/ *adj.* [F *archives* (pl.) f. L *archi(v)a* f. Gk *arkheia* public records f. *arkhē* government]

archivist /ˈɑːkɪvɪst/ *n.* a person who maintains and is in charge of archives.

archivolt /ˈɑːkɪˌvəʊlt/ *n. Archit.* **1** a band of mouldings round the lower curve of an arch. **2** the lower curve itself from impost to impost of the columns. [F *archivolte* or It. *archivolto* (as ARC, VAULT)]

archlute /ˈɑːtʃluːt, -ljuːt/ *n.* a bass lute with an extended neck and unstopped bass strings. [F *archiluth* (as ARCH-, LUTE[1])]

archon /ˈɑːkən, ˈɑːkɒn/ *n.* each of the nine chief magistrates in ancient Athens. □ **archonship** *n.* [Gk *arkhōn* ruler, = pres. part. of *arkhō* rule]

archway /ˈɑːtʃweɪ/ *n.* **1** a vaulted passage. **2** an arched entrance.

arcology /ɑːˈkɒlədʒɪ/ *n.* an ideal city that is fully integrated with its natural environment. The concept was first visualized in 1969 by the US architect Paolo Soleri (b.1919). [blend of ARCHITECTURE and ECOLOGY]

Arctic /ˈɑːktɪk/ *adj. & n.* ● *adj.* **1** of or relating to the north polar region. (*See note below.*) **2** (**arctic**) *colloq.* (esp. of weather) very cold. ● *n.* **1** the Arctic region. **2** (**arctic**) *N. Amer.* a thick waterproof overshoe. □ **Arctic skua** a migratory skua, *Stercorarius parasiticus*, breeding in northern circumpolar regions (also called (*N. Amer.*) *parasitic jaeger*). [ME f. OF *artique* f. L *ar(c)ticus* f. Gk *arktikos* f. *arktos* bear, Ursa Major]

▪ With the important exception of Greenland, most of the land in the Arctic is free from snow in the summer. Plants are able to grow (and a number of animals can live on them) even though about a metre below the surface the soil is permanently frozen. Real exploration of Arctic regions began in the 16th century when northern European nations tried to find a way to the rich trade of China and the East Indies by way of a north-east or north-west passage across the Arctic Ocean at a time when the longer but easier routes round Africa and South America were secured by Portugal and Spain. In the 19th century Fridtjof Nansen led the first expedition across Greenland and attention turned to reaching the North Pole over the floating sea-ice; in 1909 the American explorer R. E. Peary was the first person to reach the Pole; the American aviator Richard Byrd claimed to have flown over the North Pole in 1926. As air travel became more regular, flights were made over the polar ice. The last great advance in Arctic travel was the crossing of the polar ocean in 1958 by the US atomic-powered submarine *Nautilus*, which passed under the ice for 1,600 km (1,000 miles) from the Pacific to the Atlantic.

Arctic Circle the parallel of latitude 66° 33′ north of the equator. It marks the northernmost point at which the sun is visible on the northern winter solstice and the southernmost point at which the midnight sun can be seen on the northern summer solstice.

Arctic Ocean the sea that surrounds the North Pole, lying within the Arctic Circle. Much of the sea is covered with pack ice throughout the year.

Arcturus /ɑːkˈtjʊərəs/ *Astron.* the fourth brightest star in the sky, and the brightest in the constellation of Boötes. It is an orange giant. (See also CHARLES'S WAIN.) [Gk *arktos* bear + *ouros* guardian, because of its position in line with the tail of Ursa Major]

arcuate /ˈɑːkjʊət/ *adj.* shaped like a bow; curved. [L *arcuatus* past part. of *arcuare* curve f. *arcus* bow, curve]

arcus senilis /ˌɑːkʊs seˈniːlɪs/ *n.* a narrow opaque band commonly encircling the cornea in old age. [L, lit. 'senile bow']

-ard /əd/ *suffix* **1** forming nouns in depreciatory senses (*drunkard*; *sluggard*). **2** forming nouns in other senses (*bollard*; *Spaniard*; *wizard*). [ME & OF f. G *-hard* hardy (in proper names)]

Arden /ˈɑːd(ə)n/, Elizabeth (born Florence Nightingale Graham) (*c*.1880–1966), Canadian-born American businesswoman. She trained as a nurse before going to New York, where she opened her own beauty salon on Fifth Avenue in 1909. An effective use of advertising gave her brand a select and elegant image and contributed to the success of her business; she ultimately owned more than 100 beauty salons in America and Europe and her range of cosmetics comprised over 300 products.

Ardennes /ɑːˈden/ a forested upland region extending over parts of SE Belgium, NE France, and Luxembourg. It was the scene of fierce fighting in both world wars.

ardent /ˈɑːd(ə)nt/ *adj.* **1** eager, zealous; (of persons or feelings) fervent, passionate. **2** burning. □ **ardency** *n.* **ardently** *adv.* [ME f. OF *ardant* f. L *ardens -entis* f. *ardere* burn]

Ardnamurchan /ˌɑːdnəˈmɜːkən/ a peninsula on the coast of Highland Region in western Scotland. Ardnamurchan Point is the most westerly point on the British mainland.

ardour /ˈɑːdə(r)/ *n.* (*US* **ardor**) zeal, burning passion, enthusiasm. [ME f. OF f. L *ardor -oris* f. *ardere* burn]

arduous /ˈɑːdjʊəs/ *adj.* **1** (of a task etc.) hard to achieve or overcome; difficult, laborious. **2** (of an action etc.) energetic, strenuous. □ **arduously** *adv.* **arduousness** *n.* [L *arduus* steep, difficult]

are[1] 2nd sing. present & 1st, 2nd, 3rd pl. present of BE.

are[2] /ɑː(r)/ *n.* a metric unit of area equal to 100 square metres. [F f. L AREA]

area /ˈeərɪə/ *n.* **1** the extent or measure of a surface (*over a large area*; *3 acres in area*; *the area of a triangle*). **2** a region or tract (*the southern area*). **3** a space allocated for a specific purpose (*dining area*; *camping area*).

4 the scope or range of an activity or study. **5** a sunken enclosure giving access to the basement of a building. **6** (prec. by *the*) *Football* = penalty area. □ **areal** *adj.* [L, = vacant piece of level ground]

areaway /'ɛərɪəˌweɪ/ *n. N. Amer.* = AREA 5.

areca /'ærɪkə, ə'riːkə/ *n.* a tropical palm of the genus *Areca*, native to Asia. □ **areca nut** the astringent seed of an areca, *A. catechu* (also called *betel-nut*). [Port. f. Malayalam *áḍekka*]

areg *pl.* of ERG[2].

arena /ə'riːnə/ *n.* **1** the central part of an amphitheatre etc., where contests take place. **2** a scene of conflict; a sphere of action or discussion. □ **arena stage** a stage situated with the audience all round it. [L (*h*)*arena* sand, sand-strewn place of combat]

arenaceous /ˌærɪ'neɪʃəs/ *adj.* **1** (of rocks) containing sand; having a sandy texture. **2** sandlike. **3** (of plants) growing in sand. [L *arenaceus* (as ARENA, -ACEOUS)]

Arendt /'ɑːrənt/, Hannah (1906–75), German-born American philosopher and political theorist. A pupil of Heidegger, she established her reputation as a political thinker with *The Origins of Totalitarianism* (1951), one of the first works to propose that Nazism and Stalinism had common roots in the 19th century, sharing anti-Semitic, imperialist, and nationalist elements. *Eichmann in Jerusalem* (1963) aroused controversy with its suggestion that the lack of a political tradition among Jews contributed in part to their own genocide. In 1959 she became the first woman professor at Princeton University.

aren't /ɑːnt/ *contr.* **1** are not. **2** (in *interrog.*) am not (*aren't I coming too?*).

areola /æ'rɪələ/ *n.* (*pl.* **areolae** /-ˌliː/) **1** *Anat.* a circular pigmented area, esp. that surrounding a nipple. **2** any of the spaces between lines on a surface, e.g. of a leaf or an insect's wing. □ **areolar** *adj.* [L, dimin. of *area* AREA]

Areopagus /ˌærɪ'ɒpəgəs/ in ancient Greece, a hill in Athens on which was sited the highest governmental council and later a judicial court. [Gk *Areios pagos* hill of Ares]

Arequipa /ˌærɪ'kiːpə/ a city in the Andes of southern Peru; pop. (1990) 634,500.

Ares /'ɛəriːz/ *Gk Mythol.* the Greek war-god, son of Zeus and Hera. In Rome he was identified with Mars.

arête /æ'ret/ *n.* a sharp mountain ridge. [F f. L *arista* ear of corn, fishbone, spine]

argali /'ɑːgəlɪ/ *n.* (*pl.* same) a large Asiatic wild sheep, *Ovis ammon*, with massive horns. [Mongolian]

argent /'ɑːdʒənt/ *n. & adj. Heraldry* silver; silvery white. [F f. L *argentum*]

argentiferous /ˌɑːdʒ(ə)n'tɪfərəs/ *adj.* containing natural deposits of silver. [L *argentum* + -FEROUS]

Argentina /ˌɑːdʒən'tiːnə/ (also **the Argentine** /'ɑːdʒənˌtaɪn, -ˌtiːn/) a republic occupying much of the southern part of South America; pop. (1991) 32,646,000; official language, Spanish; capital, Buenos Aires. Argentina lies between the Andes foothills and the Atlantic coast, and consists mainly of extensive plains, including the treeless and semi-arid pampas. Before the arrival of Europeans, NW Argentina was part of the Inca empire, the remainder of the region being inhabited by nomadic native peoples. Colonized by the Spanish in the 16th century, Argentina declared its independence in 1816 and played a crucial role in the overthrow of European rule in the rest of South America. The country emerged as a democratic republic in the mid-19th century, but has periodically fallen under military rule. In 1982 the Argentinian claim to the Falkland Islands led to an unsuccessful war with Britain. □ **Argentine** *adj. & n.* **Argentinian** /ˌɑːdʒən'tɪnɪən/ *adj. & n.* [Sp., f. L *argentum* silver, because the region of the River Plate (Río de la Plata, = silver river) exported it]

argentine /'ɑːdʒənˌtaɪn/ *adj.* of silver; silvery. [F *argentin* f. *argent* silver]

argil /'ɑːdʒɪl/ *n.* clay, esp. that used in pottery. □ **argillaceous** /ˌɑːdʒɪ'leɪʃəs/ *adj.* [F *argile* f. L *argilla* f. Gk *argillos* f. *argos* white]

arginine /'ɑːdʒɪˌniːn, -ˌnaɪn/ *n.* an amino acid present in many animal proteins and an essential nutrient in the vertebrate diet. [G *Arginin*, of uncert. orig.]

Argive /'ɑːgaɪv/ *adj. & n.* ● *adj.* **1** of Argos in ancient Greece. **2** *literary* (esp. in Homeric contexts) Greek. ● *n.* **1** a citizen of Argos. **2** (usu. in *pl.*) *literary* a Greek. [L *Argivus* f. Gk *Argeios*]

argol /'ɑːgɒl/ *n.* crude potassium hydrogen tartrate. [ME f. AF *argoile*, of unkn. orig.]

argon /'ɑːgɒn/ *n.* an inert gaseous chemical element (atomic number 18; symbol **Ar**). One of the noble gases, argon was discovered by Lord Rayleigh and William Ramsay in 1894. It is the commonest noble gas, making up nearly one per cent of the earth's atmosphere. It is obtained by distilling liquid air, and is used in electric light bulbs to prolong the life of the filament, and also in arc welding and the growing of semiconductor crystals, where an inert atmosphere is important. It has no known chemical compounds. [Gk, neut. of *argos* idle f. *a-* not + *ergon* work]

Argonauts /'ɑːgəˌnɔːts/ *Gk Mythol.* a group of heroes who accompanied Jason on board the ship *Argo* in the quest for the Golden Fleece. Their story is one of the oldest Greek sagas, known to Homer, and may reflect early explorations in the Black Sea. [Gk, = sailors in the *Argo*]

Argos /'ɑːgɒs/ a city in the NE Peloponnese of Greece; pop. (1981) 20,702. One of the oldest cities of ancient Greece, it was in the 7th century BC the dominant city of the Peloponnese and the western Aegean.

argosy /'ɑːgəsɪ/ *n.* (*pl.* **-ies**) *poet.* a large merchant ship, orig. esp. from Ragusa (now Dubrovnik) or Venice. [prob. It. *Ragusea (nave)* Ragusan (vessel)]

argot /'ɑːgəʊ/ *n.* the jargon of a group or class, formerly esp. of criminals. [F, 19th c.: orig. unkn.]

arguable /'ɑːgjʊəb(ə)l/ *adj.* **1** that may be argued or reasonably proposed. **2** reasonable; supported by argument. □ **arguably** *adv.*

argue /'ɑːgjuː/ *v.* (**argues**, **argued**, **arguing**) **1** *intr.* (often foll. by *with*, *about*, etc.) exchange views or opinions, especially heatedly or contentiously (with a person). **2** *tr. & intr.* (often foll. by *that* + clause) indicate; maintain by reasoning. **3** *intr.* (foll. by *for*, *against*) reason (*argued against joining*). **4** *tr.* treat by reasoning (*argue the point*). **5** *tr.* (foll. by *into*, *out of*) persuade (*argued me into going*). □ **argue the toss** *colloq.* dispute a decision or choice already made. □ **arguer** *n.* [ME f. OF *arguer* f. L *argutari* prattle, frequent. of *arguere* make clear, prove, accuse]

argufy /'ɑːgjʊˌfaɪ/ *v.intr.* (**-ies**, **-ied**) *colloq.* argue excessively or tediously. [fanciful f. ARGUE: cf. SPEECHIFY]

argument /'ɑːgjʊmənt/ *n.* **1** an exchange of views, esp. a contentious or prolonged one. **2** (often foll. by *for*, *against*) a reason advanced; a reasoning process (*an argument for abolition*). **3** a summary of the subject-matter or line of reasoning of a book. **4** *Math.* an independent variable determining the value of a function. [ME f. OF f. L *argumentum* f. *arguere* (as ARGUE, -MENT)]

argumentation /ˌɑːgjʊmenˈteɪʃ(ə)n/ *n.* **1** methodical reasoning. **2** debate or argument. [F f. L *argumentatio* f. *argumentari* (as ARGUMENT, -ATION)]

argumentative /ˌɑːgjʊ'mentətɪv/ *adj.* **1** fond of arguing; quarrelsome. **2** using methodical reasoning. □ **argumentatively** *adv.* **argumentativeness** *n.* [F *argumentatif* -ive or LL *argumentativus* (as ARGUMENT, -ATIVE)]

Argus /'ɑːgəs/ *Gk Mythol.* **1** a watchman with many eyes, slain by Hermes. After Argus' death Hera took the eyes to deck the tail of the peacock. **2** the dog of Odysseus, who recognized his master on his return from Troy after an absence of twenty years. □ **Argus-eyed** vigilant. [ME f. L f. Gk *Argos*]

argus /'ɑːgəs/ *n.* a butterfly of the genus *Aricia*, with eyespots on the wings. [ARGUS 1]

Argus pheasant *n.* a large SE Asian pheasant of the genus *Argusiana* or *Rheinardia*, having a long tail with eyelike markings. [ARGUS 1]

argute /ɑː'gjuːt/ *adj. literary* **1** sharp or shrewd. **2** (of sounds) shrill. [ME f. L *argutus* past part. of *arguere*: see ARGUE]

argy-bargy /ˌɑːdʒɪ'bɑːdʒɪ/ *n. & v.joc.* ● *n.* (*pl.* **-ies**) a dispute or wrangle. ● *v.intr.* (**-ies**, **-ied**) quarrel, esp. loudly. [orig. Sc.]

Argyllshire /ɑː'gaɪlʃɪə(r)/ a former county on the west coast of Scotland. It was divided between Strathclyde and Highland regions in 1975.

Århus see AARHUS.

aria /'ɑːrɪə/ *n. Mus.* a long accompanied song for solo voice in an opera, oratorio, etc. [It.]

Ariadne /ˌærɪ'ædnɪ/ *Gk Mythol.* the daughter of King Minos of Crete and Pasiphaë. She helped Theseus to escape from the labyrinth of the Minotaur by giving him a ball of thread, which he unravelled as he went in and used to trace his way out again after killing the Minotaur. They fled together but he deserted her on the island of Naxos, where she was found by Dionysus, whom she married.

Arian /'ɛərɪən/ *n. & adj.* ● *n.* **1** *Astrol.* a person born under the sign of Aries. **2** an adherent of the doctrine of Arianism. ● *adj.* **1** *Astrol.* of or relating to a person born under the sign of Aries. **2** of or concerning Arianism.

-arian /ˈeərɪən/ *suffix* forming adjectives and nouns meaning '(one) concerned with or believing in' (*agrarian*; *antiquarian*; *humanitarian*; *vegetarian*). [L *-arius* (see -ARY¹)]

Arianism /ˈeərɪəˌnɪz(ə)m/ *n.* the principal heresy denying the divinity of Christ, named after its originator, the Alexandrian priest Arius (c.250–c.336). Arianism maintained that the son of God was created by the Father from nothing and was therefore not coeternal with the Father, nor of the same substance. The heresy was condemned by the Council of Nicaea in 325 and again at Constantinople in 381, but though driven from the Roman Empire it retained a foothold among Germanic peoples until the conversion of the Franks to Catholicism (496).

arid /ˈærɪd/ *adj.* **1 a** (of ground, climate, etc.) dry, parched. **b** too dry to support vegetation; barren. **2** uninteresting (*arid verse*). □ **aridly** *adv.* **aridness** *n.* **aridity** /əˈrɪdɪtɪ/ *n.* [F *aride* or L *aridus* f. *arere* be dry]

Ariel /ˈeərɪəl/ **1** a fairy or spirit in Shakespeare's *The Tempest*; also, an angel in Milton's *Paradise Lost* and the chief of the sylphs in Alexander Pope's *The Rape of the Lock*. **2** *Astron.* satellite I of Uranus, the twelfth closest to the planet, discovered in 1851 (diameter 1,160 km). It has cratered regions broken by fractures and faults, overlaid in parts by smooth plains. **3** a series of six American and British satellites devoted to studies of the ionosphere and X-ray astronomy, launched between 1962 and 1979.

Aries /ˈeəriːz/ *n.* **1** *Astron.* a small constellation (the Ram), said to represent the ram whose Golden Fleece was sought by Jason and the Argonauts. **2** *Astrol.* **a** the first sign of the zodiac, which the sun enters at the northern vernal equinox (about 20 Mar.). **b** a person born when the sun is in this sign. [ME f. L, = ram]

Aries, First Point of *Astron.* the point on the celestial sphere where the path of the sun (the ecliptic) crosses the celestial equator from south to north in March, marking the zero point of right ascension. It is also called the vernal equinox, and the moment that the sun reaches it is the vernal equinox of the northern hemisphere. Owing to precession of the equinoxes (see PRECESSION), it has moved from Aries into Pisces, and is now approaching Aquarius (hence the astrological Ages of Aries, Pisces, and Aquarius).

aright /əˈraɪt/ *adv.* rightly. [OE (as A², RIGHT)]

aril /ˈærɪl/ *n. Bot.* an extra seed-covering, often coloured and hairy or fleshy, e.g. the red fleshy cup around a yew seed. □ **arillate** *adj.* [mod.L *arillus*: cf. med.L *arilli* dried grape-stones]

arioso /ˌærɪˈəʊzəʊ, ˌɑːrɪ-/ *adj., adv., & n. Mus.* ● *adj. & adv.* in a melodious songlike style. ● *n. (pl. -os)* a piece of music to be performed in this way. [It. f. ARIA]

Ariosto /ˌærɪˈɒstəʊ/, Ludovico (1474–1533), Italian poet. His *Orlando Furioso* (final version 1532), about the exploits of Roland (Orlando) and other knights of Charlemagne, was the greatest of the Italian romantic epics; Spenser used its narrative form as a model for his *Faerie Queene*.

-arious /ˈeərɪəs/ *suffix* forming adjectives (*gregarious*; *vicarious*). [L *-arius* (see -ARY¹) + -OUS]

arise /əˈraɪz/ *v.intr.* (*past* **arose** /əˈrəʊz/; *past part.* **arisen** /əˈrɪz(ə)n/) **1** begin to exist; originate. **2** (usu. foll. by *from, out of*) result (*accidents can arise from carelessness*). **3** come to one's notice; emerge (*the question of payment arose*). **4** *esp. archaic & poet.* rise. [OE *ārīsan* (as A-², RISE)]

arisings /əˈraɪzɪŋz/ *n.pl.* materials forming the secondary or waste products of industrial operations.

Aristarchus¹ /ˌærɪˈstɑːkəs/ (known as Aristarchus of Samos) (3rd century BC), Greek astronomer. Founder of an important school of Hellenic astronomy, he was aware of the rotation of the earth and, by placing the sun at the centre of the universe, was able to account for the seasons. He knew that the sun must be larger than the earth and that the stars must be very distant. Many of his theories were more accurate than those of Ptolemy, which replaced them.

Aristarchus² /ˌærɪˈstɑːkəs/ (known as Aristarchus of Samothrace) (c.217–145 BC), Greek scholar. The librarian at Alexandria, he is regarded as the originator of scientific literary scholarship, and is noted for his editions of the writings of Homer and other Greek authors, as well as for commentaries and treatises on their works.

Aristides /ˌærɪˈstaɪdiːz/ (known as Aristides the Just) (5th century BC), Athenian statesman and general. In the Persian Wars he commanded the Athenian army at the battle of Plataea in 479 BC, and was subsequently prominent in founding the Athenian empire (see DELIAN LEAGUE).

Aristippus /ˌærɪˈstɪpəs/ (known as Aristippus the Elder) (late 5th century BC), Greek philosopher. He was a native of Cyrene and pupil of Socrates, and is generally considered the founder of the Cyrenaic school, holding that pleasure is the highest good and that virtue is to be equated with the ability to enjoy. His grandson Aristippus the Younger further developed his philosophy.

aristocracy /ˌærɪˈstɒkrəsɪ/ *n. (pl. -ies)* **1 a** the highest class in society; the nobility. **b** the nobility as a ruling class. **2 a** a government by the nobility or a privileged group. **b** a state governed in this way. **3** (often foll. by *of*) the best representatives or upper echelons (*aristocracy of intellect*; *aristocracy of labour*). [F *aristocratie* f. Gk *aristokratia* f. *aristos* best + *kratia* (as -CRACY)]

aristocrat /ˈærɪstəˌkræt/ *n.* a member of the aristocracy. [F *aristocrate* (as ARISTOCRATIC)]

aristocratic /ˌærɪstəˈkrætɪk/ *adj.* **1** of or relating to the aristocracy. **2 a** distinguished in manners or bearing. **b** grand; stylish. □ **aristocratically** *adv.* [F *aristocratique* f. Gk *aristokratikos* (as ARISTOCRACY)]

Aristophanes /ˌærɪˈstɒfəˌniːz/ (c.450–c.385 BC), Greek dramatist. His eleven surviving comedies, characterized by inventive situations and exuberant language, are largely occupied with topical themes; Aristophanes satirizes politicians and intellectuals (e.g. Socrates), and parodies contemporary poets such as Aeschylus and Euripides. Much use is made of political and social fantasy, as exemplified by the city of the birds ('Cloud-cuckoo-land') in the *Birds*, and by the women's sex-strike for peace in *Lysistrata*.

Aristotelian /ˌærɪstəˈtiːlɪən/ *n. & adj.* ● *n.* a disciple or student of Aristotle. ● *adj.* of or concerning Aristotle or his ideas.

Aristotle /ˈærɪstɒt(ə)l/ (384–322 BC), Greek philosopher and scientist. A pupil of Plato and tutor to Alexander the Great, in 335 BC he founded a school and library (the Lyceum) outside Athens. His surviving written works constitute a vast system of analysis, including logic, physical science, zoology, psychology, metaphysics, ethics, politics, and rhetoric. In reasoning, he established the inductive method. In metaphysics, he rejected Plato's doctrine of forms or ideals; for him form and matter were the inseparable constituents of all existing things. His empirical approach to science is most notable in the field of biology, where he analysed and described the stomach of ruminants and the development of the chick embryo. His work on the classification of animals by means of a scale ascending to man (without implying evolution) was not fully appreciated until the 19th century: Darwin acknowledged a debt to him. The influence of Aristotle in all fields has been considerable: from the 9th century it pervaded Islamic philosophy, theology, and science, and, after being lost to the West for some centuries, became the basis of scholasticism in medieval Christian thought. In astronomy, his proposal that the stars and the planets are composed of a perfect incorruptible element (ether), carried on revolving spheres centred on the earth, was a serious handicap to later thinking.

Aristotle's lantern *n. Zool.* a complex conical structure inside a sea urchin, consisting of calcareous plates, muscles, and ligaments, supporting and operating the rasping teeth.

Arita /əˈriːtə/ *n.* (usu. *attrib.*) a type of Japanese porcelain characterized by asymmetric decoration. [*Arita* in Japan]

arithmetic *n. & adj.* ● *n.* /əˈrɪθmətɪk/ **1 a** the science of numbers. **b** one's knowledge of this (*have improved my arithmetic*). **2** the use of numbers; computation (*a problem involving arithmetic*). ● *adj.* /ˌærɪθˈmetɪk/ (also **arithmetical** /ˌærɪθˈmetɪk(ə)l/) of or concerning arithmetic. □ **arithmetic mean** the central number in an arithmetic progression. **arithmetic progression 1** an increase or decrease by a constant quantity (e.g. 1, 2, 3, 4, etc., 9, 7, 5, 3, etc.). **2** a sequence of numbers showing this. □ **arithmetician** /əˌrɪθməˈtɪʃ(ə)n/ *n.* [ME f. OF *arismetique* f. L *arithmetica* f. Gk *arithmētikē* (*tekhnē*) art of counting f. *arithmos* number]

-arium /ˈeərɪəm/ *suffix* forming nouns usu. denoting a place (*aquarium*; *planetarium*). [L, neut. of adjs. in *-arius*: see -ARY¹]

Ariz. *abbr.* Arizona.

Arizona /ˌærɪˈzəʊnə/ a state of the south-western US, on the border with Mexico; pop. (1990) 3,665,230; capital, Phoenix. It became the 48th state of the US in 1912. □ **Arizonan** *n. & adj.*

Arjuna /ˈɑːdʒʊnə/ *Hinduism* a Kshatriya prince in the Mahabharata, one of the two main characters in the Bhagavadgita.

Ark. *abbr.* Arkansas.

ark /ɑːk/ *n.* **1** (also **Ark**) NOAH'S ARK 1. **2** (also **Ark**) **a** (in full **Holy**

Ark) the chest or cupboard housing the Torah scrolls in a synagogue. **b** = ARK OF THE COVENANT. **3** archaic a chest or box. □ **out of the ark** colloq. very antiquated. [OE ærc f. L arca chest]

Arkansas /'ɑːkənˌsɔː/ a state of the south central US; pop. (1990) 2,350,725; capital, Little Rock. It became the 25th state of the US in 1836.

Arkhangelsk see ARCHANGEL.

Ark of the Covenant (or **Testimony**) the wooden chest which contained the tablets of the laws of the ancient Israelites. Carried by the Israelites on their wanderings in the wilderness, it was later placed by Solomon in the Temple at Jerusalem; it was lost when Nebuchadnezzar's forces destroyed the Temple in 586 BC.

Arkwright /'ɑːkraɪt/, Sir Richard (1732–92), English inventor and industrialist. A pioneer of mechanical cotton-spinning, in 1767 he patented a water-powered spinning machine (known as a 'water frame'), the first such machine to produce yarn strong enough to be used as warp. He also improved the preparatory processes, including carding. He established spinning mills in Lancashire, Derbyshire, and Scotland, and became rich and powerful, despite disputes with rivals over patents and opposition to his mechanization of the industry.

Arles /ɑːl/ a city in SE France; pop. (1990) 52,590. It was the capital of the medieval kingdom of Arles, formed in the 10th century by the union of the kingdoms of Provence and Burgundy. The painter Van Gogh settled in the city in 1888.

Arlington /'ɑːlɪŋtən/ **1** a county in northern Virginia, forming a suburb of Washington on the right bank of the Potomac river. It is the site of the Pentagon and the Arlington National Cemetery. **2** an industrial city in northern Texas, between Dallas and Fort Worth; pop. (1990) 261,720.

Arlon /ɑːˈlɒn/ a town in SE Belgium, capital of the province of Luxembourg; pop. (1991) 23,420.

arm[1] /ɑːm/ n. **1** each of the upper limbs of the human body from the shoulder to the hand. **2 a** the forelimb of an animal. **b** the flexible limb of an invertebrate animal (e.g. an octopus). **3 a** the sleeve of a garment. **b** the side part of a chair etc., used to support a sitter's arm. **c** a thing resembling an arm in branching from a main stem (an arm of the sea). **d** a large branch of a tree. **4** a control; a means of reaching (arm of the law). □ **an arm and a leg** colloq. a large sum of money. **arm in arm** (of two or more persons) with arms linked. **arm-twisting** colloq. persuasion by the use of physical force or (esp.) moral pressure. **arm-wrestling** a trial of strength in which each party tries to force the other's arm down on to a table on which their elbows rest. **as long as your** (or **my**) **arm** colloq. very long. **at arm's length 1** as far as an arm can reach. **2** far enough to avoid undue familiarity. **in arms** (of a baby) too young to walk. **in a person's ˌarms** embraced. **on one's arm** supported by one's arm. **under one's arm** between the arm and the body. **within arm's reach** reachable without moving one's position. **with open arms** cordially. □ **armful** n. (pl. **-fuls**). **armless** adj. [OE f. Gmc]

arm[2] /ɑːm/ n. & v. ● n. **1** (usu. in pl.) **a** a weapon. **b** = FIREARM. **2** (in pl.) the military profession. **3** a branch of the military (e.g. infantry, cavalry, artillery, etc.). **4** (in pl.) heraldic devices (coat of arms). ● v.tr. & refl. **1** supply with weapons. **2** supply with tools or other requisites or advantages (armed with the truth). **3** make (a bomb etc.) able to explode. □ **arms control** international disarmament or arms limitation, esp. by mutual agreement. **arms race** (usu. prec. by the) competition between nations for superiority in the development and accumulation of weapons. **in arms** armed. **lay down one's arms** cease fighting. **take up arms** begin fighting. **under arms** ready for war or battle. **up in arms** (usu. foll. by against, about) actively rebelling. □ **armless** adj. [ME f. OF armes (pl.), armer, f. L arma arms, fittings]

Armada /ɑːˈmɑːdə/ (also **Spanish Armada**) a Spanish naval invasion force sent against England in 1588 by Philip II of Spain. The Armada, 129 ships strong and carrying almost 20,000 soldiers, was defeated in the Channel by a smaller English fleet before it could rendezvous with a Spanish army waiting in the Low Countries to be ferried across to England. The scattered survivors of the Armada tried to reach home by sailing north round Scotland, but many were lost to storms.

armada /ɑːˈmɑːdə/ n. a fleet of warships; a large fleet of vessels of any kind. [Sp. f. Rmc armata army]

armadillo /ˌɑːməˈdɪləʊ/ n. (pl. **-os**) a nocturnal insect-eating edentate mammal of the family Dasypodidae, native to Central and South America. It has large claws for digging and a body covered in bony plates, often rolling itself into a ball when threatened. [Sp. dimin. of armado armed man f. L armatus past part. of armare to arm]

Armageddon /ˌɑːməˈɡed(ə)n/ n. **1** (in the New Testament) **a** the last battle between good and evil before the Day of Judgement. **b** the place where this will be fought. **2** a bloody battle or struggle on a huge scale. [Gk f. Heb. har megiddōn hill of Megiddo: see Rev. 16:16]

Armagh /ɑːˈmɑː/ **1** one of the Six Counties of Northern Ireland, formerly an administrative area. **2** the chief town of this county; pop. (1981) 12,700. It was the seat of the kings of Ulster from the 3rd century BC to AD 333, and became the religious centre of Ireland in AD 445, when St Patrick was made archbishop of Armagh.

Armagnac /'ɑːməˌnjæk/ an area of Aquitaine in SW France, noted for its brandy.

armament /'ɑːməmənt/ n. **1** (often in pl.) military weapons and equipment, esp. guns on a warship. **2** the process of equipping for war. **3** archaic a force equipped for war. [L armamentum (as ARM[2], -MENT)]

armamentarium /ˌɑːməmenˈteərɪəm/ n. (pl. **armamentaria** /-rɪə/) **1** a set of medical equipment or drugs. **2** the resources available to a person engaged in a task. [L, = arsenal]

armature /'ɑːməˌtjʊə(r)/ n. **1 a** the rotating coil or coils of a dynamo or electric motor. **b** any moving part of an electrical machine in which a voltage is induced by a magnetic field. **2** a piece of soft iron placed in contact with the poles of a horseshoe magnet to preserve its power. Also called keeper. **3** Biol. the protective covering of an animal or plant. **4** a metal framework on which a sculpture is moulded with clay or similar material. **5** archaic arms; armour. [F f. L armatura armour (as ARM[2], -URE)]

armband /'ɑːmbænd/ n. a band worn around the upper arm to hold up a shirtsleeve or as a form of identification etc.

armchair /'ɑːmtʃeə(r), ɑːmˈtʃeə(r)/ n. **1** a comfortable, usu. upholstered, chair with side supports for the arms. **2** (attrib.) theoretical rather than active or practical; lacking first-hand experience (an armchair critic).

Armenia /ɑːˈmiːnɪə/ a landlocked country in the Caucasus of SE Europe; pop. (est. 1991) 3,360,000; official language, Armenian; capital, Yerevan. A medieval kingdom, the Armenian homeland fell under Turkish rule from the 16th century. With the decline of the Ottoman Empire Armenia was divided between Turkey, Iran, and Russia. In 1915 the Turks, fearing their Armenian subjects were sympathizing with Russia and the Allies in the First World War, forcibly deported 1,750,000 Armenians to the deserts of Syria and Mesopotamia; more than 600,000 were killed or died on forced marches. Russian Armenia was absorbed into the Soviet Union in 1922 when it became part of the Transcaucasian Soviet Federated Socialist Republic, becoming a separate constituent republic of the Soviet Union in 1936. It gained independence as a member of the Commonwealth of Independent States in 1991. Since 1988 there has been conflict with neighbouring Azerbaijan over the ethnically Armenian enclave of Nagorno-Karabakh and the predominantly Azerbaijani territory of Naxçivan.

Armenian /ɑːˈmiːnɪən/ n. & adj. ● n. **1 a** a native of Armenia. **b** a person of Armenian descent. **2** the language of Armenia. (See note below.) ● adj. of or relating to Armenia, its language, or the Christian Church established there.

▪ Armenian constitutes a separate branch of the Indo-European language family, although its vocabulary has been substantially influenced by Iranian. There are some 4 million speakers of the modern language, of whom about 3 million live in Armenia itself. Its distinctive alphabet contains thirty-eight letters and was invented in AD 400 by Christian missionaries.

Armenian Church (in full **Armenian Apostolic Orthodox Church**) an independent Christian Church established in Armenia since c.300. Though rejecting the Council of Chalcedon, the Armenian Church has been influenced by Roman and Byzantine as well as Syrian traditions. A small Armenian Catholic (Uniat) Church also exists.

armhole /'ɑːmhəʊl/ n. either of two holes in a garment through which the arms are put, usu. into a sleeve.

armiger /'ɑːmɪdʒə(r)/ n. a person entitled to heraldic arms. [L, = bearing arms, f. arma arms + gerere bear]

armillary /ɑːˈmɪlərɪ/ adj. relating to bracelets. □ **armillary sphere** hist. a representation of the celestial globe constructed from metal rings and showing the equator, the tropics, etc. [mod.L armillaris f. L armilla bracelet]

Arminian /ɑːˈmɪnɪən/ adj. & n. ● adj. of or relating to the doctrines of Jacobus Arminius (Latinized name of Jakob Hermandszoon, 1560–1609), a Dutch Protestant theologian who rejected the Calvinist doctrines of predestination and election and held that the sovereignty of God is compatible with human free will. His teachings had a considerable influence on Methodism. ● *n.* an adherent of these doctrines. □ **Arminianism** *n.*

armistice /ˈɑːmɪstɪs/ *n.* a stopping of hostilities by common agreement of the opposing sides; a truce. [F *armistice* or mod.L *armistitium*, f. *arma* arms (ARM²) + -*stitium* stoppage]

Armistice Day 11 November, the anniversary of the armistice that ended the First World War, now replaced by Remembrance Sunday in the UK and Veterans Day in the US.

armlet /ˈɑːmlɪt/ *n.* **1** a band worn round the arm. **2** a small inlet of the sea, or branch of a river.

armoire /ɑːˈmwɑː(r)/ *n.* a cupboard or wardrobe, esp. one that is ornate or antique. [F: see AUMBRY]

armor *US var. of* ARMOUR.

armorer *US var. of* ARMOURER.

armory¹ /ˈɑːmərɪ/ *n.* (*pl.* -ies) heraldry. □ **armorial** /ɑːˈmɔːrɪəl/ *adj.* [OF *armoierie:* see ARMOURY]

armory² *US var. of* ARMOURY.

armour /ˈɑːmə(r)/ *n. & v.* (*US* **armor**) ● *n.* **1** a defensive covering, usu. of metal, formerly worn to protect the body in fighting. **2 a** (in full **armour-plate**) a protective metal covering for an armed vehicle, ship, etc. **b** armoured fighting vehicles collectively. **3** a protective covering or shell on certain animals and plants. **4** heraldic devices. ● *v.tr.* (usu. as **armoured** *adj.*) provide with a protective covering, and often with guns (*armoured car; armoured train*). [ME f. OF *armure* f. L *armatura:* see ARMATURE]

armourer /ˈɑːmərə(r)/ *n.* (*US* **armorer**) **1** a maker or repairer of arms or armour. **2** an official in charge of a ship's or a regiment's arms. [AF *armurer*, OF -*urier* (as ARMOUR, -ER⁵)]

armoury /ˈɑːmərɪ/ *n.* (*US* **armory**) (*pl.* -ies) **1** a place where arms are kept; an arsenal. **2** an array of weapons, defensive resources, usable material, etc. **3** *US* a place where arms are manufactured. [ME f. OF *armoirie, armoierie* f. *armoier* to blazon f. *arme* ARM²: assim. to ARMOUR]

armpit /ˈɑːmpɪt/ *n.* **1** the hollow under the arm at the shoulder. **2** *N. Amer. colloq.* a place or part considered disgusting or contemptible (*the armpit of the world*).

armrest /ˈɑːmrest/ *n.* = ARM¹ 3b.

Armstrong¹ /ˈɑːmstrɒŋ/, Edwin Howard (1890–1954), American electrical engineer. He was the inventor of the superheterodyne radio receiver and the frequency modulation (FM) system in radio. The former involved him in bitter legal wrangles with Lee De Forest. During the 1930s Armstrong developed the FM system, which removed the static that had ruined much early broadcasting, but the radio industry was very slow to accept it. After the war Armstrong was again involved in legal battles to protect his patent, eventually committing suicide.

Armstrong² /ˈɑːmstrɒŋ/, (Daniel) Louis (known as 'Satchmo', an abbreviation of 'Satchelmouth') (1900–71), American jazz musician. He learned the cornet in the Waifs' Home in New Orleans, later switching to the trumpet. He played on Mississippi river-boats before forming his own small groups, with which he made some sixty recordings in 1925–8. He later led various big bands, toured internationally, and appeared in many films, including *The Birth of the Blues* (1941). A major influence on Dixieland jazz, he was a distinctive singer as well as a trumpet player and was noted for his talent for improvisation.

Armstrong³ /ˈɑːmstrɒŋ/, Neil (Alden) (b.1930), American astronaut. A former fighter pilot and test pilot, he began training as an astronaut in 1962, being appointed to command the Apollo 11 mission, during which he became the first man to set foot on the moon (20 July 1969).

army /ˈɑːmɪ/ *n.* (*pl.* -ies) **1** an organized force armed for fighting on land. (*See note below.*) **2** (*prec. by the*) the military profession. **3** (often foll. by *of*) a very large number (*an army of locusts; an army of helpers*). **4** an organized body regarded as fighting for a particular cause (*Salvation Army*). □ **army ant** a nomadic predatory ant of the subfamily Dorylinae, foraging in large groups. **army worm** a moth or fly larva occurring in destructive swarms. [ME f. OF *armee* f. Rmc *armata* fem. past part. of *armare* arm]

▪ In Britain, until the mid-17th century the army consisted chiefly of soldiers engaged only for the duration of particular campaigns. In 1645 Oliver Cromwell raised his New Model Army, and from then onwards a regular standing army was maintained at all times. At the Restoration in 1660 the army became dependent on the Crown, but after the Revolution of 1688 Parliament took over from the monarch the control and payment of the army.

Army List *n.* (in the UK) an official list of commissioned officers.

Arne /ɑːn/, Thomas (1710–78), English composer. He is remembered for his distinctive contribution to 18th-century theatrical music, especially with his settings of Shakespearian songs such as 'Blow, Blow Thou Winter Wind', and for his operas *Artaxerxes* (1762) and *Love in a Village* (1762). His famous song 'Rule, Britannia' (with words attributed to James Thomson) was composed for the masque *Alfred* (1740).

Arnhem /ˈɑːnəm/ a town in the eastern Netherlands, situated on the River Rhine near its junction with the IJssel, capital of the province of Gelderland; pop. (1991) 131,700. During the Second World War, in Sept. 1944, British airborne troops made a landing on nearby moorland but were eventually overwhelmed by German forces before the advancing Allied army could cross the Rhine to link up with them.

Arnhem Land a peninsula in Northern Territory, Australia, on the west of the Gulf of Carpentaria. The chief town is Nhulunbuy. In 1976 Arnhem Land was declared an Aboriginal reservation.

arnica /ˈɑːnɪkə/ *n.* **1** a composite plant of the genus *Arnica*, having erect stems bearing yellow daisy-like flower-heads, e.g. mountain tobacco. **2** a medicine prepared from this, used for bruises etc. [mod.L: orig. unkn.]

Arno /ˈɑːnəʊ/ a river which rises in the Apennines of northern Italy and flows westwards 240 km (150 miles) through Florence and Pisa to the Ligurian Sea.

Arnold /ˈɑːn(ə)ld/, Matthew (1822–88), English poet, essayist, and social critic. Author of 'The Scholar Gipsy' (1853), 'Dover Beach' (1867) and 'Thyrsis' (1867), he held the post of professor of poetry at Oxford (1857–67) and published several works of literary and social criticism, including *Culture and Anarchy* (1869). This established him as an influential social and cultural critic, who, in his views on religion, education, and the arts, criticized the Victorian age in terms of its materialism, philistinism, and complacency.

aroid /ˈeərɔɪd/ *adj. Bot.* of or relating to the family Araceae, which includes the arums, having insignificant flowers in a club-shaped spike (spadix) that is enclosed in a large bract (spathe). [ARUM + -OID]

aroma /əˈrəʊmə/ *n.* **1** a fragrance; a distinctive and pleasing smell, often of food. **2** a subtle pervasive quality. [L f. Gk *arōma -atos* spice]

aromatherapy /əˌrəʊməˈθerəpɪ/ *n.* the use of aromatic plant extracts and essential oils in massage or other treatment. □ **aromatherapist** *n.* **aromatherapeutic** /-ˌθerəˈpjuːtɪk/ *adj.*

aromatic /ˌærəˈmætɪk/ *adj. & n.* ● *adj.* **1** fragrant, spicy; (of a smell) pleasantly pungent. **2** *Chem.* (of organic compounds) having an unsaturated ring, esp. containing a benzene ring. ● *n.* an aromatic substance. □ **aromatically** *adv.* **aromaticity** /ˌærəməˈtɪsɪtɪ/ *n.* [ME f. OF *aromatique* f. LL *aromaticus* f. Gk *arōmatikos* (as AROMA, -IC)]

aromatize /əˈrəʊmətaɪz/ *v.tr.* (also **-ise**) *Chem.* convert (a compound) into an aromatic structure. □ **aromatization** /əˌrəʊmətaɪˈzeɪʃ(ə)n/ *n.*

arose *past of* ARISE.

around /əˈraʊnd/ *adv. & prep.* ● *adv.* **1** on every side; all round; round about. **2** in various places; here and there; at random (*fool around; shop around*). **3** *colloq.* **a** in existence; available (*has been around for weeks*). **b** near at hand (*it's good to have you around*). **4** approximately (*around 400 people attended*). ● *prep.* **1** on or along the circuit of. **2** on every side of; enveloping. **3** here and there in; here and there near (*chairs around the room*). **4 a** *N. Amer.* (and increasingly *Brit.*) round (*the church around the corner*). **b** (of amount, time, etc.) about; at a time near to (*come around four o'clock; happened around June*). □ **have been around** *colloq.* be widely experienced. [A² + ROUND]

arouse /əˈraʊz/ *v.tr.* **1** induce; call into existence (esp. a feeling, emotion, etc.). **2** awake from sleep. **3** stir into activity. **4** stimulate sexually. □ **arousable** *adj.* **arousal** *n.* [A² + ROUSE]

Arp /ɑːp/, Jean (also known as Hans Arp) (1887–1966), French painter, sculptor, and poet. An associate of many leading avant-garde artists of the early 20th century, he exhibited with Kandinsky's *Blaue Reiter* group in 1912, the German expressionists in 1913, and the surrealists in France in the 1920s. He co-founded the Dada movement in Zurich in 1916. His sculpture includes painted wood reliefs such as *Constellation in Five White Forms and Two Black* (1932). In the 1930s he worked in marble and bronze, making three-dimensional abstract curvilinear sculptures suggestive of organic forms.

arpeggio /ɑːˈpedʒɪˌəʊ/ *n.* (*pl.* -**os**) *Mus.* the notes of a chord played in succession, either ascending or descending. [It. f. *arpeggiare* play the harp f. *arpa* harp]

arquebus var. of HARQUEBUS.

arr. *abbr.* **1** *Mus.* arranged by. **2** arrives.

arrack /'ærək, ə'ræk/ *n.* (also **arak**) an alcoholic spirit, esp. distilled from coco sap or rice. [Arab. 'araḵ sweat, alcoholic spirit from grapes or dates]

arraign /ə'reɪn/ *v.tr.* **1** indict before a tribunal; accuse. **2** find fault with; call into question (an action or statement). □ **arraignment** *n.* [ME f. AF *arainer* f. OF *araisnier* (ult. as AD-, L *ratio -onis* reason, discourse)]

Arran /'ærən/ an island in the Firth of Clyde, in the west of Scotland.

arrange /ə'reɪndʒ/ *v.* **1** *tr.* put into the required order; classify. **2** *tr.* plan or provide for; cause to occur (*arranged a meeting*). **3** *tr.* settle beforehand the order or manner of. **4** *intr.* take measures; form plans; give instructions (*arrange to be there at eight; arranged for a taxi to come; will you arrange about the cake?*). **5** *intr.* come to an agreement (*arranged with her to meet later*). **6** *tr.* **a** *Mus.* adapt (a composition) for performance with instruments or voices other than those originally specified. **b** adapt (a play etc.) for broadcasting. **7** *tr. archaic* settle (a dispute, claim, etc.). □ **arrangeable** *adj.* **arranger** *n.* (esp. in sense 6). [ME f. OF *arangier* f. *à* to + *rangier* RANGE]

arrangement /ə'reɪndʒmənt/ *n.* **1** the act or process of arranging or being arranged. **2** the condition of being arranged; the manner in which a thing is arranged. **3** something arranged. **4** (in *pl.*) plans, measures (*make your own arrangements*). **5** *Mus.* a composition arranged for performance by different instruments or voices (see ARRANGE 6a). **6** settlement of a dispute etc. [F (as ARRANGE, -MENT)]

arrant /'ærənt/ *attrib.adj.* downright, utter, notorious (*arrant liar; arrant nonsense*). □ **arrantly** *adv.* [ME, var. of ERRANT, orig. in phrases like *arrant* (= outlawed, roving) *thief*]

Arras /'ærəs/ a town in NE France; pop. (1990) 42,700. It was a centre in medieval times for the manufacture of tapestries.

arras /'ærəs/ *n. hist.* a rich tapestry, often hung on the walls of a room, or to conceal an alcove. [ARRAS]

Arrau /æ'raʊ/, Claudio (1903–91), Chilean pianist. A child prodigy whose first public performance was at the age of 5, he became a renowned interpreter of the works of Chopin, Liszt, Beethoven, Mozart, Schumann, and Brahms. An unostentatious pianist, he built his reputation largely on the meticulous musicianship and intellectual penetration which accompanied his virtuoso technique.

array /ə'reɪ/ *n. & v.* ● *n.* **1** an imposing or well-ordered series or display. **2** an ordered arrangement, esp. of troops (*battle array*). **3** *poet.* an outfit or dress (*in fine array*). **4 a** *Math.* an arrangement of quantities or symbols in rows and columns; a matrix. **b** *Computing* an ordered set of related elements. **5** *Law* a list of jurors empanelled. ● *v.tr.* **1** deck, adorn. **2** set in order; marshal (forces). **3** *Law* empanel (a jury). [ME f. AF *araier*, OF *areer* ult. f. a Gmc root, = prepare]

arrears /ə'rɪəz/ *n.pl.* an amount still outstanding or uncompleted, esp. work undone or a debt unpaid. □ **in arrears** (or **arrear**) behindhand, esp. in payment. □ **arrearage** *n.* [ME (orig. as *adv.*) f. OF *arere* f. med.L *adretro* (as AD-, *retro* backwards): first used in phr. *in arrear*]

arrest /ə'rest/ *v. & n.* ● *v.tr.* **1 a** seize (a person) and take into custody, esp. by legal authority. **b** seize (a ship) by legal authority. **2** stop or check (esp. a process or moving thing). **3 a** attract (a person's attention). **b** attract the attention of (a person). ● *n.* **1** the act of arresting or being arrested, esp. the legal seizure of a person. **2** a stoppage or check (*cardiac arrest*). □ **arrest of judgement** *Law* the staying of proceedings, notwithstanding a verdict, on the grounds of a material irregularity in the course of the trial. **under arrest** in custody, deprived of liberty. □ **arresting** *adj.* **arrestingly** *adv.* [ME f. OF *arester* ult. f. L *restare* remain, stop]

arrestable /ə'restəb(ə)l/ *adj.* **1** susceptible of arrest. **2** *Law* (esp. of an offence) such that the offender may be arrested without a warrant.

arrester /ə'restə(r)/ *n.* (also **arrestor**) a device, esp. on an aircraft carrier, for slowing an aircraft by means of a hook and cable after landing.

arrestment /ə'restmənt/ *n.* esp. *Sc. Law* attachment of property for the satisfaction of a debt.

Arrhenius /ə'reɪnɪəs, ə'riːn-/, Svante August (1859–1927), Swedish chemist. One of the founders of modern physical chemistry, he was the first Swede to win the Nobel Prize for chemistry, which was awarded in 1903 for his work on electrolytes.

arrhythmia /ə'rɪðmɪə/ *n. Med.* deviation from the normal rhythm of the heart. [Gk *arruthmia* lack of rhythm]

arrière-pensée /ˌærɪeə'pɒnseɪ, -pɒn'seɪ/ *n.* **1** an undisclosed motive. **2** a mental reservation. [F, = behind thought]

arris /'ærɪs/ *n. Archit.* a sharp edge formed by the meeting of two flat or curved surfaces. [corrupt. f. F *areste*, mod. ARÊTE]

arrival /ə'raɪv(ə)l/ *n.* **1 a** the act or an instance of arriving. **b** an appearance on the scene. **2** a person or thing that has arrived. □ **new arrival** *colloq.* a newborn child. [ME f. AF *arrivaille* (as ARRIVE, -AL[1])]

arrive /ə'raɪv/ *v.intr.* (often foll. by *at, in*) **1** reach a destination; come to the end of a journey or a specified part of a journey (*arrived in Tibet; arrived at the station; arrived late*). **2** (foll. by *at*) reach (a conclusion, decision, etc.). **3** *colloq.* establish one's reputation or position. **4** *colloq.* (of a child) be born. **5** (of a thing) be brought (*the flowers have arrived*). **6** (of a time) come (*her birthday arrived at last*). [ME f. OF *ariver*, ult. as AD- + L *ripa* shore]

arrivisme /ˌæriːˈviːz(ə)m/ *n.* ambitious or ruthlessly self-seeking behaviour. [F (as ARRIVISTE)]

arriviste /ˌæriːˈviːst/ *n.* an ambitious or ruthlessly self-seeking person. [F f. *arriver* f. OF (as ARRIVE, -IST)]

arrogant /'ærəgənt/ *adj.* (of a person, attitude, etc.) aggressively assertive or presumptuous; haughty, overbearing. □ **arrogance** *n.* **arrogantly** *adv.* [ME f. OF (as ARROGATE, -ANT)]

arrogate /'ærəˌgeɪt/ *v.tr.* **1** (often foll. by *to* oneself) claim (power, responsibility, etc.) without justification. **2** (often foll. by *to*) attribute unjustly (to a person). □ **arrogation** /ˌærə'geɪʃ(ə)n/ *n.* [L *arrogare arrogat-* (as AD-, *rogare* ask)]

arrondissement /æ'rɒndiːsˌmɒn/ *n.* **1** a subdivision of a French department, for local government administration purposes. **2** an administrative district of a large city, esp. Paris. [F]

Arrow /'ærəʊ/, Kenneth Joseph (b.1921), American economist. He is chiefly noted for his work on general economic equilibrium and for his contribution to the study of social choice. His most startling theory, expounded in *Social Choices and Individual Values* (1951), showed the impossibility of aggregating the preferences of individuals into a single combined order of priorities for society as a whole. He shared the Nobel Prize for economics in 1972.

arrow /'ærəʊ/ *n. & v.* ● *n.* **1** a sharp pointed wooden or metal stick shot from a bow as a weapon. **2** a drawn or printed etc. representation of an arrow indicating a direction; a pointer. ● *v.* **1** *intr.* move like an arrow, move swiftly. **2** *tr.* (as **arrowed** *adj.*) provided or marked with an arrow or arrows. □ **arrow-grass** a grasslike marsh plant of the genus *Triglochin*. **arrow worm** = CHAETOGNATH. **broad arrow** a mark formerly used on British prison clothing and other government stores. □ **arrowy** *adj.* [OE *ar(e)we* f. ON *ör* f. Gmc]

arrowhead /'ærəʊˌhed/ *n.* **1** the pointed end of an arrow. **2** a water plant, *Sagittaria sagittaria*, with arrow-shaped leaves. **3** a decorative device resembling an arrowhead.

arrowroot /'ærəʊˌruːt/ *n.* a plant of the genus *Maranta*, esp. *M. arundinacea* of the West Indies, from which a starch is prepared and used for nutritional and medicinal purposes.

arroyo /ə'rɔɪəʊ/ *n.* (pl. **-os**) *N. Amer.* **1** a brook or stream. **2** a gully. [Sp.]

arse /ɑːs/ *n. & v.* (*US* **ass** /æs/) *coarse sl.* ● *n.* the buttocks. ● *v.intr.* (usu. foll. by *about, around*) play the fool. □ **arse-hole 1** the anus. **2** *offens.* a term of contempt for a person. **arse-kisser** = *arse-licker*. **arse-kissing** *n. & adj.* = *arse-licking*. **arse-licker** a sycophant. **arse-licking** *n. & adj.* obsequious(ness) for the purpose of gaining favour; toadying. [OE *ærs*]

arsenal /'ɑːsən(ə)l/ *n.* **1** a store of weapons. **2** a government establishment for the storage and manufacture of weapons and ammunition. **3** resources of anything compared with weapons (e.g. abuse), regarded collectively. [obs. F *arsenal* or It. *arzanale* f. Arab. *dārṣinā'a* f. *dār* house + *sinā'a* art, industry f. *ṣana'a* fabricate]

arsenic *n. & adj.* ● *n.* /'ɑːsənɪk, 'ɑːsnɪk/ **1** a brittle steel-grey semi-metallic chemical element (atomic number 33; symbol **As**). (*See note below.*) **2** (also **white arsenic**) a non-scientific name for arsenic trioxide, a highly poisonous white solid used in weed-killers, rat poison, etc. ● *adj.* /ɑː'senɪk/ **1** of or concerning arsenic. **2** *Chem.* containing arsenic with a valency of five. □ **red arsenic** = REALGAR. □ **arsenide** /'ɑːsəˌnaɪd/ *n.* **arsenious** /ɑː'siːnɪəs/ *adj.* [ME f. OF f. L *arsenicum* f. Gk *arsenikon* yellow orpiment, identified with *arsenikos* male, but in fact f. Arab. *az-zarnīḵ* f. *al* the + *zarnīḵ* orpiment f. Pers. f. *zar* gold]

▪ Arsenic compounds (and their poisonous properties) have been known since ancient times, and the metallic form was isolated in the Middle Ages. Arsenic occurs naturally in orpiment, realgar, and other minerals, and rarely as the free element. Arsenic is used in some specialized alloys; its toxic compounds are widely used as herbicides and pesticides.

arsenical /ɑːˈsenɪk(ə)l/ *adj. & n.* ● *adj.* of or containing arsenic. ● *n.* a drug containing arsenic.

arsine /ˈɑːsiːn/ *n. Chem.* arsenic trihydride, a colourless poisonous gas smelling slightly of garlic. [ARSENIC after AMINE]

arsis /ˈɑːsɪs/ *n. (pl.* **arses** /-siːz/) a stressed syllable or part of a metrical foot in Greek or Latin verse (opp. THESIS 3). [ME f. LL f. Gk, = lifting f. *airō* raise]

arson /ˈɑːs(ə)n/ *n.* the act of maliciously setting fire to property. □ **arsonist** *n.* [legal AF, OF, f. med.L *arsio -onis* f. L *ardere ars-* burn]

arsphenamine /ɑːsˈfenəmɪn, -ˌmiːn/ *n.* a drug formerly used in the treatment of syphilis and parasitic diseases. [ARSENIC + PHENYL + AMINE]

art[1] /ɑːt/ *n.* **1 a** human creative skill or its application. **b** work exhibiting this. **2 a** (in *pl.*; prec. by *the*) the various branches of creative activity concerned with the production of imaginative designs, sounds, or ideas, e.g. painting, music, and writing, considered collectively. **b** any one of these branches. **3** creative activity, esp. painting and drawing, resulting in visual representation (*interested in music but not art*). **4** human skill or workmanship as opposed to the work of nature (*art and nature had combined to make her a great beauty*). **5** (often foll. by *of*) a skill, aptitude, or knack (*the art of writing clearly; keeping people happy is quite an art*). **6** (in *pl.*; usu. prec. by *the*) those branches of learning (esp. languages, literature, and history) associated with creative skill as opposed to scientific, technical, or vocational skills. **7** crafty or wily behaviour; an instance of this. □ **art form 1** any medium of artistic expression. **2** an established form of composition (e.g. the novel, sonata, sonnet, etc.). **art paper** smooth-coated high-quality paper. **arts and crafts** decorative design and handicraft (see also ARTS AND CRAFTS MOVEMENT). [ME f. OF f. L *ars artis*]

art[2] /ɑːt/ *archaic* or *dial.* 2nd *sing.* present of BE.

art. /ɑːt/ *abbr.* article.

Artaud /ɑːˈtəʊ/, Antonin (1896–1948), French actor, director, and poet. Influenced by Balinese dancing and oriental drama, he sought to return drama to its symbolic and ritualistic roots, developing the concept of the non-verbal Theatre of Cruelty, which concentrated on the use of sound, mime, and lighting. He expounded his theory in a series of essays published as *Le Théâtre et son double* (1938), but his only play to be based on it was *Les Cenci* (1935). He was a significant influence on postwar experimental theatre.

Artaxerxes /ˌɑːtəˈzɜːksiːz/ the name of three kings of ancient Persia:
Artaxerxes I son of Xerxes I, reigned 464–424 BC.
Artaxerxes II son of Darius II, reigned 404–358 BC.
Artaxerxes III son of Artaxerxes II, reigned 358–338 BC.

art deco /ɑːt ˈdekəʊ/ *n.* an influential modernist style in the decorative arts of the 1920s and 1930s, characterized by geometric patterns, bright colours, and sharp edges, often involving the use of enamel, chrome, bronze, and polished stone. The term originated as an abbreviation of Exposition Internationale des Arts Décoratifs et Industriels, an exhibition held in Paris in 1925. The style, which incorporated elements of cubism and art nouveau, was used most notably in household objects such as furniture, kitchenware, and electrical appliances, and in architecture for private houses and shopfronts as well as for public buildings.

artefact /ˈɑːtɪˌfækt/ *n.* (also **artifact**) **1** a product of human art and workmanship. **2** *Archaeol.* a product of prehistoric or aboriginal workmanship as distinguished from a similar object naturally produced. **3** *Biol.* etc. a feature not naturally present, introduced during preparation or investigation (e.g. as in the preparation of a histological slide). □ **artefactual** /ˌɑːtɪˈfæktʃʊəl, -tjʊəl/ *adj.* (in senses 1 and 2). [L *arte* (ablat. of *ars* art) + *factum* (neut. past part. of *facere* make)]

artel /ɑːˈtel/ *n.* an association of craftsmen, peasants, etc., in tsarist Russia or the USSR. [Russ.]

Artemis /ˈɑːtɪmɪs/ *Gk Mythol.* a goddess, daughter of Zeus and sister of Apollo. She was a huntress and is often depicted with a bow and arrows; she is also associated with birth, fertility, and fruitfulness. She was identified with the Roman goddess Diana and with Selene, goddess of the moon; her temple at Ephesus was one of the Seven Wonders of the World.

artemisia /ˌɑːtɪˈmɪzɪə/ *n.* an aromatic or bitter-tasting composite plant of the genus *Artemisia*, which includes wormwood, mugwort, sagebrush, etc. [L f. Gk = wormwood, f. ARTEMIS (the goddess Diana), to whom it was sacred]

arterial /ɑːˈtɪərɪəl/ *adj.* **1** of or relating to an artery (*arterial blood*). **2** (esp. of a road) main, important, esp. linking large cities or towns. [F *artériel* f. *artère* artery]

arterialize /ɑːˈtɪərɪəˌlaɪz/ *v.tr.* (also **-ise**) **1** convert (blood) from venous to arterial by reoxygenation, esp. in the lungs. **2** provide with an arterial system. □ **arterialization** /ɑːˌtɪərɪəlaɪˈzeɪʃ(ə)n/ *n.*

arteriole /ɑːˈtɪərɪˌəʊl/ *n.* a small branch of an artery leading into capillaries. [F *artériole*, dimin. of *artère* ARTERY]

arteriosclerosis /ɑːˌtɪərɪəʊsklɪ(ə)ˈrəʊsɪs/ *n.* the loss of elasticity and thickening of the walls of the arteries, esp. in old age; hardening of the arteries. □ **arteriosclerotic** /-ˈrɒtɪk/ *adj.* [ARTERY + SCLEROSIS]

artery /ˈɑːtərɪ/ *n. (pl.* **-ies**) **1** any of the muscular-walled tubes forming part of the blood circulation system of the body, carrying oxygen-enriched blood from the heart (cf. VEIN *n.* 1a). **2** a main road or railway line. □ **arteritis** /ˌɑːtəˈraɪtɪs/ *n.* [ME f. L *arteria* f. Gk *artēria* prob. f. *airō* raise]

artesian well /ɑːˈtiːzɪən, ɑːˈtiːʒ(ə)n/ *n.* a well bored perpendicularly into water-bearing strata lying at an angle, so that natural pressure produces a constant supply of water with little or no pumping. [F *artésien* f. ARTOIS]

Artex /ˈɑːteks/ *n. propr.* a kind of plaster applied to give a textured finish, often in decorative patterns, to walls and ceilings. [ART[1] + TEXTURE]

artful /ˈɑːtfʊl/ *adj.* **1** (of a person or action) crafty, deceitful. **2** skilful, clever. □ **artfully** *adv.* **artfulness** *n.*

arthritis /ɑːˈθraɪtɪs/ *n.* a disease involving pain and stiffness of the joints, esp. rheumatoid arthritis and osteoarthritis. □ **arthritic** /ɑːˈθrɪtɪk/ *adj. & n.* [L f. Gk f. *arthron* joint]

arthropod /ˈɑːθrəˌpɒd/ *n. Zool.* an invertebrate animal of the phylum Arthropoda, with a segmented body, jointed limbs, and an external skeleton. Insects, crustaceans, arachnids, centipedes, and millipedes are all arthropods, and, with well over a million species, the phylum is by far the largest in the animal kingdom. [Gk *arthron* joint + *pous podos* foot]

Arthur[1] /ˈɑːθə(r)/ traditionally king of Britain, historically perhaps a 5th or 6th-century Romano-British chieftain or general. His life and court have become the focus for many romantic legends in various languages, including the exploits of adventurous knights and the quest for the Holy Grail. The stories were developed and recounted by Malory, Chrétien de Troyes, and others; the Norman writer Wace (12th century) mentions the 'Round Table', which enabled the knights to be seated in such a way that none had precedence. (See also CAMELOT.) □ **Arthurian** /ɑːˈθjʊərɪən/ *adj.*

Arthur[2] /ˈɑːθə(r)/, Chester Alan (1830–86), American Republican statesman, 21st President of the US 1881–5. He was appointed Garfield's Vice-President in 1881 and became President after Garfield's assassination. During his term of office, he was responsible for improving the strength of the US navy.

artichoke /ˈɑːtɪˌtʃəʊk/ *n.* **1** a European plant, *Cynara scolymus*, allied to the thistle. **2** (in full **globe artichoke**) the flower-head of the artichoke, the bracts of which have edible bases. **3** = JERUSALEM ARTICHOKE. [It. *articiocco* f. Arab. *al-ḵaršūfa*]

article /ˈɑːtɪk(ə)l/ *n. & v.* ● *n.* **1** (often in *pl.*) an item or commodity, usu. not further distinguished (*a collection of odd articles*). **2** a non-fictional essay, esp. one included with others in a newspaper, magazine, journal, etc. **3 a** a particular part (*an article of faith*). **b** a separate clause or portion of any document (*articles of apprenticeship*). **4** *Gram.* the definite or indefinite article. ● *v.tr.* bind by articles of apprenticeship. □ **articled clerk** a trainee solicitor. **definite article** *Gram.* the word (*the* in English) preceding a noun and implying a specific or known instance (as in *the book on the table; the art of government; the famous public school in Berkshire*). **indefinite article** *Gram.* the word (*a, an* in English) preceding a noun and implying lack of specificity (as in *bought me a book; government is an art; went to a public school*). [ME f. OF f. L *articulus* dimin. of *artus* joint]

articular /ɑːˈtɪkjʊlə(r)/ *adj.* of or relating to the joints. [ME f. L *articularis* (as ARTICLE, -AR[1])]

articulate *adj. & v.* ● *adj.* /ɑːˈtɪkjʊlət/ **1** able to speak fluently and coherently. **2** (of sound or speech) having clearly distinguishable parts. **3** having joints. ● *v.* /ɑːˈtɪkjʊˌleɪt/ **1** *tr.* **a** pronounce (words, syllables, etc.) clearly and distinctly. **b** express (an idea etc.) coherently. **2** *intr.* speak distinctly (*was quite unable to articulate*). **3** *tr.* (usu. in *passive*) connect by joints. **4** *tr.* mark with apparent joints. **5** *intr.* (often foll. by *with*) form a joint. □ **articulated lorry** *Brit.* a lorry consisting of two or more sections connected by a flexible joint. □ **articulacy** /-ləsɪ/ *n.*

articulately /-lətlɪ/ adv. **articulateness** /-lətnɪs/ n. **articulator** /-ˌleɪtə(r)/ n. [L articulatus (as ARTICLE, -ATE²)]

articulation /ɑːˌtɪkjʊˈleɪʃ(ə)n/ n. **1 a** the act of speaking. **b** articulate utterance; speech. **2 a** the act or a mode of jointing. **b** a joint. [F articulation or L articulatio f. articulare joint (as ARTICLE, -ATION)]

artifact var. of ARTEFACT.

artifice /ˈɑːtɪfɪs/ n. **1** a clever device; a contrivance. **2 a** cunning. **b** an instance of this. **3** skill, dexterity. **4** the products of human skill; man-made objects. [F f. L artificium f. ars artis art, -ficium making f. facere make]

artificer /ɑːˈtɪfɪsə(r)/ n. **1** an inventor. **2** a craftsman. **3** a skilled mechanic in the armed forces. [ME f. AF, prob. alt. of OF artificien]

artificial /ˌɑːtɪˈfɪʃ(ə)l/ adj. **1** produced by human art or effort rather than originating naturally (an artificial lake). **2** not real; imitation, fake (artificial flowers). **3** affected, insincere (an artificial smile). □ **artificial insemination** the injection of semen into the vagina or uterus other than by sexual intercourse. **artificial respiration** the restoration or initiation of breathing by manual or mechanical or mouth-to-mouth methods. **artificial silk** rayon. □ **artificially** adv. **artificiality** /-ˌfɪʃɪˈælɪtɪ/ n. [ME f. OF artificiel or L artificialis (as ARTIFICE, -AL¹)]

artificial intelligence n. the theory and development of computer programs or systems able to perform tasks that normally require human intelligence, such as decision-making and diagnosis, pattern and speech recognition, and natural language translation. Much effort has been devoted to investigating the basic processes of thought, learning, decision-making, and language use, and this is closely related to work in other fields such as physiology, neurology, psychology, linguistics, and mathematics. The more ambitious goals of creating computers and robots able to mimic human behaviour, as in science fiction, remain a long way off, and such systems may not be theoretically or practically possible.

artificial kidney n. an apparatus that can perform the functions of the human kidney (outside the body), when one or both of the organs are damaged. The first artificial kidney was demonstrated (on an animal) in 1913 and first used on a person in 1945. Also called kidney machine. (See also DIALYSIS.)

artillery /ɑːˈtɪlərɪ/ n. (pl. **-ies**) **1** large-calibre guns used in warfare on land. **2** a branch of the armed forces that uses these. □ **artillerist** n. [ME f. OF artillerie f. artiller alt. of atillier, atirier equip, arm]

artilleryman /ɑːˈtɪlərɪˌmæn/ n. (pl. **-men**) a member of the artillery.

artiodactyl /ˌɑːtɪəʊˈdæktɪl/ adj. & n. Zool. ● adj. of or relating to the order Artiodactyla, which comprises ungulate mammals with two main toes on each foot, and includes camels, pigs, and ruminants. ● n. an animal of this order. [mod.L f. Gk artios even + daktulos finger, toe]

artisan /ˌɑːtɪˈzæn/ n. a skilled (esp. manual) worker; a craftsman. □ **artisanship** n. [F f. It. artigiano, ult. f. L artitus past part. of artire instruct in the arts]

artist /ˈɑːtɪst/ n. **1** a painter. **2** a person who practises any of the arts. **3** an artiste. **4** a person who works with the dedication and attributes associated with an artist (an artist in crime). **5** colloq. a devotee; a habitual practiser of a specified (usu. reprehensible) activity (con artist). □ **artistry** n. [F artiste f. It. artista (as ART¹, -IST)]

artiste /ɑːˈtiːst/ n. a professional performer, esp. a singer or dancer. [F: see ARTIST]

artistic /ɑːˈtɪstɪk/ adj. **1** having natural skill in art. **2** made or done with art. **3** of art or artists. □ **artistically** adv.

artless /ˈɑːtlɪs/ adj. **1** guileless, ingenuous. **2** not resulting from or displaying art. **3** clumsy. □ **artlessly** adv. **artlessness** n.

art nouveau /ˌɑː nuːˈvəʊ/ n. a style of decorative art, architecture, and design that was prominent from the 1890s to the early 1900s in western Europe. It is characterized by intricate linear designs and motifs based on natural, asymmetrical forms. Typical of the style are Beardsley's drawings, Mucha's posters, Rennie Mackintosh's furniture, and the designs of Hector Guimard (1867–1942) for the Paris Métro. [F, = new art]

Artois /ɑːˈtwʌ/ a region and former province of NW France. Known in Roman times as Artesium, the area gave its name to a type of well known as the artesian well, which was first sunk there in the 12th century.

Arts and Crafts Movement n. an English decorative arts movement of the second half of the 19th century which sought to revive the ideal of craftsmanship in an age of increasing mechanization and mass production. William Morris, influenced by Jean-Jacques Rousseau and subsequently by Augustus Pugin and John Ruskin, was its most prominent member: it was Morris who translated the nostalgia for hand-crafted goods into an organized business venture; his products included hand-made textiles, wallpaper, furniture, and books. The movement had links with art nouveau and the Pre-Raphaelites and had an influence on many artists and designers, including Walter Crane, and on later groups such as Bauhaus.

artwork /ˈɑːtwɜːk/ n. the illustrations in a printed work.

arty /ˈɑːtɪ/ adj. (also esp. N. Amer. **artsy** /ˈɑːtsɪ/) (**artier, artiest**) colloq. pretentiously or affectedly artistic. □ **arty-crafty** quaintly artistic; (of furniture etc.) seeking stylistic effect rather than usefulness or comfort. □ **artiness** n.

arty-farty /ˌɑːtɪˈfɑːtɪ/ adj. (also **artsy-fartsy** /ˌɑːtsɪˈfɑːtsɪ/) colloq. pretentiously artistic.

Aruba /əˈruːbə/ an island in the Caribbean Sea, close to the Venezuelan coast; pop. (est. 1991) 60,000; capital, Oranjestad. Formerly part of the Netherlands Antilles, it separated from that group in 1986 as a step towards full independence, becoming a self-governing territory of the Netherlands.

arugula /əˈruːgjʊlə/ n. N. Amer. = ROCKET² 2. [Italian dial., ult. dimin. of Latin eruca]

arum /ˈeərəm/ n. an aroid plant of the European genus Arum, usu. stemless with arrow-shaped leaves and a white spathe, e.g. cuckoo-pint. **2** an aroid plant of certain other genera. □ **arum lily** a tall lily-like aroid plant of the African genus Zantedeschia (also called calla lily). [L f. Gk aron]

Arunachal Pradesh /ˌɑːrəˌnɑːtʃəl prəˈdeʃ/ a mountainous state in the far north-east of India, lying on the borders of Tibet to the north and Burma (Myanmar) to the east; capital, Itanagar. Formerly the North East Frontier Agency of Assam, it became a state of India in 1986.

Arunta var. of ARANDA.

arvo /ˈɑːvəʊ/ n. Austral. sl. afternoon. [abbr.]

-ary¹ /ərɪ/ suffix **1** forming adjectives (budgetary; contrary; primary; unitary). **2** forming nouns (dictionary; fritillary; granary; January). [f. L -arius connected with, or F -aire]

-ary² /ərɪ/ suffix forming adjectives (military). [F -aire or f. L -aris belonging to]

Aryabhata I /ˌærɪəˈbɑːtə/, (476–c.550), Indian astronomer and mathematician. He wrote two works, one of which is now lost. The surviving work, the Aryabhatiya (499), has sections dealing with mathematics, the measurement of time, planetary models, the sphere, and eclipses. India's first space satellite was named after him.

Aryan /ˈeərɪən/ n. & adj. ● n. **1 a** archaic a member of the peoples speaking any of the languages of the Indo-European (esp. the Indo-Iranian branch) family. (See note below.) **b** a member of a people of uncertain origin who invaded India from Persia (Iran) c.2000–1200 BC (the Vedic period). **2** archaic the parent language of the Indo-European (or esp. Indo-Iranian) family. **3** (esp. in Nazi Germany) a Caucasian not of Jewish descent. ● adj. of or relating to Aryan or the Aryans. [Skr. ārya noble]

▪ The idea of an 'Aryan race', corresponding to the parent Aryan language, was taken up by several 19th-century writers, esp. Joseph Arthur de Gobineau, who linked it with theories of racial superiority. Although the existence of any such race had been generally rejected, the idea was adopted by Hitler for political purposes and became part of the anti-Semitic doctrine of the Nazis.

aryl /ˈærail, ˈærɪl/ n. Chem. a radical derived from an aromatic hydrocarbon by removal of a hydrogen atom. [G Aryl (as AROMATIC, -YL)]

AS abbr. Anglo-Saxon.

As symb. Chem. the element arsenic.

as¹ /æz, əz/ adv., conj., & pron. ● adv. & conj. (adv. as antecedent in main sentence; conj. in relative clause expressed or implied) ... to the extent to which ... is or does etc. (I am as tall as he; am as tall as he is; am not so tall as he; (colloq.) am as tall as him; as many as six; as recently as last week; it is not as easy as you think). ● conj. (with relative clause expressed or implied) **1** (with antecedent so) expressing result or purpose (came early so as to meet us; we so arranged matters as to avoid a long wait; so good as to exceed all hopes). **2** (with antecedent adverb omitted) having concessive force (good as it is = although it is good; try as he might = although he might try). **3** (without antecedent adverb) **a** in the manner in which (do as you like; was regarded as a mistake; they rose as one man). **b** in the capacity or form of (I speak as your friend; Olivier as Hamlet; as a matter of fact). **c** during or at the time that (came up as I was speaking; fell just as I reached the door). **d** for the reason that; seeing that (as you are here, we can talk). **e** for instance (cathedral cities, as York). ● rel.pron. (with verb of relative clause expressed or implied) **1** that, who, which (*I had the same trouble as you; he is a writer, as is his wife; such money as you have; such countries as France*). **2** (with sentence as antecedent) a fact that (*he lost, as you know*). □ **as and when** to the extent and at the time that (*I'll do it as and when I want to*). **as for** with regard to (*as for you, I think you are wrong*). **as from** on and after (a specified date). **as if** (or **though**) as would be the case if (*acts as if he were in charge; as if you didn't know!; looks as though we've won*). **as it is** (or **as is**) in the existing circumstances or state. **as it were** in a way; to a certain extent (*he is, as it were, infatuated*). **as long as** see LONG[1]. **as much** see MUCH. **as of 1** = *as from*. **2** as at (a specified time). **as per** see PER. **as regards** see REGARD. **as soon as** see SOON. **as such** see SUCH. **as though** see *as if* above. **as to** with respect to; concerning (*said nothing as to money; as to you, I think you are wrong*). **as was** in the previously existing circumstances or state. **as well** see WELL[1]. **as yet** until now or a particular time in the past (usu. with neg. and with implied reserve about the future: *have received no news as yet*). [reduced form of OE *alswā* ALSO]

as[2] /æs/ n. (pl. **asses**) a Roman copper coin. [L]

as- /əs/ prefix assim. form of AD- before *s*.

ASA abbr. **1** Amateur Swimming Association. **2** American Standards Association (esp. in film-speed specification, as *200 ASA*). **3** see ADVERTISING STANDARDS AUTHORITY.

asafoetida /ˌæsəˈfiːtɪdə, -ˈfetɪdə/ n. (US **asafetida**) a resinous plant gum with a fetid ammoniac smell, formerly used in medicine, now as a herbal remedy and in Indian cooking. [ME f. med.L f. *asa* f. Pers. *azā* mastic + *fetida* (as FETID)]

Asansol /ˌæsənˈsəʊl/ an industrial city in NE India, in West Bengal, north-west of Calcutta; pop. (1991) 262,000.

Asante see ASHANTI[1].

a.s.a.p. abbr. as soon as possible.

asbestos /æzˈbestɒs, æsˈbes-/ n. **1** a fibrous silicate mineral that is incombustible. **2** this used as a heat-resistant or insulating material. □ **asbestine** /-tɪn/ adj. [ME f. OF *albeston*, ult. f. Gk *asbestos* unquenchable f. *a-* not + *sbestos* f. *sbennumi* quench]

asbestosis /ˌæzbeˈstəʊsɪs, ˌæsbe-/ n. a lung disease resulting from the inhalation of asbestos particles.

Ascalon /ˈæskəlɒn/ the ancient Greek name for ASHQELON.

ascarid /ˈæskərɪd/ n. (also **ascaris** /-rɪs/) a parasitic nematode worm of the genus *Ascaris*, e.g. the intestinal roundworm of humans and pigs. [mod.L *ascaris* f. Gk *askaris*]

ascend /əˈsend/ v. **1** intr. move upwards; rise. **2** intr. **a** slope upwards. **b** lie along an ascending slope. **3** tr. climb; go up. **4** intr. rise in rank or status. **5** tr. mount upon. **6** intr. (of sound) rise in pitch. **7** tr. go along (a river) to its source. **8** intr. *Printing* (of a letter) have part projecting upwards. □ **ascend the throne** become king or queen. [ME f. L *ascendere* (as AD-, *scandere* climb)]

ascendancy /əˈsendənsɪ/ n. (also **ascendency**) (often foll. by *over*) a superior or dominant condition or position.

ascendant /əˈsendənt/ adj. & n. ● adj. **1** rising. **2** *Astron.* rising towards the zenith. **3** *Astrol.* just above the eastern horizon. **4** predominant. ● n. *Astrol.* the point of the sun's apparent path that is ascendant at a given time (*Aries in the ascendant*). □ **in the ascendant 1** supreme or dominating. **2** rising; gaining power or authority. [ME f. OF f. L (as ASCEND, -ANT)]

ascender /əˈsendə(r)/ n. **1 a** a part of a letter that extends above the main part (as in *b* and *d*). **b** a letter having this. **2** a person or thing that ascends.

ascension /əˈsenʃ(ə)n/ n. **1** the act or an instance of ascending. **2** (**Ascension**) the ascent of Christ into heaven on the fortieth day after the Resurrection. □ **right ascension** *Astron.* the celestial coordinate equivalent to longitude, expressed in hours, minutes, and seconds east of the First Point of Aries. □ **ascensional** adj. [ME f. OF f. L *ascensio -onis* (as ASCEND, -ION)]

Ascension Day the day of Christ's Ascension (the fortieth day after Easter Day); the Thursday on which this is commemorated in the Christian Church.

Ascension Island a small island in the South Atlantic, incorporated with St Helena, with which it is a dependency of the UK; pop. (1988) 1,007. It was discovered by the Portuguese, traditionally on Ascension Day in 1501, but remained uninhabited until a small British garrison was stationed there on the arrival of Napoleon for imprisonment in 1815. It is now a British telecommunications centre and a US airbase. Of strategic importance during the Falklands War in 1982, it serves as a base for British forces and as a landing-point for aircraft travelling between Britain and the South Atlantic.

Ascensiontide /əˈsenʃ(ə)nˌtaɪd/ n. the period of ten days from Ascension Day to Whitsun Eve.

ascent /əˈsent/ n. **1** the act or an instance of ascending. **2 a** an upward movement or rise. **b** advancement or progress (*the ascent of man*). **3** a way by which one may ascend; an upward slope. [ASCEND, after *descent*]

ascertain /ˌæsəˈteɪn/ v.tr. find out as a definite fact; get to know. □ **ascertainable** adj. **ascertainment** n. [ME f. OF *acertener*, stem *acertain-* f. *à* to + CERTAIN]

ascesis /əˈsiːsɪs/ n. the practice of self-discipline. [Gk *askēsis* training f. *askeō* exercise]

ascetic /əˈsetɪk/ n. & adj. ● n. a person who practises severe self-discipline and abstains from all forms of pleasure, esp. for religious or spiritual reasons. ● adj. relating to or characteristic of ascetics or asceticism; abstaining from pleasure. □ **ascetically** adv. **asceticism** /-tɪˌsɪz(ə)m/ n. [med.L *asceticus* or Gk *askētikos* f. *askētēs* monk f. *askeō* exercise]

Ascham /ˈæskəm/, Roger (*c*.1515–68), English humanist scholar and writer. His posts included that of tutor to the future Elizabeth I and Latin secretary to Queen Mary and later to Elizabeth. He is noted for his treatise on archery, *Toxophilus* (1545), and *The Scholemaster* (1570), a practical and influential treatise on education.

ascidian /əˈsɪdɪən/ n. *Zool.* a tunicate animal of the class Ascidiacea, often found in sedentary colonies on rocks or seaweeds. Also called *sea squirt*. [mod.L *Ascidia* f. Gk *askidion* dimin. of *askos* wineskin]

ASCII /ˈæskɪ/ abbr. *Computing* American Standard Code for Information Interchange.

ascites /əˈsaɪtiːz/ n. (pl. same) *Med.* the accumulation of fluid in the abdominal cavity causing swelling. [ME f. LL f. Gk f. *askitēs* f. *askos* wineskin]

Asclepius /əˈskliːpɪəs/ *Gk Mythol.* a hero and god of healing, son of Apollo, often represented bearing a staff with a serpent coiled round it. He sometimes bears a scroll or tablet, probably representing medical learning.

ascorbic acid /əˈskɔːbɪk/ n. *Biochem.* a vitamin found in citrus fruits and green vegetables, essential in maintaining healthy connective tissue, a deficiency of which results in scurvy. Also called *vitamin C*.

Ascot /ˈæskət/ a town in Berkshire, south-west of Windsor. Its racecourse is the site of an annual race meeting, founded by Queen Anne in 1711. □ **Ascot week** this race meeting, which is held each June (also called *Royal Ascot*).

ascribe /əˈskraɪb/ v.tr. (usu. foll. by *to*) **1** attribute or impute (*ascribes his well-being to a sound constitution*). **2** regard as belonging. □ **ascribable** adj. [ME f. L *ascribere* (as AD-, *scribere* script- write)]

ascription /əˈskrɪpʃ(ə)n/ n. **1** the act or an instance of ascribing. **2** a preacher's words ascribing praise to God at the end of a sermon. [L *ascriptio -onis* (as ASCRIBE)]

asdic /ˈæzdɪk/ n. an early form of echo-sounder. (See also SONAR.) [acronym, f. Allied Submarine Detection Investigation Committee]

-ase /eɪz/ suffix *Biochem.* forming the name of an enzyme (*amylase*). [DIASTASE]

ASEAN /ˈæsɪən/ see ASSOCIATION OF SOUTH-EAST ASIAN NATIONS.

asepsis /eɪˈsepsɪs/ n. **1** the absence of harmful bacteria, viruses, or other micro-organisms. **2** a method of achieving asepsis in surgery.

aseptic /eɪˈseptɪk/ adj. **1** free from contamination caused by harmful bacteria, viruses, or other micro-organisms. **2** (of a wound, instrument, or dressing) surgically sterile or sterilized. **3** (of a surgical method etc.) aiming at the elimination of harmful micro-organisms, rather than counteraction (cf. ANTISEPTIC).

asexual /eɪˈseksjʊəl, eɪˈsekʃʊəl/ adj. **1** *Biol.* without sex or sexual organs. **2** *Biol.* (of reproduction) not involving the fusion of gametes. **3** without sexuality. □ **asexually** adv. **asexuality** /eɪˌseksjʊˈælɪtɪ, eɪˌsekʃʊ-/ n.

Asgard /ˈæzgɑːd/ *Scand. Mythol.* a region in the centre of the universe, inhabited by the gods.

ASH /æʃ/ abbr. Action on Smoking and Health.

ash[1] /æʃ/ n. **1** (often in pl.) the powdery residue left after the burning of any substance. **2** (pl.) the remains of the human body after cremation or disintegration. **3** ashlike material thrown out by a volcano. □ **ash blonde** n. **1** a very pale blonde colour. **2** a person with hair of this colour. ● adj. (often hyphenated when attrib.) very pale blonde. [OE *æsce*]

ash[2] /æʃ/ n. **1** (in full **ash tree**) a forest tree of the genus *Fraxinus*, with silver-grey bark, compound leaves, and hard, tough, pale wood.

2 the wood of this tree. **3** an Old English runic letter, = ᚫ (named from a word of which it was the first letter). □ **ash-key** the winged seed of the ash tree, growing in clusters resembling keys. **ash-plant** a sapling from an ash tree, used as a walking-stick etc. [OE *æsc* f. Gmc]

ashamed /əˈʃeɪmd/ *adj.* (usu. *predic.*) **1** (often foll. by *of* (= with regard to), *for* (= on account of), or *to* + infin.) embarrassed or disconcerted by shame (*ashamed of his aunt*; *ashamed of having lied*; *ashamed for you*; *ashamed to be seen with him*). **2** (foll. by *to* + infin.) hesitant, reluctant (but usu. not actually refusing or declining) (*am ashamed to admit that I was wrong*). □ **ashamedly** /-mɪdlɪ/ *adv.* [OE *ascamod* past part. of *ascamian* feel shame (as A-[2], SHAME)]

Ashanti[1] /əˈʃæntɪ/ (also **Asante**) a region of central Ghana. It was annexed by Britain in 1902, becoming part of the former British colony of the Gold Coast.

Ashanti[2] /əˈʃæntɪ/ *n. & adj.* ● *n.* (*pl.* same) **1** a member of the people of the Ashanti region, one of Ghana's principal ethnic groups. **2** the language of this people, a dialect of Twi. ● *adj.* of or relating to the Ashanti or their language. [Twi *Asante*]

ashcan /ˈæʃkæn/ *n.* US a dustbin.

Ashcan School a group of American realist painters active from *c.*1908 until the First World War, who painted scenes from the slums of New York. The school grew out of the group called 'the Eight' and was inspired largely by Robert Henri.

Ashcroft /ˈæʃkrɒft/, Dame Peggy (Edith Margaret Emily) (1907–91), English actress. She played a number of Shakespearian roles including Desdemona to Paul Robeson's Othello (1930) and Juliet in John Gielgud's production of *Romeo and Juliet* (1935). Other outstanding performances included the title role in Ibsen's *Hedda Gabler* (1954), for which she received a royal award. She won an Oscar for best supporting actress in the film *A Passage to India* (1984).

Ashdod /ˈæʃdɒd/ a seaport on the Mediterranean coast of Israel, situated to the south of Tel Aviv; pop. (est. 1982) 62,000.

Ashdown /ˈæʃdaʊn/, Jeremy John Durham ('Paddy') (b.1941), British Liberal Democrat politician, born in India. Formerly a Liberal MP (1983–8), he became the first leader of the Liberal Democrats (originally the Social and Liberal Democrats) in 1988.

Ashe /æʃ/, Arthur (Robert) (1943–93), American tennis player. He won the US Open championship in 1968 and Wimbledon in 1975, and was the first black male player to achieve world rankings. He died of AIDS, having contracted HIV from a blood transfusion.

ashen[1] /ˈæʃ(ə)n/ *adj.* **1** of or resembling ashes. **2** ash-coloured; grey or pale.

ashen[2] /ˈæʃ(ə)n/ *adj.* **1** of or relating to the ash tree. **2** *archaic* made of ashwood.

Asher /ˈæʃə(r)/ (in the Bible) **1** a Hebrew patriarch, son of Jacob and Zilpah (Gen. 30:12, 13). **2** the tribe of Israel traditionally descended from him.

Ashes, the a trophy for the winner of a series of test matches in a cricket season between England and Australia. The name of the trophy originated in a notice published in the *Sporting Times* on 2 Sept. 1882, a mock obituary for English cricket, which was said to have died after the sensational victory of Australia; the body would be cremated and the ashes taken to Australia. Real ashes exist, kept in an urn at Lord's, and are said to be those of a bail (or a ball) burnt at Melbourne when England won the series of 1882–3.

ashet /ˈæʃɪt/ *n.* Sc. & NZ a large plate or dish. [F *assiette*]

Ashgabat /ˈæʃgəˌbæt/ (also **Ashkhabad** /ˌæʃkəˈbæd/) the capital of the central Asian republic of Turkmenistan; pop. (1990) 407,200. Established in 1881 as a Russian fort, it was known from 1919 to 1927 as Poltoratsk.

Ashkelon see ASHQELON.

Ashkenazi /ˌæʃkəˈnɑːzɪ/ *n.* (*pl.* **Ashkenazim** /-zɪm/) a Jew of eastern or central European descent. Ashkenazim lived in France and Germany in the early Middle Ages before many migrated to Poland, Russia, and other Slavic lands around the time of the Crusades; many now live in Israel and the US. They are distinguished from the Sephardim by their customs and rites (preserving Palestinian rather than Babylonian Jewish traditions) and by their use of the Yiddish language. Over 80 per cent of Jews today are Ashkenazim. Cf. SEPHARDI. □ **Ashkenazic** *adj.* [mod.Heb., f. *Ashkenaz* (Gen. 10:3)]

Ashkenazy /ˌæʃkəˈnɑːzɪ/, Vladimir (Davidovich) (b.1937), Russian-born Icelandic pianist. A child prodigy, he made his Moscow début in 1945 and went on to win several international awards, including sharing the first prize in the 1962 Moscow Tchaikovsky Piano

Competition with John Ogdon (1937–89). Ashkenazy left the Soviet Union the following year, finally settling in Iceland in 1973. He has been acclaimed for his interpretations of composers such as Mozart, Rachmaninov, Chopin, Beethoven, and Schubert.

Ashkhabad see ASHGABAT.

ashlar /ˈæʃlə(r)/ *n.* **1** a large square-cut stone used in building. **2** masonry made of ashlars. **3** such masonry used as a facing on a rough rubble or brick wall. [ME f. OF *aisselier* f. L *axilla* dimin. of *axis* board]

ashlaring /ˈæʃlərɪŋ/ *n.* **1** ashlar masonry. **2** the short upright boarding in a garret which cuts off the acute angle between the roof and the floor.

Ashley /ˈæʃlɪ/, Laura (1925–85), Welsh fashion and textile designer. In the 1960s her clothes, in traditional floral patterns and reflecting romantic Victorian and Edwardian styles, became highly popular, as did the range of furnishing fabrics and wallpapers which her company (founded with her husband Bernard) introduced. The chain of shops under Laura Ashley's name spread through Britain and later to Europe, America, Australia, and Japan.

Ashmolean Museum /æʃˈməʊlɪən/ a museum of art and antiquities in Oxford, founded by the English antiquary Elias Ashmole (1617–92). In 1677 he deposited with Oxford University a number of items, some collected by himself, others bequeathed to him by his friend John Tradescant (1608–62), and these formed the nucleus of the collection. The museum opened in 1683 and was the first public institution of its kind in England. The collection now includes archaeological material, European works of art, and oriental works.

Ashmore and Cartier Islands /ˈæʃmɔː(r), ˈkɑːtɪˌeɪ/ an external territory of Australia in the Indian Ocean, comprising the uninhabited Ashmore Reef and Cartier Islands. The area is designated a nature reserve.

ashore /əˈʃɔː(r)/ *adv.* towards or on the shore or land (*sailed ashore*; *stayed ashore*).

ashpan /ˈæʃpæn/ *n.* a tray under a grate to catch the ash.

Ashqelon /ˈæʃkələn/ (also **Ashkelon**) an ancient Mediterranean city, situated to the south of modern Tel Aviv, in Israel. A Philistine city-state from the 12th to the 8th century BC, it was conquered by Alexander the Great in 332 BC. It was known to the Greeks as Ascalan, giving its name to the scallion, or shallot, a kind of onion. A modern city of the same name has been built close to the ancient site.

ashram /ˈæʃrəm/ *n.* (esp. in the Indian subcontinent) a place of religious retreat for Hindus; a hermitage. [Skr. *āshrama* hermitage]

ashrama /ˈæʃrəmə/ *n.* Hinduism any of the four stages of an ideal life, ascending from the status of pupil to the total renunciation of the world. [Skr.]

Ash Shariqah see SHARJAH.

Ashton /ˈæʃtən/, Sir Frederick (William Mallandaine) (1904–88), British ballet-dancer, choreographer, and director. He became chief choreographer and principal dancer of the Vic-Wells Ballet in 1935, remaining with the company when it became the Sadler's Wells and finally the Royal Ballet, of which he was director 1963–70. As a choreographer, Ashton established a lyrical and fluid style of classical ballet, creating successful new works as well as making popular adaptations of historical ballets.

ashtray /ˈæʃtreɪ/ *n.* a small receptacle for cigarette ash, stubs, etc.

Ashur see ASSUR.

Ashurbanipal /ˌæʃʊəˈbɑːnɪp(ə)l/, king of Assyria *c.*668–627 BC. The grandson of Sennacherib, he was responsible for the sacking of the Elamite capital Susa and the suppression of a revolt in Babylon. However, he is chiefly recognized for his patronage of the arts; he established a library of more than 20,000 clay tablets at Nineveh, which included literary, religious, scientific, and administrative documents.

Ash Wednesday *n.* the first day of Lent (from the custom of marking the foreheads of penitents with ashes on that day).

ashy /ˈæʃɪ/ *adj.* **1** = ASHEN[1]. **2** covered with ashes.

Asia /ˈeɪʃə, ˈeɪʒə/ the largest of the world's continents, constituting nearly one-third of the land mass, lying entirely north of the equator except for some SE Asian islands. It is connected to Africa by the Isthmus of Suez, and borders Europe (part of the same land mass) along the Ural Mountains and across the Caspian Sea. The continent is dominated by China, India, and parts of the former USSR, and contains more than half of the world's population. Some areas, particularly in the south, were colonized by European nations between the 17th and 19th centuries, emerging as independent states only after the Second World War.

Asia Minor the western peninsula of Asia, also known as Anatolia, which constitutes the bulk of modern Turkey. Located at the borders of Asia and Europe, for centuries in ancient times it served as a battlefield between East and West. The first major civilization established there was that of the Hittites in the 2nd millennium BC. Taken over the centuries by Greeks, Macedonians, and Romans, the area eventually fell to the Turks, becoming part of the Ottoman Empire from the end of the 13th century until the establishment of modern Turkey after the First World War.

Asian /'eɪʒ(ə)n, 'eɪʒ(ə)n/ n. & adj. ● n. **1** a native of Asia. **2** a person of Asian descent. ● adj. of or relating to Asia or its people, customs, or languages. [L Asianus f. Gk Asianos f. Asia]

Asian Development Bank a bank with forty-seven member countries (thirty-two are from the Asia–Pacific region) that began operations in 1966 and is located in Manila. Its aim is to promote the economic and social progress of its developing member countries.

Asiatic /,eɪʃɪ'ætɪk, ,eɪzɪ-/ n. & adj. ● n. offens. an Asian. ● adj. Asian. [L Asiaticus f. Gk Asiatikos]

A-side /'eɪsaɪd/ n. the side of a gramophone record (usu. a single) regarded as the main one.

aside /ə'saɪd/ adv. & n. ● adv. **1** to or on one side; away. **2** out of consideration (placed after noun: joking aside). ● n. **1** words spoken in a play for the audience to hear, but supposed not to be heard by the other characters. **2** an incidental remark. □ **aside from** apart from. **set aside 1** put to one side. **2** keep for a special purpose or future use. **3** reject or disregard. **4** annul. **5** remove (land) from agricultural production for fallow, forestry, or other use. **take aside** engage (a person) esp. for a private conversation. [orig. on side: see A²]

Asimov /'æzɪ,mɒf/, Isaac (1920–92), Russian-born American writer and scientist. He was a distinguished biochemist, but is more widely known as the author of many works of science fiction, books on science for non-scientists, and essays on a wide variety of subjects. Among his best-known science fiction is I, Robot (1950) and the Foundation trilogy (1951–3). Building on Karel Čapek's concept of the robot, in 1941 Asimov coined the term robotics.

asinine /'æsɪ,naɪn/ adj. **1** stupid. **2** of or concerning asses; like an ass. □ **asininity** /,æsɪ'nɪnɪtɪ/ n. [L asininus f. asinus ass]

Asir Mountains /ə'sɪə(r)/ a range of mountains in SW Saudi Arabia, running parallel to the Red Sea.

-asis /əsɪs/ suffix (usu. as **-iasis** /'aɪəsɪs/) forming the names of diseases (psoriasis; satyriasis). [L f. Gk -asis in nouns of state f. verbs in -aō]

ask /ɑːsk/ v. **1** tr. call for an answer to or about (ask her about it; ask him his name; ask a question of him). **2** tr. seek to obtain from another person (ask a favour of; ask to be allowed). **3** tr. (usu. foll. by out or over, or to (a function etc.)) invite; request the company of (must ask them over; asked her to dinner). **4** intr. (foll. by for) **a** seek to obtain, meet, or be directed to (ask for a donation; ask for the post office; asking for you). **b** invite, provoke (trouble etc.) by one's behaviour; bring upon oneself (they were asking for all they got). **5** tr. archaic require (a thing). □ **ask after** inquire about (esp. a person). **ask for it** colloq. invite trouble. **asking price** the price of an object set by the seller. **ask me another** colloq. I do not know. **for the asking** (obtainable) for nothing. **I ask you!** an exclamation of disgust, surprise, etc. **if you ask me** colloq. in my opinion. □ **asker** n. [OE āscian etc. f. WG]

askance /ə'skæns, ə'skɑːns/ adv. (also **askant** /ə'skænt, ə'skɑːnt/) sideways or squinting. □ **look askance at** regard with suspicion or disapproval. [16th c.: orig. unkn.]

askari /æ'skɑːrɪ/ n. (pl. same or **askaris**) an East African soldier or police officer. [Arab. 'askarī soldier]

askew /ə'skjuː/ adv. & predic.adj. ● adv. obliquely; awry. ● predic.adj. oblique; awry. [A² + SKEW]

aslant /ə'slɑːnt/ adv. & prep. ● adv. obliquely or at a slant. ● prep. obliquely across (lay aslant the path).

asleep /ə'sliːp/ predic.adj. & adv. **1 a** in or into a state of sleep (he fell asleep). **b** inactive, inattentive (the nation is asleep). **2** (of a limb etc.) numb. **3** euphem. dead.

ASLEF /'æzlef/ abbr. (in the UK) Associated Society of Locomotive Engineers and Firemen.

aslope /ə'sləʊp/ adv. & predic.adj. sloping; crosswise. [ME: orig. uncert.]

ASM abbr. air-to-surface missile.

Asmara /æs'mɑːrə/ (also **Asmera** /-'meərə/) the capital of Eritrea; pop. (est. 1990) 358,000.

asocial /eɪ'səʊʃ(ə)l/ adj. **1** not social; antisocial. **2** inconsiderate of or hostile to others.

Asoka /ə'səʊkə, ə'ʃəʊ-/ (died c.232 BC), emperor of India c.269–232 BC. He embarked on a campaign of conquest, but after his conversion to Buddhism (which he established as the state religion) he renounced war and sent out missionaries as far afield as Syria and Ceylon to spread his new faith.

asp /æsp/ n. **1** a small viper, Vipera aspis, native to southern Europe, resembling the adder. **2** the Egyptian cobra, Naja haje, found throughout Africa. [ME f. OF aspe or L aspis f. Gk]

asparagine /ə'spærə,dʒiːn/ n. Biochem. a hydrophilic amino acid, an amide of aspartic acid, occurring in proteins. [f. ASPARAGUS + -INE⁴]

asparagus /ə'spærəgəs/ n. **1** a liliaceous plant of the genus Asparagus. **2** one species of this, A. officinalis, with edible young shoots and leaves; this as food. □ **asparagus fern** a decorative plant, Asparagus setaceus. [L f. Gk asparagos]

aspartame /ə'spɑː,teɪm/ n. a very sweet low-calorie substance used as a sweetener instead of sugar or saccharin. [chem. name 1-methyl N-L-aspartyl-L-phenylalanine, f. ASPARTIC ACID]

aspartic acid /ə'spɑːtɪk/ n. Biochem. an acidic amino acid occurring in proteins, important in animal metabolism and as a neurotransmitter. [F aspartique, formed arbitrarily f. ASPARAGUS]

aspect /'æspekt/ n. **1 a** a particular component or feature of a matter (only one aspect of the problem). **b** a particular way in which a matter may be considered. **2 a** a facial expression; a look (a cheerful aspect). **b** the appearance of a person or thing, esp. as presented to the mind of the viewer (has a frightening aspect). **3** the side of a building or location facing a particular direction (southern aspect). **4** Gram. a verbal category or form expressing inception, duration, or completion. **5** Astrol. the relative position of planets etc. measured by angular distance. □ **aspect ratio 1** Aeron. the ratio of the span to the mean chord of an aerofoil. **2** Television the ratio of picture width to height. □ **aspectual** /æ'spektjʊəl, -tjʊəl/ adj. (in sense 4). [ME f. L aspectus f. adspicere adspect-look at (as AD-, specere look)]

Aspen /'æspən/ a resort in south central Colorado; pop. (1990) 6,850. Formerly a silver-mining town, it is now a thriving recreational centre, noted particularly for its skiing facilities.

aspen /'æspən/ n. a poplar tree, Populus tremula, with especially tremulous leaves. [earlier name asp f. OE æspe + -EN² forming adj. taken as noun]

asperity /ə'sperɪtɪ/ n. (pl. **-ies**) **1** harshness or sharpness of temper or tone. **2** roughness. **3** a rough excrescence. [ME f. OF asperité or L asperitas f. asper rough]

asperse /ə'spɜːs/ v.tr. (often foll. by with) attack the reputation of; calumniate. □ **aspersive** adj. [ME, = besprinkle, f. L aspergere aspers- (as AD-, spargere sprinkle)]

aspersion /ə'spɜːʃ(ə)n/ n. a slander, a false insinuation. □ **cast aspersions on** attack the reputation or integrity of. [L aspersio (as ASPERSE, -ION)]

asphalt /'æsfælt/ n. & v. ● n. **1** a dark bituminous kind of pitch. (See note below.) **2** a mixture of this with sand, gravel, etc., for surfacing roads etc. ● v.tr. surface with asphalt. □ **asphaltic** /æs'fæltɪk/ adj. [ME, ult. f. LL asphalton, -um, f. Gk asphalton]

- Asphalt occurs naturally in surface deposits, e.g. in Trinidad and parts of America, but commercial asphalt is mostly made in oil refineries by the distillation of crude oils.

asphodel /'æsfə,del/ n. **1 a** a liliaceous plant of the genus Asphodelus or Asphodeline, native to the Mediterranean region. **b** a related plant of the genus Narthecium or Tofieldia (bog asphodel). **2** poet. an immortal flower growing in Elysium. [L asphodelus f. Gk asphodelos: cf. DAFFODIL]

asphyxia /æs'fɪksɪə/ n. a lack of oxygen in the blood, causing unconsciousness or death; suffocation. □ **asphyxial** adj. **asphyxiant** adj. & n. [mod.L f. Gk asphuxia f. a- not + sphuxis pulse]

asphyxiate /æs'fɪksɪ,eɪt/ v.tr. cause (a person) to have asphyxia; suffocate. □ **asphyxiator** n. **asphyxiation** /-,fɪksɪ'eɪʃ(ə)n/ n.

aspic /'æspɪk/ n. a savoury meat jelly used as a garnish or to contain game, eggs, etc. [F, = ASP, from the colours of the jelly (compared to those of the asp)]

aspidistra /,æspɪ'dɪstrə/ n. a Far Eastern plant of the genus Aspidistra, with broad tapering leaves, often grown as a house plant. [mod.L f. Gk aspis -idos shield (from the shape of the leaves)]

aspirant /ˈæspɪrənt, əˈspaɪərˌ/ adj. & n. (usu. foll. by to, after, for) ● adj. aspiring. ● n. a person who aspires. [F aspirant or f. L aspirant- (as ASPIRE, -ANT)]

aspirate /ˈæspɪrət/ adj., n., & v. ● adj. **1** pronounced with an exhalation of breath. **2** blended with the sound of h. ● n. **1** a consonant pronounced in this way. **2** the sound of h. ● v. (also /ˈæspəˌreɪt/) **1 a** tr. pronounce with a breath. **b** intr. make the sound of h. **2** tr. draw (fluid) by suction from a vessel or cavity. [L aspiratus past part. of aspirare: see ASPIRE]

aspiration /ˌæspɪˈreɪʃ(ə)n/ n. **1** a strong desire to achieve an end; an ambition. **2** the act or process of drawing breath. **3** the action of aspirating. [ME f. OF aspiration or L aspiratio (as ASPIRATE, -ATION)]

aspirator /ˈæspɪˌreɪtə(r)/ n. an apparatus for aspirating fluid. [L aspirare (as ASPIRATE, -OR¹)]

aspire /əˈspaɪə(r)/ v.intr. (usu. foll. by to or after, or to + infin.) **1** have ambition or strong desire. **2** poet. rise high. [ME f. F aspirer or L aspirare f. ad to + spirare breathe]

aspirin /ˈæsprɪn/ n. a white solid, acetylsalicylic acid, used to relieve pain and reduce fever, usually in the form of tablets. Aspirin was developed in Germany in the late 1890s, originally to reduce inflammation and is now also used in the prevention of thrombosis. [G, formed as ACETYL + SPIRAEA (in an alternative name for salicylic acid) + -IN]

asquint /əˈskwɪnt/ predic.adj. & adv. (usu. look asquint). **1** to one side; from the corner of an eye. **2** with a squint. [ME perh. f. Du. schuinte slant]

Asquith /ˈæskwɪθ/, Herbert Henry, 1st Earl of Oxford and Asquith (1852–1928), British Liberal statesman, Prime Minister 1908–16. In the years before the First World War he introduced the third bill for Irish Home Rule, while also contending with the challenge posed by the women's suffrage movement and outrage from the House of Lords over Lloyd George's People's Budget (1909). In 1915 Asquith brought the Conservatives into a coalition government, but his failure to consult his colleagues divided the Liberals; he was displaced as Prime Minister by Lloyd George the following year, but retained the party leadership.

ass¹ /æs/ n. & v. ● n. **1 a** a long-eared hoofed mammal of the horse genus Equus, E. africanus of Africa and E. hemionus of Asia. **b** (in general use) a donkey. **2** a stupid person. ● v.intr. (foll. by about, around) sl. act the fool. □ **asses' bridge** = pons asinorum. **make an ass of** make (a person) look absurd or foolish. [OE assa thr. Celtic f. L asinus]

ass² US var. of ARSE.

Assad /ˈæsæd/, Hafiz al- (b.1928), Syrian Baath statesman, President since 1971. While in office, he has ensured the strengthening of Syria's oil-based economy and suppressed political opposition such as the uprising of Muslim extremists (1979–82). He supported the coalition forces during the 1991 Gulf War.

assagai var. of ASSEGAI.

assai /æˈsaɪ/ adv. Mus. very (adagio assai). [It.]

assail /əˈseɪl/ v.tr. **1** make a strong or concerted attack on. **2** make a resolute start on (a task). **3** make a strong or constant verbal attack on (was assailed with angry questions). □ **assailable** adj. [ME f. OF asaill-stressed stem of asalir f. med.L assalire f. L assilire (as AD-, salire salt- leap)]

assailant /əˈseɪlənt/ n. a person who attacks another physically or verbally. [F (as ASSAIL)]

Assam /æˈsæm/ a state in NE India; capital, Dispur. Formerly a British protectorate, it lost some territory to Pakistan and Bhutan in 1947 and was further reduced in size through the formation of the states of Arunachal Pradesh, Meghalaya, Mizoram, Manipur, and Nagaland. Most of the state lies in the valley of the Brahmaputra river, with tropical forests and heavy rainfall. It is noted for the production of tea. □ **Assamese** /ˌæsəˈmiːz/ adj. & n.

Assassin /əˈsæsɪn/ n. a member of an Ismaili Muslim sect in the time of the Crusades, allegedly responsible for a series of killings of political and religious opponents. (See NIZARI.) [F assassin or f. med.L assassinus f. Arab. ḥaššāš hashish-eater]

assassin /əˈsæsɪn/ n. a killer, esp. of a political or religious leader. □ **assassin bug** a predatory or bloodsucking heteropteran bug of the family Reduviidae. [ASSASSIN]

assassinate /əˈsæsɪˌneɪt/ v.tr. **1** kill (esp. a political or religious leader) for political or religious motives. **2** destroy or injure (esp. a person's reputation). □ **assassinator** n. **assassination** /əˌsæsɪˈneɪʃ(ə)n/ n. [med.L assassinare f. assassinus: see ASSASSIN]

assault /əˈsɔːlt, əˈsɒlt/ n. & v. ● n. **1** a violent physical or verbal attack.

2 a Law an act that threatens physical harm to a person (whether or not actual harm is done). **b** euphem. an act of rape. **3** (attrib.) relating to or used in an assault (assault craft; assault troops). **4** a vigorous start made to a lengthy or difficult task. **5** a final rush on a fortified place, esp. at the end of a prolonged attack. ● v.tr. **1** make an assault on. **2** euphem. rape. □ **assault and battery** Law a threat followed by an attack that involves physical contact (which may amount to no more than touching if done with malicious intent). **assault course** an obstacle course used in training soldiers etc. □ **assaulter** n. **assaultive** adj. [ME f. OF asaut, assauter ult. f. L (salire salt- leap)]

assay /əˈseɪ, ˈæseɪ/ n. & v. ● n. **1** the testing of a metal or ore to determine its ingredients and quality. **2** Chem. etc. the determination of the content or concentration of a substance. ● v. **1** tr. make an assay of (a metal or ore). **2** tr. Chem. etc. determine the concentration of (a dissolved substance). **3** tr. show (content) on being assayed. **4** intr. make an assay. **5** tr. archaic attempt. □ **assay office** an establishment which awards hallmarks. □ **assayer** n. [ME f. OF assaier, assai, var. of essayer, essai: see ESSAY]

assegai /ˈæsəˌgaɪ/ n. (also **assagai**) a slender iron-tipped spear of hard wood, esp. as used by southern African peoples. [obs. F azagaie or Port. azagaia f. Arab. az-zaġāyah f. al the + zaġāyah spear]

assemblage /əˈsemblɪdʒ/ n. **1** the act or an instance of bringing or coming together. **2** a collection of things or gathering of people. **3 a** the act or an instance of fitting together. **b** an object made of pieces fitted together. **4** a work of art made by grouping found or unrelated objects.

assemble /əˈsemb(ə)l/ v. **1** tr. & intr. gather together; collect. **2** tr. arrange in order. **3** tr. esp. Mech. fit together the parts of. [ME f. OF asembler ult. f. L ad to + simul together]

assembler /əˈsemblə(r)/ n. **1** a person who assembles a machine or its parts. **2** Computing **a** a program for converting instructions written in low-level symbolic code into machine code. **b** the low-level symbolic code itself; an assembly language.

assembly /əˈsemblɪ/ n. (pl. **-ies**) **1** the act or an instance of assembling or gathering together. **2 a** a group of persons gathered together, esp. as a deliberative body or a legislative council. **b** a gathering of a large part or the entire membership of a school. **3** the assembling of a machine or structure or its parts. **4** Mil. a call to assemble, given by drum or bugle. □ **assembly language** Computing the low-level symbolic code converted by an assembler. **assembly line** machinery arranged in stages by which a product is progressively assembled. **assembly room** (or **shop**) a place where a machine or its components are assembled. **assembly rooms** public rooms in which meetings or social functions are held. [ME f. OF asemblee fem. past part. of asembler: see ASSEMBLE]

assent /əˈsent/ v. & n. ● v.intr. (usu. foll. by to) **1** express agreement (assented to my view). **2** consent (assented to my request). ● n. **1** mental or inward acceptance or agreement (a nod of assent). **2** consent or sanction, esp. official. □ **royal assent** assent of the sovereign to a bill passed by Parliament. □ **assenter** n. (also **assentor**). [ME f. OF asenter, as(s)ente ult. f. L assentari (ad to, sentire think)]

assentient /əˈsenʃ(ə)nt, -ʃɪənt/ adj. & n. ● adj. assenting. ● n. a person who assents. [L assentire (as ASSENT, -ENT)]

assert /əˈsɜːt/ v. **1** tr. declare; state clearly (assert one's beliefs; assert that it is so). **2** refl. insist on one's rights or opinions; demand recognition. **3** tr. vindicate a claim to (assert one's rights). □ **assertor** n. (also **asserter**). [L asserere (as AD-, serere sert- join)]

assertion /əˈsɜːʃ(ə)n/ n. **1** a declaration; a forthright statement. **2** the act or an instance of asserting. **3** (also **self-assertion**) insistence on the recognition of one's rights or claims. [ME f. F assertion or L assertio (as ASSERT, -ION)]

assertive /əˈsɜːtɪv/ adj. **1** tending to assert oneself; forthright, positive. **2** dogmatic. □ **assertively** adv. **assertiveness** n.

asses pl. of AS², ASS¹, ASS².

assess /əˈses/ v.tr. **1 a** estimate the size or quality of. **b** estimate the value of (a property) for taxation etc. **2 a** (usu. foll. by on) fix the amount of (a tax etc.) and impose it on a person or community. **b** (usu. foll. by in, at) fine or tax (a person, community, etc.) in or at a specific amount (assessed them at £100). □ **assessable** adj. **assessment** n. [ME f. F assesser f. L assidere (as AD-, sedere sit)]

assessor /əˈsesə(r)/ n. **1** a person who assesses taxes or estimates the value of property for taxation or insurance purposes. **2** a person called upon to advise a judge, committee of inquiry, etc., on technical questions. [ME f. OF assessour f. L assessor -oris assistant-judge (as ASSESS, -OR¹): sense 1 f. med.L]

asset /'æset/ n. **1 a** a useful or valuable quality. **b** a person or thing possessing such a quality or qualities (*is an asset to the firm*). **2** (usu. in *pl.*) **a** property and possessions, esp. regarded as having value in meeting debts, commitments, etc. **b** any possession having value. □ **asset-stripper** an entrepreneur engaged in asset-stripping. **asset-stripping** the practice of taking over a company and selling off its assets to make a profit. [*assets* (taken as pl.), f. AF *asetz* f. OF *asez* enough, ult. f. L *ad* to + *satis* enough]

asseverate /ə'sevə,reɪt/ v.tr. declare solemnly or emphatically. □ **asseverative** /-rətɪv/ adj. **asseveration** /ə,sevə'reɪʃ(ə)n/ n. [L *asseverare* (as AD-, *severus* serious)]

assibilate /ə'sɪbɪ,leɪt/ v.tr. Phonet. **1** pronounce (a sound) as a sibilant or affricate ending in a sibilant. **2** alter (a syllable) to become this. □ **assibilation** /ə,sɪbɪ'leɪʃ(ə)n/ n. [L *assibilare* (as AD-, *sibilare* hiss)]

assiduity /ˌæsɪ'djuːɪtɪ/ n. (pl. **-ies**) **1** constant or close attention to what one is doing. **2** (usu. in pl.) constant attentions to another person. [L *assiduitas* (as ASSIDUOUS, -ITY)]

assiduous /ə'sɪdjʊəs/ adj. **1** persevering, hard-working. **2** attending closely. □ **assiduously** adv. **assiduousness** n. [L *assiduus* (as ASSESS)]

assign /ə'saɪn/ v. & n. ● v.tr. **1** (usu. foll. by *to*) **a** allot as a share or responsibility. **b** appoint to a position, task, etc. **2** fix (a time, place, etc.) for a specific purpose. **3** (foll. by *to*) ascribe or refer to (a reason, date, etc.) (*assigned the manuscript to 1832*). **4** (foll. by *to*) transfer formally (esp. personal property) to (another). ● n. a person to whom property or rights are legally transferred. □ **assignable** adj. **assigner** n. **assignor** /ˌæsɪ'nɔː(r)/ n. (in sense 4 of v.). [ME f. OF *asi(g)ner* f. L *assignare* mark out to (as AD-, *signum* sign)]

assignation /ˌæsɪg'neɪʃ(ə)n/ n. **1 a** an appointment to meet. **b** a secret appointment, esp. between illicit lovers. **2** the act or an instance of assigning or being assigned. [ME f. OF f. L *assignatio -onis* (as ASSIGN, -ATION)]

assignee /ˌæsaɪ'niː/ n. **1** a person appointed to act for another. **2** an assign. [ME f. OF *assigné* past part. of *assigner* ASSIGN]

assignment /ə'saɪnmənt/ n. **1** something assigned, esp. a task allotted to a person. **2** the act or an instance of assigning or being assigned. **3 a** a legal transfer. **b** the document effecting this. [ME f. OF *assignement* f. med.L *assignamentum* (as ASSIGN, -MENT)]

assimilate /ə'sɪmɪ,leɪt/ v. **1** tr. **a** absorb and digest (food etc.) into the body. **b** absorb (information etc.) into the mind. **c** absorb (people) into a larger group. **2** tr. (usu. foll. by *to*, *with*) make like; cause to resemble. **3** tr. Phonet. make (a sound) more like another in the same or next word. **4** intr. be absorbed into the body, mind, or a larger group. □ **assimilator** n. **assimilable** adj. **assimilative** /-lətɪv/ adj. **assimilatory** /-lətərɪ/ adj. **assimilation** /ə,sɪmɪ'leɪʃ(ə)n/ n. [ME f. L *assimilare* (as AD-, *similis* like)]

Assisi[1] /ə'siːzɪ/ a town in the province of Umbria in central Italy; pop. (1990) 24,790. It is famous as the birthplace of St Francis, whose tomb is located there.

Assisi[2] see CLARE OF ASSISI, ST.

Assisi[3] see FRANCIS OF ASSISI, ST.

assist /ə'sɪst/ v. & n. ● v. **1** tr. (often foll. by *in* + verbal noun) help (a person, process, etc.) (*assisted them in running the playgroup*). **2** intr. (often foll. by *in*, *at*) take part or be present (*assisted in the ceremony*). ● n. US **1** an act of helping. **2** Baseball etc. a player's action of helping to put out an opponent, score a goal, etc. □ **assisted place** (in the UK) a place in an independent school, wholly or partially subsidized by the state to enable a capable pupil of limited means to attend. **assisted suicide** the taking of lethal drugs, provided by a doctor for the purpose, by a patient considered to be incurable. □ **assistance** n. **assister** n. [ME f. F *assister* f. L *assistere* take one's stand by (as AD-, *sistere* take one's stand)]

assistant /ə'sɪstənt/ n. **1** a helper. **2** (often *attrib.*) a person who assists, esp. as a subordinate in a particular job or role. **3** = *shop assistant*. [ME *assistent* f. med.L *assistens assistent-* present (as ASSIST, -ANT)]

assize /ə'saɪz/ n. (usu. in *pl.*) hist. a court sitting at intervals in each county of England and Wales to administer the civil and criminal law. In 1972 the civil jurisdiction of assizes was transferred to the High Court and the criminal jurisdiction to the Crown Court. [ME f. OF *as(s)ise*, fem. past part. of *aseeir* sit at, f. L *assidere*: cf. ASSESS]

Assoc. abbr. (as part of a title) Association.

associable /ə'səʊʃɪəb(ə)l/ adj. (usu. foll. by *with*) capable of being connected in thought. □ **associability** /ə,səʊʃɪə'bɪlɪtɪ/ n. [F f. *associer* (as ASSOCIATE, -ABLE)]

associate v., adj., & adj. ● v. /ə'səʊʃɪ,eɪt, ə'səʊsɪ-/ **1** tr. connect in the mind (*associate holly with Christmas*). **2** tr. join or combine. **3** refl. make oneself a partner; declare oneself in agreement (*associate myself in your endeavour; did not want to associate ourselves with the plan*). **4** intr. combine for a common purpose. **5** intr. (usu. foll. by *with*) meet frequently or have dealings; be friends. ● n. /ə'səʊʃɪət, ə'səʊsɪ-/ **1** a business partner or colleague. **2** a friend or companion. **3** a subordinate member of a body, institute, etc. **4** a thing connected with another. ● adj. /ə'səʊʃɪət, ə'səʊsɪ-/ **1** joined in companionship, function, or dignity. **2** allied; in the same group or category. **3** of less than full status (*associate member*). □ **associateship** /ə'səʊʃɪət,ʃɪp, ə'səʊsɪ-/ n. **associator** /ə'səʊʃɪ,eɪtə(r), ə'səʊsɪ-/ n. [ME f. L *associatus* past part. of *associare* (as AD-, *socius* sharing, allied)]

Associated Press (abbr. **AP**) an international news agency based in New York City. Founded in 1848, it is now the largest in the world.

association /ə,səʊsɪ'eɪʃ(ə)n/ n. **1** a group of people organized for a joint purpose; a society. **2** the act or an instance of associating. **3** fellowship; human contact or cooperation. **4** a mental connection between ideas. **5** Chem. a loose aggregation of molecules. **6** Ecol. a group of associated plants. □ **associational** adj. [F *association* or med.L *associatio* (as ASSOCIATE, -ATION)]

Association football n. (also **football**; also called *soccer*) a form of football played between two teams of eleven players with a round ball which may not be handled during play except by the goalkeepers. The object of the game is to score goals, by kicking or heading the ball into the opponents' goal. Fouls and transgressions of the rules are punished by the opposition's being awarded a free kick, while a throw-in is given to return a ball played out of the playing area. A ball crossing the goal-line leads to a goal-kick for the defenders or a corner for the attacking team. The game originated in England, and is played according to rules established by the Football Association (founded 1863), which has organized the English major knock-out competition, the FA Cup, since 1872; professional clubs formed the Football League in 1888. The international game is administered by the International Football Federation (FIFA), formed in 1904, which is responsible for the World Cup. Although professional football is particularly strong in Europe and Latin America, Association football is now a truly international sport.

Association of South-East Asian Nations (abbr. **ASEAN**) a regional organization formed by Indonesia, Malaysia, the Philippines, Singapore, and Thailand through the Bangkok Declaration of 1967, designed to promote economic cooperation. Brunei joined the organization in 1984.

associative /ə'səʊʃɪətɪv, ə'səʊsɪ-/ adj. **1** of or involving association. **2** Math. & Computing involving the condition that a group of quantities connected by operators (see OPERATOR 4) gives the same result whatever their grouping, as long as their order remains the same, e.g. $(a \times b) \times c = a \times (b \times c)$.

assonance /'æsənəns/ n. the resemblance of sound between two syllables in nearby words, arising from the rhyming of two or more accented vowels, but not consonants, or the use of identical consonants with different vowels, e.g. *sonnet*, *porridge*, and *killed*, *cold*, *culled*. □ **assonant** adj. **assonate** /-,neɪt/ v.intr. [F f. L *assonare* respond to (as AD-, *sonus* sound)]

assort /ə'sɔːt/ v. **1** tr. (usu. foll. by *with*) classify or arrange in groups. **2** intr. suit; fit into; harmonize with (usu. *assort ill* or *well with*). [OF *assorter* f. *à* to + *sorte* SORT]

assortative /ə'sɔːtətɪv/ adj. assorting. □ **assortative mating** Biol. non-random mating resulting from the selection of similar partners.

assorted /ə'sɔːtɪd/ adj. **1** of various sorts put together; miscellaneous. **2** sorted into groups. **3** matched (*ill-assorted; poorly assorted*).

assortment /ə'sɔːtmənt/ n. a set of various sorts of things or people put together; a mixed collection.

ASSR abbr. hist. Autonomous Soviet Socialist Republic.

Asst. abbr. Assistant.

assuage /ə'sweɪdʒ/ v.tr. **1** calm or soothe (a person, pain, etc.). **2** appease or relieve (an appetite or desire). □ **assuagement** n. [ME f. OF *as(s)ouagier* ult. f. L *suavis* sweet]

As Sulaymaniyah see SULAYMANIYAH.

assume /ə'sjuːm/ v.tr. **1** (usu. foll. by *that* + clause) take or accept as being true, without proof, for the purpose of argument or action. **2** simulate or pretend (ignorance etc.). **3** undertake (an office or duty). **4** take or put on oneself or itself (an aspect, attribute, etc.) (*the problem assumed immense proportions*). **5** (usu. foll. by *to*) arrogate, usurp, or seize (credit, power, etc.) (*assumed to himself the right of veto*). □ **assumable**

adj. **assumedly** /-mɪdlɪ/ *adv.* [ME f. L *assumere* (as AD-, *sumere sumpt-* take)]

assumed /əˈsjuːmd/ *adj.* **1** false, adopted (*went under an assumed name*). **2** supposed, accepted (*assumed income*).

assuming /əˈsjuːmɪŋ/ *adj.* (of a person) taking too much for granted; arrogant, presumptuous.

assumption /əˈsʌmpʃ(ə)n/ *n.* **1** the act or an instance of assuming. **2 a** the act or an instance of accepting without proof. **b** a thing assumed in this way. **3** arrogance. [ME f. OF *asompsion* or L *assumptio* (as ASSUME, -ION)]

Assumption, the (in Christian thought) the doctrine that the Virgin Mary was 'assumed' (i.e. taken up and received bodily into heaven). Dating from the 4th century, it is a doctrine held by the Roman Catholic and Orthodox Churches. The feast in honour of this is held on 15 Aug.

assumptive /əˈsʌmptɪv/ *adj.* **1** taken for granted. **2** arrogant. [L *assumptivus* (as ASSUME, -IVE)]

Assur /ˈæsʊə(r)/ (also **Asur, Ashur** /ˈæʃʊə(r)/) an ancient city-state of Mesopotamia, situated on the River Tigris to the south of modern Mosul. The Assyrian empires, the first of which was established early in the 2nd millennium BC, were centred on the city.

assurance /əˈʃʊərəns/ *n.* **1** a positive declaration that a thing is true. **2** a solemn promise or guarantee. **3** insurance, esp. life insurance. (*See note below.*) **4** certainty. **5 a** self-confidence. **b** impudence. [ME f. OF *aseürance* f. *aseürer* (as ASSURE, -ANCE)]

▪ *Assurance* is the term generally used by insurance companies to denote policies where a sum is payable after a fixed number of years or on the death of the insured person, while the term *insurance* is used of policies relating to events such as fire, accident, or death within a limited period. In popular usage the word *insurance* is used in both cases.

assure /əˈʃʊə(r)/ *v.tr.* **1** (often foll. by *of*) **a** make (a person) sure; convince (*assured him of my sincerity*). **b** tell (a person) confidently (*assured him the bus went to Westminster*). **2 a** make certain of; ensure the happening etc. of (*will assure her success*). **b** make safe (against overthrow etc.). **3** confirm, encourage. **4** insure (esp. a life). **5** (as **assured** *adj.*) **a** guaranteed. **b** self-confident. **rest assured** remain confident. □ **assurer** *n.* [ME f. OF *aseürer* ult. f. L *securus* safe, SECURE]

assuredly /əˈʃʊərɪdlɪ/ *adv.* certainly.

assuredness /əˈʃʊərɪdnɪs, əˈʃʊəd-/ *n.* certainty, (self-)assurance.

Assyria /əˈsɪrɪə/ an ancient country in what is now northern Iraq. From the early part of the 2nd millennium BC Assyria was the centre of a succession of empires; it was at its peak in the 8th and late 7th centuries BC, when its rule stretched from the Persian Gulf to Egypt. It fell in 612 BC to a coalition of Medes and Chaldeans. □ **Assyrian** *adj. & n.*

Assyriology /əˌsɪrɪˈɒlədʒɪ/ *n.* the study of the language, history, and antiquities of Assyria. □ **Assyriologist** *n.*

AST *abbr.* Atlantic Standard Time.

astable /eɪˈsteɪb(ə)l/ *adj.* **1** not stable. **2** *Electr.* of or relating to a circuit which oscillates spontaneously between unstable states.

Astaire /əˈsteə(r)/, Fred (born Frederick Austerlitz) (1899–1987), American dancer, singer, and actor. He danced in music-halls from an early age, before starring in a number of film musicals, including *Top Hat* (1935), *Follow the Fleet* (1936), and *Shall We Dance?* (1937), in a successful partnership with Ginger Rogers. After his partnership with Rogers ended he continued to appear in films such as *Easter Parade* (1948) with Judy Garland.

Astarte /əˈstɑːtɪ/ *Mythol.* a Phoenician goddess of fertility and sexual love who corresponds to the goddess Ishtar and who became identified with the Egyptian Isis, the Greek Aphrodite, and others. In the Bible she is referred to as Ashtaroth or Ashtoreth and her worship is linked with that of Baal.

astatic /eɪˈstætɪk/ *adj.* **1** not static; unstable or unsteady. **2** *Physics* not tending to keep one position or direction. □ **astatic galvanometer** one in which the effect of the earth's magnetic field on the meter needle is greatly reduced. [Gk *astatos* unstable f. *a-* not + *sta-* stand]

astatine /ˈæstəˌtiːn/ *n.* a radioactive chemical element (atomic number 85; symbol **At**). One of the halogens, astatine was first produced artificially in the US in 1940, when the physicist Dale Corson and his colleagues bombarded bismuth with alpha particles. It occurs in traces in nature as a radioactive decay product. [formed as ASTATIC + -INE⁴]

aster /ˈæstə(r)/ *n.* a composite plant of the genus *Aster*, with bright daisy-like flowers, e.g. the Michaelmas daisy. □ **China aster** a related plant, *Callistephus chinensis*, cultivated for its bright and showy flowers. [L f. Gk *astēr* star]

-aster /ˈæstə(r)/ *suffix* **1** forming nouns denoting poor quality (*criticaster*; *poetaster*). **2** *Bot.* denoting incomplete resemblance (*oleaster*; *pinaster*). [L]

asterisk /ˈæstərɪsk/ *n. & v.* ● *n.* a symbol (*) used in printing and writing to mark words etc. for reference, to stand for omitted matter, etc. ● *v.tr.* mark with an asterisk. [ME f. LL *asteriscus* f. Gk *asteriskos* dimin. (as ASTER)]

asterism /ˈæstəˌrɪz(ə)m/ *n.* **1** *Astron.* a prominent pattern or group of stars, often having a popular name but smaller than a constellation. **2** a group of three asterisks (⁂) calling attention to following text. [Gk *asterismos* (as ASTER, -ISM)]

astern /əˈstɜːn/ *adv.* (often foll. by *of*) *Naut. & Aeron.* **1** aft; away to the rear. **2** backwards. [A² + STERN²]

asteroid /ˈæstəˌrɔɪd/ *n.* **1** *Astron.* a small rocky body orbiting the sun, mainly between the orbits of Mars and Jupiter. Also called *minor planet*. (*See note below.*) **2** *Zool.* a starfish. □ **asteroidal** /ˌæstəˈrɔɪd(ə)l/ *adj.* [Gk *asteroeidēs* (as ASTER, -OID)]

▪ The asteroids range from nearly 1,000 kilometres across (Ceres) to the size of dust grains. Large numbers are found in a broad belt between Mars and Jupiter, though some have eccentric orbits and may pass within a few million kilometres of earth. The first close-up photographs of asteroids (Gaspra and Ida), taken by the space probe Galileo, showed the asteroids to be irregular fragments of larger bodies. Several satellites of the planets are thought to be captured asteroids, and the majority of meteorites probably originate in the asteroid belt.

asthenia /æsˈθiːnɪə/ *n.* *Med.* loss of strength; debility. [mod.L f. Gk *astheneia* f. *asthenēs* weak]

asthenic /æsˈθenɪk/ *adj. & n.* ● *adj.* **1** of lean or long-limbed build. **2** *Med.* of or characterized by asthenia. ● *n.* a lean long-limbed person.

asthenosphere /æsˈθenəʊˌsfɪə(r)/ *n.* *Geol.* the upper layer of the earth's mantle, whose capacity for gradual flow is thought to give rise to continental drift. [f. Gk *asthenēs* weak + -SPHERE]

asthma /ˈæsmə/ *n.* a usu. allergic respiratory disease, often with paroxysms of difficult breathing. [ME f. Gk *asthma -matos* f. *azō* breathe hard]

asthmatic /æsˈmætɪk/ *adj. & n.* ● *adj.* relating to or suffering from asthma. ● *n.* a person suffering from asthma. □ **asthmatically** *adv.* [L *asthmaticus* f. Gk *asthmatikos* (as ASTHMA, -IC)]

Asti /ˈæstɪ/ *n.* (pl. **Astis**) an Italian white wine. □ **Asti spumante** /spuˈmæntɪ/ a sparkling form of this. [*Asti* in Piedmont]

astigmatism /əˈstɪgməˌtɪz(ə)m/ *n.* a defect in the eye or in a lens resulting in distorted images, as light rays are prevented from meeting at a common focus. □ **astigmatic** /ˌæstɪgˈmætɪk/ *adj.* [A-¹ + Gk *stigma -matos* point]

astilbe /əˈstɪlbɪ/ *n.* a plant of the genus *Astilbe*, of the saxifrage family, with plumelike heads of tiny white or red flowers. [mod.L f. Gk *a-* not + *stilbē* fem. of *stilbos* glittering, from the inconspicuous (individual) flowers]

astir /əˈstɜː(r)/ *predic.adj. & adv.* **1** in motion. **2** awake and out of bed (*astir early*; *already astir*). **3** excited. [A² + STIR¹ *n.*]

Aston /ˈæst(ə)n/, Francis William (1877–1945), English physicist. Aston worked in Cambridge with J. J. Thomson, inventing the mass spectrograph. With this he eventually discovered many of the 287 naturally occurring isotopes of non-radioactive elements, announcing in 1919 the whole-number rule governing their masses. He was awarded the Nobel Prize for chemistry in 1922.

astonish /əˈstɒnɪʃ/ *v.tr.* amaze; surprise greatly. □ **astonishing** *adj.* **astonishingly** *adv.* **astonishment** *n.* [obs. *astone* f. OF *estoner* f. Gallo-Roman: see -ISH²]

Astor /ˈæstə(r)/, Nancy Witcher Langhorne, Viscountess (1879–1964), American-born British Conservative politician. She became the first woman to sit in the House of Commons when she succeeded her husband as MP for Plymouth in 1919. She supported causes about which she had deep convictions, such as temperance and women's rights, rather than following the party line.

astound /əˈstaʊnd/ *v.tr.* shock with alarm or surprise; amaze. □ **astounding** *adj.* **astoundingly** *adv.* [obs. *astound* (adj.) = *astoned* past part. of obs. *astone*: see ASTONISH]

astraddle /əˈstræd(ə)l/ *adv. & predic.adj.* in a straddling position.

Astraea /æˈstriːə/ **1** *Rom. Mythol.* a goddess associated with justice. **2** *Astron.* asteroid 5, discovered in 1845 (diameter 125 km).

astragal /ˈæstrəɡ(ə)l/ *n. Archit.* a small semicircular moulding round the top or bottom of a column. [ASTRAGALUS]

astragalus /əˈstræɡələs/ *n.* (*pl.* **-li** /-ˌlaɪ/) **1** *Anat.* = TALUS[1]. **2** *Bot.* a leguminous plant of the genus *Astragalus*, e.g. the milk-vetch. [L f. Gk *astragalos* ankle-bone, moulding, a plant]

Astrakhan /ˌæstrəˈkɑːn/ a city in southern Russia, on the delta of the River Volga; pop. (est. 1989) 509,000. Astrakhan fleeces were given their name because traders from the city brought them into Russia from central Asia.

astrakhan /ˌæstrəˈkɑːn/ *n.* **1** the dark curly fleece, resembling fur, of young karakul lambs from central Asia. **2** a cloth imitating astrakhan. [ASTRAKHAN]

astral /ˈæstrəl/ *adj.* **1** of or connected with the stars. **2** consisting of stars; starry. **3** *Theosophy* relating to or arising from a supposed ethereal existence, esp. of a counterpart of the body, associated with oneself in life and surviving after death. [LL *astralis* f. *astrum* star]

astray /əˈstreɪ/ *adv. & predic.adj.* **1** in or into error or sin (esp. *lead astray*). **2** out of the right way. □ **go astray** be lost or mislaid. [ME f. OF *estraié* past part. of *estraier* ult. f. L *extra* out of bounds + *vagari* wander]

astride /əˈstraɪd/ *adv. & prep.* ● *adv.* **1** (often foll. by *of*) with a leg on each side. **2** with legs apart. ● *prep.* with a leg on each side of; extending across.

astringent /əˈstrɪndʒənt/ *adj. & n.* ● *adj.* **1** causing the contraction of body tissues. **2** checking bleeding. **3** severe, austere. ● *n.* an astringent substance or drug. □ **astringency** *n.* **astringently** *adv.* [F f. L *astringere* (as AD-, *stringere* bind)]

astro- /ˈæstrəʊ/ *comb. form* **1** relating to the stars or celestial objects. **2** relating to outer space. [Gk f. *astron* star]

astrochemistry /ˌæstrəʊˈkemɪstrɪ/ *n.* the study of molecules and radicals in interstellar space.

astrodome /ˈæstrəˌdəʊm/ *n.* **1** a domed window in an aircraft for astronomical observations. **2** esp. *US* an enclosed stadium with a domed roof.

astrohatch /ˈæstrəˌhætʃ/ *n.* = ASTRODOME 1.

astrolabe /ˈæstrəˌleɪb/ *n.* an instrument formerly used to make astronomical measurements, esp. of the altitudes of celestial bodies, and as an aid to navigation. In its earliest form (which dates from classical times) it consisted of a disc with the degrees of the circle marked round its edge, and a pivoted pointer along which a celestial body could be sighted. From late medieval times it was used by mariners for calculating latitude, until replaced by the sextant. [ME f. OF *astrelabe* f. med.L *astrolabium* f. Gk *astrolabon*, neut. of *astrolabos* star-taking]

astrology /əˈstrɒlədʒɪ/ *n.* the study of the movements and relative positions of celestial bodies interpreted as an influence on human affairs. Astrology, long seen as applied astronomy, was developed by the Greeks and reached Christian Europe via the Arabs. It was a utilitarian science linked to medicine and agriculture, as well as an ambitious philosophical system resting on the belief that the stars influence the events of the world and can be used to predict events such as wars, plagues, and the weather. The study of the movements of planets in the zodiac and the disposition of planets at a person's birth gave rise to the study of horoscopes, by which a person's life and character could be mapped with supposed scientific precision. During the Renaissance period popes such as Paul III (1468–1549) were enthusiastic patrons, and many rulers employed court astrologers for both political and medical assistance, Nostradamus (16th century) being the best known. A papal bull of 1586 condemned astrology as used in the prediction of human affairs (so-called *judicial astrology*), but its decline was slow and many leading scientists in the 17th century believed that it might have at least a residual basis of truth. Though it had lost its intellectual standing in the West by 1700, popular writers such as Old Moore (see MOORE[1]) gave it a widespread appeal which has survived to the present day. It retains a more reputable standing in many parts of the East, and has enjoyed a revival in the West in recent years. □ **astrologer** *n.* **astrologist** *n.* **astrological** /ˌæstrəˈlɒdʒɪk(ə)l/ *adj.* [ME f. OF *astrologie* f. L *astrologia* f. Gk (as ASTRO-, -LOGY)]

astronaut /ˈæstrəˌnɔːt/ *n.* a person who is trained to travel in a spacecraft. □ **astronautical** /ˌæstrəˈnɔːtɪk(ə)l/ *adj.* [ASTRO-, after *aeronaut*]

astronautics /ˌæstrəˈnɔːtɪks/ *n.* the science of space travel.

astronomical /ˌæstrəˈnɒmɪk(ə)l/ *adj.* (also **astronomic**) **1** of or relating to astronomy. **2** extremely large; too large to contemplate. □ **astronomical unit** a unit of distance in astronomy equal to the mean distance from the centre of the earth to the centre of the sun, 1.496×10^{11} metres or 92.9 million miles (symbol: **au**). **astronomical year** see YEAR *n.* 1. □ **astronomically** *adv.* [L *astronomicus* f. Gk *astronomikos*]

astronomy /əˈstrɒnəmɪ/ *n.* the scientific study of celestial objects, of space, and of the universe as a whole. People have long charted the positions and motions of the sun, moon, planets, and stars, upon which timekeeping and navigation depended. The original naked-eye observations were later refined by the use of the telescope and interpreted by the mathematics of celestial mechanics and positional astronomy. The classification and interpretation of celestial objects depend now on the use of the most sophisticated instruments, including radio telescopes, infrared detectors, and ultraviolet and X-ray satellites, while robot probes are used to visit the sun, moons, and comets of the solar system. □ **astronomer** *n.* [ME f. OF *astronomie* f. L f. Gk *astronomia* f. *astronomos* (adj.) star-arranging f. *nemō* arrange]

astrophysics /ˌæstrəʊˈfɪzɪks/ *n.* the branch of astronomy concerned with the physical nature of celestial objects, especially stars. It applies known physical laws to interpret the observations of astronomers, and reveals new laws operating in extreme conditions of temperature and density not attainable on earth. □ **astrophysical** *adj.* **astrophysicist** /-ˈfɪzɪsɪst/ *n.*

Astroturf /ˈæstrəʊˌtɜːf/ *n. propr.* an artificial grass surface, esp. for sports fields. [*Astrodome*, name of a sports stadium in Texas where it was first used, + TURF]

Asturias[1] /æˈstʊərɪˌæs/ an autonomous region and former principality of NW Spain; capital, Oviedo.

Asturias[2] /æˈstʊərɪˌæs/, Miguel Ángel (1899–1974), Guatemalan novelist and poet. He is best known for his experimental novel *The President* (1946), which deals with the disintegration of human relationships under a repressive dictatorship. Later novels, such as *Mulata* (1963), draw more extensively on his knowledge of Mayan myth and history. He was awarded the Nobel Prize for literature in 1967.

Asturias, Prince of the see PRINCE OF THE ASTURIAS.

astute /əˈstjuːt/ *adj.* **1** shrewd; sagacious. **2** crafty. □ **astutely** *adv.* **astuteness** *n.* [obs. F *astut* or L *astutus* f. *astus* craft]

Asunción /əˌsʊnsɪˈɒn/ the capital and chief port of Paraguay; pop. (1990) 729,300.

asunder /əˈsʌndə(r)/ *adv. literary* apart. [OE *on sundran* into pieces: cf. SUNDER]

Asur see ASSUR.

asura /ˈʌsʊrə/ *n.* a member of a class of divine beings in the Vedic period, which in Indian mythology tend to be evil (opposed to the devas) and in Zoroastrianism are benevolent. (Cf. DEVA.) [Skr.]

Aswan /æsˈwɑːn/ a city on the Nile in southern Egypt, 16 km (10 miles) north of Lake Nasser; pop. (est. 1986) 195,700. It is situated close to two dams across the Nile. The first was built in 1898–1902 to regulate the flooding of the Nile and the supply of water for irrigation and other purposes. It is now superseded by the High Dam, built in 1960–70, which is about 3.6 km ($2\frac{1}{4}$ miles) long and 111 m (364 ft) high. The controlled release of water from Lake Nasser behind it produces the greater part of Egypt's electricity.

asylum /əˈsaɪləm/ *n.* **1** sanctuary; protection, esp. for those pursued by the law (*seek asylum*). **2** esp. *hist.* an institution offering shelter and support to distressed or destitute individuals, esp. the mentally ill. □ **political asylum** protection given by a state to a political refugee from another country. [ME f. L f. Gk *asulon* refuge f. *a-* not + *sulon* right of seizure]

asymmetry /eɪˈsɪmɪtrɪ/ *n.* lack of symmetry. □ **asymmetric** /ˌeɪsɪˈmetrɪk/ *adj.* **asymmetrical** *adj.* **asymmetrically** *adv.* [Gk *asummetria* (as A-[1], SYMMETRY)]

asymptomatic /ˌeɪˌsɪmptəˈmætɪk/ *adj.* producing or showing no symptoms.

asymptote /ˈæsɪmpˌtəʊt/ *n.* a line that continually approaches a given curve but does not meet it at a finite distance. □ **asymptotic** /ˌæsɪmpˈtɒtɪk/ *adj.* **asymptotically** *adv.* [mod.L *asymptota* (*linea* line) f. Gk *asumptōtos* not falling together f. *a-* not + *sun* together + *ptōtos* falling f. *piptō* fall]

asynchronous /eɪˈsɪŋkrənəs/ *adj.* not synchronous. □ **asynchronously** *adv.*

asyndeton /ə'sɪndɪt(ə)n/ n. (pl. **asyndeta** /-tə/) the omission of a conjunction. □ **asyndetic** /ˌæsɪn'detɪk/ adj. [mod.L f. Gk asundeton (neut. adj.) f. a- not + sundetos bound together]

At symb. Chem. the element astatine.

at /æt, ət/ prep. **1** expressing position, exact or approximate (wait at the corner; at the top of the hill; met at Bath; is at school; at a distance). **2** expressing a point in time (see you at three; went at dawn). **3** expressing a point in a scale or range (at boiling-point; at his best). **4** expressing engagement or concern in a state or activity (at war; at work; at odds). **5** expressing a value or rate (sell at £10 each). **6 a** with or with reference to; in terms of (at a disadvantage; annoyed at losing; good at cricket; play at fighting; sick at heart; came at a run; at short notice; work at it). **b** by means of (starts at a touch; drank it at a gulp). **7** expressing: **a** motion towards (arrived at the station; went at them). **b** aim towards or pursuit of (physically or conceptually) (aim at the target; work at a solution; guess at the truth; laughed at us; has been at the milk again). □ **at all** see ALL. **at hand** see HAND. **at home** see HOME. **at it 1** engaged in an activity; working hard. **2** colloq. repeating a habitual (usu. disapproved of) activity (found them at it again). **at once** see ONCE. **at that** moreover (found one, and a good one at that). **at times** see TIME. **where it's at** sl. the fashionable scene or activity. [OE æt, rel. to L ad to]

at- /ət/ prefix assim. form of AD- before t.

Atabrine var. of ATEBRIN.

Atacama Desert /ˌætə'kɑːmə/ an arid region of western Chile, extending from the Peruvian border in the north over a distance of some 965 km (600 miles) to the south.

Atalanta /ˌætə'læntə/ Gk Mythol. a huntress, averse to marriage, who would marry only someone who could beat her in a foot-race. One of her suitors (called either Melanion or Hippomenes) managed to win the race by throwing down three golden apples given to him by Aphrodite, which were so beautiful that Atalanta stopped to pick them up.

ataractic /ˌætə'ræktɪk/ adj. & n. (also **ataraxic** /-'ræksɪk/) ● adj. calming or tranquillizing. ● n. a tranquillizing drug. [Gk ataraktos calm: cf. ATARAXY]

ataraxy /'ætəˌræksɪ/ n. (also **ataraxia** /ˌætə'ræksɪə/) calmness or tranquillity; imperturbability. [F ataraxie f. Gk ataraxia impassiveness]

Atatürk /'ætəˌtɜːk/, Kemal (born Mustafa Kemal; also called Kemal Pasha) (1881–1938), Turkish general and statesman, President 1923–38. Leader of the postwar Turkish Nationalist Party, he was elected President of a provisional government in 1920. With the official establishment of the Turkish republic in 1923, he was elected its first President, taking the name of Atatürk (= father of the Turks) in 1934. During his presidency, he introduced many political and social reforms, including the abolition of the caliphate, the adoption of the Roman alphabet for writing Turkish, and other policies designed to make Turkey a modern secular state.

atavism /'ætəˌvɪz(ə)m/ n. **1** a resemblance to remote ancestors rather than to parents in plants or animals. **2** reversion to an earlier type. □ **atavistic** /ˌætə'vɪstɪk/ adj. **atavistically** adv. [F atavisme f. L atavus great-grandfather's grandfather]

ataxy /ə'tæksɪ/ n. (also **ataxia** /-sɪə/) Med. the loss of full control of bodily movements. □ **ataxic** adj. [mod.L ataxia f. Gk f. a- not + taxis order]

ATC abbr. Brit. **1** air traffic control. **2** Air Training Corps.

ate past of EAT.

-ate¹ /ət, eɪt/ suffix **1** forming nouns denoting: **a** status or office (doctorate; episcopate). **b** state or function (curate; magistrate; mandate). **2** Chem. forming nouns denoting the salt of an acid with a corresponding name ending in -ic (chlorate; nitrate). **3** forming nouns denoting a group (electorate). **4** Chem. forming nouns denoting a product (condensate; filtrate). [from or after OF -at or é(e) or f. L -atus noun or past part.: cf. -ATE²]

-ate² /ət, eɪt/ suffix **1** forming adjectives and nouns (associate; delegate; duplicate; separate). **2** forming adjectives from Latin or English nouns and adjectives (cordate; insensate; Italianate). [from or after (F -é f.) L -atus past part. of verbs in -are]

-ate³ /ət, eɪt/ suffix forming verbs (associate; duplicate; fascinate; hyphenate; separate). [from or after (F -er f.) L -are (past part. -atus): cf. -ATE²]

A-team /'eɪtiːm/ n. a group consisting of one's best workers, advisers, etc.

Atebrin /'ætəbrɪn/ n. (also **Atabrine** /-ˌbriːn/) propr. = QUINACRINE. [orig. unkn.]

atelier /ə'telɪˌeɪ, 'ætəˌljeɪ/ n. a workshop or studio, esp. of an artist or designer. [F]

a tempo /ɑː 'tempəʊ/ adv. & adj. Mus. in the previous tempo. [It., lit. 'in time']

Aten /'ɑːt(ə)n/ (also **Aton**) Egyptian Mythol. the sun or solar disc, which became a strong monotheistic cult, particularly during the reign of Akhenaten.

Athanasian Creed /ˌæθə'neɪʃ(ə)n/ a statement of Christian faith probably written in the 5th century AD and formerly much used in the Western Church. Its attribution to St Athanasius is now generally abandoned; the author was possibly St Ambrose.

Athanasius, St /ˌæθə'neɪʃəs/ (c.296–373), Greek theologian. As bishop of Alexandria he was a consistent upholder of Christian orthodoxy, especially against Arianism. He aided the ascetic movement in Egypt and introduced knowledge of monasticism to the West. Feast day, 2 May.

Athapaskan /ˌæθə'pæskən/ n. & adj. (also **Athapascan**, **Athabaskan** /-'bæskən/) ● n. **1** a member of a widely distributed North American Indian people, speaking any of a number of closely related languages. **2** any of the languages of this people. ● adj. of or relating to the Athapaskans or their language group. [f. Lake Athabasca in western Canada f. Cree Athap-askaw grass and reeds here and there]

Atharva-veda /ə,tɜ:və'veɪdə, -'viːdə/ Hinduism a collection of hymns and ritual utterances in early Sanskrit, added at a later stage to the existing Veda material. [Skr. atharvan priest + veda VEDA]

atheism /'eɪθɪˌɪz(ə)m/ n. the theory or belief that God does not exist. □ **atheist** n. **atheistic** /ˌeɪθɪ'ɪstɪk/ adj. **atheistical** adj. [F athéisme f. Gk atheos without God f. a- not + theos god]

atheling /'æθəlɪŋ/ n. hist. a prince or lord in Anglo-Saxon England. [OE ætheling = OHG adalung f. WG, f. a base meaning 'race, family': see -ING³]

Athelstan /'æθəlstən/ (895–939), king of England 925–39. Effectively the first king of all England, Athelstan came to the thrones of Wessex and Mercia in 924 before becoming king of all England a year later. He successfully invaded both Scotland and Wales and inflicted a heavy defeat on an invading Danish army.

athematic /ˌæθɪ'mætɪk/ adj. **1** Mus. not based on the use of themes. **2** Gram. (of a verb-form) having a suffix attached to the stem without a connecting (thematic) vowel.

athenaeum /ˌæθɪ'niːəm/ n. (US **atheneum**) **1** an institution for literary or scientific study. **2** a library. [LL Athenaeum f. Gk Athēnaion temple of Athene (used as a place of teaching)]

Athenaeum, the a London club founded in 1824, originally for men of distinction in literature, art, and learning.

Athene /ə'θiːnɪ/ Gk Mythol. the patron goddess of Athens, identified with the Roman Minerva and often allegorized into a personification of wisdom; she is also called Pallas. Her statues show her as female but fully armed, and in classical times the owl is regularly associated with her. The principal story concerning her is that she sprang, fully armed and uttering her war-cry, from the head of Zeus.

Athenian /ə'θiːnɪən/ n. & adj. ● n. a native or inhabitant of ancient or modern Athens. ● adj. of or relating to Athens. [L Atheniensis f. Athenae f. Gk Athēnai Athens]

Athenian empire see DELIAN LEAGUE.

Athens /'æθɪnz/ (Greek **Athínai** /a'θine/) the capital of Greece; pop. (1991) 3,096,775. A flourishing city-state of ancient Greece from early times, Athens was an important cultural centre in the 5th century BC: this was the time of Euripides, Thucydides, and of Pericles, who commissioned many of the city's best-known buildings such as the Parthenon. Athens suffered defeat in the Peloponnesian War in 404 BC, and eventually came under Roman rule in 146 BC. It remained an important cultural centre until it fell to the Goths in AD 267. After its capture by the Turks in 1456 Athens declined to the status of a village, until chosen as the capital of a newly independent Greece in 1834 following the successful revolt against Turkish rule.

atherosclerosis /ˌæθərəʊsklɪə'rəʊsɪs/ n. a form of arteriosclerosis characterized by the degeneration of the arteries because of the build-up of fatty deposits. □ **atherosclerotic** /-'rɒtɪk/ adj. [G Atherosklerose f. Gk athērē groats + SCLEROSIS]

Atherton Tableland /'æθət(ə)n/ a plateau in the Great Dividing Range in NE Queensland, Australia.

Athínai see ATHENS.

athirst /ə'θɜːst/ predic.adj. poet. **1** (usu. foll. by for) eager (athirst for *knowledge*). **2** thirsty. [OE *ofthyrst* for *ofthyrsted* past part. of *ofthyrstan* be thirsty]

athlete /ˈæθliːt/ *n.* **1** a skilled performer in physical exercises, esp. in track and field events. **2** a healthy person with natural athletic ability. □ **athlete's foot** a fungal foot condition affecting esp. the skin between the toes. [L *athleta* f. Gk *athlētēs* f. *athleō* contend for a prize (*athlon*)]

athletic /æθˈletɪk/ *adj.* **1** of or relating to athletes or athletics (*an athletic competition*). **2** muscular or physically powerful. □ **athletically** *adv.* **athleticism** /-ˈletɪˌsɪz(ə)m/ *n.* [F *athlétique* or L *athleticus* f. Gk *athlētikos* (as ATHLETE, -IC)]

athletics /æθˈletɪks/ *n.pl.* (usu. treated as *sing.*) **1** physical exercises, esp. track and field events; the sport of taking part in such exercises or events. (*See note below.*) **2** US physical sports and games of any kind.
- The sport of athletics can be traced back at least to the Olympic Games in ancient Greece, but when they were abolished in AD 393 the sport became neglected. Evidence exists of athletic contests in England *c.*1154, but organized competitions were not held until the mid-19th century. Since the Olympics were restarted in 1896 athletics has become internationally popular and many other regular competitions are held, including a world championship. At a typical athletics meeting, running or track events include sprints of 100 and 200 metres, middle-distance races of 400, 800, and 1,500 metres, longer-distance races such as the 5,000 metres; the marathon is also run at major games such as the Olympics. Hurdles, steeplechases, relays, and walking races are also featured. Field events include the high jump, long jump, triple jump, and pole-vault, as well as throwing events such as the shot, discus, javelin, and hammer, and in major competitions the pentathlon and decathlon.

Athos, Mount /ˈæθɒs, ˈeɪθ-/ a narrow, mountainous peninsula in NE Greece, projecting into the Aegean Sea. It is inhabited by monks of the Orthodox Church who live in twenty monasteries founded in the 10th century. They forbid women and even female animals to set foot on the peninsula. □ **Athonite** /ˈæθəˌnaɪt/ *adj.* & *n.*

athwart /əˈθwɔːt/ *adv.* & *prep.* ● *adv.* **1** across from side to side (usu. obliquely). **2** perversely or in opposition. ● *prep.* **1** from side to side of. **2** in opposition to. [A² + THWART]

-atic /ˈætɪk/ *suffix* forming adjectives and nouns (*aquatic*; *fanatic*; *idiomatic*). [F *-atique* or L *-aticus*, often ult. f. Gk *-atikos*]

atilt /əˈtɪlt/ *adv.* tilted and nearly falling. [A² + TILT]

-ation /ˈeɪʃ(ə)n/ *suffix* **1** forming nouns denoting an action or an instance of it (*alteration*; *flirtation*; *hesitation*). **2** forming nouns denoting a result or product of action (*plantation*; *starvation*; *vexation*) (see also -FICATION). [from or after F *-ation* or L *-atio -ationis* f. verbs in *-are*: see -ION]

-ative /ətɪv, eɪtɪv/ *suffix* forming adjectives denoting a characteristic or propensity (*authoritative*; *imitative*; *pejorative*; *qualitative*; *talkative*). [from or after F *-atif -ative* or f. L *-ativus* f. past part. stem *-at-* of verbs in *-are* + *-ivus* (see -IVE): cf. -ATIC]

Atkinson /ˈætkɪns(ə)n/, Sir Harry (Albert) (1831–92), New Zealand statesman, Prime Minister 1876–7, 1883–4, and 1887–91. Born in Britain, he emigrated to New Zealand in 1853 and became a member of the House of Representatives in 1861, also serving as a commander in the Maori Wars in the early 1860s. During his first term as Prime Minister he passed a bill abolishing the colony's provincial governments. He later served as colonial treasurer (1879–82; 1882–3) and is chiefly remembered for the austere economic policy that he pursued throughout the 1880s to boost New Zealand's recovery from economic depression.

Atlanta /ətˈlæntə/ the state capital of Georgia in the US; pop. (1990) 394,000. Founded at the end of a railway line in 1837, the city was originally called Terminus; in 1843 it was incorporated as Marthasville, and in 1845 its name was finally changed to Atlanta.

Atlantean /ətˈlæntɪən/ *adj.* **1** of or relating to the fabled island of Atlantis. **2** *literary* of or like Atlas, esp. in physical strength. [L *Atlanteus*]

atlantes /ətˈlæntiːz/ *n.pl. Archit.* male figures carved in stone and used as columns to support the entablature of a Greek or Greek-style building. [Gk, pl. of ATLAS]

Atlantic /ətˈlæntɪk/ *n.* & *adj.* ● *n.* (**the Atlantic**) the Atlantic Ocean. ● *adj.* of, relating to, or adjoining the Atlantic. [ME f. L *Atlanticus* f. Gk *Atlantikos* (as ATLAS, -IC): orig. of the Atlas Mountains, then of the sea near the West African coast]

Atlantic, Battle of the a succession of sea operations during the Second World War in which Axis naval and air forces attempted to destroy shipping carrying supplies from North America to the UK. German U-boats, sometimes assisted by Italian submarines, were the main weapon of attack, and about 2,800 Allied, mainly British, merchant ships were lost. The situation improved for the Allies during 1943 with the deployment of more warships and aircraft to convoy protection, and the introduction of improved radar to detect U-boats on the surface, although the threat was not ended until the capture of the U-boats' bases by Allied land forces in 1944.

Atlantic Charter a declaration of eight common principles in international relations, intended to guide a postwar peace settlement, which was drawn up by Churchill and Roosevelt at a secret meeting off the coast of Newfoundland in Aug. 1941. It stipulated freely chosen governments, free trade, freedom of the seas, and disarmament of current aggressor states, and condemned territorial changes made against the wishes of local populations. In the following month other states, including the USSR, declared their support for these principles. The Atlantic Charter provided the ideological basis for the United Nations organization.

Atlantic Ocean the ocean lying between Europe and Africa to the east and North and South America to the west. It is divided by the equator into the North Atlantic and the South Atlantic oceans. A submarine ridge known as the Mid-Atlantic Ridge runs down the centre from north to south, deep basins lying on either side.

Atlantic Provinces see MARITIME PROVINCES.

Atlantic Standard Time (also **Atlantic Time**) (abbr. **AST**) the standard time in a zone including the easternmost parts of mainland Canada, four hours behind GMT.

Atlantis /ətˈlæntɪs/ a fabled island in the ocean west of the Pillars of Hercules. As described by Plato in the *Timaeus*, Atlantis was beautiful and prosperous and ruled part of Europe and Africa, but its kings were defeated by the prehistoric Athenians when it attempted to conquer the rest, and it was overwhelmed by the sea. Memories of Atlantic islands or of a great volcanic eruption such as that at Thera (Santorini) may lie behind the story.

Atlas /ˈætləs/ *Gk Mythol.* one of the Titans, who was punished for his part in their revolt against Zeus by being made to support the heavens (a popular explanation of why the sky does not fall). He became identified with the Atlas Mountains. According to a later story Perseus, with the aid of Medusa's head, turned Atlas into a mountain.

atlas /ˈætləs/ *n.* **1** a book of maps or charts. **2** *Anat.* the cervical vertebra of the backbone articulating with the skull at the neck. [ATLAS, whose picture appeared at the beginning of early atlases]

Atlas Mountains a range of mountains in North Africa extending from Morocco to Tunisia in a series of chains. These include the Anti-Atlas, High Atlas, Middle Atlas, Rif Mountains, Tell Atlas, and Sahara Atlas.

atm *abbr. Physics* atmosphere(s).

atman /ˈɑːtmæn/ *n. Hinduism* **1** the real self. **2** the supreme spiritual principle. [Skr. *ātmán* essence, breath]

atmosphere /ˈætməsˌfɪə(r)/ *n.* **1 a** the envelope of gases surrounding the earth, any other planet, or any substance. **b** the air in any particular place, esp. if unpleasant. **2 a** the pervading tone or mood of a place or situation, esp. with reference to the feelings or emotions evoked. **b** the feelings or emotions evoked by a work of art, a piece of music, etc. **c** a feeling of tension between people, caused by a disagreement etc. **3** *Physics* a unit of pressure (symbol: **atm**) equal to mean atmospheric pressure at sea level, 101,325 pascals. □ **atmospheric** /ˌætməsˈferɪk/ *adj.* **atmospherical** *adj.* **atmospherically** *adv.* [mod.L *atmosphaera* f. Gk *atmos* vapour: see SPHERE]

atmospherics /ˌætməsˈferɪks/ *n.pl.* **1** electrical disturbance in the atmosphere, esp. caused by lightning. **2** interference with telecommunications caused by this.

atoll /ˈætɒl/ *n.* a ring-shaped coral reef enclosing a lagoon. [Maldivian *atolu*]

atom /ˈætəm/ *n.* **1 a** the smallest particle of a chemical element that can exist and still retain its characteristic chemical properties. (*See note below.*) **b** (**the atom**) atoms as the source of nuclear energy. **2** (usu. with *neg.*) the least portion of a thing or quality (*not an atom of pity*). □ **atom-smasher** *colloq.* = PARTICLE ACCELERATOR. [ME f. OF *atome* f. L *atomus* f. Gk *atomos* indivisible]
- Previously believed to be hard indivisible particles (see ATOMIC THEORY), atoms have been shown in the 20th century to have an internal structure and to be capable of division and transformation. Atoms, which are roughly 10^{-8} cm in diameter, are each composed

of a positively charged nucleus of about 10^{-12} cm diameter consisting of neutrons and protons and containing 99.9 per cent of the atom's mass, surrounded by orbiting negative electrons. Each chemical element is composed of atoms with the same number of protons (see ELEMENT); nuclear transformations, which occur spontaneously in radioactive substances or are induced by collision with other particles, can change atoms of one element into another. In neutral atoms, the number of electrons equals that of the protons. Atoms are held together in molecules by an interaction or sharing of their electrons.

atom bomb n. (also **atomic bomb**) a bomb which derives its destructive power from the rapid release of nuclear energy by fission of heavy atomic nuclei, with damaging effects caused by heat, blast, and radioactivity. Suitable explosive materials are uranium-235 and plutonium-239. Small masses of such material are stable, but above a critical mass a chain reaction occurs with instantaneous release of energy. The bomb is detonated by a conventional explosive either pushing together two masses of fissile material or compressing a single mass. The first atom bomb to be used in war was exploded by the US 300 m above Hiroshima in Japan on 6 Aug. 1945, the second over Nagasaki three days later, resulting in the surrender of Japan and the end of the Second World War.

atomic /əˈtɒmɪk/ adj. **1** concerned with or using atomic energy or atom bombs. **2** of or relating to an atom or atoms. **3** Philos. (of a proposition etc.) unanalysable, ultimate. □ **atomic bomb** see ATOM BOMB. **atomic clock** a clock in which the periodic process (time-scale) is regulated by the vibrations of an atomic or molecular system, such as caesium or ammonia. **atomic energy** nuclear energy. **atomic mass** the mass of an atom measured in atomic mass units. **atomic mass unit** a unit of mass used to express atomic and molecular weights, equal to one twelfth of the mass of an atom of carbon-12 (symbol: **amu**) (also called *dalton*). **atomic number** the number of protons in the nucleus of an atom, which is characteristic of a chemical element and determines its place in the periodic table (symbol: **Z**). **atomic physics** the branch of physics concerned with the structure of the atom and the characteristics of subatomic particles. **atomic pile** a nuclear reactor. **atomic power** nuclear power. **atomic spectrum** the emission or absorption spectrum arising from electron transitions inside an atom and characteristic of the element. **atomic structure** the structure of an atom as being a central positively charged nucleus surrounded by negatively charged orbiting electrons. **atomic warfare** warfare involving the use of atom bombs. **atomic weight** = relative atomic mass. □ **atomically** adv. [mod.L atomicus (as ATOM, -IC)]

atomicity /ˌætəˈmɪsɪtɪ/ n. **1** the number of atoms in the molecules of an element. **2** the state or fact of being composed of atoms.

atomic theory 1 the theory that all matter is made up of fundamental particles called atoms. (*See note below.*) **2** the concept of an atom as being composed of subatomic particles. **3** Philos. atomism.

▪ Atomic theories have been known from antiquity, but the existence of atoms was not definitively verified until the present century. The ancient Greeks tried to explain the complexity of natural bodies and phenomena in terms of the arrangement and rearrangement of tiny indivisible particles, which differed from each other only in size, shape, and motion. The modern atomic theory — that atoms of any one element differ from those of other elements and combine in fixed proportions to form compounds — was propounded by John Dalton, who based it on his work on gases and in particular on the discovery that elements combine in definite proportions. Atoms continued to be regarded as indivisible bodies until their structure was revealed by 20th-century physicists. (See also ATOM.)

atomism /ˈætəˌmɪz(ə)m/ n. **1** Philos. the theory that all matter consists of tiny individual particles. **2** Psychol. the theory that mental states are made up of elementary units. □ **atomist** n. **atomistic** /ˌætəˈmɪstɪk/ adj.

atomize /ˈætəˌmaɪz/ v.tr. (also **-ise**) **1** reduce to atoms or fine particles. **2** fragment or divide into small units. □ **atomization** /ˌætəmaɪˈzeɪʃ(ə)n/ n.

atomizer /ˈætəˌmaɪzə(r)/ n. (also **-iser**) an instrument for emitting liquids as a fine spray.

atomy /ˈætəmɪ/ n. (pl. **-ies**) archaic **1** a skeleton. **2** an emaciated body. [ANATOMY taken as an atomy]

Aton var. of ATEN.

atonal /eɪˈtəʊn(ə)l/ adj. Mus. not written in any key or mode. □ **atonality** /ˌeɪtəʊˈnælɪtɪ/ n.

atone /əˈtəʊn/ v.intr. (usu. foll. by for) make amends; expiate for (a wrong). [back-form. f. ATONEMENT]

atonement /əˈtəʊnmənt/ n. **1** expiation; reparation for a wrong or injury. **2** the reconciliation of God and man. □ **the Atonement** the expiation by Christ of humankind's sin. **Day of Atonement** = YOM KIPPUR. [at one + -MENT, after med.L adunamentum and earlier onement f. obs. one (v.) unite]

atonic /əˈtɒnɪk/ adj. **1** without accent or stress. **2** Med. lacking bodily tone. □ **atony** /ˈætənɪ/ n.

atop /əˈtɒp/ adv. & prep. ● adv. (often foll. by of) on the top. ● prep. on the top of.

-ator /ˈeɪtə(r)/ suffix forming agent nouns, usu. from Latin words (sometimes via French) (agitator; creator; equator; escalator). See also -OR[1]. [L -ator]

-atory /ətərɪ/ suffix forming adjectives meaning 'relating to or involving (a verbal action)' (amatory; explanatory; predatory). See also -ORY[2]. [L -atorius]

ATP abbr. adenosine triphosphate.

atrabilious /ˌætrəˈbɪljəs/ adj. literary melancholy; ill-tempered. [L atra bilis black bile, transl. Gk melagkholia MELANCHOLY]

Atreus /ˈeɪtrɪəs/ Gk Mythol. the son of Pelops and father of Agamemnon and Menelaus. He quarrelled with his brother Thyestes and invited him to a banquet at which he served up the flesh of Thyestes' own children, whereupon the sun turned back on its course in horror.

atrium /ˈeɪtrɪəm/ n. (pl. **atriums** or **atria** /-trɪə/) **1 a** the central court of an ancient Roman house. **b** a usu. skylit central court rising through several storeys with galleries and rooms opening off at each level. **c** esp. US (in a modern house) a central hall or glazed court with rooms opening off it. **2** Anat. a cavity in the body, esp. one of the two upper cavities of the heart, receiving blood from the veins. □ **atrial** adj. [L]

atrocious /əˈtrəʊʃəs/ adj. **1** very bad or unpleasant (atrocious weather; their manners were atrocious). **2** extremely savage or wicked (atrocious cruelty). □ **atrociously** adv. **atrociousness** n. [L atrox -ocis cruel]

atrocity /əˈtrɒsɪtɪ/ n. (pl. **-ies**) **1** an extremely wicked or cruel act, esp. one involving physical violence or injury. **2** extreme wickedness. [F atrocité or L atrocitas (as ATROCIOUS, -ITY)]

atrophy /ˈætrəfɪ/ v. & n. ● v. (**-ies**, **-ied**) **1** intr. waste away through undernourishment, ageing, or lack of use; become emaciated. **2** tr. cause to atrophy. ● n. the process of atrophying; emaciation. [F atrophie or LL atrophia f. Gk f. a- not + trophē food]

atropine /ˈætrəˌpiːn, -pɪn/ n. a poisonous alkaloid found in deadly nightshade, used in medicine esp. to dilate the pupil of the eye, to treat intestinal spasm, and to counteract slowing of the heart. [mod.L Atropa belladonna deadly nightshade f. Gk ATROPOS]

Atropos /ˈætrəˌpɒs/ Gk Mythol. one of the three Fates. [Gk, = inflexible]

attach /əˈtætʃ/ v. **1** tr. fasten, affix, join. **2** tr. (in passive; foll. by to) be very fond of or devoted to (am deeply attached to her). **3** tr. attribute, assign (some function, quality, or characteristic) (can you attach a name to it?; attaches great importance to it). **4 a** tr. include, enclose (attach no conditions to the agreement; attach particulars). **b** intr. (foll. by to) be an attribute or characteristic (great prestige attaches to the job). **5** refl. (usu. foll. by to) (of a thing) adhere; (of a person) join, take part (the sticky stamps attached themselves to his fingers; climbers attached themselves to the expedition). **6** tr. appoint for special or temporary duties. **7** tr. Law seize (a person or property) by legal authority. □ **attachable** adj. **attacher** n. [ME f. OF estachier fasten f. Gmc: in Law sense thr. OF atachier]

attaché /əˈtæʃeɪ/ n. a person appointed to an ambassador's staff, usu. with a special sphere of activity (military attaché; press attaché). □ **attaché case** a small flat rectangular case for carrying documents etc. [F, past part. of attacher: see ATTACH]

attached /əˈtætʃt/ adj. **1** fixed, connected, enclosed. **2** (of a person) involved in a long-term relationship, esp. engagement or marriage.

attachment /əˈtætʃmənt/ n. **1** a thing attached or to be attached, esp. to a machine, device, etc., for a special function. **2** affection; devotion. **3** a means of attaching. **4** the act of attaching or the state of being attached. **5** legal seizure. **6** a temporary position in, or secondment to, an organization [ME f. F attachement f. attacher (as ATTACH, -MENT)]

attack /əˈtæk/ v. & n. ● v. **1** tr. act against with (esp. armed) force. **2** tr. seek to hurt or defeat. **3** tr. criticize adversely. **4** tr. act harmfully upon (a virus attacking the nervous system). **5** tr. vigorously apply oneself to; begin work on (attacked his meal with gusto). **6** tr. (in various games) try to score goals, points, etc. against (one's opponents). **7** intr. make an attack. **8** intr. be in a mode of attack. ● n. **1** the act or process of attacking. **2 a** an offensive operation or mode of behaviour. **b** severe criticism. **3** *Mus.* the action or manner of beginning a piece, passage, etc. **4** gusto, vigour. **5** a sudden occurrence of an illness. **6** a player or players seeking to score goals etc. □ **attacker** *n.* [F *attaque*, *attaquer* f. It. *attacco* attack, *attaccare* join (battle), attach]

attain /əˈteɪn/ *v.* **1** *tr.* arrive at; reach (a goal etc.). **2** *tr.* gain, accomplish (an aim, distinction, etc.). **3** *intr.* (foll. by *to*) arrive at by conscious development or effort. □ **attainable** *adj.* **attainableness** *n.* **attainability** /əˌteɪnəˈbɪlɪtɪ/ *n.* [ME f. AF *atain-*, *atein-*, OF *ataign-* stem of *ataindre* f. L *attingere* (as AD-, *tangere* touch)]

attainder /əˈteɪndə(r)/ *n. hist.* the forfeiture of land and civil rights suffered as a consequence of a sentence of death for treason or felony. □ **act** (or **bill**) **of attainder** an item of legislation inflicting attainder without judicial process. [ME f. AF, = OF *ateindre* ATTAIN used as noun: see -ER[6]]

attainment /əˈteɪnmənt/ *n.* **1** (often in *pl.*) something attained or achieved; an accomplishment. **2** the act or an instance of attaining.

attaint /əˈteɪnt/ *v.tr.* **1** *hist.* subject to attainder. **2 a** (of disease etc.) strike, affect. **b** taint. [ME f. obs. *attaint* (adj.) f. OF *ataint*, *ateint* past part. formed as ATTAIN: confused in meaning with TAINT]

Attalid /ˈætəlɪd/ *n. & adj.* ● *n.* a member of a Hellenistic dynasty named after Attalus I (reigned 241–197 BC), which flourished in the 3rd and 2nd centuries BC. The Attalid kings established their capital, Pergamum, as a leading cultural centre of the Greek world. ● *adj.* of or relating to this dynasty.

attar /ˈætɑː(r)/ *n.* (also **otto** /ˈɒtəʊ/) a fragrant essential oil, esp. from rose-petals. [Pers. ʿatar f. Arab. f. ʿiṭr perfume]

attempt /əˈtempt/ *v. & n.* ● *v.tr.* (often foll. by *to* + infin.) seek to achieve, complete, or master (a task, action, challenge, etc.) (*attempted the exercise*; *attempted to explain*; *attempted Everest*). ● *n.* (often foll. by *at*, *on*, or *to* + infin.) an act of attempting; an endeavour; an assault (*made an attempt at winning*; *an attempt to succeed*; *an attempt on his life*). □ **attempt the life of** *archaic* try to kill. □ **attemptable** *adj.* [OF *attempter* f. L *attemptare* (as AD-, *temptare* TEMPT)]

Attenborough[1] /ˈæt(ə)nbərə/, Sir David (Frederick) (b.1926), English naturalist and broadcaster. In 1952 he joined the BBC, where he developed the concept of filming animals in their natural habitats for the series *Zoo Quest* (1954–64). He became a household name with his documentary film series *Life on Earth* (1979), *The Living Planet* (1983), and *The Trials of Life* (1990). He is the brother of Richard Attenborough.

Attenborough[2] /ˈæt(ə)nbərə/, Richard (Samuel), Baron Attenborough of Richmond-upon-Thames (b.1923), English film actor, producer, and director. From 1942 onwards he appeared in a number of war films and comedies, and extended his repertoire into character roles such as that of Pinkie in *Brighton Rock* (1947). The films he has directed include *Oh! What a Lovely War* (1969), *Gandhi* (1982), *Cry Freedom* (1987), and *Shadowlands* (1993). He is the brother of David Attenborough.

attend /əˈtend/ *v.* **1** *tr.* **a** be present at (*attended the meeting*). **b** go regularly to (*attends the local school*). **2** *intr.* **a** be present (*many members failed to attend*). **b** be present in a serving capacity; wait. **3 a** *tr.* escort, accompany (*the king was attended by soldiers*). **b** *intr.* (foll. by *on*) wait on; serve. **4** *intr.* **a** (usu. foll. by *to*) turn or apply one's mind; focus one's attention (*attend to what I am saying*; *was not attending*). **b** (foll. by *to*) deal with, take care of (*shall attend to the matter myself*; *attend to the older people*). **5** *tr.* (usu. in *passive*) follow as a result from (*the error was attended by serious consequences*). □ **attender** *n.* [ME f. OF *atendre* f. L *attendere* (as AD-, *tendere tent-* stretch)]

attendance /əˈtendəns/ *n.* **1** the act of attending or being present. **2** the number of people present (*a high attendance*). □ **attendance allowance** (in the UK) a state benefit paid to disabled people in need of constant care at home. **attendance centre** (in the UK) a place where young offenders report by order of a court as a minor penalty. **in attendance** on hand, available for service. [ME f. OF *atendance* (as ATTEND, -ANCE)]

attendant /əˈtendənt/ *n. & adj.* ● *n.* a person employed to wait on others or provide a service (*cloakroom attendant*; *museum attendant*). ● *adj.* **1** accompanying (*attendant circumstances*). **2** waiting on; serving (*ladies attendant on the queen*). [ME f. OF (as ATTEND, -ANT)]

attendee /ˌætenˈdiː/ *n.* a person who attends (a meeting etc.).

attention /əˈten∫(ə)n/ *n. & int.* ● *n.* **1** the act or faculty of applying one's mind (*give me your attention*; *attract his attention*). **2 a** consideration (*give attention to the problem*). **b** care (*give special attention to your handwriting*). **c** notice, publicity (*only needs a bit of attention*; *labelled an attention seeker*). **3** (in *pl.*) **a** ceremonious politeness (*he paid his attentions to her*). **b** wooing,

courting (*she was the subject of his attentions*). **4** *Mil.* an erect attitude of readiness (*stand at attention*). ● *int.* (in full **stand to attention!**) *Mil.* an order to assume an attitude of attention. [ME f. L *attentio* (as ATTEND, -ION)]

attentive /əˈtentɪv/ *adj.* **1** concentrating; paying attention. **2** assiduously polite. **3** heedful. □ **attentively** *adv.* **attentiveness** *n.* [ME f. F *attentif -ive* f. attente, OF *atente*, fem. past part. of *atendre* ATTEND]

attenuate *v. & adj.* ● *v.tr.* /əˈtenjʊˌeɪt/ **1** make thin. **2** reduce in force, value, or virulence. **3** *Electr.* reduce the amplitude of (a signal or current). ● *adj.* /əˈtenjʊət/ **1** slender. **2** tapering gradually. **3** rarefied. □ **attenuated** /-ˌeɪtɪd/ *adj.* **attenuator** /-ˌeɪtə(r)/ *n.* **attenuation** /əˌtenjʊˈeɪ∫(ə)n/ *n.* [L *attenuare* (as AD-, *tenuis* thin)]

attest /əˈtest/ *v.* **1** *tr.* certify the validity of. **2** *tr.* enrol (a recruit) for military service. **3** *intr.* (foll. by *to*) bear witness to. **4** *intr.* enrol oneself for military service. □ **attestable** *adj.* **attestor** *n.* [F *attester* f. L *attestari* (as AD-, *testis* witness)]

attestation /ˌæteˈsteɪ∫(ə)n/ *n.* **1** the act of attesting. **2** a testimony. [F *attestation* or LL *attestatio* (as ATTEST, -ATION)]

Attic /ˈætɪk/ *adj. & n.* ● *adj.* of or relating to ancient Athens or Attica, or the form of Greek spoken there. ● *n.* the form of Greek used by the ancient Athenians. □ **Attic salt** (or **wit**) refined wit. [L *Atticus* f. Gk *Attikos*]

attic /ˈætɪk/ *n.* **1** the uppermost storey in a house, usu. under the roof. **2** a room in the attic area. [F *attique*, as ATTIC: orig. (Archit.) a small order above a taller one]

Attica /ˈætɪkə/ a triangular promontory of eastern Greece. With the islands in the Saronic Gulf it forms a department of Greece, of which Athens is the capital.

atticism /ˈætɪˌsɪz(ə)m/ *n.* **1** extreme elegance of speech. **2** an instance of this. [Gk *Attikismos* (as ATTIC, -ISM)]

Attila /əˈtɪlə/ (406–53), king of the Huns 434–53. Having inflicted great devastation on the Eastern Roman Empire in 445–50, Attila invaded the Western Empire but was defeated by the joint forces of the Roman army and the Visigoths at Châlons in 451. He and his army were the terror of Europe during his lifetime, earning Attila the nickname 'Scourge of God'.

Attila Line (also called *Sahin Line*) the boundary separating Greek and Turkish-occupied Cyprus, named after the Attila Plan, a secret Turkish plan of 1964 to partition the country.

attire /əˈtaɪə(r)/ *v. & n. formal* ● *v.tr.* dress, esp. in fine clothes or formal wear. ● *n.* clothes, esp. fine or formal. [ME f. OF *atir(i)er* equip f. *à tire* in order, of unkn. orig.]

Attis /ˈætɪs/ *Anatolian Mythol.* the youthful consort of Cybele. His death (after castrating himself) and his resurrection were later associated with the spring festival and with a sacrifice for the crops; his symbol was the pine tree.

attitude /ˈætɪˌtjuːd/ *n.* **1 a** a settled opinion or way of thinking. **b** behaviour reflecting this (*I don't like his attitude*). **c** *US colloq.* aggressive or uncooperative behaviour; a resentful or antagonistic manner. **d** *sl.* style, swagger. **2 a** a bodily posture. **b** a pose adopted in a painting or a play, esp. for dramatic effect (*strike an attitude*). **3** the position of an aircraft, spacecraft, etc., in relation to specified directions. □ **attitude of mind** a settled way of thinking. □ **attitudinal** /ˌætɪˈtjuːdɪn(ə)l/ *adj.* [F f. It. *attitudine* fitness, posture, f. LL *aptitudo -dinis* f. *aptus* fit]

attitudinize /ˌætɪˈtjuːdɪˌnaɪz/ *v.intr.* (also **-ise**) **1** practise or adopt attitudes, esp. for effect. **2** speak, write, or behave affectedly. [It. *attitudine* f. LL (as ATTITUDE) + -IZE]

Attlee /ˈætlɪ/, Clement Richard, 1st Earl Attlee (1883–1967), British Labour statesman, Prime Minister 1945–51. He became Labour Party leader in 1935, and deputy Prime Minister in 1942 in Churchill's coalition government. Following his party's landslide election victory in 1945, Attlee became the first Labour Prime Minister to command an absolute majority in the House of Commons. His term saw the creation of the modern welfare state and a wide programme of nationalization of major industries (including coal, gas, and electricity). Foreign policy initiatives included a progressive withdrawal from colonies and support for NATO.

attn. *abbr.* **1** attention. **2** for the attention of.

atto- /ˈætəʊ/ *comb. form Math.* denoting a factor of 10^{-18} (*attometre*). [Danish or Norw. *atten* eighteen + -O-]

attorney /əˈtɜːnɪ/ *n.* (*pl.* **-eys**) **1** a person, esp. a lawyer, appointed to act for another in business or legal matters. **2** *US* a qualified lawyer, esp. one representing a client in a lawcourt. □ **power of attorney** the authority to act for another person in legal or financial matters.

□ **attorneyship** *n.* [ME f. OF *atorné* past part. of *atorner* assign f. *à* to + *torner* turn]

Attorney-General /əˌtɜːnɪˈdʒenrəl/ *n.* the chief legal officer in England, the US, and other countries.

attract /əˈtrækt/ *v.tr.* **1** (also *absol.*) draw or bring to oneself or itself (*attracts many admirers; attracts attention*). **2** be attractive to; fascinate. **3** (of a magnet, gravity, etc.) exert a pull on (an object). □ **attractable** *adj.* **attractor** *n.* [L *attrahere* (as AD-, *trahere tract-* draw)]

attractant /əˈtræktənt/ *n. & adj.* ● *n.* a substance which attracts (esp. insects). ● *adj.* attracting.

attraction /əˈtrækʃ(ə)n/ *n.* **1 a** the act or power of attracting (*the attraction of foreign travel*). **b** a person or thing that attracts by arousing interest (*the fair is a big attraction*). **2** Physics the force by which bodies attract or approach each other (opp. REPULSION 2). **3** *Gram.* the influence exerted by one word on another which causes it to change to an incorrect form, e.g. *the wages of sin is death.* [F *attraction* or L *attractio* (as ATTRACT, -ION)]

attractive /əˈtræktɪv/ *adj.* **1** attracting or capable of attracting; interesting (*an attractive proposition*). **2** aesthetically pleasing or appealing. □ **attractively** *adv.* **attractiveness** *n.* [F *attractif -ive* f. LL *attractivus* (as ATTRACT, -IVE)]

attribute *v. & n.* ● *v.tr.* /əˈtrɪbjuːt/ (usu. foll. by *to*) **1** regard as belonging or appropriate (*a poem attributed to Shakespeare*). **2** ascribe; regard as the effect of a stated cause (*the delays were attributed to the heavy traffic*). ● *n.* /ˈætrɪˌbjuːt/ **1 a** a quality ascribed to a person or thing. **b** a characteristic quality. **2** a material object recognized as appropriate to a person, office, or status (*a large car is an attribute of seniority*). **3** *Gram.* an attributive adjective or noun. □ **attributable** /əˈtrɪbjʊtəb(ə)l/ *adj.* **attribution** /ˌætrɪˈbjuːʃ(ə)n/ *n.* [ME f. L *attribuere attribut-* (as AD-, *tribuere* assign): (n.) f. OF *attribut* or L *attributum*]

attributive /əˈtrɪbjʊtɪv/ *adj. Gram.* (of an adjective or noun) preceding the word described and expressing an attribute, as *old* in *the old dog* (but not in *the dog is old*) and *expiry* in *expiry date* (opp. PREDICATIVE). □ **attributively** *adv.* [F *attributif -ive* (as ATTRIBUTE, -IVE)]

attrit /əˈtrɪt/ *v.tr.* US colloq. wear (an enemy or opponent) down by attrition. [back-form. f. ATTRITION]

attrition /əˈtrɪʃ(ə)n/ *n.* **1 a** the act or process of gradually wearing out, esp. by friction. **b** abrasion. **2** (in Christian theology) sorrow for sin, falling short of contrition. □ **war of attrition** a war in which one side wins by gradually wearing the other down with repeated attacks etc. □ **attritional** *adj.* [ME f. LL *attritio* f. *atterere attrit-* rub]

attune /əˈtjuːn/ *v.tr.* **1** (usu. foll. by *to*) adjust (a person or thing) to a situation. **2** bring (an orchestra, instrument, etc.) into musical accord. [AT- + TUNE]

Atty. *abbr.* Attorney.

Atwood /ˈætwʊd/, Margaret (Eleanor) (b.1939), Canadian novelist, poet, critic, and short-story writer. She made her name with the novel *The Edible Woman* (1969), which was championed by the resurgent women's movement of the time. Her novels explore the question of women finding and asserting their identities, and include *The Handmaid's Tale* (1986), her dystopian vision of a patriarchal state, and *Cat's Eye* (1989).

atypical /eɪˈtɪpɪk(ə)l/ *adj.* not typical; not conforming to a type. □ **atypically** *adv.*

AU *abbr.* **1** (also **au.**) astronomical unit. **2** ångström unit.

Au *symb. Chem.* the element gold. [L *aurum*]

aubade /əʊˈbɑːd/ *n.* a poem or piece of music appropriate to the dawn or early morning. [F f. Sp. *albada* f. *alba* dawn]

auberge /əʊˈbeəʒ/ *n.* an inn. [F]

aubergine /ˈəʊbəˌʒiːn/ *n.* **1** a tropical solanaceous plant, *Solanum melongena*, having erect or spreading branches bearing white or purple egg-shaped fruit. **2** this fruit eaten as a vegetable. Also called *eggplant*. **3** the dark purple colour of this fruit. [F f. Catalan *alberginia* f. Arab. *al-bādinjān* f. Pers. *bādingān* f. Skr. *vātiṃgaṇa*]

Aubrey /ˈɔːbrɪ/, John (1626–97), English antiquarian and author. He was a pioneer of field archaeology, most of his researches being centred on the earthworks and monuments in Wiltshire (particularly Avebury and Stonehenge), and became one of the first Fellows of the Royal Society in 1663. With regard to his written works, he is chiefly remembered for the lively and anecdotal collection of biographies of eminent persons such as Milton and Bacon known as *Brief Lives*, a bowdlerized edition of which was first published in 1813.

aubrietia /ɔːˈbriːʃə/ *n.* (also **aubretia**) a dwarf perennial rock-plant

of the genus *Aubrieta*, having purple or pink flowers in spring. [alt. of mod.L *Aubrieta*, f. Claude *Aubriet*, Fr. botanist (1668–1743)]

auburn /ˈɔːb(ə)n/ *adj.* of a reddish-brown colour (usu. of a person's hair). [ME, orig. yellowish-white, f. OF *auborne, alborne*, f. L *alburnus* whitish f. *albus* white]

Aubusson /ˈəʊbjuˌsɒn/ *n.* a kind of French tapestry or carpet, principally from the 18th century. [*Aubusson* a town in central France where the tapestries were made]

AUC *abbr.* (of a date) from the foundation of the city (of Rome). [L *ab urbe condita*]

Auckland /ˈɔːklənd/ the largest city and chief seaport of New Zealand, on North Island; pop. (1990) 309,400. Established in 1840, it was the site of the first Parliament of New Zealand in 1854, remaining the capital until 1865.

au courant /ˌəʊ kuəˈrɒn/ *predic.adj.* (usu. foll. by *with, of*) knowing what is going on; well-informed. [F, = in the (regular) course]

auction /ˈɔːkʃ(ə)n/ *n. & v.* ● *n.* **1** a sale of goods, usu. in public, in which articles are sold to the highest bidder. **2** the sequence of bids made at auction bridge. ● *v.tr.* sell by auction. □ **auction bridge** see BRIDGE². **Dutch auction** a sale, usu. public, of goods in which the price is reduced by the auctioneer until a buyer is found. [L *auctio* increase, auction f. *augere auct-* increase]

auctioneer /ˌɔːkʃəˈnɪə(r)/ *n.* a person who conducts auctions professionally, by calling for bids and declaring goods sold. □ **auctioneering** *n.*

audacious /ɔːˈdeɪʃəs/ *adj.* **1** daring, bold. **2** impudent. □ **audaciously** *adv.* **audaciousness** *n.* **audacity** /ɔːˈdæsɪtɪ/ *n.* [L *audax -acis* bold f. *audere* dare]

Auden /ˈɔːd(ə)n/, W(ystan) H(ugh) (1907–73), British-born poet. *Look, Stranger!* (1936) is the collection of poems that secured his position as a leading left-wing poet. He supported the Republicans in the Spanish Civil War, and wrote *Spain* (1937). Auden also collaborated with Christopher Isherwood on several Brechtian verse dramas, notably *The Ascent of F6* (1936). After emigrating to America in 1939, he continued to publish volumes of poetry, including *The Age of Anxiety* (1947), which was awarded the Pulitzer Prize. He also worked on several opera libretti, such as Stravinsky's *The Rake's Progress* (1951).

Audenarde see OUDENARDE, BATTLE OF.

Audh see OUDH.

audible /ˈɔːdɪb(ə)l/ *adj.* capable of being heard. □ **audibly** *adv.* **audibleness** *n.* **audibility** /ˌɔːdɪˈbɪlɪtɪ/ *n.* [LL *audibilis* f. *audire* hear]

audience /ˈɔːdɪəns/ *n.* **1 a** the assembled listeners or spectators at an event, esp. a stage performance, concert, etc. **b** the people addressed by a film, book, play, etc. **2** a formal interview with a person in authority. **3** archaic a hearing (*give audience to my plea*). [ME f. OF f. L *audientia* f. *audire* hear]

audile /ˈɔːdaɪl/ *adj.* of or referring to the sense of hearing. [irreg. f. L *audire* hear, after *tactile*]

audio /ˈɔːdɪəʊ/ *n.* (usu. *attrib.*) sound or the reproduction of sound. □ **audio frequency** a frequency capable of being perceived by the human ear. **audio typist** a person who types direct from a recording. [AUDIO-]

audio- /ˈɔːdɪəʊ/ *comb. form* hearing or sound. [L *audire* hear + -O-]

audio-animatron /ˌɔːdɪəʊˈænɪməˌtrɒn/ *n.* a robot constructed using the techniques of Audio-Animatronics.

Audio-Animatronics /ˌɔːdɪəʊˌænɪməˈtrɒnɪks/ *n.pl.* (usu. treated as *sing.*) *propr.* a technique for constructing lifelike robots resembling humans, animals, etc., and programmed to perform intricate movements in synchronization with a pre-recorded soundtrack. □ **audio-animatronic** *n. & adj.* **audio-animatronically** *adv.* [AUDIO- + *animated electronics*]

audiology /ˌɔːdɪˈɒlədʒɪ/ *n.* the science of hearing. □ **audiologist** *n.*

audiometer /ˌɔːdɪˈɒmɪtə(r)/ *n.* an instrument for testing hearing.

audiophile /ˈɔːdɪəʊˌfaɪl/ *n.* a hi-fi enthusiast.

audiotape /ˈɔːdɪəʊˌteɪp/ *n. & v.* ● *n.* **1** a magnetic tape on which sound can be recorded. **b** a length of this. **2** a sound recording on tape. ● *v.tr.* record (sound, speech, etc.) on tape.

audiovisual /ˌɔːdɪəʊˈvɪzjʊəl, -ˈvɪʒʊəl/ *adj.* (esp. of teaching methods) using both sight and sound.

audit /ˈɔːdɪt/ *n. & v.* ● *n.* an official examination of accounts. ● *v.tr.* (**audited, auditing**) **1** conduct an audit of. **2** US attend (a class) informally, without working for credits. [ME f. L *auditus* hearing f. *audire audit-* hear]

audition /ɔːˈdɪʃ(ə)n/ n. & v. ● n. **1** an interview for a role as a singer, actor, dancer, etc., consisting of a practical demonstration of suitability. **2** the power of hearing or listening. ● v. **1** tr. interview (a candidate) at an audition. **2** intr. be interviewed at an audition. [F audition or L auditio f. audire audit- hear]

auditive /ˈɔːdɪtɪv/ adj. concerned with hearing. [F auditif -ive (as AUDITION, -IVE)]

auditor /ˈɔːdɪtə(r)/ n. **1** a person who conducts an audit. **2** a listener. □ **auditorial** /ˌɔːdɪˈtɔːrɪəl/ adj. [ME f. AF auditour f. L auditor -oris (as AUDITIVE, -OR¹)]

auditorium /ˌɔːdɪˈtɔːrɪəm/ n. (pl. **auditoriums** or **auditoria** /-rɪə/) the part of a theatre etc. in which the audience sits. [L neut. of auditorius (adj.): see AUDITORY, -ORIUM]

auditory /ˈɔːdɪtərɪ/ adj. **1** concerned with hearing. **2** received by the ear. [L auditorius (as AUDITOR, -ORY²)]

Audubon /ˈɔːdəb(ə)n/, John James (1785–1851), American naturalist and artist. He is chiefly remembered for his great illustrated work *The Birds of America* (1827–38), which was compiled during his travels through America. He portrayed even the largest birds life-size, and painted them not in conventionally formal postures but in dramatic and sometimes violent action. Eventually published in Britain, the book had lasting success, both artistically and as a major contribution to natural history. The National Audubon Society is a North American organization for the study and protection of birds, founded in 1886.

Auer /ˈaʊə(r)/, Carl, Baron von Welsbach (1858–1929), Austrian chemist. Working at Heidelberg under Bunsen, he discovered in 1885 that the so-called element didymium was actually a mixture of two rare-earth elements, neodymium and praseodymium. In the same year Auer patented the incandescent gas mantle for which he is remembered today. He also discovered the cerium–iron alloy that is used for flints in cigarette and gas lighters.

AUEW abbr. (in the UK) Amalgamated Union of Engineering Workers.

au fait /əʊ ˈfeɪ/ predic.adj. (usu. foll. by with) having current knowledge; conversant (fully au fait with the arrangements). □ **put** (or **make**) **au fait with** instruct in. [F]

au fond /əʊ ˈfɒn/ adv. basically; at bottom. [F]

Aug. abbr. August.

Augean /ɔːˈdʒiːən/ adj. **1** Gk Mythol. of or relating to Augeas. **2** filthy; extremely dirty. [AUGEAS]

Augeas /ɔːˈdʒiːəs/ Gk Mythol. a legendary king whose vast stables had never been cleaned. Hercules cleaned them in a day by diverting the River Alpheus to flow through them.

auger /ˈɔːɡə(r)/ n. **1** a tool resembling a large corkscrew, for boring holes in wood. **2** a similar larger tool for boring holes in the ground. [OE nafogār f. nafu NAVE², + gār pierce: for loss of n cf. ADDER]

aught¹ /ɔːt/ n. (also **ought**) archaic (usu. implying neg.) anything at all. [OE āwiht f. Gmc]

aught² var. of OUGHT².

augite /ˈɔːdʒaɪt/ n. Mineral. a complex calcium magnesium aluminous silicate occurring in many igneous rocks. [L augites f. Gk augītēs f. augē lustre]

augment v. & n. ● v.tr. & intr. /ɔːɡˈment/ make or become greater; increase. ● n. /ˈɔːɡment/ Gram. a vowel prefixed to the past tenses in the older Indo-European languages. □ **augmented interval** Mus. a perfect or major interval that is increased by a semitone. □ **augmenter** /ɔːɡˈmentə(r)/ n. [ME f. OF augment (n.), F augmenter (v.), or LL augmentum, augmentare f. L augere increase]

augmentation /ˌɔːɡmenˈteɪʃ(ə)n/ n. **1** enlargement; growth; increase. **2** Mus. the lengthening of the time-values of notes in melodic parts. [ME f. F f. LL augmentatio -onis f. augmentare (as AUGMENT)]

augmentative /ɔːɡˈmentətɪv/ adj. **1** having the property of increasing. **2** Gram. (of an affix or derived word) reinforcing the idea of the original word. [F augmentatif -ive or med.L augmentativus (as AUGMENT)]

Augrabies Falls /əˈɡrɑːbiːz/ a series of waterfalls on the Orange River in the province of Northern Cape, South Africa.

au gratin /əʊ ˈɡrætæn/ adj. Cookery cooked with a crisp brown crust usu. of breadcrumbs or melted cheese. [F f. gratter, = by grating, f. GRATE¹]

Augsburg /ˈaʊɡzbɜːɡ/ a city in southern Germany, in Bavaria; pop. (1991) 259,880. It was founded in 15 BC by the Roman emperor Augustus, who called it Augusta Vindelicorum. During the Middle Ages it was an important centre of trade between northern and southern Europe.

Augsburg Confession a statement of the Lutheran position, drawn up mainly by Melanchthon and approved by Luther before being presented to the Emperor Charles V at Augsburg on 25 June 1530.

augur /ˈɔːɡə(r)/ v. & n. ● v. **1** intr. **a** (of an event, circumstance, etc.) suggest a specified outcome (usu. augur well or ill). **b** portend, bode (all augured well for our success). **2** tr. **a** foresee, predict. **b** portend. ● n. Rom. Antiq. a religious official who observed natural signs, esp. the behaviour of birds, interpreting these as an indication of divine approval or disapproval of a proposed action. □ **augural** /ˈɔːɡjʊrəl/ adj. [L]

augury /ˈɔːɡjʊrɪ/ n. (pl. **-ies**) **1** an omen; a portent. **2** the work of an augur; the interpretation of omens. [ME f. OF augurie or L augurium f. AUGUR]

August /ˈɔːɡəst/ n. the eighth month of the year. [OE f. L AUGUSTUS]

august /ɔːˈɡʌst/ adj. inspiring reverence and admiration; venerable, impressive. □ **augustly** adv. **augustness** n. [F auguste or L augustus consecrated, venerable]

Augusta /ɔːˈɡʌstə/ **1** a resort in eastern Georgia in the US; pop. (1990) 44,640. **2** the state capital of Maine; pop. (1990) 21,320.

Augustan /ɔːˈɡʌstən/ adj. & n. ● adj. **1** connected with, occurring during, or influenced by the reign of the Roman emperor Augustus, esp. as an outstanding period of Latin literature, including writers such as Virgil, Horace, and Ovid. **2** (of a writer or of a nation's literature) refined and classical in style. (See note below.) ● n. a writer of the Augustan age. [L Augustanus f. Augustus]

▪ The Augustan age in Britain relates to a period from the late 17th to early 18th centuries, the main figures being Dryden, Pope, Addison, Swift, and Steele. These writers were admirers and conscious imitators of classical poets such as Horace and Virgil and are particularly known for their elegant style and intellectual wit, their use of satire, and their refinement of the heroic couplet. In France, in a slightly earlier period, the principal figures were Molière, Racine, and Corneille.

Augustine /ɔːˈɡʌstiːn/ n. an Augustinian friar. [ME f. OF augustin f. L Augustinus: see AUGUSTINIAN]

Augustine, St¹ /ɔːˈɡʌstɪn/ (known as St Augustine of Canterbury) (died c.604), Italian churchman. Sent from Rome by Pope Gregory the Great to refound the Church in England, he and his party landed in Kent in 597 and were favourably received by King Ethelbert (whose wife was a Christian), who was afterwards converted. Augustine founded the first church and a monastery at Canterbury, and was consecrated as its first archbishop, but failed to reach agreement with representatives of the existing Celtic Church, which still survived in Britain on bases at Iona and Lindisfarne and was at variance with Rome on questions of discipline and practice. Feast day, 26 May.

Augustine, St² /ɔːˈɡʌstɪn/ (known as St Augustine of Hippo) (354–430), Doctor of the Church. Born in North Africa of a pagan father and a Christian mother, he underwent a series of spiritual crises in his early life, described in his Confessions. While in Milan he was influenced by the bishop, Ambrose, adopting his Neoplatonic understanding of Christianity and being baptized by him in 386. Augustine henceforth lived a monastic life, becoming bishop of Hippo in North Africa in 396. His episcopate was marked by his continual opposition to the heresies of the Pelagians, Donatists, and Manichees. Of his extensive writings, perhaps his best-known work is the City of God. His theology has dominated all later Western theology, with its psychological insight, its sense of man's utter dependence on grace (expressed in his doctrine of predestination), and its conception of the Church and the sacraments. Feast day, 28 Aug.

Augustinian /ˌɔːɡəˈstɪnɪən/ adj. & n. ● adj. **1** of or relating to St Augustine of Hippo. **2** of or belonging to any of the religious orders following the Rule of St Augustine. (See note below.) ● n. **1** an adherent of the doctrines of St Augustine. **2** a member of an order of Augustinian friars. [L Augustinus Augustine]

▪ The Rule of St Augustine, a set of instructions for the religious life, was based on the writings of St Augustine of Hippo and widely disseminated after his death. The two main religious orders following this rule were the Augustinian Canons (founded in the 11th century, forming various communities across northern Italy and southern France), and the Augustinian (or Austin) Friars, a mendicant order founded c.1250. Martin Luther belonged to a congregation of the latter.

Augustus /ɔːˈɡʌstəs/ (born Gaius Octavianus; also called (until 27 BC) Octavian) (63 BC–AD 14), the first Roman emperor. Originally called Gaius Octavianus, he took the name Gaius Julius Caesar Octavianus when he was adopted by the will of his great-uncle Julius Caesar in 44

BC. He established his position as one of the triumvirate of 43 BC, gaining supreme power by his defeat of Antony in 31 BC. A constitutional settlement in 27 BC in theory restored the republic but in practice regularized his sovereignty; in the same year he was given the title Augustus (L, = venerable). His rule was marked abroad by a series of expansionist military campaigns and at home by moral and religious reforms intended to restore earlier Roman values disrupted during previous civil wars.

auk /ɔːk/ n. a diving seabird of the family Alcidae, with a heavy body, short wings, and black and white plumage, e.g. the guillemot, puffin, and razorbill. □ **great auk** a large flightless auk, *Alca impennis*, exterminated in 1844. **little auk** a small Arctic auk, *Plautus alle* (also called (*N. Amer. & Sc.*) *dovekie*). [ON *álka*]

auklet /ˈɔːklɪt/ n. any of various small auks, found mainly in the North Pacific.

auld /ɔːld/ adj. Sc. old. [OE *ald*, Anglian form of OLD]

auld lang syne /ˌɔːld læŋ ˈsaɪn/ n. times long past. [Sc., = old long since: also as the title and refrain of a song]

aumbry /ˈɔːmbrɪ/ n. (also **ambry** /ˈæm-/) (pl. **-ies**) **1** a small recess in the wall of a church. **2** hist. a small cupboard. [ME f. OF *almarie, armarie* f. L *armarium* closet, chest f. *arma* utensils]

au naturel /əʊ ˌnætjʊˈrel/ predic.adj. & adv. Cookery uncooked; (cooked) in the most natural or simplest way. [F, = in the natural state]

Aung San /aʊŋ ˈsæn/ (1914–47), Burmese nationalist leader. A leader of the radicals from his student days, during the Second World War he accepted Japanese assistance and secret military training for his supporters. Returning to Burma in 1942 he became leader of the Japanese-sponsored Burma National Army, which defected to the Allies in the closing weeks of the war in the Pacific. As leader of the postwar Council of Ministers, in Jan. 1947 he negotiated a promise of full self-government from the British; in July he and six of his colleagues were assassinated by political rivals during a meeting of the Council.

Aung San Suu Kyi /aʊŋ ˌsæn suː ˈtʃiː/ (b.1945), Burmese political leader. Daughter of Aung San, she became the co-founder and leader of the National League for Democracy (NLD), the country's main opposition party, in 1988. Although she was placed under house arrest in 1989 and not allowed to stand as a candidate, the NLD won 80 per cent of the seats in the democratic elections of 1990; the ruling military government refused to recognize the NLD's victory. A supporter of political reform through non-violent public protest and democratic processes, she was awarded the Nobel Peace Prize in 1991. She was released from house arrest in 1995.

aunt /ɑːnt/ n. **1** the sister of one's father or mother. **2** an uncle's wife. **3** colloq. an unrelated woman friend of a child or children. □ **Aunt Sally 1** a game in which players throw sticks or balls at a wooden dummy. **2** the object of an unreasonable attack. **my** (or **my sainted** etc.) **aunt** sl. an exclamation of surprise, disbelief, etc. [ME f. AF *aunte*, OF *ante*, f. L *amita*]

auntie /ˈɑːntɪ/ n. (also **aunty**) (pl. **-ies**) colloq. **1** = AUNT. **2** (**Auntie**) Brit. an institution considered to be conservative or cautious, esp. the BBC.

au pair /əʊ ˈpeə(r)/ n. a young foreign person, esp. a woman, helping with housework etc. in exchange for room, board, and pocket money, esp. as a means of learning a language. [F, = on equal terms]

aura /ˈɔːrə/ n. (pl. **aurae** /-riː/ or **auras**) **1** the distinctive atmosphere diffused by or attending a person, place, etc. **2** (in mystic or spiritualistic use) a supposed subtle emanation, visible as a sphere of white or coloured light, surrounding the body of a living creature. **3** a subtle emanation or aroma from flowers etc. **4** Med. premonitory symptom(s) in epilepsy etc. [ME f. L f. Gk, = breeze, breath]

aural[1] /ˈɔːrəl/ adj. of or relating to or received by the ear. □ **aurally** adv. [L *auris* ear]

aural[2] /ˈɔːrəl/ adj. of, relating to, or resembling an aura; atmospheric. [as AURA]

Aurangzeb /ˈɔːrəŋˌzeb, ˈaʊərəŋ-/ (1618–1707), Mogul emperor of Hindustan 1658–1707. Having usurped the throne from his father, Aurangzeb assumed the title Alamgir (Conqueror of the World). His expansionist policies increased the Mogul empire to its widest extent, and it experienced a period of great wealth and splendour, but constant rebellions and wars greatly weakened the empire and it declined sharply after his death.

aureate /ˈɔːrɪət/ adj. **1** golden, gold-coloured. **2** resplendent. **3** (of language) highly ornamented. [ME f. LL *aureatus* f. L *aureus* golden f. *aurum* gold]

Aurelian /ɔːˈriːlɪən/ (Latin name Lucius Domitius Aurelianus) (c.215–

75), Roman emperor 270–5. Originally a common soldier, he rose through the ranks and was elected emperor by the army. By a series of military campaigns, including the defeat of Queen Zenobia at Palmyra (272), he successfully quelled rebellions and repelled barbarian invaders; he also built new walls round Rome, and established the state worship of the sun. He was assassinated by his own army officers.

Aurelius /ɔːˈriːlɪəs/, Marcus (full name Caesar Marcus Aurelius Antoninus Augustus) (121–80), Roman emperor 161–80. The adopted successor of Antoninus Pius, he was occupied for much of his reign with wars against Germanic tribes invading the empire from the north. He was by nature a philosophical contemplative; his *Meditations* are a collection of aphorisms and reflections based on a Stoic outlook and written down for his own guidance.

aureole /ˈɔːrɪˌəʊl/ n. (also **aureola** /ɔːˈrɪələ/) **1** a halo or circle of light, esp. round the head or body of a portrayed religious figure. **2** a corona round the sun or moon. [ME f. L *aureola (corona)*, = golden (crown), fem. of *aureolus* f. *aureus* f. *aurum* gold: *aureole* f. OF f. L *aureola*]

aureomycin /ˌɔːrɪəʊˈmaɪsɪn/ n. an antibiotic used esp. in lung diseases. [L *aureus* golden + Gk *mukēs* fungus + -IN]

au revoir /əʊ rəˈvwɑː(r)/ int. & n. goodbye (until we meet again). [F]

Auric /ˈɔːrɪk/, Georges (1899–1983), French composer. While studying music in Paris, he met Satie and Cocteau, under whose influence he and five other composers formed the anti-romantic group Les Six. His works include operas, ballets (notably Diaghilev's *Les Matelots*, 1925), orchestral works, and songs, but he is probably best known for film music such as the scores for *The Lavender Hill Mob* (1951) and *Moulin Rouge* (1952).

auric[1] /ˈɔːrɪk/ adj. of or relating to trivalent gold. [L *aurum* gold]

auric[2] /ˈɔːrɪk/ adj. = AURAL[2].

auricle /ˈɔːrɪk(ə)l/ n. Anat. **1 a** a small muscular pouch on the surface of each atrium of the heart. **b** the atrium itself. **2** the external ear of animals. Also called *pinna*. **3** an appendage shaped like the ear. [AURICULA]

auricula /ɔːˈrɪkjʊlə/ n. a primula, *Primula auricula*, with leaves shaped like bears' ears. [L, dimin. of *auris* ear]

auricular /ɔːˈrɪkjʊlə(r)/ adj. **1** of or relating to the ear or hearing. **2** of or relating to the auricle of the heart. **3** shaped like an auricle. □ **auricularly** adv. [LL *auricularis* (as AURICULA)]

auriculate /ɔːˈrɪkjʊlət/ adj. having one or more auricles or ear-shaped appendages. [L]

auriferous /ɔːˈrɪfərəs/ adj. naturally bearing gold. [L *aurifer* f. *aurum* gold]

Auriga /ɔːˈraɪɡə/ Astron. a large northern constellation (the Charioteer), said to represent a man holding a whip. It includes the bright star Capella, several bright variable stars and star clusters, and is crossed by the Milky Way. [L]

Aurignacian /ˌɔːrɪɡˈneɪʃ(ə)n/ adj. & n. Archaeol. of, relating to, or denoting the early stages of the upper palaeolithic culture in Europe and the Near East, dated in most places to c.34,000–29,000 BP. (See also CRO-MAGNON MAN.) [F *Aurignacien* f. *Aurignac*, location of the type-site, a cave in southern France]

aurochs /ˈɔːrɒks, ˈaʊər-/ n. (pl. same) the extinct wild ox, *Bos primigenius*, ancestor of domestic cattle and formerly native to Europe and western Asia. Also called *urus*. [G f. OHG *ūrohso* f. *ūr-* urus + *ohso* ox]

Aurora /ɔːˈrɔːrə/ Rom. Mythol. goddess of the dawn, corresponding to the Greek Eos. Most of the stories about her tell of handsome men being kidnapped to live with her (see TITHONUS). [L, = dawn]

aurora /ɔːˈrɔːrə/ n. (pl. **auroras** or **aurorae** /-riː/) **1** a luminous electrical atmospheric phenomenon, usu. of streamers of light in the sky above the northern or southern magnetic pole. **2** poet. the dawn. □ **aurora australis** /ɒˈstreɪlɪs/ a southern occurrence of aurora. **aurora borealis** /ˌbɔːrɪˈeɪlɪs/ a northern occurrence of aurora. □ **auroral** adj. [L, = dawn, AURORA]

Auschwitz /ˈaʊʃvɪts/ a Nazi concentration camp in the Second World War, near the town of Oświęcim (Auschwitz) in Poland.

auscultation /ˌɔːskəlˈteɪʃ(ə)n/ n. the act of listening, esp. to sounds from the heart, lungs, etc., as a part of medical diagnosis. □ **auscultatory** /ɔːˈskʌltətərɪ/ adj. [L *auscultatio* f. *auscultare* listen to]

auspice /ˈɔːspɪs/ n. **1** (in pl.) patronage (esp. *under the auspices of*). **2** a forecast. [orig. 'observation of bird-flight in divination': F *auspice* or L *auspicium* f. *auspex* observer of birds f. *avis* bird]

auspicious /ɔː'spɪʃəs/ adj. **1** of good omen; favourable. **2** prosperous. □ **auspiciously** adv. **auspiciousness** n. [AUSPICE + -OUS]

Aussie /'ɒzɪ/ n. & adj. (also **Ossie, Ozzie**) colloq. ● n. **1** an Australian. **2** Australia. ● adj. Australian. [abbr.]

Austen /'ɒstɪn/, Jane (1775–1817), English novelist. The youngest of seven children of a Hampshire rector, she was greatly stimulated by her extended and affectionate family. Her major novels are *Sense and Sensibility* (1811), *Pride and Prejudice* (1813), *Mansfield Park* (1814), *Emma* (1815), *Northanger Abbey* (1818), and *Persuasion* (1818). They are notable for skilful characterization and penetrating social observation; Austen brings a dry wit and satirical eye to her portrayal of middle and upper-class life, capturing contemporary values and moral dilemmas.

austere /ɒ'stɪə(r)/ adj. (**austerer, austerest**) **1** severely simple. **2** morally strict. **3** harsh, stern. □ **austerely** adv. [ME f. OF f. L *austerus* f. Gk *austēros* severe]

austerity /ɒ'stɛrɪtɪ/ n. (pl. **-ies**) **1** sternness; moral severity. **2** severe simplicity, e.g. of nationwide economies. **3** (esp. in pl.) an austere practice (*the austerities of a monk's life*).

Austerlitz, Battle of /'austəlɪts, 'ɔːstə-/ a battle in 1805 near the town of Austerlitz (now in the Czech Republic), in which Napoleon defeated the Austrians and Russians.

Austin[1] /'ɒstɪn/ the state capital of Texas; pop. (1990) 465,620. First settled in 1835, it was named in 1839 after Stephen F. Austin, son of Moses Austin, leader of the first Texas colony.

Austin[2] /'ɒstɪn/, Herbert, 1st Baron Austin of Longbridge (1866–1941), British motor manufacturer. Having joined the Wolseley Sheep Shearing Machine Company in 1893, Austin persuaded the company to embark on the manufacture of cars. He produced vehicles with them until 1905, when he opened his own works near Birmingham. The output of the factory steadily increased, especially following the launch of the Austin Seven (known as 'the Baby Austin') in 1921; 300,000 models of this car were produced before 1939.

Austin[3] /'ɒstɪn/, John (1790–1859), English jurist. Regarded as the founder of analytical jurisprudence, he was greatly influenced by the utilitarianism of his friend Jeremy Bentham, as can be seen from his work *The Province of Jurisprudence Determined* (1832). An important influence on the English legal system, Austin is significant for his strict delimitation of the sphere of law and its distinction from that of morality, as well as his examination of the connotations of such common legal terms and ideas as right, duty, liberty, injury, and punishment.

Austin[4] /'ɒstɪn/, John Langshaw (1911–60), English philosopher. A lecturer and later professor of moral philosophy at Oxford University (1952–60), he was a careful and witty exponent of the linguistic school of philosophy, seeking to elucidate philosophical problems by analysis of the words in which they are expressed. Two of his courses of lectures were published posthumously in 1962: *Sense and Sensibilia* discusses perception, while *How to Do Things with Words* distinguishes 'performative' utterances (in which something is done, such as promising or making marriage vows) from utterances that convey information.

Austin[5] /'ɒstɪn, 'ɔːstɪn/ n. = AUGUSTINIAN n. [contr. of AUGUSTINE]

Austin Friars /'ɒstɪn, 'ɔːstɪn/ see AUGUSTINIAN.

austral /'ɔːstrəl, 'ɒst-/ adj. **1** southern. **2** (**Austral**) of Australia or Australasia (*Austral English*). [ME f. L *australis* f. *Auster* south wind]

Australasia /ˌɒstrə'leɪʒə, -'leɪʃə/ the region consisting of Australia, New Zealand, New Guinea, and the neighbouring islands of the Pacific. □ **Australasian** adj. & n. [f. F *Australasie*, formed as *Australia* + *Asia*]

Australia /ɒ'streɪlɪə/ an island country and continent of the southern hemisphere, in the SW Pacific, a member state of the Commonwealth; pop. (est. 1991) 17,500,000; official language, English; capital, Canberra. Much of the continent has a hot dry climate and a large part of the central area is desert or semi-desert; the most fertile areas are the eastern coastal plains and the south-western corner of Western Australia. Human habitation in Australia dates from prehistoric times, but most of the population is now of European descent, with the aboriginal peoples forming only about 1.5 per cent of the population (see ABORIGINAL n. 2). In Europe, the idea of an unknown southern land (*terra australis*) had been current since the times of the ancient Greek geographers. Pre-17th-century European sightings of Australia are claimed but not firmly attested. From 1606 onwards its western coast was explored by the Dutch, and in 1642 Tasman proved that it was an island. It was visited by an Englishman, William Dampier, in 1688 and 1699. In 1770 Captain James Cook landed at Botany Bay on the eastern side of the continent and formally took possession of New South Wales. British colonization began in 1788, as did the transportation of convicts, which was discontinued in 1868. The interior was gradually explored and opened up in the 19th century and political consolidation resulted in the declaration of a Commonwealth in 1901, when the six colonies (New South Wales, Victoria, Queensland, South Australia, Western Australia, and the offshore island of Tasmania) federated as sovereign states. Northern Territory achieved similar status in 1978. Australia played a major role on the side of the Allies in each of the two world wars. [L (*terra*) *australis* (as AUSTRAL)]

Australia Day a national public holiday in Australia, commemorating the founding on 26 Jan. 1788 of the colony of New South Wales.

Australian /ɒ'streɪlɪən/ n. & adj. ● n. **1** a native or national of Australia. **2** a person of Australian descent. ● adj. of or relating to Australia. □ **Australian bear** a koala bear. **Australian terrier** a wire-haired Australian breed of terrier. □ **Australianism** n. **Australianize** v. [F *australien* f. L (as AUSTRAL)]

Australiana /ɒ,streɪlɪ'ɑːnə/ n. pl. objects relating to or characteristic of Australia.

Australian Antarctic Territory an area of Antarctica administered by Australia, lying between longitudes 142° east and 136° east.

Australian Capital Territory a federal territory in New South Wales, Australia, consisting of two enclaves ceded by New South Wales, one in 1911 to contain Canberra, the other in 1915 containing Jervis Bay.

Australian Labor Party (abbr. **ALP**) Australia's oldest political party, founded in 1891. The party is moderate left-of-centre; it has provided three recent Australian Prime Ministers, Gough Whitlam, Bob Hawke, and Paul Keating.

Australian Rules football n. (also **Australian Rules, Australian National football**) a form of football played by teams of eighteen players on a large oval pitch with a ball similar to that used in rugby. Australian Rules football is a fast-moving, high-scoring game dating back to 1858. Players may run with the ball as long as they bounce it every nine metres. The ball may be passed forward, but in that case must be punched rather than thrown. There are four goalposts at each end: a goal, scored between the two inner posts, counts six points; a behind, scored between the outer posts, counts one point.

Austral Islands an alternative name for the TUBUAI ISLANDS.

Australoid /'ɒstrə,lɔɪd/ adj. & n. ● adj. of, relating to, or denoting the dark-skinned ethnic group comprising Australian Aboriginals and the Vedda of Sri Lanka. ● n. a member of this ethnic group. [as AUSTRALIAN]

Australopithecus /ˌɒstrələʊ'pɪθɪkəs/ n. a genus of extinct bipedal primates having apelike and human characteristics. Their fossilized remains were first found by R. Dart in South Africa in 1924, and others have since been found elsewhere in southern and East Africa. They belong to the Pliocene and Lower Pleistocene epochs, dating from *c.*4 million to 1 million years ago, and the lightly built species *A. africanus* is thought to be the immediate ancestor of the human genus *Homo*. More heavily built forms such as *A.* (or *Paranthropus*) *robustus* and *A.* (or *Zinjanthropus*) *boisei* ('Nutcracker man') appear to have been evolutionary dead ends. (See also LUCY, NUTCRACKER MAN.) □ **australopithecine** /-'pɪθɪ,siːn/ n. & adj. [mod.L f. L *australis* southern + Gk *pithēkos* ape]

Austria /'ɒstrɪə/ (called in German *Österreich*) a republic in central Europe; pop. (est. 1991) 7,700,000; official language, German; capital, Vienna. Austria is largely mountainous, with the River Danube flowing through the north-east. For a time part of the Roman Empire (the Danube forming the frontier), Austria was overrun by Germanic peoples in the 5th century AD. It was dominated from the early Middle Ages by the Habsburg family, and became the centre of a massive central European empire (see also AUSTRIA–HUNGARY). The empire's collapse in 1918 left Austria a weak and unstable country, which was incorporated within the Nazi Reich in 1938. After the Second World War the country fell under Allied military occupation, regaining its sovereignty in 1955. A referendum in 1994 approved Austria's entry into the European Community. □ **Austrian** adj. & n.

Austria–Hungary /ˌɒstrɪə'hʌŋgərɪ/ (also called *Austro-Hungarian empire*) the so-called Dual Monarchy, established by the Austrian emperor Franz Josef after Austria's defeat by Prussia in 1866, according to which Austria and Hungary became autonomous states under a common sovereign. The dualist system came under increasing pressure from the other subject nations, including Croatians, Serbs,

Slovaks, Romanians, and Czechs, and failure to resolve these nationalist aspirations was one of the causes of the First World War. The empire was dissolved by the Versailles peace settlement of 1919.

Austrian Succession, War of the a group of several related European conflicts (1740–8) triggered by the death of the Emperor Charles VI and the accession of his daughter Maria Theresa in 1740 to the Austrian throne. Austria, supported by Hungary, the Netherlands, and Great Britain, was involved in a struggle with Bavaria, Saxony, and Spain, all of which had rival claimants to the throne, and Prussia, which had territorial designs on parts of the Austrian empire. France came in on the side of Bavaria and Prussia. As a result of the war, Prussia obtained Silesia from Austria, while Maria Theresa kept her throne. (See also Pragmatic Sanction.)

Austro-[1] /ˈɒstrəʊ/ *comb. form* Austrian; Austrian and (*Austro-Hungarian*).

Austro-[2] /ˈɒstrəʊ/ *comb. form* Australian; Australian and (*Austro-Asiatic*).

Austronesian /ˌɒstrəˈniːʒ(ə)n, -ˈniːzɪən/ *adj. & n.* = Malayo-Polynesian. [G *austronesisch* f. L *australis* southern + -o- + Gk *nēsos* island]

AUT *abbr.* (in the UK) Association of University Teachers.

autarchy /ˈɔːtɑːkɪ/ *n.* (*pl.* **-ies**) **1** absolute sovereignty. **2** despotism. **3** an autarchic country or society. □ **autarchic** /ɔːˈtɑːkɪk/ *adj.* **autarchical** *adj.* [mod.L *autarchia* (as Auto-, Gk *-arkhia* f. *arkhō* rule)]

autarky /ˈɔːtɑːkɪ/ *n.* (*pl.* **-ies**) **1** self-sufficiency, esp. as an economic system. **2** a state etc. run according to such a system. □ **autarkist** *n.* **autarkic** /ɔːˈtɑːkɪk/ *adj.* **autarkical** *adj.* [Gk *autarkeia* (as Auto-, *arkeō* suffice)]

auteur /əʊˈtɜː(r)/ *n.* a film director regarded as having such a significant influence on the films that he or she directs as to be able to rank as their author. The word was first used in this sense in 1954 by François Truffaut, in re-evaluating earlier Hollywood directors. [F, lit. 'author']

authentic /ɔːˈθentɪk/ *adj.* **1 a** of undisputed origin; genuine. **b** reliable or trustworthy. **2** *Mus.* (of a mode) containing notes between the final and an octave higher (cf. PLAGAL). □ **authentically** *adv.* **authenticity** /ˌɔːθenˈtɪsɪtɪ/ *n.* [ME f. OF *autentique* f. LL *authenticus* f. Gk *authentikos* principal, genuine]

authenticate /ɔːˈθentɪˌkeɪt/ *v.tr.* **1** establish the truth or genuineness of. **2** validate. □ **authenticator** *n.* **authentication** /ɔːˌθentɪˈkeɪʃ(ə)n/ *n.* [med.L *authenticare* f. LL *authenticus*: see AUTHENTIC]

author /ˈɔːθə(r)/ *n. & v.* ● *n.* (*fem.* **authoress** /ˈɔːθrɪs, ˌɔːθəˈres/) **1** a writer, esp. of books. **2** the originator of an event, a condition, etc. (*the author of all my woes*). ● *v.tr.* be the author of. □ **authorial** /ɔːˈθɔːrɪəl/ *adj.* [ME f. AF *autour*, OF *autor* f. L *auctor* f. *augere auct-* increase, originate, promote]

authoritarian /ɔːˌθɒrɪˈteərɪən/ *adj. & n.* ● *adj.* **1** favouring, encouraging, or enforcing strict obedience to authority, as opposed to individual freedom. **2** tyrannical or domineering. ● *n.* a person favouring absolute obedience to a constituted authority. □ **authoritarianism** *n.*

authoritative /ɔːˈθɒrɪtətɪv/ *adj.* **1** being recognized as true or dependable. **2** (of a person, behaviour, etc.) commanding or self-confident. **3** official; supported by authority (*an authoritative document*). **4** having or claiming influence through recognized knowledge or expertise. □ **authoritatively** *adv.* **authoritativeness** *n.*

authority /ɔːˈθɒrɪtɪ/ *n.* (*pl.* **-ies**) **1 a** the power or right to enforce obedience. **b** (often foll. by *for*, or *to* + infin.) delegated power. **2** (esp. in *pl.*) a person or body having authority, esp. political or administrative. **3 a** an influence exerted on opinion because of recognized knowledge or expertise. **b** such an influence expressed in a book, quotation, etc. **c** a person whose opinion is accepted, esp. an expert in a subject (*an authority on vintage cars*). **4** the weight of evidence. [ME f. OF *autorité* f. L *auctoritas* f. *auctor*: see AUTHOR]

authorize /ˈɔːθəˌraɪz/ *v.tr.* (also **-ise**) **1** sanction. **2** (foll. by *to* + infin.) **a** give authority. **b** commission (a person or body) (*authorized to trade*). □ **authorization** /ˌɔːθəraɪˈzeɪʃ(ə)n/ *n.* [ME f. OF *autoriser* f. med.L *auctorizare* f. *auctor*: see AUTHOR]

Authorized Version (also called the *King James Bible*) the 1611 English translation of the Bible, ordered by James I, which was produced by about fifty scholars. It became widely popular following its publication, and although in fact never officially 'authorized' it remained for centuries the Bible of every English-speaking country.

authorship /ˈɔːθəˌʃɪp/ *n.* **1** the origin of a book or other written work (*of unknown authorship*). **2** the occupation of writing.

autism /ˈɔːtɪz(ə)m/ *n.* *Psychol.* a mental condition, usu. present from childhood, characterized by complete self-absorption and a reduced ability to respond to or communicate with the outside world. □ **autistic** /ɔːˈtɪstɪk/ *adj.* [mod.L *autismus* (as Auto-, -ISM)]

auto /ˈɔːtəʊ/ *n.* (*pl.* **-os**) US *colloq.* a motor car. [abbr. of AUTOMOBILE]

auto- /ˈɔːtəʊ/ *comb. form* (usu. **aut-** /ɔːt/ before a vowel) **1** self (*autism*). **2** one's own (*autobiography*). **3** by oneself or spontaneous (*auto-suggestion*). **4** by itself or automatic (*automobile*). [from or after Gk *auto-* f. *autos* self]

autobahn /ˈɔːtəʊˌbɑːn/ *n.* (*pl.* **autobahns** or **autobahnen** /-nən/) a German, Austrian, or Swiss motorway. [G f. *Auto* motor car + *Bahn* path, road]

autobiography /ˌɔːtəʊbaɪˈɒgrəfɪ/ *n.* (*pl.* **-ies**) **1** a personal account of one's own life, esp. for publication. **2** this as a process or literary form. □ **autobiographer** *n.* **autobiographic** /-ˌbaɪəˈgræfɪk/ *adj.* **autobiographical** *adj.* **autobiographically** *adv.*

autocade /ˈɔːtəʊˌkeɪd/ *n.* US a motorcade. [AUTOMOBILE + CAVALCADE]

autocar /ˈɔːtəʊˌkɑː(r)/ *n.* archaic a motor vehicle.

autocephalous /ˌɔːtəʊˈsefələs, -ˈkefələs/ *adj.* **1** (esp. of an Eastern Church) appointing its own head. **2** (of a bishop, church, etc.) independent. [Gk *autokephalos* (as Auto-, *kephalē* head)]

autochthon /ɔːˈtɒkθən/ *n.* (*pl.* **autochthons** or **autochthones** /-θəˌniːz/) (in *pl.*) the original or earliest known inhabitants of a country; aboriginals. □ **autochthonal** *adj.* **autochthonic** /ˌɔːtɒkˈθɒnɪk/ *adj.* [Gk, = sprung from the earth (as Auto-, *khthōn, -onos* earth)]

autochthonous /ɔːˈtɒkθənəs/ *adj.* **1** indigenous, native. **2 a** of independent or local formation. **b** *Geol.* (of a deposit) formed in its present position (cf. ALLOCHTHONOUS).

autoclave /ˈɔːtəˌkleɪv/ *n.* **1** a strong vessel used for chemical reactions at high pressures and temperatures. **2** a sterilizer using high-pressure steam. [AUTO- + L *clavus* nail or *clavis* key]

autocracy /ɔːˈtɒkrəsɪ/ *n.* (*pl.* **-ies**) **1** absolute government by one person. **2** the power exercised by such a person. **3** an autocratic country or society. [Gk *autokrateia* (as AUTOCRAT)]

autocrat /ˈɔːtəˌkræt/ *n.* **1** an absolute ruler. **2** a dictatorial person. □ **autocratic** /ˌɔːtəˈkrætɪk/ *adj.* **autocratically** *adv.* [F *autocrate* f. Gk *autokratēs* (as AUTO-, *kratos* power)]

autocross /ˈɔːtəʊˌkrɒs/ *n.* motor-racing across country or on unmade roads. [AUTOMOBILE + CROSS- 1]

Autocue /ˈɔːtəʊˌkjuː/ *n. propr.* a device, unseen by the audience, displaying a television script to a speaker or performer as an aid to memory (cf. TELEPROMPTER).

auto-da-fé /ˌɔːtəʊdɑːˈfeɪ/ *n.* (*pl.* **autos-da-fé** /ˌɔːtəʊz-/) **1** a sentence of punishment by the Spanish Inquisition. **2** the execution of such a sentence, esp. the burning of a heretic. [Port., = act of the faith]

autodidact /ˈɔːtəʊˌdaɪdækt/ *n.* a self-taught person. □ **autodidactic** /ˌɔːtəʊdaɪˈdæktɪk/ *adj.* [AUTO- + *didact* as DIDACTIC]

auto-erotism /ˌɔːtəʊˈerəˌtɪz(ə)m/ *n.* (also **auto-eroticism** /-ɪˈrɒtɪˌsɪz(ə)m/) *Psychol.* sexual excitement generated by stimulating one's own body; masturbation. □ **auto-erotic** /-ɪˈrɒtɪk/ *adj.*

autofocus /ˈɔːtəʊˌfəʊkəs/ *n.* a device for focusing a camera etc. automatically.

autogamy /ɔːˈtɒgəmɪ/ *n.* *Bot.* self-fertilization in plants. □ **autogamous** *adj.* [AUTO- + Gk *-gamia* f. *gamos* marriage]

autogenous /ɔːˈtɒdʒɪnəs/ *adj.* self-produced. □ **autogenous welding** a process of joining metal by melting the edges together, without adding material.

autogiro /ˌɔːtəʊˈdʒaɪərəʊ/ *n.* (also **autogyro**) (*pl.* **-os**) a form of aircraft with freely rotating horizontal vanes and a propeller. The autogiro differs from the later helicopter in that its wings are not powered but rotate in the slipstream, the aircraft being propelled by a conventional mounted engine. The autogiro was invented and named by the Spaniard Juan de la Cierva, whose first machine flew in 1923. [Sp. (as AUTO-, *giro* gyration)]

autograft /ˈɔːtəˌgrɑːft/ *n.* *Surgery* a graft of tissue from one point to another of the same individual's body.

autograph /ˈɔːtəˌgrɑːf/ *n. & v.* ● *n.* **1 a** a signature, esp. that of a celebrity. **b** handwriting. **2** a manuscript in an author's own handwriting. **3** a document signed by its author. ● *v.tr.* **1** sign (a photograph, autograph album, etc.). **2** write (a letter etc.) by hand. [F *autographe* or LL *autographum* f. Gk *autographon* neut. of *autographos* (as AUTO-, -GRAPH)]

autography /ɔːˈtɒgrəfɪ/ *n.* **1** writing done with one's own hand. **2** the facsimile reproduction of writing or illustration. □ **autographic** /ˌɔːtəˈɡræfɪk/ adj.

autogyro var. of AUTOGIRO.

autoharp /ˈɔːtəˌhɑːp/ n. a kind of zither with a mechanical device to allow the playing of chords.

autoimmune /ˌɔːtəʊɪˈmjuːn/ adj. Med. (of a disease) caused by antibodies or immune lymphocytes produced against substances naturally present in the body. □ **autoimmunity** n.

autointoxication /ˌɔːtəʊɪnˌtɒksɪˈkeɪʃ(ə)n/ n. Med. poisoning by a toxin formed within the body itself.

autolysis /ɔːˈtɒlɪsɪs/ n. the destruction of cells by their own enzymes. □ **autolytic** /ˌɔːtəˈlɪtɪk/ adj. [G *Autolyse* (as AUTO-, -LYSIS)]

automat /ˈɔːtəˌmæt/ n. US **1** a slot-machine that dispenses goods. **2** a cafeteria containing slot-machines dispensing food and drink. [G f. F *automate*, formed as AUTOMATON]

automate /ˈɔːtəˌmeɪt/ v.tr. convert to or operate by automation (*the ticket office has been automated*). [back-form. f. AUTOMATION]

automatic /ˌɔːtəˈmætɪk/ adj. & n. ● adj. **1** (of a machine, device, etc., or its function) working by itself, without direct human intervention. **2 a** done spontaneously, without conscious thought or intention (*an automatic reaction*). **b** necessary and inevitable (*an automatic penalty*). **3** Psychol. performed unconsciously or subconsciously. **4** (of a firearm) that continues firing until the ammunition is exhausted or the pressure on the trigger is released. **5** (of a motor vehicle or its transmission) using gears that change automatically according to speed and acceleration. ● n. **1** an automatic device, esp. a gun or transmission. **2** colloq. a vehicle with automatic transmission. □ **automatic pilot** a device for keeping an aircraft on a set course. □ **automatically** adv. **automaticity** /ˌɔːtəməˈtɪsɪtɪ/ n. [formed as AUTOMATON + -IC]

automation /ˌɔːtəˈmeɪʃ(ə)n/ n. **1** the use of automatic equipment to save mental and manual labour. **2** the automatic control of the manufacture of a product through its successive stages. [irreg. f. AUTOMATIC + -ATION]

automatism /ɔːˈtɒməˌtɪz(ə)m/ n. **1** Psychol. the performance of actions unconsciously or subconsciously; such action. **2** involuntary action. **3** unthinking routine. [F *automatisme* f. *automate* AUTOMATON]

automatize /ɔːˈtɒməˌtaɪz/ v.tr. (also **-ise**) **1** make (a process etc.) automatic. **2** subject (a business, enterprise, etc.) to automation, automate. □ **automatization** /ɔːˌtɒmətaɪˈzeɪʃ(ə)n/ n. [AUTOMATIC + -IZE]

automaton /ɔːˈtɒmət(ə)n/ n. (pl. **automata** /-tə/ or **automatons**) **1** a mechanism with concealed motive power. **2** a person who behaves mechanically, like an automaton. [L f. Gk, neut. of *automatos* acting of itself: see AUTO-]

automobile /ˈɔːtəməˌbiːl/ n. US a motor car. [F (as AUTO-, MOBILE)]

automotive /ˌɔːtəˈməʊtɪv/ adj. concerned with motor vehicles.

autonomic /ˌɔːtəˈnɒmɪk/ adj. esp. Physiol. functioning involuntarily. □ **autonomic nervous system** see NERVOUS SYSTEM. [AUTONOMY + -IC]

autonomous /ɔːˈtɒnəməs/ adj. **1** having self-government. **2** acting independently or having the freedom to do so. □ **autonomously** adv. [Gk *autonomos* (as AUTONOMY)]

autonomy /ɔːˈtɒnəmɪ/ n. (pl. **-ies**) **1** the right of self-government. **2** personal freedom. **3** freedom of the will. **4** a self-governing community. □ **autonomist** n. [Gk *autonomia* f. *autos* self + *nomos* law]

autopilot /ˈɔːtəʊˌpaɪlət/ n. an automatic pilot. [abbr.]

autopista /ˈaʊtəʊˌpiːstə/ n. a Spanish motorway. [Sp. (as AUTOMOBILE, *pista* track)]

autopsy /ˈɔːtɒpsɪ, ɔːˈtɒpsɪ/ n. (pl. **-ies**) **1** a post-mortem examination. **2** any critical analysis. **3** a personal inspection. [F *autopsie* or mod.L *autopsia* f. Gk f. *autoptēs* eye-witness]

autoradiograph /ˌɔːtəˈreɪdɪəˌɡrɑːf/ n. a photograph of an object, produced by radiation from radioactive material in the object. □ **autoradiographic** /-ˌreɪdɪəˈɡræfɪk/ adj. **autoradiography** /-ˌreɪdɪˈɒɡrəfɪ/ n.

autoroute /ˈɔːtəʊˌruːt/ n. a French motorway. [F (as AUTOMOBILE, ROUTE)]

autostrada /ˈaʊtəʊˌstrɑːdə/ n. (pl. **autostradas** or **autostrade** /-dɪ/) an Italian motorway. [It. (as AUTOMOBILE, *strada* road)]

auto-suggestion /ˌɔːtəʊsəˈdʒestʃən/ n. suggestion to oneself; the hypnotic or subconscious adoption of an idea originating within oneself.

autotelic /ˌɔːtəˈtelɪk/ adj. having or being a purpose in itself. [AUTO- + Gk *telos* end]

autotomy /ɔːˈtɒtəmɪ/ n. Zool. the casting off of a part of the body when threatened, e.g. the tail of a lizard.

autotoxin /ˌɔːtəˈtɒksɪn/ n. a poisonous substance originating within an organism. □ **autotoxic** adj.

autotrophic /ˌɔːtəˈtrəʊfɪk, -ˈtrɒfɪk/ adj. Biol. able to form complex nutritional organic substances from simple inorganic substances such as carbon dioxide (cf. HETEROTROPHIC). [AUTO- + Gk *trophos* feeder]

autotype /ˈɔːtəˌtaɪp/ n. **1** a facsimile. **2 a** a photographic printing process for monochrome reproduction. **b** a print made by this process.

autoxidation /ˌɔːˌtɒksɪˈdeɪʃ(ə)n/ n. Chem. oxidation which occurs spontaneously at ambient temperatures in the presence of oxygen.

autumn /ˈɔːtəm/ n. **1** the third season of the year, when crops and fruits are gathered and leaves fall, in the northern hemisphere from September to November and in the southern hemisphere from March to May. **2** Astron. the period from the autumnal equinox to the winter solstice. **3** a time of maturity or incipient decay. □ **autumn crocus** the meadow saffron. [ME f. OF *automne* f. L *autumnus*]

autumnal /ɔːˈtʌmn(ə)l/ adj. **1** of, characteristic of, or appropriate to autumn (*autumnal colours*). **2** occurring in autumn (*autumnal equinox*). **3** maturing or blooming in autumn. **4** past the prime of life. [L *autumnalis* (as AUTUMN, -AL[1])]

Auvergne /əʊˈveən/ a region of south central France. It was a province of the Roman Empire, and takes its name from the Arverni, a Celtic tribe who lived there in Roman times. The region is mountainous and contains the extinct volcanic cones known as the Puys.

auxanometer /ˌɔːksəˈnɒmɪtə(r)/ n. an instrument for measuring the linear growth of plants. [Gk *auxanō* increase + -METER]

auxiliary /ɔːɡˈzɪljərɪ/ adj. & n. ● adj. **1** (of a person or thing) that gives help. **2** (of services or equipment) subsidiary, additional. ● n. (pl. **-ies**) **1** an auxiliary person or thing. **2** (in pl.) Mil. auxiliary troops. **3** Gram. an auxiliary verb. □ **auxiliary troops** Mil. foreign or allied troops in a belligerent nation's service. **auxiliary verb** Gram. one used in forming tenses, moods, and voices of other verbs. [L *auxiliarius* f. *auxilium* help]

auxin /ˈɔːksɪn/ n. any of a group of plant hormones that regulate growth. [G f. Gk *auxō* increase + -IN]

AV abbr. **1** audiovisual (teaching aids etc.). **2** Authorized Version (of the Bible).

Av var. of AB[1].

avadavat /ˌævədəˈvæt/ n. (also **amadavat** /ˌæmə-/) a small brightly coloured southern Asian waxbill of the genus *Amandava*, esp. the red *A. amandava*. [AHMADABAD (where such birds were sold)]

avail /əˈveɪl/ v. & n. ● v. **1** tr. help, benefit. **2** refl. (foll. by *of*) profit by; take advantage of. **3** intr. **a** provide help. **b** be of use, value, or profit. ● n. (usu. in neg. or interrog. phrases) use, profit (*to no avail; without avail; of what avail?*). [ME f. obs. *vail* (v.) f. OF *valoir* be worth f. L *valere*]

available /əˈveɪləb(ə)l/ adj. (often foll. by *to*, *for*) **1** capable of being used; at one's disposal. **2** within one's reach. **3** (of a person) **a** free. **b** able to be contacted. □ **availably** adv. **availableness** n. **availability** /əˌveɪləˈbɪlɪtɪ/ n. [ME f. AVAIL + -ABLE]

avalanche /ˈævəˌlɑːnʃ/ n. & v. ● n. **1** a mass of snow and ice, tumbling rapidly down a mountain. **2** a sudden appearance or arrival of anything in large quantities (*faced with an avalanche of work*). ● v. **1** intr. descend like an avalanche. **2** tr. carry down like an avalanche. [F, alt. of dial. *lavanche* after *avaler* descend]

Avalon /ˈævəˌlɒn/ **1** (in Arthurian legend) the place to which Arthur was conveyed after death. **2** Welsh Mythol. the kingdom of the dead.

avant-garde /ˌævɒ̃ˈɡɑːd/ n. & adj. ● n. pioneers or innovators esp. in art and literature. ● adj. (of art, ideas, etc.) new, progressive. □ **avant-gardism** n. **avant-gardist** n. [F, = vanguard]

Avar /ˈævɑː(r)/ n. & adj. ● n. **1** a member of a people prominent in SE Europe in the 6th–9th centuries, having a kingdom that at one time extended from the Black Sea to the Adriatic. They were conquered by Charlemagne (791–9). **2** the Turkic language of these people. ● adj. of, relating to, or denoting the Avars. [Avar]

avarice /ˈævərɪs/ n. extreme greed for money or gain; cupidity. □ **avaricious** /ˌævəˈrɪʃəs/ adj. **avariciously** adv. **avariciousness** n. [ME f. OF f. L *avaritia* f. *avarus* greedy]

avast /əˈvɑːst/ int. Naut. stop, cease. [Du. *houd vast* hold fast]

avatar /ˈævəˌtɑː(r)/ n. **1** Hinduism the descent of a deity or released soul to earth in bodily form. **2** incarnation; manifestation. **3** a manifestation or phase. [Skr. *avatāra* descent f. *ava* down + *tar-* pass over]

avaunt /ə'vɔːnt/ *int. archaic* begone. [ME f. AF f. OF *avant* ult. f. L *ab* from + *ante* before]

Ave. *abbr.* Avenue.

ave /'ɑːveɪ, 'ɑːvɪ/ *int. & n.* ● *int.* **1** welcome. **2** farewell. ● *n.* **1** (**Ave**; in full **Ave Maria**) a prayer to the Virgin Mary, the opening line from Luke 1:28. Also called *Hail Mary*. **2** a shout of welcome or farewell. [ME f. L, 2nd sing. imper. of *avere* fare well]

Avebury /'eɪvbərɪ/ a village in Wiltshire, site of one of Britain's major henge monuments of the late neolithic period. The monument consists of a massive bank and ditch about 425 m (1,400 ft) across, containing the largest known stone circle, with two smaller circles and other stone settings within it. It is the centre of a complex ritual landscape that also contains a stone avenue, chambered tombs, Silbury Hill, and various other monuments.

avenge /ə'vendʒ/ *v.tr.* **1** inflict retribution on behalf of (a person, a violated right, etc.). **2** take vengeance for (an injury). □ **be avenged** avenge oneself. □ **avenger** *n.* [ME f. OF *avengier* f. *à* to + *vengier* f. L *vindicare* vindicate]

avens /'æv(ə)nz/ *n.* a rosaceous plant of the genus *Geum*. □ **mountain avens** a related alpine plant, *Dryas octopetala*. [ME f. OF *avence* (med.L *avencia*), of unkn. orig.]

aventurine /ə'ventjʊˌriːn/ *n. Mineral.* **1** brownish glass or mineral containing sparkling gold-coloured particles usu. of copper or gold. **2** a variety of spangled quartz resembling this. [F f. It. *avventurino* f. *avventura* chance (because of its accidental discovery)]

avenue /'ævəˌnjuː/ *n.* **1 a** a broad road or street, often with trees at regular intervals along its sides. **b** a tree-lined approach to a country house. **2** a way of approaching or dealing with something (*explored every avenue to find an answer*). [F, fem. past part. of *avenir* f. L *advenire* come to]

aver /ə'vɜː(r)/ *v.tr.* (**averred, averring**) *formal* assert, affirm. [ME f. OF *averer* (as AD-, L *verus* true)]

average /'ævərɪdʒ/ *n., adj., & v.* ● *n.* **1 a** the usual amount, extent, or rate. **b** the ordinary standard. **2** an amount obtained by dividing the total of given amounts by the number of amounts in the set; the mean. **3** *Law* the distribution of loss resulting from damage to a ship or cargo. ● *adj.* **1 a** usual, typical. **b** mediocre, undistinguished. **2** estimated or calculated by average. ● *v.tr.* **1** amount on average to (*the sale of the product averaged one hundred a day*). **2** do on average (*averages six hours' work a day*). **3 a** estimate the average of. **b** estimate the general standard of. □ **average adjustment** *Law* the apportionment of average. **average out** result in an average. **average out at** result in an average of. **batting average 1** *Cricket* a batsman's runs scored per completed innings. **2** *Baseball* a batter's safe hits per time at bat. **bowling average** *Cricket* a bowler's conceded runs per wicket taken. **law of averages** the principle that if one of two extremes occurs the other will also tend to so as to maintain the normal average. **on** (or **on an**) **average** as an average rate or estimate. □ **averagely** *adv.* [F *avarie* damage to ship or cargo (see sense 3), f. It. *avaria* f. Arab. *'awārīya* damaged goods f. *'awār* damage at sea, loss: -*age* after *damage*]

averment /ə'vɜːmənt/ *n.* a positive statement; an affirmation, esp. *Law* one with an offer of proof. [ME f. AF, OF *aver(r)ement* (as AVER, -MENT)]

Avernus /ə'vɜːnəs/ a lake near Naples in Italy, which fills the crater of an extinct volcano. It was described by Virgil and other Latin writers as the entrance to the underworld.

Averroës /ə'verəʊˌiːz/ (Arabic name ibn-Rushd) (c.1126–98), Spanish-born Islamic philosopher, judge, and physician. His extensive body of work includes writings on jurisprudence, science, philosophy, and religion. His most significant works were his commentaries on Aristotle, which, through a reliance on Neoplatonism, interpreted Aristotle's writings in such a way as to make them consistent with Plato's, and sought to reconcile the Greek and Arabic philosophical traditions. These commentaries exercised a strong and controversial influence on the succeeding centuries of Western philosophy and science.

averse /ə'vɜːs/ *predic.adj.* (usu. foll. by *to*; also foll. by *from*) opposed, disinclined (*was not averse to helping me*). ¶ Construction with *to* is now more common. [L *aversus* (as AVERT)]

aversion /ə'vɜːʃ(ə)n/ *n.* **1** (usu. foll. by *to*, *from*, *for*) a dislike or unwillingness (*has an aversion to hard work*). **2** an object of dislike (*my pet aversion*). □ **aversion therapy** therapy designed to make a patient averse to an existing habit. [F *aversion* or L *aversio* (as AVERT, -ION)]

avert /ə'vɜːt/ *v.tr.* (often foll. by *from*) **1** turn away (one's eyes or thoughts). **2** prevent or ward off (an undesirable occurrence). □ **avertable** *adj.* **avertible** *adj.* [ME f. L *avertere* (as AB-, *vertere vers-* turn): partly f. OF *avertir* f. Rmc]

Avesta /ə'vestə/ the sacred writings of Zoroastrianism, compiled in the 4th century. (See ZEND.) [Pers.]

Avestan /ə'vestən/ *adj. & n.* ● *adj.* of or relating to the Avesta or to the ancient Iranian language in which it is written, closely related to Vedic Sanskrit. ● *n.* the Avestan language.

avian /'eɪvɪən/ *adj.* of or relating to birds. [L *avis* bird]

aviary /'eɪvɪərɪ/ *n.* (pl. **-ies**) a large enclosure or building for keeping birds. [L *aviarium* (as AVIAN, -ARY¹)]

aviate /'eɪvɪˌeɪt/ *v.* **1** *intr.* fly in an aeroplane. **2** *tr.* pilot (an aeroplane). [back-form. f. AVIATION]

aviation /ˌeɪvɪ'eɪʃ(ə)n/ *n.* **1** the skill or practice of operating aircraft. **2** aircraft manufacture. [F f. L *avis* bird]

aviator /'eɪvɪˌeɪtə(r)/ *n.* (fem. **aviatrix** /'eɪvɪəˌtrɪks/) an airman or airwoman. [F *aviateur* f. L *avis* bird]

Avicenna /ˌævɪ'senə/ (Arabic name ibn-Sina) (980–1037), Persian-born Islamic philosopher and physician. His surviving works include treatises on philosophy, medicine, and religion. His philosophical system, while drawing heavily on Aristotle, is closer to Neoplatonism, and was the major influence on the development of 13th-century scholasticism. His *Canon of Medicine*, which combined his own knowledge with Roman and Arabic medicine, was a standard medical text in the medieval world. He also produced a philosophical encyclopedia, *The Recovery*.

aviculture /'eɪvɪˌkʌltʃə(r)/ *n.* the rearing and keeping of birds. □ **aviculturist** /ˌeɪvɪ'kʌltʃərɪst/ *n.* [L *avis* bird, after AGRICULTURE]

avid /'ævɪd/ *adj.* (usu. foll. by *of*, *for*) eager, greedy. □ **avidly** *adv.* [F *avide* or L *avidus* f. *avere* crave]

avidity /ə'vɪdɪtɪ/ *n.* **1** eagerness, greed. **2** *Biochem.* the overall strength of binding between an antibody and an antigen.

avifauna /'eɪvɪˌfɔːnə/ *n.* birds of a region or country collectively. [L *avis* bird + FAUNA]

Avignon /'ævɪːˌnjɒn/ a city on the Rhône in SE France; pop. (1990) 89,440. From 1309 until 1377 it was the residence of the popes during their exile from Rome, becoming papal property in 1348. After the papal court had returned to Rome two successive antipopes re-established a rival papal court in Avignon, which lasted until 1408. The city remained in papal hands until the French Revolution.

Ávila, Teresa of see TERESA OF ÁVILA, ST.

avionics /ˌeɪvɪ'ɒnɪks/ *n.pl.* (treated as *sing.*) electronics as applied to aviation.

avitaminosis /eɪˌvɪtəmɪ'nəʊsɪs/ *n. Med.* a condition resulting from a deficiency of one or more vitamins.

avizandum /ˌævɪ'zændəm/ *n. Sc. Law* a period of time for further consideration of a judgement. [med.L, gerund of *avizare* consider (as ADVISE)]

avocado /ˌævə'kɑːdəʊ/ *n.* (pl. **-os**) **1** (in full **avocado pear**) a pear-shaped fruit with rough leathery skin, a smooth oily edible flesh, and a large stone. Also called *alligator pear*. **2** the tropical evergreen Central American tree, *Persea americana*, of the laurel family, bearing this fruit. **3** the light green colour of the flesh of this fruit. [Sp., = advocate (substituted for Aztec *ahuacatl*)]

avocation /ˌævə'keɪʃ(ə)n/ *n.* **1** a minor occupation. **2** *colloq.* a vocation or calling. [L *avocatio* f. *avocare* call away]

avocet /'ævəˌset/ *n.* a long-legged wading bird of the genus *Recurvirostra*, with a long slender upward-curved bill and usu. black and white plumage. [F *avocette* f. It. *avosetta*]

Avogadro /ˌævə'gɑːdrəʊ/, Amedeo (1776–1856), Italian chemist and physicist. He is best known for his hypothesis (formulated in 1811), from which it became relatively simple to derive both molecular weights and a system of atomic weights.

Avogadro's hypothesis *n.* (also **Avogadro's law**) *Chem.* a law stating that equal volumes of gases at the same temperature and pressure contain equal numbers of molecules.

Avogadro's number *n.* (also **Avogadro's constant**) *Chem.* the number of molecules in one mole of a substance, equal to 6.023×10^{23}.

avoid /ə'vɔɪd/ *v.tr.* **1** refrain or keep away from (a thing, person, or action). **2** escape; evade. **3** *Law* **a** nullify (a decree or contract). **b** quash (a sentence). □ **avoidable** *adj.* **avoidably** *adv.* **avoidance** *n.* **avoider** *n.* [AF *avoider*, OF *evuider* clear out, get quit of, f. *vuide* empty, VOID]

avoirdupois /ˌævədəˈpɔɪz, ˌævwɑːˈdjuːˈpwʌ/ *n.* **1** (in full **avoirdupois weight**) a system of weights based on a pound of 16 ounces or 7,000 grains. **2** weight, heaviness. [ME f. OF *aveir de peis* goods of weight f. *aveir* f. L *habere* have + *peis* (see POISE[1])]

Avon /ˈeɪv(ə)n/ *n.* **1 a** a river of central England which rises near the Leicestershire–Northamptonshire border and flows 154 km (96 miles) south-west through Stratford to the River Severn. **b** a river of SW England which rises near the Gloucestershire–Wiltshire border and flows 121 km (75 miles) through Bath and Bristol to the River Severn. **2** a county of SW England, formed in 1974 from parts of north Somerset and Gloucestershire; county town, Bristol.

avouch /əˈvaʊtʃ/ *v.tr. & intr. archaic* or *rhet.* guarantee, affirm, confess. □ **avouchment** *n.* [ME f. OF *avochier* f. L *advocare* (as AD-, *vocare* call)]

avow /əˈvaʊ/ *v.tr.* **1** admit, confess. **2 a** *refl.* admit that one is (*avowed himself an author*). **b** (as **avowed** *adj.*) admitted (*the avowed author*). □ **avowal** *n.* **avowedly** /əˈvaʊɪdlɪ/ *adv.* [ME f. OF *avouer* acknowledge f. L *advocare* (as AD-, *vocare* call)]

avulsion /əˈvʌlʃ(ə)n/ *n.* **1** a tearing away. **2** *Law* a sudden removal of land by a flood etc. to another person's estate. [F *avulsion* or L *avulsio* f. *avellere avuls-* pluck away]

avuncular /əˈvʌŋkjʊlə(r)/ *adj.* like or of an uncle; kind and friendly, esp. towards a younger person. [L *avunculus* maternal uncle, dimin. of *avus* grandfather]

AWACS /ˈeɪwæks/ *n.* an airborne long-range radar system for detecting enemy aircraft. [abbr. of *airborne warning and control system*]

Awadh see OUDH.

await /əˈweɪt/ *v.tr.* **1** wait for. **2** (of an event or thing) be in store for (*a surprise awaits you*). [ME f. AF *awaitier*, OF *aguaitier* (as AD-, *waitier* WAIT)]

awake /əˈweɪk/ *v. & adj.* ● *v.* (*past* **awoke** /əˈwəʊk/; *past part.* **awoken** /əˈwəʊkən/) **1** *intr.* **a** cease to sleep. **b** become active. **2** *intr.* (foll. by *to*) become aware of. **3** *tr.* rouse, esp. from sleep. ● *predic.adj.* **1 a** not asleep. **b** vigilant. **2** (foll. by *to*) aware of. [OE *āwæcnan, āwacian* as A-[2], WAKE[1])]

awaken /əˈweɪkən/ *v.* **1** *tr. & intr.* = AWAKE *v.* **2** *tr.* (often foll. by *to*) make aware. [OE *onwæcnan* etc. (as A-[2], WAKEN)]

award /əˈwɔːd/ *v. & n.* ● *v.tr.* **1** give or order to be given as a payment, compensation, or prize (*awarded him a knighthood; was awarded damages*). **2** grant, assign. ● *n.* **1 a** a payment, compensation, or prize awarded. **b** the act or process of awarding. **2** a judicial decision. □ **awarder** *n.* [ME f. AF *awarder*, ult. f. Gmc: see WARD]

aware /əˈweə(r)/ *predic.adj.* **1** (often foll. by *of*, or *that* + clause) conscious; not ignorant; having knowledge. **2** well-informed. ¶ Also found in *attrib.* use in sense 2, as in *a very aware person*; this is *disp.* □ **awareness** *n.* [OE *gewær*]

awash /əˈwɒʃ/ *predic.adj.* **1** level with the surface of water, so that it just washes over. **2** carried or washed by the waves; flooded or as if flooded.

away /əˈweɪ/ *adv., adj., & n.* ● *adv.* **1** to or at a distance from the place, person, or thing in question (*go away; give away; look away; they are away; 5 miles away*). **2** towards or into non-existence (*sounds die away; explain it away; idled their time away*). **3** constantly, persistently, continuously (*work away; laugh away*). **4** without hesitation (*ask away*). ● *adj. Sport* played on an opponent's ground etc. (*away match; away win*). ● *n. Sport* an away match or win. □ **away with** (as *imper.*) take away; let us be rid of. [OE *onweg, aweg* on one's way f. A[2] + WAY]

AWB *abbr.* Afrikaner Weerstandsbeweging, the Afrikaner Resistance Movement, an extreme right-wing white political party in South Africa violently opposed to majority rule. Supporters of the party, which uses a symbol resembling a swastika, have been involved in violence against black activists and white liberals. The AWB boycotted the 1994 multiracial elections.

awe /ɔː/ *n. & v.* ● *n.* reverential fear or wonder (*stand in awe of*). ● *v.tr.* inspire with awe. □ **awe-inspiring** causing awe or wonder; amazing, magnificent. □ **awe-inspiringly** *adv.* [ME *age* f. ON *agi* f. Gmc]

aweary /əˈwɪərɪ/ *predic.adj. poet.* (often foll. by *of*) weary. [aphetic *a* + WEARY]

aweigh /əˈweɪ/ *predic.adj. Naut.* (of an anchor) clear of the sea or river bed; hanging. [A[2] + WEIGH[1]]

awesome /ˈɔːsəm/ *adj.* **1** inspiring awe. **2** *sl.* excellent, superb. □ **awesomely** *adv.* **awesomeness** *n.* [AWE + -SOME[1]]

awestricken /ˈɔːˌstrɪk(ə)n/ *adj.* (also **awestruck** /-strʌk/) struck or affected by awe.

awful /ˈɔːfʊl/ *adj.* **1** *colloq.* **a** unpleasant or horrible (*awful weather*). **b** poor in quality; very bad (*has awful writing*). **c** (*attrib.*) excessive; remarkably large (*an awful lot of money*). **2** *poet.* inspiring awe. □ **awfulness** *n.* [AWE + -FUL]

awfully /ˈɔːfəlɪ, ˈɔːflɪ/ *adv.* **1** *colloq.* in an unpleasant, bad, or horrible way (*he played awfully*). **2** *colloq.* very (*she's awfully pleased; thanks awfully*). **3** *poet.* reverently.

awhile /əˈwaɪl/ *adv.* for a short time. [OE *āne hwīle* a while]

awkward /ˈɔːkwəd/ *adj.* **1** ill-adapted for use; causing difficulty in use. **2** clumsy or bungling. **3 a** embarrassed, ill at ease (*felt awkward about it*). **b** embarrassing (*an awkward situation*). **4** difficult to deal with (*an awkward customer*). □ **the awkward age** adolescence. □ **awkwardly** *adv.* **awkwardness** *n.* [obs. *awk* backhanded, untoward (ME f. ON *afugr* turned the wrong way) + -WARD]

awl /ɔːl/ *n.* a small pointed tool used for piercing holes, esp. in leather. [OE *æl*]

awn /ɔːn/ *n. Bot.* a stiff bristle growing from the grain-sheath of grasses, or terminating a leaf etc. □ **awned** *adj.* [ME f. ON *ögn*]

awning /ˈɔːnɪŋ/ *n.* a sheet of canvas or similar material stretched on a frame and used to shade a shop window, doorway, ship's deck, or other area from the sun or rain. [17th c. (Naut.): orig. uncert.]

awoke *past* of AWAKE.

awoken *past part.* of AWAKE.

AWOL /ˈeɪwɒl/ *abbr. colloq.* (orig. *Mil.*) absent without leave.

awry /əˈraɪ/ *adv. & adj.* ● *adv.* **1** crookedly or askew. **2** improperly or amiss. ● *predic.adj.* crooked; deviant or unsound (*his theory is awry*). □ **go awry** go or do wrong. [ME f. A[2] + WRY]

axe /æks/ *n. & v.* (*US* **ax**) ● *n.* **1** a chopping-tool, usu. of iron with a steel edge and wooden handle. **2** the drastic cutting or elimination of expenditure, staff, etc. ● *v.tr.* (**axing**) **1** cut (esp. costs or services) drastically. **2** remove or dismiss. □ **axe-breaker** a hard-wooded Australian tree. **an axe to grind** private ends to serve. [OE *æx* f. Gmc]

axel /ˈæks(ə)l/ *n.* a jumping movement in skating, similar to a loop (see LOOP *n.* 7) but from one foot to the other. [*Axel* R. Paulsen, Norw. skater (1885–1938)]

axes *pl.* of AXIS[1].

axial /ˈæksɪəl/ *adj.* **1** forming or belonging to an axis. **2** round an axis (*axial rotation; axial symmetry*). □ **axially** *adv.* **axiality** /ˌæksɪˈælɪtɪ/ *n.*

axil /ˈæksɪl/ *n. Bot.* the upper angle between a leaf and the stem it springs from, or between a branch and the trunk. [L *axilla*: see AXILLA]

axilla /ækˈsɪlə/ *n.* (*pl.* **axillae** /-liː/) **1** *Anat.* the armpit. **2** *Bot.* an axil. [L, = armpit, dimin. of *ala* wing]

axillary /ækˈsɪlərɪ/ *adj.* **1** *Anat.* of or relating to the armpit. **2** *Bot.* in or growing from the axil.

axiom /ˈæksɪəm/ *n.* **1** an established or widely accepted principle. **2** esp. *Geom.* a self-evident truth. [F *axiome* or L *axioma* f. Gk *axiōma axiōmat-* f. *axios* worthy]

axiomatic /ˌæksɪəˈmætɪk/ *adj.* **1** self-evident. **2** relating to or containing axioms. □ **axiomatically** *adv.* [Gk *axiōmatikos* (as AXIOM)]

axis[1] /ˈæksɪs/ *n.* (*pl.* **axes** /-siːz/) **1 a** an imaginary line about which a body rotates or about which a plane figure is conceived as generating a solid. **b** a line which divides a regular figure symmetrically. **2** *Math.* a fixed reference line for the measurement of coordinates etc. **3** *Bot.* the central column of an inflorescence or other growth. **4** *Anat.* the second cervical vertebra. **5** *Physiol.* the central part of an organ or organism. **6 a** an agreement or alliance between two or more countries forming a centre for an eventual larger grouping of nations sharing an ideal or objective. **b** (also **Axis**) = AXIS POWERS. [L, = axle, pivot]

axis[2] /ˈæksɪs/ *n.* (*pl.* same) a white-spotted deer, *Cervus axis*, of southern Asia. Also called *chital*. [L]

Axis Powers (also **Axis**) the group of nations opposed to the Allies in the Second World War, initially comprising Germany and Italy and later extended to include Japan and other countries. Germany and Italy joined in a political association in 1936, forming a military alliance in 1939. The Axis collapsed with the fall of Mussolini and the surrender of Italy in 1943.

axle /ˈæks(ə)l/ *n.* a rod or spindle (either fixed or rotating) on which a wheel or group of wheels is fixed. [orig. *axle-tree* f. ME *axel-tre* f. ON *öxull-tré*]

Axminster /ˈæksˌmɪnstə(r)/ *n.* (in full **Axminster carpet**) a kind of machine-woven patterned carpet with a cut pile. [*Axminster* in Devon, noted since the 18th c. for the production of carpets]

axolotl /ˈæksəˌlɒt(ə)l/ *n.* an aquatic newtlike salamander, *Ambystoma mexicanum*, from Mexico, which in natural conditions retains its larval form for life but is able to breed. [Nahuatl f. *atl* water + *xolotl* servant]

axon /ˈæksɒn/ *n. Anat. & Zool.* a long threadlike part of a nerve cell, conducting impulses from the cell body. [mod.L f. Gk *axōn* axis]

axonometric /ˌæksənəˈmetrɪk/ *adj.* (of a pictorial representation) using an orthographic projection of the object on a plane inclined to each of the three principal axes of the object.

Axum see AKSUM.

ay var. of AYE[1].

Ayacucho /ˌaɪəˈkuːtʃəʊ/ a city in the Andes of south central Peru; pop. (est. 1990) 101,600. The modern city was founded by Francisco Pizarro in 1539. At the battle of Ayacucho in Dec. 1824, the Spanish forces were defeated and the independence of Peru secured.

ayah /ˈaɪə/ *n.* a nurse or maidservant, esp. of Europeans in India and other former British territories abroad. [Anglo-Ind. f. Port. *aia* nurse]

ayatollah /ˌaɪəˈtɒlə/ *n.* a Shiite religious leader in Iran. [Pers. f. Arab., = token of God]

Ayatollah Khomeini see KHOMEINI.

Ayckbourn /ˈeɪkbɔːn/, Alan (b.1939), English dramatist. *Relatively Speaking* (1967) was his first major success, and was followed by the domestic farce *Absurd Person Singular* (1973) and the trilogy *The Norman Conquests* (1974). A prolific and successful writer of comedies dealing with suburban and middle-class life, in his later plays he often explores darker themes and blurs the distinction between farce and tragedy. Other works include *Way Upstream* (1982) and *A Chorus of Disapproval* (1985). Most of his plays are premièred at Scarborough's Stephen Joseph Theatre in the Round, where he became artistic director in 1971.

aye[1] /aɪ/ *adv. & n.* (also **ay**) ● *adv.* **1** *archaic* or *dial.* yes. **2** (in voting) I assent. **3** (as **aye aye**) *Naut.* a response accepting an order. ● *n.* an affirmative answer or assent, esp. in voting. □ **the ayes have it** the affirmative votes are in the majority. [16th c.: prob. f. 1st pers. pron. expressing assent]

aye[2] /eɪ/ *adv. archaic* ever, always. □ **for aye** for ever. [ME f. ON *ei, ey* f. Gmc]

aye-aye /ˈaɪaɪ/ *n.* an arboreal nocturnal lemur, *Daubentonia madagascariensis*, native to Madagascar. [F f. Malagasy *aiay*]

Ayer /eə(r)/, Sir A(lfred) J(ules) (1910–89), English philosopher. In Vienna in 1932, he attended the meetings of the Vienna Circle, becoming a notable proponent of logical positivism; his book *Language, Truth, and Logic* (1936) was one of the most successful philosophical works of the 20th century. In moral philosophy he was disinclined to defend any specific theory of moral judgement, holding that for all practical purposes a tolerant utilitarianism was the soundest basis for private conduct and public morality.

Ayers Rock /eəz/ (called in Aboriginal *Uluru*) a red rock mass in Northern Territory, Australia, south-west of Alice Springs. The largest monolith in the world, it is 348 m (1,143 ft) high and about 9 km (6 miles) in circumference. It is named after Sir Henry Ayers, Premier of South Australia in 1872–3.

Áyios Nikólaos see AGHIOS NIKOLAOS.

Aylesbury[1] /ˈeɪlzbəri/ a town in south central England, the county town of Buckinghamshire; pop. (1985) 50,000.

Aylesbury[2] /ˈeɪlzbəri/ *n.* (pl. **Aylesburys**) a breed of large white domestic duck. [AYLESBURY[1]]

Aymara /ˈaɪməˌrɑː/ *n. & adj.* ● *n.* (pl. same or **Aymaras**) **1** a member of a native people inhabiting the plateau lands of Bolivia and Peru near Lake Titicaca. **2** the language of this people. ● *adj.* of or relating to the Aymara or their language. [Bolivian Sp.]

Ayrshire[1] /ˈeəʃɪə(r)/ *n.* a former county of SW Scotland, on the Firth of Clyde. It became a part of Strathclyde region in 1975.

Ayrshire[2] /ˈeəʃɪə(r)/ *n.* a mainly white breed of dairy cattle originating in Ayrshire.

Ayub Khan /ˌaɪjuːb ˈkɑːn/, Muhammad (1907–74), Pakistani soldier and statesman, President 1958–69. After independence he became the first Commander-in-Chief of the country's army (1951–8) and served as Minister of Defence 1954–5, taking over the presidency shortly after the declaration of martial law. His term of office saw the introduction of a new constitution and the lifting of martial law in 1962, but civil liberties were curtailed. Opposition to his foreign policy with regard to India and his increasingly repressive style of government led to widespread disorder and he was ultimately forced to resign.

ayurveda /ˌɑːjəˈveɪdə, -ˈviːdə/ *n.* a traditional Hindu system of medicine practised esp. in India and parts of the Far East, based on ideas of balance in bodily systems and placing emphasis on diet, herbal treatment, yogic breathing, etc. □ **ayurvedic** *adj.* [Skr. *āyur-veda* science of life (see VEDA)]

AZ *abbr. US* Arizona (in official postal use).

Azad Kashmir /ˌɑːzæd kæʃˈmɪə(r)/ an autonomous state in NE Pakistan, formerly part of Kashmir; administrative centre, Muzzafarabad. It was established in 1949 after Kashmir was split as a result of the partition of India. The name means 'Free Kashmir'.

azalea /əˈzeɪlɪə/ *n.* a flowering deciduous ericaceous shrub of the genus *Rhododendron*, with large pink, purple, white, or yellow flowers. [mod.L f. Gk, fem. of *azaleos* dry (from the dry soil in which it was believed to flourish)]

Azania /əˈzeɪnɪə/ an alternative name for South Africa, proposed by some supporters of majority rule for the country. □ **Azanian** *n. & adj.*

azeotrope /əˈziːətrəʊp/ *n. Chem.* a mixture of liquids in which the boiling-point remains constant during distillation, at a given pressure, without change in composition. □ **azeotropic** /əˌziːəˈtrəʊpɪk, -ˈtrɒpɪk/ *adj.* [A-[1] + Gk *zeō* boil + *tropos* turning]

Azerbaijan /ˌæzəbaɪˈdʒɑːn/ a country in the Caucasus of SE Europe, on the western shore of the Caspian Sea; pop. (est. 1991) 7,219,000; languages, Azerbaijani (official), Russian; capital, Baku. Historically, the name Azerbaijan referred to a larger Transcaucasian region which formed part of Persia. The northern part of this was ceded to Russia in the early 19th century, the southern part remaining a region of NW Iran. Russian Azerbaijan was absorbed into the Soviet Union in 1922 when it became part of the Transcaucasian Soviet Federated Socialist Republic, becoming a separate constituent republic of the Soviet Union in 1936. It gained full independence on the breakup of the USSR in 1991. Azerbaijan contains the predominantly Armenian region of Nagorno-Karabakh, over which open conflict with Armenia broke out in 1988. The Azeri autonomous republic of Naxçivan forms an enclave within the Republic of Armenia and is similarly the subject of armed conflict.

Azerbaijani /ˌæzəbaɪˈdʒɑːnɪ/ *adj. & n.* ● *n.* (pl. **Azerbaijanis**) **1** a member of a Turkic people of Azerbaijan. **2** the language of this people. ● *adj.* of or relating to the Azerbaijanis or their language.

Azeri /əˈzeərɪ/ *n. & adj.* ● *n.* (pl. **Azeris**) **1** a member of a Turkic people living mainly in Azerbaijan, Armenia, and northern Iran. **2** the language of this people. ● *adj.* of or relating to this people or their language. [Turk. *azerī* f. Pers., = fire]

azide /ˈeɪzaɪd/ *n. Chem.* any compound containing the radical –N₃. [AZO- + -IDE]

azidothymidine /ˌeɪzɪdəʊˈθaɪmɪˌdiːn, eɪˌzaɪdəʊ-/ *n.* the drug zidovudine (cf. AZT).

Azikiwe /ˌɑːzɪˈkiːweɪ/, (Benjamin) Nnamdi (b.1904), Nigerian statesman, President 1963–6. Azikiwe founded (1944) the anti-colonial National Council of Nigeria and the Cameroons, a gathering of forty political, labour, and educational groups. He was the first Governor-General of an independent Nigeria (1960–3) and its first President when it became a republic. When his civilian government was ousted by a military coup in 1966, Azikiwe joined the Biafran secessionist government. In 1978, after the reunification of Nigeria, he founded the Nigerian People's Party and was its leader until 1983.

Azilian /əˈzɪlɪən/ *adj. & n. Archaeol.* of, relating to, or denoting an early mesolithic culture in Europe, succeeding the Magdalenian and dated to 10,000–8000 BC. It is characterized by flat bone harpoons, painted pebbles, and microliths. [f. Mas d'*Azil*, the type-site, a cave in the French Pyrenees]

azimuth /ˈæzɪməθ/ *n.* **1** the angular distance from a north or south point of the horizon to the intersection with the horizon of a vertical circle passing through a given celestial object. **2** the horizontal angle or direction of a compass bearing. [ME f. OF *azimut* f. Arab. *as-sumūt* f. *al* the + *sumūt* pl. of *samt* way, direction]

azimuthal /ˌæzɪˈmjuːθəl/ *adj.* of or relating to the azimuth. □ **azimuthal projection** a map projection in which a region of the earth is projected on to a plane tangential to the surface, usually at the pole or equator.

azine /ˈeɪziːn/ *n. Chem.* any organic compound with two or more nitrogen atoms in a six-atom ring. [AZO- + -INE[4]]

azo- /ˈæzəʊ, ˈeɪzəʊ/ *prefix Chem.* containing two adjacent nitrogen atoms between carbon atoms. [F *azote* nitrogen f. Gk *azōos* without life]

azo dye /ˈeɪzəʊ/ *n.* a dye whose molecule contains two adjacent nitrogen atoms between two carbon atoms. [AZO-]

azoic /əˈzəʊɪk/ *adj.* **1** having no trace of life. **2** *Geol.* (of an age etc.) having left no organic remains. [Gk *azōos* without life]

Azores /əˈzɔːz, əˈzɔːrəʃ/ a group of volcanic islands in the Atlantic Ocean, west of Portugal, in Portuguese possession but partially autonomous; pop. (1991) 241,590; capital, Ponta Delgada.

Azov, Sea of /ˈæzɒf/ an inland sea of southern Russia and Ukraine, separated from the Black Sea by the Crimea and linked to it by a narrow strait.

Azrael /ˈæzreɪl/ *Jewish & Islamic Mythol.* the angel who severs the soul from the body at death. [Heb., = help of God]

AZT /ˌeɪzedˈtiː/ *n.* the drug zidovudine. [chem. name AZIDOTHYMIDINE]

Aztec /ˈæztek/ *n. & adj.* ● *n.* **1** a member of a native people dominant in Mexico before the Spanish conquest of the 16th century. (*See note below.*) **2** the language of the Aztecs (see also NAHUATL). ● *adj.* of or relating to the Aztecs or their language. [F *Aztèque* or Sp. *Azteca* f. Nahuatl *aztecatl* men of the north]

- The Aztecs, a Chichimec people, arrived in the central valley of Mexico after the collapse of the Toltec civilization in the 12th century, and by the 16th century they commanded a territory of tribute-paying states that covered most of the central and southern part of present-day Mexico. Their rulers (the last and most famous of whom was Montezuma) tended to be despotic, and captives taken in war were offered as sacrifices to the Aztec gods. They are also famous for their rich and elaborate civilization: when the Spaniards under Cortés arrived they found a society centred on the city of Tenochtitlán, which boasted vast pyramids, temples, and palaces with fountains.

azuki var. of ADZUKI.

azure /ˈæʒə(r), ˈæzjə(r), ˈeɪʒə(r), ˈeɪzjə(r)/ *n. & adj.* ● *n.* **1 a** a deep sky-blue colour. **b** *Heraldry* blue. **2** *poet.* the clear sky. ● *adj.* **1 a** of the colour azure. **b** *Heraldry* blue. **2** *poet.* serene, untroubled. [ME f. OF *asur, azur,* f. med.L *azzurum, azolum* f. Arab. *al* the + *lāzaward* f. Pers. *lāžward* lapis lazuli]

azygous /ˈæzɪɡəs/ *adj. & n. Anat.* ● *adj.* (of any organic structure) single, not existing in pairs. ● *n.* an organic structure occurring singly. [Gk *azugos* unyoked f. *a-* not + *zugon* yoke]

Az Zarqa see ZARQA.

Bb

B¹ /biː/ n. (also **b**) (pl. **Bs** or **B's**) **1** the second letter of the alphabet. **2** Mus. the seventh note of the diatonic scale of C major. **3** the second hypothetical person or example. **4** the second highest class or category (of roads, academic marks, etc.). **5** (usu. **b**) Algebra the second known quantity. **6** a human blood type of the ABO system. □ **B film** (or **movie**) a supporting film in a cinema programme.

B² abbr. (also **B.**) **1** black (pencil-lead). **2** Chess bishop. **3** Blessed.

B³ symb. **1** Chem. the element boron. **2** Physics magnetic flux density. **3** bel(s).

b¹ abbr. (also **b.**) **1** born. **2** Cricket **a** bowled by. **b** bye. **3** billion.

b² symb. Physics barn.

BA abbr. **1** Bachelor of Arts. **2** British Academy. **3** British Airways.

Ba symb. Chem. the element barium.

BAA abbr. British Airports Authority.

baa /baː/ v. & n. ● v.intr. (**baas, baaed** or **baa'd**) (of a sheep) bleat. ● n. (pl. **baas**) the cry of a sheep or lamb. [imit.]

Baade /ˈbaːdə/ (Wilhelm Heinrich) Walter (1893–1960), German-born American astronomer. Using cepheid variable stars, he proved that the Andromeda galaxy was much further away than had been thought, which implied that the universe was much older and more extensive than had been earlier supposed. Baade also contributed to the understanding of the life cycles of stars, identifying several radio sources optically.

Baader-Meinhof Group /ˌbaːdəˈmaɪnhɒf/ see RED ARMY FACTION.

Baal /baːl/ (also **Bel** /bel/) a male fertility god whose cult was widespread in ancient Phoenician and Canaanite lands and was strongly resisted by the Hebrew prophets. The name is found as a prefix to place-names (e.g. Baalbek) and as the last element in Phoenician names such as Hannibal and Jezebel. [Heb. ba'al lord]

Baalbek /ˈbaːlbek/ a town in eastern Lebanon, site of the ancient city of Heliopolis. Its principal monuments date from the Roman period; they include the Corinthian temples of Jupiter and Bacchus and private houses with important mosaics.

BAAS abbr. British Association for the Advancement of Science.

baas /baːs/ n. S. Afr. boss, master (often as a form of address). [Du.: cf. BOSS¹]

baasskap /ˈbaːskaːp/ n. S. Afr. domination, esp. of non-whites by whites. [Afrik. f. baas master + -skap condition]

Baath /baːθ/ (in full **Baath Party**) (also **Ba'ath**) a pan-Arab socialist party founded in Syria in 1943. Different factions of the Baath Party hold power in Syria and Iraq. □ **Baathism** n. **Baathist** adj. & n. [Arab. ba'ṯ resurrection, renaissance]

baba /ˈbaːbaː/ n. (in full **rum baba**) a small rich sponge cake, usu. soaked in rum-flavoured syrup. [F f. Pol.]

babacoote /ˈbaːbəˌkuːt/ n. = INDRI. [Malagasy babakoto]

Babbage /ˈbæbɪdʒ/, Charles (1791–1871), English mathematician, inventor, and pioneer of machine computing. His interest in the compilation of accurate mathematical and astronomical tables led to his design for a mechanical computer or 'difference engine' (in which he was assisted by Byron's daughter, Ada Lovelace), which would both perform calculations and print the results. Because of practical and financial difficulties neither this machine nor a subsequent analytical engine was finished in Babbage's lifetime (though construction of one began in the 1990s). His analysis of the postal service led to the introduction of the penny post; he also invented the heliograph and the ophthalmoscope.

Babbitt¹ /ˈbæbɪt/, Milton (Byron) (b.1916), American composer and mathematician. His compositions developed from the twelve-note system of Schoenberg and Webern; his first twelve-note work was *Composition for Orchestra* (1941). He later pioneered the use of synthesizers in composition; his works using synthesizers include *Philomel* (1964) and *Canonic Form* (1983).

Babbitt² /ˈbæbɪt/ n. a materialistic, complacent businessman. □ **Babbittry** n. [George Babbitt, a character in the novel Babbitt (1922) by Sinclair Lewis]

babbitt /ˈbæbɪt/ n. **1** (in full **babbitt metal**) any of a group of soft alloys of tin, antimony, copper, and usu. lead, used for lining bearings etc. to diminish friction. **2** a bearing-lining made of this. [Isaac Babbitt, Amer. inventor (1799–1862)]

babble /ˈbæb(ə)l/ v. & n. ● v. **1** intr. **a** talk in an inarticulate or incoherent manner. **b** chatter excessively or irrelevantly. **c** (of a stream etc.) murmur, trickle. **2** tr. repeat foolishly; divulge through chatter. ● n. **1 a** incoherent speech. **b** foolish, idle, or childish talk. **2** the murmur of voices, water, etc. **3** background disturbance caused by interference from conversations on other telephone lines. □ **babblement** n. [ME f. MLG babbelen, or imit.]

babbler /ˈbæblə(r)/ n. **1** a chatterer. **2** a person who reveals secrets. **3** a passerine bird of the large family Timaliidae, with loud chattering voices.

babe /beɪb/ n. **1** literary a baby. **2** an innocent or helpless person (babes and sucklings; babes in the wood). **3** N. Amer. sl. a young woman (often as a form of address). [ME: imit. of child's ba, ba]

babel /ˈbeɪb(ə)l/ n. **1** a confused noise, esp. of voices. **2** a noisy assembly. **3** a scene of confusion. [ME f. Heb. Bāḇel BABYLON¹ f. Akkadian bāb ili gate of God]

Babel, Tower of see TOWER OF BABEL.

Babi /ˈbaːbɪ/ n. **1** an adherent of Babism. **2** = BABISM. [Pers. Bab-ed-Din, gate (= intermediary) of the Faith]

babirusa /ˌbaːbɪˈruːsə/ n. (also **babiroussa**) a wild pig, Babyrousa babyrussa, of the Malay Archipelago, with upturned tusks. [Malay]

Babism /ˈbaːbɪz(ə)m/ n. a religion founded in 1844 as an offshoot of Islam by the Persian Mirza Ali Muhammad of Shiraz (1819/20–50, known as 'the Bab'). It holds that Muhammad was not the last of the prophets, and that a new prophet or messenger of God will come. The Baha'i faith is derived from Babism.

baboon /bəˈbuːn/ n. **1** a large Old World monkey of the genus Papio or Mandrillus, having a long doglike snout, large teeth, and naked callosities on the buttocks. **2** an ugly or uncouth person. [ME f. OF babuin or med.L babewynus, of unkn. orig.]

Babruisk /bəˈbruːɪsk/ (also **Babruysk, Bobruisk, Bobruysk**) a river port in central Belarus, on the Berezina river south-east of Minsk; pop. (1990) 222,900.

babu /ˈbaːbuː/ n. (also **baboo**) Ind. **1** a title of respect, esp. to Hindus. **2** derog. formerly, an English-writing Indian clerk. [Hindi bābū]

Babur /ˈbaːbʊə(r)/ (born Zahir al-Din Muhammad) (1483–1530), descendant of Tamerlane and the first Mogul emperor of India c.1525–30. He invaded India c.1525 and conquered the territory from the Oxus to Patna. A Muslim, he instigated the policy of religious toleration towards his non-Muslim subjects which was continued by later Mogul emperors.

babushka /bəˈbuːʃkə, Russian ˈbabuʃkə/ n. **1** (in Russia) an old woman; a grandmother. **2** a headscarf tied under the chin. [Russ., = grandmother]

Babuyan Islands /ˌbaːbʊˈjaːn/ a group of twenty-four volcanic islands lying to the north of the island of Luzon in the northern Philippines.

baby /ˈbeɪbɪ/ n. & v. ● n. (pl. **-ies**) **1** a very young child or infant, esp. one not yet able to walk. **2** an unduly childish person (is a baby about injections). **3** the youngest member of a family, team, etc. **4** (often attrib.) **a** a young or newly born animal. **b** a thing that is small of its kind (baby car; baby rose). **5** sl. a young woman; a sweetheart (often as a form of address). **6** sl. a person or thing regarded with affection or familiarity. **7** one's own responsibility, invention, concern, achievement, etc., regarded in a personal way. ● v.tr. (**-ies, -ied**) **1** treat like a baby. **2** pamper. □ **baby boom** colloq. a temporary marked increase in the birth rate. **baby boomer** a person born during a baby boom, esp. after the Second World War. **baby-bouncer** Brit. a frame supported by elastic or springs, into which a child is harnessed to exercise its limbs. **Baby Buggy** (pl. **-ies**) **1** Brit. propr. a kind of child's collapsible pushchair. **2** N. Amer. a pram. **baby carriage** N. Amer. a pram. **baby grand** the smallest size of grand piano. **baby's breath** US a scented plant, esp. Gypsophila paniculata. **baby-snatcher** colloq. **1** a person who abducts a very young child. **2** = cradle-snatcher. **baby-talk** childish talk used by or to young children. **baby-walker** a wheeled frame in which a baby learns to walk. **carry** (or **hold**) **the baby** bear unwelcome responsibility. **throw away the baby with the bath-water** reject the essential with the inessential. □ **babyhood** n. [ME, formed as BABE, -Y²]

Babygro /ˈbeɪbɪˌɡrəʊ/ n. (pl. **-os**) propr. a kind of all-in-one stretch garment for babies. [BABY + GROW]

babyish /ˈbeɪbɪʃ/ adj. **1** childish, simple. **2** immature. □ **babyishly** adv. **babyishness** n.

Babylon¹ /ˈbæbɪlən/ an ancient city in Mesopotamia, which first came to prominence in the second millennium BC under Hammurabi, who made it the capital of Babylonia. The city (of which only ruins now remain) lay on the Euphrates and was noted for its luxury, its fortifications, and particularly for the Hanging Gardens (see HANGING GARDENS OF BABYLON). [f. Gk f. Heb. Bābel (see BABEL)]

Babylon² /ˈbæbɪlən/ derog. (among blacks, esp. Rastafarians) anything which represents the degenerate or oppressive aspect of white culture, specifically: **a** white society. **b** the representatives of this, esp. the police. **c** London. [BABYLON¹]

Babylonia /ˌbæbɪˈləʊnɪə/ an ancient region of Mesopotamia, formed when the kingdoms of Akkad in the north and Sumer in the south combined in the first half of the 2nd millennium BC. In the 14th century BC Babylonia was dominated by Assyria, formerly its dependency. With the decline of Assyrian power in the 7th century BC, the Chaldeans were able to take the Babylonian throne, which they held from 625 to 539 BC. Babylonia was finally conquered by Cyrus the Great of Persia in 539 BC. □ **Babylonian** adj. & n.

Babylonian Captivity the captivity of the Israelites in Babylon, lasting from their deportation by Nebuchadnezzar in 586 BC until their release by Cyrus the Great in 539 BC.

babysit /ˈbeɪbɪˌsɪt/ v.intr. (**-sitting**; past and past part. **-sat** /-ˌsæt/) look after a child or children while the parents are out. □ **babysitter** n.

Bacall /bəˈkɔːl/ Lauren (b.1924), American actress. After little success on stage, Bacall concentrated on her film career, playing opposite Humphrey Bogart in To Have or Have Not (1944). The partnership proved compelling; the pair married in 1945 and co-starred in a succession of box-office hits, including The Big Sleep (1946) and Key Largo (1948). After a less active period, she made a stage comeback in 1967 and returned to films with Murder on the Orient Express (1974).

Bacardi /bəˈkɑːdɪ/ n. (pl. **Bacardis**) propr. a West Indian rum produced orig. in Cuba. [name of the company producing it]

baccalaureate /ˌbækəˈlɔːrɪət/ n. **1** an examination intended to qualify successful candidates for higher education, esp. in France and many international schools (International Baccalaureate). **2** the university degree of bachelor. [F baccalauréat or med.L baccalaureatus f. baccalaureus bachelor]

baccarat /ˈbækəˌrɑː/ n. a gambling card-game played by punters in turn against the banker. [F]

baccate /ˈbækeɪt/ adj. Bot. **1** bearing berries. **2** of or like a berry. [L baccatus berried f. bacca berry]

Bacchae /ˈbækiː/ n.pl. the priestesses or female devotees of the Greek god Bacchus. [L f. Gk Bakkhai]

bacchanal /ˈbækən(ə)l, -æl/ n. & adj. ● n. **1** (also /ˌbækəˈnɑːl/) a wild and drunken revelry. **2** a drunken reveller. **3** a priest, worshipper, or follower of Bacchus. ● adj. **1** of or like Bacchus or his rites. **2** riotous, roistering. [L bacchanalis f. BACCHUS god of wine f. Gk Bakkhos]

Bacchanalia /ˌbækəˈneɪlɪə/ n.pl. **1** Rom. Antiq. the festival of Bacchus. **2** (**bacchanalia**) a drunken revel. □ **Bacchanalian** adj. & n. [L, neut. pl. of bacchanalis: see BACCHANAL]

bacchant /ˈbækənt/ n. & adj. ● n. (pl. **bacchants** or **bacchantes** /bəˈkæntiːz/; fem. **bacchante** /bəˈkænti/) **1** a priest, priestess, or (usu. female) devotee of Bacchus. **2** a drunken reveller. ● adj. **1** of or like Bacchus or his rites. **2** riotous, roistering. □ **bacchantic** /bəˈkæntɪk/ adj. [F bacchante f. L bacchari celebrate Bacchanal rites]

Bacchic /ˈbækɪk/ adj. = BACCHANAL adj. [L bacchicus f. Gk bakkhikos of Bacchus]

Bacchus /ˈbækəs/ Gk Mythol. an alternative name for DIONYSUS.

baccy /ˈbækɪ/ n. Brit. colloq. tobacco. [abbr.]

Bach /bɑːx/, Johann Sebastian (1685–1750), German composer. An exceptional and prolific composer, he was known in his own lifetime chiefly as an organist; it was not until almost a century after his death that his position as an outstanding representative of the German musical baroque was fully appreciated. His compositions range from violin concertos, suites, and the six Brandenburg Concertos (1720–1) to many clavier works and more than 250 sacred cantatas. His large-scale choral works include The Passion according to St John (1723), The Passion according to St Matthew (1729), and the Mass in B minor (1733–8); through these and other liturgical works Bach expressed his devout Protestant faith in the Lutheran tradition. Of his twenty children, his eldest son Wilhelm Friedemann Bach (1710–84) became an organist and composer, Carl Philipp Emanuel Bach (1714–88) wrote much church music, more than 200 keyboard sonatas, and a celebrated treatise on clavier-playing, and Johann Christian Bach (1735–82) became music-master to the British royal family and composed thirteen operas and many instrumental works.

bachelor /ˈbætʃələ(r)/ n. **1** an unmarried man. **2** a man or woman who has taken the degree of Bachelor of Arts or Science etc. **3** hist. a young knight serving under another's banner. □ **bachelor girl** an independent unmarried young woman. **bachelor's buttons** a double-flowered form of a plant, esp. of a buttercup. □ **bachelorhood** n. **bachelorship** n. [ME & OF bacheler aspirant to knighthood, of uncert. orig.]

bacillary /bəˈsɪlərɪ/ adj. relating to or caused by bacilli.

bacilliform /bəˈsɪlɪˌfɔːm/ adj. rod-shaped.

bacillus /bəˈsɪləs/ n. (pl. **bacilli** /-laɪ/) **1** any rod-shaped bacterium. **2** (usu. in pl.) any pathogenic bacterium. [LL, dimin. of L baculus stick]

back /bæk/ n., adv., v., & adj. ● n. **1 a** the rear surface of the human body from the shoulders to the hips. **b** the corresponding upper surface of an animal's body. **c** the spine (fell and broke his back). **d** the keel of a ship. **2 a** any surface regarded as corresponding to the human back, e.g. of the head or hand, or of a chair. **b** the part of a garment that covers the back. **3 a** the less active or visible or important part of something functional, e.g. of a knife or a piece of paper (write it on the back). **b** the side or part normally away from the spectator or the direction of motion or attention, e.g. of a car, house, or room (stood at the back). **4 a** a defensive player in field games. **b** this position. **5** (**the Backs**) the grounds of Cambridge colleges which back on to the River Cam. ● adv. **1** to the rear; away from what is considered to be the front (go back a bit; ran off without looking back). **2 a** in or into an earlier or normal position or condition (came back late; went back home; ran back to the car; put it back on the shelf). **b** in return (pay back). **3** in or into the past (back in June; three years back). **4** at a distance (stand back from the road). **5** in check (hold him back). **6** (foll. by of) N. Amer. behind (was back of the house). ● v. **1** tr. **a** help with moral or financial support. **b** bet on the success of (a horse etc.). **2** tr. & intr. move, or cause (a vehicle etc.) to move, backwards. **3** tr. **a** put or serve as a back, background, or support to. **b** Mus. accompany. **4** tr. lie at the back of (a beach backed by steep cliffs). **5** intr. (of the wind) move round in an anticlockwise direction. ● adj. **1** situated behind, esp. as remote or subsidiary (backstreet; back teeth; back entrance). **2** of or relating to the past; not current (back pay; back issue). **3** reversed (back flow). □ **at a person's back** in pursuit or support. **at the back of one's mind** remembered but not consciously thought of. **back and forth** to and fro. **back bench**

a back-bencher's seat in the House of Commons. **back-bencher** a member of parliament not holding a senior office. **back-boiler** *Brit.* a boiler behind and integral with a domestic fire. **back-breaking** (esp. of manual work) extremely hard. **back country** esp. *Austral. & NZ* an area away from settled districts. **back-crawl** = BACKSTROKE. **back-cross** *Biol. v.tr.* cross a hybrid with one of its parents. ● *n.* an instance or the product of back-crossing. **back door 1** the door or entrance at the back of a building. **2** a secret or ingenious means of gaining an objective. **back-door** *attrib.adj.* (of an activity) clandestine, underhand (*back-door deal*). **back down** withdraw one's claim or point of view etc.; concede defeat in an argument etc. **back-down** *n.* an instance of backing down. **back-fill** refill an excavated hole with the material dug out of it. **back-formation 1** the formation of a word from its seeming derivative (e.g. *laze* from *lazy*). **2** a word formed in this way. **back number 1** an issue of a periodical earlier than the current one. **2** *sl.* an out-of-date person or thing. **the back of beyond** a very remote or inaccessible place. **back off 1** draw back, retreat. **2** abandon one's intention, stand, etc. **back on to** have its back adjacent to (*the house backs on to a field*). **back out** (often foll. by *of*) withdraw from a commitment. **back passage** *colloq.* the rectum. **back-pedal** (**-pedalled**, **-pedalling**; *US* **-pedaled**, **-pedaling**) **1** pedal backwards on a bicycle etc. **2** reverse one's previous action or opinion. **back-projection** the projection of a picture from behind a translucent screen for viewing or filming. **back room** (often with hyphen) *attrib.*) a place where secret work is done. **back-scattering** the scattering of radiation in a reverse direction. **back seat** an inferior position or status. **back-seat driver** a person who is eager to advise without responsibility (orig. of a passenger in a car etc.). **back slang** slang using words spelt backwards (e.g. *yob*). **back-stop** = LONGSTOP. **back talk** *US* = BACKCHAT. **back to back** with backs adjacent and opposite each other (*we stood back to back*). **back-to-back** *attrib.adj.* **1** esp. *Brit.* (of houses) with a party wall at the rear. **2** *N. Amer.* consecutive. ● *n.* a back-to-back house. **back to front 1** with the back at the front and the front at the back. **2** in disorder. **back-to-nature** (*attrib.*) denoting a movement or enthusiast for the reversion to a simpler way of life. **back up 1** give (esp. moral) support to. **2** *Computing* make a spare copy of (data, a disk, etc.). **3** (of running water) accumulate behind an obstruction. **4** reverse (a vehicle) into a desired position. **5** form a queue of vehicles etc., esp. in congested traffic. **back water** reverse a boat's forward motion using oars. **get** (or **put**) **a person's back up** annoy or anger a person. **get off a person's back** stop troubling a person. **go back on** fail to honour (a promise or commitment). **know like the back of one's hand** be entirely familiar with. **on one's back** injured or ill in bed. **on the back burner** see BURNER. **put one's back into** approach (a task etc.) with vigour. **see the back of** see SEE[1]. **turn one's back on 1** abandon. **2** disregard, ignore. **with one's back to** (or **up against**) **the wall** in a desperate situation; hard-pressed. □ **backer** *n.* (in sense 1 of *v.*). **backless** *adj.* [OE *bæc* f. Gmc]

backache /'bækeɪk/ *n.* a (usu. prolonged) pain in one's back.

backbeat /'bækbiːt/ *n. Mus.* a secondary accent on one of the normally unaccented beats of the bar, esp. in jazz and popular music.

backbite /'bækbaɪt/ *v.tr.* (*past* **-bit** /-bɪt/; *past part.* **-bitten** /-'bɪt(ə)n/) (usu. as **backbiting** *adj.*) slander; speak badly of. □ **backbiter** *n.*

backblocks /'bækblɒks/ *n.pl. Austral. & NZ* land in the remote and sparsely inhabited interior. □ **backblocker** *n.*

backboard /'bækbɔːd/ *n.* **1** a board worn to support or straighten the back. **2** a board placed at or forming the back of anything.

backbone /'bækbəʊn/ *n.* **1** the spine. **2** the main support of a structure. **3** firmness of character. **4** *US* the spine of a book.

backchat /'bæktʃæt/ *n. Brit. colloq.* the practice of replying rudely or impudently.

backcloth /'bækklɒθ/ *n.* **1** *Brit. Theatr.* a painted cloth at the back of the stage as a main part of the scenery. **2** the background to a scene or situation.

backcomb /'bækkəʊm/ *v.tr.* comb (the hair) towards the scalp to make it look thicker.

backdate /bæk'deɪt/ *v.tr.* **1** put an earlier date to (an agreement etc.) than the actual one. **2** make retrospectively valid.

backdrop /'bækdrɒp/ *n.* = BACKCLOTH.

backfire *v. & n.* ● *v.intr.* /bæk'faɪə(r)/ **1** undergo a mistimed explosion in the cylinder or exhaust of an internal-combustion engine. **2** (of a plan etc.) rebound adversely on the originator; have the opposite effect to what was intended. ● *n.* /'bækˌfaɪə(r)/ an instance of backfiring.

backflip /'bækflɪp/ *n.* a backward somersault done in the air with the arms and legs stretched out straight.

backgammon /'bækˌgæmən/ *n.* **1** a board-game for two players with pieces moved according to throws of the dice. (*See note below.*) **2** the most complete form of win in this. [BACK *adv.* + GAMMON[2]]

■ Backgammon is played on a board marked on two sides with twenty-four long triangular points in contrasting colours, twelve on each side. Players throw two dice to move their fifteen flat circular pieces around the board, the winner being the first to remove all of his or her pieces from the board. It is among the most ancient of all games, with similar games dating back as far as 3000 BC, and being played in its present form by the Romans.

background /'bækgraʊnd/ *n.* **1** part of a scene, picture, or description that serves as a setting to the chief figures or objects and foreground. **2** an inconspicuous or obscure position (*kept in the background*). **3** a person's education, knowledge, or social circumstances. **4** explanatory or contributory information or circumstances. **5** *Physics* low-intensity ambient radiation from radioisotopes present in the natural environment. **6** *Electronics* unwanted signals, such as noise in the reception or recording of sound. □ **background music** music intended as an unobtrusive accompaniment to some activity, or to provide atmosphere in a film etc.

backhand /'bækhænd/ *n. Tennis* etc. **1** a stroke played with the back of the hand turned towards the opponent. **2** (*attrib.*) of or made with a backhand (*backhand volley*).

backhanded /bæk'hændɪd/ *adj.* **1** (of a blow etc.) delivered with the back of the hand, or in a direction opposite to the usual one. **2** indirect; ambiguous (*a backhanded compliment*). **3** of or made with a backhand.

backhander /'bækˌhændə(r)/ *n.* **1 a** a backhand stroke. **b** a backhanded blow. **2** *colloq.* an indirect attack. **3** *Brit. sl.* a bribe.

backhoe /'bækhəʊ/ *n. US* a mechanical excavator which draws towards itself a bucket attached to a hinged boom.

backing /'bækɪŋ/ *n.* **1 a** a support. **b** a body of supporters. **c** material used to form a back or support. **2** musical accompaniment, esp. to a singer. □ **backing track** a recorded musical accompaniment.

backlash /'bæklæʃ/ *n.* **1** an excessive or marked adverse reaction. **2 a** a sudden recoil or reaction between parts of a mechanism. **b** excessive play between such parts.

backlist /'bæklɪst/ *n.* a publisher's list of books published before the current season and still in print.

backlit /'bæklɪt/ *adj.* (esp. in photography) illuminated from behind.

backlog /'bæklɒg/ *n.* **1** arrears of uncompleted work etc. **2** a reserve; reserves (*a backlog of goodwill*). ¶ Orig. = a large log placed at the back of a fire to sustain it.

backmarker /'bækˌmɑːkə(r)/ *n. Brit.* a competitor who has the least favourable handicap in a race etc.

backmost /'bækməʊst/ *adj.* furthest back.

backpack /'bækpæk/ *n. & v.* ● *n.* a rucksack. ● *v.intr.* travel or hike with a backpack. □ **backpacker** *n.*

backrest /'bækrest/ *n.* a support for the back.

backscratcher /'bækˌskrætʃə(r)/ *n.* **1** a rod terminating in a clawed hand for scratching one's own back. **2** a person who performs mutual services with another for gain.

backsheesh var. of BAKSHEESH.

backside /bæk'saɪd/ *n. colloq.* the buttocks.

backsight /'bæksaɪt/ *n.* **1** the sight of a rifle etc. that is nearer the stock. **2** *Surveying* a sight or reading taken backwards or towards the point of starting.

backslapping /'bækˌslæpɪŋ/ *adj.* vigorously hearty.

backslash /'bækslæʃ/ *n.* a backward-sloping diagonal line; a reverse solidus (\).

backslide /'bækslaɪd/ *v.intr.* (*past* **-slid** /-slɪd/; *past part.* **-slid** or **-slidden** /-ˌslɪd(ə)n/) relapse into bad ways or error. □ **backslider** *n.*

backspace /'bækspeɪs/ *v.intr.* move a typewriter carriage or computer cursor back one or more spaces.

backspin /'bækspɪn/ *n.* a backward spin imparted to a ball causing it to fly off at an angle on hitting a surface.

backstage /bæk'steɪdʒ/ *adv. & adj.* ● *adv.* **1** *Theatr.* out of view of the audience, esp. in the wings or dressing-rooms. **2** not known to the public. ● *adj.* that is backstage; concealed.

backstairs /'bæksteəz/ *n.pl.* **1** stairs at the back or side of a building. **2** (also **backstair**) (*attrib.*) denoting underhand or clandestine activity.

backstay /'bæksteɪ/ n. a rope etc. leading downwards and aft from the top of a mast.

backstitch /'bækstɪtʃ/ n. & v. ● n. sewing with overlapping stitches. ● v.tr. & intr. sew using backstitch.

backstreet /'bækstriːt/ n. **1** a street in a quiet part of a town, away from the main streets. **2** (attrib.) denoting illicit or illegal activity (a backstreet abortion).

backstroke /'bækstrəʊk/ n. a swimming stroke performed on the back with the arms lifted alternately out of the water in a backward circular motion and the legs extended in a kicking action.

backtrack /'bæktræk/ v.intr. **1** retrace one's steps. **2** reverse one's previous action or opinion.

backup /'bækʌp/ n. **1** moral or technical support (called for extra backup). **2** a reserve. **3** (often attrib.) Computing **a** the procedure for making security copies of data (backup facilities). **b** the copy itself (made a backup). **4** US a queue of vehicles etc., esp. in congested traffic. □ **backup light** US a reversing light.

backveld /'bækfelt, 'bækvelt/ n. S. Afr. remote country districts, esp. those strongly conservative. □ **backvelder** /-ˌfeldə(r), -ˌveldə(r)/ n. [BACK + VELD]

backward /'bækwəd/ adv. & adj. ● adv. = BACKWARDS. ¶ Backwards is now more common, esp. in literal senses. ● adj. **1** directed to the rear or starting-point (a backward look). **2** reversed. **3 a** mentally handicapped or slow. **b** slow to progress; late. **4** reluctant, shy, unassertive. **5** Cricket (of a fielding position) behind a line through the stumps at right angles to the wicket. □ **backwardness** n. [earlier abackward, assoc. with BACK]

backwardation /ˌbækwə'deɪʃ(ə)n/ n. esp. Brit. Stock Exch. the percentage paid by a person selling stock for the right of delaying the delivery of it (cf. CONTANGO).

backwards /'bækwədz/ adv. **1** away from one's front (lean backwards; look backwards). **2 a** with the back foremost (walk backwards). **b** in reverse of the usual way (count backwards; spell backwards). **3 a** into a worse state (new policies are taking us backwards). **b** into the past (looked backwards over the years). **c** (of a thing's motion) back towards the starting-point (rolled backwards). □ **backwards and forwards** in both directions alternately; to and fro. **bend** (or **fall** or **lean**) **over backwards** (often foll. by to + infin.) colloq. make every effort, esp. to be fair or helpful. **know backwards** be entirely familiar with.

backwash /'bækwɒʃ/ n. **1 a** receding waves created by the motion of a ship etc. **b** a backward current of air created by a moving aircraft. **2** repercussions.

backwater /'bækˌwɔːtə(r)/ n. **1** a place or condition remote from the centre of activity or thought. **2** stagnant water fed from a stream.

backwoods /'bækwʊdz/ n.pl. **1** remote uncleared forest land. **2** any remote or sparsely inhabited region.

backwoodsman /'bækˌwʊdzmən/ n. (pl. **-men**) **1** an inhabitant of backwoods. **2** an uncouth person. **3** Brit. a peer who very rarely attends the House of Lords.

backyard /bæk'jɑːd/ n. a yard at the back of a house etc. □ **in one's own backyard** colloq. near at hand.

baclava var. of BAKLAVA.

Bacolod /bɑː'kəʊlɒd/ a city on the NW coast of the island of Negros in the central Philippines; pop. (1990) 364,180. It is the chief city of the island and a major port.

Bacon[1] /'beɪkən/, Francis, Baron Verulam and Viscount St Albans (1561–1626), English statesman and philosopher. The pre-eminent legal figure of the late Elizabethan and early Stuart periods, he eventually rose to become Lord Chancellor under James I before falling from favour after impeachment on charges of corruption. His radical philosophical beliefs, especially as expounded in *The Advancement of Learning* (1605) and *Novum Organum* (1620), proved very influential, dominating the field for a century after his death. He advocated the inductive method and rejected the formulation of a priori hypotheses; his views were instrumental in the founding of the Royal Society in 1660.

Bacon[2] /'beɪkən/, Francis (1909–92), Irish painter. Having settled permanently in England in 1928, he first came to public prominence with the triptych *Three Studies for Figures at the Base of a Crucifixion* (1944), which depicts the Furies as monstrous semi-human figures. He often drew inspiration from photographs, film stills, or from other paintings; from the late 1950s his work chiefly depicted human figures in grotesquely distorted postures, their features blurred or erased, and set in confined interior spaces.

Bacon[3] /'beɪkən/, Roger (c.1214–94), English philosopher, scientist, and Franciscan monk. Bacon taught at Oxford and Paris; although widely acclaimed in scholarly circles he fell foul of his own order, which eventually imprisoned him as a heretic because of his interest in science. Bacon is most notable for his work in the field of optics and for emphasizing the need for an empirical approach to scientific study. He is also said to have prophesied flying machines and described the manufacture of gunpowder.

bacon /'beɪkən/ n. cured meat from the back or sides of a pig. □ **bring home the bacon** colloq. **1** succeed in one's undertaking. **2** supply material provision or support. [ME f. OF f. Frank. bako = OHG bahho ham, flitch]

Baconian /beɪ'kəʊnɪən/ adj. & n. ● adj. of or relating to the philosopher Francis Bacon (see BACON[1]), or to his inductive method of reasoning and philosophy. ● n. **1** a supporter of the view that Bacon was the author of Shakespeare's plays. **2** a follower of Bacon.

bacteria pl. of BACTERIUM.

bactericide /bæk'tɪərɪˌsaɪd/ n. a substance capable of destroying bacteria. □ **bactericidal** /-ˌtɪərɪ'saɪd(ə)l/ adj.

bacteriology /bækˌtɪərɪ'ɒlədʒɪ/ n. the study of bacteria. □ **bacteriologist** n. **bacteriological** /-ˌtɪərɪə'lɒdʒɪk(ə)l/ adj. **bacteriologically** adv.

bacteriolysis /bækˌtɪərɪ'ɒlɪsɪs/ n. the rupture of bacterial cells.

bacteriolytic /bækˌtɪərɪə'lɪtɪk/ adj. capable of lysing bacteria.

bacteriophage /bæk'tɪərɪəʊˌfeɪdʒ, -ˌfɑːʒ/ n. a virus which parasitizes a bacterium by infecting it and reproducing inside it. [BACTERIUM + Gk phagein eat]

bacteriostasis /bækˌtɪərɪəʊ'steɪsɪs/ n. the inhibition of the growth of bacteria without destroying them. [BACTERIUM + -STASIS]

bacteriostat /bæk'tɪərɪəʊˌstæt/ n. a substance which inhibits the multiplying of bacteria without destroying them. □ **bacteriostatic** /-ˌtɪərɪəʊ'stætɪk/ adj.

bacterium /bæk'tɪərɪəm/ n. (pl. **bacteria** /-rɪə/) a member of a large group of unicellular micro-organisms lacking organelles and an organized nucleus, some of which can cause disease. Bacteria are widely distributed in soil, water, and air, and on or in the tissues of plants and animals. Formerly included in the plant kingdom, they are now classified separately (as prokaryotes) from organisms with nucleated cells, to which they are thought to be ancestral. The chemical changes which many of them bring about are vital to life on earth, and include all forms of decay and the incorporation of atmospheric nitrogen into organic compounds. Classification of bacteria is problematic because they are so uniform in shape, usually spherical (cocci) or rodlike (bacilli). Multiplication is usually by simple fission, though some form spores. Because of the rapid growth rate and ease of culture of bacteria, much modern biochemical knowledge has been derived from their study. Bacteria were first observed by Antoni van Leeuwenhoek in the 17th century, and first definitely implicated in a disease (anthrax) by Robert Koch in 1876. □ **bacterial** adj. [mod.L f. Gk baktērion dimin. of baktron stick]

Bactria /'bæktrɪə/ an ancient country in central Asia, corresponding to the northern part of modern Afghanistan. Traditionally the home of Zoroaster, it was the seat of a powerful Indo-Greek kingdom in the 3rd and 2nd centuries BC.

Bactrian /'bæktrɪən/ adj. of or relating to Bactria. □ **Bactrian camel** a camel, *Camelus ferus*, native to central Asia, with two humps. [L Bactrianus f. Gk Baktrianos]

bad /bæd/ adj., n., & v. ● adj. (**worse** /wɜːs/; **worst** /wɜːst/) **1** inferior, inadequate, defective (bad work; a bad driver; bad light). **2 a** unpleasant, unwelcome (bad weather; bad news). **b** unsatisfactory, unfortunate (a bad business). **3** harmful (is bad for you). **4 a** (of food) decayed, putrid. **b** (of the atmosphere) polluted, unhealthy (bad air). **5** colloq. ill, injured (am feeling bad today; a bad leg). **6** colloq. regretful, guilty, ashamed (feels bad about it). **7** (of an unwelcome thing) serious, severe (a bad headache; a bad mistake). **8 a** morally wicked or offensive (a bad man; bad language). **b** naughty; badly behaved (a bad child). **9** worthless; not valid (a bad cheque). **10** (**badder, baddest**) esp. US sl. good, excellent. ● n. **1 a** ill fortune (take the bad with the good). **b** ruin; a degenerate condition (go to the bad). **2** the debit side of an account (£500 to the bad). **3** (as pl.; prec. by the) bad or wicked people. ● adv. N. Amer. colloq. badly (took it bad). □ **bad blood** ill feeling. **bad books** see in a person's bad books (see BOOK). **bad breath** unpleasant-smelling breath. **bad debt** a debt that is not recoverable. **bad egg** see EGG[1]. **bad faith** see FAITH. **bad form** see FORM. **a bad job** colloq. an unfortunate state of affairs. **bad-**

mannered having bad manners; rude. **bad mouth** *US sl.* malicious gossip or criticism. **bad-mouth** *v.tr. US sl.* subject to malicious gossip or criticism. **bad news** *colloq.* an unpleasant or troublesome person or thing. **from bad to worse** into an even worse state. **in a bad way** ill; in trouble (*looked in a bad way*). **not** (or **not so**) **bad** *colloq.* fairly good. **too bad** *colloq.* (of circumstances etc.) regrettable but now beyond retrieval. □ **baddish** *adj.* **badness** *n.* [ME, perh. f. OE *bæddel* hermaphrodite, womanish man: for loss of *l* cf. MUCH, WENCH]

badass /'bædæs/ *n. & adj.* (also **bad-ass**) *US sl.* ● *n.* an aggressive or formidable person; a trouble-maker. ● *adj.* **1** tough, aggressive, belligerent. **2** bad, worthless. **3** formidable, excellent.

baddy /'bædɪ/ *n.* (*pl.* **-ies**) *colloq.* a villain or criminal, esp. in a story, film, etc.

bade see BID.

Baden /'bɑːd(ə)n/ a spa town in Austria, south of Vienna; pop. (1991) 24,000. It was a royal summer retreat and fashionable resort in the 19th century.

Baden-Baden /ˌbɑːd(ə)n'bɑːd(ə)n/ a spa town in SW Germany, in the Black Forest; pop. (est. 1984) 48,700. It was a fashionable resort in the 19th century.

Baden-Powell /ˌbeɪd(ə)n'pəʊəl/, Robert (Stephenson Smyth), 1st Baron Baden-Powell of Gilwell, English soldier and founder of the Boy Scout movement. He became a national hero after his successful defence of Mafeking (1899–1900) in the Boer War. The Boy Scout movement, which he founded in 1908, and the Girl Guide movement, which he founded together with his sister Agnes and his wife Olave in 1910, grew to become important international youth movements.

Baden-Württemberg /ˌbɑːd(ə)n'vʊətəmˌbɜːɡ/ a state of western Germany; capital, Stuttgart.

Bader /'bɑːdə(r)/, Sir Douglas (Robert Steuart) (1910–82), British airman. Despite having lost both legs in a flying accident in 1931, he rejoined the RAF in 1939 and saw action as a fighter pilot during the evacuation from Dunkirk (1940) and in the Battle of Britain (1940–1), becoming a national hero. After the war he was noted for his work on behalf of disabled people.

badge /bædʒ/ *n.* **1** a distinctive emblem worn as a mark of office, membership, achievement, licensed employment, etc. **2** any feature or sign which reveals a characteristic condition or quality. [ME: orig. unkn.]

badger /'bædʒə(r)/ *n. & v.* ● *n.* **1 a** an omnivorous nocturnal Eurasian mammal, *Meles meles*, of the weasel family, living in sets. It is heavily built, with a grey back, black underparts, and a white head with two black stripes. **b** a related animal, *Taxidea taxus*, native to North America. **2** a fishing-fly, brush, etc., made of its hair. ● *v.tr.* pester, harass, tease. [16th c.: perh. f. BADGE, with ref. to its white forehead mark]

badinage /'bædɪˌnɑːʒ/ *n.* humorous or playful ridicule. [F f. *badiner* to joke]

badlands /'bædlændz/ *n.pl.* extensive uncultivable eroded tracts in arid areas, especially strikingly eroded areas of the US characterized by sharp-crested ridges and pinnacles. Such topography is found where, owing to an arid climate or over-grazing, there is little vegetation to protect the land surface from erosion, so that streams and rivers have incised it with numerous gullies and ravines. The name was originally applied to parts of South Dakota and Nebraska, which the French trappers found 'bad lands to cross'. Exposure of the rock layers has resulted in substantial finds of fossil vertebrates. [transl. F *mauvaises terres*]

badly /'bædlɪ/ *adv.* (**worse** /wɜːs/; **worst** /wɜːst/) **1** in a bad manner (*works badly*). **2** *colloq.* very much (*wants it badly*). **3** severely (*was badly defeated*).

badminton /'bædmɪntən/ *n.* **1** a game played, usu. indoors, by one or two players opposing an equivalent number across a net, using rackets and a shuttlecock. (*See note below.*) **2** a summer drink of claret, soda, and sugar. [*Badminton* in S. England]

▪ Badminton is in many ways similar to lawn tennis, but only the server can score points and the shuttlecock may not make contact with the ground. It evolved about 1870 from the ancient game of battledore and shuttlecock, when it was played at Badminton, the seat of the Duke of Beaufort in Gloucestershire (now Avon). The game was taken up by army officers, who played it in India out of doors. The first laws were drawn up in Poona in the mid-1870s.

Badon Hill, Battle of /'beɪd(ə)n/ an ancient British battle (the location of which is uncertain), in AD 516, in which the forces of King Arthur successfully defended themselves against the Saxons. Another

source implies that the battle was fought *c*.500 but does not connect it with King Arthur.

bad-tempered /ˌbæd'tempəd/ *adj.* having a bad temper; irritable; easily annoyed. □ **bad-temperedly** *adv.*

BAe /ˌbiːeɪ'iː/ *abbr.* British Aerospace.

Baedeker /'beɪdɪkə(r)/ *n.* any of various travel guidebooks published by the firm founded by the German Karl Baedeker (1801–59).

Baer /beə(r)/, Karl Ernest von (1792–1876), German biologist. His discovery that ova were particles within the ovarian follicles was his chief contribution to embryology. He also formulated the principle that in the developing embryo general characters appear before special ones, and his studies were used by Darwin in the theory of evolution.

Baeyer /'baɪə(r)/, Adolph Johann Friedrich Wilhelm von (1835–1917), German organic chemist. An able experimental chemist, he prepared the first barbiturates, and investigated dyes, synthesizing indigo and determining its structural formula. His work pioneered the study of ring structures and stimulated the synthetic dye industry. He was awarded the Nobel Prize for chemistry in 1905.

Baez /'baɪez/, Joan (b.1941), American folk-singer. From the late 1950s she was a prominent figure in the American folk revival; she is best known for her performances at civil-rights demonstrations of the early 1960s. Albums include *Any Day Now* (1968) and *Diamonds and Rust* (1975).

Baffin /'bæfɪn/, William (*c*.1584–1622), English navigator and explorer. The pilot of several expeditions in search of the North-west Passage (1612–16), he discovered the largest island of the Canadian Arctic in 1616; this and the strait between it and Greenland are named after him. The record he established for attaining the most northerly latitude was not broken until the mid-19th century.

Baffin Bay an extension of the North Atlantic between Baffin Island and Greenland, linked to the Arctic Ocean by three passages. Lying to the north of the Arctic Circle and traversed by the Labrador Current, it is largely ice-bound in winter. [BAFFIN]

Baffin Island a large island in the Canadian Arctic, situated at the mouth of Hudson Bay. It is separated from Greenland by Baffin Bay. [BAFFIN]

baffle /'bæf(ə)l/ *v. & n.* ● *v.tr.* **1** confuse or perplex (a person, one's faculties, etc.). **2 a** frustrate or hinder (plans etc.). **b** restrain or regulate the progress of (fluids, sounds, etc.). ● *n.* (also **baffle-plate**) a device used to restrain the flow of fluid, gas, etc., or to limit the emission of sound, light, etc. □ **baffle-board** a device to prevent sound from spreading in different directions, esp. round a loudspeaker cone. □ **bafflement** *n.* **baffling** *adj.* **bafflingly** *adv.* [perh. rel. to F *bafouer* ridicule, OF *beffer* mock]

baffler /'bæflə(r)/ *n.* = BAFFLE *n.*

BAFTA /'bæftə/ *abbr.* British Academy of Film and Television Arts.

bag /bæg/ *n. & v.* ● *n.* **1** a receptacle of flexible material with an opening at the top. **2 a** (usu. in *pl.*) a piece of luggage (*put the bags in the boot*). **b** a woman's handbag. **3** (in *pl.*; usu. foll. by *of*) *colloq.* a large amount; plenty (*bags of time*). **4** (in *pl.*) *Brit. colloq.* trousers. **5** *sl. derog.* a woman, esp. regarded as unattractive or unpleasant. **6** an animal's sac containing poison, honey, etc. **7** an amount of game taken by a hunter. **8** (usu. in *pl.*) baggy folds of skin under the eyes. **9** *sl.* a person's particular interest or preoccupation, esp. in a distinctive style or category of music (*his bag is Indian music*). ● *v.* (**bagged**, **bagging**) **1** *tr.* put in a bag. **2** *tr. colloq.* **a** secure; get hold of (*bagged the best seat*). **b** *colloq.* steal. **c** shoot (game). **d** (often in phr. **bags I**) *Brit. colloq.* claim on grounds of being the first to do so (*bagged first go; bags I go first*). **3 a** *intr.* hang loosely; bulge; swell. **b** *tr.* cause to do this. **4** *tr. Austral. sl.* criticize, disparage. □ **bag and baggage** with all one's belongings. **bag of bones** an emaciated person or animal. **bag lady** *N. Amer.* a homeless woman who carries her possessions around in shopping bags. **bag of nerves** a very tense or timid person. **bag** (or **whole bag**) **of tricks** *colloq.* everything; the whole lot. **in the bag** *colloq.* achieved; as good as secured. □ **bagful** *n.* (*pl.* **-fuls**). [ME, perh. f. ON *baggi*]

bagarre /bæ'ɡɑː(r)/ *n.* a scuffle or brawl. [F]

bagasse /bə'ɡæs/ *n.* the dry pulpy residue left after the extraction of juice from sugar cane, usable as fuel or to make paper etc. [F f. Sp. *bagazo*]

bagatelle /ˌbæɡə'tel/ *n.* **1** a game in which small balls are struck into numbered holes on a board, with pins as obstructions. **2** a mere trifle; a negligible amount. **3** *Mus.* a short piece of music, esp. for the piano. [F f. It. *bagatella* dimin., perh. f. *baga* BAGGAGE]

Bagehot /'bædʒət/, Walter (1826–77), English economist and journalist. He worked as a banker before becoming editor of the

Economist in 1860, a post which he held until his death. His insight into economic and political questions is shown in his books *The English Constitution* (1867), *Lombard Street* (1873), and *Economic Studies* (1880).

bagel /'beɪg(ə)l/ *n.* a hard bread roll in the shape of a ring. [Yiddish *beygel*]

baggage /'bægɪdʒ/ *n.* **1** everyday belongings packed up in suitcases etc. for travelling; luggage. **2** the portable equipment of an army. **3** *joc.* or *derog.* a girl or woman. **4** mental encumbrances. □ **baggage check** *N. Amer.* a luggage ticket. [ME f. OF *bagage* f. *baguer* tie up or *bagues* bundles: perh. rel. to BAG]

baggy /'bægɪ/ *adj.* (**baggier, baggiest**) **1** hanging in loose folds. **2** puffed out. □ **baggily** *adv.* **bagginess** *n.*

Baghdad /bæg'dæd/ the capital of Iraq, on the River Tigris; pop. (est. 1985) 4,648,600. A thriving city under the Abbasid caliphs in the 8th and 9th centuries, notably Harun ar-Rashid, it suffered repeated attacks until 1534, when it was taken by the Ottoman sultan Suleiman. It remained under Ottoman rule until the First World War. In 1920 it became the capital of the newly created state of Iraq. The discovery of oil brought prosperity to the city, but it suffered damage in the Iran–Iraq war (1980–8) and the Gulf War of 1991.

bagman /'bægmən/ *n.* (*pl.* **-men**) **1** *Brit. sl.* a travelling salesman. **2** *Austral.* a tramp. **3** *US & Austral. sl.* an agent who collects or distributes the proceeds of illicit activities.

bagnio /'ba:njəʊ/ *n.* (*pl.* **-os**) **1** a brothel. **2** an oriental prison. [It. *bagno* f. L *balneum* bath]

bagpipe /'bægpaɪp/ *n.* (usu. in *pl.*) a musical instrument with reeds that are sounded by the pressure of wind emitted from a bag squeezed by the player's arm and fed with air either by breath or by means of small bellows strapped to the waist. Bagpipes generally have at least two pipes, one (the chanter) giving the melody and any others sounding a drone or drones. They have been a popular instrument in the West from the Middle Ages and also appear, in widely varying forms, in central and eastern Europe and Asia; they are now associated especially with Scotland, Northumberland, and Ireland. □ **bagpiper** *n.*

baguette /bæ'get/ *n.* **1** a long narrow French loaf. **2** a gem cut in a long rectangular shape. **3** *Archit.* a small moulding, semicircular in section. [F f. It. *bacchetto* dimin. of *bacchio* f. L *baculum* staff]

bagworm /'bægwɜːm/ *n.* *US* a destructive caterpillar that lives in a silk case covered in plant debris.

bah /ba:/ *int.* an expression of contempt or disbelief. [prob. F]

Baha'i /bə'ha:i/ *n.* **1** a religion founded in the 19th century as a development of Babism. (See note below.) **2** an adherent of this religion. □ **Baha'ism** *n.* [Pers. *bahā'* splendour]

▪ The Baha'i faith was founded by the Persian Baha-ullah (1817–92) and his son Abdul Baha (1844–1921). The central tenet of the faith is that the essence of all religions is one; thus all religious teachers are the messengers of one God. Its adherents, who are now found in most countries of the world, work for the peace and unification of humankind, but almost from the religion's inception they have been persecuted, especially in Iran. The seat of its governing body is in Haifa in Israel, adjacent to the golden-domed shrine of the Bab (see BABISM).

Bahamas /bə'ha:məz/ a country consisting of an archipelago off the SE coast of Florida, part of the West Indies; pop. (1991) 254,680; languages, English (official), Creole; capital, Nassau. It was there that Columbus made his first landfall in the New World (12 Oct. 1492). The islands were depopulated in the 16th century as the Spaniards carried off most of the Arawak inhabitants to slavery and death. In 1648 a group of English Puritans settled there, and the islands were a British colony from the 18th century until they gained independence within the Commonwealth in 1973.

Bahamian /bə'heɪmɪən/ *n. & adj.* ● *n.* **1** a native or national of the Bahamas. **2** a person of Bahamian descent. ● *adj.* of or relating to the Bahamas.

Bahasa Indonesia /ba:'ha:sə/ *n.* the official language of Indonesia. [Indonesian *bahasa* language f. Skr. *bhāṣā* f. *bhāṣate* speech, language: see INDONESIAN]

Bahawalpur /bə'ha:wəl,pʊə(r)/ a city of central Pakistan, in Punjab province; pop. (1991) 250,000. It was formerly the capital of a princely state established by the nawabs of Bahawalpur.

Bahia /ba:'i:ə/ **1** a state of eastern Brazil, on the Atlantic coast; capital, Salvador. **2** the former name for SALVADOR.

Bahía Blanca /ba:,i:ə 'blæŋkə/ a port in Argentina serving the southern part of the country; pop. (1991) 271,500.

Bahrain /ba:'reɪn/ a sheikhdom consisting of a group of islands in the Persian Gulf; pop. (est. 1991) 518,000; official language, Arabic; capital, Manama. The islands, famous in ancient times for their pearls, were ruled by the Portuguese in the 16th century and the Persians in the 17th century. They became a British protectorate in 1861 and became independent in 1971. Bahrain's economy is dependent on the refining and export of oil, chiefly that coming by pipeline from Saudi Arabia. □ **Bahraini** *adj. & n.*

baht /ba:t/ *n.* the basic monetary unit of Thailand, equal to 100 satangs. [Thai *bāt*]

Bahutu *pl.* of HUTU.

Baikal, Lake /baɪ'ka:l/ (also **Baykal**) a large lake in southern Siberia, the largest freshwater lake in Europe and Asia and, with a depth of 1,743 m (5,714 ft), the deepest lake in the world.

Baikonur /,baɪkə'nʊə(r)/ (also **Baykonur**) a mining town in central Kazakhstan. The world's first satellite (1957) and the first manned space flight (1961) were launched from the former Soviet space centre nearby.

bail[1] /beɪl/ *n. & v.* ● *n.* **1** money etc. required as security against the temporary release of a prisoner pending trial. **2** a person or persons giving such security. ● *v.tr.* (usu. foll. by *out*) **1** release or secure the release of (a prisoner) on payment of bail. **2** (also **bale** by assoc. with *bale out* 1: see BALE[1]) release from a difficulty; come to the rescue of. □ **bail bandit** *colloq.* a person who commits a crime while on bail awaiting trial. **forfeit** (or *colloq.* **jump**) **bail** fail to appear for trial after being released on bail. **go** (or **stand**) **bail** (often foll. by *for*) act as surety (for an accused person). □ **bailable** *adj.* [ME f. OF *bail* custody, *bailler* take charge of, f. L *bajulare* bear a burden]

bail[2] /beɪl/ *n. & v.* ● *n.* **1** *Cricket* either of the two crosspieces bridging the stumps. **2** the bar on a typewriter holding the paper against the platen. **3** a bar separating horses in an open stable. **4** *Austral. & NZ* a framework for securing the head of a cow during milking. ● *v. Austral. & NZ* (usu. foll. by *up*) **1** *tr.* secure (a cow) during milking. **2** a *tr.* make (a person) hold up his or her arms to be robbed. b *intr.* surrender by throwing up one's arms. c *tr.* buttonhole (a person). [ME f. OF *bail(e)* palisade, enclosure, perh. f. L *baculum* stick: cf. BAILEY]

bail[3] /beɪl/ *v.tr.* (also **bale**) **1** (usu. foll. by *out*) scoop water out of (a boat etc.). **2** scoop (water etc.) out. □ **bail out** see *bale out* 1 (see BALE[1]). □ **bailer** *n.* [obs. *bail* (n.) bucket f. F *baille* ult. f. L *bajulus* carrier]

bailee /beɪ'li:/ *n.* *Law* a person or party to whom goods are committed for a purpose, e.g. custody or repair, without transfer of ownership. [BAIL[1] + -EE]

Bailey[1] /'beɪlɪ/ a shipping forecast area in the NE Atlantic north of Rockall and south-west of the Faeroes.

Bailey[2] /'beɪlɪ/, David (b.1938), English photographer. His career began with *Vogue* magazine in 1960, where, using a 35-mm camera and outdoor locations, he brought a new look to fashion features. He was a prominent figure of the 1960s pop culture; his idiosyncratic, jokey portraits of fellow celebrities were included in *Goodbye Baby and Amen* (1969). More recently, his subjects have included the Vietnamese boat-people and the townscape of his north London neighbourhood, contained in *David Bailey's NW1* (1982).

bailey /'beɪlɪ/ *n.* (*pl.* **-eys**) **1** the outer wall of a castle. **2** a court enclosed by it. [ME, var. of BAIL[2]]

Bailey bridge *n.* a temporary bridge of lattice steel designed for rapid assembly from prefabricated standard parts, designed by Sir Donald Bailey (1901–85) for military use in the Second World War. The bridge is built in its intended orientation beside the river and then 'launched' on rollers from one bank until it reaches the far side. Highway authorities still use such bridges for emergency work.

bailie /'beɪlɪ/ *n.* *esp. hist.* a municipal officer and magistrate in Scotland. [ME, f. OF *bailli(s)* BAILIFF]

bailiff /'beɪlɪf/ *n.* **1** a sheriff's officer who executes writs and processes and carries out distraints and arrests. **2** *Brit.* the agent or steward of a landlord. **3** *US* an official in a court of law who keeps order, looks after prisoners, etc. **4** *Brit.* (*hist.* except in formal titles) the sovereign's representative in a district, esp. the chief officer of a hundred. **5** the first civil officer in the Channel Islands. [ME f. OF *baillif* ult. f. L *bajulus* carrier, manager]

bailiwick /'beɪlɪ,wɪk/ *n.* **1** *joc.* a person's sphere of operations or particular area of interest. **2** *Law* the district or jurisdiction of a bailie or bailiff. [BAILIE + WICK[2]]

bailment /'beɪlmənt/ n. the act of delivering goods etc. for a (usu. specified) purpose.

bailor /'beɪlə(r)/ n. Law a person or party that entrusts goods to a bailee. [BAIL¹ + -OR¹]

bailout /'beɪlaʊt/ n. financial assistance given to a failing business, economy, etc., to save it from collapse.

bailsman /'beɪlzmən/ n. (pl. **-men**) a person who stands bail for another. [BAIL¹ + MAN]

Baily's beads /'beɪlɪz/ n.pl. Astron. a string of bright points seen at the edge of the darkened moon at the beginning or end of totality in an eclipse of the sun, and caused by the unevenness of the lunar topography. [F. Baily, Engl. astronomer (1774–1844), who described the phenomenon]

bain-marie /ˌbænmə'riː/ n. (pl. **bains-marie** pronunc. same) a cooking utensil consisting of a vessel of hot water in which a receptacle containing a sauce etc. can be slowly and gently heated; a double boiler. [F, transl. med.L balneum Mariae bath of Maria (an alleged Jewish alchemist)]

Bairam /baɪ'ræm, 'baɪræm/ either of two annual Muslim festivals: Greater Bairam is celebrated concurrently with the annual pilgrimage (hajj) in the twelfth month of the Muslim lunar calendar and continues for three to four days, whereas Lesser Bairam follows the month of ritual fasting (Ramadan) and lasts two to three days. [Turk. & Pers.]

Baird /beəd/, John Logie (1888–1946), Scottish pioneer of television. He started his work in the early 1920s, gave a demonstration in London in 1926, and made the first transatlantic transmission and demonstration of colour television in 1928. Baird used a mechanical system of picture scanning, which was soon displaced in television development by an electronic system developed by V. K. Zworykin and others in the 1930s.

bairn /beən/ n. Sc. & N. Engl. a child. [OE bearn]

bait /beɪt/ n. & v. ● n. **1** food used to entice a prey, esp. a fish or an animal. **2** an allurement; something intended to tempt or entice. **3** archaic a halt on a journey for refreshment or a rest. **4** var. of BATE. ● v. **1** tr. **a** harass or annoy (a person). **b** torment (a chained animal). **2** tr. put bait on (a hook, trap, etc.) to entice a prey. **3** archaic **a** tr. give food to (horses on a journey). **b** intr. stop on a journey to take food or a rest. [ME f. ON beita hunt or chase]

baize /beɪz/ n. a coarse usu. green woollen material resembling felt used as a covering or lining, esp. on the tops of billiard and card-tables. [F baies (pl.) fem. of bai chestnut-coloured (BAY⁴), treated as sing.: cf. BODICE]

Baja California /'bɑːhɑː/ (also called Lower California) a mountainous peninsula in NW Mexico, which extends southwards from the border with California and separates the Gulf of California from the Pacific Ocean. It consists of two states of Mexico: Baja California (Norte) (capital, Mexicali) and Baja California Sur (capital, La Paz).

bajra /'bɑːdʒrə/ n. Ind. pearl millet or similar grain. [Hindi]

bake /beɪk/ v. & n. ● v. **1** tr. cook (food) by dry heat in an oven or on a hot surface, without direct exposure to a flame. **b** intr. undergo the process of being baked. **2** intr. colloq. **a** (usu. as **be baking**) (of weather etc.) be very hot. **b** (of a person) become hot. **3** tr. harden (clay etc.) by heat. **b** intr. (of clay etc.) be hardened by heat. **4** a tr. (of the sun) affect by its heat, e.g. ripen (fruit). **b** intr. be affected by the sun's heat. ● n. **1** the act or an instance of baking. **2** a batch of baking. **3** US a social gathering at which baked food is eaten. □ **bake blind** see BLIND.

baked Alaska a sponge cake and ice-cream in a meringue covering.

baked beans baked haricot beans, usu. tinned in tomato sauce.

baking-powder a mixture of sodium bicarbonate, cream of tartar, etc., used instead of yeast in baking. **baking soda** sodium bicarbonate. [OE bacan]

bakehouse /'beɪkhaʊs/ n. = BAKERY.

Bakelite /'beɪkəˌlaɪt/ n. propr. a thermosetting resin or plastic made from formaldehyde and phenol, and used for cables, buttons, plates, etc. [G Bakelit f. L. H. Baekeland, its Belgian-born inventor (1863–1944)]

Baker¹ /'beɪkə(r)/, Dame Janet (Abbott) (b.1933), English operatic mezzo-soprano. From 1957 onwards she made regular appearances in opera and on the concert platform; her Covent Garden début was in 1966 in Britten's A Midsummer Night's Dream. Noted for her role as Dido in Purcell's Dido and Aeneas, she also includes in her wide repertoire interpretations of Mahler's song cycles and Bach's oratorios. She retired from the operatic stage in 1982.

Baker² /'beɪkə(r)/, Josephine (1906–75), American dancer. Appearing with the Revue Nègre in Paris at the age of 19, she caused a stir with her exotic dancing, risqué clothing, and remarkable entrances. She was a star of the Folies-Bergère in the 1930s, and became a screen idol in such films as La Sirène des tropiques (1927). She was awarded the Legion of Honour for her work with the French resistance during the Second World War. Experience of racial discrimination in the US led her to join the campaign for black civil rights in the 1950s.

baker /'beɪkə(r)/ n. a person who bakes and sells bread, cakes, etc., esp. professionally. □ **baker's dozen** thirteen (so called from the former bakers' custom of adding an extra loaf to a dozen sold; the exact reason for this is unclear). [OE bæcere]

Baker day n. colloq. a day set aside for in-service training of teachers in England and Wales. [Kenneth Baker, the Education Secretary (1986–9) who introduced the practice]

bakery /'beɪkərɪ/ n. (pl. **-ies**) a place where bread and cakes are made or sold.

Bakewell /'beɪkwel/, Robert (1725–95), English pioneer in scientific methods of livestock breeding and husbandry. He produced pedigree herds of sheep and cattle from his Leicestershire farm, while irrigation of the grassland gave four cuts a year, and feeding and selective breeding greatly increased the meat production from his animals.

Bakewell tart n. a baked open pie consisting of a pastry case lined with jam and filled with a rich almond paste. [Bakewell in Derbyshire]

baklava /'bækləvə, ˌbæklə'vɑː/ n. (also **baclava**) a rich dessert of flaky pastry, honey, and nuts. [Turk.]

baksheesh /'bækʃiːʃ/ n. (also **backsheesh**) (in some oriental countries) a small sum of money given as a gratuity or as alms. [ult. f. Pers. bakšīš f. bakšīdan give]

Bakst /bækst/, Léon (born Lev Samuilovich Rozenberg) (1866–1924), Russian painter and designer. Associated with Diaghilev's magazine The World of Art from 1899, he became one of the most influential members of the Diaghilev circle and the Ballets Russes. He designed the decor for such Diaghilev productions as Scheherazade (1910), L'Après-midi d'un faune (1912), and The Sleeping Princess (1921). His use of rich, luxuriant colour and often exotic set designs and costumes had a significant influence on fashion and the development of art deco.

Baku /bæ'kuː/ the capital of Azerbaijan, on the Caspian Sea; pop. (1990) 1,780,000. It is an industrial port and a centre of the oil industry.

Bakunin /bæ'kuːnɪn/, Mikhail (Aleksandrovich) (1814–76), Russian anarchist. After taking part in the revolutions of 1848 in France, Germany, and Poland he was exiled to Siberia, but escaped in 1861 and went to London. He participated in the First International, founded in 1864, but as a leading exponent of anarchism came into conflict with Karl Marx by calling for violent means to destroy the existing political and social order. He was expelled from the First International in 1872.

balaclava /ˌbælə'klɑːvə/ n. (in full **balaclava helmet**) a tight woollen garment covering the whole head and neck except for parts of the face, worn originally by soldiers on active service in the Crimean War.

Balaclava, Battle of /ˌbælə'klɑːvə/ a battle of the Crimean War, fought between Russia and an alliance of British, French, and Turkish forces in and around the port of Balaclava (Balaklava) in the southern Crimea in 1854. Russian forces failed to take Balaclava and the battle ended inconclusively; it is chiefly remembered as the scene of the Charge of the Light Brigade (see entry).

balalaika /ˌbælə'laɪkə/ n. a guitar-like musical instrument having a triangular body and 2–4 strings, popular in Russia and other Slav countries. [Russ.]

balance /'bæləns/ n. & v. ● n. **1** an apparatus for weighing, esp. one with a central pivot, beam, and two scales. **2 a** a counteracting weight or force. **b** (in full **balance-wheel**) the regulating device in a clock etc. **3 a** an even distribution of weight or amount. **b** stability of body or mind (regained his balance). **4** a preponderating weight or amount (the balance of opinion). **5 a** an agreement between or the difference between credits and debits in an account. **b** the difference between an amount due and an amount paid (will pay the balance next week). **c** an amount left over; the rest. **6 a** Art harmony of design and proportion. **b** Mus. the relative volume of various sources of sound (bad balance between violins and trumpets). **c** proportion. **7** (**the Balance**) the zodiacal sign or constellation Libra. ● v. **1** tr. (foll. by with, against) offset or compare (one thing) with another (must balance the advantages with the disadvantages). **2** tr. counteract, equal, or neutralize the weight or importance of. **3 a** tr. bring into or keep in equilibrium (balanced a book on her head). **b** intr. be in equilibrium (balanced on one leg). **4** tr. (usu. as **balanced** adj.) establish equal or appropriate proportions of elements in (a balanced diet; balanced opinion). **5** tr. weigh (arguments etc.) against each other. **6 a** tr. compare and esp. equalize debits and credits of (an account). **b** intr. (of an account) have credits and debits equal. □ **balance of payments** the difference in value between payments into and out of a country. **balance of power 1** a situation in which the chief states of the world have roughly equal power. **2** the power held by a small group when larger groups are of equal strength. **balance of trade** the difference in value between imports and exports. **balance sheet** a written statement of the assets and liabilities of an organization on a particular date. **balancing act** an action or activity that requires achieving a delicate balance between different situations or requirements. **in the balance** uncertain; at a critical stage. **on balance** all things considered. **strike a balance** choose a moderate course or compromise. □ **balancer** n. [ME f. OF, ult. f. LL (libra) bilanx bilancis two-scaled (balance)]

Balanchine /ˈbælənˌtʃiːn, -ˌʃiːn/, George (born Georgi Melitonovich Balanchivadze) (1904–83), Russian-born American ballet-dancer and choreographer. He worked as chief choreographer of Diaghilev's Ballets Russes during the 1920s. In 1934 he co-founded the company which later became the New York City Ballet, where he choreographed many ballets and revivals. Notable works include *The Firebird* (1949) and *A Midsummer Night's Dream* (1962).

balander /bəˈlændə(r)/ n. (also **balanda**) Austral. (Aboriginal English) a white man. [Makasarese corrupt. of *Hollander*]

balata /ˈbælətə/ n. **1** a latex-yielding tree of Central America, esp. *Manilkara bidentata*. **2** the dried sap of this used as a substitute for gutta-percha. [ult. f. Carib]

Balaton, Lake /ˈbɒlɒˌtɒn/ a large shallow lake in west central Hungary, situated in a resort and wine-producing region to the south of the Bakony mountains.

Balboa /bælˈbəʊə/, Vasco Núñez de (1475–1519), Spanish explorer. Having settled in the new Spanish colony of Hispaniola in 1501, in 1511 Balboa joined an expedition to Darien (in Panama) as a stowaway, but rose to command it after a mutiny. He founded a colony in Darien and continued to make expeditions into the surrounding areas. In 1513 he reached the western coast of the isthmus after an epic 25-day march, thereby becoming the first European to see the Pacific Ocean.

balboa /bælˈbəʊə/ n. the basic monetary unit of Panama, equal to 100 centésimos. [BALBOA]

Balbriggan /bælˈbrɪgən/ n. a knitted cotton fabric used for underwear etc. [*Balbriggan* in Ireland, where it was orig. made]

Balcon /ˈbɔːlkən/, Sir Michael (1896–1977), English film producer. He was responsible for several early Hitchcock films but is mainly remembered for his long association with Ealing Studios, during which he produced such famous comedies as *Kind Hearts and Coronets* and *Whisky Galore* (both 1949), and *The Man in the White Suit* and *The Lavender Hill Mob* (both 1952).

balcony /ˈbælkənɪ/ n. (pl. **-ies**) **1** a usu. balustraded platform on the outside of a building, with access from an upper-floor window or door. **2 a** the tier of seats in a theatre above the dress circle. **b** the upstairs seats in a cinema etc. **c** N. Amer. the dress circle in a theatre. □ **balconied** adj. [It. *balcone*]

bald /bɔːld/ adj. **1** (of a person) with the scalp wholly or partly lacking hair. **2** (of an animal, plant, etc.) not covered by the usual hair, feathers, leaves, etc. **3** colloq. with the surface worn away (a bald tyre). **4 a** blunt, unelaborated (a bald statement). **b** undisguised (the bald effrontery). **5** meagre or dull (a bald style). **6** marked with white, esp. on the face (a bald horse). □ **bald eagle** a white-headed eagle, *Haliaeetus leucocephalus*, used as the emblem of the United States. □ **balding** adj. (in senses 1–3). **baldish** adj. **baldly** adv. (in sense 4). **baldness** n. [ME *ballede*, orig. 'having a white blaze', prob. f. an OE root *ball*- 'white patch']

baldachin /ˈbɔːldəkɪn/ n. (also **baldaquin**) **1** a ceremonial canopy over an altar, throne, etc. **2** a rich brocade. [It. *baldacchino* f. *Baldacco* BAGHDAD, the brocade's place of origin]

Balder /ˈbɔːldə(r)/ Scand. Mythol. a son of Odin and god of the summer sun. He was invulnerable to all things except mistletoe, with which the god Loki, by a trick, induced the blind god Hödur to kill him.

balderdash /ˈbɔːldəˌdæʃ/ n. senseless talk or writing; nonsense. [earlier = 'mixture of drinks': orig. unkn.]

baldhead /ˈbɔːldhed/ n. a person with a bald head.

baldmoney /ˈbɔːldˌmʌnɪ/ n. (pl. **-eys**) an aromatic white-flowered umbelliferous mountain plant, *Meum athamanticum*. [ME in sense 'gentian': orig. unkn.]

baldric /ˈbɔːldrɪk/ n. hist. a belt for a sword, bugle, etc., hung from the

shoulder across the body to the opposite hip. [ME *baudry* f. OF *baudrei*: cf. MHG *balderich*, of unkn. orig.]

Baldwin¹ /ˈbɔːldwɪn/, James (Arthur) (1924–87), American novelist. His first novel, *Go Tell it on the Mountain* (1953), telling of one day in the lives of members of a Harlem church, launched Baldwin as a leading writer. *Giovanni's Room* (1956), set in Paris, dealt with sexual (especially homosexual) and racial relationships, subjects further explored in *Another Country* (1962). A civil-rights activist in the 1950s, he subsequently published several collections of essays on racial questions, including *Nobody Knows My Name* (1961) and *No Name in the Street* (1972).

Baldwin² /ˈbɔːldwɪn/, Stanley, 1st Earl Baldwin of Bewdley (1867–1947), British Conservative statesman, Prime Minister 1923–4, 1924–9, and 1935–7. His second term was marked by the return to the gold standard, the General Strike of 1926, and the annexation of Ethiopia by Fascist Italy; his last included the abdication of King Edward VIII, which he handled skilfully. Although international relations continued to deteriorate with the German occupation of the Rhineland (1936) and the outbreak of the Spanish Civil War (1936), Baldwin opposed demands for rearmament, believing that the public would not support it.

baldy /ˈbɔːldɪ/ n. & adj. (also **bally** /ˈbɔːlɪ/) Austral. colloq. ● n. **1** a Hereford. **2** a white-faced animal. ● adj. **1** Hereford (cattle). **2** (of cattle) with a white marking on the face. [Brit. dial.]

Bâle see BASLE.

bale¹ /beɪl/ n. & v. ● n. **1** a bundle of merchandise or hay etc. tightly wrapped and bound with cords or hoops. **2** the quantity in a bale as a measure, esp. *US* 500 lb of cotton. ● v.tr. make up into bales. □ **bale** (or **bail**) **out 1** (of an airman) make an emergency parachute descent from an aircraft (cf. BAIL³). **2** = BAIL¹ v. 2. [ME prob. f. MDu., ult. identical with BALL¹]

bale² /beɪl/ n. archaic or poet. evil, destruction, woe, pain, misery. [OE *b(e)alu*]

bale³ var. of BAIL³.

Balearic Islands /ˌbælɪˈærɪk/ (also **Balearics**) a group of Mediterranean islands off the east coast of Spain, forming an autonomous region of that country, with four large islands (Majorca, Minorca, Ibiza, Formentera) and seven smaller ones; capital, Palma (on Majorca). Occupied by the Romans after the destruction of Carthage, the islands were subsequently conquered by Vandals and Moors and then by Aragon in the 14th century. Tourism, fishing, and wine and fruit production are now important.

baleen /bəˈliːn/ n. whalebone. □ **baleen whale** a whale of the suborder Mysticeti, having plates of whalebone in the mouth for straining plankton from the water. [ME f. OF *baleine* f. L *balaena* whale]

baleful /ˈbeɪlfʊl/ adj. **1** (esp. of a manner, look, etc.) gloomy, menacing. **2** harmful, malignant, destructive. □ **balefully** adv. **balefulness** n. [BALE² + -FUL]

Balenciaga /bæˌlensɪˈɑːgə/, Cristóbal (1895–1972), Spanish couturier. He established his own business in Madrid in 1932, moving to Paris in 1937. His garments were noted for their simplicity, elegance, and boldness of design. In the 1950s he contributed to the move away from the tight-waisted New Look originated by Christian Dior to a looser, semi-fitted style, which culminated in 1955 with the introduction of a tunic dress, and in 1957 with a chemise ('the sack').

baler /ˈbeɪlə(r)/ n. a machine for making bales of hay, straw, metal, etc.

Balfour /ˈbælfə(r)/, Arthur James, 1st Earl of Balfour (1848–1930), British Conservative statesman, Prime Minister 1902–5. His premiership saw the formation of the Committee of Imperial Defence and the creation of the *entente cordiale* with France (1904), but the party split over the issue of tariff reform, forcing Balfour's resignation, although he remained as party leader until 1911. In 1917, in his capacity as Foreign Secretary during the First World War, Balfour issued the declaration in favour of a Jewish national home in Palestine that came to be known as the Balfour Declaration.

Bali /ˈbɑːlɪ/ a mountainous island of Indonesia, to the east of Java; chief city, Denpasar; pop. (est. 1989) 2,782,040. It is noted for its beauty and the richness of its culture.

Balinese /ˌbɑːliːˈniːz/ n. & adj. ● n. (pl. same) **1** a native of Bali. **2** the language of Bali. ● adj. of or relating to Bali or its people or language.

balk var. of BAULK.

Balkanize /ˈbɔːlkəˌnaɪz/ v.tr. (also **-ise**) divide (an area) into smaller mutually hostile states. □ **Balkanization** /ˌbɔːlkənaɪˈzeɪʃ(ə)n/ n. [BALKANS, THE]

Balkan Mountains a range of mountains stretching eastwards across Bulgaria from the Serbian frontier to the Black Sea. The highest peak is Botev Peak (2,375 m; 7,793 ft).

Balkans, the /'bɔːlkənz/ the countries occupying that part of SE Europe lying south of the Danube and Sava rivers, forming a peninsula bounded by the Adriatic and Ionian Seas in the west, the Aegean and Black Seas in the east, and the Mediterranean in the south. From the 3rd to the 7th centuries the peninsula, nominally ruled by the Byzantine emperors, was invaded by successive migrations of Slavs. In 1356 the Turkish invasions began, Constantinople falling to the Ottoman Turks in 1453. The subject peoples, though largely retaining their languages and religions, did not recover independence until the 19th century. In 1912–13 Turkey was attacked and defeated by an alliance of Balkan states, which broke up in fighting over the territory gained (see BALKAN WARS). After the First World War the peninsula was divided between Greece, Albania, Bulgaria, and Yugoslavia, with Turkey retaining only a small area including Constantinople (Istanbul). In 1991–3 the former federal republic of Yugoslavia broke up, the constituent republics of Serbia and Montenegro retaining the name 'Yugoslavia' while Slovenia, Croatia, Macedonia, and Bosnia–Herzegovina became independent against a background of civil conflict. □ **Balkan** adj. & n.

Balkan Wars two wars of 1912–13 that added to the tension in the Balkans before the First World War. In the first (1912), Bulgaria, Serbia, Greece, and Montenegro formed a league and attempted to seize the remaining European territories of the Ottoman Empire, forcing Turkey to give up Albania and Macedonia, leaving the area around Constantinople (Istanbul) as the only Ottoman territory in Europe. The second war was fought in 1913: Bulgaria disputed with Serbia, Greece, and Romania for possession of Macedonia, which was partitioned between Greece and Serbia.

Balkhash, Lake see BALQASH, LAKE.

Balkis /'bɔːlkɪs, 'bɒl-/ the name of the queen of Sheba in Arabic literature.

balky var. of BAULKY.

ball[1] /bɔːl/ n. & v. ● n. **1** a solid or hollow sphere, esp. for use in a game. **2 a** a ball-shaped object; material forming the shape of a ball (*ball of snow; ball of wool; rolled himself into a ball*). **b** a rounded part of the body (*ball of the foot*). **3** a solid non-explosive missile for a cannon etc. **4** a single delivery of a ball in cricket, baseball, etc., or passing of a ball in football. **5** (in *pl.*) *coarse sl.* **a** the testicles. **b** (usu. as an exclamation of contempt) nonsense, rubbish. **c** = *balls-up.* **d** courage, 'guts'. ● v. **1** *tr.* squeeze or wind into a ball. **2** *intr.* form or gather into a ball or balls. □ **ball-and-socket joint** a joint in which a rounded end lies in a concave cup or socket, allowing freedom of movement, esp. the shoulder or hip-joint. **ball-bearing 1** a bearing in which the two halves are separated by a ring of small metal balls which reduce friction. **2** one of these balls. **ball game 1 a** any game played with a ball. **b** *N. Amer.* a game of baseball. **2** esp. *N. Amer. colloq.* a particular affair or concern (*a whole new ball game*). **the ball is in your** etc. **court** you etc. must be next to act. **ball lightning** a rare globular form of lightning. **ball-point (pen)** a pen with a tiny ball as its writing point. **balls** (or **ball**) **up** *coarse sl.* bungle; make a mess of. **balls-up** n. *coarse sl.* a mess; a confused or bungled situation. **ball valve** a valve opened or closed by a ball which fits into a cup-shaped opening. **have the ball at one's feet** have one's best opportunity. **keep the ball rolling** maintain the momentum of an activity. **on the ball** *colloq.* alert. **play ball** *colloq.* cooperate. **start** etc. **the ball rolling** set an activity in motion; make a start. [ME f. ON *böllr* f. Gmc]

ball[2] /bɔːl/ n. **1** a formal social gathering for dancing. **2** *sl.* an enjoyable time (esp. *have a ball*). [F *bal* f. LL *ballare* to dance]

ballad /'bæləd/ n. **1** a poem or song narrating a popular story. **2** a slow sentimental or romantic song. □ **ballad metre** = *common metre.* [ME f. OF *balade* f. Prov. *balada* dancing-song f. *balar* to dance]

ballade /bæ'lɑːd/ n. **1** a poem of one or more triplets of stanzas with a repeated refrain and an envoy. **2** *Mus.* a short lyrical piece, esp. for piano. [earlier spelling and pronunc. of BALLAD]

balladeer /ˌbælə'dɪə(r)/ n. a singer or composer of ballads.

ballad opera n. a theatrical entertainment popular in early 18th-century England, in the form of a satirical play interspersed with traditional or operatic songs. The best-known example is John Gay's *The Beggar's Opera* (1728).

balladry /'bælədrɪ/ n. ballad poetry.

Ballarat /'bælə,ræt/ a mining and sheep-farming centre in Victoria, Australia; pop. (1991) 64,980. It is the site of the discovery in 1851 of the largest gold reserves in Australia.

ballast /'bæləst/ n. & v. ● n. **1** heavy material placed in a ship or the basket of a hot-air balloon etc. to secure stability. **2** coarse stone etc. used to form the bed of a railway track or road. **3** a mixture of coarse and fine aggregate for making concrete. **4** *Electr.* any device used to stabilize the current in a circuit. **5** anything that affords stability or permanence. ● *v.tr.* **1** provide with ballast. **2** afford stability or weight to. [16th c.: f. LG or Scand., of uncert. orig.]

ballboy /'bɔːlbɔɪ/ n. (*fem.* **ballgirl** /-gɜːl/) *Tennis* a boy (or girl) who retrieves balls that go out of play during a game.

ballcock /'bɔːlkɒk/ n. a floating ball on a hinged arm, whose movement up and down controls the water level in a cistern.

ballerina /ˌbælə'riːnə/ n. a female ballet-dancer. [It., fem. of *ballerino* dancing-master f. *ballare* dance f. LL: see BALL[2]]

ballet /'bæleɪ/ n. **1 a** a dramatic or representational style of dancing and mime, using set steps and techniques and usu. (esp. in classical ballet) accompanied by music. (*See note below.*) **b** a particular piece or performance of ballet. **c** the music for this. **2** a company performing ballet. □ **ballet-dancer** a dancer who specializes in ballet. □ **balletic** /bə'letɪk/ adj. [F f. It. *balletto* dimin. of *ballo* BALL[2]]

▪ Ballet developed in Renaissance Italy and flowered in France at the court of Louis XIV. It was at first a very formal spectacle which involved singing and speech as well as dance, performed by courtiers wearing flowing costumes and high heels; it was not until the early 19th century that the costumes, steps, positions, and techniques of ballet in its best-known forms developed. The great schools of romantic and classical ballet appeared in the 19th century. The former is represented by ballets such as *La Sylphide* (1832) and *Giselle* (1841), for which the technique of dancing on the tips of the toes (*sur les pointes*) was adopted. Classical ballet was perfected by the Russian Imperial Ballet under the choreographer Marius Petipa, whose *Sleeping Beauty* (1890), to music by Tchaikovsky, exemplifies the school. The early 20th century was dominated by the productions of Diaghilev's Ballets Russes, involving the collaboration of dancers, choreographers, stage designers, and composers. The work of the American dancer Isadora Duncan, who abandoned the traditional role of ballet as a narrative art for one of subjective response directly to the music, was also important (see MODERN DANCE). Classical ballet today flourishes particularly in Russia, notably in the Bolshoi and Kirov companies. A distinctively British style developed in the work of British choreographers such as Ninette de Valois with the Vic-Wells (later Sadler's Wells) Ballet and Frederick Ashton and Kenneth Macmillan (1929–92), who were particularly associated with the Royal Ballet. In the US, Diaghilev's influence was also profound, and led to the founding of such schools as the School of American Ballet (1934) and the New York City Ballet (1948), with which George Balanchine was particularly associated. Among the most famous 20th-century dancers have been Vaslav Nijinsky, Alicia Markova, Rudolf Nureyev, Margot Fonteyn, and Mikhail Baryshnikov.

balletomane /'bælɪtəʊ,meɪn/ n. a devotee of ballet. □ **balletomania** /ˌbælɪtəʊ'meɪnɪə/ n. [F]

Ballets Russes /ˌbæleɪ 'ruːs/ a ballet company formed in Paris in 1909 by the Russian impresario Sergei Diaghilev, which had a revolutionary impact in the ballet world and beyond. In place of the conventions of classical ballet was a unified whole encompassing music, dance, decor, and costume; music was commissioned from the composers Stravinsky, Satie, and Rimsky-Korsakov, and Cocteau, Bakst, and Picasso designed sets. Choreographers and dancers who worked with the company include Michel Fokine, Anna Pavlova, Vaslav Nijinsky, Léonide Massine, and later George Balanchine. The Ballets Russes performed in Europe until Diaghilev's death in 1929, but never appeared in Russia.

ballista /bə'lɪstə/ n. (*pl.* **ballistae** /-tiː/) a catapult used in ancient warfare for hurling large stones etc. [L f. Gk *ballō* throw]

ballistic /bə'lɪstɪk/ adj. **1** of or relating to projectiles. **2** moving under the force of gravity only. □ **ballistic missile** a missile which is initially powered and guided but falls under gravity on its target. **go ballistic** esp. *US* become furious. □ **ballistically** adv. [BALLISTA + -IC]

Ballistic Missile Defense Organization see STRATEGIC DEFENSE INITIATIVE.

ballistics /bə'lɪstɪks/ n.pl. (usu. treated as *sing.*) the science of projectiles and firearms.

ballocks var. of BOLLOCKS.

ballon d'essai /ˌbælɒn deˈseɪ/ *n.* (*pl.* **ballons d'essai** *pronunc.* same) an experiment to see how a new policy etc. will be received. [F, = trial balloon]

balloon /bəˈluːn/ *n. & v.* ● *n.* **1** a small inflatable rubber pouch with a neck, used as a child's toy or as decoration. **2** a large usu. round bag inflatable with hot air or gas to make it rise in the air, often carrying a basket for passengers. (*See note below.*) **3** *colloq.* a balloon shape enclosing the words or thoughts of characters in a comic strip or cartoon. **4** a large globular drinking-glass, usu. for brandy. ● *v.* **1** *intr. & tr.* swell out or cause to swell out like a balloon. **2** *intr.* travel by balloon. **3** *tr. Brit.* hit or kick (a ball etc.) high in the air. □ **when the balloon goes up** *colloq.* when the action or trouble starts. □ **balloonist** *n.* [F *ballon* or It. *ballone* large ball]

▪ A small hot-air balloon was demonstrated in Lisbon as early as 1709, but attracted little attention. Manned flight was achieved in a balloon on 21 Nov. 1783, when two Frenchmen, Pilâtre de Rozier and the Marquis d'Arlandes, were airborne for 25 minutes under a hot-air balloon built by the Montgolfier brothers, covering more than 8 km (5 miles). By the end of the 19th century flights had been made across the English Channel and the Alps, and the sport was so fashionable in London as to be rated a social grace, although it lapsed at the end of the Edwardian era with the arrival of heavier-than-air flight. Interest in ballooning has recently revived.
Anchored balloons were used by Napoleon as observation posts in some battles, while barrage balloons were arranged around cities in protective cordons to deter bombing attacks in the First and Second World Wars. Free balloons, both manned and unmanned, have had important functions in atmospheric research and weather prediction.

ballot /ˈbælət/ *n. & v.* ● *n.* **1** a process of voting, in writing and usu. secret. **2** the total of votes recorded in a ballot. **3** the drawing of lots. **4** a paper or ticket etc. used in voting. ● *v.* (**balloted, balloting**) **1** *intr.* (usu. foll. by *for*) **a** hold a ballot; give a vote. **b** draw lots for precedence etc. **2** *tr.* take a ballot of (*the union balloted its members*). □ **ballot-box** a sealed box into which voters put completed ballot-papers. **ballot-paper** a slip of paper used to register a vote. [It. *ballotta* dimin. of *balla* BALL[1]]

ballpark /ˈbɔːlpɑːk/ *n.* **1** *Amer.* a baseball ground. **2** (*attrib.*) *colloq.* approximate, rough (*a ballpark figure*). □ **in the ballpark of** *colloq.* approximately, in the region of. **in the right ballpark** *colloq.* close to one's objective; approximately correct.

ballroom /ˈbɔːlruːm, -rʊm/ *n.* a large room or hall for dancing.

ballroom dancing *n.* formal social dancing by couples or groups of couples (*formation dancing*), popular as a recreation and now also as a competitive activity. Although formal social dancing took place in some medieval courts, the ballroom dance repertoire dates only from this century: it is made up of dances such as the waltz and minuet, which developed from old European folk-dances. To these have been added Latin American dances such as the tango, samba, rumba, paso doble, and cha-cha, together with the foxtrot and quickstep, which are of 20th-century origin.

bally[1] /ˈbælɪ/ *adj. & adv. Brit. sl.* = BLOODY *adj.* 3 & *adv.* (*took the bally lot*). [alt. of BLOODY]

bally[2] var. of BALDY.

ballyhoo /ˌbælɪˈhuː/ *n.* **1** a loud noise or fuss; a confused state or commotion. **2** extravagant or sensational publicity. [19th or 20th c., orig. US (in sense 2): orig. unkn.]

Ballymena /ˌbælɪˈmiːnə/ a town in Northern Ireland, to the north of Lough Neagh, capital of a district of the same name; pop. (1981) 18,150.

ballyrag /ˈbælɪˌræg/ *v.tr.* (also **bullyrag** /ˈbʊlɪ-/) (**-ragged, -ragging**) *sl.* play tricks on; scold, harass. [18th c.: orig. unkn.]

balm /bɑːm/ *n.* **1** an aromatic ointment for anointing, soothing, or healing. **2** a fragrant and medicinal exudation from certain trees and plants. **3** a healing or soothing influence or consolation. **4** a tree that yields balm, esp. an Asian and North African tree of the genus *Commiphora*. **5** an aromatic herb, esp. one of the genus *Melissa*. **6** a pleasant perfume or fragrance. □ **balm of Gilead** (cf. Jer. 8:22) **1 a** a fragrant resin formerly much used as an unguent. **b** a tree of the genus *Commiphora* yielding such resin. **2** the balsam fir. **3** the balsam poplar. [ME f. OF *ba(s)me* f. L *balsamum* BALSAM]

balmoral /bælˈmɒrəl/ *n.* **1** a type of brimless boat-shaped cocked hat with a cockade or ribbons attached, usu. worn by certain Scottish regiments. **2** a heavy leather walking-boot with laces up the front. [BALMORAL CASTLE]

Balmoral Castle a holiday residence of the British royal family, on the River Dee in Scotland. The estate was bought in 1847 by Prince Albert, who rebuilt the castle.

balmy /ˈbɑːmɪ/ *adj.* (**balmier, balmiest**) **1** mild and fragrant; soothing. **2** yielding balm. **3** *sl.* = BARMY. □ **balmily** *adv.* **balminess** *n.*

balneology /ˌbælnɪˈɒlədʒɪ/ *n.* the scientific study of bathing and medicinal springs. □ **balneologist** *n.* **balneological** /-nɪəˈlɒdʒɪk(ə)l/ *adj.* [L *balneum* bath + -LOGY]

baloney var. of BOLONEY.

BALPA /ˈbælpə/ *abbr.* British Air Line Pilots Association.

Balqash, Lake /bælˈkɑːʃ/ (also **Balkhash**) a shallow salt lake in Kazakhstan.

balsa /ˈbɒlsə, ˈbɔːl-/ *n.* **1** (in full **balsa wood**) a type of tough lightweight wood used for making models etc. **2** the tropical American tree, *Ochroma lagopus*, from which it comes. [Sp., = raft]

balsam /ˈbɒlsəm, ˈbɔːl-/ *n.* **1** an aromatic resinous exudation, such as balm, obtained from various trees and shrubs and used as a base for certain fragrances and medical preparations. **2** an ointment, esp. one composed of a substance dissolved in oil or turpentine. **3** a tree or shrub which yields balsam. **4** a flowering plant of the genus *Impatiens* (see IMPATIENS). **5** a healing or soothing agency. □ **balsam apple** a gourdlike plant of the genus *Momordica*, having warted orange-yellow fruits. **balsam fir** a North American tree, *Abies balsamea*, which yields balsam. **balsam poplar** a North American poplar, esp. *Populus balsamifera*, yielding balsam. [OE f. L *balsamum*]

balsamic /bɒlˈsæmɪk, bɔːl-/ *adj.* **1** yielding balsam. **2** soothing; fragrant. □ **balsamic vinegar** a dark, sweet, Italian vinegar, matured in wooden barrels.

Balt /bɔːlt, bɒlt/ *n. & adj.* ● *n.* **1** a native of one of the Baltic States of Lithuania, Latvia, and Estonia. **2** *hist.* a German inhabitant of any of these states. ● *adj.* of or relating to the Balts. [L *Balthae*]

Balthasar /ˈbælθəˌzɑː(r)/ one of the three Magi (see MAGI, THE).

Balti /ˈbɔːltɪ, ˈbæltɪ/ *n.* a type of Pakistani cuisine in which the food is cooked in a small bowl-shaped frying pan. [20th c.: orig. uncert., perh. f. *Balti*, native or inhabitant of Baltistan]

Baltic /ˈbɔːltɪk, ˈbɒl-/ *n. & adj.* ● *n.* (**the Baltic**) **1** the Baltic Sea. **2** the Baltic States. ● *adj.* **1** of or relating to the Baltic. **2** of or relating to an Indo-European branch of languages comprising Old Prussian, Lithuanian, and Latvian. [med.L *Balticus* f. LL *Balthae* dwellers near the Baltic Sea]

Baltic Exchange an association of companies, with headquarters in London, whose members are engaged in numerous international trading activities, especially the chartering of vessels to carry cargo. Other activities include the sale and purchase of ships, the chartering, sale, and purchase of aircraft, and commodity trading. The origins of the Exchange can be traced to the 18th century, when shipowners and merchants met in London coffee houses, of which one of the foremost was known as 'the Virginia and Baltic' in reference to the origin of much of the trade.

Baltic Sea a sea in northern Europe. Almost landlocked, it is linked with the North Sea by the Kattegat strait and the Øresund channel.

Baltic States 1 the independent republics of Estonia, Latvia, and Lithuania. **2** the ten states bordering the Baltic Sea, members of the Council of Baltic States established in 1992: Denmark, Estonia, Finland, Germany, Latvia, Lithuania, Norway, Poland, Russia, and Sweden.

Baltimore /ˈbɔːltɪˌmɔː(r), ˈbɒl-/ a seaport in north Maryland; pop. (1990) 736,000. First settled in the 1660s, it is named after George Calvert, the first Baron Baltimore (*c.*1580–1632), who in 1632 obtained a grant of land for the colony later to become Maryland.

Baltistan /ˌbɔːltɪˈstɑːn, ˌbɒl-/ (also called *Little Tibet*) a region of the Karakoram range of the Himalayas, to the south of K2.

Baluchi /bəˈluːtʃɪ/ *n. & adj.* ● *n.* **1** a native or inhabitant of Baluchistan. **2** the Iranian language of Baluchistan. ● *adj.* of or relating to the people or language of Baluchistan. [Pers. *Balūč(ī)*]

Baluchistan /bəˌluːtʃɪˈstɑːn/ **1** a mountainous region of western Asia, which includes part of SE Iran, SW Afghanistan, and west Pakistan. **2** a province of west Pakistan; capital, Quetta.

baluster /ˈbæləstə(r)/ *n.* each of a series of often ornamental short posts or pillars supporting a rail or coping etc. ¶ Often confused with *banister*. [F *balustre* f. It. *balaustro* f. L f. Gk *balaustion* wild-pomegranate flower]

balustrade /ˌbæləˈstreɪd/ *n.* a railing supported by balusters, esp.

forming an ornamental parapet to a balcony, bridge, or terrace. □ **balustraded** *adj.* [F (as BALUSTER)]

Balzac /'bælzæk/, Honoré de (1799–1850), French novelist. He is chiefly remembered for his series of ninety-one coordinated and interconnected novels and stories known collectively as *La Comédie humaine*, which appeared in a collected edition 1842–8, and includes *Eugénie Grandet* (1833) and *Le Père Goriot* (1835). The fulfilment of his project to create an authentic and comprehensive representation of French society during the late 18th and early 19th centuries, it is a significant work of 19th-century realism. Balzac's panorama of society deals with all aspects of public and personal experience, located in rural and urban settings; recurrent themes include the role of money in shaping personal and social relations and the extent to which environment determines the individual.

bama /'bæmə, 'pæmə/ *n.* (also **pama**) *Austral.* an Aboriginal person, esp. one from northern Queensland. [f. many north Qld. languages *bama* person or man]

Bamako /'bæmə,kəʊ/ the capital of Mali, in the south of the country, on the River Niger; pop. (1987) 646,000.

bambino /bæm'bi:nəʊ/ *n.* (*pl.* **bambini** /-nɪ/) *colloq.* a young (esp. Italian) child. [It., dimin. of *bambo* silly]

bamboo /bæm'bu:/ *n.* **1** a mainly tropical giant woody grass of the subfamily Bambusidae. **2** its hollow jointed stem, used as a stick or to make furniture etc. The pulp and fibre of some species are used in paper-making, or distilled to extract substances for use in medicines and chemical reactions. [Du. *bamboes* f. Port. *mambu* f. Malay]

bamboozle /bæm'bu:z(ə)l/ *v.tr. colloq.* cheat, hoax, mystify. □ **bamboozler** *n.* [c.1700: prob. of cant. orig.]

Bamian /ˌbæmɪ'ɑ:n/ a city in central Afghanistan; pop. (1984) 8,000. Nearby are the remains of two colossal statues of Buddha and the ruins of the city of Ghulghuleh, which was destroyed by Genghis Khan *c.*1221.

ban /bæn/ *v. & n.* ● *v.tr.* (**banned**, **banning**) forbid, prohibit (an action etc.), esp. formally; refuse admittance to (a person). ● *n.* **1** a formal or authoritative prohibition (*a ban on smoking*). **2** a tacit prohibition by public opinion. **3** a sentence of outlawry. **4** *archaic* a curse or execration. [OE *bannan* summon f. Gmc]

Banaba /bə'næbə/ (also called *Ocean Island*) an island in the western Pacific, just south of the equator to the west of the Gilbert Islands. Formerly within the Gilbert and Ellice Islands, the island has been part of Kiribati since 1979.

banal /bə'nɑ:l/ *adj.* trite, feeble, commonplace. □ **banality** /-'nælɪtɪ/ *n.* (*pl.* **-ies**). **banally** /-'nɑ:llɪ/ *adv.* [orig. in sense 'compulsory', hence 'common to all', f. F f. *ban* (as BAN)]

banana /bə'nɑ:nə/ *n.* **1** a long curved fruit with soft pulpy flesh and yellow skin when ripe, growing in clusters. **2** (in full **banana-tree**) the tropical and subtropical treelike plant, *Musa sapientum*, bearing this fruit and having palmlike leaves. □ **banana republic** *derog.* a small state, esp. in Central America, economically dependent on its fruit-growing or similar trade. **banana skin 1** the skin of a banana. **2** a cause of upset or humiliation; a blunder. **banana split** a sweet dish made with split bananas, ice-cream, sauce, etc. **go bananas** *sl.* become crazy or angry. [Port. or Sp., f. a name in Guinea]

banausic /bə'nɔ:sɪk/ *adj. derog.* **1 a** uncultivated. **b** materialistic. **2** suitable only for artisans. [Gk *banausikos* for artisans]

Banbury cake /'bænbərɪ/ *n.* a flat pastry with a spicy currant filling. [*Banbury* in S. England, where it was orig. made]

banc /bæŋk/ *n.* □ **in banc** *Law* sitting as a full court. [AF (= bench) f. med.L (as BANK²)]

band¹ /bænd/ *n. & v.* ● *n.* **1** a flat, thin strip or loop of material (e.g. paper, metal, or cloth) put round something esp. to hold it together or decorate it (*headband*). **2 a** a strip of material forming part of a garment (*hatband*; *waistband*). **b** a stripe of a different colour or material on an object. **3 a** a range of frequencies or wavelengths in a spectrum (esp. of radio frequencies). **b** a range of values within a series. **4** *Mech.* a belt connecting wheels or pulleys. **5** (in *pl.*) a collar having two hanging strips, worn by some lawyers, ministers, and academics in formal dress. **6** *archaic* a thing that restrains, binds, connects, or unites; a bond. ● *v.tr.* **1** put a band on. **2** a mark with stripes. **b** (as **banded** *adj.*) *Bot. & Zool.* marked with coloured bands or stripes. **3 a** divide into, or arrange in, bands or ranges with a view to treating the bands differently. **b** group (pupils) on the basis of ability. □ **band-saw** an endless saw, consisting of a steel belt with a serrated edge running over wheels. [ME f. OF *bande, bende* (sense 6 f. ON *band*) f. Gmc]

band² /bænd/ *n. & v.* ● *n.* **1** an organized group of people having a

common object. **2 a** a group of musicians, esp. playing wind instruments (*brass band; military band*). **b** a group of musicians playing jazz, pop, or dance music. **c** *colloq.* an orchestra. **3** *N. Amer.* a herd or flock. ● *v.tr. & intr.* form into a group for a purpose (*band together for mutual protection*). [ME f. OF *bande, bander*, med.L *banda*, prob. of Gmc orig.]

Banda /'bændə/, Hastings Kamuzu (b.1906), Malawian statesman, Prime Minister 1964–94 and President 1966–94. He studied medicine in the US and practised in Britain before returning to lead his country (formerly Nyasaland) to independence. As the first President of the Republic of Malawi he created an autocratic and paternalistic one-party state; a pragmatist, he was the first black African leader to visit South Africa (1970) and later established trading links with it. Banda was defeated in Malawi's first multi-party elections in 1994.

bandage /'bændɪdʒ/ *n. & v.* ● *n.* **1** a strip of material for binding up a wound etc. **2** a piece of material used as a blindfold. ● *v.tr.* bind (a wound etc.) with a bandage. [F f. *bande* (as BAND¹)]

Band-Aid /'bændeɪd/ *n.* (also **band-aid**) **1** *propr.* **a** a type of sticking-plaster with a gauze pad. **b** a piece of this. **2** a makeshift or temporary solution.

bandanna /bæn'dænə/ *n.* a large coloured handkerchief or neckerchief, usu. of silk or cotton, and often having white spots. [prob. Port. f. Hindi]

Bandaranaike /ˌbændərə'naɪkə/, Sirimavo Ratwatte Dias (b.1916), Sinhalese stateswoman, Prime Minister of Sri Lanka 1960–5, 1970–7. The world's first woman Prime Minister, she succeeded her husband, S.W.R.D. Bandaranaike (1899–1959), after his assassination. Opposition to her policies and continuing ethnic conflict resulted in an overwhelming defeat in the 1977 elections. She was charged with misuse of power in 1980, stripped of her civil rights for six years, and expelled from Parliament. Her daughter, Chandrika Kumaratunga, became Prime Minister in 1994.

Bandar Lampung /ˌbændə 'læmpʊŋ/ a city at the southern tip of Sumatra, in Indonesia; pop. (1980) 284,275. It was created in the 1980s as a result of the amalgamation of the city of Tanjungkarang and the nearby port of Telukbetung.

Bandar Seri Begawan /ˌbændə ˌserɪ bə'gɑ:wən/ the capital of Brunei; pop. (1991) 46,000.

Banda Sea /'bændə/ a sea in eastern Indonesia, between the central and south Molucca Islands.

b. & b. *abbr.* bed and breakfast.

bandbox /'bændbɒks/ *n.* a usu. circular cardboard box for carrying hats (orig. for neckbands). □ **out of a bandbox** extremely neat. [BAND¹ + BOX¹]

bandeau /'bændəʊ/ *n.* (*pl.* **bandeaux** /-dəʊz/) a narrow band worn round the head. [F]

banderilla /ˌbændə'ri:jə/ *n.* a decorated dart thrust into a bull's neck or shoulders during a bullfight. [Sp.]

banderole /ˌbændə'rəʊl/ *n.* (also **banderol**) **1 a** a long narrow flag with a cleft end, flown at a masthead. **b** an ornamental streamer on a knight's lance. **2 a** a ribbon-like scroll. **b** a stone band resembling a banderole, bearing an inscription. [F *banderole* f. It. *banderuola* dimin. of *bandiera* BANNER]

bandicoot /'bændɪ,ku:t/ *n.* **1** a mainly insectivorous marsupial of the family Peramelidae, native to Australia and New Guinea. **2** (in full **bandicoot rat**) a destructive Asian rat of the genus *Bandicota*, esp. the large *B. indica*. [Telugu *pandikokku* pig-rat]

bandit /'bændɪt/ *n.* (*pl.* **bandits** or **banditti** /bæn'dɪtɪ/) **1** a robber or murderer, esp. a member of a gang; a gangster. **2** an outlaw. □ **banditry** *n.* [It. *bandito* (pl. *-iti*), past part. of *bandire* ban, = med.L *bannire* proclaim: see BANISH]

Bandjarmasin see BANJARMASIN.

bandmaster /'bænd,mɑ:stə(r)/ *n.* the conductor of a (esp. military or brass) band. [BAND² + MASTER]

Band of Hope *n.* an association promoting total abstinence from alcohol.

bandog /'bændɒg/ *n.* a fighting dog bred for its strength and ferocity. [orig. = dog kept on a chain, f. BAND¹ + DOG]

bandolier /ˌbændə'lɪə(r)/ *n.* (also **bandoleer**) a shoulder belt with loops or pockets for cartridges. [Du. *bandelier* or F *bandoulière*, prob. formed as BANDEROLE]

bandpass /'bænd,pɑ:s/ *n.* the range of frequencies (of sound, electrical signals, etc.) which are transmitted through a filter.

bandsman /ˈbændzmən/ n. (pl. **-men**) a player in a (esp. military or brass) band.

bandstand /ˈbændstænd/ n. a covered outdoor platform for a band to play on, usu. in a park.

Bandung /ˈbændʊŋ/ a city in Indonesia; pop. (1980) 1,462,600. Founded by the Dutch in 1810, it was the capital of the former Dutch East Indies.

bandwagon /ˈbændˌwægən/ n. orig. US a wagon used for carrying a band in a parade etc. □ **climb** (or **jump**) **on the bandwagon** join a party, cause, or group that seems likely to succeed.

bandwidth /ˈbændwɪdθ/ n. the range of frequencies within a given band (see BAND¹ n. 3a).

bandy¹ /ˈbændɪ/ adj. (**bandier, bandiest**) **1** (of the legs) curved so as to be wide apart at the knees. **2** (also **bandy-legged**) (of a person) having bandy legs. [perh. f. obs. *bandy* curved stick]

bandy² /ˈbændɪ/ v.tr. (**-ies, -ied**) **1** (often foll. by *about*) **a** pass (a story, rumour, etc.) to and fro. **b** throw or pass (a ball etc.) to and fro. **2** (often foll. by *about*) discuss disparagingly (*bandied her name about*). **3** (often foll. by *with*) exchange (blows, insults, etc.) (*don't bandy words with me*). [perh. f. F *bander* f. *bande* BAND²]

bane /beɪn/ n. **1** the cause of ruin or trouble; the curse (esp. *the bane of one's life*). **2** poet. ruin; woe. **3** archaic (except in comb.) poison (*ratsbane*). □ **baneful** adj. **banefully** adv. [OE *bana* f. Gmc]

baneberry /ˈbeɪnbərɪ/ n. (pl. **-ies**) **1** a plant of the genus *Actaea*, of the buttercup family. **2** the bitter poisonous berry of this.

Banffshire /ˈbæmfʃɪə(r)/ a former county of NE Scotland which became a part of Grampian region in 1975.

bang /bæŋ/ n., v., & adv. ● n. **1 a** a loud short sound. **b** an explosion. **c** the report of a gun. **2 a** a sharp blow. **b** the sound of this. **3** esp. N. Amer. a fringe of hair cut straight across the forehead. **4** coarse sl. **a** an act of sexual intercourse. **b** a partner in sexual intercourse. **5** sl. a drug injection (cf. BHANG). **6** N. Amer. sl. a thrill (*got a bang from going fast*). ● v. **1** tr. & intr. strike or shut noisily (*banged the door shut; banged on the table*). **2** tr. & intr. make or cause to make the sound of a blow or an explosion. **3** tr. esp. N. Amer. cut (hair) in a bang. **4** coarse sl. **a** intr. have sexual intercourse. **b** tr. have sexual intercourse with. ● adv. **1** with a bang or sudden impact. **2** colloq. exactly (*bang in the middle*). □ **bang off** Brit. sl. immediately. **bang on** Brit. colloq. adj. & adv. exactly right. ● v.intr. (foll. by *about*) talk tediously and at great length. **bang up** Brit. sl. lock up, imprison. **bang-up** N. Amer. sl. first-class, excellent (esp. *bang-up job*). **go bang 1** (of a door etc.) shut noisily. **2** explode. **3** colloq. be suddenly destroyed (*bang went their chances*). **go with a bang** go successfully. [16th c.: perh. f. Scand.]

Bangalore /ˌbæŋɡəˈlɔː(r)/ a city in south central India, capital of the state of Karnataka; pop. (1991) 2,651,000.

banger /ˈbæŋə(r)/ n. Brit. **1** sl. a sausage. **2** sl. an old car, esp. a noisy one. **3** a loud firework.

Bangkok /bæŋˈkɒk/ the capital and chief port of Thailand on the Chao Phraya waterway, 40 km (25 miles) upstream from its outlet into the Gulf of Thailand; pop. (est. 1990) 5,876,000. Originally a small port, it became capital of Thailand in 1782. Rice, rubber, and timber are among the chief exports.

Bangladesh /ˌbæŋɡləˈdeʃ/ a country of the Indian subcontinent, in the Ganges delta; pop. (est. 1991) 107,992,140; official language, Bengali; capital, Dhaka. Formerly part of British India, the region became (as East Pakistan) one of the two geographical units of Pakistan. In response to civil war an independent republic was proclaimed in East Pakistan in 1971, taking the name of Bangladesh (Bengali, = land of Bengal). It became a Commonwealth state in 1972. Cyclones in the Bay of Bengal cause repeated devastation to the country. □ **Bangladeshi** adj. & n.

bangle /ˈbæŋɡ(ə)l/ n. a rigid ornamental band worn round the arm or sometimes the ankle. [Hindi *baṅglī* glass bracelet]

bangtail /ˈbæŋteɪl/ n. & v. ● n. a horse, esp. with its tail cut straight across. ● v.tr. Austral. cut the tails of (horses or cattle) as an aid to counting or identification. □ **bangtail muster** Austral. the counting of cattle involving cutting across the tufts at the tail-ends as each is counted.

Bangui /ˈbæŋɡiː/ the capital of the Central African Republic; pop. (est. 1988) 596,800.

banian var. of BANYAN.

banish /ˈbænɪʃ/ v.tr. **1** formally expel (a person), esp. from a country.

2 dismiss from one's presence or mind. □ **banishment** n. [ME f. OF *banir* ult. f. Gmc]

banister /ˈbænɪstə(r)/ n. (also **bannister**) **1** (in pl.) the uprights and handrail at the side of a staircase. **2** (usu. in pl.) an upright supporting a handrail. ¶ Often confused with *baluster*. [earlier *barrister*, corrupt. of BALUSTER]

Banjarmasin /ˌbændʒəˈmɑːsɪn/ (also **Bandjarmasin**) a deep-water port and capital of the province of Kalimantan in Indonesia, on the island of Borneo; pop. (1980) 381,300.

banjo /ˈbændʒəʊ/ n. (pl. **-os** or **-oes**) a stringed musical instrument with a neck and head like a guitar and an open-backed body consisting of parchment stretched over a metal hoop. Thought to be West African in origin, during the late 17th century it was in use among slaves in the West Indies, and by the late 18th century was found among slaves in the US. It became the characteristic instrument of Negro minstrels, and by the 1840s had gained popularity with the white population. In the 20th century it found a place in jazz and country music. □ **banjoist** n. [US southern alt. of earlier *bandore* ult. f. Gk *pandoura* three-stringed lute]

Banjul /bænˈdʒuːl/ the capital of the Gambia; pop. (1983) 44,540. Until 1973 it was known as Bathurst.

bank¹ /bæŋk/ n. & v. ● n. **1 a** the sloping edge of land by a river. **b** the area of ground alongside a river (*had a picnic on the bank*). **2** a raised shelf of ground; a slope. **3** an elevation in the sea or a river bed. **4** the artificial slope of a road etc., enabling vehicles to maintain speed round a curve. **5** a mass of cloud, fog, snow, etc. **6** the edge of a hollow place (e.g. the top of a mine-shaft). ● v. **1** tr. & intr. (often foll. by *up*) heap or rise into banks. **2** tr. heap up (a fire) tightly so that it burns slowly. **3 a** intr. (of a vehicle or aircraft or its occupant) travel with one side higher than the other in rounding a curve. **b** tr. cause (a vehicle or aircraft) to do this. **4** tr. contain or confine within a bank or banks. **5** tr. build (a road etc.) higher at the outer edge of a bend to enable fast cornering. □ **bank swallow** N. Amer. = sand martin. **bank vole** a common Eurasian woodland vole, *Cleithrionomys glareolus*, with reddish fur. [ME f. Gmc f. ON *banki* (unrecorded: cf. Olcel. *bakki*): rel. to BENCH]

bank² /bæŋk/ n. & v. ● n. **1 a** a financial establishment which uses money deposited by customers for investment, pays it out when required, makes loans at interest, exchanges currency, etc. **b** a building in which this business takes place. **2** = *piggy bank*. **3 a** the money or tokens held by the banker in some gambling games. **b** the banker in such games. **4** a place for storing anything for future use (*blood bank; data bank*). ● v. **1** tr. deposit (money or valuables) in a bank. **2** intr. engage in business as a banker. **3** intr. (often foll. by *at, with*) keep money (at a bank). **4** intr. act as banker in some gambling games. □ **bank balance** the amount of money held in a bank account at a given moment. **bank-bill 1** Brit. a bill drawn by one bank on another. **2** N. Amer. = BANKNOTE. **bank-book** = PASSBOOK. **bank card** = cheque card. **bank manager** a person in charge of a local branch of a bank. **bank on** rely on (*I'm banking on your help*). **bank statement** a printed statement of transactions and balance issued periodically to the holder of a bank account. [F *banque* or It. *banca* f. med.L *banca, bancus*, f. Gmc: rel. to BANK¹]

bank³ /bæŋk/ n. **1** a row of similar objects, esp. of keys, lights, or switches. **2** a tier of oars. [ME f. OF *banc* f. Gmc: rel. to BANK¹, BENCH]

bankable /ˈbæŋkəb(ə)l/ adj. **1** acceptable at a bank. **2** reliable (a *bankable reputation*). **3** certain to bring profit; good for the box office (*Hollywood's most bankable stars*).

banker¹ /ˈbæŋkə(r)/ n. **1** a person who manages or owns a bank or group of banks. **2 a** a keeper of the bank or dealer in some gambling games. **b** a card-game involving gambling. **3** Brit. a result forecast identically (while other forecasts differ) in several football-pool entries on one coupon. □ **banker's card** = cheque card. **banker's order** an instruction to a bank to pay money or deliver property, signed by the owner or the owner's agent. [F *banquier* f. *banque* BANK²]

banker² /ˈbæŋkə(r)/ n. **1 a** a fishing boat off Newfoundland. **b** a Newfoundland fisherman. **2** Austral. colloq. a river flooded to the top of its banks. [BANK¹ + -ER¹]

Bank for International Settlements a bank founded in 1930 to promote the cooperation of central banks and to provide facilities for international financial operations. Its main original function was to facilitate the mobilization and transfer of German reparation payments agreed after the First World War (and cancelled in 1932). It is located at Basle in Switzerland.

bank holiday n. a weekday kept as a public holiday, when banks are officially closed. In the 19th century certain saints' days (about thirty-three per year) had been kept as holidays at the Bank of England, but in 1830 the number was reduced to eighteen and in 1834 to four. The Bank Holidays Act of 1871 formally recognized certain days as bank holidays, and the number and date of these have been altered subsequently. Although they are not statutory public holidays, bank holidays are generally observed by businesses and public organizations.

banking /'bæŋkɪŋ/ n. the business transactions of a bank.

banknote /'bæŋknəʊt/ n. a banker's promissory note, esp. from a central bank, payable to the bearer on demand, and serving as money.

Bank of England the central bank of England, founded by Act of Parliament in 1694 to raise and lend money to William III towards carrying on the war against France. It has the right of issuing legal tender, manages the national debt, and administers exchange-rate policy. The British government is its chief customer; it was nationalized in 1946.

bankroll /'bæŋkrəʊl/ n. & v. N. Amer. ● n. **1** a roll of banknotes. **2** funds. ● v.tr. colloq. support financially.

bankrupt /'bæŋkrʌpt/ adj., n., & v. ● adj. **1 a** insolvent; declared in law unable to pay debts. **b** undergoing the legal process resulting from this. **2** (often foll. by of) exhausted or drained (of some quality etc.); deficient, lacking. ● n. **1 a** an insolvent person whose estate is administered and disposed of for the benefit of the creditors. **b** an insolvent debtor. **2** a person exhausted of or deficient in a certain attribute (a moral bankrupt). ● v.tr. make bankrupt. □ **bankruptcy** /-,rʌptsɪ/ n. (pl. **-ies**). [16th c.: f. It banca rotta broken bench (as BANK[2], L rumpere rupt- break), assim. to L]

Banks /bæŋks/, Sir Joseph (1743–1820), English botanist. Banks accompanied Captain James Cook on his first voyage to the Pacific. He was president of the Royal Society for over forty years and helped to establish the Royal Botanic Gardens at Kew, both as a repository of living specimens from all over the world and as a centre for the introduction of plants to new regions. Banks also imported merino sheep from Spain and sent them on to Australia. His herbarium and library became a centre of taxonomic research, later becoming part of the British Museum.

banksia /'bæŋksɪə/ n. an evergreen flowering shrub of the genus Banksia, native to Australia. □ **banksia rose** a Chinese climbing rose with small flowers. [BANKS]

banner /'bænə(r)/ n. **1 a** a large rectangular sign bearing a slogan or design and usu. carried on two side-poles or a crossbar in a demonstration or procession. **b** a long strip of cloth etc. hung across a street or along the front of a building etc. and bearing a slogan. **2 a** flag on a pole used as the standard of a king, knight, etc., esp. in battle. **3** (attrib.) US excellent, outstanding (a banner year in sales). □ **banner headline** a large newspaper headline, esp. one across the top of the front page. **join** (or **follow**) **the banner of** adhere to the cause of. **under the banner of** associated with the cause of, esp. by the use of the same slogans as, adherence to the same principles as, etc. □ **bannered** adj. [ME f. AF banere, OF baniere f. Rmc ult. f. Gmc]

banneret /'bænərɪt/ n. hist. **1** a knight who commanded his own troops in battle under his own banner. **2** a knighthood given on the battlefield for courage. [ME & OF baneret f. baniere BANNER + -et as -ATE[1]]

Bannister /'bænɪstə(r)/, Sir Roger (Gilbert) (b.1929), British middle-distance runner and neurologist. While still a medical student, in May 1954 he became the first man to run a mile in under 4 minutes, with a time of 3 minutes 59.4 seconds. He retired from athletics in the same year and went on to a distinguished medical career.

bannister var. of BANISTER.

bannock /'bænək/ n. Sc. & N. Engl. a round flat loaf, usu. unleavened. [OE bannuc, perh. f. Celt.]

Bannockburn, Battle of /'bænək,bɜːn/ a battle which took place near Stirling in central Scotland in 1314, a decisive Scottish victory over the English. The much larger English army of Edward II, advancing to break the siege of Stirling Castle, was outmanoeuvred and defeated by Robert the Bruce.

banns /bænz/ n.pl. a notice read out on three successive Sundays in a parish church, announcing an intended marriage and giving the opportunity for objections. The custom was adopted early, but seems to have developed especially in Charlemagne's reign to aid inquiry into possible blood relationship between the parties. It was made compulsory throughout the Christian world in 1215. □ **forbid the banns** raise an objection to an intended marriage, esp. in church following the reading of the banns. [pl. of BAN]

banquet /'bæŋkwɪt/ n. & v. ● n. **1** an elaborate usu. extensive feast. **2** a dinner for many people followed by speeches in favour of a cause or in celebration of an event. ● v. (**banqueted, banqueting**) **1** intr. hold a banquet; feast. **2** tr. entertain with a banquet. □ **banqueter** n. [F, dimin. of banc bench, BANK[2]]

banquette /bæŋ'ket/ n. **1** an upholstered bench along a wall, esp. in a restaurant or bar. **2** a raised step behind a rampart. [F f. It. banchetta dimin. of banca bench, BANK[2]]

banshee /'bænʃiː/ n. Ir. & Sc. a female spirit whose wailing warns of a death in a house. [Ir. bean sídhe f. OIr. ben síde woman of the fairies]

bantam /'bæntəm/ n. **1** a small breed of domestic fowl, of which the cock is aggressive. **2** a small but aggressive person. [app. f. Banten in Java, although the fowl is not native there]

bantamweight /'bæntəm,weɪt/ n. **1** a weight in certain sports intermediate between flyweight and featherweight, in the amateur boxing scale 51–4 kg but differing for professional boxers, wrestlers, and weightlifters. **2** a sportsman of this weight.

banter /'bæntə(r)/ n. & v. ● n. good-humoured teasing. ● v. **1** tr. ridicule in a good-humoured way. **2** intr. talk humorously or teasingly. □ **banterer** n. [17th c.: orig. unkn.]

Banting /'bæntɪŋ/, Sir Frederick Grant (1891–1941), Canadian physiologist and surgeon. Banting initiated research into the secretion of the pancreas in a laboratory provided by J. J. R. Macleod. A series of experiments with dogs, carried out with C. H. Best's assistance, led to the discovery of insulin in 1921–2. They then purified the extracts of insulin and used them to treat diabetes, which had previously been an incurable and fatal disease. Banting and Macleod shared a Nobel Prize in 1923; an institute named after Banting was later opened in Toronto.

Bantu /bæn'tuː/ n. & adj. ● n. (pl. same or **Bantus**) **1** often offens. **a** a large group of Negroid peoples of central and southern Africa. **b** a member of any of these peoples. **2** the group of languages spoken by these peoples. (See note below.) ● adj. of or relating to these peoples or languages. [Bantu, = people]

▪ Bantu languages belong to the Niger-Congo language group, and there are more than 400 of them (with over 100 million speakers), of which Swahili, Xhosa, and Zulu are the most important. Their chief characteristics are the distinctiveness of high and low tones, the complex system of noun classes, often based on distinctions in meaning, and graduations of tense, often to a mouth even four degrees of past. Most Bantu languages were not written down until the 19th century. Arabs trading along the coast had brought their Arabic script, which has been used for Swahili, but elsewhere the Roman alphabet has been adopted, sometimes with additional characters. Linguistic evidence suggests that the original home of these languages may have been in the Cameroon region. Bantu-speaking people are thought to have migrated to southern Africa via the lake region of East Africa by the 3rd century AD, possibly introducing iron metallurgy to southern Africa at the time of their entry.

Bantustan /,bæntu:'stɑːn/ n. S. Afr. hist. colloq. = HOMELAND 2. [BANTU + -stan as in Hindustan]

banyan /'bænjən/ n. (also **banian**) **1** an Indian fig tree, Ficus benghalensis, the branches of which hang down and root themselves. **2** a Hindu trader. **3** a loose flannel jacket, shirt, or gown worn in India. [Port. banian f. Gujarati vāṇiyo man of trading caste, f. Skr.: applied orig. to one such tree under which banyans had built a pagoda]

banzai /bɑːn'zaɪ/ int. **1** a Japanese battle cry. **2** a form of greeting used to the Japanese emperor. [Jap., = ten thousand years (of life to you)]

baobab /'beɪəʊ,bæb/ n. an African tree, Adansonia digitata, with an enormously thick trunk and large edible pulpy fruit hanging down on stalks. [L (1592), prob. f. an Afr. language]

BAOR abbr. British Army of the Rhine.

Baotou /baʊ'təʊ/ an industrial city in Inner Mongolia, northern China, on the Yellow River; pop. (1990) 1,180,000.

bap /bæp/ n. Brit. a soft flattish bread roll. [16th c.: orig. unkn.]

baptism /'bæptɪz(ə)m/ n. **1 a** the religious rite, symbolizing admission to the Christian Church, of sprinkling the forehead with water, or (usu. only with adults) by immersion, generally accompanied by name-giving. **b** the act of baptizing or being baptized. **2** an initiation, e.g. into battle. **3** the naming of ships, church bells, etc. □ **baptism of fire 1** initiation into battle. **2** a painful new undertaking or experience. □ **baptismal** /bæp'tɪzm(ə)l/ adj. [ME f. OF ba(p)te(s)me f. eccl.L baptismus f. eccl.Gk baptismos f. baptizō BAPTIZE]

baptist /'bæptɪst/ n. **1** a person who baptizes, esp. John the Baptist.

2 (Baptist) a member of an evangelical Protestant religious denomination advocating baptism (by total immersion) only for believers, who must be old enough to hold their beliefs consciously and rationally. (*See note below.*) [ME f. OF *baptiste* f. eccl.L *baptista* f. eccl.Gk *baptistēs* f. *baptizō* BAPTIZE]

▪ Baptists form one of the largest Protestant bodies and are found throughout the world and especially in the US. Present-day membership is over 30 million. The first group was founded in Amsterdam in 1609 by the exiled English dissenter John Smyth (c.1554–1612), some of whose followers returned to London and established a Baptist Church in England. Churches arising from this were known as General Baptists, believing that anyone can be saved; those founded by a group of Calvinists, who held that salvation was only for a particular few, were known as Strict or Particular Baptists. Rigid Calvinism was gradually modified, and the two groups had largely merged by the middle of the 19th century.

baptistery /'bæptɪstrɪ/ n. (also **baptistry**) (pl. **-ies**) **1 a** the part of a church used for baptism. **b** hist. a building next to a church, used for baptism. **2** (in a Baptist chapel) a sunken receptacle used for total immersion. [ME f. OF *baptisterie* f. eccl.L *baptisterium* f. eccl.Gk *baptistērion* bathing-place f. *baptizō* BAPTIZE]

baptize /bæp'taɪz/ v.tr. (also **-ise**) **1** (also absol.) administer baptism to. **2** give a name or nickname to; christen. [ME f. OF *baptiser* f. eccl.L *baptizare* f. Gk *baptizō* immerse, baptize]

bar[1] /bɑː(r)/ n., v., & prep. ● n. **1** a long rod or piece of rigid wood, metal, etc., esp. used as an obstruction, confinement, fastening, or weapon. **2 a** something resembling a bar in being (thought of as) straight, narrow, and rigid (*bar of soap*; *bar of chocolate*). **b** a band of colour or light, esp. on a flat surface. **c** the heating element of an electric fire. **d** = CROSSBAR. **e** Brit. a metal strip below the clasp of a medal, awarded as an extra distinction. **f** a sandbank or shoal at the mouth of a harbour or an estuary. **g** Brit. a rail marking the end of each chamber in the Houses of Parliament. **h** Heraldry a narrow horizontal stripe across a shield. **3 a** a barrier of any shape. **b** a restriction (*colour bar*; *a bar to promotion*). **4 a** a counter in a public house, restaurant, or café across which alcohol or refreshments are served. **b** a room in a public house in which customers may sit and drink. **c** a public house. **d** a small shop or stall serving refreshments (*snack bar*). **e** a specialized department in a large store (*heel bar*). **5 a** an enclosure in which a prisoner stands in a lawcourt. **b** a public standard of acceptability, before which a person is said to be tried (*bar of conscience*). **c** a plea arresting an action or claim in a law case. **d** a particular court of law. **6** Mus. **a** any of the sections of usu. equal time-value into which a musical composition is divided by vertical lines across the staff. **b** = bar-line below. **7 (the Bar)** Law **a** barristers collectively. **b** the profession of barrister. ● v.tr. (**barred**, **barring**) **1 a** fasten (a door, window, etc.) with a bar or bars. **b** (usu. foll. by in, out) shut or keep in or out (*barred him in*). **2** obstruct, prevent (*bar his progress*). **3 a** (usu. foll. by from) prohibit, exclude (*bar them from attending*). **b** exclude from consideration (cf. BARRING). **4** mark with stripes. **5** Law prevent or delay (an action) by objection. ● prep. **1** except (*all were there bar a few*). **2** Horse-racing except (the horses indicated: used in stating the odds, indicating the number of horses excluded) (*33–1 bar three*). □ **bar billiards** a form of billiards in which balls are struck into holes in the table. **bar chart** (or **graph**) a chart (or graph) using bars to represent quantity. **bar-code** a machine-readable code in the form of a pattern of stripes printed on and identifying a commodity, used esp. for stock-control. **bar-line** Mus. a vertical line used to mark divisions between bars. **bar none** with no exceptions. **bar person** a barmaid or barman. **bar-room** a room with a bar selling alcoholic drinks. **bar sinister** = bend sinister (see BEND[2]). **bar tracery** Archit. tracery with strips of stone across an aperture. **be called to the Bar** Brit. be admitted as a barrister. **be called within the Bar** Brit. be appointed a Queen's Counsel. **behind bars** in prison. **the outer Bar** barristers who are not Queen's Counsels. [ME f. OF *barre*, *barrer*, f. Rmc]

bar[2] /bɑː(r)/ n. esp. Meteorol. a unit of pressure equal to 100,000 newtons per square metre, approx. one atmosphere. [Gk *baros* weight]

barathea /ˌbærə'θɪə/ n. a fine woollen cloth, sometimes mixed with silk or cotton, used esp. for coats, suits, etc. [19th c.: orig. unkn.]

Barb /bɑːb/ abbr. (also **BARB**) Broadcasters' Audience Research Board.

barb /bɑːb/ n. & v. ● n. **1** a secondary backward-facing projection from an arrow, fish-hook, etc., angled to make extraction difficult. **2** a deliberately hurtful remark. **3** a beardlike filament at the mouth of some fish, e.g. barbel and catfish. **4** any one of the fine hairlike filaments growing from the shaft of a feather, forming the vane. ● v.tr. **1** provide (an arrow, a fish-hook, etc.) with a barb or barbs. **2** (as **barbed** adj.) (of a remark etc.) deliberately hurtful. □ **barbed wire** wire bearing

sharp pointed spikes close together and used in fencing, or in warfare as an obstruction. [ME f. OF *barbe* f. L *barba* beard]

Barbados /bɑː'beɪdɒs/ the most easterly of the Caribbean islands, one of the Windward Islands group; pop. (est. 1991) 258,000; official language, English; capital, Bridgetown. The first Europeans to reach Barbados were the Spanish in 1518, and the island was largely depopulated of its native inhabitants (probably Arawaks) by the early 17th century. Barbados became a British colony in the 1630s and remained British until 1966 when it gained independence as a Commonwealth state. The economy is based on tourism, sugar, and light manufacturing industries. □ **Barbadian** adj. & n.

barbarian /bɑː'beərɪən/ n. & adj. ● n. **1** an uncultured or brutish person; a lout. **2** a member of a primitive community or tribe. ● adj. **1** rough and uncultured. **2** uncivilized. [orig. of any foreigner with a different language or customs: F *barbarien* f. *barbare* (as BARBAROUS)]

barbaric /bɑː'bærɪk/ adj. **1** brutal; cruel (*flogging is a barbaric punishment*). **2** rough and uncultured; unrestrained. **3** of or like barbarians and their art or taste; primitive. □ **barbarically** adv. [ME f. OF *barbarique* or L *barbaricus* f. Gk *barbarikos* f. *barbaros* foreign]

barbarism /'bɑːbəˌrɪz(ə)m/ n. **1 a** the absence of culture and civilized standards; ignorance and rudeness. **b** an example of this. **2** a word or expression not considered correct; a solecism. **3** anything considered to be in bad taste. [F *barbarisme* f. L *barbarismus* f. Gk *barbarismos* f. *barbarizō* speak like a foreigner f. *barbaros* foreign]

barbarity /bɑː'bærɪtɪ/ n. (pl. **-ies**) **1** savage cruelty. **2** an example of this.

barbarize /'bɑːbəˌraɪz/ v.tr. & intr. (also **-ise**) make or become barbarous. □ **barbarization** /ˌbɑːbəraɪ'zeɪʃ(ə)n/ n.

Barbarossa /ˌbɑːbə'rɒsə/ see FREDERICK I.

barbarous /'bɑːbərəs/ adj. **1** uncivilized. **2** cruel. **3** coarse and unrefined. □ **barbarously** adv. **barbarousness** n. [orig. of any foreign language or people: f. L f. Gk *barbaros* foreign]

Barbary /'bɑːbərɪ/ (also **Barbary States**) hist. the Saracen countries of North and NW Africa, together with Moorish Spain. The area was noted between the 16th and 18th centuries as a haunt of pirates. See also MAGHRIB. [ult. f. Arab. *Barbar* BERBER, applied by ancient Arab geographers to the natives of North Africa west and south of Egypt]

Barbary ape n. a tailless macaque monkey, *Macaca sylvana*, of North Africa and Gibraltar.

Barbary Coast a former name for the Mediterranean coast of North Africa from Morocco to Egypt.

barbastelle /ˌbɑːbə'stel, 'bɑːbəˌstel/ n. a bat of the genus *Barbastella*; esp. *B. barbastellus*, found rarely in western Europe. [F f. It. *barbastello*]

barbecue /'bɑːbɪˌkjuː/ n. & v. ● n. **1 a** a meal cooked on an open fire out of doors, esp. meat grilled on a metal appliance. **b** a party at which such a meal is cooked and eaten. **2 a** the metal appliance used for the preparation of a barbecue. **b** a fireplace, usu. of brick, containing such an appliance. ● v.tr. (**barbecues**, **barbecued**, **barbecuing**) cook (esp. meat) on a barbecue. □ **barbecue sauce** a highly seasoned sauce, usu. containing chillies, in which meat etc. may be cooked. [Sp. *barbacoa* f. Haitian *barbacòa* wooden frame on posts]

barbel /'bɑːb(ə)l/ n. **1** a large European freshwater fish of the genus *Barbus*, with fleshy filaments hanging from its mouth. **2** such a filament growing from the mouth of any fish. [ME f. OF f. LL *barbellus* dimin. of *barbus* barbel f. *barba* beard]

barbell /'bɑːbel/ n. an iron bar with a series of graded discs at each end, used for weightlifting exercises. [BAR[1] + BELL[1]]

Barber /'bɑːb(r)/, Samuel (1910–81), American composer. He travelled extensively in Europe and developed a style based on romanticism allied to classical forms; his music includes operas, ballets, choral works, and orchestral and chamber music. His best-known works include the *Adagio for Strings* (1936) and the opera *Vanessa* (1958).

barber /'bɑːbə(r)/ n. & v. ● n. a person who cuts men's and boys' hair and shaves or trims beards as an occupation; a men's hairdresser. Until Henry VIII's time barbers were also regular practitioners in dentistry and surgery. ● v.tr. **1** cut the hair of; shave or trim the beard of. **2** cut or trim closely (*barbered the grass*). □ **barber's pole** a spirally painted striped red and white pole hung outside barbers' shops as a business sign. [ME & AF f. OF *barbeor* f. med.L *barbator* -oris f. *barba* beard]

barberry /'bɑːbərɪ/ n. (pl. **-ies**) **1** a shrub of the genus *Berberis*, with triple spines on the branches; esp. *B. vulgaris*, with yellow flowers and red berries, often grown as hedges. **2** its berry. [ME f. OF *berberis*, of unkn. orig.: assim. to BERRY]

barber's shop quartet n. a quartet of male voices singing in close harmony, usually without accompaniment. The style appears to have originated in England in the 16th–17th centuries, when customers waiting their turn in a barber's shop would pass the time by harmonizing to a lute or guitar provided for their amusement. The style of singing had died out in England by the early part of the 18th century but was maintained in America, where it has enjoyed a 20th-century revival; it has now experienced a revival in Britain also.

barbet /'bɑːbɪt/ n. a brightly coloured fruit-eating tropical bird of the family Capitonidae, with bristles at the base of its bill. [F f. barbe beard]

barbette /bɑːˈbet/ n. a platform in a fort or ship from which guns can be fired over a parapet etc. without an embrasure. [F, dimin. of barbe beard]

barbican /'bɑːbɪkən/ n. the outer defence of a city, castle, etc., esp. a double tower above a gate or drawbridge. [ME f. OF barbacane, of unkn. orig.]

Barbican Centre an extensive arts complex in the City of London, opened in 1982. It includes two theatres, a concert hall, cinemas, and an art gallery.

barbie /'bɑːbɪ/ n. Austral. colloq. a barbecue. [abbr.]

Barbie doll /'bɑːbɪ/ n. 1 propr. a doll representing a conventionally attractive young woman. 2 colloq. a pretty but characterless or passive young woman.

Barbirolli /ˌbɑːbɪˈrɒlɪ/, Sir John (Giovanni Battista) (1899–1970), English conductor, of Franco-Italian descent. Originally a cellist, he began his conducting career in 1924. He subsequently became conductor of several major opera companies and orchestras, including Covent Garden in Britain and the New York Philharmonic in the US. In 1943 he returned to England as conductor of the Hallé Orchestra, Manchester, where he was responsible for rebuilding the orchestra's reputation; he was appointed conductor laureate there for life in 1968 in recognition of his contribution.

barbitone /'bɑːbɪˌtəʊn/ n. (US **barbital** /-t(ə)l/) a sedative drug. [as BARBITURIC ACID + -ONE, -AL²]

barbiturate /bɑːˈbɪtjʊrət, -ˌreɪt/ n. any derivative of barbituric acid used in the preparation of sedative and sleep-inducing drugs.

barbituric acid /ˌbɑːbɪˈtjʊərɪk/ n. Chem. an organic acid from which various sedatives and sleep-inducing drugs are derived. [F barbiturique f. G Barbitursäure (Säure acid) f. the name Barbara]

Barbizon School /'bɑːbɪz(ə)n/ a group of French landscape painters of the 1840s, who reacted against classical conventions and based their art on direct study of nature. Théodore Rousseau was the leader of the group, which included Charles Daubigny, Narcisse Virgile Diaz (1807–76), Jean-François Millet (1814–75), and Jules Dupré (1811–89); they took their name from a small village in the forest of Fontainebleau near Paris where Rousseau and others worked. Their fresh naturalistic approach was strongly influenced by British painters such as Constable, as well as by 17th-century Dutch traditions.

barbola /bɑːˈbəʊlə/ n. (in full **barbola work**) 1 the craft of making small models of fruit, flowers, etc. from a plastic paste. 2 articles, e.g. mirrors, decorated with such models. [arbitrarily f. barbotine clay slip for ornamenting pottery]

Barbour /'bɑːbə(r)/, John (c.1320–95), Scottish poet and prelate. He was Archdeacon of Aberdeen (1357–95), and probably taught at Oxford and Paris. The only poem ascribed to him with certainty is The Bruce, a verse chronicle relating the deeds of Robert the Bruce and his follower James Douglas, and including an account of the Battle of Bannockburn.

Barbour jacket n. propr. a type of green waxed jacket. [Barbour, name of a draper]

Barbuda an island in the West Indies: see ANTIGUA AND BARBUDA. □ **Barbudan** adj. & n.

barbule /'bɑːbjuːl/ n. a minute filament projecting from the barb of a feather. [L barbula, dimin. of barba beard]

barbwire /'bɑːbˌwaɪə(r)/ n. US = barbed wire (see BARB).

barcarole /'bɑːkəˌrəʊl/ n. (also **barcarolle** /-ˌrɒl/) 1 a song sung by Venetian gondoliers. 2 music in imitation of this. [F barcarolle f. Venetian It. barcarola boatman's song f. barca boat]

Barcelona /ˌbɑːsəˈləʊnə/ a city on the coast of NE Spain, capital of Catalonia; pop. (1991) 1,653,175. It is a large seaport and industrial city and a leading cultural centre. It was the seat of the Republican government during the Spanish Civil War.

barchan /'bɑːkən/ n. a crescent-shaped shifting sand-dune. [Turkic barkhan]

Bar-Cochba /bɑːˈkɒkbə/ Jewish leader of a rebellion in AD 132 (as designated in Christian sources; Jewish sources call him Simeon). He led the Jewish rebellion against Hadrian's intention to rebuild Jerusalem as a non-Jewish city, and claimed to be, and was accepted by some of his Jewish contemporaries as, the Messiah. A number of letters in his handwriting have been found in archaeological excavations near the Dead Sea in Israel. [Aram., = son of a star]

Barcoo /bɑːˈkuː/ adj. Austral. of or relating to a remote area of the country. □ **Barcoo rot** scurvy. **Barcoo sickness** illness marked by attacks of vomiting. **Barcoo sore** an ulcer characteristic of Barcoo rot. [river in W. Qld.]

bard¹ /bɑːd/ n. 1 a hist. a Celtic minstrel. b the winner of a prize for Welsh verse at an eisteddfod. 2 poet. a poet, esp. one treating heroic themes. □ **the Bard** (or **the Bard of Avon**) Shakespeare. □ **bardic** adj. [Gael. & Ir. bárd, Welsh bardd, f. OCelt.]

bard² /bɑːd/ n. & v. ● n. a rasher of fat bacon placed on meat or game before roasting. ● v.tr. cover (meat etc.) with bards. [F barde, orig. = horse's breastplate, ult. f. Arab.]

Bardot /bɑːˈdəʊ/, Brigitte (born Camille Javal) (b.1934), French actress. She made her film début in 1952, but it was And God Created Woman in 1956 that established her reputation as an international sex symbol. Subsequent films include Love is My Profession (1959) and A Very Private Affair (1962). After retiring from acting she became an active supporter of animal welfare.

bardy /'bɑːdɪ/ n. (pl. **-ies**) Austral. an edible wood-boring grub. [Aboriginal]

bare /beə(r)/ adj. & v. ● adj. 1 (esp. of part of the body) unclothed or uncovered (with bare head). 2 without appropriate covering or contents: a (of a tree) leafless. b unfurnished; empty (bare rooms; the cupboard was bare). c (of a floor) uncarpeted. 3 a undisguised (the bare truth). b unadorned (bare facts). 4 (attrib.) a scanty (a bare majority). b mere (bare necessities). ● v.tr. 1 uncover, unsheathe (bared his teeth). 2 reveal (bared his soul). □ **bare of** without. **with one's bare hands** without using tools or weapons. □ **bareness** n. [OE bær, barian f. Gmc]

bareback /'beəbæk/ adj. & adv. on an unsaddled horse, donkey, etc.

Barebones Parliament /'beəbəʊnz/ the nickname of Cromwell's Parliament of 1653, from one of its members, Praise-God Barbon, an Anabaptist leather-seller of Fleet Street. It replaced the Rump Parliament, but was itself dissolved within a few months.

barefaced /'beəfeɪst/ adj. undisguised; impudent (barefaced cheek). □ **barefacedly** /beəˈfeɪsɪdlɪ/ adv. **barefacedness** n.

barefoot /'beəfʊt/ adj. & adv. (also **barefooted** /beəˈfʊtɪd/) with nothing on the feet. □ **barefoot doctor** a paramedical worker with basic medical training, esp. in China.

barège /bæˈreʒ/ n. a silky gauze made from wool or other material. [F f. Barèges in SW France, where it was orig. made]

bareheaded /beəˈhedɪd/ adj. & adv. without a covering for the head.

Bareilly /bəˈreɪlɪ/ an industrial city in northern India, in Uttar Pradesh; pop. (1991) 583,000.

barely /'beəlɪ/ adv. 1 only just; scarcely (barely escaped). 2 scantily (barely furnished). 3 archaic openly, explicitly.

Barents /'bærənts/, Willem (d.1597), Dutch explorer. The leader of several expeditions in search of the North-east Passage to Asia, Barents discovered Spitsbergen and reached Novaya Zemlya, off the coast of which he died.

Barents Sea a part of the Arctic Ocean to the north of Norway and Russia, bounded to the west by Svalbard, to the north by Franz Josef Land, and to the east by Novaya Zemlya. It is named after Willem Barents.

barf /bɑːf/ v. & n. sl. ● v.intr. vomit or retch. ● n. an attack of vomiting. [20th c.: orig. unkn.]

barfly /'bɑːflaɪ/ n. (pl. **-flies**) colloq. a person who frequents bars. □ **barfly jumping** the sport of jumping at and sticking to a Velcro-covered wall while wearing a Velcro suit.

bargain /'bɑːgɪn/ n. & v. ● n. 1 a an agreement on the terms of a transaction or sale. b this seen from the buyer's viewpoint (a bad bargain). 2 something acquired or offered cheaply. ● v.intr. (often foll. by with, for) discuss the terms of a transaction (expected him to bargain, but he paid up; bargained with her; bargained for the table). □ **bargain away** part with for something worthless (had bargained away the estate). **bargain basement** the basement of a shop where bargains are displayed. **bargain for** (or *colloq.* **on**) (usu. with *neg.* actual or implied) be prepared for; expect (*didn't bargain for bad weather; more than I bargained for*). **bargain on** rely on. **drive a hard bargain** be severe or uncompromising in making a bargain. **into** (*US* **in**) **the bargain** moreover; in addition to what was expected. **make** (or **strike**) **a bargain** agree a transaction. □ **bargainer** *n.* [ME f. OF *bargaine, bargaignier,* prob. f. Gmc]

barge /bɑːdʒ/ *n. & v.* ● *n.* **1** a long flat-bottomed boat for carrying freight on canals, rivers, etc. **2** a long ornamental boat used for pleasure or ceremony. **3** a boat used by the chief officers of a warship. ● *v.intr.* **1** (often foll. by *around*) lurch or rush clumsily about. **2** (foll. by *in, into*) **a** intrude or interrupt rudely or awkwardly (*barged in while we were kissing*). **b** collide with (*barged into her*). [ME f. OF perh. f. med.L *barica* f. Gk *baris* Egyptian boat]

bargeboard /bɑːdʒbɔːd/ *n.* a board (often ornamental) fixed to the gable-end of a roof to hide the ends of the roof timbers. [perh. f. med.L *bargus* gallows]

bargee /bɑːˈdʒiː/ *n. Brit.* a person in charge of or working on a barge.

bargepole /bɑːdʒpəʊl/ *n.* a long pole used for punting barges etc. and for fending off obstacles. □ **would not touch with a bargepole** refuse to be associated or concerned with (a person or thing).

Bari /bɑːrɪ/ an industrial seaport on the Adriatic coast of SE Italy, capital of Apulia region; pop. (1990) 353,030.

barilla /bəˈrɪlə/ *n.* **1** a plant of the genus *Salsola,* found chiefly in Spain and Sicily. **2** an impure alkali made by burning either this or kelp. [Sp.]

Bariloche see SAN CARLOS DE BARILOCHE.

Barisal /ˈbʌrɪˌsʌl/ a river port in southern Bangladesh, on the Ganges delta; pop. (1991) 180,010.

barite /ˈbeəraɪt/ *n.* (also **baryte, barytes** /bəˈraɪtiːz/) a mineral form of barium sulphate. [Gk *barus* heavy + -ITE¹]

baritone /ˈbærɪˌtəʊn/ *n. & adj.* ● *n.* **1 a** the second-lowest adult male singing voice. **b** a singer with this voice. **c** a part written for it. **2 a** an instrument that is second-lowest in pitch in its family. **b** a player of such an instrument. ● *adj.* second-lowest in musical pitch. [It. *baritono* f. Gk *barutonos* f. *barus* heavy + *tonos* TONE]

barium /ˈbeərɪəm/ *n.* a soft white reactive metallic chemical element (atomic number 56; symbol **Ba**). Barium was first isolated by Sir Humphry Davy in 1808, and is one of the alkaline earth metals. It occurs in a number of minerals, notably barytes. Barium compounds are used in water purification, the glass industry, and pigments, and as an ingredient of signal flares and fireworks, giving a bright yellowish-green colour to the flame. Barium oxide is a component of high-temperature superconductors. □ **barium meal** a mixture of barium sulphate and water, which is opaque to X-rays, and is given to patients requiring radiological examination of the stomach and intestines. [BARYTA]

bark¹ /bɑːk/ *n. & v.* ● *n.* **1** the sharp explosive cry of a dog, fox, etc. **2** a sound resembling this cry. ● *v.* **1** *intr.* (of a dog, fox, etc.) give a bark. **2** *tr. & intr.* speak or utter sharply or brusquely. **3** *intr. colloq.* cough fiercely. **4** *tr. US* sell or advertise publicly by calling out; tout. □ **one's bark is worse than one's bite** one is not as ferocious as one appears. **bark up the wrong tree** be on the wrong track; make an effort in the wrong direction. [OE *beorcan*]

bark² /bɑːk/ *n. & v.* ● *n.* **1** the tough protective outer sheath of the trunks, branches, and twigs of trees or woody shrubs. **2** this material used for tanning leather or dyeing material. ● *v.tr.* **1** graze or scrape (one's shin etc.). **2** strip bark from (a tree etc.). **3** tan or dye (leather etc.) using the tannins found in bark. □ **bark beetle** a small wood-boring beetle of the family Scolytidae. [ME f. OIcel. *börkr bark-:* perh. rel. to BIRCH]

bark³ /bɑːk/ *n. poet.* a ship or boat. [= BARQUE]

barkeeper /bɑːˌkiːpə(r)/ *n.* (also **barkeep**) *N. Amer.* a person serving drinks in a bar.

barkentine var. of BARQUENTINE.

Barker /bɑːkə(r)/, George (Granville) (b.1913), English poet. A self-styled 'Augustinian anarchist', he displays in his works a penchant for puns, distortion, and abrupt changes of tone. In his *True Confession of George Barker* (1950, augmented 1965), he presented himself as both irreverent and guilt-ridden.

barker /bɑːkə(r)/ *n.* a tout at an auction, sideshow, etc., who calls out for custom to passers-by. [BARK¹ + -ER¹]

Barkly Tableland /bɑːklɪ/ a plateau region lying to the north-east of Tennant Creek in Northern Territory, Australia.

barley /bɑːlɪ/ *n.* **1** a hardy awned cereal of the genus *Hordeum,* grown for food, and to produce malt for use in beer and in malt liquors such as whisky. **2** the grain produced from this (cf. *pearl barley*). □ **barley sugar** an amber-coloured sweet made of boiled sugar, traditionally shaped as a twisted stick. **barley water** a drink made from water and a boiled barley mixture. [OE *bærlic* (adj.) f. *bære, bere* barley]

barleycorn /bɑːlɪˌkɔːn/ *n.* **1** the grain of barley. **2** a former unit of length (about a third of an inch) based on the length of a grain of barley.

barleymow /bɑːlɪˌməʊ/ *n. Brit.* a stack of barley.

barm /bɑːm/ *n.* **1** the froth on fermenting malt liquor. **2** *archaic* or *dial.* yeast or leaven. [OE *beorma*]

barmaid /bɑːmeɪd/ *n.* a woman serving behind the bar of a public house, hotel, etc.

barman /bɑːmən/ *n.* (*pl.* **-men**) a man serving behind the bar of a public house, hotel, etc.

barmbrack /bɑːmbræk/ *n.* (also **barnbrack** /bɑːnbræk/) *Ir.* soft spicy bread with currants etc. [Ir. *bairin breac* speckled cake]

Barmecide /bɑːmɪˌsaɪd/ *adj. & n.* ● *adj.* illusory, imaginary; such as to disappoint. ● *n.* a giver of benefits that are illusory or disappointing. [the name of a wealthy man in the *Arabian Nights* who gave a beggar a feast consisting of ornate but empty dishes]

bar mitzvah /bɑː ˈmɪtsvə/ *n.* **1** the religious initiation ceremony of a Jewish boy who has reached the age of 13. **2** the boy undergoing this ceremony. [Heb., = 'son of the commandment']

barmy /bɑːmɪ/ *adj.* (**barmier, barmiest**) esp. *Brit. sl.* crazy, stupid. □ **barmily** *adv.* **barminess** *n.* [earlier = frothy, f. BARM]

barn¹ /bɑːn/ *n.* **1** a large farm building for storing grain etc. **2** *derog.* a large plain or unattractive building. **3** *N. Amer.* a large shed for storing road or railway vehicles. □ **barn dance 1** an informal social gathering for country dancing, orig. in a barn. **2** a dance for a number of couples forming a line or circle, with couples moving along it in turn. **barn owl** a pale-plumaged owl, *Tyto alba,* often nesting in barns, and found throughout the world (also called *screech owl*). **barn swallow** *N. Amer.* see SWALLOW². [OE *bern, beren* f. *bere* barley + *ern, ærn* house]

barn² /bɑːn/ *n. Physics* a unit of area (symbol: **b**) equal to 10⁻²⁸ square metres, used in expressing nuclear cross-sections. [f. phrase 'as big as a barn door']

Barnabas, St /bɑːnəbəs/ a Cypriot Levite and Apostle. He introduced St Paul to the Apostles and accompanied him on the first missionary journey to Cyprus and Asia Minor, returning to Cyprus after they disagreed and separated (Acts 4–15). The traditional founder of the Cypriot Church, he is said to have been martyred in Cyprus in AD 61. Feast day, 11 June.

barnacle /bɑːnək(ə)l/ *n.* **1** a small marine crustacean of the class Cirripedia, the adults of which have valved shells and attach themselves to rocks, ships' bottoms, etc. **2** a tenacious attendant or follower who cannot easily be shaken off. □ **barnacle goose** an Arctic goose, *Branta leucopsis,* which has a black head with white cheeks, and winters in northern Europe. □ **barnacled** *adj.* [ME *bernak* (= med.L *bernaca*), of unkn. orig.]

Barnard /bɑːnɑːd/, Christiaan Neethling (b.1922), South African surgeon. A pioneer in the field of human heart transplantation, he performed the first operation of this kind in Dec. 1967.

Barnardo /bəˈnɑːdəʊ/, Thomas John (1845–1905), Irish-born doctor and philanthropist. He went to London in 1866 and while still a student of medicine, he founded the East End Mission for destitute children (1867), the first of many such homes named after him, which now cater chiefly for those with physical and mental disabilities.

Barnaul /bɑːnəˈuːl/ the capital of Altai territory on the River Ob; pop. (1990) 603,000.

barnbrack var. of BARMBRACK.

barney /bɑːnɪ/ *n.* (*pl.* **-eys**) *Brit. colloq.* a noisy quarrel. [perh. dial.]

Barnsley /bɑːnzlɪ/ a town in northern England, the administrative centre of South Yorkshire; pop. (1991) 217,300.

barnstorm /bɑːnstɔːm/ *v.intr.* **1** tour rural districts giving theatrical performances (formerly often in barns). **2** *N. Amer.* make a rapid tour holding political meetings. **3** *N. Amer. Aeron.* give informal flying exhibitions; do stunt flying. □ **barnstormer** *n.*

Barnum /bɑːnəm/, P(hineas) T(aylor) (1810–91), American showman. He became famous in the mid-19th century for his extravagant advertising and exhibition of freaks at his American Museum in New York. He billed his circus, opened in 1871, as 'The Greatest Show on Earth'; ten years later he joined forces with his former rival Anthony Bailey (1847–1906) to found the Barnum and Bailey circus.

barnyard /'bɑːnjɑːd/ n. the area around a barn; a farmyard.

Baroda /bə'rəʊdə/ **1** a former princely state of western India, now part of Gujarat. **2** the former name (until 1976) for VADODARA.

barograph /'bærə,grɑːf/ n. a barometer that records its readings on a moving chart. [Gk baros weight + -GRAPH]

barometer /bə'rɒmɪtə(r)/ n. **1** an instrument which measures atmospheric pressure, esp. for forecasting changes in the weather. (See note below.) **2** anything which reflects changes in circumstances, opinions, etc. □ **barometry** n. **barometric** /,bærə'metrɪk/ adj. **barometrical** adj.

▪ The principle of the mercury barometer, in which air pressure sustains a column of mercury, was established by Evangelista Torricelli in 1644. A successful aneroid barometer (less sensitive than the mercury type but portable and more convenient) was devised in France in the 1840s. Besides their use in meteorology, barometers are also used to measure altitude.

baron /'bærən/ n. **1 a** a member of the lowest order of the British nobility, styled Lord ——. (See note below.) **b** a similar member of a foreign nobility. **2** an important businessman or other powerful or influential person (sugar baron; newspaper baron). **3** hist. a person who held lands or property from the sovereign or a powerful overlord. □ **baron of beef** an undivided double sirloin. [ME f. AF barun, OF baron f. med.L baro, -onis man, of unkn. orig.]

▪ The use of the word as a title, as distinct from a description of a feudal relationship (see sense 3) or membership of a royal council, dates from the creation of barons by patent, which began in the reign of Richard II.

baronage /'bærənɪdʒ/ n. **1** barons or nobles collectively. **2** an annotated list of barons or peers. [ME f. OF barnage (as BARON)]

baroness /'bærənɪs/ n. **1** a woman holding the rank of baron either as a life peerage or as a hereditary rank. **2** the wife or widow of a baron. [ME f. OF baronesse (as BARON)]

baronet /'bærənɪt/ n. a member of the lowest hereditary titled British order, ranking next below a baron and above all orders of knighthood except that of the Garter. The title dates from 1611 in England and 1625 in Scotland; since 1801 new baronetcies have been of the United Kingdom. [ME f. AL baronettus (as BARON)]

baronetage /'bærənɪtɪdʒ/ n. **1** baronets collectively. **2** an annotated list of baronets.

baronetcy /'bærənɪtsɪ/ n. (pl. -ies) the domain, rank, or tenure of a baronet.

baronial /bə'rəʊnɪəl/ adj. of, relating to, or befitting barons.

barony /'bærənɪ/ n. (pl. -ies) **1** the domain, rank, or tenure of a baron. **2** (in Ireland) a division of a county. **3** (in Scotland) a large manor or estate. [ME f. OF baronie (as BARON)]

baroque /bə'rɒk/ adj. & n. ● adj. **1** highly ornate and extravagant in style, esp. of European art, architecture, and music of the 17th and 18th centuries. (See note below.) **2** of or relating to this period. ● n. **1** the baroque style. **2** baroque art collectively. [F (orig. = 'irregular pearl') f. Port. barroco, of unkn. orig.]

▪ The term is used to refer to the dominant European style of art and architecture between mannerism and rococo, characterized by dramatic and ornate detail, encompassing both grandeur and exuberance. The baroque is particularly associated with the Catholic Counter-Reformation and the reign of Louis XIV. Important figures include the Italian artists Caravaggio, Carracci, and Bernini; in northern Europe, Rubens is considered the leading exponent. In architecture, the baroque is exemplified in grand designs such as the palace at Versailles as well as in the work of Vanbrugh and Wren in England.

In music, the baroque period is generally taken as c.1600–1750. It was a time of stylistic and formal change, encompassing the formation of the orchestra and the origination of opera and the symphony, concerto, sonata, oratorio, and cantata. Major baroque composers include Vivaldi, Bach, and Handel.

baroreceptor /'bærəʊrɪ,septə(r)/ n. Physiol. a receptor sensitive to changes in pressure.

barouche /bə'ruːʃ/ n. a horse-drawn carriage with four wheels and a collapsible hood over the rear half, used esp. in the 19th century. [G (dial.) Barutsche f. It. baroccio ult. f. L birotus two-wheeled]

barque /bɑːk/ n. a sailing-ship with the rear mast fore-and-aft-rigged and the remaining (usu. two) masts square-rigged. Until the mid-19th century barques were relatively small sailing-ships, but later were built up to 3,000 tons, sometimes with four or five masts. They are now obsolete as trading vessels but some are used as training ships. The term is sometimes used in poetry to refer to any boat. [ME f. F prob. f. Prov. barca f. L barca ship's boat]

barquentine /'bɑːkən,tiːn/ n. (also **barkentine**, **barquantine**) a sailing-ship with the foremast square-rigged and the remaining (usu. two) masts fore-and-aft-rigged. [BARQUE after brigantine]

Barquisimeto /,bɑːkɪsɪ'meɪtəʊ/ a city in NW Venezuela; pop. (1991) 602,620.

Barra /'bærə/ an island of the Outer Hebrides, off the west coast of Scotland. It lies to the south of South Uist, from which it is separated by the Sound of Barra.

barrack[1] /'bærək/ n. & v. ● n. (usu. in pl., often treated as sing.) **1** a building or building complex used to house soldiers. **2** any building used to accommodate large numbers of people. **3** a large building of a bleak or plain appearance. ● v.tr. place (soldiers etc.) in barracks. □ **barrack-room lawyer** Brit. a pompously argumentative person. **barrack-square** a drill-ground near a barracks. [F baraque f. It. baracca or Sp. barraca soldier's tent, of unkn. orig.]

barrack[2] /'bærək/ v. Brit., Austral., & NZ **1** tr. (also absol.) shout or jeer at (players in a game, a performer, speaker, etc.). **2** intr. (foll. by for) (of spectators at games etc.) cheer for, encourage (a team etc.). [app. f. BORAK]

barracouta /,bærə'kuːtə/ n. (pl. same or **barracoutas**) **1** a long slender food fish, Thyrsites atun, usu. found in southern oceans. **2** NZ a small narrow loaf of bread. [var. of BARRACUDA]

barracuda /,bærə'kuːdə/ n. (pl. same or **barracudas**) a large and voracious marine fish of the family Sphyraenidae. [Amer. Sp. barracuda]

barrage /'bærɑːʒ/ n. **1** a concentrated artillery bombardment over a wide area. **2** a rapid succession of questions or criticisms. **3** an artificial barrier, esp. in a river. **4** a heat or deciding event in fencing, showjumping, etc. □ **barrage balloon** a large anchored balloon, often with netting suspended from it, used (usu. as one of a series) as a defence against low-flying aircraft. [F f. barrer (as BAR[1])]

barramundi /,bærə'mʌndɪ/ n. (pl. same or **barramundis**) an Australian freshwater fish, esp. Lates calcarifer, used as food. [Aboriginal]

Barranquilla /,bærən'kiːjə/ the chief port of Colombia; pop. (1985) 899,800. Founded in 1629, the city lies at the mouth of the Magdalena river, near the Caribbean Sea.

barrator /'bærətə(r)/ n. **1** a malicious person causing discord. **2** hist. a vexatious litigant. [ME f. AF baratour, OF barateor trickster, f. barater ult. f. Gk prattein do, perform, manage]

barratry /'bærətrɪ/ n. **1** fraud or gross negligence of a ship's master or crew at the expense of its owners or users. **2** hist. vexatious litigation or incitement to it. **3** hist. trade in the sale of Church or state appointments. □ **barratrous** adj. [ME f. OF baraterie (as BARRATOR)]

barre /bɑː(r)/ n. a horizontal bar at waist level used in dance exercises. [F]

barré /'bæreɪ/ n. Mus. a method of playing a chord on the guitar etc. with a finger laid across the strings at a particular fret, raising their pitch. [F, past part. of barrer bar]

barrel /'bærəl/ n. & v. ● n. **1** a cylindrical container usu. bulging out in the middle, traditionally made of wooden staves with metal hoops round them. **2** the contents of this. **3** a measure of capacity, usu. equal to 35 imperial gallons (about 159 litres) for oil, and 36 imperial gallons (about 164 litres) for beer etc. **4** a cylindrical tube forming part of an object such as a gun or a pen. **5** the belly and loins of a four-legged animal, e.g. a horse. ● v. (**barrelled**, **barrelling**; US **barreled**, **barreling**) **1** tr. put into a barrel or barrels. **2** intr. N. Amer. sl. drive fast. □ **barrel-chested** having a large rounded chest. **barrel of fun** (or **laughs**) (often with neg.) colloq. a source of much fun or amusement. **barrel roll** an aerobatic manoeuvre in which an aircraft follows a single turn of a spiral while rolling once about its longitudinal axis. **barrel vault** Archit. a vault forming a half cylinder. **over a barrel** colloq. in a helpless position; at a person's mercy. [ME f. OF baril perh. f. Rmc.: rel. to BAR[1]]

barrel-organ /'bærəl,ɔːgən/ n. **1** an automatic pipe-organ much used in churches in the 19th century. Projections on a cylinder (barrel) that was turned by a handle (which also worked the bellows) opened pipes to produce the required notes for a predetermined tune. **2** a 19th-century street-instrument (not of the organ type) producing notes by means of metal tongues struck by pins fixed in the barrel.

barren /'bærən/ *adj. & n.* ● *adj.* (**barrener, barrenest**) **1 a** unable to bear young. **b** unable to produce fruit or vegetation. **c** devoid of vegetation or other signs of life. **2** meagre, unprofitable. **3** dull, unstimulating. **4** (foll. by *of*) lacking in (*barren of wit*). ● *n.* a barren tract or tracts of land esp. (in *pl.*) in North America. □ **barrenly** *adv.* **barrenness** /-rənnɪs/ *n.* [ME f. AF *barai(g)ne*, OF *barhaine* etc., of unkn. orig.]

Barrett /'bærət/, Elizabeth, see BROWNING[1].

barrette /bə'ret/ *n.* a woman's bar-shaped clip or ornament for the hair. [F, dimin. of *barre* BAR[1]]

barricade /ˌbærɪ'keɪd/ *n. & v.* ● *n.* a barrier, esp. one improvised across a street etc. ● *v.tr.* block or defend with a barricade. [F f. *barrique* cask f. Sp. *barrica*, rel. to BARREL]

Barrie /'bærɪ/, Sir J(ames) M(atthew) (1860–1937), Scottish dramatist and novelist. His Scottish background provided the setting for several novels, including *The Little Minister* (1891). He abandoned fiction for the theatre in the early 1900s, gaining success with comedies such as *The Admirable Crichton* (1902) and *Dear Brutus* (1917). However, Barrie's most famous play is *Peter Pan* (1904), a fantasy for children about a boy who would not grow up. Barrie bequeathed the copyright of this to the Great Ormond Street Children's Hospital in London.

barrier /'bærɪə(r)/ *n.* **1** a fence or other obstacle that bars advance or access. **2** an obstacle or circumstance that keeps people or things apart, or prevents communication (*class barriers; a language barrier*). **3** anything that prevents progress or success. **4** a gate at a car park, railway station, etc., that controls access. **5** *colloq.* = *sound barrier*. □ **barrier cream** a cream used to protect the skin from damage or infection. **barrier reef** a coral reef separated from the shore by a broad deep channel. [ME f. AF *barrere*, OF *barriere*]

barring /'bɑːrɪŋ/ *prep.* except, not including. [BAR[1] + -ING[2]]

barrio /'bærɪəʊ/ *n.* (*pl.* **-os**) (in the US) the Spanish-speaking quarter of a town or city. [Sp., = district of a town]

barrister /'bærɪstə(r)/ *n.* (in full **barrister-at-law**) **1** *Brit.* a person called to the bar and entitled to practise as an advocate in the higher courts. **2** *US* a lawyer. [16th c.: f. BAR[1], perh. after *minister*]

barrow[1] /'bærəʊ/ *n.* **1** *Brit.* a two-wheeled handcart used esp. by street vendors. **2** = WHEELBARROW. **3** a metal frame with two wheels used for transporting luggage etc. □ **barrow boy** *Brit.* a boy who sells wares from a barrow. [OE *bearwe* f. Gmc]

barrow[2] /'bærəʊ/ *n.* a mound of earth constructed in ancient times to cover one or more burials. The earliest barrows in NW Europe date from the late 5th and 4th millenniums BC and were elongated in shape (they are often called *long barrows*). In the late 4th millennium BC round barrows came into use, and continued to be constructed intermittently up to the 10th century AD. The somewhat uniform appearance that these monuments tend to display today belies a wide range of burial practices and construction techniques. A barrow is often marked on maps of Britain as *tumulus*. [OE *beorg* f. Gmc]

Barry /'bærɪ/, Sir Charles (1795–1860), English architect. Having studied architecture during visits to Italy, France, and Greece (1817–20), he established his reputation with his Italianate design of the Travellers' Club in Pall Mall, London (1830–2). In 1836 he won a competition to design the new Houses of Parliament after the old buildings had been destroyed by fire; it was designed in the Perpendicular style, with most of the detail and internal fittings contributed by A. W. N. Pugin. Work on the building began in 1840 and continued after Charles Barry's death, his son Edward Middleton Barry (1830–80) completing the project.

Barrymore /'bærɪˌmɔː(r)/ *n.* an American family of actors. Lionel (1878–1954) withdrew from a successful career in the theatre in 1925 and devoted himself to films; these included *A Free Soul* (1931), for which he won an Oscar, and *Grand Hotel* (1932). His sister, Ethel (1879–1959), was also an actress; she gave notable stage performances in *The Second Mrs Tanqueray* (1924) and *The Corn is Green* (1942), and won an Oscar for her part in the film *None But the Lonely Heart* (1944). Their brother, John (1882–1942), was a light comedian as well as a serious actor; his most celebrated role was on stage as Hamlet, both in New York (1922) and in London (1925).

Barsac /'bɑːsæk/ *n.* a sweet white wine from the district of Barsac, department of Gironde, in western France.

Bart. /bɑːt/ *abbr.* Baronet.

bartender /'bɑːˌtendə(r)/ *n.* a person serving behind the bar of a public house.

barter /'bɑːtə(r)/ *v. & n.* ● *v.* **1** *tr.* exchange (goods or services) without using money. **2** *intr.* make such an exchange. ● *n.* trade by exchange of goods. □ **barterer** *n.* [prob. OF *barater*: see BARRATOR]

Barth[1] /bɑːθ/, John (Simmons) (b.1930), American novelist and short-story writer. He is known as a writer of complex, elaborate, experimental novels, including *The Sot-Weed Factor* (1960), a fantastic parody of an 18th-century picaresque tale, and *Giles Goat-Boy* (1966), a satirical allegory of the modern world conceived in terms of a university campus; *Letters* (1979) consists of correspondence exchanged by characters from his previous novels.

Barth[2] /bɑːt, bɑːθ/, Karl (1886–1968), Swiss Protestant theologian. Under the shadow of the First World War he was led to a radical questioning of contemporary religious thought and in 1919 published his seminal work *Epistle to the Romans*. A rebuttal of liberal 19th-century Protestant theology, the book established a neo-orthodox or theocentric approach. He emphasized the supremacy and transcendence of God and the dependence of man on divine grace, and stressed that the teachings of Christ as related in the Bible were the only route to an understanding of God. His work had its greatest impact in the 1930s, but it exerts a continuing influence on Protestant theology today.

Barthes /bɑːt/, Roland (1915–80), French writer and critic. Barthes was a leading exponent of structuralism in literary criticism and cultural analysis. In the 1950s and 1960s, he spearheaded the *nouvelle critique*, which challenged the traditional approach of literary criticism. His iconoclastic work *On Racine* (1963) is still a subject of controversy amongst literary critics. Barthes was increasingly drawn to the theory of semiotics after *Mythologies* (1957), his critique of contemporary culture, proceeding to define the theory in further detail in *Elements of Semiology* (1964). Later works, such as the essay 'The Death of the Author' (1968) and *S/Z* (1970), which stress the role of the reader in constructing a text, were influential in the development of deconstruction and post-structuralism.

Bartholomew, St /bɑː'θɒləˌmjuː/ an Apostle. He is said to have been flayed alive in Armenia, and is hence regarded as the patron saint of tanners. Feast day, 24 Aug.

bartizan /'bɑːtɪz(ə)n, ˌbɑːtɪ'zæn/ *n.* *Archit.* a battlemented parapet or an overhanging corner turret at the top of a castle or church tower. [var. of *bertisene*, erron. spelling of *bratticing*: see BRATTICE]

Bartók /'bɑːtɒk/, Béla (1881–1945), Hungarian composer. While his early work reflects the influence of Liszt, Wagner, Brahms, and Richard Strauss, he later developed an original musical language; basically homophonic and harmonically adventurous, his music is often percussive and owes much to Hungarian folk music, which he began to record, notate, and classify in 1904. His work includes six string quartets (1908–39), three piano concertos (1926; 1930–1; 1945), *Concerto for Orchestra* (1943), and an opera, *Duke Bluebeard's Castle* (1911). He emigrated to America in 1940 because of political pressures in Hungary.

Bartolommeo /ˌbɑːtɒlə'meɪəʊ/, Fra (born Baccio della Porta) (c.1472–1517), Italian painter. A Dominican friar, he worked chiefly in Florence and made visits to Venice and Rome, where he was much impressed with the work of Raphael and Michelangelo. His work particularly displays the influence of the former, as is demonstrated in the balance of his compositions and the use of rapt expressions and significant gestures. Notable works are *The Vision of St Bernard* (1507) and *The Mystic Marriage of St Catherine* (1511).

Baruch /'bɑːrʊk/ a book of the Apocrypha, attributed in the text to Baruch, the scribe of Jeremiah (Jer. 36).

baryon /'bærɪˌɒn/ *n.* *Physics* a subatomic particle that is at least as heavy as a proton, has half-integral spin, and takes part in the strong interaction (i.e. a nucleon or a hyperon). □ **baryonic** /ˌbærɪ'ɒnɪk/ *adj.* [Gk *barus* heavy]

Baryshnikov /bə'riːʃnɪˌkɒf/, Mikhail (Nikolaevich) (b.1948), American ballet-dancer, born in Latvia of Russian parents. In 1974 he defected to the West while touring with the Kirov Ballet. He then danced with the American Theater Ballet and the New York City Ballet, where roles were devised for him by Jerome Robbins (*Opus 191/The Dreamer*, 1979) and Sir Frederick Ashton (*Rhapsody*, 1980). He was artistic director of the American Theater Ballet 1980–90.

barysphere /'bærɪˌsfɪə(r)/ *n.* the dense interior of the earth, including the mantle and core, enclosed by the lithosphere. [Gk *barus* heavy + *sphaira* sphere]

baryta /bə'raɪtə/ n. barium oxide or hydroxide. □ **barytic** /-'rɪtɪk/ adj. [*barytes* (see BARITE), after *soda* etc.]

baryte (also **barytes**) var. of BARITE.

basal /'beɪs(ə)l/ adj. **1** of, at, or forming a base. **2** fundamental. □ **basal ganglia** Anat. a group of structures linked to the thalamus in the base of the brain and involved in coordination of movement. **basal metabolism** the chemical processes occurring in an organism at complete rest. [BASE¹ + -AL¹]

basalt /'bæsɔːlt/ n. **1** a dark basic volcanic rock whose strata sometimes form columns. **2** a kind of black stoneware resembling basalt. □ **basaltic** /bə'sɔːltɪk/ adj. [L *basaltes* var. of *basanites* f. Gk f. *basanos* touchstone]

bascule bridge /'bæskjuːl/ n. a type of drawbridge which is raised and lowered using counterweights. [F, earlier *bacule* see-saw f. *battre* bump + *cul* buttocks]

base¹ /beɪs/ n. & v. ● n. **1 a** a part that supports from beneath or serves as a foundation for an object or structure. **b** a notional structure or entity on which something draws or depends (*power base*). **c** (in Marxism) the economic system of a society which influences or determines that society's institutions and culture (or superstructure). **2** a principle or starting-point; a basis. **3** esp. Mil. a place from which an operation or activity is directed. **4 a** a main or important ingredient of a mixture. **b** a substance, e.g. water, in combination with which pigment forms paint etc. **5** a substance used as a foundation for make-up. **6 a** Chem. a substance capable of combining with an acid to form a salt and water. (See note below.) **b** Biochem. a purine or pyrimidine group present in a nucleotide or nucleic acid. **7** Math. a number in terms of which other numbers or logarithms are expressed (see RADIX). **8** Archit. the part of a column between the shaft and pedestal or pavement. **9** Geom. a line or surface on which a figure is regarded as standing. **10** Surveying a known line used as a geometrical base for trigonometry. **11** Electronics the middle part of a transistor separating the emitter from the collector. **12** Linguistics a root or stem as the origin of a word or a derivative. **13** Baseball etc. one of the four stations that must be reached in turn when scoring a run. **14** Bot. & Zool. the end at which an organ is attached to the trunk. **15** Heraldry the lowest part of a shield. ● v.tr. **1** (usu. foll. by *on, upon*) found or establish (*a theory based on speculation; his opinion was soundly based*). **2** (foll. by *at, in,* etc.) station (*troops were based in Malta*). □ **base camp** a camp from which expeditions set out or operations are conducted. **base hospital** esp. Austral. a hospital in a rural area, or (in warfare) removed from the field of action. **base pairing** Biochem. complementary binding by means of hydrogen bonds of a purine to a pyrimidine base in opposite strands of nucleic acids. **base rate** Brit. the interest rate set by the Bank of England, used as the basis for other banks' rates. **base unit** a unit that is defined arbitrarily and not by combinations of other units. [F *base* or L *basis* stepping f. Gk]

▪ In chemistry the term *base* is a more general one than *alkali*, which is often restricted to soluble substances that liberate hydroxide ions. Bases are generally thought of as the opposite of acids, with which they react to form salts. In chemistry bases are sometimes more widely defined as compounds which are acceptors of hydrogen ions or protons, or donors of electron pairs.

base² /beɪs/ adj. **1** lacking moral worth; cowardly, despicable. **2** menial. **3** not pure; alloyed (*base coin*). **4** (of a metal) low in value (opp. NOBLE, PRECIOUS). **5** cheap, shoddy. **6** mean, degraded. □ **basely** adv. **baseness** n. [ME in sense 'of small height', f. F *bas* f. med.L *bassus* short (in L as a cognomen)]

baseball /'beɪsbɔːl/ n. **1** a game played between two teams of nine with long straight bats and a ball, on a diamond-shaped circuit of four bases which the batsman must complete. (See note below.) **2** the ball used in this game.

▪ Baseball, played as a summer sport chiefly in the US and Canada, is thought to have developed from rounders in the 19th century; the first rules were drawn up in 1845 and the first professional league was formed in the US in 1871. The teams bat and field alternately, each team having nine innings (an inning ends when three players of the batting side are out) and aiming to score more runs than the opposition. A run is scored by hitting the ball out of reach of the fielders and running between the bases until reaching the home base. A batter is out if he or she fails to hit three consecutive balls bowled by the pitcher or if the ball is thrown to and caught by the fielder guarding the base to which the batter is running. Batters can also be caught out, or be 'tagged' or touched when running by a fielder holding the ball. The major baseball competition is the World Series.

baseboard /'beɪsbɔːd/ n. N. Amer. a skirting-board.

basehead /'beɪshed/ n. (also **base-head**) US sl. a person who habitually takes drugs such as freebase or crack.

BASE jump /beɪs/ n. & v. (also **base**) ● n. a parachute jump from a fixed point (esp. a high building or promontory) rather than an aircraft. ● v.intr. carry out such a parachute jump. □ **BASE jumper** n. [acronym f. building, antenna-tower, span, earth, denoting the types of structure used]

Basel see BASLE.

baseless /'beɪslɪs/ adj. unfounded, groundless. □ **baselessly** adv. **baselessness** n.

baseline /'beɪslaɪn/ n. **1** a line used as a base or starting-point. **2** Tennis the line marking each end of a court.

baseload /'beɪsləʊd/ n. Electr. the permanent load on power supplies etc.

baseman /'beɪsmən/ n. (pl. **-men**) Baseball a fielder stationed near a base.

basement /'beɪsmənt/ n. the lowest floor of a building, usu. at least partly below ground level. □ **basement membrane** Anat. a thin delicate membrane separating the epithelium from underlying tissue. [prob. Du., perh. f. It. *basamento* column-base]

basenji /bə'sendʒi/ n. (pl. **basenjis**) a central African breed of hunting dog, which growls and yelps but does not bark. [Bantu]

bases pl. of BASE¹, BASIS.

bash /bæʃ/ v. & n. ● v. **1** tr. **a** a strike bluntly or heavily. **b** (often foll. by *up*) colloq. attack violently. **c** (often foll. by *down, in,* etc.) damage or break by striking forcibly. **2** intr. (foll. by *into*) collide with. ● n. **1** a heavy blow. **2** sl. an attempt (*had a bash at painting*). **3** sl. a party or social event. [imit., perh. f. *bang, smash, dash,* etc.]

bashful /'bæʃfʊl/ adj. **1** shy, diffident, self-conscious. **2** sheepish. □ **bashfully** adv. **bashfulness** n. [obs. *bash* (v.), = ABASH]

Bashkir /bæʃ'kɪə(r)/ n. & adj. ● n. (pl. same or **Bashkirs**) **1** a member of a Muslim people living in the southern Urals. **2** the Turkic language of this people. ● adj. of or relating to the Bashkirs or their language.

Bashkiria /bæʃ'kɪərɪə/ (also called **Bashkir Autonomous Republic** /bæʃ'kɪər/, **Bashkortostan** /bæʃˌkɔːtə'stɑːn/) an autonomous republic in central Russia, west of the Urals; pop. (1990) 3,964,000; capital, Ufa.

basho /'bæʃəʊ/ n. (pl. same) a sumo wrestling tournament. [Jap., f. *ba* place + *shō* victory, win]

BASIC /'beɪsɪk/ n. a computer programming language using familiar English words, designed for beginners and widely used on microcomputers. [Beginner's All-purpose Symbolic Instruction Code]

basic /'beɪsɪk/ adj. & n. ● adj. **1** forming or serving as a base. **2** fundamental. **3 a** simplest or lowest in level (*basic pay; basic requirements*). **b** vulgar (*basic humour*). **4** Chem. having the properties of or containing a base. **5** Geol. (of volcanic rocks etc.) having less than 50 per cent silica. **6** of, resulting from, or designating steel-making processes involving lime-rich refractories and slags. ● n. (usu. in pl.) the fundamental facts or principles. □ **basic dye** a dye consisting of salts of organic bases. **basic industry** an industry of fundamental economic importance. **basic slag** fertilizer containing phosphates formed as a by-product during steel manufacture. **basic wage 1** a minimum wage earned before other payments (such as commission, overtime, etc.) are taken into consideration. **2** Austral. & NZ the minimum living wage, fixed by industrial tribunal. [BASE¹ + -IC]

basically /'beɪsɪkəli/ adv. **1** fundamentally, essentially. **2** (qualifying a clause) in fact, actually.

Basic English n. a simplified form of English limited to 850 selected words intended for international communication.

basicity /beɪ'sɪsɪti/ n. Chem. the number of protons with which a base will combine.

basic oxygen process n. a steel-making process in which a jet of oxygen is delivered by a lance on to a molten mixture of pig-iron and scrap steel in a retort lined with a basic refractory. Excess carbon is burnt away, producing enough heat to keep the iron molten, and the oxidized impurities escape as gases or float to the surface as lime-rich slag. The process was developed in the 1950s in Austria and is now the major means of steel production in many countries. It combines the low cost of the Bessemer process with the high-quality product of the open-hearth process, and can convert 400 tonnes of molten iron into steel in about twelve minutes. (See also STEEL.)

basidium /bə'sɪdɪəm/ n. (pl. **basidia** /-dɪə/) a microscopic spore-bearing structure produced by certain fungi. [mod.L f. Gk *basidion* dimin. of BASIS]

Basie /ˈbeɪsɪ/, Count (born William Basie) (1904–84), American jazz pianist, organist, and band-leader. He took up the piano at an early age and had lessons with Fats Waller. In 1935 he formed his own big band, which became known as the Count Basie Orchestra. One of the best-known bands of the swing era, with its strong rhythm section and employment of some of the top instrumentalists and vocalists of the day, it enjoyed great success for many years.

basil /ˈbæz(ə)l/ n. an aromatic labiate herb of the genus *Ocimum*; esp. *O. basilicum* (in full **sweet basil**), whose leaves are used as a flavouring in savoury dishes. [ME f. OF *basile* f. med.L *basilicum* f. Gk *basilikos* royal]

Basil, St /ˈbæz(ə)l/ (known as St Basil the Great) (c.330–79), Doctor of the Church. The brother of St Gregory of Nyssa, he was a staunch opponent of Arianism. He lived as a hermit until 370, when he was appointed bishop of Caesarea in Cappadocia. He put forward a monastic rule which is still the basis of monasticism in the Eastern Church. Feast day, 14 June.

basilar /ˈbæzɪlə(r)/ adj. of or at the base (esp. of the skull). [mod.L *basilaris* (as BASIS)]

Basildon /ˈbæzɪldən/ a town in SE Essex; pop. (1991) 157,500. It was developed as a new town from 1949.

basilica /bəˈzɪlɪkə/ n. **1** an ancient Roman public hall with an apse and colonnades, used as a lawcourt and place of assembly. **2** a similar building used as a Christian church. **3** a church having special privileges from the pope. □ **basilican** adj. [L f. Gk *basilikē* (*oikia, stoa*) royal (house, portico) f. *basileus* king]

Basilicata /bəˌsɪlɪˈkɑːtə/ a region of southern Italy, lying between the 'heel' of Apulia and the 'toe' of Calabria; capital, Potenza.

basilisk /ˈbæzɪlɪsk/ n. **1** a mythical reptile with a lethal breath and look. **2** a small American lizard of the genus *Basiliscus*, with a crest along its back and tail. **3** *Heraldry* a cockatrice. [ME f. L *basiliscus* f. Gk *basiliskos* kinglet, serpent]

basin /ˈbeɪs(ə)n/ n. **1 a** = wash-basin. **b** a wide round open container, esp. one for preparing food in or for holding water. **2** a hollow rounded depression. **3** any sheltered area of water where boats can moor safely. **4** a round valley. **5** an area drained by rivers and tributaries. **6** *Geol.* **a** a rock formation where the strata dip towards the centre. **b** an accumulation of rock strata formed in this dip as a result of subsidence and sedimentation. □ **basinful** n. (pl. **-fuls**). [ME f. OF *bacin* f. med.L *ba(s)cinus*, perh. f. Gaulish]

basipetal /beɪˈsɪpɪt(ə)l/ adj. *Bot.* (of each new part produced) developing nearer the base than the previous one did. □ **basipetally** adv. [BASIS + L *petere* seek]

basis /ˈbeɪsɪs/ n. (pl. **bases** /-siːz/) **1** the foundation or support of something, esp. an idea or argument. **2** the main or determining principle or ingredient (*on a purely friendly basis*). **3** the starting-point for a discussion etc. [L f. Gk, = BASE[1]]

bask /bɑːsk/ v.intr. **1** sit or lie back lazily in warmth and light (*basking in the sun*). **2** (foll. by *in*) derive great pleasure (from) (*basking in glory*). □ **basking shark** a very large shark, *Cetorhinus maximus*, which feeds on plankton and often lies near the surface. [ME, app. f. ON: rel. to BATHE]

Baskerville /ˈbæskəˌvɪl/, John (1706–75), English printer. He designed the typeface that bears his name, and from 1757 onwards produced editions of authors such as Virgil, Milton, and Horace which were notable for their quality of type and paper.

basket /ˈbɑːskɪt/ n. **1** a container made of interwoven cane etc. **2** a container resembling this. **3** the amount held by a basket. **4** the goal in basketball, or a goal scored. **5** *Econ.* a group or range (of currencies). **6** *euphem. colloq.* bastard. □ **basket case** *colloq. offens.* **1** *Med.* a person who has lost both arms and legs. **2** a thing or person regarded as hopeless, esp. a bankrupt country or a helpless or crazy person. **basket weave** a weave resembling that of a basket. □ **basketful** n. (pl. **-fuls**). [AF & OF *basket*, AL *baskettum*, of unkn. orig.]

basketball /ˈbɑːskɪtˌbɔːl/ n. **1** a game played, usu. indoors, between two teams of five players, in which the object is to score goals by making a large inflated ball drop through hooped nets (the 'basket', orig. a fruit basket) fixed high up at each end of a rectangular court. (*See note below.*) **2** the ball used in this game.

■ In basketball the players may advance while bouncing the ball or throw it to each other, but may not run or walk with the ball in their hands. The game was invented in 1891 by an American physical education instructor and quickly became established at college level.

The first professional teams formed in 1898 and basketball is now one of the major amateur and professional sports of the US. It was first included in the Olympic Games in 1936.

Basket Maker n. a member of a culture of the south-western US, forming the early stages of the Anasazi culture, from the 1st century BC until c.700 AD, so called from the basketry and other woven fragments found in early cave sites.

basketry /ˈbɑːskɪtrɪ/ n. **1** the art of making baskets. **2** baskets collectively.

basketwork /ˈbɑːskɪtˌwɜːk/ n. **1** material woven in the style of a basket. **2** the art of making this.

Basle /bɑːl/ (French **Bâle** /bɑl/, German **Basel** /ˈbɑːz(ə)l/) a commercial and industrial city on the Rhine in NW Switzerland; pop. (1991) 171,000.

basmati /bæzˈmɑːtɪ/ n. (in full **basmati rice**) a kind of rice with long thin grains and a delicate fragrance. [Hindi, = fragrant]

basophil /ˈbeɪsəfɪl/ n. (also **basophile** /-ˌfaɪl/) *Physiol.* a white blood cell that is basophilic.

basophilic /ˌbeɪsəˈfɪlɪk/ adj. *Biol.* (of a cell or its contents) readily stained with basic dyes.

Basque /bæsk, bɑːsk/ n. & adj. ● n. **1** a member of a people inhabiting the western Pyrenees in central northern Spain and the extreme south-west of France, an area often known as the Basque Country. **2** the language of this people. (*See note below.*) ● adj. of, relating to, or denoting the Basques or their language. [F f. L *Vasco -onis*]

■ Culturally the Basques are one of the most distinct groups in Europe. While they do not differ physically from other European groups, their language, Basque, is not a member of the Indo-European language family and is not demonstrably related to any other known language. An ergative language with agglutination, it is currently spoken by close on 1 million people in the Pyrenees, although it is now rarely an individual's sole language; use of a number of dialects a standard language is developing. The Basque people were largely independent until the 19th century, and retained some local autonomy in Spain until the end of the Civil War, in which most of the Basque Country opposed Franco's Nationalist forces. In the 1960s the Basque separatist movement ETA began an armed struggle against the Spanish government, which continues, although some degree of local autonomy was granted to the Basques after Franco's death. There is also a French equivalent to ETA, *Iparretarrak* ('fighters from the North Side').

basque /bæsk/ n. a close-fitting bodice with a short continuation below the waist. [BASQUE, the garment prob. originating in the Basque Country]

Basque Country (called in French *Pays Basque*) a region of the western Pyrenees in both France and Spain, the homeland of the Basque people. The region comprises the autonomous Basque Provinces of northern Spain, together with the south-western part of Aquitaine in France.

Basque Provinces an autonomous region of northern Spain, on the Bay of Biscay; capital, Vitoria. It comprises that part of the Basque Country which lies in Spain, to the south-west of the Pyrenees, and consists of the provinces of Álava, Guipúzcoa, and Vizcaya. Its chief industrial cities are Bilbao and San Sebastián.

Basra /ˈbæzrə/ an oil port of Iraq, on the Shatt al-Arab waterway; pop. (est. 1985) 616,700.

bas-relief /ˈbæsrɪˌliːf/ n. **1** = low relief (see RELIEF 6a). **2** a sculpture or carving in low relief. [earlier *basse relieve* f. It. *basso rilievo* low relief: later altered to F form]

bass[1] /beɪs/ n. & adj. ● n. **1 a** the lowest adult male singing voice. **b** a singer with this voice. **c** a part written for it. **2** the lowest part in harmonized music. **3 a** an instrument that is the lowest in pitch in its family. **b** a player of such an instrument. **4** *colloq.* **a** a bass guitar or double bass. **b** a player of this instrument. **5** the low-frequency output of a radio, record-player, etc., corresponding to the bass in music. ● adj. **1** lowest in musical pitch. **2** deep-sounding. □ **bass clef** a clef placing F below middle C on the second-highest line of the staff. **bass guitar** a four-stringed usu. electric guitar played esp. in jazz and popular music. **bass viol 1** a viola da gamba. **b** a player of this instrument. **2** *US* a double bass. □ **bassist** n. (in sense 4b). [alt. of BASE[2] after It. *basso*]

bass[2] /bæs/ n. (pl. same or **basses**) **1** the common freshwater perch. **2 a** a similar marine fish, *Dicentrarchus labrax* (family Percichthyidae), of European waters. **b** a related American fish of the genus *Morone*, of fresh and coastal waters. **3** (in full **sea-bass**) a similar marine fish of the family Serranidae, esp. one of the genus *Centropristis*. **4** (in full **black bass**) an American freshwater fish of the genus *Micropterus* (family Centrarchidae), popular with anglers. [earlier *barse* f. OE *bærs*]

bass³ /bæs/ *n.* = BAST. [alt. f. BAST]

Bassein /bæˈseɪn/ a port on the Irrawaddy delta in SW Burma (Myanmar); pop. (1983) 144,100.

Basse-Normandie /bæsˈnɔːməndɪ/ a region of NW France, on the coast of the English Channel, including the Cherbourg peninsula and the city of Caen. It was created from part of the former province of Normandy.

basset /ˈbæsɪt/ *n.* (in full **basset-hound**) a sturdy breed of hunting-dog with a long body, short legs, and big ears. [F, dimin. of *bas basse* low: see BASE²]

Basseterre /bæsˈteə(r)/ the capital of St Kitts and Nevis in the Leeward Islands, on the island of St Kitts; pop. (1980) 14,300.

Basse-Terre /bæsˈteə(r)/ **1** the main island of Guadeloupe in the West Indies. **2** its capital city, situated on the SW corner of the island; pop. (1982) 13,650.

basset-horn /ˈbæsɪtˌhɔːn/ *n.* an alto clarinet in F, with a dark tone. [G *Bassetthorn*, transl. F *cor de bassette* f. It. *corno di bassetto* f. *corno* horn + *bassetto* dimin. of *basso* BASE²]

bassinet /ˌbæsɪˈnet/ *n.* a child's wicker cradle, usu. with a hood. [F, dimin. of *bassin* BASIN]

basso /ˈbæsəʊ/ *n.* (pl. **-os** or **bassi** /-sɪ/) a singer with a bass voice. □ **basso profondo** /prɒˈfɒndəʊ/ a bass singer with an exceptionally low range. [It., = BASS¹; *profondo* deep]

bassoon /bəˈsuːn/ *n.* **1 a** a bass instrument of the oboe family, with a double reed. (*See note below*.) **b** a player of this instrument. **2** an organ stop with the quality of a bassoon. □ **bassoonist** *n.* (in sense 1b). [F *basson* f. *bas* BASS¹]

▪ The bassoon dates from around the 1660s. It has a range of about three-and-a-half octaves, and is a standard orchestral instrument. It is often used for comic effect but also has a capacity for melancholy which composers have not overlooked.

basso-relievo /ˌbæsəʊrɪˈliːvəʊ, -ˈljeɪvəʊ/ *n.* (also **basso-rilievo**) (pl. **-os**) = BAS-RELIEF. [It. *basso-rilievo*]

Bass Strait /bæs/ a channel separating Tasmania from the mainland of Australia. It is named after the English explorer George Bass, who discovered the strait in 1798.

basswood /ˈbæswʊd/ *n.* **1** the American lime, *Tilia americana*. **2** the wood of this tree. [BASS³ + WOOD]

bast /bæst/ *n.* the inner bark of lime, or other flexible fibrous bark, used as fibre in matting etc. [OE *bæst* f. Gmc]

bastard /ˈbɑːstəd/ *n. & adj.* often *offens.* ● *n.* **1** a person born of parents not married to each other. **2** *sl.* **a** an unpleasant or despicable person. **b** a person of a specified kind (*poor bastard; rotten bastard; lucky bastard*). **3** *sl.* a difficult or awkward thing, undertaking, etc. ● *adj.* (usu. *attrib.*) **1** born of parents not married to each other; illegitimate. **2** (of things): **a** unauthorized, counterfeit. **b** hybrid. □ **bastardy** *n.* (in sense 1 of *n.*). [ME f. OF f. med.L *bastardus*, perh. f. *bastum* pack-saddle]

bastardize /ˈbɑːstəˌdaɪz/ *v.tr.* (also **-ise**) **1** declare (a person) illegitimate. **2** lower in quality or character; corrupt. □ **bastardization** /ˌbɑːstədaɪˈzeɪʃ(ə)n/ *n.*

baste¹ /beɪst/ *v.tr.* moisten (meat) with gravy or melted fat during cooking. [16th c.: orig. unkn.]

baste² /beɪst/ *v.tr.* stitch loosely together in preparation for sewing; tack. [ME f. OF *bastir* sew lightly, ult. f. Gmc]

baste³ /beɪst/ *v.tr.* beat soundly; thrash. [perh. figurative use of BASTE¹]

Bastet /ˈbæstet/ *Egyptian Mythol.* a goddess usually shown as a woman with the head of a cat, wearing one gold earring (see also SEKHMET).

Bastia /ˈbæstjə/ the chief port of Corsica; pop. (1990) 38,730.

Bastille /bæˈstiːl/ a fortress in Paris built in the 14th century and used in the 17th–18th centuries as a state prison. It became a symbol of despotism and, by its storming by the mob on 14 July 1789 marked the start of the French Revolution; the anniversary of this event is kept as a national holiday in France.

bastille /bæˈstiːl/ *n. hist.* a fortress or prison. [ME f. OF *bastille* f. Prov. *bastir* build: orig. of the BASTILLE]

bastinado /ˌbæstɪˈneɪdəʊ/ *n. & v.* ● *n.* punishment by beating with a stick on the soles of the feet. ● *v.tr.* (**-oes, -oed**) punish (a person) in this way. [Sp. *bastonada* f. *bastón* BATON]

bastion /ˈbæstɪən/ *n.* **1** a projecting part of a fortification set at an angle of, or against the line of, a wall. **2** a thing serving as a defence (*bastion of freedom*). **3** a natural rock formation resembling a bastion. [F f. It. *bastione* f. *bastire* build]

basuco /bəˈzuːkəʊ/ *n.* a cheap impure form of cocaine smoked for its stimulating effect. [Colombian Sp.]

Basutoland /bəˈsuːtəʊˌlænd/ the former name (until 1966) for LESOTHO.

bat¹ /bæt/ *n. & v.* ● *n.* **1** an implement with a handle, usu. of wood and with a flat or curved surface, used for hitting balls in games. **2** a turn at using this. **3** a batsman, esp. in cricket, usu. described in some way (*an excellent bat*). **4** (usu. in *pl.*) an object like a table-tennis bat used to guide aircraft when taxiing. ● *v.* (**batted, batting**) **1** *tr.* hit with or as with a bat. **2** *intr.* take a turn at batting. □ **bat around 1** *sl.* potter aimlessly. **2** discuss (an idea or proposal). **off one's own bat** unprompted, unaided. **right off the bat** *N. Amer.* immediately. [ME f. OE *batt* club, perh. partly f. OF *batte* club f. *battre* strike]

bat² /bæt/ *n.* **1** a flying mammal of the order Chiroptera. (*See note below*.) **2** a woman, esp. regarded as unattractive or unpleasant (esp.: *old bat*). □ **have bats in the belfry** be eccentric or crazy. **like a bat out of hell** very fast. [16th c., alt. of ME *bakke* f. Scand.]

▪ Bats are the only mammals capable of true flight, using membranous wings which extend between the fingers and limbs. The fruit bats or flying foxes (suborder Megachiroptera) are larger animals confined to the tropics, generally having good eyesight and feeding mainly on fruit. The majority of bats (over 800 species) belong to the suborder Microchiroptera, being mouselike in appearance and using ultrasonic echolocation for flying and catching prey at night. Most of these are insectivorous, but there are tropical kinds that feed on fish, mammals, nectar, or blood.

bat³ /bæt/ *v.tr.* (**batted, batting**) □ **not** (or **never**) **bat an eyelid** (or **eye**) *colloq.* show no reaction or emotion. [var. of obs. *bate* flutter]

Bata /ˈbɑːtə/ a seaport in Equatorial Guinea; pop. (est. 1986) 17,000.

Batan Islands /bəˈtɑːn/ the most northerly islands of the Philippines, lying between the Babuyan Islands and Taiwan.

Batavia /bəˈteɪvɪə/ the former name (until 1949) for DJAKARTA.

batch /bætʃ/ *n. & v.* ● *n.* **1** a number of things or persons forming a group or dealt with together. **2** an instalment (*have sent off the latest batch*). **3** the loaves produced at one baking. **4** (*attrib.*) using or dealt with in batches, not as a continuous flow (*batch production*). **5** *Computing* a group of records processed as a single unit. ● *v.tr.* arrange or deal with in batches. □ **batch processing 1** the performance of an industrial process on material in batches. **2** *Computing* the processing of previously collected data or jobs in batches, esp. automatically. [ME f. OE *bæcce* f. *bacan* BAKE]

Batdambang see BATTAMBANG.

bate /beɪt/ *n.* (also **bait**) *Brit. sl.* a rage; a cross mood (*is in an awful bate*). [BAIT = state of baited person]

bateau /ˈbætəʊ/ *n.* (pl. **bateaux** /-təʊz/) a light river-boat, esp. of the flat-bottomed kind used in Canada. [F, = boat]

bated /ˈbeɪtɪd/ *adj.* □ **with bated breath** very anxiously. [past part. of obs. *bate* (v.) restrain, f. ABATE]

bateleur /ˈbætəˌlɜː(r)/ *n.* a short-tailed African eagle, *Terathopius ecaudatus*. [F, = juggler]

Bates¹ /beɪts/, Henry Walter (1825–92), English naturalist. He travelled with A. R. Wallace in Brazil, writing *The Naturalist on the River Amazons* (1863). He described the phenomenon now known as Batesian mimicry, suggesting that, by natural selection, animals which use mimicry in this way are more likely to survive.

Bates² /beɪts/, H(erbert) E(rnest) (1905–74), English novelist and short-story writer. His many short stories appeared in several collected volumes, including *The Beauty of the Dead* (1940). Of his novels, perhaps his best known is *The Darling Buds of May* (1958), which gained wide popularity in Britain when dramatized for television (1990–2).

Batesian mimicry /ˈbeɪtsɪən/ *n. Zool.* a form of mimicry in which an edible animal is protected by its resemblance to one avoided by predators (cf. MÜLLERIAN MIMICRY). [BATES¹]

Bateson /ˈbeɪts(ə)n/, William (1861–1926), English geneticist and coiner of the term *genetics* in its current sense. He decided that variation is discontinuous, publicizing Mendel's work when he came across it many years later. Bateson found that some genes were not independent of one another, though he explained this as being caused by the reduplication of gametes and did not accept the idea of chromosomes.

Bath /bɑːθ/ a spa town in the county of Avon in SW England; pop. (1991) 79,900. The town was founded by the Romans, who called it Aquae Sulis. A fashionable spa in the 18th century, Bath is still famous for its hot springs, for its Roman remains, notably the baths, and for its Regency architecture.

bath /bɑːθ/ n. & v. ● n. (pl. **baths** /bɑːðz/) **1 a** a container for liquid, usu. water, used for immersing and washing the body. **b** this with its contents (*your bath is ready*). **2** the act or process of immersing the body for washing or therapy (*have a bath; take a bath*). **3 a** a vessel containing liquid in which something is immersed, as film for developing etc. **b** this with its contents. **4** (usu. in *pl.*) a building with baths or a swimming-pool, usu. open to the public. **5** esp. *N. Amer.* a bathroom. ● v. **1** tr. wash (esp. a person) in a bath. **2** intr. take a bath. □ **bath cube** a cube of compacted bath salts. **bath salts** soluble salts used for softening or scenting bath-water. [OE *bæth* f. Gmc]

Bath, Order of the n. an order of knighthood, so called from the ceremonial bath which originally preceded installation.

Bath bun n. *Brit.* a yeast bun with currants, topped with granules of sugar. [BATH]

Bath chair n. esp. *hist.* a kind of wheelchair for invalids.

Bath chap n. a type of pickled pig's chap.

bathe /beɪð/ v. & n. ● v. **1** intr. immerse oneself in water, esp. to swim or *N. Amer.* to wash oneself. **2** tr. **a** immerse in or wash or treat with liquid esp. for cleansing or medicinal purposes. **b** *N. Amer.* wash (esp. a person) in a bath. **3** tr. (of sunlight etc.) envelop. ● n. *Brit.* an act of immersing oneself or part of the body in liquid, esp. to swim or cleanse. □ **bathing costume** (or **suit**) a garment worn for swimming. [OE *bathian* f. Gmc]

bather /ˈbeɪðə(r)/ n. **1** a person who bathes. **2** (in *pl.*) *Austral.* a bathing costume.

bathhouse /ˈbɑːθhaʊs/ n. a building with baths for public use.

batholith /ˈbæθəlɪθ/ n. a dome of igneous rock extending inwards to an unknown depth. [G f. Gk *bathos* depth + -LITH]

Bath Oliver /ˌbɑːθ ˈɒlɪvə(r)/ n. *Brit. propr.* a kind of savoury biscuit. [Dr W. *Oliver* of BATH (1695–1764), who invented it]

bathometer /bəˈθɒmɪtə(r)/ n. an instrument used to measure the depth of water. [Gk *bathos* depth + -METER]

bathos /ˈbeɪθɒs/ n. **1** an unintentional lapse in mood from the sublime to the absurd or trivial. **2** a commonplace or ridiculous feature offsetting an otherwise sublime situation; an anticlimax. □ **bathetic** /bəˈθetɪk/ adj. **bathotic** /-ˈθɒtɪk/ adj. [Gk, = depth]

bathrobe /ˈbɑːθrəʊb/ n. *US* a loose coat usu. of towelling worn esp. before and after taking a bath; a dressing gown.

bathroom /ˈbɑːθruːm, -rʊm/ n. **1 a** a room containing a bath and usu. other washing facilities. **b** bathroom fitments or units, esp. as sold together. **2** esp. *US* a room containing a lavatory.

Bathsheba /bæθˈʃiːbə, ˈbæθʃɪbə/ in the Bible, the wife of Uriah the Hittite (2 Sam. 11). She became one of the wives of David, who had caused her husband to be killed in battle, and was the mother of Solomon.

Bathurst /ˈbæθɜːst/ the former name (until 1973) for BANJUL.

bathymetry /bəˈθɪmɪtrɪ/ n. the measurement of depth in seas and lakes. □ **bathymetric** /ˌbæθɪˈmetrɪk/ adj. [f. Gk *bathus* deep + -METRY]

bathyscaphe /ˈbæθɪˌskæf/ n. a manned vessel for deep-sea diving. [Gk *bathus* deep + *skaphos* ship]

bathysphere /ˈbæθɪˌsfɪə(r)/ n. a spherical vessel for deep-sea observation. [Gk *bathus* deep + SPHERE]

batik /bəˈtiːk, ˈbætɪk/ n. a method (orig. used in Java) of producing coloured designs on textiles by applying wax to the parts to be left uncoloured; a piece of cloth treated in this way. [Javanese, = painted]

Batista /bəˈtiːstə/, Fulgencio (full name Fulgencio Batista y Zaldívar) (1901–73), Cuban soldier and statesman, President 1940–4 and 1952–9. He was instrumental in the military coup which overthrew the existing Cuban regime in 1933, and ruled Cuba indirectly through a succession of Presidents until his own election as President in 1940. He regained power after a coup in 1952; his second government was supported by the US but was notoriously corrupt and ruthless. Facing increased opposition from the guerrillas led by Fidel Castro, he was finally overthrown and fled to the Dominican Republic in 1959.

batiste /bæˈtiːst/ n. & adj. ● n. a fine linen or cotton cloth. ● adj. made of batiste. [F (earlier *batiche*), perh. rel. to *battre* BATTER¹]

Batman /ˈbætmæn/ a US cartoon character, by day the millionaire socialite Bruce Wayne but at night a cloaked and masked figure fighting crime in Gotham City (New York). First appearing in 1939 in a comic strip by artist Bob Kane (b.1916) and writer Bill Finger (1917–74), Batman has since featured in a popular 1960s TV series and two major films (1989 and 1992).

batman /ˈbætmən/ n. (pl. **-men**) *Mil.* an attendant serving an officer. [OF *bat*, *bast* f. med.L *bastum* pack-saddle + MAN]

baton /ˈbæt(ə)n/ n. **1** a thin stick used by a conductor to direct an orchestra, choir, etc. **2** *Athletics* a short stick or tube carried and passed on by the runners in a relay race. **3** a long stick carried and twirled by a drum major. **4** a staff of office or authority, esp. a Field Marshal's. **5** a police officer's truncheon. **6** *Heraldry* a narrow truncated bend. **7** a short bar replacing some figures on a clock-face etc. □ **baton round** a rubber or plastic bullet. [F *bâton*, *baston* ult. f. LL *bastum* stick]

Baton Rouge /ˌbæt(ə)n ˈruːʒ/ the state capital of Louisiana; pop. (1990) 219,530. Founded in 1719 by French settlers, the city, whose name means 'red post' in French, was named after a post serving as a boundary marker.

batrachian /bəˈtreɪkɪən/ n. & adj. ● n. any of the amphibians that discard gills and tails, esp. the frog and toad. ● adj. of or relating to the batrachians. [Gk *batrakhos* frog]

bats /bæts/ predic.adj. sl. crazy. [f. phr. *have bats in the belfry*: see BAT²]

batsman /ˈbætsmən/ n. (pl. **-men**) **1** a person who bats or is batting, esp. in cricket. **2** a signaller using bats to guide aircraft on the ground. □ **batsmanship** n. (in sense 1).

battalion /bəˈtælɪən/ n. **1** a large body of troops ready for battle, esp. an infantry unit forming part of a brigade. **2** a large body of people pursuing a common aim or sharing a major undertaking. [F *bataillon* f. It. *battaglione* f. *battaglia* BATTLE]

Battambang /ˈbætəmˌbæŋ/ (also **Batdambang**) the capital of a province of the same name in western Cambodia; pop. (1981) 551,860.

battels /ˈbæt(ə)lz/ n.pl. *Brit.* an Oxford college account for expenses, esp. for board and the supply of provisions. [perh. f. obs. *battle* (v.) fatten f. obs. *battle* (adj.) nutritious: cf. BATTEN²]

Batten /ˈbæt(ə)n/, Jean (1909–82), New Zealand aviator. She was the first woman to fly from England to Australia and back (1934–5), breaking Amy Johnson's record for the England to Australia journey by nearly five days. In 1936, she made the first direct solo flight from England to New Zealand in a time of 11 days 45 minutes.

batten¹ /ˈbæt(ə)n/ n. & v. ● n. **1** a long flat strip of squared timber or metal, esp. used to hold something in place or as a fastening against a wall etc. **2** a strip of wood used for clamping the boards of a door etc. **3** *Naut.* a strip of wood or metal for securing a tarpaulin over a ship's hatchway. ● v.tr. strengthen or fasten with battens. □ **batten down the hatches 1** *Naut.* secure a ship's tarpaulins. **2** prepare for a difficulty or crisis. [OF *batant* part. f. *batre* beat f. L *battuere*]

batten² /ˈbæt(ə)n/ v.intr. (foll. by *on*) thrive or prosper at another's expense. [ON *batna* get better f. *bati* advantage]

Battenberg /ˈbæt(ə)nˌbɜːg/ n. a kind of oblong cake, usu. of two colours of sponge and covered with marzipan. [*Battenberg* in Germany]

batter¹ /ˈbætə(r)/ v. **1 a** tr. strike repeatedly with hard blows, esp. so as to cause visible damage. **b** intr. (often foll. by *against*, *at*, etc.) strike repeated blows; pound heavily and insistently (*batter at the door*). **2** tr. (often in *passive*) **a** handle roughly, esp. over a long period. **b** censure or criticize severely. □ **battered baby** an infant that has suffered repeated violence from adults, esp. its parents. **battered wife** a wife subjected to repeated violence by her husband. **battering-ram 1** *hist.* a heavy beam, orig. with an end in the form of a carved ram's head, used in breaching fortifications. **2** a similar object used to break down doors. □ **batterer** n. **battering** n. [ME f. AF *baterer* f. OF *batre* beat f. L *battuere*]

batter² /ˈbætə(r)/ n. **1** a fluid mixture of flour, egg, and milk or water, used in cooking, esp. for pancakes and for coating food before frying. **2** *Printing* an area of damaged type. [ME f. AF *batour* f. OF *bateüre* f. *batre*: see BATTER¹]

batter³ /ˈbætə(r)/ n. *Sport* a player batting, esp. in baseball.

batter⁴ /ˈbætə(r)/ n. & v. ● n. **1** a wall etc. with a sloping face. **2** a receding slope. ● v.intr. have a receding slope. [ME: orig. unkn.]

battered /ˈbætəd/ adj. (esp. of fish) coated in batter and deep-fried.

battery /ˈbætərɪ/ n. (pl. **-ies**) **1** a device which provides a source of electric current, consisting of one or more cells in which electricity is generated by a chemical reaction. (*See note below.*) esp. *Brit.* **a** a series of cages for the intensive breeding and rearing of poultry or cattle. **3** a set of similar units of equipment, esp. connected. **4** (usu. foll. by *of*) an extensive series, sequence, or range. **5** a series of tests, esp. psychological. **6 a** a fortified emplacement for heavy guns. **b** an artillery unit of guns, men, and vehicles. **7** *Law* an act, including touching, inflicting unlawful personal violence on another person, even if no physical harm is done. **8** *Baseball* the pitcher and the catcher. [F *batterie* f. *batre, battre* strike f. L *battuere*]

▪ The first electric battery was made in 1800 by Alessandro Volta, who caused a current to pass through a wire by attaching it to two different metals (zinc and copper) in a salt solution. Dry batteries are descended from the Leclanché cell, with the electrolyte in the form of a paste or jelly instead of a liquid.

battery farming see FACTORY FARMING.

Batticaloa /ˌbætɪkəˈləʊə/ a city on the east coast of Sri Lanka; pop. (1981) 42,900.

batting /ˈbætɪŋ/ n. **1** the action of hitting with a bat. **2** cotton wadding prepared in sheets for use in quilts etc. □ **batting order** the order in which people act or take their turn, esp. of batsmen in cricket.

battle /ˈbæt(ə)l/ n. & v. ● n. **1** a prolonged fight between large organized armed forces. **2** a contest; a prolonged or difficult struggle (*life is a constant battle; a battle of wits*). ● v. **1** *intr.* struggle; fight persistently (*battled against the elements; battled for women's rights*). **2** *tr.* fight (one's way etc.). **3** *tr. N. Amer.* engage in battle with. □ **battle-cruiser** *hist.* a heavy-gunned ship faster and more lightly armoured than a battleship. **battle-cry** a cry or slogan of participants in a battle or contest. **battle fatigue** = *combat fatigue*. **battle royal 1** a battle in which several combatants or all available forces engage; a free fight. **2** a heated argument. **half the battle** the key to the success of an undertaking. □ **battler** n. [ME f. OF *bataille* ult. f. LL *battualia* gladiatorial exercises f. L *battuere* beat]

battleaxe /ˈbæt(ə)lˌæks/ n. **1** a large axe used in ancient warfare. **2** *colloq.* a formidable or domineering older woman. □ **battleaxe block** *Austral.* a battleaxe-shaped block of land, one lacking a frontage and accessible through a lane.

battlebus /ˈbæt(ə)lˌbʌs/ n. a bus or coach used as a mobile operational centre during an election campaign.

battledore /ˈbæt(ə)lˌdɔː(r)/ n. *hist.* **1 a** (in full **battledore and shuttlecock**) a game played with a shuttlecock and rackets. **b** the racket used in this. **2** a kind of wooden utensil like a paddle, formerly used in washing, baking, etc. [15th c., perh. f. Prov. *batedor* beater f. *batre* beat]

battledress /ˈbæt(ə)lˌdres/ n. the everyday uniform of a soldier.

battlefield /ˈbæt(ə)lˌfiːld/ n. (also **battleground** /-ˌɡraʊnd/) the piece of ground on which a battle is or was fought.

battlement /ˈbæt(ə)lmənt/ n. (usu. in *pl.*) **1** a parapet with recesses along the top of a wall, as part of a fortification. **2** a section of roof enclosed by this (*walking on the battlements*). □ **battlemented** adj. [OF *bataillier* furnish with ramparts + -MENT]

Battle of Aboukir Bay, Battle of Britain, etc. see ABOUKIR BAY, BATTLE OF; BRITAIN, BATTLE OF, etc.

battleship /ˈbæt(ə)lˌʃɪp/ n. a warship with the heaviest armour and the largest guns. Battleships formed the main armament of the world's major navies in the late 19th and early 20th centuries (see also DREADNOUGHT). Expensive and vulnerable to aircraft, battleships were generally discarded after the Second World War although the US retains four of the *Iowa* class. Some of these have seen intermittent action in recent years, e.g. for shore bombardment in Vietnam (1967-8), Lebanon (1983), and (with cruise missiles) the Gulf War (1991). [shortening of *line-of-battle ship* (LINE[1] 24b), orig. with ref. to the largest wooden warships]

battue /bæˈtjuː, -ˈtuː/ n. **1 a** the driving of game towards hunters by beaters. **b** a shooting-party arranged in this way. **2** wholesale slaughter. [F, fem. past part. of *battre* beat f. L *battuere*]

batty /ˈbætɪ/ adj. (**battier, battiest**) *sl.* crazy. □ **battily** adv. **battiness** n. [BAT[2] + -Y[1]]

Batwa pl. of TWA.

batwing /ˈbætwɪŋ/ attrib.adj. (esp. of a sleeve or a flame) shaped like the wing of a bat.

batwoman /ˈbætˌwʊmən/ n. (*pl.* **-women**) a female attendant serving an officer in the women's services. [as BATMAN + WOMAN]

bauble /ˈbɔːb(ə)l/ n. **1** a showy trinket or toy of little value. **2** a baton formerly used as an emblem by jesters. [ME f. OF *ba(u)bel* child's toy, of unkn. orig.]

Baucis /ˈbɔːsɪs/ *Gk Mythol.* the wife of Philemon.

baud /baʊd, bɔːd/ n. (*pl.* same or **bauds**) *Computing* etc. **1** a unit of signal transmission speed of one information unit per second. **2** (loosely) a unit of data transmission speed of one bit per second. [J. M. E. *Baudot*, Fr. engineer (1845-1903)]

Baudelaire /ˈbəʊdəˌleə(r)/, Charles (Pierre) (1821-67), French poet and critic. An associate and champion of Manet and Delacroix, he began his literary career writing art criticism and reviews, but is now largely known for *Les Fleurs du mal* (1857), a series of 101 lyrics in a variety of metres. In these he explores his sense of isolation, exile, sin, boredom, and melancholy, as well as the attraction of evil and vice and the fascination and degradation of Parisian life. He died in poverty and obscurity, and it was only in the later years of the 19th century that his importance to the symbolist movement was recognized.

Bauhaus /ˈbaʊhaʊs/ a school of design established by Walter Gropius in Weimar, Germany in 1919, which had its origins in the Weimar School of Art and Crafts. The school's style is characterized by an emphasis on functionality and simplicity of form and which played a key role in establishing the relationship between design and industrial techniques. Although some of its teachers (such as Paul Klee and Wassily Kandinsky) were painters, the school is perhaps best known for its designs of objects for utilitarian use. The socialist principles on which Bauhaus ideas rested incurred the hostility of the Nazis and, after moving to Berlin in 1932, the school was closed in 1933. The resultant movement abroad of teachers and pupils (for example, Mies van der Rohe and Josef Albers) ensured the international dissemination of its style and ideas. [G f. *Bau* building + *Haus* house]

baulk /bɔːlk, bɔːk/ v. & n. (also **balk**) ● v. **1** *intr.* **a** refuse to go on. **b** (often foll. by *at*) hesitate. **2** *tr.* **a** thwart, hinder. **b** disappoint. **3** *tr.* **a** miss, let slip (a chance etc.). **b** ignore, shirk. ● n. **1** a hindrance; a stumbling block. **2 a** a roughly squared timber beam. **b** a tie-beam of a house. **3** *Billiards* etc. the part of a billiard table between the baulk line and the bottom cushion, within the D or half-circle of which play begins. **4** *Baseball* an illegal action made by a pitcher. **5** a ridge left unploughed between furrows. □ **baulk line** *Billiards* etc. a line drawn on a billiard table parallel to the bottom cushion at a distance one-fifth of the length of the playing area. [OE *balc* f. ON *bálkr* f. Gmc]

baulky /ˈbɔːlkɪ, ˈbɔːkɪ/ adj. (also **balky**) (**-ier, -iest**) reluctant, perverse. [BAULK + -Y[1]]

bauxite /ˈbɔːksaɪt/ n. a claylike mineral containing varying proportions of alumina, the chief source of aluminium. □ **bauxitic** /bɔːkˈsɪtɪk/ adj. [F f. *Les Baux* near Arles in SE France + -ITE[1]]

bavardage /ˌbævɑːˈdɑːʒ/ n. idle gossip, chit-chat. [F, f. *bavarder* to chatter]

Bavaria /bəˈveərɪə/ (called in German *Bayern*) a state of southern Germany; capital, Munich. Formerly an independent kingdom, it joined the German Empire in 1871. □ **Bavarian** adj. & n.

bawd /bɔːd/ n. a woman who runs a brothel. [ME *bawdstrot* f. OF *baudetrot, baudestroyt* procuress]

bawdy /ˈbɔːdɪ/ adj. & n. ● adj. (**bawdier, bawdiest**) (esp. humorously) indecent; earthy, coarse. ● n. bawdy talk or writing. □ **bawdy-house** a brothel. □ **bawdily** adv. **bawdiness** n. [BAWD + -Y[1]]

bawl /bɔːl/ v. **1** *tr.* speak or call out noisily. **2** *intr.* weep loudly. □ **bawl out** *colloq.* reprimand angrily. [imit.: cf. med.L *baulare* bark, Icel. *baula* (Sw. *böla*) to low]

Bax /bæks/, Sir Arnold (Edward Trevor) (1883-1953), English composer. A lasting sympathy for the Celtic revival, particularly of Irish literature, fostered in him an enduring love of Ireland's scenery and folk-song. This is expressed musically in works such as *An Irish Elegy* (1917) for English horn, harp, and strings. *Tintagel* (1917) and his other tone poems are the best known of his works today, but he also composed seven symphonies, chamber works, songs, and choral music.

Baxter /ˈbækstə(r)/, James K(eir) (1926-72), New Zealand poet, dramatist, and critic. He published more than thirty books of poems; his early lyric poetry focuses on the New Zealand landscape and its influence on its inhabitants. A convert to Roman Catholicism, in *Jerusalem Sonnets* (1970) he develops a fluid sonnet form to express his spirituality; *Autumn Testament* (1972) reflects his humanistic socialism. His criticism includes *Aspects of Poetry in New Zealand* (1967).

bay[1] /beɪ/ n. **1** a broad inlet of the sea where the land curves inwards. **2** a recess in a mountain range. [ME f. OF *baie* f. OSp. *bahia*]

bay[2] /beɪ/ n. **1** (in full **bay laurel** or **bay tree**) a laurel, *Laurus nobilis*, having deep green leaves and purple berries. Also called *sweet bay*. **2** (in *pl.*) a wreath made of bay leaves, for a victor or poet. □ **bay leaf** the aromatic (usu. dried) leaf of the bay tree, used in cooking. **bay rum** a perfume, esp. for the hair, distilled orig. from bayberry leaves in rum (see BAYBERRY). [OF *baie* f. L *baca* berry]

bay³ /beɪ/ n. **1** a space created by a window-line projecting outwards from a wall. **2** a recess; a section of wall between buttresses or columns, esp. in the nave of a church etc. **3** a compartment (*bomb bay*). **4** an area specially allocated or marked off (*sickbay; loading bay*). **5** *Brit.* the terminus of a branch line at a railway station also having through lines, usu. at the side of an outer platform. □ **bay window** a window built into a bay. [ME f. OF *baie* f. ba(y)*er* gape f. med.L *batare*]

bay⁴ /beɪ/ adj. & n. ● adj. (esp. of a horse) dark reddish-brown. ● n. a bay horse with a black mane and tail. [OF *bai* f. L *badius*]

bay⁵ /beɪ/ v. & n. ● v. **1** *intr.* bark or howl loudly and plaintively. **2** *tr.* bay at. ● n. the sound of baying, esp. in chorus from hounds in close pursuit. □ **at bay 1** cornered, apparently unable to escape. **2** in a desperate situation. **bring to bay** gain on in pursuit; trap. **hold** (or **keep**) **at bay** hold off (a pursuer). **stand at bay** turn to face one's pursuers. [ME f. OF *bai, baïer* bark f. It. *baiare*, of imit. orig.]

Bayard /ˈbeɪɑːd/, Pierre du Terrail, Chevalier de (1473–1524), French soldier. He served under several French monarchs, including Louis XII (1462–1515), and became known as the knight 'sans peur et sans reproche' (fearless and above reproach).

bayberry /ˈbeɪbərɪ/ n. (pl. **-ies**) a North American plant, *Myrica cerifera*, having aromatic leaves and bearing berries covered in a wax coating. Also called *wax myrtle*. [BAY² + BERRY]

Bayern /ˈbaɪərn/ the German name for BAVARIA.

Bayeux Tapestry /baɪˈjɜː/ a fine example of medieval English embroidery (it is not really a tapestry), executed between 1066 and 1077, probably at Canterbury, for Odo, bishop of Bayeux and half-brother of William the Conqueror, and now exhibited at Bayeux in Normandy. It is 48 cm (19 in.) wide and, although incomplete, 70 m (230 ft) long. In seventy-nine scenes, accompanied by a Latin text and arranged like a strip cartoon, it tells the story of the Norman Conquest and the events that led up to it. It is an important historical record, relating incidents not recorded elsewhere, and is particularly valuable for the information it reveals about military tactics and equipment.

Baykal, Lake see BAIKAL, LAKE.

Baykonur see BAIKONUR.

Baylis /ˈbeɪlɪs/, Lilian Mary (1874–1937), English theatre manager. She assisted in the running of the Royal Victoria Coffee Music Hall, a temperance hall housed in the Royal Victoria Theatre (the Old Vic). Under her management from 1912, the Old Vic acquired a reputation as the world's leading house for Shakespearian productions. Her initiative in reopening the old Sadler's Wells Theatre in 1931 led to the development of the Royal Ballet and the English National Opera.

Bay of Bengal, Bay of Fundy, etc. see BENGAL, BAY OF; FUNDY, BAY OF; etc.

Bay of Pigs a bay on the SW coast of Cuba, scene of an unsuccessful attempt in 1961 by US-backed Cuban exiles to invade the country and overthrow the regime of Fidel Castro.

Bay of Plenty a region of North Island, New Zealand, extending around the bay of the same name. The port of Tauranga is situated on it.

bayonet /ˈbeɪə,net/ n. & v. ● n. **1** a stabbing blade attachable to the muzzle of a rifle. **2** an electrical or other fitting engaged by being pushed into a socket and twisted. ● v.tr. (**bayoneted, bayoneting**) stab with a bayonet. [F *baïonnette*, perh. f. *Bayonne* in SW France, where they were first made]

bayou /ˈbaɪuː/ n. a marshy offshoot of a river etc. in the southern US. [Amer. F: cf. Choctaw *bayuk*]

Bayreuth /ˈbaɪrɔɪt, baɪˈrɔɪt/ a town in Bavaria, Germany, where Wagner made his home from 1874 and where he is buried. Festivals of his operas are held regularly in a theatre specially built (1872–6) to house performances of *Der Ring des Nibelungen*.

Bay State, the Massachusetts.

bazaar /bəˈzɑː(r)/ n. **1** a market in an oriental country. **2** a fund-raising sale of goods, esp. for charity. **3** a large shop selling fancy goods etc. [Pers. *bāzār*, prob. through Turk. and It.]

bazoo /bəˈzuː/ n. *sl.* the mouth. [19th c.: orig. unkn.]

bazooka /bəˈzuːkə/ n. **1** a tubular short-range rocket-launcher used against tanks. **2** a crude trombone-like musical instrument. [app. f. *bazoo* mouth, of unkn. orig.]

BB *abbr.* double-black (pencil-lead).

BBC see BRITISH BROADCASTING CORPORATION. □ **BBC English** English as traditionally pronounced by BBC announcers.

BBFC *abbr.* British Board of Film Classification. ¶ Formerly *British Board of Film Censors*.

bbl. *abbr.* barrels (esp. of oil).

BBSRC *abbr.* (in the UK) Biotechnology and Biological Sciences Research Council.

BC *abbr.* **1** (of a date) before Christ. **2** British Columbia.

BCCI *abbr. hist.* Bank of Credit and Commerce International.

BCD /ˌbiːsiːˈdiː/ n. *Computing* a code representing decimal numbers as a string of binary digits. [abbr. for binary coded decimal]

BCE *abbr.* before the Common Era.

B-cell /ˈbiːsel/ n. *Physiol.* = B-LYMPHOCYTE.

BCG *abbr.* Bacillus Calmette-Guérin, an anti-tuberculosis vaccine.

BD *abbr.* Bachelor of Divinity.

Bde. *abbr.* Brigade.

bdellium /ˈdelɪəm/ n. **1** a tree, esp. of the genus *Commiphora*, yielding resin. **2** this fragrant resin used in perfumes. [L f. Gk *bdellion* f. Heb. *b'dhōlaḥ*]

Bdr. *abbr.* (before a name) Bombardier.

BDS *abbr.* Bachelor of Dental Surgery.

BE *abbr.* **1** Bachelor of Education. **2** Bachelor of Engineering. **3** bill of exchange.

Be *symb. Chem.* the element beryllium.

be /biː, bɪ/ v. & v.aux. (*sing. present* **am** /æm, əm/; *are* /ɑː(r), ə(r)/; *is* /ɪz/; *pl. present* **are**; *1st and 3rd sing. past* **was** /wɒz, wəz/; *2nd sing. past and pl. past* **were** /wɜː(r), wə(r), weə(r)/; *present subjunctive* **be**; *past subjunctive* **were**; *pres. part.* **being**; *past part.* **been** /biːn, bɪn/) ● v.intr. **1** (often prec. by *there*) exist, live (*I think, therefore I am; there once was a man; there is a house on the corner; there was no doubt*). **2 a** occur; take place (*dinner is at eight*). **b** occupy a position in space (*he is in the garden; she is from abroad; have you been to Paris?*). **3** remain, continue (*let it be*). **4** linking subject and predicate, expressing: **a** identity (*she is the person; today is Thursday*). **b** condition (*he is ill today*). **c** state or quality (*he is very kind; they are my friends*). **d** opinion (*I am against hanging*). **e** total (*two and two are four*). **f** cost or significance (*it is £5 to enter; it is nothing to me*). ● v.aux. **1** with a past participle to form the passive mood (*it was done; it is said; we shall be helped*). **2** with a present participle to form continuous tenses (*we are coming; it is being cleaned*). **3** with an infinitive to express duty or commitment, intention, possibility, destiny, or hypothesis (*I am to tell you; we are to wait here; he is to come at four; it was not to be found; they were never to meet again; if I were to die*). **4** *archaic* with the past participle of intransitive verbs to form perfect tenses (*the sun is set; Babylon is fallen*). □ **be about** occupy oneself with (*is about his business*). **be-all and end-all** *colloq.* (often foll. by *of*) the whole being or essence. **be at** occupy oneself with (*what is he at?; mice have been at the food*). **been** (or **been and gone**) *sl.* an expression of protest or surprise (*he's been and taken my car!*). **be off** *colloq.* go away; leave. **be that as it may** see MAY. **-to-be** of the future (in *comb.:* *bride-to-be*). [OE *beo(m)*, (*e)am, is, (e)aron*; *past f.* OE *wæs* f. *wesan* to be; there are numerous Gmc cognates]

be- /bɪ/ prefix forming verbs: **1** (from transitive verbs) **a** all over; all round (*beset; besmear*). **b** thoroughly, excessively (*begrudge; belabour*). **2** (from intransitive verbs) expressing transitive action (*bemoan; bestride*). **3** (from adjectives and nouns) expressing transitive action (*befool; befoul*). **4** (from nouns) **a** affect with (*befog*). **b** treat as (*befriend*). **c** (forming adjectives in *-ed*) having; covered with (*bejewelled; bespectacled*). [OE *be-*, weak form of *bī* BY as in *bygone, byword*, etc.]

BEA *abbr.* British Epilepsy Association.

beach /biːtʃ/ n. & v. ● n. a pebbly or sandy shore esp. of the sea between high and low-water marks. ● v.tr. **1** run or haul up (a boat etc.) on to a beach. **2** (as **beached** adj.) (of a whale etc.) stranded out of the water. □ **beach-ball** a large inflated ball for games on the beach. **beach buggy** a low wide-wheeled motor vehicle for recreational driving on sand. **beach plum 1** a maritime North American shrub, *Prunus maritima*. **2** its edible fruit. [16th c.: orig. unkn.]

Beach Boys a US pop group formed *c.*1961, known for their vocal harmonies. They made their name with simple, catchy songs like 'Surfin' USA' (1963) and 'California Girls' (1965), but the work of chief songwriter and producer Brian Wilson (b.1941) became increasingly sophisticated, culminating in the single 'Good Vibrations' and album *Pet Sounds* (both 1966).

beachcomber /ˈbiːtʃˌkəʊmə(r)/ n. **1** a vagrant who lives by searching beaches for articles of value. **2** a long wave rolling in from the sea.

beachhead /ˈbiːtʃhed/ *n. Mil.* a fortified position established on a beach by landing forces. [after *bridgehead*]

Beach-la-mar /ˌbiːtʃləˈmɑː(r)/ *n.* an English-based Creole language formerly spoken in the western Pacific. [corrupt. f. Port. *bicho do mar*: see BÊCHE-DE-MER]

beachwear /ˈbiːtʃweə(r)/ *n.* clothing suitable for wearing on the beach.

beacon /ˈbiːkən/ *n.* **1 a** a fire or light set up in a high or prominent position as a warning etc. **b** *Brit.* (now often in place-names) a hill suitable for this. **2** a visible warning or guiding point or device (e.g. a lighthouse, navigation buoy, etc.). **3** a radio transmitter whose signal helps fix the position of a ship or aircraft. **4** *Brit.* = BELISHA BEACON. [OE *bēacn* f. WG]

bead /biːd/ *n. & v.* ● *n.* **1 a** a small usu. rounded and perforated piece of glass, stone, etc., for threading with others to make jewellery, or sewing on to fabric, etc. **b** (in *pl.*) a string of beads; a rosary. **2** a drop of liquid; a bubble. **3** a small knob in the foresight of a gun. **4** the inner edge of a pneumatic tyre that grips the rim of the wheel. **5** *Archit.* **a** a moulding like a series of beads. **b** a narrow moulding with a semicircular cross-section. ● *v.* **1** *tr.* furnish or decorate with beads. **2** *tr.* string together. **3** *intr.* form or grow into beads. □ **draw a bead on** take aim at. **tell one's beads** use the beads of a rosary etc. in counting prayers. □ **beaded** *adj.* [orig. = 'prayer' (for which the earliest use of beads arose): OE *gebed* f. Gmc, rel. to BID]

beading /ˈbiːdɪŋ/ *n.* **1** decoration in the form of or resembling a row of beads, esp. lacelike looped edging. **2** *Archit.* a bead moulding. **3** the bead of a tyre.

beadle /ˈbiːd(ə)l/ *n.* **1** *Brit.* a ceremonial officer of a church, college, etc. **2** *Sc.* a church officer attending on the minister. **3** *Brit. hist.* a minor parish officer dealing with petty offenders etc. □ **beadleship** *n.* [ME f. OF *bedel* ult. f. Gmc]

beadsman /ˈbiːdzmən/ *n.* (*pl.* **-men**) *hist.* **1** a pensioner provided for by a benefactor in return for prayers. **2** an inmate of an almshouse.

beady /ˈbiːdɪ/ *adj.* (**beadier, beadiest**) **1** (of the eyes) small, round, and glittering. **2** covered with beads or drops. □ **beady-eyed** having beady eyes. □ **beadily** *adv.* **beadiness** *n.*

beagle /ˈbiːg(ə)l/ *n. & v.* ● *n.* **1** a breed of small hound with a short coat, originally used for hunting hares. **2** *hist.* an informer or spy; a constable. ● *v.intr.* (often as **beagling** *n.*) hunt with beagles. □ **beagler** *n.* [ME f. OF *beegueule* noisy person, prob. f. *beer* open wide + *gueule* throat]

Beagle Channel a channel through the islands of Tierra del Fuego at the southern tip of South America. It is named after HMS *Beagle*, the ship of Charles Darwin's voyage of 1831–6.

beak¹ /biːk/ *n.* **1 a** a bird's horny projecting jaws; a bill. **b** the similar projecting jaw of other animals, e.g. a turtle. **2** *sl.* a hooked nose. **3** *Naut. hist.* the projection at the prow of a warship. **4** a spout. □ **beaked** *adj.* **beaky** *adj.* [ME f. OF *bec* f. L *beccus*, of Celt. orig.]

beak² /biːk/ *n. Brit. sl.* **1** a magistrate. **2** a schoolmaster. [19th c.: prob. f. thieves' cant]

Beaker *attrib.adj.* of or relating to a people thought to have come to Britain from central Europe in the early Bronze Age; the term derives from the earthenware beakers found in their graves.

beaker /ˈbiːkə(r)/ *n.* **1** a tall drinking-vessel, usu. of plastic and tumbler-shaped. **2** a lipped cylindrical glass vessel for scientific experiments. **3** *archaic* or *literary* a large drinking-vessel with a wide mouth. **4** *Archaeol.* a wide-mouthed pottery vessel found in graves of the late neolithic period (3rd millennium BC) in western Europe. [ME f. ON *bikarr*, perh. f. Gk *bikos* drinking-bowl]

Beale /biːl/, Dorothea (1831–1906), English educationist. With her friend and fellow educationist Frances Buss she pioneered women's higher education in Britain. From 1858 until her death she was principal of Cheltenham Ladies' College, where she introduced a curriculum similar to that used in public schools for boys. She founded St Hilda's College in Cheltenham (1885) as the first English training college for women teachers, and established St Hilda's Hall in Oxford for women teachers in 1893. She was also a strong supporter of women's suffrage.

beam /biːm/ *n. & v.* ● *n.* **1** a long sturdy piece of squared timber or metal spanning an opening or room, usu. to support the structure above. **2 a** a ray or shaft of light. **b** a directional flow of particles or radiation. **3** a bright look or smile. **4 a** a series of radio or radar signals as a guide to a ship or aircraft. **b** the course indicated by this (*off beam*). **5** the crossbar of a balance. **6 a** a ship's breadth at its widest point.

b the width of a person's hips (esp. *broad in the beam*). **7** (in *pl.*) the horizontal cross-timbers of a ship supporting the deck and joining the sides. **8** the side of a ship (*land on the port beam*). **9** the chief timber of a plough. **10** the cylinder in a loom on which the warp or cloth is wound. **11** the main stem of a stag's antlers. **12** the lever in an engine connecting the piston-rod and crank. **13** the shank of an anchor. ● *v.* **1** *tr.* emit or direct (light, radio waves, etc.). **2** *intr.* **a** shine. **b** look or smile radiantly. **3** (usu. foll. by *up* or other adv.) (in science fiction) **a** *intr.* travel from one point to another as or along an invisible beam of energy. **b** *tr.* transport in this way. □ **beam-compass** (or **-compasses**) compasses with a beam connecting sliding sockets, used for large circles. **beam engine** a steam engine having a beam (see sense 12 of *n.*). (See also NEWCOMEN, WATT.) **a beam in one's eye** a fault that is greater in oneself than in the person one is finding fault with (see Matt. 7:3). **off** (or **off the**) **beam** *colloq.* mistaken. **on the beam** *colloq.* on the right track. **on the beam-ends** (of a ship) on its side; almost capsizing. **on one's beam-ends** near the end of one's resources. [OE *bēam* tree f. WG]

beamer /ˈbiːmə(r)/ *n. Cricket colloq.* a ball bowled at a batsman's head.

Beamon /ˈbiːmən/, Robert ('Bob'), (b.1946), American athlete. He set a world record of 8.90 metres (29 ft 2½ in.) in the long jump at the 1968 Olympic Games in Mexico City; this was not beaten until 1991.

beamy /ˈbiːmɪ/ *adj.* (of a ship) broad-beamed.

bean /biːn/ *n. & v.* ● *n.* **1 a** any kind of leguminous plant with edible usu. kidney-shaped seeds in long pods. **b** one of these seeds. **2** a similar seed of coffee and other plants. **3** *N. Amer. sl.* the head, esp. as a source of common sense. **4** (in *pl.*; with *neg.*) *US sl.* anything at all (*doesn't know beans about it*). ● *v.tr. N. Amer. sl.* hit on the head. □ **bean-counter** esp. *US colloq. derog.* an accountant (esp. one who is obsessed with returns on investments); a penny-pincher. **bean curd** jelly or paste made from beans, used esp. in Asian cookery. **bean goose** a grey goose, *Anser fabalis*, breeding in the Arctic. **bean sprout** a sprout of a bean seed, esp. of the mung bean, used as food. **full of beans** *colloq.* lively; in high spirits. **not a bean** *Brit. sl.* no money. **old bean** *Brit. sl.* a friendly form of address, usu. to a man. [OE *bēan* f. Gmc]

beanbag /ˈbiːnbæg/ *n.* **1** a small bag filled with dried beans and used esp. in children's games. **2** a large cushion filled usu. with polystyrene beads and used as a seat.

beanery /ˈbiːnərɪ/ *n.* (*pl.* **-ies**) *N. Amer. sl.* a cheap restaurant.

beanfeast /ˈbiːnfiːst/ *n.* **1** *Brit. colloq.* a celebration; a merry time. **2** an employer's annual dinner given to employees. [BEAN + FEAST, beans and bacon being regarded as an indispensable dish at an employer's annual dinner]

beanie /ˈbiːnɪ/ *n.* a small close-fitting hat worn on the back of the head. [perh. f. BEAN 'head' + -IE]

beano /ˈbiːnəʊ/ *n.* (*pl.* **-os**) *Brit. sl.* a celebration; a party. [abbr. of BEANFEAST]

beanpole /ˈbiːnpəʊl/ *n.* **1** a stick for supporting bean plants. **2** *colloq.* a tall thin person.

beanstalk /ˈbiːnstɔːk/ *n.* the stem of a bean plant.

bear¹ /beə(r)/ *v.* (*past* **bore** /bɔː(r)/; *past part.* **borne, born** /bɔːn/) ¶ In the passive *born* is used with reference to birth (e.g. *was born in July*), except for *borne* by foll. by the name of the mother (e.g. *was borne by Sarah*). **1** *tr.* carry, bring, or take (esp. visibly) (*bear gifts*). **2** *tr.* show; be marked by; have as an attribute or characteristic (*bear marks of violence*; *bears no relation to the case*; *bore no name*). **3** *tr.* **a** produce, yield (*fruit etc.*). **b** give birth to (*has borne a son*; *was born last week*). **4** *tr.* **a** sustain (a weight, responsibility, cost, etc.). **b** stand, endure (an ordeal, difficulty, etc.). **5** *tr.* (usu. with *neg.* or *interrog.*) **a** tolerate; put up with (*can't bear him*; *how can you bear it?*). **b** admit of; be fit for (*does not bear thinking about*). **6** *tr.* carry in thought or memory (*bear a grudge*). **7** *intr.* veer in a given direction (*bear left*). **8** *tr.* bring or provide (something needed) (*bear him company*). **9** *refl.* behave (in a certain way). □ **bear arms 1** carry weapons; serve as a soldier. **2** wear or display heraldic devices. **bear away** (or **off**) win (a prize etc.). **bear down** exert downward pressure. **bear down on** approach rapidly or purposefully. **bear fruit** have results. **bear a hand** help. **bear hard on** oppress. **bear in mind** take into account having remembered. **bear on** (or **upon**) be relevant to. **bear out** support or confirm (an account or the person giving it). **bear repeating** be worth repetition. **bear up 1** raise one's spirits; not despair. **2** (often foll. by *against, under*) endure, survive. **bear with** treat forbearingly; tolerate patiently. **bear witness** testify. [OE *beran* f. Gmc]

bear² /beə(r)/ *n. & v.* ● *n.* **1** a large heavy omnivorous mammal of the family Ursidae, having thick fur and a very short tail, and walking on the soles of its feet. **2** a rough, unmannerly, or uncouth person. **3** *Stock Exch.* a person who sells shares for future delivery, hoping that prices will fall and that he or she will be able to buy them at a lower price before having to deliver them. (See *note below.*) **4** = TEDDY. **5** (**the Bear**) *colloq.* Russia. ● *v. Stock Exch.* **1** *intr.* speculate for a fall in price. **2** *tr.* produce a fall in the price of (stocks etc.). □ **bear-baiting** *hist.* an entertainment involving setting dogs to attack a captive bear. **bear-hug** a tight embrace. **bear market** *Stock Exch.* a market with falling prices. **bear's breech** a kind of acanthus, *Acanthus mollis* (see BRANK-URSINE). **bear's ear** auricula. **bear's foot** stinking hellebore. **the Great Bear, the Little Bear** *Astron.* the constellations Ursa Major and Ursa Minor. **like a bear with a sore head** *Brit. colloq.* very irritable. [OE *bera* f. WG]

■ The stock-exchange usage (which dates from the 18th century) is probably derived from a proverb warning against 'selling the bear's skin before one has caught the bear'. The term *bull* (see BULL[1] *n.* 4) appeared later, and was perhaps suggested by *bear*.

bearable /ˈbeərəb(ə)l/ *adj.* that may be endured or tolerated. □ **bearably** *adv.* **bearableness** *n.* **bearability** /ˌbeərəˈbɪlɪtɪ/ *n.*

bearberry /ˈbeəbərɪ/ *n.* an evergreen ericaceous shrub or small tree of the genus *Arctostaphylos*; esp. *A. uva-ursi*, a trailing moorland plant with pinkish flowers and red berries.

beard /bɪəd/ *n. & v.* ● *n.* **1** hair growing on the chin and lower cheeks of the face. **2 a** a tuft of hair on the face or chin of a mammal, esp. a goat. **b** a growth on an animal that resembles this, e.g. near the beak of a bird. **3** the awn of a grass, sheath of barley, etc. ● *v.tr.* oppose openly; defy. □ **beardless** *adj.* [OE f. WG]

bearded /ˈbɪədɪd/ *adj.* having, or appearing to have, a beard. □ **bearded tit** a small Eurasian passerine bird, *Panurus biarmicus*, which frequents reed-beds.

beardie /ˈbɪədɪ/ *n. Brit. colloq.* a bearded man.

Beardmore Glacier /ˈbɪədmɔː(r)/ a glacier in Antarctica, flowing from the Queen Maud Mountains to the Ross Ice Shelf, at the southern edge of the Ross Sea. One of the world's largest glaciers, it is 418 km (260 miles) long.

Beardsley /ˈbɪədzlɪ/, Aubrey (Vincent) (1872–98), English artist and illustrator. His work, which was influenced by the Pre-Raphaelites and Japanese prints, first came to public notice with his illustrations for Malory's *Morte d'Arthur* (1893); in 1894 he became artistic editor of the quarterly periodical *The Yellow Book*. He produced notable illustrations for Oscar Wilde's *Salome* (1894) and for Pope's *The Rape of the Lock* (1896). He is regarded as the chief English representative of the Aesthetic Movement in art; his illustrations are characterized by linear arabesque and an emphasis on flat areas of black and white, and typify the stylistic and aesthetic considerations of art nouveau.

bearer /ˈbeərə(r)/ *n.* **1** a person or thing that bears, carries, or brings. **2** a carrier of equipment on an expedition etc. **3** a person who presents a cheque or other order to pay money. **4** (*attrib.*) payable to the possessor (*bearer stock*). **5** *hist.* (in India etc.) a personal servant.

beargarden /ˈbeəˌɡɑːd(ə)n/ *n.* a rowdy or noisy scene.

bearing /ˈbeərɪŋ/ *n.* **1** a person's bodily attitude or outward behaviour. **2** (foll. by *on, upon*) relation or relevance to (*his comments have no bearing on the subject*). **3** endurability (*beyond bearing*). **4** a part of a machine that supports a rotating or other moving part. **5** direction or position relative to a fixed point, measured esp. in degrees. **6** (in *pl.*) **a** one's position relative to one's surroundings. **b** awareness of this; a sense of one's orientation (*get one's bearings; lose one's bearings*). **7** *Heraldry* a device or charge. **8** = ball-bearing (see BALL[1]). □ **bearing-rein** a fixed rein from bit to saddle that forces a horse to arch its neck.

bearish /ˈbeərɪʃ/ *adj.* **1** like a bear, esp. in temper. **2** *Stock Exch.* causing or associated with a fall in prices.

Béarnaise sauce /ˌbeɪɑːˈneɪz/ *n.* a rich sauce thickened with egg yolks and flavoured with tarragon. [F, fem. of *béarnais* of *Béarn* in SW France]

bearskin /ˈbeəskɪn/ *n.* **1 a** the skin of a bear. **b** a wrap etc. made of this. **2** a tall furry hat worn ceremonially by some regiments.

Beas /beɪˈɑːs, biːˈɑːs/ a river of northern India which rises in the Himalayas and flows through Himachal Pradesh to join the Sutlej river in Punjab. It is one of the five rivers that gave Punjab its name. In ancient times called the Hyphasis, it marked the eastern limit of Alexander the Great's conquests.

beast /biːst/ *n.* **1** an animal other than a human being, esp. a wild quadruped. **2 a** a brutal person. **b** *colloq.* an objectionable or unpleasant person or thing (*he's a beast for not inviting her; a beast of a problem*).

3 (prec. by *the*) a human being's brutish or uncivilized characteristics (*saw the beast in him*). □ **beast of burden** an animal, e.g. a mule, used for carrying loads. **beast of prey** see PREY. [ME f. OF *beste* f. Rmc *besta* f. L *bestia*]

beastie /ˈbiːstɪ/ *n. Sc.* or *joc.* a small animal.

beastly /ˈbiːstlɪ/ *adj. & adv.* ● *adj.* (**beastlier, beastliest**) **1** *colloq.* objectionable, unpleasant. **2** like a beast; brutal. ● *adv. colloq.* very, extremely. □ **beastliness** *n.*

beat /biːt/ *v., n., & adj.* ● *v.* (*past* **beat**; *past part.* **beaten** /ˈbiːt(ə)n/) **1** *tr.* **a** strike (a person or animal) persistently or repeatedly, esp. to harm or punish. **b** strike (a thing) repeatedly, e.g. to remove dust from (a carpet etc.), to sound (a drum etc.). **2** *intr.* (foll. by *against, at, on,* etc.) **a** pound or knock repeatedly (*waves beat against the shore; beat at the door*). **b** = beat down 3. **3** *tr.* **a** overcome; surpass; win a victory over. **b** complete an activity before (another person etc.). **c** be too hard for; perplex. **4** *tr.* (often foll. by *up*) stir (eggs etc.) vigorously into a frothy mixture. **5** *tr.* (often foll. by *out*) fashion or shape (metal etc.) by blows. **6** *intr.* (of the heart, a drum, etc.) pulsate rhythmically. **7** *tr.* (often foll. by *out*) **a** indicate (a tempo or rhythm) by gestures, tapping, etc. **b** sound (a signal etc.) by striking a drum or other means (*beat a tattoo*). **8 a** *intr.* (of a bird's wings) move up and down. **b** *tr.* cause (wings) to move in this way. **9** *tr.* make (a path etc.) by trampling. **10** *tr.* strike (bushes etc.) to rouse game. **11** *intr. Naut.* sail in the direction from which the wind is blowing. ● *n.* **1 a** a main accent or rhythmic unit in music or verse (*three beats to the bar; missed a beat and came in early*). **b** the indication of rhythm by a conductor's movements (*watch the beat*). **c** the tempo or rhythm of a piece of music as indicated by the repeated fall of the main beat. **d** (in popular music) a strong rhythm. **e** (*attrib.*) characterized by a strong rhythm (*beat music*). **2 a** a stroke or blow (e.g. on a drum). **b** a measured sequence of strokes (*the beat of the waves on the rocks*). **c** a throbbing movement or sound (*the beat of his heart*). **3 a** a route or area allocated to a police officer etc. **b** a person's habitual round. **4** *Physics* a pulsation due to the combination of two sounds or electric currents of similar but not equivalent frequencies. **5** *colloq.* = BEATNIK. ● *adj.* **1** (*predic.*) *sl.* exhausted, tired out. **2** (*attrib.*) of the beat generation or its philosophy. □ **beat about** (often foll. by *for*) search (for an excuse etc.). **beat about the bush** discuss a matter without coming to the point. **beat the bounds** *Brit.* mark parish boundaries by striking certain points with rods. **beat one's breast** strike one's chest in anguish or sorrow. **beat the clock** complete a task within a stated time. **beat down 1 a** bargain with (a seller) to lower the price. **b** cause a seller to lower (the price). **2** strike (a resisting object) until it falls (*beat the door down*). **3** (of the sun, rain, etc.) radiate heat or fall continuously and vigorously. **beat the drum for** publicize, promote. **beaten at the post** defeated at the last moment. **beat in** crush. **beat it** *sl.* go away. **beat off** drive back (an attack etc.). **beat a retreat** withdraw; abandon an undertaking. **beat time** indicate or follow a musical tempo with a baton or other means. **beat a person to it** arrive or achieve something before another person. **beat up** give a beating to, esp. with punches and kicks. **beat-up** *adj. colloq.* dilapidated; in a state of disrepair. **it beats me** I do not understand (it). □ **beatable** *adj.* [OE *bēatan* f. Gmc]

beaten /ˈbiːt(ə)n/ *adj.* **1** outwitted; defeated. **2** exhausted; dejected. **3** (of gold or any other metal) shaped by a hammer. **4** (of a path etc.) well-trodden, much-used. □ **off the beaten track 1** in or into an isolated place. **2** unusual. [past part. of BEAT]

beater /ˈbiːtə(r)/ *n.* **1** an implement used for beating (esp. a carpet or eggs). **2** a person employed to rouse game for shooting. **3** a person who beats metal.

beat generation *n.* a movement of young people chiefly in the US in the 1950s and early 1960s who rejected conventional society, valuing free self-expression and favouring modern jazz and life on the road. Among writers associated with the movement were Jack Kerouac, Allen Ginsberg, and William Burroughs. The movement faded away in the early 1960s, but can be seen as anticipating the hippy movement of the later 1960s. Members were also called *beatniks* or *beats*.

beatific /ˌbiːəˈtɪfɪk/ *adj.* **1** *colloq.* blissful (*a beatific smile*). **2 a** of or relating to blessedness. **b** making blessed. □ **beatifically** *adv.* [F *béatifique* or L *beatificus* f. *beatus* blessed]

beatification /biːˌætɪfɪˈkeɪʃ(ə)n/ *n.* **1** *RC Ch.* the act of formally declaring a dead person 'blessed', often a step towards canonization. **2** making or being blessed. [F *béatification* or eccl.L *beatificatio* (as BEATIFY)]

beatify /biːˈætɪˌfaɪ/ *v.tr.* (**-ies, -ied**) **1** *RC Ch.* announce the beatification of. **2** make happy. [F *béatifier* or eccl.L *beatificare* f. L *beatus* blessed]

beating /'biːtɪŋ/ n. **1** a physical punishment or assault. **2** a defeat. □ **take some** (or **a lot of**) **beating** be difficult to surpass.

beatitude /biˈætɪˌtjuːd/ n. **1** blessedness. **2** (also **Beatitude**) (in pl.) the declarations of blessedness in Matt. 5:3–11. **3** a title of patriarchs in the Orthodox Church. [F béatitude or L beatitudo f. beatus blessed]

Beatles, the /'biːt(ə)lz/ a pop and rock group from Liverpool, England, consisting of George Harrison, John Lennon, Paul McCartney, and Ringo Starr (born Richard Starkey). The Beatles are remembered for the quality and stylistic diversity of their songs (mostly written by Lennon and McCartney), for their irreverent wit, and for the hysteria, known as 'Beatlemania', that their early performances aroused. They found success with their first single, 'Love Me Do', in 1962, and released many more hits including 'She Loves You' (1963) and 'Help' (1965) before retiring from live performances to work with producer George Martin on the more sophisticated and experimental albums *Revolver* in 1966 and *Sergeant Pepper's Lonely Hearts Club Band* in 1967. The group split up acrimoniously in 1970. In 1995 the surviving members of the group released new material based on tapes recorded by John Lennon.

beatnik /'biːtnɪk/ n. a member of the beat generation. [BEAT + -*nik* after *sputnik*, perh. infl. by US use of Yiddish -*nik* agent-suffix]

Beaton /'biːt(ə)n/, Sir Cecil (Walter Hardy) (1904–80), English photographer. During the 1930s he worked for *Vogue* magazine and quickly earned international fame for his fashion features and portraits of celebrities; he is especially remembered today for his many portraits of the British royal family. His approach to photography was essentially theatrical, with subjects often posed amid elaborate settings to create tableaux. After the Second World War he diversified into costume and set design for films, ballet, and the theatre; he won two Oscars for his design and costumes for the film *My Fair Lady* (1964).

Beatty /'biːtɪ/, David, 1st Earl Beatty of the North Sea and of Brooksby (1871–1936), British admiral. As commander of battle-cruiser squadrons during the First World War he gained victories over German cruisers off Heligoland (1914) and the Dogger Bank (1915) and played a major role in the Battle of Jutland. He was Commander-in-Chief of the Grand Fleet from 1916 and received the German naval surrender in 1918. He was First Sea Lord 1919–27 and supervised the postwar reorganization of the navy.

beau /bəʊ/ n. (pl. **beaux** or **beaus** /bəʊz, bəʊ/) **1** esp. N. Amer. an admirer; a boyfriend. **2** a fop; a dandy. [F, = handsome, f. L bellus]

Beaubourg /'bəʊbɔːɡ/ see POMPIDOU CENTRE.

Beau Brummell see BRUMMELL.

Beaufort scale /'bəʊfət/ n. Meteorol. a scale of wind speed ranging from 0 (calm) to 12 (hurricane). There are two tables for estimating wind force, for use at sea or on land respectively. [Sir Francis Beaufort, Engl. admiral (1774–1857)]

Beaufort Sea /'bəʊfət/ a part of the Arctic Ocean lying to the north of Alaska and Canada. It is named after the English admiral Sir Francis Beaufort (1774–1857).

beau geste /bəʊ 'ʒest/ n. (pl. **beaux gestes** pronunc. same) a generous or gracious act. [F, = splendid gesture]

beau idéal /ˌbəʊ iːdeɪˈɑːl/ n. the highest type of excellence or beauty. [F beau idéal = ideal beauty]

Beaujolais /'bəʊʒəˌleɪ/ n. a red or white burgundy wine from the Beaujolais district of SE France. □ **Beaujolais nouveau** Beaujolais wine sold in the first year of a vintage.

Beaumarchais /'bəʊmɑːˌʃeɪ/, Pierre Augustin Caron de (1732–99), French dramatist. An important comic dramatist, he is chiefly remembered for his comedies *The Barber of Seville* (1775) and *The Marriage of Figaro* (1784); although still popular in France, they are best known in Britain as the inspiration for operas by Rossini and Mozart.

beau monde /bəʊ 'mɒnd/ n. fashionable society. [F]

Beaumont /'bəʊmɒnt/, Francis (1584–1616), English dramatist. He was educated at Oxford and entered the Inner Temple in 1600. He became an associate of Ben Jonson and John Fletcher, and collaborated with the latter in *Philaster* (1609), *The Maid's Tragedy* (1610–11), and many other plays. *The Knight of the Burning Pestle* (c.1607) is attributed to Beaumont alone.

Beau Nash see NASH[4].

Beaune /bəʊn/ n. a red burgundy wine from the region around the town of Beaune in east central France.

beaut /bjuːt/ n. & adj. Austral. & NZ sl. ● n. an excellent or beautiful person or thing. ● adj. excellent; beautiful. [abbr. of BEAUTY]

beauteous /'bjuːtɪəs/ adj. poet. beautiful. [ME f. BEAUTY + -OUS, after bounteous, plenteous]

beautician /bjuːˈtɪʃ(ə)n/ n. **1** a person who gives beauty treatment. **2** a person who runs or owns a beauty salon.

beautiful /'bjuːtɪˌfʊl/ adj. **1** delighting the aesthetic senses (a beautiful voice). **2** pleasant, enjoyable (had a beautiful time). **3** excellent (a beautiful specimen). □ **beautifully** adv.

beautify /'bjuːtɪˌfaɪ/ v.tr. (**-ies**, **-ied**) make beautiful; adorn. □ **beautifier** n. **beautification** /ˌbjuːtɪfɪˈkeɪʃ(ə)n/ n.

beauty /'bjuːtɪ/ n. (pl. **-ies**) **1 a** a combination of qualities such as shape, colour, etc., that pleases the aesthetic senses, esp. the sight. **b** a combination of qualities that pleases the intellect or moral sense (the beauty of the argument). **2** colloq. **a** an excellent specimen (what a beauty!). **b** an attractive feature; an advantage (that's the beauty of it). **3** a beautiful woman. □ **beauty is only skin-deep** a pleasing appearance is not a guide to character. **beauty parlour** (or **salon**) an establishment in which massage, manicure, hairdressing, make-up, etc., are offered to women. **beauty queen** the woman judged most beautiful in a competition. **beauty sleep** sleep before midnight, supposed to be health-giving. **beauty spot 1** a place known for its beauty. **2** a small natural or artificial mark such as a mole on the face, considered to enhance another feature. **beauty treatment** cosmetic treatment received in a beauty parlour. [ME f. AF beuté, OF bealté, beauté, ult. f. L (as BEAU)]

Beauvoir, Simone de, see DE BEAUVOIR.

beaux pl. of BEAU.

beaux arts /bəʊ 'zɑː(r)/ n.pl. **1** fine arts. **2** (attrib.) relating to the rules and conventions of the École des Beaux-Arts in Paris (later called Académie des Beaux Arts). [F beaux-arts]

beaver[1] /'biːvə(r)/ n. & v. ● n. (pl. same or **beavers**) **1 a** a large amphibious broad-tailed rodent of the genus *Castor*, native to North America, Europe, and Asia, and able to gnaw through tree-trunks and make dams. **b** its soft light-brown fur. **c** a hat of this. **2** (in full **beaver cloth**) a heavy woollen cloth like beaver fur. **3** (**Beaver**) a boy aged six or seven who is an affiliate member of the Scout Association. ● v.intr. (usu. foll. by away) colloq. work hard. □ **beaver lamb** lambskin made to look like beaver fur. [OE be(o)for f. Gmc]

beaver[2] /'biːvə(r)/ n. in a suit of armour, the lower part of the face-guard of a helmet. The term is also used to refer to the upper part or visor, or to a single movable guard. [OF baviere bib f. baver slaver f. beve saliva f. Rmc]

beaver[3] /'biːvə(r)/ n. sl. a bearded man. [20th c.: orig. uncert.]

beaverboard /'biːvəˌbɔːd/ n. N.Amer. propr. a kind of fibreboard. [BEAVER[1] + BOARD]

Beaverbrook /'biːvəˌbrʊk/, (William) Max(well) Aitken, 1st Baron (1879–1964), Canadian-born British Conservative politician and newspaper proprietor. He made his fortune in Canadian business before coming to Britain and winning election to Parliament in 1910. However, it is for his activities as a newspaper proprietor that he is best known; he bought the *Daily Express* in 1916 and made the daily newspaper with the world's largest circulation. He launched the *Sunday Express* in 1918 and acquired the *Evening Standard* in 1923, thus consolidating his substantial newspaper empire. As Minister of Aircraft Production in Churchill's Cabinet (1940), Beaverbrook made an important contribution to victory in the Battle of Britain.

bebop /'biːbɒp/ n. a type of jazz originating in the 1940s and characterized by complex harmony and rhythms. It is associated particularly with the cerebral style of playing of Charlie Parker, Thelonious Monk, and Dizzy Gillespie. □ **bebopper** n. [imit. of the typical rhythm]

becalm /bɪˈkɑːm/ v.tr. (usu. in passive) deprive (a sailing-ship) of wind.

became past of BECOME.

because /bɪˈkɒz/ conj. for the reason that; since. □ **because of** on account of; by reason of. [ME f. BY prep. + CAUSE, after OF par cause de by reason of]

béchamel /'beʃəˌmel/ n. a kind of thick white sauce. [invented by the Marquis de Béchamel, Fr. courtier (1630–1703)]

bêche-de-mer /ˌbeʃdəˈmeə(r)/ n. (pl. same or **bêches-de-mer** pronunc. same) **1** a kind of sea cucumber eaten in China usu. in long dried strips. **2** = BEACH-LA-MAR. [pseudo-F, f. Port. bicho do mar lit. 'worm of the sea'; cf. BEACH-LA-MAR]

Bechstein /'bekstaɪn/, Friedrich Wilhelm Carl (1826–1900), German piano-builder. His name is used to designate a piano manufactured by him or by the firm which he founded in 1856.

Bechuanaland /ˌbetʃʊˈɑːnəˌlænd/ the former name (until 1966) for BOTSWANA.

beck[1] /bek/ n. N. Engl. a brook; a mountain stream. [ME f. ON bekkr f. Gmc]

beck[2] /bek/ n. poet. a gesture requesting attention, e.g. a nod, wave, etc. □ **at a person's beck and call** having constantly to obey a person's orders. [beck (v.) f. BECKON]

Beckenbauer /ˈbekənˌbaʊə(r)/, Franz (b.1945), German footballer. Under his captaincy, Bayern Munich won a number of championships and West Germany won the World Cup in 1974. After a spell in the US (1976–80) he returned to West Germany, where he was manager of the national team that won the World Cup again in 1990.

Becker /ˈbekə(r)/, Boris (b.1967), German tennis player. He became the youngest man to win the men's singles championship at Wimbledon in 1985, the first time that the title had been won by an unseeded player. He won at Wimbledon again in 1986 and 1989 and also won the US Open (1989) and the Australian Open (1991).

becket /ˈbekɪt/ n. Naut. a contrivance such as a hook, bracket, or rope-loop, for securing loose ropes, tackle, or spars. [18th c.: orig. unkn.]

Becket, St Thomas à /ˈbekɪt/ (c.1118–70), English prelate and statesman. A close and influential friend of Henry II, he served as his Chancellor and in 1162 became Archbishop of Canterbury, a position Becket accepted with reluctance, foreseeing the inevitable conflict of interests between the king and the Church. He soon found himself in open opposition to Henry, first on a matter of taxation and later over the coronation of Henry's son, and the king in anger uttered words which led four knights to assassinate Becket in his cathedral on 29 Dec. The murder aroused indignation throughout Europe, miracles were soon reported at his tomb, and Henry was obliged to do public penance there. The shrine became a major centre of pilgrimage until its destruction under Henry VIII (1538). Feast day, 29 Dec.

Beckett /ˈbekɪt/, Samuel (Barclay) (1906–89), Irish dramatist, novelist, and poet. A permanent resident in France from the mid-1930s, he is best known for his plays, especially *Waiting for Godot* (1952). A seminal work in the Theatre of the Absurd, the play was highly influential during the postwar period, especially for Beckett's use of dramatic narrative and symbolism. His later works were increasingly short and enigmatic. Beckett was awarded the Nobel Prize for literature in 1969.

Beckford /ˈbekfəd/, William (1759–1844), English writer. He inherited a large fortune from his father, which he spent lavishly. He travelled in Europe, collected works of art and curios, and commissioned the building of Fonthill Abbey in Wiltshire, a Gothic folly, where he lived in seclusion 1796–1822. He is remembered as the author of the fantastic oriental romance *Vathek* (1786, originally written in French).

Beckmann[1] /ˈbekmən/, Ernst Otto (1853–1923), German chemist. Beckmann devised a method of determining a compound's molecular weight by measuring the rise in boiling-point of a solvent containing the compound. For this he designed an accurate thermometer with an adjustable range. He also discovered a rearrangement reaction important in organic synthesis.

Beckmann[2] /ˈbekmən/, Max (1884–1950), German painter and graphic artist. Beckmann was an expressionist and his paintings typically reflect his first-hand experience of human evil during the First World War; a characteristic work is *The Night* (1919), a torture scene in which contemporary social conditions are portrayed with powerful symbolism. Beckmann was dismissed from his teaching post in Frankfurt by the Nazis in 1933; the same year he painted *Robbery of Europe*. In 1937 his work was denounced as 'degenerate' and he fled to Holland, before going to the US in 1947.

beckon /ˈbekən/ v. **1** tr. **a** attract the attention of; summon by gesture. **b** entice. **2** intr. (usu. foll. by to) make a signal to attract a person's attention; summon a person by doing this. [OE bīecnan, bēcnan ult. f. WG baukna BEACON]

becloud /bɪˈklaʊd/ v.tr. **1** obscure (becloud the argument). **2** cover with clouds.

become /bɪˈkʌm/ v. (past **became** /-ˈkeɪm/; past part. **become**) **1** intr. (foll. by complement) begin to be; come to be, turn into (became president; will become famous; tadpoles become frogs). **2** tr. **a** look well on; suit (blue becomes him). **b** befit (it ill becomes you to complain). **3** intr. (as **becoming** adj.) **a** flattering the appearance. **b** suitable; decorous. □ **become of** happen to (what will become of me?). □ **becomingly** adv. **becomingness** n. [OE becuman f. Gmc: cf. BE-, COME]

Becquerel /ˈbekəˌrel/, Antoine-Henri (1852–1908), French physicist. He shared the 1903 Nobel Prize for physics with Marie and Pierre Curie for his discovery of natural radioactivity in uranium salts, which he proceeded to investigate. Initially, the rays emitted by radioactive substances were named after him.

becquerel /ˈbekəˌrel/ n. Physics the SI unit of radioactivity (symbol: **Bq**), corresponding to one disintegration per second. [A.-H. BECQUEREL]

B.Ed. abbr. Bachelor of Education.

bed /bed/ n. & v. ● n. **1 a** a piece of furniture used for sleeping on, usu. a framework with a mattress and coverings. **b** a mattress, with or without coverings. **2** any place used by a person or animal for sleep or rest; a litter. **3 a** a garden plot, esp. one used for planting flowers. **b** a place where other things may be grown (osier bed). **4** the use of a bed: **a** colloq. for sexual intercourse (only thinks of bed). **b** for rest (needs his bed). **5** something flat, forming a support or base as in: **a** the bottom of the sea or a river. **b** the foundations of a road or railway. **c** the flat surface beneath the cloth of a billiard-table. **6** a stratum, such as a layer of oysters etc. ● v. (**bedded, bedding**) **1** tr. & intr. (usu. foll. by down) put or go to bed. **2** tr. colloq. have sexual intercourse with. **3** tr. (usu. foll. by out) plant in a garden bed. **4** tr. cover up or fix firmly in something. **5 a** tr. arrange as a layer. **b** intr. be or form a layer. □ **bed and board 1** lodging and food. **2** marital relations. **bed and breakfast 1** one night's lodging and breakfast in a hotel etc. **2** a private house offering such accommodation. **3** Stock Exch. an operation in which a shareholder sells a holding one evening while agreeing to buy it back again next morning, realizing either a gain or a loss in order to suit a tax requirement. **bed down** (esp. of a new system or arrangement) become established after an initial period or trial. **bed-head** n. the upper end of a bed. **bed-hop** v.intr. colloq. engage in a succession of casual sexual affairs. **bed of roses** a life of ease. **brought to bed** (often foll. by of) archaic delivered of a child. **get out of bed on the wrong side** be bad-tempered all day long. **go to bed 1** retire for the night. **2** (usu. foll. by with) have sexual intercourse. **3** (of a newspaper) go to press. **keep one's bed** stay in bed because of illness. **make the bed** tidy and arrange the bed for use. **make one's bed and lie in it** accept the consequences of one's acts. **put to bed 1** cause to go to bed. **2** make (a newspaper) ready for press. **take to one's bed** stay in bed because of illness. [OE bed(d), beddian f. Gmc]

bedabble /bɪˈdæb(ə)l/ v.tr. stain or splash with dirty liquid, blood, etc.

bedad /bɪˈdæd/ int. Ir. by God! [corrupt.: cf. GAD[2]]

bedaub /bɪˈdɔːb/ v.tr. smear or daub with paint etc.; decorate gaudily.

bedazzle /bɪˈdæz(ə)l/ v.tr. **1** dazzle. **2** confuse (a person). □ **bedazzlement** n.

bedbug /ˈbedbʌɡ/ n. a flat, wingless, malodorous hemipterous bug of the genus Cimex, infesting beds and unclean houses and sucking blood.

bedchamber /ˈbedˌtʃeɪmbə(r)/ n. **1** archaic a bedroom. **2** (**Bedchamber**) part of the title of some of the sovereign's attendants (Lady of the Bedchamber).

bedclothes /ˈbedkləʊðz/ n.pl. coverings for a bed, such as sheets, blankets, etc.

beddable /ˈbedəb(ə)l/ adj. colloq. sexually attractive. [BED + -ABLE]

bedder /ˈbedə(r)/ n. **1** a plant suitable for a garden bed. **2** Brit. colloq. a college bedmaker.

bedding /ˈbedɪŋ/ n. **1** a mattress and bedclothes. **2** litter for cattle, horses, etc. **3** a bottom layer. **4** Geol. the stratification of rocks, esp. when clearly visible. □ **bedding plant** a plant suitable for planting when it is in flower as part of a display in a garden bed, but discarded at the end of the season.

Bede, St /biːd/ (known as the Venerable Bede) (c.673–735), English monk, theologian, and historian. He lived and worked at the monastery in Jarrow on Tyneside. Often regarded as 'the Father of English History', he wrote a number of historical works including *The Ecclesiastical History of the English People* (completed in 731). This is considered a primary source for early English history; it has vivid descriptions and is based on careful research, separating fact from hearsay and tradition. Feast day, 27 May.

bedeck /bɪˈdek/ v.tr. adorn.

bedeguar /ˈbedɪˌɡɑː(r)/ n. a mosslike growth on rose-bushes produced by a gall wasp. [F bédégar f. Pers. bād-āwar wind-brought]

bedel /ˈbiːd(ə)l, bɪˈdel/ n. (also **bedell**) Brit. a university official with chiefly processional duties. [BEADLE]

bedevil /bɪˈdev(ə)l/ v.tr. (**bedevilled, bedevilling**; US **bedeviled, bedeviling**) **1** plague; afflict. **2** confound; confuse. **3** possess as if with a devil; bewitch. **4** treat with diabolical violence or abuse. □ **bedevilment** n.

bedew /bɪˈdjuː/ v.tr. **1** cover or sprinkle with dew or drops of water. **2** poet. sprinkle with tears.

bedfellow /ˈbedˌfeləʊ/ n. **1** a person who shares a bed. **2** an associate.

Bedford cord /ˈbedfəd/ n. a tough woven fabric having prominent ridges, similar to corduroy. [Bedford in south central England]

Bedfordshire /ˈbedfədˌʃɪə(r)/ a county of south central England; county town, Bedford.

bedight /bɪˈdaɪt/ adj. archaic arrayed; adorned. [ME past part. of bedight (v.) (as BE-, DIGHT)]

bedim /bɪˈdɪm/ v.tr. (**bedimmed**, **bedimming**) poet. make (the eyes, mind, etc.) dim.

bedizen /bɪˈdaɪz(ə)n, -ˈdɪz(ə)n/ v.tr. poet. deck out gaudily. [BE- + obs. dizen deck out]

bedjacket /ˈbedˌdʒækɪt/ n. a jacket worn when sitting up in bed.

bedlam /ˈbedləm/ n. **1** archaic a mental hospital; an asylum. (See note below.) **2** a scene of uproar and confusion (the traffic was bedlam).

■ Bedlam was originally the popular name of the hospital of St Mary of Bethlehem, founded as a priory in 1247 at Bishopsgate, London, and by the 14th century a mental hospital. In 1675 a new hospital was built in Moorfields and this in turn was replaced by a building in the Lambeth Road in 1815 (now the Imperial War Museum) and transferred to Beckenham in Kent in 1931.

bedlinen /ˈbedˌlɪnɪn/ n. sheets and pillowcases.

Bedlington terrier /ˈbedlɪŋtən/ n. a breed of terrier with a narrow head, long legs, and curly grey hair. [Bedlington in Northumberland]

bedmaker /ˈbedˌmeɪkə(r)/ n. Brit. a person employed to clean and tidy students' rooms in a college.

Bedouin /ˈbeduːɪn/ n. & adj. (also **Beduin**) ● n. (pl. same) **1** a nomadic Arab of the desert. **2** a wanderer; a nomad. ● adj. **1** of or relating to the Bedouin. **2** wandering; nomadic. [ME f. OF beduin ult. f. Arab. badawī, pl. badawīn dwellers in the desert f. badw desert]

bedpan /ˈbedpæn/ n. a receptacle used by a bedridden patient for urine and faeces.

bedplate /ˈbedpleɪt/ n. a metal plate forming the base of a machine.

bedpost /ˈbedpəʊst/ n. any of the four upright supports of a bedstead. □ **between you and me and the bedpost** colloq. in strict confidence.

bedraggle /bɪˈdræg(ə)l/ v.tr. **1** (often as **bedraggled** adj.) wet (a garment etc.) by trailing it, or so that it hangs limp. **2** (as **bedraggled** adj.) untidy; dishevelled. [BE- + DRAGGLE]

bedrest /ˈbedrest/ n. confinement of an invalid to bed.

bedridden /ˈbedˌrɪd(ə)n/ adj. **1** confined to bed by infirmity. **2** decrepit. [OE bedreda f. ridan ride]

bedrock /ˈbedrɒk/ n. **1** solid rock underlying alluvial deposits etc. **2** the underlying principles or facts of a theory, character, etc.

bedroll /ˈbedrəʊl/ n. esp. N. Amer. portable bedding rolled into a bundle, esp. a sleeping-bag.

bedroom /ˈbedruːm, -rʊm/ n. **1** a room for sleeping in. **2** (attrib.) of or referring to sexual relations (bedroom farce).

Beds. abbr. Bedfordshire.

bedside /ˈbedsaɪd/ n. **1** the space beside esp. a patient's bed. **2** (attrib.) of or relating to the side of a bed (bedside lamp). □ **bedside manner** (of a doctor) an approach or attitude to a patient.

bedsitter /ˈbedˌsɪtə(r)/ n. (also **bedsit**) colloq. = BEDSITTING ROOM. [contr.]

bedsitting room /ˈbedˌsɪtɪŋ/ n. Brit. a one-roomed unit of accommodation usu. consisting of combined bedroom and sitting-room with cooking facilities.

bedsock /ˈbedsɒk/ n. a thick sock worn in bed.

bedsore /ˈbedsɔː(r)/ n. a sore developed by an invalid because of pressure caused by lying in bed.

bedspread /ˈbedspred/ n. an often decorative cloth used to cover a bed when not in use.

bedstead /ˈbedsted/ n. the framework of a bed.

bedstraw /ˈbedstrɔː/ n. a herbaceous plant of the genus Galium, once used as straw for bedding; e.g. (in full **lady's bedstraw**) G. verum, with yellow flowers.

bedtable /ˈbedˌteɪb(ə)l/ n. a portable table or tray with legs, used by a person sitting up in bed.

bedtime /ˈbedtaɪm/ n. **1** the usual time for going to bed. **2** (attrib.) of or relating to bedtime (bedtime drink).

Beduin var. of BEDOUIN.

bedwetting /ˈbedˌwetɪŋ/ n. involuntary urination during the night. □ **bedwetter** n.

bee /biː/ n. **1 a** (also **honey bee**) a stinging hymenopterous insect of the genus Apis, which collects nectar and pollen, produces wax and honey, and lives in large communities. (See note below.) **b** a related insect of the superfamily Apoidea, either social or solitary. **2** (usu. **busy bee**) a busy person. **3** esp. US a meeting for communal work or amusement (spelling-bee). □ **bee-bread** honey or pollen used as food by bees. **bee dance** a dance performed by worker bees within the hive to inform the colony of the location of food. **bee-eater** a brightly coloured insect-eating bird of the family Meropidae, with a long slender curved bill. **a bee in one's bonnet** an obsession. **bee-keeper** a person who keeps bees. **bee-keeping** the occupation of keeping bees. **bee-master** a person who keeps bees. **bee orchid** a European orchid, Ophrys apifera, with bee-shaped flowers. **the bee's knees** sl. something outstandingly good (thinks he's the bee's knees). [OE bēo f. Gmc]

■ There are about 12,000 species of bee, most of which are solitary. The genus Apis includes four species of honey bee, including the familiar A. mellifera which was domesticated around the end of the neolithic period. The storage of honey allows honey bee colonies to survive through the winter; it remained the only sweetener for human consumption until the introduction of sugar. Human beings have also found many uses for beeswax, and bees perform a vital role in plant pollination. Social bee colonies have a complex caste structure and behaviour, based on pheromones (see SOCIAL).

Beeb /biːb/ n. (prec. by the) Brit. colloq. the BBC. [abbr.]

beech /biːtʃ/ n. **1** (in full **beech tree**) a large tree of the genus Fagus, having smooth grey bark and glossy leaves. **2** (also **beechwood**) its wood. **3** (in full **southern beech**) a similar tree of the genus Nothofagus, of the southern hemisphere. □ **beech-fern** a fern typically found in damp woods, esp. Phegopteris connectilis. **beech marten** = stone marten. [OE bēce f. Gmc]

Beecham /ˈbiːtʃəm/, Sir Thomas (1879–1961), English conductor and impresario. He was associated with most of the leading British orchestras, founding the London Philharmonic in 1932 and the Royal Philharmonic in 1947; he was also artistic director of the Royal Opera House in the 1930s. He did much to stimulate interest in new and neglected music; a champion of Delius, he was also responsible for introducing Diaghilev's Ballets Russes and the work of Sibelius and Richard Strauss to the public.

beechmast /ˈbiːtʃmɑːst/ n. (pl. same) the small rough-skinned fruit of the beech tree. [BEECH + MAST²]

beef /biːf/ n. & v. ● n. **1** the flesh of the ox, bull, or esp. the cow, for eating. **2** colloq. well-developed male muscle. **3** (pl. **beeves** /biːvz/ or US **beefs**) a cow, bull, or ox fattened for beef; its carcass. **4** (pl. **beefs**) sl. a complaint; a protest. ● v.intr. sl. complain. □ **beef tea** stewed extract of beef, given to invalids. **beef tomato** an exceptionally large and firm variety of tomato. **beef up** sl. strengthen, reinforce, augment. [ME f. AF, OF boef f. L bos bovis ox]

beefalo /ˈbiːfəˌləʊ/ n. (pl. same or **-oes**) US a cross-breed of a bison and a domestic cow. [blend of BEEF and BUFFALO]

beefburger /ˈbiːfˌbɜːgə(r)/ n. = HAMBURGER.

beefcake /ˈbiːfkeɪk/ n. sl. well-developed male muscle, esp. when displayed for admiration.

beefeater /ˈbiːfˌiːtə(r)/ n. a warder in the Tower of London; a Yeoman of the Guard. [f. obs. sense 'well-fed menial']

beefsteak /ˈbiːfsteɪk/ n. a thick slice of lean beef, esp. from the rump, usu. for grilling or frying. □ **beefsteak fungus** a red bracket fungus, Fistulina hepatica, resembling raw meat.

beefwood /ˈbiːfwʊd/ n. **1** an Australian or West Indian hardwood tree, esp. casuarina. **2** the close-grained red timber of this.

beefy /ˈbiːfɪ/ adj. (**beefier**, **beefiest**) **1** like beef. **2** solid; muscular. □ **beefily** adv. **beefiness** n.

beehive /ˈbiːhaɪv/ n. **1** an artificial habitation for bees. **2** a busy place. **3** anything resembling a wicker beehive in being domed. **4** (**the Beehive**) Astron. the star cluster Praesepe.

beeline /ˈbiːlaɪn/ n. a straight line between two places. □ **make a beeline for** hurry directly to.

Beelzebub /biːˈelzɪˌbʌb/ in the Bible (2 Kings 1), the god of the Philistine city Ekron. In the Gospels, Beelzebub was the prince of the devils, often identified with the Devil. [OE f. L f. Gk *beelzeboul* & Heb. *ba'al zᵉbûḇ* lord of the flies]

been *past part.* of BE.

beep /biːp/ *n. & v.* ● *n.* **1** the sound of a motor-car horn. **2** any similar high-pitched noise. ● *v.intr.* emit a beep. □ **beeper** *n.* [imit.]

beer /bɪə(r)/ *n.* **1 a** an alcoholic drink made from yeast-fermented malt etc., flavoured with hops. **b** a glass or can of this. **2** any of several other fermented drinks, e.g. ginger beer. □ **beer and skittles** amusement (*life is not all beer and skittles*). **beer belly** (or **gut**) *colloq.* a protruding stomach attributed to drinking large quantities of beer. **beer-cellar 1** an underground room for storing beer. **2** a basement or cellar for selling or drinking beer. **beer-engine** (or **pump**) *Brit.* a machine that draws up beer from a barrel in a cellar, usu. operated by pulling a lever on the bar. **beer garden** a garden where beer is sold and drunk. **beer hall** a large room where beer is sold and drunk. **beer-mat** a small table-mat for a beer-glass. **beer-up** *Austral. colloq.* a beer-drinking party or session. [OE *bēor* f. LL *biber* drink f. L *bibere*]

Beerbohm /ˈbɪəbəʊm/, Sir Henry Maximilian ('Max') (1872–1956), English caricaturist, essayist, and critic. A central figure of the Aesthetic Movement, from 1894 he contributed to *The Yellow Book*. He was well placed to comment on the avant-garde tendencies of the period, which he did in collections of essays and caricatures. His one completed novel, *Zuleika Dobson* (1911), is a fantasized distillation of the atmosphere of *fin-de-siècle* Oxford. From 1935 onwards he achieved success in the new medium of radio; some of his broadcasts were published in *Mainly on the Air* (1946).

beerhouse /ˈbɪəhaʊs/ *n. Brit.* a public house licensed to sell beer but not spirits.

Beersheba /bɪəˈʃiːbə/ a town in southern Israel on the northern edge of the Negev desert; pop. (1987) 114,600. In biblical times it marked the southern limit of the Hebrew kingdom of ancient Israel (Judges 20; see also DAN 3).

beery /ˈbɪərɪ/ *adj.* (**beerier, beeriest**) **1** showing the influence of drink in one's appearance or behaviour. **2** smelling or tasting of beer. □ **beerily** *adv.* **beeriness** *n.*

beestings /ˈbiːstɪŋz/ *n.pl.* (also treated as *sing.*) the first milk (esp. of a cow) after giving birth. [OE *bēsting* (implied by *bēost*), of unkn. orig.]

beeswax /ˈbiːzwæks/ *n. & v.* ● *n.* **1** the wax secreted by bees to make honeycombs. **2** this wax refined and used to polish wood. ● *v.tr.* polish (furniture etc.) with beeswax.

beeswing /ˈbiːzwɪŋ/ *n.* a filmy second crust on old port.

beet /biːt/ *n.* a plant of the genus *Beta* with an edible root (see BEETROOT, sugar beet). □ **beet sugar** sugar obtained from sugar beet. [OE *bēte* f. L *beta*, perh. of Celt. orig.]

Beethoven /ˈbeɪtəʊv(ə)n, ˈbeɪtˌhəʊ-/, Ludwig van (1770–1827), German composer. Pre-eminently an instrumental composer, he reinvigorated the forms of sonata, symphony, and concerto that had matured during the latter part of the 18th century, reshaping them and expanding their terms of reference. Despite increasing deafness, he was responsible for a prodigious musical output; his work includes nine symphonies (such as the *Eroica* of 1804, originally dedicated to his hero Napoleon), thirty-two piano sonatas, sixteen string quartets, the opera *Fidelio* (1814), and the Mass in D (the *Missa Solemnis* of 1823). In the piano sonatas of 1816–22 and the string quartets of 1824–6 the old structural forms are merely implicit; in his Ninth Symphony (1824) he broke with precedent altogether in the finale by introducing voices to sing Schiller's *Ode to Joy*. With his expansion of 18th-century forms and techniques in his earlier work, and the personal emotion and individuality of his later works, he is often seen as bridging the classical and romantic movements.

beetle¹ /ˈbiːt(ə)l/ *n. & v.* ● *n.* **1** an insect of the order Coleoptera, having the forewings modified into hardened wing-cases that cover and protect the hindwings. **2** *colloq.* any similar, usu. black, insect. **3** *sl.* a type of compact rounded Volkswagen saloon car. **4** *Brit.* a dice game in which a beetle is drawn or assembled. ● *v.intr.* (foll. by *about, away, etc.*) *Brit. colloq.* hurry, scurry. □ **beetle-crusher** *Brit. colloq.* a large boot or foot. [OE *bitula biter* f. *bītan* BITE]

beetle² /ˈbiːt(ə)l/ *n. & v.* ● *n.* **1** a tool with a heavy head and a handle, used for ramming, crushing, driving wedges, etc. **2** a machine used for heightening the lustre of cloth by pressure from rollers. ● *v.tr.* **1** ram, crush, drive, etc., with a beetle. **2** finish (cloth) with a beetle. [OE *bētel* f. Gmc]

beetle³ /ˈbiːt(ə)l/ *adj. & v.* ● *adj.* (esp. of the eyebrows) projecting, shaggy, scowling. ● *v.intr.* (usu. as **beetling** *adj.*) (of brows, cliffs, etc.) project;

overhang threateningly. □ **beetle-browed** with shaggy, projecting, or scowling eyebrows. [ME: orig. unkn.]

Beeton /ˈbiːt(ə)n/, Mrs Isabella Mary (1836–65), English author on cookery. Her best-selling *Book of Cookery and Household Management* (1861), first published serially in a women's magazine, contained over 3,000 recipes and articles, as well as sections giving advice on legal and medical matters.

beetroot /ˈbiːtruːt/ *n.* esp. *Brit.* **1** a beet, *Beta vulgaris*, with an edible spherical dark red root. **2** this root used as a vegetable.

beeves *pl.* of BEEF.

BEF see BRITISH EXPEDITIONARY FORCE.

befall /bɪˈfɔːl/ *v.* (*past* **befell** /-ˈfel/; *past part.* **befallen** /-ˈfɔːlən/) *poet.* **1** *intr.* happen (*so it befell*). **2** *tr.* happen to (a person etc.) (*what has befallen her?*). [OE *befeallan* (as BE-, *feallan* FALL)]

befit /bɪˈfɪt/ *v.tr.* (**befitted, befitting**) **1** be fitted or appropriate for; suit. **2** be incumbent on. □ **befitting** *adj.* **befittingly** *adv.*

befog /bɪˈfɒg/ *v.tr.* (**befogged, befogging**) **1** confuse; obscure. **2** envelop in fog.

befool /bɪˈfuːl/ *v.tr.* make a fool of; delude.

before /bɪˈfɔː(r)/ *conj., prep., & adv.* ● *conj.* **1** earlier than the time when (*crawled before he walked*). **2** rather than that (*would starve before he stole*). ● *prep.* **1 a** in front of (*before her in the queue*). **b** ahead of (*crossed the line before him*). **c** under the impulse of (*recoil before the attack*). **d** awaiting (*the future before them*). **2** earlier than; preceding (*Lent comes before Easter*). **3** rather than (*death before dishonour*). **4 a** in the presence of (*appear before the judge*). **b** for the attention of (*a plan put before the committee*). ● *adv.* **1 a** earlier than the time in question; already (*heard it before*). **b** in the past (*happened long before*). **2** ahead (*go before*). **3** on the front (*hit before and behind*). □ **before God** a solemn oath meaning 'as God sees me'. **before time** see TIME. [OE *beforan* f. Gmc]

before Christ *adj.* (of a date) reckoned backwards from the birth of Christ.

beforehand /bɪˈfɔːhænd/ *adv.* in anticipation; in advance; in readiness (*had prepared the meal beforehand*). □ **be beforehand with** anticipate; forestall. [ME f. BEFORE + HAND: cf. AF *avant main*]

befoul /bɪˈfaʊl/ *v.tr. poet.* **1** make foul or dirty. **2** degrade; defile (*befouled her name*).

befriend /bɪˈfrend/ *v.tr.* act as a friend to; help.

befuddle /bɪˈfʌd(ə)l/ *v.tr.* **1** make drunk. **2** confuse, bewilder. □ **befuddlement** *n.*

beg /beg/ *v.* (**begged, begging**) **1 a** *intr.* (usu. foll. by *for*) ask for (esp. food, money, etc.) (*begged for alms*). **b** *tr.* ask for (food, money, etc.) as a gift. **c** *intr.* live by begging. **2** *tr. & intr.* (usu. foll. by *for*, or *to* + infin.) ask earnestly or humbly (*begged for forgiveness; begged to be allowed out; please, I beg of you; beg your indulgence for a time*). **3** *tr.* ask formally for (*beg leave*). **4** *intr.* (of a dog etc.) sit up with the front paws raised expectantly. **5** *tr.* take or ask leave (to do something) (*I beg to differ; beg to enclose*). □ **beg one's bread** live by begging. **begging bowl 1** a bowl etc. held out for food or alms. **2** an earnest appeal for help. **beg off 1** decline to take part or attend. **2** get (a person) excused a penalty etc. **beg pardon** I beg your pardon (see PARDON *int.*). **beg the question 1** assume the truth of an argument or proposition to be proved, without arguing it. **2** *disp.* pose the question. **3** *colloq.* evade a difficulty. **go begging** (or **a-begging**) (of a chance or a thing) not be taken; be unwanted. [ME prob. f. OE *bedecian* f. Gmc: rel. to BID]

begad /bɪˈgæd/ *int. archaic colloq.* by God! [corrupt.: cf. GAD²]

began *past* of BEGIN.

begat *archaic past* of BEGET.

beget /bɪˈget/ *v.tr.* (**begetting**; *past* **begot** /-ˈgɒt/; *archaic* **begat** /-ˈgæt/; *past part.* **begotten** /-ˈgɒt(ə)n/) *literary* **1** (usu. of a father, sometimes of a father and mother) procreate. **2** give rise to; cause (*beget strife*). □ **begetter** *n.* [OE *begietan*, formed as BE- + GET = procreate]

beggar /ˈbegə(r)/ *n. & v.* ● *n.* **1** a person who begs, esp. a person who lives by begging. **2** a poor person. **3** *colloq.* a person; a fellow (*poor beggar*). ● *v.tr.* **1** reduce to poverty. **2** outshine. **3** exhaust the resources of (esp. *beggar belief; beggar description*). □ **beggar-my-neighbour 1** a card-game in which a player seeks to capture an opponent's cards. **2** (*attrib.*) (esp. of national policy) self-aggrandizing at the expense of competitors. **beggars cannot** (or **must not**) **be choosers** those without other resources must take what is offered. [ME f. BEG + -AR³]

beggarly /ˈbegəlɪ/ *adj.* **1** poverty-stricken; needy. **2** intellectually poor. **3** mean; sordid. **4** ungenerous. □ **beggarliness** *n.*

beggary /ˈbegərɪ/ *n.* extreme poverty.

Begin /'beɪgɪn/, Menachem (1913–92), Israeli statesman, Prime Minister 1977–84. A member of Irgun 1943–8, following Israel's independence he founded the Herut Party (which evolved from Irgun) and served as leader of the opposition 1948–67. In 1973 he took up the position of joint chairman of the Likud coalition, becoming Prime Minister in 1977, when the Likud party was successful at the polls. His hard line on Arab–Israeli relations, particularly with regard to retaining territories occupied by Israel during the Arab–Israeli War (1967), softened in a series of meetings with President Sadat of Egypt. The result was a peace treaty between Egypt and Israel; the two leaders shared the Nobel Peace Prize in 1978.

begin /bɪ'gɪn/ v. (**beginning**; past **began** /-'gæn/; past part. **begun** /-'gʌn/) **1** tr. perform the first part of; start (begin work; begin crying; began the book). **2** intr. come into being; arise: **a** in time (war began in 1939). **b** in space (Wales begins beyond the river). **3** tr. (usu. foll. by to + infin.) start at a certain time (then began to feel ill). **4** intr. be begun (the meeting will begin at 7). **5 a** intr. & tr. start speaking ('No,' he began). **b** intr. take the first step; be the first to do something (who wants to begin?). **6** intr. (usu. with neg.) colloq. show any attempt or likelihood (can't begin to compete). □ **begin at** start from. **begin on** (or **upon**) set to work at. **begin school** attend school for the first time. **begin with** take (a subject, task, etc.) first or as a starting-point. **to begin with** in the first place; as the first thing. [OE beginnan f. Gmc]

beginner /bɪ'gɪnə(r)/ n. a person just beginning to learn a skill etc. □ **beginner's luck** good luck supposed to attend a beginner at games etc.

beginning /bɪ'gɪnɪŋ/ n. **1** the time or place at which anything begins. **2** a source or origin. **3** the first part. □ **the beginning of the end** the first clear sign of a final result.

begone /bɪ'gɒn/ int. poet. go away at once!

begonia /bɪ'gəʊnjə/ n. a plant of the genus Begonia, with flowers having brightly coloured sepals and no petals, and often with glossy foliage. [M. Bégon, Fr. patron of science (1638–1710)]

begorra /bɪ'gɒrə/ int. Ir. by God! [corrupt.]

begot past of BEGET.

begotten past part. of BEGET.

begrime /bɪ'graɪm/ v.tr. make grimy.

begrudge /bɪ'grʌdʒ/ v.tr. **1** resent; be dissatisfied at. **2** envy (a person) the possession of. **3** be reluctant or unwilling to give (a thing to a person). □ **begrudgingly** adv.

beguile /bɪ'gaɪl/ v.tr. **1** charm; amuse. **2** divert attention pleasantly from (toil etc.). **3** (often foll. by of, out of, or into + verbal noun) delude; cheat (beguiled him into paying). □ **beguilement** n. **beguiler** n. **beguiling** adj. **beguilingly** adv. [BE- + obs. guile to deceive]

beguine /bɪ'giːn/ n. **1** a popular dance of West Indian origin. **2** its rhythm. [Amer. F f. F béguin infatuation]

begum /'beɪgəm/ n. in the Indian subcontinent: **1** a Muslim lady of high rank. **2** (**Begum**) the title of a married Muslim woman, equivalent to Mrs. [Urdu begam f. E.Turk. bīgam princess, fem. of big prince: cf. BEY]

begun past part. of BEGIN.

behalf /bɪ'hɑːf/ n. □ **on** (N. Amer. **in**) **behalf of** (or **on a person's behalf**) **1** in the interests of (a person, principle, etc.). **2** as representative of (acting on behalf of my client). [mixture of earlier phrases on his halve and bihalve him, both = on his side: see BY, HALF]

Behan /'biːən/, Brendan (Francis) (1923–64), Irish dramatist and poet. A committed supporter of Irish nationalism, he spent periods in Borstal and in prison for his involvement in terrorist activities; his period of Borstal training is described in his autobiographical novel Borstal Boy (1958). His play The Quare Fellow (1956), set in an Irish prison and evoking the horror and humour on the eve of a hanging, is based on the time he spent in Dublin's Mountjoy prison (1942–6); the work became a key text in the contemporary anti-hanging debate.

behave /bɪ'heɪv/ v. **1** intr. **a** act or react (in a specified way) (behaved well). **b** (esp. to or of a child) conduct oneself properly. **c** (of a machine etc.) work well (or in a specified way) (the computer is not behaving today). **2** refl. (esp. of or to a child) show good manners (behaved herself). □ **behave towards** treat (in a specified way). **ill-behaved** having bad manners or conduct. **well-behaved** having good manners or conduct. [BE- + HAVE]

behaviour /bɪ'heɪvjə(r)/ n. (US **behavior**) **1 a** the way one conducts oneself; manners. **b** the treatment of others; moral conduct. **2** the way in which a ship, machine, chemical substance, etc., acts or works. **3** Psychol. the response (of a person, animal, etc.) to a stimulus.

□ **behaviour therapy** Psychol. the treatment of neurotic symptoms by training the patient's reactions to stimuli (see BEHAVIOURISM). **be on one's good** (or **best**) **behaviour** behave well when being observed. [BEHAVE after demeanour and obs. haviour f. have]

behavioural /bɪ'heɪvjərəl/ adj. (US **behavioral**) of or relating to behaviour. □ **behavioural science** the scientific study of human and animal behaviour (see BEHAVIOURISM). □ **behaviouralist** n.

behaviourism /bɪ'heɪvjə,rɪz(ə)m/ n. (US **behaviorism**) Psychol. **1** the theory that human behaviour is determined by conditioning rather than by thoughts or feelings, and that psychological disorders are best treated by altering behaviour patterns. **2** such study and treatment in practice. □ **behaviourist** n. **behaviouristic** /-,heɪvjə'rɪstɪk/ adj.

behead /bɪ'hed/ v.tr. **1** cut off the head of (a person), esp. as a form of execution. **2** kill by beheading. [OE behēafdian (as BE-, hēafod HEAD)]

beheld past and past part. of BEHOLD.

behemoth /bɪ'hiːmɒθ/ n. an enormous creature or thing. [ME f. Heb. bᵉhēmôt intensive pl. of bᵉhēmāh beast, perh. f. Egyptian p-ehe-mau water-ox]

behest /bɪ'hest/ n. literary a command; an entreaty (went at his behest). [OE behǣs f. Gmc]

behind /bɪ'haɪnd/ prep., adv., & n. ● prep. **1 a** in, towards, or to the rear of. **b** on the further side of (behind the bush). **c** hidden by (something behind that remark). **2 a** in the past in relation to (trouble is behind me now). **b** late in relation to (behind schedule). **3** inferior to; weaker than (rather behind the others in his maths). **4 a** in support of (she's right behind us). **b** responsible for; giving rise to (the man behind the project; the reasons behind his resignation). **5** in the tracks of; following. ● adv. **1 a** in or to or towards the rear; further back (the street behind; glance behind). **b** on the further side (a high wall with a field behind). **2** remaining after departure (leave behind; stay behind). **3** (usu. foll. by with) **a** in arrears (behind with the rent). **b** late in accomplishing a task etc. (working too slowly and getting behind). **4** in a weak position; backward (behind in Latin). **5** following (his dog running behind). ● n. **1** colloq. the buttocks. **2** (in Australian Rules football) a kick etc. sending the ball over the behind line, scoring one point. □ **behind a person's back** without a person's knowledge. **behind line** (in Australian Rules football) the line between an inner and an outer goal post. **behind the scenes** see SCENE. **behind time** late. **behind the times** antiquated. **come from behind** win after lagging. **fall** (or **lag**) **behind** see FALL. **put behind one 1** refuse to consider. **2** get over (an unhappy experience etc.). [OE behindan, bihindan f. bi BY + hindan from behind, hinder below]

behindhand /bɪ'haɪndhænd/ adv. & predic.adj. **1** (usu. foll. by with, in) late (in discharging a duty, paying a debt, etc.). **2** out of date; behind time. [BEHIND + HAND: cf. BEFOREHAND]

Behn /beɪn/, Aphra (1640–89), English novelist and dramatist. Regarded as the first professional woman writer in England, she is best known for her philosophical novel Oroonoko, or the History of the Royal Slave (1688). Based on her trip to Suriname in 1663, the novel deplores the slave trade and Christian attitudes towards it, and encourages respect for its African hero. Of her fifteen plays, perhaps the best known is her Restoration comedy The Rover (1678).

behold /bɪ'həʊld/ v.tr. (past & past part. **beheld** /-'held/) (esp. in imper.) literary see, observe. □ **beholder** n. [OE bihaldan (as BE-, haldan hold)]

beholden /bɪ'həʊld(ə)n/ predic.adj. (usu. foll. by to) under obligation. [past part. (obs. except in this use) of BEHOLD, = bound]

behoof /bɪ'huːf/ n. (prec. by to, for, on; foll. by of) archaic benefit; advantage. [OE behōf]

behove /bɪ'həʊv/ v.tr. (US **behoove** /-'huːv/) (prec. by it as subject; foll. by to + infin.) formal **1** be incumbent on. **2** (usu. with neg.) befit (ill behoves him to protest). [OE behōfian f. behōf: see BEHOOF]

Behrens /'beərənz/, Peter (1868–1940), German architect and designer. As architect and chief designer for the electrical combine AEG from 1907, his brief included the design of the company's products and stationery in addition to their buildings, and his work was a significant influence in the development of modern architecture and industrial design. His architectural work is notable for its functional employment of modern materials such as steel and glass. Behrens was also an influential teacher and trained leading architects such as Walter Gropius and Le Corbusier.

Behring /'beərɪŋ/, Emil Adolf von (1854–1917), German bacteriologist and one of the founders of immunology. He discovered in 1890 that animals can produce substances in the blood which counteract the effects of bacterial toxins. Behring applied this knowledge to the curing of diphtheria and tetanus, injecting patients with blood serum taken from animals previously exposed to the disease. He was awarded a Nobel Prize in 1901.

Beiderbecke /'baɪdəˌbek/, Bix (born Leon Bismarck Beiderbecke) (1903–31), American jazz musician and composer. A self-taught cornettist and pianist, he was one of a handful of white musicians who profoundly influenced the development of jazz. His bell-like tone and lyrical improvisations on the cornet were his hallmarks. During a career abruptly terminated by his death from alcoholism, Beiderbecke played with Louis Armstrong and American band-leader Paul Whiteman (1890–1967).

beige /beɪʒ/ n. & adj. ● n. a pale sandy fawn colour. ● adj. of this colour. [F: orig. unkn.]

Beijing /beɪ'dʒɪŋ/ (also **Peking** /piː'kɪŋ/) the capital of China, in the north-east of the country; pop. (1990) 6,920,000. The ancient settlement was first developed as the capital of a dynastic kingdom between the 5th and 3rd centuries BC. Over the succeeding centuries it was destroyed, rebuilt, and renamed by successive conquering peoples, until Kublai Khan developed his capital on the site in the late 13th century. During the period of Mongol rule it served as the political centre for the whole of China, finally becoming the country's capital in 1421, at the start of the Ming period. Invaded once again in the 17th century by the Manchus, it became the Qing capital, surviving as the capital of the Republic of China after the revolution of 1912. At the centre of modern Beijing lies the Forbidden City, containing the former imperial palaces, to which entry was forbidden to all except the members of the imperial family and their servants. [Chin., = northern capital]

being /'biːɪŋ/ n. **1** existence. **2** the nature or essence (of a person etc.) (*his whole being revolted*). **3** a human being. **4** anything that exists or is imagined. □ **in being** existing.

Beira /'baɪrə/ a port on the coast of Mozambique, capital of the province of Sofala; pop. (1990) 299,300.

Beirut /beɪ'ruːt/ the capital and chief port of Lebanon; pop. (est. 1988) 1,500,000. The city was badly damaged during the Lebanese civil war of 1975–89.

bejabers /bɪ'dʒeɪbəz/ int. (also **bejabbers** /-'dʒæbəz/) Ir. by Jesus! [corrupt.]

Béjart /beɪ'ʒɑː(r)/, Maurice (born Maurice Jean Berger) (b.1927), French choreographer. He is chiefly identified with The Ballet of the 20th Century, the company which he founded in Brussels in 1959. His choreography is noted for its fusion of classic and modern dance. The first choreographer to stage ballet performances in a sports arena, he is remembered for his innovative productions of *The Firebird* (1970) and *Notre Faust* (1975).

bejewelled /bɪ'dʒuːəld/ adj. (US **bejeweled**) adorned with jewels.

Bekaa /bɪ'kɑː/ (also **El Beqa'a**) a fertile valley in central Lebanon between the Lebanon and Anti-Lebanon mountain ranges.

Bel /bel/ = BAAL. □ **Bel and the Dragon** a book of the Apocrypha containing additional stories of Daniel, concerned mainly with his refusal to worship Bel and his slaying of the dragon.

bel /bel/ n. a unit used in the comparison of power levels in electrical communication or intensities of sound, corresponding to an intensity ratio of 10 to 1 (cf. DECIBEL). [BELL¹]

belabour /bɪ'leɪbə(r)/ v.tr. (US **belabor**) **1 a** thrash; beat. **b** attack verbally. **2** argue or elaborate (a subject) in excessive detail. [BE- + LABOUR = exert one's strength]

Belarus /ˌbelə'ruːs/ (also **Belorussia** /ˌbeləʊ'rʌʃə/, Russian ˌbjɛlə'rusijə/; also called *White Russia*) a country in eastern Europe; pop. (est. 1991) 10,328,000; official language, Belorussian; capital, Minsk. Most of Belarus is a low-lying plain watered by the Dnieper river and its tributaries. Successively part of the grand duchy of Lithuania, Poland, and the Russian empire, it became a republic of the USSR in 1921. Belarus gained independence as a member of the Commonwealth of Independent States in 1991.

belated /bɪ'leɪtɪd/ adj. **1** coming late or too late. **2** overtaken by darkness. □ **belatedly** adv. **belatedness** n. [past part. of obs. *belate* delay (as BE-, LATE)]

Belau see PALAU.

belay /bɪ'leɪ/ v. & n. ● v. **1** tr. fix (a running rope) round a cleat, pin, rock, etc., to secure it. **2** tr. & intr. (usu. in imper.) Naut. sl. stop; enough! (esp. *belay there!*). ● n. **1** an act of belaying. **2** a spike of rock etc. used for belaying. □ **belaying-pin** a fixed wooden or iron pin used for fastening a rope round. [Du. *beleggen*]

bel canto /bel 'kæntəʊ/ n. **1** a lyrical style of operatic singing using a full rich broad tone and smooth phrasing. **2** (attrib.) (of a type of aria or voice) characterized by this type of singing. [It., = fine song]

belch /beltʃ/ v. & n. ● v. **1** intr. emit wind noisily from the stomach through the mouth. **2** tr. **a** (of a chimney, volcano, gun, etc.) send (smoke etc.) out or up. **b** utter forcibly. ● n. an act of belching. [OE *belcettan*]

beldam /'beldəm/ n. (also **beldame**) archaic **1** an old woman; a hag. **2** a virago. [ME & OF *bel* beautiful + DAM², DAME]

beleaguer /bɪ'liːgə(r)/ v.tr. **1** besiege. **2** vex; harass. [Du. *belegeren* camp round (as BE-, *leger* a camp)]

Belém /be'lem/ a city and port of northern Brazil, at the mouth of the Amazon, capital of the state of Pará; pop. (1990) 1,235,625. It is the country's chief commercial centre.

belemnite /'beləmˌnaɪt/ n. an extinct cephalopod mollusc of the mainly Mesozoic order Belemnoidea, having a bullet-shaped internal shell often found as a fossil. [mod.L *belemnites* f. Gk *belemnon* dart + -ITE¹]

bel esprit /ˌbel e'spriː/ n. (pl. **beaux esprits** /ˌbəʊz e'spriː/) a witty person. [F, lit. 'fine mind']

Belfast /bel'fɑːst/ the capital of Northern Ireland; pop. (1991) 280,970. Chartered by King James in 1613, it became a noted centre of the linen industry in the 17th century. The headquarters of the Northern Ireland Assembly are in Stormont Castle. The city suffered a dramatic population decline from the early 1970s as a result of sectarian violence by the IRA and Loyalist paramilitary groups.

belfry /'belfrɪ/ n. (pl. **-ies**) **1** a bell tower or steeple housing bells, esp. forming part of a church. **2** a space for hanging bells in a church tower. □ **bats in the belfry** see BAT². [ME f. OF *berfrei* f. Frank.: altered by assoc. with *bell*]

Belgae /'beldʒiː, 'belgaɪ/ n.pl. an ancient Celtic people inhabiting Gaul north of the Seine and Marne rivers, eventually defeated by Julius Caesar in the Gallic Wars of 58–51 BC. At the beginning of the 1st century BC some of the Belgae had crossed to southern England, where they established kingdoms around Colchester, Winchester, and Silchester; their numbers were swelled by Belgae fleeing the Romans.

Belgaum /'belgaʊm/ an industrial city in western India, in the state of Karnataka; pop. (1991) 326,000.

Belgian /'beldʒən/ n. & adj. ● n. **1** a native or national of Belgium. **2** a person of Belgian descent. ● adj. of or relating to Belgium. □ **Belgian hare** a dark red long-eared breed of domestic rabbit.

Belgian Congo the former name (1908–60) for ZAIRE.

Belgic /'beldʒɪk/ adj. **1** of the ancient Belgae of northern Gaul. **2** of the Low Countries. [L *Belgicus* f. *Belgae*]

Belgium /'beldʒəm/ (French **Belgique** /bɛlʒik/, Flemish **België** /'bɛlxiːə/) a low-lying country in western Europe on the south shore of the North Sea and English Channel; pop. (1991) 9,978,700; official languages, Flemish and French; capital, Brussels. The country takes its name from the Belgae. Flemish is mainly spoken in the north of the country, and French and Walloon in the south. Prosperous in medieval times through textile production and commerce, the region that is now Belgium was on the border of French, Dutch, and Habsburg spheres of influence in the 16th–18th centuries and, as a result, was frequently the site of military operations. After a nationalist revolt in 1830, Prince Leopold of Saxe-Coburg was elected king. Belgium's independence from the Netherlands (and its neutrality) was guaranteed by treaty in 1839. Occupied and devastated during both world wars, Belgium made a quick recovery after 1945, forming the Benelux Customs Union with the Netherlands and Luxembourg in 1948, and becoming a founder member of the EEC.

Belgorod /'belgəˌrɒd/ an industrial city in southern Russia, on the Donets river close to the border with Ukraine; pop. (1990) 306,000.

Belgrade /bel'greɪd/ (Serbian **Beograd** /bɛ'ɔgrad/) the capital of Serbia, on the River Danube; pop. (1981) 1,470,000.

Belial /'biːlɪəl/ n. the Devil. [Heb. *bᵉliyyaʿal* worthless]

belie /bɪ'laɪ/ v.tr. (**belying**) **1** give a false notion of; fail to corroborate (*its appearance belies its age*). **2 a** fail to fulfil (a promise etc.). **b** fail to justify (a hope etc.). [OE *belēogan* (as BE-, *lēogan* LIE²)]

belief /bɪ'liːf/ n. **1 a** a person's religion; religious conviction (*has no belief*). **b** a firm opinion (*my belief is that he did it*). **c** an acceptance (of a thing, fact, statement, etc.) (*belief in the afterlife*). **2** (usu. foll. by *in*) trust or confidence. □ **beyond belief** incredible. **to the best of my belief** in my genuine opinion. [ME f. OE *gelēafa* (as BELIEVE)]

believe /bɪ'liːv/ v. **1** tr. accept as true or as conveying the truth (*I believe it; don't believe him; believes what he is told*). **2** tr. think, suppose (*I believe* it's raining; *Mr Smith, I believe?*). **3** *intr.* (foll. by *in*) **a** have faith in the existence of (*believes in God*). **b** have confidence in (a remedy, a person, etc.) (*believes in alternative medicine*). **c** have trust in the advisability of (*believes in telling the truth*). **4** *intr.* have (esp. religious) faith. □ **believe one's ears** (or **eyes**) accept that what one apparently hears or sees etc. is true. **believe it or not** *colloq.* it is true though surprising. **make believe** (often foll. by *that* + clause, or *to* + infin.) pretend (*let's make believe that we're young again*). **would you believe it?** *colloq.* = believe it or not. □ **believable** *adj.* **believability** /-ˌliːvəˈbɪlɪtɪ/ *n.* [OE *belȳfan, belēfan,* with change of prefix f. *gelēfan* f. Gmc: rel. to LIEF]

believer /bɪˈliːvə(r)/ *n.* **1** an adherent of a specified religion. **2** a person who believes, esp. in the efficacy of something (*a great believer in exercise*).

Belisha beacon /bəˈliːʃə/ *n. Brit.* a flashing orange ball mounted on a striped post, marking some pedestrian crossings. [L. Hore-*Belisha* (1893–1957), Brit. Minister of Transport 1934]

belittle /bɪˈlɪt(ə)l/ *v.tr.* **1** make seem unimportant; depreciate; disparage. **2** make small; diminish in size. □ **belittlement** *n.* **belittler** *n.* **belittlingly** *adv.*

Belitung /bɪˈliːtʊŋ/ (also **Billiton** /-tɒn/) an Indonesian island in the Java Sea, between Borneo and Sumatra.

Belize /beˈliːz/ a country on the Caribbean coast of Central America; pop. (est. 1991) 190,800; languages, English (official), Creole, Spanish; capital, Belmopan. The British settled there in the 17th century, proclaiming the area (as British Honduras) a Crown Colony in 1862. It adopted the name Belize (after a river with a Mayan name meaning 'muddy water') in 1973, and in 1981 became an independent Commonwealth state. Guatemala, which bounds it on the west and south, has always claimed the territory on the basis of old Spanish treaties, although in 1992 it agreed to recognize the existence of Belize. □ **Belizian** *adj. & n.*

Belize City the principal seaport and former capital (until 1970) of Belize; pop. (1991) 46,000.

Bell[1] /bel/ Alexander Graham (1847–1922), Scottish-born American scientist and inventor. Bell studied sound waves, the mechanics of speech, and speech therapy. Having moved to the US in the early 1870s, he developed his ideas for transmitting speech electrically, and gave the first public demonstration of the telephone in 1876; he founded the Bell Telephone Company the following year. He also invented the gramophone (1897) as a successful rival to Thomas Edison's phonograph. He later carried out research in a number of other areas, including hydrofoil speedboats and aeronautics.

Bell[2] /bel/, Currer, Ellis, and Acton, the pseudonyms used by Charlotte, Emily, and Anne Brontë.

Bell[3] /bel/, Gertrude (Margaret Lowthian) (1868–1926), English archaeologist and traveller. She travelled widely as a field archaeologist in the Middle East, acquiring an extensive knowledge of the desert Arabs and local politics, and undertook liaison work with the Arabs for the British government in 1915. A supporter of Arab independence, she assisted in the negotiations for Iraq's independence (1920–1) in her capacity as Oriental Secretary to the British High Commissioner. Her writings include a description of her travels in Syria, *The Desert and the Sown* (1907).

bell[1] /bel/ *n. & v.* ● *n.* **1** a hollow usu. metal object in the shape of a deep inverted cup usu. widening at the lip, made to sound a clear musical note when struck (either externally or by means of a clapper inside). (*See note below.*) **2 a** a sound or stroke of a bell, esp. as a signal. **b** (prec. by a numeral) *Naut.* the time as indicated every half-hour of a watch by the striking of the ship's bell one to eight times. **3** anything that sounds like or functions as a bell, esp. an electronic device that rings etc. as a signal. **4 a** a bell-shaped object or part, e.g. of a musical instrument. **b** the corolla of a flower when bell-shaped. **5** (in *pl.*) *Mus.* a set of cylindrical metal tubes of different lengths, suspended in a frame and played by being struck with a hammer. ● *v.tr.* **1** provide with a bell or bells; attach a bell to. **2** (foll. by *out*) form into the shape of the lip of a bell. □ **bell-bottom 1** a marked flare below the knee (of a trouser-leg). **2** (in *pl.*) trousers with bell-bottoms. **bell-bottomed** having bell-bottoms. **bell-buoy** a buoy equipped with a warning bell rung by the motion of the sea. **bell-founder** a person who casts large bells in a foundry. **bell-glass** a bell-shaped glass cover for plants. **bell-jar** a bell-shaped glass cover or container for use in a laboratory. **bell-metal** an alloy of copper and tin for making bells (the tin content being greater than in bronze). **bell-pull** a cord or handle which rings a bell when pulled. **bell-push** a button that operates an electric bell when pushed. **bell-ringer** a person who rings church bells or handbells. **bell-ringing** this as an activity. **bells and whistles** *colloq.*

attractive but unnecessary additional features, esp. in computing. **bell-tent** a cone-shaped tent supported by a central pole. **bell-wether 1** the leading sheep of a flock, with a bell on its neck. **2** a ringleader. **clear** (or **sound**) **as a bell** perfectly clear or sound. **give a person a bell** *colloq.* telephone a person. **ring a bell** *colloq.* revive a distant recollection; sound familiar. [OE *belle:* perh. rel. to BELL[2]]

▪ Bells vary enormously in weight and size, from the huge broken bell at the Kremlin, dating from 1733 and weighing over 100 metric tons, to small handbells. They occur all over the world, have various powers assigned to them, and have long been important in ritual and religion, in the West being rung in church steeples to summon the congregation.

bell[2] /bel/ *n. & v.* ● *n.* the cry of a stag or buck at rutting-time. ● *v.intr.* make this cry. [OE *bellan* bark, bellow]

belladonna /ˌbeləˈdɒnə/ *n.* **1** deadly nightshade. **2** a drug prepared from this. □ **belladonna lily** = AMARYLLIS 1. [mod.L f. It., = fair lady, perh. from the use of its juice to make the eyes brilliant by dilating the pupils]

bellbird /ˈbelbɜːd/ *n.* a bird with a bell-like song, esp. a Central or South American cotinga of the genus *Procnias,* and a New Zealand honeyeater, *Anthornis melanura.*

bellboy /ˈbelbɔɪ/ *n.* esp. *N. Amer.* a page in a hotel or club.

belle /bel/ *n.* **1** a beautiful woman. **2** a woman recognized as the most beautiful (*the belle of the ball*). [F f. L *bella* fem. of *bellus* beautiful]

belle époque /ˌbel eˈpɒk/ *n.* the period of settled and comfortable life from the late 19th century to the war of 1914–18. [F, = fine period]

belle laide /bel ˈled/ *n.* (*pl.* **belles laides** *pronunc.* same) a fascinatingly ugly woman, esp. in France. Also called *jolie laide.* [F f. *belle* beautiful + *laide* ugly]

Bellerophon /bɪˈlerəf(ə)n/ *Gk Mythol.* a hero who slew the monster Chimera with the help of the winged horse Pegasus.

belles-lettres /bel ˈletrə/ *n.pl.* (also treated as *sing.*) writings or studies of a literary nature, esp. essays and criticisms. □ **belletrist** *n.* **belletristic** /ˌbeləˈtrɪstɪk/ *adj.* [F, = fine letters]

bellflower /ˈbelˌflaʊə(r)/ *n.* = CAMPANULA.

bellicose /ˈbelɪˌkəʊz/ *adj.* eager to fight; warlike. □ **bellicosity** /-ˈkɒsɪtɪ/ *n.* [ME f. L *bellicosus* f. *bellum* war]

belligerence /bɪˈlɪdʒərəns/ *n.* (also **belligerency** /-rənsɪ/) **1** aggressive or warlike behaviour. **2** the status of a belligerent.

belligerent /bɪˈlɪdʒərənt/ *adj. & n.* ● *adj.* **1** engaged in war or conflict. **2** given to constant fighting; pugnacious. ● *n.* a nation or person engaged in war or conflict. □ **belligerently** *adv.* [L *belligerare* wage war f. *bellum* war + *gerere* wage]

Bellingshausen Sea /ˈbelɪŋzˌhaʊz(ə)n/ a part of the SE Pacific off the coast of Antarctica, bounded to the east and south by the Antarctic Peninsula and Ellsworth Land. It is named after the Russian explorer Fabian Gottlieb von Bellingshausen (1778–1852), who in 1819–21 became the first to circumnavigate Antarctica.

Bellini[1] /beˈliːnɪ/ a family of Italian painters. Jacopo (c.1400–70) was trained by Gentile da Fabriano: his elder son Gentile (c.1429–1507) was prominent as a portraitist and narrative painter. Jacopo's younger son Giovanni (c.1430–1516) is the most famous of the family; he had a large workshop of pupils and assistants and transformed the family's native Venice into a major centre of Renaissance painting. Stylistically he was influenced by his brother-in-law Mantegna, although his painting is less severe than Mantegna's and has a serene contemplative quality. Giovanni Bellini's work in the 15th and early 16th centuries is dominated by madonnas and other sacred subjects (such as *The San Giobbe Altarpiece,* 1480). In works such as *The Agony in the Garden* (c.1460), Giovanni Bellini also made a significant contribution towards the treatment of figures within a landscape. He continued to develop as an artist, in later life painting the newly fashionable pagan themes (such as *Feast of the Gods,* 1514) and mysterious allegories, his work reflecting the influence of his pupils Giorgione and Titian.

Bellini[2] /beˈliːnɪ/, Vincenzo (1801–35), Italian composer. Of his eleven operas, the most famous are *La Sonnambula* (1831), *Norma* (1831), and *I Puritani* (1835). His work is typically dramatic and lyrical, displays a close relationship between the music and libretto, and is characterized by long, elegant melodies, such as 'Casta Diva' from *Norma.*

bellman /ˈbelmən/ *n.* (*pl.* **-men**) *hist.* a town crier.

Belloc /ˈbelɒk/ (Joseph) Hilaire (Pierre René) (1870–1953), French-born British writer, historian, and poet, of French–British descent. A devout Roman Catholic, he collaborated with his friend G. K. Chesterton in works often critical of modern industrial society and socialism, notably in *The Servile State* (1912). His writings include biographies of Napoleon and Cromwell, but he is now best known for his light verse, such as *The Bad Child's Book of Beasts* (1896) and *Cautionary Tales* (1907).

Bellow /'beləʊ/, Saul (b.1915), Canadian-born American novelist, of Russian-Jewish descent. A leading figure in mid-20th century American fiction, he has written novels as diverse as the comic *The Adventures of Augie March* (1953) and the more sombre and semi-autobiographical *Herzog* (1964). His other works include the collection of short stories *Him with His Foot in His Mouth* (1984). His fiction is both ironic and optimistic in its treatment of the human condition. He was awarded the Nobel Prize for literature in 1976.

bellow /'beləʊ/ v. & n. ● v. **1** intr. **a** emit a deep loud roar. **b** cry or shout with pain. **2** tr. utter loudly and usu. angrily. ● n. a bellowing sound. [ME: perh. rel. to BELL²]

bellows /'beləʊz/ n.pl. (also treated as *sing.*) **1** a device with an air bag that emits a stream of air when squeezed, esp.: **a** (in full **pair of bellows**) a kind with two handles used for blowing air on to a fire. **b** a kind used in a harmonium or small organ. **2** an expandable component, e.g. joining the lens to the body of e.g. a camera. [ME prob. f. OE *belga* pl. of *belig* belly]

Bell's palsy n. Med. paralysis of the facial nerve causing muscular weakness on one side of the face. [Sir Charles *Bell*, Sc. anatomist (1774–1842)]

belly /'belɪ/ n. & v. ● n. (pl. **-ies**) **1** the part of the human body below the chest, containing the stomach and bowels. **2** the stomach, esp. representing the body's need for food. **3** the front of the body from the waist to the groin. **4** the underside of a four-legged animal. **5 a** a cavity or bulging part of anything. **b** the surface of an instrument of the violin family, across which the strings are placed. ● v.tr. & intr. (**-ies, -ied**) (often foll. by *out*) swell or cause to swell; bulge. □ **belly button** *colloq.* the navel. **belly-dance** an oriental dance performed by a woman, involving voluptuous movements of the exposed belly. **belly-dancer** a woman who performs belly-dances, esp. professionally. **belly-dancing** the performance of belly-dances. **belly-landing** a crash-landing of an aircraft on the underside of the fuselage, without lowering the undercarriage. **belly-laugh** a loud unrestrained laugh. **go belly up** US colloq. fail financially. [OE *belig* (orig. = bag) f. Gmc]

bellyache /'belɪˌeɪk/ n. & v. ● n. colloq. a stomach pain. ● v.intr. sl. complain noisily or persistently. □ **bellyacher** n.

bellyband /'belɪˌbænd/ n. a band placed round a horse's belly, holding the shafts of a cart etc.

bellyflop /'belɪˌflɒp/ n. & v. colloq. ● n. a dive into water in which the body lands with the belly flat on the water. ● v.intr. (**-flopped, -flopping**) perform such a dive.

bellyful /'belɪˌfʊl/ n. (pl. **-fuls**) **1** enough to eat. **2** colloq. enough or more than enough of anything (esp. unwelcome).

Belmopan /ˌbelməʊ'pæn/ the capital of Belize. Founded in 1970, it is one of the smallest capital cities in the world; pop. (est. 1986) 3,500.

Belo Horizonte /ˌbel ɒrɪ'zɒntɪ/ a city in eastern Brazil, capital of the state of Minas Gerais; pop. (1990) 2,103,330.

belong /bɪ'lɒŋ/ v.intr. **1** (foll. by *to*) **a** be the property of. **b** be rightly assigned to as a duty, right, part, member, characteristic, etc. **c** be a member of (a club, family, group, etc.). **2** have the right personal or social qualities to be a member of a particular group (*he's nice but just doesn't belong*). **3** (foll. by *in, under*) **a** be rightly placed or classified. **b** fit a particular environment. □ **belongingness** n. [ME f. intensive BE- + *longen* belong f. OE *langian* (*gelang* at hand)]

belonging /bɪ'lɒŋɪŋ/ n. **1** (in pl.) one's movable possessions or luggage. **2** membership, relationship; esp. a person's membership of, and acceptance by, a group or society.

Belorussia see BELARUS.

Belorussian /ˌbeləʊ'rʌʃ(ə)n/ n. & adj. (also **Belarussian, Byelorussian** /ˌbjeləʊ-/) ● n. **1** a native of Belarus. **2** the East Slavonic language of Belarus. ● adj. of or relating to Belarus or its people or language.

Belostok see BIAŁYSTOK.

beloved /bɪ'lʌvɪd, also predic.adj. -'lʌvd/ adj. & n. ● adj. much loved. ● n. a much loved person. [obs. *belove* (v.)]

below /bɪ'ləʊ/ prep. & adv. ● prep. **1** lower in position (vertically, down a slope or stream, etc.) than. **2** beneath the surface of; at or to a greater depth than (*head below water; below 500 feet*). **3** lower or less than in amount or degree (*below freezing-point*). **4** lower in rank, position, or importance than. **5** unworthy of. ● adv. **1** at or to a lower point or level. **2 a** downstairs (*lives below*). **b** downstream. **3** (of a text reference)

further forward on a page or in a book (*as noted below*). **4** on the lower side (*looks similar above and below*). **5** rhet. on earth; in hell. **6** below zero; esp. below freezing-point. □ **below stairs** in the basement of a house esp. as the part occupied by servants. [ME, f. *be* BY + LOW¹]

Bel Paese /ˌbel pɑː'eɪzɪ/ n. propr. a rich white mild creamy cheese of a kind orig. made in Italy. [It., = fair country]

Belsen /'bels(ə)n/ a Nazi concentration camp in the Second World War, near the village of Belsen in NW Germany.

Belshazzar /bel'ʃæzə(r)/ (6th century BC), son of Nebuchadnezzar and last king of Babylon. According to Dan. 5, he was killed in the sack of the city and his doom was foretold by writing which appeared on the walls of his palace at a great banquet. In inscriptions and documents from Ur, however, he was perhaps the grandson of Nebuchadnezzar and the son of Nabonidos, last king of Babylon, and did not himself reign.

belt /belt/ n. & v. ● n. **1** a strip of leather or other material worn round the waist or across the chest, esp. to retain or support clothes or to carry weapons or as a safety-belt. **2** a belt worn as a sign of rank or achievement. **3 a** a circular band of material used as a driving medium in machinery. **b** a conveyor belt. **c** a flexible strip carrying machine-gun cartridges. **4** a strip of colour or texture etc. differing from that on each side. **5** a distinct region or extent (*cotton belt; commuter belt; a belt of rain*). **6** sl. a heavy blow. ● v. **1** tr. put a belt round. **2** tr. (often foll. by *on*) fasten with a belt. **3** tr. **a** beat with a belt. **b** sl. hit hard. **4** intr. sl. rush, hurry (usu. with complement: *belted along; belted home*). □ **below the belt** unfair or unfairly; disregarding the rules. **belt and braces** (of a policy etc.) of twofold security. **belt drive** a driving mechanism powered by a continuous flexible belt. **belt out** sl. sing or utter loudly and forcibly. **belt up** Brit. **1** sl. be quiet. **2** colloq. put on a seat belt. **tighten one's belt** live more frugally. **under one's belt 1** (of food) eaten. **2** securely acquired (*has a degree under her belt*). □ **belter** n. (esp. in sense of *belt out*). [OE f. Gmc f. L *balteus*]

Beltane /'belteɪn/ n. an ancient Celtic festival celebrated on May Day. [Gael. *bealltainn*]

beltman /'beltmæn/ n. (pl. **-men**) Austral. a member of a life-saving team of surfers.

beltway /'beltweɪ/ n. US a ring road.

beluga /bə'luːgə/ n. **1 a** a sturgeon, *Huso huso*, of the Caspian and Black Seas. **b** caviar obtained from it. **2** the white whale. [Russ. *beluga* f. *bely* white]

belvedere /'belvɪˌdɪə(r)/ n. a summer-house or open-sided gallery usu. at rooftop level. [It. f. *bel* beautiful + *vedere* see]

belying pres. part. of BELIE.

BEM abbr. British Empire Medal.

bemire /bɪ'maɪə(r)/ v.tr. **1** cover or stain with mud. **2** (in passive) be stuck in mud. [BE- + MIRE]

bemoan /bɪ'məʊn/ v.tr. **1** express regret or sorrow over; lament. **2** complain about. [BE- + MOAN]

bemuse /bɪ'mjuːz/ v.tr. stupefy or bewilder (a person). □ **bemusement** n. **bemusedly** /-zɪdlɪ/ adv. [BE- + MUSE²]

ben¹ /ben/ n. Sc. a high mountain or mountain peak, esp. in names (*Ben Nevis*). [Gael. *beann*]

ben² /ben/ n. Sc. an inner room, esp. of a two-roomed cottage. □ **but and ben** see BUT². [ellipt. use of *ben* (adv.), = within (OE *binnan*)]

Benares see VARANASI.

Ben Bella /ben 'belə/, (Muhammad) Ahmed (b.1916), Algerian statesman, Prime Minister 1962–3 and President 1963–5. In 1952 he founded the Front de Libération Nationale (FLN), which instigated the Algerian War of Independence (1954–62). He was elected Prime Minister of a provisional government shortly before the end of the war, becoming first President of an independent Algeria the following year. As President he initiated social and economic reform and encouraged closer links with other Arab nations. Overthrown in a military coup, he was kept under house arrest until 1979 and lived in exile until 1990, when he returned to Algeria to lead the opposition to the ruling regime.

bench /bentʃ/ n. & v. ● n. **1** a long seat of wood or stone for seating several people. **2** a working-table, e.g. for a carpenter, mechanic, or scientist. **3** (prec. by *the*) **a** the status of judge or magistrate. **b** a judge's seat in a lawcourt. **c** a lawcourt. **d** judges and magistrates collectively. **4** (often in pl.) Sport an area to the side of a pitch, with seating where coaches and players not taking part can watch the game. **5** Brit. Parl. a seat appropriated as specified (*front bench*). **6** a level ledge in masonry or an earthwork, on a hill-slope, etc. ● *v.tr.* **1** exhibit (a dog) at a show. **2** *Sport N. Amer.* withdraw (a player) from the pitch to the benches. □ **bench test** esp. *Computing n.* a test carried out on a machine, component, etc. before it is released for use. ● *v.tr.* run a bench test on. **on the bench 1** appointed a judge or magistrate. **2** *Sport* acting as substitute or reserve. [OE *benc* f. Gmc]

bencher /'bentʃə(r)/ *n. Brit.* **1** *Law* a senior member of any of the Inns of Court. **2** (in *comb.*) *Parl.* an occupant of a specified bench (*backbencher*).

benchmark /'bentʃmɑːk/ *n. & v.* ● *n.* **1** a surveyor's mark cut in a wall, pillar, building, etc., used as a reference point in measuring altitudes. **2 a** a standard or point of reference. **b** a problem designed to evaluate the performance of a computer system. ● *v.tr.* evaluate or check by comparison with a benchmark. □ **benchmark test** a test which involves comparison with a benchmark.

bend[1] /bend/ *v. & n.* ● *v.* (*past* **bent**; *past part.* **bent** except in *bended knee*) **1 a** *tr.* force or adapt (something straight) into a curve or angle. **b** *intr.* (of an object) be altered in this way. **2** *intr.* move or stretch in a curved course (*the road bends to the left*). **3** *intr. & tr.* (often foll. by *down, over,* etc.) incline or cause to incline from the vertical (*bent down to pick it up*). **4** *tr.* interpret or modify (a rule) to suit oneself. **5** *tr. & refl.* (foll. by *to, on*) direct or devote (oneself or one's attention, energies, etc.). **6** *tr.* turn (one's steps or eyes) in a new direction. **7** *tr.* (in *passive*; foll. by *on*) have firmly decided; be determined (*was bent on selling; on pleasure bent*). **8 a** *intr.* stoop or submit (*bent before his master*). **b** *tr.* force to submit. **9** *tr. Naut.* attach (a sail or cable) with a knot. ● *n.* **1** a curve in a road or other course. **2** a departure from a straight course. **3** a bent part of anything. **4** (**the bends**) *colloq.* decompression sickness. □ **bend over backwards** see BACKWARDS. **round the bend** *colloq.* crazy, insane. □ **bendable** *adj.* [OE *bendan* f. Gmc]

bend[2] /bend/ *n.* **1** *Naut.* any of various knots for tying ropes (*fisherman's bend*). **2** *Heraldry* a diagonal stripe from top right to bottom left of a shield. □ **bend sinister** *Heraldry* a diagonal stripe from top left to bottom right, as a sign of bastardy. [OE *bend* band, bond f. Gmc]

bender /'bendə(r)/ *n. sl.* **1** a wild drinking spree. **2** *offens.* a homosexual. [BEND[1] + -ER[1]]

Bendigo /'bendɪˌɡəʊ/ a former gold-mining town in the state of Victoria, Australia; pop. (1991) 57,430. Originally called Sandhurst, the town was renamed after a local boxer who had adopted the nickname of a well-known English prizefighter, William Thompson (1811–80).

bendy /'bendɪ/ *adj.* (**bendier, bendiest**) *colloq.* capable of bending; soft and flexible. □ **bendiness** *n.*

beneath /bɪ'niːθ/ *prep. & adv.* ● *prep.* **1** not worthy of; too demeaning for (*it was beneath him to reply*). **2** below, under. ● *adv.* below, under, underneath. □ **beneath contempt** see CONTEMPT. [OE *binithan, bineothan* f. *bi* BY + *nithan* etc. below f. Gmc]

benedicite /ˌbenɪ'daɪsɪtɪ/ *n.* a blessing, esp. a grace said at table in religious communities. [ME f. L, = bless ye: see BENEDICTION]

Benedict, St /'benɪdɪkt/ (*c.*480–*c.*550), Italian hermit. A hermit from the age of 14, he attracted many followers by his piety; of these he chose the most devoted to form twelve small monastic communities, ultimately establishing a monastery at Monte Cassino (*c.*540). His *Regula Monachorum* (known as the Rule of St Benedict), austere but tempered by moderation, formed the basis of Western monasticism. Feast day, 11 July (formerly 21 Mar.).

Benedictine *n. & adj.* ● *n.* **1** /ˌbenɪ'dɪktɪn/ a monk or nun of an order following the Rule of St Benedict. (*See note below.*) **2** /ˌbenɪ'dɪktiːn/ *propr.* a liqueur based on brandy, orig. made by Benedictines in France. ● *adj.* /ˌbenɪ'dɪktɪn/ of St Benedict or the Benedictines. [F *bénédictine* or mod.L *benedictinus* f. *Benedictus* Benedict]

▪ The Rule of St Benedict, established at Monte Cassino in Italy in the mid-6th century, was gradually adopted by most Western monastic houses, sometimes with their own modifications. Relaxations of discipline were followed by attempts at reform, and resulted in the formation of separate orders, such as the Cistercians and Cluniacs, which remained largely independent of any wider organization. The Benedictines were of particular importance in establishing medieval social organization in the period after the fall of the Roman Empire, preserving ideals of scholarship, and maintaining or restoring the use of art in liturgical worship.

benediction /ˌbenɪ'dɪkʃ(ə)n/ *n.* **1** the utterance of a blessing, esp. at the end of a religious service or as a special Roman Catholic service. **2** the state of being blessed. [ME f. OF f. L *benedictio -onis* f. *benedicere -dict-* bless]

benedictory /ˌbenɪ'dɪktərɪ/ *adj.* of or expressing benediction. [L *benedictorius* (as BENEDICTION)]

Benedictus /ˌbenɪ'dɪktəs/ *n.* **1** the section of the Roman Catholic Mass beginning *Benedictus qui venit in nomine Domini* (Blessed is he who comes in the name of the Lord). **2** a canticle beginning *Benedictus Dominus Deus* (Blessed be the Lord God) from Luke 1:68–79. [L, = blessed: see BENEDICTION]

benefaction /ˌbenɪ'fækʃ(ə)n/ *n.* **1** a donation or gift. **2** an act of giving or doing good. [LL *benefactio* (as BENEFIT)]

benefactor /ˌbenɪˈfæktə(r)/ *n.* (*fem.* **benefactress** /-trɪs/) a person who gives support (esp. financial) to a person or cause. [ME f. LL (as BENEFIT)]

benefice /'benɪfɪs/ *n.* **1** a living from a church office. **2** the property attached to a church office, esp. that bestowed on a rector or vicar. □ **beneficed** *adj.* [ME f. OF f. L *beneficium* favour f. *bene* well + *facere* do]

beneficent /bɪ'nefɪs(ə)nt/ *adj.* doing good; generous, actively kind. □ **beneficence** *n.* **beneficently** *adv.* [L *beneficent-* (as BENEFICE)]

beneficial /ˌbenɪ'fɪʃ(ə)l/ *adj.* **1** advantageous; having benefits. **2** improving the health. **3** *Law* relating to the use or benefit of property; having rights to this use or benefit. □ **beneficially** *adv.* [ME f. F *bénéficial* or LL *beneficialis* (as BENEFICE)]

beneficiary /ˌbenɪ'fɪʃərɪ/ *n.* (*pl.* **-ies**) **1** a person who receives benefits, esp. under a trust or life-insurance policy or will. **2** a holder of a church living. [L *beneficiarius* (as BENEFICE)]

benefit /'benɪfɪt/ *n. & v.* ● *n.* **1** a favourable or helpful factor or circumstance; advantage, profit. **2** (often in *pl.*) payment made under insurance, social security, welfare, etc. (*sickness benefit*). **3** a public performance or game of which the proceeds go to a particular player or company or charitable cause. ● *v.* (**benefited, benefiting**; *US* **benefitted, benefitting**) **1** *tr.* do good to; bring advantage to. **2** *intr.* (often foll. by *from, by*) receive an advantage or gain. □ **benefit of clergy 1** *hist.* exemption of the English tonsured clergy and nuns from the jurisdiction of the ordinary civil courts. **2** ecclesiastical sanction or approval (*marriage without benefit of clergy*). **the benefit of the doubt** a concession that a person is innocent, correct, etc., although doubt exists. **benefit society** = FRIENDLY SOCIETY. [ME f. AF *benfet*, OF *bienfet*, f. L *benefactum* f. *bene facere* do well]

Benelux /'benɪˌlʌks/ *n.* a collective name for Belgium, the Netherlands, and Luxembourg, especially with reference to their economic cooperation established in the Benelux Customs Union of 1948. [*Bel*gium + *Ne*therlands + *Lux*embourg]

Benenden /'benəndən/ an independent boarding-school for girls, in Kent in SE England. It was founded in 1923.

Beneš /'benɛʃ/, Edvard (1884–1948), Czechoslovak statesman, Prime Minister 1921–2, President 1935–8 and 1945–8. A founder (with Tomáš Masaryk) of modern Czechoslovakia, he served as Masaryk's Minister of Foreign Affairs 1919–35, during which time he championed the League of Nations (he served as its chairman six times) and established close ties with France and the Soviet Union. He resigned as President over the Munich Agreement, and during the Second World War came to London as head of the Czechoslovakian government in exile (1941–5). In 1945 he returned to his country to regain the presidency, but resigned after the 1948 Communist coup.

Benetton /'benɪˌtɒn/, Luciano (b.1935), Italian businessman, and his sister Giuliana (b.1938), Italian designer and businesswoman. Luciano established the original Benetton company with his brothers in 1965; Giuliana began by making sweaters which Luciano sold to shops in their home town of Treviso, opening the company's first shop in 1968. By the mid-1980s the firm had become a multinational clothing company with thousands of retail outlets worldwide.

benevolent /bɪ'nevələnt/ *adj.* **1** wishing to do good; actively friendly and helpful. **2** charitable (*benevolent fund; benevolent society*). □ **benevolence** *n.* **benevolently** *adv.* [ME f. OF *benivolent* f. L *bene volens -entis* well wishing f. *velle* wish]

B.Eng. *abbr.* Bachelor of Engineering.

Bengal /ben'ɡɔːl/ a region in the north-east of the Indian subcontinent, containing the Ganges and Brahmaputra river deltas. Ruled in the 18th century by a nawab dynasty, the area was taken by the British between 1757 and 1764, and became the base of British expansion in India. In 1947, with the end of British rule, the province was divided into West Bengal, which has remained a state of India, and East Bengal, now Bangladesh.

Bengal, Bay of a part of the Indian Ocean lying between India to the west and Burma (Myanmar) and Thailand to the east.

Bengali /benˈgɔːlɪ/ n. & adj. ● n. **1** a native of Bengal or Bangladesh. **2** the language of Bengal, spoken by some 160 million people. It is a descendant of Sanskrit, written in a version of the Devanagari script, and is the official language of Bangladesh. ● adj. of or relating to Bengal or its people or language.

Bengal light n. a kind of firework giving off a blue flame, used for signals.

Benghazi /benˈgɑːzɪ/ a Mediterranean port in NE Libya; pop. (1984) 485,400. It was the joint capital (with Tripoli) of Libya from 1951 to 1972.

Benguela /beŋˈgwelə, -ˈgelə/ a port and railway terminal in Angola, on the Atlantic coast; pop. (1983) 155,000. The Benguela railway line provides a link with the copper-mining regions of Zambia and Zaire.

Benguela Current a cold ocean current which flows from Antarctica northwards along the west coast of southern Africa until it meets warmer waters off the coast of Angola.

Ben-Gurion /benˈgʊərɪən/, David (1886–1973), Israeli statesman, Prime Minister 1948–53 and 1955–63. Born in Poland, he emigrated to Palestine in 1906, where he became an active Zionist. He was elected leader of the predominant socialist faction (the Mapai Party) of the Zionist movement in 1930. When the state of Israel was established in 1948, he became the country's first Prime Minister and Minister of Defence. After expulsion from the Labour Party in 1965 he formed a new party with Moshe Dayan.

benighted /bɪˈnaɪtɪd/ adj. **1** intellectually or morally ignorant. **2** overtaken by darkness. □ **benightedness** n. [obs. benight (v.)]

benign /bɪˈnaɪn/ adj. **1** gentle, mild, kindly. **2** fortunate, salutary. **3** (of climate, soil, etc.) mild, favourable. **4** Med. (of a disease, tumour, etc.) not malignant. □ **benign neglect** non-interference or neglect, intended to benefit the subject more than continual attention. □ **benignly** adv. [ME f. OF benigne f. L benignus f. bene well + -genus born]

benignant /bɪˈnɪɡnənt/ adj. **1** kindly. **2** salutary, beneficial. **3** Med. = BENIGN 4. □ **benignancy** n. **benignantly** adv. [f. BENIGN or L benignus, after malignant]

benignity /bɪˈnɪɡnɪtɪ/ n. (pl. **-ies**) **1** kindliness. **2** an act of kindness. [ME f. OF benignité or L benignitas (as BENIGN)]

Benin /beˈniːn/ a country of West Africa, immediately west of Nigeria; pop. (est. 1991) 4,883,000; languages, French (official), West African languages; capital, Porto Novo. Formerly known as Dahomey and a centre of the slave trade, the country was conquered by the French in 1893 and became part of French West Africa. In 1960 it became fully independent and in 1975 adopted the name of Benin, a former African kingdom centred on southern Nigeria, that was powerful in the 14th–17th centuries. □ **Beninese** /ˌbenɪˈniːz/ adj. & n.

Benin, Bight of a wide bay on the coast of Africa north of the Gulf of Guinea, bordered by Togo, Benin, and SW Nigeria. Lagos is its chief port.

benison /ˈbenɪz(ə)n/ n. archaic a blessing. [ME f. OF beneiçun f. L benedictio -onis]

Benjamin /ˈbendʒəmɪn/ (in the Bible) **1** a Hebrew patriarch, the youngest and favourite son of Jacob (Gen. 35:18, 42, etc.). **2** the smallest tribe of Israel, traditionally descended from him.

Bennett[1] /ˈbenɪt/, Alan (b.1934), English dramatist and actor. He achieved fame with the revue Beyond the Fringe (1960) and the satirical comedy Forty Years On (1969), lampooning the Bloomsbury Group and other cult figures. Other plays in the same vein followed, including Getting On (1972), a political satire about a Labour MP. He has also written for television, for example, the monologue series Talking Heads (1987) and the play A Question of Attribution (1991).

Bennett[2] /ˈbenɪt/, (Enoch) Arnold (1867–1931), English novelist, dramatist, and critic. He began his literary career in London writing stories for periodicals and editing the journal Woman; in Paris (1902–12) he was greatly influenced by the French realists and wrote several successful plays. However, his fame rests on the novels and stories set in the Potteries ('the Five Towns') of his youth, notably Anna of the Five Towns (1902), The Old Wives' Tale (1908), and the Clayhanger series (1902–8), in which he portrays provincial life and culture in documentary detail.

Ben Nevis /ben ˈnevɪs/ a mountain in western Scotland. Rising to 1,343 m (4,406 ft), it is the highest mountain in the British Isles.

Benoni /bəˈnəʊnɪ/ a city in South Africa, in the province of Pretoria-Witwatersrand-Vereeniging, east of Johannesburg; pop. (1980) 206,800. It is a gold-mining centre.

bent[1] /bent/ past and past part. of BEND[1] v. ● adj. **1** curved or having an angle. **2** sl. dishonest, illicit. **3** sl. a sexually deviant. **b** strange, weird, warped. **4** (foll. by on) determined to do or have. ● n. **1** an inclination or bias. **2** (foll. by for) a talent for something specified (a bent for mimicry).

bent[2] /bent/ n. **1 a** a stiff grass of the genus Agrostis. **b** a grasslike reed, rush, or sedge. **2** a stiff stalk of a grass usu. with a flexible base. **3** archaic or dial. a heath or unenclosed pasture. [ME repr. OE beonet- (in place-names), f. Gmc]

Bentham /ˈbenθəm/, Jeremy (1748–1832), English philosopher and jurist. The first major proponent of utilitarianism, as put forward in A Fragment on Government (1776) and more fully in his Introduction to the Principles of Morals and Legislation (1789), he argued that the proper object of all legislation and conduct was to secure 'the greatest happiness of the greatest number', and was concerned to reform the law by giving it a clear theoretical justification. With J. S. Mill, he co-founded the organ of the philosophical radicals, The Westminster Review, in 1824. Bentham exercised a decisive influence on 19th-century British thought, particularly in the field of political reform. □ **Benthamism** n. **Benthamite** n. & adj.

benthos /ˈbenθɒs/ n. Biol. the flora and fauna found at the bottom of a sea or lake (cf. NEKTON, PLANKTON). □ **benthic** adj. [Gk, = depth of the sea]

bentonite /ˈbentəˌnaɪt/ n. a kind of absorbent clay used esp. as a filler. [Fort Benton in Montana, US]

ben trovato /ˌben trəʊˈvɑːtəʊ/ adj. **1** well invented. **2** characteristic if not true. [It., = well found]

bentwood /ˈbentwʊd/ n. (usu. attrib.) wood that is artificially shaped for use in making furniture.

benumb /bɪˈnʌm/ v.tr. **1** make numb; deaden. **2** paralyse (the mind or feelings). [orig. = deprived, as past part. of ME benimen f. OE beniman (as BE-, niman take)]

Benxi /benˈʃiː/ a city in NE China, in the province of Liaoning; pop. (1990) 920,000.

Benz /benz, German bɛnts/, Karl Friedrich (1844–1929), German engineer and motor manufacturer. One of the pioneers of the motor car, he built the first vehicle to be driven by an internal-combustion engine in 1885, at Mannheim in Germany. The first of a series of racing cars was produced. Benz's company was merged with Daimler in 1926.

Benzedrine /ˈbenzɪˌdriːn/ n. propr. amphetamine. [BENZOIC + EPHEDRINE]

benzene /ˈbenziːn/ n. Chem. a colourless carcinogenic volatile liquid (chem. formula: C_6H_6) found in coal tar and petroleum, and used as a solvent and in making plastics etc. □ **benzene ring** a hexagonal unsaturated ring of six carbon atoms, found in benzene and many other organic compounds. □ **benzenoid** /ˈbenzɪˌnɔɪd/ adj. [BENZOIC + -ENE]

benzine /ˈbenziːn/ n. (also **benzin** /-zɪn/) a mixture of liquid hydrocarbons obtained from petroleum. [BENZOIN + -INE[4], -IN]

benzodiazepine /ˌbenzəʊdaɪˈeɪzəˌpiːn, -ˈæzəˌpiːn/ n. Chem. a heterocyclic organic compound of a kind including several tranquillizers such as Librium and Valium. [benzo- (see BENZOIC) + DI-[1] + AZO- + EPI- + -INE[4]]

benzoic /benˈzəʊɪk/ adj. Chem. containing or derived from benzoin or benzoic acid. □ **benzoic acid** a white crystalline substance (chem. formula: C_6H_5COOH) used as a food preservative. [BENZOIN + -IC]

benzoin /ˈbenzəʊɪn/ n. a fragrant gum resin obtained from various East Asian trees of the genus Styrax, and used in the manufacture of perfumes, incense, and friar's balsam. Also called gum benjamin. [earlier benjoin ult. f. Arab. lubān jāwī incense of Java]

benzol /ˈbenzɒl/ n. (also **benzole** /-zəʊl/) benzene, esp. unrefined and used as a fuel.

benzoquinone /ˌbenzəʊˈkwɪnəʊn, -kwɪˈnəʊn/ n. Chem. a yellow crystalline compound (chem. formula: $C_6H_4O_2$), related to benzene but having the hydrogen atoms on a pair of opposite carbon atoms replaced by oxygen.

benzoyl /ˈbenzəʊɪl/ n. (usu. attrib.) Chem. the radical C_6H_5CO-.

benzyl /ˈbenzaɪl, -zɪl/ n. (usu. attrib.) Chem. the radical $C_6H_5CH_2-$.

Beograd see BELGRADE.

Beowulf /ˈbeɪəˌwʊlf/ a legendary Scandinavian hero celebrated in the Old English epic poem Beowulf, the first major poem in a European vernacular language and the only complete Germanic epic that survives. The poem describes Beowulf's killing of the water-monster Grendel and its mother and his death in combat with a dragon, and includes both pagan and Christian elements. The historical events referred to in the poem belong to the first part of the 6th century, but the poem itself is generally dated to the 8th century.

bequeath /bɪˈkwiːð/ v.tr. **1** leave (a personal estate) to a person by a will. **2** hand down to posterity. □ **bequeathal** n. **bequeather** n. [OE *becwethan* (as BE-, *cwethan* say: cf. QUOTH)]

bequest /bɪˈkwest/ n. **1** the act or an instance of bequeathing. **2** a thing bequeathed. [ME f. BE- + obs. *quiste* f. OE *-cwiss*, *cwide* saying]

berate /bɪˈreɪt/ v.tr. scold, rebuke. [BE- + RATE²]

Berber /ˈbɜːbə(r)/ n. a member of any of the indigenous, mainly Muslim, Caucasian peoples of North Africa (now mainly in Morocco and Algeria), speaking related languages. Among the Berber peoples are the nomadic Tuareg, although the majority of Berbers are settled farmers or (now) migrant workers. The languages of the Berbers belong to the Afro-Asiatic family and are spoken by about 11 million people. [Arab. *Barbar* BARBARY]

Berbera /ˈbɜːbərə/ a port on the north coast of Somalia; pop. (est. 1987) 65,000.

berberis /ˈbɜːbərɪs/ n. = BARBERRY. [med.L & OF, of unkn. orig.]

berceuse /beəˈsɜːz/ n. (pl. **berceuses** pronunc. same) **1** a lullaby. **2** an instrumental piece in the style of a lullaby. [F]

Berchtesgaden /ˈbeəxtəsˌɡɑːd(ə)n/ a town in southern Germany, in the Bavarian Alps close to the border with Austria; pop. (1983) 8,186. Hitler had an alpine retreat there.

bereave /bɪˈriːv/ v.tr. (foll. by *of*) deprive of a relation, friend, etc., esp. by death. □ **bereaved** adj. **bereavement** n. [OE *berēafian* (as BE-, REAVE)]

bereft /bɪˈreft/ adj. (foll. by *of*) deprived (esp. of a non-material asset) (*bereft of hope*). [past part. of BEREAVE]

Berenice /ˌberɪˈnaɪsɪ/ (3rd century BC), Egyptian queen, wife of Ptolemy III. She dedicated her hair as a votive offering for the safe return of her husband from an expedition. The hair was stolen and (according to legend) placed in the heavens. She is commemorated in the name of the constellation Coma Berenices (*Berenice's hair*).

beret /ˈbereɪ/ n. a round flattish cap of felt or cloth. [F *béret* Basque cap f. Prov. *berret*]

Berg /beəɡ/, Alban (Maria Johannes) (1885–1935), Austrian composer. A pupil of Schoenberg, he was one of the leading exponents of twelve-note composition. He is best known for the Violin Concerto (1935), composed as a memorial after the death of the eighteen-year-old daughter of Alma and Walter Gropius, and for his two operas, *Wozzeck* (1914–21) and *Lulu* (1928–35).

berg¹ /bɜːɡ/ n. = ICEBERG. [abbr.]

berg² /bɜːɡ/ n. S. Afr. a mountain or hill. □ **berg wind** a hot dry northerly wind blowing from the interior to coastal districts. [Afrik. f. Du.]

bergamot¹ /ˈbɜːɡəˌmɒt/ n. **1** an aromatic labiate herb, esp. *Mentha citrata*. **2** an oily perfume extracted from the rind of an inedible citrus fruit that is a dwarf variety of the Seville orange. **3** the tree that produces this. [*Bergamo* in N. Italy]

bergamot² /ˈbɜːɡəˌmɒt/ n. a variety of fine pear. [F *bergamotte* f. It. *bergamotta* f. Turk. *begarmudu* prince's pear f. *beg* prince + *armud* pear + *-u* possess. suffix]

Bergen 1 /ˈbeəɡən/ a seaport in SW Norway; pop. (1991) 213,344. It is a centre of the fishing and North Sea oil industries. **2** /ˈbɛrxə/ the Flemish name for MONS.

Berger /ˈbɜːɡə(r)/, Hans (1873–1941), German psychiatrist. He attempted to correlate mental activity with brain physiology, detecting electric currents in the exposed cortex in 1924. Finding that these could also be detected through the intact skull, Berger went on to develop encephalography, which has since been used extensively to diagnose neurological conditions.

Bergerac¹ /ˈbeəʒəˌræk/ **1** a wine-producing region in the Dordogne valley in SW France. **2** a town on the Dordogne river; pop. (1990) 27,890.

Bergerac² see CYRANO DE BERGERAC.

Bergius /ˈbɜːɡɪəs/, Friedrich Karl Rudolf (1884–1949), German industrial chemist. He is best known for his process for producing petroleum and other hydrocarbons from coal dust, using hydrogen and a catalyst under high pressure. He also made a type of coal by carbonizing peat, achieved the complete hydrolysis of cellulose, and developed industrial processes for synthesizing phenol and ethylene glycol. Bergius shared the Nobel Prize for chemistry in 1931.

Bergman¹ /ˈbɜːɡmən/, (Ernst) Ingmar (b.1918), Swedish film and theatre director. His work is characterized by his use of haunting imagery and a symbolism often derived from Jungian dream analysis. He came to international fame with the film *Smiles of a Summer Night* (1955) and achieved further worldwide success with *The Seventh Seal* (1956) and *Wild Strawberries* (1957). An important theatre director, he has directed many of his players in both media.

Bergman² /ˈbɜːɡmən/, Ingrid (1915–82), Swedish actress. She made her name on stage and screen in Sweden before embarking on an international career in Hollywood in the 1930s. Although her film career was a long one, it is probably for her romantic role opposite Humphrey Bogart in *Casablanca* (1942) that she is best known. Other notable films include *For Whom the Bell Tolls* (1943) and *Anastasia* (1956); she received an Oscar for the latter, as well as for her role in *Murder on the Orient Express* (1974).

bergschrund /ˈbeəkʃrʊnt/ n. a crevasse or gap at the head of a glacier or névé. [G]

Bergson /ˈbɜːɡs(ə)n, French bɛrksɔn/, Henri (Louis) (1859–1941), French philosopher. His philosophy is dualistic, dividing the world into life (or consciousness) and matter. In his most famous work, *Creative Evolution* (1907), he attacked scientific materialism and rejected the Darwinian theory of evolution. He proposed instead that life possesses an inherent creative impulse (*élan vital*), the continuous operation of which as it seeks to impose itself upon matter leads to the production of new forms. Bergson's work influenced writers such as Marcel Proust and George Bernard Shaw; he was awarded the Nobel Prize for literature in 1927.

Beria /ˈberɪə/, Lavrenti (Pavlovich) (1899–1953), Soviet politician and head of the secret police (NKVD and MVD) 1938–53. Born in Georgia, he rose to prominence within the Soviet Communist Party under Stalin's patronage. As head of the secret police Beria was directly involved in the infamous 'purge trials' in which Stalin's opponents were eliminated; he was also responsible for the deportation of thousands to forced labour camps. After Stalin's death he was rumoured to be planning to seize power; feared by rival politicians, he was arrested. Although his fate is not certain, it was officially announced that he had been tried and shot as a traitor.

beriberi /ˌberɪˈberɪ/ n. a disease caused by a deficiency of vitamin B₁, characterized by inflammation of the nerves and often cardiac disorder. [Sinh., f. *beri* weakness]

Bering /ˈbeərɪŋ/, Vitus (Jonassen) (1681–1741), Danish navigator and explorer. At the instigation of Peter the Great he led several Russian expeditions aimed at discovering whether Asia and North America were connected by land. He sailed along the coast of Siberia and in 1741 reached Alaska from the east. On the return journey his ship was wrecked and he died on an island which now bears his name. Also named after him are the Bering Sea and Bering Strait.

Bering Sea an arm of the North Pacific lying between NE Siberia and Alaska, bounded to the south by the Aleutian Islands. It is linked to the Arctic Ocean by the Bering Strait. Both the sea and the strait are named after Vitus Bering.

Bering Strait a narrow sea passage, which separates the eastern tip of Siberia from Alaska, and links the Arctic Ocean with the Bering Sea. At its narrowest point it is about 85 km (53 miles) wide. During the Ice Age, as a result of a drop in sea levels, the area formed a bridge of land between the two continents, allowing migration of plants and animals.

Berio /ˈberɪəʊ/, Luciano (b.1925), Italian composer. A serialist, he has often adopted an experimental approach to groupings of instruments and singers, the use of electronic sound, and the combination of live and pre-recorded music. His works include *Circles* (1960), for singer, harp, and percussion, a series of *Sequences* (1958–75) for virtuoso solo instruments, and the opera *Un Re in Ascolto* (1984).

berk /bɜːk/ n. (also **burk**) Brit. sl. a fool; a stupid person. ¶ Usu. not considered *offens*. despite the etymology. [abbr. of *Berkeley* or *Berkshire Hunt*, rhyming sl. for *cunt*]

Berkeley¹ /ˈbɜːklɪ/ a city in western California, on San Francisco Bay, site of a campus of the University of California; pop. (1990) 102,724.

Berkeley² /ˈbɜːklɪ/, Busby (born William Berkeley Enos) (1895–1976), American choreographer and film director. As a leading Broadway dance director, he was introduced to films by Samuel Goldwyn, for whom he choreographed *Whoopee* (1930). In films such as the *Gold Diggers* series (1922–37) and *Babes in Arms* (1939), he quickly became famous for his spectacular and dazzling sequences in which huge casts of rhythmically moving dancers formed kaleidoscopic patterns on the screen.

Berkeley³ /ˈbɑːklɪ/, George (1685–1753), Irish philosopher and bishop. His idealist philosophy is set out in his major works *A Treatise Concerning* the *Principles of Human Knowledge* (1710) and *Three Dialogues between Hylas and Philonous* (1713). He denied the existence of matter, holding that there are only minds and mental events; material objects exist solely by being perceived. To the objection that objects would leap in and out of existence according to whether they were being looked at, he replied that God perceives everything, and that this gives objects — ideas in the mind of God — a continuous existence. He held this to be a sound argument for the existence of God.

berkelium /bɜːˈkiːlɪəm, ˈbɜːklɪəm/ *n.* an artificial radioactive metallic chemical element (atomic number 97; symbol **Bk**). A member of the actinide series, berkelium was first produced by Glenn Seaborg and his colleagues in 1949 by bombarding americium with helium ions. [f. BERKELEY[1] (where first made)]

Berks. *abbr.* Berkshire.

Berkshire /ˈbɑːkʃɪə(r)/ a county of southern England, west of London; county town, Reading.

Berlin[1] /bɜːˈlɪn/ the capital of Germany; pop. (est. 1990) 3,102,500. Founded in the 13th century, it was a seat of the Hohenzollern princely family and capital of Prussia. At the end of the Second World War the city was occupied by the Allies, and divided into two parts: *West Berlin*, comprising the American, British, and French zones, later a state of the Federal Republic of Germany, forming an enclave within the German Democratic Republic; and *East Berlin*, the zone of the city occupied by the USSR and later capital of the German Democratic Republic. Between 1961 and 1989 the Berlin Wall separated the two sectors, which were reunited in 1990; occupation formally ended in 1994.

Berlin[2] /bɜːˈlɪn/, Irving (born Israel Baline) (1888–1989), Russian-born American songwriter. He had no formal musical training, but began writing songs when he was 16; in 1911 he had a hit with 'Alexander's Ragtime Band'. Thereafter he contributed to many musical shows, revues, and films, including *Annie Get Your Gun* (1946) and *Holiday Inn* (1942); the latter contained the song 'White Christmas', sung by Bing Crosby, which proved to be one of the best-selling records of all time. Berlin also wrote 'God Bless America' (1939), which became the unofficial national anthem of the US.

Berlin airlift an operation by British and American aircraft to airlift food and supplies to Berlin in 1948–9. The city was being blockaded by Russian forces attempting to isolate it from the West and terminate the joint Allied military government of the city. After the blockade was lifted the city was formally divided into East and West Berlin.

Berliner /bɜːˈlɪnə(r)/ *n.* **1** a native or inhabitant of Berlin. **2** a lightly fried yeast bun with jam filling and vanilla icing. [G]

Berlin Wall a fortified and heavily guarded wall built in 1961 by the Communist authorities on the boundary between East and West Berlin, chiefly to curb the flow of East Germans to the West. Regarded as a symbol of the division of Europe into the Communist countries of the East and the democracies of the West, the wall was opened in Nov. 1989 after the collapse of the Communist regime in East Germany and subsequently dismantled.

Berlioz /ˈbeəlɪˌəʊz/, (Louis-)Hector (1803–69), French composer. He was one of the most original composers of his time and a major exponent of 19th-century programme music. His *Symphonie fantastique* (1830) reflects his unhappy passion for Harriet Smithson, an Irish actress. His other major works include the five-act opera *Les Troyens* (1856–9), and the cantata *La Damnation de Faust* (1846).

berm /bɜːm/ *n.* **1** a narrow path or grass strip beside a road, canal, etc. **2** a narrow ledge, esp. in a fortification between a ditch and the base of a parapet. [F *berme* f. Du. *berm*]

Bermuda /bɜːˈmjuːdə/ (also **the Bermudas**) a country consisting of about 150 small islands off the coast of North Carolina; pop. (est. 1991) 58,000; official language, English; capital, Hamilton. The islands (now a British dependency with full internal self-government) were sighted early in the 16th century by a Spaniard, Juan Bermúdez, from whom they take their name, but remained uninhabited until 1609, when an English expedition to Virginia was shipwrecked there; its leader, Sir George Somers (1554–1610), later returned from Virginia to claim the islands for Britain. □ **Bermudan** *adj. & n.* **Bermudian** *adj. & n.*

Bermuda rig *n.* a yachting rig with a high tapering mainsail.

Bermuda shorts *n.pl.* (also **Bermudas**) close-fitting shorts reaching the knees.

Bermuda Triangle an area of the western Atlantic between Bermuda and Florida, credited since the mid-19th century with a number of unexplained disappearances of ships and aircraft.

Bern see BERNE.

Bernadette, St /ˌbɜːnəˈdet/ (born Marie Bernarde Soubirous) (1844–79), French peasant girl. Her visions of the Virgin Mary in Lourdes in 1858 led to the town's establishment as a centre of pilgrimage. Bernadette later became a nun and she was canonized in 1933. Feast day, 18 Feb.

Bernadotte[1] /ˌbɜːnəˈdɒt/, Folke, Count (1895–1948), Swedish statesman. A member of the Swedish royal family, Bernadotte gained a reputation as a neutral arbiter in international disputes. As vice-president of the Swedish Red Cross during the Second World War, he arranged the exchange of prisoners of war, and in 1945 acted as an intermediary between Himmler and the Allies, conveying a German offer of capitulation. Charged with the role of UN mediator in Palestine in 1948, he was assassinated by the Stern Gang.

Bernadotte[2] /ˌbɜːnəˈdɒt/, Jean Baptiste Jules (1763–1844), French soldier. One of Napoleon's marshals, he was adopted by Charles XIII of Sweden in 1810 and became king (as Charles XIV) in 1818, thus founding Sweden's present royal house.

Bernard /beəˈnɑː(r)/, Claude (1813–78), French physiologist. Bernard used animal experiments to show the role of the pancreas in digestion, the method of regulation of body temperature, and the function of nerves supplying the internal organs. He realized that the constant composition of the body fluids was essential for the optimal functioning of the body, discovered the biological importance of glycogen, and investigated the action of curare.

Bernard, St /ˈbɜːnəd/ (*c.*996–*c.*1081), French monk. He founded two hospices to aid travellers in the Alps. The St Bernard passes, where the hospices were situated, and St Bernard dogs, once kept by the monks and trained to aid travellers, are named after him. Feast day, 28 May.

Bernard of Clairvaux, St /ˈbɜːnəd, kleəˈvəʊ/ (1090–1153), French theologian and abbot. He was the first abbot of Clairvaux in France; his monastery there became one of the chief centres of the Cistercian order. Enjoying papal favour, he was an important religious force in Europe, and the Cistercian order grew rapidly under his influence. He was noted for his asceticism, severity, and eloquence; his preaching at the council of Vézelay in 1146 instigated the Second Crusade; he had the French theologian Peter Abelard condemned for heresy. Feast day, 20 Aug.

Berne /bɜːn/ (also **Bern**) **1** a canton of Switzerland. **2** its capital, the capital of Switzerland; pop. (1990) 134,620. Founded in the late 12th century, it entered the Swiss Confederation in 1353. It was made the capital of Switzerland in 1848. □ **Bernese** *adj. & n.*

Berne Convention an international copyright agreement of 1886, later revised. The US has never been party to it.

Bernhardt /ˈbɜːnhɑːt/, Sarah (born Henriette Rosine Bernard) (1844–1923), French actress. Internationally acclaimed and regarded as the greatest tragic actress of her day, she gained her first major successes in 1872, playing Cordelia in *King Lear* and the queen in Victor Hugo's *Ruy Blas*. She was noted for her clear voice, magnetic personality, and great beauty; of all her performances, she is probably best known for her portrayal of Marguerite in *La Dame aux camélias* by Alexandre Dumas *fils*. The amputation of a leg in 1915 after an accident did not diminish her activity and she continued to act in seated roles.

Bernini /beəˈniːnɪ/, Gian Lorenzo (1598–1680), Italian sculptor, painter, and architect. An outstanding figure of the Italian baroque, Bernini is notable for the vigour, movement, and dramatic and emotional power of his works. Using a variety of materials, including stucco, stone, and marble, he fused sculpture, architecture, and painting into a decorative whole. Working chiefly in Rome, he became architect to St Peter's in 1629, for which his work included the great canopy over the high altar and the colonnade round the piazza in front of the church. One of his most famous sculptures is *The Vision of St Teresa* (1644–67) in the church of Santa Maria della Vittoria in Rome.

Bernoulli /bɜːˈnuːɪ/ a Swiss family that produced many eminent mathematicians and scientists. Jakob (Jacques or James) Bernoulli (1654–1705) made discoveries in calculus, which he used to solve minimization problems, and he contributed to geometry and the theory of probabilities. His brother Johann (Jean or John, 1667–1748) also contributed to differential and integral calculus. Both were professors of mathematics at Basle. Daniel Bernoulli (1700–82), son of Johann, was professor of mathematics at St Petersburg and then held successively the chairs of botany, physiology, and physics at Basle. Although his original studies were in medicine, his greatest contributions were to hydrodynamics and mathematical physics.

Bernstein /ˈbɜːnstiːn, -staɪn/, Leonard (1918–90), American composer, conductor, and pianist. As a conductor, he worked with the New York

Philharmonic Orchestra (1945–8 and 1957–69) and toured extensively. As a composer, he encompassed a wide range of forms and styles in his music; his large instrumental and choral works often juxtapose a romantic intensity with jazz and Latin American elements. His best-known compositions include the symphony *The Age of Anxiety* (1947–9), the musical *West Side Story* (1957), and film music such as that for *On the Waterfront* (1954).

Berra /'berə/, Yogi (born Lawrence Peter Berra) (b.1925), American baseball player. He was especially famous as a catcher with the New York Yankees in the 1950s and early 1960s, setting the record for the most home runs (313) by a catcher in the American League. He became known in the US for his pithy sayings such as (on baseball) 'You can't think and hit at the same time'. The cartoon character Yogi Bear (1958) is popularly believed to have been named after Berra, though the animators deny that this was their intention.

Berry[1] /'berɪ/ a former province of central France; chief town, Bourges.

Berry[2] /'berɪ/, Chuck (born Charles Edward Berry) (b.1931), American rock and roll singer, guitarist, and songwriter. One of the first great rock and roll stars with a large teenage following, he first had a hit with 'Maybellene' (1955); this was followed by 'Johnny B Goode' and 'Sweet Little Sixteen' (both 1958). His recording career was interrupted by a period of imprisonment (1962–4); although he continued to release albums throughout the 1970s and 1980s, his only major hit single during that time was 'My Ding A Ling' (1972). His music had a significant influence on the development of rock music; British pop and rock groups such as the Beatles and the Rolling Stones are particularly indebted to him.

berry /'berɪ/ n. & v. ● n. (pl. **-ies**) **1** a small roundish juicy fruit without a stone. **2** Bot. a fruit with its seeds enclosed in a pulp (e.g. a banana, tomato, etc.). **3** a kernel or seed (e.g. coffee bean etc.). **4** a fish egg or roe of a lobster etc. ● v.intr. (**-ies**, **-ied**) **1** (usu. as **berrying** n.) go gathering berries. **2** form a berry; bear berries. □ **berried** adj. (also in comb.). [OE *berie* f. Gmc]

berserk /bə'sɜːk, -'zɜːk/ adj. & n. ● adj. (esp. in **go berserk**) wild, frenzied; in a violent rage. ● n. (also **berserker** /-kə(r)/) an ancient Norse warrior who fought with a wild frenzy. [Icel. *berserkr* (n.) prob. f. *bern-* BEAR[2] + *serkr* coat]

berth /bɜːθ/ n. & v. ● n. **1** a fixed bunk on a ship, train, etc., for sleeping in. **2** a ship's place at a wharf. **3** room for a ship to swing at anchor. **4** adequate sea room. **5** colloq. a situation or appointment. **6** the proper place for anything. ● v. **1** tr. moor (a ship) in its berth. **2** tr. provide a sleeping place for. **3** intr. (of a ship) come to its mooring-place. □ **give a wide berth** stay away from. [prob. f. naut. use of BEAR[1] + -TH[2]]

bertha /'bɜːθə/ n. **1** a deep falling collar often of lace. **2** a small cape on a dress. [F *berthe* f. *Berthe* Bertha (the name)]

Bertillon /'beəti:ˌjɒn/, Alphonse (1853–1914), French criminologist. He devised a system of body-measurements for the identification of criminals, which was widely used in France and other countries until superseded by the technique of finger-printing at the beginning of the 20th century.

Bertolucci /ˌbeətə'luːtʃɪ/, Bernardo (b.1940), Italian film director. He made his début as a director in 1962. Critical acclaim came in 1964 with *Before the Revolution* and later with *The Spider's Stratagem* (1970), but it was with the box-office success of the sexually explicit *Last Tango in Paris* (1972) that he first gained a wide audience. His film *The Last Emperor* (1988), which dealt with the fall of the imperial dynasty in China, won nine Oscars.

Berwickshire /'berɪkˌʃɪə(r)/ a former county of SE Scotland, on the border with England. It became a part of Borders region in 1975.

Berwick-upon-Tweed /ˌberɪkəpɒn'twiːd/ a town at the mouth of the River Tweed in NE England, close to the Scottish border; pop. (1981) 13,000. It was held alternately by England and Scotland for centuries, until it was ceded to England in 1482.

beryl /'berɪl/ n. **1** a kind of transparent precious stone, esp. pale green, blue, or yellow, and consisting of beryllium aluminium silicate in a hexagonal form. **2** a mineral species which includes this, emerald, and aquamarine. [ME f. OF f. L *beryllus* f. Gk *bērullos*]

beryllium /bə'rɪlɪəm/ n. a hard grey metallic chemical element (atomic number 4; symbol **Be**). Beryllium was identified by the French chemist N. L. Vauquelin (1763–1829) in 1798. Its chief source is the mineral beryl. It is the lightest of the alkaline earth metals, and is used in the manufacture of light corrosion-resistant alloys and in windows in X-ray equipment. [BERYL]

Berzelius /bɜː'ziːlɪəs/, Jöns Jakob (1779–1848), Swedish analytical chemist. Berzelius studied about 2,000 compounds and by 1818 had determined the atomic weights of most of the then known elements. He discovered three new elements (cerium, selenium, and thorium), suggested the basic principles of modern chemical notation, and introduced the terms *isomerism, polymer, protein,* and *catalysis.*

Bes /bes/ Egyptian Mythol. a grotesque god depicted as having short legs, an obese body, and an almost bestial face, whose comic but frightening aspect had the effect of dispelling evil spirits.

Besançon /'bezæ̃ˌsɒn/ the capital of Franche-Comté in NE France; pop. (1990) 119,200.

Besant /'bez(ə)nt, bɪ'zænt/, Annie (1847–1933), English theosophist, writer, and politician. An atheist, socialist, and advocate of birth control, she became an active member of the Fabian Society in 1885. In 1889, after meeting Helena Blavatsky, she converted to theosophy. Besant was the president of the Theosophical Society from 1907 until her death, and settled in the society's headquarters in Madras in India, where she became involved in politics and pressed for Indian self-government. She established and directed the Home Rule India League (1916) and served as president of the Indian National Congress (1917–23).

beseech /bɪ'siːtʃ/ v.tr. (past and past part. **besought** /-'sɔːt/ or **beseeched**) **1** (foll. by for, or to + infin.) entreat. **2** ask earnestly for. □ **beseeching** adj. [ME f. BE- + *secan* SEEK]

beset /bɪ'set/ v.tr. (**besetting**; past and past part. **beset**) **1** attack or harass persistently (beset by worries). **2** surround or hem in (a person etc.). **3** archaic cover round with (beset with pearls). □ **besetting sin** the sin that especially or most frequently tempts one. [OE *besettan* f. Gmc]

beside /bɪ'saɪd/ prep. **1** at the side of; near. **2** compared with. **3** irrelevant to (beside the point). □ **beside oneself** overcome with worry, anger, etc. [OE *be sidan* (as BY, SIDE)]

besides /bɪ'saɪdz/ prep. & adv. ● prep. in addition to; apart from. ● adv. also; as well; moreover.

besiege /bɪ'siːdʒ/ v.tr. **1** lay siege to. **2** crowd round oppressively. **3** harass with requests. □ **besieger** n. [ME f. *assiege* by substitution of BE-, f. OF *asegier* f. Rmc]

besmear /bɪ'smɪə(r)/ v.tr. **1** smear with a greasy or sticky substance. **2** sully (a reputation etc.). [OE *bismierwan* (as BE-, SMEAR)]

besmirch /bɪ'smɜːtʃ/ v.tr. **1** soil, discolour. **2** dishonour; sully the reputation or name of. [BE- + SMIRCH]

besom /'biːz(ə)m/ n. **1** a broom made of twigs tied round a stick. **2** esp. N.Engl. derog. or joc. a woman. [OE *besema*]

besotted /bɪ'sɒtɪd/ adj. **1** infatuated. **2** foolish, confused. **3** intoxicated, stupefied. [besot (v.) (as BE-, SOT)]

besought past and past part. of BESEECH.

bespangle /bɪ'spæŋg(ə)l/ v.tr. adorn with spangles.

bespatter /bɪ'spætə(r)/ v.tr. **1** spatter (an object) all over. **2** spatter (liquid etc.) about. **3** overwhelm with abuse etc.

bespeak /bɪ'spiːk/ v.tr. (past **bespoke**; past part. **bespoken** /-'spəʊkən/ or as adj. **bespoke**) **1** engage in advance. **2** order (goods). **3** suggest; be evidence of (his gift bespeaks a kind heart). **4** literary speak to. [OE *bisprecan* (as BE-, SPEAK)]

bespectacled /bɪ'spektək(ə)ld/ adj. wearing spectacles.

bespoke /bɪ'spəʊk/ past and past part. of BESPEAK. ● adj. **1** (of goods, esp. clothing) made to order. **2** (of a tradesman) making goods to order.

bespoken past part. of BESPEAK.

besprinkle /bɪ'sprɪŋk(ə)l/ v.tr. sprinkle or strew all over with liquid etc. [ME f. BE- + *sprengen* in the same sense]

Bessarabia /ˌbesə'reɪbɪə/ a region in eastern Europe between the Dniester and Prut rivers, from 1918 to 1940 part of Romania. The major part of the region now falls in Moldova, the remainder in Ukraine. □ **Bessarabian** adj. & n.

Bessel /'bes(ə)l/, Friedrich Wilhelm (1784–1846), German astronomer and mathematician. Self-taught in navigation and astronomy, he rose to become director of the new observatory in Königsberg. He determined the positions of some 75,000 stars, and was the first to obtain accurate measurements of stellar distances using the parallax resulting from the earth's changing position. Bessel worked intensively on the orbits of planets and binary stars, developing mathematical functions that are named after him. Following a study of the orbit of Uranus he predicted the existence of an eighth planet.

Bessemer /'besɪmə(r)/, Sir Henry (1813–98), English engineer and inventor. Bessemer is best known for the steel-making process that bears his name. At the time of the Crimean War in the 1850s, his proposals for the redesign of guns received little encouragement in Britain but a great deal from Napoleon III. The material available for gun construction was inadequate, however, so Bessemer then worked on a series of experiments and patents in the search for stronger material.

Bessemer process *n.* a steel-making process in which carbon, silicon, and other impurities are removed from molten pig-iron by oxidation in a blast of air. The process was invented by Henry Bessemer, after earlier work by the iron manufacturer William Kelly (1811–88) in the US. By 1860 Bessemer had evolved his converter, a large pivoting steel vessel lined with a refractory material. When the vessel is horizontal it is loaded with molten pig iron. With the axis vertical a blast of air is blown through the converter, the reaction with oxygen removing impurities as waste gases and slag. After removal of molten slag from the surface and the addition of manganese or other additives the converter is tilted to pour out the finished steel. It was the first successful method of making steel in quantity at low cost, so enabling it to be used on a large scale, but has been replaced by more modern techniques (see STEEL).

Best[1] /best/, Charles Herbert (1899–1978), American-born Canadian physiologist. His employer, Professor J. J. R. Macleod, placed him with F. G. Banting to assist in his research on pancreatic extracts. The research team announced the discovery of insulin in 1922 (see BANTING). Unlike the others, Best was not awarded a Nobel Prize in 1923, but Banting shared half the prize money with him. He succeeded Banting as director of the Banting and Best Department of Medical Research at the University of Toronto in 1941, a post which he retained until 1967.

Best[2] /best/, George (b.1946), Northern Irish footballer. He joined Manchester United soon after leaving school and later played for Northern Ireland. An attacking winger with exceptional ball control, he was the leading scorer in the First Division in the 1967–8 season; he won a European Cup winners' medal and was named European Footballer of the Year in 1968. Unable to overcome a succession of personal and other problems in the late 1960s, his career as a top footballer came to a premature end soon after.

best /best/ *adj., adv., n., & v.* ● *adj.* (*superl.* of GOOD) of the most excellent or outstanding or desirable kind (*my best work; the best solution; the best thing to do would be to confess*). ● *adv.* (*superl.* of WELL[1]). **1** in the best manner (*does it best*). **2** to the greatest degree (*like it best*). **3** most usefully (*is best ignored*). ● *n.* **1** that which is best (*the best is yet to come*). **2** the chief merit or advantage; the best aspect or side; a person's best performance, achievement, etc. (*brings out the best in him; gave their best to the task*). **3** (foll. by *of*) a winning majority of (a certain number of games etc. played) (*the best of five*). **4** = *Sunday best*. ● *v.tr. colloq.* defeat, outwit, outbid, etc. □ **all the best** an expression used to wish a person good fortune. **as best one can** (or **may**) as effectively as possible under the circumstances. **at best** on the most optimistic view. **at one's best** in peak condition etc. **at the best of times** even in the most favourable circumstances. **be for** (or **all for**) **the best** be desirable in the end. **best boy** *esp. US* the assistant to the chief electrician of a film crew. **best end of neck** the rib end of a neck of lamb etc. for cooking. **best man** the bridegroom's chief attendant at a wedding. **the best part of** most of. **best seller 1** a book or other item that has sold in large numbers. **2** the author of such a book. **do one's best** do all one can. **get the best of** defeat, outwit. **give a person the best** admit the superiority of that person. **had best** would find it wisest to. **make the best of** derive what limited advantage one can from (something unsatisfactory or unwelcome); put up with. **to the best of one's ability, knowledge,** etc. as far as one can do, know, etc. **with the best of them** as well as anyone. [OE *betest* (adj.), *bet(o)st* (adv.), f. Gmc]

bestial /ˈbestɪəl/ *adj.* **1** brutish, cruel, savage. **2** sexually depraved; lustful. **3** of or like a beast. □ **bestialize** *v.tr.* (also **-ise**). **bestially** *adv.* [ME f. OF f. LL *bestialis* f. *bestia* beast]

bestiality /ˌbestɪˈælɪtɪ/ *n.* (*pl.* **-ies**) **1** bestial behaviour or an instance of this. **2** sexual intercourse between a person and an animal. [F *bestialité* (as BESTIAL)]

bestiary /ˈbestɪərɪ/ *n.* (*pl.* **-ies**) a moralizing medieval treatise on real and imaginary beasts. [med.L *bestiarium* f. L *bestia* beast]

bestir /bɪˈstɜː(r)/ *v.refl.* (**bestirred**, **bestirring**) exert or rouse (oneself).

bestow /bɪˈstəʊ/ *v.tr.* **1** (foll. by *on, upon*) confer (a gift, right, etc.). **2** deposit. □ **bestowal** *n.* [ME f. BE- + OE *stow* a place]

bestrew /bɪˈstruː/ *v.tr.* (*past part.* **bestrewed** or **bestrewn** /-ˈstruːn/) **1** (foll. by *with*) cover or partly cover (a surface). **2** scatter (things) about. **3** lie scattered over. [OE *bestrēowian* (as BE-, STREW)]

bestride /bɪˈstraɪd/ *v.tr.* (*past* **bestrode** /-ˈstrəʊd/; *past part.* **bestridden** /-ˈstrɪd(ə)n/) **1** sit astride on. **2** stand astride over. [OE *bestrīdan*]

bet /bet/ *v. & n.* ● *v.* (**betting**; *past* and *past part.* **bet** or **betted**) **1** *intr.* (foll. by *on* or *against* with ref. to the outcome) risk a sum of money etc. against another's with ref. to the outcome of an unpredictable event (esp. the result of a race, game, etc., or the outcome in a game of chance). **2** *tr.* risk (an amount) on such an outcome or result (*bet £10 on a horse*). **3** *tr.* risk a sum of money against (a person). **4** *tr. colloq.* feel sure (*bet they've forgotten it*). ● *n.* **1** the act of betting (*make a bet*). **2** the money etc. staked (*put a bet on*). **3** *colloq.* an opinion, esp. a quickly formed or spontaneous one (*my bet is that he won't come*). **4** *colloq.* a choice or course of action (*she's our best bet*). □ **you bet** *colloq.* you may be sure. [16th c.: perh. a shortened form of ABET]

beta /ˈbiːtə/ *n.* **1** the second letter of the Greek alphabet (*B, β*). **2** a second-class mark given for a piece of work or in an examination. **3** (**Beta**) *Astron.* the second (usu. second-brightest) star in a constellation (foll. by Latin genitive: *Beta Virginis*). **4** the second member of a series. □ **beta-blocker** *Pharm.* a drug that prevents the stimulation of increased cardiac action, used to treat angina and reduce high blood pressure. **beta decay** *Physics* a form of radioactive decay in which a beta particle is emitted. **beta particle** *Physics* a fast-moving electron emitted in radioactive decay. **beta rhythm** (or **waves**) the normal electrical activity of the brain when conscious and alert, with a frequency around 18 to 25 hertz. **beta test** *n.* a test of machinery, software, etc., carried out before its release by an assessor not involved in its development. ● *v.tr.* subject (a product) to a beta test. [ME f. L f. Gk]

betake /bɪˈteɪk/ *v.refl.* (*past* **betook** /-ˈtʊk/; *past part.* **betaken** /-ˈteɪkən/) (foll. by *to*) go to (a place or person).

betatron /ˈbiːtətrɒn/ *n. Physics* an apparatus for accelerating electrons in a circular path by magnetic induction. [BETA + -TRON]

betel /ˈbiːt(ə)l/ *n.* the leaf of the Asian evergreen climbing plant *Piper betle*, chewed in the East with parings of the areca nut. □ **betel-nut** = *areca nut*. [Port. f. Malayalam *veṭṭila*]

Betelgeuse /ˈbiːt(ə)l,dʒɜːz/ *Astron.* the tenth brightest star in the sky, in the constellation of Orion. It is a red supergiant, and variations in its brightness are associated with pulsations in its outer envelope. [F, alt. of Arab. *yad al-jauzā'* hand of the giant (Orion)]

bête noire /bet ˈnwɑː(r)/ *n.* (*pl.* ***bêtes noires*** *pronunc.* same) a person or thing one particularly dislikes or fears. [F, = black beast]

bethink /bɪˈθɪŋk/ *v.refl.* (*past* and *past part.* **bethought** /-ˈθɔːt/) (foll. by *of, how,* or *that* + clause) *formal* **1** reflect; stop to think. **2** be reminded by reflection. [OE *bithencan* f. Gmc (as BE-, THINK)]

Bethlehem /ˈbeθlɪ,hem/ a small town 8 km (5 miles) south of Jerusalem, in the West Bank; pop. (est. 1980) 14,000. It was first mentioned in Egyptian records of the 14th century BC. The native city of King David and reputed birthplace of Jesus, it contains a church built by Constantine in 330 over the supposed site of Christ's birth. St Jerome lived and worked in Bethlehem from 385, and it became a monastic centre.

Bethune /bəˈθjuːn/, Henry Norman (1890–1939), Canadian surgeon. He invented or improved a number of surgical instruments, but became disillusioned with medicine in Canada following his experiences of the surgical treatment of tuberculosis in his country. Bethune joined the Communist Party in 1935 and served in the Spanish Civil War against the Fascists, organizing the first mobile blood-transfusion service. Finally, he joined the Chinese army in their war against Japan as a surgeon, becoming a hero in the People's Republic; he died from septicaemia while in China.

betide /bɪˈtaɪd/ *v. poet.* (only in infin. and 3rd sing. subjunctive) **1** *tr.* happen to (*woe betide him*). **2** *intr.* happen (*whate'er may betide*). [ME f. obs. *tide* befall f. OE *tīdan*]

betimes /bɪˈtaɪmz/ *adv. literary* early; in good time. [ME f. obs. *betime* (as BY, TIME)]

bêtise /beˈtiːz/ *n.* **1** a foolish or ill-timed remark or action. **2** a piece of folly. [F]

Betjeman /ˈbetʃəmən/, Sir John (1906–84), English poet. His poems, as seen in collections such as *New Bats in Old Belfries* (1945), are self-deprecating, witty, and gently satirical; using traditional verse forms, they capture the spirit of his age. He also published a verse autobiography, *Summoned by Bells*, in 1960. His collection of architectural essays *In Ghastly Good Taste* (1933) reflects his interest in the preservation of Victorian and Edwardian buildings; he did much to raise public awareness of the merits of such architecture. He was appointed Poet Laureate in 1972.

betoken /bɪˈtəʊkən/ v.tr. **1** be a sign of; indicate. **2** augur. [OE (as BE-, *tācnian* signify: see TOKEN)]

betony /ˈbetənɪ/ n. (pl. **-ies**) a purple-flowered labiate plant, *Stachys officinalis*, or a plant resembling it. [ME f. OF *betoine* f. L *betonica*]

betook *past of* BETAKE.

betray /bɪˈtreɪ/ v.tr. **1** place (a person, one's country, etc.) in the hands or power of an enemy. **2** be disloyal to (another person, a person's trust, etc.). **3** reveal involuntarily or treacherously; be evidence of (*his shaking hand betrayed his fear*). **4** lead astray or into error. □ **betrayal** n. **betrayer** n. [ME f. obs. tray, ult. f. L *tradere* hand over]

betroth /bɪˈtrəʊð/ v.tr. (usu. as **betrothed** adj.) bind with a promise to marry. □ **betrothal** n. [ME f. BE- + *trouthe*, *treuthe* TRUTH, later assim. to TROTH]

better[1] /ˈbetə(r)/ adj., adv., n., & v. ● adj. **1** (compar. of GOOD) of a more excellent or outstanding or desirable kind (*a better product; it would be better to go home*). **2** (compar. of WELL[1]) partly or fully recovered from illness (*feeling better*). ● adv. (compar. of WELL[1]). **1** in a better manner (*she sings better*). **2** to a greater degree (*like it better*). **3** more usefully or advantageously (*is better forgotten*). ● n. **1** that which is better (*the better of the two; had the better of me*). **2** (usu. in pl.; prec. by my etc.) one's superior in ability or rank (*take notice of your betters*). ● v. **1** tr. improve on; surpass (*I can better his offer*). **2** tr. make better; improve. **3** refl. improve one's position etc. **4** intr. become better; improve. □ **better feelings** one's conscience. **better half** colloq. one's wife or husband. **better off** in a better (esp. financial) position. **the better part of** most of. **for better or for worse** on terms accepting all results; whatever the outcome. **get the better of** defeat, outwit; win an advantage over. **go one better 1** outbid etc. by one. **2** (*go one better than*) outdo another person. **had better** would find it wiser to. [OE *betera* f. Gmc]

better[2] /ˈbetə(r)/ n. (also **bettor**) a person who bets.

betterment /ˈbetəmənt/ n. **1** making better; improvement. **2** Econ. enhanced value (of real property) arising from local improvements.

Betterton /ˈbetət(ə)n/, Thomas (1635–1710), English actor. A leading actor of the Restoration period, he played a variety of roles including the tragic parts of Hamlet and Macbeth, and comic roles such as Sir Toby Belch in *Twelfth Night*. He also adapted the plays of John Webster, Molière, and Beaumont and Fletcher for his own productions.

Betti /ˈbetɪ/, Ugo (1892–1953), Italian dramatist, poet, and short-story writer. He practised as a judge in Rome (1930–43); his writing did not gain wide recognition until just before his death, when *Crime on Goat Island* (1950) was produced in Paris. This and his other most widely produced plays, including *Corruption in the Palace of Justice* (1949), take the form of harrowing legal examinations that result in the exposure of the real motives, evil, and guilt that lie beneath the social surfaces of the characters.

betting /ˈbetɪŋ/ n. **1** gambling by risking money on an unpredictable outcome. **2** the odds offered in this. □ **betting-shop** Brit. a bookmaker's shop or office. **what's the betting?** colloq. it is likely or to be expected (*what's the betting he'll be late?*).

bettor var. of BETTER[2].

between /bɪˈtwiːn/ prep. & adv. ● prep. **1 a** at or to a point in the area or interval bounded by two or more other points in space, time, etc. (*broke down between London and Dover; we must meet between now and Friday*). **b** along the extent of such an area or interval (*there are five shops between here and the main road; works best between five and six; the numbers between 10 and 20*). **2** separating, physically or conceptually (*the distance between here and Leeds; the difference between right and wrong*). **3 a** by combining the resources of (*great potential between them; between us we could afford it*). **b** shared by; as the joint resources of (*£5 between them*). **c** by joint or reciprocal action (*an agreement between us; sorted it out between themselves*). ¶ Use in sense 3 with reference to more than two people or things is established and acceptable (e.g. *relations between Britain, France, and Germany*). **4** to and from (*runs between London and Sheffield*). **5** taking one and rejecting the other of (*decide between eating here and going out*). ● adv. (also **in between**) at a point or in the area bounded by two or more other points in space, time, sequence, etc. (*not fat or thin but in between*). □ **between ourselves** (or **you and me**) in confidence. **between times** (or **whiles**) in the intervals between other actions; occasionally. [OE *betwēonum* f. Gmc (as BY, TWO)]

betwixt /bɪˈtwɪkst/ prep. & adv. archaic between. □ **betwixt and between** colloq. neither one thing nor the other. [ME f. OE *betwēox* f. Gmc: cf. AGAINST]

Beuthen see BYTOM.

BeV abbr. = GeV. [billion electronvolts]

Bevan /ˈbev(ə)n/, Aneurin ('Nye') (1897–1960), British Labour politician. A brilliant though often abrasive orator, he was MP for Ebbw Vale 1929–60. His most notable contribution was the creation of the National Health Service (1948) during his time as Minister of Health 1945–51. He resigned from the government in protest against the introduction of health-service charges. The leader of the left wing of the Labour Party, he was defeated by Hugh Gaitskell in the contest for the party leadership in 1955.

bevatron /ˈbevəˌtrɒn/ n. Physics a synchrotron used to accelerate protons to energies in the billion electronvolt range. [BeV + -TRON]

bevel /ˈbev(ə)l/ n. & v. ● n. **1** a slope from the horizontal or vertical in carpentry and stonework; a sloping surface or edge. **2** (in full **bevel square**) a tool for marking angles in carpentry and stonework. ● v. (**bevelled, bevelling**; US **beveled, beveling**) **1** tr. reduce (a square edge) to a sloping edge. **2** intr. slope at an angle; slant. □ **bevel gear** a gear working another gear at an angle to it by means of bevel wheels. **bevel wheel** a toothed wheel whose working face is oblique to the axis. [OF bevel (unrecorded) f. baif f. baer gape]

beverage /ˈbevərɪdʒ/ n. formal a drink (*hot beverage; alcoholic beverage*). [ME f. OF be(u)vrage, ult. f. L bibere drink]

Beveridge /ˈbevərɪdʒ/, William Henry, 1st Baron (1879–1963), British economist and social reformer, born in India. As director of the London School of Economics (1919–37), Beveridge transformed it into an institution of international repute. His most notable achievement was as chairman of the committee which prepared the Beveridge Report (Report on Social Insurance and Allied Services, 1942); this recommended the establishment of a comprehensive scheme of social insurance and formed the basis of much subsequent social legislation establishing the welfare state in the UK.

Beverly Hills /ˈbevəlɪ/ a largely residential city in California, on the north-west side of the Los Angeles conurbation; pop. (1990) 31,970. It is famous as the home of many film stars.

Bevin /ˈbevɪn/, Ernest (1881–1951), British Labour statesman and trade unionist. He was one of the founders of the Transport and General Workers' Union, serving as its first General Secretary (1921–40), and was a leading organizer of the General Strike (1926). He later entered Parliament, serving as Minister of Labour in Churchill's war Cabinet. As Foreign Secretary (1945–51), he helped form the Organization for European Economic Cooperation (1948) and NATO (1949). Unable to find a solution to the problem of Palestine, he surrendered the British mandate to the United Nations in 1947.

bevvy /ˈbevɪ/ n. (also **bevy**) (pl. **-ies**) sl. (a) drink, esp. (of) beer or other alcoholic liquor. [shortened f. BEVERAGE]

bevy /ˈbevɪ/ n. (pl. **-ies**) **1** colloq. a company or group (orig. of women). **2** a flock of quails or larks. [15th c.: orig. unkn.]

bewail /bɪˈweɪl/ v.tr. **1** greatly regret or lament. **2** wail over; mourn for. □ **bewailer** n.

beware /bɪˈweə(r)/ v. (only in imper. or infin.) **1** intr. (often foll. by of, or that, lest, etc. + clause) be cautious, take heed (*beware of the dog; told us to beware; beware that you don't fall*). **2** tr. be cautious of (*beware the Ides of March*). [BE + WARE[3]]

Bewick /ˈbjuːɪk/, Thomas (1753–1828), English artist and wood engraver. His best works are the shrewdly observed and expressive animal studies which illustrate such books as *A History of British Birds* (1797, 1804).

Bewick's swan a Eurasian race of the tundra swan, *Cygnus columbianus*, breeding in Arctic Siberia and wintering in northern Europe and central Asia. [BEWICK]

bewilder /bɪˈwɪldə(r)/ v.tr. utterly perplex or confuse. □ **bewilderedly** adv. **bewildering** adj. **bewilderingly** adv. **bewilderment** n. [BE- + obs. wilder lose one's way]

bewitch /bɪˈwɪtʃ/ v.tr. **1** enchant; greatly delight. **2** cast a spell on. □ **bewitching** adj. **bewitchingly** adv. [ME f. BE- + OE wiccian enchant f. wicca WITCH]

bey /beɪ/ n. hist. (in the Ottoman Empire) the title of a governor of a province. [Turk.]

beyond /bɪˈjɒnd/ prep., adv., & n. ● prep. **1** at or to the further side of (*beyond the river*). **2** outside the scope, range, or understanding of (*beyond repair; beyond a joke; it is beyond me*). **3** more than. ● adv. **1** at or to the further side. **2** further on. ● n. (prec. by the) the unknown after death. □ **the back of beyond** see BACK. **beyond words** inexpressible. [OE beg(e)ondan (as BY, YON, YONDER)]

bezant /'bez(ə)nt, bɪ'zænt/ n. **1** hist. a gold or silver coin orig. minted at Byzantium. **2** Heraldry a gold roundel. [ME f. OF besanz -ant f. L Byzantius Byzantine]

bezel /'bez(ə)l/ n. **1** the sloped edge of a chisel. **2** the oblique faces of a cut gem. **3 a** a groove holding a watch-glass or gem. **b** a rim holding a glass etc. cover. [OF besel (unrecorded: cf. F béseau, bizeau) of unkn. orig.]

bezique /bɪ'ziːk/ n. **1** a card-game for two with a double pack of 64 cards, including the seven to ace in each suit. **2** a combination of the queen of spades and the jack of diamonds in this game. [F bésigue, perh. f. Pers. bāzígar juggler]

bezoar /'biːzɔː(r), 'bezəʊˌɑː(r)/ n. a small stone which may form in the stomachs of certain animals, esp. ruminants, and which was once used as an antidote for various ailments. [ult. f. Pers. pādzahr antidote, Arab. bāzahr]

b.f. abbr. **1** Brit. colloq. bloody fool. **2** brought forward. **3** Printing boldface.

Bhagavadgita /ˌbʌɡəvəd'ɡiːtə/ n. the most famous religious text of Hinduism, an independent poem that was incorporated into the Mahabharata. Composed between the 2nd century BC and the 2nd century AD, it is the earliest exposition of devotional religion (see BHAKTI). The poem, presented as a dialogue between the Kshatriya prince Arjuna and his divine charioteer Krishna, stresses the importance of doing one's duty and of faith in god. [Skr. bhagavadgītā, lit. 'Song of the Lord']

bhaji /'bɑːdʒɪ/ n. (also **bhajee, bhajji**) (pl. **bhajia** or **bhajis**) **1** an Indian dish of fried vegetables. **2** a small flat cake or ball of this fried in batter (onion bhaji; spinach bhaji). [Hindi bhājī fried vegetables]

bhakti /'bʌktɪ/ n. Hinduism devotional worship directed to one supreme deity, usually Vishnu (especially in his incarnations as Rama and Krishna) or Siva, by whose grace salvation may be attained by all regardless of sex, caste, or class. It is followed by the majority of Hindus today. [Skr.]

bhang /bæŋ/ n. the leaves and flower-tops of Indian hemp used as a narcotic. [Port. bangue, Pers. & Urdu bang, later assim. to Hindi bhāṅ, f. Skr. bhaṅgā]

bhangra /'bɑːŋɡrə, 'bæŋ-/ n. a kind of pop music that combines Punjabi folk traditions with Western popular music. [Punjabi bhāṅgrā, a traditional folk-dance]

bharal /'bʌrəl/ n. (also **burhel**) a Himalayan wild sheep, Pseudois nayaur, with blue-black coat and horns curved rearward. [Hindi]

Bharat /'bʌrət/ the Hindi name for INDIA.

Bharatpur /ˌbʌrət'pʊə(r)/ a sanctuary for migratory birds, near an 18th-century fort in Rajasthan in NW India.

Bhavnagar /bʌv'nʌɡə(r)/ an industrial port in NW India, in Gujarat, on the Gulf of Cambay; pop. (1991) 401,000. It was the capital of a former Rajput princely state of the same name.

Bhojpuri /bəʊdʒ'pʊərɪ/ n. one of the Bihari group of languages, spoken by some 20 million people in western Bihar and eastern Uttar Pradesh.

Bhopal /bəʊ'pɑːl/ a city in central India, the capital of the state of Madhya Pradesh; pop. (1991) 1,604,000. In Dec. 1984 leakage of poisonous gas from an American-owned pesticide factory caused the death of about 2,500 people and thousands of injuries.

b.h.p. abbr. brake horsepower.

Bhubaneswar /ˌbʊbə'neɪʃwə(r)/ a city in eastern India, capital of the state of Orissa; pop. (1991) 412,000.

Bhutan /buː'tɑːn/ a small independent kingdom on the south-eastern slopes of the Himalayas, a protectorate of the Republic of India; pop. (est. 1991) 1,467,000; languages, Dzongkha (official), Nepali; capital, Thimphu. □ **Bhutanese** /ˌbuːtə'niːz/ adj. & n.

Bhutto[1] /'buːtəʊ/, Benazir (b.1953), Pakistani stateswoman, Prime Minister 1988–90 and since 1993. An opponent of the existing regime, she became joint leader in exile of the Pakistan People's Party (1984), returning to Pakistan in 1986 to campaign for open democratic elections. Following President Zia ul-Haq's death, she became the first woman Prime Minister of a Muslim country. She took her country back into the Commonwealth and promised radical social reform, but failed to win widespread support from other parties. She was dismissed as Prime Minister and defeated in the ensuing election, but was re-elected as head of a coalition government in 1993.

Bhutto[2] /'buːtəʊ/, Zulfikar Ali (1928–79), Pakistani statesman, President 1971–3 and Prime Minister 1973–7. As Pakistan's Foreign Minister (1963–6) he instigated a rapprochement with China and became known as an outspoken defender of his country's interests.

He formed the Pakistan People's Party in 1967, coming to power as Pakistan's first civilian President in 1971 and later (after constitutional changes) serving as Prime Minister. While in office, he did much to strengthen national morale and introduced social, constitutional, and economic reforms. He was ousted by a military coup and executed for conspiring to murder a political rival.

Bi symb. Chem. the element bismuth.

bi /baɪ/ adj. & n. sl. = BISEXUAL. [abbr.]

bi- /baɪ/ comb. form (often **bin-** before a vowel) forming nouns and adjectives meaning: **1** having two; a thing having two (bilateral; binaural; biplane). **2 a** occurring twice in every one or once in every two (bi-weekly). **b** lasting for two (biennial). **3** doubly; in two ways (biconcave). **4** Chem. a substance having a double proportion of the acid etc. indicated by the simple word (bicarbonate). **5** Bot. & Zool. (of division and subdivision) twice over (bipinnate). [L]

Biafra /bɪ'æfrə/ a state proclaimed in 1967, when part of eastern Nigeria, inhabited chiefly by the Ibo people, sought independence from the rest of the country. In the ensuing civil war the new state's troops were overwhelmed by numerically superior forces, and by 1970 it had ceased to exist. □ **Biafran** adj. & n.

Białystok /biː'ælɪˌstɒk, Polish bja'wistɔk/ (Russian **Belostok** /ˌbjila'stɔk/) an industrial city in NE Poland, close to the border with Belarus; pop. (1990) 270,568.

biannual /baɪ'ænjʊəl/ adj. occurring, appearing, etc., twice a year (cf. BIENNIAL). □ **biannually** adv.

Biarritz /bɪə'rɪts/ a seaside resort in SW France, on the Bay of Biscay; pop. (1990) 28,890.

bias /'baɪəs/ n. & v. ● n. **1** (often foll. by towards, against) a predisposition or prejudice. **2** Statistics a systematic distortion of a statistical result due to a factor not allowed for in its derivation. **3** an edge cut obliquely across the weave of a fabric. **4** Bowls **a** the irregular shape given to a bowl. **b** the oblique course this causes it to run. **5** Electr. a steady voltage, magnetic field, etc., applied to an electronic system or device. ● v.tr. (**biased, biasing; biassed, biassing**) **1** (esp. as **biased** adj.) influence (usu. unfairly); prejudice. **2** give a bias to. □ **bias binding** a strip of fabric cut obliquely and used to bind edges. **on the bias** obliquely, diagonally. [F biais, of unkn. orig.]

biathlon /baɪ'æθlɒn/ n. Sport an athletic contest in skiing and shooting or in cycling and running. □ **biathlete** n. [BI-, after PENTATHLON]

biaxial /baɪ'æksɪəl/ adj. (esp. of crystals) having two axes along which polarized light travels with equal velocity.

bib[1] /bɪb/ n. **1** a piece of cloth or plastic fastened round a child's neck to keep the clothes clean while eating. **2** the top front part of an apron, dungarees, etc. **3** an edible marine fish, Trisopterus luscus, of the cod family. Also called pout. □ **best bib and tucker** best clothes. **stick** (or **poke** etc.) **one's bib in** Austral. sl. interfere. [perh. f. BIB[2]]

bib[2] /bɪb/ v.intr. (**bibbed, bibbing**) archaic drink much or often. □ **bibber** n. [ME, perh. f. L bibere drink]

bib-cock /'bɪbkɒk/ n. a tap with a bent nozzle fixed at the end of a pipe. [perh. f. BIB[1] + COCK[1]]

bibelot /'biːbləʊ/ n. a small curio or artistic trinket. [F]

Bible /'baɪb(ə)l/ n. **1 a** the Christian scriptures consisting of the Old and New Testaments, and sometimes the Apocrypha. (See note below.) **b** the Jewish scriptures (the books of the Old Testament). **c** (**bible**) any copy of these (three bibles on the table). **d** a particular edition of the Bible (New English Bible). **2** colloq. any authoritative book (Wisden is his Bible). **3** the scriptures of any other religion. □ **Bible-basher** (or **-thumper** etc.) sl. a person given to Bible-bashing. **Bible-bashing** (or **-thumping** etc.) sl. aggressive fundamentalist preaching. **Bible belt** esp. the southern and central area of the US, noted for its adherence to fundamentalist Christianity. **Bible oath** a solemn oath taken on the Bible. [ME f. OF f. eccl.L biblia f. Gk biblia books (pl. of biblion), orig. dimin. of biblos, bublos papyrus]

▪ The Old and New Testaments have traditionally been regarded by Christians as having a unique divine authority; in addition, some Churches, including the Roman Catholic Church, accept the Apocrypha. Although there was some allegorical interpretation, even after the inception of the modern scientific movement (16th and 17th centuries) both Protestants and Catholics persevered in a belief in the literal truth of the Bible's assertions on matters not only of history, doctrine, and ethics but also of cosmology and natural science. It was not until the 19th century that critical study began to examine the historical perspective of the Bible and the circumstances and purpose of its compilation; more recently, however, many fundamentalist groups, particularly in the US, having a belief in the literal truth of the Bible, have become prominent. Though the Bible had been translated into Old English, the first medieval English translation, instituted by John Wyclif at the end of the 14th century, was opposed by the Church, and it was not until the Reformation that the Bible was again available in the vernacular: Tyndale printed the New Testament in English in 1525–6 and Miles Coverdale's translation of the Bible appeared in 1535, Martin Luther's German translation of the New Testament having appeared in 1522. The most famous English translation is probably the Authorized Version or King James Bible of 1611. This formed the basis for the Revised Version, first published in 1881. There are now translations of all or part of the Bible in more than 1,750 languages.

biblical /ˈbɪblɪk(ə)l/ *adj.* **1** of, concerning, or contained in the Bible. **2** resembling the language of the Authorized Version of the Bible. □ **biblically** *adv.*

biblio- /ˈbɪblɪəʊ/ *comb. form* denoting a book or books. [Gk f. *biblion* book]

bibliography /ˌbɪblɪˈɒgrəfɪ/ *n.* (*pl.* **-ies**) **1 a** a list of the books referred to in a scholarly work, usu. printed as an appendix. **b** a list of the books of a specific author or publisher, or on a specific subject, etc. **2 a** the history or description of books, including authors, editions, etc. **b** any book containing such information. □ **bibliographer** *n.* **bibliographic** /ˌbɪblɪəˈgræfɪk/ *adj.* **bibliographical** *adj.* **bibliographically** *adv.* [F *bibliographie* f. mod.L *bibliographia* f. Gk (as BIBLE, -GRAPHY)]

bibliomancy /ˈbɪblɪəʊˌmænsɪ/ *n.* foretelling the future by the analysis of a randomly chosen passage from a book, esp. the Bible.

bibliomania /ˌbɪblɪəʊˈmeɪnɪə/ *n.* an extreme enthusiasm for collecting and possessing books. □ **bibliomaniac** /-ˈmeɪnɪˌæk/ *n. & adj.*

bibliophile /ˈbɪblɪəʊˌfaɪl/ *n.* (also **bibliophil** /-fɪl/) a person who collects or is fond of books. □ **bibliophilic** /ˌbɪblɪəʊˈfɪlɪk/ *adj.* **bibliophily** /ˌbɪblɪˈɒfɪlɪ/ *n.* [F *bibliophile* (as BIBLIO-, -PHILE)]

bibliopole /ˈbɪblɪəʊˌpəʊl/ *n.* a seller of (esp. rare) books. [L *bibliopola* f. Gk *bibliopōlēs* f. *biblion* book + *pōlēs* seller]

Bibliothèque nationale /ˌbɪblɪəˌtek ˌnæsjɒˈnæl/ the national library of France, in Paris, which receives a copy of every book and periodical etc. published in France. Having its origins in the libraries of medieval French kings, it was expanded greatly during the long reign of Louis XIV and first came to be called by its present name in the French Revolution.

bibulous /ˈbɪbjʊləs/ *adj.* given to drinking alcoholic liquor. □ **bibulously** *adv.* **bibulousness** *n.* [L *bibulus* freely drinking f. *bibere* drink]

bicameral /baɪˈkæmərəl/ *adj.* (of a parliament or legislative body) having two chambers. □ **bicameralism** *n.* [BI- + L *camera* chamber]

bicarb /ˈbaɪkɑːb, baɪˈkɑːb/ *n. colloq.* = BICARBONATE 2. [abbr.]

bicarbonate /baɪˈkɑːbənɪt/ *n.* **1** *Chem.* any acid salt of carbonic acid. **2** (in full **bicarbonate of soda**) sodium bicarbonate used as an antacid or in baking powder.

bice /baɪs/ *n.* **1** any of various pigments made from blue or green basic copper carbonate. **2** any similar pigment made from smalt. **3** a shade of blue or green given by these. □ **blue bice** a shade of blue between ultramarine and azure derived from smalt. **green bice** a yellowish-green colour derived by adding yellow orpiment to smalt. [orig. = brownish-grey, f. OF *bis* dark grey, of unkn. orig.]

bicentenary /ˌbaɪsenˈtiːnərɪ/ *n. & adj.* ● *n.* (*pl.* **-ies**) **1** a two-hundredth anniversary. **2** a celebration of this. ● *adj.* of or concerning a bicentenary.

bicentennial /ˌbaɪsenˈtenɪəl/ *n. & adj.* ● *n.* a bicentenary. ● *adj.* **1** lasting two hundred years or occurring every two hundred years. **2** of or concerning a bicentenary.

bicephalous /baɪˈsefələs, -ˈkefələs/ *adj.* having two heads.

biceps /ˈbaɪseps/ *n.* (*pl.* same) a muscle having two heads or attachments at one end, esp. the muscle which bends the elbow. [L, = two-headed, formed as BI- + -*ceps* f. *caput* head]

bicker /ˈbɪkə(r)/ *v.intr.* **1** quarrel pettily; wrangle. **2** *poet.* **a** (of a stream, rain, etc.) patter (over stones etc.). **b** (of a flame, light, etc.) flash, flicker. □ **bickerer** *n.* [ME *biker, beker*, of unkn. orig.]

bickie /ˈbɪkɪ/ *n.* (also **bikkie**) *colloq.* a biscuit. □ **big bickies** *Austral. colloq.* a large sum of money. [abbr.]

bicolour /ˈbaɪˌkʌlə(r)/ *adj. & n.* ● *adj.* having two colours. ● *n.* a bicolour blossom or animal.

biconcave /baɪˈkɒŋkeɪv/ *adj.* (esp. of a lens) concave on both sides.

biconvex /baɪˈkɒnveks/ *adj.* (esp. of a lens) convex on both sides.

bicultural /baɪˈkʌltʃərəl/ *adj.* having or combining two cultures.

bicuspid /baɪˈkʌspɪd/ *adj. & n.* ● *adj.* having two cusps or points. ● *n.* **1** a human premolar tooth. **2** a tooth with two cusps. □ **bicuspidate** /-ˌdeɪt/ *adj.* [BI- + L *cuspis -idis* sharp point]

bicycle /ˈbaɪsɪk(ə)l/ *n. & v.* ● *n.* a vehicle of two wheels held in a frame one behind the other, propelled by pedals and steered with handlebars attached to the front wheel. (*See note below.*) ● *v.intr.* ride a bicycle. □ **bicycle chain** a chain transmitting power from the bicycle pedals to the wheels. **bicycle clip** a metal clip used to confine a cyclist's trousers at the ankle. **bicycle pump** a portable pump for inflating bicycle tyres. □ **bicycler** *n.* **bicyclist** /-klɪst/ *n.* [F f. BI- + Gk *kuklos* wheel]

■ A two-wheeled conveyance (called a velocipede) was patented in Germany in 1817: the rider sat on a bar between the wheels, and progressed by pushing the ground with each foot alternately. A system of propulsion by treadles appeared *c.*1840, and pedals were added (to the front wheel) in 1861. The early bicycle was a cumbersome and uncomfortable machine, with a heavy frame and iron tyres; it was often known as a *boneshaker*. Springs, wire-spoke wheels, and solid rubber tyres were introduced in the 1860s, and the front wheel was made much larger (the 'ordinary' bicycle or *penny farthing*). In the late 1870s 'safety' bicycles with two medium-sized wheels and with the rider's seat set further back appeared, and the successful Rover safety bicycle of 1884–5, with smaller wheels and with a toothed gearwheel connected by an endless chain with the hub of the rear wheel, was essentially the modern machine. Pneumatic tyres were introduced by John Dunlop in 1888. The next significant development did not take place until 1962, when the cross-frame bicycle was introduced, having small wheels and with one large tube as the main horizontal member. More recently the sturdy, lightweight mountain bike, with rugged tyres and often more than twenty gears, has become popular, and new designs with carbon-fibre frames are being introduced.

bid /bɪd/ *v. & n.* ● *v.* (**bidding**; *past* **bid**, *archaic* **bade** /beɪd, bæd/; *past part.* **bid**, *archaic* **bidden** /ˈbɪd(ə)n/) **1** *tr. & intr.* (*past* and *past part.* **bid**) (often foll. by *for*, *against*) **a** (esp. at an auction) offer (a certain price) (*did not bid for the vase*; *bid against the dealer*; *bid £20*). **b** offer to do work etc. for a stated price. **2** *tr. archaic* or *literary* **a** command; order (*bid the soldiers shoot*). **b** invite (*bade her start*). **3** *tr. archaic* or *literary* **a** utter (greeting or farewell) to (*I bade him welcome*). **b** proclaim (defiance etc.). **4** (*past* and *past part.* **bid**) *Cards* **a** *intr.* state before play how many tricks one intends to make. **b** *tr.* state (one's intended number of tricks). ● *n.* **1 a** (esp. at an auction) an offer (of a price) (*a bid of £5*). **b** an offer (to do work, supply goods, etc.) at a stated price; a tender. **2** *Cards* a statement of the number of tricks a player proposes to make. **3** *colloq.* an attempt; an effort (*a bid for power*). □ **bid fair to** seem likely to. **make a bid for** try to gain (*made a bid for freedom*). □ **bidder** *n.* [OE *biddan* ask f. Gmc, & OE *bēodan* offer, command]

biddable /ˈbɪdəb(ə)l/ *adj.* **1** obedient. **2** *Cards* (of a hand or suit) suitable for being bid. □ **biddability** /ˌbɪdəˈbɪlɪtɪ/ *n.*

bidden *archaic past part.* of BID.

bidding /ˈbɪdɪŋ/ *n.* **1** the offers at an auction. **2** *Cards* the act of making a bid or bids. **3** a command, request, or invitation. □ **bidding-prayer** one inviting the congregation to join in.

biddy /ˈbɪdɪ/ *n.* (*pl.* **-ies**) *sl. derog.* a woman (esp. *old biddy*). [pet-form of the name *Bridget*]

bide /baɪd/ *v.* **1** *intr. archaic* or *dial.* remain; stay. **2** *tr.* wait for (now only in *bide one's time*, await one's best opportunity). [OE *bīdan* f. Gmc]

bidet /ˈbiːdeɪ/ *n.* a low oval basin used esp. for washing the genital area. [F, = pony]

bidirectional /ˌbaɪdaɪˈrekʃən(ə)l, ˌbaɪdɪ-/ *adj.* functioning in two directions.

Biedermeier /ˈbiːdəˌmaɪə(r)/ *attrib.adj.* **1** (of styles, furnishings, etc.) characteristic of the period 1815–48 in Germany. **2** *derog.* conventional; bourgeois. [*Biedermeier* a fictitious German poet (1854)]

Bielefeld /ˈbiːləˌfelt/ an industrial city in North Rhine-Westphalia in western Germany; pop. (1991) 322,130.

biennial /baɪˈenɪəl/ *adj. & n.* ● *adj.* **1** lasting two years. **2** recurring every two years (cf. BIANNUAL). ● *n.* **1** *Bot.* a plant that takes two years to grow from seed to fruition and die (cf. ANNUAL, PERENNIAL). **2** an event celebrated or taking place every two years. □ **biennially** *adv.* [L *biennis* (as BI-, *annus* year)]

biennium /baɪˈenɪəm/ *n.* (*pl.* **bienniums** or **biennia** /-nɪə/) a period of two years. [L (as BIENNIAL)]

bier /bɪə(r)/ *n.* a movable frame on which a coffin or a corpse is placed, or taken to a grave. [OE *bēr* f. Gmc]

Bierce /bɪəs/, Ambrose (Gwinnett) (1842–*c*.1914), American writer. He served in the American Civil War (1861–5) and later became a prominent journalist in California, London, and Washington. He is best known for his realistic and sardonic short stories, strongly influenced by Edgar Allan Poe and including *Cobwebs from an Empty Skull* (1874) and *In the Midst of Life* (1898), and for the wickedly witty *The Devil's Dictionary* (1911). In 1913 he travelled to Mexico and mysteriously disappeared.

biff /bɪf/ *n. & v. sl.* ● *n.* a sharp blow. ● *v.tr.* strike (a person). [imit.]

biffin /ˈbɪfɪn/ *n. Brit.* a deep-red cooking-apple. [= *beefing* f. BEEF + -ING[1], with ref. to the colour]

bifid /ˈbaɪfɪd/ *adj.* divided by a deep cleft into two parts. [L *bifidus* (as BI-, *fidus* f. stem of *findere* cleave)]

bifocal /baɪˈfəʊk(ə)l/ *adj. & n.* ● *adj.* having two focuses, esp. of a lens with a part for distant vision and a part for near vision. ● *n.* (in *pl.*) bifocal spectacles.

BIFU /ˈbɪfuː/ *abbr.* (in the UK) Banking, Insurance, and Finance Union.

bifurcate /ˈbaɪfəˌkeɪt/ *v. & adj.* ● *v.tr. & intr.* divide into two branches; fork. ● *adj.* forked; branched. [med.L *bifurcare* f. L *bifurcus* two-forked (as BI-, *furca* fork)]

bifurcation /ˌbaɪfəˈkeɪʃ(ə)n/ *n.* **1 a** a division into two branches. **b** either or both of such branches. **2** the point of such a division.

big /bɪg/ *adj. & adv.* ● *adj.* (**bigger**, **biggest**) **1 a** of considerable size, amount, intensity, etc. (*a big mistake; a big helping*). **b** of a large or the largest size (*big toe; big drum*). **c** (of a letter) capital, upper-case. **2 a** important; significant; outstanding (*the big race; my big chance*). **b** (of a person) famous, important, esp. in a named field. **3 a** grown up (*a big boy now*). **b** elder (*big sister*). **4** *colloq.* **a** boastful (*big words*). **b** often *iron.* generous (*big of him*). **c** ambitious (*big ideas*). **5** (usu. foll. by *with*) advanced in pregnancy; fecund (*big with child; big with consequences*). ● *adv. colloq.* in a big manner, esp.: **1** effectively (*went over big*). **2** boastfully (*talk big*). **3** ambitiously (*think big*). □ **big band** a large jazz or pop orchestra. **big bud** a plant disease caused by a gall-mite. **big bug** *sl.* = BIGWIG. **big business** large-scale financial dealings, esp. when sinister or exploitative. **big deal!** *sl. iron.* I am not impressed. **big dipper** a fairground switchback. **big end** (in a motor vehicle) the end of the connecting-rod that encircles the crankpin. **big game** large animals hunted for sport. **big gun** *sl.* = BIGWIG. **big-head** *colloq.* a conceited person. **big-headed** *colloq.* conceited. **big-headedness** *colloq.* conceitedness. **big-hearted** generous. **big house 1** the principal house in a village etc. **2** *sl.* a prison. **big idea** often *iron.* an important intention or scheme. **big money** large amounts of money; high profit; high pay. **big mouth** *colloq.* loquacity; talkativeness. **big-mouth** *colloq.* a boastful or talkative person; a gossip-monger. **big name** a famous person. **big noise** (or **pot** or **shot**) *colloq.* = BIGWIG. **big smoke** *Brit. sl.* **1** London. **2** any large town. **big stick** a display of force. **the big time** *sl.* success in a profession, esp. show business. **big-timer** *sl.* a person who achieves success. **big top** the main tent in a circus. **big tree** *US* the giant sequoia (see SEQUOIA). **big wheel 1** a Ferris wheel. **2** *N. Amer. sl.* = BIGWIG. **come** (or **go**) **over big** make a great effect. **in a big way 1** on a large scale. **2** *colloq.* with great enthusiasm, display, etc. **talk big** boast. **think big** be ambitious. **too big for one's boots** (or **breeches**) *sl.* conceited. □ **biggish** *adj.* **bigness** *n.* [ME: orig. unkn.]

bigamy /ˈbɪgəmɪ/ *n.* (*pl.* **-ies**) the crime of marrying when one is lawfully married to another person. □ **bigamist** *n.* **bigamous** *adj.* [ME f. OF *bigamie* f. *bigame* bigamous f. LL *bigamus* (as BI-, Gk *gamos* marriage)]

Big Apple, the *N. Amer. sl.* New York City.

big bang *n.* (also **Big Bang**) **1** *Astron.* the explosion of dense matter which, according to a current cosmological theory, marked the origin of the universe. In the beginning a fireball of radiation at extremely high temperature and density, but occupying a tiny volume, is believed to have formed. This expanded and cooled, extremely fast at first, but more slowly as subatomic particles condensed into matter which later accumulated to form galaxies and stars. The galaxies are currently still retreating from one another. What was left of the original radiation continued to cool and has been detected as a uniform background of weak microwave radiation. (See also STEADY STATE.) **2** *Stock Exch.* (in the UK) the introduction in 1986 of major changes in the regulations and procedures for trading, especially the widening of membership, the relaxation of rules for brokers, and the introduction of computerized communications. The period around the introduction of these changes was marked by frantic trading activity in the City, and share prices rose rapidly until crashing on 'Black Monday' in Oct. 1987.

Big Ben *n.* the great clock tower of the Houses of Parliament and its bell.

Big Bend National Park a US national park in a bend of the Rio Grande, in the desert lands of southern Texas on the border with Mexico, in which were discovered, in 1975, fossil remains of the pterosaur.

Big Board *n. US colloq.* the New York Stock Exchange.

Big Brother *n.* a ruthlessly all-powerful but supposedly benevolent dictator (as in Orwell's *Nineteen Eighty-four*).

Big Chief *n.* (also **Big Daddy**) *sl.* = BIGWIG.

Big Dipper *n. N. Amer.* the Plough (see PLOUGH *n.* 4).

biggie /ˈbɪgɪ/ *n. colloq.* a big person or thing; an important event.

bighorn /ˈbɪghɔːn/ *n.* a North American sheep, *Ovis canadensis*, with large curled horns, found chiefly in the Rocky Mountains.

bight /baɪt/ *n.* **1** a curve or recess in a coastline, river, etc. **2** a loop of rope. [OE *byht*, MLG *bucht* f. Gmc: see BOW[2]]

bigot /ˈbɪgət/ *n.* an obstinate and intolerant believer in a religion, political theory, etc. □ **bigotry** *n.* [16th c. f. F: orig. unkn.]

bigoted /ˈbɪgətɪd/ *adj.* unreasonably prejudiced and intolerant.

Big Three *n.* (also **Big Four** etc.) the predominant few.

bigwig /ˈbɪgwɪg/ *n. colloq.* an important person.

Bihar /bɪˈhɑː(r)/ a state in NE India; capital, Patna.

Bihari /bɪˈhɑːrɪ/ *n.* **1** a native of Bihar. **2** a group of three closely related languages, descended from Sanskrit, spoken principally in Bihar. The three languages are Bhojpuri, Maithili, and Magahi.

bijou /ˈbiːʒuː/ *n. & adj.* ● *n.* (*pl.* **bijoux** *pronunc.* same) a jewel; a trinket. ● *attrib.adj.* small and elegant. [F]

bijouterie /biːˈʒuːtərɪ/ *n.* jewellery; trinkets. [F (as BIJOU, -ERY)]

bike /baɪk/ *n. & v.* ● *n.* **1** *colloq.* a bicycle or motorcycle. **2** *Austral. sl.* a promiscuous woman (*the town bike*). ● *v.intr.* ride a bicycle or motorcycle. [abbr.]

biker /ˈbaɪkə(r)/ *n.* a cyclist, esp. a motorcyclist.

bikie /ˈbaɪkɪ/ *n. Austral. colloq.* a member of a gang of motorcyclists.

Bikini /bɪˈkiːnɪ/ an atoll in the Marshall Islands, in the western Pacific, used by the US between 1946 and 1958 as a site for testing nuclear weapons.

bikini /bɪˈkiːnɪ/ *n.* a two-piece swimsuit for women. □ **bikini briefs** women's scanty briefs. [BIKINI, from the supposed 'explosive' effect]

bikkie var. of BICKIE.

Biko /ˈbiːkəʊ/, Stephen ('Steve') (1946–77), South African radical leader. While a medical student, he founded and became president of the South African Students Organization (1968). In 1972 he co-founded the Black People's Convention, a coalition of organizations which aimed to raise awareness of oppression in the black community and develop a sense of pride. He was banned from political activity in 1973 and his freedom of speech and association were severely restricted; detained several times in the last years of his life, he died in police custody, becoming a symbol of heroic resistance to apartheid in black townships and beyond.

bilabial /baɪˈleɪbɪəl/ *adj. Phonet.* (of a speech sound) made with closed or nearly closed lips.

bilateral /baɪˈlætərəl/ *adj.* **1** of, on, or with two sides. **2** affecting or between two parties, countries, etc. (*bilateral negotiations*). □ **bilateral symmetry** *Geom.* the property of being divisible into symmetrical halves on either side of a unique plane. □ **bilaterally** *adv.*

Bilbao /bɪlˈbɑːəʊ/ a seaport and industrial city in northern Spain; pop. (1991) 372,200.

bilberry /ˈbɪlbərɪ/ *n.* (*pl.* **-ies**) **1** a hardy dwarf ericaceous shrub of the genus *Vaccinium*, having dark blue berries; esp. *V. myrtillus* of northern Europe, which grows on heaths and mountains, and has red drooping flowers and edible berries. **2** the berry of this. [orig. uncert.: cf. Danish *bøllebær*]

bilbo /ˈbɪlbəʊ/ *n.* (*pl.* **-os** or **-oes**) *hist.* a sword noted for the temper and elasticity of its blade. [*Bilboa* = BILBAO]

bilboes /ˈbɪlbəʊz/ *n.pl. hist.* an iron bar with sliding shackles for a prisoner's ankles. [16th c.: orig. unkn.]

Bildungsroman /ˈbɪldʊŋzrəʊˌmɑːn/ *n.* a novel dealing with one person's early life and development. [G, f. *Bildung* education + *Roman* novel]

bile /baɪl/ n. **1** a bitter greenish-brown alkaline fluid which aids digestion and is secreted by the liver and stored in the gall-bladder. **2** bad temper; peevish anger. □ **bile-duct** the duct which conveys bile from the liver and the gall-bladder to the duodenum. [F f. L *bilis*]

bi-level /ˈbaɪˌlev(ə)l/ adj. & n. ● adj. **1** having or functioning on two levels; arranged on two planes. **2** US designating a style of two-storey house in which the lower storey is partially sunk below ground level, and the main entrance is between the two storeys. ● n. US a bi-level house.

bilge /bɪldʒ/ n. & v. ● n. **1 a** the almost flat part of a ship's bottom, inside or out. **b** (in full **bilge-water**) filthy water that collects inside the bilge. **2** sl. nonsense; rot (*don't talk bilge*). ● v. **1** tr. stave in the bilge of (a ship). **2** intr. spring a leak in the bilge. **3** intr. swell out; bulge. □ **bilge-keel** a plate or timber fastened under the bilge to prevent rolling. [prob. var. of BULGE]

bilharzia /bɪlˈhɑːtsɪə/ n. **1** a tropical flatworm of the genus *Schistosoma* (formerly *Bilharzia*), with larvae that are parasitic in freshwater snails and adults that are parasitic in blood vessels in the human pelvic region. Also called *schistosome*. **2** the chronic tropical disease produced by its presence. Also called *bilharziasis* or *schistosomiasis*. [f. T. *Bilharz*, Ger. physician (1825–62)]

bilharziasis /ˌbɪlhɑːˈtsaɪəsɪs/ n. = BILHARZIA 2.

biliary /ˈbɪlɪərɪ/ adj. of the bile. [F *biliaire*: see BILE, -ARY²]

bilingual /baɪˈlɪŋɡwəl/ adj. & n. ● adj. **1** able to speak two languages, esp. fluently. **2** spoken or written in two languages. ● n. a bilingual person. □ **bilingualism** n. [L *bilinguis* (as BI-, *lingua* tongue)]

bilious /ˈbɪljəs/ adj. **1** affected by a disorder of the bile. **2** bad-tempered. □ **biliously** adv. **biliousness** n. [L *biliosus* f. *bilis* bile]

bilirubin /ˌbɪlɪˈruːbɪn/ n. an orange-yellow pigment occurring in bile formed as a breakdown product of haemoglobin. [G f. L *bilis* BILE + *ruber* red]

bilk /bɪlk/ v.tr. sl. **1** cheat. **2** give the slip to. **3** avoid paying (a creditor or debt). □ **bilker** n. [orig. uncert., perh. = BAULK: earliest use (17th c.) in cribbage, = spoil one's opponent's score]

bill¹ /bɪl/ n. & v. ● n. **1 a** a printed or written statement of charges for goods supplied or services rendered. **b** the amount owed (*ran up a bill of £300*). **2** a draft of a proposed law. **3 a** a poster; a placard. **b** = HANDBILL. **4 a** a printed list, esp. a theatre programme. **b** the entertainment itself (*top of the bill*). **5** N. Amer. a banknote (*ten dollar bill*). ● v.tr. **1** put in the programme; announce. **2** (foll. by *as*) advertise. **3** send a note of charges to (*billed him for the books*). □ **bill of exchange** Econ. a written order to pay a sum of money on a given date to the drawer or to a named payee. **bill of fare 1** a menu. **2** a programme (for a theatrical event). **bill of health 1** Naut. a certificate regarding infectious disease on a ship or in a port at the time of sailing. **2** (**clean bill of health**) **a** such a certificate stating that there is no disease. **b** a declaration that a person or thing examined has been found to be free of illness or in good condition. **bill of indictment** hist. or US a written accusation as presented to a grand jury. **bill of lading** Naut. **1** a shipmaster's detailed list of the ship's cargo. **2** US = WAYBILL. **bill of sale** Econ. a certificate of transfer of personal property, esp. as a security against debt. □ **billable** adj. [ME f. AF *bille*, AL *billa*, prob. alt. of med.L *bulla* seal, sealed documents, BULL²]

bill² /bɪl/ n. & v. ● n. **1** the beak of a bird, esp. when it is slender, flattened, or weak, or belongs to a web-footed bird or a bird of the pigeon family. **2** the muzzle of a platypus. **3** a narrow promontory. **4** Naut. the point of an anchor-fluke. ● v.intr. (of doves etc.) stroke a bill with a bill. □ **bill and coo** exchange caresses. □ **billed** adj. (usu. in comb.). [OE *bile*, of unkn. orig.]

bill³ /bɪl/ n. **1** hist. a weapon like a halberd with a hook instead of a blade. **2** = BILLHOOK. [OE *bil*, ult. f. Gmc]

billabong /ˈbɪləˌbɒŋ/ n. Austral. a branch of a river forming a backwater or a stagnant pool. [Wiradhuri *bilabang* (orig. as the name of the Bell river, NSW)]

billboard /ˈbɪlbɔːd/ n. esp. US a large outdoor board for advertisements etc.

billet¹ /ˈbɪlɪt/ n. & v. ● n. **1 a** a place where troops etc. are lodged, usu. with civilians. **b** a written order requiring a householder to lodge the bearer, usu. a soldier. **2** colloq. a situation; a job. ● v.tr. (**billeted**, **billeting**) **1** (usu. foll. by *on*, *in*, *at*) quarter (soldiers etc.). **2** (of a householder) provide (a soldier etc.) with board and lodging. □ **billeter** n. **billetee** /ˌbɪlɪˈtiː/ n. [ME f. AF *billette*, AL *billetta*, dimin. of *billa* BILL¹]

billet² /ˈbɪlɪt/ n. **1** a thick piece of firewood. **2** a small metal bar. **3** Archit. each of a series of short rolls inserted at intervals in Norman decorative mouldings. [ME f. F *billette* small log, ult. prob. of Celtic orig.]

billet-doux /ˌbɪlɪˈduː/ n. (pl. **billets-doux** /-ˈduːz/) often joc. a love-letter. [F, = sweet note]

billfold /ˈbɪlfəʊld/ n. US a wallet for keeping banknotes.

billhead /ˈbɪlhed/ n. a printed account form.

billhook /ˈbɪlhʊk/ n. a sickle-shaped tool with a sharp inner edge, used for pruning, lopping, etc.

billiards /ˈbɪljədz/ n. **1** an indoor game played, usu. by two players, on an oblong cloth-covered table, with three balls struck with cues into pockets round the edge of the table. (*See note below.*) **2** (**billiard**) (in comb.) used in billiards (*billiard-ball*; *billiard-table*). [orig. pl., f. F *billard* billiards, cue, dimin. of *bille* log: see BILLET²]

■ Billiards has been played at least since the 16th–17th centuries, when it was a sport of British and French royalty. There have been public tables since the mid-17th century, although they were banned for a time in the mid-18th century because the game had acquired associations with gambling and a dissolute life. Early tables were made entirely of wood, later with an iron top, until the fabric-covered slate top was introduced c.1826. Balls were propelled with a mace (a stick with a flat square head) until the straight cue was introduced in the early 19th century. The phenomenal skill of the Australian player Walter Lindrum (1898–1960) defied competition (especially in the early 1930s), and helped to kill the game in favour of pool, where there is a greater element of chance. The most popular form of this in Britain is snooker.

Billingsgate /ˈbɪlɪŋzˌɡeɪt/ a London fish-market dating from the 16th century, known for the invective traditionally ascribed to the porters who worked there. The market moved in 1982 to the Isle of Dogs in the East End.

billion /ˈbɪljən/ n. & adj. ● n. (pl. same or (in sense 3) **billions**) (in sing. prec. by *a* or *one*) **1** a thousand million (1,000,000,000 or 10⁹). **2** (now less often, esp. Brit.) a million million (1,000,000,000,000 or 10¹²). **3** (in pl.) colloq. a very large number (*billions of years*). ● adj. that amount to a billion. □ **billionth** adj. & n. [F (as BI-, MILLION)]

billionaire /ˌbɪljəˈneə(r)/ n. a person possessing over a billion pounds, dollars, etc. [after MILLIONAIRE]

Billiton see BELITUNG.

Bill of Rights 1 a bill passed by the English Parliament in Oct. 1689 confirming the deposition of James II and the accession of William and Mary, guaranteeing the Protestant succession, and laying down principles of parliamentary supremacy. **2** the first ten amendments to the Constitution of the US, proposed by Congress in 1789 and ratified in 1791, spelling out individual rights that are regarded as inalienable. The First Amendment gives citizens freedom of religion, assembly, speech, and the press, and the right of petition. The next seven secure the rights of property and guarantee the rights of persons accused of crime. The Ninth protects rights held concurrently by the people and the Federal government, and the Tenth assures the reserved rights of the states. **3** a statement of the rights of a class of people.

billon /ˈbɪlən/ n. an alloy of gold or silver with a predominating admixture of a base metal. [F f. *bille* BILLET²]

billow /ˈbɪləʊ/ n. & v. ● n. **1** a wave. **2** a soft upward-curving flow. **3** any large soft mass. ● v.intr. move or build up in billows. □ **billowy** adj. [ON *bylgja* f. Gmc]

billposter /ˈbɪlˌpəʊstə(r)/ n. (also **billsticker** /-ˌstɪkə(r)/) a person who pastes up advertisements on hoardings. □ **billposting** n.

billy¹ /ˈbɪlɪ/ n. (pl. **-ies**) (in full **billycan**) orig. Austral. a tin or enamel cooking-pot with a lid and wire handle, for use out of doors. [perh. f. Aboriginal *billa* water]

billy² /ˈbɪlɪ/ n. (pl. **-ies**) **1** = BILLY-GOAT. **2** a bludgeon.

billycan /ˈbɪlɪˌkæn/ n. = BILLY¹.

billycart /ˈbɪlɪˌkɑːt/ n. Austral. **1** a small handcart. **2** a go-kart.

billy-goat /ˈbɪlɪˌɡəʊt/ n. a male goat. [*Billy*, pet-form of the name *William*]

billy-o /ˈbɪlɪəʊ/ n. □ **like billy-o** sl. very much, hard, strongly, etc. (*raining like billy-o*). [19th c.: orig. unkn.]

bilobate /baɪˈləʊbət/ adj. (also **bilobed** /-ˈləʊbd/) having or consisting of two lobes.

biltong /ˈbɪltɒŋ/ n. S. Afr. boneless meat salted and dried in strips. [Afrik., of uncert. orig.]

BIM abbr. British Institute of Management.

bimanal /'baɪmənəl/ adj. (also **bimanous** /-nəs/) having two hands. [BI- + L *manus* hand]

bimbo /'bɪmbəʊ/ n. (pl. **-os** or **-oes**) sl. usu. derog. **1** a person. **2** a woman, esp. a young empty-headed one. [It., = little child]

bi-media /baɪ'miːdɪə/ adj. (also **bimedia**) involving or working in two of the mass communication media, esp. radio and television.

bimetallic /ˌbaɪmɪ'tælɪk/ adj. **1** made of two metals. **2** of or relating to bimetallism. □ **bimetallic strip** a sensitive element in some thermostats made of two bands of different metals that expand at different rates when heated, causing the strip to bend and thus break the circuit. [F *bimétallique* (as BI-, METALLIC)]

bimetallism /baɪ'metəˌlɪz(ə)m/ n. a system of allowing the unrestricted currency of two metals (e.g. gold and silver) at a fixed ratio to each other, as coined money. □ **bimetallist** n.

bimillenary /ˌbaɪmɪ'lenərɪ/ adj. & n. ● adj. of or relating to a two-thousandth anniversary. ● n. (pl. **-ies**) a bimillenary year or festival.

bimodal /baɪ'məʊd(ə)l/ adj. esp. *Statistics* having two modes.

bimonthly /baɪ'mʌnθlɪ/ adj., adv., & n. ● adj. occurring twice a month or every two months. ● adv. twice a month or every two months. ● n. (pl. **-ies**) a periodical produced bimonthly. ¶ Often avoided, because of the ambiguity of meaning, in favour of *two-monthly* and *twice-monthly*.

bin /bɪn/ n. & v. ● n. a large receptacle for storage or for depositing rubbish. ● v.tr. colloq. (**binned**, **binning**) store or put in a bin; throw away. □ **bin end** one of the last bottles from a bin of wine, usu. sold at a reduced price. **bin-liner** a bag (usu. of plastic) for lining a rubbish bin. [OE *bin(n)*, *binne*]

bin- /bɪn, baɪn/ prefix var. of BI- before a vowel.

binary /'baɪnərɪ/ adj. & n. ● adj. **1 a** dual. **b** of or involving pairs. **2** of the arithmetical system using 2 as a base. ● n. (pl. **-ies**) **1** something having two parts. **2** a binary number. **3** a binary star. □ **binary code** *Computing* a coding system using the binary digits 0 and 1 to represent a letter, digit, or other character in a computer (see BCD). **binary compound** *Chem.* a compound having two elements or radicals. **binary fission** the division of a cell or organism into two parts. **binary number** (or **digit**) one of two digits (usu. 0 or 1) in a binary system of notation. **binary star** *Astron.* a system of two stars orbiting each other. **binary system** a system in which information can be expressed by combinations of the digits 0 and 1 (corresponding to 'off' and 'on' in computing). **binary tree** *Computing* a data structure in which a record is branched to the left when greater and to the right when less than the previous record. [LL *binarius* f. *bini* two together]

binate /'baɪneɪt/ adj. Bot. **1** growing in pairs. **2** composed of two equal parts. [mod.L *binatus* f. L *bini* two together]

binaural /baɪ'nɔːrəl/ adj. **1** of or used with both ears. **2** (of sound) recorded using two microphones and usu. transmitted separately to the two ears.

bind /baɪnd/ v. & n. ● v. (past and past part. **bound** /baʊnd/) (see also BOUNDEN). **1** tr. (often foll. by *to*, *on*, *together*) tie or fasten tightly. **2** tr. **a** restrain; put in bonds. **b** (as **-bound** adj.) constricted, obstructed (*snowbound*). **3** tr. esp. *Cookery* cause (ingredients) to cohere using another ingredient. **4** tr. fasten or hold together as a single mass. **5** tr. compel; impose an obligation or duty on. **6** tr. **a** edge (fabric etc.) with braid etc. **b** fix together and fasten (the pages of a book) in a cover. **7** tr. constipate. **8** tr. ratify (a bargain, agreement, etc.). **9** tr. (in passive) be required by an obligation or duty (*am bound to answer*). **10** tr. (often foll. by *up*) **a** put a bandage or other covering round. **b** fix together with something put round (*bound her hair*). **11** tr. indenture as an apprentice. **12** tr. (of snow etc.) cohere, stick. **13** intr. be prevented from moving freely. **14** intr. sl. complain. ● n. **1** colloq. **a** a nuisance; a restriction. **b** esp. *US* a tight or difficult situation. **2** = BINE. □ **be bound up with** be closely associated with. **bind over** *Law* order (a person) to do something, esp. keep the peace. **bind up** bandage. **I'll be bound** a statement of assurance, or guaranteeing the truth of something. [OE *bindan*]

binder /'baɪndə(r)/ n. **1** a cover for sheets of paper, for a book, etc. **2** a substance that acts cohesively. **3** a reaping-machine that binds grain into sheaves. **4** a bookbinder.

bindery /'baɪndərɪ/ n. (pl. **-ies**) a workshop or factory for binding books.

bindi-eye /'bɪndɪˌaɪ/ n. a small perennial Australian plant, *Calotis cuneifolia*, which has a burlike fruit. [20th c.: orig. unkn.]

binding /'baɪndɪŋ/ n. & adj. ● n. **1** something that binds, esp. the covers, glue, etc., of a book. **2** *Linguistics* the phenomenon according to which two elements in a clause may have the same reference. ● adj. (often foll. by *on*) obligatory.

bindweed /'baɪndwiːd/ n. **1** convolvulus. **2** a climbing plant, e.g. honeysuckle.

bine /baɪn/ n. **1** the twisting stem of a climbing plant, esp. the hop. **2** a flexible shoot. [orig. a dial. form of BIND]

Binet /'biːneɪ/, Alfred (1857–1911), French psychologist and pioneer of modern intelligence testing. He was requested to devise a test which would detect intellectually slow schoolchildren, and together with the psychiatrist Théodore Simon (1873–1961) he produced tests (now known as *Binet* or *Binet–Simon tests*) intended to examine general reasoning capacities rather than perceptual-motor skills. Believing that bright and dull schoolchildren were simply advanced or retarded in their mental growth, Binet devised a mental age scale which described performance in relation to the average performance of students of the same physical age.

binge /bɪndʒ/ n. & v. sl. ● n. a spree; a period of uncontrolled eating, drinking, etc. ● v.intr. (**bingeing** or **binging**) go on a spree; indulge in uncontrolled eating, drinking, etc. [prob. orig. dial., = soak]

bingle /'bɪŋg(ə)l/ n. Austral. colloq. a collision. [Brit. dial. *bing* thump, blow]

bingo /'bɪŋgəʊ/ n. & int. ● n. a game for any number of players, each having a card of squares with numbers, which are marked off as numbers are randomly drawn by a caller. The player first covering all or a set of these wins a prize. ● int. expressing sudden surprise, satisfaction, etc., as in winning at bingo. [prob. imit.: cf. dial. *bing* 'with a bang']

bingy /'bɪndʒɪ/ n. (also **bingie**) Austral. colloq. the stomach, the belly. [Dharuk *bindhi* belly]

binman /'bɪnmæn/ n. (pl. **-men**) colloq. a dustman.

binnacle /'bɪnək(ə)l/ n. Naut. a built-in housing for a ship's compass. [earlier *bittacle*, ult. f. L *habitaculum* habitation f. *habitare* inhabit]

binocular /baɪ'nɒkjʊlə(r), bɪ'nɒk-/ adj. adapted for or using both eyes. □ **binocular vision** vision in which both eyes face in the same direction, giving good three-dimensional perception where the two fields of view overlap. [BIN- + L *oculus* eye]

binoculars /bɪ'nɒkjʊləz/ n.pl. an optical instrument with a lens for each eye, for viewing distant objects.

binomial /baɪ'nəʊmɪəl/ n. & adj. ● n. **1** an algebraic expression of the sum or the difference of two terms. **2** a two-part name, esp. in taxonomy. ● adj. consisting of two terms. □ **binomial classification** a system of classification using two terms, the first one indicating the genus and the second the species. **binomial distribution** a frequency distribution of the possible number of successful outcomes in a given number of trials in each of which there is the same probability of success. **binomial theorem** a formula for finding any power of a binomial without multiplying at length. □ **binomially** adv. [F *binôme* or mod.L *binomium* (as BI-, Gk *nomos* part, portion)]

binominal /baɪ'nɒmɪn(ə)l/ adj. = BINOMIAL. [L *binominis* (as BI-, *nomen* -inis* name)]

bint /bɪnt/ n. sl. usu. offens. a girl or woman. [Arab., = daughter, girl]

binturong /'bɪntjʊˌrɒŋ/ n. a large southern Asian civet, *Arctictis binturong*, with a shaggy black coat and a prehensile tail. [Malay]

bio /'baɪəʊ/ n. & adj. ● n. **1** biology. **2** (pl. **bios**) biography. ● adj. biological. [abbr.]

bio- /'baɪəʊ/ comb. form **1** life (*biography*). **2** biological (*biomathematics*). **3** of living beings (*biophysics*). [Gk *bios* (course of) human life]

bioassay /ˌbaɪəʊə'seɪ, -'æseɪ/ n. measurement of the concentration or potency of a substance by its effect on living cells or tissues. [BIO- + ASSAY]

biochemistry /ˌbaɪəʊ'kemɪstrɪ/ n. the study of the chemical and physico-chemical processes of living organisms. □ **biochemical** adj. **biochemically** adv. **biochemist** n.

biocoenosis /ˌbaɪəʊsiː'nəʊsɪs/ n. (US **biocenosis**) (pl. **-noses** /-siːz/) **1** an association of different organisms forming a community. **2** the relationship existing between such organisms. □ **biocoenology** /-'nɒlədʒɪ/ n. **biocoenotic** /-'nɒtɪk/ adj. [mod.L f. BIO- + Gk *koinōsis* sharing f. *koinos* common]

biodegradable /ˌbaɪəʊdɪ'greɪdəb(ə)l/ adj. capable of being decomposed by bacteria or other living organisms. □ **biodegradability** /-ˌgreɪdə'bɪlɪtɪ/ n. **biodegradation** /-ˌdegrə'deɪʃ(ə)n/ n.

biodiversity /ˌbaɪəʊdaɪ'vɜːsɪtɪ/ n. diversity of plant and animal life.

bioenergetics /ˌbaɪəʊˌenə'dʒetɪks/ n. the study of the transformation of energy in living organisms.

bioengineering /ˌbaɪəʊˌendʒɪˈnɪərɪŋ/ n. **1** the application of engineering techniques to biosynthetic processes. **2** the use of artificial tissues, organs, or organ components to replace damaged or absent parts of the body, e.g. artificial limbs, heart pacemakers, etc. □ **bioengineer** n. & v.

bioethics /ˌbaɪəʊˈeθɪks/ n.pl. (treated as sing.) the ethics of medical and biological research. □ **bioethicist** /-ˈeθɪsɪst/ n.

biofeedback /ˌbaɪəʊˈfiːdbæk/ n. the technique of using the feedback of a normally automatic bodily response to a stimulus, in order to acquire voluntary control of that response.

bioflavonoid /ˌbaɪəʊˈfleɪvəˌnɔɪd/ n. Biochem. = CITRIN. [BIO- + flavonoid f. FLAVINE + -OID]

biogas /ˈbaɪəʊˌɡæs/ n. gaseous fuel, esp. methane, produced by fermentation of organic matter.

biogenesis /ˌbaɪəʊˈdʒenɪsɪs/ n. **1** the synthesis of substances by living organisms. **2** the hypothesis that a living organism arises only from another similar living organism. □ **biogenetic** /-dʒɪˈnetɪk/ adj.

biogenic /ˌbaɪəʊˈdʒenɪk/ adj. produced by living organisms.

biogeography /ˌbaɪəʊdʒɪˈɒɡrəfɪ/ n. the scientific study of the geographical distribution of plants and animals. □ **biogeographical** /-ˌdʒiːəˈɡræfɪk(ə)l/ adj.

biography /baɪˈɒɡrəfɪ/ n. (pl. -ies) **1 a** a written account of a person's life, usu. by another. **b** such writing as a branch of literature. **2** the course of a living (usu. human) being's life. □ **biographer** n. **biographic** /ˌbaɪəˈɡræfɪk/ adj. **biographical** adj. **biographically** adv. [F biographie or mod.L biographia f. med.Gk]

biohazard /ˈbaɪəʊˌhæzəd/ n. a risk to human health or the environment arising from biological work, esp. with micro-organisms.

Bioko /bɪˈəʊkəʊ/ an island of Equatorial Guinea, in the eastern part of the Gulf of Guinea. Its chief town is Malabo, the capital of Equatorial Guinea. It was known as Fernando Póo until 1973, and from 1973 to 1979 as Macias Nguema.

biological /ˌbaɪəˈlɒdʒɪk(ə)l/ adj. of or relating to biology or living organisms. □ **biological clock** an innate mechanism controlling the rhythmic physiological activities of an organism. **biological control** the control of a pest by the introduction of a natural enemy. **biological warfare** warfare involving the use of toxins or micro-organisms. □ **biologically** adv.

biology /baɪˈɒlədʒɪ/ n. **1** the science of life. (See note below.) **2** living organisms or life processes collectively. □ **biologist** n. [F biologie f. G Biologie (as BIO-, -LOGY)]

▪ Biology is an immensely broad subject, divisible into many specialized fields, and covering the morphology, physiology, anatomy, behaviour, origin, and distribution of living organisms. The word biology was coined in the 19th century when it was realized that all living things share fundamental similarities.

bioluminescence /ˌbaɪəʊˌluːmɪˈnes(ə)ns/ n. the emission of light by living organisms such as the firefly and glow-worm. □ **bioluminescent** adj.

biomass /ˈbaɪəʊˌmæs/ n. the total quantity or weight of organisms in a given area or volume. [BIO- + MASS¹]

biomathematics /ˌbaɪəʊˌmæθəˈmætɪks/ n. the science of the application of mathematics to biology.

biome /ˈbaɪəʊm/ n. **1** a large naturally occurring community of flora and fauna adapted to the particular conditions in which they occur, e.g. tundra. **2** the geographical region containing such a community. [BIO- + -OME]

biomechanics /ˌbaɪəʊmɪˈkænɪks/ n. the study of the mechanical laws relating to the movement or structure of living organisms.

biomedical /ˌbaɪəʊˈmedɪk(ə)l/ adj. of or relating to both biology and medicine. □ **biomedicine** n.

biometry /baɪˈɒmɪtrɪ/ n. (also **biometrics** /ˌbaɪəʊˈmetrɪks/) the application of statistical analysis to biological data. □ **biometric** /ˌbaɪəʊˈmetrɪk/ adj. **biometrical** adj. **biometrician** /ˌbaɪəʊmɪˈtrɪʃ(ə)n/ n. **biometricist** /ˌbaɪəʊˈmetrɪsɪst/ n.

biomorph /ˈbaɪəʊˌmɔːf/ n. a decorative form based on a living organism. □ **biomorphic** /ˌbaɪəʊˈmɔːfɪk/ adj. [BIO- + Gk morphē form]

bionic /baɪˈɒnɪk/ adj. **1** having artificial body parts or the superhuman powers resulting from these. **2** relating to bionics. □ **bionically** adv. [BIO- after ELECTRONIC]

bionics /baɪˈɒnɪks/ n.pl. (treated as sing.) the study of mechanical systems that function like living organisms or parts of living organisms.

bionomics /ˌbaɪəˈnɒmɪks/ n.pl. (treated as sing.) the study of the mode of life of organisms in their natural habitat and their adaptations to their surroundings; ecology. □ **bionomic** adj. [BIO- after ECONOMICS]

biophysics /ˌbaɪəʊˈfɪzɪks/ n.pl. (treated as sing.) the science of the physical properties of living organisms and their constituents, and the investigation of biological phenomena in general by means of the techniques of modern physics. □ **biophysical** adj. **biophysicist** n.

biopic /ˈbaɪəʊˌpɪk/ n. colloq. a biographical film.

biopsy /ˈbaɪɒpsɪ/ n. (pl. -ies) Med. the examination of tissue removed from a living body to discover the presence, cause, or extent of a disease. [F biopsie f. Gk bios life + opsis sight, after necropsy]

biorhythm /ˈbaɪəʊˌrɪðəm/ n. **1** any of the recurring cycles of biological processes thought to affect a person's emotional, intellectual, and physical activity. **2** any periodic change in the behaviour or physiology of an organism. □ **biorhythmic** /ˌbaɪəʊˈrɪðmɪk/ adj. **biorhythmically** adv.

bioscope /ˈbaɪəˌskəʊp/ n. S. Afr. sl. a cinema.

biosphere /ˈbaɪəʊˌsfɪə(r)/ n. the regions of the earth's crust and atmosphere occupied by living organisms. Also called ecosphere. [G Biosphäre (as BIO-, SPHERE)]

biosynthesis /ˌbaɪəʊˈsɪnθɪsɪs/ n. the production of organic molecules by living organisms. □ **biosynthetic** /-sɪnˈθetɪk/ adj.

biota /baɪˈəʊtə/ n. the animal and plant life of a region. [mod.L: cf. Gk biotē life]

biotechnology /ˌbaɪəʊtekˈnɒlədʒɪ/ n. the exploitation of biological processes for industrial and other purposes, esp. genetic manipulation of micro-organisms (for the production of antibiotics, hormones, etc.).

biotic /baɪˈɒtɪk/ adj. **1** relating to life or to living things. **2** of biological origin. [F biotique or LL bioticus f. Gk biōtikos f. bios life]

biotin /ˈbaɪətɪn/ n. a vitamin of the B complex, found in egg yolk, liver, and yeast, and involved in the metabolism of carbohydrates, fats, and proteins. Also called (esp. US) vitamin H. [G f. Gk bios life + -IN]

biotite /ˈbaɪəˌtaɪt/ n. Mineral. a black, dark brown, or green micaceous mineral occurring as a constituent of metamorphic and igneous rocks. [J. B. Biot, Fr. physicist (1774–1862)]

bipartisan /ˌbaɪpɑːtɪˈzæn, baɪˈpɑːtɪz(ə)n/ adj. of or involving two (esp. political) parties. □ **bipartisanship** n.

bipartite /baɪˈpɑːtaɪt/ adj. **1** consisting of two parts. **2** shared by or involving two parties. **3** Law (of a contract, treaty, etc.) drawn up in two corresponding parts or between two parties. [L bipartitus f. bipartire (as BI-, partire PART)]

biped /ˈbaɪped/ n. & adj. ● n. a two-footed animal. ● adj. two-footed. □ **bipedal** /baɪˈpiːd(ə)l, -ˈped(ə)l/ adj. [L bipes -edis as BI-, pes pedis foot)]

bipinnate /baɪˈpɪneɪt/ adj. Bot. (of a pinnate leaf) having leaflets that are further subdivided in a pinnate arrangement.

biplane /ˈbaɪpleɪn/ n. an early type of aeroplane having two sets of wings, one above the other.

bipolar /baɪˈpəʊlə(r)/ adj. having two poles or extremities. □ **bipolarity** /ˌbaɪpəʊˈlærɪtɪ/ n.

birch /bɜːtʃ/ n. & v. ● n. **1** a hardy tree or shrub of the genus Betula, having thin peeling bark, bearing catkins, and found predominantly in northern temperate regions. **2** (in full **birchwood**) the hard fine-grained pale wood of this. **3** (in full **birch-rod**) a bundle of birch twigs used for flogging. ● v.tr. beat with a birch (in sense 3). □ **birch-bark 1** the bark of Betula papyrifera used to make canoes. **2** N. Amer. such a canoe. □ **birchen** adj. [OE bi(e)rce f. Gmc]

bird /bɜːd/ n. **1** a feathered vertebrate with a beak, two wings and two feet, egg-laying and usu. able to fly. (See note below.) **2** a game bird. **3** Brit. sl. a young woman. **4** colloq. a person (a wily old bird). **5** sl. **a** a prison. **b** rhyming sl. a prison sentence (short for birdlime = time). □ **bird-bath** a basin in a garden etc. with water for birds to bathe in. **bird-call 1** a bird's natural call. **2** an instrument imitating this. **bird cherry** a wild cherry, Prunus padus. **bird-fancier** a person who knows about, collects, breeds, or deals in, birds. **a bird in the hand** something secured or certain. **the bird is** (or **has**) **flown** the prisoner, quarry, etc., has escaped. **bird-** (or **birds'-**) **nesting** hunting for birds' nests, usu. to get eggs. **bird of paradise** a bird of the family Paradiseidae, found chiefly in New Guinea, the males having brilliantly coloured and showy plumage. **bird of passage 1** a migrant. **2** any transient visitor. **bird of prey** see PREY. **bird sanctuary** an area where birds are protected and encouraged to breed. **the birds and the bees** euphem. sexual activity and reproduction. **bird's-eye** n. **1** a plant with small bright round flowers, e.g. the germander speedwell. **2** a pattern with many small spots. ● *adj.* of or having small bright round flowers (*bird's-eye primrose*). **bird's-eye view** a general view from above. **bird's-foot** (*pl.* **bird's-foots**) a plant having a part resembling the foot of a bird; esp. a vetch of the genus *Ornithopus*, having claw-shaped pods. **bird's-foot trefoil** a small leguminous plant, *Lotus corniculatus*, with yellow flowers streaked with red (also called *eggs and bacon, tom thumb*). **bird's nest soup** soup made (esp. in Chinese cookery) from the dried gelatinous coating of the nests of swifts of the genus *Aerodramus*. **birds of a feather** people of like character. **bird-strike** a collision between a bird and an aircraft. **bird table** a raised platform on which food for birds is placed. **bird-watcher** a person who observes birds in their natural surroundings. **bird-watching** this occupation. **for** (or **strictly for**) **the birds** *colloq.* trivial, uninteresting. **get the bird** *sl.* **1** be dismissed. **2** be hissed at or booed. **like a bird** without difficulty or hesitation. **a little bird** an unnamed informant. [OE *brid*, of unkn. orig.]

▪ The earliest known fossil bird, *Archaeopteryx*, clearly shows the close relationship of birds to the small dinosaurs from which they were descended (see ARCHAEOPTERYX). Modern birds, of which there are about 8,600 living species, are comparatively uniform as regards their anatomical structure. Their bones are often hollow for the sake of lightness. They are warm-blooded, and feathers probably first evolved as an insulating layer to help conserve body heat. Birds are primarily diurnal animals, with good colour vision, good hearing, and a poor sense of smell. They exhibit complex social and migratory behaviour, and show advanced care of the young, but are considered to behave more instinctively and with less ability to adapt than mammals.

birdbrain /ˈbɜːdbreɪn/ *n. colloq.* a stupid or flighty person. □ **birdbrained** *adj.*

birdcage /ˈbɜːdkeɪdʒ/ *n.* **1** a cage for birds usu. made of wire or cane. **2** an object of a similar design.

birder /ˈbɜːdə(r)/ *n. US* a bird-watcher. □ **birding** *n.*

birdie /ˈbɜːdɪ/ *n. & v.* ● *n.* **1** *colloq.* a little bird. **2** *Golf* a score of one stroke less than par at any hole. ● *v.tr.* (**birdies, birdied, birdying**) *Golf* play (a hole) in a birdie.

birdlime /ˈbɜːdlaɪm/ *n.* sticky material painted on to twigs to trap small birds.

birdseed /ˈbɜːdsiːd/ *n.* a blend of seed for feeding birds, esp. ones which are caged.

Birdseye /ˈbɜːdzaɪ/, Clarence (1886–1956), American businessman and inventor. A former fur-trader, he had observed food preservation techniques practised by local people in Labrador; this led him to develop a process of rapid freezing of foods in small packages suitable for retail selling, creating a revolution in eating habits.

birdsong /ˈbɜːdsɒŋ/ *n.* the musical cry of a bird or birds.

birefringent /ˌbaɪrɪˈfrɪndʒənt/ *adj. Physics* having two different refractive indices. □ **birefringence** *n.*

bireme /ˈbaɪriːm/ *n. hist.* an ancient Greek warship, with two files of oarsmen on each side. [L *biremis* (as BI-, *remus* oar)]

biretta /bɪˈretə/ *n.* a square usu. black cap with three flat projections on top, worn by (esp. Roman Catholic) clergymen. [It. *berretta* or Sp. *birreta* f. LL *birrus* cape]

Birgitta, St see BRIDGET, ST[2].

biriani /ˌbɪrɪˈɑːnɪ/ *n.* (also **biryani**) a dish of Indian origin made with highly seasoned rice, and meat or fish etc. [Urdu]

Birkenhead /ˌbɜːkənˈhed/ a town on the Wirral Peninsula, in Merseyside, on the River Mersey opposite Liverpool; pop. (1981) 123,884. It was formerly a centre of the shipbuilding industry.

Birmingham 1 /ˈbɜːmɪŋəm/ an industrial city in west central England; pop. (1991) 934,900. It is the administrative centre of West Midlands metropolitan county. **2** /ˈbɜːmɪŋˌhæm/ an industrial city in north central Alabama; pop. (1990) 265,968.

Biro /ˈbaɪərəʊ/ *n.* (*pl.* **-os**) *Brit. propr.* a kind of ball-point pen. [L. *Bíró*, Hungarian inventor (1899–1985)]

birth /bɜːθ/ *n. & v.* ● *n.* **1** the emergence of a (usu. fully developed) infant or other young from the body of its mother. **2** *rhet.* the beginning or coming into existence of something (*the birth of civilization; the birth of socialism*). **3 a** origin, descent, ancestry (*of noble birth*). **b** high or noble birth; inherited position. ● *v.* **1** *intr.* (esp. as **birthing** *n.*) give birth. **2** *tr. US colloq.* **a** give birth to. **b** assist (a woman) to give birth. □ **birth certificate** an official document identifying a person by name, place, and date of birth. **birth control** the control of the number of children one conceives, esp. by contraception. **birthing pool** (or **tank**) a large pool for women to give birth in. **birth pill** a contraceptive pill. **birth rate** the number of live births per thousand of population per year. **give birth** bear a child etc. **give birth to 1** produce (young) from the womb. **2** cause to begin, found. [ME f. ON *byrth* f. Gmc: see BEAR[1], -TH[2]]

birthday /ˈbɜːθdeɪ/ *n.* **1** the day on which a person etc. was born. **2** the anniversary of this. □ **birthday honours** *Brit.* titles etc. given on a sovereign's official birthday. **in one's birthday suit** *joc.* naked.

birthmark /ˈbɜːθmɑːk/ *n.* an unusual brown or red mark on one's body at or from birth.

birthplace /ˈbɜːθpleɪs/ *n.* the place where a person was born.

birthright /ˈbɜːθraɪt/ *n.* a right of possession or privilege one has from birth, esp. as the eldest son.

birthstone /ˈbɜːθstəʊn/ *n.* a gemstone popularly associated with the month of one's birth.

Birtwistle /ˈbɜːtˌwɪs(ə)l/, Sir Harrison (Paul) (b.1934), English composer and clarinettist. In 1967 he was a co-founder of the Pierrot Players and wrote much of his work for them and for the English Opera Group. His early work was influenced by Stravinsky; later compositions are more experimental.

biryani var. of BIRIANI.

Biscay, Bay of /ˈbɪskeɪ/ a part of the North Atlantic between the north coast of Spain and the west coast of France, noted for its strong currents and storms. The shipping forecast area *Biscay* extends approximately as far west as the longitude of Gijón in Spain.

biscuit /ˈbɪskɪt/ *n. & adj.* ● *n.* **1** *Brit.* a small unleavened cake, usu. flat and crisp and often sweet. **2** fired unglazed pottery. **3** a light brown colour. ● *adj.* biscuit-coloured. [ME f. OF *bescoit* etc. ult. f. L *bis* twice + *coctus* past part. of *coquere* cook]

bise /biːz/ *n.* a keen dry northerly wind in Switzerland, southern France, etc. [ME f. OF]

bisect /baɪˈsekt/ *v.tr.* divide into two (strictly equal) parts. □ **bisector** *n.* **bisection** /-ˈsekʃ(ə)n/ *n.* [BI- + L *secare sect-* cut]

bisexual /baɪˈseksjʊəl, -ˈsekʃʊəl/ *adj. & n.* ● *adj.* **1** sexually attracted by persons of both sexes. **2** *Biol.* having characteristics of both sexes. **3** of or concerning both sexes. ● *n.* a bisexual person. □ **bisexuality** /-ˌseksjʊˈælɪtɪ, -ˌsekʃʊ-/ *n.*

bish /bɪʃ/ *n. sl.* a mistake. [20th c.: orig. uncert.]

Bishkek /bɪʃˈkek/ the capital of Kyrgyzstan; pop. (1990) 625,000. From 1926 to 1991 the city was named Frunze, after the Red Army general Mikhail Vasilevich Frunze (1885–1925). Before 1926 it was known as Pishpek.

Bisho /ˈbiːʃəʊ/ a town in southern South Africa, the capital of the province of Eastern Cape, situated near the coast to the north-east of Port Elizabeth.

bishop /ˈbɪʃəp/ *n.* **1** a senior member of the Christian clergy usu. in charge of a diocese, and empowered to confer holy orders. **2** a chess piece with the top sometimes shaped like a mitre. **3** mulled and spiced wine. [OE *biscop*, ult. f. Gk *episkopos* overseer (as EPI-, *-skopos* -looking)]

bishopric /ˈbɪʃəprɪk/ *n.* **1** the office of a bishop. **2** a diocese. [OE *bisceoprīce* as BISHOP, *rīce* realm)]

Bislama /ˈbɪʃləˌmɑː/ *n.* an English-based pidgin used as a lingua franca in Fiji and as an official language in Vanuata. [alt. of BEACH-LA-MAR]

Bismarck[1] /ˈbɪzmɑːk/ the state capital of North Dakota; pop. (1990) 49,256. A terminus of the Northern Pacific Railway, it took the name of the German Chancellor in order to attract German capital for railroad building.

Bismarck[2] /ˈbɪzmɑːk/, Otto (Eduard Leopold) von, Prince of Bismarck, Duke of Lauenburg (known as 'the Iron Chancellor')(1815–98), German statesman. As Minister-President and Foreign Minister of Prussia under Wilhelm I from 1862, Bismarck was the driving force behind the unification of Germany, orchestrating wars with Denmark (1864), Austria (1866), and France (1870–1) in order to achieve this end. As Chancellor of the new German Empire (1871–90), he continued to dominate the political scene, attempting to break the influence of the Catholic Church at home (see KULTURKAMPF) while consolidating Germany's position as a European power by creating a system of alliances. Bismarck was forced to resign in 1890 after a policy disagreement with Wilhelm II.

Bismarck Sea an arm of the Pacific Ocean north-east of New Guinea and north of New Britain. In March 1943 the United States destroyed a large Japanese naval force in these waters.

bismuth /ˈbɪzməθ/ *n.* a brittle reddish-tinged metallic chemical element (atomic number 83; symbol **Bi**). The history of bismuth is obscure as medieval writers often confused it with antimony and lead. Although there are some bismuth minerals and deposits of the metal, it is usually obtained as a by-product from the smelting of the tin, lead, or copper. For a metal, bismuth has low thermal and electrical conductivity. Its main use is in specialized low-melting-point alloys; some bismuth compounds have been used medicinally. [mod.L *bisemutum*, Latinization of G *Wismut*, of unkn. orig.]

bison /ˈbaɪs(ə)n/ n. (pl. same) a wild hump-backed shaggy-haired ox of the genus *Bison*, native to North America (*B. bison*) or Europe (*B. bonasus*). [ME f. L f. Gmc]

bisque[1] /bɪsk/ n. a rich shellfish soup, made esp. from lobster. [F]

bisque[2] /bɪsk/ n. *Tennis, Croquet, & Golf* an advantage of scoring one free point, or taking an extra turn or stroke. [F]

bisque[3] /bɪsk/ n. = BISCUIT 2.

Bissagos Islands /bɪˈsɑːgəs/ a group of islands off the coast of Guinea-Bissau, West Africa.

Bissau /bɪˈsaʊ/ the capital of Guinea-Bissau; pop. (est. 1988) 125,000.

bistable /baɪˈsteɪb(ə)l/ adj. (of an electrical circuit etc.) having two stable states.

bister var. of BISTRE.

bistort /ˈbɪstɔːt/ n. a herbaceous plant, *Polygonum bistorta*, of the dock family, with a twisted root and a cylindrical spike of flesh-coloured flowers. [F *bistorte* or med.L *bistorta* f. *bis* twice + *torta* fem. past part. of *torquere* twist]

bistoury /ˈbɪstərɪ/ n. (pl. **-ies**) a surgical scalpel. [F *bistouri*, *bistorie*, orig. = dagger, of unkn. orig.]

bistre /ˈbɪstə(r)/ n. & adj. (US **bister**) ● n. **1** a brownish pigment made from the soot of burnt wood. **2** the brownish colour of this. ● adj. of this colour. [F, of unkn. orig.]

bistro /ˈbiːstrəʊ/ n. (pl. **-os**) a small restaurant. [F]

bisulphate /baɪˈsʌlfeɪt/ n. (US **bisulfate**) *Chem.* a salt or ester of sulphuric acid.

bit[1] /bɪt/ n. **1** a small piece or quantity (*a bit of cheese; give me another bit; that bit is too small*). **2** (prec. by a) **a** a fair amount (*sold quite a bit; needed a bit of persuading*). **b** *colloq.* somewhat (*am a bit tired*). **c** (foll. by of) *colloq.* rather (*a bit of an idiot*). **d** (foll. by of) *colloq.* only a little; a mere (*a bit of a boy*). **3** a short time or distance (*wait a bit; move up a bit*). **4** *US sl.* a unit of 12½ cents (used only in even multiples). □ **bit by bit** gradually. **bit of all right** *sl.* a pleasing person or thing, esp. a woman. **bit of fluff** (or **skirt** or **stuff**) see FLUFF, SKIRT, STUFF. **bit of rough** see ROUGH. **bit on the side** *sl.* **1** a sexual relationship involving infidelity to one's partner. **2** the person with whom one is unfaithful. **bit part** a minor part in a play or a film. **bits and pieces** (or **bobs**) an assortment of small items. **do one's bit** *colloq.* make a useful contribution to an effort or cause. **every bit as** see EVERY. **not a bit** (or **not a bit of it**) not at all. **to bits** into pieces. [OE *bita* f. Gmc, rel. to BITE]

bit[2] past of BITE.

bit[3] /bɪt/ n. & v. ● n. **1** a metal mouthpiece on a bridle, used to control a horse. **2** a (usu. metal) tool or piece for boring or drilling. **3** the cutting or gripping part of a plane, pincers, etc. **4** the part of a key that engages with the lock-lever. **5** the copper head of a soldering-iron. ● v.tr. **1** put a bit into the mouth of (a horse). **2** restrain. □ **take the bit between one's teeth 1** take decisive personal action. **2** escape from control. [OE *bite* f. Gmc, rel. to BITE]

bit[4] /bɪt/ n. *Computing* etc. a unit of information expressed as a choice between two possibilities (represented by 0 or 1 in binary notation). The term was coined in 1948 by C. E. Shannon. (See also INFORMATION THEORY.) [BINARY + DIGIT]

bitch /bɪtʃ/ n. & v. ● n. **1** a female dog or other canine animal. **2** *sl. offens.* a malicious or spiteful woman. **3** *sl.* a very unpleasant or difficult thing or situation. ● v. *colloq.* **1** intr. (often foll. by *about*) **a** speak scathingly. **b** complain. **2** tr. be spiteful or unfair to. **3** tr. spoil, botch. [OE *bicce*]

bitchy /ˈbɪtʃɪ/ adj. (**bitchier**, **bitchiest**) *sl.* spiteful; bad-tempered. □ **bitchily** adv. **bitchiness** n.

bite /baɪt/ v. & n. ● v. (past **bit** /bɪt/; past part. **bitten** /ˈbɪt(ə)n/) **1** tr. cut or puncture using the teeth. **2** tr. (foll. by *off*, *away*, etc.) detach with the teeth. **3** tr. (of an insect, snake, etc.) wound with a sting, fangs, etc. **4** intr. (of a wheel, screw, etc.) grip, penetrate. **5** intr. accept bait or an inducement. **6** intr. have a (desired) adverse effect. **7** tr. (in passive) **a** take in; swindle. **b** (foll. by *by*, *with*, etc.) be infected by (enthusiasm etc.). **8** tr. (as **bitten** adj.) cause a glowing or smarting pain to (*frostbitten*). **9** intr. (foll. by *at*) snap at. **10** tr. *colloq.* worry, perturb (*what's biting you?*). ● n.

1 an act of biting. **2** a wound or sore made by biting. **3 a** a mouthful of food. **b** a snack or light meal. **4** the taking of bait by a fish. **5** pungency (esp. of flavour). **6** incisiveness, sharpness. **7** = OCCLUSION 3. □ **bite back** restrain (one's speech etc.) by or as if by biting the lips. **bite** (or **bite on**) **the bullet** *sl.* behave bravely or stoically. **bite the dust** *sl.* **1** die. **2** fail; break down. **bite the hand that feeds one** hurt or offend a benefactor. **bite a person's head off** *colloq.* respond fiercely or angrily. **bite one's lip** see LIP. **bite off more than one can chew** take on a commitment one cannot fulfil. **once bitten twice shy** an unpleasant experience induces caution. **put the bite on** *US sl.* borrow or extort money from. □ **biter** n. [OE *bitan* f. Gmc]

Bithynia /bɪˈθɪnɪə/ the ancient name for the region of NW Asia Minor west of ancient Paphlagonia, bordering the Black Sea and the Sea of Marmara.

biting /ˈbaɪtɪŋ/ adj. **1** stinging; intensely cold (*a biting wind*). **2** sharp; effective (*biting wit; biting sarcasm*). □ **biting louse** a louse of the order Mallophaga, parasitic mainly on birds. □ **bitingly** adv.

bitmap /ˈbɪtmæp/ n. & v. *Computing* ● n. a representation in which each item is shown by one or more bits of information, esp. a display of the contents of a memory store. ● v.tr. provide with or represent by a bitmap.

bitten past part. of BITE.

bitter /ˈbɪtə(r)/ adj. & n. ● adj. **1** having a sharp pungent taste; not sweet. **2 a** caused by or showing mental pain or resentment (*bitter memories; bitter rejoinder*). **b** painful or difficult to accept (*bitter disappointment*). **3 a** harsh; virulent (*bitter animosity*). **b** piercingly cold. ● n. **1** *Brit.* beer strongly flavoured with hops and having a bitter taste. **2** (in pl.) liquor with a bitter flavour (esp. of wormwood) used as an additive in cocktails. □ **bitter-apple** = COLOCYNTH. **bitter orange** = SEVILLE ORANGE. **bitter pill** something unpleasant that has to be accepted. **bitter-sweet** adj. **1** sweet with a bitter after-taste. **2** arousing pleasure tinged with pain or sorrow. ● n. **1 a** a sweetness with a bitter after-taste. **b** pleasure tinged with pain or sorrow. **2** = woody nightshade (see NIGHTSHADE). **to the bitter end** to the very end in spite of difficulties. □ **bitterly** adv. **bitterness** n. [OE *biter* prob. f. Gmc: *to the bitter end* may be assoc. with a Naut. word *bitter* = 'last part of a cable': see BITTS]

bitterling /ˈbɪtəlɪŋ/ n. a small brightly coloured freshwater fish, *Rhodeus amarus*, of the carp family, found in central Europe. [BITTER + -LING[1]]

bittern[1] /ˈbɪt(ə)n/ n. a large marsh bird of the heron family, with mainly brown plumage; esp. one of the genus *Botaurus*, having a distinctive booming call. [ME f. OF *butor* ult. f. L *butio* bittern + *taurus* bull; -n perh. f. assoc. with HERON]

bittern[2] /ˈbɪt(ə)n/ n. *Chem.* the liquid remaining after the crystallization of common salt from sea water. [prob. f. BITTER adj.]

bitts /bɪts/ n.pl. *Naut.* a pair of posts on the deck of a ship, for fastening cables etc. [ME prob. f. LG: cf. LG & Du. *beting*]

bitty /ˈbɪtɪ/ adj. (**bittier**, **bittiest**) made up of unrelated bits; scrappy. □ **bittily** adv. **bittiness** n.

bitumen /ˈbɪtjʊmɪn/ n. **1** a tarlike mixture of hydrocarbons derived from petroleum naturally or by distillation, and used for road surfacing and roofing. **2** *Austral. colloq.* a tarred road. [L *bitumen -minis*]

bituminize /bɪˈtjuːmɪˌnaɪz/ v.tr. (also **-ise**) convert into, impregnate with, or cover with bitumen. □ **bituminization** /-ˌtjuːmɪnaɪˈzeɪʃ(ə)n/ n.

bituminous /bɪˈtjuːmɪnəs/ adj. of, relating to, or containing bitumen. □ **bituminous coal** a form of coal burning with a smoky flame.

bitzer /ˈbɪtsə(r)/ n. (also **bitser**) *Austral. colloq.* **1** a contraption made from previously unrelated parts. **2** a mongrel dog. [prob. abbr. of *bits and pieces*]

bivalent adj. & n. ● adj. **1** /baɪˈveɪlənt/ *Chem.* = DIVALENT. **2** /ˈbɪvələnt/ *Biol.* (of homologous chromosomes) associated in pairs. ● n. /ˈbɪvələnt/ *Biol.* a pair of homologous chromosomes.

bivalve /ˈbaɪvælv/ n. & adj. ● n. *Zool.* an aquatic mollusc of the class Bivalvia, with a laterally compressed body enclosed within two hinged shells. Also called *lamellibranch*. (See note below.) ● adj. **1** *Zool.* of or relating to this class. **2** *Biol.* having two valves, e.g. of a pea-pod.

▪ The class Bivalvia, also known as Pelecypoda or Lamellibranchia, are entirely aquatic and are usually filter-feeders. They lack a definite head, though some forms have numerous simple eyes. They may be sessile (oysters, mussels, clams), burrowers or borers (cockles, piddocks), or free-swimming (scallops).

bivouac /ˈbɪvʊˌæk/ n. & v. ● n. a temporary open encampment without tents, esp. of soldiers. ● v.intr. (**bivouacked**, **bivouacking**) camp in a bivouac, esp. overnight. [F, prob. f. Swiss G *Beiwacht* additional guard at night]

biweekly /baɪ'wiːklɪ/ *adv., adj., & n.* ● *adv.* **1** every two weeks. **2** twice a week. ● *adj.* produced or occurring biweekly. ● *n.* (*pl.* **-ies**) a biweekly periodical. ¶ See the note at *bimonthly*.

biyearly /baɪ'jɪəlɪ/ *adv. & adj.* ● *adv.* **1** every two years. **2** twice a year. ● *adj.* produced or occurring biyearly. ¶ See the note at *bimonthly*.

biz /bɪz/ *n. colloq.* business. [abbr.]

bizarre /bɪ'zɑː(r)/ *adj.* strange in appearance or effect; eccentric; grotesque. □ **bizarrely** *adv.* **bizarreness** *n.* [F, = handsome, brave, f. Sp. & Port. *bizarro* f. Basque *bizarra* beard]

bizarrerie /bɪ'zɑːrərɪ/ *n.* a bizarre quality; bizarreness. [F]

Bizerta /bɪ'zɜːtə/ (also **Bizerte**) a seaport on the northern coast of Tunisia; pop. (1984) 94,500.

Bizet /'biːzeɪ/, Georges (born Alexandre César Léopold Bizet) (1838–75), French composer. Although his first major work was the Symphony in C major (1855), this was not performed until 1933, and it was as an operatic composer that he strove to make his reputation. Critical acclaim was largely unforthcoming during his lifetime, but *The Pearl Fishers* (1863) and *Carmen* (1875) have since gained considerable popularity; the latter, based on a story by the French novelist Prosper Mérimée (1803–70), is regarded as Bizet's masterpiece.

Bjerknes /'bjɜːknəs/, Vilhelm Frimann Koren (1862–1951), Norwegian geophysicist and meteorologist. His studies of circulation and vortices led to his theory of physical hydrodynamics for atmospheric and oceanic circulation, and he developed mathematical models for weather prediction. With colleagues, he later formulated a theory which accounted for the generation of cyclones and introduced the term *front* to meteorology.

Bk *symb. Chem.* the element berkelium.

bk. *abbr.* book.

BL *abbr.* **1** *Sc. & Ir.* Bachelor of Law. **2** British Library. **3** *hist.* British Leyland. **4** bill of lading.

bl. *abbr.* **1** barrel. **2** black.

blab /blæb/ *v. & n.* ● *v.* (**blabbed, blabbing**) **1** *intr.* **a** talk foolishly or indiscreetly. **b** reveal secrets. **2** *tr.* reveal (a secret etc.) by indiscreet talk. ● *n.* a person who blabs. [ME prob. f. Gmc]

blabber /'blæbə(r)/ *n. & v.* ● *n.* (also **blabbermouth** /'blæbə,maʊθ/) a person who blabs. ● *v.intr.* (often foll. by *on*) talk foolishly or inconsequentially, esp. at length.

Black /blæk/, Joseph (1728–99), Scottish chemist. Black studied the chemistry of gases and formulated the concepts of latent heat and thermal capacity. He developed accurate techniques for following chemical reactions by weighing reactants and products. In studying the chemistry of alkalis he isolated the gas carbon dioxide, going on to investigate its chemistry in turn.

black /blæk/ *adj., n., & v.* ● *adj.* **1** very dark, having no colour from the absorption of all or nearly all incident light (like coal or soot). **2** completely dark from the absence of a source of light (*black night*). **3** (also **Black**) **a** of the human group having dark-coloured skin, esp. of African or Australian Aboriginal descent. **b** of or relating to black people (*black rights*). **4** (of the sky, a cloud, etc.) dusky; heavily overcast. **5** angry, threatening (*a black look*). **6** implying disgrace or condemnation (*in his black books*). **7** wicked, sinister, deadly (*black-hearted*). **8** gloomy, depressed, sullen (*a black mood*). **9** portending trouble or difficulty (*things looked black*). **10** (of hands, clothes, etc.) dirty, soiled. **11** (of humour or its representation) with sinister or macabre, as well as comic, import (*black comedy*). **12** (of tea or coffee) without milk. **13** *Brit.* **a** (of industrial labour or its products) boycotted, esp. by a trade union, in an industrial dispute. **b** (of a person) doing work or handling goods that have been boycotted. **14** dark in colour as distinguished from a lighter variety (*black bear*; *black pine*). ● *n.* **1** a black colour or pigment. **2** black clothes or material (*dressed in black*). **3 a** (in a game or sport) a black piece, ball, etc. **b** the player using such pieces. **4** the credit side of an account (*in the black*). **5** (also **Black**) a member of a dark-skinned race, esp. a Negro or Aborigine. ● *v.tr.* **1** make black (*blacked his face*). **2** polish with blacking. **3** *Brit.* declare (goods etc.) 'black'. □ **beyond the black stump** *Austral. colloq.* in the remote outback. **black Africa** the area of Africa, generally south of the Sahara, where blacks predominate. **black and blue** discoloured by bruises. **black and tan** a drink consisting of stout and ale. **black and white 1** recorded in writing or print (*down in black and white*). **2** (of film etc.) not in colour. **3** consisting of extremes only, oversimplified (*interpreted the problem in black and white terms*). **the black art** black

magic. **black bass** see BASS² 4. **black bean 1 a** a black variety of kidney bean. **b** a fermented soybean, used as flavouring in oriental cookery. **2** a large Australian leguminous tree, *Castanospermum australe*, with heavy seed pods. **black beetle** the common cockroach, *Blatta orientalis*. **black belt 1** a black belt worn by an expert in judo, karate, etc. **2** a person qualified to wear this. **black box 1** a flight-recorder in an aircraft. **2** any complex piece of equipment, usu. a unit in an electronic system, with contents which are mysterious to the user. **black bread** a coarse dark-coloured type of rye bread. **black bryony** see BRYONY. **black damp** = choke-damp. **black diamond** (in *pl.*) coal. **black disc** a long-playing gramophone record, as distinct from a compact disc. **black earth** = CHERNOZEM. **black economy** unofficial economic activity. **black English** the form of English spoken by many blacks, esp. as an urban dialect of the US. **black eye** bruised skin around the eye resulting from a blow. **black-eyed** (or **black-eye**) **bean** a cowpea having white seeds with a black hilum, often dried and stored prior to eating. **black-eyed Susan** a flower with a dark centre; esp. of the genus *Rudbeckia*, with yellow-coloured petals. **black-face 1** a variety of sheep with a black face. **2** the make-up used by a non-black performer playing a black role. **black flag** see FLAG¹. **black frost** see FROST. **black game** (or **grouse**) a European grouse, *Tetrao tetrix*, the male of which has black plumage and a lyre-shaped tail. **black ice** thin hard transparent ice, esp. on a road surface. **black in the face** purple in the face through strangulation, exertion, or passion. **black lead** = GRAPHITE. **black leopard** = PANTHER. **black letter** an old heavy style of type. **black light** *Physics* the invisible ultraviolet or infrared radiations of the electromagnetic spectrum. **black magic** magic involving supposed invocation of evil spirits, harmful or malevolent magic. **black mark** a mark of discredit. **black market** an illicit traffic in officially controlled or scarce commodities. **black marketeer** a person who engages in black market activities. **black mass** a travesty of the Roman Catholic mass in worship of Satan. **black nationalism** advocacy of the national civil rights of US (and occas. other) blacks. **black nightshade** see NIGHTSHADE. **black out 1 a** effect a blackout on. **b** undergo a blackout. **2** obscure windows etc. or extinguish all lights for protection esp. against an air attack. **black pepper** pepper made by grinding the whole dried berry, including the husk, of the pepper plant. **black power** a movement in support of rights and political power for blacks. **black pudding** a black sausage containing pork, dried pig's blood, suet, etc. **black sheep** *colloq.* a member of a family, group, etc. considered disreputable or a failure. **black spot 1** a place of danger or difficulty, esp. on a road (*an accident black spot*). **2** any of various diseases of plants, esp. of roses, producing black spots on leaves. **black stump** *Austral. colloq.* a mythical marker of distance in the outback. **black swan 1** something extremely rare. **2** an Australian swan, *Cygnus atratus*, with black plumage. **black tea** tea that is fully fermented before drying. **black tie 1** a black bow-tie worn with a dinner jacket. **2** *colloq.* formal evening dress. **black tracker** *Austral.* an Aborigine employed to help find persons lost or hiding in the bush. **black velvet** a drink of stout and champagne. **black-water fever** a complication of malaria, in which blood cells are rapidly destroyed, resulting in dark urine. **black widow** a venomous North American spider, *Latrodectus mactans*, of which the female devours its mate. □ **blackish** *adj.* **blackly** *adv.* **blackness** *n.* [OE *blæc*]

blackamoor /'blækə,mʊə(r), -,mɔː(r)/ *n. archaic* a black African; a very dark-skinned person. [BLACK + MOOR]

Black and Tans an armed auxiliary force recruited by the British government to fight Sinn Fein in Ireland 1920–1. They were so called because of the mixture of military (khaki) and constabulary (black) uniforms they wore. Their harsh methods caused an outcry in Britain and America.

blackball /'blækbɔːl/ *v.tr.* reject (a candidate) in a ballot (orig. by voting with a black ball).

blackberry /'blækbərɪ/ *n. & v.* ● *n.* (*pl.* **-ies**) **1** a climbing thorny rosaceous shrub, *Rubus fruticosus*, bearing white or pink flowers. See also BRAMBLE. **2** a black fleshy edible fruit of this plant. ● *v.intr.* (**-ies, -ied**) gather blackberries.

blackbird /'blækbɜːd/ *n.* **1** a common European thrush, *Turdus merula*, of which the male is black with an orange beak. **2** *US* a bird with black plumage, esp. a grackle. **3** *hist.* a kidnapped black or Polynesian on a slave ship.

blackboard /'blækbɔːd/ *n.* a board with a smooth usu. dark surface for writing on with chalk.

black body *n. Physics* a hypothetical perfect absorber and radiator of energy, with no reflecting power. An important theoretical concept, such an object emits electromagnetic radiation whose spectrum shows a peak in intensity at a wavelength inversely proportional to its absolute temperature. This effect is illustrated by a red-hot metal bar changing through yellow to white-hot as it gets hotter; real objects at high temperatures, such as stars, approximate quite closely to black bodies. The correct description of black-body radiation was an early triumph of Max Planck's quantum theory.

blackboy /ˈblækbɔɪ/ n. a tree of the genus *Xanthorrhoea*, native to Australia, with a thick dark trunk and a head of grasslike leaves. Also called *grass tree*.

blackbuck /ˈblækbʌk/ n. a small Indian gazelle, *Antilope cervicapra*, with a black back and white underbelly. Also called *sasin*.

Blackburn /ˈblækbɜːn/ an industrial town in NW England, in Lancashire; pop. (1991) 132,800.

blackcap /ˈblækkæp/ n. a European warbler, *Sylvia atricapilla*, the male of which has a black-topped head.

blackcock /ˈblækkɒk/ n. the male of the black grouse (cf. *grey-hen*).

Black Country the western part of the West Midlands in England, so named after the smoke and dust produced by the coal and iron industries of the 19th century.

blackcurrant /blækˈkʌrənt/ n. **1** a widely cultivated shrub, *Ribes nigrum*, bearing flowers in racemes. **2** the small dark edible berry of this plant.

Black Death the great epidemic of bubonic plague that killed a large proportion of the population of Europe in the mid-14th century. It originated in central Asia and China and spread rapidly through Europe, carried by the fleas of black rats, reaching England in 1348 and killing between one third and one half of the population in a matter of months. Less severe outbreaks of plague occurred at irregular intervals throughout the next few centuries. The name was introduced into English in the early 19th century.

blacken /ˈblækən/ v. **1** tr. & intr. make or become black or dark. **2** tr. speak ill of, defame (*blacken someone's character*).

Blackett /ˈblækɪt/, Patrick Maynard Stuart, Baron (1897–1974), English physicist. During the Second World War he was involved in operational research in the U-boat war and was a member of the Maud Committee which dealt with the development of the atom bomb. He modified the cloud chamber for the study of cosmic rays, being awarded the Nobel Prize for physics in 1948.

blackfellow /ˈblækˌfeləʊ/ n. hist. an Australian Aborigine.

blackfish /ˈblækfɪʃ/ n. **1** a dark-coloured fish, e.g. a North American marine fish, *Centrolophus niger*, and an Australian freshwater fish, *Gadopsis marmoratus*. **2** a salmon at spawning.

blackfly /ˈblækflaɪ/ n. (pl. **-flies**) **1** a thrips or aphid that infests plants, esp. the aphid *Aphis fabae*. **2** a biting fly of the family Simuliidae.

Blackfoot /ˈblækfʊt/ n. & adj. ● n. (pl. same or **Blackfeet**) **1** a member of an American Indian people inhabiting parts of North America. (*See note below.*) **2** The Algonquian language of the Blackfoot. ● adj. of or relating to the Blackfoot or their language.

▪ The Blackfoot confederacy is made up of three groups: the Blackfoot proper or Siksika, the Bloods, and the Peigan. Before the arrival of white settlers they lived by hunting buffalo; about 10,000 Blackfoot now live on reservations in Montana in the US and Saskatchewan and Alberta in Canada.

Black Forest (called in German *Schwarzwald*) a hilly wooded region of SW Germany, lying to the east of the Rhine valley.

Black Forest gateau n. a rich chocolate sponge cake with layers of morello cherries or cherry jam and whipped cream, originally from southern Germany.

Black Friar n. a Dominican friar, so called from the colour of the cloak worn.

blackguard /ˈblægɑːd, -gəd/ n. & v. ● n. a villain; a scoundrel; an unscrupulous, unprincipled person. ● v.tr. abuse scurrilously. □ **blackguardly** adj. [BLACK + GUARD: orig. applied collectively to menials etc.]

blackhead /ˈblækhed/ n. a black-topped pimple on the skin.

Black Hills a range of mountains in east Wyoming and west South Dakota, so called because the densely forested slopes appear dark from a distance. The highest point is Harney Peak (2,207 m, 7,242 ft); the range also includes the sculptured granite face of Mount Rushmore.

black hole n. **1** *Astron.* a region in space with a gravitational field so intense that no matter or radiation can escape. (*See note below.*) **2** a place of confinement for punishment, esp. in the armed services.

▪ Black holes are probably formed when a massive star exhausts its nuclear fuel and collapses under its own gravity. If the star is massive enough, no known force can counteract the increasing gravity, and it will collapse to a point of infinite density. Before this stage is reached, within a certain radius (the event horizon) light itself becomes trapped and the object becomes invisible. It is speculated that black holes of millions of solar masses may lurk at the centre of some galaxies, swallowing passing stars and increasing in mass. Such objects may fuel the enormous radiation of energy from quasars.

Black Hole of Calcutta see CALCUTTA, BLACK HOLE OF.

blacking /ˈblækɪŋ/ n. any black paste or polish, esp. for shoes.

blackjack[1] /ˈblækdʒæk/ n. **1** a pirates' black flag. **2** the card-game pontoon. **3** *N. Amer.* a flexible leaded bludgeon. [BLACK + JACK[1]]

blackjack[2] /ˈblækdʒæk/ n. hist. a tarred-leather vessel for alcoholic liquor. [BLACK + JACK[2]]

Black Jew see FALASHA.

blacklead /ˈblækled/ n. & v. ● n. graphite. ● v.tr. polish with graphite.

blackleg /ˈblækleg/ n. & v. ● n. (often attrib.) Brit. derog. a person who declines to take part in industrial action or who participates in strike-breaking by working for an employer whose regular workers are on strike. ● v.intr. (**-legged**, **-legging**) work as a blackleg.

blacklist /ˈblæklɪst/ n. & v. ● n. a list of persons under suspicion, in disfavour, etc. ● v.tr. put the name of (a person) on a blacklist.

blackmail /ˈblækmeɪl/ n. & v. ● n. **1 a** an extortion of payment in return for not disclosing discreditable information, a secret, etc. **b** any payment extorted in this way. **2** the use of threats or moral pressure. ● v.tr. **1** extort or try to extort money etc. from (a person) by blackmail. **2** threaten, coerce. □ **blackmailer** n. [BLACK + obs. *mail* rent, OE *māl* f. ON *mál* agreement]

Black Maria /məˈraɪə/ n. colloq. a police vehicle for transporting prisoners. [orig. US, apparently named after *Maria* Lee, a black woman who kept a boarding-house in Boston and helped the police in escorting drunk and disorderly customers to jail]

Black Monday 1 Monday 19 October 1987, when massive falls in the value of stocks on Wall Street triggered similar falls in markets around the world. **2** hist. Easter Monday (probably owing to the fact that Mondays in general were held to be unlucky).

Black Monk n. a Benedictine monk, so called from the colour of the habit worn.

Blackmore /ˈblækmɔː(r)/, Richard Doddridge (1825–1900), English novelist and poet. He published several collections of poetry before turning to fiction; his fame rests almost entirely on his popular romantic novel *Lorna Doone* (1869), set on 17th-century Exmoor.

Black Muslim n. a member of a black Islamic organization proposing a separate black nation, founded *c.*1930 in Detroit under the name Nation of Islam. The organization developed under the leadership, from 1934, of Elijah Muhammad (1897–1975), but came to national prominence under the influence of Malcolm X, who joined in 1946. After a period of dissension and conflict, the movement has recently undergone a revival since the appearance of a splinter group led by Louis Farrakhan under the original name.

blackout /ˈblækaʊt/ n. **1** a temporary or complete loss of vision, consciousness, or memory. **2** a loss of power, radio reception, etc. **3** a compulsory period of darkness as a precaution against air raids. **4** a temporary suppression of the release of information, esp. from police or government sources. **5** a sudden darkening of a theatre stage.

Black Panther n. a member of a militant political organization set up in the US in 1966 to fight for black rights. From its peak in the late 1960s it declined in the 1970s after internal conflict and the arrest of some of its leaders.

Blackpool /ˈblækpuːl/ a seaside resort in Lancashire, NW England; pop. (1991) 144,500.

Black Prince (name given to Edward, Prince of Wales and Duke of Cornwall) (1330–76), eldest son of Edward III of England. A soldier of considerable ability, he was responsible for a number of English victories during the early years of the Hundred Years War, most notably that at Poitiers in 1356. His health failed at a relatively early age and he predeceased his father, although his own son eventually came to the throne as Richard II. The name Black Prince apparently derives from the black armour he wore when fighting.

Black Rod (in full **Gentleman Usher of the Black Rod**) the chief usher of the Lord Chamberlain's department of the royal household, who is also usher to the House of Lords, so called from the ebony staff carried as his symbol of office. When the monarch is to deliver a speech in the House of Lords, Black Rod ceremonially summons the Commons to attend by knocking on their door (which has been shut against him) with his staff.

Black Sea a tideless almost landlocked sea bounded by Ukraine, Russia, Georgia, Turkey, Bulgaria, and Romania, and connected to the Mediterranean through the strait of Bosporus and the Sea of Marmara.

Blackshirt /'blækʃɜːt/ n. a member of any of various Fascist organizations. In particular, Blackshirt (*camicia nera*) was the colloquial name given to members of the Italian *Squadre d'Azione* (Action Squads), paramilitary groups founded by Mussolini in 1919 who wore black shirts and patrolled cities assaulting and intimidating suspected socialists and communists. In 1923 the Blackshirts were incorporated into the Fascist Party as a national militia. The term was also applied to the SS in Nazi Germany and to supporters of Oswald Mosley's British Union of Fascists.

blacksmith /'blæksmɪθ/ n. a smith who works in iron.

Blackstone /'blækstəʊn/, Sir William (1723–80), English jurist. His major work was the *Commentaries on the Laws of England* (1765–9), based on lectures given at Oxford University; setting out English legal structure and principles, it became a highly influential exposition of English law, forming the basis of legal education in England and the US.

blackthorn /'blækθɔːn/ n. **1** a thorny rosaceous shrub, *Prunus spinosa*, bearing white flowers before the leaves appear, and blue-black fruits. See also SLOE. **2** a cudgel or walking-stick made from its wood. □ **blackthorn winter** a spell of cold weather at the time the blackthorn flowers in early spring.

blacktop /'blæktɒp/ n. N. Amer. a type of road-surfacing material.

Black Watch (usu. prec. by *the*) the Royal Highland Regiment (so called from its dark tartan uniform).

bladder /'blædə(r)/ n. **1 a** (also **urinary bladder**) a muscular membranous bag in the body of a human or other animal, storing urine. **b** this or part of it or a similar object prepared for various uses. **c** *Anat.* an organ storing bile (*gall-bladder*), air (*swim-bladder*), or other fluid. **2** an inflated pericarp or vesicle in various plants. **3** anything inflated and hollow. [OE *blǣdre* f. Gmc]

bladderwort /'blædəwɜːt/ n. an insectivorous freshwater plant of the genus *Utricularia*, having small bladders for trapping minute aquatic animals.

bladderwrack /'blædəræk/ n. a common brown seaweed, *Fucus vesiculosus*, with fronds containing air bladders which give buoyancy.

blade /bleɪd/ n. **1 a** the flat part of a knife, chisel, etc., that forms the cutting edge. **b** = *razor-blade*. **2** the flattened functional part of an oar, spade, propeller, bat, skate, etc. **3 a** the flat, narrow, usu. pointed leaf of grass and cereals. **b** the whole of such plants before the ear is formed (*in the blade*). **c** *Bot.* the broad thin part of a leaf apart from the petiole. **4** (in full **blade-bone**) a flat bone, e.g. in the shoulder. **5** *Archaeol.* a long narrow flake (see FLAKE[1] 3). **6** *poet.* a sword. **7** *colloq.* (usu. *archaic*) a carefree young fellow. □ **bladed** adj. (also in comb.). [OE *blæd* f. Gmc]

blaeberry /'bleɪbəri/ n. (pl. **-ies**) Brit. = BILBERRY. [ME f. *blae* (Sc. and N. Engl. dial. f. ME *blo* f. ON *blár* f. Gmc: see BLUE[1]) + BERRY]

blag /blæg/ n. & v. sl. ● n. robbery, esp. with violence; theft. ● v.tr. & intr. (**blagged**, **blagging**) rob (esp. with violence); steal. □ **blagger** n. [19th c.: orig. unkn.]

blague /blɑːg/ n. humbug, claptrap. [F]

blagueur /blæ'gɜː(r)/ n. a pretentious talker. [F]

blah /blɑː/ n. (also **blah-blah**) colloq. pretentious nonsense. [imit.]

blain /bleɪn/ n. an inflamed swelling or sore on the skin. [OE *blegen* f. WG]

Blair /bleə(r)/, Anthony Charles Lynton ('Tony') (b.1953), Scottish Labour politician. A practising lawyer since 1976, he entered politics in 1983 when he became MP for Sedgefield. He was appointed shadow Home Secretary in 1992 and was elected leader of the Labour Party in 1994, following the death of the former leader John Smith (1938–94).

Blake[1] /bleɪk/, Peter (b.1932), English painter. In the late 1950s and early 1960s he was prominent in the pop art movement in the UK. His work is characterized by a nostalgia for the popular culture of the 1930s and 1940s and a combination of sophistication and naivety; his work often features collages in which painted images are juxtaposed with real objects. One of his most famous creations is the cover design for the Beatles album *Sergeant Pepper's Lonely Hearts Club Band* (1967).

Blake[2] /bleɪk/, William (1757–1827), English artist and poet. His poems mark the beginning of romanticism and a rejection of the Age of Enlightenment; they include *Songs of Innocence* (1789) and *Songs of Experience* (1794). The short poem known as 'Jerusalem', which later became a popular hymn, appears in *Milton* (1804–8). His major prose work, *The Marriage of Heaven and Hell* (c.1790–3), is a collection of paradoxical and revolutionary aphorisms. His watercolours and engravings, like his writings, were only fully appreciated after his death.

blakey /'bleɪki/ n. (also **Blakey**) (pl. **-eys**) a metal cap on the heel or toe of a shoe or boot. [*Blakey*, name of the manufacturer]

blame /bleɪm/ v. & n. ● v.tr. **1** assign fault or responsibility to. **2** (foll. by *on*) assign the responsibility for (an error or wrong) to a person etc. (*blamed his death on a poor diet*). ● n. **1** responsibility for a bad result; culpability (*shared the blame equally*; *put the blame on the bad weather*). **2** the act of blaming or attributing responsibility; censure (*she got all the blame*). □ **be to blame** (often foll. by *for*) be responsible; deserve censure (*she is not to blame for the accident*). **have only oneself to blame** be solely responsible (for something one suffers). **I don't blame you** etc. I think your etc. action was justifiable. □ **blameable** adj. [ME f. OF *bla(s)mer* (v.), *blame* (n.) f. pop.L *blastemare* f. eccl.L *blasphemare* reproach f. Gk *blasphēmeō* blaspheme]

blameful /'bleɪmfʊl/ adj. deserving blame; guilty. □ **blamefully** adv.

blameless /'bleɪmlɪs/ adj. innocent; free from blame. □ **blamelessly** adv. **blamelessness** n.

blameworthy /'bleɪmwɜːði/ adj. (of a person or action) deserving blame. □ **blameworthiness** n.

blanch /blɑːntʃ/ v. **1** tr. make white or pale by extracting colour. **2** intr. & tr. grow or make pale from shock, fear, etc. **3** tr. Cookery **a** peel (almonds etc.) by scalding. **b** immerse (vegetables or meat) briefly in boiling water. **4** tr. whiten (a plant) by depriving it of light. □ **blanch over** give a deceptively good impression of (a fault etc.) by misrepresentation. [ME f. OF *blanchir* f. *blanc* white, BLANK]

Blanchard /'blænʃɑːd/, Jean Pierre François (1753–1809), French balloonist. Together with American John Jeffries (1744–1819) he made the first crossing of the English Channel by air, flying by balloon from Dover to Calais on 7 Jan. 1785, and was the first to fly a balloon in the US. Blanchard was among the earliest to experiment with parachuting, using animals in his demonstrations; he was killed making practice jumps by parachute from a balloon.

blancmange /blə'mɒnʒ/ n. a sweet opaque gelatinous dessert made with flavoured cornflour and milk. [ME f. OF *blancmanger* f. *blanc* white (cf. BLANK) + *manger* eat f. L *manducare* MANDUCATE]

blanco /'blæŋkəʊ/ n. & v. Mil. ● n. **1** a white substance for whitening belts etc. **2** a similar coloured substance. ● v.tr. (**-oes**, **-oed**) treat with blanco. [F *blanc* white (cf. BLANK)]

bland /blænd/ adj. **1 a** mild, not irritating. **b** tasteless, unstimulating, insipid. **2** gentle in manner; suave. □ **blandly** adv. **blandness** n. [L *blandus* soft, smooth]

blandish /'blændɪʃ/ v.tr. flatter; coax; cajole. [ME f. OF *blandir* (-ISH[2]) f. L *blandiri* f. *blandus* soft, smooth]

blandishment /'blændɪʃmənt/ n. (usu. in pl.) flattery; cajolery.

blank /blæŋk/ adj., n., & v. ● adj. **1 a** (of paper) not written or printed on. **b** (of a document) with spaces left for a signature or details. **2 a** not filled; empty (*a blank space*). **b** unrelieved; plain, undecorated (*a blank wall*). **3 a** having or showing no interest or expression (*a blank face*). **b** void of incident or result. **c** puzzled, nonplussed. **d** having (temporarily) no knowledge or understanding (*my mind went blank*). **4** (with neg. import) complete, downright (*a blank refusal*; *blank despair*). **5** euphem. used in place of an adjective regarded as coarse or abusive. ● n. **1 a** a space left to be filled in a document. **b** a document having blank spaces to be filled. **2** (in full **blank cartridge**) a cartridge containing gunpowder but no bullet, used for training etc. **3** an empty space or period of time. **4 a** a coin-disc before stamping. **b** a metal or wooden block before final shaping. **5 a** a dash written instead of a word or letter, esp. instead of an obscenity. **b** euphem. used in place of a noun regarded as coarse. **6** a domino with one or both halves blank. **7** a lottery ticket that gains no prize. **8** the white centre of the target in archery etc. ● v.tr. **1** (usu. foll. by *off*, *out*) screen, obscure (*clouds blanked out the sun*). **2** (usu. foll. by *out*) cut (a metal blank). **3** N. Amer. defeat without allowing to score. □ **blank cheque 1** a cheque with the amount left for the payee to fill in. **2** colloq. unlimited freedom of action (cf. CARTE BLANCHE). **blank verse** unrhymed verse, esp. iambic pentameters. **draw a blank** elicit no response; fail. □ **blankly** adv. **blankness** n. [ME f. OF *blanc* white, ult. f. Gmc]

blanket /'blæŋkɪt/ *n., adj., & v.* ● *n.* **1** a large piece of woollen or other material used esp. as a bed-covering or to wrap up a person or an animal for warmth. **2** (usu. foll. by *of*) a thick mass or layer that covers something (*blanket of fog*; *blanket of silence*). **3** *Printing* a rubber surface transferring an impression from a plate to paper etc. in offset printing. ● *attrib.adj.* covering all cases or classes; inclusive (*blanket condemnation*; *blanket agreement*). ● *v.tr.* (**blanketed, blanketing**) **1** cover with or as if with a blanket (*snow blanketed the land*). **2** stifle; keep quiet (*blanketed all discussion*). **3** *Naut.* take wind from the sails of (another craft) by passing to windward. □ **blanket bath** a body wash given to a bedridden patient. **blanket bog** an extensive flat peat bog formed in cool regions with high rainfall or humidity. **blanket stitch** a stitch used to neaten the edges of a blanket or other material. **born on the wrong side of the blanket** illegitimate. **electric blanket** an electrically wired blanket used for heating a bed. **wet blanket** *colloq.* a gloomy person preventing the enjoyment of others. [ME f. OF *blancquet*, *blanchet* f. *blanc* white (cf. BLANK)]

blankety /'blæŋkəti/ *adj. & n.* (also **blanky** /'blæŋki/) *Brit. colloq.* = BLANK *adj. & n.* 5.

blanquette /blɒŋ'ket/ *n. Cookery* a dish consisting of white meat, e.g. veal, in a white sauce. [F (as BLANKET)]

Blantyre /blæn'taɪə(r)/ the chief commercial and industrial city of Malawi; pop. (1987) 331,600 (with Limbe, a town 8 km SE of Blantyre). The city was founded in 1876 as a Church of Scotland mission, and is named after the explorer David Livingstone's birthplace in Scotland.

blare /bleə(r)/ *v. & n.* ● *v.* **1** *tr. & intr.* sound or utter loudly. **2** *intr.* make the sound of a trumpet. ● *n.* a loud sound resembling that of a trumpet. [ME f. MDu. *blaren, bleren*, imit.]

blarney /'blɑːni/ *n. & v.* ● *n.* **1** cajoling talk; flattery. **2** nonsense. ● *v.* (**-eys, -eyed**) **1** *tr.* flatter (a person) with blarney. **2** *intr.* talk flatteringly. [*Blarney*, an Irish castle near Cork with a stone said to confer a cajoling tongue on whoever kisses it]

blasé /'blɑːzeɪ/ *adj.* **1** unimpressed or indifferent because of over-familiarity. **2** tired of pleasure; surfeited. [F]

blaspheme /blæs'fiːm/ *v.* **1** *intr.* talk profanely, making use of religious names, etc. **2** *tr.* talk profanely about; revile. □ **blasphemer** *n.* [ME f. OF *blasfemer* f. eccl.L *blasphemare* f. Gk *blasphēmeō*: cf. BLAME]

blasphemy /'blæsfəmɪ/ *n.* (*pl.* **-ies**) **1** profane talk. **2** an instance of this. □ **blasphemous** *adj.* **blasphemously** *adv.* [ME f. OF *blasfemie* f. eccl.L f. Gk *blasphēmia* slander, blasphemy]

blast /blɑːst/ *n., v., & int.* ● *n.* **1** a strong gust of wind. **2 a** a destructive wave of highly compressed air spreading outwards from an explosion. **b** such an explosion. **3** the single loud note of a wind instrument, car horn, whistle, etc. **4** *colloq.* a severe reprimand. **5** a strong current of air used in smelting etc. **6** *sl.* a party; a good time. ● *v.* **1** *tr.* blow up (rocks etc.) with explosives. **2** *tr.* **a** wither, shrivel, or blight (a plant, animal, limb, etc.). **b** destroy, ruin (*blasted her hopes*). **c** strike with divine anger; curse. **3** *intr. & tr.* make or cause to make a loud or explosive noise (*blasted away on his trumpet*). **4** *tr. colloq.* reprimand severely. **5** *colloq.* **a** *tr.* shoot; shoot at. **b** *intr.* shoot. ● *int.* expressing annoyance. □ **at full blast** *colloq.* working at maximum speed etc. **blast-hole** a hole containing an explosive charge for blasting. **blast off** (of a rocket etc.) take off from a launching site. **blast-off** *n.* **1** the launching of a rocket etc. **2** the initial thrust for this. [OE *blæst* f. Gmc]

-blast /blɑːst/ *comb. form Biol.* **1** an embryonic cell (*erythroblast*) (cf. -CYTE). **2** a germ layer of an embryo (*epiblast*). [Gk *blastos* sprout]

blasted /'blɑːstɪd/ *adj. & adv.* ● *attrib.adj.* **1** in senses of BLAST *v.* (*blasted oak*; *blasted building*). **2** damned; annoying (*that blasted dog!*). ● *adv. colloq.* damned; extremely (*it's blasted cold*).

blaster /'blɑːstə(r)/ *n.* **1** in senses of BLAST *v.* **2** *Golf* a heavy lofted club for playing from a bunker.

blast-furnace /'blɑːst,fɜːnɪs/ *n.* a smelting-furnace which uses a blast of compressed hot air. The blast-furnace is the major means of smelting iron ore to produce molten iron. It consists of a tall steel tower, lined internally with a refractory material. The furnace is charged from the top with a mixture of iron ore, coke, and limestone. Hot air is blown into the base of the tower through nozzles. The oxygen in the air reacts with carbon in the coke to produce carbon monoxide, which in turn combines with oxygen from the iron ore to produce carbon dioxide and pure iron. As the carbon dioxide passes up the tower it reacts with carbon to produce more carbon monoxide, and the cycle is repeated. The limestone combines with ash from the coke to form a liquid slag, which absorbs sulphur in the iron and collects at the bottom of the tower, floating above a pool of molten iron. The iron and the slag are tapped off separately from time to time. The exhaust gases are used to

preheat the incoming blast of air and then as fuel for raising steam to drive the air compressors or to heat the coke-ovens. The whole process is therefore a well-integrated method of producing iron continuously on a large scale.

blastula /'blæstjʊlə/ *n.* (*pl.* **blastulae** /-ˌliː/ or *US* **blastulas**) *Zool.* an embryo at an early stage of development when it is a hollow ball of cells. [mod.L f. Gk *blastos* sprout]

blatant /'bleɪt(ə)nt/ *adj.* **1** flagrant, unashamed (*blatant attempt to steal*). **2** offensively noisy or obtrusive. □ **blatancy** *n.* **blatantly** *adv.* [a word used by Spenser (1596), perh. after Sc. *blatand* = bleating]

blather /'blæðə(r)/ *n. & v.* (also **blether** /'bleðə(r)/) ● *n.* foolish chatter. ● *v.intr.* chatter foolishly. [ME *blather*, Sc. *blether*, f. ON *blathra* talk nonsense f. *blathr* nonsense]

blatherskite /'blæðəˌskaɪt/ (also **bletherskate** /'bleðəˌskeɪt/) *n.* **1** a person who blathers. **2** = BLATHER *n.* [BLATHER + *skite*, corrupt. of derog. use of SKATE²]

Blaue Reiter /ˌblaʊə 'raɪtə(r)/ a group of German expressionist painters formed in 1911, based in Munich. The painters of the group, which included Wassily Kandinsky, Jean Arp, and Paul Klee, were stylistically diverse but shared a desire to portray the spiritual side of life. [G, = the blue rider (the title of a painting by Kandinsky)]

Blavatsky /blæ'vætski/, Helena (Petrovna) (known as Madame Blavatsky; née Hahn) (1831–91), Russian spiritualist, born in Ukraine. She went to the US in 1873; in 1875 she founded the Theosophical Society in New York together with the American Henry Steel Olcott (1832–1907) (see THEOSOPHY).

blaze¹ /bleɪz/ *n. & v.* ● *n.* **1** a bright flame or fire. **2 a** a bright glaring light (*the sun set in a blaze of orange*). **b** a full light (*a blaze of publicity*). **3** a violent outburst (of passion etc.) (*a blaze of patriotic fervour*). **4 a** a glow of colour (*roses were a blaze of scarlet*). **b** a bright display (*a blaze of glory*). ● *v.intr.* **1** burn with a bright flame. **2** be brilliantly lighted. **3** be consumed with anger, excitement, etc. **4 a** show bright colours (*blazing with jewels*). **b** emit light (*stars blazing*). □ **blaze away** (often foll. by *at*) **1** fire continuously with rifles etc. **2** work enthusiastically. **blaze up 1** burst into flame. **2** burst out in anger. **like blazes** *sl.* **1** with great energy. **2** very fast. **what the blazes!** *sl.* what the hell! □ **blazingly** *adv.* [OE *blæse* torch, f. Gmc: ult. rel. to BLAZE²]

blaze² /bleɪz/ *n. & v.* ● *n.* **1** a white mark on an animal's face. **2** a mark made on a tree by slashing the bark esp. to mark a route. ● *v.tr.* mark (a tree or a path) by chipping bark. □ **blaze a trail 1** mark out a path or route. **2** be the first to do, invent, or study something; pioneer. [17th c.: ult. rel. to BLAZE¹]

blaze³ /bleɪz/ *v.tr.* proclaim as with a trumpet. □ **blaze abroad** spread (news) about. [ME f. LG or Du. *blāzen* blow, f. Gmc *blæsan*]

blazer /'bleɪzə(r)/ *n.* **1** a coloured, often striped, summer jacket worn by schoolchildren, sportsmen, etc., esp. as part of a uniform. **2** a man's plain jacket, often dark blue, not forming part of a suit. [BLAZE¹ + -ER²]

blazon /'bleɪz(ə)n/ *v. & n.* ● *v.tr.* **1** proclaim (esp. *blazon abroad*). **2** *Heraldry* **a** describe or paint (arms). **b** inscribe or paint (an object) with arms, names, etc. ● *n.* **1** *Heraldry* a correct description of armorial bearings etc. **2** a record or description, esp. of virtues etc. □ **blazonment** *n.* [ME f. OF *blason* shield, of unkn. orig.; verb also f. BLAZE³]

blazonry /'bleɪzənrɪ/ *n. Heraldry* **1 a** the art of describing or painting heraldic devices or armorial bearings. **b** such devices or bearings. **2** brightly coloured display.

bleach /bliːtʃ/ *v. & n.* ● *v.tr. & intr.* whiten by exposure to sunlight or by a chemical process. ● *n.* **1** a bleaching substance. **2** the process of bleaching. □ **bleaching-powder** calcium hypochlorite used esp. to remove colour from materials. [OE *blǣcan* f. Gmc]

bleacher /'bliːtʃə(r)/ *n.* **1 a** a person who bleaches (esp. textiles). **b** a vessel or chemical used in bleaching. **2** (usu. in *pl.*) esp. *N. Amer.* a cheap bench seat at a sports ground, usu. in an outdoor uncovered stand.

bleak¹ /bliːk/ *adj.* **1** bare, exposed; windswept. **2** unpromising; dreary (*bleak prospects*). □ **bleakly** *adv.* **bleakness** *n.* [16th c.: rel. to obs. adjs. *bleach, blake* (f. ON *bleikr*) pale, ult. f. Gmc: cf. BLEACH]

bleak² /bliːk/ *n.* a small river fish of the carp family, esp. *Alburnus alburnus*. [ME prob. f. ON *bleikja*, OHG *bleicha* f. Gmc]

blear /blɪə(r)/ *adj. & v. archaic* ● *adj.* **1** (of the eyes or the mind) dim, dull, filmy. **2** indistinct. ● *v.tr.* make dim or obscure; blur. [ME, of uncert. orig.]

bleary /'blɪərɪ/ *adj.* (**blearier, bleariest**) **1** (of the eyes or mind) dim, dull, full of sleep. **2** indistinct. □ **bleary-eyed 1** having dim sight or wits. **2** half asleep. □ **blearily** *adv.* **bleariness** *n.*

bleat /bliːt/ v. & n. ● v. **1** intr. (of a sheep, goat, or calf) make a weak, wavering cry. **2** intr. & tr. (often foll. by out) speak or say feebly, foolishly, or plaintively. ● n. **1** the sound made by a sheep, goat, etc. **2** a weak, plaintive, or foolish cry. □ **bleater** n. **bleatingly** adv. [OE blǣtan (imit.)]

bleb /bleb/ n. **1** esp. Med. a small blister on the skin. **2** a small bubble in glass or on water. [var. of BLOB]

bleed /bliːd/ v. & n. ● v. (past and past part. **bled** /bled/) **1** intr. emit blood. **2** tr. Med. draw blood from surgically or (hist.) using a leech. **3** colloq. **a** tr. extort money from. **b** intr. part with money lavishly; suffer extortion. **4** intr. (often foll. by for) suffer wounds or violent death (bled for the Revolution). **5** intr. **a** (of a plant) emit sap. **b** (of dye) come out in water. **6** tr. **a** allow (fluid or gas) to escape from a closed system through a valve etc. **b** treat (such a system) in this way. **7** Printing **a** intr. (of a printed area) be cut into when pages are trimmed. **b** tr. cut into the printed area of when trimming. **c** extend (an illustration) to the cut edge of a page. ● n. an act of bleeding (cf. NOSEBLEED). □ **one's** (or **the**) **heart bleeds** usu. iron. one is very sorrowful. [OE blēdan f. Gmc]

bleeder /'bliːdə(r)/ n. **1** coarse sl. a person (esp. as a term of contempt or disrespect) (you bleeder; lucky bleeder). **2** colloq. a haemophiliac.

bleeding /'bliːdɪŋ/ adj. & adv. Brit. coarse sl. expressing annoyance or antipathy (a bleeding nuisance). □ **bleeding heart 1** colloq. a dangerously soft-hearted person. **2** a plant with a part that is dark red; esp. Dicentra spectabilis, which has heart-shaped crimson flowers hanging from an arched stem.

bleep /bliːp/ n. & v. ● n. an intermittent high-pitched sound made electronically. ● v.intr. & tr. **1** make or cause to make such a sound, esp. as a signal. **2** alert or summon by a bleep or bleeps. [imit.]

bleeper /'bliːpə(r)/ n. a small portable electronic device which emits a bleep when the wearer is contacted.

blemish /'blemɪʃ/ n. & v. ● n. a physical or moral defect; a stain; a flaw (not a blemish on his character). ● v.tr. spoil the beauty or perfection of; stain (spots blemished her complexion). [ME f. OF ble(s)mir (-ISH²) make pale, prob. of Gmc orig.]

blench /blentʃ/ v.intr. flinch; quail. [ME f. OE blencan, ult. f. Gmc]

blend /blend/ v. & n. ● v. (past. and past. part. **blended** or poet. **blent** /blent/) **1** tr. **a** mix (esp. sorts of tea, spirits, tobacco, etc.) together to produce a desired flavour etc. **b** produce by this method (blended whisky). **2** intr. form a harmonious compound; become one. **3 a** tr. & intr. (often foll. by with) mingle or be mingled (truth blended with lies; blends well with the locals). **b** tr. (often foll. by in, with) mix thoroughly. **4** intr. (esp. of colours): **a** pass imperceptibly into each other. **b** go well together; harmonize. ● n. **1 a** a mixture, esp. of various sorts of tea, spirits, tobacco, fibres, etc. **b** a combination (of different abstract or personal qualities). **2** a portmanteau word. [ME prob. f. ON blanda mix]

blende /blend/ n. any naturally occurring metal sulphide, esp. zinc blende. [G f. blenden deceive, so called because while often resembling galena it yielded no lead]

blender /'blendə(r)/ n. **1** an implement for blending, esp. a mixing machine used in food preparation for liquidizing, chopping, or puréeing. **2** a person who blends.

Blenheim¹ /'blenɪm/ **1** a battle in 1704 in Bavaria, near the village of Blindheim, in which the English, under the Duke of Marlborough, defeated the French and the Bavarians. (See MARLBOROUGH.) **2** (also **Blenheim Palace**) the Duke of Marlborough's seat at Woodstock near Oxford, a stately home designed by Vanbrugh (1705). The house and its estate were given to the first Duke of Marlborough in honour of his victory at Blenheim.

Blenheim² /'blenɪm/ n. a breed of small spaniel with a red and white coat. [BLENHEIM¹]

Blenheim Orange n. a golden-coloured apple which ripens late in the season.

blenny /'blenɪ/ n. (pl. **-ies**) a small spiny-finned marine fish, esp. of the family Blenniidae, most of which live on the bottom in shallow water. [L blennius f. Gk blennos mucus, with ref. to its mucous coating]

blent /blent/ poet. past and past part. of BLEND.

blepharitis /ˌblefəˈraɪtɪs/ n. Med. inflammation of the eyelids. [Gk blepharon eyelid + -ITIS]

Blériot /'blerɪəʊ/, Louis (1872–1936), French aviation pioneer. Trained as an engineer, he built one of the first successful monoplanes in 1907. On 25 July 1909 he became the first to fly the English Channel (Calais to Dover) in a monoplane. Later he became an aircraft manufacturer, building more than 800 aeroplanes of forty different types between 1909 and 1914.

blesbok /'blesbɒk/ n. (also **blesbuck** /-bʌk/) a subspecies of bontebok, native to southern Africa, having small lyre-shaped horns. [Afrik. f. bles BLAZE² (from the white mark on its forehead) + bok goat]

bless /bles/ v.tr. (past and past part. **blessed**, poet. **blest** /blest/) **1** (of a priest etc.) pronounce words, esp. in a religious rite, to confer or invoke divine favour upon; ask God to look favourably on (bless this house). **2 a** consecrate (esp. bread and wine). **b** sanctify by the sign of the cross. **3** call (God) holy; adore. **4** attribute one's good fortune to (an auspicious time, one's fate, etc.); thank (bless the day I met her; bless my stars). **5** (usu. in passive; often foll. by with) make happy or successful (blessed with children; they were truly blessed). **6** euphem. curse; damn (bless the boy!). □ **(God) bless me** (or **my soul**) an exclamation of surprise, pleasure, indignation, etc. **(God) bless you!** **1** an exclamation of endearment, gratitude, etc. **2** an exclamation made to a person who has just sneezed. **I'm** (or **well, I'm**) **blessed** (or **blest**) an exclamation of surprise etc. **not have a penny to bless oneself with** be impoverished (with reference to the cross on the old silver penny). [OE blǣdsian, blēdsian, blētsian, f. blōd blood (hence mark with blood, consecrate): meaning infl. by its use at the conversion of the English to translate L benedicare praise]

blessed /'blesɪd, blest/ adj. (also poet. **blest** /blest/) **1 a** consecrated (Blessed Sacrament). **b** revered. **2** /blest/ (usu. foll. by with) often iron. fortunate (in the possession of) (blessed with good health; blessed with children). **3** euphem. cursed; damned (blessed nuisance!). **4 a** in paradise. **b** RC Ch. a title given to a dead person as an acknowledgement of his or her holy life; beatified. **5** bringing happiness; blissful (blessed ignorance). □ **blessedly** adv.

blessedness /'blesɪdnɪs/ n. **1** happiness. **2** the enjoyment of divine favour. □ **single blessedness** joc. the state of being unmarried (perversion of Shakespeare's A Midsummer Night's Dream I. i. 78).

Blessed Virgin Mary, the mother of Jesus (see MARY¹).

blessing /'blesɪŋ/ n. **1** the act of declaring, seeking, or bestowing (esp. divine) favour (sought God's blessing; mother gave them her blessing). **2** grace said before or after a meal. **3** a gift of God, nature, etc.; a thing one is glad of (what a blessing he brought it!). □ **blessing in disguise** an apparent misfortune that eventually has good results.

blest poet. var. of BLESSED.

blether var. of BLATHER.

bletherskate var. of BLATHERSKITE.

blew past of BLOW¹, BLOW³.

blewits /'bluːɪts/ n. an edible fungus of the genus Tricholoma, with a lilac stem. [prob. f. BLUE¹]

Bligh /blaɪ/, William (1754–1817), British naval officer. In 1787 he was appointed captain of HMS Bounty on a voyage to the West Indies. Two years later part of his crew mutinied and Bligh was set adrift in an open boat, arriving safely at Timor a few weeks later (see BOUNTY). He later took command of the Glatton under Admiral Nelson in the battle of Copenhagen (1801). While serving as governor of New South Wales (1805–8), he was deposed by disaffected officers, but was subsequently exonerated of blame and promoted to vice admiral.

blight /blaɪt/ n. & v. ● n. **1** a plant disease caused by mildews, rusts, smuts, fungi, or insects. **2** an insect or parasite causing such a disease. **3** any obscure force which is harmful or destructive. **4** an unsightly or neglected urban area. ● v.tr. **1** affect with blight. **2** harm, destroy. **3** spoil. [17th c.: orig. unkn.]

blighter /'blaɪtə(r)/ n. Brit. colloq. a person (esp. as a term of contempt or disparagement). [BLIGHT + -ER¹]

Blighty /'blaɪtɪ/ n. sl. **1** (used esp. by soldiers during the war of 1914–18) England; home. **2** a wound securing a return home. [Anglo-Ind. corrupt. of Hind. bilāyatī, wilāyatī foreign, European]

blimey /'blaɪmɪ/ int. (also **cor blimey** /kɔː/) Brit. sl. an expression of surprise, excitement, alarm, etc. [corrupt. of (God) blind me!]

blimp /blɪmp/ n. **1** (also (**Colonel**) **Blimp**) a proponent of reactionary Establishment opinions. **2 a** a small non-rigid airship. **b** a barrage balloon. **3** a soundproof cover for a cine-camera. □ **blimpery** n. **blimpish** adj. [20th c., of uncert. orig.: in sense 1, a pompous, obese, elderly character invented by cartoonist David Low (1891–1963), and used in anti-German or anti-government drawings before and during the war of 1939–45]

blind /blaɪnd/ adj., v., n., & adv. ● adj. **1** lacking the power of sight. **2 a** without foresight, discernment, intellectual perception, or adequate information (blind effort). **b** (often foll. by to) unwilling or unable to appreciate (a factor, circumstance, etc.) (blind to argument). **3** not governed by purpose or reason (blind forces). **4** reckless (blind hitting). **5 a** concealed (blind ditch). **b** (of a door, window, etc.) walled up. **c** closed at one end. **6** Aeron. (of flying) without direct observation, using instruments only. **7** sl. drunk. ● v. **1** tr. deprive of sight, permanently or temporarily (blinded by tears). **2** tr. (often foll. by to) rob of judgement; deceive (blinded them to the danger). **3** intr. sl. go very fast and dangerously, esp. in a motor vehicle. ● n. **1 a** a screen for a window, esp. on a roller, or with slats (roller blind; Venetian blind). **b** an awning over a shop window. **2 a** something designed or used to hide the truth; a pretext. **b** a legitimate business concealing a criminal enterprise (he's a spy, and his job is just a blind). **3** any obstruction to sight or light. **4** Brit. sl. a heavy drinking-bout. **5** Cards a stake put up by a poker player before the cards dealt are seen. **6** US = HIDE¹. ● adv. blindly (fly blind). □ **bake blind** bake (a flan case, pie base, etc.) without a filling. **blind alley 1** a cul-de-sac. **2** a course of action leading nowhere. **blind as a bat** completely blind. **blind coal** coal burning without a flame. **blind corner** a corner round which a motorist etc. cannot see. **blind date 1** a social engagement between a man and a woman who have not previously met. **2** either of the couple on a blind date. **blind drunk** extremely drunk. **blind gut** the caecum. **blind man's buff** a game in which a blindfold player tries to catch others, who push him or her about. **blind side** a direction in which one cannot see the approach of danger etc. **blind spot 1** Anat. the point of entry of the optic nerve on the retina, insensitive to light. **2** an area in which a person lacks understanding or impartiality. **3** a point of unusually weak radio reception. **blind stamping** (or **tooling**) embossing a book cover without the use of colour or gold leaf. **blind-stitch** n. sewing visible on one side only. ● v.tr. & intr. sew with this stitch. **blind to** incapable of appreciating. **blind with science** overawe with a display of (often spurious) knowledge. **go it blind** act recklessly or without proper consideration. **not a blind bit of** (or **not a blind**) sl. not the slightest; not a single (took not a blind bit of notice; not a blind word out of him). **swear blind** see SWEAR. **turn a** (or **one's**) **blind eye to** pretend not to notice. □ **blindly** adv. **blindness** n. [OE f. Gmc]

blinder /ˈblaɪndə(r)/ n. colloq. **1** an excellent piece of play in a game. **2** (in pl.) US blinkers.

blindfold /ˈblaɪndfəʊld/ v., n., adj., & adv. ● v.tr. **1** deprive (a person) of sight by covering the eyes, esp. with a tied cloth. **2** deprive of understanding; hoodwink. ● n. **1** a bandage or cloth used to blindfold. **2** any obstruction to understanding. ● adj. & adv. **1** with eyes bandaged. **2** without care or circumspection (went into it blindfold). **3** Chess without sight of board and men. [replacing (by assoc. with FOLD¹) ME blindfellen, past part. blindfelled (FELL²) strike blind]

blinding /ˈblaɪndɪŋ/ n. **1** the process of covering a newly made road etc. with grit to fill cracks. **2** such grit.

blindworm /ˈblaɪndwɜːm/ n. = SLOW-WORM.

blink /blɪŋk/ v. & n. ● v. **1** intr. shut and open the eyes quickly and usu. involuntarily. **2** intr. (often foll. by at) look with eyes opening and shutting. **3** tr. (often foll. by back) prevent (tears) by blinking. **b** (often foll. by away, from) clear (dust etc.) from the eyes by blinking. **4** tr. & (foll. by at) intr. shirk consideration of; ignore; condone. **5** intr. **a** shine with an unsteady or intermittent light. **b** cast a momentary gleam. **6** tr. blink with (eyes). ● n. **1** an act of blinking. **2** a momentary gleam or glimpse. **3** = ICEBLINK. □ **on the blink** sl. out of order, esp. intermittently. [partly var. of blenk = BLENCH, partly f. MDu. blinken shine]

blinker /ˈblɪŋkə(r)/ n. & v. ● n. **1** (usu. in pl.) either of a pair of screens attached to a horse's bridle to prevent it from seeing sideways. **2** a device that blinks, esp. a vehicle's indicator. ● v.tr. **1** obscure with blinkers. **2** (as **blinkered** adj.) having narrow and prejudiced views.

blinking /ˈblɪŋkɪŋ/ adj. & adv. Brit. sl. an intensive, esp. expressing disapproval (a blinking idiot; a blinking awful time). [BLINK + -ING² (euphem. for BLOODY)]

blip /blɪp/ n. & v. ● n. **1** a quick popping sound, as of dripping water or an electronic device. **2** a small image of an object on a radar screen. **3** a sudden small movement or fluctuation in a process or system, esp. for the worse. ● v. (**blipped**, **blipping**) **1** intr. make a blip. **2** tr. **a** rap or tap suddenly. **b** press (an accelerator) briefly. [imit.]

Bliss /blɪs/, Sir Arthur (Edward Drummond) (1891–1975), English composer. In his earlier work he showed an interest in avant-garde music, particularly that of Schoenberg and Stravinsky, composing at Elgar's behest A Colour Symphony (1922). In later years, however, he tended more to traditional forms, as in his symphony for orator, chorus, and orchestra Morning Heroes (1930), composed as a memorial to the victims of the First World War. He also wrote ballet and film scores, including the powerful music for Alexander Korda's Things to Come (1934–5).

bliss /blɪs/ n. **1 a** perfect joy or happiness. **b** enjoyment; gladness. **2 a** being in heaven. **b** a state of blessedness. [OE blīths, bliss f. Gmc blīthsjō f. blīthiz BLITHE: sense infl. by BLESS]

blissful /ˈblɪsfʊl/ adj. perfectly happy; joyful. □ **blissful ignorance** fortunate unawareness of something unpleasant. □ **blissfully** adv. **blissfulness** n.

blister /ˈblɪstə(r)/ n. & v. ● n. **1** a small bubble on the skin filled with serum and caused by friction, burning, etc. **2** a similar swelling on any other surface. **3** Med. anything applied to raise a blister. **4** sl. an annoying person. ● v. **1** tr. raise a blister on. **2** intr. come up in a blister or blisters. **3** tr. attack sharply (blistered them with his criticisms). **4** (as **blistering** adj.) causing blisters; severe; hot. □ **blister copper** copper which is almost pure. **blister gas** a poison gas causing blisters on the skin. **blister pack** a bubble pack. [ME perh. f. OF blestre, blo(u)stre swelling, pimple]

blithe /blaɪð/ adj. **1** poet. gay, joyous. **2** careless, casual (with blithe indifference). □ **blithely** adv. **blitheness** n. **blithesome** adj. [OE blīthe f. Gmc]

blithering /ˈblɪðərɪŋ/ adj. colloq. **1** senselessly talkative. **2 a** (attrib.) hopeless; contemptible (esp. in blithering idiot). **b** contemptible. [blither, var. of BLATHER + -ING²]

B.Litt. abbr. Bachelor of Letters. [L Baccalaureus Litterarum]

blitz /blɪts/ n. & v. colloq. ● n. **1 a** an intensive or sudden (esp. aerial) attack. **b** an energetic intensive attack, usu. on a specific task (must have a blitz on this room). **2** (**the Blitz**) the German air raids on London in 1940. ● v.tr. attack, damage, or destroy by a blitz. [abbr. of BLITZKRIEG]

blitzkrieg /ˈblɪtskriːg/ n. an intense military campaign intended to bring about a swift victory. [G, = lightning war]

blizzard /ˈblɪzəd/ n. a severe snowstorm with high winds. [19th-c. US, perh. imit.]

bloat /bləʊt/ v. **1** tr. & intr. inflate, swell (wind bloated the sheets; bloated with gas). **2** tr. (as **bloated** adj.) a swollen, puffed. **b** puffed up with pride or excessive wealth (bloated plutocrat). **3** tr. cure (a herring) by salting and smoking lightly. [obs. bloat swollen, soft and wet, perh. f. ON blautr soaked, flabby]

bloater /ˈbləʊtə(r)/ n. a herring cured by bloating.

blob /blɒb/ n. **1** a small roundish mass; a drop of matter. **2** a drop of liquid. **3** a spot of colour. **4** Cricket sl. a score of 0. □ **blobby** adj. [imit.: cf. BLEB]

bloc /blɒk/ n. a combination of parties, governments, groups, etc. sharing a common purpose. □ **bloc vote** = block vote. [F, = block]

Bloch /blɒx/, Ernest (1880–1959), Swiss-born American composer, of Jewish descent. Before he settled in the US in 1916, his opera Macbeth (1910) was produced in Paris to great acclaim. His musical language derives from the late 19th-century romanticism of Liszt and Richard Strauss; the influence of Jewish musical forms can be seen in the Israel Symphony (1912–16), Solomon (1916), and numerous other orchestral compositions.

block /blɒk/ n., v., & adj. ● n. **1** a solid hewn or unhewn piece of hard material, esp. of rock, stone, or wood (block of ice). **2** a flat-topped block used as a base for chopping, beheading, standing something on, hammering on, or for mounting a horse from. **3 a** a large building, esp. when subdivided (block of flats). **b** a compact mass of buildings bounded by (usu. four) streets. **4** an obstruction; anything preventing progress or normal working (a block in the pipe). **5** a chock for stopping the motion of a wheel etc. **6** a pulley or system of pulleys mounted in a case. **7** (in pl.) any of a set of solid cubes etc., used as a child's toy. **8** Printing a piece of wood or metal engraved for printing on paper or fabric. **9** a head-shaped mould used for shaping hats or wigs. **10** sl. the head (knock his block off). **11** N. Amer. **a** the area between streets in a town or suburb. **b** the length of such an area, esp. as a measure of distance (lives three blocks away). **12** a stolid, unimaginative, or hard-hearted person. **13** a large quantity or allocation of things treated as a unit, esp. shares, seats in a theatre, etc. **14** a set of sheets of paper used for writing, or esp. drawing, glued along one edge. **15** Cricket a spot on which a batsman blocks the ball before the wicket, and rests the bat before playing. **16** Athletics = starting-block. **17** Amer. Football a blocking action. **18** Austral. **a** a tract of land offered to an individual settler by a government. **b** a large area of land. **c** an urban or suburban building plot. ● v.tr. **1 a** (often foll. by up) obstruct (a passage etc.) (the road was blocked; you are blocking my view). **b** put obstacles in the way of (progress etc.). **2** restrict the use or conversion of (currency or any other asset).

3 use a block for making (a hat, wig, etc.). **4** emboss or impress a design on (a book cover). **5** *Cricket* stop (a ball) with a bat defensively. **6** *Amer. Football* intercept (an opponent) with one's body. ● *attrib.adj.* treating (many similar things) as one unit (*block booking*). □ **block and tackle** a system of pulleys and ropes, esp. for lifting. **block capitals** (or **letters**) letters printed without serifs, or written with each letter separate and in capitals. **block diagram** a diagram showing the general arrangement of parts of an apparatus. **block in 1** sketch roughly; plan. **2** confine. **block mountain** *Geol.* a mountain formed by natural faults. **block out 1 a** shut out (light, noise, etc.). **b** exclude from memory, as being too painful. **2** sketch roughly; plan. **block-ship** *Naut.* a ship used to block a channel. **block system** a system by which no railway train may enter a section that is not clear. **block tin** refined tin cast in ingots. **block up 1** confine; shut (a person etc.) in. **2** infill (a window, doorway, etc.) with bricks etc. **block vote** a vote proportional in power to the number of people a delegate represents. **mental** (or **psychological**) **block** a particular mental inability due to subconscious emotional factors. **on the block** *N. Amer.* being auctioned. **put the blocks on** prevent from proceeding. □ **blocker** *n.* [ME f. OF *bloc, bloquer* f. MDu. *blok,* of unkn. orig.]

blockade /blɒˈkeɪd/ *n. & v.* ● *n.* **1** the surrounding or blocking of a place, esp. a port, by an enemy to prevent entry and exit of supplies etc. **2** anything that prevents access or progress. **3** *N. Amer.* an obstruction by snow etc. ● *v.tr.* **1** subject to a blockade. **2** obstruct (a passage, a view, etc.). □ **blockade-runner 1** a vessel which runs or attempts to run into a blockaded port. **2** the owner, master, or one of the crew of such a vessel. **run a blockade** enter or leave a blockaded port by evading the blockading force. □ **blockader** *n.* [BLOCK + -ADE¹, prob. after *ambuscade*]

blockage /ˈblɒkɪdʒ/ *n.* **1** an obstruction. **2** a blocked state.

blockboard /ˈblɒkbɔːd/ *n.* a plywood board with a core of wooden strips.

blockbuster /ˈblɒkˌbʌstə(r)/ *n. sl.* **1** something of great power or size, esp. an epic best-selling book or successful film. **2** a heavy aerial bomb.

blockhead /ˈblɒkhed/ *n.* a stupid person. □ **blockheaded** *adj.*

blockhouse /ˈblɒkhaʊs/ *n.* **1** a reinforced concrete shelter used as an observation point etc. **2** *hist.* a one-storeyed timber building with loopholes, used as a fort. **3** a house made of squared logs.

blockish /ˈblɒkɪʃ/ *adj.* **1** resembling a block. **2** excessively dull; stupid, obtuse. **3** clumsy, rude, roughly hewn.

Bloemfontein /ˈbluːmfɒnˌteɪn/ the capital of Orange Free State and judicial capital of South Africa; pop. (1985) 233,000.

bloke /bləʊk/ *n. Brit. colloq.* a man, a fellow. □ **blokey** *adj.* [Shelta]

blond /blɒnd/ *adj. & n.* ● *adj.* **1** (of hair) light-coloured; fair. **2** (of the complexion, esp. as an indication of race) light-coloured. ● *n.* a person, esp. a woman, with fair hair and skin. □ **blondish** *adj.* **blondness** *n.* [ME f. F f. med.L *blondus, blundus* yellow, perh. of Gmc orig.]

blonde /blɒnd/ *adj. & n.* = BLOND. [F fem. of *blond;* see BLOND]

Blondin /ˈblɒndɪn/, Charles (born Jean-François Gravelet) (1824–97), French acrobat. He is renowned for walking across a tightrope suspended over Niagara Falls in 1859; on subsequent occasions he performed the same feat blindfold, with a wheelbarrow, and carrying a man on his back.

Blood /blʌd/ *n.* (*pl.* same or **Bloods**) a member of an American Indian people belonging to the Blackfoot Confederacy. [BLOOD]

blood /blʌd/ *n. & v.* ● *n.* **1** the red liquid which circulates in the arteries and veins of humans and other vertebrates. (*See below.*) **2** a corresponding fluid in invertebrates. **3** bloodshed, esp. killing. **4** passion, temperament. **5** race, descent, parentage (*of the same blood*). **6** family ties or relations (*own flesh and blood; blood is thicker than water*). **7** a dandy; a man of fashion. ● *v.tr.* **1** give (a hound) a first taste of blood. **2** initiate (a person) by experience. □ **bad blood** ill feeling. **blood-and-thunder** (*attrib.*) *colloq.* sensational, melodramatic. **blood bank** a place where supplies of blood or plasma for transfusion are stored. **blood bath** a massacre. **blood-brother** a brother by birth or by the ceremonial mingling of blood. **blood count** a count of the number of corpuscles in a specific volume of blood. **blood-curdling** horrifying. **blood donor** a person who gives blood for transfusion. **blood feud** a feud between families involving killing or injury. **blood fluke** = SCHISTOSOME. **blood group** any of the various antigenic types of human blood determining compatibility in transfusion. (See LANDSTEINER.) **blood-heat** the normal body temperature of a healthy human being, about 37 °C or 98.4 °F. **blood horse** a thoroughbred. **a person's blood is up** he or she is in a fighting mood. **blood-letting**

1 the surgical removal of some of a patient's blood. **2** bloodshed. **blood-lust** the desire for shedding blood. **blood-money 1** money paid to the next of kin of a person who has been killed. **2** money paid to a hired murderer. **3** money paid for information about a murder or murderer. **blood orange** an orange with red or red-streaked pulp. **blood-poisoning** a diseased state caused by the presence of micro-organisms or their toxins in the blood. **blood pressure** the pressure of the blood in the circulatory system, often measured for diagnosis since it is closely related to the force and rate of the heartbeat and the diameter and elasticity of the arterial walls. **blood-red** red as blood. **blood relation** (or **relative**) a relative by blood, not by marriage. **blood royal** the royal family. **blood serum** see SERUM. **blood sport** a sport involving the wounding or killing of animals, esp. hunting. **blood sugar** the concentration of glucose in the blood. **blood test** a scientific examination of blood, esp. for diagnosis. **blood transfusion** the injection of a volume of blood, previously taken from a healthy person, into a patient. **blood vessel** a vein, artery, or capillary carrying blood. **blood-wort** a plant with red roots or leaves, esp. the red-veined dock. **first blood 1** the first shedding of blood, esp. in boxing. **2** the first point gained in a contest etc. **in one's blood** inherent in one's character. **make a person's blood boil** infuriate. **make a person's blood run cold** horrify. **new** (or **fresh**) **blood** new members admitted to a group, esp. as an invigorating force. **of the blood** royal. **out for a person's blood** set on getting revenge. **taste blood** be stimulated by an early success. **young blood 1** a younger member or members of a group. **2** a rake or fashionable young man. [OE *blōd* f. Gmc]

▪ Most animals depend on a circulating fluid for transporting substances from one part of the body to another. In vertebrates this fluid is blood, which carries food, salts, oxygen, hormones, and cells and molecules of the immune system to tissues, and removes waste products; by virtue of its large volume and rapid circulation it evens out the temperature of the bodily parts. Blood consists of a mildly alkaline aqueous fluid (plasma) containing red cells (erythrocytes), white cells (leucocytes), and platelets; it is red when oxygenated and purple when deoxygenated. Red blood cells carry the protein haemoglobin, which gives blood its colour and can combine with oxygen, thus enabling the blood to carry oxygen from the lungs to the tissues. Carbon dioxide is carried in the reverse direction, mainly as bicarbonate ions. White blood cells protect the body against the invasion of foreign agents (e.g. bacteria). Platelets and other factors present in plasma are concerned in the clotting of blood, preventing haemorrhage.

blooded /ˈblʌdɪd/ *adj.* **1** (of horses etc.) of good pedigree. **2** (in *comb.*) having blood or a disposition of a specified kind (*cold-blooded; red-blooded*).

bloodhound /ˈblʌdhaʊnd/ *n.* a breed of large hound with a very keen sense of smell, used in tracking.

bloodless /ˈblʌdlɪs/ *adj.* **1** without blood. **2** unemotional; cold. **3** pale. **4** without bloodshed (*a bloodless coup*). **5** feeble; lifeless. □ **bloodlessly** *adv.* **bloodlessness** *n.*

bloodline /ˈblʌdlaɪn/ *n.* a line of descent; pedigree, descent.

bloodshed /ˈblʌdʃed/ *n.* **1** the spilling of blood. **2** slaughter.

bloodshot /ˈblʌdʃɒt/ *adj.* (of an eyeball) inflamed, tinged with blood.

bloodstain /ˈblʌdsteɪn/ *n.* a discoloration caused by blood.

bloodstained /ˈblʌdsteɪnd/ *adj.* **1** stained with blood. **2** guilty of bloodshed.

bloodstock /ˈblʌdstɒk/ *n.* thoroughbred horses.

bloodstone /ˈblʌdstəʊn/ *n.* a type of green chalcedony spotted or streaked with red, often used as a gemstone.

bloodstream /ˈblʌdstriːm/ *n.* the blood in circulation.

bloodsucker /ˈblʌdˌsʌkə(r)/ *n.* **1** an animal or insect that sucks blood, esp. a leech. **2** an extortioner. **3** a person who lives off others; a parasite. □ **bloodsucking** *adj.*

bloodthirsty /ˈblʌdˌθɜːstɪ/ *adj.* (**bloodthirstier, bloodthirstiest**) eager for bloodshed. □ **bloodthirstily** *adv.* **bloodthirstiness** *n.*

bloodworm /ˈblʌdwɜːm/ *n.* **1** the bright-red aquatic larvae of some midges of the genus *Chironomus.* **2** = TUBIFEX.

bloody /ˈblʌdɪ/ *adj., adv., & v.* ● *adj.* (**bloodier, bloodiest**) **1 a** of or like blood. **b** running or smeared with blood (*bloody bandage*). **2 a** involving, loving, or resulting from bloodshed (*bloody battle*). **b** sanguinary; cruel (*bloody butcher*). **3** *coarse sl.* expressing annoyance or antipathy, or as an intensive (*a bloody shame; a bloody sight better; not a bloody chocolate left*). **4** red. ● *adv. coarse sl.* as an intensive (*a bloody good job; I'll bloody thump* *him).* ● *v.tr.* (**-ies, -ied**) make bloody; stain with blood. □ **bloody hand** *Heraldry* the armorial device of a baronet. **bloody-minded** *colloq.* deliberately uncooperative. **bloody-mindedly** *colloq.* in a perverse or uncooperative manner. **bloody-mindedness** *colloq.* perversity, contrariness. □ **bloodily** *adv.* **bloodiness** *n.* [OE *blōdig* (as BLOOD, -Y¹)]

Bloody Assizes the trials of the supporters of the Duke of Monmouth after their defeat at the Battle of Sedgemoor, held in SW England in 1685. The government's representative, Judge Jeffreys, sentenced several hundred rebels to death and about 1,000 others to transportation to America as plantation slaves.

Bloody Mary¹ the nickname of Mary I of England (see MARY²).

Bloody Mary² *n.* a drink composed of vodka and tomato juice.

Bloody Sunday 1 in Northern Ireland, 30 Jan. 1972, when British troops shot dead thirteen marchers in Londonderry who were protesting against the government's policy of internment. **2** in Russia, 9 Jan. 1905 (22 Jan. in the New Style calendar), when troops attacked and killed hundreds of unarmed workers who had gathered in St Petersburg to present a petition to the tsar. **3** in Britain, Sunday 13 Nov. in Trafalgar Square, London, 1887, when police violently broke up a socialist demonstration against the British government's Irish policy.

bloom¹ /bluːm/ *n.* & *v.* ● *n.* **1 a** a flower, esp. one cultivated for its beauty. **b** the state of flowering (*in bloom*). **2** a state of perfection or loveliness; the prime (*in full bloom*). **3 a** (of the complexion) a flush; a glow. **b** a delicate powdery surface deposit on plums, grapes, leaves, etc., indicating freshness. **c** a cloudiness on a shiny surface. **d** (in full **algal bloom** or **water-bloom**) the rapid proliferation of microscopic algae or cyanobacteria in water, often forming a coloured scum on or near the surface. ● *v.* **1** *intr.* bear flowers; be in flower. **2** *intr.* **a** come into, or remain in, full beauty. **b** flourish; be in a healthy, vigorous state. **3** *tr. Photog.* coat (a lens) so as to reduce reflection from its surface. □ **take the bloom off** make stale. [ME f. ON *blóm, blómi* etc. f. Gmc: cf. BLOSSOM]

bloom² /bluːm/ *n.* & *v.* ● *n.* a mass of puddled iron hammered or squeezed into a thick bar. ● *v.tr.* make into bloom. [OE *blōma*]

bloomer¹ /ˈbluːmə(r)/ *n. sl.* a blunder. [= BLOOMING *error*]

bloomer² /ˈbluːmə(r)/ *n. Brit.* an oblong loaf with a rounded diagonally slashed top. [20th c.: orig. uncert.]

bloomer³ /ˈbluːmə(r)/ *n.* a plant that blooms (in a specified way) (*early autumn bloomer*).

bloomers /ˈbluːməz/ *n.pl.* **1** women's and girls' loose-fitting almost knee-length knickers. **2** *colloq.* women's knickers. **3** *hist.* women's loose-fitting trousers, gathered at the knee or (orig.) the ankle. They were named after the American social reformer Amelia Bloomer (1818–94), who advocated a costume called 'rational dress' for women, consisting of a short jacket, full skirt reaching to just below the knee, and trousers down to the ankle.

bloomery /ˈbluːmərɪ/ *n.* (*pl.* **-ies**) a factory that makes puddled iron into blooms.

Bloomfield /ˈbluːmfiːld/, Leonard (1887–1949), American linguist. One of the founders of American structural linguistics, his primary aim was to establish linguistics as an autonomous and scientific discipline; his most important work was the influential *Language* (1933).

blooming /ˈbluːmɪŋ/ *adj.* & *adv.* ● *adj.* **1** flourishing; healthy. **2** *Brit. sl.* an intensive (*a blooming miracle*). ● *adv. Brit. sl.* an intensive (*was blooming difficult*). [BLOOM¹ + -ING²: euphem. for BLOODY]

Bloomsbury /ˈbluːmzbərɪ, -brɪ/ *n.* & *adj.* ● *n.* = BLOOMSBURY GROUP. ● *adj.* **1** associated with or similar to the Bloomsbury Group. **2** intellectual; highbrow.

Bloomsbury Group a group of writers, artists, and philosophers living in or associated with Bloomsbury in London in the early 20th century. Members of the group, which included Leonard and Virginia Woolf, Lytton Strachey, Vanessa Bell (1879–1961), Duncan Grant (1885–1978), and Roger Fry (1866–1934), were critical of Victorian-age values and known for their unconventional lifestyles and attitudes. They proved a powerful force in the growth of modernism.

blooper /ˈbluːpə(r)/ *n.* esp. *N. Amer. colloq.* an embarrassing error. [imit. *bloop* + -ER¹]

blossom /ˈblɒsəm/ *n.* & *v.* ● *n.* **1** a flower or a mass of flowers, esp. of a fruit-tree. **2** the stage or time of flowering (*the cherry tree in blossom*). **3** a promising stage (*the blossom of youth*). ● *v.intr.* **1** open into flower. **2** reach a promising stage; mature, thrive. □ **blossomy** *adj.* [OE *blōstm(a)* prob. formed as BLOOM¹]

blot /blɒt/ *n.* & *v.* ● *n.* **1** a spot or stain of ink etc. **2** a moral defect in an

otherwise good character; a disgraceful act or quality. **3** any disfigurement or blemish. **4** *Biochem.* transfer from a medium used for separation to an immobilizing medium where the constituents are identified. ● *v.* (**blotted, blotting**) **1 a** *tr.* spot or stain with ink; smudge. **b** *intr.* (of a pen, ink, etc.) make blots. **2 a** use blotting-paper or other absorbent material to absorb excess ink. **b** (of blotting-paper etc.) soak up (esp. ink). **3** *tr.* disgrace (*blotted his reputation*). **4** *Biochem.* the distribution of proteins etc. on a medium on to which they have been blotted. □ **blot one's copybook** damage one's reputation. **blot on the escutcheon** a disgrace to the family name. **blot out 1 a** obliterate (writing). **b** obscure (a view, sound, etc.). **2** obliterate (from the memory) as too painful. **3** destroy. **blotting-paper** unglazed absorbent paper used for soaking up excess ink. [ME prob. f. Scand.: cf. Icel. *blettr* spot, stain]

blotch /blɒtʃ/ *n.* & *v.* ● *n.* **1** a discoloured or inflamed patch on the skin. **2** an irregular patch of ink or colour. ● *v.tr.* cover with blotches. □ **blotchy** *adj.* (**blotchier, blotchiest**). [17th c.: f. obs. *plotch* and BLOT]

blotter /ˈblɒtə(r)/ *n.* **1** a sheet or sheets of blotting-paper, usu. inserted into a frame. **2** *N. Amer.* a temporary recording-book, esp. a police charge-sheet.

blotto /ˈblɒtəʊ/ *adj. sl.* very drunk, esp. unconscious from drinking. [20th c.: perh. f. BLOT]

blouse /blaʊz/ *n.* & *v.* ● *n.* **1 a** a woman's loose, usu. lightweight, upper garment, usu. buttoned and collared. **b** the upper part of a soldier's or airman's battledress. **2** a workman's or peasant's loose linen or cotton garment, usu. belted at the waist. ● *v.* **1** *tr.* make (a bodice etc.) loose like a blouse. **2** *intr.* swell or hang loosely like a blouse. [F, of unkn. orig.]

blouson /ˈbluːzɒn/ *n.* a short blouse-shaped jacket. [F]

blow¹ /bləʊ/ *v.* & *n.* ● *v.* (*past* **blew** /bluː/; *past part.* **blown** /bləʊn/) **1 a** *intr.* (of the wind or air, or impersonally) move along; act as an air-current (*it was blowing hard*). **b** *intr.* be driven by an air-current (*waste paper blew along the gutter*). **c** *tr.* drive with an air-current (*blew the door open*). **2 a** *tr.* send out (esp. air) by breathing (*blew cigarette smoke; blew a bubble*). **b** *intr.* send a directed air-current from the mouth. **3** *tr.* & *intr.* sound or be sounded by blowing (*the whistle blew; they blew the trumpets*). **4** *tr.* **a** direct an air-current at (*blew the embers*). **b** (foll. by *off, away,* etc.) clear by means of an air-current (*blew the dust off*). **5** *tr.* (*past part.* **blowed**) (esp. in *imper.*) *sl.* curse, confound, damn (*blow it!; I'll be blowed!; let's take a taxi and blow the expense*). **6** *tr.* **a** clear (the nose) of mucus by blowing. **b** remove contents from (an egg) by blowing through it. **7 a** *intr.* puff, pant. **b** *tr.* (esp. in *passive*) exhaust of breath. **8** *sl.* **a** *tr.* depart suddenly from (*blew the town yesterday*). **b** *intr.* depart suddenly. **9** *tr.* shatter or send flying by an explosion (*the bomb blew the tiles off the roof; blew them to smithereens*). **10** *tr.* make or shape (glass or a bubble) by blowing air in. **11** *tr.* & *intr.* melt or cause to melt from overloading (*the fuse has blown*). **12** *intr.* (of a whale) eject air and water through a blowhole. **13** *tr.* break into (a safe etc.) with explosives. **14** *tr. sl.* **a** squander, spend recklessly (*blew £20 on a meal*). **b** spoil, bungle (an opportunity etc.) (*he's blown his chances of winning*). **c** reveal (a secret etc.). **15** *intr.* (of a food-tin etc.) swell and eventually burst from internal gas pressure. **16** *tr.* work the bellows of (an organ). **17** *tr.* (of flies) deposit eggs in. **18** *intr. US* & *Austral. colloq.* boast. ● *n.* **1 a** an act of blowing (e.g. one's nose, a wind instrument). **b** *colloq.* a turn or spell of playing jazz; a musical session. **2 a** a gust of wind or air. **b** exposure to fresh air. **3** = *fly-blow* (see FLY²). **4** *US* a boaster. □ **be blowed if one will** sl. be unwilling to. **blow away** sl. kill, destroy, defeat. **blow-ball** the globular seed-head of a dandelion etc. **blow-dry** *v.tr.* arrange (the hair) while drying it with a hand-held drier. ● *n.* an act of doing this. **blow-drier** (or **-dryer**) a drier used for this. **blow the gaff** reveal a secret inadvertently. **blow hot and cold** *colloq.* vacillate. **blow in 1** break inwards by an explosion. **2** *colloq.* arrive unexpectedly. **blow-job** *coarse sl.* an act or instance of fellatio or cunnilingus. **blow a kiss** kiss one's hand and wave it to a distant person. **blow a person's mind** *sl.* cause a person to have drug-induced hallucinations or a similar experience. **blow off 1** escape or allow (steam etc.) to escape forcibly. **2** *sl.* break wind noisily. **blow on** (or **upon**) **1** make stale. **2** discredit. **blow out 1 a** extinguish by blowing. **b** send outwards by an explosion. **2** (of a tyre) burst. **3** (of a fuse etc.) melt. **4** *US colloq.* a defeat convincingly; thrash. **b** render useless; break. **blow-out** *n. colloq.* **1** a burst tyre. **2** a melted fuse. **3** a huge meal. **4** *US* a resounding defeat. **blow over** (of trouble etc.) fade away without serious consequences. **blow one's own trumpet** praise oneself. **blow one's top** (*N. Amer.* **stack**) *colloq.* explode in rage. **blow up 1 a** shatter or destroy by an explosion. **b** explode, erupt. **2** *colloq.* rebuke strongly. **3** inflate (a tyre etc.). **4** *colloq.* **a** enlarge (a photograph). **b** exaggerate. **5** *colloq.* come to notice; arise. **6** *colloq.* lose one's temper. **blow-up** n. **1** colloq. an enlargement (of a photograph etc.). **2** an explosion. **blow the whistle on** see WHISTLE. [OE blāwan f. Gmc]

blow² /bləʊ/ n. **1** a hard stroke with a hand or weapon. **2** a sudden shock or misfortune. □ **at one blow** by a single stroke; in one operation. **blow-by-blow** (of a description etc.) giving all the details in sequence. **come to blows** end up fighting. **strike a blow for** (or **against**) take action on behalf of (or in opposition to). [15th c.: orig. unkn.]

blow³ /bləʊ/ v. & n. archaic ● v.intr. (past **blew** /bluː/; past part. **blown** /bləʊn/) burst into or be in flower. ● n. blossoming, bloom (in full blow). [OE blōwan f. Gmc]

blower /ˈbləʊə(r)/ n. **1** in senses of BLOW¹ v. **2** a device for creating a current of air. **3** colloq. a telephone.

blowfish /ˈbləʊfɪʃ/ n. a fish that is able to inflate its body when frightened etc., e.g. the globe-fish.

blowfly /ˈbləʊflaɪ/ n. (pl. **-flies**) a fly of the family Calliphoridae, that deposits its eggs on meat and carcases; a meat-fly, e.g. the bluebottle.

blowgun /ˈbləʊɡʌn/ n. US = BLOWPIPE.

blowhard /ˈbləʊhɑːd/ n. & adj. colloq. ● n. a boastful person. ● adj. boastful; blustering.

blowhole /ˈbləʊhəʊl/ n. **1** the nostril of a whale, on the top of its head. **2** a hole (esp. in ice) for breathing or fishing through. **3** a vent for air, smoke, etc., in a tunnel etc.

blowlamp /ˈbləʊlæmp/ n. a portable device with a very hot flame used for burning off paint, soldering, etc.

blown past part. of BLOW¹, BLOW³.

blowpipe /ˈbləʊpaɪp/ n. **1** a tube used esp. by primitive peoples for propelling arrows or darts by blowing. **2** a tube used to intensify the heat of a flame by blowing air or other gas through it at high pressure. **3** a tube used in glass-blowing.

blowtorch /ˈbləʊtɔːtʃ/ n. US = BLOWLAMP.

blowy /ˈbləʊɪ/ adj. (**blowier, blowiest**) windy, windswept.

blowzy /ˈblaʊzɪ/ adj. (**blowzier, blowziest**) **1** coarse-looking; red-faced. **2** dishevelled, slovenly. □ **blowzily** adv. **blowziness** n. [obs. blowze beggar's wench, of unkn. orig.]

blub /blʌb/ v.intr. (**blubbed, blubbing**) sl. sob. [abbr. of BLUBBER¹]

blubber¹ /ˈblʌbə(r)/ n. & v. ● n. **1** whale fat. **2** a spell of weeping. ● v. **1** intr. sob loudly. **2** tr. sob out (words). □ **blubberer** n. **blubbery** adj. [ME perh. imit. (also meanings 'foaming, bubble')]

blubber² /ˈblʌbə(r)/ adj. (of the lips) swollen, protruding. [earlier blabber, blobber, imit.]

bluchers /ˈbluːkəz/ n.pl. hist. strong leather half-boots or high shoes. [G. L. von Blücher, Prussian general (1742–1819)]

bludge /blʌdʒ/ v. & n. Austral. & NZ sl. ● v. **1** intr. shirk responsibility for hard work. **2** tr. cadge, scrounge. ● n. an easy job or assignment. □ **bludge on** impose on. [back-form. f. BLUDGER]

bludgeon /ˈblʌdʒən/ n. & v. ● n. a club with a heavy end. ● v.tr. **1** beat with a bludgeon. **2** coerce. [18th c.: orig. unkn.]

bludger /ˈblʌdʒə(r)/ n. Austral. & NZ sl. a hanger-on; a loafer. [orig. Engl. sl., = pimp, f. obs. bludgeoner f. BLUDGEON]

blue¹ /bluː/ adj., n., & v. ● adj. (**bluer, bluest**) **1** having a colour like that of a clear sky. **2** sad, depressed; (of a state of affairs) gloomy, dismal (feel blue; blue times). **3** indecent, pornographic (a blue film). **4** with bluish skin through cold, fear, anger, etc. **5** Brit. politically conservative. **6** having blue as a distinguishing colour (blue jay). ● n. **1** a blue colour or pigment. **2** blue clothes or material (dressed in blue). **3** Brit. **a** a person who has represented a university (esp. Oxford or Cambridge) in a sport. **b** this distinction. **4** Brit. a supporter of the Conservative Party. **5** a small blue-coloured butterfly of the family Lycaenidae. **6** blue powder used to whiten laundry. **7** Austral. sl. **a** an argument or row. **b** (as a nickname) a red-headed person. **8** a blue ball, piece, etc. in a game or sport. **9** (prec. by the) the clear sky. ● v.tr. (**blues, blued, bluing** or **blueing**) **1** make blue. **2** treat with laundering blue. □ **blue baby** a baby with a blue complexion from lack of oxygen in the blood due to a congenital defect of the heart or large blood vessels. **blue blood** noble birth. **blue-blooded** of noble birth. **blue book** a report issued by Parliament and the Privy Council, issued in a blue cover. **blue cheese** cheese produced with veins of blue mould, e.g. Stilton and Danish blue. **blue-chip** (attrib.) (of shares) of reliable investment, though less secure than gilt-edged stock. **blue-collar** (attrib.) of manual or unskilled work or workers. **blue dahlia** something rare or impossible. **blue ensign** see ENSIGN. **blue-eyed boy** esp. Brit. colloq.

usu. derog. a favoured person; a favourite. **blue funk** sl. a state of great terror or panic. **blue-green alga** = CYANOBACTERIUM. **blue ground** = KIMBERLITE. **blue in the face** in a state of extreme anger or exasperation. **blue line** Ice Hockey a line across the rink midway between the centre of the rink and each goal. **blue metal** broken blue stone used for road-making. **blue mould** a bluish fungus growing on food and other organic matter. **blue-pencil** (**-pencilled, -pencilling**; US **-penciled, -penciling**) censor or make cuts in (a manuscript, film, etc.). **blue ribbon 1** a high honour. **2** Brit. the ribbon of the Order of the Garter. **blue rinse** a preparation for tinting grey hair. **blue roan** see ROAN¹. **blue rock** = rock dove (see ROCK¹). **blue shift** Astron. the displacement of the spectrum to shorter wavelengths in the light coming from distant celestial objects moving towards the observer. **blue tit** a common small European tit, Parus caeruleus, with blue upperparts. **blue vitriol 1** copper sulphate crystals. **2** a bluish-grey stone used for building. **3** = BLUESTONE. **blue water** open sea. **blue whale** a rorqual, Balaenoptera musculus, the largest living mammal. **once in a blue moon** very rarely. **out of the blue** unexpectedly. □ **blueness** n. [ME f. OF bleu f. Gmc]

blue² /bluː/ v.tr. (**blues, blued, bluing** or **blueing**) sl. squander (money). [perh. var. of BLOW¹ v. 14a]

Bluebeard /ˈbluːbɪəd/ a character in a tale by Charles Perrault, who killed several wives in turn for disobeying his order to avoid a locked room, which contained the bodies of his previous wives. Local tradition in Brittany identifies him with Gilles de Rais (c.1400–40), a perpetrator of atrocities, although he had only one wife (who left him). The story was made into an opera, Duke Bluebeard's Castle (1911), by Bartók.

bluebell /ˈbluːbel/ n. **1** a liliaceous woodland plant, Hyacinthoides nonscripta, with clusters of bell-shaped blue flowers on a stem arising from a rhizome. Also called wild hyacinth or wood hyacinth. **2** Sc. = HAREBELL. **3** any other plant with blue bell-shaped flowers.

blueberry /ˈbluːbərɪ/ n. (pl. **-ies**) **1** a dwarf ericaceous shrub of the genus Vaccinium, cultivated for its edible fruit. **2** the small blue-black fruit of this plant.

bluebird /ˈbluːbɜːd/ n. a small North American songbird of the genus Sialia, of the thrush family, with sky-blue upperparts.

bluebottle /ˈbluːˌbɒt(ə)l/ n. **1** a large blowfly, Calliphora vomitoria, with a metallic-blue body. **2** Austral. a Portuguese man-of-war. **3** a dark blue cornflower. **4** Brit. colloq. a police officer.

Bluefields /ˈbluːfiːldz/ a port on the Mosquito Coast of Nicaragua, situated on an inlet of the Caribbean Sea; pop. (1985) 18,000.

bluefish /ˈbluːfɪʃ/ n. an edible blue marine fish; esp. Pomatomus saltatrix, inhabiting tropical waters and popular as a game-fish. Also called (N. Amer.) snapper.

bluegrass /ˈbluːɡrɑːs/ n. **1** a grass with bluish flowers; esp. Poa pratensis, found in Kentucky. **2** a kind of instrumental country-and-western music characterized by virtuosic playing of banjos, guitars, etc.

bluegum /ˈbluːɡʌm/ n. an Australian eucalyptus with bluish bark or leaves; esp. E. regnans, with blue-green aromatic leaves.

bluejacket /ˈbluːˌdʒækɪt/ n. a seaman in the navy.

Blue John n. a purple and white banded variety of fluorite found in Derbyshire, England.

Bluemantle /ˈbluːˌmænt(ə)l/ n. one of four pursuivants of the English College of Arms.

Blue Mountains 1 a section of the Great Dividing Range in New South Wales, Australia. Part of it is a national park and nature reserve. **2** a range of mountains in eastern Jamaica. **3** a range of mountains running from central Oregon to SE Washington state in the US.

Blue Nile one of the two principal headwaters of the Nile. Rising from Lake Tana in NW Ethiopia, it flows some 1,600 km (1,000 miles) southwards then north-westwards into Sudan, where it meets the White Nile at Khartoum.

Blue Peter n. a blue flag with a white square raised on board a ship leaving port.

blueprint /ˈbluːprɪnt/ n. & v. ● n. **1** a photographic print of the final stage of engineering or other plans in white on a blue background. **2** a detailed plan, esp. in the early stages of a project or idea. ● v.tr. US work out (a programme, plan, etc.).

Blue Riband (or **Ribbon**) a trophy for the ship making the fastest eastward sea-crossing of the Atlantic Ocean. It was originally a notional trophy for which liners unofficially but enthusiastically competed, but an actual trophy was instituted in 1935.

Blue Ridge Mountains a range of the Appalachian Mountains in the eastern US, stretching from southern Pennsylvania to northern Georgia. Mount Mitchell is the highest peak, rising to a height of 2,037 m (6,684 ft).

blues /bluːz/ *n.pl.* **1** (prec. by *the*) a bout of depression (*had a fit of the blues*). **2 a** (prec. by *the*; often treated as *sing.*) melancholic music of black American folk origin, often in a twelve-bar sequence. (*See note below.*) **b** (*pl.* same) (as *sing.*) a piece of such music (*the band played a blues*). □ **bluesy** *adj.* (in sense 2).

▪ The blues developed in the rural southern US towards the end of the 19th century, mixing work songs with African and white folk elements. The first blues was published in 1912, 'Memphis Blues' by W. C. Handy (1873–1958), an Alabaman who did much to standardize the genre's form. Early blues singers were frequently itinerant, travelling around the South and accompanying themselves on guitar or piano. Blues began to find a wider audience in the 1940s, as blacks migrated north to the cities, especially Chicago, and formed groups incorporating a rhythm section and amplified instruments. Famous blues singers of this period include Bessie Smith and Billie Holiday; later figures include Muddy Waters and B. B. King. This urban blues gave rise to rhythm and blues and rock and roll and ultimately to much of today's popular music. In the 1960s blues found a new young audience when cover versions of songs by early performers were produced by white groups such as the Rolling Stones.

bluestocking /'bluːˌstɒkɪŋ/ *n.* usu. *derog.* an intellectual or literary woman. The term is derived from a social group formed in London in 1750 by three women (Hannah More, Elizabeth Montagu, and Elizabeth Carter) for literary discussions, in which eminent men of letters often also took part. Many of these favoured informal dress, and the wearing by one man (Benjamin Stillingfleet) of grey or 'blue' worsted stockings instead of formal black silk led to the group's being known as the Blue Stocking Club.

bluestone /'bluːstəʊn/ *n.* **1** a bluish or grey building stone. **2** any of the smaller stones used in the inner parts of Stonehenge, made of dolerite and believed to have come from the Preseli Hills in South Wales.

bluet /'bluːɪt/ *n.* US a blue-flowered plant, esp. the cornflower, or the plant *Hedyotis caerulea*.

bluethroat /'bluːθrəʊt/ *n.* a Eurasian songbird, *Luscinia svecica*, with a mainly blue throat, related to the European robin.

bluey /'bluːɪ/ *n.* (*pl.* **-eys**) *Austral. colloq.* **1** a bundle carried by a bushman. **2** = BLUE¹ *n.* 7b.

bluff¹ /blʌf/ *v.* & *n.* ● *v.* **1** *intr.* make a pretence of strength or confidence to gain an advantage. **2** *tr.* mislead by bluffing. ● *n.* an act of bluffing; a show of confidence or assertiveness intended to deceive. □ **call a person's bluff** challenge a person thought to be bluffing. □ **bluffer** *n.* [19th c. (orig. in poker) f. Du. *bluffen* brag]

bluff² /blʌf/ *adj.* & *n.* ● *adj.* **1** (of a cliff, or a ship's bows) having a vertical or steep broad front. **2** (of a person or manner) blunt, frank, hearty. ● *n.* a steep cliff or headland. □ **bluffly** *adv.* (in sense 2 of *adj.*). **bluffness** *n.* (in sense 2 of *adj.*). [17th-c. Naut. word: orig. unkn.]

bluish /'bluːɪʃ/ *adj.* somewhat blue.

Blum /bluːm/, Léon (1872–1950), French statesman, Prime Minister 1936–7, 1938, 1946–7. A lawyer and literary critic, he was drawn into politics by the Dreyfus affair of 1894; he joined the Socialist Party in 1902 and became its leader in opposition in 1925. During the 1930s he led the Popular Front, being elected France's first socialist and Jewish Prime Minister in 1936. He introduced significant labour reforms, but was forced to resign the following year. Interned in Germany during the Second World War, he returned to France to head a socialist caretaker Cabinet, and retained the party leadership until his death.

Blumenbach /'bluːmənˌbɑːx/, Johann Friedrich (1752–1840), German physiologist and anatomist. He is regarded as the founder of physical anthropology, though his approach has since been much modified. He classified modern humans into five broad categories (Caucasian, Mongoloid, Malayan, Ethiopian, and American), based mainly on cranial measurements.

Blunden /'blʌndən/, Edmund (Charles) (1896–1974), English poet and critic. His prose work *Undertones of War* (1928) is a sensitive account of his experiences in the First World War. His poetry reveals his deep love of the English countryside, as can be seen in *Pastorals* (1916) and *The Waggoner and Other Poems* (1920).

blunder /'blʌndə(r)/ *n.* & *v.* ● *n.* a clumsy or foolish mistake, esp. an important one. ● *v.* **1** *intr.* make a blunder; act clumsily or ineptly. **2** *tr.*
deal incompetently with; mismanage. **3** *intr.* move about blindly or clumsily; stumble. □ **blunderer** *n.* **blunderingly** *adv.* [ME prob. f. Scand.: cf. MSw. *blundra* shut the eyes]

blunderbuss /'blʌndəˌbʌs/ *n.* a short, large-bored, muzzle-loading gun for close-range use, common in the 18th century, with a flared muzzle and a flintlock mechanism, firing many balls or slugs at one shot. [alt. of Du. *donderbus* thunder gun, assoc. with BLUNDER]

blunge /blʌndʒ/ *v.tr.* (in ceramics etc.) mix (clay etc.) with water. [after PLUNGE, BLEND]

Blunt /blʌnt/, Anthony (Frederick) (1907–83), British art historian, Foreign Office official, and Soviet spy. He was director of the Courtauld Institute of Art (1947–74), Surveyor of the King's (later Queen's) Pictures (1945–72), and one of the leading figures in establishing art history as an academic discipline in Britain. In 1965 he confessed that he had been a Soviet agent since the 1930s and had facilitated the escape of the spies Guy Burgess and Donald Maclean in 1951. These facts were made public in 1979, and he was subsequently stripped of the knighthood that he had been awarded in 1956.

blunt /blʌnt/ *adj.* & *v.* ● *adj.* **1** (of a knife, pencil, etc.) lacking in sharpness; having a worn-down point or edge. **2** (of a person or manner) direct, uncompromising, outspoken. ● *v.tr.* make blunt or less sharp. □ **bluntly** *adv.* (in sense 2 of *adj.*). **bluntness** *n.* [ME perh. f. Scand.: cf. ON *blunda* shut the eyes]

blur /blɜː(r)/ *v.* & *n.* ● *v.* (**blurred, blurring**) **1** *tr.* & *intr.* make or become unclear or less distinct. **2** *tr.* smear; partially efface. **3** *tr.* make (one's memory, perception, etc.) dim or less clear. ● *n.* something that appears or sounds indistinct or unclear. □ **blurry** *adj.* (**blurrier, blurriest**). [16th c.: perh. rel. to BLEAR]

blurb /blɜːb/ *n.* a eulogistic description of a book, esp. printed on its jacket, as promotion by its publishers. The term is said to have been coined in 1907 by the American humorist Gelett Burgess (1866–1951), who illustrated a comic book jacket with the picture of a young woman he dubbed Miss Blinda Blurb.

blurt /blɜːt/ *v.tr.* (usu. foll. by *out*) utter abruptly, thoughtlessly, or tactlessly. [prob. imit.]

blush /blʌʃ/ *v.* & *n.* ● *v.intr.* **1 a** develop a pink tinge in the face from embarrassment or shame. **b** (of the face) redden in this way. **2** feel embarrassed or ashamed. **3** be or become red or pink. ● *n.* **1** the act of blushing. **2** a pink tinge. □ **at first blush** on the first glimpse or impression. **spare a person's blushes** refrain from causing embarrassment esp. by praise. [ME f. OE *blyscan*]

blusher /'blʌʃə(r)/ *n.* a cosmetic used to give colour to the face.

bluster /'blʌstə(r)/ *v.* & *n.* ● *v.intr.* **1** behave pompously and boisterously; utter empty threats. **2** (of the wind etc.) blow fiercely. ● *n.* **1** noisily self-assertive talk. **2** empty threats. □ **blusterer** *n.* **blustery** *adj.* [16th c.: ult. imit.]

B-lymphocyte /biːˈlɪmfəˌsaɪt/ *n. Physiol.* a lymphocyte of a type that is not processed by the thymus gland, responsible for producing antibodies. Also called *B-cell*.

Blyton /'blaɪt(ə)n/, Enid (1897–1968), English writer of children's fiction. Her best-known creation for young children was the character Noddy, who first appeared in 1949; her books for older children included the series of *Famous Five* and *Secret Seven* adventure stories. In all she wrote more than 400 books, many of which were translated into other languages.

BM *abbr.* **1** British Museum. **2** Bachelor of Medicine.

BMA *abbr.* British Medical Association.

B.Mus. *abbr.* Bachelor of Music.

BMX /ˌbiːemˈeks/ *n.* **1** organized bicycle-racing on a dirt-track, esp. for youngsters. **2** a kind of bicycle used for this. **3** (*attrib.*) of or related to such racing or the equipment used (*BMX gloves*). [abbr. of bicycle moto-*cross*]

Bn. *abbr.* Battalion.

bn. *abbr.* billion.

B'nai B'rith /bəˌneɪ bəˈriːθ, ˈbrɪθ/ a Jewish organization founded in New York in 1843, which pursues educational, humanitarian, and cultural activities and attempts to safeguard the rights and interests of Jews around the world. [Heb., = sons of the covenant]

BNP see BRITISH NATIONAL PARTY.

BO *abbr. colloq.* body odour.

bo¹ /bəʊ/ *int.* = BOO. [imit.]

bo² /bəʊ/ *n.* US *colloq.* (as a form of address) pal; old chap. [19th c.: perh. f. BOY]

boa /ˈbəʊə/ n. **1** a constricting snake of the family Boidae, esp. any of the larger ones from the New World. **2** any constricting snake; an Old World python. **3** a long thin stole made of feathers or fur. □ **boa constrictor** a large boa, *Boa constrictor*, native to tropical America and the West Indies. [L]

Boadicea see BOUDICCA.

boar /bɔː(r)/ n. **1** (in full **wild boar**) the tusked wild pig, *Sus scrofa*, from which domestic pigs are descended. **2** an uncastrated male pig. **3** its flesh. **4** a male guinea pig etc. [OE *bār* f. WG]

board /bɔːd/ n. & v. ● n. **1 a** a flat thin piece of sawn timber, usu. long and narrow. **b** a piece of material resembling this, made from compressed fibres. **c** a thin slab of wood or a similar substance, often with a covering, used for any of various purposes (*chessboard; ironing-board; notice-board*). **d** thick stiff card used in bookbinding. **2** the provision of regular meals, usu. with accommodation, for payment. **3** *archaic* a table spread for a meal. **4** the directors of a company; any other specially constituted administrative body, e.g. a committee or group of councillors, examiners, etc. **5** (in *pl.*) the stage of a theatre (cf. *tread the boards*). **6** *Naut.* the side of a ship. ● v. **1** *tr.* **a** go on board (a ship, train, aircraft, etc.). **b** force one's way on board (a ship etc.) in attack. **2 a** *intr.* receive regular meals, or (esp. of a schoolchild) meals and lodging, for payment. **b** *tr.* (often foll. by *out*) arrange accommodation away from home for (esp. a child). **c** *tr.* provide (a lodger etc.) with regular meals. **3** *tr.* (usu. foll. by *up*) cover with boards; seal or close. □ **board-game** a game played on a board. **boarding-house** an unlicensed establishment providing board and lodging, esp. to holiday-makers. **boarding kennel** (often in *pl.*) a place providing board for dogs. **boarding-school** a school where pupils are resident in term-time. **board of trade** *N. Amer.* a chamber of commerce. **go by the board** be neglected, omitted, or discarded. **on board** on or on to a ship, aircraft, oil rig, etc. **take on board** consider (a new idea etc.). [OE *bord* f. Gmc]

boarder /ˈbɔːdə(r)/ n. **1** a person who boards (see BOARD v. 2a), esp. a pupil at a boarding-school. **2** a person who boards a ship, esp. an enemy.

boardroom /ˈbɔːdruːm, -rʊm/ n. a room in which a board of directors etc. meets regularly.

boardsailing /ˈbɔːdˌseɪlɪŋ/ n. = WINDSURFING. □ **boardsailor** n. (also **boardsailer**).

boardwalk /ˈbɔːdwɔːk/ n. esp. *N. Amer.* **1** a wooden walkway across sand, marsh, etc. **2** a promenade along a beach.

boart var. of BORT.

Boas /ˈbəʊæz/, Franz (1858–1942), German-born American anthropologist. A pioneer of modern anthropology, he developed the linguistic and cultural components of ethnology. Against the trend of contemporary theory he drew attention to the subjective nature of previous studies, pointing out how they were based on a Western cultural viewpoint. Instead he insisted that cultural development must be explored by means of field studies reconstructing histories of particular cultures. He did much to overturn the theory that Nordic peoples constitute an essentially superior race; his writings were burnt by the Nazis. His works include *Race, Language, and Culture* (1940).

boast /bəʊst/ v. & n. ● v. **1** *intr.* declare one's achievements, possessions, or abilities with indulgent pride and satisfaction. **2** *tr.* own or have as something praiseworthy etc. (*the hotel boasts magnificent views*). ● n. **1** an act of boasting. **2** something one is proud of. □ **boaster** n. **boastingly** *adv.* [ME f. AF *bost*, of unkn. orig.]

boastful /ˈbəʊstfʊl/ adj. **1** given to boasting. **2** characterized by boasting (*boastful talk*). □ **boastfully** *adv.* **boastfulness** n.

boat /bəʊt/ n. & v. ● n. **1** a small vessel propelled on water by an engine, oars, or sails. (See note below.) **2** (in general use) a ship of any size. **3** an elongated boat-shaped jug used for holding sauce etc. ● v.intr. travel or go in a boat, esp. for pleasure. □ **boat-hook** a long pole with a hook and a spike at one end, for moving boats. **boat-house** a shed at the edge of a river, lake, etc., for housing boats. **boat race** a race between rowing crews. (See also BOAT RACE.) **boat-train** a train scheduled to meet or go on a boat. **in the same boat** sharing the same adverse circumstances. **push the boat out** *colloq.* celebrate lavishly. □ **boatful** n. (pl. **-fuls**). [OE *bāt* f. Gmc]

▪ *Boat* is the generic name for small open craft used (for practical purposes or for pleasure) on inland waterways and close to coasts, as distinct from ships, which are sea-going. Some exceptions to this general distinction are fishing boats, gunboats, patrol boats, mail boats, some yachts, and submarines. Usage (especially by people ashore) is not always precise or consistent. (See also SHIP.)

boatel var. of BOTEL.

boater /ˈbəʊtə(r)/ n. a flat-topped hardened straw hat with a brim.

boating /ˈbəʊtɪŋ/ n. rowing or sailing in boats as a sport or form of recreation.

boatload /ˈbəʊtləʊd/ n. **1** enough to fill a boat. **2** *colloq.* a large number of people.

boatman /ˈbəʊtmən/ n. (pl. **-men**) a person who hires out boats or provides transport by boat.

boat people n.pl. refugees who have left a country by sea, in particular those Vietnamese who fled in small boats to Hong Kong, Australia, and elsewhere after the conquest of South Vietnam by North Vietnam in 1975. Some were attacked by pirates and many of those who did arrive were forcibly repatriated or held permanently in refugee camps.

Boat Race an annual rowing competition on the Thames in London between eights of Oxford and Cambridge universities. First rowed at Henley in 1829 (when Oxford won), then from Westminster to Putney in 1836 (won by Cambridge), it became an annual event in 1839. Since 1845 the race has been held over its present course, from Putney to Mortlake (6.8 km, 4.25 miles).

boatswain /ˈbəʊs(ə)n/ n. (also **bo'sun, bosun, bo's'n**) a ship's officer in charge of equipment and the crew. □ **boatswain's chair** a seat suspended from ropes for work on the side of a ship or building. [OE *bātswegen* (as BOAT, SWAIN)]

Boa Vista /ˌbəʊə ˈvɪstə/ a town in northern Brazil, capital of the state of Roraima; pop. (1990) 130,426.

bob[1] /bɒb/ v. & n. ● v.intr. (**bobbed, bobbing**) **1** move quickly up and down; dance. **2** (usu. foll. by *back, up*) **a** bounce buoyantly. **b** emerge suddenly; become active or conspicuous again after a defeat etc. **3** curtsy. **4** (foll. by *for*) try to catch with the mouth alone (fruit etc. floating or hanging). ● n. **1** a jerking or bouncing movement, esp. upward. **2** a curtsy. **3** one of several kinds of change in long peals in bell-ringing. [14th c.: prob. imit.]

bob[2] /bɒb/ n. & v. ● n. **1** a short hairstyle for women and children. **2** a weight on a pendulum, plumb-line, or kite-tail. **3** = BOB-SLED. **4** a horse's docked tail. **5** a short line at or towards the end of a stanza. **6** a knot of hair; a tassel-shaped curl. ● v. (**bobbed, bobbing**) **1** *tr.* cut (a woman's or child's hair) so that it hangs clear of the shoulders. **2** *intr.* ride on a bob-sleigh. [ME: orig. unkn.]

bob[3] /bɒb/ n. (pl. same) *Brit. colloq. hist.* a former shilling (now = 5 decimal pence). [19th c.: orig. unkn.]

bob[4] /bɒb/ n. □ **bob's your uncle** *Brit. sl.* everything is all right. [pet-form of the name *Robert*]

bobbin /ˈbɒbɪn/ n. **1 a** a cylinder or cone holding thread, yarn, wire, etc., used esp. in weaving and machine sewing. **b** a spool or reel. **2** a small bar and string for raising a door-latch. □ **bobbin-lace** lace made by hand with thread wound on bobbins. [F *bobine*]

bobbinet /ˈbɒbɪˌnɛt/ n. machine-made cotton net (imitating lace made with bobbins on a pillow). [BOBBIN + NET[1]]

bobble /ˈbɒb(ə)l/ n. & v. ● n. **1** a small woolly or tufted ball as a decoration or trimming. **2** *US* esp. *Sport* a mistake or error; a bungle. ● v.tr. & intr. *US* esp. *Sport* bungle or fumble, esp. in taking a catch. [dimin. of BOB[2]]

bobby[1] /ˈbɒbɪ/ n. (pl. **-ies**) *Brit. colloq.* a police officer. [pet-form of the name *Robert*, f. Sir *Robert* Peel (see PEEL)]

bobby[2] /ˈbɒbɪ/ n. (pl. **-ies**) (in full **bobby calf**) *Austral. & NZ* an unweaned calf slaughtered for veal. [Engl. dial.]

bobby-dazzler /ˈbɒbɪˌdæzlə(r)/ n. *colloq.* a remarkable or excellent person or thing. [dial., rel. to DAZZLE]

bobby-pin /ˈbɒbɪˌpɪn/ n. *N. Amer., Austral., & NZ* a flat hairpin. [BOB[2] + -Y[2]]

bobby socks n.pl. esp. *N. Amer.* short socks reaching just above the ankle.

bobcat /ˈbɒbkæt/ n. a small North American lynx, *Felix rufus*, with a spotted reddish-brown coat and a short tail. [BOB[2] + CAT[1]]

bobolink /ˈbɒbəˌlɪŋk/ n. a North American oriole, *Dolichonyx oryzivorus*. [orig. *Bob (o') Lincoln*: imit. of its call]

Bobruisk (also **Bobruysk**) see BABRUISK.

bob-sled /ˈbɒbslɛd/ n. & v. ● n. **1** *N. Amer.* a type of sledge consisting of two short sledges coupled together and used for tobogganing and drawing logs. **2** a mechanically braked and steered sledge used in winter sport. ● v.intr. race in a bob-sled. [BOB[2] + SLED]

bob-sledding /ˈbɒbˌslɛdɪŋ/ n. riding in a bob-sled, especially as a winter sport. Bob-sleds are normally manned by crews of two or four and are guided down a specially prepared track of solid ice with banked bends.

bob-sleigh /'bɒbsleɪ/ *n.* = BOB-SLED. □ **bob-sleighing** *n.* [BOB² + SLEIGH]

bobstay /'bɒbsteɪ/ *n.* the chain or rope holding down a ship's bowsprit. [prob. BOB¹ + STAY²]

bobtail /'bɒbteɪl/ *n.* a docked tail; a horse or a dog with a bobtail. [BOB² + TAIL¹]

bocage /bə'kɑːʒ/ *n.* the representation of silvan scenery in ceramics. [F f. OF *boscage*: see BOSCAGE]

Boccaccio /bə'kɑːtʃɪˌəʊ/, Giovanni (1313–75), Italian writer, poet, and humanist. His most famous work, the *Decameron* (1348–58), is a collection of 100 tales told by a group of ten young people living in the country to escape the Black Death. Boccaccio collected the stories, which range from the serious to the light-hearted and the bawdy, from contemporary popular fiction and transformed them into a narrative whole. He is an important figure in the history of narrative fiction and influenced many later writers, including Chaucer, Shakespeare, and Tennyson.

Boccherini /ˌbɒkə'riːnɪ/, Luigi (1743–1805), Italian composer and cellist. A prolific composer, chiefly of chamber music, he is especially known for his cello concertos and sonatas.

Boche /bɒʃ/ *n. & adj. sl. derog.* ● *n.* **1** a German, esp. a soldier. **2** (prec. by *the*) Germans, esp. German soldiers, collectively. ● *adj.* German. [F sl., orig. = rascal: applied to Germans in the First World War]

Bochum /'bəʊxʊm/ an industrial city in the Ruhr valley, North Rhine-Westphalia, Germany; pop. (1991) 398,580.

bock /bɒk/ *n.* a strong dark German beer. [F f. G abbr. of *Eimbockbier* f. *Einbeck*, a town in Lower Saxony]

BOD *abbr.* biochemical oxygen demand.

bod /bɒd/ *n. Brit. colloq.* a person. [abbr. of BODY]

bodacious /bə'deɪʃəs/ *adj. US sl.* remarkable, fabulous; excellent. [orig. uncert.: perh. f. Engl. dial. *boldacious*, blend of BOLD and AUDACIOUS]

bode /bəʊd/ *v.tr.* **1** portend, foreshow. **2** foresee, foretell (evil). □ **bode well** (or **ill**) show good (or bad) signs for the future. [OE *bodian* f. *boda* messenger]

bodega /bəʊ'diːgə/ *n.* a cellar or shop selling wine and food, esp. in a Spanish-speaking country. [Sp. f. L *apotheca* f. Gk *apothēkē* storehouse]

Bodensee /'bɔːd(ə)nˌzeː/ the German name for Lake Constance (see CONSTANCE, LAKE).

Bode's law /bəʊdz/ *n. Astron.* an empirical formula which predicts the distances from the sun of successive planets, giving reasonable figures for the first seven planets. [J. E. *Bode*, German astronomer (1747–1826), who published it]

bodge var. of BOTCH.

bodgie /'bɒdʒɪ/ *n. Austral. colloq.* **1** a youth, esp. of the 1950s, analogous to the British teddy boy. **2** something flawed or worthless. [f. Brit. dial. *bodge* work clumsily]

Bodhgaya /ˌbɒdgə'jɑː, ˌbəʊd-/ (also **Buddh Gaya** /ˌbʊd gə'jɑː/) a village in the state of Bihar, NE India, where Siddhartha Gautama (see BUDDHA) attained enlightenment. A bo tree there is said to be a descendant of the tree under which he meditated.

bodhisattva /ˌbəʊdɪ'sætvə/ *n. Buddhism* a person who is destined to become enlightened. Whereas the goal of the ancient, more orthodox schools of Buddhism (Hinayana, Theravada) is to attain nirvana oneself, that of Mahayana Buddhism is to become a bodhisattva, postponing one's own salvation in order to help others on the spiritual path. The term is applied to the Buddha before his enlightenment. [Skr., = one whose essence is perfect knowledge]

bodice /'bɒdɪs/ *n.* **1** the part of a woman's dress (excluding sleeves) which is above the waist. **2** a woman's undergarment, like a vest, for the same part of the body. [orig. *pair of bodies* = stays, corsets]

bodiless /'bɒdɪlɪs/ *adj.* **1** lacking a body. **2** incorporeal, insubstantial.

bodily /'bɒdɪlɪ/ *adj. & adv.* ● *adj.* of or concerning the body. ● *adv.* **1** with the whole bulk; as a whole (*threw them bodily*). **2** in the body; as a person.

bodkin /'bɒdkɪn/ *n.* **1** a blunt thick needle with a large eye used esp. for drawing tape etc. through a hem. **2** a long pin for fastening hair. **3** a small pointed instrument for piercing cloth, removing a piece of type for correction, etc. [ME perh. f. Celt.]

Bodleian Library /'bɒdlɪən/ (*colloq.* **Bodley** /'bɒdlɪ/) the library of Oxford University. The first library was founded in the 14th century

and benefited from the manuscript collections donated by Humphrey, Duke of Gloucester (1391–1447). It was refounded by the Diplomat and scholar Sir Thomas Bodley (1545–1613), and opened as the Bodleian Library in 1602. In 1610 the Stationers' Company agreed to give to the library a copy of every book printed in England, and the Bodleian is now one of the six copyright libraries entitled to receive on demand a copy of every book, periodical, etc. published in the UK. It also houses one of the world's most extensive collections of Western and oriental manuscripts.

Bodoni /bə'dəʊnɪ/, Giambattista (1740–1813), Italian printer. The typeface which he designed (characterized by extreme contrast between uprights and diagonals), and others based on it, are named after him.

Bodrum /'bɒdrəm/ a resort town on the Aegean coast of western Turkey, site of the ancient city of Halicarnassus.

body /'bɒdɪ/ *n. & v.* ● *n.* (*pl.* **-ies**) **1** the physical structure, including the bones, flesh, and organs, of a person or an animal, whether dead or alive. **2** the trunk apart from the head and the limbs. **3 a** the main or central part of a thing (*body of the car*). **b** the bulk or majority; the aggregate (*body of opinion*). **4 a** a group of persons regarded collectively, esp. as having a corporate function (*governing body*). **b** (usu. foll. by *of*) a collection (*body of facts*). **5** a quantity (*body of water*). **6** a piece of matter (*celestial body*). **7** *colloq.* a person. **8** a full or substantial quality of flavour, tone, etc., e.g. in wine, musical sounds, etc. **9** *colloq.* a body stocking. ● *v.tr.* (**-ies**, **-ied**) (usu. foll. by *forth*) give body or substance to. □ **body bag** a bag for carrying a corpse from the scene of warfare, an accident, etc. **body-blow** a severe setback. **body-building** the practice of strengthening the body, esp. shaping and enlarging the muscles, by exercise. **body-colour** an opaque pigment. **body double** a stand-in for a film actor during stunt or nude scenes. **body language** the process of communicating through conscious or unconscious gestures and poses. **body-line bowling** *Cricket* persistent fast bowling on the leg side threatening the batsman's body. **body louse** a variety of the louse *Pediculus humanus*, infesting human body hair. **body odour** the smell of the human body, esp. when unpleasant. **body-piercing** the piercing of a hole in some part of the body other than the ear lobe. **body politic** the nation or state as a corporate body. **body-popping** a kind of dancing with jerky robotic movements of the joints. **body scanner** a scanning X-ray machine for taking tomograms of the whole body. **body search** a search of a person's body (esp. conducted by customs officials or the police) for illicit weapons, drugs, etc. **body shop** a workshop where repairs to the bodywork of vehicles are carried out. **body stocking** a woman's undergarment, usually made of knitted cotton or nylon, which covers the torso. **body warmer** a sleeveless quilted or padded jacket worn as an outdoor garment. **body wave** a soft light permanent wave, designed to give the hair fullness. **in a body** all together. **keep body and soul together** keep alive, esp. barely. **over my dead body** *colloq.* entirely without my assent. □ **-bodied** *adj.* (in *comb.*) (*able-bodied*). [OE *bodig*, of unkn. orig.]

body-check /'bɒdɪˌtʃek/ *n. & v. Sport* ● *n.* a deliberate obstruction of one player by another. ● *v.tr.* obstruct in this way.

bodyguard /'bɒdɪˌgɑːd/ *n.* a person or group of persons escorting and protecting another person (esp. a dignitary).

body-snatcher /'bɒdɪˌsnætʃə(r)/ *n.* a person who illicitly disinterred corpses for dissection. Before the Anatomy Act (1832), there was no provision for supplying bodies to medical students for anatomical study, and disinterment was profitable though illegal. □ **body-snatching** *n.*

bodysuit /'bɒdɪˌsuːt, -ˌsjuːt/ *n.* a close-fitting one-piece stretch garment for women, used mainly for sport.

bodywork /'bɒdɪˌwɜːk/ *n.* the outer shell of a vehicle.

Boeotia /bɪ'əʊʃə/ a department of central Greece, to the north of the Gulf of Corinth, and a region of ancient Greece of which the chief city was Thebes. Hesiod, Pindar, and Plutarch all came from Boeotia. □ **Boeotian** *adj. & n.*

Boer /bɔː(r), 'bəʊə(r), bʊə(r)/ *n.* any of the early Dutch or Huguenot inhabitants of southern Africa or their descendants, esp. one of the early settlers of Transvaal or Orange Free State. Immigration took place chiefly 1652–1700, the rise in population thereafter being largely by natural increase. Calvinist by religion, the Boers developed their own fiercely self-sufficient lifestyle, many becoming wandering farmers, and by the late 18th century their own language, Afrikaans, had evolved from Dutch. Opposition to the British administration after Cape Colony became British in 1806 led eventually to the Great Trek of 1835–7 and the Boer Wars. After the defeat of the Boers in the

Second Boer War in 1902, Transvaal and Orange Free State ceased to be independent Boer republics. South Africans of Dutch descent are today usually called Afrikaners. [Du.: see BOOR]

Boer War either of two wars fought by Great Britain in southern Africa. The first (1880–1) began with the revolt of the Boer settlers in Transvaal against British rule and ended after the British defeat at Majuba Hill with the establishment of an independent Boer Republic under British suzerainty. The second (1899–1902) was caused by the Boer refusal to grant equal rights to recent British immigrants and by the imperialist ambitions of Cecil Rhodes and some Conservative politicians. In the early stages of the war the Boers gained a series of remarkable victories, but after the arrival of Roberts and Kitchener the British succeeded in capturing the Boer capital Pretoria and driving the Boer leader Kruger into exile. The second half of the war was dominated by guerrilla warfare, the British victory eventually being obtained through the use of almost half a million British and imperial troops and the employment of concentration camps to control the countryside.

Boethius /bəʊˈiːθɪəs/, Anicius Manlius Severinus (c.480–524), Roman statesman and philosopher. He is best known for *The Consolation of Philosophy*, a work written in a mixture of prose and verse while he was in prison for treason. In this he argued that the soul can attain happiness in affliction by realizing the value of goodness and meditating on the reality of God. While drawing upon Stoicism and Neoplatonism, his work echoed Christian sentiments and exercised considerable influence throughout the Middle Ages.

boffin /ˈbɒfɪn/ *n.* esp. *Brit. colloq.* a person engaged in scientific (esp. military) research. [20th c.: orig. unkn.]

Bofors gun /ˈbəʊfəz/ *n.* a type of light anti-aircraft gun. [*Bofors* in Sweden]

bog /bɒg/ *n. & v.* ● *n.* **1 a** wet spongy ground. **b** a stretch of such ground. **2** *Brit. sl.* a lavatory. ● *v.tr.* (**bogged, bogging**) (foll. by *down*; usu. in passive) impede (*was bogged down by difficulties*). □ **bog asphodel** a yellow-flowered liliaceous marsh plant, *Narthecium ossifragum*, native to Europe. **bog-bean** = BUCKBEAN. **bog cotton** = *cotton-grass*. **bog moss** = SPHAGNUM. **bog myrtle** a deciduous shrub, *Myrica gale*, which grows in damp open places and has short upright catkins and aromatic grey-green leaves (also called *sweet-gale*). **bog oak** an ancient dead oak which has been preserved in a black state in peat. **bog off** (usu. in *imper.*) *sl.* go away. **bog spavin** see SPAVIN. **bog standard** *sl.* basic, standard, unexceptional. **bog-trotter** *sl. derog.* an Irishman. □ **boggy** *adj.* (**boggier, boggiest**). **bogginess** *n.* [Ir. or Gael. *bogach* f. *bog* soft]

bogan /ˈbəʊgən/ *n. Austral. sl.* a gormless person. [20th c.: orig. uncert.]

Bogarde /ˈbəʊgɑːd/, Sir Dirk (born Derek Niven van den Bogaerde) (b.1921), British actor and writer, of Dutch descent. He became famous in the 'Doctor' series of comedy films (including *Doctor in the House*, 1953). His later films include *The Servant* (1963), *Death in Venice* (1971), and *A Bridge Too Far* (1977). He has also published a number of volumes of autobiography and several novels.

Bogart /ˈbəʊgɑːt/, Humphrey (DeForest) (1899–1957), American actor. His acting career began on the stage in 1922; his success as a ruthless gangster in the play *The Petrified Forest* was repeated in the screen version of 1936. Many memorable gangster films followed, including *They Drive by Night* (1940). His other films include *Casablanca* (1942) and *The African Queen* (1951).

bogey[1] /ˈbəʊgɪ/ *n. & v. Golf* ● *n.* (*pl.* **-eys**) **1** a score of one stroke more than par at any hole. **2** *hist.* the number of strokes that a good player should need for the course or for a hole; par. ● *v.tr.* (**-eys, -eyed**) play (a hole) in one stroke more than par. [perh. f. *Bogey* as an imaginary player]

bogey[2] /ˈbəʊgɪ/ *n.* (also **bogy**) (*pl.* **-eys** or **-ies**) **1** an evil or mischievous spirit; a devil. **2** an awkward thing or circumstance. **3** *sl.* a piece of nasal mucus. [19th c., orig. as a proper name: cf. BOGLE]

bogey[3] /ˈbəʊgɪ/ *n. & v.* (also **bogie**) *Austral.* ● *n.* a swim or bathe; a bath. ● *v.intr.* swim, bathe. [Dharuk *bugi* bathe, dive]

bogeyman /ˈbəʊgɪˌmæn/ *n.* (also **bogyman**) (*pl.* **-men**) a person (real or imaginary) causing fear or difficulty.

boggle /ˈbɒg(ə)l/ *v. colloq.* **1** *intr.* be startled or baffled (esp. *the mind boggles*). **2** *intr.* (usu. foll. by *about, at*) hesitate, demur. **3** *tr.* baffle, overwhelm mentally. [prob. f. dial. *boggle* BOGEY[2]]

bogie[1] /ˈbəʊgɪ/ *n.* esp. *Brit.* **1** a wheeled undercarriage pivoted below the end of a rail vehicle. **2** a small truck used for carrying coal, rubble, etc. [19th-c. north. dial. word: orig. unkn.]

bogie[2] var. of BOGEY[3].

bogle /ˈbəʊg(ə)l/ *n.* **1** = BOGEY[2]. **2** a phantom. **3** a scarecrow. [orig. Sc. (16th c.), prob. rel. to BOGEY[2]]

Bogotá /ˌbɒgəˈtɑː/ the capital of Colombia, situated in the eastern Andes at about 2,610 m (8,560 ft); pop. (est. 1990) 4,819,700. It was founded by the Spanish in 1538 on the site of a pre-Columbian centre of the Chibcha culture.

bogus /ˈbəʊgəs/ *adj.* sham, fictitious, spurious. □ **bogusly** *adv.* **bogusness** *n.* [19th-c. US word: orig. unkn.]

bogy var. of BOGEY[2].

bogyman var. of BOGEYMAN.

Bo Hai /bəʊ ˈhaɪ/ (also **Po Hai** /pəʊ/; also called *Gulf of Chihli*) a large inlet of the Yellow Sea, on the coast of eastern China.

bohea /bəʊˈhiː/ *n.* a black China tea, the last crop of the season and usu. regarded as of low quality. [*Bu-i* (Wuyi) Hills in China]

Bohemia /bəʊˈhiːmɪə/ a region forming the western part of the Czech Republic. Formerly a Slavonic kingdom, it fell under Austrian rule in 1526, and by the Treaty of Versailles in 1919 became a province of the newly formed Czechoslovakia. Prague was the capital of Bohemia from the 14th century.

Bohemian /bəʊˈhiːmɪən/ *n. & adj.* ● *n.* **1** a native of Bohemia, a Czech. **2** (also **bohemian**) a socially unconventional person, esp. an artist or writer. ● *adj.* **1** of, relating to, or characteristic of Bohemia or its people. **2** socially unconventional. □ **bohemianism** *n.* (in sense 2). [BOHEMIA + -AN: sense 2 f. F *bohémien* gypsy]

boho /ˈbəʊhəʊ/ *n. & adj. colloq.* ● *n.* a person who has an unconventional or Bohemian lifestyle. ● *adj.* socially unconventional, Bohemian. [abbr. of BOHEMIAN + -O]

Bohol /bəʊˈhɒl/ an island lying to the north of Mindanao in the central Philippines; chief town, Tagbilaran.

Bohr /bɔː(r)/, Niels Hendrik David (1885–1962), Danish physicist and pioneer in quantum physics. Bohr's theory of the structure of the atom incorporated quantum theory for the first time and is the basis for present-day quantum-mechanical models. In 1927 he proposed the principle of complementarity, which accounted for the paradox of regarding subatomic particles both as waves and as particles. In the 1930s Bohr fled from Nazi persecution, escaping from German-occupied Denmark in 1943 and helping to develop the atom bomb, first in Britain and then in the US. He later became concerned about the implications of atomic weapons and stressed the need to study the peaceful applications of atomic energy. Niels Bohr was awarded the 1922 Nobel Prize for physics; his son, Aage Niels Bohr (b.1922), shared the 1975 Prize for his studies of the physics of the atomic nucleus.

boil[1] /bɔɪl/ *v. & n.* ● *v.* **1** *intr.* **a** (of a liquid) start to bubble up and turn into vapour; reach a temperature at which this happens. **b** (of a vessel) contain boiling liquid (*the kettle is boiling*). **2 a** *tr.* bring (a liquid or vessel) to a temperature at which it boils. **b** *tr.* cook (food) by boiling. **c** *intr.* (of food) be cooked by boiling. **d** *tr.* subject to the heat of boiling water, e.g. to clean. **3** *intr.* **a** (of the sea etc.) undulate or seethe like boiling water. **b** (of a person or feelings) be greatly agitated, esp. by anger. **c** *colloq.* (of a person or the weather) be very hot. ● *n.* the act or process of boiling; boiling-point (*on the boil*; *bring to the boil*). □ **boil down 1** reduce volume by boiling. **2** reduce to essentials. **3** (foll. by *to*) amount to; signify basically. **boiled shirt** a dress shirt with a starched front. **boiled sweet** *Brit.* a sweet made of boiled sugar. **boil over 1** spill over in boiling. **2** lose one's temper; become over-excited. **make one's blood boil** see BLOOD. [ME f. AF *boiller*, OF *boillir*, f. L *bullire* to bubble f. *bulla* bubble]

boil[2] /bɔɪl/ *n.* an inflamed pus-filled swelling caused by infection of a hair follicle etc. [OE *býl(e)* f. WG]

Boileau /ˈbwʌləʊ/, Nicholas (full surname Boileau-Despréaux) (1636–1711), French critic and poet. Boileau is considered particularly important as one of the founders of French literary criticism, a field in which his influence has been profound. He gained wide recognition as the legislator and model for French neoclassicism with his didactic poem *Art poétique* (1674); based on Horace's *Ars Poetica*, it establishes canons of taste and defines principles of composition and criticism.

boiler /ˈbɔɪlə(r)/ *n.* **1** a fuel-burning apparatus for heating water esp. to supply a central-heating system. **2** a tank for heating water, esp. for turning it to steam under pressure. **3** a metal tub for boiling laundry etc. **4** a fowl, vegetable, etc., suitable for cooking only by boiling. □ **boiler-room** a room with a boiler and other heating equipment, esp. in the basement of a large building. **boiler suit** a one-piece suit worn as overalls for heavy manual work.

boilermaker /ˈbɔɪləˌmeɪkə(r)/ n. **1** a person who makes boilers. **2** a metalworker in heavy industry.

boiling /ˈbɔɪlɪŋ/ adj. (also **boiling hot**) colloq. very hot.

boiling-point /ˈbɔɪlɪŋˌpɔɪnt/ n. **1** the temperature at which a liquid starts to boil. **2** a state of high excitement (feelings reached boiling-point).

Boise /ˈbɔɪsɪ/ the state capital of Idaho; pop. (1990) 125,738.

boisterous /ˈbɔɪstərəs/ adj. **1** (of a person) rough; noisily exuberant. **2** (of the sea, weather, etc.) stormy, rough. □ **boisterously** adv. **boisterousness** n. [var. of ME boist(u)ous, of unkn. orig.]

Bokassa /bəˈkæsə/, Jean Bédel (b.1921), Central African Republic statesman and military leader, President 1972–6, emperor 1976–9. He led a successful coup in 1966, from which time he steadily increased his personal power, becoming President for life and later self-styled emperor of his country, which he renamed the Central African Empire. He was held responsible for many deaths and was ousted in 1979; in 1987 he was tried for his crimes and sentenced to death, but the sentence was eventually commuted.

Bokhara see BUKHORO.

Bokmål /ˈbuːkmɔːl/ n. a modified form of Danish used in Norway after its separation from Denmark (see NORWEGIAN). [Norw., f. bok book + mål language]

boko /ˈbəʊkəʊ/ n. & adj. Austral. ● n. an animal or person that is blind in one eye. ● adj. blind. [perh. f. an Aboriginal language]

bolas /ˈbəʊləs/ n. (as sing. or pl.) (esp. in South America) a missile consisting of a number of balls connected by strong cord, which when thrown entangles the limbs of the quarry. [Sp. & Port., pl. of bola ball]

bold /bəʊld/ adj. **1** confidently assertive; adventurous, courageous. **2** forthright, impudent. **3** vivid, distinct, well-marked (bold colours; a bold imagination). **4** (in full **bold-face** or **-faced**) Printing printed in a thick black typeface. □ **as bold as brass** excessively bold or self-assured. **make** (or **be**) **so bold as to** presume to; venture to. □ **boldly** adv. **boldness** n. [OE bald dangerous f. Gmc]

Boldrewood /ˈbəʊldəˌwʊd/, Rolf (pseudonym of Thomas Alexander Browne) (1826–1915), Australian novelist. His most enduring work was Robbery Under Arms (first published as a serial in 1882–3), a narration of the life and crimes of a bushranger under sentence of death.

bole[1] /bəʊl/ n. the stem or trunk of a tree. [ME f. ON bolr, perh. rel. to BAULK]

bole[2] /bəʊl/ n. fine compact earthy clay. [LL BOLUS]

bolero /bəˈleərəʊ/ n. (pl. **-os**) **1 a** a Spanish dance in simple triple time. **b** music for or in the time of a bolero. **2** (also /ˈbɒləˌrəʊ/) a woman's short open jacket. [Sp.]

boletus /bəˈliːtəs/ n. a mushroom or toadstool of the genus Boletus, with pores on the underside of the cap instead of gills.

Boleyn /bəˈlɪn/, Anne (1507–36), second wife of Henry VIII and mother of Elizabeth I. Although the king had fallen in love with Anne, and had divorced Catherine of Aragon in order to marry her (1533), she fell from favour when she failed to provide him with a male heir. She was eventually executed because of alleged infidelities.

Bolger /ˈbɒldʒə(r)/, James B(rendan) (b.1935), New Zealand statesman, Prime Minister since 1990.

Bolingbroke /ˈbɒlɪŋˌbrʊk/, the surname of Henry IV of England (see HENRY[1]).

Bolívar /ˈbɒlɪˌvɑː(r), Spanish boˈliβar/, Simón (known as 'the Liberator') (1783–1830), Venezuelan patriot and statesman. Bolívar was responsible for leading the liberation of South America from Spanish rule from 1808 onwards. Although his military career was not without its failures, and although his dream of a South American federation was never realized, he succeeded in driving the Spanish from Venezuela, Colombia, Peru, and Ecuador; Upper Peru was named Bolivia in his honour.

bolivar /ˈbɒlɪˌvɑː(r)/ n. the basic monetary unit of Venezuela, equal to 100 centimos. [BOLÍVAR]

Bolivia /bəˈlɪvɪə/ a landlocked country in South America; pop. (est. 1991) 7,356,000; languages, Spanish (official), Aymara, and Quechua; capital, La Paz; legal capital and seat of the judiciary, Sucre. Bolivia's chief topographical feature is the altiplano, the great central plateau between the two chains of the Andes. Part of the Inca empire, Bolivia became one of the most important parts of Spain's American empire following the discovery of major silver deposits soon after Pizarro's defeat of the Incas. It was freed from Spanish rule in 1825 and named after the great liberator Bolívar, but has suffered continually from political instability and has lost land (including its Pacific coast) to

surrounding countries in 19th and early 20th-century wars. □ **Bolivian** adj. & n.

boliviano /bəˌlɪvɪˈɑːnəʊ/ n. (pl. **-os**) the basic monetary unit of Bolivia (1863–1962 and since 1987), equal to 100 centavos or cents. [BOLÍVAR]

Böll /bɜːl/, Heinrich (Theodor) (1917–85), German novelist and short-story writer. His years in the German army (1938–44) provided the material for his earliest work, including the novel Adam, Where Art Thou (1951). His later work concerns aspects of postwar German society and is frequently critical of the prevailing political and business ethos; it differs stylistically from the realism of his early writing in that it uses symbolism and narrative devices such as flashback. Significant works include Billiards at Half Past Nine (1959) and The Lost Honour of Katharina Blum (1974). He was awarded the Nobel Prize for literature in 1972.

boll /bəʊl/ n. a rounded capsule containing seeds, esp. flax or cotton. □ **boll-weevil** a small North American weevil, Anthonomus grandis, whose larvae destroy cotton bolls. [ME f. MDu. bolle: see BOWL[1]]

Bollandist /ˈbɒləndɪst/ n. a member of a group of Jesuits who edit the Acta Sanctorum, a critical edition of the lives of the saints, based on early manuscripts and first edited by John Bolland (1596–1665).

bollard /ˈbɒlɑːd/ n. **1** Brit. a short metal, concrete, or plastic post in the road, esp. as part of a traffic island. **2** a short post on a quay or ship for securing a rope. [ME perh. f. ON bolr BOLE[1] + -ARD]

bollocking /ˈbɒləkɪŋ/ n. coarse sl. a severe reprimand.

bollocks /ˈbɒləks/ n. (also **ballocks**) coarse sl. **1** the testicles. **2** (often as an exclamation of contempt) nonsense, rubbish. [OE bealluc, rel. to BALL[1]]

bollocky /ˈbɒləkɪ/ adj. Austral. sl. naked.

Bologna /bəˈləʊnjə/ a city in northern Italy, capital of Emilia-Romagna region; pop. (1990) 411,800. Its university, which dates from the 11th century, is the oldest in Europe.

bologna /bəˈləʊnjə, -ˈləʊnɪ/ n. N. Amer. = BOLOGNA SAUSAGE.

Bologna sausage /bəˈləʊnjə, -ˈləʊnɪ/ n. N. Amer. a large smoked sausage made of bacon, veal, pork-suet, and other meats, and sold ready for eating.

bolometer /bəˈlɒmɪtə(r)/ n. a sensitive electrical instrument for measuring radiant energy. □ **bolometry** n. **bolometric** /ˌbəʊləˈmetrɪk/ adj. [Gk bolē ray + -METER]

boloney /bəˈləʊnɪ/ n. (also **baloney**) (pl. **-eys**) sl. **1** humbug, nonsense. **2** = BOLOGNA SAUSAGE. [20th c.: alt. of BOLOGNA]

Bolshevik /ˈbɒlʃəˌvɪk/ n. & adj. ● n. **1** hist. a member of the revolutionary faction of the Social Democratic party in Russia which, from 1903, favoured revolutionary tactics. (See note below.) **2** a Russian communist. **3** (in general use) any revolutionary socialist. ● adj. **1** of, relating to, or characteristic of the Bolsheviks. **2** communist. □ **Bolshevism** n. **Bolshevist** n. [Russ., = a member of the majority, from the fact that this faction formed the majority group of the Russian Social Democratic Party in 1903, f. bol'she greater]

▪ Led by Lenin, the Bolsheviks opposed cooperation with moderate reformers and advocated the instigation of revolution by a small political élite prepared to shape the ideas of the working class. After the Russian Revolution in 1917 they eventually succeeded in seizing complete control of the country from the various other revolutionary groups. In Mar. 1918 they were renamed the (Russian) Communist Party.

Bolshie /ˈbɒlʃɪ/ adj. & n. (also **Bolshy**) sl. ● adj. (usu. **bolshie**) **1** uncooperative, rebellious, awkward; bad-tempered. **2** left-wing, socialist. ● n. (pl. **-ies**) a Bolshevik. □ **bolshiness** n. (in sense 1 of adj.). [abbr.]

Bolshoi Ballet /ˈbɒlʃɔɪ/ a Moscow ballet company, one of the most prestigious in the world. It is generally regarded as dating from 1776 and since 1825 has been established at the Bolshoi Theatre, where it staged the first production of Swan Lake, with music by Tchaikovsky, in 1877. After the Russian Revolution in 1917 the company was reorganized and new Soviet ballets such as The Red Poppy (1927) were introduced, but the Bolshoi has since reverted to a more traditional repertoire. The company's first appearances in the West were in London in 1956 and New York in 1959. [Russ. bol'shoi large, f. Bolshoi Teatr, where it performs]

bolster[1] /ˈbəʊlstə(r)/ n. & v. ● n. **1** a long thick pillow. **2** a pad or support, esp. in a machine. **3** Building a short timber cap over a post to increase the bearing of the beams it supports. ● v.tr. (usu. foll. by up) **1** encourage, reinforce (bolstered our morale). **2** support with a bolster; prop up. □ **bolsterer** n. [OE f. Gmc]

bolster² /ˈbəʊlstə(r)/ n. a chisel for cutting bricks. [20th c.: orig. uncert.]

bolt¹ /bəʊlt/ n., v., & adv. ● n. **1** a sliding bar and socket used to fasten or lock a door, gate, etc. **2** a large usu. metal pin with a head, usu. riveted or used with a nut, to hold things together. **3** a discharge of lightning. **4** an act of bolting (cf. sense 4 of v.); a sudden escape or dash for freedom. **5** an arrow for shooting from a crossbow. **6** a roll of fabric (orig. as a measure). ● v. **1** tr. fasten or lock with a bolt. **2** tr. (foll. by in, out) keep (a person etc.) from leaving or entering by bolting a door. **3** tr. fasten together with bolts. **4** intr. a dash suddenly away, esp. to escape. **b** (of a horse) suddenly gallop out of control. **5** tr. gulp down (food) unchewed; eat hurriedly. **6** intr. (of a plant) run to seed. ● adv. (usu. in **bolt upright**) rigidly, stiffly. □ **a bolt from the blue** a complete surprise. **bolt-hole 1** a means of escape. **2** a secret refuge. **shoot one's bolt** do all that is in one's power. □ **bolter** n. (in sense 4 of v.). [OE bolt arrow]

bolt² /bəʊlt/ v.tr. (also **boult**) sift (flour etc.). [ME f. OF bulter, buleter, of unkn. orig.]

Bolton /ˈbəʊlt(ə)n/ a town in NW England, in Greater Manchester; pop. (1991) 253,300. Between the 16th and 19th centuries Bolton was a centre of the textile industries, moving in the 18th century from the production of wool to that of cotton.

Boltzmann /ˈbɒltsmən/, Ludwig (1844–1906), Austrian physicist. Boltzmann made contributions to the kinetic theory of gases, statistical mechanics, and thermodynamics. He recognized the importance of the electromagnetic theory of J. C. Maxwell, but had difficulty in getting his own work on statistical mechanics accepted until the discoveries in atomic physics at the turn of the century. He derived the Maxwell–Boltzmann equation for the distribution of energy among colliding atoms, correlating entropy with probability when he brought thermodynamics and molecular physics together.

bolus /ˈbəʊləs/ n. (pl. **boluses**) **1** a soft ball, esp. of chewed food. **2** a large pill. [LL f. Gk bōlos clod]

Bolzano /bɒlˈtsɑːnəʊ/ a city in NE Italy, capital of the Trentino-Alto Adige region; pop. (1990) 100,380.

bomb /bɒm/ n. & v. ● n. **1 a** a container with explosive, incendiary material, smoke, or gas etc., designed to explode on impact or by means of a time-mechanism or remote-control device. **b** an ordinary object fitted with an explosive device (letter-bomb). **2** (prec. by the) the atomic or hydrogen bomb considered as a weapon with supreme destructive power. **3** Brit. sl. a large sum of money (cost a bomb). **4** a mass of solidified lava thrown from a volcano. **5** US colloq. a bad failure (esp. a theatrical one). **6** sl. a drugged cigarette. ● v. **1** tr. attack with bombs; drop bombs on. **2** tr. (foll. by out) drive (a person etc.) out of a building or refuge by using bombs. **3** intr. throw or drop bombs. **4** intr. esp. US sl. fail badly. **5** intr. (usu. foll. by along, off) colloq. move or go very quickly. **6** tr. US sl. criticize fiercely. □ **bomb-bay** a compartment in an aircraft used to hold bombs. **bomb-disposal** the defusing or removal and detonation of an unexploded bomb. **bomb-sight** a device in an aircraft for aiming bombs. **bomb-site** an area where buildings have been destroyed by bombs. **bomb squad** a division of a police force investigating crimes involving bombs. **go down a bomb** colloq., often iron. be very well received. **like a bomb** Brit. colloq. **1** often iron. very successfully. **2** very fast. [F bombe f. It. bomba f. L bombus f. Gk bombos hum]

bombard /bɒmˈbɑːd/ v.tr. **1** attack with a number of bombs, shells, etc. **2** (often foll. by with) subject to persistent questioning, abuse, etc. **3** Physics direct a stream of high-speed particles at (a substance). □ **bombardment** n. [F bombarder f. bombarde f. med.L bombarda a stone-throwing engine: see BOMB]

bombarde /ˈbɒmbɑːd/ n. Mus. a medieval alto-pitched shawm. [OF bombarbe, prob. f. L bombus: see BOMB]

bombardier /ˌbɒmbəˈdɪə(r)/ n. **1** (in the British army) a non-commissioned officer in the artillery. **2** (in North America) a member of a bomber crew responsible for sighting and releasing bombs. [F (as BOMBARD)]

bombardon /bɒmˈbɑːd(ə)n, ˈbɒmbəd(ə)n/ n. Mus. **1** a type of valved bass tuba. **2** an organ stop imitating this. [It. bombardone f. bombardo bassoon]

bombasine var. of BOMBAZINE.

bombast /ˈbɒmbæst/ n. pompous or extravagant language. □ **bombastic** /bɒmˈbæstɪk/ adj. **bombastically** adv. [earlier bombace cotton wool f. F f. med.L bombax -acis alt. f. bombyx; see BOMBAZINE]

Bombay /bɒmˈbeɪ/ a city and port on the west coast of India, capital of the state of Maharashtra; pop. (1991) 9,990,000. In 1995 the city's official name was changed to the Hindi form Mumbai.

Bombay duck n. a dried fish, esp. bummalo, usu. eaten with curried dishes. [corrupt. of bombil (see BUMMALO), after BOMBAY]

bombazine /ˈbɒmbəˌziːn/ (also **bombasine**) n. a twilled dress-material of worsted with or without an admixture of silk or cotton, esp., when black, formerly used for mourning. [F bombasin f. med.L bombacinum f. LL bombycinus silken f. bombyx -ycis silk or silkworm f. Gk bombux]

bombe /bɒmb/ n. Cookery a dome-shaped dish or confection, often frozen. [F, = BOMB]

bomber /ˈbɒmə(r)/ n. **1** an aircraft equipped to carry and drop bombs. **2** a person using bombs, esp. illegally. □ **bomber jacket** a short leather or cloth jacket tightly gathered at the waist and cuffs.

bombora /bɒmˈbɔːrə/ n. Austral. a dangerous sea area where waves break over a submerged reef. [Aboriginal, perh. Dharuk bumbora]

bombproof /ˈbɒmpruːf/ adj. strong enough to resist the effects of blast from a bomb.

bombshell /ˈbɒmʃel/ n. **1** an overwhelming surprise or disappointment. **2** an artillery bomb. **3** sl. a very attractive woman (blonde bombshell).

Bon, Cape /bɒn/ a peninsula of NE Tunisia, extending into the Mediterranean Sea, noted for its resorts and its wine.

bona fide /ˌbəʊnə ˈfaɪdɪ/ adj. & adv. ● adj. genuine; sincere. ● adv. genuinely; sincerely. [L, ablat. sing. of BONA FIDES]

bona fides /ˌbəʊnə ˈfaɪdiːz/ n. **1** esp. Law an honest intention; sincerity. **2** (as pl.) colloq. documentary evidence of acceptability (his bona fides are in order). [L, = good faith]

Bonaire /bɒˈneə(r)/ one of the two principal islands of the Netherlands Antilles (the other is Curaçao); chief town, Kralendijk.

bonanza /bəˈnænzə/ n. & adj. ● n. **1** a source of wealth or prosperity. **2** a large output (esp. of a mine). **3 a** prosperity; good luck. **b** a run of good luck. ● adj. greatly prospering or productive. [orig. US f. Sp., = fair weather, f. L bonus good]

Bonaparte /ˈbəʊnəˌpɑːt/ (Italian **Buonaparte** /ˌbwonaˈparte/) a Corsican family including the three French rulers named Napoleon.

bona vacantia /ˌbəʊnə vəˈkæntɪə/ n.pl. Law goods without an apparent owner, and to which the Crown has right. [L, = ownerless goods]

Bonaventura, St /ˌbɒnəvenˈtjʊərə/ (born Giovanni di Fidanza; known as 'the Seraphic Doctor') (1221–74), Franciscan theologian. Appointed minister general of his order in 1257, he was made cardinal bishop of Albano in 1273. He wrote the official biography of St Francis and had a lasting influence as a spiritual writer. Feast day, 15 (formerly 14) July.

bon-bon /ˈbɒnbɒn/ n. a piece of confectionery; a sweet. [F f. bon good f. L bonus]

bonce /bɒns/ n. Brit. **1** sl. the head. **2** a large marble. [19th c.: orig. unkn.]

Bond /bɒnd/, James, a suave British secret agent in the spy novels of Ian Fleming, known also by his code name 007. The character has been popularized since the 1960s by a series of films.

bond /bɒnd/ n. & v. ● n. **1 a** a thing that ties another down or together. **b** (usu. in pl.) a thing restraining bodily freedom (broke his bonds). **2** (often in pl.) **a** a uniting force (sisterly bond). **b** a restraint; a responsibility (bonds of duty). **3** a binding engagement; an agreement (his word is his bond). **4 a** Commerce a certificate issued by a government or a public company promising to repay borrowed money at a fixed rate of interest at a specified time; a debenture. **b** Insurance an insurance policy held by a travel agent, tour operator, airline, etc., which protects travellers' holidays and money from the company's bankruptcy. **5** adhesiveness. **6** Law a deed by which a person is bound to make payment to another. **7** Chem. a linkage between atoms in a chemical compound. **8** Building the laying of bricks in one of various patterns in a wall in order to ensure strength (English bond; Flemish bond). ● v. **1** tr. **a** lay (bricks) overlapping. **b** bind together (resin with fibres, etc.). **2** intr. adhere; hold together. **3** tr. connect with a bond. **4** tr. place (goods) in bond. **5** intr. become emotionally attached. **6** tr. join or hold by an emotional or psychological bond. □ **bond paper** high-quality writing-paper. **bond-washing** dividend-stripping. **in bond** (of goods) stored in a bonded warehouse until the importer pays the duty owing (see BONDED). [ME var. of BAND¹]

bondage /ˈbɒndɪdʒ/ n. **1** serfdom; slavery. **2** subjection to constraint, influence, obligation, etc. **3** sado-masochistic practices, including the use of physical restraints or mental enslavement. [ME f. AL *bondagium*: infl. by BOND]

bonded /'bɒndɪd/ *adj.* **1** (of goods) placed in bond. **2** (of material) reinforced by or cemented to another. **3 a** (of a debt) secured by bonds. **b** *Insurance* (of a travel company etc.) protected by a bond (see BOND *n.* 4b). □ **bonded warehouse** a customs-controlled warehouse for the retention of imported goods until the duty owed is paid.

Bondi /'bɒndaɪ/ a coastal resort in New South Wales, Australia, a suburb of Sydney. It is noted for its popular beach.

bondi /'bɒndaɪ/ *n. Austral.* a heavy club with a knob on the end. □ **give a person bondi** attack savagely. [Wiradhuri *bundi*]

bondsman /'bɒndzmən/ *n.* (*pl.* **-men**) **1** a slave. **2** a person in thrall to another. [var. of *bondman* (f. archaic *bond* in serfdom or slavery) as though f. *bond's* genitive of BOND]

Bône /bəʊn/ the former name for ANNABA.

bone /bəʊn/ *n. & v.* ● *n.* **1** any of the pieces of hard tissue making up the skeleton in vertebrates. (*See note below.*) **2** (in *pl.*) **a** the skeleton, esp. as remains after death. **b** the body, esp. as a seat of intuitive feeling (*felt it in my bones*). **3 a** the material of which bones consist. **b** a similar substance such as ivory, dentine, or whalebone. **4** a thing made of bone. **5** (in *pl.*) the essential part of a thing (*the bare bones*). **6** (in *pl.*) **a** dice. **b** castanets. **7** a strip of stiffening in a corset etc. ● *v.* **1** *tr.* take out the bones from (meat or fish). **2** *tr.* stiffen (a garment) with bone etc. **3** *tr. Brit. sl.* steal. □ **bone china** fine china made of clay mixed with the ash from bones. **bone-dry** completely dry. **bone idle** (or **lazy**) utterly idle or lazy. **bone-meal** crushed or ground bones used esp. as a fertilizer. **bone of contention** a source or ground of dispute. **bone-setter** a person who sets broken or dislocated bones, esp. without being a qualified surgeon. **bone spavin** see SPAVIN. **bone up** (often foll. by *on*) *colloq.* study (a subject) intensively. **close to** (or **near**) **the bone 1** near the permitted limit (esp. of good taste or decency). **2** destitute; hard up. **have a bone to pick** (usu. foll. by *with*) have a cause for dispute (with another person). **make no bones about 1** admit or allow without fuss. **2** not hesitate or scruple. **point the bone** (usu. foll. by *at*) *Austral.* **1** wish bad luck on. **2** cast a spell on in order to kill. **to the bone 1** to the bare minimum. **2** penetratingly. **work one's fingers to the bone** work very hard, esp. thanklessly. □ **boneless** *adj.* [OE *bān* f. Gmc]

▪ Bone is a living tissue, composed of special cells which secrete around themselves a material consisting of calcium salts (which provide hardness and strength in compression) and collagen fibres (which provide tensile strength). The material of bone varies considerably in density and compactness, that near the surface of a bone generally being more compact. Many bones have a central cavity containing marrow, a tissue which is the source of most blood cells and also stores fats. The calcium salts in bone are crucial in the regulation of the level of calcium throughout the body.

bonefish /'bəʊnfɪʃ/ *n. N. Amer.* a large game-fish with many small bones, esp. *Albula vulpes*.

bonehead /'bəʊnhed/ *n. sl.* a stupid person. □ **boneheaded** *adj.*

boner /'bəʊnə(r)/ *n. sl.* a stupid mistake. [BONE + -ER¹]

boneshaker /'bəʊnˌʃeɪkə(r)/ *n.* **1** a decrepit or uncomfortable old vehicle. **2** an old type of bicycle with solid tyres.

bonfire /'bɒnˌfaɪə(r)/ *n.* a large open-air fire for burning rubbish, as part of a celebration, or as a signal. □ **make a bonfire of** destroy by burning. [earlier *bonefire* f. BONE (bones being the chief material formerly used) + FIRE]

Bonfire Night (also called *Guy Fawkes Night*) (in the UK) 5 November, on which fireworks are set off and an effigy of Guy Fawkes burnt in memory of the Gunpowder Plot.

bongo¹ /'bɒŋgəʊ/ *n.* (*pl.* **-os** or **-oes**) either of a pair of small long-bodied drums usu. held between the knees and played with the fingers. [Amer. Sp. *bongó*]

bongo² /'bɒŋgəʊ/ *n.* (*pl.* same or **-os**) a forest antelope, *Tragelaphus euryceros*, native to central Africa, having a chestnut-red coat with narrow white vertical stripes. [cf. central African language *mbangani*, Lingala *mongu*]

Bonhoeffer /'bɒnˌhɜːfə(r)/, Dietrich (1906–45), German Lutheran theologian and pastor. Originally influenced by the neo-orthodox theology of Karl Barth, in his later writings he moved away from biblical orthodoxy and concentrated on a Christianity informed by contemporary social and political issues. He was an active opponent of Nazism both before and during the Second World War, becoming

involved in the German resistance movement. Arrested in 1943, he was sent to Buchenwald concentration camp and later executed.

bonhomie /ˌbɒnɒ'miː/ *n.* geniality; good-natured friendliness. [F f. *bonhomme* good fellow]

bonhomous /'bɒnəməs/ *adj.* full of bonhomie.

Boniface, St /'bɒnɪˌfeɪs/ (born Wynfrith; known as 'the Apostle of Germany') (680–754), Anglo-Saxon missionary. Sent to Frisia and Germany to spread the Christian faith, he laid the foundations of a settled ecclesiastical organization there and made many converts. He was appointed Primate of Germany in 732, and in 741 was given authority to reform the whole Frankish Church. The first papal legate north of the Alps, Boniface greatly assisted the spread of papal influence. He was martyred in Frisia. Feast day, 5 June.

Bonington /'bɒnɪŋtən/, Christian John Storey ('Chris') (b.1934), English mountaineer. He made the first British ascent of the north face of the Eiger in 1962. He was the leader of two expeditions to Mount Everest: one by a hitherto unclimbed route in 1975 and the other in 1985, when he reached the summit.

bonito /bə'niːtəʊ/ *n.* (*pl.* **-os**) a tuna striped like mackerel and common in tropical seas. [Sp.]

bonk /bɒŋk/ *v. & n.* ● *v.* **1** *tr.* hit resoundingly. **2** *intr.* bang; bump. **3** *sl.* **a** *intr.* have sexual intercourse. **b** *tr.* have sexual intercourse with. ● *n.* **1** an abrupt heavy sound of impact; a bump. **2** *sl.* an act of sexual intercourse. [imit.: cf. BANG, BUMP, CONK²]

bonkers /'bɒŋkəz/ *predic.adj. sl.* crazy. [20th c.: orig. unkn.]

bon mot /bɒn 'məʊ/ *n.* (*pl.* **bons mots** pronunc. same) a witty saying. [F]

Bonn /bɒn/ a city in the state of North Rhine-Westphalia in Germany; pop. (1991) 296,240. From 1949 until the reunification of Germany in 1990 Bonn was the capital of the Federal Republic of Germany.

Bonnard /'bɒnɑː(r)/, Pierre (1867–1947), French painter and graphic artist. A member of a group of painters called the Nabi Group, he produced ornamental screens and lithographs before concentrating on painting from about 1905. His works continue and develop the impressionist tradition; notable for their rich, glowing colour harmonies, they mostly depict domestic interior scenes, nudes, and landscapes.

bonne bouche /bɒn 'buːʃ/ *n.* (*pl.* **bonnes bouches** pronunc. same) a titbit, esp. to end a meal with. [F f. *bonne* fem. good + *bouche* mouth]

bonnet /'bɒnɪt/ *n.* **1 a** a woman's or child's hat tied under the chin and usu. with a brim framing the face. **b** a soft round brimless hat like a beret worn by men and boys in Scotland (cf. TAM-O'-SHANTER). **c** *colloq.* any hat. **2** *Brit.* a hinged cover over the engine of a motor vehicle. **3** the ceremonial feathered head-dress of an American Indian. **4** the cowl of a chimney etc. **5** a protective cap in various machines. **6** *Naut.* additional canvas laced to the foot of a sail. □ **bonnet monkey** an Indian macaque, *Macaca radiata*, with a bonnet-like tuft of hair. □ **bonneted** *adj.* [ME f. OF *bonet* short for *chapel de bonet* cap of some kind of material (med.L *bonetus*)]

bonnethead /'bɒnɪtˌhed/ *n.* a hammerhead shark, *Sphyrna tiburo*, with a relatively narrow head. Also called *shovelhead*.

Bonnie Prince Charlie see STUART².

bonny /'bɒnɪ/ *adj.* (also **bonnie**) (**bonnier**, **bonniest**) esp. *Sc. & N. Engl.* **1 a** physically attractive. **b** healthy-looking. **2** good, fine, pleasant. □ **bonnily** *adv.* **bonniness** *n.* [16th c.: perh. f. F *bon* good]

bonsai /'bɒnsaɪ/ *n.* (*pl.* same) **1** the art of cultivating ornamental artificially dwarfed varieties of trees and shrubs. **2** a tree or shrub grown by this method. [Jap.]

bonspiel /'bɒnspiːl/ *n.* esp. *Sc.* a curling-match. [16th c.: perh. f. LG]

bontebok /'bɒntɪˌbɒk/ *n.* (also **bontbok** /'bɒntbɒk/) (*pl.* same or **-boks**) a large chestnut antelope, *Damaliscus dorcas*, native to southern Africa, having a white tail and a white patch on its head and rump. [Afrik. f. *bont* spotted + *bok* BUCK¹]

bonus /'bəʊnəs/ *n.* **1** an unsought or unexpected extra benefit. **2 a** a usu. seasonal gratuity to employees beyond their normal pay. **b** an extra dividend or issue paid to the shareholders of a company. **c** a distribution of profits to holders of an insurance policy. [L *bonus*, *bonum* good (thing)]

bon vivant /ˌbɒn viː'vɒn/ *n.* (*pl.* **bons vivants** pronunc. same) a person indulging in good living; a gourmand. [F, lit. 'one who lives well' f. *vivre* to live]

bon viveur /ˌbɒn viː'vɜː(r)/ *n.* (*pl.* **bons viveurs** pronunc. same) = BON VIVANT. [pseudo-F, after BON VIVANT]

bon voyage /ˌbɒn vwaɪˈɑːʒ/ *int. & n.* an expression of good wishes to a departing traveller. [F]

bony /ˈbəʊnɪ/ *adj.* (**bonier**, **boniest**) **1** (of a person) thin with prominent bones. **2** having many bones. **3** of or like bone. **4** (of a fish) having bones rather than cartilage. □ **boniness** *n.*

bonze /bɒnz/ *n.* a Japanese or Chinese Buddhist priest. [F *bonze* or Port. *bonzo* prob. f. Jap. *bonzō*, *bonsō* priest]

bonzer /ˈbɒnzə(r)/ *adj. Austral. sl.* excellent, first-rate. [perh. f. BONANZA]

boo /buː/ *int., n., & v.* ● *int.* **1** an expression of disapproval or contempt. **2** a sound, made esp. to a child, intended to surprise. ● *n.* an utterance of *boo*, esp. as an expression of disapproval or contempt made to a performer etc. ● *v.* (**boos**, **booed**) **1** *intr.* utter a boo or boos. **2** *tr.* jeer at (a performer etc.) by booing. □ **can't** (or **wouldn't**) **say boo to a goose** is very shy or timid. [imit.]

boob[1] /buːb/ *n. & v. sl.* ● *n.* **1** *Brit.* an embarrassing mistake. **2** a simpleton. ● *v.intr. Brit.* make an embarrassing mistake. [abbr. of BOOBY]

boob[2] /buːb/ *n. sl.* a woman's breast. □ **boob tube** *sl.* **1** a woman's low-cut close-fitting usu. strapless top. **2** (usu. prec. by *the*) *N. Amer.* television; one's television set. [earlier *bubby*, *booby*, of uncert. orig.]

booboo /ˈbuːbuː/ *n. sl.* a mistake. [BOOB[1]]

boobook /ˈbuːbʊk/ *n.* (in full **boobook owl**) a brown spotted owl, *Ninox novaeseelandiae*, native to Australia and New Zealand. Also called *mopoke*, *morepork*. [imit. of its call]

booby /ˈbuːbɪ/ *n.* (*pl.* **-ies**) **1** a stupid or childish person. **2** a seabird of the genus *Sula*, related to the gannet. □ **booby-hatch** esp. *US sl.* a mental hospital. **booby prize** a prize given to the least successful competitor in any contest. **booby trap 1** a trap intended as a practical joke, e.g. an object placed on top of a door ajar ready to fall on the next person to pass through. **2** *Mil.* an apparently harmless explosive device intended to kill or injure anyone touching it. **booby-trap** *v.tr.* place a booby trap or traps in or on. [prob. f. Sp. *bobo* (in both senses) f. L *balbus* stammering]

boodle /ˈbuːd(ə)l/ *n. sl.* money, esp. when gained or used dishonestly, e.g. as a bribe. [Du. *boedel* possessions]

boofhead /ˈbʊfhed/ *n. Austral. sl.* a fool. [prob. f. obs. *bufflehead* fool]

boogie /ˈbuːgɪ/ *v. & n.* ● *v.intr.* (**boogies**, **boogied**, **boogying**) *sl.* dance to pop music. ● *n.* **1** = BOOGIE-WOOGIE. **2** *sl.* a dance to pop music. [BOOGIE-WOOGIE]

boogie-woogie /ˌbuːgɪˈwuːgɪ/ *n.* a style of playing blues or jazz on the piano, marked by a persistent bass rhythm. [20th c.: orig. unkn.]

boohoo /buːˈhuː/ *int. & v.* ● *int.* expressing weeping. ● *v.intr.* (**boohoos**, **boohooed**) (esp. of a child) weep loudly. [imitative]

book /bʊk/ *n. & v.* ● *n.* **1 a** a written or printed work consisting of pages glued or sewn together along one side and bound in covers. **b** a literary composition intended for publication (*is working on her book*). **2** a bound set of blank sheets for writing or keeping records in. **3** a set of tickets, stamps, matches, cheques, samples of cloth, etc., bound up together. **4** (in *pl.*) a set of records or accounts. **5** a main division of a literary work, or of the Bible (*the Book of Deuteronomy*). **6** (in full **book of words**) **a** a libretto, script of a play, etc. **b** a set of rules or regulations. **7** *colloq.* a magazine. **8** a telephone directory (*his number's in the book*). **9** a record of bets made and money paid out at a race meeting by a bookmaker. **10** a set of six tricks collected together in a card-game. **11** an imaginary record or list (*the book of life*). ● *v.* **1** *tr.* **a** engage (a seat etc.) in advance; make a reservation of. **b** engage (a guest, supporter, etc.) for some occasion. **2** *tr.* **a** take the personal details of (an offender or rule-breaker). **b** enter in a book or list. **3** *tr.* make a reservation for (a person). **4** *intr.* make a reservation (*no need to book*). □ **book club** a society which sells its members selected books on special terms. **book-end** a usu. ornamental prop used to keep a row of books upright esp. one of a pair. **book in** esp. *Brit.* register one's arrival at a hotel etc. **book learning** mere theory. **book-louse** (*pl.* **-lice**) a minute insect of the order Psocoptera, often damaging to books. **book-plate** a decorative label stuck in the front of a book bearing the owner's name. **book-rest** an adjustable support for an open book on a table. **book token** *Brit.* a voucher which can be exchanged for books to a specified value. **book up 1** buy tickets in advance for a theatre, concert, holiday, etc. **2** (as **booked up**) with all places reserved. **book value** the value of a commodity as entered in a firm's books (opp. *market value*). **bring to book** call to account. **closed** (or **sealed**) **book** a subject of which one is ignorant. **go by the book** proceed according to the rules. **the good Book** the Bible. **in a person's bad** (or **good**) **books** in disfavour (or favour) with a person. **in my book** in my opinion. **make**

a book take bets and pay out winnings at a race meeting. **not in the book** disallowed. **on the books** contained in a list of members etc. **suits my book** is convenient to me. **take a leaf out of a person's book** imitate a person. **throw the book at** *colloq.* charge or punish to the utmost. [OE *bōc*, *bōcian*, f. Gmc, usu. taken to be rel. to BEECH (the bark of which was used for writing on)]

bookbinder /ˈbʊkˌbaɪndə(r)/ *n.* a person who binds books professionally. □ **bookbinding** *n.*

bookcase /ˈbʊkkeɪs/ *n.* a set of shelves for books in the form of a cabinet.

Booker Prize /ˈbʊkə(r)/ a literary prize awarded annually for a novel published by a British or Commonwealth citizen during the previous year. It was founded in 1969 and financed by the multinational company Booker McConnell.

bookie /ˈbʊkɪ/ *n. colloq.* = BOOKMAKER.

booking /ˈbʊkɪŋ/ *n.* the act or an instance of booking or reserving a seat, a room in a hotel, etc.; a reservation (see BOOK *v.* 1). □ **booking-clerk** an official selling tickets at a railway station. **booking-hall** (or **-office**) *Brit.* a room or area at a railway station in which tickets are sold.

bookish /ˈbʊkɪʃ/ *adj.* **1** studious; fond of reading. **2** acquiring knowledge from books rather than practical experience. **3** (of a word, language, etc.) literary; not colloquial. □ **bookishness** *n.*

bookkeeper /ˈbʊkˌkiːpə(r)/ *n.* a person who keeps accounts for a trader, a public office, etc. □ **bookkeeping** *n.*

bookland /ˈbʊklænd/ *n. hist.* an area of common land granted by charter to a private owner.

booklet /ˈbʊklɪt/ *n.* a small book consisting of a few sheets usu. with paper covers.

bookmaker /ˈbʊkˌmeɪkə(r)/ *n.* a person who takes bets, esp. on horse-races, calculates odds, and pays out winnings. □ **bookmaking** *n.*

bookman /ˈbʊkmən/ *n.* (*pl.* **-men**) a literary man, esp. a reviewer.

bookmark /ˈbʊkmɑːk/ *n.* (also **bookmarker** /-ˌmɑːkə(r)/) a strip of leather, card, etc., used to mark one's place in a book.

bookmobile /ˈbʊkməˌbiːl/ *n. N. Amer.* a mobile library. [after AUTOMOBILE]

Book of Changes see I CHING.

Book of Common Prayer the official service book of the Church of England. It was compiled through the efforts of Thomas Cranmer and others as a simplified and condensed English version of the Latin service books used by the medieval Church and was first issued in 1549. After the book had been in turn revised and suppressed under different monarchs, a version came out in 1662, which remained almost unchanged until the 20th century. Measures of 1965 and 1974 authorized the use also of alternative services and in 1980 the Alternative Service Book presented these alternative services (in modern English) in a canonical form.

book of hours *n.* in the Christian Church, a book containing the prayers or offices to be said at the canonical hours of the day, used today in the Roman Catholic Church and particularly popular in the Middle Ages. Many medieval examples still exist, small in size and often lavishly decorated.

bookseller /ˈbʊkˌselə(r)/ *n.* a dealer in books.

bookshop /ˈbʊkʃɒp/ *n.* a shop where books are sold.

bookstall /ˈbʊkstɔːl/ *n.* a stand for selling books, newspapers, etc., esp. out of doors or at a station.

bookstore /ˈbʊkstɔː(r)/ *n. N. Amer.* = BOOKSHOP.

booksy /ˈbʊksɪ/ *adj. colloq.* having literary or bookish pretensions.

bookwork /ˈbʊkwɜːk/ *n.* the study of books (as opposed to practical work).

bookworm /ˈbʊkwɜːm/ *n.* **1** *colloq.* a person devoted to reading. **2** the larva of a moth or beetle which feeds on paper and glue used in books.

Boole /buːl/, George (1815–64), English mathematician. Professor at Cork in Ireland from 1849, Boole wrote important works on differential equations and other branches of mathematics, but is remembered chiefly for the algebraic description of reasoning now known as Boolean algebra. The study of mathematical or symbolic logic developed mainly from his ideas.

Boolean /ˈbuːlɪən/ *adj.* denoting a system of algebraic notation used to represent logical propositions by means of the binary digits 0 (false) and 1 (true). [BOOLE]

boom[1] /buːm/ n. & v. ● n. a deep resonant sound. ● v.intr. make or speak with a boom. [imit.]

boom[2] /buːm/ n. & v. ● n. a period of prosperity or sudden activity in commerce. ● v.intr. (esp. of commercial ventures) be suddenly prosperous or successful. □ **boom town** a town undergoing sudden growth due to a boom. □ **boomlet** n. [19th-c. US word, prob. f. BOOM[1]]

boom[3] /buːm/ n. **1** Naut. a pivoted spar to which the foot of a sail is attached, allowing the angle of the sail to be changed. **2** a long pole over a television or film set, carrying a microphone or other equipment. **3** a floating barrier across the mouth of a harbour or river. [Du., = BEAM]

boomer /'buːmə(r)/ n. **1** a large male kangaroo. **2** a North American mountain beaver, *Aplodontia rufa*. **3** a large wave.

boomerang /'buːmə,ræŋ/ n. & v. ● n. **1** a curved flat hardwood missile used by Australian Aboriginals to kill prey, and often of a kind able to return in flight to the thrower. **2** a plan or scheme that recoils on its originator. ● v.intr. **1** act as a boomerang. **2** (of a plan or action) backfire. [Dharuk *umarin*[y]]

boomslang /'buːmslæŋ/ n. a large venomous tree-snake, *Dispholidus typus*, native to southern Africa. [Afrik. f. *boom* tree + *slang* snake]

boon[1] /buːn/ n. **1** an advantage; a blessing. **2** archaic **a** a thing asked for; a request. **b** a gift; a favour. [ME, orig. = prayer, f. ON *bón* f. Gmc]

boon[2] /buːn/ adj. close, intimate, favourite (usu. *boon companion*). [ME (orig. = jolly, congenial) f. OF *bon* f. L *bonus* good]

boondock /'buːndɒk/ n. (usu. in *pl.*) N. Amer. sl. rough or isolated country. [Tagalog *bundok* mountain]

boondoggle /'buːn,dɒg(ə)l/ n. & v. US sl. ● n. a trivial or useless undertaking; a dishonest undertaking, a fraud. ● v.intr. take part in a trivial, useless, or dishonest undertaking. [20th c.: orig. uncert.]

Boone /buːn/, Daniel (c.1734–1820), American pioneer. Moving west from his native Pennsylvania, Boone made trips into the unexplored area of Kentucky from 1767 onwards, organizing settlements and successfully defending them against hostile Indians. He later moved further west to Missouri, being granted land there in 1799. As a hunter, trail-blazer, and fighter against the Indians he became a legend in his own lifetime.

boong /bʊŋ/ n. Austral. sl. offens. an Aborigine. [20th c.: orig. uncert.]

boonies /'buːnɪz/ n.pl. (prec. by *the*) N. Amer. sl. = BOONDOCK.

boor /bʊə(r)/ n. **1** a rude, ill-mannered person. **2** a clumsy person. **3** a rustic, a yokel. □ **boorish** adj. **boorishly** adv. **boorishness** n. [LG *būr* or Du. *boer* (cf. BOER) farmer: cf. BOWER[3]]

boost /buːst/ v. & n. colloq. ● v.tr. **1 a** promote or increase the reputation of (a person, scheme, commodity, etc.) by praise or advertising; push; increase or assist (*boosted his spirits*; *boost sales*). **b** push from below; assist (*boosted me up the tree*). **2 a** raise the voltage in (an electric circuit etc.). **b** amplify (a radio signal). ● n. **1** an act, process, or result of boosting; a push (*asked for a boost up the hill*). **2 a** an advertisement campaign. **b** the resulting advance in value, reputation, etc. [19th-c. US word: orig. unkn.]

booster /'buːstə(r)/ n. **1** a device for increasing electrical power or voltage. **2** an auxiliary engine or rocket used to give initial acceleration. **3** Med. a dose of an immunizing agent increasing or renewing the effect of an earlier one. **4** colloq. a person who boosts by helping or encouraging.

boot[1] /buːt/ n. & v. ● n. **1** an outer covering for the foot, esp. of leather, reaching above the ankle, often to the knee. **2** Brit. the luggage compartment of a motor car, usu. at the rear. **3** colloq. a firm kick. **4** (prec. by *the*) colloq. dismissal, esp. from employment (*gave them the boot*). **5** a covering to protect the lower part of a horse's leg. **6** hist. an instrument of torture encasing and crushing the foot. **7** derog. a person. ● v.tr. **1** kick, esp. hard. **2** (often foll. by *out*) dismiss (a person) forcefully. **3** (usu. foll. by *up*) put (a computer) in a state of readiness (cf. BOOTSTRAP 2). □ **the boot is on the other foot** (or **leg**) the truth or responsibility is the other way round; the tables have now been turned. **die with one's boots on** (of a soldier etc.) die fighting. **put the boot in 1** kick brutally. **2** act decisively against a person. **you bet your boots** sl. it is quite certain. □ **booted** adj. [ME f. ON *bóti* or f. OF *bote*, of unkn. orig.]

boot[2] /buːt/ n. □ **to boot** as well; to the good; in addition. [orig. = 'advantage': OE *bōt* f. Gmc]

bootblack /'buːtblæk/ n. N. Amer. a person who polishes boots and shoes.

bootboy /'buːtbɔɪ/ n. **1** a boy employed to clean shoes. **2** a hooligan typically wearing heavy boots.

bootee /buː'tiː/ n. **1** a soft shoe, esp. a woollen one, worn by a baby. **2** a woman's short boot.

Boötes /bəʊ'əʊtiːz/ Astron. a northern constellation (the Herdsman), said to represent a man holding the leash of two dogs (Canes Venatici) while driving a bear (Ursa Major). It contains the bright star Arcturus. [Gk, = ox-driver]

Booth /buːð/, William (1829–1912), English religious leader, founder and first general of the Salvation Army. A Methodist revivalist preacher, he worked actively to improve the condition of the poor, attending to both their physical and spiritual needs. In 1865, assisted by his wife Catherine (1829–90), he established a mission in the East End of London which became the Salvation Army in 1878. Booth's family had a close involvement with the movement, which continued into the 20th century; his eldest son, William Branwell Booth (1856–1929), succeeded his father as general in 1912, while a granddaughter, Catherine Branwell Booth (1884–1987), was a commissioner in the Army. (See also SALVATION ARMY.)

booth /buːð, buː:θ/ n. **1** a small temporary roofed structure of canvas, wood, etc., used esp. as a market stall, for puppet shows, etc. **2** an enclosure or compartment for various purposes, e.g. telephoning or voting. **3** a set of a table and benches in a restaurant or bar. [ME f. Scand.]

Boothia, Gulf of /'buːθɪə/ a gulf in the Canadian Arctic, between the Boothia Peninsula and Baffin Island, in the Northwest Territories. Both the gulf and the peninsula were named in honour of Sir Felix Booth (1775–1850), patron of the expedition to the Arctic (1829–33) led by Sir John Ross.

Boothia Peninsula /'buːθɪə/ a peninsula of northern Canada, in the Northwest Territories, situated between Victoria Island and Baffin Island.

bootjack /'buːtdʒæk/ n. a device for holding a boot by the heel to ease withdrawal of the leg.

bootlace /'buːtleɪs/ n. a cord or leather thong for lacing boots.

bootleg /'buːtleg/ adj., v., & n. ● adj. **1** (esp. of liquor) smuggled; illicitly sold. **2** (of musical or other recordings) made and distributed without authorization. ● v.tr. (-**legged**, -**legging**) **1** make, distribute, or smuggle (illicit goods, esp. alcohol). **2** make and distribute (musical or other recordings) without authorization. ● n. **1** smuggled or illicit alcohol. **2** a bootleg recording. □ **bootlegger** n. [f. the smugglers' practice of concealing bottles in their boots]

bootless /'buːtlɪs/ adj. archaic unavailing, useless. [OE *bōtlēas* (as BOOT[2], LESS)]

bootlicker /'buːt,lɪkə(r)/ n. colloq. a person who behaves obsequiously or servilely; a toady. □ **bootlick** v.intr. **bootlicking** n. & adj.

bootmaker /'buːt,meɪkə(r)/ n. a maker or manufacturer of boots and shoes.

boots /buːts/ n. Brit. a hotel employee who cleans boots and shoes, carries luggage, etc.

bootstrap /'buːtstræp/ n. **1** a loop at the back of a boot used to pull it on. **2** Computing a technique of loading a program into a computer by means of a few initial instructions which enable the introduction of the rest of the program from an input device. □ **pull oneself up by one's bootstraps** better oneself by one's own efforts.

booty /'buːtɪ/ n. **1** plunder gained esp. in war or by piracy. **2** colloq. something gained or won. [ME f. MLG *būte*, *buite* exchange, of uncert. orig.]

booze /buːz/ n. & v. colloq. ● n. **1** alcoholic drink. **2** the drinking of this (*on the booze*). ● v.intr. drink alcoholic liquor, esp. excessively or habitually. □ **booze-up** sl. a drinking-bout. [earlier *bouse*, *bowse*, f. MDu. *būsen* drink to excess]

boozer /'buːzə(r)/ n. colloq. **1** a person who drinks alcohol, esp. to excess. **2** Brit. a public house.

boozy /'buːzɪ/ adj. (**boozier**, **booziest**) colloq. intoxicated; addicted to drink. □ **boozily** adv. **booziness** n.

bop[1] /bɒp/ n. & v. colloq. ● n. **1** = BEBOP. **2 a** a spell of dancing, esp. to pop music. **b** an organized social occasion for this. ● v.intr. (**bopped**, **bopping**) dance, esp. to pop music. □ **bopper** n. [abbr. of BEBOP]

bop[2] /bɒp/ v. & n. colloq. ● v.tr. (**bopped**, **bopping**) hit, punch lightly. ● n. a light blow or hit. [imit.]

bo-peep /bəʊ'piːp/ n. a game of hiding and suddenly reappearing, played with a young child. [BO[1] + PEEP[1]]

Bophuthatswana /ˌbəʊpuːtətˈswɑːnə/ a former homeland established in South Africa for the Tswana people, now part of North-West Province and Eastern Transvaal. (See also HOMELAND.)

bora[1] /ˈbɔːrə/ n. a strong cold dry north-east wind blowing in the upper Adriatic. [It. dial. f. L *boreas* north wind: see BOREAL]

bora[2] /ˈbɔːrə/ n. *Austral.* an Aboriginal rite in which boys are initiated into manhood. [Aboriginal]

Bora-Bora /ˌbɔːrəˈbɔːrə/ an island of the Society Islands group in French Polynesia.

boracic /bəˈræsɪk/ adj. of borax; containing boron. □ **boracic acid** = boric acid. [med.L *borax -acis*]

borage /ˈbɒrɪdʒ/ n. a plant of the genus *Borago*; esp. *B. officinalis*, with bright blue flowers and hairy leaves sometimes used as flavouring. [OF *bourrache* f. med.L *borrago* perh. f. Arab. *'abū ḥurāš* father of roughness (with ref. to the leaves)]

borak /ˈbɔːræk/ n. *Austral.* & *NZ sl.* banter, ridicule, nonsense. □ **poke borak at** make fun of. [Austral. pidgin, f. Aboriginal]

borane /ˈbɔːreɪn/ n. *Chem.* a hydride of boron.

Borås /bʊˈrɔːs/ an industrial city in SW Sweden; pop. (1990) 101,770.

borate /ˈbɔːreɪt/ n. a salt or ester of boric acid.

borax /ˈbɔːræks/ n. **1** the mineral salt sodium borate, occurring in alkaline deposits as an efflorescence or as crystals. **2** the purified form of this salt, used in making glass and china, and as an antiseptic. [ME f. OF *boras* f. med.L *borax* f. Arab. *būrak* f. Pers. *būrah*]

Borazon /ˈbɔːrəˌzɒn/ n. *propr.* a hard form of boron nitride, used as an abrasive. [BORON + AZO- + -ON]

borborygmus /ˌbɔːbəˈrɪɡməs/ n. (pl. **borborygmi** /-maɪ/) a rumbling of gas in the intestines. □ **borborygmic** adj. [mod.L f. Gk]

Bordeaux[1] /bɔːˈdəʊ/ a port of SW France on the River Garonne, capital of Aquitaine; pop. (1990) 213,270. It is a centre of the wine trade.

Bordeaux[2] /bɔːˈdəʊ/ n. (pl. same /-ˈdəʊz/) any of various red, white, or rosé wines from the district of Bordeaux in SW France. □ **Bordeaux mixture** a fungicide for vines, fruit-trees, etc., composed of equal quantities of copper sulphate and calcium oxide in water.

bordello /bɔːˈdeləʊ/ n. (pl. **-os**) esp. *N. Amer.* a brothel. [ME (f. It. *bordello*) f. OF *bordel* cabin, dimin. of *borde* ult. f. Frank.: see BOARD]

Border /ˈbɔːdə(r)/, Allan (Robert) (b.1955), Australian cricketer. A batsman and occasional spin bowler, he first played for his country in 1978 and became captain six years later. At the time of his retirement from international cricket in 1994 he had made 156 test match appearances (93 as captain) and scored 11,174 runs (all three figures being world records).

border /ˈbɔːdə(r)/ n. & v. ● n. **1** the edge or boundary of anything, or the part near it. **2 a** the line separating two political or geographical areas, esp. countries. **b** the district on each side of this. **c** (**the Border**) a particular boundary and its adjoining districts, esp. between Scotland and England (usu. **the Borders**), or Northern Ireland and the Republic of Ireland (see also BORDERS). **3** a distinct edging round anything, esp. for strength or decoration. **4** a long narrow bed of flowers or shrubs in a garden (*herbaceous border*). ● v. **1** tr. be a border to. **2** tr. provide with a border. **3** intr. (usu. foll. by *on*, *upon*) **a** adjoin; come close to being. **b** approximate, resemble. [ME f. OF *bordure*: cf. BOARD]

Border collie n. a common working sheepdog originating near the border between England and Scotland.

borderer /ˈbɔːdərə(r)/ n. a person who lives near a border, esp. that between Scotland and England.

borderland /ˈbɔːdəˌlænd/ n. **1** the district near a border. **2** an intermediate condition between two extremes. **3** an area for debate.

borderline /ˈbɔːdəˌlaɪn/ n. & adj. ● n. **1** the line dividing two (often extreme) conditions. **2** a line marking a boundary. ● adj. **1** on the borderline. **2** verging on an extreme condition; only just acceptable.

Borders /ˈbɔːdəz/ an administrative region of southern Scotland; administrative centre, Newtown St Boswells.

Border terrier n. a breed of small terrier with rough hair, originating in the Cheviot Hills.

Bordet /ˈbɔːdeɪ/, Jules (1870–1961), Belgian bacteriologist and immunologist. He discovered the heat-sensitive complement system found in blood serum, and demonstrated its role in antibody–antigen reactions and bacterial lysis. He also isolated a number of pathogenic bacteria, and developed a vaccine for whooping cough. Bordet was awarded a Nobel Prize in 1919.

bordure /ˈbɔːdjʊə(r)/ n. *Heraldry* a border round the edge of a shield. [ME form of BORDER]

bore[1] /bɔː(r)/ v. & n. ● v. **1** tr. make a hole in, esp. with a revolving tool. **2** tr. hollow out (a tube etc.). **3** tr. **a** make (a hole) by boring or excavation. **b** make (one's way) through a crowd etc. **4** intr. (of an athlete, racehorse, etc.) push another competitor out of the way. **5** intr. drill a well (for oil etc.). ● n. **1** the hollow of a firearm barrel or of a cylinder in an internal-combustion engine. **2** the diameter of this; the calibre. **3** = BOREHOLE. [OE *borian* f. Gmc]

bore[2] /bɔː(r)/ n. & v. ● n. a tiresome or dull person or thing. ● v.tr. weary by tedious talk or dullness. □ **bore a person to tears** weary (a person) in the extreme. [18th c.: orig. unkn.]

bore[3] /bɔː(r)/ n. (in full **tidal bore**) a large wave that travels up a narrow estuary or tidal river, caused by constriction of the rising tide, esp. a spring tide. Also called *eagre*. [ME, perh. f. ON *bára* wave]

bore[4] *past* of BEAR[1].

boreal /ˈbɔːrɪəl/ adj. **1** of the north or northern regions. **2** of the north wind. [ME f. F *boréal* or LL *borealis* f. L *Boreas* f. Gk *Boreas* god of the north wind]

boredom /ˈbɔːdəm/ n. the state of being bored; ennui.

borehole /ˈbɔːhəʊl/ n. **1** a deep narrow hole, esp. one made in the earth to find water, oil, etc. **2** *Austral.* a water-hole for cattle.

borer /ˈbɔːrə(r)/ n. **1** a worm, mollusc, insect, or insect larva which bores into wood, other plant material, or rock. **2** a tool for boring.

Borg /bɔːɡ/, Björn (Rune) (b.1956), Swedish tennis player. His first major titles were the Italian and French championships (1974); he then went on to win five consecutive men's singles titles at Wimbledon (1976–80), beating the record of three consecutive wins held by Fred Perry.

Borges /ˈbɔːxes/, Jorge Luis (1899–1986), Argentinian poet, short-story writer, and essayist. His first three collections of poetry (1923–9) explore the themes of time and identity that are later treated in his fiction. His first volume of short stories, *A Universal History of Infamy* (1935, revised 1954), recounting the lives of real and fictitious criminals and exploring the relationships between fiction, truth, and identity, is regarded as a founding work of magic realism. His fiction also includes *The Aleph and Other Stories* (1949).

boric /ˈbɔːrɪk/ adj. of or containing boron. □ **boric acid** an acid derived from borax, used as a mild antiseptic and in the manufacture of heat-resistant glass and enamels.

boring /ˈbɔːrɪŋ/ adj. that makes one bored; uninteresting, tedious, dull. □ **boringly** adv. **boringness** n.

Boris Godunov /ˈbɒrɪs/ see GODUNOV.

Born /bɔːn/, Max (1882–1970), German theoretical physicist and a founder of quantum mechanics. In 1921 he was appointed professor of theoretical physics at Göttingen, but in 1933 he had to flee the Nazi regime and settled in Britain. After retirement from Edinburgh University he returned to Göttingen to write mainly on the philosophy of physics and the social responsibility of scientists. He provided a link between wave mechanics and quantum theory by postulating a probabilistic interpretation of Schrödinger's wave equation, for which he was awarded the Nobel Prize for physics in 1954. He also wrote popular textbooks on optics and atomic physics.

born /bɔːn/ adj. **1** existing as a result of birth. **2 a** being such or likely to become such by natural ability or quality (*a born leader*). **b** (usu. foll. by *to* + infin.) having a specified destiny or prospect (*born lucky*; *born to be king*; *born to lead men*). **3** (in *comb.*) of a certain status by birth (*French-born*; *well-born*). □ **born-again** (attrib.) converted (esp. to fundamentalist Christianity). **born and bred** by birth and upbringing. **in all one's born days** *colloq.* in one's life so far. **not born yesterday** *colloq.* not stupid; shrewd. [past part. of BEAR[1]]

borne /bɔːn/ **1** *past part.* of BEAR[1]. **2** (in *comb.*) carried or transported by (*airborne*).

borné /ˈbɔːneɪ/ adj. **1** narrow-minded; of limited ideas. **2** having limitations. [F]

Borneo /ˈbɔːnɪˌəʊ/ a large island of the Malay Archipelago, comprising Kalimantan (a region of Indonesia), Sabah and Sarawak (states of Malaysia), and Brunei. □ **Bornean** adj.

Bornholm /ˈbɔːnhəʊm/ a Danish island in the Baltic Sea, south-east of Sweden.

Bornholm disease n. a viral infection with fever and pain in the muscles of the ribs. It was named after the Danish island of Bornholm from which a series of early epidemics were described.

boro- /ˈbɔːrəʊ/ *comb. form* indicating salts containing boron.

Borobudur /ˌbɒrəʊbʊˈdʊə(r)/ a Buddhist monument in central Java, built *c*.800, abandoned *c*.1000, restored in 1907-11 and again in the 1980s. It consists of five square successively smaller terraces, one above the other, surmounted by three concentric galleries and a stupa. Illustrations on the terrace walls show the life of the Buddha and successive stages towards perfection.

Borodin /ˈbɒrədɪn/, Aleksandr (Porfirevich) (1833-87), Russian composer. He began to compose music at the age of 9, but trained as a chemist before undertaking formal musical studies in 1862. A member of the group known as 'The Five' or 'The Mighty Handful' (the others were Mily (Alekseevich) Balakirev (1837-1910), César (Antonovich) Cui (1835-1918), Mussorgsky, and Rimsky-Korsakov), he composed symphonies, string quartets, songs, and piano music, but is best known for the epic opera *Prince Igor* (completed after his death by Rimsky-Korsakov and Glazunov).

Borodino, Battle of /ˌbɒrəˈdiːnəʊ/ a battle in 1812 at Borodino, a village about 110 km (70 miles) west of Moscow, at which Napoleon's forces defeated the Russian army under Prince Kutuzov (1745-1813). This allowed the French to advance to Moscow, but the heavy losses that they suffered at Borodino contributed to their eventual defeat.

boron /ˈbɔːrɒn/ *n.* a non-metallic chemical element (atomic number 5; symbol **B**). Boron was first isolated by Sir Humphry Davy in 1807, although borax, its chief source, had been known to alchemists. The element is usually prepared as an amorphous brown powder, but when very pure it forms hard, shiny, black crystals with semiconducting properties. The element has some specialized uses, e.g. in alloy steels; several compounds such as borax and boric acid are commercially important. [f. BORAX after CARBON]

boronia /bəˈrəʊnɪə/ *n.* a sweet-scented shrub of the genus *Boronia*, native to Australia. [F. *Borone*, It. botanist (1769-94)]

borosilicate /ˌbɔːrəʊˈsɪlɪˌkeɪt/ *n.* any of many substances containing boron, silicon, and oxygen generally used in glazes and enamels and in the production of glass.

borough /ˈbʌrə/ *n.* **1** *Brit.* **a** a town (as distinct from a city) with a corporation and privileges granted by a royal charter. **b** *hist.* a town sending representatives to Parliament. **2** an administrative division of London. **3** a municipal corporation in certain US states. **4** each of five divisions of New York City. **5** (in Alaska) a county. [OE *burg, burh* f. Gmc: cf. BURGH]

Borovets /ˈbɒrəˌvets/ a ski resort in the Rila Mountains of western Bulgaria.

Borromini /ˌbɒrəˈmiːnɪ/, Francesco (1599-1667), Italian architect. He worked in Rome for most of his life, first as a mason at St Peter's and subsequently on the Palazzo Barberini (1620-31) as the chief assistant to his rival Bernini. Borromini's own buildings include the churches of S. Carlo alle Quattro Fontane (1641) and S. Ivo della Sapienza (1643-60). A leading figure of the Italian baroque, he used subtle architectural forms, innovative spatial composition, and much sculptural decoration. His work was central to the development of the baroque in Italy and an important influence on architecture in Austria and southern Germany.

Borrow /ˈbɒrəʊ/, George (Henry) (1803-81), English writer. His travels in England, Europe, Russia, and the Far East provided material for his picaresque narrative *Lavengro* (1851) and its sequel *The Romany Rye* (1857). In these books he combines fiction with a factual account of his travels. He also wrote *Wild Wales* (1862).

borrow /ˈbɒrəʊ/ *v.* **1 a** *tr.* acquire temporarily with the promise or intention of returning. **b** *intr.* obtain money in this way. **2** *tr.* use (an idea, invention, etc.) originated by another; plagiarize. **3** *intr. Golf* **a** play the ball uphill so that it rolls back towards the hole. **b** allow for the wind or a slope. □ **borrow pit** a pit resulting from the excavation of material for use in embanking etc. **borrowed time** an unexpected extension esp. of life. □ **borrower** *n.* **borrowing** *n.* [OE *borgian* give a pledge]

borsch var. of BORTSCH.

Borstal /ˈbɔːst(ə)l/ *n. Brit. hist.* an institution for reforming and training young offenders. ¶ Now replaced by *detention centre* and *youth custody centre.* [*Borstal* in S. England, where the first of these was established]

bort /bɔːt/ *n.* (also **boart**) **1** an inferior or malformed diamond, used for cutting. **2** fragments of diamonds produced in cutting. [Du. *boort*]

bortsch /bɔːtʃ/ *n.* (also **borsch** /bɔːʃ/) a highly seasoned Russian or Polish soup with various ingredients including beetroot and cabbage and served with sour cream. [Russ. *borshch*]

borzoi /ˈbɔːzɔɪ/ *n.* a Russian breed of large wolfhound with a narrow head and silky, usu. white, coat. [Russ. f. *borzy* swift]

Bosanquet /ˈbəʊs(ə)n,ket/, Bernard James Tindall (1877-1936), English all-round cricketer. As a bowler for Middlesex and England he is credited with inventing the googly, which in Australia was named the bosie after him.

boscage /ˈbɒskɪdʒ/ *n.* (also **boskage**) **1** masses of trees or shrubs. **2** wooded scenery. [ME f. OF *boscage* f. Gmc: cf. BUSH¹]

Bosch /bɒʃ/, Hieronymus (*c*.1450-1516), Dutch painter. His highly detailed works are typically crowded with creatures of fantasy, half-human half-animal, and grotesque demons, and are interspersed with human figures in non-realistic settings that are often representations of sin and folly. His style is distinctly unlike that of mainstream Dutch painting of the period, and the elements of the grotesque and fantasy in his work prefigure the style of the surrealists.

Bose¹ /bəʊs/, Sir Jagdis Chandra (1858-1937), Indian physicist and plant physiologist. He investigated the properties of very short radio waves, wireless telegraphy, and radiation-induced fatigue in inorganic materials. His physiological work involved comparative measurements of the responses of plants exposed to stress.

Bose² /bəʊs/, Satyendra Nath (1894-1974), Indian physicist. He contributed to statistical mechanics, quantum statistics, and unified field theory, and derived Planck's black-body radiation law without reference to classical electrodynamics. With Einstein, he described fundamental particles which later came to be known as bosons.

bosh /bɒʃ/ *n. & int. sl.* nonsense; foolish talk. [Turk. *boş* empty]

bosie /ˈbəʊzɪ/ *n. Austral. Cricket* = GOOGLY. [BOSANQUET + -IE]

Boskop /ˈbɒskɒp/ a town in South Africa, in North-West Province, where a skull fossil was found in 1913. The fossil is undated and morphologically shows no primitive features. At the time of discovery, this find was regarded as representative of a distinct 'Boskop race' but is now thought to be related to the San-Nama (Bushman-Hottentot) types.

bosky /ˈbɒskɪ/ *adj. literary* wooded, bushy. [ME *bosk* thicket]

bo's'n var. of BOATSWAIN.

Bosnia /ˈbɒznɪə/ **1** a region in the Balkans forming the larger, northern part of Bosnia-Herzegovina. **2** = BOSNIA-HERZEGOVINA. □ **Bosnian** *adj. & n.*

Bosnia-Herzegovina /ˌbɒznɪəˌheətsəˈgɒvɪnə, -gəˈviːnə/ (also **Bosnia and Herzegovina**) a country in the Balkans, formerly a constituent republic of Yugoslavia; pop. (est. 1991) 4,365,000; capital, Sarajevo. Settled by Slavs in the 7th century, Bosnia and Herzegovina were conquered by the Turks in 1463. The province of Bosnia-Herzegovina came under Austrian control in 1878 and was annexed in 1908, an event which contributed towards the outbreak of the First World War. In 1918, it became part of the Kingdom of Serbs, Croats, and Slovenes, which changed its name to Yugoslavia in 1929. In 1992 Bosnia-Herzegovina followed Slovenia and Croatia in declaring independence, but ethnic conflict amongst Muslims, Serbs, and Croats created a state of civil war and UN peacekeeping forces were deployed to prevent further advances by Bosnian Serb forces, which controlled more than half of Bosnian territory. In 1994 Bosnian Muslims and Croats reached an accord establishing a federation. In 1995, after negotiations in the US, the presidents of Bosnia, Croatia, and Serbia signed a peace agreement which effectively divided Bosnia into separate self-governing parts, the Muslim-Croat Federation and the Bosnian Serb Republic. (See also HERZEGOVINA.)

bosom /ˈbʊz(ə)m/ *n. & adj.* ● *n.* **1 a** a person's breast or chest, esp. a woman's. **b** *colloq.* each of a woman's breasts. **c** the enclosure formed by a person's breast and arms. **2** the seat of the emotions; an emotional centre, esp. as the source of an enfolding relationship (*in the bosom of one's family*). **3** the part of a woman's dress covering the breast. ● *adj.* (esp. in **bosom friend**) close, intimate. [OE *bōsm* f. Gmc]

bosomy /ˈbʊzəmɪ/ *adj.* (of a woman) having large breasts.

boson /ˈbəʊzɒn/ *n. Physics* a subatomic particle, such as a photon, which has zero or integral spin and follows the statistical description given by S. N. Bose and Einstein (cf. FERMION). [BOSE² + -ON]

Bosporus /ˈbɒspərəs/ (also **Bosphorus** /ˈbɒsfə-/) a strait connecting the Black Sea with the Sea of Marmara, and separating Europe from the Anatolian peninsula of western Asia. Istanbul is located at its south end. It is spanned by two long suspension bridges. [Gk *bos* ox, cow + *poros* passage, crossing; the name is linked with the story of Io]

BOSS /bɒs/ *abbr.* Bureau of State Security, the former South African intelligence and security organization. It was replaced by the National

Intelligence Service after evidence of corruption was revealed in 1978. Its activities abroad against opponents of apartheid caused much resentment in the countries where it operated.

boss[1] /bɒs/ *n. & v. colloq.* ● *n.* **1** a person in charge; an employer, manager, or overseer. **2** *US* a person who controls or dominates a political organization. ● *v.tr.* **1** (usu. foll. by *about, around*) treat domineeringly; give constant peremptory orders to. **2** be the master or manager of. [orig. US: f. Du. *baas* master]

boss[2] /bɒs/ *n.* **1** a round knob, stud, or other protuberance, esp. on the centre of a shield or in ornamental work. **2** *Archit.* a piece of ornamental carving etc. covering the point where the ribs in a vault or ceiling cross. **3** *Geol.* a large mass of igneous rock. **4** *Mech.* an enlarged part of a shaft. [ME f. OF *boce* f. Rmc]

bossa nova /ˌbɒsə ˈnəʊvə/ *n.* **1** a dance like the samba, originating in Brazil. **2** a piece of music for this or in its rhythm. [Port., = new style]

boss-eyed /ˈbɒsaɪd/ *adj. Brit. colloq.* **1** having only one good eye; cross-eyed. **2** crooked; out of true. [dial. *boss* miss, bungle]

boss-shot /ˈbɒsʃɒt/ *n. Brit. dial. & sl.* **1** a bad shot or aim. **2** an unsuccessful attempt. [as BOSS-EYED]

bossy /ˈbɒsɪ/ *adj.* (**bossier**, **bossiest**) *colloq.* domineering; tending to boss. □ **bossy-boots** *colloq.* a domineering person. □ **bossily** *adv.* **bossiness** *n.*

Boston /ˈbɒstən/ the state capital of Massachusetts; pop. (1990) 574,280. It was founded *c.*1630 by the Massachusetts Bay Company under its governor, John Winthrop (1588–1649), and named after Boston in Lincolnshire. It was an early centre of New England Puritanism and was the scene of disturbances leading to the War of American Independence at the end of the 18th century. □ **Bostonian** /bɒsˈtəʊnɪən/ *n. & adj.*

Boston Tea Party a violent demonstration in 1773 by American colonists prior to the War of American Independence. Dressed as Mohawk Indians, colonists boarded vessels moored in the harbour of Boston, Massachusetts, and threw the cargoes of tea into the water in protest at the imposition of a tax on tea by the British Parliament, in which the colonists had no representation.

bosun (also **bo'sun**) var. of BOATSWAIN.

Boswell /ˈbɒzwəl/, James (1740–95), Scottish author and biographer. After travels in Europe, where he met Jean-Jacques Rousseau and Voltaire, he practised law, although his ambitions were directed towards literature and politics. He first met Samuel Johnson in London (1763) and Boswell's *Journal of a Tour to the Hebrides* (1785) describes their travels together in 1773. He is now best known for his celebrated biography *The Life of Samuel Johnson* (1791), which gives a vivid and intimate portrait of Johnson and an invaluable panorama of the age and its personalities.

Bosworth Field /ˈbɒzwəθ/ (also **Battle of Bosworth**) a battle of the Wars of the Roses fought in 1485 near Market Bosworth in Leicestershire. Henry Tudor defeated and killed the Yorkist king Richard III, enabling him to take the throne as Henry VII.

bot /bɒt/ *n.* (also **bott**) the parasitic larva of a bot-fly, infesting the digestive organs of horses, sheep, etc. □ **bot-fly** a dipterous fly of the genus *Oestrus*. [prob. of LG orig.]

bot. *abbr.* **1** bottle. **2** botanic; botanical; botany. **3** bought.

botanize /ˈbɒtəˌnaɪz/ *v.intr.* (also **-ise**) study plants, esp. in their habitat.

Botany /ˈbɒtənɪ/ *n.* (in full **Botany wool**) merino wool, esp. from Australia. [BOTANY BAY]

botany /ˈbɒtənɪ/ *n.* **1** the scientific study of plants. (*See note below.*) **2** the plant life of a particular area or time. □ **botanist** *n.* **botanic** /bəˈtænɪk/ *adj.* **botanical** *adj.* **botanically** *adv.* [*botanic* f. F *botanique* or LL *botanicus* f. Gk *botanikos* f. *botanē* plant]

■ The study of plants has a long history: Theophrastus (*c.*300 BC) wrote about their form and function, and they remained of interest thereafter, especially because of their medicinal properties. John Ray produced the first systematic account of English flora, and Linnaeus devised a system for naming and classifying plants which is still used (in modified form) today. Study of plant anatomy developed in the 17th century, and of plant physiology a century later, with important advances such as the discovery of photosynthesis. Modern botany has contributed much to agriculture and horticulture with discoveries such as plant hormones, systematic breeding techniques, and advances in plant biology and pathology.

Botany Bay an inlet of the Tasman Sea in New South Wales, Australia, just south of Sydney. It was the site of Captain James Cook's landing in 1770 and of an early British penal settlement. Cook named the bay after the large variety of plants collected there by his companion, Sir Joseph Banks.

botch /bɒtʃ/ *v. & n.* (also **bodge** /bɒdʒ/) ● *v.tr.* **1** bungle; do badly. **2** patch or repair clumsily. ● *n.* bungled or spoilt work (*made a botch of it*). □ **botcher** *n.* [ME: orig. unkn.]

botel /bəʊˈtel/ *n.* (also **boatel**) a waterside hotel with facilities for mooring boats. [blend of BOAT and HOTEL]

both /bəʊθ/ *adj., pron., & adv.* ● *adj. & pron.* the two, not only one (*both boys; both the boys; both of the boys; the boys are both here*). ¶ Widely used with *of*, esp. when followed by a pronoun (e.g. *both of us*) or a noun implying separate rather than collective consideration, e.g. *both of the boys* suggests *each boy* rather than the two together. ● *adv.* with equal truth in two cases (*both the boy and his sister are here; are both here and hungry*). □ **both ways** = *each way.* **have it both ways** alternate between two incompatible points of view to suit the needs of the moment. [ME f. ON *báthir*]

Botha[1] /ˈbəʊtə/, Louis (1862–1919), South African soldier and statesman, first Prime Minister of the Union of South Africa 1910–19. One of the most successful Boer leaders in the Boer War, Botha became Commander-in-Chief in 1900 and waged guerrilla warfare against the British forces. As Transvaal's first Prime Minister he played a leading role in the National Convention (1908–9), which was responsible for drafting the constitution for the Union of South Africa; he became its Prime Minister a year later. Botha supported the Allies in the First World War, gaining recognition for his annexation of German South West Africa in 1915.

Botha[2] /ˈbəʊtə/, P(ieter) W(illem) (b.1916), South African statesman, Prime Minister 1978–84, State President 1984–9. He joined the National Party in 1936 and was involved in party organization, particularly in Cape Province, until his election as Prime Minister in 1978. He abolished the office of Prime Minister in 1984, replacing it with that of State President. An authoritarian leader, he continued to enforce apartheid, but in response to pressure introduced limited reforms, including a new constitution (1984) giving certain classes of non-whites a degree of political representation. His resistance to more radical change ultimately led to his fall from power.

Botham /ˈbəʊθəm/, Ian (Terence) (b.1955), English all-round cricketer. He made his county début for Somerset in 1974 and his test début against Australia in 1977 and was admired for his bold attacking style of batting and skill as a medium-fast bowler. In 1978 he became the first player to score 100 runs and take eight wickets in a single test match; in 1982 he also achieved the record of 3,000 runs and 250 wickets in test matches overall. He retired from first-class cricket in 1993.

bother /ˈbɒðə(r)/ *v., n., & int.* ● *v.* **1** *tr.* **a** give trouble to; worry, disturb. **b** *refl.* (often foll. by *about*) be anxious or concerned. **2** *intr.* **a** (often foll. by *about*, or *to* + infin.) worry or trouble oneself (*don't bother about that; didn't bother to tell me*). **b** (foll. by *with*) be concerned. ● *n.* **1 a** a person or thing that bothers or causes worry. **b** a minor nuisance. **2 a** a trouble, worry, fuss. **b** a state of worry. ● *int.* esp. *Brit.* expressing annoyance or impatience. □ **cannot be bothered** will not make the effort needed. [Ir. *bodhraim* deafen]

botheration /ˌbɒðəˈreɪʃ(ə)n/ *n. & int. colloq.* = BOTHER *n. & int.*

bothersome /ˈbɒðəsəm/ *adj.* causing bother; troublesome.

Bothnia, Gulf of /ˈbɒθnɪə/ a northern arm of the Baltic Sea, between Sweden and Finland.

Bothwell /ˈbɒθwel/, 4th Earl of (title of James Hepburn) (*c.*1536–78), Scottish nobleman and third husband of Mary, Queen of Scots. Mary's chief adviser, he was implicated in the murder of Lord Darnley (1567). He was tried for the crime but acquitted and he married Mary later the same year. The marriage was annulled in 1570 and Bothwell was imprisoned in Denmark, where he died.

bothy /ˈbɒθɪ/ *n.* (also **bothie**) (*pl.* **-ies**) *Sc.* a small hut or cottage, esp. one for housing labourers or for use as a mountain refuge. [18th c.: orig. unkn.: perh. rel. to BOOTH]

bo tree /bəʊ/ *n.* the Indian fig tree, *Ficus religiosa*, regarded as sacred by Buddhists. Also called *peepul, pipal*. [repr. Sinh. *bōgaha* tree of knowledge (Buddha's enlightenment having occurred beneath such a tree)]

botryoidal /ˌbɒtrɪˈɔɪd(ə)l/ *adj. Mineral.* shaped like a cluster of grapes. [Gk *botruoeidēs* f. *botrus* bunch of grapes]

Botswana /bɒtˈswɑːnə/ a landlocked country in southern Africa; pop. (est. 1991) 1,300,000; official languages, Setswana and English;

capital, Gaborone. Much of Botswana is an arid tableland, with the Kalahari Desert occupying the western half of the country. Inhabited by Sotho people and, in the Kalahari, San (Bushmen), the area was made the British Protectorate of Bechuanaland in 1885. It became an independent republic within the Commonwealth in 1966, adopting the name Botswana. □ **Botswanan** adj. & n.

bott var. of BOT.

Botticelli /ˌbɒtɪˈtʃɛlɪ/, Sandro (born Alessandro di Mariano Filipepi) (1445–1510), Italian painter. A pupil of Filippo Lippi, he had his own studio in Florence and enjoyed the patronage of the Medici. His reputation rests largely on his mythological paintings such as *Primavera* (c.1478) and *The Birth of Venus* (c.1480). His work was neglected until the second half of the 19th century, when it was re-evaluated and championed by John Ruskin and Walter Pater; it had a significant influence on the Pre-Raphaelites, especially Burne-Jones.

bottle /ˈbɒt(ə)l/ n. & v. ● n. **1** a container, usu. of glass or plastic and with a narrow neck, for storing liquid. **2** the amount that will fill a bottle. **3** a baby's feeding-bottle. **4** = *hot-water bottle*. **5** a metal cylinder for liquefied gas. **6** *Brit. sl.* courage, confidence. ● v.tr. **1** put into bottles or jars. **2** preserve (fruit etc.) in jars. **3** (foll. by *up*) **a** conceal or restrain for a time (esp. a feeling). **b** keep (an enemy force etc.) contained or entrapped. **4** (as **bottled** adj.) *sl.* drunk. □ **bottle bank** a place where used bottles may be deposited for recycling. **bottle-brush 1** a cylindrical brush for cleaning inside bottles. **2** a plant with a flower of this shape, esp. an Australian shrub such as banksia. **bottle-green** a dark shade of green. **bottle party** a party to which guests bring bottles of drink. **bottle tree** an Australian tree of the genus *Brachychiton*, with a swollen bottle-shaped trunk. **hit the bottle** *sl.* drink heavily. **on the bottle** *sl.* drinking (alcoholic drink) heavily. [ME f. OF *botele*, *botaille* f. med.L *butticula* dimin. of LL *buttis* BUTT⁴]

bottle-feed /ˈbɒt(ə)lˌfiːd/ v.tr. (past and past part. **-fed** /-ˌfed/) feed (a baby) with milk by means of a bottle.

bottleneck /ˈbɒt(ə)lˌnek/ n. **1** a narrow place at which the flow of traffic, production, etc., is constricted. **2** an obstruction to the even flow of production etc. **3** *Mus.* the neck of a bottle, worn on the finger by a guitarist to produce sliding effects on the strings over the fingerboard. □ **bottleneck blues** a style of blues played on a bottleneck guitar. **bottleneck guitar** a guitar played with a bottleneck, or the style of music so produced.

bottlenose /ˈbɒt(ə)lˌnəʊz/ n. a swollen nose. □ **bottlenose dolphin** a dolphin, *Tursiops truncatus*, with a bottle-shaped snout. □ **bottlenosed** adj.

bottler /ˈbɒtlə(r)/ n. **1** a person who bottles drinks etc. **2** *Austral. & NZ sl.* an excellent person or thing.

bottle-washer /ˈbɒt(ə)lˌwɒʃə(r)/ n. (esp. in phr. **chief cook and bottle-washer**) *colloq.* a menial, a factotum.

bottom /ˈbɒtəm/ n., adj., & v. ● n. **1 a** the lowest point or part (*bottom of the stairs*). **b** the part on which a thing rests (*bottom of a saucepan*). **c** the underneath part (*scraped the bottom of the car*). **d** the furthest or inmost part (*bottom of the garden*). **2** *colloq.* **a** the buttocks. **b** the seat of a chair etc. **3 a** the less honourable, important, or successful end of a table, a class, etc. (*at the bottom of the list of requirements*). **b** a person occupying this place (*he's always bottom of the class*). **4** the ground under the water of a lake, river, etc. (*swam until he touched the bottom*). **5** the basis; the origin (*he's at the bottom of it*). **6** the essential character; reality. **7** *Naut.* **a** the keel or hull of a ship. **b** a ship, esp. as a cargo-carrier. **8** staying power; endurance. **9** = *bottom gear* (see GEAR). ● adj. lowest; last; at or forming the bottom (*bottom button*; *got the bottom score*). ● v. **1** tr. put a bottom to (a chair, saucepan, etc.). **2** intr. (of a ship) reach or touch the bottom. **3** tr. find the extent or real nature of; work out. **4** tr. (usu. foll. by *on*) base (an argument etc.) (*reasoning bottomed on logic*). **5** tr. touch the bottom or lowest point of. □ **at bottom** basically, essentially. **be at the bottom of** have caused. **bet one's bottom dollar** *sl.* stake all; be very confident. **bottom dog** = UNDERDOG. **bottom drawer** *Brit.* linen etc. stored by a woman in preparation for her marriage. **bottom falls (or drops) out** collapse occurs. **bottom gear** see GEAR. **bottom line** *colloq.* the underlying or ultimate truth; the ultimate, esp. financial, criterion. **bottom out** reach the lowest level. **bottoms up!** a call to drain one's glass. **bottom-up 1** upside-down. **2** (of institutions etc.) non-hierarchical. **3** (of knowledge etc.) proceeding from detail to general theory. **get to the bottom of** fully investigate and explain. **knock the bottom out of** prove (a thing) worthless. □ **bottommost** /ˈbɒtəmˌməʊst/ adj. [OE *botm* f. Gmc]

bottomless /ˈbɒtəmlɪs/ adj. **1** without a bottom. **2** (of a supply etc.) inexhaustible.

bottomry /ˈbɒtəmrɪ/ n. & v. *Naut.* ● n. a system of using a ship as security against a loan to finance a voyage, the lender losing his or her money if the ship sinks. ● v.tr. (**-ies**, **-ied**) pledge (a ship) in this way. [BOTTOM = ship + -RY, after Du. *bodemerij*]

botulism /ˈbɒtjʊˌlɪz(ə)m/ n. poisoning caused by a toxin produced by the bacillus *Clostridium botulinum*, which grows in poorly preserved food. [G *Botulismus* f. L *botulus* sausage]

Boucher /ˈbuːʃeɪ/, François (1703–70), French painter and decorative artist. One of the foremost artists of the rococo style in France, Boucher reflected the elegant social life of the period in his work, and enjoyed both royal and aristocratic patronage. His output ranged from large decorative paintings of mythological scenes to popular engravings, *fêtes galantes*, and tapestry design. Significant paintings include *The Rising of the Sun* (1753) and *Summer Pastoral* (1749).

Boucher de Perthes /ˌbuːʃeɪ də ˈpeət/, Jacques (1788–1868), French archaeologist. He discovered some of the first evidence of man-made stone tools near the bones of extinct (Pleistocene) animals in the valley of the River Somme in northern France. In the decade following 1837 he argued that these tools (and their makers) belonged to a remote pre-Celtic 'antediluvian' age, but it was not until the 1850s, when geologists supported his claims, that his findings were accepted.

bouclé /ˈbuːkleɪ/ n. **1** a looped or curled yarn (esp. wool). **2** a fabric, esp. knitted, made of this. [F, = buckled, curled]

Boudicca /bəʊˈdɪkə, ˈbuːdɪkə/ (also **Boadicea** /ˌbəʊdɪˈsiːə/) (d. AD 62), a queen of the Britons, ruler of the Iceni tribe in eastern England. When Rome broke the treaty made with King Prasutagus (her husband) after his death in AD 60, Boudicca led her forces in revolt against the Romans and sacked Colchester, St Albans, and London before being completely defeated by the Roman governor Suetonius Paulinus. She committed suicide soon after her defeat, but her name became a symbol of native resistance to the Roman occupation.

boudoir /ˈbuːdwɑː(r)/ n. a woman's small private room or bedroom. [F, lit. 'sulking-place' f. *bouder* sulk]

bouffant /ˈbuːfɒn/ adj. (of a dress, hair, etc.) puffed out. [F]

Bougainville[1] /ˈbuːɡənˌvɪl/ a volcanic island in the South Pacific, the largest of the Solomon Islands. It is named after Louis de Bougainville, who visited it in 1768.

Bougainville[2] /ˈbuːɡənˌvɪl/, Louis Antoine de (1729–1811), French explorer. After a distinguished early military career, Bougainville joined the French navy and led the first successful French circumnavigation of the globe 1766–9, visiting many of the islands of the South Pacific and compiling an invaluable scientific record of his findings. The island of Bougainville is named after him, as is the plant bougainvillea.

bougainvillea /ˌbuːɡənˈvɪlɪə/ n. a widely cultivated tropical plant of the genus *Bougainvillea*, native to South America, with large colourful bracts (usu. purple, red, or white) almost concealing the flowers. [BOUGAINVILLE[2]]

bough /baʊ/ n. a branch of a tree, esp. a main one. [OE *bōg*, *bōh* f. Gmc]

bought past and past part. of BUY.

boughten /ˈbɔːt(ə)n/ adj. N. Amer. or dial. bought at a shop, not home-made. [var. of past part. of BUY]

bougie /ˈbuːʒiː/ n. **1** *Med.* a thin flexible surgical instrument for exploring, dilating, etc. the passages of the body. **2** a wax candle. [F f. Arab. *Bijāya* Algerian town with a wax trade]

bouillabaisse /ˌbuːjəˈbes/ n. a rich, spicy fish-stew, orig. from Provence. [F]

bouilli /ˈbuːjɪ/ n. stewed or boiled meat. [F, past part. of *bouillir* = boiled]

bouillon /ˈbuːjɒn/ n. thin soup; broth. [F f. *bouillir* to boil]

boulder /ˈbəʊldə(r)/ n. a large stone worn smooth by erosion. □ **boulder-clay** *Geol.* a mixture of boulders etc. formed by deposition from massive bodies of melting ice, giving distinctive glacial formations. [short for *boulderstone*, ME f. Scand.]

boule[1] /buːl/ n. (also **boules** pronunc. same) a French form of bowls, played on rough ground with usu. metal balls. [F, = BOWL[2]]

boule[2] /ˈbuːlɪ/ n. a legislative body of ancient or modern Greece. [Gk *boulē* senate]

boule[3] var. of BUHL.

boules var. of BOULE[1].

boulevard /ˈbuːləˌvɑːd, ˈbuːlvɑː(r)/ n. **1** a broad tree-lined avenue. **2** esp. *US* a broad main road. [F f. G *Bollwerk* BULWARK, orig. of a promenade on a demolished fortification]

Boulez /'buːlez/, Pierre (b.1925), French composer and conductor. His first publicly acclaimed work was *Le Marteau sans maître* (1954, revised 1957); other more recent compositions include *Répons* (1981–6). He was principal conductor with the New York Philharmonic Orchestra (1971–8). His works explore and develop serialism and aleatory music, making use of traditional and, in particular, electronic instruments.

boulle var. of BUHL.

Boulogne /buˈlɔɪn/ (also **Boulogne-sur-Mer** /-sjuəˈmeə(r)/) a ferry port and fishing town in northern France; pop. (1990) 44,240.

Boult /bəult/, Sir Adrian (Cedric) (1889–1983), English conductor. Noted especially for his championship of English composers, he was music director of the BBC (1930–49) and trained and conducted the BBC Symphony Orchestra (1931–50). He was also principal conductor of the London Philharmonic Orchestra (1950–7) and continued to conduct until 1981.

boult var. of BOLT².

Boulting /'bəultɪŋ/, John (1913–85) and Roy (b.1913), English film producers and directors. Twin brothers, they worked together, interchanging responsibilities as film producer and director. They made a number of memorable films, including *Brighton Rock* (1947) and *Seven Days to Noon* (1950). From the late 1950s their main output was comedy and farce, including *Private's Progress* (1956) and *I'm All Right Jack* (1959).

Boulton /'bəult(ə)n/, Matthew (1728–1809), English engineer and manufacturer. With his partner James Watt he pioneered the manufacture of steam engines, which they began to produce in 1774. Under Boulton's influence the engines began to enjoy widespread commercial success.

bounce /bəuns/ v. & n. ● v. **1 a** intr. (of a ball etc.) rebound. **b** tr. cause to rebound. **c** tr. & intr. bounce repeatedly. **2** intr. sl. (of a cheque) be returned by a bank when there are insufficient funds to meet it. **3** intr. **a** (foll. by *about, up*) (of a person, dog, etc.) jump or spring energetically. **b** (foll. by *in, out*, etc.) rush noisily, angrily, enthusiastically, etc. (*bounced into the room*; *bounced out in a temper*). **4** tr. (usu. foll. by *into* + verbal noun) colloq. hustle, persuade (*bounced him into signing*). **5** intr. colloq. talk boastfully. **6** tr. sl. eject forcibly (from a dancehall, club, etc.). ● n. **1 a** a rebound. **b** the power of rebounding (*this ball has a good bounce*). **2** colloq. **a** swagger, self-confidence (*has a lot of bounce*). **b** liveliness. **c** resilience. **3** (often prec. by *the*) sl. dismissal or ejection. □ **bounce back** regain one's good health, spirits, prosperity, etc. [ME *bunsen* beat, thump, (perh. imit.), or f. LG *bunsen*, Du. *bons* thump]

bouncer /'baʊnsə(r)/ n. **1** sl. a person employed to eject troublemakers from a dancehall, club, etc. **2** Cricket a ball bowled fast and short so as to rise high after pitching.

bouncing /'baʊnsɪŋ/ adj. **1** (esp. of a baby) big and healthy. **2** boisterous.

bouncy /'baʊnsɪ/ adj. (**bouncier, bounciest**) **1** (of a ball etc.) that bounces well. **2** cheerful and lively. **3** resilient, springy (*a bouncy sofa*). □ **bouncily** adv. **bounciness** n.

bound¹ /baʊnd/ v. & n. ● v.intr. **1 a** spring, leap (*bounded out of bed*). **b** walk or run with leaping strides. **2** (of a ball etc.) recoil from a wall or the ground; bounce. ● n. **1** a springy movement upwards or outwards; a leap. **2** a bounce. □ **by leaps and bounds** see LEAP. [F *bond*, *bondir* (orig. of sound) f. LL *bombitare* f. L *bombus* hum]

bound² /baʊnd/ n. & v. (usu. in pl.) **1** a limitation; a restriction (*beyond the bounds of possibility*). **2** a border of a territory; a boundary. ● v.tr. **1** (esp. in passive; foll. by *by*) set bounds to; limit (*views bounded by prejudice*). **2** be the boundary of. □ **out of bounds 1** outside the part of a school etc. in which one is allowed to be. **2** beyond what is acceptable; forbidden. [ME f. AF *bounde*, OF *bodne*, *bonde*, etc., f. med.L *bodina*, earlier *butina*, of unkn. orig.]

bound³ /baʊnd/ adj. **1** (usu. foll. by *for*) ready to start or having started (*bound for stardom*). **2** (in comb.) moving in a specified direction (*northbound*; *outward bound*). [ME f. ON *búinn* past part. of *búa* get ready: -d euphonic, or partly after BIND]

bound⁴ /baʊnd/ past and past part. of BIND. □ **bound to** certain to (*he's bound to come*).

boundary /'baʊndərɪ, -drɪ/ n. (pl. -**ies**) **1** a real or notional line marking the limits of an area, territory, etc.; the limit itself or the area near it (*the fence is the boundary*; *boundary between liberty and licence*). **2** Cricket a hit crossing the limits of the field, scoring four or six runs. □ **boundary layer** a layer of more or less stationary fluid immediately surrounding an immersed and moving object. **boundary rider** *Austral. & NZ* a person employed to ride round the fences etc. of a cattle or sheep station and keep them in good order. **boundary umpire** (in Australian Rules football) an umpire on the sidelines who signals when the ball is out. [dial. *bounder* f. BOUND² + -ER¹ perh. after *limitary*]

bounden /'baʊndən/ adj. archaic obligatory. □ **bounden duty** solemn responsibility. [archaic past part. of BIND]

bounder /'baʊndə(r)/ n. colloq. or joc. a cad; an ill-bred person. [BOUND¹ + -ER¹]

boundless /'baʊndlɪs/ adj. unlimited; immense (*boundless enthusiasm*). □ **boundlessly** adv. **boundlessness** n.

bounteous /'baʊntɪəs/ adj. poet. **1** generous, liberal. **2** freely given (*bounteous affection*). □ **bounteously** adv. **bounteousness** n. [ME f. OF *bontif* f. *bonté* BOUNTY after *plenteous*]

bountiful /'baʊntɪˌfʊl/ adj. **1** = BOUNTEOUS. **2** ample. □ **bountifully** adv. [BOUNTY + -FUL]

Bounty /'baʊntɪ/ a ship of the British navy which was bound from Tahiti to the Cape of Good Hope and the West Indies when on 28 Apr. 1789 part of the crew mutinied against their commander, Captain Bligh. Bligh and eighteen companions, set adrift in an open boat, succeeded in reaching Timor in the East Indies, nearly 6,400 km (4,000 miles) away. The mutineers, led by Fletcher Christian (c.1764–c.1794), returned to Tahiti, from where some of them went on to Pitcairn Island and founded a settlement which was not discovered until 1808.

bounty /'baʊntɪ/ n. (pl. -**ies**) **1** liberality; generosity. **2** a gift or reward, made usu. by the state, esp.: **a** a sum paid for a valiant act. **b** a sum paid to encourage a trading enterprise etc. **c** a sum paid to army or navy recruits on enlistment. □ **bounty-hunter** a person who pursues a criminal or seeks an achievement for the sake of the reward. **king's** (or **queen's**) **bounty** hist. a grant made to the mother of three or more children born at once. [ME f. OF *bonté* f. L *bonitas -tatis* f. *bonus* good]

bouquet /buːˈkeɪ, bəʊˈkeɪ/ n. **1** a bunch of flowers, esp. for carrying at a wedding or other ceremony. **2** the scent of wine etc. **3** a favourable comment; a compliment. □ **bouquet garni** /ˈɡɑːnɪ/ a bunch of herbs used for flavouring stews etc. [F f. dial. var. of OF *bos*, *bois* wood]

Bourbaki /buəˈbɑːkɪ/, Nicolas. A pseudonym under which a group of mathematicians, mainly French, have attempted to give a complete account of the foundations of pure mathematics. Volumes in different areas of mathematics have been published since 1939. The group was named, humorously, after a defeated French general of the Franco-Prussian War (1870–1).

Bourbon¹ /'buəb(ə)n, 'bɔːbɒn/ the surname of a branch of the royal family of France, who became the ruling monarchs when Henry IV succeeded to the throne in 1589 and reached the peak of their power under Louis XIV in the late 17th century. The ruling line was terminated when Louis XVI was executed in 1793 but renewed with the Restoration of 1814. The last Bourbon king of France was Louis Philippe, and the French monarchy came to an end when he was overthrown in 1848. Members of this family have also been kings of Spain (1700–1931, with a few interruptions, and since 1975) and of Naples and Sicily (1735–1860).

Bourbon² /'buəb(ə)n, 'bɔːbɒn/ n. **1** (in full **Bourbon biscuit**) a chocolate-flavoured biscuit with chocolate-cream filling. **2** US a reactionary. [BOURBON¹]

bourbon /'bɜːb(ə)n, 'buəb(ə)n/ n. an American whisky distilled from maize and rye. [*Bourbon* County, Kentucky, where it was first made]

Bourbonnais /'buəbɒˌneɪ/ a former duchy and province of central France; chief town, Moulins. It forms part of the Auvergne and Centre regions.

bourdon /'buəd(ə)n/ n. Mus. **1** a low-pitched stop in an organ or harmonium. **2** the lowest bell in a peal of bells. **3** the drone pipe of a bagpipe. [F, = bagpipe-drone, f. Rmc. imit.]

bourgeois /'buəʒwʌ/ adj. & n. often derog. ● adj. **1 a** conventionally middle-class. **b** humdrum, unimaginative. **c** selfishly materialistic. **2** upholding the interests of the capitalist class; non-communist. ● n. (pl. same) a bourgeois person. [F: see BURGESS]

bourgeoisie /ˌbuəʒwʌˈziː/ n. **1** the capitalist class. **2** the middle class. [F]

Bourgogne see BURGUNDY.

Bourguiba /buəˈɡiːbə/, Habib ibn Ali (b.1903), Tunisian nationalist and statesman, President 1957–87. Having negotiated the settlement that led to his country's autonomy, he was its first Prime Minister after independence in 1956 and was chosen as its first President when the country became a republic in 1957. A moderate socialist, he embarked on a reform programme intended to improve Tunisia's economy and to establish democratic government. He was deposed following continuing political unrest.

Bourke-White /bɜːkˈwaɪt/, Margaret (1906–71), American photojournalist. As a staff photographer with *Life* magazine in 1937 she took photographs of the effects of the Depression among the rural poor in the southern US. During the Second World War she was the first female photographer to be attached to the US armed forces, at the end of the war accompanying the Allied forces when they entered the concentration camps. Later assignments included the Korean War (1950–3) and work in India and South Africa.

bourn[1] /bɔːn, bʊən/ n. a small stream. [ME: S. Engl. var. of BURN[2]]

bourn[2] /bɔːn, bʊən/ n. (also **bourne**) *archaic* **1** a goal; a destination. **2** a limit. [F *borne* f. OF *bodne* BOUND[2]]

Bournemouth /ˈbɔːnməθ/ a resort on the south coast of England, in Dorset; pop. (1991) 154,400.

bourrée /ˈbʊreɪ/ n. **1** a lively French dance like a gavotte. **2** the music for this dance. [F]

bourse /bʊəs/ n. **1** (**Bourse**) the Paris equivalent of the Stock Exchange. **2** a money market. [F, = purse, f. med.L *bursa*: cf. PURSE]

boustrophedon /ˌbaʊstrəˈfiːd(ə)n, ˌbuːstrə-/ adj. & adv. (of written words) from right to left and from left to right in alternate lines. [Gk (adv.) = as an ox turns in ploughing f. *bous* ox + *-strophos* turning]

bout /baʊt/ n. (often foll. by *of*) **1 a** a limited period (of intensive work or exercise). **b** a drinking session. **c** a period (of illness) (*a bout of flu*). **2 a** a wrestling or boxing-match. **b** a trial of strength. [16th c.: app. the same as obs. *bought* bending]

boutique /buːˈtiːk/ n. a small shop or department of a store, selling (esp. fashionable) clothes or accessories. [F, = small shop, f. L (as BODEGA)]

boutonnière /ˌbuːtɒnɪˈeə(r)/ n. a spray of flowers worn in a buttonhole. [F]

Bouvet Island /ˈbuːveɪ/ an uninhabited Norwegian island in the South Atlantic, named after the French navigator François Lozier-Bouvet (1705–86), who visited it in 1739.

bouzouki /buːˈzuːkɪ/ n. a Greek form of mandolin. [mod.Gk]

bovate /ˈbəʊveɪt/ n. *hist.* a measure of land, as much as one ox could plough in a year, varying from 10 to 18 acres. [med.L *bovata* f. L *bos bovis* ox]

bovine /ˈbəʊvaɪn/ adj. **1** of or relating to cattle. **2** stupid, dull. □ **bovine spongiform encephalopathy** see BSE. □ **bovinely** adv. [LL *bovinus* f. L *bos bovis* ox]

Bovril /ˈbɒvrɪl/ n. *propr.* a concentrated essence of beef diluted with hot water to make a drink. [L *bos bovis* ox, cow]

bovver /ˈbɒvə(r)/ n. *Brit. sl.* deliberate troublemaking. □ **bovver boot** a heavy laced boot worn typically by skinheads. **bovver boy** a violent hooligan. [cockney pronunc. of BOTHER]

Bow /bəʊ/, Clara (1905–65), American actress. One of the most popular stars and sex symbols of the 1920s, she was known as the 'It Girl'; her best-known roles were in the silent films *It* (1927) and *The Wild Party* (1929).

bow[1] /bəʊ/ n. & v. ● n. **1 a** a slip-knot with a double loop. **b** a ribbon, shoelace, etc., tied with this. **c** a decoration (on clothing, or painted etc.) in the form of a bow. **2** a device for shooting arrows with a taut string joining the ends of a curved piece of wood etc. **3 a** a rod with horsehair stretched along its length, used for playing the violin, cello, etc. **b** a single stroke of a bow over strings. **4 a** a shallow curve or bend. **b** a rainbow. **5** = *saddle-bow*. **6** a metal ring forming the handle of scissors, a key, etc. **7** *US* the side-piece of a spectacle-frame. **8** *Archery* = BOWMAN[1]. ● v.tr. (also *absol.*) use a bow on (a violin etc.) (*he bowed vigorously*). □ **bow-compass** (or **-compasses**) compasses with jointed legs. **bow-legged** having bandy legs. **bow-legs** bandy legs. **bow-saw** *Carpentry* a narrow saw stretched like a bowstring on a light frame. **bow-tie** a necktie in the form of a bow (sense 1). **bow-window** a curved bay window. **have two** (or **many**) **strings to one's bow** have two (or many) alternative resources. [OE *boga* f. Gmc: cf. BOW[2]]

bow[2] /baʊ/ v. & n. ● v. **1** intr. incline the head or trunk, esp. in greeting or assent or acknowledgement of applause. **2** intr. submit (*bowed to the inevitable*). **3** tr. cause to incline or submit (*bowed his head; bowed his will to hers*). **4** tr. express (thanks, assent, etc.) by bowing (*bowed agreement to the plan*). **5** tr. (foll. by *in*, *out*) usher or escort obsequiously (*bowed us out of the restaurant*). ● n. an inclining of the head or body in greeting, assent, or in the acknowledgement of applause, etc. □ **bow and scrape** be obsequious; fawn. **bow down 1** bend or kneel in submission or reverence (*bowed down before the king*). **2** (usu. in *passive*) make stoop; crush (*was bowed down by care*). **bowing acquaintance** a person one acknowledges but does not know well enough to speak to. **bow out 1** make one's exit (esp. formally). **2** retreat, withdraw; retire gracefully. **make one's bow** make a formal exit or entrance. **take a bow** acknowledge applause. [OE *būgan*, f. Gmc: cf. BOW[1]]

bow[3] /baʊ/ n. *Naut.* **1** (often in *pl.*) the fore-end of a boat or a ship. **2** = BOWMAN[2]. □ **bow wave** a wave set up at the bows of a moving ship or in front of a body moving in air. **on the bow** within 45° of the point directly ahead. **shot across the bows** a warning. [LG *boog*, Du. *boeg*, ship's bow, orig. shoulder: see BOUGH]

bowdlerize /ˈbaʊdlə̩raɪz/ v.tr. (also **-ise**) expurgate (a book etc.). □ **bowdlerism** n. **bowdlerization** /ˌbaʊdlə̩raɪˈzeɪʃ(ə)n/ n. [T. *Bowdler* (1754–1825), expurgator of Shakespeare]

bowel /ˈbaʊəl/ n. **1** (often in *pl.*) the part of the alimentary canal below the stomach; the intestine. **2** (in *pl.*) the depths; the innermost parts (*the bowels of the earth*). □ **bowel movement 1** discharge from the bowels; defecation. **2** the faeces discharged from the body. [ME f. OF *buel* f. L *botellus* little sausage]

Bowen /ˈbəʊɪn/, Elizabeth (Dorothea Cole) (1899–1973), British novelist and short-story writer, born in Ireland. Her writing is distinguished by delicate characterization and acute observation, especially of emotions and the relationships between her chiefly upper-middle-class characters. Among her best-known novels are *The Death of the Heart* (1938) and *The Heat of the Day* (1949).

bower[1] /ˈbaʊə(r)/ n. & v. ● n. **1 a** a secluded place, esp. in a garden, enclosed by foliage; an arbour. **b** a summer-house. **2** *poet.* an inner room; a boudoir. ● v.tr. *poet.* embower. □ **bowery** adj. [OE *būr* f. Gmc]

bower[2] /ˈbaʊə(r)/ n. (in full **bower-anchor**) either of two anchors carried at a ship's bow. □ **best bower** the starboard bower. **bower-cable** the cable attached to a bower-anchor. **small bower** the port bower. [BOW[3] + -ER[1]]

bower[3] /ˈbaʊə(r)/ n. either of two cards at euchre and similar games. □ **left bower** the jack of the same colour as the right bower. **right bower** the jack of trumps. [G *Bauer* peasant, jack at cards, rel. to Du. *boer* (cf. BOER): see BOOR]

bowerbird /ˈbaʊə̩bɜːd/ n. **1** a bird of the family Ptilonorhynchidae, native to Australia and New Guinea, the males of which construct elaborate bowers of feathers, grasses, shells, etc. during courtship. **2** a person who collects bric-à-brac.

bowery /ˈbaʊərɪ/ n. (also **Bowery**) (pl. **-ies**) *US* a district known as a resort of drunks and down-and-outs. [orig. the Bowery, a street in New York City, f. Du. *bouwerij* farm]

bowfin /ˈbəʊfɪn/ n. a large voracious freshwater fish, *Amia calva*, native to North America. [BOW[1] + FIN]

bowhead /ˈbəʊhed/ n. a kind of right whale, *Balaena mysticetus*, found in Arctic waters.

Bowie /ˈbəʊɪ/, David (born David Robert Jones) (b.1947), English rock singer, songwriter, and actor. His first hit single 'Space Oddity' (1969) was followed by a number of other hit singles and albums such as *Hunky Dory* (1971), *Ziggy Stardust* (1972), and *Aladdin Sane* (1973); later albums include *Let's Dance* (1983). He became known for his theatrical performances and unconventional stage personae, involving the use of elaborate costumes and make-up. He also acted in a number of films, especially in the 1980s.

bowie /ˈbəʊɪ/ n. (in full **bowie knife**) a long knife with a blade double-edged at the point, used as a weapon by American pioneers. It is named after the American folk hero James ('Jim') Bowie (1796–1836), a colonel in the Texan forces during the war with Mexico, who is said to have used such a knife for hunting.

bowl[1] /bəʊl/ n. **1 a** a usu. round deep basin used for food or liquid. **b** the quantity (of soup etc.) a bowl holds. **c** the contents of a bowl. **2 a** any deep-sided container shaped like a bowl (*lavatory bowl*). **b** the bowl-shaped part of a tobacco-pipe, spoon, balance, etc. **3** esp. *US* a bowl-shaped region or building, esp. an amphitheatre (*Hollywood Bowl*). □ **bowlful** n. (pl. **-fuls**) [OE *bolle*, *bolla*, f. Gmc]

bowl[2] /bəʊl/ n. & v. ● n. **1 a** a wooden or hard rubber ball, slightly asymmetrical so that it runs on a curved course, used in the game of bowls. **b** a wooden ball or disc used in playing skittles. **c** a large ball with indents for gripping, used in tenpin bowling. **2** (in *pl.*; usu. treated as *sing.*) **a** a game played with bowls (sense 1a) esp. on grass. (See BOWLS.) **b** tenpin bowling. (See BOWLING 1.) **c** skittles. **3** a spell or turn of bowling in cricket. ● v. **1 a** tr. roll (a ball, a hoop, etc.) along the ground. **b** intr. play bowls or skittles. **2** tr. (also *absol.*) *Cricket* etc. **a** deliver (a ball, an over, etc.) (*bowled six overs*; *bowled well*). **b** (often foll. by *out*, *for*) dismiss (a batsman or a side) by knocking down the wicket with a ball (*bowled him out*; *bowled for six*). **c** (often foll. by *down*) knock (a wicket) over. **3** *intr.* (often foll. by *along*) go along rapidly by revolving, esp. on wheels (*the cart bowled along the road*). □ **bowl out** *Cricket* etc. dismiss (a batsman or a side). **bowl over 1** knock down. **2** *colloq.* **a** impress greatly. **b** overwhelm (*bowled over by her energy*). [ME & F *boule* f. L *bulla* bubble]

bowler[1] /ˈbəʊlə(r)/ *n.* **1** *Cricket* etc. a member of the fielding side who bowls or is bowling. **2** a player at bowls.

bowler[2] /ˈbəʊlə(r)/ *n.* (in full **bowler hat**) a man's hard felt hat with a round dome-shaped crown. □ **bowler-hat** (**-hatted**, **-hatting**) *sl.* retire (a person) from the army etc. (*he's been bowler-hatted*). [William Bowler, Engl. hatter, who designed it in 1850]

bowline /ˈbəʊlɪn/ *n. Naut.* **1** a rope attaching the weather side of a square sail to the bow. **2** a simple knot for forming a non-slipping loop at the end of a rope. [ME f. MLG *bōlīne* (as BOW[3], LINE[1])]

bowling /ˈbəʊlɪŋ/ *n.* **1** (also **tenpin bowling**) an indoor sport in which heavy bowls are rolled down a long narrow lane with the aim of knocking down a cluster of wooden pegs (*pins*). (*See note below.*) **2** playing at bowls; the action of rolling a ball etc. **3** *Cricket* the action of delivering the ball. □ **bowling-alley 1** a long enclosure for skittles or tenpin bowling. **2** a building containing these. **bowling-crease** *Cricket* the line from behind which a bowler delivers the ball. **bowling-green** a smooth green used for playing bowls.

 ▪ Bowling as an indoor game is of great antiquity. Similar games were played in ancient Egypt. In Germany, congregations played at bowling in church cloisters (c.1400) and from there the game spread to Britain and the rest of Europe. It was taken to America by Dutch settlers in the 17th century, when the target consisted of nine pins; a tenth pin was later added, and it is in this form that bowling has became a popular social activity, especially in the US. In 1895 the American Bowling Congress codified the rules and standardized the equipment.

bowls /bəʊlz/ *n.* a game played with heavy wooden bowls, the object of which is to propel one's bowl so that it comes to rest as close as possible to a previously bowled small white ball (the *jack*) without touching it. Bowls is played chiefly out of doors (though indoor bowls is also popular) on a closely trimmed lawn (the *green*). An intentional imbalance (*bias*) of balls was introduced in the 16th century, greatly increasing the tactical scope of the game by allowing balls to run in curved paths round obstructions. Bowls is one of the oldest games known in Britain, being first recorded in the late 13th century; it is popularly associated with the story of Sir Francis Drake refusing to allow the sighting of the Spanish Armada to interrupt his game of bowls. The modern rules for bowls were drawn up in the mid-19th century.

bowman[1] /ˈbəʊmən/ *n.* (*pl.* **-men**) an archer.

bowman[2] /ˈbaʊmən/ *n.* (*pl.* **-men**) the rower nearest the bow of esp. a racing boat.

bowser /ˈbaʊzə(r)/ *n.* **1** a tanker used for fuelling aircraft etc., or for supplying water. **2** *Austral.* & *NZ propr.* a petrol pump. [orig. unkn.]

bowshot /ˈbəʊʃɒt/ *n.* the distance to which a bow can send an arrow.

bowsprit /ˈbəʊsprɪt/ *n. Naut.* a spar running out from a ship's bow to which the forestays are fastened. [ME f. Gmc (as BOW[3], SPRIT)]

Bow Street Runners /bəʊ/ (also **Bow Street Officers**) the popular name for London's police force during the first half of the 19th century, coined because the main police court was situated in Bow Street. They were formed in 1749 by Henry Fielding.

bowstring /ˈbəʊstrɪŋ/ *n.* & *v.* ● *n.* the string of an archer's bow. ● *v.tr.* strangle with a bowstring (a former Turkish method of execution).

bow-wow /ˈbaʊwaʊ/ *int.* & *n.* ● *int.* an imitation of a dog's bark. ● *n.* **1** *colloq.* a dog. **2** a dog's bark. [imit.]

bowyang /ˈbəʊjæŋ/ *n. Austral.* & *NZ* either of a pair of bands or straps worn round the trouser-legs below the knee. [dial. *bowy-yangs* etc.]

bowyer /ˈbəʊjə(r)/ *n.* a maker or seller of archers' bows.

box[1] /bɒks/ *n.* & *v.* ● *n.* **1** a container, usu. with flat sides and of firm material such as wood or card, esp. for holding solids. **2 a** the amount that will fill a box. **b** = *Christmas box*. **3** a separate compartment for any of various purposes, e.g. for a small group in a theatre, for witnesses in a lawcourt, for horses in a stable or vehicle. **4** an enclosure or receptacle for a special purpose (often in *comb.*: *money box*; *telephone box*). **5** a facility at a newspaper office for receiving replies to an advertisement. **6** (prec. by *the*) *colloq.* television; one's television set (*what's on the box?*). **7** an enclosed area or space. **8** a space or area of

print on a page, enclosed by a border. **9** *Brit.* a small country house for use when shooting, fishing, or for other sporting activity. **10** a protective casing for a piece of mechanism. **11** a light shield for protecting the genitals in sport, esp. in cricket. **12** (prec. by *the*) *Football colloq.* the penalty area. **13** *Baseball* the area occupied by the batter or the pitcher. **14** a coachman's seat. ● *v.tr.* **1** put in or provide with a box. **2** (foll. by *in*, *up*) confine; restrain from movement. **3** (foll. by *up*) *Austral.* & *NZ* mix up (different flocks of sheep). □ **box camera** a simple box-shaped hand camera. **box the compass** *Naut.* recite the points of the compass in the correct order. **box girder** a hollow girder square in cross-section. **box junction** *Brit.* a road area at a junction marked with a yellow grid, which a vehicle should enter only if its exit from it is clear. **box kite** a kite in the form of a long box open at each end. **box number** a number by which replies are made to a private advertisement in a newspaper. **box office 1** an office for booking seats and buying tickets at a theatre, cinema, etc. **2** the commercial aspect of the arts and entertainment (often *attrib.*: *a box-office failure*). **box pleat** a pleat consisting of two parallel creases forming a raised band. **box spanner** a spanner with a box-shaped end fitting over the head of a nut. **box spring** each of a set of vertical springs housed in a frame, e.g. in a mattress. □ **boxful** *n.* (*pl.* **-fuls**). **boxlike** *adj.* [OE f. LL *buxis* f. L PYXIS]

box[2] /bɒks/ *v.* & *n.* ● *v.* **1** *tr.* fight (an opponent) at boxing. **b** *intr.* practise boxing. **2** slap (esp. a person's ears). ● *n.* a slap with the hand, esp. on the ears. □ **box clever** *colloq.* act in a clever or effective way. [ME: orig. unkn.]

box[3] /bɒks/ *n.* **1** (in full **box tree**) an evergreen shrub or small tree of the genus *Buxus*; esp. *B. sempervirens*, a slow-growing tree with small dark leaves which is often used in hedging. **2** its wood, used for carving, turning, engraving, etc. **3** an Australasian tree with similar wood or foliage, esp. a eucalyptus. □ **box elder** the American ash-leaved maple, *Acer negundo*. [OE f. L *buxus*, Gk *puxos*]

Box and Cox /bɒks, kɒks/ *n.* & *v.* ● *n.* (often *attrib.*) two persons sharing accommodation, office facilities, etc., and using them at different times. ● *v.intr.* share accommodation, duties, etc. by a strictly timed arrangement. [title of a farce (1847) by J. M. Morton (1811–91); the names of two characters to whom a landlady lets the same room without their knowing it, one being out at work all night and the other all day]

boxcar /ˈbɒkskɑː(r)/ *n. N. Amer.* an enclosed railway goods wagon, usu. with sliding doors on the sides.

Boxer /ˈbɒksə(r)/ *n.* a member of a fiercely nationalistic Chinese secret society that flourished in the late 19th century. Assisted by the Dowager Empress, in 1899 the society led a Chinese uprising against Western domination and besieged the foreign legations in Beijing. The uprising was eventually crushed by a combined European force, aided by Japan and the US, and the foreign powers took their opportunity to strengthen their hold over the Chinese. [transl. Chin. *yì hé quán*, lit. 'righteous harmony fists']

boxer /ˈbɒksə(r)/ *n.* **1** a person who practises boxing, esp. for sport. **2** a breed of dog with a smooth brown coat and puglike face. □ **boxer shorts** men's underpants similar to shorts worn in boxing, with a shallow curved slit at each side.

boxing /ˈbɒksɪŋ/ *n.* the practice of fighting with the fists, esp. in padded gloves in a roped square, as a sport. (*See note below.*) □ **boxing glove** a heavily padded glove used in boxing.

 ▪ Boxing is a sport of great antiquity. It was popular in ancient Greece, being mentioned in Homer and depicted in Minoan wall-paintings. The Romans added metal weights to the leather thongs with which the boxers' fists were bound. In the 18th century, prize fighting with bare fists was popular, often for large amounts of money with huge side bets. The modern rules of boxing were codified by the Marquess of Queensberry (1844–1900) in 1865 and are named the Queensberry Rules after him. They include the use of boxing gloves, the division of fights into a set number of short, usually three-minute rounds, banning of gouging and wrestling, and a count of ten seconds before a boxer who has been knocked down is declared the loser. Contestants are divided into categories according to their weight. Professional contests generally last for ten or twelve rounds; amateur contests are normally fought over three rounds, a ruling which tends to emphasize skill and speed more than strength and stamina.

Boxing Day /ˈbɒksɪŋ/ the first day (strictly the first weekday) after Christmas. [from the custom of giving tradesmen or money: see BOX[1] *n.* 2b, *Christmas box*]

boxing weight *n.* each of a series of fixed weight-ranges at which boxers are matched. The British professional scale is as follows: flyweight, bantamweight, featherweight, lightweight, light welterweight, welterweight, light middleweight, middleweight, light heavyweight, heavyweight. (See separate entries.) The weights and divisions are modified in the amateur scale.

boxroom /ˈbɒksruːm, -rʊm/ n. Brit. a small room or large cupboard esp. for storing boxes, cases, etc.

boxwood /ˈbɒkswʊd/ n. **1** the wood of the box tree used esp. by engravers for the fineness of its grain and for its hardness. **2** = BOX³ 1.

boxy /ˈbɒksɪ/ adj. (**boxier, boxiest**) **1** resembling a box; (of a room or space) very cramped. **2** (of recorded sound) restricted in tone.

boy /bɔɪ/ n. & int. ● n. **1** a male child or youth. **2** a young man, esp. regarded as not yet mature. **3** a male servant, attendant, etc. **4** (**the boys**) colloq. a group of men mixing socially. ● int. (also **oh boy**) (often prec. by the) expressing pleasure, surprise, etc. □ **boy scout** = SCOUT¹ 4. **boys in blue** Brit. policemen. □ **boyhood** n. **boyish** adj. **boyishly** adv. **boyishness** n. [ME = servant, perh. ult. f. L boia fetter]

boyar /bəʊˈjɑː(r), ˈbɔɪə(r)/ n. hist. a member of the old aristocracy in Russia. [Russ. boyarin grandee]

Boyce /bɔɪs/, William (1711–79), English composer and organist. His compositions include songs, overtures, church anthems and services, and eight symphonies; one of his most famous songs is 'Hearts of Oak'. He is noted also for his Cathedral Music (1760–73), an anthology of English sacred music of the 16th to 18th centuries.

Boycott /ˈbɔɪkɒt/, Geoffrey (b.1940), English cricketer. His career with Yorkshire began in 1962 and he first played for England two years later. Often controversial, he was an opening batsman of great concentration. He was captain of his county 1971–5, and had scored more than 150 centuries by the time he retired from first-class cricket in 1986.

boycott /ˈbɔɪkɒt/ v. & n. ● v.tr. **1** combine in refusing social or commercial relations with (a person, group, country, etc.) usu. as punishment or coercion. (See note below.) **2** refuse to handle (goods) to this end. ● n. such a refusal.

 ▪ The term comes from the name of the Irish land-agent Capt. C. C. Boycott (1832–97). He was asked by the Irish Land League in 1880 to reduce rents after a bad harvest, and when he refused the tenants avoided any communication with him. (See also LAND LEAGUE.)

Boyd /bɔɪd/, Arthur (Merric Bloomfield) (b.1920), Australian painter, potter, etcher, and ceramic artist. His first paintings were influenced by expressionism and surrealism; later works, such as The Mockers (1945), show the influence of Rembrandt, Bosch, and Bruegel. He became famous for his large ceramic totem pole erected in Melbourne and for his series of twenty pictures inspired by his travels among the Aboriginals of central Australia.

Boyer /ˈbɔɪeɪ/, Charles (1897–1977), French-born American actor. Before going to Hollywood in the 1930s, he enjoyed a successful stage career in France. He played romantic leading roles in films such as Mayerling (1936); his other memorable films include Gaslight (1944) and Barefoot in the Park (1968).

boyfriend /ˈbɔɪfrend/ n. a person's regular male companion or lover.

Boyle /bɔɪl/, Robert (1627–91), Irish-born scientist. A founder member of the Royal Society, Boyle advanced a corpuscular view of matter which was a precursor of the modern theory of chemical elements and a cornerstone of his mechanical philosophy which became very influential. He is best known for his experiments with the air pump, assisted initially by Robert Hooke, which led to the law named after him.

Boyle's law n. Chem. a law stating that the pressure of a given mass of gas is inversely proportional to its volume at constant temperature. [BOYLE]

Boyne, Battle of the /bɔɪn/ a battle fought near the River Boyne in Ireland in 1690, in which the Protestant army of William of Orange, the newly crowned William III, defeated the Catholic army (including troops from both France and Ireland) led by the recently deposed James II. The battle is celebrated annually (on 12 July) in Northern Ireland as a victory for the Protestant cause.

boyo /ˈbɔɪəʊ/ n. (pl. **-os**) Welsh & Ir. colloq. boy, fellow (esp. as a form of address).

Boys' Brigade the oldest of the national organizations for boys in Britain, founded in 1883 with the aim of promoting 'Christian manliness', discipline, and self-respect. Companies are now also found in the US and in Commonwealth countries; each is connected with a church.

boysenberry /ˈbɔɪz(ə)nbərɪ/ n. (pl. **-ies**) **1** a hybrid of several species of bramble. **2** the large red edible fruit of this plant. [R. Boysen, 20th-c. Amer. horticulturalist]

Boz /bɒz/ the pseudonym used by Charles Dickens in his Pickwick Papers and contributions to the Morning Chronicle.

bozo /ˈbəʊzəʊ/ n. (pl. **-os**) US sl. a stupid or insignificant person. [20th c.: orig. uncert.]

BP abbr. **1** boiling-point. **2** blood pressure. **3** before the present (era). **4** British Petroleum. **5** British Pharmacopoeia.

Bp. abbr. Bishop.

BPC abbr. British Pharmaceutical Codex.

B.Phil. abbr. Bachelor of Philosophy.

bpi abbr. Computing bits per inch.

bps abbr. Computing bits per second.

Bq abbr. becquerel.

BR abbr. British Rail.

Br symb. Chem. the element bromine.

Br. abbr. **1** British. **2** Brother.

bra /brɑː/ n. (pl. **bras**) an undergarment worn by women to support the breasts. □ **braless** adj. [abbr. of BRASSIÈRE]

Brabant /brəˈbænt/ a former duchy in western Europe, lying between the Meuse and Scheldt rivers. Its capital was Brussels. It is now divided into two provinces in two countries: North Brabant in the Netherlands, of which the capital is 's-Hertogenbosch; and Brabant in Belgium, of which the capital remains Brussels.

Brabham /ˈbræbəm/, Sir John Arthur ('Jack') (b.1926), Australian motor-racing driver. After making his Grand Prix début in Australia, he moved to the UK, where he won his first Formula One world championship in 1959, repeating his success again in 1960 and in 1966.

brace /breɪs/ n. & v. ● n. **1 a** a device that clamps or fastens tightly. **b** a strengthening piece of iron or timber in building. **2** (in pl.) Brit. straps supporting trousers from the shoulders. **3** a wire device for straightening the teeth. **4** (pl. same) a pair (esp. of game). **5** a rope attached to the yard of a ship for trimming the sail. **6 a** a connecting mark { or } used in printing. **b** Mus. a similar mark connecting staves to be performed at the same time. ● v.tr. **1** fasten tightly, give firmness to. **2** make steady by supporting. **3** (esp. as **bracing** adj.) invigorate, refresh. **4** (often refl.) prepare for a difficulty, shock, etc. □ **brace and bit** a revolving tool with a D-shaped central handle for boring. □ **bracingly** adv. **bracingness** n. [ME f. OF brace two arms, bracier embrace, f. L bra(c)chia arms]

bracelet /ˈbreɪslɪt/ n. **1** an ornamental band, hoop, or chain worn on the wrist or arm. **2** sl. a handcuff. [ME f. OF, dimin. of bracel f. L bracchiale f. bra(c)chium arm]

bracer /ˈbreɪsə(r)/ n. colloq. a tonic.

brachial /ˈbreɪkɪəl/ adj. **1** of or relating to the arm (brachial artery). **2** like an arm. [L brachialis f. bra(c)chium arm]

brachiate /ˈbreɪkɪˌeɪt/ v. & adj. ● v.intr. (of certain apes and monkeys) move by using the arms to swing from branch to branch. ● adj. Biol. **1** having arms. **2** having paired branches on alternate sides. □ **brachiator** n. **brachiation** /ˌbreɪkɪˈeɪʃ(ə)n/ n. [L bra(c)chium arm]

brachiopod /ˈbreɪkɪəˌpɒd, ˈbræk-/ n. Zool. a marine invertebrate of the phylum Brachiopoda, having a two-valved shell and a ciliated feeding arm. Brachiopods are common as fossils. Also called lamp shell. [mod.L f. Gk brakhiōn arm + pous podos foot]

brachiosaurus /ˌbreɪkɪəˈsɔːrəs, ˌbræk-/ n. (also **brachiosaur** /ˈbreɪkɪəˌsɔː(r), ˈbræk-/ a huge long-necked plant-eating dinosaur of the genus Brachiosaurus, of the Jurassic period, with forelegs longer than its hind legs. It is the tallest known dinosaur. [mod.L f. Gk brakhiōn arm + sauros lizard]

brachistochrone /brəˈkɪstəˌkrəʊn/ n. Math. a curve between two points along which a body can move in a shorter time than for any other curve. [Gk brakhistos shortest + khronos time]

brachy- /ˈbrækɪ/ comb. form short. [Gk brakhus short]

brachycephalic /ˌbrækɪsɪˈfælɪk, ˌbrækɪkɪ-/ adj. Anthropol. having a short rounded skull whose width is at least 80 per cent of its length, i.e. having a cephalic index of 80 or more. □ **brachycephalous** /-ˈsefələs, -ˈkef-/ adj. **brachycephaly** n. [BRACHY- + Gk kephalē head]

brachylogy /brəˈkɪlədʒɪ/ n. (pl. **-ies**) **1** over-conciseness of expression. **2** an instance of this.

brack /bræk/ n. Ir. cake or bread containing dried fruit etc. [abbr. of BARMBRACK]

bracken /ˈbrækən/ n. **1** a large coarse fern; esp. *Pteridium aquilinum*, which is abundant on heaths and moorlands, and in woods. **2** a mass of such ferns. Also called *brake*. [north. ME f. ON]

bracket /ˈbrækɪt/ n. & v. ● n. **1** a right-angled or other support attached to and projecting from a vertical surface. **2** a shelf fixed with such a support to a wall. **3** each of a pair of marks () [] {} <> used to enclose words or figures. **4** a group classified as containing similar elements or falling between given limits (*income bracket*). **5** Mil. the distance between two artillery shots fired either side of the target to establish range. ● v.tr. (**bracketed**, **bracketing**) **1 a** couple (names etc.) with a bracket. **b** imply a connection or equality between. **2 a** enclose in brackets as parenthetic or spurious. **b** Math. enclose in brackets as having specific relations to what precedes or follows. **3** Mil. establish the range of (a target) by firing two preliminary shots one short of and the other beyond it. □ **bracket fungus** a fungus forming shelflike projections on tree-trunks etc. [F *braguette* or Sp. *bragueta* codpiece, dimin. of F *brague* f. Prov. *braga* f. L *braca*, pl. *bracae* breeches]

brackish /ˈbrækɪʃ/ adj. (of water etc.) slightly salty. □ **brackishness** n. [obs. *brack* (adj.) f. MLG, MDu. *brac*]

bract /brækt/ n. a modified and often brightly coloured leaf, with a flower or an inflorescence in its axil. □ **bracteal** adj. **bracteate** /ˈbræktɪɪt/ adj. [L *bractea* thin plate, gold-leaf]

brad /bræd/ n. a thin, flat, small-headed nail. [var. of ME *brod* goad, pointed instrument, f. ON *broddr* spike]

bradawl /ˈbrædɔːl/ n. a small tool with a pointed end for boring holes by hand. [BRAD + AWL]

Bradbury[1] /ˈbrædbəri/, Malcolm (Stanley) (b.1932), English novelist, critic, and academic. His first three novels (including *The History Man*, 1975) are satires of university campus life; *Rates of Exchange* (1983) recounts the experiences of an academic on a lecture tour of an eastern European country. His critical works include studies of Evelyn Waugh (1962) and Saul Bellow (1982).

Bradbury[2] /ˈbrædbəri/, Ray (Douglas) (b.1920), American writer of science fiction. He is best known for his collections of short stories, such as *The Martian Chronicles* (1950) and *The Golden Apples of the Sun* (1953). His novels include *Fahrenheit 451* (1951), a depiction of a future totalitarian state.

Bradford /ˈbrædfəd/ an industrial city in West Yorkshire, England; pop. (1990) 449,100. It was noted for the production of woollen cloth until the decline of the textile industry in the 1970s.

Bradley /ˈbrædlɪ/, James (1693–1762), English astronomer. Bradley was appointed Savilian professor of astronomy at Oxford in 1721 and Astronomer Royal in 1742. His attempt to measure the distance of stars by means of stellar parallax resulted in his discovery of the aberration of light, which he ascribed correctly to the combined effect of the velocity of light and the earth's annual orbital motion. He also observed the oscillation of the earth's axis, which he termed *nutation*. His star catalogue was published posthumously.

Bradman /ˈbrædmən/, Sir Donald George ('Don') (b.1908), Australian cricketer. An outstanding batsman who dominated the sport in his day, he began his career in 1927 with New South Wales and played for his country from 1928 until his retirement in 1948. He scored 117 centuries in first-class cricket, twenty-nine of them in test matches; he holds the record for the highest Australian test score against England (334 at Leeds in 1930). His test match batting average of 99.94 is well above that of any other cricketer of any era.

Bradshaw /ˈbrædʃɔː/ n. a timetable of all passenger trains in Britain, issued 1839–1961. It was named after its first publisher, the printer and engraver George Bradshaw (1801–53).

bradycardia /ˌbrædɪˈkɑːdɪə/ n. Med. abnormally slow heart-action. [Gk *bradus* slow + *kardia* heart]

brae /breɪ/ n. Sc. a steep bank or hillside. [ME f. ON *brá* eyelash]

brag /bræg/ v. & n. ● v. (**bragged**, **bragging**) **1** intr. talk boastfully. **2** tr. boast about. ● n. **1** a card-game like poker. **2** a boastful statement; boastful talk. □ **bragger** n. **braggingly** adv. [ME, orig. adj., = spirited, boastful: orig. unkn.]

Braga /ˈbrɑːɡə/ a city in northern Portugal, capital of a mountainous district of the same name; pop. (est. 1987) 63,030. Known to the Romans as Bracara Augusta, it was capital of the Roman province of Lusitania.

Braganza[1] /brəˈɡænzə/ (Portuguese **Bragança** /brɐˈɣɐ̃sɐ/) a city in NE Portugal, capital of a mountainous district of the same name; pop. (est. 1987) 13,900. It was the original seat of the Braganza dynasty.

Braganza[2] /brəˈɡænzə/ the dynasty that ruled Portugal from 1640

until the end of the monarchy in 1910 and Brazil (on its independence from Portugal) from 1822 until the formation of a republic in 1889.

Bragg /bræɡ/, Sir William Henry (1862–1942), English physicist, a founder of solid-state physics. His early work was concerned with X-rays and ionizing radiation, but in 1912 he began to collaborate with his son, Sir (William) Lawrence Bragg (1890–1971), in developing the technique of X-ray diffraction for determining the atomic structure of crystals; for this they shared the 1915 Nobel Prize for physics. He later established a research school for crystallography at University College, London, and became director of the Royal Institution in 1923. His son was appointed head of the same establishment in 1953. Their diffraction studies of organic crystals were of fundamental importance in molecular biology.

braggadocio /ˌbræɡəˈdəʊtʃɪˌəʊ, -ˈdəʊʃɪˌəʊ/ n. empty boasting; a boastful manner of speech and behaviour. [*Braggadochio*, a braggart in Spenser's *Faerie Queene*, f. BRAG or BRAGGART + It. augment. suffix *-occio*]

braggart /ˈbræɡət/ n. & adj. ● n. a person given to bragging. ● adj. boastful. [F *bragard* f. *braguer* BRAG]

Brahe /ˈbrɑːhɪ, ˈbrɑːə/, Tycho (1546–1601), Danish astronomer. A nobleman, he built his own observatory, which he equipped with new instruments for determining planetary motions and star positions with great precision, making due allowance for atmospheric refraction. He was regarded as the leading observational astronomer of his day; his observations of comets, published in the book *De Nova Stella* (1577), demonstrated that they followed regular sun-centred paths, but despite this he adhered to a geocentric picture of the orbits of the planets. A 'new star' which he observed in 1572 is now known to have been a supernova. From the 1590s, he corresponded with Kepler, who later assisted him in his work.

Brahma /ˈbrɑːmə/ Hinduism the creator god, who forms a triad with Vishnu and Siva. Brahma was an important god of late Vedic religion, but has been little worshipped since the 5th century AD and has only one major temple dedicated to him in India today. [Skr. *brahma* (nom. of *brahman*) priest]

brahma /ˈbrɑːmə/ n. = BRAHMAPUTRA. [abbr.]

Brahman /ˈbrɑːmən/ n. **1** Hinduism the supreme being of the Upanishads, often identified with the atman or inner core of the individual. It is the eternal and conscious ground of the universe, the source of dharma, and the special sphere of the priestly brahman class. A neuter term, Brahman was personified in early Hindu mythology as the male creator god Brahma. **2** (also **brahman**) a member of the highest of the four great Hindu classes or varnas, whose members are traditionally eligible for the priesthood, versed in sacred knowledge (i.e. the Veda). **3** (usu. **Brahmin**) = ZEBU. □ **Brahmanic** /brɑːˈmænɪk/ adj. **Brahmanical** adj. [Skr. *brāhmaṇa* f. *brahman* priest]

Brahmana /ˈbrɑːmənə/ any of the lengthy commentaries on the Vedas, composed in Sanskrit c.900–700 BC and containing expository material relating to Vedic sacrificial ritual. [as BRAHMAN]

Brahmanism /ˈbrɑːməˌnɪz(ə)m/ n. (also **Brahminism**) the complex sacrificial religion that emerged in post-Vedic India (c.900 BC) under the influence of the dominant priesthood (Brahmans), an early stage in the development of Hinduism. It was largely as a reaction to Brahman orthodoxy that religions such as Buddhism and Jainism were formed.

Brahmaputra /ˌbrɑːməˈpuːtrə/ a river of southern Asia, rising in the Himalayas and flowing 2,900 km (1,800 miles) through Tibet, NE India, and Bangladesh, to join the Ganges at its delta on the Bay of Bengal.

brahmaputra /ˌbrɑːməˈpuːtrə/ n. an Asian breed of large domestic fowl. [BRAHMAPUTRA, from where it was brought]

Brahmin /ˈbrɑːmɪn/ n. **1** = BRAHMAN. **2** US a socially or intellectually superior person. [var. of BRAHMAN]

Brahms /brɑːmz/, Johannes (1833–97), German composer and pianist. He lived for most of the last thirty-five years of his life in Vienna and owed much of his early success to the friendship and patronage of Schumann, whom he first met in 1853. Firmly opposed to the 'New German' school of Liszt and the young Wagner, he eschewed programme music and opera and concentrated his energies on 'pure' and traditional forms. He wrote four symphonies, two piano concertos, a violin concerto (1879) and the Double Concerto (1887), chamber and piano music, choral works including the *German Requiem* (1857–68), and nearly 200 songs.

braid /breɪd/ n. & v. ● n. **1** a woven band of silk or thread used for edging or trimming. **2** a length of entwined hair. ● v.tr. **1** plait or intertwine (hair or thread). **2** trim or decorate with braid. [OE *bregdan* f. Gmc]

braiding /ˈbreɪdɪŋ/ n. **1** various types of braid collectively. **2** braided work.

Brăila /brəˈiːlə/ an industrial city and port on the Danube, in eastern Romania; pop. (1989) 242,600.

Braille /breɪl/ n. & v. ● n. a system of writing and printing for the blind, in which characters are represented by patterns of raised dots. ● v.tr. print or transcribe in Braille. [Louis *Braille*, Fr. teacher of the blind (1809–52), who invented the system]

Brain /breɪn/, Dennis (1921–57), English French-horn player. His mastery of the instrument's entire range, together with his subtleties of phrasing and variety of tone, inspired many composers, including Benjamin Britten and Paul Hindemith, to write works for him. He was killed in a car accident.

brain /breɪn/ n. & v. ● n. **1** an organ of soft nervous tissue, situated in the skull in vertebrates, which functions as the coordinating centre of sensation and nervous activity, and in humans is the seat of consciousness and intellectual activity. (*See note below.*) **2** (in *pl.*) the substance of the brain, esp. as food. **3 a** a person's intellectual capacity (*has a poor brain*). **b** (often in *pl.*) intelligence; high intellectual capacity (*has a brain; has brains*). **c** *colloq.* a clever person. **4** (usu. in *pl.*; prec. by *the*) *colloq.* **a** the cleverest person in a group. **b** a person who originates a complex plan or idea (*the brains behind the robbery*). **5** an electronic device with functions comparable to those of a brain. ● v.tr. **1** dash out the brains of. **2** strike hard on the head. □ **brain-coral** a stony coral of the genus *Meandrina*, forming a rounded mass with a surface like the convolutions of the brain. **brain-dead 1** suffering from brain death. **2** *sl. offens.* feeble-minded, stupid, dull. **brain death** irreversible brain damage causing the end of independent respiration, regarded as indicative of death. **brain drain** *colloq.* the loss of academic or skilled personnel by emigration. **brain fever** inflammation of the brain. **brain-pan** *colloq.* the skull. **brain stem** the central trunk of the brain, upon which the cerebrum and cerebellum are set, and which continues downwards to form the spinal cord. **brains** (US **brain**) **trust** a group of experts who give impromptu answers to questions, usu. publicly. **brain-teaser** (or **-twister**) *colloq.* a puzzle or problem. **on the brain** *colloq.* obsessively in one's thoughts. [OE *brægen* f. WG]

▪ The human brain consists of three main parts. (i) The forebrain, greatly developed into the cerebrum, consists of two hemispheres joined by a bridge of nerve fibres, and is responsible for the exercise of thought and control of speech. Its wrinkled surface or cortex is coated with nerve cells, which appear grey in contrast to the white of the deeper nerve-fibres. Embedded in it are the thalamus and the hypothalamus, the latter being concerned with the regulation of temperature and water balance, with hunger, thirst, emotions, and the sex drive. The left half of the cerebrum is associated with the right side of the body, regulating its posture and movements and appreciation of bodily sensations, and the right half with the left side. The functions of most of the cerebral cortex, the nature of the physical embodiment of memory and of consciousness itself, remain unknown. (ii) The midbrain, the upper part of the tapering brain-stem, contains cells concerned in eye-movements. (iii) The hindbrain, the lower part of the brain stem, contains cells responsible for breathing and for regulating heart action, the flow of digestive juices, etc. The cerebellum, which lies behind the brain stem, plays an important role in the execution of highly skilled movements. The brain of other vertebrates is similar but less developed in some respects; in invertebrates a collection of ganglia has an analogous function.

brainbox /ˈbreɪnbɒks/ n. *colloq.* a clever person.

brainchild /ˈbreɪntʃaɪld/ n. (pl. **-children**) *colloq.* an idea, plan, or invention regarded as the result of a person's mental effort.

Braine /breɪn/, John (Gerard) (1922–86), English novelist. His first novel, *Room at the Top* (1957), was an instant success, its opportunistic hero being hailed as a representative example of an 'angry young man'. Braine's later novels express less radical views and include *Finger of Fire* (1977) and *One and Last Love* (1981).

brainless /ˈbreɪnlɪs/ adj. stupid, foolish.

brainpower /ˈbreɪnpaʊə(r)/ n. mental ability or intelligence.

brainstorm /ˈbreɪnstɔːm/ n. & v. ● n. **1** a violent or excited outburst often as a result of a sudden mental disturbance. **2** *colloq.* mental confusion. **3** esp. *N. Amer.* a brainwave. **4** a concerted intellectual treatment of a problem by discussing spontaneous ideas about it. ● v.intr. & tr. engage in a brainstorming session; discuss (an idea, problem, etc.) by brainstorming. □ **brainstorming** n. (in sense 4).

brainwash /ˈbreɪnwɒʃ/ v.tr. subject (a person) to a prolonged process by which ideas other than and at variance with those already held are implanted in the mind. □ **brainwashing** n.

brainwave /ˈbreɪnweɪv/ n. **1** (usu. in *pl.*) an electrical impulse in the brain. **2** *colloq.* a sudden bright idea.

brainy /ˈbreɪnɪ/ adj. (**brainier**, **brainiest**) intellectually clever or active. □ **brainily** adv. **braininess** n.

braise /breɪz/ v.tr. fry lightly and then stew slowly with a little liquid in a closed container. [F *braiser* f. *braise* live coals]

brake[1] /breɪk/ n. & v. ● n. **1** (often in *pl.*) a device for checking the motion of a mechanism, esp. a wheel or vehicle, or for keeping it at rest. **2** anything that has the effect of hindering or impeding (*shortage of money was a brake on their enthusiasm*). ● v. **1** *intr.* apply a brake. **2** *tr.* retard or stop with a brake. □ **brake block 1** a block (usu. of hardened rubber) which is applied to a bicycle wheel as a brake. **2** a block used to hold a brake shoe. **brake drum** a cylinder attached to a wheel on which the brake shoe presses to brake. **brake fluid** fluid used in a hydraulic brake system. **brake horsepower** the power of an engine reckoned in terms of the force needed to brake it. **brake lining** a strip of fabric attached to a brake shoe to increase its friction. **brake shoe** a block which presses on the brake drum when brakes are applied. **brake van** *Brit.* a railway coach or vehicle from which the train's brakes can be controlled. [prob. obs. *brake* in sense 'machine-handle, bridle']

brake[2] /breɪk/ n. a large estate car. [var. of BREAK[2]]

brake[3] /breɪk/ n. & v. ● n. **1** a toothed instrument used for crushing flax and hemp. **2** (in full **brake harrow**) a heavy kind of harrow for breaking up large lumps of earth. ● v.tr. crush (flax or hemp) by beating it. [ME, rel. to BREAK[1]]

brake[4] /breɪk/ n. **1** a thicket. **2** brushwood. [ME f. OF *bracu*, MLG *brake* branch, stump]

brake[5] /breɪk/ n. bracken. [ME, perh. shortened f. BRACKEN, *-en* being taken as a pl. ending]

brake[6] archaic past of BREAK[1].

brakeman /ˈbreɪkmən/ n. (pl. **-men**) **1** *US* an official on a train, responsible for maintenance on a journey. **2** a person in charge of brakes. [BRAKE[1] + MAN]

brakesman /ˈbreɪksmən/ n. (pl. **-men**) *Brit.* = BRAKEMAN 2.

Bramah /ˈbræmə/, Joseph (1748–1814), English inventor. One of the most influential engineers of the Industrial Revolution, Bramah is best known for his hydraulic press, used for heavy forging; he also patented a successful lock. His other inventions included milling and planing machines and other machine tools, a beer-engine, a machine for numbering banknotes, and a water-closet.

Bramante /bræˈmæntɪ/, Donato (di Angelo) (1444–1514), Italian architect. Strongly influenced by the architecture of ancient Rome, his work often typifies the Renaissance spirit striving for the ideal of classical perfection as exemplified by the Tempietto in the cloister of S. Pietro in Montorio (1502). As architect to Pope Julius II he designed works at the Vatican as well as the new St Peter's (begun in 1506); his floor-plan for the latter, in the form of a Greek cross crowned with a central dome, was the starting-point for subsequent work on the basilica.

bramble /ˈbræmb(ə)l/ n. & v. ● n. **1** a thorny rosaceous shrub bearing fleshy red or black berries, esp. the blackberry bush, *Rubus fructicosus*. **2** the edible berry of this. **3** a rosaceous shrub with similar foliage, esp. the dog rose, *Rosa canina*. ● v.intr. (esp. as **brambling** n.) gather blackberries. □ **brambly** adj. [OE *brǣmbel* (earlier *brǣmel*): see BROOM]

brambling /ˈbræmblɪŋ/ n. a northern Eurasian finch, *Fringilla montifringilla*, which has brightly coloured plumage with a white rump, and is a winter visitor to Britain. [orig. uncert.: perh. var. of BRANDLING]

Bramley /ˈbræmlɪ/ n. (pl. **-eys**) (in full **Bramley's seedling**) a large green variety of cooking apple. [Matthew *Bramley*, Engl. butcher in whose garden it is said to have first grown *c.*1850]

bran /bræn/ n. grain husks separated from the flour. □ **bran-tub** *Brit.* a lucky dip with prizes concealed in bran. [ME f. OF. of unkn. orig.]

Branagh /ˈbrænə/, Kenneth (Charles) (b.1960), English actor, producer, and director. In 1984 he joined the Royal Shakespeare Company, where he attracted critical acclaim for roles such as Henry V; in 1989 he played the same part in the film of *Henry V*, which he also directed. In 1987 he founded the Renaissance Theatre Company, both acting in and directing some of its productions.

branch /brɑːntʃ/ n. & v. ● n. **1** a limb extending from a tree or bough.

2 a lateral extension or subdivision, esp. of a river, road, or railway. **3** a conceptual extension or subdivision, as of a family, knowledge, a subject, etc. **4** a local division or office etc. of a large business, as of a bank, library, etc. ● *v.intr.* (often foll. by *off*) **1** diverge from the main part. **2** divide into branches. **3** (of a tree) bear or send out branches. □ **branch out** extend one's field of interest. □ **branched** *adj.* **branchlet** *n.* **branchlike** *adj.* **branchy** *adj.* [ME f. OF *branche* f. LL *branca* paw]

branchia /ˈbræŋkɪə/ *n.pl.* (also **branchiae** /-kɪˌiː/) gills. □ **branchial** *adj.* **branchiate** /-kɪˌeɪt/ *adj.* [L *branchia*, pl. -*ae*, f. Gk *bragkhia* pl.]

Brancusi /brænˈkuːzɪ/, Constantin (1876–1957), Romanian sculptor, who spent much of his working life in France. His sculpture represents an attempt to move away from a representational art and to capture the essence of forms by reducing them to their ultimate, almost abstract, simplicity. His subjects, often executed in marble and polished bronze, were frequently repeated and refined several times in his quest for simplicity, as can be seen in his series of twenty-eight 'bird' sculptures (*c.*1911–36). Brancusi's influence on modern sculpture is particularly apparent in the work of Jacob Epstein and Barbara Hepworth.

brand /brænd/ *n. & v.* ● *n.* **1 a** a particular make of goods. **b** an identifying trademark, label, etc. **2** (usu. foll. by *of*) a special or characteristic kind (*brand of humour*). **3** an identifying mark burned on livestock or (formerly) prisoners etc. with a hot iron. **4** an iron used for this. **5** a piece of burning, smouldering, or charred wood. **6** a stigma; a mark of disgrace. **7** *poet.* **a** a torch. **b** a sword. **8** a kind of blight, leaving leaves with a burnt appearance. ● *v.tr.* **1** mark with a hot iron. **2** stigmatize; mark with disgrace (*they branded him a liar; was branded for life*). **3** impress unforgettably on one's mind. **4** assign a trademark or label to. □ **brand leader** the leading or best-selling product of its type. **brand name 1** a trade or proprietary name. **2** a product with a brand name. **brand-new** completely or obviously new. □ **brander** *n.* [OE f. Gmc]

Brandenburg /ˈbrændənˌbɜːɡ/ a state of NE Germany; capital, Potsdam. The city of Berlin lies in the centre of the state, but is administratively independent of it. The modern state corresponds to the western part of the former Prussian electorate, of which the eastern part was ceded to Poland after the Second World War.

Brandenburg Gate one of the city gates of Berlin, the only one that survives. It was built 1788–91 in neoclassical style. After the construction of the Berlin Wall in 1961 it stood in East Berlin, a conspicuous symbol of a divided city. It was reopened in Dec. 1989.

brandish /ˈbrændɪʃ/ *v.tr.* wave or flourish as a threat or in display. □ **brandisher** *n.* [OF *brandir* ult. f. Gmc, rel. to BRAND]

brandling /ˈbrændlɪŋ/ *n.* a red earthworm, *Eisenia foetida*, with rings of a brighter colour, which is often found in manure and used as bait. [BRAND + -LING[1]]

Brando /ˈbrændəʊ/, Marlon (b.1924), American actor. An exponent of method acting, he first attracted critical acclaim in the stage production of *A Streetcar Named Desire* (1947); he starred in the film version four years later. Other notable early films included *The Wild One* (1953) and *On the Waterfront* (1954), for which he won an Oscar. His later career included memorable roles in *The Godfather* (1972) and *Apocalypse Now* (1979).

Brands Hatch /brændz ˈhætʃ/ a motor-racing circuit near Farningham in Kent.

Brandt /brænt/, Willy (Herbert Ernst Karl Frahm) (1913–92), German statesman, Chancellor of West Germany 1969–74. He was mayor of West Berlin 1957–66 and in 1964 became chairman of the West German Social Democratic Party. He achieved international recognition for his policy of détente and the opening of relations with the former Communist countries of the Eastern bloc (see OSTPOLITIK). A pragmatist, he encouraged the negotiation of joint economic projects and a policy of non-aggression. He also chaired the Brandt Commission on the state of the world economy, whose report was published in 1980. He was awarded the Nobel Peace Prize in 1971.

brandy /ˈbrændɪ/ *n.* (pl. **-ies**) a strong alcoholic spirit distilled from wine or fermented fruit juice. □ **brandy-ball** *Brit.* a kind of brandy-flavoured sweet. **brandy butter** a rich sweet hard sauce made with brandy, butter, and sugar. **brandy-snap** a crisp rolled gingerbread wafer usu. filled with cream. [earlier *brand(e)wine* f. Du. *brandewijn* burnt (distilled) wine]

brank-ursine /bræŋkˈɜːsaɪn/ *n.* an acanthus, *Acanthus mollis* or *A. spinosus*, which has three-lobed flowers and spiny leaves, and is used as

a motif for the Corinthian capital. See also *bear's breech* (BEAR[2]). [F *branche ursine*, med.L *branca ursina* bear's claw: see BRANCH, URSINE]

Branson /ˈbræns(ə)n/, Richard (b.1950), English businessman. He set up his own record company, Virgin Records, in 1969, and later influenced the opening up of air routes with his company Virgin Atlantic Airways, established in 1984. He is also famous for making the fastest transatlantic crossing by boat in 1986 and the first by hot-air balloon in 1987.

brant *US* var. of BRENT.

Braque /bræk/, Georges (1882–1963), French painter. A co-founder of the cubist movement with Picasso in 1908, Braque was influenced by Cézanne's geometrical simplification of forms. His collages, which introduced commercial lettering and fragmented objects into pictures to contrast the real with the 'illusory' painted image, were the first stage in the development of synthetic cubism. He continued to paint during the postwar period in an essentially cubist style, gradually introducing brighter colours into his work, and becoming noted for his treatment of still life.

brash[1] /bræʃ/ *adj.* **1** vulgarly or ostentatiously self-assertive. **2** hasty, rash. **3** impudent. □ **brashly** *adv.* **brashness** *n.* [orig. dial., perh. f. RASH[1]]

brash[2] /bræʃ/ *n.* **1** loose broken rock or ice. **2** clippings from hedges, shrubs, etc. [18th c.: orig. unkn.]

brash[3] /bræʃ/ *n.* an eruption of fluid from the stomach. [16th c., perh. imit.]

Brasil see BRAZIL.

Brasilia /brəˈzɪljə/ the capital, since 1960, of Brazil; pop. (1990) 1,841,000. Designed by Lúcio Costa in 1956, the city was located in the centre of the country with the intention of drawing people away from the crowded coastal areas.

Braşov /ˈbræʃɒv/ (Hungarian **Brassó** /ˈbrɒʃʃoː/) a city in Romania; pop. (1989) 352,640. It belonged to Hungary until after the First World War, and was ceded to Romania in 1920. From 1950 to 1960 it was known in Romanian as Oraşul Stalin, meaning 'Stalin City'. Formerly a centre for expatriate Germans, it is known in German as Kronstadt.

brass /brɑːs/ *n. & adj.* ● *n.* **1** a yellow alloy of copper and zinc. **2 a** an ornament or other decorated piece of brass. **b** brass objects collectively. **3** *Mus.* brass wind instruments (including trumpet, horn, trombone, and tuba) forming a band or a section of an orchestra, group, etc. **4** *Brit. sl.* money. **5** (in full **horse-brass**) a round flat brass ornament for the harness of a draught-horse. **6** (in full **top brass**) *colloq.* persons in authority or of high (esp. military) rank. **7** an inscribed or engraved memorial tablet of brass. **8** *colloq.* effrontery (*then had the brass to demand money*). **9** a brass block or die used for making a design on a book binding. ● *adj.* made of brass. □ **brass band** a group of musicians playing brass instruments, sometimes also with percussion. **brassed off** *sl.* fed up. **brass hat** *Brit. colloq.* an officer of high rank, usu. one with gold braid on the cap. **brass neck** *Brit. colloq.* cheek, effrontery; = BRASS 8 above. **brass-rubbing 1** the rubbing of heelball etc. over paper laid on an engraved brass to take an impression of its design. **2** an impression obtained by this. **brass tacks** *sl.* actual details; real business (*get down to brass tacks*). **cold enough to freeze the balls off a brass monkey** *coarse sl.* extremely cold. **not have a brass farthing** *colloq.* have no money or assets at all. [OE *bræs*, of unkn. orig.]

brassard /ˈbræsɑːd/ *n.* a band worn on the sleeve, esp. with a uniform. [F *bras* arm + -ARD]

brasserie /ˈbræsərɪ/ *n.* a restaurant, orig. one serving beer with food. [F, = brewery]

Brassey /ˈbræsɪ/, Thomas (1805–70), English engineer and railway contractor. He built his first railway in England in the 1830s and went on to contract work for railways in Europe, India, South America, and Australia. He built more than 10,000 km (6,500 miles) of railways worldwide and during one period employed an estimated 75,000 workers.

brassica /ˈbræsɪkə/ *n.* a cruciferous plant of the genus *Brassica*, having tap roots and erect branched stems. The genus includes cabbage, Brussels sprout, broccoli, mustard, rape, cauliflower, kohlrabi, calabrese, kale, swede, and turnip. [L, = cabbage]

brassie /ˈbræsɪ/ *n.* (also **brassy**) (pl. **-ies**) a wooden-headed golf club with a brass sole.

brassière /ˈbræzɪə(r), ˈbræsɪˌeə(r)/ *n.* = BRA. [F, = child's vest]

Brassó see BRAŞOV.

brassy[1] /ˈbrɑːsɪ/ *adj.* (**brassier**, **brassiest**) **1** of or like brass, esp.

in colour. **2** impudent. **3** pretentious, showy. **4** loud and blaring. □ **brassily** adv. **brassiness** n.

brassy[2] var. of BRASSIE.

brat /bræt/ n. usu. derog. a child, esp. an ill-behaved one. □ **brat pack** a rowdy or ostentatious group of young celebrities, esp. film stars. □ **bratty** adj. [perh. abbr. of Sc. bratchet infant, or f. dial. word brat rough garment]

Bratislava /ˌbrætɪˈslɑːvə/ (called in German Pressburg; Hungarian Pozsony) the capital of Slovakia, a port on the Danube; pop. (1991) 441,450. From 1526 to 1784 it was the capital of Hungary.

brattice /ˈbrætɪs/ n. a wooden partition or shaft-lining in a coal mine. [ME ult. f. OE brittisc BRITISH]

bratwurst /ˈbrɑːtvʊəst, -vɜːst/ n. a type of small German pork sausage. [G f. braten fry, roast + Wurst sausage]

Braun[1] /braʊn/, Karl Ferdinand (1850–1918), German physicist. Braun contributed to wireless telegraphy and to the development of the cathode ray tube. He discovered the rectification properties of certain crystals and invented the coupled system of radio transmission. His demonstration that a beam of electrons could be deflected by a voltage difference between deflector plates in an evacuated tube, or by a magnetic field, led to the Braun tube, the forerunner of the cathode ray tube. He shared the Nobel Prize for physics in 1909.

Braun[2] /braʊn/, Wernher Magnus Maximilian von (1912–77), German-born American rocket engineer. An enthusiast for space travel from boyhood, Braun began to develop rockets in the 1930s. His work received support from the German army, his team being responsible for the V-2 rockets used by Germany in the Second World War. After the war he moved to the US, leading the efforts which eventually resulted in successful launches of satellites, interplanetary missions, and the landing of men on the moon in 1969.

Braunschweig see BRUNSWICK.

bravado /brəˈvɑːdəʊ/ n. a bold manner or a show of boldness intended to impress. [Sp. bravata f. bravo: cf. BRAVE, -ADO]

brave /breɪv/ adj., n., & v. ● adj. **1** able or ready to face and endure danger or pain. **2** formal splendid, spectacular (make a brave show). ● n. a North American Indian warrior. ● v.tr. defy; encounter bravely. □ **brave it out** behave defiantly under suspicion or blame; see a thing through to the end. □ **bravely** adv. **braveness** n. [ME f. F, ult. f. L barbarus BARBAROUS]

bravery /ˈbreɪvərɪ/ n. **1** brave conduct. **2** a brave nature. [F braverie or It. braveria (as BRAVE)]

bravo[1] /brɑːˈvəʊ/ int. & n. ● int. expressing approval of a performer etc. ● n. (pl. **-os**) a cry of bravo. [F f. It.]

bravo[2] /ˈbrɑːvəʊ/ n. (pl. **-oes** or **-os**) a hired ruffian or killer. [It.: see BRAVE]

bravura /brəˈvʊərə, -ˈvjʊərə/ n. (often attrib.) **1** a brilliant or ambitious action or display. **2 a** a style of (esp. vocal) music requiring exceptional ability. **b** a passage of this kind. **3** bravado. [It.]

braw /brɔː/ adj. Sc. fine, good. [var. of brawf BRAVE]

brawl /brɔːl/ n. & v. ● n. a noisy quarrel or fight. ● v.intr. **1** quarrel noisily or roughly. **2** (of a stream) run noisily. □ **brawler** n. [ME prob. ult. rel. to BRAY[1]]

brawn /brɔːn/ n. **1** muscularity, physical strength (all brawn and no brain). **2** muscle; lean flesh. **3** Brit. a jellied preparation of the chopped meat from a boiled pig's head. [ME f. AF braun, OF braon f. Gmc]

brawny /ˈbrɔːnɪ/ adj. (**brawnier**, **brawniest**) muscular, strong. □ **brawniness** n.

bray[1] /breɪ/ n. & v. ● n. **1** the cry of a donkey. **2** a sound like this cry, e.g. that of a harshly played brass instrument, a laugh, etc. ● v. **1** intr. make a braying sound. **2** tr. utter harshly. [ME f. OF braire, perh. ult. f. Celt.]

bray[2] /breɪ/ v.tr. archaic pound or crush to small pieces, esp. with a pestle and mortar. [ME f. AF braier, OF breier f. Gmc]

Bray, Vicar of /breɪ/ the hero of an 18th-century song who kept his benefice from Charles II's reign to George I's by changing his beliefs to suit the times. The song is apparently based on an anecdote of an unidentified vicar of Bray, Berks., in Thomas Fuller's Worthies of England (1662).

braze[1] /breɪz/ v. & n. ● v.tr. solder with an alloy of brass and zinc at a high temperature. ● n. **1** a brazed joint. **2** the alloy used for brazing. [F braser solder f. braise live coals]

braze[2] /breɪz/ v.tr. **1** a make of brass. **b** cover or ornament with brass. **2** make hard like brass. [OE bræsen f. bræs BRASS]

brazen /ˈbreɪz(ə)n/ adj. & v. ● adj. **1** (also **brazen-faced**) flagrant and

shameless; insolent. **2** of or like brass. **3** harsh in sound. ● v.tr. (foll. by out) face or undergo defiantly. □ **brazen it out** be defiantly unrepentant under censure. □ **brazenly** adv. **brazenness** /ˈbreɪzənnɪs/ n. [OE bræsen f. bræs brass]

brazier[1] /ˈbreɪzɪə(r), ˈbreɪzə(r)/ n. a portable heater consisting of a pan or stand for holding lighted coals. [F brasier f. braise hot coals]

brazier[2] /ˈbreɪzɪə(r), ˈbreɪzə(r)/ n. a worker in brass. □ **braziery** n. [ME prob. f. BRASS + -IER, after glass, glazier]

Brazil /brəˈzɪl/ (Portuguese **Brasil** /braˈzil/) the largest country in South America; pop. (est. 1991) 150,400,000; official language, Portuguese; capital, Brasilia. Brazil is the fifth largest country in the world, occupying almost half of the continent and containing most of the Amazon basin with its tropical rainforests. The region was inhabited largely by Tupi and Guarani peoples, who now form a very small proportion of the population, restricted mainly to Amazonia. Brazil was colonized by the Portuguese, who imported large numbers of slaves from West Africa between the 16th and 19th centuries, largely to work on sugar plantations. Brazil was proclaimed an independent empire in 1822, becoming a republic after the overthrow of the monarchy in 1889. In 1930 Getúlio Vargas seized power, ruling as a dictator for fifteen years. The country is named after brazil wood, of which it was a major source. □ **Brazilian** adj. & n.

brazil /brəˈzɪl/ n. (also **Brazil**) **1** (in full **brazil nut**) a large three-sided nut with an edible kernel, obtained from the South American forest tree Bertholletia excelsa. **2** (in full **brazil wood**) a hard red wood from a tropical South American tree of the genus Caesalpinia, yielding dyes. [Sp. & Port. brasil, med. L brasilium (orig. unkn.: perh. a corruption of an oriental name for the wood)]

Brazzaville /ˈbræzəˌvɪl/ the capital and a major port of the Republic of the Congo; pop. (est. 1990) 760,300. It was founded in 1880 by the French explorer Savorgnan de Brazza (1852–1905) and was capital of French Equatorial Africa from 1910 to 1958.

BRCS abbr. British Red Cross Society.

breach /briːtʃ/ n. & v. ● n. **1** (often foll. by of) the breaking of or failure to observe a law, contract, etc. **2 a** a breaking of relations; an estrangement. **b** a quarrel. **3 a** a broken state. **b** a gap, esp. one made by artillery in fortifications. ● v.tr. **1** break through; make a gap in. **2** break (a law, contract, etc.). □ **breach of the peace** an infringement or violation of the public peace by any disturbance or riot etc. **breach of promise** the breaking of a promise, esp. a promise to marry. **stand in the breach** bear the brunt of an attack. **step into the breach** give help in a crisis, esp. by replacing someone who has dropped out. [ME f. OF breche, ult. f. Gmc]

bread /bred/ n. & v. ● n. **1** a staple food consisting of baked dough made from flour usu. leavened with yeast and moistened. **2 a** a necessary food. **b** (also **daily bread**) one's livelihood. **3** sl. money. ● v.tr. coat with breadcrumbs for cooking. □ **bread and butter 1** bread spread with butter. **2 a** one's livelihood. **b** routine work to ensure an income. **bread-and-butter letter** a letter of thanks for hospitality. **bread and circuses** the public provision of subsistence and entertainment. **bread and wine** the Eucharist. **bread basket 1** a basket for bread or rolls. **2** sl. the stomach. **3** a region etc. that supplies cereals to another. **bread bin** a container for keeping bread in. **bread sauce** a white sauce thickened with breadcrumbs. **cast one's bread upon the waters** do good without expecting gratitude or reward. **know which side one's bread is buttered** know where one's advantage lies. **take the bread out of a person's mouth** take away a person's living, esp. by competition etc. [OE brēad f. Gmc]

breadboard /ˈbredbɔːd/ n. **1** a board for cutting bread on. **2** a board for making an experimental model of an electric circuit.

breadcrumb /ˈbredkrʌm/ n. **1** a small fragment of bread. **2** (in pl.) bread crumbled for use in cooking.

breadfruit /ˈbredfruːt/ n. **1** an evergreen tree, Artocarpus altilis, which is native to islands in the Pacific, and bears edible usu. seedless fruit. **2** this fruit, which when roasted has a texture like new bread.

breadline /ˈbredlaɪn/ n. **1** subsistence level (esp. on the breadline). **2** N. Amer. a queue of people waiting to receive free food.

breadth /bredθ, bretθ/ n. **1** the distance or measurement from side to side of a thing; broadness. **2** a piece (of cloth etc.) of standard or full breadth. **3** extent, distance, room. **4** (usu. foll. by of) capacity to respect other opinions; freedom from prejudice or intolerance (esp. breadth of mind or view). **5** Art unity of the whole, achieved by the disregard of unnecessary details. □ **breadthways** adv. **breadthwise** adv. [obs. brede, OE brǣdu, f. Gmc, rel. to BROAD]

breadwinner /ˈbredˌwɪnə(r)/ n. a person who earns the money to support a family.

break[1] /breɪk/ v. & n. ● v. (past **broke** /brəʊk/ or archaic **brake** /breɪk/; past part. **broken** /ˈbrəʊkən/ or archaic **broke**) **1** tr. & intr. **a** separate into pieces under a blow or strain; shatter. **b** make or become inoperative, esp. from damage (the toaster has broken). **c** break a bone in or dislocate (part of the body). **d** break the skin of (the head or crown). **2 a** tr. cause or effect an interruption in (broke our journey; the spell was broken; broke the silence). **b** intr. have an interval between spells of work (let's break now; we broke for tea). **3** tr. fail to observe or keep (a law, promise, etc.). **4 a** tr. & intr. make or become subdued or weakened; yield or cause to yield (broke his spirit; he broke under the strain). **b** tr. weaken the effect of (a fall, blow, etc.). **c** tr. = break in 3c. **d** tr. defeat, destroy (broke the enemy's power). **e** tr. defeat the object of (a strike, e.g. by engaging other personnel). **5** tr. surpass (a record). **6** intr. (foll. by with) **a** quarrel or cease association with (another person etc.). **b** repudiate, depart from (a tradition, practice, etc.) **7** tr. **a** be no longer subject to (a habit). **b** (foll. by of) cause (a person) to be free of a habit (broke them of their addiction). **8** tr. & intr. reveal or be revealed; (cause to) become known (broke the news; the story broke on Friday). **9** intr. **a** (of the weather) change suddenly, esp. after a fine spell. **b** (of waves) curl over and dissolve into foam. **c** (of the day) dawn. **d** (of clouds) move apart; show a gap. **e** (of a storm) begin violently. **10** tr. Electr. disconnect (a circuit) (opp. MAKE v. 23). **11** intr. (of the voice) change with emotion. **b** (of a boy's voice) change in register etc. at puberty. **12** tr. **a** (often foll. by up) divide (a set etc.) into parts, e.g. by selling to different buyers. **b** change (a banknote etc.) for coins. **13** tr. ruin (an individual or institution) financially (see also BROKE adj.). **14** tr. penetrate (e.g. a safe) by force. **15** tr. decipher (a code). **16** tr. make a way, path, etc.) by separating obstacles. **17** intr. burst forth (the sun broke through the clouds). **18** Mil. a tr. (of troops) disperse in confusion. **b** tr. make a rupture in (ranks). **19 a** intr. (usu. foll. by free, loose, out, etc.) escape from constraint by a sudden effort. **b** tr. escape or emerge from (prison, bounds, cover, etc.). **20** tr. Tennis etc. win a game against (an opponent's service). **21** intr. Boxing etc. (of two fighters, usu. at the referee's command) come out of a clinch. **22** tr. Mil. demote (an officer). **23** intr. esp. Stock Exch. (of prices) fall sharply. **24** intr. Cricket (of a bowled ball) change direction on bouncing. **25** intr. Billiards etc. make the first stroke at the beginning of a game. **26** tr. unfurl (a flag etc.). **27** tr. Phonet. subject (a vowel) to fracture. **28** tr. disprove (an alibi). ● n. **1 a** an act or instance of breaking. **b** a point where something is broken; a gap. **2** an interval, an interruption; a pause in work; a holiday. **3** a sudden dash (esp. to escape). **4** colloq. **a** a piece of good luck; a fair chance. **b** (also **bad break**) an unfortunate remark or action, a blunder. **c** US (in pl., prec. by the) fate. **5** Cricket a change in direction of a bowled ball on bouncing. **6** Billiards etc. **a** a series of points scored during one turn. **b** the opening shot that disperses the balls. **7** Mus. (in jazz) a short unaccompanied passage for a soloist, usu. improvised. **8** Electr. a discontinuity in a circuit. □ **bad break** colloq. **1** a piece of bad luck. **2** a mistake or blunder. **break away** make or become free or separate (see also BREAKAWAY). **break the back of 1** do the hardest or greatest part of (a problem etc.). **2** overburden (a person) physically or mentally; crush, defeat. **break bulk** see BULK. **break crop** a crop grown to avoid the continual growing of cereals. **break-dancing** an energetic and acrobatic style of street-dancing, developed by US blacks. **break down 1** a fail in mechanical action; cease to function. **b** (of human relationships etc.) fail, collapse. **c** (of health) fail, deteriorate; (of a person) fail in (esp. mental) health. **d** be overcome by emotion; collapse in tears. **2 a** demolish, destroy. **b** overcome (resistance). **c** force (a person) to yield under pressure. **3** analyse into components (see also BREAKDOWN). **break even** emerge from a transaction etc. with neither profit nor loss. **break a person's heart** see HEART. **break the ice 1** begin to overcome formality or shyness, esp. between strangers. **2** make a start. **break in 1** enter premises by force, esp. with criminal intent. **2** interrupt. **3 a** accustom to a habit etc. **b** wear etc. until comfortable. **c** tame or discipline (an animal); accustom (a horse) to saddle and bridle etc. **4** Austral. & NZ bring (virgin land) into cultivation. **break-in** n. an illegal forced entry into premises, esp. with criminal intent. **breaking and entering** (formerly) the illegal entering of a building with intent to commit a felony. **breaking-point** the point of greatest strain, at which a thing breaks or a person gives way. **break in on** disturb; interrupt. **break into 1** enter forcibly or violently. **2 a** suddenly begin, burst forth with (a song, laughter, etc.). **b** suddenly change one's pace for (a faster one) (broke into a gallop). **3** interrupt. **break-line** Printing the last line of a paragraph (usu. not of full length). **break new ground** innovate; start on something new. **break of day** dawn. **break off 1** detach by breaking. **2** bring to an end. **3** cease talking etc. **break open** open forcibly. **break out 1** escape by force,

esp. from prison. **2** begin suddenly; burst forth (then violence broke out). **3** (foll. by in) become covered in (a rash etc.). **4** exclaim. **5** release (a run-up flag). **6** US **a** open up (a receptacle) and remove its contents. **b** remove (articles) from a place of storage. **break-out** n. a forcible escape. **break point 1** a place or time at which an interruption or change is made. **2** (usu. **breakpoint**) Computing a place in a computer program where the sequence of instructions is interrupted, esp. by another program. **3** Tennis **a** the state of the game when the player or side receiving service needs only one more point to win the game. **b** this point. **4** = breaking-point. **break ship** fail to rejoin one's ship after absence on leave. **break step** get out of step. **break up 1** break into small pieces. **2** disperse; disband. **3** end the school term. **4 a** terminate a relationship; disband. **b** cause to do this. **5** (of the weather) change suddenly (esp. after a fine spell). **6** esp. US **a** upset or be upset. **b** excite or be excited. **c** convulse or be convulsed (see also BREAKUP). **break wind** release gas from the anus. **break one's word** see WORD. [OE brecan f. Gmc]

break[2] /breɪk/ n. **1** a carriage-frame without a body, for breaking in young horses. **2** = BRAKE[2]. [perh. = brake framework: 17th c., of unkn. orig.]

breakable /ˈbreɪkəb(ə)l/ adj. & n. ● adj. that may or is apt to be broken easily. ● n. (esp. in pl.) a breakable thing.

breakage /ˈbreɪkɪdʒ/ n. **1** an act or instance of breaking. **2 a** a broken thing. **b** damage caused by breaking.

breakaway /ˈbreɪkəˌweɪ/ n. **1** the act or an instance of breaking away or seceding. **2** (attrib.) that breaks away or has broken away; separate. **3** Austral. a stampede, esp. at the sight or smell of water. **4** a false start in a race. **5** Rugby each of the two flank forwards on the outsides of the second row of a scrum formation.

breakdown /ˈbreɪkdaʊn/ n. **1 a** a mechanical failure. **b** a loss of (esp. mental) health and strength. **2 a** a collapse or disintegration (breakdown of communication). **b** physical or chemical decomposition. **3** a detailed analysis (of statistics, chemical components, etc.).

breaker /ˈbreɪkə(r)/ n. **1** a person or thing that breaks something, esp. disused machinery. **2** a person who breaks in a horse. **3** a heavy wave that breaks.

breakfast /ˈbrekfəst/ n. & v. ● n. the first meal of the day. ● v.intr. have breakfast. □ **breakfast television** early-morning television. **have a person for breakfast** sl. get the better of (a person) in a dispute etc.; defeat easily. □ **breakfaster** n. [BREAK[1] interrupt + FAST[2]]

breakneck /ˈbreɪknek/ attrib.adj. (of speed) dangerously fast.

Breakspear /ˈbreɪkspɪə(r)/, Nicholas, see ADRIAN IV.

breakthrough /ˈbreɪkθruː/ n. **1** a major advance or discovery. **2** an act of breaking through an obstacle etc.

breakup /ˈbreɪkʌp/ n. **1** the disintegration or collapse of a thing. **2** a dispersal.

breakwater /ˈbreɪkˌwɔːtə(r)/ n. a barrier built out into the sea to break the force of waves.

Bream /briːm/, Julian (Alexander) (b.1933), English guitarist and lute-player. He made his London début in 1950 and benefited from an early involvement with Andrés Segovia. He formed the Julian Bream Consort for the performance of early consort music and has revived and edited much early music. Britten, Walton, and others composed works for him.

bream[1] /briːm/ n. (pl. same) **1** a yellowish deep-bodied freshwater fish, Abramis brama, of the carp family. **2** (in full **sea bream**) a similarly shaped sparid marine fish; esp. one of the genus Pagellus or Spondyliosoma, of the NE Atlantic. [ME f. OF bre(s)me f. WG]

bream[2] /briːm/ v.tr. Naut. hist. clean (a ship's bottom) by burning and scraping. [prob. f. LG: rel. to BROOM]

breast /brest/ n. & v. ● n. **1 a** either of two milk-secreting organs on the upper front of a woman's body. **b** the corresponding usu. rudimentary part of a man's body. **2 a** the upper front part of a human body; the chest. **b** the corresponding part of an animal. **3** the part of a garment that covers the breast. **4** the breast as a source of nourishment or emotion. ● v.tr. **1** face, meet in full opposition (breast the wind). **2** contend with, face (prepared to breast the difficulties of the journey). **3** reach the top of (a hill). □ **breast-feed** (past and past part. **-fed**) **1** tr. feed (a baby) from the breast. **2** intr. (of a baby) feed from the breast. **breast-high** as high as the breast; submerged to the breast. **breast-pin** a brooch etc. worn on the breast. **breast-stroke** a stroke made while swimming on the breast by extending arms forward and sweeping them back in unison. **breast the tape** see TAPE. **make a**

clean breast of confess fully. □ **breasted** adj. (also in comb.). [OE *brēost* f. Gmc]

breastbone /'brestbəʊn/ n. a thin flat vertical bone and cartilage in the chest connecting the ribs.

breastplate /'brestpleɪt/ n. a piece of armour covering the breast.

breastsummer /'bresəmə(r)/ n. Archit. a beam across a broad opening, sustaining a superstructure. [BREAST + SUMMER²]

breastwork /'brestwɜːk/ n. a low temporary defence or parapet.

breath /breθ/ n. **1 a** the air taken into or expelled from the lungs. **b** one respiration of air. **c** an exhalation of air that can be seen, smelt, or heard (*breath steamed in the cold air; bad breath*). **2 a** a slight movement of air; a breeze. **b** a whiff of perfume etc. **3** a whisper, a murmur (esp. of a scandalous nature). **4** the power of breathing; life (*is there breath in him?*). □ **below** (or **under**) **one's breath** in a whisper. **breath of fresh air 1** a small amount of or a brief time in the fresh air. **2** a refreshing change. **breath of life** a necessity. **breath test** Brit. a test of a person's alcohol consumption, using a breathalyser. **breath-test** Brit. administer a breath test to. **catch one's breath 1** cease breathing momentarily in surprise, suspense, etc. **2** rest after exercise to restore normal breathing. **draw breath** breathe; live. **hold one's breath 1** cease breathing temporarily. **2** colloq. wait in eager anticipation. **in the same breath** (esp. of saying two contradictory things) within a short time. **out of breath** gasping for air, esp. after exercise. **take breath** pause for rest. **take a person's breath away** astound; surprise; awe; delight. **waste one's breath** talk or give advice without effect. [OE *brǣth* f. Gmc]

Breathalyser /'breθəˌlaɪzə(r)/ n. (also **Breathalyzer**) Brit. propr. an instrument for measuring the amount of alcohol in the breath, esp. of a driver. □ **breathalyse** v.tr. (also **-lyze**). [BREATH + ANALYSE + -ER¹]

breathe /briːð/ v. **1** intr. take air into and expel it from the lungs. **2** intr. be or seem alive (*is she breathing?*). **3** tr. utter; say (esp. quietly) (*breathed her forgiveness*). **b** express; display (*breathed defiance*). **4** intr. take breath, pause. **5** tr. send out or take in (as if) with breathed air (*breathed enthusiasm into them; breathed whisky*). **6** intr. (of wine, fabric, etc.) be exposed to fresh air. **7** intr. **a** sound, speak (esp. quietly). **b** (of wind) blow softly. **8** tr. allow (a horse etc.) to breathe; give rest after exertion. □ **breathe again** (or **freely**) recover from a shock, fear, etc., and be at ease. **breathe down a person's neck** follow or check up on a person, esp. menacingly. **breathe new life into** revitalize, refresh. **breathe one's last** die. **breathe upon** tarnish, taint. **not breathe a word** keep silent. **not breathe a word of** keep quite secret. [ME f. BREATH]

breather /'briːðə(r)/ n. **1** colloq. a brief pause for rest. **2** a safety-vent in the crankcase of a motor vehicle etc.

breathing /'briːðɪŋ/ n. **1** the process of taking air into and expelling it from the lungs. **2** a sign in Greek indicating that an initial vowel or rho is aspirated (**rough breathing**) or not aspirated (**smooth breathing**). □ **breathing-space** time to breathe; a pause.

breathless /'breθlɪs/ adj. **1** panting, out of breath. **2** (as if) holding the breath because of excitement, suspense, etc. (*a state of breathless expectancy*). **3** unstirred by wind; still. □ **breathlessly** adv. **breathlessness** n.

breathtaking /'breθˌteɪkɪŋ/ adj. astounding; awe-inspiring. □ **breathtakingly** adv.

breathy /'breθɪ/ adj. (**breathier**, **breathiest**) (of a singing voice etc.) containing the sound of breathing. □ **breathily** adv. **breathiness** n.

breccia /'bretʃɪə/ n. & v. Geol. ● n. a rock of angular stones etc. cemented by finer material. ● v.tr. form into breccia. □ **brecciate** /-tʃɪˌeɪt/ v.tr. **brecciation** /ˌbretʃɪ'eɪʃ(ə)n/ n. [It., = gravel, f. Gmc, rel. to BREAK¹]

Brecht /brext/, Bertolt (1898–1956), German dramatist, producer, and poet. His interest in combining music and drama led to a number of successful collaborations with Kurt Weill, the first of these being *The Threepenny Opera* (1928), an adaptation of John Gay's *The Beggar's Opera*. In his later drama, which was written in exile after Hitler's rise to power and includes *Mother Courage* (1941) and *The Caucasian Chalk Circle* (1948), Brecht experimented with his ideas of a Marxist 'epic theatre' with its 'alienation effect', whereby theatrical illusion is controlled and the audience is confronted with the real political issues at stake.

Breconshire /'brekən.ʃɪə(r)/ (also **Brecknockshire** /'breknɒk.ʃɪə(r)/) a former county of south central Wales. It was divided between Powys and Gwent in 1974.

bred past and past part. of BREED.

Breda /'breɪdə, Dutch bre'daː/ a manufacturing town in the SW Netherlands; pop. (1991) 124,800. It is noted for the Compromise of

Breda of 1566, a protest against Spanish rule over the Netherlands; the 1660 manifesto of Charles II (who lived there in exile), stating his terms for accepting the throne of Britain; and the Treaty of Breda, which ended the Anglo-Dutch war of 1665–7.

breech /briːtʃ/ n. & v. ● n. **1 a** the part of a cannon behind the bore. **b** the back part of a rifle or gun barrel. **2** archaic the buttocks. ● v.tr. archaic put (a boy) into breeches after being in petticoats since birth. □ **breech birth** (or **delivery**) the delivery of a baby which is so positioned in the uterus that its buttocks or feet are delivered first. **breech-block** a metal block which closes the breech aperture in a gun. **breech-loader** a gun loaded at the breech, not through the muzzle. **breech-loading** (of a gun) loaded at the breech, not through the muzzle. [OE *brōc* (as sing. in sense of BREECHES), pl. *brēc* (treated as sing. in ME), f. Gmc]

breeches /'brɪtʃɪz/ n.pl. (also **pair of breeches** sing.) **1** short trousers, esp. fastened below the knee, now used esp. for riding or in court costume. **2** colloq. trousers, knickerbockers, or underpants. □ **breeches buoy** a lifebuoy suspended from a rope which has canvas breeches for the user's legs. [pl. of BREECH]

Breeches Bible the Geneva Bible of 1560 with *breeches* in Gen. 3:7 for the garments made by Adam and Eve.

breed /briːd/ v. & n. ● v. (past and past part. **bred** /bred/) **1** tr. & intr. bear, generate (offspring); reproduce. **2** tr. & intr. propagate or cause to propagate; raise (livestock). **3** tr. **a** yield, produce; result in (*war breeds famine*). **b** spread (*discontent bred by rumour*). **4** intr. arise; spread (*disease breeds in the Tropics*). **5** tr. bring up; train (*bred to the law; Hollywood breeds stars*). **6** tr. Physics create (fissile material) by nuclear reaction. ● n. **1** a stock of animals or plants within a species, having a similar appearance, and usu. developed by deliberate selection. **2** a race; a lineage. **3** a sort, a kind. □ **bred and born** = born and bred. **bred in the bone** hereditary. **breeder reactor** a nuclear reactor that can create more fissile material than it consumes. **breed in** mate with or marry near relations. □ **breeder** n. [OE *brēdan*: rel. to BROOD]

breeding /'briːdɪŋ/ n. **1** the process of developing or propagating (animals, plants, etc.). **2** generation; childbearing. **3** the result of training or education; behaviour. **4** good manners (as produced by an aristocratic heredity) (*has no breeding*).

breeks /briːks/ n.pl. Sc. var. of BREECHES.

breeze¹ /briːz/ n. & v. ● n. **1** a gentle wind. **2** Meteorol. a wind between force 2 and force 6 on the Beaufort scale (4–31 m.p.h. or 1.6–13.8 metres per second). **3** a wind blowing from land at night or sea during the day. **4** esp. Brit. colloq. a quarrel or display of temper. **5** esp. US colloq. an easy task. ● v.intr. (foll. by in, out, along, etc.) colloq. come or go in a casual or light-hearted manner. [prob. f. OSp. & Port. *briza* north-east wind]

breeze² /briːz/ n. small cinders. □ **breeze-block** a lightweight building block, esp. one made from breeze mixed with sand and cement. [F *braise* live coals]

breeze³ /briːz/ n. a gadfly or cleg. [OE *briosa*, of unkn. orig.]

breezy /'briːzɪ/ adj. (**breezier**, **breeziest**) **1 a** windswept. **b** pleasantly windy. **2** colloq. lively; jovial. **3** colloq. careless (*with breezy indifference*). □ **breezily** adv. **breeziness** n.

Bregenz /'breɪgents/ a city in western Austria, on the eastern shores of Lake Constance; pop. (1991) 27,240. It is the capital of the state of Vorarlberg.

Bremen /'breɪmən/ **1** a state of NE Germany. Divided into two parts, which centre on the city of Bremen and the port of Bremerhaven, it is surrounded by the state of Lower Saxony. **2** its capital, an industrial city linked by the River Weser to the port of Bremerhaven and the North Sea; pop. (1989) 537,600.

bremsstrahlung /'bremzˌʃtrɑːlʊŋ/ n. (often attrib.) Physics electromagnetic radiation produced by the acceleration or esp. the deceleration of a charged particle after passing through the electric and magnetic fields of a nucleus. [G, = braking radiation]

Bren /bren/ n. (in full **Bren gun**) a lightweight quick-firing machine-gun. [BRNO (where orig. made) + Enfield, England (where later made)]

Brendan, St /'brendən/ (c.486–c.575), Irish abbot. The legend of the 'Navigation of St Brendan' (c.1050), describing his voyage with a band of monks to a promised land (possibly Orkney or the Hebrides), was widely popular in the Middle Ages.

Brenner Pass /'brenə(r)/ an Alpine pass at the border between Austria and Italy, on the route between Innsbruck and Bolzano, at an altitude of 1,371 m (4,450 ft).

brent /brent/ n. (US **brant** /brænt/) (in full **brent-goose**) a small black, grey, and white migratory goose, *Branta bernicla*, which breeds in the Arctic. [16th c.: orig. unkn.]

Brescia /ˈbreʃə/ an industrial city in Lombardy, in northern Italy; pop. (1990) 196,770.

Breslau see WROCŁAW.

Bresson /ˈbresɒn/, Robert (b.1907), French film director. An influential though not a prolific director, he is noted for his austere intellectual style and meticulous attention to detail. His most notable films, most of which feature unknown actors, include *Diary of a Country Priest* (1951), *The Trial of Joan of Arc* (1962), and *The Devil, Probably* (1977).

Brest /brest/ **1** a port and naval base on the Atlantic coast of Brittany, in NW France; pop. (1990) 153,100. **2** (Polish **Brześć nad Bugiem** /ˌbʒɛʃtʃ nad ˈbuɡjɛm/) a river port and industrial city in Belarus, situated close to the border with Poland; pop. (1990) 268,800. The peace treaty between Germany and Russia was signed there in March 1918. It was known until 1921 as Brest-Litovsk.

Bretagne see BRITTANY.

brethren see BROTHER.

Breton[1] /ˈbret(ə)n, French brətɔ̃/ *n. & adj.* ● *n.* **1** a native of Brittany. **2** the language of the Bretons, belonging to the Brythonic branch of the Celtic language group. (*See note below.*) ● *adj.* of or relating to Brittany or its people or language. [OF, = BRITON]

▪ Breton is the only Celtic language now spoken on the European mainland, representing the modern development of the language brought from Cornwall in the 5th and 6th centuries by Britons fleeing from the Saxon invaders. Until the 20th century Breton was widely spoken in Brittany, but official encouragement of the use of French has contributed to its decline. Recently attempts have been made to revive its use in Brittany, but although it is taught in some schools its use is rare in the younger generation; some 500,000 are thought to use it on an everyday basis.

Breton[2] /ˈbret(ə)n, French brətɔ̃/, André (1896–1966), French poet, essayist, and critic. Influenced by the work of Sigmund Freud and the poet Paul Valéry, Breton was first involved with Dadaism, co-founding the movement's review *Littérature* in 1919. When Dada collapsed in the early 1920s, Breton launched the surrealist movement; its chief theorist, he first outlined the movement's philosophy in his manifesto of 1924. His creative writing is characterized by surrealist techniques such as 'automatic' writing and the startling juxtaposition of images, as in his famous poetic novel *Nadja* (1928).

bretzel var. of PRETZEL.

Breughel see BRUEGEL.

Breuil /brɜːɪ/, Henri (Édouard Prosper) (1877–1961), French archaeologist. He is noted for his work on palaeolithic cave-paintings, in particular those at Altamira in Spain, which he was able to authenticate. He also made detailed studies of examples in the Dordogne region of France and in southern Africa.

breve /briːv/ *n.* **1** *Mus.* a note, now rarely used, having the time value of two semibreves. **2** a written or printed mark (˘) indicating a short or unstressed vowel. **3** *hist.* an authoritative letter from a sovereign or pope. [ME var. of BRIEF]

brevet /ˈbrevɪt/ *n. & v.* ● *n.* (often *attrib.*) a document conferring a privilege from a sovereign or government, esp. a rank in the army, without the appropriate pay (*was promoted by brevet; brevet major*). ● *v.tr.* (**breveted, breveting** or **brevetted, brevetting**) confer brevet rank on. [ME f. OF dimin. of *bref* BRIEF]

breviary /ˈbriːvɪərɪ/ *n.* (*pl.* **-ies**) *RC Ch.* a book containing the service for each day, to be recited by those in orders. [L *breviarium* summary f. *breviare* abridge: see ABBREVIATE]

brevity /ˈbrevɪtɪ/ *n.* **1** economy of expression; conciseness. **2** shortness (of time etc.) (*the brevity of happiness*). [AF *breveté*, OF *brieveté* f. *bref* BRIEF]

brew /bruː/ *v. & n.* ● *v.* **1** *tr.* **a** make (beer etc.) by infusion, boiling, and fermentation. **b** make (tea etc.) by infusion or (punch etc.) by mixture. **2** *intr.* undergo either of these processes (*the tea is brewing*). **3** *intr.* (of trouble, a storm, etc.) gather force; threaten (*mischief was brewing*). **4** *tr.* bring about; set in train; concoct (*brewed their fiendish scheme*). ● *n.* **1** an amount (of beer etc.) brewed at one time (*this year's brew*). **2** what is brewed (esp. with regard to its quality) (*a good strong brew*). **3** the action or process of brewing. □ **brew up 1** make tea. **2** = BREW *v.* 2 above. **3** = BREW *v.* 4 above. **brew-up** *n.* an instance of making tea. □ **brewer** *n.* [OE *brēowan* f. Gmc]

brewery /ˈbruːərɪ/ *n.* (*pl.* **-ies**) a place where beer etc. is brewed commercially.

Brewster /ˈbruːstə(r)/, Sir David (1781–1868), Scottish physicist. He is

best known for his work on the laws governing the polarization of light, and for his invention of the kaleidoscope. Brewster also worked extensively on the optical classification of crystals and minerals, and on the use of spectroscopy for chemical analysis.

Brezhnev /ˈbreʒnef/, Leonid (Ilich) (1906–82), Soviet statesman, General Secretary of the Communist Party of the USSR 1966–82 and President 1977–82. Born in Russia, he held offices within the Soviet Communist Party (CPSU) before and after the Second World War, rising to become Chairman of the Presidium in 1960. In 1964 he and Kosygin forced Khrushchev to resign and Brezhnev eventually became General Secretary of the Party. His period in power was marked by intensified persecution of dissidents at home and by attempted détente followed by renewed cold war in 1968. He was largely responsible for the decision to invade Czechoslovakia in 1968.

Briansk see BRYANSK.

briar[1] var. of BRIER[1].

briar[2] var. of BRIER[2].

bribe /braɪb/ *v. & n.* ● *v.tr.* (often foll. by *to* + infin.) persuade (a person etc.) to act esp. illegally or dishonestly in one's favour by a gift of money, services, etc. (*bribed the guard to release the suspect*). ● *n.* money or services offered in the process of bribing. □ **bribable** *adj.* **briber** *n.* **bribery** *n.* [ME f. OF *briber, brimber* beg, of unkn. orig.]

bric-à-brac /ˈbrɪkəˌbræk/ *n.* (also **bric-a-brac, bricabrac**) miscellaneous, often old, ornaments, trinkets, furniture, etc., of no great value. [F f. obs. *à bric et à brac* at random]

brick /brɪk/ *n., v., & adj.* ● *n.* **1 a** a small, usu. rectangular, block of fired or sun-dried clay, used in building. **b** the material used to make these. **c** a similar block of concrete etc. **3** *Brit.* a child's toy building-block. **3** a brick-shaped solid object (*a brick of ice-cream*). **4** *sl.* a generous or loyal person. ● *v.tr.* (foll. by *in, up*) close or block with brickwork. ● *adj.* **1** built of brick (*brick wall*). **2** of a dull red colour. □ **bang** (or **knock** or **run**) **one's head against a brick wall** be continually rebuffed, try repeatedly to no avail. **brick-field** a place where bricks are made. **brick-red** *n. & adj.* ● *n.* a dull red typical of bricks. ● *adj.* of this colour. **like a ton of bricks** *colloq.* with crushing weight, force, or authority. **see through a brick wall** have miraculous insight. □ **bricky** *adj.* [ME f. MLG, MDu. *bri(c)ke*, of unkn. orig.]

brickbat /ˈbrɪkbæt/ *n.* **1** a piece of brick, esp. when used as a missile. **2** an uncomplimentary remark.

brickfielder /ˈbrɪkˌfiːldə(r)/ *n.* *Austral.* a hot, dry north wind.

brickie /ˈbrɪkɪ/ *n.* *sl.* a bricklayer.

bricklayer /ˈbrɪkˌleɪə(r)/ *n.* a worker who builds with bricks. □ **bricklaying** *n.*

brickwork /ˈbrɪkwɜːk/ *n.* **1** work executed in brick. **2** building with bricks; bricklaying.

brickyard /ˈbrɪkjɑːd/ *n.* a place where bricks are made.

bridal /ˈbraɪd(ə)l/ *adj.* of or concerning a bride or a wedding. □ **bridally** *adv.* [orig. as noun, = wedding-feast, f. OE *brȳd-ealu* f. *brȳd* BRIDE + *ealu* ale-drinking]

bride /braɪd/ *n.* a woman on her wedding day and for some time before and after it. □ **bride-price** money or goods given to a bride's family by the bridegroom's, esp. in tribal societies. [OE *brȳd* f. Gmc]

Bride, St see BRIDGET, ST[1].

bridegroom /ˈbraɪdɡruːm, -ɡrʊm/ *n.* a man on his wedding day and for some time before and after it. [OE *brȳdguma* (as BRIDE, *guma* man, assim. to GROOM)]

bridesmaid /ˈbraɪdzmeɪd/ *n.* a girl or unmarried woman attending a bride on her wedding day. [earlier *bridemaid*, f. BRIDE + MAID]

bridewell /ˈbraɪdwəl, -wel/ *n.* *archaic* a prison; a reformatory. [St *Bride's* (or *Bridget's*) *Well* in London, between Fleet Street and the Thames, where such a building (formerly a royal palace) stood]

Bridge /brɪdʒ/, Frank (1879–1941), English composer, conductor, and violist. His compositions include chamber music, songs, and orchestral works, among them *The Sea* (1910–11). His later works, such as the string trio *Rhapsody* (1928) and *Oration* (for cello and orchestra, 1930), show stylistic elements akin to those of Schoenberg. Benjamin Britten was one of his pupils.

bridge[1] /brɪdʒ/ *n. & v.* ● *n.* **1 a** a structure carrying a road, path, railway, etc., across a stream, ravine, road, railway, etc. **b** anything providing a connection between different things (*English is a bridge between nations*). **2** the superstructure on a ship from which the captain and officers direct operations. **3** the upper bony part of the nose. **4** *Mus.* an upright piece of wood on a violin etc. over which the strings are stretched.

5 = BRIDGEWORK. **6** *Billiards* etc. **a** a long stick with a structure at the end which is used to support a cue for a difficult shot. **b** a support for a cue formed by a raised hand. **7** = *land-bridge*. ● *v.tr.* **1 a** be a bridge over (*a fallen tree bridges the stream*). **b** make a bridge over; span. **2** span as if with a bridge (*bridged their differences with understanding*). □ **bridge of asses** = *pons asinorum*. **bridge of boats** a bridge formed by mooring boats together abreast across a river etc. **bridge passage** *Mus.* a transitional section between main themes. **bridging loan** a loan from a bank etc. to cover the short interval between buying a house etc. and selling another. **cross a** (or **that**) **bridge when one comes to it** deal with a problem when and if it arises. □ **bridgeable** *adj.* [OE *brycg* f. Gmc]

▪ The main types of bridge are the girder, arch, suspension, cantilever, and drawbridge or bascule. The simplest and oldest type consists of a straight beam or girder placed across a span; variants of this survive in the Dartmoor clapper bridges, and in China, where granite slabs of up to 200 metric tons in weight and 21 m (70 ft) long were used. Although the masonry arch was developed in China in the 4th millennium BC, it was the Romans who first used the semicircular arch extensively, building many bridges, such as the Pont du Gard aqueduct, that remain in use today. With increasing confidence, medieval and Renaissance engineers built arches with lower rise than the semicircle and coped with the resulting increased thrust on the abutment; these can be seen, for example, in the remaining elliptical arches of the bridge at Avignon (1171) and the segmented arches of the Ponte Vecchio in Florence (1345). In 1779 the Coalbrookdale bridge in Shropshire heralded a new era of iron and steel arch bridges: several, including the Sydney Harbour Bridge (1932), have spans of over 500 m (1640 ft).
The suspension bridge in its primitive form, widespread geographically, consists of three ropes hanging from anchorages on each side, forming walkway and handrailing. The Lan Chin bridge in China (AD 65) has iron chains supporting a wooden deck, but in the West iron chains were not used until the early 19th century, as in Telford's Menai Bridge (1819–26). Modern suspension bridges use parallel stranded wire ropes for the cables; famous examples include the Golden Gate Bridge near San Francisco (1937) and the Humber Bridge in England, which on its completion in 1980 had the world's longest single span (1410 m, 4626 ft).
The cantilever bridge, a refinement of the beam, has a central span supported on the ends of side-spans which are cantilevered out over piers, as in the Forth railway bridge in Scotland (1882–9) and the Quebec bridge (completed in 1917).
The medieval drawbridge, lowered and raised by ropes or chains to allow or prevent passage, finds its modern equivalent in designs using counterweights and gears to allow a bridge to be raised, e.g. to allow the passage of ships. London's Tower Bridge uses two bascules to span 61 m (200 ft).

bridge² /brɪdʒ/ *n.* a card-game derived from whist, played by two partnerships of two players who at the beginning of each hand bid for the right to name the trump suit, the highest bid also representing a contract to make a specified number of tricks. (*See note below.*) □ **bridge roll** a small soft bread roll. [19th c.: prob. of eastern Mediterranean origin]

▪ In bridge, the members of each pair sit opposite one another; one member of the partnership which made the highest bid has to play both his or her own hand as well as the exposed cards of his or her partner. The modern form of bridge, *contract bridge*, was developed mainly by the American player Ely Culbertson (1891–1955) and his associates on the basis of an earlier form, known as *auction bridge*, popular with the British in India in the 19th century. The principal features of contract bridge are a complex bidding system, including many conventions for indicating the strength and distribution of a hand, and a scoring system for the card play in which only tricks that were bid as well as won count towards game. The main form of competitive bridge for tournaments is *duplicate bridge*, in which the same pre-arranged hands are played independently by several pairs or teams.

bridgehead /ˈbrɪdʒhed/ *n. Mil.* a fortified position held on the enemy's side of a river or other obstacle.

Bridge of Sighs a 16th-century enclosed bridge in Venice between the Doges' Palace and the state prison, originally crossed by prisoners on their way to torture or execution. The name is also used for two similarly constructed bridges in Oxford and Cambridge.

Bridges /ˈbrɪdʒɪz/, Robert (Seymour) (1844–1930), English poet and literary critic. His long philosophical poem *The Testament of Beauty*

(1929), written in the Victorian tradition, was instantly popular; he was Poet Laureate 1913–30. He also wrote two important critical essays, *Milton's Prosody* (1893) and *John Keats* (1895). He made an important contribution to literature in publishing his friend Gerard Manley Hopkins's poems in 1918.

Bridget, St¹ /ˈbrɪdʒɪt/ (known as St Bridget of Ireland) (also **Bride** /ˈbriːdə, braɪd/, **Brigid** /ˈbrɪdʒɪd/) (6th century), Irish abbess. She was venerated in Ireland as a virgin saint and noted in miracle stories for her compassion; her cult soon spread over most of western Europe. It has been suggested that she may represent the Irish goddess Brig. Feast day, 23 July.

Bridget, St² /ˈbrɪdʒɪt/ (known as St Bridget of Sweden) (also **Birgitta** /bɪəˈɡɪtə/) (c.1303–73), Swedish nun and visionary. She experienced her first vision of the Virgin Mary at the age of 7. After her husband's death she was inspired by further visions to devote herself to religion and she founded the Order of Bridgettines (c.1346) at Vadstena in Sweden. Feast day, 1 Feb.

Bridgetown /ˈbrɪdʒtaʊn/ the capital of Barbados, a port on the south coast; pop. (est. 1988) 102,000.

bridgework /ˈbrɪdʒwɜːk/ *n. Dentistry* a dental structure used to cover a gap, joined to and supported by the teeth on either side.

Bridgman /ˈbrɪdʒmən/, Percy Williams (1882–1961), American physicist. He worked mainly on the properties of liquids and solids under very high pressures, and designed an apparatus that achieved a fluid pressure of 30,000 atmospheres. His techniques were later used in making artificial diamonds and other minerals, and he became involved in the Manhattan Project. Bridgman also published several major contributions to the philosophy of science, including *The Logic of Modern Physics* (1927). He was awarded the Nobel Prize for physics in 1946.

bridie /ˈbraɪdɪ/ *n. Sc.* a meat pasty. [orig. uncert., perh. f. obs. *bride's pie*]

bridle /ˈbraɪd(ə)l/ *n. & v.* ● *n.* **1 a** the headgear used to control a horse, consisting of buckled leather straps, a metal bit, and reins. **b** a restraining device or influence (*put a bridle on your tongue*). **2** *Naut.* a mooring-cable. **3** *Anat.* a ligament checking the motion of a part. ● *v.* **1** *tr.* put a bridle on (a horse etc.). **2** *tr.* bring under control; curb. **3** *intr.* (often foll. by *at* or *up at*) express offence, resentment, etc., esp. by throwing up the head and drawing in the chin. □ **bridle-path** (or **-road** or **-way**) a rough path or road fit only for riders or walkers, not vehicles. [OE *brīdel*]

bridoon /brɪˈduːn/ *n.* the snaffle and rein of a military bridle. [F *bridon* f. *bride* bridle]

Brie /briː/ *n.* a kind of soft cheese. [*Brie* in northern France]

brief /briːf/ *adj., n., & v.* ● *adj.* **1** of short duration; fleeting. **2** concise in expression. **3** abrupt, brusque (*was rather brief with me*). **4** (of clothes) lacking in substance, scanty (*wearing a brief skirt*). ● *n.* **1** (in *pl.*) **a** women's brief pants. **b** men's brief underpants. **2** *Law* **a** a summary of the facts and legal points of a case drawn up for counsel. **b** a piece of work for a barrister. **3** instructions given for a task, operation, etc. (orig. a bombing plan given to an aircrew). **4** *RC Ch.* a letter from the pope to a person or community on a matter of discipline. **5** *US* a short account or summary; a synopsis. ● *v.tr.* **1** *Brit. Law* instruct (a barrister) by brief. **2** instruct (an employee, a participant, etc.) in preparation for a task; inform or instruct thoroughly in advance (*briefed him for the interview*) (cf. DEBRIEF). □ **be brief** use few words. **hold a brief for 1** argue in favour of. **2** be retained as counsel for. **in brief** in short. **watching brief 1** a brief held by a barrister following a case for a client not directly involved. **2** a state of interest maintained in a proceeding not directly or immediately concerning one. □ **briefly** *adv.* **briefness** *n.* [ME f. AF *bref*, OF *brief*, f. L *brevis* short]

briefcase /ˈbriːfkeɪs/ *n.* a flat rectangular case for carrying documents etc.

briefing /ˈbriːfɪŋ/ *n.* **1** a meeting for giving information or instructions. **2** the information or instructions given; a brief. **3** the action of informing or instructing.

briefless /ˈbriːflɪs/ *adj. Law* (of a barrister) having no clients.

brier¹ /ˈbraɪə(r)/ *n.* (also **briar**) a prickly bush, esp. of a wild rose. □ **brier-rose** the dog rose, *Rosa canina*. **sweet-brier** a wild rose, *Rosa eglanteria*, with small fragrant leaves and flowers. □ **briery** *adj.* [OE *brǣr, brēr*, of unkn. orig.]

brier² /ˈbraɪə(r)/ *n.* (also **briar**) **1** a white tree heather, *Erica arborea*, native to southern Europe. Also called *tree heath*. **2** a tobacco pipe made from its root. [19th-c. *bruyer* f. F *bruyère* heath]

Brig. *abbr.* Brigadier.

brig[1] /brɪg/ *n.* **1** a two-masted square-rigged ship, with an additional lower fore-and-aft sail on the gaff and a boom to the mainmast. **2** *orig. US sl.* a prison, esp. on a warship. [abbr. of BRIGANTINE]

brig[2] /brɪg/ *n. Sc. & N. Engl.* var. of BRIDGE[1].

brigade /brɪˈgeɪd/ *n. & v.* ● *n.* **1** *Mil.* **a** a subdivision of an army. **b** a British infantry unit consisting usu. of three battalions and forming part of a division. **c** a corresponding armoured unit. **2** an organized or uniformed band of workers (*fire brigade*). **3** *colloq.* a group of people with a characteristic in common (*the couldn't-care-less brigade*). ● *v.tr.* form into a brigade. [F f. It. *brigata* company f. *brigare* be busy with f. *briga* strife]

brigadier /ˌbrɪgəˈdɪə(r)/ *n. Mil.* **1** an officer commanding a brigade. **2 a** a staff officer of similar standing, above a colonel and below a major-general. **b** the titular rank granted to such an officer. □ **brigadier general** an officer ranking next above colonel in the US army, air force, and marine corps. [F (as BRIGADE, -IER)]

brigalow /ˈbrɪgəˌloʊ/ *n. Austral.* an acacia tree, esp. *Acacia harpophylla.* [Aboriginal]

brigand /ˈbrɪgənd/ *n.* a member of a robber band living by pillage and ransom, usu. in wild terrain. □ **brigandage** *n.* **brigandry** *n.* [ME f. OF f. It. *brigante* f. *brigare*: see BRIGADE]

brigantine /ˈbrɪgənˌtiːn/ *n.* a sailing-ship with two masts, the foremast square-rigged, used esp. in the 18th and 19th centuries for short coastal and trading voyages. The name comes from the fact that they were favourite vessels of brigands or pirates. [OF *brigandine* or It. *brigantino* f. *brigante* BRIGAND]

Briggs /brɪgz/, Henry (1561–1630), English mathematician. He was renowned for his work on logarithms, in which he introduced the decimal base, made the thousands of calculations necessary for the tables, and popularized their use. Briggs also devised the usual method used for long division.

Bright /braɪt/, John (1811–89), English Liberal politician and reformer. A noted orator, Bright was the leader, along with Richard Cobden, of the campaign to repeal the Corn Laws. He was also a vociferous opponent of the Crimean War (1854) and was closely identified with the 1867 Reform Act.

bright /braɪt/ *adj., adv., & n.* ● *adj.* **1** emitting or reflecting much light; shining. **2** (of colour) intense, vivid. **3** clever, talented, quick-witted (*a bright idea; a bright child*). **4 a** (of a person) cheerful, vivacious. **b** (of prospects, the future, etc.) promising, hopeful. ● *adv.* esp. *poet.* brightly (*the moon shone bright*). ● *n.* (in *pl.*) **1** bright colours. **2** *N. Amer.* headlights switched to full beam. □ **bright and early** very early in the morning. **bright-eyed and bushy-tailed** *colloq.* alert and sprightly. **the bright lights** the glamour and excitement of the city. **look on the bright side** be optimistic. □ **brightish** *adj.* **brightly** *adv.* **brightness** *n.* [OE *beorht*, (adv.) *beorhte*, f. Gmc]

brighten /ˈbraɪt(ə)n/ *v.tr. & intr.* (often foll. by *up*) **1** make or become brighter. **2** make or become more cheerful or hopeful.

Brighton /ˈbraɪt(ə)n/ a resort town on the south coast of England, in East Sussex; pop. (1991) 133,400. It was patronized by the Prince of Wales (later George IV) from *c.*1780 to 1827, and is noted for its Regency architecture.

Bright's disease *n.* = NEPHRITIS. [Richard *Bright*, English physician (1759–1858), who established its nature]

Brigid, St see BRIDGET, ST[1].

brill[1] /brɪl/ *n.* a European flatfish, *Scophthalmus rhombus*, resembling a turbot. [15th c.: orig. unkn.]

brill[2] /brɪl/ *adj. colloq.* = BRILLIANT *adj.* 4. [abbr.]

brilliance /ˈbrɪlɪəns/ *n.* (also **brilliancy** /-ənsɪ/) **1** great brightness; sparkling or radiant quality. **2** outstanding talent or intelligence.

brilliant /ˈbrɪlɪənt/ *adj. & n.* ● *adj.* **1** very bright; sparkling. **2** outstandingly talented or intelligent. **3** showy; outwardly impressive. **4** *colloq.* excellent, superb. ● *n.* a diamond of the finest cut with many facets. □ **brilliantly** *adv.* [F *brillant* part. of *briller* shine f. It. *brillare*, of unkn. orig.]

brilliantine /ˈbrɪlɪənˌtiːn/ *n.* **1** an oily liquid dressing for making the hair glossy. **2** *US* a lustrous dress fabric. [F *brillantine* (as BRILLIANT)]

brim /brɪm/ *n. & v.* ● *n.* **1** the edge or lip of a cup or other vessel, or of a hollow. **2** the projecting edge of a hat. ● *v.tr. & intr.* (**brimmed**, **brimming**) fill or be full to the brim. □ **brim over** overflow. □ **brimless** *adj.* **brimmed** *adj.* (usu. in *comb.*). [ME *brimme*, of unkn. orig.]

brim-full /brɪmˈfʊl/ *adj.* (also **brimful**) (often foll. by *of*) filled to the brim.

brimstone /ˈbrɪmstəʊn/ *n.* **1** *archaic* the element sulphur. **2** a butterfly, *Gonepteryx rhamni*, or moth, *Opisthograptis luteolata*, having yellow wings. [ME prob. f. OE *bryne* burning + STONE]

brindled /ˈbrɪnd(ə)ld/ *adj.* (also **brindle**) brownish or tawny with streaks of other colour (esp. of domestic animals). [earlier *brinded, brended* f. *brend*, perh. of Scand. orig.]

Brindley /ˈbrɪndlɪ/, James (1716–72), pioneer British canal builder. He began with the Bridgwater canal near Manchester, which included an aqueduct that was a wonder of the age. He designed some 600 km (375 miles) of waterway, connecting most of the major rivers of England. Brindley believed in building contour canals with the minimum of locks, embankments, cuttings, or tunnels, at the expense of greater lengths, and such canals have proved to be the longest to survive.

brine /braɪn/ *n. & v.* ● *n.* **1** water saturated or strongly impregnated with salt. **2** sea water. ● *v.tr.* soak in or saturate with brine. □ **brine shrimp** a small crustacean of the genus *Artemia*, inhabiting salt lakes, salt-pans, etc., and used by aquarists as fish food. [OE *brīne*, of unkn. orig.]

bring /brɪŋ/ *v.tr.* (*past* and *past part.* **brought** /brɔːt/) **1 a** come conveying esp. by carrying or leading. **b** come with. **2** cause to come or be present (*what brings you here?*). **3** cause or result in (*war brings misery*). **4** be sold for; produce as income. **5 a** prefer (a charge). **b** initiate (legal action). **6** cause to become or to reach a particular state (*brings me alive; brought them to their senses; cannot bring myself to agree*). **7** adduce (evidence, an argument, etc.). □ **bring about 1** cause to happen. **2** turn (a ship) around. **bring-and-buy sale** *Brit.* a charity sale at which participants bring items for sale and buy what is brought by others. **bring back** call to mind. **bring down 1** cause to fall. **2** lower (a price). **3** *sl.* make unhappy or less happy. **4** *colloq.* damage the reputation of; demean. **bring forth 1** produce, emit, cause. **2** *archaic* give birth to. **bring forward 1** move to an earlier date or time. **2** transfer from the previous page or account. **3** draw attention to; adduce. **bring home to** cause to realize fully (*brought home to me that I was wrong*). **bring the house down** receive rapturous applause. **bring in 1** introduce (legislation, a custom, fashion, topic, etc.). **2** yield as income or profit. **bring into play** cause to operate; activate. **bring low** overcome, humiliate. **bring off** achieve successfully. **bring on 1** cause to happen or appear. **2** accelerate the progress of. **bring out 1** emphasize; make evident. **2** publish. **bring over** convert to one's own side. **bring round 1** restore to consciousness. **2** persuade. **bring through** aid (a person) through adversity, esp. illness. **bring to 1** restore to consciousness (*brought him to*). **2** check the motion of. **bring to bear** (usu. foll. by *on*) direct and concentrate (forces). **bring to mind** recall; cause one to remember. **bring to pass** cause to happen. **bring under** subdue. **bring up 1** rear (a child). **2** vomit, regurgitate. **3** call attention to; broach. **4** (*absol.*) stop suddenly. **bring upon oneself** be responsible for (something one suffers). □ **bringer** *n.* [OE *bringan* f. Gmc]

brinjal /ˈbrɪndʒəl/ *n.* (in India and Africa) an aubergine. [ult. Port. *berinjela* formed as AUBERGINE]

Brink /brɪŋk/, André (b.1935), South African novelist, short-story writer, and dramatist. Brink, who writes in Afrikaans and translates his work into English, gained international recognition with his seventh novel *Looking on Darkness* (1973), an open criticism of apartheid which became the first novel in Afrikaans to be banned by the South African government. Subsequent novels include *A Dry White Season* (1979) and *A Chain of Voices* (1982).

brink /brɪŋk/ *n.* **1** the extreme edge of land before a precipice, river, etc., esp. when a sudden drop follows. **2** the furthest point before something dangerous or exciting is discovered. □ **on the brink of** about to experience or suffer; in imminent danger of. [ME f. ON: orig. unkn.]

brinkmanship /ˈbrɪŋkmənˌʃɪp/ *n.* the art or policy of pursuing a dangerous course to the brink of catastrophe before desisting, esp. in politics.

briny /ˈbraɪnɪ/ *adj. & n.* ● *adj.* of brine or the sea; salty. ● *n.* (prec. by *the*) *Brit. sl.* the sea. □ **brininess** *n.*

brio /ˈbriːəʊ/ *n.* dash, vigour, vivacity. [It.]

brioche /ˈbriːɒʃ/ *n.* a small rounded sweet roll made with a light yeast dough. [F]

briquette /brɪˈket/ *n.* (also **briquet**) a block of compressed coal dust used as fuel. [F *briquette*, dimin. of *brique* brick]

Brisbane /ˈbrɪzbən/ the capital of Queensland, Australia; pop. (1990)

1,273,500. Founded in 1824 as a penal colony, it is named after the Scottish soldier and patron of astronomy Sir Thomas Brisbane (1773–1860), governor of New South Wales from 1821 to 1825.

brisk /brɪsk/ *adj. & v.* ● *adj.* **1** quick, lively, keen (*a brisk pace; brisk trade*). **2** enlivening, fresh, keen (*a brisk wind*). **3** curt, peremptory (*a brisk manner*). ● *v.tr. & intr.* (often foll. by *up*) make or grow brisk. □ **brisken** *v.tr. & intr.* **briskly** *adv.* **briskness** *n.* [prob. F *brusque* BRUSQUE]

brisket /ˈbrɪskɪt/ *n.* an animal's breast, esp. as a joint of meat. [AF f. OF *bruschet*, perh. f. ON]

brisling /ˈbrɪzlɪŋ, ˈbrɪslɪŋ/ *n.* (*pl.* same or **brislings**) a small herring or sprat. [Norw. & Danish, = sprat]

bristle /ˈbrɪs(ə)l/ *n. & v.* ● *n.* **1** a short stiff hair, esp. one of those on an animal's back. **2** such hairs, or a man-made substitute, used in clumps to make a brush. ● *v.* **1** *a intr.* (of the hair) stand upright, esp. in anger or pride. **b** *tr.* make (the hair) do this. **2** *intr.* show irritation or defensiveness. **3** *intr.* (usu. foll. by *with*) be covered or abundant (in) (*bristling with weapons*). [ME *bristel, brestel* f. OE *byrst*]

bristlecone pine /ˈbrɪs(ə)lˌkəʊn/ *n.* a very long-lived shrubby pine, *Pinus aristata*, of western North America, used in dendrochronology.

bristletail /ˈbrɪs(ə)lˌteɪl/ *n.* a small primitive wingless insect of the order Thysanura or Diplura, with three or two terminal bristles respectively, e.g. the silverfish.

bristly /ˈbrɪslɪ/ *adj.* (**bristlier, bristliest**) full of bristles; rough, prickly.

Bristol /ˈbrɪst(ə)l/ a city in SW England, the county town of Avon; pop. (1991) 370,300. Situated on the River Avon about 10 km (6 miles) from the Bristol Channel, it has been a leading port since the 12th century.

Bristol board *n.* a kind of fine smooth pasteboard for drawing on.

Bristol Channel a wide inlet of the Atlantic between South Wales and the south-western peninsula of England, narrowing into the estuary of the River Severn.

Bristol fashion *n.* (functioning as *predic.adj.*) (in full **shipshape and Bristol fashion**) orig. *Naut.* with all in good order.

bristols /ˈbrɪst(ə)lz/ *n.pl. Brit. sl.* a woman's breasts. [rhyming sl. f. *Bristol cities* = titties]

Brit /brɪt/ *n. colloq.* a British person. [abbr.]

Brit. *abbr.* **1** British. **2** Britain.

Britain /ˈbrɪt(ə)n/ the island containing England, Wales, and Scotland, and including the small adjacent islands (see also GREAT BRITAIN). The name is broadly synonymous with Great Britain, but the longer form is more usual for the political unit. After the Old English period the name was only a historical term until the mid-16th century, when it was used politically in connection with the efforts made to unite England and Scotland. In 1604 James I was proclaimed 'King of Great Britain', and this name was adopted for the United Kingdom at the Union of the English and Scottish Parliaments in 1707. For a time after this the terms South Britain and North Britain, for England and Scotland respectively, were often used in Acts of Parliament. [13th c. *Bretayne* f. OF *Bretaigne* f. L *Britannia* (OE *Breoton* and variants)]

Britain, Battle of a series of air battles fought over Britain (Aug.–Oct. 1940), in which the RAF successfully resisted raids by the numerically superior German air force. The heavy losses sustained and the failure to establish control in the air led Hitler to abandon plans to invade Britain, although the Germans continued to bomb British cities by night for several months afterwards.

Britannia /brɪˈtænjə/ the personification of Britain, esp. as a helmeted woman with shield and trident. Such a figure had appeared on Roman coins of the time of Hadrian or earlier; she was revived with the name Britannia on the coinage of Charles II, one of whose mistresses, the Duchess of Richmond, is said to have been the model for the figure. Since 1682 the Royal Navy has always had a ship of this name, the latest being the royal yacht (1900). [L f. Gk *Brettania* f. *Brettanoi* Britons]

Britannia metal *n.* a silvery alloy of tin, antimony, and copper.

Britannia silver *n.* silver that is at least 95.8 per cent pure.

Britannic /brɪˈtænɪk/ *adj.* (esp. in **His** (or **Her**) **Britannic Majesty**) of Britain. [L *Britannicus* (as BRITANNIA)]

Briticism /ˈbrɪtɪˌsɪz(ə)m/ *n.* (also **Britishism** /-ˌʃɪz(ə)m/) an idiom used in Britain but not in other English-speaking countries. [BRITISH, after GALLICISM]

British /ˈbrɪtɪʃ/ *adj. & n.* ● *adj.* **1** of or relating to Great Britain or the United Kingdom, or to its people or language. **2** of the Commonwealth or (formerly) the British Empire (*British subject*). ● *n.* (prec. by *the*; treated as *pl.*) the British people. □ **British English** English as used in Great

Britain, as distinct from that used elsewhere. **British thermal unit** the amount of heat needed to raise 1 lb of water at maximum density by one degree Fahrenheit, equivalent to 1.055×10^3 joules. □ **Britishness** *n.* [OE *Brettisc* etc. f. Bret f. L *Britto* or OCelt.]

British Academy an institution founded in 1901 for the promotion of historical, philosophical, and philological studies.

British Antarctic Territory that part of Antarctica claimed by Britain. Designated in 1962 from territory that was formerly part of the Falkland Islands Dependencies, it includes some 388,500 sq. km (150,058 sq. miles) of the continent of Antarctica as well as the South Orkney Islands and South Shetland Islands in the South Atlantic.

British Broadcasting Corporation (abbr. **BBC**) a public corporation having the monopoly of radio and television broadcasting in Britain until the introduction of the first commercial TV station in 1954. The BBC was established in 1927 by royal charter to carry on work previously performed by the private British Broadcasting Company. It is financed by the sale of television viewing licences rather than by revenue from advertising and has an obligation to remain impartial in its reporting. An influential figure was the first director-general, John Reith, under whom the world's first regular television service began in 1936 and the corporation's ethos of public-service broadcasting was instituted. Today the BBC operates two television channels (BBC1 and BBC2) and a radio network within the UK, and also runs the World Service, as well as exporting drama, documentary, and comedy programmes around the world.

British Columbia a province on the west coast of Canada; pop. (1991) 3,218,500; capital, Victoria. Formed in 1866 by the union of Vancouver Island (a former British colony) and the mainland area, then called New Caledonia, the province includes the Queen Charlotte Islands.

British Commonwealth see COMMONWEALTH, THE 1.

British Council an organization established in 1934 with the aims of promoting a wider knowledge of Britain and the English language abroad and of developing closer cultural relations with other countries. Most of its funds are provided by Parliament.

British Empire British overseas possessions, from the late 17th to the mid-20th century, acquired for commercial, strategic, and territorial reasons. The colonization of North America started in the early 17th century, although the colonies south of Canada were lost in the War of American Independence. British domination of India began under the auspices of the East India Company, also in the 17th century, while a series of small colonies, mostly in the West Indies, were gained in colonial wars with France during the late 17th–early 19th centuries. Australia, New Zealand, and various possessions in the Far East (notably Hong Kong) were added in the 19th century, while large areas of Africa came under British control in the last few decades of the century, at the height of the imperialist age. The movement of the British colonies towards independence began in the mid-19th century with the granting of self-government to Canada, Australia, New Zealand, and South Africa. This trend was accelerated by the two world wars, with most of the remaining colonies gaining independence in the decade and a half following the end of the Second World War.

Britisher /ˈbrɪtɪʃə(r)/ *n.* a British subject, esp. of British descent. ¶ Not used in British English.

British Expeditionary Force (abbr. **BEF**) any of the British forces made available by the army reform of 1908 for service overseas against foreign countries. Such a force was sent to France in 1914 at the outbreak of the First World War. The force sent to France early in the Second World War (1939) was evacuated from Dunkirk in 1940.

British Honduras the former name (until 1973) for BELIZE.

British India that part of the Indian subcontinent administered by the British from 1765, when the East India Company acquired control over Bengal, until 1947, when India became independent and Pakistan was created. By 1850 British India was coterminous with India's boundaries in the west and north and by 1885 it included Burma in the east. The period of British rule was known as the Raj. See also INDIA.

British Indian Ocean Territory a British dependency in the Indian Ocean, comprising the islands of the Chagos Archipelago and (until 1976) some other groups now belonging to the Seychelles. Ceded to Britain by France in 1814, the islands were administered from Mauritius until the designation of a separate dependency in 1965. There are no permanent inhabitants, but British and US naval personnel occupy the island of Diego Garcia.

British Isles a group of islands lying off the coast of NW Europe, from which they are separated by the North Sea and the English Channel. They include Britain, Ireland, the Isle of Man, the Hebrides, the Orkney Islands, the Shetland Islands, the Scilly Isles, and the Channel Islands.

Britishism var. of BRITICISM.

British Legion see ROYAL BRITISH LEGION.

British Library the national library of Britain, containing the former library departments of the British Museum. As the principal copyright library it receives a copy of every book, periodical, etc., published in the UK. It was established separately from the British Museum in 1972. The library is in the process of moving from dispersed sites around London to a new building in St Pancras.

British Lions see LIONS.

British Museum a national museum of antiquities etc. in Bloomsbury, London. It was established with public funds in 1753 to house the library and collections of Sir Hans Sloane and manuscripts from other sources, and now has one of the world's finest collections, particularly of Egyptian, Assyrian, Greek, Roman, and oriental antiquities. Among its holdings are Magna Carta, the Elgin Marbles, and the Rosetta Stone. During the 19th century the natural history collections grew so extensively that in 1881 they were moved to a separate building in South Kensington; in 1963 the Natural History Museum was made completely independent. The library departments, which had been enriched by George II's donation of the royal library and the purchase of George IV's books, were transferred in 1972 to the British Library.

British National Party (abbr. **BNP**) an extreme right-wing political party in Britain supporting racial discrimination and strongly opposing immigration. The party arose in the 1980s as a breakaway group from the National Front, and in 1993 became the first openly racist party to win a seat in a British council election.

British Somaliland /səˈmɑːlɪˌlænd/ a former British protectorate established on the Somali coast of East Africa in 1884. In 1960 it united with former Italian territory to form the independent republic of Somalia.

British Summer Time (abbr. **BST**) time as advanced one hour ahead of Greenwich Mean Time for daylight saving between March and October.

British Virgin Islands see VIRGIN ISLANDS.

Briton /ˈbrɪt(ə)n/ n. **1** one of the people of southern Britain before the Roman conquest. **2** a native or national of Great Britain or (formerly) of the British Empire. [ME & OF Breton f. L Britto -onis f. OCelt.]

Brittany /ˈbrɪtənɪ/ (French **Bretagne** /brəˈtaɲ/) a region and former duchy of NW France, forming a peninsula between the Bay of Biscay and the English Channel. It was occupied in the 5th and 6th centuries by Britons (known as Bretons) fleeing the Saxon invasions of Britain. Brittany was an independent duchy from 1196 until 1532, when it was incorporated into France.

Britten /ˈbrɪt(ə)n/, (Edward) Benjamin, Lord Britten of Aldeburgh (1913–76), English composer, pianist, and conductor. Chiefly known for his operas, he made settings of the work of a varied range of writers, including George Crabbe (*Peter Grimes*, 1945), Shakespeare (*A Midsummer Night's Dream*, 1960), and Thomas Mann (*Death in Venice*, 1973). His many choral works include the *War Requiem* (1962), based on Wilfred Owen's war poems. In 1948, with the tenor Peter Pears (1910–86), he founded the Aldeburgh Festival, which became one of Britain's major music festivals. He was made a life peer in 1976, the first composer to be so honoured.

brittle /ˈbrɪt(ə)l/ adj. & n. ● adj. **1** hard and fragile; apt to break. **2** frail, weak; unstable. ● n. a brittle sweet made from nuts and set melted sugar. □ **brittle-bone disease** = OSTEOPOROSIS. **brittle-star** an echinoderm of the class Ophiuroidea, with long flexible arms radiating from a small central body. □ **brittlely** adv. **brittleness** n. [ME ult. f. a Gmc root rel. to OE brēotan break up]

Brittonic /brɪˈtɒnɪk/ adj. & n. = BRYTHONIC. [L Britto -onis Briton]

Brno /ˈbɜːnəʊ/ an industrial city in the Czech Republic; pop. (1991) 388,000. It is the capital of Moravia.

bro. abbr. brother.

broach /brəʊtʃ/ v. & n. ● v.tr. **1** raise (a subject) for discussion. **2** pierce (a cask) to draw liquor. **3** open and start using the contents of (a box, bale, bottle, etc.). **4** begin drawing (liquor). ● n. **1** a bit for boring. **2** a roasting-spit. □ **broach spire** an octagonal church spire rising from a square tower without a parapet. [ME f. OF broche (n.), brocher (v.) ult. f. L brocc(h)us projecting]

broad /brɔːd/ adj. & n. ● adj. **1** large in extent from one side to the other; wide. **2** (following a measurement) in breadth (*2 metres broad*).

3 spacious or extensive (*broad acres; a broad plain*). **4** full and clear (*broad daylight*). **5** explicit, unmistakable (*broad hint*). **6** general; not taking account of detail (*broad intentions; a broad inquiry; in the broadest sense of the word*). **7** (of speech) markedly regional (*broad Scots*). **8** chief or principal (*the broad facts*). **9** tolerant, liberal; widely inclusive (*take a broad view*). **10** somewhat coarse (*broad humour*). ● n. **1** the broad part of something (*broad of the back*). **2** N. Amer. sl. a woman. □ **broad arrow** see ARROW. **broad bean 1** a kind of bean, Vicia faba, with pods containing large edible flat seeds. **2** one of these seeds. **broad-brush** as if painted with a broad brush; general; lacking in detail. **broad gauge** a railway track with a gauge wider than the standard one. **broad pennant** a short swallow-tailed pennant distinguishing the commodore's ship in a squadron. **broad spectrum** (of a medicinal substance) effective against a large variety of micro-organisms. □ **broadness** n. **broadways** adv. **broadwise** adv. [OE brād f. Gmc]

broadcast /ˈbrɔːdkɑːst/ v., n., adj., & adv. ● v. (past **broadcast** or **broadcasted**; past part. **broadcast**) **1** tr. a transmit (programmes or information) by radio or television. **b** disseminate (information) widely. **2** intr. undertake or take part in a radio or television transmission. **3** tr. a scatter (seed etc.) over a large area, esp. by hand. ● n. a radio or television programme or transmission. ● adj. **1** transmitted by radio or television. **2 a** scattered widely. **b** (of information etc.) widely disseminated. ● adv. over a large area. □ **broadcaster** n. **broadcasting** n. [BROAD + CAST v.]

Broad Church n. **1** a group within the Anglican Church favouring a liberal interpretation of doctrine. **2** any group allowing its members a wide range of opinion.

broadcloth /ˈbrɔːdklɒθ/ n. a fine cloth of wool, cotton, or silk. [orig. with ref. to width and quality]

broaden /ˈbrɔːd(ə)n/ v.tr. & intr. make or become broader.

broadleaved /ˈbrɔːdliːvd/ adj. (of a tree) having relatively broad flat leaves rather than needles; non-coniferous.

broadloom /ˈbrɔːdluːm/ adj. (esp. of carpet) woven in broad widths.

broadly /ˈbrɔːdlɪ/ adv. in a broad manner; widely (*grinned broadly*). □ **broadly speaking** disregarding minor exceptions.

broad-minded /brɔːdˈmaɪndɪd/ adj. tolerant or liberal in one's views. □ **broad-mindedly** adv. **broad-mindedness** n.

Broadmoor /ˈbrɔːdmʊə(r), -mɔː(r)/ a special hospital in Berkshire, England, for the secure holding of patients regarded as both mentally ill and potentially dangerous. The hospital was established in 1863.

Broads, the a network of shallow freshwater lakes, traversed by slow-moving rivers, in an area of Norfolk and Suffolk, England. They were formed by the gradual natural flooding of medieval peat diggings.

broadsheet /ˈbrɔːdʃiːt/ n. **1** a large sheet of paper printed on one side only, esp. with information. **2** a newspaper with a large format.

broadside /ˈbrɔːdsaɪd/ n. **1** the firing of all guns from one side of a ship. **2** a vigorous verbal onslaught. **3** the side of a ship above the water between the bow and quarter. □ **broadside on** sideways on.

broadsword /ˈbrɔːdsɔːd/ n. a sword with a broad blade, for cutting rather than thrusting.

broadtail /ˈbrɔːdteɪl/ n. **1** the karakul sheep. **2** the fleece or wool from its lamb.

Broadway /ˈbrɔːdweɪ/ a street traversing the length of Manhattan, New York. It is famous for its theatres, and its name has become synonymous with show business. It is also known as the Great White Way, in reference to its brilliant street illuminations.

broadway /ˈbrɔːdweɪ/ n. a large open or main road.

brocade /brəˈkeɪd, brəʊˈkeɪd/ n. & v. ● n. a rich fabric with a silky finish woven with a raised pattern, and often with gold or silver thread. ● v.tr. weave with this design. [Sp. & Port. brocado f. It. broccato f. brocco twisted thread]

broccoli /ˈbrɒkəlɪ/ n. **1** a variety of cabbage, similar to the cauliflower, with a loose cluster of greenish flower buds. **2** the flower-stalk and head used as a vegetable. [It., pl. of broccolo dimin. of brocco sprout]

broch /brɒk, brɒx/ n. (in Scotland) a prehistoric circular stone tower. [ON borg castle]

brochette /brɒˈʃet/ n. a skewer on which chunks of meat are cooked, esp. over an open fire. [F, dimin. of broche BROACH]

brochure /ˈbrəʊʃə(r), brəʊˈʃjʊə(r)/ n. a pamphlet or leaflet, esp. one giving descriptive information. [F, lit. 'stitching', f. brocher stitch]

brock /brɒk/ n. (esp. in rural use) a badger. [OE broc(c), of Celt. orig.: cf. Welsh broch, Ir., Gael. broc]

Brocken /ˈbrɒkən/ a mountain in the Harz Mountains of north central Germany, rising to 1,143 m (3,747 ft). It is noted for the phenomenon of the Brocken spectre and for witches' revels which reputedly took place there on Walpurgis night.

Brocken spectre *n.* a magnified shadow of the spectator, often encircled by rainbow-like bands, thrown on a bank of cloud in high mountains when the sun is low. This phenomenon was first observed on the Brocken.

brocket /'brɒkɪt/ *n.* a small deer of the genus *Mazama*, native to Central and South America, having short straight antlers. [ME f. AF *broque* (= *broche* BROACH)]

broderie anglaise /ˌbrəʊdərɪ ɒŋˈgleɪz/ *n.* open embroidery on white linen or cambric, esp. in floral patterns. [F, = English embroidery]

Brodsky /'brɒdskɪ/, Joseph (born Iosif Aleksandrovich Brodsky) (1940–96), Russian-born American poet. He wrote both in Russian and in English, and his poetry was preoccupied with themes of loss and exile. Brodsky was most famous for his collection *The End of a Beautiful Era* (1977). He was awarded the Nobel Prize for literature in 1987.

brogue[1] /brəʊg/ *n.* **1** a strong outdoor shoe with ornamental perforated bands. **2** a rough shoe of untanned leather. [Gael. & Ir. *bróg* f. ON *brók*]

brogue[2] /brəʊg/ *n.* a marked accent, esp. Irish. [18th c.: orig. unkn.: perh. allusively f. BROGUE[1]]

broil[1] /brɔɪl/ *v.* esp. US **1** *tr.* cook (meat) on a rack or a gridiron. **2** *tr. & intr.* make or become very hot, esp. from the sun. [ME f. OF *bruler* burn f. Rmc]

broil[2] /brɔɪl/ *n.* a row; a tumult. [obs. *broil* to muddle: cf. EMBROIL]

broiler /'brɔɪlə(r)/ *n.* **1** a young chicken raised for broiling or roasting. **2** a gridiron etc. for broiling. **3** *colloq.* a very hot day. □ **broiler house** a building for rearing broiler chickens in close confinement.

broke /brəʊk/ *past* of BREAK[1]. ● *predic.adj. colloq.* having no money; financially ruined. □ **go for broke** *sl.* risk everything in an all-out effort. [(adj.) archaic past part. of BREAK[1]]

broken /'brəʊkən/ *past part.* of BREAK[1]. ● *adj.* **1** that has been broken; out of order. **2** (of a person) reduced to despair; defeated. **3** (of a language or of speech) spoken falteringly and with many mistakes, as by a foreigner (*broken English*). **4** disturbed, interrupted (*broken time*). **5** uneven (*broken ground*). **6** (of an animal) trained to obey; tamed. **7** transgressed, not observed (*broken rules*). □ **broken chord** *Mus.* a chord in which the notes are played successively. **broken-down 1** worn out by age, use, or ill-treatment. **2** not functioning. **broken-hearted** overwhelmed with sorrow or grief. **broken-heartedness** grief. **broken home** a family in which the parents are divorced or separated. **broken reed** a person who has become unreliable or ineffective. **broken wind** a chronic disabling condition of a horse caused by ruptured air-cells in the lungs. **broken-winded** *adj.* suffering from this. □ **brokenly** *adv.* **brokenness** /-kənnɪs/ *n.*

Broken Hill 1 a town in New South Wales, Australia; pop. (est. 1987) 24,170. It is a centre of lead, silver, and zinc mining. **2** the former name (1904–65) for KABWE.

broker /'brəʊkə(r)/ *n.* **1** an agent who buys and sells for others; a middleman. **2** a member of the Stock Exchange dealing in stocks and shares. ¶ In the UK from Oct. 1986 officially called **broker-dealer** and entitled to act as agent and principal in share dealings. **3** *Brit.* an official appointed to sell or appraise distrained goods. [ME f. AF *brocour*, of unkn. orig.]

brokerage /'brəʊkərɪdʒ/ *n.* **1** the action or service of a broker. **2** a company providing such a service. **3** a broker's fee or commission.

broking /'brəʊkɪŋ/ *n.* the trade or business of a broker.

brolga /'brɒlgə/ *n. Austral.* a large Australian crane, *Grus rubicunda*, with a booming call. [Aboriginal]

brolly /'brɒlɪ/ *n.* (*pl.* **-ies**) *Brit.* **1** *colloq.* an umbrella. **2** *sl.* a parachute. [abbr.]

Bromberg /'brɒmbɛrk/ the German name for BYDGOSZCZ.

brome /brəʊm/ *n.* an oatlike grass of the genus *Bromus*, having slender stems with flowering spikes. [mod.L *Bromus* f. Gk *bromos* oat]

bromelia /brəʊˈmiːlɪə/ *n.* (also **bromeliad** /-lɪəd/) a New World plant of the family Bromeliaceae, esp. of the genus *Bromelia*, having short stems with rosettes of stiff usu. spiny leaves, e.g. pineapple. [O. *Bromel*, Swedish botanist (1639–1705)]

bromic /'brəʊmɪk/ *adj. Chem.* of or containing bromine. □ **bromic acid** a strong acid (chem. formula: $HBrO_3$) used as an oxidizing agent. □ **bromate** *n.*

bromide /'brəʊmaɪd/ *n.* **1** *Chem.* a compound of bromine with another

element or group. **2** *Pharm.* a preparation of usu. potassium bromide, used as a sedative. **3** a trite remark. **4** *Printing* a reproduction or proof on bromide paper. □ **bromide paper** photographic printing paper coated with silver bromide emulsion.

bromine /'brəʊmiːn/ *n.* a dark red fuming liquid chemical element (atomic number 35; symbol **Br**). One of the halogens, bromine was discovered by the French chemist Antoine Balard in 1826. It is toxic, with a choking irritating smell. Bromine occurs chiefly as salts in sea water and brines; bromine compounds have many uses in medicine, photography, and chemical industry. [F *brome* f. Gk *brōmos* stink]

bromism /'brəʊmɪz(ə)m/ *n. Med.* a condition of torpor and weakness due to excessive intake of bromides.

bromo- /'brəʊməʊ/ *comb. form Chem.* bromine.

bronc /brɒŋk/ *n. US colloq.* = BRONCO. [abbr.]

bronchi *pl.* of BRONCHUS.

bronchial /'brɒŋkɪəl/ *adj.* of or relating to the bronchi or bronchioles. □ **bronchial tree** the branching system of bronchi and bronchioles conducting air from the windpipe into the lungs.

bronchiole /'brɒŋkɪəʊl/ *n.* any of the minute divisions of a bronchus. □ **bronchiolar** /ˌbrɒŋkɪˈəʊlə(r)/ *adj.*

bronchitis /brɒŋˈkaɪtɪs/ *n.* inflammation of the mucous membrane in the bronchial tubes. □ **bronchitic** /-ˈkɪtɪk/ *adj. & n.*

broncho- /'brɒŋkəʊ/ *comb. form* bronchi.

bronchocele /'brɒŋkəˌsiːl/ *n.* a goitre.

bronchodilator /ˌbrɒŋkəʊdaɪˈleɪtə(r)/ *n.* a substance which causes widening of the bronchi, used esp. to alleviate asthma.

bronchopneumonia /ˌbrɒŋkəʊnjuːˈməʊnɪə/ *n.* inflammation of the lungs, arising in the bronchi or bronchioles.

bronchoscope /'brɒŋkəˌskəʊp/ *n.* a usu. fibre-optic instrument for inspecting the bronchi. □ **bronchoscopy** /brɒŋˈkɒskəpɪ/ *n.*

bronchus /'brɒŋkəs/ *n.* (*pl.* **bronchi** /-kaɪ/) any of the major air passages of the lungs, esp. either of the two main divisions of the windpipe. [LL f. Gk *brogkhos* windpipe]

bronco /'brɒŋkəʊ/ *n.* (*pl.* **-os**) a wild or half-tamed horse of the western US. □ **bronco-buster** *US sl.* a person who breaks in horses. [Sp., = rough]

Brontë /'brɒnteɪ/, Charlotte (1816–55), Emily (1818–48), and Anne (1820–49), English novelists. Motherless and largely educated at home, the three sisters led a lonely childhood in the village of Haworth in a remote part of Yorkshire. Apart from work as governesses and, for Emily and Charlotte, a visit to Brussels, their experience of the outside world was unusually limited. All died young, Emily of tuberculosis after the publication (but before the success) of her masterpiece *Wuthering Heights* (1847). Anne also died of tuberculosis, after publishing *Agnes Grey* (1845) and *The Tenant of Wildfell Hall* (1847). Charlotte died during pregnancy, when she was already famous for her romantic tour-de-force *Jane Eyre* (1847) and for *Shirley* (1849) and *Villette* (1853). Their works were published under the pseudonyms Currer, Ellis, and Acton Bell. Of their early poetry, collected in *Poems by Currer, Ellis and Acton Bell* (1846), Emily's has been the most highly regarded.

brontosaurus /ˌbrɒntəˈsɔːrəs/ *n.* (also **brontosaur** /'brɒntəˌsɔː(r)/) = APATOSAURUS. [mod.L f. Gk *brontē* thunder + *sauros* lizard]

Bronx, the /brɒŋks/ a borough in the north-east of New York City. It is named after a Dutch settler, Jonas Bronck, who purchased land there in 1641.

bronze /brɒnz/ *n., adj., & v.* ● *n.* **1** an alloy of copper with up to one-third tin. (See *note below.*) **2** its brownish colour. **3** a thing made of bronze, esp. as a work of art. ● *adj.* made of or coloured like bronze. ● *v.* **1** *tr.* give a surface of bronze or resembling bronze to. **2** *tr. & intr.* make or become brown; tan. □ **bronze medal** a medal usu. awarded to a competitor who comes third (esp. in sport). □ **bronzy** *adj.* [F f. It. *bronzo*, prob. f. Pers. *birinj* brass]

▪ Bronze was first smelted in the Near East, the Aegean, and the Balkans in the late 4th and early 3rd millennium BC. It is harder than pure copper and therefore superior for making weapons and tools. Until the introduction of iron, bronze remained the sole metal for utilitarian purposes, and afterwards it continued in general use to the end of antiquity for sculpture, many domestic objects, and (after the 5th century BC) for small-denomination coins. In more recent times it has continued to be used for coinage (the bronze coin was introduced into Britain in 1860); most so-called 'copper' coins are in fact bronze.

Bronze Age n. a period of prehistory when certain weapons and tools came to be made of bronze rather than stone, following the Stone Age and preceding the Iron Age (see PREHISTORY). It began in the Near East and SE Europe in the late 4th and early 3rd millennium BC and is associated with the first European civilizations, the spread of the wheel, and the establishment of far-reaching trade networks. It is equated with the beginnings of urban life in China (beginning c.2000 BC), but in America developed only in the final stages of some of the Meso-American civilizations (AD c.1000). In NW Europe the Bronze Age was unaccompanied by developments in civilized life but merely followed on from the Stone Age; in Africa and Australasia it did not appear at all. It ended in most areas with the general use of iron technology, in the 8th century BC in northern Europe and in the 5th century AD in China; in Greece and other Aegean countries it ended c.1200 BC with the start of a dark age.

brooch /brəʊtʃ/ n. an ornament fastened to clothing with a hinged pin. [ME broche = BROACH n.]

brood /bruːd/ n. & v. ● n. **1** the young of an animal (esp. a bird) produced at one hatching or birth. **2** colloq. the children in a family. **3** a group of related things. **4** bee or wasp larvae. **5** (attrib.) kept for breeding (brood-mare). ● v. **1** intr. (often foll. by on, over, etc.) worry or ponder (esp. resentfully). **2 a** intr. sit as a hen on eggs to hatch them. **b** tr. sit on (eggs) to hatch them. **3** intr. (usu. foll. by over) (of silence, a storm, etc.) hang or hover closely. □ **broodingly** adv. [OE brōd f. Gmc]

brooder /'bruːdə(r)/ n. **1** a heated house for chicks, piglets, etc. **2** a person who broods.

broody /'bruːdɪ/ adj. (**broodier**, **broodiest**) **1** (of a hen) wanting to brood. **2** sullenly thoughtful or depressed. **3** colloq. (of a woman) wanting to have a baby. □ **broodily** adv. **broodiness** n.

Brook /brʊk/, Peter (Stephen Paul) (b.1925), English theatre director. Appointed co-director of the Royal Shakespeare Company in 1962, he earned critical acclaim with King Lear (1963) and A Midsummer Night's Dream (1970). In 1971 he founded the International Centre for Theatre Research in Paris, developing new acting techniques drawn from mime and other cultures.

brook¹ /brʊk/ n. a small stream. □ **brooklet** /-lɪt/ n. [OE brōc, of unkn. orig.]

brook² /brʊk/ v.tr. (usu. with neg.) literary tolerate, allow. [OE brūcan f. Gmc]

Brooke /brʊk/, Rupert (Chawner) (1887–1915), English poet. His works include 'Tiara Tahiti' and other poems, but he is most famous for his wartime poetry 1914 and Other Poems (1915) and for his lighter verse, such as 'The Old Vicarage, Grantchester'. He died of blood-poisoning while on naval service in the Mediterranean.

Brooklands /'brʊkləndz/ a motor-racing circuit near Weybridge in Surrey, England, opened in 1907. During the Second World War the course was converted for aeroplane manufacture. Part of the site is currently being restored as a museum of aviation and motor vehicles.

brooklime /'brʊklaɪm/ n. a speedwell, Veronica beccabunga, growing in wet ground. [f. BROOK¹ + lime alt. f. OE hleomoce name of the plant]

Brooklyn /'brʊklɪn/ a borough of New York City, at the south-western corner of Long Island. The Brooklyn Bridge (1869–83) links Long Island with lower Manhattan.

Brookner /'brʊknə(r)/, Anita (b.1928), English novelist and art historian. She has written books on Jacques-Louis David (1981) and Watteau (1968). Her career as a novelist began in 1981 with A Start in Life; three years later she won the Booker Prize for Hotel du Lac (1984). Her novels are characterized by their pervading atmosphere of melancholy and their use of allusion.

brookweed /'brʊkwiːd/ n. a small white-flowered plant, Samolus valerandi, of the primrose family, growing in wet ground.

broom /bruːm/ n. **1** a long-handled brush of bristles, twigs, etc. for sweeping (orig. one made of twigs of broom). **2** a leguminous shrub of the genus Cytisus or Genista, usu. with yellow flowers, esp. C. scoparius. □ **new broom** a newly appointed person eager to make changes. [OE brōm]

broomrape /'bruːmreɪp/ n. a parasitic plant of the genus Orobanche, with tubular flowers on a leafless brown stem, and living on the roots of broom and other plants. [BROOM + L rapum tuber]

broomstick /'bruːmstɪk/ n. the handle of a broom, esp. as allegedly ridden through the air by witches.

Bros. abbr. Brothers (esp. in the name of a firm).

brose /brəʊz/ n. esp. Sc. a dish of oatmeal with boiling water or milk poured on it. [Sc. form of brewis broth: ME f. OF bro(u)ez, ult. f. Gmc]

broth /brɒθ/ n. **1 a** a thin soup of meat or fish stock. **b** unclarified meat or fish stock. **2** Biol. meat stock as a nutrient medium for bacteria. [OE f. Gmc: rel. to BREW]

brothel /'brɒθəl/ n. a house etc. where prostitution takes place. [orig. brothel-house f. ME brothel worthless man, prostitute, f. OE brēothan go to ruin]

brother /'brʌðə(r)/ n. **1** a man or boy in relation to other sons and daughters of his parents. **2 a** (often as a form of address) a close male friend or associate. **b** a male fellow member of a trade union etc. **3** (pl. also **brethren** /'breðrɪn/) **a** a member of a male religious order, esp. a monk. **b** a fellow member of the Christian Church, a religion, or (formerly) a guild etc. **4** a fellow human being. □ **brother german** see GERMAN. **brother-in-law** (pl. **brothers-in-law**) **1** the brother of one's wife or husband. **2** the husband of one's sister. **3** the husband of one's sister-in-law. **brother uterine** see UTERINE 2. □ **brotherless** adj. **brotherly** adj. & adv. **brotherliness** n. [OE brōthor f. Gmc]

brotherhood /'brʌðəˌhʊd/ n. **1 a** the relationship between brothers. **b** brotherly friendliness; companionship. **2 a** an association, society, or community of people linked by a common interest, religion, trade, etc. **b** its members collectively. **3** N. Amer. a trade union. **4** community of feeling between all human beings. [ME, alt. f. brotherred (f. BROTHER as KINDRED) after words in -HOOD, -HEAD]

brougham /'bruːəm, bruːm/ n. hist. **1** a horse-drawn closed carriage with a driver perched outside in front. **2** a motor car with an open driver's seat. [Lord Brougham (1778–1868)]

brought past and past part. of BRING.

brouhaha /'bruːhaːˌhaː/ n. commotion, sensation; hubbub, uproar. [F]

Brouwer /'braʊə(r)/, Adriaen (c.1605–38), Flemish painter. He was based in Haarlem, where he probably studied with Frans Hals; he provides an important link between Dutch and Flemish genre painting. His most typical works represent peasant scenes in taverns; they are characterized by a delicate use of colour, which contrasts with the coarseness of the subject-matter.

brow /braʊ/ n. **1** the forehead. **2** (usu. in pl.) an eyebrow. **3** the summit of a hill or pass. **4** the edge of a cliff etc. **5** colloq. intellectual level. □ **browed** adj. (in comb.). [OE brū f. Gmc]

browbeat /'braʊbiːt/ v.tr. (past **-beat**; past part. **-beaten** /-ˌbiːt(ə)n/) intimidate with stern looks and words. □ **browbeater** n.

Brown¹ /braʊn/, Sir Arthur Whitten (1886–1948), Scottish aviator. He made the first transatlantic flight in 1919 with Sir John William Alcock.

Brown² /braʊn/, James (b.1928), American soul and funk singer and songwriter. Influenced by gospel and early rhythm and blues, he became known as 'Soul Brother Number One' in the 1960s, playing a leading role in the development of funk and having a significant influence on many other areas of popular music. His many hits include 'Papa's Got a Brand New Bag' (1965) and 'Sex Machine' (1970).

Brown³ /braʊn/, John (1800–59), American abolitionist. The leader of an unsuccessful uprising in Virginia in 1859, he was captured and executed after raiding the government arsenal at Harpers Ferry, intending to arm runaway black slaves and start a revolt. Although the revolt never materialized, Brown became a hero of the American abolitionists in the subsequent American Civil War. He is commemorated in the popular marching-song 'John Brown's Body'.

Brown⁴ /braʊn/, Lancelot (known as Capability Brown) (1716–83), English landscape gardener. He evolved an English style of landscape parks, made to look natural by serpentine waters, clumps of trees, and other artifices. Famous examples of his work are to be found at Blenheim Palace in Oxfordshire, Chatsworth House in Derbyshire, and Kew Gardens. He earned his nickname by telling his patrons that their estates had 'great capabilities'.

brown /braʊn/ adj., n., & v. ● adj. **1** having the colour produced by mixing red, yellow, and black, as of dark wood or rich soil. **2** dark-skinned or suntanned. **3** (of bread) made from a dark flour such as wholemeal or wheatmeal. **4** (of species or varieties) distinguished by brown coloration. ● n. **1** a brown colour or pigment. **2** brown clothes or material (dressed in brown). **3** (in a game or sport) a brown ball, piece, etc. ● v.tr. & intr. make or become brown by cooking, sunburn, etc. □ **brown ale** a dark, mild, bottled beer. **brown bag** esp. US a plain brown paper bag in which a lunch is packed and carried to work etc. **brown-bagger** esp. US a person who takes a packed lunch to work etc. **brown bear** a large brown bear, Ursus arctos, found in parts of Eurasia and North America. **brown coal** = LIGNITE. **brown dwarf** Astron. a celestial body intermediate in size between a giant planet and a small star, believed to emit mainly infrared radiation. **browned off** *Brit. sl.* fed up, disheartened. **brown fat** a dark-coloured adipose tissue with a rich supply of blood vessels. **brown holland** see HOLLAND. **brown-nose** *US coarse sl.* toady; be servile. **brown-noser** *US coarse sl.* a toady, a yes-man. **brown owl 1** an owl with brown plumage, esp. the tawny owl. **2 (Brown Owl)** an adult leader of a Brownie Guides pack. **brown rice** unpolished rice with only the husk of the grain removed. **brown sugar** unrefined or partially refined sugar. **brown trout** a common European trout, *Salmo trutta*, esp. a small dark non-migratory race found in smaller rivers and pools. **fire into the brown** *Brit.* fire at a mass of flying game birds; aim in the general direction of a large target. **in a brown study** see STUDY. □ **brownish** *adj.* **brownness** /ˈbraʊnnɪs/ *n.* **browny** *adj.* [OE *brūn* f. Gmc]

Browne /braʊn/, Sir Thomas (1605–82), English author and physician. He achieved prominence with his *Religio Medici* (1642), a confession of Christian faith, drawing together a collection of imaginative and erudite opinions on a vast number of subjects more or less connected with religion. *Hydriotaphia; Urn Burial* (1658), a study of burial customs, is a notable example of Browne's elaborately ornate language.

brownfield /ˈbraʊnfiːld/ *n.* (*attrib.*) (of a piece of land) having formerly been the site of commercial or industrial activity, but now cleared and available for redevelopment.

Brownian motion /ˈbraʊnɪən/ *n.* (also **Brownian movement**) *Physics* the erratic random movement of microscopic particles in a fluid, as a result of continuous bombardment from molecules of the surrounding medium. [Robert *Brown*, Sc. botanist (1773–1858), who discovered the phenomenon]

Brownie /ˈbraʊnɪ/ *n.* **1** (in full **Brownie Guide**) a member of the junior branch of the Guides Association. **2** (**brownie**) **a** a small square of rich, usu. chocolate, cake with nuts. **b** *Austral. & NZ* a sweet currant-bread. **3** (**brownie**) a benevolent elf said to haunt houses and do household work secretly. □ **Brownie point** *colloq.* a notional credit for something done to please or win favour.

Browning[1] /ˈbraʊnɪŋ/, Elizabeth Barrett (1806–61), English poet. After first becoming known with *The Seraphim* (1838), she established her reputation with *Poems* (1844), which was so well received that she was seriously considered as a possible successor to Wordsworth as Poet Laureate. In 1845 Robert Browning began his passionate correspondence with her. The pair met and were secretly married the following year, eloping to Italy to escape the wrath of Barrett's domineering father. She is best known for the love poems *Sonnets from the Portuguese* (1850), the experimental verse novel *Aurora Leigh* (1857), and the posthumous *Last Poems* (1862).

Browning[2] /ˈbraʊnɪŋ/, Robert (1812–89), English poet. In 1842 he established his name as a poet with the publication of *Dramatic Lyrics*, containing such poems as 'The Pied Piper of Hamelin' and 'My Last Duchess'. *Dramatic Romances and Lyrics* (1845), which included 'Home Thoughts from Abroad', built on this success. In 1846 he eloped to Italy with Elizabeth Barrett, and a highly creative period followed: *Men and Women* (1855) and *The Ring and the Book* (1868–9), a series of dramatic monologues, were among the important works completed during this time.

browning /ˈbraʊnɪŋ/ *n. Brit.* browned flour or any other additive used to colour gravy.

Browning gun *n.* a type of water-cooled automatic machine-gun, named after the US designer J. M. Browning (1855–1926).

Brownshirt /ˈbraʊnʃɜːt/ *n.* a member of an early Nazi militia, the Storm Troops (German *Sturmabteilung*, abbr. SA), founded by Adolf Hitler in Munich in 1921, who wore brown uniforms reminiscent of those of Mussolini's Blackshirts. Their violent intimidation of political opponents and of Jews played a key role in Hitler's rise to power. By 1933 they numbered some 2 million, double the size of the army, which was hostile to them. On 29–30 June 1934 Hitler had more than seventy members of the SA, including their leader Ernst Röhm, summarily executed by the SS on the 'night of the long knives', and greatly reduced the Brownshirts' power.

brownstone /ˈbraʊnstəʊn/ *n. US* **1** a kind of reddish-brown sandstone used for building. **2** a building faced with this.

browse /braʊz/ *v. & n.* ● *v.* **1** *intr. & tr.* read or survey desultorily. **2** *intr.* (often foll. by *on*) feed on leaves, twigs, or scanty vegetation. **3** *tr.* crop and eat. ● *n.* **1** twigs, young shoots, etc., as fodder for cattle. **2** an act of browsing. □ **browser** *n.* [(n.) f. earlier *brouse* f. OF *brost* young shoot, prob. f. Gmc; (v.) f. F *broster*]

BRS *abbr.* British Road Services.

Bruce[1] /bruːs/, Robert the, see ROBERT I.

Bruce[2] /bruːs/, James ('the Abyssinian') (1730–94), Scottish explorer. After serving as consul-general at Algiers 1763–5, Bruce set off from Cairo in 1768 on an expedition to Ethiopia, becoming the first European to discover the source of the Blue Nile in 1770. His *Travels to Discover the Sources of the Nile* (1790), containing an account of his expedition, was dismissed by his contemporaries as fabrication.

brucellosis /ˌbruːsəˈləʊsɪs/ *n.* a disease caused by bacteria of the genus *Brucella*, affecting esp. cattle and causing undulant fever in humans. [mod.L *Brucella* f. Sir David *Bruce*, Sc. physician (1855–1931) + -OSIS]

brucite /ˈbruːsaɪt/ *n.* a white, grey, or greenish mineral form of magnesium hydroxide. [Archibald *Bruce*, Amer. mineralogist (1777–1818)]

Bruckner /ˈbrʊknə(r)/, Anton (1824–96), Austrian composer and organist. After dividing his time for several years between teaching and organ-playing, he turned to composition, writing ten symphonies between 1863 and his death, together with four masses and a *Te Deum* (1884). Bruckner was often persuaded by well-meaning friends to alter his scores; for the most part, though, editors have been able to trace his original intentions.

Bruegel /ˈbrɔɪɡ(ə)l/ (also **Breughel**, **Brueghel**), Pieter (c.1525–69), Flemish artist. Known as Pieter Bruegel the Elder, he joined the Antwerp guild in 1551, but produced the bulk of his famous work in Brussels, where he moved in 1563. He worked successfully in a variety of genres, including landscapes, religious allegories, and satires of peasant life. His major works include *The Procession to Calvary* (1564), *The Blind Leading the Blind* (1568), and *The Peasant Dance* (1568). Both of his sons also worked as painters, chiefly in Antwerp. Pieter ('Hell') Bruegel the Younger (1564–1638) is known primarily as a very able copyist of his father's work; he is also noted for his paintings of devils (hence his diabolic nickname). Jan ('Velvet') Bruegel (1568–1623) was a celebrated painter of flower, landscape, and mythological pictures.

Bruges /bruːʒ/ (Flemish **Brugge** /ˈbryxə/) a city in NW Belgium, capital of the province of West Flanders; pop. (1991) 117,000. A centre of the Flemish textile trade until the 15th century, it is a well-preserved medieval city surrounded by canals.

Brugge see BRUGES.

Bruin /ˈbruːɪn/ *n.* a personal name used for a bear. [ME f. Du., = BROWN: used as a name in the *Roman de Renart* (see REYNARD)]

bruise /bruːz/ *n. & v.* ● *n.* **1** an injury appearing as an area of discoloured skin on a human or animal body, caused by a blow or impact which ruptures underlying blood vessels. **2** a similar area of damage on a fruit etc. ● *v.* **1** *tr.* **a** inflict a bruise on. **b** hurt mentally. **2** *intr.* be susceptible to bruising (*bruises easily*). **3** *tr.* crush or pound. [ME f. OE *brÿsan* crush, reinforced by AF *bruser*, OF *bruisier* break]

bruiser /ˈbruːzə(r)/ *n. colloq.* **1** a large tough-looking person. **2** a professional boxer.

bruit /bruːt/ *v. & n.* ● *v.tr.* (often foll. by *abroad*, *about*) spread (a report or rumour). ● *n. archaic* a report or rumour. [F, = noise f. *bruire* roar]

Brum /brʌm/ *n. colloq.* Birmingham (in England). [abbr. of BRUMMAGEM]

brumby /ˈbrʌmbɪ/ *n.* (*pl.* **-ies**) *Austral.* a wild or unbroken horse. [19th c.: orig. unkn.]

brume /bruːm/ *n. literary* mist, fog. [F f. L *bruma* winter]

Brummagem /ˈbrʌmədʒəm/ *adj.* **1** cheap and showy (*Brummagem goods*). **2** counterfeit. [dial. form of BIRMINGHAM 1, with ref. to counterfeit coins and plated goods once made there]

Brummell /ˈbrʌm(ə)l/, George Bryan (known as Beau Brummell) (1778–1840), English dandy. He was the arbiter of British fashion for the first decade and a half of the 19th century, owing his social position to his close friendship with the Prince of Wales (later George IV). Brummell quarrelled with his patron and fled to France to avoid his creditors in 1816, eventually dying penniless in a mental asylum in Caen.

Brummie /ˈbrʌmɪ/ *n. & adj.* (also **Brummy**) *colloq.* ● *n.* (*pl.* **-ies**) a native of Birmingham. ● *adj.* of or characteristic of a Brummie (*a Brummie accent*). [BRUM]

brunch /brʌntʃ/ *n. & v.* ● *n.* a late-morning meal eaten as the first meal of the day. ● *v.intr.* eat brunch. [blend of BREAKFAST + LUNCH]

Brundtland /ˈbrʌntlænd/, Gro Harlem (b.1939), Norwegian Labour stateswoman. After serving as Environment Minister in the Labour government, in February 1981 she became Norway's first woman Prime Minister, but only held office until October of that year. During her second premiership (1986–89), she chaired the World Commission on Environment and Development (known as the Brundtland

Commission), which produced the report *Our Common Future* in 1987. During her third term of office (1990–), Norway applied to join the EEC.

Brunei /bruːˈnaɪ/ (official name **Brunei Darussalam**) a small oil-rich sultanate on the NW coast of Borneo, pop. (est. 1991) 264,000; languages, Malay (official), English (official), Chinese; capital, Bandar Seri Begawan. By the early 16th century Brunei's power extended over all of Borneo and parts of the Philippines, but declined as Portuguese and Dutch influence grew, and in 1888 it was placed under British protection; it became a fully independent Commonwealth state in 1984. □ **Bruneian** /bruːˈnaɪən/ adj. & n.

Brunel[1] /bruˈnel/, Isambard Kingdom (1806–59), English engineer. Son of Sir Marc Isambard Brunel, he was equally versatile, designing the famous Clifton suspension bridge in Bristol (1829–30), then in 1833 becoming chief engineer of the Great Western Railway, for which he surveyed more than a thousand miles of line (originally laid with broad gauge track, which he favoured) and designed many engineering works. He turned to steamship construction with the *Great Western* (1838), the first successful transatlantic steamship. His *Great Eastern* (1858) remained the world's largest ship until 1899. A little-known but remarkable achievement was Brunel's design in 1855 of a prefabricated hospital for the Crimean War.

Brunel[2] /bruˈnel/, Sir Marc Isambard (1769–1849), French-born English engineer. He came to England in 1799 and persuaded the government to adopt his designs for mass-production machinery at Portsmouth dockyard, an early example of automation. He also designed other machines for woodworking, boot-making, knitting, and printing. A versatile civil engineer, he built bridges, landing stages, and the first tunnelling shield, which he used to construct the first tunnel under the Thames (1825–43).

Brunelleschi /ˌbruːnəˈleskɪ/, Filippo (born Filippo di Ser Brunnellesco) (1377–1446), Italian architect. The most famous Florentine architect of the 15th century, he is especially noted for the dome of Florence cathedral (1420–61), which he raised, after the fashion of ancient Roman construction, without the use of temporary supports. It is the largest dome in the world in diameter and inspired Michelangelo's design for St Peter's in Rome. Brunelleschi revived Roman architectural forms and is also often credited with the Renaissance 'discovery' of perspective.

brunette /bruːˈnet/ n. & adj. ● n. a woman with dark brown hair. ● adj. (of a woman) having dark brown hair. [F, fem. of *brunet*, dimin. of *brun* BROWN]

Brunhild /ˈbruːnhɪlt/ in the Nibelungenlied, the wife of Gunther, who instigated the murder of Siegfried. In the Norse versions she is a Valkyrie whom Sigurd (the counterpart of Siegfried) wins by penetrating the wall of fire behind which she lies in an enchanted sleep.

Bruno /ˈbruːnəʊ/, Giordano (1548–1600), Italian philosopher. After a period in the Dominican order, he became a follower of the magical tradition of Hermes Trismegistus. He was a supporter of the heliocentric Copernican view of the solar system, envisaging an infinite universe of numerous worlds moving in space. He was tried by the Inquisition for heresy and later burned at the stake.

Bruno, St /ˈbruːnəʊ/ (c.1032–1101), German-born French churchman. After withdrawing to the mountains of Chartreuse in 1084, he founded the Carthusian order at La Grande Chartreuse in SE France in the same year. Feast day, 6 Oct.

Brunswick /ˈbrʌnzwɪk/ (German **Braunschweig** /ˈbraʊnʃvaɪk/) **1** a former duchy and state of Germany, mostly incorporated into Lower Saxony. **2** the capital of this former duchy, an industrial city in Lower Saxony, Germany; pop. (1991) 259,130.

brunt /brʌnt/ n. the chief or initial impact of an attack, task, etc. (esp. *bear the brunt of*). [ME: orig. unkn.]

brush /brʌʃ/ n. & v. ● n. **1** an implement with bristles, hair, wire, etc. varying in firmness set into a block or projecting from the end of a handle, for any of various purposes, esp. cleaning or scrubbing, painting, arranging the hair, etc. **2** the application of a brush; brushing. **3 a** (usu. foll. by *with*) a short esp. unpleasant encounter (*a brush with the law*). **b** a skirmish. **4 a** the bushy tail of a fox. **b** a brushlike tuft. **5** *Electr.* **a** a piece of carbon or metal serving as an electrical contact esp. with a moving part. **b** (in full **brush discharge**) a brushlike discharge of sparks. **6** esp. *N. Amer. & Austral.* **a** undergrowth, thicket; small trees and shrubs. **b** *US* such wood cut in faggots. **c** land covered with brush. **d** *Austral.* dense forest. **7** *Austral. & NZ sl.* a girl or young woman. ● v. **1** *tr.* **a** sweep or scrub or put in order with a brush. **b** treat (a surface) with a brush so as to change its nature or appearance. **2** *tr.* **a** remove (dust etc.) with a brush. **b** apply (a liquid preparation) to a surface with a brush. **3** *tr. & intr.* graze or touch in passing. **4** *intr.* perform a brushing action or motion. □ **brush aside** dismiss or dispose of (a person, idea, etc.) curtly or lightly. **brushed aluminium** aluminium treated so that it has a lustreless surface. **brushed fabric** fabric brushed so as to raise the nap. **brush off** dismiss abruptly. **brush-off** n. a rebuff; an abrupt dismissal. **brush over** paint lightly. **brush turkey** *Austral.* a large mound-building megapode, *Alectura lathami*. **brush up 1** clean up or smarten. **2** revive one's former knowledge of (a subject). **brush-up** n. the process of cleaning up. □ **brushlike** adj. **brushy** adj. [ME f. OF *brosse*]

brushless /ˈbrʌʃlɪs/ adj. not requiring the use of a brush.

brushwood /ˈbrʌʃwʊd/ n. **1** cut or broken twigs etc. **2** undergrowth; a thicket.

brushwork /ˈbrʌʃwɜːk/ n. **1** manipulation of the brush in painting. **2** a painter's style in this.

brusque /brʊsk, bruːsk, brʌsk/ adj. abrupt or offhand in manner or speech. □ **brusquely** adv. **brusqueness** n. **brusquerie** /ˈbrʊskəˌriː/ n. [F f. It. *brusco* sour]

Brussels /ˈbrʌs(ə)lz/ (French **Bruxelles** /brysɛl/, Flemish **Brussel** /ˈbrys(ə)l/) the capital of Belgium and of the Belgian province of Brabant; pop. (1991) 954,000. The headquarters of the European Commission is located there.

Brussels carpet n. a carpet with a wool pile and a stout linen back.

Brussels lace n. an elaborate needlepoint or pillow lace.

Brussels sprout n. **1** a variety of cabbage producing many small compact buds borne close together along a tall single stem. **2** one of these buds used as a vegetable.

brut /bruːt/ adj. (of wine) unsweetened, very dry. [F]

brutal /ˈbruːt(ə)l/ adj. **1** savagely or coarsely cruel. **2** harsh, merciless. □ **brutally** adv. **brutality** /bruːˈtælɪtɪ/ n. (pl. **-ies**). [F *brutal* or med.L *brutalis* f. *brutus* BRUTE]

brutalism /ˈbruːtəˌlɪz(ə)m/ n. **1** brutality. **2** (also **Brutalism, New Brutalism**) a heavy plain style of architecture etc., in particular a style of architecture in the 1950s and 1960s that evolved from the work of Le Corbusier and Mies van der Rohe, stressing functionality and typically making use of steel and exposed concrete. □ **Brutalist** n. & adj. (in sense 2).

brutalize /ˈbruːtəˌlaɪz/ v.tr. (also **-ise**) **1** make brutal. **2** treat brutally. □ **brutalization** /ˌbruːtəlaɪˈzeɪʃ(ə)n/ n.

brute /bruːt/ n. & adj. ● n. **1 a** a brutal or violent person or animal. **b** *colloq.* an unpleasant person. **2** an animal as opposed to a human being. ● adj. (usu. *attrib.*) **1** not possessing the capacity to reason. **2 a** animal-like, cruel. **b** stupid, sensual. **3** unthinking, merely material (*brute force*; *brute matter*). □ **brutehood** n. **brutish** adj. **brutishly** adv. **brutishness** n. [F f. L *brutus* stupid]

Brutus[1] /ˈbruːtəs/ legendary Trojan hero, great-grandson of Aeneas and supposed ancestor of the British people. He is said to have brought a group of Trojans to England and founded Troynovant or New Troy (later called London), becoming the progenitor of a line of kings. His story is told by the chronicler Geoffrey of Monmouth.

Brutus[2] /ˈbruːtəs/, Lucius Junius, legendary founder of the Roman Republic. Traditionally he led a popular uprising, after the rape of Lucretia, against the king (his uncle) and drove him from Rome. He and the father of Lucretia were elected as the first consuls of the Republic (509 BC).

Brutus[3] /ˈbruːtəs/, Marcus Junius (85–42 BC), Roman senator. With Cassius he was a leader of the conspirators who assassinated Julius Caesar in the name of the Republic in 44. He and Cassius were defeated by Caesar's supporters, Antony and Octavian, at the battle of Philippi in 42, after which he committed suicide.

Bruxelles see BRUSSELS.

bruxism /ˈbrʌksɪz(ə)m/ n. involuntary or habitual grinding or clenching of the teeth. [Gk *brukhein* gnash the teeth]

Bryansk /brɪˈænsk/ (also **Briansk**) an industrial city in European Russia, south-west of Moscow, on the Desna river; pop. (1990) 456,000.

bryology /braɪˈɒlədʒɪ/ n. the study of bryophytes. □ **bryologist** n. **bryological** /ˌbraɪəˈlɒdʒɪk(ə)l/ adj. [Gk *bruon* moss]

bryony /ˈbraɪənɪ/ n. (pl. **-ies**) a climbing plant of the genus *Bryonia*, of the gourd family; esp. *B. dioica* (in full **white bryony**), with greenish-white flowers and red berries. □ **black bryony** a climbing plant,

Tamus communis, of the yam family, with glossy leaves and poisonous red berries. [L *bryonia* f. Gk *bruōnia*]

bryophyte /'braɪəˌfaɪt/ *n. Bot.* a cryptogamous plant of the division Bryophyta, which comprises the mosses and liverworts. Bryophytes bear structures resembling roots, stems, and leaves, but they have no specialized vascular (conducting) tissue, and are confined to damp places. □ **bryophytic** /ˌbraɪə'fɪtɪk/ *adj.* [mod.L Bryophyta f. Gk *bruon* moss + *phuton* plant]

bryozoan /ˌbraɪə'zəʊən/ *n. & adj. Zool.* ● *n.* an aquatic invertebrate animal of the phylum Bryozoa, forming mosslike colonies attached to rocks, seaweeds, etc. Also called *polyzoan*. ● *adj.* of or relating to the phylum Bryozoa. □ **bryozoology** /-ˌzuː'ɒlədʒɪ, -ˌzəʊ'ɒl-/ *n.* [Gk *bruon* moss + *zōia* animals]

Brythonic /brɪ'θɒnɪk/ *n. & adj.* ● *n.* the southern group of the Celtic languages, comprising Welsh, Cornish, and Breton. Also called *Brittonic.* (See note below.) ● *adj.* of or relating to Brythonic. [Welsh *Brython* Britons f. OCelt.]

▪ The Brythonic languages constituted the vernaculars spoken in Britain before and during the Roman occupation. After Britain was invaded by Germanic-speaking peoples in the 5th–7th centuries, the languages of the Britons died out except in the remote, mountainous west, where they survived as Welsh and Cornish. The latter was carried by British emigrants across the English Channel, where it survives as the Breton language in Brittany. Brythonic is also known as *P-Celtic.* (Cf. GOIDELIC.)

Brześć nad Bugiem see BREST 2.

BS *abbr.* **1** US Bachelor of Science. **2** Bachelor of Surgery. **3** Blessed Sacrament. **4** British Standard(s).

B.Sc. *abbr.* Bachelor of Science.

BSE *abbr.* bovine spongiform encephalopathy, a usu. fatal disease of cattle, believed to be caused by either prions or virinos. BSE affects the central nervous system and causes agitated behaviour and staggering. Popularly known as 'mad cow disease', it was identified in Britain in 1986, and appears to be related to scrapie, CJD, and kuru.

BSI *abbr.* British Standards Institution.

B-side /'biːsaɪd/ *n.* the side of a gramophone record (usu. a single) regarded as less important.

BST *abbr.* **1** see BRITISH SUMMER TIME. **2** British Standard Time, one hour ahead of Greenwich Mean Time (in use continuously 1968–71). **3** bovine somatotrophin, a growth hormone produced naturally by cows and introduced into cattle-feed to boost milk production.

BT *abbr.* British Telecom.

Bt. *abbr.* Baronet.

B.th.u. *abbr.* (also **B.t.u., BTU, B.Th.U.**) British thermal unit(s).

bu. *abbr.* bushel(s).

bub /bʌb/ *n. N. Amer. colloq.* a boy or a man, often used as a form of address. [earlier *bubby*, perh. a childish form of BROTHER or f. G *Bube* boy]

bubal /'bjuːb(ə)l/ *n.* = HARTEBEEST. [L *bubalus* f. Gk *boubalos* oxlike antelope]

bubble /'bʌb(ə)l/ *n. & v.* ● *n.* **1 a** a thin sphere of liquid enclosing air etc. **b** an air-filled cavity in a liquid or a solidified liquid such as glass or amber. **c** (in *pl.*) froth, foam. **2** the sound or appearance of bubbling; an agitated or bubbling motion. **3** a transparent domed cavity. **4** a visionary or unrealistic project or enterprise (*the South Sea Bubble*). ● *v.intr.* **1** rise in or send up bubbles. **2** make the sound of rising or bursting bubbles. □ **bubble and squeak** *Brit.* cooked cabbage fried with cooked potatoes or meat. **bubble bath 1** a preparation for adding to bath water to make it foam. **2** a bath with this added. **bubble car** *Brit.* a small motor car with a transparent canopy. **bubble chamber** *Physics* an apparatus designed to make the tracks of ionizing particles visible as a row of bubbles in a liquid. **bubble gum** chewing-gum that can be blown into bubbles. **bubble memory** *Computing* a type of memory which stores data as a pattern of magnetized regions in a thin layer of magnetic material. **bubble over** (often foll. by *with*) be exuberant with laughter, excitement, anger, etc. **bubble pack** a small package enclosing goods in a transparent material on a backing. [ME: prob. imit.]

bubbly /'bʌblɪ/ *adj. & n.* ● *adj.* (**bubblier, bubbliest**) **1** having or resembling bubbles. **2** exuberant, vivacious. ● *n. colloq.* champagne. □ **bubbly-jock** *Sc.* a turkeycock.

Buber /'buːbə(r)/, Martin (1878–1965), Israeli religious philosopher, born in Austria. A supporter of Hasidism and a committed Zionist, he

settled in Palestine in 1938 after fleeing the Nazis. His most famous work *I and Thou* (1923) sums up much of his religious philosophy, comparing mutual and reciprocal relationships with objective or utilitarian ones. His other important publications include *Between Man and Man* (1946) and *Eclipse of God* (1952).

bubo /'bjuːbəʊ/ *n.* (*pl.* **-oes**) a swollen inflamed lymph node in the armpit or groin. [med.L *bubo -onis* swelling f. Gk *boubōn* groin]

bubonic plague /bjuː'bɒnɪk/ *n.* a contagious bacterial disease characterized by fever, delirium, and the formation of buboes. Epidemics of bubonic plague, transmitted to humans by rats' fleas, broke out in Europe in the Middle Ages. The first killed millions in the mid-14th century (see BLACK DEATH), and the disease re-emerged periodically for several centuries thereafter, the last serious outbreak in Britain being in 1665–6 (see GREAT PLAGUE). It is still endemic in parts of Asia.

buccal /'bʌk(ə)l/ *adj. Anat.* **1** of or relating to the cheek. **2** of or in the mouth. [L *bucca* cheek]

buccaneer /ˌbʌkə'nɪə(r)/ *n. & v.* ● *n.* **1** a pirate, orig. off the Spanish-American coasts. **2** an unscrupulous adventurer. ● *v.intr.* be a buccaneer. □ **buccaneering** *n. & adj.* **buccaneerish** *adj.* [F *boucanier* f. *boucaner* cure meat on a barbecue f. *boucan* f. Tupi *mukem*]

buccinator /'bʌksɪˌneɪtə(r)/ *n. Anat.* a flat thin cheek muscle. [L f. *buccinare* blow a trumpet (*buccina*)]

Bucephalus /bjuː'sefələs/ the favourite horse of Alexander the Great, who tamed the horse as a boy and took it with him on his campaigns until its death, after a battle, in 326 BC.

Buchan[1] /'bʌkən/, Alexander (1829–1907), Scottish meteorologist. He proposed that at certain times in the year the temperature regularly deviated from the normal, though it is now thought that such cold spells are probably distributed at random. As well as writing a standard textbook on meteorology, he produced maps and tables of atmospheric circulation, and of ocean currents and temperatures, based largely on information gathered on the voyage of HMS *Challenger* in 1872–6.

Buchan[2] /'bʌkən/, John, 1st Baron Tweedsmuir (1875–1940), Scottish novelist. Although he wrote several non-fictional works, he is remembered for his adventure stories, which often feature recurring heroes such as Richard Hannay. These include *The Thirty-Nine Steps* (1915), *Greenmantle* (1916), and *The Three Hostages* (1924). Buchan was also active in public life, serving as MP for the Scottish Universities (1927–35) and as Governor-General of Canada (1935–40).

Buchanan /bjuː'kænən/, James (1791–1868), American Democratic statesman, 15th President of the US 1857–61. He consistently leaned towards the pro-slavery side in the developing dispute over slavery. Towards the end of his term the issue grew more fraught and he retired from politics in 1861.

Bucharest /ˌbuːkə'rest/ (Romanian **Bucureşti** /ˌbuku'reʃtj/) the capital of Romania and former capital of Wallachia; pop. (1989) 2,318,900. It was founded in the 14th century on the trade route between Europe and Constantinople.

Buchenwald /'bʊkənˌvɑːlt, 'buːxən-/ a Nazi concentration camp in the Second World War, near the village of Buchenwald in eastern Germany.

Buchner /'bʌknə(r), 'buːxnə(r)/, Eduard (1860–1917), German organic chemist. He discovered that intact yeast cells were not necessary for alcoholic fermentation, which could be carried out by an active extract that he called *zymase*. Buchner investigated the chain of reactions involved in fermentation, identifying various other enzymes. He won the Nobel Prize for chemistry in 1907.

buck[1] /bʌk/ *n. & v.* ● *n.* **1** the male of various animals, esp. the deer, hare, or rabbit. **2** *archaic* a fashionable young man. **3** (*attrib.*) **a** *sl.* male (*buck antelope*). **b** US Mil. *sl.* of the lowest rank (*buck private*). ● *v.* **1** *intr.* (of a horse) jump upwards with back arched and feet drawn together. **2** *tr.* **a** (usu. foll. by *off*) throw (a rider or burden) in this way. **b** esp. *N. Amer.* oppose, resist (*tried to buck the trend*). **3** *tr. & intr.* (usu. foll. by *up*) *colloq.* **a** make or become more cheerful. **b** make or become more vigorous or lively (*needs to buck up his ideas*). **4** *tr.* (as **bucked** *adj.*) *colloq.* encouraged, elated. □ **buck fever** US nervousness when called on to act. **buckhorn** horn of male deer as a material for knife-handles etc. **buckhound** a small kind of staghound. **buck rarebit** Welsh rarebit with a poached egg on top. **buck-tooth** an upper tooth that projects. [OE *buc* male deer, *bucca* male goat, f. ON]

buck[2] /bʌk/ *n. N. Amer. & Austral. etc. sl.* a dollar. □ **big bucks** a great deal of money. **a fast buck** easy money. [19th c.: orig. unkn.]

buck[3] /bʌk/ *n. sl.* an article placed as a reminder before a player whose turn it is to deal at poker. □ **pass the buck** *colloq.* shift responsibility (to another). [19th c.: orig. unkn.]

buck[4] /bʌk/ *n.* **1** *N. Amer.* a saw-horse; = SAWBUCK 1. **2** a vaulting-horse. [Du. (*zaag*)*boc*]

buck[5] /bʌk/ *n.* the body of a cart. [perh. f. obs. *bouk* belly, f. OE *būc* f. Gmc]

buckbean /'bʌkbiːn/ *n.* a bog plant, *Menyanthes trifoliata*, with white or pinkish hairy flowers. Also called *bog-bean*.

buckboard /'bʌkbɔːd/ *n.* US a horse-drawn vehicle with the body formed by a plank fixed to the axles. [BUCK[5] + BOARD]

bucket /'bʌkɪt/ *n. & v.* ● *n.* **1 a** a roughly cylindrical open container usu. of metal or plastic with a handle, used for carrying, drawing, or holding water etc. **b** the amount contained in this (*need three buckets to fill the bath*). **2** (in *pl.*) large quantities of liquid, esp. rain or tears (*wept buckets*). **3** a compartment on the outer edge of a water-wheel. **4** the scoop of a dredger or a grain-elevator. ● *v.* (**bucketed**, **bucketing**) **1** *intr.* (usu. foll. by *down*) (of liquid, esp. rain) pour heavily. **2** *intr. & tr.* (often foll. by *along*) *Brit.* move or drive jerkily or bumpily. □ **bucket seat** a seat with a rounded back to fit one person, esp. in a car. **bucket-shop 1** an office for gambling in stocks, speculating on markets, etc. **2** *colloq.* a travel agency specializing in cheap air tickets. □ **bucketful** *n.* (*pl.* **-fuls**). [ME & AF *buket*, *buquet*, perh. f. OE *būc* pitcher]

buckeye /'bʌkaɪ/ *n.* **1** a tree or shrub of the genus *Aesculus*, an American horse chestnut, with large sticky buds and showy red or white flowers. **2** the shiny brown fruit of this plant.

Buckingham Palace /'bʌkɪŋəm/ the London residence of the British sovereign since 1837, adjoining St James's Park, Westminster. It was built for the Duke of Buckingham in the early 18th century, bought by George III in 1761, and redesigned by John Nash for George IV c.1821–30; the façade facing the Mall was redesigned in 1913.

Buckinghamshire /'bʌkɪŋəm,ʃɪə(r)/ a county of central England; county town, Aylesbury.

Buckland /'bʌklənd/, William (1784–1856), English geologist. He taught at Oxford, later becoming dean of Westminster. He helped to redefine geology, linking type of deposit with local dynamic conditions and using the associated fossils to establish former habitats and climate. He was the first to describe and name a dinosaur (*Megalosaurus*), in 1824. Buckland supported the idea of a past catastrophic event, first interpreting this as being the biblical flood and later moving to the idea of an ice age.

buckle /'bʌk(ə)l/ *n. & v.* ● *n.* **1** a flat often rectangular frame with a hinged pin, used for joining the ends of a belt, strap, etc. **2** a similarly shaped ornament, esp. on a shoe. ● *v.* **1** *tr.* (often foll. by *up*, *on*, etc.) fasten with a buckle. **2** *tr. & intr.* (often foll. by *up*) give way or cause to give way under longitudinal pressure; crumple up. □ **buckle down** make a determined effort. **buckle to** (or **down to**) prepare for, set about (work etc.). **buckle to** get to work, make a vigorous start. [ME f. OF *boucle* f. L *buccula* cheek-strap of a helmet f. *bucca* cheek: sense 2 of *v.* f. F *boucler* bulge]

buckler /'bʌklə(r)/ *n.* **1** *hist.* a small round shield held by a handle. **2** a fern of the genus *Dryopteris*, having buckler-shaped indusia. Also called *shield-fern*. [ME f. OF *bocler* lit. 'having a boss' f. *boucle* BOSS[2]]

Buckley's /'bʌklɪz/ *n.* (in full **Buckley's chance**) *Austral. & NZ colloq.* little or no chance. The origins of the expression are uncertain; possibly it derives from the name of William Buckley (1780–1856), an escaped convict who lived for thirty-two years with Aboriginals in south Victoria.

buckling /'bʌklɪŋ/ *n.* a smoked herring. [G *Bückling* bloater]

buckminsterfullerene /,bʌkmɪnstə'fʊlə,riːn/ *n.* *Chem.* a form of carbon with a molecule of 60 atoms joined in a polyhedron of 20 hexagons and 12 pentagons. (See also FULLERENE.) [R. *Buckminster Fuller* (FULLER[1])]

bucko /'bʌkəʊ/ *n. & adj.* *Naut. sl.* ● *n.* (*pl.* **-oes**) a swaggering or domineering person. ● *adj.* blustering, swaggering, bullying. [BUCK[1] + -o]

buckram /'bʌkrəm/ *n. & adj.* ● *n.* **1** a coarse linen or other cloth stiffened with gum or paste, and used as interfacing or in bookbinding. **2** *archaic* stiffness in manner. ● *adj.* *archaic* starchy; formal. □ **men in buckram** non-existent persons, figments (Shakespeare's *1 Henry IV* II. iv. 210–50). [ME f. AF *bukeram*, OF *boquerant*, perh. f. BUKHORO]

Bucks. *abbr.* Buckinghamshire.

Buck's Fizz /bʌks/ *n.* a cocktail of champagne or sparkling white wine and orange juice. [*Buck's Club* in London + FIZZ]

buckshee /bʌk'ʃiː/ *adj. & adv.* *Brit. sl.* free of charge. [corrupt. of BAKSHEESH]

buckshot /'bʌkʃɒt/ *n.* coarse lead shot.

buckskin /'bʌkskɪn/ *n.* **1 a** the skin of a buck (male deer). **b** leather made from a buck's skin. **2** a thick smooth cotton or woollen cloth.

buckthorn /'bʌkθɔːn/ *n.* a thorny shrub of the family Rhamnaceae, esp. *Rhamnus cathartica*, which has berries formerly used as a cathartic.

buckwheat /'bʌkwiːt/ *n.* a cereal plant of the genus *Fagopyrum*, esp. *F. esculentum*, which has seeds used for fodder and for flour to make bread and pancakes. [MDu. *boecweite* beech wheat, its grains being shaped like beechmast]

buckyball /'bʌkɪ,bɔːl/ *n.* *Chem. colloq.* (usu. in *pl.*) a molecule of a fullerene. [f. BUCKMINSTERFULLERENE + BALL[1]]

bucolic /bju:'kɒlɪk/ *adj. & n.* ● *adj.* of or concerning shepherds or pastoral life; rural. ● *n.* **1** (usu. in *pl.*) a pastoral poem or poetry. **2** a peasant. □ **bucolically** *adv.* [L *bucolicus* f. Gk *boukolikos* f. *boukolos* herdsman f. *bous* ox]

Bucureşti see BUCHAREST.

bud[1] /bʌd/ *n. & v.* ● *n.* **1 a** an immature knoblike shoot from which a stem, leaf, or flower develops. **b** a flower or leaf that is not fully open. **2** *Biol.* an asexual outgrowth from a parent organism that separates to form a new individual. **3** anything still undeveloped. ● *v.* (**budded**, **budding**) **1** *intr. Bot. & Zool.* form a bud or buds. **2** *intr.* begin to grow or develop (*a budding cricketer*). **3** *tr.* graft a bud (of a plant) on to another plant. □ **in bud** having newly formed buds. [ME: orig. unkn.]

bud[2] /bʌd/ *n. N. Amer. colloq.* (as a form of address) = BUDDY. [abbr.]

Budapest /,bu:də'pest/ the capital of Hungary; pop. (1990) 2,000,000. The city was formed in 1873 by the union of the hilly city of Buda on the right bank of the River Danube with the low-lying city of Pest on the left.

Buddha /'bʊdə/ a title given to successive teachers (past and future) of Buddhism, although it usually denotes the founder of Buddhism, Siddhartha Gautama (c.563 BC–c.480 BC). Although born an Indian prince (in what is now Nepal), he renounced his kingdom, wife, and child to become an ascetic, taking religious instruction until he attained enlightenment (nirvana) through meditation beneath a bo tree in the village of Bodhgaya. He then taught all who wanted to learn, regardless of sex, class, or caste, until his death. [Skr., = enlightened, past part. of *budh* know]

Buddh Gaya see BODHGAYA.

Buddhism /'bʊdɪz(ə)m/ *n.* a widespread Asian religion or philosophy, founded by Siddhartha Gautama (see BUDDHA), in NE India in the 5th century BC as a reaction against the sacrificial religion of orthodox Brahmanism. It is a religion without a god, of which the central doctrine is that of karma. The basic teachings of Buddhism are contained in the 'four noble truths': that all existence is suffering, that the cause of suffering is desire, that freedom from suffering is nirvana, and that nirvana may be attained through the 'eightfold path' that combines ethical conduct, mental discipline (in particular the practice of meditation), and wisdom. There are two major traditions or 'vehicles', Theravada (developed from Hinayana) and Mahayana. □ **Buddhist** *n. & adj.* **Buddhistic** /bʊ'dɪstɪk/ *adj.* **Buddhistical** *adj.*

buddleia /'bʌdlɪə/ *n.* a shrub of the genus *Buddleia*, with spikes of fragrant lilac, yellow, or white flowers attractive to butterflies. [Adam *Buddle*, Engl. botanist (d.1715)]

buddy /'bʌdɪ/ *n. & v. N. Amer. colloq.* ● *n.* (*pl.* **-ies**) **1** (often as a form of address) a close friend or mate. **2** a person who befriends and provides support for someone in need, esp. a person with Aids. ● *v.intr.* (**-ies**, **-ied**) (often foll. by *up*) become friendly. □ **buddy movie** a film featuring friendship between two individuals, esp. men. [perh. corrupt. of *brother*, or var. of BUTTY[1]]

Budge /bʌdʒ/, John Donald ('Don') (b.1915), American tennis player. He was the first to win the four major singles championships — Australia, France, Britain, and the US — in one year (1938). In both 1937 and 1938 he won the Wimbledon singles, men's doubles, and mixed doubles.

budge /bʌdʒ/ *v.* (usu. with *neg.*) **1** *intr.* **a** make the slightest movement. **b** change one's opinion (*he's stubborn, he won't budge*). **2** *tr.* cause or compel to budge (*nothing will budge him*). □ **budge up** (or **over**) make room for another person by moving. [F *bouger* stir ult. f. L *bullire* boil]

budgerigar /'bʌdʒərɪ,gɑː(r)/ *n.* a small green Australian parrot, *Melopsittacus undulatus*, which is bred in several colour varieties that are often kept as cage-birds. [Aboriginal]

budget /'bʌdʒɪt/ n. & v. • n. **1** the amount of money needed or available (for a specific item etc.) (a budget of £200; mustn't exceed the budget). **2 a** (**the Budget**) Brit. the usu. annual estimate of national revenue and expenditure. **b** a similar estimate of revenue or income and expenditure made by a company, family, private individual, etc. **3** (attrib.) inexpensive. **4** archaic a quantity of material etc., esp. written or printed. • v.tr. & intr. (**budgeted, budgeting**) (often foll. by for) allow or arrange for in a budget (have budgeted for a new car; can budget £60). □ **budget account** (or **plan**) a bank account, or account with a store, into which one makes regular, usu. monthly, payments to cover bills. **on a budget** with a restricted amount of money. □ **budgetary** adj. [ME = pouch, f. OF bougette dimin. of bouge leather bag f. L bulga (f. Gaulish) knapsack: cf. BULGE]

budgie /'bʌdʒɪ/ n. colloq. = BUDGERIGAR. [abbr.]

Budweis /'buːtvaɪs/ the German name for ČESKÉ BUDĚJOVICE.

Buenaventura /ˌbweɪnəvenˈtʊərə/ the chief Pacific port of Colombia; pop. (1985) 122,500.

Buenos Aires /ˌbweɪnəs ˈaɪriːz/ the capital city and chief port of Argentina, on the River Plate; pop. (1991) 2,961,000.

buff[1] /bʌf/ adj., n., & v. • adj. of a yellowish-beige colour (buff envelope). • n. **1** a yellowish-beige colour. **2** colloq. an enthusiast, esp. for a particular hobby (railway buff). **3** (prec. by the) colloq. the human skin unclothed. **4 a** a velvety dull-yellow ox-leather. **b** (attrib.) (of a garment etc.) made of this (buff gloves). **5** (**the Buffs**) the former East Kent Regiment (from the colour of its uniform facings). • v.tr. **1** polish (metal, fingernails, etc.). **2** make (leather) velvety like buff, by removing the surface. □ **in the buff** colloq. naked. [orig. sense 'buffalo', prob. f. F buffle; sense 2 of n. orig. f. buff uniforms formerly worn by New York volunteer firemen, applied to enthusiastic fire-watchers]

buff[2] /bʌf/ n. a blow. □ **blind man's buff** see BLIND. [OF bufe BUFFET[2]]

Buffalo /'bʌfəˌləʊ/ an industrial city in New York State; pop. (1990) 328,120. Situated at the eastern end of Lake Erie, it is a major port of the St Lawrence Seaway.

buffalo /'bʌfəˌləʊ/ n. & v. • n. (pl. same or **-oes**) **1 a** an Asiatic ox of the genus Bubalus, with backswept horns; esp. B. arnee (in full **water buffalo**), which is domesticated and used as a draught animal. **b** (in full **Cape buffalo**), a powerful wild ox, Syncerus caffer, found in eastern and southern Africa. **2** a North American bison (see BISON). • v.tr. (**-oes, -oed**) N. Amer. sl. overawe, outwit. □ **buffalo grass 1** a grass, Buchloe dactyloides, of the North American plains. **2** a grass, Stenotaphrum secundatum, of Australia and New Zealand. [prob. f. Port. bufalo f. LL bufalus f. L bubalus f. Gk boubalos antelope, wild ox]

Buffalo Bill (born William Frederick Cody) (1846–1917), American showman. A former US army scout and dispatch-bearer, Cody gained his nickname for killing 4,280 buffalo in eight months to feed the Union Pacific Railroad workers. He subsequently devoted his life to show business, particularly his Wild West Show, which travelled all over Europe and the US. These dramatics more than any real frontier exploits made Cody a national figure and his death in 1917 was widely seen as symbolizing the end of an era.

buffer[1] /'bʌfə(r)/ n. & v. • n. **1 a** a device that protects against or reduces the effect of an impact. **b** Brit. such a device (usu. either of a pair) on the end of a railway vehicle or at the end of a track. **2** Chem. a substance that acts to minimize the change in the hydrogen ion concentration of a solution when an acid or alkali is added. **3** Computing a temporary memory area or queue for data to aid its transfer between devices or programs operating at different speeds etc. • v.tr. **1** act as a buffer to. **2** Chem. treat with a buffer. □ **buffer state** a small state situated between two larger ones potentially hostile to one another and regarded as reducing the likelihood of open hostilities. **buffer stock** a reserve of commodity to offset price fluctuations. [prob. f. obs. buff (v.), imit. of the sound of a soft body struck]

buffer[2] /'bʌfə(r)/ n. Brit. sl. a silly or incompetent old man (esp. old buffer). [18th c.: prob. formed as BUFFER[1] or with the sense 'stutterer']

buffet[1] /'bʊfeɪ, 'bʌfeɪ/ n. **1** a room or counter where light meals or snacks may be bought (station buffet). **2** a meal consisting of several dishes set out from which guests serve themselves (buffet lunch). **3** (also /'bʌfɪt/) a sideboard or recessed cupboard for china etc. □ **buffet car** Brit. a railway coach serving light meals or snacks. [F f. OF bufet stool, of unkn. orig.]

buffet[2] /'bʌfɪt/ v. & n. • v. (**buffeted, buffeting**) **1** tr. **a** strike or knock repeatedly (wind buffeted the trees). **b** strike, esp. repeatedly, with the hand or fist. **2** tr. (of fate etc.) treat badly; plague (cheerful though buffeted by misfortune). **3 a** intr. struggle; fight one's way (through difficulties

etc.). **b** tr. contend with (waves etc.). • n. **1** a blow, esp. of the hand or fist. **2** a shock. [ME f. OF dimin. of bufe blow]

buffeting /'bʌfɪtɪŋ/ n. **1** a beating; repeated blows. **2** Aeron. an irregular oscillation, caused by air eddies, of part of an aircraft.

bufflehead /'bʌf(ə)lˌhed/ n. a black and white diving duck, Bucephala albeola, native to North America, with a large head. [obs. buffle buffalo + HEAD]

buffo /'bʊfəʊ/ n. & adj. • n. (pl. **-os**) a comic actor, esp. in Italian opera. • adj. comic, burlesque. [It.]

Buffon /'buːfɒn/, Georges-Louis Leclerc, Comte de (1707–88), French naturalist. He was one of the founders of palaeontology and suggested that the earth was much older than was generally accepted. Buffon saw all life as the physical property of matter, thereby stressing the unity of all living species and minimizing the apparent differences between animals and plants. He produced a remarkable compilation of the animal kingdom, the Histoire Naturelle, which was begun in 1749 and had reached thirty-six volumes by the time of his death.

buffoon /bəˈfuːn/ n. **1** a ludicrous person. **2** a jester; a mocker. □ **buffoonery** n. **buffoonish** adj. [F bouffon f. It. buffone f. med.L buffo clown f. Rmc]

bug /bʌg/ n. & v. • n. **1 a** an insect of the order Hemiptera, having the mouthparts adapted for piercing and sucking. **b** colloq. any small insect. **2** colloq. a micro-organism, esp. a bacterium, or a disease caused by it. **3** a concealed microphone. **4** an error in a computer program or system etc. **5** colloq. **a** an obsession, enthusiasm, etc. **b** an enthusiast. • v. (**bugged, bugging**) **1** tr. conceal a microphone in (esp. a building or room). **2** tr. sl. annoy, bother. **3** intr. **a** esp. US sl. (often foll. by out) leave quickly. **b** (foll. by off) go away. □ **bug-eyed** with bulging eyes. [17th c.: orig. unkn.]

bugaboo /'bʌgəˌbuː/ n. a bogey (see BOGEY[2]) or bugbear. [prob. f. dial. orig.: cf. Welsh bwcibo the Devil, bwci hobgoblin]

Buganda /buːˈgændə, bjuː-/ a former kingdom of East Africa, on the north shore of Lake Victoria, now part of Uganda.

bugbear /'bʌgbeə(r)/ n. **1** a cause of annoyance or anger; a bête noire. **2** an object of baseless fear. **3** archaic a sort of hobgoblin or other being invoked to intimidate children. [obs. bug + BEAR[1]]

bugger /'bʌgə(r)/ n., v., & int. coarse sl. (except in sense 2 of n. and 3 of v.) • n. **1 a** an unpleasant or awkward person or thing (the bugger won't fit). **b** a person of a specified kind (he's a miserable bugger; you clever bugger!). **2** a person who commits buggery. • v. **1** tr. as an exclamation of annoyance (bugger the thing!). **2** (often foll. by up) Brit. **a** ruin; spoil (really buggered it up; no good, it's buggered). **b** (esp. as **buggered** adj.) exhaust, tire out. **3** commit buggery with. • int. expressing annoyance. □ **bugger about** (or **around**) **1** (often foll. by with) fool about. **2** mislead; persecute; make things awkward for. **bugger-all** nothing. **bugger off** (often in imper.) go away. [ME f. MDu. f. OF bougre, orig. 'heretic' f. med.L Bulgarus Bulgarian (member of the Greek Orthodox Church)]

buggery /'bʌgərɪ/ n. **1** anal intercourse. **2** = BESTIALITY 2. [ME f. MDu. buggerie f. OF bougerie: see BUGGER]

buggy[1] /'bʌgɪ/ n. (pl. **-ies**) **1** a light, horse-drawn, esp. two-wheeled, vehicle for one or two people. **2** a small, sturdy, esp. open, motor vehicle (beach buggy; dune buggy). **3 a** N. Amer. a pram. **b** a child's collapsible pushchair (cf. Baby Buggy 1). [18th c.: orig. unkn.]

buggy[2] /'bʌgɪ/ adj. (**buggier, buggiest**) infested with bugs.

bugle[1] /'bjuːg(ə)l/ n. & v. • n. (also **bugle-horn**) a brass instrument like a small trumpet, used esp. for military signals and in fox-hunting etc. • v. **1** intr. sound a bugle. **2** tr. sound (a note, a call, etc.) on a bugle. □ **bugler** n. [ME, orig. = 'buffalo', f. OF f. L buculus dimin. of bos ox]

bugle[2] /'bjuːg(ə)l/ n. a creeping labiate plant, Ajuga reptans, with blue flowers. [ME f. LL bugula]

bugle[3] /'bjuːg(ə)l/ n. a tube-shaped bead sewn on a dress etc. for ornament. [16th c.: orig. unkn.]

bugloss /'bjuːglɒs/ n. **1** a bristly plant related to borage, esp. of the genus Anchusa, with bright blue tubular flowers. **2** = viper's bugloss (see VIPER). [F buglosse or L buglossus f. Gk bouglōssos ox-tongued]

buhl /buːl/ n. (also **boule, boulle**) **1** pieces of brass, tortoiseshell, etc., cut to make a pattern and used as decorative inlays esp. on furniture. **2** work inlaid with buhl. **3** (attrib.) inlaid with buhl. [(buhl Germanized) f. A. C. Boulle, Fr. cabinet-maker (1642–1732)]

build /bɪld/ v. & n. • v.tr. (past and past. part. **built** /bɪlt/) **1 a** construct (a house, vehicle, fire, road, model, etc.) by putting parts or material together. **b** commission, finance, and oversee the building of (*the council has built two new schools*). **2 a** (often foll. by *up*) establish, develop, make, or accumulate gradually (*built the business up from nothing*). **b** (often foll. by *on*) base (hopes, theories, etc.) (*ideas built on a false foundation*). **3** (as **built** *adj.*) having a specified build (*sturdily built; brick-built*). ● *n.* **1** the proportions of esp. the human body (*a slim build*). **2** a style of construction; a make (*the build of his suit was pre-war*). □ **build in** incorporate as part of a structure. **build in** (or **round** or **up**) surround with houses etc.; block up. **build on** add (an extension etc.). **build up 1** increase in size or strength. **2** praise; boost. **3** gradually become established. **build-up** *n.* **1** a favourable description in advance; publicity. **2** a gradual approach to a climax or maximum (*the build-up was slow but sure*). **3** an accumulation or increase. **built-in 1** (esp. of the fittings of a house) constructed to form an integral part of a larger unit (*built-in cupboard*). **2** inherent, integral, innate (*built-in obsolescence; built-in integrity*). **built on sand** unstable. **built-up 1** (of a locality) densely covered by houses etc. **2** increased in height etc. by the addition of parts. **3** composed of separately prepared parts. [OE *byldan* f. *bold* dwelling f. Gmc]

builder /ˈbɪldə(r)/ *n.* **1** a contractor for building houses etc.; a master builder. **2** a person engaged as a bricklayer etc. on a building site. □ **builders' merchant** a supplier of materials to builders.

building /ˈbɪldɪŋ/ *n.* **1** a permanent fixed structure forming an enclosure and providing protection from the elements etc. (e.g. a house, school, factory, or stable). **2** the constructing of such structures. □ **building line** a limit or boundary between a house and a street beyond which the owner may not build. **building site** an area before or during the construction of a house etc.

building society *n.* a society which pays interest on funds invested and lends money to people buying houses etc. Building societies date from the first part of the 19th century, when they consisted of a group of people contributing to a fund which was used to purchase a house for each of its members, the society being dissolved when this had been achieved. Later in the 19th century societies were established on a permanent basis. Since 1986 building societies have been able to widen the range of services they provide and they now compete in many areas with the high-street banks.

built *past* and *past part.* of BUILD.

Bujumbura /ˌbuːdʒəmˈbʊərə/ the capital of Burundi, at the north-eastern end of Lake Tanganyika; pop. (1989) 241,000. It was known as Usumbura until 1962.

Bukharin /buˈkɑːrɪn/, Nikolai (Ivanovich) (1888–1938), Russian revolutionary activist and theorist. Editor of *Pravda* (1918–29), a member of the Politburo (1924–9), and chairman of Comintern from 1926, he initially supported Stalin but was later denounced by him and expelled from the Politburo. After working as editor of the official government newspaper *Izvestia* (1934–7), he became one of the victims of Stalin's purges and was arrested, convicted, and shot. Bukharin was a model for Rubashov, hero of Arthur Koestler's novel *Darkness at Noon* (1940).

Bukhoro /buːˈkɔːrə/ (also **Bukhara** /bəˈkɑːrə/, **Bokhara**) a city in the central Asian republic of Uzbekistan; pop. (1990) 246,200. Situated in a large cotton-growing district, it is one of the oldest trade centres in central Asia, and is noted for the production of karakul fleeces.

Bukovina /ˌbʊkəˈviːnə/ a region of SE Europe in the Carpathians, divided between Romania and Ukraine. Formerly a province of Moldavia, it was ceded to Austria by the Turks in 1775 and became an Austrian duchy in 1849. After the First World War it was made part of Romania. The northern part was occupied by the Soviets in the Second World War, and became incorporated into the Ukrainian SSR.

Bulawayo /ˌbʊləˈweɪəʊ/ an industrial city in western Zimbabwe; pop. (1982) 414,800.

bulb /bʌlb/ *n.* **1 a** a fleshy-leaved storage organ of some plants (e.g. lily, onion) sending roots downwards and leaves upwards. **b** a plant grown from this, e.g. a daffodil. **2** = *light bulb* (see LIGHT¹). **3** any object or part shaped like a bulb. [L *bulbus* f. Gk *bolbos* onion]

bulbil /ˈbʌlbɪl/ *n. Bot.* **1** a small bulb which grows among the leaves or flowers of a plant. **2** a small bulb formed at the side of an ordinary bulb. [mod. L *bulbillus* dimin. of *bulbus* BULB]

bulbous /ˈbʌlbəs/ *adj.* **1** shaped like a bulb; fat or bulging. **2** having a bulb or bulbs. **3** (of a plant) growing from a bulb.

bulbul /ˈbʊlbʊl/ *n.* **1** an Asian or African songbird of the family Pycnonotidae, of dull plumage with contrasting bright patches. **2** a singer or poet. [Pers. f. Arab., of imit. orig.]

Bulganin /bʊlˈɡɑːnɪn/, Nikolai (Aleksandrovich) (1895–1975), Soviet statesman, Chairman of the Council of Ministers (Premier) 1955–8. Born in Russia, he succeeded Stalin as Minister of Defence in 1946, and was appointed Vice-Premier in the government of Georgi Malenkov (1903–88) in 1953. Following Malenkov's resignation in 1955, Bulganin became Premier, sharing power with Khrushchev, who replaced him.

Bulgar /ˈbʌlɡɑː(r)/ *n.* **1** a member of an ancient Turkic tribe that conquered the Slavs of the lower Danube area in the 7th century AD and settled in what is now Bulgaria, adopting a Slavonic language (Bulgarian) from the people they had conquered. **2** a Bulgarian. [med.L *Bulgarus* f. OBulg. *Blŭgarinŭ*]

bulgar /ˈbʌlɡə(r)/ *n.* (also **bulghur**, **bulgur**) a cereal food of whole wheat partially boiled then dried, eaten esp. in Turkey. [Turk.]

Bulgaria /bʌlˈɡeərɪə/ a country in SE Europe on the western shores of the Black Sea; pop. (est. 1991) 8,798,000; official language, Bulgarian; capital, Sofia. It was settled by Slavonic peoples during the Dark Ages, and conquered by the Bulgars, from whom it takes its name, in the 7th century. Part of the Ottoman Empire from the 14th century, Bulgaria remained under Turkish rule until the Russo-Turkish Wars of the late 19th century. Becoming fully independent (as a monarchy) in 1908, it fought on the German side in both world wars. Bulgaria was occupied by the Soviets after the Second World War, and a Communist state was set up which was one of the most consistently pro-Soviet members of the Warsaw Pact. A multi-party democratic system was introduced in 1989.

Bulgarian /bʌlˈɡeərɪən/ *n. & adj.* ● *n.* **1 a** a native or national of Bulgaria. **b** a person of Bulgarian descent. **2** the official language of Bulgaria, belonging to the Slavonic group of languages, within which it is most closely related to Serbo-Croat. ● *adj.* of or relating to Bulgaria or its people or language. [med.L *Bulgaria* f. *Bulgarus*: see BULGAR]

bulge /bʌldʒ/ *n. & v.* **1 a** a convex part of an otherwise flat or flatter surface. **b** an irregular swelling; a lump. **2** *colloq.* a temporary increase in quantity or number (*baby bulge*). **3** *Naut.* the bilge of a ship. **4** *Mil.* a salient. ● *v.* **1** *intr.* swell outwards. **2** *intr.* be full or replete. **3** *tr.* swell (a bag, cheeks, etc.) by stuffing. □ **have** (or **get**) **the bulge on** *sl.* have or get an advantage over. □ **bulgingly** *adv.* **bulgy** *adj.* [ME f. OF *boulge, bouge* f. L *bulga*: see BUDGET]

Bulge, Battle of the in the Second World War, a German counter-offensive in the Ardennes aimed at preventing an Allied invasion of Germany, late 1944–early 1945. The Germans drove a salient or 'bulge' about 60 miles (110 km) deep in the front line, but were later forced to retreat.

bulgur (also **bulghur**) var. of BULGAR.

bulimarexia /bjuːˌlɪməˈreksɪə, buː-/ *n.* esp. *US* = BULIMIA 2. □ **bulimarexic** *adj. & n.* [BULIMIA + ANOREXIA]

bulimia /bjuːˈlɪmɪə, buː-/ *n. Med.* **1** insatiable overeating. **2** (in full **bulimia nervosa**) an emotional disorder in which bouts of extreme overeating are followed by depression and self-induced vomiting, purging, or fasting. □ **bulimic** *adj. & n.* [mod.L f. Gk *boulimia* f. *bous* ox + *limos* hunger]

bulk /bʌlk/ *n. & v.* ● *n.* **1** (usu. prec. by *the* and foll. by *of*; the verb agrees with the complement) the greater part or number (*the bulk of the applicants are women*). **2** a bodily frame of large proportions (*jacket barely covered his bulk*). **3 a** size, magnitude, or volume (*its bulk is not enormous*). **b** a large mass or shape. **c** great quantity or volume (often *attrib.: a bulk supplier*). **4** roughage in food. **5** *Naut.* a ship's cargo, esp. unpackaged. ● *v.* **1** *intr.* seem in respect of size or importance (*bulks large in his reckoning*). **2** *tr.* (often foll. by *out*) make (a book, a textile yarn, etc.) seem thicker by suitable treatment (*bulked it with irrelevant stories*). **3** *tr.* combine (consignments etc.). □ **break bulk** begin unloading (cargo). **bulk-buy** buy in bulk; engage in bulk-buying. **bulk-buying 1** buying in large amounts at a discount. **2** the purchase by one buyer of all or most of a producer's output. **in bulk 1** in large quantities, usu. at a lower price. **2** (of a cargo) loose, not packaged. [sense 'cargo' f. ON *búlki*; sense 'mass' etc. perh. alt. f. obs. *bouk* (cf. BUCK⁵)]

bulkhead /ˈbʌlkhed/ *n.* an upright partition separating the compartments in a ship, aircraft, vehicle, etc. [*bulk stall* f. ON *bálkr* + HEAD]

bulky /ˈbʌlkɪ/ *adj.* (**bulkier**, **bulkiest**) **1** taking up much space, large. **2** awkwardly large, unwieldy. □ **bulkily** *adv.* **bulkiness** *n.*

bull¹ /bʊl/ *n., adj., & v.* ● *n.* **1 a** an uncastrated male bovine animal. **b** a male of the whale, elephant, and other large animals. **2** (**the Bull**) the zodiacal sign or constellation Taurus. **3** *Brit.* the bull's-eye of a target. **4** *Stock Exch.* a person who buys shares hoping to sell them at a higher price later (cf. BEAR[2] n. 3). ● attrib.adj. like that of a bull (bull neck). ● v. **1** tr. & intr. act or treat violently. **2** Stock Exch. **a** intr. speculate for a rise. **b** tr. raise price of (stocks, etc.). □ **bull ant** Austral. = bulldog ant. **bull at a gate** a hasty or rash person. **bull-fiddle** US colloq. a double bass. **bull-horn** a megaphone. **bull in a china shop** a reckless or clumsy person. **bull market** a market with shares rising in price. **bull-nose** (or **-nosed**) with a rounded end. **bull session** N. Amer. an informal group discussion. **bull's-eye 1** the centre of a target; a shot that hits this. **2** a large hard peppermint-flavoured sweet. **3** a hemisphere or thick disc of glass in a ship's deck or side to admit light. **4** a small circular window. **5 a** a hemispherical lens. **b** a lantern fitted with this. **6** a boss of glass at the centre of a blown glass sheet. **bull terrier** a breed of short-haired dog that is a cross between a bulldog and a terrier. **take the bull by the horns** face danger or challenge boldly. [ME f. ON boli = MLG, MDu bulle]

bull[2] /bʊl/ n. a papal edict. [ME f. OF bulle f. L bulla rounded object, in med.L 'seal']

bull[3] /bʊl/ n. **1** (also **Irish bull**) an expression containing a contradiction in terms or implying ludicrous inconsistency. **2** sl. (cf. BULLSHIT) **a** nonsense. **b** trivial or insincere talk or writing. **c** unnecessary routine tasks or discipline. **d** US a bad blunder. [17th c.: orig. unkn.]

bullace /'bʊlɪs/ n. a thorny rosaceous shrub, Prunus insititia, bearing globular yellow or purple-black fruits, of which the damson is the cultivated form. [ME f. OF buloce, beloce]

bulldog /'bʊldɒg/ n. **1** a sturdy powerful breed of dog with a large head and smooth hair. **2** a tenacious and courageous person. □ **bulldog ant** Austral. a large ant with a powerful sting. **bulldog clip** a strong sprung clip for papers.

bulldoze /'bʊldəʊz/ v.tr. **1** clear with a bulldozer. **2** colloq. **a** intimidate, force. **b** make (one's way) forcibly. [orig. (US) in sense 2a, f. BULL[1]: second element uncertain]

bulldozer /'bʊl,dəʊzə(r)/ n. **1** a powerful tractor with a broad curved upright blade at the front for clearing ground. **2** a forceful and domineering person.

bullet /'bʊlɪt/ n. a small round or cylindrical missile usu. with a pointed end, fired from a rifle, revolver, etc. □ **bullet-headed** having a round head. **bullet train** a high-speed streamlined passenger train, esp. in Japan. [F boulet, boulette dimin. of boule ball f. L bulla bubble]

bulletin /'bʊlɪtɪn/ n. **1** a short official statement of news. **2** a regular list of information etc. issued by an organization or society. □ **bulletin-board 1** US a notice-board. **2** a system for storing information on particular topics in a computer network so that users can access and add to it from remote terminals. [F f. It. bullettino dimin. of bulletta passport, dimin. of bulla seal, BULL[2]]

bulletproof /'bʊlɪt,pruːf/ adj. (of a material) designed to resist the penetration of bullets.

bullfighting /'bʊl,faɪtɪŋ/ n. a sport of baiting and (usu.) killing a bull as a public spectacle, taking place in an outdoor arena. In the most familiar form, the bull is tormented and wounded by mounted picadors and has darts stuck into its neck before a matador plays it with a red cape and eventually tries to kill it by plunging a sword between its shoulder and shoulder-blade. Bullfighting is the national spectator sport of Spain, and is found also in some parts of Latin America and southern France. An early type of bullfighting was practised by Minoans, Greeks, and Romans, and the sport seems to have been introduced into Spain in about the 11th century. □ **bullfight** n. **bullfighter** n.

bullfinch /'bʊlfɪntʃ/ n. a stocky Eurasian finch of the genus Pyrrhula, with a short stout bill; esp. P. pyrrhula, which is mainly grey with a pink breast.

bullfrog /'bʊlfrɒg/ n. a large frog, Rana catesbiana, native to North America, with a deep croak.

bullhead /'bʊlhed/ n. a small freshwater fish with a large head, esp. one of the family Cottidae such as the miller's thumb.

bull-headed /bʊl'hedɪd/ adj. obstinate; impetuous; blundering. □ **bull-headedly** adv. **bull-headedness** n.

bullion /'bʊlɪən/ n. a metal (esp. gold or silver) in bulk before coining, or valued by weight. [AF = mint, var. of OF bouillon ult. f. L bullire boil]

bullish /'bʊlɪʃ/ adj. **1** like a bull, esp. in temper. **2 a** Stock Exch. causing or associated with a rise in prices. **b** colloq. optimistic.

bullock /'bʊlək/ n. & v. ● n. a castrated male of domestic cattle, raised for beef. ● v.intr. (often foll. by at) Austral. colloq. work very hard. [OE bulluc, dimin. of BULL[1]]

bullocky /'bʊləki/ n. Austral. & NZ colloq. a bullock-driver.

bullring /'bʊlrɪŋ/ n. an arena for bullfights.

Bull Run a small river in east Virginia, scene of two Confederate victories, in 1861 and 1862, during the American Civil War.

bullshit /'bʊlʃɪt/ n. & v. coarse sl. ● n. **1** (often as int.) nonsense, rubbish. **2** trivial or insincere talk or writing. ● v.intr. & tr. (**-shitted**, **-shitting**) talk nonsense or as if one has specialized knowledge (to). □ **bullshitter** n. [BULL[1] + SHIT]

bulltrout /'bʊltraʊt/ n. Brit. a sea trout.

bully[1] /'bʊli/ n. & v. ● n. (pl. **-ies**) a person who uses strength or power to coerce others by fear. ● v.tr. (**-ies**, **-ied**) **1** persecute or oppress by force or threats. **2** (foll. by into + verbal noun) pressure or coerce (a person) to do something (bullied him into agreeing). □ **bully-boy** a hired ruffian. [orig. as a term of endearment, prob. f. MDu. boele lover]

bully[2] /'bʊli/ adj. & int. colloq. ● esp. N. Amer. adj. very good; first-rate. ● int. (foll. by for) expressing admiration or approval, or iron. (bully for them!). [perh. f. BULLY[1]]

bully[3] /'bʊli/ n. & v. (in full **bully off**) ● n. (pl. **-ies**) the start of play in hockey in which two opponents strike each other's sticks three times and then go for the ball. ● v.intr. (**-ies**, **-ied**) start play in this way. [19th c.: perh. f. bully scrum in Eton football, of unkn. orig.]

bully[4] /'bʊli/ n. (in full **bully beef**) corned beef. [F bouilli boiled beef f. bouillir BOIL[1]]

bullyrag var. of BALLYRAG.

bully tree n. = BALATA. [corrupt.]

bulrush /'bʊlrʌʃ/ n. **1** = reed-mace. **2** a rushlike water-plant, Scirpus lacustris, used for weaving. **3** Bibl. a papyrus plant. [perh. f. BULL[1] = large, coarse, as in bullfrog, bulltrout, etc.]

Bultmann /'bʊltmən/, Rudolf (Karl) (1884–1976), German Lutheran theologian. He held that the Gospels were a patchwork of traditional elements and insisted on the need for demythologizing the whole Gospel story. He emphasized what he saw as its 'existential' rather than its historical significance, asserting that faith in Christ rather than belief in him was the key. His important works include The History of the Synoptic Tradition (1921), Jesus Christ and Mythology (1960), and Existence and Faith (1964).

bulwark /'bʊlwək/ n. & v. ● n. **1** a defensive wall, esp. of earth; a rampart; a mole or breakwater. **2** a person, principle, etc., that acts as a defence. **3** (usu. in pl.) a ship's side above deck. ● v.tr. serve as a bulwark to; defend, protect. [ME f. MLG, MDu. bolwerk: see BOLE[1], WORK]

Bulwer-Lytton /,bʊlwə'lɪt(ə)n/ see LYTTON.

bum[1] /bʌm/ n. Brit. sl. the buttocks. □ **bum-bag** colloq. a small pouch for money and other valuables, on a belt worn round the waist or hips. **bum-bailiff** hist. a bailiff empowered to collect debts or arrest debtors for non-payment. **bum-boat** a small boat plying with provisions etc. for ships. **bum-fluff** sl. an adolescent's first beard growth. **bum-sucker** sl. a toady. **bum-sucking** sl. toadying. [ME bom, of unkn. orig.]

bum[2] /bʌm/ n., v., & adj. sl. ● n. a habitual loafer or tramp; a lazy dissolute person. ● v. (**bummed**, **bumming**) **1** intr. (often foll. by about, around) loaf or wander around; be a bum. **2** tr. get by begging; cadge. ● attrib.adj. **1** of poor quality; bad, worthless. **2** false, fabricated. □ **bum rap** N. Amer. sl. imprisonment on a false charge. **bum's rush** (prec. by the) N. Amer. sl. forcible ejection. **bum steer** N. Amer. sl. false information. **on the bum** vagrant, begging. [orig. N. Amer., prob. abbr. or back-form. f. BUMMER]

bumble /'bʌmb(ə)l/ v.intr. **1** (foll. by on) speak in a rambling incoherent way. **2** (often as **bumbling** adj.) move or act ineptly; blunder. **3** make a buzz or hum. □ **bumbler** n. [BOOM[1] + -LE[4]: partly f. bumble = blunderer]

bumble-bee /'bʌmb(ə)l,biː/ n. a large social bee of the genus Bombus, with a loud hum. Also called humble-bee. [as BUMBLE]

bumboy /'bʌmbɔɪ/ n. coarse sl. a young male homosexual, esp. a prostitute.

bumf /bʌmf/ n. (also **bumph**) Brit. colloq. **1** usu. derog. papers, documents. **2** lavatory paper. [abbr. of bum-fodder]

bumiputra /,buːmɪ'puːtrə/ n. & adj. ● n. (pl. same or **bumiputras**) a Malaysian of indigenous Malay origin. ● adj. of or relating to the bumiputra. [Malay f. Skr., = son of the soil]

bummalo /'bʌmə,ləʊ/ n. (pl. same) a small fish, Harpodon nehereus, of southern Asian coasts, dried and used as food (see BOMBAY DUCK). [perh. f. Marathi bombīl(a)]

bummer /'bʌmə(r)/ n. US sl. **1** an idler; a loafer. **2** an unpleasant occurrence. [19th c.: perh. f. G Bummler]

bump /bʌmp/ n., v., & adv. ● n. **1** a dull-sounding blow or collision. **2** a swelling or dent caused by this. **3** an uneven patch on a road, field, etc. **4** *hist.* a prominence on the skull formerly thought to indicate a particular mental faculty. **5** (in narrow-river races where boats make a spaced start one behind another) the point at which a boat begins to overtake (and usu. touches) the boat ahead, thereby defeating it. **6** *Aeron.* **a** an irregularity in an aircraft's motion. **b** a rising air current causing this. **7** (**the bumps**) *Brit.* (on a person's birthday) the act of lifting a person by the arms and legs and letting him or her down on to the ground, once for each year of age. ● *v.* **1** *tr.* hit or come against with a bump. **b** *intr.* (of two objects) collide. **2** *intr.* (foll. by *against*, *into*) hit with a bump; collide with. **3** *tr.* (often foll. by *against*, *on*) hurt or damage by striking (*bumped my head on the ceiling*; *bumped the car while parking*). **4** *intr.* (usu. foll. by *along*) move or travel with much jolting (*we bumped along the road*). **5** *tr.* (in a boat race) gain a bump against. **6** *tr.* *N. Amer.* displace, esp. by seniority. ● *adv.* with a bump; suddenly; violently. □ **bump into** *colloq.* meet by chance. **bump off** *sl.* murder. **bump-start** = *push-start*. **bump up** *colloq.* increase (prices etc.). [16th c., imit.: perh. f. Scand.]

bumper /'bʌmpə(r)/ *n.* **1** a horizontal bar or strip fixed across the front or back of a motor vehicle to reduce damage in a collision or as a trim. **2** (usu. *attrib.*) an unusually large or fine example (*a bumper crop*). **3** *Cricket* = BOUNCER 2. **4** a brim-full glass of wine etc. □ **bumper car** = DODGEM.

bumph var. of BUMF.

bumpkin /'bʌmpkɪn/ *n.* a rustic or socially inept person. [perh. Du. *boomken* little tree or MDu. *bommekijn* little barrel]

bumptious /'bʌmpʃəs/ *adj.* offensively self-assertive or conceited. □ **bumptiously** *adv.* **bumptiousness** *n.* [BUMP, after FRACTIOUS]

bumpy /'bʌmpɪ/ *adj.* (**bumpier**, **bumpiest**) **1** having many bumps (*a bumpy road*). **2** affected by bumps (*a bumpy ride*). □ **bumpily** *adv.* **bumpiness** *n.*

bun /bʌn/ *n.* **1** a small usu. sweetened bread roll or cake, often with dried fruit. **2** *Sc.* a rich fruit cake or currant bread. **3** hair worn in a tight coil at the back of the head. **4** (in *pl.*) *N. Amer. sl.* the buttocks. □ **bun fight** *Brit. sl.* a tea party. **have a bun in the oven** *sl.* be pregnant. **hot cross bun** a bun marked with a cross, traditionally eaten on Good Friday. [ME: orig. unkn.]

Buna /'bjuːnə, 'buːnə/ *n. propr.* a synthetic rubber made by polymerization of butadiene. [G (as BUTADIENE, *Na* chem. symbol for sodium)]

Bunbury /'bʌnbərɪ/ a seaport and resort to the south of Perth in Western Australia; pop. (est. 1989) 26,400.

bunch /bʌntʃ/ *n. & v.* ● *n.* **1** a cluster of things growing or fastened together (*bunch of grapes*; *bunch of keys*). **2** a collection; a set or lot (*best of the bunch*). **3** *colloq.* a group; a gang. **4 a** (*pl.*) *Brit.* a hairstyle (esp. of girls) in which the hair is drawn back into two pony-tails, one at each side of the head. **b** (in *sing.*) either of these pony-tails. ● *v.* **1** *tr.* make into a bunch or bunches; gather into close folds. **2** *intr.* form into a group or crowd. □ **bunch grass** *N. Amer.* a grass that grows in clumps, esp. of the genus *Poa* or *Festuca*. **bunch of fives** see FIVE. □ **bunchy** *adj.* [ME: orig. unkn.]

bunco /'bʌŋkəʊ/ *n. & v. US sl.* ● *n.* (*pl.* **-os**) a swindle, esp. by card-sharping or a confidence trick. ● *v.tr.* (**-oes**, **-oed**) swindle, cheat. [perh. f. Sp. *banca* a card-game]

buncombe var. of BUNKUM.

Bundesbank /'bʊndəsˌbæŋk/ the central bank of Germany, established in 1875. Its headquarters are in Frankfurt. [G f. *Bund* federation + BANK²]

Bundesrat /'bʊndəsˌrɑːt/ *n.* the upper House of Parliament in Germany or in Austria. [G f. *Bund* federation + *Rat* council]

Bundestag /'bʊndəsˌtɑːg/ *n.* the lower House of Parliament in Germany. [G f. *Bund* federation + *tagen* confer]

bundle /'bʌnd(ə)l/ *n. & v.* ● *n.* **1** a collection of things tied or fastened together. **2** a set of nerve fibres etc. banded together. **3** *sl.* a large amount of money. ● *v.* **1** *tr.* (usu. foll. by *up*) tie in or make into a bundle (*bundled up my squash kit*). **2** *tr.* (usu. foll. by *into*) throw or push, esp. quickly or confusedly (*bundled the papers into the drawer*). **3** *tr.* (usu. foll. by *out*, *off*, *away*, etc.) send (esp. a person) away hurriedly or unceremoniously (*bundled them off the premises*). **4** *tr.* (foll. by *with*) (usu. as **bundled** *adj.*) *Computing* sell as a package with (*training is bundled with the software*). **5** *intr.* sleep clothed with another person, esp. a fiancé(e), as a local custom. □ **be a bundle of nerves** (or **prejudices** etc.) be extremely nervous (or prejudiced etc.). **bundle up** dress warmly or cumbersomely. **go a bundle on** *sl.* be very fond of. □ **bundler** *n.* [ME, perh. f. OE *byndelle* a binding, but also f. LG, Du *bundel*]

bung¹ /bʌŋ/ *n. & v.* ● *n.* a stopper for closing a hole in a container, esp. a cask. ● *v.tr.* **1** stop with a bung. **2** *Brit. sl.* throw, toss. □ **bunged up** closed, blocked. **bung-hole** a hole for filling or emptying a cask etc. [MDu. *bonghe*]

bung² /bʌŋ/ *adj. Austral. & NZ sl.* dead; ruined, useless. □ **go bung 1** die. **2** fail; go bankrupt. [Aboriginal]

bung³ /bʌŋ/ *n. sl.* a bribe. [20th c.: orig. unkn.]

bungalow /'bʌŋgəˌləʊ/ *n.* a one-storeyed house. [Gujarati *bangalo* f. Hind. *baṅglā* belonging to Bengal]

bungee /'bʌndʒɪ/ *n.* (in full **bungee cord**, **rope**) elasticated cord or rope used for securing baggage and in bungee jumping. □ **bungee jumping** the sport of jumping from a height while secured by a bungee from the ankles or a harness. [20th c.: orig. unkn.]

bungle /'bʌŋg(ə)l/ *v. & n.* ● *v.* **1** *tr.* blunder over, mismanage, or fail at (a task). **2** *intr.* work badly or clumsily. ● *n.* a bungled attempt; bungled work. □ **bungler** *n.* [imit.: cf. BUMBLE]

Bunin /'buːnɪn/, Ivan (Alekseevich) (1870–1953), Russian poet and prose-writer. An opponent of modernism, he made peasant life and love the most prominent themes in his prose works, which include the novel *The Village* (1910), the short-story collection *The Gentleman from San Francisco* (1914), and an autobiography, *The Well of Days* (1910). He opposed the October Revolution and left Russia in 1918, eventually reaching France and remaining in permanent exile. In 1933 he became the first Russian to be awarded the Nobel Prize for literature.

bunion /'bʌnjən/ *n.* a swelling on the foot, esp. at the first joint of the big toe. [OF *buignon* f. *buigne* bump on the head]

bunk¹ /bʌŋk/ *n.* a sleeping-berth, esp. a shelflike bed against a wall, e.g. in a ship. □ **bunk-bed** each of two or more beds one above the other, forming a unit. **bunk-house** a house where workers etc. are lodged. [18th c.: orig. unkn.]

bunk² /bʌŋk/ *v. & n. Brit.* ● *v.tr.* (also *absol.*; often foll. by *off*) play truant from (school etc.). ● *n.* (in **do a bunk**) leave or abscond hurriedly. [19th c.: orig. unkn.]

bunk³ /bʌŋk/ *n. sl.* nonsense, humbug. [abbr. of BUNKUM]

bunker /'bʌŋkə(r)/ *n. & v.* ● *n.* **1** a large container or compartment for storing fuel. **2** a reinforced underground shelter, esp. for use in wartime. **3** a hollow filled with sand, used as an obstacle in a golf course. ● *v.tr.* **1** fill the fuel bunkers of (a ship etc.). **2** (usu. in *passive*) **a** trap in a bunker (in sense 3). **b** bring into difficulties. [19th c.: orig. unkn.]

Bunker Hill the first pitched battle (1775) of the War of American Independence (actually fought on Breed's Hill near Boston, Massachusetts). Although the British were able to drive the Americans from their positions, the good performance of the untrained American irregulars gave considerable impetus to the Revolution.

bunkum /'bʌŋkəm/ *n.* (also **buncombe**) nonsense; humbug. [orig. *buncombe* f. *Buncombe* County in North Carolina, mentioned in a nonsense speech by its Congressman, *c*.1820]

bunny /'bʌnɪ/ *n.* (*pl.* **-ies**) **1** a child's name for a rabbit. **2** *Austral. sl.* a victim or dupe. **3** (in full **bunny girl**) a club hostess, waitress, etc., wearing a skimpy costume with ears and a tail suggestive of a rabbit. [dial. *bun* rabbit]

Bunsen /'bʌns(ə)n, German 'bʊnz(ə)n/, Robert Wilhelm Eberhard (1811–99), German chemist. Bunsen was a pioneer of chemical spectroscopy and photochemistry. During his early research he lost the use of his right eye in an explosion. With G. Kirchhoff he developed spectroscopy, using it to detect new elements (caesium and rubidium) and to determine the composition of substances and of the sun and stars. Bunsen designed numerous items of chemical apparatus, besides the burner (developed in 1855) for which he is best known.

Bunsen burner *n.* a small adjustable gas burner used in scientific work as a source of heat. [BUNSEN]

bunt¹ /bʌnt/ *n.* the baggy centre of a fishing-net, sail, etc. [16th c.: orig. unkn.]

bunt² /bʌnt/ *n.* a disease of wheat caused by the fungus *Tilletia caries*. [18th c.: orig. unkn.]

bunt³ /bʌnt/ *v. & n. Baseball* ● *v.tr.* (of a batter) make (the ball) rebound from the bat without swinging. ● *n.* an instance of this. [19th c.: cf. BUTT¹]

buntal /'bʌnt(ə)l/ *n.* the straw from a talipot palm. [Tagalog]

Bunter /'bʌntə(r)/, Billy, a schoolboy character, noted for his fatness and gluttony, in stories by Frank Richards (pseudonym of Charles Hamilton, 1876–1961).

bunting[1] /ˈbʌntɪŋ/ n. a seed-eating bird of the family Emberizidae, related to the finches, with streaky sparrow-like plumage, e.g. the yellowhammer. [ME: orig. unkn.]

bunting[2] /ˈbʌntɪŋ/ n. **1** flags and other decorations. **2** a loosely woven fabric used for these. [18th c.: orig. unkn.]

buntline /ˈbʌntlaɪn/ n. a line for confining the bunt (see BUNT[1]) when furling a sail.

Buñuel /buːˈnwel/, Luis (1900–83), Spanish film director. Profoundly influenced by surrealism, he wrote and directed his first film Un Chien andalou (1928) jointly with Salvador Dali. Remarkable for their shocking and terrifying images, his films often attacked the Establishment (in particular the Church); after the banning of his early work, he left Spain and worked on dubbing American films for fifteen years before re-establishing his reputation in Mexico with Los Olvidados (1950). Among the notable films that followed were Belle de jour (1967) and The Discreet Charm of the Bourgeoisie (1972).

bunya /ˈbʌnjə/ n. (also **bunya bunya**) Austral. a tall coniferous tree, Araucaria bidwillii, bearing large edible cones. [Aboriginal]

Bunyan /ˈbʌnjən/, John (1628–88), English writer. He served with the Parliamentary army during the Civil War, an experience which possibly inspired his allegory The Holy War (1682). In 1653 he joined the Nonconformist Church at Bedford, where he preached and clashed with the Quakers. He was put under arrest in 1660 for unlicensed preaching and spent most of the next twelve years in prison, where he wrote his spiritual autobiography Grace Abounding (1666) and a number of other works; during a later period of imprisonment, he began his major work The Pilgrim's Progress (1678–84), an allegory recounting the spiritual journey of its hero Pilgrim.

bunyip /ˈbʌnjɪp/ n. Austral. **1** a fabulous monster inhabiting swamps and lagoons. **2** an impostor. [Aboriginal]

Buonaparte see BONAPARTE.

Buonarroti /ˌbwɒnəˈrɒti/, Michelangelo, see MICHELANGELO.

buoy /bɔɪ/ n. & v. ● n. **1** an anchored float serving as a navigation mark or to show reefs etc. **2** a lifebuoy. ● v.tr. **1** (usu. foll. by up) **a** keep afloat. **b** sustain the courage or spirits of (a person etc.); uplift, encourage. **2** (often foll. by out) mark with a buoy or buoys. [ME prob. f. MDu. bo(e)ye, ult. f. L boia collar f. Gk boeiai ox-hides]

buoyancy /ˈbɔɪənsi/ n. **1** the capacity to be or remain buoyant. **2** resilience; recuperative power. **3** cheerfulness. □ **buoyancy aid** a sleeveless jacket lined with buoyant material, worn for water sports.

buoyant /ˈbɔɪənt/ adj. **1 a** able or apt to keep afloat or rise to the top of a liquid or gas. **b** (of a liquid or gas) able to keep something afloat. **2** light-hearted, resilient. □ **buoyantly** adv. [F buoyant or Sp. boyante part. of boyar float f. boya BUOY]

BUPA /ˈbuːpə/ abbr. British United Provident Association, a private health insurance organization.

bur /bɜː(r)/ n. (also **burr**) **1 a** a prickly clinging seed-case or flower-head. **b** a plant producing these. **2** a person hard to shake off. **3** = BURR n. 2. □ **bur oak** a North American oak, Quercus macrocarpa with large fringed acorn-cups. **bur walnut** walnut wood containing knots, used as veneer. [ME: cf. Danish burre bur, burdock, Sw. kard-borre burdock]

burb /bɜːb/ n. (usu. in pl.) N. Amer. colloq. a suburb. [abbr.]

Burbage /ˈbɜːbɪdʒ/, Richard (c.1567–1619), English actor. He was the creator of most of Shakespeare's great tragic roles – Hamlet, Othello, Lear, and Richard III. He was also associated with the building of the Globe Theatre.

Burbank /ˈbɜːbæŋk/ a city in southern California, on the north side of the Los Angeles conurbation; pop. (1990) 93,640. It is a centre of the film and television industries.

Burberry /ˈbɜːbəri/ n. propr. a distinctive type of raincoat. [Burberry's, name of the manufacturer]

burble /ˈbɜːb(ə)l/ v. & n. ● v.intr. **1** speak ramblingly; make a murmuring noise. **2** Aeron. (of an airflow) break up into turbulence. ● n. **1** a murmuring noise. **2** rambling speech. □ **burbler** n. [19th c.: imit.]

burbot /ˈbɜːbət/ n. an elongated freshwater fish, Lota lota, of the carp family, with a broad head and barbels. [ME: cf. OF barbote]

burden /ˈbɜːd(ə)n/ n. & v. (also archaic **burthen** /ˈbɜːð(ə)n/) ● n. **1** a load, esp. a heavy one. **2** an oppressive duty, obligation, expense, emotion, etc. **3** the bearing of loads (beast of burden). **4** a ship's carrying-capacity, tonnage. **5 a** the refrain or chorus of a song. **b** the chief theme or gist of a speech, book, poem, etc. ● v.tr. load with a burden; encumber,

oppress. □ **burden of proof** the obligation to prove one's case. □ **burdensome** adj. **burdensomeness** n. [OE byrthen: rel. to BIRTH]

burdock /ˈbɜːdɒk/ n. a coarse composite plant of the genus Arctium, with large docklike leaves and prickly fruits. [BUR + DOCK[3]]

bureau /ˈbjʊərəʊ/ n. (pl. **bureaux** or **bureaus** /-rəʊz/) **1 a** Brit. a writing-desk with drawers and usu. an angled top opening downwards to form a writing surface. **b** N. Amer. a chest of drawers. **2 a** an office or department for transacting specific business. **b** a government department. [F, = desk, orig. its baize covering, f. OF burel f. bure, buire dark brown ult. f. Gk purros red]

bureaucracy /bjʊəˈrɒkrəsi/ n. (pl. **-ies**) **1 a** government by central administration. **b** a state or organization so governed. **2** the officials of such a government, esp. regarded as oppressive and inflexible. **3** conduct typical of such officials. [F bureaucratie: see BUREAU]

bureaucrat /ˈbjʊərəˌkræt/ n. **1** an official in a bureaucracy. **2** an inflexible or insensitive administrator. □ **bureaucratic** /ˌbjʊərəˈkrætɪk/ adj. **bureaucratically** adv. [F bureaucrate (as BUREAUCRACY)]

bureaucratize /bjʊəˈrɒkrəˌtaɪz/ v.tr. (also **-ise**) govern by or transform into a bureaucratic system. □ **bureaucratization** /-ˌrɒkrətaɪˈzeɪʃ(ə)n/ n.

burette /bjʊəˈret/ n. (US **buret**) a graduated glass tube with a tap at the lower end, for measuring small volumes of liquid in chemical analysis. [F]

burg /bɜːg/ n. N. Amer. colloq. a town or city. [see BOROUGH]

burgage /ˈbɜːgɪdʒ/ n. hist. (in England and Scotland) tenure of land in a town on a yearly rent. [ME f. med.L burgagium f. burgus BOROUGH]

Burgas /bʊəˈgæs/ an industrial port and resort in Bulgaria, on the coast of the Black Sea; pop. (1990) 226,120.

burgee /bɜːˈdʒiː/ n. a flag bearing the colours or emblem of a sailing club. [18th c.: perh. = (ship)owner, ult. F bourgeois: see BURGESS]

Burgenland /ˈbʊəgənˌlænd/ a state of eastern Austria; capital, Eisenstadt.

burgeon /ˈbɜːdʒən/ v. & n. literary ● v.intr. **1** begin to grow rapidly; flourish. **2** put out young shoots; bud. ● n. a bud or young shoot. [ME f. OF bor-, burjon ult. f. LL burra wool]

burger /ˈbɜːgə(r)/ n. **1** colloq. a hamburger. **2** (in comb.) a certain kind of hamburger or variation of it (beefburger; nutburger). [abbr.]

Burgess[1] /ˈbɜːdʒɪs/, Anthony (pseudonym of John Anthony Burgess Wilson) (1917–93), English novelist and critic. His experiences as an education officer in Malaya and Borneo (1954–60) inspired his first novels The Malayan Trilogy (1956–9). One of his best-known novels is A Clockwork Orange (1962), a disturbing, futuristic vision of juvenile delinquency, violence, and high technology. His many other works include the best-selling novel Earthly Powers (1980).

Burgess[2] /ˈbɜːdʒɪs/, Guy (Francis de Moncy) (1911–63), British Foreign Office official and spy. Acting as a Soviet agent from the 1930s, he worked for MI5 while ostensibly employed by the BBC. After the war, he served the Foreign Office and became Second Secretary at the British Embassy in Washington, DC, under Kim Philby. Charged with espionage in 1951, he fled to the USSR with Donald Maclean.

burgess /ˈbɜːdʒɪs/ n. **1** Brit. an inhabitant of a town or borough, esp. of one with full municipal rights. **2** Brit. hist. a member of parliament for a borough, corporate town, or university. **3** US a borough magistrate or governor. [ME f. OF burgeis ult. f. LL burgus BOROUGH]

burgh /ˈbʌrə/ n. hist. a Scottish borough or chartered town. ¶ This status was abolished in 1975. □ **burghal** /ˈbɜːg(ə)l/ adj. [Sc. form of BOROUGH]

burgher /ˈbɜːgə(r)/ n. **1** a citizen or freeman, esp. of a foreign town. **2** S. Afr. hist. a citizen of a Boer republic. **3** a descendant of a Dutch or Portuguese colonist in Sri Lanka. [G Burger or Du. burger f. Burg, burg BOROUGH]

Burghley /ˈbɜːli/, William Cecil, 1st Baron (1520–98), English statesman. Secretary of State to Queen Elizabeth I 1558–72 and Lord High Treasurer 1572–98, Burghley was the queen's most trusted councillor and minister and the driving force behind many of her government's policies.

burglar /ˈbɜːglə(r)/ n. a person who commits burglary. □ **burglarious** /bɜːˈgleərɪəs/ adj. [legal AF burgler, rel. to OF burgier pillage]

burglarize /ˈbɜːgləˌraɪz/ v.tr. & intr. (also **-ise**) US = BURGLE.

burglary /ˈbɜːgləri/ n. (pl. **-ies**) **1** entry into a building illegally with intent to commit theft, do bodily harm, or do damage. **2** an instance of this. ¶ Before 1968 in English law a crime under statute and in common law; after 1968 a statutory crime only (cf. HOUSEBREAKING). [legal AF *burglarie*: see BURGLAR]

burgle /'bɜːg(ə)l/ v. **1** tr. commit burglary on (a building or person). **2** intr. commit burglary. [back-form. f BURGLAR]

burgomaster /'bɜːgəˌmɑːstə(r)/ n. the mayor of a Dutch or Flemish town. [Du. *burgemeester* f. *burg* BOROUGH: assim. to MASTER]

Burgos /'bʊəgɒs/ a town in northern Spain; pop. (1991) 169,280. The capital of the former kingdom of Castile during the 11th century, it became politically significant again during the Spanish Civil War, when Franco made it the official seat of his Nationalist government (1936–9).

Burgoyne /bɜː'gɔɪn/, John ('Gentleman Johnny') (1722–92), English general and dramatist. He is largely remembered for surrendering to the Americans at Saratoga (1777) in the War of American Independence. His plays include the comedies *The Maid of the Oaks* (1774) and *The Heiress* (1786).

burgrave /'bɜːgreɪv/ n. hist. the ruler of a town or castle. [G *Burggraf* f. *Burg* BOROUGH + *Graf* COUNT²]

Burgundy /'bɜːgəndɪ/ (French **Bourgogne** /burgɔɲ/) a region and former duchy of east central France, centred on Dijon. Under a series of strong dukes Burgundy achieved considerable independence from imperial control in the later Middle Ages, before being absorbed by France when King Louis XI claimed the duchy in 1477. The region is noted for its wine. □ **Burgundian** /bɜː'gʌndɪən/ adj. & n.

burgundy /'bɜːgəndɪ/ n. (pl. **-ies**) **1 a** a wine (usu. red) from Burgundy. **b** a similar wine from another place. **2** the red colour of burgundy.

burhel var. of BHARAL.

burial /'berɪəl/ n. **1 a** the burying of a dead body. **b** a funeral. **2** Archaeol. a grave or its remains. □ **burial-ground** a cemetery. [ME, erron. formed as sing. of OE *byrgels* f. Gmc: rel. to BURY]

burin /'bjʊərɪn/ n. **1** a steel tool for engraving on copper or wood. **2** Archaeol. a flint tool with a chisel point. [F]

burk var. of BERK.

burka /'bɜːkə/ n. a long enveloping garment worn in public by Muslim women. [Hind. f. Arab. *burḳa*']

Burke¹ /bɜːk/, Edmund (1729–97), British man of letters and Whig politician. Burke was a prolific writer on the issues of political emancipation and moderation, supporting proposals for relaxing the laws against Roman Catholics in Britain and protesting against the harsh handling of the American colonies. He was a fierce opponent of the radical excesses of the French Revolution, calling on European leaders to resist the new regime in the influential *Reflections on the Revolution in France* (1790).

Burke² /bɜːk/, John (1787–1848), Irish genealogical and heraldic writer. He compiled *Burke's Peerage* (1826), the first reference guide of peers and baronets in alphabetical order.

Burke³ /bɜːk/, Robert O'Hara (1820–61), Irish explorer. He emigrated to Australia in 1853 and led a successful expedition from south to north across Australia in the company of William Wills and two other men — the first white men to make this journey. On the return journey, however, Burke, Wills, and a third companion died of starvation.

Burke⁴ /bɜːk/, William (1792–1829), Irish body-snatcher operating in Edinburgh, he was hanged for a series of fifteen or more murders carried out for profit. His accomplice, fellow-Irishman William Hare (fl.1820s), gave King's evidence and was never heard of again.

Burkina /bɜː'kiːnə/ (official name **Burkina Faso** /'fæsəʊ/) a landlocked country in western Africa, in the Sahel, known until 1984 as Upper Volta; pop. (est. 1991) 9,271,000; official language, French; capital, Ouagadougou. A French protectorate from 1898, it became an autonomous republic within the French Community in 1958 and a fully independent republic in 1960. □ **Burkinan** adj. & n.

Burkitt's lymphoma /'bɜːkɪts/ n. Med. a malignant tumour of the lymphatic system, caused by the Epstein–Barr virus, esp. affecting children in central Africa. [D. P. *Burkitt*, Brit. surgeon (1911–93)]

burl /bɜːl/ n. **1** a knot or lump in wool or cloth. **2** N. Amer. a flattened knotty growth on a tree. [ME f. OF *bourle* tuft of wool, dimin. of *bourre* coarse wool f. LL *burra* wool]

burlap /'bɜːlæp/ n. **1** coarse canvas esp. of jute used for sacking etc. **2** a similar lighter material for use in dressmaking or furnishing. [17th c.: orig. unkn.]

burlesque /bɜː'lesk/ n., adj., & v. ● n. **1 a** a comic imitation, esp. in parody of a dramatic or literary work. **b** a performance or work of this kind.

c bombast, mock-seriousness. **2** US a variety show, often including striptease. ● adj. of or in the nature of burlesque. ● v.tr. (**burlesques, burlesqued, burlesquing**) make or give a burlesque of. □ **burlesquer** n. [F f. It. *burlesco* f. *burla* mockery]

Burlington /'bɜːlɪŋtən/ a city in southern Canada, on Lake Ontario south-west of Toronto; pop. (1991) 129,600.

burly /'bɜːlɪ/ adj. (**burlier, burliest**) (of a person) of stout sturdy build; big and strong. □ **burliness** n. [ME *borli* prob. f. an OE form = 'fit for the bower' (BOWER¹)]

Burma /'bɜːmə/ (official name *Myanmar*) a country in SE Asia, on the Bay of Bengal; pop. (est. 1991) 42,528,000; official language, Burmese; capital, Rangoon. An independent empire centred on the city of Pagan in the 11th–13th centuries, Burma fell to the Mongols and was generally split into small rival states until unified once again in 1757. As a result of the Burmese Wars of 1823–86, Burma was gradually annexed by the British, remaining under British administration until the Second World War. Occupied by the Japanese from 1942 to 1945, the country became an independent republic in 1948, choosing not to join the Commonwealth. In 1962 an army coup led by Ne Win overthrew the government and established an authoritarian state, which continued to maintain a policy of neutrality and limited foreign contact. The official name of the country was changed to the Union of Myanmar in 1989. In the late 1980s the pro-democracy movement gathered momentum, but following widespread demonstrations against the government in 1988 the army formally imposed military rule. The National League for Democracy (NLD) won the election held in May 1990, even though its leader Aung San Suu Kyi was under house arrest; however, the military regime did not relinquish power.

Burman /'bɜːmən/ adj. & n. (pl. **Burmans**) = BURMESE.

Burma Road a route linking Lashio in Burma to Kunming in China, covering 1,154 km (717 miles). Completed in 1939, it was built by the Chinese in response to the Japanese occupation of the Chinese coast, to serve as a supply route to the interior. In 1942 the Japanese seized Lashio, closing the supply route at its source, but the Allies then constructed a route linking Ledo in India to a part of the Burma Road still in Chinese hands.

Burmese /bɜː'miːz/ n. & adj. ● n. (pl. same) **1 a** a native or national of Burma (Myanmar) in SE Asia. **b** a person of Burmese descent. **2** a member of the largest ethnic group of Burma. **3** the official language of Burma, the first language of about three-quarters of its population and a lingua franca for most of the remainder. (*See note below.*) **4** (in full **Burmese cat**) a breed of short-coated domestic cat. ● adj. of or relating to Burma or its people or language.

▪ Burmese is a tonal language belonging to the Sino-Tibetan group. It uses an alphabet derived from the ancient Indian script in which Pali, the language of Theravada Buddhist texts, was written.

burn¹ /bɜːn/ v. & n. ● v. (past and past part. **burnt** /bɜːnt/ or **burned** /bɜːnd/) **1** tr. & intr. be or cause to be consumed or destroyed by fire. **2** intr. **a** blaze or glow with fire. **b** be in the state characteristic of fire. **3** tr. & intr. be or cause to be injured or damaged by fire or great heat or by radiation. **4** tr. & intr. use or be used as a source of heat, light, or other energy. **5** tr. & intr. char or scorch in cooking (*burned the vegetables; the vegetables are burning*). **6** tr. produce (a hole, a mark, etc.) by fire or heat. **7** tr. **a** subject (clay, chalk, etc.) to heat for a purpose. **b** harden (bricks) by fire. **c** make (lime or charcoal) by heat. **8** tr. colour, tan, or parch with heat or light (*we were burnt brown by the sun*). **9** tr. & intr. put or be put to death by fire. **10** tr. **a** cauterize, brand. **b** (foll. by *in*) imprint by burning. **11** tr. & intr. make or be hot, give or feel a sensation or pain of or like heat; smart or cause to smart. **12 a** tr. & intr. (often foll. by *with*) make or be passionate; feel or cause to feel great emotion (*burn with shame*). **b** intr. (usu. foll. by *to* + infin.) desire passionately, long. **13** intr. sl. drive fast. **14** tr. US sl. anger, infuriate. **15** intr. (foll. by *into*) (of acid etc.) gradually penetrate (into) causing disintegration. ● n. **1** a mark or injury caused by burning. **2** the ignition of a rocket engine in flight, giving extra thrust. **3** N. Amer., Austral., & NZ **a** the clearing of vegetation by burning. **b** an area so cleared. **4** sl. a cigarette. **5** sl. a car race. □ **burn one's boats** (or **bridges**) commit oneself irrevocably. **burn the candle at both ends** exhaust one's strength or resources by undertaking too much. **burn down 1 a** destroy (a building) by burning. **b** (of a building) be destroyed by fire. **2** burn less vigorously as fuel fails. **burn one's fingers** suffer for meddling or rashness. **burn a hole in one's pocket** (of money) be quickly spent. **burning-glass** a lens for concentrating the sun's rays on an object to burn it. **burn low** (of fire) be nearly out. **burn the midnight oil** read or work late into the night. **burn out 1** be reduced to nothing by burning. **2** fail or cause to fail by burning. **3** (usu. refl.) suffer physical or emotional exhaustion.

4 consume the contents of by burning. **5** make (a person) homeless by burning his or her house. **burn-out** n. **1** physical or emotional exhaustion, esp. caused by stress. **2** depression, disillusionment. **burnt ochre** (or **sienna** or **umber**) a pigment darkened by burning. **burnt offering 1** an offering burnt on an altar as a sacrifice. **2** joc. overcooked or charred food. **burnt-out** physically or emotionally exhausted. **burn up 1** get rid of by fire. **2** begin to blaze. **3** sl. be or make furious. **4** colloq. traverse at high speed. **have money to burn** have more money than one needs. [OE birnan, bærnan f. Gmc]

burn[2] /bɜːn/ n. Sc. & N.Engl. a small stream. [OE burna etc. f. Gmc]

Burne-Jones /bɜːnˈdʒəʊnz/, Sir Edward (Coley) (1833–98), English painter and designer. A founder member of William Morris's business venture, he created many tapestry and stained-glass window designs for Morris & Company. He was preoccupied with medieval and literary themes, and invented an escapist, dreamlike world in his paintings; his work is regarded, with that of Morris, as typical of the later Pre-Raphaelite style. Major works include the tapestry The Adoration of the Magi in Exeter College Chapel, Oxford, and the paintings The Golden Stairs (1880) and The Mirror of Venus (1898–9).

burner /ˈbɜːnə(r)/ n. the part of a gas cooker, lamp, etc. that emits and shapes the flame. □ **on the back** (or **front**) **burner** colloq. receiving little (or much) attention.

burnet /ˈbɜːnɪt/ n. **1** a rosaceous plant of the genus Sanguisorba, with globular pinkish flower-heads. **2** a day-flying moth of the family Zygaenidae, often with crimson spots on greenish-black wings. [obs. burnet (adj.) dark brown f. OF burnete]

Burnett /bɜːˈnet/, Frances (Eliza) Hodgson (1849–1924), British-born American novelist. She is remembered chiefly for her novels for children, including Little Lord Fauntleroy (1886), The Little Princess (1905), and The Secret Garden (1911).

Burney /ˈbɜːnɪ/, Frances ('Fanny') (1752–1840), English novelist. Her first novel, the satire Evelina (1778), brought her fame and the patronage of Dr Johnson. Her second novel Cecilia (1782) was also a success. However, her father exhorted her to accept a post serving Queen Charlotte at court in 1786. Unhappy, Burney sought permission to retire in 1791. Two years later she married General Alexandre d'Arblay (1753–1818), with whom she was interned in France by Napoleon, finally returning in 1812.

burning /ˈbɜːnɪŋ/ adj. **1** ardent, intense (burning desire). **2** hotly discussed, exciting; vital, urgent (burning question). **3** flagrant (burning shame). **4** that burns; on fire; very hot. □ **burningly** adv.

burning bush n. **1** a shrub with red fruits or red autumn leaves. **2** fraxinella. ['the bush that burned and was not consumed' in Exod. 3:2]

burnish /ˈbɜːnɪʃ/ v.tr. polish by rubbing. □ **burnisher** n. [ME f. OF burnir = brunir f. brun BROWN]

burnous /bɜːˈnuːs/ n. an Arab or Moorish hooded cloak. [F f. Arab. burnus f. Gk birros cloak]

Burns /bɜːnz/, Robert (1759–96), Scottish poet. He developed an inclination for literature at an early age; his Poems, Chiefly in the Scottish Dialect (1786) was an immediate success. The satire 'The Jolly Beggars' (1786) and the narrative poem 'Tam o' Shanter' (1791) are among his most important poems. Approached in 1786 to collect old Scottish songs for The Scots Musical Museum (1787–1803), he responded with over 200 songs, including the famous lyrics 'Auld Lang Syne' and 'Ye Banks and Braes'. Burns was a firm patriot, and his popularity with the Scots is reaffirmed annually in the Burns Night celebrations held worldwide on his birthday, 25 Jan.

burnt past and past part. of BURN[1].

burp /bɜːp/ v. & n. colloq. ● v. **1** intr. belch. **2** tr. make (a baby) belch, usu. by patting its back. ● n. a belch. □ **burp gun** US sl. an automatic pistol. [imit.]

burpee /ˈbɜːpiː/ n. a physical exercise consisting of a squat thrust made from, and ending in, a standing position. [Royal H. Burpee, Amer. psychologist (b.1897)]

Burr /bɜː(r)/, Aaron (1756–1836), American Democratic Republican statesman. After losing the presidential election to Jefferson in 1800, Burr was elected Vice-President. He was defeated in the contest for the governorship of New York in 1804, largely through the campaign of his rival Alexander Hamilton. Later the same year Burr killed Hamilton in a duel. He subsequently plotted to invade Mexico in order to form an independent administration there; it was also alleged that he intended to annex Mexico to the western US and establish a separate empire. Burr was tried for treason and acquitted in 1807.

burr /bɜː(r)/ n. & v. ● n. **1 a** a whirring sound. **b** a rough sounding of the letter r. **2** (also **bur**) **a** a rough edge left on cut or punched metal or paper. **b** a surgeon's or dentist's small drill. **3 a** a siliceous rock used for millstones. **b** a whetstone. **4** = BUR 1, 2. **5** the coronet of a deer's antler. ● v. **1** tr. pronounce with a burr. **2** intr. speak indistinctly. **3** intr. make a whirring sound. [var. of BUR]

Burra /ˈbʌrə/, Edward (1905–76), English painter. He was fascinated by low-life and seedy subjects, as paintings such as Harlem (1934) attest. Usually he worked in watercolour, but on a large scale and using layers of pigment. In the mid-1930s he became fascinated with the bizarre and fantastic (Dancing Skeletons, 1934), and in the 1950s and 1960s he turned to landscape.

burrawang /ˈbʌrəˌwæŋ/ n. Austral. **1** a cycad of the genus Macrozamia. **2** the nut produced by this tree. [Mount Budawang in New South Wales]

burrito /bəˈriːtəʊ/ n. (pl. **-os**) US a tortilla rolled round a savoury filling. [Amer. Sp., dimin. of burro BURRO]

burro /ˈbʌrəʊ/ n. (pl. **-os**) N. Amer. a small donkey used as a pack-animal. [Sp.]

Burroughs[1] /ˈbʌrəʊz/, Edgar Rice (1875–1950), American novelist and writer of science fiction. He is remembered principally for his adventure stories about Tarzan, who first featured in Tarzan of the Apes (1914).

Burroughs[2] /ˈbʌrəʊz/, William S(eward) (b.1914), American novelist. In the 1940s he became addicted to heroin and also became associated with figures who were later prominent members of the beat generation. His best-known writing deals in a unique, surreal style with life as a drug addict (Junkie, 1953; The Naked Lunch, 1959).

burrow /ˈbʌrəʊ/ n. & v. ● n. a hole or tunnel dug by a small animal, esp. a rabbit, as a dwelling. ● v. **1** intr. make or live in a burrow. **2** tr. (often foll. by in, under, etc.) make (a hole etc.) by digging. **3** intr. hide oneself. **4** intr. (foll. by into) investigate, search. □ **burrower** n. [ME, app. var. of BOROUGH]

Bursa /ˈbɜːsə/ a city in NW Turkey, capital of a province of the same name; pop. (1990) 834,580. Captured by the Turks in 1326, it was the capital of the Ottoman Empire from then until 1402.

bursa /ˈbɜːsə/ n. (pl. **bursae** /-siː/ or **bursas**) Anat. a fluid-filled sac or saclike cavity, esp. one countering friction at a joint. □ **bursal** adj. [med.L f. Gk: cf. PURSE]

bursa of Fabricius /fəˈbrɪʃəs/ n. Zool. a glandular sac opening into a bird's cloaca, producing B-cells. [Latinized form of Girolamo Fabrici, It. anatomist (1533–1619)]

bursar /ˈbɜːsə(r)/ n. **1** a treasurer, esp. the person in charge of the funds and other property of a college. **2** the holder of a bursary. □ **bursarship** n. [F boursier or (in sense 1) med.L bursarius f. bursa bag]

bursary /ˈbɜːsəri/ n. (pl. **-ies**) **1** a grant, esp. a scholarship. **2** the post or room of a bursar. □ **bursarial** /bɜːˈseərɪəl/ adj. [med.L bursaria (as BURSAR)]

bursitis /bɜːˈsaɪtɪs/ n. Med. inflammation of a bursa.

burst /bɜːst/ v. & n. ● v. (past and past part. **burst**) **1 a** intr. break suddenly and violently apart by expansion of contents or internal pressure. **b** tr. cause to do this. **c** tr. cause (a container etc.) to split or puncture. **2 a** tr. open forcibly. **b** intr. come open or be opened forcibly. **3 a** intr. (usu. foll. by in, out) make one's way suddenly, dramatically, or by force. **b** tr. break away from or through (the river burst its banks). **4** tr. & intr. be full to or be full to overflowing. **5** intr. appear or come suddenly (burst into flame; burst upon the view; sun burst out). **6** intr. (foll. by into) suddenly begin to shed or utter (esp. burst into tears or laughter or song). **7** intr. be as if about to burst because of effort, excitement, etc. **8** tr. suffer bursting of (burst a blood vessel). **9** tr. separate (continuous stationery) into single sheets. ● n. **1** the act of or an instance of bursting; a split. **2** a sudden issuing forth (burst of flame). **3** a sudden outbreak (burst of applause). **4 a** a short sudden effort; a spurt. **b** a gallop. **5** an explosion. □ **burst out 1** suddenly begin (burst out laughing). **2** exclaim. □ **burster** n. [OE berstan f. Gmc]

burstproof /ˈbɜːstpruːf/ adj. (of a door lock) able to withstand a violent impact.

burthen archaic var. of BURDEN.

Burton[1] /ˈbɜːt(ə)n/, Richard (born Richard Jenkins) (1925–84), Welsh actor. He played a number of Shakespearian roles and performed in the radio adaptation of Dylan Thomas's Under Milk Wood (1954), before becoming well known in films such as The Spy Who Came in from the Cold (1966) and Who's Afraid of Virginia Woolf? (1966). He often co-starred with Elizabeth Taylor (to whom he was twice married).

Burton[2] /ˈbɜːt(ə)n/, Sir Richard (Francis) (1821–90), English explorer, anthropologist, and translator. He joined the Indian Army in 1842; subsequent travels took him to Mecca disguised as a Pathan, and to Brazil, Damascus, and Trieste as consul. With John Hanning Speke, he became the first European to discover Lake Tanganyika (1858). As a translator, he is best known for his unexpurgated versions of the *Arabian Nights* (1885–8), the *Kama Sutra* (1883), *The Perfumed Garden* (1886), and other works of Arabian erotica. His interest in sexual behaviour and deviance, as well as his detailed ethnographical notes, led him to risk prosecution many times under the Obscene Publications Act of 1857.

burton[1] /ˈbɜːt(ə)n/ *n.* □ **go for a burton** *Brit. sl.* be lost or destroyed or killed. [20th c.: perh. *Burton* ale f. BURTON-UPON-TRENT]

burton[2] /ˈbɜːt(ə)n/ *n.* (more fully **burton-tackle**) a light two-block tackle for hoisting. [ME *Breton tackles*: see BRETON[1]]

Burton-upon-Trent /ˌbɜːt(ə)nəpɒnˈtrent/ a town in west central England, in Staffordshire, situated on the River Trent to the north-east of Birmingham; pop. (1981) 59,600. The town has long been noted for its breweries.

Burundi /bʊˈrʊndɪ/ a central African country on the east side of Lake Tanganyika, to the south of Rwanda; pop. (est. 1991) 5,800,000; official languages, French and Kirundi; capital, Bujumbura. Inhabited mainly by Hutu and Tutsi peoples, the area formed part of German East Africa from the 1890s until the First World War, after which it was administered by Belgium. As in Rwanda, Belgium ruled through the Tutsi kings, institutionalizing the traditional power of the minority Tutsi in the region. The country became an independent monarchy in 1962 and a republic in 1966. Following coups in 1976 and 1987, multi-party elections in 1993 resulted in the country being led for the first time by a member of the Hutu majority. However, he was assassinated within months, sparking ethnic violence in which at least 150,000 people died; more fighting followed the death in 1994 of the country's next leader, together with the President of Rwanda. □ **Burundian** *adj. & n.*

bury /ˈberɪ/ *v.tr.* (**-ies, -ied**) **1 a** place (a dead body) in the earth, in a tomb, or in the sea. **b** lose by death (*has buried three husbands*). **2 a** put under ground (*bury alive*). **b** hide (treasure, a bone, etc.) in the earth. **c** cover up; submerge. **3 a** put out of sight (*buried his face in his hands*). **b** consign to obscurity (*the idea was buried after brief discussion*). **c** put away; forget. **4** involve deeply (*buried himself in his work; was buried in a book*). □ **bury the hatchet** cease to quarrel. **bury one's head in the sand** ignore unpleasant realities, refuse to face facts. **burying-beetle** a sexton beetle. **burying-ground** (or **-place**) a cemetery. [OE *byrgan* f. WG: cf. BURIAL]

Buryatia /bʊəˈjaːtɪə/ (also **Buryat Republic** /bʊəˈjaːt/) an autonomous republic in SE Russia, between Lake Baikal and the Mongolian border; pop. (1990) 1,049,000; capital, Ulan-Ude.

bus /bʌs/ *n. & v.* ● *n.* (pl. **buses** or *US* **busses**) **1** a large passenger vehicle, esp. one serving the public on a fixed route. (See note below.) **2** *colloq.* a motor car, aeroplane, etc. **3** *Computing* a defined set of conductors carrying data and control signals within a computer. ● *v.* (**buses** or **busses, bussed, bussing**) **1** *intr.* go by bus. **2** *tr. N. Amer.* transport by bus, esp. to promote racial integration. □ **bus lane** a part of a road's length marked off mainly for use by buses. **bus shelter** a shelter from rain etc. beside a bus-stop. **bus station** a centre, esp. in a town, where (esp. long-distance) buses depart and arrive. **bus-stop** a regular stopping-place for buses, esp. one marked by a sign. [abbr. of OMNIBUS]

▪ The first regular service of horse-drawn buses for public transportation appeared in Paris in 1827 and was soon imitated in other cities. Later versions were motor-driven, first by petrol engines and then from c.1930 onwards by diesel engines. From early days passengers were carried on the roof as well as inside the vehicle, and from this practice there developed the form of the double-decker bus with upper and lower sections.

busbar /ˈbʌsbɑː(r)/ *n. Electr.* a system of conductors in a generating or receiving station on which power is concentrated for distribution.

busby /ˈbʌzbɪ/ *n.* (pl. **-ies**) (not in official use) a tall fur hat worn by hussars etc. [18th c.: orig. unkn.]

Bush /bʊʃ/, George (Herbert Walter) (b.1924), American Republican statesman, 41st President of the US 1989–93. He was director of the CIA from 1976 to 1977, and President Reagan's Vice-President from 1981 to 1988. In 1989 Bush was elected President. While in office he negotiated further arms reductions with the Soviet Union and organized international action to liberate Kuwait following the Iraqi invasion in 1990.

bush[1] /bʊʃ/ *n.* **1** a shrub or clump of shrubs with stems of moderate length. **2** a thing resembling this, esp. a clump of hair or fur. **3** (esp. in Australia and Africa) a wild uncultivated district; woodland or forest. **4** *hist.* a bunch of ivy as a vintner's sign. ● **bush basil** a culinary herb, *Ocimum minimum.* **bush jacket** a light cotton jacket with a belt. **bush lawyer 1** *Austral. & NZ* a person claiming legal knowledge without qualifications for it. **2** *NZ* a bramble. **bush league** *US* a minor league of a professional sport. **bush-league** *US colloq.* inferior, minor, unsophisticated. **bush-leaguer** *US* **1** a person who plays in a bush league. **2** *colloq.* an unimportant or second-rate person. **bush sickness** a disease of animals due to a lack of cobalt in the soil. **bush telegraph** a rapid informal spreading of information, a rumour, etc.; the network by which this takes place. **go bush** *Austral.* leave one's usual surroundings; run wild. [ME f. OE & ON, ult. f. Gmc]

bush[2] /bʊʃ/ *n. & v.* ● *n.* **1** a metal lining for a round hole enclosing a revolving shaft etc. **2** a sleeve providing electrical insulation. ● *v.tr.* provide with a bush. [MDu. *busse* BOX[1]]

bushbaby /ˈbʊʃˌbeɪbɪ/ *n.* a small tree-dwelling nocturnal primate of the family Lorisidae, native to Africa, with very large eyes; a galago.

bushbuck /ˈbʊʃbʌk/ *n.* a small antelope, *Tragelaphus scriptus*, of southern Africa, having a chestnut coat with white stripes. [BUSH[1] + BUCK[1], after Du. *boschbok* f. *bosch* bush]

bushed /bʊʃt/ *adj. colloq.* **1** tired out. **2** *Austral. & NZ* **a** lost in the bush. **b** bewildered.

bushel /ˈbʊʃ(ə)l/ *n.* a measure of capacity for corn, fruit, liquids, etc. (*Brit.* 8 gallons, or 36.4 litres; *US* 64 US pints). The bushel formerly had a great variety of other values, varying not only from place to place but in the same place according to the commodity in question; often it was a weight (not a measure) of so many pounds. □ **hide one's light under a bushel** see HIDE[1]. □ **bushelful** *n.* (pl. **-fuls**). [ME f. OF *buissiel* etc., perh. of Gaulish orig.]

bushfire /ˈbʊʃˌfaɪə(r)/ *n.* a fire in a forest or in scrub often spreading widely.

bushido /buːˈʃiːdəʊ/ *n.* the code of honour and morals evolved by the Japanese samurai. [Jap. f. *bushi* samurai + *dō* way]

bushing /ˈbʊʃɪŋ/ *n.* = BUSH[2] *n.*

Bushman /ˈbʊʃmən/ *n. & adj.* ● *n.* (pl. **-men**) **1** a member of any of various aboriginal peoples living in southern Africa, particularly in the Kalahari Desert. Traditionally nomadic hunter-gatherers, many now perform paid work for farmers. Also called *San.* **2** any of the languages of these peoples; San. ● *adj.* of or relating to the Bushmen or their languages. [as BUSHMAN, after Du. *boschjesman* f. *bosch* bush]

bushman /ˈbʊʃmən/ *n.* (pl. **-men**) *Austral.* a person who lives or travels in the bush.

bushmaster /ˈbʊʃˌmɑːstə(r)/ *n.* a venomous viper, *Lachesis muta*, of Central and South America. [perh. f. Du. *bosmeester*]

bushranger /ˈbʊʃˌreɪndʒə(r)/ *n. hist.* an Australian outlaw living in the bush.

bushveld /ˈbʊʃfelt, ˈbʊʃvelt/ *n.* open country consisting largely of bush. [BUSH[1] + VELD, after Afrik. *bosveld*]

bushwhack /ˈbʊʃwæk/ *v.* **1** *intr. US, Austral., & NZ* **a** clear woods and bush country. **b** live or travel in bush country. **2** *tr. US* ambush.

bushwhacker /ˈbʊʃˌwækə(r)/ *n.* **1** *US, Austral., & NZ* **a** a person who clears woods and bush country. **b** a person who lives or travels in bush country. **2** *US* a guerrilla fighter (orig. in the American Civil War).

bushy[1] /ˈbʊʃɪ/ *adj.* (**bushier, bushiest**) **1** growing thickly like a bush (*bushy eyebrows*). **2** having many bushes. **3** covered with bush. □ **bushily** *adv.* **bushiness** *n.*

bushy[2] /ˈbʊʃɪ/ *n.* (pl. **-ies**) *Austral. & NZ colloq.* a person who lives in the bush (as distinct from in a town).

busily see BUSY.

business /ˈbɪznɪs/ *n.* **1** one's regular occupation, profession, or trade. **2** a thing that is one's concern. **3 a** a task or duty. **b** a reason for coming (*what is your business?*). **4** serious work or activity (*get down to business*). **5** *derog.* **a** an affair, a matter (*sick of the whole business*). **b** a structure (*a lath-and-plaster business*). **6** a thing or series of things needing to be dealt with (*the business of the day*). **7** trade, relations; dealings, esp. of a commercial nature (*good stroke of business*). **8** a commercial house or firm. **9** *Theatr.* action on stage. **10** a difficult matter (*what a business it is!; made a great business of it*). **11** (**the business**) *colloq.* exactly what is required; an exemplary person or thing. □ **business card** a card printed with one's name and professional details. **the business end** *colloq.* the functional part of a tool or device. **business park** an area designed to accommodate businesses and light industry. **business person** a businessman or businesswoman. **business studies** training in economics, management, etc. **has no business to** has no right to. **in business 1** trading or dealing. **2** able to begin operations. **in the business of 1** engaged in. **2** intending to (we are not in the business of surrendering). **like nobody's business** colloq. extraordinarily. **make it one's business to** undertake to. **mind one's own business** not meddle. **on business** with a definite purpose, esp. one relating to one's regular occupation. **send a person about his or her business** dismiss a person; send a person away. [OE bisignis (as BUSY, -NESS)]

businesslike /'bɪznɪsˌlaɪk/ adj. efficient, systematic, practical.

businessman /'bɪznɪsmən/ n. (pl. **-men**; fem. **businesswoman**, pl. **-women**) a man or woman engaged in trade or commerce, esp. at a senior level.

busk /bʌsk/ v.intr. perform (esp. music) in the street or other public places for donations of money. □ **busker** n. **busking** n. [busk peddle etc. (perh. f. obs. F busquer seek)]

buskin /'bʌskɪn/ n. **1** a thick-soled laced boot worn by an ancient Athenian tragic actor to gain height. **2** (usu. prec. by the) tragic drama; its style or spirit. **3** hist. a calf or knee-high boot of cloth or leather worn in the Middle Ages. □ **buskined** adj. [prob. f. OF bouzequin, var. of bro(u)sequin, of unkn. orig.]

busman /'bʌsmən/ n. (pl. **-men**) the driver of a bus. □ **busman's holiday** leisure time spent in an activity similar to one's regular work.

Busoni /buːˈsəʊniː/, Ferruccio (Benvenuto) (1866–1924), Italian composer, conductor, and pianist. A child prodigy, Busoni began performing at the age of 9 and went on to become an international concert pianist. As a composer he is best known for his works for piano, and for his unfinished opera Doktor Faust (1925). His later music anticipated some of the harmonic and rhythmic developments of Webern, Bartók, and Messiaen.

Buss /bʌs/, Frances Mary (1827–94), English educationist. At the age of 18 she was in charge of her own school, which in 1850 became known as the North London Collegiate School for Ladies. She was to remain headmistress there until 1894 and was the first to use the title headmistress. In 1886 she co-founded a training college for women teachers in Cambridge. With her friend Dorothea Beale, she also campaigned for higher education for women.

buss /bʌs/ n. & v. archaic or N. Amer. colloq. ● n. a kiss. ● v.tr. kiss. [earlier bass (n. & v.): cf. F baiser f. L basiare]

bust[1] /bʌst/ n. **1 a** the human chest, esp. that of a woman; the bosom. **b** the circumference of a woman's body at bust level (a 36-inch bust). **2** a sculpture of a person's head, shoulders, and chest. [F buste f. It. busto, of unkn. orig.]

bust[2] /bʌst/ v., n., & adj. colloq. ● v. (past and past part. **busted** or **bust**) **1** tr. & intr. burst, break. **2** tr. esp. US reduce (a soldier etc.) to a lower rank; dismiss. **3** tr. esp. US a raid, search. **b** arrest. ● n. **1** a sudden failure; a bankruptcy. **2** esp. US **a** a police raid. **b** an arrest. **3** a drinking-bout. **4** esp. US a punch; a hit. **5** a worthless thing. **6** a bad hand at cards. ● adj. (also **busted**) **1** broken, burst, collapsed. **2** bankrupt. □ **bust a gut** make every possible effort. **bust up 1** bring or come to collapse; explode. **2** (of esp. a married couple) separate. **bust-up** n. **1** a quarrel. **2** a collapse; an explosion. **go bust** become bankrupt; fail. [orig. a (dial.) pronunc. of BURST]

bustard /'bʌstəd/ n. a large swift-running terrestrial bird of the family Otididae, with long neck and stout body. Bustards are the heaviest flying birds, and are found mainly in Africa. [ME f. OF bistarde f. L avis tarda 'slow bird' (perh. a perversion of a foreign word)]

bustee /'bʌstiː/ n. Ind. a shanty town; a slum. [Hind. bastī dwelling]

buster /'bʌstə(r)/ n. **1** sl. mate; fellow (used esp. as a disrespectful form of address). **2** a violent gale.

bustier /'bʌstɪˌeɪ/ n. a women's strapless close-fitting bodice. [F]

bustle[1] /'bʌs(ə)l/ v. & n. ● v. **1** intr. (often foll. by about) **a** work etc. showily, energetically, and officiously. **b** move hurriedly (bustled about the kitchen banging saucepans). **2** tr. make (a person) hurry or work hard (bustled him into his overcoat). **3** intr. (as **bustling** adj.) colloq. full of activity. ● n. excited activity; a fuss. [perh. f. buskle frequent. of busk prepare]

bustle[2] /'bʌs(ə)l/ n. hist. a pad or frame worn under a skirt and puffing it out behind. [18th c.: orig. unkn.]

busty /'bʌstɪ/ adj. (**bustier**, **bustiest**) (of a woman) having a prominent bosom. □ **bustiness** n.

busy /'bɪzɪ/ adj., v., & n. ● adj. (**busier**, **busiest**) **1** (often foll. by in, with, at, or pres. part.) occupied or engaged in work etc. with the attention concentrated (busy at their needlework; he was busy packing). **2** full of activity or detail; fussy (a busy evening; a picture busy with detail). **3** employed continuously; unresting (busy as a bee). **4** meddlesome; prying. **5** esp. N. Amer. (of a telephone line) engaged. ● v.tr. (**-ies**, **-ied**) (often refl.) keep busy; occupy (the work busied him for many hours; busied herself with the accounts). ● n. (pl. **-ies**) sl. a detective; a police officer. □ **busy Lizzie** an East African plant, Impatiens walleriana, with abundant, red, pink, or white flowers, often grown as a bedding plant or house plant. □ **busily** /'bɪzɪlɪ/ adv. **busyness** /'bɪzɪnɪs/ n. (cf. BUSINESS). [OE bisig]

busybody /'bɪzɪˌbɒdɪ/ n. (pl. **-ies**) **1** a meddlesome person. **2** a mischief-maker.

but[1] /bʌt, bət/ conj., prep., adv., pron., n., & v. ● conj. **1 a** nevertheless, however (tried hard but did not succeed; I am old, but I am not weak). **b** on the other hand; on the contrary (I am old but you are young). **2** (prec. by can etc.; in neg. or interrog.) except, other than, otherwise than (cannot choose but do it; what could we do but run?). **3** without the result that (it never rains but it pours). **4** prefixing an interruption to the speaker's train of thought (the weather is ideal — but is that a cloud on the horizon?). ● prep. except; apart from; other than (everyone went but me; nothing but trouble). ● adv. **1** only; no more than; only just (we can but try; is but a child; had but arrived; did it but once). **2** introducing emphatic repetition; definitely (wanted to see nobody, but nobody). **3** Austral., NZ, & Sc. though, however (didn't like it, but). ● rel.pron. who not; that not (there is not a man but feels pity). ● n. an objection (ifs and buts). ● v.tr. (in phr. **but me no buts**) do not raise objections. □ **but for** without the help or hindrance etc. of (but for you I'd be rich by now). **but one** (or **two** etc.) excluding one (or two etc.) from the number (next door but one; last but one). **but that** (prec. by neg.) that (I don't deny but that it's true). **but that** (or colloq. **what**) other than that; except that (who knows but that it is true?). **but then** (or **yet**) however, on the other hand (I won, but then the others were beginners). [OE be-ūtan, būtan, būta outside, without]

but[2] /bʌt/ n. Sc. an outer room, esp. of a two-roomed cottage. □ **but and ben** a two-roomed cottage; one's humble home (see BEN[2]). [BUT[1] = outside]

butadiene /ˌbjuːtəˈdaɪiːn/ n. Chem. a colourless gaseous hydrocarbon (chem. formula: C_4H_6) used in the manufacture of synthetic rubbers. [BUTANE + DI-[1] + -ENE]

butane /'bjuːteɪn/ n. Chem. a gaseous hydrocarbon of the alkane series used in liquefied form as fuel (chem. formula: C_4H_{10}). [BUTYL + -ANE[2]]

butch /bʊtʃ/ adj. & n. sl. ● adj. masculine; tough-looking. ● n. **1** (often attrib.) **a** a mannish woman. **b** a mannish lesbian. **2** a tough, usu. muscular, youth or man. [perh. abbr. of BUTCHER]

butcher /'bʊtʃə(r)/ n. & v. ● n. **1 a** a person whose trade is dealing in meat. **b** a person who slaughters animals for food. **2** a person who kills or has people killed indiscriminately or brutally. ● v.tr. **1** slaughter or cut up (an animal) for food. **2** kill (people) wantonly or cruelly. **3** ruin (esp. a job or a piece of music) through incompetence. □ **the butcher, the baker, the candlestick-maker** people of all kinds or trades. **butcher-bird 1** a shrike of the family Laniidae. **2** a similar bird of the Australasian family Cracticidae. **butcher's** rhyming sl. a look (short for butcher's hook). **butcher's-broom** a low evergreen liliaceous shrub, Ruscus aculeatus, with stiff flat shoots that look like spine-tipped leaves. **butcher's meat** slaughtered fresh meat excluding game, poultry, and bacon. [ME f. OF bo(u)chier f. buc BUCK[1]]

butchery /'bʊtʃərɪ/ n. (pl. **-ies**) **1** wanton or cruel slaughter (of people). **2** the butcher's trade. **3** a slaughterhouse. [ME f. OF boucherie (as BUTCHER)]

Bute /bjuːt/, 3rd Earl of (title of John Stuart) (1713–92), Scottish courtier and Tory statesman, Prime Minister 1762–3. His influence with the king ensured his appointment as Premier, but he was widely disliked and soon fell out of favour with the king.

Buthelezi /ˌbuːtəˈleɪzɪ/, Chief Mangosuthu (Gatsha) (b.1928), South African politician. In 1953 he was appointed assistant to the Zulu king Cyprian, a position he held until 1968. He was elected leader of Zululand (later KwaZulu) in 1970 and was responsible for the revival of the Inkatha movement, of which he became leader in 1975. He was appointed Minister of Home Affairs in Nelson Mandela's Cabinet (1994).

butle var. of BUTTLE.

Butler[1] /'bʌtlə(r)/, Reginald Cotterell ('Reg') (1913–81), English sculptor. As a conscientious objector, he worked as a blacksmith during the Second World War. In 1947 he took up sculpture as an assistant to Henry Moore, working mainly in forged or cast metal. Butler suddenly became prominent in 1953, when he won an international competition for a monument (never built) to the Unknown Political Prisoner.

Butler[2] /ˈbʌtlə(r)/, Samuel ('Hudibras') (1612–80), English poet. His reputation rests on his three-part satirical poem *Hudibras* (1663–78), a mock romance parodying the Puritan sects and the Civil War, which rapidly became the most popular poem in England at the time. It was highly approved by Charles II, who granted the author a pension, although Butler is said to have died in penury.

Butler[3] /ˈbʌtlə(r)/, Samuel (1835–1902), English novelist. Emigrating in 1859, he became a successful sheep farmer in New Zealand, before returning to England in 1864. Turning to literature, he published his satirical anti-utopian novel *Erewhon* (1872) and its sequel *Erewhon Revisited* (1901), both of which challenged aspects of Darwinism. His semi-autobiographical *The Way of All Flesh* (1903) parodies child–parent relations and the effects of inherited family traits.

butler /ˈbʌtlə(r)/ *n.* the principal manservant of a household, usu. in charge of the wine cellar, pantry, etc. [ME f. AF *buteler*, OF *bouteillier*: see BOTTLE]

butt[1] /bʌt/ *v. & n.* ● *v.* **1** *tr. & intr.* push with the head or horns. **2 a** *intr.* (usu. foll. by *against, upon*) touch with one end flat, meet end to end with, abut. **b** *tr.* (usu. foll. by *against*) place (timber etc.) with the end flat against a wall etc. ● *n.* **1** a push with the head. **2** a join of two edges. □ **butt in** interrupt, meddle. **butt out** esp. *US* leave off (doing something); stop interfering. [ME f. AF *buter*, OF *boter* f. Gmc: infl. by BUTT[2] and ABUT]

butt[2] /bʌt/ *n.* **1** (often foll. by *of*) an object of ridicule etc. (*the butt of his jokes; made him their butt*). **2 a** a mound behind a target. **b** (in *pl.*) a shooting-range. **c** a target. **3** a grouse-shooter's stand screened by low turf or a stone wall. [ME f. OF *but* goal, of unkn. orig.]

butt[3] /bʌt/ *n.* **1** (also **butt-end**) the thicker end, esp. of a tool or a weapon (*gun butt*). **2** (also **butt-end**) **a** the stub of a cigar or a cigarette. **b** the remaining part. **3** esp. *N. Amer. sl.* the buttocks. **4** (also **butt-end**) the square end of a plank meeting a similar one. **5** the trunk of a tree, esp. the part just above the ground. □ **butt weld** a weld in which the pieces are joined end to end. [Du. *bot* stumpy]

butt[4] /bʌt/ *n.* a cask, esp. as a measure of wine or ale. [AL *butta, bota*, AF *but*, f. OF *bo(u)t* f. LL *buttis*]

butt[5] /bʌt/ *n.* a flatfish (e.g. a sole, plaice, or turbot). [MLG, MDu. *but* flatfish]

butte /bjuːt/ *n.* esp. *N. Amer.* a high isolated steep-sided hill. [F, = mound]

butter /ˈbʌtə(r)/ *n. & v.* ● *n.* **1 a** a pale yellow edible fatty substance made by churning cream and used as a spread or in cooking. **b** a substance of a similar consistency or appearance (*peanut butter*). **2** excessive flattery. ● *v.tr.* spread, cook, or serve with butter (*butter the bread; buttered carrots*). □ **butter-and-eggs** a plant with two shades of yellow in the flower, esp. yellow toadflax. **butter-bean 1** a flat, dried, white lima bean. **2** a yellow-podded bean. **butter-cream** (or **-icing**) a mixture of butter, icing sugar, etc. used as a filling or a topping for a cake. **butter-fingers** *colloq.* a person prone to drop things, esp. one who fails to hold a catch. **butter-knife** a blunt knife used for cutting butter at table. **butter muslin** a thin, loosely woven cloth with a fine mesh, orig. for wrapping butter. **butter-nut 1** a North American tree, *Juglans cinerea*. **2** the oily nut of this tree. **butter up** *colloq.* flatter excessively. **look as if butter wouldn't melt in one's mouth** seem demure or innocent, probably deceptively. [OE *butere* f. L *butyrum* f. Gk *bouturon*]

butterball /ˈbʌtəˌbɔːl/ *n.* **1** a piece of butter shaped into a ball. **2** *N. Amer.* = BUFFLEHEAD. **3** *N. Amer. sl.* a plump person or animal.

butterbur /ˈbʌtəˌbɜː(r)/ *n.* a waterside composite plant of the genus *Petasites*, with pale purple flowers and large soft leaves formerly used to wrap butter.

buttercup /ˈbʌtəˌkʌp/ *n.* a herbaceous plant of the genus *Ranunculus* (family Ranunculaceae), having bright yellow cup-shaped flowers, found commonly in grassland.

butterfat /ˈbʌtəˌfæt/ *n.* the essential fats of pure butter.

butterfish /ˈbʌtəˌfɪʃ/ *n.* = GUNNEL[1].

butterfly /ˈbʌtəˌflaɪ/ *n.* (*pl.* **-flies**) **1** a lepidopterous insect with large, often showy, wings, and a long coiled tongue. (*See note below.*) **2** a showy or frivolous person. **3** (in *pl.*) *colloq.* a nervous sensation felt in the stomach. **4** (in full **butterfly stroke**) a stroke in swimming, with both arms raised and lifted forwards together. □ **butterfly bush** a buddleia, esp. *Buddleia davidii*. **butterfly fish** a fish resembling a butterfly in shape or colour; esp. *Blennius ocellatus*, which has a broad dorsal fin with eye-spots, or a brightly coloured fish of the family

Chaetodontidae. **butterfly net** a fine net on a ring attached to a pole, used for catching butterflies. **butterfly nut** a kind of wing-nut. **butterfly valve** a valve with hinged semicircular plates. [OE *buttorflēoge* (as BUTTER, FLY[2])]

■ The butterflies are distinguished from moths in most instances by their day-flying behaviour, clubbed or dilated antennae, thin bodies, and the usually erect position of the wings when at rest. However, they are not taxonomically distinct, and there are several groups intermediate between butterflies and moths. (See also LEPIDOPTERA.)

buttermilk /ˈbʌtəˌmɪlk/ *n.* a slightly acid liquid left after churning butter.

butterscotch /ˈbʌtəˌskɒtʃ/ *n.* a brittle sweet made from butter, brown sugar, etc. [SCOTCH]

butterwort /ˈbʌtəˌwɜːt/ *n.* an insectivorous bog plant of the genus *Pinguicula*, esp. *P. vulgaris*, with violet-like flowers and sticky fleshy leaves that trap small insects.

buttery[1] /ˈbʌtərɪ/ *n.* (*pl.* **-ies**) a room, esp. in a college, where provisions are kept and sold to students etc. [ME f. AF *boterie* butt-store (as BUTT[4])]

buttery[2] /ˈbʌtərɪ/ *adj.* like, containing, or spread with butter. □ **butteriness** *n.*

buttle /ˈbʌt(ə)l/ *v.intr.* (also **butle**) *joc.* work as a butler. [back-form. f. BUTLER]

buttock /ˈbʌtək/ *n.* (usu. in *pl.*) **1** either of the two fleshy protuberances on the lower rear part of the human body. **2** the corresponding part of an animal. [*butt* ridge + -OCK]

button /ˈbʌt(ə)n/ *n. & v.* ● *n.* **1** a small disc or knob sewn on to a garment, either to fasten it by being pushed through a buttonhole, or as an ornament or badge. **2** a knob on a piece of esp. electrical equipment which is pressed to operate it. **3 a** a small disc-shaped object (*chocolate buttons*). **b** (*attrib.*) anything resembling a button (*button nose*). **4 a** a bud. **b** a button mushroom. **5** *Fencing* a terminal knob on a foil making it harmless. ● *v.* **1** *tr. & intr.* = *button up* 1. **2** *tr.* supply with buttons. □ **buttonball tree** (or **buttonwood**) the North American plane tree, *Platanus occidentalis.* **button chrysanthemum** a variety of chrysanthemum with small spherical flowers. **buttoned up** *colloq.* **1** formal and inhibited in manner. **2** silent. **button one's lip** *sl.* remain silent. **button mushroom** a young unopened mushroom. **button-through** (of a dress) fastened with buttons from neck to hem like a coat. **button up 1** fasten with buttons. **2** *colloq.* complete (a task etc.) satisfactorily. **3** *colloq.* become silent. **not worth a button** worthless. **on the button** esp. *N. Amer. sl.* precisely. □ **buttoned** *adj.* **buttonless** *adj.* **buttony** *adj.* [ME f. OF *bouton*, ult. f. Gmc]

buttonhole /ˈbʌt(ə)nˌhəʊl/ *n. & v.* ● *n.* **1** a slit made in a garment to receive a button for fastening. **2** a flower or spray worn in a lapel buttonhole. ● *v.tr.* **1** *colloq.* accost and detain (a reluctant listener). **2** make buttonholes in. □ **buttonhole stitch** a looped stitch used for edging buttonholes.

buttonhook /ˈbʌt(ə)nˌhʊk/ *n.* a hook formerly used esp. for pulling the buttons on tight boots into place for fastening.

buttons /ˈbʌt(ə)nz/ *n. colloq.* a liveried page-boy. [from the rows of buttons on his jacket]

buttress /ˈbʌtrɪs/ *n. & v.* ● *n.* **1 a** a projecting support of stone or brick etc. built against a wall. **b** a source of help or encouragement (*she was a buttress to him in his trouble*). **2** a projecting portion of a hill or mountain. ● *v.tr.* (often foll. by *up*) **1** support with a buttress. **2** support by argument etc. (*claim buttressed by facts*). [ME f. OF (*ars*) *bouterez* thrusting (arch) f. *bouteret* f. *bouter* BUTT[1]]

butty[1] /ˈbʌtɪ/ *n.* (*pl.* **-ies**) **1** *colloq.* or *dial.* a mate; a companion. **2** *hist.* a middleman negotiating between a mine-owner and the miners. **3** a barge or other craft towed by another. □ **butty-gang** a gang of men contracted to work on a large job and sharing the profits equally. [19th c.: perh. f. BOOTY in phr. *play booty* join in sharing plunder]

butty[2] /ˈbʌtɪ/ *n.* (also **buttie**) (*pl.* **-ies**) *N. Engl.* **1** a sandwich (*bacon butty*). **2** a slice of bread and butter. [BUTTER + -Y[2]]

butyl /ˈbjuːtaɪl, -tɪl/ *n. Chem.* the univalent alkyl radical $-C_4H_9$. □ **butyl rubber** a synthetic rubber used in the manufacture of tyre inner tubes. [*butyric* (see BUTYRIC ACID) + -YL]

butyric acid /bjuːˈtɪrɪk/ *n. Chem.* either of two colourless syrupy liquid organic acids found in rancid butter or arnica oil. □ **butyrate** /ˈbjuːtɪˌreɪt/ *n.* [L *butyrum* BUTTER + -IC]

buxom /ˈbʌksəm/ *adj.* (esp. of a woman) plump and healthy-looking; large and shapely; busty. □ **buxomness** *n.* [earlier sense *pliant*: ME f. stem of OE *būgan* BOW[2] + -SOME[1]]

Buxtehude /ˈbʊkstəˌhuːdə/, Dietrich (*c.*1637–1707), Danish organist and composer. He worked as an organist in Lübeck from 1668 until his death, turning the traditional Sunday evening concerts there into celebrated occasions for the performance of his vocal, organ, and chamber music. His skill as an organist inspired Bach to walk more than 200 miles from Anstadt to hear him play. His toccatas, preludes, fugues, and choral variations give some idea of his mastery of the instrument as well as of his gifts as a composer.

buy /baɪ/ *v. & n.* ● *v.* (**buys, buying**; *past* and *past part.* **bought** /bɔːt/) **1** *tr.* **a** obtain in exchange for money etc. **b** (often in *neg.*) serve to obtain (*money can't buy happiness*). **2** *tr.* **a** procure (the loyalty etc.) of a person by bribery, promises, etc. **b** win over (a person) in this way. **3** *tr.* get by sacrifice, great effort, etc. (*dearly bought; bought with our sweat*). **4** *tr. sl.* accept, believe in, approve of (*it's a good scheme, I'll buy it; he bought it, he's so gullible*). **5** *absol.* be a buyer for a store etc. (*buys for Selfridges; are you buying or selling?*). ● *n. colloq.* a purchase (*that sofa was a good buy*). □ **best buy** the purchase giving the best value in proportion to its price; a bargain. **buy-back** the buying-back or repurchase of goods, shares, etc., often by contractual agreement. **buy in 1** buy a stock of. **2** withdraw (an item) at auction because of failure to reach the reserve price. **buy-in** *esp. US* **1** the purchase of shares on the Stock Exchange, esp. after the non-delivery of similar shares bought. **2** the buying-back of a company's own shares. **buy into** obtain a share in (an enterprise) by payment. **buy it** (usu. in *past*) *sl.* be killed. **buy off** get rid of (a claim, a claimant, a blackmailer) by payment. **buy oneself out** obtain one's release (esp. from the armed services) by payment. **buy out** pay (a person) to give up an ownership, interest, etc. **buy-out** *n.* the purchase of a controlling share in a company etc. **buy over** bribe. **buy time** delay an event, conclusion, etc., temporarily. **buy up 1** buy as much as possible of. **2** absorb (another firm etc.) by purchase. [OE *bycgan* f. Gmc]

buyer /ˈbaɪə(r)/ *n.* **1** a person employed to select and purchase stock for a large store etc. **2** a purchaser, a customer. □ **buyer's** (or **buyers'**) **market** an economic position in which goods are plentiful and cheap and buyers have the advantage.

buzz /bʌz/ *n. & v.* ● *n.* **1** the hum of a bee etc. **2** the sound of a buzzer. **3 a** a confused low sound as of people talking; a murmur. **b** a stir; hurried activity (*a buzz of excitement*). **c** *colloq.* a rumour. **4** *sl.* a telephone call. **5** *sl.* a thrill; a euphoric sensation. ● *v.* **1** *intr.* make a humming sound. **2 a** *tr. & intr.* signal or signal to with a buzzer. **b** *tr. sl.* telephone. **3** *intr.* **a** (often foll. by *about*) move or hover busily. **b** (of a place) have an air of excitement or purposeful activity. **4** *tr. colloq.* throw hard. **5** *tr. Aeron. colloq.* fly fast and very close to (another aircraft, the ground, etc.). □ **buzz off** *sl.* go or hurry away. **buzz-saw** *N. Amer.* a circular saw. **buzz-word** *sl.* **1** a fashionable piece of esp. technical jargon. **2** a catchword; a slogan. [imit.]

buzzard /ˈbʌzəd/ *n.* **1** a predatory bird of the hawk family, esp. the genus *Buteo*, with broad wings adapted for soaring flight. **2** *N. Amer.* a vulture, esp. the turkey vulture. [ME f. OF *busard, buson* f. L *buteo -onis* falcon]

buzzer /ˈbʌzə(r)/ *n.* **1** an electrical device, similar to a bell, that makes a buzzing noise. **2** a whistle or hooter.

BVM *abbr.* Blessed Virgin Mary.

b/w *abbr.* black and white (television etc., as opposed to colour).

bwana /ˈbwɑːnə/ *n. Afr.* master, sir. [Swahili]

BWI *abbr. hist.* British West Indies.

BWR *abbr.* boiling-water (nuclear) reactor.

by /baɪ/ *prep., adv., & n.* ● *prep.* **1** near, beside, in the region of (*stand by the door; sit by me; path by the river*). **2** through the agency, means, instrumentality, or causation of (*by proxy; bought by a millionaire; a poem by Donne; went by bus; succeeded by persisting; divide four by two*). **3** not later than; as soon as (*by next week; by now; by the time he arrives*). **4 a** past, beyond (*drove by the church; came by us*). **b** passing through; via (*went by Paris*). **5** in the circumstances of (*by day; by daylight*). **6** to the extent of (*missed by a foot; better by far*). **7** according to; using as a standard or unit (*judge by appearances; paid by the hour*). **8** with the succession of (*worse by the minute; day by day; one by one*). **9** concerning; in respect of (*did our duty by them; Smith by name; all right by me*). **10** used in mild oaths (orig. = as surely as one believes in) (*by God; by gum; swear by all that is sacred*). **11** placed between specified lengths in two directions (*three feet by two*). **12** avoiding, ignoring (*pass by him; passed us by*). **13** (esp. in names of compass points) inclining to (*north by east: between north and north-north-east; north-east by north: between north-east and north-north-east*). ● *adv.* **1** near (*sat by, watching; lives close by*). **2** aside; in reserve (*put £5 by*). **3** past (*they marched by*). ● *n.* (*pl.* **byes**) = BYE[1]. □ **by and by** before long; eventually. **by and large** on the whole,

everything considered. **by the by** (or **bye**) incidentally, parenthetically. **by oneself 1 a** unaided. **b** without prompting. **2** alone; without company. [OE *bī, bi, be* f. Gmc]

by- /baɪ/ *prefix* (also **bye-**) subordinate, incidental, secondary (*by-product; byroad*).

Byatt /ˈbaɪət/, A(ntonia) S(usan) (b.1936), English novelist and literary critic. Her fiction is noted for its use of literary and historical allusion and pastiche. Major novels include *The Virgin in the Garden* (1978), set largely in the coronation year of 1953, with complex allegorical references to Spenser, Shakespeare, and Raleigh, and *Possession* (1990), a satire of the literary biography industry, which won the Booker Prize. She is the elder sister of the novelist Margaret Drabble.

Byblos /ˈbɪblɒs/ an ancient Mediterranean seaport, situated on the site of modern Jebeil, to the north of Beirut in Lebanon. An important trading centre with strong links with Egypt, it became a thriving Phoenician city in the 2nd millennium BC. It was particularly noted for the export of papyrus and cedar wood.

by-blow /ˈbaɪbləʊ/ *n.* **1** a side-blow not at the main target. **2** a man's illegitimate child.

Bydgoszcz /ˈbɪdɡɒʃtʃ/ (called in German *Bromberg*) an industrial river port in north central Poland; pop. (1990) 381,530. It was ruled by Prussia from 1792 to 1919. A monument commemorates the massacre of 20,000 of its citizens by Nazis in Sept. 1939.

bye[1] /baɪ/ *n.* **1** *Cricket* a run scored from a ball that passes the batsman without being hit, counted as an extra. **2** the status of an unpaired competitor in a sport, who proceeds to the next round as if having won. **3** *Golf* one or more holes remaining unplayed after the match has been decided. □ **by the bye** = *by the by*. **leg-bye** *Cricket* a run scored from a ball that touches the batsman, counted as an extra. [BY as noun]

bye[2] /baɪ/ *int. colloq.* = GOODBYE. [abbr.]

bye- *prefix* var. of BY-.

bye-bye[1] /ˈbaɪbaɪ, bəˈbaɪ/ *int. colloq.* = GOODBYE. [childish corrupt.]

bye-bye[2] /ˈbaɪbaɪ/ *n.* (also **bye-byes** /-baɪz/) (a child's word for) sleep. [ME, f. the sound used in lullabies]

by-election /ˈbaɪɪˌlekʃ(ə)n/ *n.* the election of an MP in a single constituency to fill a vacancy arising during a government's term of office.

Byelorussian var. of BELORUSSIAN.

by-form /ˈbaɪfɔːm/ *n.* a collateral form of a word etc.

bygone /ˈbaɪɡɒn/ *adj. & n.* ● *adj.* past, antiquated (*bygone years*). ● *n.* (in *pl.*) past offences (*let bygones be bygones*).

by-law /ˈbaɪlɔː/ *n.* (also **bye-law**) **1** *Brit.* a regulation made by a local authority or corporation. **2** a rule made by a company or society for its members. [ME prob. f. obs. *byrlaw* local custom (ON *býjar* genitive sing. of *býr* town, but assoc. with BY)]

byline /ˈbaɪlaɪn/ *n.* **1** a line in a newspaper etc. naming the writer of an article. **2** a secondary line of work. **3** a goal-line or touch-line.

byname /ˈbaɪneɪm/ *n.* a sobriquet; a nickname.

BYOB *abbr.* bring your own bottle.

bypass /ˈbaɪpɑːs/ *n. & v.* ● *n.* **1** a road passing round a town or its centre to provide an alternative route for through traffic. **2 a** a secondary channel or pipe etc. to allow a flow when the main one is closed or blocked. **b** an alternative passage for the circulation of blood during a surgical operation on the heart. ● *v.tr.* **1** avoid; go round. **2** provide with a bypass.

bypath /ˈbaɪpɑːθ/ *n.* **1** a secluded path. **2** a minor or obscure branch of a subject.

byplay /ˈbaɪpleɪ/ *n.* a secondary action or sequence of events, esp. in a play.

by-product /ˈbaɪˌprɒdʌkt/ *n.* **1** an incidental or secondary product made in the manufacture of something else. **2** a secondary result.

Byrd[1] /bɜːd/, Richard E(velyn) (1888–1957), American explorer, naval officer, and aviator. He claimed to have made the first aeroplane flight over the North Pole in 1926, although his actual course has been disputed. He was the first to fly over the South Pole in 1929 and led further scientific expeditions to the Antarctic in 1933–4 and 1939–41.

Byrd[2] /bɜːd/, William (1543–1623), English composer. Joint-organist of the Chapel Royal with Tallis, he is often held to be one of the finest Tudor composers. As a Roman Catholic under the Anglican Elizabeth I, he wrote for both Churches and is most famous for his Latin masses for three, four, and five voices and his Anglican Great Service. He composed a great quantity of music for virginals in addition to much consort music, including more than forty consort songs.

Byrds, the /bɜːdz/ an influential US pop and rock group led by Roger McGuinn (b.1942), known for their pure vocal harmonies and twelve-string guitar sound. Beginning as a Beatles-influenced group playing a pleasantly melodic mixture of original, traditional, and Bob Dylan songs (notably 'Mr Tambourine Man', 1965), they later added Eastern, jazz, and experimental elements, before becoming one of the first rock groups to play country music, on their *Sweetheart of the Rodeo* album (1968).

byre /'baɪə(r)/ *n.* a cowshed. [OE *bȳre*: perh. rel. to BOWER[1]]

byroad /'baɪrəʊd/ *n.* a minor road.

Byron /'baɪərən/, George Gordon, 6th Baron (1788–1824), English poet. His first literary success was *Childe Harold's Pilgrimage* (1812–18). In 1815 there were rumours of an incestuous relationship with his half-sister, his wife left him, and debts associated with his ancestral home increased. Ostracized and embittered, he left England permanently and stayed with Shelley in Geneva, finally settling in Italy. In *Beppo* (1818) he found a new ironic colloquial voice, which he fully developed in his epic satire *Don Juan* (1819–24). Though criticized on moral grounds, Byron's poetry exerted considerable influence on the romantic movement, particularly on the Continent. In 1824 he joined the fight for Greek independence, but died of malaria before seeing serious battle.

Byronic /baɪ'rɒnɪk/ *adj.* **1** characteristic of Lord Byron or his romantic poetry. **2** (of a man) handsomely dark, mysterious, or moody.

byssinosis /ˌbɪsɪ'nəʊsɪs/ *n. Med.* a lung disease caused by prolonged inhalation of textile fibre dust. [mod.L f. Gk *bussinos* made of byssus + -OSIS]

byssus /'bɪsəs/ *n. hist.* (*pl.* **byssuses** or **byssi** /-saɪ/) **1** a fine textile fibre and fabric of flax. **2** *Zool.* a tuft of tough silky filaments by which some molluscs adhere to rocks etc. [ME f. L f. Gk *bussos*]

bystander /'baɪˌstændə(r)/ *n.* a person who stands by but does not take part; a mere spectator.

byte /baɪt/ *n. Computing* a group of binary digits (usu. eight), operated on as a unit. [20th c.: perh. based on BIT[4] and BITE]

Bytom /'bɪtəm/ (German **Beuthen** /'bɔɪt(ə)n/) a city in southern Poland, north-west of Katowice; pop. (1990) 231,200. Formerly part of Prussia, it was incorporated into Poland in 1945.

byway /'baɪweɪ/ *n.* **1** a byroad or bypath. **2** a minor activity.

byword /'baɪwɜːd/ *n.* **1** a person or thing cited as a notable example (*is a byword for luxury*). **2** a familiar saying; a proverb.

Byzantine /bɪ'zæntaɪn, baɪ-, 'bɪz(ə)nˌtiːn, -ˌtaɪn/ *adj. & n.* ● *adj.* **1** of Byzantium or the Eastern Roman Empire. **2** (of a political situation etc.): **a** extremely complicated. **b** inflexible. **c** carried on by underhand methods. **3** *Archit. & Art* of a highly decorated style developed in the Eastern Empire. (*See note below.*) ● *n.* a citizen of Byzantium or the Eastern Roman Empire. □ **Byzantinism** *n.* **Byzantinist** *n.* [F *byzantin* or L *Byzantinus* f. *Byzantium*, later Constantinople and now Istanbul]

▪ Byzantine art and architecture spread beyond the Byzantine Empire into Italy, and long remained influential in Russia. Pictorial art, largely in the form of religious wall-paintings and icons, was characterized by clearly defined rules, heavily stylized figures depicted frontally and without depth, and much use of gold and other rich colours. The artists, generally monks, were expected to remain anonymous. Byzantine churches were often (with the exception of St Sophia in Constantinople, now Istanbul) comparatively small, but frequently featured sumptuous decoration, with interiors that made much use of marble panelling, mosaics, and wall-paintings. Domes supported by pendentives were a common feature.

Byzantine Empire the empire in SE Europe and Asia Minor formed from the eastern part of the Roman Empire (cf. EASTERN EMPIRE). The Roman Empire was divided in AD 395 by the Emperor Theodosius between his sons; Constantinople (Byzantium) became the capital of the Eastern Empire in 476, when Rome fell to the Ostrogoths. Justinian (emperor 527–65) reconquered North Africa and part of Italy, but by 750 the Byzantine Empire had been pushed back, mainly by Muslim Arab conquests, to the Balkans and Asia Minor. In 1054, theological and political differences between Constantinople and Rome led to a schism between Eastern and Western Christianity (see GREAT SCHISM 1). After the strong reign of Alexius Comnenus (1081–1118), the empire gradually declined under his dynastic successors. The loss of Constantinople to the Ottoman Turks in 1453 was the end of the empire, although its rulers held Trebizond (Trabzon) until 1461.

Byzantium /bɪ'zæntɪəm, baɪ-/ an ancient Greek city, founded in the 7th century BC, at the southern end of the Bosporus, site of the modern city of Istanbul. It was rebuilt by Constantine the Great in AD 324–30 as Constantinople.

Cc

C¹ /siː/ n. (also **c**) (pl. **Cs** or **C's**) **1** the third letter of the alphabet. **2** *Mus.* the first note of the diatonic scale of C major (the major scale having no sharps or flats). **3** the third hypothetical person or example. **4** the third highest class or category (of academic marks etc.). **5** (usu. **c**) *Algebra* the third known quantity. **6** (as a Roman numeral) 100. **7** (**C**) the name of a computer programming language.

C² *abbr.* (also **C.**) **1** Cape. **2** Conservative. **3** Command Paper (second series, 1870–99). **4** Celsius, Centigrade.

C³ *symb.* **1** *Chem.* the element carbon. **2** (also ©) copyright. **3** coulomb(s). **4** capacitance.

c¹ *abbr.* **1** century; centuries. **2** chapter. **3** cent(s). **4** cold. **5** (usu. **c.**) *circa*, approximately. **6** colt. **7** *Cricket* caught by.

c² *symb.* (usu. **c**) *Physics* the speed of light in a vacuum.

c/- *abbr. Austral. & NZ* care of.

CA *abbr.* **1** *US* California (in official postal use). **2** *Sc. & Can.* chartered accountant.

Ca *symb. Chem.* the element calcium.

ca. *abbr. circa*, about.

CAA *abbr.* (in the UK) Civil Aviation Authority.

Caaba var. of KAABA.

CAB *abbr.* **1** Citizens' Advice Bureau. **2** *US* Civil Aeronautics Board. **3** see COMMON AGRICULTURAL BUREAUX.

cab /kæb/ n. **1** the driver's compartment in a lorry, train, or crane. **2** a taxi. **3** *hist.* a hackney carriage. [abbr. of CABRIOLET]

cabal /kəˈbæl/ n. & v. ● n. **1** a secret intrigue. **2** a political clique or faction. **3** *hist.* a committee of five ministers under Charles II, whose surnames happened to begin with C, A, B, A, and L. ● *v.intr.* (**caballed**, **caballing**) (often foll. by *together, against*) plot, intrigue. [F *cabale* f. med.L *cabala*, CABBALA]

cabala var. of CABBALA.

caballero /ˌkæbəˈljeərəʊ/ n. (pl. **-os**) a Spanish gentleman. [Sp.: see CAVALIER]

cabana /kəˈbɑːnə/ n. *US* a hut or shelter at a beach or swimming-pool. [Sp. *cabaña* f. LL (as CABIN)]

cabaret /ˈkæbəˌreɪ/ n. **1** an entertainment in a nightclub or restaurant while guests eat or drink at tables. **2** such a nightclub etc. [F, = wooden structure, tavern]

cabbage /ˈkæbɪdʒ/ n. **1 a** a cultivated variety of *Brassica oleracea*, with thick green or purple leaves forming a round heart or head. **b** these leaves usu. eaten as a vegetable. **2** *colloq. derog.* a person who is inactive or lacks interest. □ **cabbage palm** a palm tree that resembles a cabbage in some way, esp. one of the genus *Roystonea* or *Sabal*, with edible buds, or of the genus *Livistona*, with large leaves. **cabbage rose** a double rose with a large round compact flower. **cabbage tree 1** a palmlike tree, *Cordyline australis* (agave family), native to New Zealand and widely grown for its sugary sap or for ornament (also called *ti*). **2** = *cabbage palm*. **cabbage white** a white butterfly, *Pieris brassicae*, whose caterpillars feed on cabbage leaves. □ **cabbagy** *adj.* [earlier *cabache*, *-oche* f. OF (Picard) *caboche* head, OF *caboce*, of unkn. orig.]

cabbala /kəˈbɑːlə, ˈkæbələ/ n. (also **cabala**, **kabbala**) **1** the ancient Jewish tradition of mystic interpretation of the Bible, first transmitted orally and using esoteric methods (including ciphers). It reached the height of its influence in the later Middle Ages and remains significant in Hasidism. **2** mystic interpretation; any esoteric doctrine or occult lore. □ **cabbalism** n. **cabbalist** n. **cabbalistic** /ˌkæbəˈlɪstɪk/ *adj.* [med.L f. rabbinical Heb. *ḳabbālâ* tradition]

cabby /ˈkæbɪ/ n. (also **cabbie**) (pl. **-ies**) *colloq.* a taxi-driver. [CAB + -Y²]

caber /ˈkeɪbə(r)/ n. a roughly trimmed tree-trunk used in the Scottish Highland sport of tossing the caber. This involves holding the caber upright and running forward to toss it so that it lands on one end and falls forward. There is no standard size for the caber, but once tossed in a contest a caber must never be shortened; that used at the Braemar games is 5.79 m (19 ft) long and weighs 54.5 kg (120 lb). [Gael. *cabar* pole]

Cabernet /ˈkæbəˌneɪ/ n. **1** (in full **Cabernet Franc** /frɒŋk/ or **Cabernet Sauvignon** /ˈsəʊvɪˌnjɒn/) a variety of black grape used in wine-making. **2** a vine on which these grow. **3** a wine made from these grapes. [F]

cabin /ˈkæbɪn/ n. & v. ● n. **1** a small shelter or house, esp. of wood. **2** a room or compartment in an aircraft or ship for passengers or crew. **3** a driver's cab. ● *v.tr.* (**cabined**, **cabining**) confine in a small place, cramp. □ **cabin-boy** a boy who waits on a ship's officers or passengers. **cabin class** the intermediate class of accommodation in a ship. **cabin crew** the crew members on an aeroplane attending to passengers and cargo. **cabin cruiser** a large motor boat with living accommodation. [ME f. OF *cabane* f. Prov. *cabana* f. LL *capanna, cavanna*]

Cabinda /kəˈbɪndə/ n. **1** an enclave of Angola at the mouth of the River Congo, separated from the rest of Angola by a wedge of Zaire territory. **2** its capital; pop. (1991) 163,000.

cabinet /ˈkæbɪnɪt/ n. **1 a** a cupboard or case with drawers, shelves, etc., for storing or displaying articles. **b** a piece of furniture housing a radio or television set etc. **2** (**Cabinet**) the committee of senior ministers responsible for controlling government policy. (See note below.) **3** *archaic* a small private room. □ **cabinet-maker** a skilled joiner. **Cabinet minister** *Brit.* a member of the Cabinet. **cabinet photograph** a photograph of about 6 by 4 inches. **cabinet pudding** a steamed pudding with dried fruit. [CABIN + -ET¹, infl. by F *cabinet*]

▪ Although the monarchs of England always had advisers, the first Cabinet (or Cabinet Council) developed after the Restoration in 1660. It consisted of the most trusted members of the Privy Council and met with the monarch in a 'cabinet' or private room, taking decisions without consulting the full Privy Council. In the time of Queen Anne it became the main machinery of executive government, and the Privy Council was relegated to a formal and judicial role. From about 1717 the monarch ceased to attend, but it was not until after the Reform Act of 1832 that the royal power was dissolved and Cabinets came to depend for their existence and policies upon the support of a majority in the House of Commons.

cable /ˈkeɪb(ə)l/ n. & v. ● n. **1** a thick rope of wire or hemp. **2** an encased group of insulated wires for transmitting electricity or electrical signals. **3** a cablegram. **4 a** *Naut.* the chain of an anchor. **b** a measure of 200 yards. **5** (in full **cable stitch**) a knitted stitch resembling twisted rope. **6** *Archit.* a rope-shaped ornament. ● v. **1 a** *tr.* transmit (a message) by cablegram. **b** *tr.* inform (a person) by cablegram. **c** *intr.* send a cablegram. **2** *tr.* furnish or fasten with a cable or cables. **3** *Archit. tr.* furnish with cables. □ **cable-car 1** a small cabin (often one of a series) suspended on an endless cable, by which it is drawn up and down a mountainside etc. **2** a carriage drawn along a cable railway. **cable-laid** (of rope) having three triple strands. **cable railway** a railway along which carriages are drawn by an endless cable. **cable television** a broadcasting system with signals transmitted by cable to subscribers' sets. [ME f. OF *chable*, ult. f. LL *capulum* halter f. Arab. *ḥabl*]

cablegram /ˈkeɪb(ə)lˌɡræm/ n. a telegraph message sent by undersea cable etc.

cableway /ˈkeɪb(ə)lˌweɪ/ n. a transporting system with a usu. elevated cable.

cabman /ˈkæbmən/ n. (pl. **-men**) hist. the driver of a cab.

cabochon /ˈkæbəˌʃɒn/ n. (often attrib.) a gem polished but not faceted. □ **en cabochon** (of a gem) treated in this way. [F dimin. of *caboche*: see CABBAGE]

caboodle /kəˈbuːd(ə)l/ n. □ **the whole (kit and) caboodle** sl. the whole lot (of persons or things). [19th c. US: perh. f. phr. *kit and boodle*]

caboose /kəˈbuːs/ n. **1** a kitchen on a ship's deck. **2** N. Amer. a guard's van; a car on a freight train for workmen etc. [Du. *cabūse*, of unkn. orig.]

Cabora Bassa /kəˌbɔːrə ˈbæsə/ a lake on the Zambezi river in western Mozambique. Its waters are impounded by a dam and massive hydroelectric complex supplying power mainly to Maputo and South Africa.

Cabot /ˈkæbət/, John (Italian name Giovanni Caboto) (c.1450–c.1498), Italian explorer and navigator. He and his son Sebastian (c.1475–1557) sailed from Bristol in 1497 with letters patent from Henry VII of England in search of Asia, but in fact discovered the mainland of North America. The site of their arrival is uncertain (it may have been Cape Breton Island, Newfoundland, or Labrador. John Cabot returned to Bristol and undertook a second expedition in 1498. Sebastian made further voyages of exploration after his father's death, most notably to Brazil and the River Plate (1526).

cabotage /ˈkæbəˌtɑːʒ, -tɪdʒ/ n. **1** Naut. coastal navigation and trade. **2** esp. Aeron. the reservation to a country of (esp. air) traffic operation within its territory. [F f. *caboter* to coast, perh. f. Sp. *cabo* CAPE²]

cabotin /ˈkæbəˌtæn/ n. (fem. **cabotine** /-ˌtiːn/) a second-rate actor; a strolling player. [F, = strolling player, perh. formed as CABOTAGE, from the resemblance to vessels travelling from port to port]

cabriole /ˈkæbrɪˌəʊl/ n. a kind of curved leg characteristic of Chippendale and Queen Anne furniture. [F f. *cabrioler, caprioler* f. It. *capriolare* to leap in the air; from the resemblance to a leaping animal's foreleg: see CAPRIOLE]

cabriolet /ˌkæbrɪəʊˈleɪ/ n. **1** a light two-wheeled carriage with a hood, drawn by one horse. **2** a motor car with a folding top. [F f. *cabriole* goat's leap (cf. CAPRIOLE), applied to the carriage's motion]

ca'canny /kɑːˈkænɪ/ n. colloq. **1** the practice of going slow at work; a trade union policy of limiting output. **2** extreme caution. [Sc., = proceed warily: see CALL v. 16, CANNY]

cacao /kəˈkɑːəʊ, -ˈkeɪəʊ/ n. (pl. **-os**) **1** a seed pod from which cocoa and chocolate are made. **2** a small widely cultivated evergreen tree, *Theobroma cacao*, bearing these. [Sp. f. Nahuatl *cacauatl* (*uatl* tree)]

cachalot /ˈkæʃəˌlɒt, -ˌləʊt/ n. a sperm whale. [F f. Sp. & Port. *cachalote*, of unkn. orig.]

cache /kæʃ/ n. & v. ● n. **1** a hiding-place for treasure, provisions, ammunition, etc. **2** what is hidden in a cache. **3** (in full **cache memory**) Computing an auxiliary memory from which high-speed retrieval is possible. ● v.tr. put in a cache. [F f. *cacher* to hide]

cachectic /kəˈkektɪk/ adj. Med. relating to or having the symptoms of cachexia.

cachet /ˈkæʃeɪ/ n. **1** a distinguishing mark or seal. **2** prestige. **3** Med. a flat capsule enclosing a dose of unpleasant-tasting medicine. [F f. *cacher* press ult. f. L *coactare* constrain]

cachexia /kəˈkeksɪə/ n. (also **cachexy** /-sɪ/) Med. a condition of wasting associated with chronic disease, esp. cancer. [F *cachexie* or LL *cachexia* f. Gk *kakhexia* f. *kakos* bad + *hexis* habit]

cachinnate /ˈkækɪˌneɪt/ v.intr. literary laugh loudly. □ **cachinnatory** adj. **cachinnation** /ˌkækɪˈneɪʃ(ə)n/ n. [L *cachinnare cachinnat-*]

cacholong /ˈkæʃəˌlɒŋ/ n. a kind of opal. [F f. Mongolian *kashchilon* beautiful stone]

cachou /ˈkæʃuː/ n. **1** a lozenge to sweeten the breath. **2** var. of CATECHU. [F f. Port. *cachu* f. Malay *kāchu*: cf. CATECHU]

cachucha /kəˈtʃuːtʃə/ n. a Spanish solo dance. [Sp.]

cacique /kəˈsiːk/ n. **1** a West Indian or American Indian native chief. **2** a political boss in Spain or Latin America. [Sp., of Carib orig.]

cack-handed /kækˈhændɪd/ adj. colloq. **1** awkward, clumsy. **2** left-handed. □ **cack-handedly** adv. **cack-handedness** n. [dial. *cack* excrement]

cackle /ˈkæk(ə)l/ n. & v. ● n. **1** a clucking sound as of a hen or a goose. **2** a loud silly laugh. **3** noisy inconsequential talk. ● v. **1** intr. emit a cackle. **2** intr. talk noisily and inconsequentially. **3** tr. utter or express with a cackle. □ **cut the cackle** colloq. stop talking aimlessly and come to the point. [ME prob. f. MLG, MDu. *kākelen* (imit.)]

cacodemon /ˌkækəˈdiːmən/ n. (also **cacodaemon**) **1** an evil spirit. **2** a malignant person. [Gk *kakodaimōn* f. *kakos* bad + *daimōn* spirit]

cacodyl /ˈkækəˌdaɪl, -ˌdɪl/ n. Chem. **1** a malodorous, toxic, spontaneously flammable liquid, tetramethyldiarsine (chem. formula: $((CH_3)_2As)_2$). **2** (usu. attrib.) the radical $-As(CH_3)_2$ derived from this. □ **cacodylic** /ˌkækəˈdaɪlɪk, -ˈdɪlɪk/ adj. [Gk *kakōdēs* stinking f. *kakos* bad]

cacoethes /ˌkækəʊˈiːθiːz/ n. an urge to do something inadvisable. [L f. Gk *kakoēthes* neut. adj. f. *kakos* bad + *ēthos* disposition]

cacography /kəˈkɒɡrəfɪ/ n. **1** bad handwriting. **2** bad spelling. □ **cacographer** n. **cacographic** /ˌkækəˈɡræfɪk/ adj. **cacographical** adj. **cacographically** adv. [Gk *kakos* bad, after *orthography*]

cacology /kəˈkɒlədʒɪ/ n. **1** bad choice of words. **2** bad pronunciation. [LL *cacologia* f. Gk *kakologia* vituperation f. *kakos* bad]

cacomistle /ˈkækəˌmɪs(ə)l/ n. a raccoon-like American animal of the genus *Bassariscus*, with a dark-ringed tail; esp. *B. astutus*, found from the north-western US to Mexico. [Amer. Sp. *cacomixtle* f. Nahuatl *tlacomiztli*]

cacophony /kəˈkɒfənɪ/ n. (pl. **-ies**) **1** a harsh discordant mixture of sound. **2** dissonance; discord. □ **cacophonous** adj. [F *cacophonie* f. Gk *kakophōnia* f. *kakophōnos* f. *kakos* bad + *phōnē* sound]

cactus /ˈkæktəs/ n. (pl. **cacti** /-taɪ/ or **cactuses**) a succulent plant of the family Cactaceae, with a thick fleshy stem, usu. spines but no leaves, and often brilliantly coloured flowers. They are native to the arid areas of the New World. □ **cactus dahlia** a dahlia with quilled petals resembling a cactus flower. □ **cactaceous** /kækˈteɪʃəs/ adj. [L f. Gk *kaktos* cardoon]

cacuminal /kæˈkjuːmɪn(ə)l/ adj. Phonet. pronounced with the tip of the tongue curled up towards the hard palate. [L *cacuminare* make pointed f. *cacumen -minis* tree-top]

CAD abbr. computer-aided design.

cad /kæd/ n. a person (esp. a man) who behaves dishonourably. □ **caddish** adj. **caddishly** adv. **caddishness** n. [abbr. of CADDIE in earlier sense 'odd job man']

cadastral /kəˈdæstrəl/ adj. of or showing the extent, value, and ownership of land, esp. for taxation. [F f. *cadastre* register of property f. Prov. *cadastro* f. It. *catasto* f. earlier *catastico* f. late Gk *katastikhon* list, register f. *kata stikhon* line by line]

cadaver /kəˈdeɪvə(r), -ˈdɑːvə(r)/ n. esp. Med. a corpse. □ **cadaveric** /-ˈdævərɪk/ adj. [ME f. L f. *cadere* fall]

cadaverous /kəˈdævərəs/ adj. **1** corpselike. **2** deathly pale. [L *cadaverosus* (as CADAVER)]

Cadbury /ˈkædbərɪ/, George (1839–1922), English cocoa and chocolate manufacturer and social reformer. He and his brother Richard (1835–99) took over their father's business in 1861 and established Cadbury Brothers. Committed Quakers, they greatly improved working conditions, and in 1879 George Cadbury moved the works to a new factory on a rural site (which he called Bournville) outside Birmingham, where he subsequently built a housing estate intended primarily for his workers.

caddie /ˈkædɪ/ n. & v. (also **caddy**) ● n. (pl. **-ies**) a person who assists a golfer during a match, by carrying clubs etc. ● v.intr. (**caddies, caddied, caddying**) act as caddie. □ **caddie car** (or **cart**) a light two-wheeled trolley for transporting golf clubs during a game. [orig. Sc. f. F CADET]

caddis-fly /ˈkædɪsˌflaɪ/ n. (pl. **-flies**) a small mothlike insect of the order Trichoptera, having aquatic larvae that often make a cylindrical protective case of debris. [17th c.: orig. unkn.]

caddish see CAD.

caddis-worm /ˈkædɪsˌwɜːm/ n. (also **caddis**) a larva of the caddis-fly, living in water and making protective cylindrical cases of sticks, leaves, etc., and used as fishing-bait. [as CADDIS-FLY]

caddy¹ /ˈkædɪ/ n. (pl. **-ies**) a small container, esp. a box for holding tea. [earlier *catty* weight of 1⅓ lb, f. Malay *kātī*]

caddy² var. of CADDIE.

cadence /ˈkeɪd(ə)ns/ n. **1** a fall in pitch of the voice, esp. at the end of a phrase or sentence. **2** intonation, tonal inflection. **3** *Mus.* the close of a musical phrase. **4** rhythm; the measure or beat of sound or movement. □ **cadenced** *adj.* [ME f. OF f. It. *cadenza*, ult. f. L *cadere* fall]

cadential /kəˈdenʃ(ə)l/ *adj.* of a cadence or cadenza.

cadenza /kəˈdenzə/ *n.* in music, a virtuoso solo passage inserted into a movement in a concerto or other work. Originally, a cadenza was a flourish inserted by a singer into an aria; in the late 17th and 18th centuries such insertions were improvised by the performer, but in the 19th century composers increasingly wrote out instrumental cadenzas. [It.: see CADENCE]

cadet /kəˈdet/ *n.* **1** a young trainee in the armed services or police force. **2** *NZ* an apprentice in sheep-farming. **3** a younger son. □ **cadetship** *n.* [F f. Gascon dial. *capdet*, ult. f. L *caput* head]

cadge /kædʒ/ *v.* **1** *tr.* get or seek by begging. **2** *intr.* beg. □ **cadger** *n.* [19th c., earlier = ? bind, carry: orig. unkn.]

cadi /ˈkɑːdɪ, ˈkeɪdɪ/ *n.* (also **kadi**) (*pl.* **-is**) a judge in a Muslim country. [Arab. *ḳāḍī* f. *ḳaḍā* to judge]

Cadiz /kəˈdɪz/ (Spanish **Cádiz** /ˈkaðiθ/) a city and port on the coast of SW Spain; pop. (1991) 156,560. It was important in the 16th to 18th centuries as the headquarters of the Spanish fleet. In 1587 Sir Francis Drake burnt the ships of Philip II at anchor there.

Cadmean /kædˈmiːən/ *adj.* = PYRRHIC[1]. [L *Cadmeus* f. Gk *Kadmeios* f. *Kadmos* CADMUS]

cadmium /ˈkædmɪəm/ *n.* a silvery-white metallic chemical element (atomic number 48; symbol **Cd**. (*See note below.*) □ **cadmium cell** *Electr.* a standard primary cell. **cadmium yellow** an intense yellow pigment containing cadmium sulphide, used in paints. [obs. *cadmia* calamine f. L *cadmia* f. Gk *kadm(e)ia (gē)* Cadmean (earth), f. CADMUS]

▪ Cadmium was discovered by the German chemist Friedrich Stromeyer in 1817. It chiefly occurs associated with zinc, which it resembles, and is obtained as a by-product of zinc smelting. It is used as a component in low-melting-point alloys, as a corrosion-resistant coating on other metals, and in the manufacture of pigments. Its compounds are very toxic.

Cadmus /ˈkædməs/ *Gk Mythol.* the brother of Europa and traditional founder of Thebes in Boeotia. He killed a dragon which guarded a spring, and when (on Athene's advice) he sowed the dragon's teeth there came up a harvest of armed men; he disposed of the majority by setting them to fight one another, and the survivors formed the ancestors of the Theban nobility.

cadre /ˈkɑːdə(r), ˈkɑːdrə/ *n.* **1** a basic unit, esp. of servicemen, forming a nucleus for expansion when necessary. **2** also /ˈkeɪdə(r)/ **a** a group of activists in a communist or other revolutionary party. **b** a member of such a group. [F f. It. *quadro* f. L *quadrus* square]

caduceus /kəˈdjuːsɪəs/ *n.* (*pl.* **caducei** /-sɪˌaɪ/) an ancient Greek or Roman herald's wand, esp. as carried by the messenger-god Hermes or Mercury. [L f. Doric Gk *karuk(e)ion* f. *kērux* herald]

caducous /kəˈdjuːkəs/ *adj. Biol.* (of organs and parts) easily detached or shed at an early stage. □ **caducity** /-ˈdjuːsɪtɪ/ *n.* [L *caducus* falling f. *cadere* fall]

caecilian /siːˈsɪlɪən/ *n.* (also **coecilian**) *Zool.* a burrowing wormlike amphibian of the order Gymnophiona, having poorly developed eyes and no limbs. [L *caecilia* kind of lizard]

caecitis /sɪˈkaɪtɪs/ *n.* (*US* **cecitis**) *Med.* inflammation of the caecum.

caecum /ˈsiːkəm/ *n.* (*US* **cecum**) (*pl.* **-ca** /-kə/) *Anat.* a blind-ended pouch at the junction of the small and large intestines. □ **caecal** *adj.* [L for *intestinum caecum* f. *caecus* blind, transl. Gk *tuphlon enteron*]

Caedmon /ˈkædmən/ (7th century), English poet and monk. According to Bede he was an illiterate herdsman, who received in a vision the power of song and was called to sing in praise of the Creation. He then joined the monastery at Whitby and later wrote English poetry inspired by biblical themes. The only authentic fragment of his work is a song in praise of the Creation, quoted by Bede.

Caen /kɒn/ an industrial city and river port in Normandy in northern France, on the River Orne, capital of the region of Basse-Normandie; pop. (1990) 115,620. It is the burial place of William the Conqueror. The town was the scene of fierce fighting between the Germans and the Allies in June and July 1944.

Caernarfon /kəˈnɑːv(ə)n/ (also **Caernarvon**) a town in NW Wales on the shore of the Menai Strait, the administrative centre of Gwynedd; pop. (1981) 9,400. Its 13th-century castle was the birthplace of Edward II.

Caernarvonshire /kəˈnɑːv(ə)nˌʃɪə(r)/ a former county of NW Wales. It became a part of Gwynedd in 1974.

Caerns. *abbr.* Caernarvonshire.

Caerphilly /keəˈfɪlɪ, kə-/ *n.* a kind of mild white cheese orig. made in Caerphilly in South Wales.

Caesar[1] /ˈsiːzə(r)/ *n.* **1** the title of the Roman emperors, esp. from Augustus to Hadrian. **2** an autocrat. **3** *Med. sl.* a Caesarean section; a case of this. □ **Caesar's wife** a person required to be above suspicion. [L, family name of Gaius Julius *Caesar* (see JULIUS CAESAR)]

Caesar[2] see JULIUS CAESAR.

Caesarea /ˌsiːzəˈrɪə/ an ancient port on the Mediterranean coast of Israel. Founded in 22 BC by Herod the Great on the site of a Phoenician harbour and named in honour of the Roman emperor Augustus Caesar, Caesarea became one of the principal cities of Roman Palestine. It later declined as its harbour silted up.

Caesarea Mazaca /ˈmæzəkə/ the former name for KAYSERI.

Caesarean /sɪˈzeərɪən/ *adj. & n.* (also **Caesarian**, *US* **Cesarean**) ● *adj.* **1** of Caesar or the Caesars. **2** (of a birth) effected by Caesarean section. ● *n.* a Caesarean section. □ **Caesarean section** an operation for delivering a child by cutting into the mother's womb through the wall of the abdomen (Julius Caesar supposedly having been born this way). [L *Caesarianus*]

Caesarea Philippi /ˈfɪlɪˌpaɪ, fɪˈlɪpaɪ/ a city in ancient Palestine, on the site of the present-day village of Baniyas in the Golan Heights. It was the site of a Hellenistic shrine to the god Pan and then of a temple built towards the end of the 1st century BC by Herod the Great and named in honour of the Roman emperor Augustus Caesar.

Caesar salad *n.* a salad dish consisting of cos lettuce and croutons served with a dressing of olive oil, lemon juice, raw egg, Worcester sauce, and seasoning. [*Caesar* Cardini, Mexican restaurateur, who invented it in 1924]

caesious /ˈsiːzɪəs/ *adj. Bot.* bluish or greyish-green. [L *caesius*]

caesium /ˈsiːzɪəm/ *n.* (*US* **cesium**) a soft silvery reactive metallic chemical element (atomic number 55; symbol **Cs**). (*See note below.*) □ **caesium clock** an atomic clock that uses the vibrations of caesium atoms as a time standard. [as CAESIOUS (from its spectrum lines)]

▪ Discovered spectroscopically by Robert Bunsen and Gustav Kirchhoff in 1860, caesium is a member of the alkali-metal group. It is a rare element, occurring in traces in some rocks and minerals; it has few commercial uses. It melts at 28.5°C and so is liquid in a warm room. A specified transition of the caesium-133 atom is used in defining the second as a unit of time (see SECOND[2]).

caesura /sɪˈzjʊərə/ *n.* (*pl.* **caesuras**) *Prosody* **1** (in Greek and Latin verse) a break between words within a metrical foot. **2** (in modern verse) a pause near the middle of a line. □ **caesural** *adj.* [L f. *caedere caes-* cut]

CAF *abbr. US* cost and freight.

cafard /kɑːˈfɑː(r)/ *n.* melancholia. [F, = cockroach, hypocrite]

café /ˈkæfeɪ, ˈkæfɪ/ *n.* (also **cafe**; also *joc.* /kæf, keɪf/) **1** a small coffee-house or teashop; a simple restaurant. **2** *US* a bar. □ **café au lait** /əʊ ˈleɪ/ **1** coffee with milk. **2** the colour of this. **café noir** /nwɑː(r)/ black coffee. **café society** the regular patrons of fashionable restaurants and nightclubs. [F, = coffee, coffee-house]

cafeteria /ˌkæfɪˈtɪərɪə/ *n.* a restaurant in which customers collect their meals on trays at a counter and usu. pay before sitting down to eat. [Amer. Sp. *cafetería* coffee-shop]

cafetière /ˌkæfɪˈtjeə(r)/ *n.* a coffee-pot with a plunger that pushes the grounds to the bottom. [F]

caff /kæf/ *n. Brit. sl.* = CAFÉ. [abbr.]

caffeine /ˈkæfiːn/ *n.* an alkaloid drug with stimulant action, found in tea leaves and coffee beans. [F *caféine* f. *café* coffee]

caftan /ˈkæftæn/ *n.* (also **kaftan**) **1** a long usu. belted tunic worn by men in countries of the Near East. **2 a** a woman's long loose dress. **b** a loose shirt or top. [Turk. *ḳaftān*, partly through F *cafetan*]

Cagayan Islands /ˌkɑːɡəˈjɑːn/ a group of seven small islands in the Sulu Sea in the western Philippines.

Cage /keɪdʒ/, John (Milton) (1912–92), American composer, pianist, and writer. A pupil of Schoenberg, he is notable for his experimental approach to music, including the use of silence and the role of chance (in which he was a pioneer: see ALEATORY MUSIC). He experimented with musical instruments, inventing the 'prepared piano' (with pieces of metal, rubber, etc., inserted between the strings to alter the tone), and also used electronic instruments and various sound-effects.

cage /keɪdʒ/ n. & v. ● n. **1** a structure of bars or wires, esp. for confining animals or birds. **2** any similar open framework, esp. an enclosed platform or lift in a mine or the compartment for passengers in a lift. **3** colloq. a camp for prisoners of war. ● v.tr. place or keep in a cage. □ **cage-bird** a bird of the kind customarily kept in a cage. [ME f. OF f. L cavea]

cagey /ˈkeɪdʒɪ/ adj. (also **cagy**) (**cagier**, **cagiest**) colloq. cautious and uncommunicative; wary. □ **cagily** adv. **caginess** n. (also **cageyness**). [20th-c. US: orig. unkn.]

Cagliari /kæˈljɑːrɪ, Italian ˈkaʎʎari/ the capital of Sardinia, a port on the south coast; pop. (1990) 211,720.

Cagney /ˈkægnɪ/, James (1899–1986), American actor. He is chiefly remembered for his parts as a gangster in films such as *The Public Enemy* (1931) and *Angels with Dirty Faces* (1938). Also a skilled dancer and comedian, he received an Oscar for his part in the musical *Yankee Doodle Dandy* (1942).

cagoule /kəˈguːl/ n. a hooded thin windproof and waterproof garment, usu. pulled on over the head. [F]

cahoots /kəˈhuːts/ n.pl. □ **in cahoots** (often foll. by *with*) sl. in collusion. [19th c.: orig. uncert.]

CAI abbr. computer-assisted (or -aided) instruction.

caiman var. of CAYMAN.

Cain /keɪn/ in the Bible, the eldest son of Adam and murderer of his brother Abel (Gen. 4:1–16). □ **raise Cain** colloq. make a disturbance; create trouble.

Caine /keɪn/, Sir Michael (born Maurice Micklewhite) (b.1933), English film actor. He had his first major role in *Zulu* (1963), and soon established a reputation for laconic, anti-heroic roles, e.g. as a spy in *The Ipcress File* (1965) and as a streetwise cockney in *Alfie* (1966). Since then, he has appeared in a wide variety of films, such as *Educating Rita* (1983) and *Hannah and Her Sisters* (1986), for which he won an Oscar.

Cainozoic var. of CENOZOIC.

caique /kaɪˈiːk/ n. **1** a light rowing-boat on the Bosporus. **2** an eastern Mediterranean sailing-ship. [F f. It. *caicco* f. Turk. *kayík*]

cairn /keən/ n. **1** a mound of rough stones built as a monument or landmark. **2** (in full **cairn terrier**) a breed of small terrier with short legs, a longish body, and a shaggy coat (perhaps so called from its being used to hunt among cairns). [Gael. *carn*]

cairngorm /ˈkeəngɔːm/ n. a yellow or wine-coloured semi-precious form of quartz. [CAIRNGORM MOUNTAINS (where first found)]

Cairngorm Mountains (also **Cairngorms**) a mountain range in northern Scotland. Its highest peak, Ben Macdhui (1,309 m; 4,296 ft) is the second-highest mountain in the British Isles. [Gael. *carn gorm* blue cairn]

Cairo /ˈkaɪrəʊ/ (Arabic **Al Qahira** /æl ˈkɑːhiːrɑː/, also **El Qahira**) the capital of Egypt, a port on the Nile near the head of its delta; pop. (est. 1991) 13,300,000. Founded by the Fatimid dynasty in 969, it was later fortified by the Crusaders by Saladin, whose citadel, built c.1179, still survives.

caisson /ˈkeɪs(ə)n, kəˈsuːn/ n. **1** a watertight chamber in which underwater construction work can be done. **2** a floating vessel used as a floodgate in docks. **3** an ammunition chest or wagon. □ **caisson disease** = decompression sickness. [F (f. It. *cassone*) assim. to *caisse* CASE²]

Caithness /ˈkeɪθnes/ a former county in the extreme north-east of Scotland. It became part of Highland Region in 1975.

caitiff /ˈkeɪtɪf/ n. & adj. poet. or archaic ● n. a base or despicable person; a coward. ● adj. base, despicable, cowardly. [ME f. OF *caitif*, *chaitif* ult. f. L *captivus* CAPTIVE]

cajole /kəˈdʒəʊl/ v.tr. (often foll. by *into*, *out of*) persuade by flattery, deceit, etc. □ **cajolement** n. **cajolery** /-ˈdʒəʊlərɪ/ n. [F *cajoler*]

Cajun /ˈkeɪdʒən/ n. a member of any of the largely self-contained communities in the bayou areas of southern Louisiana, which were formed by descendants of the French Canadians driven out of Nova Scotia by the British in the 18th century. The Cajuns speak an archaic form of French and are particularly known for their lively folk music and spicy cooking. [alt. of ACADIAN]

cake /keɪk/ n. & v. ● n. **1 a** a mixture of flour, butter, eggs, sugar, etc., baked in the oven. **b** a quantity of this baked in a flat round or ornamental shape and often iced and decorated. **2 a** other food in a flat round shape (*fish cake*). **b** = cattle-cake. **3** a flattish compact mass (*a cake of soap*). **4** Sc. & N. Engl. thin oaten bread. ● v. **1** tr. & intr. form into a compact mass. **2** tr. (usu. foll. by *with*) cover (with a hard or sticky mass) (*boots caked with mud*). □ **cake-hole** sl. the mouth. **cakes and ale**

merrymaking. **have one's cake and eat it** colloq. enjoy both of two mutually exclusive alternatives. **a piece of cake** colloq. something easily achieved. **sell** (or **go**) **like hot cakes** colloq. be sold (or go) quickly; be popular. **a slice of the cake** participation in benefits. [ME f. ON *kaka*]

cakewalk /ˈkeɪkwɔːk/ n. **1** a dance developed from an American black contest in graceful walking with a cake as a prize. **2** colloq. an easy task. **3** a form of fairground entertainment consisting of a promenade moved by machinery.

CAL abbr. computer-assisted learning.

Cal abbr. large calorie(s) or kilocalorie(s).

Cal. abbr. California.

cal abbr. small calorie(s).

Calabar /ˈkæləˌbɑː(r)/ a seaport in Nigeria; pop. (1983) 126,000.

Calabar bean n. a poisonous seed of the tropical African climbing plant *Physostigma venosum*, yielding a medicinal extract.

calabash /ˈkæləˌbæʃ/ n. **1 a** an evergreen tree, *Crescentia cujete*, native to tropical America, bearing fruit in the form of large gourds. **b** a gourd from this tree. **2** the shell of this or a similar gourd used as a vessel for water, to make a tobacco pipe, etc. [F *calebasse* f. Sp. *calabaza* perh. f. Pers. *karbuz* melon]

calaboose /ˌkæləˈbuːs/ n. US a prison. [Black F *calabouse* f. Sp. *calabozo* dungeon]

calabrese /ˌkæləˈbriːz, -ˈbreɪseɪ/ n. a large succulent variety of sprouting broccoli. [It., = Calabrian]

Calabria /kəˈlæbrɪə/ a region of SW Italy, forming the 'toe' of the Italian peninsula; capital, Reggio di Calabria. The name was formerly applied by the Byzantines to the eastern promontory forming the 'heel', but was transferred to the 'toe' in the west when the area was seized by the Lombards AD c.700. □ **Calabrian** adj. & n.

Calais /ˈkæleɪ/ a ferry port in northern France; pop. (1990) 75,840. Captured by Edward III in 1347 after a long siege, it remained an English possession until it was retaken by the French in 1558. The French terminal of the Channel Tunnel is at Fréthun, to the south of Calais.

calamanco /ˌkæləˈmæŋkəʊ/ n. (pl. **-oes**) hist. a glossy woollen cloth chequered on one side. [16th c.: orig. unkn.]

calamander /ˈkæləˌmændə(r)/ n. a fine-grained red-brown ebony streaked with black, from the Asian tree *Diospyros qualsita*, used in furniture. [19th c.: orig. unkn.: perh. connected with Sinh. word for the tree *kalu-madiriya*]

calamary /ˈkæləˌmɑːrɪ/ n. (pl. **-ies**) a squid with a horny penlike internal shell, esp. one of the genus *Loligo*. [med.L *calamarium* pen-case f. L *calamus* pen]

calamine /ˈkæləˌmaɪn/ n. **1** a pink powder consisting of zinc carbonate and ferric oxide, used as a lotion or ointment. **2** a zinc mineral, usu. zinc carbonate. [ME f. F f. med.L *calamina* alt. f. L *cadmia*: see CADMIUM]

calamint /ˈkæləˌmɪnt/ n. an aromatic labiate herb or shrub of the genus *Calamintha*; esp. *C. officinalis*, with purple or lilac flowers. [ME f. OF *calament* f. med.L *calamentum* f. LL *calaminthe* f. Gk *kalaminthē*]

calamity /kəˈlæmɪtɪ/ n. (pl. **-ies**) **1** a disaster, a great misfortune. **2 a** adversity. **b** deep distress. □ **calamitous** adj. **calamitously** adv. [ME f. F *calamité* f. L *calamitas -tatis*]

Calamity Jane (born Martha Jane Cannary) (c.1852–1903), American frontierswoman. A colourful character noted for her skill at shooting and riding, she dressed as a man and lived for a time in Wyoming, where she became known for her wild behaviour and heavy drinking. She later joined Buffalo Bill's Wild West Show.

calando /kæˈlændəʊ/ adv. Mus. gradually decreasing in speed and volume. [It., = slackening]

calash /kəˈlæʃ/ n. hist. **1 a** a light low-wheeled carriage with a removable folding hood. **b** the folding hood itself. **2** Can. a two-wheeled horse-drawn vehicle. **3** a woman's hooped silk hood. [F *calèche* f. G *Kalesche* f. Pol. *kolaska* or Czech *kolesa*]

calc- /kælk/ comb. form lime or calcium. [G *Kalk* f. L CALX]

calcaneus /kælˈkeɪnɪəs/ n. (also **calcaneum** /-nɪəm/) (pl. **calcanei** /-nɪˌaɪ/ or **calcanea** /-nɪə/) Anat. the bone forming the heel. [L]

calcareous /kælˈkeərɪəs/ adj. (also **calcarious**) of or containing calcium carbonate; chalky. [L *calcarius* (as CALX)]

calceolaria /ˌkælsɪəˈleərɪə/ n. a South American plant of the genus *Calceolaria*, of the figwort family, with slipper-shaped flowers. [mod.L f. L *calceolus* dimin. of *calceus* shoe + *-aria* fem. = -ARY¹]

calceolate /ˈkælsɪˌleɪt/ *adj. Bot.* slipper-shaped.

calces *pl.* of CALX.

calcicole /ˈkælsɪˌkəʊl/ *adj. & n. Bot.* ● *adj.* growing best in calcareous soil. ● *n.* a calcicole plant. [f. CALX + *colere* inhabit]

calciferol /kælˈsɪfəˌrɒl/ *n.* one of the D vitamins, routinely added to dairy products, essential for the deposition of calcium in bones. Also called *ergocalciferol, vitamin D₂*. [CALCIFEROUS + -OL]

calciferous /kælˈsɪfərəs/ *adj.* yielding calcium salts, esp. calcium carbonate. [L CALX lime + -FEROUS]

calcifuge /ˈkælsɪˌfjuːdʒ/ *adj. & n. Bot.* ● *adj.* not suited to calcareous soil. ● *n.* a calcifuge plant. [f. CALX + -FUGE]

calcify /ˈkælsɪˌfaɪ/ *v.tr. & intr.* (**-ies, -ied**) **1** harden or become hardened by deposition of calcium salts; petrify. **2** convert or be converted to calcium carbonate. □ **calcific** /kælˈsɪfɪk/ *adj.* **calcification** /ˌkælsɪfɪˈkeɪʃ(ə)n/ *n.*

calcine /ˈkælsɪn, -saɪn/ *v.* **1** *tr.* **a** reduce, oxidize, or desiccate by strong heat. **b** burn to ashes; consume by fire; roast. **c** reduce to calcium oxide by roasting or burning. **2** *tr.* consume or purify as if by fire. **3** *intr.* undergo any of these. □ **calcination** /ˌkælsɪˈneɪʃ(ə)n/ *n.* [ME f. OF *calciner* or med.L *calcinare* f. LL *calcina* lime f. L CALX]

calcite /ˈkælsaɪt/ *n.* natural crystalline calcium carbonate. [G *Calcit* f. L CALX lime]

calcium /ˈkælsɪəm/ *n.* a soft grey metallic chemical element (atomic number 20; symbol **Ca**). (*See note below.*) □ **calcium carbonate** a white insoluble solid occurring naturally as chalk, limestone, marble, and calcite, forming mollusc shells and stony corals, and used in the manufacture of lime and cement. **calcium hydroxide** a white crystalline powder used in the manufacture of plaster and cement; slaked lime. **calcium oxide** quicklime (see LIME¹ 1). **calcium phosphate** the main constituent of animal bones and used as bone ash fertilizer. **calcium sulphate** a white crystalline solid occurring as anhydrite and gypsum. [L CALX lime]

▪ Calcium was first isolated by Sir Humphry Davy in 1808, and belongs to the alkaline earth metals. A common element in the earth's crust, its compounds occur naturally in limestone, fluorite, gypsum, etc. The metal now has a number of specialized uses. Calcium is essential to life: many physiological processes depend on the movement of calcium ions, and calcium salts are an essential constituent of bone, teeth, and shells.

calcrete /ˈkælkriːt/ *n. Geol.* a conglomerate formed by the cementation of sand and gravel with calcium carbonate. [CALC- + *concrete*]

calcspar /ˈkælkspɑː(r)/ *n.* = CALCITE. [CALC- + SPAR³]

calculable /ˈkælkjʊləb(ə)l/ *adj.* able to be calculated or estimated. □ **calculably** *adv.* **calculability** /ˌkælkjʊləˈbɪlɪti/ *n.*

calculate /ˈkælkjʊˌleɪt/ *v.* **1** *tr.* ascertain or determine beforehand, esp. by mathematics or by reckoning. **2** *tr.* plan deliberately. **3** *intr.* (foll. by *on, upon*) rely on; make an essential part of one's reckoning (*calculated on a quick response*). **4** *tr. US colloq.* suppose, believe. □ **calculative** /-lətɪv/ *adj.* [LL *calculare* (as CALCULUS)]

calculated /ˈkælkjʊˌleɪtɪd/ *adj.* **1** (of an action) done with awareness of the likely consequences. **2** (foll. by *to* + infin.) designed or suitable; intended. □ **calculatedly** *adv.*

calculating /ˈkælkjʊˌleɪtɪŋ/ *adj.* (of a person) shrewd, scheming. □ **calculatingly** *adv.*

calculation /ˌkælkjʊˈleɪʃ(ə)n/ *n.* **1** the act or process of calculating. **2** a result obtained by calculating. **3** a reckoning or forecast. [ME f. OF f. LL *calculatio* (as CALCULATE)]

calculator /ˈkælkjʊˌleɪtə(r)/ *n.* **1** a device (esp. a small electronic one) used for making mathematical calculations. **2** a person or thing that calculates. **3** a set of tables used in calculation. [ME f. L (as CALCULATE)]

calculus /ˈkælkjʊləs/ *n.* (*pl.* **calculuses** or **calculi** /-ˌlaɪ/) **1** *Math.* **a** a particular method of calculation or reasoning (*calculus of probabilities*). **b** (also **infinitesimal calculus**) the part of mathematics concerned with the integration and differentiation of functions. (*See note below.*) **2 a** *Med.* a stone or concretion of minerals formed within the body, esp. in the kidney or gall-bladder. **b** *Dentistry* = TARTAR 1. □ **calculous** *adj.* (in sense 2). [L, = small stone used in reckoning on an abacus]

▪ Calculus emerged in the 17th century from the work of Leibniz, Newton, and others, who used methods based on the summation of infinitesimal differences in order to find rates of change of varying quantities. It was developed for its main applications in mechanics and in geometry, where it provides techniques for finding tangents of curves and areas of curvilinear figures.

Calcutta /kælˈkʌtə/ a port and industrial centre in eastern India, capital of the state of West Bengal and the second largest city in India; pop. (1991) 10,916,000. It is situated on the Hooghly river near the Bay of Bengal. Founded c.1690 by the East India Company, Calcutta was the capital of India from 1772 to 1912.

Calcutta, Black Hole of a dungeon in Fort William, Calcutta, where, following the capture of Calcutta by Siraj-ud-Dawlah (c.1729–57), nawab of Bengal, in 1756, perhaps as many as 146 English prisoners were confined in a narrow cell 6 m (20 ft) square for the night of 20 June, only twenty-three of them still being alive the next morning.

caldarium /kælˈdeərɪəm/ *n.* (*pl.* **caldariums** or **caldaria** /-rɪə/) *Rom. Antiq.* in a Roman bath, the room containing the first, hot, bath. (See also ROMAN BATHS.) [L f. *calidus* hot]

Caldecott /ˈkɔːldɪˌkɒt/, Randolph (1846–86), English graphic artist and watercolour painter. He is best known for his illustrations for children's books, such as those for *The House that Jack Built* (1878). A medal awarded annually for the illustration of American children's books is named after him.

Calder /ˈkɔːldə(r)/, Alexander (1898–1976), American sculptor and painter. He was one of the first artists to introduce movement into sculpture, from the early 1930s, making mobiles incorporating abstract forms and often using wire. His static sculptures (e.g. *The Red Crab*, 1962) are known by contrast as 'stabiles'.

caldera /kælˈdeərə/ *n. Geol.* a large volcanic crater whose breadth greatly exceeds that of the vent or vents within it. [Sp. f. LL *caldaria* boiling-pot]

Calderón de la Barca /ˌkɔːldəˌrɒn deɪ læ ˈbɑːkə, ˌkɒl-/, Pedro (1600–81), Spanish dramatist and poet. He wrote some 120 plays, more than 70 of them religious dramas for outdoor performance on the festival of Corpus Christi. His secular dramas include *El Alcalde de Zalamea* (c.1643).

caldron *var.* of CAULDRON.

Caledonian /ˌkælɪˈdəʊnɪən/ *adj. & n.* ● *adj.* **1** of or relating to Scotland or (in Roman times) northern Britain. **2** *Geol.* of a mountain-forming period in Europe in the Palaeozoic era. ● *n.* a Scotsman. [L *Caledonia* Roman name for part of northern Britain]

Caledonian Canal a system of lochs and canals crossing Scotland from Inverness on the east coast to Fort William on the west. Built by Thomas Telford, it was opened in 1822. It traverses the Great Glen, part of its length being formed by Loch Ness.

calefacient /ˌkælɪˈfeɪʃ(ə)nt/ *n. & adj. Med.* ● *n.* a substance producing or causing a sensation of warmth. ● *adj.* having this property. [L *calefacere* f. *calere* be warm + *facere* make]

calendar /ˈkælɪndə(r)/ *n. & v.* ● *n.* **1** a system by which the beginning, length, and subdivisions of the year are fixed. (See GREGORIAN CALENDAR, JULIAN CALENDAR.) **2** a chart or series of pages showing the days, weeks, and months of a particular year, or giving special seasonal information. **3** a timetable or programme of appointments, special events, etc. **4** a list or register, esp. of canonized saints, cases for trial, etc. ● *v.tr.* register or enter in a calendar or timetable etc. □ **calendar month** see MONTH 1. **calendar year** see YEAR 2. □ **calendric** /kəˈlendrɪk/ *adj.* **calendrical** *adj.* [ME f. AF *calender*, OF *calendier* f. L *calendarium* account-book (as CALENDS)]

calender /ˈkælɪndə(r)/ *n. & v.* ● *n.* a machine in which cloth, paper, etc., is pressed by rollers to glaze or smooth it. ● *v.tr.* press in a calender. [F *calendre(r)*, of unkn. orig.]

calends /ˈkælɪndz/ *n.pl.* (also **kalends**) the first of the month in the ancient Roman calendar. [ME f. OF *calendes* f. L *kalendae*]

calendula /kəˈlendjʊlə/ *n.* a composite plant of the genus *Calendula*, with large yellow or orange flowers, e.g. marigold. [mod.L dimin. of *calendae* (as CALENDS), perh. = little clock]

calenture /ˈkæləntʃə(r)/ *n. hist.* a tropical delirium of sailors, who imagined the sea to be green fields. [F f. Sp. *calentura* fever f. *calentar* be hot ult. f. L *calere* be warm]

calf¹ /kɑːf/ *n.* (*pl.* **calves** /kɑːvz/) **1** a young bovine animal, used esp. of domestic cattle. **2** the young of other animals, e.g. elephant, deer, and whale. **3** *Naut.* a floating piece of ice detached from an iceberg. □ **calf-love** romantic attachment or affection between adolescents. **in** (or **with**) **calf** (of a cow) pregnant. □ **calflike** *adj.* [OE *cælf* f. WG]

calf² /kɑːf/ *n.* (*pl.* **calves** /kɑːvz/) the fleshy hind part of the human leg below the knee. □ **-calved** /kɑːvd/ *adj.* (in *comb.*). [ME f. ON *kálfi*, of unkn. orig.]

calfskin /ˈkɑːfskɪn/ *n.* calf-leather, esp. in bookbinding and shoemaking.

Calgary /ˈkælgərɪ/ a city in southern Alberta, SW Canada; pop. (1991) 710,680. It is on the edge of an agricultural area to the east of the Rocky Mountains. An annual rodeo known as the Calgary Stampede, inaugurated in 1912, is held there.

Cali /ˈkɑːlɪ/ an industrial city in western Colombia; pop. (est. 1992) 1,624,000.

calibrate /ˈkælɪˌbreɪt/ *v.tr.* **1** mark (a gauge) with a standard scale of readings. **2** correlate the readings of (an instrument) with a standard. **3** determine the calibre of. **4** determine the correct capacity or value of. □ **calibrator** *n.* [CALIBRE + -ATE³]

calibration /ˌkælɪˈbreɪʃ(ə)n/ *n.* **1** the act or process of calibrating an instrument, gauge, etc. **2** each of a set of graduations on an instrument etc.

calibre /ˈkælɪbə(r)/ *n.* (*US* **caliber**) **1 a** the internal diameter of a gun or tube. **b** the diameter of a bullet or shell. **2** strength or quality of character; ability, importance (*we need someone of your calibre*). □ **calibred** *adj.* (also in *comb.*). [F *calibre* or It. *calibro*, f. Arab. *ḳālib* mould]

caliche /kəˈliːtʃɪ/ *n.* **1** a mineral deposit of gravel, sand, and nitrates, esp. Chile saltpetre, found in dry areas of America. **2** = CALCRETE. [Amer. Sp.]

calico /ˈkælɪˌkəʊ/ *n.* & *adj.* ● *n.* (*pl.* **-oes** or *US* **-os**) **1** a cotton cloth, esp. plain white or unbleached. **2** *N. Amer.* a printed cotton fabric. ● *adj.* **1** made of calico. **2** *N. Amer.* multicoloured, piebald. [earlier *calicut* f. CALICUT]

Calicut /ˈkælɪˌkʌt/ a seaport in the state of Kerala in SW India, on the Malabar Coast; pop. (1991) 420,000. The cotton fabric known as calico originated there. In the 17th and 18th centuries Calicut became a centre of the textile trade with Europe.

Calif. *abbr.* California.

California /ˌkælɪˈfɔːnɪə/ a state of the US, on the Pacific coast; pop. (1990) 29,760,000; capital, Sacramento. Formerly part of Mexico, it was ceded to the US in 1847, having briefly been an independent republic, and became the 31st state of the US in 1850. Large numbers of settlers were attracted to California in the 19th century, especially during the gold rushes of the 1840s; it is now the most populous state. □ **Californian** *adj.* & *n.*

California, Gulf of an arm of the Pacific Ocean separating the Baja California peninsula from mainland Mexico.

Californian poppy *n.* a plant of the poppy family, *Eschscholtzia californica*, grown for its brilliant yellow or orange flowers.

californium /ˌkælɪˈfɔːnɪəm/ *n.* an artificial radioactive metallic chemical element (atomic number 98; symbol **Cf**). A member of the actinide series, californium was first produced by Glenn Seaborg and his colleagues in 1950 by bombarding curium with helium ions. It is used in industry and medicine as a source of neutrons. [CALIFORNIA, where it was first made]

Caligula /kəˈlɪgjʊlə/ (born Gaius Julius Caesar Germanicus) (AD 12–41), Roman emperor 37–41. Brought up in a military camp, he gained the nickname 'Caligula' (= little boot) as an infant on account of the miniature military boots he wore. Caligula's brief reign as emperor, which began when he succeeded Tiberius and ended with his assassination, became notorious for its tyrannical excesses.

caliper var. of CALLIPER.

caliph /ˈkeɪlɪf, ˈkæl-/ *n. hist.* the chief Muslim civil and religious ruler, regarded as the successor of Muhammad. The first caliph was Abu Bakr (573–634), who was chosen by the small Muslim community following the death of Muhammad in 632 to spread Islam, a task that included military leadership. He and the following three caliphs had had personal links with Muhammad, but subsequently the caliphate became a hereditary position with the establishment of the Umayyad and Abbasid dynasties (respectively *c.*660–750 and from 750). The latter ruled in Baghdad until 1258 and then in Egypt until the Ottoman conquest (1517), though by the 11th century most of the caliph's authority had passed to a hierarchy of officials. The title was then held by the Ottoman sultans until the nationalist revolution of 1922, and the caliphate was abolished by Atatürk in 1924. □ **caliphate** *n.* [ME f. OF *caliphe* f. Arab. *ḳalīfa* successor]

calisthenics var. of CALLISTHENICS.

calk *US* var. of CAULK.

call /kɔːl/ *v.* & *n.* ● *v.* **1** *intr.* **a** (often foll. by *out*) cry, shout; speak loudly. **b** (of a bird or animal) emit its characteristic note or cry. **2** *tr.* communicate or converse with by telephone or radio. **3** *tr.* **a** bring to one's presence by calling; summon (*will you call the children?*). **b** arrange for (a person or thing) to come or be present (*called a taxi*). **4** *intr.* (often

foll. by *at, in, on*) pay a brief visit (*called at the house; called in to see you; come and call on me*). **5** *tr.* **a** order to take place; fix a time for (*called a meeting*). **b** direct to happen; announce (*call a halt*). **6 a** *intr.* require one's attention or consideration (*duty calls*). **b** *tr.* urge, invite, nominate (*call to the bar*). **7** *tr.* name; describe as (*call her Amanda*). **8** *tr.* consider; regard or estimate as (*I call that silly*). **9** *tr.* rouse from sleep (*call me at 8*). **10** *intr.* guess the outcome of tossing a coin etc. **11** *intr.* (foll. by *for*) order, require, demand (*called for silence*). **12** *tr.* (foll. by *over*) read out (a list of names to determine those present). **13** *intr.* (foll. by *on, upon*) invoke; appeal to; request or require (*called on us to be quiet*). **14** *tr. Cricket* (of an umpire) disallow a ball from (a bowler). **15** *tr. Cards* specify (a suit or contract) in bidding. **16** *tr. Sc.* drive (an animal, vehicle, etc.). **17** *tr.* (in country dancing etc.) announce (the next figure or set of steps) by shouting rhythmically. ● *n.* **1** a shout or cry; an act of calling. **2 a** the characteristic cry of a bird or animal. **b** an imitation of this. **c** an instrument for imitating it. **3** a brief visit (*paid them a call*). **4 a** an act of telephoning. **b** a telephone conversation. **5 a** an invitation or summons to appear or be present. **b** an appeal or invitation (from a specific source or discerned by a person's conscience etc.) to follow a certain profession, set of principles, etc. **6** (foll. by *for*, or *to* + infin.) a duty, need, or occasion (*no call to be rude; no call for violence*). **7** (foll. by *for, on*) a demand (*not much call for it these days; a call on one's time*). **8** a signal on a bugle etc.; a signalling-whistle. **9** *Stock Exch.* an option of buying stock at a fixed price at a given date. **10** *Cards* **a** a player's right or turn to make a bid. **b** a bid made. **11** (in country dancing etc.) a direction called to the dancers. □ **at call** = *on call*. **call away** divert, distract. **call-box** a public telephone box or kiosk. **call-boy** a theatre attendant who summons actors when needed on stage. **call down 1** invoke. **2** reprimand. **call forth** elicit. **call-girl** a prostitute who accepts appointments by telephone. **call in 1** withdraw from circulation. **2** seek the advice or services of. **calling-card** *N. Amer.* = *visiting-card*. **call in** (or **into**) **question** dispute; doubt the validity of. **call into play** give scope for; make use of. **call a person names** abuse a person verbally. **call off 1** cancel (an arrangement etc.). **2** order (an attacker or pursuer) to desist. **call of nature** a need to urinate or defecate. **call out 1** summon (troops etc.) to action. **2** order (workers) to strike. **call-out** an instance of being called out, esp. in order to rescue, do repairs, etc. **call-over 1** a roll-call. **2** reading aloud of a list of betting prices. **call the shots** (or **tune**) be in control; take the initiative. **call-sign** (or **-signal**) a broadcast signal identifying the radio transmitter used. **call to account** see ACCOUNT. **call to mind** recollect; cause one to remember. **call to order 1** request to be orderly. **2** declare (a meeting) open. **call up 1** reach by telephone. **2** imagine, recollect. **3** summon, esp. to serve in the army. **call-up** *n.* the act or process of calling up or being called up (sense 3). **on call 1** (of a doctor etc.) available if required but not formally on duty. **2** (of money lent) repayable on demand. **within call** near enough to be summoned by calling. [OE *ceallian* f. ON *kalla*]

calla /ˈkælə/ *n.* **1** (in full **calla lily**) = *arum lily*. **2** an aquatic plant, *Calla palustris*, of the arum family. [mod.L]

Callaghan /ˈkæləhən/, (Leonard) James, Baron Callaghan of Cardiff (b.1912), British Labour statesman, Prime Minister 1976–9. He became Prime Minister following Harold Wilson's resignation; the leader of a minority government, he was forced in 1977 to negotiate an agreement with the Liberal Party (known as the Lib–Lab Pact) to stay in power. After widespread strikes in the so-called 'winter of discontent' (1978–9), Callaghan received a vote of no confidence; the Labour Party was defeated by the Conservatives in the subsequent election.

Callao /kæˈjɑːəʊ/ the principal seaport of Peru; pop. (1990) 588,600.

Callas /ˈkæləs/, Maria (born Maria Cecilia Anna Kalageropoulos) (1923–77), American-born operatic soprano, of Greek parentage. She was a coloratura soprano whose bel canto style of singing especially suited her to early Italian opera; a number of works by Rossini, Bellini, and Donizetti were revived for her.

caller /ˈkɔːlə(r)/ *n.* **1** a person who calls, esp. a person who pays a visit or makes a telephone call. **2** *Austral.* a racing commentator.

calligraphy /kəˈlɪgrəfɪ/ *n.* **1** handwriting, esp. when fine or pleasing. **2** the art of handwriting. □ **calligrapher** *n.* **calligraphist** *n.* **calligraphic** /ˌkælɪˈgræfɪk/ *adj.* [Gk *kalligraphia* f. *kallos* beauty]

Callil /kəˈlɪl/, Carmen (Thérèse) (b.1938), Australian publisher. She settled in Britain and worked with various publishers before founding her own feminist publishing house, Virago, in 1972. Virago books, with their distinctive jacket design, soon established a strong market identity, bringing overlooked women writers of the past to a wider public and also publishing new female authors.

Callimachus /kəˈlɪməkəs/ (*c.*305–*c.*240 BC), Greek poet and scholar.

As head of the library at Alexandria, he compiled a critical catalogue of the existing Greek literature of his day. As a poet he is best known for his short or episodic poetry, especially hymns and epigrams.

calling /'kɔ:lɪŋ/ n. **1** a profession or occupation. **2** an inwardly felt call or summons; a vocation.

Calliope /kə'laɪəpɪ/ Gk & Rom. Mythol. the Muse of epic poetry. [Gk, = beautiful-voiced]

calliope /kə'laɪəpɪ/ n. an American keyboard instrument resembling an organ, with a set of steam whistles producing musical notes. [CALLIOPE]

calliper /'kælɪpə(r)/ n. & v. (also **caliper**) ● n. **1** (in pl.) (also **calliper compasses**) compasses with bowed legs for measuring the diameter of convex bodies, or with out-turned points for measuring internal dimensions. **2** (in full **calliper splint**) a metal splint to support the leg. ● v.tr. measure with callipers. [app. var. of CALIBRE]

callisthenics /ˌkælɪs'θenɪks/ n.pl. (also **calisthenics**) gymnastic exercises to achieve bodily fitness and grace of movement. □ **callisthenic** adj. [Gk kallos beauty + sthenos strength]

Callisto /kə'lɪstəʊ/ **1** Gk Mythol. a nymph who was changed into a she-bear by Zeus (see URSA MAJOR). **2** Astron. satellite IV of Jupiter, the eighth closest to the planet, and one of the Galilean moons (diameter 4,800 km). It is probably composed mainly of ice; the surface is dark and heavily cratered.

callop /'kæləp/ n. Austral. a gold-coloured freshwater fish, Plectroplites ambiguus, used as food. Also called golden perch. [perh. Aboriginal (S. Australia)]

callosity /kə'lɒsɪtɪ/ n. (pl. **-ies**) a hard thick area of skin usu. occurring in parts of the body subject to pressure or friction. [F callosité or L callositas (as CALLOUS)]

callous /'kæləs/ adj. & n. ● adj. **1** unfeeling, insensitive. **2** (of skin) hardened or hard. ● n. = CALLUS 1. □ **calloused** adj. **callously** adv. (in sense 1 of adj.). **callousness** n. [ME f. L callosus (as CALLOUS) or F calleux]

callow /'kæləʊ/ adj. inexperienced, immature. □ **callowly** adv. **callowness** n. [OE calu]

calluna /kə'lu:nə/ n. the common heather, Calluna vulgaris, native to Europe and North Africa. [mod.L f. Gk kallunō beautify f. kallos beauty]

callus /'kæləs/ n. **1** a hard thick area of skin or tissue. **2** a hard tissue formed round bone ends after a fracture. **3** Bot. a new protective tissue formed over a wound. [L]

calm /kɑ:m/ adj., n., & v. ● adj. **1** tranquil, quiet, windless (a calm sea; a calm night). **2** (of a person or disposition) settled; not agitated (remained calm throughout the ordeal). **3** self-assured, confident (his calm assumption that we would wait). ● n. **1** a state of being calm; stillness, serenity. **2** a period without wind or storm. **3** (in pl.) an area, esp. of the sea, with predominantly calm weather. **4** Meteorol. a condition of no wind, force 0 on the Beaufort scale. ● v.tr. & intr. (often foll. by down) make or become calm. □ **calmly** adv. **calmness** n. [ME ult. f. LL cauma f. Gk kauma heat]

calmative /'kælmətɪv, 'kɑ:m-/ adj. & n. Med. ● adj. tending to calm or sedate. ● n. a calmative drug etc.

calomel /'kælə,mel/ n. mercury chloride (chem. formula: Hg_2Cl_2), a white powder formerly used as a purgative. [mod.L perh. f. Gk kalos beautiful + melas black]

Calor gas /'kælə(r)/ n. propr. liquefied butane gas stored under pressure in containers for domestic use and used as a substitute for mains gas. [L calor heat]

caloric /kə'lɒrɪk, 'kælərɪk/ adj. & n. ● adj. of heat or calories. ● n. hist. a supposed material form or cause of heat. [F calorique f. L calor heat]

calorie /'kælərɪ/ n. (also **calory**) (pl. **-ies**) a unit of heat energy: **1** (in full **small calorie**) the energy needed to raise the temperature of 1 gram of water through 1 °C, now usu. defined as 4.1868 joules (abbr.: **cal**). **2** (in full **large calorie**) the energy needed to raise the temperature of 1 kilogram of water through 1 °C, often used to measure the energy value of foods (abbr.: **Cal**). (Also called kilocalorie.) [F, arbitrarily f. L calor heat + -ie]

calorific /ˌkælə'rɪfɪk/ adj. producing heat. □ **calorific value** the amount of heat produced by a specified quantity of fuel, food, etc. □ **calorifically** adv. [L calorificus f. calor heat]

calorimeter /ˌkælə'rɪmɪtə(r)/ n. an instrument for measuring quantity of heat, esp. to find calorific values. □ **calorimetry** n. **calorimetric** /ˌkælərɪ'metrɪk/ adj. [L calor heat + -METER]

calque /kælk/ n. Philol. = loan-translation. [F, = copy, tracing f. calquer trace ult. f. L calcare tread]

caltrop /'kæltrəp/ n. (also **caltrap**) **1** hist. a four-spiked iron ball thrown on the ground to impede cavalry horses. **2** Heraldry a representation of this. **3** a creeping plant of the genus Tribulus, with woody carpels usu. having hard spines. [(sense 3) OE calcatrippe f. med.L calcatrippa: (senses 1–2) ME f. OF chauchetrape f. chauchier tread, trappe trap: ult. the same word]

calumet /'kælju,met/ n. a North American Indian peace-pipe. [F, ult. f. L calamus reed]

calumniate /kə'lʌmnɪ,eɪt/ v.tr. slander. □ **calumniator** n. **calumniatory** /-nɪətərɪ/ adj. **calumniation** /-,lʌmnɪ'eɪʃ(ə)n/ n. [L calumniari]

calumny /'kæləmnɪ/ n. & v. ● n. (pl. **-ies**) **1** slander; malicious representation. **2** an instance of this. ● v.tr. (**-ies**, **-ied**) slander. □ **calumnious** /kə'lʌmnɪəs/ adj. [L calumnia]

calvados /'kælvə,dɒs/ n. an apple brandy. [Calvados in France]

Calvary /'kælvərɪ/ n. **1** the place (just outside ancient Jerusalem) where Christ was crucified. **2** a representation of the Crucifixion. [ME f. LL calvaria skull, transl. Gk GOLGOTHA (Matt. 27:33)]

calve /kɑ:v/ v. **1 a** intr. give birth to a calf. **b** tr. (esp. in passive) give birth to (a calf). **2** tr. (also absol.) (of an iceberg) break off or shed (a mass of ice). [OE calfian]

calves pl. of CALF[1], CALF[2].

Calvin[1] /'kælvɪn/, John (1509–64), French Protestant theologian and reformer. He began his theological career in France, but was forced to flee to Basle in Switzerland after embracing Protestantism in the early 1530s. At Basle he published (1536) the first edition of his Institutes of the Christian Religion, the first systematic account of reformed Christian doctrine and a work which he continued to revise and extend throughout his lifetime. He attempted a re-ordering of society on reformed Christian principles, with strong and sometimes ruthless control over the private lives of citizens. From 1541 he lived in Geneva, where he established the first Presbyterian government. Through his Institutes and an extensive body of commentary on the Scriptures, he exerted an important influence on the development of Protestant thought; his theology was further developed by his followers, notably Theodore Beza (1519–1605). (See also CALVINISM.)

Calvin[2] /'kælvɪn/, Melvin (b.1911), American biochemist. He investigated the metabolic pathways involved in photosynthesis, discovering the cycle of reactions which is named after him, and attempting to duplicate them synthetically. He was awarded the Nobel Prize for chemistry in 1961.

Calvin cycle n. Biochem. a cycle of reactions, forming an important part of photosynthesis, whereby plants synthesize glucose from atmospheric carbon dioxide. [CALVIN[2]]

Calvinism /'kælvɪ,nɪz(ə)m/ n. the Protestant theological system of John Calvin and his successors. In Calvinism, Martin Luther's doctrine of justification by faith alone becomes an overriding emphasis on the grace of God, which culminates in the central place given to the doctrine of predestination: certain souls (known as the elect) are predestined by God to salvation, and the rest to damnation. Calvinism was adopted particularly in Scotland, Switzerland, and the Netherlands, by the Puritans, and by American Congregationalists and Presbyterians; in the 20th century Calvinist ideas have undergone something of a revival in the work of Karl Barth. □ **Calvinist** n. & adj. **Calvinistic** /ˌkælvɪ'nɪstɪk/ adj. **Calvinistical** adj. [F calvinisme or mod.L calvinismus]

Calvino /kæl'vi:nəʊ/, Italo (1923–87), Italian novelist and short-story writer, born in Cuba. His first novel The Path to the Nest of Spiders (1947) is considered a significant example of neo-realism, whereas his later works increasingly use fantasy, allegory, and innovative narrative structures and have been associated with magic realism; his later novels include Invisible Cities (1972) and If on a Winter's Night a Traveller (1979).

calx /kælks/ n. (pl. **calces** /'kælsi:z/) **1** a powdery metallic oxide formed when an ore or mineral has been heated. **2** quicklime (see LIME[1] 1). [L calx calcis lime prob. f. Gk khalix pebble, limestone]

Calypso /kə'lɪpsəʊ/ Gk Mythol. a nymph who kept Odysseus on her island, Ogygia, for seven years. [Gk, = she who conceals]

calypso /kə'lɪpsəʊ/ n. (pl. **-os**) **1** a kind of West Indian music in syncopated orig. African rhythm, usu. involving improvisation on a topical theme. **2** a song in this style. [20th c.: orig. unkn.]

calyx /'keɪlɪks, 'kæl-/ n. (also **calix**) (pl. **calyces** /-lɪ,si:z/ or **calyxes**) **1** Bot. the sepals collectively, forming the protective layer of a flower in bud. **2** Biol. any cuplike cavity or structure. [L f. Gk kalux case of bud, husk: cf. kaluptō hide]

calzone /kæl'tsəʊnɪ/ *n*. (*pl*. **calzoni** *pronunc*. same or **calzones**) a type of pizza consisting of a folded half-moon of dough stuffed with mozzarella cheese, prosciutto, etc., and baked or fried. [It. dial., prob. rel. to *calzone* trouser-leg]

cam /kæm/ *n*. a projection on a rotating part in machinery, shaped to impart reciprocal or variable motion to the part in contact with it. [Du. *kam* comb: cf. Du. *kamrad* cog-wheel]

camaraderie /ˌkæməˈrɑːdərɪ/ *n*. mutual trust and sociability among friends. [F]

Camargue, the /kæˈmɑːg/ a region of the Rhône delta in SE France, characterized by numerous shallow salt lagoons. The region is known for its white horses and as a nature reserve.

camarilla /ˌkæməˈrɪlə/ *n*. a cabal or clique. [Sp., dimin. of *camara* chamber]

Camb. *abbr*. Cambridge.

Cambay, Gulf of /kæmˈbeɪ/ (also **Gulf of Khambat** /kɑːmˈbɑːt/) an inlet of the Arabian Sea on the Gujarat coast of western India, north of Bombay.

camber /ˈkæmbə(r)/ *n*. & *v*. ● *n*. **1** the slightly convex or arched shape of the surface of a road, ship's deck, aircraft wing, etc. **2** the slight sideways inclination of the front wheel of a motor vehicle. ● *v*. **1** *intr*. (of a surface) have a camber. **2** *tr*. give a camber to; build with a camber. [F *cambre* arched f. L *camurus* curved inwards]

Camberwell Beauty /ˈkæmbəˌwel/ *n*. a butterfly, *Nymphalis antiopa*, with deep purple yellow-bordered wings. Also called (*N. Amer.*) *mourning cloak*. [Camberwell in London]

cambium /ˈkæmbɪəm/ *n*. (*pl*. **cambia** /-bɪə/ or **cambiums**) *Bot*. a cellular plant tissue responsible for the increase in girth of stems and roots. □ **cambial** *adj*. [med.L, = change, exchange]

Cambodia /kæmˈbəʊdɪə/ a country in SE Asia between Thailand and southern Vietnam; pop. (est. 1991) 8,660,000; official language, Khmer; capital, Phnom Penh. Formerly part of the Khmer empire, the country was made a French protectorate in 1863 and remained under French influence until it became fully independent in 1953. It was bombed and then invaded by US forces during the Vietnam War, and then embroiled in a civil war in 1970–5, which was won by the Khmer Rouge; more than 2 million Cambodians died before the regime was toppled by a Vietnamese invasion in 1979. The country has continued to be plagued by intermittent fighting, although the Vietnamese withdrew in 1989; the monarchy was restored in 1993. The country was known officially as the Khmer Republic from 1970 to 1975 and as Kampuchea from 1976 to 1989.

Cambodian /kæmˈbəʊdɪən/ *n*. & *adj*. ● *n*. **1 a** a native or national of Cambodia. **b** a person of Cambodian descent. **2** the Khmer language. ● *adj*. of or relating to Cambodia or its people or the Khmer language.

Cambrian /ˈkæmbrɪən/ *adj*. & *n*. ● *adj*. **1** Welsh. **2** *Geol*. of or relating to the first period of the Palaeozoic era. (*See note below.*) ● *n*. *Geol*. this period or the corresponding geological system. [L *Cambria* var. of *Cumbria* f. Welsh *Cymry* Welshman or *CYMRU* Wales]

■ The Cambrian period lasted from about 590 to 505 million years BP, between the end of the Precambrian era and the beginning of the Ordovician period. It was a time of widespread seas, and is the first period in which fossils (notably trilobites) can be used in geological dating. Rocks of this period were first recognized in Wales (see SEDGWICK).

cambric /ˈkæmbrɪk, ˈkeɪm-/ *n*. a fine white linen or cotton fabric. [*Kamerijk*, Flem. form of *Cambrai* in northern France, where it was orig. made]

Cambridge /ˈkeɪmbrɪdʒ/ **1** a city in eastern England, the county town of Cambridgeshire; pop. (1991) 101,000. Cambridge University is located there. **2** a city in eastern Massachusetts, forming part of the conurbation of Boston; pop. (1990) 95,800. Harvard University and the Massachusetts Institute of Technology are located there.

Cambridge blue *n*. & *adj*. ● *n*. **1** a pale blue. **2** a blue (BLUE[1] *n*. 3a) of Cambridge University. ● *adj*. of this colour.

Cambridgeshire[1] /ˈkeɪmbrɪdʒˌʃɪə(r)/ a county of eastern England; county town, Cambridge.

Cambridgeshire[2] /ˈkeɪmbrɪdʒˌʃɪə(r)/ *n*. a handicap horse-race run annually at Newmarket in Suffolk, England, in early October, inaugurated in 1839.

Cambridge University a university at Cambridge in England, first established when a group of students migrated from Oxford to Cambridge in 1209 and formally founded in 1230. The university comprises a federation of thirty-one colleges, the oldest of which,

Peterhouse, was founded in 1284, another nine following in the 14th and 15th centuries. Colleges for women were founded in the mid-19th century.

Cambs. *abbr*. Cambridgeshire.

Cambyses /kæmˈbaɪsiːz/ (d.522 BC), son of Cyrus, king of Persia 529–522 BC. He is chiefly remembered for his conquest of Egypt in 525 BC.

camcorder /ˈkæmˌkɔːdə(r)/ *n*. a combined video camera and video recorder. [*camera* + *recorder*]

came past of COME.

camel /ˈkæm(ə)l/ *n*. **1** a large cud-chewing mammal of the genus *Camelus*, with a long neck, slender cushion-footed legs, and either one or two humps on its back (*Arabian camel, Bactrian camel*). **2** a fawn colour. **3** an apparatus for providing additional buoyancy to ships etc. □ **camel** (or **camel's**) **-hair 1** the hair of a camel. **2 a** a fine soft hair used in artists' brushes. **b** a fabric made of this. [OE f. L *camelus* f. Gk *kamēlos*, of Semitic orig.]

cameleer /ˌkæməˈlɪə(r)/ *n*. a camel-driver.

camellia /kəˈmiːlɪə, -ˈmelɪə/ *n*. an evergreen East Asian shrub of the genus *Camellia*, of the tea family, with shiny leaves and showy flowers. [J. *Camellus* or *Kamel*, 17th-c. Jesuit botanist]

camelopard /ˈkæmələˌpɑːd, kəˈmel-/ *n*. *archaic* a giraffe. [L *camelopardus* f. Gk *kamēlopardalis* (as CAMEL, PARD)]

Camelot /ˈkæmɪˌlɒt/ (in Arthurian legend) the place where King Arthur held his court, variously identified as Caerleon in Wales, Camelford in Cornwall, Cadbury Castle in Somerset, and (by Thomas Malory) Winchester in Hampshire.

camelry /ˈkæməlrɪ/ *n*. (*pl*. **-ies**) troops mounted on camels.

Camembert /ˈkæməmˌbeə(r)/ *n*. a kind of soft creamy cheese. [*Camembert* in northern France, where it was orig. made]

cameo /ˈkæmɪəʊ/ *n*. (*pl*. **-os**) **1 a** a small piece of onyx or other hard stone carved in relief with a background of a different colour. **b** a similar relief design using other materials. **2 a** a short descriptive literary sketch or acted scene. **b** a small character part in a play or film, usu. brief and played by a distinguished actor. [ME f. OF *camahieu* and med.L *cammaeus*]

camera /ˈkæmrə, ˈkæmərə/ *n*. **1** an apparatus for recording visual images on light-sensitive film or plates, consisting essentially of a lightproof box into which light, focused by a lens, can be briefly admitted by a shutter mechanism. (*See note below.*) (See also PHOTOGRAPHY.) **2** an apparatus for forming an optical image and converting it into a video signal for transmission or storage. □ **camera-ready** *Printing* (of copy) in a form suitable for immediate photographic reproduction. **in camera 1** *Law* in a judge's private room. **2** privately; not in public. **on camera** (esp. of an actor) being filmed or televised at a particular moment. [orig. = chamber f. L *camera* f. Gk *kamara* vault etc.]

■ Cameras were developed during the 19th century and originally used metal and then glass plates with a sensitized coating. Later, flexible roll-film of coated celluloid was used, enabling cameras to be made much smaller and more portable, and the tendency towards miniaturization has continued. Reflex cameras originally used two lenses, one to give an image for viewing on a ground-glass screen, the other to focus the image on the film; single-lens reflex cameras achieve a similar result by a complex arrangement of reflecting mirrors. Developments in photoelectric devices have enabled highly automated systems of focusing and exposure to be introduced. In cine cameras, the film is moved rapidly past a rotating shutter, though such cameras have largely been superseded for domestic use by the video camcorder. (See also TELEVISION.)

camera lucida /ˈluːsɪdə/ *n*. (*pl*. **camera lucidas**) an instrument, esp. attached to a microscope, by which an image of an object is reflected by a prism on to a flat surface on which it can be traced. [L, = bright chamber]

cameraman /ˈkæmrəmən, ˈkæmərə-/ *n*. (*pl*. **-men**) a person who operates a camera professionally, esp. in film-making or television.

camera obscura /əbˈskjʊərə/ *n*. (*pl*. **camera obscuras**) a darkened box or room with a lens and an aperture for projecting an image of a distant object on to a screen within. Aristotle was aware of the principle on which this depends, and early astronomers such as Johannes Kepler used a form of it to watch and record phenomena such as eclipses. Its use by artists became a craze in the 18th century, and camera obscuras were used by Canaletto and others in topographical painting. [L, = dark chamber]

camerawork /ˈkæmrəˌwɜːk, ˈkæmərə-/ n. the technique of using cameras in films or television.

Cameron /ˈkæmərən/, Julia Margaret (1815–79), English photographer. Although she did not take up photography until the age of 48, she quickly gained acclaim for her portraits of prominent figures such as Tennyson, Darwin, and Carlyle. She is credited with being the first to use soft-focus techniques; her work often reflects the influence of contemporary painting, especially that of the Pre-Raphaelites.

Cameron Highlands a hill resort region in Pahang, Malaysia, named after the surveyor William Cameron, who mapped the area in 1885.

Cameroon /ˌkæməˈruːn/ (French **Cameroun** /kamrun/) a country on the west coast of Africa between Nigeria and Gabon; pop. (est. 1991) 12,081,000; languages, French (official), English (official), many local languages, pidgin; capital, Yaoundé. The territory, largely inhabited by Bantu-speaking peoples, became a German protectorate in 1884; from 1916 it was administered by France and Britain, latterly under League of Nations (later UN) trusteeship. In 1960 French Cameroons became an independent republic, to be joined in 1961 by part of the British Cameroons, the remainder became part of Nigeria. Cameroon joined the Commonwealth in 1995. □ **Cameroonian** adj. & n.

camiknickers /ˈkæmɪˌnɪkəz/ n.pl. Brit. a one-piece undergarment worn by women. [CAMISOLE + KNICKERS]

camisole /ˈkæmɪˌsəʊl/ n. a woman's under-bodice, usu. embroidered. [F f. It. camiciola or Sp. camisola: see CHEMISE]

Camões /kæˈmɔɪnʃ/, Luis (Vaz) de (also **Camoëns** /ˈkæməʊˌens/) (c.1524–80), Portuguese poet. Little is known of his early life; in 1553 he travelled to India and the Far East, returning to Portugal in 1570. His most famous work The Lusiads (1572) describes Vasco da Gama's voyage and discovery of the sea route to India and celebrates the golden age of Portuguese discovery; combining classical elements with a strong sense of Portuguese history and identity, it became established from the 17th century as Portugal's national epic.

camomile /ˈkæməˌmaɪl/ n. (also **chamomile**) an aromatic composite plant with daisy-like flowers, esp. Chamaemulum nobile. □ **camomile tea** an infusion of the dried flowers of C. nobile, used as a tonic or a drink. [ME f. OF camomille f. LL camomilla or chamomilla f. Gk khamaimēlon earth-apple (from the apple-smell of its flowers)]

Camorra /kəˈmɒrə/ a secret criminal society that evolved and grew powerful in Naples and Neapolitan emigrant communities in the 19th century. Although suppressed in the early 20th century it has remained active, and some members have moved to the US and formed links with the Mafia. [It. camorra smock-frock]

camouflage /ˈkæməˌflɑːʒ/ n. & v. ● n. **1 a** the disguising of military vehicles, aircraft, ships, artillery, and installations by painting them or covering them to make them blend with their surroundings. **b** such a disguise. **2** the natural colouring of an animal which enables it to blend in with its surroundings. **3** a misleading or evasive precaution or expedient. ● v.tr. hide or disguise by means of camouflage. [F f. camoufler disguise f. It. camuffare disguise, deceive]

camp[1] /kæmp/ n. & v. ● n. **1 a** a place where troops are lodged or trained. **b** the military life (court and camp). **2** temporary overnight lodging in tents etc. in the open. **3 a** a temporary accommodation of various kinds, usu. consisting of huts or tents, for detainees, homeless persons, and other emergency use. **b** a complex of buildings for holiday accommodation, usu. with extensive recreational facilities. **4** an ancient fortified site or its remains. **5** the adherents of a particular party or doctrine regarded collectively (the Labour camp was jubilant). **6** S. Afr. a portion of veld fenced off for pasture on farms. **7** Austral. & NZ an assembly place of sheep or cattle. ● v.intr. **1** set up or spend time in a camp (in senses 1 and 2 of n.). **2** (often foll. by out) lodge in temporary quarters or in the open. **3** Austral. & NZ (of sheep or cattle) flock together esp. for rest. **4** (usu. foll. by on, on to) reserve (a telephone call) for connection to an engaged telephone as soon as it becomes free. □ **camp-bed** a folding portable bed of a kind used in camping. **camp-fire** an open-air fire in a camp etc. **camp-follower 1** a civilian worker in a military camp. **2** a disciple or adherent. **camp on** (or **camp on busy**) a facility allowing a telephone call to be camped on to an engaged telephone. **camp-site** a place for camping. □ **camping** n. [F f. It. campo f. L campus level ground]

camp[2] /kæmp/ adj., n., & v. colloq. ● adj. **1** affected, effeminate. **2** homosexual. **3** done in an exaggerated way for effect. ● n. a camp manner or style. ● v.intr. & tr. behave or do in a camp way. □ **camp it**

up overact; behave affectedly. □ **campy** adj. (**campier**, **campiest**). **campily** adv. **campiness** n. [20th c.: orig. uncert.]

campaign /kæmˈpeɪn/ n. & v. ● n. **1** an organized course of action for a particular purpose, esp. to arouse public interest (e.g. before a political election). **2 a** a series of military operations in a definite area or to achieve a particular objective. **b** military service in the field (on campaign). ● v.intr. conduct or take part in a campaign. □ **campaigner** n. [F campagne open country f. It. campagna f. LL campania]

Campaign for Nuclear Disarmament (abbr. **CND**) a British organization which campaigns for the abolition of nuclear weapons worldwide and calls for unilateral disarmament by the West. It was founded in 1958 under the presidency of Bertrand Russell. After a period of dormancy in the late 1960s and 1970s, the campaign was revived in 1979 to oppose the siting of US cruise missiles in Britain, drawing huge crowds to demonstrations in London. With the improvement in East–West relations and the breakup of the Soviet Union, the organization has had a lower public profile.

Campania /kæmˈpeɪnɪə/ a region of west central Italy; capital, Naples. □ **Campanian** n. & adj.

campanile /ˌkæmpəˈniːlɪ/ n. a bell-tower (usu. free-standing), esp. in Italy. [It. f. campana bell]

campanology /ˌkæmpəˈnɒlədʒɪ/ n. **1** the study of bells. **2** the art or practice of bell-ringing. □ **campanologist** n. **campanological** /-nəˈlɒdʒɪk(ə)l/ adj. [mod.L campanologia f. LL campana bell]

campanula /kæmˈpænjʊlə/ n. a plant of the genus Campanula, with bell-shaped usu. blue, purple, or white flowers. Also called bellflower. [mod.L dimin. of L campana bell]

campanulate /kæmˈpænjʊlət/ adj. esp. Bot. bell-shaped.

Campbell[1] /ˈkæmb(ə)l/, Donald (Malcolm) (1921–67), English motor-racing driver and holder of world speed records. Following in the footsteps of his father Sir Malcolm Campbell, he broke a number of world speed records in boats and cars named Bluebird. In 1964 he achieved a speed of 276.33 m.p.h. (445 k.p.h.) on water and 403 m.p.h. (649 k.p.h.) on land, both in Australia. He was killed in an attempt to achieve a water speed of 300 m.p.h. (483 k.p.h.) on Coniston Water in England.

Campbell[2] /ˈkæmb(ə)l/, Sir Malcolm (1885–1948), English motor-racing driver and holder of world speed records. In 1935 he became the first man to exceed a land speed of 300 m.p.h. (483 k.p.h.), driving his car Bluebird on Bonneville Flats in Utah in the US, a record which was not broken until 1950. He also achieved a water-speed record of 141.74 m.p.h. (228 k.p.h.) in his boat of the same name in 1939.

Campbell[3] /ˈkæmb(ə)l/, Mrs Patrick (née Beatrice Stella Tanner) (1865–1940), English actress. Renowned for her wit and beauty, she created the part of Paula in Pinero's The Second Mrs Tanqueray (1893) and also gave notable performances in roles ranging from Shakespeare to Ibsen. George Bernard Shaw wrote the part of Eliza Doolittle in Pygmalion (1914) for her and they exchanged letters over a long period.

Campbell[4] /ˈkæmb(ə)l/, (Ignatius) Roy(ston Dunnachie) (1901–57), South African poet. His works include The Flaming Terrapin (1924), an allegorical narrative of the Flood, and The Wayzgoose (1928), a satire on South African life. His first autobiography Broken Record (1934) and the long poem Flowering Rifle (1939) show strong right-wing sympathies; he fought for Franco's side in the Spanish Civil War.

Campbell[5] /ˈkæmb(ə)l/, Thomas (1777–1844), Scottish poet. He published The Pleasures of Hope (1799) and Gertrude of Wyoming (1809) among other volumes of verse, and is now chiefly remembered for his patriotic lyrics such as 'The Battle of Hohenlinden' and 'Ye Mariners of England'.

Campbell-Bannerman /ˌkæmb(ə)lˈbænəmən/, Sir Henry (1836–1908), British Liberal statesman, Prime Minister 1905–8. He was first elected to Parliament as MP for the Stirling burghs in 1868 and became leader of his party in 1899. His premiership, which ended with his resignation only a few days before his death, saw the grant of self-government to the defeated Boer republics of Transvaal (1906) and the Orange River Colony (1907), the passing of the important Trade Disputes Act (1906), which exempted trade unions from certain liabilities in connection with strikes, and the entente with Russia (1907).

Camp David the retreat of the president of the US, in the Appalachian Mountains in Maryland. President Carter hosted talks there between the leaders of Israel and Egypt which resulted in the Camp David agreements (1978) and the Egypt–Israel peace treaty of 1979.

Campeachy wood /kæm'pi:tʃɪ/ n. = LOGWOOD. [CAMPECHE, from where it was first exported]

Campeche /kæm'peɪtʃɪ/ **1** a state of SE Mexico, on the Yucatán Peninsula. **2** its capital, a seaport on the Gulf of Mexico; pop. (1990) 172,200.

camper /'kæmpə(r)/ n. **1** a person who camps out or lives temporarily in a tent, hut, etc., esp. on holiday. **2** a large motor vehicle with accommodation for camping out.

camphor /'kæmfə(r)/ n. a white translucent crystalline volatile substance with aromatic smell and bitter taste, used to make celluloid and in medicine. □ **camphoric** /kæm'fɒrɪk/ adj. [ME f. OF camphore or med.L camphora f. Arab. kāfūr f. Skr. karpūram]

camphorate /'kæmfəˌreɪt/ v.tr. impregnate or treat with camphor.

Campinas /kæm'pi:nəs/ a city in SE Brazil, north-west of São Paulo; pop. (1991) 835,000. It is a major centre of the coffee trade.

campion /'kæmpɪən/ n. **1** a plant of the genus Silene, of the pink family, with usu. pink or white notched flowers. **2** a similar cultivated plant of the genus Lychnis. [perh. f. obs. campion f. OF, = CHAMPION: transl. of Gk lukhnis stephanōmatikē a plant used for (champions') garlands]

Campion, St Edmund /'kæmpɪən/ (1540–81), English Jesuit priest and martyr. He was ordained a deacon in the Church of England in 1569 but because of his Roman Catholic sympathies he went abroad, becoming a Catholic and a Jesuit priest (1573). He was a member of the first Jesuit mission to England (1580). In 1581 he was arrested, charged with conspiracy against the Crown, tortured, and executed. Feast day, 1 Dec.

Campobasso /ˌkæmpəʊˈbæsəʊ/ a city in central Italy, capital of Molise region; pop. (1990) 51,300.

Campo Grande /ˌkæmpuː ˈgrændɪ/ a city in SW Brazil, capital of the state of Mato Grosso do Sul; pop. (1991) 489,000.

campus /'kæmpəs/ n. (pl. **campuses**) **1** the grounds and buildings of a university or college. **2** university or college life. [L, = field]

campylobacter /'kæmpɪləʊˌbæktə(r)/ n. a curved or spiral bacterium of the genus Campylobacter, esp. as a cause of food poisoning. [mod.L f. Gk kampulos bent + BACTERIUM]

CAMRA /'kæmrə/ abbr. Campaign for Real Ale.

camshaft /'kæmʃɑ:ft/ n. a shaft with one or more cams attached to it.

Camulodunum /ˌkæmjʊləʊˈdu:nəm, -'dju:nəm/ the Roman name for COLCHESTER.

Camus /'kæmju:/, Albert (1913–60), French novelist, dramatist, and essayist. He joined the French resistance during the Second World War and became co-editor (with Jean-Paul Sartre) of the left-wing daily Combat (1944–7). His essay The Myth of Sisyphus (1942) and his first novel The Outsider (1942) gained him international respect; each conveys his conception of the absurdity of human existence and aligns him closely with Sartre's existentialism. Other notable works include the novel The Plague (1947) and the essay The Rebel (1951). He was awarded the Nobel Prize for literature in 1957.

camwood /'kæmwʊd/ n. a red African wood, formerly that of the tree Baphia nitida, now usu. of the African padouk, Pterocarpus soyauxii, native to West Africa. [perh. f. Temne]

Can. abbr. Canada; Canadian.

can[1] /kæn, kən/ v.aux. (3rd sing. present **can**; past **could** /kʊd/) (foll. by infin. without to, or absol.; present and past only in use) **1 a** be able to; know how to (I can run fast; can he?; can you speak German?). **b** potentially capable of (you can do it if you try). **c** (in past) colloq. feel inclined to (I could murder him). **2** be permitted to (can we go to the party?). □ **can-do** attrib.adj. designating a determination or willingness to achieve something (the American can-do philosophy). [OE cunnan know]

can[2] /kæn/ n. & v. ● n. **1** a vessel for holding liquids. **2** a usu. cylindrical metal container in which food or drink is hermetically sealed to enable storage over long periods. (See note below.) **3** ((in pl.)) sl. headphones. **4** (prec. by the) sl. **a** prison (sent to the can). **b** US lavatory. **5** US sl. the buttocks. ● v.tr. (**canned, canning**) **1** put or preserve in a can. **2** record on film or tape for future use. □ **can of worms** colloq. a complicated matter, likely to prove problematic or scandalous. **can-opener** a device for opening cans (in sense 2 of n.). **in the can** colloq. completed, ready (orig. of filmed or recorded material). □ **canner** n. [OE canne]

▪ Storage of food in cans or tins was introduced in the 19th century. In 1810 a French confectioner, Nicholas Appert, began to supply preserved food to the French army. Food was heated in glass jars which were then sealed until use. The reason for its effectiveness was not known until the work of Louis Pasteur: heat destroyed the micro-organisms present in food, and sealing prevented others from entering. Tin-coated steel containers, still the most common type, were developed by Bryan Donkin and introduced in the 1830s. Lids were soldered until the early 20th century, when a method was devised of sealing the cover to the can. Can-openers were not invented until the mid-19th century; the first cans were opened with a hammer and chisel.

Cana /'keɪnə/ an ancient small town in Galilee, where Christ is said to have performed his first miracle by changing water into wine during a marriage feast (John 2:1–11).

Canaan /'keɪnən/ n. **1** the land, later known as Palestine, which the Israelites gradually conquered and occupied during the latter part of the 2nd millennium BC. In the Bible it is the land promised by God to Abraham and his descendants (Gen. 12:7). **2** a promised land. **3** heaven. [eccl.L f. eccl.Gk Khanaan f. Heb. kⁿna'an]

Canaanite /'keɪnəˌnaɪt/ adj. & n. ● adj. of or relating to Canaan, its people, or its culture. ● n. a native or inhabitant of Canaan.

Canada /'kænədə/ the second largest country in the world, covering the entire northern half of North America with the exception of Alaska; pop. (1991) 26,832,400; official languages, English and French; capital, Ottawa. Native peoples had inhabited all of Canada for several thousand years; they now form about two per cent of the population. The first European visitors were Vikings, who briefly settled in Newfoundland in about 1000. John Cabot reached the east coast in 1497, Jacques Cartier explored the St Lawrence in 1535, and Samuel de Champlain founded Quebec in 1608 and penetrated the interior (1609–16). Eastern Canada was colonized by the French in the 17th century, with the British emerging as the ruling colonial power in 1763, after the Seven Years War. Canada became a federation of provinces with dominion status in 1867, and the last step in attaining legal independence from the UK was taken with the signing of the Constitution Act of 1982. Canada is a member of the Commonwealth, and played a major role on the side of the Allies in each of the two world wars. Throughout the late 19th and early 20th centuries the vast Canadian West was gradually opened up, but the country remains sparsely populated for its size. French-speakers are largely concentrated in Quebec, the focal point for the French-Canadian nationalist movement, with minorities in other provinces.

Canada balsam n. a yellow resin obtained from the balsam fir and used for mounting preparations on microscope slides (its refractive index being similar to that of glass).

Canada goose n. a wild brownish-grey North American goose, Branta canadensis, with white cheeks and breast and a loud honking call, widely introduced in Britain etc.

Canadian /kə'neɪdɪən/ n. & adj. ● n. **1** a native or national of Canada. **2** a person of Canadian descent. ● adj. of or relating to Canada.

Canadian Pacific Railway the first transcontinental railway in Canada, completed in 1885.

Canadian pondweed n. an invasive American aquatic plant, Elodea canadensis, of the frogbit family, naturalized in Europe and grown in aquariums.

Canadian Shield (also called Laurentian Plateau) a large plateau which occupies over two-fifths of the land area of Canada and is drained by rivers flowing into Hudson Bay.

canaille /kæ'naɪ/ n. the rabble; the populace. [F f. It. canaglia pack of dogs f. cane dog]

canal /kə'næl/ n. **1** an artificial waterway for inland navigation or irrigation. (See note below.) **2** a tubular duct in a plant or animal, for carrying food, liquid, or air. **3** Astron. hist. any of a number of spurious linear markings formerly reported as seen by telescope on the planet Mars, named canali 'channels' by G. V. Schiaparelli. □ **canal boat** a long narrow boat for use on canals. [ME f. OF (earlier chanel) f. L canalis or It. canale]

▪ Examples of irrigation canals date from the 5th millennium BC in Mesopotamia. Early canals are also to be found in Egypt and in China, where the Grand Canal, 1,700 km (1,060 miles) in length, was begun in the 4th–5th centuries BC and completed in 1327. Many notable canals are designed to shorten sea passages: in France the Canal du Midi joined the Atlantic and Mediterranean in 1681, the Suez Canal joined the latter to the Red Sea in 1869, and in 1914 the Panama Canal joined the Atlantic and Pacific Oceans. In England canals were the arteries of the Industrial Revolution, greatly reducing the cost of transport to below that of land transport over inadequate roads.

James Brindley and Thomas Telford were the great canal builders between 1760 and 1840, by which time a network of some 6,800 km (4,250 miles) of canal was in existence. Thereafter canal use declined in Britain, owing to the expansion of the railways and the restricted size of the early canals, but in Europe and North America canal use has grown in route length and in size of vessels, while in the late 20th century canals in Britain have become increasingly popular recreational waterways. The Rhine, Main, and Danube were linked in 1992, creating a canal connecting the North Sea with the Black Sea, while the Great Lakes are accessible to ocean-going vessels through the St Lawrence Seaway.

Canaletto /ˌkænəˈletəʊ/ (born Giovanni Antonio Canale) (1697–1768), Italian painter. Working chiefly in his native city of Venice, he was especially popular with the English aristocracy, who commissioned his paintings of Venetian festivals and scenery as mementoes of their grand tour. His early work is dramatic and freely handled, reflecting his training as a theatrical scene painter, but from c.1730 he changed to the more topographically precise style for which he is mainly remembered, aided by his use of the camera obscura. His works include *Scene in Venice: The Piazzetta Entrance to the Grand Canal* (c.1726–8).

canalize /ˈkænəˌlaɪz/ *v.tr.* (also **-ise**) **1** make a canal through. **2** convert (a river) into a canal. **3** provide with canals. **4** give the desired direction or purpose to. □ **canalization** /ˌkænəlaɪˈzeɪʃ(ə)n/ *n.* [F *canaliser:* see CANAL]

Canal Zone see PANAMA CANAL.

canapé /ˈkænəpɪ, -ˌpeɪ/ *n.* **1** a small piece of bread or pastry with a savoury on top, often served as an hors-d'œuvre. **2** a sofa. [F]

canard /kəˈnɑːd, ˈkænɑːd/ *n.* **1** an unfounded rumour or story. **2** an extra surface attached to an aeroplane forward of the main lifting surface, for extra stability or control. [F, = duck]

Canarese var. of KANARESE.

canary /kəˈneərɪ/ *n.* (*pl.* **-ies**) **1** a small finch of the genus *Serinus;* esp. *S. canaria,* a songbird native to the Canary Islands, with mainly yellow plumage, frequently kept as a pet. **2** *hist.* a sweet wine from the Canary Islands. □ **canary-coloured** canary yellow in colour. **canary creeper** a climbing plant, *Tropaeolum peregrinum,* having flowers with bright yellow deeply toothed petals which give the appearance of a small bird in flight. **canary grass** a Mediterranean plant *Phalaris canariensis,* grown as a crop plant for bird seed. **canary yellow** bright yellow. [CANARY ISLANDS]

Canary Islands /kəˈneərɪ/ (also **Canaries**) a group of islands in the Atlantic Ocean, off the NW coast of Africa, forming an autonomous region of Spain; capital, Las Palmas; pop. (est. 1989) 1,557,530. It includes the islands of Tenerife, Gomera, La Palma, Hierro, Gran Canaria, Fuerteventura, and Lanzarote. The name is said to derive from the Latin *canis,* meaning 'dog', one of the islands being noted in Roman times for large dogs.

canasta /kəˈnæstə/ *n.* **1** a card-game using two packs and resembling rummy, the aim being to collect sets (or melds) of cards. **2** a set of seven cards in this game. [Sp., = basket]

canaster /kəˈnæstə(r)/ *n.* tobacco made from coarsely broken dried leaves. [orig. the container: Sp. *canastro* ult. f. Gk *kanastron*]

Canaveral, Cape /kəˈnævərəl/ a cape on the east coast of Florida, known as Cape Kennedy from 1963 until 1973. It is the location of the John F. Kennedy Space Center, from which the Apollo space missions were launched.

Canberra /ˈkænbərə/ the capital of Australia and seat of the federal government, in Australian Capital Territory, an enclave of New South Wales; pop. (est. 1990) 310,000.

cancan /ˈkænkæn/ *n.* a lively stage-dance with high kicking, performed by women in long skirts and petticoats. It was invented in 1830 as a variant of the quadrille, but soon appeared in the French music-halls, where it developed an increasingly uninhibited emphasis on the throwing up of the legs of the dancers and the display of their underwear, which led to its eventual banning. The music most often associated with it is from Offenbach's operettas, notably *Orpheus in the Underworld.* [F, redupl. of *canard* duck]

cancel /ˈkæns(ə)l/ *v. & n.* ● *v.* (**cancelled, cancelling;** *US* **canceled, canceling**) **1** *tr.* **a** withdraw or revoke (a previous arrangement). **b** discontinue (an arrangement in progress). **2** *tr.* obliterate or delete (writing etc.). **3** *tr.* mark or pierce (a ticket, stamp, etc.) to invalidate it. **4** *tr.* annul; make void; abolish. **5** (often foll. by *out*) **a** *tr.* (of one factor or circumstance) neutralize or counterbalance (another). **b** *intr.* (of two factors or circumstances) neutralize each other. **6** *tr. Math.* strike out (an

equal factor) on each side of an equation or from the numerator and denominator of a fraction. ● *n.* **1** a countermand. **2** the cancellation of a postage stamp. **3** *Printing* a new page or section inserted in a book to replace the original text, usu. to correct an error. **4** *Mus. US =* NATURAL *n.* 3a. □ **canceller** *n.* [ME f. F *canceller* f. L *cancellare* f. *cancelli* crossbars, lattice]

cancellate /ˈkænsɪlət/ *adj.* (also **cancellated** /-ˌleɪtɪd/) *Biol.* **1** marked with crossing lines. **2** = CANCELLOUS. [L *cancelli* lattice]

cancellation /ˌkænsɪˈleɪʃ(ə)n/ *n.* **1** the act or an instance of cancelling or being cancelled. **2** something that has been cancelled, esp. a booking or reservation. [L *cancellatio* (as CANCEL)]

cancellous /ˈkænsɪləs/ *adj.* (of a bone) having pores. [L *cancelli* lattice]

Cancer /ˈkænsə(r)/ *n.* **1** *Astron.* a constellation (the Crab), said to represent a crab crushed under the foot of Hercules. It is most noted for the globular star cluster of Praesepe or the Beehive. **2** *Astrol.* **a** the fourth sign of the zodiac, which the sun enters at the northern summer solstice (about 21 June). **b** a person born when the sun is in this sign. □ **Tropic of Cancer** see TROPIC. □ **Cancerian** /kænˈsɪərɪən/ *n. & adj.* [ME f. L, = crab, cancer, after Gk *karkinos*]

cancer /ˈkænsə(r)/ *n.* **1** a malignant tumour or growth of body cells which tends to spread and may recur if removed; disease in which such a growth is formed. (*See note below.*) **2** an evil influence or corruption spreading uncontrollably. □ **cancer stick** *sl.* a cigarette. □ **cancerous** *adj.* [as CANCER]

▪ Cancer is a disorder of the processes of growth, development, and repair, in which there is uncontrolled growth of cells which differ morphologically and biochemically from the normal cells of the tissue of origin. The normal control mechanisms do not operate, and cancer cells tend both to infiltrate neighbouring tissues and to spread to distant parts of the body, forming secondary growths or metastases. Many kinds of cancer are known, affecting different parts of the body; some are known to be caused by external factors such as exposure to certain chemicals, ionizing radiation, and pathogenic organisms. Various forms of therapy such as surgery, chemotherapy, and radiation treatment have been used with differing degrees of success.

cancroid /ˈkæŋkrɔɪd/ *adj. & n.* ● *adj.* **1** crablike. **2** resembling cancer. ● *n.* a disease resembling cancer.

Cancún /kænˈkuːn/ a resort in SE Mexico, on the NE coast of the Yucatán Peninsula; pop. (1980) 27,500.

candela /kænˈdiːlə, -ˈdeɪlə/ *n. Physics* the SI base unit of luminous intensity (symbol: **cd**). One candela is $\frac{1}{60}$ of the luminous intensity per square centimetre, in the perpendicular direction, of the surface of a black body at the temperature of solidification of platinum at standard atmospheric pressure. [L, = candle]

candelabrum /ˌkændɪˈlɑːbrəm/ *n.* (also **candelabra** /-brə/) (*pl.* **candelabra;** *US* **candelabrums, candelabras**) a large branched candlestick or lamp-holder. □ **candelabrum tree** a tropical East African tree, *Euphorbia candelabrum,* with foliage shaped like a candelabrum. [L f. *candela* CANDLE]

candescent /kænˈdes(ə)nt/ *adj.* glowing with or as with white heat. □ **candescence** *n.* [L *candere* be white]

candid /ˈkændɪd/ *adj.* **1** frank; not hiding one's thoughts. **2** (of a photograph) taken informally, usu. without the subject's knowledge. □ **candid camera** a small camera for taking candid photographs. □ **candidly** *adv.* **candidness** *n.* [F *candide* or L *candidus* white]

candida /ˈkændɪdə/ *n.* a yeastlike parasitic fungus of the genus *Candida;* esp. *C. albicans,* that causes thrush. [mod.L fem. of L *candidus:* see CANDID]

candidate /ˈkændɪdət, -ˌdeɪt/ *n.* **1** a person who seeks or is nominated for an office, award, etc. **2** a person or thing likely to gain some distinction or position. **3** a person entered for an examination. □ **candidacy** /-dəsɪ/ *n.* **candidature** /-dətʃə(r)/ *n. Brit.* [F *candidat* or L *candidatus* white-robed (Roman candidates wearing white)]

candle /ˈkænd(ə)l/ *n. & v.* ● *n.* **1** a cylinder or block of wax or tallow with a central wick, for giving light when burning. **2** (also **international candle**) a unit of luminous intensity superseded by the candela. (Cf. CANDLEPOWER.) ● *v.tr.* test (an egg) for fertility by holding it to the light. □ **cannot hold a candle to** cannot be compared with; is much inferior to. **not worth the candle** not justifying the cost or trouble. □ **candler** *n.* [OE *candel* f. L *candela* f. *candere* shine]

candlelight /ˈkænd(ə)lˌlaɪt/ *n.* **1** light provided by candles. **2** dusk. □ **candlelit** *adj.*

Candlemas /ˈkænd(ə)lməs, -ˌmæs/ *n.* a Christian feast with blessing of candles (2 Feb.), commemorating the purification of the Virgin Mary and the presentation of Christ in the Temple. [OE *Candelmæsse* (as CANDLE, MASS[2])]

candlepower /ˈkænd(ə)lˌpaʊə(r)/ *n.* luminous intensity, orig. as expressed in candles.

candlestick /ˈkænd(ə)lˌstɪk/ *n.* a holder for one or more candles.

candlewick /ˈkænd(ə)lˌwɪk/ *n.* **1** a thick soft cotton yarn. **2** material made from this, usu. with a tufted pattern.

Candolle /kænˈdɒl/, Augustin Pyrame de (1778–1841), Swiss botanist. His prolific writings in taxonomy and botany were highly influential, particularly his belief that taxonomy should be based on morphological characters, and his scheme of classification prevailed for many years. Candolle also contributed to agronomy, and was a pioneer in plant geography and the linking of soil type with vegetation.

candour /ˈkændə(r)/ *n.* (*US* **candor**) candid behaviour or action; frankness. [F *candeur* or L *candor* whiteness]

C. & W. *abbr.* country-and-western.

candy /ˈkændɪ/ *n. & v.* ● *n.* (*pl.* **-ies**) **1** (in full **sugar-candy**) sugar crystallized by repeated boiling and slow evaporation. **2** *N. Amer.* sweets; a sweet. ● *v.tr.* (**-ies, -ied**) (usu. as **candied** *adj.*) preserve by coating and impregnating with a sugar syrup (*candied fruit*). [F *sucre candi* candied sugar f. Arab. *ḳand* sugar]

candyfloss /ˈkændɪˌflɒs/ *n. Brit.* a fluffy mass of spun sugar wrapped round a stick.

candystripe /ˈkændɪˌstraɪp/ *n.* a pattern consisting of alternate stripes of white and a colour (usu. pink). □ **candystriped** *adj.*

candytuft /ˈkændɪˌtʌft/ *n.* a cruciferous plant of the genus *Iberis*, native to western Europe, with white, pink, or purple flowers in tufts. [obs. *Candy* (Candia Crete) + TUFT]

cane /keɪn/ *n. & v.* ● *n.* **1 a** the hollow jointed stem of giant reeds or grasses (*bamboo cane*). **b** the solid stem of slender palms (*malacca cane*). **2** = *sugar cane*. **3** a raspberry-cane. **4** material of cane used for wickerwork etc. **5 a** a cane used as a walking-stick or a support for a plant or an instrument of punishment. **b** any slender walking-stick. ● *v.tr.* **1** beat with a cane. **2** weave cane into (a chair etc.). □ **cane-brake** *N. Amer.* a tract of land overgrown with canes. **cane chair** a chair with a seat made of woven cane strips. **cane-sugar** sugar obtained from sugar-cane. **cane toad** a large brown toad, *Bufo marinus*, native to tropical America and introduced elsewhere originally for pest control. **cane-trash** see TRASH *n.* 3. □ **caner** *n.* (in sense 2 of *v*). **caning** *n.* [ME f. OF f. L *canna* f. Gk *kanna*]

Canes Venatici /ˌkeɪniːz vɪˈnætɪˌsaɪ/ *Astron.* a small northern constellation (the Hunting Dogs), said to represent two dogs (Asterion and Chara) held on a leash by Boötes. It contains a fine globular cluster and a spiral galaxy, the Whirlpool Galaxy. [L]

canine /ˈkeɪnaɪn, ˈkæn-/ *adj. & n.* ● *adj.* **1** of a dog or dogs. **2** *Zool.* of or belonging to the family Canidae, which includes dogs, wolves, jackals, foxes, etc. ● *n.* **1** a dog. **2** (in full **canine tooth**) a pointed tooth between the incisors and premolars. [ME f. *canin -ine* or f. L *caninus* f. *canis* dog]

Canis Major /ˈkeɪnɪs/ *Astron.* a small constellation (the Great Dog), said to represent one of the dogs following Orion. It contains the brightest star, Sirius. [L]

Canis Minor /ˈkeɪnɪs/ *Astron.* a small constellation (the Little Dog), said to represent one of the dogs following Orion. It contains the bright star Procyon. [L]

canister /ˈkænɪstə(r)/ *n.* **1** a small container, usu. of metal and cylindrical, for storing tea etc. **2 a** a cylinder of shot, tear-gas, etc., that explodes on impact. **b** such cylinders collectively. [L *canistrum* f. Gk f. *kanna* CANE]

canker /ˈkæŋkə(r)/ *n. & v.* ● *n.* **1 a** a destructive fungal disease of trees and plants. **b** an open wound in the stem of a tree or plant. **2** *Zool.* an ulcerous ear disease of animals, esp. cats and dogs. **3** *Med.* an ulceration esp. of the lips. **4** a corrupting influence. ● *v.tr.* **1** consume with canker. **2** corrupt. **3** (as **cankered** *adj.*) soured, malignant, crabbed. □ **cankerous** *adj.* [OE *cancer* & ONF *cancre*, OF *chancre* f. L *cancer* crab]

cankerworm /ˈkæŋkəˌwɜːm/ *n. US* a caterpillar of a wingless moth that consumes the buds and leaves of shade and fruit trees.

Canmore /ˈkænmɔː(r)/, the nickname of Malcolm III of Scotland (see MALCOLM).

canna /ˈkænə/ *n.* a tropical plant of the genus *Canna*, with bright flowers and ornamental leaves. [L: see CANE]

cannabinol /kəˈnæbɪˌnɒl, ˈkænəbɪ-/ *n. Chem.* a crystalline phenol, derivatives of which (esp. tetrahydrocannabinol) are the active principles of cannabis. [CANNABIS + -OL]

cannabis /ˈkænəbɪs/ *n.* **1** a hemp plant of the genus *Cannabis*, esp. Indian hemp (see HEMP 1). **2** a preparation of parts of this used as an intoxicant or hallucinogen. □ **cannabis resin** a sticky product containing the active principles of cannabis, esp. from the flowering tops of the female cannabis plant. [L f. Gk]

canned /kænd/ *adj.* **1** pre-recorded (*canned laughter; canned music*). **2** supplied in a can (*canned beer*). **3** *sl.* drunk.

cannel /ˈkæn(ə)l/ *n.* (in full **cannel coal**) a bituminous coal burning with a bright flame. [16th c.: orig. N.Engl.]

cannelloni /ˌkænəˈləʊnɪ/ *n.pl.* tubes or rolls of pasta stuffed with a meat or vegetable mixture. [It. f. *cannello* stalk]

cannelure /ˈkænəˌljʊə(r)/ *n.* the groove round a bullet etc. [F f. *canneler* f. *canne* reed, CANE]

cannery /ˈkænərɪ/ *n.* (*pl.* **-ies**) a factory where food is canned.

Cannes /kæn/ a resort on the Mediterranean coast of France; pop. (1990) 69,360. An international film festival is held there annually.

cannibal /ˈkænɪb(ə)l/ *n. & adj.* ● *n.* **1** a person who eats human flesh. **2** an animal that feeds on flesh of its own species. ● *adj.* of or like a cannibal. □ **cannibalistic** /ˌkænɪbəˈlɪstɪk/ *adj.* **cannibalistically** *adv.* [orig. pl. *Canibales* f. Sp.: var. of *Caribes* Caribs (see CARIB)]

cannibalism /ˈkænɪbəˌlɪz(ə)m/ *n.* the eating of the flesh of members of one's own species. Archaeological evidence suggests that cannibalism among humans has occurred since palaeolithic times in many places throughout the world, although now, if it exists at all, it is limited to isolated parts of Melanesia and South America. Cannibalism is associated with ritual, religious, or magical beliefs far more than it is with hunger, and occasions such as when the starving survivors of an air crash have resorted to cannibalism are extremely rare. Anthropologists distinguish two categories of cannibalism: *endocannibalism*, in which the remains of relatives or other members of one's own group are consumed, generally out of respect and reverence; and *exocannibalism*, in which the remains of one's enemies are consumed, as a form of ritualized vengeance or with the aim of absorbing the vitality or other qualities of vanquished foes. The particular organs or portions consumed have frequently had ritual significance.

cannibalize /ˈkænɪbəˌlaɪz/ *v.tr.* (also **-ise**) use (a machine etc.) as a source of spare parts for others. □ **cannibalization** /ˌkænɪbəlaɪˈzeɪʃ(ə)n/ *n.*

cannikin /ˈkænɪkɪn/ *n.* a small can. [Du. *kanneken* (as CAN[2], -KIN)]

Canning /ˈkænɪŋ/, George (1770–1827), British Tory statesman, Prime Minister 1827. Foreign Secretary from 1807, he resigned in 1809 after a disagreement with his rival Castlereagh over a disastrous expedition in the Napoleonic Wars, but returned to office following Castlereagh's suicide in 1822. During this ministry he presided over a reversal of Britain's hitherto conservative foreign policy, being particularly responsible for the support of nationalist movements in various parts of Europe. He succeeded Lord Liverpool as Prime Minister in 1827, but died shortly afterwards.

Cannizzaro /ˌkænɪˈtsɑːrəʊ/, Stanislao (1826–1910), Italian chemist. He is chiefly remembered for his revival of Avogadro's hypothesis, using it to distinguish clearly between atoms and molecules, and introducing the unified system of atomic and molecular weights. Cannizzaro also discovered a reaction (named after him) in which an aldehyde is converted into an acid and an alcohol in the presence of a strong alkali.

cannon /ˈkænən/ *n. & v.* ● *n.* **1** *hist.* (*pl.* usu. same) a large heavy gun installed on a carriage or mounting. The cannon dates (in Europe) from the early 14th century. **2** an automatic aircraft gun firing shells. **3** *Billiards* the hitting of two balls successively by the cue-ball. **4** *Mech.* a hollow cylinder moving independently on a shaft. **5** (in full **cannon-bit**) a smooth round bit for a horse. ● *v.intr.* **1** (usu. foll. by *against, into*) collide heavily or obliquely. **2** *Billiards* make a cannon shot. □ **cannon-ball** *hist.* a large usu. metal ball fired by a cannon. **cannon-bone** the tubular bone between the hock and fetlock of a horse. **cannon-fodder** soldiers regarded merely as material to be expended in war. [F *canon* f. It. *cannone* large tube f. *canna* CANE; in billiards sense f. older CAROM]

cannonade /ˌkænəˈneɪd/ *n. & v.* ● *n.* a period of continuous heavy gunfire. ● *v.tr.* bombard with a cannonade. [F f. It. *cannonata*]

cannot /ˈkænɒt, kæˈnɒt/ *v.aux.* can not.

cannula /ˈkænjʊlə/ *n.* (*pl.* **cannulae** /-liː/ or **cannulas**) *Surgery* a small tube for inserting into the body to allow fluid to enter or escape. [L, dimin. of *canna* cane]

cannulate /ˈkænjʊˌleɪt/ *v.tr. Surgery* introduce a cannula into.

canny /ˈkænɪ/ *adj.* (**cannier, canniest**) **1 a** shrewd, worldly-wise. **b** thrifty. **c** circumspect. **2** sly, drily humorous. **3** *Sc. & N. Engl.* pleasant, agreeable. □ **cannily** *adv.* **canniness** *n.* [CAN¹ (in sense 'know') + -Y¹]

canoe /kəˈnuː/ *n. & v.* ● *n.* a small narrow boat with pointed ends usu. propelled by paddling. (*See note below.*) ● *v.intr.* (**canoes, canoed, canoeing**) travel in a canoe. □ **canoeist** *n.* [Sp. and Haitian *canoa*]

■ The term was originally applied to a small open boat used by West Indian aboriginals, which was hollowed out of a single tree-trunk. It was extended to embrace similar craft, in which paddles were the motive force, all over the world: some of these (particularly among the Pacific islands) were remarkably large vessels in which two banks of paddlers, up to twenty or thirty a side, were used. In the West canoeing was popularized as a sport in the mid-19th century, and today fibre-glass or wood canoes are commonly used for recreation.

canon /ˈkænən/ *n.* **1 a** a general law, rule, principle, or criterion. **b** a Church decree or law. **2 a** a member of a cathedral chapter. **b** (*fem.* **canoness**) a member of certain RC orders. **3 a** a collection or list of sacred books etc. accepted as genuine. **b** the recognized genuine works of a particular author; a list of these. **4** the part of the Roman Catholic mass containing the words of consecration. **5** *Mus.* a piece with different parts taking up the same theme successively, either at the same or at a different pitch. □ **canon law** ecclesiastical law. **canon regular** (or **regular canon**) see REGULAR *adj.* 9b. [OE f. L f. Gk *kanōn*, in ME also f. AF & OF *canun, -on*; in sense 2 ME f. OF *canonie* f. eccl.L *canonicus*: cf. CANONICAL]

cañon var. of CANYON.

canonic /kəˈnɒnɪk/ *adj.* = CANONICAL *adj.* [OE f. OF *canonique* or L *canonicus* f. Gk *kanonikos* (as CANON)]

canonical /kəˈnɒnɪk(ə)l/ *adj. & n.* ● **1 a** according to or ordered by canon law. **b** included in the canon of Scripture. **2** authoritative, standard, accepted. **3** of a cathedral chapter or a member of it. **4** *Mus.* in canon form. ● *n.* (in *pl.*) the canonical dress of the clergy. □ **canonical hours** *Eccl.* the times fixed for a formal set of prayers or for the celebration of marriage. □ **canonically** *adv.* [med.L *canonicalis* (as CANONIC)]

canonicate /kəˈnɒnɪkət/ *n.* = CANONRY.

canonicity /ˌkænəˈnɪsɪtɪ/ *n.* the status of being canonical. [L *canonicus* canonical]

canonist /ˈkænənɪst/ *n.* an expert in canon law. [ME f. F *canoniste* or f. med.L *canonista*: see CANON]

canonize /ˈkænəˌnaɪz/ *v.tr.* (also **-ise**) **1 a** declare officially to be a saint, usu. with a ceremony. **b** regard as a saint. **2** admit to the canon of Scripture. **3** sanction by Church authority. □ **canonization** /ˌkænənaɪˈzeɪʃ(ə)n/ *n.* [ME f. med.L *canonizare*: see CANON]

canonry /ˈkænənrɪ/ *n.* (*pl.* **-ies**) the office or benefice of a canon.

canoodle /kəˈnuːd(ə)l/ *v.intr. colloq.* kiss and cuddle amorously. [19th-c. US: orig. unkn.]

Canopic jar /kəˈnəʊpɪk/ *n.* (also **Canopic vase**) each of a set of (usually four) urns for containing the different organs of an embalmed body in an ancient Egyptian burial. The lids, originally plain, were later modelled as the human, falcon, dog, and jackal heads of the four sons of Horus, protectors of the jars. [L *Canopicus* f. *Canopus* in ancient Egypt]

Canopus /kəˈnəʊpəs/ **1** *Gk Mythol.* the pilot of the fleet of King Menelaus in the Trojan War. **2** *Astron.* the second brightest star in the sky, and the brightest in the constellation of Carina. It is a supergiant, visible only to observers in the southern hemisphere.

canopy /ˈkænəpɪ/ *n. & v.* ● *n.* (*pl.* **-ies**) **1 a** a covering hung or held up over a throne, bed, person, etc. **b** the sky. **c** an overhanging shelter. **2** *Archit.* a rooflike projection over a niche etc. **3** the uppermost layers of foliage etc. in a forest. **4 a** the expanding part of a parachute. **b** the cover of an aircraft's cockpit. ● *v.tr.* (**-ies, -ied**) supply or be a canopy to. [ME f. med.L *canopeum* f. L *conopeum* f. Gk *kōnōpeion* couch with mosquito-curtains f. *kōnōps* gnat]

canorous /kəˈnɔːrəs/ *adj.* melodious, resonant. [L *canorus* f. *canere* sing]

Canova /kəˈnəʊvə/, Antonio (1757–1822), Italian sculptor. Canova's inventive approach to classical models has led to him being regarded as a leading exponent of neoclassicism. His most famous works range from classical subjects such as *Cupid and Psyche* (1792) and *The Three Graces* (1813–16) to funeral monuments and life-size busts; highly

regarded in his day, he executed commissions for papal and royal monuments and for Napoleon and his family.

canst /kænst/ *archaic 2nd person sing.* of CAN¹.

Cant. *abbr.* (in the Bible) Canticles.

cant¹ /kænt/ *n. & v.* ● *n.* **1** insincere pious or moral talk. **2** ephemeral or fashionable catchwords. **3** language peculiar to a class, profession, sect, etc.; jargon. ● *v.intr.* use cant. □ **canting arms** *Heraldry* arms containing an allusion to the name of the bearer. [earlier of musical sound, of intonation, and of beggars' whining; perh. from the singing of religious mendicants: prob. f. L *canere* sing]

cant² /kænt/ *n. & v.* ● *n.* **1 a** a slanting surface, e.g. of a bank. **b** a bevel of a crystal etc. **2** an oblique push or movement that upsets or partly upsets something. **3** a tilted or sloping position. ● *v.* **1** push or pitch out of level; tilt. **2** *intr.* take or lie in a slanting position. **3** *tr.* impart a bevel to. **4** *intr. Naut.* swing round. □ **cant-dog** (or **-hook**) an iron hook at the end of a long handle, used for rolling logs. [ME f. MLG *kant, kante*, MDu. *cant*, point, side, edge, ult. f. L *cant(h)us* iron tire]

can't /kɑːnt/ *contr.* can not.

Cantab. /ˈkæntæb/ *abbr.* of Cambridge University. [L *Cantabrigiensis*]

cantabile /kænˈtɑːbɪlɪ/ *adv., adj., & n. Mus.* ● *adv. & adj.* in a smooth singing style. ● *n.* a cantabile passage or movement. [It., = singable]

Cantabria /kænˈtæbrɪə/ an autonomous region of northern Spain; capital, Santander. □ **Cantabrian** *adj. & n.*

Cantabrigian /ˌkæntəˈbrɪdʒɪən/ *adj. & n.* ● *adj.* of Cambridge or Cambridge University. ● *n.* **1** a member of Cambridge University. **2** a native of Cambridge. [L *Cantabrigia* Cambridge]

cantal /ˈkænt(ə)l/ *n.* a type of hard strong French cheese. [*Cantal*, a department of Auvergne, France]

cantaloup /ˈkæntəˌluːp/ *n.* (also **cantaloupe**) a small round ribbed variety of melon with orange flesh. [F *cantaloup* f. *Cantaluppi* near Rome, where it was first grown in Europe]

cantankerous /kænˈtæŋkərəs/ *adj.* bad-tempered, quarrelsome. □ **cantankerously** *adv.* **cantankerousness** *n.* [perh. f. Ir. *cant* outbidding + *rancorous*]

cantata /kænˈtɑːtə/ *n.* a short narrative or descriptive musical composition with vocal solos and usu. orchestral and choral accompaniment. Although the cantata in the early 17th century was of the nature of an extended song, the form is known principally from the choral cantatas written by Bach and his contemporaries for church use, frequently based on hymn tunes and harmonized to allow the congregation to join in the singing at the end. The cantata declined after 1750, but isolated examples appear in the 20th century, for example those by Britten and Stravinsky. [It. *cantata* (*aria*) sung (air) f. *cantare* sing]

canteen /kænˈtiːn/ *n.* **1 a** a restaurant for employees in an office or factory etc. **b** a shop selling provisions or liquor in a barracks or camp. **2** a case or box of cutlery. **3** a soldier's or camper's water-flask or set of eating or drinking utensils. [F *cantine* f. It. *cantina* cellar]

canter /ˈkæntə(r)/ *n. & v.* ● *n.* a gentle gallop. ● *v.* **1** *intr.* (of a horse or its rider) go at a canter. **2** *tr.* make (a horse) canter. □ **in a canter** easily (*win in a canter*). [short for *Canterbury pace*, from the supposed easy pace of medieval pilgrims to Canterbury]

Canterbury /ˈkæntəbərɪ/ a city in Kent, SE England; pop. (1981) 39,700. St Augustine established a church and monastery there in 597 and it became a place of medieval pilgrimage. It is the seat of the Archbishop of Canterbury.

canterbury /ˈkæntəbərɪ/ *n.* (*pl.* **-ies**) a piece of furniture with partitions for holding music etc. [CANTERBURY]

Canterbury, Archbishop of the archbishop of the southern province of England, first peer of the realm, and Primate of All England of the Anglican Church. The Archbishop of Canterbury is appointed by the Prime Minister, and crowns the sovereign. The office was created by St Augustine, who was the first archbishop (601–4). The official residence is Lambeth Palace in London.

Canterbury bell *n.* a cultivated campanula with large flowers. [after the bells of Canterbury pilgrims' horses: see CANTER]

Canterbury Plains a region on the central east coast of South Island, New Zealand.

cantharides /kænˈθærɪˌdiːz/ *n.pl.* a preparation made from dried bodies of the beetle *Lytta vesicatoria*, causing blistering of the skin and formerly used in medicine and as an aphrodisiac. See also *Spanish fly.* [L f. Gk *kantharis* Spanish fly]

canthus /ˈkænθəs/ n. (pl. **canthi** /-θaɪ/) the outer or inner corner of the eye, where the upper and lower lids meet. [L f. Gk *kanthos*]

canticle /ˈkæntɪk(ə)l/ n. **1** a hymn or chant with usu. a biblical text, forming a regular part of a church service. **2** (in pl.: **Canticles**) see SONG OF SOLOMON. [ME f. OF *canticle* (var. of *cantique*) or L *canticulum* dimin. of *canticum* f. *canere* sing]

cantilena /ˌkæntɪˈliːnə/ n. Mus. a simple or sustained melody. [It.]

cantilever /ˈkæntɪˌliːvə(r)/ n. & v. ● n. **1** a long bracket or beam etc. projecting from a wall to support a balcony etc. **2** a beam or girder fixed at only one end. ● v. **1** intr. & tr. project as a cantilever. **2** tr. support by a cantilever or cantilevers. □ **cantilever bridge** a bridge made of cantilevers projecting from the piers and connected by girders. [17th c.: orig. unkn.]

cantillate /ˈkæntɪˌleɪt/ v.tr. & intr. chant or recite with musical tones. □ **cantillation** /ˌkæntɪˈleɪʃ(ə)n/ n. [L *cantillare* sing low: see CHANT]

cantina /kænˈtiːnə/ n. a bar-room or wine-shop. [Sp. & It.]

canto /ˈkæntəʊ/ n. (pl. **-os**) a division of a long poem. [It., = song, f. L *cantus*]

Canton see GUANGZHOU.

canton n. & v. ● n. **1** /ˈkæntɒn/ **a** a subdivision of a country. **b** a state of the Swiss Confederation. **2** /ˈkæntən/ Heraldry a square division, less than a quarter, in the upper (usu. dexter) corner of a shield. ● v.tr. **1** /kænˈtuːn/ put (troops) into quarters. **2** /kænˈtɒn/ divide into cantons. □ **cantonal** /ˈkæntən(ə)l, kænˈtɒn(ə)l/ adj. [OF, = corner (see CANT²): (v.) also partly f. F *cantonner*]

Cantonese /ˌkæntəˈniːz/ adj. & n. ● adj. of Canton or the Cantonese dialect of Chinese. ● n. (pl. same) **1** a native of Canton. **2** the dialect of Chinese spoken in SE China and Hong Kong.

cantonment /kænˈtuːnmənt/ n. **1** a lodging assigned to troops. **2** hist. a permanent military station in India. [F *cantonnement*: see CANTON]

Cantor /ˈkæntɔː(r)/, Georg (1845–1918), Russian-born German mathematician. Cantor's work on numbers laid the foundations of the theory of sets. He introduced the concept of transfinite numbers, and his work stimulated 20th-century exploration of number theory and the logical foundation of mathematics.

cantor /ˈkæntɔː(r)/ n. **1** the leader of the singing in church; a precentor. **2** the precentor in a synagogue. [L, = singer f. *canere* sing]

cantorial /kænˈtɔːrɪəl/ adj. **1** of or relating to the cantor. **2** of the north side of the choir in a church, the side on which the cantor sits (cf. DECANAL).

cantoris /kænˈtɔːrɪs/ adj. Mus. to be sung by the cantorial side of the choir in antiphonal singing (cf. DECANI). [L, genitive of CANTOR]

cantrail /ˈkæntreɪl/ n. Brit. a timber etc. support for the roof of a railway carriage. [CANT² + RAIL¹]

cantrip /ˈkæntrɪp/ n. Sc. **1** a witch's trick. **2** a piece of mischief; a playful act. [18th c.: orig. unkn.]

Canuck /kəˈnʌk/ n. & adj. Can. sl. ● n. **1** a Canadian, orig. esp. a French Canadian. **2** a Canadian horse or pony. ● adj. Canadian, orig. esp. French Canadian. [app. f. CANADA]

Canute /kəˈnjuːt/ (also **Cnut, Knut**) (d.1035), son of Sweyn I, Danish king of England 1017–35, Denmark 1018–35, and Norway 1028–35. After Edmund Ironside's murder in 1016, Canute became king of England, ending a prolonged struggle for the throne. As king, he presided over a period of relative peace. He is most commonly remembered for demonstrating to fawning courtiers his inability to stop the rising tide; this demonstration of the limited powers of human beings, even if they be kings, has become distorted in folklore so as to suggest that Canute really expected to succeed in turning back the tide.

canvas /ˈkænvəs/ n. & v. ● n. **1 a** a strong coarse kind of cloth made from hemp or flax or other coarse yarn and used for sails and tents etc. and as a surface for oil-painting. **b** a piece of this. **2** a painting on canvas, esp. in oils. **3** an open kind of canvas used as a basis for tapestry and embroidery. **4** sl. the floor of a boxing or wrestling ring. **5** a racing-boat's covered end. ● v.tr. (**canvassed, canvassing**; US **canvased, canvasing**) cover with canvas. □ **by a canvas** (in boat-racing) by a small margin (win by a canvas). **canvas-back** a wild duck Aythya valisineria, of North America, with back feathers the colour of unbleached canvas. **under canvas 1** in a tent or tents. **2** with sails spread. [ME & ONF *canevas*, ult. f. L CANNABIS]

canvass /ˈkænvəs/ v. & n. ● v. **1 a** intr. solicit votes. **b** tr. solicit votes from (electors in a constituency). **2** tr. **a** ascertain opinions of. **b** seek custom from. **c** discuss thoroughly. **3** tr. Brit. propose (an idea or plan etc.). **4** intr.

US check the validity of votes. ● n. the process of or an instance of canvassing, esp. of electors. □ **canvasser** n. [orig. = toss in a sheet, agitate, f. CANVAS]

canyon /ˈkænjən/ n. (also **cañon**) a deep gorge, often with a stream or river. [Sp. *cañón* tube, ult. f. L *canna* CANE]

canzonetta /ˌkænzəˈnetə/ n. (also **canzonet** /-ˈnet/) **1** a short light song. **2** a kind of madrigal. [It., dimin. of *canzone* song f. L *cantio -onis* f. *canere* sing]

caoutchouc /ˈkaʊtʃʊk/ n. raw rubber. [F f. Carib *cahuchu*]

CAP see COMMON AGRICULTURAL POLICY.

cap /kæp/ n. & v. ● n. **1 a** a soft brimless head-covering, usu. with a peak. **b** a head-covering worn in a particular profession (nurse's cap). **c** esp. Brit. a cap awarded as a sign of membership of a sports team. **d** an academic mortarboard or soft hat. **e** a special hat as part of Highland costume. **2 a** a cover like a cap in shape or position (kneecap; toecap). **b** a device to seal a bottle or protect the point of a pen, lens of a camera, etc. **3 a** = Dutch cap. **b** = percussion cap. **4** = CROWN n. 9b. ● v.tr. (**capped, capping**) **1 a** put a cap on. **b** cover the top or end of. **c** set a limit to (rate-capping). **2 a** esp. Brit. award a sports cap to. **b** Sc. & NZ confer a university degree on. **3 a** lie on top of; form the cap of. **b** surpass, excel. **c** improve on (a story, quotation, etc.), esp. by producing a better or more apposite one. □ **cap in hand** humbly. **cap of liberty** Rom. Hist. a conical cap given to slaves on emancipation, often used as a republican symbol. **cap of maintenance** a cap or hat worn as a symbol of official dignity or carried before the sovereign etc. **cap rock** a hard rock or stratum overlying a deposit of oil, gas, coal, etc. **cap sleeve** a sleeve extending only a short distance from the shoulder. **the cap fits** (said of a generalized comment) it seems to be true of the person concerned. **set one's cap at** try to attract as a suitor. □ **capful** n. (pl. **-fuls**). **capping** n. [OE *cæppe* f. LL *cappa*, perh. f. L *caput* head]

cap. abbr. **1** capital. **2** capital letter. **3** chapter. [L *capitulum* or *caput*]

capability /ˌkeɪpəˈbɪlɪtɪ/ n. (pl. **-ies**) **1** (often foll. by of, for, to) ability, power; the condition of being capable. **2** an undeveloped or unused faculty.

Capability Brown see BROWN⁴.

Capablanca /ˌkæpəˈblæŋkə/, José Raúl (1888–1942), Cuban chess player. As world champion 1921–7 he made a considerable impact on the game, particularly on opening theory.

capable /ˈkeɪpəb(ə)l/ adj. **1** competent, able, gifted. **2** (foll. by of) **a** having the ability, fitness, or necessary quality for. **b** susceptible of admitting of (explanation or improvement etc.). □ **capably** adv. [F f. LL *capabilis* f. L *capere* hold]

capacious /kəˈpeɪʃəs/ adj. roomy; able to hold much. □ **capaciously** adv. **capaciousness** n. [L *capax -acis* f. *capere* hold]

capacitance /kəˈpæsɪt(ə)ns/ n. Electr. **1** the ability of a system to store an electric charge. **2** the ratio of the change in an electric charge in a system to the corresponding change in its electric potential (symbol: **C**). [CAPACITY + -ANCE]

capacitate /kəˈpæsɪˌteɪt/ v.tr. **1** (usu. foll. by for, or to + infin.) render capable. **2** make legally competent.

capacitor /kəˈpæsɪtə(r)/ n. Electr. a device of one or more pairs of conductors separated by insulators used to store an electric charge.

capacity /kəˈpæsɪtɪ/ n. (pl. **-ies**) **1 a** the power of containing, receiving, experiencing, or producing (capacity for heat, pain, etc.). **b** the maximum amount that can be contained or produced etc. **c** the volume, e.g. of the cylinders in an internal-combustion engine. **d** (attrib.) fully occupying the available space, resources, etc. (a capacity audience). **2 a** mental power. **b** a faculty or talent. **3** a position or function (in a civil capacity; in my capacity as a critic). **4** legal competence. **5** Electr. capacitance. □ **measure of capacity** a measure used for vessels and liquids or grains etc. **to capacity** fully; using all resources (working to capacity). □ **capacitative** /-tətɪv/ adj. (also **capacitive**) (in sense 5). [ME f. F f. L *capacitas -tatis* (as CAPACIOUS)]

caparison /kəˈpærɪs(ə)n/ n. & v. ● n. **1** (usu. in pl.) a horse's trappings. **2** equipment, finery. ● v.tr. put caparisons on; adorn richly. [obs. F *caparasson* f. Sp. *caparazón* saddle-cloth f. *capa* CAPE¹]

cape¹ /keɪp/ n. **1** a sleeveless cloak. **2** a short sleeveless cloak as a fixed or detachable part of a longer cloak or coat. [F f. Prov., Sp. *capa* f. LL *cappa* CAP]

cape² /keɪp/ n. a headland or promontory. [ME f. OF *cap* f. Prov. *cap* ult. f. L *caput* head]

Cape, the 1 the Cape of Good Hope. **2** the former Cape Province of South Africa.

Cape Agulhas, Cape Bon, etc. see AGULHAS, CAPE; BON, CAPE, etc.

Cape Breton Island /'bret(ə)n/ an island that forms the north-eastern part of the province of Nova Scotia, Canada. Believed to have been visited by John Cabot in 1497, it was a French possession until 1763 and a colony of Britain from 1784 to 1820.

Cape Cod a sandy peninsula in SE Massachusetts, forming a wide curve enclosing Cape Cod Bay. The Pilgrim Fathers landed on the northern tip of Cape Cod in Nov. 1620.

Cape Colony an early name (1814–1910) for the former CAPE PROVINCE.

Cape coloured adj. & n. ● adj. of the coloured (see COLOURED 2) population of the former Cape Province in South Africa, now largely in the province of Western Cape. ● n. a member of this population.

Cape doctor S. Afr. colloq. a strong south-east wind.

Cape Dutch n. archaic = AFRIKAANS.

Cape gooseberry n. **1** an edible soft roundish yellow berry enclosed in a lantern-like husk. **2** the plant, *Physalis peruviana*, bearing these.

Cape Johnson Depth the deepest point of the Philippine Trench, off the east coast of the Philippines, dropping to 10,497 m (34,440 ft) below sea level. It is named after the USS *Cape Johnson*, which took soundings there in 1945.

Čapek /'tʃæpek/, Karel (1890–1938), Czech novelist and dramatist. He wrote several plays with his brother Josef (1887–1945), including *The Insect Play* (1921), a satire on human society and totalitarianism. Čapek's best-known independent work was *R.U.R.* (*Rossum's Universal Robots*) (1920), a cautionary drama about the dangers of mechanization set 'on a remote island in 1950–60'. The title introduced the word *robot* to the English language.

capelin /'kæplɪn/ n. (also **caplin**) a small smeltlike fish, *Mallotus villosus*, of the North Atlantic, used as food and as bait for catching cod etc. Also called *ice-fish*. [F f. Prov. *capelan*: see CHAPLAIN]

Capella /kə'pelə/ Astron. the sixth brightest star in the sky, and the brightest in the constellation of Auriga, overhead in winter to observers in the northern hemisphere. It is a yellow giant. [L, = little goat, f. *caper* goat]

Cape of Good Hope a mountainous promontory south of Cape Town, South Africa, near the southern extremity of Africa. Sighted towards the end of the 15th century by Bartolomeu Dias, it was rounded for the first time by Vasco da Gama in 1497.

Cape Province a former province of South Africa, containing the Cape of Good Hope. Seized by the British in 1806, the area became a British colony in 1814. It was known as Cape Colony, or the Cape, from then until 1910, when it joined the Union of South Africa. In 1994 it was divided into the provinces of Northern Cape, Western Cape, and Eastern Cape.

caper[1] /'keɪpə(r)/ v. & n. ● v.intr. jump or run about playfully. ● n. **1** a playful jump or leap. **2 a** a fantastic proceeding; a prank. **b** sl. any activity or occupation. □ **cut a caper** (or **capers**) act friskily. □ **caperer** n. [abbr. of CAPRIOLE]

caper[2] /'keɪpə(r)/ n. **1** a bramble-like southern European shrub, *Capparis spinosa*. **2** (in pl.) its flower buds cooked and pickled for use as flavouring esp. for a savoury sauce. [ME *capres* & F *câpres* f. L *capparis* f. Gk *kapparis*, treated as pl.: cf. CHERRY, PEA]

capercaillie /ˌkæpə'keɪlɪ/ n. (also **capercailzie** /-'keɪlzɪ/) a large grouse, *Tetrao urogallus*, native to the coniferous forests of northern Europe. [Gael. *capull coille* horse of the wood]

capeskin /'keɪpskɪn/ n. a soft leather made from South African sheepskin.

Capet /'kæpɪt, kæ'pet, French kapɛ/, Hugh (or Hugo) (938–96), king of France 987–96. His election as king in 987 marked the foundation of the Capetian dynasty.

Capetian /kə'pi:ʃ(ə)n/ adj. & n. ● adj. of or relating to a dynasty of kings of France founded by Hugh Capet in 987 in succession to the Carolingian dynasty. It survived until 1328, giving way to the House of Valois. The extinction of the direct line of Capetians gave rise to Edward III's claim to the French throne and the start of the Hundred Years War. ● n. a member of this dynasty.

Cape Town the legislative capital of South Africa and administrative capital of the province of Western Cape; pop. (1985) 776,600. It is situated at the foot of Table Mountain.

Cape Verde Islands /vɜːd/ a country consisting of a group of islands in the Atlantic off the coast of Senegal, named after the most westerly cape of Africa; pop. (est. 1991) 383,000; languages, Portuguese (official), Creole; capital, Praia. Previously uninhabited, the islands were settled by the Portuguese from the 15th century. They remained a Portuguese colony until 1975, when an independent republic was established. □ **Cape Verdean** /'vɜːdɪən/ adj. & n.

capias /'keɪpɪˌæs, 'keɪp-/ n. Law a writ ordering the arrest of the person named. [L, = you are to seize, f. *capere* take]

capillarity /ˌkæpɪ'lærɪtɪ/ n. = capillary action. [F *capillarité* (as CAPILLARY)]

capillary /kə'pɪlərɪ/ adj. & n. ● adj. **1** of or like a hair. **2** (of a tube) of hairlike internal diameter. **3** of one of the delicate ramified blood vessels intervening between arteries and veins. ● n. (pl. **-ies**) **1** a capillary tube. **2** a capillary blood vessel. □ **capillary action** the tendency of liquid in a capillary tube or absorbent material to rise or fall as a result of surface tension (also called *capillarity*). [L *capillaris* f. *capillus* hair]

capital[1] /'kæpɪt(ə)l/ n., adj., & int. ● n. **1** the most important town or city of a country or region, usu. its seat of government and administrative centre. **2 a** the money or other assets with which a company starts in business. **b** accumulated wealth, esp. as used in further production. **c** money invested or lent at interest. **3** capitalists generally. **4** a capital letter. ● adj. **1 a** principal; most important; leading. **b** colloq. excellent, first-rate. **2 a** involving or punishable by death (*capital punishment; a capital offence*). **b** (of an error etc.) vitally harmful; fatal. **3** (of letters of the alphabet) large in size and of the form used to begin sentences and names etc. ● int. expressing approval or satisfaction. □ **capital gain** a profit from the sale of investments or property. **capital gains tax** a tax levied on capital gains. **capital goods** goods, esp. machinery, plant, etc., used or to be used in producing commodities (opp. *consumer goods*). **capital levy 1** the appropriation by the state of a fixed proportion of the wealth in the country. **2** a wealth tax. **capital sum** a lump sum of money, esp. payable to an insured person. **capital territory** a territory containing the capital city of a country. **capital transfer tax** hist. (in the UK) a tax levied on the transfer of capital by gift or bequest etc., replaced in 1986 by *inheritance tax*. **make capital out of** use to one's advantage. **with a capital —** emphatically such (*art with a capital A*). □ **capitally** adv. [ME f. OF f. L *capitalis* f. *caput -itis* head]

capital[2] /'kæpɪt(ə)l/ n. Archit. the head or cornice of a pillar or column. [ME f. OF *capitel* f. LL *capitellum* dimin. of L *caput* head]

capitalism /'kæpɪtəˌlɪz(ə)m/ n. **1 a** an economic system in which the production and distribution of goods depend on invested private capital and profit-making. **b** the possession of capital or wealth. **2** Polit. the dominance of private owners of capital and production for profit.

capitalist /'kæpɪtəlɪst/ n. & adj. ● n. **1** a person using or possessing capital; a rich person. **2** an advocate of capitalism. ● adj. of or favouring capitalism. □ **capitalistic** /ˌkæpɪtə'lɪstɪk/ adj. **capitalistically** adv.

capitalize /'kæpɪtəˌlaɪz/ v. (also **-ise**) **1** tr. **a** convert into or provide with capital. **b** calculate or realize the present value of an income. **c** reckon (the value of an asset) by setting future benefits against the cost of maintenance. **2** tr. **a** write (a letter of the alphabet) as a capital. **b** begin (a word) with a capital letter. **3** intr. (foll. by *on*) use to one's advantage; profit from. □ **capitalization** /ˌkæpɪtəlaɪ'zeɪʃ(ə)n/ n. [F *capitaliser* (as CAPITAL[1])]

capital punishment n. infliction of death by an authorized public authority as punishment for a crime. It was recognized by ancient legal systems, methods of execution varying from drowning in Babylon, stoning among the Hebrews, to strangulation, exposure to wild beasts, or crucifixion in Rome. In medieval Europe hanging and beheading were the usual methods; religious heretics were burnt at the stake. In more recent times, hanging was used in Britain, the guillotine in France, and the garrotte in Spain; in the US the chief methods of execution have been the electric chair and the gas chamber. In the 19th century in Britain the death penalty, previously available for a wide range of offences (some quite trivial), was restricted to cases of treason and murder, and in 1965 it was abolished for murder. Many countries, particularly in the West, have abolished the penalty, but it is retained in Russia and China and its use is becoming more common again in some states of the US. Regular calls for its restoration are made in Britain, despite the danger of executing a wrongly convicted person and the lack of conclusive evidence that it functions as a deterrent to criminals.

capitation /ˌkæpɪ'teɪʃ(ə)n/ n. **1** a tax or fee at a set rate per person. **2** the levying of such a tax or fee. □ **capitation grant** a grant of a sum calculated from the number of people to be catered for, esp. in education. [F *capitation* or LL *capitatio* poll-tax f. *caput* head]

Capitol, the /'kæpɪt(ə)l/ **1** the temple of Jupiter on the Capitoline Hill in ancient Rome. **2** (also **Capitol Hill**) the seat of the US Congress in Washington, DC. Its site was chosen by George Washington, who laid the first stone in 1793. [OF f. L *caput* head]

Capitol Hill see CAPITOL, THE 2.

capitular /kə'pɪtjʊlə(r)/ *adj.* **1** of or relating to a cathedral chapter. **2** *Anat.* of or relating to a terminal protuberance of a bone. [LL *capitularis* f. L *capitulum* CHAPTER]

capitulary /kə'pɪtjʊlərɪ/ *n.* (*pl.* **-ies**) *hist.* a collection of ordinances, esp. of the Frankish kings. [LL *capitularius* (as CAPITULAR)]

capitulate /kə'pɪtjʊˌleɪt/ *v.intr.* surrender, esp. on stated conditions. □ **capitulator** *n.* **capitulatory** /-lətərɪ/ *adj.* [med.L *capitulare* draw up under headings f. L *caput* head]

capitulation /kə,pɪtjʊ'leɪʃ(ə)n/ *n.* **1** the act of capitulating; surrender. **2** a statement of the main divisions of a subject. **3** an agreement or set of conditions.

capitulum /kə'pɪtjʊləm/ *n.* (*pl.* **capitula** /-lə/) *Bot.* a dense flat cluster of flowers or florets, as in the daisy family. [L, dimin. of *caput* head]

caplin var. of CAPELIN.

cap'n /'kæp(ə)n/ *n. colloq.* captain. [contr.]

capo[1] /'kæpəʊ/ *n.* (in full **capo tasto** /'tæstəʊ/) (*pl.* **capos** or **capo tastos**) *Mus.* a device secured across the neck of a fretted instrument to raise equally the tuning of all strings by the required amount. [It. *capo tasto* head stop]

capo[2] /'kæpəʊ/ *n.* (*pl.* **capos**) esp. *US* the head of a crime syndicate or one of its branches. [It., head, captain]

Capo di Monte /ˌkæpəʊ dɪ 'mɒntɪ/ *n.* a type of porcelain first produced at the Capo di Monte palace near Naples in the mid-18th century. It is usually in the form of tableware or figures (frequently of peasants or characters from the *commedia dell'arte*), and is generally white with richly coloured rococo decoration.

capon /'keɪp(ə)n/ *n.* a domestic cock castrated and fattened for eating. □ **caponize** *v.tr.* (also **-ise**). [OE f. AF *capun*, OF *capon*, ult. f. L *capo -onis*]

Capone /kə'pəʊn/, Alphonse ('Al') (1899–1947), Italian-born American gangster. He was notorious for his domination of organized crime in Chicago in the 1920s; his earnings from liquor, prostitution, gambling, and other rackets were estimated to be $30 million per year. Although indirectly responsible for many murders, including those of the St Valentine's Day Massacre (1929), he was never tried for any of them; it was for federal income-tax evasion that he was eventually imprisoned in 1931.

caponier /ˌkæpə'nɪə(r)/ *n.* a covered passage across a ditch round a fort. [Sp. *caponera*, lit. 'capon-pen']

capot /kə'pɒt/ *n. & v.* ● *n.* (in piquet) the winning of all the tricks by one player. ● *v.tr.* (**capotted, capotting**) score a capot against (an opponent). [F]

Capote /kə'pəʊtɪ/, Truman (born Truman Streckfus Persons) (1924–84), American writer. His works range from the light-hearted novella *Breakfast at Tiffany's* (1958) to the grim and meticulous re-creation of a brutal multiple murder in *In Cold Blood* (1966).

capote /kə'pəʊt/ *n. hist.* a long cloak with a hood, formerly worn by soldiers and travellers etc. [F, dimin. of *cape* CAPE[1]]

Cappadocia /ˌkæpə'dəʊʃə/ an ancient region of central Asia Minor, between Lake Tuz and the Euphrates, north of Cilicia. It was an important centre of early Christianity. □ **Cappadocian** *adj. & n.*

cappuccino /ˌkæpʊ'tʃiːnəʊ/ *n.* (*pl.* **-os**) coffee with milk made frothy with pressurized steam. [It., = CAPUCHIN]

Capra /'kæprə/, Frank (1897–1991), Italian-born American film director. His reputation rests on the film comedies which he made in the 1930s and early 1940s, such as *It Happened One Night* (1934), *Mr Deeds Goes to Town* (1936), and *Arsenic and Old Lace* (1944). He won six Oscars for his films.

Capri /kə'priː, Italian 'kaːpri/ an island off the west coast of Italy, south of Naples.

capriccio /kə'prɪtʃɪ,əʊ/ *n.* (*pl.* **-os**) **1** a lively and usu. short musical composition. **2** a painting etc. representing a fantasy or a mixture of real and imaginary features. [It., = sudden start, orig. 'horror']

capriccioso /kə,prɪtʃɪ'əʊsəʊ/ *adv., adj., & n. Mus.* ● *adv. & adj.* in a free and impulsive style. ● *n.* (*pl.* **-os**) a capriccioso passage or movement. [It., = capricious]

caprice /kə'priːs/ *n.* **1 a** an unaccountable or whimsical change of mind or conduct. **b** a tendency to this. **2** a work of lively fancy in painting, drawing, or music; a capriccio. [F f. It. CAPRICCIO]

capricious /kə'prɪʃəs/ *adj.* **1** guided by or given to caprice. **2** irregular, unpredictable. □ **capriciously** *adv.* **capriciousness** *n.* [F *capricieux* f. It. CAPRICCIOSO]

Capricorn /'kæprɪˌkɔːn/ *n.* **1** (usu. **Capricornus** /-'kɔːnəs/) *Astron.* a constellation (the Goat), said to represent a goat with a fish's tail. It has few bright stars. **2** *Astrol.* **a** the tenth sign of the zodiac, which the sun enters at the northern winter solstice (about 21 Dec.). **b** a person born when the sun is in this sign. □ **Tropic of Capricorn** see TROPIC. □ **Capricornian** /ˌkæprɪ'kɔːnɪən/ *n. & adj.* [ME f. OF *capricorne* f. L *capricornus* f. *caper -pri* goat + *cornu* horn]

caprine /'kæpraɪn/ *adj.* of or like a goat. [ME f. L *caprinus* f. *caper -pri* goat]

capriole /'kæprɪˌəʊl/ *n. & v.* ● *n.* **1** a leap or caper. **2** a trained horse's high leap and kick without advancing. ● *v.* **1** *intr.* (of a horse or its rider) perform a capriole. **2** *tr.* make (a horse) capriole. [F f. It. *capriola* leap, ult. f. *caper -pri* goat]

Capris /kə'priːz/ *n.pl.* (also **Capri pants**) women's close-fitting tapered trousers. [CAPRI]

Caprivi Strip /kə'priːvɪ/ a narrow strip of Namibia, which extends towards Zambia from the north-eastern corner of Namibia and reaches the Zambezi river. It was part of German South West Africa until after the First World War, having been ceded by Britain in 1893. It is named after Leo Graf von Caprivi, German imperial Chancellor 1890–4.

caps. /kæps/ *abbr.* capital letters.

Capsian /'kæpsɪən/ *adj. & n. Archaeol.* of, relating to, or denoting a palaeolithic culture of North Africa and southern Europe (c.8000–4500 BC) noted for its microliths. [f. L *Capsa* = GAFSA in Tunisia, the type-site]

capsicum /'kæpsɪkəm/ *n.* **1** a solanaceous plant of the genus *Capsicum*, having edible capsular fruits containing many seeds; esp. *C. annuum*, which yields several varieties of pepper, esp. green pepper and red pepper (cf. PIMIENTO). Also called *sweet pepper*. **2** the fruit of these plants, which vary in size, colour, and pungency. [mod.L, perh. f. L *capsa* box]

capsid[1] /'kæpsɪd/ *n. & adj. Zool.* = MIRID. [mod.L *Capsus* a genus of bug]

capsid[2] /'kæpsɪd/ *n.* the protein coat or shell of a virus. [F *capside* f. L *capsa* box]

capsize /kæp'saɪz/ *v.* **1** *tr.* upset or overturn (a boat). **2** *intr.* be capsized. □ **capsizal** *n.* [cap- as in Prov. *capvirar*, F *chavirer*: -size unexpl.]

capstan /'kæpstən/ *n.* **1** a thick revolving cylinder with a vertical axis, for winding an anchor cable or a halyard etc. **2** a motor-driven revolving spindle on a tape recorder that guides the tape past the head at a constant speed. □ **capstan lathe** a lathe with a revolving tool-holder. [Prov. *cabestan*, ult. f. L *capistrum* halter f. *capere* seize]

capstone /'kæpstəʊn/ *n.* coping; a coping-stone.

capsule /'kæpsjuːl/ *n.* **1** a small soluble case of gelatine enclosing a dose of medicine and swallowed with it. **2** a detachable compartment of a spacecraft or nose-cone of a rocket. **3** an enclosing membrane in the body. **4** a top or cover for a bottle, esp. the foil or plastic covering the cork in a wine bottle. **5 a** a dry fruit that releases its seeds when ripe. **b** the spore-producing part of mosses and liverworts. **6** *Biol.* an enveloping layer surrounding certain bacteria. **7** (*attrib.*) concise; highly condensed (*a capsule history of jazz*). □ **capsular** *adj.* **capsulate** *adj.* [F f. L *capsula* f. *capsa* CASE[2]]

capsulize /'kæpsjʊˌlaɪz/ *v.tr.* (also **-ise**) put (information etc.) in compact form.

Capt. *abbr.* Captain.

captain /'kæptɪn/ *n. & v.* ● *n.* **1 a** a chief or leader. **b** the leader of a team, esp. in sports. **c** a powerful or influential person (*captain of industry*). **2 a** the person in command of a merchant or passenger ship. **b** the pilot of a civil aircraft. **3** (as a title **Captain**) **a** an army or *US* air force officer next above lieutenant. **b** a navy officer in command of a warship; one ranking below commodore or rear admiral and above commander. **c** (in the US and elsewhere) a police officer in charge of a precinct, ranking below chief officer. **4 a** a foreman. **b** a head boy or girl in a school. **c** *Scouting* the adult leader of a company of Guides. **d** *N. Amer.* a supervisor of waiters or bellboys. **5 a** a great soldier or strategist. **b** an experienced commander. ● *v.tr.* be captain of; lead. □ **captain-general** (in the UK) an honorary officer, esp. of artillery. □ **captaincy** *n.* (*pl.* **-ies**). **captainship** *n.* [ME & OF *capitain* f. LL *capitaneus* chief f. L *caput capit-* head]

Captain Cooker n. NZ a wild boar (descended from introduced pigs). [Cook¹]

Captain of the Fleet n. (in the Royal Navy) a staff officer in charge of maintenance.

caption /ˈkæpʃ(ə)n/ n. & v. ● n. **1** a title or brief explanation appended to an illustration, cartoon, etc. **2** wording appearing on a cinema or television screen as part of a film or broadcast. **3** the heading of a chapter or article etc. **4** Law a certificate attached to or written on a document. ● v.tr. provide with a caption. [ME f. L captio f. capere capt- take]

captious /ˈkæpʃəs/ adj. given to finding fault or raising petty objections. □ **captiously** adv. **captiousness** n. [ME f. OF captieux or L captiosus (as CAPTION)]

captivate /ˈkæptɪˌveɪt/ v.tr. **1** overwhelm with charm or affection. **2** fascinate. □ **captivating** adj. **captivatingly** adv. **captivation** /ˌkæptɪˈveɪʃ(ə)n/ n. [LL captivare take captive (as CAPTIVE)]

captive /ˈkæptɪv/ n. & adj. ● n. a person or animal that has been taken prisoner or confined. ● adj. **1 a** taken prisoner. **b** kept in confinement or under restraint. **2 a** unable to escape. **b** in a position of having to comply (captive audience; captive market). **3** of or like a prisoner (captive state). □ **captive balloon** a balloon held by a rope from the ground. [ME f. L captivus f. capere capt- take]

captivity /kæpˈtɪvɪtɪ/ n. (pl. **-ies**) **1** the condition or circumstances of being a captive. **2** a period of captivity.

Captivity, the see BABYLONIAN CAPTIVITY.

captor /ˈkæptə(r), -tɔː(r)/ n. a person who holds captive or captures a person etc. [L (as CAPTIVE)]

capture /ˈkæptʃə(r)/ v. & n. ● v.tr. **1 a** take prisoner; seize as a prize. **b** obtain by force or trickery. **2** portray in permanent form (could not capture the likeness). **3** Physics absorb (a subatomic particle). **4** (in board games) make a move that secures the removal of (an opposing piece) from the board. **5** (of a stream) divert the upper course of (another stream) by encroaching on its basin. **6** cause (data) to be stored in a computer. **7** Astron. (of a star or planet) bring (an object) permanently within its gravitational influence. ● n. **1** the act of capturing. **2** a thing or person captured. □ **capturer** n. [F f. L captura f. capere capt- take]

Capuchin /ˈkæpjʊˌtʃɪn/ n. **1** a friar of a branch of the Franciscan order that observes a strict rule drawn up in 1529. (See also FRANCISCAN.) **2** a cloak and hood formerly worn by women. **3** (**capuchin**) **a** a South American monkey of the genus Cebus, with cowl-like hair on the head. **b** a variety of pigeon with head and neck feathers resembling a cowl. [F f. It. cappuccino f. cappuccio cowl f. cappa CAPE¹]

capybara /ˌkæpɪˈbɑːrə/ n. a very large semiaquatic rodent, Hydrochoerus hydrochaeris, native to South America. [Tupi]

car /kɑː(r)/ n. (in full **motor car**) **1** a usu. four-wheeled road vehicle able to carry a small number of people and powered by an internal-combustion engine. (See note below.) **2** (in comb.) **a** a wheeled vehicle, esp. of a specified kind (tramcar). **b** a railway carriage of a specified type (dining-car). **3** N. Amer. any railway carriage or van. **4** the passenger compartment of a lift, cableway, balloon, etc. **5** poet. a chariot. □ **car bomb** a terrorist bomb concealed in or under a parked car. **car boot sale** an outdoor sale at which participants sell unwanted possessions from the boots of their cars or from tables set up nearby. **car coat** a short coat designed esp. for car drivers. **car park** esp. Brit. an area for parking cars. **car phone** a radio-telephone for use in a motor vehicle. **car wash 1** an establishment containing equipment for washing vehicles automatically. **2** the equipment itself. □ **carful** n. (pl. **-fuls**). [ME f. AF & ONF carre ult. f. L carrum, carrus, of OCelt. orig.]

▪ The first self-propelled road vehicle was a steam-driven carriage designed and built in France by Nicholas-Joseph Cugnot (1725–1804) in 1769; other pioneers included Richard Trevithick. In Britain development clashed with the interests of the railway promoters, who used their influence to secure a highly restrictive law (1865) which required every power-driven road vehicle to have a man walking 100 yards in front with a red flag by day and a red lantern by night; the provisions of this law were not abolished until 1896. On the Continent, however, progress was being made with the internal-combustion engine, which Daimler fitted to a cycle and Benz to a three-wheeled vehicle (1885) that carried two passengers. Early cars were very small two-seat carts, with no roofs, poor springs, and wooden- or wire-spoked wheels. The car became a significant social phenomenon when mass production was introduced in the early 20th century, by pioneers such as Henry Ford. Cars are now ubiquitous in industrialized countries and exercise an overriding influence on the development of land. Alternatives to oil-based fuels,

such as electric batteries or solar power, have not yet been sufficiently viable to challenge the dominance of the internal-combustion engine.

carabineer /ˌkærəbɪˈnɪə(r)/ n. (also **carabinier**) hist. **1** a soldier whose principal weapon is a carbine. **2** (**the Carabineers**) a former regiment of dragoon guards, now incorporated in the Royal Scots Dragoon Guards. [F carabinier f. carabine CARBINE]

carabiniere /ˌkærəbɪˈnjeərɪ/ n. (pl. **carabinieri** pronunc. same) an Italian gendarme. [It.]

caracal /ˈkærəˌkæl/ n. a lynx, Felis caracal, native to North Africa and SW Asia. [F or Sp. f. Turk. karakulak f. kara black + kulak ear]

Caracalla /ˌkærəˈkælə/ (born Septimius Bassanius; later called Marcus Aurelius Severus Antoninus Augustus) (188–217), Roman emperor 211–17. He spent much of his reign campaigning in Germany and in the East, where he hoped to repeat the conquests of Alexander the Great, but was assassinated in Mesopotamia. By an edict of 212 he granted Roman citizenship to all free inhabitants of the Roman Empire. His name was derived from that of a kind of tunic he characteristically wore.

caracara /ˌkærəˈkɑːrə/ n. a bird of prey found mainly in tropical America, related to falcons but somewhat resembling a vulture. [Sp. or Port. caracará, f. Tupi-Guarani]

Caracas /kəˈrækəs/ the capital of Venezuela; pop. (1991) 1,824,890. It was the birthplace of Simón Bolívar.

caracole /ˈkærəˌkəʊl/ n. & v. ● n. a horse's half-turn to the right or left. ● v. **1** intr. (of a horse or its rider) perform a caracole. **2** tr. make (a horse) caracole. [F]

Caratacus see CARATACUS.

caracul var. of KARAKUL.

carafe /kəˈræf, -ˈrɑːf/ n. a glass container for water or wine, esp. at a table or bedside. [F f. It. caraffa, ult. f. Arab. ġarrāfa drinking-vessel]

Carajás /ˌkærəˈʒɑːs/ a mining region in north Brazil, the site of one of the richest deposits of iron ore in the world.

carambola /ˌkærəmˈbəʊlə/ n. **1** a small tree, Averrhoa carambola, native to SE Asia, bearing golden-yellow ribbed fruit. **2** this fruit. Also called star fruit. [Port., prob. of Indian or E. Indian orig.]

caramel /ˈkærəˌmel, -m(ə)l/ n. **1 a** a sugar or syrup heated until it turns brown, then used as a flavouring or to colour spirits etc. **b** a kind of soft toffee made with sugar, butter, etc., melted and further heated. **2** the light-brown colour of caramel. [F f. Sp. caramelo]

caramelize /ˈkærəməˌlaɪz/ v. (also **-ise**) **1 a** tr. convert (sugar or syrup) into caramel. **b** intr. (of sugar or syrup) be converted into caramel. **2** tr. coat or cook (food) with caramelized sugar or syrup. □ **caramelization** /ˌkærəmələˈzeɪʃ(ə)n/ n.

carangid /kəˈrændʒɪd/ adj. & n. Zool. ● adj. of or relating to the large fish family Carangidae, which includes scads, pompanos, etc. ● n. a fish of this family. [f. mod.L Caranx, genus name]

carapace /ˈkærəˌpeɪs/ n. the hard upper shell of a tortoise or a crustacean. [F f. Sp. carapacho]

carat /ˈkærət/ n. **1** a unit of weight for precious stones, now equivalent to 200 milligrams. **2** (US **karat**) a measure of purity of gold, pure gold being 24 carats. [F f. It. carato f. Arab. ḳīrāṭ weight of four grains, f. Gk keration fruit of the carob (dimin. of keras horn)]

Caratacus /kəˈrætəkəs/ (also **Caractacus** /-ˈræktəkəs/) (1st century AD), British chieftain, son of Cunobelinus (see CYMBELINE). He took part in the resistance to the Roman invasion of AD 43, and when defeated fled to Yorkshire, where he was handed over to the Romans in AD 51.

Caravaggio /ˌkærəˈvædʒɪˌəʊ/, Michelangelo Merisi da (c.1571–1610), Italian painter. An important figure in the transition from late mannerism to baroque, he reinvigorated religious art and had a far-reaching influence on later artists. The characteristic features of his work include naturalistic realism (achieved partly by the use of ordinary people as models for biblical characters) and dramatic use of light and shade.

caravan /ˈkærəˌvæn/ n. & v. ● n. **1 a** Brit. a vehicle equipped for living in and usu. towed by a motor vehicle or a horse. **b** US a covered motor vehicle equipped for living in. **2** a company of merchants or pilgrims etc. travelling together, esp. across a desert in Asia or North Africa. **3** a covered cart or carriage. ● v.intr. (**caravanned**, **caravanning**) travel or live in a caravan. □ **caravan site** (or **park**) a place where caravans are parked as dwellings, often with special amenities. □ **caravanner** n. [F caravane f. Pers. kārwān]

caravanette /ˌkærəvæˈnet/ n. a motor vehicle with a caravan-like rear compartment for eating, sleeping, etc.

caravanserai /ˌkærəˈvænsəraɪ, -ˌraɪ/ n. an Eastern inn with a central court where caravans (see CARAVAN 2) may rest. [Pers. *kārwānsarāy* f. *sarāy* palace]

caravel /ˈkærəˌvel/ n. (also **carvel** /ˈkaːv(ə)l/) *hist.* a small light fast ship, chiefly Spanish and Portuguese of the 15th–17th centuries. [F *caravelle* f. Port. *caravela* f. Gk *karabos* horned beetle, light ship]

caraway /ˈkærəˌweɪ/ n. an umbelliferous plant, *Carum carvi*, bearing clusters of tiny white flowers. □ **caraway seed** its fruit used as flavouring and as a source of oil. [prob. OSp. *alcarahueya* f. Arab. *alkarāwiyā*, perh. f. Gk *karon*, *kareon* cumin]

carb /kaːb/ n. *colloq.* a carburettor. [abbr.]

carbamate /ˈkaːbəˌmeɪt/ n. *Chem.* a salt or ester of an amide of carbonic acid. Some carbamates are important as insecticides. [CARBONIC + AMIDE]

carbide /ˈkaːbaɪd/ n. *Chem.* a binary compound of carbon.

carbie /ˈkaːbɪ/ n. (also **carby**) *Austral. colloq.* a carburettor. [abbr.]

carbine /ˈkaːbaɪn/ n. a short firearm, usu. a rifle, orig. for cavalry use. [F *carabine* (this form also earlier in Engl.), weapon of the *carabin* mounted musketeer]

carbo- /ˈkaːbəʊ/ *comb. form* carbon (*carbohydrate*; *carbolic*; *carboxyl*).

carbohydrate /ˌkaːbəˈhaɪdreɪt/ n. *Biochem.* any of a large group of energy-producing organic compounds containing carbon, hydrogen, and oxygen, e.g. starch, glucose, and other sugars.

carbolic /kaːˈbɒlɪk/ n. (in full **carbolic acid**) phenol, esp. when used as a disinfectant. □ **carbolic soap** soap containing this. [CARBO- + -OL + -IC]

carbon /ˈkaːb(ə)n/ n. **1** a non-metallic chemical element (atomic number 6; symbol C). (*See note below.*) **2 a** = *carbon copy*. **b** = *carbon paper*. **3** a rod of carbon in an arc lamp. □ **carbon black** a fine carbon powder made by burning hydrocarbons in insufficient air. **carbon copy 1** a copy made with carbon paper. **2** a person or thing identical or similar to another (*is a carbon copy of his father*). **carbon cycle 1** *Biol.* the continuous transfer of carbon compounds from the atmosphere to living organisms through plant photosynthesis, and back to the atmosphere by respiration and decay. **2** *Physics* a cycle of thermonuclear reactions believed to occur in stars, in which carbon nuclei are repeatedly formed and broken down in the conversion of hydrogen into helium. **carbon dating** the determination of the age of an organic object from the ratio of isotopes which changes as carbon-14 decays (see RADIOCARBON). **carbon dioxide** a colourless odourless gas (chem. formula: CO_2) occurring naturally in the atmosphere and formed by respiration. **carbon disulphide** a colourless liquid (chem. formula: CS_2) used as a solvent. **carbon fibre** a thin strong crystalline filament of carbon used as strengthening material in resins, ceramics, etc. **carbon-14** a long-lived radioactive carbon isotope of mass 14, used in radiocarbon dating and as a tracer in biochemistry. **carbon microphone** a microphone depending for its action on the varying electrical resistance of carbon granules. **carbon monoxide** a colourless odourless toxic gas (chem. formula: CO) formed by the incomplete burning of carbon. **carbon paper** a thin carbon-coated paper used for making (esp. typed) copies. **carbon steel** a steel with properties dependent on the percentage of carbon present. **carbon tax** a tax on the carbon emissions that result from burning fossil fuels (e.g. in motor vehicles) because of their contribution to the greenhouse effect. **carbon tetrachloride** a colourless volatile liquid (chem. formula: CCl_4) used as a solvent. **carbon-12** the principal stable isotope of carbon, of mass 12. [F *carbone* f. L *carbo -onis* charcoal]

▪ Pure carbon exists in two allotropic forms, diamond and graphite. The element also occurs in impure form as charcoal, soot, coke, coal, and related materials; molecules consisting of hollow cages of carbon atoms (fullerenes) have recently been discovered. Carbon has many economic uses, but its unique significance is that its compounds form the physical basis of all living organisms. Carbon has the property of linking with itself and other elements to form molecules consisting of rings or long chains of atoms, and an immense variety of compounds exist. The chemistry of such compounds found in living things or their remains was once thought to obey laws different from those governing other substances, whence the original separation of organic chemistry as a distinct subject.

carbonaceous /ˌkaːbəˈneɪʃəs/ adj. **1** consisting of or containing carbon. **2** of or like coal or charcoal.

carbonade /ˌkaːbəˈneɪd/ n. a rich beef stew made with onions and beer. [F]

carbonado /ˌkaːbəˈneɪdəʊ/ n. (pl. **-os**) a dark opaque or impure kind of diamond used as an abrasive, for drills etc. [Port.]

carbonate /ˈkaːbəˌneɪt/ n. & v. ● n. *Chem.* a salt of carbonic acid. ● *v.tr.* **1** impregnate with carbon dioxide; aerate. **2** convert into a carbonate. □ **carbonation** /ˌkaːbəˈneɪʃ(ə)n/ n. [F *carbonat* f. mod.L *carbonatum* (as CARBON)]

carbonic /kaːˈbɒnɪk/ adj. *Chem.* containing carbon. □ **carbonic acid** a very weak acid formed from carbon dioxide dissolved in water. **carbonic acid gas** *archaic* carbon dioxide.

carboniferous /ˌkaːbəˈnɪfərəs/ adj. & n. ● adj. **1** producing coal. **2** (**Carboniferous**) *Geol.* of or relating to a period of the Palaeozoic era. (*See note below.*) ● n. (**Carboniferous**) *Geol.* this period or the corresponding geological system.

▪ The Carboniferous period lasted from about 360 to 286 million years BP, between the Devonian and the Permian. During this period seed-bearing plants appeared, corals were widespread, and extensive limestone deposits were formed. Rivers formed deltas, and luxuriant vegetation developed on coastal swamps. This vegetation was later drowned and buried under mud and sand, and subsequently became coal.

carbonize /ˈkaːbəˌnaɪz/ v.tr. (also **-ise**) **1** convert into carbon by heating. **2** reduce to charcoal or coke. **3** coat with carbon. □ **carbonization** /ˌkaːbənaɪˈzeɪʃ(ə)n/ n.

carbonyl /ˈkaːbəˌnaɪl, -nɪl/ n. (used *attrib.*) *Chem.* the divalent radical :CO. □ **carbonyl chloride** = PHOSGENE.

carborundum /ˌkaːbəˈrʌndəm/ n. a compound of carbon and silicon used esp. as an abrasive. [CARBON + CORUNDUM]

carboxyl /kaːˈbɒksaɪl, -sɪl/ n. *Chem.* the univalent acid radical (–COOH), present in most organic acids. [CARBON + OXYGEN + -YL]

carboxylic acid /ˌkaːbɒkˈsɪlɪk/ n. *Chem.* an organic acid containing the carboxyl group. □ **carboxylate** /kaːˈbɒksɪˌleɪt/ n.

carboy /ˈkaːbɔɪ/ n. a large globular glass bottle usu. protected by a frame, for containing liquids. [Pers. *ḳarāba* large glass flagon]

carbuncle /ˈkaːbʌŋk(ə)l/ n. **1** a severe abscess in the skin. **2** a bright red gem. □ **carbuncular** /kaːˈbʌŋkjʊlə(r)/ adj. [ME f. OF *charbucle* etc. f. L *carbunculus* small coal f. *carbo* coal]

carburation /ˌkaːbjʊˈreɪʃ(ə)n/ n. the process of charging air with a spray of liquid hydrocarbon fuel, esp. in an internal-combustion engine. [as CARBURET]

carburet /ˌkaːbjʊˈret/ v.tr. (**carburetted**, **carburetting**; US **carbureted**, **carbureting**) combine (a gas etc.) with carbon. [earlier *carbure* f. F f. L *carbo* (as CARBON)]

carburettor /ˌkaːbjʊˈretə(r), ˌkaːbəˈretə(r)/ n. (also **carburetter**, US **carburetor** /ˈkaːbjʊˌretə(r)/) an apparatus for carburation of petrol and air in an internal-combustion engine. [as CARBURET + -OR[1]]

carburize /ˈkaːbjʊˌraɪz/ v.tr. (also **-ise**) add carbon to (iron). □ **carburization** /ˌkaːbjʊraɪˈzeɪʃ(ə)n/ n. [f. CARBURET]

carby var. of CARBIE.

carcajou /ˈkaːkəˌdʒuː, -ˌʒuː/ n. N. Amer. = WOLVERINE. [F, app. Algonquian (cf. KINKAJOU)]

carcass /ˈkaːkəs/ n. (also **carcase**) **1** the dead body of an animal, esp. a trunk for cutting up as meat. **2** the bones of a cooked bird. **3** *derog.* the human body, living or dead. **4** the skeleton or framework of a building, ship, etc. **5** worthless remains. □ **carcass meat** raw meat as distinct from canned or tinned meat. [ME f. AF *carcois* (OF *charcois*) & f. F *carcasse*: ult. orig. unkn.]

Carcassonne /ˈkaːkəˌsɒn/ a walled city in SW France; pop. (1990) 45,000. It is noted for its medieval fortifications.

Carchemish /ˈkaːkɪmɪʃ/ an ancient city on the upper Euphrates, north-east of Aleppo. It was a Hittite stronghold, annexed by Sargon II of Assyria in 717 BC.

carcinogen /kaːˈsɪnədʒən/ n. any substance that produces cancer. [as CARCINOMA + -GEN]

carcinogenesis /ˌkaːsɪnəˈdʒenɪsɪs/ n. the production of cancer.

carcinogenic /ˌkaːsɪnəˈdʒenɪk/ adj. tending to cause cancer. □ **carcinogenicity** /-dʒəˈnɪsɪtɪ/ n.

carcinoma /ˌkaːsɪˈnəʊmə/ n. (pl. **carcinomas** or **carcinomata** /kaːsɪˈnəʊmətə/) *Med.* a cancer, esp. one arising in epithelial tissue. □ **carcinomatous** adj. [L f. Gk *karkinōma* f. *karkinos* crab]

Card. *abbr.* Cardinal.

card[1] /kɑːd/ n. & v. ● n. **1** thick stiff paper or thin pasteboard. **2 a** a flat piece of this, esp. for writing or printing on. **b** = POSTCARD. **c** a card used to send greetings, issue an invitation, etc. (*birthday card*). **d** = visiting-card. **e** = business card. **f** a ticket of admission or membership. **3 a** = PLAYING CARD. **b** a similar card in a set designed for particular games, e.g. happy families. **c** (in *pl.*) card-playing; a card-game. **4** (in *pl.*) *colloq.* an employee's documents, esp. for tax and national insurance, held by the employer. **5 a** a programme of events at a race-meeting etc. **b** Cricket a score-card. **c** a list of holes on a golf course, on which a player's scores are entered. **6** *colloq.* a person, esp. an odd or amusing one (*what a card!; a knowing card*). **7** a plan or expedient (*sure card*). **8** a printed or written notice, set of rules, etc., for display. **9** a small rectangular piece of plastic issued by a bank, building society, etc., with personal (often machine-readable) data on it, used chiefly to obtain cash or credit (*cheque card; credit card; do you have a card?*). **10** *Electr.* a printed circuit board. ● *v.tr.* **1** fix to a card. **2** write on a card, esp. for indexing. □ **ask for** (or **get**) **one's cards** ask (or be told) to leave one's employment. **card-carrying** being a registered member of an organization, esp. a political party or trade union. **card-game** a game in which playing cards are used. **card index** an index in which each item is entered on a separate card. **card-index** *v.tr.* make a card index of. **card-playing** the playing of card-games. **card-sharp** (or **-sharper**) a swindler at card-games. **card-table** a table for card-playing, esp. a folding one. **card up one's sleeve** a plan in reserve; a hidden advantage. **card vote** a block vote, esp. in trade-union meetings. **on** (*N. Amer.* **in**) **the cards** possible or likely. **put** (or **lay**) **one's cards on the table** reveal one's resources, intentions, etc. [ME f. OF *carte* f. L *charta* f. Gk *khartēs* papyrus-leaf]

card[2] /kɑːd/ n. & v. ● n. a toothed instrument, wire brush, etc., for raising a nap on cloth or for disentangling fibres before spinning. ● *v.tr.* brush, comb, cleanse, or scratch with a card. □ **carding-wool** short-stapled wool. □ **carder** n. [ME f. OF *carde* f. Prov. *carda* f. *cardar* tease, comb, ult. f. L *carere* card]

cardamom /ˈkɑːdəməm/ n. (also **cardamum**) **1** an aromatic SE Asian plant, *Elettaria cardamomum*. **2** the seed-capsules of this used as a spice. [L *cardamomum* or F *cardamome* f. Gk *kardamōmon* f. *kardamon* cress + *amōmon* a spice plant]

Cardamom Mountains a range of mountains in western Cambodia, rising to a height of 1,813 m (5,886 ft) at its highest point.

cardan joint /ˈkɑːd(ə)n/ n. Engin. a universal joint. [G. *Cardano*, It. mathematician (1501–76)]

cardan shaft /ˈkɑːd(ə)n/ n. Engin. a shaft with a universal joint at one or both ends. [as CARDAN JOINT]

cardboard /ˈkɑːdbɔːd/ n. & adj. ● n. pasteboard or stiff paper, esp. for making cards or boxes. ● adj. **1** made of cardboard. **2** flimsy, insubstantial. □ **cardboard city** a place where homeless people gather at night using cardboard boxes etc. for shelter.

cardholder /ˈkɑːdˌhəʊldə(r)/ n. the holder of a membership card, credit card, etc.

cardiac /ˈkɑːdɪˌæk/ adj. & n. ● adj. **1** of or relating to the heart. **2** of or relating to the part of the stomach nearest the oesophagus. ● n. colloq. a person with heart disease. [F *cardiaque* or L *cardiacus* f. Gk *kardiakos* f. *kardia* heart]

cardie var. of CARDY.

Cardiff /ˈkɑːdɪf/ the capital of Wales and county town of South Glamorgan, a seaport on the Bristol Channel; pop. (1991) 272,600.

cardigan /ˈkɑːdɪgən/ n. a knitted jacket fastening down the front, usu. with long sleeves. The cardigan is named after the 7th Earl of Cardigan (James Thomas Brudenell, 1797–1868), who led the disastrous Charge of the Light Brigade in the Crimean War.

Cardiganshire /ˈkɑːdɪgənˌʃɪə(r)/ a former county of SW Wales. It became part of Dyfed in 1974.

Cardin /ˈkɑːdæn/, Pierre (b.1922), French couturier. He was the first designer in the field of *haute couture* to show a collection of clothes for men as well as women (1960). He is also noted for his ready-to-wear clothes and accessories.

cardinal /ˈkɑːdɪn(ə)l/ n. & adj. ● n. **1** (as a title **Cardinal**) a member of the Sacred College of the Roman Catholic Church. Cardinals hold the highest rank next to the pope, who is chosen from their number. They characteristically wear a red cassock and hat. **2** an American songbird of the subfamily Cardinalinae, related to the buntings; esp. the common *Cardinalis cardinalis*, the male of which has scarlet plumage. **3** *hist.* a woman's cloak, orig. of scarlet cloth with a hood. ● adj. **1** chief, fundamental; on which something hinges. **2** of deep scarlet (like a

cardinal's cassock). □ **cardinal-flower** the scarlet lobelia. **cardinal humour** see HUMOUR n. 5. **cardinal numbers** those denoting quantity (one, two, three, etc.), as opposed to ordinal numbers (first, second, third, etc.). **cardinal points** the four main points of the compass (north, south, east, west). **cardinal virtues** the chief moral attributes (esp. justice, prudence, temperance, and fortitude). □ **cardinalate** /-ˌleɪt/ n. (in sense 1 of n.). **cardinally** adv. **cardinalship** n. (in sense 1 of n.). [ME f. OF f. L *cardinalis* f. *cardo* -*inis* hinge: in Engl. first applied to the four virtues on which conduct 'hinges']

cardio- /ˈkɑːdɪəʊ/ comb. form heart (*cardiogram; cardiology*). [Gk *kardia* heart]

cardiogram /ˈkɑːdɪəʊˌgræm/ n. a record of muscle activity within the heart, made by a cardiograph.

cardiograph /ˈkɑːdɪəʊˌgrɑːf/ n. **1** an instrument for recording heart muscle activity. **2** an electrocardiograph. □ **cardiographer** /ˌkɑːdɪˈɒgrəfə(r)/ n. **cardiography** n.

cardioid /ˈkɑːdɪˌɔɪd/ n. & adj. ● n. **1** Math. a heart-shaped curve traced by a point on the circumference of a circle as it rolls around another identical circle. **2** a directional microphone with a pattern of sensitivity of this shape. ● adj. **1** of the shape of a cardioid. **2** (of a microphone) having a cardioid pattern of sensitivity. [Gk *kardioeidēs* heart-shaped, f. *kardia* heart]

cardiology /ˌkɑːdɪˈɒlədʒɪ/ n. the branch of medicine concerned with diseases and abnormalities of the heart. □ **cardiologist** n.

cardiomyopathy /ˌkɑːdɪəʊmaɪˈɒpəθɪ/ n. Med. chronic disease of the heart muscle.

cardiopulmonary /ˌkɑːdɪəʊˈpʌlmənərɪ/ adj. Med. of or relating to the heart and the lungs.

cardiovascular /ˌkɑːdɪəʊˈvæskjʊlə(r)/ adj. of or relating to the heart and blood vessels.

cardoon /kɑːˈduːn/ n. a thistle-like plant, *Cynara cardunculus*, allied to the globe artichoke, with leaves used as a vegetable. [F *cardon* ult. f. L *cardu(u)s* thistle]

cardphone /ˈkɑːdfəʊn/ n. a public telephone operated by the insertion of a prepaid plastic machine-readable card instead of money.

cardy /ˈkɑːdɪ/ n. (also **cardie**) (pl. **-ies**) colloq. a cardigan. [abbr.]

care /keə(r)/ n. & v. ● n. **1** worry, anxiety. **2** an occasion for this. **3** serious attention; heed, caution, pains (*assembled with care; handle with care*). **4 a** protection, charge. **b** Brit. = child care. **5** a thing to be done or seen to. ● *v.intr.* **1** (usu. foll. by about, for, whether, if, etc.) feel concern, liking, regard, or interest. **2** (foll. by for, about, and usu. with neg. expressed or implied) have a preference or liking for (an activity etc.) (*don't care for jazz*). **3** (foll. by for or to + infin.) wish or be willing (*would you care to try one?; would you care for a cup of coffee?*). □ **care for** provide for; look after. **care-label** a label attached to clothing, with instructions for washing etc. **care of** at the address of (*sent it care of his sister*). **have a care** take care; be careful. **I** (etc.) **couldn't** (US **could**) **care less** colloq. an expression of complete indifference. **in care** Brit. (of a child) taken into the care of a local authority. **take care 1** be careful. **2** (foll. by to + infin.) not fail or neglect. **take care of 1** look after; keep safe. **2** deal with. **3** dispose of. [OE *caru, carian,* f. Gmc]

careen /kəˈriːn/ v. **1** tr. turn (a ship) on one side for cleaning, caulking, or repair. **2 a** intr. tilt; lean over. **b** tr. cause to do this. **3** intr. N. Amer. swerve about; career. ¶ Sense 3 is infl. by career (v.). □ **careenage** n. [earlier as noun, = careened position of ship, f. F *carène* f. It. *carena* f. L *carina* keel]

career /kəˈrɪə(r)/ n. & v. ● n. **1 a** one's advancement through life, esp. in a profession. **b** the progress through history of a group or institution. **2** a profession or occupation, esp. as offering advancement. **3** (attrib.) **a** pursuing or wishing to pursue a career (*career woman*). **b** working permanently in a specified profession (*career diplomat*). **4** swift course; impetus (*in full career*). ● *v.intr.* **1** move or swerve about wildly. **2** go swiftly. □ **career structure** a recognized pattern of advancement within a profession. [F *carrière* f. It. *carriera* ult. f. L *carrus* CAR]

careerist /kəˈrɪərɪst/ n. a person predominantly or overly concerned with personal advancement in a career. □ **careerism** n.

carefree /ˈkeəfriː/ adj. free from anxiety or responsibility; light-hearted. □ **carefreeness** n.

careful /ˈkeəfʊl/ adj. **1** painstaking, thorough. **2** cautious. **3** done with care and attention. **4** (usu. foll. by that + clause, or to + infin.) taking care; not neglecting. **5** (foll. by *for, of*) concerned for; taking care of. □ **carefully** *adv.* **carefulness** *n.* [OE *carful* (as CARE, -FUL)]

careless /'keəlɪs/ *adj.* **1** not taking care or paying attention. **2** unthinking, insensitive. **3** done without care; inaccurate. **4** lighthearted. **5** (foll. by *of*) not concerned about; taking no heed of. **6** effortless; casual. □ **carelessly** *adv.* **carelessness** *n.* [OE *carlēas* (as CARE, -LESS)]

carer /'keərə(r)/ *n.* a person who cares for a sick or elderly person, esp. a relative at home.

caress /kə'res/ *v. & n.* ● *v.tr.* **1** touch or stroke gently or lovingly; kiss. **2** treat fondly or kindly. ● *n.* a loving or gentle touch or kiss. □ **caressingly** *adv.* [F *caresse* (n.), *caresser* (v.), f. It. *carezza* ult. f. L *carus* dear]

caret /'kærət/ *n.* a mark (^, ⁁) indicating a proposed insertion in printing or writing. [L, = is lacking]

caretaker /'keə.teɪkə(r)/ *n.* **1** a person employed to look after something, esp. a house in the owner's absence, or *Brit.* a public building. **2** (*attrib.*) exercising temporary authority (*caretaker government*).

careworn /'keəwɔːn/ *adj.* showing the effects of prolonged worry.

Carey /'keərɪ/, George (Leonard) (b.1935), English Anglican churchman, Archbishop of Canterbury since 1991. He was formerly a theology lecturer and comes from a broadly evangelical background. The controversial introduction of women priests into the Church of England was finally approved under his leadership.

carfare /'kɑːfeə(r)/ *n. N. Amer.* a passenger's fare for travel by public transport.

cargo /'kɑːɡəʊ/ *n.* (*pl.* **-oes** or **-os**) **1 a** goods carried on a ship or aircraft. **b** a load of such goods. **2** *US* **a** goods carried in a motor vehicle. **b** a load of such goods. □ **cargo cult** (orig. in the Pacific Islands) a belief in the forthcoming arrival of ancestral spirits bringing cargoes of food and other goods. [Sp. *cargo* = OF *charge* f. LL *car(ri)care*: see CHARGE]

carhop /'kɑːhɒp/ *n. US colloq.* a waiter at a drive-in restaurant.

Caria /'keərɪə/ an ancient region of SW Asia Minor, south of the Maeander river and north-west of Lycia. □ **Carian** *adj. & n.*

cariama var. of SERIEMA.

Carib /'kærɪb/ *n. & adj.* ● *n.* **1** a member of the pre-Columbian native peoples of the Lesser Antilles and parts of the neighbouring South American coast or their descendants. (*See note below.*) **2** the language of the Caribs. ● *adj.* of or relating to the Caribs or their language.

▪ The Caribs forced the Arawak peoples out of the Antilles to South America and were still expanding at the time of the Spanish conquest, but they were all but wiped out by the Spaniards, who alleged that they were savage cannibals (the word is derived from their name). Today in the West Indies only a few hundred remain, on the island of Dominica. In mainland South America Carib-speaking groups occupy territory in the north-eastern and Amazon regions, living in small autonomous communities.

Caribbean /.kærɪ'bɪən, kə'rɪbɪən/ *n. & adj.* ● *n.* the Caribbean Sea. ● *adj.* **1** of or relating to this sea or region. **2** of the Caribs or their language or culture.

Caribbean Community and Common Market (abbr. **CARICOM**) an organization established in 1973 to promote cooperation in economic affairs and social services etc. and to coordinate foreign policy among its members, all of which are independent states of the Caribbean region. The headquarters are at Georgetown, Guyana.

Caribbean Sea the part of the Atlantic Ocean lying between the Antilles and the mainland of Central and South America.

caribou /'kærɪ.buː/ *n.* (*pl.* same) *N. Amer.* a reindeer. [Canad. F f. Micmac, lit. 'snow-shoveller']

caricature /'kærɪkə.tjʊə(r)/ *n. & v.* ● *n.* **1** a grotesque usu. comic representation of a person by exaggeration of characteristic traits, in a picture, writing, or mime. **2** a ridiculously poor or absurd imitation or version. ● *v.tr.* make or give a caricature of. □ **caricatural** *adj.* **caricaturist** *n.* [F f. It. *caricatura* f. *caricare* load, exaggerate: see CHARGE]

CARICOM /'kærɪ.kɒm/ see CARIBBEAN COMMUNITY AND COMMON MARKET.

caries /'keərɪːz, -rɪ.iːz/ *n.* (*pl.* same) decay and crumbling of a tooth or bone. [L]

carillon /kə'rɪljən, 'kærɪljən/ *n.* **1** a set of bells sounded either from a keyboard or mechanically. **2** a tune played on bells. **3** an organ-stop imitating a peal of bells. [F f. OF *quarregnon* peal of four bells, alt. of Rmc *quaternio* f. L *quattuor* four]

carina /kə'riːnə/ *n. Biol.* a keel-shaped structure, esp. the ridge of a bird's breastbone. □ **carinal** *adj.* [L, = keel]

carinate /'kærɪ.neɪt/ *adj.* (of a bird) having a keeled breastbone (opp. RATITE). [L *carinatus* keeled f. *carina* keel]

caring /'keərɪŋ/ *adj.* **1** compassionate. **2** involving the care of the sick, elderly, or disabled.

Carinthia /kə'rɪnθɪə/ (German **Kärnten** /'kɛrnt(ə)n/) an Alpine state of southern Austria; capital, Klagenfurt. □ **Carinthian** *n. & adj.*

carioca /.kærɪ'ɒkə/ *n.* **1 a** a Brazilian dance resembling the samba. **b** the music for this. **2** a native of Rio de Janeiro. [Port.]

cariogenic /.keərɪəʊ'dʒenɪk/ *adj.* causing caries.

carious /'keərɪəs/ *adj.* (of bones or teeth) decayed. [L *cariosus*]

carjacking /'kɑː.dʒækɪŋ/ *n.* the hijacking of a car. □ **carjack** *v.tr.* **carjacker** *n.*

carking /'kɑːkɪŋ/ *adj. archaic* burdensome (*carking care*). [part. of obs. *cark* (v.) f. ONF *carkier* f. Rmc, rel. to CHARGE]

carl /kɑːl/ *n. Sc.* a man; a fellow. [OE f. ON *karl*, rel. to CHURL]

carline /'kɑːlɪn/ *n.* a thistle-like composite plant of the genus *Carlina*, esp. *C. vulgaris*. [F f. med.L *carlina* perh. for *cardina* (L *carduus* thistle), assoc. with *Carolus Magnus* Charlemagne]

Carlisle /kɑː'laɪl/ a town in NW England, the county town of Cumbria; pop. (1991) 99,800. Founded as a Roman settlement close to Hadrian's Wall, Carlisle served for centuries as a fortress town in defence of the north-western borders of England against the Scots.

carload /'kɑːləʊd/ *n.* **1** a quantity that can be carried in a car. **2** *US* the minimum quantity of goods for which a lower rate is charged for transport.

Carlovingian see CAROLINGIAN.

Carlow /'kɑːləʊ/ **1** a county of the Republic of Ireland, in the southeast, in the province of Leinster. **2** its county town; pop. (1991) 11,275.

Carlyle /kɑː'laɪl/, Thomas (1795–1881), Scottish historian and political philosopher. He worked as a teacher before starting to write articles for the *Edinburgh Encyclopedia* and critical works on German literature in the 1820s. His first major philosophical work was *Sartor Resartus* (1833–4), which dealt with social values and is written in a mannered prose style. He established his reputation as a historian with his history of the French Revolution (1837), the first volume of which he was forced to rewrite after the manuscript was accidentally burnt. Many of his works on history, politics, and social philosophy advocate a benevolent autocracy rather than democracy; *On Heroes, Hero Worship, and the Heroic in History* (1841) was the starting-point for his development of these ideas. Carlyle's influence on the development of social and political ideas in Britain during the 19th century was considerable.

carman /'kɑːmæn/ *n. US* **1** the driver of a van. **2** a carrier.

Carmarthen /kə'mɑːð(ə)n/ a town in SW Wales, the administrative centre of Dyfed; pop. (1991) 54,800.

Carmarthenshire /kə'mɑːð(ə)n.ʃɪə(r)/ a former county of South Wales. It became part of Dyfed in 1974.

Carmel, Mount /'kɑːm(ə)l/ a group of mountains near the Mediterranean coast in NW Israel, sheltering the port of Haifa. Caves on the south-western slopes have provided evidence of human occupation dating from the palaeolithic to the mesolithic periods. The Carmelite order was founded on Mount Carmel during the Crusades.

Carmelite /'kɑːmɪ.laɪt/ *n.* a friar or nun of the Christian order of Our Lady of Mount Carmel. (*See note below.*) ● *adj.* of or relating to the Carmelites. [F *Carmelite* or med.L *carmelita* f. CARMEL, MOUNT]

▪ The male order of Carmelites was founded at Mount Carmel in Palestine by St Berthold *c.*1154; members are also called 'White Friars' from the colour of their cloaks. The corresponding order of nuns was established in 1452. Towards the end of the 16th century, after discipline in the order had become relaxed, the 'discalced' reform movement was instituted by St Teresa of Ávila and St John of the Cross, leading to the formation of a 'barefoot' order of Carmelite nuns and friars (in reality they wore sandals rather than shoes). The Carmelite order is contemplative and has produced great mystics.

Carmichael /kɑː'maɪk(ə)l/, Hoagy (born Howard Hoagland Carmichael) (1899–1981), American jazz pianist, composer, and singer. His best-known songs include 'Stardust' (1929), 'Two Sleepy People' (1938), and 'In the Cool, Cool, Cool of the Evening' (1951).

carminative /'kɑːmɪnətɪv/ *adj. & n.* ● *adj.* relieving flatulence. ● *n.* a carminative drug. [F *carminatif -ive* or med.L *carminare* heal (by incantation): see CHARM]

carmine /'kɑːmaɪn/ *adj. & n.* ● *adj.* of a vivid crimson colour. ● *n.* **1** this colour. **2** a vivid crimson pigment made from cochineal. [F *carmin* or med.L *carminium* perh. f. *carmesinum* crimson + *minium* cinnabar]

Carnaby Street /'kɑːnəbɪ/ a street in the West End of London. It became famous in the 1960s as a centre of the popular fashion industry.

Carnac /'kɑːnæk/ the site in NW France, in Brittany near the Atlantic coast, of a group of stone monuments dating from the neolithic period. There are nearly 3,000 stones, which include single standing stones (menhirs), dolmens, and long avenues of grey monoliths arranged in order of height so that they decrease steadily from about 3–3.7 metres (10–12 ft) to 1–1.2 metres (3–4 ft); some of these avenues end in semicircular or rectangular enclosures of standing stones.

carnage /'kɑːnɪdʒ/ *n.* great slaughter, esp. of human beings in battle. [F f. It. *carnaggio* f. med.L *carnaticum* f. L *caro carnis* flesh]

carnal /'kɑːn(ə)l/ *adj.* **1** of the body or flesh; worldly. **2** sensual, sexual. □ **carnal knowledge** esp. *Law* sexual intercourse. □ **carnalize** *v.tr.* (also **-ise**) **carnally** *adv.* **carnality** /kɑː'nælɪtɪ/ *n.* [ME f. LL *carnalis* f. *caro carnis* flesh]

Carnap /'kɑːnæp/, Rudolf (1891–1970), German-born American philosopher. One of the originators of logical positivism, he was a founder and the most influential member of the Vienna Circle, noted for his contributions to logic, the analysis of language, the theory of probability, and the philosophy of science. His emphasis on scientific method in philosophy and the need to verify statements through observation marked a turning-point in philosophical enquiry and the rejection of traditional metaphysics. His major works include *The Logical Structure of the World* (1928) and *The Logical Foundations of Probability* (1950).

carnassial /kɑː'næsɪəl/ *adj. & n.* ● *adj.* (of a carnivore's upper premolar and lower molar teeth) adapted for shearing flesh. ● *n.* such a tooth. Also called *sectorial*. [F *carnassier* carnivorous]

carnation[1] /kɑː'neɪʃ(ə)n/ *n.* **1** a cultivated variety of the clove pink, with variously coloured showy flowers (see CLOVE[2]). **2** this flower. [orig. uncert.: in 16th c. varying with *coronation*]

carnation[2] /kɑː'neɪʃ(ə)n/ *n. & adj.* ● *n.* a rosy pink colour. ● *adj.* of this colour. [F f. It. *carnagione* ult. f. L *caro carnis* flesh]

carnauba /kɑː'naʊbə, -'nɔːbə, -'naʊbə/ *n.* **1** a fan palm, *Copernicia cerifera*, native to NE Brazil. Also called *wax palm*. **2** (in full **carnauba wax**) the yellowish leaf-wax of this tree used as a polish etc. [Port.]

Carné /kɑː'neɪ/, Marcel (b.1909), French film director. He held a dominant position among film-makers of the 1930s and 1940s, gaining his reputation in particular for the films which he made with the poet and scriptwriter Jacques Prévert (1900–77). Characterized by a fatalistic outlook and a masterly evocation of atmosphere, they include *Quai des brumes* (1938), *Le Jour se lève* (1939), and *Les Enfants du paradis* (1945).

Carnegie /kɑː'neɪgɪ, 'kɑːnəgɪ/, Andrew (1835–1919), Scottish-born American industrialist and philanthropist. He built up a considerable fortune in the steel industry in the US, then retired from business in 1901 and devoted his wealth to charitable purposes on both sides of the Atlantic, supporting many educational institutions, libraries, and the arts. One of his most notable achievements was the creation of the Carnegie Peace Fund to promote international peace.

carnelian var. of CORNELIAN.

carnet /'kɑːneɪ/ *n.* **1** a customs permit to take a motor vehicle across a frontier for a limited period. **2** a permit allowing use of a camp-site. [F, = notebook]

carnival /'kɑːnɪv(ə)l/ *n.* **1 a** a period of festivity, esp. occurring at a regular date; a festival, usu. involving a procession. **b** the festivities usual during the period before Lent in Roman Catholic countries. **2** merrymaking, revelry. **3** *N. Amer.* a travelling funfair or circus. [It. *carne-, carnovale* f. med.L *carnelevarium* etc. Shrovetide f. L *caro carnis* flesh + *levare* put away]

carnivore /'kɑːnɪˌvɔː(r)/ *n.* **1 a** *Zool.* a flesh-eating mammal of the order Carnivora. (*See note below.*) **b** any other flesh-eating animal. **2** a flesh-eating plant. [F, f. as CARNIVOROUS]

▪ The order Carnivora consists of predatory mammals with powerful jaws and teeth that are adapted for stabbing, tearing, and eating flesh. They have developed a great variety of hunting strategies, and there are a number of families comprising the dogs, bears, raccoons, weasels, civets, mongooses, hyenas, and cats.

carnivorous /kɑː'nɪvərəs/ *adj.* **1** (of an animal) feeding on flesh. **2** (of

a plant) digesting trapped insects or other animal substances. **3** of or relating to the order Carnivora. □ **carnivorously** *adv.* **carnivorousness** *n.* [L *carnivorus* f. *caro carnis* flesh + -VOROUS]

Carnot /'kɑːnəʊ/, Nicolas Léonard Sadi (1796–1832), French scientist. An army officer for most of his life, Carnot became interested in the principles of operation of steam engines, and analysed the efficiency of such engines using the notion of a cycle of reversible temperature and pressure changes of the gases. Carnot's work was recognized after his death as being of crucial importance to the theory of thermodynamics.

carnotite /'kɑːnəˌtaɪt/ *n. Mineral.* a lemon-yellow radioactive mineral containing uranium, vanadium, potassium, and some other elements. [M. A. *Carnot*, Fr. inspector of mines (1839–1920)]

carob /'kærəb/ *n.* **1** (in full **carob tree**) an evergreen leguminous tree, *Ceratonia siliqua*, native to the Mediterranean, bearing edible pods. **2** its bean-shaped edible seed pod, sometimes used as a substitute for chocolate. Also called *locust bean*. [obs. F *carobe* f. med.L *carrubia, -um* f. Arab. ḵarrūba]

carol /'kærəl/ *n. & v.* ● *n.* a joyous song, esp. a Christmas hymn. ● *v.* (**carolled, carolling**; *US* **caroled, caroling**) **1** *intr.* sing carols, esp. outdoors at Christmas. **2** *tr. & intr.* sing joyfully. □ **caroler** *n.* (also **caroller**). [ME f. OF *carole, caroler*, of unkn. orig.]

Caroline /'kærəˌlaɪn/ *adj.* **1** (also **Carolean** /ˌkærə'liən/) of the time of Charles I or II of England. **2** = CAROLINGIAN *adj.* 2. [L *Carolus* Charles]

Caroline Islands (also **Carolines**) a group of islands in the western Pacific Ocean, north of the equator, forming the Federated States of Micronesia.

Carolingian /ˌkærə'lɪndʒɪən/ *adj. & n.* (also **Carlovingian** /ˌkɑːlə'vɪndʒɪən/) ● *adj.* **1** of or relating to a Frankish dynasty that ruled in western Europe from 750 to 987, many of whose members were called Charles. Most notable was Charlemagne (= Charles the Great), whose father Pepin III (*c*.714–68) founded the dynasty. **2** of a style of script developed in France at the time of Charlemagne. ● *n.* **1** a member of the Carolingian dynasty. **2** the Carolingian style of script. [F *carlovingien* f. *Karl* Charles after *mérovingien* (see MEROVINGIAN): re-formed after L *Carolus*]

Carolingian Renaissance a period marked by achievements in art, architecture, learning, and music during the reign of the Frankish emperor Charlemagne and his successors. Credit for stimulating this renaissance is traditionally given to Charlemagne's adviser Alcuin.

carom /'kærəm/ *n. & v. N. Amer. Billiards* ● *n.* a cannon. ● *v.intr.* **1** make a carom. **2** (usu. foll. by *off*) strike and rebound. [abbr. of *carambole* f. Sp. *carambola*]

carotene /'kærəˌtiːn/ *n.* an orange or red plant pigment found in carrots, etc., acting as a source of vitamin A. [G *Carotin* f. L *carota* CARROT]

carotenoid /kə'rɒtɪˌnɔɪd/ *n.* any of a group of fat-soluble yellow, orange, red, or brown pigments giving characteristic colour to plant organs, e.g. ripe tomatoes, carrots, autumn leaves, etc.; esp. carotene.

Carothers /kə'rʌðəz/, Wallace Hume (1896–1937), American industrial chemist. He took up the study of long-chain molecules, now called polymers, and developed the first successful synthetic rubber, neoprene, and the first synthetic fibre able to be spun from a melt, Nylon 6.6. He committed suicide at the age of 41 before nylon had been commercially exploited.

carotid /kə'rɒtɪd/ *n. & adj.* ● *n.* each of the two main arteries carrying blood to the head and neck. ● *adj.* of or relating to either of these arteries. [F *carotide* or mod.L *carotides* f. Gk *karōtides* (pl.) f. *karoō* stupefy (compression of these arteries being thought to cause stupor)]

carouse /kə'raʊz/ *v. & n.* ● *v.intr.* **1** have a noisy or lively drinking-party. **2** drink heavily. ● *n.* a noisy or lively drinking-party. □ **carousal** *n.* **carouser** *n.* [orig. as *adv.* = right out, in phr. *drink carouse* f. G *gar aus trinken*]

carousel /ˌkærə'sel, -'zel/ *n.* (*US* **carrousel**) **1** *US* a merry-go-round or roundabout. **2** a rotating delivery or conveyor system, esp. for passengers' luggage at an airport. **3** *hist.* a kind of equestrian tournament. [F *carrousel* f. It. *carosello*]

carp[1] /kɑːp/ *n.* (pl. same) a freshwater fish of the family Cyprinidae; esp. *Cyprinus carpio*, often bred for use as food. [ME f. OF *carpe* f. Prov. or f. LL *carpa*]

carp[2] /kɑːp/ *v.intr.* (usu. foll. by *at*) find fault; complain pettily. □ **carper** *n.* [obs. ME senses 'talk, say, sing' f. ON *karpa* to brag: mod. sense (16th c.) from or infl. by L *carpere* pluck at, slander]

Carpaccio /kɑː'pætʃɪˌəʊ/, Vittore (*c*.1455–1525), Italian painter. He is noted especially for his paintings of Venice and for his lively narrative cycle of paintings *Scenes from the Life of St Ursula* (1490–5).

carpal /ˈkɑːp(ə)l/ *adj. & n.* ● *adj.* of or relating to the bones in the wrist. ● *n.* any of the bones forming the wrist. [CARPUS + -AL¹]

Carpathian Mountains /kɑːˈpeɪθɪən/ (also **Carpathians**) a mountain system extending south-eastwards from southern Poland and the Czech Republic into Romania.

carpel /ˈkɑːp(ə)l/ *n. Bot.* the female reproductive organ of a flower, consisting of a stigma, style, and ovary. □ **carpellary** *adj.* [F *carpelle* or mod.L *carpellum* f. Gk *karpos* fruit]

Carpentaria, Gulf of /ˌkɑːpənˈteərɪə/ a large bay on the north coast of Australia, between Arnhem Land and the Cape York Peninsula.

carpenter /ˈkɑːpɪntə(r)/ *n. & v.* ● *n.* a person skilled in woodwork, esp. of a structural kind (cf. JOINER). ● *v.* **1** *intr.* do carpentry. **2** *tr.* make by means of carpentry. **3** *tr.* (often foll. by *together*) construct; fit together. □ **carpenter ant** a large ant of the genus *Camponotus*, that bores into wood to nest. **carpenter bee** a solitary bee that bores into wood. [ME & AF; OF *carpentier* f. LL *carpentarius* f. *carpentum* wagon f. Gaulish]

carpentry /ˈkɑːpɪntrɪ/ *n.* **1** the work or occupation of a carpenter. **2** woodwork made by a carpenter. [ME f. OF *carpenterie* f. L *carpentaria*: see CARPENTER]

carpet /ˈkɑːpɪt/ *n. & v.* ● *n.* **1 a** a thick fabric for covering a floor or stairs. **b** a piece of this fabric. **c** a rug, esp. a large oriental rug. **2** an expanse or layer resembling a carpet in being smooth, soft, bright, or thick (*carpet of snow*). ● *v.tr.* (**carpeted, carpeting**) **1** cover with or as with a carpet. **2** *colloq.* reprimand, reprove. □ **carpet-bag** a travelling-bag of a kind orig. made of carpet-like material. **carpet-bagger 1** esp. *US* a political candidate in an area where the candidate has no local connections (orig. a northerner in the southern US after the Civil War). **2** an unscrupulous opportunist. **carpet beetle** a small beetle of the genus *Anthrenus*, whose larvae are destructive to carpets and other textiles. **carpet bombing** intensive bombing. **carpet slipper** a kind of slipper with the upper made orig. of carpet-like material. **carpet-sweeper** a household implement with a revolving brush or brushes for sweeping carpets. **on the carpet 1** *colloq.* being reprimanded. **2** under consideration. **sweep under the carpet** conceal (a problem or difficulty) in the hope that it will be forgotten. [ME f. OF *carpite* or med.L *carpita*, f. obs. It. *carpita* woollen counterpane, ult. f. L *carpere* pluck, pull to pieces]

carpeting /ˈkɑːpɪtɪŋ/ *n.* **1** material for carpets. **2** carpets collectively.

carpology /kɑːˈpɒlədʒɪ/ *n.* the study of the structure of fruit and seeds. [Gk *karpos* fruit]

carport /ˈkɑːpɔːt/ *n.* a shelter with a roof and open sides for a car, usu. beside a house.

carpus /ˈkɑːpəs/ *n.* (*pl.* **carpi** /-paɪ/) the small bones between the forelimb and metacarpus in terrestrial vertebrates, forming the wrist in humans. [mod.L f. Gk *karpos* wrist]

Carr /kɑː(r)/, Emily (1871–1945), Canadian painter and writer. Her paintings, inspired by the wilderness of British Columbia, often drew on the motifs of American Indian folk art. From 1927 she came into contact with the Group of Seven and produced such expressionist works as *Forest Landscape II* and *Sky* (both 1934–5).

carr /kɑː(r)/ *n.* **1** a marsh or fen, esp. overgrown with shrubs or reclaimed as meadowland. **2** a marshy copse, esp. of alders. [ON *kjarr* brushwood]

Carracci /kəˈrɑːtʃɪ/ a family of Italian painters. Ludovico (1555–1619) is remembered chiefly as a distinguished teacher; with his cousins he established an academy at Bologna which was responsible for training many important painters. His cousin Annibale (1560–1609) is the most famous of the family, especially for his work in Rome, such as the ceiling of the Farnese Gallery (1597–1600). Taking the art of Raphael and Michelangelo as a starting-point, he developed a style which proved to be a foundation of the Italian baroque. He is also remembered for his invention of the caricature. Annibale's brother Agostino (1557–1602) was chiefly an engraver, but he also worked with his brother in the Farnese Gallery.

carrack /ˈkærək/ *n. hist.* a large merchant ship of northern and southern Europe in the 14th–17th centuries, the forerunner and first example of the larger three-masted ship which dominated naval architecture until the general introduction of steam propulsion in the mid-19th century. (See also GALLEON.) [ME f. F *caraque* f. Sp. *carraca* f. Arab. *karākir*]

carrageen /ˈkærəgiːn/ *n.* (also **carragheen**) an edible red seaweed, *Chondrus crispus*, of the northern hemisphere. See also *Irish moss*. [orig.

uncert.: perh. f. Ir. *cosáinín carraige* carrageen, lit. 'little stem of the rock']

carrageenan /ˌkærəˈgiːnən/ *n. Biochem.* a mixture of polysaccharides extracted from red and purple seaweeds, used as a thickening or emulsifying agent in food products. [f. CARRAGEEN]

Carrara /kəˈrɑːrə/ a town in Tuscany in NW Italy, famous for the white marble quarried there since Roman times; pop. (1990) 68,480.

Carrel /kəˈrel, ˈkærəl/, Alexis (1873–1944), French surgeon and biologist. He developed improved techniques for suturing arteries and veins, and carried out some of the first organ transplants. He also succeeded in keeping organs alive outside the body by perfusion, using a glass pump devised with the aid of Charles Lindbergh. Carrel, who spent much of his career in the US, received a Nobel Prize in 1912.

carrel /ˈkærəl/ *n.* **1** a small cubicle for a reader in a library. **2** *hist.* a small enclosure or study in a cloister. [OF *carole*, med.L *carola*, of unkn. orig.]

Carreras /kəˈreərəs/, José (b.1946), Spanish operatic tenor. Since his operatic début in his native city of Barcelona (1970), he has given many performances in opera houses worldwide. Noted for his soft voice in the upper register, he has had great success in the operas of Verdi, Puccini, and Donizetti; his repertoire also includes traditional Catalan songs and popular light music, such as the musical *West Side Story*.

carriage /ˈkærɪdʒ/ *n.* **1** *Brit.* a railway passenger vehicle. **2** a wheeled passenger vehicle, esp. one with four wheels and pulled by horses. **3 a** the conveying of goods. **b** the cost of this (*carriage paid*). **4** the part of a machine (e.g. a typewriter) that carries other parts into the required position. **5** a gun-carriage. **6** a manner of carrying oneself; one's bearing or deportment. □ **carriage and pair** a carriage with two horses pulling it. **carriage clock** a portable clock in a rectangular case with a handle on top. **carriage-dog** a Dalmatian dog. [ME f. ONF *cariage* f. *carier* CARRY]

carriageway /ˈkærɪdʒweɪ/ *n. Brit.* the part of a road intended for vehicles.

carrick bend /ˈkærɪk/ *n. Naut.* a kind of knot used to join ropes. [BEND²: *carrick* perh. f. CARRACK]

Carrick-on-Shannon /ˌkærɪkɒnˈʃænən/ the county town of Leitrim in the Republic of Ireland, on the River Shannon; pop. (1991) 6,168.

carrier /ˈkærɪə(r)/ *n.* **1** a person or thing that carries. **2** a person or company undertaking to convey goods or passengers for payment. **3** = *carrier bag*. **4** part of a bicycle etc. for carrying luggage or a passenger. **5** a person or animal that may transmit a disease or a hereditary characteristic without suffering from or displaying it. **6** = *aircraft-carrier*. **7** a substance used to support or convey a pigment, a catalyst, radioactive material, etc. **8** *Physics* a mobile electron or hole that carries a charge in a semiconductor. **9** *Biochem.* a molecule that transfers another molecule or ion, esp. across a cell membrane. □ **carrier bag** *Brit.* a disposable plastic or paper bag with handles. **carrier pigeon** a pigeon trained to carry messages tied to its neck or leg. **carrier wave** a high-frequency electromagnetic wave modulated in amplitude or frequency to convey a signal.

carriole /ˈkærɪəʊl/ *n.* **1** a Canadian sledge. **2** *hist.* a small open carriage for one. **3** *hist.* a covered light cart. [F f. It. *carriuola*, dimin. of *carro* CAR]

carrion /ˈkærɪən/ *n. & adj.* ● *n.* **1** dead putrefying flesh. **2** something vile or filthy. ● *adj.* rotten, loathsome. □ **carrion crow** a common Eurasian crow, *Corvus corone*, that has uniformly black plumage in its typical form. **carrion flower** = STAPELIA. [ME f. AF & ONF *caroine*, *-oigne*, OF *charoigne* ult. f. L *caro* flesh]

Carroll /ˈkærəl/, Lewis (pseudonym of Charles Lutwidge Dodgson) (1832–98), English writer. He worked as a mathematics lecturer at Oxford University 1855–81; after a boat trip with Alice Liddell, the daughter of the dean of his college, he was inspired to write *Alice's Adventures in Wonderland* (1865) and *Through the Looking Glass* (1871). Both books tell the story of a child's fantastic dream adventures; illustrated by John Tenniel, they became classics of children's literature. Carroll also wrote nonsense verse, notably *The Hunting of the Snark* (1876), and experimented in portrait photography.

carrot /ˈkærət/ *n.* **1 a** an umbelliferous plant, *Daucus carota*, with a tapering orange-coloured root. **b** this root as a vegetable. **2** a means of enticement or persuasion. **3** (in *pl.*) *sl.* a red-haired person. □ **carroty** *adj.* [F *carotte* f. L *carota* f. Gk *karōton*]

carrousel *US* var. of CAROUSEL.

carry /ˈkærɪ/ *v. & n.* ● *v.* (**-ies, -ied**) **1** *tr.* support or hold up, esp. while moving. **2** *tr.* convey with one from one place to another. **3** *tr.* have on one's person (*carry a watch*). **4** *tr.* conduct or transmit (*pipe carries water*;

wire carries electric current). **5** *tr.* take (a process etc.) to a specified point (*carry into effect*; *carry a joke too far*). **6** *tr.* (foll. by *to*) continue or prolong (*carry modesty to excess*). **7** *tr.* involve, imply; have as a feature or consequence (*carries a two-year guarantee*; *principles carry consequences*). **8** *tr.* (in reckoning) transfer (a figure) to a column of higher value. **9** *tr.* hold in a specified way (*carry oneself erect*). **10** *tr.* **a** (of a newspaper or magazine) publish; include in its contents, esp. regularly. **b** (of a radio or television station) broadcast, esp. regularly. **11** *tr.* (of a retailing outlet) keep a regular stock of (particular goods for sale) (*have stopped carrying that brand*). **12** *intr.* **a** (of sound, esp. a voice) be audible at a distance. **b** (of a missile) travel, penetrate. **13** *tr.* (of a gun etc.) propel to a specified distance. **14** *tr.* **a** win victory or acceptance for (a proposal etc.). **b** win acceptance from (*carried the audience with them*). **c** win, capture (a prize, a fortress, etc.). **d** *US* gain (a state or district) in an election. **e** *Golf* cause the ball to pass beyond (a bunker etc.). **15** *tr.* **a** endure the weight of; support (*columns carry the dome*). **b** be the chief cause of the effectiveness of; be the driving force in (*you carry the sales department*). **16** *tr.* be pregnant with (*is carrying twins*). **17** *tr.* **a** (of a motive, money, etc.) cause or enable (a person) to go to a specified place. **b** (of a journey) bring (a person) to a specified point. ● *n.* (*pl.* **-ies**) **1** an act of carrying. **2** *Golf* the distance a ball travels before reaching the ground. **3** a portage between rivers etc. **4** the range of a gun etc. □ **carry-all 1** a light carriage (cf. CARRIOLE). **2** *US* a car with seats placed sideways. **3** *N. Amer.* a large bag or case. **carry all before one** succeed spectacularly; overcome all opposition. **carry away 1** remove. **2** inspire; affect emotionally or spiritually. **3** deprive of self-control (*got carried away*). **4** *Naut.* **a** lose (a mast etc.) by breakage. **b** break off or away. **carry back** take (a person) back in thought to a past time. **carry one's bat** *Cricket* be not out at the end of a side's completed innings. **carry the can** *colloq.* bear the responsibility or blame. **carry conviction** be convincing. **carry-cot** a portable cot for a baby. **carry the day** be victorious or successful. **carry forward** transfer to a new page or account. **carrying-on** (or **carryings-on**) = *carry-on*. **carrying-trade** the conveying of goods from one country to another by water or air as a business. **carry it off** (or **carry it off well**) do well under difficulties. **carry off 1** take away, esp. by force. **2** win (a prize). **3** (esp. of a disease) kill. **4** render acceptable or passable. **carry on 1** continue (*carry on eating*; *carry on, don't mind me*). **2** engage in (a conversation or a business). **3** *colloq.* behave strangely or excitedly. **4** (often foll. by *with*) *colloq.* flirt or have a love affair. **5** advance (a process) by a stage. **carry-on** *n. Brit. sl.* **1** a state of excitement or fuss. **2** a questionable piece of behaviour. **3** a flirtation or love affair. **carry out** put (ideas, instructions, etc.) into practice. **carry-out** *attrib.adj. & n.* esp. *Sc. & US* = *take-away*. **carry over 1** = *carry forward*. **2** postpone (work etc.). **3** *Stock Exch.* keep over to the next settling-day. **carry-over** *n.* **1** something carried over. **2** *Stock Exch.* postponement to the next settling-day. **carry through 1** complete successfully. **2** bring safely out of difficulties. **carry weight** be influential or important. **carry with one** bear in mind. [ME f. AF & ONF *carier* (as CAR)]

carrying capacity *n.* **1** the quantity of people, things, etc., which can be conveyed by a vehicle, container, etc. **2** *Ecol.* the number of people, animals, or crops which a region can support without environmental degradation.

carse /kɑːs/ *n. Sc.* fertile lowland beside a river. [ME, perh. f. *carrs* swamps]

carsick /ˈkɑːsɪk/ *adj.* affected with nausea caused by the motion of a car. □ **carsickness** *n.*

Carson /ˈkɑːs(ə)n/, Rachel (Louise) (1907–64), American zoologist. Remembered as a pioneer ecologist and popularizer of science, she wrote *The Sea Around Us* (1951) and *Silent Spring* (1963), an attack on the indiscriminate use of pesticides and weed-killers.

Carson City the state capital of Nevada; pop. (1990) 40,440.

cart /kɑːt/ *n. & v.* ● *n.* **1** a strong vehicle with two or four wheels for carrying loads, usu. drawn by a horse. **2** a light vehicle for pulling by hand. **3** a light vehicle with two wheels for driving in, drawn by a single horse. **4** *US* (in full **shopping cart**) a supermarket trolley. ● *v.tr.* **1** convey in or as in a cart. **2** *sl.* carry (esp. a cumbersome thing) with difficulty or over a long distance (*carted it all the way home*). □ **cart-horse** a thickset horse suitable for heavy work. **cart-load 1** an amount filling a cart. **2** a large quantity of anything. **cart off** remove, esp. by force. **cart-track** a track or road too rough for ordinary vehicles. **cart-wright** a maker of carts. **in the cart** *sl.* in trouble or difficulty. **put the cart before the horse 1** reverse the proper order or procedure. **2** take an effect for a cause. □ **carter** *n.* **cartful** *n.* (*pl.* **-fuls**). [ME f. ON *kartr* cart (= OE *cræt*), prob. infl. by AF & ONF *carete* dimin. of *carre* CAR]

cartage /ˈkɑːtɪdʒ/ *n.* the price paid for carting.

Cartagena /ˌkɑːtəˈdʒiːnə/ **1** a port in SE Spain; pop. (1991) 172,150. Originally named Mastia, it was refounded as Carthago Nova (New Carthage) by Hasdrubal (son-in-law of Hamilcar) in *c.*225 BC, as a base for the Carthaginian conquest of Spain. It has a fine natural harbour and has been a naval port since the 16th century. **2** a port, resort, and oil-refining centre in NW Colombia, on the Caribbean Sea; pop. (1985) 529,600. Founded as a Spanish city in 1533, it was frequently attacked by British privateers, including Sir Francis Drake.

carte var. of QUART 4.

carte blanche /kɑːt ˈblɒnʃ/ *n.* full discretionary power given to a person. [F, = blank paper]

cartel /kɑːˈtel/ *n.* **1** an informal association of manufacturers or suppliers to maintain prices at a high level and control production, marketing arrangements, etc. **2** a political combination between parties. □ **cartelize** /ˈkɑːtəˌlaɪz/ *v.tr. & intr.* (also **-ise**). [G *Kartell* f. F *cartel* f. It. *cartello* dimin. of *carta* CARD[1]]

Carter[1] /ˈkɑːtə(r)/, Angela (1940–92), English novelist and short-story writer. Her fiction is characterized by fantasy, black humour, and eroticism, while her second novel *The Magic Toyshop* (1967) established her as a major exponent of magic realism. Later novels, such as *The Passion of New Eve* (1977), offer a strong feminist perspective on capitalism and Western society. Her exploration of the symbolic function of myth and folklore in the unconscious is reflected in short stories such as 'The Company of Wolves' (1979), which formed the basis for a film in 1984.

Carter[2] /ˈkɑːtə(r)/, Elliott (Cook) (b.1908), American composer. He is noted for his innovative approach to metre and eclectic choice of sources as diverse as modern jazz and Renaissance madrigals.

Carter[3] /ˈkɑːtə(r)/, James Earl ('Jimmy') (b.1924), American Democratic statesman, 39th President of the US 1977–81. A progressive and reformist governor of Georgia (1970–4), he was elected President on a manifesto of civil rights and economic reform. Although his administration was notable for achieving the Panama Canal Treaty (1977) and the Camp David agreements (1978), it was dogged in the last few years by Carter's inability to resolve the crisis caused by the seizure of American hostages in Iran.

Cartesian /kɑːˈtiːzɪən, -ˈtiːʒ(ə)n/ *adj. & n.* ● *adj.* of or relating to Descartes or his work in philosophy, science, and mathematics. ● *n.* a follower of Descartes. □ **Cartesian coordinates** a system for describing the position of a point by reference to its perpendicular distance from two or three axes intersecting at right angles (x, y, and z). **Cartesian diver** a toy device that rises and falls in liquid when the vessel containing it is subjected to varying pressure. □ **Cartesianism** *n.* [mod.L *Cartesianus* f. *Cartesius*, Latinized name of DESCARTES]

Carthage /ˈkɑːθɪdʒ/ an ancient city on the coast of North Africa near present-day Tunis, founded by the Phoenicians traditionally in 814 BC. Carthage became a major force in the Mediterranean, with interests throughout North Africa, Spain, and Sicily that brought it into conflict with Greece until the 3rd century BC and then with Rome in the Punic Wars; the Romans finally destroyed it in 146 BC.

Carthaginian /ˌkɑːθəˈdʒɪnɪən/ *adj. & n.* ● *adj.* of or relating to Carthage or its inhabitants. ● *n.* a native or inhabitant of Carthage.

Carthusian /kɑːˈθjuːzɪən/ *n.* ● a Christian monk or nun of a strictly contemplative order founded at Chartreuse in SE France by St Bruno in 1084, leading a hermitic way of life of great austerity and self-denial. ● *adj.* of or relating to the Carthusians. [med.L *Carthusianus* f. L *Cart(h)usia* Chartreuse]

Cartier /ˈkɑːtɪˌeɪ/, Jacques (1491–1557), French explorer. The first to establish France's claim to North America, he made three voyages to Canada between 1534 and 1541, sailing up the St Lawrence River as far as present-day Montreal and building a fort at Cap Rouge (a few miles upstream of what is now Quebec City).

Cartier-Bresson /ˌkɑːtɪeɪˈbresɒn/, Henri (b.1908), French photographer and film director. Intent on capturing the 'decisive moment' of a scene or event, he travelled widely, recording the lives of ordinary people without artificial composition and establishing a reputation as a humane and perceptive observer. His collections of photographs include *The Decisive Moment* (1952). Between 1936 and 1939 he worked as assistant to the film director Jean Renoir, also making his own documentary film about the Spanish Civil War, *Return to Life* (1937).

Cartier Islands see ASHMORE AND CARTIER ISLANDS.

cartilage /ˈkɑːtɪlɪdʒ/ n. **1** the firm flexible semi-opaque connective tissue which forms the infant skeleton and in adults is mainly replaced by bone except in the articulating surfaces of joints and in structures such as the trachea, larynx, etc. **2** a structure made of this. □ **cartilaginoid** /ˌkɑːtɪˈlædʒɪˌnɔɪd/ adj. [F f. L cartilago -ginis]

cartilaginous /ˌkɑːtɪˈlædʒɪnəs/ adj. consisting of or resembling cartilage. □ **cartilaginous fish** a fish of the class Selachii, with a skeleton of cartilage rather than bone, esp. a shark or a ray.

Cartland /ˈkɑːtlənd/, Dame (Mary) Barbara (Hamilton) (b.1901), English writer. A prolific author, she specializes in light romantic fiction; her popular romances include *Bride to a Brigand* (1983) and *A Secret Passage to Love* (1992).

cartogram /ˈkɑːtəˌgræm/ n. a map with diagrammatic statistical information. [F cartogramme f. carte map, card]

cartography /kɑːˈtɒgrəfɪ/ n. the science or practice of map-drawing. □ **cartographer** n. **cartographic** /ˌkɑːtəˈgræfɪk/ adj. **cartographical** adj. [F cartographie f. carte map, card]

cartomancy /ˈkɑːtəˌmænsɪ/ n. fortune-telling by interpreting a random selection of playing cards. [F cartomancie f. carte CARD[1]]

carton /ˈkɑːt(ə)n/ n. a light box or container, esp. one made of cardboard. [F (as CARTOON)]

cartoon /kɑːˈtuːn/ n. & v. ● n. **1** a humorous drawing in a newspaper, magazine, etc., esp. as a topical comment. **2** a sequence of drawings, often with speech indicated, telling a story (*strip cartoon*). **3** a filmed sequence of drawings using the technique of animation. **4** a full-size drawing on stout paper as an artist's preliminary design for a painting, tapestry, mosaic, etc. ● v. **1** tr. draw a cartoon of. **2** intr. draw cartoons. □ **cartoonish** adj. **cartoonist** n. **cartoony** adj. [It. cartone f. carta CARD[1]]

cartouche /kɑːˈtuːʃ/ n. **1 a** Archit. a scroll-like ornament, e.g. the volute of an Ionic capital. **b** a tablet imitating, or a drawing of, a scroll with rolled-up ends, used ornamentally or bearing an inscription. **c** an ornate frame. **2** Archaeol. an oval ring enclosing Egyptian hieroglyphs, usu. representing the name and title of a king. [F, = cartridge, f. It. cartoccio f. carta CARD[1]]

cartridge /ˈkɑːtrɪdʒ/ n. **1** a case containing a charge of propelling explosive for firearms or blasting, with a bullet or shot if intended for small arms. **2** a spool of film, magnetic tape, etc., in a sealed container ready for insertion. **3** a component carrying the stylus on the pick-up head of a record-player. **4** an ink-container for insertion in a pen or in a laser or ink-jet printer. □ **cartridge-belt** a belt with pockets or loops for cartridges (in sense 1). **cartridge paper** thick rough paper used for cartridges, for drawing, and for strong envelopes. [corrupt. of CARTOUCHE (but recorded earlier)]

cartwheel /ˈkɑːtwiːl/ n. **1** the (usu. spoked) wheel of a cart. **2** a circular sideways handspring with the arms and legs extended.

Cartwright /ˈkɑːtraɪt/, Edmund (1743-1823), English engineer, inventor of the power loom. Initially a clergyman, he became interested in textile machinery, and despite financial failures continued to innovate, developing machines for wool-combing and rope-making and an engine which used alcohol rather than steam. His achievements were recognized eventually in 1809 by the government, which voted him an award of £10,000.

caruncle /ˈkærəŋk(ə)l, kəˈrʌŋ-/ n. **1** Zool. a fleshy excrescence, e.g. a turkeycock's wattles or the red prominence at the inner angle of the eye. **2** Bot. an outgrowth from a seed near the micropyle. □ **caruncular** /kəˈrʌŋkjʊlə(r)/ adj. [obs. F f. L caruncula f. caro carnis flesh]

Caruso /kəˈruːsəʊ/, Enrico (1873-1921), Italian operatic tenor. His voice combined a brilliant upper register with a baritone-like warmth. He appeared in both French and Italian opera and had his greatest successes in operas by Verdi, Puccini, and Jules Massenet (1842-1912). The first major tenor to be recorded on gramophone records, he became a household name even among those who never attended operatic performances.

carve /kɑːv/ v. **1** tr. produce or shape (a statue, representation in relief, etc.) by cutting into a hard material (*carved a figure out of rock*; *carved it in wood*). **2** tr. **a** cut patterns, designs, letters, etc. in (hard material). **b** (foll. by *into*) form a pattern, design, etc., from (*carved it into a bust*). **c** (foll. by *with*) cover or decorate (material) with figures or designs cut in it. **3** tr. (also absol.) cut (meat etc.) into slices for eating. □ **carve out 1** take from a larger whole. **2** establish (a career etc.) purposefully (*carved out a name for themselves*). **carve up 1** divide into several pieces; subdivide (territory etc.). **2** drive aggressively into the path of (another

vehicle). **carve-up** n. sl. a sharing-out, esp. of spoils. **carving knife** a knife with a long blade, for carving meat. [OE ceorfan cut f. WG]

carvel /ˈkɑːv(ə)l/ n. var. of CARAVEL. □ **carvel-built** (of a boat) made with planks flush, not overlapping (cf. CLINKER-BUILT). [as CARAVEL]

carven /ˈkɑːv(ə)n/ archaic past part. of CARVE.

Carver /ˈkɑːvə(r)/ n. US a chair with arms, a rush seat, and a back having vertical and horizontal spindles. [John Carver (1576-1621), first governor of the Plymouth colony, for whom a prototype was allegedly made]

carver /ˈkɑːvə(r)/ n. **1** a person who carves. **2 a** a carving knife. **b** (in pl.) a knife and fork for carving. **3** Brit. the principal chair, with arms, in a set of dining-chairs, intended for the person who carves. ¶ To be distinguished (in sense 3) from CARVER.

carvery /ˈkɑːvərɪ/ n. (pl. **-ies**) a buffet or restaurant with joints of meat displayed, and carved as required, in front of customers.

carving /ˈkɑːvɪŋ/ n. a carved object, esp. as a work of art.

Cary /ˈkeərɪ/, (Arthur) Joyce (Lunel) (1888-1957), English novelist. A number of his novels are set in Africa, where he had served briefly in the colonial service. His major works constitute two trilogies; the first is concerned with art and includes *The Horse's Mouth* (1944), a memorable portrait of an outrageous artist, while the second deals with political life and includes *Not Honour More* (1955).

caryatid /ˌkærɪˈætɪd/ n. (pl. **caryatides** /-ˌdiːz/ or **caryatids**) Archit. a pillar in the form of a draped female figure, supporting an entablature. [F caryatide f. It. cariatide or L f. Gk karuatis -idos priestess at Caryae (Karuai) in Laconia]

caryopsis /ˌkærɪˈɒpsɪs/ n. (pl. **caryopses** /-ˈɒpsiːz/) Bot. a dry one-seeded indehiscent fruit, as in wheat and maize. [mod.L f. Gk karuon nut + opsis appearance]

Casablanca /ˌkæsəˈblæŋkə/ the largest city of Morocco, a seaport on the Atlantic coast; pop. (1982) 2,139,200. It was founded by the Portuguese in 1515.

Casals /kəˈsælz/, Pablo (or Pau) (1876-1973), Spanish cellist, conductor, and composer. The foremost cellist of his time, he was noted especially for his performances of Bach suites and the Dvořák Cello Concerto. He refused to perform in Hitler's Germany and went into voluntary exile in 1939 during the Franco regime, living first in the French Pyrenees and then in Puerto Rico. His compositions include the oratorio *The Manger* (1943-60).

Casanova /ˌkæsəˈnəʊvə/, Giovanni Jacopo (full surname Casanova de Seingalt) (1725-98), Italian adventurer. He is famous for his memoirs (first published in French 1828-38), describing his adventures in Europe and especially his sexual encounters.

casbah var. of KASBAH.

cascade /kæsˈkeɪd/ n. & v. ● n. **1** a small waterfall, esp. forming one in a series or part of a large broken waterfall. **2** a succession of electrical devices or stages in a process. **3** a quantity of material etc. draped in descending folds. **4** a process of disseminating information from senior to junior levels in an organization. ● v.intr. fall in or like a cascade. [F f. It. cascata f. cascare to fall ult. f. L casus: see CASE[1]]

Cascade Range a range of mountains in western North America, extending from southern British Columbia through Washington and Oregon to northern California. Its highest peak, Mount Rainier, rises to a height of 4,395 m (14,410 ft). The range also includes the active volcano Mount St Helens.

cascara /kæsˈkɑːrə/ n. (in full **cascara sagrada** /səˈgrɑːdə/) the bark of a Californian buckthorn, *Rhamnus purshiana*, used as a purgative. [Sp., = sacred bark]

case[1] /keɪs/ n. **1** an instance of something occurring. **2** a state of affairs, hypothetical or actual. **3 a** an instance of a person receiving professional guidance or treatment, e.g. from a doctor or social worker. **b** this person or the circumstances involved. **4** a matter under official investigation, esp. by the police. **5** Law **a** a cause or suit for trial. **b** a statement of the facts in a cause *sub judice*, drawn up for a higher court's consideration (*judge states a case*). **c** a cause that has been decided and may be cited (*leading case*). **6 a** the sum of the arguments on one side, esp. in a lawsuit (*that is our case*). **b** a set of arguments, esp. in relation to persuasiveness (*have a good case*; *have a weak case*). **c** a valid set of arguments (*have no case*). **7** Gram. **a** the relation of a word to other words in a sentence. **b** a form of a noun, adjective, or pronoun expressing this. **8** colloq. a comical person. **9** one's position or circumstances. □ **as the case may be** according to the situation. **case history** information about a person for use in professional treatment, e.g. by a doctor. **case-law** the law as established by the outcome of former cases (cf. *common law, statute law*). **case-load** the cases with which a doctor etc. is concerned at one time. **case-study 1** an attempt to understand a person, institution, etc., from collected information. **2** a record of such an attempt. **3** the use of a particular instance as an exemplar of general principles. **in any case** whatever the truth is; whatever may happen. **in case 1** in the event that; if. **2** lest; in provision against a stated or implied possibility (*take an umbrella in case it rains; took it in case*). **in case of** in the event of. **in the case of** as regards. **in no case** under no circumstances. **in that case** if that is true; should that happen. **is** (or **is not**) **the case** is (or is not) so. [ME f. OF *cas* f. L *casus* fall f. *cadere cas-* to fall]

case[2] /keɪs/ *n. & v.* ● *n.* **1** a container or covering serving to enclose or contain. **2** a container with its contents. **3** the outer protective covering of a watch, book, seed-vessel, sausage, etc. **4** an item of luggage, esp. a suitcase. **5** *Printing* a partitioned receptacle for type. **6** a glass box for showing specimens, curiosities, etc. ● *v.tr.* **1** enclose in a case. **2** (foll. by *with*) surround. **3** *sl.* reconnoitre (a house etc.), esp. with a view to robbery. □ **case-bound** (of a book) in a hard cover. **case-harden 1** harden the surface of, esp. give a steel surface to (iron) by carbonizing. **2** make callous. **case-knife** a knife carried in a sheath. **case-shot 1** bullets in an iron case fired from a cannon. **2** shrapnel. **lower case** small letters. **upper case** capitals. [ME f. OF *casse, chasse*, f. L *capsa* f. *capere* hold]

casebook /ˈkeɪsbʊk/ *n.* a book containing a record of legal or medical cases.

casein /ˈkeɪsɪn, ˈkeɪsiːn/ *n.* the main protein in milk, which occurs in coagulated form in cheese and is used in plastics, adhesives, paint, etc. [L *caseus* cheese]

casemate /ˈkeɪsmeɪt/ *n.* **1** a chamber in the thickness of the wall of a fortress, with embrasures. **2** an armoured enclosure for guns on a warship. [F *casemate* & It. *casamatta* or Sp. *-mata*, f. *camata*, perh. f. Gk *khasma -atos* gap]

Casement /ˈkeɪsmənt/, Sir Roger (David) (1864–1916), Irish nationalist. He served with the British consular service in Africa until his retirement in 1912, when he joined the Irish nationalist cause. Shortly after the outbreak of the First World War he visited Germany to seek support for an Irish uprising. He was captured on his return to Ireland before the Easter Rising of 1916, and subsequently hanged by the British for treason. His diaries reveal his homosexuality and were used to discredit a campaign for his reprieve; they were not made available to the public until 1959.

casement /ˈkeɪsmənt/ *n.* **1** a window or part of a window hinged vertically to open like a door. **2** *poet.* a window. [ME f. AL *cassimentum* f. *cassa* CASE[2]]

casework /ˈkeɪswɜːk/ *n.* social work concerned with individuals, esp. involving understanding of the client's family and background. □ **caseworker** *n.*

Cash /kæʃ/, Johnny (b.1932), American country music singer and songwriter. The poverty and hardship of his childhood are reflected in his early songs, which tend to feature outlaws, prisoners, or characters who are unlucky in life or love. He formed a brief association with Bob Dylan in the 1960s; during the 1970s he turned increasingly to gospel music. His most famous hits include 'I Walk the Line' (1956) and 'A Boy Named Sue' (1969).

cash[1] /kæʃ/ *n. & v.* ● *n.* **1** money in coins or notes, as distinct from cheques or orders. **2** (also **cash down**) money paid as full payment at the time of purchase, as distinct from credit. **3** *colloq.* wealth. ● *v.tr.* give or obtain cash for (a note, cheque, etc.). □ **cash and carry 1** a system of wholesaling in which goods are paid for in cash and taken away by the purchaser. **2** a store where this system operates. **cashbook** a book in which receipts and payments of cash are recorded. **cash cow** *colloq.* a business, or part of one, that provides a steady cash flow. **cash crop** a crop produced for sale, not for use as food etc. **cash desk** a counter or compartment in a shop where goods are paid for. **cash dispenser** an automatic machine from which customers of a bank etc. may withdraw cash, esp. by using a cashcard. **cash flow** the movement of money into and out of a business, as a measure of profitability, or as affecting liquidity. **cash in 1** obtain cash for. **2** (usu. foll. by *on*) *colloq.* profit (from); take advantage (of). **3** pay into a bank etc. **4** (in full **cash in one's chips** or **checks**) *colloq.* die. **cash on delivery** a system of paying the carrier for goods when they are delivered. **cash register** a machine in a shop etc. with a drawer for money, recording the amount of each sale, totalling receipts, etc. **cash up** *Brit.* count and check cash takings at the end of a day's trading.

□ **cashable** *adj.* **cashless** *adj.* [obs. F *casse* box or It. *cassa* f. L *capsa* CASE[2]]

cash[2] /kæʃ/ *n.* (*pl.* same) *hist.* any of various small coins of China or the East Indies. [ult. f. Port. *ca(i)xa* f. Tamil *kāsu* f. Skr. *karsha*]

cashcard /ˈkæʃkɑːd/ *n.* a plastic card (see CARD[1] *n.* 9) which enables the holder to draw money from a cash dispenser.

cashew /ˈkæʃuː, kæˈʃuː/ *n.* **1** a bushy evergreen tree, *Anacardium occidentale*, native to Central and South America, bearing kidney-shaped nuts attached to fleshy fruits. **2** (in full **cashew nut**) the edible nut of this tree. □ **cashew apple** the edible fleshy fruit of this tree. [Port. f. Tupi *(a)caju*]

cashier[1] /kæˈʃɪə(r)/ *n.* a person dealing with cash transactions in a shop, bank, etc. [Du. *cassier* or F *caissier* (as CASH[1])]

cashier[2] /kæˈʃɪə(r)/ *v.tr.* dismiss from service, esp. from the armed forces with disgrace. [Flem. *kasseren* disband, revoke, f. F *casser* f. L *quassare* QUASH]

cashmere /ˈkæʃmɪə(r)/ *n.* **1** a fine soft wool, esp. that of a breed of goat from the Himalayas. **2** a material made from this. [KASHMIR]

cashpoint /ˈkæʃpɔɪnt/ *n.* = *cash dispenser*.

casing /ˈkeɪsɪŋ/ *n.* **1** a protective or enclosing cover or shell. **2** the material for this.

casino /kəˈsiːnəʊ/ *n.* (*pl.* **-os**) a public room or building for gambling. [It., dimin. of *casa* house f. L *casa* cottage]

cask /kɑːsk/ *n.* **1** a large barrel-like container made of wood, metal, or plastic, esp. one for alcoholic liquor. **2** its contents. **3** its capacity. [F *casque* or Sp. *casco* helmet]

casket /ˈkɑːskɪt/ *n.* **1** a small often ornamental box or chest for jewels, letters, etc. **2 a** a small wooden box for cremated ashes. **b** *US* a coffin, esp. a rectangular one. [perh. f. AF form of OF *cassette* f. It. *cassetta* dimin. of *cassa* f. L *capsa* CASE[2]]

Caslon /ˈkæzlən/, William (1692–1766), English typographer. He established a type foundry (continued by his son William, 1720–78) which supplied printers on the Continent as well as in England. His name is applied to the types cut by the Caslons or to later type styles modelled on the same characteristics.

Caspar /ˈkæspə(r), -pɑː(r)/ one of the three Magi (see MAGI, THE).

Caspian Sea /ˈkæspɪən/ a large landlocked salt lake, bounded by Russia, Kazakhstan, Turkmenistan, Azerbaijan, and Iran. It is the world's largest body of inland water. Its surface lies 28 m (92 ft) below sea level.

casque /kæsk/ *n.* **1** *hist.* or *poet.* a helmet. **2** *Zool.* a helmet-like structure, e.g. the process on the bill of the cassowary. [F f. Sp. *casco*]

Cassandra /kəˈsændrə/ *Gk Mythol.* a daughter of the Trojan king Priam, who was loved by Apollo and given by him the gift of prophecy. When she cheated him, however, he turned this into a curse by causing her prophecies, though true, to be disbelieved.

cassata /kəˈsɑːtə/ *n.* a type of ice-cream containing candied or dried fruit and nuts. [It.]

cassation /kəˈseɪʃ(ə)n/ *n. Mus.* an informal instrumental composition of the 18th century, similar to a divertimento and orig. often for outdoor performance. [It. *cassazione*]

cassava /kəˈsɑːvə/ *n.* **1 a** a plant of the genus *Manihot*, of the spurge family; esp. the cultivated *bitter cassava* (*M. esculenta*) and *sweet cassava* (*M. dulcis*), which have starchy tuberous roots. **b** the roots themselves. **2** a starch or flour obtained from these roots. See also TAPIOCA, MANIOC. [earlier *cas(s)avi* etc., f. Taino *casavi*, infl. by F *cassave*]

Cassegrain telescope /ˈkæsɪˌɡreɪn/ *n.* a reflecting telescope in which light reflected from a convex secondary mirror is brought to a focus just behind a hole in the primary mirror. [N. *Cassegrain*, 17th-c. Fr. scientist]

casserole /ˈkæsəˌrəʊl/ *n. & v.* ● *n.* **1** a covered dish, usu. of earthenware or glass, in which food is cooked, esp. slowly in the oven. **2** food cooked in a casserole. ● *v.tr.* cook in a casserole. [F f. *cassole* dimin. of *casse* f. Prov. *casa* f. LL *cattia* ladle, pan f. Gk *kuathion* dimin. of *kuathos* cup]

cassette /kəˈset/ *n.* a sealed case containing a length of tape, ribbon, etc., ready for insertion in a machine, esp.: **1** a length of magnetic tape wound on to spools, ready for insertion in a tape recorder. **2** a length of photographic film, ready for insertion in a camera. [F, dimin. of *casse* CASE[2]]

cassia /ˈkæsɪə, ˈkæʃə/ *n.* **1** a tree of the genus *Cassia*, bearing leaves from which senna is extracted. **2** the cinnamon-like bark of this tree used as a spice. [L f. Gk *kasia* f. Heb. *k'ṣī'āh* bark like cinnamon]

Cassini /kəˈsiːnɪ/, Giovanni Domenico (1625–1712), Italian-born

French astronomer. He helped to establish the Paris Observatory and became its director, specializing in the solar system. He determined the rotational periods of Jupiter and Saturn, calculated the movements of the Galilean moons of Jupiter, discovered four of the moons of Saturn, and described a gap in the rings of Saturn (now known as Cassini's division). Cassini also contributed to geodesy, but wrongly believed the shape of the earth to be a prolate spheroid, a theory supported by the three generations of his descendants who succeeded him at the Paris Observatory.

Cassiopeia /ˌkæsɪəˈpiːə/ **1** *Gk Mythol.* the wife of Cepheus, king of Ethiopia, and mother of Andromeda. She boasted that she herself (or, in some versions, her daughter) was more beautiful than the nereids, thus incurring the wrath of Poseidon. (See also ANDROMEDA.) **2** *Astron.* a constellation near the north celestial pole, recognizable by the conspicuous 'W' pattern of its brightest stars. It contains a supernova remnant which is the strongest radio source in the sky.

cassis /kæˈsiːs/ *n.* a syrupy usu. alcoholic blackcurrant flavouring for drinks etc. [F, = blackcurrant]

cassiterite /kəˈsɪtəˌraɪt/ *n.* a naturally occurring ore of tin dioxide, from which tin is extracted. Also called *tinstone*. [Gk *kassiteros* tin]

Cassius /ˈkæsɪəs/, Gaius (full name Gaius Cassius Longinus) (d.42 BC), Roman general. With Brutus he was one of the leaders of the conspiracy in 44 BC to assassinate Julius Caesar. He and Brutus were defeated by Caesar's supporters, Antony and Octavian, at the battle of Philippi in 42 BC, in the course of which he committed suicide.

cassock /ˈkæsək/ *n.* a full length usu. black or red garment worn by clergy, members of choirs, etc. □ **cassocked** *adj.* [F *casaque* long coat f. It. *casacca* horseman's coat, prob. f. Turkic: cf. COSSACK]

cassoulet /ˈkæsʊˌleɪ/ *n.* a ragout of meat and beans. [F, dimin. of dial. *cassolo* stew-pan]

cassowary /ˈkæsəˌweərɪ/ *n.* (*pl.* **-ies**) a large flightless Australasian bird of the genus *Casuarius*, with heavy body, stout legs, a wattled neck, and a bony crest on its forehead. [Malay *kasuārī, kasawārī*]

cast /kɑːst/ *v. & n.* ● *v.* (*past* and *past part.* **cast**) **1** *tr.* throw, esp. deliberately or forcefully. **2** *tr.* (often foll. by *on*, *over*) **a** direct or cause to fall (one's eyes, a glance, light, a shadow, a spell, etc.). **b** express (doubts, aspersions, etc.). **3** *tr.* throw out (a net, fishing line, etc.) into the water. **4** *tr.* let down (an anchor or sounding-lead). **5** *tr.* **a** throw off, get rid of. **b** shed (skin etc.), esp. in the process of growth. **c** (of a horse) lose (a shoe). **6** *tr.* record, register, or give (a vote). **7** *tr.* **a** shape (molten metal or plastic material) in a mould. **b** make (a product) in this way. **8** *tr. Printing* make (type). **9** *tr.* **a** (usu. foll. by *as*) assign (an actor) to play a particular character. **b** allocate roles in (a play, film, etc.). **10** *tr.* (foll. by *in*, *into*) arrange or formulate (facts etc.) in a specified form. **11** *tr. & intr.* reckon, add up, calculate (accounts or figures). **12** *tr.* calculate and record details of (a horoscope). ● *n.* **1 a** the throwing of a missile etc. **b** the distance reached by this. **2** a throw or a number thrown at dice. **3** a throw of a net, sounding-lead, or fishing-line. **4 a** that which is cast in fishing, esp. the extremity of a fishing line with its hook or fly. **b** a place for casting. **5 a** an object of metal, clay, etc., made in a mould. **b** a moulded mass of solidified material, esp. plaster protecting a broken limb. **6** the actors taking part in a play, film, etc. **7** form, type, or quality (*cast of features; cast of mind*). **8** a tinge or shade of colour. **9 a** (in full **cast in the eye**) a slight squint. **b** a twist or inclination. **10 a** a mass of earth excreted by a worm. **b** a mass of indigestible food regurgitated by a hawk, owl, etc. **11** the form into which any work is thrown or arranged. **12 a** a wide area covered by a dog or pack to find a trail. **b** *Austral. & NZ* a wide sweep made by a sheepdog in mustering sheep. □ **cast about** (or **around** or **round**) make an extensive search (actually or mentally) (*cast about for a solution*). **cast adrift** leave to drift. **cast ashore** (of waves etc.) throw to the shore. **cast aside** give up using; abandon. **cast away 1** reject. **2** (in *passive*) be shipwrecked (cf. CASTAWAY). **cast one's bread upon the waters** see BREAD. **cast down** depress, deject (cf. DOWNCAST). **casting vote** a deciding vote usu. given by the chairperson when the votes on two sides are equal (from an obsolete sense of *cast* = turn the scale). **cast iron** a hard brittle alloy of iron, carbon (in greater proportion than in steel), and silicon, cast in a mould. **cast-iron** *adj.* **1** made of cast iron. **2** hard, unchallengeable, unchangeable. **cast loose** detach; detach oneself. **cast lots** see LOT. **cast-net** a net thrown out and immediately drawn in. **cast off 1** abandon. **2** *Knitting* take the stitches off the needle by looping each over the next to finish the edge. **3** *Naut.* **a** set a ship free from a quay etc. **b** loosen and throw off (rope etc.). **4** *Printing* estimate the space that will be taken in print by manuscript copy. **cast-off** *adj.* abandoned, discarded. ● *n.* a cast-off thing, esp. a garment. **cast on** *Knitting* make the first row of loops on the needle.

cast out expel. **cast up 1** (of the sea) deposit on the shore. **2** add up (figures etc.). [ME f. ON *kasta*]

Castalia /kæˈsteɪlɪə/ a spring on Mount Parnassus in central Greece, sacred in antiquity to Apollo and the Muses. □ **Castalian** *adj.*

castanet /ˌkæstəˈnet/ *n.* (usu. in *pl.*) a small concave piece of hardwood, ivory, etc., in pairs held in the hands and clicked together by the fingers as a rhythmic accompaniment, esp. by Spanish dancers. [Sp. *castañeta* dimin. of *castaña* f. L *castanea* chestnut]

castaway /ˈkɑːstəˌweɪ/ *n. & adj.* ● *n.* a shipwrecked person. ● *adj.* **1** shipwrecked. **2** cast aside; rejected.

caste /kɑːst/ *n.* **1** any of the Hindu hereditary classes. (*See note below.*) **2** a more or less exclusive social class. **3** a system of such classes. **4** the position it confers. **5** *Zool.* a form of social insect having a particular function. (See SOCIAL.) □ **caste mark** a symbol on the forehead denoting a person's caste. **lose caste** descend in the social order. [Sp. and Port. *casta* lineage, race, breed, fem. of *casto* pure, CHASTE]

▪ The term occurs first in Spanish, but was applied by the Portuguese in the 16th century to the rigid social divisions found in the Indian subcontinent. Although the caste system applies in particular to Hindus, it extends also to other religious groups such as Christians and Muslims. Castes are associated with different levels of purity and pollution. They are traditionally defined by occupation but are also linked with geographical location and (esp. dietary) customs; members marry within their caste and have strictly defined social relations with members of other castes. Hinduism recognizes four broad classes or varnas (Brahman, Kshatriya, Vaisya, and Sudra: priests, warriors, merchants and farmers, and labourers), based on the varnas of the ancient Aryans. Within each of these are many castes; the lowest caste, that of the Harijans or untouchables, is outside the varna system, and its members often suffer extreme discrimination, despite recent attempts at positive discrimination in their favour.

casteism /ˈkɑːstiːˌɪz(ə)m/ *n.* often *derog.* the caste system.

Castel Gandolfo /ˌkæstel ɡænˈdɒlfəʊ/ the summer residence of the pope, situated on the western edge of Lake Albano, 16 km (10 miles) south-east of Rome.

castellan /ˈkæstələn/ *n. hist.* the governor of a castle. [ME f. ONF *castelain* f. med.L *castellanus*: see CASTLE]

castellated /ˈkæstəˌleɪtɪd/ *adj.* **1** having battlements. **2** castle-like. □ **castellation** /ˌkæstəˈleɪʃ(ə)n/ *n.* [med.L *castellatus*: see CASTLE]

caster /ˈkɑːstə(r)/ *n.* **1** var. of CASTOR[1]. **2** a person who casts. **3** a machine for casting type.

castigate /ˈkæstɪˌɡeɪt/ *v.tr.* rebuke or punish severely. □ **castigator** *n.* **castigatory** *adj.* **castigation** /ˌkæstɪˈɡeɪʃ(ə)n/ *n.* [L *castigare* reprove f. *castus* pure]

Castile /kæˈstiːl/ a region of central Spain, on the central plateau of the Iberian peninsula. It was formerly an independent Spanish kingdom. The marriage of Isabella of Castile to Ferdinand of Aragon in 1469 linked these two powerful kingdoms and led eventually to the unification of Spain.

Castile soap *n.* a fine hard white or mottled soap made with olive oil and soda.

Castilian /kəˈstɪlɪən/ *n. & adj.* ● *n.* **1** a native of Castile in Spain. **2** the language of Castile, standard spoken and literary Spanish. ● *adj.* of or relating to Castile or Castilian.

Castilla-La Mancha /kæˌstiːjələˈmæntʃə/ an autonomous region of central Spain; capital, Toledo.

Castilla-León /kæˌstiːjəlerˈɒn/ an autonomous region of northern Spain; capital, Valladolid.

casting /ˈkɑːstɪŋ/ *n.* an object made by casting, esp. of molten metal.

castle /ˈkɑːs(ə)l/ *n. & v.* ● *n.* **1 a** a large fortified building or group of buildings; a stronghold. Although castles are characteristic of the Middle Ages, the term is also applied in proper names to ancient British or Roman earthworks. **b** a formerly fortified mansion. **2** *Chess* = ROOK[2]. ● *v. Chess* **1** *intr.* make a special move (once only in a game on each side) in which the king is moved two squares along the back rank and the nearer rook is moved to the square passed over by the king. **2** *tr.* move (the king) by castling. □ **castles in the air** (or **in Spain**) a visionary unattainable scheme; a day-dream. □ **castled** *adj.* [AF & ONF *castel, chastel* f. L *castellum* dimin. of *castrum* fort]

Castlebar /ˌkɑːs(ə)lˈbɑː(r)/ the county town of Mayo, in the Republic of Ireland; pop. (est. 1991) 6,070. In 1798 a French force routed the English in an engagement known as the 'Races of Castlebar'.

Castlereagh /ˈkɑːs(ə)lˌreɪ/, Robert Stewart, Viscount (1769–1822), British Tory statesman. Born in Ulster, he began his political career as a Whig in the Irish Parliament and continued to concern himself with Irish affairs after becoming a Tory in 1795. He became Foreign Secretary in 1812, and in this capacity represented his country at the Congress of Vienna (1814–15), playing a central part in reviving the Quadruple Alliance. He committed suicide, apparently as a result of mental strain owing to pressure of work.

Castor /ˈkɑːstə(r)/ **1** *Gk Mythol.* twin brother of Pollux (see DIOSCURI). **2** *Astron.* the second brightest star in the constellation of Gemini, close to Pollux. It is a multiple star system, the three components visible in a moderate telescope being close binaries.

castor[1] /ˈkɑːstə(r)/ n. (also **caster**) **1** a small swivelled wheel (often one of a set) fixed to a leg (or the underside) of a piece of furniture. **2** a small container with holes in the top for sprinkling the contents. □ **castor action** swivelling of vehicle wheels to ensure stability. **castor sugar** finely granulated white sugar. [orig. a var. of CASTER (in the general sense)]

castor[2] /ˈkɑːstə(r)/ n. an oily substance secreted by beavers and used in medicine and perfumes. [F or L f. Gk *kastōr* beaver]

castor oil n. **1** an oil from the seeds of the shrub *Ricinus communis*, used as a purgative and lubricant. **2** (in full **castor oil plant**) this plant, with large divided spiky leaves. □ **castor oil bean** (or **castor bean**) the poisonous seed of the castor oil plant, containing the highly poisonous substance ricin. [18th c.: orig. uncert.: perh. so called as having succeeded CASTOR[2] in the medical sense]

castrate /kæˈstreɪt/ v.tr. **1** remove the testicles of; geld. **2** deprive of vigour. □ **castrator** n. **castration** /-ˈstreɪʃ(ə)n/ n. [L *castrare*]

castrato /kæˈstrɑːtəʊ/ n. (pl. **castrati** /-tɪ/) *hist.* a male singer castrated in boyhood so as to retain a soprano or alto voice. Castrati were in great demand in Italy for opera during the 17th and 18th centuries, and parts were written for the castrato voice, which had a special brilliance, as late as 1824. Castrati sang in the Vatican chapel until the late 19th century, the practice of castration being banned in 1903. The last castrato died in 1922. [It., past part. of *castrare*: see CASTRATE]

Castries /kæsˈtriːs/ the capital of the Caribbean island of St Lucia, a seaport on the NW coast; pop. (1988) 52,900.

Castro /ˈkæstrəʊ/, Fidel (b.1927), Cuban statesman, Prime Minister 1959–76 and President since 1976. He forced President Batista from power in 1959, setting up a Communist regime which he has led ever since. The abortive US-backed invasion attempt by Cuban exiles at the Bay of Pigs in 1961 boosted his popularity, as did his successful survival of the Cuban Missile Crisis of 1962. He became leader of the Non-Aligned Movement in 1979, in spite of Cuba's reliance on the USSR for economic aid. Since the collapse of the Soviet bloc Castro has strictly maintained Communism in Cuba.

casual /ˈkæʒʊəl, ˈkæzjʊ-/ adj. & n. ● adj. **1** accidental; due to chance. **2** not regular or permanent; temporary, occasional (*casual work; a casual affair*). **3 a** unconcerned, uninterested (*was very casual about it*). **b** made or done without great care or thought (*a casual remark*). **c** acting carelessly or unmethodically. **4** (of clothes) informal. ● n. **1** a casual worker. **2** (usu. in *pl.*) casual clothes or shoes. □ **casually** adv. **casualness** n. [ME f. OF *casuel* & L *casualis* f. *casus* CASE[1]]

casualty /ˈkæʒʊəltɪ, ˈkæzjʊ-/ n. (pl. **-ies**) **1** a person killed or injured in a war or accident. **2** a thing lost or destroyed. **3** = casualty department. **4** an accident, mishap, or disaster. □ **casualty department** (or **ward**) the part of a hospital where casualties are treated. [ME f. med.L *casualitas* (as CASUAL), after ROYALTY etc.]

casuarina /ˌkæsjʊəˈriːnə/ n. a tree of the genus *Casuarina*, native to Australia and SE Asia, having tiny scale leaves on slender jointed branches, resembling gigantic horsetails. [mod.L *casuarius* cassowary (from the resemblance between branches and feathers)]

casuist /ˈkæzjʊɪst, ˈkæzjʊ-/ n. **1** a person, esp. a theologian, who resolves problems of conscience, duty, etc., often with clever but false reasoning. **2** a sophist or quibbler. □ **casuistry** n. **casuistic** /ˌkæzjʊˈɪstɪk, ˌkæzjʊ-/ **casuistical** adj. **casuistically** adv. [F *casuiste* f. Sp. *casuista* f. L *casus* CASE[1]]

casus belli /ˌkɑːzəs ˈbelɪ, ˌkeɪsəs/ n. an act or situation provoking or justifying war. [L]

CAT /kæt/ abbr. **1** computer-assisted (or -aided) testing. **2** *Med.* computerized axial tomography (*CAT scanner*).

cat[1] /kæt/ n. & v. ● n. **1** a small soft-furred flesh-eating domesticated mammal, *Felis catus*, usu. having a short snout and retractile claws, and kept as a pet or for catching mice. (*See note below.*) **2 a** a wild

member of the family Felidae, which includes the lion, tiger, leopard, etc. **b** = WILDCAT n. 1. **3** a catlike animal of any other family (*civet cat*). **4** *colloq.* a malicious or spiteful woman. **5** *sl.* a person (of a specified kind); a jazz enthusiast. **6** *Naut.* = CATHEAD. **7** = *cat-o'-nine-tails*. **8** a short tapered stick in the game of tipcat. ● v.tr. (also *absol.*) (**catted, catting**) *Naut.* raise (an anchor) from the surface of the water to the cathead. □ **cat-and-dog** (of a relationship etc.) full of quarrels. **cat burglar** a burglar who enters by climbing to an upper storey. **cat flap** (or **door**) a small swinging flap in an outer door, for a cat to pass in and out. **the cat has got his** (or **her** etc.) **tongue** *colloq.* he (or she etc.) refuses to speak. **cat-ice** thin ice unsupported by water. **cat-o'-nine-tails** *hist.* a rope whip with nine knotted lashes for flogging sailors, soldiers, or criminals. **cat's cradle** a child's game in which a loop of string is held between the fingers and patterns are formed. **Cat's-eye** *Brit. propr.* one of a series of reflector studs set into a road. **cat's-eye** a precious stone of Sri Lanka and the Malabar Coast. **cat's-foot** a small composite plant of the genus *Antennaria*, having soft woolly leaves and growing on the surface of the ground. **cat's-tail** = *reed-mace*. **cat's whisker** a fine adjustable wire in a crystal radio receiver. **cat's whiskers** (or **pyjamas**) *sl.* an excellent person or thing. **let the cat out of the bag** reveal a secret, esp. involuntarily. **like a cat on hot bricks** (or **on a hot tin roof**) very agitated or agitatedly. **put** (or **set**) **the cat among the pigeons** cause trouble. **rain cats and dogs** *colloq.* rain very hard. [OE *catt*(e) f. LL *cattus*]

▪ The cat family is characterized by muscular limbs and sharp retractile claws, suitable for climbing and for catching prey by ambush. The cat was probably domesticated in ancient Egypt, its ancestor being the North African form of the wild cat *Felis silvestris*. Cats were held in great reverence by the Egyptians, who mummified them in great numbers. The members of the cat family are notable for their uniformity, with all but two species usually placed in the genera *Felis* (the smaller cats) or *Panthera* (the great cats).

cat[2] /kæt/ n. a catalytic converter. [abbr.]

cata- /ˈkætə/ prefix (usu. **cat-** before a vowel or h) **1** down, downwards (*catadromous*). **2** wrongly, badly (*catachresis*). **3** completely (*catalogue*). **4** against, alongside (*catapult*). [Gk *kata* down]

catabolism /kəˈtæbəˌlɪz(ə)m/ n. (also **katabolism**) *Biochem.* the breakdown of complex molecules in living organisms to form simpler ones with the release of energy; destructive metabolism (opp. ANABOLISM). □ **catabolic** /ˌkætəˈbɒlɪk/ adj. [Gk *katabolē* descent f. *kata* down + *bolē* f. *ballō* throw]

catachresis /ˌkætəˈkriːsɪs/ n. (pl. **catachreses** /-siːz/) an incorrect use of words. □ **catachrestic** /-ˈkriːstɪk, -ˈkrestɪk/ adj. [L f. Gk *katakhrēsis* f. *khraomai* use]

cataclasis /ˌkætəˈkleɪsɪs/ n. (pl. **cataclases** /-siːz/) *Geol.* the natural process of fracture, shearing, or breaking up of rocks. □ **cataclastic** /-ˈklæstɪk/ adj. [mod.L f. Gk *kataklasis* breaking down]

cataclasm /ˈkætəˌklæz(ə)m/ n. a violent break; a disruption. [Gk *kataklasma* (as CATA-, *klaō* to break)]

cataclysm /ˈkætəˌklɪz(ə)m/ n. **1 a** a violent, esp. social or political, upheaval or disaster. **b** a great change. **2** a great flood or deluge. □ **cataclysmal** /ˌkætəˈklɪzm(ə)l/ adj. **cataclysmic** adj. **cataclysmically** adv. [F *cataclysme* f. L *cataclysmus* f. Gk *kataklusmos* f. *klusmos* flood f. *kluzō* wash]

catacomb /ˈkætəˌkuːm, -ˌkəʊm/ n. (often in *pl.*) **1** an underground cemetery, esp. a Roman subterranean gallery with recesses for tombs. (*See note below.*) **2** a similar underground construction; a cellar. [F *catacombes* f. LL *catacumbas*, of unkn. orig.]

▪ Catacombs, notably those under the city of Rome, were used by the early Christians for the reception of their dead. Since Roman legislation regarded every burial place as sacrosanct, Christians were able to hide in them during the era of the persecutions, and their violation was extremely rare. Having been long covered up and forgotten, the catacombs around Rome were fortuitously discovered in 1578; the stucco paintings which often covered the walls are the earliest surviving examples of Christian art.

catadioptric /ˌkætədaɪˈɒptrɪk/ adj. involving both the reflection and refraction of light. The word is chiefly used to describe a compact telescope or photographic lens using a spherical mirror, with a lens or plate to correct the resulting spherical aberration. [f. CATA- + DIOPTRIC]

catadromous /kəˈtædrəməs/ adj. (of a fish, e.g. the eel) that swims down rivers to the sea to spawn (cf. ANADROMOUS). [Gk *katadromos* f. *kata* down + *dromos* running]

catafalque /ˈkætəˌfælk/ n. a decorated wooden framework for supporting the coffin of a distinguished person during a funeral or while lying in state. [F f. It. *catafalco*, of unkn. orig.: cf. SCAFFOLD]

Catalan /ˈkætəˌlæn/ n. & adj. ● n. **1** a native of Catalonia in Spain. **2** a Romance language most closely related to Castilian Spanish and to Provençal. (*See note below.*) ● adj. of or relating to Catalonia or its people or to Catalan. [F f. Sp.]

▪ Traditionally Catalan is the language of Catalonia, but it is also spoken in Andorra (where it has official status), the Balearic Islands, and some parts of southern France. Since the death of Franco in 1975, Catalan has been accorded parity with Castilian Spanish in Catalonia.

catalase /ˈkætəˌleɪz/ n. Biochem. an enzyme that catalyses the reduction of hydrogen peroxide. [CATALYSIS]

catalepsy /ˈkætəˌlepsɪ/ n. a state of trance or seizure with loss of sensation and consciousness accompanied by rigidity of the body. □ **cataleptic** /ˌkætəˈleptɪk/ adj. & n. [F *catalepsie* or LL *catalepsia* f. Gk *katalēpsis* (as CATA-, *lēpsis* seizure)]

catalogue /ˈkætəˌlɒg/ n. & v. (US **catalog**) ● n. **1** a complete list of items (e.g. articles for sale, books held by a library), usu. in alphabetical or other systematic order and often with a description of each. **2** an extensive list (*a catalogue of crimes*). **3** US a university course-list etc. ● v.tr. (**catalogues, catalogued, cataloguing**; US **catalogs, cataloged, cataloging**) **1** make a catalogue of. **2** enter in a catalogue. □ **cataloguer** n. (US **cataloger**). [F f. LL *catalogus* f. Gk. *katalogos* f. *katalegō* enrol (as CATA-, *legō* choose)]

catalogue raisonné /ˌkætəˌlɒg ˌreɪzɒˈneɪ/ n. a descriptive catalogue with explanations or comments. [F, = explained catalogue]

Catalonia /ˌkætəˈləʊnɪə/ (Spanish **Cataluña** /ˌkataˈluɲa/) an autonomous region of NE Spain; capital, Barcelona. The region has a strong separatist tradition; the normal language for everyday purposes is Catalan, which has also won acceptance in recent years for various official purposes.

catalpa /kəˈtælpə/ n. a tree of the genus *Catalpa*, with large heart-shaped leaves, trumpet-shaped flowers, and long pods. [Creek]

catalyse /ˈkætəˌlaɪz/ v.tr. (US **catalyze**) Chem. produce (a reaction) by catalysis. [as CATALYSIS after *analyse*]

catalysis /kəˈtælɪsɪs/ n. (pl. **catalyses** /-ˌsiːz/) Chem. & Biochem. the acceleration of a chemical or biochemical reaction by a catalyst. [Gk *katalusis* dissolution (as CATA-, *luō* set free)]

catalyst /ˈkætəlɪst/ n. **1** Chem. a substance that, without itself undergoing any permanent chemical change, increases the rate of a reaction. **2** a person or thing that precipitates a change. [as CATALYSIS after *analyst*]

catalytic /ˌkætəˈlɪtɪk/ adj. Chem. relating to or involving catalysis. □ **catalytic converter** a device incorporated in the exhaust system of a motor vehicle, with a catalyst for converting pollutant gases into less harmful ones. **catalytic cracker** a device for cracking (see CRACK v. 9) petroleum oils by catalysis.

catalyze US var. of CATALYSE.

catamaran /ˌkætəməˈræn/ n. **1** a boat with twin hulls in parallel. **2** a raft of yoked logs or boats. **3** colloq. a quarrelsome woman. [Tamil *kaṭṭumaram* tied wood]

catamite /ˈkætəˌmaɪt/ n. **1** a boy kept for homosexual practices. **2** the passive partner in sodomy. [L *catamitus* through Etruscan f. Gk *Ganumēdēs* GANYMEDE]

catamountain /ˈkætəˌmaʊntɪn/ n. (also **catamount**) **1** a lynx, leopard, puma, or similar cat. **2** a wild quarrelsome person. [ME f. *cat of the mountain*]

catananche /ˌkætəˈnæŋkɪ/ n. a composite plant of the genus *Catananche*, with blue or yellow flowers. [mod.L f. L *catanancē* plant used in love-potions f. Gk *katanagkē* (as CATA-, *anagkē* compulsion)]

Catania /kaˈtaːnɪə/ a seaport situated at the foot of Mount Etna, on the east coast of Sicily; pop. (1990) 364,180.

cataplexy /ˈkætəˌpleksɪ/ n. sudden temporary paralysis due to fright etc. □ **cataplectic** /ˌkætəˈplektɪk/ adj. [Gk *kataplēxis* stupefaction]

catapult /ˈkætəˌpʌlt/ n. & v. ● n. **1** a forked stick etc. with elastic for shooting stones. **2** hist. a military machine worked by a lever and ropes for hurling large stones etc. **3** a mechanical device for launching a glider, an aircraft from the deck of a ship, etc. ● v. **1** tr. a hurl from or launch with a catapult. **b** fling forcibly. **2** intr. leap or be hurled forcibly. [F *catapulte* or L *catapulta* f. Gk *katapeltēs* (as CATA-, *pallō* hurl)]

cataract /ˈkætəˌrækt/ n. **1 a** a large waterfall or cascade. **b** a downpour; a rush of water. **2** Med. a condition in which the eye-lens

becomes progressively opaque resulting in blurred vision. [L *cataracta* f. Gk *katarrhaktēs* down-rushing; in med. sense prob. f. obs. sense 'portcullis']

catarrh /kəˈtɑː(r)/ n. **1** inflammation of the mucous membrane of the nose, air passages, etc. **2** a watery discharge in the nose or throat due to this. □ **catarrhal** adj. [F *catarrhe* f. LL *catarrhus* f. Gk *katarrheō* flow down]

catarrhine /ˈkætəˌraɪn/ adj. & n. Zool. ● adj. (of primates) having the nostrils close together and directed downwards, characteristic of Old World monkeys, apes, and humans (cf. PLATYRRHINE). ● n. such an animal. (See PRIMATE.) [CATA- + *rhis rhinos* nose]

catastrophe /kəˈtæstrəfɪ/ n. **1** a great and usu. sudden disaster. **2** the denouement of a drama. **3** a disastrous end; ruin. **4** an event producing a subversion of the order of things. □ **catastrophe theory** a branch of mathematics concerned with systems showing abrupt discontinuous change. □ **catastrophic** /ˌkætəˈstrɒfɪk/ adj. **catastrophically** adv. [L *catastropha* f. Gk *katastrophē* (as CATA-, *strophē* turning f. *strephō* turn)]

catastrophism /kəˈtæstrəˌfɪz(ə)m/ n. Geol. the theory that changes in the earth's crust have occurred in sudden violent and unusual events. □ **catastrophist** n.

catatonia /ˌkætəˈtəʊnɪə/ n. **1** Med. abnormal motor behaviour (esp. with episodes of catalepsy or overactivity), associated esp. with a form of schizophrenia. **2** catalepsy. [G *Katatonie* (as CATA-, TONE)]

catatonic /ˌkætəˈtɒnɪk/ adj. & n. ● adj. **1** of or relating to catatonia. **2** (usu. foll by *with*) made rigid or powerless (by an emotion etc.) (*catatonic with rage*). ● n. a person affected with catatonia.

catawba /kəˈtɔːbə/ n. **1** a US variety of grape. **2** a white wine made from this. [River *Catawba* in S. Carolina]

catbird /ˈkætbɜːd/ n. a bird with a mewing call like that of a cat; esp. a slaty-black North American mockingbird, *Dumetella carolinensis*.

catboat /ˈkætbəʊt/ n. a sailing-boat with a single mast placed well forward and carrying only one sail. [perh. f. *cat* a former type of coaler in NE England + BOAT]

catcall /ˈkætkɔːl/ n. & v. ● n. a shrill whistle of disapproval made at meetings etc. ● v. **1** intr. make a catcall. **2** tr. make a catcall at.

catch /kætʃ/ v. & n. ● v. (past and past part. **caught** /kɔːt/) **1** tr. **a** lay hold of so as to restrain or prevent from escaping; capture in a trap, in one's hands, etc. **b** (also **catch hold of**) get into one's hands so as to retain, operate, etc. (*caught hold of the handle*). **2** tr. detect or surprise (a person, esp. in a wrongful or embarrassing act) (*caught me in the act; caught him smoking*). **3** tr. **a** intercept and hold (a moving thing) in the hands etc. (*failed to catch the ball; a bowl to catch the drips*). **b** Cricket dismiss (a batsman) by catching the ball before it reaches the ground. **4** tr. **a** contract (a disease) by infection or contagion. **b** acquire (a quality or feeling) from another's example (*caught her enthusiasm*). **5** tr. **a** reach in time and board (a train, bus, etc.). **b** be in time to see etc. (a person or thing about to leave or finish) (*if you hurry you'll catch them; caught the end of the performance*). **6** tr. **a** apprehend with the senses or the mind (esp. a thing occurring quickly or briefly) (*didn't catch what he said*). **b** (of an artist etc.) reproduce faithfully. **7** tr. intr. become fixed or entangled; be checked (*the bolt began to catch*). **b** tr. cause to do this (*caught her tights on a nail*). **c** tr. (often foll. by *on*) hit, deal a blow to (*caught him on the nose; caught his elbow on the table*). **8** tr. draw the attention of; captivate (*caught his eye; caught her fancy*). **9** intr. begin to burn. **10** tr. (often foll. by *up*) reach or overtake (a person etc. ahead). **11** tr. check suddenly (*caught his breath*). **12** tr. (foll. by *at*) grasp or try to grasp. ● n. **1 a** an act of catching. **b** Cricket a chance or act of catching the ball. **2 a** an amount of a thing caught, esp. of fish. **b** a thing or person caught or worth catching, esp. in marriage. **3 a** a question, trick, etc., intended to deceive, incriminate, etc. **b** an unexpected or hidden difficulty or disadvantage. **4** a device for fastening a door or window etc. **5** Mus. a round, esp. with words arranged to produce a humorous effect. □ **catch-all** (often attrib.) a thing designed to be all-inclusive. **catch-as-catch-can** a style of wrestling with few holds barred. **catch at a straw** see STRAW. **catch crop** a crop grown between two staple crops (in position or time). **catch one's death of cold** see DEATH. **catch fire** see FIRE. **catch it** sl. be punished or in trouble. **catch me!** etc. (often foll. by *pres. part.*) colloq. you may be sure I etc. shall not. **catch on** colloq. **1** (of a practice, fashion, etc.) become popular. **2** (of a person) understand what is meant. **catch out 1** detect in a mistake etc. **2** take unawares; cause to be bewildered or confused. **3** = sense 3b of v. **catch-phrase** a phrase in frequent use. **catch the sun 1** be in a sunny position. **2** become sunburnt. **catch up 1 a** (often foll. by *with*) reach a person etc. ahead (*he caught up in the end; he caught us up; he* *caught up with us).* **b** (often foll. by *with, on*) make up arrears (of work etc.) (*must catch up with my correspondence*). **2** snatch or pick up hurriedly. **3** (often in *passive*) **a** involve; entangle (*caught up in suspicious dealings*). **b** fasten up (*hair caught up in a ribbon*). □ **catchable** *adj.* [ME f. AF & ONF *cachier*, OF *chacier*, ult. f. L *captare* try to catch]

catcher /ˈkætʃə(r)/ *n.* **1** a person or thing that catches. **2** *Baseball* a fielder who stands behind the batter.

catchfly /ˈkætʃflaɪ/ *n.* (*pl.* **-flies**) a plant of the genus *Silene* or *Lychnis*, of the pink family, with a sticky stem.

catching /ˈkætʃɪŋ/ *adj.* **1 a** (of a disease) infectious. **b** (of a practice, habit, etc.) likely to be imitated. **2** attractive; captivating.

catchline /ˈkætʃlaɪn/ *n. Printing* a short line of type esp. at the head of copy or as a running headline.

catchment /ˈkætʃmənt/ *n.* the collection of rainfall. □ **catchment area 1** the area from which rainfall flows into a river etc. **2** the area served by a school, hospital, etc.

catchpenny /ˈkætʃˌpenɪ/ *adj.* intended merely to sell quickly; superficially attractive.

catch-22 /ˌkætʃtwentɪˈtuː/ *n.* (often *attrib.*) a dilemma or circumstance from which there is no escape because of two mutually incompatible conditions, both of which are necessary. The phrase comes from the title of a novel by Joseph Heller (1961), set in the US air force in the Second World War: the hero, an American bombardier, wishes to avoid combat duty, to do which he has to be adjudged insane; but since anyone wishing to avoid combat duty is obviously sane, he must therefore be fit for duty.

catchup var. of KETCHUP.

catchweight /ˈkætʃweɪt/ *adj. & n.* ● *adj.* unrestricted as regards weight. ● *n.* unrestricted weight, as a weight category in sports.

catchword /ˈkætʃwɜːd/ *n.* **1** a word or phrase in common (often temporary) use; a topical slogan. **2** a word so placed as to draw attention. **3** *Theatr.* an actor's cue. **4** *Printing* the first word of a page given at the foot of the previous one.

catchy /ˈkætʃɪ/ *adj.* (**catchier, catchiest**) **1** (of a tune) easy to remember; attractive. **2** that snares or entraps; deceptive. **3** (of the wind etc.) fitful, spasmodic. □ **catchily** *adv.* **catchiness** *n.* [CATCH + -Y¹]

cate /keɪt/ *n.* archaic (usu. in *pl.*) choice food, delicacies. [obs. *acate* purchase f. AF *acat*, OF *achat* f. *acater, achater* buy: see CATER]

catechetical /ˌkætɪˈketɪk(ə)l/ *adj.* (also **catechetic**) **1** of or by oral teaching. **2** according to the catechism of a Church. **3** consisting of or proceeding by question and answer. □ **catechetically** *adv.* **catechetics** *n.* [eccl.Gk *katēkhētikos* f. *katēkhētēs* oral teacher: see CATECHIZE]

catechism /ˈkætɪˌkɪz(ə)m/ *n.* **1 a** a summary of the principles of a religion in the form of questions and answers. **b** a book containing this. **2** a series of questions put to anyone. □ **catechismal** /ˌkætɪˈkɪzm(ə)l/ *adj.* [eccl.L *catechismus* (as CATECHIZE)]

catechist /ˈkætɪkɪst/ *n.* a religious teacher, esp. one using a catechism.

catechize /ˈkætɪkaɪz/ *v.tr.* (also **-ise**) **1** instruct by means of question and answer, esp. from a catechism. **2** put questions to; examine. □ **catechizer** *n.* [LL *catechizare* f. eccl.Gk *katēkhizō* f. *katēkheō* make hear (as CATA-, *ēkheō* sound)]

catecholamine /ˌkætɪˈkəʊləˌmiːn/ *n. Biochem.* any of a class of aromatic amine compounds, many of which are neurotransmitters (e.g. adrenalin, dopamine). [from *catechol* the parent compound (orig. extracted from CATECHU) + AMINE]

catechu /ˈkætɪˌtʃuː/ *n.* (also **cachou** /ˈkæʃuː/) gambier or similar vegetable extract, containing tannin. [mod.L f. Malay *kachu*]

catechumen /ˌkætɪˈkjuːmən/ *n.* a Christian convert under instruction before baptism. [ME f. OF *catechumene* or eccl.L *catechumenus* f. Gk *katēkheō*: see CATECHIZE]

categorical /ˌkætɪˈɡɒrɪk(ə)l/ *adj.* (also **categoric**) unconditional, absolute; explicit, direct (*a categorical refusal*). □ **categorical imperative** (in ethics) an unconditional moral obligation derived from pure reason; the bidding of conscience as ultimate moral law. □ **categorically** *adv.* [F *catégorique* or LL *categoricus* f. Gk *katēgorikos*: see CATEGORY]

categorize /ˈkætɪɡəˌraɪz/ *v.tr.* (also **-ise**) place in a category or categories. □ **categorization** /ˌkætɪɡəraɪˈzeɪʃ(ə)n/ *n.*

category /ˈkætɪɡərɪ/ *n.* (*pl.* **-ies**) **1** a class or division. **2** *Philos.* **a** one of a possibly exhaustive set of classes among which all things might be distributed. **b** one of the a priori conceptions applied by the mind to

sense-impressions. **c** any relatively fundamental philosophical concept. □ **categorial** /ˌkætɪˈɡɔːrɪəl/ *adj.* [F *catégorie* or LL *categoria* f. Gk *katēgoria* statement f. *katēgoros* accuser]

catena /kəˈtiːnə/ *n.* (*pl.* **catenae** /-niː/ or **catenas**) **1** a connected series of patristic comments on Scripture. **2** a series or chain. [L, = chain: orig. *catena patrum* chain of the Fathers (of the Church)]

catenary /kəˈtiːnərɪ/ *n. & adj.* ● *n.* (*pl.* **-ies**) a curve formed by a uniform chain hanging freely from two points not in the same vertical line. ● *adj.* of or resembling such a curve. □ **catenary bridge** a suspension bridge hung from such chains. [L *catenarius* f. *catena* chain]

catenate /ˈkætɪˌneɪt/ *v.tr.* connect like links of a chain. □ **catenation** /ˌkætɪˈneɪʃ(ə)n/ *n.* [L *catenare catenat-* (as CATENARY)]

cater /ˈkeɪtə(r)/ *v.intr.* **1** (often foll. by *for*) provide food. **2** (foll. by *for, to*) provide what is desired or needed by. **3** (foll. by *to*) pander to (esp. low tastes). [obs. noun *cater* (now *caterer*), f. *acater* f. AF *acatour* buyer f. *acater* buy f. Rmc]

cateran /ˈkætərən/ *n. Sc. hist.* a Highland irregular fighting man; a marauder. [ME f. med.L *cateranus* & Gael. *ceathairne* peasantry]

cater-cornered /ˈkætəˌkɔːnəd/ *adj. & adv.* (also **cater-corner, catty-cornered** /ˈkætɪ-/) *N. Amer.* ● *adj.* placed or situated diagonally. ● *adv.* diagonally. [dial. adv. *cater* diagonally (cf. obs. *cater* the four on dice f. F *quatre* f. L *quattuor* four)]

caterer /ˈkeɪtərə(r)/ *n.* a person who supplies food for social events, esp. professionally.

catering /ˈkeɪtərɪŋ/ *n.* the profession or work of a caterer.

caterpillar /ˈkætəˌpɪlə(r)/ *n.* **1 a** the larva of a butterfly or moth. **b** (in general use) any similar larva of various insects. **2** (also **Caterpillar**) **a** (in full **caterpillar track** or **tread**) *propr.* an endless articulated steel band passing round the wheels of a tractor etc. for travel on rough ground. **b** a vehicle with these tracks, e.g. a tractor or tank. [perh. AF var. of OF *chatepelose* lit. 'hairy cat', infl. by obs. *piller* ravager]

caterwaul /ˈkætəˌwɔːl/ *v. & n.* ● *v.intr.* make the shrill screaming noise characteristic of a cat on heat. ● *n.* a caterwauling noise. [ME f. CAT¹ + -*waul* etc. imit.]

catfish /ˈkætfɪʃ/ *n.* **1** a mainly freshwater fish of the order Siluriformes, having barbels round the mouth and a scaleless body. **2** a wolf-fish.

catgut /ˈkætɡʌt/ *n.* a material used for the strings of musical instruments and surgical sutures, made of the dried twisted intestines of the sheep, horse, or ass (but not the cat).

Cath. abbr. **1** Cathedral. **2** Catholic.

Cathar /ˈkæθə(r)/ *n.* (*pl.* **Cathars** or **Cathari** /-rɪ/) a member of a heretical Christian sect in the Middle Ages professing a form of Manichaean dualism (see MANICHAEISM). The name appears to have been used first in Germany and Italy in the 12th century, although similar sects were active from the 10th century in the Balkans and then in France, where they were identified with the Albigenses. The sect was ruthlessly suppressed and had disappeared by the beginning of the 15th century. □ **Catharism** *n.* **Catharist** *n.* [med.L *Cathari* (pl.) f. Gk *katharoi* pure]

catharsis /kəˈθɑːsɪs/ *n.* (*pl.* **catharses** /-siːz/) **1** an emotional release in drama or art. **2** *Psychol.* the process of freeing repressed emotion by association with the cause, and elimination by abreaction. **3** *Med.* purgation. [mod.L f. Gk *katharsis* f. *kathairō* cleanse: sense 1 f. Aristotle's *Poetics*, with ref. to tragedy]

cathartic /kəˈθɑːtɪk/ *adj. & n.* ● *adj.* **1** effecting catharsis. **2** purgative. ● *n.* a cathartic drug. □ **cathartically** *adv.* [LL *catharticus* f. Gk *kathartikos* (as CATHARSIS)]

Cathay /kæˈθeɪ/ (also **Khitai** /kɪˈtaɪ/) the name by which China was known to medieval Europe. [med.L *Cataya* f. Turkic *Khitāy*]

cathead /ˈkæthed/ *n. Naut.* a horizontal beam from each side of a ship's bow for raising and carrying the anchor.

cathectic see CATHEXIS.

cathedral /kəˈθiːdr(ə)l/ *n.* the principal church of a diocese, containing the bishop's throne. □ **cathedral city** a city in which there is a cathedral. [ME (as adj.) f. OF *cathedral* or f. LL *cathedralis* f. L f. Gk *kathedra* seat]

Cather /ˈkæðə(r)/, Willa (Sibert) (1876–1947), American novelist and short-story writer. The state of Nebraska, where she was brought up, provides the setting for some of her best writing; early major novels include *O Pioneers!* (1913), *The Song of the Lark* (1915), and *My Antonia* (1918), while among later works is the best-selling *Death Comes for the Archbishop* (1927), a celebration of the Catholic Church in New Mexico.

Catherine II /ˈkæθrɪn/, (known as Catherine the Great) (1729–96), empress of Russia, reigned 1762–96. A German princess, she was made empress following a plot which deposed her husband Peter III (1728–62). Her attempted social and political reforms were impeded by entrenched aristocratic interests, and in later years her reign became increasingly conservative. Under Catherine Russia played an important part in European affairs, participating in the three partitions of Poland and forming close links with Prussia and Austria, while to the south and east further territorial advances were made at the expense of the Turks and Tartars.

Catherine, St (known as St Catherine of Alexandria) (died *c*.307), early Christian martyr. Traditionally, she opposed the persecution of Christians under the Roman emperor Maxentius, debated with fifty scholars sent to undermine her position, and refused to recant or to marry the emperor. She is then said to have been tortured on a spiked wheel and beheaded when it broke. The Catherine wheel subsequently became her emblem. Feast day, 25 Nov.

Catherine de' Medici (1519–89), queen of France. The wife of Henry II of France, Catherine ruled as regent 1560–74 during the minority reigns of their three sons, Francis II (reigned 1559–60), Charles IX (reigned 1560–74), and Henry III (reigned 1574–89). She proved unable or unwilling to control the confused situation during the French Wars of Religion, and it was on her instigation that Huguenots were killed in the Massacre of St Bartholomew (1572).

Catherine of Aragon (1485–1536), first wife of Henry VIII, youngest daughter of Ferdinand and Isabella of Castile, mother of Mary I. Originally married to Henry's elder brother Arthur in 1501, she was widowed six months later, and married Henry in 1509. She gave birth to five children, but all except her daughter Mary died in infancy. Concerned about Catherine's failure to produce a male heir, and attracted to Anne Boleyn, Henry attempted to divorce his wife on the debatable grounds that her marriage to his brother made the marriage illegal. Catherine was sent into retirement, but neither she nor the pope accepted the annulment of the marriage, which led eventually to England's break with the Roman Catholic Church.

Catherine wheel *n.* **1** a firework in the form of a flat coil which spins when fixed and lit. **2** a circular window with radial divisions. [mod.L *Catharina* f. Gk *Aikaterina* Catherine (see CATHERINE, ST)]

catheter /ˈkæθɪtə(r)/ *n. Med.* a tube for insertion into a body cavity for introducing or removing fluid. [LL f. Gk *kathetēr* f. *kathiēmi* send down]

catheterize /ˈkæθɪtəˌraɪz/ *v.tr.* (also **-ise**) *Med.* insert a catheter into.

cathetometer /ˌkæθɪˈtɒmɪtə(r)/ *n.* a telescope mounted on a graduated scale along which it can slide, used for accurate measurement of small vertical distances. [L *cathetus* f. Gk *kathetos* perpendicular line (as CATHETER + -METER)]

cathexis /kəˈθeksɪs/ *n.* (pl. **cathexes** /-siːz/) *Psychol.* concentration of mental energy in one channel. □ **cathectic** /-ˈθektɪk/ *adj.* [Gk *kathexis* retention]

cathode /ˈkæθəʊd/ *n.* (also **kathode**) (opp. ANODE) *Electr.* **1** a negatively charged electrode. **2** the terminal by which electric current leaves a device. □ **cathodal** /kəˈθəʊd(ə)l/ *adj.* [Gk *kathodos* descent f. *kata* down + *hodos* way]

cathode ray *n.* a beam of electrons emitted from the cathode of a high-vacuum discharge tube. The study of high-voltage electrical discharges through gases at low pressure in the late 19th century revealed that a highly energetic radiation was being emitted from the region of the cathode or negative terminal of the tube. J. J. Thomson in 1897 established that cathode rays consisted of streams of high-velocity electrons. Cathode-ray tubes are used widely in television receivers and other visual display units. The modern television tube, for example, is an advanced form of cathode-ray tube producing electrons by evaporation from a heated cathode. The electrons are accelerated and focused by the anode, and fall on a fluorescent screen where they produce pinpoints of light. The electron beam is made to follow a scanning pattern which covers the whole screen. Application of a positive or negative potential (e.g. a video signal) to a metal grid or mesh inserted between cathode and anode has the effect of varying the intensity of the beam so as to build up an image on the screen.

cathode-ray tube /ˌkæθəʊdˈreɪ/ *n.* (abbr. **CRT**) a high-vacuum discharge tube in which cathode rays are made to produce a luminous image on a fluorescent screen.

cathodic /kəˈθɒdɪk/ *adj.* of or relating to a cathode. □ **cathodic protection** protection of a metal structure from corrosion under water by making it act as an electrical cathode (cf. *sacrificial anode*).

catholic /ˈkæθəlɪk, ˈkæθlɪk/ *adj. & n.* ● *adj.* **1** of interest or use to all; universal. **2** all-embracing; of wide sympathies or interests (*has catholic tastes*). **3** (**Catholic**) **a** of the Roman Catholic religion. **b** including all Christians. **c** including all of the Western Church. ● *n.* (**Catholic**) a Roman Catholic. □ **catholically** *adv.* **catholicly** *adv.* **Catholicism** /kəˈθɒlɪˌsɪz(ə)m/ *n.* **catholicity** /ˌkæθəˈlɪsɪtɪ/ *n.* [ME f. OF *catholique* or LL *catholicus* f. Gk *katholikos* universal f. *kata* in respect of + *holos* whole]

Catholic Emancipation the granting of full political and civil liberties to Roman Catholics in Britain and Ireland. Although religious toleration had been practised since the late 17th century, Catholics were barred from holding public office and restricted in many other ways under the Penal Laws and Test Acts. Many of the restrictions were lifted in 1791, and heavy pressure was applied from Ireland after the Act of Union (1801) for the admission of Catholics to Parliament; the issue was fiercely debated, but emancipation was strongly opposed by George III. The election to Parliament of Daniel O'Connell in 1828 led to the passing of the Catholic Emancipation Act of 1829, which repealed most of the discriminatory laws. (See also PENAL LAWS, TEST ACTS.)

catholicize /kəˈθɒlɪˌsaɪz/ *v.tr. & intr.* (also **-ise**) **1** make or become catholic. **2** (**Catholicize**) make or become a Roman Catholic.

Catholic League see HOLY LEAGUE.

Catiline /ˈkætɪˌlaɪn/ (Latin name Lucius Sergius Catilina) (*c*.108–62 BC), Roman nobleman and conspirator. Repeatedly thwarted in his ambition to be elected consul, in 63 BC he planned an uprising in Italy. His fellow-conspirators in Rome were successfully suppressed and executed on the initiative of the consul Cicero, and Catiline died in a battle in Etruria.

cation /ˈkætˌaɪən/ *n.* a positively charged ion; an ion that is attracted to the cathode in electrolysis (opp. ANION). [CATA- + ION]

cationic /ˌkætaɪˈɒnɪk/ *adj.* **1** of a cation or cations. **2** having an active cation.

catkin /ˈkætkɪn/ *n.* a spike of usu. downy or silky male or female flowers hanging from a willow, hazel, etc. [obs. Du. *katteken* kitten]

catlick /ˈkætlɪk/ *n. colloq.* a perfunctory wash.

catlike /ˈkætlaɪk/ *adj.* **1** like a cat. **2** stealthy.

catmint /ˈkætmɪnt/ *n.* a downy labiate plant, *Nepeta cataria*, with purple-spotted white flowers and a mintlike smell attractive to cats. Also called *catnip*.

catnap /ˈkætnæp/ *n. & v.* ● *n.* a short sleep. ● *v.intr.* (**-napped, -napping**) have a catnap.

catnip /ˈkætnɪp/ *n.* = CATMINT. [CAT¹ + dial. *nip* catmint, var. of dial. *nep*]

Cato /ˈkeɪtəʊ/, Marcus Porcius (known as Cato the Elder or Cato the Censor) (234–149 BC), Roman statesman, orator, and writer. An implacable enemy of Carthage, he fought in the second Punic War as a young man and continually warned against the Carthaginian threat when a senator. As censor in 184 BC he engaged in a vigorous programme of moral and social reform, and attempted to stem the growing influence of Greek culture on Roman life. His many writings include a lost history of Rome and an extant work on agriculture. His great-grandson, Cato the Younger (95–46 BC), was an opponent of the dictatorial ambitions of Julius Caesar.

catoptric /kəˈtɒptrɪk/ *adj.* of or relating to a mirror, a reflector, or reflection. □ **catoptrics** *n.* [Gk *katoptrikos* f. *katoptron* mirror]

Catskill Mountains /ˈkætskɪl/ (also **Catskills**) a range of mountains in the state of New York, part of the Appalachian system.

cat's-paw /ˈkætspɔː/ *n.* **1** a person used as a tool by another (from the fable of the monkey who used the paw of his friend the cat to rake roasted chestnuts out of the fire). **2** a slight breeze rippling the surface of water.

catsuit /ˈkætsuːt, -sjuːt/ *n.* a close-fitting garment with trouser legs, covering the body from neck to feet.

catsup /ˈkætsəp/ *esp. US* var. of KETCHUP.

cattery /ˈkætərɪ/ *n.* (pl. **-ies**) a place where cats are boarded or bred.

cattish /ˈkætɪʃ/ *adj.* = CATTY. □ **cattishly** *adv.* **cattishness** *n.*

cattle /ˈkæt(ə)l/ *n.pl.* **1** large cud-chewing horned animals of the genus *Bos*; esp. domesticated bovine animals, *B. taurus*, kept for milk or meat. (See note below.) **2** *archaic* livestock. □ **cattle-cake** *Brit.* a concentrated food for cattle, in cake form. **cattle-dog** *Austral. & NZ* a dog trained to work with cattle. **cattle-duff** *Austral.* to steal cattle. **cattle-duffer** *Austral.* a cattle thief. **cattle-grid** *Brit.* a grid covering a ditch, allowing vehicles to pass over but not cattle, sheep, etc. **cattle-guard** *US* = *cattle-grid.* **cattle-plague** rinderpest. **cattle-stop** *NZ* = *cattle-grid.* [ME & AF *catel* f. OF *chatel* CHATTEL]

▪ Cattle were domesticated from the wild ox or aurochs in neolithic times, certainly before 3000 BC and possibly as early as 7000 BC. They were first kept for meat and hides, and their early use as draught animals transformed agriculture. Milk yields were slowly increased by selective breeding, and later breeds of dairy cattle became perhaps the most important of domestic animals. The Brahman or zebu was developed in India, where it is held sacred by Hindus, and has been extensively cross-bred with western breeds.

cattleman /ˈkæt(ə)lmən/ n. (pl. **-men**) N. Amer. a person who tends or rears cattle.

cattleya /ˈkætlɪə/ n. an epiphytic orchid of the genus *Cattleya*, with handsome violet, pink, or yellow flowers. [mod.L f. William *Cattley*, Engl. patron of botany (d.1832)]

catty /ˈkætɪ/ adj. (**cattier, cattiest**) **1** sly, spiteful; deliberately hurtful in speech. **2** catlike. □ **cattily** adv. **cattiness** n.

catty-cornered var. of CATER-CORNERED.

Catullus /kəˈtʌləs/, Gaius Valerius (c.84–c.54 BC), Roman poet. His one book of verse contains poems in a variety of metres on a range of subjects; he is best known for his poems to a married woman addressed as 'Lesbia', although he also wrote a number of longer mythological pieces. His importance for later Latin poetry lies both in the impetus he gave to the development of the love-elegy and in his cultivation of an Alexandrian refinement and learning.

catwalk /ˈkætwɔːk/ n. **1** a narrow footway along a bridge, above a theatre stage, etc. **2** a narrow platform or gangway used in fashion shows etc.

Caucasian /kɔːˈkeɪz(ə)n, -ˈkeɪzɪən/ adj. & n. ● adj. **1** of or relating to the white or light-skinned division of humankind, so called because the German physiologist Blumenbach believed that it originated in the Caucasus region of SE Europe. **2** of or relating to the Caucasus. **3** of or relating to a group of languages spoken in the region of the Caucasus. (*See note below.*) ● n. a Caucasian or white person.

▪ Thirty-eight Caucasian languages are known, only a few of which have been committed to writing; the best-known and most widely spoken is Georgian. The languages are divided into four families, not all of which are necessarily related to each other; they are characterized by ergativity and agglutination.

Caucasoid /ˈkɔːkəˌsɔɪd/ adj. of or relating to the Caucasian division of humankind.

Caucasus, the /ˈkɔːkəsəs/ (also **Caucasia** /kɔːˈkeɪʒə, -ˈkeɪzɪə/) a mountainous region of SE Europe, lying between the Black Sea and the Caspian Sea, in Georgia, Armenia, Azerbaijan, and SE Russia. Its highest peak, Mount Elbrus, rises to 5,642 m (18,481 ft).

Cauchy /ˈkaʊʃi, French koʃi/, Augustin Louis, Baron (1789–1857), French mathematician. His numerous textbooks and writings introduced new standards of criticism and rigorous argument in calculus, from which grew the field of mathematics known as analysis. He transformed the theory of complex functions by discovering his integral theorems and introducing the calculus of residues. Cauchy also founded the modern theory of elasticity, produced fundamental new ideas about the solution of differential equations, and contributed substantially to the founding of group theory.

caucus /ˈkɔːkəs/ n. **1** US **a** a meeting of the members of a political party, esp. in the Senate etc., to decide policy. **b** a bloc of such members. **c** this system as a political force. **2** often derog. (esp. in the UK) **a** a usu. secret meeting of a group within a larger organization or party. **b** such a group. [18th-c. US, perh. f. Algonquian *cau'-cau-as'u* adviser]

caudal /ˈkɔːd(ə)l/ adj. **1** of or like a tail. **2** of the posterior part of the body. □ **caudally** adv. [mod.L *caudalis* f. L *cauda* tail]

caudate /ˈkɔːdeɪt/ adj. having a tail. [see CAUDAL]

caudillo /kaʊˈdiːjəʊ/ n. (pl. **-os**) (in Spanish-speaking countries) a military or political leader. 'Caudillo' was the title taken by General Franco in his rule of Spain 1939–75. [Sp. f. LL *capitellum* dimin. of *caput* head]

caught past and past part. of CATCH.

caul /kɔːl/ n. **1 a** the inner membrane enclosing a foetus; the amnion. **b** part of this occasionally found on a child's head at birth, thought to bring good luck. The superstition existed among many sailors at least until the early 20th century that possession of the caul of a newborn child was a sure protection against death by drowning. **2** hist. **a** a woman's close-fitting indoor head-dress. **b** the plain back part of a woman's indoor head-dress. **3** the omentum. [ME perh. f. OF *cale* small cap]

cauldron /ˈkɔːldrən/ n. (also **caldron**) a large deep bowl-shaped vessel for boiling over an open fire; an ornamental vessel resembling this. [ME f. AF & ONF *caudron*, ult. f. L CALDARIUM]

cauliflower /ˈkɒlɪˌflaʊə(r)/ n. **1** a variety of cabbage with a large immature flower-head of small usu. creamy-white flower-buds. **2** the flower-head eaten as a vegetable. □ **cauliflower cheese** a savoury dish of cauliflower in a cheese sauce. **cauliflower ear** an ear thickened by repeated blows, esp. in boxing. [earlier *cole-florie* etc. f. obs. F *chou fleuri* flowered cabbage, assim. to COLE and FLOWER]

caulk /kɔːk/ v.tr. (US **calk**) **1** stop up (the seams of a boat etc.) with oakum etc. and waterproofing material, or by driving plate-junctions together. **2** make (esp. a boat) watertight by this method. □ **caulker** n. [OF dial. *cauquer* tread, press with force, f. L *calcare* tread f. *calx* heel]

causal /ˈkɔːz(ə)l/ adj. **1** of, forming, or expressing a cause or causes. **2** relating to, or of the nature of, cause and effect. □ **causally** adv. [LL *causalis*: see CAUSE]

causality /kɔːˈzælɪtɪ/ n. **1** the relation of cause and effect. **2** the principle that everything has a cause.

causation /kɔːˈzeɪʃ(ə)n/ n. **1** the act of causing or producing an effect. **2** = CAUSALITY. [F *causation* or L *causatio* pretext etc., in med.L the action of causing, f. *causare* CAUSE]

causative /ˈkɔːzətɪv/ adj. **1** acting as cause. **2** (foll. by *of*) producing; having an effect. **3** Gram. expressing cause. □ **causatively** adv. [ME f. OF *causatif* or f. LL *causativus*: see CAUSATION]

cause /kɔːz/ n. & v. ● n. **1 a** that which produces an effect, or gives rise to an action, phenomenon, or condition. **b** a person or thing that occasions something. **c** a reason or motive; a ground that may be held to justify something (*no cause for complaint*). **2** a reason adjudged adequate (*show cause*). **3** a principle, belief, or purpose which is advocated or supported (*faithful to the cause*). **4 a** a matter to be settled at law. **b** an individual's case offered at law (*plead a cause*). **5** the side taken by any party in a dispute. ● v.tr. **1** be the cause of, produce, make happen (*caused a commotion*). **2** (foll. by *to* + infin.) induce (*caused me to smile; caused it to be done*). □ **cause and effect 1** a cause and the effect it produces. **2** the operation or relation of a cause and its effect. **in the cause of** to maintain, defend, or support (*in the cause of justice*). **make common cause with** join the side of. □ **causable** adj. **causeless** adj. **causer** n. [ME f. OF f. L *causa*]

'cause /kɒz, kəz/ conj. & adv. colloq. = BECAUSE. [abbr.]

cause célèbre /ˌkɔːz seˈlɛbrə/ n. (pl. **causes célèbres** pronunc. same) **1** a lawsuit that attracts much attention. **2** an issue that gives rise to widespread public discussion. [F]

causerie /ˈkəʊzərɪ/ n. (pl. **causeries** pronunc. same) an informal article or talk, esp. on a literary subject. [F f. *causer* talk]

causeway /ˈkɔːzweɪ/ n. **1** a raised road or track across low or wet ground or a stretch of water. **2** a raised path by a road. [earlier *cauce, causeway* f. ONF *caucié* ult. f. L CALX lime, limestone]

causey /ˈkɔːzɪ/ n. archaic or dial. = CAUSEWAY.

caustic /ˈkɔːstɪk/ adj. & n. ● adj. **1** that burns or corrodes organic tissue. **2** sarcastic, biting. **3** Chem. strongly alkaline. **4** Physics formed by the intersection of reflected or refracted parallel rays from a curved surface. ● n. **1** a caustic substance. **2** Physics a caustic surface or curve. □ **caustic potash** potassium hydroxide. **caustic soda** sodium hydroxide. □ **caustically** adv. **causticity** /kɔːˈstɪsɪtɪ/ n. [L *causticus* f. Gk *kaustikos* f. *kaustos* burnt f. *kaiō* burn]

cauterize /ˈkɔːtəˌraɪz/ v.tr. (also **-ise**) Med. burn or coagulate (tissue) with a heated instrument or caustic substance, esp. to stop bleeding. □ **cauterization** /ˌkɔːtəraɪˈzeɪʃ(ə)n/ n. [F *cautériser* f. LL *cauterizare* f. Gk *kautēriazō* f. *kautērion* branding-iron f. *kaiō* burn]

cautery /ˈkɔːtərɪ/ n. (pl. **-ies**) Med. **1** an instrument or caustic for cauterizing. **2** the operation of cauterizing. [L *cauterium* f. Gk *kautērion*: see CAUTERIZE]

caution /ˈkɔːʃ(ə)n/ n. & v. ● n. **1** attention to safety; prudence, carefulness. **2 a** esp. Brit. a warning, esp. a formal one in law. **b** a formal warning and reprimand. **3** colloq. an amusing or surprising person or thing. ● v.tr. **1** (often foll. by *against*, or *to* + infin.) warn or admonish. **2** esp. Brit. issue a caution to. □ **caution money** Brit. a sum deposited as security for good conduct. [ME f. OF f. L *cautio -onis* f. *cavere caut-* take heed]

cautionary /ˈkɔːʃənərɪ/ adj. that gives or serves as a warning (*a cautionary tale*).

cautious /ˈkɔːʃəs/ adj. careful, prudent; attentive to safety. □ **cautiously** adv. **cautiousness** n. [ME f. OF f. L: see CAUTION]

Cauvery /ˈkɔːvərɪ/ (also **Kaveri**) a river in south India which rises in north Kerala and flows 765 km (475 miles) eastwards to the Bay of Bengal, south of Pondicherry. It is held sacred by Hindus.

cavalcade /ˌkævəlˈkeɪd/ n. a procession or formal company of riders, motor vehicles, etc. [F f. It. *cavalcata* f. *cavalcare* ride ult. f. L *caballus* pack-horse]

cavalier /ˌkævəˈlɪə(r)/ n. & adj. ● n. **1** *hist.* (**Cavalier**) a supporter of Charles I in the English Civil War. The term originally had pejorative connotations, derived from the supposedly overenthusiastic attitude of the king's supporters towards the prospect of war. **2** a courtly gentleman, esp. as a lady's escort. **3** *archaic* a horseman. ● adj. offhand, supercilious, blasé. □ **cavalierly** adv. [F f. It. *cavaliere*: see CHEVALIER]

cavalry /ˈkævəlrɪ/ n. (pl. **-ies**) (usu. treated as pl.) soldiers on horseback or in armoured vehicles. □ **cavalry twill** a strong fabric in a double twill. [F *cavallerie* f. It. *cavalleria* f. *cavallo* horse f. L *caballus*]

cavalryman /ˈkævəlrɪmən/ n. (pl. **-men**) a soldier of a cavalry regiment.

Cavan /ˈkæv(ə)n/ **1** a county of the Republic of Ireland, part of the old province of Ulster. **2** its county town; pop. (1991) 3,330.

cavatina /ˌkævəˈtiːnə/ n. **1** a short simple song. **2** a similar piece of instrumental music, usu. slow and emotional. [It.]

cave[1] /keɪv/ n. & v. ● n. **1** a large hollow in the side of a cliff, hill, etc., or underground. **2** *Brit. hist.* a dissident political group. ● *v.intr.* explore caves, esp. interconnecting or underground. □ **cave-bear** an extinct kind of large bear, whose bones have been found in caves. **cave-dweller** = CAVEMAN. **cave in 1 a** (of a wall, earth over a hollow, etc.) subside, collapse. **b** cause (a wall, earth, etc.) to do this. **2** yield or submit under pressure; give up. **cave-in** n. a collapse, submission, etc. □ **cavelike** adj. **caver** n. [ME f. OF f. L *cava* f. *cavus* hollow: *cave in* prob. f. E. Anglian dial. *calve in*]

cave[2] /ˈkeɪvɪ/ int. Brit. school sl. look out! (as a warning cry). □ **keep cave** act as lookout. [L, = beware]

caveat /ˈkævɪˌæt/ n. **1** a warning or proviso. **2** *Law* a process in court to suspend proceedings. [L, = let a person beware]

caveat emptor /ˌkævɪˌæt ˈemptɔː(r)/ n. the principle that the buyer alone is responsible if dissatisfied. [L, = let the buyer beware]

Cavell /ˈkæv(ə)l/, Edith (Louisa) (1865–1915), English nurse. In charge of the Berkendael Medical Institute in Brussels during the First World War, she helped many Allied soldiers to escape from occupied Belgium. She was arrested by the Germans and brought before a military tribunal, where she openly admitted her actions and was sentenced to death. Her execution provoked widespread condemnation and she became famous as a heroine of the Allied cause.

caveman /ˈkeɪvmæn/ n. (pl. **-men**) **1** a prehistoric man living in a cave. **2** a primitive or crude man.

Cavendish /ˈkæv(ə)ndɪʃ/, Henry (1731–1810), English chemist and physicist. Pursuing his research in his own private laboratory, Cavendish identified hydrogen as a separate gas, studied carbon dioxide, and determined their densities relative to atmospheric air. He also established that water was a compound, and determined the density of the earth. The full extent of his discoveries in electrostatics was not known until his manuscripts were published in 1879: he had anticipated Coulomb, Ohm, and Faraday, deduced the inverse square law of electrical attraction and repulsion, and discovered specific inductive capacity. The Cavendish Laboratory at Cambridge was named after him.

cave-painting /ˈkeɪvˌpeɪntɪŋ/ n. a picture of an animal etc. on the interior of a cave, especially one by prehistoric peoples. Among the finest prehistoric cave-paintings are those at Altamira in Spain.

cavern /ˈkæv(ə)n/ n. **1** a cave, esp. a large or dark one. **2** a dark cavelike place, e.g. a room. □ **cavernous** adj. **cavernously** adv. [ME f. OF *caverne* or f. L *caverna* f. *cavus* hollow]

caviar /ˈkævɪˌɑː(r)/ n. (also **caviare**) the pickled roe of sturgeon or other large fish, eaten as a delicacy. [early forms repr. It. *caviale*, F *caviar*, prob. f. med.Gk *khaviari*]

cavil /ˈkævɪl/ v. & n. ● *v.intr.* (**cavilled**, **cavilling**; US **caviled**, **caviling**) (usu. foll. by *at*, *about*) make petty objections; carp. ● n. a trivial objection. □ **caviller** n. [F *caviller* f. L *cavillari* f. *cavilla* mockery]

caving /ˈkeɪvɪŋ/ n. exploring caves as a sport or pastime.

cavitation /ˌkævɪˈteɪʃ(ə)n/ n. **1** the formation of a cavity in a structure. **2** the formation of bubbles or cavities in a liquid caused by the movement of a propeller etc. through it.

cavity /ˈkævɪtɪ/ n. (pl. **-ies**) **1** a hollow within a solid body. **2** a decayed part of a tooth. □ **cavity wall** a wall formed from two skins of brick or blockwork with a space between. [F *cavité* or LL *cavitas* f. L *cavus* hollow]

cavort /kəˈvɔːt/ *v.intr. colloq.* caper excitedly; gambol, prance. [US, perh. f. CURVET]

Cavour /kəˈvʊə(r)/, Camillo Benso, Conte di (1810–61), Italian statesman. He was the driving force behind the unification of Italy under Victor Emmanuel II, king of the kingdom of Sardinia. In 1847 Cavour founded the newspaper *Il Risorgimento* to further the cause of unification. As Premier of Piedmont (1852–59; 1860–1), he obtained international support by forming an alliance with France and participating in the Crimean and Franco-Austrian wars. In 1861 he saw Victor Emmanuel crowned king of a united Italy, and became Italy's first Premier.

cavy /ˈkeɪvɪ/ n. (pl. **-ies**) a South American rodent of the family Caviidae, having a sturdy body and vestigial tail, esp. a guinea pig. [mod.L *cavia* f. Galibi *cabiai*]

caw /kɔː/ n. & v. ● n. the harsh cry of a rook, crow, etc. ● *v.intr.* utter this cry. [imit.]

Cawley /ˈkɔːlɪ/, Evonne (Fay) (née Goolagong) (b.1951), Australian tennis player. She won two Wimbledon singles titles (1971; 1980) and was three times Australian singles champion (1974–6).

Cawnpore see KANPUR.

Caxton /ˈkækstən/, William (c.1422–91), the first English printer. Having learned the art of printing on the Continent, Caxton printed his first English text in 1474 and went on to produce about eighty other texts, including editions of Malory's *Le Morte d'Arthur*, Chaucer's *Canterbury Tales*, and his own translations of French romances.

cay /keɪ/ n. a low insular bank or reef of coral, sand, etc. (cf. KEY[2]). [Sp. *cayo* shoal, reef f. F *quai*: see QUAY]

Cayenne /keɪˈen/ the capital and chief port of French Guiana; pop. (1990) 41,600.

cayenne /keɪˈen, ˈkeɪen/ n. (in full **cayenne pepper**) a pungent red powder prepared from ground dried chillies and used for seasoning. [Tupi *kyynha* assim. to CAYENNE]

Cayley[1] /ˈkeɪlɪ/, Arthur (1821–95), English mathematician and barrister. Cayley wrote almost a thousand mathematical papers in algebra and geometry. These include articles on determinants, the newly developing group theory, and the algebra of matrices. He also studied dynamics and physical astronomy. The Cayley numbers, a generalization of complex numbers, are named after him.

Cayley[2] /ˈkeɪlɪ/, Sir George (1773–1857), British engineer, the father of British aeronautics. He is best known for his understanding of the principles of flight, his model gliders, and the first man-carrying glider flight in 1853. Cayley's research, inventions, and designs covered schemes and devices for land reclamation, artificial limbs, theatre architecture, railways, lifeboats, finned projectiles, optics, electricity, hot-air engines, and what was later called the Caterpillar tractor. He was a founder of the original Polytechnic Institution, and was an MP for a time.

cayman /ˈkeɪmən/ n. (also **caiman**) a crocodilian of the family Alligatoridae, esp. the genus *Caiman*, native to Central and South America, and smaller than alligators. [Sp. & Port. *caiman*, f. Carib *acayuman*]

Cayman Islands /ˈkeɪmən/ (also **Caymans**) a group of three islands in the Caribbean Sea, south of Cuba; pop. (est. 1992) 27,000; official language, English; capital, George Town. The Cayman Islands are a British dependency. Columbus, discovering the uninhabited islands in 1503, named them Las Tortugas (Sp., = the turtles) because of their abundance of turtles. A British colony was established in the late 17th century.

Cayuga /keɪˈjuːgə, kaɪ-/ n. & adj. ● n. (pl. same or **Cayugas**) **1** a member of an Iroquois people, one of the five comprising the original Iroquois confederacy, formerly inhabiting part of New York State. **2** the language of this people. ● adj. of or relating to the Cayugas or their language. [Iroquoian]

CB abbr. **1** citizens' band. **2** (in the UK) Companion of the Order of the Bath.

Cb symb. US Chem. the element columbium.

CBC abbr. Canadian Broadcasting Corporation.

CBE abbr. Commander of the Order of the British Empire.

CBI see CONFEDERATION OF BRITISH INDUSTRY.

CBS abbr. (in the US) Columbia Broadcasting System.

CC *abbr.* **1** *Brit.* **a** City Council. **b** County Council. **c** County Councillor. **2** Cricket Club. **3** Companion of the Order of Canada.

cc *abbr.* (also **c.c.**) **1** cubic centimetre(s). ¶ Often in descriptions of vehicle engine capacity. **2** carbon copy.

CCD *abbr. Electronics* charge-coupled device, a high-speed semiconductor device used esp. in image detection.

CD *abbr.* **1** compact disc. **2** Civil Defence. **3** *Corps Diplomatique.*

Cd *symb. Chem.* the element cadmium.

Cd. *abbr.* Command Paper (1900–18).

cd *abbr.* candela.

CDC *abbr.* **1** (in the US) Centers for Disease Control. **2** see COMMONWEALTH DEVELOPMENT CORPORATION.

Cdr. *abbr. Mil.* Commander.

Cdre. *abbr.* Commodore.

CD-ROM /ˌsiːdiːˈrɒm/ *abbr.* compact disc read-only memory (for retrieval of text or data on a VDU screen).

CDT *abbr.* **1** Central Daylight Time, one hour ahead of Central Standard Time. **2** *Education* craft, design, and technology.

CD-video /ˌsiːdiːˈvɪdɪəʊ/ *n.* a system of simultaneously reproducing high-quality sound and video pictures from a compact disc.

CE *abbr.* **1** Church of England. **2** civil engineer. **3** Common Era.

Ce *symb. Chem.* the element cerium.

ceanothus /ˌsiːəˈnəʊθəs/ *n.* a North American shrub of the genus *Ceanothus*, of the buckthorn family, with small blue or white flowers. [mod.L f. Gk *keanōthos* kind of thistle]

Ceará /ˌseɪəˈrɑː/ a state in NE Brazil, on the Atlantic coast; capital, Fortaleza.

cease /siːs/ *v. & n.* ● *v.tr. & intr.* stop; bring or come to an end (*ceased breathing*). ● *n.* (in **without cease**) unendingly. □ **cease fire** *Mil.* stop firing. **cease-fire** *n.* **1** the order to do this. **2** a period of truce; a suspension of hostilities. [ME f. OF *cesser*, L *cessare* frequent. of *cedere cess-* yield]

ceaseless /ˈsiːslɪs/ *adj.* without end; not ceasing. □ **ceaselessly** *adv.*

Ceauşescu /tʃaʊˈʃɛskuː/, Nicolae (1918–89), Romanian Communist statesman, first President of the Socialist Republic of Romania 1974–89. Noted for his independence of the USSR, for many years he fostered his own personality cult, making his wife Elena his deputy and appointing many other members of his family to high office. His regime became increasingly totalitarian, repressive, and corrupt; a popular uprising in Dec. 1989 resulted in its downfall and in the arrest, summary trial, and execution of Ceauşescu and his wife.

Cebu /sɪˈbuː/ **1** an island of the south central Philippines. **2** its chief city and port; pop. (1990) 610,000. Ferdinand Magellan landed there in 1521.

Cecil /ˈses(ə)l, ˈsɪs-/, William, see BURGHLEY.

Cecilia, St /sɪˈsiːljə/ (2nd or 3rd century), Roman martyr. According to legend, she took a vow of celibacy but was forced to marry a young Roman; she converted her husband to Christianity and both were martyred. She is frequently pictured playing the organ and is the patron saint of church music. Feast day, 22 Nov.

cecitis *US var. of* CAECITIS.

cecum *US var. of* CAECUM.

cedar /ˈsiːdə(r)/ *n.* **1** a spreading evergreen conifer of the genus *Cedrus*, bearing tufts of small needles and cones of papery scales; esp. the *Cedar of Lebanon, C. libani,* native to Asia Minor. **2** a similar conifer yielding timber. **3** (in full **cedar wood**) the fragrant durable wood of any cedar tree. □ **cedarn** *adj. poet.* [ME f. OF *cedre* f. L *cedrus* f. Gk *kedros*]

cede /siːd/ *v.tr.* give up one's rights to or possession of. [F *céder* or L *cedere* yield]

cedi /ˈsiːdɪ/ *n.* the basic monetary unit of Ghana, equal to 100 pesewas. [W. Afr. language]

cedilla /sɪˈdɪlə/ *n.* **1** a mark written under the letter *c*, esp. in French, to show that it is sibilant (as in *façade*). **2** a similar mark under *s* in Turkish and other oriental languages. [Sp. *cedilla* dimin. of *zeda* f. Gk *zēta* letter Z]

Ceefax /ˈsiːfæks/ *n. Brit. propr.* a teletext service provided by the BBC. [repr. pronunc. of *seeing + facsimile*]

CEGB *abbr.* (in the UK) Central Electricity Generating Board.

ceilidh /ˈkeɪlɪ/ *n. orig. Ir. & Sc.* an informal gathering for conversation, music, dancing, songs, and stories. [Gael.]

ceiling /ˈsiːlɪŋ/ *n.* **1 a** the upper interior surface of a room or other similar compartment. **b** the material forming this. **2** an upper limit on prices, wages, performance, etc. **3** *Aeron.* the maximum altitude a given aircraft can reach. **4** *Naut.* the inside planking of a ship's bottom and sides. [ME *celynge, siling*, perh. ult. f. L *caelum* heaven or *celare* hide]

celadon /ˈselədɒn/ *n. & adj.* ● *n.* **1** a willow-green colour. **2** a grey-green glaze used on some pottery. **3** Chinese pottery glazed in this way. ● *adj.* of a grey-green colour. [F, f. the name of a character in *L'Astrée* by Honoré d'Urfé (1567–1625)]

celandine /ˈseləndaɪn/ *n.* **1** (in full **greater celandine**) a yellow-flowered plant, *Chelidonium majus*, of the poppy family. **2** (in full **lesser celandine**) a small yellow-flowered plant, *Ranunculus ficaria*, of the buttercup family. [ME and OF *celidoine* ult. f. Gk *khelidōn* swallow: the flowering of the plant was associated with the arrival of swallows]

-cele /siːl/ *comb. form* (also **-coele**) *Med.* swelling, hernia (*gastrocele*). [Gk *kēlē* tumour]

celeb /sɪˈleb/ *n. colloq.* a celebrity, a star.

Celebes /ˈselɪbiːz/ the former name for SULAWESI.

Celebes Sea a part of the western Pacific between the Philippines and Sulawesi, bounded to the west by Borneo. It is linked to the Java Sea by the Makassar Strait.

celebrant /ˈselɪbrənt/ *n.* a person who performs a rite, esp. a priest at the Eucharist. [F *célébrant* or L *celebrare celebrant-*: see CELEBRATE]

celebrate /ˈselɪbreɪt/ *v.* **1** *tr.* mark (a festival or special event) with festivities etc. **2** *tr.* perform publicly and duly (a religious ceremony etc.). **3 a** *tr.* officiate at (the Eucharist). **b** *intr.* officiate, esp. at the Eucharist. **4** *intr.* engage in festivities, usu. after a special event etc. **5** *tr.* (esp. as **celebrated** *adj.*) honour publicly, make widely known. □ **celebrator** *n.* **celebratory** *adj.* **celebration** /ˌselɪˈbreɪʃ(ə)n/ *n.* [L *celebrare* f. *celeber -bris* frequented, honoured]

celebrity /sɪˈlebrɪtɪ/ *n.* (*pl.* **-ies**) **1** a well-known person. **2** fame. [F *célébrité* or L *celebritas f. celeber*: see CELEBRATE]

celeriac /sɪˈlerɪˌæk/ *n.* a variety of celery with a swollen turnip-like stem-base used as a vegetable. [CELERY: *-ac* is unexplained]

celerity /sɪˈlerɪtɪ/ *n. archaic or literary* swiftness (esp. of a living creature). [ME f. OF *célérité* f. L *celeritas -tatis* f. *celer* swift]

celery /ˈselərɪ/ *n.* an umbelliferous plant, *Apium graveolens*, with closely packed succulent leaf-stalks used as a vegetable. □ **celery pine** an Australasian tree, *Phyllocladus trichomanoides*, with branchlets like celery leaves. [F *céleri* f. It. dial. *selleri* f. L *selinum* f. Gk *selinon* parsley]

celesta /sɪˈlestə/ *n. Mus.* a small keyboard instrument resembling a glockenspiel, with hammers striking steel plates suspended over wooden resonators, giving an ethereal bell-like sound. [pseudo-L f. F *céleste*: see CELESTE]

celeste /sɪˈlest/ *n. Mus.* **1** an organ and harmonium stop with a soft tremulous tone. **2** = CELESTA. [F *céleste* heavenly f. L *caelestis* f. *caelum* heaven]

celestial /sɪˈlestɪəl/ *adj.* **1** heavenly; divinely good or beautiful; sublime. **2** of the sky, or of outer space as observed in astronomy etc. □ **celestial body** the sun, the moon, a planet, a star, etc. **celestial coordinate** *Astron.* either of two coordinates, declination and right ascension, that define a position on the celestial sphere. **celestial equator** *Astron.* the great circle of the sky in the plane perpendicular to the earth's axis. **celestial horizon** see HORIZON 1c. **celestial mechanics** *Astron.* the mathematical description of the positions and motions of celestial bodies on the celestial sphere. **celestial navigation** navigation by the stars etc. **celestial sphere** *Astron.* the abstract sphere of unit radius on which the positions of celestial objects are projected to form a map of the heavens, whose poles and equator are projections of the corresponding terrestrial features. □ **celestially** *adv.* [ME f. OF f. med.L *caelestialis* f. L *caelestis*: see CELESTE]

celestial pole *n. Astron.* the point on the celestial sphere directly above the earth's geographic pole (north or south), around which the stars and planets appear to rotate during the course of the night. The north celestial pole is currently within one degree of the star Polaris, but because of precession it appears to trace out a circle on the celestial sphere over a period of some 26,000 years.

celiac *US var. of* COELIAC.

celibate /ˈselɪbət/ *adj. & n.* ● *adj.* **1** committed to abstention from sexual relations and from marriage, esp. for religious reasons. **2** abstaining from sexual relations. ● *n.* a celibate person. □ **celibacy** *n.* [F *célibat* or L *caelibatus* unmarried state f. *caelebs -ibis* unmarried]

cell /sel/ *n.* **1** a small room, esp. in a prison or monastery. **2** a small compartment, e.g. in a honeycomb. **3** a small group as a nucleus of political activity, esp. of a subversive kind. **4** *hist.* a small monastery or nunnery dependent on a larger one. **5** *Biol.* **a** the structural and functional unit of which organisms consist. (*See note below.*) **b** an enclosed cavity in an organism etc. **6** *Electr.* a vessel for containing electrodes within an electrolyte for current-generation or electrolysis. □ **cell wall** *Biol.* the rigid layer that encloses a plant or bacterial cell. □ **celled** *adj.* (also in *comb.*). [ME f. OF *celle* or f. L *cella* storeroom etc.]

▪ Nearly all organisms are composed of cells (the simplest consist of a single cell), which are microscopic structures bounded by a membrane and capable of metabolism, self-repair, and reproduction. There are two basic types of cell: those of prokaryotes (bacteria and blue-green algae), which have nuclei and other distinct structures, and those of eukaryotes (animals and plants), which are larger and more complex and have the genetic material (DNA) contained in a membrane-bound nucleus. The cytoplasm of eukaryotic cells contains various specialized structures (organelles), such as mitochondria. The membrane of a cell regulates the exchange of materials with its environment. In large organisms such as humans there are many specialized types of cell lacking some of the features described above; for example, red blood cells have no nuclei. Most plant cells have a thick wall of cellulose outside the cell membrane, a large fluid-filled cavity within the cytoplasm, and (in green plants) structures that contain chlorophyll.

cellar /ˈselə(r)/ *n. & v.* ● *n.* **1** a room below ground level in a house, used for storage, esp. of wine or coal. **2** a stock of wine in a cellar (*has a good cellar*). ● *v.tr.* store or put in a cellar. [ME f. AF *celer*, OF *celier* f. LL *cellarium* storehouse]

cellarage /ˈselərɪdʒ/ *n.* **1** cellar accommodation. **2** the charge for the use of a cellar or storehouse.

cellarer /ˈselərə(r)/ *n.* a monastic officer in charge of wine.

cellaret /ˌseləˈret/ *n.* a case or sideboard for holding wine bottles in a dining-room.

Cellini /tʃɪˈliːniː/, Benvenuto (1500–71), Italian goldsmith and sculptor. His work is characterized by its elaborate virtuosity; the salt-cellar of gold and enamel which he made while working for Francis I of France is an outstanding example and typifies the late Renaissance style. The latter part of his life was spent in Florence, where he cast the bronze *Perseus* (1545–54), regarded as his masterpiece. His autobiography is famous for its racy style and its vivid picture of Italian Renaissance life.

Cellnet /ˈselnet/ *n. propr.* a cellular telephone service.

cello /ˈtʃeləʊ/ *n.* (pl. **-os**) a large bass to tenor instrument of the violin family, played while being held upright on the floor between the legs of the seated player. It is a standard member of the orchestra and of string quartets, while its four-octave range and deep, expressive tone make it an ideal solo instrument, for which there is a large repertoire. □ **cellist** *n.* [abbr. of VIOLONCELLO]

Cellophane /ˈseləˌfeɪn/ *n. propr.* a thin transparent wrapping material made from viscose, first produced in Switzerland in 1908. [CELLULOSE + *-phane* (cf. DIAPHANOUS)]

cellphone /ˈselfəʊn/ *n.* a small portable radio-telephone having access to a cellular radio system.

cellular /ˈseljʊlə(r)/ *adj.* **1** of or having small compartments or cavities. **2** of open texture; porous. **3** *Physiol.* of or consisting of cells. **4** designating or relating to a mobile telephone system that uses a number of short-range radio stations to cover the area it serves, the signal being automatically switched from one station to another as the user travels about. □ **cellular blanket** a blanket of open texture. **cellular plant** a plant with no distinct stem, leaves, etc. **cellular radio** a system of mobile radio-telephone transmission with an area divided into 'cells' each served by its own small transmitter. □ **cellulous** *adj.* **cellulate** /-lət/ *adj.* **cellularity** /ˌseljʊˈlærɪtɪ/ *n.* **cellulation** /-ˈleɪʃ(ə)n/ *n.* [F *cellulaire* f. mod.L *cellularis*: see CELLULE]

cellule /ˈseljuːl/ *n. Biol.* a small cell or cavity. [F *cellule* or L *cellula* dimin. of *cella* CELL]

cellulite /ˈseljʊˌlaɪt/ *n.* a lumpy form of fat, esp. on the hips and thighs of women, causing puckering of the skin. [F (as CELLULE)]

cellulitis /ˌseljʊˈlaɪtɪs/ *n.* inflammation of subcutaneous connective tissue.

celluloid /ˈseljʊˌlɔɪd/ *n.* **1** a transparent flammable plastic made from camphor and cellulose nitrate. **2** cinema film. [CELLULOSE + -OID]

cellulose /ˈseljʊˌləʊz, -ˌləʊs/ *n.* **1** *Biochem.* a carbohydrate forming the main constituent of plant-cell walls. (*See note below.*) **2** (in general use) a paint or lacquer consisting of esp. cellulose acetate or nitrate in

solution. □ **cellulose acetate** the cellulose ester of acetic acid, widely used as an artificial fibre or plastic. **cellulose nitrate** = NITROCELLULOSE. □ **cellulosic** /ˌseljʊˈləʊsɪk/ *adj.* [F (as CELLULE)]

▪ Cellulose, the main structural material of plants, has a molecular structure consisting of long unbranched chains of glucose molecules. It is a major constituent of wood, from which it is produced industrially. Paper and plant-based textile fibres such as cotton consist largely of cellulose, and in chemically modified forms it is used to make rayon, some plastics, and other products. It is important in the human diet as a constituent of dietary fibre.

celom *US* var. of COELOM.

Celsius¹ /ˈselsɪəs/, Anders (1701–44), Swedish astronomer, best known for his thermometer scale. He was professor of astronomy at Uppsala, and joined an expedition to measure a meridian in the north, which verified Newton's theory that the earth is flattened at the poles. In 1742 he advocated a metric temperature scale with 100° as the freezing-point of water and 0° as the boiling-point; however, the thermometer which was actually introduced at the Uppsala Observatory had its scale reversed.

Celsius² /ˈselsɪəs/ *adj.* of or denoting a temperature on the Celsius scale. □ **Celsius scale** a scale of temperature on which water freezes at 0° and boils at 100° under standard conditions (also called *centigrade*). [CELSIUS¹]

Celt /kelt/ *n.* a member of a group of peoples who occupied a large part of western Europe in the Iron Age, including the pre-Roman inhabitants of Britain and Gaul and their descendants. The culture of the Celts can be traced back to the Bronze Age of the upper Danube in the 13th century BC, with successive stages represented by the urnfield and Hallstatt cultures. Spreading over western and central Europe from perhaps as early as 900 BC, they reached the height of their power in the La Tène period of the 5th–1st centuries BC; they sacked Rome in 390 BC before being conquered by the Romans and by Germanic tribes and confined largely to Gaul and the British Isles. Today their languages survive in Ireland, Scotland, Wales, and Brittany. The Celts made significant developments in agriculture and metal-working: they cultivated fields on a regular basis with ox-drawn ploughs in place of manual implements, and have left much fine jewellery and metalwork. [L *Celtae* (pl.) f. Gk *Keltoi*]

celt /kelt/ *n. Archaeol.* a stone or metal prehistoric implement with a chisel edge. [med.L *celtes* chisel]

Celtic /ˈkeltɪk, ˈsel-/ *adj. & n.* ● *adj.* of or relating to the Celts or their languages. ● *n.* a branch of the Indo-European language family, today spoken by some in parts of the British Isles and in Brittany. (*See note below.*) □ **Celtic cross** a Latin cross with a circle round the centre. □ **Celticism** /-tɪˌsɪz(ə)m/ *n.* [L *celticus* (as CELT) or F *celtique*]

▪ The Celtic languages are divided into three groups, Continental Celtic (including the dead languages of pre-Roman and early Roman Europe, such as Gaulish), Goidelic (including Irish, Scottish Gaelic, and Manx), and Brythonic (including Welsh, Cornish, and Breton). A characteristic feature of Celtic languages is lenition, in which the initial consonant of a word is pronounced differently depending on its syntactic role and the preceding word. Thus, the vocative of the Gaelic name *Màiri* /ˈmaːrɪ/ is *A Mhàiri* /ə ˈvaːrɪ/. (See also BRYTHONIC, GOIDELIC.)

Celtic Church the Christian Church in the British Isles from its foundation in the 2nd or 3rd century until its assimilation into the Roman Catholic Church. It was largely driven out of England by the Saxons in the 5th century, surviving in Scotland, Wales, and Ireland, from where it attempted to convert the invaders by means of missionaries. St Augustine's expedition from Rome in 597 refounded the Church in England but failed to reconcile the differences that had developed between the Celtic and Roman Churches; these were resolved by the Synod of Whitby in 664, a date which effectively marked the end of the Celtic Church in England, although it continued to exist in Wales, Scotland, and Ireland until about the 12th century.

Celtic Sea the part of the Atlantic Ocean between southern Ireland and SW England.

cembalo /ˈtʃembəˌləʊ/ *n.* (pl. **-os**) a harpsichord. □ **cembalist** *n.* [abbr. of CLAVICEMBALO]

cement /sɪˈment/ *n. & v.* ● *n.* **1** a powdery substance made by calcining lime and clay, mixed with water to form mortar or used in concrete (see also PORTLAND CEMENT). **2** any similar substance that hardens and fastens on setting. **3** a uniting factor or principle. **4** a substance for filling cavities in teeth. **5** (also **cementum**) *Anat.* a thin layer of bony material that fixes teeth to the jaw. ● *v.tr.* **1 a** unite with or as with cement. **b** establish or strengthen (a friendship etc.). **2** apply cement to. **3** line or cover with cement. □ **cement-mixer** a machine (usu. with a revolving drum) for mixing cement with water. □ **cementer** n. [ME f. OF *ciment* f. L *caementum* quarry stone f. *caedere* hew]

cementation /ˌsiːmenˈteɪʃ(ə)n/ n. **1** the act or process of cementing or being cemented. **2** the heating of iron with charcoal powder to form steel.

cemetery /ˈsemɪtərɪ/ n. (pl. **-ies**) a burial ground, esp. one not in a churchyard. [LL *coemeterium* f. Gk *koimētērion* dormitory f. *koimaō* put to sleep]

C.Eng. abbr. Brit. chartered engineer.

cenobite US var. of COENOBITE.

cenotaph /ˈsenəˌtɑːf/ n. a tomblike monument, esp. a war memorial, to a dead person whose body is elsewhere. [F *cénotaphe* f. LL *cenotaphium* f. Gk *kenos* empty + *taphos* tomb]

Cenotaph, the a monument, designed by Sir Edwin Lutyens, erected in 1919–20 in Whitehall, London, as a memorial to the British servicemen who died in the First World War. An inscription now commemorates also those who died in the Second World War.

Cenozoic /ˌsiːnəˈzəʊɪk/ adj. & n. (also **Cainozoic** /ˌkaɪnə-/) Geol. of, relating to, or denoting the most recent geological era, following the Mesozoic and lasting from about 65 million years ago to the present day. The Cenozoic includes the Tertiary and Quaternary periods, and has seen the rapid evolution of mammals. [Gk *kainos* recent + *zōē* life]

censer /ˈsensə(r)/ n. a vessel in which incense is burnt, esp. during a religious procession or ceremony. [ME f. AF *censer*, OF *censier* aphetic of *encensier* f. *encens* INCENSE[1]]

censor /ˈsensə(r)/ n. & v. ● n. **1** an official authorized to examine printed matter, films, news, etc., before public release, and to suppress any parts on the grounds of obscenity, a threat to security, etc. **2** Rom. Hist. either of two annual magistrates responsible for holding censuses and empowered to supervise public morals. **3** Psychol. an impulse which is said to prevent certain ideas and memories from emerging into consciousness. ● v.tr. **1** act as a censor of. **2** make deletions or changes in. ¶ As a verb, often confused with *censure*. □ **censorship** n. **censorial** /senˈsɔːrɪəl/ adj. [L f. *censere* assess: in sense 3 mistransl. G *Zensur* censorship]

censorious /senˈsɔːrɪəs/ adj. severely critical; fault-finding; quick or eager to criticize. □ **censoriously** adv. **censoriousness** n. [L *censorius*: see CENSOR]

censure /ˈsenʃə(r)/ v. & n. ● v.tr. criticize harshly; reprove. ¶ Often confused with *censor*. ● n. harsh criticism; expression of disapproval. □ **censurable** adj. [ME f. OF f. L *censura* f. *censere* assess]

census /ˈsensəs/ n. (pl. **censuses**) the official count of a population or of a class of things, often with various statistics noted. [L f. *censere* assess]

cent /sent/ n. **1 a** a monetary unit in various countries, equal to one-hundredth of a dollar or other decimal currency unit. **b** a coin of this value. **2** colloq. a very small sum of money. **3** see PER CENT. [F *cent* or It. *cento* or L *centum* hundred]

cent. abbr. century.

centaur /ˈsentɔː(r)/ n. Gk Mythol. a member of a tribe of creatures with the head, arms, and torso of a man and the body and legs of a horse. They are often portrayed as wild and drunken, although some, notably Chiron, were learned teachers of humans. [ME f. L *centaurus* f. Gk *kentauros*, of unkn. orig.]

Centaurus /senˈtɔːrəs/ n. Astron. a large southern constellation (the Centaur), said to represent the figure of a centaur. It lies in the Milky Way and is very rich in stars, containing the stars Alpha and Proxima Centauri, and one of the finest globular clusters. [L]

centaury /ˈsentɔːrɪ/ n. (pl. **-ies**) a plant of the genus *Centaurium*, of the gentian family; esp. C. *erythraea*, with pink flowers, formerly used in medicine. [LL *centaurea* ult. f. Gk *kentauros* CENTAUR: from the legend that it was discovered by the centaur Chiron]

centavo /senˈtɑːvəʊ/ n. a small coin of Spain, Portugal, Mexico, Brazil, and other (chiefly Latin American) countries, worth one-hundredth of the standard unit. [Sp. f. L *centum* hundred]

CENTCOM, Centcom /ˈsentkɒm/ see CENTRAL COMMAND.

centenarian /ˌsentɪˈneərɪən/ n. & adj. ● n. a person a hundred or more years old. ● adj. a hundred or more years old.

centenary /senˈtiːnərɪ/ n. & adj. ● n. (pl. **-ies**) **1** a hundredth anniversary. **2** a celebration of this. ● adj. **1** of or relating to a centenary.

2 occurring every hundred years. [L *centenarius* f. *centeni* a hundred each f. *centum* a hundred]

centennial /senˈtenɪəl/ adj. & n. ● adj. **1** lasting for a hundred years. **2** occurring every hundred years. ● n. = CENTENARY n. [L *centum* a hundred, after BIENNIAL]

center US var. of CENTRE.

centerboard US var. of CENTREBOARD.

centerfold US var. of CENTREFOLD.

centering US var. of CENTRING.

centesimal /senˈtesɪm(ə)l/ adj. reckoning or reckoned by hundredths. □ **centesimally** adv. [L *centesimus* hundredth f. *centum* hundred]

centésimo /senˈtesɪˌməʊ/ n. (pl. **-os**) a monetary unit of Uruguay (equal to one-hundredth of a peso) and Panama (equal to one-hundredth of a balboa). [Sp.]

centi- /ˈsentɪ/ comb. form **1** one-hundredth, esp. of a unit in the metric system (*centigram*; *centilitre*). **2** hundred (abbr. **c**). [L *centum* hundred]

centigrade /ˈsentɪˌɡreɪd/ adj. **1** = CELSIUS[2]. **2** having a scale of a hundred degrees. ¶ In sense 1 *Celsius* is usually preferred in technical use. [F f. L *centum* hundred + *gradus* step]

centigram /ˈsentɪˌɡræm/ n. (also **centigramme**) a metric unit of mass, equal to one-hundredth of a gram (symbol: **cg**).

centilitre /ˈsentɪˌliːtə(r)/ n. (US **centiliter**) a metric unit of capacity, equal to one-hundredth of a litre (symbol: **cl**).

centime /ˈsɒntiːm/ n. **1** a monetary unit in various countries, equal to one-hundredth of a franc or other decimal currency unit. **2** a monetary unit equal to one-hundredth of a gourde in Haiti or of a dirham in Morocco. [F f. L *centum* a hundred]

centimetre /ˈsentɪˌmiːtə(r)/ n. (US **centimeter**) a metric unit of length, equal to one-hundredth of a metre (symbol: **cm**). □ **centimetre-gram-second** denoting a system of measure using these as basic units of length, mass, and time (abbr. **cgs**).

centimo /ˈsentɪˌməʊ/ n. (pl. **-os**) a monetary unit of Spain and a number of Latin American countries, equal to one-hundredth of the basic unit. [Sp.]

centipede /ˈsentɪˌpiːd/ n. a predatory arthropod of the class Chilopoda, with an elongated flattened body of many segments, most of which bear a pair of legs. [F *centipède* or L *centipeda* f. *centum* hundred + *pes pedis* foot]

cento /ˈsentəʊ/ n. (pl. **-os**) a composition made up of quotations from other authors. [L, = patchwork garment]

central /ˈsentr(ə)l/ adj. **1** of, at, or forming the centre. **2** from the centre. **3** chief, essential, most important. □ **central heating** a method of warming a building by pipes, radiators, etc., fed from a central source of heat. **central locking** a locking system in motor vehicles whereby the locks of several doors can be operated from a single lock. **central nervous system** see NERVOUS SYSTEM. **central processor** (or **processing unit**) (abbr. **CPU**) the part of a computer in which the control and execution of operations occur, consisting mainly of a microprocessor. **central reservation** see RESERVATION 5. □ **centrally** adv. **centrality** /senˈtrælɪtɪ/ n. [F *central* or L *centralis* f. *centrum* CENTRE]

Central African Republic a country of central Africa; pop. (est. 1991) 3,113,000; languages, French (official), Sango; capital, Bangui. Formerly the French colony of Ubanghi Shari, it became a republic within the French Community in 1958 and a fully independent state in 1960. In 1976 its President, Jean Bédel Bokassa, declared himself emperor and changed the country's name to Central African Empire, but it reverted to the name of Republic after he was ousted in 1979.

Central America the southernmost part of North America linking the continent to South America and consisting of the countries of Guatemala, Belize, Honduras, El Salvador, Nicaragua, Costa Rica, and Panama.

Central American Common Market an economic organization comprising Guatemala, Honduras, El Salvador, Nicaragua, and Costa Rica, set up in 1960 to reduce trade barriers, stimulate exports, and encourage industrialization. It lost impetus in the 1970s and by the mid-1980s was in a state of suspension.

central bank n. a national bank that provides financial and banking services for its country's government and commercial banking system, as well as implementing the government's monetary policy and issuing currency. Major central banks include the Bank of England in the UK, the Federal Reserve Bank of the US, the *Bundesbank* in Germany, and France's *Banque de France*.

Central Command (abbr. **CENTCOM** or **Centcom**) a US military strike force consisting of units from the army, air force, and navy, established in 1979 (as the Rapid Deployment Force) to operate in the Middle East and North Africa. Led by Norman Schwarzkopf, it was involved in the Gulf War of 1991.

Central Criminal Court see OLD BAILEY.

Central Intelligence Agency (abbr. **CIA**) a Federal agency in the US, established in 1947, responsible for coordinating government intelligence activities. It was developed for overseas use in the cold war, but has since figured in domestic operations. Abroad, the agency has been involved, amongst other operations, in the Bay of Pigs incident in Cuba (1961) and the coup against President Allende of Chile (1973).

centralism /ˈsentrəˌlɪz(ə)m/ n. a system that centralizes (esp. an administration) (see also *democratic centralism*). □ **centralist** n.

centralize /ˈsentrəˌlaɪz/ v. (also **-ise**) **1** tr. & intr. bring or come to a centre. **2** tr. **a** concentrate (administration) at a single centre. **b** subject (a state) to this system. □ **centralization** /ˌsentrəlaɪˈzeɪʃ(ə)n/ n.

Central Park a large public park in the city of New York, in the centre of Manhattan.

Central Powers a Triple Alliance of Germany, Austria–Hungary, and Italy, between 1882 and 1914.

Central Region a local government region in central Scotland; administrative centre, Stirling.

Central Standard Time (also **Central Time**) (abbr. **CST**) the standard time in a zone including parts of central Canada and the US, six hours behind GMT.

Centre /ˈsɒntrə/ a region of central France, including the cities of Orleans, Tours, Chartres, and Bourges.

centre /ˈsentə(r)/ n. & v. (US **center**) ● n. **1** the middle point, esp. of a line, circle, or sphere, equidistant from the ends or from any point on the circumference or surface. **2** a pivot or axis of rotation. **3 a** a place or group of buildings forming a central point in a district, city, etc., or a main area for an activity (*shopping centre*; *town centre*). **b** (with preceding word) a piece or set of equipment for a number of connected functions (*music centre*). **4** a point of concentration or dispersion; a nucleus or source. **5** a political party or group holding moderate opinions. **6** the filling in a chocolate etc. **7** *Sport* **a** the middle player in a line or group in some field games. **b** a kick or hit from the side to the centre of the pitch. **8** (in a lathe etc.) a conical adjustable support for the workpiece. **9** (*attrib.*) of or at the centre. ● v. **1** intr. (foll. by *in*, *on*; *disp.* foll. by *round*) have as its main centre. **2** tr. place in the centre. **3** tr. mark with a centre. **4** tr. (foll. by *in*, *on* etc.) concentrate or focus (a thing) in, around, etc. **5** tr. *Sport* kick or hit (the ball) from the side to the centre of the pitch. □ **centre-bit** a boring-tool with a centre point and side cutters.

centre forward *Sport* the middle player or position in a forward line.
centre half *Sport* the middle player or position in a defensive line.
centre of attention 1 a person or thing that draws general attention.
2 *Physics* the point to which bodies tend by gravity. **centre of gravity** a point from which the weight of a body or system may be considered to act (in uniform gravity the same as the centre of mass). **centre of mass** a point representing the mean position of the matter in a body or system. **centre-piece 1** an ornament for the middle of a table. **2** a principal item. **centre spread** the two facing middle pages of a newspaper etc. □ **centred** adj. (often in *comb.*). **centremost** adj. **centric** /ˈsentrɪk/ adj. **centrical** adj. **centricity** /senˈtrɪsɪtɪ/ n. [ME f. OF *centre* or L *centrum* f. Gk *kentron* sharp point]

centreboard /ˈsentəˌbɔːd/ n. (US **centerboard**) a board for lowering through a boat's keel to prevent leeway.

centrefold /ˈsentəˌfəʊld/ n. (US **centerfold**) **1** a printed and usu. illustrated sheet folded to form the centre spread of a magazine etc. **2** a model, usu. naked or scantily clad, pictured on such a spread.

centreing var. of CENTRING.

-centric /ˈsentrɪk/ comb. form forming adjectives with the sense 'having a (specified) centre' (*anthropocentric*; *eccentric*). [after concentric etc. f. Gk *kentrikos*: see CENTRE]

centrifugal /senˈtrɪfjʊg(ə)l, ˌsentrɪˈfjuːg(ə)l/ adj. moving or tending to move outwards from a centre (cf. CENTRIPETAL). □ **centrifugally** adv. [mod.L *centrifugus* f. L *centrum* centre + *fugere* flee]

centrifugal force n. an apparent force that acts outwards on a body moving about a centre. Centrifugal force is actually a manifestation of inertia. A body in uniform circular motion, e.g. at the end of a string fixed at the other end, is constantly accelerating because its direction of motion is always changing. The force causing this acceleration is

actually centripetal and is manifested as tension in the string. If the string is cut (and the centripetal force ceases) the body will tend through inertia to move in a straight line at a tangent to its previous circular path. It is possible to assign a magnitude to the centrifugal force equal to that of the centripetal force.

centrifuge /ˈsentrɪˌfjuːdʒ/ n. & v. ● n. a container designed to rotate rapidly so as to apply centrifugal force to its contents, usu. in order to separate liquids from solids or other liquids (e.g. cream from milk). (*See note below.*) ● v.tr. **1** subject to the action of a centrifuge. **2** separate by centrifuge. □ **centrifugation** /ˌsentrɪfjʊˈgeɪʃ(ə)n/ n.

■ Simple centrifuges consist of a mechanically driven bowl or cylinder which turns inside a stationary casing; most have buckets fixed at the ends of a precisely balanced rotor. In all centrifuges the components are impelled to the periphery, those of lesser density collecting nearer the middle or else passing through a filter and being removed. Ultracentrifuges, very fast machines able to create forces many thousands of times that of gravity, are used in biochemistry to precipitate colloidal materials and large molecules from solution. Centrifugal separation of gases has been used to separate uranium isotopes in the production of nuclear fuels.

centring /ˈsentrɪŋ/ (also **centreing** /ˈsentər-/, US **centering**) n. a temporary frame used to support an arch, dome, etc., while under construction.

centriole /ˈsentrɪˌəʊl/ n. *Biol.* a minute organelle usu. within a centrosome involved esp. in the development of spindles in cell division. [med.L *centriolum* dimin. of *centrum* centre]

centripetal /senˈtrɪpɪt(ə)l, ˌsentrɪˈpiː t(ə)l/ adj. moving or tending to move towards a centre (cf. CENTRIFUGAL). □ **centripetal force** the force acting on a body causing it to move about a centre. □ **centripetally** adv. [mod.L *centripetus* f. L *centrum* centre + *petere* seek]

centrist /ˈsentrɪst/ n. *Polit.* often *derog.* a person who holds moderate views. □ **centrism** n.

centromere /ˈsentrəˌmɪə(r)/ n. *Biol.* the point on a chromosome by which it is attached to the spindle during cell division. [L *centrum* centre + Gk *meros* part]

centrosome /ˈsentrəˌsəʊm/ n. *Biol.* a distinct part of the cytoplasm in a cell, usu. near the nucleus, that contains the centriole. [G *Centrosoma* f. L *centrum* centre + Gk *sōma* body]

centuple /ˈsentjʊp(ə)l, senˈtjuː p-/ n., adj., & v. ● n. a hundredfold amount. ● adj. increased a hundredfold. ● v.tr. multiply by a hundred; increase a hundredfold. [F *centuple* or eccl.L *centuplus*, *centuplex* f. L *centum* hundred]

centurion /senˈtjʊərɪən/ n. the commander of a century in the ancient Roman army. [ME f. L *centurio -onis* (as CENTURY)]

century /ˈsentʃərɪ, ˈsentjʊrɪ/ n. (pl. **-ies**) **1 a** a period of one hundred years. **b** any of the centuries reckoned from the birth of Christ (*twentieth century* = 1901–2000; *fifth century* BC = 500–401 BC). ¶ In modern use often reckoned as (e.g.) 1900–1999. **2 a** a score etc. of a hundred in a sporting event, esp. a hundred runs by one batsman in cricket. **b** a group of a hundred things. **3 a** a company in the ancient Roman army, orig. of 100 men. **b** an ancient Roman political division for voting. □ **century plant** the American aloe (see AGAVE). [L *centuria* f. *centum* hundred]

cep /sep/ n. an edible fungus, *Boletus edulis*, with a stout stalk, a smooth brown cap, and pores rather than gills. [F *cèpe* f. Gascon *cep* f. L *cippus* stake]

cephalic /sɪˈfælɪk, kɪ-/ adj. of or in the head. □ **cephalic index** *Anthropol.* a number expressing the ratio of a head's greatest breadth and length. [F *céphalique* f. L *cephalicus* f. Gk *kephalikos* f. *kephalē* head]

-cephalic /sɪˈfælɪk, kɪ-/ comb. form = -CEPHALOUS.

cephalization /ˌsefəlaɪˈzeɪʃ(ə)n, ˌkef-/ n. (also **-isation**) *Zool.* the concentration of sense organs, nervous control, etc., at the anterior end of the body, forming a head and brain, both during evolution and in the course of an embryo's development.

Cephalonia /ˌsefəˈləʊnɪə, ˌkef-/ (Greek **Kefallinía** /ˌkɛfaliˈnia/) a Greek island in the Ionian Sea; pop. (1981) 31,300.

cephalopod /ˈsefələˌpɒd, ˈkef-/ n. & adj. *Zool.* ● n. a marine mollusc of the class Cephalopoda, having a distinct head with large eyes, and a ring of tentacles around the beaked mouth. (*See note below.*) ● adj. of or relating to this class. [mod.L f. Gk *kephalē* head + *pous podos* foot]

■ There are about 700 species of cephalopod, including octopuses, squid, cuttlefish, and nautiluses, but fossil species are much more numerous and include the ammonites and belemnites. They have large brains and are the most intelligent of all invertebrates. They are also notable for their well-developed eyes, which are quite similar to the vertebrate eye, and their ability to swim by a form of jet propulsion. The rare giant squid is the world's largest invertebrate.

cephalothorax /ˌsefələʊˈθɔːræks, ˌkef-/ n. (pl. **-thoraces** /-ˈθɔːrəˌsiːz/ or **-thoraxes**) Anat. the fused head and thorax of a spider, crab, or other arthropod.

-cephalous /ˈsefələs, ˈkef-/ comb. form -headed (brachycephalous; dolichocephalic). [Gk kephalē head]

cepheid /ˈsiːfiːd, ˈsef-/ n. (in full **cepheid variable**) Astron. a variable star with a regular cycle of brightness, caused by pulsations of the surface layers. A precise relationship exists between the total luminosity of the star and the period of pulsation, allowing its distance to be inferred. [f. delta Cephei, the original example, f. CEPHEUS]

Cepheus /ˈsiːfiːəs/ **1** Gk Mythol. a king of Ethiopia, the husband of Cassiopeia. **2** Astron. a constellation near the north celestial pole, with no bright stars.

ceramic /sɪˈræmɪk, kɪ-/ adj. & n. ● adj. **1** made of (esp.) clay and permanently hardened by heat (a ceramic bowl). **2** of or relating to ceramics (the ceramic arts). ● n. **1** a ceramic article or product. **2** a substance, esp. clay, used to make ceramic articles. [Gk keramikos f. keramos pottery]

ceramicist /sɪˈræmɪsɪst, kɪ-/ n. a person who makes ceramics.

ceramics /sɪˈræmɪks, kɪ-/ n.pl. **1** ceramic products collectively (exhibition of ceramics). **2** (usu. treated as sing.) the art of making ceramic articles.

Ceram Sea /ˈseɪrəm/ (also **Seram Sea**) the part of the western Pacific Ocean at the centre of the Molucca Islands.

cerastes /sɪˈræstiːz/ n. a North African viper of the genus Cerastes; esp. C. cerastes, the horned viper, having a sharp upright spike over each eye. [L f. Gk kerastēs f. keras horn]

cerastium /sɪˈræstɪəm/ n. a plant of the genus Cerastium, of the pink family, with white flowers and often horn-shaped capsules. [mod.L f. Gk kerastes horned f. keras horn]

Cerberus /ˈsɜːbərəs/ Gk Mythol. a monstrous watchdog with three (or in some accounts fifty) heads, which guarded the entrance to Hades. Cerberus could be appeased with a cake, as by Aeneas, or lulled to sleep (as by Orpheus) with lyre music; one of the twelve labours of Hercules was to bring him up from the underworld.

cercaria /sɜːˈkeərɪə/ n. (pl. **cercariae** /-rɪˌiː/) Zool. & Med. a free-swimming larval stage with a propelling tail, in which a parasitic fluke passes from an intermediate host (esp. a snail) to another intermediate host or to the final vertebrate host (cf. MIRACIDIUM). [mod.L, irreg. f. Gk kerkos tail]

cercus /ˈsɜːkəs/ n. (pl. **cerci** /-kaɪ/) Zool. either of a pair of small appendages at the end of the abdomen of some insects and other arthropods. [mod.L f. Gk kerkos tail]

cere /sɪə(r)/ n. a waxy fleshy covering at the base of the upper beak in some birds. [L cera wax]

cereal /ˈsɪərɪəl/ n. & adj ● n. **1** (usu. in pl.) **a** any kind of grain used for food. **b** any grass producing this, e.g. wheat, maize, rye, etc. **2** a breakfast food made from a cereal and requiring no cooking. ● adj. of or relating to cereal or products of it. [L cerealis f. CERES]

cerebellum /ˌserɪˈbeləm/ n. (pl. **cerebellums** or **cerebella** /-lə/) the part of the brain at the back of the skull in vertebrates, which coordinates and regulates muscular activity. □ **cerebellar** adj. [L dimin. of CEREBRUM]

cerebral /ˈserɪbrəl/ adj. **1** of the brain. **2** intellectual rather than emotional. **3** = CACUMINAL. □ **cerebral hemisphere** each of the two halves of the vertebrate cerebrum. **cerebral palsy** Med. a condition marked by weakness and impaired coordination of the limbs, caused esp. by damage to the brain before or at birth. □ **cerebrally** adv. [L cerebrum brain]

cerebration /ˌserɪˈbreɪʃ(ə)n/ n. working of the brain. □ **unconscious cerebration** action of the brain with results reached without conscious thought. □ **cerebrate** /ˈserɪˌbreɪt/ v.intr.

cerebro- /ˈserɪbrəʊ/ comb. form brain (cerebrospinal).

cerebrospinal /ˌserɪbrəʊˈspaɪn(ə)l/ adj. Anat. of the brain and spine. □ **cerebrospinal fluid** a clear watery fluid which fills the space between the arachnoid membrane and the pia mater.

cerebrovascular /ˌserɪbrəʊˈvæskjʊlə(r)/ adj. of the brain and its blood vessels.

cerebrum /ˈserɪbrəm/ n. (pl. **cerebra** /-brə/) the principal part of the

brain in vertebrates, located in the front area of the skull, which integrates complex sensory and neural functions. [L, = brain]

cerecloth /ˈsɪəklɒθ/ n. hist. waxed cloth used as a waterproof covering or (esp.) as a shroud. [earlier cered cloth f. cere to wax f. L cerare f. cera wax]

cerement /ˈsɪəmənt/ n. (usu. in pl.) literary **1** cerecloth for wrapping the dead. **2** clothes. [first used by Shakespeare in Hamlet: app. f. CERECLOTH]

ceremonial /ˌserɪˈməʊnɪəl/ adj. & n. ● adj. **1** with or concerning ritual or ceremony. **2** formal (a ceremonial bow). ● n. **1** a system of rites etc. to be used esp. at a formal or religious occasion. **2** the formalities or behaviour proper to any occasion (with all due ceremonial). **3** RC Ch. a book containing an order of ritual. □ **ceremonialism** n. **ceremonialist** n. **ceremonially** adv. [LL caerimonialis (as CEREMONY)]

ceremonious /ˌserɪˈməʊnɪəs/ adj. **1** excessively polite; punctilious. **2** having or showing a fondness for ritualistic observance or formality. □ **ceremoniously** adv. **ceremoniousness** n. [F cérémonieux or LL caerimoniosus (as CEREMONY)]

ceremony /ˈserɪmənɪ/ n. (pl. **-ies**) **1** a formal religious or public occasion, esp. celebrating a particular event or anniversary. **2** formalities, esp. of an empty or ritualistic kind (ceremony of exchanging compliments). **3** excessively polite behaviour (bowed low with great ceremony). □ **stand on ceremony** insist on the observance of formalities. **without ceremony** informally. [ME f. OF ceremonie or L caerimonia religious worship]

Cerenkov radiation /tʃɪˈreŋkɒf/ n. (also **Cherenkov**) Physics the electromagnetic radiation emitted by particles moving in a medium at speeds faster than that of light in the same medium. This is the cause of light emission seen when a source of gamma rays is placed in a suitable liquid (or solid). With the aid of photomultipliers, the radiation can be used to measure radioactivity. [CHERENKOV]

Ceres /ˈsɪəriːz/ **1** Rom. Mythol. the corn-goddess, commonly identified by the Romans with Demeter. **2** Astron. the first asteroid to be discovered, found by G. Piazzi of Palermo on 1 Jan. 1801. With a diameter of 913 km it is also much the largest, and its surface appears to be rich in carbon.

ceresin /ˈserɪsɪn/ n. a hard whitish wax used with or instead of beeswax. [mod.L ceres f. L cera wax + -IN]

cerise /səˈriːz, -ˈriːs/ adj. & n. ● adj. of a light clear red. ● n. this colour. [F, = cherry]

cerium /ˈsɪərɪəm/ n. a silvery-white metallic chemical element (atomic number 58; symbol Ce). The most abundant of the lanthanide elements, cerium was first isolated in 1875, although as a mixture with other rare earths it had been identified by Berzelius and others at the beginning of the 19th century. It is the main component of the alloy misch metal, and the dioxide is used for polishing glass. [named after the asteroid CERES, discovered at about the same time]

cermet /ˈsɜːmet/ n. a heat-resistant material made of ceramic and sintered metal. [ceramic + metal]

CERN /sɜːn/ abbr. European Council (or Organization) for Nuclear Research, an organization of European countries established in 1954 for research into the fundamental structure of matter. The organization is now called the European Laboratory for Particle Physics. Its headquarters are at Meyrin, near Geneva. [tr. F Conseil Européen (later Organisation Européene) pour la Recherche Nucléaire]

cero- /ˈsɪərəʊ/ comb. form wax (cf. CEROGRAPHY, CEROPLASTIC). [L cera or Gk kēros wax]

cerography /sɪəˈrɒgrəfɪ/ n. the technique of engraving or designing on or with wax.

ceroplastic /ˌsɪərəʊˈplæstɪk/ adj. **1** modelled in wax. **2** of or concerning wax-modelling.

cert /sɜːt/ n. sl. (esp. **dead cert**) **1** an event or result regarded as certain to happen. **2** a horse strongly tipped to win. [abbr. of CERTAIN, CERTAINTY]

cert. /sɜːt/ abbr. **1** a certificate. **2** certified.

certain /ˈsɜːt(ə)n, -tɪn/ adj. & pron. ● adj. **1 a** (often foll. by of, or that + clause) confident, convinced (certain that I put it here). **b** (often foll. by that + clause) indisputable; known for sure (it is certain that he is guilty). **2** (often foll. by to + infin.) **a** that may be relied on to happen (it is certain to rain). **b** destined (certain to become a star). **3** definite, unfailing, reliable (a certain indication of the coming storm; his touch is certain). **4** (of a person, place, etc.) that might be specified, but is not (a certain lady; of a certain age). **5** some though not much (a certain reluctance). **6** (of a person, place, etc.) existing, though probably unknown to the reader or hearer (a certain John Smith). ● pron. (as pl.) some but not all (certain of them were *wounded*). □ **for certain** without doubt. **make certain** = *make sure* (see SURE). [ME f. OF ult. f. L *certus* settled]

certainly /ˈsɜːt(ə)nlɪ, -tɪnlɪ/ *adv.* **1** undoubtedly, definitely. **2** confidently. **3** (in affirmative answer to a question or command) yes; by all means.

certainty /ˈsɜːt(ə)ntɪ, -tɪntɪ/ *n.* (*pl.* **-ies**) **1 a** an undoubted fact. **b** a certain prospect (*his return is a certainty*). **2** (often foll. by *of*, or *that* + clause) an absolute conviction (*has a certainty of his own worth*). **3** (often foll. by *to* + infin.) a thing or person that may be relied on (*a certainty to win the Derby*). □ **for a certainty** beyond the possibility of doubt. [ME f. AF *certainté*, OF *-eté* (as CERTAIN)]

Cert. Ed. *abbr.* (in the UK) Certificate in Education.

certifiable /ˌsɜːtɪˈfaɪəb(ə)l/ *adj.* **1** able or needing to be certified. **2** *colloq.* insane.

certificate *n. & v.* ● *n.* /səˈtɪfɪkət/ a formal document attesting a fact, esp. birth, marriage, or death, a medical condition, a level of achievement, a fulfilment of requirements, ownership of shares, etc. ● *v.tr.* /səˈtɪfɪˌkeɪt/ (esp. as **certificated** *adj.*) provide with or license or attest by a certificate. □ **certification** /ˌsɜːtɪfɪˈkeɪʃ(ə)n/ *n.* [F *certificat* or med.L *certificatum* f. *certificare*: see CERTIFY]

Certificate of Secondary Education *n.* (abbr. **CSE**) *hist.* **1** an examination set for secondary-school pupils in England and Wales. **2** the certificate gained by passing it. ¶ Replaced in 1988 by the *General Certificate of Secondary Education.*

certify /ˈsɜːtɪˌfaɪ/ *v.tr.* (**-ies, -ied**) **1** make a formal statement of; attest; attest to (*certified that he had witnessed the crime*). **2** declare by certificate (that a person is qualified or competent) (*certified as a trained bookkeeper*). **3** officially declare insane (*he should be certified*). □ **certified cheque** a cheque the validity of which is guaranteed by a bank. **certified mail** *US* = *recorded delivery* (see RECORD). **certified milk** milk guaranteed free from the tuberculosis bacillus. **certified public accountant** *US* a member of an officially accredited professional body of accountants. [ME f. OF *certifier* f. med.L *certificare* f. L *certus* certain]

certiorari /ˌsɜːtɪɔːˈreəraɪ/ *n. Law* a writ from a higher court requesting the records of a case tried in a lower court. [LL passive of *certiorare* inform f. *certior* compar. of *certus* certain]

certitude /ˈsɜːtɪˌtjuːd/ *n.* a feeling of absolute certainty or conviction. [ME f. LL *certitudo* f. *certus* certain]

cerulean /səˈruːlɪən/ *adj. & n. literary* ● *adj.* deep blue like a clear sky. ● *n.* this colour. [L *caeruleus* sky-blue f. *caelum* sky]

cerumen /səˈruːmen/ *n.* the yellow waxy substance in the outer ear. □ **ceruminous** /-mɪnəs/ *adj.* [mod.L f. L *cera* wax]

ceruse /ˈsɪəruːs, sɪˈruːs/ *n.* white lead. [ME f. OF f. L *cerussa*, perh. f. Gk *kēros* wax]

Cervantes /sɜːˈvæntiːz/, Miguel de (1547–1616) (full surname Cervantes Saavedra), Spanish novelist and dramatist. His most famous work is *Don Quixote* (1605–15), a satire on chivalric romances. It tells the story of an amiable knight who imagines himself called upon to roam the world in search of adventure on his horse Rosinante, accompanied by the shrewd squire Sancho Panza. The character of Don Quixote had widespread appeal in many other countries and continues to inspire innumerable imitations; his name has passed into the English language as the adjective *quixotic.*

cervelat /ˈsɜːvəˌlɑː, -ˌlæt/ *n.* a kind of smoked pork sausage. [obs. F f. It. *cervellata*]

cervical /sɜːˈvaɪk(ə)l, ˈsɜːvɪk(ə)l/ *adj. Anat.* **1** of or relating to the neck (*cervical vertebrae*). **2** of or relating to the cervix. □ **cervical screening** examination of a large number of apparently healthy women for cervical cancer. **cervical smear** a specimen of cellular material from the neck of the womb that is smeared on a microscope slide and examined for cancerous cells. [F *cervical* or mod.L *cervicalis* f. L *cervix -icis* neck]

cervine /ˈsɜːvaɪn/ *adj.* of or like a deer. [L *cervinus* f. *cervus* deer]

cervix /ˈsɜːvɪks/ *n.* (*pl.* **cervices** /-vɪˌsiːz/) *Anat.* **1** the neck. **2** any necklike structure, esp. the neck of the womb. [L]

Cesarean (also **Cesarian**) *US* var. of CAESAREAN.

Cesarewitch /sɪˈzærɪwɪtʃ/ *n.* a handicap horse-race run annually at Newmarket in Suffolk, England, in late October, inaugurated in 1839 and named in honour of the state visit of the Russian prince who became Alexander II. [Russ. *tsesarevich* heir to the throne]

cesium *US* var. of CAESIUM.

České Budějovice /ˌtʃeskeɪ ˌbʊdjeˈjəʊvɪtsə/ a city in the south of the Czech Republic, on the River Vltava; pop. (1991) 173,400. The city,

known in German as Budweis, is noted for the production of Budvar, or Budweiser, beer.

cess[1] /ses/ *n.* (also **sess**) *Sc., Ir.,* & *Ind.* etc. a tax, a levy. [properly *sess* for obs. *assess* n.: see ASSESS]

cess[2] /ses/ *n. Ir.* □ **bad cess to** may evil befall (*bad cess to their clan*). [perh. f. CESS[1]]

cessation /seˈseɪʃ(ə)n/ *n.* **1** a ceasing (*cessation of the truce*). **2** a pause (*resumed fighting after the cessation*). [ME f. L *cessatio* f. *cessare* CEASE]

cesser /ˈsesə(r)/ *n. Law* a coming to an end; a cessation (of a term, a liability, etc.). [AF & OF = CEASE]

cession /ˈseʃ(ə)n/ *n.* **1** (often foll. by *of*) the ceding or giving up (of rights, property, and esp. of territory by a state). **2** the territory etc. so ceded. [ME f. OF *cession* or L *cessio* f. *cedere cess-* go away]

cessionary /ˈseʃənərɪ/ *n.* (*pl.* **-ies**) *Law* = ASSIGN *n.*

cesspit /ˈsespɪt/ *n.* **1** a pit for the disposal of refuse. **2** = CESSPOOL. [*cess* in CESSPOOL + PIT[1]]

cesspool /ˈsespuːl/ *n.* **1** an underground container for the temporary storage of liquid waste or sewage. **2** a centre of corruption, depravity, etc. [perh. alt., after POOL[1], f. earlier *cesperalle*, f. *suspiral* vent, water-pipe, f. OF *souspirail* air-hole f. L *suspirare* breathe up, sigh (as SUB-, *spirare* breathe)]

cestode /ˈsestəʊd/ *n.* (also **cestoid** /-tɔɪd/) a parasitic flatworm of the class Cestoda, which comprises the tapeworms. [L *cestus* f. Gk *kestos* girdle]

CET *abbr.* Central European Time.

cetacean /sɪˈteɪʃ(ə)n/ *n. & adj. Zool.* ● *n.* a marine mammal of the order Cetacea, which comprises the whales and dolphins. (See note below.) ● *adj.* of or relating to cetaceans. □ **cetaceous** *adj.* [mod.L f. L *cetus* f. Gk *kētos* whale]

▪ The cetaceans are large marine mammals with a streamlined hairless body, a dorsal blowhole for breathing, forelimbs modified as flippers, and no hindlimbs. There are two major groups, the toothed whales (suborder Odontoceti), which include the sperm whales, killer whales, dolphins, and porpoises, and the baleen whales (suborder Mysticeti), which feed by filter-feeding and include the blue whale, the largest of all animals.

cetane /ˈsiːteɪn/ *n. Chem.* a colourless liquid hydrocarbon of the alkane series used in standardizing ratings of diesel fuel. □ **cetane number** a measure of the ignition properties of diesel fuel. [f. SPERMACETI after methane etc.]

ceteris paribus /ˌsetərɪs ˈpærɪˌbʊs/ *adv.* other things being equal. [L]

Cetus /ˈsiːtəs/ *Astron.* a large northern constellation (the Whale), said to represent the sea monster which threatened Andromeda. It contains the variable star Mira, but no other bright stars. [L]

Ceuta /ˈseɪʊtə/ a Spanish enclave on the coast of North Africa, in Morocco; pop. (est. 1989) 68,000 (with Melilla). Held by Spain since 1580, it consists of a free port and a military post and overlooks the Mediterranean approach to the Strait of Gibraltar.

Cévennes /seɪˈven/ a mountain range on the south-eastern edge of the Massif Central in France.

Ceylon /sɪˈlɒn/ the former name (until 1972) for SRI LANKA.

Ceylon moss *n.* a red seaweed, *Gracilaria lichenoides*, of the Indian subcontinent, used as the main source of agar.

Cézanne /seɪˈzæn/, Paul (1839–1906), French painter. Although his early work is associated with impressionism, it is with post-impressionism that he is most closely identified. From the 1880s his work is dominated by the increasing use of simplified geometrical forms (the cylinder, the sphere, and the cone), which he regarded as being the structural basis of nature; this later work was an important influence on the development of cubism. For this reason he is regarded as an important influence on the development of cubism. He is especially known for his landscapes, many of which depict the Mont Sainte-Victoire in Provence, for still lifes such as *Still Life with Cupid* (1895), and for his *Bathers* paintings (1890–1905).

CF *abbr. Brit.* Chaplain to the Forces.

Cf *symb. Chem.* the element californium.

cf. *abbr.* compare. [L *confer* imper. of *conferre* compare]

c.f. *abbr.* carried forward.

CFC *abbr. Chem.* chlorofluorocarbon, any of a class of usu. gaseous compounds of carbon, hydrogen, chlorine, and fluorine. CFCs are non-flammable, non-toxic, and unreactive synthetic compounds which have been used since the 1930s as working fluids in refrigerators and propellants for aerosol sprays. They have now been shown to be harmful to the earth's ozone layer, as well as being major contributors to the greenhouse effect, and their use is being reduced. CFC molecules which have been released into the environment are broken down by the sun's ultraviolet radiation in the upper atmosphere, forming chlorine which reacts with ozone.

CFE *abbr.* College of Further Education.

cg *abbr.* centigram(s).

CGS *abbr.* Chief of General Staff.

cgs *abbr.* centimetre-gram-second.

CH *abbr.* (in the UK) Companion of Honour.

ch. *abbr.* **1** church. **2** chapter. **3** chestnut.

cha var. of CHAR³.

Chablis /ˈʃæblɪ/ *n.* (*pl.* same /-lɪz/) a dry white burgundy wine. [*Chablis* in E. France]

Chabrol /ʃæˈbrɒl/, Claude (b.1930), French film director. He gained recognition in the late 1950s as one of the directors of the *nouvelle vague*, making his directorial début with *Le Beau serge* (1958). An admirer of the work of Alfred Hitchcock, he has directed many films that combine mystery and suspense with studies of personal relationships and social situations. Notable films include *Les Biches* (1968) and *Le Boucher* (1970).

cha-cha /ˈtʃɑːtʃɑː/ *n. & v.* (also **cha-cha-cha** /ˌtʃɑːtʃɑːˈtʃɑː/) ● *n.* **1** a ballroom dance with a Latin American rhythm. **2** music for or in the rhythm of a cha-cha. ● *v.intr.* (**cha-chas, cha-chaed** /-tʃɑːd/ or **cha-cha'd, cha-chaing** /-tʃɑːɪŋ/) dance the cha-cha. [Amer. Sp.]

Chaco see GRAN CHACO.

chaconne /ʃəˈkɒn/ *n. Mus.* **1 a** a musical form consisting of variations on a ground bass. **b** a musical composition in this style. **2** *hist.* a dance performed to this music. [F f. Sp. *chacona*]

Chaco War /ˈtʃɑːkəʊ/ a war in 1932–5 between Bolivia and Paraguay, a boundary dispute triggered by the discovery of oil in the northern part of the Gran Chaco. Paraguay eventually forced Bolivia to sue for peace and gained most of the disputed territory, but casualties on both sides were heavy.

Chad /tʃæd/ a landlocked country in northern central Africa; pop. (est. 1991) 5,828,000; official languages, French and Arabic; capital, N'Djamena. Much of the country lies in the Sahel and, in the north, the Sahara Desert. The population includes several ethnic groups with a variety of languages and religions. French expeditions entered the region in 1890, and by 1913 the country was organized as a French colony. It became autonomous within the French Community in 1958, and fully independent as a republic in 1960. Civil war between the Muslim north and the southern-based government broke out in 1965, and conflict involving various factions continued intermittently; in 1993 a national conference was established to prepare for democratic elections. □ **Chadian** *adj. & n.*

Chad, Lake a shallow lake on the borders of Chad, Niger, and Nigeria in north central Africa. Its size varies seasonally from *c.*10,360 sq. km (4,000 sq. miles) to *c.*25,900 sq. km (10,000 sq. miles).

Chadic /ˈtʃædɪk/ *n. & adj.* ● *n.* a group of Afro-Asiatic languages spoken in the region of Lake Chad, of which the most important is Hausa. ● *adj.* of or relating to this group of languages.

chador /ˈtʃʌdə(r)/ *n.* (also **chadar, chuddar**) a large piece of cloth worn in some countries by Muslim women, wrapped around the body to leave only the face exposed. [Pers. *chador*, Hindi *chador*]

Chadwick /ˈtʃædwɪk/, Sir James (1891–1974), English physicist. He became assistant director at the Cavendish Laboratory, Cambridge, where he studied the artificial disintegration of elements such as beryllium when bombarded by alpha particles. This led to the discovery of the neutron, for which he received the 1935 Nobel Prize for physics. During World War Two he was involved with the atom bomb project, and afterwards stressed the importance of university research into nuclear physics.

chaetognath /ˈkiːtəɡˌnæθ/ *n. Zool.* a dart-shaped worm of the phylum Chaetognatha, usu. living among marine plankton, and having a head with external grasping spines. Also called *arrow worm*. [mod.L *Chaetognatha* f. Gk *khaitē* long hair + *gnathos* jaw]

chafe /tʃeɪf/ *v. & n.* ● *v.* **1** *tr. & intr.* make or become sore or damaged by rubbing. **2** *tr.* rub (esp. the skin to restore warmth or sensation). **3** *tr. & intr.* make or become annoyed; fret (*was chafed by the delay*). ● *n.* **1 a** an act of chafing. **b** a sore resulting from this. **2** a state of annoyance. [ME f. OF *chaufer* ult. f. L *calefacere* f. *calere* be hot + *facere* make]

chafer /ˈtʃeɪfə(r)/ *n.* a large slow-moving strong-flying beetle of the family Scarabaeidae, e.g. the cockchafer. Chafers are plant-eaters, and several kinds are serious pests. [OE *ceafor, cefer* f. Gmc]

chaff /tʃɑːf/ *n. & v.* ● *n.* **1** the husks of corn or other seed separated by winnowing or threshing. **2** chopped hay and straw used as fodder. **3** light-hearted joking; banter. **4** worthless things; rubbish. **5** strips of metal foil released in the atmosphere to obstruct radar detection. ● *v.tr.* **1** tease; banter. **2** chop (straw etc.). □ **chaff-cutter** a machine for chopping fodder. **separate the wheat from the chaff** distinguish good from bad. □ **chaffy** *adj.* [OE *ceaf, cæf* prob. f. Gmc: sense 3 of *n.* & 1 of *v.* perh. f. CHAFE]

chaffer /ˈtʃæfə(r)/ *v. & n.* ● *v.intr.* haggle; bargain. ● *n.* bargaining; haggling. □ **chafferer** *n.* [ME f. OE *ceapfaru* f. *ceap* bargain + *faru* journey]

chaffinch /ˈtʃæfɪntʃ/ *n.* a common European finch, *Fringilla coelebs*, the male of which has a blue-grey head with pinkish cheeks and breast. [OE *ceaffinc*: see CHAFF, FINCH]

chafing-dish /ˈtʃeɪfɪŋˌdɪʃ/ *n.* **1** a cooking pot with an outer pan of hot water, used for keeping food warm. **2** a dish with a spirit-lamp etc. for cooking at table. [obs. sense of CHAFE = warm]

Chagall /ʃəˈɡɑːl/, Marc (1887–1985), Russian-born French painter and graphic artist. Working chiefly in Paris, he was associated with the avant-garde circle of Delaunay, Modigliani, and Chaim Soutine. Inspired by Russian folk art and the work of the fauves (see FAUVISM), he used rich emotive colour and dream imagery, as can be seen in paintings such as *Maternity* (1913); his early work had a significant influence on the development of surrealism. Other achievements included theatre design (the costumes and decor for Stravinsky's *The Firebird*, 1945), stained-glass windows, and murals.

Chagas' disease /ˈʃɑːɡəs/ *n.* (also **Chagas's disease**) a disease caused by trypanosomes transmitted by blood-sucking hemipteran bugs, endemic in South America and characterized by damage to the heart muscle and central nervous system. [Carlos Chagas, Brazilian physician (1879–1934)]

Chagos Archipelago /ˈtʃɑːɡəs/ an island group in the Indian Ocean, formerly a dependency of Mauritius and now forming the British Indian Ocean Territory.

chagrin /ˈʃæɡrɪn, ʃəˈɡriːn/ *n. & v.* ● *n.* acute vexation or mortification. ● *v.tr.* affect with chagrin. [F *chagrin(er)*, of uncert. orig.]

Chain /tʃeɪn/, Sir Ernst Boris (1906–79), German-born British biochemist. (See FLOREY.)

chain /tʃeɪn/ *n. & v.* ● *n.* **1 a** a connected flexible series of esp. metal links as decoration or for a practical purpose. **b** something resembling this (*formed a human chain*). **2** (in *pl.*) **a** fetters used to confine prisoners. **b** any restraining force. **3** a sequence, series, or set (*chain of events; mountain chain*). **4** a group of associated hotels, shops, etc. **5** a badge of office in the form of a chain worn round the neck (*mayoral chain*). **6 a** a jointed measuring-line consisting of linked metal rods. (See also GUNTER'S CHAIN.) **b** the length of this (66 ft). **7** *Chem.* a group of (esp. carbon) atoms bonded in sequence in a molecule. **8** a figure in a quadrille or similar dance. **9** (in *pl.*) *Naut.* channels (see CHANNEL²). **10** (also **chain-shot**) *hist.* two cannon-balls or half balls joined by a chain and used in sea battles for bringing down a mast etc. ● *v.tr.* **1** (often foll. by *up*) secure or confine with a chain. **2** confine or restrict (a person) (*is chained to the office*). □ **chain-armour** armour made of interlaced rings. **chain bridge** a suspension bridge on chains. **chain drive** a system of transmission by endless chains. **chain-gang** a team of convicts chained together and forced to work in the open air. **chain-gear** a gear transmitting motion by means of an endless chain. **chain letter** one of a sequence of letters requesting the recipient to send copies to a specific number of other people, who are told to do similarly. **chain-link** made of wire in a diamond-shaped mesh (*chain-link fencing*). **chain-mail** = *chain-armour*. **chain reaction 1** *Physics* a self-sustaining nuclear reaction, esp. one in which a neutron from a fission reaction initiates a series of these reactions. **2** *Chem.* a self-sustaining molecular reaction in which intermediate products initiate further reactions. **3** a series of events, each caused by the previous one. **chain-saw** a motor-driven saw with teeth on an endless chain. **chain-smoke** smoke (cigarettes) continuously, esp. by lighting the next cigarette from the stub of the last one smoked. **chain-smoker** a person who chain-smokes. **chain stitch 1** an ornamental embroidery or crochet stitch resembling chains. **2** a stitch made by a sewing machine using a single thread that is hooked through its own loop on the underside of the fabric sewn. **chain store** one of a series of shops owned by one firm and selling the same sort of goods. **chain-wale** = CHANNEL².

chain-wheel a wheel transmitting power by a chain fitted to its edges. [ME f. OF cha(e)ine f. L catena]

chair /tʃeə(r)/ n. & v. ● n. **1** a separate seat for one person, of various forms, usu. having a back and four legs. **2 a** a professorship (offered the chair in physics). **b** a seat of authority, esp. on a board of directors. **c** a mayoralty. **3 a** a chairperson. **b** the seat or office of a chairperson (will you take the chair?; I'm in the chair). **4** US = electric chair. **5** an iron or steel socket holding a railway rail in place. **6** hist. = SEDAN 1. ● v.tr. **1** act as chairman of or preside over (a meeting). **2** Brit. carry (a person) aloft in a chair or in a sitting position, in triumph. **3** install in a chair, esp. as a position of authority. □ **chair-bed** a chair that unfolds into a bed. **chair-borne** colloq. (of an administrator) not active. **chair-lift** a series of chairs on an endless cable for carrying passengers up and down a mountain etc. **take a chair** sit down. [ME f. AF chaere, OF chaiere f. L cathedra f. Gk kathedra: see CATHEDRAL]

chairlady /'tʃeəˌleɪdɪ/ n. (pl. -ies) = chairwoman (see CHAIRMAN).

chairman /'tʃeəmən/ n. (pl. -men; fem. **chairwoman**, pl. -women) **1** a person chosen to preside over a meeting. **2 a** the permanent president of a committee, a board of directors, a firm, etc. **b** (**Chairman**) the leading figure in the Communist Party of China having the dominant role in the country's political affairs (or formerly in the USSR and the Soviet bloc). **3** the master of ceremonies at an entertainment etc. **4** hist. either of two sedan-bearers. ¶ In senses 1, 2a, and 3 now sometimes avoided for the more neutral chair or chairperson. □ **chairmanship** n.

chairperson /'tʃeəˌpɜːs(ə)n/ n. a chairman or chairwoman (used as a neutral alternative).

chaise /ʃeɪz/ n. **1** esp. hist. a horse-drawn carriage for one or two people, esp. one with an open top and two wheels. **2** = post-chaise (see POST²). [F var. of chaire, formed as CHAIR]

chaise longue /ʃeɪz 'lɒŋ/ n. (pl. **chaise longues** or **chaises longues** pronunc. same) a sofa with a backrest at only one end. [F, lit. 'long chair']

Chaka see SHAKA.

chakra /'tʃʌkrə/ n. each of the centres of spiritual power in the human body recognized in yoga. [Skr. cakra (cogn. with WHEEL n.; in earlier senses, a discus or mystic circle depicted in the hands of Hindu deities)]

chalaza /kə'leɪzə/ n. (pl. **chalazae** /-ziː/) either of two twisted membranous strips joining the yolk to the ends of an egg. [mod.L f. Gk, = hailstone]

Chalcedon /'kælsɪˌdɒn, kæl'siːd(ə)n/ (Turkish **Kadiköy** /kɑ'dikœj/) a former city on the Bosporus in Asia Minor, now a district of Istanbul. The site was quarried for building materials, including chalcedony, during the construction of Constantinople by the Romans (AD 324–30). □ **Chalcedonian** /ˌkælsɪ'dəʊnɪən/ adj.

Chalcedon, Council of the fourth ecumenical council of the Christian Church, held at Chalcedon in 451. It condemned the Monophysite position and affirmed the dual but united nature of Christ as god and man.

chalcedony /kæl'sedənɪ/ n. a type of quartz occurring in several different forms, e.g. onyx, agate, tiger's eye, etc. □ **chalcedonic** /ˌkælsɪ'dɒnɪk/ adj. [ME f. L c(h)alcedonius f. Gk khalkēdōn: see CHALCEDON]

Chalcis /'kælsɪs/ (Greek **Khalkís** /xal'kis/) the chief town of the island of Euboea, on the coast opposite mainland Greece; pop. (1981) 44,800.

chalcolithic /ˌkælkə'lɪθɪk/ adj. Archaeol. of, relating to, or denoting a short period (also called the Copper Age) following the end of the Stone Age in some areas, during which both stone and copper implements were used, before bronze technology was introduced. (See also COPPER AGE.) [Gk khalkos copper + lithos stone]

chalcopyrite /ˌkælkə'paɪraɪt/ n. a yellow mineral of copper-iron sulphide, which is the principal ore of copper. Also called copper pyrites. [Gk khalkos copper + PYRITE]

Chaldea /kæl'diːə/ an ancient country in what is now southern Iraq inhabited by the Chaldeans, forming the southern part of Babylonia from c.800 to 539 BC.

Chaldean /kæl'diːən/ n. & adj. ● n. **1 a** a member of a Semitic people originating in Arabia, who settled in what became Chaldea c.800 BC and ruled Babylonia 625–539 BC. They were famous as astronomers. **b** the Semitic language of the Chaldeans. **2** an astrologer. **3** a member of a Syrian Uniat (formerly Nestorian) Church in Iran and elsewhere. ● adj. **1** of or relating to ancient Chaldea or its people or language. **2** of or relating to astrology. **3** of or relating to the East Syrian Uniat Church. [L Chaldaeus f. Gk Khaldaios f. Assyr. Kaldu]

Chaldee /kæl'diː/ n. **1** the language of the Chaldeans. **2** a native of ancient Chaldea. **3** the Aramaic language as used in Old Testament books. [ME, repr. L Chaldaei (pl.) (as CHALDEAN)]

chalet /'ʃæleɪ/ n. **1** a small suburban house or bungalow, esp. with an overhanging roof. **2** a small, usu. wooden, hut or house on a beach or in a holiday camp. **3** a Swiss cowherd's hut, or wooden cottage, with overhanging eaves. [Swiss F]

Chaliapin /ʃæ'ljɑːpɪn/, Fyodor (Ivanovich) (1873–1938), Russian operatic bass. He made his début in St Petersburg, later going to Moscow, where he excelled in Russian opera, most notably in the title role of Mussorgsky's Boris Godunov. He left Russia after the Revolution and developed a successful international career, appearing in a wide range of operas in many different countries.

chalice /'tʃælɪs/ n. **1** literary a goblet. **2** a wine-cup used in the Communion service. [ME f. OF f. L calix -icis cup]

chalk /tʃɔːk/ n. & v. ● n. **1** a white soft earthy limestone (calcium carbonate) formed from the skeletal remains of sea creatures. **2 a** a similar substance (calcium sulphate), sometimes coloured, used for writing or drawing. **b** a piece of this (a box of chalks). **3** a series of strata consisting mainly of chalk. **4** = French chalk. ● v.tr. **1** rub, mark, draw, or write with chalk. **2** (foll. by up) **a** write or record with chalk. **b** register (a success etc.). **c** charge (to an account). □ **as different as chalk and (or from) cheese** fundamentally different. **by a long chalk** Brit. by far (from the use of chalk to mark the score in games). **chalk and talk** traditional teaching (employing blackboard, chalk, and interlocution). **chalk out** sketch or plan a thing to be accomplished. **chalk-pit** a quarry in which chalk is dug. **chalk-stone** a concretion of urates like chalk in tissues and joints esp. of hands and feet. **chalk-stripe** a pattern of thin white stripes on a dark background. **chalk-striped** having chalk-stripes. [OE cealc ult. f. WG f. L CALX]

chalkboard /'tʃɔːkbɔːd/ n. N. Amer. = BLACKBOARD.

chalkie /'tʃɔːkɪ/ n. Austral. colloq. a schoolteacher.

chalky /'tʃɔːkɪ/ adj. (**chalkier**, **chalkiest**) **1 a** abounding in chalk. **b** white as chalk. **2** like or containing chalk stones. □ **chalkiness** n.

challah /'hɑːlə, xɑː'lɑː/ n. (also **challa**, **halla**) (pl. **challahs** or **chalot** /xɑː'ləʊt/) a loaf of white leavened bread, often plaited in form, traditionally baked to celebrate the Jewish sabbath. [Heb. ḥalah, perh. f. ḥll to pierce or Akkadian ellu pure]

challenge /'tʃælɪndʒ/ n. & v. ● n. **1 a** a summons to take part in a contest or a trial of strength etc., esp. to a duel. **b** a summons to prove or justify something. **2** a demanding or difficult task (rose to the challenge of the new job). **3** an act of disputing or denying a statement, claim, etc. **4** Law an objection made to a jury member. **5** a call to respond. **6** a sentry's call for a password etc. **6** an invitation to a sporting contest, esp. one issued to a reigning champion. **7** Med. a test of immunity after immunization treatment. ● v.tr. **1** (often foll. by to + infin.) **a** invite to take part in a contest, game, debate, duel, etc. **b** invite to prove or justify something. **2** dispute, deny (I challenge that remark). **3 a** stretch, stimulate (challenges him to produce his best). **b** (as **challenging** adj.) demanding; stimulatingly difficult. **4** (of a sentry) call to respond. **5** claim (attention, etc.). **6** Law object to (a jury member, evidence, etc.). **7** Med. test by a challenge. □ **challengeable** adj. **challenger** n. **challengingly** adv. [ME f. OF c(h)alenge, c(h)alenger f. L calumnia calumniari calumny]

challenged /'tʃælɪndʒd/ adj. (in comb.) euphem. disabled or handicapped (physically challenged).

Challenger Deep /'tʃælɪndʒə(r)/ the deepest part (11,034 m, 36,201 ft) of the Mariana Trench in the North Pacific, discovered by HMS Challenger II in 1948.

challis /'ʃælɪs, 'ʃælɪ/ n. a lightweight soft clothing fabric. [perh. f. a surname]

chalybeate /kə'lɪbɪət/ adj. (of mineral water etc.) impregnated with iron salts. [mod.L chalybeatus f. L chalybs f. Gk khalups -ubos steel]

chamaephyte /'kæmɪˌfaɪt/ n. a plant whose buds are on or near the ground. [Gk khamai on the ground + -PHYTE]

chamber /'tʃeɪmbə(r)/ n. **1 a** a hall used by a legislative or judicial body. **b** the body that meets in it. **c** any of the houses of a parliament (second chamber). **2** (in pl.) Brit. **a** rooms used by a barrister or group of barristers, esp. in the Inns of Court. **b** a judge's room used for hearing cases not needing to be taken in court. **3** poet. or archaic a room, esp. a bedroom. **4** a large underground cavity; a cave. **5** (attrib.) Mus. of or for a small group of instruments; relating to chamber music (chamber orchestra). **6** an enclosed space in machinery etc. (esp. the part of a gun-bore that contains the charge). **7 a** a cavity in a plant or in the body of an animal. **b** a compartment in a structure. **8** = chamber-pot.

□ **chamber of commerce** an association to promote local commercial interests. **chamber-pot** a receptacle for urine etc., used in a bedroom. [ME f. OF *chambre* f. L CAMERA]

chambered /ˈtʃeɪmbəd/ *adj.* (of a tomb) containing a burial chamber.

Chamberlain[1] /ˈtʃeɪmbəlɪn/, (Arthur) Neville (1869–1940), British Conservative statesman, Prime Minister 1937–40. The son of Joseph Chamberlain, as Prime Minister of a coalition government he pursued a policy of appeasement towards Germany, Italy, and Japan; in 1938 he signed the Munich Agreement ceding the Sudetenland to Germany, which he claimed would mean 'peace in our time'. Although the policy was primarily intended to postpone war until Britain had rearmed, it caused increasing discontent in his own party; he was forced to abandon it and prepare for war when Hitler invaded the rest of Czechoslovakia in 1939. Chamberlain's war leadership proved inadequate and he was replaced by Winston Churchill.

Chamberlain[2] /ˈtʃeɪmbəlɪn/, Joseph (1836–1914), British Liberal statesman. He became a Liberal MP in 1876, but left the party in 1886 because of Gladstone's support of Irish Home Rule. The leader of the Liberal Unionists from 1891, in the coalition government of 1895 he served as Colonial Secretary, in which post he played a leading role in the handling of the Second Boer War.

Chamberlain[3] /ˈtʃeɪmbəlɪn/, Owen (b.1920), American physicist. He worked on the Manhattan Project during the war, after which he investigated subatomic particles using a bevatron accelerator. He and E. G. Segrè discovered the antiproton in 1955, and four years later they shared the Nobel Prize for physics.

chamberlain /ˈtʃeɪmbəlɪn/ *n.* **1** an officer managing the household of a sovereign or a great noble. **2** the treasurer of a corporation etc. □ **chamberlainship** *n.* [ME f. OF *chamberlain* etc. f. Frank. f. L *camera* CAMERA]

chambermaid /ˈtʃeɪmbəˌmeɪd/ *n.* **1** a housemaid at a hotel etc. **2** *US* a housemaid.

chamber music *n.* music played by a small ensemble, with one player to a part, in a room rather than a concert hall. The madrigals and consorts of the 16th century can be classed as chamber music, but the form became established with the development of the string quartet.

Chamber of Deputies *n.* the lower legislative assembly in some parliaments.

Chambers /ˈtʃeɪmbəz/, Sir William (1723–96), Scottish architect. Travels in the Far East and studies in France and Italy helped to mould his eclectic but conservative neoclassical style. His most notable buildings include Somerset House in London (1776) and the pagoda in Kew Gardens (1757–62).

Chambertin /ˈʃɒmbəˌtæn/ *n.* a high-quality dry red burgundy wine. [*Gevrey Chambertin* region in E. France]

chamber tomb *n.* a tomb (esp. a megalithic structure) containing a chamber or chambers for deposition of the dead.

Chambéry /ˈʃɒmbərɪ/ a town in eastern France; pop. (1990) 55,600.

chambray /ˈʃæmbreɪ/ *n.* a linen-finished gingham cloth with a white weft and a coloured warp. [irreg. f. *Cambrai*: see CAMBRIC]

chambré /ˈʃɒmbreɪ/ *adj.* (of red wine) brought to room temperature. [F, past part. of *chambrer* f. *chambre* room: see CHAMBER]

chameleon /kəˈmiːlɪən/ *n.* **1 a** a slow-moving lizard of the family Chamaeleontidae, with a grasping tail, long tongue, protruding eyes which can swivel independently, and the ability to change colour; esp. *Chamaeleo chamaeleon*, of the Mediterranean area. **b** *US* an anole. **2** a variable or inconstant person. □ **chameleonic** /-ˌmiːlɪˈɒnɪk/ *adj.* [ME f. L f. Gk *khamaileōn* f. *khamai* on the ground + *leōn* lion]

chamfer /ˈtʃæmfə(r)/ *v. & n.* ● *v.tr.* bevel symmetrically (a right-angled edge or corner). ● *n.* a bevelled surface at an edge or corner. [back-form. f. *chamfering* f. F *chamfrain* f. *chant* edge (CANT[2]) + *fraint* broken f. OF *fraindre* break f. L *frangere*]

chamois *n.* (*pl.* same) **1** /ˈʃæmwɑ/ (*pl.* -wɑz) an agile goat antelope, *Rupicapra rupicapra*, native to the mountains of Europe and Asia. **2** /ˈʃæmɪ, ˈʃæmwɑ/ (in full **chamois leather**) a soft pliable leather from sheep, goats, deer, etc. **b** a piece of this for polishing etc. [F: cf. Gallo-Roman *camox*]

chamomile var. of CAMOMILE.

Chamonix /ˈʃæmɒˌniː/ (in full **Chamonix-Mont-Blanc** /mɒnˈblɒ̃k/) a ski resort at the foot of Mont Blanc, in the Alps of eastern France; pop. (1982) 9,255.

champ[1] /tʃæmp/ *v. & n.* ● *v.* **1** *tr.* & *intr.* munch or chew noisily. **2** *tr.* (of a horse etc.) work (the bit) noisily between the teeth. **3** *intr.* fret with impatience (*is champing to be away*). ● *n.* a chewing noise or motion. □ **champ at the bit** be restlessly impatient. [prob. imit.]

champ[2] /tʃæmp/ *n. colloq.* a champion. [abbr.]

Champagne /ʃæmˈpeɪn/ a region and former province of NE France, which now corresponds to the Champagne-Ardenne administrative region. The province, powerful during the 12th and 13th centuries, was united with France in 1314. The region is noted for the white sparkling wine first produced there in about 1700.

champagne /ʃæmˈpeɪn/ *n.* **1 a** a white sparkling wine from Champagne. **b** (loosely) a similar wine from elsewhere. ¶ Use in sense b is strictly incorrect. **2** a pale cream or straw colour. [CHAMPAGNE]

Champagne-Ardenne /ʃæmˌpeɪnɑːˈden/ a region of NE France, comprising part of the Ardennes forest and the vine-growing area of Champagne.

champaign /ʃæmˈpeɪn/ *n. literary* **1** open country. **2** an expanse of open country. [ME f. OF *champagne* f. LL *campania*: cf. CAMPAIGN]

champers /ˈʃæmpəz/ *n. colloq.* champagne.

champerty /ˈtʃæmpətɪ/ *n.* (*pl.* -**ies**) *Law* an illegal agreement in which a person not naturally interested in a lawsuit finances it with a view to sharing the disputed property. □ **champertous** *adj.* [ME f. AF *champartie* f. OF *champart* feudal lord's share of produce, f. L *campus* field + *pars* part]

champion /ˈtʃæmpɪən/ *n., v., adj., & adv.* ● *n.* **1** (often *attrib.*) a person (esp. in a sport or game), an animal, plant, etc., that has defeated or surpassed all rivals in a competition etc. **2 a** a person who fights or argues for a cause or on behalf of another person. **b** *hist.* a knight etc. who fought in single combat on behalf of a king etc. ● *v.tr.* support the cause of, defend, argue in favour of. ● *adj. colloq.* or *dial.* first-class, splendid. ● *adv. colloq.* or *dial.* splendidly, well. [ME f. OF f. med.L *campio -onis* fighter f. L *campus* field]

Champion of England *n.* (also **King's** or **Queen's Champion**) (in the UK) a hereditary official who ceremonially offers to defend a new monarch's title to the crown at his or her coronation.

championship /ˈtʃæmpɪənˌʃɪp/ *n.* **1** (often in *pl.*) a contest for the position of champion in a sport etc. **2** the position of champion over all rivals. **3** the advocacy or defence (of a cause etc.).

Champlain /ʃæmˈpleɪn/, Samuel de (1567–1635), French explorer and colonial statesman. He made his first voyage to Canada in 1603, and between 1604 and 1607 explored the eastern coast of North America. In 1608 he was sent to establish a settlement at Quebec, where he developed alliances with the native peoples for trade and defence. He was appointed Lieutenant-Governor in 1612; much of his subsequent career was spent exploring the Canadian interior. After capture and imprisonment by the English (1629–32), he returned to Canada for a final spell as governor (1633–5).

Champlain, Lake a lake in North America, situated to the east of the Adirondack Mountains. It forms part of the border between New York State and Vermont and its northern tip extends into Quebec. It is named after Samuel de Champlain, who reached it in 1609.

champlevé /ˌʃɒmləˈveɪ/ *n. & adj.* ● *n.* a type of enamelwork in which hollows made in a metal surface are filled with coloured enamels. ● *adj.* of or relating to *champlevé* (cf. CLOISONNÉ). [F, = raised field]

Champollion /ʃɒmˈpɒljɒn/, Jean-François (1790–1832), French Egyptologist. A pioneer in the study of ancient Egypt, he is best known for his success in deciphering some of the hieroglyphic inscriptions on the Rosetta Stone in 1822.

Champs Élysées /ˌʃɒmz eɪˈliːzeɪ/ an avenue in Paris, leading from the Place de la Concorde to the Arc de Triomphe. It is noted for its fashionable shops and restaurants.

chance /tʃɑːns/ *n., adj., & v.* ● *n.* **1 a** a possibility (*just a chance we will catch the train*). **b** (often in *pl.*) probability (*the chances are against it*). **2** a risk (*have to take a chance*). **3 a** an accidental occurrence (*just a chance that they met*). **b** the absence of design or discoverable cause (*here merely because of chance*). **4** an opportunity (*didn't have a chance to speak to him*). **5** the way things happen; fortune; luck (*we'll just leave it to chance*). **6** (often **Chance**) the course of events regarded as a power; fate (*blind Chance rules the universe*). **7** *Sport* an opportunity for scoring a goal, dismissing a batsman, etc. ● *adj.* fortuitous, accidental (*a chance meeting*). ● *v.* **1** *tr. colloq.* risk (*we'll chance it and go*). **2** *intr.* (often foll. by *that* + clause, or *to* + infin.) happen without intention (*it chanced that I found it; I chanced to find it*). □ **by any chance** as it happens; perhaps. **by chance** without design; unintentionally. **chance one's arm** make an attempt though unlikely to succeed. **chance on** (or **upon**) happen to find, meet, etc. **game of chance** a game decided by luck, not skill. **the off chance** the slight possibility. **on the chance** (often foll. by *of*, or *that* + clause) in view of the possibility. **stand a chance** have a prospect of success etc. **take a chance** (or **chances**) behave riskily; risk failure. **take a** (or **one's**) **chance on** (or **with**) consent to take the consequences of; trust to luck. [ME f. AF *ch(e)aunce*, OF *chëance chëoir* fall ult. f. L *cadere*]

chancel /ˈtʃɑːns(ə)l/ *n.* the part of a church near the altar, reserved for the clergy, the choir, etc., usu. enclosed by a screen or separated from the nave by steps. [ME f. OF f. L *cancelli* lattice]

chancellery /ˈtʃɑːnsələrɪ/ *n.* (*pl.* **-ies**) **1 a** the position, office, staff, department, etc., of a chancellor. **b** the official residence of a chancellor. **2** *US* an office attached to an embassy or consulate. [ME f. OF *chancellerie* (as CHANCELLOR)]

chancellor /ˈtʃɑːnsələ(r)/ *n.* **1** a state or legal official of various kinds. **2** (**Chancellor**) **a** the head of the government in some European countries, e.g. Germany. **b** = CHANCELLOR OF THE EXCHEQUER. **3** the non-resident honorary head of a university. **4** a bishop's law officer. **5** *US* the president of a chancery court. □ **chancellorship** *n.* [OE f. AF *c(h)anceler*, OF *-ier* f. LL *cancellarius* porter, secretary, f. *cancelli* lattice]

Chancellor of the Duchy of Lancaster *n.* (in the UK) a member of the government legally representing the Crown as Duke of Lancaster, often a Cabinet Minister employed on non-departmental work.

Chancellor of the Exchequer *n.* the Finance Minister of the United Kingdom, who prepares the nation's annual budgets. The office dates from the reign of Henry III and was originally that of assistant to the treasurer of the Exchequer. (See also EXCHEQUER, TREASURY.)

Chancellor of the Garter *n.* (in the UK) a government officer who seals commissions etc.

chance-medley /tʃɑːnsˈmedlɪ/ *n.* (*pl.* **-eys**) **1** *Law* a fight, esp. homicidal, beginning unintentionally. **2** inadvertency. [AF *chance medlee* (see MEDDLE) mixed chance]

chancery /ˈtʃɑːnsərɪ/ *n.* (*pl.* **-ies**) **1** (**Chancery**) *Law* the Lord Chancellor's court, a division of the High Court of Justice. **2** *hist.* the records office of an order of knighthood. **3** *hist.* the court of a bishop's Chancellor. **4** an office attached to an embassy or consulate. **5** a public record office. **6** *US* a court of equity. □ **in chancery** *sl.* (of a boxer or wrestler) with his or her head held, contrary to the rules, between the opponent's arm and body and unable to avoid blows. [ME, contracted f. CHANCELLERY]

Chan Chan /tʃæn ˈtʃæn/ the capital of the pre-Inca civilization of the Chimu. Its extensive adobe ruins are situated on the coast of north Peru.

Chan-chiang see ZHANJIANG.

chancre /ˈʃæŋkə(r)/ *n.* a painless ulcer developing in venereal disease etc. [F f. L CANCER]

chancroid /ˈʃæŋkrɔɪd/ *n.* ulceration of lymph nodes in the groin, from venereal disease.

chancy /ˈtʃɑːnsɪ/ *adj.* (**chancier**, **chanciest**) subject to chance; uncertain; risky. □ **chancily** *adv.* **chanciness** *n.*

chandelier /ˌʃændɪˈlɪə(r)/ *n.* an ornamental branched hanging support for several candles or electric light bulbs. [F (*chandelle* f. as CANDLE)]

Chandigarh /ˌtʃʌndɪˈɡɑː(r)/ **1** a Union Territory of NW India, created in 1966. **2** a city in this territory; pop. (1991) 503,000. The present city was designed in 1950 by Le Corbusier as a new capital for the Punjab, the old capital, Lahore, having been lost to Pakistan in 1947. Chandigarh is the capital both of the modern Indian state of Punjab, which lies to the west of it, and of Haryana, which lies to the south of it.

Chandler /ˈtʃɑːndlə(r)/, Raymond (Thornton) (1888–1959), American novelist. Of British descent and educated largely in England (1896–1912), Chandler is particularly remembered as the creator of the private detective Philip Marlowe and as one of the exponents of the tough, realistic style of hard-boiled detective fiction. His novels, which include *The Big Sleep* (1939), *Farewell, My Lovely* (1940), and *The Long Goodbye* (1953), are written in an ironic, terse, and fast-moving style; many were made into films, especially of the *film noir* genre, and Chandler himself worked on the screenplays.

chandler /ˈtʃɑːndlə(r)/ *n.* a dealer in candles, oil, soap, paint, groceries, etc. □ **corn chandler** a dealer in corn. **ship** (or **ship's**) **chandler** a dealer in cordage, canvas, etc. [ME f. AF *chaundeler*, OF *chandelier* (as CANDLE)]

chandlery /ˈtʃɑːndlərɪ/ *n.* the goods sold by a chandler.

Chandrasekhar /ˌtʃʌndrəˈsiːkə(r)/, -ˈseɪkə(r)/, Subrahmanyan (1910–95), Indian-born American astronomer. He worked on stellar evolution, suggesting the process whereby some stars eventually collapse to form a dense white dwarf. He demonstrated that for this to happen the star's mass must not exceed 1.44 solar masses (the Chandrasekhar limit): stars above this mass collapse further to form neutron stars.

Chanel /ʃəˈnel/, Coco (born Gabrielle Bonheur Chanel) (1883–1971), French couturière. She opened her first Paris couture house in 1924, quickly achieving success with her simple but elegant designs. Loose and comfortable, her garments were a radical departure from the dominant stiff corseted styles of the day. She also manufactured her own range of perfumes, costume jewellery, and textiles.

Changan /tʃæŋˈɑːn/ the former name (202 BC until AD 618) for XIAN.

Chang-chiakow see ZHANGJIAKOU.

Changchun /tʃæŋˈtʃʊn/ an industrial city in NE China, capital of Jilin province; pop. (1990) 2,070,000.

change /tʃeɪndʒ/ *n. & v.* ● *n.* **1 a** the act or an instance of making or becoming different. **b** an alteration or modification (*the change in her expression*). **2 a** money given in exchange for money in larger units or a different currency. **b** money returned as the balance of that given in payment. **c** = *small change*. **3** a new experience; variety (*fancied a change*; *for a change*). **4 a** the substitution of one thing for another; an exchange (*change of scene*). **b** a set of clothes etc. put on in place of another. **5** (in full **change of life**) *colloq.* the menopause. **6** (usu. in *pl.*) any of the different orders in which a peal of bells can be rung. **7** (**Change**) (also **'Change**) *hist.* a place where merchants etc. met to do business. **8** (of the moon) arrival at a fresh phase, esp. at the new moon. ● *v.* **1** *tr. & intr.* undergo, show, or subject to change; make or become different (*the wig changed his appearance*; *changed from an introvert into an extrovert*). **2** *tr.* **a** take or use another instead of; go from one to another (*change one's socks*; *changed his doctor*; *changed trains*). **b** (usu. foll. by *for*) give up or get rid of in exchange (*changed the car for a van*). **3** *tr.* **a** give or get change in smaller denominations for (*can you change a ten-pound note?*). **b** (foll. by *for*) exchange (a sum of money) for (*changed his dollars for pounds*). **4** *tr. & intr.* put fresh clothes or coverings on (*changed the baby as he was wet*; *changed into something loose*). **5** *tr.* (often foll. by *with*) give and receive, exchange (*changed places with him*; *we changed places*). **6** *intr.* change trains etc. (*changed at Crewe*). **7** *intr.* (of the moon) arrive at a fresh phase, esp. become new. □ **change colour** blanch or flush. **change down** engage a lower gear in a vehicle. **change gear** engage a different gear in a vehicle. **change hands 1** pass to a different owner. **2** substitute one hand for the other. **change one's mind** adopt a different opinion or plan. **change of air** a different climate; variety. **change of heart** a conversion to a different view. **change over** change from one system or situation to another. **change-over** *n.* such a change. **change step** alter one's step so that the opposite leg is the one that marks time when marching etc. **change the subject** begin talking of something different, esp. to avoid embarrassment. **change one's tune 1** voice a different opinion from that expressed previously. **2** change one's style of language or manner, esp. from an insolent to a respectful tone. **change up** engage a higher gear in a vehicle. **get no change out of** *colloq.* **1** fail to get information from. **2** fail to get the better of (in business etc.). **ring the changes** (**on**) vary the ways of expressing, arranging, or doing something. □ **changeful** *adj.* **changer** *n.* [ME f. AF *chaunge*, OF *change*, *changer* f. LL *cambiare*, L *cambire* barter, prob. of Celt. orig.]

changeable /ˈtʃeɪndʒəb(ə)l/ *adj.* **1** irregular, inconstant. **2** that can change or be changed. □ **changeably** *adv.* **changeableness** *n.* **changeability** /ˌtʃeɪndʒəˈbɪlɪtɪ/ *n.* [ME f. OF, formed as CHANGE]

changeless /ˈtʃeɪndʒlɪs/ *adj.* unchanging. □ **changelessly** *adv.* **changelessness** *n.*

changeling /ˈtʃeɪndʒlɪŋ/ *n.* a child believed to have been substituted for another by stealth, esp. by fairies.

Changsha /tʃæŋˈʃɑː/ the capital of Hunan province in east central China; pop. (1990) 1,300,000.

Chania /kɑːˈnjɑː/ (Greek **Khaniá** /xaˈnjɑː/) a port on the north coast of Crete, capital of the island from 1841 to 1971; pop. (1981) 47,340.

channel[1] /ˈtʃæn(ə)l/ *n. & v.* ● *n.* **1 a** a length of water wider than a strait, joining two larger areas, esp. seas. **b** (**the Channel**) the English Channel between Britain and France. **2** a medium of communication; an agency for conveying information (*through the usual channels*). **3** *Broadcasting* **a** a band of frequencies used in radio and television transmission, esp. as used by a particular station. **b** a service or station using this. **4** the course in which anything moves; a direction. **5 a** a natural or artificial hollow bed of water. **b** the navigable part of a waterway. **6** a tubular passage for liquid. **7** *Electronics* a lengthwise strip on recording tape etc. **8** a groove or a flute, esp. in a column. ● *v.tr.* (**channelled, channelling**; *US* **channeled, channeling**) **1** guide, direct (*channelled them through customs*). **2** form channels in; groove. [ME f. OF *chanel* f. L *canalis* CANAL]

channel² /ˈtʃæn(ə)l/ *n. Naut.* any of the broad thick planks projecting horizontally from a ship's side abreast of the masts, used to widen the basis for the shrouds. [for *chain-wale*: cf. *gunnel* for *gunwale*]

Channel Country an area of SW Queensland and NE South Australia, watered intermittently by natural channels, where rich grasslands produced by the summer rains provide grazing for cattle.

Channel Four (in the UK) a commercial television company that aims to show material of minority or specialist interest in addition to popular programmes. It began broadcasting in 1982.

Channel Islands a group of islands in the English Channel off the NW coast of France, of which the largest are Jersey, Guernsey, and Alderney; pop. (1981) 128,900. They are the only portions of the former dukedom of Normandy that still owe allegiance to England, to which they have been attached since the Norman Conquest in 1066.

channelize /ˈtʃænəˌlaɪz/ *v.tr.* (also **-ise**) convey in, or as if in, a channel; guide.

Channel Tunnel a railway tunnel under the English Channel, linking the coasts of England and France, begun in 1987 and opened in 1994. Terminals near Folkestone in England and Calais in France enable motor vehicles to be loaded on to trains for the journey through the 49-km (31-mile) tunnel. Several abortive schemes for constructing a tunnel under the Channel had been put forward (and in 1882 actually begun), the earliest by a French engineer in 1802.

chanson de geste /ˌʃɒnsɒn də ˈʒest/ *n.* (*pl.* **chansons** pronunc. same) any of a group of French historical verse romances, mostly connected with Charlemagne, composed in the 11th–13th centuries. Perhaps the best-known today is the *Chanson de Roland* (12th century). [F, = song of heroic deeds]

chant /tʃɑːnt/ *n. & v.* ● *n.* **1 a** a spoken singsong phrase, esp. one performed in unison by a crowd etc. **b** a repetitious singsong way of speaking. **2** *Mus.* **a** a short musical passage in two or more phrases used for singing unmetrical words, e.g. psalms, canticles. **b** the psalm or canticle so sung. **c** a song, esp. monotonous or repetitive. **3** a musical recitation, esp. of poetry. ● *v.tr. & intr.* **1** talk or repeat monotonously (*a crowd chanting slogans*). **2** sing or intone (a psalm etc.). [ME (orig. as verb) f. OF *chanter* sing f. L *cantare* frequent. of *canere cant-* sing]

chanter /ˈtʃɑːntə(r)/ *n. Mus.* the melody-pipe with finger-holes, of a bagpipe.

chanterelle /ˌtʃæntəˈrel/ *n.* an edible fungus, *Cantharellus cibarius*, with a yellow funnel-shaped cap and smelling of apricots. [F f. mod.L *cantharellus* dimin. of *cantharus* f. Gk *kantharos* a kind of drinking-vessel]

chanteuse /ʃɒnˈtɜːz/ *n.* a female singer of popular songs. [F]

chanticleer /ˌtʃæntɪˈklɪə(r), ˌtʃɑːn-, ˌʃæn-, ˌʃɑːn-/ *n. literary* a name given to a domestic cock, esp. in fairy tales etc. [ME f. OF *chantecler* (as CHANT, CLEAR): used as a name in the *Roman de Renart* (see REYNARD)]

Chantilly /ʃænˈtɪli/ *n.* **1** a delicate kind of bobbin-lace. **2** sweetened or flavoured whipped cream. [*Chantilly* near Paris]

chantry /ˈtʃɑːntri/ *n.* (*pl.* **-ies**) **1** an endowment for a priest or priests to celebrate masses for the founder's soul. **2** the priests, chapel, altar, etc., so endowed. [ME f. AF *chaunterie*, OF *chanterie* f. *chanter* CHANT]

chanty var. of SHANTY².

Chanukkah var. of HANUKKAH.

Chanute /ʃəˈnuːt/, Octave (1832–1910), French-born American aviation pioneer. Educated as a railway engineer, he built his first glider in 1896 and later produced others, of which the most successful was a biplane which made over 700 flights. His encouragement of the Wright brothers and of the serious study of aeronautics greatly assisted them in making the world's first controlled powered flight.

chaology /keɪˈɒlədʒi/ *n. Math.* = CHAOS THEORY.

Chao Phraya /ˌtʃaʊ prəˈjɑː/ a major waterway of central Thailand, formed by the junction of the Ping and Nan rivers.

Chaos /ˈkeɪɒs/ *Gk Mythol.* the first created being, from which came the primeval deities Gaia (Earth), Tartarus, Erebus (Darkness), and Nyx (Night). [Gk, = gaping void]

chaos /ˈkeɪɒs/ *n.* **1 a** utter confusion, disorder. **b** chaotic behaviour

(CHAOTIC 2). **2** the formless matter supposed to have existed before the creation of the universe. [F or L f. Gk *khaos*: *-otic* after *erotic* etc.]

chaos theory *n.* the mathematical study of complex systems whose development is highly sensitive to slight changes in conditions, so that small events can give rise to strikingly great consequences. For example, a tiny disruption of the atmosphere in the Arctic Circle might lead to the development of a hurricane in the tropics. Such behaviour is in practice unpredictable beyond a short time-scale. Chaos theory, which has applications in physics, biology, ecology, economics, and other fields, has two main aspects. On the one hand, processes that seem random or irregular may actually follow discoverable laws. On the other hand, some processes that used to be thought predictable, e.g. the movements of bodies in the solar system, have been shown to be chaotic in the long term.

chaotic /keɪˈɒtɪk/ *adj.* **1** characterized by chaos, disorderly, confused. **2** (of a system or its behaviour) so unpredictable as to appear random, owing to great sensitivity to small changes in conditions. (See also CHAOS THEORY.) □ **chaotically** *adv.*

chap¹ /tʃæp/ *v. & n.* ● *v.* (**chapped, chapping**) **1** *intr.* (esp. of the skin; also of dry ground etc.) crack in fissures, esp. because of exposure and dryness. **2** *tr.* (of the wind, cold, etc.) cause to chap. ● *n.* (usu. in *pl.*) **1** a crack in the skin. **2** an open seam. [ME, perh. rel. to MLG, MDu. *kappen* chop off]

chap² /tʃæp/ *n. colloq.* a man; a boy; a fellow. [abbr. of CHAPMAN]

chap³ /tʃæp/ *n.* the lower jaw or half of the cheek, esp. of a pig as food. □ **chap-fallen** dispirited, dejected (with the lower jaw hanging). [16th c.: var. of CHOP², of unkn. orig.]

chap. *abbr.* chapter.

chaparejos /ˌʃæpəˈreɪɒs, ˌtʃæp-, -ˈreɪhəʊs/ *n.pl. US* a cowboy's leather protection for the front of the legs. [Mex. Sp. *chaparreras* from *chaparro* dwarf evergreen oak]

chaparral /ˌʃæpəˈræl, ˌtʃæp-/ *n. US* dense tangled brushwood; undergrowth. □ **chaparral cock** = ROADRUNNER. [Sp. f. *chaparra* evergreen oak]

chapatti /tʃəˈpɑːti, -ˈpæti/ *n.* (also **chapati, chupatty**) (*pl.* **-is** or **chupatties**) (in Indian cookery) a flat thin cake of unleavened wholemeal bread. [Hindi *capātī*]

chap-book /ˈtʃæpbʊk/ *n. hist.* a small pamphlet containing tales, ballads, tracts, etc., hawked by chapmen. [19th c.: see CHAPMAN]

chape /tʃeɪp/ *n.* **1** the metal cap of a scabbard-point. **2** the back-piece of a buckle attaching it to a strap etc. **3** a sliding loop on a belt or strap. [ME f. OF, = cope, hood, formed as CAP]

chapeau-bras /ˌʃæpəʊˈbrɑː/ *n.* (*pl.* **chapeaux-bras** pronunc. same) a three-cornered flat silk hat often carried under the arm. [F f. *chapeau* hat + *bras* arm]

chapel /ˈtʃæp(ə)l/ *n.* **1 a** a place for private Christian worship in a large church or esp. a cathedral, with its own altar and dedication (*Lady chapel*). **b** a place of Christian worship attached to a private house or institution. **2** *Brit.* **a** a place of worship for nonconformist bodies. **b** (*predic.*) an attender at or believer in nonconformist worship (*they are strictly chapel*). **c** a chapel service. **d** attendance at a chapel. **3** an Anglican church subordinate to a parish church. **4** *Printing* **a** the members or branch of a printers' trade union at a specific place of work. The name reflects the early connection of printing with the production of religious texts. **b** a meeting of such a union branch. □ **chapel of ease** an Anglican chapel for the convenience of remote parishioners. **chapel of rest** an undertaker's mortuary. **chapel royal** a chapel in a royal palace. **father of chapel** (or **the chapel**) the shop steward of a printers' chapel. [ME f. OF *chapele* f. med.L *cappella* dimin. of *cappa* cloak: the first chapel was a sanctuary in which St Martin's sacred cloak (*cappella*) was preserved]

Chapel Royal the body of clergy, singers, and musicians employed by the English monarch for religious services, known to have existed as early as 1135 and now based at St James's Palace, London. Among members of the Chapel Royal have been Thomas Tallis, William Byrd, and Henry Purcell.

chapelry /ˈtʃæp(ə)lri/ *n.* (*pl.* **-ies**) a district served by an Anglican chapel.

chaperon /ˈʃæpəˌrəʊn/ *n. & v.* (also **chaperone**) ● *n.* **1** a person, esp. an older woman, who ensures propriety by accompanying a young unmarried woman on social occasions. **2** a person who takes charge of esp. young people in public. ● *v.tr.* act as a chaperon to. □ **chaperonage** /-rənɪdʒ/ *n.* [F, = hood, chaperon, dimin. of *chape* cope, formed as CAP]

chaplain /ˈtʃæplɪn/ n. a member of the clergy attached to a private chapel, institution, ship, regiment, etc. □ **chaplaincy** n. (pl. **-ies**). [ME f. AF & OF c(h)apelain f. med.L cappellanus, orig. custodian of the cloak of St Martin: see CHAPEL]

chaplet /ˈtʃæplɪt/ n. **1** a garland or circlet for the head. **2** a string of 55 beads (one-third of the rosary number) for counting prayers, or as a necklace. **3** a bead-moulding. □ **chapleted** adj. [ME f. OF chapelet, ult. f. LL cappa CAP]

Chaplin /ˈtʃæplɪn/, Sir Charles Spencer ('Charlie') (1889–1977), English film actor and director. Moving to Hollywood in 1914 he made many short comedies, mostly playing a little bowler-hatted tramp, a character which remained his trademark for more than twenty-five years. A master of mime who combined pathos with slapstick clowning, he was best suited to the silent medium; his most successful films include The Kid (1921) and The Gold Rush (1925). The director of all his films, he combined speech and mime in Modern Times (1936), while The Great Dictator (1940), a satire on Hitler, was his first proper sound film and his last appearance in his familiar bowler-hatted role.

Chapman /ˈtʃæpmən/, George (c.1560–1634), English poet and dramatist. Although acclaimed as a dramatist in his day, he is now chiefly known for his translations of Homer; twelve books of the Iliad were published in 1611 and the complete Iliad and Odyssey in 1616. They are commemorated in Keats's sonnet 'On First Looking into Chapman's Homer' (1817).

chapman /ˈtʃæpmən/ n. (pl. **-men**) hist. a pedlar. [OE cēapman f. cēap barter]

chappal /ˈtʃæp(ə)l/ n. an Indian sandal, usu. of leather. [Hindi]

Chappaquiddick Island /ˌtʃæpəˈkwɪdɪk/ a small island in the western Atlantic, off the coast of Martha's Vineyard in Massachusetts. It was the scene of a car accident in 1969 involving Senator Edward Kennedy, the brother of President John F. Kennedy, in which a Democratic Party worker, Mary Jo Kopechne, drowned.

Chappell /ˈtʃæp(ə)l/, Gregory Stephen ('Greg') (b.1948), Australian cricketer. After making his first-class début for South Australia in 1966, he played county cricket in England in 1968 and 1969, and was first selected to play for his country in 1970. He captained Australia from 1975 to 1984, becoming the first Australian to score more than 7,000 runs in test matches.

chappie /ˈtʃæpɪ/ n. colloq. = CHAP².

chappy /ˈtʃæpɪ/ adj. full of chaps or cracks; chapped (chappy knuckles).

chaps /tʃæps, ʃæps/ n. = CHAPAREJOS. [abbr.]

chapstick /ˈtʃæpstɪk/ n. US a cylinder of a cosmetic substance used to prevent chapping of the lips.

chapter /ˈtʃæptə(r)/ n. **1** a main division of a book. **2** a period of time (in a person's life, a nation's history, etc.). **3** a series or sequence (a chapter of misfortunes). **4 a** the canons of a cathedral or other religious community or knightly order. **b** a meeting of these. **5** an Act of Parliament numbered as part of a session's proceedings. **6** N. Amer. a local branch of a society. □ **chapter and verse** an exact reference or authority. **chapter house 1** a building used for the meetings of a chapter. **2** US the place where a college fraternity or sorority meets. [ME f. OF chapitre f. L capitulum dimin. of caput -itis head]

char¹ /tʃɑː(r)/ v.tr. & intr. (**charred, charring**) **1** make or become black by burning; scorch. **2** burn or be burnt to charcoal. [app. back-form. f. CHARCOAL]

char² /tʃɑː(r)/ n. & v. Brit. colloq. ● n. = CHARWOMAN. ● v.intr. (**charred, charring**) work as a charwoman. [earlier chare f. OE cerr a turn, cierran to turn]

char³ /tʃɑː(r)/ n. (also **cha** /tʃɑː/) Brit. colloq. tea. [Chin. cha]

char⁴ /tʃɑː(r)/ n. (also **charr**) (pl. same) a small trout of the genus Salvelinus, esp. S. alpinus, of Arctic waters and northern lakes. [17th c.: orig. unkn.]

charabanc /ˈʃærəˌbæŋ/ n. Brit. hist. a long vehicle, originally horse-drawn and open, later an early form of motor coach, with seating on transverse benches facing forward. [F char à bancs carriage with benches]

characin /ˈkærəsɪn/ n. a freshwater fish of the family Characidae, mainly of South and Central America, including piranhas and tetras. [mod.L Characinus genus name, f. Gk kharax a kind of fish]

character /ˈkærɪktə(r)/ n. & v. ● n. **1** the collective qualities or characteristics, esp. mental and moral, that distinguish a person or thing. **2 a** moral strength (has a weak character). **b** reputation, esp. good reputation. **3 a** a person in a novel, play, etc. **b** a part played by an actor; a role. **4** colloq. a person, esp. an eccentric or outstanding individual (he's a real character). **5 a** a printed or written letter, symbol, or distinctive mark, especially one denoting a sound, syllable, or idea, as in the Chinese alphabet. **b** Computing any of a group of symbols representing a letter etc. **6** a written description of a person's qualities; a testimonial. **7** a characteristic (esp. of a biological species). ● v.tr. archaic inscribe; describe. □ **character actor** an actor who specializes in playing eccentric or unusual persons. **character assassination** a malicious attempt to harm or destroy a person's good reputation. **in** (or **out of**) **character** consistent (or inconsistent) with a person's character. □ **characterful** adj. **characterfully** adv. **characterless** adj. [ME f. OF caractere f. L character f. Gk kharaktēr stamp, impress]

characteristic /ˌkærɪktəˈrɪstɪk/ adj. & n. ● adj. typical, distinctive (with characteristic expertise). ● n. **1** a characteristic feature or quality. **2** Math. the whole number or integral part of a logarithm. □ **characteristic curve** a graph showing the relationship between two variable but interdependent quantities. **characteristic radiation** radiation the wavelengths of which are peculiar to the element which emits them. □ **characteristically** adv. [F caractéristique or med.L characterizare f. Gk kharaktērizō]

characterize /ˈkærɪktəˌraɪz/ v.tr. (also **-ise**) **1 a** describe the character of. **b** (foll. by as) describe as. **2** be characteristic of. **3** impart character to. □ **characterization** /ˌkærɪktəraɪˈzeɪʃ(ə)n/ n. [F caractériser or med.L characterizare f. Gk kharaktērizō]

charade /ʃəˈrɑːd/ n. **1 a** (usu. in pl., treated as sing.) a game of guessing a word from a written or acted clue given for each syllable and for the whole. **b** one such clue. **2** an absurd pretence. [F f. mod.Prov. charrado conversation f. charra chatter]

charas /ˈtʃɑːrəs/ n. a narcotic resin made from the flower-heads of hemp; cannabis resin. [Hindi]

charbroil /ˈtʃɑːbrɔɪl/ v.tr. grill (meat etc.) on a rack over charcoal. [CHARCOAL + BROIL¹]

charcoal /ˈtʃɑːkəʊl/ n. **1 a** an amorphous form of carbon consisting of a porous black residue from wood, bones, etc., heated in the absence of air. **b** (usu. in pl.) a piece of this used for drawing. **2** a drawing in charcoal. **3** (in full **charcoal grey**) a dark grey colour. □ **charcoal biscuit** a biscuit containing wood-charcoal to aid digestion. [ME COAL = charcoal: first element perh. chare turn (cf. CHAR¹, CHAR²)]

Charcot /ˈʃɑːkəʊ/, Jean-Martin (1825–93), French neurologist. Working at the Salpêtrière clinic in Paris, he established links between various neurological conditions and particular lesions in the central nervous system. He described several such diseases, some of which are named after him, and he is regarded as one of the founders of modern neurology. Charcot's work on hysteria was taken up by his pupil Sigmund Freud.

charcuterie /ʃɑːˈkuːtərɪ/ n. **1** cold cooked meats. **2** a shop selling these. [F]

chard /tʃɑːd/ n. a kind of beet, Beta vulgaris, with edible broad white leaf-stalks and green blades. Also called seakale beet. [F carde, and chardon thistle: cf. CARDOON]

Chardonnay /ˈʃɑːdɒˌneɪ/ n. **1** a variety of white grape used for making champagne and other wines. **2** the vine on which this grape grows. **3** a wine made from Chardonnay grapes. [F]

Charente /ʃæˈrɒnt/ a river of western France, which rises in the Massif Central and flows 360 km (225 miles) westwards to enter the Bay of Biscay at Rochefort.

charge /tʃɑːdʒ/ v. & n. ● v. **1** tr. ask (an amount) as a price (charges £5 a ticket). **b** ask (a person) for an amount as a price (you forgot to charge me). **2** tr. **a** (foll. by to, up to) debit the cost of to (a person or account) (charge it to my account; charge it up to me). **b** debit (a person or an account) (bought a new car and charged the company). **3** tr. (often foll. by with) accuse (of an offence) (charged him with theft). **b** (foll. by that + clause) make an accusation that. **4** tr. (foll. by to + infin.) instruct or urge. **5** (foll. by with) **a** tr. entrust with. **b** refl. undertake. **6 a** intr. make a rushing attack; rush headlong. **b** tr. make a rushing attack on; throw oneself against. **7** (often foll. by up) **a** tr. give an electric charge to (a body); store energy in (a battery etc.). **b** intr. (of a battery etc.) receive and store energy. **8** tr. (often foll. by with) load or fill (a vessel, gun, etc.) to the full or proper extent. **9** tr. (usu. as **charged** adj.) **a** (foll. by with) saturated with (air charged with vapour). **b** (usu. foll. by with) pervaded (with strong feelings etc.) (atmosphere charged with emotion; a charged atmosphere). ● n. **1 a** a price asked for goods or services. **b** a financial liability or commitment. **2** an accusation, esp. against a prisoner brought to trial. **3 a** a task, duty, or commission. **b** care, custody, responsible possession. **c** a person or thing entrusted; a minister's congregation. **4 a** an impetuous rush or attack, esp. in a battle. **b** the signal for this. **5** the appropriate amount of material to be put into a receptacle, mechanism, etc., at one time, esp. of explosive for a gun. **6 a** a property of matter that is a consequence of the interaction between its constituent particles and exists in a positive or negative form, causing electrical phenomena. **b** the quantity of this carried by a body. **c** energy stored chemically for conversion into electricity. **d** the process of charging a battery. **7** an exhortation; directions, orders. **8** a burden or load. **9** *Heraldry* a device; a bearing. □ **charge account** *N. Amer.* a credit account at a shop etc. **charge card** a credit card, esp. for use at a particular store. **charge conjugation** *Physics* replacement of a particle by its antiparticle. **charge-coupled device** see CCD. **charge-hand** *Brit.* a worker, ranking below a foreman, in charge of others on a particular job. **charge-nurse** *Brit.* a male nurse in charge of a ward (the male equivalent of a nursing sister). **charge-sheet** *Brit.* a record of cases and charges made at a police station. **free of charge** free, without charge. **give a person in charge** hand a person over to the police. **in charge** having command. **lay to a person's charge** accuse a person of. **put a person on a charge** charge a person with a specified offence. **return to the charge** begin again, esp. in argument. **take charge** (often foll. by *of*) assume control or direction. □ **chargeable** *adj.* [ME f. OF *charger* f. LL *car(r)icare* load f. L *carrus* CAR]

chargé d'affaires /ˌʃɑːʒeɪ dæˈfeə(r)/ *n.* (also **chargé**) (*pl.* **chargés** *pronunc.* same) **1** an ambassador's deputy. **2** an envoy to a minor country. [F, = in charge (of affairs)]

Charge of the Light Brigade a British cavalry charge in 1854 during the Battle of Balaclava in the Crimean War. A misunderstanding between the commander of the Light Brigade, Lord Cardigan, and his superiors, Lords Raglan and Lucan, led to the British cavalry being committed to an attack up a valley strongly held on three sides by the Russians. Immortalized in verse by Tennyson, the charge destroyed some of the finest light cavalry in the world to very little military purpose.

charger[1] /ˈtʃɑːdʒə(r)/ *n.* **1 a** a cavalry horse. **b** *poet.* any horse. **2** an apparatus for charging a battery. **3** a person or thing that charges.

charger[2] /ˈtʃɑːdʒə(r)/ *n. archaic* a large flat dish. [ME f. AF *chargeour*]

chariot /ˈtʃærɪət/ *n. & v.* ● *n.* **1** *hist.* **a** a two-wheeled vehicle drawn by horses, used in ancient warfare and racing. Chariots were known in Mesopotamia from the end of the 3rd millennium BC and spread from there throughout Europe and Asia. **b** a four-wheeled carriage with back seats only. **2** *poet.* a stately or triumphal vehicle. ● *v.tr. literary* convey in or as in a chariot. [ME f. OF, augment. of *char* CAR]

charioteer /ˌtʃærɪəˈtɪə(r)/ *n.* a chariot-driver.

charisma /kəˈrɪzmə/ *n.* (*pl.* **charismata** /-mətə/) **1 a** the ability to inspire followers with devotion and enthusiasm. **b** an attractive aura; great charm. **2** a divinely conferred power or talent. [eccl.L f. Gk *kharisma* f. *kharis* favour, grace]

charismatic /ˌkærɪzˈmætɪk/ *adj. & n.* ● *adj.* **1** having charisma; inspiring enthusiasm. **2** (of Christian worship) characterized by spontaneity, ecstatic utterances, etc. ● *n.* a person who claims divine inspiration; an adherent of charismatic worship. □ **charismatic movement** a Pentecostal movement within Roman Catholic, Anglican, and other Christian Churches. □ **charismatically** *adv.*

charitable /ˈtʃærɪtəb(ə)l/ *adj.* **1** generous in giving to those in need. **2** of, relating to, or connected with a charity or charities. **3** apt to judge favourably of persons, acts, and motives. □ **charitableness** *n.* **charitably** *adv.* [ME f. OF f. *charité* CHARITY]

charity /ˈtʃærɪtɪ/ *n.* (*pl.* **-ies**) **1** a giving voluntarily to those in need. **b** the help, esp. money, so given. **2** an institution or organization for helping those in need. **3 a** kindness, benevolence. **b** tolerance in judging others. **c** love of one's fellow humans. [OE f. OF *charité* f. L *caritas -tatis* f. *carus* dear]

Charity Commission (in the UK) a board established to control charitable trusts.

charivari /ˌʃɑːrɪˈvɑːrɪ/ *n.* (also **shivaree** /ˌʃɪvəˈriː/) **1** a serenade of banging saucepans etc. to a newly married couple. **2** a medley of sounds; a hubbub. [F, = serenade with pans, trays, etc., to an unpopular person]

charlady /ˈtʃɑːˌleɪdɪ/ *n.* (*pl.* **-ies**) = CHARWOMAN.

charlatan /ˈʃɑːlət(ə)n/ *n.* a person falsely claiming a special knowledge or skill. □ **charlatanism** *n.* **charlatanry** *n.* [F f. It. *ciarlatano* f. *ciarlare* babble]

Charlemagne /ˈʃɑːləˌmeɪn/ (Latin *Carolus Magnus* Charles the Great)

(742–814), king of the Franks 768–814 and Holy Roman emperor (as Charles I) 800–14. He created an empire by conquering and Christianizing the Lombards (774), Saxons (772–7; 782–5), and Avars (791–9), and restoring areas of Italy to the pope. His coronation by Pope Leo III in Rome on Christmas Day, 800, is taken as having inaugurated the Holy Roman Empire. He gave government new moral drive and religious responsibility, and encouraged commerce and agriculture. A well-educated man, he promoted the arts and education, and under Alcuin his principal court at Aachen became a major centre of learning. The political cohesion of his empire did not last, but the influence of his scholars persisted in the Carolingian Renaissance.

Charleroi /ˈʃɑːləˌrwʌ/ an industrial city in SW Belgium; pop. (1991) 206,200.

Charles[1] /tʃɑːlz/ the name of two kings of England, Scotland, and Ireland:

Charles I (1600–49), son of James I, reigned 1625–49. His reign was dominated by the deepening religious and constitutional crisis that eventually resulted in the English Civil War. His attempt to rule without Parliament (1629–40) eventually failed when he was obliged to recall Parliament to fund his war with Scotland; disputes with the new Parliament led to civil war in 1642. Charles surrendered to the Scots in 1646 and was handed over to Parliament in 1647. He escaped and made an alliance with the Scots in return for religious concessions, but the Royalist forces were defeated in 1648 and the Parliamentary army demanded Charles' death. He was tried by a special Parliamentary court and beheaded.

Charles II (1630–85), son of Charles I, reigned 1660–85. After his father's execution in 1649 Charles was declared king in Scotland and then crowned there in 1651, but was forced into exile the same year, when his army attempted to invade England and was defeated by Cromwell's forces at Worcester. He remained in exile on the Continent for nine years before he was restored after the collapse of Cromwell's regime. Charles displayed considerable adroitness in handling the difficult constitutional situation, but continuing religious and political strife dogged his reign. Although he failed to produce a legitimate heir, he moved to ensure the Protestant succession by arranging the marriage of his niece Mary to William of Orange.

Charles[2] /tʃɑːlz/ the name of four kings of Spain, notably:

Charles I (1500–58), son of Philip I, reigned 1516–56, Holy Roman emperor (as Charles V) 1519–56. He united the Spanish and imperial thrones when he inherited the latter in 1519. His reign was characterized by the struggle against Protestantism in Germany by rebellion in Castile, and by war with France (1521–44). Exhausted by these struggles, Charles handed Naples, the Netherlands, and Spain over to his son Philip II and the imperial crown (1556) to his brother Ferdinand, and retired to a monastery in Spain.

Charles II (1661–1700), reigned 1665–1700. The last Habsburg to be king of Spain, he inherited a kingdom already in a decline which he was unable to halt. Childless, he chose Philip of Anjou, grandson of Louis XIV of France, as his successor; this ultimately gave rise to the War of the Spanish Succession.

Charles IV (1748–1819), reigned 1788–1808. He was dominated by his wife Maria Luisa and her lover Manuel de Godoy (Prime Minister from 1792). During the Napoleonic Wars he suffered the loss of the Spanish fleet, destroyed along with that of France at Trafalgar in 1805. Following the French invasion of Spain in 1807, Charles was forced to abdicate. He died in exile in Rome.

Charles[3] /tʃɑːlz/ the name of seven Holy Roman emperors, notably:

Charles V Charles I of Spain (see CHARLES[2]).

Charles VI (1685–1740), Holy Roman emperor, reigned 1711–40. His claim to the Spanish throne instigated the War of the Spanish Succession, but he was ultimately unsuccessful. He became emperor on the death of his elder brother; with no surviving male heirs, he drafted the Pragmatic Sanction in an attempt to ensure that his daughter Maria Theresa succeeded to the Habsburg dominions. The failure for this to be accepted by the whole of Europe triggered a struggle for power on Charles's death and the War of the Austrian Succession.

Charles VII /tʃɑːlz, French ʃɑrl/ (1403–61), king of France 1422–61. At the time of his accession to the throne, much of northern France was under English occupation, including Reims, where he should have been crowned. After the intervention of Joan of Arc, however, the French experienced a dramatic military revival. Charles was crowned at Reims, and his reign eventually saw the defeat of the English and the end of the Hundred Years War. He modernized the administration of the army and did much to lay the foundations of French power in the following decades.

Charles XII /tʃɑːlz/ (also **Karl XII** /kɑːl/) (1682–1718), king of Sweden 1697–1718. Three years after his succession, he embarked on the Great Northern War against the encircling powers of Denmark, Poland-Saxony, and Russia. In the early years he won a series of victories, but in 1709 he embarked on an expedition deep into Russia which ended in the destruction of his army at Poltava and the internment of Charles until 1715. He resumed his military career after his return but was killed while besieging a fortress in Norway.

Charles, Prince /ˈtʃɑːlz/, Charles Philip Arthur George, Prince of Wales (b.1948), heir apparent to Elizabeth II. Educated at Gordonstoun School in Scotland and Trinity College, Cambridge, he was invested as Prince of Wales in 1969. He served in the Royal Navy 1971–6, and married Lady Diana Spencer (see DIANA, PRINCESS) in 1981; the couple publicly announced their separation in 1993. They have two children, Prince William Arthur Philip Louis (b.1982) and Prince Henry Charles Albert David (b.1984).

Charles Martel /mɑːˈtel/ (c.688–741), Frankish ruler of the eastern part of the Frankish kingdom from 715 and the whole kingdom from 719. He earned his nickname *Martel* ('the hammer') from his victory at Poitiers in 732, which effectively checked the Muslim advance into Europe. His rule marked the beginning of Carolingian power; Charlemagne was his grandson.

Charles's Law /ˈtʃɑːlzɪz/ (also **Charles' Law** /tʃɑːlz/) n. *Chem.* a law stating that the volume of an ideal gas at constant pressure is directly proportional to the absolute temperature. [J. A. C. *Charles*, Fr. scientist (1746–1823)]

Charles's Wain /ˈtʃɑːlzɪz/ n. *Astron.* the Plough in Ursa Major. The name *Charles* refers to Charlemagne: it was originally called the 'Wain of Arcturus', i.e. King Arthur, with whom Charlemagne was associated in legend.

Charleston /ˈtʃɑːlstən/ **1** the state capital of West Virginia; pop. (1990) 57,290. **2** a city and port in South Carolina; pop. (1990) 80,410. The bombardment in 1861 of Fort Sumter, in the harbour, by Confederate troops marked the beginning of the American Civil War.

charleston /ˈtʃɑːlstən/ n. & v. (also **Charleston**) ● n. a lively American dance of the 1920s with side-kicks from the knee. ● v.intr. dance the charleston. [CHARLESTON 2]

charley horse /ˈtʃɑːlɪ/ n. N. Amer. sl. stiffness or cramp in an arm or leg. [19th c.: orig. uncert.]

charlie /ˈtʃɑːlɪ/ n. Brit. sl. **1** a fool. **2** (in pl.) a woman's breasts. [dimin. of the name *Charles*]

charlock /ˈtʃɑːlɒk/ n. a wild mustard, *Sinapis arvensis*, with yellow flowers. Also called *field mustard*. [OE *cerlic*, of unkn. orig.]

Charlotte /ˈʃɑːlət/ a commercial city and transportation centre in southern North Carolina; pop. (1990) 395,930. The city is named after the wife of King George III.

charlotte /ˈʃɑːlət/ n. a pudding made of stewed fruit with a casing or layers or covering of bread, sponge cake, biscuits, or breadcrumbs (*apple charlotte*). □ **charlotte russe** /ruːs/ custard etc. enclosed in sponge cake or a casing of sponge fingers. [F]

Charlotte Amalie /əˈmɑːlɪə/ the capital of the Virgin Islands, on the island of St Thomas; pop. (1985) 52,660. The town is named after the wife of King Christian V of Denmark.

Charlotte Dundas /dʌnˈdæs/ the first vessel to use steam propulsion commercially, built on the River Clyde. The engine drove a single paddle-wheel and the ship made her first voyage in 1802.

Charlottetown /ˈʃɑːlət ˌtaʊn/ the capital and chief port of Prince Edward Island, Canada; pop. (1991) 33,150.

Charlton[1] /ˈtʃɑːlt(ə)n/, John ('Jack') (b.1935), English footballer and manager, brother of Bobby Charlton. A rugged defender, he played for Leeds United (1952–73) and was a member of the England side that won the World Cup in 1966. He later managed Middlesbrough, Sheffield Wednesday, and Newcastle United before taking over the management of the Republic of Ireland national team (1986).

Charlton[2] /ˈtʃɑːlt(ə)n/, Sir Robert ('Bobby') (b.1937), English footballer, brother of Jack Charlton. An outstanding striker, he played for Manchester United (1954–73) and for England (1957–73); he scored forty-nine goals for his country and was a member of the side that won the World Cup in 1966.

charm /tʃɑːm/ n. & v. ● n. **1 a** the power or quality of giving delight, arousing admiration, or influencing people. **b** fascination, attractiveness. **c** (usu. in pl.) an attractive or enticing quality. **2** a trinket on a bracelet etc. **3 a** an object, act, or word(s) supposedly having occult or magic power; a spell. **b** a thing worn to avert evil etc.; an amulet. **4** *Physics* a property of matter manifested by certain quarks. ● v.tr. **1** delight, captivate (*charmed by the performance*). **2** influence or protect as if by magic (*leads a charmed life*). **3 a** gain by charm (*charmed agreement out of him*). **b** influence by charm (*charmed her into consenting*). **4** cast a spell on, bewitch. □ **charm-bracelet** a bracelet hung with small trinkets. **charm offensive** the deliberate use of charm or cooperation in order to achieve a (usu. political) goal. **like a charm** perfectly, wonderfully. □ **charmer** n. [ME f. OF *charme*, *charmer* f. L *carmen* song]

charmeuse /ʃɑːˈmɜːz/ n. a soft smooth silky dress-fabric. [F, fem. of *charmeur* charmer (as CHARM)]

charming /ˈtʃɑːmɪŋ/ adj. **1** delightful, attractive, pleasing. **2** (often as int.) iron. expressing displeasure or disapproval. □ **charmingly** adv.

charmless /ˈtʃɑːmlɪs/ adj. lacking charm; unattractive, ungracious. □ **charmlessly** adv. **charmlessness** n.

charnel-house /ˈtʃɑːn(ə)l ˌhaʊs/ n. a house or vault in which dead bodies or bones are piled. [ME & OF *charnel* burying-place f. med.L *carnale* f. LL *carnalis* CARNAL]

Charolais /ˈʃærəˌleɪ/ n. (also **Charollais**) (pl. same) a breed of large white beef cattle. [Monts du *Charollais* in E. France]

Charon /ˈkeərən/ **1** *Gk Mythol.* an aged ferryman who, for a fee of one obol, ferried the souls of the dead across the rivers Styx and Acheron to Hades. It was usual for the Greeks to place a coin in the mouth of the dead for this fee. **2** *Astron.* the satellite of Pluto, probably composed mainly of ice, discovered in 1978. As its diameter of 1,190 km is more than half that of Pluto they should properly be regarded as a binary system.

charpoy /ˈtʃɑːpɔɪ/ n. Ind. a light bedstead. [Hind. *chārpāi*]

charr var. of CHARR[4].

chart /tʃɑːt/ n. & v. ● n. **1** a geographical map or plan, esp. for navigation by sea or air. **2** a sheet of information in the form of a table, graph, or diagram. **3** (usu. in pl.) a listing of the currently most popular records, esp. pop singles. ● v.tr. make a chart of, map. [F *charte* f. L *charta* CARD[1]]

chartbuster /ˈtʃɑːtˌbʌstə(r)/ n. colloq. a best-selling popular song, record, etc.

charter /ˈtʃɑːtə(r)/ n. & v. ● n. **1 a** a written grant of rights, by the sovereign or legislature, esp. the creation of a borough, company, university, etc. **b** a written constitution or description of an organization's functions etc. **2** a contract to hire an aircraft, ship, etc., for a special purpose. **3** = CHARTER-PARTY. ● v.tr. **1** grant a charter to. **2** hire (an aircraft, ship, etc.). □ **chartered accountant**, **engineer**, **librarian**, **surveyor**, etc. Brit. a member of a professional body that has a royal charter. **chartered libertine** a person allowed to do as he or she pleases. **charter flight** a flight by a chartered aircraft. **charter-member** an original member of a society, corporation, etc. □ **charterer** n. [ME f. OF *chartre* f. L *chartula* dimin. of *charta* CARD[1]]

Charter Mark n. (in the UK) an award granted to institutions for exceptional public service under the terms of the Citizen's Charter.

charter-party /ˈtʃɑːtəˌpɑːtɪ/ n. (pl. **-ies**) a deed between a ship-owner and a merchant for the hire of a ship and the delivery of cargo. [F *charte partie* f. med.L *charta partita* divided charter, indenture]

Chartism /ˈtʃɑːtɪz(ə)m/ n. a popular movement in Britain for electoral and social reform 1837–48, the principles of which were set out in a manifesto called *The People's Charter*. This called for universal suffrage for men, equal electoral districts, voting by secret ballot, abolition of the property qualification for Parliament, payment of MPs, and annual parliaments. Despite mass support, petitions sent to Parliament in 1839 and 1842 (another was drawn up in 1848), and some rioting the movement collapsed, but most of its demands later became law. □ **Chartist** n.

Chartres /ˈʃɑːtrə/ a city in northern France; pop. (1990) 41,850. It is noted for its Gothic cathedral.

chartreuse /ʃɑːˈtrɜːz/ n. **1** a pale green or yellow liqueur of brandy and aromatic herbs etc. **2** the pale yellow or pale green colour of this. **3** a dish of fruit enclosed in jelly etc. [La Grande *Chartreuse* (Carthusian monastery near Grenoble)]

charwoman /ˈtʃɑːˌwʊmən/ n. (pl. **-women**) a woman employed as a cleaner in houses or offices.

chary /ˈtʃeərɪ/ adj. (**charier**, **chariest**) **1** cautious, wary (*chary of employing such people*). **2** sparing; ungenerous (*chary of giving praise*). **3** shy. □ **charily** adv. [OE *cearig*]

Charybdis /kəˈrɪbdɪs/ *Gk Mythol.* a dangerous whirlpool in a narrow channel of the sea, opposite the cave of the sea monster Scylla. It was later identified with the Strait of Messina, although there is no whirlpool there.

Chas. /tʃæz/ *abbr.* Charles.

chase[1] /tʃeɪs/ *v. & n.* ● *v.* **1** *tr.* pursue in order to catch. **2** *tr.* (foll. by *from, out of, to,* etc.) drive. **3** *intr.* **a** (foll. by *after*) hurry in pursuit of (a person). **b** (foll. by *round* etc.) *colloq.* act or move about hurriedly. **4** *tr.* (usu. foll. by *up*) *colloq.* pursue (overdue work, payment, etc. or the person responsible for it). **5** *tr. colloq.* **a** try to attain. **b** court persistently and openly. ● *n.* **1** pursuit. **2** unenclosed hunting-land. **3** (prec. by *the*) hunting, esp. as a sport. **4** an animal etc. that is pursued. **5** = STEEPLECHASE. □ **go and chase oneself** (usu. in *imper.*) *colloq.* depart. [ME f. OF *chace chacier,* ult. f. L *capere* take]

chase[2] /tʃeɪs/ *v.tr.* emboss or engrave (metal). [app. f. earlier *enchase* f. F *enchasser* enshrine, set (as EN-[1], CASE[2])]

chase[3] /tʃeɪs/ *n. Printing* a metal frame holding composed type. [F *châsse* f. L *capsa* CASE[2]]

chase[4] /tʃeɪs/ *n.* **1** the part of a gun enclosing the bore. **2** a trench or groove cut to receive a pipe etc. [F *chas* enclosed space f. Prov. *ca(u)s* f. med.L *capsum* thorax]

chaser /ˈtʃeɪsə(r)/ *n.* **1** a person or thing that chases. **2** a horse for steeplechasing. **3** *colloq.* a drink taken after another of a different kind, e.g. spirits after beer. **4** *US colloq.* (in full **skirt-chaser**) an amorous pursuer of women.

Chasid, Chasidism var. of HASID, HASIDISM.

chasm /ˈkæz(ə)m/ *n.* **1** a deep fissure or opening in the earth, rock, etc. **2** a wide difference of feeling, interests, etc.; a gulf. **3** *archaic* a hiatus. □ **chasmic** /ˈkæzmɪk/ *adj.* [L *chasma* f. Gk *khasma* gaping hollow]

chasse /ʃæs/ *n.* a liqueur taken after coffee etc. [F f. *chasser* CHASE[1]]

chassé /ˈʃæseɪ/ *n. & v.* ● *n.* a gliding step in dancing. ● *v.intr.* (**chasséd; chasséing**) make this step. [F, = chasing]

chasseur /ʃæˈsɜː(r)/ *n.* (*pl.* same) **1** *hist.* a soldier (esp. French) equipped and trained for rapid movement. **2** a hotel messenger, esp. in France. **3** a huntsman. □ **chasseur sauce** a rich sauce with wine and mushrooms for meat or poultry. **chicken** (or **beef** etc.) **chasseur** a dish of chicken (or beef etc.) cooked in chasseur sauce. [F, f. *chasser* CHASE[1]]

chassis /ˈʃæsɪ/ *n.* (*pl.* same /-sɪz/) **1** the base-frame of a motor vehicle, carriage, etc. **2** a frame to carry radio etc. components. [F *châssis* ult. f. L *capsa* CASE[2]]

chaste /tʃeɪst/ *adj.* **1** abstaining from extramarital, or from all, sexual intercourse. **2** (of behaviour, speech, etc.) pure, virtuous, decent. **3** (of artistic etc. style) simple, unadorned. □ **chaste-tree** an ornamental shrub, *Vitex agnus-castus,* of the verbena family, with blue or white flowers. □ **chastely** *adv.* **chasteness** *n.* [ME f. OF f. L *castus*]

chasten /ˈtʃeɪs(ə)n/ *v.tr.* **1** (esp. as **chastening, chastened** *adjs.*) subdue, restrain (*a chastening experience; chastened by his failure*). **2** discipline, punish. **3** moderate. □ **chastener** *n.* [obs. *chaste* (v.) f. OF *chastier* f. L *castigare* CASTIGATE]

chastise /tʃæsˈtaɪz/ *v.tr.* **1** rebuke or reprimand severely. **2** punish, esp. by beating. □ **chastisement** *n.* **chastiser** *n.* [ME, app. irreg. formed f. obs. verbs *chaste, chasty:* see CHASTEN]

chastity /ˈtʃæstɪtɪ/ *n.* **1** being chaste. **2** sexual abstinence; virginity. **3** simplicity of style or taste. □ **chastity belt** *hist.* a garment designed to prevent the woman wearing it from having sexual intercourse. [ME f. OF *chasteté* f. L *castitas -tatis* f. *castus* CHASTE]

chasuble /ˈtʃæzjʊb(ə)l/ *n.* a loose sleeveless usu. ornate outer vestment worn by a priest celebrating mass or the Eucharist. [ME f. OF *chesible,* later *-uble,* ult. f. L *casula* hooded cloak, little cottage, dimin. of *casa* cottage]

chat[1] /tʃæt/ *v. & n.* ● *v.intr.* (**chatted, chatting**) talk in a light familiar way. ● *n.* **1** informal conversation or talk. **2** an instance of this. □ **chat show** *Brit.* a television or radio programme in which celebrities are interviewed informally. **chat up** *Brit. colloq.* chat to, esp. flirtatiously or with an ulterior motive. [ME: shortening of CHATTER]

chat[2] /tʃæt/ *n.* a small bird with a harsh call, esp. a stonechat or whinchat, or certain American or Australian warblers. [prob. imit.]

château /ˈʃætəʊ/ *n.* (*pl.* **châteaux** /-təʊz/) a large French country house or castle, often giving its name to wine made in its neighbourhood. [F f. OF *chastel* CASTLE]

Chateaubriand /ˌʃætəʊˈbriːɒn/, François-René, Vicomte de (1768–1848), French writer and diplomat. An important figure in early French romanticism, he established his literary reputation with *Atala* (1801), but *Le Génie du Christianisme* (1802), which contributed to the post-Revolution religious revival in France, won him his greatest fame. A supporter of the royalist cause during the French Revolution, he lived in exile in England (1793–1800), where he published his *Essai sur les révolutions* (1797). His autobiography *Mémoires d'outre-tombe* (1849–50) gives an eloquent account of his life against a background of political upheaval.

chateaubriand /ˌʃætəʊˈbriːɒn/ *n.* a thick fillet of beef steak. [CHATEAUBRIAND]

chatelaine /ˈʃætəˌleɪn/ *n.* **1** the mistress of a large house. **2** *hist.* a set of short chains attached to a woman's belt, for carrying keys etc. [F *châtelaine,* fem. of *-ain* lord of a castle, f. med.L *castellanus* CASTELLAN]

Chatham /ˈtʃætəm/, 1st Earl of, see PITT[1].

Chatham Islands a group of two islands, Pitt Island and Chatham Island, in the SW Pacific to the east of New Zealand.

chatline /ˈtʃætlaɪn/ *n.* a telephone service which sets up conference calls, esp. among young people.

chattel /ˈtʃæt(ə)l/ *n.* (usu. in *pl.*) **1** *Law* any property other than freehold land, including tangible goods (*chattels personal*) and leasehold interests (*chattels real*). **2** (in general use) a personal possession. □ **chattel mortgage** *N. Amer.* the conveyance of chattels by mortgage as security for a debt. **goods and chattels** personal possessions. [ME f. OF *chatel:* see CATTLE]

chatter /ˈtʃætə(r)/ *v. & n.* ● *v.intr.* **1** talk quickly, incessantly, trivially, or indiscreetly. **2** (of a bird, monkey, etc.) emit short quick sounds. **3** (of the teeth) click repeatedly together (usu. from cold). **4** (of a tool) clatter from vibration. ● *n.* **1** chattering talk or sounds. **2** the vibration of a tool. □ **chattering classes** *Brit. colloq. derog.* members of the educated middle class who are given to debating social, political, or cultural issues; the intelligentsia. □ **chatterer** *n.* **chattery** *adj.* [ME: imit.]

chatterbox /ˈtʃætəˌbɒks/ *n.* a talkative person.

Chatterton /ˈtʃætət(ə)n/, Thomas (1752–70), English poet. He is chiefly remembered for his fabricated poems professing to be the work of Thomas Rowley, an imaginary 15th-century monk. Poverty and lack of recognition drove Chatterton to suicide at the age of 17. First published in 1777, the Rowley poems were eventually proved spurious in Skeat's 1871 edition. Chatterton's tragic life was much romanticized by Keats and Wordsworth.

chatty /ˈtʃætɪ/ *adj.* (**chattier, chattiest**) **1** fond of chatting; talkative. **2** resembling chat; informal and lively (*a chatty letter*). □ **chattily** *adv.* **chattiness** *n.*

Chaucer /ˈtʃɔːsə(r)/, Geoffrey (c.1342–1400), English poet. Born in London, he served at court and in a number of official posts before going on a series of diplomatic missions to Spain, Italy, France, and Flanders in the 1370s. From the late 1380s, particularly with the death of his wife in 1387, he suffered a number of periods of misfortune and financial insecurity. His early work (including *The Book of the Duchess,* 1369) shows the influence of French literature; he also translated part of the *Roman de la rose* during this period. During the middle period of his career, in which he wrote the long narrative poem *Troilus and Criseyde* (1385) and *The Parlement of Fowles,* his work was particularly influenced by Italian poets of his day, such as Petrarch and Boccaccio. It was not until the later part of his life that he wrote his most famous work, the *Canterbury Tales* (c.1387–1400), a cycle of linked tales told by a group of pilgrims who meet in a London tavern before their pilgrimage to the shrine of St Thomas à Becket in Canterbury. In this work in particular Chaucer demonstrated skills of characterization, humour, versatility, and a distinctive narrative style that established him as the traditional starting-point for English literature and as the first great English poet; in addition Chaucer's vernacular work helped to establish the East Midland dialect of English of his day as the standard literary language.

Chaucerian /tʃɔːˈsɪərɪən/ *adj. & n.* ● *adj.* of or relating to the English poet Chaucer or his style. ● *n.* a student of Chaucer.

chaud-froid /ʃəʊˈfrwʌ/ *n.* a dish of cold cooked meat or fish in jelly or sauce. [F f. *chaud* hot + *froid* cold]

chauffeur /ˈʃəʊfə(r), ʃəʊˈfɜː(r)/ *n. & v.* ● *n.* (*fem.* **chauffeuse** /ʃəʊˈfɜːz/) a person employed to drive a private or hired motor car. ● *v.tr.* drive (a car or a passenger) as a chauffeur. [F, = stoker]

Chauliac /ˈʃəʊlɪˌæk/, Guy de (c.1300–68), French physician. Probably the most influential surgeon of the Middle Ages, he was private physician to three successive popes in Avignon from 1342. In his

Chirurgia Magna (1363) Chauliac was the first to describe many surgical techniques, and this remained the standard work in Europe until at least the 17th century.

chaulmoogra /tʃɔːlˈmuːɡrə/ *n.* a tropical Asian tree of the family Flacourtiaceae; esp. *Hydnocarpus kurzii*, with seeds yielding an oil formerly used to treat skin diseases. [Bengali]

chautauqua /tʃɔːˈtɔːkwə, ʃɔː-/ *n.* N. Amer. hist. a summer school or similar educational course. [*Chautauqua* in New York State]

chauvinism /ˈʃəʊvɪˌnɪz(ə)m/ *n.* **1** exaggerated or aggressive patriotism. **2** excessive or prejudiced support or loyalty for one's cause or group or sex (*male chauvinism*). [*Chauvin*, a Napoleonic veteran popularized as a character in the Cogniards' *Cocarde Tricolore* (1831)]

chauvinist /ˈʃəʊvɪnɪst/ *n.* **1** a person exhibiting chauvinism. **2** (in full **male chauvinist**) a man showing excessive loyalty to men and prejudice against women. □ **chauvinistic** /ˌʃəʊvɪˈnɪstɪk/ *adj.* **chauvinistically** *adv.*

Chavín /tʃɑːˈviːn/ a civilization that flourished in Peru 1000–200 BC and united a large part of the country's coastal region in a common culture. The unifying force was probably religious rather than political; characteristic remains have been found of fanged figures, presumably gods.

Ch.B. *abbr.* Bachelor of Surgery. [L *Chirurgiae Baccalaureus*]

cheap /tʃiːp/ *adj. & adv.* ● *adj.* **1** low in price; worth more than its cost (*a cheap holiday*; *cheap labour*). **2** charging low prices; offering good value (*a cheap restaurant*). **3** of poor quality; inferior (*cheap housing*). **4 a** costing little effort or acquired by discreditable means and hence of little worth (*cheap popularity*; *a cheap joke*). **b** contemptible; despicable (*a cheap criminal*). ● *adv.* cheaply (*got it cheap*). □ **cheap and cheerful** inexpensive but not unattractive. **cheap and nasty** low cost and bad quality. **dirt cheap** very cheap. **feel cheap** feel ashamed or contemptible. **on the cheap** cheaply. □ **cheapish** *adj.* **cheaply** *adv.* **cheapness** *n.* [obs. phr. *good cheap* f. *cheap* a bargain f. OE *cēap* barter, ult. f. L *caupo* innkeeper]

cheapen /ˈtʃiːp(ə)n/ *v.tr. & intr.* make or become cheap or cheaper; depreciate, degrade.

cheapjack /ˈtʃiːpdʒæk/ *n. & adj.* ● *n.* a seller of inferior goods at low prices. ● *adj.* inferior, shoddy. [CHEAP + JACK¹]

cheapo /ˈtʃiːpəʊ/ *adj. sl.* cheap.

cheapskate /ˈtʃiːpskeɪt/ *n. colloq.* a stingy or parsimonious person; a miser. [CHEAP + SKATE³ (19th c.: orig. unkn.)]

cheat /tʃiːt/ *v. & n.* ● *v.* **1** *tr.* **a** (often foll. by *into, out of*) deceive or trick (*cheated into parting with his savings*). **b** (foll. by *of*) deprive of (*cheated of a chance to reply*). **2** *intr.* gain unfair advantage by deception or breaking rules, esp. in a game or examination. **3** *tr.* avoid (something undesirable) by luck or skill (*cheated the bad weather*). **4** *tr. archaic* divert attention from, beguile (*time, tedium, etc.*). ● *n.* **1** a person who cheats. **2** a trick, fraud, or deception. **3** an act of cheating. □ **cheat on** *colloq.* be sexually unfaithful to. □ **cheatingly** *adv.* [ME *chete* f. *achete*, var. of ESCHEAT]

cheater /ˈtʃiːtə(r)/ *n.* **1** a person who cheats. **2** (in *pl.*) US sl. spectacles.

Cheboksary /ˌtʃebək'sɑːrɪ/ a city in west central Russia, on the River Volga, west of Kazan, capital of the autonomous republic of Chuvashia; pop. (1990) 429,000.

Chechen Republic /ˈtʃetʃen/ (also **Chechnya** /ˌtʃetʃˈnjɑː/) an autonomous republic in the Caucasus in SW Russia, on the border with Georgia; pop. (1990) 1,290,000; capital, Grozny. The republic declared itself independent of Russia in 1991; Russian troops invaded the republic in 1994.

check¹ /tʃek/ *v., n., & int.* ● *v.* **1** *tr.* (also *absol.*) **a** examine the accuracy, quality, or condition of. **b** (often foll. by *that* + clause) make sure; verify; establish to one's satisfaction (*checked that the doors were locked*; *checked the train times*). **2** *tr.* **a** stop or slow the motion of; curb, restrain (*progress was checked by bad weather*). **b** *colloq.* find fault with; rebuke. **3** *tr. Chess* move a piece into a position that directly threatens (the opposing king). **4** *intr.* US agree or correspond when compared. **5** *tr.* US mark with a tick etc. **6** *tr.* N. Amer. deposit (luggage etc.) for storage or dispatch. **7** *intr.* (of hounds) pause to ensure or regain scent. ● *n.* **1** a means or act of testing or ensuring accuracy, quality, satisfactory condition, etc. **2 a** a stopping or slowing of motion; a restraint on action. **b** a rebuff or rebuke. **c** a person or thing that restrains. **3** (also as *int.*) *Chess* **a** the exposure of a king to direct attack from an opposing piece. **b** an announcement of this by the attacking player. **4** US a bill in a restaurant. **5** esp. N. Amer. a token of identification for left luggage etc. **6** US Cards a counter used in various games. **7** a temporary loss of the scent in hunting. **8** a crack or flaw in timber. **9** N. Amer. colloq. = TICK¹ *n.*

3. ● *int.* US expressing assent or agreement. □ **check in 1** arrive or register at a hotel, airport, etc. **2** record the arrival of. **check-in** *n.* the act or place of checking in. **check into** register one's arrival at (a hotel etc.). **check-list** a list for reference and verification. **check-nut** = *lock-nut.* **check off** mark on a list etc. as having been examined or dealt with. **check on** examine carefully or in detail; ascertain the truth about; keep a watch on (a person, work done, etc.). **check out 1** (often foll. by *of*) leave a hotel etc. with due formalities. **2** *colloq.* investigate; examine for authenticity or suitability. **check over** examine for errors; verify. **check-rein** a rein attaching one horse's rein to another's bit, or preventing a horse from lowering its head. **checks and balances** measures designed to counterbalance damaging or unwanted influences. **check sum** *Electronics* a digit representing the sum of the digits in a digital signal and transmitted with it as a check against errors. **check through** inspect or examine exhaustively; verify successive items of. **check up** ascertain, verify, make sure. **check-up** *n.* a thorough (esp. medical) examination. **check up on** = *check on.* **check-valve** a valve allowing flow in one direction only. **in check** under control, restrained. □ **checkable** *adj.* [ME f. OF *eschequier* play chess, give check to, and OF *eschec*, ult. f. Pers. *šāh* king]

check² /tʃek/ *n.* **1** a pattern of small squares. **2** fabric having this pattern. **3** (*attrib.*) so patterned. [ME, prob. f. CHEQUER]

check³ US var. of CHEQUE.

checked /tʃekt/ *adj.* having a check pattern.

checker¹ /ˈtʃekə(r)/ *n.* **1** a person or thing that verifies or examines, esp. in a factory etc. **2** US a cashier in a supermarket etc.

checker² /ˈtʃekə(r)/ *n.* **1** var. of CHEQUER. **2** US **a** (in *pl.*, usu. treated as *sing.*) the game of draughts. **b** = CHECKERMAN.

checkerberry /ˈtʃekəbərɪ/ *n.* (*pl.* **-ies**) **1** a low-growing North American ericaceous shrub, *Gaultheria procumbens*, with aromatic leaves and white flowers. Also called *wintergreen.* **2** the edible red fruit of this plant. [*checkers* berries of service tree]

checkerboard /ˈtʃekəˌbɔːd/ *n.* N. Amer. = DRAUGHTBOARD.

checkerman /ˈtʃekəˌmæn/ *n.* (*pl.* **-men**) each of the pieces in a game of draughts.

checking account /ˈtʃekɪŋ/ *n.* US a current account at a bank. [CHECK³]

checkmate /ˈtʃekmeɪt/ *n. & v.* ● *n.* **1** (also as *int.*) *Chess* **a** check from which a king cannot escape. **b** an announcement of this. **2** a final defeat or deadlock. ● *v.tr.* **1** *Chess* put into checkmate. **2** defeat; frustrate. [ME f. OF *eschec mat* f. Pers. *šāh māt* the king is dead]

checkout /ˈtʃekaʊt/ *n.* **1** an act of checking out. **2** a point at which goods are paid for in a supermarket etc.

checkpoint /ˈtʃekpɔɪnt/ *n.* a place, esp. a barrier or manned entrance, where documents, vehicles, etc., are inspected.

checkroom /ˈtʃekruːm, -rʊm/ *n.* N. Amer. **1** a cloakroom in a hotel or theatre. **2** an office for left luggage etc.

Cheddar /ˈtʃedə(r)/ *n.* a kind of firm smooth yellow cheese originally made in Cheddar but now widely imitated. [*Cheddar*, a village in SW England]

cheek /tʃiːk/ *n. & v.* ● *n.* **1 a** the side of the face below the eye. **b** the side-wall of the mouth. **2 a** impertinent speech. **b** impertinence; cool confidence (*had the cheek to ask for more*). **3** *sl.* either buttock. **4 a** either of the side-posts of a door etc. **b** either of the jaws of a vice. **c** either of the side-pieces of various parts of machines arranged in lateral pairs. ● *v.tr.* speak impertinently to. □ **cheek-bone** the bone below the eye. **cheek by jowl** close together; intimate. **turn the other cheek** accept attack etc. meekly; refuse to retaliate. [OE *cē(a)ce, cēoce*]

cheeky /ˈtʃiːkɪ/ *adj.* (**cheekier, cheekiest**) impertinent, impudent. □ **cheekily** *adv.* **cheekiness** *n.*

cheep /tʃiːp/ *n. & v.* ● *n.* the weak shrill cry of a young bird. ● *v.intr.* make such a cry. [imit.: cf. PEEP²]

cheer /tʃɪə(r)/ *n. & v.* ● *n.* **1** a shout of encouragement or applause. **2** mood, disposition (*full of good cheer*). **3** cheerfulness, joy. **4** (in *pl.*; as *int.*) *Brit. colloq.* **a** expressing good wishes on parting. **b** expressing good wishes before drinking. **c** expressing gratitude. ● *v.* **1** *tr.* applaud with shouts. **b** (usu. foll. by *on*) urge or encourage with shouts. **2** *intr.* shout for joy. **3** *tr.* gladden; comfort. □ **cheer-leader** a person who leads cheers of applause etc. **cheer up** make or become less depressed. **three cheers** three successive hurrahs for a person or thing honoured. [ME f. AF *chere* face etc., OF *chiere* f. LL *cara* face f. Gk *kara* head]

cheerful /ˈtʃɪəfʊl/ adj. **1** in good spirits, noticeably happy (a cheerful disposition). **2** bright, pleasant (a cheerful room). **3** willing, not reluctant. □ **cheerfully** adv. **cheerfulness** n.

cheerio /ˌtʃɪrɪˈəʊ/ int. Brit. colloq. expressing good wishes on parting; goodbye.

cheerless /ˈtʃɪəlɪs/ adj. gloomy, dreary, miserable. □ **cheerlessly** adv. **cheerlessness** n.

cheerly /ˈtʃɪəlɪ/ adv. & adj. ● adv. esp. Naut. heartily, with a will. ● adj. archaic cheerful.

cheery /ˈtʃɪərɪ/ adj. (**cheerier, cheeriest**) lively; in good spirits; genial, cheering. □ **cheerily** adv. **cheeriness** n.

cheese[1] /tʃiːz/ n. **1 a** a food made from the pressed curds of milk. **b** a complete cake of this with rind. **2** a conserve having the consistency of soft cheese (lemon cheese). **3** a round flat object, e.g. the heavy flat wooden disc used in skittles. □ **cheese-cutter 1** a knife with a broad curved blade. **2** a device for cutting cheese by pulling a wire through it. **cheese-fly** (pl. **-flies**) a small black fly, Piophila casei, breeding in cheese. **cheese-head** the squat cylindrical head of a screw etc. **cheese-mite** a mite of the genus Tyroglyphus, feeding on cheese. **cheese-paring** adj. stingy. ● n. stinginess. **cheese plant** = Swiss cheese plant. **cheese-skipper** = cheese-fly. **cheese straw** a thin cheese-flavoured strip of pastry. **hard cheese** sl. bad luck. [OE cēse etc. ult. f. L caseus]

cheese[2] /tʃiːz/ v.tr. Brit. sl. (as **cheesed** adj.) (often foll. by off) bored, fed up. □ **cheese it** stop it, leave off. [19th c.: orig. unkn.]

cheese[3] /tʃiːz/ n. (also **big cheese**) sl. an important person. [perh. f. Hind. chīz thing]

cheeseboard /ˈtʃiːzbɔːd/ n. **1** a board from which cheese is served. **2** a selection of cheeses.

cheeseburger /ˈtʃiːzˌbɜːgə(r)/ n. a hamburger with cheese in or on it.

cheesecake /ˈtʃiːzkeɪk/ n. **1** a tart filled with sweetened curds etc. **2** sl. the portrayal of women in a sexually attractive manner.

cheesecloth /ˈtʃiːzklɒθ/ n. thin loosely woven cloth, used orig. for wrapping cheese.

cheesemonger /ˈtʃiːzˌmʌŋgə(r)/ n. a dealer in cheese, butter, etc.

cheesewood /ˈtʃiːzwʊd/ n. **1** an Australian tree of the genus Pittosporum. **2** its hard yellowish wood.

cheesy /ˈtʃiːzɪ/ adj. (**cheesier, cheesiest**) **1** like cheese in taste, smell, appearance, etc. **2** sl. inferior; cheap and nasty; disagreeable. □ **cheesiness** n.

cheetah /ˈtʃiːtə/ n. a slender small-headed feline, Acinonyx jubatus, native to the plains of Africa and SW Asia, with a leopard-like spotted coat. The cheetah is unique among cats in having non-retractile claws, and it is the fastest-running animal. [Hindi cītā, perh. f. Skr. citraka speckled]

chef /ʃef/ n. a (usu. male) cook, esp. the chief cook in a restaurant etc. [F, = head]

chef-d'œuvre /ʃeɪˈdɜːvrə/ n. (pl. **chefs-d'œuvre** pronunc. same) a masterpiece. [F]

Chefoo /tʃiːˈfuː/ the former name for YANTAI.

cheiro- comb. form var. of CHIRO-.

Cheka /ˈtʃekə/ an organization set up in 1917 under the Soviet regime for the investigation of counter-revolutionary activities. It helped to stabilize Lenin's regime by removing real and alleged enemies of the Soviet state, many being executed at its headquarters, the Lubyanka prison in Moscow. In 1922 the Cheka was abolished and immediately replaced by the OGPU. [Russ. abbr., = Extraordinary Commission (for combating counter-revolution, sabotage, and speculation)]

Chekhov /ˈtʃekɒf/, Anton (Pavlovich) (1860–1904), Russian dramatist and short-story writer. Chekhov studied medicine in Moscow, and combined his medical practice with writing short humorous stories for journals. He is best known as the author of such plays as The Seagull (1895), Uncle Vanya (1900), The Three Sisters (1901), and The Cherry Orchard (1904). First produced at the Moscow Art Theatre under Konstantin Stanislavsky, they established the theatre's reputation and style. Chekhov's work portrays upper-class life in pre-revolutionary Russia with a blend of naturalism and symbolism and almost imperceptible shifts from comedy to tragedy. He had a considerable influence on 20th-century drama; George Bernard Shaw paid tribute to him in Heartbreak House (1919). □ **Chekhovian** /tʃeˈkəʊvɪən/ adj.

Chekiang see ZHEJIANG.

chela[1] /ˈkiːlə/ n. (pl. **chelae** /-liː/) a prehensile claw of crabs, lobsters, scorpions, etc. [mod.L f. L chele, or Gk khēlē claw]

chela[2] /ˈtʃeɪlə/ n. (pl. **chelas**) esp. Hinduism a disciple, a pupil. [Hindi, = servant]

chelate /ˈkiːleɪt/ n., adj., & v. ● n. Chem. a usu. organometallic compound containing a bonded ring of atoms including a metal atom. ● adj. **1** Chem. involving or able to form a chelate. **2** Zool. & Anat. of or having chelae. ● v.intr. Chem. form a chelate. □ **chelation** /kiːˈleɪʃ(ə)n/ n. [CHELA[1] + -ATE[2]]

chelicera /tʃeˈlɪsərə/ n. (pl. **-rae** /-riː/) Zool. (in a spider or similar arthropod) each of a pair of appendages modified as pincer-like jaws, often used to inject venom. [Gk khēlē claw + keras horn]

chelicerate /keˈlɪsəˌreɪt/ adj. & n. Zool. ● adj. of or relating to the subphylum Chelicerata, which comprises arthropods with a pair of chelicerae. ● n. a chelicerate arthropod, e.g. an arachnid, horseshoe crab, or sea-spider. [CHELICERA]

Chellean /ˈʃelɪən/ adj. Archaeol. = ABBEVILLIAN. [F chelléen f. Chelles near Paris]

Chelmsford /ˈtʃelmzfəd/ a cathedral city in SE England, the county town of Essex; pop. (1991) 152,418.

chelonian /kɪˈləʊnɪən/ n. & adj. Zool. ● n. a reptile of the order Chelonia, including turtles, terrapins, and tortoises, having a shell of bony plates covered with horny scales. ● adj. of or relating to this order. [mod.L Chelonia f. Gk khelōnē tortoise]

Chelsea /ˈtʃelsɪ/ a residential district of London, on the north bank of the River Thames.

Chelsea bun n. a kind of currant bun in the form of a flat spiral.

Chelsea pensioner n. an inmate of the Chelsea Royal Hospital for old or disabled soldiers.

Chelsea ware n. any of various soft-paste porcelains made at Chelsea in the 18th century.

Cheltenham /ˈtʃelt(ə)nəm/ a town in western England, in Gloucestershire. Noted for its saline springs, it became a fashionable spa town in the 19th century.

Chelyabinsk /tʃɪlˈjɑːbɪnsk/ an industrial city in southern Russia on the eastern slopes of the Ural Mountains; pop. (1990) 1,148,000.

chemi- comb. form var. of CHEMO-.

chemical /ˈkemɪk(ə)l/ adj. & n. ● adj. of, made by, or employing chemistry or chemicals. ● n. a substance obtained or used in chemistry. □ **chemical bond** the force holding atoms together in a molecule or crystal. **chemical engineer** a person engaged in chemical engineering, esp. professionally. **chemical engineering** the design, manufacture, and operation of industrial chemical plants. **chemical reaction** a process that involves change in the structure of atoms, molecules, or ions. **chemical warfare** warfare using poison gas and other chemicals. **chemical weapon** a weapon depending for its effect on the release of a toxic or noxious substance. **fine chemicals** chemicals of high purity usu. used in small amounts. **heavy chemicals** bulk chemicals used in industry and agriculture. □ **chemically** adv. [chemic alchemic f. F chimique or mod.L chimicus, chymicus, f. med.L alchymicus: see ALCHEMY]

chemico- /ˈkemɪkəʊ/ comb. form chemical; chemical and (chemico-physical).

chemiluminescence /ˌkemɪˌluːmɪˈnes(ə)ns/ n. the emission of light during a chemical reaction. □ **chemiluminescent** adj. [G Chemilumineszenz (as CHEMI-, LUMINESCENCE)]

chemin de fer /ʃə‚mæn də ˈfeə(r)/ n. a form of baccarat. [F, = railway, lit. 'road of iron']

chemise /ʃəˈmiːz/ n. a woman's loose-fitting under-garment or dress hanging straight from the shoulders. [ME f. OF f. LL camisia shirt]

chemisorption /ˌkemɪˈsɔːpʃ(ə)n, -ˈzɔːpʃ(ə)n/ n. adsorption by chemical bonding. [CHEMI- + ADSORPTION (see ADSORB)]

chemist /ˈkemɪst/ n. **1** Brit. **a** a dealer in medicinal drugs, usu. also selling other medical goods and toiletries. **b** an authorized dispenser of medicines. **2** a person practising or trained in chemistry. [earlier chymist f. F chimiste f. mod.L chimista f. alchimista ALCHEMIST (see ALCHEMY)]

chemistry /ˈkemɪstrɪ/ n. (pl. **-ies**) **1** the study of the elements and the compounds they form and the reactions they undergo. **2** Chem. the chemical composition and properties of a substance. **3** any complex (esp. emotional) change or process (the chemistry of fear). **4** colloq. **a** a person's personality or temperament. **b** the attraction or interaction between people.

Chemnitz /ˈkemnɪts/ an industrial city in eastern Germany, on the Chemnitz river; pop. (est. 1990) 310,000. Between 1953 and 1990 it was called Karl-Marx-Stadt.

chemo /ˈkiːməʊ/ n. colloq. chemotherapy.

chemo- /ˈkiːməʊ/ comb. form (also **chemi-** /ˈkemɪ/) chemical.

chemoreceptor /ˈkiːməʊˌseptə(r)/ n. Physiol. a sensory organ responsive to chemical stimuli.

chemosynthesis /ˌkiːməˈsɪnθɪsɪs/ n. the synthesis of organic compounds by energy derived from chemical reactions.

chemotaxis /ˌkiːməˈtæksɪs/ n. Biol. movement of a motile cell or organism towards or away from an increasing concentration of a particular substance. □ **chemotactic** adj.

chemotherapy /ˌkiːməˈθerəpɪ/ n. the treatment of disease, esp. cancer, by use of chemical substances. □ **chemotherapist** n.

chemurgy /ˈkemɜːdʒɪ/ n. US the chemical and industrial use of organic raw materials. □ **chemurgic** /kemˈɜːdʒɪk/ adj. [CHEMO-, after metallurgy]

Chenab /tʃɪˈnɑːb/ a river of northern India and Pakistan, which rises in the Himalayas and flows through Himachal Pradesh and Jammu and Kashmir, to join the Sutlej river in Punjab. It is one of the five rivers that gave Punjab its name.

Chen-chiang see ZHENJIANG.

Chengchow see ZHENGZHOU.

Chengdu /tʃeŋˈduː/ the capital of Sichuan province in west central China; pop. (1990) 2,780,000.

chenille /ʃəˈniːl/ n. **1** a tufty velvety cord or yarn, used in trimming furniture etc. **2** fabric made from this. [F, = hairy caterpillar f. L canicula dimin. of canis dog]

cheongsam /ˌtʃɪɒŋˈsæm/ n. a Chinese woman's garment with a high neck and slit skirt. [Chin.]

Cheops /ˈkiːɒps/ (Egyptian name Khufu) (c.2613–c.2494 BC), Egyptian pharaoh of the 4th dynasty. He commissioned the building of the Great Pyramid at Giza.

cheque /tʃek/ n. (US **check**) **1** a written order to a bank to pay the stated sum from the drawer's account. **2** the printed form on which such an order is written. **3** Austral. the total sum received by a rural worker at the end of a seasonal contract. □ **cheque-book** a book of forms for writing cheques. **cheque-book journalism** the payment of large sums for exclusive rights to material for (esp. personal) newspaper stories. **cheque card** a card issued by a bank to guarantee the honouring of cheques up to a stated amount. [special use of CHECK¹ to mean 'device for checking the amount of an item']

chequer /ˈtʃekə(r)/ n. & v. (also **checker**) ● n. **1** (often in pl.) a pattern of squares often alternately coloured. **2** (in pl.) (usu. as **checkers**) US the game of draughts. ● v.tr. **1** mark with chequers. **2** variegate; break the uniformity of. **3** (as **chequered** adj.) with varied fortunes (a chequered career). □ **chequer-board 1** a chessboard. **2** a pattern resembling it. **chequered flag** in motor-racing, a flag with a black and white chequered pattern, displayed to drivers at the moment of finishing a race. [ME f. EXCHEQUER]

Chequers /ˈtʃekəz/ a Tudor mansion in the Chilterns near Princes Risborough, Bucks., which serves as a country seat of the Prime Minister in office. It was presented to the British nation for this purpose in 1917 by Lord and Lady Lee of Fareham.

Cher /ʃeə(r)/ a river of central France, which rises in the Massif Central, flowing 350 km (220 miles) northwards to meet the Loire near Tours.

Cherbourg /ˈʃɜːbʊəɡ, French ʃɛrbur/ a seaport and naval base in Normandy, northern France; pop. (1990) 28,770.

Cherenkov /tʃɪˈreŋkɒf/, Pavel (Alekseevich) (also **Cerenkov**) (1904–90), Soviet physicist. From the 1930s he investigated the effects of high-energy particles, in particular, the blue light emitted from water containing a radioactive substance. He suggested the cause of this radiation (an example of what is now called *Cerenkov radiation*), and shared the 1958 Nobel Prize for physics for this discovery.

Cherenkov radiation var. of CERENKOV RADIATION.

Cherepovets /ˌtʃerɪpəˈvjets/ a city in NW Russia, on the Rybinsk reservoir; pop. (1990) 313,000.

cherish /ˈtʃerɪʃ/ v.tr. **1** protect or tend (a child, plant, etc.) lovingly. **2** hold dear, cling to (hopes, feelings, etc.). [ME f. OF cherir f. cher f. L carus dear]

Cherkassy /tʃɜːˈkæsɪ/ a port in central Ukraine, on the River Dnieper; pop. (1990) 297,000.

Cherkessk /tʃəˈkesk/ a city in the Caucasus in southern Russia, capital of the republic of Karachai-Cherkessia; pop. (1990) 113,000.

Chernenko /tʃəˈnjenkəʊ/, Konstantin (Ustinovich) (1911–85), Soviet statesman, General Secretary of the Communist Party of the USSR and President 1984–5. Born in Siberia, he became a full member of the Politburo in 1978 and was a close associate of Brezhnev from this time. Chernenko succeeded Andropov in the presidency, but died after only thirteen months in office.

Chernigov /tʃəˈniːɡɒf/ a port in northern Ukraine, on the River Desna; pop. (1990) 301,000.

Chernivtsi /tʃəˈnɪvtsɪ, Ukrainian ˌtʃɛrɲiwˈtsi/ (Russian **Chernovtsy** /ˌtʃernəvˈtsɪ/) a city in western Ukraine, in the foothills of the Carpathians, close to the border with Romania; pop. (est. 1990) 257,000. It was part of Romania between 1918 and 1940.

Chernobyl /tʃəˈnɒbɪl, -ˈnəʊbɪl/ a town near Kiev in Ukraine where, in April 1986, an accident at a nuclear power station resulted in a serious escape of radioactive material and the subsequent contamination of Ukraine, Belarus, and other parts of Europe.

Chernoreche /ˌtʃɜːnəˈretʃjə/ a former name (until 1919) for DZERZHINSK.

chernozem /ˈtʃɜːnəʊˌzem/ n. a fertile black soil rich in humus, found in temperate regions, esp. southern Russia. Also called *black earth*. [Russ. f. cherny black + zemlya earth]

Cherokee /ˈtʃerəˌkiː/ n. & adj. ● n. **1** a member of an American Indian people formerly inhabiting much of the southern US and now living in reservations in Oklahoma and North Carolina. **2** the language of this people. ● adj. of or relating to the Cherokees or their language. [Cherokee tsaliki]

Cherokee rose n. a fragrant white rose, Rosa laevigata, of the southern US.

cheroot /ʃəˈruːt/ n. a cigar with both ends open. [F cheroute f. Tamil shuruṭṭu roll]

cherry /ˈtʃerɪ/ n. & adj. ● n. (pl. **-ies**) **1 a** a small soft round stone-fruit. **b** a tree of the genus Prunus, bearing this fruit or grown for its ornamental flowers. **2** (in full **cherry wood**) the wood of a cherry. **3** US sl. **a** a virginity. **b** a virgin. **4** a bright deep red colour. ● adj. of a bright deep red colour. □ **cherry brandy** a dark red liqueur of brandy in which cherries have been steeped. **cherry-laurel** Brit. a small evergreen tree, Prunus laurocerasus, with white flowers and cherry-like fruits. **cherry-pick** cream off; selectively choose (the best or best part of). **cherry-picker** colloq. a crane for raising and lowering people. **cherry-pie 1** a pie made with cherries. **2** a garden heliotrope. **cherry plum 1** a tree, Prunus cerasifera, native to SW Asia, with solitary white flowers and red fruits. **2** the fruit of this tree. **cherry tomato** a miniature tomato with a strong flavour. [ME f. ONF cherise (taken as pl.: cf. PEA) f. med.L ceresia perh. f. L f. Gk kerasos]

Chersonese /ˈkɜːsəˌniːs/ an ancient region corresponding to the Thracian or Gallipoli peninsula on the north side of the Hellespont. The word was occasionally also applied to other peninsulas. [L chersonesus peninsula f. Gk khersonēsos f. khersos dry + nēsos island]

chert /tʃɜːt/ n. a flintlike form of quartz composed of chalcedony. □ **cherty** adj. [17th c.: orig. unkn.]

cherub /ˈtʃerəb/ n. **1** (pl. **cherubim** /-bɪm/) an angelic being of the second order of the celestial hierarchy (see ORDER n. 19). **2 a** a representation of a winged child or the head of a winged child. **b** a beautiful or innocent child. □ **cherubic** /tʃɪˈruːbɪk/ adj. **cherubically** adv. [ME f. OE cherubin and f. Heb. kᵉrūḇ, pl. kᵉrūḇīm]

Cherubini /ˌkeruˈbiːnɪ/, (Maria) Luigi (Carlo Zenobio Salvatore) (1760–1842), Italian composer. Born in Florence, he spent most of his composing career in Paris (where he became director of the Conservatoire) and is principally known for his church music and operas. He is now chiefly remembered for his opera The Water-Carrier (1800), of which the overture is sometimes still performed.

chervil /ˈtʃɜːvɪl/ n. an umbelliferous plant, Anthriscus cerefolium, with small white flowers, used as a herb for flavouring soup, salads, etc. [OE cerfille f. L chaerephylla f. Gk khairephullon]

Cherwell /ˈtʃɑːwel/, Frederick Alexander Lindemann, 1st Viscount (1886–1957), German-born British physicist. He studied a wide variety of subjects, and a number of theories and items are named after him. These include a theory of specific heat, a theory of the upper atmosphere, a formula concerning the melting-point of crystals, an electrometer, and a glass for transmitting X-rays. He was Churchill's adviser on scientific and aeronautical matters during the war.

Ches. abbr. Cheshire.

Chesapeake Bay /ˈtʃesəˌpiːk/ a large inlet of the North Atlantic on the US coast, extending 320 km (200 miles) northwards through the states of Virginia and Maryland.

Cheshire[1] /ˈtʃeʃə(r)/ a county of west central England; county town, Chester.

Cheshire[2] /ˈtʃeʃə(r)/, (Geoffrey) Leonard (1917–92), British airman and philanthropist. He served as a fighter pilot in the Second World War, was awarded the VC in 1944 for his one hundred bombing missions, and was an official observer of the atom bomb dropped on Nagasaki in 1945. From the late 1940s he founded the Cheshire Foundation Homes for the disabled and incurably sick; these spread to forty-five countries. Cheshire married the philanthropist Sue Ryder in 1959.

Cheshire[3] /ˈtʃeʃə(r)/ n. a crumbly cheese originally made in Cheshire.

Cheshire cat n. a cat depicted with a broad fixed grin, as popularized through Lewis Carroll's *Alice's Adventures in Wonderland* (1865). Its origin is uncertain.

Chesil Beach /ˈtʃez(ə)l/ (also **Chesil Bank**) a shingle beach in southern England, on the Dorset coast. Separated from the mainland by a tidal lagoon, it is over 25 km (17 miles) long.

chess /tʃes/ n. a game of skill played between two persons on a chequered board divided into sixty-four squares. Each player has sixteen 'men' (king, queen, two bishops, two knights, two castles or rooks, eight pawns), which are moved according to strict rules in simulation of a battle where the object is to manoeuvre the opponent's king into a position (*checkmate*) from which escape is impossible. Many sequences of moves are named after the great players who originated them. The game seems to be a descendant (5th century) of an earlier Indian game and to have reached Persia and Arab countries and spread from there until by the 13th century it was known all over western Europe. [ME f. OF *esches*, pl. of *eschec* (CHECK[1])]

chessboard /ˈtʃesbɔːd/ n. a chequered board of sixty-four squares on which chess and draughts are played.

chessman /ˈtʃesmæn/ n. (pl. **-men**) any of the thirty-two pieces and pawns with which chess is played.

chest /tʃest/ n. **1** a large strong box, esp. for storage or transport. **2 a** the part of a human or animal body enclosed by the ribs. **b** the circumference of the body at chest level. **c** the front surface of the body from neck to waist. **3** a small cabinet for medicines etc. **4 a** the treasury or financial resources of an institution. **b** the money available from it. □ **chest of drawers** a piece of furniture consisting of a set of drawers in a frame. **chest-voice** the lowest register of the voice in singing or speaking. **get a thing off one's chest** *colloq.* disclose a fact, secret, etc., to relieve one's anxiety about it. **play (one's cards, a thing,** etc.) **close to one's chest** *colloq.* be cautious or secretive about. □ **-chested** adj. (in comb.). [OE *cest, cyst* f. Gmc f. L f. Gk *kistē*]

Chester /ˈtʃestə(r)/ a town in western England, the county town of Cheshire; pop. (1991) 115,000.

Chesterfield /ˈtʃestəˌfiːld/ a town in Derbyshire; pop. (1991) 99,700. It is noted for its 14th-century church, which, as a result of the warping of timber, has a twisted spire.

chesterfield /ˈtʃestəˌfiːld/ n. **1** a sofa with arms and back of the same height and curved outwards at the top. **2** a man's plain overcoat usu. with a velvet collar. [19th-c. Earl of *Chesterfield*]

Chesterton /ˈtʃestət(ə)n/, G(ilbert) K(eith) (1874–1936), English essayist, novelist, and critic. He first came to prominence as a journalist for *The Speaker*, in which, with Hilaire Belloc, he took an anti-imperialist platform on the Boer War question. His best-known novel is his 'Merry England' fantasy *The Napoleon of Notting Hill* (1904), but he is also widely remembered for his creation of the character Father Brown, a priest with a talent for crime detection, who first appears in *The Innocence of Father Brown* (1911). Chesterton became a Roman Catholic in 1922; his other writings include biographies of St Francis of Assisi and St Thomas Aquinas.

chestnut /ˈtʃesnʌt/ n. & adj. ● n. **1 a** a glossy hard brown edible nut. **b** the tree *Castanea sativa*, bearing these nuts enclosed in a spiny fruit. Also called *Spanish chestnut* or *sweet chestnut*. **2** any other tree of the genus *Castanea*. **3** = *horse chestnut*. **4** (in full **chestnut-wood**) the heavy wood of any chestnut tree. **5** a horse of a reddish-brown or yellowish-brown colour. **6** *colloq.* a stale joke or anecdote. **7** a small hard patch on a horse's leg. **8** a reddish-brown colour. ● adj. of the colour chestnut. □ **liver chestnut** a dark kind of chestnut horse. [obs. *chesten* f. OF *chastaine* f. L *castanea* f. Gk *kastanea*]

chesty /ˈtʃestɪ/ adj. (**chestier, chestiest**) **1** *Brit. colloq.* inclined to or

symptomatic of chest disease. **2** *colloq.* having a large chest or prominent breasts. **3** *US sl.* arrogant. □ **chestily** adv. **chestiness** n.

Chesvan see HESVAN.

Chetnik /ˈtʃetnɪk/ n. a member of a guerrilla force in the Balkans. During the Second World War the term referred to the group of royalist Serbs led by Dragoljub Mihajlović against the Germans and the Communist partisans, while it has recently been used of Serbs fighting in the former Yugoslavia. [Serbo-Croat *četnik* f. *četa* band, troop]

Chetumal /ˌtʃetuːˈmaːl/ a port in SE Mexico, on the Yucatán Peninsula at the border with Belize, capital of the state of Quintana Roo; pop. (1981) 40,000.

cheval-glass /ʃəˈvælglɑːs/ n. a tall mirror swung on an upright frame. [F *cheval* horse, frame]

Chevalier /ʃəˈvælˌeɪ/, Maurice (1888–1972), French singer and actor. He gained an international reputation in the Paris music-halls of the 1920s, particularly in the Folies-Bergère, where he regularly partnered the French dancer Mistinguett (1874–1956). He went on to star in successful Hollywood musicals such as *Innocents of Paris* (1929), *Love Me Tonight* (1932), and *Gigi* (1958).

chevalier /ˌʃevəˈlɪə(r)/ n. **1 a** a member of certain orders of knighthood, and of modern French orders, as the Legion of Honour. **b** *archaic* or *hist.* a knight. **2** *hist.* the title of the Old and Young Pretenders. **3** a chivalrous man; a cavalier. [ME f. AF *chevaler*, OF *chevalier* f. med.L *caballarius* f. L *caballus* horse]

chevet /ʃəˈveɪ/ n. *Archit.* the apsidal end of a church, sometimes with an attached group of apses. [F, = pillow, f. L *capitium* f. *caput* head]

Cheviot /ˈtʃevɪət, ˈtʃiːv-/ n. **1** a breed of large sheep with short thick wool. **2** (**cheviot**) the wool or cloth obtained from this breed. [CHEVIOT HILLS]

Cheviot Hills /ˈtʃevɪət, ˈtʃiːv-/ (also **Cheviots**) a range of hills on the border between England and Scotland.

chèvre /ˈʃevrə/ n. a variety of goat's-milk cheese. [F, = goat, she-goat]

chevron /ˈʃevrən/ n. **1** a badge in a V shape on the sleeve of a uniform indicating rank or length of service. **2** *Heraldry & Archit.* a bent bar of an inverted V shape. **3** any V-shaped line or stripe. [ME f. OF ult. f. L *caper* goat: cf. L *capreoli* pair of rafters]

chevrotain /ˈʃevrəˌteɪn/ (also **chevrotin** /-tɪn/) n. a small deerlike animal of the family Tragulidae, native to Africa and SE Asia, having small tusks. Also called *mouse deer*. [F, dimin. of OF *chevrot* dimin. of *chèvre* goat]

chevy var. of CHIVVY.

chew /tʃuː/ v. & n. ● v.tr. (also *absol.*) work (food etc.) between the teeth; crush or indent with the teeth. ● n. **1** an act of chewing. **2** something for chewing, esp. a chewy sweet. □ **chew the cud** reflect, ruminate. **chew the fat** (or **rag**) *sl.* **1** chat. **2** grumble. **chewing-gum** flavoured gum, esp. chicle, for chewing. **chew on 1** work continuously between the teeth (*chewed on a piece of string*). **2** think about; meditate on. **chew out** *N. Amer. colloq.* reprimand. **chew over 1** discuss, talk over. **2** think about; meditate on. □ **chewable** adj. **chewer** n. [OE *cēowan*]

chewy /ˈtʃuːɪ/ adj. (**chewier, chewiest**) **1** needing much chewing. **2** suitable for chewing. □ **chewiness** n.

Cheyenne[1] /ʃaɪˈæn, -ˈen/ the state capital of Wyoming; pop. (1990) 50,000.

Cheyenne[2] /ʃaɪˈæn, -ˈen/ n. & adj. ● n. (pl. same) **1** a member of an American Indian people formerly living between the Missouri and Arkansas rivers, now living on reservations in Montana and Oklahoma. **2** the language of these people. ● adj. of or relating to the Cheyenne or their language. [Canad. F f. Dakota *Sahiyena*]

Cheyne–Stokes respiration /tʃeɪnˈstəʊks/ n. *Med.* a breathing cycle with a gradual decrease of movement to a complete stop, followed by a gradual increase. [J. *Cheyne*, Sc. physician (1777–1836), and W. *Stokes*, Ir. physician (1804–78)]

chez /ʃeɪ/ prep. at the house or home of. [F f. OF *chiese* f. L *casa* cottage]

chi /kaɪ/ n. the twenty-second letter of the Greek alphabet (X, χ). □ **chi-rho** a monogram of chi and rho as the first two letters of Greek *Khristos* Christ. **chi-square test** *Statistics* a method of comparing observed and expected values of a variable quantity. [ME f. Gk *khi*]

chiack /ˈtʃaɪæk/ v. & n. (also **chyack**) *Austral. & NZ* ● v.tr. jeer, taunt. ● n. jeering, banter. □ **chiacking** n. [19th c.: orig. unkn.]

Chiang Kai-shek /ˌtʃjæŋ kaɪˈʃek/ (also **Jiang Jie Shi** /ˌdʒjæŋ dʒjiː ˈʃiː/) (1887–1975), Chinese statesman and general, President of China 1928–31 and 1943–9 and of Taiwan 1950–75. A prominent general in the army of Sun Yat-sen, in 1925 he became leader of the Kuomintang when Sun Yat-sen died, and launched a military campaign to unite China. In the 1930s he concentrated more on defeating the Chinese Communists than on resisting the invading Japanese, but he proved unable to establish order and was defeated by the Communists after the end of the Second World War. Forced to abandon mainland China in 1949, he set up a separate Nationalist Chinese State in Taiwan.

Chiangmai /tʃjæŋˈmaɪ/ a city in NW Thailand; pop. (1990) 164,900.

Chianti /kɪˈænti/ n. (pl. **Chiantis**) a dry red Italian wine. [*Chianti*, an area in Tuscany, Italy]

Chiapas /tʃɪˈɑːpəs/ a state of southern Mexico, on the border with Guatemala; capital, Tuxtla Gutiérrez.

chiaroscuro /kɪˌɑːrəˈskʊərəʊ/ n. **1** the treatment of light and shade in drawing and painting. **2** the use of contrast in literature etc. **3** (*attrib.*) half-revealed. [It. f. *chiaro* CLEAR + *oscuro* dark, OBSCURE]

chiasma /kaɪˈæzmə/ n. (pl. **chiasmata** /-mətə/) Biol. the point at which paired chromosomes remain in contact after crossing over during meiosis. [mod.L f. Gk *chiasma* a cross-shaped mark]

chiasmus /kaɪˈæzməs/ n. inversion in the second of two parallel phrases of the order followed in the first (e.g. *to stop too fearful and too faint to go*). □ **chiastic** /-ˈæstɪk/ adj. [mod.L f. Gk *khiasmos* crosswise arrangement f. *khiazō* mark with letter CHI]

Chiba /tʃiːbə/ a city in Japan, on the island of Honshu, east of Tokyo; pop. (1990) 829,470.

Chibcha /tʃɪbtʃə/ n. & adj. ● n. (pl. same) **1** a member of a native people of Colombia. (*See note below.*) **2** the language of this people. ● adj. of or relating to the Chibcha or their language. □ **Chibchan** adj. & n. [Amer. Sp. f. Chibcha *zipa* chief, hereditary leader]

▪ The Chibcha's ancient civilization, flourishing when the Spaniards first encountered them in 1537, was destroyed by the Europeans; their language died out in the 18th century, and the Chibcha became assimilated into the rest of the population.

chibouk /tʃɪˈbuːk/ n. (also **chibouque**) a long Turkish tobacco-pipe. [Turk. *çubuk* tube]

chic /ʃiːk/ adj. & n. ● adj. (**chic-er**, **chic-est**) stylish, elegant (in dress or appearance). ● n. stylishness, elegance. □ **chicly** adv. [F]

Chicago /ʃɪˈkɑːgəʊ/ a city in Illinois, on Lake Michigan; pop. (1990) 2,783,730. Selected in 1848 as a terminal for the new Illinois and Michigan canal, Chicago developed during the 19th century as a major grain market and food-processing centre. □ **Chicagoan** n. & adj.

chicane /ʃɪˈkeɪn/ n. & v. ● n. **1** chicanery. **2** an artificial barrier or obstacle on a motor-racing track. **3** Bridge a hand without trumps, or without cards of one suit. ● v. archaic **1** intr. use chicanery. **2** tr. (usu. foll. by *into*, *out of*, etc.) cheat (a person). [F *chicane*(r) quibble]

chicanery /ʃɪˈkeɪnəri/ n. (pl. **-ies**) **1** clever but misleading talk; a false argument. **2** trickery, deception. [F *chicanerie* (as CHICANE)]

Chicano /tʃɪˈkɑːnəʊ/ n. (pl. **-os**) a North American of Mexican origin. [Sp. *mejicano* Mexican]

Chichén Itzá /tʃɪˌtʃen ɪtˈsɑː/ a site in northern Yucatán, Mexico, which was the centre of the Mayan empire after AD 918, with elaborate ceremonial buildings centred on a sacred well.

Chichester[1] /tʃɪtʃɪstə(r)/ a city in southern England, the county town of West Sussex; pop. (1981) 27,200.

Chichester[2] /tʃɪtʃɪstə(r)/, Sir Francis (Charles) (1901–72), English yachtsman. In 1960 he won the first solo transatlantic yacht race in his boat *Gipsy Moth III*. In 1966–7 he sailed alone round the world in *Gipsy Moth IV*, taking 107 days to sail from Plymouth to Sydney, and 119 days to make the return voyage. He was knighted on his return.

chichi /ʃiːʃiː/ adj. & n. ● adj. **1** (of a thing) frilly, showy. **2** (of a person or behaviour) fussy, affected. ● n. **1** over-refinement, pretentiousness, fussiness. **2** a frilly, showy, or pretentious object. [F]

Chichimec /ˌtʃiːtʃɪˈmek/ n. & adj. ● n. (pl. same or **Chichimecs**) **1** a member of a group of native peoples dominant in central Mexico from the 10th to the 16th centuries. (*See note below.*) **2** the language of this people. ● adj. of or relating to the Chichimec or their language. [Sp. f. Nahuatl]

▪ The Chichimec comprised a group of peoples, including the Toltec and the Aztecs. The Toltec were amongst the first of the Chichimec peoples to invade the central valley of Mexico from the north-west in *c.*900, and dominated the region until overthrown by other invading Chichimec peoples in the 12th century. (See AZTEC, TOLTEC.)

chick[1] /tʃɪk/ n. **1** a young bird, esp. one newly hatched. **2** sl. **a** a young woman. **b** a child. [ME: shortening of CHICKEN]

chick[2] /tʃɪk/ n. Ind. a screen for a doorway etc., made from split bamboo and twine. [Hindi *chik*]

chickadee /ˈtʃɪkədiː/ n. N. Amer. a small bird of the tit family. [imit.]

chicken /tʃɪkɪn/ n., adj., & v. ● n. **1** a domestic fowl, esp. a young bird. **2 a** a domestic fowl prepared as food. **b** its flesh. **3** a youthful person (usu. with *neg.*: *is no chicken*). **4** colloq. a children's pastime testing courage, usu. recklessly. ● adj. colloq. cowardly. ● v.intr. (foll. by *out*) colloq. withdraw from or fail in some activity through fear or lack of nerve. □ **chicken-and-egg problem** (or **dilemma** etc.) the unresolved question as to which of two things caused the other. **chicken brick** an earthenware container in two halves for roasting a chicken in its own juices. **chicken cholera** see CHOLERA. **chicken-feed 1** food for poultry. **2** colloq. an unimportant amount, esp. of money. **chicken-hearted** (or **-livered**) easily frightened; lacking nerve or courage. **chicken-wire** a light wire netting with a hexagonal mesh. [OE *cīcen*, *cȳcen* f. Gmc]

chickenpox /ˈtʃɪkɪnˌpɒks/ n. Med. an infectious disease, esp. of children, with a rash of small blisters and caused by the varicella-zoster virus, which also causes shingles. Also called *varicella*.

chick pea n. **1** a leguminous plant, *Cicer arietinum*, with short swollen pods containing yellow beaked seeds. See also GRAM[2]. **2** this seed used as a vegetable. [orig. *ciche pease* f. L *cicer*: see PEASE]

chickweed /ˈtʃɪkwiːd/ n. a small white-flowered plant of the pink family; esp. *Stellaria media*, a garden weed with slender stems and tiny flowers.

chicle /ˈtʃɪk(ə)l, tʃiːkliː/ n. the milky juice of the sapodilla tree, used in the manufacture of chewing-gum. [Amer. Sp. f. Nahuatl *tzietli*]

chicory /ˈtʃɪkəri/ n. (pl. **-ies**) **1** a blue-flowered composite plant, *Cichorium intybus*, cultivated for its salad leaves and its root. **2** its root, roasted and ground for use with or instead of coffee. **3** N. Amer. = ENDIVE. [ME f. obs. F *cicorée* endive f. med.L *cic(h)orea* f. L *cichorium* f. Gk *kikhorion*: cf. SUCCORY]

chide /tʃaɪd/ v.tr. & intr. (past **chided** or **chid** /tʃɪd/; past part. **chided** or **chidden** /ˈtʃɪd(ə)n/) esp. archaic or literary scold, rebuke. □ **chider** n. **chidingly** adv. [OE *cīdan*, of unkn. orig.]

chief /tʃiːf/ n. & adj. ● n. **1 a** a leader or ruler. **b** the head of a tribe, clan, etc. **2** the head of a department; the highest official. **3** Heraldry the upper third of a shield. ● adj. (usu. attrib.) **1** first in position, importance, influence, etc. (*chief engineer*). **2** prominent, leading. □ **-in-chief** supreme (*commander-in-chief*). □ **chiefdom** n. [ME f. OF *ch(i)ef* ult. f. L *caput* head]

Chief Constable n. the head of the police force of a county or other region.

chiefly /ˈtʃiːfli/ adv. above all; mainly but not exclusively.

Chief of Staff n. the senior staff officer of a service or command.

chieftain /ˈtʃiːftən/ n. (fem. **chieftainess** /-tənɪs/) the leader of a tribe, clan, etc. □ **chieftaincy** /-tənsɪ/ n. (pl. **-ies**). **chieftainship** n. [ME f. OF *chevetaine* f. LL *capitaneus* CAPTAIN: assim. to CHIEF]

chiffchaff /ˈtʃɪftʃæf/ n. a common small European warbler, *Phylloscopus collybita*, with a distinctive repetitive song. [imit.]

chiffon /ˈʃɪfɒn/ n. & adj. ● n. a light diaphanous fabric of silk, nylon, etc. ● adj. **1** made of chiffon. **2** (of a pie-filling, dessert, etc.) light-textured. [F f. *chiffe* rag]

chiffonier /ˌʃɪfəˈnɪə(r)/ n. **1** a movable low cupboard with a sideboard top. **2** US a tall chest of drawers. [F *chiffonnier*, *-ière* rag-picker, chest of drawers for odds and ends]

Chifley /ˈtʃɪflɪ/, Joseph Benedict (1885–1951), Australian Labor statesman, Prime Minister 1945–9. He entered Parliament in 1928; after the Second World War he became Prime Minister on the death of John Curtin. During his term of office he continued to fulfil Labor's nationalization and welfare programme; he also initiated Australia's immigration policy and the Snowy Mountains hydroelectric scheme. He was defeated in the 1949 election but remained leader of the Labor Party until his death.

chigger var. of JIGGER[2].

chignon /ˈʃiːnjɒn/ n. a coil or mass of hair at the back of a woman's head. [F, orig. = nape of the neck]

chigoe /ˈtʃɪgəʊ/ n. = JIGGER[2] 1. [Carib]

Chihli, Gulf of /ˈtʃiːliː/ an alternative name for Bo Hai.

Chihuahua /tʃɪˈwɑːwə/ **1** a state of northern Mexico. **2** its capital, a principal city of north central Mexico; pop. (1990) 530,490.

chihuahua /tʃɪˈwɑːwə/ n. a breed of very small dog with smooth hair and large eyes, originating in Mexico. [CHIHUAHUA]

chilblain /ˈtʃɪlbleɪn/ n. a painful itching swelling of the skin usu. on a hand, foot, etc., caused by exposure to cold and by poor circulation. □ **chilblained** adj. [CHILL + BLAIN]

child /tʃaɪld/ n. (pl. **children** /ˈtʃɪldrən/) **1 a** a young human being below the age of puberty. **b** an unborn or newborn human being. **2** one's son or daughter (at any age). **3** (foll. by of) a descendant, follower, adherent, or product of (children of Israel; child of God; child of nature). **4** a childish person. □ **child abuse** maltreatment of a child, esp. by physical violence or sexual molestation. **child allowance 1** hist. (in the UK) a tax allowance granted to parents of dependent children. **2** = child benefit. **child benefit** (in the UK) regular payment by the state to the parents of a child up to a certain age. **child care** the care of a child or children, esp. by a crèche or child-minder while parents are working, or by a local authority when a normal home life is lacking. **child-centred** (of education) concerned with developing the esp. creative potential of the individual child, rather than with external standards, performance in tests, etc. **child-minder** a person who looks after children for payment, strictly speaking a person registered with the local authority to give paid daytime care in his or her own home for children under eight. **child's play** an easy task. □ **childless** adj. **childlessness** n. [OE cild]

childbed /ˈtʃaɪldbed/ n. archaic = CHILDBIRTH.

childbirth /ˈtʃaɪldbɜːθ/ n. the act of giving birth to a child.

Childe /tʃaɪld/ n. archaic a youth of noble birth (Childe Harold). [var. of CHILD]

Childermas /ˈtʃɪldəˌmæs/ n. archaic the feast of the Holy Innocents, 28 Dec. [OE cildramæsse f. cildra genitive pl. of cild CHILD + mæsse MASS²]

Childers /ˈtʃɪldəz/, (Robert) Erskine (1870–1922), Irish writer and political activist, born in England. His fame as a writer stems from his novel The Riddle of the Sands (1903), in which two amateur yachtsmen discover German preparations for an invasion of England. A supporter of Irish Home Rule from 1910, he settled in Ireland in 1920, became a Sinn Fein MP in 1921, and, in the same year, Minister of Propaganda. In 1922 he was court-martialled and shot for his involvement in the civil war following the establishment of the Irish Free State. His son Erskine Hamilton Childers (1905–74) was President of Ireland 1973–4.

childhood /ˈtʃaɪldhʊd/ n. the state or period of being a child. □ **second childhood** a person's dotage. [OE cildhād]

childish /ˈtʃaɪldɪʃ/ adj. **1** of, like, or proper to a child. **2** immature, silly. □ **childishly** adv. **childishness** n.

childlike /ˈtʃaɪldlaɪk/ adj. having the good qualities of a child as innocence, frankness, etc.

childproof /ˈtʃaɪldpruːf/ adj. that cannot be damaged or operated by a child.

children pl. of CHILD.

Children's Crusade a movement in 1212 in which tens of thousands of children (mostly from France and Germany) embarked on a crusade to the Holy Land. Most of the children never reached their destination; arriving at French and Italian ports, many were sold into slavery.

Chile /ˈtʃɪli/ a country occupying a long coastal strip down the southern half of the west of South America, between the Andes and the Pacific Ocean; pop. (est. 1991) 13,360,000; official language, Spanish; capital, Santiago. Most of Chile was part of the Inca empire and became part of Spanish Peru after Pizarro's conquest, although the native peoples of the south resisted both imperial powers quite successfully. Chilean independence was proclaimed in 1810 by Bernardo O'Higgins and finally achieved in 1818 with help from Argentina. After the overthrow of the Marxist democrat Salvador Allende in 1973, Chile was ruled by the right-wing military dictatorship of General Pinochet until a democratically elected President took office in 1990. □ **Chilean** adj. & n.

chile var. of CHILLI.

Chile pine n. a monkey-puzzle tree.

Chile saltpetre n. (also **Chile nitre**) naturally occurring sodium nitrate.

chili var. of CHILLI.

chiliad /ˈkɪliˌæd/ n. **1** a thousand. **2** a thousand years. [LL chilias chiliad- f. Gk khilias -ados]

chiliasm /ˈkɪliˌæz(ə)m/ n. the doctrine of or belief in Christ's prophesied reign of 1,000 years on earth (see MILLENNIUM, MILLENARIANISM). [Gk khiliasmos: see CHILIAD]

chiliast /ˈkɪliˌæst/ n. a believer in chiliasm. □ **chiliastic** /ˌkɪliˈæstɪk/ adj. [LL chiliastes: see CHILIAD, CHILIASM]

chill /tʃɪl/ n., v., & adj. ● n. **1 a** an unpleasant cold sensation; lowered body temperature. **b** a feverish cold (catch a chill). **2** unpleasant coldness (of air, water, etc.). **3 a** a depressing influence (cast a chill over). **b** a feeling of fear or dread accompanied by coldness. **4** coldness of manner. ● v. **1** tr. & intr. make or become cold. **2** tr. **a** depress, dispirit. **b** horrify, frighten. **3** tr. cool (food or drink); preserve by cooling. **4** tr. harden (molten metal) by contact with cold material. **5** esp. US sl. **a** (often foll. by out) relax, take it easy. **b** (often foll. by with) pass time idly; hang around. ● adj. literary chilly. □ **take the chill off** warm slightly. □ **chillingly** adv. **chillness** n. **chillsome** adj. literary. [OE cele, ciele, etc.: in mod. use the verb is the oldest (ME), and is of obscure orig.]

chiller /ˈtʃɪlə(r)/ n. **1** = spine-chiller. **2** a cold cabinet or refrigerator, esp. in a shop, garage, etc.

chilli /ˈtʃɪli/ n. (also **chile**, US **chili**) (pl. **-ies**) **1** (in full **chilli pepper**) a small hot-tasting (dried) pod, usu. red or green, of a capsicum, Capsicum annuum, used in sauces, relishes, etc. **2** = chilli powder. **3** esp. US = chilli con carne. □ **chilli con carne** /kɒn ˈkɑːni/ a stew of chilli-flavoured minced beef and beans. **chilli powder** hot cayenne. **chilli sauce** a hot sauce made with tomatoes, chillies, and spices. [Sp. chile, chili, f. Nahuatl chilli]

chilly /ˈtʃɪli/ adj. (**chillier**, **chilliest**) **1** (of the weather or an object) somewhat cold. **2** (of a person or animal) feeling somewhat cold; sensitive to the cold. **3** unfriendly; unemotional. □ **chilliness** n.

Chilpancingo /ˌtʃɪlpænˈsɪŋɡəʊ/ a city in SW Mexico, capital of the state of Guerrero; pop. (1980) 120,000.

Chiltern Hills /ˈtʃɪlt(ə)n/ (also **Chilterns**) a range of chalk hills in southern England, north of the River Thames and west of London.

Chiltern Hundreds /ˈtʃɪlt(ə)n/ a district (formerly called a hundred) which includes part of the Chilterns in southern England and is Crown property. Stewardship of the district is legally an office of profit under the Crown, the holding of which disqualifies a person from being an MP. An MP wishing to resign from his or her seat is said to 'apply for the Chiltern Hundreds'.

chimaera var. of CHIMERA.

Chimborazo /ˌtʃɪmbəˈrɑːzəʊ/ the highest peak of the Andes in Ecuador, rising to 6,310 m (20,487 ft).

chime¹ /tʃaɪm/ n. & v. ● n. **1 a** a set of attuned bells. **b** the series of sounds given by this. **c** (usu. in pl.) a set of attuned bells as a door bell. **2** agreement, correspondence, harmony. ● v. **1 a** intr. (of bells) ring. **b** tr. sound (a bell or chime) by striking. **2** tr. show (the hour) by chiming. **3** intr. (usu. foll. by together, with) be in agreement, harmonize. □ **chime in 1** interject a remark. **2** join in harmoniously. **3** (foll. by with) agree with. □ **chimer** n. [ME, prob. f. chym(b)e bell f. OE cimbal f. L cymbalum f. Gk kumbalon CYMBAL]

chime² /tʃaɪm/ n. (also **chimb**) the projecting rim at the end of a cask. [ME: cf. MDu., MLG kimme]

chimera /kaɪˈmɪərə, kɪ-/ (also **chimaera**) n. **1** Gk Mythol. a fire-breathing female monster with a lion's head, a goat's body, and a serpent's tail, killed by Bellerophon. **2** a fantastic or grotesque product of the imagination; a bogey. **3** a fabulous beast with parts taken from various animals. **4** Biol. **a** an organism containing genetically different tissues, formed by the fusion of two early embryos, grafting, mutation, etc. **b** a nucleic acid formed by laboratory manipulation. **5** a cartilaginous fish of the family Chimaeridae, usu. having erect pointed fins and a long tail. □ **chimeric** /-ˈmerɪk/ adj. **chimerical** adj. **chimerically** adv. [L f. Gk khimaira she-goat, chimera]

chimney /ˈtʃɪmni/ n. & v. ● n. (pl. **-eys**) **1** a vertical channel conducting smoke or combustion gases etc. up and away from a fire, furnace, engine, etc. **2** the part of this which projects above a roof. **3** a glass tube protecting the flame of a lamp. **4** a narrow vertical crack in a rock-face, often used by mountaineers to ascend. ● v.tr. (esp. as **chimneyed** adj.) provide with a chimney or chimneys. □ **chimney-breast** a projecting interior wall surrounding a chimney. **chimney-piece** an ornamental structure around an open fireplace; a mantelpiece. **chimney-pot** an earthenware or metal pipe at the top of a chimney, narrowing the aperture and increasing the up draught. **chimney-stack 1** a number of chimneys grouped in one structure. **2** = sense 2. **chimney-sweep** a person whose job is removing soot from inside chimneys. [ME f. OF cheminée f. LL caminata having a fire-place, f. L caminus f. Gk kaminos oven]

chimp /tʃɪmp/ n. colloq. = CHIMPANZEE. [abbr.]

chimpanzee /ˌtʃɪmpənˈziː/ n. either of two apes of the genus Pan, native to central and West Africa: P. troglodytes, which resembles man more closely than does any other ape, and *P. paniscus* (the **pygmy chimpanzee**). [F *chimpanzé* f. Kongo]

Chimu /tʃiːˈmuː/ *n. & adj.* ● *n.* (*pl.* same or **Chimus**) **1** a member of a native people of Peru, who developed the largest and most important civilization before the Incas. (*See note below.*) **2** the language of this people. ● *adj.* of or relating to the Chimu or their language. [Sp., f. Amer. Indian]

▪ The Chimu were conquered by the Incas in the mid-15th century, passing on much of their culture, engineering skills, and social organization to the conquerors. The Chimu language died out in the 19th century.

Chin var. of JIN.

Ch'in var. of QIN.

chin /tʃɪn/ *n.* the front of the lower jaw. □ **chin-strap** a strap for fastening a hat etc. under the chin. **chin up** *colloq.* cheer up. **chin-wag** *sl.* a talk or chat. ● *v.intr.* (**-wagged, -wagging**) have a gossip. **keep one's chin up** *colloq.* remain cheerful, esp. in adversity. **take on the chin 1** suffer a severe blow from (a misfortune etc.). **2** endure courageously. □ **-chinned** *adj.* (in comb.). [OE *cin*(n) f. Gmc]

China /ˈtʃaɪnə/ (official name **People's Republic of China**) a country in eastern Asia, the third largest and most populous in the world; pop. (est. 1991) 1,151,200,000; language, Chinese (of which Mandarin is the official form); capital, Beijing. Chinese civilization stretches back until at least the 3rd millennium BC, the country being ruled by a series of dynasties, until the Qing (or Manchu) dynasty was overthrown by Sun Yat-sen in 1911, China being proclaimed a republic the following year. The country was stricken by civil war (1927–37 and 1946–9) and by Japanese invasion, and soon after the Second World War the corrupt and ineffective Kuomintang government of Chiang Kai-shek was overthrown by the Communists, the People's Republic of China being declared in 1949. The Communists have remained in power, and in June 1989 a pro-democracy movement was ruthlessly repressed. For many centuries China remained generally closed to Western economic or political penetration, both under its old imperial rulers and its new Communist ones. (See also TAIWAN.)

china /ˈtʃaɪnə/ *n. & adj.* ● *n.* **1** a kind of fine white or translucent ceramic ware, porcelain, etc. **2** things made from ceramic, esp. household tableware. **3** *rhyming sl.* one's 'mate', i.e. husband or wife (short for *china plate*). ● *adj.* made of china. □ **china clay** kaolin. [orig. *China ware* (from CHINA: name f. Pers. *chīnī*)]

China, Republic of the official name for TAIWAN.

Chinagraph /ˈtʃaɪnəˌɡrɑːf/ *n. propr.* a waxy coloured pencil used to write on china, glass, etc.

Chinaman /ˈtʃaɪnəmən/ *n.* (*pl.* **-men**) **1** *archaic* or *derog.* (now usu. *offens.*) a native of China. **2** *Cricket* a ball bowled by a left-handed bowler that spins from off to leg.

China Sea the part of the Pacific Ocean off the coast of China, divided by the island of Taiwan into the *East China Sea* in the north and the *South China Sea* in the south.

China syndrome *n.* an imaginary sequence of events following the meltdown of a nuclear reactor, in which the core melts through its containment structure and deep into the earth. [CHINA, as being on the opposite side of the earth from a reactor in the US]

China tea *n.* smoke-cured tea from a small-leaved tea plant grown in China.

Chinatown /ˈtʃaɪnəˌtaʊn/ *n.* a district of any non-Chinese town, esp. a city or seaport, in which the population is predominantly Chinese.

chinch /tʃɪntʃ/ *n.* (in full **chinch-bug**) *US* **1** a North American plant bug, *Blissus leucopterus*, that destroys the shoots of grasses and grains. **2** *N. Amer.* a bedbug. [Sp. *chinche* f. L *cimex -icis*]

chincherinchee /ˌtʃɪntʃərɪnˈtʃiː/ *n.* a white-flowered bulbous liliaceous plant, *Ornithogalum thyrsoides*, native to southern Africa. [imit. of the squeaky rubbing of its stalks]

chinchilla /tʃɪnˈtʃɪlə/ *n.* **1 a** a small rodent of the genus *Chinchilla*, native to South America, having soft silver-grey fur and a bushy tail. **b** its highly valued fur. **2** a breed of cat or rabbit. [Sp. prob. f. Aymara or Quechua]

chin-chin /tʃɪnˈtʃɪn/ *int. Brit. colloq.* a toast; a greeting or farewell. [Chin. *qingqing* (pr. ch-)]

Chindit /ˈtʃɪndɪt/ *n. hist.* a member of the Allied forces behind the Japanese lines in Burma in 1943–5. [Burmese *chinthé*, a mythical creature]

Chindwin /tʃɪnˈdwɪn/ a river which rises in northern Burma

(Myanmar) and flows southwards for 885 km (550 miles) to meet the Irrawaddy, of which it is the principal tributary.

chine[1] /tʃaɪn/ *n. & v.* ● *n.* **1 a** a backbone, esp. of an animal. **b** a joint of meat containing all or part of this. **2** a ridge or arête. ● *v.tr.* cut (meat) across or along the backbone. [ME f. OF *eschine* f. L *spina* SPINE]

chine[2] /tʃaɪn/ *n.* a deep narrow ravine in the Isle of Wight or Dorset. [OE *cinu* chink etc. f. Gmc]

chine[3] /tʃaɪn/ *n.* the join between the side and the bottom of a ship etc. [var. of CHINE[2]]

Chinese /tʃaɪˈniːz/ *adj. & n.* ● *adj.* **1** of or relating to China. **2** of Chinese descent. ● *n.* **1** the Chinese language. (*See note below.*) **2** (*pl.* same) **a** a native or national of China. **b** a person of Chinese descent. □ **Chinese cabbage** = Chinese leaf. **Chinese gooseberry** = kiwi fruit. **Chinese lantern 1** a collapsible paper lantern. **2** a solanaceous plant, *Physalis alkekengi*, bearing white flowers and globular orange fruits enclosed in an orange-red papery calyx. **Chinese leaf** a lettuce-like cabbage, *Brassica chinensis*. **Chinese puzzle** a very intricate puzzle or problem. **Chinese water chestnut** see *water chestnut* 2. **Chinese white** zinc oxide as a white pigment.

▪ Chinese is a member of the Sino-Tibetan language group and is an isolating tonal language with no inflections, declensions, or conjugations. It is the world's most commonly spoken first language, with an estimated one billion speakers in China and elsewhere. Of the many dialects, the Mandarin group and Cantonese are the most widespread, and a form of Mandarin is used as the official language of China. Chinese script is logographic: many of the characters were in origin pictographs, with each sign standing for an object, but they gradually developed into a system of non-pictorial ideographs which can also represent both abstract concepts and the sounds of syllables. Despite its complexity, the script has advantages; for example, it makes written communication possible between people speaking mutually incomprehensible dialects. Examples of Chinese writing date back well beyond 1000 BC. Traditionally Chinese books were arranged in vertical columns and read from right to left, but now they are usually composed horizontally from left to right. Until the beginning of the 20th century the greater part of written Chinese was in a style imitative of the Chinese classics (largely written before 200 BC), but a reform movement was started to make the literature available to the masses, and many simplified characters were introduced. Since 1958 the Pinyin system of spelling, using the Roman alphabet, has been adopted for international and educational use.

Ch'ing var. of QING.

Chin Hills /tʃɪn/ a range of hills in western Burma (Myanmar), close to the borders with India and Bangladesh.

Chink /tʃɪŋk/ *n. sl. offens.* a Chinese. [abbr.]

chink[1] /tʃɪŋk/ *n.* **1** an unintended crack that admits light or allows an attack; a flaw. **2** a narrow opening; a slit. [16th c.: rel. to CHINE[2]]

chink[2] /tʃɪŋk/ *v. & n.* ● *v.* **1** *intr.* make a slight ringing sound, as of glasses or coins striking together. **2** *tr.* cause to make this sound. ● *n.* this sound. [imit.]

Chinkiang see ZHENJIANG.

Chinky /ˈtʃɪŋkɪ/ *n.* (*pl.* **-ies**) *sl. offens.* **1** a Chinese. **2** a Chinese restaurant.

chinless /ˈtʃɪnlɪs/ *adj. colloq.* weak or feeble in character. □ **chinless wonder** *Brit.* an ineffectual esp. upper-class person.

chino /ˈtʃiːnəʊ/ *n.* esp. *US* (*pl.* **-os**) **1** a cotton twill fabric, usu. khaki-coloured. **2** (in *pl.*) a garment, esp. trousers, made from this. [Amer. Sp., = toasted]

Chino- /ˈtʃaɪnəʊ/ *comb. form* = SINO-.

chinoiserie /ʃiːˈnwɑːzərɪ/ *n.* **1** the imitation or evocation of Chinese motifs and techniques in Western art and architecture, particularly that of the 18th century. **2** an object or objects in this style. [F]

Chinook /ʃɪˈnuːk, tʃɪ-, -ˈnʊk/ *n. & adj.* ● *n.* (*pl.* same or **Chinooks**) **1** a member of an American Indian people originally inhabiting the region around the Columbus river in Oregon, and now living on reservations. **2** the language of this people. ● *adj.* of or relating to the Chinook or their language. □ **Chinook jargon** a pidgin composed of elements from Chinook, Nootka, English, French, and elsewhere, used formerly in the Pacific north-west of North America. [Salish *tsinúk*]

chinook /ʃɪˈnuːk, tʃɪ-, -ˈnʊk/ *n.* **1** a warm dry wind which blows east of the Rocky Mountains. **2** a warm wet southerly wind west of the Rocky Mountains. □ **chinook salmon** a large salmon, *Oncorhynchus tshawytscha*, of the North Pacific. [CHINOOK]

chintz /tʃɪnts/ *n. & adj.* ● *n.* a printed multicoloured cotton fabric with a glazed finish. ● *adj.* made from or upholstered with this fabric. [earlier *chints* (pl.) f. Hindi *chīṇṭ* f. Skr. *citra* variegated]

chintzy /ˈtʃɪntsɪ/ *adj.* (**chintzier, chintziest**) **1** like chintz. **2** gaudy, cheap. **3** characteristic of the decor associated with chintz soft furnishings. □ **chintzily** *adv.* **chintziness** *n.*

chionodoxa /ˌkaɪənəˈdɒksə/ *n.* a liliaceous plant of the genus *Chionodoxa*, having early-blooming blue flowers. Also called *glory-of-the-snow.* [mod.L f. Gk *khiōn* snow + *doxa* glory]

Chios /ˈkaɪɒs/ (Greek **Khios** /ˈxiɔs/) a Greek island in the Aegean Sea; pop. (1991) 52,690.

chip /tʃɪp/ *n. & v.* ● *n.* **1** a small piece removed by or in the course of chopping, cutting, or breaking, esp. from hard material such as wood or stone. **2** the place where such a chip has been made. **3 a** (usu. in *pl.*) a strip of potato, deep fried. **b** (in *pl.*) *N. Amer.* potato crisps. **4** a counter used in some gambling games to represent money. **5** *Electronics* = MICROCHIP. **6 a** thin strip of wood, straw, etc., used for weaving hats, baskets, etc. **b** a basket made from these. **7** *Football* etc. & *Golf* a short shot, kick, or pass with the ball describing an arc. ● *v.* (**chipped, chipping**) **1** *tr.* (often foll. by *off, away*) cut or break (a piece) from a hard material. **2** *intr.* (foll. by *at, away at*) cut pieces off (a hard material) to alter its shape, break it up, etc. **3** *intr.* (of stone, china, etc.) be susceptible to being chipped; be apt to break at the edge (*will chip easily*). **4** *tr.* (also *absol.*) *Football* etc. & *Golf* strike or kick (the ball) with a chip (cf. sense 7 of *n.*). **5** *tr.* (usu. as **chipped** *adj.*) cut (potatoes) into chips. □ **chip heater** *Austral.* & *NZ* a domestic water-heater that burns wood chips. **chip in** *colloq.* **1** interrupt or contribute abruptly to a conversation (*chipped in with a reminiscence*). **2** contribute (money or resources). **a chip off the old block** a child who resembles a parent, esp. in character. **a chip on one's shoulder** *colloq.* a disposition or inclination to feel resentful or aggrieved. **chip shot** = sense 7 of *n.* **have had one's chips** *Brit. colloq.* be unable to avoid defeat, punishment, etc. **in the chips** *sl.* moneyed, affluent. **when the chips are down** *colloq.* when it comes to the point. [ME f. OE *cipp, cyp* beam]

chipboard /ˈtʃɪpbɔːd/ *n.* a rigid sheet or panel made from compressed wood chips and resin.

chipmunk /ˈtʃɪpmʌŋk/ *n.* a North American ground squirrel of the genus *Tamias*, having alternate light and dark stripes running down the body. [Algonquian]

chipolata /ˌtʃɪpəˈlɑːtə/ *n. Brit.* a small thin sausage. [F f. It. *cipollata* a dish of onions f. *cipolla* onion]

Chippendale /ˈtʃɪp(ə)nˌdeɪl/ *adj.* **1** (of furniture) designed or made by the English cabinet-maker Thomas Chippendale (1718–79). **2** in the ornately elegant style of Chippendale's furniture.

chipper /ˈtʃɪpə(r)/ *adj. esp. N. Amer. colloq.* **1** cheerful. **2** smartly dressed. [perh. f. N.Engl. dial. *kipper* lively]

chipping /ˈtʃɪpɪŋ/ *n.* **1** a small fragment of stone, wood, etc. **2** (in *pl.*) these used as a surface for roads, roofs, etc.

chippy /ˈtʃɪpɪ/ *n.* (also **chippie**) (*pl.* **-ies**) *Brit. colloq.* **1** a fish-and-chip shop. **2** a carpenter.

Chips /tʃɪps/ *n. Naut. sl.* a ship's carpenter.

Chirac /ˈʃɪræk/, Jacques (René) (b.1932), French statesman, Prime Minister 1974–6 and 1986–8 and President since 1995. He was elected mayor of Paris in 1977, a position he held for eighteen years. The founder and leader of the right-wing RPR (Rally for the Republic) Party, Chirac headed the right's coalition in the National Assembly during the socialist government of 1981–6. When his coalition was victorious in the 1986 National Assembly elections, he was appointed Prime Minister by the socialist President François Mitterrand. After an unsuccessful bid for the presidency in 1988, Chirac was elected to succeed Mitterrand as President in 1995.

chiral /ˈkaɪrəl/ *adj. Chem.* (of an optically active compound) asymmetric and not superposable on its mirror image. □ **chirality** /ˌkaɪəˈrælɪtɪ/ *n.* [Gk *kheir* hand]

Chirico /ˈkɪrɪˌkəʊ/, Giorgio de (1888–1978), Greek-born Italian painter. After 1910 he started painting disconnected and unsettling dream images, a style that became known as 'metaphysical painting'. His work exerted a significant influence on surrealism and he participated in the surrealist's Paris exhibition of 1925. Major works include *The Uncertainty of the Poet* (1913), portraying a bust with bananas, an arcade, and a distant train.

chiro- /ˈkaɪrəʊ/ (also **cheiro-**) *comb. form* of the hand. [Gk *kheir* hand]

chirography /ˌkaɪəˈrɒɡrəfɪ/ *n.* handwriting, calligraphy.

chiromancy /ˈkaɪrəʊˌmænsɪ/ *n.* palmistry.

Chiron /ˈkaɪərɒn/ **1** *Gk Mythol.* a learned centaur who acted as teacher to Jason, Achilles, and many other heroes. **2** *Astron.* asteroid 2060, discovered in 1977, which is unique in having an orbit lying mainly between the orbits of Saturn and Uranus. It is now believed to have a diameter of 370 km.

chiropody /kɪˈrɒpədɪ/ *n.* the treatment of the feet and their ailments. □ **chiropodist** *n.* [CHIRO- + Gk *pous podos* foot]

chiropractic /ˌkaɪərəʊˈpræktɪk/ *n.* the diagnosis and manipulative treatment of mechanical disorders of the joints, esp. of the spinal column. □ **chiropractor** /ˈkaɪərəʊˌpræktə(r)/ *n.* [CHIRO- + Gk *praktikos*: see PRACTICAL]

chiropteran /ˌkaɪəˈrɒptərən/ *n. & adj. Zool.* ● *n.* a flying mammal of the order Chiroptera, which comprises the bats. ● *adj.* of or relating to bats. □ **chiropterous** *adj.* [mod.L f. CHIRO- + Gk *pteron* wing]

chirp /tʃɜːp/ *v. & n.* ● *v.* **1** *intr.* (usu. of small birds, grasshoppers, etc.) utter a short sharp high-pitched note. **2** *tr. & intr.* (esp. of a child) speak or utter in a lively or jolly way. ● *n.* a chirping sound. □ **chirper** *n.* [ME, earlier *chirk, chirt*: imit.]

chirpy /ˈtʃɜːpɪ/ *adj. colloq.* (**chirpier, chirpiest**) cheerful, lively. □ **chirpily** *adv.* **chirpiness** *n.*

chirr /tʃɜː(r)/ *v. & n.* (also **churr**) ● *v.intr.* (esp. of insects) make a prolonged low trilling sound. ● *n.* this sound. [imit.]

chirrup /ˈtʃɪrəp/ *v. & n.* ● *v.intr.* (**chirruped, chirruping**) (esp. of small birds) chirp, esp. repeatedly; twitter. ● *n.* a chirruping sound. □ **chirrupy** *adj.* [trilled form of CHIRP]

chisel /ˈtʃɪz(ə)l/ *n. & v.* ● *n.* a hand-tool with a squared bevelled blade for shaping wood, stone, or metal. ● *v.* **1** *tr.* (**chiselled, chiselling**; US **chiseled, chiseling**) cut or shape with a chisel. **2** *tr.* (as **chiselled** *adj.*) (of facial features) clean-cut, fine. **3** *tr. & intr. sl.* cheat, swindle. □ **chiseller** *n.* [ME f. ONF ult. f. LL *cisorium* f. L *caedere caes-* cut]

Chişinău /ˌkɪʃɪˈnaʊ/ (Russian **Kishinev** /ˌkɪʃɪˈnjɔf/) the capital of Moldova; pop. (est. 1989) 665,000.

chit¹ /tʃɪt/ *n.* **1** *derog.* or *joc.* a young, small, or frail girl or woman (esp. *a chit of a girl*). **2** a young child. [ME, = whelp, cub, kitten, perh. = dial. *chit* sprout]

chit² /tʃɪt/ *n.* **1** a note of requisition; a note of a sum owed, esp. for food or drink. **2** esp. *Brit.* a note or memorandum. [earlier *chitty*: Anglo-Ind. f. Hindi *citṭhī* pass f. Skr. *citra* mark]

chital /ˈtʃiːt(ə)l/ *n.* = AXIS². [Hindi *cītal*]

chit-chat /ˈtʃɪttʃæt/ *n. & v. colloq.* ● *n.* light conversation; gossip. ● *v.intr.* (**-chatted, -chatting**) talk informally; gossip. [redupl. of CHAT¹]

chitin /ˈkaɪtɪn/ *n. Chem.* a polysaccharide forming the major constituent in the exoskeleton of arthropods and in the cell walls of fungi. □ **chitinous** *adj.* [F *chitine* irreg. f. Gk *khitōn*: see CHITON]

chiton /ˈkaɪt(ə)n/ *n.* **1** a long woollen tunic worn by ancient Greeks. **2** a marine mollusc of the class Amphineura, having a shell of overlapping plates. [Gk *khitōn* tunic]

Chittagong /ˈtʃɪtəˌɡɒŋ/ a seaport in SE Bangladesh, on the Bay of Bengal; pop. (1991) 1,566,070.

chitterling /ˈtʃɪtəlɪŋ/ *n.* (usu. in *pl.*) the smaller intestines of pigs etc., esp. as cooked for food. [ME: orig. uncert.]

chivalrous /ˈʃɪvəlrəs/ *adj.* **1** (usu. of a male) gallant, honourable, courteous. **2** involving or showing chivalry. □ **chivalrously** *adv.* [ME f. OF *chevalerous*: see CHEVALIER]

chivalry /ˈʃɪvəlrɪ/ *n.* **1** the medieval knightly system with its religious, moral, and social code. **2** the combination of qualities expected of an ideal knight, esp. courage, honour, courtesy, justice, and readiness to help the weak. **3** a man's courteous behaviour, esp. towards women. **4** *archaic* knights, noblemen, and horsemen collectively. □ **chivalric** *adj.* [ME f. OF *chevalerie* etc. f. med.L *caballerius* for LL *caballarius* horseman: see CAVALIER]

chive /tʃaɪv/ *n.* a small allium, *Allium schoenoprasum*, having purple-pink flowers and dense tufts of long tubular leaves which are used as a herb. [ME f. OF *cive* f. L *cepa* onion]

chivvy /ˈtʃɪvɪ/ *v.tr.* (also **chivy, chevy** /ˈtʃevɪ/) (**-ies, -ied**) hurry (a person) up; harass, nag; pursue. [*chevy* (n. & v.), prob. f. the ballad of *Chevy Chase*, a place on the Scottish border]

Chkalov /ˈtʃkɑːlɒf/ the former name (1938–57) for ORENBURG.

chlamydia /kləˈmɪdɪə/ *n.* (*pl.* **chlamydiae** /-dɪˌiː/) a small parasitic bacterium of the genus *Chlamydia*, only able to multiply within the cells of a host, some of which cause diseases such as trachoma, psittacosis, and non-specific urethritis. [mod.L f. Gk *khlamus -udos* cloak]

chlamydomonas /ˌklæmɪdə'məʊnəs/ n. a motile unicellular green alga of the genus *Chlamydomonas*. [mod.L (as CHLAMYDIA)]

chlor- var. of CHLORO-.

chloracne /klɔː'ræknɪ/ n. Med. a skin disease resembling severe acne, caused by exposure to chlorinated chemicals. [CHLORINE + ACNE]

chloral /'klɔːrəl/ n. **1** a colourless liquid aldehyde used in making DDT. **2** (in full **chloral hydrate**) Pharm. a colourless crystalline solid made from chloral and used as a sedative. [CHLOR- + -AL²]

chloramphenicol /ˌklɔːræm'fenɪˌkɒl/ n. Pharm. an antibiotic prepared from the bacterium *Streptomyces venezuelae* or produced synthetically, used esp. against typhoid fever. [CHLORO- + AMIDE + PHENO- + NITRO- + GLYCOL]

chlorate /'klɔːreɪt/ n. Chem. a salt of chloric acid.

chlorella /klɔː'relə/ n. a non-motile unicellular green alga of the genus *Chlorella*. [mod.L, dimin. of Gk *khlōros* green]

chloric acid /'klɔːrɪk/ n. Chem. a colourless liquid acid with strong oxidizing properties. [CHLORO- + -IC]

chloride /'klɔːraɪd/ n. Chem. **1** any compound of chlorine with another element or group. **2** any bleaching agent containing chloride. [CHLORO- + -IDE]

chlorinate /'klɔːrɪˌneɪt/ v.tr. **1** impregnate or treat with chlorine. **2** Chem. cause to react or combine with chlorine. □ **chlorinator** n.

chlorination /ˌklɔːrɪ'neɪʃ(ə)n/ n. **1** the treatment of water with chlorine to disinfect it. **2** Chem. a reaction in which chlorine is introduced into a compound.

chlorine /'klɔːriːn/ n. a toxic irritant pale green gaseous chemical element (atomic number 17; symbol **Cl**). A member of the halogen group, chlorine was isolated by C. W. Scheele in 1774 and recognized as an element by Sir Humphry Davy in 1810. It occurs in nature mainly as sodium chloride in sea water and salt deposits. The gas was used as a poison gas in the First World War. Chlorine is added to water supplies as a disinfectant, and is a constituent of many bleaches, antiseptics, insecticides, and other commercial substances. [Gk *khlōros* green + -INE⁴]

chlorite¹ /'klɔːraɪt/ n. Mineral. a dark green mineral found in many rocks, consisting of a basic aluminosilicate of magnesium, iron, etc. □ **chloritic** /klɔː'rɪtɪk/ adj. [L *chloritis* f. Gk *khlōritis* green precious stone, f. as CHLORO-]

chlorite² /'klɔːraɪt/ n. Chem. a salt of chlorous acid.

chloro- /'klɔːrəʊ, 'klɒrəʊ/ comb. form (also **chlor-** esp. before a vowel) **1** Bot. & Mineral. green. **2** Chem. chlorine. [Gk *khlōros* green: in sense 2 f. CHLORINE]

chlorofluorocarbon /ˌklɔːrəʊˌflʊərəʊ'kɑːb(ə)n/ n. see CFC.

chloroform /'klɒrəˌfɔːm, 'klɔːrə-/ n. & v. n. a colourless volatile sweet-smelling liquid (chem. formula: CHCl₃) used as a solvent and formerly used as a general anaesthetic. ● v.tr. render (a person) unconscious with this. [F *chloroforme* formed as CHLORO- + *formyle*: see FORMIC ACID]

Chloromycetin /ˌklɔːrəʊmaɪ'siːtɪn/ n. propr. = CHLORAMPHENICOL. [CHLORO- + Gk *mukēs -ētos* fungus]

chlorophyll /'klɒrəfɪl/ n. the green pigment found in most plants, responsible for light absorption to provide energy for photosynthesis. (See PHOTOSYNTHESIS.) □ **chlorophyllous** /ˌklɔːrəʊ'fɪləs/ adj. [F *chlorophylle* f. Gk *phullon* leaf: see CHLORO-]

chloroplast /'klɔːrəʊˌplæst/ n. Bot. a plastid containing chlorophyll, found in plant cells undergoing photosynthesis. [G: (as CHLORO-, PLASTID)]

chlorosis /klə'rəʊsɪs, klɔː'rəʊ-/ n. **1** Med. a severe form of anaemia from iron deficiency esp. in young women, causing a greenish complexion (cf. GREENSICK). **2** Bot. a reduction or loss of the normal green coloration of leaves. □ **chlorotic** /-'rɒtɪk/ adj. [CHLORO- + -OSIS]

chlorous acid /'klɔːrəs/ n. Chem. a pale yellow liquid acid (chem. formula: HClO₂) with oxidizing properties. [CHLORO- + -OUS]

chlorpromazine /klɔː'prəʊməˌziːn/ n. Pharm. a drug used as a sedative and to control nausea and vomiting. [F (as CHLORO-, PROMETHAZINE)]

Ch.M. abbr. Master of Surgery. [L *Chirurgiae Magister*]

choc /tʃɒk/ n. & adj. colloq. chocolate. □ **choc-ice** a bar of ice-cream covered with a thin coating of chocolate. [abbr.]

chocho /'tʃəʊtʃəʊ/ n. (pl. **-os**) West Indies = CHOKO.

chock /tʃɒk/ n., v, & adv. ● n. a block or wedge of wood to check motion, esp. of a cask or a wheel. ● v.tr. **1** fit or make fast with chocks. **2** (usu. foll. by up) Brit. cram full. ● adv. as closely or tightly as possible. [prob. f. OF *couche*, *coche*, of unkn. orig.]

chock-a-block /'tʃɒkəˌblɒk/ adj. & adv. crammed close together; crammed full (*a street chock-a-block with cars*). [orig. Naut., with ref. to tackle with the two blocks run close together]

chocker /'tʃɒkə(r)/ adj. Brit. sl. fed up, disgusted. [CHOCK-A-BLOCK]

chock-full /'tʃɒkfʊl/ adj. & adv. crammed full (*chock-full of rubbish*). [CHOCK + FULL¹: ME *chokkefulle* (rel. to CHOKE¹) is doubtful]

chocoholic /ˌtʃɒkə'hɒlɪk/ n. & adj. (also **chocaholic**) ● n. a person who is addicted to or very fond of chocolate. ● adj. of or relating to chocoholics; addicted to or very fond of chocolate.

chocolate /'tʃɒklət, 'tʃɒkələt/ n. & adj. ● n. **1 a** a food preparation in the form of a paste or solid block made from roasted and ground cacao seeds, usually sweetened. **b** a sweet made of or coated with this. **c** a drink made with chocolate. **2** a deep brown colour. ● adj. **1** made from or of chocolate. **2** chocolate-coloured. □ **chocolate-box 1** a decorated box filled with chocolates. **2** (attrib.) stereotypically pretty or romantic. □ **chocolatey** adj. (also **chocolaty**). [F *chocolat* or Sp. *chocolate* f. Aztec *chocolatl*]

Choctaw /'tʃɒktɔː/ n. & adj. ● n. (pl. same or **Choctaws**) **1 a** a member of an American Indian people originally inhabiting Mississippi and Alabama. **b** the Muskogean language of this people. **2** (in skating) a step from one edge of a skate to the other edge of the other skate in the opposite direction. ● adj. of or relating to the Choctaw or their language. [Choctaw]

choice /tʃɔɪs/ n. & adj. ● n. **1 a** the act or an instance of choosing. **b** a thing or person chosen (*not a good choice*). **2** a range from which to choose. **3** (usu. foll. by of) the élite, the best. **4** the power or opportunity to choose (*what choice have I?*). ● adj. of superior quality; carefully chosen. □ **choicely** adv. **choiceness** n. [ME f. OF *chois* f. *choisir* CHOOSE]

choir /'kwaɪə(r)/ n. **1** a regular group of singers, esp. taking part in church services. **2** the part of a cathedral or large church between the altar and the nave, used by the choir and clergy. **3** a company of singers, birds, angels etc. (*a heavenly choir*). **4** Mus. a group of instruments of one family playing together. □ **choir organ** the softest of three parts making up a large organ having its row of keys the lowest of the three. **choir-stall** = STALL¹ n. 3a. [ME f. OF *quer* f. L *chorus*: see CHORUS]

choirboy /'kwaɪəˌbɔɪ/ n. (fem. **choirgirl**) a boy or girl who sings in a church or cathedral choir.

choke¹ /tʃəʊk/ v. & n. ● v. **1** tr. hinder or impede the breathing of (a person or animal) esp. by constricting the windpipe or (of gas, smoke, etc.) by being unbreathable. **2** intr. suffer a hindrance or stoppage of breath. **3** tr. & intr. make or become speechless from emotion. **4** tr. retard the growth of or kill (esp. plants) by the deprivation of light, air, nourishment, etc. **5** tr. (often foll. by back) suppress (feelings) with difficulty. **6** tr. block or clog (a passage, tube, etc.). **7** tr. (as **choked** adj.) colloq. disgusted, disappointed. **8** tr. enrich the fuel mixture in (an internal-combustion engine) by reducing the intake of air. ● n. **1** the valve in the carburettor of an internal-combustion engine that controls the intake of air, esp. to enrich the fuel mixture. **2** Electr. an inductance coil used to smooth the variations of an alternating current or to alter its phase. □ **choke-chain** a chain looped round a dog's neck to exert control by pressure on its windpipe when the dog pulls. **choke-cherry** an astringent North American cherry, *Prunus virginiana*. **choke-damp** carbon dioxide in mines, wells, etc. **choke down** swallow with difficulty. **choke up** block (a channel etc.). [ME f. OE *ācēocian* f. *cēoce*, *cēce* CHEEK]

choke² /tʃəʊk/ n. the centre part of an artichoke. [prob. confusion of the ending of *artichoke* with CHOKE¹]

chokeberry /'tʃəʊkˌbærɪ/ n. (pl. **-ies**) Bot. **1** a North American rosaceous shrub of the genus *Aronia*. **2** its red, black, or purple astringent fruit.

choker /'tʃəʊkə(r)/ n. **1** a close-fitting necklace or ornamental neckband. **2** a clerical or other high collar.

choko /'tʃəʊkəʊ/ n. (pl. **-os**) Austral. & NZ a succulent green pear-shaped vegetable like a cucumber in flavour. [Brazilian Ind. *chocho*]

choky¹ /'tʃəʊkɪ/ n. (also **chokey**) (pl. **-ies** or **-eys**) Brit. sl. prison. [orig. Anglo-Ind., f. Hindi *caukī* shed]

choky² /'tʃəʊkɪ/ adj. (**chokier, chokiest**) tending to choke or to cause choking.

cholangiography /ˌkɒlændʒɪ'ɒgrəfɪ/ n. Med. X-ray examination of the bile ducts, used to find the site and nature of any obstruction. [CHOLE- + Gk *aggeion* vessel + -GRAPHY]

chole- /'kɒlɪ/ comb. form (also **chol-** esp. before a vowel) Med. & Chem. bile. [Gk *kholē* gall, bile]

cholecalciferol /ˌkɒlɪkæl'sɪfəˌrɒl/ n. one of the D vitamins, produced by the action of sunlight on a cholesterol derivative widely distributed in the skin, a deficiency of which results in rickets in children and osteomalacia in adults. Also called *vitamin D₃*. [CHOLE- + CALCIFEROL]

cholecystography /ˌkɒlɪsɪˈstɒɡrəfɪ/ *n. Med.* X-ray examination of the gall-bladder, esp. used to detect the presence of any gallstones. [CHOLE- + CYSTO- + -GRAPHY]

choler /ˈkɒlə(r)/ *n.* **1** *hist.* one of the four humours, bile. **2** *poet.* or *archaic* anger, irascibility. [ME f. OF *colere* bile, anger f. L *cholera* f. Gk *kholera* diarrhoea, in LL = bile, anger, f. Gk *kholē* bile]

cholera /ˈkɒlərə/ *n. Med.* a dangerous infectious disease of the small intestine caused by the bacterium *Vibrio cholerae*, resulting in severe vomiting and diarrhoea leading to dehydration. Cholera is endemic in parts of Asia, although there were outbreaks in western Europe in the 19th century. It is a mainly water-borne disease, and can be contained by purification of water supplies and by vaccination. □ **chicken** (or **fowl**) **cholera** an infectious disease of fowls. □ **choleraic** /ˌkɒləˈreɪk/ *adj.* [ME f. L f. Gk *kholera*: see CHOLER]

choleric /ˈkɒlərɪk/ *adj.* irascible, angry. □ **cholerically** *adv.* [ME f. OF *cholerique* f. L *cholericus* f. Gk *kholerikos*: see CHOLER]

cholesterol /kəˈlestəˌrɒl/ *n. Biochem.* a sterol found in most body tissues, including the blood, where high concentrations may involve a risk of arteriosclerosis. [*cholesterin* f. Gk *kholē* bile + *stereos* stiff]

choli /ˈtʃəʊlɪ/ *n.* (*pl.* **cholis**) a type of short-sleeved bodice worn by Indian women. [Hindi *colī*]

choliamb /ˈkəʊlɪˌæmb/ *n. Prosody* = SCAZON. □ **choliambic** /ˌkəʊlɪˈæmbɪk/ *adj.* [LL *choliambus* f. Gk *khōliambos* f. *khōlos* lame: see IAMBUS]

choline /ˈkəʊliːn, -lɪn/ *n. Biochem.* a basic nitrogenous organic compound occurring widely in living matter. [G *Cholin* f. Gk *kholē* bile]

cholla /ˈtʃɔɪə/ *n.* an opuntia cactus of Mexico and the south-western US. [Mex. Sp., lit. 'skull, head']

chomp /tʃɒmp/ *v.tr.* = CHAMP¹. [imit.]

Chomsky /ˈtʃɒmskɪ/, (Avram) Noam (b.1928), American theoretical linguist and political activist. His theory of transformational grammar is set out in *Syntactic Structures* (1957). Language is seen as a formal system possessing an underlying 'deep structure', from which the 'surface structure' is derived by a series of transformations. A grammar is seen as a set of rules capable of generating all and only the grammatical sentences of a language when applied to the lexicon of that language. A distinction is made between a speaker's linguistic competence, which is idealized, and actual performance; the theory sets out to account only for the former. Chomsky has revised the theory since 1957, most notably in *Lectures on Government and Binding* (1981). In addition, Chomsky was a leading opponent of the American involvement in the Vietnam War, and later also voiced his doubts about the American role in the Gulf War of 1991.

chondrite /ˈkɒndraɪt/ *n.* a stony meteorite containing small mineral granules. [G *Chondrit* f. Gk *khondros* granule]

chondrocranium /ˌkɒndrəʊˈkreɪnɪəm/ *n. Anat.* the embryonic skull composed of cartilage and later replaced by bone. [Gk *khondros* grain, cartilage]

Chongjin /tʃʌŋˈdʒɪn/ a port on the NE coast of North Korea; pop. (est. 1984) 754,100.

Chongqing /tʃʊŋˈtʃɪŋ/ (also **Chungking** /-ˈkɪŋ/) a city in Sichuan province in central China; pop. (1990) 2,960,000. It was the capital of China from 1938 to 1946, a period when the present capital Beijing, and Nanjing, another former capital, were held by the Japanese.

choo-choo /ˈtʃuːtʃuː/ *n.* (also **choo-choo train**) *colloq.* (esp. as a child's word) a railway train or locomotive, esp. a steam engine. [imit.]

chook /tʃʊk/ *n.* (also **chookie** /ˈtʃʊkɪ/) *Austral. & NZ colloq.* **1** a chicken or fowl. **2** *sl.* an older woman. [Engl. dial. *chuck* chicken]

choose /tʃuːz/ *v.* (*past* **chose** /tʃəʊz/; *past part.* **chosen** /ˈtʃəʊz(ə)n/) **1** *tr.* select out of a greater number. **2** *intr.* (usu. foll. by *between, from*) take or select one or another. **3** *tr.* (usu. foll. by *to* + infin.) decide, be determined (*chose to stay behind*). **4** *tr.* (foll. by complement) select as (*was chosen king*). **5** *tr.* (in Christian theology) (esp. as **chosen** *adj.*) destine to be saved (*God's chosen people*). □ **cannot choose but** *archaic* must. **nothing** (or **little**) **to choose between them** they are equivalent. □ **chooser** *n.* [OE *cēosan* f. Gmc]

choosy /ˈtʃuːzɪ/ *adj.* (**choosier, choosiest**) *colloq.* fastidious. □ **choosily** *adv.* **choosiness** *n.*

chop¹ /tʃɒp/ *v. & n.* ● *v.tr.* (**chopped, chopping**) **1** (usu. foll. by *off, down,* etc.) cut or fell by a blow, usu. with an axe. **2** (often foll. by *up*) cut (esp. meat or vegetables) into small pieces. **3** strike (esp. a ball) with a short

heavy edgewise blow. **4** *Brit. colloq.* dispense with; shorten or curtail. ● *n.* **1** a cutting blow, esp. with an axe. **2** a thick slice of meat (esp. pork or lamb) usu. including a rib. **3** a short heavy edgewise stroke or blow in tennis, cricket, boxing, etc. **4** the broken motion of water, usu. owing to the action of the wind against the tide. **5** (prec. by *the*) *Brit. sl.* **a** a dismissal from employment. **b** the action of killing or being killed. [ME, var. of CHAP¹]

chop² /tʃɒp/ *n.* (usu. in *pl.*) the jaw of an animal etc. [16th-c. var. (occurring earlier) of CHAP³, of unkn. orig.]

chop³ /tʃɒp/ *v.intr.* (**chopped, chopping**) □ **chop and change** vacillate; change direction frequently. **chop logic** argue pedantically. [ME, = barter, exchange, perh. rel. to *chap* f. OE *cēapian* (as CHEAP)]

chop⁴ /tʃɒp/ *n. Brit. archaic* a trademark; a brand of goods. □ **not much chop** esp. *Austral. & NZ* no good. [orig. in India & China, f. Hindi *chāp* stamp]

chop-chop /tʃɒpˈtʃɒp/ *adv. & int.* (pidgin English) quickly, quick. [f. Chin. dial. *k'wâi-k'wâi*]

Chopin /ˈʃəʊpæn/, Frédéric (François) (Polish name Fryderyk Franciszek Szopen) (1810–49), Polish-born French composer and pianist. A concert pianist from the age of 8, he wrote almost exclusively for the piano. Inspired by Polish folk music, his mazurkas (fifty-five in all) and polonaises (some thirteen) were to become his trademarks. He is also noted for short piano pieces full of poetic intensity; good examples are his nineteen nocturnes and twenty-four preludes. As well as writing for solo piano, he wrote two piano concertos (1829; 1830). For some years Chopin was the lover of the French writer George Sand, but their affair ended two years before his death from tuberculosis.

chopper /ˈtʃɒpə(r)/ *n.* **1 a** *Brit.* a short axe with a large blade. **b** a butcher's cleaver. **2** *colloq.* a helicopter. **3** *Electr.* a device for regularly interrupting an electric current or light-beam. **4** *colloq.* a type of bicycle or motorcycle with high handlebars. **5** (in *pl.*) *Brit. sl.* teeth. **6** *US sl.* a machine-gun.

choppy /ˈtʃɒpɪ/ *adj.* (**choppier, choppiest**) (of the sea, the weather, etc.) fairly rough. □ **choppily** *adv.* **choppiness** *n.* [CHOP¹ + -Y¹]

chopstick /ˈtʃɒpstɪk/ *n.* each of a pair of small thin sticks of wood or ivory etc., held both in one hand as eating utensils by the Chinese, Japanese, etc. [pidgin Engl. f. *chop* = quick + STICK¹ equivalent of Cantonese *k'wâi-tsze* nimble ones]

chopsuey /tʃɒpˈsuːɪ/ *n.* (*pl.* **-eys**) a Chinese-style dish of meat stewed and fried with bean sprouts, bamboo shoots, onions, and served with rice. [Cantonese *shap sui* mixed bits]

choral /ˈkɔːrəl/ *adj.* of, for, or sung by a choir or chorus. □ **choral ode** a song of the chorus in a Greek drama. **choral society** a group which meets regularly to sing choral music. □ **chorally** *adv.* [med.L *choralis* f. L *chorus*: see CHORUS]

chorale /kɔːˈrɑːl/ *n.* (also **choral**) **1** a stately and simple hymn tune; a harmonized version of this. **2** esp. *US* a choir or choral society. [G *Choral(gesang)* f. med.L *cantus choralis*]

chord¹ /kɔːd/ *n. Mus.* a group of (usu. three or more) notes sounded together, as a basis of harmony. □ **chordal** *adj.* [orig. *cord* f. ACCORD: later confused with CHORD²]

chord² /kɔːd/ *n.* **1** *Math. & Aeron.* etc. a straight line joining the ends of an arc, the wings of an aeroplane, etc. **2** *Anat.* = CORD. **3** *poet.* the string of a harp etc. **4** *Engin.* one of the two principal members, usu. horizontal, of a truss. □ **strike a chord 1** recall something to a person's memory. **2** elicit sympathy. **touch the right chord** appeal skilfully to the emotions. □ **chordal** *adj.* [16th-c. refashioning of CORD after L *chorda*]

chordate /ˈkɔːdeɪt/ *n. & adj. Zool.* ● *n.* an animal of the phylum Chordata, possessing a notochord at some stage during its development. The vertebrates constitute the bulk of the phylum, but it also includes the tunicates and lancelets. ● *adj.* of or relating to the chordates. [mod.L *chordata* f. L *chorda* CHORD² after *Vertebrata* etc.]

chore /tʃɔː(r)/ *n.* a tedious or routine task, esp. domestic. [orig. dial. & US form of CHAR²]

chorea /kɒˈrɪə/ *n. Med.* a disorder characterized by jerky involuntary movements affecting esp. the shoulders, hips, and face. □ **Huntington's chorea** chorea accompanied by a progressive dementia. **Sydenham's chorea** chorea, esp. in children, as one of the manifestations of rheumatic fever (also called *St Vitus's dance*). [L f. Gk *khoreia* (as CHORUS)]

choreograph /ˈkɒrɪəˌɡrɑːf/ *v.tr.* compose the choreography for (a ballet etc.). □ **choreographer** /ˌkɒrɪˈɒɡrəfə(r)/ *n.* [back-form. f. CHOREOGRAPHY]

choreography /ˌkɒrɪˈɒɡrəfɪ/ n. **1** the design or arrangement of a ballet or other staged dance. **2** the sequence of steps and movements in dance. **3** the written notation for this. □ **choreographic** /ˌkɒrɪəˈɡræfɪk/ adj. **choreographically** adv. [Gk khoreia dance + -GRAPHY]

choreology /ˌkɒrɪˈɒlədʒɪ/ n. the study and description of the movements of dancing. □ **choreologist** n.

choriambus /ˌkɒrɪˈæmbəs/ n. (pl. **choriambi** /-baɪ/) Prosody a metrical foot consisting of two short (unstressed) syllables between two long (stressed) ones. □ **choriambic** adj. [LL Gk khoriambos f. khoreios of the dance + IAMBUS]

choric /ˈkɒrɪk/ adj. of, like, or for a chorus in drama or recitation. [LL choricus f. Gk khorikos (as CHORUS)]

chorine /ˈkɔːriːn/ n. US a chorus girl. [CHORUS + -INE³]

chorion /ˈkɔːrɪən/ n. Anat. the outermost membrane surrounding an embryo of a reptile, bird, or mammal. □ **chorionic** /ˌkɔːrɪˈɒnɪk/ adj. [Gk khorion]

chorister /ˈkɒrɪstə(r)/ n. **1** a member of a choir, esp. a choirboy or choirgirl. **2** US the leader of a church choir. [ME, ult. f. OF cueriste f. quer CHOIR]

chorography /kəˈrɒɡrəfɪ/ n. the systematic description of regions or districts. □ **chorographer** n. **chorographic** /ˌkɒrəˈɡræfɪk/ adj. [F chorographie or L f. Gk khōrographia f. khōra region]

choroid /ˈkɔːrɔɪd/ adj. & n. ● adj. like a chorion in shape or vascularity. ● n. (in full **choroid coat** or **membrane**) a layer of the eyeball between the retina and the sclera. [Gk khoroeidēs for khorioeidēs: see CHORION]

chortle /ˈtʃɔːt(ə)l/ v. & n. ● v.intr. chuckle gleefully. ● n. a gleeful chuckle. [portmanteau word coined by Lewis Carroll, prob. f. CHUCKLE + SNORT]

chorus /ˈkɔːrəs/ n. & v. ● n. (pl. **choruses**) **1** a group (esp. a large one) of singers; a choir. **2** a piece of music composed for a choir. **3** the refrain or the main part of a popular song, in which a chorus participates. **4** any simultaneous utterance by many persons etc. (a chorus of disapproval followed). **5** a group of singers and dancers performing in concert in a musical comedy, opera, etc. **6** Gk Antiq. **a** in Greek tragedy, a group of performers who comment together in voice and movement on the main action. **b** an utterance of the chorus. **7** esp. in Elizabethan drama, a character who speaks the prologue and other linking parts of the play. **8** the part spoken by this character. ● v.tr. & intr. (**chorused**, **chorusing**) (of a group) speak or utter simultaneously. □ **chorus girl** a young woman who sings or dances in the chorus of a musical comedy etc. **in chorus** (uttered) together; in unison. [L f. Gk khoros]

chose past of CHOOSE.

chosen past part. of CHOOSE.

Chou var. of ZHOU.

Chou En-lai see ZHOU ENLAI.

chough /tʃʌf/ n. a small black crow of the genus Pyrrhocorax, with a curved pointed bill; esp. P. pyrrhocorax, with a red bill and legs. [ME, prob. orig. imit.]

choux pastry /ʃuː/ n. very light pastry enriched with eggs. [F, pl. of chou cabbage, rosette]

chow /tʃaʊ/ n. **1** sl. food. **2** offens. a Chinese. **3** a Chinese breed of dog resembling a Pomeranian, with long hair and a bluish-black tongue. [shortened f. CHOW-CHOW]

chow-chow /ˈtʃaʊtʃaʊ/ n. **1** = CHOW. **2** a Chinese preserve of ginger, orange-peel, etc., in syrup. **3** a mixed vegetable pickle. [pidgin Engl.]

chowder /ˈtʃaʊdə(r)/ n. N. Amer. a soup or stew usu. of fresh fish, clams, or corn with bacon, onions, etc. [perh. F chaudière pot: see CAULDRON]

chow mein /tʃaʊ ˈmeɪn/ n. a Chinese-style dish of fried noodles with shredded meat or shrimps etc. and vegetables. [Chin. chao mian fried flour]

Chr. abbr. (in the Bible) Chronicles.

chrestomathy /krɛˈstɒməθɪ/ n. (pl. **-ies**) a selection of passages used esp. to help in learning a language. [F chrestomathie or Gk khrēstomatheia f. khrēstos useful + -matheia learning]

Chrétien /ˈkreɪtjɛn/, (Joseph-Jacques) Jean (b.1934), Canadian Liberal statesman, Prime Minister since 1993. He became an MP in 1963 and held a series of ministerial posts in the Pearson and Trudeau administrations before serving briefly as deputy Prime Minister in 1984. As leader of the Liberal Party from 1990, he led the party to victory in the 1993 elections. A Quebecker, Chrétien has always been committed to keeping Quebec within the Canadian federation.

Chrétien de Troyes /də ˈtrwʌ/ (12th century), French poet. The author of courtly romances, including some of the earliest on Arthurian themes, he is also thought to have written a romance about the legendary knight Tristram. Of his four extant volumes of romances, Lancelot (c.1177–81) is the most famous, while his unfinished Perceval (1181–90) contains the first reference in literature to the Holy Grail.

chrism /ˈkrɪz(ə)m/ n. a consecrated oil or unguent used esp. for anointing in Catholic and Greek Orthodox rites. [OE crisma f. eccl.L f. Gk khrisma anointing]

chrisom /ˈkrɪz(ə)m/ n. hist. **1** = CHRISM. **2** (in full **chrisom-cloth**) a white robe put on a child at baptism, and used as its shroud if it died within the month. [ME, alt. of CHRISM (through pronunc. with two syllables)]

Christ /kraɪst/ n. & int. ● n. **1** the title, also now treated as a name, given to Jesus of Nazareth, believed by Christians to have fulfilled the Old Testament prophecies of a coming Messiah. **2** the Messiah as prophesied in the Old Testament. **3** an image or picture of Jesus. ● int. sl. expressing surprise, anger, etc. □ **Christhood** n. **Christlike** adj. **Christly** adj. [OE Crīst f. L Christus f. Gk khristos anointed one f. khriō anoint: transl. Heb. māšīaḥ MESSIAH]

Christadelphian /ˌkrɪstəˈdɛlfɪən/ n. a member of a Christian sect founded in America in 1848, claiming to return to the beliefs and practices of the earliest disciples. The core of the Christadelphian faith is that Christ will return in power to set up a worldwide theocracy beginning at Jerusalem, and that belief in this is necessary for salvation. [CHRIST + Gk adelphos brother]

Christchurch /ˈkraɪsttʃɜːtʃ/ a city on South Island, New Zealand; pop. (1990) 303,400. The city was founded in 1850 by English Anglican colonists and named after the university college of Christ Church in Oxford.

christen /ˈkrɪs(ə)n/ v.tr. **1** give a Christian name to at baptism as a sign of admission to a Christian Church. **2** give a name to (a thing), esp. formally or with a ceremony. **3** colloq. use for the first time. □ **christener** n. **christening** n. [OE crīstnian make Christian]

Christendom /ˈkrɪs(ə)ndəm/ n. Christians worldwide, regarded as a collective body. [OE crīstendōm f. cristen CHRISTIAN + -DOM]

Christian /ˈkrɪstɪən, ˈkrɪstʃən/ adj. & n. ● adj. **1** of Christ's teaching or religion. **2** believing in or following the religion of Jesus Christ. **3** showing the qualities associated with Christ's teaching. **4** colloq. (of a person) kind, fair, decent. ● n. **1** a person who has received Christian baptism. **b** an adherent of Christ's teaching. **2** a person exhibiting Christian qualities. □ **Christian era** the era reckoned from the traditional date of Christ's birth. **Christian name** a forename, esp. as given at baptism. □ **Christianly** adv. **Christianize** v.tr. & intr. (also **-ise**). **Christianization** /ˌkrɪstɪənaɪˈzeɪʃ(ə)n, ˌkrɪstʃə-/ n. [Christianus f. Christus CHRIST]

Christian Aid a charity supported by most of the Christian Churches in the UK and operating chiefly in developing countries, where it works for disaster relief and supports development projects. It was founded in 1945.

Christiania /ˌkrɪstɪˈɑːnɪə/ (also **Kristiania**) the former name (1624–1924) for OSLO.

Christianity /ˌkrɪstɪˈænɪtɪ/ n. **1** the Christian religion; its beliefs and practices. (See note below.) **2** being a Christian; Christian quality or character. **3** = CHRISTENDOM. [ME cristianite f. OF crestienté f. crestien CHRISTIAN]

● Christianity is today the world's most widespread religion, with more than a billion members. It is divided into many denominations and sects, the main divisions being the Roman Catholic, Protestant, and Eastern Orthodox Churches. Christians in general believe in one God in three Persons (the Father, the Son, and the Holy Spirit) and that Jesus is the son of God who rose from the dead after being crucified; a Christian hopes to attain eternal life through faith in Jesus Christ and should try to live by his teachings. At first Christians were members of a Jewish sect who believed that Jesus of Nazareth was the Messiah (or 'Christ', = anointed one). Largely owing to the missionary efforts of the Apostle St Paul, it quickly became an independent, mainly Gentile, organization. In the early centuries Christians experienced intermittent persecution in the Roman Empire, though there was no clear legal basis for this until the reign of the Emperor Decius (AD 250). By the 3rd century Christianity was widespread throughout the Roman Empire; in 313 Constantine ended persecution and in 380 Theodosius I recognized it as the state religion. There were frequent disputes between Christians, mainly over the status of Christ and the nature of the Trinity, and later over Church organization. Division between East and West, in origin largely cultural and linguistic, intensified, culminating in the Great Schism of 1054. In the West the organization of the Church, focused on the Roman papacy, was fragmented by the Reformation of the 16th century, which gave rise to the Protestant Churches. In the 20th century the ecumenical movement has sought to reconcile these ancient divisions.

Christian Science *n.* the doctrine of the Church of Christ, Scientist, a Christian sect founded in Boston in 1879 by Mary Baker Eddy. Members believe that God and the mind are the only ultimate reality, and that matter and evil have no existence; illness and sin, they believe, are illusions that can be overcome by prayer and faith. As a consequence they generally refuse medical treatment. The movement has flourished in America and spread to many other countries; it has its own daily newspaper, *The Christian Science Monitor*. □ **Christian Scientist** *n.*

Christie[1] /ˈkrɪstɪ/, Dame Agatha (1890–1976), English writer of detective fiction. Many of her novels feature the Belgian Hercule Poirot or the resourceful Miss Marple, her two most famous and successful creations. Among her best-known detective stories are *Murder on the Orient Express* (1934) and *Death on the Nile* (1937). Her novels are characterized by brisk, humorous dialogue and ingenious plots. She also wrote plays; *The Mousetrap* (1952) has had a record run of more than forty years on the London stage.

Christie[2] /ˈkrɪstɪ/, Linford (b.1960), Jamaican-born British sprinter. Having won Olympic silver medals in the 100 metres and 4 × 100 metres relay events in 1988, Christie took the individual 100-metre title in 1992. This was followed by the world championship title at this distance in 1993, and in 1994 he took the European 100-metre championship title for the third consecutive time.

Christie[3] /ˈkrɪstɪ/ *n.* (also **Christy**) (*pl.* **-ies**) *Skiing* a sudden turn in which the skis are kept parallel, used for changing direction fast or stopping short. [abbr. of *Christiania* (now Oslo) in Norway]

Christingle /ˈkrɪstɪŋ(ə)l/ *n.* a lighted candle set in an orange, symbolizing Christ as the light of the world, received at a Christingle service. [f. CHRIST; origin of second element unkn.]

Christingle service *n.* a children's Advent service, originally in the Moravian Church and recently popularized outside it, at which each participant is given a Christingle.

Christmas /ˈkrɪsməs/ *n. & int.* ● *n.* (*pl.* **Christmases**) **1** (also **Christmas Day**) the annual festival of Christ's birth, celebrated by most Christian Churches on 25 Dec. and marked especially by present-giving, family reunion, and eating and drinking. (*See note below.*) **2** the season in which this occurs; the time immediately before and after 25 Dec. ● *int. sl.* expressing surprise, dismay, etc. □ **Christmas box** a present or gratuity given at Christmas esp. to tradesmen and employees. **Christmas cake** *Brit.* a rich fruit cake usu. covered with marzipan and icing and eaten at Christmas. **Christmas card** a card sent with greetings at Christmas. The Christmas card as we know it dates from the mid-19th century. **Christmas Eve** the day or the evening before Christmas Day. **Christmas pudding** *Brit.* a rich boiled pudding eaten at Christmas, made with flour, suet, dried fruit, etc. **Christmas rose** a white-flowered winter-blooming evergreen hellebore, *Helleborus niger*. [OE *Crīstes mæsse* (MASS[2])]

■ The festival of Christ's birth has been celebrated in the Western Church on 25 December from about the end of the 4th century; in the East it was originally held on 6 January in conjunction with the Epiphany. There is no biblical or other direct evidence of the season of Christ's birth, and the date of Christmas may have been chosen to facilitate the conversion of followers of older religions, many of which held festivals around this time. Many of the customs now associated with Christmas were taken over from the Roman Saturnalia and from pre-Christian festivals celebrating the winter solstice (21 December).

Christmas Island 1 an island in the Indian Ocean 350 km (200 miles) south of Java, administered as an external territory of Australia since 1958; pop. (1991) 1,275. **2** the former name (until 1981) for KIRITIMATI.

Christmassy /ˈkrɪsməsɪ/ *adj. colloq.* characteristic of or suitable for Christmas. [CHRISTMAS + -Y[1]]

Christmas tree *n.* an evergreen (usu. spruce) or artificial tree set up with decorations at Christmas. The Christmas tree originated in Germany, where it was known from the 16th century, and spread

elsewhere from there. It became fashionable in Britain after Prince Albert introduced it into his family.

Christo- /ˈkrɪstəʊ/ *comb. form* Christ.

Christology /krɪˈstɒlədʒɪ/ *n.* the branch of theology relating to Christ.

Christopher, St /ˈkrɪstəfə(r)/ a legendary Christian martyr, adopted as the patron saint of travellers. According to tradition he was martyred in Asia Minor. He is represented as a giant who carried travellers across a river; it is said that he once carried a child whose weight bowed him down, as the child was Christ and his weight that of the world. His feast day (25 July) was dropped from the Roman calendar in 1969. [Gk, = one who bore Christ]

Christ's Hospital a boys' school founded in London in 1552 for poor children, now a public school at Horsham, Sussex. Pupils wear a distinctive uniform of long dark blue belted gowns and yellow stockings.

Christy var. of CHRISTIE[1].

chroma /ˈkrəʊmə/ *n.* purity or intensity of colour. [Gk *khrōma* colour]

chromate /ˈkrəʊmeɪt/ *n. Chem.* a salt or ester of chromic acid.

chromatic /krəˈmætɪk/ *adj.* **1** of or produced by colour; in (esp. bright) colours. **2** *Mus.* **a** of or having notes not belonging to a diatonic scale. **b** (of a scale) ascending or descending by semitones. □ **chromatic aberration** *Optics* the failure of different wavelengths of electromagnetic radiation to come to the same focus after refraction. **chromatic semitone** *Mus.* an interval between a note and its flat or sharp. □ **chromatically** *adv.* **chromaticism** /-tɪˌsɪz(ə)m/ *n.* [F *chromatique* or L *chromaticus* f. Gk *khrōmatikos* f. *khrōma -atos* colour]

chromaticity /ˌkrəʊməˈtɪsɪtɪ/ *n.* the quality of colour regarded independently of brightness.

chromatid /ˈkrəʊmətɪd/ *n. Biol.* either of two threadlike strands into which a chromosome divides longitudinally during cell division. [Gk *khrōma -atos* colour + -ID[2]]

chromatin /ˈkrəʊmətɪn/ *n. Biochem.* the material in a cell nucleus that stains with basic dyes and consists of protein, RNA, and DNA, of which eukaryotic chromosomes are composed. [G: see CHROMATID]

chromato- /ˈkrəʊmətəʊ/ *comb. form* (also **chromo-** /ˈkrəʊməʊ/) colour (*chromatopsia*). [Gk *khrōma -atos* colour]

chromatography /ˌkrəʊməˈtɒgrəfɪ/ *n. Chem.* a technique for the chemical separation of a mixture by passing it in solution or suspension through a medium in which the components move at different rates. □ **chromatographic** /-mətəʊˈgræfɪk/ *adj.* **chromatograph** /krəʊˈmætəˌgrɑːf/ *n.* [G *Chromatographie* (as CHROMATO-, -GRAPHY)]

chromatopsia /ˌkrəʊməˈtɒpsɪə/ *n. Med.* abnormally coloured vision. [CHROMATO- + Gk -*opsia* seeing]

chrome /krəʊm/ *n.* **1** chromium, esp. as plating. **2** (in full **chrome yellow**) a yellow pigment obtained from lead chromate. □ **chrome leather** leather tanned with chromium salts. **chrome steel** a hard fine-grained steel containing much chromium and used for tools etc. [F, = chromium, f. Gk *khrōma* colour]

chrome-moly /krəʊmˈmɒlɪ/ *n.* a corrosion-resistant alloy steel of high tensile strength containing chromium and molybdenum.

chromic /ˈkrəʊmɪk/ *adj. Chem.* of or containing trivalent chromium. □ **chromic acid** an acid that exists only in solution or in the form of chromate salts.

chrominance /ˈkrəʊmɪnəns/ *n. Television* the colorimetric difference between a given colour and a standard colour of equal luminance. [Gk *khrōma* colour + LUMINANCE]

chromite /ˈkrəʊmaɪt/ *n.* **1** *Mineral.* a black mineral of chromium and iron oxides, which is the principal ore of chromium. **2** *Chem.* a salt of bivalent chromium.

chromium /ˈkrəʊmɪəm/ *n.* a hard white metallic chemical element (atomic number 24; symbol **Cr**). A transition metal, chromium was identified as an element by the French chemist N. L. Vauquelin in 1798. The chief ore is chromite. Chromium is often plated on to other metals for decorative purposes and to prevent corrosion, and it is used in many alloys, notably stainless steel. Chromium compounds, many of which are brightly coloured, are used as pigments, in dyeing, and in the tanning of leather. □ **chromium steel** = chrome steel. [f. F CHROME]

chromium-plate /ˌkrəʊmɪəmˈpleɪt/ *n. & v.* ● *n.* an electrolytically deposited protective coating of chromium. ● *v.tr.* **1** coat with this. **2** (as **chromium-plated** *adj.*) pretentiously decorative.

chromo /ˈkrəʊməʊ/ *n.* (*pl.* **-os**) *Austral. sl.* a prostitute. [abbr. of CHROMOLITHOGRAPH, with ref. to her make-up]

chromo-[1] /'krəʊməʊ/ *comb. form Chem.* chromium.

chromo-[2] *comb. form var. of* CHROMATO-.

chromodynamics /ˌkrəʊmədaɪˈnæmɪks/ *n.* (in full **quantum chromodynamics**) *Physics* a theory in which the strong interaction between subatomic particles is described in terms of the exchange of gluons by quarks. [CHROMO-[2] colour (see COLOUR *n.* 11) after *thermodynamics*]

chromolithograph /ˌkrəʊməʊˈlɪθəˌɡrɑːf/ *n. & v.* ● *n.* a coloured picture printed by lithography. ● *v.tr.* print or produce by this process. □ **chromolithographer** /-ˈθɒɡrəfə(r)/ *n.* **chromolithography** *n.* **chromolithographic** /-ˌlɪθəˈɡræfɪk/ *adj.*

chromosome /'krəʊməˌsəʊm/ *n. Biol.* a rodlike or threadlike structure found in the nuclei of most cells and carrying the genetic information in the form of genes. Chromosomes, only found in eukaryotic cells, can normally be seen as separate structures only when the cell is dividing. A normal undivided chromosome contains a single DNA double helix (associated with protein) in which the genes are arranged in linear order. The number of chromosomes per cell varies between species. Humans have twenty-three pairs of which one pair determines the sex of the individual: a female carries two similar X chromosomes in each cell while a male has one X and one smaller Y chromosome. (See also DNA.) □ **chromosome map** a plan showing the relative positions of genes along the length of a chromosome. □ **chromosomal** /ˌkrəʊməˈsəʊm(ə)l/ *adj.* [G *Chromosom* (as CHROMO-[2], -SOME[3])]

chromosphere /'krəʊməˌsfɪə(r)/ *n. Astron.* the region immediately above the photosphere of the sun which, together with the corona, constitutes its outer atmosphere. Material in this region is at temperatures of 10,000–20,000°C, and is subject to magnetic forces that may occasionally propel it outwards for thousands of kilometres as solar flares. □ **chromospheric** /ˌkrəʊməˈsferɪk/ *adj.* [CHROMO-[2] + SPHERE]

Chron. *abbr.* (in the Bible) Chronicles.

chronic /'krɒnɪk/ *adj.* **1** persisting for a long time (usu. of an illness or a personal or social problem). **2** having a chronic complaint. **3** *colloq. disp.* habitual, inveterate (*a chronic liar*). **4** *Brit. colloq.* very bad; intense, severe. □ **chronic fatigue syndrome** = *myalgic encephalomyelitis*. □ **chronically** *adv.* **chronicity** /krɒˈnɪsɪti/ *n.* [F *chronique* f. L *chronicus* (in LL of disease) f. Gk *khronikos* f. *khronos* time]

chronicle /'krɒnɪk(ə)l/ *n. & v.* ● *n.* **1** a register of events in order of their occurrence. **2** a narrative, a full account. ● *v.tr.* record (events) in the order of their occurrence. □ **chronicler** *n.* [ME f. AF *cronicle* ult. f. L *chronica* f. Gk *khronika* annals: see CHRONIC]

Chronicles /'krɒnɪk(ə)lz/ either of two books of the Bible, recording the history of Israel and Judah from the Creation until the return from Exile (536 BC).

chrono- /'krɒnə, 'krəʊnə/ *comb. form* time. [Gk *khronos* time]

chronograph /'krɒnəˌɡrɑːf, 'krəʊnə-, -ˌɡræf/ *n.* **1** an instrument for recording time with extreme accuracy. **2** a stopwatch. □ **chronographic** /ˌkrɒnəˈɡræfɪk, ˌkrəʊnə-/ *adj.*

chronological /ˌkrɒnəˈlɒdʒɪk(ə)l/ *adj.* **1** (of a number of events) arranged or regarded in the order of their occurrence. **2** of or relating to chronology. □ **chronologically** *adv.*

chronology /krəˈnɒlədʒi/ *n.* (*pl.* **-ies**) **1** the study of historical records to establish the dates of past events. **2 a** the arrangement of events, dates, etc. in the order of their occurrence. **b** a table or document displaying this. □ **chronologist** *n.* **chronologize** *v.tr.* (also **-ise**). [mod.L *chronologia* (as CHRONO-, -LOGY)]

chronometer /krəˈnɒmɪtə(r)/ *n.* a time-measuring instrument, esp. one keeping accurate time in spite of movement or variations in temperature, humidity, and air pressure. The main stimulus to the development of chronometers was the need for an accurate method of determining longitude at sea. Longitude is calculated by comparing local time with Greenwich Mean Time; the former can be found by astronomical observation, but without an accurate timekeeper it is not easy to ascertain Greenwich time. By 1785 French and English clockmakers had evolved a chronometer accurate to better than one second a day, employing special balance-wheels and springs. Modern chronometers use a quartz crystal kept in oscillation at a constant frequency by electronic means. The advent of radio time signals did away with the need for expensive marine chronometers.

chronometry /krəˈnɒmɪtri/ *n.* the science of accurate time-measurement. □ **chronometric** /ˌkrɒnəˈmetrɪk, ˌkrəʊnə-/ *adj.* **chronometrical** *adj.* **chronometrically** *adv.*

chrysalis /'krɪsəlɪs/ *n.* (*pl.* **chrysalises** or **chrysalides** /krɪˈsælɪˌdiːz/) **1 a** a quiescent pupa of a butterfly or moth. **b** the hard outer case enclosing it. **2** a preparatory or transitional state. [L f. Gk *khrusallis -idos* f. *khrusos* gold]

chrysanth /krɪˈsænθ/ *n. colloq.* any of the autumn-blooming cultivated varieties of chrysanthemum. [abbr.]

chrysanthemum /krɪˈsænθəməm/ *n.* **1** a composite plant of the genus *Chrysanthemum*, e.g. the corn marigold, *C. segetum*. **2** a composite plant now of the genus *Dendranthema*, cultivated for its brightly coloured flowers. [L f. Gk *khrusanthemon* f. *khrusos* gold + *anthemon* flower]

chryselephantine /ˌkrɪselɪˈfæntaɪn/ *adj.* (of ancient Greek sculpture) overlaid with gold and ivory. [Gk *khruselephantinos* f. *khrusos* gold + *elephas* ivory]

chrysoberyl /'krɪsəˌberɪl/ *n.* a yellowish-green gem consisting of a beryllium salt. [L *chrysoberyllus* f. Gk *khrusos* gold + *bērullos* beryl]

chrysolite /'krɪsəˌlaɪt/ *n.* a precious stone, a yellowish-green or brownish variety of olivine. [ME f. OF *crisolite* f. med.L *crisolitus* f. L *chrysolithus* f. Gk *khrusolithos* f. *khrusos* gold + *lithos* stone]

chrysoprase /'krɪsəˌpreɪz/ *n.* **1** an apple-green variety of chalcedony containing nickel and used as a gem. **2** (in the New Testament) prob. a golden-green variety of beryl. [ME f. OF *crisopace* f. L *chrysopassus* var. of L *chrysoprasus* f. Gk *khrusoprasos* f. *khrusos* gold + *prason* leek]

Chrysostom, St John /'krɪsəstəm/ (*c.*347–407), Doctor of the Church, bishop of Constantinople. His name (Gk, = golden-mouthed) is a tribute to the eloquence of his preaching. As patriarch of Constantinople, he attempted to reform the corrupt state of the court, clergy, and people; this offended many, including the Empress Eudoxia (d.404), who banished him in 403. Feast day, 27 Jan.

chrysotile /'krɪsəˌtaɪl/ *n. Mineral.* a form of asbestos. [Gk *khrusos* gold + *tilos* fibre]

chthonic /'kθɒnɪk, 'θɒn-/ (also **chthonian** /'kθəʊnɪən, 'θəʊ-/) *adj.* of, relating to, or inhabiting the underworld. [Gk *khthōn* earth]

chub /tʃʌb/ *n.* a thick-bodied coarse-fleshed river fish, *Leuciscus cephalus*. [15th c.: orig. unkn.]

chubby /'tʃʌbi/ *adj.* (**chubbier, chubbiest**) plump and rounded (esp. of a person or a part of the body). □ **chubbily** *adv.* **chubbiness** *n.* [CHUB]

Chubu /'tʃuːbuː/ a mountainous region of Japan, on the island of Honshu; capital, Nagoya.

chuck[1] /tʃʌk/ *v. & n.* ● *v.tr.* **1** *colloq.* fling or throw carelessly or with indifference. **2** (often foll. by *in, up*) *colloq.* give up; reject, abandon; jilt (*chucked in my job; chucked her boyfriend*). **3** touch playfully, esp. under the chin. ● *n.* **1** a playful touch under the chin. **2** a toss. **3** (prec. by *the*) *sl.* dismissal (*he got the chuck*). □ **chucker-out** *colloq.* a person employed to expel troublesome people from a gathering etc. **chuck it** *sl.* stop, desist. **chuck off** *Austral. & NZ sl.* sneer, scoff. **chuck out** *colloq.* **1** expel (a person) from a gathering etc. **2** get rid of, discard. [16th c., perh. f. F *chuquer, choquer* to knock]

chuck[2] /tʃʌk/ *n. & v.* ● *n.* **1** a cut of beef between the neck and the ribs. **2** a device for holding a workpiece in a lathe or a tool in a drill. ● *v.tr.* fix (wood, a tool, etc.) to a chuck. [var. of CHOCK]

chuck[3] /tʃʌk/ *n. N. Amer. colloq.* food. □ **chuck-wagon 1** a provision-cart on a ranch etc. **2** a roadside eating-place. [19th c.: perh. f. CHUCK[2]]

chuckle /'tʃʌk(ə)l/ *v. & n.* ● *v.intr.* laugh quietly or inwardly. ● *n.* a quiet or suppressed laugh. □ **chuckler** *n.* [*chuck* cluck]

chucklehead /'tʃʌk(ə)lˌhed/ *n. colloq.* a stupid person. □ **chuckleheaded** *adj.* [*chuckle* clumsy, prob. rel. to CHUCK[2]]

chuddar var. of CHADOR.

chuff /tʃʌf/ *v.intr.* (of a steam engine etc.) work with a regular sharp puffing sound. [imit.]

chuffed /tʃʌft/ *adj. Brit. sl.* delighted. [dial. *chuff* pleased]

chug /tʃʌɡ/ *v. & n.* ● *v.intr.* (**chugged, chugging**) **1** emit a regular muffled explosive sound, as of an engine running slowly. **2** move with this sound. ● *n.* a chugging sound. [imit.]

Chugoku /tʃuːˈɡəʊkuː/ a region of Japan, on the island of Honshu; capital, Hiroshima.

chukar /'tʃʊkɑː(r)/ *n.* a red-legged partridge, *Alectoris chukar*, native to India. [Hindi *cakor*]

Chukchi Sea /'tʃʊktʃi/ part of the Arctic Ocean lying between North America and Asia and to the north of the Bering Strait.

chukker /'tʃʌkə(r)/ *n.* (also **chukka** /'tʃʌkə/) each of the periods of play into which a game of polo is divided. □ **chukka boot** an ankle-high leather boot as worn for polo. [Hindi *cakkar* f. Skr. *cakra* wheel]

chum[1] /tʃʌm/ n. & v. ● n. colloq. (esp. among schoolchildren) a close friend. ● v.intr. (often foll. by *with*) share rooms. □ **chum up** (often foll. by *with*) become a close friend (of). □ **chummy** adj. (**chummier, chummiest**). **chummily** adv. **chumminess** n. [17th c.: prob. short for *chamber-fellow*]

chum[2] /tʃʌm/ n. & v. US ● n. **1** refuse from fish. **2** chopped fish used as bait. ● v. **1** intr. fish using chum. **2** tr. bait (a fishing place) using chum. [19th c.: orig. unkn.]

chump /tʃʌmp/ n. **1** colloq. a foolish person. **2** Brit. the thick end, esp. of a loin of lamb or mutton (*chump chop*). **3** a short thick block of wood. **4** Brit. sl. the head. □ **off one's chump** Brit. sl. crazy. [18th c.: blend of CHUNK and LUMP[1]]

chunder /tʃʌndə(r)/ v.intr. & n. Austral. sl. vomit. [20th c.: orig. unkn.]

Chungking see CHONGQING.

Chung-shan see ZHONGSHAN.

chunk /tʃʌŋk/ n. **1** a thick solid slice or piece of something firm or hard. **2** a substantial amount or piece. [prob. var. of CHUCK[2]]

chunky /tʃʌŋkɪ/ adj. (**chunkier, chunkiest**) **1** containing or consisting of chunks. **2** short and thick; small and sturdy. **3** (of clothes) made of a thick material. □ **chunkiness** n.

Chunnel /tʃʌn(ə)l/ n. colloq. = CHANNEL TUNNEL.

chunter /tʃʌntə(r)/ v.intr. Brit. colloq. mutter, grumble. [prob. imit.]

chupatty var. of CHAPATTI.

Chuquisaca /ˌtʃuːkiːˈsaːkə/ the former name (1539–1840) for SUCRE[1].

church /tʃɜːtʃ/ n. & v. ● n. **1** a building for public (usu. Christian) worship. **2** a meeting for public worship in such a building (*go to church; met after church*). **3** (**Church**) the body of all Christians. **4** (**Church**) the clergy or clerical profession (*went into the Church*). **5** (**Church**) an organized Christian group or society of any time, country, or distinct principles of worship (*the primitive Church; Church of Scotland; High Church*). **6** (**Church**) institutionalized religion as a political or social force (*Church and State*). ● v.tr. archaic take (esp. a woman after childbirth) to church for a service of thanksgiving. □ **church school** a school founded by or associated with the Church of England. [OE *cirice, circe,* etc. f. med. Gk *kurikon* f. Gk *kuriakon (dōma)* Lord's (house) f. *kurios* Lord: cf. KIRK]

Church Army a voluntary Anglican organization of lay workers, founded in 1882 on the model of the Salvation Army, for evangelistic purposes. Its activities include welfare work.

Church Commissioners a body managing the finances of the Church of England.

churchgoer /tʃɜːtʃˌɡəʊə(r)/ n. a person who goes to church, esp. regularly. □ **churchgoing** n. & adj.

Churchill[1] /tʃɜːtʃɪl/, Caryl (b.1938), English dramatist. She is best known for the satire *Serious Money* (1986); written in rhyming couplets, it deals with 1980s speculators and the ethics of high finance. Her other plays include *Top Girls* (1982) and *Mad Forest* (1990).

Churchill[2] /tʃɜːtʃɪl/, Sir Winston (Leonard Spencer) (1874–1965), British Conservative statesman, Prime Minister 1940–5 and 1951–5. He served as Home Secretary (1910–11) under the Liberals and First Lord of the Admiralty (1911–15), but lost this post after the unsuccessful Allied attack on the Turks in the Dardanelles. He was only to regain it in 1939 under Neville Chamberlain, whom he replaced as Prime Minister in May 1940. Serving as war leader of a coalition government until 1945, Churchill demonstrated rare qualities of leadership and outstanding gifts as an orator. Part of his contribution to victory was to forge and maintain the Alliance, especially with the US, which defeated the Axis Powers. After the victory he was defeated in the general election of 1945; elected Prime Minister for a second term in 1951, he retired from the premiership in 1955, but remained an MP until 1964. His writings include *The Second World War* (1948–53) and *A History of the English-Speaking Peoples* (1956–8); he was awarded the Nobel Prize for literature in 1953.

churchman /tʃɜːtʃmən/ n. (pl. **-men**) **1** a member of the clergy or of a church. **2** a supporter of the church.

Church of England the English branch of the Western Christian Church, which rejects the pope's authority and has the monarch as its titular head and nominator of its bishops and archbishops. A synod held at Whitby in 664 resolved the earlier conflict between the indigenous Celtic Church, dominated by missionaries from Ireland and Scotland, and the Roman customs, introduced by St Augustine's

mission (597), in favour of the latter. The English Church remained part of the Western Catholic Church until the Reformation of the 16th century: against a background of religious dissatisfaction, growing national self-awareness, and Continental Protestantism, Henry VIII failed to obtain a divorce from Catherine of Aragon and subsequently repudiated papal supremacy, bringing the English Church under the control of the Crown. The Church achieved its definitive form under Elizabeth I, when the Book of Common Prayer became its service-book and the Thirty-nine Articles its statement of doctrine. A chief aim of the Church of England has been, while rejecting the claims of Rome, to maintain its continuity with earlier tradition.

Church of Scotland the national (Presbyterian) Christian Church in Scotland. At the Reformation the Calvinist party in Scotland, under John Knox, reformed the established Church and organized it on Presbyterian lines (1560). During the next century there were repeated attempts by the Stuart monarchs to impose episcopalianism, and the Church of Scotland was not finally established as Presbyterian until 1690. Like many Protestant Churches it has had a complicated history of schism and reunification.

Church Slavonic n. (also **Old Slavonic** or **Old Church Slavonic**) the earliest written Slavonic language, surviving as a liturgical language in the Orthodox Church. It was originally a South Slavonic dialect from the region of Macedonia, used in the 9th century by St Cyril and his brother St Methodius for their missionary purposes in the Slav countries of Moravia and Pannonia. Throughout the Middle Ages it was the language of culture for many Orthodox peoples of eastern Europe, playing a role similar to that of Latin in the West. Two different alphabets were used, Glagolitic and Cyrillic.

Churchward /tʃɜːtʃwəd/, George Jackson (1857–1933), English railway engineer. He spent most of his working life at the Swindon works of the Great Western Railway, rising to take effective control of rolling stock from 1899 to 1921. Churchward made Swindon the most modern locomotive works in the country, and is particularly remembered for the standard four-cylinder 4-6-0 locomotives that he introduced in 1903–11. These formed the basis of many subsequent designs. Churchward also became the first mayor of Swindon; he was killed while crossing a railway line.

churchwarden /tʃɜːtʃˈwɔːd(ə)n/ n. **1** either of two elected lay representatives of a parish, assisting with routine administration. **2** a long-stemmed clay pipe.

churchwoman /tʃɜːtʃˌwʊmən/ n. (pl. **-women**) **1** a woman member of the clergy or of a church. **2** a woman supporter of the Church.

churchy /tʃɜːtʃɪ/ adj. **1** obtrusively or intolerantly devoted to the Church or opposed ·to religious dissent. **2** like a church. □ **churchiness** n.

churchyard /tʃɜːtʃjɑːd/ n. the enclosed ground around a church, esp. as used for burials.

churinga /tʃʌˈrɪŋɡə/ n. (pl. same or **churingas**) a sacred object, esp. an amulet, among the Australian Aboriginals. [Aboriginal]

churl /tʃɜːl/ n. **1** an ill-bred person. **2** archaic a peasant; a person of low birth. **3** archaic a surly or mean person. [OE *ceorl* f. a WG root, = man]

churlish /tʃɜːlɪʃ/ adj. surly; mean. □ **churlishly** adv. **churlishness** n. [OE *cierlisc, ceorlisc* f. *ceorl* CHURL]

churn /tʃɜːn/ n. & v. ● n. **1** Brit. a large milk-can. **2** a machine for making butter by agitating milk or cream. ● v. **1** tr. agitate (milk or cream) in a churn. **2** tr. produce (butter) in this way. **3** tr. (usu. foll. by *up*) cause distress to; upset, agitate. **4** intr. (of a liquid) seethe, foam violently (*the churning sea*). **5** tr. agitate or move (liquid) vigorously, causing it to foam. □ **churn out** produce routinely or mechanically, esp. in large quantities. [OE *cyrin* f. Gmc]

churr var. of CHIRR.

churrascaria /tʃʊˌræskəˈrɪə/ n. a restaurant specializing in churrasco. [Sp.]

churrasco /tʃʊˈræskəʊ/ n. a South American dish consisting of steak barbecued over a wood or charcoal fire. [Sp., prob. f. Sp. dial. *churrascar* burn]

Churrigueresque /ˌtʃʊərɪɡəˈresk/ adj. Archit. of the lavishly ornamented late Spanish baroque style. Churrigueresque architecture takes its name from the Spanish architect José de Churriguera (1650–1725).

chute[1] /ʃuːt/ n. **1** a sloping channel or slide, with or without water, for conveying things to a lower level. **2** a slide into a swimming-pool. **3** US a cataract or cascade of water; a steep descent in a river bed producing a swift current. [F *chute* fall (of water etc.), f. OF *cheoite* fem. past part. of *cheoir* fall f. L *cadere*; in some senses = SHOOT]

chute[2] /ʃuːt/ *n. colloq.* parachute. □ **chutist** *n.* [abbr.]

chutney /ˈtʃʌtnɪ/ *n.* (*pl.* **-eys**) a pungent orig. Indian condiment made of fruits or vegetables, vinegar, spices, sugar, etc. [Hindi *caṭnī*]

chutzpah /ˈxʊtspə/ *n. sl.* shameless audacity; cheek. [Yiddish]

Chuvashia /tʃuːˈvɑːʃɪə/ an autonomous republic in European Russia, east of Nizhni Novgorod; pop. (1990) 1,340,000; capital, Cheboksary.

chyack var. of CHIACK.

chyle /kaɪl/ *n. Biochem.* lymph which contains products of food digestion, esp. fats, and thus appears milky. □ **chylous** *adj.* [LL *chylus* f. Gk *khulos* juice]

chyme /kaɪm/ *n. Biochem.* the acidic semisolid and partly digested food produced in the stomach by gastric activity. □ **chymous** *adj.* [LL *chymus* f. Gk *khumos* juice]

chypre /ˈʃiːprə/ *n.* a heavy perfume made from sandalwood. [F, = CYPRUS, perh. where it was first made]

CI *abbr.* **1** Channel Islands. **2** *hist.* Order of the Crown of India.

Ci *abbr.* curie.

CIA *abbr.* (in the US) Central Intelligence Agency.

ciabatta /tʃəˈbɑːtə/ *n.* (*pl.* **ciabattas**, **ciabatte** /-tɪ/) a loaf of moist aerated Italian bread made with olive oil. [It., lit. 'slipper' (f. its shape)]

ciao /tʃaʊ/ *int. colloq.* **1** goodbye. **2** hello. [It.]

ciborium /sɪˈbɔːrɪəm/ *n.* (*pl.* **ciboria** /-rɪə/) **1** *Eccl.* a vessel with an arched cover used to hold the Eucharist. **2** *Archit.* **a** a canopy. **b** a shrine with a canopy. [med.L f. Gk *kibōrion* seed-vessel of the water-lily, a cup made from it]

cicada /sɪˈkɑːdə, -ˈkeɪdə/ *n.* (also **cicala** /-ˈkɑːlə/) a large homopterous bug of the family Cicadidae, with large transparent wings held rooflike over the back. The males make a loud shrill chirping sound at night. [L *cicada*, It. f. L *cicala*, It. *cigala*]

cicatrice /ˈsɪkətrɪs/ *n.* (also **cicatrix** /-trɪks/) (*pl.* **cicatrices** /ˌsɪkəˈtraɪsiːz/) **1** any mark left by a healed wound; a scar. **2** *Bot.* **a** a mark on a stem etc. left when a leaf or other part becomes detached. **b** a scar on the bark of a tree. □ **cicatricial** /ˌsɪkəˈtrɪʃ(ə)l/ *adj.* [ME f. OF *cicatrice* or L *cicatrix -icis*]

cicatrize /ˈsɪkəˌtraɪz/ *v.* (also **-ise**) **1** *tr.* heal (a wound) by scar formation. **2** *intr.* (of a wound) heal by scar formation. □ **cicatrization** /ˌsɪkətraɪˈzeɪʃ(ə)n/ *n.* [F *cicatriser*: see CICATRICE]

cicely /ˈsɪsəlɪ/ *n.* (*pl.* **-ies**) an umbelliferous plant, esp. sweet cicely. [app. f. L *seselis* f. Gk, assim. to the woman's forename]

Cicero /ˈsɪsəˌrəʊ/, Marcus Tullius (106–43 BC), Roman statesman, orator, and writer. A supporter of Pompey against Julius Caesar, in the *Philippics* (43 BC) he attacked Mark Antony, who had him put to death. As an orator and writer, Cicero established a model for Latin prose; his surviving works include many speeches, treatises on rhetoric, philosophical works, and books of letters, recording his personal and political interests and activities.

cicerone /ˌtʃɪtʃəˈrəʊnɪ, ˌsɪsə-/ *n.* (*pl.* **ciceroni** *pronunc.* same) a guide who gives information about antiquities, places of interest, etc. to sightseers. [It., f. CICERO, app. with ref. to his learning or eloquence]

Ciceronian /ˌsɪsəˈrəʊnɪən/ *adj.* (of language) eloquent, classical, or rhythmical, in the style of Cicero.

cichlid /ˈsɪklɪd/ *n. Zool.* a tropical freshwater fish of the family Cichlidae, many of which are kept in aquariums. [mod.L *Cichlidae* f. Gk *kikhlē* a kind of fish]

CID *abbr.* (in the UK) Criminal Investigation Department.

Cid, El /sɪd/ (also **the Cid**) (born Rodrigo Díaz de Vivar), Count of Bivar (c.1043–99), Spanish soldier. A champion of Christianity against the Moors, he began his long fighting career in 1065; in 1094 he captured Valencia, which he went on to rule. He is immortalized in the Spanish *Poema del Cid* (12th century) and in Corneille's play *Le Cid* (1637).

-cide /saɪd/ *suffix* forming nouns meaning: **1** a person or substance that kills (*regicide*; *insecticide*). **2** the killing of (*infanticide*; *suicide*). [F f. L *-cida* (sense 1), *-cidium* (sense 2), *caedere* kill]

cider /ˈsaɪdə(r)/ *n.* (also **cyder**) **1** *Brit.* an alcoholic drink made from fermented apple juice. **2** *N. Amer.* an unfermented drink made from apple juice. □ **cider-press** a press for crushing apples to make cider. [ME f. OF *sidre*, ult. f. Heb. *šēkār* strong drink]

ci-devant /ˌsiːdəˈvɒn/ *adj. & adv.* that has been (with person's earlier name or status); former or formerly. [F, = heretofore]

CIE *abbr. hist.* Companion (of the Order) of the Indian Empire.

c.i.f. *abbr.* cost, insurance, freight (as being included in a price).

cig /sɪg/ *n. colloq.* cigarette, cigar. [abbr.]

cigala /sɪˈgɑːlə/ *n.* = CICADA. [F *cigale*, It. & Prov. *cigala* f. L *cicada*]

cigar /sɪˈgɑː(r)/ *n.* a cylinder of tobacco rolled in tobacco leaves for smoking. [F *cigare* or Sp. *cigarro*]

cigarette /ˌsɪgəˈret/ *n.* (*US* also **cigaret**) **1** a thin cylinder of finely cut tobacco rolled in paper for smoking. **2** a similar cylinder containing a narcotic, herbs, or a medicated substance. □ **cigarette card** a small picture card of a kind formerly included in a packet of cigarettes. **cigarette-end** the unsmoked remainder of a cigarette. [F, dimin. of *cigare* CIGAR]

cigarillo /ˌsɪgəˈrɪləʊ/ *n.* (*pl.* **-os**) a small cigar. [Sp., dimin. of *cigarro* CIGAR]

ciggy /ˈsɪgɪ/ *n.* (*pl.* **-ies**) *colloq.* cigarette. [abbr.]

CIGS *abbr. hist.* Chief of the Imperial General Staff.

cilia *pl.* of CILIUM.

ciliary /ˈsɪlɪərɪ/ *adj.* **1** *Biol.* of or relating to cilia. **2** *Anat.* **a** of or relating to the eyelids or eyelashes. **b** of or denoting the part of the eye connecting the iris to the choroid, and the muscle in it which controls the shape of the lens.

ciliate /ˈsɪlɪˌeɪt, -lɪət/ *adj. & n.* ● *adj.* having cilia. ● *n. Zool.* a protozoan of the phylum Ciliophora, bearing cilia.

cilice /ˈsɪlɪs/ *n.* **1** haircloth. **2** a garment of this. [F f. L *cilicium* f. Gk *kilikion* f. *Kilikia* CILICIA, orig. made of Cilician goat's hair]

Cilicia /sɪˈlɪʃə/ an ancient region on the coast of SE Asia Minor, corresponding to the present-day province of Adana, Turkey.

Cilician /sɪˈlɪʃ(ə)n/ *adj.* of or relating to Cilicia. □ **Cilician Gates** a mountain pass in the Taurus Mountains of southern Turkey, historically linking Anatolia with the Mediterranean coast.

cilium /ˈsɪlɪəm/ *n.* (*pl.* **cilia** /-lɪə/) **1** *Anat. & Biol.* a short minute hairlike vibrating structure on the surface of some cells, usu. found in large numbers that (in stationary cells) create currents in the surrounding fluid or (in some small organisms) provide propulsion. **2** an eyelash. □ **ciliated** /-lɪˌeɪtɪd/ *adj.* **ciliation** /ˌsɪlɪˈeɪʃ(ə)n/ *n.* [L, = eyelash]

cill var. of SILL.

cimbalom /ˈsɪmbələm/ *n.* a dulcimer. [Hungarian f. It. *cembalo*]

Cimmerian /sɪˈmɪərɪən/ *n. & adj.* ● *n.* **1** a member of an ancient nomadic people, the earliest known inhabitants of the Crimea, who overran Asia Minor in the 7th century BC. They conquered Phrygia c.676 BC and terrorized Ionia, but were gradually destroyed by epidemics and in wars with Lydia and Assyria. **2** *Gk Mythol.* a member of a people who lived in perpetual mist and darkness, near the land of the dead. ● *adj.* of or relating to the Cimmerians. [L *Cimmerius* f. Gk *Kimmerios*, Assyr. *Gimirri* (the 'Gomer' of Gen. 10:2, Ezek. 38:6)]

C.-in-C. *abbr.* commander-in-chief.

cinch /sɪntʃ/ *n. & v.* ● *n.* **1** *colloq.* **a** a sure thing; a certainty. **b** an easy task. **2** a firm hold. **3** esp. *US* a girth for a saddle or pack, used in Mexico and the western US. ● *v.tr.* **1 a** tighten as with a cinch (*cinched at the waist with a belt*). **b** secure a grip on. **2** *sl.* make certain of. **3** fix (a saddle etc) securely by means of a girth; put a cinch on (a horse). [Sp. *cincha*]

cinchona /sɪŋˈkəʊnə/ *n.* **1 a** an evergreen tree or shrub of the South American genus *Cinchona*, of the madder family, having fragrant flowers. **b** (in full **cinchona bark**) this bark, which contains quinine. **2** any drug obtained from this bark, formerly used as a tonic and to stimulate the appetite. □ **cinchonic** /-ˈkɒnɪk/ *adj.* **cinchonine** /ˈsɪŋkəˌniːn/ *n.* [mod.L f. Countess of Chinchón (d.1641), introducer of drug into Spain]

Cincinnati /ˌsɪnsɪˈnætɪ/ an industrial city in Ohio, on the Ohio River; pop. (1990) 364,000.

cincture /ˈsɪŋktʃə(r)/ *n.* **1** *literary* a girdle, belt, or border. **2** *Archit.* a ring at either end of a column-shaft. [L *cinctura* f. *cingere* cinct- gird]

cinder /ˈsɪndə(r)/ *n.* **1** the residue of coal or wood etc. that has stopped giving off flames but still has combustible matter in it. **2** slag. **3** (in *pl.*) ashes. □ **burnt to a cinder** made useless by burning. □ **cindery** *adj.* [OE *sinder* f. Gmc, assim. to the unconnected F *cendre* and L *cinis* ashes: cf. SINTER]

Cinderella /ˌsɪndəˈrelə/ a character in many traditional European fairy tales (including one by Charles Perrault), a girl neglected or exploited as a servant by her family but enabled by a fairy godmother to attend a royal ball. She meets and captivates Prince Charming but has to flee at midnight, leaving the prince to identify her by the glass slipper which she leaves behind.

cine- /'sɪnɪ/ *comb. form* cinematographic (*cine-camera; cinephotography*). [abbr.]

cineaste /'sɪnɪˌæst/ *n.* (also **cineast**) a cinema enthusiast. [F *cinéaste* (as CINE-): cf. ENTHUSIAST]

cinema /'sɪnɪmə, -ˌmɑː/ *n.* **1** *Brit.* a theatre where motion-picture films are shown. **2 a** films collectively. **b** the production of films as an art or industry; cinematography. (*See note below.*) □ **cinema organ** *Mus.* a kind of organ with extra stops and special effects. [F *cinéma*: see CINEMATOGRAPH]

▪ Photographic motion pictures projected on to a screen became available to the general public from about 1895, in the form of short silent comedies, dramas, or documentaries shown either as sideshows at fairgrounds or as items in music-hall programmes. From 1900 to 1914 the film industry was international, led by France, Italy, Britain, and the US, and films increased in length from a few minutes to two hours. During the First World War the demand for films grew at a time when European producers were least able to meet it. The American industry took advantage of this and the US became the foremost film-making country, Hollywood in California being the chief centre of production. In the 1920s the US consolidated this position, developing the star-system and film publicity simultaneously, and the cinema became the people's entertainment. Sound films evolved in the late 1920s and (because of the language barrier) forced national film industries to develop independently. In 1932 a three-colour process known as Technicolor was developed. After the Second World War the increasing popularity of television seriously threatened the prosperity of the film industry, which fought back with the introduction of new techniques, such as wide screens and stereo sound, and ever more expensive and technologically sophisticated productions. The growth in the 1970s of the video industry, rather than killing off cinema as predicted by many, appears to have had the opposite effect of stimulating greater interest in films.

CinemaScope /'sɪnɪməˌskəʊp/ *n. propr.* a cinematographic process in which special lenses are used to compress a wide image into a standard frame and then expand it again during projection. The resulting image is almost two and a half times as wide as it is high. The process was copyrighted by Twentieth Century-Fox in 1952, and similar processes were later adopted by other studios.

cinematheque /ˌsɪnɪmə'tek/ *n.* **1** a film library or archive. **2** a small cinema. [F *cinémathèque*, f. as CINEMA, after *bibliothèque* library]

cinematic /ˌsɪnɪ'mætɪk/ *adj.* **1** having the qualities characteristic of the cinema. **2** of or relating to the cinema. □ **cinematically** *adv.*

cinematograph /ˌsɪnɪ'mætəˌgrɑːf/ (also **kinematograph** /ˌkɪnɪ-/) *n.* an apparatus for showing motion-picture films. [F *cinématographe* f. Gk *kinēma -atos* movement f. *kineō* move]

cinematography /ˌsɪnɪmə'tɒgrəfɪ/ *n.* the art of making motion-picture films. □ **cinematographer** *n.* **cinematographic** /-ˌmætə'græfɪk/ *adj.* **cinematographically** *adv.*

cinéma-vérité /ˌsɪnɛmɑː'verɪˌteɪ, -ˌverɪ'teɪ/ *n.* the art or process of making realistic (esp. documentary) films which avoid artificiality and overt artistic effect. This style of film-making developed in France and the US in the late 1950s and early 1960s, having been made possible by the production for television of hand-held 16 mm cameras and portable sound equipment. In order to achieve immediacy and truth there was no script and no director, and the film crew consisted only of a cameraman and a sound-man. [F, = cinema truth]

cineraria /ˌsɪnə'reərɪə/ *n.* a composite plant, *Pericallis cruenta*, having bright flowers and ash-coloured down on its leaves. It is native to the Canary Islands, but various hybrids are cultivated. [mod.L, fem. of L *cinerarius* of ashes f. *cinis -eris* ashes, from the ash-coloured down on the leaves]

cinerarium /ˌsɪnə'reərɪəm/ *n.* (*pl.* **cinerariums**) a place where a cinerary urn is deposited. [LL, neut. of *cinerarius*: see CINERARIA]

cinerary /'sɪnərərɪ/ *adj.* of ashes. □ **cinerary urn** an urn for holding the ashes after cremation. [L *cinerarius*: see CINERARIA]

cinereous /sɪ'nɪərɪəs/ *adj.* (esp. of a bird or plumage) ash-grey. [L *cinereus* f. *cinis -eris* ashes]

ciné-vérité /ˌsɪneɪ'verɪˌteɪ, -ˌverɪ'teɪ/ *n. Cinematog.* = CINÉMA-VÉRITÉ.

Cingalese /ˌsɪŋɡə'liːz/ *adj. & n.* (*pl.* same) *archaic* Sinhalese. [F *cing(h)alais*: see SINHALESE]

cingulum /'sɪŋɡjʊləm/ *n.* (*pl.* **cingula** /-lə/) *Anat.* a girdle, belt, or analogous structure, esp. a ridge surrounding the base of the crown of a tooth. [L, = belt]

cinnabar /'sɪnəˌbɑː(r)/ *n.* **1** a bright red mineral form of mercuric sulphide from which mercury is obtained. **2** vermilion. **3** a day-flying moth, *Tyria jacobaeae*, with bright red wing markings. [ME f. L *cinnabaris* f. Gk *kinnabari*, of oriental orig.]

cinnamon /'sɪnəmən/ *n.* **1** an aromatic spice from the peeled, dried, and rolled bark of a SE Asian tree. **2** a tree of the genus *Cinnamomum*; esp. *C. zeylanicum*, which yields the spice. **3** yellowish-brown. [ME f. OF *cinnamome* f. L *cinnamomum* f. Gk *kinnamōmon*, and L *cinnamon* f. Gk *kinnamon*, f. Semitic (cf. Heb. *ḳinnāmôn*)]

cinque /sɪŋk/ *n.* (also **cinq**) the five on dice. [ME f. OF *cinc, cink*, f. L *quinque* five]

cinquecento /ˌtʃɪŋkwɪ'tʃentəʊ/ *n.* the style of Italian art and literature of the 16th century, with a reversion to classical forms. □ **cinquecentist** *n.* [It., = 500, used with ref. to the years 1500–99]

cinquefoil /'sɪŋkfɔɪl/ *n.* **1** a herbaceous rosaceous plant of the genus *Potentilla*, with compound leaves of five leaflets. (See also POTENTILLA.) **2** *Archit.* a five-cusped ornament in a circle or arch. [ME f. L *quinquefolium* f. *quinque* five + *folium* leaf]

Cinque Ports /sɪŋk/ *n.pl.* a group of medieval ports in SE England (originally five: Dover, Hastings, Hythe, Romney, and Sandwich; Rye and Winchelsea were added later), formerly allowed various trading privileges in exchange for providing the bulk of England's navy. The origins of the association are unknown, but it existed long before its first real charter was granted by Edward I. Most of the old privileges were abolished in the 19th century and the Wardenship of the Cinque Ports is now a purely honorary post. [ME f. OF *cink porz*, L *quinque portus* five ports]

Cintra see SINTRA.

CIO see CONGRESS OF INDUSTRIAL ORGANIZATIONS.

cion *US* var. of SCION 1.

cipher /'saɪfə(r)/ *n. & v.* (also **cypher**) ● *n.* **1 a** a secret or disguised way of writing. **b** a thing written in this way. **c** the key to it. **2** the arithmetical symbol (0) denoting no amount but used to occupy a vacant place in decimal etc. numeration (as in 12.05). **3** a person or thing of no importance. **4** the interlaced initials of a person or company etc.; a monogram. **5** any Arabic numeral. **6** continuous sounding of an organ-pipe, caused by a mechanical defect. ● *v.* **1** *tr.* put into secret writing, encipher. **2 a** *tr.* (usu. foll. by *out*) work out by arithmetic, calculate. **b** *intr. archaic* do arithmetic. [ME, f. OF *cif(f)re*, ult. f. Arab. *ṣifr* ZERO]

cipolin /'sɪpəlɪn/ *n.* an Italian white-and-green marble. [F *cipolin* or It. *cipollino* f. *cipolla* onion]

circa /'sɜːkə/ *prep.* (often preceding a date) approximately. [L]

circadian /sɜː'keɪdɪən/ *adj. Physiol.* occurring or recurring about once per day. [irreg. f. L *circa* about + *dies* day]

Circassian /sɜː'kæsɪən/ *n. & adj.* ● *n.* **1** a member of a group of peoples of the Caucasus. The majority of Circassians are Sunni Muslims, many of whom live in traditionally organized pastoral societies. **2** the Caucasian language of these peoples. ● *adj.* of or relating to these peoples. [*Circassia* district in northern Caucasus (Russ. *Cherkes* tribe calling themselves Adygei)]

Circe /'sɜːsɪ/ *Gk Mythol.* an enchantress who lived with her wild animals on the island of Aeaea. When Odysseus visited the island his companions were changed into pigs by her potions, but he protected himself with the mythical herb *moly* and forced her to restore his men into human form.

circinate /'sɜːsɪˌneɪt/ *adj. Bot. & Zool.* rolled up with the apex in the centre, e.g. of young fronds of ferns. [L *circinatus* past part. of *circinare* make round f. *circinus* pair of compasses]

circle /'sɜːk(ə)l/ *n. & v.* ● *n.* **1 a** a round plane figure whose circumference is everywhere equidistant from its centre. **b** the line enclosing a circle. **2** a roundish enclosure or structure. **3** a ring. **4** a curved upper tier of seats in a theatre etc. (*dress circle*). **5** a circular route. **6** *Archaeol.* a group of (usu. large embedded) stones arranged in a circle. **7** *Hockey* = *striking-circle*. **8** persons grouped round a centre of interest. **9** a set or class or restricted group (*literary circles; not done in the best circles*). **10** a period or cycle (*the circle of the year*). **11** (in full **vicious circle**) **a** an unbroken sequence of reciprocal cause and effect. **b** an action and reaction that intensify each other (cf. *virtuous circle*). **c** the fallacy of proving a proposition from another which depends on the first for its own proof. ● *v.* **1** *intr.* (often foll. by *round, about*) move in a circle. **2** *tr.* **a** revolve round. **b** form a circle round. □ **circle back** move in a wide loop towards the starting-point. **come**

full circle return to the starting-point. **go round in circles** make no progress despite effort. **great circle** see GREAT. **run round in circles** *colloq.* be fussily busy with little result. **small circle** see SMALL. [ME f. OF *cercle* f. L *circulus* dimin. of *circus* ring]

circlet /'sɜːklɪt/ n. **1** a small circle. **2** a circular band, esp. of gold or jewelled etc., as an ornament.

circs /sɜːks/ n.pl. *colloq.* circumstances. [abbr.]

circuit /'sɜːkɪt/ n. **1 a** a line or course enclosing an area; the distance round; the circumference. **b** the area enclosed. **2** *Electr.* **a** the path of a current. **b** the apparatus through which a current passes. **3 a** the journey of a judge in a particular district to hold courts. **b** this district. **c** the lawyers following a circuit. **4** a chain of theatres or cinemas etc. under a single management. **5** *Brit.* a motor-racing track. **6 a** a sequence of sporting events (*the US tennis circuit*). **b** a sequence of athletic exercises. **7** a roundabout journey. **8 a** a group of local Methodist churches forming a minor administrative unit. **b** the journey of an itinerant minister within this. **9** an itinerary or route followed by an entertainer, politician, etc.; a sphere of operation (*election circuit; cabaret circuit*). □ **circuit board** a thin rigid board bearing the components of an electric circuit; esp. a printed circuit. **circuit-breaker** an automatic device for stopping the flow of current in an electric circuit. [ME f. OF, f. L *circuitus* f. CIRCUM- + *ire it-* go]

circuitous /sɜːˈkjuːɪtəs/ adj. **1** indirect (and usu. long). **2** going a long way round. □ **circuitously** adv. **circuitousness** n. [med.L *circuitosus* f. *circuitus* CIRCUIT]

circuitry /'sɜːkɪtrɪ/ n. (pl. **-ies**) **1** a system of electric circuits. **2** the equipment forming this.

circular /'sɜːkjʊlə(r)/ adj. & n. ● adj. **1 a** having the form of a circle. **b** moving or taking place along a circle; indirect, circuitous (*circular tour*). **2** *Logic* (of reasoning) depending on a vicious circle. **3** (of a letter or advertisement etc.) for distribution to a large number of people. ● n. a circular letter, leaflet, etc. □ **circular saw** a power saw with a rapidly rotating toothed disc. □ **circularly** adv. **circularity** /ˌsɜːkjʊˈlærɪtɪ/ n. [ME f. AF *circuler*, OF *circulier, cerclier* f. LL *circularis* f. L *circulus* CIRCLE]

circularize /'sɜːkjʊləˌraɪz/ v.tr. (also **-ise**) **1** distribute circulars to. **2** *US* seek opinions of (people) by means of a questionnaire. □ **circularization** /ˌsɜːkjʊləraɪˈzeɪʃ(ə)n/ n.

circulate /'sɜːkjʊˌleɪt/ v. **1** intr. go round from one place or person etc. to the next and so on; be in circulation. **2** tr. **a** cause to go round; put into circulation. **b** give currency to (a report etc.). **c** circularize. **3** intr. be actively sociable at a party, gathering, etc. □ **circulating library** a small library with books lent to a group of subscribers in turn. **circulating medium** notes or gold etc. used in exchange. □ **circulative** adj. **circulator** n. [L *circulare circulat-* f. *circulus* CIRCLE]

circulation /ˌsɜːkjʊˈleɪʃ(ə)n/ n. **1 a** movement to and fro, or from and back to a starting-point, esp. of a fluid in a confined area or circuit. **b** the movement of blood to and from the tissues of the body. **c** a similar movement of sap etc. **2 a** the transmission or distribution (of news or information or books etc.). **b** the number of copies sold, esp. of journals and newspapers. **3 a** currency, coin, etc. **b** the movement or exchange of this in a country etc. □ **in** (or **out of**) **circulation** participating (or not participating) in activities etc. [F *circulation* or L *circulatio* f. *circulare* CIRCULATE]

circulatory /ˌsɜːkjʊˈleɪtərɪ, 'sɜːkjʊlətərɪ/ adj. of or relating to the circulation of blood or sap.

circum- /'sɜːkəm/ *comb. form* round, about, around, used: **1** adverbially (*circumambient; circumfuse*). **2** prepositionally (*circumlunar; circumocular*). [from or after L *circum* prep. = round, about]

circumambient /ˌsɜːkəmˈæmbɪənt/ adj. (esp. of air or another fluid) surrounding. □ **circumambience** n. **circumambiency** n.

circumambulate /ˌsɜːkəmˈæmbjʊˌleɪt/ v.tr. & intr. walk round or about. □ **circumambulatory** adj. **circumambulation** /-ˌæmbjʊˈleɪʃ(ə)n/ n. [CIRCUM- + *ambulate* f. L *ambulare* walk]

circumcircle /'sɜːkəmˌsɜːk(ə)l/ n. *Geom.* a circle touching all the vertices of a triangle or polygon.

circumcise /'sɜːkəmˌsaɪz/ v.tr. **1** cut off the foreskin of, as a Jewish or Muslim rite or a surgical operation. **2** cut off the clitoris (and sometimes the labia) of, usu. as a religious rite. **3** *Bibl.* purify (the heart etc.). [ME f. OF f. L *circumcidere circumcis-* (as CIRCUM-, *caedere* cut)]

circumcision /ˌsɜːkəmˈsɪʒ(ə)n/ n. **1** the act or rite of circumcising or being circumcised. **2** (**Circumcision**) *Eccl.* the feast of the Circumcision of Christ, 1 Jan. [ME f. OF *circoncision* f. LL *circumcisio -onis* (as CIRCUMCISE)]

circumference /səˈkʌmfərəns/ n. **1** the enclosing boundary, esp. of a circle or other figure enclosed by a curve. **2** the distance round. □ **circumferential** /ˌsɜːkəmfəˈrenʃ(ə)l/ adj. **circumferentially** adv. [ME f. OF *circonference* f. L *circumferentia* (as CIRCUM-, *ferre* bear)]

circumflex /'sɜːkəmˌfleks/ n. & adj. ● n. (in full **circumflex accent**) a mark (ˆ or ˜) placed over a vowel in some languages to indicate a contraction, length, or a special quality. ● adj. *Anat.* curved, bending round something else (*circumflex nerve*). [L *circumflexus* (as CIRCUM-, *flectere flex-* bend), transl. Gk *perispōmenos* drawn around]

circumfluent /səˈkʌmflʊənt/ adj. flowing round, surrounding. □ **circumfluence** n. [L *circumfluere* (as CIRCUM-, *fluere* flow)]

circumfuse /ˌsɜːkəmˈfjuːz/ v.tr. pour round or about. [CIRCUM- + L *fundere fus-* pour]

circumjacent /ˌsɜːkəmˈdʒeɪs(ə)nt/ adj. situated around. [L *circumjacere* (as CIRCUM-, *jacēre* lie)]

circumlocution /ˌsɜːkəmləˈkjuːʃ(ə)n/ n. **1 a** a roundabout expression. **b** evasive talk. **2** the use of many words where fewer would do; verbosity. □ **circumlocutional** adj. **circumlocutionary** adj. **circumlocutionist** n. **circumlocutory** /-ˈlɒkjʊtərɪ/ adj. [ME f. F *circumlocution* or L *circumlocutio* (as CIRCUM-, LOCUTION), transl. Gk PERIPHRASIS]

circumlunar /ˌsɜːkəmˈluːnə(r)/ adj. moving or situated around the moon.

circumnavigate /ˌsɜːkəmˈnævɪˌgeɪt/ v.tr. sail round (esp. the world). □ **circumnavigator** n. **circumnavigation** /-ˌnævɪˈgeɪʃ(ə)n/ n. [L *circumnavigare* (as CIRCUM-, NAVIGATE)]

circumpolar /ˌsɜːkəmˈpəʊlə(r)/ adj. **1** *Geog.* around or near one of the earth's poles. **2** *Astron.* (of a star, constellation, motion etc.) above the horizon at all times in a given latitude.

circumscribe /'sɜːkəmˌskraɪb/ v.tr. **1** (of a line etc.) enclose or outline. **2** lay down the limits of; confine, restrict. **3** *Geom.* draw (a figure) round another, touching it at points but not cutting it (cf. INSCRIBE 4). □ **circumscriber** n. **circumscribable** /ˌsɜːkəmˈskraɪbəb(ə)l/ adj. **circumscription** /-ˈskrɪpʃ(ə)n/ n. [L *circumscribere* (as CIRCUM-, *scribere script-* write)]

circumsolar /ˌsɜːkəmˈsəʊlə(r)/ adj. moving or situated around or near the sun.

circumspect /'sɜːkəmˌspekt/ adj. wary, cautious; taking everything into account. □ **circumspectly** adv. **circumspection** /ˌsɜːkəmˈspekʃ(ə)n/ n. [ME f. L *circumspicere circumspect-* (as CIRCUM-, *specere spect-* look)]

circumstance /'sɜːkəmstəns/ n. **1 a** a fact, occurrence, or condition, esp. (in pl.) the time, place, manner, cause, occasion, etc., or surroundings of an act or event. **b** (in pl., or sing. as non-count noun) the external conditions that affect or might affect an action (*a victim of circumstance(s)*). **2** (often foll. by that + clause) an incident, occurrence, or fact, as needing consideration (*the circumstance that he left early*). **3** (in pl.) one's state of financial or material welfare (*in reduced circumstances*). **4** ceremony, fuss (*pomp and circumstance*). **5** full detail in a narrative (*told it with much circumstance*). □ **in** (or **under**) **the** (or **these**) **circumstances** the state of affairs being what it is. **in** (or **under**) **no circumstances** not at all; never. □ **circumstanced** adj. [ME f. OF *circonstance* or L *circumstantia* (as CIRCUM-, *stantia* f. *sto* stand)]

circumstantial /ˌsɜːkəmˈstænʃ(ə)l/ adj. **1** given in full detail (*a circumstantial account*). **2** (of evidence, a legal case, etc.) tending to establish a conclusion by inference from known facts hard to explain otherwise. **3 a** depending on circumstances. **b** adventitious, incidental. □ **circumstantially** adv. **circumstantiality** /-ˌstænʃɪˈælɪtɪ/ n. [L *circumstantia:* see CIRCUMSTANCE]

circumterrestrial /ˌsɜːkəmtəˈrestrɪəl/ adj. moving or situated around the earth.

circumvallate /ˌsɜːkəmˈvæleɪt/ v.tr. surround with or as with a rampart. [L *circumvallare circumvallat-* (as CIRCUM-, *vallare* f. *vallum* rampart)]

circumvent /ˌsɜːkəmˈvent/ v.tr. **1 a** evade (a difficulty); find a way round. **b** baffle, outwit. **2** entrap (an enemy) by surrounding. □ **circumvention** /-ˈvenʃ(ə)n/ n. [L *circumvenire circumvent-* (as CIRCUM-, *venire* come)]

circumvolution /ˌsɜːkəmvəˈluːʃ(ə)n, -ˈljuːʃ(ə)n/ n. **1** rotation. **2** the winding of one thing round another. **3** a sinuous movement. [ME f. L *circumvolvere circumvolut-* (as CIRCUM-, *volvere* roll)]

circus /'sɜːkəs/ n. (pl. **circuses**) **1** a travelling show of performing animals, acrobats, clowns, etc. (*See note below.*) **2** *colloq.* **a** a scene of lively action; a disturbance. **b** a group of people in a common activity, esp.

sport. **3** *Brit.* an open space in a town, where several streets converge (*Piccadilly Circus*). **4** a circular hollow surrounded by hills. **5** *Rom. Antiq.* **a** a rounded or oval arena with tiers of seats, for chariot racing, gladiatorial combats, and other sports and contests. **b** a performance given there (*bread and circuses*). **6** (**the Circus**) *sl.* the British Secret Service. [L, = ring]

▪ The modern circus dates from the late 18th century, when horse-riding displays were given in London in an arena that also had a stage for singing, dancing, and pantomime. This form of entertainment spread to other parts of Britain, and to Europe, Russia, and the US, some permanent shows being mounted and others travelling around in caravans and wagons (later by train). From the early 19th century they were often combined with the travelling menageries and wild animal performances that had become very popular. In the second half of the 19th century acts involving acrobatics became popular after the invention of the flying trapeze by Jules Léotard (1830–70) and the spectacular tightrope walk across Niagara Falls by Charles Blondin in 1859. In Europe circuses were generally restricted to a single ring, while in the US circuses became bigger and more spectacular, often with three rings, the best-known being the Barnum and Bailey circus, which started in 1881. After losing its popularity in Britain in the early 20th century, the circus enjoyed something of a revival through large and elaborate productions such as those of Sir C. B. Cochran and Bertram Mills (1873–1938), but traditional circuses are now less common in Britain and their use of animals has led to protests by animal rights activists and bans by some local authorities. In Russia and China, however, the circus has continued to flourish.

ciré /ˈsɪəreɪ/ *n. & adj.* ● *n.* a fabric with a smooth shiny surface obtained esp. by waxing and heating. ● *adj.* having such a surface. [F, = waxed]

Cirencester /ˈsaɪərənˌsestə(r)/ a town in Gloucestershire; pop. (1981) 14,000. It was a major town in Roman Britain, when it was known as Corinium Dobunorum.

cire perdue /ˌsɪə peəˈdjuː/ *n. Art* a method of bronze-casting using a clay core and a wax coating placed in a mould: the wax is melted in the mould and bronze poured into the space left, producing a hollow bronze figure when the core is discarded. [F, = lost wax]

cirque /sɜːk/ *n.* **1** *Geol.* a deep bowl-shaped hollow at the head of a valley or on a mountainside. **2** *poet.* **a** a ring. **b** an amphitheatre or arena. [F f. L CIRCUS]

cirrhosis /sɪˈrəʊsɪs/ *n. Med.* a chronic disease of the liver marked by the degeneration of cells, inflammation, and the fibrous thickening of tissue, as a result of alcoholism, hepatitis, etc. □ **cirrhotic** /-ˈrɒtɪk/ *adj.* [mod.L f. Gk *kirrhos* tawny]

cirriped /ˈsɪrɪˌped/ *n.* (also **cirripede** /-ˌpiːd/) *Zool.* a marine crustacean of the class Cirripedia, which comprises the barnacles. [mod.L *Cirripedia* f. L *cirrus* curl (from the form of the legs) + *pes pedis* foot]

cirro- /ˈsɪrəʊ/ *comb. form* cirrus (cloud).

cirrocumulus /ˌsɪrəʊˈkjuːmjʊləs/ *n.* (often *attrib.*) *Meteorol.* a cloud type forming a broken layer of small fleecy clouds at high altitude, as in a mackerel sky.

cirrostratus /ˌsɪrəʊˈstrɑːtəs, -ˈstreɪtəs/ *n.* (often *attrib.*) *Meteorol.* a cloud type forming a thin, fairly uniform layer at high altitude.

cirrus /ˈsɪrəs/ *n.* (*pl.* **cirri** /-raɪ/) **1** (often *attrib.*) *Meteorol.* a cloud type forming wispy filamentous tufts at high altitude. **2** *Bot.* a tendril. **3** *Zool.* a long slender appendage or filament. □ **cirrose** *adj.* **cirrous** *adj.* [L, = curl]

CIS *abbr.* Commonwealth of Independent States.

cis- /sɪs/ *prefix* (opp. TRANS- or ULTRA-). **1** on this side of; on the side nearer to the speaker or writer (*cisatlantic*). **2** *Rom. Antiq.* on the Roman side of (*cisalpine*). **3** (of time) closer to the present (*cis-Elizabethan*). **4** *Chem.* (of an isomer) having two atoms or groups on the same side of a given plane in the molecule. [L *cis* on this side of]

cisalpine /sɪsˈælpaɪn/ *adj.* on the southern side of the Alps.

Cisalpine Gaul see GAUL[1].

cisatlantic /ˌsɪsətˈlæntɪk/ *adj.* on this side of the Atlantic.

cisco /ˈsɪskəʊ/ *n.* (*pl.* **-oes**) a freshwater whitefish of the genus *Coregonus*, native to North America. [19th c.: orig. unkn.]

Ciskei /sɪsˈkaɪ/ a former homeland established in South Africa for the Xhosa people, now part of the province of Eastern Cape. (See also HOMELAND.)

cislunar /sɪsˈluːnə(r)/ *adj.* between the earth and the moon.

cispontine /sɪsˈpɒntaɪn/ *adj.* on the north side of the Thames in London. [CIS- (orig. the better-known side) + L *pons pont-* bridge]

cissy var. of SISSY.

cist[1] /sɪst, kɪst/ *n.* (also **kist** /kɪst/) *Archaeol.* a coffin or burial-chamber made from stone or a hollowed tree. [Welsh, = CHEST]

cist[2] /sɪst/ *n. Gk Antiq.* a box used for sacred utensils. [L *cista* f. Gk *kistē* box]

Cistercian /sɪˈstɜːʃ(ə)n/ *n. & adj.* ● *n.* a member of a monastic order founded in 1098 at Cîteaux near Dijon in France. (*See note below.*) ● *adj.* of or relating to the Cistercians. [F *cistercien* f. L *Cistercium* Cîteaux]

▪ Founded for strict observance of the Rule of St Benedict, Cistercian houses for both monks and nuns spread throughout Europe in the 12th–13th centuries. The monks are now divided into two observances, the strict observance (following the original rule), whose adherents are known popularly as Trappists, and the common observance, which has certain relaxations.

cistern /ˈsɪstən/ *n.* **1** a tank for storing water, esp. one in a roof-space supplying taps or as part of a flushing lavatory. **2** an underground reservoir for rainwater. [ME f. OF *cisterne* f. L *cisterna* (as CIST[2])]

cistus /ˈsɪstəs/ *n.* an evergreen Mediterranean shrub of the genus *Cistus*, with large white or red flowers. See also *rock rose*. [mod.L f. Gk *kistos*]

citadel /ˈsɪtəd(ə)l, -ˌdel/ *n.* **1** a fortress, usu. on high ground protecting or dominating a city. **2** a meeting-hall of the Salvation Army. [F *citadelle* or It. *citadella*, ult. f. L *civitas -tatis* city]

citation /saɪˈteɪʃ(ə)n/ *n.* **1** the citing of a book or other source; a passage cited. **2** a mention in an official dispatch. **3** a note accompanying an award, describing the reasons for it.

cite /saɪt/ *v.tr.* **1** adduce as an instance. **2** quote (a passage, book, or author) in support of an argument etc. **3** mention in an official dispatch. **4** summon to appear in a lawcourt. □ **citable** *adj.* [ME f. F f. L *citare* f. *ciere* set moving]

citified /ˈsɪtɪˌfaɪd/ *adj.* (also **cityfied**) usu. *derog.* city-like or urban in appearance or behaviour.

citizen /ˈsɪtɪz(ə)n/ *n.* **1** a member of a state or Commonwealth, either native or naturalized (*British citizen*). **2** (usu. foll. by *of*) **a** an inhabitant of a city. **b** a freeman of a city. **3** *US* a civilian. □ **citizen of the world** a person who is at home anywhere; a cosmopolitan. **citizen's arrest** an arrest by an ordinary person without a warrant, allowable in certain cases. **citizens' band** a system of local intercommunication by individuals on special radio frequencies. □ **citizenhood** *n.* **citizenry** *n.* **citizenship** *n.* [ME f. AF *citesein*, OF *citeain* ult. f. L *civitas -tatis* city: cf. DENIZEN]

Citizens' Advice Bureau *n.* (abbr. **CAB**) (in the UK) an office at which the public can receive free advice and information on civil matters.

Citizen's Charter *n. Polit.* a document setting out the rights of citizens, esp. a British government document of 1991 designed to guarantee that public services meet certain standards of performance and to give the public rights of redress when the standards are not met.

Citlaltépetl /ˌsiːtlælˈteɪpet(ə)l/ (called in Spanish *Pico de Orizaba*) the highest peak in Mexico, in the east of the country, north of the city of Orizaba. It rises to a height of 5,699 m (18,503 ft) and is an extinct volcano. Its Aztec name means 'star mountain'.

citole /sɪˈtəʊl/ *n. hist.* a musical instrument akin to the lute, a precursor of the cittern. [ME f. OF: rel. to CITTERN with dimin. suffix]

citrate /ˈsɪtreɪt/ *n. Chem.* a salt or ester of citric acid.

citric /ˈsɪtrɪk/ *adj.* derived from citrus fruit. □ **citric acid** a sharp-tasting water-soluble organic acid found in the juice of lemons and other sour fruits used as a flavouring for carbonated drinks etc. [F *citrique* f. L *citrus* citron]

citrin /ˈsɪtrɪn/ *n. Biochem.* a group of substances occurring mainly in citrus fruits and blackcurrants, and formerly thought to be a vitamin. Also called *bioflavonoid*.

citrine /ˈsɪtrɪn/ *adj. & n.* ● *adj.* lemon-coloured. ● *n.* a transparent yellow variety of quartz. Also called *false topaz*. [ME f. OF *citrin* (as CITRUS)]

citron /ˈsɪtrən/ *n.* **1** a shrubby tree, *Citrus medica*, bearing large lemon-like fruits with thick fragrant peel. **2** this fruit. [F f. L CITRUS, after *limon* lemon]

citronella /ˌsɪtrəˈnelə/ *n.* **1** a fragrant grass of the genus *Cymbopogon*, native to southern Asia. **2** the scented oil from this, used in insect repellent, and perfume and soap manufacture. [mod.L, formed as CITRON + dimin. suffix]

citrus /ˈsɪtrəs/ n. (pl. **citruses**) **1** a tree of the genus *Citrus*, of the rue family, having fruits with juicy flesh and pulpy rind. Selective breeding has produced the lemon, lime, orange, tangerine, grapefruit, and various hybrids, which are cultivated in warm countries. **2** (in full **citrus fruit**) a fruit from such a tree. □ **citrous** adj. [L, = citron-tree or thuja]

cittern /ˈsɪt(ə)n/ n. hist. a wire-stringed lutelike instrument usu. played with a plectrum. [L *cithara*, Gk *kithara* a kind of harp, assim. to GITTERN]

city /ˈsɪtɪ/ n. (pl. **-ies**) **1 a** a large town. **b** Brit. (strictly) a town created a city by charter and containing a cathedral. **c** US a municipal corporation occupying a definite area. **2** (**the City**) **a** the part of London governed by the Lord Mayor and the Corporation. **b** the business part of this. **c** commercial circles; high finance. **3** (attrib.) of a city or the City. □ **city desk** a department of a newspaper etc. dealing with business news or N. Amer. with local news. **city father** (usu. in pl.) a person concerned with or experienced in the administration of a city. **city hall** N. Amer. municipal offices or officers. **city manager** N. Amer. an official directing the administration of a city. **city page** Brit. the part of a newspaper or magazine dealing with the financial and business news. **city slicker** usu. derog. **1** a smart and sophisticated city-dweller. **2** a plausible rogue as usu. found in cities. **city-state** esp. hist. a city that with its surrounding territory forms an independent state. □ **cityward** adj. & adv. **citywards** adv. [ME f. OF cité f. L civitas -tatis f. civis citizen]

City Company n. a corporation descended from an ancient trade-guild.

City editor n. **1** the editor dealing with financial news in a newspaper or magazine. **2** (**city editor**) N. Amer. the editor dealing with local news.

cityfied var. of CITIFIED.

City of London see CITY 2a.

cityscape /ˈsɪtɪˌskeɪp/ n. **1** a view of a city (actual or depicted). **2** city scenery.

City Technology College n. (abbr. **CTC**) (in the UK) any of a number of schools established in 1988 and designed to teach (esp. technological subjects) in inner-city areas, taking pupils aged 11–18.

Ciudad Bolívar /sjuːˈdæd bɒˈliːvɑː(r)/ a city in SE Venezuela, on the Orinoco river; pop. (1991) 225,850. Formerly called Angostura, its name was changed in 1846 to honour the country's liberator, Simón Bolívar.

Ciudad Trujillo /sjuːˌdæd truːˈhiːjəʊ/ the former name (1936–61) for SANTO DOMINGO.

Ciudad Victoria /sjuːˌdæd vɪkˈtɔːrɪə/ a city in NE Mexico, capital of the state of Tamaulipas; pop. (1990) 207,830.

civet /ˈsɪvɪt/ n. **1** (in full **civet-cat**) a slender catlike mammal of the family Viverridae; esp. *Viverra civetta* of central Africa, having well developed anal scent glands. **2** a strong musky perfume obtained from the secretions of this animal's scent glands. [F *civette* f. It. *zibetto* f. med.L *zibethum* f. Arab. *azzabād* f. *al* the + *zabād* this perfume]

civic /ˈsɪvɪk/ adj. **1** of a city; municipal. **2** of or proper to citizens (*civic virtues*). **3** of citizenship, civil. □ **civic centre** Brit. the area where municipal offices and other public buildings are situated; the buildings themselves. □ **civically** adv. [F *civique* or L *civicus* f. *civis* citizen]

civics /ˈsɪvɪks/ n.pl. (usu. treated as sing.) the study of the rights and duties of citizenship.

civil /ˈsɪv(ə)l, -vɪl/ adj. **1** of or belonging to citizens. **2** of ordinary citizens and their concerns, as distinct from military or naval or ecclesiastical matters. **3** polite, obliging, not rude. **4** Law relating to civil law (see below), not criminal or political matters (*civil court; civil lawyer*). **5** (of the length of a day, year, etc.) fixed by custom or law, not natural or astronomical. **6** occurring within a community or among fellow citizens; internal (*civil unrest*). □ **civil aviation** non-military, esp. commercial, aviation. **civil commotion** a riot or similar disturbance. **civil defence** the organization and training of civilians for the protection of lives and property during and after attacks in wartime. **civil disobedience** the refusal to comply with certain laws or to pay taxes etc. as a peaceful form of political protest. **civil engineer** an engineer who designs or maintains roads, bridges, dams, etc. **civil engineering** this work. **civil law 1** law concerning private rights (opp. *criminal law*). **2** hist. Roman or non-ecclesiastical law, as of Canada, France, or Germany (opp. *common law*). **civil libertarian** an advocate of increased civil liberty. **civil liberty** (often in pl.) freedom

of action and speech subject to the law. **civil marriage** a marriage solemnized as a civil contract without religious ceremony. **civil rights** the rights of citizens to political and social freedom and equality. **civil servant** a member of the civil service. **civil state** being single or married or divorced etc. **civil war** war between citizens of the same country. **civil year** see YEAR 2. □ **civilly** adv. [ME f. OF f. L *civilis* f. *civis* citizen]

civilian /sɪˈvɪljən/ n. & adj. ● n. a person not in the armed services or the police force. ● adj. of or for civilians.

civilianize /sɪˈvɪljəˌnaɪz/ v.tr. (also **-ise**) make civilian in character or function. □ **civilianization** /-ˌvɪljənaɪˈzeɪʃ(ə)n/ n.

civility /sɪˈvɪlɪtɪ/ n. (pl. **-ies**) **1** politeness. **2** an act of politeness. [ME f. OF *civilité* f. L *civilitas -tatis* (as CIVIL)]

civilization /ˌsɪvɪlaɪˈzeɪʃ(ə)n, ˌsɪvɪlɪ-/ n. (also **-isation**) **1** an advanced stage or system of social development. **2** those peoples of the world regarded as having this. **3** a people or nation (esp. of the past) regarded as an element of social evolution (*ancient civilizations; the Inca civilization*). **4** making or becoming civilized.

civilize /ˈsɪvɪˌlaɪz/ v.tr. (also **-ise**) **1** bring to an advanced stage or system of social development. **2** enlighten; refine and educate. □ **civilizable** adj. **civilizer** n. [F *civiliser* (as CIVIL)]

Civil List n. (in the UK) an annual allowance voted by Parliament for the royal family's household expenses.

civil service n. the body of full-time officers employed by a state in the administration of civil (non-military) affairs. The Roman Empire had such a service, and there was one in China from the 7th century; in the 19th century many European countries copied Napoleon's system of an organized hierarchy. The term *civil service* was originally applied to the part of the service of the British East India Company carried on by staff who did not belong to the army or navy. In the UK civil servants do not change with the government and are expected to be politically neutral, but in the US, France, and many other countries political appointments are made to the higher ranks.

civvy /ˈsɪvɪ/ n. & adj. sl. ● n. (pl. **-ies**) **1** (in pl.) civilian clothes. **2** a civilian. ● adj. civilian. [abbr.]

Civvy Street n. sl. civilian life. [abbr.]

CJ abbr. Chief Justice.

CJD abbr. Creutzfeldt–Jakob disease.

Cl symb. Chem. the element chlorine.

cl abbr. **1** centilitre(s). **2** class.

clack /klæk/ v. & n. ● v.intr. **1** make a sharp sound as of boards struck together. **2** chatter, esp. loudly. ● n. **1** a clacking sound. **2** clacking talk. □ **clacker** n. [ME, = to chatter, prob. f. ON *klaka*, of imit. orig.]

Clactonian /klækˈtəʊnɪən/ adj. & n. Archaeol. of, relating to, or denoting the lower palaeolithic culture represented by flint implements found at Clacton, Essex, dated c.250,000–200,000 BP.

clad[1] /klæd/ adj. **1** clothed (often in comb.: *leather-clad; scantily clad*). **2** provided with cladding. [past part. of CLOTHE]

clad[2] /klæd/ v.tr. (**cladding**; past and past part. **cladded** or **clad**) provide with cladding. [app. f. CLAD[1]]

cladding /ˈklædɪŋ/ n. a covering or coating on a structure or material etc.

clade /kleɪd/ n. Biol. a group of organisms evolved from a common ancestor. [Gk *klados* branch]

cladistics /kləˈdɪstɪks/ n.pl. (usu. treated as sing.) Biol. a method of classification of animals and plants on the basis of those shared characteristics that are assumed to indicate common ancestry. □ **cladistic** adj. **cladism** /ˈklædɪz(ə)m/ n. [as CLADE + -IST + -ICS]

cladode /ˈkleɪdəʊd/ n. Bot. a flattened leaflike stem. [Gk *kladōdēs* many-shooted f. *klados* shoot]

cladogram /ˈkleɪdəʊˌgræm/ n. Biol. a branching diagram showing the cladistic relationship between a number of species.

claim /kleɪm/ v. & n. ● v.tr. **1 a** (often foll. by *that* + clause) demand as one's due or property. **b** (usu. absol.) submit a request for payment under an insurance policy. **2 a** represent oneself as having or achieving (*claim victory; claim accuracy*). **b** (foll. by *to* + infin.) profess (*claimed to be the owner*). **c** assert, contend (*claim that one knows*). **3** have as an achievement or a consequence (*could then claim five wins; the fire claimed many victims*). **4** (of a thing) deserve (one's attention etc.). ● n. **1 a** a demand or request for something considered one's due (*lay claim to; put in a claim*). **b** an application for compensation under the terms of an insurance policy. **2** (foll. by *to, on*) a right or title to a thing (*his only claim to fame; have many claims on my time*). **3** a contention or assertion.

4 a thing claimed. **5** a statement of the novel features in a patent. **6** *Mining* a piece of land allotted or taken. □ **no claim** (or **claims**) **bonus** a reduction of an insurance premium after an agreed period without a claim under the terms of the policy. □ **claimable** *adj.* **claimer** *n.* [ME f. OF *claime* f. *clamer* call out f. L *clamare*]

claimant /ˈkleɪmənt/ *n.* a person making a claim, esp. in a lawsuit or for a state benefit.

Clair /kleə(r)/, René (born René Lucien Chomette) (1898–1981), French film director. He made a number of silent films, notably *Un Chapeau de paille d'Italie* (1927); his early sound films included *Sous les toits de Paris* (1930), *Le Million* (1931), and *À nous la liberté* (1931). He later achieved wider success with *Les Belles de nuit* (1952). Clair was always involved in the scriptwriting for his films, which typically contain elements of surrealism, underpinned by satire.

clairaudience /kleərˈɔːdɪəns/ *n.* the supposed faculty of perceiving, as if by hearing, what is inaudible. □ **clairaudient** *adj. & n.* [F *clair* CLEAR, + AUDIENCE, after CLAIRVOYANCE]

clairvoyance /kleəˈvɔɪəns/ *n.* **1** the supposed faculty of perceiving things or events in the future or beyond normal sensory contact. **2** exceptional insight. [F *clairvoyance* f. *clair* CLEAR + *voir* voy- see]

clairvoyant /kleəˈvɔɪənt/ *n. & adj.* ● *n.* (*fem.* **clairvoyante** *pronunc.* same) a person having clairvoyance. ● *adj.* having clairvoyance. □ **clairvoyantly** *adv.*

clam /klæm/ *n. & v.* ● *n.* **1** a bivalve mollusc; esp. two edible North American species, the *hard* or *round clam* (*Venus mercenaria*), and the *soft* or *long clam* (*Mya arenaria*). **2** *colloq.* a shy or withdrawn person. ● *v.intr.* (**clammed**, **clamming**) **1** dig for clams. **2** (foll. by *up*) *colloq.* refuse to talk. [16th c.: app. f. *clam* a clamp]

clamant /ˈkleɪmənt/ *adj. literary* noisy; insistent; urgent. □ **clamantly** *adv.* [L *clamare* clamant- cry out]

clamber /ˈklæmbə(r)/ *v. & n.* ● *v.intr.* climb with hands and feet, esp. with difficulty or laboriously. ● *n.* a difficult climb. [ME, prob. f. *clamb*, obs. past tense of CLIMB]

clammy /ˈklæmɪ/ *adj.* (**clammier**, **clammiest**) **1** unpleasantly damp and sticky or slimy. **2** (of weather) cold and damp. □ **clammily** *adv.* **clamminess** *n.* [ME f. *clam* to daub]

clamour /ˈklæmə(r)/ *n. & v.* (US **clamor**) ● *n.* **1** loud or vehement shouting or noise. **2** a protest or complaint; an appeal or demand. ● *v.* **1** *intr.* make a clamour. **2** *tr.* utter with a clamour. □ **clamorous** *adj.* **clamorously** *adv.* **clamorousness** *n.* [ME f. OF f. L *clamor -oris* f. *clamare* cry out]

clamp[1] /klæmp/ *n. & v.* ● *n.* **1** a device, esp. a brace or band of iron etc., for strengthening other materials or holding things together. **2** (in full **wheel clamp**) a device for immobilizing an illegally parked car. ● *v.tr.* **1** strengthen or fasten with a clamp. **2** place or hold firmly. **3** (in full **wheel clamp**) immobilize (an illegally parked car) by fixing a clamp to one of its wheels. □ **clamp down 1** (often foll. by *on*) be rigid in enforcing a rule etc. **2** (foll. by *on*) try to suppress. **clamp-down** *n.* an act of clamping down or trying to suppress. [ME prob. f. MDu., MLG *klamp(e)*]

clamp[2] /klæmp/ *n.* **1** a heap of potatoes or other root vegetables stored under straw or earth. **2** a pile of bricks for firing. **3** a pile of turf or peat or garden rubbish etc. [16th c.: prob. f. Du. *klamp* heap (in sense 2 related to CLUMP)]

clan /klæn/ *n.* **1** a group of people with a common ancestor, esp. in the Scottish Highlands. **2** a large family as a social group. **3** a group with a strong common interest. **4 a** a genus, species, or class. **b** a family or group of animals, e.g. elephants. [ME f. Gael. *clann* f. L *planta* sprout]

clandestine /klænˈdestɪn/ *adj.* surreptitious; secret. □ **clandestinely** *adv.* **clandestinity** /ˌklændesˈtɪnɪtɪ/ *n.* [F *clandestin* or L *clandestinus* f. *clam* secretly]

clang /klæŋ/ *n. & v.* ● *n.* a loud resonant metallic sound as of a bell or hammer etc. ● *v.* **1** *intr.* make a clang. **2** *tr.* cause to clang. [imit.: infl. by L *clangere* resound]

clanger /ˈklæŋə(r)/ *n. sl.* a mistake or blunder. □ **drop a clanger** commit a conspicuous indiscretion.

clangour /ˈklæŋgə(r)/ *n.* (US **clangor**) **1** a prolonged or repeated clanging noise. **2** an uproar or commotion. □ **clangorous** *adj.* **clangorously** *adv.* [L *clangor* noise of trumpets etc.]

clank /klæŋk/ *n. & v.* ● *n.* a sound as of heavy pieces of metal meeting or a chain rattling. ● *v.* **1** *intr.* make a clanking sound. **2** *tr.* cause to clank. □ **clankingly** *adv.* [imit.: cf. CLANG, CLINK[1], Du. *klank*]

clannish /ˈklænɪʃ/ *adj. usu. derog.* **1** (of a family or group) tending to

hold together. **2** of or like a clan. □ **clannishly** *adv.* **clannishness** *n.*

clanship /ˈklænʃɪp/ *n.* **1** a patriarchal system of clans. **2** loyalty to one's clan.

clansman /ˈklænzmən/ *n.* (*pl.* **-men**; *fem.* **clanswoman**, *pl.* **-women**) a member or fellow member of a clan.

clap[1] /klæp/ *v. & n.* ● *v.* (**clapped**, **clapping**) **1 a** *intr.* strike the palms of one's hands together as a signal or repeatedly as applause. **b** *tr.* strike (the hands) together in this way. **2** *tr.* applaud or show one's approval of (esp. a person) in this way. **3** *tr.* (of a bird) flap (its wings) audibly. **4** *tr.* put or place quickly or with determination (*clapped him in prison; clap a tax on whisky*). **5** *tr.* (foll. by *on*) slap (a person) encouragingly on (the back, shoulder, etc.). ● *n.* **1** the act of clapping, esp. as applause. **2** an explosive sound, esp. of thunder. **3** a slap, a pat. □ **clap eyes on** *colloq.* see. **clap on the back** = *slap on the back.* **clapped out** *Brit. sl.* worn out (esp. of machinery etc.); exhausted. [OE *clappian* throb, beat, of imit. orig.]

clap[2] /klæp/ *n.* coarse *sl.* venereal disease, esp. gonorrhoea. [OF *clapoir* venereal bubo]

clapboard /ˈklæpbɔːd, ˈklæbəd/ *n. N. Amer.* = WEATHERBOARD. [anglicized f. LG *klappholt* cask-stave]

clapper /ˈklæpə(r)/ *n.* the tongue or striker of a bell. □ **like the clappers** *Brit. sl.* very fast or hard.

clapperboard /ˈklæpəˌbɔːd/ *n. Cinematog.* a device of hinged boards struck together to synchronize the starting of picture and sound machinery in filming.

clapper bridge *n.* a rough bridge consisting of a series of slabs or planks resting on piles of stones. (See BRIDGE[1].) [perh. f. L *claperius* heap of stones (orig. unkn.)]

Clapton /ˈklæptən/, Eric (b.1945), English blues and rock guitarist, singer, and composer. He played in the Yardbirds and then formed his own group, Cream (1966–8), whose lengthy improvisations and experimental harmonies were influential in the development of rock music. The song 'Layla' (1972) is perhaps his best known; since then he has developed a restrained style that displays less of his instrumental virtuosity but has brought immense commercial success.

claptrap /ˈklæptræp/ *n.* **1** insincere or pretentious talk, nonsense. **2** language used or feelings expressed only to gain applause. [CLAP[1] + TRAP[1]]

claque /klæk, klɑːk/ *n.* a group of people hired to applaud in a theatre etc. [F f. *claquer* to clap]

claqueur /klæˈkɜː(r), klɑːˈkɜː(r)/ *n.* a member of a claque. [F (as CLAQUE)]

clarabella /ˌklærəˈbelə, ˌklɑːrə-/ *n.* an organ-stop with the quality of a flute. [fem. forms of L *clarus* clear and *bellus* pretty]

Clare[1] /kleə(r)/ a county of the Republic of Ireland, on the west coast in the province of Munster; county town, Ennis.

Clare[2] /kleə(r)/, John (1793–1864), English poet. After working as a day labourer and gardener, he published *Poems Descriptive of Rural Life and Scenery* (1820). This was followed by *The Shepherd's Calendar* (1827), which has received renewed critical attention in recent years, along with *The Rural Muse* (1835) and other later poems. Clare's poetry is a simple man's celebration of the natural world; it is notable for its use of the poet's own dialect and grammar. In 1837 he was certified insane and spent the rest of his life in an asylum.

clarence /ˈklærəns/ *n. hist.* a four-wheeled closed carriage with seats for four inside and two on the box. [Duke of *Clarence*, afterwards William IV]

Clarenceux /ˈklærənˌsuː/ *n. Heraldry* (in the UK) the title given to the second King of Arms, with jurisdiction south of the River Trent. [ME f. AF f. Duke of *Clarence* f. *Clare* in Suffolk]

Clarendon /ˈklærəndən/, Earl of (title of Edward Hyde) (1609–74), English statesman and historian, chief adviser to Charles II 1660–7. He shifted his allegiance from the Roundheads to Charles I on the outbreak of the Civil War, becoming royal adviser and accompanying the king to Oxford. He was Chancellor of Oxford University (1660–7) and author of the prestigious *History of the Rebellion and Civil Wars in England*, which he began writing in 1646, but which was published posthumously (1702–4). A number of University and public buildings in Oxford are named after him.

Clare of Assisi, St (1194–1253), Italian saint and abbess. She joined St Francis in 1212 and together they founded the order of Poor Ladies of San Damiano, more commonly known as the 'Poor Clares', of which she was appointed abbess. She was canonized two years after her death and in 1958 Pope Pius XII declared her the patron saint of television, alluding to a story of her miraculously experiencing the Christmas midnight mass being held in the Church of St Francis in Assisi when on her deathbed. Feast day, 11 (formerly 12) Aug.

claret /ˈklærət/ n. & adj. ● n. **1** red wine, esp. from Bordeaux. **2** a deep purplish-red. **3** archaic sl. blood. ● adj. claret-coloured. [ME f. OF (vin) claret f. med.L claratum (vinum) f. L clarus clear]

clarify /ˈklærɪˌfaɪ/ v. (-ies, -ied) **1** tr. & intr. make or become clearer. **2** tr. **a** free (liquid, butter, etc.) from impurities. **b** make transparent. **c** purify. □ **clarifier** n. **clarification** /ˌklærɪfɪˈkeɪʃ(ə)n/ n. **clarificatory** /-ˈkeɪtərɪ/ n. [ME f. OF clarifier f. L clarus clear]

clarinet /ˌklærɪˈnet/ n. **1 a** a woodwind instrument with a single-reed mouthpiece. (See note below.) **b** a player of this instrument. **2** an organ-stop with a quality resembling that of a clarinet. □ **clarinettist** n. (US **clarinetist**). [F clarinette, dimin. of clarine a kind of bell]

■ The clarinet is a cylindrical tube, usu. of dark wood, with a flared end, having holes stopped by keys. It has a range of just over three octaves, but the highest notes have a squeaky, harsh sound and are generally used only in jazz. The clarinet dates from the early 18th century, and has been a standard orchestral instrument since the end of that century.

clarion /ˈklærɪən/ n. & adj. ● n. **1** a clear rousing sound. **2** hist. a shrill narrow-tubed war trumpet. **3** an organ-stop with the quality of a clarion. ● adj. clear and loud. [ME f. med.L clario -onis f. L clarus clear]

clarity /ˈklærɪtɪ/ n. the state or quality of being clear, esp. of sound or expression. [ME f. L claritas f. clarus clear]

Clark /klɑːk/, William (1770–1838), American explorer. With Meriwether Lewis, he jointly commanded the Lewis and Clark expedition (1804–6) across the North American continent. (See LEWIS[6].)

Clarke[1] /klɑːk/, Arthur C(harles) (b.1917), English writer of science fiction. Originally a scientific researcher, he conceived the idea of communications satellites. He is better known as the writer of such novels as *Earthlight* (1955) and *The Fountains of Paradise* (1979). He also co-wrote (with Stanley Kubrick, the film's director) the screenplay for the film *2001: A Space Odyssey* (1968); Clarke and Kubrick collaborated in writing the novel of the same title, published in the year of the film's release.

Clarke[2] /klɑːk/, Marcus (Andrew Hislop) (1846–81), British-born Australian writer. In 1863 he emigrated to Australia, where he worked on a sheep station before becoming a journalist. His fame is based on his novel *For the Term of his Natural Life* (1874) about an Australian penal settlement, as well as his shorter stories of Australian life, such as *Old Tales of a Young Country* (1871).

clarkia /ˈklɑːkɪə/ n. a plant of the genus *Clarkia*, native to America, with showy white, pink, or purple flowers. [mod.L f. CLARK]

clary /ˈkleərɪ/ n. (pl. -ies) an aromatic labiate herb of the genus *Salvia*. [ME f. obs. F clarie repr. med.L sclarea]

clash /klæʃ/ n. & v. ● n. **1 a** a loud jarring sound as of metal objects being struck together. **b** a collision, esp. with force. **2 a** a conflict or disagreement. **b** a discord of colours etc. ● v. **1 a** intr. make a clashing sound. **b** tr. cause to clash. **2** intr. collide; coincide awkwardly. **3** intr. (often foll. by with) **a** come into conflict or be at variance. **b** (of colours) be discordant. □ **clasher** n. [imit.: cf. clack, clang, crack, crash]

clasp /klɑːsp/ n. & v. ● n. **1 a** a device with interlocking parts for fastening. **b** a buckle or brooch. **c** a metal fastening on a book-cover. **2 a** an embrace; a person's reach. **b** a grasp or handshake. **3** a bar of silver on a medal-ribbon with the name of the battle etc. at which the wearer was present. ● v. **1** tr. fasten with or as with a clasp. **2 a** grasp, hold closely. **b** embrace, encircle. **3** intr. fasten a clasp. □ **clasp hands** shake hands with fervour or affection. **clasp one's hands** interlace one's fingers. **clasp-knife** a folding knife, usu. with a catch holding the blade when open. □ **clasper** n. [ME: orig. unkn.]

clasper /ˈklɑːspə(r)/ n. (in pl.) the appendages of some male fish and insects used to hold the female in copulation.

class /klɑːs/ n., v., & adj. ● n. **1** any set of persons or things grouped together, or graded or differentiated from others esp. by quality (first class; economy class). **2 a** a division or order of society (upper class; professional classes). **b** a caste system, a system of social classes. **c** (the classes) archaic the rich or educated. **3** colloq. distinction or high quality in appearance, behaviour, etc.; stylishness. **4 a** a group of students or pupils taught together. **b** the occasion when they meet. **c** their course of instruction. **5** N. Amer. all the college or school students of the same standing or graduating in a given year (the class of 1990). **6** (in conscripted armies) all the recruits of a given year (the 1950 class).

7 Brit. a division of candidates according to merit in an examination. **8** Biol. a grouping of organisms, the next major rank below a division or phylum. ● v.tr. assign to a class or category. ● adj. colloq. classy, stylish. □ **class-conscious** aware of and reacting to social divisions or one's place in a system of social class. **class-consciousness** this awareness. **class-list** Brit. a list of candidates in an examination with the class achieved by each. **class war** conflict between social classes. **in a class of** (or on) **its** (or one's) **own** unequalled. **no class** colloq. a lack of quality or distinction. [L classis assembly]

classic /ˈklæsɪk/ adj. & n. ● adj. **1 a** of the first class; of acknowledged excellence. **b** remarkably typical; outstandingly important (a classic case). **c** having enduring worth; timeless. **2 a** of ancient Greek and Latin literature, art, or culture. **b** (of style in art, music, etc.) simple, harmonious, well-proportioned; in accordance with established forms. **3** having literary or historic associations (classic ground). **4** (of clothes) made in a simple elegant style not much affected by changes in fashion. ● n. **1** a classic writer, artist, work, or example. **2 a** an ancient Greek or Latin writer. **b** (in pl.) ancient Greek and Latin literature and history. **c** archaic a scholar of ancient Greek and Latin. **3** a follower of classic models. **4** a garment in classic style. **5** (in pl.) Brit. the classic races. □ **classic races** Brit. the five main flat races of the horse-racing season, namely the Two Thousand and the One Thousand Guineas, the Derby, the Oaks, and the St Leger. [F classique or L classicus f. classis class]

classical /ˈklæsɪk(ə)l/ adj. **1 a** of ancient Greek or Latin literature, art, or culture. **b** (of language) having the form used by the ancient standard authors (classical Latin; classical Hebrew). **c** based on the study of ancient Greek and Latin (a classical education). **d** learned in classical studies. **2** (also **Classical**) **a** denoting or relating to classical music. **b** of the period from c.1750–1800 (cf. ROMANTIC). **3 a** (also **Classical**) in or following the restrained style of classical antiquity; characteristic of classicism (cf. ROMANTIC). **b** (of a form or period of art etc.) representing an exemplary standard; having a long-established worth. **4** Physics relating to the concepts which preceded relativity and quantum theory. □ **classicism** n. **classicalist** n. **classically** adv. **classicality** /ˌklæsɪˈkælɪtɪ/ n. [L classicus (as CLASSIC)]

classical music n. serious or conventional music following long-established principles rather than a folk, jazz, or popular tradition. It is associated with acoustic instruments, in particular the orchestra, and the sonata form; however, modern experimental composers such as Karlheinz Stockhausen and John Cage, using electronic instruments and other devices, are generally considered to be working within classical music. The term is used more specifically with reference to music written c.1750–1800, as opposed to baroque and romantic music, and is exemplified by the work of Haydn, Mozart, and the young Beethoven. During this period the orchestra, the chamber group, and the various compositional forms such as symphony, concerto, and sonata became standardized.

classicism /ˈklæsɪˌsɪz(ə)m/ n. **1** (also **Classicism**) the following of a classic style. (See note below.) **2 a** classical scholarship. **b** the advocacy of a classical education. **3** an ancient Greek or Latin idiom. □ **classicist** n.

■ In the arts, classicism denotes a style modelled on the qualities traditionally associated with the art of ancient Greece and Rome, for example harmony, restraint, and adherence to recognized standards of form and craftsmanship. The term is frequently opposed to romanticism. The painters and sculptors of Renaissance Italy such as Michelangelo and Raphael exemplify classicism, as does the 16th-century architect Palladio. In France, classicists included the painter Nicholas Poussin and the dramatists Racine and Corneille; in England the style was favoured by the Augustan poets John Dryden and Alexander Pope. The classicism of the later 18th century is frequently referred to as neoclassicism.

classicize /ˈklæsɪˌsaɪz/ v. (also **-ise**) **1** tr. make classic. **2** intr. imitate a classical style.

classified /ˈklæsɪˌfaɪd/ adj. & n. ● adj. **1** arranged in classes or categories. **2** (of information etc.) designated as officially secret. **3** Brit. (of a road) assigned to a category according to its importance. **4** (of newspaper advertisements) arranged in columns according to various categories. ● n. (in pl.) classified advertisements.

classify /ˈklæsɪˌfaɪ/ v.tr. (-ies, -ied) **1 a** arrange in classes or categories. **b** assign (a thing) to a class or category. **2** designate as officially secret or not for general disclosure. □ **classifiable** adj. **classifier** n. **classification** /ˌklæsɪfɪˈkeɪʃ(ə)n/ n. **classificatory** /-ˈkeɪtərɪ/ adj. [back-form. f. classification f. F (as CLASS)]

classism /'klɑ:sɪz(ə)m/ n. discrimination on the grounds of social class. □ **classist** adj. & n.

classless /'klɑ:slɪs/ adj. making or showing no distinction of classes (classless society; classless accent). □ **classlessness** n.

classmate /'klɑ:smeɪt/ n. a fellow member of a class, esp. at school.

classroom /'klɑ:sru:m, -rʊm/ n. a room in which a class of students is taught, esp. in a school.

classy /'klɑ:sɪ/ adj. (**classier, classiest**) colloq. superior, stylish. □ **classily** adv. **classiness** n.

clastic /'klæstɪk/ adj. Geol. designating a rock composed of broken pieces of older rocks. [F clastique f. Gk klastos broken in pieces]

clathrate /'klæθreɪt/ n. Chem. a solid in which one component is enclosed in the structure of another. [L clathratus f. clathri lattice-bars f. Gk klēthra]

clatter /'klætə(r)/ n. & v. ● n. 1 a rattling sound as of many hard objects struck together. 2 noisy talk. ● v. 1 intr. a make a clatter. b fall or move etc. with a clatter. 2 tr. cause (plates etc.) to clatter. [OE, of imit. orig.]

Claude Lorraine /,klɔːd lə'reɪn/ (also **Lorrain**) (born Claude Gellée) (1600–82), French painter. By 1630 he had achieved great fame in Italy, where he served his apprenticeship as a landscape painter. His paintings were so much in demand that he recorded them in the form of sketches in his Liber Veritatis (c.1635) to guard against forgeries. Although the influence of the late mannerists is evident in his early landscapes, his mature works concentrate on the poetic power of light and atmosphere. He was particularly admired in England, where he inspired the painters J. M. W. Turner and Richard Wilson (1714–82).

claudication /,klɔ:dɪ'keɪʃ(ə)n/ n. Med. a cramping pain, esp. in the leg, caused by arterial obstruction; limping. [L claudicare limp f. claudus lame]

Claudius /'klɔ:dɪəs/ (full name Tiberius Claudius Drusus Nero Germanicus) (10 BC–AD 54), Roman emperor 41–54. He spent his early life engaged in historical study, prevented from entering public life by his physical infirmity; he was proclaimed emperor after the murder of Caligula. His reign was noted for its restoration of order after Caligula's decadence and for its expansion of the Roman Empire, in particular his invasion of Britain in the year 43, in which he personally took part. His fourth wife, Agrippina (AD 15–59), is said to have killed him with a dish of poisoned mushrooms.

clause /klɔ:z/ n. 1 Gram. a distinct part of a sentence, including a subject and predicate. 2 a single statement in a treaty, law, bill, or contract. □ **clausal** adj. [ME f. OF f. L clausula conclusion f. claudere claus- shut]

Clausewitz /'klaʊzə,vɪts/, Karl von (1780–1831), Prussian general and military theorist. A Chief of Staff in the Prussian army (1815) and later a general (1818), he went on to write the detailed study On War (1833), which had a marked influence on strategic studies in the 19th and 20th centuries.

Clausius /'klaʊzɪəs/, Rudolf (1822–88), German physicist, one of the founders of modern thermodynamics. He was the first, in 1850, to formulate the second law of thermodynamics, thereby reconciling apparent conflicts between the work of Nicolas Carnot and James Joule. He later developed the concept of (and coined the term) entropy, which was his greatest contribution to physics. Clausius also carried out pioneering work on the kinetic theory of gases, in which he introduced the idea of the mean path length and effective radius of a moving molecule.

claustral /'klɔ:str(ə)l/ adj. 1 of or associated with the cloister; monastic. 2 narrow-minded. [ME f. LL claustralis f. claustrum CLOISTER]

claustrophobia /,klɒstrə'fəʊbɪə, ,klɔ:strə-/ n. an abnormal fear of confined places. □ **claustrophobe** /'klɒstrə,fəʊb, 'klɔ:strə-/ n. [mod.L f. L claustrum: see CLOISTER]

claustrophobic /,klɒstrə'fəʊbɪk, ,klɔ:strə-/ adj. 1 suffering from claustrophobia. 2 inducing claustrophobia. □ **claustrophobically** adv.

clavate /'kleɪveɪt/ adj. Bot. club-shaped. [mod.L clavatus f. L clava club]

clave[1] /kleɪv, klɑ:v/ n. Mus. a hardwood stick used in pairs to make a hollow sound when struck together. [Amer. Sp. f. Sp., = keystone, f. L clavis key]

clave[2] archaic past of CLEAVE[2].

clavicembalo /,klævɪ'tʃembə,ləʊ/ n. (pl. **-os**) a harpsichord. [It.]

clavichord /'klævɪ,kɔ:d/ n. a small keyboard instrument with a very soft tone, developed in the 14th century and in use from the early 15th. The strings are struck by brass blades, called tangents, to produce a clear and quiet tone, for domestic music-making. It was a favourite instrument of both J. S. Bach and his son C. P. E. Bach. [ME f. med.L clavichordium f. L clavis key, chorda string (see CHORD)]

clavicle /'klævɪk(ə)l/ n. the collar-bone. □ **clavicular** /klə'vɪkjʊlə(r)/ adj. [L clavicula dimin. of clavis key (from its shape)]

clavier /klə'vɪə(r), 'klævɪə(r)/ n. Mus. 1 any keyboard instrument. 2 its keyboard. [F clavier or G Klavier f. med.L claviarius, orig. = key-bearer, f. L clavis key]

claviform /'klævɪ,fɔ:m/ adj. club-shaped. [L clava club]

claw /klɔ:/ n. & v. ● n. 1 a a pointed horny nail on an animal's or bird's foot. b a foot with such a nail or nails. 2 the pincer of a crab or other crustacean. 3 a device for grappling, holding, etc. ● v. 1 a tr. & intr. scratch, maul, or pull (a person or thing) with claws. b intr. (often foll. by at) grasp, clutch, or scrabble at as with claws. 2 tr. & intr. Sc. scratch gently. 3 intr. Naut. beat to windward. □ **claw back 1** regain laboriously or gradually. 2 recover (money paid out) from another source (e.g. taxation). **claw-back** n. 1 the act of clawing back. 2 money recovered in this way. **claw-hammer** a hammer with one side of the head forked for extracting nails. □ **clawed** adj. (also in comb.). **clawless** adj. [OE clawu, clawian]

Clay /kleɪ/, Cassius, see MUHAMMAD ALI[2].

clay /kleɪ/ n. 1 a stiff sticky earth, used for making bricks, pottery, ceramics, etc. 2 poet. the substance of the human body. 3 (in full **clay pipe**) a tobacco-pipe made of clay. □ **clay-pan** Austral. a natural hollow in clay soil, retaining water after rain. **clay pigeon** a breakable disc thrown up from a trap as a target for shooting. □ **clayey** adj. **clayish** adj. **claylike** adj. [OE clæg f. WG]

claymore /'kleɪmɔ:(r)/ n. 1 hist. a a Scottish two-edged broadsword. b a broadsword, often with a single edge, having a hilt with a basketwork design. 2 a type of anti-personnel mine. [Gael. claidheamh mór great sword]

-cle /k(ə)l/ suffix forming (orig. diminutive) nouns (article; particle). [as -CULE]

clean /kli:n/ adj., adv., v., & n. ● adj. 1 (often foll. by of) free from dirt or contaminating matter, unsoiled. 2 clear; unused or unpolluted; preserving what is regarded as the original state (clean air; clean page). 3 free from obscenity or indecency. 4 a attentive to personal hygiene and cleanliness. b (of children and animals) toilet-trained or house-trained. 5 complete, clear-cut, unobstructed, even. 6 a (of a ship, aircraft, or car) streamlined, smooth. b well-formed, slender and shapely (clean-limbed; the car has clean lines). 7 adroit, skilful (clean fielding). 8 (of a nuclear weapon) producing relatively little fallout. 9 a free from ceremonial defilement or from disease. b (of food) not prohibited. 10 a free from any record of a crime, offence, etc. (a clean driving-licence). b sl. not carrying a weapon or incriminating material; free from suspicion. c observing the rules of a sport or game; fair (a clean fight). 11 (of a taste, smell, etc.) sharp, fresh, distinctive. 12 (of timber) free from knots. ● adv. 1 completely, outright, simply (clean bowled; cut clean through; clean forgot). 2 in a clean manner. ● v. 1 tr. (also foll. by of) & intr. make or become clean. 2 tr. eat all the food on (one's plate). 3 tr. remove the innards of (fish or fowl). 4 intr. make oneself clean. ● n. the act or process of cleaning or being cleaned (give it a clean). □ **clean bill of health** see bill of health 2 (see BILL[1]). **clean break** a quick and final separation. **clean-cut** sharply outlined. **clean down** clean by brushing or wiping. **clean hands** freedom from guilt. **clean-living** of upright character. **clean out 1** clean or clear thoroughly. 2 sl. empty or deprive (esp. of money). **clean-shaven** without beard, whiskers, or moustache. **clean sheet** (or **slate**) freedom from commitments or imputations; the removal of these from one's record. **clean up 1 a** clear (a mess) away. b (also absol.) put (things) tidy. c make (oneself) clean. 2 restore order or morality to. 3 sl. a acquire as gain or profit. b make a gain or profit. **clean-up** n. an act of cleaning up. **come clean** colloq. own up; confess everything. **make a clean breast of** see BREAST. **make a clean job of** colloq. do thoroughly. **make a clean sweep of** see SWEEP. □ **cleanable** adj. **cleanish** adj. **cleanness** /'kli:nnɪs/ n. [OE clæne (adj. & adv.), clēne (adv.), f. WG]

cleaner /'kli:nə(r)/ n. 1 a person employed to clean the interior of a building. 2 (usu. in pl.) a commercial establishment for cleaning clothes. 3 a device or substance for cleaning. □ **take to the cleaners** sl. 1 defraud or rob (a person) of all his or her money. 2 criticize severely.

cleanly[1] /'kli:nlɪ/ adv. 1 in a clean way. 2 efficiently; without difficulty. [OE clænlīce: see CLEAN, -LY[2]]

cleanly[2] /ˈklɛnlɪ/ *adj.* (**cleanlier, cleanliest**) habitually clean; with clean habits. □ **cleanliness** *n.* [OE *clǣnlic*: see CLEAN, -LY[1]]

cleanse /klɛnz/ *v.tr.* **1** usu. *formal* make clean. **2** (often foll. by *of*) purify from sin or guilt. **3** *archaic* cure (a leper etc.). □ **cleansing cream** cream for removing unwanted matter from the face, hands, etc. **cleansing department** *Brit.* a local service of refuse collection etc. □ **cleanser** *n.* [OE *clǣnsian* (see CLEAN)]

cleanskin /ˈkliːnskɪn/ *n. Austral.* **1** an unbranded animal. **2** *sl.* a person free from blame, without a police record, etc.

clear /klɪə(r)/ *adj., adv., & v.* ● *adj.* **1** free from dirt or contamination. **2** (of weather, the sky, etc.) not dull or cloudy. **3 a** transparent. **b** lustrous, shining; free from obscurity. **c** (of the complexion) fresh and unblemished. **4** (of soup) not containing solid ingredients. **5** (of a fire) burning with little smoke. **6 a** distinct, easily perceived by the senses. **b** unambiguous, easily understood (*make a thing clear; make oneself clear*). **c** manifest; not confused or doubtful (*clear evidence*). **7** that discerns or is able to discern readily and accurately (*clear thinking; clear-sighted*). **8** (usu. foll. by *about, on,* or *that* + clause) confident, convinced, certain. **9** (of a conscience) free from guilt. **10** (of a road etc.) unobstructed, open. **11 a** net, without deduction (*a clear £1,000*). **b** complete (*three clear days*). **12** (often foll. by *of*) free, unhampered; unencumbered by debt, commitments, etc. **13** (foll. by *of*) not obstructed by. ● *adv.* **1** clearly (*speak loud and clear*). **2** completely (*he got clear away*). **3** apart, out of contact (*keep clear; stand clear of the doors*). **4** (foll. by *to*) *US* all the way. ● *v.* **1** *tr. & intr.* make or become clear. **2 a** *tr.* (often foll. by *of*) free from prohibition or obstruction. **b** *tr. & intr.* make or become empty or unobstructed. **c** *tr.* free (land) for cultivation or building by cutting down trees etc. **d** *tr.* cause people to leave (a room etc.). **3** *tr.* (often foll. by *of*) show or declare (a person) to be innocent (*cleared them of complicity*). **4** *tr.* approve (a person) for special duty, access to information, etc. **5** *tr.* pass over or by safely or without touching, esp. by jumping. **6** *tr.* make (an amount of money) as a net gain or to balance expenses. **7** *tr.* pass (a cheque) through a clearing-house. **8** *tr.* pass through (a customs office etc.). **9** *tr.* remove (an obstruction, an unwanted object, etc.) (*clear them out of the way*). **10** *tr.* (also *absol.*) *Football* send (the ball) out of one's defensive zone. **11** *intr.* (often foll. by *away, up*) (of physical phenomena) disappear, gradually diminish (*mist cleared by lunchtime; my cold has cleared up*). **12** *tr.* (often foll. by *off*) discharge (a debt). □ **clear the air 1** make the air less sultry. **2** disperse an atmosphere of suspicion, tension, etc. **clear away 1** remove completely. **2** remove the remains of a meal from the table. **clear-cut** sharply defined. **clear the decks** prepare for action. **clear off 1** get rid of. **2** *colloq.* go away. **clear out 1** empty. **2** remove. **3** *colloq.* go away. **clear one's throat** cough slightly to make one's voice clear. **clear up 1** tidy up. **2** solve (a mystery etc.); remove (a difficulty etc.). **3** (of weather) become fine. **4** disappear. **clear the way 1** remove obstacles. **2** stand aside. **clear a thing with** get approval or authorization for a thing from (a person). **in clear** not in cipher or code. **in the clear** free from suspicion or difficulty. **out of a clear sky** as a complete surprise. □ **clearable** *adj.* **clearer** *n.* **clearly** *adv.* **clearness** *n.* [ME f. OF *cler* f. L *clarus*]

clearance /ˈklɪərəns/ *n.* **1 a** the removal of obstructions etc., esp. removal of buildings, persons, etc., so as to clear land. **b** the removal of contents, esp. of a house. **2** clear space allowed for the passing of two objects or two parts in machinery etc. **3** special authorization or permission (esp. for an aircraft to take off or land, or for access to information etc.). **4 a** the clearing of a person, ship, etc., by customs. **b** a certificate showing this. **5** the clearing of cheques. **6** *Football* a kick sending the ball out of a defensive zone. **7** making clear. □ **clearance order** an order for the demolition of buildings. **clearance sale** *Brit.* a sale to dispose of superfluous stock.

clearcole /ˈklɪəkəʊl/ *n. & v.* ● *n.* a mixture of size and whiting or white lead, used as a primer for distemper. ● *v.tr.* paint with clearcole. [F *claire colle* clear glue]

clearing /ˈklɪərɪŋ/ *n.* **1** in senses of CLEAR *v.* **2** an area in a forest cleared for cultivation. □ **clearing bank** *Brit.* a bank which is a member of a clearing-house. **clearing-house 1** a bankers' establishment where cheques and bills from member banks are exchanged, so that only the balances need be paid in cash. **2** an agency for collecting and distributing information etc.

clearstory *US* var. of CLERESTORY.

clearway /ˈklɪəweɪ/ *n. Brit.* a main road (other than a motorway) on which vehicles are not normally permitted to stop.

clearwing /ˈklɪəwɪŋ/ *n.* a day-flying moth of the family Sesiidae, with largely transparent wings and often resembling a wasp etc.

cleat /kliːt/ *n.* **1** a piece of metal, wood, etc., bolted on for fastening ropes to, or to strengthen woodwork etc. **2** a projecting piece on a spar, gangway, boot, etc., to give footing or prevent a rope from slipping. **3** a wedge. [OE]

cleavage /ˈkliːvɪdʒ/ *n.* **1** the hollow between a woman's breasts, esp. as exposed by a low-cut garment. **2 a** a division or splitting. **b** *Biol.* cell division, esp. of a fertilized egg cell. **3** the splitting of rocks, crystals, etc., in a preferred direction.

cleave[1] /kliːv/ *v.* (*past* **clove** /kləʊv/ *or* **cleft** /klɛft/ *or* **cleaved**; *past part.* **cloven** /ˈkləʊv(ə)n/ *or* **cleft** *or* **cleaved**) *literary* **1 a** *tr.* chop or break apart, split, esp. along the grain or the line of cleavage. **b** *intr.* come apart in this way. **2** *tr.* make one's way through (air or water). □ **cleavable** *adj.* [OE *clēofan* f. Gmc]

cleave[2] /kliːv/ *v.intr.* (*past* **cleaved** *or archaic* **clave** /kleɪv/) (foll. by *to*) *literary* stick fast; adhere. [OE *cleofian, clifian* f. WG: cf. CLAY]

cleaver /ˈkliːvə(r)/ *n.* **1** a tool for cleaving, esp. a heavy chopping tool used by butchers. **2** a person who cleaves.

cleavers /ˈkliːvəz/ *n.* (also **clivers** /ˈklɪvəz/) (treated as *sing.* or *pl.*) a climbing plant, *Galium aparine*, having hooked bristles on its stem that catch on clothes etc. Also called *goosegrass*. [OE *clife*, formed as CLEAVE[2]]

Cleese /kliːz/, John (Marwood) (b.1939), English comic actor and writer. He began his television writing and acting career in the 1960s, gaining widespread fame with *Monty Python's Flying Circus* (1969–74). He also co-wrote and starred in the situation comedy *Fawlty Towers* (1975–9). John Cleese has also appeared in films, notably *A Fish Called Wanda* (1988), which he wrote.

clef /klɛf/ *n. Mus.* any of several symbols placed at the beginning of a staff, indicating the pitch of the notes written on it. [F f. L *clavis* key]

cleft[1] /klɛft/ *adj.* split, partly divided. □ **cleft lip** a congenital split in the upper lip. **cleft palate** a congenital split in the roof of the mouth. **in a cleft stick** in a difficult position, esp. one allowing neither retreat nor advance. [past part. of CLEAVE[1]]

cleft[2] /klɛft/ *n.* a split or fissure; a space or division made by cleaving. [OE (rel. to CLEAVE[1]): assim. to CLEFT[1]]

cleg /klɛg/ *n. Brit.* a horsefly. [ON *kleggi*]

Cleisthenes /ˈklaɪsθəˌniːz/ (*c.*570 BC–*c.*508 BC), Athenian statesman. From 525 to 524 he served as principal archon of Athens. In 508 he sought to undermine the power of the nobility by forming an alliance with the Popular Assembly, passing laws to assign political responsibility according to citizenship of a locality as opposed to membership of a clan or kinship group. His reforms consolidated the Athenian democratic process begun by Solon and later influenced the policies of Pericles.

cleistogamic /ˌklaɪstəˈɡæmɪk/ *adj. Bot.* (of a flower) permanently closed and self-fertilizing. [Gk *kleistos* closed + *gamos* marriage]

clematis /ˈklɛmətɪs, kləˈmeɪtɪs/ *n.* an erect or climbing ranunculaceous plant of the genus *Clematis*, bearing white, pink, or purple flowers and feathery seeds; e.g. traveller's joy, *C. vitalba*. [L f. Gk *klēmatis* f. *klēma* vine branch]

Clemenceau /ˈklɛmənˌsəʊ/, Georges (Eugène Benjamin) (1841–1929), French statesman, Prime Minister 1906–9 and 1917–20. A radical politician and journalist, he persistently opposed the government during the early years of the First World War, before becoming Premier and seeing France through to victory in 1918. He presided at the Versailles peace talks, where he pushed hard for a punitive settlement with Germany, but failed to obtain all that he demanded (notably the River Rhine as a frontier). He founded the newspaper *L'Aurore*.

Clemens /ˈklɛmənz/, Samuel Langhorne, see TWAIN.

clement /ˈklɛmənt/ *adj.* **1** mild (*clement weather*). **2** merciful. □ **clemency** *n.* [ME f. L *clemens -entis*]

Clement, St /ˈklɛmənt/ (known as St Clement of Rome) (1st century AD), pope *c.*88–*c.*97. Probably the third bishop of Rome after St Peter, he was the author of an epistle written *c.*96 to the Church at Corinth, insisting that certain deposed presbyters be reinstated. In later tradition he became the subject of a variety of legends; one held that he was martyred. Feast day, 23 Nov.

clementine /ˈklɛmənˌtiːn, -ˌtaɪn/ *n.* a small citrus fruit, thought to be a hybrid between a tangerine and sweet orange. [F *clémentine*]

Clement of Alexandria, St /ˈklɛmənt/ (Latin name Titus Flavius Clemens) (*c.*150–*c.*215), Greek theologian. He was head of the catechetical school at Alexandria (*c.*190–202), but was forced to flee from Roman imperial persecution and was succeeded by his pupil Origen. Clement's main contribution to theological scholarship was to relate the ideas of Greek philosophy to the Christian faith. Feast day, 5 Dec.

clench /klentʃ/ v. & n. ● v.tr. **1** close (the teeth or fingers) tightly. **2** grasp firmly. **3** = CLINCH v. 4. ● n. **1** a clenching action. **2** a clenched state. [OE f. Gmc: cf. CLING]

Cleopatra /klɪəˈpætrə/ (also **Cleopatra VII**) (69–30 BC), queen of Egypt 47–30. The last Ptolemaic ruler, she was restored to the throne by Julius Caesar, having been ousted by the guardians of her father Ptolemy Auletes (d.51). After her brief liaison with Caesar she forged a longer political and romantic alliance with Mark Antony, by whom she had three children. Their ambitions for the expansion of the Egyptian empire ultimately brought them into conflict with Rome, and she and Antony were defeated by Octavian at the battle of Actium in 31. She is reputed to have committed suicide by allowing herself to be bitten by an asp.

Cleopatra's Needles a pair of granite obelisks erected at Heliopolis by Tuthmosis III c.1475 BC. They were moved to Alexandria in 12 BC and taken from Egypt in 1878, one being set up on the Thames Embankment in London and the other in Central Park, New York. They have no known historical connection with Cleopatra.

clepsydra /ˈklɛpsɪdrə/ n. an ancient time-measuring device worked by a flow of water. [L f. Gk klepsudra f. kleptō steal + hudōr water]

clerestory /ˈklɪəstərɪ, -ˌstɔːrɪ/ n. (US **clearstory**) (pl. **-ies**) **1** an upper row of windows in a cathedral or large church, above the level of the aisle roofs. **2** US a raised section of the roof of a railway carriage, with windows or ventilators. [ME f. CLEAR + STOREY]

clergy /ˈklɜːdʒɪ/ n. (pl. **-ies**) (usu. prec. by the) **1** the body of all persons ordained for religious duties in the Christian churches. **2** a number of such persons (ten clergy were present). [ME, partly f. OF clergé f. eccl.L cléricatus, partly f. OF clergie f. clerc CLERK]

clergyman /ˈklɜːdʒɪmən/ n. (pl. **-men**) a member of the clergy, esp. of the Church of England.

cleric /ˈklɛrɪk/ n. a member of the clergy. [(orig. adj.) f. eccl.L f. Gk klērikos f. klēros lot, heritage, as in Acts 1:17]

clerical /ˈklɛrɪk(ə)l/ adj. **1** of the clergy or clergymen. **2** of or done by a clerk or clerks. □ **clerical collar** a stiff upright white collar fastening at the back, as worn by the clergy in some Churches. **clerical error** an error made in copying or writing out. □ **clericalism** n. **clericalist** n. **clerically** adv. [eccl.L clericalis (as CLERIC)]

clerihew /ˈklɛrɪˌhjuː/ n. a short comic or nonsensical verse, usu. in two rhyming couplets with lines of unequal length and referring to a famous person. [E. Clerihew Bentley, Engl. writer (1875–1956), its inventor]

clerk /klɑːk/ n. & v. ● n. **1** a person employed in an office, bank, shop, etc., to keep records, accounts, etc. **2** a secretary, agent, or record-keeper of a local council (town clerk), court, etc. **3** a lay officer of a church (parish clerk), college chapel, etc. **4** a senior official in Parliament. **5** N. Amer. an assistant in a shop or hotel. **6** archaic a clergyman. ● v.intr. work as a clerk. □ **clerk in holy orders** formal a clergyman. **clerk of the course** the judges' secretary etc. in horse- or motor-racing. **clerk of the works** (or **of works**) an overseer of building works etc. □ **clerkdom** n. **clerkess** n. Sc. **clerkish** adj. **clerkly** adj. **clerkship** n. [OE cleric, clerc, & OF clerc, f. eccl.L cléricus CLERIC]

Clermont-Ferrand /ˌkleəmɒnfeˈrɒn/ an industrial city in central France, capital of the Auvergne region, at the centre of the Massif Central; pop. (1990) 140,170.

Cleveland[1] /ˈkliːvlənd/ **1** a county on the North Sea coast of NE England; administrative centre, Middlesbrough. It was formed in 1974 from parts of Durham and North Yorkshire. **2** a major port and industrial city in NE Ohio, situated on Lake Erie; pop (1990) 505,600.

Cleveland[2] /ˈkliːvlənd/, (Stephen) Grover (1837–1908), American Democratic statesman, 22nd and 24th President of the US 1885–9 and 1893–7. His first term was marked by efforts to reverse the heavily protective import tariff, and his second by his application of the Monroe doctrine to Britain's border dispute with Venezuela (1895).

clever /ˈklevə(r)/ adj. (**cleverer, cleverest**) **1 a** skilful, talented; quick to understand and learn. **b** colloq. showing good sense or wisdom, wise. **2** adroit, dexterous. **3** (of the doer or the thing done) ingenious, cunning. □ **clever Dick** (or **clogs** etc.) colloq. a person who is or purports to be smart or knowing. □ **cleverly** adv. **cleverness** n. [ME, = adroit: perh. rel. to CLEAVE², with sense 'apt to seize']

clevis /ˈklevɪs/ n. **1** a U-shaped piece of metal at the end of a beam for attaching tackle etc. **2** a connection in which a bolt holds one part that fits between the forked ends of another. [16th c.: rel. to CLEAVE¹]

clew /kluː/ n. & v. ● n. **1** Naut. **a** a lower or after corner of a sail. **b** a set of small cords suspending a hammock. **2** archaic **a** a ball of thread or yarn, esp. with reference to the legend of Theseus and the labyrinth. **b** = CLUE n. ● v.tr. Naut. **1** (foll. by up) draw the lower ends of (a sail) to the upper yard or the mast ready for furling. **2** (foll. by down) let down (a sail) by the clews in unfurling. [OE cliwen, cleowen]

clianthus /klɪˈænθəs/ n. a leguminous plant of the genus Clianthus, native to Australia and New Zealand, bearing drooping clusters of red pealike flowers. [mod.L, app. f. Gk klei-, kleos glory + anthos flower]

cliché /ˈkliːʃeɪ/ n. (also **cliche**) **1** a hackneyed phrase or opinion. **2** Brit. Printing a metal casting of a stereotype or electrotype. [F f. clicher to stereotype]

clichéd /ˈkliːʃeɪd/ adj. (also **cliched**) hackneyed; full of clichés.

click /klɪk/ n. & v. ● n. **1** a slight sharp sound as of a switch being operated. **2** Phonet. a sharp non-vocal suction, used as a speech sound in some languages. **3** a catch in machinery acting with a slight sharp sound. **4** an action causing a horse's hind foot to touch the shoe of a fore foot. ● v. **1** a intr. make a click. **b** tr. cause (one's tongue, heels, etc.) to click. **2** intr. colloq. **a** become clear or understandable (often prec. by it as subject: when I saw them it all clicked). **b** be successful, secure one's object. **c** (foll. by with) become friendly, esp. with a person of the opposite sex. **d** come to an agreement. **3** intr. & tr. (often foll. by on) Computing press (one of the buttons on a mouse); select (an item represented on the screen, a particular function, etc.) by so doing. □ **click beetle** a beetle of the family Elateridae, springing up with a click when lying on its back (also called skipjack, elater). □ **clicker** n. [imit.: cf. Du. klikken, F cliquer]

client /ˈklaɪənt/ n. **1** a person using the services of a lawyer, architect, social worker, or other professional person. **2** a customer. **3** Rom. Hist. a plebeian under the protection of a patrician. **4** archaic a dependant or hanger-on. □ **clientship** n. [ME f. L cliens -entis f. cluere hear, obey]

clientele /ˌkliːɒnˈtel/ n. **1** clients collectively. **2** customers, esp. of a shop. **3** the patrons of a theatre etc. [L clientela clientship & F clientèle]

Clifden nonpareil /ˈklɪfdən/ n. a large Eurasian noctuid moth, Catocala fraxini, with blue and black hindwings. [Clifden former name of Cliveden, village in England + NONPAREIL]

cliff /klɪf/ n. a steep rock-face, esp. at the edge of the sea. □ **cliff-hanger** a story etc. with a strong element of suspense; a suspenseful ending to an episode of a serial. **cliff-hanging** full of suspense. □ **clifflike** adj. **cliffy** adj. [OE clif f. Gmc]

climacteric /klaɪˈmæktərɪk, ˌklaɪmækˈterɪk/ n. & adj. ● n. **1** Med. the period of life when fertility and sexual activity are in decline. **2** a supposed critical period in life (esp. occurring at intervals of seven years). ● adj. **1** Med. occurring at the climacteric. **2** constituting a crisis; critical. [F climatérique or L climactericus f. Gk klimaktērikos f. klimaktēr critical period f. klimax -akos ladder, climax]

climactic /klaɪˈmæktɪk/ adj. of or forming a climax. □ **climactically** adv. [CLIMAX + -IC, perh. after SYNTACTIC or CLIMACTERIC]

climate /ˈklaɪmɪt/ n. **1** the prevailing weather conditions of an area. **2** a region with particular weather conditions. **3** the prevailing trend of opinion or public feeling. □ **climatic** /klaɪˈmætɪk/ adj. **climatical** adj. **climatically** adv. [ME f. OF climat or LL clima climat- f. Gk klima f. klinō slope]

climatology /ˌklaɪməˈtɒlədʒɪ/ n. the scientific study of climate. □ **climatologist** n. **climatological** /-təˈlɒdʒɪk(ə)l/ adj.

climax /ˈklaɪmæks/ n. & v. ● n. **1** the event or point of greatest intensity or interest; a culmination or apex. **2** a sexual orgasm. **3** Rhet. **a** a series arranged in order of increasing importance etc. **b** the last term in such a series. **4** Ecol. a state of equilibrium reached by a plant community. ● v.tr. & intr. colloq. bring or come to a climax. [LL f. Gk klimax -akos ladder, climax]

climb /klaɪm/ v. & n. ● v. **1** tr. & intr. (often foll. by up) ascend, mount, go or come up, esp. by using one's hands. **2** intr. (of a plant) grow up a wall, tree, trellis, etc. by clinging with tendrils or by twining. **3** intr. move along by grasping or clinging; go with effort, clamber (climbed across the ditch; climbed into bed). **4** intr. make progress in social rank, intellectual or moral strength, etc. **5** intr. (of an aircraft, the sun, etc.) go upwards. **6** intr. slope upwards. ● n. **1** an ascent by climbing. **2 a** a place, esp. a hill, climbed or to be climbed. **b** a recognized route up a mountain etc. □ **climb down 1** descend with the help of one's hands. **2** withdraw from a position taken up in argument, negotiation, etc. **climb-down** n. such a withdrawal. **climbing-frame** a structure of joined bars etc. for children to climb on. **climbing-iron** a set of spikes attachable to a boot for climbing trees or ice slopes. **climbing perch** a southern

Asian freshwater fish, *Anabas testudinens*, able to breathe air and move over land (also called *anabas*). **climb into** put on (clothes). □ **climbable** *adj.* [OE *climban* f. WG, rel. to CLEAVE[2]]

climber /ˈklaɪmə(r)/ *n.* **1** a mountaineer. **2** a climbing plant. **3** a person with strong social etc. aspirations.

clime /klaɪm/ *n. literary* **1** a region. **2** a climate. [LL *clima*: see CLIMATE]

clinch /klɪntʃ/ *v. & n.* ● *v.* **1** *tr.* confirm or settle (an argument, bargain, etc.) conclusively. **2** *intr. Boxing & Wrestling* (of participants) become too closely engaged. **3** *intr. colloq.* embrace. **4** *tr.* secure (a nail or rivet) by driving the point sideways when through. **5** *tr. Naut.* fasten (a rope) with a particular half hitch. ● *n.* **1 a** a clinching action. **b** a clinched state. **2** *colloq.* an (esp. amorous) embrace. **3** *Boxing & Wrestling* an action or state in which participants become too closely engaged. [16th-c. var. of CLENCH]

clincher /ˈklɪntʃ(ə)r/ *n. colloq.* a remark or argument that settles a matter conclusively.

clincher-built var. of CLINKER-BUILT.

cline /klaɪn/ *n.* **1** *Biol.* a graded sequence of differences within a species etc., esp. between different parts of its geographical range. **2** a continuum with an infinite number of gradations. □ **clinal** *adj.* [Gk *klinein* to slope]

cling /klɪŋ/ *v. & n.* ● *v.intr.* (*past* and *past part.* **clung** /klʌŋ/) **1** (foll. by *to*) adhere, stick, or hold on (by means of stickiness, suction, grasping, or embracing). **2** (foll. by *to*) remain persistently or stubbornly faithful (to a friend, habit, idea, etc.). **3** maintain one's grasp; keep hold; resist separation. ● *n.* = CLINGSTONE. □ **cling film** a very thin clinging transparent plastic film, used as a covering esp. for food. **cling together** remain in one body or in contact. □ **clinger** *n.* **clingingly** *adv.* [OE *clingan* f. Gmc: cf. CLENCH]

clingstone /ˈklɪŋstəʊn/ *n.* a variety of peach or nectarine in which the flesh adheres to the stone (cf. FREESTONE 2).

clingy /ˈklɪŋɪ/ *adj.* (**clingier, clingiest**) **1** (of clothes) close-fitting. **2** (of a child) timid and tending to cling to a parent. □ **clinginess** *n.*

clinic /ˈklɪnɪk/ *n.* **1** *Brit.* a private or specialized hospital. **2** a place or occasion for giving specialist medical treatment or advice (*eye clinic*; *fertility clinic*). **3** a gathering at a hospital bedside for the teaching of medicine or surgery. **4** *N. Amer.* a conference or short course on a particular subject (*golf clinic*). □ **clinician** /klɪˈnɪʃ(ə)n/ *n.* [F *clinique* f. Gk *klinikē* (*tekhnē*) clinical, lit. 'bedside (art)']

clinical /ˈklɪnɪk(ə)l/ *adj.* **1** *Med.* **a** of or for the treatment of patients. **b** taught or learned at the hospital bedside. **2** dispassionate, coldly detached. **3** (of a room, building, etc.) bare, functional. □ **clinical death** death judged by observation of a person's condition. **clinical medicine** medicine dealing with the observation and treatment of patients. **clinical thermometer** a thermometer with a small range, for taking a person's temperature. □ **clinically** *adv.* [L *clinicus* f. Gk *klinikos* f. *klinē* bed]

clink[1] /klɪŋk/ *n. & v.* ● *n.* a sharp ringing sound. ● *v.* **1** *intr.* make a clink. **2** *tr.* cause (glasses etc.) to clink. [ME, prob. f. MDu. *klinken*; cf. CLANG, CLANK]

clink[2] /klɪŋk/ *n.* (often prec. by *in*) *sl.* prison. [16th c.: orig. unkn.]

clinker[1] /ˈklɪŋkə(r)/ *n.* **1** a mass of slag or lava. **2** a stony residue from burnt coal. [earlier *clincard* etc. f. obs. Du. *klinkaerd* f. *klinken* CLINK[1]]

clinker[2] /ˈklɪŋkə(r)/ *n.* **1** *Brit. sl.* something excellent or outstanding. **2** *N. Amer. sl.* a mistake or blunder. [CLINK[1] + -ER[1]]

clinker-built /ˈklɪŋkəˌbɪlt/ *adj.* (also **clincher-built** /ˈklɪntʃə-/) (of a boat) having external planks overlapping downwards and secured with clinched copper nails. [*clink* N.Engl. var. of CLINCH + -ER[1]]

clinkstone /ˈklɪŋkstəʊn/ *n. Mineral.* a kind of feldspar that rings like iron when struck.

clinometer /klaɪˈnɒmɪtə(r)/ *n. Surveying* an instrument for measuring slopes. [Gk *klinō* to slope + -METER]

clint /klɪnt/ *n. Geol.* a mass of rock between fissures (grikes) in a limestone pavement. [Danish, Sw. *klint* f. OSw. *klinter* rock]

Clinton /ˈklɪntən/, William Jefferson ('Bill') (b.1946), American Democratic statesman, 42nd President of the US (since 1993). He was governor of Arkansas 1979–81 and 1983–92; during his presidential campaign he pledged to reduce the large US budget deficit, introduce radical health-care reforms, and end discrimination against homosexuals in the armed forces. Problems at home during the early part of his presidency included contested allegations of financial and sexual misconduct and difficulty in obtaining congressional support for proposed measures, while foreign policy was concerned mainly with conflicts in Bosnia, Somalia, and Haiti.

Clio /ˈklaɪəʊ/ *Gk & Rom. Mythol.* the Muse of history. [Gk *kleiō* celebrate]

cliometrics /ˌklaɪə'metrɪks/ *n.pl.* (usu. treated as *sing.*) a method of historical research making much use of statistical information and methods. [CLIO + METRIC + -ICS]

clip[1] /klɪp/ *n. & v.* ● *n.* **1** a device for holding things together or for attachment to an object as a marker, esp. a paper-clip or a device worked by a spring. **2** a piece of jewellery fastened by a clip. **3** a set of attached cartridges for a firearm. ● *v.tr.* (**clipped, clipping**) **1** fix with a clip. **2** grip tightly. **3** surround closely. □ **clip-on** attached by a clip. [OE *clyppan* embrace f. WG]

clip[2] /klɪp/ *v. & n.* ● *v.tr.* (**clipped, clipping**) **1** cut with shears or scissors, esp. cut short or trim (hair, wool, etc.). **2** trim or remove the hair or wool of (a person or animal). **3** *colloq.* hit smartly. **4 a** curtail, diminish, cut short. **b** omit (a letter etc.) from a word; omit letters or syllables of (words pronounced). **5** *Brit.* remove a small piece of (a ticket) to show that it has been used. **6** cut (an extract) from a newspaper etc. **7** *sl.* swindle, rob. **8** pare the edge of (a coin). ● *n.* **1** an act of clipping, esp. shearing or hair-cutting. **2** *colloq.* a smart blow, esp. with the hand. **3** a short sequence from a motion picture. **4** the quantity of wool clipped from a sheep, flock, etc. **5** *colloq.* speed, esp. rapid. □ **clip-joint** *sl.* a nightclub etc. charging exorbitant prices. **clip a person's wings** prevent a person from pursuing ambitions or acting effectively. □ **clippable** *adj.* [ME f. ON *klippa*, prob. imit.]

clipboard /ˈklɪpbɔːd/ *n.* a small board with a spring clip for holding papers etc. and providing support for writing.

clip-clop /ˈklɪpklɒp/ *n. & v.* ● *n.* a sound such as the beat of a horse's hooves. ● *v.intr.* (**-clopped, -clopping**) make such a sound. [imit.]

clipper /ˈklɪpə(r)/ *n.* **1** (usu. in *pl.*) any of various instruments for clipping hair, fingernails, hedges, etc. **2** a fast sailing-ship, esp. one with raking bows and masts. (*See note below.*) **3** a fast horse.

▪ The term *clipper* was first applied to the speedy schooners built in Virginia and Maryland in the early 19th century; it is said to have arisen because such ships could 'clip' the passage-time of the regular packet ships. The hull design was long and low with a very sharply raked stem, and was later combined with the three-masted square rig, making the clippers of the mid-19th century among the finest productions of the age of sail. The first British clippers were built for the tea trade, a profitable cargo, with the first arrivals in London from China each year commanding the highest prices; the best known of the tea-clippers is the *Cutty Sark*, now preserved on the dockside at Greenwich in London. The opening of the Suez Canal in 1869 removed the need for fast ships that could travel round the Cape of Good Hope, and heralded the end of the clipper era.

clippie /ˈklɪpɪ/ *n. Brit. colloq.* a bus conductress.

clipping /ˈklɪpɪŋ/ *n.* a piece clipped or cut from something, esp. from a newspaper, hedge, etc.

clique /kliːk/ *n.* a small exclusive group of people. □ **cliquey** *adj.* **cliquish** *adj.* **cliquishness** *n.* [F f. *cliquer* CLICK]

C.Lit. *abbr.* (in the UK) Companion of Literature.

clitic /ˈklɪtɪk/ *n.* (often *attrib.*) *Gram.* an enclitic or proclitic. □ **cliticization** /ˌklɪtɪkaɪˈzeɪʃ(ə)n/ *n.*

clitoridectomy /ˌklɪtərɪˈdektəmɪ/ *n.* (pl. **-ies**) the surgical removal of the clitoris.

clitoris /ˈklɪtərɪs, ˈklaɪ-/ *n.* a small erectile part of the female genitals at the anterior end of the vulva. □ **clitoral** *adj.* [mod.L f. Gk *kleitoris*]

Clive /klaɪv/, Robert, 1st Baron Clive of Plassey (known as Clive of India) (1725–74), British general and colonial administrator. In 1743 he joined the East India Company in Madras, becoming governor of Fort St David in 1755. Following the Black Hole of Calcutta incident, he commanded the forces that recaptured Calcutta from Siraj-ud-Dawlah (c.1729–57), nawab of Bengal in 1757. Clive's victory at Plassey later that year made him virtual ruler of Bengal, helping the British to gain an important foothold in India. After returning to Britain 1760–5, he served as governor of Bengal until 1767, restructuring the administration of the colony and restoring discipline to the East India Company, whose reputation had been called into question. Clive was subsequently implicated in the company's corruption scandals; although officially exonerated, he committed suicide.

clivers var. of CLEAVERS.

Cllr. *abbr. Brit.* Councillor.

cloaca /kləʊˈeɪkə/ *n.* (pl. **cloacae** /-ˈeɪsiː/) **1** *Zool.* the genital and excretory cavity at the end of the intestinal canal in birds, reptiles, etc. **2** a sewer. □ **cloacal** adj. [L, = sewer]

cloak /kləʊk/ n. & v. ● n. **1** an outdoor over-garment, usu. sleeveless, hanging loosely from the shoulders. **2** a covering (cloak of snow). **3** (in pl.) = CLOAKROOM. ● v.tr. **1** cover with a cloak. **2** conceal, disguise. □ **cloak-and-dagger** involving intrigue and espionage. **under the cloak of** using as a pretext. [ME f. OF cloke, dial. var. of cloche bell, cloak (from its bell shape) f. med.L clocca bell: see CLOCK[1]]

cloakroom /'kləʊkruːm, -rʊm/ n. **1** a room where outdoor clothes or luggage may be left by visitors, clients, etc. **2** Brit. euphem. a lavatory.

clobber[1] /'klɒbə(r)/ n. Brit. sl. clothing or personal belongings. [19th c.: orig. unkn.]

clobber[2] /'klɒbə(r)/ v.tr. sl. **1** hit repeatedly; beat up. **2** defeat. **3** criticize severely. [20th c.: orig. unkn.]

cloche /klɒʃ, kləʊʃ/ n. **1** a small translucent cover for protecting or forcing outdoor plants. **2** (in full **cloche hat**) a woman's close-fitting bell-shaped hat. [F, = bell, f. med.L clocca: see CLOCK[1]]

clock[1] /klɒk/ n. & v. ● n. **1** an instrument for measuring time, driven mechanically or electrically and indicating hours, minutes, etc., by hands on a dial or by displayed figures. (See note below.) **2 a** any measuring device resembling a clock. **b** colloq. a speedometer, taximeter, or stopwatch. **3** time taken as an element in competitive sports etc. (ran against the clock). **4** Brit. sl. a person's face. **5** a downy seedhead, esp. that of a dandelion. ● v.tr. **1** colloq. **a** (often foll. by up) attain or register (a stated time, distance, or speed, esp. in a race). **b** time (a race) with a stopwatch. **2** Brit. sl. hit, esp. on the head. **3** sl. take note of (a person or thing); watch, stare at. □ **clock golf** a game in which a golf ball is putted into a hole from successive points in a circle. **clock in** (or **on**) register one's arrival at work, esp. by means of an automatic recording clock. **clock off** (or **out**) register one's departure similarly. **clock radio** a combined radio and alarm clock. **round the clock** all day and (usu.) night. **watch the clock** = CLOCK-WATCH. [ME f. MDu., MLG klocke f. med.L clocca bell, perh. f. Celt.]

▪ The first true mechanical clock dates from c.1280. Such clocks were large weight-driven structures fitted into towers; they had no dial or hands, but sounded a signal which alerted a keeper to toll a bell. In the 14th century public striking-clocks appeared, using a foliot or weighted arm as an oscillating flywheel controlled by a toothed wheel and an escapement mechanism. The first spring-driven clocks appeared c.1500, and led to the development of watches. The 17th century saw the introduction of the anchor escapement and the balance-wheel, followed by the balance-spring, whose frequency of oscillation was controlled by a spiral spring of the now familiar type. In 1657 Christiaan Huygens first used the swinging pendulum, a great advance because it is free from the friction of gears to which the foliot was subject. Electrically driven clocks were introduced by 1840, and the quartz clock in 1927–30, using a quartz crystal maintained in oscillation at a fixed high frequency. Most accurate of all is the caesium clock, depending on the vibration of atoms of caesium. (See also WATCH, WATER-CLOCK.)

clock[2] /klɒk/ n. an ornamental pattern on the side of a stocking or sock near the ankle. [16th c.: orig. unkn.]

clock-watch /'klɒkwɒtʃ/ v.intr. work over-anxiously to time, esp. so as not to exceed minimum working hours. □ **clock-watcher** n. **clock-watching** n.

clockwise /'klɒkwaɪz/ adj. & adv. in a curve corresponding in direction to the movement of the hands of a clock.

clockwork /'klɒkwɜːk/ n. **1** a mechanism like that of a mechanical clock, with a spring and gears. **2** (attrib.) **a** driven by clockwork. **b** regular, mechanical. □ **like clockwork** smoothly, regularly, automatically.

clod /klɒd/ n. **1** a lump of earth, clay, etc. **2** colloq. a silly or foolish person. **3** meat cut from the neck of an ox. [ME: var. of CLOT]

cloddish /'klɒdɪʃ/ adj. loutish, foolish, clumsy. □ **cloddishly** adv. **cloddishness** n.

clodhopper /'klɒdˌhɒpə(r)/ n. colloq. **1** (usu. in pl.) a large heavy shoe. **2** = CLOD 2.

clodhopping /'klɒdˌhɒpɪŋ/ adj. colloq. = CLODDISH.

clodpoll /'klɒdpɒl/ n. colloq. = CLOD 2.

clog /klɒg/ n. & v. ● n. **1** a shoe with a thick wooden sole. **2** archaic an encumbrance or impediment. **3** a block of wood to impede an animal's movement. ● v. (**clogged, clogging**) **1** (often foll. by up) **a** tr. obstruct, esp. by accumulation of glutinous matter. **b** intr. become obstructed. **2** tr. impede, hamper. **3** tr. & intr. (often foll. by up) fill with glutinous or

choking matter. □ **clog-dance** a dance performed in clogs. [ME: orig. unkn.]

cloggy /'klɒgɪ/ adj. (**cloggier, cloggiest**) **1** lumpy, knotty. **2** sticky.

cloisonné /'klwazɒˌneɪ/ n. & adj. ● n. **1** an enamel finish produced by forming areas of different colours separated by strips of wire placed edgeways on a metal backing. **2** this process. ● adj. (of enamel) made by this process. [F f. cloison compartment]

cloister /'klɔɪstə(r)/ n. & v. ● n. **1** a covered walk, often with a wall on one side and a colonnade open to a quadrangle on the other, esp. in a convent, monastery, college, or cathedral. **2** monastic life or seclusion. **3** a convent or monastery. ● v.tr. seclude or shut up, usu. in a convent or monastery. □ **cloistral** adj. [ME f. OF cloistre f. L claustrum, clostrum lock, enclosed place f. claudere claus- CLOSE[2]]

cloistered /'klɔɪstəd/ adj. **1** secluded, sheltered. **2** monastic.

clomp var. of CLUMP v. 2.

clone /kləʊn/ n. & v. ● n. **1 a** a group of cells or organisms produced asexually from one stock or ancestor. **b** one such organism. **2** a person or thing regarded as identical with another. ● v.tr. propagate as a clone. □ **clonal** adj. [Gk klōn twig, slip]

clonk /klɒŋk/ n. & v. ● n. an abrupt heavy sound of impact. ● v. **1** intr. make such a sound. **2** tr. colloq. hit. [imit.]

Clonmel /klɒn'mel/ the county town of Tipperary, in the Republic of Ireland; pop. (1991) 14,500.

clonus /'kləʊnəs/ n. Physiol. a spasm with alternate muscular contractions and relaxations. □ **clonic** /'klɒnɪk/ adj. [Gk klonos turmoil]

clop /klɒp/ n. & v. ● n. the sound made by a horse's hooves. ● v.intr. (**clopped, clopping**) make this sound. [imit.]

cloqué /'kləʊkeɪ/ n. a fabric with an irregularly raised surface. [F, = blistered]

close[1] /kləʊs/ adj., adv., & n. ● adj. **1** (often foll. by to) situated at only a short distance or interval. **2 a** having a strong or immediate relation or connection (close friend; close relative). **b** in intimate friendship or association (were very close). **c** corresponding almost exactly (close resemblance). **d** fitting tightly (close cap). **e** (of hair etc.) short, near the surface. **3** in or almost in contact (close combat; close proximity). **4** dense, compact, with no or only slight intervals (close texture; close writing; close formation; close thicket). **5** in which competitors are almost equal (close contest; close election). **6** leaving no gaps or weaknesses, rigorous (close reasoning). **7** concentrated, searching (close examination; close attention). **8** (of air etc.) stuffy or humid. **9 a** closed, shut. **b** shut up, under secure confinement. **10** limited or restricted to certain persons etc. (close corporation; close scholarship). **11 a** hidden, secret, covered. **b** secretive. **12** (of a danger etc.) directly threatening, narrowly avoided (that was close). **13** niggardly. **14** (of a vowel) pronounced with a relatively narrow opening of the mouth. **15** narrow, confined, contracted. **16** under prohibition. ● adv. **1** (often foll. by by, on, to, upon) at only a short distance or interval (they live close by; close to the church). **2** closely, in a close manner (shut close). ● n. **1** an enclosed space. **2** Brit. a street closed at one end. **3** Brit. the precinct of a cathedral. **4** Brit. a school playing-field or playground. **5** Sc. an entry from the street to a common stairway or to a court at the back. □ **at close quarters** very close together. **close-fisted** niggardly. **close-fitting** (of a garment) fitting close to the body. **close-grained** without gaps between fibres etc. **close harmony** harmony in which the notes of the chord are close together. **close-hauled** (of a ship) with the sails hauled aft to sail close to the wind. **close-knit** tightly bound or interlocked; closely united in friendship. **close-mouthed** reticent. **close season** Brit. the season when something, esp. the killing of game etc., is illegal. **close-set** set close together; separated only by a small interval or intervals. **close shave** colloq. a narrow escape. **close to the wind** see SAIL. **close-up 1** a photograph etc. taken at close range and showing the subject on a large scale. **2** an intimate description. **go close** (of a racehorse) win or almost win. □ **closely** adv. **closeness** n. **closish** adj. [ME f. OF clos f. L clausum enclosure & clausus past part. of claudere shut]

close[2] /kləʊz/ v. & n. ● v. **1 a** tr. shut (a lid, box, door, room, house, etc.). **b** intr. become shut (the door closed slowly). **c** tr. block up. **2 a** tr. & intr. bring or come to an end. **b** intr. finish speaking (closed with an expression of thanks). **c** tr. settle (a bargain etc.). **3 a** tr. end the day's business. **b** tr. end the day's business at (a shop, office, etc.). **4** tr. & intr. bring or come closer or into contact (close ranks). **5** tr. make (an electric circuit etc.) continuous. **6** intr. (foll. by with) express agreement (with an offer, terms, or the person offering them). **7** intr. (often foll. by with) come within striking distance; grapple. **8** intr. (foll. by on) (of a hand, box, etc.) grasp or entrap. ● n. **1** a conclusion, an end. **2** Mus. a cadence. □ **close down**

1 discontinue (or cause to discontinue) business, esp. permanently. **2** *Brit.* (of a broadcasting station) end transmission esp. until the next day. **close one's eyes 1** (foll. by *to*) pay no attention. **2** die. **close in 1** enclose. **2** come nearer. **3** (of days) get successively shorter with the approach of the winter solstice. **close off** prevent access to by blocking or sealing the entrance. **close out** *N. Amer.* discontinue, terminate, dispose of (a business). **close up 1** (often foll. by *to*) move closer. **2** shut, esp. temporarily. **3** block up. **4** (of an aperture) grow smaller. **5** coalesce. **closing-time** the time at which a public house, shop, etc., ends business. □ **closable** *adj.* **closer** *n.* [ME f. OF *clos-* stem of *clore* f. L *claudere* shut]

closed /kləʊzd/ *adj.* **1** not giving access; shut. **2** (of a shop etc.) having ceased business temporarily. **3** (of a society, system, etc.) self-contained; not communicating with others. **4** (of a sport etc.) restricted to specified competitors etc. □ **closed book** see BOOK. **closed-circuit** (of television) transmitted by wires to a restricted set of receivers. **closed-end** having a predetermined extent (cf. *open-ended*). **closed season** *N. Amer.* = *close season* (see CLOSE¹). **closed shop 1** a place of work etc. where all employees must belong to an agreed trade union. **2** this system. **closed syllable** a syllable ending in a consonant.

closed universe *n. Astron.* the condition in which there is sufficient matter in the universe to halt the expansion driven by the big bang and cause eventual re-collapse. Current observations suggest that the amount of visible matter is only a tenth of that required for closure, but apparent uncertainties in the masses of galaxies leave open the possibility of there being large quantities of dark matter.

closet /ˈklɒzɪt/ *n. & v.* ● *n.* **1** a small or private room. **2** a cupboard or recess. **3** = *water-closet.* **4** (*attrib.*) secret, covert (*closet homosexual*). ● *v.tr.* (**closeted, closeting**) shut away, esp. in private conference or study. □ **closet play** a play to be read rather than acted. **come out of the closet** stop hiding something about oneself, esp. one's homosexuality. [ME f. OF, dimin. of *clos:* see CLOSE¹]

Closet, Clerk of the *n.* (in the UK) the sovereign principal chaplain.

clostridial /klɒˈstrɪdɪəl/ *adj. Med.* of, relating to, or caused by rod-shaped bacteria of the genus *Clostridium,* many of which cause disease (e.g. tetanus, botulism). [mod.L f. Gk *klōstēr* spindle]

closure /ˈkləʊʒə(r)/ *n. & v.* ● *n.* **1** the act or process of closing. **2** a closed condition. **3** something that closes or seals, e.g. a cap or tie. **4** in a legislative assembly, a procedure for ending a debate and taking a vote. ● *v.tr.* apply the closure to (a motion, speakers, etc.) in a legislative assembly. [ME f. OF f. LL *clausura* f. *claudere* *claus-* CLOSE²]

clot /klɒt/ *n. & v.* ● *n.* **1 a** a thick mass of coagulated liquid, esp. of blood. **b** a mass of material stuck together. **2** *Brit. colloq.* a silly or foolish person. ● *v.tr. & intr.* (**clotted, clotting**) form into clots. □ **clotted cream** esp. *Brit.* thick cream obtained by slow scalding. [OE *clot(t)* f. WG]

cloth /klɒθ/ *n.* (*pl.* **cloths** /klɒθs, klɒðz/) **1** woven or felted material. **2** a piece of this. **3** a piece of cloth for a particular purpose; a tablecloth, dishcloth, etc. **4** woollen woven fabric as used for clothes. **5 a** a profession or status, esp. of the clergy, as shown by clothes (*respect due to his cloth*). **b** (prec. by *the*) the clergy. □ **cloth-cap** relating to or associated with the working class. **cloth-eared** *colloq.* somewhat deaf or insensitive to sound. **cloth of gold** (or **silver**) tissue of gold (or silver) threads interwoven with silk or wool. **man of the cloth** see MAN. [OE *clāth,* of unkn. orig.]

clothe /kləʊð/ *v.tr.* (*past* and *past part.* **clothed** or *archaic* or *literary* **clad**) **1** put clothes on; provide with clothes. **2** cover as with clothes or a cloth. **3** (foll. by *with*) invest or provide (with qualities etc.). [OE: rel. to CLOTH]

clothes /kləʊðz/ *n.pl.* **1** garments worn to cover the body and limbs. **2** bedclothes. □ **clothes-horse 1** a frame for airing washed clothes. **2** *colloq.* an affectedly fashionable person. **clothes-line** a rope or wire etc. on which washed clothes are hung to dry. **clothes-moth** a small moth of the family Tineidae, with a larva destructive to wool, fur, etc. **clothes-peg** *Brit.* a clip or forked device for securing clothes to a clothes-line. **clothes-pin** *N. Amer.* a clothes-peg. [OE *clāthas* pl. of *clāth* CLOTH]

clothier /ˈkləʊðɪə(r)/ *n.* a seller of men's clothes. [ME *clother* f. CLOTH]

clothing /ˈkləʊðɪŋ/ *n.* clothes collectively.

Clotho /ˈkləʊθəʊ/ *Gk Mythol.* one of the three Fates. [Gk, = she who spins]

cloture /ˈkləʊtʃə(r), ˈkləʊtjʊə(r)/ *n. & v. US* ● *n.* the closure of a debate. ● *v.tr.* closure. [F *clôture* f. OF CLOSURE]

clou /klu:/ *n.* **1** the point of greatest interest; the chief attraction. **2** the central idea. [F, = nail]

cloud /klaʊd/ *n. & v.* ● *n.* **1** a visible mass of condensed watery vapour in the atmosphere high above the general level of the ground. **2** a mass of smoke or dust. **3** (foll. by *of*) a great number of insects, birds, etc., moving together. **4 a** a state of gloom, trouble, or suspicion. **b** a frowning or depressed look (*a cloud on his brow*). **5** a local dimness or a vague patch of colour in or on a liquid or a transparent body. **6** an unsubstantial or fleeting thing. **7** obscurity. ● *v.* **1** *tr.* cover or darken with clouds or gloom or trouble. **2** *intr.* (often foll. by *over, up*) become overcast or gloomy. **3** *tr.* make unclear. **4** *tr.* variegate with vague patches of colour. □ **cloud-castle** a day-dream. **cloud chamber** *Physics* a device containing vapour for tracking the paths of charged particles, X-rays, and gamma rays. **clouded leopard** a large spotted arboreal feline, *Neofelis nebulosa,* native to SE Asia. **clouded yellow** an orange or yellow and black butterfly of the genus *Crocias.* **cloud-hopping** movement of an aircraft from cloud to cloud esp. for concealment. **cloud-land** a utopia or fairyland. **in the clouds 1** unreal, imaginary, mystical. **2** (of a person) abstracted, inattentive. **on cloud nine** (or **seven**) *colloq.* extremely happy. **under a cloud** out of favour, discredited, under suspicion. **with one's head in the clouds** day-dreaming, unrealistic. □ **cloudless** *adj.* **cloudlessly** *adv.* **cloudlet** *n.* [OE *clūd* mass of rock or earth, prob. rel. to CLOD]

cloudberry /ˈklaʊdbərɪ/ *n.* (*pl.* **-ies**) a dwarf mountain bramble, *Rubus chamaemorus,* with a white flower and an orange-coloured fruit.

cloudburst /ˈklaʊdbɜːst/ *n.* a sudden violent rainstorm.

cloud-cuckoo-land /klaʊdˈkuːkuːˌlænd/ *n.* a fanciful or ideal place. [transl. Gk *Nephelokokkugia* f. *nephelē* cloud + *kokkux* cuckoo (in Aristophanes' *Birds*)]

cloudscape /ˈklaʊdskeɪp/ *n.* **1** a picturesque grouping of clouds. **2** a picture or view of clouds. [CLOUD *n.,* after *landscape*]

cloudy /ˈklaʊdɪ/ *adj.* (**cloudier, cloudiest**) **1 a** (of the sky) covered with clouds, overcast. **b** (of weather) characterized by clouds. **2** not transparent; unclear. □ **cloudily** *adv.* **cloudiness** *n.*

Clouet /kluːˈeɪ/, Jean (*c.*1485–1541) and his son François (*c.*1516–72), French painters. Jean worked as court painter to Francis I (1494–1547); the monarch's portrait in the Louvre is attributed to him. François succeeded his father as court painter, and is chiefly known for his undated portraits of Elizabeth of Austria (now in the Louvre) and Mary, Queen of Scots (now in the Wallace Collection in London).

Clough /klʌf/, Arthur Hugh (1819–61), English poet. He is especially remembered for his longer poems: *The Bothie of Tober-na-Vuolich* (1848) is about a student reading party in Scotland; *Amours de Voyage* (1858) concerns a traveller's spiritual crisis in Rome; *Dipsychus* (1865) is a dialogue reminiscent of Faust and set in Venice.

clough /klʌf/ *n. dial.* a steep valley usu. with a torrent bed; a ravine. [OE *clōh* f. Gmc]

clout /klaʊt/ *n. & v.* ● *n.* **1** a heavy blow. **2** *colloq.* influence, power of effective action, esp. in politics or business. **3** *dial.* a piece of cloth or clothing (*cast not a clout*). **4** *Archery hist.* a piece of canvas on a frame, used as a mark. **5** a nail with a large flat head. **6** a patch. ● *v.tr.* **1** hit hard. **2** mend with a patch. [OE *clūt,* rel. to CLEAT, CLOT]

clove¹ /kləʊv/ *n.* **1 a** a dried flower-bud of a tropical plant, *Eugenia aromatica,* used as a pungent aromatic spice. **b** this plant. **2** (in full **clove gillyflower** or **clove pink**) a clove-scented pink, *Dianthus caryophyllus,* the ancestor of the carnation and other double pinks. [ME f. OF *clou (de girofle)* nail (of gillyflower), from its shape: see GILLYFLOWER]

clove² /kləʊv/ *n.* any of the small bulbs making up a compound bulb of garlic, shallot, etc. [OE *clufu,* rel. to CLEAVE¹]

clove³ *past* of CLEAVE¹.

clove hitch *n.* a knot by which a rope is secured by passing it twice round a spar or rope that it crosses at right angles. [old past part. of CLEAVE¹, as showing parallel separate lines]

cloven /ˈkləʊv(ə)n/ *adj.* split, partly divided. □ **cloven hoof** (or **foot**) the divided hoof of ruminant quadrupeds (e.g. oxen, sheep, goats); also ascribed to the god Pan, and so to the Devil. **show the cloven hoof** reveal one's evil nature. □ **cloven-footed** /ˌkləʊv(ə)nˈfʊtɪd/ *adj.* **cloven-hoofed** /-ˈhuːft/ *adj.* [past part. of CLEAVE¹]

clover /ˈkləʊvə(r)/ *n.* a leguminous fodder plant of the genus *Trifolium,* having dense flower-heads and leaves each consisting of usu. three leaflets. □ **clover leaf** a junction of roads intersecting at different levels with connecting sections forming the pattern of a four-leaved clover. **in clover** in ease and luxury. [OE *clāfre* f. Gmc]

Clovis¹ /ˈkləʊvɪs/ *adj. & n. Archaeol.* of, relating to, or denoting a prehistoric culture and its remains first found at a site near the city of Clovis in eastern New Mexico, referring in particular to a type of heavy leaf-shaped stone projectile point often found in conjunction with bones of mammoths. The points precede the Folsom type, and date from the 10th millennium BC and earlier. (See also FOLSOM.)

Clovis[2] /'kləʊvɪs/ (465–511), king of the Franks 481–511. He succeeded his father Childeric (d.481) as king of the Salian Franks at Tournai, and extended Merovingian rule to Gaul and Germany after victories at Soissons (486) and Cologne (496). After his conversion to Christianity, he championed orthodoxy against the Arian Visigoths, finally defeating them in the battle of Poitiers (507). He made Paris his capital.

clown /klaʊn/ n. & v. ● n. **1** a comic entertainer, esp. in a pantomime or circus, usu. with traditional costume and make-up. (See GRIMALDI[2].) **2** a silly, foolish, or playful person. **3** archaic a rustic. ● v. **1** intr. (often foll. by about, around) behave like a clown; act foolishly or playfully. **2** tr. perform (a part, an action, etc.) like a clown. □ **clownery** n. **clownish** adj. **clownishly** adv. **clownishness** n. [16th c.: perh. of LG orig.]

cloy /klɔɪ/ v.tr. (usu. foll. by with) satiate or sicken with an excess of sweetness, richness, etc. □ **cloyingly** adv. [ME f. earlier accloy f. OF encloer f. med.L inclavare f. L clavus nail]

cloze /kləʊz/ n. the exercise of supplying a word that has been omitted from a passage as a test of readability or comprehension (usu. attrib.: cloze test). [CLOSURE]

club /klʌb/ n. & v. ● n. **1** a heavy stick with a thick end, used as a weapon etc. **2** a stick used in a game, esp. a stick with a head used in golf. **3 a** a playing card of a suit denoted by a black trefoil. **b** (in pl.) this suit. **4** an association of persons united by a common interest, usu. meeting periodically for a shared activity (tennis club; yacht club). **5 a** an organization or premises offering members social amenities, meals and temporary residence, etc. **b** a nightclub. **6** a commercial organization offering subscribers special deals (book club). **7** a group of persons, nations, etc., having something in common. **8** = CLUBHOUSE. **9** a structure or organ, esp. in a plant, with a knob at the end. ● v. (**clubbed, clubbing**) **1** tr. beat with or as with a club. **2** intr. (foll. by together) combine with others for joint action, esp. making up a sum of money for a purpose. **3** tr. contribute (money etc.) to a common stock. **4** intr. colloq. go out to nightclubs (went clubbing every weekend). □ **club-class** a class of fare on aircraft etc. designed for the business traveller. **club-foot** a congenitally deformed foot. **club-footed** having a club-foot. **club-man** (pl. -**men**) a member of one or more clubs (in sense 5 of n.). **club-root** a disease of cabbages etc. with swelling at the base of the stem. **club sandwich** a sandwich with two layers of filling between three slices of toast or bread. **in the club** Brit. sl. pregnant. **on the club** colloq. receiving relief from the funds of a benefit society. □ **clubber** n. [ME f. ON klubba assim. form of klumba club, rel. to CLUMP]

clubbable /'klʌbəb(ə)l/ adj. sociable; fit for membership of a club. □ **clubbableness** n. **clubbability** /ˌklʌbə'bɪlɪtɪ/ n.

clubby /'klʌbɪ/ adj. (**clubbier, clubbiest**) esp. US sociable; friendly.

clubhouse /'klʌbhaʊs/ n. the premises used by a club.

clubland /'klʌblænd/ n. Brit. an area where many clubs are, esp. St James's in London.

clubmoss /'klʌbmɒs/ n. a small creeping plant of the pteridophyte family Lycopodiaceae, with needle-like or scalelike leaves and upright spikes of spore-cases.

cluck /klʌk/ n. & v. ● n. **1** a guttural cry like that of a hen. **2** sl. a silly or foolish person (dumb cluck). ● v.intr. emit a cluck or clucks. [imit.]

clucky /'klʌkɪ/ adj. (of a hen) sitting on eggs.

clue /kluː/ n. & v. ● n. **1** a fact or idea that serves as a guide, or suggests a line of inquiry, in a problem or investigation. **2** a piece of evidence etc. in the detection of a crime. **3** a verbal formula serving as a hint as to what is to be inserted in a crossword. **4 a** the thread of a story. **b** a train of thought. ● v.tr. (**clues, clued, cluing** or **clueing**) provide a clue to. □ **clue in** (or **up**) sl. inform. **not have a clue** colloq. be ignorant or incompetent. [var. of CLEW]

clueless /'kluːlɪs/ adj. colloq. ignorant, stupid. □ **cluelessly** adv. **cluelessness** n.

Cluj–Napoca /kluːʒ'nɑːpɒkə/ (also **Cluj**) a city in west central Romania; pop. (1989) 317,900. Founded (as Klausenburg) by German colonists in the 12th century, by the 19th century, when Cluj belonged to Hungary, the city had become a noted centre of learning and cultural capital of Transylvania; it is called Kolozsvár in Hungarian. Cluj became Cluj–Napoca in the mid-1970s, combining its former name with that of a nearby ancient settlement, believed to be Roman.

clump /klʌmp/ n. & v. ● n. **1** (foll. by of) a cluster of plants, esp. trees or shrubs. **2** an agglutinated mass of blood cells etc. **3** a thick extra sole on a boot or shoe. ● v. **1 a** intr. form a clump. **b** tr. heap or plant together. **2** intr. (also **clomp** /klɒmp/) walk with heavy tread. **3** tr. colloq. hit. □ **clumpy** adj. (**clumpier, clumpiest**). [MLG klumpe, MDu. klompe: see CLUB]

clumsy /'klʌmzɪ/ adj. (**clumsier, clumsiest**) **1** awkward in movement or shape; ungainly. **2** difficult to handle or use. **3** tactless. □ **clumsily** adv. **clumsiness** n. [obs. clumse be numb with cold (prob. f. Scand.)]

clung past and past part. of CLING.

Cluniac /'kluːnɪˌæk/ adj. & n. ● adj. of or relating to a monastic order founded at Cluny in eastern France in 910. The order was formed with the object of returning to the strict Benedictine rule, and became centralized and influential in the 11th–12th centuries. ● n. a monk of this order.

clunk /klʌŋk/ n. & v. ● n. a dull sound as of thick pieces of metal meeting. ● v.intr. make such a sound. [imit.]

cluster /'klʌstə(r)/ n. & v. ● n. **1** a close group or bunch of similar things growing together. **2** a close group or swarm of people, animals, faint stars, gems, etc. **3** a group of successive consonants or vowels. ● v. **1** tr. bring into a cluster or clusters. **2** intr. be or come into a cluster or clusters. **3** intr. (foll. by round, around) gather, congregate. □ **cluster bomb** an anti-personnel bomb spraying pellets on impact. **cluster fly** a dipterous fly, Pollenia rudis, which gathers around buildings in autumn and has larvae that are parasitic on earthworms. **cluster pine** a Mediterranean pine, Pinus pinaster, with clustered cones (also called pinaster). [OE clyster: cf. CLOT]

clustered /'klʌstəd/ adj. **1** growing in or brought into a cluster. **2** Archit. (of pillars, columns, or shafts) several close together, or disposed round or half detached from a pier.

clutch[1] /klʌtʃ/ v. & n. ● v. **1** tr. seize eagerly; grasp tightly. **2** intr. (foll. by at) try, esp. desperately, to seize or grasp. ● n. **1 a** a tight grasp. **b** (foll. by at) an act of grasping. **2** (in pl.) grasping hands, esp. as representing a cruel or relentless grasp or control. **3 a** (in a motor vehicle) a device for connecting and disconnecting the engine to the transmission. **b** the pedal operating this. **c** an arrangement for connecting or disconnecting working parts of a machine. □ **clutch bag** a slim flat handbag without handles. [ME clucche, clicche f. OE clyccan crook, clench, f. Gmc]

clutch[2] /klʌtʃ/ n. **1** a set of eggs for hatching. **2** a brood of chickens. [18th c.: prob. S.Engl. var. of cletch f. cleck to hatch f. ON klekja, assoc. with CLUTCH[1]]

Clutha /'kluːθə/ a gold-bearing river at the southern end of South Island, New Zealand. It flows 338 km (213 miles) to the Pacific Ocean.

clutter /'klʌtə(r)/ n. & v. ● n. **1** a crowded and untidy collection of things. **2** an untidy state. ● v.tr. (often foll. by up, with) crowd untidily, fill with clutter. [partly var. of clotter coagulate, partly assoc. with CLUSTER, CLATTER]

Clwyd /'kluːɪd/ a county in NE Wales: county town, Mold.

Clyde /klaɪd/ a river in western central Scotland which flows 170 km (106 miles) from the Southern Uplands to the Firth of Clyde. It flows through Glasgow, and was formerly famous for the shipbuilding industries along its banks.

Clyde, Firth of the estuary of the River Clyde in western Scotland. Opening on to the North Channel to the south, the firth separates southern Scotland to the east from the southern extremities of the Highlands to the north-west.

Clydesdale /'klaɪdzdeɪl/ n. **1** a breed of draught horse. **2** a breed of small smooth-haired terrier. [orig. bred in the valley of the CLYDE]

clypeus /'klɪpɪəs/ n. (pl. **clypei** -pɪˌaɪ/) Zool. a broad plate at the front of an insect's head. □ **clypeal** adj. **clypeate** adj. [L, = round shield]

clyster /'klɪstə(r)/ n. & v. archaic ● n. an enema. ● v.tr. treat with an enema. [ME f. OF clystere or f. L f. Gk klustēr syringe f. kluzō wash out]

Clytemnestra /ˌklaɪtɪm'nestrə/ Gk Mythol. sister of Helen and the Dioscuri, wife of Agamemnon. She conspired with her lover Aegisthus to murder Agamemnon on his return from the Trojan War, and was murdered in retribution by her son Orestes and her daughter Electra.

CM abbr. Member of the Order of Canada.

Cm symb. Chem. the element curium.

Cm. abbr. Brit. Command Paper (sixth series, 1986–).

cm abbr. centimetre(s).

Cmd. abbr. Brit. Command Paper (fourth series, 1918–56).

Cmdr. abbr. Commander.

Cmdre. abbr. Commodore.

CMEA *abbr.* Council for Mutual Economic Assistance (see COMECON).

CMG *abbr.* (in the UK) Companion (of the Order) of St Michael and St George.

Cmnd. *abbr. Brit.* Command Paper (fifth series, 1956–86).

CMV *abbr. Med.* cytomegalovirus.

CNAA *abbr.* Council for National Academic Awards.

CND see CAMPAIGN FOR NUCLEAR DISARMAMENT.

cnr. *abbr.* corner.

CNS *abbr.* central nervous system.

CN Tower /si:'en/ a tower in Toronto, Canada, the tallest self-supporting man-made structure in the world. Completed in 1976, it stands 553 m (1,815 ft) high including a 100 m (328 ft) communications mast. [Canadian National (Railways)]

Cnut see CANUTE.

CO *abbr.* **1** Commanding Officer. **2** conscientious objector. **3** *US* Colorado (in official postal use).

Co *symb. Chem.* the element cobalt.

Co. *abbr.* **1** company. **2** county. □ **and Co.** /kəʊ/ *colloq.* and the rest of them; and similar things.

c/o *abbr.* care of.

co- /kəʊ/ *prefix* **1** added to: **a** nouns, with the sense 'joint, mutual, common' (*co-author; coequality*). **b** adjectives and adverbs, with the sense 'jointly, mutually' (*co-belligerent; coequal; coequally*). **c** verbs, with the sense 'together with another or others' (*cooperate; co-author*). **2** *Math.* **a** of the complement of an angle (*cosine*). **b** the complement of (*co-latitude; coset*). [orig. a form of COM-]

coacervate /kəʊ'æsə,veɪt/ *n. Chem.* a viscous liquid phase which may separate from a colloidal solution on addition of a third component. [L *coacervat- coacervare,* f. *acervare* f. *acervus* heap]

coach /kəʊtʃ/ *n. & v.* ● *n.* **1** a single-decker bus, usu. comfortably equipped for longer journeys. **2** a railway carriage. **3** a horse-drawn carriage, usu. closed, esp. a state carriage or a stagecoach. **4 a** an instructor or trainer in sport. **b** a private tutor. **5** *US* economy-class seating in an aircraft. ● *v.* **1** *tr.* **a** train or teach (a pupil, sports team, etc.) as a coach. **b** give hints to; prime with facts. **2** *intr.* travel by stagecoach (*in the old coaching days*). □ **coach-built** (of motor-car bodies) individually built by craftsmen. **coach-house** an outhouse for carriages. **coach station** a stopping-place for a number of coaches, usu. with buildings and amenities. [F *coche* f. Hungarian *kocsi* (adj.) f. *Kocs* in Hungary]

coachload /'kəʊtʃləʊd/ *n.* a number of people, esp. holiday-makers, taken by coach.

coachman /'kəʊtʃmən/ *n.* (*pl.* **-men**) the driver of a horse-drawn carriage.

coachwood /'kəʊtʃwʊd/ *n. Austral.* a tree, esp. *Ceratopetalum apetalum*, with close-grained wood suitable for cabinet-making.

coachwork /'kəʊtʃwɜːk/ *n.* the bodywork of a road or rail vehicle.

coadjutor /kəʊ'ædʒʊtə(r)/ *n.* an assistant, esp. an assistant bishop. [ME f. OF *coadjuteur* f. LL *coadjutor* (as CO-, *adjutor* f. *adjuvare* -jut- help)]

coagulant /kəʊ'ægjʊlənt/ *n.* a substance that produces coagulation.

coagulate /kəʊ'ægjʊ,leɪt/ *v.tr. & intr.* **1** change from a fluid to a solid or semisolid state. **2** clot, curdle. **3** set, solidify. □ **coagulator** *n.* **coagulable** /-ləb(ə)l/ *adj.* **coagulative** /-lətɪv/ *adj.* [ME f. L *coagulare* f. *coagulum* rennet]

coagulation /kəʊ,ægjʊ'leɪʃ(ə)n/ *n.* the process by which a liquid changes to a semisolid mass. [as COAGULATE]

coagulum /kəʊ'ægjʊləm/ *n.* (*pl.* **coagula** /-lə/) a mass of coagulated matter. [L: see COAGULATE]

Coahuila /,kəʊə'wiːlə/ a state of northern Mexico, on the border with the US; capital, Saltillo.

coal /kəʊl/ *n. & v.* ● *n.* **1 a** a hard black or blackish rock, mainly carbonized plant matter, found in underground seams and used as a fuel and in the manufacture of gas, tar, etc. (*See note below.*) **b** *Brit.* a piece of this for burning. **2** a red-hot piece of coal, wood, etc. in a fire. ● *v.* **1** *intr.* take in a supply of coal. **2** *tr.* put coal into (an engine, fire, etc.). □ **coal-black** completely black. **coal-fired** heated or driven by coal. **coal gas** mixed gases extracted from coal and used for lighting and heating. **coal-hole** *Brit.* a compartment or small cellar for storing coal. **coal measures** a series of rocks formed by seams of coal with intervening strata. **coal mine** a mine in which coal is dug. **coal miner** a person who works in a coal mine. **coal-mining** digging for

coal. **coal oil** *N. Amer.* petroleum or paraffin. **coal-sack 1** a sack for carrying coal. **2** *Astron.* a dark nebula in the Milky Way, esp. one near the Southern Cross. **coal-scuttle** a container for coal to supply a domestic fire. **coal-seam** a stratum of coal suitable for mining. **coals to Newcastle** something brought or sent to a place where it is already plentiful. **coal tar** a thick black oily liquid distilled from coal and used as a source of benzene. **coal-tit** (or **cole-tit**) a small greyish titmouse, *Parus ater*, with a black head (also called *coalmouse*). **haul** (or **call**) **over the coals** reprimand. □ **coaly** *adj.* [OE *col* f. Gmc]

■ Coal is the most abundant form of solid fuel, formed from the remains of trees and other plant material, mostly during the Carboniferous period. Types of coal vary from black anthracite (almost pure carbon) through bituminous coal, which contains tarry substances (mostly hydrocarbons which form gases when heated), to lignite, a soft brown coal of later formation and with lower calorific value. Coal occurs in layers or seams typically a few metres thick. Layers near the surface are mined by opencast methods, removing and later replacing the covering soil; deeper layers are mined by sinking shafts and excavating the coal layers by human labour or, increasingly, by machine, leaving pillars to support the ground above. Coal is widely used as a fuel in industry for the production of heat and electric power, and for producing coke and (formerly) gas. The pre-eminent fuel of the Industrial Revolution, coal has lost much of its importance to oil and natural gas; however, these are limited in quantity, whereas coal deposits are sufficient for several hundred years.

coaler /'kəʊlə(r)/ *n.* a ship etc. transporting coal.

coalesce /,kəʊə'les/ *v.intr.* **1** come together and form one whole. **2** combine in a coalition. □ **coalescence** *n.* **coalescent** *adj.* [L *coalescere* (as CO-, *alescere alit-* grow f. *alere* nourish)]

coalface /'kəʊlfeɪs/ *n.* an exposed surface of coal in a mine.

coalfield /'kəʊlfiːld/ *n.* an extensive area with strata containing coal.

coalfish /'kəʊlfɪʃ/ *n.* = SAITHE.

coalition /,kəʊə'lɪʃ(ə)n/ *n.* **1** *Polit.* a temporary alliance for combined action, esp. of distinct parties forming a government, or of states. **2** fusion into one whole. □ **coalitionist** *n.* [med.L *coalitio* (as COALESCE)]

coalman /'kəʊlmən/ *n.* (*pl.* **-men**) a man who carries or delivers coal.

coalmouse /'kəʊlmaʊs/ *n.* (also **colemouse**) (*pl.* **-mice**) = coal-tit. [OE *colmāse* f. *col* COAL + *māse* as TITMOUSE]

Coalport /'kəʊlpɔːt/ *n.* a kind of china and porcelain produced at Coalport, a town in Shropshire, from the late 18th century. It is largely tableware and is frequently decorated with floral designs.

coaming /'kəʊmɪŋ/ *n.* a raised border round the hatches etc. of a ship to keep out water. [17th c.: orig. unkn.]

coarse /kɔːs/ *adj.* **1 a** rough or loose in texture or grain; made of large particles. **b** (of a person's features) rough or large. **2** lacking refinement or delicacy; crude, obscene (*coarse humour*). **3** rude, uncivil. **4** inferior, common. □ **coarse fish** *Brit.* any freshwater fish other than salmon and trout. □ **coarsely** *adv.* **coarseness** *n.* **coarsish** *adj.* [ME: orig. unkn.]

coarsen /'kɔːs(ə)n/ *v.tr. & intr.* make or become coarse.

coast /kəʊst/ *n. & v.* ● *n.* **1 a** the border of the land near the sea; the seashore. **b** (**the Coast**) *US* the Pacific coast of the US. **2 a** a run, usu. downhill, on a bicycle without pedalling or in a motor vehicle without using the engine. **b** *N. Amer.* a toboggan slide or slope. ● *v.intr.* **1** ride or move, usu. downhill, without use of power, free-wheel. **2** make progress without much effort. **3** *N. Amer.* slide down a hill on a toboggan. **4 a** sail along the coast. **b** trade between ports on the same coast. □ **the coast is clear** there is no danger of being observed or caught. **coast-to-coast** across an island or continent. □ **coastal** *adj.* [ME f. OF *coste, costeier* f. L *costa* rib, flank, side]

coaster /'kəʊstə(r)/ *n.* **1** a ship that travels along the coast from port to port. **2** a small tray or mat for a bottle or glass. **3** *N. Amer.* **a** a toboggan for coasting. **b** a roller-coaster.

coastguard /'kəʊstgɑːd/ *n.* **1** an organization keeping watch on the coasts and on local shipping to save life, prevent smuggling, etc. **2** a member of this.

coastline /'kəʊstlaɪn/ *n.* the line of the seashore, esp. with regard to its shape (*a rugged coastline*).

coastwise /'kəʊstwaɪz/ *adj. & adv.* along, following, or connected with the coast.

coat /kəʊt/ *n. & v.* ● *n.* **1** an outer garment with sleeves and often extending below the hips; an overcoat or jacket. **2 a** an animal's fur, hair, etc. **b** *Physiol.* a structure, esp. a membrane, enclosing or lining an organ. **c** a skin, rind, or husk. **d** a layer of a bulb etc. **3 a** a layer or covering. **b** a covering of paint etc. laid on a surface at one time. ● *v.tr.* **1** (usu. foll. by *with, in*) apply a coat of paint etc. to; provide with a layer or covering. **b** (as **coated** *adj.*) covered with. **2** (of paint etc.) form a covering to. □ **coat armour** coats of arms. **coat dress** a woman's tailored dress resembling a coat. **coat-hanger** see HANGER¹ **2. coat of arms** the heraldic bearings or shield of a person, family, or corporation. **coat of mail** a jacket of mail armour (see MAIL²). **on a person's coat-tails** undeservedly benefiting from another's success. □ **coated** *adj.* (also in *comb.*). [ME f. OF *cote* f. Rmc f. Frank., of unkn. orig.]

coatee /kəʊˈtiː/ *n.* **1** a woman's or infant's short coat. **2** *archaic* a close-fitting short coat.

coati /kəʊˈɑːtɪ/ *n.* (*pl.* **coatis**) a raccoon-like flesh-eating mammal of the genera *Nasua* or *Nasuella*, native to Central and South America, with a long flexible snout and a long usu. ringed tail. [Tupi f. *cua* belt + *tim* nose]

coatimundi /kəʊˌɑːtɪˈmʌndɪ/ *n.* (*pl.* **coatimundis**) = COATI. [as COATI + Tupi *mondi* solitary]

coating /ˈkəʊtɪŋ/ *n.* **1** a thin layer or covering of paint etc. **2** material for making coats.

Coats Land /kəʊts/ a region of Antarctica, to the east of the Antarctic Peninsula. Bordering the Weddell Sea, it lies south of South Georgia and the South Sandwich Islands.

co-author /kəʊˈɔːθə(r)/ *n. & v.* ● *n.* a joint author. ● *v.tr.* be a joint author of.

coax¹ /kəʊks/ *v.tr.* **1** (usu. foll. by *into*, or *to* + infin.) persuade (a person) gradually or by flattery. **2** (foll. by *out of*) obtain (a thing from a person) by coaxing. **3** manipulate (a thing) carefully or slowly. □ **coaxer** *n.* **coaxingly** *adv.* [16th c.: f. 'make a *cokes* of' f. obs. *cokes* simpleton, of unkn. orig.]

coax² /ˈkəʊæks/ *n. colloq.* coaxial cable. [abbr.]

coaxial /kəʊˈæksɪəl/ *adj.* **1** having a common axis. **2** *Electr.* (of a cable or line) transmitting by means of two concentric conductors separated by an insulator. □ **coaxially** *adv.*

cob¹ /kɒb/ *n.* **1** a roundish lump of coal etc. **2** *Brit.* a domed loaf of bread. **3** *Brit.* = corn-cob (see CORN¹). **4** (in full **cob-nut**) a large hazelnut. **5** a sturdy short-legged horse for riding. **6** a male swan. [ME: orig. unkn.]

cob² /kɒb/ *n.* a material for walls, made from compressed earth, clay, or chalk reinforced with straw. [17th c.: orig. unkn.]

cobalt /ˈkəʊbɔːlt, -bɒlt/ *n. & adj.* ● *n.* a hard silvery-white metallic chemical element (atomic number 27; symbol **Co**). (*See note below.*) ● *adj.* = cobalt blue. □ **cobalt blue 1** a pigment containing a cobalt salt. **2** the deep blue colour of this. □ **cobaltic** /kəʊˈbɔːltɪk/ *adj.* **cobaltous** /-ˈbɔːltəs/ *adj.* [G *Kobalt* etc., prob. = KOBOLD goblin or demon of the mines, cobalt ore being viewed by medieval German miners as a worthless nuisance (cf. NICKEL)]

■ Cobalt was known to Paracelsus, though its modern discovery is usually credited to the Swedish chemist Georg Brandt in 1733. It is chiefly obtained as a by-product from nickel and copper ores. Cobalt is a transition metal similar in many respects to nickel. Its main use is as a component of magnetic alloys and those designed for use at high temperatures. Cobalt compounds have been used since ancient times to colour ceramics, and they are also used as catalysts.

cobber /ˈkɒbə(r)/ *n. Austral. & NZ colloq.* a companion or friend. [19th c.: perh. rel. to Engl. dial. *cob* take a liking to]

Cobbett /ˈkɒbɪt/, William (1763–1835), English writer and political reformer. He started his political life as a Tory, but later became a radical; the change is reflected in *Cobbett's Political Register*, a periodical that he founded in 1802 and continued for the rest of his life. Cobbett was one of the leaders of the campaign for political and social reform after 1815, although he had already spent two years in prison for his outspoken criticism of flogging in the army (1810–12). A prolific writer, he published more than forty works in his lifetime, including *Rural Rides* (1830).

cobble¹ /ˈkɒb(ə)l/ *n. & v.* ● *n.* **1** (in full **cobblestone**) a small rounded stone of a size used for paving. **2** (in *pl.*) *Brit.* coal in lumps of this size. ● *v.tr.* pave with cobbles. [ME *cobel(-ston)*, f. COB¹]

cobble² /ˈkɒb(ə)l/ *v.tr.* **1** mend or patch up (esp. shoes). **2** (often foll. by *together*) join or assemble roughly. [back-form. f. COBBLER]

cobbler /ˈkɒblə(r)/ *n.* **1** a person who mends shoes, esp. professionally. **2** an iced drink of wine etc., sugar, and lemon (*sherry cobbler*). **3 a** a pie topped with scones. **b** esp. *US* a fruit pie with a rich thick crust. **4** (in

pl.) *Brit. sl.* nonsense. **5** *Austral. & NZ sl.* the last sheep to be shorn. [ME, of unkn. orig.: sense 4 f. rhyming sl. *cobbler's awls* = balls: sense 5 with pun on LAST³]

Cobden /ˈkɒbdən/, Richard (1804–65), English political reformer. A Manchester industrialist, Cobden was one of the leading spokesmen of the free-trade movement in Britain. From 1838, together with John Bright, he led the Anti-Corn Law League in its successful campaign for the repeal of the Corn Laws (1846).

COBE /ˈkəʊbɪ/ Cosmic Background Explorer, a NASA satellite launched in 1989 to map the background microwave radiation in a search for evidence of the big bang. [acronym]

co-belligerent /ˌkəʊbɪˈlɪdʒərənt/ *n. & adj.* ● *n.* any of two or more nations engaged in war as allies. ● *adj.* of or as a co-belligerent. □ **co-belligerence** *n.* **co-belligerency** *n.*

coble /ˈkəʊb(ə)l/ *n.* a flat-bottomed fishing-boat in Scotland and NE England. [OE, perh. f. Celt.]

COBOL /ˈkəʊbɒl/ *n. Computing* a programming language designed for use in commerce. [common business oriented language]

cobra /ˈkəʊbrə, ˈkɒb-/ *n.* a venomous African or Asian snake, esp. of the genus *Naja*, which can dilate the neck like a hood when excited. [Port. f. L *colubra* snake]

cobweb /ˈkɒbweb/ *n.* **1 a** a fine network of threads spun by a spider from a liquid extruded from its spinnerets, used to trap insects etc. **b** the thread of this. **2** anything compared with a cobweb, esp. in flimsiness of texture. **3** a trap or insidious entanglement. **4** (in *pl.*) a state of lethargy; fustiness (esp. *blow* or *clear away the cobwebs*). □ **cobwebbed** *adj.* **cobwebby** *adj.* [ME *cop(pe)web* f. obs. *coppe* spider]

coca /ˈkəʊkə/ *n.* **1** a South American shrub, *Erythroxylum coca*. **2** its dried leaves, containing cocaine, chewed as a stimulant. [Sp. f. Aymara *kuka* or Quechua *koka*]

Coca-Cola /ˌkəʊkəˈkəʊlə/ *n. propr.* a carbonated non-alcoholic drink sometimes flavoured with cola seeds.

cocaine /kəˈkeɪn, kəʊˈkeɪn/ *n.* a drug derived from coca or prepared synthetically, used as a local anaesthetic and as a stimulant. [COCA + -INE⁴]

coccidiosis /ˌkɒksɪdɪˈəʊsɪs/ *n.* a disease of birds and mammals caused by parasitic protozoa, esp. of the genus *Eimeria*, affecting the intestine. [*coccidium* (mod.L f. Gk *kokkis* dimin. of *kokkos* berry) + -OSIS]

coccus /ˈkɒkəs/ *n.* (*pl.* **cocci** /ˈkɒksaɪ, ˈkɒkaɪ/) *Biol.* any spherical or roughly spherical bacterium. □ **coccal** *adj.* **coccoid** *adj.* [mod.L f. Gk *kokkos* berry]

coccyx /ˈkɒksɪks/ *n.* (*pl.* **coccyges** /kɒkˈsaɪdʒiːz/ or **coccyxes**) *Anat.* the small triangular bone at the base of the spinal column in humans and some apes. □ **coccygeal** /kɒkˈsɪdʒɪəl/ *adj.* [L f. Gk *kokkux -ugos* cuckoo (from being shaped like its bill)]

Cochabamba /ˌkɒtʃəˈbæmbə/ a city in Bolivia, situated at the centre of a rich agricultural region; pop. (est. 1988) 403,600.

Cochin /ˈkəʊtʃɪn, ˈkɒtʃɪn/ a seaport and naval base on the Malabar Coast of SW India, in the state of Kerala; pop. (1991) 504,000.

cochin /ˈkəʊtʃɪn, ˈkɒtʃɪn/ *n.* (in full **cochin-china**) an Asian breed of fowl with feathery legs. [COCHIN-CHINA]

Cochin-China /ˌkəʊtʃɪnˈtʃaɪnə, ˌkɒtʃɪn-/ the former name for the southern region of what is now Vietnam. Part of French Indo-China from 1862, in 1946 it became a French overseas territory, then merged officially with Vietnam in 1949.

cochineal /ˌkɒtʃɪˈniːl/ *n.* **1** a scarlet dye used esp. for colouring food. **2** the dried bodies of the female of a scale insect, *Dactylopius coccus*, native to Mexico, yielding this. [F *cochenille* or Sp. *cochinilla* f. L *coccinus* scarlet f. Gk *kokkos* berry]

cochlea /ˈkɒklɪə/ *n.* (*pl.* **cochleae** /-lɪˌiː/) the spiral cavity of the internal ear, where sound vibrations are converted into nervous impulses. □ **cochlear** *adj.* [L, = snail-shell, f. Gk *kokhlias*]

Cochran¹ /ˈkɒkrən/, Sir C(harles) B(lake) (1872–1951), English theatrical producer. Agent for Houdini and the French dancer Mistinguett (1874–1956), he is most famous for the musical revues which he produced from 1918 onwards at the London Pavilion, including Noël Coward's *Bitter Sweet* (1929) and *Cavalcade* (1931).

Cochran² /ˈkɒkrən/, Jacqueline (1910–80), American aviator. She made aviation history when she became the first woman to enter the trans-American Bendix Cup Air Race in 1935 and the first to win it in 1938. She achieved more than 200 speed, distance, and altitude records in her flying career, more than any of her contemporaries. Her achievements included becoming the first woman to break the sound barrier (1953), logging record speeds for men and women for the 15 km, 100 km, and 500 km distances (1953), and breaking the women's jet speed record (1964) when she flew at 1,429 m.p.h. (2,300 k.p.h.).

cock[1] /kɒk/ n. & v. ● n. **1 a** a male bird, esp. of a domestic fowl. **b** a male lobster, crab, or salmon. **c** = WOODCOCK. **2** Brit. sl. (usu. **old cock** as a form of address) a friend; a fellow. **3** coarse sl. the penis. **4** Brit. sl. nonsense (cf. POPPYCOCK). **5 a** a firing lever in a gun which can be raised to be released by the trigger. **b** the cocked position of this (at full cock). **6** a tap or valve controlling flow. ● v.tr. **1** raise or make upright or erect. **2** turn or move (the eye or ear) attentively or knowingly. **3** set aslant, or turn up the brim of (a hat). **4** raise the cock of (a gun). □ **at half cock** only partly ready. **cock-a-doodle-doo** a cock's crow. **cock-and-bull story** an absurd or incredible account. **cock crow** dawn. **cocked hat 1** a brimless triangular hat pointed at the front, back, and top. **2** hist. a hat with a wide brim permanently turned up towards the crown (e.g. a tricorn). **cock-fight** a fight between cocks as sport. **cock-fighting** this sport. **cock-of-the-rock** (pl. **cocks-of-the-rock**) a South American bird, a cotinga of the genus Rupicola, having bright orange or red plumage and a prominent crest. **cock-of-the-walk** a dominant or arrogant person. **cock-of-the-wood** **1** a capercaillie. **2** N. Amer. a red-crested woodpecker. **cock-shy 1 a** a target for throwing with sticks, stones, etc. **b** a throw at this. **2** an object of ridicule or criticism. **cock a snook** see SNOOK[1]. **cock sparrow 1** a male sparrow. **2** a lively quarrelsome person. **cock up** Brit. sl. bungle; make a mess of. **cock-up** n. Brit. sl. a muddle or mistake. **knock into a cocked hat** defeat utterly. [OE cocc and OF coq prob. f. med.L coccus]

cock[2] /kɒk/ n. & v. ● n. a small pile of hay, straw, etc. with vertical sides and a rounded top. ● v.tr. pile into cocks. [ME, perh. of Scand. orig.]

cockade /kɒˈkeɪd/ n. a rosette etc. worn in a hat as a badge of office or party, or as part of a livery. □ **cockaded** adj. [F cocarde orig. in bonnet à la coquarde, f. fem. of obs. coquard saucy f. coq COCK[1]]

cock-a-hoop /ˌkɒkəˈhuːp/ adj. & adv. ● adj. exultant; crowing boastfully. ● adv. exultantly. [16th c.: orig. in phr. set cock a hoop denoting some action preliminary to hard drinking]

cock-a-leekie /ˌkɒkəˈliːkɪ/ n. (also **cocky-leeky** /ˌkɒkɪ-/) a soup traditionally made in Scotland with boiling fowl and leeks. [COCK[1] + LEEK]

cockalorum /ˌkɒkəˈlɔːrəm/ n. colloq. a self-important little man. [18th c.: arbitrarily f. COCK[1]]

cockatiel /ˌkɒkəˈtiːl/ n. (also **cockateel**) Austral. a small delicately coloured crested parrot, Nymphicus hollandicus. [Du. kaketielje]

cockatoo /ˌkɒkəˈtuː/ n. **1** an Australasian parrot of the family Cacatuidae, having a powerful beak and an erectile crest. **2** Austral. & NZ colloq. a small farmer. [Du. kaketoe f. Malay kakatua, assim. to COCK[1]]

cockatrice /ˈkɒkətrɪs, -ˌtraɪs/ n. **1** = BASILISK 1. **2** Heraldry a fabulous animal, a cock with a serpent's tail. [ME f. OF cocatris f. L calcare tread, track, rendering Gk ikhneumōn tracker: see ICHNEUMON]

cockboat /ˈkɒkbəʊt/ n. a small ship's-boat. [obs. cock small boat (f. OF coque) + BOAT]

cockchafer /ˈkɒkˌtʃeɪfə(r)/ n. a large night-flying chafer, Melolontha melolontha, which feeds on leaves and whose larva feeds on the roots of crops etc. Also called May-bug. [perh. f. COCK[1] as expressing size or vigour + CHAFER]

Cockcroft /ˈkɒkkrɒft/, Sir John Douglas (1897–1967), English physicist. He joined the Cavendish Laboratory, Cambridge, and in 1932 succeeded (with E. T. S. Walton) in splitting the atom by means of artificially accelerated protons. With their high-energy particle accelerator they bombarded lithium atoms with protons and produced alpha particles. This demonstrated the transmutation of elements and Einstein's theory of the equivalence of mass and energy, and ushered in the whole field of nuclear and particle physics which relied on particle accelerators. The two shared the 1951 Nobel Prize for physics. After the war, Cockcroft became the first director of the Atomic Energy Research Establishment at Harwell.

cocker /ˈkɒkə(r)/ n. (in full **cocker spaniel**) a breed of small spaniel with a silky coat. [as COCK[1], from use in hunting woodcocks etc.]

cockerel /ˈkɒkrəl/ n. a young cock. [ME: dimin. of COCK[1]]

Cockerell /ˈkɒkərəl/, Sir Christopher Sydney (b.1910), English engineer. He was a boat-designer, and in 1955 he took out the first patent on the vessel that later came to be known as the hovercraft.

cock-eyed /ˈkɒkaɪd/ adj. colloq. **1** crooked, askew, not level. **2** (of a scheme etc.) absurd, not practical. **3** drunk. **4** squinting. [19th c.: app. f. COCK[1] + EYE]

cockle[1] /ˈkɒk(ə)l/ n. **1 a** a small edible bivalve mollusc of the genus Cardium, having a chubby ribbed shell. **b** its shell. **2** (in full **cockleshell**) a small shallow boat. □ **warm the cockles of one's heart** make one contented; be satisfying. [ME f. OF coquille shell ult. f. Gk kogkhulion f. kogkhē CONCH]

cockle[2] /ˈkɒk(ə)l/ n. **1** (in full **corn-cockle**) a pink-flowered plant, Agrostemma githago, of the pink family, formerly common in cornfields. **2** a disease of wheat that turns the grains black. [OE coccul, perh. ult. f. LL coccus]

cockle[3] /ˈkɒk(ə)l/ v. & n. ● v. **1** intr. pucker, wrinkle. **2** tr. cause to cockle. ● n. a pucker or wrinkle in paper, glass, etc. [F coquiller blister (bread in cooking) f. coquille: see COCKLE[1]]

cocklebur /ˈkɒk(ə)lˌbɜː(r)/ n. a composite weed of the genus Xanthium, with fruit covered in hooked bristles. [COCKLE[2] + BUR]

cockney /ˈkɒknɪ/ n. & adj. ● n. (pl. **-eys**) **1 a** a native of London, especially the East End; traditionally a cockney is someone born within the sound of Bow Bells. **b** the dialect or accent typical of this area. A distinctive feature is its use of rhyming slang. **2** Austral. a young snapper fish, Chrysophrys auratus. Also called squire. ● adj. of or characteristic of cockneys or their dialect or accent. □ **cockneyism** n. [ME cokeney cock's egg, later derog. for 'townsman']

cockpit /ˈkɒkpɪt/ n. **1 a** a compartment for the pilot (or the pilot and crew) of an aircraft or spacecraft. **b** a similar compartment for the driver in a racing car. **c** a space for the helmsman in some small yachts. **2** an arena of war or other conflict. **3** a place where cock-fights are held. [orig. in sense 3, f. COCK[1] + PIT[1]]

cockroach /ˈkɒkrəʊtʃ/ n. a flat-bodied scavenging beetle-like insect of the order Dictyoptera; esp. Blatta orientalis and Periplaneta americana, which infest kitchens, warehouses, etc. [Sp. cucaracha, assim. to COCK[1], ROACH[1]]

cockscomb /ˈkɒkskəʊm/ n. **1** the crest or comb of a cock. **2** a garden plant, Celosia cristata, with a terminal plume of tiny white or red flowers.

cocksfoot /ˈkɒksfʊt/ n. a pasture grass of the genus Dactylis, with broad leaves and green or purplish spikes.

cocksure /kɒkˈʃʊə(r), -ˈʃɔː(r)/ adj. **1** presumptuously or arrogantly confident. **2** (foll. by of, about) absolutely sure. □ **cocksurely** adv. **cocksureness** n. [cock = God + SURE, later assoc. with COCK[1]]

cocktail /ˈkɒkteɪl/ n. **1** a usu. alcoholic drink made by mixing various spirits, fruit juices, etc. **2** a dish of mixed ingredients (fruit cocktail; prawn cocktail). **3** any (esp. unpleasant or dangerous) mixture or concoction. □ **cocktail dress** a usu. short evening dress suitable for wearing at a drinks party. **cocktail stick** a small pointed stick for serving an olive, cherry, small sausage, etc. [orig. unkn.: cf. earlier sense 'docked horse' f. COCK[1]: the connection is unclear]

cocky[1] /ˈkɒkɪ/ adj. (**cockier**, **cockiest**) **1** conceited, arrogant. **2** saucy, impudent. □ **cockily** adv. **cockiness** n. [COCK[1] + -Y[1]]

cocky[2] /ˈkɒkɪ/ n. (pl. **-ies**) Austral. & NZ colloq. = COCKATOO 2. [abbr.]

cocky-leeky var. of COCK-A-LEEKIE.

coco /ˈkəʊkəʊ/ n. (also **cocoa**) (pl. **cocos** or **cocoas**) the coconut palm. [abbr. of COCONUT, though in earlier use as a separate word]

cocoa /ˈkəʊkəʊ/ n. **1** a powder made from crushed cacao seeds, often with other ingredients. **2** a drink made from this. The Aztecs were the first to make such a drink, and in the 16th century the Spaniards brought it to Europe, where it was enjoyed as an expensive luxury. □ **cocoa bean** a cacao seed. **cocoa butter** a fatty substance obtained from cocoa beans and used for confectionery, cosmetics, etc. [alt. of CACAO]

coco-de-mer /ˌkəʊkəʊdəˈmeə(r)/ n. a tall palm tree, Lodoicea maldivica, of the Seychelles, with gigantic nuts that look like double coconuts. [F]

coconut /ˈkəʊkənʌt/ n. (also **cocoanut**) **1 a** the large ovate brown seed of a tall tropical palm, Cocos nucifera, with a hard shell lined with edible white flesh enclosing a milky juice. **b** (in full **coconut palm** or **tree**) the palm itself. **c** the flesh of a coconut. **2** sl. the human head. □ **coconut butter** a solid oil obtained from the lining of the coconut, and used in soap, candles, ointment, etc. **coconut ice** a sweet of sugar and desiccated coconut. **coconut matting** a matting made of fibre from coconut husks. **coconut shy** a fairground sideshow where balls are thrown to dislodge coconuts. **double coconut** the very large nut of the coco-de-mer. [f. Sp. and Port. coco = grimace (the base of the shell resembling a face) + NUT]

cocoon /kəˈkuːn/ n. & v. ● n. **1 a** a silky case spun by many insect larvae for protection as pupae. **b** a similar structure made by other animals.

2 a thing that encloses like a cocoon; a protective covering, esp. to prevent corrosion of metal equipment. ● *v.* **1** *tr.* (usu. as **cocooned** *adj.*) wrap (as) in a cocoon; protect, enclose. **2** *tr.* spray with a protective coating. **3** *intr.* form a cocoon. [F *cocon* f. mod. Prov. *coucoun* dimin. of *coca* shell]

Cocos Islands /'kəʊkəs/ (also called *Keeling Islands*) a group of twenty-seven small coral islands in the Indian Ocean, administered as an external territory of Australia since 1955. The islands (previously uninhabited) were discovered in 1609 by Captain William Keeling of the East India Company; pop. (1990) 603.

cocotte /kə'kɒt/ *n.* **1** a heatproof dish or small casserole in which food can be both cooked and served, often as an individual portion. **2** *archaic* a fashionable prostitute. [F]

Cocteau /'kɒktəʊ/, Jean (1889–1963), French dramatist, novelist, and film director. His plays, noted for their striking blend of poetry, irony, and fantasy, include *La Machine infernale* (1934), based on the Oedipus legend, and *Orphée* (1926), which he made into a film in 1950. Among his other major films are *Le Sang d'un poète* (1930) and *La Belle et la bête* (1946), both of which mingle myth and reality. Also a prolific novelist, he is best known for *Les Enfants terribles* (1929).

COD *abbr.* **1 a** cash on delivery. **b** *US* collect on delivery. **2** Concise Oxford Dictionary.

cod[1] /kɒd/ *n.* (*pl.* same) a marine fish of the family Gadidae, having barbels on the chin; esp. the large *Gadus morrhua*, of the North Atlantic, an important food fish. □ **cod-liver oil** an oil pressed from the fresh liver of cod, which is rich in vitamins D and A. [ME: orig. unkn.]

cod[2] /kɒd/ *n. & v. Brit. sl.* ● *n.* **1** a parody. **2** a hoax. **3** (*attrib.*) = MOCK *adj.* ● *v.* (**codded**, **codding**) **1 a** *intr.* perform a hoax. **b** *tr.* play a trick on; fool. **2** *tr.* parody. [19th c.: orig. unkn.]

cod[3] /kɒd/ *n. sl.* nonsense. [abbr. of CODSWALLOP]

coda /'kəʊdə/ *n.* **1** *Mus.* the concluding passage of a piece or movement, usu. forming an addition to the basic structure. **2** *Ballet* the concluding section of a dance. **3** a concluding event or series of events. [It. f. L *cauda* tail]

coddle /'kɒd(ə)l/ *v.tr.* **1 a** treat as an invalid; protect attentively. **b** (foll. by *up*) strengthen by feeding. **2** cook (an egg) in water below boiling-point. □ **coddler** *n.* [prob. dial. var. of *caudle* invalids' gruel]

code /kəʊd/ *n. & v.* ● *n.* **1** a system of words, letters, figures, or symbols, used to represent others for secrecy or brevity. **2** a system of prearranged signals, esp. used to ensure secrecy in transmitting messages. **3** *Computing* a piece of program text. **4 a** a systematic collection of statutes, a body of laws so arranged as to avoid inconsistency and overlapping. **b** a set of rules on any subject. **5 a** the prevailing morality of a society or class (*code of honour*). **b** a person's standard of moral behaviour. ● *v.* **1** *tr.* put (a message, program, etc.) into code. **2** *intr.* (foll. by *for*) *Biochem.* be the genetic code for (an amino acid etc.). □ **code-book** a list of symbols etc. used in a code. **code-breaking** cryptanalysis. **code-name** (or **-number**) a word or symbol (or number) used for secrecy or convenience instead of the usual name. □ **coder** *n.* [ME f. OF f. L CODEX]

codeine /'kəʊdiːn/ *n. Pharm.* an alkaloid derived from morphine and used to relieve pain. [Gk *kōdeia* poppy-head + -INE[4]]

codependency /ˌkəʊdɪ'pendənsɪ/ *n.* emotional dependency on supporting or caring for another person or other people. □ **codependent** *adj. & n.* [CO- + DEPENDENCY]

co-determination /ˌkəʊdɪˌtɜːmɪ'neɪʃ(ə)n/ *n.* cooperation between management and workers in decision-taking. [CO- + DETERMINATION, after G *Mitbestimmung*]

codex /'kəʊdeks/ *n.* (*pl.* **codices** /'kəʊdɪˌsiːz, 'kɒd-/ or **codexes**) **1** a kind of ancient manuscript text in book form, which between the 1st and 4th centuries AD gradually replaced the continuous roll previously used for written documents. **2** a collection of pharmaceutical descriptions of drugs. [L, = block of wood, tablet, book]

codfish /'kɒdfɪʃ/ *n.* = COD[1].

codger /'kɒdʒə(r)/ *n.* (usu. in **old codger**) *colloq.* a person, esp. an old or strange one. [perh. var. of *cadger*: see CADGE]

codices *pl.* of CODEX.

codicil /'kəʊdɪsɪl, 'kɒd-/ *n.* an addition explaining, modifying, or revoking a will or part of one. □ **codicillary** /ˌkəʊdɪ'sɪlərɪ, ˌkɒd-/ *adj.* [L *codicillus*, dimin. of CODEX]

codicology /ˌkəʊdɪ'kɒlədʒɪ/ *n.* the study of manuscripts. □ **codicological** /-kə'lɒdʒɪk(ə)l/ *adj.* **codicologically** *adv.* [F *codicologie* f. L *codex codicis*: see CODEX]

codify /'kəʊdɪˌfaɪ, 'kɒd-/ *v.tr.* (**-ies**, **-ied**) arrange (laws etc.) systematically into a code. □ **codifier** *n.* **codification** /ˌkəʊdɪfɪ'keɪʃ(ə)n, ˌkɒd-/ *n.*

codling[1] /'kɒdlɪŋ/ *n.* (also **codlin** /-lɪn/) **1** a variety of cooking-apple, having a long tapering shape. **2** (in full **codling moth**) a small moth, *Carpocapsa pomonella*, the larva of which feeds on apples. □ **codlings-and-cream** the great hairy willowherb, *Epilobium hirsutum*. [ME f. AF *quer de lion* lion-heart]

codling[2] /'kɒdlɪŋ/ *n.* a small codfish.

codomain /ˌkəʊdəʊˌmeɪn/ *n. Math.* a set that includes all the possible expressions of a given function. [CO- 2 + DOMAIN]

codon /'kəʊdɒn/ *n. Biochem.* a sequence of three nucleotides, forming a unit of genetic code in a DNA or RNA molecule. [CODE + -ON]

codpiece /'kɒdpiːs/ *n. hist.* an appendage like a small bag or flap at the front of a man's breeches. [ME, f. *cod* scrotum + PIECE]

co-driver /kəʊ'draɪvə(r)/ *n.* a person who shares the driving of a vehicle with another, esp. in a race, rally, etc.

codswallop /'kɒdzˌwɒləp/ *n. Brit. sl.* nonsense. [20th c.: orig. unkn.]

Cody /'kəʊdɪ/, William Frederick, see BUFFALO BILL.

Coe /kəʊ/, Sebastian (b.1956), British middle-distance runner and Conservative politician. He won a gold medal in the 1,500 metres at the Olympic Games in 1980 and 1984. In 1981 he created new world records in the 800 metres, 1,000 metres, and the mile. On retiring from athletics, Coe became an MP in 1992.

coecilian var. of CAECILIAN.

coed *n. & adj. colloq.* ● *n.* /'kəʊed/ **1** a coeducational system or institution. **2** esp. *US* a female student at a coeducational institution. ● *adj.* /kəʊ'ed/ coeducational. [abbr.]

coeducation /ˌkəʊedjʊ'keɪʃ(ə)n/ *n.* the education of pupils of both sexes together. □ **coeducational** *adj.*

coefficient /ˌkəʊɪ'fɪʃ(ə)nt/ *n.* **1** *Math.* a quantity placed before and multiplying an algebraic expression (e.g. 4 in $4x^y$). **2** *Physics* a multiplier or factor that measures some property (*coefficient of expansion*). [mod.L *coefficiens* (as CO-, EFFICIENT)]

coelacanth /'siːləˌkænθ/ *n.* a large bony marine fish, *Latimeria chalumnae*, having a trilobed tail fin and fleshy pectoral fins, thought to be related to the ancestors of the first land vertebrates. It was known only from fossils until one was caught off South Africa in 1938; since then other specimens have been obtained near the Comoro Islands in the Indian Ocean. [mod.L *Coelacanthus* f. Gk *koilos* hollow + *akantha* spine]

-coele *comb. form* var. of -CELE.

coelenterate /siː'lentəˌreɪt/ *n. Zool.* an aquatic invertebrate animal of the phylum Coelenterata (now usually called Cnidaria), having radial symmetry and stinging tentacles. They have either a free-swimming medusa form like the jellyfish, or a sessile polyp form like the corals and sea anemones, or both in alternating generations. [mod.L *Coelenterata* f. Gk *koilos* hollow + *enteron* intestine]

coeliac /'siːlɪæk/ *adj.* (*US* **celiac**) *Med.* of or affecting the abdomen. □ **coeliac disease** a digestive disease of the small intestine brought on by hypersensitivity to dietary gluten. [L *coeliacus* f. Gk *koiliakos* f. *koilia* belly]

coelom /'siːləm, -ləʊm/ *n.* (*US* **celom**) (*pl.* **-oms** or **-omata** /siː'ləʊmətə/) *Zool.* the principal body cavity in animals, between the intestinal canal and the body wall. □ **coelomate** /-ˌmeɪt/ *adj. & n.* [Gk *koilōma* cavity]

coelostat /'siːləˌstæt/ *n. Astron.* an instrument with a rotating mirror that continuously reflects the light from the same area of sky allowing the path of a celestial object to be monitored. [L *caelum* sky + -STAT]

coenobite /'siːnəˌbaɪt/ *n.* (*US* **cenobite**) a member of a monastic community. □ **coenobitic** /ˌsiːnə'bɪtɪk/ *adj.* **coenobitical** *adj.* [OF *cenobite* or eccl.L *coenobita* f. LL *coenobium* f. Gk *koinobion* convent f. *koinos* common + *bios* life]

coenzyme /kəʊ'enzaɪm/ *n. Biochem.* a non-proteinaceous compound that assists in the action of an enzyme. □ **coenzyme A** a coenzyme derived from pantothenic acid, important in many biochemical reactions, esp. respiration.

coequal /kəʊ'iːkwəl/ *adj. & n. archaic* or *literary* ● *adj.* equal with one another. ● *n.* an equal. □ **coequally** *adv.* **coequality** /ˌkəʊɪ'kwɒlɪtɪ/ *n.* [ME f. L or eccl.L *coaequalis* (as CO-, EQUAL)]

coerce /kəʊ'ɜːs/ *v.tr.* (often foll. by *into*) persuade or restrain (an unwilling person) by force (*coerced you into signing*). □ **coercible** *adj.* [ME f. L *coercere* restrain (as CO-, *arcere* restrain)]

coercion /kəʊˈɜːʃ(ə)n/ n. **1** the act or process of coercing. **2** government by force. □ **coercive** /-ˈɜːsɪv/ adj. **coercively** adv. **coerciveness** n. [OF cohercion, -tion f. L coer(c)tio, coercitio -onis (as COERCE)]

coercivity /ˌkəʊəˈsɪvɪti/ n. Physics the resistance of a magnetic material to changes in magnetization, esp. measured as the field intensity necessary to demagnetize it when fully magnetized.

Coetzee /kʊtˈsɪə/, J(ohn) M(axwell) (b.1940), South African novelist. In his major works, such as the two Dusklands novellas (1974), and the novels In the Heart of the Country (1977) and the Booker Prize-winning The Life and Times of Michael K (1983), he explores the psychology and mythology of colonialism and racial domination. His recent output includes White Writing (1988), a collection of critical essays, and the novel Age of Iron (1990).

coeval /kəʊˈiːv(ə)l/ adj. & n. ● adj. **1** having the same age or date of origin. **2** living or existing at the same epoch. **3** having the same duration. ● n. a coeval person, a contemporary. □ **coevally** adv. **coevality** /ˌkəʊiːˈvælɪti/ n. [LL coaevus (as CO-, L aevum age)]

coexist /ˌkəʊɪɡˈzɪst/ v.intr. (often foll. by with) **1** exist together (in time or place). **2** (esp. of nations) exist in mutual tolerance though professing different ideologies etc. □ **coexistence** n. **coexistent** adj. [LL coexistere (as CO-, EXIST)]

coextensive /ˌkəʊɪkˈstensɪv/ adj. extending over the same space or time.

C. of E. abbr. Church of England.

coffee /ˈkɒfɪ/ n. **1 a** a drink made from the roasted and ground beanlike seeds of a tropical shrub of the genus Coffea. **b** a cup of this. **2 a** the shrub that yields these seeds, one or more of which are contained in each berry. **b** these seeds raw, or roasted and ground. **3 a** pale brown colour, as of coffee mixed with milk. □ **coffee bar** a bar or café serving coffee and light refreshments from a counter. **coffee bean** the beanlike seeds of the coffee shrub. **coffee-cup** a small cup for serving coffee. **coffee-essence** a concentrated extract of coffee usu. containing chicory. **coffee-house** a place serving coffee and other refreshments. **coffee-mill** a small machine for grinding roasted coffee beans. **coffee-morning** a morning gathering at which coffee is served, often in aid of charity. **coffee nibs** coffee beans removed from their shells. **coffee-shop** a small informal restaurant, often in a hotel or department store. **coffee-table** a small low table. **coffee-table book** a large lavishly illustrated book. [ult. f. Turk. kahveh f. Arab. ḳahwa, the drink]

coffer /ˈkɒfə(r)/ n. **1** a box, esp. a large strongbox for valuables. **2** (in pl.) a treasury or store of funds. **3** a sunken panel in a ceiling etc. □ **coffer-dam** a watertight enclosure pumped dry to permit work below the waterline on building bridges etc., or for repairing a ship. □ **coffered** adj. [ME f. OF coffre f. L cophinus f. Gk kophinos basket]

coffin /ˈkɒfɪn/ n. & v. ● n. **1** a long narrow usu. wooden box in which a corpse is buried or cremated. **2** the part of a horse's hoof below the coronet. ● v.tr. (**coffined**, **coffining**) put in a coffin. □ **coffin-bone** a bone in a horse's hoof. **coffin corner** Amer. Football the corner between the goal-line and sideline. **coffin-joint** the joint at the top of a horse's hoof. **coffin-nail** sl. a cigarette. [ME f. OF cof(f)in little basket etc. f. L cophinus: see COFFER]

coffle /ˈkɒf(ə)l/ n. a line of animals, slaves, etc., fastened together. [Arab. ḳāfila caravan]

co-founder /kəʊˈfaʊndə(r)/ n. a joint founder. □ **co-found** v.tr.

cog /kɒɡ/ n. **1** each of a series of projections on the edge of a wheel or bar transferring motion by engaging with another series. **2** an unimportant member of an organization etc. □ **cog-wheel** a wheel with cogs. □ **cogged** adj. [ME: prob. f. Scand. orig.]

cogent /ˈkəʊdʒənt/ adj. (of arguments, reasons, etc.) convincing, compelling. □ **cogency** n. **cogently** adv. [L cogere compel (as CO-, agere act- drive)]

cogitable /ˈkɒdʒɪtəb(ə)l/ adj. able to be grasped by the mind; conceivable. [L cogitabilis (as COGITATE)]

cogitate /ˈkɒdʒɪteɪt/ v.tr. & intr. ponder, meditate. □ **cogitator** n. **cogitative** /-tətɪv/ adj. **cogitation** /ˌkɒdʒɪˈteɪʃ(ə)n/ n. [L cogitare think (as CO-, AGITATE)]

cogito /ˈkəʊɡɪˌtəʊ, ˈkɒdʒɪ-/ n. Philos. the principle establishing the existence of a being from the fact of its thinking or awareness. [L, = I think, in Fr. philosopher Descartes's formula (1641) cogito, ergo sum I think, therefore I am]

cognac /ˈkɒnjæk/ n. a high-quality brandy, properly that distilled in Cognac in western France.

cognate /ˈkɒɡneɪt/ adj. & n. ● adj. **1** related to or descended from a common ancestor (cf. AGNATE). **2** Philol. (of a word) having the same linguistic family or derivation (as another); representing the same original word or root (e.g. English father, German Vater, Latin pater). ● n. **1** a relative. **2** Philol. a cognate word. □ **cognate object** Gram. an object that is related in origin and sense to the verb governing it (as in live a good life). □ **cognately** adv. **cognateness** n. [L cognatus (as CO-, natus born)]

cognition /kɒɡˈnɪʃ(ə)n/ n. **1** Philos. knowing, perceiving, or conceiving as an act or faculty distinct from emotion and volition. **2** a result of this; a perception, sensation, notion, or intuition. □ **cognitional** adj. **cognitive** /ˈkɒɡnɪtɪv/ adj. [L cognitio (as CO-, gnoscere gnit- apprehend)]

cognizable /ˈkɒɡnɪzəb(ə)l, ˈkɒn-/ adj. (also **-isable**) **1** perceptible, recognizable; clearly identifiable. **2** Law within the jurisdiction of a court. □ **cognizably** adv. [COGNIZANCE + -ABLE]

cognizance /ˈkɒɡnɪz(ə)ns, ˈkɒn-/ n. (also **cognisance**) **1** knowledge or awareness; perception, notice. **2** the sphere of one's observation or concern. **3** Law the right of a court to deal with a matter. **4** Heraldry a distinctive device or mark. □ **have cognizance of** know, esp. officially. **take cognizance of** attend to; take account of. [ME f. OF conoisance ult. f. L cognoscent- f. cognitio: see COGNITION]

cognizant /ˈkɒɡnɪz(ə)nt, ˈkɒn-/ adj. (also **cognisant**) (foll. by of) having knowledge or being aware of.

cognomen /kɒɡˈnəʊmen/ n. (pl. **cognomens** or **cognomina** /-ˈnɒmɪnə/) **1** a nickname. **2** an ancient Roman's third personal name (as in Marcus Tullius Cicero) or personal epithet (as in Publius Cornelius Scipio Africanus). [L]

cognoscente /ˌkɒnjəˈʃentɪ/ n. (pl. **cognoscenti** pronunc. same) (usu. in pl.) a connoisseur. [obs. It.]

cohabit /kəʊˈhæbɪt/ v.intr. (**cohabited**, **cohabiting**) live together, esp. as husband and wife without being married to one another. □ **cohabitant** n. **cohabiter** n. **cohabitation** /-ˌhæbɪˈteɪʃ(ə)n/ n. **cohabitee** /-ˈtiː/ n. [L cohabitare (as CO-, habitare dwell)]

cohere /kəʊˈhɪə(r)/ v.intr. **1** (of parts or a whole) stick together, remain united. **2** (of reasoning etc.) be logical or consistent. [L cohaerere cohaes- (as CO-, haerere stick)]

coherent /kəʊˈhɪərənt/ adj. **1** (of a person) able to speak intelligibly and articulately. **2** (of speech, an argument, etc.) logical and consistent; easily followed. **3** cohering; sticking together. **4** Physics (of waves) having a constant phase relationship. □ **coherence** n. **coherency** n. **coherently** adv. [L cohaerere cohaerent- (as COHERE)]

cohesion /kəʊˈhiːʒ(ə)n/ n. **1 a** the act or condition of sticking together. **b** a tendency to cohere. **2** Chem. the force with which molecules cohere. □ **cohesive** /-ˈhiːsɪv/ adj. **cohesively** adv. **cohesiveness** n. [L cohaes- (see COHERE) after adhesion]

Cohn /kəʊn/, Ferdinand Julius (1828–98), German botanist, a founder of bacteriology. Noted for his studies of algae, bacteria, and other micro-organisms, Cohn was the first to devise a systematic classification of bacteria into genera and species. It was Cohn who recognized the importance of the work of Robert Koch on anthrax.

coho /ˈkəʊhəʊ/ n. (also **cohoe**) (pl. **-os** or **-oes**) a salmon, Oncorhynchus kisutch, of the North Pacific. Also called silver salmon. [19th c.: orig. unkn.]

cohort /ˈkəʊhɔːt/ n. **1** an ancient Roman military unit, equal to one-tenth of a legion. **2** a band of warriors. **3 a** persons banded or grouped together, esp. in a common cause. **b** a group of persons with a common statistical characteristic. **4** N. Amer. a companion or colleague. [ME f. F cohorte or L cohors cohort- enclosure, company]

COHSE /ˈkəʊzi/ abbr. hist. (in the UK) Confederation of Health Service Employees, merged with NALGO and NUPE to form UNISON in 1993.

COI abbr. (in the UK) Central Office of Information.

coif /kɔɪf/ n. **1** a close-fitting cap, now esp. as worn by nuns under a veil. **2** hist. a protective metal skullcap worn under armour. [ME f. OF coife f. LL cofia helmet]

coiffeur /kwʌˈfɜː(r)/ n. (fem. **coiffeuse** /-ˈfɜːz/) a hairdresser. [F]

coiffure /kwʌˈfjʊə(r)/ n. the way hair is arranged; a hairstyle. □ **coiffured** adj. [F]

coign /kɔɪn/ n. □ **coign of vantage** a favourable position for observation or action. [earlier spelling of COIN in the sense 'cornerstone']

coil¹ /kɔɪl/ n. & v. ● n. **1** anything arranged in a joined sequence of concentric circles. **2** a length of rope, a spring, etc., arranged in this way. **3** a single turn of something coiled, e.g. a snake. **4** a lock of hair twisted and coiled. **5** an intrauterine contraceptive device in the form of a coil. **6** *Electr.* a device consisting of a coiled wire for converting low voltage to high voltage, esp. for transmission to the sparking plugs of an internal-combustion engine. **7** a piece of wire, piping, etc., wound in circles or spirals. **8** a roll of postage stamps. ● *v.* **1** *tr.* arrange in a series of concentric loops or rings. **2** *tr. & intr.* twist or be twisted into a circular or spiral shape. **3** *intr.* move sinuously. [OF *coillir* f. L *colligere* COLLECT[1]]

coil[2] /kɔɪl/ *n.* □ **this mortal coil** the difficulties of earthly life (with ref. to Shakespeare's *Hamlet* III. i. 67). [16th c.: orig. unkn.]

Coimbatore /ˌkɔɪmbəˈtɔː(r)/ a city in the state of Tamil Nadu, in southern India; pop. (1991) 853,000.

Coimbra /kəʊˈɪmbrə/ a university city in central Portugal; pop. (est. 1987) 71,780.

coin /kɔɪn/ *n. & v.* ● *n.* **1** a piece of flat usu. round metal stamped and issued by authority as money. **2** (*collect.*) metal money. ● *v.tr.* **1** make (coins) by stamping. **2** make (metal) into coins. **3** invent or devise (esp. a new word or phrase). □ **coin-box 1** a telephone operated by inserting coins. **2** the receptacle for these. **coin money** (or **coin it, coin it in**) make much money quickly. **coin-op** a launderette etc. with automatic machines operated by inserting coins. **to coin a phrase** *iron.* introducing a banal remark or cliché. [ME f. OF, = stamping-die, f. L *cuneus* wedge]

coinage /ˈkɔɪnɪdʒ/ *n.* **1** the act or process of coining. Coinage is reputed to be the invention of the Lydians of Asia Minor, probably in the mid-7th century BC. **2 a** coins collectively. **b** a system or type of coins in use (*decimal coinage; bronze coinage*). **3** an invention, esp. of a new word or phrase. [ME f. OF *coigniage*]

coincide /ˌkəʊɪnˈsaɪd/ *v.intr.* **1** occur at or during the same time. **2** occupy the same portion of space. **3** (often foll. by *with*) be in agreement; have the same view. [med.L *coincidere* (as CO-, INCIDENT)]

coincidence /kəʊˈɪnsɪd(ə)ns/ *n.* **1 a** occurring or being together. **b** an instance of this. **2** a remarkable concurrence of events or circumstances without apparent causal connection. **3** *Physics* the presence of ionizing particles etc. in two or more detectors simultaneously, or of two or more signals simultaneously in a circuit. [med.L *coincidentia* (as COINCIDE)]

coincident /kəʊˈɪnsɪd(ə)nt/ *adj.* **1** occurring together in space or time. **2** (foll. by *with*) in agreement; harmonious. □ **coincidently** *adv.*

coincidental /kəʊˌɪnsɪˈdent(ə)l/ *adj.* **1** in the nature of or resulting from a coincidence. **2** happening or existing at the same time. □ **coincidentally** *adv.*

coiner /ˈkɔɪnə(r)/ *n.* **1** a person who coins money, esp. *Brit.* the maker of counterfeit coin. **2** a person who invents or devises something (esp. a new word or phrase).

Cointreau /ˈkwɑːntrəʊ/ *n. propr.* an orange-flavoured liqueur that is a colourless form of curaçao. [F]

coir /ˈkɔɪə(r)/ *n.* fibre from the outer husk of the coconut, used for ropes, matting, in potting compost, etc. [Malayalam *kāyar* cord f. *kāyaru* be twisted]

coition /kəʊˈɪʃ(ə)n/ *n. Med.* = COITUS. [L *coitio* f. *coire coit-* go together]

coitus /ˈkəʊɪtəs/ *n. Med.* sexual intercourse. □ **coitus interruptus** /ˌɪntəˈrʌptəs/ sexual intercourse in which the penis is withdrawn before ejaculation. □ **coital** *adj.* [L (as COITION)]

Coke /kəʊk/ *n. propr.* Coca-Cola. [abbr.]

coke[1] /kəʊk/ *n. & v.* ● *n.* **1** a solid substance left after the gases have been extracted from coal. **2** a residue left after the incomplete combustion of petrol etc. ● *v.tr.* convert (coal) into coke. [prob. f. N.Engl. dial. *colk* core, of unkn. orig.]

coke[2] /kəʊk/ *n. sl.* cocaine. [abbr.]

Col. *abbr.* **1** Colonel. **2** Epistle to the Colossians (New Testament).

col /kɒl/ *n.* **1** a depression in the summit-line of a chain of mountains, generally affording a pass from one slope to another. **2** *Meteorol.* a low-pressure region between anticyclones. [F, = neck, f. L *collum*]

col. *abbr.* column.

col- /kɒl, kəl/ *prefix* assim. form of COM- before *l*.

cola /ˈkəʊlə/ *n.* (also **kola**) **1** a small tree of the genus *Cola*, native to West Africa, bearing seeds containing caffeine. **2** a carbonated drink usu. flavoured with these seeds. □ **cola nut** a seed of the tree. [Temne]

colander /ˈkʌləndə(r)/ *n.* a perforated vessel used to strain off liquid in cookery. [ME, ult. f. L *colare* strain]

co-latitude /kəʊˈlætɪˌtjuːd/ *n. Astron.* the complement of the latitude, the difference between it and 90°.

Colbert /ˈkɒlbeə(r)/, Jean Baptiste (1619–83), French statesman, chief minister to Louis XIV 1665–83. A vigorous reformer, he put order back into the country's finances, boosted industry and commerce, and established the French navy as one of the most formidable in Europe. His reforms, however, could not keep pace with the demands of Louis's war policies and extensive royal building programme, and by the end of Louis's reign the French economy was again experiencing severe problems.

Colchester /ˈkəʊltʃɪstə(r)/ a town in Essex; pop. (1981) 82,000. A prominent town in Roman Britain, when it was known as Camulodunum, Colchester is noted for its Roman ruins.

colchicine /ˈkɒltʃɪˌsiːn, ˈkɒlkɪ-/ *n. Pharm.* a yellow alkaloid obtained from colchicum, used in the treatment of gout, in plant breeding to induce mutations, and in biological research to block cell division.

colchicum /ˈkɒltʃɪkəm, ˈkɒlkɪ-/ *n.* **1** a liliaceous plant of the genus *Colchicum*; esp. meadow saffron. **2** its dried corm or seed. [L f. Gk *kolkhikon* of Kolkhis, COLCHIS]

Colchis /ˈkɒlkɪs/ (Greek **Kolkhis** /ˈkɒlxis/) an ancient region south of the Caucasus mountains at the eastern end of the Black Sea. In classical mythology it was the goal of Jason's expedition for the Golden Fleece.

cold /kəʊld/ *adj., n., & adv.* ● *adj.* **1** of or at a low or relatively low temperature, esp. when compared with the human body. **2** not heated; cooled after being heated. **3** (of a person) feeling cold. **4** lacking ardour, friendliness, or affection; undemonstrative, apathetic. **5** depressing, dispiriting, uninteresting (*cold facts*). **6 a** dead. **b** *colloq.* unconscious. **7** *colloq.* at one's mercy (*had me cold*). **8** sexually frigid. **9** (of soil) slow to absorb heat. **10** (of a scent in hunting) having become weak. **11** (in children's games) far from finding or guessing what is sought. **12** without preparation or rehearsal. ● *n.* **1 a** the prevalence of a low temperature, esp. in the atmosphere. **b** cold weather; a cold environment (*went out into the cold*). **2** a viral infection in which the mucous membrane of the nose and throat becomes inflamed, causing running at the nose, sneezing, sore throat, etc. ● *adv.* **1** unrehearsed, without preparation. **2** esp. *US sl.* completely, entirely (*was stopped cold mid-sentence*). □ **catch a cold 1** become infected with a cold virus. **2** encounter trouble or difficulties. **cold call** sell goods or services by making unsolicited calls on prospective customers by telephone or in person. **cold cathode** *Electr.* a cathode that emits electrons without being heated. **cold chisel** a chisel suitable for cutting metal. **cold comfort** poor or inadequate consolation. **cold cream** ointment for cleansing and softening the skin. **cold cuts** slices of cold cooked meats. **cold feet** *colloq.* loss of nerve or confidence. **cold frame** an unheated frame with a glass top for growing small plants. **cold front** the forward edge of an advancing mass of cold air. **cold shoulder** a show of intentional unfriendliness. **cold-shoulder** *v.tr.* be deliberately unfriendly to. **cold sore** inflammation and blisters in and around the mouth, caused by infection with herpes simplex virus. **cold storage 1** storage in a refrigerator or other cold place for preservation. **2** a state in which something (esp. an idea) is put aside temporarily. **cold sweat** a state of sweating induced by fear or illness. **cold table** a selection of dishes of cold food. **cold turkey** *sl.* **1** abrupt withdrawal from addictive drugs; the symptoms of this. **2** *US* blunt statements or blunt treatment. **cold wave 1** a temporary spell of cold weather over a wide area. **2** a kind of permanent wave for the hair using chemicals and without heat. **in cold blood** without feeling or passion; deliberately, ruthlessly. **out in the cold** ignored, neglected. **throw** (or **pour**) **cold water on** be discouraging or depreciatory about. □ **coldish** *adj.* **coldly** *adv.* **coldness** *n.* [OE *cald* f. Gmc, rel. to L *gelu* frost]

cold-blooded /kəʊldˈblʌdɪd/ *adj.* **1** having a body temperature varying with that of the environment (e.g. of fish); poikilothermic. **2 a** callous; deliberately cruel. **b** without excitement or sensibility, dispassionate. □ **cold-bloodedly** *adv.* **cold-bloodedness** *n.*

cold fusion *n. Physics* nuclear fusion at room temperature, seen as a potential source of pollution-free energy. In March 1989 it was announced by Stanley Pons of the US and Martin Fleischmann of Britain, and independently by Stephen E. Jones and colleagues in the US, that they had achieved cold fusion. They passed an electric current through a rod of palladium wrapped with platinum wire and placed in a tube of deuterium, claiming an output of four times more energy than they had put in, together with the production of neutrons and tritium. Although research into possible energy generation by methods like this continues, most scientists are sceptical of the claim that nuclear fusion is involved.

cold-hearted /kəʊldˈhɑːtɪd/ *adj.* lacking affection or warmth; unfriendly. □ **cold-heartedly** *adv.* **cold-heartedness** *n.*

Colditz /'kəʊldɪts/ a medieval castle near Leipzig, eastern Germany, used as a top-security camp for Allied prisoners in the Second World War.

cold-short /'kəʊldʃɔːt/ adj. (of a metal) brittle in its cold state. [Sw. kallskör f. kall cold + skör brittle: assim. to SHORT]

cold start n. & v. ● n. **1** the starting of an engine or machine at the ambient temperature. **2** the starting of a process, enterprise, etc. without prior preparation. ● v.tr. (**cold-start**) start something at the ambient temperature, or without prior preparation.

cold war n. a state of hostility between nations without actual fighting, consisting of threats, violent propaganda, subversive political activities, etc. The term is used specifically to refer to the hostility between the Soviet bloc countries and the Western powers which began after the Second World War with the Soviet takeover of the countries of eastern Europe. In 1949 NATO was formed, and was followed by the establishment of COMECON (also 1949) and later, the Warsaw Pact (1955). Over the following decades, the cold war spread to all parts of the world, with the development of the nuclear-arms race in the 1950s. After the resumption of START talks in 1985, agreement for limited arms control in 1987, and withdrawal of Soviet forces from Afghanistan in 1988–9, the cold war was formally and officially ended in November 1990 by a declaration of friendship and a treaty agreeing a great reduction of conventional armaments in Europe.

cole /kəʊl/ n. (usu. in comb.) **1** cabbage. **2** = RAPE[2]. [ME f. ON kál f. L caulis stem, cabbage]

-cole /kəʊl/ comb. form forming nouns and adjectives denoting living or growing in or on (calcicole). [f. L -colus inhabiting, f. colere inhabit]

colemouse var. of COALMOUSE.

Coleoptera /ˌkɒlɪ'ɒptərə/ n.pl. Zool. an order of insects comprising the beetles and weevils, which have the forewings modified into hardened wing-cases that cover and protect the hindwings, and biting mouthparts. □ **coleopteran** n. & adj. **coleopterist** n. **coleopterous** adj. [mod.L f. Gk koleopteros f. koleon sheath + pteron wing]

coleoptile /ˌkɒlɪ'ɒptaɪl/ n. Bot. a sheath protecting a young shoot tip in grasses. [Gk koleon sheath + ptilon feather]

Coleraine /kəʊl'reɪn/ a town in the north of Northern Ireland, on the River Bann in County Londonderry; pop. (1981) 16,000.

Coleridge /'kəʊləˌrɪdʒ/, Samuel Taylor (1772–1834), English poet, critic, and philosopher. His Lyrical Ballads (1798), written with William Wordsworth, marked the start of English romanticism. The collection included Coleridge's famous poem 'The Rime of the Ancient Mariner'. His other well-known poems include the ballad 'Christabel' (1816) and the opium fantasy 'Kubla Khan' (1816). Coleridge's opium addiction is also recorded in his pessimistic poem 'Dejection: an Ode' (1802). During the latter part of his life he wrote little poetry, but contributed significantly to critical and philosophical literature.

coleseed /'kəʊlsiːd/ n. = RAPE[2].

coleslaw /'kəʊlslɔː/ n. a dressed salad of sliced raw cabbage, carrot, onion, etc. [Du. koolsla: see COLE, SLAW]

cole-tit var. of coal-tit.

Colette /kɒ'let/ (born Sidonie Gabrielle Claudine) (1873–1954), French novelist. Her Claudine series (1900–3) was published by her husband, the novelist Henri Gauthier-Villars (1859–1931), who caused a scandal by inserting salacious passages. Colette made her name as a serious writer with her novels Chéri (1920) and La Fin de Chéri (1926), telling of a passionate relationship between a young man and an older woman. She was awarded the Legion of Honour in 1953.

coleus /'kəʊlɪəs/ n. a labiate plant of the genus Solenostemon, having variegated coloured leaves. [mod.L f. Gk koleon sheath]

coley /'kəʊlɪ/ n. (pl. **-eys**) = SAITHE. [perh. f. coalfish]

colic /'kɒlɪk/ n. a severe spasmodic abdominal pain. □ **colicky** adj. [ME f. F colique f. LL colicus: see COLON[2]]

Colima /kɒ'liːmə/ **1** a state of SW Mexico, on the Pacific coast. **2** the capital city of this state; pop. (est. 1984) 58,000.

coliseum /ˌkɒlɪ'siːəm/ n. (also **colosseum**) a large stadium or amphitheatre. [med.L, neut. of colosseus gigantic (as COLOSSUS)]

colitis /kə'laɪtɪs/ n. Med. inflammation of the lining of the colon.

Coll /kɒl/ an island in the Inner Hebrides, west of Scotland.

Coll. abbr. College.

collaborate /kə'læbəˌreɪt/ v.intr. (often foll. by with) **1** work jointly, esp. in a literary or artistic production. **2** cooperate traitorously with an enemy. □ **collaborator** n. **collaborative** /-rətɪv/

adj. **collaboratively** adv. **collaboration** /-ˌlæbə'reɪʃ(ə)n/ n. **collaborationist** n. & adj. [L collaborare collaborat- (as COM-, laborare work)]

collage /'kɒlaːʒ, kə'laːʒ/ n. **1** a form of art in which various materials (e.g. photographs, pieces of paper, fabric, or wood) are arranged and glued to a backing. **2** a work of art done in this way. **3** a collection of unrelated things. □ **collagist** n. [F, = gluing]

collagen /'kɒlədʒən/ n. Biochem. the major structural protein found in animal connective tissue, yielding gelatin on boiling. [F collagène f. Gk kolla glue + -gène f. -GEN]

collapsar /kə'læpsaː(r)/ n. Astron. an old star that has gravitationally collapsed to form a white dwarf, neutron star, or black hole. [f. COLLAPSE after pulsar, quasar]

collapse /kə'læps/ n. & v. ● n. **1** the tumbling down or falling in of a structure; folding up; giving way. **2** a sudden failure of a plan, undertaking, etc. **3** a physical or mental breakdown. ● v. **1 a** intr. undergo or experience a collapse. **b** tr. cause to collapse. **2** intr. colloq. lie or sit down and relax, esp. after prolonged effort (collapsed into a chair). **3 a** intr. (of furniture etc.) be foldable into a small space. **b** tr. fold (furniture) in this way. □ **collapsible** adj. **collapsibility** /-ˌlæpsɪ'bɪlɪtɪ/ n. [L collapsus past part. of collabi (as COM-, labi slip)]

collar /'kɒlə(r)/ n. & v. ● n. **1** the part of a shirt, dress, coat, etc., that goes round the neck, either upright or turned over. **2** a band of linen, lace, etc., completing the upper part of a costume. **3** a band of leather or other material put round an animal's (esp. a dog's) neck. **4** a restraining or connecting band, ring, or pipe in machinery. **5** a coloured marking resembling a collar round the neck of a bird or animal. **6** Brit. a piece of meat rolled up and tied. ● v.tr. **1** seize (a person) by the collar or neck. **2** capture, apprehend. **3** colloq. accost. **4** sl. take, esp. illicitly. □ **collar-beam** a horizontal beam connecting two rafters and forming with them an A-shaped roof-truss. **collar-bone** either of two bones joining the breastbone and the shoulder-blades. **collared dove** a grey-brown dove, Streptopelia decaocto, having distinct neck-markings. □ **collared** adj. (also in comb.). **collarless** adj. [ME f. AF coler, OF colier, f. L collare f. collum neck]

collard /'kɒlaːd/ n. (also **collards**, **collard greens**) US & dial. a variety of cabbage without a distinct heart. [reduced form of colewort, f. COLE + WORT]

collate /kə'leɪt/ v.tr. **1** analyse and compare (texts, statements, etc.) to identify points of agreement and difference. **2** verify the order of (sheets of a book) by their signatures. **3** assemble (information) from different sources. **4** (often foll. by to) Eccl. appoint (a clergyman) to a benefice. □ **collator** n. [L collat- past part. stem of conferre compare]

collateral /kə'lætərəl/ n. & adj. ● n. **1** security pledged as a guarantee for repayment of a loan. **2** a person having the same descent as another but by a different line. ● adj. **1** descended from the same stock but by a different line. **2** side by side; parallel. **3 a** additional but subordinate. **b** contributory. **c** connected but aside from the main subject, course, etc. □ **collateral damage** destruction or injury beyond that intended or expected, esp. in the vicinity of a military target. □ **collaterally** adv. **collaterality** /-ˌlætə'rælɪtɪ/ n. [ME f. med.L collateralis (as COM-, LATERAL)]

collation /kə'leɪʃ(ə)n/ n. **1** the act or an instance of collating. **2** RC Ch. a light meal allowed during a fast. **3** a light informal meal. [ME f. OF f. L collatio -onis (see COLLATE): sense 2 f. Cassian's Collationes Patrum (= Lives of the Fathers) read by Benedictines and followed by a light meal]

colleague /'kɒliːg/ n. a fellow official or worker, esp. in a profession or business. [F collègue f. L collega (as COM-, legare depute)]

collect[1] /kə'lekt/ v., adj., & adv. ● v. **1** tr. & intr. bring or come together; assemble, accumulate. **2** tr. systematically seek and acquire (books, stamps, etc.), esp. as a continuing hobby. **3 a** tr. obtain (taxes, contributions, etc.) from a number of people. **b** intr. colloq. receive money. **4** tr. call for; fetch (went to collect the laundry). **5 a** refl. regain control of oneself esp. after a shock. **b** tr. concentrate (one's energies, thoughts, etc.). **c** tr. (as **collected** adj.) calm and cool; not perturbed or distracted. **6** tr. infer, gather, conclude. ● adj. & adv. US to be paid for by the receiver (of a telephone call, parcel, etc.). □ **collectable** adj. **collectedly** adv. [F collecter or med.L collectare f. L collectus past part. of colligere (as COM-, legere pick)]

collect[2] /'kɒlekt/ n. a short prayer of the Anglican and Roman Catholic Church, esp. one assigned to a particular day or season. [ME f. OF collecte f. L collecta fem. past part. of colligere: see COLLECT[1]]

collectible /kə'lektɪb(ə)l/ adj. & n. ● adj. worth collecting. ● n. an item sought by collectors.

collection /kəˈlekʃ(ə)n/ n. **1** the act or process of collecting or being collected. **2** a group of things collected together (e.g. works of art, literary items, or specimens), esp. systematically. **3** (foll. by of) an accumulation; a mass or pile (a collection of dust). **4 a** the collecting of money, esp. in church or for a charitable cause. **b** the amount collected. **5** the regular removal of mail, esp. from a postbox, for dispatch. **6** (in pl.) Brit. college examinations held at the end of a term, esp. at Oxford University. [ME f. OF f. L collectio -onis (as COLLECT[1])]

collective /kəˈlektɪv/ adj. & n. ● adj. **1** formed by or constituting a collection. **2** taken as a whole; aggregate (our collective opinion). **3** of or from several or many individuals; common. ● n. **1 a** = collective farm. **b** any cooperative enterprise. **c** its members. **2** Gram. = collective noun. □ **collective bargaining** negotiation of wages etc. by an organized body of employees. **collective farm** a jointly operated esp. state-owned amalgamation of several smallholdings. **collective memory** Anthropol. the memory of a group of people, often passed from one generation to the next. **collective noun** Gram. a noun that is grammatically singular and denotes a collection or number of individuals (e.g. assembly, family, troop). **collective ownership** ownership of land, means of production, etc., by all for the benefit of all. **collective unconscious** Psychol. (in Jungian theory) the part of the unconscious mind derived from ancestral memory and experience common to all humankind, as distinct from the personal unconscious. □ **collectively** adv. **collectiveness** n. **collectivity** /ˌkɒlekˈtɪvɪti/ n. [F collectif or L collectivus (as COLLECT[1])]

collectivism /kəˈlektɪˌvɪz(ə)m/ n. the theory and practice of the collective ownership of land and the means of production. □ **collectivist** n. **collectivistic** /-ˌlektɪˈvɪstɪk/ adj.

collectivize /kəˈlektɪˌvaɪz/ v.tr. (also **-ise**) organize on the basis of collective ownership. □ **collectivization** /-ˌlektɪvaɪˈzeɪʃ(ə)n/ n.

collector /kəˈlektə(r)/ n. **1** a person who collects, esp. things of interest as a hobby. **2** a person who collects money etc. due (tax-collector; ticket-collector). **3** Electronics the region in a transistor that absorbs carriers of a charge. □ **collector's item** (or **piece**) a valuable object, esp. one of interest to collectors. [ME f. AF collectour f. med.L collector (as COLLECT[1])]

colleen /kɒˈliːn/ n. Ir. a girl. [Ir. cailín, dimin. of caile country-woman]

college /ˈkɒlɪdʒ/ n. **1** an establishment for further or higher education, sometimes part of a university. **2** an establishment for specialized professional education (business college; college of music; naval college). **3** the buildings or premises of a college (lived in college). **4** the students and teachers in a college. **5** Brit. a public school. **6** an organized body of persons with shared functions and privileges (College of Physicians). □ **college of education** a training college for schoolteachers. **college pudding** Brit. a small baked or steamed suet pudding with dried fruit. □ **collegial** /kəˈliːdʒəl/ adj. [ME f. OF college or L collegium f. collega (as COLLEAGUE)]

College of Arms (also called Heralds' College) an English heraldic corporation responsible for recording lineage, establishing the right to bear arms, and granting arms, formed in 1484. It comprises three Kings of Arms, six heralds, and four pursuivants.

College of Cardinals (also called Sacred College) the body of cardinals of the Roman Catholic Church, founded in the 11th century and since 1179 responsible for the election of the pope. It also takes on government of the Church during vacancies in the papacy.

collegian /kəˈliːdʒən/ n. a member of a college. [med.L collegianus (as COLLEGE)]

collegiate /kəˈliːdʒət/ adj. **1** of the nature of, constituted as, or belonging to, a college; corporate. **2** (of a university) composed of different colleges. □ **collegiate church 1** a church endowed for a chapter of canons but without a bishop's see. **2** US & Sc. a church or group of churches established under a joint pastorate. □ **collegiately** adv. [LL collegiatus (as COLLEGE)]

Collembola /kəˈlembələ/ n.pl. Zool. an order of wingless insects comprising the springtails. □ **collembolan** n. & adj. [mod.L f. Gk kolla glue + embolon peg]

collenchyma /kɒˈleŋkɪmə/ n. Bot. a tissue of cells with thick cellulose cell walls, strengthening young stems etc. [Gk kolla glue + egkhuma infusion]

Colles' fracture /ˈkɒlɪs/ n. a fracture of the lower end of the radius with a backward displacement of the hand. [A. Colles, Ir. surgeon (1773–1843)]

collet /ˈkɒlɪt/ n. **1** a flange or socket for setting a gem in jewellery. **2** Engin. a segmented band or sleeve put round a shaft or spindle and

tightened to grip it. **3** a small collar in a clock etc. to which the inner end of a balance spring is attached. [F, dimin. of COL]

collide /kəˈlaɪd/ v.intr. (often foll. by with) **1** come into abrupt or violent impact. **2** be in conflict. [L collidere collis- (as COM-, laedere strike, damage)]

collider /kəˈlaɪdə(r)/ n. Physics an accelerator in which two beams of particles are made to collide.

collie /ˈkɒli/ n. a breed of sheepdog, orig. Scottish, with a long pointed nose and usu. dense long hair. [perh. f. coll COAL (as being orig. black)]

collier /ˈkɒlɪə(r)/ n. **1** a coal miner. **2 a** a coal-ship. **b** a member of its crew. [ME, f. COAL + -IER]

colliery /ˈkɒlɪəri/ n. (pl. **-ies**) a coal mine and its associated buildings.

colligate /ˈkɒlɪˌɡeɪt/ v.tr. bring into connection (esp. isolated facts by a generalization). □ **colligation** /ˌkɒlɪˈɡeɪʃ(ə)n/ n. [L colligare colligat- (as COM-, ligare bind)]

collimate /ˈkɒlɪˌmeɪt/ v.tr. **1** adjust the line of sight of (a telescope etc.). **2** make (telescopes or rays) accurately parallel. □ **collimation** /ˌkɒlɪˈmeɪʃ(ə)n/ n. [L collimare, erron. for collineare align (as COM-, linea line)]

collimator /ˈkɒlɪˌmeɪtə(r)/ n. **1** a device for producing a parallel beam of rays or radiation. **2** a small fixed telescope used for adjusting the line of sight of an astronomical telescope, etc.

collinear /kəˈlɪnɪə(r)/ adj. Geom. (of points) lying in the same straight line. □ **collinearity** /-ˌlɪnɪˈærɪti/ n.

Collins[1] /ˈkɒlɪnz/, Joan (Henrietta) (b.1933), English actress. She established a reputation as a sex symbol in her nine films of the 1950s, including the romantic comedy Our Girl Friday (1953). She continued to act in glamorous roles throughout the 1960s, 1970s, and 1980s, including that of Alexis in the US television soap opera Dynasty (1981–9). She is the sister of the novelist Jackie Collins.

Collins[2] /ˈkɒlɪnz/, Michael (1890–1922), Irish nationalist leader and politician. He took part in the Easter Rising in 1916. Elected to Parliament as a member of Sinn Fein in 1919, he became Minister of Finance in the provisional government, at the same time directing the Irish Republican Army's guerrilla campaign against the British. He was one of the negotiators of the Anglo-Irish Treaty of 1921, and commanded the Irish Free State forces in the civil war that followed partition. On the death of Arthur Griffith in 1922, he became head of state, but was shot in an ambush ten days later.

Collins[3] /ˈkɒlɪnz/, (William) Wilkie (1824–89), English novelist. Although he wrote in a number of genres, he is chiefly remembered as the writer of the first full-length detective stories in English, notably The Woman in White (1860) and The Moonstone (1868). These are striking for their use of multiple narrators and the complexity of their plots.

Collins[4] /ˈkɒlɪnz/ n. an iced drink made of gin or whisky etc. with soda, lemon or lime juice, and sugar. [20th c.: orig. unkn.]

collision /kəˈlɪʒ(ə)n/ n. **1** a violent impact of a moving body, esp. a vehicle or ship, with another or with a fixed object. **2** the clashing of opposed interests or considerations. **3** Physics the action of particles striking or coming together. □ **collision course** a course or action that is bound to cause a collision or conflict. □ **collisional** adj. [ME f. LL collisio (as COLLIDE)]

collocate /ˈkɒləˌkeɪt/ v.tr. **1** place together or side by side. **2** arrange; set in a particular place. **3** (often foll. by with) Linguistics juxtapose (a word etc.) with another. □ **collocation** /ˌkɒləˈkeɪʃ(ə)n/ n. [L collocare collocat- (as COM-, locare to place)]

collocutor /ˈkɒləˌkjuːtə(r), kəˈlɒkjʊtə(r)/ n. a person who takes part in a conversation. [LL f. colloqui (as COM-, loqui locut- talk)]

collodion /kəˈləʊdɪən/ n. a syrupy solution of cellulose nitrate in a mixture of alcohol and ether, used in photography and surgery. [Gk kollōdēs gluelike f. kolla glue]

collogue /kəˈləʊɡ/ v.intr. (**collogues, collogued, colloguing**) (foll. by with) talk confidentially. [prob. alt. of obs. colleague conspire, by assoc. with L colloqui converse]

colloid /ˈkɒlɔɪd/ n. **1** Chem. a non-crystalline substance consisting of ultramicroscopic particles, often large single molecules, usu. dispersed through a second substance, as in gels, sols, and emulsions. **2** Med. a substance of a homogeneous gelatinous consistency. □ **colloidal** /kəˈlɔɪd(ə)l/ adj. [Gk kolla glue + -OID]

collop /ˈkɒləp/ n. a slice, esp. of meat or bacon; an escalope. [ME, = fried bacon and eggs, of Scand. orig.]

colloquial /kəˈləʊkwɪəl/ adj. belonging to or proper to ordinary or familiar conversation, not formal or literary. □ **colloquially** *adv.* [L *colloquium* COLLOQUY]

colloquialism /kə'ləʊkwɪəˌlɪz(ə)m/ *n.* **1** a colloquial word or phrase. **2** the use of colloquialisms.

colloquium /kə'ləʊkwɪəm/ *n.* (*pl.* **colloquiums** or **colloquia** /-kwɪə/) an academic conference or seminar. [L: see COLLOQUY]

colloquy /'kɒləkwɪ/ *n.* (*pl.* **-quies**) **1** *formal* **a** the act of conversing. **b** a conversation. **2** *Eccl.* a gathering for discussion of theological questions. [L *colloquium* (as COM-, *loqui* speak)]

collotype /'kɒləˌtaɪp/ *n. Printing* **1** a thin sheet of gelatin exposed to light, treated with reagents, and used to make high quality prints by lithography. **2** a print made by this process. [Gk *kolla* glue + TYPE]

collude /kə'lu:d, -'lju:d/ *v.intr.* come to an understanding or conspire together, esp. for a fraudulent purpose. □ **colluder** *n.* [L *colludere collus-* (as COM-, *ludere lus-* play)]

collusion /kə'lu:ʒ(ə)n, -'lju:ʒ(ə)n/ *n.* **1** a secret understanding, esp. for a fraudulent purpose. **2** *Law* such an understanding between ostensible opponents in a lawsuit. □ **collusive** /-'lu:sɪv, -'lju:sɪv/ *adj.* **collusively** *adv.* [ME f. OF *collusion* or L *collusio* (as COLLUDE)]

collyrium /kə'lɪrɪəm/ *n.* (*pl.* **collyria** /-rɪə/) a medicated eye-lotion. [L f. Gk *kollurion* poultice f. *kollura* coarse bread-roll]

collywobbles /'kɒlɪˌwɒb(ə)lz/ *n.pl. colloq.* **1** a rumbling or pain in the stomach. **2** a feeling of strong apprehension. [fanciful, f. COLIC + WOBBLE]

Colo. *abbr.* Colorado.

colobus /'kɒləbəs/ *n.* any leaf-eating monkey of the genus *Colobus*, native to Africa, having shortened thumbs. [mod.L f. Gk *kolobos* docked]

colocynth /'kɒləsɪnθ/ *n.* (also **coloquintida** /ˌkɒlə'kwɪntɪdə/) **1 a** a plant, *Citrullus colocynthis*, of the gourd family, bearing a pulpy fruit. **b** this fruit. **2** a bitter purgative drug obtained from the fruit. [L *colocynthis* f. Gk *kolokunthis*]

Cologne /kə'ləʊn/ (German **Köln** /kœln/) an industrial and university city in western Germany, in North Rhine-Westphalia; pop. (1991) 956,690. Founded by the Romans, Cologne is situated on the west bank of the River Rhine. It is renowned for its medieval cathedral. The archbishop of Cologne was one of the most powerful German secular princes in the Middle Ages.

cologne /kə'ləʊn/ *n.* (in full **cologne water**) eau-de-Cologne or a similar scented toilet water. [COLOGNE]

Colombia /kə'lɒmbɪə/ a country in the extreme NW of South America, having a coastline on both the Atlantic and the Pacific Ocean; pop. (est. 1991) 32,873,000; official language, Spanish; capital, Bogotá. The Pacific coastal plain is humid and swampy, and most of the population is concentrated in the temperate valleys of the Andes. The economy is mainly agricultural, coffee being the chief export. Mineral resources are rich and include gold, silver, platinum (one of the world's richest deposits), emeralds, and salt, as well as oil, coal, and natural gas. Inhabited by the Chibcha and other native peoples, Colombia was conquered by the Spanish in the early 16th century, and under Spanish rule the capital Bogotá developed such a reputation for intellectual and social life as to be called 'the Athens of South America'. Like the rest of Spain's South American empire, Colombia achieved independence in the early 19th century, although the resulting Republic of Great Colombia lasted only until 1830, when first Venezuela and then Ecuador broke away to become independent states in their own right. The remaining state was known as New Granada, changing its name to Colombia in 1863. The country was stricken by civil war between 1949 and 1953, and since then has struggled with endemic poverty and social problems. It is notorious for its drugs traffic. □ **Colombian** *adj.* & *n.*

Colombo /kə'lʌmbəʊ/ the capital and chief port of Sri Lanka; pop. (1990) 615,000.

Colón /kɒ'lɒn/ the chief port of Panama, at the Caribbean end of the Panama Canal; pop. (1990) 140,900. Founded in 1850 by the American William Aspinwall (1807–55), after whom it was originally named, it was renamed in 1903 after Christopher Columbus, Colón being the Spanish form of his surname.

colon¹ /'kəʊlən, -lɒn/ *n.* a punctuation mark (:), used esp. to introduce a quotation or a list of items or to separate clauses when the second expands or illustrates the first; also between numbers in a statement of proportion (as in 10:1) and in biblical references (as in Exodus 3:2). [L f. Gk *kōlon* limb, clause]

colon² /'kəʊlən, -lɒn/ *n. Anat.* the greater part of the large intestine,

from the caecum to the rectum. □ **colonic** /kə'lɒnɪk/ *adj.* [ME, ult. f. Gk *kolon*]

colón /kɒ'lɒn/ *n.* (*pl.* **colones** /-'lɒnez/) the basic monetary unit of Costa Rica and El Salvador, equal to 100 centimos in Costa Rica and 100 centavos in El Salvador. [*Colón*, Sp. form of the name *Columbus*: see COLUMBUS²]

colonel /'kɜ:n(ə)l/ *n.* **1** an army officer in command of a regiment, immediately below a brigadier in rank. **2** *US* an officer of corresponding rank in the air force. **3** = *lieutenant colonel*. □ **Colonel Blimp** see BLIMP *n.* 1. □ **colonelcy** *n.* (*pl.* **-ies**). [obs. F *coronel* f. It. *colonnello* f. *colonna* COLUMN]

colonial /kə'ləʊnɪəl/ *adj.* & *n.* ● *adj.* **1 a** of, relating to, or characteristic of a colony or colonies, esp. of a British Crown Colony. **b** of colonialism. **2** (esp. of architecture or furniture) built or designed in, or in a style characteristic of, the period of the British colonies in America before independence. ● *n.* **1** a native or inhabitant of a colony. **2** a house built in colonial style. □ **colonial goose** *Austral.* & *NZ* a boned and stuffed roast leg of mutton. □ **colonially** *adv.*

colonialism /kə'ləʊnɪəˌlɪz(ə)m/ *n.* **1** a policy of acquiring or maintaining colonies. **2** *derog.* this policy regarded as the esp. economic exploitation of weak or backward peoples by a larger power. □ **colonialist** *n.*

colonist /'kɒlənɪst/ *n.* a settler in or inhabitant of a colony.

colonize /'kɒləˌnaɪz/ *v.* (also **-ise**) **1** *tr.* establish a colony or colonies in (a country or area). **b** settle as colonists. **2** *intr.* establish or join a colony. **3** *tr. US Polit.* plant voters in (a district) for party purposes. **4** *tr. Biol.* (of plants and animals) become established (in an area). □ **colonizer** *n.* **colonization** /ˌkɒlənaɪ'zeɪʃ(ə)n/ *n.*

colonnade /ˌkɒlə'neɪd/ *n. Archit.* a row of columns, esp. supporting an entablature or roof. □ **colonnaded** *adj.* [F f. *colonne* COLUMN]

colony /'kɒlənɪ/ *n.* (*pl.* **-ies**) **1 a** a group of settlers in a new country (whether or not already inhabited) fully or partly subject to the mother country. **b** the settlement or its territory. **2 a** people of one nationality or race or occupation in a city, esp. if living more or less in isolation or in a special quarter. **b** a separate or segregated group (*nudist colony*). **3** *Biol.* a community of animals or plants of one kind forming a physically connected structure or living close together. [ME f. L *colonia* f. *colonus* farmer f. *colere* cultivate]

colophon /'kɒləˌfɒn, -f(ə)n/ *n.* **1** a publisher's device or imprint, esp. on the title-page. **2** a tailpiece in a manuscript or book, often ornamental, giving the writer's or printer's name, the date, etc. [LL f. Gk *kolophōn* summit, finishing touch]

colophony /kə'lɒfənɪ/ *n.* = ROSIN. [L *colophonia* (resin) from Colophon in Asia Minor]

coloquintida var. of COLOCYNTH.

color *US* var. of COLOUR etc.

Colorado /ˌkɒlə'rɑ:dəʊ/ **1** a river which rises in the Rocky Mountains of northern Colorado and flows generally south-westwards for 2,333 km (1,468 miles) to the Gulf of California. Approximately 447 km (278 miles) of its course is through the Grand Canyon in Arizona. **2** a state in the central US; pop. (1990) 3,294,400; capital, Denver. Named from the Colorado river which rises there, Colorado extends from the Great Plains in the east to the Rocky Mountains in the west. Part of it was acquired by the Louisiana Purchase in 1803 and the rest ceded by Mexico in 1848. It became the 38th state in 1876.

Colorado beetle *n.* a yellow and black striped leaf-beetle, *Leptinotarsa decemlineata*, the larva of which is highly destructive to the potato plant. [COLORADO 2]

coloration /ˌkʌlə'reɪʃ(ə)n/ *n.* (also **colouration**) **1** colouring; a scheme or method of applying colour. **2** the natural (esp. variegated) colour of living things or animals. [F *coloration* or LL *coloratio* f. *colorare* COLOUR]

coloratura /ˌkɒlərə'tʊərə/ *n.* **1** elaborate ornamentation of a vocal melody. **2** a singer (esp. a soprano) skilled in coloratura singing. [It. f. L *colorare* COLOUR]

colorific /ˌkʌlə'rɪfɪk/ *adj.* **1** producing colour. **2** highly coloured. [F *colorifique* or mod.L *colorificus* (as COLOUR)]

colorimeter /ˌkʌlə'rɪmɪtə(r)/ *n.* an instrument for measuring the intensity of colour. □ **colorimetry** *n.* **colorimetric** /-rɪ'metrɪk/ *adj.* [L *color* COLOUR + -METER]

colossal /kə'lɒs(ə)l/ *adj.* **1** of immense size; huge, gigantic. **2** *colloq.* remarkable, splendid. **3** *Archit.* (of an order) having more than one storey of columns. **4** *Sculpture* (of a statue) about twice life size. □ **colossally** *adv.* [F f. *colosse* COLOSSUS]

colosseum *n.* var. of COLISEUM.

Colosseum, the the name since medieval times of the *Amphitheatrum Flavium*, a vast amphitheatre in Rome begun by Vespasian AD *c*.75 and continued and completed by Titus and Domitian. It was capable of holding 50,000 people, with seating in three tiers and standing-room above; an elaborate system of staircases served all parts. The timber-floored arena was the scene of gladiatorial combats, fights between men and beasts, and large-scale mock battles.

Colossians, Epistle to the /kə'lɒʃ(ə)nz/ a book of the New Testament, an epistle of St Paul to the Church at Colossae in Phrygia, Asia Minor.

colossus /kə'lɒsəs/ *n.* (*pl.* **colossi** /-saɪ/ or **colossuses**) **1** a statue much bigger than life size. **2** a gigantic person, animal, building, etc. **3** an imperial power personified. [L f. Gk *kolossos*]

Colossus of Rhodes a huge bronze statue of the sun-god Helios, one of the Seven Wonders of the World, said by Pliny the Elder to have been over 30.5 m (100 ft) high. Built *c*.292–280 BC, it stood beside the harbour entrance at Rhodes for about fifty years but was destroyed in an earthquake in 224 BC.

colostomy /kə'lɒstəmɪ/ *n.* (*pl.* **-ies**) a surgical operation in which the colon is shortened to remove a damaged part and its cut end directed to an opening in the abdominal wall. [as COLON[2] + Gk *stoma* mouth]

colostrum /kə'lɒstrəm/ *n.* the first secretion from the mammary glands occurring after giving birth, rich in antibodies. [L]

colotomy /kə'lɒtəmɪ/ *n.* (*pl.* **-ies**) *Surgery* an incision in the colon. [as COLON[2] + -TOMY]

colour /'kʌlə(r)/ *n. & v.* ● *n.* (*US* **color**) **1 a** the sensation produced on the eye by rays of light when resolved into different wavelengths, as by a prism, selective reflection, etc. (*black* being the effect produced by no light or by a surface reflecting no rays, and *white* the effect produced by rays of unresolved light). (*See note below.*) **b** perception of colour; a system of colours. **2** one, or any mixture, of the constituents into which light can be separated as in a spectrum or rainbow, sometimes including (loosely) black and white. **3** a colouring substance, esp. paint. **4** the use of all colours, not only black and white, as in photography and television. **5 a** pigmentation of the skin, esp. when dark. **b** this as a ground for prejudice or discrimination. **6** ruddiness of complexion (*a healthy colour*). **7** (in *pl.*) appearance or aspect (*see things in their true colours*). **8** (in *pl.*) **a** *Brit.* a coloured ribbon or uniform etc. worn to signify membership of a school, club, team, etc. **b** the flag of a regiment or ship. **c** a national flag. **9** quality, mood, or variety in music, literature, speech, etc.; distinctive character or timbre. **10** a show of reason; a pretext (*lend colour to; under colour of*). **11** *Physics* a quantum number assigned to quarks and gluons in the theory of quantum chromodynamics. ● *v.* **1** *tr.* apply colour to, esp. by painting or dyeing or with coloured pens or pencils. **2** *tr.* influence (*an attitude coloured by experience*). **3** *tr.* **a** misrepresent, exaggerate, esp. with spurious detail (*a highly coloured account*). **b** disguise. **4** *intr.* take on colour; blush. □ **colour bar** the denial of services and facilities to non-white people. **colour-blind** unable to distinguish certain colours. **colour-blindness** the condition of being colour-blind. **colour code** use of colours as a standard means of identification. **colour-code** *v.tr.* identify by means of a colour code. **colour-fast** dyed in colours that will not fade or be washed out. **colour-fastness** the condition of being colour-fast. **colour scheme** an arrangement or planned combination of colours esp. in interior design. **colour-sergeant** the senior sergeant of an infantry company. **colour supplement** *Brit.* a magazine with coloured illustrations, issued as a supplement to a newspaper. **colour temperature** *Astron. & Photog.* the temperature at which a black body would emit radiation of the same colour as a given object. **colour wash** coloured distemper. **colour-wash** *v.tr.* paint with coloured distemper. **person of colour** a non-white person. **Queen's** (or **King's** or **regimental**) **colour** a flag carried by a regiment. **show one's true colours** reveal one's true character or intentions. **under false colours** falsely, deceitfully. **with flying colours** see FLYING. [ME f. OF *color, colorer* f. L *color, colorare*]

▪ Newton showed that white light can be split by a prism into its constituent colours, ranging from violet light, which has the highest frequency (shortest wavelength), to red light, which has the lowest frequency (longest wavelength). Colour perceived depends on the relative intensities of the mixture of frequencies present in the light received. Opaque objects appear coloured according to the wavelengths they reflect (other wavelengths being absorbed).

Coloured lights can be mixed to produce other colours. Red, blue, and green lights — the primary colours — combine to produce white light, and in unequal combinations can produce almost any colour sensation. In the case of pigments or dyes the situation is different; combination is subtractive, and the colour of a mixture of two pigments depends on the wavelengths absorbed by neither. For pigments (as in printing, photography, etc.) magenta, yellow, and cyan are used as primary colours.

colourable /'kʌlərəb(ə)l/ *adj.* (*US* **colorable**) **1** specious, plausible. **2** counterfeit. □ **colourably** *adv.*

colourant /'kʌlərənt/ *n.* (*US* **colorant**) a colouring substance.

colouration var. of COLORATION.

coloured /'kʌləd/ *adj. & n.* (*US* **colored**) ● *adj.* **1** having colour(s). **2 a** wholly or partly of non-white descent. **b** *S. Afr.* of mixed white and non-white descent. **c** of or relating to coloured people (*a coloured audience*). ● *n.* **1 a** often *offens.* a coloured person. **b** *S. Afr.* a person of mixed descent speaking Afrikaans or English as the mother tongue. **2** (in *pl.*) coloured clothing etc. for washing.

colour-field painting /'kʌlə,fi:ld/ *n.* a style of American abstract painting prominent from the late 1940s to the 1960s which features large expanses of unmodulated colour covering the greater part of the canvas. Barnett Newman was considered its chief exponent, and other painters associated with it include Mark Rothko and Clyfford Still (1904–80) (see also ABSTRACT EXPRESSIONISM).

colourful /'kʌlə,fʊl/ *adj.* (*US* **colorful**) **1** having much or varied colour; brightly coloured. **2** full of interest; vivid, lively. □ **colourfully** *adv.* **colourfulness** *n.*

colouring /'kʌlərɪŋ/ *n.* (*US* **coloring**) **1** the process of or skill in using colour(s). **2** the style in which a thing is coloured, or in which an artist uses colour. **3** facial complexion.

colourist /'kʌlərɪst/ *n.* (*US* **colorist**) a person who uses colour, esp. in art.

colourless /'kʌləlɪs/ *adj.* (*US* **colorless**) **1** without colour. **2** lacking character or interest. **3** dull or pale in hue. **4** neutral, impartial, indifferent. □ **colourlessly** *adv.*

coloury /'kʌlərɪ/ *adj.* (*US* **colory**) having a distinctive colour, esp. as indicating good quality.

colposcopy /kɒl'pɒskəpɪ/ *n. Med.* examination of the vagina and the neck of the womb. □ **colposcope** /'kɒlpə,skəʊp/ *n.* [Gk *kolpos* womb + -SCOPY]

Colt[1] /kəʊlt/, Samuel (1814–62), American inventor. He is remembered chiefly for the automatic pistol named after him. Colt patented his invention in 1836, but had to wait ten years before its adoption by the US army after the outbreak of the war with Mexico. The revolver was highly influential in the 19th-century development of small arms.

Colt[2] /kəʊlt/ *n. propr.* a type of repeating pistol. [COLT[1]]

colt /kəʊlt/ *n.* **1** a young uncastrated male horse, usu. less than four years old. **2** *Sport* a young or inexperienced player; a member of a junior team. □ **colthood** *n.* **coltish** *adj.* **coltishly** *adv.* **coltishness** *n.* [OE, = young ass or camel]

colter *US* var. of COULTER.

Coltrane /kɒl'treɪn/, John (William) (1926–67), American jazz musician. Once established as a jazz saxophonist, he played in groups led by Dizzy Gillespie and Miles Davis. In 1960 he formed his own quartet and soon became a leading figure in avant-garde jazz, bridging the transition between the harmonically dense jazz of the 1950s and the free jazz that was evolving in the 1960s.

coltsfoot /'kəʊltsfʊt/ *n.* (*pl.* **coltsfoots**) a composite plant, *Tussilago farfara*, with large heart-shaped leaves and yellow flowers.

colubrid /'kɒljʊbrɪd/ *n. & adj. Zool.* ● *n.* a snake of the large family Colubridae, to which most non-venomous snakes belong. ● *adj.* of or relating to this family. [mod.L *Colubridae*: see COLUBRINE]

colubrine /'kɒljʊ,braɪn/ *adj.* **1** snakelike. **2** *Zool.* of the subfamily Colubrinae of colubrid snakes with solid teeth. [L *colubrinus* f. *coluber* snake]

colugo /kə'lu:gəʊ/ *n.* (*pl.* **-os**) = flying lemur. [18th c.: orig. unkn.]

Columba, St /kə'lʌmbə/ (*c*.521–97), Irish abbot and missionary. After founding several churches and monasteries in his own country, he established the monastery at Iona in *c*.563, led a number of missions to mainland Scotland from there, and converted the Picts to Christianity. He is considered one of the leading figures of the Celtic missionary tradition in the British Isles (see CELTIC CHURCH) and contributed significantly to the literature of Celtic Christianity. Feast day, 9 June.

Columbia /kəˈlʌmbɪə/ **1** a river in NW North America which rises in the Rocky Mountains of south-east British Columbia, Canada, and flows 1,953 km (1,230 miles) generally southwards into the US, where it turns westwards to form the Washington–Oregon border and enters the Pacific south of Seattle. **2** the state capital of South Carolina; pop. (1990) 98,000.

Columbia, District of see DISTRICT OF COLUMBIA.

Columbia University a university in New York City, one of the most prestigious in the US. It was founded in 1754.

Columbine /ˈkɒləmˌbaɪn/ a character in Italian *commedia dell'arte*, the mistress of Harlequin. She appears in the harlequinade of English pantomime as a short-skirted dancer. [F *Colombine* f. It. *Colombina* f. *colombino* dovelike]

columbine /ˈkɒləmˌbaɪn/ n. a plant of the genus *Aquilegia*; esp. *A. vulgaris*, having purple-blue flowers. (See AQUILEGIA.) [ME f. OF *colombine* f. med.L *colombina herba* dovelike plant f. L *columba* dove (from the supposed resemblance of the flower to a cluster of 5 doves)]

columbite /kəˈlʌmbaɪt/ n. US Chem. an ore of iron and niobium found in America. [*Columbia*, a poetic name for America, + -ITE[1]]

columbium /kəˈlʌmbɪəm/ n. US Chem. hist. = NIOBIUM.

Columbus[1] /kəˈlʌmbəs/ the state capital of Ohio; pop. (1990) 632,900.

Columbus[2] /kəˈlʌmbəs/, Christopher (Spanish name Cristóbal Colón) (1451–1506), Italian-born Spanish explorer. A Genoese by birth, Columbus persuaded the rulers of Spain, Ferdinand and Isabella, to sponsor an expedition to sail westwards across the Atlantic in search of Asia and prove that the world was round. Sailing with three small ships in 1492, he discovered the New World (in fact various Caribbean islands). He made three further voyages to the New World between 1493 and 1504, in 1498 discovering the South American mainland and finally exploring the coast of Mexico.

column /ˈkɒləm/ n. **1** Archit. an upright cylindrical pillar often slightly tapering and usu. supporting an entablature or arch, or standing alone as a monument. **2** a structure or part shaped like a column. **3** a vertical cylindrical mass of liquid or vapour. **4 a** a vertical division of a page, chart, etc., containing a sequence of figures or words. **b** the figures or words themselves. **5** a part of a newspaper regularly devoted to a particular subject (*gossip column*). **6 a** Mil. an arrangement of troops in successive lines, with a narrow front. **b** Naut. a similar arrangement of ships. □ **column-inch** a quantity of print (esp. newsprint) occupying a one-inch length of a column. **dodge the column** colloq. shirk one's duty; avoid work. □ **columned** adj. **columnar** /kəˈlʌmnə(r)/ adj. [ME f. OF *columpne* & L *columna* pillar]

columnist /ˈkɒləmnɪst, -mɪst/ n. a journalist contributing regularly to a newspaper.

colure /kəˈljʊə(r)/ n. Astron. either of two great circles intersecting at right angles at the celestial poles and passing through the ecliptic at either the equinoxes or the solstices. [ME f. LL *colurus* f. Gk *kolouros* truncated]

colza /ˈkɒlzə/ n. = RAPE[2]. [F *kolza*(t) f. LG *kōlsāt* (as COLE, SEED)]

COM /kɒm/ abbr. computer output on microfilm or microfiche.

com- /kɒm, kəm, kʌm/ prefix (also **co-, col-, con-, cor-**) with, together, jointly, altogether. ¶ com- is used before *b*, *m*, *p*, and sometimes before vowels and *f*; co- esp. before vowels, *h*, and gn; col- before *l*, cor- before *r*, and con- before other consonants. [L *com-, cum* with]

coma[1] /ˈkəʊmə/ n. (pl. **comas**) a prolonged deep unconsciousness, caused esp. by severe injury or excessive use of drugs. [med.L f. Gk *kōma* deep sleep]

coma[2] /ˈkəʊmə/ n. (pl. **comae** /-miː/) **1** Astron. a spherical cloud of gas and dust surrounding the nucleus of a comet. **2** Bot. a tuft of silky hairs at the end of some seeds. [L f. Gk *komē* hair of head]

Coma Berenices /ˌkəʊmə ˌberɪˈnaɪsiːz/ Astron. a small constellation (Berenice's Hair), said to represent the tresses of Queen Berenice (see BERENICE). It contains a large number of galaxies. [L]

Comanche /kəˈmæntʃɪ/ n. & adj. ● n. (pl. same or **Comanches**) **1** a member of an American Indian people of the south-western US, about 4,000 of whom now live in Oklahoma. **2** the Shoshonean language of this people. (*See note below.*) ● adj. of or relating to this people or their language. [Sp.]

▪ In the 18th and 19th centuries the Comanche were a powerful nomadic people, who were among the first to acquire horses (from the Spanish) and who resisted white settlers fiercely.

Comaneci /ˌkɒməˈnetʃ/, Nadia (b.1961), Romanian-born American gymnast. She became the first competitor in the history of the Olympic Games to be awarded the maximum score (10.00) when she won three gold medals for her performances at the Montreal Olympics in 1976. She emigrated to the US in 1989.

comatose /ˈkəʊməˌtəʊz, -ˌtəʊs/ adj. **1** in a coma. **2** drowsy, sleepy, lethargic.

comb /kəʊm/ n. & v. ● n. **1** a toothed strip of rigid material for tidying and arranging the hair, or for keeping it in place. **2** a part of a machine having a similar design or purpose. **3 a** the red fleshy crest of a fowl, esp. a cock. **b** an analogous growth in other birds. **4** a honeycomb. ● v.tr. **1** arrange or tidy (the hair) by drawing a comb through. **2** curry (a horse). **3** dress (wool or flax) with a comb. **4** search (a place) thoroughly. □ **comb-jelly** = CTENOPHORE. **comb out 1** tidy and arrange (hair) with a comb. **2** remove with a comb. **3** search or attack systematically. **4** search out and get rid of (anything unwanted). □ **combed** adj. [OE *camb* f. Gmc]

combat /ˈkɒmbæt, ˈkʌm-/ n. & v. ● n. **1** a fight, an armed encounter or conflict; fighting, battle. **2** a struggle, contest, or dispute. ● v. (**combated, combating**) **1** intr. engage in combat. **2** tr. engage in combat with. **3** tr. oppose; strive against. □ **combat fatigue** mental disorder caused by stress in wartime combat. **single combat** a duel. [F *combat* f. *combattre* f. LL (as COM-, L *batuere* fight)]

combatant /ˈkɒmbət(ə)nt, ˈkʌm-/ n. & adj. ● n. a person engaged in fighting. ● adj. **1** fighting. **2** for fighting.

combative /ˈkɒmbətɪv, ˈkʌm-/ adj. ready or eager to fight; pugnacious. □ **combatively** adv. **combativeness** n.

combe var. of COOMB.

comber[1] /ˈkəʊmə(r)/ n. **1** a person or thing that combs, esp. a machine for combing cotton or wool very fine. **2** a long curling wave; a breaker.

comber[2] /ˈkəʊmə(r)/ n. a fish of the sea-perch family, *Serranus cabrilla*. [18th c.: orig. unkn.]

combination /ˌkɒmbɪˈneɪʃ(ə)n/ n. **1** the act or an instance of combining; the process of being combined. **2** a combined state (*in combination with*). **3** a combined set of things or people. **4** a sequence of numbers or letters used to open a combination lock. **5** Brit. a motorcycle with side-car attached. **6** (in pl.) Brit. a single undergarment for the body and legs. **7** a group of things chosen from a larger number without regard to their arrangement. **8** a united action. **b** Chess a coordinated and effective sequence of moves. **9** Chem. a union of substances in a compound with new properties. □ **combination lock** a lock that can be opened only by a specific sequence of movements. □ **combinational** adj. **combinative** /ˈkɒmbɪnətɪv/ adj. **combinatory** /-nətərɪ/ adj. [obs. F *combination* or LL *combinatio* (as COMBINE)]

Combination Act any of the British laws of 1799–1800 making illegal the confederacy of persons to further their own interests, affect the rate of wages, etc. Formulated in the wake of naval mutinies and the Irish rebellion to prevent revolutionary ideas from spreading to England after the French Revolution, the laws were supposed to apply to employers and workers alike, but in fact the employers were allowed to combine freely and restrictions were enforced against workers' unions; most of the legislation was repealed in 1824.

combinatorial /ˌkɒmbɪnəˈtɔːrɪəl/ adj. Math. relating to combinations of items.

combine v. & n. ● v. /kəmˈbaɪn/ **1** tr. & intr. join together; unite for a common purpose. **2** tr. possess (qualities usually distinct) together (*combines charm and authority*). **3 a** intr. coalesce in one substance. **b** tr. cause to do this. **c** intr. form a chemical compound. **4** intr. cooperate. **5** /ˈkɒmbaɪn/ tr. harvest (crops etc.) by means of a combine harvester. ● n. /ˈkɒmbaɪn/ a combination of esp. commercial interests to control prices etc. □ **combine harvester** a mobile machine that reaps and threshes in one operation. **combining form** Gram. a linguistic element used in combination with another element to form a word (e.g. *Anglo-* = English, *bio-* = life, *-graphy* = writing). ¶ In this dictionary, *combining form* is used of an element that contributes to the particular sense of words (as with both elements of *biography*), as distinct from a prefix or suffix that adjusts the sense of or determines the function of words (as with *un-, -able*, and *-ation*). □ **combinable** /kəmˈbaɪnəb(ə)l/ adj. [ME f. OF *combiner* or LL *combinare* (as COM-, L *bini* two)]

combing /ˈkəʊmɪŋ/ n. (in pl.) hairs combed off. □ **combing wool** long-stapled wool, suitable for combing and making into worsted.

combo /ˈkɒmbəʊ/ n. (pl. **-os**) sl. a small jazz or dance band. [abbr. of COMBINATION + -o]

combs /kɒmz/ n.pl. colloq. combinations (see COMBINATION 6).

combust /kəm'bʌst/ v.tr. subject to combustion. [obs. *combust* (adj.) f. L *combustus* past part. (as COMBUSTION)]

combustible /kəm'bʌstɪb(ə)l/ adj. & n. ● adj. **1** capable of or used for burning. **2** excitable; easily irritated. ● n. a combustible substance. □ **combustibility** /-ˌbʌstɪ'bɪlɪtɪ/ n. [F *combustible* or med.L *combustibilis* (as COMBUSTION)]

combustion /kəm'bʌstʃən/ n. **1** burning; consumption by fire. **2** *Chem.* the development of light and heat from the chemical combination of a substance with oxygen. □ **combustion chamber** a space in which combustion takes place, e.g. of gases in a boiler-furnace or fuel in an internal-combustion engine. □ **combustive** /-'bʌstɪv/ adj. [ME f. F *combustion* or LL *combustio* f. L *comburere combust-* burn up]

come /kʌm/ v. & n. ● v.intr. (*past* **came** /keɪm/; *past part.* **come**) **1** move, be brought towards, or reach a place thought of as near or familiar to the speaker or hearer (*come and see me; shall we come to your house?; the books have come*). **2** reach or be brought to a specified situation or result (*you'll come to no harm; have come to believe it; has come to be used wrongly; came into prominence*). **3** reach or extend to a specified point (*the road comes within a mile of us*). **4** traverse or accomplish (with complement: *have come a long way*). **5 a** get to be in a certain condition (*how did you come to break your leg?*). **b** (of time) arrive in due course (*the day soon came*). **6** take or occupy a specified position in space or time (*it comes on the third page; Nero came after Claudius; it does not come within the scope of the inquiry*). **7** become perceptible or known (*the church came into sight; the news comes as a surprise; it will come to me*). **8** be available (*the dress comes in three sizes; this model comes with optional features*). **9** become (with complement: *the handle has come loose; won't come clean*). **10** (foll. by *of, from*) **a** be descended from (*comes of a rich family*). **b** be the result of (*that comes of complaining*). **11** (foll. by *from*) **a** originate in; have as its source. **b** have as one's home. **12** *colloq.* play the part of; behave like (with complement: *don't come the bully with me*). **13** *sl.* have a sexual orgasm. **14** (in *subjunctive*) *colloq.* when a specified time is reached (*come next month*). **15** (also **come, come!**) (as *int.*) expressing caution or reserve (*come, it cannot be that bad*). ● n. *sl.* semen ejaculated at a sexual orgasm. □ **as ... as they come** typically or supremely so (*is as tough as they come*). **come about 1** happen; take place. **2** *Naut.* tack. **come across 1** meet or find by chance (*came across an old jacket*). **2** *informal* be effective or understood; give a specified impression. **3** (foll. by *with*) *sl.* hand over what is wanted. **come again** *colloq.* **1** make a further effort. **2** (as *imper.*) what did you say? **come along 1** make progress; move forward. **2** (as *imper.*) hurry up. **come and go 1** pass to and fro; be transitory. **2** pay brief visits. **come apart** fall or break into pieces, disintegrate. **come at 1** reach, discover; get access to. **2** attack (*came at me with a knife*). **come-at-able** /-'ætəb(ə)l/ adj. reachable, accessible. **come away 1** become detached or broken off (*came away in my hands*). **2** (foll. by *with*) be left with a feeling, impression, etc. (*came away with many misgivings*). **come back 1** return. **2** recur to one's memory. **3** become fashionable or popular again. **4** *N. Amer.* reply, retort. **come before** be dealt with by (a judge etc.). **come between 1** interfere with the relationship of. **2** separate; prevent contact between. **come by 1** pass; go past. **2** call on a visit (*why not come by after work?*). **3** acquire, obtain; attain (*came by a new camera*). **come clean** see CLEAN. **come down 1** come to a place or position regarded as lower. **2** lose position or wealth (*has come down in the world*). **3** be handed down by tradition or inheritance. **4** be reduced; show a downward trend (*prices are coming down*). **5** (foll. by *against, in favour of*, etc.) reach a decision or recommendation (*the report came down against change*). **6** (foll. by *to*) signify or betoken basically; be dependent on (a factor) (*it comes down to who is willing to go*). **7** (foll. by *on*) criticize harshly; rebuke, punish. **8** (foll. by *with*) begin to suffer from (a disease). **come for 1** come to collect or receive. **2** attack (*came for me with a hammer*). **come forward 1** advance. **2** offer oneself for a task, post, etc. **come-hither** attrib.adj. *colloq.* (of a look or manner) enticing, flirtatious. **come in 1** enter a house or room. **2 a** take a specified position in a race etc. (*came in third*). **b** *colloq.* win. **3** become fashionable or seasonable. **4 a** have a useful role or function. **b** (with complement) prove to be (*came in very handy*). **c** have a part to play (*where do I come in?*). **5** be received (*more news has just come in*). **6** begin speaking, esp. in radio transmission. **7** be elected; come to power. **8** *Cricket* begin an innings. **9** (foll. by *for*) receive; be the object of (usu. something unwelcome) (*came in for much criticism*). **10** (foll. by *on*) join (an enterprise etc.). **11** (of a tide) turn to high tide. **12** (of a train, ship, or aircraft) approach its destination. **13** return to base (*come in, number 9*). **come into 1** see senses 2, 7 of v. **2** receive, esp. as heir. **come near** see NEAR. **come of age** see AGE. **come off 1** *colloq.* (of an action) succeed; be accomplished. **2** (with complement) fare; turn out (*came off badly; came off the winner*). **3** *coarse sl.* have a sexual

orgasm. **4** be detached or detachable (from). **5** fall (from). **6** be reduced or subtracted from (*£5 came off the price*). **come off it** (as *imper.*) *colloq.* an expression of disbelief or refusal to accept another's opinion, behaviour, etc. **come on 1** continue to come. **2** advance, esp. to attack. **3** make progress; thrive (*is really coming on*). **4** (foll. by *to* + *infin.*) begin (*it came on to rain*). **5** appear on the stage, field of play, etc. **6** be heard or seen on television, on the telephone, etc. **7** arise to be discussed. **8** (as *imper.*) expressing encouragement. **9** = *come upon*. **come-on** n. *sl.* a lure or enticement. **come out 1 a** emerge; become known (*it came out that he had left*). **b** end, turn out. **2** appear or be published (*comes out every Saturday*). **3 a** declare oneself as being for or against something; make a decision (*came out in favour of joining*). **b** openly declare something (esp. that one is a homosexual). **4** *Brit.* go on strike. **5 a** be satisfactorily visible in a photograph etc., or present in a specified way (*the dog didn't come out; he came out badly*). **b** (of a photograph) be produced satisfactorily or in a specified way (*only three have come out; they all came out well*). **6** attain a specified result in an examination etc. **7** (of a stain etc.) be removed. **8** make one's début in society. **9** (foll. by *in*) be covered with (*came out in spots*). **10** (of a problem) be solved. **11** (foll. by *with*) declare openly; disclose. **come out of the closet** see CLOSET. **come over 1** come from some distance or nearer to the speaker (*came over from Paris; come over here a moment*). **2** change sides or one's opinion. **3 a** (of a feeling etc.) overtake or affect (a person). **b** *colloq.* feel suddenly (*came over faint*). **4** appear or sound in a specified way (*you came over very well; the ideas came over clearly*). **5** affect or influence (*I don't know what came over me*). **come round 1** pay an informal visit. **2** recover consciousness. **3** be converted to another person's opinion. **4** (of a date or regular occurrence) recur; be imminent again. **come through 1** be successful; survive. **2** be received by telephone. **3** survive or overcome (a difficulty) (*came through the ordeal*). **come to 1** recover consciousness. **2** *Naut.* bring a vessel to a stop. **3** reach in total; amount to. **4** *refl.* **a** recover consciousness. **b** stop being foolish. **5** have as a destiny; reach (*what is the world coming to?*). **6** be a question of (*when it comes to wine, he is an expert*). **come to hand** become available; be recovered. **come to light** see LIGHT[1]. **come to nothing** have no useful result in the end; fail. **come to pass** happen, occur. **come to rest** cease moving. **come to one's senses** see SENSE. **come to that** *colloq.* in fact; if that is the case. **come under 1** be classified as or among. **2** be subject to (influence or authority). **come up 1** come to a place or position regarded as higher. **2** attain wealth or position (*come up in the world*). **3 a** (of an issue, problem, etc.) arise; present itself; be mentioned or discussed. **b** (of an event etc.) approach, occur, happen (*coming up next on BBC1*). **4** (often foll. by *to*) **a** approach a person, esp. to talk. **b** approach or draw near to a specified time, event, etc. (*is coming up to eight o'clock*). **5** (foll. by *to*) match (a standard etc.). **6** (foll. by *with*) produce (an idea etc.), esp. in response to a challenge. **7** (of a plant etc.) spring up out of the ground. **8** become brighter (e.g. with polishing); shine more brightly. **come up against** be faced with or opposed by. **come upon 1** meet or find by chance. **2** attack by surprise. **come what may** no matter what happens. **have it coming to one** *colloq.* be about to get one's deserts. **how come?** *colloq.* how did that happen? **if it comes to that** in that case. **to come** future; in the future (*the year to come; many problems were still to come*). [OE *cuman* f. Gmc]

comeback /'kʌmbæk/ n. **1** a return to a previous (esp. successful) state. **2** *sl.* a retaliation or retort. **3** *Austral.* a sheep bred from crossbred and purebred parents for both wool and meat.

Comecon /'kɒmɪˌkɒn/ abbr. (also **CMEA**) Council for Mutual Economic Assistance, an economic association of east European countries, founded in 1949 and analogous to the European Economic Community. With the collapse of Communism in eastern Europe, the association was dissolved in 1991.

comedian /kə'miːdɪən/ n. **1** a humorous entertainer on stage, television, etc. **2** an actor in comedy. **3** *sl.* a buffoon; a foolish person. [F *comédien* f. *comédie* COMEDY]

Comédie Française /ˌkɒmeɪˌdiː frɒn'seɪz/ the French national theatre (used for both comedy and tragedy), in Paris, founded in 1680 by Louis XIV and reconstituted by Napoleon I in 1803. It is organized as a cooperative society in which each actor holds a share or part-share. [F, = French comedy]

comedienne /kəˌmiːdɪ'en/ n. a female comedian. [F fem. (as COMEDIAN)]

comedist /'kɒmɪdɪst/ n. a writer of comedies.

comedo /'kɒmɪˌdəʊ/ n. (pl. **comedones** /ˌkɒmɪ'dəʊniːz/) *Med.* a blackhead. [L, = glutton f. *comedere* eat up]

comedown /'kʌmdaʊn/ n. **1** a loss of status; decline or degradation. **2** a disappointment.

comedy /ˈkɒmɪdɪ/ n. (pl. **-ies**) **1 a** a play, film, etc., of an amusing or satirical character, usu. with a happy ending. **b** the dramatic genre consisting of works of this kind. (*See note below.*) **2** an amusing or farcical incident or series of incidents in everyday life. **3** humour, esp. in a work of art etc. □ **comedy of manners** see MANNER. □ **comedic** /kəˈmiːdɪk/ adj. [ME f. OF comedie f. L comoedia f. Gk kōmōidia f. kōmōidos comic poet f. kōmos revel]

▪ Comedy tends to remain closer to everyday life than tragedy, with which it is often contrasted, and to explore everyday situations and common human failings rather than exalted acts and passions. Forms include romantic comedy, farce, slapstick, burlesque, and satire. The term was also applied in the Middle Ages to narrative poems with happy endings, such as Dante's *Divine Comedy*. Comedy in Europe dates back to ancient Athens, where the plays of Aristophanes (5th century BC) mix political and social satire with fantasy. Menander's New Comedy (4th century BC), in which young lovers undergo endless vicissitudes in the company of stock fictional characters, was developed in the Roman comedy of Plautus and Terence and eventually by Shakespeare. The great period of European comedy is often regarded as the 17th century, when Shakespeare and Ben Jonson were succeeded by Molière in France and by the Restoration comedy of such writers as William Congreve, Aphra Behn, and William Wycherley. 18th-century comedy is exemplified by Oliver Goldsmith and Richard Brinsley Sheridan, while Oscar Wilde and George Bernard Shaw typify the 19th century. In the 20th century absurdist writers such as Samuel Beckett and Eugène Ionesco and black humorists such as Joe Orton have been influential; film and television performers have included Charlie Chaplin, the Marx Brothers, and Woody Allen. In the 20th century the term has also come to encompass the work of stand-up comics.

comely /ˈkʌmlɪ/ adj. (**comelier, comeliest**) (usu. of a woman) pleasant to look at. □ **comeliness** n. [ME cumelich, cumli prob. f. becumelich f. BECOME]

comer /ˈkʌmə(r)/ n. **1** a person who comes, esp. as an applicant, participant, etc. (*offered the job to the first comer*). **2** colloq. a person likely to be a success. □ **all comers** any applicants (with reference to a position, or esp. a challenge to a champion, that is unrestricted in entry).

comestible /kəˈmestɪb(ə)l/ n. (usu. in pl.) formal or joc. food. [ME f. F f. med.L comestibilis f. L comedere comest- eat up]

comet /ˈkɒmɪt/ n. Astron. a luminous object seen in the night sky, originally considered a supernatural omen, but now known to be a ball of ice and dust orbiting the sun. The nucleus generates a coma and a tail of gas and dust particles as it approaches the sun. Originating in the remotest regions of the solar system, most comets follow regular orbits and appear in the inner solar system as periodic comets, some of which break up and can be the origin of annual meteor showers. Sometimes one is captured by the gravitational field of a large planet, especially Jupiter, and then goes into a new orbit around the planet (see SHOEMAKER–LEVY 9, COMET). (See also HALLEY'S COMET.) □ **cometary** adj. [ME f. OF comete f. L cometa f. Gk komētēs long-haired (star)]

comeuppance /kʌmˈʌp(ə)ns/ n. colloq. one's deserved fate or punishment (*got his comeuppance*). [COME + UP + -ANCE]

comfit /ˈkʌmfɪt/ n. archaic a sweet consisting of a nut, seed, etc., coated in sugar. [ME f. OF confit f. L confectum past part. of conficere prepare: see CONFECTION]

comfort /ˈkʌmfət/ n. & v. ● n. **1** consolation; relief in affliction. **2 a** a state of physical well-being; being comfortable (*live in comfort*). **b** (usu. in pl.) things that make life easy or pleasant (*has all the comforts*). **3** a cause of satisfaction (*a comfort to me that you are here*). **4** a person who consoles or helps one (*he's a comfort to her in her old age*). **5** US a warm quilt. ● v.tr. **1** soothe in grief; console. **2** make comfortable (*comforted by the warmth of the fire*). □ **comfort station** US euphem. a public lavatory. [ME f. OF confort(er) f. LL confortare strengthen (as COM-, L fortis strong)]

comfortable /ˈkʌmftəb(ə)l, -fətəb(ə)l/ adj. & n. ● adj. **1 a** such as to avoid hardship or trouble and give comfort or ease (*a comfortable pair of shoes*). **b** (of a person) relaxing to be with, congenial. **2** free from discomfort; at ease (*I'm quite comfortable thank you*). **3** colloq. having an adequate standard of living; free from financial worry. **4 a** having an easy conscience (*did not feel comfortable about refusing him*). **b** colloq. complacent, placidly self-satisfied. **5** with a wide margin (*a comfortable win*). ● n. US a warm quilt. □ **comfortableness** n. **comfortably** adv. [ME f. AF confortable (as COMFORT)]

comforter /ˈkʌmfətə(r)/ n. **1** a person who comforts. **2** a baby's

dummy. **3** archaic a woollen scarf. **4** N. Amer. a warm quilt. [ME f. AF confortour, OF -eor (as COMFORT)]

comfortless /ˈkʌmfətlɪs/ adj. **1** dreary, cheerless. **2** without comfort.

comfrey /ˈkʌmfrɪ/ n. (pl. **-eys**) a plant of the genus Symphytum, of the borage family; esp. *S. officinale*, which has large hairy leaves and clusters of usu. white or purple bell-shaped flowers. [ME f. AF cumfrie, ult. f. L conferva (as COM-, fervere boil)]

comfy /ˈkʌmfɪ/ adj. (**comfier, comfiest**) colloq. comfortable. □ **comfily** adv. **comfiness** n. [abbr.]

comic /ˈkɒmɪk/ adj. & n. ● adj. **1** (often attrib.) of, or in the style of, comedy (*a comic actor; comic opera*). **2** causing or meant to cause laughter; funny (*comic to see his struggles*). ● n. **1** a professional comedian. **2 a** a children's periodical, mainly in the form of comic strips. **b** a similar publication intended for adults. □ **comic opera 1** an opera with much spoken dialogue, usu. with humorous treatment. **2** this genre of opera. **comic strip** a horizontal series of drawings in a comic, newspaper, etc., telling a story. [L comicus f. Gk kōmikos f. kōmos revel]

comical /ˈkɒmɪk(ə)l/ adj. funny; causing laughter. □ **comically** adv. **comicality** /ˌkɒmɪˈkælɪtɪ/ n. [COMIC]

coming /ˈkʌmɪŋ/ adj. & n. ● attrib.adj. **1** approaching, next (*in the coming week; this coming Sunday*). **2** of potential importance (*a coming man*). ● n. arrival; approach. □ **coming and going** (or **comings and goings**) activity, esp. intense. **not know if one is coming or going** be confused from being very busy.

Comino /kɒˈmiːnəʊ/ the smallest of the three main islands of Malta;

COMINT /ˈkɒmɪnt/ abbr. communications intelligence.

Comintern /ˈkɒmɪnˌtɜːn/ n. the Third International (see INTERNATIONAL), a Communist organization (1919–43). [Russ. Komintern f. Russ. forms of communist, international]

comitadji /ˌkɒmɪˈtædʒɪ/ n. (also **komitadji, komitaji**) a member of an irregular band of soldiers in the Balkans. [Turk. komitacı, lit. 'member of a (revolutionary) committee']

comity /ˈkɒmɪtɪ/ n. (pl. **-ies**) **1** courtesy, civility; considerate behaviour towards others. **2 a** an association of nations etc. for mutual benefit. **b** (in full **comity of nations**) the mutual recognition by nations of the laws and customs of others. [L comitas f. comis courteous]

comma /ˈkɒmə/ n. **1** a punctuation mark (,) indicating a pause between parts of a sentence, or dividing items in a list, string of figures, etc. **2** Mus. a definite minute interval or difference of pitch. **3** (in full **comma butterfly**) an orange-brown butterfly, Polygonia c-album, with ragged edges to the wings and a white comma-shaped mark on the underside of the hindwing. □ **comma bacillus** Biol. a comma-shaped bacillus causing cholera. [L f. Gk komma clause]

command /kəˈmɑːnd/ v. & n. ● v.tr. **1** (also absol.; often foll. by to + infin., or that + clause) give formal order or instructions to (*commands us to obey; commands that it be done*). **2** (also absol.) have authority or control over. **3 a** (often refl.) restrain, master. **b** gain the use of; have at one's disposal or within reach (skill, resources, etc.) (*commands an extensive knowledge of history; commands a salary of £40,000*). **4** deserve and get (sympathy, respect, etc.). **5** Mil. dominate (a strategic position) from a superior height; look down over. ● n. **1** an authoritative order; an instruction. **2** mastery, control, possession (*a good command of languages; has command of the resources*). **3** the exercise or tenure of authority, esp. naval or military (*has command of this ship*). **4** Mil. **a** a body of troops etc. under a commander (*Bomber Command*). **b** a district under a commander (*Western Command*). **5** Computing **a** an instruction causing a computer to perform one of its basic functions. **b** a signal initiating such an operation. □ **at command** ready to be used at will. **at** (or **by**) a **person's command** in pursuance of a person's bidding. **command economy** = planned economy (see PLAN). **command module** the control compartment in a spacecraft. **command performance** (in the UK) a theatrical or film performance given by royal command. **command post** the headquarters of a military unit. **in command of** commanding; having under control. **under command of** commanded by. **word of command 1** Mil. an order for a movement in a drill etc. **2** a prearranged spoken signal for the start of an operation. [ME f. AF comaunder, OF comander f. LL commandare COMMEND]

commandant /ˌkɒmənˈdænt, -ˈdɑːnt/ n. a commanding officer, esp. of a particular force, military academy, etc. □ **Commandant-in-Chief** the supreme commandant. □ **commandantship** n. [F commandant, or It. or Sp. commandante (as COMMAND)]

commandeer /ˌkɒmənˈdɪə(r)/ v.tr. **1** seize (men or goods) for military purposes. **2** take possession of without authority. [S. Afr. Du. kommanderen f. F commander COMMAND]

commander /kəˈmɑːndə(r)/ n. **1** a person who commands, esp.: **a** a naval officer next in rank below captain. **b** = *wing commander*. **2** an officer in charge of a London police district. **3** (in full **knight commander**) a member of a higher class in some orders of knighthood. **4** a large wooden mallet. □ **commander-in-chief** (*pl.* **commanders-in-chief**) the supreme commander, esp. of a nation's forces. □ **commandership** n. [ME f. OF *comandere*, *-eör* f. Rmc (as COMMAND)]

Commander of the Faithful n. a title of a caliph.

commanding /kəˈmɑːndɪŋ/ adj. **1** dignified, exalted, impressive. **2** (of a hill or other high point) giving a wide view. **3** (of an advantage, a position, etc.) controlling; dominating (*has a commanding lead*). □ **commandingly** adv.

commandment /kəˈmɑːndmənt/ n. a divine command. [ME f. OF *comandement* (as COMMAND)]

commando /kəˈmɑːndəʊ/ n. (*pl.* **-os**) *Mil.* **1** (often **Commando**) **a** a unit of British amphibious shock troops; a member of such a unit. (*See note below.*) **b** a similar unit or member of such a unit elsewhere. **2 a** a party of men called out for military service. **b** a body of troops. **3** *hist.* (in the Boer War) a unit of the Boer army composed of the militia of an electoral district **4** (*attrib.*) of or concerning a commando (*a commando operation*). [Port. f. *commandar* COMMAND]

■ The commandos were British troops raised jointly by the army and the Royal Marines during the Second World War. They were trained originally (in 1940) as shock troops for repelling the threatened German invasion of England, later for carrying out raids on enemy-held territory.

Command Paper n. (in the UK) a paper laid before Parliament (in practice, by the government) by command of the Crown.

comme ci, comme ça /kɒm ˌsiː kɒm ˈsæ/ adv. & adj. so so; middling or middlingly. [F, = like this, like that]

commedia dell'arte /kɒˌmeɪdɪə delˈɑːteɪ/ n. an improvised kind of popular comedy developed in 16th-century Italy and developed throughout Europe until the end of the 18th century. It featured stock character types (e.g. Harlequin, Columbine, Pantaloon), played by professional actors, who adapted their comic dialogue and action according to a few basic plots (commonly love intrigues) and to topical issues. It had an enormous influence on European drama, as seen in the work of Ben Jonson and Molière, and many of its characters continued into other theatrical forms, such as the harlequinade and the Punch and Judy show. [It., = comedy of art]

comme il faut /ˌkɒm iːl ˈfəʊ/ adj. & adv. ● *predic.adj.* (esp. of behaviour, etiquette, etc.) proper, correct. ● adv. properly, correctly. [F, = as is necessary]

commemorate /kəˈmeməˌreɪt/ v.tr. **1** celebrate in speech or writing. **2 a** preserve in memory by some celebration. **b** (of a stone, plaque, etc.) be a memorial of. □ **commemorator** n. **commemorative** /-rətɪv/ adj. [L *commemorare* (as COM-, *memorare* relate f. *memor* mindful)]

commemoration /kəˌmeməˈreɪʃ(ə)n/ n. **1** an act of commemorating. **2** a service or part of a service in memory of a person, an event, etc. [ME f. F *commemoration* or L *commemoratio* (as COMMEMORATE)]

commence /kəˈmens/ v.tr. & intr. formal begin. [ME f. OF *com(m)encier* f. Rmc (as COM-, L *initiare* INITIATE)]

commencement /kəˈmensmənt/ n. formal **1** a beginning. **2** esp. *US* a ceremony of degree conferment. [ME f. OF (as COMMENCE)]

commend /kəˈmend/ v.tr. **1** (often foll. by *to*) entrust, commit (*commends his soul to God*). **2** praise (*commends her singing voice*). **3** (often *refl.*) recommend (*the idea naturally commends itself*; *has much to commend it*). □ **commend me to** archaic remember me kindly to. **highly commended** (of a competitor etc.) just missing the top places. [ME f. L *commendare* (as COM-, *mendare* = *mandare* entrust: see MANDATE)]

commendable /kəˈmendəb(ə)l/ adj. praiseworthy. □ **commendably** adv. [ME f. OF f. L *commendabilis* (as COMMEND)]

commendation /ˌkɒmenˈdeɪʃ(ə)n/ n. **1** an act of commending or recommending (esp. a person to another's favour). **2** praise. [ME f. OF f. L *commendatio* (as COMMEND)]

commendatory /kəˈmendətərɪ/ adj. commending, recommending. [LL *commendatorius* (as COMMEND)]

commensal /kəˈmens(ə)l/ adj. & n. ● adj. **1** *Biol.* of, relating to, or exhibiting commensalism. **2** (of a person) eating at the same table as another. ● n. **1** *Biol.* a commensal organism. **2** a person who eats at the same table as another. □ **commensality** /ˌkɒmənˈsælɪtɪ/ n. [ME f. F *commensal* or med.L *commensalis* (in sense 2) (as COM-, *mensa* table)]

commensalism /kəˈmensəˌlɪz(ə)m/ n. *Biol.* an association between two organisms in which one benefits and the other derives neither benefit nor harm.

commensurable /kəˈmenʃərəb(ə)l, -ˈmensərəb(ə)l/ adj. **1** (often foll. by *with*, *to*) measurable by the same standard. **2** (foll. by *to*) proportionate to. **3** *Math.* (of numbers) in a ratio equal to the ratio of integers. □ **commensurably** adv. **commensurability** /-ˌmenʃərə'bɪlɪtɪ, -ˌmensjərə-/ n. [LL *commensurabilis* (as COM-, MEASURE)]

commensurate /kəˈmenʃərət, -ˈmensjərət/ adj. **1** (usu. foll. by *with*) having the same size, duration, etc.; coextensive. **2** (often foll. by *to*, *with*) proportionate. □ **commensurately** adv. [LL *commensuratus* (as COM-, MEASURE)]

comment /ˈkɒment/ n. & v. ● n. **1 a** a remark, esp. critical; an opinion (*passed a comment on her hat*). **b** commenting; criticism (*his behaviour aroused much comment*; *an hour of news and comment*). **2 a** an explanatory note (e.g. on a written text). **b** written criticism or explanation (e.g. of a text). **3** (of a play, book, etc.) a critical illustration; a parable (*his art is a comment on society*). ● v.intr. **1** (often foll. by *on*, *upon*, or *that* + clause) make (esp. critical) remarks (*commented on her choice of friends*). **2** (often foll. by *on*, *upon*) write explanatory notes. □ **no comment** colloq. I decline to answer your question. □ **commenter** n. [ME f. L *commentum* contrivance (in LL also = interpretation), neut. past part. of *comminisci* devise, or F *commenter* (v.)]

commentary /ˈkɒməntərɪ/ n. (*pl.* **-ies**) **1** a set of explanatory or critical notes on a text etc. **2** a descriptive spoken account (esp. on radio or television) of an event or a performance as it happens. [L *commentarius*, *-ium* adj. used as noun (as COMMENT)]

commentate /ˈkɒmənˌteɪt/ v.intr. act as a commentator. [back-form. f. COMMENTATOR]

commentator /ˈkɒmənˌteɪtə(r)/ n. **1** a person who provides a commentary on an event etc. **2** the writer of a commentary. **3** a person who writes or speaks on current events. [L f. *commentari* frequent. of *comminisci* devise]

commerce /ˈkɒmɜːs/ n. **1** financial transactions, esp. the buying and selling of merchandise, on a large scale. **2** social intercourse (*the daily commerce of gossip and opinion*). **3** archaic sexual intercourse. [F *commerce* or L *commercium* (as COM-, *mercium* f. *merx mercis* merchandise)]

commercial /kəˈmɜːʃ(ə)l/ adj. & n. ● adj. **1** of, engaged in, or concerned with, commerce. **2** having profit as a primary aim rather than artistic etc. value; philistine. **3** (of chemicals) supplied in bulk more or less unpurified. ● n. **1** a television or radio advertisement. **2** archaic a commercial traveller. □ **commercial art** art used in advertising, selling, etc. **commercial broadcasting** television or radio broadcasting in which programmes are financed by advertisements. **commercial traveller** a firm's travelling salesman or saleswoman who visits shops to get orders. **commercial vehicle** a vehicle used for carrying goods or fare-paying passengers. □ **commercialism** n. **commercially** adv. **commerciality** /-ˌmɜːʃɪˈælɪtɪ/ n.

commercialize /kəˈmɜːʃəˌlaɪz/ v.tr. (also **-ise**) **1** exploit or spoil for the purpose of gaining profit. **2** make commercial. □ **commercialization** /-ˌmɜːʃəlaɪˈzeɪʃ(ə)n/ n.

commère /ˈkɒmeə(r)/ n. Brit. a female compère. [F, fem. of COMPÈRE]

Commie /ˈkɒmɪ/ n. sl. derog. a Communist. [abbr.]

commination /ˌkɒmɪˈneɪʃ(ə)n/ n. **1** the threatening of divine vengeance. **2 a** the recital of divine threats against sinners in the Anglican Liturgy for Ash Wednesday. **b** the service that includes this. [ME f. L *comminatio* f. *comminari* threaten]

comminatory /ˈkɒmɪnətərɪ/ adj. threatening, denunciatory. [med.L *comminatorius* (as COMMINATION)]

commingle /kəˈmɪŋg(ə)l/ v.tr. & intr. literary mingle together.

comminute /ˈkɒmɪˌnjuːt/ v.tr. **1** reduce to small fragments. **2** divide (property) into small portions. □ **comminuted fracture** a fracture producing multiple bone splinters. □ **comminution** /ˌkɒmɪˈnjuːʃ(ə)n/ n. [L *comminuere comminut-* (as COM-, *minuere* lessen)]

commis /ˈkɒmɪ, ˈkɒmɪs/ n. (*pl.* same /ˈkɒmɪ, ˈkɒmɪz/) a junior waiter or chef. [orig. = deputy, clerk, f. F, past part. of *commettre* entrust (as COMMIT)]

commiserate /kəˈmɪzəˌreɪt/ v. **1** intr. (usu. foll. by *with*) express or feel pity. **2** tr. archaic express or feel pity for (*commiserate you on your loss*). □ **commiserator** n. **commiserative** /-rətɪv/ adj. **commiseration** /-ˌmɪzəˈreɪʃ(ə)n/ n. [L *commiserari* (as COM-, *miserari* pity f. *miser* wretched)]

commissar /ˈkɒmɪˌsɑː(r)/ n. hist. **1** an official of the Soviet Communist Party responsible for political education and organization. **2** the head of a government department in the USSR before 1946. [Russ. *komissar* f. F *commissaire* (as COMMISSARY)]

commissariat /ˌkɒmɪˈseərɪət, -ˈsærɪˌæt/ *n.* **1** esp. *Mil.* **a** a department for the supply of food etc. **b** the food supplied. **2** *hist.* a government department of the USSR before 1946. [F *commissariat* & med.L *commissariatus* (as COMMISSARY)]

commissary /ˈkɒmɪsərɪ, kəˈmɪs-/ *n.* (*pl.* **-ies**) **1** a deputy or delegate. **2** a representative or deputy of a bishop. **3** *Mil.* an officer responsible for the supply of food etc. to soldiers. **4** *US* **a** a restaurant in a film studio etc. **b** the food supplied. **5** *US Mil.* a store for the supply of food etc. to soldiers. □ **commissaryship** *n.* **commissarial** /ˌkɒmɪˈseərɪəl/ *adj.* [ME f. med.L *commissarius* person in charge (as COMMIT)]

commission /kəˈmɪʃ(ə)n/ *n. & v.* ● *n.* **1 a** the authority to perform a task or certain duties. **b** a person or group entrusted esp. by a government with such authority (*set up a commission to look into it*). **c** an instruction, command, or duty given to such a group or person (*their commission was to simplify the procedure; my commission was to find him*). **2** an order for something, esp. a work of art, to be produced specially. **3** *Mil.* **a** a warrant conferring the rank of officer in the army, navy, or air force. **b** the rank so conferred. **4 a** the authority to act as agent for a company etc. in trade. **b** a percentage paid to the agent from the profits of goods etc. sold, or business obtained (*his salary is low, but he gets 20 per cent commission*). **c** the pay of a commissioned agent. **5** the act of committing (a crime, sin, etc.). **6** the office or department of a commissioner. ● *v.tr.* **1** authorize or empower by a commission. **2 a** give (an artist etc.) a commission for a piece of work. **b** order (a work) to be written (*commissioned a new concerto*). **3 a** give (an officer) the command of a ship. **b** prepare (a ship) for active service. **4** bring (a machine, equipment, etc.) into operation. □ **commission-agent** a bookmaker. **commission of the peace 1** Justices of the Peace. **2** the authority given to them. **in commission** (of a warship etc.) manned, armed, and ready for service. **out of commission** (esp. of a ship) not in service, not in working order. [ME f. OF f. L *commissio -onis* (as COMMIT)]

commissionaire /kəˌmɪʃəˈneə(r)/ *n.* esp. *Brit.* a uniformed door-attendant at a theatre, cinema, etc. [F (as COMMISSIONER)]

commissioner /kəˈmɪʃənə(r)/ *n.* **1** a person appointed by a commission to perform a specific task, e.g. the head of the London police, a delegate to the General Assembly of the Church of Scotland, etc. **2** a person appointed as a member of a government commission (*Charity Commissioner; Civil Service Commissioner*). **3** a representative of the supreme authority in a district, department, etc. [ME f. med.L *commissionarius* (as COMMISSION)]

Commissioner for Oaths *n.* a solicitor authorized to administer an oath to a person making an affidavit.

commissure /ˈkɒmɪˌsjʊə(r)/ *n.* **1** a junction, joint, or seam. **2** *Anat.* **a** the joint between two bones. **b** a band of nerve tissue connecting the hemispheres of the brain, the two sides of the spinal cord, etc. **c** the line where the upper and lower lips, or eyelids, meet. **3** *Bot.* a joint etc. between different parts of a plant. □ **commissural** /kəˈmɪsjʊərəl, ˌkɒmɪˈsjʊərəl/ *adj.* [ME f. L *commissura* junction (as COMMIT)]

commit /kəˈmɪt/ *v.tr.* (**committed, committing**) **1** (usu. foll. by *to*) entrust or consign for: **a** safe keeping (*I commit him to your care*). **b** treatment, usu. destruction (*committed the book to the flames*). **c** official custody as a criminal or as insane (*you could be committed for such behaviour*). **2** perpetrate, do (esp. a crime, sin, or blunder). **3** pledge, involve, or bind (esp. oneself) to a certain course or policy (*does not like committing herself; committed by the vow he had made*). **4** (as **committed** *adj.*) (often foll. by *to*) **a** morally dedicated or politically aligned (*a committed Christian; committed to the cause; a committed socialist*). **b** obliged (to take certain action) (*felt committed to staying there*). **5** *Polit.* refer (a bill etc.) to a committee. □ **commit to memory** learn (a thing) so as to be able to recall it. **commit to prison** consign officially to custody, esp. on remand. □ **committable** *adj.* **committer** *n.* [ME f. L *committere* join, entrust (as COM-, *mittere miss-* send)]

commitment /kəˈmɪtmənt/ *n.* **1** an engagement or (esp. financial) obligation that restricts freedom of action. **2** the process or an instance of committing oneself; a pledge or undertaking. **3** dedication, application.

committal /kəˈmɪt(ə)l/ *n.* **1** the act of committing a person to an institution, esp. prison or a mental hospital. **2** the burial of a body.

committee /kəˈmɪtɪ/ *n.* **1 a** a body of persons appointed for a specific function by, and usu. out of, a larger body. **b** such a body appointed by Parliament etc. to consider the details of proposed legislation. **c** (**Committee**) the whole House of Commons when sitting as a

committee. **2** /ˌkɒmɪˈtiː/ *Law* a person entrusted with the charge of another person or another person's property. □ **committee-man** (*pl.* **-men**, *fem.* **committee-woman**, *pl.* **-women**) a member of a committee, esp. a habitual member of committees. **committee stage** *Brit.* the third of five stages of a bill's progress through Parliament when it may be considered in detail and amendments made. **select committee** a small parliamentary committee appointed for a special purpose. **standing committee** a committee that is permanent during the existence of the appointing body. [COMMIT + -EE]

Committee of Public Safety (during the French Revolution) a governing body set up in France in Apr. 1793. Consisting of nine (later twelve) members, it was at first dominated by Danton but later in the year came under the influence of Robespierre and initiated the Terror. The Committee's power ended with the fall of Robespierre in 1794, although it was not dissolved until the following year.

commix /kɒˈmɪks/ *v.tr. & intr. archaic or poet.* mix. □ **commixture** *n.* [ME: back-form. f. *commixt* past part. f. L *commixtus* (as COM-, MIXED)]

Commo /ˈkɒməʊ/ *n.* (*pl.* **-os**) esp. *Austral. & NZ sl.* a Communist. [abbr.]

commode /kəˈməʊd/ *n.* **1** a chest of drawers. **2** (also **night-commode**) **a** a bedside table with a cupboard containing a chamber-pot. **b** a chamber-pot concealed in a chair with a hinged cover. **3** = CHIFFONIER. [F, *adj.* (as noun) f. L *commodus* convenient (as COM-, *modus* measure)]

commodious /kəˈməʊdɪəs/ *adj.* **1** roomy and comfortable. **2** *archaic* convenient. □ **commodiously** *adv.* **commodiousness** *n.* [F *commodieux* or f. med.L *commodiosus* f. L *commodus* (as COMMODE)]

commodity /kəˈmɒdɪtɪ/ *n.* (*pl.* **-ies**) **1** an article or raw material that can be bought and sold, esp. a product as opposed to a service. **2** a useful thing. [ME f. OF *commodité* or f. L *commoditas* (as COMMODE)]

commodore /ˈkɒməˌdɔː(r)/ *n.* **1** a naval officer above a captain and below a rear admiral. **2** the commander of a squadron or other division of a fleet. **3** the president of a yacht-club. **4** the senior captain of a shipping line. [prob. f. Du. *komandeur* f. F *commandeur* COMMANDER]

Commodore-in-Chief /ˌkɒməˌdɔːrɪnˈtʃiːf/ *n.* the supreme officer in the air force.

common /ˈkɒmən/ *adj. & n.* ● *adj.* (**commoner, commonest**) **1 a** occurring often (*a common mistake*). **b** occurring too frequently, overused, trite. **c** ordinary; of ordinary qualities; without special rank or position (*no common mind; common soldier; the common people*). **2 a** shared by, coming from, or done by, more than one (*common knowledge; by common consent; our common benefit*). **b** belonging to, open to, or affecting, the whole community or the public (*common land*). **3** *derog.* low-class; vulgar; inferior (*a common little man*). **4** of the most familiar type (*common cold; common nightshade*). **5** *Math.* belonging to two or more quantities (*common denominator; common factor*). **6** *Gram.* (of gender) referring to individuals of either sex (e.g. *teacher*). **7** *Prosody* (of a syllable) that may be either short or long. **8** *Mus.* having two or four beats, esp. four crotchets, in a bar. **9** *Law* (of a crime) of lesser importance (cf. GRAND *adj.* 8, PETTY 4). ● *n.* **1** a piece of open public land, esp. in a village or town. **2** *sl.* = *common sense* (*use your common*). **3** *Eccl.* a service used for each of a group of occasions. **4** (in full **right of common**) *Law* a person's right over another's land, e.g. for pasturage. □ **common carrier** a person or firm undertaking to transport any goods or person in a specified category. **common chord** *Mus.* any note with its major or minor third and perfect fifth. **common crier** see CRIER. **common denominator** see DENOMINATOR. **common ground** a point or argument accepted by both sides in a dispute. **common gull** a grey and white gull, *Larus canus*, which has usu. greenish legs and is widespread in the northern hemisphere (also called (*N.Amer.*) *mew gull*). **common jury** a jury with members of no particular social standing (cf. *special jury*). **common law** law derived from custom and judicial precedent rather than statutes (cf. *case-law* (see CASE[1]), *statute law*). **common-law husband** (or **wife**) **1** a partner in a marriage recognized in some jurisdictions as valid by common law, though not brought about by a civil or ecclesiastical ceremony. **2** *colloq.* a partner in a relationship in which a man and woman cohabit for a period long enough to suggest stability. **common market** an organization promoting free trade among a group of countries (see also COMMON MARKET). **common metre** a hymn stanza of four lines with 8, 6, 8, and 6 syllables. **common noun** (or **name**) *Gram.* a name denoting a class of objects or a concept as opposed to a particular individual (e.g. *boy, chocolate, beauty*). **common or garden** *colloq.* ordinary. **common-room 1** a room in some colleges, schools, etc., which members may use for relaxation or work. **2** the members who use this. **common**

salt see SALT *n.* 1. **common seal 1** a seal with a mottled grey coat, *Phoca vitulina*, found in the North Atlantic and North Pacific. **2** the official seal of a corporate body. **common sense** sound practical sense, esp. in everyday matters (often hyphenated when *attrib.: a common-sense approach*). **common soldier** see SOLDIER *n.* 2. **common stock** *US* = *ordinary shares*. **common weal** public welfare. **common year** see YEAR 2. **in common 1** in joint use; shared. **2** of joint interest (*have little in common*). **in common with** in the same way as. **least** (or **lowest**) **common denominator, multiple** see DENOMINATOR, MULTIPLE. **out of the common** unusual. □ **commonly** *adv.* **commonness** /ˈkɒmənnɪs/ *n.* [ME f. OF *comun* f. L *communis*]

commonable /ˈkɒmənəb(ə)l/ *adj.* **1** (of an animal) that may be pastured on common land. **2** (of land) that may be held in common. [obs. *common* to exercise right of common + -ABLE]

commonage /ˈkɒmənɪdʒ/ *n.* **1** right of common (see COMMON *n.* 4). **2 a** land held in common. **b** the state of being held in common. **3** the common people; commonalty.

Common Agricultural Bureaux (abbr. **CAB**, now called **CAB International**) an international institution, founded in 1929, consisting of four institutes and eleven bureaux whose purpose is to provide a scientific information service and mutual assistance in agricultural science to the countries contributing to its funds.

Common Agricultural Policy (abbr. **CAP**) the system in the EC, set out in the Treaty of Rome (1957), for establishing common prices for most agricultural products within the European Community, a single fund for price supports, and levies on imports. It also lays down a common policy for the export of agricultural products to countries outside the Community.

commonality /ˌkɒməˈnælɪtɪ/ *n.* (*pl.* **-ies**) **1** the sharing of an attribute. **2** a common occurrence. **3** = COMMONALTY. [var. of COMMONALTY]

commonalty /ˈkɒmənəltɪ/ *n.* (*pl.* **-ies**) **1** the common people. **2** the general body (esp. of humankind). **3** a corporate body. [ME f. OF *comunalté* f. med.L *communalitas -tatis* (as COMMON)]

commoner /ˈkɒmənə(r)/ *n.* **1** one of the common people, as opposed to the aristocracy. **2** a person who has the right of common. **3** a student at a British university who does not have a scholarship. [ME f. med.L *communarius* f. *communa* (as COMMUNE¹)]

Common Era the Christian era.

commonhold /ˈkɒmənˌhəʊld/ *n.* a type of property ownership for blocks of flats, whereby each flat is owned on a freehold basis by its occupant, but responsibilities and services that relate to the whole block are shared among all the occupants. □ **commonholder** *n.*

Common Market the European Economic Community.

commonplace /ˈkɒmənˌpleɪs/ *adj. & n.* ● *adj.* lacking originality; trite. ● *n.* **1 a** an everyday saying; a platitude (*uttered a commonplace about the weather*). **b** an ordinary topic of conversation. **2** anything usual or trite. **3** a notable passage in a book etc. copied into a commonplace-book. □ **commonplace-book** a book into which notable extracts from other works are copied for personal use. □ **commonplaceness** *n.* [transl. L *locus communis* = Gk *koinos topos* general theme]

Common Prayer the Church of England liturgy originally set forth in the *Book of Common Prayer* of Edward VI (1549).

commons /ˈkɒmənz/ *n.pl.* **1** (**the Commons**) = HOUSE OF COMMONS. **2 a** the common people. **b** (prec. by *the*) the common people regarded as a part of a political, esp. British, system. **3** provisions shared in common; daily fare. □ **short commons** insufficient food. [ME pl. of COMMON]

commonsensical /ˌkɒmənˈsensɪk(ə)l/ *adj.* possessing or marked by common sense. [*common sense* (see COMMON)]

Common Serjeant *n.* a circuit judge of the Central Criminal Court with duties in the City of London.

commonweal /ˈkɒmənˌwiːl/ *n. archaic* **1** = *common weal*. **2** = COMMONWEALTH.

commonwealth /ˈkɒmənˌwelθ/ *n.* **1** an independent state or community, esp. a democratic republic. **2** such a community or organization of shared interests in a non-political field (*the commonwealth of learning*). [COMMON + WEALTH]

Commonwealth, the 1 (in full **the Commonwealth of Nations**) an international association of states that were once part of the British Empire, and their dependencies. (*See note below.*) **2** the period of republican government in Britain between the execution of Charles I in 1649 and the Restoration of Charles II in 1660. **3** a part of the title of Puerto Rico and some of the states of the US. **4** the title of the federated Australian states.

▪ In 1931 a group of dependent states, originally colonies of the British Empire, was recognized by the Statute of Westminster as having special status and autonomy; it was called the *British Commonwealth of Nations*. In 1946 the group was renamed the *Commonwealth of Nations*, and in 1947 the first countries with majority non-European populations (India, Pakistan, and Ceylon) became members. A small number of countries, including the Republic of Ireland (1948), South Africa (1961), Pakistan (1972), and Fiji (1987), later withdrew their membership; Pakistan rejoined in 1989, and South Africa in 1994. Leaders of Commonwealth states meet in conference every two years, but decisions are not binding on member states, and there is no longer a system of preferential tariffs for members. The symbolic head of the Commonwealth is the British monarch.

Commonwealth Day the name since 1959 of a day each year celebrating the Commonwealth, now the second Monday in March. Originally called Empire Day, it was instituted to commemorate assistance given to Britain by the colonies during the Boer War (1899–1902).

Commonwealth Development Corporation (abbr. **CDC**) an organization set up in 1948 to assist commercial and industrial development in British dependent territories and in certain cases any Commonwealth or developing country elsewhere.

Commonwealth Games an amateur sports competition held every four years between member countries of the Commonwealth. Some of the sports included are athletics, gymnastics, swimming, boxing, cycling, and shooting. The first Commonwealth Games (then called the British Empire Games) were held in Ontario, Canada, in 1930; women's events were first held in 1934.

Commonwealth of Independent States (abbr. **CIS**) a confederation of independent states, formerly constituent republics of the Soviet Union, established in 1991 following a meeting in the Belorussian city of Brest at which the USSR was dissolved. The member states are Armenia, Belarus, Kazakhstan, Kyrgyzstan, Moldova, Russia, Tajikistan, Turkmenistan, Ukraine, and Uzbekistan (Azerbaijan left in 1992). The administrative headquarters of the CIS is at Minsk in Belarus.

commotion /kəˈməʊʃ(ə)n/ *n.* **1 a** a confused and noisy disturbance or outburst. **b** loud and confusing noise. **2** a civil insurrection. [ME f. OF *commotion* or L *commotio* (as COM-, MOTION)]

communal /ˈkɒmjʊn(ə)l/ *adj.* **1** relating to or benefiting a community; for common use (*communal baths*). **2** (of conflicts etc.) between different esp. ethnic or religious communities (*communal violence*). **3** of a commune, esp. the Paris Commune. □ **communally** *adv.* **communality** /ˌkɒmjʊˈnælɪtɪ/ *n.* [F f. LL *communalis* (as COMMUNE¹)]

communalism /ˈkɒmjʊnəˌlɪz(ə)m/ *n.* **1** a principle of political organization based on federated communes. **2** the principle of communal ownership etc. □ **communalist** *n.* **communalistic** /ˌkɒmjʊnəˈlɪstɪk/ *adj.*

communalize /ˈkɒmjʊnəˌlaɪz/ *v.tr.* (also **-ise**) make communal. □ **communalization** /ˌkɒmjʊnəlaɪˈzeɪʃ(ə)n/ *n.*

communard /ˈkɒmjʊˌnɑːd/ *n.* **1** a member of a commune. **2** (also **Communard**) *hist.* a supporter of the Paris Commune. [F (as COMMUNE¹)]

commune¹ /ˈkɒmjuːn/ *n.* **1 a** a group of people, not necessarily related, sharing living accommodation, goods, etc., esp. as a political act. **b** a communal settlement esp. for the pursuit of shared interests. **2 a** the smallest French territorial division for administrative purposes. **b** a similar division elsewhere. [F f. med.L *communia* neut. pl. of L *communis* common]

commune² /kəˈmjuːn/ *v.intr.* **1** (usu. foll. by *with*) **a** speak confidentially and intimately (*communed together about their loss; communed with his heart*). **b** feel in close touch (with nature etc.) (*communed with the hills*). **2** *US* receive Holy Communion. [ME f. OF *comuner* share f. *comun* COMMON]

Commune, the (also **Commune of Paris, Paris Commune**) **1** a group which seized the municipal government of Paris in the French Revolution and in this capacity played a leading part in the Reign of Terror until suppressed in 1794. **2** a municipal government organized on communalistic principles and containing socialists and revolutionaries, elected in Paris in 1871 after the Franco-Prussian War and the collapse of the Second Empire. It was soon brutally suppressed by government troops from Versailles.

communicable /kəˈmjuːnɪkəb(ə)l/ *adj.* **1** (esp. of a disease) able to be passed on. **2** *archaic* communicative. □ **communicably** *adv.*

communicability /-ˌmjuːnɪkəˈbɪlɪtɪ/ n. [ME f. OF *communicable* or LL *communicabilis* (as COMMUNICATE)]

communicant /kəˈmjuːnɪkənt/ n. **1** a person who receives Holy Communion, esp. regularly. **2** a person who imparts information. [L *communicare communicant-* (as COMMON)]

communicate /kəˈmjuːnɪˌkeɪt/ v. **1** tr. **a** transmit or pass on by speaking or writing (*communicated his ideas*). **b** transmit (heat, motion, etc.). **c** pass on (an infectious illness). **d** impart (feelings etc.) non-verbally (*communicated his affection*). **2** intr. (often foll. by *with*) be in communication; succeed in conveying information, evoking understanding etc. (*he communicates well*). **3** intr. (often foll. by *with*) share a feeling or understanding; relate socially. **4** intr. (often foll. by *with*) (of a room etc.) have a common door (*my room communicates with yours*). **5** tr. administer Holy Communion to. **b** intr. receive Holy Communion. □ **communicator** n. **communicatory** adj. [L *communicare communicat-* (as COMMON)]

communication /kəˌmjuːnɪˈkeɪʃ(ə)n/ n. **1 a** the act of imparting, esp. news. **b** an instance of this. **c** the information etc. communicated. **2** a means of connecting different places, such as a door, passage, road, or railway. **3** social intercourse (*it was difficult to maintain communication in the uproar*). **4** (in pl.) the science and practice of transmitting information esp. by electronic or mechanical means. **5** (in pl.) Mil. the means of transport between a base and the front. **6** a paper read to a learned society. □ **communication cord** Brit. a cord or chain in a railway carriage that may be pulled to stop the train in an emergency. **communication** (or **communications**) **satellite** an artificial satellite used to relay telephone circuits or broadcast programmes. **communication theory** the study of the principles and methods by which information is conveyed.

communicative /kəˈmjuːnɪkətɪv/ adj. **1** open, talkative, informative. **2** ready to communicate. □ **communicatively** adv. [LL *communicativus* (as COMMUNICATE)]

communion /kəˈmjuːnɪən/ n. **1** a sharing, esp. of thoughts etc.; fellowship (*their minds were in communion*). **2** participation; a sharing in common (*communion of interests*). **3** (**Communion**, **Holy Communion**) **a** the Eucharist. **b** participation in the Communion service. **c** (attrib.) of or used in the Communion service (*Communion table*; *Communion cloth*; *Communion rail*). **4** fellowship, esp. between branches of the Catholic Church. **5** a body or group within the Christian faith (*the Methodist communion*). □ **Communion in both kinds** a Eucharist in which both the consecrated elements of bread and wine are administered. **communion of saints** fellowship between Christians living and dead. [ME f. OF *communion* or L *communio* f. *communis* common]

communiqué /kəˈmjuːnɪˌkeɪ/ n. an official communication, esp. a news report. [F, = communicated]

communism /ˈkɒmjʊˌnɪz(ə)m/ n. **1** a political theory derived from Marxism, advocating a society in which all property is publicly owned and each person is paid and works according to his or her needs and abilities. **2** (usu. **Communism**) **a** the communistic form of society established in the former USSR and elsewhere. (*See note below.*) **b** any movement or political doctrine advocating communism. **3** = COMMUNALISM. [F *communisme* f. *commun* COMMON]

■ Although it has taken different forms in different countries, the most familiar form of Communism is that established by the Bolsheviks after the Russian Revolution of 1917. It has generally been defined in terms of the system practised by the former USSR and its allies in eastern Europe, in China since 1949, and in certain developing countries such as Cuba, Vietnam, and North Korea. Perhaps the most important political force in the early and mid-20th centuries, Communism embraced a revolutionary ideology based on the overthrow of the capitalist system and, in theory at least, on the notion of constant progress towards the perfect stateless society. While Marx envisaged the state 'withering away' after takeover by the proletariat, the Communist countries of the 20th century, however, saw the state grow to control all aspects of society. The end of 1989 saw the collapse of bureaucratic Communism in eastern Europe against a background of its failure to meet people's economic expectations, a shift to more democracy in political life, under Soviet leader Mikhail Gorbachev, and growing nationalist aspirations in the Soviet republics, which led to the breakup of the USSR .

Communism Peak one of the principal peaks in the Pamir Mountains of Tajikistan, rising to 7,495 m (24,590 ft). Known as Mount Garmo until 1933 and Stalin Peak until 1962, it was the highest mountain in the Soviet Union.

communist /ˈkɒmjʊnɪst/ n. & adj. ● n. **1** a person advocating or practising communism. **2** (**Communist**) a member of a Communist Party. ● adj. of or relating to communism (*a communist play*). □ **communistic** /ˌkɒmjʊˈnɪstɪk/ adj. [COMMUNISM]

communitarian /kəˌmjuːnɪˈteərɪən/ n. & adj. ● n. a member of a communistic community. ● adj. of or relating to such a community. [COMMUNITY + -ARIAN after *unitarian* etc.]

community /kəˈmjuːnɪtɪ/ n. (pl. **-ies**) **1 a** all the people living in a specific locality, including its inhabitants. **2 a** body of people having a religion, a profession, etc., in common (*the immigrant community*). **3** fellowship of interests etc.; similarity (*community of intellect*). **4** a monastic, socialistic, etc. body practising common ownership. **5** joint ownership or liability (*community of goods*). **6** (prec. by *the*) the public. **7** a body of nations unified by common interests. **8** Ecol. a group of animals or plants living or growing together in the same area. □ **community centre** a place providing social etc. facilities for a neighbourhood. **community charge** hist. (in the UK) a short-lived tax levied locally on every adult in a community replaced in 1993 by the council tax. **community chest** a fund for charity and welfare work in a community. **community home** Brit. a centre for housing young offenders and other juveniles in need of custodial care. **community policeman** a policeman involved in community policing. **community policing** policing by officers who have personal knowledge of and involvement in the community they police. **community service 1** work in the community (esp. voluntary and unpaid); an act of community service. **2** work stipulated under a community service order. **community service order** an order for a convicted offender to perform a period of unpaid work in the community. **community singing** singing by a large crowd or group, esp. of old popular songs or hymns. **community spirit** a feeling of belonging to a community, expressed in mutual support etc. **community worker** a person whose job is to help a community promote its own welfare. [ME f. OF *comuneté* f. L *communitas -tatis* (as COMMON)]

communize /ˈkɒmjʊˌnaɪz/ v.tr. (also **-ise**) **1** make (land etc.) common property. **2** make (a person etc.) communistic. □ **communization** /ˌkɒmjʊnaɪˈzeɪʃ(ə)n/ n. [L *communis* COMMON]

commutable /kəˈmjuːtəb(ə)l/ adj. **1** convertible into money; exchangeable. **2** Law (of a punishment) able to be commuted. **3** within commuting distance. □ **commutability** /-ˌmjuːtəˈbɪlɪtɪ/ n. [L *commutabilis* (as COMMUTE)]

commutate /ˈkɒmjʊˌteɪt/ v.tr. Electr. **1** regulate the direction of (an alternating current), esp. to make it a direct current. **2** reverse the direction of (an electric current). [L *commutare commutat-* (as COMMUTE)]

commutation /ˌkɒmjʊˈteɪʃ(ə)n/ n. **1** the act or process of commuting or being commuted (in legal and exchange senses). **2** Electr. the act or process of commutating or being commutated. **3** Math. the reversal of the order of two quantities. □ **commutation ticket** US a season ticket. [F *commutation* or L *commutatio* (as COMMUTE)]

commutative /kəˈmjuːtətɪv/ adj. **1** relating to or involving substitution. **2** Math. unchanged in result by the interchange of the order of quantities. [F *commutatif* or med.L *commutativus* (as COMMUTE)]

commutator /ˈkɒmjʊˌteɪtə(r)/ n. **1** Electr. a device for reversing electric current. **2** an attachment connected with the armature of a dynamo which directs and makes continuous the current produced.

commute /kəˈmjuːt/ v. **1** intr. travel to and from one's daily work, usu. in a city, esp. by car or train. **2** tr. (usu. foll. by *to*) Law change (a judicial sentence etc.) to another less severe. **3** tr. (often foll. by *into*, *for*) **a** change (one kind of payment) for another. **b** make a payment etc. to change (an obligation etc.) for another. **4** tr. **a** exchange; interchange (two things). **b** change (to another thing). **5** tr. Electr. commutate. **6** intr. Math. have a commutative relation. **7** intr. US buy and use a season ticket. [L *commutare commutat-* (as COM-, *mutare* change)]

commuter /kəˈmjuːtə(r)/ n. a person who travels some distance to work, esp. in a city, usu. by car or train.

Como, Lake /ˈkəʊməʊ/ a lake in the foothills of the Alps in northern Italy.

Comodoro Rivadavia /ˌkɒməˌdɔːrəʊ ˌriːvəˈdɑːvɪə/ a port in Argentina situated on the Atlantic coast of Patagonia; pop. (1991) 124,000.

Comorin, Cape /ˈkɒmərɪn/ a cape at the southern tip of India, in the state of Tamil Nadu.

Comoros /ˈkɒməˌrəʊz/ a country consisting of a group of islands in the Indian Ocean north of Madagascar; pop. (est. 1991) 492,000;

languages, French (official), Arabic (official), Comoran Swahili; capital, Moroni. The islands were first visited by the English at the end of the 16th century. At that time and for long afterwards Arab influence was dominant. In the mid-19th century they came under French protection, until in 1974 all but one of the four major islands voted for independence. □ **Comoran** adj. & n.

comose /ˈkəʊməʊz, -məʊs/ adj. Bot. (of seeds etc.) having hairs, downy. [L comosus (as COMA²)]

comp /kɒmp/ n. & v. colloq. ● n. **1** a competition. **2** Printing a compositor. **3** Mus. an accompaniment. ● v. **1** Mus. **a** tr. accompany. **b** intr. play an accompaniment. **2** Printing **a** intr. work as a compositor. **b** tr. work as a compositor on. [abbr.]

compact¹ adj., v., & n. ● adj. /kəmˈpækt, ˈkɒmpækt/ **1** closely or neatly packed together. **2** (of a piece of equipment, a room, etc.) well-fitted and practical though small. **3** (of style etc.) condensed; brief. **4** (esp. of the human body) small but well-proportioned. **5** (foll. by of) composed or made up of. ● v.tr. /kəmˈpækt/ **1** join or press firmly together. **2** condense. **3** (usu. foll. by of) compose; make up. ● n. /ˈkɒmpækt/ **1** a small flat and usu. decorated case for face-powder, a mirror, etc. **2** an object formed by compacting powder. **3** N. Amer. a medium-sized motor car. □ **compact disc** /ˈkɒmpækt/ a disc on which information or sound is recorded digitally and reproduced by reflection of laser light. (See also SOUND RECORDING.) □ **compaction** /kəmˈpækʃ(ə)n/ n. **compactly** /-ˈpæktlɪ/ adv. **compactness** /-ˈpæktnɪs/ n. **compactor** /-ˈpæktə(r)/ n. [ME f. L compingere compact- (as COM-, pangere fasten)]

compact² /ˈkɒmpækt/ n. an agreement or contract between two or more parties. [L compactum f. compacisci compact- (as COM-, pacisci covenant): cf. PACT]

compages /kəmˈpeɪdʒiːz/ n. (pl. same) **1** a framework; a complex structure. **2** something resembling a compages in complexity etc. [L compages (as COM-, pages f. pangere fasten)]

compander /kəmˈpændə(r)/ n. (also **compandor**) a device that improves the quality of reproduced or transmitted sound by compressing the range of amplitudes of the signal before transmission and then expanding it on reproduction or reception. □ **compand** v.t. [COMPRESSOR + EXPANDER (see EXPAND)]

companion¹ /kəmˈpænjən/ n. & v. ● n. **1 a** (often foll. by in, of) a person who accompanies, associates with, or shares with, another (a companion in adversity; they were close companions). **b** a person, esp. an unmarried or widowed woman, employed to live with and assist another. **2** a handbook or reference book on a particular subject (A Companion to North Wales). **3** a thing that matches another (the companion of this book-end is over there). **4** (**Companion**) a member of the lowest grade of some orders of knighthood (Companion of the Bath). **5** Astron. a star etc. that accompanies another. **6** equipment or a piece of equipment that combines several uses. ● v. **1** tr. accompany. **2** intr. (often foll. by with) literary be a companion. □ **companion in arms** a fellow-soldier. **companion-set** a set of fireside implements on a stand. [ME f. OF compaignon ult. f. L panis bread]

companion² /kəmˈpænjən/ n. Naut. **1** a raised frame with windows let into the quarterdeck of a ship to allow light into the cabins etc. below. **2** = companion-way. □ **companion-hatch** a wooden covering over a companion-way. **companion hatchway** an opening in a deck leading to a cabin. **companion ladder** a ladder from a deck to a cabin. **companion-way** a staircase to a cabin. [obs. Du. kompanje quarterdeck f. OF compagne f. It. (camera della) compagna pantry, prob. ult. rel. to COMPANION¹]

companionable /kəmˈpænjənəb(ə)l/ adj. agreeable as a companion; sociable. □ **companionableness** n. **companionably** adv.

companionate /kəmˈpænjənɪt/ adj. **1** well-suited; (of clothes) matching. **2** of or like a companion.

Companion of Honour n. (in the UK) a member of an order of knighthood founded in 1917.

Companion of Literature n. (in the UK) a holder of an honour awarded by the Royal Society of Literature and founded in 1961.

companionship /kəmˈpænjənˌʃɪp/ n. good fellowship; friendship.

company /ˈkʌmpənɪ/ n. & v. ● n. (pl. **-ies**) **1 a** a number of people assembled; a crowd; an audience (addressed the company). **b** guests or a guest (am expecting company). **2** a state of being a companion or fellow; companionship, esp. of a specific kind (enjoys low company; do not care for his company). **3 a** a commercial business. **b** (usu. **Co.**) the partner or partners not named in the title of a firm (Smith and Co.). **4** a troupe of actors or entertainers. **5** Mil. a subdivision of an infantry battalion usu. commanded by a major or a captain. **6** a group of Guides. ● v. (**-ies**,

-ied) **1** tr. archaic accompany. **2** intr. (often foll. by with) literary be a companion. □ **company officer** a captain or a lower commissioned officer. **company sergeant-major** see SERGEANT. **err** (or **be**) **in good company** discover that one's companions, or better people, have done the same as oneself. **good** (or **bad**) **company 1** a pleasant (or dull) companion. **2** a suitable (or unsuitable) associate or group of friends. **in company** not alone. **in company with** together with. **keep company** (often foll. by with) associate habitually. **keep** (archaic **bear**) **a person company** accompany a person; be sociable. **part company** (often foll. by with) cease to associate. **ship's company** the entire crew. [ME f. AF compainie, OF compai(g)nie f. Rmc (as COMPANION¹)]

comparable /ˈkɒmpərəb(ə)l, disp. kəmˈpær-/ adj. **1** (often foll. by with) able to be compared. **2** (often foll. by to) fit to be compared; worth comparing. ¶ Use with to and with corresponds to the senses at compare; to is more common. □ **comparably** adv. **comparableness** n. **comparability** /ˌkɒmpərəˈbɪlɪtɪ/ n. [ME f. OF f. L comparabilis (as COMPARE)]

comparative /kəmˈpærətɪv/ adj. & n. ● adj. **1** perceptible by comparison; relative (in comparative comfort). **2** estimated by comparison (the comparative merits of the two ideas). **3** of or involving comparison (esp. of a science or subject of study). **4** Gram. (of an adjective or adverb) expressing a higher degree of a quality, but not the highest possible (e.g. braver, more fiercely) (cf. POSITIVE adj. 3b, SUPERLATIVE adj. 2). ● n. Gram. **1** the comparative expression or form of an adjective or adverb. **2** a word in the comparative. □ **comparatively** adv. [ME f. L comparativus (as COMPARE)]

comparator /kəmˈpærətə(r)/ n. Engin. a device for comparing a product, an output, etc., with a standard, esp. an electronic circuit comparing two signals.

compare /kəmˈpeə(r)/ v. & n. ● v. **1** tr. (usu. foll. by to) express similarities in; liken (compared the landscape to a painting). **2** tr. (often foll. by to, with) estimate the similarity or dissimilarity of; assess the relation between (compared radio with television; that lacks quality compared to this). ¶ In current use to and with are generally interchangeable, but with often implies a greater element of formal analysis, as in compared my account with yours. **3** intr. (often foll. by with) bear comparison (compares favourably with the rest). **4** intr. (often foll. by with) be equal or equivalent to. **5** tr. Gram. form the comparative and superlative degrees of (an adjective or an adverb). ● n. literary comparison (beyond compare; without compare; has no compare). □ **compare notes** exchange ideas or opinions. [ME f. OF comparer f. L comparare (as COM-, parare f. par equal)]

comparison /kəmˈpærɪs(ə)n/ n. **1** the act or an instance of comparing. **2** a simile or semantic illustration. **3** capacity for being likened; similarity (there's no comparison). **4** (in full **degrees of comparison**) Gram. the positive, comparative, and superlative forms of adjectives and adverbs. □ **bear** (or **stand**) **comparison** (often foll. by with) be able to be compared favourably. **beyond comparison 1** totally different in quality. **2** greatly superior; excellent. **in comparison with** compared to. [ME f. OF comparesoun f. L comparatio -onis (as COMPARE)]

compartment /kəmˈpɑːtmənt/ n. & v. ● n. **1** a space within a larger space, separated from the rest by partitions, e.g. in a railway carriage, wallet, desk, etc. **2** Naut. a watertight division of a ship. **3** an area of activity etc. kept apart from others in a person's mind. ● v.tr. put into compartments. □ **compartmentation** /-ˌpɑːtmənˈteɪʃ(ə)n/ n. [F compartiment f. It. compartimento f. LL compartiri (as COM-, partiri share)]

compartmental /ˌkɒmpɑːtˈment(ə)l/ adj. consisting of or relating to compartments or a compartment. □ **compartmentally** adv.

compartmentalize /ˌkɒmpɑːtˈmentəˌlaɪz/ v.tr. (also **-ise**) divide (esp. an activity, time, etc.) into compartments or categories. □ **compartmentalization** /-ˌmentəlaɪˈzeɪʃ(ə)n/ n.

compass /ˈkʌmpəs/ n. & v. ● n. **1** an instrument showing the direction of magnetic north and bearings from it. (See note below.) **2** (usu. in pl.) an instrument for taking measurements and describing circles, with two arms connected at one end by a movable joint. **3** a circumference or boundary. **4** area, extent; scope (e.g. of knowledge or experience) (beyond my compass). **5** the range of tones of a voice or a musical instrument. ● v.tr. literary **1** hem in. **2** grasp mentally. **3** contrive, accomplish. **4** go round. □ **compass card** a circular rotating card showing the thirty-two principal bearings, forming the indicator of a magnetic compass. **compass rose** a circle of the principal directions marked on a chart. **compass-saw** a saw with a narrow blade, for cutting curves. **compass window** a bay window with a semicircular curve. □ **compassable** adj. [ME f. OF compas ult. f. L passus PACE¹]

∎ The simplest form of compass, the magnetic compass, depends on the property of a piece of magnetic material such as iron aligning itself in the north–south direction in the earth's magnetic field. Its use in navigation at sea was reported from China *c.*1100, western Europe 1187, Arabia *c.*1220, and Scandinavia *c.*1300, although it probably dates from much earlier. In simple compasses the compass card is fixed and a magnetic needle swings round above it; in all aircraft and ships' magnetic compasses, however, the card is fixed on top of the needle and swings round with it. In ships the compass casing is swung on gimbals so that it remains face upwards in spite of the rocking of the ship. The magnetic compass, which points to the magnetic north (whose position varies over the years), is subject to a number of errors (deviations can be caused by adjacent metal fitments etc.). Since the early 20th century the gyrocompass has become standard equipment for ships and aircraft, although a magnetic compass may be fitted as a stand-by.

compassion /kəmˈpæʃ(ə)n/ *n.* pity inclining one to help or be merciful. □ **compassion fatigue** indifference to human suffering as a result of overexposure to charitable appeals, news reports, etc.[ME f. OF f. eccl.L *compassio -onis* f. *compati* (as COM-, *pati pass-* suffer)]

compassionate /kəmˈpæʃənət/ *adj.* sympathetic, pitying. □ **compassionate leave** *Brit.* leave granted on grounds of bereavement etc. □ **compassionately** *adv.* [obs. F *compassioné* f. *compassioner* feel pity (as COMPASSION)]

compatible /kəmˈpætəb(ə)l/ *adj.* & *n.* ● *adj.* **1** (often foll. by *with*) **a** able to coexist; well-suited; mutually tolerant (*a compatible couple*). **b** consistent (*their views are not compatible with their actions*). **2** (of equipment, machinery, etc.) capable of being used in combination. ● *n.* (usu. in comb.) *Computing* a piece of equipment that can use software etc. designed for another brand of the same equipment (*IBM compatibles*). □ **compatibly** *adv.* **compatibility** /-ˌpætəˈbɪlɪtɪ/ *n.* [F f. med.L *compatibilis* (as COMPASSION)]

compatriot /kəmˈpætrɪət, -ˈpeɪtrɪət/ *n.* a fellow-countryman. □ **compatriotic** /-ˌpætrɪˈɒtɪk, -ˌpeɪtrɪ-/ *adj.* [F *compatriote* f. LL *compatriota* (as COM-, *patriota* PATRIOT)]

compeer /ˈkɒmpɪə(r), kəmˈpɪə(r)/ *n.* **1** an equal, a peer. **2** a comrade. [ME f. OF *comper* (as COM-, PEER²)]

compel /kəmˈpel/ *v.tr.* (**compelled**, **compelling**) **1** (usu. foll. by *to* + infin.) force, constrain (*compelled them to admit it*). **2** bring about (an action) by force (*compel submission*). **3** (as **compelling** *adj.*) rousing strong interest, attention, conviction, or admiration. **4** *archaic* drive forcibly. □ **compellingly** *adv.* [ME f. L *compellere compuls-* (as COM-, *pellere* drive)]

compellable /kəmˈpeləb(ə)l/ *adj. Law* (of a witness etc.) that may be made to attend court or give evidence.

compendious /kəmˈpendɪəs/ *adj.* (esp. of a book etc.) comprehensive but fairly brief. □ **compendiously** *adv.* **compendiousness** *n.* [ME f. OF *compendieux* f. L *compendiosus* brief (as COMPENDIUM)]

compendium /kəmˈpendɪəm/ *n.* (*pl.* **compendiums** or **compendia** /-dɪə/) **1** esp. *Brit.* a usu. one-volume handbook or encyclopedia. **2 a** a summary or abstract of a larger work. **b** an abridgement. **3 a** (in full **compendium of games**) a collection of games in a box. **b** any collection or mixture. **4** a package of writing paper, envelopes, etc. [L, = what is weighed together, f. *compendere* (as COM-, *pendere* weigh)]

compensate /ˈkɒmpenˌseɪt/ *v.* **1** *tr.* (often foll. by *for*) recompense (a person) (*compensated him for his loss*). **2** *intr.* (usu. foll. by *for* a thing, *to* a person) make amends (*compensated for the insult*; *will compensate to her in full*). **3** *tr.* counterbalance; make up for, make amends for. **4** *tr. Mech.* provide (a pendulum etc.) with extra or less weight etc. to neutralize the effects of temperature etc. **5** *intr. Psychol.* offset a disability or frustration by development in another direction. □ **compensator** *n.* **compensative** /ˈkɒmpenˌseɪtɪv, kəmˈpensətɪv/ *adj.* **compensatory** /ˈkɒmpenˌseɪtərɪ, kəmˈpensətərɪ/ *adj.* [L *compensare* (as COM-, *pensare* frequent. of *pendere pens-* weigh)]

compensation /ˌkɒmpenˈseɪʃ(ə)n/ *n.* **1 a** the act of compensating. **b** the process of being compensated. **2** something, esp. money, given as a recompense. **3** *Psychol.* **a** an act of compensating. **b** the result of compensating. **4** *N. Amer.* a salary or wages. □ **compensation pendulum** *Physics* a pendulum designed to neutralize the effects of temperature variation. □ **compensational** *adj.* [ME f. OF f. L *compensatio* (as COMPENSATE)]

compère /ˈkɒmpeə(r)/ *n.* & *v. Brit.* ● *n.* a person who introduces and links the artistes in a variety show etc.; a master of ceremonies. ● *v.* **1** *tr.* act as a compère to. **2** *intr.* act as compère. [F, = godfather f. Rmc (as COM-, L *pater* father)]

compete /kəmˈpiːt/ *v.intr.* **1** (often foll. by *with*, *against* a person, *for* a

thing) strive for superiority or supremacy (*competed with his brother*; *compete against each other*; *compete for the job*). **2** (often foll. by *in*) take part (in a contest etc.) (*competed in the hurdles*). [L *competere competit-*, in late sense 'strive after or contend for (something)' (as COM-, *petere* seek)]

competence /ˈkɒmpɪt(ə)ns/ *n.* (also **competency** /-sɪ/) **1 a** (often foll. by *for*, or *to* + infin.) ability; the state of being competent. **b** an area in which a person is competent; a skill. **2** an income large enough to live on, usu. unearned. **3** *Law* the legal capacity (of a court, a magistrate, etc.) to deal with a matter.

competent /ˈkɒmpɪt(ə)nt/ *adj.* **1 a** (usu. foll. by *to* + infin. or *for*) properly qualified or skilled (*not competent to drive*); adequately capable, satisfactory. **b** effective (*a competent batsman*). **2** *Law* (of a judge, court, or witness) legally qualified or qualifying. □ **competently** *adv.* [ME f. OF *competent* or L *competent-* (as COMPETE)]

competition /ˌkɒmpəˈtɪʃ(ə)n/ *n.* **1 a** (often foll. by *for*) competing, esp. in an examination, in trade, etc. **b** *Biol.* the interaction between two or more organisms, populations, or species that share a limited environmental resource. **2** an event or contest in which people compete. **3 a** the people competing against a person. **b** the opposition they represent. [LL *competitio* rivalry (as COMPETITIVE)]

competitive /kəmˈpetɪtɪv/ *adj.* **1** involving, offered for, or by competition. **2** (of prices etc.) low enough to compare well with those of rival traders. **3** (of a person) having a strong urge to win; keen to compete. □ **competitively** *adv.* **competitiveness** *n.* [*competit-*, past part. stem of L *competere* COMPETE]

competitor /kəmˈpetɪtə(r)/ *n.* a person who competes; a rival, esp. in business or commerce. [F *compétiteur* or L *competitor* (as COMPETE)]

compilation /ˌkɒmpɪˈleɪʃ(ə)n/ *n.* **1 a** the act of compiling. **b** the process of being compiled. **2** something compiled, esp. a book etc. composed of separate articles, stories, etc. [ME f. OF f. L *compilatio -onis* (as COMPILE)]

compile /kəmˈpaɪl/ *v.tr.* **1 a** collect (material) into a list, volume, etc. **b** make up (a volume etc.) from such material. **2** accumulate (a large number of). **3** *Computing* produce (a machine-coded form of a high-level program). [ME f. OF *compiler* or its apparent source L *compilare* plunder, plagiarize]

compiler /kəmˈpaɪlə(r)/ *n.* **1** *Computing* a program for translating a high-level programming language into machine code. **2** a person who compiles.

complacency /kəmˈpleɪsənsɪ/ *n.* (also **complacence**) tranquil pleasure, self-satisfaction, esp. when uncritical or unwarranted. [med.L *complacentia* f. L *complacere* (as COM-, *placere* please)]

complacent /kəmˈpleɪs(ə)nt/ *adj.* smugly self-satisfied, calmly content. ¶ Sometimes confused with *complaisant*. □ **complacently** *adv.* [L *complacere*: see COMPLACENCY]

complain /kəmˈpleɪn/ *v.intr.* **1** (often foll. by *about*, *at*, or *that* + clause) express dissatisfaction (*complained at the state of the room*; *is always complaining*). **2** (foll. by *of*) **a** announce that one is suffering from (an ailment) (*complained of a headache*). **b** state a grievance concerning (*complained of the delay*). **3** make a mournful sound; groan, creak under a strain. □ **complainer** *n.* **complainingly** *adv.* [ME f. OF *complaindre* (stem *complaign-*) f. med.L *complangere* bewail (as COM-, *plangere planct-* lament)]

complainant /kəmˈpleɪnənt/ *n. Law* a plaintiff in certain lawsuits.

complaint /kəmˈpleɪnt/ *n.* **1** an act of complaining. **2** a grievance. **3** an ailment or illness. **4** *Law* the plaintiff's initial pleading in a civil action. [ME f. OF *complainte* f. *complaint* past part. of *complaindre*: see COMPLAIN]

complaisant /kəmˈpleɪz(ə)nt/ *adj.* **1** politely deferential. **2** willing to please; acquiescent. ¶ Sometimes confused with *complacent*. □ **complaisance** *n.* [F f. *complaire* (stem *complais-*) acquiesce to please, f. L *complacere*: see COMPLACENCY]

compleat *archaic* var. of COMPLETE.

complement *n.* & *v.* ● *n.* /ˈkɒmplɪmənt/ **1 a** something that completes. **b** one of a pair, or one of two things that go together. **2** (often **full complement**) the full number needed to man a ship, fill a conveyance, etc. **3** *Gram.* a word or phrase added to a verb to complete the predicate of a sentence. **4** *Biochem.* a group of proteins in the blood, which by combining with an antigen-antibody complex can produce various effects, esp. the lysis of bacteria. **5** *Math.* any element not belonging to a specified set or class. **6** *Geom.* the amount by which an angle is less than 90° (cf. SUPPLEMENT *n.* 5). ● *v.tr.* /ˈkɒmplɪˌment/ **1** complete. **2** form a complement to (*the scarf complements her dress*).

☐ **complemental** /ˌkɒmplɪˈment(ə)l/ *adj.* [ME f. L *complementum* (as COMPLETE)]

complementarity /ˌkɒmplɪmenˈtærɪtɪ/ *n.* (*pl.* **-ies**) **1** a complementary relationship or situation. **2** *Physics* the concept that a single model may not be adequate to explain atomic systems in different experimental conditions.

complementary /ˌkɒmplɪˈmentərɪ/ *adj.* **1** completing; forming a complement. **2** (of two or more things) complementing each other. ☐ **complementary angle** either of two angles making up 90°. **complementary colour** a colour that combined with a given colour makes white or black. **complementary medicine** alternative medicine. ☐ **complementarily** *adv.* **complementariness** *n.*

complete /kəmˈpliːt/ *adj. & v.* ● *adj.* **1** having all its parts; entire (*the set is complete*). **2** finished (*my task is complete*). **3** of the maximum extent or degree (*a complete surprise; a complete stranger*). **4** (also **compleat** after Sir Izaak Walton's *Compleat Angler*) usu. *joc.* accomplished (*the complete horseman*). ● *v.tr.* **1** finish. **2 a** make whole or perfect. **b** make up the amount of (*completes the quota*). **3** fill in the answers to (a questionnaire etc.). **4** (usu. *absol.*) *Law* conclude a sale of property. ☐ **complete with** having (as an important accessory) (*comes complete with instructions*). ☐ **completely** *adv.* **completeness** *n.* **completion** /-ˈpliːʃ(ə)n/ *n.* [ME f. OF *complet* or L *completus* past part. of *complere* fill up]

completist /kəmˈpliːtɪst/ *n. & adj.* ● *n.* an obsessive or indiscriminate collector. ● *adj.* obsessed with or intent on completeness as an end in itself.

complex /ˈkɒmpleks/ *n. & adj.* ● *n.* **1** a building, a series of rooms, a network, etc. made up of related parts (*the arts complex*). **2** *Psychol.* a related group of usu. repressed feelings or thoughts which cause abnormal behaviour or mental states (*inferiority complex; Oedipus complex*). **3** (in general use) a preoccupation or obsession (*has a complex about punctuality*). **4** *Chem.* a compound in which molecules or ions form coordinate bonds to a metal atom or ion. ● *adj.* **1** consisting of related parts; composite. **2** complicated (*a complex problem*). **3** *Math.* containing real and imaginary parts (cf. IMAGINARY). ☐ **complex sentence** a sentence containing a subordinate clause or clauses. ☐ **complexly** *adv.* **complexity** /kəmˈpleksɪtɪ/ *n.* (*pl.* **-ies**). [F *complexe* or L *complexus* past part. of *complectere* embrace, assoc. with *complexus* plaited]

complexion /kəmˈplekʃ(ə)n/ *n.* **1** the natural colour, texture, and appearance, of the skin, esp. of the face. **2** an aspect; a character (*puts a different complexion on the matter*). ☐ **complexioned** *adj.* (also in *comb.*) [ME f. OF L *complexio -onis* (as COMPLEX): orig. = combination of supposed qualities determining the nature of a body]

compliance /kəmˈplaɪəns/ *n.* **1** the act or an instance of complying; obedience to a request, command, etc. **2** *Mech.* **a** the capacity to yield under an applied force. **b** the degree of such yielding. **3** unworthy acquiescence. ☐ **in compliance with** according to (a wish, command, etc.).

compliant /kəmˈplaɪənt/ *adj.* disposed to comply; yielding, obedient. ☐ **compliantly** *adv.*

complicate /ˈkɒmplɪˌkeɪt/ *v.tr. & intr.* **1** (often foll. by *with*) make or become difficult, confused, intricate, or complex. **2** (as **complicated** *adj.*) complex; intricate. ☐ **complicatedly** *adv.* **complicatedness** *n.* [L *complicare complicat-* (as COM-, *plicare* fold)]

complication /ˌkɒmplɪˈkeɪʃ(ə)n/ *n.* **1 a** an involved or confused condition or state. **b** a complicating circumstance; a difficulty. **2** *Med.* a secondary disease or condition aggravating a previous one. [F *complication* or LL *complicatio* (as COMPLICATE)]

complicity /kəmˈplɪsɪtɪ/ *n.* partnership in a crime or wrongdoing. [*complice* (see ACCOMPLICE) + -ITY]

compliment *n. & v.* ● *n.* /ˈkɒmplɪmənt/ **1 a** a spoken or written expression of praise. **b** an act or circumstance implying praise (*their success was a compliment to their efforts*). **2** (in *pl.*) **a** formal greetings, esp. as a written accompaniment to a gift etc. (*with the compliments of the management*). **b** praise (*my compliments to the cook*). ● *v.tr.* /ˈkɒmplɪˌment/ **1** (often foll. by *on*) congratulate; praise (*complimented him on his roses*). **2** (often foll. by *with*) present as a mark of courtesy (*complimented her with his attention*). ☐ **compliments of the season** greetings appropriate to the time of year, esp. Christmas. **compliments slip** a printed slip of paper sent with a gift etc., esp. from a business firm. **pay a compliment to** praise. **return the compliment 1** give a compliment in return for another. **2** retaliate or recompense in kind. [F *complimenter* f. It. *complimento* ult. f. L (as COMPLEMENT)]

complimentary /ˌkɒmplɪˈmentərɪ/ *adj.* **1** expressing a compliment;

praising. **2** (of a ticket for a play etc.) given free of charge, esp. as a mark of favour. ☐ **complimentarily** *adv.*

compline /ˈkɒmplɪn, -plaɪn/ *n. Eccl.* **1** the last of the canonical hours of prayer. **2** the service taking place during this. [ME f. OF *complie*, fem. past part. of obs. *complir* complete, ult. f. L *complere* fill up]

comply /kəmˈplaɪ/ *v.intr.* (**-ies, -ied**) (often foll. by *with*) act in accordance (with a wish, command, etc.) (*complied with the conditions; had no choice but to comply*). [It. *complire* f. Catalan *complir*, Sp. *cumplir* f. L *complere* fill up]

compo /ˈkɒmpəʊ/ *n. & adj.* ● *n.* (*pl.* **-os**) a composition of plaster etc., e.g. stucco. ● *adj.* = COMPOSITE *adj.* ☐ **compo rations** a large pack of food designed to last for several days. [abbr.]

component /kəmˈpəʊnənt/ *n. & adj.* ● *n.* **1** a part of a larger whole, esp. part of a motor vehicle. **2** *Math.* one of two or more vectors equivalent to a given vector. ● *adj.* being part of a larger whole (*assembled the component parts*). ☐ **componential** /ˌkɒmpəˈnenʃ(ə)l/ *adj.* [L *componere component-* (as COM-, *ponere* put)]

comport /kəmˈpɔːt/ *v.refl.* usu. *literary* conduct oneself; behave. ☐ **comport with** suit, befit. ☐ **comportment** *n.* [L *comportare* (as COM-, *portare* carry)]

compos /ˈkɒmpɒs/ *adj.* = COMPOS MENTIS.

compose /kəmˈpəʊz/ *v.* **1 a** *tr.* construct or create (a work of art, esp. literature or music). **b** *intr.* compose music (*gave up composing in 1917*). **2** *tr.* constitute; make up (*six tribes which composed the German nation*). **3** *tr.* put together to form a whole, esp. artistically; order; arrange (*composed the group for the photographer*). **4** *tr.* (often *refl.*) calm; settle (*compose your expression; composed himself to wait*). **b** (as **composed** *adj.*) calm, settled. **5** *tr.* settle (a dispute etc.). **6** *tr. Printing* **a** set up (type) to form words and blocks of words. **b** set up (a manuscript etc.) in type. ☐ **composed of** made up of, consisting of (*a flock composed of sheep and goats*). ☐ **composedly** /-zɪdlɪ/ *adv.* [F *composer*, f. L *componere* (as COM-, *ponere* put)]

composer /kəmˈpəʊzə(r)/ *n.* a person who composes (esp. music).

composite /ˈkɒmpəzɪt, -ˌzaɪt/ *adj., n., & v.* ● *adj.* **1** made up of various parts; blended. **2** (esp. of a synthetic building material) made up of recognizable constituents. **3** (**Composite**) *Archit.* of the fifth classical order of architecture, consisting of elements of the Ionic and Corinthian orders. (See also ORDER *n.* 11.) **4** *Bot.* of the plant family Compositae. (*See note below.*) ● *n.* **1** a thing made up of several parts or elements. **2** a synthetic building material. **3** *Bot.* a plant of the family Compositae. (*See note below.*) **4** *Polit.* a resolution composed of two or more related resolutions. ● *v.tr. Polit.* amalgamate (two or more similar resolutions). ☐ **compositely** *adv.* **compositeness** *n.* [F f. L *compositus* past part. of *componere* (as COM-, *ponere posit-* put)]

▪ The composites constitute a very large family of flowering plants, having 'flowers' that actually consist of many small florets, each of which is a reduced flower. The florets are either all strap-shaped (as in the dandelion), all disc-shaped (as in the tansy), or both together (as in the daisy). Many composites are highly successful as weeds, while others are cultivated for ornamental use.

composition /ˌkɒmpəˈzɪʃ(ə)n/ *n.* **1 a** the act of putting together; formation or construction. **b** something so composed; a mixture. **c** the constitution of such a mixture; the nature of its ingredients (*the composition is two parts oil to one part vinegar*). **2 a** a literary or musical work. **b** the act or art of producing such a work. **c** an essay, esp. written by a schoolchild or student. **d** an artistic arrangement (of parts of a picture, subjects for a photograph, etc.). **3** mental constitution; character (*jealousy is not in his composition*). **4** (often *attrib.*) a compound artificial substance, esp. one serving the purpose of a natural one. **5** *Printing* the setting-up of type. **6** *Gram.* the formation of words into a compound word. **7** *Law* **a** a compromise, esp. a legal agreement to pay a sum in lieu of a larger sum, or other obligation (*made a composition with his creditors*). **b** a sum paid in this way. **8** *Math.* the combination of functions in a series. ☐ **compositional** *adj.* **compositionally** *adv.* [ME f. OF, f. L *compositio -onis* (as COMPOSITE)]

compositor /kəmˈpɒzɪtə(r)/ *n. Printing* a person who sets up type for printing. [ME f. AF *compositour* f. L *compositor* (as COMPOSITE)]

compos mentis /ˌkɒmpɒs ˈmentɪs/ *adj.* having control of one's mind; sane. [L]

compossible /kəmˈpɒsɪb(ə)l/ *adj. formal* (often foll. by *with*) able to coexist. [OF f. med.L *compossibilis* (as COM-, POSSIBLE)]

compost /ˈkɒmpɒst/ *n. & v.* ● *n.* **1 a** a mixed manure, esp. of organic origin. **b** a loam soil or other medium with added compost, used for growing plants. **2** a mixture of ingredients (*a rich compost of lies and* *innuendo).* ● *v.tr.* **1** treat (soil) with compost. **2** make (manure, vegetable matter, etc.) into compost. □ **compost heap** (or **pile**) a layered structure of garden refuse, soil, etc., which decays to become compost. [ME f. OF *composte* f. L *compos(i)tum* (as COMPOSITE)]

composure /kəm'pəʊʒə(r)/ *n.* a tranquil manner; calmness. [COMPOSE + -URE]

compote /'kɒmpəʊt, -pɒt/ *n.* fruit preserved or cooked in syrup. [F f. OF *composte* (as COMPOSITE)]

compound[1] *n., adj., & v.* ● *n.* /'kɒmpaʊnd/ **1** a mixture of two or more things, qualities, etc. **2** (also **compound word**) a word made up of two or more existing words. **3** *Chem.* a substance formed from two or more elements chemically united in fixed proportions. ● *adj.* /'kɒmpaʊnd/ **1 a** made up of several ingredients. **b** consisting of several parts. **2** combined; collective. **3** *Zool.* consisting of individual organisms. **4** *Biol.* consisting of several or many parts. ● *v.* /kəm'paʊnd/ **1** *tr.* mix or combine (ingredients, ideas, motives, etc.) (*grief compounded with fear*). **2** *tr.* increase or complicate (difficulties etc.) (*anxiety compounded by discomfort*). **3** *tr.* make up or concoct (a composite whole). **4** *tr.* (also *absol.*) settle (a debt, dispute, etc.) by concession or special arrangement. **5** *tr. Law* **a** condone (a liability or offence) in exchange for money etc. **b** forbear from prosecuting (a felony) from private motives. **6** *intr.* (usu. foll. by *with, for*) *Law* come to terms with a person, for forgoing a claim etc. for an offence. **7** *tr.* combine (words or elements) into a word. □ **compound eye** an eye consisting of numerous visual units, as found in insects and crustaceans (cf. *simple eye*). **compound fracture** a fracture with the skin pierced by a bone end and involving a risk of infection. **compound interest** interest payable on capital and its accumulated interest (cf. *simple interest*). **compound interval** *Mus.* an interval exceeding one octave. **compound leaf** a leaf consisting of several or many leaflets. **compound sentence** a sentence with more than one subject or predicate. **compound time** *Mus.* music having more than one group of simple-time units in each bar. □ **compoundable** /kəm'paʊndəb(ə)l/ *adj.* [ME *compoun(e)* f. OF *compondre* f. L *componere* (as COM-, *ponere* put: *-d* as in *expound*)]

compound[2] /'kɒmpaʊnd/ *n.* **1** a large open enclosure for housing workers etc., esp. miners in South Africa. **2** an enclosure, esp. in India, China, etc., in which a factory or a house stands (cf. KAMPONG). **3** a large enclosed space in a prison or prison camp. **4** = POUND[3]. [Port. *campon* or Du. *kampong* f. Malay]

comprador /ˌkɒmprə'dɔ:(r)/ *n.* (also **compradore**) **1** *hist.* a Chinese business agent of a foreign company. **2** an agent of a foreign power. [Port. *comprador* buyer f. LL *comparator* f. L *comparare* purchase]

comprehend /ˌkɒmprɪ'hend/ *v.tr.* **1** grasp mentally; understand (a person or a thing). **2** include; take in. [ME f. OF *comprehender* or L *comprehendere comprehens-* (as COM-, *prehendere* grasp)]

comprehensible /ˌkɒmprɪ'hensɪb(ə)l/ *adj.* **1** that can be understood; intelligible. **2** that can be included or contained. □ **comprehensibly** *adv.* **comprehensibility** /-ˌhensɪ'bɪlɪtɪ/ *n.* [F *compréhensible* or L *comprehensibilis* (as COMPREHEND)]

comprehension /ˌkɒmprɪ'henʃ(ə)n/ *n.* **1 a** the act or capability of understanding, esp. writing or speech. **b** an extract from a text set as an examination, with questions designed to test understanding of it. **2** inclusion. **3** *Eccl. hist.* the inclusion of Nonconformists in the Anglican Church. [F *compréhension* or L *comprehensio* (as COMPREHENSIBLE)]

comprehensive /ˌkɒmprɪ'hensɪv/ *adj. & n.* ● *adj.* **1** complete; including all or nearly all elements, aspects, etc. (*a comprehensive grasp of the subject*). **2** of or relating to understanding (*the comprehensive faculty*). **3** (of motor-vehicle insurance) providing complete protection. ● *n.* (in full **comprehensive school**) *Brit.* a secondary school catering for children of all abilities from a given area. □ **comprehensively** *adv.* **comprehensiveness** *n.* [F *compréhensif -ive* or LL *comprehensivus* (as COMPREHENSIBLE)]

compress *v. & n.* ● *v.tr.* /kəm'pres/ **1** squeeze together. **2** bring into a smaller space or shorter extent. ● *n.* /'kɒmpres/ a pad of lint etc. pressed on to part of the body to relieve inflammation, stop bleeding, etc. □ **compressed air** air at more than atmospheric pressure. □ **compressible** /kəm'presɪb(ə)l/ *adj.* **compressibility** /-ˌpresɪ'bɪlɪtɪ/ *n.* **compressive** /-'presɪv/ *adj.* [ME f. OF *compresser* or LL *compressare* frequent. of L *comprimere* compress- (as COM-, *premere* press)]

compression /kəm'preʃ(ə)n/ *n.* **1** the act of compressing or being compressed. **2** the reduction in volume (causing an increase in pressure) of the fuel mixture in an internal-combustion engine before ignition. **3** *Electronics* the process of reducing the amount of space taken up by data that is being stored or transmitted. □ **compression ratio** the ratio of the maximum to minimum volume in the cylinder of an internal-combustion engine. [F f. L *compressio* (as COMPRESS)]

compressor /kəm'presə(r)/ *n.* an instrument or device for compressing, esp. a machine used for increasing the pressure of air or other gases.

comprise /kəm'praɪz/ *v.tr. & intr.* **1** *tr.* consist of, be made up of; contain (*the book comprises 350 pages*). **2** *tr.* make up, compose (*the essays comprise his total work*). **3** *tr.* (in passive, foll by *of*) consist of (*the army was comprised of volunteers*). **4** *intr.* (foll. by *of*) *disp.* consist of (*the region comprises of three separate districts*). □ **comprisable** *adv.* ¶ The use of *comprise* in senses 2 and 3 is still regarded as non-standard by some people, while the intransitive use in sense 4, a construction formed by analogy with *consist of*, is especially frowned upon; *consist of, be composed of*, or, simply, *comprise* (without *of*, as in sense 1) provide more generally acceptable alternatives. [ME f. F, fem. past part. of *comprendre* COMPREHEND]

compromise /'kɒmprə.maɪz/ *n. & v.* ● *n.* **1** the settlement of a dispute by mutual concession (*reached a compromise by bargaining*). **2** (often foll. by *between*) an intermediate state between conflicting opinions, actions, etc., reached by mutual concession or modification (*a compromise between ideals and material necessity*). ● *v.* **1 a** *intr.* settle a dispute by mutual concession (*compromised over the terms*). **b** *tr. archaic* settle (a dispute) by mutual concession. **2** *tr.* bring into disrepute or danger esp. by indiscretion or folly. □ **compromiser** *n.* **compromisingly** *adv.* [ME f. OF *compromis* f. LL *compromissum* neut. past part. of *compromittere* (as COM-, *promittere* PROMISE)]

compte rendu /ˌkɒmt rɒn'dju:/ *n.* (*pl.* **comptes rendus** pronunc. same) a report; a review; a statement. [F]

Comptometer /kɒmp'tɒmɪtə(r)/ *n. propr.* an early type of calculating-machine. [app. f. F *compte* COUNT[1] + -METER]

Compton /'kɒmptən/, Arthur Holly (1892–1962), American physicist. He observed that the wavelength of X-rays increased when scattered by electrons (later known as the Compton effect). This demonstrated the dual particle and wave properties of electromagnetic radiation and matter, predicted by quantum theory. Compton shared the 1927 Nobel Prize for physics, and during the war he developed plutonium production for the Manhattan Project.

Compton-Burnett /ˌkɒmptənbɜː'net, -'bɜːnɪt/, Dame Ivy (1884–1969), English novelist. Her works include *Brothers and Sisters* (1929), *A Family and a Fortune* (1939), and *Manservant and Maidservant* (1947). Her novels typically portray life at the turn of the (nineteenth) century and are characterized by ironic wit and an emphasis on dialogue.

comptroller /kən'trəʊlə(r)/ *n.* a controller (used in the title of some financial officers) (*Comptroller and Auditor General*). [var. of CONTROLLER, by erron. assoc. with COUNT[1], L *computus*]

compulsion /kəm'pʌlʃ(ə)n/ *n.* **1** a constraint; an obligation. **2** *Psychol.* an irresistible urge to a form of behaviour, esp. against one's conscious wishes. □ **under compulsion** because one is compelled. [ME f. F f. LL *compulsio -onis* (as COMPEL)]

compulsive /kəm'pʌlsɪv/ *adj.* **1** compelling. **2** resulting or acting from, or as if from, compulsion (*a compulsive gambler*). **3** *Psychol.* resulting or acting from compulsion against one's conscious wishes. **4** holding one's attention (*the film made compulsive viewing*). □ **compulsively** *adv.* **compulsiveness** *n.* [med.L *compulsivus* (as COMPEL)]

compulsory /kəm'pʌlsərɪ/ *adj.* **1** required by law or a rule (*it is compulsory to keep dogs on leads*). **2** essential; necessary. □ **compulsory purchase** the enforced purchase of land or property by a local authority etc., for public use. □ **compulsorily** *adv.* **compulsoriness** *n.* [med.L *compulsorius* (as COMPEL)]

compunction /kəm'pʌŋkʃ(ə)n/ *n.* (usu. with *neg.*) **1** the pricking of the conscience; remorse. **2** slight regret; a scruple (*without compunction; have no compunction in refusing him*). □ **compunctious** *adj.* **compunctiously** *adv.* [ME f. OF *componction* f. eccl.L *compunctio -onis* f. L *compungere compunct-* (as COM-, *pungere* prick)]

compurgation /ˌkɒmpɜː'geɪʃ(ə)n/ *n. Law hist.* an acquittal from a charge or accusation obtained by the oaths of witnesses. □ **compurgatory** /kəm'pɜːgətərɪ/ *adj.* [med.L *compurgatio* f. L *compurgare* (as COM-, *purgare* purify)]

compurgator /'kɒmpɜːˌgeɪtə(r)/ *n. Law hist.* a witness who swore to the innocence or good character of an accused person.

computation /ˌkɒmpjʊ'teɪʃ(ə)n/ *n.* **1** the act or an instance of reckoning; calculation. **2** the use of a computer. **3** a result obtained by calculation. □ **computational** *adj.*

compute /kəm'pju:t/ *v.* **1** *tr.* (often foll. by *that* + clause) reckon or calculate (a number, an amount, etc.). **2** *intr.* make a reckoning, esp.

using a computer. □ **computable** *adj.* **computability** /-ˌpjuːtəˈbɪlɪtɪ/ *n.* [F *computer* or L *computare* (as COM-, *putare* reckon)]

computer /kəmˈpjuːtə(r)/ *n.* **1** an electronic device for storing and processing data according to instructions given to it in a variable program. (*See note below.*) **2** a person who computes or makes calculations. □ **computer-literate** able to use computers; familiar with the operation of computers. **computer science** the study of the principles and use of computers. **computer virus** a hidden code within a computer program intended to corrupt a system or destroy data stored in it.

▪ Calculating machines and other mechanical computing devices date from the 17th century and important early work was carried out, for example, by Charles Babbage and Ada Lovelace, but the electronic digital computer, one of the most revolutionary inventions of the 20th century, differs fundamentally from these earlier devices in its ability to carry out different tasks according to its programming, without any physical change. Computers are also much faster in operation and can handle data of greater quantity and complexity. Computers depend on the property of electronic switches to store information in binary form, the digits 0 and 1 being represented by the 'off' and 'on' positions. The first digital computers were developed in the US in the 1940s. In the rapid development of computers since then, several generations have been identified, distinguished by progressively reduced size and increased speed and capacity: the first generation, broadly in the period 1945–55, used vacuum tubes; in second-generation machines, roughly 1955–60, solid-state devices (transistors) replaced valves; the third generation, 1960–70, saw the introduction of integrated circuits and operating systems; the fourth generation, since 1970, employs large-scale integrated-circuit technology and very large rapid-access memory; fifth-generation computers are expected to begin to use artificial intelligence.

computerate /kəmˈpjuːtərət/ *adj.* = computer-literate. [COMPUTER + -ATE², after LITERATE]

computerize /kəmˈpjuːtəˌraɪz/ *v.tr.* (also **-ise**) **1** equip with a computer; install a computer in. **2** store, perform, or produce by computer. □ **computerization** /-ˌpjuːtəraɪˈzeɪʃ(ə)n/ *n.*

comrade /ˈkɒmreɪd, -rɪd/ *n.* **1** (also **comrade-in-arms**) **a** (usu. of males) a workmate, friend, or companion. **b** fellow soldier etc. **2** *Polit.* a fellow socialist or communist (often as a form of address). □ **comradely** *adj.* **comradeship** *n.* [earlier *cama- camerade* f. F *camerade, camarade* (orig. fem.) f. Sp. *camarada* room-mate (as CHAMBER)]

Comsat /ˈkɒmsæt/ *n. propr.* a communications satellite. [abbr.]

Comte /kɒmt/, Auguste (1798–1857), French philosopher, one of the founders of sociology. In his historical study of the progress of the human mind, he discerned three phases: the theological, the metaphysical, and the positive. He argued that only the last phase survives in mature sciences. Comte's positivist philosophy attempted to define the laws of social evolution and to found a genuine social science that could be used for social reconstruction. Major works include his *Cours de philosophie positive* (1830–42) and *Système de politique positive* (1851–4). (See also SOCIOLOGY, POSITIVISM.)

con¹ /kɒn/ *n. & v. sl.* ● *n.* a confidence trick. ● *v.tr.* (**conned, conning**) swindle; deceive (*conned him into thinking he had won*). □ **con man** = confidence man. [abbr.]

con² /kɒn/ *n., prep., & adv.* ● *n.* (usu. in *pl.*) a reason against. ● *prep. & adv.* against (cf. PRO²). [L *contra* against]

con³ /kɒn/ *n. sl.* a convict. [abbr.]

con⁴ /kɒn/ *v.tr.* (US **conn**) (**conned, conning**) *Naut.* direct the steering of (a ship). [app. weakened form of obs. *cond, condie,* f. F *conduire* f. L *conducere* CONDUCT]

con⁵ /kɒn/ *v.tr.* (**conned, conning**) *archaic* (often foll. by *over*) study, learn by heart (*conned his part well*). [ME *cunn-, con,* forms of CAN¹]

con- /kɒn, kən/ *prefix* assim. form of COM- before *c, d, f, g, j, n, q, s, t, v,* and sometimes before vowels.

conacre /ˈkɒnˌeɪkə(r)/ *n. Ir.* the letting by a tenant of small portions of land prepared for crops or grazing. [CORN¹ + ACRE]

Conakry /ˈkɒnəˌkriː/ the capital and chief port of Guinea; pop. (1983) 705,300.

con amore /ˌkɒn æˈmɔːrɪ/ *adv.* **1** with devotion or zeal. **2** *Mus.* tenderly. [It., = with love]

Conan Doyle /ˈkəʊnən/ see DOYLE.

conation /kəˈneɪʃ(ə)n/ *n. Philos. & Psychol.* **1** the desire to perform an action. **2** voluntary action; volition. □ **conative** /ˈkɒnətɪv, ˈkəʊn-/ *adj.* [L *conatio* f. *conari* try]

con brio /kɒn ˈbriːəʊ/ *adv. Mus.* with vigour. [It.]

concatenate /kənˈkætɪˌneɪt/ *v. & adj.* ● *v.tr.* link together (a chain of events, things, etc.). ● *adj.* joined; linked. □ **concatenation** /-ˌkætɪˈneɪʃ(ə)n/ *n.* [LL *concatenare* (as COM-, *catenare* f. *catena* chain)]

concave /ˈkɒŋkeɪv/ *adj.* having an outline or surface curved like the interior of a circle or sphere (cf. CONVEX). □ **concavely** *adv.* **concavity** /kɒŋˈkævɪtɪ/ *n.* [L *concavus* (as COM-, *cavus* hollow), or through F *concave*]

conceal /kənˈsiːl/ *v.tr.* **1** (often foll. by *from*) keep secret (*concealed her motive from him*). **2** not allow to be seen; hide (*concealed the letter in her pocket*). □ **concealer** *n.* **concealment** *n.* [ME f. OF *conceler* f. L *concelare* (as COM-, *celare* hide)]

concede /kənˈsiːd/ *v.tr.* **1 a** (often foll. by *that* + clause) admit (a defeat etc.) to be true (*conceded that his work was inadequate*). **b** admit defeat in. **2** (often foll. by *to*) grant, yield, or surrender (a right, a privilege, points or a start in a game, etc.). **3** *Sport* allow an opponent to score (a goal) or to win (a match), etc. □ **conceder** *n.* [F *concéder* or L *concedere concess-* (as COM-, *cedere* yield)]

conceit /kənˈsiːt/ *n.* **1** personal vanity; pride. **2** *literary* **a** a far-fetched comparison, esp. as a stylistic affectation; a convoluted or unlikely metaphor. **b** a fanciful notion. [ME f. CONCEIVE after *deceit, deceive,* etc.]

conceited /kənˈsiːtɪd/ *adj.* vain, proud. □ **conceitedly** *adv.* **conceitedness** *n.*

conceivable /kənˈsiːvəb(ə)l/ *adj.* capable of being grasped or imagined; understandable. □ **conceivably** *adv.* **conceivability** /-ˌsiːvəˈbɪlɪtɪ/ *n.*

conceive /kənˈsiːv/ *v.* **1** *intr.* become pregnant. **2** *tr.* become pregnant with (a child). **3** *tr.* (often foll. by *that* + clause) **a** imagine, fancy, think (*can't conceive that he could be guilty*). **b** (usu. in *passive*) formulate, express (a belief, a plan, etc.). □ **conceive of** form in the mind; imagine. [ME f. OF *conceiv-* stressed stem of *concevoir* f. L *concipere concept-* (as COM-, *capere* take)]

concelebrate /kənˈselɪˌbreɪt/ *v.intr. RC Ch.* **1** (of priests) celebrate the mass together. **2** (esp. of a newly ordained priest) celebrate the mass with the ordaining bishop. □ **concelebrant** /-brənt/ *n.* **concelebration** /-ˌselɪˈbreɪʃ(ə)n/ *n.* [L *concelebrare* (as COM-, *celebrare* CELEBRATE)]

concentrate /ˈkɒnsənˌtreɪt/ *v. & n.* ● *v.* **1** *intr.* (often foll. by *on, upon*) focus all one's attention or mental ability. **2** *tr.* bring together (troops, power, attention, etc.) to one point; focus. **3** *tr.* increase the strength of (a dissolved substance, liquid etc.) by removing water or other diluting agent. **4** *tr.* (as **concentrated** *adj.*) intense, strong. ● *n.* **1** a concentrated substance. **2** a concentrated form of esp. food. □ **concentratedly** *adv.* **concentrative** *adj.* **concentrator** *n.* [after *concentre* f. F *concentrer* (as CON- + CENTRE)]

concentration /ˌkɒnsənˈtreɪʃ(ə)n/ *n.* **1** the act or power of focusing all one's attention or mental ability; fixed attention (*broke my concentration; needs to develop concentration*). **2** the act of concentrating or bringing together. **3** something concentrated (*a concentration of resources*). **4** *Chem.* **a** the strengthening of a solution by removal of solvent. **b** the strength of a solution, esp. the amount of solute per unit volume of a solution.

concentration camp *n.* **1** a camp for the detention or extermination of political prisoners, internees, etc., esp. in Nazi Germany. Several hundred camps were established by the Nazis in Germany and occupied Europe 1933–45, perhaps the most infamous being Dachau, Buchenwald, and Belsen in Germany and Auschwitz and Treblinka in Poland. Over 6 million people died or were murdered in the camps, mainly Jews but also religious dissidents, homosexuals, gypsies, and the disabled. A system of concentration camps was also instituted in the USSR (see GULAG). **2** any of the camps (instituted by Lord Kitchener) where non-combatants of a district accommodated during the Second Boer War of 1899–1902. A similar type of camp had been introduced in Cuba in 1896–7 during the struggle for independence from Spanish rule.

concentre /kənˈsentə(r)/ *v.tr. & intr.* (US **concenter**) bring or come to a common centre. [F *concentrer*: see CONCENTRATE]

concentric /kənˈsentrɪk/ *adj.* (often foll. by *with*) (esp. of circles) having a common centre (opp. ECCENTRIC *adj.* 2b). □ **concentrically** *adv.* **concentricity** /ˌkɒnsenˈtrɪsɪtɪ/ *n.* [ME f. OF *concentrique* or med.L *concentricus* (as COM-, *centricus* as CENTRE)]

Concepción /kɒnˌsepsɪˈɒn/ an industrial city in south central Chile; pop. (est. 1987) 294,000.

concept /ˈkɒnsept/ *n.* **1** a general notion; an abstract idea (*the concept* *of evolution).* **2** *colloq.* an idea or invention to help sell or publicize a commodity (*a new concept in swimwear*). **3** *Philos.* an idea or mental picture of a group or class of objects formed by combining all their aspects. [LL *conceptus* f. *concept-*: see CONCEIVE]

conception /kənˈsepʃ(ə)n/ *n.* **1 a** the act or an instance of conceiving; the process of being conceived. **b** the faculty of conceiving in the mind; apprehension, imagination. **2** an idea or plan, esp. as being new or daring (*the whole conception showed originality*). **3** (usu. foll. by *of*) understanding, ability to imagine (*has no conception of what it entails*). □ **conceptional** *adj.* [ME f. OF f. L *conceptio -onis* (as CONCEPT)]

conceptive /kənˈseptɪv/ *adj.* **1** conceiving mentally. **2** of conception. [L *conceptivus* (as CONCEPTION)]

conceptual /kənˈseptʃʊəl, -tjʊəl/ *adj.* of mental conceptions or concepts. □ **conceptually** *adv.* [med.L *conceptualis* (*conceptus* as CONCEPT)]

conceptual art *n.* art in which the idea or concept is considered more important than the finished product, if such exists. The notion goes back to Marcel Duchamp but became a recognized phenomenon in the 1960s. Typical manifestations include maps, diagrams, photography, and video.

conceptualism /kənˈseptʃʊəˌlɪz(ə)m, kənˈseptjʊ-/ *n. Philos.* the theory that universals exist, but only as concepts in the mind. □ **conceptualist** *n.*

conceptualize /kənˈseptʃʊəˌlaɪz, kənˈseptjʊ-/ *v.tr.* (also **-ise**) form a concept or idea of. □ **conceptualization** /-ˌseptʃʊəlaɪˈzeɪʃ(ə)n, -ˌseptjʊ-/ *n.*

conceptus /kənˈseptəs/ *n. Med.* the product of conception in the womb, esp. the embryo in the early stages of pregnancy. [L, = conception, embryo]

concern /kənˈsɜːn/ *v. & n.* ● *v.tr.* **1 a** be relevant or important to (*this concerns you*). **b** relate to; be about. **2** (usu. *refl.*; often foll. by *with, in, about,* or *to* + infin.) interest or involve oneself (*don't concern yourself with my problems*). **3** worry, cause anxiety to (*it concerns me that he is always late*). ● *n.* **1 a** anxiety, worry (*felt a deep concern*). **b** solicitous regard; care, consideration. **2 a** a matter of interest or importance to one (*no concern of mine*). **b** (usu. in *pl.*) affairs; private business (*chatted about departmental concerns*). **3** a business, a firm (*quite a prosperous concern*). **4** *colloq.* a complicated or awkward thing (*have lost the whole concern*). □ **have a concern in** have an interest or share in. **have no concern with** have nothing to do with. **to whom it may concern** to those who have a proper interest in the matter (as an address to the reader of a testimonial, reference, etc.). [F *concerner* or LL *concernere* (as COM-, *cernere* sift, discern)]

concerned /kənˈsɜːnd/ *adj.* **1** involved, interested (*the people concerned; concerned with proving his innocence*). **2** (often foll. by *that, about, at, for,* or *to* + infin.) troubled, anxious (*concerned about him; concerned to hear that*). □ **as** (or **so**) **far as I am concerned** as regards my interests. **be concerned** (often foll. by *in*) take part. **I am not concerned** it is not my business. □ **concernedly** /-nɪdlɪ/ *adv.* **concernedness** /-nɪdnɪs/ *n.*

concerning /kənˈsɜːnɪŋ/ *prep.* about, regarding.

concernment /kənˈsɜːnmənt/ *n. formal* **1** an affair or business. **2** importance. **3** (often foll. by *with*) a state of being concerned; anxiety.

concert *n. & v.* ● *n.* /ˈkɒnsət/ **1** a musical performance of usu. several separate compositions. **2** agreement, accordance, harmony. **3** a combination of voices or sounds. ● *v.tr.* /kənˈsɜːt/ arrange (by mutual agreement or coordination). □ **concert-goer** a person who often goes to concerts. **concert grand** the largest size of grand piano, used for concerts. **concert-master** esp. *US* the leading first-violin player in some orchestras. **concert overture** *Mus.* a piece like an overture but intended for independent performance. **concert performance** *Mus.* a performance (of an opera etc.) without scenery, costumes, or action. **in concert 1** (often foll. by *with*) acting jointly and accordantly. **2** (*predic.*) (of a musician) in a performance. [F *concert* (n.), *concerter* (v.) f. It. *concertare* harmonize]

concerted /kənˈsɜːtɪd/ *adj.* **1** combined together; jointly arranged or planned (*a concerted effort*). **2** *Mus.* arranged in parts for voices or instruments.

concertina /ˌkɒnsəˈtiːnə/ *n. & v.* ● *n.* a musical instrument, usu. polygonal in form, held in the hands and stretched and squeezed like bellows, having reeds and a set of buttons at each end to control the valves. ● *v.tr. & intr.* (**concertinas, concertinaed** /-nəd/ or **concertina'd, concertinaing** /-nəɪŋ/) compress or collapse in folds

like those of a concertina (*the car concertinaed into the bridge*). [CONCERT + -INA]

concertino /ˌkɒntʃəˈtiːnəʊ/ *n.* (*pl.* **-os**) *Mus.* **1** a simple or short concerto. **2** a solo instrument or solo instruments playing in a concerto. [It., dimin. of CONCERTO]

concerto /kənˈtʃeətəʊ, -ˈtʃɜːtəʊ/ *n.* (*pl.* **-os** or **concerti** /-tɪ/) *Mus.* a composition for a solo instrument or instruments accompanied by an orchestra. (*See note below.*) □ **concerto grosso** /ˈɡrɒsəʊ, ˈɡrəʊs-/ (*pl.* **concerti grossi** /-sɪ/ or **concerto grossos**) a composition for a group of solo instruments accompanied by an orchestra. [It. (see CONCERT): *grosso* big]

▪ The concerto for an individual player was developed by J. S. Bach in his harpsichord concertos from the concerti grossi of Corelli and his contemporaries. Handel's organ concertos were an important development, being among the first to provide a cadenza in which the soloist could display skill by extemporization. By this time the concerto had usually a three-movement form, and the classical composers continued this structure. Mozart composed nearly fifty concertos for various instrumental combinations, and established the modern style. Since the 19th century it has been usual for the composer to write out a cadenza.

concert pitch *n.* **1** *Mus.* the standard pitch, internationally agreed in 1960, whereby the A above middle C = 440 Hz, the previous standard being slightly lower at between 435 (European) and 439 (British). This note is generally given out by a tuning-fork or, in an orchestra, the oboe, one of the instruments least affected by temperature change. **2** a state of unusual readiness, efficiency, and keenness (for action etc.).

concession /kənˈseʃ(ə)n/ *n.* **1 a** the act or an instance of conceding something asked or required (*made the concession that we were right*). **b** a thing conceded. **2** a reduction in price for a certain category of person. **3 a** the right to use land or other property, granted esp. by a government or local authority, esp. for a specific use. **b** the right, given by a company, to sell goods, esp. in a particular territory. **c** the land or property used or given. □ **concessionary** *adj.* **concessional** *adj.* [F *concession* f. L *concessio* (as CONCEDE)]

concessionaire /kənˌseʃəˈneə(r)/ *n.* (also **concessionnaire**) the holder of a concession or grant, esp. for the use of land or trading rights. [F *concessionnaire* (as CONCESSION)]

concessive /kənˈsesɪv/ *adj.* **1** of or tending to concession. **2** *Gram.* **a** (of a preposition or conjunction) introducing a phrase or clause which might be expected to preclude the action of the main clause, but does not (e.g. *in spite of, although*). **b** (of a phrase or clause) introduced by a concessive preposition or conjunction. [LL *concessivus* (as CONCEDE)]

conch /kɒŋk, kɒntʃ/ *n.* (*pl.* **conchs** /kɒŋks/ or **conches** /ˈkɒntʃɪz/) **1 a** a thick heavy spiral shell, occasionally bearing long projections, of various marine gastropod molluscs of the family Strombidae. **b** one of these gastropods. **2** *Archit.* the domed roof of a semicircular apse. **3** = CONCHA. [L *concha* shell f. Gk *kogkhē*]

concha /ˈkɒŋkə/ *n.* (*pl.* **conchae** /-kiː/) *Anat.* any part resembling a shell, esp. the depression in the external ear leading to its central cavity. [L: see CONCH]

conchie /ˈkɒntʃɪ/ *n.* (also **conchy**) (*pl.* **-ies**) *sl. derog.* a conscientious objector. [abbr.]

conchoidal /kɒŋˈkɔɪd(ə)l/ *adj. Mineral.* (of a solid fracture etc.) resembling the surface of a bivalve shell.

conchology /kɒŋˈkɒlədʒɪ/ *n. Zool.* the scientific study of molluscs and their shells. □ **conchologist** *n.* **conchological** /ˌkɒŋkəˈlɒdʒɪk(ə)l/ *adj.* [Gk *kogkhē* shell + -LOGY]

conchy var. of CONCHIE.

concierge /ˌkɒnsɪˈeəʒ/ *n.* (esp. in France) a door-keeper or porter of a block of flats etc. [F, prob. ult. f. L *conservus* fellow slave]

conciliar /kənˈsɪlɪə(r)/ *adj.* of or concerning a council, esp. an ecclesiastical council. [med.L *consiliarius* counsellor]

conciliate /kənˈsɪlɪˌeɪt/ *v.tr.* **1** make calm and amenable; pacify. **2** gain (esteem or goodwill). **3** *archaic* reconcile, make compatible. □ **conciliator** *n.* **conciliative** /-ˈsɪlɪətɪv/ *adj.* **conciliatory** /-ˈsɪlɪətərɪ/ *adj.* **conciliatoriness** *n.* [L *conciliare* combine, gain (*concilium* COUNCIL)]

conciliation /kənˌsɪlɪˈeɪʃ(ə)n/ *n.* the use of conciliating measures; reconcilement. [L *conciliatio* (as CONCILIATE)]

concinnity /kənˈsɪnɪtɪ/ *n.* elegance or neatness of literary style. [L *concinnitas* f. *concinnus* well-adjusted]

concise /kənˈsaɪs/ *adj.* (of speech, writing, style, or a person) brief but comprehensive in expression. □ **concisely** *adv.* **conciseness** *n.* [F *concis* or L *concisus* past part. of *concidere* (as COM-, *caedere* cut)]

concision /kənˈsɪʒ(ə)n/ n. (esp. of literary style) the quality or state of being concise. [ME f. L concisio (as CONCISE)]

conclave /ˈkɒŋkleɪv/ n. **1** a private meeting. **2** RC Ch. **a** the assembly of cardinals for the election of a pope. **b** the meeting-place for a conclave. [ME f. OF f. L conclave lockable room (as COM-, clavis key)]

conclude /kənˈkluːd/ v. **1** tr. & intr. bring or come to an end. **2** tr. (often foll. by from, or that + clause) infer (from given premisses) (what did you conclude?; concluded from the evidence that he had been mistaken). **3** tr. settle, arrange (a treaty etc.). **4** intr. (usu. foll. by to + infin.) esp. US decide. [ME f. L concludere (as COM-, claudere shut)]

conclusion /kənˈkluːʒ(ə)n/ n. **1** a final result; a termination. **2** a judgement reached by reasoning. **3** the summing-up of an argument, article, book, etc. **4** a settling; an arrangement (the conclusion of peace). **5** Logic a proposition that is reached from given premisses; the third and last part of a syllogism. □ **in conclusion** lastly, to conclude. **try conclusions with** engage in a trial of skill etc. with. [ME f. OF conclusion or L conclusio (as CONCLUDE)]

conclusive /kənˈkluːsɪv/ adj. decisive, convincing. □ **conclusively** adv. **conclusiveness** n. [LL conclusivus (as CONCLUSION)]

concoct /kənˈkɒkt/ v.tr. **1** make by mixing ingredients (concocted a stew). **2** invent (a story, a lie, etc.). □ **concocter** n. **concoctor** n. **concoction** /-ˈkɒkʃ(ə)n/. [L concoquere concoct- (as COM-, coquere cook)]

concomitance /kənˈkɒmɪt(ə)ns/ n. (also **concomitancy** /-sɪ/) **1** coexistence. **2** (in Christian theology) the doctrine of the coexistence of the body and blood of Christ both in the bread and in the wine of the Eucharist. [med.L concomitantia (as CONCOMITANT)]

concomitant /kənˈkɒmɪt(ə)nt/ adj. & n. ● adj. going together; associated (concomitant circumstances). ● n. an accompanying thing. □ **concomitantly** adv. [LL concomitari (as COM-, comitari f. L comes -mitis companion)]

Concord /ˈkɒŋkɔːd/ **1** the state capital of New Hampshire; pop. (1990) 36,000. **2** a town in NE Massachusetts; pop. (1990) 17,080. Battles there and at Lexington in Apr. 1775 marked the start of the War of American Independence.

concord /ˈkɒŋkɔːd/ n. **1** agreement or harmony between people or things. **2** a treaty. **3** Mus. a chord that is pleasing or satisfactory in itself. **4** Gram. agreement between words in gender, number, etc. [ME f. OF concorde f. L concordia f. concors of one mind (as COM-, cors f. cor cordis heart)]

concordance /kənˈkɔːd(ə)ns/ n. **1** agreement. **2** a book containing an alphabetical list of the important words used in a book or by an author, usu. with citations of the passages concerned. [ME f. OF f. med.L concordantia (as CONCORDANT)]

concordant /kənˈkɔːd(ə)nt/ adj. **1** (often foll. by with) agreeing, harmonious. **2** Mus. in harmony. □ **concordantly** adv. [ME f. OF f. L concordare f. concors (as CONCORD)]

concordat /kənˈkɔːdæt/ n. an agreement, esp. between the Roman Catholic Church and a state. [F concordat or L concordatum neut. past part. of concordare (as CONCORDANCE)]

Concorde /ˈkɒŋkɔːd/ a supersonic airliner, the only one to have entered operational service, able to cruise at twice the speed of sound. Concorde was produced through Anglo-French cooperation and made its maiden flight in 1969. It has been in commercial service since 1976, with British Airways and Air France.

concourse /ˈkɒŋkɔːs/ n. **1** a crowd. **2** a coming together; a gathering (a concourse of ideas). **3** an open central area in a large public building, a railway station, etc. [ME f. OF concours f. L concursus (as CONCUR)]

concrescence /kənˈkres(ə)ns/ n. Biol. coalescence; growing together. □ **concrescent** adj. [CON-, after excrescence etc.]

concrete /ˈkɒŋkriːt/ adj., n., & v. ● adj. **1 a** existing in a material form; real. **b** specific, definite (concrete evidence; a concrete proposal). **2** Gram. (of a noun) denoting a material object as opposed to an abstract quality, state, or action. ● n. (often attrib.) a composition of gravel, sand, cement, and water, used for building. (See note below.) ● v. **1** tr. **a** cover with concrete. **b** embed in concrete. **2 a** tr. & intr. form into a mass; solidify. **b** tr. make concrete instead of abstract. □ **concrete jungle** a densely built-up urban area with impersonal modern buildings and few trees, grassed areas, etc. **concrete-mixer** a machine, usu. with a revolving drum, used for mixing concrete. **concrete music** see MUSIQUE CONCRÈTE. **concrete poetry** poetry in which part of the meaning is conveyed visually, by means of patterns of words or letters and other typographical devices. **in the concrete** in reality or in practice. □ **concretely** adv. **concreteness** n. [F concret or L concretus past part. of concrescere (as COM-, crescere cret- GROW)]

■ Concrete was used by the Romans as a building material, but its modern use dates from the 19th century with the development of Portland cement. Reinforced concrete, strengthened by embedded metal rods, bars, or mesh, was developed in France and England in the mid-19th century, and prestressed concrete was invented at the end of the century. In prestressed concrete the reinforcing metal bars are stretched while the concrete is wet and released when it has set round them, thereby compressing the concrete longitudinally and reducing its tendency to bend under a load; it has been used with notable success in the building of bridges.

concretion /kənˈkriːʃ(ə)n/ n. **1 a** a hard solid concreted mass. **b** the forming of this by coalescence. **2** Med. a stony mass formed within the body. **3** Geol. a small round mass of rock particles embedded in limestone or clay. □ **concretionary** adj. [F f. L concretio (as CONCRETE)]

concretize /ˈkɒŋkrɪˌtaɪz/ v.tr. (also **-ise**) make concrete instead of abstract. □ **concretization** /ˌkɒŋkrɪtaɪˈzeɪʃ(ə)n/ n.

concubinage /kənˈkjuːbɪnɪdʒ/ n. **1** the cohabitation of a man and woman not married to each other. **2** the state of being or having a concubine. [ME f. F (as CONCUBINE)]

concubine /ˈkɒŋkjʊˌbaɪn/ n. **1** a woman who lives with a man without being his wife; a kept mistress. **2** (in polygamous societies) a secondary wife. □ **concubinary** /kənˈkjuːbɪnərɪ/ adj. [ME f. OF f. L concubina (as COM-, cubina f. cubare lie)]

concupiscence /kənˈkjuːpɪs(ə)ns/ n. formal sexual desire. □ **concupiscent** adj. [ME f. OF f. LL concupiscentia f. L concupiscere begin to desire (as COM-, inceptive f. cupere desire)]

concur /kənˈkɜː(r)/ v.intr. (**concurred, concurring**) **1** happen together; coincide. **2** (often foll. by with) **a** agree in opinion. **b** express agreement. **3** combine together for a cause; act in combination. [L concurrere (as COM-, currere run)]

concurrent /kənˈkʌrənt/ adj. **1** (often foll. by with) **a** existing or in operation at the same time (served two concurrent sentences). **b** existing or acting together. **2** Geom. (of three or more lines) meeting at or tending towards one point. **3** agreeing, harmonious. □ **concurrence** n. **concurrently** adv.

concuss /kənˈkʌs/ v.tr. **1** subject to concussion. **2** shake violently. **3** archaic intimidate. □ **concussive** adj. [L concutere concuss- (as COM-, cutere = quatere shake)]

concussion /kənˈkʌʃ(ə)n/ n. **1** Med. temporary unconsciousness or incapacity due to a blow on the head. **2** violent shaking; shock. [L concussio (as CONCUSS)]

condemn /kənˈdem/ v.tr. **1** express utter disapproval of; censure (was condemned for his irresponsible behaviour). **2 a** find guilty; convict. **b** (usu. foll. by to) sentence to (a punishment, esp. death). **c** bring about the conviction of (his looks condemn him). **3** pronounce (a building etc.) unfit for use or habitation. **4** (usu. foll. by to) doom or assign (to something unwelcome or painful) (condemned to spending hours at the kitchen sink). **5** declare (smuggled goods, property, etc.) to be forfeited. □ **condemned cell** a cell for a prisoner who has been condemned to death. □ **condemnable** /-ˈdemnəb(ə)l/ adj. **condemnatory** /-ˈdemnətərɪ/ adj. **condemnation** /ˌkɒndemˈneɪʃ(ə)n/ n. [ME f. OF condem(p)ner f. L condemnare (as COM-, damnare DAMN)]

condensate /kənˈdenseɪt, ˈkɒndənˌseɪt/ n. a substance produced by condensation.

condensation /ˌkɒndənˈseɪʃ(ə)n/ n. **1** the act of condensing. **2** any condensed material (esp. water on a cold surface). **3** an abridgement. **4** Chem. the combination of molecules with the elimination of water or other small molecules. □ **condensation trail** = vapour trail. [LL condensatio (as CONDENSE)]

condense /kənˈdens/ v. **1** tr. make denser or more concentrated. **2** tr. express in fewer words; make concise. **3** tr. & intr. reduce or be reduced from a gas or vapour to a liquid (or solid). □ **condensed milk** milk thickened by evaporation and sweetened. □ **condensable** adj. [F condenser or L condensare (as COM-, densus thick)]

condenser /kənˈdensə(r)/ n. **1** an apparatus or vessel for condensing vapour. **2** Electr. = CAPACITOR. **3** a lens or system of lenses for concentrating light. **4** a person or thing that condenses.

condescend /ˌkɒndɪˈsend/ v.intr. **1** (usu. foll. by to + infin.) often iron. be gracious enough (to do a thing) (condescended to attend the meeting). **2** (foll. by to) behave as if one is on equal terms with (an inferior), usu. while maintaining an attitude of superiority (condescended to the visitors). **3** (as **condescending** adj.) patronizing. □ **condescendingly** adv. [ME f. OF condescendre f. eccl.L condescendere (as COM-, DESCEND)]

condescension /ˌkɒndɪˈsenʃ(ə)n/ n. a patronizing manner. [obs. F f. eccl.L *condescensio* (as CONDESCEND)]

condign /kənˈdaɪn/ adj. (of a punishment etc.) severe and well-deserved. □ **condignly** adv. [ME f. OF *condigne* f. L *condignus* (as COM-, *dignus* worthy)]

condiment /ˈkɒndɪmənt/ n. a seasoning or relish for food. [ME f. L *condimentum* f. *condire* pickle]

condition /kənˈdɪʃ(ə)n/ n. & v. ● n. **1** a stipulation; something upon the fulfilment of which something else depends. **2 a** the state of being or fitness of a person or thing (*arrived in bad condition; not in a condition to be used*). **b** an ailment or abnormality (*a heart condition*). **3** (in pl.) circumstances, esp. those affecting the functioning or existence of something (*working conditions are good*). **4** archaic social rank (*all sorts and conditions of men*). **5** Gram. a conditional clause. **6** US a subject in which a student must pass an examination within a stated time to maintain a provisionally granted status. ● v.tr. **1 a** bring into a good or desired state or condition. **b** make fit (esp. dogs or horses). **2** teach or accustom to adopt certain habits etc. (*conditioned by society*). **3** govern, determine (*his behaviour was conditioned by his drunkenness*). **4 a** impose conditions on. **b** be essential to (*the two things condition each other*). **5** test the condition of (textiles etc.). **6** US subject (a student) to re-examination. □ **conditioned reflex** a reflex response to a non-natural stimulus, established by training. **in** (or **out of**) **condition** in good (or bad) condition. **in no condition to** certainly not fit to. **on condition that** with the stipulation that. [ME f. OF *condicion* (n.), *condicionner* (v.) or med.L *condicionare* f. L *condicio -onis* f. *condicere* (as COM-, *dicere* say)]

conditional /kənˈdɪʃən(ə)l/ adj. & n. ● adj. **1** (often foll. by on) dependent; not absolute; containing a condition or stipulation (*a conditional offer*). **2** Gram. (of a clause, mood, etc.) expressing a condition. ● n. Gram. **1** a conditional clause etc. **2** the conditional mood. □ **conditional discharge** Law an order made by a criminal court whereby an offender will not be sentenced for an offence unless a further offence is committed within a stated period. □ **conditionally** adv. **conditionality** /-ˌdɪʃəˈnælɪtɪ/ n. [ME f. OF *condicionel* or f. LL *conditionalis* (as CONDITION)]

conditioner /kənˈdɪʃənə(r)/ n. an agent that brings something into better condition, esp. a substance applied to the hair.

condo /ˈkɒndəʊ/ n. (pl. **-os**) N. Amer. colloq. a condominium. [abbr.]

condolatory /kənˈdəʊlətərɪ/ adj. expressing condolence. [CONDOLE, after *consolatory* etc.]

condole /kənˈdəʊl/ v.intr. (foll. by with) express sympathy with a person over a loss, grief, etc. ¶ Often confused with *console*. [LL *condolere* (as COM-, *dolere* suffer)]

condolence /kənˈdəʊləns/ n. (often in pl.) an expression of sympathy (*sent my condolences*).

condom /ˈkɒndɒm/ n. a rubber sheath worn on the penis or (usu. **female condom**) in the vagina during sexual intercourse as a contraceptive or to prevent infection. [18th c.: orig. unkn.]

condominium /ˌkɒndəˈmɪnɪəm/ n. **1** the joint control of a state's affairs by other states. **2** N. Amer. a building containing flats which are individually owned. [mod.L (as COM-, *dominium* DOMINION)]

condone /kənˈdəʊn/ v.tr. **1** forgive or overlook (an offence or wrongdoing). **2** approve or sanction, usu. reluctantly. **3** (of an action) atone for (an offence); make up for. □ **condoner** n. **condonation** /ˌkɒndəˈneɪʃ(ə)n/ n. [L *condonare* (as COM-, *donare* give)]

condor /ˈkɒndɔː(r)/ n. a very large American mountain-dwelling vulture. There are two species, the Andean condor (*Vultur gryphus*), of South America, having black plumage with a white neck ruff and a fleshy wattle on the forehead, and the endangered Californian condor (*Gymnogyps californianus*), restricted to California and now the subject of a captive-breeding recovery programme. [Sp. f. Quechua *cuntur*]

condottiere /ˌkɒndɒˈtjeɑrɪ/ n. (pl. **condottieri** pronunc. same) hist. a leader or a member of a troop of mercenaries in Italy etc. [It. f. *condotto* troop under contract (*condotta*) (as CONDUCT)]

conduce /kənˈdjuːs/ v.intr. (foll. by to) (usu. of an event or attribute) lead or contribute to (a result). [L *conducere* conduct- (as COM-, *ducere* duct- lead)]

conducive /kənˈdjuːsɪv/ adj. (often foll. by to) contributing or helping (towards something) (*not a conducive atmosphere for negotiation; good health is conducive to happiness*).

conduct n. & v. ● n. /ˈkɒndʌkt/ **1** behaviour (esp. in its moral aspect). **2** the action or manner of directing or managing (business, war, etc.). **3** Art mode of treatment, execution. **4** leading, guidance. ● v. /kənˈdʌkt/ **1** tr. lead or guide (a person or persons). **2** tr. direct or manage (business

etc.). **3** tr. (also absol.) be the conductor of (an orchestra, choir, etc.). **4** tr. transmit (heat, electricity, etc.) by conduction; serve as a channel for. **5** refl. behave (*conducted himself appropriately*). □ **conducted tour** a tour led by a guide on a fixed itinerary. **conduct sheet** a record of a person's offences and punishments. □ **conductible** /kənˈdʌktɪb(ə)l/ adj. **conductibility** /-ˌdʌktɪˈbɪlɪtɪ/ n. [ME f. L *conductus* (as COM-, *ducere* duct- lead): (v.) f. OF *conduite* past part. of *conduire*]

conductance /kənˈdʌktəns/ n. Physics the power of a specified material to conduct electricity.

conduction /kənˈdʌkʃ(ə)n/ n. **1 a** the transmission of heat through a substance from a region of higher temperature to a region of lower temperature. **b** the transmission of electricity through a substance by the application of an electric field. **2** the transmission of impulses along nerves. **3** the conducting of liquid through a pipe etc. [F *conduction* or L *conductio* (as CONDUCT)]

conductive /kənˈdʌktɪv/ adj. having the property of conducting (esp. heat, electricity, etc.). □ **conductive education** a system of education for children and adults with motor disorders. □ **conductively** adv.

conductivity /ˌkɒndʌkˈtɪvɪtɪ/ n. the conducting power of a specified material.

conductor /kənˈdʌktə(r)/ n. **1** a person who directs the performance of an orchestra or choir etc. **2** (fem. **conductress** /-trɪs/) **a** a person who collects fares in a bus etc. **b** US an official in charge of a train. **3** Physics **a** a thing that conducts or transmits heat or electricity, esp. regarded in terms of its capacity to do this (*a poor conductor*). **b** = *lightning conductor*. **4** a guide or leader. **5** a manager or director. □ **conductor rail** a rail transmitting current to an electric train etc. □ **conductorship** n. [ME f. F *conducteur* f. L *conductor* (as CONDUCT)]

conductus /kənˈdʌktəs/ n. (pl. **conducti** /-taɪ/) a musical composition of the 12th–13th centuries, with Latin text. [med.L: see CONDUIT]

conduit /ˈkɒndɪt, -djʊɪt/ n. **1** a channel or pipe for conveying liquids. **2 a** a tube or trough for protecting insulated electric wires. **b** a length or stretch of this. [ME f. OF *conduit* f. med.L *conductus* CONDUCT n.]

condyle /ˈkɒndɪl/ n. Anat. a rounded process at the end of some bones, forming an articulation with another bone. □ **condylar** adj. **condyloid** adj. [F f. L *condylus* f. Gk *kondulos* knuckle]

cone /kəʊn/ n. & v. ● n. **1** a solid figure with a circular (or other curved) plane base, tapering to a point. **2** a thing of a similar shape, solid or hollow, e.g. as used to mark off areas of roads. **3** the dry fruit of a conifer. **4** an ice-cream cornet. **5** any of the minute cone-shaped structures in the retina. **6** a conical mountain esp. of volcanic origin. **7** (in full **cone-shell**) a marine gastropod mollusc of the family Conidae, with a conical shell. **8** Pottery a ceramic pyramid, melting at a known temperature, used to indicate the temperature of a kiln. ● v.tr. **1** shape like a cone. **2** (foll. by off) Brit. mark off (a road etc.) with cones. [F *cône* f. L *conus* f. Gk *kōnos*]

Conegliano /ˌkɒneˈljɑːnəʊ/, Emmanuele, see DA PONTE.

coney var. of CONY.

Coney Island /ˈkəʊnɪ/ a resort and amusement park on the Atlantic coast in Brooklyn, New York City, on the south shore of Long Island.

confab /ˈkɒnfæb/ n. & v. colloq. ● n. a conversation, a chat. ● v.intr. (**confabbed**, **confabbing**) = CONFABULATE. [abbr.]

confabulate /kənˈfæbjʊˌleɪt/ v.intr. **1** converse, chat. **2** Psychol. fabricate imaginary experiences as compensation for the loss of memory. □ **confabulatory** /-lətərɪ/ adj. **confabulation** /-ˌfæbjʊˈleɪʃ(ə)n/ n. [L *confabulari* (as COM-, *fabulari* f. *fabula* tale)]

confect /kənˈfekt/ v.tr. literary make by putting together ingredients. [L *conficere* confect- put together (as COM-, *facere* make)]

confection /kənˈfekʃ(ə)n/ n. **1** a dish or delicacy made with sweet ingredients. **2** mixing, compounding. **3** a fashionable or elaborate article of women's dress. [ME f. OF f. L *confectio -onis* (as CONFECT)]

confectioner /kənˈfekʃənə(r)/ n. a maker or retailer of confectionery.

confectionery /kənˈfekʃənərɪ/ n. sweets and other confections.

confederacy /kənˈfedərəsɪ/ n. (pl. **-ies**) **1** a league or alliance, esp. of confederate states. **2** a league for an unlawful or evil purpose; a conspiracy. **3** the condition or fact of being confederate; alliance; conspiracy. [ME, AF, OF *confederacie* (as CONFEDERATE)]

Confederacy, the see CONFEDERATE STATES.

confederate adj., n., & v. ● adj. /kənˈfedərət/ esp. Polit. allied; joined by an agreement or treaty. ● n. /kənˈfedərət/ **1** an ally, esp. (in a bad sense) an accomplice. **2** (**Confederate**) a supporter of the Confederate states. ● v. /kənˈfedəˌreɪt/ (often foll. by with) **1** tr. bring (a person, state, or oneself) into alliance. **2** *intr.* come into alliance. [LL *confoederatus* (as COM-, FEDERATE)]

Confederate states (also **the Confederacy**) the eleven southern states (Alabama, Arkansas, Florida, Georgia, Louisiana, Mississippi, North Carolina, South Carolina, Tennessee, Texas, Virginia) which seceded from the United States in 1860–1 and formed a confederacy of their own, thus precipitating the American Civil War. The Confederate states were defeated in 1865, after which they were reunited with the US.

confederation /kənˌfedəˈreɪʃ(ə)n/ *n.* **1** a union or alliance of states etc. **2** the act or an instance of confederating; the state of being confederated. [F *confédération* (as CONFEDERATE)]

Confederation of British Industry (abbr. **CBI**) an organization founded in 1965 (combining earlier associations) to promote the prosperity of British business. The organization, which has a membership of about 50,000 companies, provides its members with a wide range of services and practical advice and voices the views of the management side of industry in the UK.

confer /kənˈfɜː(r)/ *v.* (**conferred, conferring**) **1** *tr.* (often foll. by *on, upon*) grant or bestow (a title, degree, favour, etc.). **2** *intr.* (often foll. by *with*) converse, consult. □ **conferrable** *adj.* [L *conferre* (as COM-, *ferre* bring)]

conferee /ˌkɒnfəˈriː/ *n.* **1** a person on whom something is conferred. **2** a participant in a conference.

conference /ˈkɒnfərəns/ *n. & v.* ● *n.* **1** consultation, discussion. **2** a meeting for discussion, esp. a regular one held by an association or organization. **3** an annual assembly of the Methodist Church. **4** an association in commerce, sport, etc. **5** the linking of several telephones, computer terminals, etc., so that each user may communicate with the others simultaneously; (often *attrib.*: *conference call*). ● *v.intr.* (usu. as **conferencing** *n.*) take part in a conference or conference call. □ **in conference** engaged in discussion. [F *conférence* or med.L *conferentia* (as CONFER)]

Conference on Disarmament a committee with forty nations as members that seeks to negotiate multilateral disarmament. It was constituted in 1962 as the Committee on Disarmament (with eighteen nations as members) and adopted its present title in 1984. It meets in Geneva.

conferment /kənˈfɜːmənt/ *n.* **1** the conferring of a degree, honour, etc. **2** an instance of this.

conferral /kənˈfɜːrəl/ *n.* esp. *US* = CONFERMENT.

confess /kənˈfes/ *v.* **1 a** *tr.* (also *absol.*) acknowledge or admit (a fault, wrongdoing, etc.). **b** *intr.* (foll. by *to*) admit to (*confessed to having lied*). **2** *tr.* admit reluctantly (*confessed it would be difficult*). **3 a** *tr.* (also *absol.*) declare (one's sins) to a priest. **b** *tr.* (of a priest) hear the confession of. **c** *refl.* declare one's sins to a priest. [ME f. OF *confesser* f. Rmc f. L *confessus* past part. of *confiteri* (as COM-, *fateri* declare, avow)]

confessant /kənˈfes(ə)nt/ *n.* a person who confesses to a priest.

confessedly /kənˈfesɪdlɪ/ *adv.* by one's own or general admission.

confession /kənˈfeʃ(ə)n/ *n.* **1 a** confessing or acknowledgement of a fault, wrongdoing, a sin to a priest, etc. **b** an instance of this. **c** a thing confessed. **2** (in full **confession of faith**) **a** a declaration of one's religious beliefs. **b** a statement of one's principles. □ **confessionary** *adj.* [ME f. OF f. L *confessio -onis* (as CONFESS)]

confessional /kənˈfeʃən(ə)l/ *n. & adj.* ● *n.* an enclosed stall in a church in which a priest hears confessions. ● *adj.* **1** of or relating to confession. **2** denominational. [F f. It. *confessionale* f. med.L, neut. of *confessionalis* (as CONFESSION)]

confessor /kənˈfesə(r)/ *n.* **1** a person who makes a confession. **2** (also /ˈkɒnfesə(r)/) a priest who hears confessions and gives spiritual counsel. **3** a person who avows a religion in the face of its suppression, but does not suffer martyrdom. [ME f. AF *confessur*, OF *-our*, f. eccl.L *confessor* (as CONFESS)]

confetti /kənˈfetɪ/ *n.* small pieces of coloured paper thrown by wedding guests at the bride and groom. [It., = sweetmeats f. L (as COMFIT)]

confidant /ˈkɒnfɪˌdɒnt/ *n.* (*fem.* **confidante** *pronunc.* same) a person trusted with knowledge of one's private affairs. [18th-c. for earlier CONFIDENT *n.*, prob. to represent the pronunc. of F *confidente* (as CONFIDE)]

confide /kənˈfaɪd/ *v.* **1** *tr.* (usu. foll. by *to*) tell (a secret etc.) in confidence. **2** *tr.* (foll. by *to*) entrust (an object of care, a task, etc.) to. **3** *intr.* (foll. by *in*) **a** have trust or confidence in. **b** talk confidentially to. □ **confidingly** *adv.* [L *confidere* (as COM-, *fidere* trust)]

confidence /ˈkɒnfɪd(ə)ns/ *n.* **1** firm trust (*have confidence in his ability*). **2 a** a feeling of reliance or certainty. **b** a sense of self-reliance; boldness. **3 a** something told confidentially. **b** the telling of private matters with mutual trust. □ **confidence level** *Statistics* the probability that the value of a parameter falls within a specified range of values. **confidence man** a man who robs or swindles by means of a confidence trick. **confidence trick** (*US* **game**) a swindle in which the victim is persuaded to trust the swindler in some way. **in confidence** as a secret. **in a person's confidence** trusted with a person's secrets. **take into one's confidence** confide in. [ME f. L *confidentia* (as CONFIDE)]

confident /ˈkɒnfɪd(ə)nt/ *adj. & n.* ● *adj.* **1** feeling or showing confidence; self-assured, bold (*spoke with a confident air*). **2** (often foll. by *of,* or *that* + clause) assured, trusting (*confident of your support; confident that he will come*). ● *n. archaic* = CONFIDANT. □ **confidently** *adv.* [F f. It. *confidente* (as CONFIDE)]

confidential /ˌkɒnfɪˈdenʃ(ə)l/ *adj.* **1** spoken or written in confidence. **2** entrusted with secrets (*a confidential secretary*). **3** confiding. □ **confidentially** *adv.* **confidentiality** /-ˌdenʃɪˈælɪtɪ/ *n.*

configuration /kənˌfɪgjʊˈreɪʃ(ə)n, -ˌfɪgəˈreɪʃ(ə)n/ *n.* **1 a** an arrangement of parts or elements in a particular form or figure. **b** the form, shape, or figure resulting from such an arrangement. **2** *Astron. & Astrol.* the relative position of planets etc. **3** *Psychol.* = GESTALT. **4** *Physics* the distribution of electrons among the energy levels of an atom, or of nucleons among the energy levels of a nucleus, as specified by quantum numbers. **5** *Chem.* the fixed three-dimensional relationship of the atoms in a molecule. **6** *Computing* **a** the interrelating or interconnecting of a computer system or elements of it so that it will accommodate a particular specification. **b** an instance of this. □ **configurational** *adj.* **configure** /kənˈfɪgə(r)/ *v.tr.* (in senses 1, 2, 6). [LL *configuratio* f. L *configurare* (as COM-, *figurare* fashion)]

confine *v. & n.* ● *v.tr.* /kənˈfaɪn/ (often foll. by *in, to, within*) **1** keep or restrict (within certain limits etc.). **2** hold captive; imprison. ● *n.* /ˈkɒnfaɪn/ (usu. in *pl.*) a limit or boundary (*within the confines of the town*). □ **be confined** be in childbirth. [(v.) f. F *confiner*, (n.) ME f. F *confins* (pl.), f. L *confinia* (as COM-, *finia* neut. pl. f. *finis* end, limit)]

confinement /kənˈfaɪnmənt/ *n.* **1** the act or an instance of confining; the state of being confined. **2** the time of a woman's giving birth.

confirm /kənˈfɜːm/ *v.tr.* **1** provide support for the truth or correctness of; make definitely valid (*confirmed my suspicions; confirmed his arrival time*). **2** ratify (a treaty, possession, title, etc.); make formally valid. **3** (foll. by *in*) encourage (a person) in (an opinion etc.). **4** establish more firmly (power, possession, etc.). **5** administer the religious rite of confirmation to. □ **confirmative** /-ˈfɜːmətɪv/ *adj.* **confirmatory** /-ˈfɜːmətərɪ/ *adj.* [ME f. OF *confermer* f. L *confirmare* (as COM-, FIRM¹)]

confirmand /ˈkɒnfəˌmænd/ *n.* *Eccl.* a person who is to be or has just been confirmed.

confirmation /ˌkɒnfəˈmeɪʃ(ə)n/ *n.* **1 a** the act or an instance of confirming; the state of being confirmed. **b** an instance of this. **2 a** a religious rite confirming a baptized person, esp. at the age of discretion, as a member of the Christian Church. **b** in Judaism, the ceremony of bar mitzvah. [ME f. OF f. L *confirmatio -onis* (as CONFIRM)]

confirmed /kənˈfɜːmd/ *adj.* firmly settled in some habit or condition (*confirmed in his ways; a confirmed bachelor*).

confiscate /ˈkɒnfɪˌskeɪt/ *v.tr.* **1** take or seize by authority. **2** appropriate to the public treasury (by way of a penalty). □ **confiscator** *n.* **confiscable** /kənˈfɪskəb(ə)l/ *adj.* **confiscatory** /-ˈfɪskətərɪ/ *adj.* **confiscation** /ˌkɒnfɪˈskeɪʃ(ə)n/ *n.* [L *confiscare* (as COM-, *fiscare* f. *fiscus* treasury)]

conflagration /ˌkɒnfləˈɡreɪʃ(ə)n/ *n.* a great and destructive fire. [L *conflagratio* f. *conflagrare* (as COM-, *flagrare* blaze)]

conflate /kənˈfleɪt/ *v.tr.* blend or fuse together (esp. two variant texts into one). □ **conflation** /-ˈfleɪʃ(ə)n/ *n.* [L *conflare* (as COM-, *flare* blow)]

conflict *n. & v.* ● *n.* /ˈkɒnflɪkt/ **1 a** a state of opposition or hostilities. **b** a fight or struggle. **2** (often foll. by *of*) **a** the clashing of opposed principles etc. **b** an instance of this. **3** *Psychol.* **a** the opposition of incompatible wishes or needs in a person. **b** an instance of this. **c** the distress resulting from this. ● *v.intr.* /kənˈflɪkt/ **1** clash; be incompatible. **2** (often foll. by *with*) struggle or contend. **3** (as **conflicting** *adj.*) contradictory. □ **in conflict** conflicting. □ **confliction** /kənˈflɪkʃ(ə)n/ *n.* **conflictual** /-ˈflɪktʃʊəl, -ˈflɪktjʊ-/ *adj.* [ME f. L *confligere* conflict- (as COM-, *fligere* strike)]

confluence /ˈkɒnflʊəns/ *n.* **1** a place where two rivers meet. **2 a** a coming together. **b** a crowd of people. [L *confluere* (as COM-, *fluere* flow)]

confluent /ˈkɒnfluənt/ adj. & n. ● adj. flowing together, uniting. ● n. a stream joining another.

conflux /ˈkɒnflʌks/ n. = CONFLUENCE. [LL confluxus (as CONFLUENCE)]

conform /kənˈfɔːm/ v. **1** intr. comply with rules or general custom. **2** intr. & tr. (often foll. by to) be or make accordant or suitable. **3** tr. (often foll. by to) form according to a pattern; make similar. **4** intr. (foll. by to, with) comply with; be in accordance with. □ **conformer** n. [ME f. OF conformer f. L conformare (as COM-, FORM)]

conformable /kənˈfɔːməb(ə)l/ adj. **1** (often foll. by to) similar. **2** (often foll. by with) consistent. **3** (often foll. by to) adapted. **4** tractable, submissive. **5** Geol. (of strata in contact) lying in the same direction. □ **conformably** adv. **conformability** /-ˌfɔːməˈbɪlɪtɪ/ n. [med.L conformabilis (as CONFORM)]

conformal /kənˈfɔːm(ə)l/ adj. (of a map) showing any small area in its correct shape. □ **conformally** adv. [LL conformalis (as CONFORM)]

conformance /kənˈfɔːməns/ n. (often foll. by to, with) = CONFORMITY 1, 2.

conformation /ˌkɒnfɔːˈmeɪʃ(ə)n/ n. **1** the way in which a thing is formed; shape, structure. **2** (often foll. by to) adjustment in form or character; adaptation. **3** Chem. a spatial arrangement of atoms in a molecule, able to convert freely to other arrangements esp. by rotation. [L conformatio (as CONFORM)]

conformist /kənˈfɔːmɪst/ n. & adj. ● n. **1** a person who conforms to an established practice; a conventional person. **2** Brit. a person who conforms to the practices of the Church of England. ● adj. (of a person) conforming to established practices; conventional. □ **conformism** n.

conformity /kənˈfɔːmɪtɪ/ n. **1** (often foll. by to, with) action or behaviour in accordance with established practice; compliance. **2** (often foll. by to, with) correspondence in form or manner; likeness, agreement. **3** Brit. compliance with the practices of the Church of England. [ME f. OF conformité or LL conformitas (as CONFORM)]

confound /kənˈfaʊnd/ v. & int. ● v.tr. **1** throw into perplexity or confusion. **2** mix up; confuse (in one's mind). **3** archaic defeat, overthrow. ● int. expressing annoyance (confound you!). [ME f. AF conf(o)undre, OF confondre f. L confundere mix up (as COM-, fundere fuspour)]

confounded /kənˈfaʊndɪd/ adj. colloq. damned (a confounded nuisance!). □ **confoundedly** adv.

confraternity /ˌkɒnfrəˈtɜːnɪtɪ/ n. (pl. **-ies**) a brotherhood, esp. religious or charitable. [ME f. OF confraternité f. med.L confraternitas (as COM-, FRATERNITY)]

confrère /ˈkɒnfreə(r)/ n. a fellow member of a profession, scientific body, etc. [ME f. OF f. med.L confrater (as COM-, frater brother)]

confront /kənˈfrʌnt/ v.tr. **1** a face in hostility or defiance. **b** face up to and deal with (a problem, difficulty, etc.). **2** (of a difficulty etc.) present itself to (countless obstacles confronted us). **3** (foll. by with) **a** bring (a person) face to face with (a circumstance), esp. by way of accusation (confronted them with the evidence). **b** set (a thing) face to face with (another) for comparison. **4** meet or stand facing. □ **confrontation** /ˌkɒnfrʌnˈteɪʃ(ə)n/ n. **confrontational** adj. [F confronter f. med.L confrontare (as COM-, frontare f. frons frontis face)]

Confucianism /kənˈfjuːʃəˌnɪz(ə)m/ n. a system of philosophical and ethical teachings founded by Confucius in China in the 6th century BC and developed by Mencius in the 4th century BC, one of the two major Chinese ideologies (the other being Taoism). The basic concepts are ethical ones, for example love for one's fellows and filial piety, and the ideal of the superior man; traditional ideas such as yin and yang have also been incorporated into Confucianism. The publication in AD 1190 of the four great Confucian texts revitalized Confucianism throughout China; a second series of texts, the 'five classics', includes the I Ching. □ **Confucian** adj. & n. **Confucianist** n. & adj.

Confucius /kənˈfjuːʃəs/ (Latinized name of K'ung Fu-tzu = 'Kong the master') (551–479 BC), Chinese philosopher. He spent much of his life as a moral teacher of a group of disciples, at first working for the government and later taking up the role of an itinerant sage. His ideas about the importance of practical moral values formed the basis of the philosophy of Confucianism. His teachings were collected by his pupils after his death in the Analects; later collections of Confucianist writings are probably only loosely based on his work.

confusable /kənˈfjuːzəb(ə)l/ adj. that is able or liable to be confused. □ **confusability** /-ˌfjuːzəˈbɪlɪtɪ/ n.

confuse /kənˈfjuːz/ v.tr. **1** a disconcert, perplex, bewilder. **b** embarrass. **2** mix up in the mind; mistake (one for another). **3** make indistinct (that point confuses the issue). **4** (as **confused** adj.) **a** mentally decrepit

or senile. **b** puzzled, perplexed. **5** (often as **confused** adj.) make muddled or disorganized, throw into disorder (a confused jumble of clothes). □ **confusedly** /-zɪdlɪ/ adv. **confusing** adj. **confusingly** adv. [19th-c. back-form. f. confused (14th c.) f. OF confus f. L confusus: see CONFOUND]

confusion /kənˈfjuːʒ(ə)n/ n. **1 a** the act of confusing (the confusion of fact and fiction). **b** an instance of this; a misunderstanding (confusions arise from a lack of communication). **2 a** the result of confusing; a confused state; embarrassment, disorder (thrown into confusion by his words; trampled in the confusion of battle). **b** (foll. by of) a disorderly jumble (a confusion of ideas). **3 a** civil commotion (confusion broke out at the announcement). **b** an instance of this. [ME f. OF confusion or L confusio (as CONFUSE)]

confute /kənˈfjuːt/ v.tr. **1** prove (a person) to be in error. **2** prove (an argument) to be false. □ **confutation** /ˌkɒnfjʊˈteɪʃ(ə)n/ n. [L confutare restrain]

conga /ˈkɒŋgə/ n. & v. ● n. **1** a Latin American dance of African origin, usu. with several persons in a single line, one behind the other. **2** (also **conga drum**) a tall, narrow, low-toned drum beaten with the hands. ● v.intr. (**congas**, **congaed** /-gəd/ or **conga'd**, **congaing** /-gəɪŋ/) perform the conga. [Amer. Sp. f. Sp. conga (fem.) of the Congo]

congé /ˈkɒnʒeɪ/ n. an unceremonious dismissal; leave-taking. [F: earlier congee, ME f. OF congié f. L commeatus leave of absence f. commeare go and come (as COM-, meare go): now usu. treated as mod. F]

congeal /kənˈdʒiːl/ v.tr. & intr. **1** make or become semi-solid by cooling. **2** (of blood etc.) coagulate. □ **congealable** adj. **congealment** n. [ME f. OF congeler f. L congelare (as COM-, gelare f. gelu frost)]

congelation /ˌkɒndʒɪˈleɪʃ(ə)n/ n. **1** the process of congealing. **2** a congealed state. **3** a congealed substance. [ME f. OF congelation or L congelatio (as CONGEAL)]

congener /ˈkɒndʒɪnə(r)/ n. a thing or person of the same kind or category as another, esp. animals or plants of a specified genus (the raspberry and the blackberry are congeners). [L (as CON-, GENUS)]

congeneric /ˌkɒndʒɪˈnerɪk/ adj. **1** of the same genus, kind, or race. **2** allied in nature or origin; akin. □ **congenerous** /kənˈdʒenərəs/ adj.

congenial /kənˈdʒiːnɪəl/ adj. **1** pleasant, agreeable; friendly (congenial environment). **2** (often foll. by to) suited or agreeable to the nature of anything (congenial to my mood). □ **congenially** adv. **congeniality** /-ˌdʒiːnɪˈælɪtɪ/ n. [CON- + GENIAL[1]]

congenital /kənˈdʒenɪt(ə)l/ adj. **1** (esp. of a disease, defect, etc.) existing from birth. **2** that is (or as if) such from birth (a congenital liar). □ **congenitally** adv. [L congenitus (as COM-, genitus past part. of gigno beget)]

conger /ˈkɒŋgə(r)/ n. (in full **conger eel**) a large marine eel of the family Congridae; esp. Conger conger, which is caught for food. [ME f. OF congre f. L conger, congrus, f. Gk goggros]

congeries /kənˈdʒɪəriːz, -ˈdʒerɪˌiːz/ n. (pl. same) a disorderly collection; a mass or heap. [L, formed as CONGEST]

congest /kənˈdʒest/ v.tr. (esp. as **congested** adj.) affect with congestion; obstruct, block (congested streets; congested lungs). □ **congestive** adj. [L congerere congest- (as COM-, gerere bring)]

congestion /kənˈdʒestʃən/ n. abnormal accumulation, crowding, or obstruction, esp. of traffic etc. or of blood or mucus in a part of the body. [F f. L congestio -onis (as CONGEST)]

conglomerate adj., n., & v. ● adj. /kənˈglɒmərət/ **1** gathered into a rounded mass. **2** Geol. (of rock) made up of small stones held together (cf. AGGLOMERATE). ● n. /kənˈglɒmərət/ **1** a number of things or parts forming a heterogeneous mass. **2** a group or corporation formed by the merging of separate and diverse firms. **3** Geol. conglomerate rock. ● v.tr. & intr. /kənˈglɒməˌreɪt/ collect into a coherent mass. □ **conglomeration** /-ˌglɒməˈreɪʃ(ə)n/ n. [L conglomeratus past part. of conglomerare (as COM-, glomerare f. glomus -eris ball)]

Congo /ˈkɒŋgəʊ/ **1** (also called Zaire River) a major river of central Africa, which rises as the Lualaba, to the south of Kisangani in northern Zaire, and flows 4,630 km (2,880 miles) in a great curve westwards, turning south-westwards to form the border between the Congo and Zaire, before emptying into the Atlantic. **2** (also **the Congo**) an equatorial country in Africa, with a short Atlantic coastline; pop. (est. 1991) 2,351,000; languages, French (official), Kikongo, and other Bantu languages; capital, Brazzaville. The River Congo and its tributary the Ubanghi form most of the country's eastern boundary (with Zaire). Largely occupied by Bantu-speaking peoples, the region was colonized in the 19th century by France. It became known as Middle Congo, and formed part of the larger territory of French Congo (later, French Equatorial Africa). After becoming independent in 1960, the Congo was the scene of civil war for nearly two decades.

Congolese /ˌkɒŋɡəˈliːz/ *adj. & n.* ● *adj.* of or relating to Congo or its people. ● *n.* (*pl.* same) **1** a native or inhabitant of Congo. **2** the language of the Congolese.

congou /ˈkɒŋɡuː, -ɡəʊ/ *n.* a variety of black China tea. [Cantonese *kungfûch'a*, Mandarin *gōngfu chá* tea made for refined tastes]

congrats /kənˈɡræts/ *n.pl. & int. colloq.* congratulations. [abbr.]

congratulate /kənˈɡrætjʊˌleɪt/ *v.tr. & refl.* (often foll. by *on, upon*) **1** *tr.* express pleasure at the happiness or good fortune or excellence of (a person) (*congratulated them on their success*). **2** *refl.* think oneself fortunate or clever. □ **congratulant** *adj. & n.* **congratulator** *n.* **congratulatory** /-lətərɪ/ *adj.* [L *congratulari* (as COM-, *gratulari* show joy f. *gratus* pleasing)]

congratulation /kənˌɡrætjʊˈleɪʃ(ə)n/ *n.* **1** congratulating. **2** (also as *int.*; usu. in *pl.*) an expression of this (*congratulations on winning!*). [L *congratulatio* (as CONGRATULATE)]

congregant /ˈkɒŋɡrɪɡənt/ *n.* a member of a congregation (esp. Jewish). [L *congregare* (as CONGREGATE)]

congregate /ˈkɒŋɡrɪˌɡeɪt/ *v.intr. & tr.* collect or gather into a crowd or mass. [ME f. L *congregare* (as COM-, *gregare* f. *grex gregis* flock)]

congregation /ˌkɒŋɡrɪˈɡeɪʃ(ə)n/ *n.* **1** the process of congregating; collection into a crowd or mass. **2** a crowd or mass gathered together. **3 a** a body assembled for religious worship. **b** a body of persons regularly attending a particular church etc. **c** *RC Ch.* a body of persons obeying a common religious rule. **d** *RC Ch.* any of several permanent committees of the Roman Catholic College of Cardinals. **4** (**Congregation**) *Brit.* (in some universities) a general assembly of resident senior members. [ME f. OF *congregation* or L *congregatio* (as CONGREGATE)]

congregational /ˌkɒŋɡrɪˈɡeɪʃ(ə)l/ *adj.* **1** of a congregation. **2** (**Congregational**) of or adhering to Congregationalism.

Congregationalism /ˌkɒŋɡrɪˈɡeɪʃənəˌlɪz(ə)m/ *n.* a system of Christian ecclesiastical organization in England and elsewhere, whereby individual churches are largely self-governing. Originally known as Independents, Congregationalists formed the backbone of Cromwell's army, but were persecuted under the 1662 Act of Uniformity. The independence of Congregational churches did not prevent them from forming associations for mutual support; in 1832 these associations combined, and in 1972 the Congregational Church in England and Wales combined with the Presbyterian Church of England to form the United Reformed Church. □ **Congregationalist** *n.* **Congregationalize** *v.tr.* (also **-ise**).

Congregation of the Mission see LAZARIST.

congress /ˈkɒŋɡres/ *n.* **1** a formal meeting of delegates for discussion. **2** (**Congress**) a national legislative body, esp. that of the US. (*See note below.*) **3** a society or organization. **4** coming together, meeting. □ **congressional** /kənˈɡreʃ(ə)l/ *adj.* [L *congressus* f. *congredi* (as COM-, *gradi* walk)]

▪ The US Congress, established by the Constitution of 1787, is composed of two houses, the upper, or Senate, and the lower, or House of Representatives. The Senate is made up of two members for each state, each member sitting for six years and one-third of the members coming up for re-election every two years; the House of Representatives is composed of 435 members (re-elected every two years) divided between the states on the basis of population. Congress meets at the Capitol in Washington, DC. New legislation may be put forward by any member of Congress and may be suggested by the President; the President may also veto bills, a veto that can be overridden only by a two-thirds majority in each house.

congressman /ˈkɒŋɡresmən/ *n.* (*pl.* **-men**; *fem.* **congresswoman**, *pl.* **-women**) a member of the US Congress.

Congress of Industrial Organizations (abbr. **CIO**) a federation of North American trade unions, organized largely by industry rather than craft, founded in 1935 as the Committee on Industrial Organization. It changed its name in 1937 on being expelled from the American Federation of Labor, with which it merged in 1955 to form the AFL–CIO.

Congreve /ˈkɒŋɡriːv/, William (1670–1729), English dramatist. A close associate of Swift, Pope, and Steele, he wrote plays such as *Love for Love* (1695) and *The Way of the World* (1700), which epitomize the wit and satire of Restoration comedy.

congruence /ˈkɒŋɡrʊəns/ *n.* (also **congruency** /-sɪ/) **1** agreement, consistency. **2** *Geom.* the state of being congruent. [ME f. L *congruentia* (as CONGRUENT)]

congruent /ˈkɒŋɡrʊənt/ *adj.* **1** (often foll. by *with*) suitable, agreeing. **2** *Geom.* (of figures) coinciding exactly when superimposed. □ **congruently** *adv.* [ME f. L *congruere* agree]

congruous /ˈkɒŋɡrʊəs/ *adj.* (often foll. by *with*) suitable, agreeing; fitting. □ **congruously** *adv.* **congruity** /kənˈɡruːɪtɪ/ *n.* [L *congruus* (as CONGRUENT)]

conic /ˈkɒnɪk/ *adj. & n.* ● *adj.* of a cone. ● *n.* **1** a conic section. **2** (in *pl.*) the study of conic sections. □ **conic section** a figure formed by the intersection of a cone and a plane. [mod.L *conicus* f. Gk *kōnikos* (as CONE)]

conical /ˈkɒnɪk(ə)l/ *adj.* cone-shaped. □ **conical projection** (also **conic projection**) a map projection in which a spherical surface is projected on to a cone, usually with its vertex above the pole. □ **conically** *adv.*

conidium /kəˈnɪdɪəm/ *n.* (*pl.* **conidia** /-dɪə/) a spore produced asexually by various fungi. [mod.L dimin. f. Gk *konis* dust]

conifer /ˈkɒnɪfə(r), ˈkəʊn-/ *n.* a tree of the order Coniferales, typically bearing cones and needle-like leaves (often evergreen), e.g. a pine, yew, cedar, or redwood. □ **coniferous** /kəˈnɪfərəs/ *adj.* [L (as CONE, -FEROUS)]

coniform /ˈkəʊnɪˌfɔːm/ *adj.* cone-shaped. [L *conus* cone + -FORM]

coniine /ˈkəʊnɪˌiːn/ *n.* a poisonous alkaloid found in hemlock, that paralyses the nerves. [L *conium* f. Gk *kōneion* hemlock]

conjectural /kənˈdʒektərəl/ *adj.* based on, involving, or given to conjecture. □ **conjecturally** *adv.* [F f. L *conjecturalis* (as CONJECTURE)]

conjecture /kənˈdʒektʃə(r)/ *n. & v.* ● *n.* **1 a** the formation of an opinion on incomplete information; guessing. **b** an opinion or conclusion reached in this way. **2 a** (in textual criticism) the guessing of a reading not in the text. **b** a proposed reading. ● *v.* **1** *tr. & intr.* guess. **2** *tr.* (in textual criticism) propose (a reading). □ **conjecturable** *adj.* [ME f. OF *conjecture* or L *conjectura* f. *conjicere* (as COM-, *jacere* throw)]

conjoin /kənˈdʒɔɪn/ *v.tr. & intr.* join, combine. [ME f. OF *conjoign-* pres. stem of *conjoindre* f. L *conjungere* (as COM-, *jungere junct-* join)]

conjoint /kənˈdʒɔɪnt/ *adj.* associated, conjoined. □ **conjointly** *adv.* [ME f. OF, past part. (as CONJOIN)]

conjugal /ˈkɒndʒʊɡ(ə)l/ *adj.* of marriage or the relation between husband and wife. □ **conjugal rights** those rights (esp. to sexual relations) regarded as exercisable in law by each partner in a marriage. □ **conjugally** *adv.* **conjugality** /ˌkɒndʒʊˈɡælɪtɪ/ *n.* [L *conjugalis* f. *conjux* consort (as COM-, *-jux -jugis* f. root of *jungere* join)]

conjugate *v., adj., & n.* ● *v.* **1** *tr. Gram.* give the different forms of (a verb). **2** *intr.* usu. *Biol.* **a** unite sexually. **b** (of gametes) become fused. **3** *intr. Chem.* (of protein) combine with non-protein. ● *adj.* /ˈkɒndʒʊɡət/ **1** joined together, esp. as a pair. **2** *Gram.* derived from the same root. **3** *Biol.* fused. **4** *Chem.* (of an acid or base) related by loss or gain of a proton. **5** *Math.* joined in a reciprocal relation, esp. having the same real parts, and equal magnitudes but opposite signs of imaginary parts. ● *n.* /ˈkɒndʒʊɡət/ a conjugate word or thing. [L *conjugare* yoke together (as COM-, *jugare* f. *jugum* yoke)]

conjugation /ˌkɒndʒʊˈɡeɪʃ(ə)n/ *n.* **1** *Gram.* a system of verbal inflection. **2** the act or an instance of conjugating. **b** an instance of this. **3** *Biol.* **a** the fusion of two gametes in reproduction. **b** the temporary union of two unicellular organisms for the exchange of genetic material. □ **conjugational** *adj.* [L *conjugatio* (as CONJUGATE)]

conjunct /kənˈdʒʌŋkt/ *adj.* joined together; combined; associated. [ME f. L *conjunctus* (as CONJOIN)]

conjunction /kənˈdʒʌŋkʃ(ə)n/ *n.* **1 a** the action of joining; the condition of being joined. **b** an instance of this. **2** *Gram.* a word used to connect clauses or sentences or words in the same clause (e.g. *and*, *but*, *if*). **3 a** a combination (of events or circumstances). **b** a number of associated persons or things. **4** *Astron. & Astrol.* the alignment of two bodies in the solar system so that they have the same longitude as seen from the earth. □ **in conjunction with** together with. □ **conjunctional** *adj.* [ME f. OF *conjonction* f. L *conjunctio -onis* (as CONJUNCT)]

conjunctiva /ˌkɒndʒʌŋkˈtaɪvə, kənˈdʒʌŋktɪvə/ *n.* (*pl.* **conjunctivas**) *Anat.* the mucous membrane that covers the front of the eye and lines the inside of the eyelids. □ **conjunctival** *adj.* [med.L (*membrana*) *conjunctiva* (as CONJUNCTIVE)]

conjunctive /kənˈdʒʌŋktɪv/ *adj. & n.* ● *adj.* **1** serving to join; connective. **2** *Gram.* of the nature of a conjunction. ● *n. Gram.* a conjunctive word. □ **conjunctively** *adv.* [LL *conjunctivus* (as CONJOIN)]

conjunctivitis /kən,dʒʌŋktɪ'vaɪtɪs/ n. inflammation of the conjunctiva.

conjuncture /kən'dʒʌŋktʃə(r)/ n. a combination of events; a state of affairs. [obs. F f. It. *congiuntura* (as CONJOIN)]

conjuration /ˌkɒndʒʊ'reɪʃ(ə)n/ n. an incantation; a magic spell. [ME f. OF f. L *conjuratio -onis* (as CONJURE)]

conjure /'kʌndʒə(r)/ v. **1** intr. perform tricks which are seemingly magical, esp. by rapid movements of the hands. **2** tr. (usu. foll. by *out of, away, to,* etc.) cause to appear or disappear as if by magic (*conjured a rabbit out of a hat; conjured them to a desert island; his pain was conjured away*). **3** tr. call upon (a spirit) to appear. **4** intr. perform marvels. **5** tr. /kən'dʒʊə(r)/ (often foll. by *to* + infin.) appeal solemnly to (a person). □ **conjure up 1** bring into existence or cause to appear as if by magic. **2** cause to appear to the eye or mind; evoke. [ME f. OF *conjurer* plot, exorcise f. L *conjurare* band together by oath (as COM-, *jurare* swear)]

conjuror /'kʌndʒərə(r)/ n. (also **conjurer**) a performer of conjuring tricks. [CONJURE + -ER[1] & AF *conjurour* (OF -*eor*) f. med.L *conjurator* (as CONJURE)]

conk[1] /kɒŋk/ v.intr. (usu. foll. by *out*) colloq. **1** (of a machine etc.) break down. **2** (of a person) become exhausted and give up; faint; die. [20th c.: orig. unkn.]

conk[2] /kɒŋk/ n. & v. sl. ● n. **1 a** the nose. **b** the head. **2** a punch on the nose or head; a blow. ● v.tr. punch on the nose; hit on the head etc. [19th c.: perh. = CONCH]

conker /'kɒŋkə(r)/ n. **1** the hard fruit of a horse chestnut. **2** (in *pl.*) Brit. a children's game played with conkers on strings, one hit against another to try to break it. [dial. *conker* snail-shell (orig. used in the game), assoc. with CONQUER]

con moto /kɒn 'məʊtəʊ/ adv. Mus. with movement. [It., = with movement]

Conn. abbr. Connecticut.

conn US var. of CON[4].

Connacht /'kɒnɔːt/ (also **Connaught**) a province of the Republic of Ireland, in the west on the Atlantic coast.

connate /'kɒneɪt/ adj. **1** existing in a person or thing from birth; innate. **2** formed at the same time. **3** allied, congenial. **4** Bot. (of organs) congenitally united so as to form one part. **5** Geol. (of water) trapped in sedimentary rock during its deposition. [LL *connatus* past part. of *connasci* (as COM-, *nasci* be born)]

connatural /kə'nætʃrəl/ adj. **1** (often foll. by *to*) innate; belonging naturally. **2** of like nature. □ **connaturally** adv. [LL *connaturalis* (as COM-, NATURAL)]

Connaught see CONNACHT.

connect /kə'nekt/ v. **1 a** tr. (often foll. by *to, with*) join (one thing with another) (*connected the hose to the tap*). **b** tr. join (two things) (*a track connected the two villages*). **c** intr. be joined or joinable (*the two parts do not connect*). **2** tr. (often foll. by *with*) associate mentally or practically (*did not connect the two ideas; never connected her with the theatre*). **3** intr. (foll. by *with*) (of a train etc.) be synchronized at its destination with another train etc., so that passengers can transfer (*the train connects with the boat*). **4** tr. put into communication by telephone. **5 a** tr. (usu. in *passive*; foll. by *with*) unite or associate with others in relationships etc. (*am connected with the royal family*). **b** intr. form a logical sequence; be meaningful. **6** intr. colloq. hit or strike effectively. □ **connecting-rod** the rod between the piston and the crankpin etc. in an internal-combustion engine or between the wheels of a locomotive. □ **connectable** adj. **connector** n. [L *connectere connex-* (as COM-, *nectere* bind)]

connected /kə'nektɪd/ adj. **1** joined in sequence. **2** (of ideas etc.) coherent. **3** related or associated. □ **well-connected** associated, esp. by birth, with persons of good social position. □ **connectedly** adv. **connectedness** n.

Connecticut /kə'netɪkət/ **1** a state in the north-eastern US, on the Atlantic coast; capital, Hartford; pop. (1990) 3,287,100. A Puritan settlement in the 17th century, it was one of the original thirteen states of the Union and ratified the draft US Constitution in 1788. **2** the longest river in New England, rising in northern New Hampshire and flowing south for 655 km (407 miles) to enter Long Island Sound.

connection /kə'nekʃ(ə)n/ n. (also Brit. **connexion**) **1 a** the act of connecting; the state of being connected. **b** an instance of this. **2** the point at which two things are connected (*broke at the connection*). **3 a** a thing or person that connects; a link, a relationship or association (*a radio formed the only connection with the outside world; cannot see the connection between the two ideas*). **b** a telephone link (*got a bad connection*). **4** arrangement or opportunity for catching a connecting train etc.;

the train etc. itself (*missed the connection*). **5** Electr. **a** the linking up of an electric current by contact. **b** a device for effecting this. **6** (often in *pl.*) a relative or associate, esp. one with influence (*has connections in the Home Office; heard it through a business connection*). **7** a relation of ideas; a context (*in this connection I have to disagree*). **8** sl. a supplier of narcotics. **9** a religious body, esp. Methodist. □ **in connection with** with reference to. **in this** (or **that**) **connection** with reference to this (or that). □ **connectional** adj. [L *connexio* (as CONNECT): spelling -*ct*- after CONNECT]

connective /kə'nektɪv/ adj. & n. ● adj. serving or tending to connect. ● n. something that connects. □ **connective tissue** Anat. tissue connecting and binding other tissues and organs, containing relatively few cells in a non-living matrix, e.g. bone, cartilage, blood, and (esp.) fibrous tissue rich in collagen.

Connemara /ˌkɒnɪ'mɑːrə/ a mountainous coastal region of Galway, in the west of the Republic of Ireland.

Connery /'kɒnərɪ/, Sean (born Thomas Connery) (b.1930), Scottish actor. He is best known for his portrayal of James Bond in the films of Ian Fleming's spy thrillers. He played the part seven times from his first performance in *Dr No* (1962) to his last in *Never Say Never Again* (1984). He has since appeared in other films, such as *The Name of the Rose* (1986).

conning tower /'kɒnɪŋ/ n. **1** the superstructure of a submarine from which steering, firing, etc., are directed on or near the surface, and which contains the periscope. **2** the armoured pilot-house of a warship. [CON[4] + -ING[1]]

connivance /kə'naɪv(ə)ns/ n. **1** (often foll. by *at, in*) conniving (*connivance in the crime*). **2** tacit permission (*done with his connivance*). [F *connivence* or L *conniventia* (as CONNIVE)]

connive /kə'naɪv/ v.intr. **1** (foll. by *at*) disregard or tacitly consent to (a wrongdoing). **2** (usu. foll. by *with*) conspire. □ **conniver** n. **conniving** adj. [F *conniver* or L *connivere* shut the eyes (to)]

connoisseur /ˌkɒnə'sɜː(r)/ n. (often foll. by *of, in*) an expert judge in matters of taste (*a connoisseur of fine wine*). □ **connoisseurship** n. [F, obs. spelling of *connaisseur* f. pres. stem of *connaître* know + -*eur* -OR[1]: cf. *reconnoitre*]

Connolly /'kɒnəlɪ/, Maureen Catherine (known as 'Little Mo') (1934–69), American tennis player. She was 16 when she first won the US singles title and 17 when she took the Wimbledon title; she retained these titles for a further two years each. In 1953 she became the first woman to win the grand slam. She lost a mere four matches after the age of 16, being forced to retire in 1954 after a riding accident.

Connors /'kɒnəz/, James Scott ('Jimmy') (b.1952), American tennis player. He established himself as one of the world's top players when he defeated the Australian Ken Rosewall (b.1934) in the 1974 Wimbledon and US Open championships. He won the US title again in 1976, 1978, 1982, and 1983 and Wimbledon in 1982.

connotation /ˌkɒnə'teɪʃ(ə)n/ n. **1** that which is implied by a word etc. in addition to its literal or primary meaning (*a letter with sinister connotations*). **2** the act of connoting or implying.

connote /kə'nəʊt/ v.tr. **1** (of a word etc.) imply in addition to the literal or primary meaning. **2** (of a fact) imply as a consequence or condition. **3** mean, signify. □ **connotative** /'kɒnə,teɪtɪv, kə'nəʊtətɪv/ adj. [med.L *connotare* mark in addition (as COM-, *notare* f. *nota* mark)]

connubial /kə'njuːbɪəl/ adj. of or relating to marriage or the relationship of husband and wife. □ **connubially** adv. **connubiality** /-ˌnjuːbɪ'ælɪtɪ/ n. [L *connubialis* f. *connubium* (*nubium* f. *nubere* marry)]

conoid /'kəʊnɔɪd/ adj. & n. ● adj. (also **conoidal** /kəʊ'nɔɪd(ə)l/) cone-shaped. ● n. a cone-shaped object.

conquer /'kɒŋkə(r)/ v.tr. **1 a** overcome and control (an enemy or territory) by military force. **b** absol. be victorious. **2** overcome (a habit, emotion, disability, etc.) by effort (*conquered his fear*). **3** climb (a mountain) successfully. □ **conquerable** adj. [ME f. OF *conquerre* f. Rmc f. L *conquirere* (as COM-, *quaerere* seek, get)]

conqueror /'kɒŋkərə(r)/ n. **1** a person who conquers. **2** Brit. = CONKER. [ME f. AF *conquerour* (OF -*eor*) f. *conquerre* (as CONQUER)]

conquest /'kɒŋkwest/ n. **1** the act or an instance of conquering; the state of being conquered. **2** a conquered territory. **b** something won. **3** a person whose affection or favour has been won. □ **make a conquest of** win the affections of. [ME f. OF *conquest(e)* f. Rmc (as CONQUER)]

Conquest, the see NORMAN CONQUEST.

conquistador /kɒn'kwɪstə,dɔː(r)/ n. (pl. **conquistadores**

/-ˌkwɪstəˈdɔːrez/ or **conquistadors**) one of the Spanish soldiers and adventurers who conquered South and Central America in the 16th century. While the initial object of most of their expeditions was the search for the fabled riches of the area, they ended by overthrowing the Aztec, Mayan, and Inca civilizations and establishing Spanish colonies. The best-known conquistadores are Hernando Cortés, Francisco Pizarro, and Diego Velázquez de Cuéllar. [Sp., = conqueror]

Conrad /ˈkɒnræd/, Joseph (born Józef Teodor Konrad Korzeniowski) (1857–1924), Polish-born British novelist. Conrad's long career at sea (1874–94) inspired many of his most famous works, including his novel *Lord Jim* (1900). Much of Conrad's work, including his story *Heart of Darkness* (1902) and the novel *Nostromo* (1904), explores the darker side of human nature; however, the novel that initially brought him fame and success was *Chance* (1913), in which there is a romantic theme. Although French was his first foreign language, Conrad wrote in English and became a British citizen in 1886. His work had a significant influence on modernist fiction.

con-rod /ˈkɒnrɒd/ n. colloq. connecting-rod. [abbr.]

Cons. abbr. Conservative.

consanguineous /ˌkɒnsæŋˈɡwɪnɪəs/ adj. descended from the same ancestor; akin. [L consanguineus (as COM-, sanguis -inis blood)]

consanguinity /ˌkɒnsæŋˈɡwɪnɪtɪ/ n. relationship by descent from the same ancestor; blood relationship. [L consanguinitas, f. consanguineus (see CONSANGUINEOUS)]

conscience /ˈkɒnʃ(ə)ns/ n. a moral sense of right and wrong esp. as felt by a person or collectively by a group and affecting behaviour (*my conscience won't allow me to do that*). □ **bad** (or **guilty**) **conscience** a conscience troubled by feelings of guilt or wrongdoing. **case of conscience** a matter in which one's conscience has to decide a conflict of principles. **clear** (or **good**) **conscience** a conscience untroubled by feelings of guilt or wrongdoing. **conscience clause** a clause in a law, ensuring respect for the consciences of those affected. **conscience money** a sum paid to relieve one's conscience, esp. about a payment previously evaded. **conscience-stricken** (or **-struck**) made uneasy by a bad conscience. **for conscience** (or **conscience's**) **sake** to satisfy one's conscience. **freedom of conscience** a system allowing all citizens freedom of choice in matters of religion, moral issues, etc. **in all conscience** colloq. by any reasonable standard; by all that is fair. **on one's conscience** causing one feelings of guilt. **prisoner of conscience** a person imprisoned by a state for holding political or religious views it does not tolerate. □ **conscienceless** adj. [ME f. OF f. L conscientia f. conscire be privy to (as COM-, scire know)]

conscientious /ˌkɒnʃɪˈenʃəs/ adj. (of a person or conduct) governed by a sense of duty; diligent and scrupulous. □ **conscientious objector** a person who for reasons of conscience objects to conforming to a requirement, esp. that of military service. □ **conscientiously** adv. **conscientiousness** n. [F consciencieux f. med.L conscientiosus (as CONSCIENCE)]

conscious /ˈkɒnʃəs/ adj. & n. ● adj. **1** awake and aware of one's surroundings and identity. **2** (usu. foll. by of, or that + clause) aware, knowing (*conscious of his inferiority*). **3** (of actions, emotions, etc.) realized or recognized by the doer; intentional (*made a conscious effort not to laugh*). **4** (in comb.) aware of; concerned with (*fashion-conscious*). ● n. (prec. by the) the conscious mind. □ **consciously** adv. [L conscius knowing with others or in oneself f. conscire (as COM-, scire know)]

consciousness /ˈkɒnʃəsnɪs/ n. **1** the state of being conscious (*lost consciousness during the fight*). **2 a** awareness, perception (*had no consciousness of being ridiculed*). **b** (in comb.) awareness of (*class-consciousness*). **3** the totality of a person's thoughts, feelings, and sensations or of a class of these (*moral consciousness*). □ **consciousness-raising** the activity of increasing esp. social or political sensitivity or awareness.

conscribe /kənˈskraɪb/ v.tr. = CONSCRIPT v. [L conscribere (as CONSCRIPTION)]

conscript v. & n. ● v.tr. /kənˈskrɪpt/ enlist by conscription. ● n. /ˈkɒnskrɪpt/ a person enlisted by conscription. [(v.) back-form. f. CONSCRIPTION: (n.) f. F conscrit f. L conscriptus (as CONSCRIPTION)]

conscription /kənˈskrɪpʃ(ə)n/ n. compulsory enlistment for state service, esp. military service. [F f. LL conscriptio levying of troops f. L conscribere conscript- write)]

consecrate /ˈkɒnsɪˌkreɪt/ v.tr. **1** make or declare sacred; dedicate formally to a religious or divine purpose. **2** (in Christian belief) make (bread and wine) into the body and blood of Christ. **3** (foll. by to) devote (one's life etc.) to (a purpose). **4** ordain (esp. a bishop) to a sacred office.

□ **consecrator** n. **consecratory** adj. **consecration** /ˌkɒnsɪˈkreɪʃ(ə)n/ n. [ME f. L consecrare (as COM-, secrare = sacrare dedicate f. sacer sacred)]

consecution /ˌkɒnsɪˈkjuːʃ(ə)n/ n. **1** logical sequence (in argument or reasoning). **2** sequence, succession (of events etc.). [L consecutio f. consequi consecut- overtake (as COM-, sequi pursue)]

consecutive /kənˈsekjʊtɪv/ adj. **1 a** following continuously. **b** in unbroken or logical order. **2** Gram. expressing consequence. □ **consecutive intervals** Mus. intervals of the same kind (esp. fifths or octaves), occurring in succession between two voices or parts in harmony. □ **consecutively** adv. **consecutiveness** n. [F consécutif -ive f. med.L consecutivus (as CONSECUTION)]

consensual /kənˈsensjʊəl, -ˈsenʃʊəl/ adj. of or by consent or consensus. □ **consensually** adv. [L consensus (see CONSENSUS) + -AL¹]

consensus /kənˈsensəs/ n. (often foll. by of) **1 a** general agreement (of opinion, testimony, etc.). **b** an instance of this. **2** (attrib.) majority view, collective opinion (*consensus politics*). [L, = agreement (as CONSENT)]

consent /kənˈsent/ v. & n. ● v.intr. (often foll. by to) express willingness, give permission, agree. ● n. voluntary agreement, permission, compliance. □ **age of consent** the age at which consent to sexual intercourse is valid in law. **consenting adult** an adult who consents to something, esp. a homosexual act. [ME f. OF consentir f. L consentire (as COM-, sentire sens- feel)]

consentient /kənˈsenʃ(ə)nt/ adj. **1** agreeing, united in opinion. **2** concurrent. **3** (often foll. by to) consenting. [L consentient- (as CONSENT)]

consequence /ˈkɒnsɪkwəns/ n. **1** the result or effect of an action or condition. **2 a** importance (*it is of no consequence*). **b** social distinction (*persons of consequence*). **3** (in pl., often treated as sing.) a game in which a narrative (usu. describing the meeting of a man and woman, and its consequences) is made up by the players, each ignorant of what has already been contributed. □ **in consequence** as a result. **take the consequences** accept the results of one's choice or action. [ME f. OF f. L consequentia (as CONSEQUENT)]

consequent /ˈkɒnsɪkwənt/ adj. & n. ● adj. **1** (often foll. by on, upon) following as a result or consequence. **2** logically consistent. ● n. **1** a thing that follows another. **2** Logic the second part of a conditional proposition, dependent on the antecedent. [ME f. OF f. L consequi (as CONSECUTION)]

consequential /ˌkɒnsɪˈkwenʃ(ə)l/ adj. **1** following as a result or consequence. **2** resulting indirectly (*consequential damage*). **3** important; significant. □ **consequentially** adv. **consequentiality** /-ˌkwenʃɪˈælɪtɪ/ n. [L consequentia]

consequently /ˈkɒnsɪkwəntlɪ/ adv. & conj. as a result; therefore.

conservancy /kənˈsɜːvənsɪ/ n. (pl. **-ies**) **1** Brit. a commission etc. controlling a port, river, etc. (*Thames Conservancy*). **2** a body concerned with the preservation of natural resources. **3** conservation; official preservation (of forests etc.). [18th-c. alt. of obs. conservacy f. AF conservacie f. AL conservatia f. L conservatio (as CONSERVE)]

conservation /ˌkɒnsəˈveɪʃ(ə)n/ n. preservation, esp. of the natural environment. □ **conservation area** an area containing a noteworthy environment and specially protected by law against undesirable changes. **conservation of energy** (or **mass** or **momentum** etc.) Physics the principle that the total quantity of energy etc. of any system not subject to external action remains constant. □ **conservational** adj. [ME f. OF conservation or L conservatio (as CONSERVE)]

conservationist /ˌkɒnsəˈveɪʃənɪst/ n. a supporter or advocate of environmental conservation.

conservative /kənˈsɜːvətɪv/ adj. & n. ● adj. **1 a** averse to rapid change. **b** (of views, taste, etc.) moderate, avoiding extremes (*conservative in his dress*). **2** (of an estimate etc.) purposely low; moderate, cautious. **3** (**Conservative**) of or characteristic of Conservatives or the Conservative Party. **4** tending to conserve. ● n. **1** a conservative person. **2** (**Conservative**) a supporter or member of the Conservative Party. □ **conservative surgery** surgery that seeks to preserve tissues as far as possible. □ **conservatism** n. **conservatively** adv. **conservativeness** n. [ME f. LL conservativus (as CONSERVE)]

Conservative Judaism n. a Jewish movement chiefly in North America seeking to preserve Jewish tradition and ritual while at the same time making some modifications to strict Orthodox Judaism. The movement has its origin in the ideas of a group of 19th-century Jewish intellectuals, in particular Zachariah Frankel (1801–75).

Conservative Party a political party disposed to maintain existing institutions and promote private enterprise. The modern Conservative

Party in Britain (full name Conservative and Unionist Party) emerged from the old Tory Party under Peel in the 1830s and 1840s. Under Disraeli it was the party committed to traditional institutions, social reform, and the defence of the British Empire. After the First World War the Conservatives benefited from the decline of their traditional opponents, the Liberals, and dominated the political scene until defeated in the general election of 1945 by the Labour Party. During the postwar period the Conservatives were in power 1951–64, 1970–4, and from 1979. Under Margaret Thatcher and then John Major the party won an unprecedented four general elections and was able to enact a radical right-wing economic and social programme, including privatization of nationalized industries, reduction in direct taxation (but increased indirect taxation), dismantling of trade-union and wages policy legislation, deregulation of money markets, and reduction of local government control especially in health and education.

conservatoire /kən'sɜ:və,twɑ:(r)/ n. a (usu. European) school of music or other arts. [F f. It. *conservatorio* (as CONSERVATORY)]

conservator /'kɒnsə,veɪtə(r), kən'sɜ:vətə(r)/ n. a person who preserves something; an official custodian (of a museum etc.). [ME f. AF *conservatour*, OF *-ateur* f. L *conservator -oris* (as CONSERVE)]

conservatorium /kən,sɜ:və'tɔ:rɪəm/ n. *Austral.* = CONSERVATOIRE.

conservatory /kən'sɜ:vətərɪ/ n. (pl. **-ies**) **1** a greenhouse for tender plants; a room, esp. one attached to and communicating with a house, designed for the growing or displaying of plants. **2** esp. *N. Amer.* = CONSERVATOIRE. [LL *conservatorium* (as CONSERVE): sense 2 through It. *conservatorio*]

conserve /kən'sɜ:v/ v. & n. ● v.tr. **1** store up; keep from harm or damage, esp. for later use. **2** *Physics* maintain a quantity of (heat etc.). **3** preserve (food, esp. fruit), usu. with sugar. ● n. (also /'kɒnsɜ:v/) **1** fruit etc. preserved in sugar. **2** fresh fruit jam. [ME f. OF *conserver* f. L *conservare* (as COM-, *servare* keep)]

consider /kən'sɪdə(r)/ v.tr. (often *absol.*) **1 a** contemplate mentally, esp. in order to reach a conclusion; give attention to. **b** examine the merits of (a course of action, a candidate, claim, etc.). **2** look attentively at. **3** take into account; show regard for. **4** (foll. by *that* + clause) have the opinion. **5** (foll. by complement) believe; regard as (*consider it to be genuine; consider it settled*). **6** (as **considered** *adj.*) formed after careful thought (*a considered opinion*). □ **all things considered** taking everything into account. [ME f. OF *considerer* f. L *considerare* examine]

considerable /kən'sɪdərəb(ə)l/ adj. **1** enough in amount or extent to merit consideration, much; a lot of (*considerable pain*). **2** notable, important (*considerable achievement*). □ **considerably** adv.

considerate /kən'sɪdərət/ adj. **1** thoughtful towards other people; careful not to cause hurt or inconvenience. **2** *archaic* careful. □ **considerately** adv.

consideration /kən,sɪdə'reɪʃ(ə)n/ n. **1** the act of considering; careful thought. **2** thoughtfulness for others; being considerate. **3** a fact or a thing taken into account in deciding or judging something. **4** compensation; a payment or reward. **5** *Law* (in a contractual agreement) anything given or promised or forborne by one party in exchange for the promise or undertaking of another. **6** *archaic* importance or consequence. □ **in consideration of** in return for; on account of. **take into consideration** include as a factor, reason, etc.; make allowance for. **under consideration** being considered. [ME f. OF f. L *consideratio -onis* (as CONSIDER)]

considering /kən'sɪdərɪŋ/ prep., conj., & adv. ● prep. & conj. in view of; taking into consideration (*considering their youth; considering that it was snowing*). ● adv. *colloq.* all in all; taking everything into account (*not so bad, considering*).

consign /kən'saɪn/ v.tr. (often foll. by *to*) **1** hand over; deliver to a person's possession or trust. **2** assign; commit decisively or permanently (*consigned it to the dustbin; consigned to years of misery*). **3** transmit or send (goods), usu. by a public carrier. □ **consignor** n.

consignee /,kɒnsaɪ'ni:/ n. [ME f. F *consigner* or L *consignare* mark with a seal (as COM-, SIGN)]

consignment /kən'saɪnmənt/ n. **1** the act or an instance of consigning; the process of being consigned. **2** a batch of goods consigned.

consist /kən'sɪst/ v.intr. **1** (foll. by *of*) be composed of; have ingredients or elements as specified. **2** (foll. by *in, of*) have its essential features as specified (*its beauty consists in the use of colour*). **3** (usu. foll. by *with*) harmonize; be consistent. [L *consistere* exist (as COM-, *sistere* stop)]

consistency /kən'sɪstənsɪ/ n. (also **consistence**) (pl. **-ies** or **-es**)

1 a the degree of firmness with which a substance holds together. **b** the degree of density, esp. of thick liquids. **2** the state of being consistent; conformity with other or earlier attitudes, practice, etc. [F *consistence* or LL *consistentia* (as CONSIST)]

consistent /kən'sɪstənt/ adj. (usu. foll. by *with*) **1** compatible or in harmony; not contradictory. **2** (of a person) constant to the same principles of thought or action. □ **consistently** adv. [L *consistere* (as CONSIST)]

consistory /kən'sɪstərɪ/ n. (pl. **-ies**) **1** *RC Ch.* the council of cardinals (with or without the pope). **2** (in full **consistory court**) (in the Church of England) a court presided over by a bishop, for the administration of ecclesiastical law in a diocese. **3** (in other Churches) a local administrative body. □ **consistorial** /,kɒnsɪ'stɔ:rɪəl/ adj. [ME f. AF *consistorie*, OF *-oire* f. LL *consistorium* (as CONSIST)]

consociation /kən,səʊsɪ'eɪʃ(ə)n/ n. **1** close association, esp. of Churches or religious communities. **2** *Ecol.* a closely related subgroup of plants having one dominant species. [L *consociatio, -onis* f. *consociare* (as COM-, *socius* fellow)]

consolation /,kɒnsə'leɪʃ(ə)n/ n. **1** the act of an instance of consoling; the state of being consoled. **2** a consoling thing, person, or circumstance. □ **consolation prize** a prize given to a competitor who just fails to win a main prize. □ **consolatory** /kən'sɒlətərɪ/ adj. [ME f. OF, f. L *consolatio -onis* (as CONSOLE¹)]

console¹ /kən'səʊl/ v.tr. comfort, esp. in grief or disappointment. ¶ Often confused with *condole*. □ **consolable** adj. **consoler** n. **consolingly** adv. [F *consoler* f. L *consolari*]

console² /'kɒnsəʊl/ n. **1** a panel or unit accommodating a set of switches, controls, etc. **2** a cabinet for television or radio equipment etc. **3** *Mus.* a cabinet with the keyboards, stops, pedals, etc., of an organ. **4** an ornamented bracket supporting a shelf etc. □ **console table** a table supported by a bracket against a wall. [F, perh. f. *consolider* (as CONSOLIDATE)]

consolidate /kən'sɒlɪ,deɪt/ v. **1** tr. & intr. make or become strong or solid. **2** tr. reinforce or strengthen (one's position, power, etc.). **3** tr. combine (territories, companies, debts, etc.) into one whole. □ **consolidator** n. **consolidatory** adj. **consolidation** /-,sɒlɪ'deɪʃ(ə)n/ n. [L *consolidare* (as COM-, *solidare* f. *solidus* solid)]

Consolidated Fund in Britain, the Exchequer account at the Bank of England into which public monies (such as tax receipts) are paid and from which the main payments are made, other than those dependent on periodic parliamentary approval. These payments include interest on the national debt, grants to the royal family, and payments on the civil list. The fund was established by William Pitt the Younger in 1786.

consols /'kɒnsɒlz/ n.pl. British government securities without redemption date and with fixed annual interest. [abbr. of *consolidated annuities*]

consommé /kən'sɒmeɪ/ n. a clear soup made with meat stock. [F, past part. of *consommer* f. L *consummare* (as CONSUMMATE)]

consonance /'kɒnsənəns/ n. **1** agreement, harmony. **2** *Prosody* a recurrence of similar-sounding consonants. **3** *Mus.* a harmonious combination of notes; a harmonious interval. [ME f. OF *consonance* or L *consonantia* (as CONSONANT)]

consonant /'kɒnsənənt/ n. & adj. ● n. **1** a speech sound in which the breath is at least partly obstructed, and which to form a syllable must be combined with a vowel. **2** a letter or letters representing this. ● adj. (foll. by *with, to*) **1** consistent; in agreement or harmony. **2** similar in sound. **3** *Mus.* making a concord. □ **consonantly** adv. **consonantal** /,kɒnsə'nænt(ə)l/ adj. [ME f. F f. L *consonare* (as COM-, *sonare* sound f. *sonus*]

con sordino /kɒn sɔ:'di:nəʊ/ adv. *Mus.* with the use of a mute. [It.]

consort¹ /'kɒnsɔ:t/ n. *Mus.* a group of musicians who regularly perform together, esp. one specializing in English music of about 1570 to 1720. □ **consort song** a song for solo voice and a consort usually of viols. [earlier form of CONCERT]

consort² n. & v. ● n. /'kɒnsɔ:t/ **1** a wife or husband, esp. of royalty (*prince consort*). **2** a ship sailing with another. ● v. /kən'sɔ:t/ **1** intr. (usu. foll. by *with, together*) **a** keep company; associate. **b** harmonize. **2** tr. class or bring together. [ME f. F f. L *consors* sharer, comrade (as COM-, *sors sortis* lot, destiny)]

consortium /kən'sɔ:tɪəm/ n. (pl. **consortia** /-tɪə/ or **consortiums**) **1** an association, esp. of several business companies. **2** *Law* the right of association with a husband or wife (*loss of consortium*). [L, = partnership (as CONSORT²)]

conspecific /ˌkɒnspɪˈsɪfɪk/ adj. Biol. of the same species.

conspectus /kənˈspektəs/ n. **1** a general or comprehensive survey. **2** a summary or synopsis. [L f. conspicere conspect- (as COM-, spicere look at)]

conspicuous /kənˈspɪkjʊəs/ adj. **1** clearly visible; striking to the eye (a conspicuous hole in his shirt). **2** remarkable; noteworthy (conspicuous success). **3** (of expenditure etc.) lavish, with a view to enhancing one's prestige. □ **conspicuously** adv. **conspicuousness** n. [L conspicuus (as CONSPECTUS)]

conspiracy /kənˈspɪrəsɪ/ n. (pl. -ies) **1** a secret plan to commit a crime or do harm, often for political ends; a plot. **2** the act of conspiring. □ **conspiracy of silence** an agreement to say nothing. **conspiracy theory** a belief that some covert but influential agency or organization is responsible for an unexplained event. [ME f. AF conspiracie, alt. form of OF conspiration f. L conspiratio -onis (as CONSPIRE)]

conspirator /kənˈspɪrətə(r)/ n. a person who takes part in a conspiracy. □ **conspiratorial** /-ˌspɪrəˈtɔːrɪəl/ adj. **conspiratorially** adv. [ME f. AF conspiratour, OF -teur (as CONSPIRE)]

conspire /kənˈspaɪə(r)/ v.intr. **1** combine secretly to plan and prepare an unlawful or harmful act. **2** (often foll. by against, or to + infin.) (of events or circumstances) seem to be working together, esp. disadvantageously. [ME f. OF conspirer f. L conspirare agree, plot (as COM-, spirare breathe)]

Constable /ˈkʌnstəb(ə)l/, John (1776–1837), English painter. Among his best-known works are early paintings like Flatford Mill (1817) and The Hay Wain (1821), both inspired by the landscape of his native Suffolk. By the late 1820s and early 1830s, Constable had become fascinated by the painting of changing weather patterns, and in works such as Sketch for 'Hadleigh Castle' (c.1828–9) and The Valley Farm (1835) he focused on the transient effects of clouds and light, so breaking new ground in landscape painting. He proved a great influence on French painters, especially Eugène Delacroix and the Barbizon School.

constable /ˈkʌnstəb(ə)l/ n. **1** Brit. **a** a policeman or policewoman. **b** (also **police constable**) a police officer of the lowest rank. **2** the governor of a royal castle. **3** hist. the principal officer in a royal household. [ME f. OF conestable f. LL comes stabuli count of the stable]

constabulary /kənˈstæbjʊlərɪ/ n. & adj. ● n. (pl. -ies) an organized body of police; a police force. ● attrib.adj. of or concerning the police force. [med.L constabularius (as CONSTABLE)]

Constance, Lake /ˈkɒnstəns/ (called in German Bodensee) a lake in SE Germany on the north side of the Swiss Alps, at the meeting-point of Germany, Switzerland, and Austria, forming part of the course of the River Rhine.

constancy /ˈkɒnstənsɪ/ n. **1** the quality of being unchanging and dependable; faithfulness. **2** firmness, endurance. [L constantia (as CONSTANT)]

constant /ˈkɒnstənt/ adj. & n. ● adj. **1** continuous (needs constant attention). **2** occurring frequently (receive constant complaints). **3** (often foll. by to) unchanging, faithful, dependable. ● n. **1** anything that does not vary. **2** Math. a component of a relationship between variables that does not change its value. **3** Physics a number expressing a relation, property, etc., and remaining the same in all circumstances. **b** such a number that remains the same for a substance in the same conditions. □ **constantly** adv. [ME f. OF f. L constare (as COM-, stare stand)]

Constanţa /kɒnˈstæntsə/ (also **Constanza**) the chief port of Romania, on the Black Sea; pop. (1989) 315,920. Founded in the 7th century BC by the Greeks, it was under Roman rule from 72 BC. Formerly called Tomis, it was renamed after Constantine I in the 4th century.

constantan /ˈkɒnstənˌtæn/ n. an alloy of copper and nickel used in electrical equipment. [CONSTANT + -AN]

Constantine[1] /ˈkɒnstənˌtaɪn/ a city in NE Algeria; pop. (1989) 449,000. In ancient times it was the capital of the Roman province of Numidia. It was destroyed in 311 but rebuilt soon afterwards by Constantine the Great and given his name.

Constantine[2] /ˈkɒnstənˌtaɪn/ (known as Constantine the Great) (c.274–337), Roman emperor. The years from 305 until Constantine became sole emperor in 324 were marked by civil wars and continuing rivalry for the imperial throne. Constantine was the first Roman emperor to be converted to Christianity; he issued a decree of toleration towards Christians in the empire in 313. In 324 he made Christianity a state religion, although paganism was also tolerated. In 330 he moved his capital from Rome to Byzantium, renaming it Constantinopolis (Constantinople). His reign was marked by

increasing imperial control of the Eastern Church and much church building, especially at the holy sites in Palestine. In the Orthodox Church he is venerated as a saint.

Constantinople /ˌkɒnstæntɪˈnəʊp(ə)l/ the former name for Istanbul from AD 330 (when it was given its name by Constantine the Great) to the capture of the city by the Turks in 1453. Constantinople is the anglicized form of Constantinopolis, 'city of Constantine'.

Constanza see CONSTANŢA.

constellate /ˈkɒnstəˌleɪt/ v.tr. **1** form into (or as if into) a constellation. **2** adorn as with stars.

constellation /ˌkɒnstəˈleɪʃ(ə)n/ n. **1** Astron. a group of stars forming a recognizable pattern in the sky. (See note below.) **2** a group of associated persons, ideas, etc. [ME f. OF f. LL constellatio -onis (as COM-, stella star)]

▪ The constellations are traditionally identified by some imaginative name describing their apparent form or identifying them with a mythological figure. They have been placed on a formal basis by modern astronomers, who divide the sky into eighty-eight constellations with defined boundaries. Many of the currently recognized constellations are identical with those of the ancient Egyptians. (See also ASTERISM 1.)

consternate /ˈkɒnstəˌneɪt/ v.tr. (usu. in passive) dismay; fill with anxiety. [L consternare (as COM-, sternere throw down)]

consternation /ˌkɒnstəˈneɪʃ(ə)n/ n. anxiety or dismay causing mental confusion. [F consternation or L consternatio (as CONSTERNATE)]

constipate /ˈkɒnstɪˌpeɪt/ v.tr. (esp. as **constipated** adj.) affect with constipation. [L constipare (as COM-, stipare press)]

constipation /ˌkɒnstɪˈpeɪʃ(ə)n/ n. **1** a condition with hardened faeces and difficulty in emptying the bowels. **2** a restricted state. [ME f. OF constipation or LL constipatio (as CONSTIPATE)]

constituency /kənˈstɪtjʊənsɪ/ n. (pl. -ies) **1** a body of voters in a specified area who elect a representative member to a legislative body. **2** the area represented in this way. **3** a body of customers, supporters, etc.

constituent /kənˈstɪtjʊənt/ adj. & n. ● adj. **1** composing or helping to make up a whole. **2** able to make or change a (political etc.) constitution (constituent assembly). **3** appointing or electing. ● n. **1** a member of a constituency (esp. political). **2** a component part. **3** Law a person who appoints another as agent. [L constituent- partly through F -ant (as CONSTITUTE)]

constitute /ˈkɒnstɪˌtjuːt/ v.tr. **1** be the components or essence of; make up, form. **2 a** be equivalent or tantamount to (this constitutes an official warning). **b** formally establish (does not constitute a precedent). **3** give legal or constitutional form to; establish by law. [L constituere (as COM-, statuere set up)]

constitution /ˌkɒnstɪˈtjuːʃ(ə)n/ n. **1** the act or method of constituting; the composition (of something). **2 a** the body of fundamental principles or established precedents according to which a state or other organization is acknowledged to be governed. **b** a (usu. written) record of this. (See also CONSTITUTION, THE.) **3** a person's physical state as regards vitality, health, strength, etc. **4** a person's mental or psychological make-up. **5** hist. a decree or ordinance. [ME f. OF constitution or L constitutio (as CONSTITUTE)]

Constitution, the (in full **Constitution of the USA**) the basic written set of principles and precedents of federal government in the US. The Constitution, consisting of seven main articles, was drawn up in 1787, largely based on the earlier Articles of Confederation (1781–7); after ratification by the states of the Union, it came into operation in 1789. Since 1789 twenty-six amendments have been added, including, in particular, the Bill of Rights (1791), which contains the first ten amendments. The underlying principle and aim of the Constitution is the preservation of individual citizens' and states' rights while ensuring an effective federal government; the division into legislative (Congress), judicial (Supreme Court), and executive (President) arms of government was designed to limit overall federal powers.

constitutional /ˌkɒnstɪˈtjuːʃ(ə)n(ə)l/ adj. & n. ● adj. **1** of, consistent with, authorized by, or limited by a political constitution (a constitutional monarchy). **2** inherent in, stemming from, or affecting the physical or mental constitution. ● n. usu. archaic or joc. a walk taken regularly to maintain or restore good health. □ **constitutionalize** v.tr. (also -ise). **constitutionally** adv. **constitutionality** /-ˌtjuːʃəˈnælɪtɪ/ n.

constitutionalism /ˌkɒnstɪˈtjuːʃənəˌlɪz(ə)m/ n. **1** a constitutional system of government. **2** the adherence to or advocacy of such a system. □ **constitutionalist** n.

constitutive /ˈkɒnstɪˌtjuːtɪv/ *adj.* **1** able to form or appoint. **2** component. **3** essential. □ **constitutively** *adv.* [LL *constitutivus* (as CONSTITUTE)]

constrain /kənˈstreɪn/ *v.tr.* **1** compel; urge irresistibly or by necessity. **2 a** confine forcibly; imprison. **b** restrict severely as regards action, behaviour, etc. **3** bring about by compulsion. **4** (as **constrained** *adj.*) forced, embarrassed (*a constrained voice; a constrained manner*). □ **constrainedly** /-nɪdlɪ/ *adv.* [ME f. OF *constraindre* f. L *constringere* (as COM-, *stringere* strict- tie)]

constraint /kənˈstreɪnt/ *n.* **1** the act or result of constraining or being constrained; restriction of liberty. **2** something that constrains; a limitation on motion or action. **3** the restraint of natural feelings or their expression; a constrained manner. [ME f. OF *constreinte*, fem. past part. (as CONSTRAIN)]

constrict /kənˈstrɪkt/ *v.tr.* **1** make narrow or tight; compress. **2** *Biol.* cause (organic tissue) to contract. □ **constrictive** *adj.* **constriction** /-ˈstrɪkʃ(ə)n/ *n.* [L (as CONSTRAIN)]

constrictor /kənˈstrɪktə(r)/ *n.* **1** any snake (esp. a boa) that kills by coiling round its prey and compressing it. **2** *Anat.* any muscle that compresses or contracts an organ or part of the body. [mod.L (as CONSTRICT)]

construct *v. & n.* ● *v.tr.* /kənˈstrʌkt/ **1** make by fitting parts together; build, form (something physical or abstract). **2** *Geom.* draw or delineate, esp. accurately to given conditions (*construct a triangle*). ● *n.* /ˈkɒnstrʌkt/ **1** a thing constructed, esp. by the mind. **2** *Linguistics* a group of words forming a phrase. □ **constructor** /kənˈstrʌktə(r)/ *n.* [L *construere construct-* (as COM-, *struere* pile, build)]

construction /kənˈstrʌkʃ(ə)n/ *n.* **1** the act or a mode of constructing. **2** a thing constructed. **3** an interpretation or explanation (*they put a generous construction on his act*). **4** *Gram.* an arrangement of words according to syntactical rules. □ **constructional** *adj.* **constructionally** *adv.* [ME f. OF f. L *constructio -onis* (as CONSTRUCT)]

constructive /kənˈstrʌktɪv/ *adj.* **1 a** of construction; tending to construct. **b** tending to form a basis for ideas (opp. DESTRUCTIVE) (*constructive criticism*). **2** helpful, positive (*a constructive approach*). **3** derived by inference; not expressed (*constructive permission*). **4** belonging to the structure of a building. □ **constructively** *adv.* **constructiveness** *n.* [LL *constructivus* (as CONSTRUCT)]

constructivism /kənˈstrʌktɪˌvɪz(ə)m/ *n.* a 20th-century artistic style and movement, characterized by work which is generally abstract and three-dimensional and uses machine-age materials such as sheet metal, glass, perspex, wire, and tubing. The original movement in the 1920s stemmed from the art and writings of Russian artists such as Naum Gabo and Antoine Pevsner, and is often called *Russian constructivism*; it had a significant impact on Bauhaus and the De Stijl group. In subsequent and recent years, the term acquired a wider, generalized currency, and was loosely associated with many aspects of modern architecture and design; these later developments are often referred to as *international* or *European constructivism*. □ **constructivist** *n. & adj.* [Russ. *konstruktivizm* (as CONSTRUCT)]

construe /kənˈstruː/ *v.tr.* (**construes**, **construed**, **construing**) **1** interpret (words or actions) (*their decision can be construed in many ways*). **2** (often foll. by *with*) combine (words) grammatically (*'rely' is construed with 'on'*). **3** analyse the syntax of (a sentence). **4** translate word for word. □ **construable** *adj.* **construal** *n.* [ME f. L *construere* CONSTRUCT]

consubstantial /ˌkɒnsəbˈstænʃ(ə)l/ *adj.* (in Christian theology) of the same substance (esp. of the three persons of the Trinity). □ **consubstantiality** /-ˌstænʃɪˈælɪtɪ/ *n.* [ME f. eccl.L *consubstantialis*, transl. Gk *homoousios* (as COM-, SUBSTANTIAL)]

consubstantiation /ˌkɒnsəbˌstænʃɪˈeɪʃ(ə)n/ *n.* (in Christian theology) the presence in the Eucharist, after consecration of the elements, of the real substances of the body and blood of Christ coexisting with those of the bread and wine. The doctrine asserting this, associated especially with Lutheran belief, was formulated in opposition to the doctrine of transubstantiation. [mod.L *consubstantiatio*, after *transubstantiatio* TRANSUBSTANTIATION]

consuetude /ˈkɒnswɪˌtjuːd/ *n.* a custom, esp. one having legal force in Scotland. □ **consuetudinary** /ˌkɒnswɪˈtjuːdɪnərɪ/ *adj.* [ME f. OF *consuetude* or L *consuetudo -dinis* f. *consuetus* accustomed]

consul /ˈkɒns(ə)l/ *n.* **1** an official appointed by a state to live in a foreign city and protect the state's citizens and interests there. **2** *hist.* either of two annually elected chief magistrates in ancient Rome. **3** any of the three chief magistrates of the French republic (1799–1804).

consulship *n.* **consular** /ˈkɒnsjʊlə(r)/ *adj.* [ME f. L, rel. to *consulere* take counsel]

consulate /ˈkɒnsjʊlət/ *n.* **1** the building officially used by a consul. **2** the office, position, or period of office of consul. **3** *hist.* government by consuls. **4** *hist.* the period of office of a consul. **5** *hist.* (**Consulate**) the government of France by three consuls (1799–1804). [ME f. L *consulatus* (as CONSUL)]

consult /kənˈsʌlt/ *v.* **1** *tr.* seek information or advice from (a person, book, watch, etc.). **2** *intr.* (often foll. by *with*) refer to a person for advice, an opinion, etc. **3** *tr.* seek permission or approval from (a person) for a proposed action. **4** *tr.* take into account; consider (feelings, interests, etc.). □ **consultative** /-tətɪv/ *adj.* [F *consulter* f. L *consultare* frequent. of *consulere* consult- take counsel]

consultancy /kənˈsʌltənsɪ/ *n.* (*pl.* **-ies**) the professional practice or position of a consultant.

consultant /kənˈsʌlt(ə)nt/ *n.* **1** a person providing professional advice etc., esp. for a fee. **2** a senior specialist in a branch of medicine responsible for patients in a hospital. [prob. F (as CONSULT)]

consultation /ˌkɒnsəlˈteɪʃ(ə)n/ *n.* **1** a meeting arranged to consult (esp. with a consultant). **2** the act or an instance of consulting. **3** a conference. [ME f. OF *consultation* or L *consultatio* (as CONSULTANT)]

consulting /kənˈsʌltɪŋ/ *attrib.adj.* giving professional advice to others working in the same field or subject (*consulting physician*).

consumable /kənˈsjuːməb(ə)l/ *adj. & n.* ● *adj.* that can be consumed; intended for consumption. ● *n.* (usu. in *pl.*) a commodity that is eventually used up, worn out, or eaten.

consume /kənˈsjuːm/ *v.tr.* **1** eat or drink. **2** completely destroy; reduce to nothing or to tiny particles (*fire consumed the building*). **3** engage the full attention of, engross; dominate (often foll. by *with* or *by*: *consumed with rage*). **4** use up (time, energy, etc.). □ **consumingly** *adv.* [ME f. L *consumere* (as COM-, *sumere sumpt-* take up): partly through F *consumer*]

consumer /kənˈsjuːmə(r)/ *n.* **1** a person who consumes, esp. a person who uses a product. **2** a purchaser of goods or services. □ **consumer durable** a household product with a relatively long useful life (e.g. a radio or washing-machine). **consumer goods** goods put to use by consumers, not used in producing other goods (opp. *capital goods* (see CAPITAL¹)). **consumer research** investigation of purchasers' needs and opinions. **consumer society** a society in which the marketing of goods and services is an important social and economic activity.

consumerism /kənˈsjuːməˌrɪz(ə)m/ *n.* **1** the protection or promotion of consumers' interests in relation to the producer. **2** (often *derog.*) a preoccupation with consumer goods and their acquisition. □ **consumerist** *adj. & n.*

consummate *v. & adj.* ● *v.tr.* /ˈkɒnsəˌmeɪt/ **1** complete; make perfect. **2** complete (a marriage) by sexual intercourse. ● *adj.* /kənˈsʌmɪt, ˈkɒnsəmɪt/ complete, perfect; fully skilled (*a consummate general*). □ **consummately** /kənˈsʌmɪtlɪ, ˈkɒnsəmɪtlɪ/ *adv.* **consummator** /ˈkɒnsəˌmeɪtə(r)/ *n.* [L *consummare* (as COM-, *summare* complete f. *summus* utmost]

consummation /ˌkɒnsəˈmeɪʃ(ə)n/ *n.* **1** completion, esp. of a marriage by sexual intercourse. **2** a desired end or goal; perfection. [ME f. OF *consommation* or L *consummatio* (as CONSUMMATE)]

consumption /kənˈsʌmpʃ(ə)n/ *n.* **1** the act or an instance of consuming; the process of being consumed. **2** any disease causing wasting of tissues, esp. pulmonary tuberculosis. **3** an amount consumed. **4** the purchase and use of goods etc. **5** use by a particular person or group (*a film unfit for children's consumption*). [ME f. OF *consomption* f. L *consumptio* (as CONSUME)]

consumptive /kənˈsʌmptɪv/ *adj. & n.* ● *adj.* **1** of or tending to consumption. **2** tending to or affected with pulmonary tuberculosis. ● *n.* a consumptive patient. □ **consumptively** *adv.* [med.L *consumptivus* (as CONSUMPTION)]

cont. *abbr.* **1** contents. **2** continued.

contact /ˈkɒntækt/ *n. & v.* ● *n.* **1** the state or condition of touching, meeting, or communicating. **2** a person who is or may be communicated with for information, supplies, assistance, etc. **3** *Electr.* **a** a connection for the passage of a current. **b** a device for providing this. **4** a person likely to carry a contagious disease through being associated with an infected person. **5** (usu. in *pl.*) *colloq.* a contact lens. ● *v.tr.* (also /kənˈtækt/) **1** get into communication with (a person). **2** begin correspondence or personal dealings with. □ **contact lens** a small lens placed directly on the eyeball to correct the vision. **contact print** a photographic print made by placing a negative directly on sensitized paper etc. and illuminating it. **contact sport** a sport in which participants necessarily come into bodily contact with one another. □ **contactable** /'kɒntæktəb(ə)l, kən'tækt-/ adj. [L contactus f. contingere (as COM-, tangere touch)]

contagion /kən'teɪdʒən/ n. **1 a** the communication of disease from one person to another by bodily contact. **b** a contagious disease. **2** a contagious or harmful influence. **3** moral corruption, esp. when tending to be widespread. [ME f. L contagio (as COM-, tangere touch)]

contagious /kən'teɪdʒəs/ adj. **1 a** (of a person) likely to transmit disease by contact. **b** (of a disease) transmitted in this way. **2** (of emotions, reactions, etc.) likely to affect others (contagious enthusiasm). □ **contagious abortion** brucellosis in cattle, as producing abortion. □ **contagiously** adv. **contagiousness** n. [ME f. LL contagiosus (as CONTAGION)]

contain /kən'teɪn/ v.tr. **1** hold or be capable of holding within itself; include, comprise. **2** (of measures) consist of or be equal to (a gallon contains eight pints). **3** prevent (an enemy, difficulty, etc.) from moving or extending. **4** control or restrain (oneself, one's feelings, etc.). **5** (of a number) be divisible by (a factor) without a remainder. □ **containable** adj. [ME f. OF contenir f. L continere content- (as COM-, tenere hold)]

container /kən'teɪnə(r)/ n. **1** a vessel, box, etc., for holding particular things. **2** a large boxlike receptacle of standard design for the transport of goods, esp. one readily transferable from one form of transport to another. □ **container port** a port specializing in handling goods stored in containers. **container ship** (or **train**) a ship (or train) designed to carry goods stored in containers.

containerize /kən'teɪnə,raɪz/ v.tr. (also **-ise**) **1** pack in or transport by container. **2** adapt to transport by container. □ **containerization** /-,teɪnəraɪ'zeɪʃ(ə)n/ n.

containment /kən'teɪnmənt/ n. the action or policy of preventing the expansion of a hostile country or influence.

contaminate /kən'tæmɪ,neɪt/ v.tr. **1** pollute, esp. with radioactivity. **2** infect; corrupt. □ **contaminant** n. **contaminator** n. **contamination** /-,tæmɪ'neɪʃ(ə)n/ n. [L contaminare (as COM-, tamen- rel. to tangere touch)]

contango /kən'tæŋgəʊ/ n. (pl. **-os**) Brit. Stock Exch. **1** the postponement of the transfer of stock from one account day to the next. **2** a percentage paid by the buyer for such a postponement. □ **contango day** the eighth day before settling day. [19th c.: prob. an arbitrary formation]

conte /kɒnt/ n. **1** a short story (as a form of literary composition). **2** a medieval narrative tale. [F]

contemn /kən'tem/ v.tr. literary despise; treat with disregard. □ **contemner** /-'temə(r), -'temnə(r)/ n. [ME f. OF contemner or L contemnere (as COM-, temnere tempt- despise)]

contemplate /'kɒntəm,pleɪt/ v. **1** tr. survey with the eyes or in the mind. **2** tr. regard (an event) as possible. **3** tr. intend; have as one's purpose (we contemplate leaving tomorrow). **4** intr. meditate. □ **contemplator** n. **contemplation** /,kɒntəm'pleɪʃ(ə)n/ n. [L contemplari (as COM-, templum place for observations)]

contemplative /kən'templətɪv/ adj. & n. ● adj. of or given to (esp. religious) contemplation; meditative. ● n. a person whose life is devoted to religious contemplation. □ **contemplatively** adv. [ME f. OF contemplatif -ive, or L contemplativus (as CONTEMPLATE)]

contemporaneous /kən,tempə'reɪnɪəs/ adj. (usu. foll. by with) **1** existing or occurring at the same time. **2** of the same period. □ **contemporaneously** adv. **contemporaneousness** n. **contemporaneity** /-pərə'ni:ɪtɪ/ n. [L contemporaneus (as COM-, temporaneus f. tempus -oris time)]

contemporary /kən'tempərərɪ/ adj. & n. ● adj. **1** living or occurring at the same time. **2** approximately equal in age. **3** following modern ideas or fashion in style or design. ● n. (pl. **-ies**) **1** a person or thing living or existing at the same time as another. **2** a person of roughly the same age as another. □ **contemporarily** adv. **contemporariness** n. [med.L contemporarius (as CONTEMPORANEOUS)]

contempt /kən'tempt/ n. **1** a feeling that a person or a thing is beneath consideration or worthless, or deserving scorn or extreme reproach. **2** the condition of being held in contempt. **3** (in full **contempt of court**) disobedience to or disrespect for a court of law and its officers. □ **beneath contempt** utterly despicable. **hold in contempt** despise. [ME f. L contemptus (as CONTEMN)]

contemptible /kən'temptɪb(ə)l/ adj. deserving contempt; despicable. □ **contemptibly** adv. **contemptibility** /-,temptɪ'bɪlɪtɪ/ n. [ME f. OF or LL contemptibilis (as CONTEMN)]

contemptuous /kən'temptjʊəs/ adj. (often foll. by of) showing contempt; scornful; insolent. □ **contemptuously** adv.

contemptuousness n. [med.L contemptuosus f. L contemptus (as CONTEMPT)]

contend /kən'tend/ v. **1** intr. (usu. foll. by with) strive, fight. **2** intr. (usu. foll. by for, with) compete (contending for the title; contending emotions). **3** tr. (usu. foll. by that + clause) assert, maintain. □ **contender** n. (esp. in sense 2) [OF contendre or L contendere (as COM-, tendere tent- stretch, strive)]

content[1] /kən'tent/ adj., v., & n. ● predic.adj. **1** satisfied; adequately happy; in agreement. **2** (foll. by to + infin.) willing. ● v.tr. make content; satisfy. ● n. a contented state; satisfaction. □ **to one's heart's content** to the full extent of one's desires. [ME f. OF f. L contentus satisfied, past part. of continere (as CONTAIN)]

content[2] /'kɒntent/ n. **1** (usu. in pl.) what is contained in something, esp. in a vessel, book, or house. **2** the amount of a constituent contained (low sodium content). **3 a** the substance or material dealt with (in a speech, work of art, etc.) as distinct from its form or style. **b** meaning or significance. **4** the capacity or volume of a thing. **5** (in pl.) (in full **table of contents**) a list of the titles of chapters etc. given at the front of a book, periodical, etc. [ME f. med.L contentum (as CONTAIN)]

contented /kən'tentɪd/ adj. (often foll. by with, or to + infin.) **1** happy, satisfied. **2** (foll. by with) willing to be content (was contented with the outcome). □ **contentedly** adv. **contentedness** n.

contention /kən'tenʃ(ə)n/ n. **1** a dispute or argument; rivalry. **2** a point contended for in an argument (it is my contention that you are wrong). □ **in contention** competing, esp. with a good chance of success. [ME f. OF contention or L contentio (as CONTEND)]

contentious /kən'tenʃəs/ adj. **1** argumentative, quarrelsome. **2** likely to cause an argument; disputed, controversial. □ **contentiously** adv. **contentiousness** n. [ME f. OF contentieux f. L contentiosus (as CONTENTION)]

contentment /kən'tentmənt/ n. a satisfied state; tranquil happiness.

conterminous /kɒn'tɜːmɪnəs/ adj. (often foll. by with) **1** having a common boundary. **2** coextensive in space, time, or meaning. □ **conterminously** adv. [L conterminus (as COM-, terminus boundary)]

contessa /kɒn'tesə/ n. an Italian countess. [It. f. LL comitissa: see COUNTESS]

contest n. & v. ● n. /'kɒntest/ **1** a process of contending; a competition. **2** a dispute; a controversy. ● v.tr. /kən'test/ **1** challenge or dispute (a decision etc.). **2** debate (a point, statement, etc.). **3** contend or compete for (a prize, parliamentary seat, etc.); compete in (an election). □ **contestable** /kən'testəb(ə)l/ adj. **contester** /-'testə(r)/ n. [L contestari (as COM-, testis witness)]

contestant /kən'testənt/ n. a person who takes part in a contest or competition.

contestation /,kɒnte'steɪʃ(ə)n/ n. **1** a disputation. **2** an assertion contended for. [L contestatio partly through F (as CONTEST)]

context /'kɒntekst/ n. **1** the parts of something written or spoken that immediately precede and follow a word or passage and clarify its meaning. **2** the circumstances relevant to something under consideration. □ **in context** with the surrounding words or circumstances. **out of context** without the surrounding words or circumstances and so not fully understandable. [ME f. L contextus (as COM-, texere text- weave)]

contextual /kən'tekstʃʊəl, -tjʊəl/ adj. of or belonging to the context; depending on the context. □ **contextually** adv.

contextualize /kən'tekstʃʊə,laɪz, kən'tekstjʊ-/ v.tr. (also **-ise**) place in a context; study in context. □ **contextualization** /-,tekstʃʊəlaɪ'zeɪʃ(ə)n, -,tekstjʊ-/ n.

contiguity /,kɒntɪ'gju:ɪtɪ/ n. **1** being contiguous; proximity; contact. **2** Psychol. the proximity of ideas or impressions in place or time, as a principle of association.

contiguous /kən'tɪgjʊəs/ adj. (usu. foll. by with, to) touching, esp. along a line; in contact. □ **contiguously** adv. [L contiguus (as COM-, tangere touch)]

continent[1] /'kɒntɪnənt/ n. **1** any of the main continuous expanses of land (Europe, Asia, Africa, North and South America, Australia, Antarctica). **2** (**the Continent**) Brit. the mainland of Europe as distinct from the British Isles. **3** continuous land; a mainland. [L terra continens (see CONTAIN) continuous land]

continent[2] /'kɒntɪnənt/ adj. **1** able to control movements of the bowels and bladder. **2** exercising self-restraint, esp. sexually. □ **continence** n. **continently** adv. [ME f. L (as CONTAIN)]

continental /ˌkɒntɪˈnent(ə)l/ adj. & n. ● adj. **1** of or characteristic of a continent. **2** (**Continental**) Brit. of, relating to, or characteristic of mainland Europe. ● n. an inhabitant of mainland Europe. □ **continental breakfast** a light breakfast of coffee, rolls, etc. **continental climate** a climate having wide variations of temperature. **continental divide** a watershed separating two river systems of a continent (see also CONTINENTAL DIVIDE). **continental quilt** Brit. a duvet. **continental shelf** an area of relatively shallow seabed between the shore of a continent and the deeper ocean. **continental slope** the relatively steep slope between the outer edge of the continental shelf and the ocean bed. □ **continentally** adv.

Continental Army US hist. the army raised (with George Washington as commander) by the Continental Congress of 1775.

Continental Celtic see CELTIC.

Continental Congress US hist. each of the three congresses held by the American colonies in revolt against British rule in 1774, 1775, and 1776 respectively. The second Congress, convened in the wake of the battles at Lexington and Concord, created a Continental Army under George Washington, which fought and eventually won the American War of Independence.

Continental Divide (also called Great Divide) the main series of mountain ridges in North America, chiefly the crests of the Rocky Mountains, which form a watershed separating the rivers flowing eastwards into the Atlantic Ocean or the Gulf of Mexico from those flowing westwards into the Pacific.

continental drift n. the gradual movement of the continents across the earth's surface through geological time. The German meteorologist Alfred Wegener, beginning in 1910, was the first serious proponent of continental drift. Wegener's theory was based mainly on similarities in rock formations and in the flora and fauna of different continents, as well as the correspondence of coastlines, e.g. of South America and Africa, but his ideas made little headway, mainly because he could not suggest a plausible physical mechanism. In the 1960s information from the floors of the oceans, seismic studies, palaeomagnetic surveys, and radiometric dating of rocks confirmed the reality of continental drift, and led to the theory of plate tectonics. It is believed that the continental crust of the earth once formed a single supercontinent called Pangaea, which broke up to form Gondwana and Laurasia; these further split to form the present-day continents. The motive force is believed to come from convection currents in the earth's mantle, but this remains poorly understood. Modern satellite surveying techniques enable continental drift to be directly measured and motions of the order of a few centimetres per year are detected.

contingency /kənˈtɪndʒənsɪ/ n. (pl. **-ies**) **1** a future event or circumstance regarded as likely to occur, or as influencing present action. **2** something dependent on another uncertain event or occurrence. **3** uncertainty of occurrence. **4 a** one thing incident to another. **b** an incidental expense etc. □ **contingency fund** a fund to cover incidental or unforeseen expenses. **contingency plan** a plan designed to take account of a possible future event or circumstance. [earlier contingence f. LL contingentia (as CONTINGENT)]

contingent /kənˈtɪndʒənt/ adj. & n. ● adj. **1** (usu. foll. by on, upon) conditional, dependent (on an uncertain event or circumstance). **2** associated. **3** (usu. foll. by to) incidental. **4 a** that may or may not occur. **b** fortuitous; occurring by chance. **5** true only under existing or specified conditions. ● n. a body forming part of a larger group. □ **contingently** adv. [L contingere (as COM-, tangere touch)]

continual /kənˈtɪnjʊəl/ adj. constantly or frequently recurring; always happening. □ **continually** adv. [ME f. OF continuel f. continuer (as CONTINUE)]

continuance /kənˈtɪnjʊəns/ n. **1** a state of continuing in existence or operation. **2** the duration of an event or action. **3** US Law an adjournment. [ME f. OF (as CONTINUE)]

continuant /kənˈtɪnjʊənt/ n. & adj. Phonet. ● n. a speech sound in which the vocal tract is only partly closed, allowing the breath to pass through and the sound to be prolonged (as with f, r, s, v). ● adj. of or relating to such a sound. [F continuant and L continuare (as CONTINUE)]

continuation /kənˌtɪnjʊˈeɪʃ(ə)n/ n. **1** the act or an instance of continuing; the process of being continued. **2** a part that continues something else. **3** Brit. Stock Exch. the carrying over of an account to the next settling day. □ **continuation day** Stock Exch. = contango day. [ME f. OF f. L continuatio -onis (as CONTINUE)]

continuative /kənˈtɪnjʊətɪv/ adj. tending or serving to continue. [LL continuativus (as CONTINUATION)]

continue /kənˈtɪnjuː/ v. (**continues**, **continued**, **continuing**) **1** tr. (often foll. by verbal noun, or to + infin.) persist in, maintain, not stop (an action etc.). **2 a** tr. (also absol.) resume or prolong (a narrative, journey, etc.). **b** intr. recommence after a pause (the concert will continue shortly). **3** tr. be a sequel to. **4** intr. **a** remain in existence or unchanged. **b** (with complement) remain in a specified state (the weather continued fine). **5** tr. US Law adjourn (proceedings). □ **continuable** adj. **continuer** n. [ME f. OF continuer f. L continuare make or be CONTINUOUS]

continuity /ˌkɒntɪˈnjuːɪtɪ/ n. (pl. **-ies**) **1 a** the state of being continuous. **b** an unbroken succession. **c** a logical sequence. **2** the detailed and self-consistent scenario of a film or broadcast. **3** the linking of broadcast items. □ **continuity girl** (or **man**) Cinematog. the person responsible for agreement of detail between different sessions of filming. [F continuité f. L continuitas -tatis (as CONTINUOUS)]

continuo /kənˈtɪnjʊəʊ/ n. (pl. **-os**) Mus. an accompaniment providing a bass line and harmonies which are indicated by figures, often played on a keyboard instrument. [basso continuo (It., = continuous bass)]

continuous /kənˈtɪnjʊəs/ adj. **1** unbroken, uninterrupted, connected throughout in space or time. **2** Gram. = PROGRESSIVE adj. 7. □ **continuous assessment** the evaluation of a pupil's progress throughout a course of study, as well as or instead of by examination. **continuous creation** the creation of the universe or the matter in it regarded as a continuous process. **continuous stationery** a continuous ream of paper, usu. perforated to form single sheets. □ **continuously** adv. **continuousness** n. [L continuus uninterrupted f. continere (as COM-, tenere hold)]

continuum /kənˈtɪnjʊəm/ n. (pl. **continua** /-jʊə/) anything seen as having a continuous, not discrete, structure (space–time continuum). [L, neut. of continuus: see CONTINUOUS]

contort /kənˈtɔːt/ v.tr. twist or force out of normal shape. [L contorquere contort- (as COM-, torquere twist)]

contortion /kənˈtɔːʃ(ə)n/ n. **1** the act or process of twisting. **2** a twisted state, esp. of the face or body. [L contortio (as CONTORT)]

contortionist /kənˈtɔːʃənɪst/ n. an entertainer who adopts contorted postures.

contour /ˈkɒntʊə(r)/ n. & v. ● n. **1** an outline, esp. representing or bounding the shape or form of something. **2** the outline of a natural feature, e.g. a coast or mountain mass. **3** (in full **contour line**) a line on a map joining points of equal altitude. **4** a line separating differently coloured parts of a design. ● v.tr. **1** mark with contour lines. **2** carry (a road or railway) round the side of a hill. □ **contour feather** any of the feathers which form the outline of a bird's plumage. **contour map** a map marked with contour lines. **contour ploughing** ploughing along lines of constant altitude to minimize soil erosion. [F f. It. contorno f. contornare draw in outline (as COM-, tornare turn)]

Contra /ˈkɒntrə/ n. a member of a guerrilla force in Nicaragua which opposed the left-wing Sandinista government 1979–90, and was supported by the US for much of that time. It was officially disbanded in 1990, after the Sandinistas' electoral defeat. [abbr. of Sp. contrarevolucionario counter-revolutionary]

contra- /ˈkɒntrə/ comb. form **1** against, opposite (contradict). **2** Mus. (of instruments, organ-stops, etc.) pitched an octave below (contra-bassoon). [L contra against]

contraband /ˈkɒntrəˌbænd/ n. & adj. ● n. **1** goods that have been smuggled, or imported or exported illegally. **2** prohibited trade; smuggling. **3** (in full **contraband of war**) goods forbidden to be supplied by neutrals to belligerents. ● adj. **1** forbidden to be imported or exported (at all or without payment of duty). **2** concerning traffic in contraband (contraband trade). □ **contrabandist** n. [Sp. contrabanda f. It. (as CONTRA-, bando proclamation)]

contrabass /ˈkɒntrəˌbeɪs/ n. Mus. a double bass. [It. (basso BASS¹)]

contraception /ˌkɒntrəˈsepʃ(ə)n/ n. the intentional prevention of conception and pregnancy; the use of contraceptive methods. Although some birth control methods date back to ancient times, the widespread practice of contraception is a 20th-century phenomenon. The commonest barrier method is the condom or sheath, worn on the penis; rubber condoms are now mass-produced very cheaply, and are also used to protect against sexually transmitted diseases. Other barrier methods, used by women, include the diaphragm or cap, often used in conjunction with spermicidal agents; intrauterine devices, such as the coil, act by preventing the fertilized ovum from implanting in the uterus. Hormonal methods, introduced in the 1960s, use synthetic sex hormones which prevent ovulation. These are commonly taken as a daily pill, although longer-acting injections are also used.

Other forms of contraception include male and female sterilization, coitus interruptus, and abstinence from intercourse close to the time of ovulation (the rhythm method). Artificial contraception is contrary to the teaching of some religious groups, notably the Roman Catholic Church. [CONTRA- + CONCEPTION]

contraceptive /ˌkɒntrə'septɪv/ adj. & n. ● adj. preventing pregnancy. ● n. a contraceptive device or drug.

contract n. & v. ● n. /'kɒntrækt/ **1** a written or spoken agreement between two or more parties, intended to be enforceable by law. **2** a document recording this. **3** marriage regarded as a binding commitment. **4** Bridge etc. an undertaking to win the number of tricks bid. **5** sl. an arrangement for someone to be killed, usu. by a hired assassin. ● v. /kən'trækt/ **1** tr. & intr. **a** make or become smaller. **b** draw together (muscles, the brow, etc.) or, be drawn together. **2 a** intr. (usu. foll. by with) make a contract. **b** intr. (usu. foll. by for, or to + infin.) enter formally into a business or legal arrangement. **c** tr. (often foll. by out) arrange (work) to be done by contract. **3** intr. place under a contract. **4** tr. catch or develop (a disease). **5** tr. form or develop (a friendship, habit, etc.). **6** tr. enter into (marriage). **7** tr. incur (a debt etc.). **8** tr. shorten (a word) by combination or elision. □ **contract bridge** see BRIDGE². **contract in** (or **out**) (also refl.) Brit. choose to be involved in (or withdraw or remain out of) a scheme or commitment. □ **contractive** /kən'træktɪv/ adj. [earlier as adj., = contracted: OF, f. L contractus (as COM-, trahere tract- draw)]

contractable /kən'træktəb(ə)l/ adj. (of a disease) that can be contracted.

contractible /kən'træktɪb(ə)l/ adj. that can be shrunk or drawn together.

contractile /kən'træktaɪl/ adj. capable of or producing contraction. □ **contractility** /ˌkɒntræk'tɪlɪtɪ/ n.

contraction /kən'trækʃ(ə)n/ n. **1** the act of contracting. **2** (often in pl.) Med. a shortening of the uterine muscles occurring at intervals during childbirth. **3** shrinking, diminution. **4 a** a shortening of a word by combination or elision. **b** a contracted word or group of words. [F f. L contractio -onis (as CONTRACT)]

contractor /kən'træktə(r)/ n. a person who undertakes a contract, esp. to provide materials, conduct building operations, etc. [LL (as CONTRACT)]

contractual /kən'træktʃʊəl, -'træktjʊəl/ adj. of or in the nature of a contract. □ **contractually** adv.

contradict /ˌkɒntrə'dɪkt/ v.tr. **1** deny or express the opposite of (a statement). **2** deny or express the opposite of a statement made by (a person). **3** be in opposition to or in conflict with (new evidence contradicted our theory). □ **contradictor** n. [L contradicere contradict- (as CONTRA-, dicere say)]

contradiction /ˌkɒntrə'dɪkʃ(ə)n/ n. **1 a** statement of the opposite; denial. **b** an instance of this. **2** inconsistency. □ **contradiction in terms** a self-contradictory statement or group of words. [ME f. OF f. L contradictio -onis (as CONTRADICT)]

contradictory /ˌkɒntrə'dɪktərɪ/ adj. **1** expressing a denial or opposite statement. **2** (of statements etc.) mutually opposed or inconsistent. **3** (of a person) inclined to contradict. **4** Logic (of two propositions) so related that one and only one must be true. □ **contradictorily** adv. **contradictoriness** n. [ME f. LL contradictorius (as CONTRADICT)]

contradistinction /ˌkɒntrədɪ'stɪŋkʃ(ə)n/ n. a distinction made by contrasting.

contradistinguish /ˌkɒntrədɪ'stɪŋgwɪʃ/ v.tr. (usu. foll. by from) distinguish two things by contrasting them.

contraflow /'kɒntrəˌfləʊ/ n. Brit. a flow (esp. of road traffic) alongside, and in a direction opposite to, an established or usual flow, esp. as a temporary or emergency arrangement.

contrail /'kɒntreɪl/ n. a condensation trail, esp. from an aircraft. [abbr.]

contraindicate /ˌkɒntrə'ɪndɪˌkeɪt/ v.tr. Med. act as an indication against (the use of a particular substance or treatment). □ **contraindication** /-ˌɪndɪ'keɪʃ(ə)n/ n.

contralto /kən'træltəʊ/ n. (pl. -os) **1 a** the lowest female singing voice. **b** a singer with this voice. **2** a part written for contralto. [It. (as CONTRA-, ALTO)]

contraposition /ˌkɒntrəpə'zɪʃ(ə)n/ n. **1** opposition or contrast. **2** Logic conversion of a proposition from all A is B to all not-B is not-A. □ **contrapositive** /-'pɒzɪtɪv/ adj. & n. [LL contrapositio (as CONTRA-, ponere posit- place)]

contraption /kən'træpʃ(ə)n/ n. often derog. or joc. a machine or device, esp. a strange or cumbersome one. [19th c.: perh. f. CONTRIVE, INVENTION: assoc. with TRAP¹]

contrapuntal /ˌkɒntrə'pʌnt(ə)l/ adj. Mus. of or in counterpoint. □ **contrapuntally** adv. **contrapuntist** n. [It. contrappunto counterpoint]

contrariety /ˌkɒntrə'raɪɪtɪ/ n. **1** opposition in nature, quality, or action. **2** disagreement, inconsistency. [ME f. OF contrarieté f. LL contrarietas -tatis (as CONTRARY)]

contrariwise /kən'treərɪˌwaɪz/ adv. **1** on the other hand. **2** in the opposite way. **3** perversely. [ME f. CONTRARY + -WISE]

contrary /'kɒntrərɪ/ adj., n., & adv. ● adj. **1** (usu. foll. by to) opposed in nature or tendency. **2** /kən'treərɪ/ colloq. perverse, self-willed. **3** (of a wind) unfavourable, impeding. **4** mutually opposed. **5** opposite in position or direction. ● n. (pl. **-ies**) (prec. by the) the opposite. ● adv. (foll. by to) in opposition or contrast (contrary to expectations it rained). □ **on the contrary** intensifying a denial of what has just been implied or stated. **to the contrary** to the opposite effect (can find no indication to the contrary). □ **contrarily** /'kɒntrərɪlɪ/ (/kən'treərɪlɪ/ in sense 2 of adj.) adv. **contrariness** /'kɒntrərɪnɪs/ (/kən'treərɪnɪs/ in sense 2 of adj.) n. [ME f. AF contrarie, OF contraire, f. L contrarius f. contra against]

contrast n. & v. ● n. /'kɒntrɑːst/ **1 a** a juxtaposition or comparison showing striking differences. **b** a difference so revealed. **2** (often foll. by to) a thing or person having qualities noticeably different from another. **3 a** the degree of difference between tones in a television picture or a photograph. **b** the change of apparent brightness or colour of an object caused by the juxtaposition of other objects. ● v. /kən'trɑːst/ (often foll. by with) **1** tr. distinguish or set together so as to reveal a contrast. **2** intr. have or show a contrast. □ **contrastingly** /kən'trɑːstɪŋlɪ/ adv. **contrastive** /-'trɑːstɪv/ adj. [F contraste, contraster, f. It. contrasto f. med.L contrastare (as CONTRA-, stare stand)]

contrasty /'kɒntrɑːstɪ/ adj. (of a television picture, a photograph, etc.) showing a high degree of contrast.

contra-suggestible /ˌkɒntrəsə'dʒestɪb(ə)l/ adj. Psychol. tending to respond to a suggestion by believing or doing the contrary.

contrate wheel /'kɒntreɪt/ n. = crown wheel. [med.L & Rmc contrata: see COUNTRY]

contravene /ˌkɒntrə'viːn/ v.tr. **1** infringe (a law or code of conduct). **2** (of things) conflict with. □ **contravener** n. [LL contravenire (as CONTRA-, venire vent- come)]

contravention /ˌkɒntrə'venʃ(ə)n/ n. **1** infringement. **2** an instance of this. □ **in contravention of** infringing, violating (a law etc.). [F f. med.L contraventio (as CONTRAVENE)]

contretemps /'kɒntrəˌtɒm/ n. (pl. same) **1** an awkward or unfortunate occurrence; an unexpected mishap. **2** colloq. a dispute or disagreement. [F]

contribute /kən'trɪbjuːt, disp. 'kɒntrɪˌbjuːt/ v. (often foll. by to) **1** tr. give (money, an idea, help, etc.) towards a common purpose (contributed £5 to the fund). **2** intr. help to bring about a result etc. (contributed to their downfall). **3** tr. (also absol.) supply (an article etc.) for publication with others in a journal etc. □ **contributive** /kən'trɪbjʊtɪv/ adj. [L contribuere contribut- (as COM-, tribuere bestow)]

contribution /ˌkɒntrɪ'bjuːʃ(ə)n/ n. **1** the act of contributing. **2** something contributed, esp. money. **3** an article etc. contributed to a publication. [ME f. OF contribution or LL contributio (as CONTRIBUTE)]

contributor /kən'trɪbjʊtə(r)/ n. a person who contributes (esp. an article or literary work).

contributory /kən'trɪbjʊtərɪ/ adj. & n. ● adj. **1** that contributes. **2** operated by means of contributions (contributory pension scheme). ● n. Brit. Law a person liable to contribute towards the payment of a wound-up company's debts. □ **contributory negligence** Law negligence on the part of the injured party through failure to take precautions against an accident. [med.L contributorius (as CONTRIBUTE)]

contrite /'kɒntraɪt, kən'traɪt/ adj. **1** completely penitent. **2** feeling remorse or penitence; affected by guilt. **3** (of an action) showing a contrite spirit. □ **contritely** adv. **contriteness** n. [ME f. OF contrit f. L contritus bruised (as COM-, terere trit- rub)]

contrition /kən'trɪʃ(ə)n/ n. the state of being contrite; thorough penitence. [ME f. OF f. LL contritio -onis (as CONTRITE)]

contrivance /kən'traɪv(ə)ns/ n. **1** something contrived, esp. a mechanical device or a plan. **2** an act of contriving, esp. deceitfully. **3** inventive capacity.

contrive /kən'traɪv/ v.tr. **1** devise; plan or make resourcefully or with skill. **2** (often foll. by *to* + infin.) manage (*contrived to make matters worse*). □ **contrivable** *adj.* **contriver** *n.* [ME f. OF *controver* find, imagine f. med.L *contropare* compare]

contrived /kən'traɪvd/ *adj.* planned so carefully as to seem unnatural; artificial, forced (*the plot seemed contrived*).

control /kən'trəʊl/ *n.* & *v.* ● *n.* **1** the power of directing, command (*under the control of*). **2** the power of restraining, esp. self-restraint. **3** a means of restraint; a check. **4** (usu. in *pl.*) a means of regulating prices etc. **5** (usu. in *pl.*) switches and other devices by which a machine, esp. an aircraft or vehicle, is controlled (also *attrib.*: *control panel*; *control room*). **6 a** a place where something is controlled or verified. **b** a person or group that controls something. **7** a standard of comparison for checking the results of a survey or experiment. **8** a member of an intelligence organization who personally directs the activities. ● *v.tr.* (**controlled, controlling**) **1** have control or command of; dominate. **2** exert control over; regulate. **3** hold in check; restrain (*told him to control himself*). **4** serve as control to. **5** check, verify. □ **control group** a group forming the standard of comparison in an experiment. **controlling interest** a means of determining the policy of a business etc., esp. by ownership of a majority of the stock. **control rod** a rod of neutron-absorbing material used to vary the output power of a nuclear reactor. **control tower** a tall building at an airport etc. from which air traffic is controlled. **in control** (often foll. by *of*) directing an activity. **out of control** no longer subject to proper direction or restraint. **under control** being controlled; in order. □ **controllable** *adj.* **controllably** *adv.* **controllability** /-ˌtrəʊlə'bɪlɪtɪ/ *n.* [ME f. AF *controreller* keep a copy of a roll of accounts, f. med.L *contrarotulare* (as CONTRA-, *rotulus* ROLL *n.*): (n.) perh. f. F *contrôle*]

controller /kən'trəʊlə(r)/ *n.* **1** a person or thing that controls. **2** a person in charge of expenditure, esp. a steward or comptroller. □ **controllership** *n.* [ME *counterroller* f. AF *contrerollour* (as CONTROL)]

controversial /ˌkɒntrə'vɜːʃ(ə)l/ *adj.* **1** causing or subject to controversy; disputed, esp. publicly. **2** of or relating to controversy. **3** given to controversy, prone to argue. □ **controversialism** *n.* **controversialist** *n.* **controversially** *adv.* [LL *controversialis* (as CONTROVERSY)]

controversy /'kɒntrə'vɜːsɪ, *disp.* kən'trɒvəsɪ/ *n.* (*pl.* **-ies**) **1** disagreement on a matter of opinion. **2** a prolonged argument or dispute, esp. when conducted publicly. [ME f. L *controversia* (as CONTROVERT)]

controvert /'kɒntrə'vɜːt, ˌkɒntrə'vɜːt/ *v.tr.* **1** dispute, deny. **2** argue about; discuss. □ **controvertible** /ˌkɒntrə'vɜːtɪb(ə)l/ *adj.* [orig. past part.; f. F *controvers(e)* f. L *controversus* (as CONTRA-, *vertere vers-* turn)]

contumacious /ˌkɒntjʊ'meɪʃəs/ *adj.* insubordinate; stubbornly or wilfully disobedient, esp. to a court order. □ **contumaciously** *adv.* [L *contumax*, perh. rel. to *tumere* swell]

contumacy /'kɒntjʊməsɪ/ *n.* stubborn refusal to obey or comply. [L *contumacia* f. *contumax*: see CONTUMACIOUS]

contumelious /ˌkɒntjʊ'miːlɪəs/ *adj.* reproachful, insulting, or insolent. □ **contumeliously** *adv.* [ME f. OF *contumelieus* f. L *contumeliosus* (as CONTUMELY)]

contumely /'kɒntjuːmlɪ/ *n.* **1** insolent or reproachful language or treatment. **2** disgrace. [ME f. OF *contumelie* f. L *contumelia* (as COM-, *tumere* swell)]

contuse /kən'tjuːz/ *v.tr.* injure without breaking the skin; bruise. □ **contusion** *n.* [L *contundere contus-* (as COM-, *tundere* thump)]

conundrum /kə'nʌndrəm/ *n.* **1** a riddle, esp. one with a pun in its answer. **2** a hard or puzzling question. [16th c.: orig. unkn.]

conurbation /ˌkɒnɜː'beɪʃ(ə)n/ *n.* an extended urban area, esp. one consisting of several towns and merging suburbs. [CON- + L *urbs urbis* city + -ATION]

conure /'kɒnjʊə(r)/ *n.* a medium-sized tropical American parrot of the genus *Pyrrhura*, with mainly green plumage and a long gradated tail. [mod.L *conurus* f. Gk *kōnos* cone + *oura* tail]

convalesce /ˌkɒnvə'les/ *v.intr.* recover one's health after illness or medical treatment. [ME f. L *convalescere* (as COM-, *valere* be well)]

convalescent /ˌkɒnvə'les(ə)nt/ *adj.* & *n.* ● *adj.* **1** recovering from an illness. **2** of or for persons in convalescence. ● *n.* a convalescent person. □ **convalescence** *n.*

convection /kən'vekʃ(ə)n/ *n.* **1** transference of heat in a gas or liquid by upward movement of the heated and less dense medium. **2** *Meteorol.* the transfer of heat by the upward flow of hot air or downward flow of cold air. □ **convection current** circulation that results from

convection. □ **convectional** *adj.* **convective** /-'vektɪv/ *adj.* [LL *convectio* f. L *convehere convect-* (as COM-, *vehere vect-* carry)]

convector /kən'vektə(r)/ *n.* a heating appliance that circulates warm air by convection.

convenance /'kɒnvəˌnɒns/ *n.* (usu. in *pl.*) conventional propriety. [F f. *convenir* be fitting (as CONVENE)]

convene /kən'viːn/ *v.* **1** *tr.* summon or arrange (a meeting etc.). **2** *intr.* assemble. **3** *tr.* summon (a person) before a tribunal. □ **convenable** *adj.* **convener** *n.* **convenor** *n.* [ME f. L *convenire convent-* assemble, agree, fit (as COM-, *venire* come)]

convenience /kən'viːnɪəns/ *n.* & *v.* ● *n.* **1** the quality of being convenient; suitability. **2** freedom from difficulty or trouble; material advantage (*for convenience*). **3** an advantage (*a great convenience*). **4** a useful thing, esp. an installation or piece of equipment. **5** *Brit.* a lavatory, esp. a public one. ● *v.tr.* afford convenience to; suit, accommodate. □ **at one's convenience** at a time or place that suits one. **at one's earliest convenience** as soon as one can. **convenience food** food, esp. complete meals, sold in convenient form and requiring very little preparation. **convenience store** a shop with extended opening hours, stocking a wide range of goods. **make a convenience of** take advantage of (a person) insensitively. [ME f. L *convenientia* (as CONVENE)]

convenient /kən'viːnɪənt/ *adj.* **1** (often foll. by *for, to*) **a** serving one's comfort or interests; suitable. **b** free of trouble or difficulty. **2** available or occurring at a suitable time or place (*will try to find a convenient moment*). **3** within easy reach; easily accessible (*convenient for the shops*). □ **conveniently** *adv.* [ME (as CONVENE)]

convent /'kɒnv(ə)nt, -vent/ *n.* **1** a religious (usu. Christian) community, esp. of nuns, under vows. **2** the premises occupied by this. **3** (in full **convent school**) a school attached to and run by a convent. [ME f. AF *covent*, OF *convent* f. L *conventus* assembly (as CONVENE)]

conventicle /kən'ventɪk(ə)l/ *n.* esp. *hist.* **1** a secret or unlawful religious meeting, esp. of dissenters. **2** a building used for this. [ME f. L *conventiculum* (place of) assembly, dimin. of *conventus* (as CONVENE)]

convention /kən'venʃ(ə)n/ *n.* **1 a** general agreement, esp. agreement on social behaviour etc. by implicit consent of the majority. **b** a custom or customary practice, esp. an artificial or formal one. **2 a** a formal assembly or conference for a common purpose. **b** *US* an assembly of the delegates of a political party to select candidates for office. **c** *hist.* a meeting of Parliament without a summons from the sovereign. **3 a** a formal agreement. **b** an agreement between states, esp. one less formal than a treaty. **4** *Cards* an accepted method of play (in leading, bidding, etc.) used to convey information to a partner. **5** the act of convening. [ME f. OF f. L *conventio -onis* (as CONVENE)]

conventional /kən'venʃən(ə)l/ *adj.* **1** depending on or according with convention. **2** (of a person) attentive to social conventions. **3** usual; of agreed significance. **4** not spontaneous or sincere or original. **5** (of weapons or power) non-nuclear. **6** *Art* following tradition rather than nature. □ **conventionalism** *n.* **conventionalist** *n.* **conventionalize** *v.tr.* (also **-ise**). **conventionally** *adv.* **conventionality** /-ˌvenʃə'nælɪtɪ/ *n.* (*pl.* **-ies**). [F *conventionnel* or LL *conventionalis* (as CONVENTION)]

conventioneer /kən,venʃə'nɪə(r)/ *n.* *US* a person attending a convention.

conventual /kən'ventʃʊəl, -tjʊəl/ *adj.* & *n.* ● *adj.* **1** of or belonging to a convent. **2** (**Conventual**) of the less strict branch of the Franciscans, living in large convents. (See also FRANCISCAN.) ● *n.* **1** a member or inmate of a convent. **2** (**Conventual**) a Conventual Franciscan. [ME f. med.L *conventualis* (as CONVENT)]

converge /kən'vɜːdʒ/ *v.intr.* **1** come together as if to meet or join. **2** (of lines) tend to meet at a point. **3** (foll. by *on, upon*) approach from different directions. **4** *Math.* (of a series) approximate in the sum of its terms towards a definite limit. [LL *convergere* (as COM-, *vergere* incline)]

convergent /kən'vɜːdʒənt/ *adj.* **1** converging; tending to converge. **2** *Biol.* (of unrelated organisms) having the tendency to become similar while evolving to fill a particular niche. **3** *Psychol.* (of thought) tending to reach only the most rational result. □ **convergence** *n.* **convergency** *n.*

conversant /kən'vɜːs(ə)nt, 'kɒnvəs-/ *adj.* (foll. by *with*) well experienced or acquainted with a subject, person, etc. □ **conversance** *n.* **conversancy** *n.* [ME f. OF, pres. part. of *converser* CONVERSE[1]]

conversation /ˌkɒnvə'seɪʃ(ə)n/ *n.* **1** the informal exchange of ideas, information, etc., by spoken words. **2** an instance of this. □ **conversation piece 1** a small genre painting of a group of figures.

2 a thing that serves as a topic of conversation because of its unusualness etc. **conversation stopper** colloq. an unexpected remark, esp. one that cannot readily be answered. [ME f. OF f. L conversatio -onis (as CONVERSE¹)]

conversational /ˌkɒnvəˈseɪʃən(ə)l/ adj. **1** of or in conversation. **2** fond of or good at conversation. **3** colloquial. □ **conversationally** adv.

conversationalist /ˌkɒnvəˈseɪʃənəlɪst/ n. a person who is good at or fond of conversing.

conversazione /ˌkɒnvəˌsætsɪˈəʊnɪ/ n. (pl. **conversaziones** or **conversazioni** pronunc. same) a social gathering held by a learned or art society. [It. f. L (as CONVERSATION)]

converse¹ v. & n. ● v.intr. /kənˈvɜːs/ (often foll. by with) engage in conversation (conversed with him about various subjects). ● n. /ˈkɒnvɜːs/ archaic conversation. □ **converser** /kənˈvɜːsə(r)/ n. [ME f. OF converser f. L conversari keep company (with), frequent. of convertere (CONVERT)]

converse² /ˈkɒnvɜːs/ adj. & n. ● adj. opposite, contrary, reversed. ● n. **1** something that is opposite or contrary. **2** a statement formed from another statement by the transposition of certain words, e.g. some philosophers are men from some men are philosophers. **3** Math. a theorem whose hypothesis and conclusion are the conclusion and hypothesis of another. □ **conversely** /ˈkɒnvɜːslɪ, kənˈvɜːslɪ/ adv. [L conversus, past part. of convertere (CONVERT)]

conversion /kənˈvɜːʃ(ə)n/ n. **1** the act or an instance of converting or the process of being converted, esp. in belief or religion. **2 a** an adaptation of a building for new purposes. **b** a converted building. **3** transposition, inversion. **4** (in Christian theology) the turning of sinners to God. **5** the transformation of fertile into fissile material in a nuclear reactor. **6** Rugby the scoring of points by a successful kick at goal after scoring a try. **7** Psychol. the change of an unconscious conflict into a physical disorder or disease. [ME f. OF f. L conversio -onis (as CONVERT)]

convert v. & n. ● v. /kənˈvɜːt/ **1** tr. (usu. foll. by into) change in form, character, or function. **2** tr. cause (a person) to change beliefs, opinion, party, etc. **3** tr. change (money, stocks, units in which a quantity is expressed, etc.) into others of a different kind. **4** tr. make structural alterations in (a building) to serve a new purpose. **5** tr. (also absol.) **a** Rugby score extra points from (a try) by a successful kick at goal. **b** Amer. Football complete (a touchdown) by kicking a goal or crossing the goal-line. **6** intr. be converted or convertible (the sofa converts into a bed). **7** tr. Logic interchange the terms of (a proposition). ● n. /ˈkɒnvɜːt/ (often foll. by to) a person who has been converted to a different belief, opinion, etc. □ **convert to one's own use** wrongfully make use of (another's property). [ME f. OF convertir ult. f. L convertere convers- turn about (as COM-, vertere turn)]

converter /kənˈvɜːtə(r)/ n. (also **convertor**) **1** a person or thing that converts. **2 a** an electrical apparatus for the interconversion of alternating current and direct current. **b** Electronics an apparatus for converting a signal from one frequency to another. **3** a reaction vessel used in making steel. □ **converter reactor** a nuclear reactor that converts fertile material into fissile material.

convertible /kənˈvɜːtɪb(ə)l/ adj. & n. ● adj. **1** that may be converted. **2** (of currency etc.) that may be converted into other forms, esp. into gold or US dollars. **3** (of a car) having a folding or detachable roof. **4** (of terms) synonymous. ● n. a car with a folding or detachable roof. □ **convertibly** adv. **convertibility** /-ˌvɜːtɪˈbɪlɪtɪ/ n. [OF f. L convertibilis (as CONVERT)]

convex /ˈkɒnveks/ adj. having an outline or surface curved like the exterior of a circle or sphere (cf. CONCAVE). □ **convexly** adv. **convexity** /kɒnˈveksɪtɪ/ n. [L convexus vaulted, arched]

convey /kənˈveɪ/ v.tr. **1** transport or carry (goods, passengers, etc.). **2** communicate (an idea, meaning, etc.). **3** Law transfer the title to (property). **4** transmit (sound, smell, etc.). □ **conveyable** adj. [ME f. OF conveier f. med.L conviare (as COM-, L via way)]

conveyance /kənˈveɪəns/ n. **1 a** the act or process of carrying. **b** the communication (of ideas etc.). **c** transmission. **2** a means of transport; a vehicle. **3** Law **a** the transfer of property from one owner to another. **b** a document effecting this. □ **conveyancer** n. (in sense 3). **conveyancing** n. (in sense 3).

conveyor /kənˈveɪə(r)/ n. (also **conveyer**) a person or thing that conveys. □ **conveyor belt** an endless moving belt for conveying articles or materials, esp. in a factory.

convict v. & n. ● v.tr. /kənˈvɪkt/ **1** (often foll. by of) prove to be guilty (of a crime etc.). **2** declare guilty by the verdict of a jury or the decision of a judge. ● n. /ˈkɒnvɪkt/ **1** a person found guilty of a criminal offence.

2 a person serving a prison sentence, esp. (hist.) in a penal colony. [ME f. L convincere convict- (as COM-, vincere conquer): noun f. obs. convict convicted]

conviction /kənˈvɪkʃ(ə)n/ n. **1 a** the act or process of proving or finding guilty. **b** an instance of this (has two previous convictions). **2 a** the action or resulting state of being convinced. **b** a firm belief or opinion. **c** an act of convincing. [L convictio (as CONVICT)]

convince /kənˈvɪns/ v.tr. **1** (often foll. by of, or that + clause) persuade (a person) to believe or realize. **2** (as **convinced** adj.) firmly persuaded (a convinced pacifist). □ **convincer** n. **convincible** adj. [L (as CONVICT)]

convincing /kənˈvɪnsɪŋ/ adj. **1** able to or such as to convince. **2** leaving no margin of doubt, substantial (a convincing victory). □ **convincingly** adv.

convivial /kənˈvɪvɪəl/ adj. **1** fond of good company; sociable and lively. **2** festive (a convivial atmosphere). □ **convivially** adv. **conviviality** /-ˌvɪvɪˈælɪtɪ/ n. [L convivialis f. convivium feast (as COM-, vivere live)]

convocation /ˌkɒnvəˈkeɪʃ(ə)n/ n. **1** the act of calling together. **2** a large formal gathering of people, esp.: **a** Brit. a provincial synod of the Anglican clergy of Canterbury or York. **b** Brit. a legislative or deliberative assembly of a university. □ **convocational** adj. [ME f. L convocatio (as CONVOKE)]

convoke /kənˈvəʊk/ v.tr. formal call (people) together to a meeting etc.; summon to assemble. [L convocare convocat- (as COM-, vocare call)]

convoluted /ˈkɒnvəˌluːtɪd/ adj. **1** coiled, twisted. **2** complex, intricate. □ **convolutedly** adv. [past part. of convolute f. L convolutus (as COM-, volvere volut- roll)]

convolution /ˌkɒnvəˈluːʃ(ə)n/ n. **1** coiling, twisting. **2** a coil or twist. **3** complexity. **4** Anat. a sinuous fold in the surface of the brain. □ **convolutional** adj. [med.L convolutio (as CONVOLUTED)]

convolve /kənˈvɒlv/ v.tr. & intr. (esp. as **convolved** adj.) roll together; coil up. [L convolvere (as CONVOLUTED)]

convolvulus /kənˈvɒlvjʊləs/ n. (pl. **convolvuluses**) a twining plant of the genus Convolvulus, with trumpet-shaped flowers, e.g. bindweed. [L]

convoy /ˈkɒnvɔɪ/ n. & v. ● n. **1** a group of ships travelling together or under escort. **2** a supply of provisions etc. under escort. **3** a group of vehicles travelling on land together or under escort. **4** the act of travelling or moving in a group or under escort. ● v.tr. **1** (of a warship) escort (a merchant or passenger vessel). **2** escort, esp. with armed force. □ **in convoy** under escort with others; as a group. [OF convoyer var. of conveier CONVEY]

convulsant /kənˈvʌls(ə)nt/ adj. & n. Pharm. ● adj. producing convulsions. ● n. a drug that may produce convulsions. [F f. convulser (as CONVULSE)]

convulse /kənˈvʌls/ v.tr. **1** (usu. in passive) affect with convulsions. **2** colloq. cause to laugh uncontrollably. **3** shake violently; agitate, disturb. [L convellere convuls- (as COM-, vellere pull)]

convulsion /kənˈvʌlʃ(ə)n/ n. **1** (usu. in pl.) violent irregular motion of a limb or limbs or the body caused by involuntary contraction of muscles, esp. as a disorder of infants. **2** a violent natural disturbance, esp. an earthquake. **3** violent social or political agitation. **4** (in pl.) colloq. uncontrollable laughter. □ **convulsionary** adj. [F convulsion or L convulsio (as CONVULSE)]

convulsive /kənˈvʌlsɪv/ adj. **1** characterized by or affected with convulsions. **2** producing convulsions. □ **convulsively** adv.

cony /ˈkəʊnɪ/ n. (also **coney**) (pl. **-ies** or **-eys**) **1 a** Heraldry or dial. a rabbit. **b** its fur. **2** (in the Bible) a hyrax. [ME cunin(g) f. AF coning, OF conin, f. L cuniculus]

coo /kuː/ n., v., & int. ● n. a soft murmuring sound like that of a dove or pigeon. ● v. (**coos, cooed**) **1** intr. make the sound of a coo. **2** intr. & tr. talk or say in a soft or amorous voice. ● int. Brit. sl. expressing surprise or incredulity. □ **cooingly** adv. [imit.]

cooee /ˈkuːiː/ n., int., & v. colloq. ● n. & int. a sound used to attract attention, esp. at a distance. ● v.intr. (**cooees, cooeed, cooeeing**) make this sound. □ **within cooee** (or **a cooee**) **of** Austral. & NZ colloq. very near to. [imit. of a signal used by Australian Aboriginals and copied by settlers]

Cook¹ /kʊk/, Captain James (1728–79), English explorer. He conducted an expedition to the Pacific (1768–71) in his ship Endeavour, charting the coasts of New Zealand and New Guinea as well as exploring the east coast of Australia and claiming it for Britain. He returned to the Pacific in 1772–5 to search for the fabled Antarctic continent, landing at Tahiti, the New Hebrides, and New Caledonia. Cook's final voyage (1776–9) to discover a passage round North America from the Pacific side ended in disaster when he was killed in a skirmish with native peoples in Hawaii.

Cook[2] /kʊk/, Thomas (1808–92), English founder of the travel firm Thomas Cook. In 1841 he organized the first publicly advertised excursion train in England, carrying 570 passengers from Leicester to Loughborough and back, to attend a temperance meeting, for the price of one shilling. The success of this induced him to organize further excursions both in Britain and abroad and to lay the foundations for the tourist and travel-agent industry of the 20th century.

cook /kʊk/ v. & n. ● v. **1** tr. prepare (food) by heating it. **2** intr. (of food) undergo cooking. **3** tr. (esp. as **cook the books**) colloq. falsify (accounts etc.); alter to produce a desired result. **4** tr. sl. ruin, spoil. **5** tr. (esp. as **cooked** adj.) Brit. sl. fatigue, exhaust. **6** tr. & intr. US colloq. do or proceed successfully. **7** intr. (as **be cooking**) colloq. be happening or about to happen (went to find out what was cooking). ● n. a person who cooks, esp. professionally or in a specified way (a good cook). □ **cook-chill** attrib.adj. designating food which has been cooked and refrigerated by the manufacturer ready for reheating by the consumer. **cook a person's goose** ruin a person's chances. **cook up** colloq. invent or concoct (a story, excuse, etc.). □ **cookable** adj. & n. [OE cōc f. pop.L cocus for L coquus]

Cook, Mount (called in Maori Aorangi) the highest peak in New Zealand, in the Southern Alps on South Island, rising to a height of 3,764 m (12,349 ft). It is named after Captain James Cook.

cookbook /ˈkʊkbʊk/ n. a cookery book.

Cooke /kʊk/, Sir William Fothergill (1806–79), English inventor. He became interested in the application of electric telegraphy to alarm systems and railway signalling, and went into partnership with Charles Wheatstone in the 1830s. They took out a joint patent for a railway alarm system in 1837, when they set up the first practical telegraph between two stations in London, and they progressively improved the system over the next few years. The patents were later acquired by the Electro-Telegraph Company, which Cooke formed in 1848.

cooker /ˈkʊkə(r)/ n. **1 a** Brit. a stove used for cooking food. **b** (usu. in comb.) a container in which food is cooked. **2** Brit. a fruit etc. (esp. an apple) that is more suitable for cooking than for eating raw.

cookery /ˈkʊkərɪ/ n. (pl. **-ies**) **1** the art or practice of cooking. **2** US a place or establishment for cooking. □ **cookery book** Brit. a book containing recipes and other information about cooking.

cookhouse /ˈkʊkhaʊs/ n. **1** a camp kitchen. **2** an outdoor kitchen in warm countries. **3** a ship's galley.

cookie /ˈkʊkɪ/ n. **1** N. Amer. a small sweet biscuit. **2** sl. a person. **3** Sc. a plain bun. □ **the way the cookie crumbles** esp. N. Amer. colloq. how things turn out; the unalterable state of affairs. [Du. koekje dimin. of koek cake]

cooking /ˈkʊkɪŋ/ n. **1** the art or process by which food is cooked. **2** (attrib.) suitable for or used in cooking (cooking apple; cooking utensils).

Cook Islands a group of fifteen islands in the SW Pacific Ocean between Tonga and French Polynesia, which have the status of a self-governing territory in free association with New Zealand; pop. (1986) 17,185; languages, English (official), Rarotongan; capital, Avarua, on Rarotonga. The islands were named after Captain James Cook, who visited them in 1773.

cookout /ˈkʊkaʊt/ n. N. Amer. a gathering with an open-air cooked meal; a barbecue.

Cookson /ˈkʊks(ə)n/, Dame Catherine (Anne) (b.1906), English writer. A prolific author of light romantic fiction, she is best known for the Mary Ann series (1956–67), the Mallen trilogy (1973–4), and the Tilly Trotter series (1980–2).

Cook Strait the strait separating the North and South Islands of New Zealand. It was named after Captain James Cook, who visited it in 1770.

cookware /ˈkʊkweə(r)/ n. utensils for cooking, esp. dishes, pans, etc.

cool /kuːl/ adj., n., & v. ● adj. **1** of or at a fairly low temperature, fairly cold (a cool day; a cool bath). **2** suggesting or achieving coolness (cool colours; cool clothes). **3 a** (of a person) calm, unexcited. **b** (of an act) done without emotion. **4** lacking zeal or enthusiasm. **5** unfriendly; lacking cordiality (got a cool reception). **6** (of jazz playing) restrained, relaxed. **7** calmly audacious (a cool customer). **8** (prec. by a) colloq. (usu. as an intensive; esp. of large sums of money) not less than; a full (cost me a cool thousand). **9** colloq. **a** excellent, marvellous. **b** suave, stylish. ● n. **1** coolness. **2** cool air; a cool place. **3** sl. calmness, composure (keep one's cool; lose one's cool). ● v. (often foll. by down, off) **1** tr. & intr. make or become cool. **2** intr. (of anger, emotions, etc.) lessen, become calmer. □ **cool-**

bag (or **-box**) an insulated container for keeping food cool. **cool-headed** not easily excited. **cool one's heels** see HEEL[1]. **cooling-off period** an interval to allow for a change of mind before commitment to action. **cooling tower** a tall structure for cooling hot water before reuse, esp. in industry. **cool it** sl. relax, calm down. □ **coolish** adj. **coolly** /ˈkuːllɪ/ adv. **coolness** n. [OE cōl, cōlian, f. Gmc: cf. COLD]

coolabah /ˈkuːləˌbɑː/ n. (also **coolibah** /-lɪˌbɑː/) Austral. a gum-tree; esp. Eucalyptus microtheca, found by watercourses. [Aboriginal]

coolant /ˈkuːlənt/ n. **1** a cooling agent, esp. fluid, to remove heat from an engine, nuclear reactor, etc. **2** a fluid used to lessen the friction of a cutting tool. [COOL + -ANT after lubricant]

cooler /ˈkuːlə(r)/ n. **1** a vessel in which a thing is cooled. **2** N. Amer. a refrigerator. **3** a long drink, esp. a spritzer. **4** sl. prison or a prison cell.

Coolgardie safe /kuːlˈgɑːdɪ/ n. Austral. a food safe cooled by strips of wetted fabric. [Coolgardie, a town in Western Australia]

coolibah var. of COOLABAH.

Coolidge /ˈkuːlɪdʒ/, (John) Calvin (1872–1933), American Republican statesman, 30th President of the US 1923–9. Highly popular personally, he was seen as an embodiment of thrift, caution, and honesty in a decade when corruption in public life was common, even in his own administration. He was committed to reducing income taxes and the national debt, and was noted for his policy of non-interference in foreign affairs, which culminated in the signing of the Kellogg Pact in 1928.

coolie /ˈkuːlɪ/ n. (also **cooly**) (pl. **-ies**) **1** an unskilled native labourer in countries of the Far East. **2** derog. usu. offens. a person from the Indian subcontinent; a person of Indian descent. □ **coolie hat** a broad conical hat as worn in Asia by coolies. [perh. f. Kulī, an aboriginal tribe of Gujarat]

coomb /kuːm/ n. (also **combe**) Brit. **1** a valley or hollow on the side of a hill. **2** a short valley running up from the coast. [OE cumb: cf. CWM]

coon /kuːn/ n. **1** N. Amer. a raccoon. **2** sl. offens. a black person. [abbr.]

coon-can /ˈkuːnkæn/ n. a simple card-game like rummy (orig. Mexican). [perh. f. Sp. con quién with whom?]

coonskin /ˈkuːnskɪn/ n. **1** the skin of a raccoon. **2** a cap etc. made of this.

coop /kuːp/ n. & v. ● n. **1** a cage or pen for confining poultry. **2** a small place of confinement, esp. a prison. **3** Brit. a basket used in catching fish. ● v.tr. **1** put or keep (a fowl) in a coop. **2** (often foll. by up, in) confine (a person) in a small space. [ME cupe basket f. MDu., MLG kūpe, ult. f. L cupa cask]

co-op /ˈkəʊɒp/ n. colloq. **1** a cooperative society or shop. **2** a cooperative business or enterprise. [abbr.]

Cooper[1] /ˈkuːpə(r)/, Gary (born Frank James Cooper) (1901–61), American actor. His performance in such westerns as The Virginian (1929) and High Noon (1952) established his reputation in tough cowboy roles. He also starred in other films, including For Whom the Bell Tolls (1943).

Cooper[2] /ˈkuːpə(r)/, James Fenimore (1789–1851), American novelist. He is renowned for his tales of American Indians and frontier life, including The Last of the Mohicans (1826), The Prairie (1827), and The Deerslayer (1841). He also wrote novels inspired by his early career at sea, as well as historical studies.

cooper /ˈkuːpə(r)/ n. & v. ● n. a maker or repairer of casks, barrels, etc. ● v.tr. make or repair (a cask). [ME f. MDu., MLG kūper f. kūpe COOP]

cooperage /ˈkuːpərɪdʒ/ n. **1** the work or establishment of a cooper. **2** money payable for a cooper's work.

cooperate /kəʊˈɒpəˌreɪt/ v.intr. (also **co-operate**) **1** (often foll. by with) work or act together; assist. **2** (of things) concur in producing an effect. □ **cooperant** adj. **cooperator** n. [eccl.L cooperari (as CO-, operari f. opus operis work)]

cooperation /kəʊˌɒpəˈreɪʃ(ə)n/ n. (also **co-operation**) **1** the process of working together to the same end; assistance. **2** Econ. the formation and operation of cooperatives. [ME f. L cooperatio (as COOPERATE): partly through F coopération]

cooperative /kəʊˈɒpərətɪv/ adj. & n. (also **co-operative**) ● adj. **1** of or affording cooperation. **2** willing to cooperate. **3** Econ. (of a farm, shop, or other business, or a society owning such businesses) owned and run jointly by its members, with profits shared among them. (See note below.) ● n. a cooperative farm or society or business. □ **cooperatively** adv. **cooperativeness** n. [LL cooperativus (as COOPERATE)]

▪ The modern cooperative movement began with the Rochdale Society of Equitable Pioneers, formed in 1844 in order to reduce what was seen as the exploitation of consumers by the capitalist system. The ideas of Welsh social reformer Robert Owen were strongly influential in the early cooperatives. (See also COOPERATIVE WHOLESALE SOCIETY.) Today cooperatives in agriculture and manufacturing are common in developing countries; cooperatives in industrial market economies include the kibbutzim of Israel and agricultural distribution cooperatives in France. In Communist countries, for example in China and the former USSR, cooperatives have usually formed the basis of the centrally controlled network for organizing farm produce.

Cooperative Party a British political party formed in 1917 to represent the cooperative movement in local and national government. One of the party's candidates was elected at the 1918 general election and by 1945 the party had twenty-three MPs. Since then the Cooperative Party has effectively been assimilated by the Labour Party, with which it always had close links.

Cooperative Wholesale Society (abbr. **CWS**) a British cooperative society, the largest in the world, formed in 1863. It acts as a manufacturer, wholesaler, and banker for the numerous cooperative retail societies by which it is owned and controlled.

co-opt /kəʊˈɒpt/ v.tr. **1** appoint to membership of a body by invitation of the existing members. **2** absorb into a larger (esp. political) group; take over, adopt (an idea etc.). □ **co-optive** adj. **co-option** /-ˈɒpʃ(ə)n/ n. **co-optation** /ˌkəʊɒpˈteɪʃ(ə)n/ n. [L cooptare (as CO-, optare choose)]

coordinate v., adj., & n. (also **co-ordinate**) ● v. /kəʊˈɔːdɪˌneɪt/ **1** tr. bring (various parts, movements, etc.) into a proper or required relation to ensure harmony or effective operation etc. **2** intr. work or act together effectively. **3** tr. make coordinate; organize, classify. ● adj. /kəʊˈɔːdɪnət/ **1** equal in rank or importance. **2** in which the parts are coordinated; involving coordination. **3** Gram. (of parts of a compound sentence) equal in status (cf. SUBORDINATE). **4** Chem. denoting a type of covalent bond in which one atom provides both the shared electrons. ● n. /kəʊˈɔːdɪnət/ **1** Math. each of a system of magnitudes used to fix the position of a point, line, or plane. **2** a person or thing equal in rank or importance. **3** (in pl.) matching items of clothing. □ **coordinative** /-ˌneɪtɪv/ adj. **coordinator** /-ˌneɪtə(r)/ n. [CO- + L ordinare ordinat- f. ordo -inis order]

coordination /kəʊˌɔːdɪˈneɪʃ(ə)n/ n. **1** the harmonious or effective working together of different parts. **2** the arrangement of parts etc. into an effective relation. **3** Chem. the formation of a coordinate bond.

coot /kuːt/ n. **1** a black aquatic bird of the genus Fulica, of the rail family; esp. F. atra, with the upper mandible extended backwards to form a white plate on the forehead. **2** colloq. a stupid person. [ME, prob. f. LG]

cootie /ˈkuːtɪ/ n. sl. a body louse. [perh. f. Malay kutu a biting parasite]

cop[1] /kɒp/ n. & v. sl. ● n. **1** a police officer. **2** Brit. a capture or arrest (it's a fair cop). ● v.tr. (**copped**, **copping**) **1** catch or arrest (an offender). **2** receive, suffer. **3** take, seize. □ **cop it 1** get into trouble; be punished. **2** be killed. **cop out 1** withdraw; give up an attempt. **2** go back on a promise. **3** escape. **cop-out** n. **1** a cowardly or feeble evasion. **2** an escape; a way of escape. **cop-shop** a police station. **not much** (or **no**) **cop** Brit. of little or no value or use. [perh. f. obs. cap arrest f. OF caper seize f. L capere: (n.) cf. COPPER[2]]

cop[2] /kɒp/ n. (in spinning) a conical ball of thread wound on a spindle. [OE cop summit]

Copacabana Beach /ˌkɒpəkəˈbænə/ a resort on the Atlantic coast of Brazil near Rio de Janeiro.

copacetic /ˌkəʊpəˈsɛtɪk, -ˈsiːtɪk/ adj. N. Amer. sl. excellent; in good order. [20th c.: orig. unkn.]

copaiba /kəˈpaɪbə/ n. an aromatic oil or resin from a South American leguminous tree of the genus Copaifera, used in medicine and perfumery. [Sp. & Port. f. Guarani cupauba]

copal /ˈkəʊp(ə)l/ n. a resin from a tropical tree, used for varnish. [Sp. f. Aztec copalli incense]

Copán /kəʊˈpæn/ an ancient Mayan city, in western Honduras near the Guatemalan frontier, which flourished from the 4th to the 10th centuries AD. It was the southernmost point of the Mayan empire.

copartner /kəʊˈpɑːtnə(r)/ n. a partner or associate, esp. when sharing equally. □ **copartnership** n.

cope[1] /kəʊp/ v.intr. **1** (foll. by with) deal effectively or contend successfully with a person or task. **2** manage successfully; deal with a situation or problem (found they could no longer cope). [ME f. OF coper, colper f. cop, colp blow f. med.L colpus f. L colaphus f. Gk kolaphos blow with the fist]

cope[2] /kəʊp/ n. & v. ● n. **1** Eccl. a long cloaklike vestment worn by a

priest or bishop in ceremonies and processions. **2** esp. poet. anything likened to a cope. ● v.tr. cover with a cope or coping. [ME ult. f. LL cappa CAP, CAPE[1]]

copeck /ˈkəʊpek, ˈkɒp-/ n. (also **kopeck, kopek**) a monetary unit of Russia and some other countries of the former USSR, equal to one-hundredth of a rouble. [Russ. kopeika dimin. of kop'e lance (from the figure of Ivan IV bearing a lance instead of a sword in 1535)]

Copenhagen /ˌkəʊpənˈheɪgən/ (Danish **København** /ˌkøbənˈhaʊn/) the capital and chief port of Denmark, a city occupying the eastern part of Zealand and northern part of the island of Amager; pop. (1990) 466,700. Based on a fortified town built in the 12th century, Copenhagen became the capital of Denmark and seat of the Danish royal family in the mid-15th century. During the Reformation and later during wars with Sweden in the late 17th century, the city was often sacked. It was also the site of the battle of Copenhagen in 1801, at which Nelson was victorious over the Danes.

copepod /ˈkəʊpɪˌpɒd/ n. Zool. a small aquatic crustacean of the class Copepoda, without a carapace and with paddle-like feet. Many are minute and form a major component of marine plankton, though some are ectoparasites. [Gk kōpē oar-handle + pous podos foot]

coper /ˈkəʊpə(r)/ n. a horse-dealer. [obs. cope buy, f. MDu., MLG kōpen, G kaufen: rel. to CHEAP]

Copernican /kəˈpɜːnɪkən/ adj. hist. of or relating to Copernicus or his theories. □ **Copernican system** the theory that the sun is the centre of the solar system, with the planets orbiting round it (cf. Ptolemaic system). [COPERNICUS]

Copernicus /kəˈpɜːnɪkəs/, Nicolaus (Latinized name of Mikołaj Kopernik) (1473–1543), Polish astronomer. Copernicus, canon of the cathedral at Frauenberg, first published his astronomical theories in outline in 1530, and more fully in De Revolutionibus Orbium Coelestium (1543). In order to avoid the complex system of epicycles required to explain planetary motions in Ptolemaic theory, he proposed a simpler model in which the planets orbited in perfect circles around the sun. His work ultimately led to the overthrow of the established geocentric cosmology.

copestone /ˈkəʊpstəʊn/ n. **1** = coping-stone. **2** a finishing touch. [COPE[2] + STONE]

copiable /ˈkɒpɪəb(ə)l/ adj. that can or may be copied.

copier /ˈkɒpɪə(r)/ n. a machine or person that copies (esp. documents).

copilot /ˈkəʊˌpaɪlət/ n. a second pilot in an aircraft.

coping /ˈkəʊpɪŋ/ n. the top (usu. sloping) course of masonry in a wall or parapet. □ **coping-stone** a stone used in a coping. [COPE[2]]

coping saw n. a D-shaped saw for cutting curves in wood. [cope cut wood f. OF coper: see COPE[1]]

copious /ˈkəʊpɪəs/ adj. **1** abundant, plentiful. **2** producing much. **3** providing much information. **4** profuse in speech. □ **copiously** adv. **copiousness** n. [ME f. OF copieux or f. L copiosus f. copia plenty]

copita /kəˈpiːtə/ n. **1** a tulip-shaped sherry-glass. **2** a glass of sherry. [Sp., dimin. of copa cup]

coplanar /kəʊˈpleɪnə(r)/ adj. Geom. in the same plane. □ **coplanarity** /ˌkəʊpləˈnærɪtɪ/ n.

Copland /ˈkəʊplənd/, Aaron (1900–90), American composer, pianist, and conductor, of Lithuanian descent. He worked to establish a distinctive American style in music, borrowing from jazz in his Music for the Theater (1925), from Shaker music in Appalachian Spring (1944), and from other folk and traditional music in the ballet score Rodeo (1942). The first American composer to earn a living from music, he contributed to the professional standing of composers nationally by establishing the American Composers' Alliance (1937).

Copley /ˈkɒplɪ/, John Singleton (1738–1815), American painter. A distinguished colonial portraitist, he sailed for England in 1774 and subsequently settled there. He made his mark with such paintings as The Death of Chatham (1779–80) and The Death of Major Peirson (1783), which are among the first large-scale paintings of contemporary events.

copolymer /kəʊˈpɒlɪmə(r)/ n. Chem. a polymer with units of more than one kind. □ **copolymerize** v.tr. & intr. (also **-ise**). **copolymerization** /-ˌpɒlɪməraɪˈzeɪʃ(ə)n/ n. (also **-isation**).

copper[1] /ˈkɒpə(r)/ n., adj., & v. ● n. **1** a red-brown metallic chemical element (atomic number 29; symbol **Cu**). (See note below.) **2** (usu. in pl.) a bronze coin, esp. one of little value. **3** a large metal vessel for boiling esp. laundry. **4** a small butterfly of the genus Lycaena, with reddish-orange wings; esp. (in full small copper) the common European L.

phlaeas. ● *adj.* made of or coloured like copper. ● *v.tr.* cover (a ship's bottom, a pan, etc.) with copper. ☐ **copper beech** a variety of beech with copper-coloured leaves. **copper belt** a copper-mining area of central Africa. **copper-bit** a soldering tool pointed with copper. **copper-bottomed 1** (esp. of a ship or pan) having a bottom sheathed with copper. **2** (esp. financially) genuine or reliable. **copper pyrites** = CHALCOPYRITE. **copper sulphate** a blue crystalline solid used in electroplating, textile dyeing, etc. **copper vitriol** copper sulphate. [OE *copor, coper,* ult. f. L *cyprium aes* Cyprus metal, since the Roman supply of metal came almost entirely from Cyprus]

▪ A transition metal, copper is found in the native state as well as in the form of ores. It was the earliest metal to be used by humans, first by itself and then later alloyed with tin to form bronze. A ductile easily worked metal, it is a very good conductor of heat and electricity. Copper is a component of many alloys, but it is still used mainly in its pure state, especially for electrical wiring. Copper compounds are used in the production of green pigments, insecticides, and fungicides. The symbol Cu is from the Latin word for copper *cuprum*.

copper² /ˈkɒpə(r)/ *n. Brit. sl.* a police officer. [COP¹ + -ER¹]

Copper Age a prehistoric period (also called *chalcolithic*) when some weapons and tools were made of copper. The Copper Age is sometimes regarded as coming between the Stone Age and the Bronze Age, and sometimes as forming the earliest phase of the Bronze Age. (See also CHALCOLITHIC.)

copperas /ˈkɒpərəs/ *n.* hydrated ferrous sulphate, forming green crystals. [ME *coperose* f. OF *couperose* f. med.L *cup(e)rosa*: perh. orig. *aqua cuprosa* copper water]

Copperbelt, the /ˈkɒpəˌbelt/ a region of central Zambia with rich deposits of copper, cobalt, and uranium, which are all mined there; chief town, Ndola.

copperhead /ˈkɒpəˌhed/ *n.* **1** a venomous pit viper, *Agkistrodon contortrix*, native to North America. **2** a venomous Australian snake, *Denisonia superba*, of the cobra family.

copperplate /ˈkɒpəˌpleɪt/ *n. & adj.* ● *n.* **1 a** a polished copper plate for engraving or etching. **b** a print made from this. **2** an ornate style of handwriting resembling that orig. used in engravings. ● *adj.* of or in copperplate writing.

coppersmith /ˈkɒpəˌsmɪθ/ *n.* a person who works in copper.

coppery /ˈkɒpərɪ/ *adj.* of or like copper, esp. in colour.

coppice /ˈkɒpɪs/ *n. & v.* ● *n.* an area of undergrowth and small trees, grown for periodic cutting. ● *v.* **1** *tr.* cut back (young trees) periodically to stimulate growth of shoots. **2** *intr.* (of a tree) produce new shoots from a stump. ☐ **coppiced** *adj.* [OF *copeïz* ult. f. med.L *colpus* blow: see COPE¹]

Coppola /ˈkɒpələ/, Francis Ford (b.1939), American film director, writer, and producer. Coppola's reputation rests chiefly on *The Godfather* (1972) and its two sequels, a film trilogy charting the fortunes of a New York Mafia family over several generations; it earned him three Oscars as writer and director. Other films include *Apocalypse Now* (1979), a retelling of Joseph Conrad's story *Heart of Darkness* in the context of the Vietnam War, *The Cotton Club* (1984), and *Bram Stoker's Dracula* (1993).

copra /ˈkɒprə/ *n.* the dried kernels of the coconut. [Port. f. Malayalam *koppara* coconut]

co-precipitation /ˌkəʊprɪˌsɪpɪˈteɪʃ(ə)n/ *n. Chem.* the simultaneous precipitation of more than one compound from a solution.

copro- /ˈkɒprəʊ/ *comb. form* dung, faeces. [Gk *kopros* dung]

coprocessor /kəʊˈprəʊsesə(r)/ *n. Computing* a microprocessor which provides additional functions to supplement a computer's primary processor.

co-production /ˌkəʊprəˈdʌkʃ(ə)n/ *n.* a production of a play, broadcast, etc., jointly by more than one company.

coprolite /ˈkɒprəˌlaɪt/ *n. Archaeol.* fossil dung or a piece of it.

coprophagous /kɒˈprɒfəgəs/ *adj. Zool.* dung-eating. [COPRO-]

coprophilia /ˌkɒprəˈfɪlɪə/ *n.* an abnormal interest in faeces and defecation.

coprosma /kəˈprɒzmə/ *n.* a small evergreen tree or shrub of the genus *Coprosma*, native to Australasia. [mod.L f. Gk *kopros* dung + *osmē* smell]

copse /kɒps/ *n.* **1** = COPPICE. **2** (in general use) a small wood. [shortened f. COPPICE]

copsewood /ˈkɒpswʊd/ *n.* undergrowth.

Copt /kɒpt/ *n.* **1** a native Egyptian in the Hellenistic and Roman periods. **2** a native Christian of the independent Egyptian Church.

[F *Copte* or mod.L *Coptus* f. Arab. *al-ḳibṭ, al-ḳubṭ* Copts f. Coptic *Gyptios* f. Gk *Aiguptios* Egyptian]

Coptic /ˈkɒptɪk/ *n. & adj.* ● *n.* the language of the Copts. (*See note below.*) ● *adj.* of or relating to the Copts.

▪ Coptic represents the final stage of ancient Egyptian. It has an alphabet largely based on the Greek but with some letters borrowed from Egyptian demotic, and now survives only as the liturgical language of the Coptic Church. In the 3rd century AD Coptic was the prevailing language of Christian Egypt, but after the Arab conquest in 642 it began to give way to Arabic, dying out as a spoken language in the 17th century.

Coptic Church the native Christian Church in Egypt, traditionally founded by St Mark. It became isolated from much of the rest of the Christian world in 451 when it adhered to the Monophysite doctrine condemned by the Council of Chalcedon, and its numbers declined as the conquest of Egypt by Muslim Arabs in the 7th century was followed by centuries of persecution. Coptic Christians now make up about 5 per cent of Egypt's population.

copula /ˈkɒpjʊlə/ *n.* (*pl.* **copulas**) *Logic & Gram.* a connecting word, esp. a part of the verb *be* connecting a subject and predicate. ☐ **copular** *adj.* [L (as CO-, *apere* fasten)]

copulate /ˈkɒpjʊˌleɪt/ *v.intr.* (often foll. by *with*) have sexual intercourse. ☐ **copulatory** /-lətərɪ/ *adj.* [L *copulare* fasten together (as COPULA)]

copulation /ˌkɒpjʊˈleɪʃ(ə)n/ *n.* **1** sexual union. **2** a grammatical or logical connection. [ME f. OF f. L *copulatio* (as COPULATE)]

copulative /ˈkɒpjʊlətɪv/ *adj.* **1** serving to connect. **2** *Gram.* **a** (of a word) that connects words or clauses linked in sense (cf. DISJUNCTIVE *adj.* 2). **b** connecting a subject and predicate. **3** relating to sexual union. ☐ **copulatively** *adv.* [ME f. OF *copulatif -ive* or LL *copulativus* (as COPULATE)]

copy /ˈkɒpɪ/ *n. & v.* ● *n.* (*pl.* **-ies**) **1** a thing made to imitate or be identical to another. **2** a single specimen of a publication or issue (*ordered twenty copies*). **3 a** a matter to be printed. **b** material for a newspaper or magazine article (*scandals make good copy*). **c** the text of an advertisement. **4 a** a model to be copied. **b** a page written after a model (of penmanship). ● *v.* (**-ies, -ied**) **1** *tr.* **a** make a copy of. **b** (often foll. by *out*) transcribe. **2** *intr.* make a copy rather than produce something original, esp. clandestinely. **3** *tr.* (foll. by *to*) send a copy of (a letter) to a third party. **4** *tr.* do the same as; imitate. ☐ **copy-edit** edit (copy) for printing. **copy editor** a person who edits copy for printing. **copy-typist** a person who makes typewritten transcripts of documents. [ME f. OF *copie, copier,* ult. f. L *copia* abundance (in med.L = transcript)]

copybook /ˈkɒpɪˌbʊk/ *n.* **1** a book containing models of handwriting for learners to imitate. **2** (*attrib.*) **a** tritely conventional. **b** accurate, exemplary.

copycat /ˈkɒpɪˌkæt/ *n. colloq.* (esp. as a child's word) a person who copies another, esp. slavishly.

copydesk /ˈkɒpɪˌdesk/ *n.* the desk at which copy is edited for printing.

copyhold /ˈkɒpɪˌhəʊld/ *n. Brit. hist.* **1** tenure of land based on manorial records. **2** land held in this way.

copyholder /ˈkɒpɪˌhəʊldə(r)/ *n.* **1** *Brit. hist.* a person who held land in copyhold. **2** a clasp for holding copy while it is keyboarded etc.

copyist /ˈkɒpɪɪst/ *n.* **1** a person who makes (esp. written) copies. **2** an imitator. [earlier *copist* f. F *copiste* or med.L *copista* (as COPY)]

copyreader /ˈkɒpɪˌriːdə(r)/ *n.* a person who reads and edits copy for a newspaper or book. ☐ **copyread** *v.tr.*

copyright /ˈkɒpɪˌraɪt/ *n., adj., & v.* ● *n.* the exclusive legal right, given to the originator or his or her assignee for a fixed number of years, to print, publish, perform, film, or record literary, artistic, or musical material and to authorize others to do the same. (*See note below.*) ● *adj.* (of such material) protected by copyright. ● *v.tr.* secure copyright for (material).

▪ Protection of authors' rights was made necessary by the invention of printing (15th century). Rulers issued monopoly rights to individuals or guilds, and at first the only protection available to the author was against unauthorized publication; once published the use to which the material could be put was out of the author's control. The first English statute recognizing the author's rights was passed in 1710 and gave protection for twenty-eight years only; in 1790 a similar copyright law was passed in the US. In Britain, copyright now generally subsists for the author's lifetime plus fifty years; in the US protection lasts for twenty-eight years, renewable for a second twenty-eight-year term. Legislation has recently been introduced to protect rights in computer programs and computer-stored material. International copyright protection was recognized as a growing problem in the 19th century, and in 1886 fourteen countries adopted an international copyright agreement known as the Berne Convention. In 1955 this was supplemented by the Universal Copyright Convention, to which the United States and certain other countries were signatories.

copyright library *n.* a library entitled to a free copy of each book, periodical, etc. published in the specified country. The copyright libraries in the British Isles are the British Library, the Bodleian Library, Cambridge University Library, the National Library of Wales, the National Library of Scotland, and the library of Trinity College, Dublin; each may demand a copy of each book published in the UK, with the exception of the British Library, which receives books automatically.

copywriter /ˈkɒpɪˌraɪtə(r)/ *n.* a person who writes or prepares copy (esp. of advertising material) for publication. □ **copywriting** *n.*

coq au vin /ˌkɒk əʊ ˈvæn/ *n.* a casserole of chicken pieces cooked in wine. [F]

coquetry /ˈkɒkɪtrɪ, ˈkəʊk-/ *n.* (*pl.* **-ies**) **1** coquettish behaviour. **2** a coquettish act. **3** trifling with serious matters. [F *coquetterie* f. *coqueter* (as COQUETTE)]

coquette /kɒˈket, kəʊˈket/ *n.* **1** a woman who flirts. **2** a crested hummingbird of the genus *Lophornis*. □ **coquettish** *adj.* **coquettishly** *adv.* **coquettishness** *n.* [F, fem. of *coquet* wanton, dimin. of *coq* cock]

coquina /kəˈkiːnə/ *n.* a soft limestone of broken shells, used in road-making in the West Indies and Florida. [Sp., = cockle]

coquito /kəˈkiːtəʊ/ *n.* (*pl.* **-os**) a palm tree, *Jubaea chilensis*, native to Chile, yielding sweet sap used for making palm wine, and fibre. [Sp., dimin. of *coco* coconut]

Cor. *abbr.* **1** Epistle to the Corinthians (New Testament). **2** *US* corner.

cor /kɔː(r)/ *int. Brit. sl.* expressing surprise, excitement, alarm, etc. □ **cor blimey** see BLIMEY. [corrupt. of *God*]

cor- /kər/ *prefix* assim. form of COM- before *r*.

coracle /ˈkɒrək(ə)l/ *n.* a small boat, occasionally circular but more often rectangular with rounded corners, constructed of wickerwork and made watertight originally with animal hides but later with pitch or some other watertight material. Coracles were used for river and coastal transport by the ancient Britons and are still employed by fishermen on the rivers and lakes of Wales and Ireland. [Welsh *corwgl* (*corwg* = Ir. *currach* boat: cf. CURRACH)]

coracoid /ˈkɒrəˌkɔɪd/ *n.* (in full **coracoid process**) a short projection from the shoulder-blade in vertebrates. [mod.L *coracoides* f. Gk *korakoeidēs* raven-like f. *korax -akos* raven]

coral /ˈkɒrəl/ *n. & adj.* ● *n.* **1 a** a hard calcareous substance secreted by certain marine coelenterates as an external skeleton and often forming large reefs. **b** any similar substance secreted by other marine organisms, esp. (in full *red coral*) a tough red substance produced by gorgonians of the genus *Corallium* and used in jewellery. **2** *Zool.* a marine coelenterate of the class Anthozoa, with a calcareous, horny, or soft skeleton, and often colonial, esp. (*stony* or *true coral*) of the reef-forming family Madreporaria. (*See note below.*) **3** the yellowish or reddish-pink colour of some corals. **4** the unfertilized roe of a lobster or scallop. ● *adj.* **1** of the colour of coral. **2** made of coral. □ **coral island** (or **reef**) one formed by the growth of red coral. **coral rag** limestone containing beds of petrified corals. **coral-root 1** a cruciferous woodland plant, *Cardamine bulbifera*, with purple flowers and scaly rhizomes. **2** (in full **coral-root orchid**) a brown saprophytic orchid of the genus *Corallorhiza*. **coral-snake** a brightly coloured snake of the cobra family, esp. a venomous one of the American genus *Micrurus*, with red, black, and yellow bands. [ME f. OF f. L *corallum* f. Gk *korallion*, prob. of Semitic orig.]

▪ The corals constitute several orders of solitary and colonial anthozoan coelenterates. The order Alcyonaria contains the soft corals (e.g. dead man's fingers), while the order Gorgonacea contains the horny corals or gorgonians. The massive stony corals of the order Madreporaria secrete a calcareous skeleton, and because they contain symbiotic photosynthetic algae they inhabit shallow waters, where they tend to form reefs, atolls, and islands. They are therefore vitally important for the habitats they provide, and numerous fossil forms are known.

coralline /ˈkɒrəˌlaɪn/ *n. & adj.* ● *n.* **1** a seaweed of the genus *Corallina*, having a calcareous jointed stem. **2** (in general use) a plantlike compound organism. ● *adj.* **1** of the colour of red coral. **2** of or like coral. [F *corallin* & It. *corallina* f. LL *corallinus* (as CORAL)]

corallite /ˈkɒrəˌlaɪt/ *n.* **1** the coral skeleton of a marine polyp. **2** fossil coral. [L *corallum* CORAL]

coralloid /ˈkɒrəˌlɔɪd/ *adj. & n.* ● *adj.* like or akin to coral. ● *n.* a coralloid organism.

Coral Sea a part of the western Pacific lying between Australia, New Guinea, and Vanuatu. It was the scene of a naval battle between US and Japanese carriers in 1942, in which the US forestalled Japanese moves against Port Moresby and the Solomon Islands.

coram populo /ˌkɔːrəm ˈpɒpjʊˌləʊ/ *adv.* in public. [L, = in the presence of the people]

cor anglais /ˌkɔːr ˈɒŋgleɪ/ *n.* (*pl.* **cors anglais** *pronunc.* same) *Mus.* **1** an alto woodwind instrument of the oboe family. **2** a player of this instrument. **3** an organ stop with the quality of a cor anglais. [F, = English horn]

corbel /ˈkɔːb(ə)l/ *n. & v. Archit.* ● *n.* **1** a projection of stone, timber, etc., jutting out from a wall to support a weight. **2** a short timber laid longitudinally under a beam to help support it. ● *v.tr. & intr.* (**corbelled**, **corbelling**; *US* **corbeled**, **corbeling**) (foll. by *out*, *off*) support or project on corbels. □ **corbel-table** a projecting course resting on corbels. [obs. ME *corbel* raven f. OF, dimin. of *corp*: see CORBIE]

corbie /ˈkɔːbɪ/ *n. Sc.* **1** a raven. **2** a carrion crow. □ **corbie-steps** the steplike projections on the sloping sides of a gable. [ME f. OF *corb*, *corp* f. L *corvus* crow]

Corcovado /ˌkɔːkəˈvɑːdəʊ/ a peak rising to 711 m (2,310 ft) on the south side of Rio de Janeiro. A gigantic statue of Christ, 40 m (131 ft) high, named 'Christ the Redeemer', stands on its summit.

Corcyra /kɔːˈsaɪərə/ the ancient Greek name for CORFU.

cord /kɔːd/ *n. & v.* ● *n.* **1 a** a long thin string or rope made from several twisted strands. **b** a piece of this. **2** *Anat.* a structure in the body resembling a cord (*spinal cord*). **3 a** a ribbed fabric, esp. corduroy. **b** (in *pl.*) corduroy trousers. **c** a cordlike rib on fabric. **4** an electric flex. **5** a measure of cut wood (usu. 128 cu.ft, 3.62 cubic metres). **6** a moral or emotional tie (*cords of affection*). ● *v.tr.* **1** fasten or bind with cord. **2** (as **corded** *adj.*) **a** (of cloth) ribbed. **b** provided with cords. **c** (of muscles) standing out like taut cords. □ **cordlike** *adj.* [ME f. OF *corde* f. L *chorda* f. Gk *khordē* gut, string of musical instrument]

cordage /ˈkɔːdɪdʒ/ *n.* cords or ropes, esp. in the rigging of a ship. [ME f. F (as CORD)]

cordate /ˈkɔːdeɪt/ *adj.* heart-shaped. [mod.L *cordatus* f. L *cor cordis* heart]

Corday /kɔːˈdeɪ/, Charlotte (full name Marie Anne Charlotte Corday d'Armont) (1768–93), French political assassin. The daughter of an impoverished nobleman and royalist, she became involved with the Girondists and in 1793 assassinated the revolutionary leader Jean Paul Marat in his bath. The Revolutionary Tribunal found her guilty of treason and she was guillotined four days later.

Cordelier /ˌkɔːdɪˈlɪə(r)/ *n.* a Franciscan Observant (wearing a knotted cord round the waist). [ME f. OF f. *cordele* dimin. of *corde* CORD]

cordial /ˈkɔːdɪəl/ *adj. & n.* ● *adj.* **1** heartfelt, sincere. **2** warm, friendly. ● *n.* **1** a fruit-flavoured drink. **2** a comforting or pleasant-tasting medicine. □ **cordially** *adv.* **cordiality** /ˌkɔːdɪˈælɪtɪ/ *n.* [ME f. med.L *cordialis* f. L *cor cordis* heart]

cordillera /ˌkɔːdɪˈljeərə/ *n.* a system or group of usu. parallel mountain ranges together with intervening plateaux etc., esp. of the Andes and in Central America and Mexico. [Sp. f. *cordilla* dimin. of *cuerda* CORD]

cordite /ˈkɔːdaɪt/ *n.* a smokeless explosive made from cellulose nitrate and nitroglycerine. [CORD (from its appearance) + -ITE[1]]

cordless /ˈkɔːdlɪs/ *adj.* (of an electrical appliance, telephone, etc.) working from an internal source of energy etc. (esp. a battery) and without a connection to a mains supply or central unit.

Cordoba /ˈkɔːdəbə/ (also **Cordova** /-dəvə/; Spanish **Córdoba** /ˈkorðoβa/) **1** a city in Andalusia, southern Spain; pop. (1991) 309,200. Founded by the Carthaginians, it was under Moorish rule from 711 to 1236. As capital of the most powerful of the Arab states in Spain, it was a centre of learning and culture, earning the title of 'the Athens of the West', and was renowned for its architecture, particularly the Great Mosque. **2** a city in central Argentina; pop. (1990) 1,198,000.

cordoba /ˈkɔːdəbə/ *n.* the basic monetary unit of Nicaragua, equal to 100 centavos. [F. Fernández de *Córdoba* (*fl.* 1524), Sp. governor of Nicaragua]

cordon /ˈkɔːd(ə)n/ *n. & v.* ● *n.* **1** a line or circle of police, soldiers, guards, etc., esp. preventing access to or from an area. **2 a** an ornamental cord or braid. **b** the ribbon of a knightly order. **3** a fruit-tree trained to grow as a single stem. **4** *Archit.* a string-course. ● *v.tr.* (often foll. by *off*) enclose or separate with a cordon of police etc. [It. *cordone* augmentative of *corda* CORD, & F *cordon* (as CORD)]

cordon bleu /ˌkɔːdɒn ˈblɜː/ *n. & adj.* ● *n.* a first-class cook; cooking of the highest class. (*See note below.*) ● *adj.* (of cooking) first class.

▪ The term, French for 'blue ribbon', referred originally to the ribbon worn by Knights-grand-cross of the French order of the Holy Ghost, the highest order of chivalry under the Bourbon kings; the term was later extended to other first-class distinctions.

cordon sanitaire /ˌkɔːdɒn ˌsænɪˈteə(r)/ *n.* **1** a guarded line between infected and uninfected districts. **2** any measure designed to prevent communication or the spread of undesirable influences. [F]

Cordova see CORDOBA.

cordovan /ˈkɔːdəv(ə)n/ *n.* a kind of soft leather. [Sp. *cordovan* f. Cordova (CORDOBA) where it was orig. made]

corduroy /ˈkɔːdəˌrɔɪ, -djuˌrɔɪ/ *n.* **1** a thick cotton fabric with velvety ribs. **2** (in *pl.*) corduroy trousers. □ **corduroy road** a road made of tree-trunks laid across a swamp. [18th c.: prob. f. CORD ribbed fabric + obs. *duroy* coarse woollen fabric]

cordwainer /ˈkɔːdˌweɪnə(r)/ *n. Brit. archaic* a shoemaker (usu. in names of guilds etc.). [obs. *cordwain* CORDOVAN]

cordwood /ˈkɔːdwʊd/ *n.* wood that is or can easily be measured in cords.

CORE /kɔː(r)/ *abbr.* (in the US) Congress of Racial Equality.

core /kɔː(r)/ *n. & v.* ● *n.* **1** the horny central part of various fruits, containing the seeds. **2 a** the central or most important part of anything (also *attrib.*: *core curriculum*). **b** the central part, of different character from the surroundings. **3** the central region of the earth. **4** the central part of a nuclear reactor, containing the fissile material. **5** a magnetic structural unit in a computer, storing one bit of data (see BIT⁴). **6** the inner strand of an electric cable, rope, etc. **7** a piece of soft iron forming the centre of an electromagnet or an induction coil. **8** an internal mould filling a space to be left hollow in a casting. **9** an internal part cut out (esp. of rock etc. in boring). **10** *Archaeol.* a piece of flint from which flakes or blades have been removed. ● *v.tr.* remove the core from. □ **core dump** *Computing* a dump of the contents of main memory, usually done as an aid to debugging. **core memory** *Computing* the memory of a computer consisting of many cores. **core time** (in a flexitime system) the central part of the working day, when all employees must be present. □ **corer** *n.* (usu. in *comb.*). [ME: orig. unkn.]

corelation var. of CORRELATION.

co-religionist /ˌkəʊrɪˈlɪdʒənɪst/ *n.* (US **coreligionist**) an adherent of the same religion.

corella /kəˈrelə/ *n.* a long-billed white cockatoo, *Cacatua tenuirostris*, native to Australia. [Wiradhwi]

Corelli¹ /kəˈrelɪ/, Arcangelo (1653–1713), Italian violinist and composer. His best-known works are his trio and solo sonatas for the violin (1681; 1685; 1689; 1694; 1700), and his concerti grossi (published posthumously in 1714), especially the 'Christmas' concerto, with its pastorale on the Nativity. Corelli's innovative use of harmony and attention to melody had an important influence on composers abroad, particularly Purcell, J. S. Bach, and Handel.

Corelli² /kəˈrelɪ/, Marie (pseudonym of Mary Mackay) (1855–1924), English writer of romantic fiction. After the success of her first novel, *A Romance of Two Worlds* (1886), the sales of her novels *Thelma* (1887), *Barabbas* (1893), and *The Sorrows of Satan* (1895) broke all existing records for book sales. Her popular success was not matched by critical acclaim, however; although briefly championed by Oscar Wilde and William Gladstone, she had little enduring literary success.

coreopsis /ˌkɔːrɪˈɒpsɪs/ *n.* a composite plant of the genus *Coreopsis*, having rayed usu. yellow flowers. [mod.L f. Gk *koris* bug + *opsis* appearance, with ref. to the shape of the seed]

co-respondent /ˌkəʊrɪˈspɒndənt/ *n.* (US **corespondent**) a person cited in a divorce case as having committed adultery with the respondent.

corf /kɔːf/ *n.* (*pl.* **corves** /kɔːvz/) *Brit.* **1** a basket in which fish are kept alive in the water. **2** a small wagon, formerly a large basket, used in mining. [MDu., MLG *korf*, OHG *chorp*, *korb* f. L *corbis* basket]

Corfu /kɔːˈfuː, -ˈfjuː/ (called in Greek *Kérkira*) a Greek island, one of the largest of the Ionian Islands, off the west coast of mainland Greece. It was known in ancient times as Corcyra.

corgi /ˈkɔːgɪ/ *n.* (*pl.* **corgis**) (in full **Welsh corgi**) a short-legged breed of dog with a foxlike head. [Welsh f. *cor* dwarf + *ci* dog]

coriaceous /ˌkɒrɪˈeɪʃəs/ *adj.* like leather; leathery. [LL *coriaceus* f. *corium* leather]

coriander /ˌkɒrɪˈændə(r)/ *n.* **1** an umbelliferous plant, *Coriandrum sativum*, with leaves used for flavouring and small round aromatic fruits. **2** (also **coriander seed**) the dried fruit of this plant used for flavouring curries etc. [ME f. OF *coriandre* f. L *coriandrum* f. Gk *koriannon*]

Corinth /ˈkɒrɪnθ/ (Greek **Kórinthos** /ˈkɔrɪnθɒs/) a city on the north coast of the Peloponnese, Greece; pop. (1981) 22,700. The modern city, built in 1858, is a little to the north-east of the site of an ancient city of the same name, which was a prominent city-state in ancient Greece and is associated with the teaching of St Paul.

Corinth, Gulf of (also called *Gulf of Lepanto*) an inlet of the Ionian Sea extending between the Peloponnese and central Greece.

Corinth, Isthmus of a narrow neck of land linking the Peloponnese with central Greece and separating the Gulf of Corinth from the Saronic Gulf.

Corinth Canal a man-made shipping channel across the narrowest part of the Isthmus of Corinth (a distance of 6.4 km, or 4 miles). Opened in 1893, it links the Gulf of Corinth and the Saronic Gulf.

Corinthian /kəˈrɪnθɪən/ *adj. & n.* ● *adj.* **1** of or relating to ancient Corinth in southern Greece. **2** *Archit.* of an order characterized by ornate decoration and flared capitals with rows of acanthus leaves, used esp. by the Romans. **3** amateur (in sport). **4** *archaic* profligate. ● *n.* **1** a native of Corinth. **2** a wealthy amateur of sport. [L *Corinthius* f. Gk *Korinthios* + -AN]

Corinthians, Epistle to the either of two books of the New Testament, epistles of St Paul to the Church at Corinth.

Coriolanus /ˌkɒrɪəˈleɪnəs/, Gaius (or Gnaeus) Marcius (5th century BC), Roman general. He earned his name by capturing the Volscian town of Corioli. He is said to have been banished from Rome in 491 BC after opposing the distribution of corn to the starving people and being charged with tyrannical conduct. He joined forces with the Volscians; according to legend, he led a Volscian army against Rome in 491 BC, and was turned back only by the pleas of his mother Veturia and his wife Volumnia. He was subsequently put to death by the Volscians.

Coriolis effect /ˌkɒrɪˈəʊlɪs/ *n. Physics* an effect whereby a body moving relative to a rotating frame of reference is accelerated in a direction perpendicular both to its direction of motion and to the axis of rotation of the frame. The effect helps to explain global wind patterns (rotating clockwise in the northern hemisphere, anticlockwise in the southern) and the trajectories of rockets over the earth's surface. [G. G. de *Coriolis*, Fr. scientist (1792–1843)]

corium /ˈkɔːrɪəm/ *n. Anat.* the dermis. [L, = skin]

Cork /kɔːk/ **1** a county of the Republic of Ireland, on the south coast in the province of Munster. **2** its county town, a port on the River Lee; pop. (1991) 127,000.

cork /kɔːk/ *n. & v.* ● *n.* **1** the buoyant light-brown material obtained from the cork-oak, being the cork (sense 4) of that tree. **2** a bottle-stopper of cork or other material. **3** a float of cork used in fishing etc. **4** *Bot.* a protective layer of dead cells immediately below the bark of woody plants. **5** (*attrib.*) made of cork. ● *v.tr.* (often foll. by *up*) **1** stop or confine. **2** restrain (feelings etc.). **3** blacken with burnt cork. □ **cork-oak** an evergreen Mediterranean oak, *Quercus suber*, which yields cork. **cork-tipped** *Brit.* (of a cigarette) having a filter of corklike material. □ **corklike** *adj.* [ME f. Du. & LG *kork* f. Sp. *alcorque* cork sole, perh. f. Arab.]

corkage /ˈkɔːkɪdʒ/ *n.* a charge made by a restaurant or hotel for serving wine etc. when brought in by customers.

corked /kɔːkt/ *adj.* **1** stopped with a cork. **2** (of wine) spoilt by a decayed cork. **3** blackened with burnt cork.

corker /ˈkɔːkə(r)/ *n. sl.* an excellent or astonishing person or thing.

corking /ˈkɔːkɪŋ/ *adj. sl.* strikingly large or splendid.

corkscrew /ˈkɔːkskruː/ *n. & v.* ● *n.* **1** a spirally twisted steel device for extracting corks from bottles. **2** (often *attrib.*) a thing with a spiral shape. ● *v.tr. & intr.* move spirally; twist.

corkwood /ˈkɔːkwʊd/ *n.* **1** a shrub or tree yielding a light porous wood, esp. *Leitneria floridana*, of the south-eastern US, and *Eritelea arborescens*, of New Zealand. **2** this wood.

corky /ˈkɔːkɪ/ *adj.* **1** corklike. **2** (of wine) corked.

corm /kɔːm/ *n. Bot.* an underground swollen stem base of some plants, e.g. the crocus. [mod.L *cormus* f. Gk *kormos* trunk with boughs lopped off]

cormorant /'kɔːmərənt/ n. a diving seabird of the family Phalacrocoracidae, with a long neck and bill, and mainly dark plumage; esp. *Phalacrocorax carbo*, which has lustrous black plumage. [ME f. OF *cormaran* f. med.L *corvus marinus* sea-raven: for ending *-ant* cf. *peasant, tyrant*]

corn[1] /kɔːn/ n. & v. ● n. **1 a** any cereal before or after harvesting, esp. the chief crop of a region: wheat, oats, or (in North America and Australia) maize. **b** a grain or seed of a cereal plant. **2** *colloq.* something corny or trite. ● *v. tr.* (as **corned** *adj.*) sprinkled or preserved with salt or brine (*corned beef*). □ **corn circle** = CROP CIRCLE. ● **corn-cob** the cylindrical centre of the maize ear to which rows of grains are attached. **corn-cob pipe** a tobacco-pipe made from a corn-cob. **corn-cockle** see COCKLE[2]. **corn dolly** a symbolic or decorative figure made of plaited straw. **corn exchange** a place for trade in corn. **corn-factor** *Brit.* a dealer in corn. **corn marigold** a yellow-flowered composite plant, *Chrysanthemum segetum*, growing amongst corn. **corn on the cob** maize cooked and eaten from the corn-cob. **corn-salad** = *lamb's lettuce* (see LAMB). **corn spurrey** see SPURREY. **corn-whiskey** *US* whisky distilled from maize. [OE f. Gmc: rel. to L *granum* grain]

corn[2] /kɔːn/ n. a small area of horny usu. tender skin esp. on the toes, extending into subcutaneous tissue. [ME f. AF f. L *cornu* horn]

cornbrash /'kɔːnbræʃ/ n. Geol. Brit. an earthy limestone layer of the Jurassic period. [CORN[1] + BRASH[2]]

corncrake /'kɔːnkreɪk/ n. a bird of the rail family, *Crex crex*, inhabiting grassland and nesting on the ground, with a rasping double call.

cornea /'kɔːnɪə/ n. the transparent layer covering the front of the eye, over the iris and pupil. □ **corneal** *adj.* [med.L *cornea tela* horny tissue, f. L *corneus* horny f. *cornu* horn]

Corneille /kɔːˈneɪ/, Pierre (1606–84), French dramatist. He worked as a magistrate 1624–30 before moving to Paris in the early 1630s. He is generally regarded as the founder of classical French tragedy; his plays in this genre include *Le Cid* (1637), *Cinna* (1641), and *Polyeucte* (1643); the newly founded Académie française criticized *Le Cid* for moral laxity and performances of it were subsequently banned. He also wrote comedies such as *Mélite* (his first play, 1629) and *Le Menteur* (1642).

cornel /'kɔːn(ə)l/ n. a shrub or tree of the genus *Cornus*, which includes the dogwoods; esp. a dwarf kind, *C. suecica*. [ME f. L *cornus*]

cornelian /kɔːˈniːlɪən/ n. (also **carnelian** /kɑːˈniːl-/) **1** a dull red or reddish-white variety of chalcedony. **2** this colour. [ME f. OF *corneline*; *car-* after L *caro carnis* flesh]

corneous /'kɔːnɪəs/ adj. hornlike, horny. [L *corneus* f. *cornu* horn]

corner /'kɔːnə(r)/ n. & v. ● n. **1** a place where converging sides or edges meet. **2** a projecting angle, esp. where two streets meet. **3** the internal space or recess formed by the meeting of two sides, esp. of a room. **4** a difficult position, esp. one from which there is no escape (*driven into a corner*). **5** a secluded or remote place. **6** a region or quarter, esp. a remote one (*from the four corners of the earth*). **7** the action or result of buying or controlling the whole available stock of a commodity, thereby dominating the market. **8** *Boxing & Wrestling* **a** an angle of the ring, esp. one where a contestant rests between rounds. **b** a contestant's supporters offering assistance at the corner between rounds. **9** *Football & Hockey* a free kick or hit from a corner of the pitch after the ball has been kicked over the goal-line by a defending player. **10** a triangular cut of gammon or ham. ● *v.* **1** *tr.* force (a person or animal) into a difficult or inescapable position. **2** *tr.* establish a corner in (a commodity). **b** dominate (dealers or the market) in this way. **3** *intr.* (esp. of or in a vehicle) go round a corner. □ **corner shop** a small local shop, esp. at a street corner. **just round** (or **around**) **the corner** *colloq.* very near, imminent. [ME f. AF ult. f. L *cornu* horn]

cornerback /'kɔːnəbæk/ n. Amer. Football a back lining up just behind and to the outside of the linebackers.

cornerstone /'kɔːnəˌstəʊn/ n. **1 a** a stone in a projecting angle of a wall. **b** a foundation-stone. **2** an indispensable part or basis of something.

cornerwise /'kɔːnəˌwaɪz/ adv. diagonally.

cornet[1] /'kɔːnɪt/ n. **1** *Mus.* **a** a brass instrument resembling a trumpet but shorter and wider. **b** a player of this instrument. **c** an organ stop with the quality of a cornet. **d** *Brit.* a conical wafer for holding ice-cream. □ **cornetist** /kɔːˈnetɪst, 'kɔːnɪtɪst/ n. **cornettist** /kɔːˈnetɪst/ n. [ME f. OF ult. f. L *cornu* horn]

cornet[2] /'kɔːnɪt/ n. Brit. hist. the fifth commissioned officer in a cavalry troop, who carried the colours. □ **cornetcy** n. (pl. **-ies**). [earlier sense 'pennon, standard' f. F *cornette* dimin. of *corne* ult. f. L *cornua* horns]

cornett /'kɔːnɪt, kɔːˈnet/ n. Mus. = CORNETTO. [var. of CORNET[1]]

cornetto /kɔːˈnetəʊ/ n. (pl. **cornetti** /-tɪ/) in early music, a wooden wind instrument with finger-holes and a cup-shaped mouthpiece. [It., dimin. of *corno* horn (as CORNET[1])]

cornfield /'kɔːnfiːld/ n. a field in which corn is being grown.

cornflake /'kɔːnfleɪk/ n. **1** (in pl.) a breakfast cereal of toasted flakes made from maize flour. **2** a flake of this cereal.

cornflour /'kɔːnˌflaʊə(r)/ n. **1** a fine-ground maize flour. Also called *cornstarch*. **2** a flour of rice or other grain.

cornflower /'kɔːnˌflaʊə(r)/ n. a composite plant of the genus *Centaurea*, growing among corn; esp. *C. cyanus*, with deep blue flowers.

cornice /'kɔːnɪs/ n. **1** *Archit.* **a** an ornamental moulding round the wall of a room just below the ceiling. **b** a horizontal moulded projection crowning a building or structure, esp. the uppermost member of the entablature of an order, surmounting the frieze. **2** *Mountaineering* an overhanging mass of hardened snow at the edge of a precipice. □ **corniced** *adj.* [F *corniche* etc. f. It. *cornice*, perh. f. L *cornix -icis* crow]

corniche /'kɔːnɪʃ, kɔːˈniːʃ/ n. (in full **corniche road**) **1** a road cut into the edge of a cliff etc. **2** a coastal road with wide views. [F: see CORNICE]

Cornish /'kɔːnɪʃ/ adj. & n. ● adj. of or relating to Cornwall. ● n. the ancient Celtic language of Cornwall, belonging to the Brythonic branch of the Celtic language group. It was formerly spoken in Cornwall but gradually died out in the 17th–18th centuries, though attempts have been made to revive it. □ **Cornish cream** clotted cream. **Cornish pasty** a pasty containing seasoned meat and vegetables. □ **Cornishman** n. (pl. **-men**).

Corn Laws a series of 19th-century laws introduced in an attempt to maintain the prosperity enjoyed by British agriculture during the Napoleonic Wars. The original Corn Law (1815) allowed foreign grain to be imported only after the price of home-grown wheat had risen above 80 shillings a quarter (= 8 bushels), but this had the unintended effect of forcing bread prices so high that both consumer and producer suffered. A sliding scale of import duties was introduced in 1828, but opposition to the Corn Laws continued to mount and they were eventually repealed by Robert Peel in 1846, an act which split the Conservative Party. (See also ANTI-CORN-LAW LEAGUE.)

cornstarch /'kɔːnstɑːtʃ/ n. = CORNFLOUR.

cornstone /'kɔːnstəʊn/ n. Brit. Geol. a mottled red and green limestone usu. formed under arid conditions, esp. in the Devonian period.

cornucopia /ˌkɔːnjʊˈkəʊpɪə/ n. **1 a** a symbol of plenty consisting of a goat's horn overflowing with flowers, fruit, and corn. **b** an ornamental vessel shaped like this. **2** an abundant supply. □ **cornucopian** *adj.* [LL f. L *cornu copiae* horn of plenty]

Cornwall /'kɔːnwəl/ a county occupying the extreme south-western peninsula of England; county town, Truro.

corny /'kɔːnɪ/ adj. (**cornier, corniest**) **1** *colloq.* **a** trite. **b** feebly humorous. **c** sentimental. **2** old-fashioned; out of date. **2** of or abounding in corn. □ **cornily** adv. **corniness** n. [CORN[1] + -Y[1]: sense 1 f. sense 'rustic']

corolla /kəˈrɒlə/ n. Bot. a whorl or whorls of petals forming the inner envelope of a flower. [L, dimin. of *corona* crown]

corollary /kəˈrɒlərɪ/ n. & adj. ● n. (pl. **-ies**) **1 a** a proposition that follows from (and is often appended to) one already proved. **b** an immediate deduction. **2** (often foll. by *of*) a natural consequence or result. ● adj. **1** supplementary, associated. **2** (often foll. by *to*) forming a corollary. [ME f. L *corollarium* money paid for a garland, gratuity: neut. adj. f. COROLLA]

Coromandel Coast /ˌkɒrəˈmænd(ə)l/ the southern part of the east coast of India, from Point Calimere to the mouth of the Krishna River. It takes its name from *Cholamandalaru* 'country of the Cholas', an ancient Dravidian people.

corona[1] /kəˈrəʊnə/ n. (pl. **coronae** /-niː/) **1 a** a small circle of light round the sun or moon. **b** *Astron.* the outermost layers of the sun's atmosphere. (*See note below.*) **2** a circular chandelier hung from a roof. **3** *Anat.* a crown or crownlike structure. **4** *Bot.* a crownlike outgrowth from the inner side of a corolla. **5** *Archit.* a broad vertical face of a cornice, usu. of considerable projection. **6** *Electr.* the glow around a conductor at high potential. [L, = crown]

■ The sun's corona is normally visible only during a total solar eclipse, when it is seen as a pearly halo surrounding the darkened disc of the moon, extending to several times the radius of the sun. It is caused by light from the photosphere being scattered by very hot high-energy electrons and slower dust particles. There is evidence of such regions in other stars.

corona[2] /kə'rəʊnə/ n. a long straight-sided cigar. [formerly proprietary name, f. Sp. *La Corona* lit. 'the crown']

Corona Borealis /kə,rəʊnə ,bɔːrɪ'eɪlɪs/ *Astron.* a northern constellation (the Northern Crown), in which the main stars form a small but prominent arc. [L]

coronach /'kɒrənək, -nəx/ n. *Sc. & Ir.* a funeral-song or dirge. [Ir. *coranach*, Gael. *corranach* f. *comh-* together + *rànach* outcry]

coronagraph /kə'rəʊnə,grɑːf/ n. an instrument for observing the sun's corona, esp. other than during a solar eclipse.

coronal[1] /kə'rəʊn(ə)l, 'kɒrən(ə)l/ adj. **1** *Astron. & Bot.* of a corona. **2** *Anat.* of the crown of the head. □ **coronal bone** the frontal bone of the skull. **coronal plane** an imaginary plane dividing the body into dorsal and ventral parts. **coronal suture** a transverse suture of the skull separating the frontal bone from the parietal bones. [F *coronal* or L *coronalis* (as CORONA[1])]

coronal[2] /'kɒrən(ə)l/ n. **1** a circlet (esp. of gold or gems) for the head. **2** a wreath or garland. [ME, app. f. AF f. *corone* CROWN]

coronary /'kɒrənəri/ adj. & n. ● adj. *Anat.* resembling or encircling like a crown, esp. denoting or involving the arteries which supply blood to the heart. ● n. (pl. **-ies**) **1** = *coronary thrombosis*. **2** a heart attack. □ **coronary artery** an artery supplying blood to the heart. **coronary thrombosis** *Med.* a blockage of the blood flow caused by a blood clot in a coronary artery. [L *coronarius* f. *corona* crown]

coronation /,kɒrə'neɪʃ(ə)n/ n. the ceremony of crowning a sovereign or a sovereign's consort. [ME f. OF f. med.L *coronatio -onis* f. *coronare* to crown f. CORONA[1]]

Coronation stone = *Stone of Scone* (see SCONE).

coroner /'kɒrənə(r)/ n. **1** an officer of a county, district, or municipality, holding inquests on deaths thought to be violent or accidental, and inquiries in cases of treasure trove. **2** *hist.* an officer charged with maintaining the rights of the private property of the Crown. □ **coronership** n. [ME f. AF *cor(o)uner* f. *coro(u)ne* CROWN]

coronet /'kɒrənɪt, -,net/ n. **1** a small crown (esp. as worn, or used as a heraldic device, by a peer or peeress). **2** a circlet of precious materials, esp. as a woman's head-dress or part of one. **3** a garland for the head. **4** the lowest part of a horse's pastern. **5** a ring of bone at the base of a deer's antler. □ **coroneted** adj. [OF *coronet(t)e* dimin. of *corone* CROWN]

Corot /'kɒrəʊ/, (Jean-Baptiste) Camille (1796–1875), French landscape painter. Trained in the neoclassical tradition, he worked in an essentially classical style despite his contact with the Barbizon School and his preference for taking preliminary studies outdoors. One of his most famous paintings is *La Danse des nymphes* (1850). Corot was a major influence on the impressionists, notably Camille Pissarro.

corozo /kə'rəʊzəʊ/ n. (pl. **-os**) a South American palm tree, esp. the ivory-nut palm, *Phytelephas macrocarpa*, and the American oil-palm, *Elaeis oleifera*. □ **corozo nut** = *ivory-nut*. [Sp.]

Corp. abbr. **1** Corporal. **2** US Corporation.

corpora pl. of CORPUS.

corporal[1] /'kɔːpərəl, -prəl/ n. **1** a non-commissioned army or air-force officer ranking next below sergeant. **2** (in full **ship's corporal**) *Brit.* an officer under the master-at-arms, attending to police matters. **3** *N. Amer.* = FALLFISH. [obs. F, var. of *caporal* f. It. *caporale* prob. f. L *corporalis* (as CORPORAL[2]), confused with It. *capo* head]

corporal[2] /'kɔːpərəl, -prəl/ adj. of or relating to the human body (cf. CORPOREAL). □ **corporal punishment** punishment inflicted on the body, esp. by beating. □ **corporally** adv. [ME f. OF f. L *corporalis* f. *corpus -oris* body]

corporal[3] /'kɔːpərəl, -prəl/ n. a cloth on which the vessels containing the consecrated elements are placed during the celebration of the Eucharist. [OE f. OF *corporal* or med.L *corporale pallium* body cloth (as CORPORAL[2])]

corporality /,kɔːpə'rælɪti/ n. (pl. **-ies**) **1** material existence. **2** a body. [ME f. LL *corporalitas* (as CORPORAL[2])]

corporate /'kɔːpərət/ adj. & n. ● adj. **1** forming a corporation (*corporate body*; *body corporate*). **2** forming one body of many individuals. **3** of or belonging to a corporation or group (*corporate responsibility*). **4** corporative. ● n. *Commerce* a large industrial corporation. □ **corporate raider** *Stock Exch.* a person who makes a practice of taking over companies against their wishes or interests for personal profit or self-interest. □ **corporately** adv. **corporatism** n. [L *corporare corporat-* form into a body (*corpus -oris*)]

corporation /,kɔːpə'reɪʃ(ə)n/ n. **1** a group of people authorized to act as an individual and recognized in law as a single entity, esp. in business. **2** the municipal authorities of a borough, town, or city. **3** *joc.* a protruding stomach. [LL *corporatio* (as CORPORATE)]

corporative /'kɔːpərətɪv/ adj. **1** of a corporation. **2** governed by or organized in corporations, esp. of employers and employed. □ **corporativism** n.

corporeal /kɔː'pɔːrɪəl/ adj. **1** bodily, physical, material, esp. as distinct from spiritual (cf. CORPORAL[2]). **2** *Law* consisting of material objects. □ **corporeally** adv. **corporeality** /-,pɔːrɪ'ælɪti/ n. [LL *corporealis* f. L *corporeus* f. *corpus -oris* body]

corporeity /,kɔːpə'riːɪti/ n. **1** the quality of being or having a material body. **2** bodily substance. [F *corporéité* or med.L *corporeitas* f. L *corporeus* (as CORPOREAL)]

corposant /'kɔːpəz,ænt/ n. a luminous electrical discharge seen esp. on a ship or aircraft during a storm (cf. ST ELMO'S FIRE). [OSp., Port., It. *corpo santo* holy body]

corps /kɔː(r)/ n. (pl. **corps** /kɔːz/) **1** *Mil.* **a** a body of troops with special duties (*intelligence corps*; *Royal Army Medical Corps*). **b** a main subdivision of an army in the field, consisting of two or more divisions. **2** a body of people engaged in a special activity (*diplomatic corps*; *press corps*). [F (as CORPSE)]

corps de ballet /,kɔː də 'bæleɪ/ n. the company of ensemble dancers in a ballet. [F]

corps d'élite /,kɔː deɪ'liːt/ n. a select group. [F]

corps diplomatique /,kɔː ,dɪpləmæ'tiːk/ n. a diplomatic corps. [F]

corpse /kɔːps/ n. a dead (usu. human) body. □ **corpse-candle 1** a lambent flame seen in a churchyard or over a grave, regarded as an omen of death. **2** a lighted candle placed beside a corpse before burial. [ME *corps*, var. spelling of *cors* (CORSE), f. OF *cors* f. L *corpus* body]

corpulent /'kɔːpjʊlənt/ adj. bulky in body, fat. □ **corpulence** n. **corpulency** n. [ME f. L *corpulentus* f. *corpus* body]

corpus /'kɔːpəs/ n. (pl. **corpora** /-pərə/ or **corpuses**) **1** a body or collection of writings, texts, spoken material, etc. **2** *Anat.* a distinctive structure in an animal body. [ME f. L, = body]

corpus callosum /,kɔːpəs kə'ləʊsəm/ n. *Anat.* a broad band of nerve fibres joining the two hemispheres of the brain. [mod.L f. CORPUS + *callosus, -um* tough]

Corpus Christi[1] /,kɔːpəs 'krɪstɪ/ a city and port in southern Texas; pop. (1990) 257,400. It is situated on Corpus Christi Bay, an inlet of the Gulf of Mexico.

Corpus Christi[2] /,kɔːpəs 'krɪstɪ/ a festival commemorating the Christian Eucharist, originating in the 13th century and observed on the Thursday after Trinity Sunday. In medieval times it was the occasion when the guilds of many towns performed religious plays. [L, = Body of Christ]

corpuscle /'kɔːpʌs(ə)l/ n. **1** a minute body or cell in an organism, esp. (in pl.) the red or white cells in the blood of vertebrates. **2** *hist.* a minute particle seen as the basic constituent of matter, light, etc. □ **corpuscular** /kɔː'pʌskjʊlə(r)/ adj. [L *corpusculum* (as CORPUS)]

corpus delicti /,kɔːpəs dɪ'lɪktaɪ/ n. *Law* the facts and circumstances constituting a breach of a law. [L, = body of offence]

corpus luteum /,kɔːpəs 'luːtɪəm/ n. (pl. **corpora lutea** /,kɔːpərə 'luːtɪə/) *Anat.* a hormone-secreting structure developed in the ovary after discharge of the ovum, degenerating after a few days unless pregnancy has begun. [mod.L f. CORPUS + *luteus, -um* yellow]

corral /kɒ'rɑːl/ n. & v. ● n. **1** *N. Amer.* a pen for cattle, horses, etc. **2** an enclosure for capturing wild animals. **3** esp. *US hist.* a defensive enclosure of wagons in an encampment. ● v.tr. (**corralled, corralling**) **1** put or keep in a corral. **2** form (wagons) into a corral. **3** *N. Amer. colloq.* acquire. [Sp. *corral*, Port. *curral* (as KRAAL)]

corrasion /kə'reɪʒ(ə)n/ n. *Geol.* erosion of the earth's surface by rock material being carried over it by water, ice, etc. [L *corradere corras-* scrape together (as COM-, *radere* scrape)]

correct /kə'rekt/ adj. & v. ● adj. **1** true, right, accurate. **2** (of conduct, manners, etc.) proper, right. **3** in accordance with good standards of taste etc. ● v.tr. & intr. **1** *tr.* set right; amend (an error, omission, etc., or the person responsible for it). **2** *tr.* mark the errors in (written or printed work etc.). **3** *tr.* substitute the right thing for (the wrong one). **4** *tr.* **a** admonish or rebuke (a person). **b** punish (a person or fault). **5** *tr.* counteract (a harmful quality). **6** *tr.* adjust (an instrument etc.) to function accurately or accord with a standard. **7** *intr.* (foll. by *for*) make necessary adjustments for; set right. □ **correctable** adj. **correctly** adv. **correctness** n. [ME (adj. through F) f. L *corrigere correct-* (as COM-, *regere* guide)]

correction /kəˈrekʃ(ə)n/ n. **1 a** the act or process of correcting. **b** an instance of this. **2** a thing substituted for what is wrong. **3** archaic punishment. □ **correction fluid** a usu. white liquid that can be painted over an error. **house of correction** hist. an institution where vagrants and minor offenders were confined and set to work. □ **correctional** adj. [ME f. OF f. L correctio -onis (as CORRECT)]

correctitude /kəˈrektɪˌtjuːd/ n. correctness, esp. conscious correctness of conduct. [19th c., f. CORRECT + RECTITUDE]

corrective /kəˈrektɪv/ adj. & n. ● adj. serving or tending to correct or counteract something undesired or harmful. ● n. a corrective measure or thing. □ **correctively** adv. [F correctif -ive or LL correctivus (as CORRECT)]

corrector /kəˈrektə(r)/ n. a person who corrects or points out faults. [ME f. AF correctour f. L corrector (as CORRECT)]

Correggio /kɒˈredʒɪˌəʊ/, Antonio Allegri da (born Antonio Allegri) (c.1494–1534), Italian painter. He is best known for his series of frescos in Parma churches, especially S. Giovanni Evangelista (1520–3) and Parma cathedral (c.1526–30). His treatment of the cupolas shows the influence of Mantegna in its use of extreme foreshortening to give the illusion of great height. His devotional painting The Mystic Marriage of St Catherine (1520–6) and such mythological paintings as Jupiter and Io (c.1530) epitomize his soft and sensual style of painting, which influenced the rococo of the 18th century.

correlate /ˈkɒrəˌleɪt/ v. & n. ● v. **1** intr. (foll. by with, to) have a mutual relation. **2** tr. (usu. foll. by with) bring into a mutual relation. ● n. either of two related or complementary things (esp. so related that one implies the other). [back-form. f. CORRELATION, CORRELATIVE]

correlation /ˌkɒrəˈleɪʃ(ə)n/ n. (also **corelation** /ˌkəʊrɪˈleɪ-/) **1** a mutual relation between two or more things. **2 a** interdependence of variable quantities. **b** a quantity measuring the extent of this. **3** the act of correlating. □ **correlational** adj. [med.L correlatio (as CORRELATIVE)]

correlative /kəˈrelətɪv/ adj. & n. ● adj. **1** (often foll. by with, to) having a mutual relation. **2** Gram. (of words) corresponding to each other and regularly used together (as neither and nor). ● n. a correlative word or thing. □ **correlatively** adv. **correlativity** /-ˌreləˈtɪvɪtɪ/ n. [med.L correlativus (as COM-, RELATIVE)]

correspond /ˌkɒrɪˈspɒnd/ v.intr. **1 a** (usu. foll. by to) be analogous or similar. **b** (usu. foll. by to) agree in amount, position, etc. **c** (usu. foll. by with, to) be in harmony or agreement. **2** (usu. foll. by with) communicate by interchange of letters. □ **corresponding member** an honorary member of a learned society etc. with no voice in the society's affairs. □ **correspondingly** adv. [F correspondre f. med.L correspondere (as COM-, RESPOND)]

correspondence /ˌkɒrɪˈspɒndəns/ n. **1** (usu. foll. by with, to, between) agreement, similarity, or harmony. **2 a** communication by letters. **b** letters sent or received. □ **correspondence college** (or **school**) a college conducting correspondence courses. **correspondence column** the part of a newspaper etc. that contains letters from readers. **correspondence course** a course of study conducted by post. [ME f. OF f. med.L correspondentia (as CORRESPOND)]

correspondent /ˌkɒrɪˈspɒndənt/ n. & adj. ● n. **1** a person who writes letters to a person or a newspaper, esp. regularly. **2** a person employed to contribute material for publication in a periodical or for broadcasting (our chess correspondent; the BBC's Moscow correspondent). **3** a person or firm having regular business relations with another, esp. in another country. ● adj. (often foll. by to, with) archaic corresponding. [ME f. OF correspondant or med.L (as CORRESPOND)]

corrida /kɒˈriːdə/ n. **1** a bullfight. **2** bullfighting. [Sp. corrida de toros running of bulls]

corridor /ˈkɒrɪˌdɔː(r)/ n. **1** a passage from which doors lead into rooms (orig. an outside passage connecting parts of a building, now usu. a main passage in a large building). **2** a passage in a railway carriage from which doors lead into compartments. **3** a strip of the territory of one state passing through that of another, esp. securing access to the sea. **4** (in full **air corridor**) a route to which aircraft are restricted, esp. over a foreign country. □ **corridors of power** places where covert influence is said to be exerted in government. [F f. It. corridore corridor for corridojo running-place f. correre run, by confusion with corridore runner]

corrie /ˈkɒrɪ/ n. esp. Sc. a circular hollow on a mountainside; a cirque. [Gael. coire cauldron]

corrigendum /ˌkɒrɪˈgendəm, -ˈdʒendəm/ n. (pl. **corrigenda** /-də/) a thing to be corrected, esp. an error in a printed book. [L, neut. gerundive of corrigere: see CORRECT]

corrigible /ˈkɒrɪdʒɪb(ə)l/ adj. **1** capable of being corrected. **2** (of a person) submissive; open to correction. □ **corrigibly** adv. [ME f. F f. med.L corrigibilis (as CORRECT)]

corroborate /kəˈrɒbəˌreɪt/ v.tr. confirm or give support to (a statement or belief, or the person holding it), esp. in relation to witnesses in a law court. □ **corroborator** n. **corroboratory** /-rətərɪ/ adj. **corroborative** /-rətɪv/ adj. **corroboration** /-ˌrɒbəˈreɪʃ(ə)n/ n. [L corroborare strengthen (as COM-, roborare f. robur -oris strength)]

corroboree /kəˈrɒbərɪ/ n. **1** a festive or warlike dance-drama with song of Australian Aboriginals. **2** a noisy party. [Dharuk garabari a style of dancing]

corrode /kəˈrəʊd/ v. **1 a** tr. wear away, esp. by chemical action. **b** intr. be worn away; decay. **2** tr. destroy gradually (optimism corroded by recent misfortunes). □ **corrodible** adj. [ME f. L corrodere corros- (as COM-, rodere gnaw)]

corrosion /kəˈrəʊʒ(ə)n/ n. **1** the process of corroding, esp. of a rusting metal. **2 a** damage caused by corroding. **b** a corroded area.

corrosive /kəˈrəʊsɪv/ adj. & n. ● adj. tending to corrode or consume. ● n. a corrosive substance. □ **corrosive sublimate** mercuric chloride, a poisonous acrid substance used as a fungicide, antiseptic, etc. (chem. formula: $HgCl_2$). □ **corrosively** adv. **corrosiveness** n. [ME f. OF corosif -ive (as CORRODE)]

corrugate /ˈkɒrʊˌgeɪt/ v. **1** tr. (esp. as **corrugated** adj.) form into alternate ridges and grooves, esp. to strengthen (corrugated iron; corrugated paper). **2** tr. & intr. contract into wrinkles or folds. □ **corrugation** /ˌkɒrʊˈgeɪʃ(ə)n/ n. [L corrugare (as COM-, rugare f. ruga wrinkle)]

corrugator /ˈkɒrʊˌgeɪtə(r)/ n. Anat. either of two muscles that contract the brow in frowning. [mod.L (as CORRUGATE)]

corrupt /kəˈrʌpt/ adj. & v. ● adj. **1** morally depraved; wicked. **2** influenced by or using bribery or fraudulent activity. **3** (of a text, language, etc.) harmed (esp. made suspect or unreliable) by errors or alterations. **4** rotten. ● v. **1** tr. & intr. make or become corrupt or depraved. **2** tr. affect or harm by errors or alterations. **3** tr. infect, taint. □ **corrupt practices** fraudulent activity, esp. at elections. □ **corrupter** n. **corruptible** adj. **corruptive** adj. **corruptly** adv. **corruptibility** /-ˌrʌptɪˈbɪlɪtɪ/ n. [ME f. OF corrupt or L corruptus past part. of corrumpere corrupt- (as COM-, rumpere break)]

corruption /kəˈrʌpʃ(ə)n/ n. **1** moral deterioration, esp. widespread. **2** use of corrupt practices, esp. bribery or fraud. **3 a** irregular alteration (of a text, language, etc.) from its original state. **b** an irregularly altered form of a word. **4** decomposition, esp. of a corpse or other organic matter. [ME f. OF corruption or L corruptio (as CORRUPT)]

corsac /ˈkɔːsæk/ n. (also **corsak**) a fox, Vulpes corsac, of central Asia. [Turkic]

corsage /kɔːˈsɑːʒ/ n. **1** a small bouquet worn by a woman. **2** the bodice of a woman's dress. [ME f. OF f. cors body: see CORPSE]

corsair /ˈkɔːseə(r)/ n. **1** a pirate ship. **2** a pirate. **3** hist. a privateer, esp. of the Barbary Coast. [F corsaire f. med.L cursarius f. cursus inroad f. currere run]

corsak var. of CORSAC.

Corse see CORSICA.

corse /kɔːs/ n. archaic a corpse. [var. of CORPSE]

corselet var. of CORSLET, CORSELETTE.

corselette /ˈkɔːslɪt, ˈkɔːsəˌlet/ n. (also **corselet**) a woman's foundation garment combining corset and brassière.

corset /ˈkɔːsɪt/ n. & v. **1** a closely fitting undergarment worn by women to support the abdomen. **2** a similar garment worn by men and women because of injury, weakness, or deformity. ● v.tr. (**corseted**, **corseting**) **1** provide with a corset. **2** control closely. □ **corseted** adj. **corsetry** n. [ME f. OF, dimin. of cors body: see CORPSE]

corsetière /ˌkɔːsɪˈtjeə(r)/ n. a woman who makes or fits corsets. [F, fem. of corsetier (as CORSET, -IER)]

Corsica /ˈkɔːsɪkə/ (French **Corse** /kɔːs/) a mountainous island off the west coast of Italy, forming an administrative region of France; pop. (1990) 249,740; chief towns, Bastia (northern department) and Ajaccio (southern department). It was the birthplace of Napoleon I.

Corsican /ˈkɔːsɪkən/ adj. & n. ● adj. of or relating to Corsica. ● n. **1** a native of Corsica. **2** the Italian dialect of Corsica.

corslet /ˈkɔːslɪt/ n. (also **corselet**) **1** a garment (usu. tight-fitting) covering the trunk but not the limbs. **2** hist. a piece of armour covering the trunk. [OF corselet, dimin. formed as CORSET]

Cort /kɔːt/, Henry (1740–1800), English ironmaster. Initially a supplier of wrought iron for naval and ordnance use, he set up his own forge, and patented a process for producing iron bars by passing iron through grooved rollers to avoid the laborious business of hammering. He later patented the puddling process for refining molten pig-iron, which gave Britain a lead in the industry and earned Cort the nickname 'the Great Finer'.

cortège /kɔː'teɪʒ/ *n.* **1** a procession, esp. for a funeral. **2** a train of attendants. [F]

Cortes /'kɔːtez, Spanish 'kortes/ *n.* the legislative assembly of Spain and formerly of Portugal. [Sp. & Port., pl. of *corte* COURT]

Cortés /'kɔːtez, Spanish kor'tes/, Hernando (also **Cortez**) (1485–1547), first of the Spanish conquistadores. Cortés overthrew the Aztec empire with a comparatively small army of adventurers; he conquered its capital city, Tenochtitlán, in 1519 and deposed the emperor, Montezuma. In 1521 he destroyed Tenochtitlán completely and established Mexico City as the new capital of Mexico (then called New Spain), serving briefly as governor of the colony.

cortex /'kɔːteks/ *n.* (*pl.* **cortices** /-tɪˌsiːz/) **1** *Anat.* the outer part of an organ, esp. of the brain (*cerebral cortex*) or kidneys (*renal cortex*). **2** *Bot.* **a** an outer layer of tissue immediately below the epidermis. **b** bark. □ **cortical** /-tɪk(ə)l/ *adj.* [L *cortex, -icis* bark]

Corti /'kɔːtɪ/ *n.* □ **organ of Corti** *Anat.* a structure in the inner ear of mammals, responsible for converting sound signals into nerve impulses. [Alfonso *Corti*, It. anatomist (1822–76)]

corticate /'kɔːtɪˌkeɪt/ *adj.* (also **corticated**) **1** having bark or rind. **2** barklike. [L *corticatus* (as CORTEX)]

corticosteroid /ˌkɔːtɪkəʊ'stɪərɔɪd/ *n. Biochem.* **1** any of a group of steroid hormones produced in the adrenal cortex and concerned with regulation of salts and carbohydrates, inflammation, and sexual physiology. **2** an analogous synthetic steroid.

corticotrophic hormone /ˌkɔːtɪkəʊ'trəʊfɪk, -'trɒfɪk/ *adj.* (also **corticotropic** /-'trəʊpɪk, -'trɒpɪk/) = ADRENOCORTICOTROPHIC HORMONE.

corticotrophin /ˌkɔːtɪkəʊ'trəʊfɪn/ *n.* (also **corticotropin** /-'trəʊpɪn/) = ADRENOCORTICOTROPHIN.

cortisol /'kɔːtɪˌsɒl/ *n.* = HYDROCORTISONE.

cortisone /'kɔːtɪˌzəʊn/ *n. Biochem.* a steroid hormone produced in the adrenal cortex or synthetically, used medicinally esp. against inflammation and allergy. [chem. name 17-hydroxy-11-dehydro*corticosterone*]

corundum /kə'rʌndəm/ *n.* extremely hard crystallized alumina, used esp. as an abrasive. Some varieties of it, e.g. ruby and sapphire, are used for gemstones. [Tamil *kurundam* f. Skr. *kuruvinda* ruby]

Corunna /kə'rʌnə/ (Spanish **La Coruña** /la ko'ruɲa/) a port in NW Spain; pop. (1991) 251,300. The Armada set sail from Corunna in 1588, and the town was sacked by Francis Drake in 1589. It was the site of a battle in 1809 in the Peninsular War, at which British forces under Sir John Moore defeated the French. Moore, who was killed in the battle, was buried in the city.

coruscate /'kɒrəˌskeɪt/ *v.intr.* **1** give off flashing light; sparkle. **2** be showy or brilliant. □ **coruscation** /ˌkɒrə'skeɪʃ(ə)n/ *n.* [L *coruscare* glitter]

corvée /'kɔːveɪ, kɔː'veɪ/ *n.* **1** *hist.* a day's work of unpaid labour due to a lord from a vassal. **2** labour exacted in lieu of paying taxes. **3** an onerous task. [ME f. OF ult. f. L *corrogare* ask for, collect (as COM-, *rogare* ask)]

corves *pl.* of CORF.

corvette /kɔː'vet/ *n. Naut.* **1** a small naval escort-vessel. **2** *hist.* a flush-decked warship with one tier of guns. [F f. MDu. *korf* kind of ship + dimin. -ETTE]

corvid /'kɔːvɪd/ *n. & adj. Zool.* ● *n.* a bird of the family Corvidae, which comprises the crows. ● *adj.* of or relating to this family. [mod.L *Corvidae* f. L *corvus* crow]

corvine /'kɔːvaɪn/ *adj.* of or resembling a raven or crow. [L *corvinus* f. *corvus* raven]

corybantic /ˌkɒrɪ'bæntɪk/ *adj.* wild, frenzied. [*Corybantes* priests of Cybele performing wild dances (L f. Gk *Korubantes*)]

corymb /'kɒrɪmb, -rɪm/ *n. Bot.* a flat-topped cluster of flowers with the flower-stalks proportionally longer lower down the stem. □ **corymbose** /kə'rɪmbəʊs/ *adj.* [F *corymbe* or L *corymbus* f. Gk *korumbos* cluster]

coryphée /'kɒrɪˌfeɪ/ *n.* a leading dancer in a *corps de ballet*. [F f. Gk *koruphaios* leader of a chorus f. *koruphē* head]

coryza /kə'raɪzə/ *n. Med.* **1** a catarrhal inflammation of the mucous membrane in the nose; a cold in the head. **2** a disease with this as a symptom. [L f. Gk *koruza* running at the nose]

Cos see KOS.

cos[1] /kɒs/ *n.* a variety of lettuce with crisp narrow leaves forming a long upright head. [Cos]

cos[2] /kɒs, kɒz/ *abbr.* cosine.

cos[3] /kɒz, kəz/ *conj. & adv.* (also **'cos**) *colloq.* because. [abbr.]

Cosa Nostra /ˌkəʊzə 'nɒstrə/ *n.* the American branch of the Mafia. [It., = our affair]

cosec /'kəʊsek/ *abbr.* cosecant.

cosecant /kəʊ'siːkənt/ *n. Math.* the ratio of the hypotenuse (in a right-angled triangle) to the side opposite an acute angle; the reciprocal of sine. [mod.L *cosecans* and F *cosécant* (as CO-, SECANT)]

coseismal /kəʊ'saɪzm(ə)l/ *adj. & n.* ● *adj.* of or relating to points of simultaneous arrival of an earthquake wave. ● *n.* a straight line or a curve connecting these points. [CO- + SEISMAL (see SEISMIC)]

coset /'kəʊset/ *n. Math.* a set composed of all the products obtained by multiplying on the right or on the left each element of a subgroup in turn by one particular element of the group containing the subgroup. [CO- + SET[2]]

cosh[1] /kɒʃ/ *n. & v. Brit. colloq.* ● *n.* a heavy blunt weapon. ● *v.tr.* hit with a cosh. [19th c.: orig. unkn.]

cosh[2] /kɒʃ, kɒs'eɪtʃ/ *abbr. Math.* hyperbolic cosine.

co-signatory /kəʊ'sɪɡnətərɪ/ *n. & adj.* (US **cosignatory**) ● *n.* (*pl.* **-ies**) a person or state signing a treaty etc. jointly with others. ● *adj.* signing jointly.

Cosimo de' Medici /'kɒzɪˌməʊ/ (known as Cosimo the Elder) (1389–1464), Italian statesman and banker. He laid the foundations for the Medici family's power in Florence, becoming the city's ruler in 1434. He used his considerable wealth to promote the arts and learning, and funded the establishment of public buildings such as the Medici Library.

cosine /'kəʊsaɪn/ *n. Math.* the ratio of the side adjacent to an acute angle (in a right-angled triangle) to the hypotenuse. [mod.L *cosinus* (as CO-, SINE)]

cosmetic /kɒz'metɪk/ *adj. & n.* ● *adj.* **1** intended to adorn or beautify the body, esp. the face. **2** intended to improve only appearances; superficially improving or beneficial (*a cosmetic change*). **3** (of surgery or a prosthetic device) imitating, restoring, or enhancing the normal appearance. ● *n.* a cosmetic preparation, esp. for the face. □ **cosmetically** *adv.* [F *cosmétique* f. Gk *kosmētikos* f. *kosmeō* adorn f. *kosmos* order, adornment]

cosmic /'kɒzmɪk/ *adj.* **1** of the universe or cosmos, esp. as distinct from the earth. **2** of or for space travel. □ **cosmic dust** small particles of matter distributed throughout space. **cosmic rays** (or **radiation**) radiations from space etc. that reach the earth from all directions, usu. with high energy and penetrative power. **cosmic string** see STRING *n.* 11b. □ **cosmical** *adj.* **cosmically** *adv.*

cosmogony /kɒz'mɒɡənɪ/ *n.* (*pl.* **-ies**) **1** the origin of the universe. **2** a theory about this. □ **cosmogonist** *n.* **cosmogonic** /ˌkɒzmə'ɡɒnɪk/ *adj.* **cosmogonical** *adj.* [Gk *kosmogonia* f. *kosmos* world + *-gonia* -begetting]

cosmography /kɒz'mɒɡrəfɪ/ *n.* (*pl.* **-ies**) a description or mapping of general features of the universe. □ **cosmographer** *n.* **cosmographic** /ˌkɒzmə'ɡræfɪk/ *adj.* **cosmographical** *adj.* [ME f. F *cosmographie* or f. LL f. Gk *kosmographia* (as COSMOS[1], -GRAPHY)]

cosmology /kɒz'mɒlədʒɪ/ *n.* **1** the study of the creation and development of the universe. (*See note below.*) **2** an account or theory of the origin of the universe. □ **cosmologist** *n.* **cosmological** /ˌkɒzmə'lɒdʒɪk(ə)l/ *adj.* [F *cosmologie* or mod.L *cosmologia* (as COSMOS[1], -LOGY)]

▪ Theories of the origin of the universe belonged essentially to the realms of mythology and religion until the 20th century, when observational evidence for processes of large-scale change in the universe became available: in particular, the recognition of the true nature of galaxies as external star systems like the Milky Way, and of the expansion of the visible universe. This led to the big bang theory which dominates modern cosmology, and which brings together observational astronomy and particle physics. The rival steady-state theory, which accommodated expansion within an unchanging infinite universe, has not been supported by recent discoveries.

cosmonaut /ˈkɒzməˌnɔːt/ n. a Russian astronaut. [Russ. *kosmonavt*, as COSMOS[1], after *astronaut*]

cosmopolis /kɒzˈmɒpəlɪs/ n. a cosmopolitan city. [Gk *kosmos* world + *polis* city]

cosmopolitan /ˌkɒzməˈpɒlɪt(ə)n/ adj. & n. ● adj. **1 a** of or from or knowing many parts of the world. **b** consisting of people from many or all parts. **2** free from national limitations or prejudices. **3** (of a plant or animal) widely distributed. ● n. **1** a cosmopolitan person. **2** a widely distributed animal or plant. □ **cosmopolitanism** n. **cosmopolitanize** v.tr. & intr. (also **-ise**). [COSMOPOLITE + -AN]

cosmopolite /kɒzˈmɒpəˌlaɪt/ n. & adj. ● n. **1** a cosmopolitan person. **2** = COSMOPOLITAN n. 2. ● adj. free from national attachments or prejudices. [F f. Gk *kosmopolitēs* f. *kosmos* world + *politēs* citizen]

cosmos[1] /ˈkɒzmɒs/ n. **1** the universe, esp. as a well-ordered whole. **2 a** an ordered system of ideas etc. **b** a sum total of experience. [Gk *kosmos*]

cosmos[2] /ˈkɒzmɒs/ n. a composite plant of the genus *Cosmos*, bearing single dahlia-like blossoms of various colours. [mod.L f. Gk *kosmos* in sense 'ornament']

COSPAR /ˈkəʊspɑː(r)/ abbr. Committee on Space Research.

Cossack /ˈkɒsæk/ n. & adj. ● n. **1** a member of a people living in southern Russia, Ukraine, and Siberia, noted from late medieval times for their horsemanship and military skill. (*See note below.*) **2** a member of a Cossack military unit. ● adj. of, relating to, or characteristic of the Cossacks. [F *cosaque* f. Russ. *kazak* f. Turkic, = vagabond]

- The Cossacks had their origins in the 15th century when refugees from religious persecution, outlaws, adventurers, and escaped serfs banded together in settled groups for mutual protection. In Imperial Russia the Cossacks were allowed considerable autonomy for their communities in return for service in protecting the frontiers; later they were organized into military units which were frequently used against insurrectionaries. After the Russian Revolution the Cossack communities lost their autonomy, although the military units were revived in the Second World War. With the collapse of Soviet rule Cossack groups have reasserted their identity in both Russia and Ukraine.

cosset /ˈkɒsɪt/ v.tr. (**cosseted, cosseting**) pamper. [dial. *cosset* = pet lamb, prob. f. AF *coscet, cozet* f. OE *cotsǣta* cottager (as COT[2], SIT)]

cossie /ˈkɒzɪ/ n. (also **cozzie**) esp. *Austral. sl.* a swimming-costume. [abbr.]

Cossyra /kəˈsaɪərə/ the Roman name for PANTELLERIA.

cost /kɒst/ v. & n. ● v. (*past* and *past part.* **cost**) **1** tr. be obtainable for (a sum of money); have as a price (*what does it cost?; it cost me £50*). **2** tr. involve as a loss or sacrifice (*it cost them much effort; it cost him his life*). **3** tr. (*past* and *past part.* **costed**) fix or estimate the cost or price of. **4** colloq. **a** tr. be costly to (*it'll cost you*). **b** intr. be costly. ● n. **1** what a thing costs; the price paid or to be paid. **2** a loss or sacrifice; an expenditure of time, effort, etc. **3** (in pl.) legal expenses, esp. those allowed in favour of the winning party or against the losing party in a suit. □ **at all costs** (or **at any cost**) no matter what the cost or risk may be. **at cost** at the initial cost; at cost price. **at the cost of** at the expense of losing or sacrificing. **cost accountant** an accountant who records costs and (esp. overhead) expenses in a business concern. **cost-benefit** designating or relating to a process that assesses the relation between the cost of an operation and the value of the resulting benefits (*cost-benefit analysis*). **cost** (or **costing**) **clerk** a clerk who records costs and expenses in a business concern. **cost a person dear** (or **dearly**) involve a person in a high cost or a heavy penalty. **cost-effective** effective or productive in relation to its cost. **cost of living** the level of prices esp. of the basic necessities of life. **cost-plus** designating or relating to a pricing system in which a fixed profit factor is added to cost incurred. **cost price** the price paid for a thing by a person who later sells it. **cost push** *Econ.* factors other than demand that cause inflation. **to a person's cost** at a person's expense; with loss or disadvantage to a person. [ME f. OF *coster, couster, coust* ult. f. L *constare* stand firm, stand at a price (as COM-, *stare* stand)]

Costa /ˈkɒstə/, Lúcio (1902–63), French-born Brazilian architect, town planner, and architectural historian. He headed the group that designed the Ministry of Education in Rio de Janeiro (1937–43) and achieved a worldwide reputation with his plan for Brazil's new capital Brasília, which was chosen by an international jury in 1956.

Costa Blanca /ˌkɒstə ˈblæŋkə/ a resort region on the Mediterranean coast of SE Spain. [Sp., = white coast]

Costa Brava /ˌkɒstə ˈbrɑːvə/ a resort region to the north of Barcelona, on the Mediterranean coast of NE Spain. [Sp., = wild coast]

Costa del Sol /ˌkɒstə del ˈsɒl/ a resort region on the Mediterranean coast of southern Spain. Marbella and Torremolinos are the principal resort towns. [Sp., = coast of the sun]

costal /ˈkɒst(ə)l/ adj. of the ribs. [F f. mod.L *costalis* f. L *costa* rib]

co-star /ˈkəʊstɑː(r)/ n. & v. ● n. a cinema or stage star appearing with another or others of equal importance. ● v. (**-starred, -starring**) **1** intr. take part as a co-star. **2** tr. (of a production) include as a co-star.

costard /ˈkɒstəd/ n. *Brit.* **1** a large ribbed variety of apple. **2** archaic joc. the head. [ME f. AF *coste* rib f. L *costa*]

Costa Rica /ˌkɒstə ˈriːkə/ a republic in Central America on the Isthmus of Panama; pop. (est. 1991) 2,875,000; official language, Spanish; capital, San José. Inhabited by Carib and other native peoples, the territory was colonized by Spain in the early 16th century. Costa Rica achieved independence in 1823 and emerged as a separate country in 1838 after fourteen years within the United Provinces of Central America. In 1948 the army was abolished, the President declaring it unnecessary as the country loved peace. □ **Costa Rican** adj. & n.

costate /ˈkɒsteɪt/ adj. ribbed; having ribs or ridges. [L *costatus* f. *costa* rib]

coster /ˈkɒstə(r)/ n. *Brit.* = COSTERMONGER. [abbr.]

costermonger /ˈkɒstəˌmʌŋɡə(r)/ n. *Brit.* a person who sells fruit, vegetables, etc., in the street from a barrow. [COSTARD + MONGER]

costive /ˈkɒstɪv/ adj. **1** constipated. **2** niggardly. □ **costively** adv. **costiveness** n. [ME f. OF *costivé* f. L *constipatus*: see CONSTIPATE]

costly /ˈkɒstlɪ/ adj. (**costlier, costliest**) **1** costing much; expensive. **2** of great value. □ **costliness** n.

costmary /ˈkɒstˌmeərɪ/ n. (pl. **-ies**) an aromatic composite plant, *Balsamita major*, formerly used in medicine and for flavouring ale. [OE *cost* f. L *costum* f. Gk *kostos* f. Arab. *ḳusṭ* an aromatic plant + (*St*) *Mary* (with whom it was associated in medieval times)]

costume /ˈkɒstjuːm/ n. & v. ● n. **1** a style or fashion of dress, esp. that of a particular place, time, or class. **2** a set of clothes. **3** clothing for a particular activity (*swimming-costume*). **4** an actor's clothes for a part. **5** a woman's matching jacket and skirt. ● v.tr. provide with a costume. □ **costume jewellery** artificial jewellery worn to adorn clothes. **costume play** (or **piece** or **drama**) a play or television drama in which the actors wear historical costume. [F f. It. f. L *consuetudo* CUSTOM]

costumier /kɒˈstjuːmɪə(r)/ n. (also **costumer** /-mə(r)/) a person who makes or deals in costumes, esp. for theatrical use. [F *costumier* (as COSTUME)]

cosy /ˈkəʊzɪ/ adj., n., & v. (*US* **cozy**) ● adj. (**cosier, cosiest**) **1** comfortable and warm; snug. **2** derog. complacent; self-serving. **3** warm and friendly. ● n. (pl. **-ies**) **1** a cover to keep something hot, esp. a teapot or a boiled egg. **2** a canopied corner seat for two. ● v.tr. (**-ies, -ied**) (often foll. by *along*) colloq. reassure, esp. deceptively. □ **cosy up to** *US* colloq. ingratiate oneself with. **2** snuggle up to. □ **cosily** adv. **cosiness** n. [18th c. f. Sc., of unkn. orig.]

cot[1] /kɒt/ n. **1** *Brit.* a small bed with high sides, esp. for a baby or very young child. **2** a hospital bed. **3** *US* a small folding bed. **4** *Ind.* a light bedstead. **5** *Naut.* a kind of swinging bed hung from deck beams, formerly used by officers. □ **cot-case** a person too ill to leave his or her bed. **cot-death** *Brit.* the unexplained death of a baby while sleeping. [Anglo-Ind., f. Hindi *khāṭ* bedstead, hammock]

cot[2] /kɒt/ n. & v. ● n. **1** a small shelter; a cote (*bell-cot*; *sheep-cot*). **2** *poet.* a cottage. ● v.tr. (**cotted, cotting**) put (sheep) in a cot. [OE f. Gmc, rel. to COTE]

cot[3] /kɒt/ abbr. *Math.* cotangent.

cotangent /kəʊˈtændʒənt/ n. *Math.* the ratio of the side adjacent to an acute angle (in a right-angled triangle) to the opposite side.

cote /kəʊt/ n. a shelter, esp. for animals or birds; a shed or stall (*sheep-cote*). [OE f. Gmc, rel. to COT[2]]

Côte d'Ivoire /kot divwar/ the French name for IVORY COAST.

coterie /ˈkəʊtərɪ/ n. **1** an exclusive group of people sharing interests. **2** a select circle in society. [F, orig. = association of tenants, ult. f. MLG *kote* COTE]

coterminous /kəʊˈtɜːmɪnəs/ adj. (often foll. by *with*) having the same boundaries or extent (in space, time, or meaning). [CO- + TERMINUS + -OUS]

coth /kɒθ/ abbr. *Math.* hyperbolic cotangent.

co-tidal line /kəʊˈtaɪd(ə)l/ n. a line on a map connecting points at which tidal levels (as high tide or low tide) occur simultaneously.

cotillion /kə'tɪljən/ n. **1** a French dance with elaborate steps, figures, and ceremonial. **2** US **a** a ballroom dance resembling a quadrille. **b** a formal ball. [F *cotillon* petticoat, dimin. of *cotte* f. OF *cote* COAT]

cotinga /kə'tɪŋgə/ n. a tropical American passerine bird of the family Cotingidae, often with brilliant plumage. [F f. Tupi *cutinga*]

Cotman /'kɒtmən/, John Sell (1782–1842), English painter. His main importance is as a watercolourist and landscape painter; he is regarded as one of the leading figures of the Norwich School. His early watercolours, including *Greta Bridge* (1805), are notable for their bold configurations of light and shade and have been compared in their flat areas of colour to Japanese painting. In the 1820s he developed a more richly coloured style, as in *The Drop Gate* (c.1826). He also completed a number of distinctive etchings, but these have not enjoyed the same popularity as his watercolours.

cotoneaster /kə,təʊnɪ'æstə(r)/ n. a rosaceous Eurasian shrub of the genus *Cotoneaster*, with red or black berries. [mod.L f. L *cotoneum* QUINCE + -ASTER]

Cotonou /,kɒtə'nuː/ the largest city, chief port, and chief commercial and political centre of Benin, on the coast of West Africa; pop. (1982) 487,000.

Cotopaxi /,kɒtə'pæksɪ/ the highest active volcano in the world, rising to 5,896 m (19,142 ft) in the Andes of central Ecuador. Its name is Quechuan and means 'shining peak'.

Cotswold Hills /'kɒtswəʊld/ (also **Cotswolds**) a range of limestone hills in SW England, largely in the county of Gloucestershire.

cotta /'kɒtə/ n. *Eccl.* a surplice-like garment usu. reaching just above the waist. [It., formed as COAT]

cottage /'kɒtɪdʒ/ n. **1** a small simple house, esp. in the country. **2** a dwelling forming part of a farm establishment, used by a worker. **3** *sl.* a public toilet. □ **cottage cheese** soft white cheese made from curds of skimmed milk without pressing. **cottage hospital** *Brit.* a small hospital not having resident medical staff. **cottage industry** a business activity partly or wholly carried on at home. **cottage loaf** a loaf formed of two round masses, the smaller on top of the larger. **cottage pie** *Brit.* a dish of minced meat topped with browned mashed potato. □ **cottagey** adj. [ME f. AF, formed as COT², COTE]

cottager /'kɒtɪdʒə(r)/ n. a person who lives in a cottage.

cottaging /'kɒtɪdʒɪŋ/ n. *sl.* homosexual behaviour in public toilets. [COTTAGE 3]

cottar /'kɒtə(r)/ n. (also **cotter**) **1** *Sc. & hist.* a farm-labourer or tenant occupying a cottage in return for labour as required. **2** *Ir. hist.* = COTTIER. [COT² + -ER¹ (Sc. *-ar*)]

Cottbus /'kɒtbʊs/ an industrial city in SE Germany, in Brandenburg, on the River Spree; pop. (1991) 123,320.

cotter /'kɒtə(r)/ n. **1** a bolt or wedge for securing parts of machinery etc. **2** (in full **cotter pin**) a split pin that opens after passing through a hole. [17th c. (rel. to earlier *cotterel*): orig. unkn.]

cottier /'kɒtɪə(r)/ n. *Brit.* **1** a cottager. **2** *hist.* an Irish peasant under cottier tenure. □ **cottier tenure** *hist.* the letting of land in small portions at a rent fixed by competition. [ME f. OF *cotier* f. med.L *cotarius*: see COTERIE]

cotton /'kɒt(ə)n/ n. & v. ● n. **1** a soft white fibrous substance covering the seeds of certain plants, esp. of the genus *Gossypium*. **2 a** (in full **cotton plant**) a tropical or subtropical plant of the genus *Gossypium* (mallow family), grown for the fibre or seeds. **b** this plant as a crop. **3** thread or cloth made from the fibre. **4** (*attrib.*) made of cotton. ● v.intr. (foll. by *to*) be attracted by (a person). □ **cotton-cake** compressed cotton seed used as food for cattle. **cotton candy** *N. Amer.* candyfloss. **cotton-gin** a machine for separating cotton from its seeds. **cotton-grass** a sedge of the genus *Eriophorum*, having fruiting heads with long white silky hairs, growing in bogs. **cotton on** (often foll. by *to*) *colloq.* begin to understand. **cotton-picking** *N. Amer. sl.* unpleasant, wretched. **cotton waste** refuse yarn used to clean machinery etc. **cotton wool 1** esp. *Brit.* fluffy wadding of a kind orig. made from raw cotton. **2** US raw cotton. □ **cottony** adj. [ME f. OF *coton* f. Arab. *ḳuṭn*]

cottontail /'kɒt(ə)n,teɪl/ n. a rabbit of the genus *Sylvilagus*, native to America, having a mainly white fluffy tail.

cottonwood /'kɒt(ə)n,wʊd/ n. **1** a North American poplar with seeds covered in white cottony hairs. **2** a downy-leaved shrub, *Bedfordia salicina*, native to Australia.

cotyledon /,kɒtɪ'liːd(ə)n/ n. **1** *Bot.* an embryonic leaf in seed-bearing plants. **2** a succulent plant of the southern African genus *Cotyledon* or the European genus *Umbilicus*, esp. navelwort. □ **cotyledonary** adj.

cotyledonous adj. [L, = pennywort, f. Gk *kotulēdōn* cup-shaped cavity f. *kotulē* cup]

coucal /'kuːkæl/ n. a ground-nesting bird of the genus *Centropus*, related to the cuckoos, found widely in the Old World. [F, perh. f. *coucou* cuckoo + *alouette* lark]

couch¹ /kaʊtʃ/ n. & v. ● n. **1** an upholstered piece of furniture for several people; a sofa. **2** a long padded seat with a headrest at one end, esp. one on which a psychoanalyst's subject or doctor's patient reclines during examination. ● v. **1** tr. (foll. by *in*) express in words of a specified kind (*couched in simple language*). **2** tr. lay on or as on a couch. **3** intr. **a** (of an animal) lie, esp. in its lair. **b** lie in ambush. **4** tr. lower (a spear etc.) to the position for attack. **5** tr. *Med.* treat (a cataract) by displacing the lens of the eye. □ **couch potato** *sl.* a person who likes lazing at home, esp. watching television. [ME f. OF *couche, coucher* f. L *collocare* (as COM-, *locare* place)]

couch² /kuːtʃ, kaʊtʃ/ n. (in full **couch grass**) a grass of the genus *Agropyron*; esp. *A. repens*, which has long creeping roots. [var. of QUITCH]

couchant /'kaʊtʃənt/ adj. (placed after noun) *Heraldry* (of an animal) lying with the body resting on the legs and the head raised. [F, pres. part. of *coucher*: see COUCH¹]

couchette /kuː'ʃet/ n. **1** a railway carriage with seats convertible into sleeping-berths. **2** a berth in this. [F, = little bed, dimin. of *couche* COUCH¹]

coudé /kuː'deɪ/ adj. & n. ● adj. of or relating to an astronomical telescope in which the rays are bent to a focus at a fixed point off the axis. ● n. such a telescope. [F, past part. of *couder* bend at right angles f. *coude* elbow formed as CUBIT]

Couéism /'kuːeɪ,ɪz(ə)m/ n. a system of psychotherapy using usu. optimistic auto-suggestion. [Émile *Coué*, Fr. psychologist (1857–1926)]

cougar /'kuːgə(r)/ n. *N. Amer.* a puma. [F, repr. Guarani *guaçu ara*]

cough /kɒf/ v. & n. ● v.intr. **1** expel air from the lungs with a sudden sharp sound produced by abrupt opening of the glottis, to remove an obstruction or congestion. **2** (of an engine, gun, etc.) make a similar sound. **3** *sl.* confess. ● n. **1** an act of coughing. **2** a condition of the respiratory organs causing coughing. **3** a tendency to cough. □ **cough drop** (or **sweet**) a medicated lozenge for relieving a cough. **cough mixture** a liquid medicine for relieving a cough. **cough out 1** eject by coughing. **2** say with a cough. **cough up 1** = *cough out.* **2** *sl.* bring out or give (money or information) reluctantly. [ME *coghe, cowhe*, rel. to MDu. *kuchen*, MHG *kûchen*, of imit. orig.]

could past of CAN¹.

couldn't /'kʊd(ə)nt/ contr. could not.

coulée /'kuːleɪ, -lɪ/ n. *Geol.* **1** a stream of molten or solidified lava. **2** *N. Amer. dial.* a deep ravine. [F, fem. past part. of *couler* flow, f. L *colare* strain, filter]

coulis /'kuːlɪ/ n. (pl. same) a light sauce of puréed fruit or vegetables, thin enough to pour. [F f. *couler* to run]

coulisse /kuː'liːs/ n. **1** (usu. in pl.) *Theatr.* a piece of side scenery or a space between two of these; the wings. **2** a place of informal discussion or negotiation. [F f. *coulis* sliding: see PORTCULLIS]

couloir /'kuːlwɑː(r)/ n. a steep narrow gully on a mountainside. [F f. *couler* glide: see COULÉE]

Coulomb /'kuːlɒm/, Charles-Augustin de (1736–1806), French military engineer. He conducted research on structural mechanics, elasticity, friction, electricity, and magnetism. He is best known for Coulomb's Law, established with a sensitive torsion balance in 1785, according to which the forces between two electrical charges are proportional to the product of the sizes of the charges and inversely proportional to the square of the distance between them. Coulomb's verification of the inverse square law of electrostatic force enabled the quantity of electric charge to be defined.

coulomb /'kuːlɒm/ n. *Electr.* the SI unit of electric charge (symbol: **C**), equal to the quantity of electricity conveyed in one second by a current of one ampere. [COULOMB]

coulometry /kuː'lɒmɪtrɪ/ n. *Chem.* a method of chemical analysis by measurement of the number of coulombs used in electrolysis. □ **coulometric** /,kuːlə'metrɪk/ adj.

coulter /'kəʊltə(r)/ n. (US **colter**) a vertical cutting blade fixed in front of a ploughshare. [OE f. L *culter*]

coumarin /'kuːmərɪn/ n. *Chem.* an aromatic compound found in many plants and formerly used for flavouring food. [F *coumarine* f. Tupi *cumarú* substance from tonka beans]

coumarone /'kuːmə,rəʊn/ n. *Chem.* an organic liquid compound present in coal tar and used in paints and varnishes. □ **coumarone resin** a thermoplastic resin formed by polymerization of coumarone. [COUMARIN + -ONE]

council /'kaʊns(ə)l/ n. **1 a** an advisory, deliberative, or administrative body of people formally constituted and meeting regularly. **b** a meeting of such a body. **2 a** the elected local administrative body of a parish, district, town, city, or administrative county and its paid officers and workforce. **b** (attrib.) (esp. of housing) provided by a local council (council flat; council estate). **3** a body of persons chosen as advisers (Privy Council). **4** an ecclesiastical assembly (ecumenical council). □ **council-chamber** a room in which a council meets. **council-house 1** a house owned and let by a local council. **2** a building in which a council meets. **council of war 1** an assembly of officers called in a special emergency. **2** any meeting held to plan a response to an emergency. **council tax** (in the UK) a tax levied by local authorities, based on the estimated value of a property, replacing the community charge from 1993. [ME f. AF cuncile f. L concilium convocation, assembly f. calare summon: cf. COUNSEL]

councillor /'kaʊnsələ(r)/ n. an elected member of a council, esp. a local council. □ **councillorship** n. [ME, alt. of COUNSELLOR: assim. to COUNCIL]

councilman /'kaʊns(ə)lmən/ n. (pl. **-men**; fem. **councilwoman**, pl. **-women**) esp. US a member of a council; a councillor.

Council of Chalcedon, Council of Europe, etc. see CHALCEDON, COUNCIL OF; EUROPE, COUNCIL OF, etc.

counsel /'kaʊns(ə)l/ n. & v. ● n. **1** advice, esp. formally given. **2** consultation, esp. to seek or give advice. **3** (pl. same) a barrister or other legal adviser; a body of these advising in a case. **4** a plan of action. ● v.tr. (**counselled, counselling**; US **counseled, counseling**) **1** (often foll. by to + infin.) advise (a person). **2 a** give advice to (a person) on social or personal problems, esp. professionally. **b** assist or guide (a person) in resolving personal difficulties. **3** (often foll. by that) recommend (a course of action). □ **counsel of despair** action to be taken when all else fails. **counsel of perfection 1** advice that is ideal but not feasible. **2** advice guiding towards moral perfection. **keep one's own counsel** not confide in others. **take counsel** (usu. foll. by with) consult. [ME f. OF c(o)unseil, conseiller f. L consilium consultation, advice]

counselling /'kaʊnsəlɪŋ/ n. (US **counseling**) **1** the act or process of giving counsel. **2** the process of assisting and guiding clients, esp. by a trained person on a professional basis, to resolve esp. personal, social, or psychological problems and difficulties (cf. COUNSEL v. 2).

counsellor /'kaʊnsələ(r)/ n. (US **counselor**) **1** a person who gives counsel; an adviser. **2** a person trained to give guidance on personal, social, or psychological problems (marriage guidance counsellor). **3** a senior officer in the diplomatic service. **4 a** (also **counselor-at-law**) US a barrister. **b** (also **counsellor-at-law**) Ir. an advising barrister. [ME f. OF conseiller (f. L consiliarius), conseillour, -eur (f. L consiliator): see COUNSEL]

Counsellor of State n. (in the UK) a temporary regent during a sovereign's absence.

count[1] /kaʊnt/ v. & n. ● v. **1** tr. determine the total number or amount of, esp. by assigning successive numbers (count the stations). **2** intr. repeat numbers in ascending order; conduct a reckoning. **3 a** tr. (often foll. by in) include in one's reckoning or plan (you can count me in; fifteen people, counting the guide). **b** intr. be included in a reckoning or plan. **4** tr. consider (a thing or a person) to be (lucky etc.) (count no man happy until he is dead). **5** intr. (often foll. by for) have value; matter (his opinion counts for a great deal). ● n. **1 a** the act of counting; a reckoning (after a count of fifty). **b** the sum total of a reckoning (blood count; pollen count). **2** Law each charge in an indictment (guilty on ten counts). **3** a count of up to ten seconds by a referee when a boxer is knocked down. **4** Polit. the act of counting the votes after a general or local election. **5** one of several points under discussion. **6** the measure of the fineness of a yarn expressed as the weight of a given length or the length of a given weight. **7** Physics the number of ionizing particles detected by a counter. □ **count against** be reckoned to the disadvantage of. **count one's blessings** be grateful for what one has. **count one's chickens** be over-optimistic or hasty in anticipating good fortune. **count the cost** consider the risks before taking action; calculate the damage resulting from an action. **count the days** (or **hours** etc.) be impatient. **count down** recite numbers backwards to zero, esp. as part of a rocket-launching procedure. **counting-house** a place where accounts are kept. **count noun** a countable noun (see COUNTABLE 2). **count on** (or **upon**) depend on, rely on; expect confidently. **count out 1** count

while taking from a stock. **2** complete a count of ten seconds over (a fallen boxer etc.), indicating defeat. **3** (in children's games) select (a player) for dismissal or a special role by use of a counting rhyme etc. **4** colloq. exclude from a plan or reckoning (I'm too tired, count me out). **5** Brit. Polit. procure the adjournment of (the House of Commons) when fewer than 40 members are present. **count up** find the sum of. **keep count** take note of how many there have been etc. **lose count** fail to take note of how many there have been etc.; forget the number noted in counting. **not counting** excluding from the reckoning. **out for the count 1** Boxing defeated by being unable to rise within ten seconds. **2 a** defeated or demoralized. **b** sound asleep; unconscious. **take the count** Boxing be defeated. [ME f. OF co(u)nter, co(u)nte f. LL computus, computare COMPUTE]

count[2] /kaʊnt/ n. a foreign noble corresponding to an earl. □ **countship** n. [OF conte f. L comes comitis companion]

countable /'kaʊntəb(ə)l/ adj. **1** that can be counted. **2** Gram. (of a noun) that can form a plural or be used with the indefinite article (e.g. book, kindness).

countdown /'kaʊntdaʊn/ n. **1 a** the act of counting down, esp. at the launching of a rocket etc. **b** the procedures carried out during this time. **2** the final moments before any significant event.

countenance /'kaʊntɪnəns/ n. & v. ● n. **1 a** the face. **b** the facial expression. **2** composure. **3** moral support. ● v.tr. **1** give approval to (an act etc.) (cannot countenance this breach of the rules). **2** (often foll. by in) encourage (a person or a practice). □ **change countenance** alter one's expression as an effect of emotion. **keep one's countenance** maintain composure, esp. by refraining from laughter. **keep a person in countenance** support or encourage a person. **lose countenance** become embarrassed. **out of countenance** disconcerted. [ME f. AF c(o)untenance, OF contenance bearing f. contenir: see CONTAIN]

counter[1] /'kaʊntə(r)/ n. **1 a** a long flat-topped fitment in a shop, bank, etc., across which business is conducted with customers. **b** a similar structure used for serving food etc. in a cafeteria or bar. **2 a** a small disc used for keeping the score etc. in table-games. **b** a token representing a coin. **c** something used in bargaining; a pawn (a counter in the struggle for power). **3 a** an apparatus used for counting. **b** Physics an apparatus used for counting individual ionizing particles etc. **4** a person who counts something. □ **over the counter** by ordinary retail purchase. **under the counter** attrib.adj. (esp. of illicit goods) obtained surreptitiously. [AF count(e)our, OF conteo(i)r, f. med.L computatorium (as COMPUTE)]

counter[2] /'kaʊntə(r)/ v., adv., adj., & n. ● v. **1** tr. a oppose, contradict (countered our proposal with their own). **b** meet by a countermove. **2** intr. **a** make a countermove. **b** make an opposing statement ('I shall!' he countered). **3** intr. Boxing give a return blow while parrying. ● adv. **1** in the opposite direction (ran counter to the fox). **2** contrary (his action was counter to my wishes). ● adj. **1** opposed; opposite. **2** duplicate; serving as a check. ● n. **1** a parry; a countermove. **2** something opposite or opposed. □ **act** (or **go**) **counter to** disobey (instructions etc.). **go** (or **hunt** or **run**) **counter** run or ride against the direction taken by a quarry. **run counter to** act contrary to. [ME f. OF countre f. L contra against: see COUNTER-]

counter[3] /'kaʊntə(r)/ n. **1** the part of a horse's breast between the shoulders and under the neck. **2** the curved part of the stern of a ship. **3** Printing a part of a printing type etc. that is completely enclosed by an outline (e.g. the loop of P). [17th c.: orig. unkn.]

counter[4] /'kaʊntə(r)/ n. the back part of a shoe or a boot round the heel. [abbr. of counterfort buttress]

counter- /'kaʊntə(r)/ comb. form denoting: **1** retaliation, opposition, or rivalry (counter-demonstration; counter-inflationary). **2** opposite direction (counter-current). **3** correspondence, duplication, or substitution (counterpart; countersign). [from or after AF countre-, OF contre f. L contra against]

counteract /ˌkaʊntər'ækt/ v.tr. **1** hinder or oppose by contrary action. **2** neutralize. □ **counteractive** adj. **counteraction** /-'ækʃ(ə)n/ n.

counter-attack /'kaʊntərəˌtæk/ n. & v. ● n. an attack in reply to an attack by an enemy or opponent. ● v.tr. & intr. attack in reply.

counter-attraction /'kaʊntərəˌtrækʃ(ə)n/ n. **1** a rival attraction. **2** the attraction of a contrary tendency.

counterbalance n. & v. ● n. /'kaʊntəˌbæləns/ **1** a weight balancing another. **2** an argument, force, etc., balancing another. ● v.tr. /ˌkaʊntə'bæləns/ act as a counterbalance to.

counterblast /ˈkaʊntəˌblɑːst/ n. (often foll. by *to*) an energetic or violent verbal or written reply to an argument etc.

counterchange /ˌkaʊntəˈtʃeɪndʒ/ v. **1** tr. change (places or parts); interchange. **2** tr. *literary* chequer, esp. with contrasting colours etc. **3** intr. change places or parts. [F *contrechanger* (as COUNTER-, CHANGE)]

countercharge n. & v. ● n. /ˈkaʊntəˌtʃɑːdʒ/ a charge or accusation in return for one received. ● v.tr. /ˌkaʊntəˈtʃɑːdʒ/ make a countercharge against.

countercheck n. & v. ● n. /ˈkaʊntəˌtʃek/ **1 a** a restraint that opposes something. **b** a restraint that operates against another. **2** a second check, esp. for security or accuracy. **3** *archaic* a retort. ● v.tr. /ˌkaʊntəˈtʃek/ make a countercheck on.

counter-claim n. & v. ● n. /ˈkaʊntəˌkleɪm/ **1** a claim made against another claim. **2** *Law* a claim made by a defendant in a suit against the plaintiff. ● v.tr. & intr. /ˌkaʊntəˈkleɪm/ make a counter-claim (for).

counter-clockwise /ˌkaʊntəˈklɒkwaɪz/ adv. & adj. N. Amer. = ANTICLOCKWISE.

counter-culture /ˈkaʊntəˌkʌltʃə(r)/ n. a way of life etc. opposed to that usually considered normal.

counter-espionage /ˌkaʊntərˈespɪəˌnɑːʒ/ n. action taken to frustrate enemy spying.

counterfeit /ˈkaʊntəˌfɪt, -ˌfiːt/ adj., n., & v. ● adj. **1** (of a coin, writing, etc.) made in imitation; not genuine; forged. **2** (of a claimant etc.) pretended. ● n. a forgery; an imitation. ● v.tr. **1 a** imitate fraudulently (a coin, handwriting, etc.); forge. **b** make an imitation of. **2** simulate (feelings etc.) (*counterfeited interest*). **3** resemble closely. □ **counterfeiter** n. [ME f. OF *countrefet, -fait*, past part. of *contrefaire* f. Rmc]

counterfoil /ˈkaʊntəˌfɔɪl/ n. the part of a cheque, receipt, etc., retained by the person issuing it and containing details of the transaction.

counter-intelligence /ˌkaʊntərɪnˈtelɪdʒəns/ n. = COUNTER-ESPIONAGE.

counterirritant /ˌkaʊntərˈɪrɪt(ə)nt/ n. **1** *Med.* something used to produce surface irritation of the skin, thereby counteracting more painful symptoms. **2** anything resembling a counterirritant in its effects. □ **counterirritation** /-ˌɪrɪˈteɪʃ(ə)n/ n.

countermand v. & n. ● v.tr. /ˌkaʊntəˈmɑːnd/ **1** *Mil.* **a** revoke (an order or command). **b** recall (forces etc.) by a contrary order. **2** cancel an order for (goods etc.). ● n. /ˈkaʊntəˌmɑːnd/ an order revoking a previous one. [ME f. OF *contremander* f. med.L *contramandare* (as CONTRA-, *mandare* order)]

countermarch v. & n. ● v.intr. & tr. /ˌkaʊntəˈmɑːtʃ/ esp. *Mil.* march or cause to march in the opposite direction, e.g. with the front marchers turning and marching back through the ranks. ● n. /ˈkaʊntəˌmɑːtʃ/ an act of countermarching.

countermeasure /ˈkaʊntəˌmeʒə(r)/ n. an action taken to counteract a danger, threat, etc.

countermine n. & v. ● n. /ˈkaʊntəˌmaɪn/ **1** *Mil.* **a** a mine dug to intercept another dug by an enemy. **b** a submarine mine sunk to explode an enemy's mines. **2** a counterplot. ● v.tr. /ˌkaʊntəˈmaɪn/ make a countermine against.

countermove n. & v. ● n. /ˈkaʊntəˌmuːv/ a move or action in opposition to another. ● v.intr. /ˌkaʊntəˈmuːv/ make a countermove. □ **countermovement** n.

counter-offensive /ˈkaʊntərəˌfensɪv/ n. **1** *Mil.* an attack made from a defensive position in order to effect an escape. **2** any attack made from a defensive position.

counterpane /ˈkaʊntəˌpeɪn/ n. a bedspread. [alt. (with assim. to *pane* in obs. sense 'cloth') f. obs. *counterpoint* f. OF *contrepointe* alt. f. *cou(l)tepointe* f. med.L *culcita puncta* quilted mattress]

counterpart /ˈkaʊntəˌpɑːt/ n. **1 a** a person or thing extremely like another. **b** a person or thing forming a natural complement or equivalent to another. **2** *Law* one of two copies of a legal document. □ **counterpart funds** a sum of money in a local currency equivalent to goods or services received from abroad.

counterplot n. & v. ● n. /ˈkaʊntəˌplɒt/ a plot intended to defeat another plot. ● v. /ˌkaʊntəˈplɒt/ (**-plotted, -plotting**) **1** intr. make a counterplot. **2** tr. make a counterplot against.

counterpoint /ˈkaʊntəˌpɔɪnt/ n. & v. ● n. **1** *Mus.* **a** the art or technique of setting, writing, or playing a melody or melodies in conjunction with another, according to fixed rules. **b** a melody played in conjunction with another. **2** a contrasting argument, plot, idea, or literary theme, used to set off the main element. ● v.tr. **1** *Mus.* add

counterpoint to. **2** set (an argument, plot, etc.) in contrast to (a main element). □ **strict counterpoint** an academic exercise in writing counterpoint, not necessarily intended as a composition. [OF *contrepoint* f. med.L *contrapunctum* pricked or marked opposite, i.e. to the original melody (as CONTRA-, *pungere punct-* prick)]

counterpoise n. & v. ● n. /ˈkaʊntəˌpɔɪz/ **1** a force etc. equivalent to another on the opposite side. **2** a state of equilibrium. **3** a counterbalancing weight. ● v.tr. /ˌkaʊntəˈpɔɪz/ **1** counterbalance. **2** compensate. **3** bring into or keep in equilibrium. [ME f. OF *contrepeis, -pois, contrepeser* (as COUNTER-, *peis, pois* f. L *pensum* weight: cf. POISE[1])]

counter-productive /ˌkaʊntəprəˈdʌktɪv/ adj. having the opposite of the desired effect.

Counter-Reformation the reform of the Roman Catholic Church in Europe from the mid-16th to the mid-17th century. Though stimulated by Protestant opposition, reform movements within the Roman Catholic Church had begun almost simultaneously with the Reformation itself. Measures to oppose the spread of the Reformation were resolved on at the Council of Trent (1545–63). The Jesuit order became the spearhead of the movement, both within Europe and as a missionary force, while Spain, the strongest military power of the day, extended the Inquisition to countries under its influence. Although most of northern Europe remained Protestant, southern Germany and Poland were brought back to the Roman Catholic Church.

counter-reformation /ˌkaʊntəˌrefəˈmeɪʃ(ə)n/ n. a reformation running counter to another. (See also COUNTER-REFORMATION.)

counter-revolution /ˌkaʊntəˌrevəˈluːʃ(ə)n/ n. a revolution opposing a former one or reversing its results. □ **counter-revolutionary** adj. & n. (pl. **-ies**).

counterscarp /ˈkaʊntəˌskɑːp/ n. *Mil.* the outer wall or slope of a ditch in a fortification. [F *contrescarpe* f. It. *contrascarpa* (as CONTRA-, SCARP)]

countershaft /ˈkaʊntəˌʃɑːft/ n. an intermediate shaft driven by a main shaft and transmitting motion to a particular machine etc.

countersign v. & n. ● v.tr. /ˌkaʊntəˈsaɪn/ **1** add a signature to (a document already signed by another). **2** ratify. ● n. /ˈkaʊntəˌsaɪn/ **1** a watchword or password spoken to a person on guard. **2** a mark used for identification etc. □ **counter-signature** /ˌkaʊntəˈsɪgnətʃə(r)/ n. [F *contresigner* (v.), *contresigne* (n.) f. It. *contrasegno* (as COUNTER-, SIGN)]

countersink /ˈkaʊntəˌsɪŋk/ v.tr. (past and past part. **-sunk** /-ˌsʌŋk/) **1** enlarge and bevel the rim of (a hole) so that a screw or bolt can be inserted flush with the surface. **2** sink (a screw etc.) in such a hole.

counterstroke /ˈkaʊntəˌstrəʊk/ n. a blow given in return for another.

counter-tenor /ˈkaʊntəˌtenə(r)/ n. *Mus.* **1 a** a male alto singing voice. **b** a singer with this voice. **2** a part written for counter-tenor. [ME f. F *contre-teneur* f. obs. It. *contratenore* (as CONTRA-, TENOR)]

counter-transference /ˌkaʊntəˈtrænsfərəns, -ˈtrɑːnsfərəns/ n. *Psychol.* **1** the redirection towards a patient of childhood emotions felt by an analyst (see TRANSFERENCE 2). **2** any emotion felt by an analyst towards a patient.

countervail /ˌkaʊntəˈveɪl/ v. **1** tr. counterbalance. **2** tr. & intr. (often foll. by *against*) oppose forcefully and usu. successfully. □ **countervailing duty** a tax put on imports to offset a subsidy in the exporting country or a tax on similar goods not from abroad. [ME f. AF *contrevaloir* f. L *contra valere* be of worth against]

countervalue /ˈkaʊntəˌvæljuː/ n. an equivalent value, esp. in military strategy.

counterweight /ˈkaʊntəˌweɪt/ n. a counterbalancing weight.

countess /ˈkaʊntɪs/ n. **1** the wife or widow of a count or an earl. **2** a woman holding the rank of count or earl. [ME f. OF *contesse, cuntesse,* f. LL *comitissa* fem. of *comes* COUNT[2]]

countless /ˈkaʊntlɪs/ adj. too many to be counted.

Count Palatine n. **1** (in the later Roman Empire) a count attached to the imperial palace and having supreme judicial authority in certain cases. **2** (under the German emperors) a count having supreme jurisdiction in his fief. **3** (in England) (also called *Earl Palatine*) the proprietor of a County Palatine, now applied to the earldom of Chester and duchy of Lancaster, dignities which are attached to the Crown. [see PALATINE[1]]

countrified /ˈkʌntrɪˌfaɪd/ adj. (also **countryfied**) often *derog.* rural or rustic, esp. of manners, appearance, etc. [past part. of *countrify* f. COUNTRY]

country /ˈkʌntrɪ/ n. (pl. **-ies**) **1 a** the territory of a nation with its own government; a state. **b** a territory possessing its own language, people, culture, etc. **2** (often *attrib.*) rural districts as opposed to towns or the capital (*a cottage in the country; a country town*). **3** the land of a person's birth or citizenship; a fatherland or motherland. **4 a** a territory, esp. an area of interest or knowledge. **b** a region associated with a particular person, esp. a writer (*Hardy country*). **5** *Brit.* a national population, esp. as voters (*the country won't stand for it*). **6** = COUNTRY MUSIC. □ **across country** not keeping to roads. **country and western** see COUNTRY MUSIC. **country club** a sporting and social club in a rural setting. **country cousin** often *derog.* a person with a countrified appearance or manners. **country dance** a traditional sort of dance, esp. English, with couples facing each other in long lines. **country gentleman** a gentleman with landed property. **country house** a usu. large house in the country, often the seat of a country gentleman. **country party** a political party supporting agricultural interests. **country rock 1** *Geol.* the rock which encloses a mineral deposit or an igneous intrusion. **2** a blend of country music with rock music. **country seat** a large country house belonging to an aristocratic family. **country-wide** extending throughout a nation. **go** (or **appeal) to the country** *Brit.* test public opinion after an adverse or doubtful vote in the House of Commons or at the end of a government's term of office, by dissolving Parliament and holding a general election. **in the country** *Cricket sl.* far from the wickets; in the deep field. **line of country** a subject about which a person is knowledgeable. **unknown country** an unfamiliar place or topic. [ME f. OF *cuntree*, f. med.L *contrata* (*terra*) (land) lying opposite (CONTRA-)]

countryfied var. of COUNTRIFIED.

countryman /'kʌntrɪmən/ *n.* (*pl.* **-men**; *fem.* **countrywoman**, *pl.* **-women**) **1** a person living in a rural area. **2 a** (also **fellow-countryman**) a person of one's own country or district. **b** (often in *comb.*) a person from a specified country or district (*north-countryman*).

country music *n.* (also called *country and western*) a form of popular music which originated among poor white people in the rural southern US, deriving ultimately from the folk music of early British settlers. Country music is a mixture of ballads and dance tunes played characteristically on fiddle, banjo, guitar, and the mournful pedal steel guitar; its lyrics deal with recurrent themes such as love and separation, often with a strong, clear narrative. The earliest country-music recordings date from the 1920s, when they quickly became popular; early stars included Jimmie Rodgers (1897-1933), Hank Williams (1923-53), and Patsy Cline (1932-1963). Since then country music has become an industry centred on Nashville, Tennessee, with internationally known performers such as Dolly Parton, Willie Nelson (b.1933), and Emmylou Harris (b.1949).

countryside /'kʌntrɪˌsaɪd/ *n.* **1 a** a rural area. **b** rural areas in general. **2** the inhabitants of a rural area.

county /'kaʊntɪ/ *n. & adj.* ● *n.* (*pl.* **-ies**) **1 a** a territorial division of some countries, forming the chief unit of local administration. **b** *US* a political and administrative division of a state. **2** often *iron.* the people of a county, esp. the leading families. ● *adj.* having the social status or characteristics of county families. □ **county borough** *hist.* a large borough ranking as a county for administrative purposes. **county corporate** *hist.* a city or town ranking as an administrative county. **county council** the elected governing body of an administrative county. **county court** a judicial court for civil cases (in the US for civil and criminal cases). **county cricket** cricket matches between teams representing counties. **county family** an aristocratic family with an ancestral seat in a county. **county town** (*US* **seat**) the administrative capital of a county. [ME f. AF *counté*, OF *conté*, *cunté*, f. L *comitatus* (as COUNT²)]

County Palatine *n.* (in England) a county of which the earl formerly had royal privileges, with exclusive jurisdiction (now Cheshire and Lancashire, formerly also Durham, Pembroke, Ely, etc.). [see PALATINE¹]

coup /kuː/ *n.* (*pl.* **coups** /kuːz/) **1** a notable or successful stroke or move. **2** = COUP D'ÉTAT. **3** *Billiards* a direct pocketing of the ball. [F f. med.L *colpus* blow: see COPE¹]

coup de grâce /ˌkuː də ˈɡrɑːs/ *n.* (*pl.* **coups de grâce** *pronunc.* same) a finishing stroke, esp. to kill a wounded animal or person. [F, lit. 'stroke of grace']

coup de main /ˌkuː də ˈmæn/ *n.* (*pl.* **coups de main** *pronunc.* same) a sudden vigorous attack. [F, lit. 'stroke of the hand']

coup d'état /ˌkuː deɪˈtɑː/ *n.* (*pl.* **coups d'état** *pronunc.* same) a violent or illegal seizure of power. [F, lit. 'stroke of the state']

coup d'œil /ˌkuː ˈdɜːɪ/ *n.* (*pl.* **coups d'œil** *pronunc.* same) **1** a comprehensive glance. **2** a general view. [F, lit. 'stroke of the eye']

coupe /kuːp/ *n.* **1** a shallow glass or dish used for serving fruit, ice-cream, etc. **2** fruit, ice-cream, etc. served in this. [F, = goblet]

coupé /'kuːpeɪ/ *n.* (*US* **coupe** /kuːp/) **1** a car with a hard roof, two (or four) seats, and usu. a sloping rear. **2** *hist.* a four-wheeled enclosed carriage for two passengers and a driver. [F, past part. of *couper* cut (formed as COUP)]

Couperin /'kuːpəˌræn/, François (1668-1733), French composer, organist, and harpsichordist. A composer at the court of Louis XIV, he is principally known for his harpsichord works, particularly those 220 pieces contained in his four books (1713; 1716-17; 1722; 1730). These pieces are characterized by extensive ornamentation and a blend of Italian and French styles; Couperin's music, as well as his treatise *L'Art de toucher le clavecin* (1716), later proved a significant influence on J. S. Bach.

couple /'kʌp(ə)l/ *n. & v.* ● *n.* **1** (usu. foll. by *of*; often as *sing.*) **a** two (*a couple of girls*). **b** about two (*a couple of hours*). **2** (often as *sing.*) **a** a married or engaged pair; a set of two people who live together or are habitual companions. **b** a pair of partners in a dance, a game, etc. **c** a pair of rafters. **3** (*pl.* **couple**) a pair of hunting dogs (*six couple of hounds*). **4** (in *pl.*) a pair of joined collars used for holding hounds together. **5** *Mech.* a pair of equal and parallel forces acting in opposite directions, and tending to cause rotation about an axis perpendicular to the plane containing them. ● *v.* **1** *tr.* fasten or link together; connect (esp. railway carriages). **2** *tr.* (often foll. by *together*, *with*) associate in thought or speech (*papers coupled their names; couple our congratulations with our best wishes*). **3** *tr. & intr.* (often foll. by *with*, *up* (*with*)) bring or come together as partners. **4** *intr.* copulate. **5** *tr. Physics* connect (oscillators) with a coupling. [ME f. OF *cople, cuple, copler, cupler* f. L *copulare*, L COPULA]

coupler /'kʌplə(r)/ *n.* **1** *Mus.* **a** a device in an organ for connecting two manuals, or a manual with pedals, so that they both sound when only one is played. **b** (also **octave coupler**) a similar device for connecting notes with their octaves above or below. **2** anything that connects two things, esp. a transformer used for connecting electric circuits.

couplet /'kʌplɪt/ *n. Prosody* two successive lines of verse, usu. rhyming and of the same length. [F dimin. of *couple*, formed as COUPLE]

coupling /'kʌplɪŋ/ *n.* **1 a** a link connecting railway carriages etc. **b** a device for connecting parts of machinery. **2** *Physics* a connection between two systems, causing one to oscillate when the other does so. **3** *Mus.* **a** the arrangement of items on a gramophone record. **b** each such item. **4** (an act of) sexual intercourse.

coupon /'kuːpɒn/ *n.* **1** a form in a newspaper, magazine, etc., which may be filled in and sent as an application for a purchase, information, etc. **2** *Brit.* an entry form for a football pool or other competition. **3** a voucher given with a retail purchase, a certain number of which entitle the holder to a discount etc. **4 a** a detachable ticket entitling the holder to a ration of food, clothes, etc., esp. in wartime. **b** a similar ticket entitling the holder to payment, goods, services, etc. [F, = piece cut off f. *couper* cut: see COUPÉ]

courage /'kʌrɪdʒ/ *n.* the ability to disregard fear; bravery. □ **have the courage of one's convictions** have the courage to act on one's beliefs. **lose courage** become less brave. **pluck up** (or **take) courage** muster one's courage. **take one's courage in both hands** nerve oneself to a venture. [ME f. OF *corage*, f. L *cor* heart]

courageous /kəˈreɪdʒəs/ *adj.* brave, fearless. □ **courageously** *adv.* **courageousness** *n.* [ME f. AF *corageous*, OF *corageus* (as COURAGE)]

courante /kʊˈrɒnt/ *n.* **1** *hist.* a running or gliding dance. **2** *Mus.* the music used for this, esp. as a movement of a suite. [F, fem. pres. part. (as noun) of *courir* run f. L *currere*]

Courbet /'kʊəbeɪ/, Gustave (1819-77), French painter. A leader of the 19th-century realist school of painting, he favoured an unidealized choice of subject-matter that did not exclude the ugly or vulgar. Important works include *Burial at Ornans* (1850) and *Painter in his Studio* (1855).

courgette /kʊəˈʒet/ *n.* a small green variety of vegetable marrow. Also called *zucchini*. [F, dimin. of *courge* gourd]

courier /'kʊrɪə(r)/ *n.* **1** a person employed, usu. by a travel company, to guide and assist a group of tourists. **2** a special messenger. [ME f. obs. F, f. It. *corriere*, & f. OF *coreor*, both f. L *currere* run]

Courrèges /kʊəˈreɪʒ/, André (b.1923), French fashion designer. After training with Balenciaga (1952-60), he opened a fashion house in Paris in 1961. He became known for his futuristic and youth-oriented styles, introducing the miniskirt in 1964 and promoting unisex fashion with his trouser suits for women. He briefly ceased trading in 1966 in protest at the mass-produced imitations of his designs, but reopened the following year with both ready-made and couture clothes.

course /kɔːs/ *n. & v.* ● *n.* **1** a continuous onward movement or progression. **2 a** a line along which a person or thing moves; a direction taken (*has changed course; the course of the winding river*). **b** a correct or intended direction or line of movement. **c** the direction taken by a ship or aircraft. **3 a** the ground on which a race (or other sport involving extensive linear movement) takes place. **b** a series of fences, hurdles, or other obstacles to be crossed in a race etc. **4 a** a series of lectures, lessons, etc., in a particular subject. **b** a book for such a course (*A Modern French Course*). **5** any of the successive parts of a meal. **6** *Med.* a sequence of medical treatment etc. (*prescribed a course of antibiotics*). **7 a** line of conduct (*disappointed by the course he took*). **8** *Archit.* a continuous horizontal layer of brick, stone, etc., in a building. **9** a channel in which water flows. **10** the pursuit of game (esp. hares) with hounds, esp. greyhounds, by sight rather than scent. **11** *Naut.* a sail on a square-rigged ship (*fore course; main course*). ● *v.* **1** *intr.* (esp. of liquid) run, esp. fast (*blood coursed through his veins*). **2** *tr.* (also *absol.*) **a** use (hounds) to hunt. **b** pursue (hares etc.) in hunting. □ **the course of nature** ordinary events or procedure. **in course of** in the process of. **in the course of** during. **in the course of time** as time goes by; eventually. **a matter of course** the natural or expected thing. **of course** naturally; as is or was to be expected; admittedly. **on** (or **off**) **course** following (or deviating from) the desired direction or goal. **run** (or **take**) **its course** (esp. of an illness) complete its natural development. □ **courser** *n.* (in sense 2 of *v.*). [ME f. OF *cours* f. L *cursus* f. *currere* *curs-* run]

courser[1] /ˈkɔːsə(r)/ *n. poet.* a swift horse. [ME f. OF *corsier* f. Rmc]

courser[2] /ˈkɔːsə(r)/ *n.* a fast-running long-legged plover-like bird of the genus *Cursorius*, native to open arid areas of Africa and Asia. [LL *cursorius* adapted for running]

court /kɔːt/ *n. & v.* ● *n.* **1** (in full **court of law**) **a** an assembly of judges or other persons acting as a tribunal in civil and criminal cases. **b** = COURTROOM. **2 a** an enclosed quadrangular area for games, which may be open or covered (*tennis-court; squash-court*). **b** an area marked out for lawn tennis etc. (*hit the ball out of court*). **3 a** a small enclosed street in a town, having a yard surrounded by houses, and adjoining a larger street. **b** *Brit.* = COURTYARD. **c** (**Court**) the name of a large house, block of flats, street, etc. (*Grosvenor Court*). **d** (at Cambridge University) a college quadrangle. **e** a subdivision of a building, usu. a large hall extending to the ceiling with galleries and staircases. **4 a** the establishment, retinue, and courtiers of a sovereign. **b** a sovereign and his or her councillors, constituting a ruling power. **c** a sovereign's residence. **d** an assembly held by a sovereign; a state reception. **5** attention paid to a person whose favour, love, or interest is sought (*paid court to her*). **6 a** the qualified members of a company or a corporation. **b** (in some Friendly Societies) a local branch. **c** a meeting of a court. ● *v.tr.* **1** a try to win the affection or favour of (a person). **b** pay amorous attention to (*courting couples*). **2** seek to win (applause, fame, etc.). **3** invite (misfortune) by one's actions (*you are courting disaster*). □ **court-card** a playing card that is a king, queen, or jack (orig. *coat-card*). **court circular** *Brit.* a daily report of royal court affairs, published in some newspapers. **court dress** formal dress worn at a royal court. **court-house 1** a building in which a judicial court is held. **2** US a building containing the administrative offices of a county. **Court leet** see LEET[1]. **court of record** a court whose proceedings are recorded and available as evidence of fact. **court of summary jurisdiction** a court having the authority to use summary proceedings and arrive at a judgement or conviction. **court order** a direction issued by a court or a judge, usu. requiring a person to do or not do something. **court plaster** *hist.* sticking-plaster for cuts etc. (formerly used by ladies at court for face-patches). **court roll** *hist.* a manorial-court register of holdings. **court shoe** a woman's light, usu. high-heeled, shoe with a low-cut upper. **court tennis** N. Amer. = REAL TENNIS. **go to court** take legal action. **in court** appearing as a party or an advocate in a court of law. **out of court 1** (of a plaintiff) not entitled to be heard. **2** (of a settlement) arranged before a hearing or judgement can take place. **3** not worthy of consideration (*that suggestion is out of court*). [ME f. AF *curt*, OF *cort*, ult. f. L *cohors*, *-hortis* yard, retinue: (v.) after OIt. *corteare*, OF *courtoyer*]

Courtauld /ˈkɔːtəʊld/, Samuel (1876–1947), English industrialist. He was a director of his family's silk firm and one of the earliest British collectors of French impressionist and post-impressionist paintings. He presented his collection to the University of London, endowed the Courtauld Institute of Art, and bequeathed to it his house in Portman Square, London.

court bouillon /ˌkʊə buːˈjɒn/ *n.* stock usu. made from wine, vegetables, etc., often used in fish dishes. [F f. *court* short + BOUILLON]

courteous /ˈkɜːtɪəs/ *adj.* polite, kind, or considerate in manner; well-mannered. □ **courteously** *adv.* **courteousness** *n.* [ME f. OF *corteis*, *curteis* f. Rmc (as COURT): assim. to words in -OUS]

courtesan /ˌkɔːtɪˈzæn/ *n. literary* **1** a prostitute, esp. one with wealthy or upper-class clients. **2** the mistress of a wealthy man. [F *courtisane* f. It. *cortigiana*, fem. of *cortigiano* courtier f. *corte* COURT]

courtesy /ˈkɜːtɪsɪ/ *n.* (pl. **-ies**) **1** courteous behaviour; good manners. **2** a courteous act. **3** *archaic* = CURTSY. □ **by courtesy** by favour, not by right. **by courtesy of** with the formal permission of (a person etc.). **courtesy light** a light in a car that is switched on by opening a door. **courtesy title** a title held by courtesy, usu. having no legal validity, e.g. a title given to the heir of a duke etc. [ME f. OF *curtesie*, *co(u)rtesie* f. *curteis* etc. COURTEOUS]

courtier /ˈkɔːtɪə(r)/ *n.* a person who attends or frequents a sovereign's court. [ME f. AF *courte(i)our*, f. OF f. *cortoyer* be present at court]

courtly /ˈkɔːtlɪ/ *adj.* (**courtlier**, **courtliest**) **1** polished or refined in manners. **2** obsequious. **3** punctilious. □ **courtly love** the conventional medieval tradition of knightly love for a lady, and the etiquette used in its (esp. literary) expression. □ **courtliness** *n.*

court martial *n. & v.* ● *n.* (pl. **courts martial**) a judicial court for trying members of the armed services. ● *v.tr.* (**court-martial**) (**-martialled**, **-martialling**; US **-martialed**, **-martialing**) try by a court martial.

Court of Appeal (in the UK) a court of law hearing appeals against judgements in the Crown Court, High Court, County Court, etc.

Court of Arches the ecclesiastical court of appeal for the province of Canterbury, so known because it was formerly held at the church of St Mary-le-Bow, famous for its arched crypt.

Court of Protection (in the UK) the department of the Supreme Court attending to the affairs of the mentally unfit.

Court of Session the supreme civil court in Scotland.

Court of St James's the British sovereign's court.

Courtrai see KORTRIJK.

courtroom /ˈkɔːtruːm, -rʊm/ *n.* the place or room in which a court of law meets.

courtship /ˈkɔːtʃɪp/ *n.* **1 a** courting with a view to marriage. **b** the courting behaviour of male animals, birds, etc. **c** a period of courting. **2** an attempt, often protracted, to gain advantage by flattery, attention, etc.

courtyard /ˈkɔːtjɑːd/ *n.* an area enclosed by walls or buildings, often opening off a street.

couscous /ˈkuːskuːs/ *n.* **1** a type of North African pasta in granules made from crushed durum wheat. **2** a dish of this, often with meat or fruit added. [F f. Arab. *kuskus* f. *kaskasa* to pound]

cousin /ˈkʌz(ə)n/ *n.* **1** (also **first cousin**, **cousin-german**) the child of one's uncle or aunt. **2** (usu. in *pl.*) applied to the people of kindred races or nations (*our American cousins*). **3** *hist.* a title formerly used by a sovereign in addressing another sovereign or a noble of his or her own country. □ **second cousin** a child of one's parent's first cousin. □ **cousinhood** *n.* **cousinly** *adj.* **cousinship** *n.* [ME f. OF *cosin*, *cusin*, f. L *consobrinus* mother's sister's child]

Cousteau /ˈkuːstəʊ/, Jacques-Yves (b.1910), French oceanographer and film director. A naval officer interested in underwater exploration, he began using a camera under water in 1939 and devised the scuba apparatus. Cousteau has made three feature films and several popular series for television. He has turned increasingly to biological research and marine conservation issues.

couth /kuːθ/ *adj. joc.* cultured; well-mannered. [back-form. as antonym of UNCOUTH]

couture /kuːˈtjʊə(r)/ *n.* the design and manufacture of fashionable clothes; = HAUTE COUTURE. [F, = sewing, dressmaking]

couturier /kuːˈtjʊərɪˌeɪ/ *n.* (fem. **couturière** /-ˌtjʊərɪˈeə(r)/) a fashion designer or dressmaker. [F]

couvade /kuːˈvɑːd/ *n.* a custom in some cultures by which a man takes to his bed and goes through certain rituals when his child is being born. [F f. *couver* hatch f. L *cubare* lie down]

couvert /kuːˈveə(r)/ *n.* = COVER *n.* 6. [F]

couverture /ˌkuːvəˈtjʊə(r)/ *n.* chocolate for covering sweets, cakes, etc. [F, = covering]

covalency /kəʊˈveɪlənsɪ/ *n. Chem.* **1** the linking of atoms by a covalent bond. **2** the number of pairs of electrons an atom can share with another.

covalent /kəʊˈveɪlənt/ *adj. Chem.* of, relating to, or characterized by covalency. □ **covalent bond** *Chem.* a bond formed by sharing of electrons usu. in pairs by two atoms in a molecule. □ **covalence** *n.* **covalently** *adv.*

cove[1] /kəʊv/ *n. & v.* ● *n.* **1** a small, esp. sheltered, bay or creek. **2** a sheltered recess. **3** *Archit.* a concave arch or arched moulding, esp. one formed at the junction of a wall with a ceiling. ● *v.tr. Archit.* **1** provide (a room, ceiling, etc.) with a cove. **2** slope (the sides of a fireplace) inwards. [OE *cofa* chamber f. Gmc]

cove[2] /kəʊv/ *n. Brit. sl.* a fellow; a chap. [16th-c. cant: orig. unkn.]

coven /ˈkʌv(ə)n/ *n.* a meeting or company of witches. [var. of *covent*; see CONVENT]

covenant /ˈkʌvənənt/ *n. & v.* ● *n.* **1** an agreement; a contract. **2** *Law* **a** a contract drawn up under a seal, esp. undertaking to make regular payments to a charity. **b** a clause of a covenant. **3** (**Covenant**) *Bibl.* the agreement between God and the Israelites (see ARK OF THE COVENANT). ● *v.tr. & intr.* agree, esp. by legal covenant. □ **land of the Covenant** Canaan. □ **covenantor** *n.* **covenantal** /ˌkʌvəˈnænt(ə)l/ *adj.* [ME f. OF, pres. part. of *co(n)venir*, formed as CONVENE]

covenanted /ˈkʌvənəntɪd/ *adj.* bound by a covenant.

covenanter /ˈkʌvənəntə(r)/ *n.* **1** a person who covenants. **2** (**Covenanter**) *hist.* an adherent of the National Covenant or the Solemn League and Covenant in 17th-century Scotland, in support of Presbyterianism (see SOLEMN LEAGUE AND COVENANT).

Covent Garden /ˈkɒv(ə)nt/ a district in central London, originally the convent garden of the Abbey of Westminster. It was the site for 300 years of London's chief fruit and vegetable market, which in 1974 was moved to Nine Elms, Battersea. The first Covent Garden Theatre was opened in 1732 and such famous plays as Goldsmith's *She Stoops to Conquer* (1773) and Sheridan's *The Rivals* (1775) were first performed there. The theatre was several times destroyed and reconstructed, and since 1946 has been the home of the national opera and ballet companies, based at the Royal Opera House (built 1888).

Coventry /ˈkɒv(ə)ntrɪ/ an industrial city in the West Midlands, central England; pop. (1991) 292,600. Its cathedral, built in 1443, was badly damaged during the Second World War and was replaced by a new cathedral, consecrated in 1962. □ **send a person to Coventry** refuse to associate with or speak to a person.

cover /ˈkʌvə(r)/ *v. & n.* ● *v.tr.* **1 a** (often foll. by *with*) protect or conceal by means of a cloth, lid, etc. **b** prevent the perception or discovery of, conceal (*cover my embarrassment*). **2** extend over; occupy the whole surface of (*covered in dirt; covered with writing*). **b** (often foll. by *with*) strew thickly or thoroughly (*covered the floor with straw*). **c** lie over; be a covering to (*the blanket scarcely covered him*). **3 a** protect; clothe. **b** (as **covered** *adj.*) wearing a hat; having a roof. **4** include; comprise; deal with (*the talk covered recent discoveries*). **5** travel (a specified distance) (*covered sixty miles*). **6** *Journalism* **a** report (events, a meeting, etc.). **b** investigate as a reporter. **7** be enough to defray (expenses, a bill, etc.) (*£20 should cover it*). **8 a** *refl.* take precautionary measures so as to protect oneself (*had covered myself by saying I might be late*). **b** (*absol.*; foll. by *for*) deputize or stand in for (a colleague etc.) (*will you cover for me?*). **9** *Mil.* **a** aim a gun etc. at. **b** (of a fortress, guns, etc.) command (a territory). **c** stand behind (a person in the front rank). **d** protect (an exposed person etc.) by being able to return fire. **10 a** esp. *Cricket* stand behind (another player) to stop any missed balls. **b** (in team games) mark (a corresponding player of the other side). **11** (also *absol.*) (in some card-games) play a card higher than (one already played to the same trick). **12** make a cover version of (a song etc.). **13** (of a stallion, a bull, etc.) copulate with. ● *n.* **1** something that covers or protects, esp.: **a** a lid. **b** the binding of a book. **c** either board of this. **d** an envelope or the wrapper of a parcel (*under separate cover*). **e** the outer case of a pneumatic tyre. **f** (in *pl.*) bedclothes. **2 a** a hiding-place; a shelter. **3** woods or undergrowth sheltering game or covering the ground (see COVERT). **4 a** a pretence; a screen (*under cover of humility*). **b** a spy's pretended identity or activity, intended as concealment. **c** *Mil.* a supporting force protecting an advance party from attack. **5 a** funds, esp. obtained by insurance, to meet a liability or secure against a contingent loss. **b** the state of being protected (*third-party cover*). **6** a place setting at table, esp. in a restaurant. **7** (in full **cover version**) a recording of a previously recorded song etc., esp. one made to take advantage of the original's success. **8** *Cricket* **a** = *cover-point*. **b** (**the covers**) the general area of cover-point and extra cover. □ **break cover** (of an animal, esp. game, or a hunted person) leave a place of shelter, esp. vegetation. **cover charge** an extra charge levied per head in a restaurant, nightclub, etc. **cover crop** a crop grown for the protection and enrichment of the soil. **cover-drive** *Cricket* a drive past cover-point. **cover girl** a

female model whose picture appears on magazine covers etc. **cover in** provide with a roof etc. **covering letter** (or **note**) an explanatory letter sent with an enclosure. **cover note** *Brit.* a temporary certificate of current insurance. **cover-point** *Cricket* a fielding position a little in front of the batsman on the off side and half way to the boundary. **cover slip** a small thin piece of glass used to protect a specimen on a microscope slide. **cover story** a news story in a magazine, that is illustrated or advertised on the front cover. **cover one's tracks** conceal evidence of what one has done. **cover up** **1** completely cover or conceal. **2** conceal (circumstances etc., esp. illicitly) (also *absol.*: *refused to cover up for them*). **cover-up** *n.* an act of concealing circumstances, esp. illicitly. **from cover to cover** from beginning to end of a book etc. **take cover** use a natural or prepared shelter against an attack. □ **coverable** *adj.* [ME f. OF *covrir, cuvrir* f. L *cooperire* (as CO-, *operire opert-* cover)]

coverage /ˈkʌvərɪdʒ/ *n.* **1** an area or an amount covered. **2** *Journalism* the amount of press etc. publicity received by a particular story, person, etc. **3** a risk covered by an insurance policy. **4** an area reached by a particular broadcasting station or advertising medium.

coverall /ˈkʌvərɔːl/ *n. & adj.* ● *n.* **1** something that covers entirely. **2** (usu. in *pl.*) a full-length protective outer garment often zipped up the front. ● *attrib.adj.* covering entirely (*a coverall term*).

Coverdale /ˈkʌvədeɪl/, Miles (1488–1568), English biblical scholar. He translated the first complete printed English Bible (1535), published in Zurich while he was in exile for preaching against confession and images. He also edited the Great Bible, brought out in 1539 by the printer Richard Grafton (*c.*1513–*c.*1572).

covering /ˈkʌvərɪŋ/ *n.* something that covers, esp. a bedspread, blanket, etc., or clothing.

coverlet /ˈkʌvəlɪt/ *n.* a bedspread. [ME f. AF *covrelet, -lit* f. OF *covrir* cover + *lit* bed]

covert /ˈkʌvət/ *adj. & n.* ● *adj.* (also /ˈkəʊvɜːt/) secret or disguised (*a covert glance; covert operations*). ● *n.* **1** a shelter, esp. a thicket hiding game. **2** a feather covering the base of a bird's flight-feather. □ **covert coat** a short, light, overcoat worn for shooting, riding, etc. □ **covertly** *adv.* **covertness** *n.* [ME f. OF *covert* past part. of *covrir* COVER]

coverture /ˈkʌvətjʊə(r), -tʃə(r)/ *n.* **1** covering; shelter. **2** *Law hist.* the position of a married woman, considered to be under her husband's protection. [ME f. OF (as COVERT)]

covet /ˈkʌvɪt/ *v.tr.* (**coveted, coveting**) desire greatly (esp. something belonging to another person) (*coveted her friend's earrings*). □ **covetable** *adj.* [ME f. OF *cu-, coveitier* f. Rmc]

covetous /ˈkʌvɪtəs/ *adj.* (usu. foll. by *of*) **1** greatly desirous (esp. of another person's property). **2** grasping, avaricious. □ **covetously** *adv.* **covetousness** *n.* [ME f. OF *coveitous* f. Gallo-Roman]

covey /ˈkʌvɪ/ *n.* (*pl.* **-eys**) **1** a brood or flock of partridges. **2** a small party or group of people or things. [ME f. OF *covee* f. Rmc f. L *cubare* lie]

covin /ˈkʌvɪn/ *n.* **1** *Law* a conspiracy to commit a crime etc. against a third party. **2** *archaic* fraud, deception. [ME f. OF *covin(e)* f. med.L *convenium -ia* f. *convenire*: see CONVENE]

coving /ˈkəʊvɪŋ/ *n.* = COVE[1] 3.

cow[1] /kaʊ/ *n.* **1 a** a fully grown female of a bovine animal, esp. of the genus *Bos*, used as a source of milk and beef. **b** a domestic bovine animal (regardless of sex or age). **2** the female of other large animals, esp. the elephant, whale, and seal. **3** *sl. derog.* **a** a woman, esp. a coarse or unpleasant one. **b** *Austral. & NZ* an unpleasant person, thing, situation, etc. □ **cow-fish 1** a manatee or small cetacean. **2** a marine fish, *Lactoria diaphana*, covered in hard bony plates and having hornlike spines over the eyes and on other parts of the body. **cow-heel** the foot of a cow or an ox stewed to a jelly. **cow-lick** a projecting lock of hair. **cow-parsley** an umbelliferous hedgerow plant, *Anthriscus sylvestris*, having lacelike umbels of flowers (also called *Queen Anne's lace*). **cow-pat** a flat round piece of cow-dung. **cow-tree** a tree, *Brosimum galactodendron*, native to South America, yielding a milklike juice which is used as a substitute for cow's milk. **cow-wheat** a plant of the genus *Melampyrum*, of the figwort family; esp. *M. pratense*, growing on heathland. **till the cows come home** *colloq.* for an indefinitely long time. [OE *cū* f. Gmc, rel. to L *bos*, Gk *bous*]

cow[2] /kaʊ/ *v.tr.* (usu. in *passive*) intimidate or dispirit (*cowed by ill-treatment*). [prob. f. ON *kúga* oppress]

cowage /ˈkaʊɪdʒ/ *n.* (also **cowhage**) a leguminous climbing plant, *Mucuna pruriens*, having hairy pods which cause stinging and itching. [Hindi *kawāñch*]

Coward /ˈkaʊəd/, Sir Noel (Pierce) (1899–1973), English dramatist, actor, and composer. He is remembered for witty, satirical plays such as *Hay Fever* (1925) and *Private Lives* (1930), as well as revues and musicals, including *Cavalcade* (1931), *Words and Music* (1932), and *Sigh No More* (1945). Individual songs from some of these, including 'Mad Dogs and Englishmen' (1932), became famous in their own right. Coward was also active in film-making. Among his successes were the war film *In Which We Serve* (1942), of which he was producer, co-director, writer, and star, and *Brief Encounter* (1945), which he wrote and produced.

coward /ˈkaʊəd/ *n. & adj.* ● *n.* a person who is easily frightened or intimidated by danger or pain. ● *adj. poet.* easily frightened. [ME f. OF *cuard, couard* ult. f. L *cauda* tail]

cowardice /ˈkaʊədɪs/ *n.* a lack of bravery. [ME f. OF *couardise* (as COWARD)]

cowardly /ˈkaʊədlɪ/ *adj. & adv.* ● *adj.* **1** of or like a coward; lacking courage. **2** (of an action) done against a person who cannot retaliate. ● *adv. archaic* like a coward; with cowardice. □ **cowardliness** *n.*

cowbane /ˈkaʊbeɪn/ *n.* = water hemlock.

cowbell /ˈkaʊbel/ *n.* **1** a bell worn round a cow's neck for easy location of the animal. **2** a similar bell used as a percussion instrument.

cowberry /ˈkaʊbərɪ/ *n.* (*pl.* **-ies**) **1** an evergreen ericaceous dwarf shrub, *Vaccinium vitis-idaea*, bearing dark red berries. **2** the berry of this plant.

cowbird /ˈkaʊbɜːd/ *n.* a North American oriole that often eats insects stirred up by grazing cattle, esp. the brown-plumaged *Molothrus ater*, which lays its eggs in the nests of other birds.

cowboy /ˈkaʊbɔɪ/ *n.* **1** (*fem.* **cowgirl** /-gɜːl/) a (usu. mounted) person who herds and tends cattle, esp. in the western US. **2** this as a conventional figure in American folklore, esp. in films. **3** *colloq.* an unscrupulous or reckless person in business, esp. an unqualified one.

cowcatcher /ˈkaʊˌkætʃə(r)/ *n. N. Amer.* a peaked metal frame at the front of a locomotive for pushing aside obstacles on the line.

cower /ˈkaʊə(r)/ *v.intr.* **1** crouch or shrink back, esp. in fear; cringe. **2** stand or squat in a bent position. [ME f. MLG *kūren* lie in wait, of unkn. orig.]

Cowes /kaʊz/ a town on the Isle of Wight, southern England; pop. (1981) 16,300. It is internationally famous as a yachting centre.

cowhage var. of COWAGE.

cowherd /ˈkaʊhɜːd/ *n.* a person who tends cattle.

cowhide /ˈkaʊhaɪd/ *n.* **1 a** a cow's hide. **b** leather made from this. **2** a leather whip made from cowhide.

cowhouse /ˈkaʊhaʊs/ *n.* a shed or shelter for cows.

cowl /kaʊl/ *n.* **1 a** the hood of a monk's habit. **b** a loose hood. **c** a monk's hooded habit. **d** *Relig.* a cloak with wide sleeves worn by members of Benedictine orders. **2** the hood-shaped covering of a chimney or ventilating shaft. **3** the removable cover of a vehicle or aircraft engine. □ **cowled** *adj.* (in sense 1). [OE *cugele, cūle* f. eccl.L *cuculla* f. L *cucullus* hood of a cloak]

cowling /ˈkaʊlɪŋ/ *n.* = COWL 3.

cowman /ˈkaʊmən/ *n.* (*pl.* **-men**) **1** = COWHERD. **2** *US* = COWBOY 1.

co-worker /kəʊˈwɜːkə(r)/ *n.* a person who works in collaboration with another.

cowpea /ˈkaʊpiː/ *n.* **1** a leguminous plant, *Vinca unguiculata*, grown for fodder in southern Europe and the southern US. **2** its seed, eaten as a pulse (cf. *black-eyed bean*).

Cowper /ˈkuːpə(r)/, William (1731–1800), English poet. He wrote *Olney Hymns* (1779) with the evangelical minister John Newton (1725–1807), contributing 'Oh! for a Closer Walk with God' amongst other well-known hymns. His famous comic ballad *John Gilpin* appeared in 1782. Cowper is best known, however, for his long poem *The Task* (1785), notable for its intimate sketches of rural life.

cowpoke /ˈkaʊpəʊk/ *n. N. Amer. colloq.* = COWBOY 1.

cowpox /ˈkaʊpɒks/ *n.* a disease of cows, of which the virus was formerly used in vaccination against smallpox. Also called *vaccinia*.

cowpuncher /ˈkaʊˌpʌntʃə(r)/ *n. N. Amer. colloq.* = COWBOY 1.

cowrie /ˈkaʊrɪ/ *n.* (also **cowry**) (*pl.* **-ies**) **1** a gastropod mollusc of the family Cypraeidae, having a smooth, glossy, and often brightly coloured shell. **2** its shell, formerly used as money in parts of Africa and southern Asia. [Urdu & Hindi *kaurī*]

cowshed /ˈkaʊʃed/ *n.* **1** a shed for cattle that are not at pasture. **2** a milking-shed.

cowslip /ˈkaʊslɪp/ *n.* **1** a spring-flowering primula, *Primula veris*, with fragrant yellow flowers on tall stems, growing in pastures. **2** *US* a

marsh marigold. [OE *cūslyppe* f. *cū* cow¹ + *slyppe* slimy substance, i.e. cow-dung]

Cox /kɒks/ *n.* (in full **Cox's orange pippin**) a variety of eating-apple with a red-tinged green skin. [Richard *Cox*, Engl. amateur fruit grower (c.1776–1845), who raised it]

cox /kɒks/ *n. & v.* ● *n.* a coxswain, esp. of a racing-boat. ● *v.* **1** *intr.* act as a cox (*coxed for Cambridge*). **2** *tr.* act as cox for (*coxed the winning boat*). [abbr.]

coxa /ˈkɒksə/ *n.* (*pl.* **coxae** /-siː/) **1** *Anat.* the hip-bone or hip-joint. **2** *Zool.* the first or basal segment of the leg in insects etc. □ **coxal** *adj.* [L]

coxcomb /ˈkɒkskəʊm/ *n.* an ostentatiously conceited man; a dandy. □ **coxcombry** /-kəmrɪ/ *n.* (*pl.* **-ies**). [var. of COCKSCOMB]

Cox's Bazar /ˌkɒksɪz bəˈzɑː(r)/ a port and resort town on the Bay of Bengal, near Chittagong, southern Bangladesh; pop. (1981) 29,600.

coxswain /ˈkɒks(ə)n/ *n. & v.* ● *n.* **1** the steersman of a ship's boat, lifeboat, racing-boat, etc. **2** the senior petty officer in a small ship. ● *v.* **1** *intr.* act as a coxswain. **2** *tr.* act as a coxswain of. □ **coxswainship** *n.* [ME f. *cock* (see COCKBOAT) + SWAIN: cf. BOATSWAIN]

Coy. *abbr. Mil.* Company.

coy /kɔɪ/ *adj.* (**coyer, coyest**) **1** archly or affectedly shy. **2** irritatingly reticent (*always coy about her age*). **3** modest or shy, esp. in sexual matters. □ **coyly** *adv.* **coyness** *n.* [ME f. OF *coi, quei* f. L *quietus* QUIET]

coyote /kɔɪˈəʊtɪ, ˈkɔɪəʊt/ *n.* (*pl.* same or **coyotes**) a wolflike wild dog, *Canis latrans*, native to North America. [Mex. Sp. f. Aztec *coyotl*]

coypu /ˈkɔɪpuː/ *n.* (*pl.* **coypus**) an aquatic beaver-like rodent, *Myocastor coypus*, native to South America and kept in captivity for its fur (nutria). The coypu has become naturalized in some parts of Europe and the US. [Araucanian]

coz /kʌz/ *n. archaic* cousin. [abbr.]

cozen /ˈkʌz(ə)n/ *v. literary* **1** *tr.* (often foll. by *of, out of*) cheat, defraud. **2** *tr.* (often foll. by *into*) beguile; persuade. **3** *intr.* act deceitfully. □ **cozenage** *n.* [16th-c. cant, perh. rel. to COUSIN]

Cozumel /ˌkɒzəˈmel/ a resort island in the Caribbean, off the NE coast of the Yucatán Peninsula of Mexico.

cozy *US* var. of COSY.

cozzie var. of COSSIE.

CP *abbr.* **1** Communist Party. **2** (in South Africa) Conservative Party.

cp. *abbr.* compare.

c.p. *abbr.* candlepower.

CPA *abbr. US* certified public accountant.

Cpl. *abbr.* Corporal.

CPO *abbr.* Chief Petty Officer.

CPR *abbr.* Canadian Pacific Railway.

CPRE *abbr.* Council for the Protection of Rural England.

CPS *abbr.* (in the UK) Crown Prosecution Service.

cps *abbr.* (also **c.p.s.**) **1** *Computing* characters per second. **2** cycles per second.

CPSA *abbr.* (in the UK) Civil and Public Services Association.

CPU *abbr. Computing* central processing unit.

CPVE *abbr.* (in the UK) Certificate of Pre-Vocational Education, a qualification introduced in 1986 for students aged 16 or over who complete a one-year course of preparation for work or for further vocational study or training.

CR *abbr.* Community of the Resurrection.

Cr *symb. Chem.* the element chromium.

Cr. *abbr.* **1** Councillor. **2** creditor.

crab¹ /kræb/ *n.* **1 a** a ten-legged crustacean with the first pair of legs modified as pincers. **b** the flesh of a crab, esp. *Cancer pagurus*, as food. **2** (**the Crab**) the zodiacal sign or constellation Cancer. **3** (in full **crab louse**) (often in *pl.*) a parasitic louse, *Phthirus pubis*, infesting hairy parts of the body and causing extreme irritation. **4** a machine for hoisting heavy weights. □ **catch a crab** *Rowing* effect a faulty stroke in which the oar is jammed under water or misses the water altogether. **crab-grass** *US* a creeping grass; esp. *Digitaria sanguinalis*, widespread as a weed in warmer parts of the world. **crab-pot** a wicker trap for crabs. □ **crablike** *adj.* [OE *crabba*, rel. to ON *krafla* scratch]

crab² /kræb/ *n.* **1** (in full **crab-apple**) a small sour apple-like fruit. **2** (in full **crab tree** or **crab-apple tree**) a tree that bears this fruit, esp. the European wild apple, *Malus sylvestris*. **3** a sour person. [ME, perh. alt. (after CRAB¹ or CRABBED) of earlier *scrab*, prob. of Scand. orig.]

crab[3] /kræb/ v. (**crabbed**, **crabbing**) colloq. **1** tr. & intr. criticize adversely or captiously; grumble. **2** tr. act so as to spoil (the mistake crabbed his chances). [orig. of hawks fighting, f. MLG krabben]

Crabbe /kræb/, George (1754–1832), English poet. Crabbe's name is associated with grimly realistic narrative poems, such as The Village (1783) and The Borough (1810); the latter was based on his native Aldeburgh in Suffolk, and included tales of Peter Grimes and Ellen Orford. These later provided the subject-matter for Benjamin Britten's opera Peter Grimes (1945).

crabbed /ˈkræbɪd/ adj. **1** irritable or morose. **2** (of handwriting) ill-formed and hard to decipher. **3** perverse or cross-grained. **4** difficult to understand. □ **crabbedly** adv. **crabbedness** n. [ME f. CRAB[1], assoc. with CRAB[2]]

crabby /ˈkræbɪ/ adj. (**crabbier**, **crabbiest**) = CRABBED 1, 3. □ **crabbily** adv. **crabbiness** n.

Crab Nebula Astron. an irregular patch of luminous gas in the constellation of Taurus, believed to be the remnant of a supernova explosion seen by Chinese astronomers in 1054. At its centre is the first pulsar to be observed visually, and the nebula is a strong source of high-energy radiation.

crabwise /ˈkræbwaɪz/ adv. & attrib.adj. (of movement) sideways or backwards like a crab.

crack /kræk/ n., v., & adj. ● n. **1 a** a sudden sharp or explosive noise (the crack of a whip; a rifle crack). **b** (in a voice) a sudden harshness or change in pitch. **2** a sharp blow (a crack on the head). **3 a** a narrow opening formed by a break (entered through a crack in the wall). **b** a partial fracture, with the parts still joined (the teacup has a crack in it). **c** a chink (looked through the crack formed by the door; a crack of light). **4** colloq. a mischievous or malicious remark or aside (a nasty crack about my age). **5** colloq. an attempt (I'll have a crack at it). **6** the exact moment (the crack of dawn). **7** colloq. a first-rate player, horse, etc. **8** dial. colloq. conversation; good company; fun (only went there for the crack). **9** sl. a potent hard crystalline form of cocaine broken into small pieces and inhaled or smoked for its stimulating effect. ● v. **1** tr. & intr. break without a complete separation of the parts (cracked the window; the cup cracked on hitting the floor). **2** intr. & tr. make or cause to make a sudden sharp or explosive sound. **3** intr. & tr. break or cause to break with a sudden sharp sound. **4** intr. & tr. give way or cause to give way (under torture etc.); yield. **5** intr. (of the voice, esp. of an adolescent boy or a person under strain) become dissonant; break. **6** tr. colloq. find a solution to (a problem, code, etc.). **7** tr. say (a joke etc.) in a jocular way. **8** tr. colloq. hit sharply or hard (cracked her head on the ceiling). **9** tr. Chem. decompose (heavy oils) by heat and pressure with or without a catalyst to produce lighter hydrocarbons (such as petrol). **10** tr. break (wheat) into coarse pieces. ● attrib.adj. colloq. excellent; first-rate (a crack regiment; a crack shot). □ **crack a bottle** open a bottle, esp. of wine, and drink it. **crack-brained** colloq. crazy. **crack a crib** sl. break into a house. **crack-down** colloq. severe measures (esp. against law-breakers etc.). **crack down on** colloq. take severe measures against. **crack-jaw** colloq. ● adj. (of a word) difficult to pronounce. ● n. such a word. **crack of doom** a thunder-peal announcing the Day of Judgement. **crack on** colloq. (often foll. by with) proceed briskly or vigorously. **crack up** colloq. **1** collapse under strain. **2** laugh. **3** repute (not all it's cracked up to be). **crack-up** n. colloq. **1** a mental breakdown. **2** a car crash. **crack-willow** a willow, Salix fragilis, with brittle branches. **fair crack of the whip** colloq. a fair chance to participate etc. **get cracking** colloq. begin promptly and vigorously. **have a crack at** colloq. attempt. [OE cracian resound]

cracked /krækt/ adj. **1** having cracks. **2** (predic.) sl. crazy. □ **cracked wheat** wheat that has been crushed into small pieces.

cracker /ˈkrækə(r)/ n. **1** a paper cylinder both ends of which are pulled at Christmas etc. making a sharp noise and releasing a small toy etc. **2** a firework exploding with a sharp noise. **3** (usu. in pl.) an instrument for cracking (nutcrackers). **4 a** a thin dry biscuit often eaten with cheese. **b** a light crisp made of rice or tapioca flour. **5** Brit. sl. a notable or attractive person. **6** N. Amer. offens. = poor white. □ **cracker-barrel** US (of philosophy etc.) homespun; unsophisticated.

crackerjack /ˈkrækəˌdʒæk/ adj. & n. US sl. ● adj. exceptionally fine or expert. ● n. an exceptionally fine thing or person.

crackers /ˈkrækəz/ predic.adj. Brit. sl. crazy.

cracking /ˈkrækɪŋ/ adj. & adv. sl. ● adj. **1** outstanding; very good (a cracking performance). **2** (attrib.) fast and exciting (a cracking speed). ● adv. outstandingly (a cracking good time).

crackle /ˈkræk(ə)l/ v. & n. ● v.intr. make a repeated slight cracking sound (radio crackled; fire was crackling). ● n. **1** such a sound. **2 a** a paintwork,

china, or glass decorated with a pattern of minute surface cracks. **b** the smooth surface of such paintwork etc. □ **crackly** adj. [CRACK + -LE[4]]

crackling /ˈkræklɪŋ/ n. **1** the crisp skin of roast pork. **2** joc. or offens. attractive women regarded collectively as objects of sexual desire. □ **bit of crackling** joc. or offens. an attractive woman.

cracknel /ˈkrækn(ə)l/ n. a light crisp biscuit. [ME f. F craquelin f. MDu. krākelinc f. krāken CRACK]

crackpot /ˈkrækpɒt/ n. & adj. sl. ● n. an eccentric or impractical person. ● adj. mad, unworkable (a crackpot scheme).

cracksman /ˈkræksmən/ n. (pl. **-men**) sl. a burglar, esp. a safe-breaker.

cracky /ˈkrækɪ/ adj. covered with cracks.

Cracow /ˈkrækaʊ/ (Polish **Kraków** /ˈkrakuf/) an industrial and university city in southern Poland, on the River Vistula; pop. (1990) 750,540. It was the capital of Poland from 1320 until replaced by Warsaw in 1609. The city's many fine medieval buildings survived the Second World War largely unscathed.

-cracy /krəsɪ/ comb. form denoting a particular form of government, rule, or influence (aristocracy; bureaucracy). [from or after F -cratie f. med.L -cratia f. Gk -kratia f. kratos strength, power]

cradle /ˈkreɪd(ə)l/ n. & v. ● n. **1 a** a child's bed or cot, esp. one mounted on rockers. **b** a place in which a thing begins, esp. a civilization etc., or is nurtured in its infancy (cradle of choral singing; cradle of democracy). **2** a framework resembling a cradle, esp.: **a** that on which a ship, a boat, etc., rests during construction or repairs. **b** that on which a worker is suspended to work on a ceiling, a ship, the vertical side of a building, etc. **c** the part of a telephone on which the receiver rests when not in use. ● v.tr. **1** contain or shelter as if in a cradle (cradled his head in her arms). **2** place in a cradle. □ **cradle-snatcher** sl. a person amorously attached to a much younger person. **cradle-song** a lullaby. **from the cradle** from infancy. **from the cradle to the grave** from infancy till death (esp. of state welfare). [OE cradol, perh. rel. to OHG kratto basket]

cradling /ˈkreɪdlɪŋ/ n. Archit. a wooden or iron framework, esp. one used as a structural support in a ceiling.

craft /krɑːft/ n. & v. ● n. **1** skill, esp. in practical arts. **2 a** (esp. in comb.) a trade or an art (statecraft; handicraft; priestcraft; the craft of pottery). **b** the members of a craft. **3** (pl. **craft**) **a** a boat or vessel. **b** an aircraft or spacecraft. **4** cunning or deceit. **5** (**the Craft**) the brotherhood of Freemasons. ● v.tr. make in a skilful way (crafted a poem; a well-crafted piece of work). □ **craft-brother** a fellow worker in the same trade. **craft-guild** hist. a guild of workers of the same trade. [OE cræft]

craftsman /ˈkrɑːftsmən/ n. (pl. **-men**; fem. **craftswoman**, pl. **-women**) **1** a skilled and usu. time-served worker. **2** a person who practises a handicraft. **3** a private soldier in the Royal Electrical and Mechanical Engineers. □ **craftsmanship** n. [ME, orig. craft's man]

crafty /ˈkrɑːftɪ/ adj. (**craftier**, **craftiest**) cunning, artful, wily. □ **craftily** adv. **craftiness** n. [OE cræftig]

crag[1] /kræg/ n. Brit. a steep or rugged rock. [ME, of Celt. orig.]

crag[2] /kræg/ n. Geol. rock consisting of a shelly sand. [18th c.: perh. f. CRAG[1]]

craggy /ˈkrægɪ/ adj. (**craggier**, **craggiest**) **1** (esp. of a person's face) rugged; rough-textured. **2** (of a landscape) having crags. □ **craggily** adv. **cragginess** n.

cragsman /ˈkrægzmən/ n. (pl. **-men**) a skilled climber of crags.

Craiova /krəˈjəʊvə/ a city in SW Romania; pop. (1989) 300,030.

crake /kreɪk/ n. **1** a bird of the rail family, esp. one of the shorter-billed kinds of the genus Porzana, or the corncrake. **2** the cry of a corncrake. [ME f. ON kráka (imit.): cf. CROAK]

cram /kræm/ v. (**crammed**, **cramming**) **1** tr. **a** fill to bursting; stuff (the room was crammed). **b** (foll. by in, into) force (a thing) into (cram the sandwiches into the bag). **2** tr. & intr. prepare for an examination by intensive study. **3** tr. (often foll. by with) feed (poultry etc.) to excess. **4** tr. & intr. colloq. eat greedily. □ **cram-full** as full as possible. **cram in** push in to bursting point (crammed in another five minutes' work). [OE crammian f. Gmc]

crambo /ˈkræmbəʊ/ n. a game in which a player gives a word or verse-line to which each of the others must find a rhyme. [earlier crambe, app. allusive f. L crambe repetita cabbage served up again]

crammer /ˈkræmə(r)/ n. a person or institution that crams pupils for examinations.

cramp /kræmp/ n. & v. ● n. **1 a** a painful involuntary contraction of a muscle or muscles due to cold, exertion, etc. **b** = writer's cramp (see WRITER). **2** (also **cramp-iron**) a metal bar with bent ends for holding masonry etc. together. **3** a portable tool for holding two planks etc. together; a clamp. **4** a restraint. ● *v.tr.* **1** affect with cramp. **2** confine narrowly. **3** restrict (energies etc.). **4** (as **cramped** *adj.*) **a** (of handwriting) small and difficult to read. **b** (of a room etc.) uncomfortably crowded, lacking space. **5** fasten with a cramp. □ **cramp a person's style** prevent a person from acting freely or naturally. **cramp up** confine narrowly. [ME f. OF *crampe* f. MDu., MLG *krampe*, OHG *krampfo* f. adj. meaning 'bent': cf. CRIMP]

crampon /ˈkræmpən/ *n.* (usu. in *pl.*) **1** an iron plate with spikes fixed to a boot for walking on ice, climbing, etc. **2** a metal hook for lifting timber, rock, etc.; a grappling-iron. [ME f. F (as CRAMP)]

cran /kræn/ *n. Sc.* a measure for fresh herrings (37½ gallons). [= Gael. *crann*, of uncert. orig.]

Cranach /ˈkrænək/, Lucas (known as Cranach the Elder) (1472–1553), German painter. A member of the Danube School, he is noted for his early religious pictures, in which landscape plays a prominent part, as in *The Rest on the Flight into Egypt* (1504). He also painted portraits, including several of his friend Martin Luther, and is regarded as the originator of the full-length secular portrait as a subject in its own right. His son Lucas (known as Cranach the Younger) (1515–86) continued working in the same tradition.

cranage /ˈkreɪnɪdʒ/ *n.* **1** the use of a crane or cranes. **2** the money paid for this.

cranberry /ˈkrænbəri/ *n.* (*pl.* **-ies**) **1** an evergreen ericaceous shrub of the genus *Vaccinium*, esp. *V. macrocarpon* of America and *V. oxycoccos* of Europe, yielding small red acid berries. **2** a berry from this used for a sauce and in cooking. [17th c.: named by Amer. colonists f. G *Kranbeere*, LG *kranebere* crane-berry]

Crane /kreɪn/, Stephen (1871–1900), American writer. His reputation rests on his novel *The Red Badge of Courage* (1895), a study of an inexperienced soldier and his reactions to the ordeal of battle in the American Civil War. It was hailed as a masterpiece of psychological realism, even though Crane himself had no personal experience of war. In 1895 he came to England, where he developed a close friendship with Joseph Conrad, to whose work his own has been compared.

crane /kreɪn/ *n. & v.* ● *n.* **1** a machine for moving heavy objects, usu. by suspending them from a projecting arm or beam. **2** a tall bird of the family Gruidae, with long legs, long neck, and straight bill; esp. the grey *Grus grus* of Europe. **3** a moving platform supporting a television camera or cine-camera. ● *v.tr.* **1** (also *absol.*) stretch out (one's neck) in order to see something. **2** *tr.* move (an object) by a crane. □ **crane-fly** (*pl.* **-flies**) a large fly of the family Tipulidae, having very long legs (also called *daddy-long-legs*). [OE *cran* (in sense 2), rel. to L *grus*, Gk *geranos*]

cranesbill /ˈkreɪnzbɪl/ *n.* a plant of the genus *Geranium*, with usu. purple, violet, or pink five-petalled flowers. [CRANE + BILL² (with ref. to a long spur on the fruit)]

cranial /ˈkreɪnɪəl/ *adj.* of or relating to the skull. □ **cranial index** the ratio of the width and length of a skull. **cranial nerve** *Anat.* each of twelve pairs of nerves arising directly from the brain, not from the spinal cord. [CRANIUM + -AL¹]

craniate /ˈkreɪnɪət/ *adj. & n.* ● *adj.* having a skull. ● *n.* a craniate animal. [mod.L *craniatus* f. CRANIUM]

cranio- /ˈkreɪnɪəʊ/ *comb. form* cranium.

craniology /ˌkreɪnɪˈɒlədʒɪ/ *n.* the scientific study of the shape and size of the human skull. □ **craniologist** *n.* **craniological** /-nɪəˈlɒdʒɪk(ə)l/ *adj.*

craniometry /ˌkreɪnɪˈɒmɪtrɪ/ *n.* the scientific measurement of skulls. □ **craniometric** /-nɪəˈmetrɪk/ *adj.*

craniotomy /ˌkreɪnɪˈɒtəmɪ/ *n.* (*pl.* **-ies**) **1** surgical removal of a portion of the skull. **2** surgical perforation of the skull of a dead foetus to ease delivery.

cranium /ˈkreɪnɪəm/ *n.* (*pl.* **craniums** or **crania** /-nɪə/) *Anat.* **1** the skull. **2** the part of the skeleton that encloses the brain. [ME f. med.L f. Gk *kranion* skull]

crank¹ /kræŋk/ *n. & v.* ● *n.* **1** part of an axle or shaft bent at right angles for interconverting reciprocal and circular motion. **2** an elbow-shaped connection in bell-hanging. ● *v.tr.* **1** cause to move by means of a crank. **2 a** bend into a crank-shape. **b** furnish or fasten with a crank. □ **crank up 1** start (a car engine) by turning a crank. **2** *sl.* increase (speed etc.) by intensive effort. [OE *cranc*, app. f. *crincan*, rel. to *cringan* fall in battle, orig. 'curl up']

crank² /kræŋk/ *n.* **1 a** an eccentric person, esp. one obsessed by a particular theory (*health-food crank*). **b** *N. Amer.* a bad-tempered person.

2 *literary* a fanciful turn of speech (*quips and cranks*). [back-form. f. CRANKY]

crank³ /kræŋk/ *adj. Naut.* liable to capsize. [f. Sc. or dial. *crank* weak, shaky, or CRANK¹]

crankcase /ˈkræŋkkeɪs/ *n.* a case enclosing a crankshaft.

crankpin /ˈkræŋkpɪn/ *n.* a pin by which a connecting-rod is attached to a crank.

crankshaft /ˈkræŋkʃɑːft/ *n.* a shaft driven by a crank.

cranky /ˈkræŋkɪ/ *adj.* (**crankier, crankiest**) **1** *colloq.* eccentric, esp. obsessed with a particular theory (*cranky ideas about women*). **2** (esp. of a machine) working badly; shaky. **3** esp. *N. Amer.* ill-tempered or crotchety. □ **crankily** *adv.* **crankiness** *n.* [perh. f. obs. *crank* rogue feigning sickness]

Cranmer /ˈkrænmə(r)/, Thomas (1489–1556), English Protestant cleric and martyr. After helping to negotiate Henry VIII's divorce from Catherine of Aragon, he was appointed Archbishop of Canterbury in 1532. The first Protestant to hold this position, he was largely responsible for English liturgical reform, particularly under Edward VI, and for the compilation of the Book of Common Prayer (1549). After the accession of the Catholic Mary Tudor, Cranmer was tried for treason and heresy and burnt at the stake in Oxford.

crannog /ˈkrænəg/ *n.* an ancient lake-dwelling in Scotland or Ireland. [Ir. f. *crann* tree, beam]

cranny /ˈkrænɪ/ *n.* (*pl.* **-ies**) a chink, a crevice, a crack. □ **crannied** /-nɪd/ *adj.* [ME f. OF *crané* past part. of *craner* f. *cran* f. pop.L *crena* notch]

crap¹ /kræp/ *n., v., & adj. coarse sl.* ● *n.* **1** (often as *int.*) nonsense, rubbish (*he talks crap*). **2** faeces. ● *v.intr.* (**crapped, crapping**) defecate. ● *adj.* of bad quality, useless, inferior. □ **crap out** *US* **1** be unsuccessful. **2** withdraw from a game etc. [earlier senses 'chaff, refuse from fat-boiling': ME f. Du. *krappe*]

crap² /kræp/ *n. N. Amer.* a losing throw of 2, 3, or 12 in craps. □ **crap game** a game of craps. [formed as CRAPS]

crape /kreɪp/ *n.* **1** crêpe, usu. of black silk or imitation silk, formerly used for mourning clothes. **2** a band of this formerly worn round a person's hat etc. as a sign of mourning. □ **crape fern** a New Zealand fern, *Leptopteris superba*, with tall dark green fronds. **crape hair** artificial hair used in stage make-up. □ **crapy** *adj.* [earlier *crispe, crespe* f. F *crespe* CRÊPE]

crappy /ˈkræpɪ/ *adj.* (**crappier, crappiest**) *coarse sl.* **1** rubbishy, cheap. **2** disgusting.

craps /kræps/ *n.pl. US* a gambling game played with dice. □ **shoot craps** play craps. [19th c.: perh. f. *crab* lowest throw at dice]

crapulent /ˈkræpjʊlənt/ *adj.* **1** given to indulging in alcohol. **2** drunk. **3** resulting from drunkenness. □ **crapulence** *n.* **crapulous** *adj.* [LL *crapulentus* very drunk f. L *crapula* inebriation f. Gk *kraipalē* drunken headache]

craquelure /ˈkrækəˌljʊə(r)/ *n.* a network of fine cracks in a painting or its varnish. [F]

crash¹ /kræʃ/ *v., n., & adv.* ● *v.* **1** *intr. & tr.* make or cause to make a loud smashing noise (*the cymbals crashed; crashed the plates together*). **2** *tr. & intr.* throw, drive, move, or fall with a loud smashing noise. **3** *intr. & tr.* **a** collide or cause (a vehicle) to collide violently with another vehicle, obstacle, etc.; overturn at high speed. **b** fall or cause (an aircraft) to fall violently on to the land or the sea (*crashed the plane; the airman crashed into the sea*). **4** *intr.* (usu. foll. by *into*) collide violently (*crashed into the window*). **5** *intr.* undergo financial ruin. **6** *tr. colloq.* enter without permission (*crashed the cocktail party*). **7** *intr. colloq.* be heavily defeated (*crashed to a 4–0 defeat*). **8** *intr. Computing* (of a machine or system) fail suddenly. **9** *tr. colloq.* pass (a red traffic light etc.). **10** *intr.* (often foll. by *out*) *sl.* go to sleep, esp. in an improvised setting. ● *n.* **1 a** a loud and sudden smashing noise (*a thunder crash; the crash of crockery*). **b** a breakage (esp. of crockery, glass, etc.). **2 a** a violent collision, esp. of one vehicle with another or with an object. **b** the violent fall of an aircraft on to the land or sea. **3** ruin, esp. financial. **4** *Computing* a sudden failure which puts a system out of action. **5** (*attrib.*) done rapidly or urgently (*a crash course in first aid; a crash diet*). ● *adv.* with a crash (*the window went crash*). □ **crash barrier** a barrier intended to prevent a car from leaving the road etc. **crash-dive** *v.* **1** *intr.* **a** (of a submarine or its pilot) dive hastily and steeply in an emergency. **b** (of an aircraft or its pilot) dive and crash. **2** *tr.* cause to crash-dive. ● *n.* such a dive. **crash-halt** a sudden stop by a vehicle. **crash-helmet** a helmet worn esp. by a motorcyclist to protect the head in a crash. **crash-land 1** *intr.* (of an aircraft or its pilot) land hurriedly with a crash, usu. without lowering the undercarriage. **2** *tr.* cause (an aircraft) to crash-land.

crash landing a hurried landing with a crash. **crash pad** *sl.* a place to sleep, esp. in an emergency. **crash-stop** = *crash-halt.* **crash-tackle** *Football* a vigorous tackle. [ME: imit.]

crash[2] /kræʃ/ *n.* a coarse plain linen, cotton, etc., fabric. [Russ. *krashenina* coloured linen]

crashing /'kræʃɪŋ/ *adj. colloq.* overwhelming (*a crashing bore*).

crasis /'kreɪsɪs/ *n.* (*pl.* **crases** /-siːz/) the contraction of two adjacent vowels in ancient Greek into one long vowel or diphthong. [Gk *krasis* mixture]

crass /kræs/ *adj.* **1** grossly stupid (*a crass idea*). **2** gross (*crass stupidity*). **3** *literary* thick or gross. □ **crassitude** *n.* **crassly** *adv.* **crassness** *n.* [L *crassus* solid, thick]

Crassus /'kræsəs/, Marcus Licinius ('Dives') (*c.*115–53 BC), Roman politician. He defeated Spartacus in 71 BC, though Pompey claimed credit for the victory. Crassus joined Caesar and Pompey in the First Triumvirate in 60. In 55 he was made consul and given a special command in Syria, where he hoped to regain a military reputation equal to that of his allies by a victory over the Parthians, but after some successes he was defeated and killed.

-crat /kræt/ *comb. form* a member or supporter of a particular form of government or rule (*autocrat; democrat*). [from or after F *-crate:* see *-CRACY*]

cratch /krætʃ/ *n.* a rack used for holding food for farm animals out of doors. [ME f. OF *creche* f. Gmc: rel. to CRIB]

crate /kreɪt/ *n. & v.* ● *n.* **1** a large wickerwork basket or slatted wooden case etc. for packing esp. fragile goods for transportation. **2** *sl.* an old aeroplane or other vehicle. ● *v.tr.* pack in a crate. □ **crateful** *n.* (*pl.* **-fuls**). [ME, perh. f. Du. *krat* basket etc.]

crater /'kreɪtə(r)/ *n. & v.* ● *n.* **1** the mouth of a volcano. **2** a bowl-shaped cavity, esp. that made by the explosion of a shell or bomb. **3** *Astron.* a hollow with a raised rim on the surface of a planet or moon, caused by the impact of a meteorite. **4** *Antiq.* a large ancient Greek bowl, used for mixing wine. ● *v.tr.* form a crater in. □ **craterous** *adj.* [L f. Gk *kratēr* mixing-bowl: see CRASIS]

Crater Lake a lake filling a volcanic crater in the Cascade mountains of SW Oregon. With a depth of more than 600 m (1,968 ft) it is the deepest lake in the US.

-cratic /'krætɪk/ *comb. form* (also **-cratical** /'krætɪk(ə)l/) denoting a particular kind of government or rule (*autocratic; democratic*). □ **-cratically** *comb. form* (*adv.*) [from or after F *-cratique:* see -CRACY]

craton /'kreɪtɒn/ *n. Geol.* a large stable block of the earth's crust.

cravat /krə'væt/ *n.* **1** a scarf worn by men inside an open-necked shirt. **2** *hist.* a necktie. □ **cravatted** *adj.* [F *cravate* f. G *Krawat, Kroat* f. Serbo-Croat *Hrvat* Croat]

crave /kreɪv/ *v.* **1** *tr.* **a** long for (*craved affection*). **b** beg for (*craves a blessing*). **2** *intr.* (foll. by *for*) long for; beg for (*craved for comfort*). □ **craver** *n.* [OE *crafian*, rel. to ON *krefja*]

craven /'kreɪv(ə)n/ *adj. & n.* ● *adj.* (of a person, behaviour, etc.) cowardly; abject. ● *n.* a cowardly person. □ **cravenly** *adv.* **cravenness** /-v(ə)nnɪs/ *n.* [ME *cravand* etc. perh. f. OF *cravanté* defeated, past part. of *cravanter* ult. f. L *crepare* burst; assim. to -EN[3]]

craving /'kreɪvɪŋ/ *n.* (usu. foll. by *for*) a strong desire or longing.

craw /krɔː/ *n.* the crop of a bird or insect. □ **stick in one's craw** be unacceptable. [ME, rel. to MDu. *crāghe*, MLG *krage*, MHG *krage* neck, throat]

crawfish /'krɔːfɪʃ/ *n. & v.* ● *n.* (*pl.* same) esp. N. Amer. = CRAYFISH. ● *v.intr.* (often foll. by *out*) *US colloq.* retreat; back out. [var. of CRAYFISH]

Crawford[1] /'krɔːfəd/, Joan (born Lucille le Sueur) (1908–77), American actress. For over forty years she ranked among Hollywood's leading film stars. A dancer in her early films, she later played the female lead in films such as *Rain* (1932) and *Mildred Pierce* (1945), as well as mature roles, such as her part in the horror film *Whatever Happened to Baby Jane?* (1962).

Crawford[2] /'krɔːfəd/, Osbert Guy Stanhope (1886–1957), British archaeologist. He pioneered the use of aerial photography in the detection of previously unlocated or buried archaeological sites and monuments. He also made an important contribution to the cartographic representation of archaeology when he worked as the Ordnance Survey's first archaeology officer (1920–40).

crawl /krɔːl/ *v. & n.* ● *v.intr.* **1** move slowly, esp. on hands and knees. **2** (of an insect, snake, etc.) move slowly with the body close to the ground etc. **3** walk or move slowly (*the train crawled into the station*). **4** (often foll. by *to*) *colloq.* behave obsequiously or ingratiatingly in the hope of advantage. **5** (often foll. by *with*) be covered or filled with crawling or moving things, or with people etc. compared to this. **6** (esp. of the skin) feel a creepy sensation. **7** swim with a crawl stroke. ● *n.* **1** an act of crawling. **2** a slow rate of movement. **3** a high-speed swimming stroke with alternate overarm movements and rapid straight-legged kicks. **4** (usu. in *comb.*) *colloq.* a leisurely journey between places of interest (*church-crawl*). **b** = *pub-crawl.* □ **crawlingly** *adv.* **crawly** *adj.* (in senses 5, 6 of *v.*). [ME: orig. unkn.: cf. Sw. *kravla*, Danish *kravle*]

crawler /'krɔːlə(r)/ *n.* **1** *sl.* a person who behaves obsequiously in the hope of advantage. **2** anything that crawls, e.g. an insect or a slow-moving road vehicle. **3** a tractor moving on a caterpillar track. **4** (usu. in *pl.*) esp. *US* a baby's overall for crawling in; rompers.

cray /kreɪ/ *n. Austral. & NZ* = CRAYFISH.

crayfish /'kreɪfɪʃ/ *n.* (*pl.* same) **1** a small lobster-like freshwater crustacean, esp. of the genus *Astacus.* **2** = *spiny lobster.* (See also CRAWFISH.) [ME f. OF *crevice, crevis,* ult. f. OHG *krebiz* CRAB[1]: assim. to FISH[1]]

crayon /'kreɪən, 'kreɪɒn/ *n. & v.* ● *n.* **1** a stick or pencil of coloured chalk, wax, etc., used for drawing. **2** a drawing made with this. ● *v.tr.* draw with crayons. [F f. *craie* f. L *creta* chalk]

craze /kreɪz/ *v. & n.* ● *v.* **1** *tr.* (usu. as **crazed** *adj.*) make insane (*crazed with grief*). **2 a** *tr.* produce fine surface cracks on (pottery glaze etc.). **b** *intr.* develop such cracks. ● *n.* **1 a** a usu. temporary enthusiasm (*a craze for hula hoops*). **b** the object of this. **2** an insane fancy or condition. [ME, orig. = break, shatter, perh. f. ON]

crazy /'kreɪzɪ/ *adj.* (**crazier, craziest**) **1** *colloq.* (of a person, an action, etc.) insane or mad; foolish. **2** *colloq.* extremely enthusiastic. **3** (usu. foll. by *about*) *sl.* **a** exciting, unrestrained. **b** excellent. **4** (*attrib.*) (of paving, a quilt, etc.) made of irregular pieces fitted together. **5** *archaic* (of a ship, building, etc.) unsound, shaky. □ **crazy bone** *US* the funny bone. **like crazy** *colloq.* = like mad (see MAD). □ **crazily** *adv.* **craziness** *n.*

Crazy Horse (Sioux name Ta-Sunko-Witko) (*c.*1849–77), Sioux chief. In 1876 he led a successful rearguard action of Sioux and Cheyenne warriors against invading US army forces in Montana. Shortly afterwards he and his men joined Sitting Bull at Little Bighorn, where Crazy Horse played an important strategic and military role in the massacre of US forces under General Custer. He surrendered in 1877 and was killed in custody in Nebraska a few months later.

CRC *abbr.* camera-ready copy.

creak /kriːk/ *n. & v.* ● *n.* a harsh scraping or squeaking sound. ● *v.intr.* **1** make a creak. **2 a** move with a creaking noise. **b** move stiffly and awkwardly. **c** show weakness or frailty under strain. □ **creakingly** *adv.* [ME, imit.: cf. CRAKE, CROAK]

creaky /'kriːkɪ/ *adj.* (**creakier, creakiest**) **1** liable to creak. **2 a** stiff or frail (*creaky joints*). **b** (of a practice, institution, etc.) decrepit, dilapidated, outmoded. □ **creakily** *adv.* **creakiness** *n.*

cream /kriːm/ *n., v., & adj.* ● *n.* **1 a** the fatty content of milk which gathers at the top and can be made into butter by churning. **b** this eaten (often whipped) with a dessert, as a cake-filling, etc. (*strawberries and cream; cream gateau*). **2** the part of a liquid that gathers at the top. **3** (usu. prec. by *the*) the best or choicest part of something, esp.: **a** point of an anecdote. **b** an élite group of people (*the cream of the nation*). **4** a creamlike preparation, esp. a cosmetic (*hand cream*). **5** a very pale yellow or off-white colour. **6 a** a dish or sweet like or made with cream. **b** a soup or sauce containing milk or cream. **c** a full-bodied mellow sweet sherry. **d** a biscuit with a creamy sandwich filling. **e** a chocolate-covered usu. fruit-flavoured fondant. ● *v.* **1** *tr.* (usu. foll. by *off*) **a** take the cream from (milk). **b** take the best or a specified part from (*creamed off the brightest pupils*). **2** *tr.* work (butter etc.) to a creamy consistency. **3** *tr.* treat (the skin etc.) with cosmetic cream. **4** *tr.* add cream to (coffee etc.). **5** *intr.* (of milk or any other liquid) form a cream or scum. **6** *tr. US colloq.* defeat (esp. in a sporting contest). ● *adj.* pale yellow; off-white. □ **cream bun** (or **cake**) a bun or cake filled or topped with cream. **cream cheese** a soft rich cheese made from unskimmed milk and cream. **cream-coloured** pale yellowish-white. **cream cracker** *Brit.* a crisp dry unsweetened biscuit usu. eaten with cheese. **cream-laid** (or **-wove**) laid (or wove) cream-coloured paper. **cream of tartar** crystallized potassium hydrogen tartrate, used in medicine, baking powder, etc. **cream puff 1** a cake made of puff pastry filled with cream. **2** *colloq.* an ineffectual or effeminate person. **cream soda** a carbonated vanilla-flavoured soft drink. **cream tea** afternoon tea with scones, jam, and cream. [ME f. OF *cre(s)me* f. LL *cramum* (perh. f. Gaulish) & eccl.L *chrisma* CHRISM]

creamer /'kriːmə(r)/ *n.* **1** a flat dish used for skimming the cream off milk. **2** a machine used for separating cream from milk. **3** a cream or milk substitute for adding to coffee or tea. **4** *US* a jug for cream.

creamery /'kriːmərɪ/ *n.* (*pl.* **-ies**) **1** a factory producing butter and cheese. **2** a shop where milk, cream, etc., are sold; a dairy. [CREAM, after F *crémerie*]

creamy /'kriːmɪ/ *adj.* (**creamier**, **creamiest**) **1** like cream in consistency or colour. **2** rich in cream. □ **creaminess** *n.*

crease¹ /kriːs/ *n. & v.* ● *n.* **1 a** a line in paper etc. caused by folding. **b** a fold or wrinkle. **2** *Cricket* a line marking the position of the bowler or batsman (see POPPING-CREASE, *bowling-crease*). **3** an area near the goal in ice hockey or lacrosse into which the puck or the ball must precede the players. ● *v.* **1** *tr.* make creases in (material). **2** *intr.* become creased (*linen creases badly*). **3** *tr. & intr. sl.* make or become incapable through laughter. **4** *tr. esp. US* (often foll. by *up*) *sl.* **a** tire out. **b** stun or kill. [earlier *creast* = CREST ridge in material]

crease² var. of KRIS.

create /kriː'eɪt/ *v.* **1** *tr.* **a** (of natural or historical forces) bring into existence; cause (*poverty creates resentment*). **b** (of a person or persons) make or cause (*create a diversion; create a good impression*). **2** *tr.* originate (*an actor creates a part*). **3** *tr.* invest (a person) with a rank (*created him a lord*). **4** *intr. Brit. sl.* make a fuss; grumble. □ **creatable** *adj.* [ME f. L *creare*]

creatine /'kriːətɪn/ *n. Biochem.* a product of protein metabolism found (esp. as the phosphate) in the muscles etc. of vertebrates. [Gk *kreas* meat + -INE⁴]

creation /kriː'eɪʃ(ə)n/ *n.* **1 a** the act of creating. **b** an instance of this. **2 a** (usu. **the Creation**) the creating of the universe regarded as an act of God. **b** (usu. **Creation**) everything so created; the universe. **3** a product of human intelligence, esp. of imaginative thought or artistic ability. **4 a** the act of investing with a title or rank. **b** an instance of this. □ **creation science** the reinterpretation of scientific knowledge in accord with belief in the literal truth of the Bible, esp. regarding the origin of matter, life, and humankind (cf. CREATIONISM). [ME f. OF f. L *creatio -onis* (as CREATE)]

creationism /kriː'eɪʃəˌnɪz(ə)m/ *n.* the belief that the universe and living organisms originated from specific acts of divine creation, as in the biblical account, rather than by processes such as evolution; creation science. □ **creationist** *n.*

creative /kriː'eɪtɪv/ *adj.* **1** inventive and imaginative. **2** creating or able to create. □ **creatively** *adv.* **creativeness** *n.* **creativity** /ˌkriːeɪ'tɪvɪtɪ/ *n.*

creator /kriː'eɪtə(r)/ *n.* **1** a person who creates. **2** (as **the Creator**) God. [ME f. OF *creat(o)ur* f. L *creator -oris* (as CREATE)]

creature /'kriːtʃə(r)/ *n.* **1 a** an animal, as distinct from a human being. **b** any living being (*we are all God's creatures*). **2** a person of a specified kind (*poor creature*). **3** a person owing status to and obsequiously subservient to another. **4** anything created; a creation. □ **creature comforts** material comforts such as good food, warmth, etc. **creature of habit** a person set in an unvarying routine. □ **creaturely** *adj.* [ME f. OF f. LL *creatura* (as CREATE)]

crèche /kreʃ, kreɪʃ/ *n.* **1** a day nursery for babies and young children. **2** *US* a representation of a Nativity scene. [F (as CRATCH)]

Crécy, Battle of /'kresɪ/ a battle between the English and the French in 1346 near the village of Crécy-en-Ponthieu in Picardy in northern France, at which the forces of Edward III defeated those of Philip VI. It was the first great English victory of the Hundred Years War.

cred /kred/ *n. sl.* = CREDIBILITY 2. [abbr.]

credal see CREED.

credence /'kriːd(ə)ns/ *n.* **1** belief. **2** (in full **credence table**) a small side-table, shelf, or niche which holds the elements of the Eucharist before they are consecrated. □ **give credence to** believe. **letter of credence** a letter of introduction, esp. of an ambassador. [ME f. OF f. med.L *credentia* f. *credere* believe]

credential /krɪ'denʃ(ə)l/ *n.* (usu. in *pl.*) **1** evidence of a person's achievements or trustworthiness, usu. in the form of certificates, references, etc. **2** a letter or letters of introduction. [med.L *credentialis* (as CREDENCE)]

credenza /krɪ'denzə/ *n.* a sideboard or cupboard. [It. f. med.L (as CREDENCE)]

credibility /ˌkredɪ'bɪlɪtɪ/ *n.* **1** the condition of being credible or believable. **2** reputation; status; acceptability among one's peers. □ **credibility gap** an apparent difference between what is said and what is true.

credible /'kredɪb(ə)l/ *adj.* **1** (of a person or statement) believable or worthy of belief. **2** (of a threat etc.) convincing. □ **credibly** *adv.* [ME f. L *credibilis* f. *credere* believe]

credit /'kredɪt/ *n. & v.* ● *n.* **1** (usu. of a person) a source of honour, pride, etc. (*is a credit to the school*). **2** the acknowledgement of merit (*must give him credit for consistency*). **3** a good reputation (*his credit stands high*). **4** a belief or trust (*I place credit in that*). **b** something believable or trustworthy (*that statement has credit*). **5 a** a person's financial standing; the sum of money at a person's disposal in a bank etc. **b** the power to obtain goods etc. before payment (based on the trust that payment will be made). **6** (usu. in *pl.*) an acknowledgement of a contributor's services to a film, television programme, etc. **7** a grade above a pass in an examination. **8** a reputation for solvency and honesty in business. **9 a** (in bookkeeping) the acknowledgement of being paid by an entry on the credit side of an account. **b** the sum entered. **c** the credit side of an account. **10 a** a certificate or other acknowledgement of a student's completion of a course. **b** a unit of study counting towards a degree etc. ● *v.tr.* (**credited**, **crediting**) **1** believe (*cannot credit it*). **2** (foll. by *to*, *with*) **a** enter on the credit side of an account (*credited £20 to him; credited him with £20*). **b** ascribe a good quality or achievement to (*the goal was credited to Shearer; credit me with common sense*). □ **credit account** *Brit.* an account with a shop etc. for obtaining goods or services before payment. **credit card** a card from a bank etc. authorizing the obtaining of goods on credit. **credit note** a note given by a shop etc. in return for goods returned, stating the value of goods owed to the customer. **credit rating** an estimate of a person's suitability to receive commercial credit. **credit sale** the sale of goods on credit. **credit title** a person's name appearing at the beginning or end of a film or broadcast etc. as an acknowledgement. **credit transfer** a transfer of money from one bank account to another. **do credit to** (or **do a person credit**) enhance the reputation of. **get credit for** be given credit for. **give a person credit for 1** enter (a sum) to a person's credit. **2** ascribe (a good quality) to a person. **give credit to** believe. **letter of credit** a letter from a banker authorizing a person to draw money up to a specified amount, usu. from another bank. **on credit** with an arrangement to pay later. **to one's credit** in one's praise, commendation, or defence (*to his credit, he refused the offer*). [F *crédit* f. It. *credito* or L *creditum* f. *credere* credit- believe, trust]

creditable /'kredɪtəb(ə)l/ *adj.* (often foll. by *to*) bringing credit or honour. □ **creditably** *adv.* **creditability** /ˌkredɪtə'bɪlɪtɪ/ *n.*

creditor /'kredɪtə(r)/ *n.* **1** a person to whom a debt is owing. **2** a person or company that gives credit for money or goods (cf. DEBTOR). [ME f. AF *creditour* (OF *-eur*) f. L *creditor -oris* (as CREDIT)]

creditworthy /'kredɪtˌwɜːðɪ/ *adj.* considered suitable to receive commercial credit. □ **creditworthiness** *n.*

credo /'kreɪdəʊ, 'kriːdəʊ/ *n.* (*pl.* **-os**) **1** (**Credo**) a statement of belief; a creed, esp. the Apostles' or Nicene Creed beginning in Latin with *credo*. **2** a musical setting of the Nicene Creed. [ME f. L, = I believe]

credulous /'kredjʊləs/ *adj.* **1** too ready to believe; gullible. **2** (of behaviour) showing such gullibility. □ **credulously** *adv.* **credulousness** *n.* **credulity** /krɪ'djuːlɪtɪ/ *n.* [L *credulus* f. *credere* believe]

Cree /kriː/ *n. & adj.* ● *n.* (*pl.* same or **Crees**) **1** a member of an American Indian people living in central Canada and around Hudson Bay. **2** the Algonquian language of this people. ● *adj.* of or relating to this people or their language. [Canad. F *Cris* (earlier *Cristinaux*) f. Algonquian]

creed /kriːd/ *n.* **1** a set of principles or opinions, esp. as a philosophy of life (*his creed is moderation in everything*). **2** a brief formal summary of Christian belief, esp. (often **the Creed**): **a** = APOSTLES' CREED. **b** = NICENE CREED (cf. ATHANASIAN CREED). □ **credal** /'kriːd(ə)l/ *adj.* **creedal** *adj.* [OE *crēda* f. L CREDO]

Creek /kriːk/ *n. & adj.* ● *n.* **1 a** (*pl.* same) a member of an American Indian people now settled in Oklahoma but originally from Alabama and Georgia. **b** the Muskogean language of this people. **2** a confederacy of several American Indian peoples, of whom the Creek proper were the most numerous. ● *adj.* of or relating to these peoples or their language. [CREEK (prob. as orig. often living by waterways)]

creek /kriːk/ *n.* **1** *Brit.* **a** a small bay or harbour on a sea-coast. **b** a narrow inlet on a sea-coast or in a river-bank. **2** *N. Amer., Austral., & NZ* a stream or brook. □ **up shit creek** *coarse sl.* = *up the creek* 1. **up the creek** *sl.* **1** in difficulties or trouble. **2** crazy. [ME *crike* f. ON *kriki* nook (or partly f. OF *crique* f. ON), & ME *crēke* f. MDu. *krēke* (or f. *crike* by lengthening); ult. orig. unkn.]

creel /kriːl/ *n.* **1** a large wicker basket for fish. **2** an angler's fishing-basket. [ME, orig. Sc.: ult. orig. unkn.]

creep /kriːp/ v. & n. ● v.intr. (past and past part. **crept** /krept/) **1** move with the body prone and close to the ground; crawl. **2** (often foll. by in, out, up, etc.) come, go, or move slowly and stealthily or timidly (crept out without being seen). **3** enter slowly (into a person's affections, life, awareness, etc.) (a feeling crept over her; crept into her heart). **4** colloq. act abjectly or obsequiously in the hope of advancement. **5** (of a plant) grow along the ground or up a wall by means of tendrils. **6** (as **creeping** adj.) developing slowly and steadily (creeping inflation). **7** (of the flesh) feel as if insects etc. were creeping over it, as a result of fear, horror, etc. **8** (of materials) undergo creep. ● n. **1 a** the act of creeping. **b** an instance of this. **2** (in pl.; prec. by the) colloq. a nervous feeling of revulsion or fear (gives me the creeps). **3** sl. an unpleasant person. **4** the gradual downward movement of disintegrated rock due to gravitational forces etc. (soil creep). **5** a gradual deformation of materials under stress. **6** a low arch under a railway embankment, road, etc. □ **creeping barrage** an artillery barrage moving ahead of advancing troops. **creeping Jenny** = MONEYWORT. **creeping Jesus** sl. an abject or hypocritical person. **creep up on** approach (a person) stealthily or unnoticed. [OE crēopan f. Gmc]

creeper /ˈkriːpə(r)/ n. **1** a climbing or creeping plant. **2** a bird that climbs, esp. a treecreeper. **3** sl. a soft-soled shoe.

creepy /ˈkriːpɪ/ adj. (**creepier**, **creepiest**) **1** colloq. having or producing a creeping of the flesh (I feel creepy; a creepy film). **2** given to creeping. □ **creepily** adv. **creepiness** n. [CREEP]

creepy-crawly /ˌkriːpɪˈkrɔːlɪ/ n. & adj. Brit. colloq. ● n. (pl. **-ies**) an insect, worm, or other small crawling creature. ● adj. creeping and crawling.

creese var. of KRIS.

cremate /krɪˈmeɪt/ v.tr. burn (a corpse etc.) to ashes. □ **cremator** n. [L cremare burn]

cremation /krɪˈmeɪʃ(ə)n/ n. the practice of disposing of a corpse by burning. Cremation appears to have been rare in prehistory, but in the ancient civilized world it was the normal custom except in Egypt, Judaea, and China. Belief in the resurrection of the body made the practice repugnant to the early Christians, and by the 5th century Christian influence had caused it to be abandoned throughout the Roman Empire. It was revived in the West in the 19th century; in the East it has remained the most general method of disposal of the dead.

crematorium /ˌkremaˈtɔːrɪəm/ n. (pl. **crematoria** /-rɪə/ or **crematoriums**) a place for cremating corpses in a furnace. [mod.L (as CREMATE, -ORY¹)]

crematory /ˈkremətərɪ/ adj. & n. ● adj. of or relating to cremation. ● n. (pl. **-ies**) N. Amer. = CREMATORIUM.

crème /krem/ n. **1** = CREAM n. 6a. **2** a creamy liqueur (crème de cassis). □ **crème brûlée** /bruːˈleɪ/ a pudding of cream or custard topped with caramelized sugar. **crème caramel** a custard coated with caramel. **crème de la crème** /də læ ˈkrem/ the best part; the élite. **crème de menthe** /də ˈmɒnθ/ a peppermint-flavoured liqueur. [F, = cream]

crème fraîche /krem ˈfreɪʃ/ n. a type of heavy slightly soured cream used esp. in making sauces etc. [F, lit. 'fresh cream']

Cremona /krɪˈməʊnə/ a city in Lombardy, in northern Italy; pop. (1990) 75,160. Between the 16th and the 18th century the city was home to three renowned families of violin-makers: the Amati, the Guarneri, and the Stradivari.

crenate /ˈkriːneɪt/ adj. Bot. & Zool. having a notched edge or rounded teeth. □ **crenated** adj. **crenation** /krɪˈneɪʃ(ə)n/ n. **crenature** /ˈkrenəˌtjʊə(r), ˈkriːnə-/ n. [mod.L crenatus f. pop.L crena notch]

crenel /ˈkren(ə)l/ n. (also **crenelle** /krɪˈnel/) an indentation or gap in the parapet of a tower, castle, etc., orig. for shooting through etc. [ME f. OF crenel, ult. f. pop.L crena notch]

crenellate /ˈkrenəˌleɪt/ v.tr. (also **crenelate**) provide (a tower etc.) with battlements or loopholes. □ **crenellation** /ˌkrenəˈleɪʃ(ə)n/ n. [F créneler (as CRENEL)]

Creole /ˈkriːəʊl/ n. & adj. ● n. **1 a** a descendant of European (esp. Spanish) settlers in the West Indies or Central or South America. **b** a white descendant of French settlers in the southern US. **c** a person of mixed European and black descent. **2** a mother tongue formed from the contact of a European language (esp. English, French, or Portuguese) with another (esp. African) language. ● adj. **1** of or relating to a Creole or Creoles. **2** (usu. **creole**) of Creole origin or production (creole cooking). [F créole, criole f. Sp. criollo, prob. f. Port. crioulo home-born slave f. criar breed f. L creare CREATE]

creolize /ˈkriːəˌlaɪz/ v.tr. (also **-ise**) form a Creole from (another language). □ **creolization** /ˌkriːəlaɪˈzeɪʃ(ə)n/ n.

creosote /ˈkriːəˌsəʊt/ n. & v. ● n. **1** (in full **creosote oil**) a dark brown oil distilled from coal tar, used as a wood-preservative. **2** a colourless oily fluid distilled from wood, used as an antiseptic. ● v.tr. treat with creosote. □ **creosote bush** a shrub, Larrea tridentata, native to arid regions of Mexico and the western US, with leaves smelling of creosote. [G Kreosote f. Gk kreas flesh + sōtēr preserver, with ref. to its antiseptic properties]

crêpe /kreɪp/ n. **1** a fine often gauzelike fabric with a wrinkled surface. **2** a thin pancake, usu. with a savoury or sweet filling. **3** (also **crêpe rubber**) a very hard-wearing wrinkled sheet rubber used for the soles of shoes etc. □ **crêpe de Chine** /də ˈʃiːn/ a fine silk crêpe. **crêpe paper** thin crinkled paper. **crêpe Suzette** /suːˈzet/ a small dessert pancake flamed in alcohol at the table. □ **crêpey** adj. **crêpy** adj. [F f. OF crespe curled f. L crispus]

crepitate /ˈkrepɪˌteɪt/ v.intr. **1** make a crackling sound. **2** Zool. (of a beetle) eject pungent fluid with a sharp report. □ **crepitant** adj. [L crepitare frequent. of crepare creak]

crepitation /ˌkrepɪˈteɪʃ(ə)n/ n. **1** Med. = CREPITUS. **2** the action or sound of crackling or rattling.

crepitus /ˈkrepɪtəs/ n. Med. **1** a grating noise from the ends of a fractured bone rubbing together. **2** a similar sound heard from the chest in pneumonia etc. [L f. crepare rattle]

crept past and past part. of CREEP.

crepuscular /krɪˈpʌskjʊlə(r)/ adj. **1 a** of twilight. **b** dim. **2** Zool. appearing or active in twilight. [L crepusculum twilight]

Cres. abbr. Crescent.

cresc. abbr. (also **cres.**) Mus. = CRESCENDO.

crescendo /krɪˈʃendəʊ/ n., adv., adj., & v. ● n. (pl. **-os**) **1** Mus. **a** a gradual increase in loudness. **b** a passage to be played with such an increase. **2** a progress towards a climax (a crescendo of emotions). **b** disp. a climax (reached a crescendo then died away). ● adv. & adj. Mus. with a gradual increase in loudness. ● v.intr. (**-oes**, **-oed**) increase gradually in loudness or intensity. [It., part. of crescere increase f. L crescere grow]

crescent /ˈkrez(ə)nt, ˈkres(ə)nt/ n. & adj. ● n. **1** the curved sickle shape of the waxing or waning moon. **2** anything of this shape, esp. Brit. a street forming an arc. **3 a** the crescent-shaped emblem of Islam or Turkey. (See note below.) **b** (**the Crescent**) the world or power of Islam. ● adj. **1** poet. increasing. **2** crescent-shaped. □ **crescentic** /krɪˈsentɪk/ adj. [ME f. AF cressaunt, OF creissant, f. L crescere grow]

▪ Regarded in the West as the quintessential emblem of the Muslim East from the mid-15th century, the crescent was a symbol of the Byzantines, and had been used by the Ottoman Turks even before their conquest of the Byzantine Empire in 1453. The new moon itself is of great importance in Islam, for its appearance defines the first and last days of Ramadan and the start of the annual pilgrimage. Today the crescent is the central motif on the national flags of many Muslim countries, such as Turkey, Tunisia, Pakistan, and Algeria.

cresol /ˈkriːsɒl/ n. Chem. any of three isomeric phenols (chem. formula: $(CH_3)C_6H_4OH$) present in creosote and used as disinfectants. [CREOSOTE + -OL]

cress /kres/ n. a cruciferous plant usu. with pungent edible leaves, e.g. watercress. [OE cresse f. WG]

cresset /ˈkresɪt/ n. hist. a metal container filled with fuel, lighted and usu. mounted on a pole for illumination. [ME f. OF cresset, craisset, f. craisse = graisse GREASE]

Cressida /ˈkresɪdə/ (in medieval legends of the Trojan War) the daughter of Calchas, a priest. She was faithless to her lover Troilus, a son of Priam.

crest /krest/ n. & v. ● n. **1 a** a comb or tuft of feathers, fur, etc. on a bird's or animal's head. **b** something resembling this, esp. a plume of feathers on a helmet. **c** a helmet; the top of a helmet. **2** the top of something, esp. of a mountain, wave, roof, etc. **3** Heraldry **a** a device above the shield and helmet of a coat of arms. **b** such a device reproduced on writing paper or on a seal, signifying a family. **4 a** a line along the top of the neck of some animals. **b** the hair growing from this; a mane. **5** Anat. a ridge along the surface of a bone. ● v. **1** tr. reach the crest of (a hill, wave, etc.). **2** tr. **a** provide with a crest. **b** serve as a crest to. **3** intr. (of a wave) form into a crest. □ **on the crest of a wave** at the most favourable moment in one's progress. □ **crested** adj. (also in comb.). **crestless** adj. [ME f. OF creste f. L crista tuft]

Cresta Run /ˈkrestə/ a hazardously winding, steeply banked channel of ice built each year at St Moritz, Switzerland, as a tobogganing course, on which competitors race on light toboggans in a characteristic head-first position. A run down the Cresta valley was first built in 1884.

crestfallen /ˈkrestˌfɔːlən/ adj. **1** dejected, dispirited. **2** with a fallen or drooping crest.

cresyl /ˈkriːsaɪl, -sɪl/ n. (usu. attrib.) Chem. a radical (CH₃)C₆H₄O–, derived from a cresol.

cretaceous /krɪˈteɪʃəs/ adj. & n. ● adj. **1** of the nature of chalk. **2** (**Cretaceous**) Geol. of or relating to the last period of the Mesozoic era. (See note below.) ● n. (**Cretaceous**) Geol. this era or the corresponding geological system. [L cretaceus f. creta chalk]

▪ The Cretaceous lasted from about 144 to 65 million years ago, between the Jurassic and Tertiary periods. The climate was warm and the sea level rose; the period is characterized especially in NW Europe by the deposition of chalk. The first flowering plants emerged and the domination of the dinosaurs continued, although they died out quite abruptly towards the end of it. A widespread layer of sediment dating from the Cretaceous–Tertiary (K/T) boundary has been shown since 1980 to be enriched in iridium and other elements and to contain minerals showing evidence of thermal shock and carbon deposits indicative of extensive fires. This appears to indicate the catastrophic impact of one or more large meteorites, and geologists have identified a formation at Chicxulub in the Yucatán Peninsula, Mexico, as a probable impact site. A resulting drastic climate change has been suggested as the cause of the extinction of dinosaurs and many other organisms at this time, but this remains controversial.

Crete /kriːt/ (Greek **Kríti** /ˈkriti/) a Greek island in the eastern Mediterranean; pop. (1991) 536,980; capital, Heraklion. It is noted for the remains of the Minoan civilization which flourished there in the 2nd millennium BC. It fell to Rome in 67 BC and was subsequently ruled by Byzantines, Venetians, and Turks. Crete played an important role in the Greek struggle for independence from the Turks in the late 19th and early 20th centuries, becoming administratively part of an independent Greece in 1913. □ **Cretan** adj. & n.

cretic /ˈkriːtɪk/ n. Prosody a foot containing one short or unstressed syllable between two long or stressed ones. [L Creticus f. Gk Krētikos (as CRETAN)]

cretin /ˈkretɪn/ n. **1** Med. a person who is deformed and mentally retarded as the result of a thyroid deficiency. **2** colloq. a stupid person. □ **cretinism** n. **cretinize** v.tr. (also **-ise**). **cretinous** adj. [F crétin f. Swiss F. creitin, crestin f. L Christianus CHRISTIAN]

cretonne /kreˈtɒn, ˈkretɒn/ n. (often attrib.) a heavy cotton fabric with a usu. floral pattern printed on one or both sides, used for upholstery. [F, of unknown origin]

Creutzfeldt–Jakob disease /ˌkrɔɪtsfeltˈjækɒb/ n. (abbr. **CJD**) Med. a degenerative disease affecting nerve cells in the brain, causing mental, physical, and sensory disturbances such as dementia and seizures. It is thought to be caused by either prions or virinos, and hence to be related to kuru, scrapie, and BSE. [Hans G. Creutzfeldt (1885–1964) and Alfons Jakob (1884–1931), German physicians]

crevasse /krəˈvæs/ n. **1** a deep open crack, esp. in a glacier. **2** US a breach in a river levee. [F f. OF crevace: see CREVICE]

crevice /ˈkrevɪs/ n. a narrow opening or fissure, esp. in a rock or building etc. [ME f. OF crevace f. crever burst f. L crepare]

crew¹ /kruː/ n. & v. ● n. (often treated as pl.) **1 a** a body of people manning a ship, aircraft, train, etc. **b** such a body as distinguished from the captain or officers. **c** a body of people working together; a team. **2** colloq. a company of people; a gang (a motley crew). ● v. **1** tr. supply or act as a crew or member of a crew for. **2** intr. act as a crew or member of a crew. □ **crew cut** a very short haircut. **crew neck** a close-fitting round neckline, esp. on a sweater. [ME f. OF creüe increase, fem. past part. of croistre grow f. L crescere]

crew² past of CROW².

Crewe /kruː/ a town and major railway junction in Cheshire; pop. (1981) 47,800.

crewel /ˈkruːəl/ n. a thin worsted yarn used for tapestry and embroidery. □ **crewel-work** a design worked in crewel on linen or cloth. [ME crule etc., of unkn. orig.]

crewman /ˈkruːmən/ n. (pl. **-men**) a member of a crew.

crib /krɪb/ n. & v. ● n. **1 a** a child's bed with barred or latticed sides; a cot. **b** a model of the Nativity of Christ, with a manger as a bed. **2** a barred container or rack for animal fodder. **3** colloq. **a** a translation of a text for the (esp. surreptitious) use of students. **b** plagiarized work etc. **4** a small house or cottage. **5** a framework lining the shaft of a mine. **6** colloq. **a** cribbage. **b** a set of cards given to the dealer at cribbage by all the players. **7** heavy crossed timbers used in foundations in loose soil etc. **8** sl. a brothel. **9** Austral. & NZ a light meal; food. ● v.tr. (also absol.) (**cribbed, cribbing**) **1** colloq. copy (another person's work) unfairly or without acknowledgement. **2** confine in a small space. **3** colloq. pilfer, steal. **4** colloq. grumble. □ **crib-biting** a horse's habit of biting the manger while noisily breathing in and swallowing. □ **cribber** n. [OE crib(b)]

cribbage /ˈkrɪbɪdʒ/ n. a card-game for two, three, or four players, in which the dealer may score from the cards in the crib (see CRIB n. 6b). According to John Aubrey, the game was invented by the English poet Sir John Suckling; it seems to have been developed from an older game called Noddy. □ **cribbage-board** a board with pegs and holes used for scoring at cribbage. [17th c.: orig. unkn.]

cribo /ˈkrɪbəʊ, ˈkraɪbəʊ/ n. (pl. **-os**) = indigo snake. [19th c.: orig. unkn.]

cribriform /ˈkrɪbrɪˌfɔːm/ adj. Anat. & Bot. having numerous small holes. [L cribrum sieve + -FORM]

cribwork /ˈkrɪbwɜːk/ n. = CRIB n. 7.

Crichton /ˈkraɪt(ə)n/, James ('the Admirable Crichton') (1560–c.1585), Scottish adventurer. Crichton was an accomplished swordsman, poet, and scholar. He travelled a great deal in Europe, serving in the French army and later making a considerable impression on French and Italian universities with his intellect and skills as a polyglot orator.

Crick /krɪk/, Francis Harry Compton (b.1916), English biophysicist. Together with J. D. Watson he proposed the double helix structure of the DNA molecule, thus broadly explaining how genetic information is carried in living organisms and how genes replicate. He has since worked on the triplet code of bases in DNA, the processes of transcription into RNA and translation into amino acids, and the structure of other macromolecules and of viruses. He shared a Nobel Prize with Watson and M. H. F. Wilkins in 1962.

crick /krɪk/ n. & v. ● n. a sudden painful stiffness in the neck or the back etc. ● v.tr. produce a crick in (the neck etc.). [ME: orig. unkn.]

cricket¹ /ˈkrɪkɪt/ n. & v. ● n. an open-air game played on a pitch usu. of grass with ball, bats, and two wickets, between teams of eleven players each. (See note below.) ● v.intr. (**cricketed, cricketing**) play cricket. □ **cricket-bag** a long bag used for carrying a cricketer's bat, clothing, etc. **not cricket** Brit. colloq. underhand or dishonourable behaviour, infringing the tradition of fair play. □ **cricketer** n. [16th c.: orig. uncert.]

▪ Cricket was first played in England in Tudor times, and the laws were first drawn up in 1744. It has spread as a major sport in many of the territories formerly under British rule; full international or test matches are played between teams representing England, Australia, South Africa, the West Indies, New Zealand, India, Pakistan, Sri Lanka, and Zimbabwe. The object of the game is to score more runs than the opposition. Two members of the batting side are on the field at once, one or other of them defending his or her wicket against balls bowled (in sequences or overs of six) by a member of the fielding side and attempting to strike the ball out of reach of the fielders so as to score runs by running between the wickets before the ball is returned. A hit of the ball to the boundary scores four or six runs. Batsmen may be dismissed in several ways, of which the common ones are bowled, caught, run out, stumped, and leg before wicket (l.b.w.). The full game with two innings per side can last three, four, or five days, although shorter single-innings matches are usual at amateur level and have become popular at professional level since the 1960s.

cricket² /ˈkrɪkɪt/ n. a grasshopper-like insect of the order Orthoptera and esp. of the family Gryllidae, the male of which produces a chirping sound (traditionally regarded as cheerful), esp. (in full **house cricket**) Acheta domestica, formerly common indoors near fireplaces, ovens, etc. [ME f. OF criquet f. criquer creak etc. (imit.)]

cricoid /ˈkraɪkɔɪd/ adj. & n. ● adj. ring-shaped. ● n. (in full **cricoid cartilage**) Anat. the ring-shaped cartilage of the larynx. [mod.L cricoides f. Gk krikoeidēs f. krikos ring]

cri de cœur /ˌkriː də ˈkɜː(r)/ n. (pl. **cris de cœur** pronunc. same) a passionate appeal, complaint, or protest. [F, = cry from the heart]

cried past and past part. of CRY.

crier /ˈkraɪə(r)/ n. (also **cryer**) **1** a person who cries. **2** an officer who makes public announcements in a court of justice. □ **town** (or **common**) **crier** hist. an officer employed by a town council etc. to make public announcements in the streets or market-place. [ME f. AF criour, OF criere f. crier CRY]

crikey /ˈkraɪki/ int. Brit. sl. an expression of astonishment. [euphem. for CHRIST]

crim /krɪm/ n. & adj. Austral. sl. = CRIMINAL. [abbr.]

crime /kraɪm/ n. & v. ● n. **1 a** an act or omission regarded as a serious offence against an individual or the state, and punishable by law. **b** such illegal acts as a whole (*resorted to crime*). **2** an evil act (*a crime against humanity*). **3** colloq. a shameful act (*a crime to tease them*). **4** a soldier's offence against military regulations. ● v.tr. Mil. charge with or convict of an offence. □ **crime-sheet** Mil. a record of a defendant's offences. **crime wave** a sudden increase in crime. **crime-writer** a writer of detective fiction or thrillers. [ME f. OF f. L *crimen -minis* judgement, offence]

Crimea, the /kraɪˈmɪə/ (also **Crimea**) a peninsula of Ukraine lying between the Sea of Azov and the Black Sea. It was the scene of the Crimean War in the 1850s. The majority of the population is Russian. □ **Crimean** adj.

Crimean War /kraɪˈmɪən/ a war (1853–6) between Russia and an alliance of Great Britain, France, Sardinia, and Turkey. Russian aggression against Turkey led to war in 1853, with Turkey's European allies intervening to destroy Russian naval power in the Black Sea in 1854. The main theatre of the war was the Crimea, where an Anglo-French army eventually captured the fortress city of Sebastopol in 1855 after a lengthy siege; a peace treaty was signed early in 1856. Casualties were very heavy, and in Britain the war was chiefly remembered for the deficiencies in the British army's medical services exposed by the work of Florence Nightingale and others.

crime passionnel /ˌkriːm ˌpæsjɒˈnel/ n. (pl. **crimes passionnels** pronunc. same) a crime, esp. murder, committed in a fit of sexual jealousy. [F, = crime of passion]

criminal /ˈkrɪmɪn(ə)l/ n. & adj. ● n. a person who has committed a crime or crimes. ● adj. **1** of, involving, or concerning crime (*criminal records*). **2** having committed (and usu. been convicted of) a crime. **3** Law relating to or expert in criminal law rather than civil or political matters (*criminal code*; *criminal lawyer*). **4** colloq. scandalous, deplorable. □ **criminal law** law concerned with punishment of offenders (opp. civil law). **criminal libel** see LIBEL. □ **criminally** adv. **criminality** /ˌkrɪmɪˈnælɪtɪ/ n. [ME f. LL *criminalis* (as CRIME)]

criminalistic /ˌkrɪmɪnəˈlɪstɪk/ adj. relating to criminals or their habits.

criminalistics /ˌkrɪmɪnəˈlɪstɪks/ n.pl. esp. US forensic science.

criminalize /ˈkrɪmɪnəˌlaɪz/ v.tr. (also **-ise**) **1** turn (an activity) into a crime by making it illegal. **2** turn (a person) into a criminal, esp. by making his or her activities illegal. □ **criminalization** /ˌkrɪmɪnəlaɪˈzeɪʃ(ə)n/ n.

criminology /ˌkrɪmɪˈnɒlədʒɪ/ n. the branch of knowledge concerned with the study of crime. □ **criminologist** n. **criminological** /-nəˈlɒdʒɪk(ə)l/ adj. [L *crimen -minis* CRIME + -OLOGY]

crimp /krɪmp/ v. & n. ● v.tr. **1** compress into small folds or ridges; frill. **2** make narrow wrinkles or flutings in; corrugate. **3** make waves in (the hair) with a hot iron. ● n. a crimped thing or form. □ **put a crimp in** N. Amer. sl. thwart; interfere with. □ **crimper** n. **crimpy** adj. **crimpily** adv. **crimpiness** n. [ME, prob. ult. f. OHG *krimphan*]

Crimplene /ˈkrɪmpliːn/ n. propr. a synthetic crease-resistant fibre and fabric.

crimson /ˈkrɪmz(ə)n/ adj., n., & v. ● adj. of a rich deep red inclining to purple. ● n. this colour. ● v.tr. & intr. make or become crimson. [ME *cremesin*, *crimesin*, ult. f. Arab. *ķirmizī* KERMES]

cringe /krɪndʒ/ v. & n. ● v.intr. **1** shrink back in fear or apprehension; cower. **2** (often foll. by to) behave obsequiously. ● n. the act or an instance of cringing. □ **cringer** n. [ME *crenge*, *crenche*, OE *cringan*, *crincan*: see CRANK[1]]

cringle /ˈkrɪŋɡ(ə)l/ n. Naut. an eye of rope containing a thimble for another rope to pass through. [LG *kringel* dimin. of *kring* ring f. root of CRANK[1]]

crinkle /ˈkrɪŋk(ə)l/ n. & v. ● n. a wrinkle or crease in paper, cloth, etc. ● v. **1** intr. form crinkles. **2** tr. form crinkles in. □ **crinkle-cut** (of vegetables) cut into pieces with wavy edges. □ **crinkly** adj. [ME f. OE *crincan*: see CRANK[1]]

crinoid /ˈkrɪnɔɪd/ n. & adj. Zool. ● n. an echinoderm of the class Crinoidea, usu. sedentary and with feathery arms, e.g. sea lilies and feather stars. ● adj. of or relating to this class. □ **crinoidal** /krɪˈnɔɪd(ə)l/ adj. [Gk *krinoeidēs* lily-like f. *krinon* lily]

crinoline /ˈkrɪnəlɪn/ n. **1** a stiffened or hooped petticoat formerly worn to make a long skirt stand out. **2** a stiff fabric of horsehair etc. used for linings, hats, etc. [F f. L *crinis* hair + *linum* thread]

Crippen /ˈkrɪpɪn/, Hawley Harvey (known as Doctor Crippen) (1862–1910), American-born British murderer. Crippen poisoned his wife at their London home and sailed to Canada with his former secretary. His arrest in Canada was achieved through the intervention of radio-telegraphy, the first case of its use in apprehending a criminal. Crippen was later hanged.

cripple /ˈkrɪp(ə)l/ n. & v. ● n. a person who is permanently lame. ● v.tr. **1** make a cripple of; lame. **2** disable, impair. **3** weaken or damage (an institution, enterprise, etc.) seriously (*crippled by the loss of funding*). □ **crippledom** n. **crippler** n. [OE *crypel*, rel. to CREEP]

cris var. of KRIS.

crisis /ˈkraɪsɪs/ n. (pl. **crises** /-siːz/) **1 a** a decisive moment. **b** a time of danger or great difficulty. **2** the turning-point, esp. of a disease. [L f. Gk *krisis* decision f. *krinō* decide]

crisp /krɪsp/ adj., n., & v. ● adj. **1** hard but brittle. **2 a** (of air) bracing. **b** (of a style or manner) lively, brisk and decisive. **c** (of features etc.) neat and clear-cut. **d** (of paper) stiff and crackling. **e** (of hair) closely curling. ● n. (in full **potato crisp**) Brit. a thin fried slice of potato eaten as a snack or appetizer. ● v.tr. & intr. **1** make or become crisp. **2** curl in short stiff folds or waves. □ **burn to a crisp** make inedible or useless by burning. □ **crisply** adv. **crispness** n. [OE f. L *crispus* curled]

crispate /ˈkrɪspeɪt/ adj. **1** crisped. **2** Bot. & Zool. having a wavy margin. [L *crispare* curl]

crispbread /ˈkrɪspbred/ n. **1** a thin crisp biscuit of crushed rye etc. **2** these collectively (*a packet of crispbread*).

crisper /ˈkrɪspə(r)/ n. a compartment in a refrigerator for storing fruit and vegetables.

crispy /ˈkrɪspɪ/ adj. (**crispier**, **crispiest**) **1** crisp, brittle. **2** curly. **3** brisk. □ **crispiness** n.

criss-cross /ˈkrɪskrɒs/ n., adj., adv., & v. ● n. **1** a pattern of crossing lines. **2** the crossing of lines or currents etc. ● adj. crossing in cross lines (*criss-cross marking*). ● adv. crosswise; at cross purposes. ● v. **1** intr. **a** intersect repeatedly. **b** move crosswise. **2** tr. mark or make with a criss-cross pattern. [15th c., f. *Christ's cross*: later treated as redupl. of CROSS]

crista /ˈkrɪstə/ n. (pl. **cristae** /-tiː/) **1** Anat. & Zool. a ridge or crest. **2** Biol. an infold of the inner membrane of a mitochondrion. □ **cristate** /-teɪt/ adj. [L]

cristobalite /krɪˈstəʊbəˌlaɪt/ n. Mineral. a principal form of silica, occurring as opal. [G *Cristobalit* f. Cerro San *Cristóbal* in Mexico]

crit /krɪt/ n. colloq. **1** = CRITICISM 2. **2** = CRITIQUE. **3** Physics critical mass. [abbr.]

criterion /kraɪˈtɪərɪən/ n. (pl. **criteria** /-rɪə/) a principle or standard that a thing is judged by. □ **criterial** adj. [Gk *kritērion* means of judging (cf. CRITIC)]

critic /ˈkrɪtɪk/ n. **1** a person who censures. **2** a person who reviews or judges the merits of literary, artistic, or musical works etc., esp. regularly or professionally. **3** a person engaged in textual criticism. [L *criticus* f. Gk *kritikos* f. *kritēs* judge f. *krinō* judge, decide]

critical /ˈkrɪtɪk(ə)l/ adj. **1 a** making or involving adverse or censorious comments or judgements. **b** expressing or involving criticism. **2** skilful at or engaged in criticism. **3** providing textual criticism (*a critical edition of Milton*). **4 a** of or at a crisis; involving risk or suspense (*in a critical condition*; *a critical operation*). **b** decisive, crucial (*of critical importance*; *at the critical moment*). **5 a** Math. & Physics marking a transition from one state etc. to another (*critical angle*). **b** Physics (of a nuclear reactor) maintaining a self-sustaining chain reaction. □ **critical apparatus** = APPARATUS 4. **critical mass** Physics the amount of fissile material needed to maintain a nuclear chain reaction. **critical path** the sequence of stages determining the minimum time needed for an operation. **critical temperature** Chem. the temperature above which a gas cannot be liquefied by pressure. □ **critically** adv. **criticalness** n. **criticality** /ˌkrɪtɪˈkælɪtɪ/ n. (in sense 5). [L *criticus*: see CRITIC]

criticaster /ˌkrɪtɪˈkæstə(r)/ n. a minor or inferior critic.

criticism /ˈkrɪtɪˌsɪz(ə)m/ n. **1 a** finding fault; censure. **b** a statement or remark expressing this. **2 a** the work of a critic. **b** an article, essay, etc., expressing or containing an analytical evaluation of something. □ **the higher criticism** criticism dealing with the origin and character etc. of texts, esp. of biblical writings. **the lower criticism** textual criticism of the Bible. [CRITIC or L *criticus* + -ISM]

criticize /ˈkrɪtɪˌsaɪz/ v.tr. (also **-ise**) (also absol.) **1** find fault with; censure. **2** discuss critically. □ **criticizable** adj. **criticizer** n.

critique /krɪˈtiːk/ n. & v. ● n. a critical essay or analysis; an instance or the process of formal criticism. ● *v.tr.* (**critiques**, **critiqued**, **critiquing**) discuss critically. [F f. Gk *kritikē tekhnē* critical art]

critter /'krɪtə(r)/ *n.* **1** *dial.* or *joc.* a creature. **2** *derog.* a person. [var. of CREATURE]

croak /krəʊk/ *n. & v.* ● *n.* **1** a deep hoarse sound as of a frog or a raven. **2** a sound resembling this. ● *v.* **1** *intr.* utter a croak. **b** *tr.* utter with a croak or in a dismal manner. **2** *sl.* **a** *intr.* die. **b** *tr.* kill. [ME: imit.]

croaker /'krəʊkə(r)/ *n.* **1** an animal that croaks. **2** a prophet of evil.

croaky /'krəʊkɪ/ *adj.* (**croakier**, **croakiest**) (of a voice) croaking; hoarse. □ **croakily** *adv.*

Croat /'krəʊæt/ *n. & adj.* = CROATIAN. [mod.L *Croatae* f. Serbo-Croat *Hrvat*]

Croatia /krəʊ'eɪʃə/ (Croatian **Hrvatska** /'hrvɑːtskə/) a country in SE Europe, formerly a constituent republic of Yugoslavia; pop. (est. 1991) 4,760,000; language, Croatian; capital, Zagreb. The interior of Croatia, which extends as far east as the Danube, is separated from a long Adriatic coastline by the Dinaric Alps. The Croats migrated to this area in the 6th–7th centuries, later forming a kingdom which was united with the Hungarian crown in 1102. Apart from a period of Turkish rule in the 16th–17th centuries, Croatia largely remained linked with Hungary until 1918, when it joined the Kingdom of the Serbs, Croats, and Slovenes (later Yugoslavia). After a period in the Second World War as a Nazi puppet state (1941–5), Croatia became part of Yugoslavia once more and remained a constituent republic until it declared itself independent in 1991. The secession of Croatia led to conflict between Croats and the Serb minority.

Croatian /krəʊ'eɪʃ(ə)n/ *n. & adj.* ● *n.* **1 a** a native or national of Croatia. **b** a person of Croatian descent. **2** the language of the Croatians, a form of Serbo-Croat written in the Roman alphabet. (See also SERBO-CROAT.) ● *adj.* of or relating to Croatia, the Croatians, or their language.

croc /krɒk/ *n. colloq.* a crocodile. [abbr.]

Croce /'krəʊtʃeɪ/, Benedetto (1866–1952), Italian philosopher. The author of some seventy books, he founded and edited the influential review *Critica* (1903), in which he sought to revitalize Italian thought. In a series of works (1902–17) Croce presented his philosophical system ('Philosophy of the Spirit'), which is notable for its denial of the physical reality of a work of art and its identification of philosophical endeavour with a methodological approach to history. Croce served as Minister of Education 1920–1, but opposed Mussolini and all forms of totalitarianism; he returned to political life and helped to rebuild democracy in Italy after the fall of Mussolini.

croceate /'krəʊsɪˌeɪt/ *adj.* saffron-coloured. [L *croceus* f. CROCUS]

crochet /'krəʊʃeɪ, -ʃɪ/ *n. & v.* ● *n.* **1** a handicraft in which yarn is made up into a patterned fabric by means of a hooked needle. **2** work made in this way. ● *v.* (**crocheted** /-ʃeɪd/; **crocheting** /-ʃeɪɪŋ/) **1** *tr.* make by crocheting. **2** *intr.* do crochet. □ **crocheter** /-ʃeɪə(r)/ *n.* [F, dimin. of *croc* hook]

crocidolite /krəʊ'sɪdəˌlaɪt/ *n.* a fibrous blue or green silicate of iron and sodium; blue asbestos. [Gk *krokis -idos* nap of cloth]

crock¹ /krɒk/ *n. & v. colloq.* ● *n.* **1** an inefficient, broken-down, or worn-out person. **2** a worn-out vehicle, ship, etc. ● *v.* **1** *intr.* (foll. by *up*) break down, collapse. **2** *tr.* (often foll. by *up*) disable, cause to collapse. [orig. Sc., perh. f. Flem.]

crock² /krɒk/ *n.* **1** an earthenware pot or jar. **2** a broken piece of earthenware. [OE *croc(ca)*]

crockery /'krɒkərɪ/ *n.* earthenware or china dishes, plates, etc. [obs. *crocker* potter: see CROCK²]

crocket /'krɒkɪt/ *n. Archit.* a small carved ornament (usu. a bud or curled leaf) on the inclined side of a pinnacle etc. [ME f. var. of OF *crochet*: see CROCHET]

Crockett /'krɒkɪt/, David ('Davy') (1786–1836), American frontiersman, soldier, and politician. He was a member of the House of Representatives 1827–35 and cultivated the image of a rough backwoods legislator. On leaving politics he returned to the frontier, where he took up the cause of Texan independence and was killed at the siege of the Alamo.

Crockford /'krɒkfəd/ Crockford's Clerical Directory, a reference book of Anglican clergy in the British Isles first issued in 1860. It takes its name from John Crockford (1823–65), its first publisher.

crocodile /'krɒkəˌdaɪl/ *n.* **1 a** a large crocodilian of the family Crocodylidae, esp. the genus *Crocodylus*, found in tropical and subtropical regions. Its long snout is narrower than that of the alligator, and when the jaws are closed the fourth tooth on each side of the lower jaw projects outside the snout. **b** leather from its skin, used to make bags, shoes, etc. **2** *Brit. colloq.* a line of schoolchildren etc. walking in pairs. □ **crocodile clip** a clip with teeth for gripping. **crocodile tears** insincere grief (from the belief that crocodiles wept while devouring or alluring their prey). [ME f. OF *cocodrille* f. med.L *cocodrillus* f. L *crocodilus* f. Gk *krokodilos* f. *krokē* pebble + *drilos* worm]

crocodilian /ˌkrɒkə'dɪlɪən/ *n. & adj.* ● *n.* a large predatory amphibious reptile of the order Crocodilia, which includes crocodiles, alligators, caymans, and gharials. They have a long snout, strong jaws, short legs with claws and webbed toes, and a heavy powerful tail. ● *adj.* of or relating to the crocodilians.

crocus /'krəʊkəs/ *n.* (*pl.* **crocuses** or **croci** /-kaɪ/) **1** a small spring-flowering plant of the genus *Crocus*, of the iris family, growing from a corm and having bright yellow, purple, or white flowers. **2** (in full **autumn crocus**) the meadow saffron. [ME, = saffron, f. L f. Gk *krokos* crocus, of Semitic orig.]

Croesus /'kriːsəs/ (6th century BC), last king of Lydia *c*.560–546 BC. He subjugated the Greek cities on the coast of Asia Minor, and was proverbial for his great wealth. His empire, with its capital at Sardis, was overthrown by the Persian king Cyrus the Great. At this point Croesus' fate becomes the theme of legend; Cyrus is said to have cast him on a pyre from which he was saved by the miraculous intervention of the god Apollo.

croft /krɒft/ *n. & v. Brit.* ● *n.* **1** an enclosed piece of (usu. arable) land. **2** a small rented farm in Scotland or northern England. ● *v.intr.* farm a croft; live as a crofter. [OE: orig. unkn.]

crofter /'krɒftə(r)/ *n. Brit.* a person who rents a smallholding, esp. a joint tenant of a divided farm in parts of Scotland.

Crohn's disease /krəʊnz/ *n. Med.* a chronic inflammatory disease of the intestines, esp. the colon and ileum, causing ulcers and fistulae. [Burrill B. *Crohn*, Amer. pathologist (1884–1983)]

croissant /'krwʌsɒn/ *n.* a crescent-shaped roll made of rich yeast pastry. [F, formed as CRESCENT]

Cro-Magnon man /krəʊ'mænjɒn, -'mægnən/ *n.* the earliest form of modern human, associated with the Aurignacian flint industry. Remains were first found in a rock shelter in the Cro-Magnon hill near Les Eyzies, Dordogne, France, in 1868, and were the first evidence of modern humans existing in the upper palaeolithic. Their appearance in western Europe heralded the apparent decline and disappearance of the existing Neanderthal populations and their middle palaeolithic industries. The group persisted into the neolithic period, and some authorities consider that it survived to historic times in the earliest inhabitants (now extinct) of the Canary Islands. (See also HOMO.)

Cromarty Firth an inlet of the Moray Firth on the coast of Highland Region, northern Scotland. The shipping forecast area *Cromarty* extends far beyond this, covering Scottish coastal waters roughly from Aberdeen in the south to John o'Groats in the north.

Crome /krəʊm/, John (1768–1821), English painter. Founder and leading member of the Norwich School, he was influenced by Dutch artists such as Hobbema and Ruisdael. He later developed a distinctive romantic style of his own, exemplified in such landscapes as *Slate Quarries* and *Moonrise on the Marshes of the Yare* (both undated).

cromlech /'krɒmlek/ *n.* **1** (esp. in Wales) a dolmen; a megalithic tomb. **2** (in Brittany) a stone circle. [Welsh f. *crom* fem. of *crwm* arched + *llech* flat stone, in sense 2 via F f. Breton *krommlec'h*]

Crompton¹ /'krɒmptən/, Richmal (pseudonym of Richmal Crompton Lamburn) (1890–1969), English writer. A classics teacher who wrote stories for magazines, she made her name with *Just William* (1922), a collection of stories for children about a mischievous schoolboy, William Brown. She published a further thirty-seven collections based on the same character, as well as some fifty books for adults, during her writing career.

Crompton² /'krɒmptən/, Samuel (1753–1827), English inventor. Famed for his invention of the spinning mule, he lacked the means to obtain a patent and sold his rights to a Bolton industrialist for £67. The House of Commons subsequently gave him £5,000 in compensation.

Cromwell¹ /'krɒmwel/, Oliver (1599–1658), English general and statesman. He was the driving force in the revolutionary opposition to Charles I in the English Civil War, and was the leader of the Parliamentary forces (or Roundheads), winning decisive battles at Marston Moor and Naseby. After helping to arrange the trial and execution of Charles I, he returned to military command to suppress resistance to the Commonwealth in Ireland and Scotland, finally defeating a Scottish army at Worcester (1651). He styled himself Lord

Protector of the Commonwealth (1653–8); although he called and dissolved a succession of Parliaments, he refused Parliament's offer of the crown in 1657. His rule was notable for its puritan reforms in the Church of England and for the establishment of the Commonwealth as the major Protestant power in the world.

Cromwell[2] /'krɒmwel/, Thomas (*c.*1485–1540), English statesman, chief minister to Henry VIII 1531–40. After serving Cardinal Wolsey from 1514, he succeeded him as the king's chief adviser. He presided over the king's divorce from Catherine of Aragon (1533) and his break with the Roman Catholic Church, as well as the dissolution of the monasteries and a series of administrative measures, such as the Act of Supremacy (1534), designed to strengthen the Crown. He fell from favour over Henry's marriage to Anne of Cleves and was executed on a charge of treason.

crone /krəʊn/ *n.* **1** a withered old woman. **2** an old ewe. [ME, ult. f. ONF *carogne* CARRION]

Cronin /'krəʊnɪn/, A(rchibald) J(oseph) (1896–1981), Scottish novelist. His novels often reflect his early experiences as a doctor; they include *The Citadel* (1937), telling of the struggles of an idealistic young doctor, and *The Stars Look Down* (1935), about a mining community. Cronin's Scottish medical stories were successfully adapted for radio and television as *Dr Finlay's Casebook* in the 1960s and 1990s.

cronk /krɒŋk/ *adj. Austral. colloq.* **1** unsound; liable to collapse. **2 a** fraudulent. **b** (of a horse) dishonestly run, unfit. [19th c.: cf. CRANK[3]]

Cronus /'krɒnəs, 'krəʊn-/ (also **Kronos**) *Gk Mythol.* the youngest son of Uranus (Heaven) and Gaia (Earth) and leader of his brothers, the Titans. He overthrew and castrated his father, and then married his sister Rhea, who gave birth to many of the future gods, including Zeus. Because he was fated to be overcome by one of his male children, Cronus swallowed all of them as soon as they were born, but when Zeus was born, Rhea deceived him with a stone wrapped in swaddling clothes and hid the baby away in Crete. Cronus swallowed the stone, and Zeus eventually dethroned him as ruler of the universe.

crony /'krəʊnɪ/ *n.* (*pl.* **-ies**) (often *derog.*) a close friend or companion. [17th-c. *chrony*, university sl. f. Gk *khronios* long-standing f. *khronos* time]

crook /krʊk/ *n., v., & adj.* ● *n.* **1** the hooked staff of a shepherd or bishop. **2 a** a bend, curve, or hook. **b** anything hooked or curved. **3** *colloq.* **a** a rogue; a swindler. **b** a professional criminal. ● *v.tr. & intr.* bend, curve. ● *adj.* **1** crooked. **2** *Austral. & NZ colloq.* **a** unsatisfactory, out of order; (of a person) unwell, injured. **b** unpleasant. **c** dishonest, unscrupulous. **d** bad-tempered, irritable, angry. □ **crook-back** a hunchback. **crook-backed** hunchbacked. **go crook** (usu. foll. by *at, on*) *Austral. & NZ colloq.* lose one's temper; become angry. □ **crookery** *n.* [ME f. ON *krókr* hook]

crooked /'krʊkɪd/ *adj.* (**crookeder, crookedest**) **1 a** not straight or level; bent, curved, twisted. **b** deformed, bent with age. **2** *colloq.* not straightforward; dishonest. **3** /krʊkt/ *Austral. & NZ sl.* = CROOK *adj.* 2. **4** (foll. by *on*) *Austral. sl.* hostile to. □ **crookedly** *adv.* **crookedness** *n.* [ME f. CROOK, prob. after ON *krókóttr*]

Crookes /krʊks/, Sir William (1832–1919), English physicist and chemist. Crookes combined private experimental research with business; he also edited several photographic and scientific journals. In 1861, shortly after the spectroscopic discoveries of Robert Bunsen and Gustav Kirchhoff, he discovered the element thallium. This led him indirectly to the invention of the radiometer in 1875. He later developed a vacuum tube (the precursor of the X-ray tube) and in 1903 invented the spinthariscope. His interest in spiritualism and psychic research led him into controversy.

croon /kru:n/ *v. & n.* ● *v.tr. & intr.* hum or sing in a low subdued voice, esp. in a sentimental manner. ● *n.* such singing. □ **crooner** *n.* [ME (orig. Sc. & N.Engl.) f. MDu. & MLG *krōnen* groan, lament]

crop /krɒp/ *n. & v.* ● *n.* **1 a** the produce of cultivated plants, esp. cereals. **b** the season's total yield of this (*a good crop*). **2** a group or an amount produced or appearing at one time (*this year's crop of students*). **3** (in full **hunting crop**) the stock or handle of a whip. **4 a** a style of hair cut very short. **b** the cropping of hair. **5** *Zool.* **a** the thin-walled pouch in a bird's gullet, where food is stored or prepared for digestion before it is passed to the stomach. **b** a similar organ in other animals. **6** the entire tanned hide of an animal. **7** a piece cut off or out of something. ● *v.* (**cropped, cropping**) **1** *tr.* **a** cut off. **b** (of animals) bite off and eat (the tops of plants). **2** *tr.* cut (hair, cloth, edges of a book, etc.) short. **3** *tr.* gather or reap (produce). **4** *tr.* (foll. by *with*) sow or plant (land) with a crop. **5** *intr.* (of land) bear a crop. □ **crop-dusting** the spraying of powdered insecticide or fertilizer on crops, esp. from the air. **crop-eared** having the ears (esp. of animals) or hair cut short. **crop-full** having a full crop or stomach. **crop out** *Geol.* appear at the surface.

crop-over a West Indian celebration marking the end of the sugar-cane harvest. **crop up 1** (of a subject, circumstance, etc.) appear or come to one's notice unexpectedly. **2** *Geol.* appear at the surface. [OE *crop(p)*]

crop circle *n.* a circular area in a standing crop, esp. wheat, where the plants have been flattened to the ground, sometimes associated with straight lines and complex patterns. Suggested causes have ranged from whirlwinds and fungi to alien spaceships, though it now seems that many were made by hoaxers.

cropper /'krɒpə(r)/ *n.* a crop-producing plant of specified quality (*a good cropper; a heavy cropper*). □ **come a cropper** *sl.* **1** fall heavily. **2** fail badly.

croquet /'krəʊkeɪ, -kɪ/ *n. & v.* ● *n.* **1** a game played on a lawn, with wooden balls which are driven through a series of square-topped hoops by mallets. It owes its origin to *paille maille* or *Pell-Mell*, a game known in France from the 16th century. **2** the act of croqueting a ball. ● *v.tr.* (**croqueted** /-keɪd/; **croqueting** /-keɪɪŋ/) drive away (one's opponent's ball in croquet) by placing one's own against it and striking one's own. [perh. dial. form of F CROCHET hook]

croquette /krə'ket/ *n.* a fried breaded roll or ball of mashed potato or minced meat etc. [F f. *croquer* crunch]

crore /krɔ:(r)/ *n. Ind.* **1** ten million. **2** one hundred lakhs (of rupees, units of measurement, persons, etc.). [Hindi *k(a)rōr*, ult. f. Skr. *koṭi* apex]

Crosby /'krɒzbɪ/, Bing (born Harry Lillis Crosby) (1904–77), American singer and actor. His songs include 'Pennies from Heaven', 'Blue Skies', and in particular 'White Christmas' (from the film *Holiday Inn*, 1942), which has sold more than 30 million copies. He also starred in the series of *Road* films (1940–62) with Bob Hope and Dorothy Lamour (b.1914).

crosier /'krəʊzɪə(r), 'krəʊʒə(r)/ *n.* (also **crozier**) **1** a hooked staff carried by a bishop as a symbol of pastoral office. **2** a crook. [orig. = bearer of a crook, f. OF *crocier* & OF *croisier* f. *crois* CROSS]

cross /krɒs/ *n., v., & adj.* ● *n.* **1** an upright post with a transverse bar, as used in antiquity for crucifixion. **2 a** (**the Cross**) in Christianity, the cross on which Christ was crucified. **b** a representation of this as an emblem of Christianity. **c** = *sign of the cross.* **3** a staff surmounted by a cross and borne before an archbishop or in a religious procession. **4 a** a thing or mark shaped like a cross, esp. a figure made by two short intersecting lines (+ or x). **b** a monument in the form of a cross, esp. one in the centre of a town or on a tomb. **5** a cross-shaped decoration indicating rank in some orders of knighthood or awarded for personal valour. **6 a** an intermixture of animal breeds or plant varieties. **b** an animal or plant resulting from this. **7** (foll. by *between*) a mixture or compromise of two things. **8 a** a crosswise movement, e.g. of an actor on stage. **b** *Football* etc. a pass of the ball across the field of play, esp. from a wing to the centre. **c** *Boxing* a blow with a crosswise movement of the fist. **9** a trial or affliction; something to be endured (*bear one's crosses*). ● *v.* **1** *tr.* (often foll. by *over*; also *absol.*) go across or to the other side of (a road, river, sea, etc.). **2 a** *intr.* intersect or be across one another (*the roads cross near the bridge*). **b** *tr.* cause to do this; place crosswise (*cross one's legs*). **3** *tr.* **a** draw a line or lines across. **b** *Brit.* mark (a cheque) with two parallel lines, and often an annotation, to indicate that it must be paid into a named bank account. **4** *tr.* (foll. by *off, out, through*) cancel or obliterate or remove from a list with lines drawn across. **5** *tr.* (often *refl.*) make the sign of the cross on or over. **6** *intr.* **a** pass in opposite or different directions. **b** (of letters between two correspondents) each be dispatched before receipt of the other. **c** (of telephone lines) become wrongly interconnected so that intrusive calls can be heard. **7** *tr.* **a** cause to interbreed. **b** cross-fertilize (plants). **8** *tr.* thwart or frustrate (*crossed in love*). **9** *tr. sl.* cheat. **10** *tr. & intr. Football* etc. pass (the ball) across the field of play, esp. from a wing to the centre. ● *adj.* **1** (often foll. by *with*) peevish, angry. **2** (usu. *attrib.*) transverse; reaching from side to side. **3** (usu. *attrib.*) intersecting. **4** (usu. *attrib.*) contrary, opposed, reciprocal. □ **as cross as two sticks** extremely angry or peevish. **at cross purposes** misunderstanding or conflicting with one another. **cross one's fingers** (or **keep one's fingers crossed**) **1** put one finger across another as a sign of hoping for good luck. **2** trust in good luck. **cross the floor** join the opposing side in a debating-assembly. **cross one's heart** make a solemn pledge, esp. by crossing one's front. **cross one's mind** (of a thought etc.) occur to one, esp. transiently. **cross a person's palm** (usu. foll. by *with*) pay a person for a favour. **cross the path of 1** meet with (a person). **2** thwart. **cross swords** (often foll. by *with*) encounter in opposition; have an argument or dispute. **cross wires** (or **get one's wires crossed**) **1** become wrongly connected by telephone. **2** have a misunderstanding. **on the**

cross 1 diagonally. **2** *sl.* fraudulently, dishonestly. □ **crossly** *adv.* **crossness** *n.* [OE *cros* f. ON *kross* f. OIr. *cros* f. L *crux cruc-*]

cross- /krɒs/ *comb. form* **1** denoting movement or position across something (*cross-channel*; *cross-country*). **2** denoting interaction (*cross-breed*; *cross-cultural*; *cross-fertilize*). **3 a** passing from side to side; transverse (*crossbar*; *cross-current*). **b** having a transverse part (*crossbow*). **4** describing the form or figure of a cross (*cross-keys*; *crossroads*).

crossbar /ˈkrɒsbɑː(r)/ *n.* a horizontal bar, esp. held on a pivot or between two upright bars etc., e.g. of a bicycle or of a football goal.

cross-bedding /ˈkrɒsˌbedɪŋ/ *n. Geol.* lines of stratification crossing the main rock strata. Also called *false bedding*.

cross-bench /ˈkrɒsbentʃ/ *n. Brit.* a seat in Parliament (now only the House of Lords) occupied by a member not taking the whip from a political party. □ **cross-bencher** *n.*

crossbill /ˈkrɒsbɪl/ *n.* a finch of the genus *Loxia*, having a bill with crossed mandibles with which it opens pine cones.

crossbones /ˈkrɒsbəʊnz/ *n.* a representation of two crossed thigh-bones. □ **skull and crossbones** see SKULL.

crossbow /ˈkrɒsbəʊ/ *n. esp. hist.* a weapon for shooting bolts, consisting of a bow fixed across a usu. wooden stock, with a groove for the bolt and a mechanism for drawing and releasing the string. □ **crossbowman** *n.* (*pl.* **-men**).

cross-breed *n. & v.* ● *n.* /ˈkrɒsbriːd/ **1** a breed of animals or plants produced by crossing. **2** an individual animal or plant of a cross-breed. ● *v.tr.* /krɒsˈbriːd/ (*past and past part.* **-bred** /-ˈbred/) produce by crossing.

cross-check *v. & n.* ● *v.tr.* /krɒsˈtʃek/ check by a second or alternative method, or by several methods. ● *n.* /ˈkrɒstʃek/ an instance of cross-checking.

cross-country /krɒsˈkʌntri/ *adj., adv., & n.* ● *adj. & adv.* **1** across fields or open country. **2** not keeping to main or direct roads. ● *n.* (*pl.* **-ies**) a cross-country race.

cross-cut *adj. & n.* ● *adj.* /krɒsˈkʌt/ cut across the main grain or axis. ● *n.* /ˈkrɒskʌt/ a diagonal cut, path, etc. □ **cross-cut saw** a saw for cutting the grain of wood.

cross-dating /krɒsˈdeɪtɪŋ/ *n. Archaeol.* dating by correlation with another site or level.

crosse /krɒs/ *n.* a stick with a triangular net at the end for conveying the ball in lacrosse. [F f. OF *croce, croc* hook]

cross-examine /ˌkrɒsɪɡˈzæmɪn/ *v.tr.* examine (esp. a witness in a lawcourt) to check or extend testimony already given. □ **cross-examiner** *n.* **cross-examination** /-ˌzæmɪˈneɪʃ(ə)n/ *n.*

cross-eyed /krɒsˈaɪd/ *adj.* (as a disorder) having one or both eyes turned permanently inwards towards the nose.

cross-fade /krɒsˈfeɪd/ *v. & n. Radio etc.* ● *v. intr.* /krɒsˈfeɪd/ fade in one sound as another is faded out. ● *n.* /ˈkrɒsfeɪd/ an act of cross-fading.

cross-fertilize /krɒsˈfɜːtɪˌlaɪz/ *v.tr.* (also **-ise**) **1** fertilize (an animal or plant) from another of the same species. **2** improve or strengthen by the interchange of ideas etc. □ **cross-fertilization** /-ˌfɜːtɪlaɪˈzeɪʃ(ə)n/ *n.*

crossfire /ˈkrɒsˌfaɪə(r)/ *n.* **1** firing in two crossing directions simultaneously. **2 a** attack or criticism from several sources at once. **b** a lively or combative exchange of views etc.

cross-grain /ˈkrɒsɡreɪn/ *n.* a grain in timber, running across the regular grain.

cross-grained /krɒsˈɡreɪnd/ *adj.* **1** (of timber) having a cross-grain. **2** perverse, intractable.

cross-hair /ˈkrɒshpeə(r)/ *n.* a fine wire at the focus of an optical instrument for use in measurement.

cross-hatch /krɒsˈhætʃ/ *v.tr.* shade with intersecting sets of parallel lines.

cross-head /ˈkrɒshed/ *n.* **1** a bar between the piston-rod and connecting-rod in a steam engine. **2** = CROSS-HEADING.

cross-heading /ˈkrɒsˌhedɪŋ/ *n.* a heading to a paragraph printed across a column in the body of an article in a newspaper etc.

crossing /ˈkrɒsɪŋ/ *n.* **1** a place where things (esp. roads) cross. **2** a place at which one may cross a street etc. (*pedestrian crossing*). **3** a journey across water (*had a smooth crossing*). **4** the intersection of a church nave and transepts. **5** *Biol.* mating. □ **crossing over** *Biol.* an exchange of genes between homologous chromosomes (cf. RECOMBINATION).

cross-legged /krɒsˈleɡd, -ˈleɡɪd/ *adj.* with one leg crossed over the other.

cross-link /ˈkrɒslɪŋk/ *n.* (also **cross-linkage**) *Chem.* a bond between chains of atoms in a polymer etc.

crossmatch /krɒsˈmætʃ/ *v.tr. Med.* test the compatibility of (a donor's and a recipient's blood or tissues). □ **crossmatching** *n.*

crossover /ˈkrɒsˌəʊvə(r)/ *n. & adj.* ● *n.* **1** a point or place of crossing from one side to the other. **2** the process of crossing over, esp. from one style or genre to another. ● *adj.* **1** having a part that crosses over. **2** that crosses over, esp. from one style or genre to another.

crosspatch /ˈkrɒspætʃ/ *n. colloq.* a bad-tempered person. [CROSS *adj.* 1 + obs. *patch* fool, clown]

crosspiece /ˈkrɒspiːs/ *n.* a transverse beam or other component of a structure etc.

cross-ply /ˈkrɒsplaɪ/ *adj.* (of a tyre) having fabric layers with cords lying crosswise.

cross-pollinate /krɒsˈpɒlɪˌneɪt/ *v.tr.* pollinate (a plant) with pollen from another plant. □ **cross-pollination** /-ˌpɒlɪˈneɪʃ(ə)n/ *n.*

cross-question /krɒsˈkwestʃən/ *v.tr.* = CROSS-EXAMINE.

cross-refer /ˌkrɒsrɪˈfɜː(r)/ *v.intr.* (**-referred, -referring**) refer from one part of a book, article, etc., to another.

cross-reference /krɒsˈrefərəns/ *n. & v.* ● *n.* a reference from one part of a book, article, etc., to another. ● *v.tr.* provide with cross-references.

crossroad /ˈkrɒsrəʊd/ *n.* **1** (usu. in *pl.*) an intersection of two or more roads. **2** *US* a road that crosses a main road or joins two main roads. □ **at the crossroads** at a critical point in one's life.

cross-ruff /ˈkrɒsrʌf/ *n. & v. Bridge etc.* ● *n.* the alternate trumping of partners' leads. ● *v.intr.* play in this way.

cross-section /ˈkrɒsˌsekʃən/ *n.* **1 a** a cutting of a solid at right angles to an axis. **b** a plane surface produced in this way. **c** a representation of this. **2** a representative sample, esp. of people. **3** *Physics* a quantity, expressed as an area, representing the probability of interaction between subatomic particles. □ **cross-sectional** *adj.*

cross stitch /ˈkrɒsstɪtʃ/ *n.* **1** a stitch formed of two stitches crossing each other. **2** needlework done using this stitch. ● *v.tr.* (**cross-stitch**) sew or embroider with cross stitches.

crosstalk /ˈkrɒstɔːk/ *n.* **1** unwanted transfer of signals between communication channels. **2** *Brit.* witty talk; repartee.

cross-trees /ˈkrɒstriːz/ *n.pl. Naut.* a pair of horizontal timbers at the top of a lower mast, supporting the topmast.

cross-voting /ˈkrɒsˌvəʊtɪŋ/ *n.* voting for a party not one's own, or for more than one party.

crosswalk /ˈkrɒswɔːk/ *n. N. Amer. & Austral.* a pedestrian crossing.

crossways /ˈkrɒsweɪz/ *adv.* = CROSSWISE.

crosswind /ˈkrɒswɪnd/ *n.* a wind blowing across one's direction of travel.

crosswise /ˈkrɒswaɪz/ *adj. & adv.* **1** in the form of a cross; intersecting. **2** transverse or transversely.

crossword /ˈkrɒswɜːd/ *n.* (also **crossword puzzle**) a puzzle consisting of a grid of squares and blanks into which words crossing vertically and horizontally have to be filled from clues. Invention of the crossword is attributed to a journalist, Arthur Wynne, whose puzzle (called a 'word-cross') appeared in a Sunday newspaper, the *New York World*, on 21 Dec. 1913.

crostini /krɒˈstiːni/ *n.pl.* small pieces of toasted or fried bread served with a topping as a starter. [It., pl. of *crostino* little crust]

crotch /krɒtʃ/ *n.* a place where something forks, esp. the legs of the human body or a garment (cf. CRUTCH). [perh. = ME & OF *croc(he)* hook, formed as CROOK]

crotchet /ˈkrɒtʃɪt/ *n.* **1** *Mus.* a note having the time value of a quarter of a semibreve and usu. representing one beat, drawn as a large dot with a stem. Also called (*US*) *quarter note*. **2** a whimsical fancy. **3** a hook. [ME f. OF *crochet* dimin. of *croc* hook (see CROTCH)]

crotchety /ˈkrɒtʃɪti/ *adj.* peevish, irritable. □ **crotchetiness** *n.* [CROTCHET 2 + -Y[1]]

croton /ˈkrəʊt(ə)n/ *n.* **1** a strong-scented herbaceous plant, shrub, or tree of the genus *Croton*, of the spurge family, producing a capsule-like fruit. **2** an evergreen tree or shrub of the genus *Codiaeum*, of the spurge family; esp. *C. variegatum*, with glossy coloured leaves. □ **croton oil** a powerful purgative obtained from the fruit of *Croton tiglium*. [mod.L f. Gk *krotōn* sheep-tick, croton (from the shape of its seeds)]

crouch /kraʊtʃ/ *v. & n.* ● *v.intr.* (often foll. by *down*) lower the body with the limbs close to the chest, esp. for concealment, or (of an animal) before pouncing; be in this position. ● *n.* an act of crouching; a crouching position. [ME, perh. f. OF *crochir* be bent f. *croc* hook: cf. CROOK]

croup[1] /kruːp/ *n.* an inflammation of the larynx and trachea in children, with a hard cough and difficulty in breathing. □ **croupy** *adj.* [*croup* to croak (imit.)]

croup[2] /kruːp/ *n.* the rump or hindquarters esp. of a horse. [ME f. OF *croupe*, rel. to CROP]

croupier /ˈkruːpɪə(r), -pɪˌeɪ/ *n.* **1** the person in charge of a gaming-table, raking in and paying out money etc. **2** the assistant chairperson at a public dinner, seated at the foot of the table. [F, orig. = rider on the croup: see CROUP[2]]

crouton /ˈkruːtɒn/ *n.* a small piece of fried or toasted bread served with soup or used as a garnish. [F *croûton* f. *croûte* CRUST]

crow[1] /krəʊ/ *n.* **1** a large bird of the genus *Corvus*, with a powerful beak, harsh call, and usu. glossy black plumage; esp. (in full **carrion crow**) the common *C. corone* of Europe. **2** any bird of the family Corvidae, which includes the magpie, jays, etc., as well as the black-plumaged members (raven, rook, jackdaw, and chough). **3** *sl. derog.* a woman, esp. an old or ugly one. □ **as the crow flies** in a straight line. **crow-bill** a forceps for extracting bullets etc. **crow's-foot** (*pl.* **-feet**) **1** (usu. in *pl.*) a wrinkle at the outer corner of a person's eye. **2** *Mil.* a caltrop. **crow's-nest** a barrel or platform fixed at the masthead of a sailing vessel as a shelter for a lookout man. **crow steps** corbie-steps. **eat crow** *N. Amer.* submit to humiliation. [OE *crāwe* ult. f. WG]

crow[2] /krəʊ/ *v.* ● *v.intr.* **1** (*past* **crowed** or **crew** /kruː/) (of a cock) utter its characteristic loud cry. **2** (of a baby) utter happy cries. **3** (usu. foll. by *over*) express unrestrained gleeful satisfaction. ● *n.* **1** the cry of a cock. **2** a happy cry of a baby. [OE *crāwan*, of imit. orig.]

crowbar /ˈkrəʊbɑː(r)/ *n.* an iron bar with a flattened end, used as a lever.

crowberry /ˈkrəʊbərɪ/ *n.* (*pl.* **-ies**) **1 a** an evergreen ericaceous dwarf shrub, *Empetrum nigrum*, bearing black berries. **b** the flavourless edible berry of this plant. **2** *US* a cranberry.

crowd /kraʊd/ *n. & v.* ● *n.* **1** a large number of people gathered together, usu. without orderly arrangement. **2** a mass of spectators; an audience. **3** *colloq.* a particular company or set of people (*met the crowd from the sales department*). **4** (prec. by *the*) the mass or multitude of people (*go along with the crowd*). **5** a large number (of things). **6** actors representing a crowd. ● *v.* **1 a** *intr.* come together in a crowd. **b** *tr.* cause to do this. **c** *intr.* force one's way. **2** *tr.* **a** (foll. by *into*) force or compress into a confined space. **b** (often foll. by *with*; usu. in *passive*) fill or make abundant with (*was crowded with tourists*). **3** *tr.* **a** (of a number of people) come aggressively close to. **b** *colloq.* harass or pressure (a person). □ **crowd out** exclude by crowding. □ **crowdedness** *n.* [OE *crūdan* press, drive]

crowfoot /ˈkrəʊfʊt/ *n.* a aquatic plant of the genus *Ranunculus*, with white buttercup-like flowers held above the water.

crown /kraʊn/ *n. & v.* ● *n.* **1** a monarch's ornamental and usu. jewelled head-dress. **2** (**the Crown**) **a** the monarch, esp. as head of state. **b** the power or authority residing in the monarchy. **3 a** a wreath of leaves or flowers etc. worn on the head, esp. as an emblem of victory. **b** an award or distinction gained by a victory or achievement, esp. in sport. **4** a crown-shaped thing, esp. a device or ornament. **5** the top part of a thing, esp. of the head or a hat. **6 a** the highest or central part of an arched or curved thing (*crown of the road*). **b** a thing that completes or forms the summit. **7** the part of a plant just above and below the ground. **8** the upper part of a cut gem above the girdle. **9 a** the part of a tooth projecting from the gum. **b** an artificial replacement or covering for this. **10 a** a former British coin equal to five shillings (25p). **b** a foreign coin with a name meaning 'crown', esp. the krona or krone. **11** a former size of paper, 504 x 384 mm. ● *v.tr.* **1** put a crown on (a person or a person's head). **2** invest (a person) with a royal crown or authority. **3** be a crown to; encircle or rest on the top of. **4 a** (often as **crowning** *adj.*) be or cause to be the consummation, reward, or finishing touch to (*the crowning glory*). **b** bring (efforts) to a happy issue. **5** fit a crown to (a tooth). **6** *sl.* hit on the head. **7** *Draughts* promote (a piece in draughts) to king. □ **crown cap** a cork-lined metal cap for a bottle. **crown glass** glass which contains no lead or iron, formerly blown for use in windows and now used as an optical glass of low dispersion. **crown green** a kind of bowling-green rising towards the middle. **crown imperial** a tall fritillary, *Fritillaria imperialis*, with a flower-cluster at the top of the stalk. **crown jewels** the regalia and other jewellery worn by the sovereign on certain state occasions. **crown of thorns** a large spiky starfish of the genus *Acanthaster*, feeding on coral. **Crown prince** a male heir to a sovereign throne. **Crown princess 1** the wife of a Crown prince. **2** a female heir to a

sovereign throne. **crown roast** a roast of rib-pieces of pork or lamb arranged like a crown. **crown saw** a cylinder with a toothed edge for making a circular hole. **crown wheel** a wheel with teeth set at right angles to its plane, esp. in the gears of motor vehicles. [ME f. AF *corune*, OF *corone* f. L *corona*]

Crown Colony *n.* a British colony controlled by the Crown.

Crown Court *n.* a court of criminal jurisdiction in England and Wales.

Crown Derby *n.* a soft-paste porcelain made at Derby and often marked with a crown above the letter 'D'.

Crown Office (in the UK) an office of the Supreme Court transacting common-law business of Chancery.

Crozet Islands /krəʊˈzeɪ/ a group of five small islands in the southern Indian Ocean, under French administration.

crozier var. of CROSIER.

CRT *abbr.* cathode-ray tube.

cru /kruː/ *n.* **1** a French vineyard or wine-producing region. **2** the grade of wine produced from it. [F f. *crû* grown]

cruces *pl.* of CRUX.

crucial /ˈkruːʃl/ *adj.* **1** decisive, critical. **2** *colloq. disp.* very important. **3** *sl.* excellent. □ **crucially** *adv.* **cruciality** /ˌkruːʃɪˈælɪtɪ/ *n.* (*pl.* **-ies**). [F f. L *crux crucis* cross]

crucian /ˈkruːʃn/ *n.* (more fully **crucian carp**) a yellow cyprinoid fish, *Carassius carassius*, allied to the goldfish. [LG *karusse* etc.]

cruciate /ˈkruːʃɪeɪt/ *adj. esp. Anat.* cross-shaped. □ **cruciate ligament** either of a pair of ligaments in the knee which cross each other and connect the femur and tibia. [mod.L *cruciatus* f. L (as CRUCIBLE)]

crucible /ˈkruːsɪb(ə)l/ *n.* **1** a melting-pot for metals etc. **2** a severe test or trial. [ME f. med.L *crucibulum* night-lamp, crucible, f. L *crux crucis* cross]

crucifer /ˈkruːsɪfə(r)/ *n.* **1** *Bot.* a cruciferous plant. **2** a person carrying a processional cross or crucifix.

cruciferous /kruːˈsɪfərəs/ *adj. Bot.* of the plant family Cruciferae, having flowers with four petals arranged in a cross. The family includes many vegetables, such as the brassicas and mustards. [LL *crucifer* (as CRUCIAL, -FEROUS)]

crucifix /ˈkruːsɪfɪks/ *n.* a model or image of a cross with a figure of Christ on it. [ME f. OF f. eccl.L *crucifixus* f. L *cruci fixus* fixed to a cross]

crucifixion /ˌkruːsɪˈfɪkʃ(ə)n/ *n.* **1 a** a crucifying or being crucified. **b** an instance of this. (*See note below.*) **2** (**Crucifixion**) **a** the crucifixion of Christ. **b** a representation of this. [eccl.L *crucifixio* (as CRUCIFIX)]

▪ Crucifixion was a form of capital punishment used by various ancient peoples including the Persians, Carthaginians, and Romans; it was normally confined to slaves and other persons with no civil rights. The condemned person was first flogged and then made to carry a crosspiece to the place of execution, where a stake had been fixed in the ground. The prisoner was fastened to the beam by nails or cords, and the beam was drawn up and fixed to the stake so that the feet were clear of the ground; sometimes the feet were fastened to the upright. Death apparently resulted from exhaustion, perhaps caused by the difficulty of breathing when the body's weight is suspended by the arms in this way; it could be hastened by breaking the legs. The penalty was abolished in the Roman Empire by the emperor Constantine (337).

cruciform /ˈkruːsɪfɔːm/ *adj.* cross-shaped (esp. of a church with transepts). [L *crux crucis* cross + -FORM]

crucify /ˈkruːsɪfaɪ/ *v.tr.* (**-ies**, **-ied**) **1** put to death by fastening to a cross. (*See* CRUCIFIXION.) **2 a** cause extreme pain to. **b** persecute, torment. **c** *sl.* defeat thoroughly in an argument, match, etc. □ **crucifier** *n.* [ME f. OF *crucifier* f. LL *crucifigere* (as CRUCIFIX)]

cruck /krʌk/ *n. Brit. hist.* either of a pair of curved timbers extending to the ground in the framework of a type of medieval house-roof. [var. of CROOK]

crud /krʌd/ *n. sl.* **1** unwanted impurities, grease, etc.; dirt. **2** an unpleasant person. **3** nonsense. □ **cruddy** *adj.* (**cruddier**, **cruddiest**). [var. of CURD]

crude /kruːd/ *adj. & n.* ● *adj.* **1 a** in the natural or raw state; not refined (*crude oil*). **b** rough, unpolished; lacking finish. **2 a** (of an action or statement or manners) rude, blunt. **b** offensive, indecent (*a crude gesture*). **3 a** (of statistics) not adjusted or corrected by reference to modifying circumstances. **b** rough (*a crude estimate*). ● *n.* natural unrefined mineral oil. □ **crudely** *adv.* **crudeness** *n.* **crudity** *n.* [ME f. L *crudus* raw, rough]

crudités /ˈkruːdɪˌteɪ/ *n.pl.* an hors-d'œuvre of mixed raw vegetables often served with a sauce into which they are dipped. [F]

cruel /ˈkruːəl/ *adj. & v.* ● *adj.* (**crueller, cruellest** or **crueler, cruelest**) **1** indifferent to or gratified by another's suffering. **2** causing pain or suffering, esp. deliberately. ● *v.tr.* (**cruelled, cruelling**) *Austral. sl.* thwart, spoil. □ **cruelly** *adv.* **cruelness** *n.* [ME f. OF f. L *crudelis*, rel. to *crudus* (as CRUDE)]

cruelty /ˈkruːəltɪ/ *n.* (*pl.* **-ies**) **1** a cruel act or attitude; indifference to another's suffering. **2** a succession of cruel acts; a continued cruel attitude (*suffered much cruelty*). **3** *Law* physical or mental harm inflicted (whether or not intentional), esp. as a ground for divorce. □ **cruelty-free** (of cosmetics etc.) produced without involving any cruelty to animals in the development or manufacturing process. [OF *cruaulté* ult. f. L *crudelitas*]

cruet /ˈkruːɪt/ *n.* **1** a small container for salt, pepper, oil, or vinegar for use at table. **2** (in full **cruet-stand**) a stand holding cruets. **3** *Eccl.* a small container for the wine and water in the celebration of the Eucharist. [ME through AF f. OF *crue* pot f. OS *krūka*: rel. to CROCK²]

Crufts /krʌfts/ an annual dog-show held in London, first organized in 1886 by British dog-breeder Charles Cruft (1852–1938).

Cruikshank /ˈkrʊkʃæŋk/, George (1792–1878), English painter, illustrator, and caricaturist. The most eminent political cartoonist of his day, he was known for exposing the private life of the Prince Regent. His later work includes illustrations for Charles Dickens's *Sketches by Boz* (1836), as well as a series of etchings supporting the temperance movement.

cruise /kruːz/ *v. & n.* ● *v.* **1** *intr.* make a journey by sea calling at a series of ports usu. according to a predetermined plan, esp. for pleasure. **2** *intr.* sail about without a precise destination. **3** *intr.* **a** (of a motor vehicle or aircraft) travel at a moderate or economical speed. **b** (of a vehicle or its driver) travel at random, esp. slowly. **4** *intr.* achieve an objective, win a race etc., with ease. **5** *intr. & tr. sl.* walk or drive about (the streets etc.) in search of a sexual (esp. homosexual) partner. ● *n.* a cruising voyage, esp. as a holiday. □ **cruise missile** one able to fly at a low altitude and guide itself by reference to the features of the region it traverses. **cruising speed** a comfortable and economical speed for a motor vehicle, below its maximum speed. [prob. f. Du. *kruisen* f. *kruis* CROSS]

cruiser /ˈkruːzə(r)/ *n.* **1** a warship of high speed and medium armament. **2** = *cabin cruiser*. **3** *N. Amer.* a police patrol car. [Du. *kruiser* (as CRUISE)]

cruiserweight /ˈkruːzəˌweɪt/ *n.* esp. *Brit.* = *light heavyweight* (see HEAVYWEIGHT).

cruller /ˈkrʌlə(r)/ *n. N. Amer.* a small cake made of a rich dough twisted or curled and fried in fat. [prob. f. Du. *krullen* curl]

crumb /krʌm/ *n. & v.* ● *n.* **1 a** a small fragment, esp. of bread. **b** a small particle (*a crumb of comfort*). **2** the soft inner part of a loaf of bread. **3** *sl.* an objectionable person. ● *v.tr.* **1** cover with breadcrumbs. **2** break into crumbs. [OE *cruma*]

crumble /ˈkrʌmb(ə)l/ *v. & n.* ● *v.* **1** *tr. & intr.* break or fall into crumbs or fragments. **2** *intr.* (of power, a reputation, etc.) gradually disintegrate. ● *n.* **1** *Brit.* a mixture of flour and fat, rubbed to the texture of breadcrumbs and cooked as a topping for fruit etc. (*apple crumble*; *vegetable crumble*). **2** a crumbly or crumbled substance. [ME f. OE, formed as CRUMB]

crumbly /ˈkrʌmblɪ/ *adj. & n.* ● *adj.* (**crumblier, crumbliest**) consisting of, or apt to fall into, crumbs or fragments. ● *n.* (*pl.* **-ies**) *sl. offens.* an old person. □ **crumbliness** *n.*

crumbs /krʌmz/ *int. Brit. sl.* expressing dismay or surprise. [euphem. for *Christ*]

crumby /ˈkrʌmɪ/ *adj.* (**crumbier, crumbiest**) **1** like or covered in crumbs. **2** = CRUMMY.

crumhorn /ˈkrʌmhɔːn/ *n.* (also **krummhorn**) a medieval and Renaissance wind instrument with a double reed and a curved end. It has seven finger-holes with three extension keys for low notes. [G *Krummhorn* f. *krumm* crooked + *Horn* HORN]

crummy /ˈkrʌmɪ/ *adj.* (**crummier, crummiest**) *colloq.* dirty, squalid; inferior, worthless. □ **crummily** *adv.* **crumminess** *n.* [var. of CRUMBY]

crump /krʌmp/ *n. & v. sl.* ● *n.* the sound of a bursting bomb or shell. ● *v.intr.* make this sound. [imit.]

crumpet /ˈkrʌmpɪt/ *n.* **1** a soft flat cake of a yeast mixture cooked on a griddle and eaten toasted and buttered. **2** *Brit. joc.* or *offens.* **a** a sexually attractive person, esp. a woman. **b** women regarded collectively, esp. as objects of sexual desire. **3** *archaic sl.* the head. [17th c.: orig. uncert.]

crumple /ˈkrʌmp(ə)l/ *v. & n.* ● *v.* **1** *tr. & intr.* (often foll. by *up*) **a** crush or become crushed into creases. **b** ruffle, wrinkle. **2** *intr.* (often foll. by *up*) collapse, give way. ● *n.* a crease or wrinkle. □ **crumple zone** a part of a motor vehicle, esp. the extreme front and rear, designed to crumple easily in a crash and absorb impact. □ **crumply** *adj.* [obs. *crump* (v. & adj.) (make or become) curved]

crunch /krʌntʃ/ *v. & n.* ● *v.* **1** *tr.* **a** crush noisily with the teeth. **b** grind (gravel, dry snow, etc.) under foot, wheels, etc. **2** *intr.* (often foll. by *up*, *through*) make a crunching sound in walking, moving, etc. ● *n.* **1** crunching; a crunching sound. **2** *colloq.* a decisive event or moment (*when it came to the crunch*). [earlier *cra(u)nch*, assim. to *munch*]

crunchy /ˈkrʌntʃɪ/ *adj.* (**crunchier, crunchiest**) that can be or has been crunched or crushed into small pieces; hard and crispy. □ **crunchily** *adv.* **crunchiness** *n.*

crupper /ˈkrʌpə(r)/ *n.* **1** a strap buckled to the back of a saddle and looped under the horse's tail to hold the harness back. **2** the hindquarters of a horse. [ME f. OF *cropiere* (cf. CROUP²)]

crural /ˈkrʊərəl/ *adj. Anat.* of the leg. [F *crural* or L *cruralis* f. *crus cruris* leg]

crusade /kruːˈseɪd/ *n. & v.* ● *n.* **1** (usu. **Crusade**) any of several military expeditions made by western European Christians in the 11th–13th centuries to recover the Holy Land from the Saracen Muslims. (*See note below.*) **2** a vigorous campaign in favour of a cause. ● *v.intr.* engage in a crusade. □ **crusader** *n.* [earlier *croisade* (F f. *croix* cross) or *crusado* (Sp. f. *cruz* cross)]

▪ The First Crusade (1096–9) resulted in the capture of Jerusalem and the establishment of Crusader states in the Holy Land, but the second (1147–9) failed to stop a Muslim resurgence, and Jerusalem fell to Saladin in 1187. The third (1189–92) recaptured some lost ground but not Jerusalem, while the fourth (1202–4) was diverted against the Byzantine Empire, which was fatally weakened by the resultant sack of Constantinople. The fifth (1217–21) was delayed in Egypt, where it accomplished nothing, and although the sixth (1228–9) resulted in the return of Jerusalem to Christian hands the city was lost to the Turks in 1244. The seventh (1248–54) ended in disaster in Egypt, while the eighth and last (1270–1) petered out when its leader, Louis IX of France, died on his way east. Although undertaken in a religious cause, the Crusades were carried on like most other medieval wars and were generally badly organized and indecisive.

cruse /kruːz/ *n. archaic* an earthenware pot or jar. [OE *crūse*, of unkn. orig.]

crush /krʌʃ/ *v. & n.* ● *v.tr.* **1** compress with force or violence, so as to break, bruise, etc. **2** reduce to powder by pressure. **3** crease or crumple by rough handling. **4** defeat or subdue completely (*crushed by my reply*). ● *n.* **1** an act of crushing. **2** a crowded mass of people. **3** a drink made from the juice of crushed fruit. **4** *colloq.* **a** (usu. foll. by *on*) a (usu. passing) infatuation. **b** the object of an infatuation (*who's the latest crush?*). □ **crush bar** a place in a theatre for audiences to buy drinks in the intervals. **crush barrier** a barrier, esp. a temporary one, for restraining a crowd. □ **crushable** *adj.* **crusher** *n.* **crushing** *adj.* **crushingly** *adv.* [ME f. AF *cruissir*, *corussier*, OF *croissir*, *cruissir*, gnash (teeth), crack, f. Rmc]

crust /krʌst/ *n. & v.* ● *n.* **1 a** the hard outer part of a loaf of bread. **b** a piece of this with some soft bread attached. **c** a hard dry scrap of bread. **d** esp. *Austral. sl.* a livelihood (*what do you do for a crust?*). **2** the pastry covering of a pie. **3** a hard casing of a softer thing, e.g. a harder layer over soft snow. **4** *Geol.* the outer rocky portion of the earth, esp. the part overlying the mantle and separated from it by the moho. **5 a** a coating or deposit on the surface of anything. **b** a hard dry formation on the skin, a scab. **6** a deposit of tartar formed in bottles of old wine. **7 a** *sl.* impudence (*you have a crust!*). **b** a superficial hardness of manner. ● *v.tr. & intr.* **1** cover or become covered with a crust. **2** form into a crust. □ **crustal** *adj.* (in sense 4 of *n.*). [ME f. OF *crouste* f. L *crusta* rind, shell]

Crustacea /krʌˈsteɪʃə/ *n.pl. Zool.* a subphylum of arthropods comprising the crabs, lobsters, shrimps, etc., many of which have a hard calcareous carapace and numerous legs. The great majority are aquatic, and include numerous minute and larval forms which are a major constituent of plankton. □ **crustacean** *n. & adj.* **crustaceous** /-ʃəs/ *adj.* **crustaceology** /-ˌsteɪʃɪˈɒlədʒɪ/ *n.* [mod.L *crustaceus* f. *crusta*: see CRUST]

crusted /ˈkrʌstɪd/ *adj.* **1 a** having a crust. **b** (of wine) having deposited a crust. **2** antiquated, venerable (*crusted prejudice*).

crustose /ˈkrʌstəʊz, -təʊs/ *adj.* (esp. of a lichen) forming or resembling a crust. [L *crustosus* f. *crusta* crust]

crusty /ˈkrʌstɪ/ *adj. & n.* ● *adj.* (**crustier, crustiest**) **1** having a crisp crust (*a crusty loaf*). **2** irritable, curt. **3** hard, crustlike. ● *n. sl.* a person of unkempt appearance, often with matted hair or dreadlocks, esp. a New Age traveller. □ **crustily** *adv.* **crustiness** *n.*

crutch /krʌtʃ/ *n.* **1** a support for a lame person, usu. with a crosspiece at the top fitting under the armpit (*pair of crutches*). **2** any support or prop. **3** the crotch of the human body or garment. [OE *cryc(c)* f. Gmc]

Crutched Friars /krʌtʃt, 'krʌtʃɪd/ an order of mendicant friars established in Italy by 1169, which spread to England, France, and the Low Countries in the 13th century and was suppressed in 1656. The word *crutched* meant 'cross-bearing', and referred to the cross worn on top of their staves, and later on the front of their habits.

Crux /krʌks/ (in full **Crux Australis**) *Astron.* the smallest constellation (the Cross, also known as the Southern Cross), but the most familiar one to observers in the southern hemisphere. It contains the bright star Acrux, and the star cluster of the Jewel Box. [L]

crux /krʌks/ *n.* (*pl.* **cruxes** or **cruces** /'kruːsiːz/) **1** the decisive point at issue. **2** a difficult matter; a puzzle. [L, = cross]

Cruyff /krɔɪf/, Johan (b.1947), Dutch footballer. An attacking midfielder, he played for the Amsterdam team Ajax (1965–73), and was a member of the team that won three consecutive European Cup Finals (1971–3). Cruyff's exceptional skills were recognized when he was made European Footballer of the Year in both 1973 and 1974. He later played for Barcelona (1973–8) and captained the Netherlands in their unsuccessful World Cup Final against West Germany (1974).

cruzado /kruːˈzaːdəʊ/ *n.* (*pl.* **-os**) the chief monetary unit of Brazil between 1988 and 1990. [Port. *cruzado, crusado,* = marked with the cross]

cruzeiro /kruːˈzeərəʊ/ *n.* (*pl.* **-os**) a former monetary unit of Brazil. [Port., = large cross]

cry /kraɪ/ *v. & n.* ● *v.* (**cries, cried**) **1** *intr.* (often foll. by *out, for*) make a loud or shrill sound, esp. to express pain, grief, etc., or to appeal for help. **2 a** *intr.* shed tears; weep. **b** *tr.* shed (tears). **3** *tr.* (often foll. by *out*) say or exclaim loudly or excitedly. **4** *intr.* (of an animal, esp. a bird) make a loud call. **5** *tr.* (of a hawker etc.) proclaim (wares etc.) in the street. ● *n.* (*pl.* **cries**) **1** a loud inarticulate utterance of grief, pain, fear, joy, etc. **2** a loud excited utterance of words. **3** an urgent appeal or entreaty. **4** a spell of weeping. **5 a** public demand; a strong movement of opinion. **b** a watchword or rallying call. **6** the natural utterance of an animal, esp. of hounds on the scent. **7** the street-call of a hawker etc. □ **cry-baby** a person, esp. a child, who sheds tears frequently. **cry down** disparage, belittle. **cry one's eyes** (or **heart**) **out** weep bitterly. **cry for the moon** ask for what is unattainable. **cry from the heart** a passionate appeal or protest. **cry off** *colloq.* withdraw from a promise or undertaking. **cry out for** demand as a self-evident requirement or solution. **cry over spilt milk** see MILK. **cry stinking fish** disparage one's own efforts, products, etc. **cry up** praise, extol. **cry wolf** see WOLF. **a far cry from 1** a long way from. **2** a very different thing from. **for crying out loud** *colloq.* an exclamation of surprise or annoyance. **in full cry** (of hounds) in keen pursuit. [ME f. OF *crier,* cri f. L *quiritare* wail]

cryer var. of CRIER.

crying /'kraɪɪŋ/ *attrib.adj.* (of an injustice or other evil) flagrant, demanding redress (*a crying need; a crying shame*).

cryo- /'kraɪəʊ/ *comb. form* (extreme) cold. [Gk *kruos* frost]

cryobiology /ˌkraɪəʊbaɪˈɒlədʒɪ/ *n.* the biology of organisms below their normal temperatures. □ **cryobiologist** *n.* **cryobiological** /-ˌbaɪəˈlɒdʒɪk(ə)l/ *adj.*

cryogen /'kraɪəʊdʒən/ *n.* a freezing-mixture; a substance used to produce very low temperatures.

cryogenics /ˌkraɪəʊˈdʒɛnɪks/ *n.* **1** the branch of physics dealing with the production and effects of very low temperatures. (*See note below.*) **2** = CRYONICS. □ **cryogenic** *adj.*

■ Although absolute zero is in principle unattainable, modern cryogenic techniques can reduce temperatures to a tiny fraction of 1 kelvin. Low temperature environments are increasingly important in science and in industry. At very low temperatures many elements and alloys become superconductors, a phenomenon used, for example, in the production of powerful electromagnets. At very low temperatures also the thermal 'noise' produced by vibrating atoms is greatly reduced and electrical devices, especially detectors of electromagnetic signals, become much more sensitive and discriminating.

cryolite /'kraɪəʊˌlaɪt/ *n. Mineral.* a lustrous sodium aluminium fluoride mineral, used as a flux in the smelting of aluminium. [CRYO- (deposits were discovered in Greenland)]

cryonics /kraɪˈɒnɪks/ *n. Med.* the practice or technique of deep-freezing the bodies of those who have died of an incurable disease, in the hope of a future cure. □ **cryonic** *adj.* [contraction of CRYOGENICS]

cryoprotectant /ˌkraɪəʊprəˈtɛktənt/ *n. Biochem.* a substance that prevents the freezing of tissues; also, one which prevents damage to cells etc. during freezing.

cryopump /'kraɪəʊˌpʌmp/ *n.* a vacuum pump using liquefied gases.

cryostat /'kraɪəʊˌstæt/ *n.* an apparatus for maintaining a very low temperature.

cryosurgery /ˌkraɪəʊˈsɜːdʒərɪ/ *n.* surgery using the local application of intense cold for anaesthesia or therapy.

crypt /krɪpt/ *n.* **1** an underground room or vault, esp. one beneath a church, used usu. as a burial place. **2** *Anat.* a small tubular gland, pit, or recess. [ME f. L *crypta* f. Gk *kruptē* f. *kruptos* hidden]

cryptanalysis /ˌkrɪptəˈnælɪsɪs/ *n.* the art or process of deciphering encoded text by analysis; code-breaking. □ **cryptanalyst** /krɪptˈænəlɪst/ *n.* **cryptanalytic** /ˌkrɪptænəˈlɪtɪk/ *adj.* **cryptanalytical** *adj.* [CRYPTO- + ANALYSIS]

cryptic /'krɪptɪk/ *adj.* **1 a** obscure in meaning. **b** (of a crossword clue etc.) indirect; indicating the solution in a way that is not obvious. **c** secret, mysterious, enigmatic. **2** *Zool.* (of coloration etc.) serving for concealment. □ **cryptically** *adv.* [LL *crypticus* f. Gk *kruptikos* (as CRYPTO-)]

crypto /'krɪptəʊ/ *n.* (*pl.* **-os**) *colloq.* a person having a secret allegiance to a political creed etc., esp. communism. [as CRYPTO-]

crypto- /'krɪptəʊ/ *comb. form* concealed, secret (*crypto-communist*). [Gk *kruptos* hidden]

cryptocrystalline /ˌkrɪptəʊˈkrɪstəˌlaɪn/ *adj.* having a crystalline structure visible only when magnified.

cryptogam /'krɪptəˌgæm/ *n. Bot.* a plant that has no true flowers or seeds, e.g. ferns, mosses, algae, and fungi. □ **cryptogamic** /ˌkrɪptəˈgæmɪk/ *adj.* **cryptogamous** /krɪpˈtɒgəməs/ *adj.* [F *cryptogame* f. mod.L *cryptogamae (plantae)* formed as CRYPTO- + Gk *gamos* marriage]

cryptogram /'krɪptəˌgræm/ *n.* a text written in cipher.

cryptography /krɪpˈtɒgrəfɪ/ *n.* the art of writing or solving codes. □ **cryptographer** *n.* **cryptographic** /ˌkrɪptəˈgræfɪk/ *adj.* **cryptographically** *adv.*

cryptomeria /ˌkrɪptəˈmɪərɪə/ *n.* a tall evergreen tree, *Cryptomeria japonica*, native to China and Japan, with long curved spirally arranged leaves and short cones. Also called *Japanese cedar*. [CRYPTO- + Gk *meros* part (because the seeds are enclosed by scales)]

cryptozoology /ˌkrɪptəʊzuːˈɒlədʒɪ, -zəʊˈɒlədʒɪ/ *n.* the search for and study of animals whose existence is disputed or unsubstantiated, such as the Loch Ness monster. □ **cryptozoologist** *n.* **cryptozoological** /-ˌzuːəˈlɒdʒɪk(ə)l, -ˌzəʊə-/ *adj.*

crystal /'krɪst(ə)l/ *n. & adj.* ● *n.* **1 a** a clear transparent mineral, esp. rock crystal. **b** a piece of this. **2** (in full **crystal glass**) a highly transparent glass; flint glass. **b** articles made of this. **3** the glass over a watch-face. **4** *Electronics* a crystalline piece of semiconductor. **5** *Chem.* **a** an aggregation of atoms or molecules with a regular internal structure and the external form of a solid enclosed by symmetrically arranged plane faces. **b** a solid whose constituent particles are symmetrically arranged. ● *adj.* (usu. *attrib.*) made of, like, or clear as crystal. □ **crystal ball** a glass globe used in crystal-gazing. **crystal class** *Crystallog.* each of thirty-two categories of crystals classified according to the elements of symmetry they possess. **crystal clear** unclouded, transparent; readily understood. **crystal-gazing** the process of concentrating one's gaze on a crystal ball supposedly in order to obtain a picture of future events etc. **crystal lattice** *Crystallog.* the regular repeating pattern of atoms or molecules in a crystalline substance. **crystal set** a simple early form of radio receiving apparatus with a crystal touching a metal wire as the rectifier. **crystal system** *Crystallog.* each of seven distinct symmetrical forms (cubic, tetragonal, orthorhombic, trigonal, hexagonal, monoclinic, and triclinic) into which crystals can be classified according to the relations of their axes. [OE f. OF *cristal* f. L *crystallum* f. Gk *krustallos* ice, crystal]

crystalline /'krɪstəˌlaɪn/ *adj.* **1** of, like, or clear as crystal. **2** *Chem. & Mineral.* having the structure and form of a crystal. □ **crystalline lens** a transparent lens enclosed in a membranous capsule behind the iris of the eye. □ **crystallinity** /ˌkrɪstəˈlɪnɪtɪ/ *n.* [ME f. OF *cristallin* f. L *crystallinus* f. Gk *krustallinos* (as CRYSTAL)]

crystallite /'krɪstəˌlaɪt/ *n.* **1** a small crystal. **2** an individual perfect crystal or grain in a metal etc. **3** *Bot.* a region of cellulose etc. with a crystal-like structure.

crystallize /ˈkrɪstəˌlaɪz/ v. (also **-ise**) **1** tr. & intr. form or cause to form crystals. **2** (often foll. by *out*) **a** intr. (of ideas or plans) become definite. **b** tr. make definite. **3** tr. & intr. coat or impregnate or become coated or impregnated with sugar (*crystallized fruit*). □ **crystallizable** adj. **crystallization** /ˌkrɪstəlaɪˈzeɪʃ(ə)n/ n.

crystallography /ˌkrɪstəˈlɒɡrəfɪ/ n. the science of crystal form and structure. □ **crystallographer** n. **crystallographic** /-ləˈɡræfɪk/ adj.

crystalloid /ˈkrɪstəˌlɔɪd/ adj. & n. ● adj. **1** crystal-like. **2** having a crystalline structure. ● n. a substance that in solution is able to pass through a semi-permeable membrane (cf. COLLOID).

Crystal Palace a large building of prefabricated iron and glass, resembling a giant greenhouse, designed by Joseph Paxton for the Great Exhibition of 1851 in Hyde Park, London, and re-erected at Sydenham near Croydon; it was accidentally burnt down in 1936.

CS abbr. **1** Civil Service. **2** chartered surveyor. **3** Court of Session.

Cs symb. Chem. the element caesium.

c/s abbr. cycles per second.

csardas /ˈtʃɑːdɑːʃ/ n. (also **czardas**) (pl. same) a Hungarian dance with a slow start and a quick wild finish. [Hungarian *csárdás* f. *csárda* inn]

CSC abbr. **1** Civil Service Commission. **2** Conspicuous Service Cross.

CSE abbr. see CERTIFICATE OF SECONDARY EDUCATION.

CS gas /siːˈes/ n. a gas causing tears and choking, used to control riots etc. [B. B. Corson (b.1896) & R. W. Stoughton (1906–57), Amer. chemists]

CSI abbr. Companion of the Order of the Star of India.

CSIRO abbr. Commonwealth Scientific and Industrial Research Organization.

CSM abbr. (in the UK) Company Sergeant-Major.

CST abbr. Central Standard Time.

CSU abbr. (in the UK) Civil Service Union.

CT abbr. US Connecticut (in official postal use).

ct. abbr. **1** carat. **2** cent.

CTC 1 (in the UK) Cyclists' Touring Club. **2** see CITY TECHNOLOGY COLLEGE.

ctenoid /ˈtiːnɔɪd, ˈten-/ adj. Zool. (of fish scales) having marginal projections like the teeth of a comb (cf. PLACOID). [Gk *kteis ktenos* comb]

ctenophore /ˈtiːnəˌfɔː(r), ˈtenə-/ n. Zool. a marine coelenterate animal of the phylum Ctenophora, having a jellyfish-like body bearing rows of cilia, e.g. sea gooseberries. Also called *comb-jelly*. [mod.L *ctenophorus* (as CTENOID)]

Ctesiphon /ˈtesɪf(ə)n/ an ancient city on the Tigris near Baghdad, capital of the Parthian kingdom from *c.*224 and then of Persia under the Sassanian dynasty. It was taken by the Arabs in 636 and destroyed in the 8th century.

CU abbr. Cambridge University.

Cu symb. Chem. the element copper.

cu. abbr. cubic.

cub /kʌb/ n. & v. ● n. **1** the young of a fox, bear, lion, etc. **2** an ill-mannered young man. **3** (**Cub**) (in full **Cub Scout**) a member of the junior branch of the Scout Association. **4** (in full **cub reporter**) colloq. a young or inexperienced newspaper reporter. **5** US an apprentice. ● v.tr. (**cubbed**, **cubbing**) **1** (also absol.) give birth to (cubs). **2** hunt fox cubs. □ **cubhood** n. [16th c.: orig. unkn.]

Cuba /ˈkjuːbə/ a Caribbean country, the largest and furthest west of the islands of the West Indies, situated at the mouth of the Gulf of Mexico; pop. (est. 1990) 10,800,000; official language, Spanish; capital, Havana. The island was visited by Columbus in 1492, and colonized by Spain, the native inhabitants being virtually exterminated in the process. Cuba became nominally independent, although effectively under American control, after the Spanish-American War of 1898. Several periods of dictatorship followed the granting of full autonomy in 1934, and Cuba was taken over by a Communist revolution in 1959, although the US retained a naval base at Guantánamo Bay. Since then Cuba has been under the presidency of Fidel Castro. Having defeated an attempted invasion of US-backed Cuban exiles at the Bay of Pigs (1961), Castro's regime made advances in health care and education, but the country has suffered under a US trade embargo. Cuba was for many years heavily dependent on the USSR, and after the collapse of the Soviet Union and the Eastern bloc much of the country's trade was curtailed, with severe consequences for the economy.

Cuban /ˈkjuːb(ə)n/ adj. & n. ● adj. of or relating to Cuba or its people.

● n. a native or national of Cuba. □ **Cuban heel** a moderately high straight heel of a man's or woman's shoe.

Cubango see OKAVANGO.

Cuban Missile Crisis an international crisis in October 1962, the closest approach to nuclear war at any time between the US and the USSR. The crisis began when the US discovered that Soviet nuclear missiles had been placed on Cuba. President John F. Kennedy demanded their removal and announced a naval blockade of the island; the Soviet leader Khrushchev acceded to the US demands a week later.

cubby /ˈkʌbɪ/ n. (pl. **-ies**) (in full **cubby-hole**) **1** a very small room. **2** a snug or confined space. [dial. *cub* stall, pen, of LG orig.]

cube /kjuːb/ n. & v. ● n. **1** a solid contained by six equal squares. **2** a cube-shaped block. **3** Math. the product of a number multiplied by its square. ● v.tr. **1** find the cube of (a number). **2** cut (food for cooking etc.) into small cubes. □ **cube root** the number which produces a given number when cubed. [F *cube* or L *cubus* f. Gk *kubos*]

cubeb /ˈkjuːbeb/ n. **1** a SE Asian shrub, *Piper cubeba*, of the pepper family, bearing pungent berries. **2** this berry crushed for use in medicated cigarettes. [ME f. OF *cubebe*, *quibibe* ult. f. Arab. *kobāba*, *kubāba*]

cubic /ˈkjuːbɪk/ adj. **1** cube-shaped. **2** of three dimensions. **3** involving the cube (and no higher power) of a number (*cubic equation*). **4** Crystallog. having three equal axes at right angles. □ **cubic content** the volume of a solid expressed in cubic metres. **cubic metre** etc. the volume of a cube whose edge is one metre etc. [F *cubique* or L *cubicus* f. Gk *kubikos* (as CUBE)]

cubical /ˈkjuːbɪk(ə)l/ adj. cube-shaped. □ **cubically** adv.

cubicle /ˈkjuːbɪk(ə)l/ n. **1** a small partitioned space, screened for privacy. **2** a small separate sleeping-compartment. [L *cubiculum* f. *cubare* lie down]

cubiform /ˈkjuːbɪˌfɔːm/ adj. cube-shaped.

cubism /ˈkjuːbɪz(ə)m/ n. an early 20th-century movement in art, esp. painting, in which perspective with a single viewpoint was abandoned and use was made of simple geometric shapes, interlocking planes, and, later, collage. (*See note below.*) □ **cubist** n. & adj. [F *cubisme* (as CUBE)]

▪ Cubism was a reaction against traditional modes of representation and perspective and impressionist concerns with light and colour. The style, created by Picasso and Braque and first named by the French critic Louis Vauxcelles in 1908, was inspired by the later work of Cézanne and by African sculpture. In its first phase, known as *analytical cubism*, the artists restricted their colour range and subject-matter and depicted objects as a series of planes, as they would be seen from a variety of different viewpoints. In the later phase, *synthetic cubism* (1912 onwards), they experimented with collages, sticking pieces of newspaper, matchboxes, etc., on to the canvas and combining them with drawing or painting. Other artists associated with cubism include Juan Gris (1887–1927) and Fernand Léger (1881–1955).

cubit /ˈkjuːbɪt/ n. an ancient measure of length, approximately equal to the length of a forearm. [ME f. L *cubitum* elbow, cubit]

cubital /ˈkjuːbɪt(ə)l/ adj. **1** Anat. of the forearm. **2** Zool. of the corresponding part in animals. [ME f. L *cubitalis* (as CUBIT)]

cuboid /ˈkjuːbɔɪd/ adj. & n. ● adj. cube-shaped; like a cube. ● n. **1** Geom. a rectangular parallelepiped. **2** (in full **cuboid bone**) Anat. the outer bone of the tarsus. □ **cuboidal** /kjuːˈbɔɪd(ə)l/ adj. [mod.L *cuboides* f. Gk *kuboeidēs* (as CUBE)]

cucking-stool /ˈkʌkɪŋˌstuːl/ n. hist. a chair on which disorderly women were ducked as a punishment. [ME f. obs. *cuck* defecate]

cuckold /ˈkʌkəʊld/ n. & v. ● n. the husband of an adulteress. ● v.tr. (of a man) have a sexual relationship with the wife of. □ **cuckoldry** n. [ME *cukeweld*, *cokewold*, f. OF *cucu* cuckoo]

cuckoo /ˈkʊkuː/ n. & adj. ● n. a long-tailed, often parasitic songbird of the family Cuculidae; esp. the migratory European bird *Cuculus canorus*, which has a characteristic call and deposits its eggs in the nests of small birds, who hatch and rear the chicks as their own. ● predic.adj. sl. crazy, foolish. □ **cuckoo bee** a non-social bee, esp. one of the genus *Psithyrus* (resembling a bumble-bee), that lays its eggs in the nest of another species. **cuckoo clock** a clock that strikes the hour with a sound like a cuckoo's call, usu. with the emergence on each note of a mechanical cuckoo. **cuckoo flower 1** a cruciferous meadow plant, *Cardamine pratensis*, with pale lilac flowers (also called *lady's smock*). **2** = *ragged robin*. **cuckoo in the nest** an unwelcome intruder. **cuckoo-pint** a wild aroid plant, *Arum maculatum*, with arrow-shaped leaves and scarlet berries (also called *lords and ladies, wild arum*). **cuckoo-spit** froth exuded by larvae of leafhoppers of the family Cercopidae on leaves, stems, etc. **cuckoo wasp** a non-social wasp, e.g. a ruby-tailed wasp, that lays its eggs in the nest of another species. [ME f. OF *cucu*, imit.]

cucumber /ˈkjuːˌkʌmbə(r)/ *n.* **1** a long green fleshy fruit, used in salads. **2** the climbing plant, *Cucumis sativus*, of the gourd family, yielding this fruit. [ME f. OF *co(u)combre* f. L *cucumer*]

cucurbit /kjuːˈkɜːbɪt/ *n.* = GOURD 1b. □ **cucurbitaceous** /-ˌkɜːbɪˈteɪʃəs/ *adj.* [L *cucurbita*]

cud /kʌd/ *n.* half-digested food returned from the first stomach of ruminants to the mouth for further chewing. [OE *cwidu, cudu* what is chewed, corresp. to OHG *kuti, quiti* glue]

cuddle /ˈkʌd(ə)l/ *v. & n.* ● *v.* **1** *tr.* hug, embrace, fondle. **2** *intr.* **a** (often foll. by *up*) nestle together; lie close and snug. **b** kiss and fondle amorously. ● *n.* a prolonged and fond hug. □ **cuddlesome** *adj.* [16th c.: perh. f. dial. *couth* snug]

cuddly /ˈkʌdlɪ/ *adj.* (**cuddlier, cuddliest**) tempting to cuddle; given to cuddling.

cuddy /ˈkʌdɪ/ *n.* (*pl.* **-ies**) *Sc.* **1** a donkey. **2** a stupid person. [perh. a pet-form of the name *Cuthbert*]

cudgel /ˈkʌdʒəl/ *n. & v.* ● *n.* a short thick stick used as a weapon. ● *v.tr.* (**cudgelled, cudgelling**; *US* **cudgeled, cudgeling**) beat with a cudgel. □ **cudgel one's brains** think hard about a problem. **take up the cudgels** (often foll. by *for*) make a vigorous defence. [OE *cycgel*, of unkn. orig.]

Cudlipp /ˈkʌdlɪp/, Hugh, Baron Cudlipp of Aldingbourne (b.1913), British newspaper editor. The features editor on the *Daily Mirror* from 1935 and (after the Second World War) its editorial director, he was a pioneer of tabloid journalism. He conceived the formula of the sensationalized presentation of sex and crime together with populist politics that later proved a basis for other tabloids; this dramatically increased the paper's circulation. He gives an account of his career in *Publish and Be Damned* (1953).

cudweed /ˈkʌdwiːd/ *n.* a composite plant of the genus *Gnaphalium* or *Filago*, with hairy or downy foliage.

cue[1] /kjuː/ *n. & v.* ● *n.* **1 a** the last words of an actor's speech serving as a signal to another actor to enter or speak. **b** a similar signal to a singer or player etc. **2 a** a stimulus to perception etc. **b** a signal for action. **c** a hint on how to behave in particular circumstances. **3** a facility for cueing audio equipment by allowing the tape to be played at high speed during fast forward wind and stopped when the desired place is reached. ● *v.tr.* (**cues, cued, cueing** or **cuing**) **1** give a cue to. **2** put (a piece of audio equipment, esp. a record-player or tape recorder) in readiness to play a particular part of the recorded material. □ **cue-bid** *Bridge* an artificial bid to show a particular card etc. in the bidder's hand. **cue in 1** insert a cue for. **2** give information to. **on cue** at the correct moment. **take one's cue from** follow the example or advice of. [16th c.: orig. unkn.]

cue[2] /kjuː/ *n. & v. Billiards* etc. ● *n.* a long straight tapering rod for striking the ball. ● *v.* (**cues, cued, cueing** or **cuing**) **1** *tr.* strike (a ball) with a cue. **2** *intr.* use a cue. □ **cue-ball** the ball that is to be struck with the cue. □ **cueist** *n.* [var. of QUEUE]

Cuenca /ˈkwɛŋkə/ a city in the Andes in southern Ecuador; pop. (1990) 332,920. Founded in 1557, it is known as the 'marble city' because of its many fine buildings.

Cuernavaca /ˌkweənəˈvækə/ a resort town in central Mexico, at an altitude of 1,542 m (5,060 ft), capital of the state of Morelos; pop. (est. 1990) 400,000.

cuesta /ˈkwɛstə/ *n. Geog.* a gentle slope, esp. one ending in a steep drop. [Sp., = slope, f. L *costa*: see COAST]

cuff[1] /kʌf/ *n.* **1 a** the end part of a sleeve. **b** a separate band of linen worn round the wrist so as to appear under the sleeve. **c** the part of a glove covering the wrist. **2** *N. Amer.* a trouser turn-up. **3** (in *pl.*) *colloq.* handcuffs. □ **cuff-link** a device of two joined studs etc. to fasten the sides of a cuff together. **off the cuff** *colloq.* without preparation, extempore. □ **cuffed** *adj.* (also in *comb.*). [ME: orig. unkn.]

cuff[2] /kʌf/ *v. & n.* ● *v.tr.* strike with an open hand. ● *n.* such a blow. [16th c.: perh. imit.]

Cufic var. of KUFIC.

Cuiabá /ˌkuːjəˈbɑː/ **1** a river port in west central Brazil, on the Cuiabá river, capital of the state of Mato Grosso; pop. (1990) 389,070. **2** a river of western Brazil, which rises in the Mato Grosso plateau and flows for

483 km (300 miles) to join the São Lourenço river near the border with Bolivia.

cui bono? /kwiː ˈbɒnəʊ, ˈbəʊn-/ who stands, or stood, to gain? (with the implication that this person is responsible). [L, = to whom (is it) a benefit?]

cuirass /kwɪˈræs/ *n.* **1** *hist.* a piece of armour consisting of breastplate and back-plate fastened together. **2** a device for artificial respiration. [ME f. OF *cuirace*, ult. f. LL *coriaceus* f. *corium* leather]

cuirassier /ˌkwɪərəˈsɪə(r)/ *n. hist.* a cavalry soldier wearing a cuirass. [F (as CUIRASS)]

cuish var. of CUISSE.

cuisine /kwɪˈziːn/ *n.* a style or method of cooking, esp. of a particular country or establishment. [F f. L *coquina* f. *coquere* to cook]

cuisse /kwɪs/ *n.* (also **cuish** /kwɪʃ/) (usu. in *pl.*) *hist.* thigh armour. [ME, f. OF *cuisseaux* pl. of *cuissel* f. LL *coxale* f. *coxa* hip]

Culbertson /ˈkʌlbəts(ə)n/, Ely (1891-1955), American bridge player. An authority on contract bridge, he revolutionized the game by formalizing a system of bidding. This, together with his other activities in the early 1930s, such as well-publicized challenge matches with high stakes, helped to establish this form of the game in preference to auction bridge.

Culdee /kʌlˈdiː/ *n.* any of various Irish and Scottish monks in the 8th-12th centuries, who lived as hermits, usually in groups of thirteen on the analogy of Christ and his Apostles. They died out as the Celtic Church of which they were a part was brought under Roman Catholic rule. [f. Ir. *céile Dé* client of God]

cul-de-sac /ˈkʌldəˌsæk, ˈkʊl-/ *n.* (*pl.* **culs-de-sac** pronunc. same) **1** a street or passage closed at one end. **2** a route or course leading nowhere; a position from which one cannot escape. **3** *Anat.* = DIVERTICULUM. [F, = sack-bottom]

-cule /kjuːl/ *suffix* forming (orig. diminutive) nouns (*molecule*). [F *-cule* or L *-culus*]

Culiacán Rosales /ˌkʊljəˌkɑːn rəʊˈzɑːles/ a city in NW Mexico, capital of the state of Sinaloa; pop. (1990) 662,110.

culinary /ˈkʌlɪnərɪ/ *adj.* of or for cooking or the kitchen. □ **culinarily** *adv.* [L *culinarius* f. *culina* kitchen]

cull /kʌl/ *v. & n.* ● *v.tr.* **1** select, choose, or gather from a large quantity or amount (*knowledge culled from books*). **2** pick or gather (flowers, fruit, etc.). **3 a** select (animals) according to quality, esp. poor surplus specimens for killing. **b** reduce the numbers of (an animal population) by selective slaughter. ● *n.* **1** an act of culling. **2** an animal or animals culled. □ **culler** *n.* [ME f. OF *coillier* etc. ult. f. L *colligere* COLLECT[1]]

cullet /ˈkʌlɪt/ *n.* recycled waste or broken glass used in glass-making. [var. of COLLET]

Culloden, Battle of /kəˈlɒd(ə)n/ the final engagement of the Jacobite uprising of 1745-6, fought on a moor near Inverness in NE Scotland, the last pitched battle on British soil. The small and poorly supplied Jacobite army of the Young Pretender, Charles Edward Stuart, was crushed by the Hanoverian army under the Duke of Cumberland, and a ruthless pursuit after the battle effectively prevented any chance of saving the Jacobite cause.

culm[1] /kʌlm/ *n.* **1** coal-dust, esp. of anthracite. **2** *Geol.* strata under coal measures, esp. in SW England. [ME, prob. rel. to COAL]

culm[2] /kʌlm/ *n. Bot.* the stem of a plant, esp. of grasses. □ **culmiferous** /-ˈmɪfərəs/ *adj.* [L *culmus* stalk]

culminant /ˈkʌlmɪnənt/ *adj.* **1** at or forming the top. **2** *Astron.* on the meridian. [as CULMINATE + -ANT]

culminate /ˈkʌlmɪˌneɪt/ *v.* **1** *intr.* (usu. foll. by *in*) reach its highest or final point (*the antagonism culminated in war*). **2** *tr.* bring to its highest or final point. **3** *intr. Astron.* (of a star etc.) be on the meridian. □ **culmination** /ˌkʌlmɪˈneɪʃ(ə)n/ *n.* [LL *culminare culminat-* f. *culmen* summit]

culottes /kjuːˈlɒts/ *n.pl.* women's (usu. short) trousers cut to resemble a skirt. [F, = knee-breeches]

culpable /ˈkʌlpəb(ə)l/ *adj.* deserving blame. □ **culpably** *adv.* **culpability** /ˌkʌlpəˈbɪlɪtɪ/ *n.* [ME f. OF *coupable* f. L *culpabilis* f. *culpare* f. *culpa* blame]

Culpeper /ˈkʌlˌpepə(r)/, Nicholas (1616-54), English herbalist. He is chiefly remembered for his *Complete Herbal* (1653), which popularized herbalism and, despite embracing ideas of astrology and the doctrine of signatures (see SIGNATURE 2), was important in the development of botany and pharmacology. Culpeper was unpopular among his contemporaries, partly because of his unauthorized and critical translation of the London *Pharmacopoeia* (1649).

culprit /ˈkʌlprɪt/ n. a person accused of or guilty of an offence. [17th c.: orig. in the formula *Culprit, how will you be tried?*, said by the Clerk of the Crown to a prisoner pleading Not Guilty: perh. abbr. of AF *Culpable: prest d'averrer* etc. (You are) guilty: (I am) ready to prove etc.]

cult /kʌlt/ n. **1** a system of religious worship esp. as expressed in ritual. **2 a** devotion or homage to a person or thing (*the cult of aestheticism*). **b** a popular fashion esp. followed by a specific section of society. **3** (*attrib.*) denoting a person or thing popularized in this way (*cult film; cult figure*). □ **cultic** adj. **cultism** n. **cultist** n. [F *culte* or L *cultus* worship f. *colere cult-* inhabit, till, worship]

cultivar /ˈkʌltɪˌvɑː(r)/ n. Bot. a cultivated plant variety produced by selective breeding. [CULTIVATE + VARIETY]

cultivate /ˈkʌltɪˌveɪt/ v.tr. **1 a** prepare and use (soil etc.) for crops or gardening. **b** break up (the ground) with a cultivator. **2 a** raise or produce (crops). **b** = CULTURE v. **3 a** (often as **cultivated** adj.) apply oneself to improving or developing (the mind, manners, etc.). **b** pay attention to or nurture (a person or a person's friendship); ingratiate oneself with (a person). □ **cultivatable** adj. **cultivable** /-vəb(ə)l/ adj. **cultivation** /ˌkʌltɪˈveɪʃ(ə)n/ n. [med.L *cultivare* f. *cultiva (terra)* arable (land) (as CULT]

cultivator /ˈkʌltɪˌveɪtə(r)/ n. **1** a mechanical implement for breaking up the ground and uprooting weeds. **2** a person or thing that cultivates.

cultural /ˈkʌltʃərəl/ adj. of or relating to the cultivation of the mind or manners, esp. through artistic or intellectual activity. □ **culturally** adv.

Cultural Revolution a political upheaval in China, 1966–8, intended to bring about a return to revolutionary Maoist beliefs. Largely carried forward by the Red Guard, the Cultural Revolution resulted in attacks on intellectuals and what were seen as bourgeois elements, a large-scale purge in party posts, and the appearance of a personality cult around the Chinese leader Mao Zedong, who had been in semi-retirement since 1959. It led to considerable economic dislocation and was gradually brought to a halt by premier Zhou Enlai.

culture /ˈkʌltʃə(r)/ n. & v. ● n. **1 a** the arts and other manifestations of human intellectual achievement regarded collectively (*a city lacking in culture*). **b** a refined understanding of this; intellectual development (*a person of culture*). **2 a** the customs, civilization, and achievements of a particular time or people (*studied Chinese culture*). **b** the way of life of a particular society or group. **3** improvement by mental or physical training. **4 a** the cultivation of plants; the rearing of bees, silkworms, etc. **b** the cultivation of the soil. **5** Biol. **a** the cultivation of bacteria, tissue cells, etc. in an artificial nutrient medium. **b** a growth of cells etc. so obtained. ● v.tr. Biol. maintain (bacteria, tissue cells, etc.) in conditions suitable for growth. □ **culture shock** the feeling of disorientation experienced by a person suddenly subjected to an unfamiliar culture or way of life. **culture vulture** colloq. a person eager to acquire culture. **the two cultures** the arts and science. [ME f. F *culture* or L *cultura* (as CULT): (v.) f. obs. F *culturer* or med.L *culturare*]

cultured /ˈkʌltʃəd/ adj. having refined taste and manners and a good education. □ **cultured pearl** a pearl formed by an oyster after the insertion of a foreign body into its shell.

cultus /ˈkʌltəs/ n. a system of religious worship; a cult. [L: see CULT]

culverin /ˈkʌlvərɪn/ n. hist. **1** a long cannon. **2** a small firearm. [ME f. OF *coulevrine* f. *couleuvre* snake ult. f. L *colubra*]

culvert /ˈkʌlvət/ n. **1** an underground channel carrying water across a road etc. **2** a channel for an electric cable. [18th c.: orig. unkn.]

cum /kʌm/ prep. (usu. in *comb.*) with, combined with, also used as (*a bedroom-cum-study*). [L]

cumber /ˈkʌmbə(r)/ v. & n. ● v.tr. literary hamper, hinder, inconvenience. ● n. a hindrance, obstruction, or burden. [ME, prob. f. ENCUMBER]

Cumberland[1] /ˈkʌmbələnd/ a former county of NW England. In 1974 it was united with Westmorland and part of Lancashire to form the county of Cumbria.

Cumberland[2] /ˈkʌmbələnd/, William Augustus, Duke of (1721–65), third son of George II, English military commander. He gained great notoriety (and his nickname 'the Butcher') for the severity of his suppression of the Jacobite clans in the aftermath of his victory at the Battle of Culloden (1746).

Cumbernauld /ˌkʌmbəˈnɔːld/ a town in Strathclyde region, central Scotland; pop. (1981) 47,900. It was built as a new town in 1955.

cumbersome /ˈkʌmbəsəm/ adj. inconvenient in size, weight, or shape; unwieldy. □ **cumbersomely** adv. **cumbersomeness** n. [ME f. CUMBER + -SOME[1]]

cumbia /ˈkʊmbɪə/ n. **1** a kind of dance music of Colombian origin, similar to salsa. **2** a dance performed to this music. [adapted f. Colombian Sp., perh. f. Sp. *cumbé*]

Cumbria /ˈkʌmbrɪə/ a county of NW England; county town, Carlisle. Cumbria was an ancient British kingdom, and the name continued to be used for the hilly north-western region of England containing the Lake District and much of the northern Pennines. The county of Cumbria was formed in 1974, largely from the former counties of Westmorland and Cumberland. □ **Cumbrian** adj. & n. [med. L f. Welsh *Cymry* Welshman]

cumbrous /ˈkʌmbrəs/ adj. = CUMBERSOME. □ **cumbrously** adv. **cumbrousness** n. [CUMBER + -OUS]

cum grano salis /kʊm ˌɡrɑːnəʊ ˈsɑːlɪs/ adv. with a grain of salt (cf. *take with a pinch of salt* (see SALT)). [L]

cumin /ˈkʌmɪn/ n. (also **cummin**) **1** an umbelliferous plant, *Cuminum cyminum*, bearing aromatic seeds. **2** these seeds used as flavouring, esp. ground and used in curry powder. [ME f. OF *cumin, comin* f. L *cuminum* f. Gk *kuminon*, prob. of Semitic orig.]

cummerbund /ˈkʌməˌbʌnd/ n. a waist sash. [Hind. & Pers. *kamar-band* loin-band]

cummin var. of CUMIN.

cummings /ˈkʌmɪŋz/, e(dward) e(stlin) (1894–1962), American poet and novelist. His novel *The Enormous Room* (1922), an account of his brief internment in a French detention camp in 1917, won him an international reputation and introduced many of the themes of his subsequent work. He is now chiefly remembered for his poems, which are characterized by their experimental typography (most notably in the avoidance of capital letters), technical skill, frank vocabulary, and the sharpness of his satire. His many volumes of poetry include *95 Poems* (1956).

cumquat var. of KUMQUAT.

cumulate v. & adj. ● v.tr. & intr. /ˈkjuːmjʊˌleɪt/ accumulate, amass; combine. ● adj. /ˈkjuːmjʊlət/ heaped up, massed. □ **cumulation** /ˌkjuːmjʊˈleɪʃ(ə)n/ n. [L *cumulare* f. *cumulus* heap]

cumulative /ˈkjuːmjʊlətɪv/ adj. **1 a** increasing or increased in amount, force, etc., by successive additions (*cumulative evidence*). **b** formed by successive additions (*learning is a cumulative process*). **2** Stock Exch. (of shares) entitling holders to arrears of interest before any other distribution is made. □ **cumulative error** an error that increases with the size of the sample revealing it. **cumulative voting** a system in which each voter has as many votes as there are candidates and may give all to one candidate. □ **cumulatively** adv. **cumulativeness** n.

cumulo- /ˈkjuːmjʊləʊ/ comb. form cumulus (cloud).

cumulonimbus /ˌkjuːmjʊləʊˈnɪmbəs/ n. (often attrib.) (pl. **cumulonimbuses** or **cumulonimbi** /-baɪ/) Meteorol. a cloud type like cumulus but forming towering masses, such as occur in thunderstorms.

cumulus /ˈkjuːmjʊləs/ n. (often attrib.) (pl. **cumuli** /-ˌlaɪ/) Meteorol. a cloud type consisting of rounded masses heaped on each other above a horizontal base at fairly low altitude. □ **cumulous** adj. [L, = heap]

Cunard /kjuːˈnɑːd/, Sir Samuel (1787–1865), Canadian-born British shipowner. One of the pioneers of the regular transatlantic passenger service, he founded the steamship company which still bears his name with the aid of a contract to carry the mail between Britain and Canada. The first such voyage for the company was made in 1840.

cuneate /ˈkjuːnɪət/ adj. wedge-shaped. [L *cuneus* wedge]

cuneiform /ˈkjuːnɪˌfɔːm/ adj. & n. ● adj. **1** wedge-shaped. **2** of, relating to, or using an ancient system of writing with wedge-shaped marks impressed on soft clay with a straight length of reed, bone, wood, or metal, or incised into stone etc. (*See note below.*) ● n. cuneiform writing. [F *cunéiforme* or mod.L *cuneiformis* f. L *cuneus* wedge]

▪ Cuneiform writing was used (though perhaps not invented) by the Sumerians in the 3rd millennium BC and with the dissemination of their civilization came to be used in modified forms for a number of languages in the Near East until towards the end of the 1st millennium BC. Cuneiform scripts remained undeciphered until the 19th century.

Cunene /kjuːˈneɪnə/ a river of Angola, which rises near the city of Huambo and flows 250 km (156 miles) southwards as far as the frontier with Namibia, which then follows it westwards to the Atlantic.

cunjevoi /ˈkʌndʒɪˌvɔɪ/ n. Austral. **1** a tall plant, *Alocasia macrorrhiza*, of the arum family. **2** a sea squirt. [Aboriginal]

cunnilingus /ˌkʌnɪˈlɪŋgəs/ n. (also **cunnilinctus** /-ˈlɪŋktəs/) oral stimulation of the female genitals. [L f. *cunnus* vulva + *lingere* lick]

cunning /ˈkʌnɪŋ/ adj. & n. ● adj. (**cunninger, cunningest**) **1 a** skilled in ingenuity or deceit. **b** selfishly clever or crafty. **2** ingenious (*a cunning device*). **3** N. Amer. attractive, quaint. ● n. **1** craftiness; skill in deceit. **2** skill, ingenuity. □ **cunning man** (or **woman**) hist. a person possessing magical knowledge or skill. □ **cunningly** adv. **cunningness** n. [ME f. ON *kunnandi* knowing f. *kunna* know: cf. CAN¹]

Cunningham /ˈkʌnɪŋəm/, Merce (b.1919), American dancer and choreographer. While a dancer with the Martha Graham Dance Company (1939–45), he began to experiment with choreography, collaborating with the composer John Cage in solo performances in 1944. He formed his own company in 1953 and explored new abstract directions for modern dance. Cunningham's works include *Suite for Five* (1956) and *Travelogue* (1977).

Cunobelinus see CYMBELINE.

cunt /kʌnt/ n. coarse sl. **1** the female genitals. **2** offens. an unpleasant or stupid person. [ME f. Gmc]

CUP abbr. Cambridge University Press.

cup /kʌp/ n. & v. ● n. **1** a small bowl-shaped container, usu. with a handle for drinking from. **2 a** its contents (*a cup of tea*). **b** = CUPFUL. **c** esp. N. Amer. (in cookery) a standard measure of liquid or dry capacity equal to half a US pint (0.237 litre or 8.33 British fluid oz). **3** a cup-shaped thing, esp. the calyx of a flower or the socket of a bone. **4** flavoured wine, cider, etc., usu. chilled. **5** an ornamental cup-shaped trophy as a prize for victory or prowess, esp. in a sports contest. **6** one's fate or fortune (*a bitter cup*). **7** either of the two cup-shaped parts of a brassière. **8** the chalice used or the wine taken at the Eucharist. **9** Golf the hole on a putting-green or the metal container in it. ● v.tr. (**cupped, cupping**) **1** form (esp. one's hands) into the shape of a cup. **2** take or hold as in a cup. **3** hist. bleed (a person) by using a glass in which a partial vacuum is formed by heating. □ **cup-cake** a small cake baked in a cup-shaped foil or paper container and often iced. **cup lichen** a lichen, *Cladonia pyxidata*, with cup-shaped processes arising from the thallus. **one's cup of tea** colloq. what interests or suits one. **cup-tie** a match in a competition for a cup. **in one's cups** colloq. while drunk; drunk. [OE *cuppe* f. med.L *cuppa* cup, prob. differentiated from L *cupa* tub]

cupbearer /ˈkʌpˌbeərə(r)/ n. a person who serves wine, esp. an officer of a royal or noble household.

cupboard /ˈkʌbəd/ n. a recess or piece of furniture with a door and (usu.) shelves, in which things are stored. □ **cupboard love** a display of affection meant to secure some gain. [ME f. CUP + BOARD]

cupel /ˈkjuːp(ə)l/ n. & v. ● n. a small flat porous vessel used in assaying gold or silver in the presence of lead. ● v.tr. (**cupelled, cupelling**; US **cupeled, cupeling**) assay or refine in a cupel. □ **cupellation** /ˌkjuːpəˈleɪʃ(ə)n/ n. [F *coupelle* f. LL *cupella* dimin. of *cupa*: see CUP]

Cup Final n. the final match in a competition for a cup.

cupful /ˈkʌpfʊl/ n. (pl. **-fuls**) **1 a** the amount held by a cup. **b** esp. N. Amer. (in cookery) a half-pint or 8-ounce measure. **2** the contents of a full cup (*drank a cupful of water*). ¶ A *cupful* is a measure, and so *three cupfuls* is a quantity regarded in terms of a cup; *three cups full* denotes the actual cups, as in *three cups full of water*. Sense 2 is an intermediate use.

Cupid /ˈkjuːpɪd/ n. **1** Rom. Mythol. the god of love, identified by the Romans with Eros. He is often pictured as a beautiful naked boy with wings, carrying a bow and arrows, with which he wounds his victims. **2** (also **cupid**) a representation of Cupid. □ **Cupid's bow** the upper lip when shaped like the double-curved bow carried by Cupid. [ME f. L *Cupido* f. *cupere* desire]

cupidity /kjuːˈpɪdɪtɪ/ n. greed for gain; avarice. [ME f. OF *cupidité* or L *cupiditas* f. *cupidus* desirous]

cupola /ˈkjuːpələ/ n. **1** Archit. **a** a rounded dome forming a roof or ceiling. **b** a small rounded dome adorning a roof. **2** a revolving dome protecting mounted guns on a warship or in a fort. **3** (in full **cupola-furnace**) a furnace for melting metals. □ **cupolaed** /-ləd/ adj. [It. f. LL *cupula* dimin. of *cupa* cask]

cuppa /ˈkʌpə/ n. (also **cupper** /ˈkʌpə(r)/) Brit. colloq. **1** a cup of. **2** a cup of tea. [corruption]

cuprammonium /ˌkjuːprəˈməʊnɪəm/ n. Chem. a complex ion of divalent copper and ammonia, solutions of which dissolve cellulose. [LL *cuprum* copper + AMMONIUM]

cupreous /ˈkjuːprɪəs/ adj. of or like copper. [LL *cupreus* f. *cuprum* copper]

cupric /ˈkjuːprɪk/ adj. Chem. of copper, esp. divalent copper. [LL *cuprum* copper]

cupriferous /kjuːˈprɪfərəs/ adj. yielding or containing copper. [LL *cuprum* copper + -IFEROUS]

cupro- /ˈkjuːprəʊ/ comb. form copper (*cupro-nickel*).

cupro-nickel /ˌkjuːprəʊˈnɪk(ə)l/ n. an alloy of copper and nickel, esp. in the proportions 3:1 as used in 'silver' coins.

cuprous /ˈkjuːprəs/ adj. Chem. of copper, esp. monovalent copper. [LL *cuprum* copper]

cupule /ˈkjuːpjuːl/ n. Bot. & Zool. a cup-shaped organ, receptacle, etc. [LL *cupula* CUPOLA]

cur /kɜː(r)/ n. **1** a worthless or snappy dog. **2** colloq. a contemptible man. [ME, prob. orig. in *cur-dog*, perh. f. ON *kurr* grumbling]

curable /ˈkjʊərəb(ə)l/ adj. that can be cured. □ **curability** /ˌkjʊərəˈbɪlɪtɪ/ n. [CURE]

Curaçao /ˌkjʊərəˈsaʊ, -ˈseɪəʊ/ the largest island of the Netherlands Antilles, situated in the Caribbean Sea 60 km (37 miles) north of the Venezuelan coast; pop. (est. 1990) 144,960; chief town, Willemstad.

curaçao /ˌkjʊərəˈsaʊ, -ˈseɪəʊ/ n. (also **curaçoa** /-ˈsəʊə/) (pl. **-os** or **curaçoas**) a liqueur of spirits flavoured with the peel of bitter oranges, e.g. Cointreau. [CURAÇAO, which produces these oranges]

curacy /ˈkjʊərəsɪ/ n. (pl. **-ies**) a curate's office or the tenure of it.

curare /kjʊəˈrɑːrɪ/ n. a resinous bitter substance prepared from South American plants of the genera *Strychnos* and *Chondodendron*. It is able to paralyse the motor nerves, and is traditionally used by American Indians to poison arrows and blowpipe darts; it was formerly used as a muscle relaxant in surgery. [Carib]

curassow /ˈkjʊərəˌsəʊ/ n. a game bird of the family Cracidae, native to Central and South America, resembling a turkey. [anglicized f. CURAÇAO]

curate /ˈkjʊərət/ n. **1** a member of the clergy engaged as assistant to a parish priest. **2** archaic an ecclesiastical pastor. □ **curate-in-charge** a curate appointed to take charge of a parish in place of a priest. **curate's egg** a thing that is partly good and partly bad. [ME f. med.L *curatus* f. L *cura* CURE]

curative /ˈkjʊərətɪv/ adj. & n. ● adj. tending or able to cure (esp. disease). ● n. a curative medicine or agent. [F *curatif -ive* f. med.L *curativus* f. L *curare* CURE]

curator /kjʊəˈreɪtə(r)/ n. a keeper or custodian of a museum or other collection. □ **curatorship** n. **curatorial** /ˌkjʊərəˈtɔːrɪəl/ adj. [ME f. AF *curatour* (of *-eur*) or L *curator* (as CURATIVE)]

curb /kɜːb/ n. & v. ● n. **1** a check or restraint. **2** a strap etc. fastened to the bit and passing under a horse's lower jaw, used as a check. **3** an enclosing border or edging such as the frame round the top of a well or a fender round a hearth. **4** = KERB. ● v.tr. **1** restrain. **2** put a curb on (a horse). □ **curb roof** a roof of which each face has two slopes, the lower one steeper. [ME f. OF *courber* f. L *curvare* bend, CURVE]

curcuma /ˈkɜːkjʊmə/ n. **1** the spice turmeric. **2** a tuberous plant of the genus *Curcuma*, yielding this and other commercial substances. [med.L f. Arab. *kurkum* saffron f. Skr. *kuṇkuma*]

curd /kɜːd/ n. **1** (often in pl.) a coagulated substance formed by the action of acids or rennet on milk, which may be made into cheese or eaten as food. **2** a fatty substance found between flakes of boiled salmon flesh. **3** the edible head of a cauliflower. □ **curd cheese** a soft smooth cheese made from skimmed milk curds. □ **curdy** adj. [ME: orig. unkn.]

curdle /ˈkɜːd(ə)l/ v.tr. & intr. make into or become curds, (of milk) turn sour; coagulate, congeal. □ **make one's blood curdle** fill one with horror. □ **curdler** n. [frequent. form of CURD (as verb)]

cure /kjʊə(r)/ v. & n. ● v. **1** tr. (often foll. by *of*) restore (a person or animal) to health; relieve (*was cured of pleurisy*). **2** tr. eliminate (a disease, evil, etc.). **3** tr. preserve (meat, fruit, tobacco, or skins) by salting, drying, etc. **4** tr. **a** vulcanize (rubber). **b** harden (concrete or plastic). **5** intr. effect a cure. **6** intr. undergo a process of curing. ● n. **1** restoration to health. **2** a thing that effects a cure. **3** a course of medical or healing treatment. **4 a** the process of curing rubber or plastic. **b** (with qualifying adj.) the degree of this. **5 a** the office or function of a curate. **b** a parish or other sphere of spiritual ministration. □ **cure-all** a panacea; a universal remedy. □ **curer** n. [ME f. OF *curer* f. L *curare* take care of f. *cura* care]

curé /ˈkjʊəreɪ/ n. a parish priest in France etc. [F f. med.L *curatus*: see CURATE]

curettage /kjʊəˈretɪdʒ, ˌkjʊərɪˈtɑːʒ/ *n. Med.* the use of or an operation involving the use of a curette. [F (as CURETTE)]

curette /kjʊəˈret/ *n. & v. Med.* ● *n.* a surgeon's small instrument for scraping. ● *v.tr. & intr.* clean or scrape with a curette. [F, f. *curer* cleanse (as CURE)]

curfew /ˈkɜːfjuː/ *n.* **1 a** a regulation restricting or forbidding the public circulation of people, esp. requiring people to remain indoors between specified hours, usu. at night. **b** the hour designated as the beginning of such a restriction. **c** a daily signal indicating this. **2** *hist.* **a** a medieval regulation requiring people to extinguish fires at a fixed hour in the evening. **b** the hour for this. **c** the bell announcing it. **3** the ringing of a bell at a fixed evening hour. [ME f. AF *coeverfu*, OF *cuevrefeu* f. the stem of *couvrir* COVER + *feu* fire]

Curia /ˈkjʊərɪə/ *n.* (also **curia**) the papal court; the government departments of the Vatican. □ **Curial** *adj.* [L: orig. a division of an ancient Roman tribe, the Senate house at Rome, a feudal court of justice]

Curie /ˈkjʊərɪ/, Marie (1867–1934), Polish-born French physicist, and Pierre (1859–1906), French physicist, pioneers of radioactivity. Pierre's early researches were on piezoelectricity and on the effects of temperature on magnetism. The two scientists married in 1895. Working together on the mineral pitchblende, they discovered the elements polonium and radium, for which they shared the 1903 Nobel Prize for physics with A.-H. Becquerel. Marie succeeded to her husband's chair of physics at the Sorbonne after his accidental death, receiving another Nobel Prize (for chemistry) in 1911 for her isolation of radium. She also studied radioactive decay and the applications of radioactivity to medicine, pioneered mobile X-ray units, headed the French Radiological Service during the First World War, and afterwards worked in the new Radium Institute. She died of leukaemia, undoubtedly caused by prolonged exposure to radioactive materials.

curie /ˈkjʊərɪ/ *n. Physics* **1** a unit of radioactivity, corresponding to 3.7×10^{10} disintegrations per second (symbol: **Ci**). **2** a quantity of radioactive substance having this activity. [Pierre *Curie* (see CURIE)]

curio /ˈkjʊərɪəʊ/ *n.* (pl. **-os**) a rare or unusual object or person. [19th-c. abbr. of CURIOSITY]

curiosa /ˌkjʊərɪˈəʊsə/ *n.pl.* **1** curiosities. **2** erotic or pornographic books. [neut. pl. of L *curiosus*: see CURIOUS]

curiosity /ˌkjʊərɪˈɒsɪtɪ/ *n.* (pl. **-ies**) **1** an eager desire to know; inquisitiveness. **2** strangeness. **3** a strange, rare, or interesting object. [ME f. OF *curiouseté* f. L *curiositas -tatis* (as CURIOUS)]

curious /ˈkjʊərɪəs/ *adj.* **1** eager to learn; inquisitive. **2** strange, surprising, odd. **3** *euphem.* (of books etc.) erotic, pornographic. □ **curiously** *adv.* **curiousness** *n.* [ME f. OF *curios* f. L *curiosus* careful f. *cura* care]

Curitiba /ˌkʊərɪˈtiːbə/ a city in southern Brazil, capital of the state of Paraná; pop. (1990) 1,248,400.

curium /ˈkjʊərɪəm/ *n.* an artificial radioactive metallic chemical element (atomic number 96; symbol **Cm**). A member of the actinide series, curium was first produced by Glenn Seaborg and his colleagues in 1944 by bombarding plutonium with helium ions. [Marie and Pierre *Curie* (see CURIE)]

curl /kɜːl/ *v. & n.* ● *v.* **1** *tr. & intr.* (often foll. by *up*) bend or coil into a spiral; form or cause to form curls. **2** *intr.* move in a spiral form (*smoke curling upwards*). **3 a** *intr.* (of the upper lip) be raised slightly on one side as an expression of contempt or disapproval. **b** *tr.* cause (the lip) to do this. **4** *intr.* play curling. ● *n.* **1** a lock of curled hair. **2** anything spiral or curved inwards. **3 a** a curling movement or act. **b** the state of being curled. **4** a disease of plants in which the leaves are curled up. □ **curl up 1** lie or sit with the knees drawn up. **2** *colloq.* writhe with embarrassment, horror, or amusement. **make a person's hair curl** *colloq.* shock or horrify a person. **out of curl** lacking energy. [ME; earliest form *crolled*, *crulled* f. obs. adj. *crolle*, *crulle* curly f. MDu. *krul*]

curler /ˈkɜːlə(r)/ *n.* **1** a pin or roller etc. for curling the hair. **2** a player in the game of curling.

curlew /ˈkɜːljuː/ *n.* a wading bird of the genus *Numenius*, with a long slender down-curved bill; esp. the common European *N. arquatus*, which has a call resembling its name. [ME f. OF *courlieu*, *courlis* orig. imit., but assim. to *courliu* courier f. *courre* run + *lieu* place]

curlicue /ˈkɜːlɪˌkjuː/ *n.* a decorative curl or twist. [CURLY + CUE² (= pigtail) or Q¹]

curling /ˈkɜːlɪŋ/ *n.* **1** in senses of CURL *v.* **2** a game played on ice, especially in Scotland, in which large flat rounded stones are hurled along a defined space (the *rink*) towards a mark (the *tee*). (*See note below.*)

□ **curling-tongs** (or **-iron** or **-pins**) a heated device for twisting the hair into curls.

■ Curling may have originated in the Low Countries (two of Breughel's landscapes (16th century) show a similar game being played on frozen ponds), but it developed in Scotland, from where it spread to other countries, such as Canada, where the climatic conditions are suitable.

curly /ˈkɜːlɪ/ *adj.* (**curlier, curliest**) **1** having or arranged in curls. **2** moving in curves. □ **curly kale** see KALE. □ **curliness** *n.*

curmudgeon /kəˈmʌdʒən/ *n.* a bad-tempered or miserly person. □ **curmudgeonly** *adj.* [16th c.: orig. unkn.]

currach /ˈkʌrə/ *n.* (also **curragh**) *Ir.* a coracle. [Ir.: cf. CORACLE]

Curragh, the /ˈkʌrə/ a plain in County Kildare in the Republic of Ireland, noted for the breeding of racehorses. The Irish Derby is run annually on its racecourse.

currajong var. of KURRAJONG.

currant /ˈkʌrənt/ *n.* **1** a dried fruit of a small seedless variety of grape originally grown in the eastern Mediterranean region and much used in cookery. **2 a** a shrub of the genus *Ribes* producing red, white, or black berries. **b** the fruit of these shrubs. □ **flowering currant** an ornamental currant, *Ribes sanguineum*, native to North America. [ME *raysons of coraunce* f. AF, = grapes of Corinth (the orig. source)]

currawong /ˈkʌrəˌwɒŋ/ *n. Austral.* a large Australian woodland songbird of the genus *Strepera*, resembling the magpie and having a resonant call. [Aboriginal]

currency /ˈkʌrənsɪ/ *n.* (pl. **-ies**) **1 a** the money in general use in a country. **b** any other commodity used as a medium of exchange. **2** the condition of being current; prevalence (e.g. of words or ideas). **3** the time during which something is current. **4** *Austral. hist.* a native-born Australian person (cf. STERLING *n.* 2).

current /ˈkʌrənt/ *adj. & n.* ● *adj.* **1** belonging to the present time; happening now (*current events; the current week*). **2** (of money, opinion, a rumour, a word, etc.) in general circulation or use. ● *n.* **1** a body of water, air, etc., moving in a definite direction, esp. through a stiller surrounding body. **2 a** a flow of electricity; an ordered movement of electrically charged particles. **b** a quantity representing the rate of this. **3** (usu. foll. by *of*) a general tendency or course (of events, opinions, etc.). □ **current account** a bank account from which money may be drawn without notice. **pass current** be generally accepted as true or genuine. [ME f. OF *corant* f. L *currere* run]

currently /ˈkʌrəntlɪ/ *adv.* at the present time; now.

curricle /ˈkʌrɪk(ə)l/ *n. hist.* a light open two-wheeled carriage drawn by two horses abreast. [L *curriculum*: see CURRICULUM]

curriculum /kəˈrɪkjʊləm/ *n.* (pl. **curricula** /-lə/) **1** the subjects that are studied or prescribed for study in a school (*not part of the school curriculum*). **2** any programme of activities. □ **curricular** *adj.* [L, = course, race-chariot, f. *currere* run]

curriculum vitae /ˈviːtaɪ/ *n.* (pl. **curricula vitae** or **vitarum** /vɪˈtɑːrəm/) a brief account of one's education, qualifications, and previous occupations. [L, = course of life]

currier /ˈkʌrɪə(r)/ *n.* a person who dresses and colours tanned leather. [ME f. OF *corier*, f. L *coriarius* f. *corium* leather]

currish /ˈkɜːrɪʃ/ *adj.* **1** like a cur; snappish. **2** ignoble. □ **currishly** *adv.* **currishness** *n.*

curry¹ /ˈkʌrɪ/ *n. & v.* ● *n.* (pl. **-ies**) a dish of meat, vegetables, etc., cooked in a sauce of hot-tasting spices, usu. served with rice. ● *v.tr.* (**-ies, -ied**) prepare or flavour with a sauce of hot-tasting spices (*curried eggs*). □ **curry-powder** a preparation of turmeric and other spices for making curry. [Tamil]

curry² /ˈkʌrɪ/ *v.tr.* (**-ies, -ied**) **1** groom (a horse) with a curry-comb. **2** treat (tanned leather) to improve its properties. **3** thrash. □ **curry-comb** a hand-held metal serrated device for grooming horses. **curry favour** ingratiate oneself. [ME f. OF *correier* ult. f. Gmc]

curse /kɜːs/ *n. & v.* ● *n.* **1** a solemn utterance intended to invoke a supernatural power to inflict destruction or punishment on a person or thing. **2** the evil supposedly resulting from a curse. **3** a violent exclamation of anger; a profane oath. **4** a thing that causes evil or harm. **5** (prec. by *the*) *colloq.* menstruation. **6** a sentence of excommunication. ● *v.* **1** *tr.* **a** utter a curse against. **b** (in *imper.*) may God curse. **2** *tr.* (usu. in *passive*; foll. by *with*) afflict with (*cursed with blindness*). **3** *intr.* utter expletive curses; swear. **4** *tr.* excommunicate. □ **curser** *n.* [OE *curs, cursian*, of unkn. orig.]

cursed /'kɜːsɪd, kɜːst/ *adj.* damnable, abominable. □ **cursedly** /-sɪdlɪ/ *adv.* **cursedness** /-sɪdnɪs/ *n.*

cursillo /kʊə'sɪləʊ/ *n.* (*pl.* **-os**) *RC Ch.* a short informal spiritual retreat by a group of devotees esp. in Latin America. [Sp., = little course]

cursive /'kɜːsɪv/ *adj. & n.* ● *adj.* (of writing) done with joined characters. ● *n.* cursive writing (cf. PRINT v. 4, UNCIAL). □ **cursively** *adv.* [med.L (*scriptura*) *cursiva* f. L *currere curs-* run]

cursor /'kɜːsə(r)/ *n.* **1** *Math.* etc. a transparent slide engraved with a hairline and forming part of a slide-rule. **2** *Computing* a movable indicator on a VDU screen identifying a particular position in the display, esp. the position that the program will operate on with the next keystroke. [L, = runner (as CURSIVE)]

cursorial /kɜː'sɔːrɪəl/ *adj. Zool.* having limbs adapted for running. [as CURSOR + -IAL]

cursory /'kɜːsərɪ/ *adj.* hasty, hurried (*a cursory glance*). □ **cursorily** *adv.* **cursoriness** *n.* [L *cursorius* of a runner (as CURSOR)]

curst /kɜːst/ *archaic var.* of CURSED.

curt /kɜːt/ *adj.* noticeably or rudely brief. □ **curtly** *adv.* **curtness** *n.* [L *curtus* cut short, abridged]

curtail /kɜː'teɪl/ *v.tr.* **1** cut short; reduce; terminate esp. prematurely (*curtailed his visit to America*). **2** (foll. by *of*) *archaic* deprive of. □ **curtailment** *n.* [obs. *curtal* horse with docked tail f. F *courtault* f. *court* short f. L *curtus*: assim. to *tail*]

curtain /'kɜːt(ə)n/ *n. & v.* ● *n.* **1** a piece of cloth etc. hung up as a screen, usu. movable sideways or upwards, esp. at a window or between the stage and auditorium of a theatre. **2** *Theatr.* **a** the rise or fall of the stage curtain at the beginning or end of an act or scene. **b** = *curtain-call*. **3** a partition or cover. **4** (in *pl.*) *sl.* the end. ● *v.tr.* **1** furnish or cover with a curtain or curtains. **2** (foll. by *off*) shut off with a curtain or curtains. □ **curtain-call** *Theatr.* an audience's summons to actor(s) to take a bow after the fall of the curtain. **curtain-fire** *Mil.* a concentration of rapid and continuous fire. **curtain lecture** a wife's private reproof to her husband, orig. behind bed-curtains. **curtain-raiser 1** *Theatr.* a piece prefaced to the main performance. **2** a preliminary event. **curtain-wall 1** *Fortification* the plain wall of a fortified place, connecting two towers etc. **2** *Archit.* a piece of plain wall not supporting a roof. [ME f. OF *cortine* f. LL *cortina* transl. Gk *aulaia* f. *aulē* court]

curtana /kɜː'teɪnə, -'tɑːnə/ *n. Brit.* an unpointed sword borne before English sovereigns at their coronation, as an emblem of mercy. [ME f. AL *cortana* (*spatha* sword) f. AF *curtain*, OF *cortain* name of Roland's similar sword f. *court* short (as CURT)]

curtilage /'kɜːtɪlɪdʒ/ *n.* an area attached to a dwelling-house and forming one enclosure with it. [ME f. AF *curtilage*, OF *co(u)rtillage* f. *co(u)rtil* small court f. *cort* COURT]

Curtin /'kɜːtɪn/, John (Joseph Ambrose) (1885–1945), Australian Labor statesman, Prime Minister 1941–5. He led the Labor party from 1935 to 1945. As Premier during the Second World War, he mobilized Australian resources to meet the danger of Japanese invasion, laid down the groundwork for the postwar economy, and introduced various welfare measures. Curtin died while in office.

Curtiss /'kɜːtɪs/, Glenn (Hammond) (1878–1930), American air pioneer and aircraft designer. He began by building and selling bicycles; from 1901 he built motorcycles, setting motorcycle speed records in 1905 and 1907. He built his first aeroplane in 1909, and invented the aileron and demonstrated the first practical seaplane two years later. In 1908 Curtiss made the first public American flight of 1.0 km (0.6 miles), and won the James Gordon Bennett Cup in 1909 for a flight in his own aeroplane at 46.6 m.p.h.

curtsy /'kɜːtsɪ/ *n. & v.* (also **curtsey**) ● *n.* (*pl.* **-ies** or **-eys**) a woman's or girl's formal greeting or salutation made by bending the knees and lowering the body. ● *v.intr.* (**-ies**, **-ied** or **-eys**, **-eyed**) make a curtsy. [var. of COURTESY]

curule /'kjʊəruːl/ *adj. Rom. Hist.* designating or relating to the authority exercised by the senior Roman magistrates, chiefly the consul and praetor, who were entitled to use the *sella curulis* ('curule seat' or seat of office). [L *curulis* f. *currus* chariot (in which the chief magistrate was conveyed to the seat of office)]

curvaceous /kɜː'veɪʃəs/ *adj. colloq.* (esp. of a woman) having a shapely curved figure.

curvature /'kɜːvətʃə(r)/ *n.* **1** the act or state of curving. **2** a curved form. **3** *Geom.* **a** the deviation of a curve from a straight line, or of a curved surface from a plane. **b** the quantity expressing this. [OF f. L *curvatura* (as CURVE)]

curve /kɜːv/ *n. & v.* ● *n.* **1** a line or surface having along its length a regular deviation from being straight or flat, as exemplified by the surface of a sphere or lens. **2** a curved form or thing. **3** a curved line on a graph. **4** *Baseball* a ball caused to deviate by the pitcher's spin. ● *v.tr. & intr.* bend or shape so as to form a curve. □ **curved** *adj.* [orig. as *adj.* (in *curve line*) f. L *curvus* bent: (v.) f. L *curvare*]

curvet /kɜː'vet/ *n. & v.* ● *n.* a horse's leap with the forelegs raised together and the hind legs raised with a spring before the forelegs reach the ground. ● *v.intr.* (**curvetted**, **curvetting** or **curveted**, **curveting**) (of a horse or rider) make a curvet. [It. *corvetta* dimin. of *corva* CURVE]

curvi- /'kɜːvɪ/ *comb. form* curved. [L *curvus* curved]

curviform /'kɜːvɪˌfɔːm/ *adj.* having a curved shape.

curvilinear /ˌkɜːvɪ'lɪnɪə(r)/ *adj.* contained by or consisting of curved lines. □ **curvilinearly** *adv.* [CURVI- after *rectilinear*]

curvy /'kɜːvɪ/ *adj.* (**curvier**, **curviest**) **1** having many curves. **2** (of a woman's figure) shapely. □ **curviness** *n.*

cuscus[1] /'kʌskəs/ *n.* the aromatic fibrous root of an Indian grass, *Vetiveria zizanoides*, used for making fans etc. [Pers. *ḵaškaš*]

cuscus[2] /'kʌskəs/ *n.* a nocturnal phalanger (marsupial), esp. of the genus *Phalanger*, native to New Guinea and northern Australia. [F *couscous* f. Du. *koeskoes* f. native Moluccas name]

cusec /'kjuːsek/ *n.* a unit of flow (esp. of water) equal to one cubic foot per second. [abbr.]

Cush /kʊʃ/ **1** in the Bible, the eldest son of Ham and grandson of Noah (Gen. 10:6). **2** the southern part of ancient Nubia, first mentioned in Egyptian records of the Middle Kingdom (see EGYPT). In the Bible it is the country of the descendants of Cush.

cush /kʊʃ/ *n.* esp. *Billiards colloq.* a cushion. [abbr.]

cushat /'kʌʃət/ *n. Sc.* a woodpigeon. [OE *cūscute*, of unkn. orig.]

cush-cush /'kʊʃkʊʃ/ *n.* a yam, *Dioscorea trifida*, native to South America. [perh. ult. of Afr. orig.]

Cushing[1] /'kʊʃɪŋ/, Harvey Williams (1869–1939), American surgeon. He introduced techniques that greatly increased the likelihood of success in neurosurgical operations, and described the hormonal disorder that was later named after him.

Cushing[2] /'kʊʃɪŋ/, Peter (1913–94), English actor. Cushing was known particularly for his roles in horror films, especially those made by Hammer films; these include *Dracula* (1958) and *Frankenstein Must Be Destroyed* (1969). Making his début in 1939, he appeared in more than a hundred films; he played Sherlock Holmes in *The Hound of the Baskervilles* (1959), and also appeared in *Star Wars* (1977).

Cushing's disease *n. Med.* Cushing's syndrome accompanied and caused by a tumour of the pituitary gland. [CUSHING[1]]

Cushing's syndrome *n. Med.* a metabolic disorder caused by overactivity of the adrenal cortex and often involving obesity and hypertension. [CUSHING[1]]

cushion /'kʊʃ(ə)n/ *n. & v.* ● *n.* **1** a bag of cloth etc. stuffed with a mass of soft material, used as a soft support for sitting or leaning on etc. **2** a means of protection against shock. **3** the elastic lining of the sides of a billiard-table, from which the ball rebounds. **4** a body of air supporting a hovercraft etc. **5** the frog of a horse's hoof. ● *v.tr.* **1** provide or protect with a cushion or cushions. **2** provide with a defence; protect. **3** mitigate the adverse effects of (*cushioned the blow*). **4** quietly suppress. **5** place or bounce (the ball) against the cushion in billiards. □ **cushiony** *adj.* [ME f. OF *co(i)ssin*, *cu(i)ssin* f. Gallo-Roman f. L *culcita* mattress, cushion]

Cushitic /kʊ'ʃɪtɪk/ *adj.* denoting a group of East African languages belonging to the Afro-Asiatic family, spoken mainly in Ethiopia and Somalia. [CUSH 2 + -ITE[1] + -IC]

cushy /'kʊʃɪ/ *adj.* (**cushier**, **cushiest**) *colloq.* **1** (of a job etc.) easy and pleasant. **2** *US* (of a seat, surroundings, etc.) soft, comfortable. □ **cushiness** *n.* [Anglo-Ind. f. Hind. *ḵhūsh* pleasant]

cusp /kʌsp/ *n.* **1** an apex or peak. **2** the horn of a crescent moon etc. **3** *Astrol.* the initial point of an astrological sign or house. **4** *Archit.* a projecting point between small arcs in Gothic tracery. **5** *Geom.* the point at which two arcs meet from the same direction terminating with a common tangent. **6** *Bot.* a pointed end, esp. of a leaf. **7** *Anat.* a cone-shaped prominence on the surface of a tooth, esp. a molar or premolar. **8** *Anat.* a pocket or fold in a valve of the heart. □ **cusped** *adj.*

cuspidal *adj.* **cuspate** /'kʌspeɪt/ *adj.* [L *cuspis, -idis* point, apex]

cuspidor /'kʌspɪˌdɔː(r)/ *n. N. Amer.* a spittoon. [Port., = spitter f. *cuspir* spit f. L *conspuere*]

cuss /kʌs/ *n. & v. colloq.* ● *n.* **1** a curse. **2** usu. *derog.* a person; a creature. ● *v.tr. & intr.* curse. □ **cuss-word** a swear-word. [var. of CURSE]

cussed /ˈkʌsɪd/ *adj. colloq.* awkward and stubborn. □ **cussedly** *adv.* **cussedness** *n.* [var. of CURSED]

custard /ˈkʌstəd/ *n.* **1** a dish made with milk and eggs, usu. sweetened. **2** a sweet sauce made with milk and flavoured cornflour. □ **custard-apple** a West Indian fruit, borne by trees of the genus *Annona*, with a sweet custard-like pulp. **custard-pie 1** a pie containing custard, commonly thrown in slapstick comedy. **2** (*attrib.*) denoting slapstick comedy. **custard powder** a preparation of cornflour etc. for making custard. [ME, earlier *crusta(r)de* f. AF f. OF *crouste* CRUST]

Custer /ˈkʌstə(r)/, George (Armstrong) (1839–76), American cavalry general. He served with distinction in the American Civil War but led his men to their deaths in a clash (popularly known as Custer's Last Stand) with the Sioux at Little Bighorn in Montana. Controversy over his conduct in the final battle still continues.

custodian /kʌˈstəʊdɪən/ *n.* a guardian or keeper, esp. of a public building etc. □ **custodianship** *n.* [CUSTODY + -AN, after *guardian*]

custody /ˈkʌstədɪ/ *n.* **1** guardianship; protective care. **2** imprisonment. □ **take into custody** arrest. □ **custodial** /kʌˈstəʊdɪəl/ *adj.* [L *custodia* f. *custos -odis* guardian]

custom /ˈkʌstəm/ *n.* **1 a** the usual way of behaving or acting (*a slave to custom*). **b** a particular established way of behaving (*our customs seem strange to foreigners*). **2** *Law* established usage having the force of law. **3** business patronage; regular dealings or customers (*lost a lot of custom*). **4** (in *pl.*; also treated as *sing.*) **a** a duty levied on certain imported and exported goods. **b** the official department that administers this. **c** the area at a port, frontier, etc., where customs officials deal with incoming goods, baggage, etc. □ **custom-built** (or **-made** etc.) made to a customer's order. **custom-house** the office at a port or frontier etc. at which customs duties are levied. **customs union** a group of states with an agreed common tariff, and usu. free trade with each other. [ME and OF *custume* ult. f. L *consuetudo -dinis*: see CONSUETUDE]

customary /ˈkʌstəmərɪ/ *adj. & n.* ● *adj.* **1** usual; in accordance with custom. **2** *Law* in accordance with custom. ● *n.* (*pl.* **-ies**) *Law* a book etc. listing the customs and established practices of a community. □ **customarily** *adv.* **customariness** *n.* [med.L *custumarius* f. *custuma* f. AF *custume* (as CUSTOM)]

customer /ˈkʌstəmə(r)/ *n.* **1** a person who buys goods or services from a shop or business. **2** a person one has to deal with (*an awkward customer*). [ME f. AF *custumer* (as CUSTOMARY), or f. CUSTOM + -ER¹]

customize /ˈkʌstəˌmaɪz/ *v.tr.* (also **-ise**) make to order or modify according to individual requirements.

cut /kʌt/ *v. & n.* ● *v.* (**cutting**; past and past part. **cut**) **1** *tr.* (also *absol.*) penetrate or wound with a sharp-edged instrument (*cut his finger; the knife won't cut*). **2** *tr. & intr.* (often foll. by *into*) divide or be divided with a knife etc. (*cut the bread; cut the cloth into metre lengths*). **3** *tr.* **a** trim or reduce the length of (hair, a hedge, etc.) by cutting. **b** detach all or the significant part of (flowers, corn, etc.) by cutting. **c** reduce the length of (a book, film, etc.). **4** *tr.* (foll. by *loose, open,* etc.) make loose, open, etc. by cutting. **5** *tr.* (esp. as **cutting** *adj.*) cause sharp physical or mental pain to (*a cutting remark; a cutting wind; was cut to the quick*). **6** *tr.* (often foll. by *down*) **a** reduce (wages, prices, time, etc.). **b** reduce or cease (services etc.). **7** *tr.* **a** shape or fashion (a coat, gem, key, record, etc.) by cutting. **b** make (a path, tunnel, etc.) by removing material. **8** *tr.* perform, execute, make (*cut a caper; cut a sorry figure; cut a deal*). **9** *tr.* (also *absol.*) cross, intersect (*the line cuts the circle at two points; the two lines cut*). **10** *intr.* (foll. by *across, through,* etc.) pass or traverse, esp. in a hurry or as a shorter way (*cut across the grass*). **11** *tr.* **a** ignore or refuse to recognize (a person). **b** renounce (a connection). **12** *tr.* esp. *N. Amer.* deliberately fail to attend (a class etc.). **13** *Cards* **a** *tr.* divide (a pack) into two parts. **b** *intr.* select a dealer etc. by dividing the pack. **14** *Cinematog.* **a** *tr.* edit (a film or tape). **b** *intr.* (often in *imper.*) stop filming or recording. **c** *intr.* (foll. by *to*) go quickly to (another shot). **15** *tr.* switch off (an engine etc.). **16** *tr.* **a** hit (a ball) with a chopping motion. **b** *Golf* slice (the ball). **17** *tr. N. Amer.* dilute, adulterate. **18** *tr.* (as **cut** *adj.*) *Brit. sl.* drunk. **19** *intr. Cricket* (of the ball) turn sharply on pitching. **20** *intr. sl.* run. **21** *tr.* castrate. ● *n.* **1** an act of cutting. **2** a division or wound made by cutting. **3** a stroke with a knife, sword, whip, etc. **4 a** a reduction (in prices, wages, etc.). **b** a cessation (of a power supply etc.). **5** an excision of part of a play, film, book, etc. **6** a wounding remark or act. **7** the way or style in which a garment, the hair, etc., is cut. **8** a piece of meat etc. cut from a carcass. **9** *colloq.* commission; a share of profits. **10** *Tennis & Cricket* etc. a stroke made by cutting. **11** ignoring of or refusal to recognize a person. **12 a** an engraved block for printing. **b** a woodcut or other print. **13** a railway cutting. **14** a new channel made for a river. □ **a cut above** *colloq.* noticeably superior to. **be cut out** (foll. by *for,* or *to* + infin.) be suited (*was not cut out to be a teacher*). **cut across 1** transcend or take no account of (normal limitations etc.) (*their concern cuts across normal rivalries*). **2** see sense 10 of *v.* **cut-and-come-again** *n.* a green vegetable etc. that can be frequently cut or picked. ● *adj.* inexhaustible. **cut and dried 1** completely decided; prearranged; inflexible. **2** (of opinions etc.) ready-made, lacking freshness. **cut and paste** (esp. *attrib.*) **1** a method of compiling a book, article, etc. from extracts of others or without independent research. **2** *Computing* a method or facility for taking pieces of text from one point and inserting them at another. **cut and run** *sl.* run away. **cut and thrust 1** a lively interchange of argument etc. **2** the use of both the edge and the point of a sword. **cut back 1** reduce (expenditure etc.). **2** prune (a tree etc.). **3** *Cinematog.* repeat part of a previous scene for dramatic effect. **cut-back** *n.* an instance or the act of cutting back, esp. a reduction in expenditure. **cut both ways 1** serve both sides of an argument etc. **2** (of an action) have both good and bad effects. **cut one's coat according to one's cloth 1** adapt expenditure to resources. **2** limit ambition to what is feasible. **cut a corner** go across and not round it. **cut corners** do a task etc. perfunctorily or incompletely, esp. to save time. **cut a dash** see DASH. **cut dead** completely refuse to recognize (a person). **cut down 1 a** bring or throw down by cutting. **b** kill by means of a sword or disease. **2** see sense 6 of *v.* **3** reduce the length of (*cut down the trousers to make shorts*). **4** (often foll. by *on*) reduce one's consumption (*tried to cut down on beer*). **cut a person down to size** *colloq.* ruthlessly expose the limitations of a person's importance, ability, etc. **cut one's eye-teeth** attain worldly wisdom. **cut glass** glass with patterns and designs cut on it. **cut in 1** interrupt. **2** pull in too closely in front of another vehicle (esp. having overtaken it). **3** give a share of profits etc. to (a person). **4** connect (a source of electricity). **5** join in a card-game by taking the place of a player who cuts out. **6** interrupt a dancing couple to take over from one partner. **cut into 1** make a cut in (*they cut into the cake*). **2** interfere with and reduce (*travelling cuts into my free time*). **cut it fine** see FINE¹. **cut it out** (usu. in *imper.*) *sl.* stop doing that (esp. quarrelling). **cut the knot** solve a problem in an irregular but efficient way. **cut-line 1** a caption to an illustration. **2** the line in squash above which a served ball must strike the wall. **cut loose 1** begin to act freely. **2** see sense 4 of *v.* **cut one's losses** (or **a loss**) abandon an unprofitable enterprise before losses become too great. **cut the mustard** *N. Amer. sl.* reach the required standard. **cut no ice** *sl.* **1** have no influence or importance. **2** achieve little or nothing. **cut off 1** remove (an appendage) by cutting. **2 a** (often in *passive*) bring to an abrupt end or (esp. early) death. **b** intercept, interrupt; prevent from continuing (*cut off supplies; cut off the gas*). **c** disconnect (a person engaged in a telephone conversation) (*was suddenly cut off*). **3 a** prevent from travelling or venturing out (*was cut off by the snow*). **b** (as **cut off** *adj.*) isolated, remote (*felt cut off in the country*). **4 a** disinherit (*was cut off without a penny*). **b** sever a relationship (*was cut off from the children*). **cut-off** *n.* **1** the point at which something is cut off. **2** a device for stopping a flow. **3** *US* a short cut. **4** (in *pl.*) shorts, esp. made from cut-down jeans. **cut out 1** remove from the inside by cutting. **2** make by cutting from a larger whole. **3** omit; leave out. **4** *colloq.* stop doing or using (something) (*managed to cut out chocolate; let's cut out the arguing*). **5** cease or cause to cease functioning (*the engine cut out*). **6** outdo or supplant (a rival). **7** *US* detach (an animal) from the herd. **8** *Cards* be excluded from a card-game as a result of cutting the pack. **9** *colloq.* prepare, plan (*has his work cut out*). **cut-out 1** a figure cut out of paper etc. **2** a device for automatic disconnection, the release of exhaust gases, etc. **cut-out box** *US* = *fuse-box* (see FUSE¹). **cut-price** (or **-rate**) selling or sold at a reduced price. **cut short 1** interrupt; terminate prematurely (*cut short his visit*). **2** make shorter or more concise. **cut one's teeth on** acquire initial practice or experience from (something). **cut a tooth** have it appear through the gum. **cut up 1** cut into pieces. **2** destroy utterly. **3** (usu. in *passive*) distress greatly (*was very cut up about it*). **4** criticize severely. **5** *N. Amer. derog.* behave in a comical or unruly manner. **cut up rough** *Brit. sl.* show anger or resentment. **cut up well** *sl.* bequeath a large fortune. **have one's work cut out** see WORK. [ME *cutte, kitte, kette,* perh. f. OE *cyttan* (unrecorded)]

cutaneous /kjuːˈteɪnɪəs/ *adj.* of the skin. [mod.L *cutaneus* f. L *cutis* skin]

cutaway /ˈkʌtəˌweɪ/ *adj.* **1** (of a diagram etc.) with some parts left out to reveal the interior. **2** (of a coat) with the front below the waist cut away.

cute /kjuːt/ *adj. colloq.* **1 a** attractive, pretty; quaint. **b** affectedly attractive. **2** clever, ingenious. □ **cutely** *adv.* **cuteness** *n.* [shortening of ACUTE]

Cuthbert, St /ˈkʌθbət/ (d.687), English monk. He travelled in the north of England as a missionary before living as a hermit on Farne Island and later becoming bishop of Lindisfarne. Feast day, 20 Mar.

cuticle /ˈkjuːtɪk(ə)l/ n. **1 a** the dead skin at the base of a fingernail or toenail. **b** the epidermis or other superficial skin. **2** *Bot. & Zool.* the outer layer of an organism, esp. a non-cellular protective layer covering the epidermis of a plant or insect. □ **cuticular** /kjuːˈtɪkjʊlə(r)/ *adj.* [L *cuticula*, dimin. of *cutis* skin]

cutie /ˈkjuːtɪ/ n. *sl.* an attractive young woman.

cutis /ˈkjuːtɪs/ n. *Anat.* the true skin or dermis, underlying the epidermis. [L, = skin]

cutlass /ˈkʌtləs/ n. a short sword with a slightly curved blade, esp. of the type formerly used by sailors. [F *coutelas* ult. f. L *cultellus*: see CUTLER]

cutler /ˈkʌtlə(r)/ n. a person who makes or deals in cutlery. [ME f. AF *cotillere*, OF *coutelier* f. *coutel* f. L *cultellus* dimin. of *culter* COULTER]

cutlery /ˈkʌtlərɪ/ n. knives, forks, and spoons for use at table. [OF & F *coutel(l)erie* (as CUTLER)]

cutlet /ˈkʌtlɪt/ n. **1** a neck-chop of mutton or lamb. **2** a small piece of veal etc. for frying. **3** a flat cake of minced meat or nuts and breadcrumbs etc. [F *côtelette*, OF *costelet* dimin. of *coste* rib f. L *costa*]

cutpurse /ˈkʌtpɜːs/ n. *archaic* a pickpocket; a thief.

cutter /ˈkʌtə(r)/ n. **1** a person or thing that cuts. **2** a tailor etc. who takes measurements and cuts cloth. **3** *Naut.* **a** a small fast sailing-ship. **b** a small boat carried by a large ship. **4** *Cricket* a ball turning sharply on pitching. **5** *N. Amer.* a light horse-drawn sleigh.

cutthroat /ˈkʌtθrəʊt/ n. & adj. ● n. **1** a violent criminal; a murderer. **2** (in full **cutthroat razor**) a razor having a long blade set in a handle and usu. folding like a penknife. **3** a trout, *Salmo clarki*, with a red mark under the jaw. ● adj. **1** (of competition) ruthless and intense. **2** (of a person) murderous. **3** (of a card-game) three-handed.

cutting /ˈkʌtɪŋ/ n. & adj. ● n. **1** a piece cut from a newspaper etc. **2** a piece cut from a plant for propagation. **3** an excavated channel through high ground for a railway or road. ● adj. that cuts (see CUT v. 5). □ **cutting edge 1** an edge that cuts. **2** the forefront or vanguard (of a movement, industry, etc.). **cutting-edge** *attrib.adj.* pioneering, innovative. □ **cuttingly** *adv.*

cuttle /ˈkʌt(ə)l/ n. = CUTTLEFISH. □ **cuttle-bone** the internal shell of the cuttlefish crushed and used for polishing teeth etc. or as a supplement to the diet of a cage-bird. [OE *cudele*, ME *codel*, rel. to *cod* bag, with ref. to its ink-bag]

cuttlefish /ˈkʌt(ə)lˌfɪʃ/ n. a marine cephalopod mollusc of the genera *Sepia* or *Sepiola*, having ten arms and ejecting a black fluid when threatened or pursued.

cutty /ˈkʌtɪ/ adj. & n. *Sc. & N. Engl.* ● adj. cut short; abnormally short. ● n. (pl. **-ies**) a short tobacco pipe. □ **cutty-stool** *hist.* a stool on which offenders had to sit to be publicly rebuked during a church service.

Cutty Sark /ˌkʌtɪ ˈsɑːk/ the only survivor of the British tea-clippers, launched in 1869 and now preserved as a museum ship at Greenwich, London. The name comes from the poem 'Tam o' Shanter' by Robert Burns, about a Scottish farmer chased by a young witch who wore only her 'cutty sark' (= short shift); the ship's figurehead is a representation of the witch with her arm outstretched to catch the tail of the horse on which the farmer was escaping.

cutwater /ˈkʌtˌwɔːtə(r)/ n. **1** the forward edge of a ship's prow. **2** a wedge-shaped projection from the pier of a bridge.

cutworm /ˈkʌtwɜːm/ n. a caterpillar that eats through the stems of young plants level with the ground.

cuvée /kjuːˈveɪ/ n. a blend or batch of wine. [F, = vatful f. *cuve* cask f. L *cupa*]

cuvette /kjuːˈvet/ n. **1** a shallow vessel for liquid. **2** *Chem.* a straight-sided transparent vessel for holding a liquid sample in a spectrophotometer. [F, dimin. of *cuve* cask f. L *cupa*]

Cuvier /ˈkuːvɪˌeɪ/, Georges Léopold Chrétien Frédéric Dagobert, Baron (1769–1832), French naturalist. Cuvier carried out a study of fossil elephants which in effect founded the science of palaeontology. Pioneering also in comparative anatomy, he was the first to classify the lower invertebrates. He realized that each species could be derived from another by small changes in structure, which proved crucial in the emergence of evolutionary theory. However, he believed resolutely in the creationist view and quarrelled publicly with the early proponents of evolution, notably Lamarck.

Cuzco /ˈkʊskəʊ/ a city in the Andes in southern Peru; pop. (1990) 275,000. It was the capital of the Inca empire until the Spanish conquest in 1533.

c.v. *abbr.* curriculum vitae.

CVO *abbr.* Commander of the Royal Victorian Order.

CVS *abbr.* chorionic villus sample, a test on a pregnant woman to detect any chromosomal abnormalities in the foetus.

Cwlth. *abbr.* Commonwealth.

cwm /kuːm/ n. **1** (in Wales) = COOMB. **2** *Geog.* a cirque. [Welsh]

Cwmbran /kʊmˈbrɑːn/ a town in SE Wales, the administrative centre of Gwent; pop. (1981) 44,800.

c.w.o. *abbr.* cash with order.

CWS see COOPERATIVE WHOLESALE SOCIETY.

cwt. *abbr.* hundredweight.

-cy /sɪ/ *suffix* (see also -ACY, -ANCY, -CRACY, -ENCY, -MANCY). **1** denoting state or condition (*bankruptcy*; *idiocy*). **2** denoting rank or status (*captaincy*). [from or after L *-cia*, *-tia*, Gk *-k(e)ia*, *-t(e)ia*]

cyan /ˈsaɪæn/ adj. & n. ● adj. greenish-blue. ● n. a greenish-blue colour. [Gk *kuan(e)os* dark blue]

cyanamide /saɪˈænəˌmaɪd/ n. *Chem.* **1** a weakly acidic colourless crystalline compound (chem. formula: CH_2N_2). **2** a salt of this, esp. the calcium salt used as a fertilizer. [CYANOGEN + AMIDE]

cyanic acid /saɪˈænɪk/ n. *Chem.* a very unstable acid (chem. formula: HOCN). (See also FULMINIC ACID, ISOCYANIC ACID.) □ **cyanate** /ˈsaɪəˌneɪt/ n. [CYANOGEN]

cyanide /ˈsaɪəˌnaɪd/ n. any of the highly poisonous salts or esters of hydrocyanic acid, esp. the potassium salt used in the extraction of gold and silver. [CYANOGEN + -IDE]

cyanobacterium /ˌsaɪənəʊbækˈtɪərɪəm/ n. (pl. **cyanobacteria** /-rɪə/) a prokaryotic micro-organism of the division Cyanobacteria. They contain chlorophyll and are capable of photosynthesis, and some kinds are responsible for releasing toxins into water. Also called *blue-green alga*. [CYANOGEN + BACTERIUM]

cyanocobalamin /ˌsaɪəˌnəʊkəˈbæləmɪn/ n. a vitamin of the B complex, found in foods of animal origin such as liver, fish, and eggs, a deficiency of which can cause pernicious anaemia. Also called *vitamin B_{12}*. [CYANOGEN + *cobalamin* f. COBALT + VITAMIN]

cyanogen /saɪˈænədʒən/ n. *Chem.* a colourless highly poisonous gas (chem. formula: C_2N_2), made as an intermediate in fertilizer manufacture. [F *cyanogène* f. Gk *kuanos* dark blue mineral, as being a constituent of Prussian blue]

cyanosis /ˌsaɪəˈnəʊsɪs/ n. *Med.* a bluish discoloration of the skin due to the presence of oxygen-deficient blood. □ **cyanotic** /-ˈnɒtɪk/ *adj.* [mod.L f. Gk *kuanōsis* blueness (as CYANOGEN)]

Cybele /ˈsɪbɪlɪ/ *Mythol.* a mother goddess worshipped especially in Phrygia and later in Greece (where she was associated with Demeter), Rome, and the Roman provinces, with her consort Attis.

cybernation /ˌsaɪbəˈneɪʃ(ə)n/ n. control by machines. □ **cybernate** /ˈsaɪbəˌneɪt/ v.tr. [f. CYBERNETICS + -ATION]

cybernetics /ˌsaɪbəˈnetɪks/ n.pl. (usu. treated as *sing.*) the science of communications and automatic control systems in both machines and living things. □ **cybernetic** *adj.* **cybernetician** /-neˈtɪʃ(ə)n/ n. **cyberneticist** /-ˈnetɪsɪst/ n. [Gk *kubernētēs* steersman]

cyberpunk /ˈsaɪbəˌpʌŋk/ n. **1** a style of science fiction featuring urban counter-culture in a world of high technology and virtual reality. **2** *colloq.* a computer hacker. [CYBERNETICS + PUNK]

cyberspace /ˈsaɪbəˌspeɪs/ n. the notional environment in which electronic communication occurs; virtual reality.

cyborg /ˈsaɪbɔːg/ n. a person whose physical abilities are extended beyond normal human limitations by machine technology (as yet undeveloped). [CYBERNETIC + ORGANISM]

cycad /ˈsaɪkæd/ n. *Bot.* a palmlike gymnosperm of the order Cycadales, including many fossil forms, inhabiting tropical and subtropical regions and often growing to a great height. [mod.L *cycas*, *cycad-* f. supposed Gk *kukas*, scribal error for *koikas*, pl. of *koix* Egyptian palm]

Cyclades /ˈsɪkləˌdiːz/ (Greek **Kikládhes** /kiˈklaðes/) a large group of islands in the southern Aegean Sea, regarded in antiquity as circling around the sacred island of Delos. The Cyclades form a department of modern Greece. □ **Cycladic** /saɪˈklædɪk, sɪ-/ *adj.* [L, f. Gk *kuklos* circle) f. Gk *Kuklades* f. *kuklos* circle (of islands)]

cyclamate /ˈsaɪkləˌmeɪt, ˈsɪk-/ n. a salt of cyclohexylsulphamic acid, esp. one formerly used as a sweetening agent. [chem. name *cyclohexylsulphamate*]

cyclamen /ˈsɪkləmən/ n. **1** a European plant of the genus Cyclamen, of the primrose family, having pink, red, or white flowers with reflexed petals, often grown in pots. **2** the shade of colour of the red or pink cyclamen flower. [med.L f. Gk kuklaminos, perh. f. kuklos circle, with ref. to its bulbous roots]

cycle /ˈsaɪk(ə)l/ n. & v. ● n. **1 a** a recurrent round or period (of events, phenomena, etc.). **b** the time needed for one such round or period. **2 a** Physics etc. a recurrent series of operations or states. **b** Electr. = HERTZ. **3** a series of songs, poems, etc., usu. on a single theme. **4** a bicycle, tricycle, or similar machine. ● v.intr. **1** ride a bicycle etc. **2** move in cycles. □ **cycle-track** (or **-way**) a path or road for bicycles. [ME f. OF, or f. LL cyclus f. Gk kuklos circle]

cyclic /ˈsaɪklɪk/ adj. **1 a** recurring in cycles. **b** belonging to a chronological cycle. **2** Chem. with constituent atoms forming a ring. **3** of a cycle of songs etc. **4** Bot. (of a flower) with its parts arranged in whorls. **5** Math. of a circle or cycle. [F cyclique or L cyclicus f. Gk kuklikos (as CYCLE)]

cyclical /ˈsaɪklɪk(ə)l, ˈsɪk-/ adj. = CYCLIC 1. □ **cyclically** adv.

cycling /ˈsaɪklɪŋ/ n. the sport or recreation of riding bicycles. Cycling as a sport began in 1868, when a 1200 m race was ridden in a park near Paris. Racing can take place on an oval track, on roads (usually over long distances broken into day-long stages), or across country (CYCLO-CROSS). The sport spread to Britain and America, but has remained more popular in Continental Europe; the sport's major competition, the Tour de France, was instituted in 1903.

cyclist /ˈsaɪklɪst/ n. a rider of a cycle, esp. a bicycle.

cyclo- /ˈsaɪkləʊ/ comb. form circle, cycle, or cyclic (cyclometer; cyclorama). [Gk kuklos circle]

cycloalkane /ˌsaɪkləʊˈælkeɪn/ n. Chem. a saturated cyclic hydrocarbon.

cyclo-cross /ˈsaɪkləʊˌkrɒs/ n. cross-country racing on bicycles, a sport mainly confined to Europe (especially Luxembourg, the Low Countries, and Russia) but practised also in the US, especially around Chicago.

cyclograph /ˈsaɪkləˌɡrɑːf/ n. an instrument for tracing circular arcs.

cyclohexane /ˌsaɪkləʊˈhekseɪn/ n. Chem. a colourless liquid cycloalkane (chem. formula: C_6H_{12}) used as a solvent and paint remover. □ **cyclohexyl** n.

cycloid /ˈsaɪklɔɪd/ n. Math. a curve traced by a point on a circle when the circle is rolled along a straight line. □ **cycloidal** /saɪˈklɔɪd(ə)l/ adj. [Gk kukloeidēs (as CYCLE, -OID)]

cyclometer /saɪˈklɒmɪtə(r)/ n. **1** an instrument for measuring circular arcs. **2** an instrument for measuring the distance traversed by a bicycle etc.

cyclone /ˈsaɪkləʊn/ n. **1** Meteorol. a weather system with low barometric pressure at its centre, around which air circulates in an anticlockwise (northern hemisphere) or clockwise (southern hemisphere) direction; a depression. **2** (often **tropical cyclone**) a violent and persistent storm system, usu. formed in the tropics, in which winds revolve around an area of low atmospheric pressure; a hurricane or typhoon. □ **cyclonic** /saɪˈklɒnɪk/ adj. **cyclonically** adv. [prob. repr. Gk kuklōma wheel, coil of a snake]

cycloparaffin /ˌsaɪkləʊˈpærəfɪn/ n. Chem. = CYCLOALKANE.

Cyclopean /ˌsaɪkləˈpiːən, saɪˈkləʊpɪən/ adj. (also **Cyclopian**) **1** (of ancient masonry) made with massive irregular blocks. **2** of or resembling a Cyclops.

cyclopedia /ˌsaɪkləˈpiːdɪə/ n. (also **cyclopaedia**) an encyclopedia. □ **cyclopedic** adj. [shortening of ENCYCLOPEDIA]

cyclopropane /ˌsaɪkləʊˈprəʊpeɪn/ n. Chem. a colourless gaseous cycloalkane (chem. formula: C_3H_6) used as a general anaesthetic.

Cyclops /ˈsaɪklɒps/ n. (pl. **Cyclops** or **Cyclopes** /saɪˈkləʊpiːz/) Gk Mythol. a member of a race of savage one-eyed giants who lived as shepherds. In the Odyssey, Odysseus escaped from the Cyclops Polyphemus by putting out his eye while he slept. Elsewhere they are portrayed as three one-eyed giants who made thunderbolts. [L f. Gk Kuklōps f. kuklos circle + ōps eye]

cyclops /ˈsaɪklɒps/ n. (pl. **cyclops** or **cyclopes** /saɪˈkləʊpiːz/) Zool. a minute predatory copepod crustacean, esp. of the genus Cyclops, with a single central eye. [CYCLOPS]

cyclorama /ˌsaɪkləʊˈrɑːmə/ n. a circular panorama, curved wall, or cloth at the rear of a stage, esp. one used to represent the sky. □ **cycloramic** /-ˈræmɪk/ adj.

cyclostome /ˈsaɪkləˌstəʊm/ n. Zool. an eel-like jawless vertebrate of the subclass Cyclostomata, having a large sucking mouth, e.g. a lamprey. □ **cyclostomate** /saɪˈklɒstəmət/ adj. [CYCLO- + Gk stoma mouth]

cyclostyle /ˈsaɪkləˌstaɪl/ n. & v. ● n. an apparatus for printing copies of writing from a stencil. ● v.tr. print or reproduce with this.

cyclothymia /ˌsaɪkləʊˈθaɪmɪə/ n. Psychol. a disorder characterized by the occurrence of marked swings of mood from cheerfulness to misery. □ **cyclothymic** adj. [CYCLO- + Gk thumos temper]

cyclotron /ˈsaɪkləˌtrɒn/ n. Physics an apparatus in which charged particles are accelerated by an alternating electric field while following an outward spiral or circular path in a magnetic field. (See also PARTICLE ACCELERATOR.)

cyder var. of CIDER.

cygnet /ˈsɪɡnɪt/ n. a young swan. [ME f. AF cignet dimin. of OF cigne swan f. med.L cycnus f. Gk kuknos]

Cygnus /ˈsɪɡnəs/ n. Astron. a prominent northern constellation (the Swan), said to represent a flying swan, as the form adopted by Zeus on one occasion. It contains the bright star Deneb, the strong X-ray source Cygnus X-1 (which probably contains a black hole), and some rich star fields of the Milky Way. [L, = swan]

cylinder /ˈsɪlɪndə(r)/ n. **1 a** a uniform solid or hollow body with straight sides and a circular section. **b** a thing of this shape, e.g. a container for liquefied gas. **2** a cylinder-shaped part of various machines, esp. a piston-chamber in an engine. **3** Printing a metal roller. □ **cylinder saw** = crown saw. □ **cylindrical** /sɪˈlɪndrɪk(ə)l/ adj. **cylindrically** adv. [L cylindrus f. Gk kulindros f. kulindō roll]

cylinder seal n. a small cylindrical object engraved with a design, used in antiquity as a mark of property and later as a signature to authenticate clay documents, particularly in Mesopotamia from the late 4th to the 1st millennium BC.

cyma /ˈsaɪmə/ n. (pl. **cymas** or **cymae** /-miː/) **1** Archit. an ogee moulding of a cornice. **2** = CYME. [mod.L f. Gk kuma wave, wavy moulding]

cymbal /ˈsɪmb(ə)l/ n. a percussion instrument consisting of a concave brass or bronze plate, struck with another or with a stick to make a ringing sound. The cymbal is an ancient instrument, having been used at least since the time of the Assyrians. The earliest cymbals were clashed together as a pair, in combination with a drum, as a rhythm instrument. The paired cymbal still finds a place in the orchestra and in marching bands, but the single suspended cymbal is commoner today, especially as part of the drum-kit. □ **cymbalist** n. [ME f. L cymbalum f. Gk kumbalon f. kumbē cup]

Cymbeline /ˈsɪmbəˌliːn/ (also **Cunobelinus** /ˌkjuːnəʊbəˈlaɪnəs/) (died c.42 AD), British chieftain. He was a powerful ruler whose tribe occupied a wide area from Northamptonshire to SE England. He made Camulodunum (Colchester) his capital, and established a mint there. He was the subject of a medieval fable used by Shakespeare for his play Cymbeline.

cymbidium /sɪmˈbɪdɪəm/ n. a tropical epiphytic orchid of the genus Cymbidium, with a recess in the flower-lip. [mod.L f. Gk kumbē cup]

cymbiform /ˈsɪmbɪˌfɔːm/ adj. Anat. & Bot. boat-shaped. [L cymba f. Gk kumbē boat + -FORM]

cyme /saɪm/ n. Bot. an inflorescence in which the primary axis bears a single terminal flower that develops first, the system being continued by the axes of secondary and higher orders each with a flower (cf. RACEME). □ **cymose** /ˈsaɪməʊz, -məʊs/ adj. [F, var. of cime summit, ult. f. Gk kuma wave]

Cymric /ˈkɪmrɪk/ adj. Welsh. [Welsh CYMRU]

Cymru /ˈkʌmrɪ/ the Welsh name for WALES.

Cynewulf /ˈkɪnɪˌwʊlf/ (late 8th–9th centuries), Anglo-Saxon poet. Of the many poems that have been attributed to him in the past, modern scholarship restricts attribution to four: Juliana, Elene, The Fates of the Apostles, and Christ II. Each of these is inscribed with his name in runes in Anglo-Saxon collections.

Cynic /ˈsɪnɪk/ n. a member of an ancient Greek sect of philosophers founded by Antisthenes, a pupil of Socrates, who were characterized by an ostentatious contempt for ease, wealth, and the enjoyments of life. The most famous was Diogenes, a pupil of Antisthenes (c.445–c.365 BC), who carried these principles to an extreme of asceticism. The movement flourished in the 3rd century BC and revived in the 1st century AD, when Cynic beggar philosophers became a common sight in the Roman Empire. [L Cynicus f. Gk kunikos f. kuōn kunos dog, nickname of Diogenes]

cynic /ˈsɪnɪk/ n. & adj. ● n. a cynical person, someone with little faith in human sincerity etc. ● *adj.* = CYNICAL. □ **cynicism** /'sɪnɪˌsɪz(ə)m/ *n.* [CYNIC]

cynical /'sɪnɪk(ə)l/ *adj.* **1** of or characteristic of a cynic; distrustful or incredulous of human sincerity or goodness. **2** (of behaviour etc.) disregarding accepted standards, unscrupulous. **3** sneering, mocking. □ **cynically** *adv.*

cynocephalus /ˌsaɪnəʊ'sefələs, -'kefələs/ *n.* **1** a fabled dog-headed man. **2** *Zool.* a flying lemur. [Gk *kunokephalos* f. *kuōn kunos* dog + *kephalē* head]

cynosure /'saɪnəˌzjʊə(r), 'sɪnə-/ *n.* **1** a centre of attraction or admiration. **2** a guiding star. [F *cynosure* or L *cynosura* f. Gk *kunosoura* dog's tail, Ursa Minor f. *kuōn kunos* dog + *oura* tail]

cypher var. of CIPHER.

cy pres /si: 'preɪ/ *adv. & adj.* Law as near as possible to the testator's or donor's intentions when these cannot be precisely followed. [AF, = *si près* so near]

cypress /'saɪprəs/ *n.* **1** an evergreen coniferous tree of the genus *Cupressus* or *Chamaecyparis*, having flattened shoots with scalelike leaves; esp. *Cupressus sempervirens*, with dark foliage and hard wood. **2** this tree, or branches from it, as a symbol of mourning. [ME f. OF *cipres* f. LL *cypressus* f. Gk *kuparissos*]

Cyprian /'sɪprɪən/ *n. & adj.* = CYPRIOT. [L *Cyprius* of Cyprus]

Cyprian, St /'sɪprɪən/ (d.258), Carthaginian bishop and martyr. The author of a work on the nature of true unity in the Church in its relation to the episcopate, he was martyred in the reign of the Roman emperor Valerian. Feast day, 16 or 26 Sept.

cyprinoid /'sɪprɪˌnɔɪd/ *adj. & n.* ● *adj.* of or like a carp. ● *n.* a carp or related fish. [L *cyprinus* f. Gk *kuprinos* carp]

Cypriot /'sɪprɪət/ *n. & adj.* (also **Cypriote** /-rɪˌəʊt/) ● *n.* **1** a native or national of Cyprus. **2** the dialect of Greek used in Cyprus. ● *adj.* **1** of or relating to Cyprus or its people or dialect. **2** denoting an ancient syllabic script related to the Minoan and Mycenaean scripts and used to write the Cypriot dialect esp. from the 6th to the 3rd centuries BC. [Gk *Kupriōtes* f. *Kupros* Cyprus]

cypripedium /ˌsɪprɪ'pi:dɪəm/ *n.* an orchid of the genus *Cypripedium*, esp. the lady's slipper. [mod.L f. Gk *Kupris* Aphrodite + *pedilon* slipper]

Cyprus /'saɪprəs/ an island in the eastern Mediterranean about 80 km (50 miles) south of the Turkish coast; pop. (est. 1991) 708,000; official languages, Greek and Turkish; capital, Nicosia. The largest Mediterranean island, Cyprus was colonized from Greece in the 14th century BC, and in classical times was noted for copper (which is named after it) and the cult of Aphrodite. Placed at the crossroads of a number of ancient civilizations, Cyprus was successively subject to Assyrian, Egyptian, Persian, Ptolemaic, and Roman overlords; in medieval times it was ruled by Byzantines, Arabs, Franks, and Venetians, until it was conquered in 1571 by the Turks. In 1878 it was placed under British administration, becoming a Crown Colony in 1925. The island's recent history has been dominated by tension between the Greek Cypriots (some of whom favour enosis or union with Greece) and the Turkish Cypriots. After a period of virtual civil war, Cyprus became an independent Commonwealth republic in 1960. Tensions continued, and in 1974 Turkish forces intervened, taking over the northern part of the island. This became the Turkish Federated State of Cyprus in 1975, and proclaimed itself the independent Turkish Republic of Northern Cyprus in 1983, but has not received international recognition.

cypsela /'sɪpsɪlə/ *n.* (*pl.* **cypselae** /-ˌliː/) *Bot.* a dry single-seeded fruit formed from a double ovary of which only one develops into a seed, characteristic of the daisy family Compositae. [mod.L f. Gk *kupselē* hollow vessel]

Cyrano de Bergerac /ˌsɪrənəʊ də 'beəʒə,ræk/, Savinien (1619–55), French soldier, duellist, and writer. He wrote comedies and satire, but is now chiefly remembered for the large number of duels that he fought (many on account of his proverbially large nose); this aspect of his life is immortalized in a play by Edmond Rostand (*Cyrano de Bergerac*, 1897).

Cyrenaic /ˌsaɪrɪ'neɪɪk/ *adj.* of or denoting the hedonistic school of philosophy founded *c.*400 BC by Aristippus the Elder, a pupil of Socrates. Its ethical doctrines anticipated those of the Epicureans.

Cyrenaica /ˌsaɪrɪ'neɪkə/ a region of NE Libya, bordering on the Mediterranean Sea, settled by the Greeks *c.*640 BC.

Cyrene /saɪ'riːniː/ an ancient Greek city in North Africa, near the coast in Cyrenaica. From the 4th century BC it was a great intellectual centre,

with a noted medical school, and gave its name to the Cyrenaic school of philosophy.

Cyril, St /'sɪrɪl/ (826–69), Greek missionary. The invention of the Cyrillic alphabet is ascribed to him. He and his brother St Methodius (*c.*815–85) became known as the 'Apostles to the Slavs'. Sent to Moravia, they taught in the vernacular, which they adopted also for the liturgy, and circulated a Slavonic version of the Scriptures. Feast day (in the Eastern Church) 11 May; (in the Western Church) 14 Feb.

Cyrillic /sɪ'rɪlɪk/ *n.* the alphabet used today for writing Russian, Bulgarian, Serbian, Ukrainian, and some other Slavonic languages. It was derived from Greek uncials, and was reputedly introduced by St Cyril.

Cyril of Alexandria, St /'sɪrɪl/ (d.444), Doctor of the Church and patriarch of Alexandria. A champion of orthodox thought on the person of Christ, he is best known for his vehement opposition to the views of Nestorius (whose condemnation he secured at the Council of Ephesus in 431). His extensive writings include a series of theological treatises, sermons, and letters. Feast day, 9 Feb.

Cyrus[1] /'saɪərəs/ (known as Cyrus the Great) (died *c.*530 BC), father of Cambyses, king of Persia 559–530 BC and founder of the Achaemenid dynasty. He became ruler of the Median empire after the capture of King Astyages in 550 BC, and went on to conquer Asia Minor, Babylonia, Syria, Palestine, and most of the Iranian plateau. He is said to have ruled his empire with wisdom and moderation, maintaining good relations with the Jews (whom he freed from the Babylonian Captivity) and the Phoenicians.

Cyrus[2] /'saɪərəs/ (known as Cyrus the Younger) (d.401 BC), Persian prince, second son of Darius II. On the death of his father (405 BC), Cyrus led an army of mercenaries against his elder brother, who had succeeded to the throne as Artaxerxes II; his campaign is recounted by the historian Xenophon, who had enlisted in his army. Cyrus was killed in battle north of Babylon.

cyst /sɪst/ *n.* **1** *Med.* **a** a sac or cavity of abnormal character, containing fluid. **b** a structure containing a larva of a parasitic worm etc. **2** *Biol.* **a** a hollow organ, bladder, etc., in an animal or plant, containing a liquid secretion. **b** a cell or cavity enclosing reproductive bodies, an embryo, etc. [LL *cystis* f. Gk *kustis* bladder]

cysteine /'sɪstɪˌiːn, -tɪn/ *n.* *Biochem.* a sulphur-containing amino acid, essential in the human diet and a constituent of many enzymes. [CYSTINE + -*eine* (var. of -INE[4])]

cystic /'sɪstɪk/ *adj.* **1** of the urinary bladder. **2** of the gall-bladder. **3** of the nature of a cyst. □ **cystic fibrosis** *Med.* a hereditary disease affecting the exocrine glands and usu. resulting in respiratory infections. [F *cystique* or mod.L *cysticus* (as CYST)]

cystine /'sɪstiːn/ *n.* *Biochem.* an organic base which is a naturally occurring dimer of cysteine. [Gk *kustis* bladder (because first found in urinary calculi) + -INE[4]]

cystitis /sɪ'staɪtɪs/ *n.* an inflammation of the urinary bladder, esp. in women, often caused by infection and usu. accompanied by frequent painful urination.

cysto- /'sɪstəʊ/ *comb. form* the urinary bladder (*cystoscope*; *cystotomy*). [Gk *kustē, kustis* bladder]

cystoscope /'sɪstəˌskəʊp/ *n.* an instrument inserted in the urethra for examining the urinary bladder. □ **cystoscopic** /ˌsɪstə'skɒpɪk/ *adj.* **cystoscopy** /sɪ'stɒskəpɪ/ *n.*

cystotomy /sɪ'stɒtəmɪ/ *n.* (*pl.* -ies) a surgical incision into the urinary bladder.

-cyte /saɪt/ *comb. form* Biol. a mature cell (*leucocyte*) (cf. -BLAST). [Gk *kutos* vessel]

cyto- /'saɪtəʊ/ *comb. form* Biol. cells or a cell. [as -CYTE]

cytochrome /'saɪtəʊˌkrəʊm/ *n.* *Biochem.* a compound consisting of a protein linked to a haem group, which is involved in electron transfer reactions.

cytogenetics /ˌsaɪtəʊdʒɪ'netɪks/ *n.* the study of inheritance in relation to the structure and function of cells. □ **cytogeneticist** /-'netɪsɪst/ *n.* **cytogenetic** *adj.* **cytogenetical** *adj.* **cytogenetically** *adv.*

cytology /saɪ'tɒlədʒɪ/ *n.* the study of cells. □ **cytologist** *n.* **cytological** /ˌsaɪtə'lɒdʒɪk(ə)l/ *adj.* **cytologically** *adv.*

cytomegalovirus /ˌsaɪtəʊ'megələʊˌvaɪərəs/ *n.* (abbr. **CMV**) *Med.* a kind of herpesvirus which may cause nerve damage in babies or in people with weakened immune systems.

cytoplasm /ˈsaɪtəʊˌplæz(ə)m/ *n.* the protoplasmic content of a cell excluding its nucleus. □ **cytoplasmic** /ˌsaɪtəʊˈplæzmɪk/ *adj.*

cytosine /ˈsaɪtəʊˌsiːn/ *n. Biochem.* a pyrimidine found in all living tissue as a component base of DNA and RNA.

cytoskeleton /ˈsaɪtəʊˌskelɪt(ə)n/ *n. Biol.* a network of protein filaments and tubules giving shape and coherence to a living cell.

cytotoxic /ˌsaɪtəʊˈtɒksɪk/ *adj.* toxic to cells.

czar var. of TSAR.

czardas var. of CSARDAS.

czarevich var. of TSAREVICH.

czarina var. of TSARINA.

Czech /tʃek/ *n. & adj.* ● *n.* **1** a native or national of the Czech Republic or (formerly) Czechoslovakia. **2** the Slavonic language spoken in the Czech Republic or (formerly) Czechoslovakia. ● *adj.* **1** of or relating to the Czech Republic or (formerly) Czechoslovakia. **2** of or relating to the Czech language. [Pol. spelling of Bohemian *Čech*]

Czechoslovakia /ˌtʃekəsləˈvækɪə/ a former country in central Europe in the period 1918–1992, now divided between the Czech Republic and Slovakia; capital, Prague. Czechoslovakia was created out of the northern part of the old Austro-Hungarian empire at the end of the First World War. It incorporated the Czech-speaking Bohemians and Moravians in the west with the Slovaks of Slovakia in the east. The country was crushed by Nazi Germany, which annexed the Sudetenland in 1938 and the rest of Bohemia and Moravia in 1939. After the Second World War, power was seized by the Communists and Czechoslovakia remained under Soviet domination, an attempt at liberalization being crushed by Soviet military intervention in 1968. Communist control collapsed in Dec. 1989, and democratic reforms were introduced. In Aug. 1990 the official name of the country was changed to the Czech and Slovak Federal Republic, and the two parts separated on 1 Jan. 1993. □ **Czechoslovakian** *adj. & n.* **Czechoslovak** /-ˈsləʊvæk/ *n. & adj.*

Czech Republic a country in central Europe; pop. (1991) 10,298,700; official language, Czech; capital, Prague. Formerly one of the two constituent republics of Czechoslovakia, the Czech Republic became independent on the partition of that country on 1 Jan. 1993. It comprises the former provinces of Bohemia, Silesia, and Moravia.

Czerny /ˈtʃeənɪ/, Karl (1791–1857), Austrian pianist, teacher, and composer. A pupil of Beethoven and the teacher of Liszt, he was active at a time when the piano was undergoing important structural developments; the bulk of his output is made up of more than 1,000 exercises and studies for this instrument.

Częstochowa /ˌtʃenstəˈkəʊvə/ an industrial city in south central Poland; pop. (1990) 258,000. It is famous for the statue of the black Madonna in its church.

Dd

D¹ /diː/ *n.* (also **d**) (*pl.* **Ds** or **D's**) **1** the fourth letter of the alphabet. **2** *Mus.* the second note of the diatonic scale of C major. **3** (as a Roman numeral) 500. **4** = DEE. **5** the fourth highest class or category (of academic marks etc.).

D² *abbr.* (also **D.**) **1** *US* Democrat. **2** dimension (3-D).

D³ *symb.* **1** *Chem.* the isotope deuterium. **2** electric flux density. **3** dextrorotatory.

d¹ *abbr.* (also **d.**) **1** died. **2** departs. **3** delete. **4** daughter. **5** diameter. **6** depth.

d² *symb.* **1** deci-. **2** *Brit.* (pre-decimal) penny. [sense 2 from Latin *denarius* silver coin]

'd /d/ *abbr.* (usu. after pronouns) had, would (*I'd*; *he'd*).

DA *abbr.* **1** *US* District Attorney. **2** *sl.* = *duck's arse* (see DUCK¹).

D/A *abbr. Computing* digital to analogue.

da *abbr.* deca-.

dab¹ /dæb/ *v.* & *n.* ● *v.* (**dabbed**, **dabbing**) **1** *tr.* press (a surface) briefly with a cloth, sponge, etc., without rubbing, esp. in cleaning or to apply a substance. **2** *tr.* press (a sponge etc.) lightly on a surface. **3** *tr.* (foll. by *on*) apply (a substance) by dabbing a surface. **4** *intr.* (usu. foll. by *at*) aim a feeble blow; tap. **5** *tr.* strike lightly; tap. ● *n.* **1** a brief application of a cloth, sponge, etc., to a surface without rubbing. **2** a small amount of something applied in this way (*a dab of paint*). **3** a light blow or tap. **4** (in *pl.*) *Brit. sl.* fingerprints. □ **dabber** *n.* [ME, imit.]

dab² /dæb/ *n.* a small flatfish of the genus *Limanda*. [15th c.: orig. unkn.]

dab³ /dæb/ *adj.* esp. *Brit. colloq.* □ **dab hand** (usu. foll. by *at*) a person especially skilled (in) (*a dab hand at cooking*). [17th c.: orig. unkn.]

dabble /ˈdæb(ə)l/ *v.* **1** *intr.* (usu. foll. by *in*, *at*) take a casual or superficial interest or part (in a subject or activity). **2** *intr.* move the feet, hands, etc., about in (usu. a small amount of) liquid. **3** *tr.* wet partly or intermittently; moisten, stain, splash. □ **dabbling duck** a duck that habitually feeds in shallow usu. fresh water by dabbling or upending, esp. one of the tribe Anatini, which includes the mallard, teal, etc. □ **dabbler** *n.* [16th c.: f. Du. *dabbelen* or DAB¹]

dabchick /ˈdæbtʃɪk/ *n.* = *little grebe*. [16th c., in earlier forms *dap-*, *dop-*: perh. rel. to OE *dūfedoppa*, DEEP, DIP]

da capo /dɑː ˈkɑːpəʊ/ *adv. Mus.* (a direction at the end of a piece of music to repeat) from the beginning. [It.]

Dacca see DHAKA.

dace /deɪs/ *n.* (*pl.* same) a slender freshwater fish related to the carp, esp. the European *Leuciscus leuciscus*. [OF *dars*: see DART]

dacha /ˈdætʃə/ *n.* a country house or cottage in Russia. [Russ., orig. = grant (of land)]

Dachau /ˈdæxaʊ, ˈdækaʊ/ a Nazi concentration camp in southern Bavaria, Germany, from 1933 to 1945.

dachshund /ˈdækshʊnd/ *n.* a breed of dog with short legs and a long body, originating in Germany. [G, = badger-dog]

Dacia /ˈdeɪʃə, ˈdeɪsɪə/ an ancient country of SE Europe in what is now NW Romania. It was annexed by Trajan in AD 106 as a province of the Roman Empire. □ **Dacian** *adj.*

dacite /ˈdeɪsaɪt/ *n. Geol.* a volcanic rock similar to andesite. [f. DACIA + -ITE¹]

dacoit /dəˈkɔɪt/ *n.* (in India and Burma) a member of a band of armed robbers. [Hindi *ḍakait* f. *ḍākā* gang-robbery]

dactyl /ˈdæktɪl/ *n.* a metrical foot (‾ ˘ ˘) consisting of one long (or stressed) syllable followed by two short (or unstressed). [ME f. L *dactylus* f. Gk *daktulos* finger, the three bones corresponding to the three syllables]

dactylic /dækˈtɪlɪk/ *adj.* & *n.* ● *adj.* of or using dactyls. ● *n.* (usu. in *pl.*) dactylic verse. [L *dactylicus* f. Gk *daktulikos* (as DACTYL)]

dad /dæd/ *n. colloq.* father. [perh. imit. of a child's *da*, *da* (cf. DADDY)]

Dada /ˈdɑːdɑː/ *n.* an early 20th-century international movement in art, literature, music, and film, which revolted against and mocked artistic and social conventions and laid emphasis on the illogical and absurd. The movement was started in Zurich in 1916 by the poet Tristan Tzara and others and soon spread to New York, Paris, and Cologne; the name *dada*, a French word for a child's hobby-horse, is said to have been chosen at random from a dictionary. Typical products include montage, collage, and the ready-made, which all emphasize the anti-rational and the arbitrariness of creative form. The leading figures were Jean Arp, André Breton, Max Ernst, Man Ray, and Marcel Duchamp. □ **Dadaism** *n.* **Dadaist** *n.* & *adj.* **Dadaistic** /ˌdɑːdɑːˈɪstɪk/ *adj.*

daddy /ˈdædɪ/ *n.* (*pl.* **-ies**) *colloq.* **1** father. **2** (usu. foll. by *of*) the oldest or supreme example (*had a daddy of a headache*). □ **daddy-long-legs 1** = *crane-fly*. **2** *US* = HARVESTMAN. [DAD + -Y²]

dado /ˈdeɪdəʊ/ *n.* (*pl.* **-os**) **1** the lower part of the wall of a room when visually distinct from the upper part. **2** the plinth of a column. **3** the cube of a pedestal between the base and the cornice. [It., = DIE²]

Dadra and Nagar Haveli /ˈdɑːdrə, ˌnɑːɡə həˈveɪlɪ/ a Union Territory in western India, on the Arabian Sea; pop. (1991) 138,500; capital, Silvassa.

Daedalus /ˈdiːdələs/ *Gk Mythol.* a craftsman who is said to have built the labyrinth for Minos, king of Crete. Minos imprisoned him and his son Icarus, but they escaped on wings which Daedalus made. Icarus was killed when he flew too near the sun, but Daedalus reached Sicily safely. Daedalus was considered the inventor of carpentry and is credited with the creation of many works of craftsmanship.

daemon var. of DEMON¹ 5.

daemonic var. of DEMONIC.

daff /dæf/ *n. colloq.* a daffodil. [abbr.]

daffodil /ˈdæfədɪl/ *n.* **1** a bulbous plant of the genus *Narcissus*, with yellow flowers having a large trumpet-shaped corona; esp. the common *N. pseudonarcissus*. **2** a pale-yellow colour. [earlier *affodil*, as ASPHODEL]

daffy /ˈdæfɪ/ *adj.* (**daffier**, **daffiest**) *sl.* = DAFT. □ **daffily** *adv.* **daffiness** *n.* [*daff* simpleton + -Y¹]

daft /dɑːft/ *adj.* esp. *Brit. colloq.* **1** silly, foolish, crazy. **2** (foll. by *about*) fond of; infatuated with. □ **daftly** *adv.* **daftness** *n.* [ME *daffte* = OE *gedæfte* mild, meek, f. Gmc]

dag /dæg/ *n.* & *v. Austral.* & *NZ* ● *n.* **1** (usu. in *pl.*) a lock of wool clotted with dung on the hinder parts of a sheep. **2** *sl.* an eccentric or noteworthy person; a character (*he's a bit of a dag*). ● *v.tr.* (**dagged**, **dagging**) remove dags from (a sheep). □ **rattle one's dags** *sl.* hurry up. □ **dagger** *n.* [orig. Engl. dial.]

da Gama /də ˈɡɑːmə/, Vasco (c.1469–1524), Portuguese explorer. He led the first European expedition round the Cape of Good Hope in 1497, sighting and naming Natal on Christmas Day before crossing the Indian Ocean and arriving in Calicut in 1498. The Portuguese king Manuel I (1469–1521) ennobled him on his return and chose him to lead a second expedition to Calicut in 1502. He forced the raja of Calicut (who had massacred Portuguese settlers from an earlier expedition) to make peace, also establishing colonies on the coast of Mozambique. After twenty years in retirement da Gama was sent east again in 1524 as a viceroy to India, but he died soon after his arrival.

Dagestan /ˌdægɪ'stɑːn/ an autonomous republic in SW Russia, on the western shore of the Caspian Sea; pop. (1990) 1,823,000; capital, Makhachkala.

dagga /'dægə/ n. S. Afr. **1** hemp used as a narcotic. **2** a plant of the genus *Leonotis* used similarly. [Afrik. f. Hottentot *dachab*]

dagger /'dægə(r)/ n. **1** a short stabbing-weapon with a pointed and edged blade. **2** *Printing* = OBELUS. □ **at daggers drawn** in bitter enmity. **look daggers at** glare angrily or venomously at. [ME, perh. f. obs. *dag* pierce, infl. by OF *dague* long dagger]

daggy /'dægɪ/ adj. Austral. & NZ sl. **1** dowdy, scruffy. **2** unfashionable. [DAG + -Y[1]]

dago /'deɪgəʊ/ n. (pl. **-os**) sl. offens. a foreigner, esp. a Spaniard, Portuguese, or Italian. [Sp. *Diego* = James]

Dagon /'deɪgɒn/ (in the Bible) a national deity of the ancient Philistines, represented as a fish-tailed man. [Heb. *Dāgōn*]

Daguerre /də'geə(r)/, Louis-Jacques-Mandé (1789–1851), French physicist, painter, and inventor of the first practical photographic process. While working as a painter he co-invented the diorama, and he later went into partnership with Joseph-Nicéphore Niépce (1765–1833) to improve the latter's heliography process. Daguerre greatly reduced the exposure time, and in 1839 he presented his daguerreotype process to the French Academy of Sciences. (See also PHOTOGRAPHY.)

daguerreotype /də'gerəʊˌtaɪp/ n. **1** a photograph taken by an early photographic process employing an iodine-sensitized silvered plate and mercury vapour. **2** this process. [DAGUERRE]

dah /dɑː/ n. Telegraphy (in the Morse system) = DASH (cf. DIT). [imit.]

Dahl /dɑːl/, Roald (1916–90), British writer, of Norwegian descent. The short-story collection *Tales of the Unexpected* (1979) is characteristic of his fiction and drama, which typically include macabre plots and unexpected outcomes. Dahl is also widely known for his stories and poems for children; they include *Charlie and the Chocolate Factory* (1964), *The BFG* (1982), and *Revolting Rhymes* (1982).

dahlia /'deɪlɪə/ n. a composite garden plant of the genus *Dahlia*, of Mexican origin, cultivated for its many-coloured single or double flowers. [A. *Dahl*, Swedish botanist (1751–89)]

Dahomey /də'həʊmɪ/ the former name (until 1975) of BENIN.

Dáil /dɔɪl/ (in full **Dáil Éireann** /'eərən/) the lower House of Parliament in the Republic of Ireland, composed of 166 members (called Teachtí Dála) elected on a basis of proportional representation. It was first established in 1919, when Irish republicans (whose representatives had been elected to Westminster in the 1918 election) proclaimed an Irish state. [Ir., = assembly (of Ireland)]

daily /'deɪlɪ/ adj., adv., & n. ● adj. **1** done, produced, or occurring every day or every weekday. **2** constant, regular. ● adv. **1** every day; from day to day. **2** constantly. ● n. (pl. **-ies**) colloq. **1** a daily newspaper. **2** Brit. a charwoman or domestic help working daily. □ **daily bread** necessary food; a livelihood. **daily dozen** Brit. colloq. regular exercises, esp. on rising. [ME f. DAY + -LY[1], -LY[2]]

Daimler /'deɪmlə(r)/, Gottlieb (1834–1900), German engineer and motor manufacturer. An employee of Nikolaus Otto, he produced a small engine using the Otto cycle in 1884 and made it propel a bicycle in 1886, using petrol vapour. He founded the Daimler motor company in 1890.

daimon /'daɪməʊn/ n. = DEMON[1] 5. □ **daimonic** /daɪ'mɒnɪk/ adj. [Gk, = deity]

dainty /'deɪntɪ/ adj. & n. ● adj. (**daintier, daintiest**) **1** delicately pretty. **2** delicate of build or in movement. **3** (of food) choice. **4** fastidious; having delicate taste and sensibility. ● n. (pl. **-ies**) a choice morsel; a delicacy. □ **daintily** adv. **daintiness** n. [AF *dainté*, OF *daintié, deintié* f. L *dignitas -tatis* f. *dignus* worthy]

daiquiri /'dækərɪ, 'daɪk-/ n. (pl. **daiquiris**) a cocktail of rum, lime juice, etc. [*Daiquiri* in Cuba]

Dairen /daɪ'ren/ the former name for DALIAN.

dairy /'deərɪ/ n. (pl. **-ies**) **1** a building or room for the storage, processing, and distribution of milk and its products. **2** a shop where milk and milk products are sold. **3** (attrib.) **a** of, containing, or concerning milk and its products (and sometimes eggs). **b** used for

dairy products (*dairy cow*). [ME *deierie* f. *deie* maidservant f. OE *dæge* kneader of dough]

dairying /'deərɪɪŋ/ n. the business of producing, storing, and distributing milk and its products.

dairymaid /'deərɪˌmeɪd/ n. a woman employed in a dairy.

dairyman /'deərɪmən/ n. (pl. **-men**) **1** a man dealing in dairy products. **2** a man employed in a dairy.

dais /'deɪs/ n. a low platform, usu. at the upper end of a hall and used to support a table, lectern, etc. [ME f. OF *deis* f. L *discus* disc, dish, in med.L = table]

daisy /'deɪzɪ/ n. (pl. **-ies**) **1 a** a small plant, *Bellis perennis* (family Compositae), bearing flowers each with a yellow disc and white rays. **b** any other composite plant with daisy-like flowers (*ox-eye daisy*; *Michaelmas daisy*; *Shasta daisy*). **2** sl. a first-rate specimen of anything. □ **daisy-chain** a string of daisies threaded together. **daisy-cutter** *Cricket* a ball bowled so as to roll along the ground. **daisy wheel** *Computing* a disc of spokes extending radially from a central hub, each terminating in a printing character, used as a printer in word processors and typewriters. **pushing up the daisies** sl. dead and buried. [OE *dæges ēage* day's eye, the flower opening in the morning]

Dak. abbr. Dakota.

Dakar /'dækɑː(r)/ the capital of Senegal, a port on the Atlantic coast of West Africa; pop. (est. 1985) 1,382,000.

Dakota[1] /də'kəʊtə/ a former territory of the US, organized in 1889 into the states of North Dakota and South Dakota.

Dakota[2] /də'kəʊtə/ see SIOUX.

dal var. of DHAL.

Dalai Lama /ˌdælaɪ 'lɑːmə/ the spiritual head of Tibetan Buddhism and, until the establishment of Chinese Communist rule, the spiritual and temporal ruler of Tibet. The title is applied to a series of lamas, each of whom is believed to be the reincarnation of the bodhisattva Avalokitesvara. When the incumbent lama dies, the lamas search for a child who gives evidence that the soul of the deceased has entered into him; when found, the child succeeds to the office. The present Dalai Lama, the fourteenth incarnation, escaped to India in 1959 following the invasion of Tibet by the Chinese in the early 1950s; he was awarded the Nobel Peace Prize in 1989. [Mongolian *dalai* ocean; see LAMA]

dalasi /də'lɑːsɪ/ n. (pl. same or **dalasis**) the basic monetary unit of the Gambia. [name of an earlier local coin]

Dalcroze see JAQUES-DALCROZE.

Dale /deɪl/, Sir Henry Hallett (1875–1968), English physiologist and pharmacologist. He worked first on the physiological action of ergot, going on to investigate the role of histamine in anaphylactic shock and allergy. He discovered the role of acetylcholine as a natural neurotransmitter, which led to a clearer understanding of the chemical transmission of nerve impulses. Dale later held the most senior posts in British medical research, and shared a Nobel Prize in 1936.

dale /deɪl/ n. a valley, esp. in northern England. [OE *dæl* f. Gmc]

dalek /'dɑːlek/ n. a member of a race of hostile alien machine-organisms which appeared in the BBC television science-fiction serial *Dr Who* from 1963. [invented word, named after an encyclopedia volume covering the alphabetical sequence *dal–lek*]

dalesman /'deɪlzmən/ n. (pl. **-men**) an inhabitant of the dales in Northern England.

Dalhousie /dæl'haʊzɪ/, 1st Marquis of (title of James Andrew Broun Ramsay) (1812–60), British colonial administrator. As Governor-General of India (1847–56) he was responsible for a series of reforms and innovations, notably the introduction of railways and telegraphic communications, as well as the drafting of legislation against slavery, suttee, and female infanticide. He considerably expanded British territory in India, taking Punjab and Pegu and annexing Oudh and Nagpur.

Dali /'dɑːlɪ/, Salvador (1904–89), Spanish painter. He joined the surrealists in 1928 and became one of the most prominent members of the movement. Like many surrealists, he was much influenced by Sigmund Freud's writings on dreams; many of his paintings portray subconscious or dream images painted with almost photographic realism against backgrounds of arid Catalan landscapes. Expelled from the surrealist group in 1938 because of his repudiation of its Marxist politics, he settled in the US; after becoming a Roman Catholic he devoted the latter part of his life to symbolic religious paintings. His most famous pictures include *The Persistence of Memory* (1931) and *Christ of St John of the Cross* (1951). He also produced surrealist writings and collaborated with Luis Buñuel in the production of *Un chien andalou* (1928) and other films.

Dalian /ˌdɑːliˈæn/ a port and shipbuilding centre on the Liaodong Peninsula in NE China, now part of the urban complex of Luda. It was formerly called Dairen.

Dalit /ˈdɑːlɪt/ *n.* (in the Indian subcontinent) a member of the caste of Harijans or untouchables (see UNTOUCHABLE). [Skr., = depressed]

Dallapiccola /ˌdæləˈpɪkələ/, Luigi (1904–75), Italian composer. Influenced by the serial technique of Schoenberg and Webern, he further developed the twelve-note system in a variety of compositions, including songs, opera, ballet music, and a piano concerto. Among his best-known works is *Songs of Prison* (1938–41).

Dallas /ˈdæləs/ a city in NE Texas, noted as a centre of the oil industry; pop. (1990) 1,006,900. President John F. Kennedy was assassinated there in Nov. 1963.

dalliance /ˈdælɪəns/ *n.* **1** a leisurely or frivolous passing of time. **2** the act or an instance of light-hearted flirting; a casual love affair. [DALLY + -ANCE]

dally /ˈdælɪ/ *v.intr.* (**-ies**, **-ied**) **1** delay; waste time, esp. frivolously. **2** (often foll. by *with*) play about; flirt, treat frivolously (*dallied with her affections*). □ **dally away** waste or fritter (one's time, life, etc.). [ME f. OF *dalier* chat]

Dalmatia /dælˈmeɪʃə/ an ancient region in what is now SW Croatia, comprising mountains and a narrow coastal plain along the Adriatic, together with offshore islands. It once formed part of the Roman province of Illyricum.

Dalmatian /dælˈmeɪʃ(ə)n/ *adj. & n.* ● *adj.* of or relating to Dalmatia. ● *n.* a breed of large dog having short white hair with dark spots.

dalmatic /dælˈmætɪk/ *n.* a wide-sleeved long loose vestment open at the sides, worn by deacons and bishops and by a monarch at his or her coronation. [ME f. OF *dalmatique* or LL *dalmatica* (*vestis* robe) of Dalmatia]

Dalriada /dælˈrɪədə/ an ancient Gaelic kingdom in northern Ireland whose people (the *Scots*: see SCOT 2) established a colony in SW Scotland from about the late 5th century. By the 9th century Irish Dalriada had declined but the people of Scottish Dalriada gradually acquired dominion over the whole of Scotland, giving that country its present name.

dal segno /dæl ˈseɪnjəʊ/ *adv. Mus.* (a direction in a piece of music to repeat) from the point marked by a sign. [It., = from the sign]

Dalton /ˈdɔːlt(ə)n/, John (1766–1844), English chemist, the father of modern atomic theory. His study of gases led to his formulation of the law of partial pressures (Dalton's law), according to which the total pressure of a mixture of gases is equal to the sum of the pressures that each gas would exert separately. He then worked on the solubility of gases, producing fundamental work on atomic theory. He defined an atom as the smallest part of a substance that could participate in a chemical reaction, argued that elements are composed of atoms and that elements combine in definite proportions, and produced the first table of comparative atomic weights. He also gave the first detailed description of colour-blindness, based on his own inability to distinguish green from red.

dalton /ˈdɔːlt(ə)n/ *n. Chem.* = atomic mass unit. [DALTON]

daltonism /ˈdɔːltəˌnɪz(ə)m/ *n.* colour-blindness, esp. a congenital inability to distinguish between red and green. [F *daltonisme* f. DALTON]

dam[1] /dæm/ *n. & v.* ● *n.* **1** a barrier constructed to hold back water and raise its level, forming a reservoir or preventing flooding. (*See note below.*) **2** a barrier constructed in a stream by a beaver. **3** anything functioning like a dam does. **4** a causeway. ● *v.tr.* (**dammed, damming**) **1** provide or confine with a dam. **2** (often foll. by *up*) block up; hold back; obstruct. [ME f. MLG, MDu.]

▪ Some of the earliest dams, dating from *c.*2500 BC, are to be found in Egypt, Iran, Iraq, and China, made of earth or masonry. Earth dams, with a waterproof core of clay to prevent seepage, are still found, although where site conditions are suitable concrete is now used. Gravity dams rely on the weight of the dam for stability; arch dams, with a curved structure having its convex face upstream, are supported by abutments at either side of the structure. The main uses of dams are to generate electricity by means of water turbines, to store water for irrigation and industrial or domestic water supply, and to supply canals with water. It is estimated that at present more than one-tenth of the stream-flow of the world's rivers is regulated by dams; by the year 2000 two-thirds could be regulated.

dam[2] /dæm/ *n.* the female parent of an animal, esp. a mammal. [ME: var. of DAME]

damage /ˈdæmɪdʒ/ *n. & v.* ● *n.* **1** harm or injury impairing the value or usefulness of something, or the health or normal function of a person. **2** (in *pl.*) *Law* a sum of money claimed or awarded in compensation for a loss or an injury. **3** the loss of what is desirable. **4** (prec. by *the*) *sl.* cost (*what's the damage?*). ● *v.tr.* **1** inflict damage on. **2** (esp. as **damaging** *adj.*) detract from the reputation of (*a most damaging admission*). □ **damagingly** *adv.* [ME f. OF *damage* (n.), *damagier* (v.), f. *dam(me)* loss f. L *damnum* loss, damage]

Daman and Diu /dəˈmɑːn, ˈdiːuː/ a Union Territory in India, on the west coast north of Bombay; pop. (1991) 101,400; capital, Daman. It consists of the district of Daman and the island of Diu, and until 1987 was administered with Goa.

Damara /dəˈmɑːrə/ *n. & adj.* ● *n.* (*pl.* same or **Damaras**) a member of a people inhabiting mountainous parts of SW Africa and speaking the Nama language. ● *adj.* of or relating to the Damara. [Nama]

Damaraland /dəˈmɑːrəˌlænd/ a plateau region of central Namibia inhabited chiefly by the Damara and Herero peoples.

damascene /ˈdæməˌsiːn/ *v., n., & adj.* ● *v.tr.* decorate (metal, esp. iron or steel) by etching or inlaying esp. with gold or silver, or with a watered pattern produced in welding. ● *n.* a design or article produced in this way. ● *adj.* of, relating to, or produced by this process. [*Damascene* of Damascus, f. L *Damascenus* f. Gk *Damaskēnos*]

Damascus /dəˈmɑːskəs, -ˈmæskəs/ the capital of Syria since the country's independence in 1946; pop. (est. 1987) 1,292,000. It has existed as a city for over 4,000 years.

damask /ˈdæməsk/ *n., adj., & v.* ● *n.* **1 a** a figured woven fabric (esp. silk or linen) with a pattern visible on both sides. **b** twilled table linen with woven designs shown by the reflection of light. **2** a tablecloth made of this material. **3** *hist.* steel with a watered pattern produced in welding. ● *adj.* **1** made of or resembling damask. **2** coloured like a damask rose, velvety pink or vivid red. ● *v.tr.* **1** weave with figured designs. **2** = DAMASCENE *v.* **3** ornament. □ **damask rose** an old sweet-scented variety of rose, with very soft velvety petals, used to make attar. [ME, ult. f. L DAMASCUS]

dame /deɪm/ *n.* **1** (**Dame**) **a** (in the UK) the title given to a woman with the rank of Knight Commander or holder of the Grand Cross in the Orders of Chivalry. **b** a woman holding this title. **2** *Brit.* a comic middle-aged female character in modern pantomime, usu. played by a man. **3** *archaic* a mature woman. **4** *N. Amer. sl.* a woman. □ **dame-school** *hist.* a primary school of a kind kept by elderly women. [ME f. OF f. L *domina* mistress]

damfool /ˈdæmfuːl/ *adj. colloq.* foolish, stupid. [DAMN + FOOL[1]]

Damietta /ˌdæmɪˈetə/ (Arabic **Dumyat** /dʊmˈjaːt/) **1** the eastern branch of the Nile delta. **2** a port at the mouth of this; pop. (est. 1986) 121,200.

dammar /ˈdæmə(r)/ *n.* **1** an eastern Asian and Australasian coniferous tree, esp. of the genus *Agathis* or *Shorea*, yielding a resin used in varnish-making. **2** this resin. [Malay *damar*]

dammit /ˈdæmɪt/ *int.* damn it.

damn /dæm/ *v., n., adj., & adv.* ● *v.tr.* **1** (often *absol.* or as *int.* of anger or annoyance, = may God damn) curse (a person or thing). **2** doom to hell; cause the damnation of. **3** condemn, censure (*a review damning the performance*). **4 a** (often as **damning** *adj.*) (of a circumstance, piece of evidence, etc.) show or prove to be guilty; bring condemnation upon (*evidence against them was damning*). **b** be the ruin of. ● *n.* **1** an uttered curse. **2** *sl.* a negligible amount (*not worth a damn*). ● *adj. & adv. colloq.* = DAMNED. □ **damn all** *colloq.* nothing at all. **damn well** *colloq.* (as an emphatic) simply (*damn well do as I say*). **damn with faint praise** commend so unenthusiastically as to imply disapproval. **I'm** (or **I'll be) damned if** *colloq.* I certainly do not, will not, etc. **not give a damn** see GIVE. **well I'm** (or **I'll be) damned** *colloq.* exclamation of surprise, dismay, etc. □ **damningly** *adv.* [ME f. OF *damner* f. L *damnare* inflict loss on f. *damnum* loss]

damnable /ˈdæmnəb(ə)l/ *adj.* hateful, annoying. □ **damnably** *adv.* [ME f. OF *damnable* (as DAMN)]

damnation /dæmˈneɪʃ(ə)n/ *n. & int.* ● *n.* condemnation to eternal punishment, esp. in hell. ● *int.* expressing anger or annoyance. [ME f. OF *damnation* (as DAMN)]

damnatory /ˈdæmnətərɪ/ *adj.* conveying or causing censure or damnation. [L *damnatorius* (as DAMN)]

damned /dæmd/ *adj. & adv. colloq.* ● *adj.* damnable, infernal, unwelcome. ● *adv.* extremely (*damned hot*; *damned lovely*). □ **damned well** (as an emphatic) simply (*you've damned well got to*). **do one's damnedest** do one's utmost.

damnify /'dæmnɪ̩faɪ/ *v.tr.* (**-ies, -ied**) *Law* cause injury to. □ **damnification** /ˌdæmnɪfɪ'keɪʃ(ə)n/ *n.* [OF *damnifier* etc. f. LL *damnificare* injure (as DAMN)]

Damocles /'dæmə̩kliːz/ a legendary courtier who extravagantly praised the happiness of Dionysius I, ruler of Syracuse. To show him how precarious this happiness was, Dionysius seated him at a banquet with a sword hung by a single hair over his head. □ **sword of Damocles** an imminent danger at a time of apparent well-being.

Damon /'deɪmən/ a legendary Syracusan of the 4th century BC whose friend Pythias (also called Phintias) was sentenced to death by Dionysius I. Damon stood bail for Pythias, who returned from settling his affairs just in time to save him; Pythias was then reprieved. □ **Damon and Pythias** (or **Phintias**) faithful friends.

damp /dæmp/ *adj., n., & v.* ● *adj.* slightly wet; moist. ● *n.* **1** diffused moisture in the air, on a surface, or in a solid, esp. as a cause of inconvenience or danger. **2** dejection; discouragement. **3** = FIREDAMP. ● *v.tr.* **1** make damp; moisten. **2** (often foll. by *down*) **a** take the force or vigour out of (*damp one's enthusiasm*). **b** make flaccid or spiritless. **c** make (a fire) burn less strongly by reducing the flow of air to it. **3** reduce or stop the vibration of (esp. the strings of a musical instrument). **4** quieten. □ **damp** (or **damp-proof**) **course** a layer of waterproof material in the wall of a building near the ground, to prevent rising damp. **damp off** (of a plant) die from a fungus attack in damp conditions. **damp squib** an unsuccessful attempt to impress etc. □ **dampish** *adj.* **damply** *adv.* **dampness** *n.* [ME f. MLG, = vapour etc., OHG *dampf* steam f. WG]

dampen /'dæmpən/ *v.* **1** *tr. & intr.* make or become damp. **2** *tr.* make less forceful or vigorous; stifle, choke. □ **dampener** *n.*

damper /'dæmpə(r)/ *n.* **1** a person or thing that discourages, or tempers enthusiasm. **2** a device that reduces shock or noise. **3** a metal plate in a flue to control the draught, and so the rate of combustion. **4** *Mus.* a pad silencing a piano string except when removed by means of a pedal or by the note's being struck. **5** esp. *Austral. & NZ* an unleavened loaf or cake of flour and water baked in wood ashes. □ **put a damper on** take the vigour or enjoyment out of.

Dampier /'dæmpɪə(r)/, William (1652–1715), English explorer and adventurer. He is notable for having sailed round the world twice. In 1683 he set out on a privateering expedition from Panama, crossing the Pacific to the Philippines, China, and Australia before eventually reaching England again in 1691. In 1699 he was commissioned by the British government to explore the NW coast of Australia and circumnavigated the globe again, despite being shipwrecked on Ascension Island on the way home.

damsel /'dæmz(ə)l/ *n. archaic* or *literary* a young unmarried woman. [ME f. OF *dam(e)isele* ult. f. L *domina* mistress]

damselfish /'dæmz(ə)lˌfɪʃ/ *n.* a small brightly coloured tropical marine fish of the family Pomacentridae; esp. *Chromis chromis*, found in or near coral reefs.

damselfly /'dæmz(ə)lˌflaɪ/ *n.* (*pl.* **-flies**) an insect of the order Odonata, like a slender dragonfly but with its wings folded over the body when resting.

damson /'dæmz(ə)n/ *n. & adj.* ● *n.* **1** (in full **damson plum**) **a** a small dark purple plumlike fruit. **b** the small deciduous tree, *Prunus institia*, bearing this. **2** a dark purple colour. ● *adj.* damson-coloured. □ **damson cheese** a solid preserve of damsons and sugar. [ME *damacene, -scene, -sene* f. L *damascenum* (*prunum* plum) of Damascus]

Dan /dæn/ (in the Bible) **1** a Hebrew patriarch, son of Jacob and Bilhah (Gen. 30:6). **2** the tribe of Israel traditionally descended from him. **3** an ancient town in the north of Canaan, where the tribe of Dan settled. It marked the northern limit of the ancient Hebrew kingdom of Israel (Judges 20; see also BEERSHEBA).

Dan. *abbr.* (in the Bible) Daniel.

dan¹ /dæn/ *n.* **1** any of ten degrees of advanced proficiency in judo. **2** a person who has achieved any of these. [Jap.]

dan² /dæn/ *n.* (in full **dan buoy**) a small buoy used as a marker in deep-sea fishing, or to mark the limits of an area cleared by minesweepers. [17th c.: orig. unkn.]

Dana¹ /'deɪnə/, James Dwight (1813–95), American naturalist, geologist, and mineralogist. At the age of 24 he produced *A System of Mineralogy* and founded a classification of minerals based on chemistry and physics; revised editions of this still appear under his name. His view of the earth as a unit, with its physical features changing and developing progressively, was an evolutionary one, but he did not accept Darwin's theory of evolution until the last edition of his *Manual of Geology*, published shortly before his death.

Dana² /'deɪnə/, Richard Henry (1815–82), American adventurer, lawyer, and writer. He wrote a faithful and lively account of his voyage from Boston round Cape Horn to California, published anonymously as *Two Years before the Mast* (1840). In professional life he was an expert in maritime law and editor of an international law journal.

Danae /'dæneɪˌiː/ *Gk Mythol.* the daughter of Acrisius, king of Argos. An oracle foretold that she would bear a son who would kill her father. In an attempt to evade this Acrisius imprisoned Danae in a tower, but Zeus visited her in the form of a shower of gold and she conceived Perseus, who after many adventures killed Acrisius by accident.

Danaids /'dæneɪˌɪdz/ *Gk Mythol.* the daughters of Danaus, king of Argos, who were compelled to marry the sons of his brother Aegyptus but murdered their husbands on the wedding night, except for one, Hypermnestra, who helped her husband to escape. After their deaths the Danaids, with the exception of Hypermnestra, were punished in Hades by being set to fill a leaky jar with water.

Danakil /'dænəˌkɪl, də'nɑːkɪ(ə)l/ *n. & adj.* = AFAR. [Arab *danākil* pl. of *danākilī*]

Danakil Depression a long low-lying desert region of NE Ethiopia and northern Djibouti, between the Red Sea and the Great Rift Valley.

Da Nang /dɑːˈnæŋ/ a port and city (formerly called Tourane) in central Vietnam, on the South China Sea; pop. (est. 1991) 400,000. During the Vietnam War it was used as a US military base.

dance /dɑːns/ *v. & n.* ● *v.* **1** *intr.* move about rhythmically alone or with a partner or in a set, usu. in fixed steps or sequences to music, for pleasure or as entertainment. **2** *intr.* move in a lively way; skip or jump about. **3** *tr.* **a** perform (a specified dance or form of dancing). **b** perform (a specified role) in a ballet etc. **4** *intr.* move up and down (on water, in the field of vision, etc.). **5** *tr.* move (esp. a child) up and down; dandle. ● *n.* **1 a** a piece of dancing; a sequence of steps in dancing. **b** a special form of this. **2** a single round or turn of a dance. **3** a social gathering for dancing, a ball. **4** a piece of music for dancing to or in a dance rhythm. **5** a dancing or lively motion. □ **dance attendance on** follow or wait on (a person) obsequiously. **dance to a person's tune** accede obsequiously to a person's demands and wishes. **lead a person a dance** (or **merry dance**) *Brit.* cause a person much trouble in following a course one has instigated. □ **danceable** *adj.* [ME f. OF *dance, danse* (n.), *dancer, danser* (v.), f. Rmc, of unkn. orig.]

dancehall /'dɑːnshɔːl/ *n.* a public hall for dancing.

dance of death *n.* a allegorical representation of death, popular in the Middle Ages, in which skeletal figures lead people of all types in a dance to the grave. The inclusion of popes and kings as well as beggars and paupers emphasized the equality of all before death. The earliest painting of the dance of death (*c.*1425) was a mural in a Parisian cemetery; the best-known example is the series of forty-one woodcuts by Holbein (*c.*1523–6), based on a mural in Basle.

dancer /'dɑːnsə(r)/ *n.* **1** a person who performs a dance. **2** a person whose profession is dancing.

d. and c. *n. Med.* dilatation (of the cervix) and curettage (of the uterus), performed after a miscarriage or for the removal of cysts, tumours, etc.

dandelion /'dændɪˌlaɪən/ *n.* a composite plant of the genus *Taraxacum*, with jagged leaves and a large bright yellow flower on a hollow stalk, followed by a globular head of seeds with downy tufts; esp. the common *T. officinale*. □ **dandelion clock** the downy seed-head of a dandelion. **dandelion coffee** dried and powdered dandelion roots; a drink made from this. [F *dent-de-lion* transl. med.L *dens leonis* lion's tooth]

dander /'dændə(r)/ *n. colloq.* temper, anger, indignation. □ **get one's dander up** lose one's temper; become angry. [19th c.: orig. uncert.]

dandify /'dændɪˌfaɪ/ *v.tr.* (**-ies, -ied**) cause to resemble a dandy, dress up very smartly.

dandle /'dænd(ə)l/ *v.tr.* **1** dance (a child) on one's knees or in one's arms. **2** pamper, pet. [16th c.: orig. unkn.]

Dandong /dæn'dʊŋ/ a port in Liaoning province, NE China, near the mouth of the Yalu river, on the border with North Korea; pop. (1984) 560,000. It was formerly called Antung.

dandruff /'dændrʌf/ *n.* dead skin in small scales among the hair. [16th c.: *-ruff* perh. rel. to ME *rove* scurfiness f. ON *hrufa* or MLG, MDu. *rōve*]

dandy /'dændɪ/ n. & adj. ● n. (pl. **-ies**) **1** a man unduly devoted to style, smartness, and fashion in dress and appearance. **2** colloq. an excellent thing. ● adj. (**dandier, dandiest**) esp. N. Amer. colloq. very good of its kind; splendid, first-rate. □ **dandy-brush** a brush for grooming a horse. **dandy roll** (or **roller**) a device for solidifying, and impressing a watermark in, paper during manufacture. □ **dandyish** adj. **dandyism** n. [18th c.: perh. orig. = Andrew, in Jack-a-dandy]

Dane /deɪn/ n. **1** a native or national of Denmark. **2** hist. a Viking invader of England in the 9th–11th centuries. [ME f. ON Danir (pl.), LL Dani]

Danegeld /'deɪngeld/ n. a land-tax levied in Anglo-Saxon England during the reign of King Ethelred (especially 991–1016) to buy off Danish invaders, or finance forces to oppose them. The term also applies to taxes collected for national defence by the Norman kings until 1162. [OE (as DANE) + ON gjald payment]

Danelaw /'deɪnlɔː/ the part of northern and eastern England occupied or subjugated by Danes from the late 9th century and administered according to their laws until after the Norman Conquest. [OE Dena lagu Danes' law]

danger /'deɪndʒə(r)/ n. **1** liability or exposure to harm; risk, peril. **2** a thing that causes or is likely to cause harm. **3** an unwelcome possibility. **4** the status of a railway signal directing a halt or caution. □ **danger list** a list of those dangerously ill, esp. in a hospital. **danger man** a man perceived as posing a particular threat, esp. an outstanding sportsman. **danger money** extra payment for dangerous work. **in danger of** likely to incur or to suffer from. [earlier sense 'jurisdiction, power': ME f. OF dangier ult. f. L dominus lord]

dangerous /'deɪndʒərəs/ adj. involving or causing danger. □ **dangerously** adv. **dangerousness** n. [ME f. AF dangerous, daungerous, OF dangereus (as DANGER)]

dangle /'dæŋg(ə)l/ v. **1** intr. be loosely suspended, so as to be able to sway to and fro. **2** tr. hold or carry loosely suspended. **3** tr. hold out (a hope, temptation, etc.) enticingly. **4** intr. Gram. (as **dangling** adj.) (of a participle in an absolute clause or phrase) with the subject omitted, e.g. of having in having said that, the problem can be solved. □ **dangler** n. **dangly** adj. [16th c. (imit.): cf. Sw. dangla, Danish dangle]

Daniel /'dænjəl/ **1** a Hebrew prophet (6th century BC), who spent his life as a captive at the court of Babylon. In the Bible he interpreted the dreams of Nebuchadnezzar and was delivered by God from the lions' den in which he had been thrown as the result of a trick; in the apocryphal Book of Susanna he is portrayed as a wise judge (Sus. 45–64). **2** a book of the Bible containing his prophecies. It was probably written at the outbreak of persecution of the Jews under Seleucid rule c.167 BC.

Daniell cell /'dænjəl/ n. Physics & Chem. a primary voltaic cell with a copper anode and a zinc-amalgam cathode giving a standard electromotive force when either copper sulphate or sulphuric acid is used as the electrolyte. [John Daniell, Brit. chemist (1790–1845), its inventor]

Danish /'deɪnɪʃ/ adj. & n. ● adj. of or relating to Denmark or its people or language. ● n. **1** the Scandinavian language spoken in Denmark and also in Greenland and the Faeroes. **2** (prec. by the; treated as pl.) the Danish people. **3** colloq. = DANISH PASTRY. [ME f. AF danes, OF daneis f. med.L Danensis (as DANE)]

Danish blue n. a soft salty white cheese with blue veins.

Danish pastry n. a cake of sweetened yeast pastry topped with icing, fruit, nuts, etc.

dank /dæŋk/ adj. disagreeably damp and cold. □ **dankly** adv. **dankness** n. [ME prob. f. Scand.: cf. Sw. dank marshy spot]

Danmark see DENMARK.

d'Annunzio /dɑː'nʊntsɪˌəʊ/, Gabriele (1863–1938), Italian novelist, dramatist, and poet. The publication of his 'Romances of the Rose' trilogy, including The Child of Pleasure (1890), established d'Annunzio as a leading Italian literary figure of the fin de siècle. His strong admiration for Nietzsche is particularly evident in the second of these novels, The Triumph of Death (1894).

Dano- /'deɪnəʊ/ comb. form **1** Danish and. **2** Denmark or the Danes in connection with. [L Danus Danish]

Dano-Norwegian /ˌdeɪnəʊnɔː'wiːdʒən/ n. a form of the Norwegian language (see NORWEGIAN).

danse macabre /ˌdɒns mə'kɑːbrə/ n. = DANCE OF DEATH. [F (as DANCE, MACABRE)]

danseur /dɒn'sɜː(r)/ n. (fem. **danseuse** /-'sɜːz/) a ballet-dancer. [F, = dancer]

Dante /'dæntɪ/ (full name Dante Alighieri) (1265–1321), Italian poet. His early work consisted mainly of courtly love poetry; his first book, Vita nuova (c.1290–4), consists of thirty-one poems linked by a prose narrative and tells of his love for Beatrice Portinari (c.1265–90). However, Dante's international renown and reputation as the founding figure of Italian literature rests on The Divine Comedy (c.1309–20), an epic poem telling of his spiritual journey, in the form of an imagined visit to Hell and Purgatory with Virgil as guide and finally to Paradise with Beatrice, now a blessed spirit, as guide. Dante also wrote scholarly treatises on a number of subjects, including philosophy, science, and politics; his political activity led to his spending part of his life in exile from his native Florence. His innovative use of Italian did much to establish a vernacular literature; his Latin treatise De Vulgari Eloquentia (c.1303) promoted vernacular Italian as a literary language fit to replace Latin.

danthonia /dæn'θəʊnɪə/ n. Austral. & NZ a tufted pasture grass of the genus Danthonia. [mod.L f. E. Danthoine 19th-c. Fr. botanist]

Danton /'dæntən/, French dɑ̃tɔ̃/, Georges (Jacques) (1759–94), French revolutionary. A noted orator, he won great popularity in the early days of the French Revolution. He served as Minister of Justice (1792–4) in the new republic and was a founder member of the Committee of Public Safety (1793). Initially an ally of Robespierre and the Jacobins, he later revolted against their radicalism and the severity of the Revolutionary Tribunal, only to be arrested and executed on Robespierre's orders.

Danube /'dænjuːb/ (German **Donau** /'dɔːnaʊ/) a river which rises in the Black Forest in SW Germany and flows about 2,850 km (1,770 miles) into the Black Sea. It is the second longest river in Europe after the Volga; the cities of Vienna, Budapest, and Belgrade are situated on it.

Danube School a group of landscape painters working in the Danube region in the early 16th century, whose work is significant in that they chose to paint landscape for its own sake. Its members included Altdorfer and Cranach the Elder (at the time of his early work).

Danubian /dæ'njuːbɪən/ adj. of or relating to the River Danube.

Danubian principalities the former European principalities of Moldavia and Wallachia. In 1861 they united to form the state of Romania.

Danzig see GDAŃSK.

dap /dæp/ v. (**dapped, dapping**) **1** intr. fish by letting the bait bob on the water. **2** tr. & intr. dip lightly. **3** tr. & intr. bounce on the ground. [cf. DAB[1]]

Daphne /'dæfnɪ/ Gk Mythol. a nymph who was turned into a laurel-bush to save her from the amorous pursuit of Apollo.

daphne /'dæfnɪ/ n. a flowering shrub of the genus Daphne, e.g. the spurge laurel or the mezereon. [ME, = laurel, f. Gk daphnē]

daphnia /'dæfnɪə/ n. a minute freshwater crustacean of the genus Daphnia, enclosed in a transparent carapace and with long antennae and prominent eyes. Dried daphnia are often used to feed fish in aquaria. Also called freshwater flea or water flea. [mod.L f. DAPHNE]

Daphnis /'dæfnɪs/ Gk Mythol. a Sicilian shepherd who, according to one version of the legend, was struck with blindness for his infidelity to the nymph Echenaïs. He consoled himself with pastoral poetry, of which he was the inventor.

Da Ponte /dɑː 'pɒntɪ/, Lorenzo (born Emmanuele Conegliano) (1749–1838), Italian poet and librettist. He became poet to the Court Opera in Vienna in 1784 and wrote the libretti for Mozart's Marriage of Figaro (1786), Don Giovanni (1787), and Così fan tutte (1790).

dapper /'dæpə(r)/ adj. **1** neat and precise, esp. in dress or movement. **2** sprightly. □ **dapperly** adv. **dapperness** n. [ME f. MLG, MDu. dapper strong, stout]

dapple /'dæp(ə)l/ v. & n. ● v. **1** tr. mark with spots or rounded patches of colour or shade. **2** intr. become marked in this way. ● n. **1** a dappled effect. **2** a dappled animal, esp. a horse. □ **dapple grey 1** (of an animal's coat) grey or white with darker spots. **2** a horse of this colour. [ME dappled, dappeld, (adj.), of unkn. orig.]

Dapsang /dʌp'sʌŋ/ an alternative name for the mountain K2.

dapsone /'dæpsəʊn/ n. Pharm. a bacteriostatic sulphur compound used esp. to treat leprosy. [chem. name di(para-aminophenyl)sulphone]

Daqing /dɑː'tʃɪŋ/ (also **Taching** /tɑː-/) a major industrial city in NE China, in Heilongjiang province; pop. (est. 1984) 802,100.

DAR see DAUGHTERS OF THE AMERICAN REVOLUTION.

darbies /'dɑːbɪz/ n.pl. Brit. sl. handcuffs. [allusive use of *Father Darby's bands*, some rigid form of agreement for debtors (16th c.)]

Darby and Joan /'dɑːbɪ, dʒəʊn/ a devoted old married couple. □ **Darby and Joan club** Brit. a club for people over 60. [18th c.: cf. a poem of 1735 in the *Gentleman's Magazine* containing the lines 'Old Darby, with Joan by his side, You've often regarded with wonder: He's dropsical, she is sore-eyed, Yet they're never happy asunder']

Dard /dɑːd/ n. & adj. ● n. **1** a member of a group of peoples inhabiting eastern Afghanistan, northern Pakistan, and Kashmir. **2** a group of Indic languages spoken by these peoples. ● adj. of or relating to the Dards or their languages. □ **Dardic** adj. & n.

Dardanelles /ˌdɑːdə'nelz/ a narrow strait between Europe and Asiatic Turkey (called the Hellespont in classical times), linking the Sea of Marmara with the Aegean Sea. It is 60 km (38 miles) long. In 1915, during the First World War, it was the scene of an unsuccessful attack on Turkey by Allied troops, with Australian and New Zealand contingents playing a major part (see GALLIPOLI).

dare /deə(r)/ v. & n. ● v.tr. (*3rd sing. present usu.* **dare** before an expressed or implied infinitive without *to*; *past* **dared** or *archaic* or *dial.* **durst**) **1** (foll. by infin. with or without *to*) venture (to); have the courage or impudence (to) (*dare he do it?*; *if they dare to come*; *how dare you?*; *I dare not speak*; *I do not dare to jump*). **2** (usu. foll. by *to* + infin.) defy or challenge (a person) (*I dare you to own up*). **3** *literary* attempt; take the risk of (*dare all things*; *dared their anger*). ● n. **1** an act of daring. **2** a challenge, esp. to prove courage. □ **I dare say 1** (often foll. by *that* + clause) it is probable. **2** probably; I grant that much (*I dare say, but you are still wrong*). □ **darer** n. [OE *durran* with Gmc cognates: cf. Skr. *dhṛsh*, Gk *tharseō* be bold]

daredevil /'deə,dev(ə)l/ n. & adj. ● n. a recklessly daring person. ● adj. recklessly daring. □ **daredevilry** n.

Dar es Salaam /ˌdɑːr es sə'lɑːm/ the chief port and former capital of Tanzania; pop. (1988) 1,360,850. It was founded in 1866 by the sultan of Zanzibar, who built his summer palace there. Its Arabic name means 'haven of peace'.

Darfur /dɑː'fʊə(r)/ a region in the west of Sudan. Until 1874 it was an independent kingdom.

darg /dɑːg/ n. Sc., N. Engl., & Austral. **1** a day's work. **2** a definite amount of work; a task. [ME f. *daywerk* or *daywark* day-work]

Dari /'dɑːri/ n. the form of Persian spoken in Afghanistan. [Pers.]

Darien /'deəriən, 'dær-/ a sparsely populated province of eastern Panama. The name was formerly applied to the whole of the Isthmus of Panama. At the end of the 17th century an unsuccessful attempt was made by Scottish settlers to establish a colony in the tropical wilderness of this region, with the aim of controlling trade between the Atlantic and Pacific Oceans.

Darien, Gulf of part of the Caribbean Sea between Panama and Colombia.

daring /'deərɪŋ/ n. & adj. ● n. adventurous courage. ● adj. adventurous, bold; prepared to take risks. □ **daringly** adv.

dariole /'dærɪ,əʊl/ n. a savoury or sweet dish cooked and served in a small mould usu. shaped like a flowerpot. [ME f. OF]

Darius I /də'raɪəs/ (known as Darius the Great) (*c.*550–486 BC), king of Persia 521–486 BC. His reign divided the empire into provinces governed by satraps, allowing each province its own government while maintaining some centralizing authority. He developed commerce, building a network of roads, exploring the Indus valley, and connecting the Nile with the Red Sea by canal. After suppressing a revolt of the Greek cities in Ionia (499–494 BC), he invaded Greece to punish the mainland Greeks for their interference, but was defeated at Marathon (490 BC).

Darjeeling[1] /dɑː'dʒiːlɪŋ/ (also **Darjiling**) a hill station at an altitude of 2,150 m (7,054 ft) in West Bengal, NE India, near the Sikkim border; pop. (1991) 73,090. The surrounding area produces a high-quality tea.

Darjeeling[2] /dɑː'dʒiːlɪŋ/ n. the high-quality tea grown in the area around Darjeeling.

dark /dɑːk/ adj. & n. ● adj. **1** with little or no light. **2** of a deep or sombre colour. **3** (of a person) with deep brown or black hair, complexion, or skin. **4** gloomy, depressing, dismal (*dark thoughts*). **5** evil, sinister (*dark deeds*). **6** sullen, angry (*a dark mood*). **7** remote, secret, mysterious, little-known (*the dark and distant past*; *keep it dark*). **8** ignorant, unenlightened. **9** (of a theatre) not in use, closed. ● n. **1** absence of light. **2** nightfall (*don't go out after dark*). **3** a lack of knowledge. **4** a dark area or colour, esp. in painting (*the skilled use of lights and darks*). □ **the dark ages** (or **age**) any period of supposed unenlightenment (see DARK AGES). **dark**

glasses spectacles with dark-tinted lenses; sunglasses. **dark horse** a little-known person who unexpectedly becomes successful or prominent. **dark matter** *Astron.* hypothetical non-luminous material in the universe, which has yet to be detected but whose existence is predicted by some cosmological theories. **dark star** *Astron.* an invisible star known to exist from reception of physical data other than light. **in the dark 1** with little or no light. **2** lacking information. □ **darkish** adj. **darkly** adv. **darkness** n. **darksome** adj. poet. [OE *deorc* prob. f. Gmc]

Dark Ages the period in the West between the fall of the Roman Empire and the high Middle Ages, *c.*500–1100, so called because it has been judged a time of relative unenlightenment and obscurity. Germanic tribes swept through Europe and North Africa, often attacking and destroying towns; some formed their own kingdoms, e.g. the Ostrogoths and Vandals in Italy, the Franks in France and western Germany, and the Angles and Saxons in England. The period was a time of political fragmentation with a lack of major cities and centres of learning, although the negative connotations of the term disguise some real achievements and imply an underestimation of classical influences on the newly settled Germanic peoples; the period saw the foundation of Christian monasteries, which kept scholarship alive, while learning was encouraged at the courts of Charlemagne and Alfred the Great.

The term has also been applied to a similar period in the history of Greece and the Aegean from the end of the Bronze Age until the beginning of the historical period, when the region seems to have been heavily depopulated, there was no building of palaces and fortresses, and the art of writing was apparently lost. However, the term glosses over some important factors, such as the growth of a strong oral tradition, which resulted in the emergence of the Homeric poems.

Dark Continent, the Africa, esp. in the time before it was fully explored by Europeans.

darken /'dɑːkən/ v. **1** tr. make dark or darker. **2** intr. become dark or darker. □ **never darken a person's door** keep away permanently. □ **darkener** n.

Darkhan /dɑː'kɑːn/ an industrial and mining city in northern Mongolia, established in 1961; pop. (1990) 80,100.

darkie var. of DARKY.

darkling /'dɑːklɪŋ/ adj. & adv. poet. in the dark; in the night.

darkroom /'dɑːkruːm, -rʊm/ n. a room for photographic work, with normal light excluded.

darky /'dɑːki/ n. (also **darkie**) (pl. **-ies**) sl. offens. a black person.

Darling /'dɑːlɪŋ/, Grace (1815–42), English heroine. The daughter of a lighthouse keeper on the Farne Islands off the coast of Northumberland, she became a national heroine when in September 1838 she and her father rowed through a storm to rescue the survivors of the wrecked *Forfarshire*.

darling /'dɑːlɪŋ/ n. & adj. ● n. **1** a beloved or lovable person or thing. **2** a favourite. **3** colloq. a pretty or endearing person or thing. ● adj. **1** beloved, lovable. **2** favourite. **3** colloq. charming or pretty. [OE *dēorling* (as DEAR, -LING[1])]

Darling River a river of SE Australia, flowing 2,757 km (1,712 miles) in a generally south-westward course to join the Murray River.

Darlington /'dɑːlɪŋtən/ an industrial town in county Durham, NE England; pop. (1991) 96,700.

Darmstadt /'dɑːmʃtæt/ an industrial town in Hesse, western Germany; pop. (1991) 140,040.

darn[1] /dɑːn/ v. & n. ● v.tr. **1** mend (esp. knitted material, or a hole in it) by interweaving yarn across the hole with a needle. **2** embroider with a large running stitch. ● n. a darned area in material. □ **darning needle 1** a long needle with a large eye, used in darning. **2** US a dragonfly. [16th c.: perh. f. obs. *dern* hide]

darn[2] /dɑːn/ v.tr., int., adj., & adv. (US dial. **durn** /dɜːn/) colloq. = DAMN (in imprecatory senses). [corrupt. of DAMN]

darned /dɑːnd/ adj. & adv. (US **durned** /dɜːnd/) colloq. = DAMNED.

darnel /'dɑːn(ə)l/ n. a grass of the genus *Lolium*; esp. *L. temulentum*, growing chiefly in Mediterranean countries as a weed among cereal crops, which it is liable to contaminate with ergot. [ME: cf. Walloon *darnelle*]

darner /'dɑːnə(r)/ n. a person or thing that darns, esp. a darning needle.

darning /ˈdɑːnɪŋ/ n. **1** the action of a person who darns. **2** things to be darned.

Darnley /ˈdɑːnlɪ/, Lord (title of Henry Stewart or Stuart) (1545–67), Scottish nobleman, second husband of Mary, Queen of Scots and father of James I of England. He was implicated in the murder of his wife's secretary Rizzio in 1566, and was later killed in a mysterious gunpowder explosion in Edinburgh.

Dart /dɑːt/, Raymond Arthur (1893–1988), Australian-born South African anthropologist and anatomist. While based in Johannesburg he became interested in fossil primates. He obtained the skull of a juvenile hominid from a limestone quarry at Taung near Kimberley in 1924, and a year later coined the genus name *Australopithecus* for it. It was assigned to the species *A. africanus*, and was the first of many specimens of *Australopithecus* obtained in Africa.

dart /dɑːt/ n. & v. ● n. **1 a** a small pointed missile used as a weapon. **b** a similar missile, usu. with a feather or plastic flight used in the game of darts. **2** a sudden rapid movement. **3** *Zool.* a dartlike structure, such as an insect's sting or the calcareous projections of a snail (used during copulation). **4** a tapering tuck stitched in a garment. ● v. **1** *intr.* (often foll. by *out, in, past,* etc.) move or go suddenly or rapidly (*darted into the shop*). **2** *tr.* throw (a missile). **3** *tr.* direct suddenly (a glance etc.). [ME f. OF *darz, dars,* f. Frank.]

dartboard /ˈdɑːtbɔːd/ n. a circular board marked with numbered segments, used as a target in darts.

darter /ˈdɑːtə(r)/ n. **1** a diving fish-eating bird of the genus *Anhinga*, with a long thin neck (cf. ANHINGA, *snake bird*). **2** a small quick-moving freshwater fish of the family Percidae, native to North America.

Dartmoor /ˈdɑːtmʊə(r), -mɔː(r)/ **1** a moorland district in Devon that was a royal forest in Saxon times, now a national park. **2** a prison near Princetown in this district, originally built to hold French prisoners of war from the Napoleonic Wars.

Dartmoor pony n. a breed of small pony with a long shaggy coat.

Dartmouth /ˈdɑːtməθ/ a port in Devon; pop. (1981) 6,210. It is the site of the Royal Naval College.

darts /dɑːts/ n.pl. (usu. treated as *sing.*) an indoor game in which darts are thrown at a circular target to score points. Darts is a predominantly British game; lightweight darts with a plastic or feather flight are thrown at a dartboard, which is marked with concentric circles and a bull's-eye in the centre and divided by radiating lines into numbered sectors. The game dates from the time of George III or earlier, although the present system of numbering the sectors was not introduced until 1896. It has long been associated with inns and public houses; recently, however, darts competitions have been televised.

Darwin[1] /ˈdɑːwɪn/ the capital of Northern Territory, Australia; pop. (1990) 73,300.

Darwin[2] /ˈdɑːwɪn/, Charles (Robert) (1809–82), English natural historian and geologist, proponent of the theory of evolution by natural selection. Grandson of the eminent physician and biologist Erasmus Darwin (1731–1802), he failed to complete his medical training and narrowly achieved a theological degree. Darwin took the post of unpaid naturalist on HMS *Beagle* for her voyage around the southern hemisphere (1831–6), during which he collected the material which became the basis for his ideas on natural selection. On his return he made his name as a geologist, in particular with his accounts of the formation of coral reefs and atolls. In 1858, he and A. R. Wallace agreed to publish simultaneously their similar thoughts on evolution, to the consternation of theologians. He went on to write an extensive series of books, monographs, and papers; *On the Origin of Species* (1859) and *The Descent of Man* (1871) changed our concepts of nature and of humanity's place within it.

Darwinian /dɑːˈwɪnɪən/ adj. & n. ● adj. of or relating to Darwin's theory of the evolution of species by the action of natural selection. ● n. an adherent of this theory. □ **Darwinism** /ˈdɑːwɪˌnɪz(ə)m/ n. **Darwinist** n. & adj.

Darwin's finches n.pl. a diverse group of buntings found only in the Galapagos Islands, evolved from a common ancestor into thirteen species with a wide variety of bills adapted for different feeding methods. One of them (the woodpecker finch, *Camarhynchus pallidus*) uses a twig as a tool to extract insects from holes. Darwin used them as an example of speciation by adaptive radiation.

dash /dæʃ/ v. & n. ● v. **1** *intr.* rush hastily or forcefully (*dashed up the stairs*). **2** *tr.* strike or fling with great force, esp. so as to shatter (*dashed it to the ground; the cup was dashed from my hand*). **3** *tr.* frustrate, daunt, dispirit (*dashed their hopes*). **4** *tr. colloq.* (esp. **dash it** or **dash it all**) =

DAMN v. **1.** ● n. **1** a rush or onset; a sudden advance (*made a dash for shelter*). **2** a horizontal stroke in writing or printing to mark a pause or break in sense or to represent omitted letters or words. **3** impetuous vigour or the capacity for this. **4** showy appearance or behaviour, stylishness. **5** *N. Amer.* a sprinting-race. **6** the longer signal of the two used in Morse code (cf. DOT[1] n. 3). **7** a slight admixture, esp. of a liquid. **8** = DASHBOARD. □ **cut a dash** make a brilliant show. **dash down** (or **off**) write or finish hurriedly. [ME, prob. imit.]

dashboard /ˈdæʃbɔːd/ n. **1** the surface below the windscreen of a motor vehicle or aircraft, containing instruments and controls. **2** *hist.* a board of wood or leather in front of a carriage, to keep out mud.

dashiki /dɑːˈʃiːkɪ/ n. a loose brightly coloured shirt of West African origin, also worn by some blacks in America. [prob. f. Yoruba f. Hausa]

dashing /ˈdæʃɪŋ/ adj. **1** spirited, lively. **2** stylish. □ **dashingly** adv. **dashingness** n.

dashpot /ˈdæʃpɒt/ n. a device for damping shock or vibration.

dassie /ˈdæsɪ, ˈdɑːsɪ/ n. S. Afr. **1** a rock hyrax, *Procavia capensis*, of southern Africa. **2** a small coastal fish *Diplodus sargus*, with a black tail-spot. [Afrik. f. Du. *dasje* dimin. of *das* badger]

dastardly /ˈdæstədlɪ/ adj. cowardly, despicable. □ **dastardliness** n. [*dastard* base coward, prob. f. *dazed* past part. + -ARD, or obs. *dasart* dullard, DOTARD]

dasyure /ˈdæsɪˌjʊə(r)/ n. a small catlike flesh-eating marsupial of the genus *Dasyurus*. [F f. mod.L *dasyurus* f. Gk *dasus* rough + *oura* tail]

DAT abbr. digital audio tape.

data /ˈdeɪtə, ˈdɑːtə/ n.pl. (also treated as *sing.*, as in *that is all the data we have,* although the singular form is strictly *datum*) **1** known facts or things used as a basis for inference or reckoning. **2** quantities or characters operated on by a computer etc. □ **data bank 1** a store or source of data. **2** = DATABASE. **data capture** the action or process of entering data into a computer. **data processing** a series of operations on data, esp. by a computer, to retrieve or classify etc. information. **data processor** a machine, esp. a computer, that carries out data processing. **data protection** legal control over access to data stored in computers. [pl. of DATUM]

database /ˈdeɪtəˌbeɪs/ n. a structured set of data held in a computer, esp. one that is accessible in various ways.

datable /ˈdeɪtəb(ə)l/ adj. (often foll. by *to*) capable of being dated (to a particular time).

DataGlove /ˈdeɪtəˌglʌv/ n. a device worn like a glove, which allows the manual manipulation of images in virtual reality.

date[1] /deɪt/ n. & v. ● n. **1** a day of the month, esp. specified by a number. **2** a particular day or year, esp. when a given event occurred. **3** a statement (usu. giving the day, month, and year) in a document or inscription etc., of the time of composition or publication. **4** the period to which a work of art etc. belongs. **5** the time when an event happens or is to happen. **6** *colloq.* **a** an appointment or social engagement, esp. with a person of the opposite sex. **b** esp. *N. Amer.* a person with whom one has a social engagement. ● v. **1** *tr.* mark with a date. **2** *tr.* **a** assign a date to (an object, event, etc.). **b** (foll. by *to*) assign to a particular time, period, etc. **3** *intr.* (often foll. by *from, back to,* etc.) have its origins at a particular time. **4** *intr.* be recognizable as from a past or particular period; become evidently out of date (*a design that does not date*). **5** *tr.* indicate or expose as being out of date (*that hat really dates you*). **6** *colloq.* **a** *tr.* make an arrangement with (a person) to meet socially. **b** *intr.* meet socially by agreement (*they are now dating regularly*). □ **date line** a line at the head of a dispatch or special article in a newspaper showing the date and place of writing. **date rape** the rape of a girl or woman by a person with whom she is on a date. **date-stamp** n. **1** an adjustable rubber stamp etc. used to record a date. **2** the impression made by this. ● v.tr. mark with a date-stamp. **out of date** (*attrib.* **out-of-date**) old-fashioned, obsolete. **to date** until now. **up to date** (*attrib.* **up-to-date**) meeting or according to the latest requirements, knowledge, or fashion; modern. [ME f. OF f. med.L *data*, fem. past part. of *dare* give: from the L formula used in dating letters, *data* (*epistola*) (letter) given or delivered (at a particular time or place)]

date[2] /deɪt/ n. **1** a dark oval single-stoned fruit. **2** (in full **date-palm**) the tall palm tree *Phoenix dactylifera*, native to western Asia and North Africa, bearing this fruit. [ME f. OF f. L *dactylus* f. Gk *daktulos* finger, from the shape of its leaf]

dateless /ˈdeɪtlɪs/ adj. **1** having no date. **2** of immemorial age. **3** not likely to become out of date.

Date Line (in full **International Date Line**) an imaginary north–south line through the Pacific Ocean, east of which the date is a day earlier than it is to the west. It lies mainly along the meridian furthest from Greenwich (i.e. longitude 180°), with diversions to pass around some island groups. It was adopted worldwide in 1884, at the same time as the Greenwich meridian.

dative /ˈdeɪtɪv/ n. & adj. Gram. ● n. the case of nouns and pronouns (and words in grammatical agreement with them) indicating an indirect object or recipient. ● adj. of or in the dative. □ **datival** /deɪˈtaɪv(ə)l/ adj. [ME f. L (casus) dativus f. dare dat- give]

Datong /dɑːˈtʊŋ/ a city in northern China in Shanxi province; pop. (1990) 1,090,000. Nearby are the Yungang caves, which contain the earliest examples of Buddhist stone carvings in China.

datum /ˈdeɪtəm, ˈdɑːtəm/ n. (pl. **data**: see DATA) **1** a piece of information. **2** a thing known or granted; an assumption or premiss from which inferences may be drawn (see sense-datum). **3** a fixed starting-point of a scale etc. (datum-line). [L, = thing given, neut. past part. of dare give]

datura /dəˈtjʊərə/ n. a poisonous plant of the genus Datura, e.g. the thorn apple. [mod.L f. Hindi dhatura]

daub /dɔːb/ v. & n. ● v.tr. **1** spread (paint, plaster, or some other thick substance) crudely or roughly on a surface. **2** coat or smear (a surface) with paint etc. **3 a** (also absol.) paint crudely or unskilfully. **b** lay (colours) on crudely and clumsily. ● n. **1** paint or other substance daubed on a surface. **2** plaster, clay, etc., for coating a surface, esp. mixed with straw and applied to laths or wattles to form a wall. **3** a crude painting. [ME f. OF dauber f. L dealbare whitewash f. albus white]

daube /dəʊb/ n. a stew of braised meat (usu. beef) with wine etc. [F]

dauber /ˈdɔːbə/ n. a person or implement that daubs, esp. in painting. □ **get one's dauber down** US sl. become dispirited or depressed.

Daubigny /ˈdəʊbiːˌnjiː/, Charles François (1817–78), French landscape painter. He was a member of the Barbizon School and is often regarded as a linking figure between this group of painters and the impressionists. His landscapes frequently feature stretches of water, for example The Banks of the Oise (1872).

Daudet /ˈdəʊdeɪ/, Alphonse (1840–97), French novelist and dramatist. He is best known for his sketches of life in his native Provence, particularly the Lettres de mon moulin (1869), and as the creator of Tartarin, a caricature of the Frenchman of the south of France, whose comic exploits are first related in Tartarin de Tarascon (1872).

daughter /ˈdɔːtə(r)/ n. **1** a girl or woman in relation to either or both of her parents. **2** a female descendant. **3** (foll. by of) a female member of a family, nation, etc. **4** (foll. by of) a woman who is regarded as the spiritual descendant of, or as spiritually attached to, a person or thing. **5** a product or attribute personified as a daughter in relation to its source (Fortune and its daughter Confidence). **6** Physics a nuclide formed by the radioactive decay of another. **7** Biol. a cell etc. formed by the division etc. of another. □ **daughter-in-law** (pl. **daughters-in-law**) the wife of one's son. □ **daughterhood** n. **daughterly** adj. [OE dohtor f. Gmc]

Daughters of the American Revolution a US patriotic society whose aims include encouraging education and the study of US history and which tends to be politically conservative. Membership is limited to female descendants of soldiers and others of the revolutionary period who aided the cause of independence. It was first organized in 1890.

Daumier /ˈdəʊmiˌeɪ/, Honoré (1808–78), French painter and lithographer. From the 1830s he worked as a cartoonist for periodicals such as Charivari, where he produced over 4,000 lithographs sharply satirizing French society and politics. His later oil paintings, such as Don Quixote (1868), deal with their subjects in a powerfully realistic and unromanticized manner.

daunt /dɔːnt/ v.tr. discourage, intimidate. □ **daunting** adj. **dauntingly** adv. [ME f. AF daunter, OF danter, donter f. L domitare frequent. of domare tame]

dauntless /ˈdɔːntlɪs/ adj. intrepid, persevering, fearless. □ **dauntlessly** adv. **dauntlessness** n.

dauphin /ˈdɔːfɪn/ n. the title borne by the eldest son of kings of France from 1349 to 1830. Originally a family name, it became the title of the lords of an area of SE France which was afterwards called Dauphiné. In 1349 the future Charles V acquired the lands and the title; when king he ceded them to his eldest son in 1368, establishing the practice of passing both title and lands to the Crown prince. [ME f. F, ult. f. L delphinus DOLPHIN, as a family name]

Dauphiné /ˈdəʊfiˌneɪ/ a region and former province of SE France (see DAUPHIN). Its capital was Grenoble.

Davao /dɑːˈvaːəʊ/ a seaport in the southern Philippines, on the island of Mindanao; pop. (1990) 850,000. Founded in 1849, it is the largest city on the island and the third largest city in the Philippines.

Davenport /ˈdæv(ə)nˌpɔːt/ n. **1** Brit. an ornamental writing-desk with drawers and a sloping surface for writing. **2** US a large heavily upholstered sofa. [prob. f. Capt. Davenport, for whom early examples of this type of desk were made in the late 18th c.]

David[1] /ˈdeɪvɪd/ (died c.962 BC), king of Judah and Israel c.1000–c.962 BC. In the biblical account he was the youngest son of Jesse, and was made a military commander by Saul after slaying the Philistine Goliath. On Saul's death he became king of Judah and later of the whole of Israel, making Jerusalem his capital. He is traditionally regarded as the author of the Psalms, but it is unlikely that more than a fraction of the psalter is his work.

David[2] /ˈdeɪvɪd/ the name of two kings of Scotland:

David I (c.1084–1153), sixth son of Malcolm III, reigned 1124–53. Much of his youth was spent at the English court, after his sister Matilda (1080–1118) married King Henry I of England in 1100. After succeeding his brother Alexander (c.1080–1124) as king of Scotland, David established a strong administration on the Norman model, bringing many retainers with him from England, encouraging the development of trade, and introducing legal reforms. In 1136, after Henry's death, David invaded England in support of his niece Matilda's claim to the throne, but was decisively defeated at the Battle of the Standard in Yorkshire in 1138.

David II (1324–71), son of Robert the Bruce, reigned 1329–71. His long reign witnessed a renewal of fighting between England and Scotland, with Edward III taking advantage of the Scottish king's minority to introduce Edward de Baliol (c.1283–1364) as an English puppet in his place. After returning from exile in France (1334–41), David was defeated by the English at Neville's Cross (1346) and spent eleven years in prison. His death without issue in 1371 left the throne to the Stuarts.

David[3] /ˈdeɪvɪd/, Elizabeth (1913–92), British cookery writer. She played a leading role in introducing Mediterranean cuisine to Britain in the 1950s and 1960s. Her best-selling books include French Provincial Cooking (1960).

David[4] /dæˈviːd/, Jacques-Louis (1748–1825), French painter. He is famous for neoclassical paintings such as The Oath of the Horatii (1784) and The Intervention of the Sabine Women (1799). He became actively involved in the French Revolution, voting for the death of Louis XVI and supporting Robespierre. Famous works of the revolutionary era include The Dead Marat (1793), which treats contemporary events with a grandeur hitherto reserved for history painting. Imprisoned after the fall of Robespierre, he returned to prominence under Napoleon.

David, St (also **Dewi** /ˈdewiː/) (6th century), Welsh monk. Since the 12th century he has been regarded as the patron saint of Wales. Little is known of his life, but it is generally accepted that he transferred the centre of Welsh ecclesiastical administration from Caerleon to Mynyw, now St David's. He also established a number of monasteries in England and Wales and many churches in South Wales. Feast day, 1 Mar.

Davies[1] /ˈdeɪvɪs/, Sir Peter Maxwell (b.1934), English composer and conductor. In 1967, with Harrison Birtwistle, he co-founded the Pierrot Players (later the Fires of London ensemble), for whom he composed many of his works. These include Taverner (1970) and Eight Songs for a Mad King (1969). Davies's work is influenced particularly by serialism and early English music.

Davies[2] /ˈdeɪvɪs/, (William) Robertson (1913–95), Canadian novelist, dramatist, and journalist. He won international recognition with his Deptford trilogy of novels. The first two of these, Fifth Business (1970) and The Manticore (1972), show the influence on Davies of Jungian psychology; the third novel, World of Wonders (1975), also explores psychological and spiritual issues.

Davies[3] /ˈdeɪvɪs/, W(illiam) H(enry) (1871–1940), English poet. He emigrated to the US and lived as a vagrant and jobbing labourer there, writing The Autobiography of a Super-Tramp (1908) about his experiences. His poems often focus on the natural world; collections include The Soul's Destroyer and Other Poems (1905), which earned him the patronage of George Bernard Shaw.

da Vinci, Leonardo, see LEONARDO DA VINCI.

Davis[1] /ˈdeɪvɪs/, Bette (born Ruth Elizabeth Davis) (1908–89), American actress. She established her Hollywood career playing a number of strong, independent female characters in such films as Dangerous

(1935), *Jezebel* (1938), and *All About Eve* (1950). Her flair for suggesting the macabre and menacing emerged in later films, such as the melodrama *Whatever Happened to Baby Jane?* (1962) and the thriller *Murder with Mirrors* (1984).

Davis[2] /'deɪvɪs/, Joe (1901–78) and Fred (b.1913), English billiards and snooker players. Joe Davis was the dominant figure in snooker for many years, holding the world championship from 1927 until his retirement in 1946. He was also world billiards champion (1928–32). His brother Fred Davis was world snooker champion (1948–9; 1951–6) and world billiards champion (1980).

Davis[3] /'deɪvɪs/, Miles (Dewey) (1926–91), American jazz trumpeter, composer, and band-leader. In the 1950s he played and recorded arrangements by Gil Evans (1912–88) in a new style which became known as 'cool' jazz, heard on albums such as *Kind of Blue* (1959). For much of the time between 1955 and 1960 his quintet included John Coltrane. In the 1960s Davis pioneered the fusion of jazz and rock, on albums such as *In a Silent Way* (1969).

Davis[4] /'deɪvɪs/, Steve (b.1957), English snooker player. He won a number of national and international events in the 1980s, becoming UK Professional Champion (1980–1; 1984–7) and World Professional Champion (1981; 1983–4; 1987–9).

Davis Cup an annual lawn tennis championship for men, first held in 1900, between teams from different countries. It was named after the American D. F. Davis (1879–1945), the doubles champion who donated the trophy.

Davisson /'deɪvɪs(ə)n/, Clinton Joseph (1881–1958), American physicist. Following work in support of a patent suit, and after a laboratory accident, he and L. H. Germer (1896–1971) discovered electron diffraction, thus confirming de Broglie's theory of the wave nature of electrons. He shared the Nobel Prize for physics for this in 1937, and later worked on the application of electron waves to crystal physics and electron microscopy.

Davis Strait a sea passage 645 km (400 miles) long separating Greenland from Baffin Island and connecting Baffin Bay with the Atlantic Ocean. The strait is named after the English explorer John Davis (1550–1605), who sailed through it in 1587.

davit /'dævɪt, 'deɪvɪt/ n. a small crane on board a ship, esp. one of a pair for suspending or lowering a lifeboat. [AF & OF *daviot* dimin. of *Davi* David]

Davos /dɑ:'vəʊs/ a resort and winter-sports centre in eastern Switzerland; pop. (1990) 10,500.

Davy /'deɪvɪ/, Sir Humphry (1778–1829), English chemist, a pioneer of electrochemistry. After an apprenticeship with an apothecary-surgeon, he discovered nitrous oxide (laughing gas) and was invited to join the Royal Institution. Using electrolytic decomposition, he discovered the elements sodium, potassium, magnesium, calcium, strontium, and barium. Davy identified and named the element chlorine after he had demonstrated that oxygen was not a necessary constituent of acids, determined the properties of iodine, and demonstrated that diamond was a form of carbon. He appointed Faraday as his assistant.

Davy Jones n. *Naut. sl.* the evil spirit of the sea. □ **Davy Jones's locker** the bottom of the sea, esp. regarded as the grave of those who are drowned or buried at sea. [18th c.: orig. unkn.]

Davy lamp n. an early type of safety lamp for miners, invented by Humphry Davy in 1815. The lamp was enclosed in wire gauze, the effect of which was to quench any flame which might emerge from inside. This reduced the risk of igniting the methane found in underground collieries.

daw /dɔ:/ n. = JACKDAW. [ME: cf. OHG *tāha*]

dawdle /'dɔ:d(ə)l/ v. & n. ● v. **1** *intr.* **a** walk slowly and idly. **b** delay; waste time. **2** *tr.* (foll. by *away*) waste (time). ● n. an act or instance of dawdling. □ **dawdler** n. [perh. rel. to dial. *daddle*, *doddle* idle, dally]

Dawkins /'dɔ:kɪnz/, Richard (b.1941), English biologist. Dawkins's book *The Selfish Gene* (1976) did much to popularize the theory of sociobiology. In *The Blind Watchmaker* (1986), Dawkins discussed evolution by natural selection and suggested that the theory could answer the fundamental question of why life exists.

dawn /dɔ:n/ n. & v. ● n. **1** the first light of day; daybreak. **2** the beginning or incipient appearance of something. ● v.intr. **1** (of a day) begin; grow light. **2** begin to appear or develop. **3** (often foll. by *on*, *upon*) begin to become evident or understood (by a person). □ **dawn chorus** the singing of many birds at the break of day. **dawn redwood** a Chinese deciduous coniferous tree, *Metasequoia glyptostroboides*, of a genus first known only from fossils. [orig. as verb: back-form. f. *dawning*, ME f. earlier *dawing* after Scand. (as DAY)]

dawning /'dɔ:nɪŋ/ n. **1** daybreak. **2** the first beginning of something.

Day /deɪ/, Doris (born Doris Kappelhoff) (b.1924), American actress and singer. She became a star in the 1950s with roles in a number of films, at first musicals but later also comedies and light romances. Her films include *Calamity Jane* (1953) and *Pillow Talk* (1959).

day /deɪ/ n. **1** the time between sunrise and sunset. **2 a** a period of twenty-four hours as a unit of time, esp. from midnight to midnight, corresponding to a complete revolution of the earth on its axis. **b** a corresponding period on other planets (*Martian day*). **3** daylight (*clear as day*). **4** the time in a day during which work is normally done (*an eight-hour day*). **5 a** (also *pl.*) a period of the past or present (*the modern day*; *the old days*). **b** (prec. by *the*) the present time (*the issues of the day*). **6** the lifetime of a person or thing, esp. regarded as useful or productive (*have had my day*; *in my day things were different*). **7** a point of time (*will do it one day*). **8 a** the date of a particular festival or event (*graduation day*; *pay day*; *Christmas day*). **9** a particular date; a date agreed on. **10** a day's endeavour, or the period of an endeavour, esp. as bringing success (*win the day*). □ **all in a** (or **the**) **day's work** part of normal routine. **at the end of the day** in the final reckoning, when all is said and done. **call it a day** end a period of activity, esp. resting content that enough has been done. **day after day** without respite. **day and night** all the time. **day-boy** (or **-girl**) *Brit.* a boy or girl who goes daily from home to school, esp. a school that also has boarders. **day by day** gradually. **day care 1** the supervision of young children during the working day. **2** the care provided in a day centre. **day centre** a place providing care for the elderly or handicapped during the day. **day-dream** n. a pleasant fantasy or reverie. ● v.intr. indulge in this. **day-dreamer** a person who indulges in day-dreams. **day in, day out** routinely, constantly. **day labourer** an unskilled labourer hired by the day. **day lily** a plant of the genus *Hemerocallis*, whose flowers last only a day. **day nursery** a nursery where children are looked after during the working day. **day off** a day's holiday from work. **day of reckoning** see RECKONING. **day of rest** a day off from work, esp. the sabbath. **day out** a trip or excursion for a day. **day owl** an owl that hunts by day, esp. the short-eared owl. **day release** *Brit.* a system of allowing employees days off work for education. **day return** a fare or ticket at a reduced rate for a journey out and back in one day. **day-room** a room, esp. a communal room in an institution, used during the day. **day-school** a school for pupils living at home. **day-to-day** mundane, routine. **day-trip** a trip or excursion completed in one day. **day-tripper** a person who goes on a day-trip. **from day one** *colloq.* from the beginning. **not one's day** a day of successive misfortunes for a person. **on one's day** at one's peak of capability. **one of these days** before very long. **one of those days** a day when things go badly. **some day** at some point in the future. **that will be the day** *colloq.* that will never happen. **this day and age** the present time or period. [OE *dæg* f. Gmc]

Dayak var. of DYAK.

Dayan /daɪ'æn/, Moshe (1915–81), Israeli statesman and general. He fought in the British army in the Second World War, in which he lost an eye. After commanding Israeli forces at the time of the Suez crisis he entered Parliament, originally representing the Labour Party but later forming an independent group with David Ben-Gurion. Dayan became Minister of Defence in 1967 and oversaw Israel's victory in the Six Day War, but resigned in 1974 following criticisms of the country's state of readiness at the start of the Yom Kippur War. As Foreign Minister (1977–9) he played a prominent role in negotiations towards the Israeli–Egyptian peace treaty of 1979 (see CAMP DAVID).

daybed /'deɪbed/ n. a bed for daytime rest.

daybook /'deɪbʊk/ n. an account-book in which a day's transactions are entered, for later transfer to a ledger.

daybreak /'deɪbreɪk/ n. the first appearance of light in the morning.

Day-Glo /'deɪɡləʊ/ n. & adj. ● n. *propr.* a make of fluorescent paint or other colouring. ● adj. coloured with or like this. [DAY + GLOW]

Day-Lewis /deɪ'lu:ɪs/, C(ecil) (1904–72), English poet and critic. During the 1930s he was associated with a group of left-wing poets which included W. H. Auden and Stephen Spender, and his early volumes of verse, such as *Transitional Poems* (1929), reflect the influence of radical and revolutionary ideas. After 1940, however, he became increasingly a figure of the Establishment; he published several works of criticism, including *The Poetic Image* (1947), and further collections of verse, such as *The Whispering Roots* (1970). He was Poet Laureate 1968–72.

daylight /'deɪlaɪt/ n. **1** the light of day. **2** dawn (before daylight). **3 a** openness, publicity. **b** open knowledge. **4** a visible gap or interval, e.g. between boats in a race. **5** (usu. in pl.) archaic sl. one's eyes or internal organs. □ **beat, scare,** etc. **the (living) daylights out of** colloq. beat, scare, etc., severely. **daylight robbery** colloq. blatant overcharging. **daylight saving** the achieving of longer evening daylight, esp. in summer, by setting the time an hour ahead of the standard time. **daylight time** esp. N. Amer. time as adjusted for daylight saving (cf. summer time). **see daylight** begin to understand what was previously obscure.

daylong /'deɪlɒŋ/ adj. lasting for a (whole) day.

Day of Atonement see YOM KIPPUR.

Day of Judgement = JUDGEMENT DAY.

daysack /'deɪsæk/ n. a small rucksack for use on one-day hikes.

dayside /'deɪsaɪd/ n. **1** US staff, esp. of a newspaper, who work during the day. **2** Astron. the side of a planet that faces the sun.

daytime /'deɪtaɪm/ n. the part of the day when there is natural light.

Dayton /'deɪt(ə)n/ a city in western Ohio; pop. (1990) 182,000. It was the home of the aviation pioneers the Wright brothers and is still a centre of aerospace research.

daywork /'deɪwɜːk/ n. work paid for according to the time taken.

daze /deɪz/ v. & n. ● v.tr. stupefy, bewilder. ● n. a state of confusion or bewilderment (in a daze). □ **dazedly** /-zɪdlɪ/ adv. [ME dased past part., f. ON dasathr weary]

dazzle /'dæz(ə)l/ v. & n. ● v. **1** tr. blind temporarily or confuse the sight of by an excess of light. **2** tr. impress or overpower (a person) with knowledge, ability, or any brilliant display or prospect. **3** intr. archaic (of eyes) be dazzled. ● n. bright confusing light. □ **dazzlement** n. **dazzler** n. **dazzling** adj. **dazzlingly** adv. [ME, f. DAZE + -LE⁴]

dB abbr. decibel(s).

DBE abbr. (in the UK) Dame Commander of the Order of the British Empire.

DBS abbr. **1** direct-broadcast satellite. **2** direct broadcasting by satellite.

DC abbr. **1** (also **d.c.**) direct current. **2** District of Columbia. **3** da capo. **4** District Commissioner.

DCB abbr. (in the UK) Dame Commander of the Order of the Bath.

DCL abbr. Doctor of Civil Law.

DCM abbr. (in the UK) Distinguished Conduct Medal.

DCMG abbr. (in the UK) Dame Commander of the Order of St Michael and St George.

DCVO abbr. (in the UK) Dame Commander of the Royal Victorian Order.

DD abbr. Doctor of Divinity.

D-Day /'diːdeɪ/ **1** the day (6 June 1944) in the Second World War on which Allied forces invaded German-occupied northern France. Landings on beaches in Normandy by British, US, and Canadian forces under the overall command of General Dwight D. Eisenhower established a bridgehead from which the Allies were able to launch an offensive which drove the Germans out of France. **2** the day on which an important operation is to begin or a change to take effect. [D for day + DAY]

DDC see DEWEY DECIMAL CLASSIFICATION.

ddI abbr. Pharm. a drug intended for use against the Aids virus. [chem. name dideoxyinosine]

DDR hist. German Democratic Republic. [Deutsche Demokratische Republik]

DDT abbr. dichlorodiphenyltrichloroethane, a synthetic organic compound used as an insecticide. DDT is a typical member of a class of pesticides introduced in the 1940s, based on chlorinated aromatic hydrocarbons. Once applied DDT tends to persist in the environment and to become concentrated in animals at the head of the food chain, with damaging effects especially on their reproductive success. Its use is now banned in many countries.

DE abbr. US Delaware (in official postal use).

de- /dɪ, diː/ prefix **1** forming verbs and their derivatives: **a** down, away (descend; deduct). **b** completely (declare; denude; deride). **2** added to verbs and their derivatives to form verbs and nouns implying removal or reversal (decentralize; de-ice; demoralization). [from or after L de (adv. & prep.) = off, from: sense 2 through OF des- f. L dis-]

deaccession /ˌdiːæk'seʃ(ə)n/ v.tr. (of a museum, library, etc.) sell or otherwise dispose of (a work etc. in the collection).

deacon /'diːkən/ n. & v. ● n. **1** (in episcopal Churches) a minister of the third order, below bishop and priest. **2** (in Nonconformist Churches) a lay officer attending to a congregation's secular affairs. **3** (in the early Church) an appointed minister of charity. ● v.tr. appoint or ordain as a deacon. □ **deaconship** n. **deaconate** /-kənɪt/ n. [OE diacon f. eccl.L diaconus f. Gk diakonos servant]

deaconess /ˌdiːkə'nes, 'diːkənɪs/ n. a woman in the early Church and in some modern Churches with functions analogous to a deacon's. [DEACON, after LL diaconissa]

deactivate /diː'æktɪ,veɪt/ v.tr. make inactive or less reactive. □ **deactivator** n. **deactivation** /-ˌæktɪ'veɪʃ(ə)n/ n.

dead /ded/ adj., adv., & n. ● adj. **1** no longer alive. **2 a** benumbed; affected by loss of sensation (my fingers are dead). **b** colloq. extremely tired or unwell. **3** (foll. by to) unappreciative or unconscious of; insensitive to. **4** no longer effective or in use; obsolete, extinct. **5** (of a match, of coal, etc.) no longer burning; extinguished. **6** inanimate. **7 a** lacking force or vigour; dull, lustreless, muffled. **b** (of sound) not resonant. **c** (of sparkling wine etc.) no longer effervescent. **8 a** quiet; lacking activity (the dead season). **b** motionless, idle. **9 a** (of a microphone, telephone, etc.) not transmitting any sound, esp. because of a fault. **b** (of a circuit, conductor, etc.) carrying or transmitting no current; not connected to a source of electricity (a dead battery). **10** (of the ball in a game) out of play. **11** abrupt, complete, exact, unqualified, unrelieved (come to a dead stop; a dead faint; a dead calm; in dead silence; a dead certainty). **12** without spiritual life. ● adv. **1** absolutely, exactly, completely (dead on target; dead level; dead tired). **2** colloq. very, extremely (dead good; dead easy). ● n. (prec. by the) **1** (treated as pl.) those who have died. **2** a time of silence or inactivity (the dead of night). □ **dead-and-alive** Brit. (of a place, person, activity, etc.) dull, monotonous; lacking interest. **dead as the dodo** see DODO. **dead as a doornail** see DOORNAIL. **dead-ball line** Rugby a line behind the goal-line beyond which the ball is considered out of play. **dead bat** Cricket a bat held loosely so that it imparts no motion to the ball when struck. **dead beat** **1** colloq. exhausted. **2** Physics (of an instrument) without recoil. **dead-beat** n. **1** colloq. an idle, feckless, or disreputable person. **2** US sl. a person constantly in debt. **dead centre 1** the exact centre. **2** the position of a crank etc. in line with the connecting-rod and not exerting torque. **dead cert** see CERT. **dead duck** colloq. an unsuccessful or useless person or thing. **dead end 1** a closed end of a road, passage, etc. **2** (often with hyphen) attrib.) a situation offering no prospects of progress or advancement. **dead-eye** Naut. a round flat three-holed block for extending shrouds. **dead from the neck up** colloq. stupid. **dead hand** an oppressive persisting influence, esp. posthumous control. **dead heat 1** a race in which two or more competitors finish exactly level. **2** the result of such a race. **dead-heat** v.intr. run a dead heat. **dead language** a language no longer commonly spoken, e.g. Latin. **dead letter 1** a law or practice no longer observed or recognized. **2** an unclaimed or undelivered letter. **dead lift** the exertion of one's utmost strength to lift something. **dead loss 1** colloq. a useless person or thing. **2** a complete loss. **dead man's fingers 1** the early purple orchid, Orchis mascula. **2** a soft coral of the genus Alcyonium, with spongy lobes. **3** the finger-like divisions of a lobster's or crab's gills. **dead man's handle** (or **pedal** etc.) a controlling device on an electric or diesel train, allowing power to be connected only as long as the operator presses on it. **dead march** a funeral march. **dead men** colloq. bottles after the contents have been drunk. **dead-nettle** a labiate plant of Lamium or a related genus, having nettle-like leaves but without stinging hairs. **dead-on** exactly right. **dead reckoning** Naut. calculation of a ship's position from the log, compass, etc., when observations are impossible. **dead ringer** see be a ringer for (see RINGER). **dead set** a determined attack or attempt. **dead shot** a person who is extremely accurate with a gun etc. **dead time** Physics the period after the recording of a pulse etc. when the detector is unable to record another. **dead to the world** colloq. fast asleep; unconscious. **dead weight** (or **dead-weight**) **1 a** an inert mass. **b** a heavy weight or burden. **2** a debt not covered by assets. **3** the total weight carried on a ship. **dead wood** colloq. one or more useless people or things. **make a dead set at** see SET². **wouldn't be seen dead with** (or **in** etc.) colloq. shall have nothing to do with; shall refuse to wear etc. □ **deadness** n. [OE dēad f. Gmc, rel. to DIE¹]

deadbolt /'dedbəʊlt/ n. esp. US a bolt engaged by turning a knob or key, rather than by spring action.

deaden /'ded(ə)n/ v. **1** tr. & intr. deprive of or lose vitality, force, brightness, sound, feeling, etc. **2** tr. (foll. by to) make insensitive. □ **deadener** n.

deadeye /'dedaɪ/ n. **1** Naut. a circular wooden block with a groove round the circumference to take a lanyard, used singly or in pairs to tighten a shroud. **2** US colloq. an expert marksman.

deadfall /ˈdedfɔːl/ n. N. Amer. a trap in which a raised weight is made to fall on and kill esp. large game.

deadhead /ˈdedhed/ n. & v. ● n. **1** a faded flower-head. **2** a passenger or member of an audience who has made use of a free ticket. **3** a useless or unenterprising person. ● v. **1** tr. remove deadheads from (a plant). **2** intr. US (of a driver etc.) complete a journey with an empty train, bus, etc.

deadlight /ˈdedlaɪt/ n. **1** Naut. a shutter inside a porthole. **2** US a skylight that cannot be opened.

deadline /ˈdedlaɪn/ n. **1** a time-limit for the completion of an activity etc. **2** hist. a line beyond which prisoners were not allowed to go. Such a line was marked round a military prison at Andersonville, Georgia, US, c.1864; a prisoner going beyond it was liable to be shot down.

deadlock /ˈdedlɒk/ n. & v. ● n. **1** a situation, esp. one involving opposing parties, in which no progress can be made. **2** a type of lock requiring a key to open or close it. ● v.tr. & intr. bring or come to a standstill.

deadly /ˈdedlɪ/ adj. & adv. ● adj. (**deadlier, deadliest**) **1 a** causing or able to cause fatal injury or serious damage. **b** poisonous (deadly snake). **2** intense, extreme (deadly dullness). **3** (of an aim etc.) extremely accurate or effective. **4** deathlike (deadly pale; deadly faintness; deadly gloom). **5** colloq. dreary, dull. **6** implacable. ● adv. **1** like death; as if dead (deadly faint). **2** extremely, intensely (deadly serious). □ **deadly nightshade** see NIGHTSHADE. **deadly sin** in Christian tradition, a sin regarded as leading to damnation for a person's soul (see also SEVEN DEADLY SINS). □ **deadliness** n. [OE dēadlic, dēadlīce (as DEAD, -LY¹)]

deadpan /ˈdedpæn/ adj., adv., & v. ● adj. & adv. with a face or manner totally lacking expression or emotion. ● v. (**deadpanned, deadpanning**) **1** intr. speak or behave in a deadpan manner. **2** tr. say or behave towards in a deadpan manner.

Dead Sea a salt lake or inland sea in the Jordan valley, on the Israel–Jordan border. Its surface is 400 m (1,300 ft) below sea level.

Dead Sea scrolls a collection of Hebrew and Aramaic manuscripts discovered in pottery storage jars in caves near Qumran, at the north-eastern end of the Dead Sea, between 1947 and 1956. They are thought to belong to the library of a splinter Jewish sect, generally equated with the Essenes, and it is presumed that the scrolls were hidden shortly before the Jewish revolt against Roman rule AD 66–70. They include texts of many books of the Hebrew Bible (the Old Testament), as well as commentaries, psalms, and other works; they are some 1,000 years older than previously known versions.

deadstock /ˈdedstɒk/ n. farm machinery. [in contrast to livestock]

de-aerate /diːˈeəreɪt/ v.tr. remove air from. □ **de-aeration** /ˌdiːeəˈreɪʃ(ə)n/ n.

deaf /def/ adj. **1** wholly or partly without hearing (deaf in one ear). **2** (foll. by to) refusing to listen or comply. **3** insensitive to harmony, rhythm, etc. (tone-deaf). □ **deaf-aid** Brit. a hearing-aid. **deaf-and-dumb alphabet** (or **language** etc.) = sign language. ¶ Sign language is preferred in official use. **deaf as a post** completely deaf. **deaf mute** a deaf and dumb person. **fall on deaf ears** be ignored. **turn a deaf ear** (usu. foll. by to) be unresponsive. □ **deafness** n. [OE dēaf f. Gmc]

deafen /ˈdef(ə)n/ v.tr. **1** (often as **deafening** adj.) overpower with sound. **2** deprive of hearing by noise, esp. temporarily. □ **deafeningly** adv.

deal¹ /diːl/ v. & n. ● v. (past and past part. **dealt** /delt/) **1** intr. **a** (foll. by with) take measures concerning (a problem, person, etc.), esp. in order to put something right. **b** (foll. by with) do business with; associate with. **c** (foll. by with) discuss or treat (a subject). **d** (often foll. by by) behave in a specified way towards a person (dealt honourably by them). **2** intr. (foll. by in) sell or be concerned with commercially (deals in insurance). **3** tr. (often foll. by out, round) distribute or apportion to several people etc. **4** tr. (also absol.) distribute (cards) to players for a game or round. **5** tr. cause to be received; administer (deal a heavy blow). **6** tr. assign as a share or deserts to a person (Providence dealt them much happiness). **7** tr. (foll. by in) colloq. include (a person) in an activity (you can deal me in). ● n. **1** (usu. **a good** or **great deal**) colloq. **a** a large amount (a good deal of trouble). **b** to a considerable extent (is a great deal better). **2** a business arrangement; a transaction. **3** a specified form of treatment given or received (gave them a rough deal; got a fair deal). **4 a** the distribution of cards by dealing. **b** a player's turn to do this (it's my deal). **c** the round of play following this. **d** a set of hands dealt to players. □ **it's a deal** colloq. expressing assent to an agreement. [OE dǣl, dǣlan, f. Gmc]

deal² /diːl/ n. **1** fir or pine timber, esp. sawn into boards of a standard size. **2 a** a board of this timber. **b** such boards collectively. [ME f. MLG, MDu. dele plank f. Gmc]

dealer /ˈdiːlə(r)/ n. **1** a person or business dealing in (esp. retail) goods (contact your dealer; car-dealer; a dealer in tobacco). **2** the player dealing at cards. **3** a jobber on the Stock Exchange. ¶ In the UK from Oct. 1986 the name has been merged with **broker** (see BROKER 2, JOBBER 1). □ **dealership** n. (in sense 1).

dealings /ˈdiːlɪŋz/ n.pl. contacts or transactions, esp. in business.

dealt past and past part. of DEAL¹.

Dean¹ /diːn/, Christopher, see TORVILL AND DEAN.

Dean² /diːn/, James (born James Byron) (1931–55), American actor. He starred in only three films, East of Eden (1955), Rebel Without a Cause (1955; released posthumously), and Giant (1956), before dying in a car accident. However, he became a cult figure closely identified with the title role of Rebel Without a Cause, symbolizing for many the disaffected youth of the postwar era.

dean¹ /diːn/ n. **1 a** the head of the chapter of a cathedral or collegiate church. **b** (usu. **rural dean**) Brit. a member of the clergy exercising supervision over a group of parochial clergy within a division of an archdeaconry. **2 a** a college or university official, esp. one of several fellows of a college, with disciplinary and advisory functions. **b** the head of a university faculty or department or of a medical school. **3** = DOYEN. [ME f. AF deen, OF deien, f. LL decanus f. decem ten; orig. = chief of a group of ten]

dean² var. of DENE¹.

deanery /ˈdiːnərɪ/ n. (pl. **-ies**) **1** a dean's house or office. **2** Brit. the group of parishes presided over by a rural dean.

Dean of Faculty n. the president of the Faculty of Advocates in Scotland.

dear /dɪə(r)/ adj., n., adv., & int. ● adj. **1 a** beloved or much esteemed. **b** as a merely polite or ironic form (my dear man). **2** used as a formula of address, esp. at the beginning of letters (Dear Sir). **3** (often foll. by to) precious; much cherished. **4** (usu. in superl.) earnest, deeply felt (my dearest wish). **5 a** high-priced relative to its value. **b** having high prices. **c** (of money) available as a loan only at a high rate of interest. ● n. (esp. as a form of address) dear person. ● adv. at a high price or great cost (buy cheap and sell dear; will pay dear). ● int. expressing surprise, dismay, pity, etc. (dear me!; oh dear!; dear, dear!). □ **Dear John** (**letter**) colloq. a letter from a woman terminating a personal relationship. **for dear life** see LIFE. □ **dearness** n. [OE dēore f. Gmc]

dearie /ˈdɪərɪ/ n. (esp. as a form of address) usu. joc. or iron. my dear. □ **dearie me!** int. expressing surprise, dismay, etc.

dearly /ˈdɪəlɪ/ adv. **1** affectionately, fondly (loved him dearly). **2 a** earnestly; keenly. **b** very much, greatly (would dearly love to go). **3** at a high price or great cost.

dearth /dɜːθ/ n. scarcity or lack. [ME, formed as DEAR]

deasil /ˈdes(ə)l/ adv. esp. Sc. in the direction of the sun's apparent course (considered as lucky); clockwise (opp. WIDDERSHINS). [Gael. deiseil]

death /deθ/ n. **1** the final cessation of vital functions in an organism; the ending of life. **2** the event that terminates life. **3 a** the fact or process of being killed or killing (stone to death; fight to the death). **b** the fact or state of being dead (eyes closed in death; their deaths caused rioting). **4 a** the destruction or permanent cessation of something (was the death of our hopes). **b** colloq. something terrible or appalling. **5** (usu. **Death**) a personification of death, esp. as a destructive power, usu. represented by a skeleton. **6** a lack of religious faith or spiritual life. □ **as sure as death** quite certain. **at death's door** close to death. **be in at the death 1** be present when an animal is killed, esp. in hunting. **2** witness the (esp. sudden) ending of an enterprise etc. **be the death of 1** cause the death of. **2** be very harmful to. **catch one's death of cold** colloq. catch a serious chill. **death adder** a venomous snake of the genus Acanthophis, esp. A. antarcticus of Australia. **death camp** a prison camp in which many people die or are put to death. **death cap** a poisonous toadstool, Amanita phalloides. **death cell** a prison cell for a person condemned to death. **death certificate** an official statement of the cause and date and place of a person's death. **death duty** Brit. hist. a tax levied on property after the owner's death. ¶ Replaced in 1975 by capital transfer tax, which was replaced in 1986 by inheritance tax. **death grant** Brit. a state grant towards funeral expenses. **death-knell 1** the tolling of a bell to mark a person's death. **2** an event that heralds the end or destruction of something. **death-mask** a cast taken of a dead person's face. **death penalty** punishment by being put to death, capital punishment. **death rate** the number of deaths per thousand of population per year. **death-rattle** a gurgling sound sometimes heard in a dying person's throat. **death-roll 1** those killed in an accident, battle, etc. **2** a list of these. **death row** (esp. with reference to the US) a prison block or section for prisoners sentenced to death. **death's head** a human skull as an emblem of mortality. **death's head moth** (or **hawkmoth**) a large dark hawkmoth, *Acherontia atropos*, with skull-like markings on the back of the thorax. **death squad** an armed paramilitary group formed to kill political enemies etc. **death tax** *US* a tax on property payable on the owner's death. **death-toll** the number of people killed in an accident, battle, etc. **death-trap** *colloq.* a dangerous or unhealthy building, vehicle, etc. **death-warrant 1** an order for the execution of a condemned person. **2** anything that causes the end of an established practice etc. **death-wish** *Psychol.* a desire (usu. unconscious) for the death of oneself or another. **do to death 1** *archaic* kill. **2** overdo. **fate worse than death** *archaic* or *joc.* being raped or seduced. **like death warmed up** *sl.* very tired or ill. **put to death** kill or cause to be killed, execute. **to death** to the utmost, extremely (*bored to death*; *worked to death*). □ **deathless** *adj.* **deathlessness** *n.* **deathlike** *adj.* [OE *dēath* f. Gmc: rel. to DIE¹]

deathbed /'deθbed/ *n.* a bed as the place where a person is dying or has died.

deathblow /'deθbləʊ/ *n.* **1** a blow or other action that causes death. **2** an event or circumstance that abruptly ends an activity, enterprise, etc.

deathly /'deθlɪ/ *adj. & adv.* ● *adj.* suggestive of death (*deathly silence*). ● *adv.* in a deathly way (*deathly pale*).

Death Valley a deep arid desert basin in SE California and SW Nevada, the hottest and driest part of North America and one of the hottest places on earth.

death-watch beetle /'deθwɒtʃ/ *n.* (also **death-watch**) a small beetle, *Xestobium rufovillosum*, whose larvae bore in old wood and are very destructive to old buildings. The adult makes a sound like a watch ticking, once supposed to portend death.

deattribute /ˌdiːə'trɪbjuːt/ *v.tr.* cease to attribute (a work of art) to a particular artist. □ **deattribution** /ˌdiːætrɪ'bjuːʃ(ə)n/ *n.*

deb /deb/ *n. colloq.* a débutante. [abbr.]

débâcle /deɪ'bɑːk(ə)l/ *n.* (also **debacle**) **1 a** an utter defeat or failure; a fiasco or disaster. **b** a sudden collapse or downfall. **2** a confused rush or rout; a stampede. **3 a** a breakup of ice in a river, with resultant flooding. **b** a sudden rush of water carrying along blocks of stone and other debris. [F f. *débâcler* unbar]

debag /diː'bæg/ *v.tr.* (**debagged, debagging**) *Brit. sl.* remove the trousers of (a person), esp. as a joke.

debar /dɪ'bɑː(r), diː'bɑː(r)/ *v.tr.* (**debarred, debarring**) (foll. by *from*) exclude from admission or from a right; prohibit from an action (*was debarred from entering*). □ **debarment** *n.* [ME f. F *débarrer*, OF *desbarrer* (as DE-, BAR¹)]

debark¹ /diː'bɑːk, dɪ'bɑːk/ *v.tr. & intr.* land from a ship. □ **debarkation** /ˌdiːbɑː'keɪʃ(ə)n/ *n.* [F *débarquer* (as DE-, BARK³)]

debark² /diː'bɑːk/ *v.tr.* remove the bark from (a tree).

debase /dɪ'beɪs/ *v.tr.* **1** lower in quality, value, or character. **2** depreciate (coin) by alloying etc. □ **debasement** *n.* **debaser** *n.* [DE- + obs. *base* for ABASE]

debatable /dɪ'beɪtəb(ə)l/ *adj.* **1** questionable; subject to dispute. **2** capable of being debated. □ **debatably** *adv.* [OF *debatable* or AL *debatabilis* (as DEBATE)]

debate /dɪ'beɪt/ *v. & n.* ● *v.* **1** *tr.* (also *absol.*) discuss or dispute about (an issue, proposal, etc.), esp. formally in a legislative assembly, public meeting, etc. **2 a** *tr.* consider or ponder (a matter). **b** *intr.* consider different sides of a question. ● *n.* **1** a formal discussion on a particular màtter, esp. in a legislative assembly etc. **2** debating, discussion (*open to debate*). □ **debating point** an inessential matter used to gain advantage in a debate. □ **debater** *n.* [ME f. OF *debatre*, *debat* f. Rmc (as DE-, BATTLE)]

debauch /dɪ'bɔːtʃ/ *v. & n.* ● *v.tr.* **1** corrupt morally. **2** make intemperate or sensually indulgent. **3** deprave or debase (taste or judgement). **4** (as **debauched** *adj.*) dissolute, licentious. **5** *archaic* seduce (a woman). ● *n.* **1** a bout of sensual indulgence. **2** debauchery. □ **debaucher** *n.* [F *débaucher*, OF *desbaucher*, of unkn. orig.]

debauchee /ˌdɪbɔː'tʃiː, ˌdeb-/ *n.* a person addicted to excessive sensual indulgence. [F *débauché* past part.: see DEBAUCH, -EE]

debauchery /dɪ'bɔːtʃərɪ/ *n.* excessive licentiousness or sensual indulgence.

de Beauvoir /də 'bəʊvwɑː(r)/, Simone (1908–86), French existentialist philosopher, novelist, and feminist. While studying philosophy at the

Sorbonne in 1929 she began her lifelong association with Jean-Paul Sartre; they became leading exponents of existentialism and founded the review *Les Temps modernes* (1945). De Beauvoir is regarded as an important figure in the 'second wave' of feminism, bringing the ideas of psychology, myth, political theory, and history to bear on the issue in her best-known work, *The Second Sex* (1949). Her fiction includes *The Blood of Others* (1944).

debenture /dɪ'bentʃə(r)/ *n.* **1** *Brit.* an acknowledgement of indebtedness, esp. a bond of a company or corporation acknowledging a debt and providing for payment of interest at fixed intervals. **2** *US* (in full **debenture bond**) a fixed-interest bond of a company or corporation, backed by general credit rather than specified assets. □ **debenture stock** *Brit.* stock comprising debentures, with only the interest secured. [ME f. L *debentur* are owing f. *debere* owe: assim. to -URE]

debilitate /dɪ'bɪlɪˌteɪt/ *v.tr.* enfeeble, enervate. □ **debilitatingly** *adv.* **debilitative** /-tətɪv/ *adj.* **debilitation** /-ˌbɪlɪ'teɪʃ(ə)n/ *n.* [L *debilitare* (as DEBILITY)]

debility /dɪ'bɪlɪtɪ/ *n.* feebleness, esp. of health. [ME f. OF *debilité* f. L *debilitas -tatis* f. *debilis* weak]

debit /'debɪt/ *n. & v.* ● *n.* **1** an entry in an account recording a sum owed. **2** the sum recorded. **3** the total of such sums. **4** the debit side of an account. ● *v.tr.* (**debited, debiting**) **1** (foll. by *against, to*) enter (an amount) on the debit side of an account (*debited £500 against me*). **2** (foll. by *with*) enter (a person) on the debit side of an account (*debited me with £500*). □ **debit card** a card allowing the holder to transfer money direct from his or her own bank account to another when making a purchase etc. [F *débit* f. L *debitum* DEBT]

debonair /ˌdebə'neə(r)/ *adj.* **1** carefree, cheerful, self-assured. **2** having pleasant manners, urbane. □ **debonairly** *adv.* [ME f. OF *debonaire* = *de bon aire* of good stock]

Deborah /'debərə, 'debrə/ a biblical prophet and leader who inspired the Israelite army to defeat the Canaanites (Judges 4–5). The 'Song of Deborah', a song of victory attributed to her, is thought to be one of the oldest sections of the Bible.

debouch /dɪ'baʊtʃ, -'buːʃ/ *v.intr.* **1** (of troops or a stream) issue from a ravine, wood, etc., into open ground. **2** (often foll. by *into*) (of a river, road, etc.) merge into a larger body or area. □ **debouchment** *n.* [F *déboucher* (as DE-, *bouche* mouth)]

Debrecen /'debrəˌtsen/ an industrial and commercial city in eastern Hungary; pop. (est. 1989) 220,000.

Debrett /dɪ'bret/, John (*c.*1750–1822), English publisher. He compiled *The Peerage of England, Scotland and Ireland* (first issued in 1803 and until 1971 issued annually), which is regarded as the authority on the British nobility; it is now published every five years.

debridement /dɪ'briːdmənt/ *n. Med.* the removal of damaged tissue or foreign matter from a wound etc. [F, lit. 'unbridling']

debrief /diː'briːf/ *v.tr.* interrogate (a person, e.g. a diplomat or pilot) about a completed mission or undertaking. □ **debriefing** *n.*

debris /'debriː, 'deɪb-/ *n.* **1** scattered fragments, esp. of something wrecked or destroyed. **2** *Geol.* an accumulation of loose material, e.g. from rocks or plants. [F *débris* f. obs. *débriser* break down (as DE-, *briser* break)]

de Broglie /də 'brəʊglɪ, French də brɔj/, Louis-Victor, Prince (1892–1987), French physicist. He was the first to suggest that subatomic particles can also have the properties of waves, and his name is now applied to such a wave. He further developed the study of wave mechanics, which was fundamental to the subsequent development of quantum mechanics. He was awarded the Nobel Prize for physics in 1929.

de Broglie wavelength *n. Physics* the wavelength representing the wavelike properties of a particle in wave mechanics.

debt /det/ *n.* **1** something that is owed, esp. money. **2** a state of obligation to pay something owed (*in debt*; *out of debt*; *get into debt*). □ **debt-collector** a person who is employed to collect debts for creditors. **debt of honour** a debt not legally recoverable, esp. a sum lost in gambling. **in a person's debt** under an obligation to a person. [ME *det(te)* f. OF *dette* (later *debte*) ult. f. L *debitum* past part. of *debere* owe]

debtor /'detə(r)/ *n.* a person who owes a debt, esp. of money. [ME f. OF *det(t)or, -our* f. L *debitor* (as DEBT)]

debug /diː'bʌg/ *v.tr.* (**debugged, debugging**) **1** trace and remove concealed listening devices from (a room etc.). **2** identify and remove defects from (a machine, computer program, etc.). **3** remove bugs from, delouse.

debugger /diːˈbʌɡə(r)/ n. Computing a program for debugging other programs.

debunk /diːˈbʌŋk/ v.tr. colloq. **1** show the good reputation or aspirations of (a person, institution, etc.) to be spurious. **2** expose the falseness of (a claim etc.). □ **debunker** n.

debus /diːˈbʌs/ v.tr. & intr. (**debussed, debussing**) esp. Mil. unload (personnel or stores) or alight from a motor vehicle.

Debussy /dəˈbuːsɪ, French dəbysi/, (Achille) Claude (1862–1918), French composer and critic. Debussy carried the ideas of impressionist art and symbolist poetry into music, using melodies based on the whole-tone scale and delicate harmonies exploiting overtones. His orchestral tone-poem *Prélude à l'après-midi d'un faune* (1894) and his books of piano preludes and studies are outstanding examples of the delicate, suggestive character of his music. Debussy was to have a profound influence on later composers such as Berg, Bartók, and Boulez.

début /ˈdeɪbjuː, -buː/ n. (also **debut**) **1** the first public appearance of a performer on stage etc., or the opening performance of a show etc. **2** the first appearance of a débutante in society. [F f. *débuter* lead off]

débutante /ˈdebjʊˌtɒnt, ˈdeɪb-/ n. (also **debutante**) **1** a (usu. wealthy) young woman making her social début. **2** a female performer making her first public appearance. [F, fem. part. of *débuter*: see DÉBUT]

Debye /dɪˈbaɪ/, Peter Joseph William (1884–1966), Dutch-born American chemical physicist. Debye made substantial contributions in several fields. He established the existence of permanent electric dipole moments in many molecules, demonstrated the use of these to determine molecular size and shape, modified Einstein's theory of specific heats as applied to solids, pioneered the use of X-ray scattering to determine crystal structure, and solved problems of electrolytic conductance. He received the Nobel Prize for chemistry in 1936.

Dec. abbr. December.

dec. abbr. **1** deceased. **2** declared.

deca- /ˈdekə/ comb. form (also **dec-** before a vowel) **1** having ten. **2** tenfold. **3** ten, esp. of a metric unit (*decagram; decalitre*). [Gk *deka* ten]

decade /ˈdekeɪd, -ˈkeɪd/ n. **1** a period of ten years. **2** a set, series, or group of ten. [ME f. F *décade* f. LL *decas -adis* f. Gk *deka* ten]

decadence /ˈdekəd(ə)ns/ n. **1** moral or cultural deterioration, esp. after a peak or culmination of achievement. **2** decadent behaviour; a state of decadence. [F *décadence* f. med.L *decadentia* f. *decadere* DECAY]

decadent /ˈdekəd(ə)nt/ adj. & n. ● adj. **1 a** in a state of moral or cultural deterioration; showing or characterized by decadence. **b** of a period of decadence. **2** self-indulgent. ● n. a decadent person. □ **decadently** adv. [F *décadent* (as DECADENCE)]

Decaf /ˈdiːkæf/ n. & adj. (also **decaff**) ● n. propr. decaffeinated coffee. ● adj. (**decaf**) decaffeinated.

decaffeinate /diːˈkæfɪˌneɪt/ v.tr. **1** remove the caffeine from. **2** reduce the quantity of caffeine in (usu. coffee).

decagon /ˈdekəɡən/ n. a plane figure with ten sides and angles. □ **decagonal** /dɪˈkæɡən(ə)l/ adj. [med.L *decagonum* f. Gk *dekagōnon* (as DECA-, -GON)]

decagynous /deˈkædʒɪnəs/ adj. Bot. having ten pistils. [mod.L *decagynus* (as DECA-, Gk *gūne* woman)]

decahedron /ˌdekəˈhiːdrən/ n. a solid figure with ten faces. □ **decahedral** adj. [DECA- + -HEDRON after POLYHEDRON]

decal /ˈdiːkæl/ n. = DECALCOMANIA 2. [abbr.]

decalcify /diːˈkælsɪˌfaɪ/ v.tr. (**-ies, -ied**) remove lime or calcareous matter from (a bone, tooth, etc.). □ **decalcifier** n. **decalcification** /-ˌkælsɪfɪˈkeɪʃ(ə)n/ n.

decalcomania /diːˌkælkəˈmeɪnɪə/ n. **1** a process of transferring designs from specially prepared paper to the surface of glass, porcelain, etc. **2** a picture or design used in or made by this process. [F *décalcomanie* f. *décalquer* transfer]

decalitre /ˈdekəˌliːtə(r)/ n. (US **decaliter**) a metric unit of capacity, equal to 10 litres.

Decalogue /ˈdekəˌlɒɡ/ n. the Ten Commandments. [ME f. F *décalogue* or eccl.L *decalogus* f. Gk *dekalogos* (after *hoi deka logoi* the Ten Commandments)]

Decameron /dɪˈkæmərən/ a work by Boccaccio, written between 1348 and 1358, containing a hundred tales supposedly told in ten days by a party of ten young people who had fled from the Black Death in

Florence. The work was influential on later writers such as Chaucer and Shakespeare. [It. f. Gk f. *deka* ten + *hēmera* day]

decametre /ˈdekəˌmiːtə(r)/ n. (US **decameter**) a metric unit of length, equal to 10 metres.

decamp /dɪˈkæmp, diːˈkæmp/ v.intr. **1** break up or leave a camp. **2** depart suddenly; abscond. □ **decampment** n. [F *décamper* (as DE-, CAMP¹)]

decanal /dɪˈkeɪn(ə)l, ˈdekən(ə)l/ adj. **1** of a dean or deanery. **2** of the south side of a choir, the side on which the dean sits (cf. CANTORIAL). [med.L *decanalis* f. LL *decanus* DEAN¹]

decandrous /deˈkændrəs/ adj. Bot. having ten stamens. [DECA- + Gk *andr-* man (= male organ)]

decani /dɪˈkeɪnaɪ/ adj. Mus. to be sung by the decanal side in antiphonal singing (cf. CANTORIS). [L, genitive of *decanus* DEAN¹]

decant /dɪˈkænt/ v.tr. **1** gradually pour off (liquid, esp. wine or a solution) from one container to another, esp. without disturbing the sediment. **2** empty out; move as if by pouring. [med.L *decanthare* (as DE-, L *canthus* f. Gk *kanthos* canthus, used of the lip of a beaker)]

decanter /dɪˈkæntə(r)/ n. a stoppered glass container into which wine or spirits are decanted.

decapitate /dɪˈkæpɪˌteɪt/ v.tr. **1** behead (esp. as a form of capital punishment). **2** cut the head or end from. □ **decapitator** n. **decapitation** /-ˌkæpɪˈteɪʃ(ə)n/ n. [LL *decapitare* (as DE-, *caput -itis* head)]

decapod /ˈdekəˌpɒd/ n. Zool. **1** a crustacean of the chiefly marine order Decapoda, characterized by five pairs of legs (the first pair of which are often modified as pincers), e.g. shrimps, crabs, and lobsters. **2** a mollusc of the class Cephalopoda with ten tentacles, e.g. squids and cuttlefish. □ **decapodan** /dɪˈkæpəd(ə)n/ adj. [F *décapode* f. Gk *deka* ten + *pous podos* foot]

decarbonize /diːˈkɑːbəˌnaɪz/ v.tr. (also **-ise**) remove carbon or carbonaceous deposits from (an internal-combustion engine etc.). □ **decarbonization** /-ˌkɑːbənaɪˈzeɪʃ(ə)n/ n.

decastyle /ˈdekəˌstaɪl/ n. & adj. Archit. ● n. a ten-columned portico. ● adj. having ten columns. [Gk *dekastulos* f. *deka* ten + *stulos* column]

decasyllable /ˈdekəˌsɪləb(ə)l/ n. a metrical line of ten syllables. □ **decasyllabic** /ˌdekəsɪˈlæbɪk/ adj. & n.

decathlon /dɪˈkæθlən/ n. an athletic event taking place over two days, in which each competitor takes part in the same prescribed ten events. These are, in order: 100 metres sprint, long jump, shot-put, high jump, 400 metres, 110 metres hurdles, discus, pole vault, javelin, and 1,500 metres. The decathlon is an Olympic event. □ **decathlete** /-liːt/ n. [DECA- + Gk *athlon* contest]

decay /dɪˈkeɪ/ v. & n. ● v. **1 a** intr. rot, decompose. **b** tr. cause to rot or decompose. **2** intr. & tr. decline or cause to decline in quality, power, wealth, energy, beauty, etc. **3** intr. Physics **a** (usu. foll. by *to*) (of a substance etc.) undergo change by radioactivity. **b** undergo a gradual decrease in magnitude of a physical quantity. ● n. **1** a rotten or ruinous state; a process of wasting away. **2** decline in health, quality, etc. **3** Physics **a** change into another substance etc. by radioactivity. **b** a decrease in the magnitude of a physical quantity, esp. the intensity of radiation or amplitude of oscillation. **4** decayed tissue. [ME f. OF *decair* f. Rmc (as DE-, L *cadere* fall)]

Deccan /ˈdekən/ a triangular plateau in southern India, bounded by the Malabar Coast in the west, the Coromandel Coast in the east, and by the Vindhaya mountains in the north.

decease /dɪˈsiːs/ n. & v. formal esp. Law ● n. death. ● v.intr. die. [ME f. OF *deces* f. L *decessus* f. *decedere* (as DE-, *cedere cess-* go)]

deceased /dɪˈsiːst/ adj. & n. formal ● adj. dead. ● n. (usu. prec. by *the*) a person who has died, esp. recently.

decedent /dɪˈsiːd(ə)nt/ n. N. Amer. Law a deceased person. [L *decedere* die: see DECEASE]

deceit /dɪˈsiːt/ n. **1** the act or process of deceiving or misleading, esp. by concealing the truth. **2** a dishonest trick or stratagem. **3** a tendency to deceive or mislead. [ME f. OF f. past part. of *deceveir* f. L *decipere* deceive (as DE-, *capere* take)]

deceitful /dɪˈsiːtfʊl/ adj. **1** (of a person) using deceit, esp. habitually. **2** (of an act, practice, etc.) intended to deceive. □ **deceitfully** adv. **deceitfulness** n.

deceive /dɪˈsiːv/ v. **1** tr. make (a person) believe what is false, mislead purposely. **2** tr. be unfaithful to, esp. sexually. **3** intr. use deceit. **4** tr. archaic disappoint (esp. hopes). □ **be deceived** be mistaken or deluded. **deceive oneself** persist in a mistaken belief. □ **deceivable** adj.

deceiver n. [ME f. OF *deceivre* or *deceiv-* stressed stem of *deceveir* (as DECEIT)]

decelerate /diːˈseləˌreɪt/ v. **1** intr. & tr. begin or cause to begin to reduce speed. **2** tr. make slower (*decelerated motion*). □ **decelerator** n. **deceleration** /-ˌseləˈreɪʃ(ə)n/ n. **decelerometer** /-ˈrɒmɪtə(r)/ n. [DE-, after ACCELERATE]

December /dɪˈsembə(r)/ n. the twelfth month of the year. [ME f. OF *decembre* f. L *December* f. *decem* ten: orig. the tenth month of the Roman year]

Decembrist /dɪˈsembrɪst/ n. a member of a group of Russian revolutionaries who in Dec. 1825 led an unsuccessful revolt against Tsar Nicholas I. The leaders were executed and later came to be regarded as martyrs by the Left. [transl. Russ. *dekabrist*]

decency /ˈdiːsənsɪ/ n. (pl. **-ies**) **1** generally accepted standards of behaviour or propriety. **2** avoidance of obscenity. **3** (in pl.) the requirements of correct behaviour. [L *decentia* f. *decere* be fitting]

decennial /dɪˈsenɪəl/ adj. **1** lasting ten years. **2** recurring every ten years. □ **decennially** adv. [L *decennis* of ten years f. *decem* ten + *annus* year]

decent /ˈdiːs(ə)nt/ adj. **1 a** conforming with generally accepted standards of behaviour or propriety. **b** avoiding obscenity. **2** respectable. **3** acceptable, passable; good enough. **4** Brit. kind, obliging, generous (*was decent enough to apologize*). □ **decently** adv. [F *décent* or L *decere* be fitting]

decentralize /diːˈsentrəˌlaɪz/ v.tr. (also **-ise**) **1** transfer (powers etc.) from a central to a local authority. **2** reorganize (a centralized institution, organization, etc.) on the basis of greater local autonomy. □ **decentralist** n. & adj. **decentralization** /-ˌsentrəlaɪˈzeɪʃ(ə)n/ n.

deception /dɪˈsepʃ(ə)n/ n. **1** the act or an instance of deceiving; the process of being deceived. **2** a thing that deceives; a trick or sham. [ME f. OF or LL *deceptio* f. *decipere* (as DECEIT)]

deceptive /dɪˈseptɪv/ adj. apt to deceive; easily mistaken for something else or as having a different quality. □ **deceptively** adv. **deceptiveness** n. [OF *deceptif -ive* or LL *deceptivus* (as DECEPTION)]

decerebrate /diːˈserɪbrət/ adj. having had the cerebrum removed.

deci- /ˈdesɪ/ comb. form one-tenth, esp. of a unit in the metric system (*decilitre*; *decimetre*). [L *decimus* tenth]

decibel /ˈdesɪˌbel/ n. a unit (one-tenth of a bel) used in the comparison of two power levels relating to electrical signals or sound intensities, one of the pair usually being taken as a standard (symbol: **dB**).

decide /dɪˈsaɪd/ v. **1 a** intr. (often foll. by *on*, *about*) come to a resolution as a result of consideration. **b** tr. (usu. foll. by *to* + infin., or *that* + clause) have or reach as one's resolution about something (*decided to stay*; *decided that we should leave*). **2** tr. **a** cause (a person) to reach a resolution (*was unsure about going but the weather decided me*). **b** resolve or settle (a question, dispute, etc.). **3** intr. (usu. foll. by *between*, *for*, *against*, *in favour of*, or *that* + clause) give a judgement concerning a matter. □ **decidable** adj. [ME f. F *décider* or f. L *decidere* (as DE-, *cædere* cut)]

decided /dɪˈsaɪdɪd/ adj. **1** (usu. attrib.) definite, unquestionable (*a decided difference*). **2** (of a person, esp. as a characteristic) having clear opinions, resolute, not vacillating. □ **decidedness** n.

decidedly /dɪˈsaɪdɪdlɪ/ adv. undoubtedly, undeniably.

decider /dɪˈsaɪdə(r)/ n. **1** a game, race, etc., to decide between competitors finishing equal in a previous contest. **2** any person or thing that decides.

deciduous /dɪˈsɪdjʊəs/ adj. **1** (of a tree) shedding its leaves annually. **2** (of leaves, horns, teeth, etc.) shed periodically. **3** (of an ant etc.) shedding its wings after copulation. **4** fleeting, transitory. □ **deciduousness** n. [L *deciduus* f. *decidere* f. *cadere* fall]

decigram /ˈdesɪˌɡræm/ n. (also **decigramme**) a metric unit of mass, equal to 0.1 gram.

decile /ˈdesɪl, -saɪl/ n. Statistics any of the nine values of a random variable which divide a frequency distribution into ten groups, each containing one-tenth of the total population. [F *décile*, ult. f. L *decem* ten]

decilitre /ˈdesɪˌliːtə(r)/ n. (US **deciliter**) a metric unit of capacity, equal to 0.1 litre.

decimal /ˈdesɪm(ə)l/ adj. & n. ● adj. **1** (of a system of numbers, weights, measures, etc.) based on the number ten, in which the smaller units are related to the principal units as powers of ten (units, tens, hundreds, thousands, etc.). **2** of tenths or ten; reckoning or proceeding by tens. ● n. a decimal fraction. □ **decimal fraction** a fraction whose denominator is a power of ten, esp. when expressed positionally by

units to the right of a decimal point. **decimal place** the position of a digit to the right of a decimal point. **decimal point** a full point or dot placed before the numerator in a decimal fraction. **decimal scale** a scale with successive places denoting units, tens, hundreds, etc. □ **decimally** adv. [mod.L *decimalis* f. L *decimus* tenth]

decimalization /ˌdesɪməlaɪˈzeɪʃ(ə)n/ n. the conversion of a system of weights, measures, or (especially) coinage to a decimal system. Decimal currency was first used in 1785, when the US standardized the dollar and cent, and was adopted in France after the Revolution; in the UK the old pound of 240 pence or 20 shillings was replaced by a pound divided into 100 pence in Feb. 1971. (See also METRIC SYSTEM.)

decimalize /ˈdesɪməˌlaɪz/ v.tr. (also **-ise**) **1** express as a decimal. **2** convert to a decimal system (esp. of coinage).

decimate /ˈdesɪˌmeɪt/ v.tr. **1** destroy, kill, or remove a large proportion of. **2** orig. Mil. kill or remove one in every ten of. □ **decimator** n. **decimation** /ˌdesɪˈmeɪʃ(ə)n/ n. [L *decimare* take the tenth man f. *decimus* tenth]

decimetre /ˈdesɪˌmiːtə(r)/ n. (US **decimeter**) a metric unit of length, equal to 0.1 metre.

decipher /dɪˈsaɪfə(r)/ v.tr. **1** convert (a text written in cipher) into an intelligible script or language. **2** determine the meaning of (anything obscure or unclear). □ **decipherable** adj. **decipherment** n.

decision /dɪˈsɪʒ(ə)n/ n. **1** the act or process of deciding. **2** a conclusion or resolution reached, esp. as to future action, after consideration (*have made my decision*). **3** (often foll. by *of*) **a** the settlement of a question. **b** a formal judgement. **4** a tendency to decide firmly; resoluteness. [ME f. OF *decision* or L *decisio* (as DECIDE)]

decisive /dɪˈsaɪsɪv/ adj. **1** that decides an issue; conclusive. **2** (of a person, esp. as a characteristic) able to decide quickly and effectively. □ **decisively** adv. **decisiveness** n. [F *décisif -ive* f. med.L *decisivus* (as DECIDE)]

Decius /ˈdiːsɪəs/, Gaius Messius Quintus Trajanus (c.201–51), Roman emperor 249–51. He was the first Roman emperor to promote systematic persecution of the Christians in the empire; popular protest eventually forced him to reverse this policy shortly before the end of his reign. He resisted a Gothic invasion of Moesia in 249, but was defeated and killed. □ **Decian** adj.

deck /dek/ n. & v. ● n. **1 a** a platform in a ship covering all or part of the hull's area at any level and serving as a floor. **b** the accommodation on a particular deck of a ship. **2** anything compared to a ship's deck, e.g. the floor or compartment of a bus. **3** a component or unit that comprises the mechanism for recording discs, tapes, etc., in sound-reproduction equipment. **4 a** N. Amer. a pack of cards. **b** US sl. a packet of narcotics. **5** sl. the ground. **6** any floor or platform, esp. the floor of a pier or a platform for sunbathing. ● v.tr. **1** (often foll. by *out*) decorate, adorn. **2** furnish with or cover as a deck. **3** sl. knock (a person) to the ground; floor. □ **below deck** (or **decks**) in or into the space below the main deck. **deck-chair** a folding chair of wood and canvas, of a kind used on deck on passenger ships. **deck-hand** a person employed in cleaning and odd jobs on a ship's deck. **deck quoits** a game played, esp. on ships, in which rope quoits are aimed at a peg. **deck tennis** a game played, esp. on ships, in which a quoit of rope, rubber, etc., is tossed to and fro over a net. **on deck 1** in the open air on a ship's main deck. **2** esp. US ready for action, work, etc. [ME, = covering f. MDu. *dec* roof, cloak]

-decker /ˈdekə(r)/ comb. form having a specified number of decks or layers (*double-decker*).

deckle /ˈdek(ə)l/ n. a device in a paper-making machine for limiting the size of the sheet. □ **deckle edge** the rough uncut edge formed by a deckle. **deckle-edged** having a deckle edge. [G *Deckel* dimin. of *Decke* cover]

declaim /dɪˈkleɪm/ v. **1** intr. & tr. speak or utter rhetorically or affectedly. **2** intr. practise oratory or recitation. **3** intr. (foll. by *against*) protest forcefully. **4** intr. deliver an impassioned (rather than reasoned) speech. □ **declaimer** n. [ME f. F *déclamer* or f. L *declamare* (as DE-, CLAIM)]

declamation /ˌdekləˈmeɪʃ(ə)n/ n. **1** the act or art of declaiming. **2** a rhetorical exercise or set speech. **3** an impassioned speech; a harangue. □ **declamatory** /dɪˈklæmətərɪ/ adj. [F *déclamation* or L *declamatio* (as DECLAIM)]

declarant /dɪˈkleərənt/ n. a person who makes a legal declaration. [F *déclarant* part. of *déclarer* (as DECLARE)]

declaration /ˌdekləˈreɪʃ(ə)n/ n. **1** the act or process of declaring. **2 a** a formal, emphatic, or deliberate statement or announcement. **b** a statement asserting or protecting a legal right. **3** a written public announcement of intentions, terms of an agreement, etc. **4** *Cricket* an act of declaring an innings closed. **5** *Cards* **a** the naming of trumps. **b** an announcement of a combination held. **6** *Law* **a** a plaintiff's statement of claim. **b** an affirmation made instead of taking an oath. **7** (in full **declaration of the poll**) a public official announcement of the votes cast for candidates in an election. [ME f. L *declaratio* (as DECLARE)]

Declaration of Independence a document drawn up declaring the US to be independent of the British Crown, signed on 4 July 1776 by the Congressional representatives of thirteen states (Thomas Jefferson, Benjamin Franklin, John Adams, Roger Sherman (1721–93), and Robert Livingston (1746–1813)) (see THIRTEEN COLONIES).

Declaration of Indulgence any of various proclamations made by the two Stuart kings Charles II and James II (esp. those of 1662, 1672, and 1687–8), which aimed to dispense with repressive legislation against religious nonconformists.

Declaration of Rights a statute passed by the English Parliament in 1689, later incorporated in the Bill of Rights, which established the joint monarchy of William and Mary and which was designed to ensure that the Crown would not act without Parliament's consent.

declarative /dɪ'klærətɪv/ *adj. & n.* ● *adj.* **1 a** of the nature of, or making, a declaration. **b** *Gram.* (of a sentence) that takes the form of a simple statement. **2** *Computing* designating high-level programming languages which can be used to solve problems without requiring the programmer to specify an exact procedure to be followed. ● *n.* **1** a declaratory statement or act. **2** *Gram.* a declarative sentence. □ **declaratively** *adv.* [OF *déclaratif*-*ive* or Latin *declarativus* (as DECLARE)]

declare /dɪ'kleə(r)/ *v.* **1** *tr.* announce openly or formally (*declare war*; *declare a dividend*). **2** *tr.* pronounce (a person or thing) to be something (*declared him to be an impostor*; *declared it invalid*). **3** *tr.* (usu. foll. by *that* + clause) assert emphatically; state explicitly. **4** *tr.* acknowledge possession of (dutiable goods, income, etc.). **5** *tr.* (as **declared** *adj.*) who admits to be such (*a declared atheist*). **6** *tr.* (also *absol.*) *Cricket* close (an innings) voluntarily before all the wickets have fallen. **7** *tr.* *Cards* **a** (also *absol.*) name (the trump suit). **b** announce that one holds (certain combinations of cards etc.). **8** *tr.* (of things) make evident, prove (*your actions declare your honesty*). **9** *intr.* (foll. by *for*, *against*) take the side of one party or another. □ **declare oneself** reveal one's intentions or identity. **well, I declare** (or **I do declare**) an exclamation of incredulity, surprise, or vexation. □ **declarable** *adj.* **declarer** *n.* **declaredly** /-rɪdlɪ/ *adv.* **declaratory** /-'klærətərɪ/ *adj.* [ME f. L *declarare* (as DE-, *clarare* f. *clarus* clear)]

déclassé /deɪ'klæseɪ/ *adj.* (*fem.* **déclassée** *pronunc.* same) that has fallen in social status. [F]

declassify /di:'klæsɪˌfaɪ/ *v.tr.* (**-ies**, **-ied**) declare (information etc.) to be no longer secret. □ **declassification** /-ˌklæsɪfɪ'keɪʃ(ə)n/ *n.*

declension /dɪ'klenʃ(ə)n/ *n.* **1** *Gram.* **a** the variation of the form of a noun, pronoun, or adjective, by which its grammatical case, number, and gender are identified. **b** the class in which a noun etc. is put according to the exact form of this variation. **2** deterioration, declining. □ **declensional** *adj.* [OF *declinaison* f. *decliner* DECLINE after L *declinatio*: assim. to ASCENSION etc.]

declination /ˌdeklɪ'neɪʃ(ə)n/ *n.* **1** a downward bend or turn. **2** *Astron.* the celestial coordinate equivalent to latitude, expressed as angular distance north or south of the celestial equator. **3** *Physics* the angular deviation of a compass needle from true north. **4** *US* a formal refusal. □ **declination axis** *Astron.* the axis of an equatorial mount at right angles to the polar axis, about which a telescope is turned to alter declination. □ **declinational** *adj.* [ME f. L *declinatio* (as DECLINE)]

decline /dɪ'klaɪn/ *v. & n.* ● *v.* **1** *tr.* deteriorate; lose strength or vigour; decrease. **2 a** *tr.* reply with formal courtesy that one will not accept (an invitation, honour, etc.). **b** *tr.* refuse, esp. formally and courteously (*declined to be made use of*; *declined doing anything*). **c** *tr.* turn away from (a challenge, battle, discussion, etc.). **d** *intr.* give or send a refusal. **3** *intr.* slope downwards. **4** *intr.* bend down, droop. **5** *tr.* *Gram.* state the forms of (a noun, pronoun, or adjective) corresponding to cases, number, and gender. **6** *intr.* (of a day, life, etc.) draw to a close. **7** *intr.* decrease in price etc. **8** *tr.* bend down. ● *n.* **1** a gradual loss of vigour or excellence (*on the decline*). **2** decay, deterioration. **3** setting; the last part of the course (of the sun, of life, etc.). **4** a fall in price. **5** *archaic* tuberculosis or a similar wasting disease. □ **declining years** old age. **on the decline** in a declining state. □ **declinable** *adj.* **decliner** *n.* [ME f. OF *decliner* f. L *declinare* (as DE-, *clinare* bend)]

declivity /dɪ'klɪvɪtɪ/ *n.* (*pl.* **-ies**) a downward slope, esp. a piece of sloping ground. □ **declivitous** *adj.* [L *declivitas* f. *declivis* (as DE-, *clivus* slope)]

declutch /di:'klʌtʃ/ *v.intr.* disengage the clutch of a motor vehicle. □ **double-declutch** release and re-engage the clutch twice when changing gear.

deco /'dekəʊ/ *n.* (also **Deco**) (usu. *attrib.*) = ART DECO. [F *décoratif* DECORATIVE]

decoct /dɪ'kɒkt/ *v.tr.* extract the essence from by decoction. [ME f. L *decoquere* boil down]

decoction /dɪ'kɒkʃ(ə)n/ *n.* **1** concentration of, or extraction of the essence of, a substance by boiling in water etc. **2** the extracted liquor resulting from this. [ME f. OF *decoction* or LL *decoctio* (as DE-, L *coquere* *coct-* boil)]

decode /di:'kəʊd/ *v.tr.* convert (a coded message) into intelligible language. □ **decodable** *adj.*

decoder /di:'kəʊdə(r)/ *n.* **1** a person or thing that decodes. **2** an electronic device for analysing signals and feeding separate amplifier-channels.

decoke *v. & n. Brit.* ● *v.tr.* /di:'kəʊk/ remove carbon or carbonaceous material from (an internal-combustion engine). ● *n.* /'di:kəʊk/ the process of decoking.

decollate /dɪ'kɒleɪt, 'dekəˌleɪt/ *v.tr. formal* **1** behead. **2** truncate. □ **decollation** /ˌdi:kɒ'leɪʃ(ə)n, ˌdekə-/ *n.* [L *decollare* *decollat-* (as DE-, *collum* neck)]

décolletage /ˌdeɪkɒl'tɑ:ʒ/ *n.* a low neckline of a woman's dress etc. [F (as DE-, *collet* collar of a dress)]

décolleté /deɪ'kɒlteɪ/ *adj. & n.* ● *adj.* (also **décolletée**) **1** (of a dress etc.) having a low neckline. **2** (of a woman) wearing a dress etc. with a low neckline. ● *n.* a low neckline. [F (as DÉCOLLETAGE)]

decolonize /di:'kɒləˌnaɪz/ *v.tr.* (also **-ise**) free (a colony) from dependent status. □ **decolonization** /-ˌkɒlənaɪ'zeɪʃ(ə)n/ *n.*

decolorize /di:'kʌləˌraɪz/ *v.* (also **-ise**) **1** *tr.* remove the colour from. **2** *intr.* lose colour. □ **decolorization** /-ˌkʌləraɪ'zeɪʃ(ə)n/ *n.*

decommission /ˌdi:kə'mɪʃ(ə)n/ *v.tr.* **1** close down (a nuclear reactor etc.). **2** take (a ship) out of service.

decompose /ˌdi:kəm'pəʊz/ *v.* **1** *intr.* decay, rot. **2** *tr.* separate (a substance, light, etc.) into its elements or simpler constituents. **3** *intr.* disintegrate; break up. □ **decomposition** /-ˌkɒmpə'zɪʃ(ə)n/ *n.* [F *décomposer* (as DE-, COMPOSE)]

decomposer /ˌdi:kəm'pəʊzə(r)/ *n. Ecol.* a thing, esp. a living organism, that performs decomposition.

decompress /ˌdi:kəm'pres/ *v.tr.* subject to decompression; relieve or reduce the compression on.

decompression /ˌdi:kəm'preʃ(ə)n/ *n.* **1** release from compression. **2** a gradual reduction of air pressure on a person who has been subjected to high pressure (esp. under water). □ **decompression chamber** an enclosed space for subjecting a person to decompression. **decompression sickness** a condition which results from too rapid decompression (and consequent formation of nitrogen bubbles in the tissues) and may involve pain in the joints, numbness, nausea, paralysis, etc. (also called *caisson disease*, *the bends*).

decompressor /ˌdi:kəm'presə(r)/ *n.* a device for reducing pressure in the engine of a motor vehicle.

decongestant /ˌdi:kən'dʒestənt/ *adj. & n.* ● *adj.* that relieves (esp. nasal) congestion. ● *n.* a medicinal agent that relieves nasal congestion.

deconsecrate /di:'kɒnsɪˌkreɪt/ *v.tr.* transfer (esp. a building) from sacred to secular use. □ **deconsecration** /-ˌkɒnsɪ'kreɪʃ(ə)n/ *n.*

deconstruct /ˌdi:kən'strʌkt/ *v.tr.* subject to deconstruction. □ **deconstructive** *adj.* [back-form. f. DECONSTRUCTION]

deconstruction /ˌdi:kən'strʌkʃ(ə)n/ *n.* an approach to and method of critical analysis whose philosophy and practice attempts to challenge certain perceived assumptions of the Western philosophical tradition. Deconstruction was initiated by the French philosopher Jacques Derrida in the late 1960s and subsequently taken up by mainly US literary critics such as Paul de Man and J. Hillis Miller. Derrida claimed that Western philosophy had become rooted in a tradition which sought truth and certainty of meaning by privileging certain types of interpretation and repressing others; he emphasized, on the other hand, the instability and deferral of meaning in language and the limitlessness (or impossibility) of interpretation and coined his word *différance* in reference to this. Although deconstruction seemed to favour a challenge to traditional hierarchies and centres of power and thus has formed a strand in political movements such as feminism, its emphasis on the internal workings of language and texts and on the dissociation of text and external world has led to it being criticized for its insularity and lack of political engagement. □ **deconstructionism** n. **deconstructionist** adj. & n. [F *déconstruction* (as DE-, CONSTRUCTION)]

decontaminate /ˌdiːkənˈtæmɪˌneɪt/ v.tr. remove contamination from; free from (the risk of) disease, radioactivity, etc. □ **decontamination** /-ˌtæmɪˈneɪʃ(ə)n/ n.

decontrol /ˌdiːkənˈtrəʊl/ v. & n. ● v.tr. (**decontrolled, decontrolling**) release (a commodity etc.) from controls or restrictions, esp. those imposed by the state. ● n. the act of decontrolling.

decor /ˈdeɪkɔː(r), ˈdek-/ n. (also **décor**) **1** the furnishing and decoration of a room etc. **2** the decoration and scenery of a stage. [F f. *décorer* (as DECORATE)]

decorate /ˈdekəˌreɪt/ v.tr. **1** provide with adornments. **2** provide (a room or building) with new paint, wallpaper, etc. **3** serve as an adornment to. **4** confer an award or distinction on. [L *decorare decorat-* f. *decus -oris* beauty]

Decorated style n. *Archit.* the second stage of English Gothic (14th century), marked by increasing use of decoration and geometrical tracery.

decoration /ˌdekəˈreɪʃ(ə)n/ n. **1** the process or art of decorating. **2** a thing that decorates or serves as an ornament. **3** a medal etc. conferred and worn as an honour. **4** (in pl.) flags etc. put up on an occasion of public celebration. [F *décoration* or LL *decoratio* (as DECORATE)]

Decoration Day (in the US) Memorial Day.

decorative /ˈdekərətɪv/ adj. serving to decorate. □ **decoratively** adv. **decorativeness** n. [F *décoratif* (as DECORATE)]

decorator /ˈdekəˌreɪtə(r)/ n. a person who decorates, esp. a person who paints or papers houses professionally.

decorous /ˈdekərəs/ adj. respecting good taste or propriety; dignified and decent. □ **decorously** adv. **decorousness** n. [L *decorus* seemly]

decorticate /diːˈkɔːtɪˌkeɪt/ v.tr. **1** remove the bark, rind, or husk from. **2** remove the outside layer from (the kidney, brain, etc.). [L *decorticare decorticat-* (as DE-, *cortex -icis* bark)]

decortication /diːˌkɔːtɪˈkeɪʃ(ə)n/ n. **1** the removal of the outside layer from an organ (e.g. the kidney) or structure. **2** an operation removing the blood clot and scar tissue formed after bleeding in the chest cavity.

decorum /dɪˈkɔːrəm/ n. **1 a** seemliness, propriety. **b** behaviour required by politeness or decency. **2** a particular requirement of this kind. **3** etiquette. [L, neut. of *decorus* seemly]

découpage /ˌdeɪkuːˈpɑːʒ/ n. the decoration of surfaces with paper cut-outs. [F, = the action of cutting out]

decouple /diːˈkʌp(ə)l/ v.tr. **1** *Electr.* make the interaction between (oscillators etc.) so weak that there is little transfer of energy between them. **2** separate, disengage, dissociate.

decoy n. & v. ● n. /ˈdiːkɔɪ, dɪˈkɔɪ/ **1 a** a person or thing used to lure an animal or person into a trap or danger. **b** a bait or enticement. **2** a pond from which narrow netted channels lead, into which wild duck may be enticed for capture. ● v.tr. /dɪˈkɔɪ/ (often foll. by *into, out of*) allure or entice, esp. by means of a decoy. [17th c.: perh. f. Du. *de kooi* the decoy f. DE + *kooi* f. L *cavea* cage]

decrease v. & n. ● v.tr. & intr. /dɪˈkriːs/ make or become smaller or fewer. ● n. /ˈdiːkriːs/ **1** the act or an instance of decreasing. **2** the amount by which a thing decreases. □ **decreasingly** /dɪˈkriːsɪŋlɪ/ adv. [ME f. OF *de(s)creiss-*, pres. stem of *de(s)creistre* ult. f. L *decrescere* (as DE-, *crescere* cret- grow)]

decree /dɪˈkriː/ n. & v. ● n. **1** an official order issued by a legal authority. **2** a judgement or decision of certain lawcourts, esp. in matrimonial cases. ● v.tr. (**decrees, decreed, decreeing**) ordain by decree. □ **decree absolute** a final order for divorce, enabling either party to remarry. **decree nisi** a provisional order for divorce, made absolute unless cause to the contrary is shown within a fixed period. [ME f. OF *decré* f. L *decretum* neut. past part. of *decernere* decide (as DE-, *cernere* sift)]

decrement /ˈdekrɪmənt/ n. **1** *Physics* the ratio of the amplitudes in successive cycles of a damped oscillation. **2** the amount lost by diminution or waste. **3** the act of decreasing. [L *decrementum* (as DECREASE)]

decrepit /dɪˈkrepɪt/ adj. **1** weakened or worn out by age and infirmity. **2** worn out by long use; dilapidated. □ **decrepitude** n. [ME f. L *decrepitus* (as DE-, *crepitus* past part. of *crepare* creak)]

decrepitate /dɪˈkrepɪˌteɪt/ v. **1** tr. roast or calcine (a mineral or salt) until it stops crackling. **2** intr. crackle under heat. □ **decrepitation** /-ˌkrepɪˈteɪʃ(ə)n/ n. [prob. mod.L *decrepitare* f. DE- + L *crepitare* crackle]

decrescendo /ˌdiːkrɪˈʃendəʊ, ˌdeɪkrɪ-/ adv., adj., & n. (pl. **-os**) *Mus.* = DIMINUENDO. [It., part. of *decrescere* DECREASE]

decrescent /dɪˈkres(ə)nt/ adj. (usu. of the moon) waning, decreasing. [L *decrescere*: see DECREASE]

decretal /dɪˈkriːt(ə)l/ n. **1** a papal decree. **2** (in pl.) a collection of these, forming part of canon law. [ME f. med.L *decretale* f. LL (*epistola*) *decretalis* (letter) of decree f. L *decernere*: see DECREE]

decriminalize /diːˈkrɪmɪnəˌlaɪz/ v.tr. (also **-ise**) cease to treat (an action etc.) as criminal. □ **decriminalization** /-ˌkrɪmɪnəlaɪˈzeɪʃ(ə)n/ n.

decry /dɪˈkraɪ/ v.tr. (**-ies, -ied**) disparage, belittle. □ **decrial** n. **decrier** n. [after F *décrier*: cf. *cry down*]

decrypt /diːˈkrɪpt/ v.tr. decipher (a cryptogram), with or without knowledge of its key. □ **decryption** /-ˈkrɪpʃ(ə)n/ n. [DE- + CRYPTOGRAM]

decumbent /dɪˈkʌmb(ə)nt/ adj. *Bot.* & *Zool.* (of a plant, shoot, or bristles) lying along the ground or a surface. [L *decumbere decumbent-* lie down]

decurve /diːˈkɜːv/ v.tr. & intr. *Zool.* & *Bot.* (esp. as **decurved** adj.) curve or bend down (a *decurved bill*). □ **decurvature** n.

decussate /diːˈkʌseɪt/ adj. & v. ● adj. **1** X-shaped. **2** *Bot.* with pairs of opposite leaves etc. each at right angles to the pair below. ● v.tr. & intr. **1** arrange or be arranged in a decussate form. **2** intersect. □ **decussation** /ˌdiːkʌˈseɪʃ(ə)n/ n. [L *decussatus* past part. of *decussare* divide in a cross shape f. *decussis* the numeral ten or the shape X f. *decem* ten]

dedans /dəˈdɒn/ n. (in real tennis) **1** the open gallery at the end of the service side of a court. **2** the spectators watching a match (from this area). [F, = inside]

Dedekind /ˈdeɪdəˌkɪnt/, Richard (1831–1916), German mathematician. One of the founders of abstract algebra and modern mathematics, Dedekind was professor at Brunswick from 1862 until his death. His analysis of the properties of real numbers solved the question of what numbers are, and supplied a satisfactory foundation on which analysis could be based. He is remembered also for his theory of rings of algebraic integers, which cast the theory of algebraic numbers into its general modern form. He introduced collections of numbers as entities of interest in their own right, whose relationships to each other may be studied by means of set theory.

dedicate /ˈdedɪˌkeɪt/ v.tr. **1** (foll. by *to*) devote (esp. oneself) to a special task or purpose. **2** (foll. by *to*) address (a book, piece of music, etc.) as a compliment to a friend, patron, etc. **3** (often foll. by *to*) devote (a building etc.) to a deity or a sacred person or purpose. **4** (as **dedicated** adj.) **a** (of a person) devoted to an aim or vocation; having single-minded loyalty or integrity. **b** (of equipment, esp. a computer) designed for a specific purpose. □ **dedicatedly** adv. **dedicative** adj. **dedicator** n. **dedicatory** adj. **dedicatee** /ˌdedɪkəˈtiː/ n. [L *dedicare* (DE-, *dicare* declare, dedicate)]

dedication /ˌdedɪˈkeɪʃ(ə)n/ n. **1** the act or an instance of dedicating; the quality or process of being dedicated. **2** the words with which a book etc. is dedicated. **3** a dedicatory inscription. [ME f. OF *dedicacion* or L *dedicatio* (as DEDICATE)]

deduce /dɪˈdjuːs/ v.tr. **1** (often foll. by *from*) infer; draw as a logical conclusion. **2** *archaic* trace the course or derivation of. □ **deducible** adj. [L *deducere* (as DE-, *ducere duct-* lead)]

deduct /dɪˈdʌkt/ v.tr. (often foll. by *from*) subtract, take away, withhold (an amount, portion, etc.). [L (as DEDUCE)]

deductible /dɪˈdʌktɪb(ə)l/ adj. & n. ● adj. that may be deducted, esp. from tax to be paid or taxable income. ● n. US = EXCESS n. 6. □ **deductibility** /dɪˌdʌktɪˈbɪlɪtɪ/ n.

deduction /dɪˈdʌkʃ(ə)n/ n. **1 a** the act of deducting. **b** an amount deducted. **2 a** the inferring of particular instances from a general law (cf. INDUCTION). **b** a conclusion deduced. [ME f. OF *deduction* or L *deductio* (as DEDUCE)]

deductive /dɪˈdʌktɪv/ adj. of or reasoning by deduction. □ **deductively** adv. [med.L *deductivus* (as DEDUCE)]

de Duve /də ˈduːv/, Christian René (b.1917), British-born Belgian biochemist. He was a pioneer in the study of cell biology, and his research suggested the existence of organelles that contain and isolate a cell's digestive enzymes. In 1955, with the aid of electron microscopy, he proved the existence of lysosomes. De Duve shared a Nobel Prize in 1974.

Dee /diː/ **1** a river in NE Scotland, which rises in the Grampian

Mountains and flows eastwards past Balmoral Castle to the North Sea at Aberdeen. **2** a river which rises in North Wales and flows past Chester and on into the Irish Sea.

dee /diː/ *n.* **1** the letter D. **2 a** a thing shaped like this. **b** *Physics* either of two hollow semicircular electrodes in a cyclotron. [the name of the letter]

deed /diːd/ *n. & v.* ● *n.* **1** a thing done intentionally or consciously. **2** a brave, skilful, or conspicuous act. **3** actual fact or performance (*kind in word and deed; in deed and not in name*). **4** *Law* a written or printed document often used for a legal transfer of ownership and bearing the disposer's signature. ● *v.tr. US* convey or transfer by legal deed. □ **deed-box** a strong box for keeping deeds and other documents. **deed of covenant** an agreement to pay a specified amount regularly to a charity etc., enabling the recipient to recover the tax paid by the donor on an equivalent amount of income. **deed poll** a deed made and executed by one party only, esp. to change one's name. [OE *dēd* f. Gmc: cf. DO¹]

deejay /ˈdiːdʒeɪ/ *n. sl.* a disc jockey. [repr. pronunc. of DJ]

deem /diːm/ *v.tr.* *formal* regard, consider, judge (*deem it my duty; was deemed sufficient*). [OE *dēman* f. Gmc, rel. to DOOM]

de-emphasize /diːˈemfəˌsaɪz/ *v.tr.* (also **-ise**) **1** remove emphasis from. **2** reduce emphasis on.

deemster /ˈdiːmstə(r)/ *n.* a judge in the Isle of Man. [DEEM + -STER]

deep /diːp/ *adj., n., & adv.* ● *adj.* **1 a** extending far down from the top (*deep hole; deep water*). **b** extending far in from the surface or edge (*deep wound; deep plunge; deep shelf; deep border*). **2** (*predic.*) **a** extending to or lying at a specified depth (*water six feet deep; ankle-deep in mud*). **b** in a specified number of ranks one behind another (*soldiers drawn up six deep*). **3** situated far down or back or in (*hands deep in his pockets*). **4** coming or brought from far down or in (*deep breath; deep sigh*). **5** low-pitched, full-toned, not shrill (*deep voice; deep note; deep bell*). **6** intense, vivid, extreme (*deep disgrace; deep sleep; deep colour; deep secret*). **7** heartfelt, absorbing (*deep affection; deep feelings; deep interest*). **8** (*predic.*) fully absorbed or overwhelmed (*deep in a book; deep in debt*). **9** profound, penetrating, not superficial; difficult to understand (*deep thinker; deep thought; deep insight; deep learning*). **10** *Cricket* distant from the batsman (*deep mid-off*). **11** *Football* distant from the front line of one's team. **12** *colloq.* cunning or secretive (*a deep one*). ● *n.* **1** (prec. by *the*) poet. the sea. **2** a deep part of the sea. **3** an abyss, pit, or cavity. **4** (prec. by *the*) *Cricket* the parts of the field distant from the batsman. **5** a deep state (*deep of the night*). **6** *poet.* a mysterious region of thought or feeling. ● *adv.* deeply; far down or in (*dig deep; read deep into the night*). □ **deep breathing** breathing with long breaths, esp. as a form of exercise. **deep-drawn** (of metal etc.) shaped by forcing through a die when cold. **deep-fry** (**-fries**, **-fried**) fry (food) in an amount of fat or oil sufficient to cover it. **deep kiss** a kiss with contact between tongues. **deep-laid** (of a scheme) secret and elaborate. **deep mourning** mourning expressed by wearing only black clothes. **deep-mouthed** (esp. of a dog) having a deep voice. **deep-rooted** (esp. of convictions) firmly established. **deep sea** the deeper parts of the ocean. **deep-seated** (of emotion, disease, etc.) firmly established, profound. **deep space** the regions beyond the solar system or the earth's atmosphere. **deep structure** (in transformational grammar) the representation of the fundamental abstract grammatical or semantic relationships of the elements of a phrase or sentence (opp. *surface structure*). **deep therapy** curative treatment with short-wave X-rays of high penetrating power. **go off** (or **go in off**) **the deep end** *colloq.* give way to anger or emotion. **in deep water** (or **waters**) in trouble or difficulty. **jump** (or **be thrown**) **in at the deep end** face a difficult problem, undertaking, etc., with little experience of it. □ **deeply** *adv.* **deepness** *n.* [OE *dēop* (adj.), *dīope, dēope* (adv.), f. Gmc: rel. to DIP]

deepen /ˈdiːp(ə)n/ *v.tr. & intr.* make or become deep or deeper.

deepening /ˈdiːp(ə)nɪŋ/ *n.* the act or process of making deeper, esp. the implementation of measures (such as economic and monetary union) to deepen and strengthen the ties among EC countries.

deep-freeze /diːpˈfriːz/ *n. & v.* ● *n.* **1** a refrigerator in which food can be quickly frozen and kept for long periods at a very low temperature. **2** a suspension of activity. ● *v.tr.* (**-froze** /-ˈfrəʊz/, **-frozen** /-ˈfrəʊz(ə)n/) freeze or store (food) in a deep-freeze.

Deep South the southernmost parts of the US, esp. the states bordering the Gulf of Mexico.

deer /dɪə/ *n.* (*pl.* same) a hoofed grazing or browsing animal of the family Cervidae, the males of which usu. have branching antlers. Deer range in size from the tiny brockets to the huge moose or elk. They are characterized by the bony antlers borne by the males of most species,

which are grown annually, used for fighting in the rut, and then cast. The reindeer, the only species to have been fully domesticated, is unusual in that the female also bears antlers. The musk deer lacks antlers and belongs to a different family. □ **deer fly** a bloodsucking fly of the genus *Chrysops*. **deer-forest** an extensive area of wild and often treeless land reserved for the stalking of deer. **deer-hound** a large rough-haired greyhound. **deer-lick** a spring or damp spot impregnated with salt etc. where deer come to lick. **deer mouse** a mouse of the American genus *Peromyscus*, esp. the common *P. maniculatus*. [OE *dēor* animal, deer]

deerskin /ˈdɪəskɪn/ *n. & adj.* ● *n.* leather from a deer's skin. ● *adj.* made from a deer's skin.

deerstalker /ˈdɪəˌstɔːkə(r)/ *n.* **1** a soft cloth cap with peaks in front and behind and ear-flaps often joined at the top. **2** a person who stalks deer.

de-escalate /diːˈeskəˌleɪt/ *v.tr.* reduce the level or intensity of. □ **de-escalation** /-ˌeskəˈleɪʃ(ə)n/ *n.*

def /def/ *adj.* esp. *US sl.* excellent. [corrupt. of DEATH or shortened f. DEFINITIVE]

deface /dɪˈfeɪs/ *v.tr.* **1** spoil the appearance of; disfigure. **2** make illegible. □ **defacement** *n.* **defacer** *n.* [ME f. F *défacer* f. OF *desfacier* (as DE-, FACE)]

de facto /diː ˈfæktəʊ, deɪ/ *adv., adj., & n.* ● *adv.* in fact, whether by right or not. ● *adj.* that exists or is such in fact (*a de facto ruler*). ● *n.* (in full **de facto wife** or **husband**) a person living with another as if married. [L]

defalcate /ˈdiːfælˌkeɪt/ *v.intr. formal* misappropriate property in one's charge, esp. money. □ **defalcator** *n.* [med.L *defalcare* lop (as DE-, L *falx -cis* sickle)]

defalcation /ˌdiːfælˈkeɪʃ(ə)n/ *n.* **1** *Law* **a** a misappropriation of money. **b** an amount misappropriated. **2** *formal* a shortcoming. **3** *formal* defection. [ME f. med.L *defalcatio* (as DEFALCATE)]

de Falla, Manuel, see FALLA.

defame /dɪˈfeɪm/ *v.tr.* attack the good reputation of; speak ill of. □ **defamer** *n.* **defamatory** /-ˈfæmətərɪ/ *adj.* **defamation** /ˌdefəˈmeɪʃ(ə)n, ˌdiːfə-/ *n.* [ME f. OF *diffamer* etc. f. L *diffamare* spread evil report (as DIS-, *fama* report)]

defat /diːˈfæt/ *v.tr.* (**defatted**, **defatting**) remove fat or fats from.

default /dɪˈfɔːlt, -ˈfɒlt/ *n. & v.* ● *n.* **1** failure to fulfil an obligation, esp. to pay money, appear in a lawcourt, or act in some way. **2** lack, absence. **3** a preselected option adopted by a computer program when no alternative is specified by the user or programmer. ● *v.* **1** *intr.* fail to fulfil an obligation, esp. to pay money or to appear in a lawcourt. **2** *tr. Law* declare (a party) in default and give judgement against that party. □ **by default 1** because of inaction. **2** because of a lack of opposition. **go by default** fail by default. **in default of** because of the absence of. **judgement by default** judgement given for the plaintiff on the defender's failure to plead. [ME f. OF *defaut(e)* f. *defaillir* fail f. Rmc (as DE-, L *fallere* deceive): cf. FAIL]

defaulter /dɪˈfɔːltə(r), -ˈfɒltə(r)/ *n.* a person who defaults, esp. *Brit.* a soldier guilty of a military offence.

defeasance /dɪˈfiːz(ə)ns/ *n.* esp. *Law* the act or process of rendering null and void. [ME f. OF *defesance* f. *de(s)faire* undo (as DE-, *faire* make f. L *facere*)]

defeasible /dɪˈfiːzɪb(ə)l/ *adj.* esp. *Law* **1** capable of annulment. **2** liable to forfeiture. □ **defeasibly** *adv.* **defeasibility** /-ˌfiːzɪˈbɪlɪtɪ/ *n.* [AF (as DEFEASANCE)]

defeat /dɪˈfiːt/ *v. & n.* ● *v.tr.* **1** overcome in a battle or other contest. **2** frustrate, baffle. **3** reject (a motion etc.) by voting. **4** *Law* annul. ● *n.* the act or an instance of defeating or being defeated. [ME f. OF *deffait*, *desfait* past part. of *desfaire* f. med.L *disfacere* (as DIS-, L *facere* do)]

defeatism /dɪˈfiːtɪz(ə)m/ *n.* **1** an excessive readiness to accept defeat. **2** conduct conducive to this. □ **defeatist** *n. & adj.* [F *défaitisme* f. *défaite* DEFEAT]

defecate /ˈdefɪˌkeɪt/ *v.intr.* discharge faeces from the body. □ **defecation** /ˌdefɪˈkeɪʃ(ə)n/ *n.* [earlier as adj., = purified, f. L *defaecare* (as DE-, *faex faecis* dregs)]

defect *n. & v.* ● *n.* /ˈdiːfekt, dɪˈfekt/ **1** lack of something essential or required; imperfection. **2** a shortcoming or failing. **3** a blemish. **4** the amount by which a thing falls short. ● *v.intr.* /dɪˈfekt/ abandon one's country, cause, etc., in favour of another, esp. flee to a country with an opposing ideology. □ **defector** /dɪˈfektə(r)/ *n.* [L *defectus* f. *deficere* desert, fail (as DE-, *facere* do)]

defection /dɪˈfekʃ(ə)n/ n. the abandonment of one's country, cause, etc., for another. [L *defectio* (as DEFECT)]

defective /dɪˈfektɪv/ adj. & n. ● adj. **1** having a defect or defects; incomplete, imperfect, faulty. **2** archaic or offens. mentally handicapped. **3** (usu. foll. by in) lacking, deficient. **4** Gram. not having all the usual inflections. ● n. archaic or offens. a mentally handicapped person. □ **defectively** adv. **defectiveness** n. [ME f. OF *defectif -ive* or LL *defectivus* (as DEFECT)]

defence /dɪˈfens/ n. (US **defense**) **1** the act of defending from or resisting attack. **2 a** a means of resisting attack. **b** a thing that protects. **c** the military resources of a country. **3** (in pl.) fortifications. **4 a** justification, vindication. **b** a speech or piece of writing used to this end. **5 a** the defendant's case in a lawsuit. **b** the counsel for the defendant. **6 a** the action or role of defending one's goal etc. against attack. **b** the players in a team who perform this role. □ **defence mechanism 1** the body's reaction against disease organisms. **2** a usu. unconscious mental process to avoid conscious conflict or anxiety. □ **defenceless** adj. **defencelessly** adv. **defencelessness** n. [ME f. OF *defens(e)* f. LL *defensum, -a,* past part. of *defendere*: see DEFEND]

defenceman /dɪˈfensmən/ n. (US **defenseman**) (pl. **-men**) (in ice hockey and lacrosse) a player in a defensive position.

defend /dɪˈfend/ v.tr. (also absol.) **1** (often foll. by against, from) resist an attack made on; protect (a person or thing) from harm or danger. **2** support or uphold by argument; speak or write in favour of. **3** conduct the case for (a defendant in a lawsuit). **4** compete to retain (a title) in a contest. **5** (in various sports and games) try to prevent the opposition scoring goals, points, etc.; resist attacks. □ **defendable** adj. **defender** n. [ME f. OF *defendre* f. L *defendere*: cf. OFFEND]

defendant /dɪˈfendənt/ n. a person etc. sued or accused in a court of law. [ME f. OF, part. of *defendre*: see DEFEND]

Defender of the Faith a title conferred on Henry VIII by Pope Leo X in 1521 in recognition of his treatise defending the seven sacraments against Luther. It was recognized by Parliament as an official title of the English monarch in 1544, and has been borne by all subsequent sovereigns. [transl. L *Fidei Defensor*]

defenestration /diːˌfenɪˈstreɪʃ(ə)n/ n. formal or joc. the action of throwing (esp. a person) out of a window. □ **defenestrate** /-ˈfenɪˌstreɪt/ v.tr. [mod.L *defenestratio* (as DE-, L *fenestra* window)]

defense US var. of DEFENCE.

defensible /dɪˈfensɪb(ə)l/ adj. **1** justifiable; supportable by argument. **2** that can be easily defended militarily. □ **defensibly** adv. **defensibility** /-ˌfensɪˈbɪlɪti/ n. [ME f. LL *defensibilis* (as DEFEND)]

defensive /dɪˈfensɪv/ adj. **1** done or intended for defence or to defend. **2** (of a person or attitude) concerned to challenge criticism. **3** Sport primarily concerned to protect one's goal, wicket, etc.; cautious. □ **defensive end** Amer. Football either of the two defensive players who line up next to the nose tackle. **on the defensive 1** expecting criticism. **2** in an attitude or position of defence. □ **defensively** adv. **defensiveness** n. [ME f. F *défensif -ive* f. med.L *defensivus* (as DEFEND)]

defer[1] /dɪˈfɜː(r)/ v.tr. (**deferred, deferring**) **1** put off to a later time; postpone. **2** US postpone the conscription of (a person). □ **deferred payment** payment by instalments. □ **deferment** n. **deferrable** adj. **deferral** n. [ME, orig. the same as DIFFER]

defer[2] /dɪˈfɜː(r)/ v.intr. (**deferred, deferring**) (foll. by to) yield or make concessions in opinion or action. □ **deferrer** n. [ME f. F *déférer* f. L *deferre* (as DE-, *ferre* bring)]

deference /ˈdefərəns/ n. **1** courteous regard, respect. **2** compliance with the advice or wishes of another (*pay deference to*). □ **in deference to** out of respect for. [F *déférence* (as DEFER[2])]

deferential /ˌdefəˈrenʃ(ə)l/ adj. showing deference; respectful. □ **deferentially** adv. [DEFERENCE, after PRUDENTIAL etc.]

defiance /dɪˈfaɪəns/ n. **1** open disobedience; bold resistance. **2** a challenge to fight or maintain a cause, assertion, etc. □ **in defiance of** disregarding; in conflict with. [ME f. OF (as DEFY)]

defiant /dɪˈfaɪənt/ adj. **1** showing defiance. **2** openly disobedient. □ **defiantly** adv.

defibrillation /diːˌfɪbrɪˈleɪʃ(ə)n/ n. Med. the stopping of the fibrillation of the heart. □ **defibrillator** /-ˈfɪbrɪˌleɪtə(r)/ n.

deficiency /dɪˈfɪʃənsi/ n. (pl. **-ies**) **1** the state or condition of being deficient. **2** (usu. foll. by of) a lack or shortage. **3** a thing lacking. **4** the amount by which a thing, esp. revenue, falls short. □ **deficiency disease** a disease caused by the lack of some essential or important element in the diet.

deficient /dɪˈfɪʃ(ə)nt/ adj. **1** (usu. foll. by in) incomplete; not having enough of a specified quality or ingredient. **2** insufficient in quantity, force, etc. **3** (in full **mentally deficient**) archaic or offens. having a mental handicap. □ **deficiently** adv. [L *deficiens* part. of *deficere* (as DEFECT)]

deficit /ˈdefɪsɪt/ n. **1** the amount by which a thing (esp. a sum of money) is too small. **2** an excess of liabilities over assets in a given period, esp. a financial year (opp. SURPLUS). □ **deficit financing** financing of (esp. state) spending by borrowing. **deficit spending** spending, esp. by the state, financed by borrowing. [F *déficit* f. L *deficit* 3rd sing. pres. of *deficere* (as DEFECT)]

defier /dɪˈfaɪə(r)/ n. a person who defies.

defilade /ˌdefɪˈleɪd/ v. & n. ● v.tr. secure (a fortification) against enfilading fire. ● n. this precaution or arrangement. [DEFILE[2] + -ADE[1]]

defile[1] /dɪˈfaɪl/ v.tr. **1** make dirty; pollute, foul. **2** corrupt. **3** desecrate, profane. **4** deprive (esp. a woman) of virginity; euphem. rape. **5** make ceremonially unclean. □ **defilement** n. **defiler** n. [ME *defoul* f. OF *defouler* trample down, outrage (as DE-, *fouler* tread, trample) altered after obs. *befile* f. OE *befȳlan* (BE-, *fūl* FOUL)]

defile[2] n. & v. ● n. /ˈdiːfaɪl, dɪˈfaɪl/ **1** a narrow way through which troops can march only in file. **2** a gorge. ● v.intr. /dɪˈfaɪl/ march in file. [F *défiler* and *défilé* past part. (as DE-, FILE[2])]

define /dɪˈfaɪn/ v.tr. **1** give the exact meaning of (a word etc.). **2** describe or explain the scope of (*define one's position*). **3** make clear, esp. in outline (*well-defined image*). **4** mark out the boundary or limits of. **5** (of properties) make up the total character of. □ **definable** adj. **definer** n. [ME f. OF *definer* ult. f. L *definire* (as DE-, *finire* finish, f. *finis* end)]

definite /ˈdefɪnɪt/ adj. & n. ● adj. **1** having exact and discernible limits. **2** clear and distinct; not vague. **3** certain, sure (*a definite offer; are you definite he was there?*). ¶ See the note at *definitive*. ● n. a definite thing, esp. Gram. a noun denoting a definite thing or object. □ **definite article** see ARTICLE. **definite integral** see INTEGRAL. □ **definiteness** n. [L *definitus* past part. of *definire* (as DEFINE)]

definitely /ˈdefɪnɪtli/ adv. & int. ● adv. **1** in a definite manner. **2** certainly; without doubt (*they were definitely there*). ● int. colloq. yes, certainly.

definition /ˌdefɪˈnɪʃ(ə)n/ n. **1 a** the act or process of defining. **b** a statement of the meaning of a word or the nature of a thing. **2 a** the degree of distinctness in outline of an object or image (esp. of an image produced by a lens or shown in a photograph or on a cinema or television screen). **b** making or being distinct in outline. □ **definitional** adj. **definitionally** adv. [ME f. OF f. L *definitio* (as DEFINE)]

definitive /dɪˈfɪnɪtɪv/ adj. **1** (of an answer, treaty, verdict, etc.) decisive, unconditional, final. ¶ Often confused in this sense with *definite*, which does not have connotations of authority and conclusiveness: *a definite no* is a firm refusal, whereas *a definitive no* is an authoritative judgement or decision that something is not the case. **2** (of an edition of a book etc.) most authoritative. **3** Philately (of a series of stamps) for permanent use, not commemorative etc. □ **definitively** adv. [ME f. OF *definitif -ive* f. L *definitivus* (as DEFINE)]

deflagrate /ˈdefləˌɡreɪt, ˈdiːf-/ v.tr. & intr. burn away with sudden flame. □ **deflagrator** n. **deflagration** /ˌdefləˈɡreɪʃ(ə)n, ˌdiːf-/ n. [L *deflagrare* (as DE-, *flagrare* blaze)]

deflate /dɪˈfleɪt/ v. **1 a** tr. let air or gas out of (a tyre, balloon, etc.). **b** intr. be emptied of air or gas. **2 a** tr. cause to lose confidence or conceit. **b** intr. lose confidence. **3** Econ. **a** tr. subject (a currency or economy) to deflation. **b** intr. pursue a policy of deflation. **4** tr. reduce the importance of, depreciate. □ **deflator** n. [DE- + INFLATE]

deflation /dɪˈfleɪʃ(ə)n/ n. **1** the act or process of deflating or being deflated. **2** Econ. reduction of the amount of money in circulation to increase its value as a measure against inflation. **3** Geol. the removal of particles of rock etc. by the wind. □ **deflationary** adj. **deflationist** n. & adj.

deflect /dɪˈflekt/ v. **1** tr. & intr. bend or turn aside from a straight course or intended purpose. **2** (often foll. by from) **a** tr. cause to deviate. **b** intr. deviate. [L *deflectere* (as DE-, *flectere flex-* bend)]

deflection /dɪˈflekʃ(ə)n/ n. (also **deflexion**) **1** the act or process of deflecting or being deflected. **2** a lateral bend or turn; a deviation. **3** Physics the displacement of a pointer on an instrument from its zero position. [LL *deflexio* (as DEFLECT)]

deflector /dɪˈflektə(r)/ n. a thing that deflects, esp. a device for deflecting a flow of air etc.

defloration /ˌdiːflɔːˈreɪʃ(ə)n/ n. deflowering. [ME f. OF or f. LL *defloratio* (as DEFLOWER)]

deflower /dɪˈflaʊə(r)/ v.tr. **1** deprive (esp. a woman) of virginity.

2 ravage, spoil. **3** strip of flowers. [ME f. OF *deflourer*, *des-*, ult. f. LL *deflorare* (as DE-, L *flos floris* flower)]

defocus /diːˈfəʊkəs/ *v.tr. & intr.* (**defocused**, **defocusing** or **defocussed**, **defocussing**) put or go out of focus.

Defoe /dɪˈfəʊ/, Daniel (1660–1731), English novelist and journalist. After a varied career of political journalism and secret service work, Defoe wrote *Robinson Crusoe* (1719) when he was nearly 60. Loosely based on the true story of the sailor Alexander Selkirk, it has a claim to being the first English novel. Its tale of the shipwrecked Crusoe battling for survival in virtual solitude is one of the most familiar and resonant myths of modern literature. Other major works include the novel *Moll Flanders* (1722) and the historical fiction *A Journal of the Plague Year* (1722).

defoliate /diːˈfəʊlɪˌeɪt/ *v.tr.* remove leaves from, esp. as a military tactic. □ **defoliant** *n. & adj.* **defoliator** *n.* **defoliation** /-ˌfəʊlɪˈeɪʃ(ə)n/ *n.* [LL *defoliare* f. *folium* leaf]

De Forest /də ˈfɒrɪst/, Lee (1873–1961), American physicist and electrical engineer. His triode valve (patented in 1907) became the basic device for the large-scale amplification of signals, and was crucial to the development of radio communication, television, and computers. De Forest was one of the pioneers of radio broadcasting, successfully transmitting a live broadcast in 1910. He later developed a method of representing sound waves visually for recording and reproducing sound on cinema film.

deforest /diːˈfɒrɪst/ *v.tr.* clear of forests or trees.

deforestation /diːˌfɒrɪˈsteɪʃ(ə)n/ *n.* the deliberate destruction of natural forests. The felling and clearing of forests, mainly for agriculture and logging, has long been associated with human occupation in temperate regions, but has accelerated rapidly in the late 20th century, especially in the tropics. The annual rate of destruction is estimated to be about 20 million hectares; over 50 per cent of the world's tropical forest has been destroyed in the last hundred years. Besides the loss of habitat and the threat to wildlife, deforestation often has other undesirable results such as soil erosion and disturbance to water supplies. Of most concern are effects on the atmosphere. Photosynthesis by forest trees is very important in removing carbon dioxide from the atmosphere, and deforestation therefore directly exacerbates the problem of global warming. Deforestation is part of the wider problem of the economic relationship between the developing countries and the developed world, and successful international action to control it has been limited.

deform /dɪˈfɔːm/ *v.* **1** *tr.* make ugly, deface. **2** *tr.* put out of shape, misshape. **3** *intr.* undergo deformation; be deformed. □ **deformable** *adj.* [ME f. OF *deformer* etc. f. med.L *difformare* ult. f. L *deformare* (as DE-, *formare* f. *forma* shape)]

deformation /ˌdiːfɔːˈmeɪʃ(ə)n/ *n.* **1** disfigurement. **2** *Physics* **a** (often foll. by *of*) change in shape. **b** a quantity representing the amount of this change. **3** an altered form of a word (e.g. *dang* for *damn*). □ **deformational** *adj.* [ME f. OF *deformation* or L *deformatio* (as DEFORM)]

deformed /dɪˈfɔːmd/ *adj.* (of a person or limb) misshapen.

deformity /dɪˈfɔːmɪtɪ/ *n.* (pl. **-ies**) **1** the state of being deformed; ugliness, disfigurement. **2** a malformation, esp. of body or limb. **3** a moral defect; depravity. [ME f. OF *deformité* etc. f. L *deformitas -tatis* f. *deformis* (as DE-, *forma* shape)]

defraud /dɪˈfrɔːd/ *v.tr.* (often foll. by *of*) cheat by fraud. □ **defrauder** *n.* [ME f. OF *defrauder* or L *defraudare* (as DE-, FRAUD)]

defray /dɪˈfreɪ/ *v.tr.* provide money to pay (a cost or expense). □ **defrayable** *adj.* **defrayal** *n.* **defrayment** *n.* [F *défrayer* (as DE-, obs. *frai(t)* cost, f. med.L *fredum*, *-us* fine for breach of the peace)]

defrock /diːˈfrɒk/ *v.tr.* deprive (a person, esp. a priest) of ecclesiastical status. [F *défroquer* (as DE-, FROCK)]

defrost /diːˈfrɒst/ *v.* **1** *tr.* **a** free (the interior of a refrigerator) of excess frost, usu. by turning it off for a period. **b** remove frost or ice from (esp. the windscreen of a motor vehicle). **2** *tr.* unfreeze (frozen food). **3** *intr.* become unfrozen. □ **defroster** *n.*

deft /deft/ *adj.* neatly skilful or dexterous; adroit. □ **deftly** *adv.* **deftness** *n.* [ME, var. of DAFT in obs. sense 'meek']

defunct /dɪˈfʌŋkt/ *adj.* **1** no longer existing. **2** no longer used or in fashion. **3** dead or extinct. [L *defunctus* dead, past part. of *defungi* (as DE-, *fungi* perform)]

defuse /diːˈfjuːz/ *v.tr.* **1** remove the fuse from (an explosive device). **2** reduce the tension or potential danger in (a crisis, difficulty, etc.).

defy /dɪˈfaɪ/ *v.tr.* (**-ies**, **-ied**) **1** resist openly; refuse to obey. **2** (of a thing) present insuperable obstacles to (*defies solution*). **3** (foll. by *to* + infin.) challenge (a person) to do or prove something. **4** *archaic* challenge to combat. [ME f. OF *defier* f. Rmc (as DIS-, L *fidus* faithful)]

deg. *abbr.* degree.

dégagé /deɪˈɡɑːʒeɪ/ *adj.* (fem. **dégagée** pronunc. same) unconstrained, relaxed; unconcerned, detached. [F, past part. of *dégager* set free]

Degas /ˈdeɪɡɑː/, French /dəɡa/, (Hilaire Germain) Edgar (1834–1917), French painter and sculptor. Having been a pupil of Ingres, he became associated with impressionism; he took part in most of the impressionist exhibitions, including the first in 1874. Unlike other impressionist painters, who often concentrated on landscape, Degas is best known for his paintings of ballet-dancers; in his sculpture, to which he turned in later life, he also concentrated on the human form.

degas /diːˈɡæs/ *v.tr.* (**degassed**, **degassing**) remove unwanted gas from.

de Gaulle /də ˈɡəʊl/, Charles (André Joseph Marie) (1890–1970), French general and statesman, head of government 1944–6, President 1959–69. He served in the French army during the First World War, and during the Second World War was a member of the Cabinet at the time of France's surrender in June 1940. He escaped to Britain, where he was an instigator of the resistance and organized the Free French movement. Following the war he became interim President of the new French Republic, but later resigned. Having been asked to form a government, he became President in 1959 and went on to establish the presidency as a democratically elected office (1962). He resigned in 1969 after proposed constitutional changes were rejected by the electorate. In addition to extricating France from the Algerian crisis and strengthening the French economy, he is remembered for his assertive foreign policy (including withdrawing French forces from NATO and blocking Britain's entry to the EEC) and for quelling the student uprisings and strikes of May 1968.

degauss /diːˈɡaʊs/ *v.tr.* neutralize the magnetism in (a thing) by encircling it with a current-carrying conductor. □ **degausser** *n.* [DE- + GAUSS]

degenerate *adj., n., & v.* ● *adj.* /dɪˈdʒenərət/ **1** having lost the qualities that are normal and desirable or proper to its kind; fallen from former excellence; debased, corrupt. **2** *Biol.* having a simplified structure. ● *n.* /dɪˈdʒenərət/ a degenerate person or animal. ● *v.intr.* /dɪˈdʒenəˌreɪt/ become degenerate. □ **degeneracy** /-rəsɪ/ *n.* **degenerately** /-rətlɪ/ *adv.* [L *degeneratus* past part. of *degenerare* (as DE-, *genus -eris* race)]

degenerate art *n.* a term coined by the Nazi regime in Germany in the 1930s to describe contemporary (esp. experimental and modernist) art, which was considered to be incompatible with Nazi ideology. An exhibition of so-called degenerate art was held in Munich in 1937, which included the work of Matisse, Picasso, Klee, and Van Gogh.

degeneration /dɪˌdʒenəˈreɪʃ(ə)n/ *n.* **1 a** the process of becoming degenerate. **b** the state of being degenerate. **2** *Med.* morbid deterioration of tissue or change in its structure. [ME f. F *dégénération* or f. LL *degeneratio* (as DEGENERATE)]

degenerative /dɪˈdʒenərətɪv/ *adj.* **1** of or tending to degeneration. **2** (of disease) characterized by progressive often irreversible deterioration.

degrade /dɪˈɡreɪd/ *v.* **1** *tr.* reduce to a lower rank, esp. as a punishment. **2** *tr.* bring into dishonour or contempt. **3** lower in character or quality; debase. **4** *tr. Chem.* reduce to a simpler molecular structure. **5** *tr. Physics* reduce (energy) to a less convertible form. **6** *tr. Geol.* wear down (rocks etc.) by disintegration. **7** *intr.* degenerate. **8** *intr. Chem.* disintegrate. □ **degradable** *adj.* **degrader** *n.* **degradative** /-dətɪv/ *adj.* **degradability** /-ˌɡreɪdəˈbɪlɪtɪ/ *n.* **degradation** /ˌdeɡrəˈdeɪʃ(ə)n/ *n.* [ME f. OF *degrader* f. eccl.L *degradare* (as DE-, L *gradus* step)]

degrading /dɪˈɡreɪdɪŋ/ *adj.* that debases or lowers in dignity; humiliating; causing a loss of self-respect. □ **degradingly** *adv.*

degrease /diːˈɡriːs/ *v.tr.* remove unwanted grease or fat from. □ **degreaser** *n.*

degree /dɪˈɡriː/ *n.* **1** a stage in an ascending or descending scale, series, or process. **2** a stage in intensity or amount (*to a high degree*; *in some degree*). **3** relative condition (*each is good in its degree*). **4** *Math.* a unit of measurement of angles, one-ninetieth of a right angle or the angle subtended by one-three-hundred-and-sixtieth of the circumference of a circle (symbol: °). **5** *Physics* a unit in a scale of temperature, hardness, etc. (symbol: ° or **deg.**). **6** *Med.* each of a number of grades, usu. three, used to classify burns according to their severity (often *attrib.*: *third-degree burns*). **7** an academic rank conferred by a college or university after examination or after completion of a course, or conferred as an honour on a distinguished person. **8** a grade of crime or criminality (*murder in the first degree*). **9** a step in direct genealogical descent. **10** social or official rank. **11** *Math.* the highest power of unknowns or variables in an equation etc. (*equation of the third degree*). **12** a masonic rank. **13** a thing placed like a step in a series; a tier or row. **14** *Mus.* the classification of a note by its position in the scale. □ **by degrees** a little at a time; gradually. **degree of freedom 1** *Physics* the independent direction in which motion can occur. **2** *Chem.* the number of independent factors required to specify a system at equilibrium. **3** *Statistics* the number of independent values or quantities which can be assigned to a statistical distribution. **degrees of comparison** see COMPARISON. **forbidden** (or **prohibited**) **degrees** a number of degrees of descent within which marriage between two related persons is forbidden. **to a degree** *colloq.* considerably. □ **degreeless** *adj.* [ME f. OF *degré* f. Rmc (as DE-, L *gradus* step)]

degressive /dɪ'ɡresɪv/ *adj.* **1** (of taxation) at successively lower rates on low amounts. **2** reducing in amount. [L *degredi* (as DE-, *gradi* walk)]

de haut en bas /də ˌəʊt ɒn 'bɑ:/ *adv.* in a condescending or superior manner. [F, = from above to below]

de Havilland /də 'hævɪlənd/, Sir Geoffrey (1882–1965), English aircraft designer and manufacturer. Having built the BE series of fighters in the First World War, he started his own company in 1920. He designed and built many famous aircraft, including the Moth series, the Mosquito of the Second World War, and some of the first jet aircraft, and also produced the Gipsy series of aircraft engines.

dehire /di:'haɪə(r)/ *v.tr.* *US colloq.* dismiss (a person) from employment; sack.

dehisce /dɪ'hɪs/ *v.intr.* gape or burst open (esp. of a pod or seed-vessel or of a cut or wound). □ **dehiscence** *n.* **dehiscent** *adj.* [L *dehiscere* (as DE-, *hiscere* inceptive of *hiare* gape)]

de Hooch /də 'hu:tʃ, Dutch də 'hoːx/, Pieter (also **de Hoogh**) (c.1629–c.1684), Dutch genre painter. He is noted for his depictions of domestic interior and courtyard scenes, which he painted while living in Delft. His best works, such as *Interior with a Woman Peeling Apples* (1663), are characterized by their sensitive handling of light and tranquil atmosphere.

dehorn /di:'hɔ:n/ *v.tr.* remove the horns from (an animal).

dehumanize /di:'hju:mənaɪz/ *v.tr.* (also **-ise**) **1** deprive of human characteristics. **2** make impersonal or machine-like. □ **dehumanization** /-ˌhju:mənaɪ'zeɪʃ(ə)n/ *n.*

dehumidify /ˌdi:hju:'mɪdɪfaɪ/ *v.tr.* (**-ies**, **-ied**) reduce the degree of humidity of; remove moisture from (a gas, esp. air). □ **dehumidifier** *n.* **dehumidification** /-ˌmɪdɪfɪ'keɪʃ(ə)n/ *n.*

dehydrate /ˌdi:haɪ'dreɪt/ *v.* **1** *tr.* **a** remove water from (esp. foods for preservation and storage in bulk). **b** make dry, esp. make (the body) deficient in water. **c** render lifeless or uninteresting. **2** *intr.* lose water. □ **dehydrator** *n.* **dehydration** /-'dreɪʃ(ə)n/ *n.*

dehydrogenate /ˌdi:haɪ'drɒdʒɪ,neɪt/ *v.tr.* *Chem.* remove a hydrogen atom or atoms from (a compound). □ **dehydrogenation** /-ˌdrɒdʒɪ'neɪʃ(ə)n/ *n.*

Deianira /ˌdi:ə'naɪərə/ *Gk Mythol.* the wife of Hercules, who was tricked into smearing poison on a garment, which caused his death.

de-ice /di:'aɪs/ *v.tr.* **1** remove ice from. **2** prevent the formation of ice on.

de-icer /di:'aɪsə(r)/ *n.* a device or substance for de-icing, esp. a windscreen or ice on an aircraft.

deicide /'di:ɪ,saɪd, 'deɪɪ-/ *n.* **1** the killer of a god. **2** the killing of a god. [eccl.L *deicida* f. L *deus* god + -CIDE]

deictic /'daɪktɪk/ *adj. & n.* *Philol. & Gram.* ● *adj.* pointing, demonstrative. ● *n.* a deictic word. [Gk *deiktikos* f. *deiktos* capable of proof f. *deiknumi* show]

deify /'di:ɪ,faɪ, 'deɪɪ-/ *v.tr.* (**-ies**, **-ied**) **1** make a god of. **2** regard or worship as a god. □ **deification** /ˌdi:ɪfɪ'keɪʃ(ə)n, ˌdeɪɪ-/ *n.* [ME f. OF *deifier* f. eccl.L *deificare* f. *deus* god]

Deighton /'deɪt(ə)n/, Len (b.1929), English writer. His reputation is based on his spy thrillers, several of which have been adapted as films and for television. The best known include his first novel *The Ipcress File* (1962) and the trilogy *Berlin Game*, *Mexico Set*, and *London Match* (1983–5).

deign /deɪn/ *v.* **1** *intr.* (foll. by *to* + infin.) think fit, condescend. **2** *tr.* (usu. with *neg.*) *archaic* condescend to give (an answer etc.). [ME f. OF *degnier*, *deigner*, *daigner* f. L *dignare*, *-ari* deem worthy f. *dignus* worthy]

Dei gratia /ˌdeɪɪ 'ɡrɑ:tɪə, 'ɡrɑ:ʃɪə/ *adv.* by the grace of God. [L]

Deimos /'deɪmɒs/ **1** *Gk Mythol.* one of the sons of Ares. **2** *Astron.* the outer of the two small satellites of Mars, discovered in 1877. It is 15 km long and 12 km across and is heavily cratered.

deinonychus /daɪ'nɒnɪkəs/ *n.* a bipedal carnivorous dinosaur of the genus *Deinonychus*, of the Cretaceous period, with an enlarged claw on each hind foot. [mod.L f. Gk *deinos* terrible + *onukh- onux* claw]

deinothere /'daɪnə,θɪə(r)/ *n.* an extinct elephant-like animal of the genus *Deinotherium*, having downward curving tusks. [mod.L *dinotherium* f. Gk *deinos* terrible + *thērion* wild beast]

deinstitutionalize /ˌdi:ɪnstɪ'tju:ʃənə,laɪz/ *v.tr.* (also **-ise**) **1** (usu. as **deinstitutionalized** *adj.*) remove from an institution or from the effects of institutional life. **2** make less institutional; reorganize on non-institutional or more individualistic lines. □ **deinstitutionalization** /-ˌtju:ʃənəlaɪ'zeɪʃ(ə)n/ *n.*

deionize /di:'aɪə,naɪz/ *v.tr.* (also **-ise**) remove the ions or ionic constituents from (water, air, etc.). □ **deionizer** *n.* **deionization** /-ˌaɪənaɪ'zeɪʃ(ə)n/ *n.*

deipnosophist /daɪp'nɒsəfɪst/ *n.* a person skilled in dining and table talk. [Gk *deipnosophistēs* (in pl. as title of a work by Athenaeus (3rd c.) describing long discussions at a banquet) f. *deipnon* dinner + *sophistēs* wise man: see SOPHIST]

Deirdre /'dɪədrɪ/ *Ir. Mythol.* a tragic heroine of whom it was prophesied that her beauty would bring banishment and death to heroes. King Conchobar of Ulster wanted to marry her, but she fell in love with Naoise, son of Usnach, who with his brothers carried her off to Scotland. They were lured back by Conchobar and treacherously slain, and Deirdre took her own life. The legend has been dramatized by George William Russell, J. M. Synge, and W. B. Yeats.

deism /'di:ɪz(ə)m, 'deɪɪz(ə)m/ *n.* belief in a supreme being or creator, specifically one that created but does not intervene in the universe. The term originally meant belief in a god known through natural religion and reason rather than revelation or teaching, and in this sense is applied with reference to a group in England in the 17th–18th centuries, strongly influenced by Locke's empiricism, who attacked the practice of orthodox Christianity. Never widely accepted in England, the concept of deism had a greater influence in France (where Voltaire and Jean-Jacques Rousseau were among its adherents), Germany, and the US. Cf. THEISM. □ **deist** *n.* **deistic** /di:'ɪstɪk, deɪ-/ *adj.* **deistical** *adj.* [L *deus* god + -ISM]

deity /'di:ɪtɪ, *disp.* 'deɪɪtɪ/ *n.* (*pl.* **-ies**) **1** a god or goddess. **2** divine status, quality, or nature. **3** (**the Deity**) the Creator, God. [ME f. OF *deité* f. eccl.L *deitas -tatis* transl. Gk *theotēs* f. *theos* god]

déjà vu /ˌdeɪʒɑ: 'vu:/ *n.* **1** *Psychol.* an illusory feeling of having already experienced a present situation. **2** the feeling that something similar has already been experienced; tedious familiarity. [F, = already seen]

deject /dɪ'dʒekt/ *v.tr.* (usu. as **dejected** *adj.*) make sad or dispirited; depress. □ **dejectedly** *adv.* [ME f. L *dejicere* (DE-, *jacere* throw)]

dejection /dɪ'dʒekʃ(ə)n/ *n.* a dejected state; low spirits. [ME f. L *dejectio* (as DEJECT)]

de jure /di: 'dʒʊərɪ, deɪ 'jʊəreɪ/ *adj. & adv.* ● *adj.* rightful. ● *adv.* rightfully; by right. [L]

Dekker /'dekə(r)/, Thomas (c.1570–1632), English dramatist and novelist. He is chiefly known for his two-part tragicomedy *The Honest Whore* (1604; 1630), the first part of which he wrote jointly with Thomas Middleton. He is also remembered for the revenge tragedy *The Witch of Edmonton* (1623), in which he collaborated with John Ford (c.1586–c.1640) and William Rowley (c.1585–1626).

dekko /'dekəʊ/ *n.* (*pl.* **-os**) *Brit. sl.* a look or glance (*took a quick dekko*). [Hindi *dekho*, imper. of *dekhnā* look]

de Klerk /də 'kleək/, F(rederik) W(illem) (b.1936), South African statesman, State President 1989–94. Becoming State President only months after assuming leadership of the National Party, he instigated significant political reforms designed to bring about the dismantling of apartheid in South Africa. After freeing Nelson Mandela and other ANC leaders in 1990 he lifted the ban on membership of the ANC and opened the negotiations with black political leaders that led to the country's first true democratic elections in 1994. After the ANC's electoral victory de Klerk was given a Cabinet post as a Deputy State President. In 1993 he shared the Nobel Peace Prize with Nelson Mandela.

de Kooning /də 'ku:nɪŋ/, Willem (b.1904), Dutch-born American painter. He and Jackson Pollock are generally regarded as the leading exponents of abstract expressionism. De Kooning's work in this genre from the late 1940s onwards usually retained figurative elements, whether represented or merely hinted at, as in *Painting* (1948). The female form became a central theme in his later work, notably in the *Women* series (1950–3).

Del. *abbr.* Delaware.

de la Beche /ˌdə læ ˈbiːtʃ/, Sir Henry Thomas (1796–1855), English geologist. He travelled extensively to study geology in the field, and produced the first geological description and map of Jamaica. Having already surveyed Devon, he was involved in the establishment of the Geological Survey of Great Britain in 1835, directing it from then until his death.

Delacroix /ˌdeləˈkrwʌ/, (Ferdinand Victor) Eugène (1798–1863), French painter. The chief painter of the French romantic school, he is known for his use of vivid colour, free drawing, and exotic, violent, or macabre subject-matter. *The Massacre at Chios* (1824) was an early example of his work and attracted much criticism when it was first exhibited. In his later work Delacroix experimented with complementary colours, purifying his palette to exclude black and earth colours and thus anticipating impressionist methods.

de la Mare /ˌdə læ ˈmeə(r)/, Walter (John) (1873–1956), English poet and novelist. Essentially a lyric poet, he had his first major success with *The Listeners* (1912). His many volumes of verse for children include *Peacock Pie* (1913) and *Tom Tiddler's Ground* (1932).

delate /dɪˈleɪt/ *v.tr. archaic* **1** inform against; impeach (a person). **2** report (an offence). □ **delator** *n.* **delation** /-ˈleɪʃ(ə)n/ *n.* [L *delat-* (as DE-, *lat-* past part. stem of *ferre* carry)]

Delaunay /dəˈlɔːneɪ/, Robert (1885–1941), French painter. For most of his career he experimented with the abstract qualities of colour, notably in his Eiffel Tower series (1910–12), and was one of the founder members of Orphism together with his wife, Sonia Delaunay-Terk. He was influenced by early cubism and painted some of the first purely abstract pictures, including his *Formes circulaires cosmiques* series.

Delaunay-Terk /dəˌlɔːneɪˈteək/, Sonia (1885–1979), Russian-born French painter and textile designer. She and her husband Robert Delaunay were among the founders of the movement of Orphism. She created abstract paintings based on harmonies of form and colour, her interest in the use of colour being reflected in her fabric and tapestry designs of the 1920s; her work in this field had a significant impact on international fashion.

Delaware[1] /ˈdeləˌweə(r)/ **1** a river of the north-eastern US. Rising in the Catskill Mountains in New York State, it flows some 450 km (280 miles) southwards to northern Delaware, where it meets the Atlantic at Delaware Bay. For much of its length it forms the eastern border of Pennsylvania. **2** a state of the US on the Atlantic coast, one of the original thirteen states of the Union (1787); pop. (1990) 666,168; capital, Dover.

Delaware[2] /ˈdeləˌweə(r)/ *n. & adj.* ● *n.* (*pl.* **Delawares** or same) **1** a member of an American Indian people formerly inhabiting the Delaware river valley of New Jersey and eastern Pennsylvania. **2** either of two Algonquian languages spoken by this people. ● *adj.* of or relating to the Delawares or their languages.

delay /dɪˈleɪ/ *v. & n.* ● *v.* **1** *tr.* postpone; defer. **2** *tr.* make late (*was delayed at the traffic lights*). **3** *intr.* loiter; be late (*don't delay!*). ● *n.* **1** the act or an instance of delaying; the process of being delayed. **2** time lost by inaction or the inability to proceed. **3** a hindrance. □ **delayed-action** (*attrib.*) (esp. of a bomb, camera, etc.) operating some time after being primed or set. **delaying action** *n.* **1** *Mil.* an action fought to delay the advance of an enemy. **2** any action taken to delay an unwelcome event or process. **delay line** *Electr.* a device producing a desired delay in the transmission of a signal. □ **delayer** *n.* [ME f. OF *delayer* (v.), *delai* (n.), prob. f. *des-* DIS- + *laier* leave: see RELAY]

dele /ˈdiːlɪ/ *v. & n. Printing* ● *v.tr.* (**deled** /-lɪd/, **deleing** /-lɪŋ/) delete or mark for deletion (a letter, word, etc., struck out of a text). ● *n.* a sign marking something to be deleted; a deletion. [L, imper. of *delere*: see DELETE]

delectable /dɪˈlektəb(ə)l/ *adj. esp. joc. or literary* **1** delightful, pleasant. **2** (of food) delicious. □ **delectably** *adv.* **delectability** /-ˌlektəˈbɪlɪtɪ/ *n.* [ME f. OF f. L *delectabilis* f. *delectare* DELIGHT]

delectation /ˌdiːlekˈteɪʃ(ə)n/ *n. esp. joc. or literary* pleasure, enjoyment (*sang for his delectation*). [ME f. OF (as DELECTABLE)]

delegacy /ˈdelɪɡəsɪ/ *n.* (*pl.* **-ies**) **1** a system of delegating. **2** **a** an appointment as a delegate. **b** a body of delegates; a delegation.

delegate *n. & v.* ● *n.* /ˈdelɪɡət/ **1** an elected representative sent to a conference. **2** a member of a committee. **3** a member of a deputation. ● *v.tr.* /ˈdelɪˌɡeɪt/ **1** (often foll. by *to*) **a** commit (authority, power, etc.) to an agent or deputy. **b** entrust (a task) to another person. **2** send or

authorize (a person) as a representative; depute. □ **delegable** /-ɡəb(ə)l/ *adj.* **delegator** /-ˌɡeɪtə(r)/ *n.* [ME f. L *delegatus* (as DE-, *legare* depute)]

delegation /ˌdelɪˈɡeɪʃ(ə)n/ *n.* **1** a body of delegates; a deputation. **2** the act or process of delegating or being delegated. [L *delegatio* (as DELEGATE)]

de Lenclos, Ninon, see LENCLOS.

delete /dɪˈliːt/ *v.tr.* remove or obliterate (written or printed matter), esp. by striking out. □ **deletion** /-ˈliːʃ(ə)n/ *n.* [L *delere delet-* efface]

deleterious /ˌdelɪˈtɪərɪəs/ *adj.* harmful (to the mind or body). □ **deleteriously** *adv.* [med.L *deleterius* f. Gk *dēlētērios* noxious]

Delfont /ˈdelfɒnt/, Bernard, Baron Delfont of Stepney (born Boris Winogradsky) (1909–94), British impresario, born in Russia. After emigrating to Britain with his brother (see GRADE) in 1912, he pursued a successful career in theatrical management; from the early 1940s onwards he presented more than 200 shows in London's West End.

Delft /delft/ a town in the Netherlands, in the province of South Holland; pop. (1991) 89,400. Since the 17th century the town has been noted for its pottery. It was also the home of the painters Pieter de Hooch and Jan Vermeer.

delft /delft/ *n.* (also **delftware** /ˈdelftweə(r)/) glazed, usu. blue and white, earthenware (as) made in Delft.

Delhi /ˈdelɪ/ a Union Territory in north central India, containing the cities of Old and New Delhi; pop. (1991) 7,175,000. *Old Delhi*, a walled city on the River Jumna, was made the capital of the Mogul empire in 1638 by Shah Jahan (1592–1666), who built within the city the Red Fort containing the imperial palace. *New Delhi*, the capital of India, was built in 1912–29 to replace Calcutta as the capital of British India.

deli /ˈdelɪ/ *n.* (*pl.* **delis**) *colloq.* a delicatessen shop. [abbr.]

Delian /ˈdiːlɪən/ *adj. & n.* ● *adj.* of or relating to Delos. ● *n.* a native or inhabitant of Delos.

Delian League an alliance of ancient Greek city-states that joined in 478–447 BC against the Persians, with its headquarters on Delos. The alliance was dominated by Athens (it is also known as the *Athenian empire*): after the end of the war with Persia the treasury was moved from Delos to Athens. The league was disbanded on the defeat of Athens in the Peloponnesian War (404 BC), but again united under Athens' leadership against Spartan aggression in 377–338 BC.

deliberate *adj. & v.* ● *adj.* /dɪˈlɪbərət/ **1 a** intentional (*a deliberate foul*). **b** fully considered; not impulsive (*made a deliberate choice*). **2** slow in deciding; cautious (*a ponderous and deliberate mind*). **3** (of movement etc.) leisurely and unhurried. ● *v.* /dɪˈlɪbəˌreɪt/ **1** *intr.* think carefully; take counsel (*the jury deliberated for an hour*). **2** *tr.* consider, discuss carefully, debate (*deliberated the question*). □ **deliberately** /-rətlɪ/ *adv.* **deliberateness** /-rətnɪs/ *n.* **deliberator** /-ˌreɪtə(r)/ *n.* [L *deliberatus* past part. of *deliberare* (as DE-, *librare* weigh f. *libra* balance)]

deliberation /dɪˌlɪbəˈreɪʃ(ə)n/ *n.* **1** careful consideration. **2 a** the discussion of reasons for and against. **b** a debate or discussion. **3 a** caution and care. **b** (of action or behaviour) purposefulness. **c** (of movement) slowness or ponderousness. [ME f. OF f. L *deliberatio -onis* (as DELIBERATE)]

deliberative /dɪˈlɪbərətɪv/ *adj.* of, or appointed for the purpose of, deliberation or debate (*a deliberative assembly*). □ **deliberatively** *adv.* **deliberativeness** *n.* [F *délibératif -ive* or L *deliberativus* (as DELIBERATE)]

Delibes /dəˈliːb/, (Clément Philibert) Léo (1836–91), French composer and organist. He wrote a number of light operas such as *Lakmé* (1883), but his best-known works are the ballets *Coppélia* (1870) and *Sylvia* (1876).

delicacy /ˈdelɪkəsɪ/ *n.* (*pl.* **-ies**) **1** (esp. in craftsmanship or artistic or natural beauty) fineness or intricacy of structure or texture; gracefulness. **2** susceptibility to injury or disease; weakness. **3** the quality of requiring discretion or sensitivity (*a situation of some delicacy*). **4** a choice or expensive food. **5 a** consideration for the feelings of others. **b** avoidance of vulgarity or offensiveness. **6** (esp. in a person, a sense, or an instrument) accuracy of perception; sensitiveness. [ME f. DELICATE + -ACY]

delicate /ˈdelɪkət/ *adj.* **1 a** fine or intricate in texture or structure; soft, slender, or slight. **b** of exquisite quality or workmanship. **c** (of colour) subtle or subdued; not bright. **d** subtle, hard to appreciate. **2 a** (of a person) easily injured; susceptible to illness. **b** (of a thing) easily spoiled or damaged. **3 a** requiring sensitivity or careful handling; tricky (*a delicate situation*). **b** (of an instrument) highly sensitive. **4** deft (*a delicate touch*). **5** (of a person) avoiding the vulgar or offensive. **6** (esp. of actions) considerate. **7** (of food) dainty; suitable for an invalid. □ **in a delicate condition** *archaic* pregnant. □ **delicately** *adv.* **delicateness** *n.* [ME f. OF *delicat* or L *delicatus*, of unkn. orig.]

delicatessen /ˌdelɪkə'tes(ə)n/ *n.* **1** a shop selling cooked meats, cheeses, and unusual or foreign prepared foods. **2** (often *attrib.*) such foods collectively (*a delicatessen counter*). [G *Delikatessen* or Du. *delicatessen* f. F *délicatesse* f. *délicat* (as DELICATE)]

delicious /dɪ'lɪʃəs/ *adj.* **1** highly delightful and enjoyable to the taste or sense of smell. **2** (of a joke etc.) very witty. □ **deliciously** *adv.* **deliciousness** *n.* [ME f. OF f. LL *deliciosus* f. L *deliciae* delight]

delict /dɪ'lɪkt, 'diːlɪkt/ *n. archaic* a violation of the law; an offence. [L *delictum* neut. past part. of *delinquere* offend (as DE-, *linquere* leave)]

delight /dɪ'laɪt/ *v. & n.* ● *v.* **1** *tr.* (often foll. by *with*, or *that* + clause, or *to* + infin.) please greatly (*the gift delighted them; was delighted with the result; was delighted that you won; would be delighted to help*). **2** *intr.* (often foll. by *in*, or *to* + infin.) take great pleasure; be highly pleased (*he delighted in her success; they delight to humour him*). ● *n.* **1** great pleasure. **2** something giving pleasure (*her singing is a delight*). □ **delighted** *adj.* **delightedly** *adv.* [ME f. OF *delitier, delit,* f. L *delectare* frequent. of *delicere*: alt. after *light* etc.]

delightful /dɪ'laɪtfʊl/ *adj.* causing delight; very pleasant, charming. □ **delightfully** *adv.* **delightfulness** *n.*

Delilah /dɪ'laɪlə/ (in the Bible) a woman who betrayed Samson to the Philistines (Judges 16) by revealing to them that the secret of his strength lay in his long hair. (See SAMSON.)

delimit /dɪ'lɪmɪt/ *v.tr.* (**delimited, delimiting**) **1** determine the limits of. **2** fix the territorial boundary of. □ **delimiter** *n.* **delimitation** /-ˌlɪmɪ'teɪʃ(ə)n/ *n.* [F *délimiter* f. L *delimitare* (as DE-, *limitare* f. *limes -itis* boundary)]

delimitate /dɪ'lɪmɪˌteɪt/ *v.tr.* = DELIMIT.

delineate /dɪ'lɪnɪˌeɪt/ *v.tr.* portray by drawing etc. or in words (*delineated her character*). □ **delineator** *n.* **delineation** /-ˌlɪnɪ'eɪʃ(ə)n/ *n.* [L *delineare delineat-* (as DE-, *lineare* f. *linea* line)]

delinquency /dɪ'lɪŋkwənsɪ/ *n.* (*pl.* **-ies**) **1** a minor crime in general, esp. that of young people (*juvenile delinquency*). **b** a crime, usu. not of a serious kind; a misdeed. **2** wickedness (*moral delinquency; an act of delinquency*). **3** neglect of one's duty. [eccl. L *delinquentia* f. L *delinquens* part. of *delinquere* (as DELICT)]

delinquent /dɪ'lɪŋkwənt/ *n. & adj.* ● *n.* an offender, a person guilty of delinquency (*juvenile delinquent*). ● *adj.* **1** guilty of a minor crime or a misdeed. **2** failing in one's duty. **3** *US* in arrears. □ **delinquently** *adv.*

deliquesce /ˌdelɪ'kwes/ *v.intr.* **1** become liquid, melt. **2** *Chem.* dissolve in water absorbed from the air. □ **deliquescence** *n.* **deliquescent** *adj.* [L *deliquescere* (as DE-, *liquescere* inceptive of *liquere* be liquid)]

delirious /dɪ'lɪrɪəs/ *adj.* **1** affected with delirium; temporarily or apparently mad; raving. **2** wildly excited, ecstatic. **3** (of behaviour) betraying delirium or ecstasy. □ **deliriously** *adv.*

delirium /dɪ'lɪrɪəm/ *n.* **1** an acutely disordered state of mind involving incoherent speech, hallucinations, and frenzied excitement, occurring in metabolic disorders, intoxication, fever, etc. **2** great excitement, ecstasy. □ **delirium tremens** /'triːmenz/ a psychosis of chronic alcoholism involving tremors and hallucinations. [L f. *delirare* be deranged (as DE-, *lira* ridge between furrows)]

Delius /'diːlɪəs/, Frederick (1862–1934), English composer, of German and Scandinavian descent. He is best known for pastoral works such as *Brigg Fair* (1907) and *On Hearing the First Cuckoo* (1912), but also wrote songs, concertos, and much choral and theatre music, including operas, often showing his deep interest in German and Scandinavian music and culture. Delius settled in France in the 1890s; blinded and paralysed by illness in 1928 he dictated his last works to an amanuensis, Eric Fenby (b.1906).

deliver /dɪ'lɪvə(r)/ *v.tr.* **1 a** distribute (letters, parcels, ordered goods, etc.) to the addressee or the purchaser. **b** (often foll. by *to*) hand over (*delivered the boy safely to his teacher*). **2** (often foll. by *from*) save, rescue, or set free (*delivered him from his enemies*). **3 a** give birth to (*delivered a girl*). **b** (in *passive*; often foll. by *of*) give birth (*was delivered of a child*). **c** assist at the birth of (*delivered six babies that week*). **d** assist in giving birth (*delivered the patient successfully*). **4 a** (often *refl.*) utter or recite (an opinion, a speech, etc.) (*delivered himself of the observation; delivered the sermon well*). **b** (of a judge) pronounce (a judgement). **5** (often foll. by *up, over*) abandon, surrender; hand over (*delivered his soul up to God*). **6** present or render (an account). **7** launch or aim (a blow, a ball, or an attack). **8** *Law* hand over formally (esp. a sealed deed to a grantee). **9** (*absol.*) *colloq.* = deliver the goods. **10** *US* cause (voters etc.) to support a

candidate. □ **deliver the goods** *colloq.* carry out one's part of an agreement. □ **deliverable** *adj.* **deliverer** *n.* [ME f. OF *delivrer* f. Gallo-Roman (as DE-, LIBERATE)]

deliverance /dɪ'lɪvərəns/ *n.* **1 a** the act or an instance of rescuing; the process of being rescued. **b** a rescue. **2** a formally expressed opinion. [ME f. OF *delivrance* (as DELIVER)]

delivery /dɪ'lɪvərɪ/ *n.* (*pl.* **-ies**) **1 a** the delivering of letters etc. **b** a regular distribution of letters etc. (*two deliveries a day*). **c** something delivered. **2 a** the process of childbirth. **b** an act of this. **3** deliverance. **4 a** an act of throwing, esp. of a cricket ball. **b** the style of such an act (*a good delivery*). **5** the act of giving or surrendering (*delivery of the town to the enemy*). **6 a** the uttering of a speech etc. **b** the manner or style of such a delivery (*a measured delivery*). **7** *Law* **a** the formal handing over of property. **b** the transfer of a deed to a grantee or a third party. □ **take delivery of** receive (something purchased). [ME f. AF *delivree* fem. past part. of *delivrer* (as DELIVER)]

dell /del/ *n.* a small usu. wooded hollow or valley. [OE f. Gmc]

Della Cruscan /ˌdelə 'krʌskən/ *adj. & n.* ● *adj.* **1** of or relating to the Academy della Crusca in Florence, an institution concerned with the purity of the Italian language. **2** of or concerning a late 18th-century school of English poets with an artificial style modelled on that of purist Italian writers. ● *n.* a member of the Academy della Crusca or the Della Cruscan school of English poets. [It. (*Accademia*) *della Crusca* (Academy) of the bran (with ref. to sifting)]

della Francesca see PIERO DELLA FRANCESCA.

della Quercia /ˌdelə 'kwɜːʃə/, Jacopo (*c.*1374–1438), Italian sculptor. He is noted for his tomb of Ilaria del Carretto in Lucca cathedral (*c.*1406) and, above all, for the biblical reliefs on the portal of San Petronio in Bologna (1425–35), which are notable for the sense of depth they achieve. He became chief architect of the cathedral in his native Siena in 1435.

della Robbia /ˌdelə 'rɒbɪə/, a family of Italian sculptors and ceramicists. Luca (1400–82) is the most famous, and is particularly well known for his relief panels in Florence cathedral. He invented vitreous glazes to colour terracotta figures, thus making it possible for polychromatic sculpture to be used in outdoor settings without suffering from the effects of damp. His nephew Andrea (1434–1525) carried on the family business of glazed terracotta production.

delocalize /diː'ləʊkəˌlaɪz/ *v.tr.* (also **-ise**) **1 a** detach or remove (a thing) from its place. **b** not limit to a particular location. **2** (as **delocalized** *adj.*) *Chem.* (of electrons) shared among more than two atoms in a molecule. □ **delocalization** /-ˌləʊkəlaɪ'zeɪʃ(ə)n/ *n.*

Delors /də'lɔː(r)/, Jacques (Lucien Jean) (b.1925), French socialist politician, president of the European Commission 1985–94. During his presidency he pressed for closer European union and oversaw the introduction of a single market within the European Community, which came into effect on 1 Jan. 1993.

Delos /'diːlɒs/ (Greek **Dhílos** /'ðilɒs/) a small Greek island in the Aegean Sea, regarded as the centre of the Cyclades. It is now virtually uninhabited, but thrived in classical times: considered to be sacred to Apollo, it was, according to legend, the birthplace of both Apollo and Artemis.

delouse /diː'laʊs/ *v.tr.* rid (a person or animal) of lice.

Delphi /'delfɪ, -faɪ/ (Greek **Dhelfoí** /ðɛl'fi/) one of the most important religious sanctuaries of the ancient Greek world, dedicated to Apollo and situated on the lower southern slopes of Mount Parnassus above the Gulf of Corinth. Thought of as the navel of the earth, it was the seat of the Delphic Oracle, whose often riddling responses to a wide range of religious, political, and moral questions were delivered in a state of ecstasy by the Pythia, the priestess of Apollo.

Delphic /'delfɪk/ *adj.* (also **Delphian** /-fɪən/) **1** (of an utterance, prophecy, etc.) obscure, ambiguous, or enigmatic. **2** of or concerning the ancient Greek oracle at Delphi.

delphinium /del'fɪnɪəm/ *n.* (*pl.* **delphiniums**) a ranunculaceous garden plant of the genus *Delphinium*, with tall spikes of usu. blue flowers. Also called *larkspur*. [mod.L f. Gk *delphinion* larkspur f. *delphin* dolphin]

delphinoid /'delfɪˌnɔɪd/ *adj. & n. Zool.* ● *adj.* **1** of the division Delphinoidea, which includes dolphins, porpoises, grampuses, etc. **2** dolphin-like. ● *n.* **1** a member of the Delphinoidea division. **2** a dolphin-like animal. [Gk *delphinoeidēs* f. *delphin* dolphin]

del Sarto, Andrea, see SARTO.

delta /'deltə/ *n.* **1** a triangular tract of deposited earth, alluvium, etc., at the mouth of a river, formed by its diverging outlets. **2 a** the fourth letter of the Greek alphabet (*Δ*, *δ*). **b** a fourth-class mark given for a piece of work or in an examination. **3** *Astron.* the fourth (usu. fourth brightest) star in a constellation (foll. by Latin genitive: *Delta Cephei*). **4** *Math.* an increment of a variable. □ **delta connection** *Electr.* a triangular arrangement of three-phase windings with circuit wire from each angle. **delta rays** *Physics* rays of low penetrative power consisting of slow electrons ejected from an atom by the impact of ionizing radiation. **delta rhythm** (or **wave**) low-frequency electrical activity of the brain during sleep. **delta wing** the triangular swept-back wing of an aircraft. □ **deltaic** /del'teɪɪk/ *adj.* [ME f. Gk f. Phoenician *daleth*]

deltiology /ˌdeltɪ'ɒlədʒɪ/ *n.* the collecting and study of postcards. □ **deltiologist** *n.* [Gk *deltion* dimin. of *deltos* writing-tablet + -LOGY]

deltoid /'deltɔɪd/ *adj. & n.* ● *adj.* triangular; like a river delta. ● *n.* (in full **deltoid muscle**) a thick triangular muscle covering the shoulder joint and used for raising the arm away from the body. [F *deltoïde* or mod.L *deltoides* f. Gk *deltoeidēs* (as DELTA, -OID)]

delude /dɪ'luːd, -'ljuːd/ *v.tr.* deceive or mislead (*deluded by false optimism*). □ **deluder** *n.* [ME f. L *deludere* mock (as DE-, *ludere lus-* play)]

deluge /'deljuːdʒ/ *n. & v.* ● *n.* **1** a great flood. **2** (**the Deluge**) the biblical Flood. **3** a great outpouring (of words, paper, etc.). **4** a heavy fall of rain. ● *v.tr.* **1** flood. **2** inundate with a great number or amount (*deluged with complaints*). [ME f. OF f. L *diluvium*, rel. to *lavare* wash]

delusion /dɪ'luːʒ(ə)n, -'ljuːʒ(ə)n/ *n.* **1** a false belief or impression. **2** *Psychol.* this as a symptom or form of mental disorder. □ **delusions of grandeur** a false idea of oneself as being important, noble, famous, etc. □ **delusional** *adj.* [ME f. LL *delusio* (as DELUDE)]

delusive /dɪ'luːsɪv, -'ljuːsɪv/ *adj.* **1** deceptive or unreal. **2** disappointing. □ **delusively** *adv.* **delusiveness** *n.*

delusory /dɪ'luːsərɪ, -'ljuːsərɪ/ *adj.* = DELUSIVE. [LL *delusorius* (as DELUSION)]

delustre /diː'lʌstə(r)/ *v.tr.* (US **deluster**) remove the lustre from (a textile).

de luxe /də 'lʌks, 'lʊks/ *adj.* **1** luxurious or sumptuous. **2** of a superior kind. [F, = of luxury]

delve /delv/ *v.* **1** *intr.* (often foll. by *in*, *into*) **a** search energetically (*delved into his pocket*). **b** make a laborious search in documents etc.; research (*delved into his family history*). **2** *tr. & intr. poet.* dig. □ **delver** *n.* [OE *delfan* f. WG]

Dem. *abbr.* US Democrat.

demagnetize /diː'mægnɪˌtaɪz/ *v.tr.* (also **-ise**) remove the magnetic properties of. □ **demagnetizer** *n.* **demagnetization** /-ˌmægnɪtaɪ'zeɪʃ(ə)n/ *n.*

demagogue /'deməˌgɒg/ *n.* (US **demagog**) **1** a political agitator appealing to the basest instincts of a mob. **2** *hist.* a leader of the people, esp. in ancient times. □ **demagogy** *n.* **demagogic** /ˌdemə'gɒgɪk/ *adj.* **demagoguery** /-'gɒgərɪ/ *n.* [Gk *dēmagōgos* f. *dēmos* the people + *agōgos* leading]

de Maintenon see MAINTENON.

demand /dɪ'mɑːnd/ *n. & v.* ● *n.* **1** an insistent and peremptory request, made as of right. **2** *Econ.* the desire of purchasers or consumers for a commodity (*no demand for solid tyres these days*). **3** an urgent claim (*care of her mother makes demands on her*). ● *v.tr.* **1** (often foll. by *of*, *from*, or *to* + infin., or *that* + clause) ask for (something) insistently and urgently, as of right (*demanded to know*; *demanded five pounds from him*; *demanded that his wife be present*). **2** require or need (*a task demanding skill*). **3** insist on being told (*demanded her business*). **4** (as **demanding** *adj.*) **a** making demands, esp. persistently (*a demanding child*). **b** requiring skill, effort, etc. (*a demanding but worthwhile job*). □ **demand feeding** the practice of feeding a baby when it cries for a feed rather than at set times. **demand-led** *Econ.* motivated or propelled by demand, esp. from consumers. **demand note 1** a written request for payment. **2** US a bill payable at sight. **demand pull** *Econ.* available money as a factor causing economic inflation. **in demand** sought after. **on demand** as soon as a demand is made (*a cheque payable on demand*). □ **demander** *n.* **demandingly** *adv.* [ME f. OF *demande* (n.), *demander* (v.) f. L *demandare* entrust (as DE-, *mandare* order: see MANDATE)]

demantoid /dɪ'mæntɔɪd/ *n.* a lustrous green garnet. [G]

demarcation /ˌdiːmɑː'keɪʃ(ə)n/ *n.* **1** the act of marking a boundary or limits. **2** the trade-union practice of strictly assigning specific jobs to different unions. □ **demarcation dispute** an inter-union dispute about who does a particular job. □ **demarcate** /'diːmɑːˌkeɪt/ *v.tr.* **demarcator** *n.* [Sp. *demarcación* f. *demarcar* mark the bounds of (as DE-, MARK¹)]

démarche /deɪ'mɑːʃ/ *n.* a political step or initiative. [F f. *démarcher* take steps (as DE-, MARCH¹)]

dematerialize /ˌdiːmə'tɪərɪəˌlaɪz/ *v.tr. & intr.* (also **-ise**) make or become non-material or spiritual (esp. of psychic phenomena etc.). □ **dematerialization** /-ˌtɪərɪəlaɪ'zeɪʃ(ə)n/ *n.*

de Maupassant, Guy, see MAUPASSANT.

deme /diːm/ *n.* **1 a** a political division of Attica in ancient Greece. **b** an administrative division in modern Greece. **2** *Biol.* a local population of closely related plants or animals. [Gk *dēmos* the people]

demean¹ /dɪ'miːn/ *v.tr.* (usu. *refl.*) lower the dignity or status of (*would not demean myself to take it*). [DE- + MEAN², after *debase*]

demean² /dɪ'miːn/ *v.refl.* (with *adv.*) behave (*demeaned himself well*). [ME f. OF *demener* f. Rmc (as DE-, L *minare* drive animals f. *minari* threaten)]

demeanour /dɪ'miːnə(r)/ *n.* (US **demeanor**) outward behaviour or bearing. [DEMEAN², prob. after obs. *havour* behaviour]

de' Medici¹ see MEDICI.

de' Medici², Catherine, see CATHERINE DE' MEDICI.

de' Medici³, Cosimo, see COSIMO DE' MEDICI.

de' Medici⁴, Giovanni, the name of Pope Leo X (see LEO²).

de' Medici⁵, Lorenzo, see LORENZO DE' MEDICI.

de Médicis, Marie, see MARIE DE MÉDICIS.

dement /dɪ'ment/ *n. archaic* a demented person. [orig. adj. f. F *dément* or L *demens* (as DEMENTED)]

demented /dɪ'mentɪd/ *adj.* mad; crazy. □ **dementedly** *adv.* **dementedness** *n.* [past part. of *dement* verb f. OF *dementer* or f. LL *dementare* f. *demens* out of one's mind (as DE-, *mens mentis* mind)]

démenti /deɪ'mɒntɪ/ *n.* (pl. **démentis** *pronunc.* same) an official denial of a rumour etc. [F f. *démentir* accuse of lying]

dementia /dɪ'menʃə/ *n. Med.* a chronic or persistent disorder of the mental processes marked by memory disorders, personality changes, impaired reasoning, etc., due to brain disease or injury. □ **dementia praecox** /'priːkɒks/ schizophrenia. [L f. *demens* (as DEMENTED)]

Demerara /ˌdemə'reərə, -'rɑːrə/ **1** a river of northern Guyana. Rising in the Guiana Highlands, it flows about 320 km (200 miles) northwards to the Atlantic. **2** a former Dutch colony in South America, now part of Guyana.

demerara /ˌdemə'reərə, -'rɑːrə/ *n.* (in full **demerara sugar**) light-brown cane-sugar coming orig. and chiefly from Demerara.

demerge /diː'mɜːdʒ/ *v.tr.* separate (a business) from another. □ **demerger** *n.*

demerit /diː'merɪt/ *n.* **1** a quality or action deserving blame; a fault. **2** *N. Amer.* a mark awarded against an offender. □ **demeritorious** /-ˌmerɪ'tɔːrɪəs/ *adj.* [ME f. OF *de(s)merite* or L *demeritum* neut. past part. of *demereri* deserve]

demersal /dɪ'mɜːs(ə)l/ *adj.* (of a fish etc.) being or living near the sea-bottom (cf. PELAGIC). [L *demersus* past part. of *demergere* (as DE-, *mergere* plunge)]

demesne /dɪ'miːn, -'meɪn/ *n.* **1 a** a sovereign's or state's territory; a domain. **b** land attached to a mansion etc. **c** landed property; an estate. **2** (usu. foll. by *of*) a region or sphere. **3** *Law hist.* possession (of real property) as one's own. □ **held in demesne** (of an estate) occupied by the owner, not by tenants. [ME f. AF, OF *demeine* (later AF *demesne*) belonging to a lord f. L *dominicus* (as DOMINICAL)]

Demeter /dɪ'miːtə(r)/ *Gk Mythol.* the corn-goddess, daughter of Cronus and Rhea and mother of Persephone. She is identified with Ceres and also Cybele; her symbol is often an ear of corn. The Eleusinian mysteries were held in honour of her. (See also PERSEPHONE.) [Gk f. *mētēr* mother]

demi- /'demɪ/ *prefix* **1** half; half-size. **2** partially or imperfectly such (*demigod*). [ME f. F f. med.L *dimedius* half, for L *dimidius*]

demigod /'demɪˌgɒd/ *n.* (fem. **-goddess** /-ˌgɒdɪs/) **1 a** a partly divine being. **b** the offspring of a god or goddess and a mortal. **2** *colloq.* a person of compelling beauty, powers, or personality.

demijohn /'demɪˌdʒɒn/ *n.* a bulbous narrow-necked bottle holding from 3 to 10 gallons and usu. in a wicker cover. [prob. corrupt. of F *dame-jeanne* Lady Jane, assim. to DEMI- + the name *John*]

demilitarize /diː'mɪlɪtəˌraɪz/ *v.tr.* (also **-ise**) remove a military organization or forces from (a frontier, a zone, etc.). □ **demilitarization** /-ˌmɪlɪtəraɪ'zeɪʃ(ə)n/ *n.*

de Mille /də 'mɪl/, Cecil B(lount) (1881–1959), American film producer and director. He founded the Jesse L. Lasky Feature Play Company (later Paramount) with Samuel Goldwyn in 1913 and chose the then little-known Los Angeles suburb of Hollywood as a location for their first film, *The Squaw Man* (1914); the success of this first American feature film helped to establish Hollywood as the world's prime centre of film-making. As a film-maker de Mille is particularly known for lavish biblical spectacles such as *The Ten Commandments* (1923; remade by de Mille in 1956) and *Samson and Delilah* (1949).

demi-mondaine /ˌdemɪˈmɒndeɪn/ *n.* a woman of a *demi-monde*.

demi-monde /ˈdemɪˌmɒnd/ *n.* **1 a** *hist.* a class of women in 19th-century France considered to be of doubtful social standing and morality. **b** a similar class of women in any society, esp. kept women or prostitutes. **2** any group considered to be on the fringes of respectable society. [F, = half-world]

demineralize /diːˈmɪnərəˌlaɪz/ *v.tr.* (also **-ise**) remove salts from (sea water etc.). □ **demineralization** /-ˌmɪnərəlaɪˈzeɪʃ(ə)n/ *n.*

demi-pension /ˌdemɪˈpɒnsjɒn/ *n.* hotel accommodation with bed, breakfast, and one main meal per day. [F (as DEMI-, PENSION²)]

demirep /ˈdemɪˌrep/ *n.* *archaic* a woman of doubtful sexual reputation. [abbr. of *demi-reputable*]

demise /dɪˈmaɪz/ *n. & v.* ● *n.* **1** death (*left a will on her demise; the demise of the agreement*). **2** *Law* conveyance or transfer (of property, a title, etc.) by demising. ● *v.tr.* *Law* **1** convey or grant (an estate) by will or lease. **2** transmit (a title etc.) by death. [AF use of past part. of OF *de(s)mettre* DISMISS, in refl. abdicate]

demisemiquaver /ˌdemɪˈsemɪˌkweɪvə(r)/ *n.* *Mus.* a note having the time value of half a semiquaver and represented by a large dot with a three-hooked stem. Also called *thirty-second note*.

demist /diːˈmɪst/ *v.tr.* clear mist from (a windscreen). □ **demister** *n.*

demit /dɪˈmɪt/ *v.tr.* (**demitted, demitting**) (often *absol.*) resign or abdicate (an office etc.). □ **demission** /-ˈmɪʃ(ə)n/ *n.* [F *démettre* f. L *demittere* (as DE-, *mittere* miss- send)]

demitasse /ˈdemɪˌtæs/ *n.* **1** a small coffee-cup. **2** its contents. [F, = half-cup]

demiurge /ˈdemɪˌɜːdʒ/ *n.* **1** (in the philosophy of Plato) the creator of the universe. **2** (in Gnosticism etc.) a heavenly being subordinate to the Supreme Being. □ **demiurgic** /ˌdemɪˈɜːdʒɪk/ *adj.* [eccl.L f. Gk *dēmiourgos* craftsman f. *dēmios* public f. *dēmos* people + *-ergos* working]

demo /ˈdemoʊ/ *n.* (*pl.* **-os**) *colloq.* **1** = DEMONSTRATION 2, 3. **2** (*attrib.*) demonstrating the capabilities of computer software, a group of musicians, etc. [abbr.]

demob /diːˈmɒb/ *v. & n.* *Brit. colloq.* ● *v.tr.* (**demobbed, demobbing**) demobilize. ● *n.* demobilization. [abbr.]

demobilize /diːˈmoʊbɪˌlaɪz/ *v.tr.* (also **-ise**) disband (troops, ships, etc.). □ **demobilization** /-ˌmoʊbɪlaɪˈzeɪʃ(ə)n/ *n.* [F *démobiliser* (as DE-, MOBILIZE)]

democracy /dɪˈmɒkrəsɪ/ *n.* (*pl.* **-ies**) **1 a** a system of government by the whole population, usu. through elected representatives. **b** a state so governed. **c** any organization governed on democratic principles. **2** an egalitarian and tolerant form of society. **3** *US* **a** the principles of the Democratic Party. **b** its members. [F *démocratie* f. LL *democratia* f. Gk *dēmokratia* f. *dēmos* the people + -CRACY]

democrat /ˈdeməˌkræt/ *n.* **1** an advocate of democracy. **2** (**Democrat**) (in the US) a member of the Democratic Party. [F *démocrate* (as DEMOCRACY), after *aristocrate*]

democratic /ˌdeməˈkrætɪk/ *adj.* **1** of, like, practising, advocating, or constituting democracy or a democracy. **2** favouring social equality. □ **democratic centralism** an organizational system in which policy is decided centrally and is binding on all members. □ **democratically** *adv.* [F *démocratique* f. med.L *democraticus* f. Gk *dēmokratikos* f. *dēmokratia* DEMOCRACY]

Democratic Party one of the two main US political parties (the other being the Republican Party). The name dates from *c.*1828, but the party existed earlier as the Democratic Republican Party and under other names; it claims Thomas Jefferson as its founder. Traditionally associated with the South, it has, since Roosevelt and especially in the 1960s, followed a broadly liberal programme, tending to support social reform and minority rights.

Democratic Republican Party *hist.* a US political party founded in 1792 by Thomas Jefferson, a forerunner of the modern Democratic Party.

democratize /dɪˈmɒkrəˌtaɪz/ *v.tr.* (also **-ise**) make (a state, institution, etc.) democratic. □ **democratization** /-ˌmɒkrətaɪˈzeɪʃ(ə)n/ *n.*

Democritus /dɪˈmɒkrɪtəs/ (*c.*460–*c.*370 BC), Greek philosopher. He wrote widely on a variety of subjects and further developed the atomic theory originated by his teacher, Leucippus (5th century BC); this theory explained natural phenomena in terms of the arrangement and rearrangement of atoms moving in a void (see also ATOMIC THEORY). Democritus was nicknamed 'the laughing philosopher', reputedly because he showed amusement at human nature.

démodé /deɪˈmoʊdeɪ/ *adj.* out of fashion. [F, past part. of *démoder* (as DE-, *mode* fashion)]

demodulate /diːˈmɒdjʊˌleɪt/ *v.tr.* *Physics* **1** extract (a modulating signal) from its carrier. **2** separate a modulating signal from. □ **demodulator** *n.* **demodulation** /-ˌmɒdjʊˈleɪʃ(ə)n/ *n.*

demography /dɪˈmɒɡrəfɪ/ *n.* the study of the statistics of births, deaths, disease, etc., as illustrating the conditions of life in communities. □ **demographer** *n.* **demographic** /ˌdeməˈɡræfɪk/ *adj.* **demographical** *adj.* **demographically** *adv.* [Gk *dēmos* the people + -GRAPHY]

demoiselle /ˌdemwʌˈzel/ *n.* **1** a demoiselle crane. **2 a** a damselfly. **b** a damselfish. **3** *archaic* a young woman. □ **demoiselle crane** a small crane, *Anthropoides virgo*, native to Asia and North Africa. [F, = DAMSEL]

demolish /dɪˈmɒlɪʃ/ *v.tr.* **1** pull down (a building). **b** completely destroy or break. **2** overthrow (an institution). **3** refute (an argument, theory, etc.). **4** *joc.* eat up completely and quickly. □ **demolisher** *n.* **demolition** /ˌdeməˈlɪʃ(ə)n/ *n.* **demolitionist** *n.* [F *démolir* f. L *demoliri* (as DE-, *moliri moliti-* construct f. *moles* mass)]

demon¹ /ˈdiːmən/ *n.* **1 a** an evil spirit or devil, esp. one thought to possess a person. **b** the personification of an evil passion. **2** a malignant supernatural being; the Devil. **3** (often *attrib.*) a forceful, fierce, or skilful performer (*a demon on the tennis-court; a demon player*). **4** a cruel or destructive person. **5** (also **daemon**) **a** an inner or attendant spirit; genius (*the demon of creativity*). **b** a divinity or supernatural being in ancient Greece. □ **demon bowler** *Cricket* a very fast bowler. **a demon for work** *colloq.* a person who works strenuously. [ME f. med.L *demon* f. L *daemon* f. Gk *daimōn* deity]

demon² /ˈdiːmən/ *n.* *Austral. sl.* a police officer. [app. f. VAN DIEMEN'S LAND after DEMON¹]

demonetize /diːˈmʌnɪˌtaɪz/ *v.tr.* (also **-ise**) withdraw (a coin etc.) from use as money. □ **demonetization** /-ˌmʌnɪtaɪˈzeɪʃ(ə)n/ *n.* [F *démonétiser* (as DE-, L *moneta* MONEY)]

demoniac /dɪˈmoʊnɪˌæk/ *adj. & n.* ● *adj.* = DEMONIC. ● *n.* a person supposedly possessed by an evil spirit. □ **demoniacal** /ˌdiːməˈnaɪək(ə)l/ *adj.* **demoniacally** *adv.* [ME f. OF *demoniaque* f. eccl.L *daemoniacus* f. *daemonium* f. Gk *daimonion* dimin. of *daimōn*: see DEMON¹]

demonic /dɪˈmɒnɪk/ *adj.* (also **daemonic**) **1** fiercely energetic or frenzied. **2 a** supposedly possessed by an evil spirit. **b** of or concerning such possession. **3** of or like demons. **4** having or seeming to have supernatural genius or power. [LL *daemonicus* f. Gk *daimonikos* (as DEMON¹)]

demonism /ˈdiːməˌnɪz(ə)m/ *n.* belief in the power of demons.

demonize /ˈdiːməˌnaɪz/ *v.tr.* (also **-ise**) **1** make into or like a demon. **2** represent as a demon. □ **demonization** /ˌdiːmənaɪˈzeɪʃ(ə)n/ *n.*

demonolatry /ˌdiːməˈnɒlətrɪ/ *n.* the worship of demons.

demonology /ˌdiːməˈnɒlədʒɪ/ *n.* the study of demons etc. □ **demonologist** *n.*

demonstrable /ˈdemənstrəb(ə)l, dɪˈmɒnstrəb(ə)l/ *adj.* capable of being shown or logically proved. □ **demonstrably** *adv.* **demonstrability** /ˌdemənstrəˈbɪlɪtɪ/ *n.* [ME f. L *demonstrabilis* (as DEMONSTRATE)]

demonstrate /ˈdemənˌstreɪt/ *v.* **1** *tr.* show evidence of (feelings etc.). **2** *tr.* describe and explain (a scientific proposition, machine, etc.) by experiment, practical use, etc. **3** *tr.* **a** logically prove the truth of. **b** be proof of the existence of. **4** *intr.* take part in or organize a public demonstration. **5** *intr.* act as a demonstrator. [L *demonstrare* (as DE-, *monstrare* show)]

demonstration /ˌdemənˈstreɪʃ(ə)n/ *n.* **1** (foll. by *of*) **a** the outward showing of feeling etc. **b** an instance of this. **2** a show of public opinion on a political or moral question, usu. in the form of a mass march or meeting. **3 a** a practical exhibition or explanation of something, designed to teach or inform. **b** a practical display of a piece of equipment to show its capabilities. **4** proof provided by logic, argument, etc. **5** *Mil.* a show of military force. □ **demonstrational** *adj.* [ME f. OF *demonstration* or L *demonstratio* (as DEMONSTRATE)]

demonstrative /dɪˈmɒnstrətɪv/ *adj. & n.* ● *adj.* **1** given to or marked by an open expression of feeling, esp. of affection (*a very demonstrative person*). **2** (usu. foll. by *of*) logically conclusive; giving proof (*the work is demonstrative of their skill*). **3 a** serving to point out or exhibit.

b involving esp. scientific demonstration (*demonstrative technique*). **4** *Gram.* (of an adjective or pronoun) indicating the person or thing referred to (e.g. *this, that, those*). ● *n. Gram.* a demonstrative adjective or pronoun. □ **demonstratively** *adv.* **demonstrativeness** *n.* [ME f. OF *demonstratif -ive* f. L *demonstrativus* (as DEMONSTRATION)]

demonstrator /ˈdemənˌstreɪtə(r)/ *n.* **1** a person who takes part in a political demonstration etc. **2 a** a person who demonstrates, esp. machines, equipment, etc., to prospective customers. **b** a machine etc., esp. a car, used for such demonstrations. **3** a person who teaches by demonstration, esp. in a laboratory etc. [L (as DEMONSTRATE)]

de Montespan, Marquise de, see MONTESPAN.

de Montfort, Simon, see MONTFORT¹.

demoralize /dɪˈmɒrəˌlaɪz/ *v.tr.* (also **-ise**) **1** destroy the morale of; make hopeless. **2** *archaic* corrupt the morals of. □ **demoralizing** *adj.* **demoralizingly** *adv.* **demoralization** /-ˌmɒrəlaɪˈzeɪʃ(ə)n/ *n.* [F *démoraliser* (as DE-, MORAL)]

Demosthenes /dɪˈmɒsθəˌniːz/ (384–322 BC), Athenian orator and statesman. He is best known for his political speeches on the need to defend Athens against the pretensions of Philip II of Macedon, which are known as the *Philippics*. Demosthenes was at the forefront of the campaign to unite the Greek city-states militarily against Macedon; the Greeks were defeated at the battle of Chaeronea in 338 BC, and Demosthenes committed suicide after the failure of an Athenian revolt against Macedon.

demote /dɪˈməʊt/ *v.tr.* reduce to a lower rank or class. □ **demotion** /-ˈməʊʃ(ə)n/ *n.* [DE- + PROMOTE]

demotic /dɪˈmɒtɪk/ *adj. & n.* ● *adj.* **1** (esp. of language) popular, colloquial, or vulgar. **2 a** of or relating to the everyday spoken form of modern Greek, containing many Turkish and other elements (cf. KATHAREVOUSA). **b** of or relating to a popular simplified form of ancient Egyptian writing, a cursive script based partially on hieratic, which dates from *c*.650 BC and was gradually replaced by Greek in the Ptolemaic period. ● *n.* demotic language, esp. that of ancient Egypt or modern Greece. [Gk *dēmotikos* f. *dēmotēs* one of the people (*dēmos*)]

demotivate /diːˈməʊtɪˌveɪt/ *v.tr.* (also *absol.*) cause to lose motivation; discourage. □ **demotivation** /-ˌməʊtɪˈveɪʃ(ə)n/ *n.*

demount /diːˈmaʊnt/ *v.tr.* **1** take (apparatus, a gun, etc.) from its mounting. **2** dismantle for later reassembly. □ **demountable** *adj. & n.* [F *démonter*: cf. DISMOUNT]

Dempsey /ˈdempsɪ/, William Harrison ('Jack') (1895–1983), American boxer. He was world heavyweight champion 1919–26, and during this time drew extremely large audiences to boxing; his defence of the title in 1921 was the first fight at which a million dollars was taken at the gate.

demulcent /dɪˈmʌls(ə)nt/ *adj. & n.* ● *adj.* soothing. ● *n.* an agent that forms a protective film soothing irritation or inflammation in the mouth. [L *demulcere* (as DE-, *mulcere* soothe)]

demur /dɪˈmɜː(r)/ *v. & n.* ● *v.intr.* (**demurred, demurring**) **1** (often foll. by *to, at*) raise scruples or objections. **2** *Law* put in a demurrer. ● *n.* (also **demurral** /-ˈmʌrəl/) (usu. in *neg.*) **1** an objection (*agreed without demur*). **2** the act or process of objecting. [ME f. OF *demeure* (n.), *demeurer* (v.) f. Rmc (as DE-, L *morari* delay)]

demure /dɪˈmjʊə(r)/ *adj.* (**demurer, demurest**) **1** composed, quiet, and reserved; modest. **2** affectedly shy and quiet; coy. **3** decorous (*a demure high collar*). □ **demurely** *adv.* **demureness** *n.* [ME, perh. f. AF *demuré* f. OF *demoré* past part. of *demorer* remain, stay (as DEMUR): infl. by OF *meür* f. L *maturus* ripe]

demurrable /dɪˈmʌrəb(ə)l/ *adj.* esp. *Law* open to objection.

demurrage /dɪˈmʌrɪdʒ/ *n.* **1 a** a rate or amount payable to a shipowner by a charterer for failure to load or discharge a ship within the time agreed. **b** a similar charge on railway trucks or goods. **2** a detention or delay caused in this way. [OF *demo(u)rage* f. *demorer* (as DEMUR)]

demurrer /dɪˈmʌrə(r)/ *n.* an objection, esp. *Law* a pleading that an opponent's point is irrelevant. [AF (infin. as noun), = DEMUR]

demy /dɪˈmaɪ/ *n.* *Printing* a size of paper, 564 × 444 mm. [ME, var. of DEMI-]

demystify /diːˈmɪstɪˌfaɪ/ *v.tr.* (**-ies, -ied**) **1** clarify (obscure beliefs or subjects etc.), remove the mystery from. **2** reduce or remove the irrationality in (a person). □ **demystification** /-ˌmɪstɪfɪˈkeɪʃ(ə)n/ *n.*

demythologize /ˌdiːmɪˈθɒləˌdʒaɪz/ *v.tr.* (also **-ise**) **1** remove mythical elements from (a legend, story of famous person's life, etc.). **2** (in Christian theology) reinterpret what some consider to be the mythological elements in (the Bible).

den /den/ *n.* **1** a wild animal's lair. **2** a place of crime or vice (*den of iniquity; opium den*). **3 a** a small private room or place for pursuing a hobby etc. **b** a hide-out or secret place for children. [OE *denn* f. Gmc, rel. to DENE¹]

denarius /dɪˈneərɪəs/ *n.* (*pl.* **denarii** /-rɪˌaɪ/) an ancient Roman silver coin. [L, = (coin) of ten asses (as DENARY: see AS²)]

denary /ˈdiːnərɪ/ *adj.* of ten; decimal. □ **denary scale** = *decimal scale.* [L *denarius* containing ten (*deni* by tens)]

denationalize /diːˈnæʃ(ə)nəˌlaɪz/ *v.tr.* (also **-ise**) **1** transfer (a nationalized industry or institution etc.) from public to private ownership. **2 a** deprive (a nation) of its status or characteristics as a nation. **b** deprive (a person) of nationality or national characteristics. □ **denationalization** /-ˌnæʃ(ə)nəlaɪˈzeɪʃ(ə)n/ *n.* [F *dénationaliser* (as DE-, NATIONAL)]

denaturalize /diːˈnætʃrəˌlaɪz/ *v.tr.* (also **-ise**) **1** change the nature or properties of; make unnatural. **2** deprive of the rights of citizenship. **3** = DENATURE *v.* 1. □ **denaturalization** /-ˌnætʃrəˈzeɪʃ(ə)n/ *n.*

denature /diːˈneɪtʃə(r)/ *v.tr.* **1** change the properties of (a protein etc.) by heat, acidity, etc. **2** make (alcohol) unfit for drinking, esp. by the addition of another substance. □ **denaturant** *n.* **denaturation** /-ˌnætʃəˈreɪʃ(ə)n/ *n.* [F *dénaturer* (as DE-, NATURE)]

Denbighshire /ˈdenbɪˌʃɪə(r)/ a former county of North Wales. It was divided between Clwyd and Gwynedd in 1974.

dendrite /ˈdendraɪt/ *n.* **1 a** a stone or mineral with natural treelike or mosslike markings. **b** such marks on stones or minerals. **2** *Chem.* a crystal with branching treelike growth. **3** *Zool. & Anat.* a branching process of a nerve cell conducting signals to a cell body. [F f. Gk *dendritēs* (adj.) f. *dendron* tree]

dendritic /denˈdrɪtɪk/ *adj.* **1** of or like a dendrite. **2** treelike in shape or markings. □ **dendritically** *adv.*

dendrochronology /ˌdendrəʊkrəˈnɒlədʒɪ/ *n.* a system of dating using the annual growth rings of trees to assign dates to timber; the study of these growth rings. Trees add a ring of growth each year, and variations in climate affect the width of the rings and give rise to recognizable sequences. By matching these from a tree of known date with those from an older tree overlapping in age, and combining a number of such overlaps, a master plot of tree-ring patterns can be built up covering several thousand years. Timber of unknown date from within the area of this plot can then be dated exactly by matching its rings, and this has also enabled radiocarbon dates to be more accurately calibrated. □ **dendrochronologist** *n.* **dendrochronological** /-ˌkrɒnəˈlɒdʒɪk(ə)l/ *adj.* [Gk *dendron* tree + CHRONOLOGY]

dendrogram /ˈdendrəˌɡræm/ *n.* *Biol.* a branched diagram showing the relationship between kinds of organism. [Gk *dendron* tree + -GRAM]

dendroid /ˈdendrɔɪd/ *adj.* tree-shaped. [Gk *dendrōdēs* treelike + -OID]

dendrology /denˈdrɒlədʒɪ/ *n.* the scientific study of trees. □ **dendrologist** *n.* **dendrological** /ˌdendrəˈlɒdʒɪk(ə)l/ *adj.* [Gk *dendron* tree + -LOGY]

dene¹ /diːn/ *n.* (also **dean**) *Brit.* **1** a narrow wooded valley. **2** a vale (esp. as the ending of place-names). [OE *denu*, rel. to DEN]

dene² /diːn/ *n.* *Brit.* a bare sandy tract, or a low sand-hill, by the sea. [orig. unkn.: cf. DUNE]

Deneb /ˈdeneb/ *Astron.* the brightest star in the constellation of Cygnus, a yellow supergiant. [Arab., = tail (of the swan)]

de-net /diːˈnet/ *v.tr.* (also **denet**) sell (a book) at a price lower than that fixed under the terms of the net book agreement.

Deneuve /dəˈnɜːv/, Catherine (born Catherine Dorléac) (b.1943), French actress. She is best known for her roles in such films as *Repulsion* (1965) and Luis Buñuel's *Belle de jour* (1967).

dengue /ˈdeŋɡɪ/ *n.* an infectious viral disease of the tropics, transmitted by mosquitoes and causing a fever and acute pains in the joints. [West Indian Sp., f. Swahili *denga, dinga*, with assim. to Sp. *dengue* fastidiousness, with ref. to the stiffness of the patient's neck and shoulders]

Deng Xiaoping /ˌdeŋ ʃaʊˈpɪŋ, dʌŋ/ (also **Teng Hsiao-p'ing** /ˌteŋ/) (b.1904), Chinese Communist statesman, Vice-Premier 1973–6 and 1977–80; Vice-Chairman of the Central Committee of the Chinese Communist Party 1977–80. Discredited during the Cultural Revolution, he was reinstated in 1977, becoming the most prominent exponent of economic modernization, improving relations with the West, and taking a firm stance in relation to the Soviet Union. Since then, despite the announcement of his retirement in 1989, he has been regarded as the effective leader of China. In 1989 his orders led to the massacre of some 2,000 pro-democracy demonstrators in Beijing's Tiananmen Square.

Den Haag see HAGUE, THE.

deniable /dɪˈnaɪəb(ə)l/ adj. that may be denied. □ **deniability** /-ˌnaɪəˈbɪlɪtɪ/ n.

denial /dɪˈnaɪəl/ n. **1** the act or an instance of denying. **2** a refusal of a request or wish. **3** a statement that a thing is not true; a rejection (denial of the accusation). **4** a disavowal of a person as one's leader etc. **5** = SELF-DENIAL. **6** Psychol. the (usu. unconscious) suppression of a painful or unacceptable emotion or truth.

denier /ˈdenjə(r)/ n. a unit of weight by which the fineness of silk, rayon, or nylon yarn is measured. [orig. the name of a small coin: ME f. OF f. L denarius]

denigrate /ˈdenɪˌɡreɪt/ v.tr. defame or disparage the reputation of (a person); blacken. □ **denigrator** n. **denigratory** adj. **denigration** /ˌdenɪˈɡreɪʃ(ə)n/ n. [L denigrare (as DE-, nigrare f. niger black)]

denim /ˈdenɪm/ n. **1** (often attrib.) a usu. blue hard-wearing cotton twill fabric used for jeans etc. (a denim skirt). **2** (in pl.) colloq. jeans etc. made of this. [for serge de Nim f. NÎMES]

De Niro /də ˈnɪərəʊ/, Robert (b.1943), American actor. Since his first major film success, in Mean Streets (1972), he has starred in many films, often playing gangsters and other tough characters and frequently working with director Martin Scorsese. He won Oscars for The Godfather Part II (1974) and Raging Bull (1980). More recently De Niro made his début as a director with A Bronx Tale (1994), in which he also acted.

Denis /dəˈniː/, Maurice (1870–1943), French painter, designer, and art theorist. A founder member and leading theorist of the Nabi Group, he wrote many works on art, including Théories (1913) and Nouvelles Théories (1921). As a painter he is best known for his group portrait Hommage à Cézanne (1900) and his religious paintings. He founded the Ateliers d'Art Sacré in 1919 as part of a programme to revive religious painting.

Denis, St /ˈdenɪs/ (also **Denys**; Roman name Dionysius) (died c.250), Italian-born French bishop, patron saint of France. According to tradition he was one of a group of seven missionaries sent from Rome to convert Gaul; he became bishop of Paris, and was martyred in the reign of the emperor Valerian. He was later confused with Dionysius the Areopagite. Feast day, 9 Oct.

denitrify /diːˈnaɪtrɪfaɪ/ v.tr. (-ies, -ied) remove the nitrates or nitrites from (soil etc.). □ **denitrification** /-ˌnaɪtrɪfɪˈkeɪʃ(ə)n/ n.

denizen /ˈdenɪz(ə)n/ n. **1** (usu. foll. by of) an inhabitant or occupant. **2** a foreigner admitted to certain rights in his or her adopted country. **3** a naturalized foreign word, animal, or plant. □ **denizenship** n. [ME f. AF deinzein f. OF deinz within f. L de from + intus within + -ein f. L -aneus: see -ANEOUS]

Denmark /ˈdenmɑːk/ (Danish **Danmark** /ˈdanmarɡ/) a Scandinavian country consisting of the greater part of the Jutland peninsula and several neighbouring islands, between the North Sea and the Baltic; pop. (est. 1991) 5,100,000; official language, Danish; capital, Copenhagen. Denmark emerged as a separate country during the Viking period of the 10th and 11th centuries. In the 14th century Denmark and Norway were united under a Danish king, the union being joined between 1389–97 and 1523 by Sweden. Norway was ceded to Sweden after the Napoleonic Wars. Although neutral, Denmark was occupied by Germany for much of the Second World War. Denmark joined the EEC in 1973.

denominate /dɪˈnɒmɪˌneɪt/ v.tr. **1** give a name to. **2** call or describe (a person or thing) as. [L denominare (as DE-, NOMINATE)]

denomination /dɪˌnɒmɪˈneɪʃ(ə)n/ n. **1** a Church or religious sect. **2** a class of units within a range or sequence of numbers, weights, money, etc. (money of small denominations). **3 a** a name or designation, esp. a characteristic or class name. **b** a class or kind having a specific name. **4** the rank of a playing card within a suit, or of a suit relative to others. [ME f. OF denomination or L denominatio (as DENOMINATE)]

denominational /dɪˌnɒmɪˈneɪʃən(ə)l/ adj. **1** belonging to or of the nature of a denomination. **2** (of education, a school, etc.) founded on the principles of a Church or religious sect, sectarian. □ **denominationalism** n. **denominationally** adv.

denominative /dɪˈnɒmɪnətɪv/ adj. serving as or giving a name. [LL denominativus (as DENOMINATION)]

denominator /dɪˈnɒmɪˌneɪtə(r)/ n. Math. the number below the line in a vulgar fraction; a divisor. □ **common denominator 1** a common multiple of the denominators of several fractions. **2** a common feature of members of a group. **least** (or **lowest**) **common denominator** the lowest common multiple as above. [F dénominateur or med.L denominator (as DE-, NOMINATE)]

de nos jours /ˌdə nəʊ ˈʒʊə(r)/ adj. (placed after noun) of the present time. [F, = of our days]

denote /dɪˈnəʊt/ v.tr. **1** be a sign of; indicate (the arrow denotes direction). **2** (usu. foll. by that + clause) mean, convey. **3** stand as a name for; signify. □ **denotative** /-tətɪv/ adj. **denotation** /ˌdiːnəʊˈteɪʃ(ə)n/ n. [F dénoter or f. L denotare (as DE-, notare mark f. nota NOTE)]

denouement /deɪˈnuːmɒn/ n. (also **dénouement**) **1** the final unravelling of a plot or complicated situation. **2** the final scene in a play, novel, etc., in which the plot is resolved. [F dénouement f. dénouer unknot (as DE-, L nodare f. nodus knot)]

denounce /dɪˈnaʊns/ v.tr. **1** accuse publicly; condemn (denounced him as a traitor). **2** inform against (denounced her to the police). **3** give notice of the termination of (an armistice, treaty, etc.). □ **denouncement** n. **denouncer** n. [ME f. OF denoncier f. L denuntiare (as DE-, nuntiare make known f. nuntius messenger)]

de nouveau /də ˈnuːvəʊ/ adv. starting again; anew. [F]

de novo /diː ˈnəʊvəʊ, deɪ/ adv. starting again; anew. [L]

Denpasar /denˈpɑːsɑː(r)/ the chief city of the island of Bali, a seaport on the south coast; pop. (1980) 261,200.

dense /dens/ adj. **1** closely compacted in substance; thick (dense fog). **2** crowded together (the population is less dense on the outskirts). **3** colloq. stupid. □ **densely** adv. **denseness** n. [F dense or L densus]

densitometer /ˌdensɪˈtɒmɪtə(r)/ n. an instrument for measuring the photographic density of an image on a film or photographic print.

density /ˈdensɪtɪ/ n. (pl. **-ies**) **1** the degree of compactness of a substance. **2** Physics degree of consistency measured by the quantity of mass per unit volume. **3** the opacity of a photographic image. **4** a crowded state. **5** colloq. stupidity. [F densité or L densitas (as DENSE)]

dent /dent/ n. & v. ● n. **1** a slight mark or hollow in a surface made by, or as if by, a blow with a hammer etc. **2** a noticeable effect (lunch made a dent in our funds). ● v.tr. **1** mark with a dent. **2** have (esp. an adverse) effect on (the news dented our hopes). [ME, prob. f INDENT[1]]

dental /ˈdent(ə)l/ adj. **1** of the teeth; of or relating to dentistry. **2** Phonet. (of a consonant) produced with the tongue-tip against the upper front teeth (as th) or the ridge of the teeth (as n, s, t). □ **dental floss** a thread of floss silk etc. used to clean between the teeth. **dental formula** Zool. a formula expressing the number and kinds of teeth possessed by a mammal. **dental mechanic** a person who makes and repairs artificial teeth. **dental surgeon** a dentist. □ **dentalize** v.tr. (also **-ise**). [LL dentalis f. L dens dentis tooth]

dentalium /denˈteɪlɪəm/ n. (pl. **dentalia** /-lɪə/) **1** a tusk shell, esp. of the genus Dentalium. **2** this shell used as an ornament or as a form of currency. [mod.L f. LL dentalis: see DENTAL]

dentate /ˈdenteɪt/ adj. Bot. & Zool. toothed; with toothlike notches; serrated. [L dentatus f. dens dentis tooth]

denticle /ˈdentɪk(ə)l/ n. Zool. a small tooth or toothlike projection, scale, etc. □ **denticulate** /denˈtɪkjʊlət/ adj. [ME f. L denticulus dimin. of dens dentis tooth]

dentifrice /ˈdentɪfrɪs/ n. a paste or powder for cleaning the teeth. [F f. L dentifricium f. dens dentis tooth + fricare rub]

dentil /ˈdentɪl/ n. Archit. each of a series of small rectangular blocks as a decoration under the moulding of a cornice in classical architecture. [obs. F dentille dimin. of dent tooth f. L dens dentis]

dentilingual /ˌdentɪˈlɪŋɡwəl/ adj. Phonet. formed by the teeth and the tongue.

dentine /ˈdentiːn/ n. (US **dentin** /-tɪn/) a hard dense bony tissue forming the bulk of a tooth. □ **dentinal** /-tɪn(ə)l/ adj. [L dens dentis tooth + -INE[4]]

dentist /ˈdentɪst/ n. a person who is qualified to treat the diseases and conditions that affect the mouth, jaws, teeth, and their supporting tissues, esp. the repair and extraction of teeth and the insertion of artificial ones. [F dentiste f. dent tooth]

dentistry /ˈdentɪstrɪ/ n. the profession or practice of a dentist. The Etruscans were practising dental surgery by the 9th century BC, and ancient Hindu writings describe extraction, scaling, and filling, and the fitting of artificial teeth. In medieval Europe dentistry was the province of barbers and barber-surgeons. Dentistry was taught in some hospitals from the early 19th century, and the first separate dental school was started at Baltimore in the US in 1839. Some instruments, notably the forceps, are of great antiquity, but many, such as the dental drill, date from the early 19th century.

dentition /den'tɪʃ(ə)n/ n. **1** the type, number, and arrangement of teeth in a species etc. **2** the cutting of teeth; teething. [L *dentitio* f. *dentire* to teethe]

denture /'dentʃə(r)/ n. a removable artificial replacement for one or more teeth carried on a removable plate or frame. [F f. *dent* tooth]

denuclearize /di:'nju:klɪə‚raɪz/ v.tr. (also **-ise**) remove nuclear armaments from (a country etc.). □ **denuclearization** /-‚nju:klɪəraɪ'zeɪʃ(ə)n/ n.

denude /dɪ'nju:d/ v.tr. **1** make naked or bare. **2** (foll. by *of*) **a** strip of clothing, a covering, etc. **b** deprive of a possession or attribute. **3** *Geol.* lay (rock or a formation etc.) bare by removing what lies above. □ **denudative** /-dətɪv/ adj. **denudation** /‚di:nju:'deɪʃ(ə)n/ n. [L *denudare* (as DE-, *nudus* naked)]

denumerable /dɪ'nju:mərəb(ə)l/ adj. *Math.* countable by correspondence with the infinite set of integers. □ **denumerably** adv. **denumerability** /-‚nju:mərə'bɪlɪtɪ/ n. [LL *denumerare* (as DE-, *numerare* NUMBER)]

denunciation /dɪ‚nʌnsɪ'eɪʃ(ə)n/ n. **1** the act of denouncing (a person, policy, etc.); public condemnation. **2** an instance of this. □ **denunciate** /-'nʌnsɪ‚eɪt/ v.tr. **denunciator** n. **denunciative** /-'nʌnsɪətɪv/ adj. **denunciatory** /-'nʌnsɪətərɪ/ adj. [F *dénonciation* or L *denunciatio* (as DENOUNCE)]

Denver /'denvə(r)/ the state capital of Colorado; pop. (1990) 467,600. Situated at an altitude of 1,608 m (5,280 ft) on the eastern side of the Rocky Mountains, Denver was developed in the 1870s as a silver-mining town.

deny /dɪ'naɪ/ v.tr. (**-ies**, **-ied**) **1** declare untrue or non-existent (*denied the charge; denied that it is so; denied having lied*). **2** repudiate or disclaim (*denied his faith; denied his signature*). **3** (foll. by *to*) refuse (a person or thing, or something to a person) (*this was denied to me; denied him the satisfaction*). **4** refuse access to (a person sought) (*denied him his son*). □ **deny oneself** be abstinent. □ **denier** n. [ME f. OF *denier* f. L *denegare* (as DE-, *negare* say no)]

Denys, St see DENIS, ST.

deoch an doris /‚dɒx ən 'dɒrɪs, ‚dɒk/ n. (also **doch an dorris**) *Sc.* & *Ir.* a drink taken at parting; a stirrup-cup. [Gael. *deoch an doruis* drink at the door]

deodar /'di:ə‚dɑ:(r)/ n. the Himalayan cedar *Cedrus deodara*, the tallest of the cedar family, with drooping branches bearing large barrel-shaped cones. [Hindi *dě' odār* f. Skr. *deva-dāru* divine tree]

deodorant /di:'əʊdərənt/ n. (often *attrib.*) a substance sprayed or rubbed on to the body or sprayed into the air to remove or conceal unwanted or unpleasant smells (*a roll-on deodorant; has a deodorant effect*). [as DEODORIZE + -ANT]

deodorize /di:'əʊdə‚raɪz/ v.tr. (also **-ise**) remove or destroy the (usu. unpleasant) smell of. □ **deodorizer** n. **deodorization** /-‚əʊdəraɪ'zeɪʃ(ə)n/ n. [DE- + L *odor* smell]

Deo gratias /‚deɪəʊ 'grɑ:tɪəs, 'grɑ:ʃɪəs/ int. thanks be to God. [L, = (we give) thanks to God]

deontic /di:'ɒntɪk/ adj. *Philos.* of or relating to duty and obligation as ethical concepts. [Gk *deont-* part. stem of *dei* it is right]

deontology /‚di:ɒn'tɒlədʒɪ/ n. *Philos.* the study of duty. □ **deontologist** n. **deontological** /-tə'lɒdʒɪk(ə)l/ adj.

Deo volente /‚deɪəʊ və'lenteɪ/ adv. God willing; if nothing prevents it. [L]

deoxygenate /di:'ɒksɪdʒə‚neɪt/ v.tr. remove oxygen, esp. free oxygen, from. □ **deoxygenation** /-‚ɒksɪdʒə'neɪʃ(ə)n/ n.

deoxyribonucleic acid /di:‚ɒksɪ‚raɪbəʊnju:'kli:ɪk, -'kleɪɪk/ n. *Biochem.* see DNA. [DE- + OXYGEN + RIBONUCLEIC ACID]

deoxyribose /di:‚ɒksɪ'raɪbəʊz, -bəʊs/ n. a sugar related to ribose and found in the nucleosides of DNA. [DE- + OXY-[2] + RIBOSE]

dep. abbr. **1** departs. **2** deputy.

Depardieu /dəpardjø/, Gérard (b.1948), French actor. He made his screen début in 1965. His international reputation is based on the many films he has made since the early 1980s; these include *Danton* (1982), *Jean de Florette* (1986), and *Cyrano de Bergerac* (1990).

depart /dɪ'pɑ:t/ v. **1** intr. **a** (usu. foll. by *from*) go away; leave (*the train departs from this platform*). **b** (usu. foll. by *for*) start; set out (*trains depart for Crewe every hour*). **2** intr. (usu. foll. by *from*) diverge; deviate (*departs from standard practice*). **3 a** intr. leave by death; die. **b** tr. *formal or literary*

leave by death (*departed this life*). [ME f. OF *departir* ult. f. L *dispertire* divide]

departed /dɪ'pɑ:tɪd/ adj. & n. ● adj. bygone (*departed greatness*). ● n. (prec. by *the*) *euphem.* a particular dead person or dead people (*we are here to mourn the departed*).

department /dɪ'pɑ:tmənt/ n. **1** a separate part of a complex whole, esp.: **a** a branch of municipal or state administration (*Housing Department; Department of Social Security*). **b** a branch of study and its administration at a university, school, etc. (*the physics department*). **c** a specialized section of a large store (*hardware department*). **2** *colloq.* an area of special expertise. **3** an administrative district in France and other countries. □ **department store** a large shop stocking many varieties of goods in different departments. [F *département* (as DEPART)]

departmental /‚di:pɑ:t'ment(ə)l/ adj. of or belonging to a department. □ **departmental store** = department store. □ **departmentalism** n. **departmentally** adv. **departmentalize** v.tr. (also **-ise**). **departmentalization** /-‚mentəlaɪ'zeɪʃ(ə)n/ n.

departure /dɪ'pɑ:tʃə(r)/ n. **1** the act or an instance of departing. **2** (often foll. by *from*) a deviation (from the truth, a standard, etc.). **3** (often *attrib.*) the starting of a train, an aircraft, etc. (*the departure was late; departure lounge*). **4** a new course of action or thought (*driving a car is rather a departure for him*). **5** *Naut.* the amount of a ship's change of longitude. [OF *departeüre* (as DEPART)]

depasture /diː'pɑ:stʃə(r)/ v. **1 a** tr. (of cattle) graze upon. **b** intr. graze. **c** tr. put (cattle) to graze. **2** tr. (of land) provide pasturage for (cattle). □ **depasturage** /-rɪdʒ/ n.

dépaysé /deɪ'peɪ‚zeɪ/ adj. (fem. **dépaysée** pronunc. same) removed from one's habitual surroundings. [F, = removed from one's own country]

depend /dɪ'pend/ v.intr. **1** (often foll. by *on*, *upon*) be controlled or determined by (*success depends on hard work; it depends on whether they agree; it depends how you tackle the problem*). **2** (foll. by *on*, *upon*) **a** be unable to do without (*depends on her mother*). **b** rely on (*I'm depending on you to come*). **3** (foll. by *on*, *upon*) be grammatically dependent on. **4** (often foll. by *from*) *archaic or poet.* hang down. □ **depending on** according to. **depend upon it!** you may be sure! **it** (or **it all** or **that**) **depends** expressing uncertainty or qualification in answering a question (*Will they come? It depends*). [ME f. OF *dépendre* ult. f. L *dependere* (as DE-, *pendere* hang)]

dependable /dɪ'pendəb(ə)l/ adj. reliable. □ **dependably** adv. **dependability** /-‚pendə'bɪlɪtɪ/ n.

dependant /dɪ'pendənt/ n. (US **dependent**) **1** a person who relies on another, esp. for financial support. **2** a subordinate, a servant. [F *dépendant* pres. part. of *dépendre* (as DEPEND)]

dependence /dɪ'pendəns/ n. **1** the state of being dependent, esp. on financial or other support. **2** reliance; trust; confidence (*shows great dependence on his judgement*). [F *dépendance* (as DEPEND)]

dependency /dɪ'pendənsɪ/ n. (pl. **-ies**) **1** a country or province controlled by another. **2** anything subordinate or dependent. **3** = DEPENDENCE 1. □ **dependency culture** a way of life etc. characterized by dependence on state benefits.

dependent /dɪ'pendənt/ adj. & n. ● adj. **1** (usu. foll. by *on*) depending, conditional, or subordinate. **2** unable to do without (esp. a drug). **3** maintained at another's cost. **4** *Math.* (of a variable) having a value determined by that of another variable. **5** *Gram.* (of a clause, phrase, or word) subordinate to a sentence or word. ● n. US var. of DEPENDANT. □ **dependently** adv. [ME, earlier *-ant* = DEPENDANT]

depersonalization /di:‚pɜ:sənəlaɪ'zeɪʃ(ə)n/ n. (also **-isation**) esp. *Psychol.* the loss of one's sense of identity.

depersonalize /di:'pɜ:sənə‚laɪz/ v.tr. (also **-ise**) **1** make impersonal. **2** deprive of personality.

depict /dɪ'pɪkt/ v.tr. **1** represent in a drawing or painting etc. **2** portray in words; describe (*the play depicts him as vain and petty*). □ **depicter** n. **depiction** /-'pɪkʃ(ə)n/ n. [L *depingere depict-* (as DE-, *pingere* paint)]

depilate /'depɪ‚leɪt/ v.tr. remove the hair from. □ **depilation** /‚depɪ'leɪʃ(ə)n/ n. [L *depilare* (as DE-, *pilare* f. *pilus* hair)]

depilatory /dɪ'pɪlətərɪ/ adj. & n. ● adj. that removes unwanted hair. ● n. (pl. **-ies**) a depilatory substance.

de Pisan /də 'pi:zæn/, Christine (also **de Pizan**) (c.1364–c.1430), Italian writer, resident in France from 1369. After the death of her husband when she was 25, she turned to writing for her living, becoming the first professional woman writer in France. She is best known for her works in defence of women's virtues and achievements; these include *Epistre au dieu d'amour* (1399), *La Cité des dames* (1405; *Book of the City of Ladies*), which celebrates the lives of famous women throughout history, and *Le Livre des trois vertus* (1406), a treatise on women's education. Other works include a biography of Charles V of France (1404).

deplane /diːˈpleɪn/ v. esp. US **1** intr. disembark from an aeroplane. **2** tr. remove from an aeroplane.

deplete /dɪˈpliːt/ v.tr. (esp. in passive) **1** reduce in numbers or quantity (depleted forces). **2** empty out; exhaust (their energies were depleted). □ **depletion** /-ˈpliːʃ(ə)n/ n. [L deplere (as DE-, plere plet- fill)]

deplorable /dɪˈplɔːrəb(ə)l/ adj. **1** exceedingly bad (a deplorable meal). **2** that can be deplored. □ **deplorably** adv.

deplore /dɪˈplɔː(r)/ v.tr. feel or express strong disapproval of. □ **deploringly** adv. [F déplorer or It. deplorare f. L deplorare (as DE-, plorare bewail)]

deploy /dɪˈplɔɪ/ v. **1** Mil. **a** tr. cause (troops) to spread out from a column into a line. **b** intr. (of troops) spread out in this way. **2** tr. bring (arguments, forces, etc.) into effective action. □ **deployment** n. [F déployer f. L displicare (as DIS-, plicare fold) & LL deplicare explain]

deplume /diːˈpluːm/ v.tr. **1** strip of feathers, pluck. **2** deprive of honours etc. [ME f. F déplumer or f. med.L deplumare (as DE-, L pluma feather)]

depolarize /diːˈpəʊləˌraɪz/ v.tr. (also **-ise**) Physics reduce or remove the polarization of. □ **depolarization** /-ˌpəʊləraɪˈzeɪʃ(ə)n/ n.

depoliticize /ˌdiːpəˈlɪtɪˌsaɪz/ v.tr. (also **-ise**) **1** make (a person, an organization, etc.) non-political. **2** remove from political activity or influence. □ **depoliticization** /-ˌlɪtɪsaɪˈzeɪʃ(ə)n/ n.

depolymerize /diːˈpɒlɪməˌraɪz/ v.tr. & intr. (also **-ise**) Chem. break down into monomers or other smaller units. □ **depolymerization** /-ˌpɒlɪməraɪˈzeɪʃ(ə)n/ n.

deponent /dɪˈpəʊnənt/ adj. & n. ● adj. Gram. (of a verb, esp. in Latin or Greek) passive or middle in form but active in meaning. ● n. **1** Gram. a deponent verb. **2** Law **a** a person making a deposition under oath. **b** a witness giving written testimony for use in court etc. [L deponere (as DE-, ponere posit- place): adj. from the notion that the verb had laid aside the passive sense]

depopulate /diːˈpɒpjʊˌleɪt/ v. **1** tr. reduce the population of. **2** intr. decline in population. □ **depopulation** /-ˌpɒpjʊˈleɪʃ(ə)n/ n. [L depopulari (as DE-, populari lay waste f. populus people)]

deport /dɪˈpɔːt/ v.tr. **1** remove (an immigrant or foreigner) forcibly to another country; banish. **2** exile (a native) to another country. □ **deportable** adj. **deportation** /ˌdiːpɔːˈteɪʃ(ə)n/ n. [OF deporter and (sense 1) F déporter (as DE-, L portare carry)]

deportee /ˌdiːpɔːˈtiː/ n. a person who has been or is being deported.

deportment /dɪˈpɔːtmənt/ n. bearing, demeanour, or manners, esp. of a cultivated kind. [F déportement (as DEPORT)]

depose /dɪˈpəʊz/ v. **1** tr. remove from office, esp. dethrone. **2** intr. (usu. foll. by to, or that + clause) Law bear witness, esp. on oath in court. [ME f. OF deposer f. L deponere: see DEPONENT, POSE[1]]

deposit /dɪˈpɒzɪt/ n. & v. ● n. **1 a** Brit. a sum of money kept in an account in a bank. **b** anything stored or entrusted for safe keeping, usu. in a bank. **2 a** a sum payable as a first instalment on an item bought on hire purchase or a mortgage, or as a pledge for a contract. **b** a returnable sum payable on the short-term hire of a car, boat, etc. **3** Brit. Polit. a sum of money deposited by an election candidate and forfeited if he or she fails to receive a certain proportion of the votes. **4 a** a natural layer of sand, rock, coal, etc. **b** a layer of precipitated matter on a surface, e.g. fur on a kettle. ● v.tr. (**deposited**, **depositing**) **1 a** put or lay down in a (usu. specified) place (deposited the book on the floor). **b** (of water, wind, etc.) leave (matter etc.) lying in a displaced position. **2 a** store or entrust for keeping. **b** pay (a sum of money) into a bank account, esp. a deposit account. **3** pay (a sum) as a first instalment or as a pledge for a contract. □ **deposit account** Brit. a bank account that pays interest but from which money cannot usu. be withdrawn without notice or loss of interest. **on deposit** (of money) placed in a deposit account. [L depositum (n.), med.L depositare f. L deponere deposit- (as DEPONENT)]

depositary /dɪˈpɒzɪtəri/ n. (pl. **-ies**) a person to whom something is entrusted; a trustee. [LL depositarius (as DEPOSIT)]

deposition /ˌdiːpəˈzɪʃ(ə)n, ˌdep-/ n. **1** the act or an instance of deposing, esp. a monarch; dethronement. **2** Law **a** the process of giving sworn evidence; allegation. **b** an instance of this. **c** evidence given under oath; a testimony. **3** the act or an instance of depositing. **4** (**the Deposition**) **a** the taking down of the body of Christ from the Cross. **b** a representation of this. [ME f. OF f. L depositio -onis f. deponere: see DEPOSIT]

depositor /dɪˈpɒzɪtə(r)/ n. a person who deposits money, property, etc.

depository /dɪˈpɒzɪtəri/ n. (pl. **-ies**) **1 a** a storehouse, a repository. **b** a store of wisdom, knowledge, etc. (the book is a depository of wit). **2** = DEPOSITARY. [LL depositorium (as DEPOSIT)]

depot /ˈdepəʊ/ n. **1** a storehouse. **2** Mil. **a** a storehouse for equipment etc. **b** the headquarters of a regiment. **3 a** a building for the servicing, parking, etc., of esp. buses, trains, or goods vehicles. **b** N. Amer. a railway or bus station. [F dépôt, OF depost f. L (as DEPOSIT)]

deprave /dɪˈpreɪv/ v.tr. pervert or corrupt, esp. morally. □ **depravation** /ˌdeprəˈveɪʃ(ə)n/ n. [ME f. OF depraver or L depravare (as DE-, pravare f. pravus crooked)]

depravity /dɪˈprævɪti/ n. (pl. **-ies**) **1 a** moral corruption; wickedness. **b** an instance of this; a wicked act. **2** (in Christian theology) the innate corruptness of human nature. [DE- + obs. pravity f. L pravitas (as DEPRAVE)]

deprecate /ˈdeprɪˌkeɪt/ v.tr. **1** express disapproval of or a wish against; deplore (deprecate hasty action). ¶ Often confused with depreciate. **2** plead earnestly against. **3** archaic pray against. □ **deprecator** n. **deprecatory** adj. **deprecatingly** adv. **deprecative** /-kətɪv/ adj. **deprecation** /ˌdeprɪˈkeɪʃ(ə)n/ n. [L deprecari (as DE-, precari pray)]

depreciate /dɪˈpriːʃɪˌeɪt, dɪˈpriːsɪ-/ v. **1** tr. & intr. diminish in value (the car has depreciated). **2** tr. disparage; belittle (they are always depreciating his taste). **3** tr. reduce the purchasing power of (money). □ **depreciatory** /-ˈpriːʃɪətəri, -ˈpriːsɪə-/ adj. [LL depretiare (as DE-, pretiare f. pretium price)]

depreciation /dɪˌpriːʃɪˈeɪʃ(ə)n, dɪˌpriːsɪ-/ n. **1** the amount of wear and tear (of a property etc.) for which a reduction may be made in a valuation, an estimate, or a balance sheet. **2** the action or process of diminishing in value, esp. a fall in the exchange rate of a currency. **3** the act or an instance of depreciating; belittlement.

depredation /ˌdeprɪˈdeɪʃ(ə)n/ n. (usu. in pl.) **1** despoiling, ravaging, or plundering. **2** an instance or instances of this. [F déprédation f. LL depraedatio (as DE-, praedatio -onis f. L praedari plunder)]

depredator /ˈdeprɪˌdeɪtə(r)/ n. a despoiler or pillager. □ **depredatory** /ˈdeprɪˌdeɪtəri, dɪˈpredɪtəri/ adj. [LL depraedator (as DEPREDATION)]

depress /dɪˈpres/ v.tr. **1** push or pull down; lower (depressed the lever). **2** make dispirited or dejected. **3** Econ. reduce the activity of (esp. trade). **4** (as **depressed** adj.) **a** dispirited or miserable. **b** Psychol. suffering from depression. □ **depressed area** an area suffering from economic depression. □ **depressible** adj. **depressing** adj. **depressingly** adv. [ME f. OF depresser f. LL depressare (as DE-, pressare frequent. of premere press)]

depressant /dɪˈpres(ə)nt/ adj. & n. ● adj. **1** that depresses. **2** Med. sedative. ● n. **1** Med. an agent, esp. a drug, that sedates. **2** an influence that depresses.

depression /dɪˈpreʃ(ə)n/ n. **1 a** Psychol. a state of extreme dejection or excessive melancholy; a mood of hopelessness and feelings of inadequacy, often with physical symptoms. **b** a reduction in vitality, vigour, or spirits. **2** a long period of financial and industrial decline; a slump. (See also DEPRESSION, THE.) **3** Meteorol. a region of low barometric pressure, esp. = CYCLONE 1. **4** a sunken place or hollow on a surface. **5 a** a lowering or sinking (often foll. by of: depression of freezing-point). **b** pressing down. **6** Astron. & Geog. the angular distance of an object below the horizon or a horizontal plane. [ME f. OF or L depressio (as DE-, premere press- press)]

Depression, the, (also called the Great Depression) the severe economic depression of 1929–34. The Wall Street stock market crash of 1929 is often seen as its starting-point, although an earlier agricultural crisis caused by overproduction was also a factor. The sudden loss of confidence in the US and consequent withdrawal of funds resulted in widespread business failures and the calling in of loans made to Europe. A major Austrian bank collapsed, followed by others, and the Depression spread across Europe also. Unemployment reached 13.7 million in the US, 5.6 million in Germany, and 2.8 million in Britain. The most dramatic results of the Depression were the introduction of extensive state-controlled economic planning, such as the American New Deal, and the rise of right-wing movements such as the German Nazi Party.

depressive /dɪˈpresɪv/ adj. & n. ● adj. **1** tending to depress. **2** Psychol. involving or characterized by depression. ● n. Psychol. a person suffering or with a tendency to suffer from depression. [F dépressif -ive or med.L depressivus (as DEPRESSION)]

depressor /dɪˈpresə(r)/ n. **1** Anat. **a** (in full **depressor muscle**) a muscle that causes the lowering of some part of the body. **b** a nerve that lowers blood pressure. **2** Surgery an instrument for pressing down an organ etc. [L (as DEPRESSION)]

depressurize /diːˈpreʃəˌraɪz/ v.tr. (also **-ise**) cause an appreciable drop in the pressure of the gas inside (a container), esp. to the ambient level. □ **depressurization** /-ˌpreʃəraɪˈzeɪʃ(ə)n/ n.

deprivation /ˌdeprɪˈveɪʃ(ə)n/ n. **1** (usu. foll. by of) the act or an instance of depriving; the state of being deprived (deprivation of liberty; suffered many deprivations). **2** archaic **a** deposition from esp. an ecclesiastical office. **b** an instance of this. [med.L deprivatio (as DEPRIVE)]

deprive /dɪˈpraɪv/ v.tr. **1** (usu. foll. by of) strip, dispossess; debar from enjoying (illness deprived him of success). **2** (as **deprived** adj.) **a** (of a child etc.) suffering from the effects of a poor or loveless home. **b** (of an area) with inadequate housing, facilities, employment, etc. **3** archaic depose (esp. a clergyman) from office. □ **deprivable** adj. **deprival** n. [ME f. OF depriver f. med.L deprivare (as DE-, L privare deprive)]

de profundis /ˌdeɪ prəˈfʊndɪs/ adv. & n. ● adv. from the depths (of sorrow etc.). ● n. a cry from the depths. [opening L words of Ps. 130]

Dept. abbr. Department.

depth /depθ/ n. **1 a** deepness (the depth is not great at the edge). **b** the measurement from the top down, from the surface inwards, or from the front to the back (depth of the drawer is 12 inches). **2** difficulty; abstruseness. **3 a** sagacity; wisdom. **b** intensity of emotion etc. (the poem has little depth). **4** an intensity of colour, darkness, etc. **5** (in pl.) **a** deep water, a deep place; an abyss. **b** a low, depressed state. **c** the lowest or inmost part (the depths of the country). **6** the middle (in the depth of winter). □ **depth-charge** (or **-bomb**) a bomb capable of exploding under water, esp. for dropping on a submerged submarine etc. **depth psychology** psychoanalysis to reveal hidden motives etc. **in depth** comprehensively, thoroughly, or profoundly. **in-depth** adj. thorough; done in depth. **out of one's depth 1** in water over one's head. **2** engaged in a task or on a subject too difficult for one. [ME (as DEEP, -TH²)]

depthless /ˈdepθlɪs/ adj. **1** extremely deep; fathomless. **2** shallow, superficial.

depurate /ˈdepjʊˌreɪt/ v.tr. & intr. make or become free from impurities. □ **depurator** n. **depuration** /ˌdepjʊˈreɪʃ(ə)n/ n. **depurative** /dɪˈpjʊərətɪv/ adj. & n. [med.L depurare (as DE-, purus pure)]

deputation /ˌdepjʊˈteɪʃ(ə)n/ n. a group of people appointed to represent others, usu. for a specific purpose; a delegation. [ME f. LL deputatio (as DEPUTE)]

depute v. & n. ● v.tr. /dɪˈpjuːt/ (often foll. by to) **1** appoint as a deputy. **2** delegate (a task, authority, etc.) (deputed the leadership to her). ● n. /ˈdepjuːt/ Sc. a deputy. [ME f. OF député past part. of deputer f. L deputare regard as, allot (as DE-, putare think)]

deputize /ˈdepjʊˌtaɪz/ v.intr. (also **-ise**) (usu. foll. by for) act as a deputy or understudy.

deputy /ˈdepjʊtɪ/ n. (pl. **-ies**) **1** a person appointed or delegated to act for another or others (also attrib.: deputy manager). **2** Polit. a parliamentary representative in certain countries, e.g. France. **3** a coal mine official responsible for safety. □ **by deputy** by proxy. **deputy lieutenant** Brit. the deputy of the Lord Lieutenant of a county. □ **deputyship** n. [ME var. of DEPUTE n.]

De Quincey /də ˈkwɪnsɪ/, Thomas (1785–1859), English essayist and critic. After first taking opium for toothache at Oxford, he became a lifelong addict. He achieved fame with his Confessions of an English Opium Eater (1822), a study of his addiction and its psychological effects, ranging from euphoria to nightmares. His writing had an important influence on Baudelaire and Poe.

deracinate /diːˈræsɪˌneɪt/ v.tr. literary **1** tear up by the roots. **2** obliterate, expunge. □ **deracination** /-ˌræsɪˈneɪʃ(ə)n/ n. [F déraciner (as DE-, racine f. LL radicina dimin. of radix root)]

derail /dɪˈreɪl/ v.tr. (usu. in passive) cause (a train etc.) to leave the rails. □ **derailment** n. [F dérailler (as DE-, RAIL¹)]

derailleur /dɪˈreɪlə(r), -ˈreɪljə(r)/ n. a bicycle gear in which the ratio is changed by switching the line of the chain while pedalling so that it jumps to a different sprocket. [F dérailleur (as DERAIL)]

Derain /dəˈræn/, André (1880–1954), French painter. He was one of the exponents of fauvism, joining with Vlaminck and Matisse in treating colour as an independent decorative and expressive element; this is particularly evident in his paintings of Hyde Park and the Thames (1906–7). He was also influenced by early cubism and the post-impressionist painting of Cézanne. In addition to painting, he designed theatre sets and costumes, notably for the Ballets Russes.

derange /dɪˈreɪndʒ/ v.tr. **1** throw into confusion; disorganize; cause to act irregularly. **2** (esp. as **deranged** adj.) make insane (deranged by the

tragic events). **3** disturb; interrupt. □ **derangement** n. [F déranger (as DE-, rang RANK¹)]

derate /diːˈreɪt/ v. **1** tr. remove part or all of the burden of rates from. **2** intr. diminish or remove rates.

deration /diːˈræʃ(ə)n/ v.tr. free (food etc.) from rationing.

Derbent /dəˈbent/ a city in southern Russia, in Dagestan on the western shore of the Caspian Sea; pop. (1985) 80,000. The city was formerly on an important land route for trade between Europe and Asia.

Derby¹ /ˈdɑːbɪ/ a city in Derbyshire, England, on the River Derwent; pop. (1991) 214,000.

Derby² /ˈdɑːbɪ/, 14th Earl of (title of Edward George Geoffrey Smith Stanley) (1799–1869), British Conservative statesman, Prime Minister 1852, 1858–9, and 1866–8. He led the protectionists in the House of Lords in their opposition to Peel's attempted repeal of the Corn Laws in 1846, and in his last term as Prime Minister he carried the second Reform Act (1867) through Parliament.

Derby³ /ˈdɑːbɪ/ n. (pl. **-ies**) **1 a** an annual flat horse-race for three-year-olds, founded in 1780 by the 12th Earl of Derby. The race is run on Epsom Downs in England in late May or early June. **b** a similar race elsewhere (Kentucky Derby). **2** (in full **local Derby**) a match between two teams from the same district. **3** any important sporting contest. □ **Derby Day** the day on which the Derby is run.

derby /ˈdɑːbɪ/ n. (pl. **-ies**) N. Amer. a bowler hat. [DERBY³]

Derbyshire /ˈdɑːbɪˌʃɪə(r)/ a county of north central England; county town, Matlock.

derecognize /diːˈrekəgˌnaɪz/ v.tr. (also **-ise**) cease to recognize the status of (esp. a trade union).

deregister /diːˈredʒɪstə(r)/ v.tr. remove from a register. □ **deregistration** /-ˌredʒɪˈstreɪʃ(ə)n/ n.

de règle /də ˈreglə/ predic.adj. customary; proper. [F, = of rule]

deregulate /diːˈregjʊˌleɪt/ v.tr. remove regulations or restrictions from.

deregulation /diːˌregjʊˈleɪʃ(ə)n/ n. the policy or process of removing or reducing controls and restrictions, specifically the removal of government control over the operation of markets in order to increase competition. In Britain deregulation became a major political policy issue in the 1980s; important events include the deregulation of the Stock Exchange (see BIG BANG 2) and the deregulation of bus services.

derelict /ˈderɪlɪkt/ adj. & n. ● adj. **1** abandoned, ownerless (esp. of a ship at sea or an empty decrepit property). **2** (esp. of property) ruined; dilapidated. **3** N. Amer. negligent (of duty etc.). ● n. **1** a social outcast; a person without a home, a job, or property. **2** abandoned property, esp. a ship. [L derelictus past part. of derelinquere (as DE-, relinquere leave)]

dereliction /ˌderɪˈlɪkʃ(ə)n/ n. **1** (usu. foll. by of) **a** neglect; failure to carry out one's obligations (dereliction of duty). **b** an instance of this. **2** the act or an instance of abandoning; the fact or process of being abandoned. **3 a** the retreat of the sea exposing new land. **b** the land so exposed. [L derelictio (as DERELICT)]

derequisition /diːˌrekwɪˈzɪʃ(ə)n/ v.tr. return (requisitioned property) to its former owner.

derestrict /ˌdiːrɪˈstrɪkt/ v.tr. **1** remove restrictions from. **2** remove speed restrictions from (a road, area, etc.). □ **derestriction** /-ˈstrɪkʃ(ə)n/ n.

deride /dɪˈraɪd/ v.tr. laugh scornfully at; ridicule, mock. □ **derider** n. **deridingly** adv. [L deridere (as DE-, ridere ris- laugh)]

de rigueur /də rɪˈgɜː(r)/ predic.adj. required by custom or etiquette (evening dress is de rigueur). [F, = of strictness]

derision /dɪˈrɪʒ(ə)n/ n. ridicule; mockery (bring into derision). □ **hold** (or **have**) **in derision** archaic mock at. □ **derisible** /-ˈrɪzɪb(ə)l/ adj. [ME f. OF f. LL derisio -onis (as DERIDE)]

derisive /dɪˈraɪsɪv/ adj. scoffing; ironical; scornful (derisive cheers). □ **derisively** adv. **derisiveness** n.

derisory /dɪˈraɪsərɪ/ adj. **1** so small or unimportant as to be ridiculous (derisory offer; derisory costs). **2** = DERISIVE. [LL derisorius (as DERISION)]

derivation /ˌderɪˈveɪʃ(ə)n/ n. **1** the act or an instance of deriving or obtaining from a source; the fact or process of being derived. **2 a** the formation of a word from another word or from a root. **b** a derivative. **c** the tracing of the origin of a word. **d** a statement or account of this. **3** extraction, descent. **4** Math. a sequence of statements showing that a formula, theorem, etc., is a consequence of previously accepted statements. □ **derivational** adj. [F dérivation or L derivatio (as DERIVE)]

derivative /dɪˈrɪvətɪv/ *adj. & n.* ● *adj.* derived from another source; not original (*his music is derivative and uninteresting*). ● *n.* **1** something derived from another source, esp.: **a** a word derived from another or from a root (e.g. *quickly* from *quick*). **b** *Chem.* a chemical compound that is derived from another. **2** *Math.* an expression which represents the rate of change of a function with respect to a particular variable. □ **derivatively** *adv.* [F *dérivatif* *-ive* f. L *derivativus* (as DERIVE)]

derive /dɪˈraɪv/ *v.* **1** *tr.* (usu. foll. by *from*) get, obtain, or form (*derived satisfaction from work*). **2** *intr.* (foll. by *from*) arise from, originate in, be descended or obtained from (*happiness derives from many things*). **3** *tr.* gather or deduce (*derived the information from the clues*). **4** *tr.* **a** trace the descent of (a person). **b** show the origin of (a thing). **5** *tr.* (usu. foll. by *from*) show or state the origin or formation of (a word etc.) (*derived the word from Latin*). **6** *tr. Math.* obtain (a function) by differentiation. □ **derivable** *adj.* [ME f. OF *deriver* or f. L *derivare* (as DE-, *rivus* stream)]

derm (also **derma**) var. of DERMIS.

Dermaptera /dɜːˈmæptərə/ *n.pl. Zool.* an order of insects comprising the earwigs. □ **dermapteran** *n. & adj.* **dermapterous** *adj.* [mod.L f. Gk *derma* skin + *pteron* wing]

dermatitis /ˌdɜːməˈtaɪtɪs/ *n.* inflammation of the skin. [Gk *derma* *-atos* skin + -ITIS]

dermatoglyphics /ˌdɜːmətəʊˈglɪfɪks/ *n.* the science or study of skin markings or patterns, esp. of the fingers, hands, and feet. □ **dermatoglyphic** *adj.* **dermatoglyphically** *adv.* [as DERMATITIS + Gk *gluphē* carving: see GLYPH]

dermatology /ˌdɜːməˈtɒlədʒɪ/ *n.* the study of the diagnosis and treatment of skin disorders. □ **dermatologist** *n.* **dermatological** /-tə'lɒdʒɪk(ə)l/ *adj.* [as DERMATITIS + -LOGY]

dermis /ˈdɜːmɪs/ *n.* (also **derm** /dɜːm/ or **derma** /ˈdɜːmə/) **1** (in general use) the skin. **2** *Anat.* the true skin, the thick layer of living connective tissue below the epidermis. □ **dermal** *adj.* **dermic** *adj.* [mod.L, after EPIDERMIS]

dernier cri /ˌdeənɪ ˌeɪ 'kriː/ *n.* the very latest fashion. [F, = last cry]

derogate /ˈderəˌgeɪt/ *v.intr.* (foll. by *from*) *formal* **1 a** take away a part from; detract from (a merit, a right, etc.). **b** disparage. **2** deviate from (correct behaviour etc.). □ **derogative** /dɪˈrɒgətɪv/ *adj.* [L *derogare* (as DE-, *rogare* ask)]

derogation /ˌderəˈgeɪʃ(ə)n/ *n.* **1** (foll. by *of*) a lessening or impairment of (a law, authority, position, dignity, etc.). **2** deterioration; debasement. [ME f. F *dérogation* or L *derogatio* (as DEROGATE)]

derogatory /dɪˈrɒgətərɪ/ *adj.* (often foll. by *to*) involving disparagement or discredit; insulting, depreciatory (*made a derogatory remark*; *derogatory to my position*). □ **derogatorily** *adv.* [LL *derogatorius* (as DEROGATE)]

derrick /ˈderɪk/ *n.* **1** a kind of crane for moving or lifting heavy weights, having a movable pivoted arm. **2** the framework over an oil well or similar excavation, holding the drilling machinery. [obs. senses *hangman*, *gallows*, f. the name of a London hangman *c.*1600]

Derrida /ˈderɪdə/, Jacques (b.1930), French philosopher and critic. His radical critique of traditional Western philosophy and literary analysis led to the emergence of the school of deconstruction in Paris in the late 1960s. His work, which includes *Of Grammatology* (1967) and *Writing and Difference* (1967), rejects earlier structuralist assumptions about the relationship between language and meaning and between text and the objective world, often concentrating on ambiguity and contradiction in meaning. (See also DECONSTRUCTION.) □ **Derridean** /ˌderɪˈdɪən/ *adj. & n.*

derrière /ˌderɪˈeə(r)/ *n. colloq. euphem.* the buttocks. [F, = behind]

derring-do /ˌderɪŋˈduː/ *n. literary or joc.* heroic courage or action. [ME, = *daring to do*, misinterpreted by Spenser and by Scott]

derringer /ˈderɪndʒə(r)/ *n.* a small large-bore pistol. [Henry *Deringer*, Amer. inventor (1786–1868)]

derris /ˈderɪs/ *n.* **1** a tropical leguminous climbing plant of the genus *Derris*, bearing leathery pods. **2** an insecticide made from the powdered root of some kinds of derris. [mod.L f. Gk, = leather covering (with ref. to its pod)]

Derry /ˈderɪ/ *n.* □ see LONDONDERRY.

derry /ˈderɪ/ *n.* □ **have a derry on** *Austral. & NZ colloq.* be prejudiced against (a person). [app. f. the song-refrain *derry down*]

derv /dɜːv/ *n. Brit.* diesel oil for road vehicles. [f. diesel-engined road-vehicle]

dervish /ˈdɜːvɪʃ/ *n.* a Muslim (specifically Sufi) religious man who has taken vows of poverty and austerity. Dervishes first appeared in the 12th century; they were noted for their ecstatic rituals and use of hypnotic trance-states (such as dancing or ritual chanting), and were known as *dancing*, *whirling*, or *howling dervishes* according to the practice of their order. The order of whirling dervishes, founded in Anatolia in the 13th century by the poet and mystic Jalal ad-Din ar-Rumi, was dissolved in 1925 by order of Kemal Atatürk. [Turk. *derviş* f. Pers. *darvēsh* poor, a mendicant]

DES *abbr.* (in the UK) Department of Education and Science.

de Sade, Marquis, see SADE.

desalinate /diːˈsælɪˌneɪt/ *v.tr.* remove salt from (esp. sea water). □ **desalination** /-ˌsælɪˈneɪʃ(ə)n/ *n.*

desalt /diːˈsɔːlt, -ˈsɒlt/ *v.tr.* = DESALINATE.

desaparecido /ˌdezəˌpærəˈsiːdəʊ/ *n.* (*pl.* **-os**) (esp. in South America) a person who has disappeared, presumed killed by members of the armed forces or police. [Sp., lit. 'disappeared']

descale /diːˈskeɪl/ *v.tr.* remove the scale from.

descant *n. & v.* ● *n.* /ˈdeskænt/ **1** *Mus.* an independent treble melody usu. sung or played above a basic melody, esp. of a hymn tune. **2** *poet.* a melody; a song. ● *v.intr.* /dɪsˈkænt/ **1** (foll. by *on*, *upon*) talk lengthily and prosily, esp. in praise of. **2** *Mus.* sing or play a descant. □ **descant recorder** the most common size of recorder, with a range of two octaves. [ME f. OF *deschant* f. med.L *discantus* (as DIS-, *cantus* song, CHANT)]

Descartes /ˈdeɪkɑːt/, René (1596–1650), French philosopher, mathematician, and man of science, often called the father of modern philosophy. Aiming to reach totally secure foundations for knowledge, he began by attacking all his beliefs with sceptical doubts. What was left was the certainty of his own conscious experience, and with it of his existence: '*Cogito, ergo sum*' (I think, therefore I am). From this certainty he argued for the existence of God (as the first cause) and the reality of the physical world, and developed a dualistic theory of mind (conscious experience) and matter. His approach was of fundamental importance in the development of modern philosophy, particularly epistemology. In mathematics he developed the use of coordinates to locate a point in two or three dimensions: this enabled the techniques of algebra and calculus to be used to solve geometrical problems. Descartes suppressed his heretical doctrines of the earth's rotation and the infinity of the universe; fragments of his work in this field were published after his death. From 1628 to 1649 he lived in Holland, then departed for Sweden at the invitation of Queen Christina (1626–89). (See also CARTESIAN.)

descend /dɪˈsend/ *v.* **1** *tr. & intr.* go or come down (a hill, stairs, etc.). **2** *intr.* (of a thing) sink, fall (*rain descended heavily*). **3** *intr.* slope downwards, lie along a descending slope (*fields descended to the beach*). **4** *intr.* (usu. foll. by *on*) **a** make a sudden attack. **b** make an unexpected and usu. unwelcome visit (*hope they don't descend on us at the weekend*). **5** *intr.* (usu. foll. by *from*, *to*) (of property, qualities, rights, etc.) be passed by inheritance (*the house descends from my grandmother*; *the property descended to me*). **6** *intr.* **a** sink in rank, quality, etc. **b** (foll. by *to*) degrade oneself morally to (an unworthy act) (*descend to violence*). **7** *intr. Mus.* (of sound) become lower in pitch. **8** *intr.* (usu. foll. by *to*) proceed (in discourse or writing): **a** in time (to a subsequent event etc.). **b** from the general (to the particular) (*now let's descend to details*). **9** *tr.* go along (a river etc.) to the sea etc. **10** *intr. Printing* (of a letter) have its tail below the line. □ **be descended from** have as an ancestor. □ **descendent** *adj.* [ME f. OF *descendre* f. L *descendere* (as DE-, *scandere* climb)]

descendant /dɪˈsendənt/ *n.* (often foll. by *of*) a person or thing descended from another (*a descendant of Charles I*). [F, part. of *descendre* (as DESCEND)]

descender /dɪˈsendə(r)/ *n. Printing* a part of a letter that extends below the line.

descendible /dɪˈsendɪb(ə)l/ *adj.* **1** (of a slope etc.) that may be descended. **2** *Law* capable of descending by inheritance. [OF *descendable* (as DESCEND)]

descent /dɪˈsent/ *n.* **1 a** the act of descending. **b** an instance of this. **c** a downward movement. **2 a** a way or path etc. by which one may descend. **b** a downward slope. **3 a** being descended; lineage, family origin (*traces his descent from William the Conqueror*). **b** the transmission of qualities, property, privileges, etc., by inheritance. **4 a** a decline; a fall. **b** a lowering (of pitch, temperature, etc.). **5** a sudden violent attack. [ME f. OF *descente* f. *descendre* DESCEND]

descramble /diːˈskræmb(ə)l/ *v.tr.* **1** convert or restore (a signal) to intelligible form. **2** counteract the effects of (a scrambling device). **3** recover an original signal from (a scrambled signal). (Cf. UN-SCRAMBLE.) □ **descrambler** *n.*

describe /dɪˈskraɪb/ *v.tr.* **1 a** state the characteristics, appearance, etc., of, in spoken or written form (*described the landscape*). **b** (foll. by *as*) assert to be; call (*described him as a habitual liar*). **2 a** mark out or draw (esp. a geometrical figure) (*described a triangle*). **b** move in (a specified way, esp. a curve) (*described a parabola through the air*). □ **describable** *adj.* **describer** *n.* [L *describere* (as DE-, *scribere* script- write)]

description /dɪˈskrɪpʃ(ə)n/ *n.* **1 a** the act or an instance of describing; the process of being described. **b** a spoken or written representation (of a person, object, or event). **2** a sort, kind, or class (*no food of any description*). □ **answers** (or **fits**) **the description** has the qualities specified. [ME f. OF f. L *descriptio -onis* (as DESCRIBE)]

descriptive /dɪˈskrɪptɪv/ *adj.* **1** serving or seeking to describe (*a descriptive writer*). **2** describing or classifying without expressing feelings or judging (*a purely descriptive account*). **3** *Linguistics* describing a language without comparing, endorsing, or condemning particular usage, vocabulary, etc. (opp. PRESCRIPTIVE 2). **4** *Gram.* (of an adjective) describing the noun, rather than its relation, position, etc., e.g. *blue* as distinct from *few*. □ **descriptively** *adv.* **descriptiveness** *n.* [LL *descriptivus* (as DESCRIBE)]

descriptor /dɪˈskrɪptə(r)/ *n. Linguistics* a word or expression etc. used to describe or identify. [L, = describer (as DESCRIBE)]

descry /dɪˈskraɪ/ *v.tr.* (**-ies, -ied**) *literary* catch sight of; discern (*descried him in the crowd*; *descries no glimmer of light in her situation*). [ME (earlier senses 'proclaim, DECRY') f. OF *descrier*: prob. confused with var. of obs. *descrive* f. OF *descrivre* DESCRIBE]

desecrate /ˈdesɪˌkreɪt/ *v.tr.* **1** violate (a sacred place or thing) with violence, profanity, etc. **2** deprive (a church, a sacred object, etc.) of sanctity; deconsecrate. □ **desecrator** *n.* **desecration** /ˌdesɪˈkreɪʃ(ə)n/ *n.* [DE- + CONSECRATE]

deseed /diːˈsiːd/ *v.tr.* remove the seeds from (a plant, vegetable, etc.).

desegregate /diːˈsegrɪˌgeɪt/ *v.tr.* abolish racial segregation in (schools etc.) or of (people etc.). □ **desegregation** /-ˌsegrɪˈgeɪʃ(ə)n/ *n.*

deselect /ˌdiːsɪˈlekt/ *v.tr. Polit.* decline to select or retain as a constituency candidate in an election. □ **deselection** /-ˈlekʃ(ə)n/ *n.*

desensitize /diːˈsensɪˌtaɪz/ *v.tr.* (also **-ise**) reduce or destroy the sensitiveness of (photographic materials, an allergic person, etc.). □ **desensitizer** *n.* **desensitization** /-ˌsensɪtaɪˈzeɪʃ(ə)n/ *n.*

desert[1] /dɪˈzɜːt/ *v.* **1** *tr.* abandon, give up, leave (*deserted the sinking ship*). **2** *tr.* forsake or abandon (a cause or a person, people, etc., having claims on one) (*deserted his wife and children*). **3** *tr.* fail (*his presence of mind deserted him*). **4** *intr. Mil.* run away (esp. from military service). **5** *tr.* (as **deserted** *adj.*) empty, abandoned (*a deserted house*). □ **deserter** *n.* (in sense 4 of *v.*). **desertion** /-ˈzɜːʃ(ə)n/ *n.* [F *déserter* f. LL *desertare* f. L *desertus* (as DESERT[2])]

desert[2] /ˈdezət/ *n. & adj.* ● *n.* **1 a** a dry barren often sand-covered area of land, characteristically desolate, waterless, and without vegetation. **2** an uninteresting or barren subject, period, etc. (*a cultural desert*). ● *adj.* **1** uninhabited, desolate. **2** uncultivated, barren. □ **desert boot** a suede etc. boot reaching to or extending just above the ankle. **desert island** a remote (usu. tropical) island presumed to be uninhabited. [ME f. OF f. L *desertus*, eccl.L *desertum* (n.), past part. of *deserere* leave, forsake]

desert[3] /dɪˈzɜːt/ *n.* **1** (in *pl.*) **a** acts or qualities deserving reward or punishment. **b** such reward or punishment (*has got his deserts*). **2** the fact of being worthy of reward or punishment; deservingness. [ME f. OF f. *deservir* DESERVE]

desertification /dɪˌzɜːtɪfɪˈkeɪʃ(ə)n/ *n.* the process of making or becoming a desert.

desert rat *n. Brit. colloq.* a soldier of the 7th British armoured division (with the jerboa as a badge) in the North African desert campaign of 1941–2.

deserve /dɪˈzɜːv/ *v.tr.* **1** (often foll. by *to* + infin.) show conduct or qualities worthy of (reward, punishment, etc.) (*deserves to be imprisoned*; *deserves a prize*). **2** (as **deserved** *adj.*) rightfully merited or earned (*a deserved win*). □ **deserve well** (or **ill**) **of** be worthy of good (or bad) treatment at the hands of (*deserves well of the electorate*). □ **deservedly** /-vɪdlɪ/ *adv.* **deservedness** /-vɪdnɪs/ *n.* [ME f. OF *deservir* f. L *deservire* (as DE-, *servire* serve)]

deserving /dɪˈzɜːvɪŋ/ *adj.* meritorious. □ **deserving of** showing conduct or qualities worthy of (praise, blame, help, etc.). □ **deservingly** *adv.* **deservingness** *n.*

desex /diːˈseks/ *v.tr.* **1** castrate or spay (an animal). **2** deprive of sexual qualities or attraction.

desexualize /diːˈseksjʊəˌlaɪz, diːˈsekʃʊə-/ *v.tr.* (also **-ise**) deprive of sexual character or of the distinctive qualities of a sex.

déshabillé var. of DISHABILLE.

De Sica /dəˈsiːkə/, Vittorio (1901–74), Italian film director and actor. After acting in more than 150 films, in 1940 he turned to directing, becoming a key figure in Italian neo-realist cinema. His celebrated films in this genre include *Shoeshine* (1946) and the Oscar-winning *Bicycle Thieves* (1948). During the 1960s he made a number of successful films starring Sophia Loren, notably *Two Women* (1960), which won an Oscar.

desiccant /ˈdesɪkənt/ *n. Chem.* a hygroscopic substance used as a drying agent.

desiccate /ˈdesɪˌkeɪt/ *v.tr.* remove the moisture from, dry (esp. food for preservation) (*desiccated coconut*). □ **desiccative** /-kətɪv/ *adj.* **desiccation** /ˌdesɪˈkeɪʃ(ə)n/ *n.* [L *desiccare* (as DE-, *siccus* dry)]

desiccator /ˈdesɪˌkeɪtə(r)/ *n.* **1** an apparatus for desiccating. **2** *Chem.* an apparatus containing a drying agent to remove the moisture from specimens.

desiderate /dɪˈzɪdəˌreɪt/ *v.tr. archaic* feel to be missing; regret the absence of; wish to have. [L *desiderare* (as DE-, *siderare* as in CONSIDER)]

desiderative /dɪˈzɪdərətɪv/ *adj. & n.* ● *adj.* **1** *Gram.* (of a verb, conjugation, etc.) formed from another verb etc. and denoting a desire to perform the action of that verb etc. **2** desiring. ● *n. Gram.* a desiderative verb, conjugation, etc. [LL *desiderativus* (as DESIDERATE)]

desideratum /dɪˌzɪdəˈrɑːtəm, dɪˌsɪd-/ *n.* (*pl.* **desiderata** /-tə/) something lacking but needed or desired. [L neut. past part.: see DESIDERATE]

design /dɪˈzaɪn/ *n. & v.* ● *n.* **1 a** a preliminary plan or sketch for the making or production of a building, machine, garment, etc. **b** the art of producing these. **2** a scheme of lines or shapes forming a pattern or decoration. **3** a plan, purpose, or intention. **4 a** the general arrangement or layout of a product, a work of art, printed material, etc. **b** the action or act of planning and creating such an arrangement etc. **c** an established version of a product (*one of our most popular designs*). ● *v.* **1** *tr.* produce a design for (a building, machine, picture, garment, etc.). **2** *tr.* intend, plan, or purpose (*the remark was designed to offend*; *a course designed for beginners*; *designed an attack*). **3** *absol.* be a designer. □ **argument from design** (in Christian theology) the argument that God's existence is provable by the evidence of design in the universe. **by design** on purpose. **have designs on** plan to harm, appropriate, or attract. [F *désigner* appoint or obs. F *desseing* ult. f. L *designare* DESIGNATE]

designate *v. & adj.* ● *v.tr.* /ˈdezɪgˌneɪt/ **1** (often foll. by *as*) appoint to an office or function (*designated him as postmaster general*; *designated his own successor*). **2** specify or particularize (*receives guests at designated times*). **3** (often foll. by *as*) describe as; entitle, style. **4** serve as the name or distinctive mark of (*English uses French words to designate ballet steps*). ● *adj.* /ˈdezɪgnət/ (placed after noun) appointed to an office but not yet installed (*bishop designate*). □ **designator** /-ˌneɪtə(r)/ *n.* [L *designare*, past part. *designatus* (as DE-, *signare* f. *signum* mark)]

designation /ˌdezɪgˈneɪʃ(ə)n/ *n.* **1** a name, description, or title. **2** the act or process of designating. [ME f. OF *designation* or L *designatio* (as DESIGNATE)]

designedly /dɪˈzaɪnɪdlɪ/ *adv.* by design; on purpose.

designer /dɪˈzaɪnə(r)/ *n.* **1** a person who makes artistic designs or plans for construction, e.g. for clothing, machines, theatre sets; a draughtsman. **2** (*attrib.*) (of clothing etc.) bearing the name or label of a famous designer; prestigious.

designer drug *n.* a drug synthesized to imitate a legally restricted or prohibited drug without being subject to such restriction. The term is sometimes also used to describe any fashionable or new drug, even when illegal, such as Ecstasy.

designing /dɪˈzaɪnɪŋ/ *adj.* crafty, artful, or scheming. □ **designingly** *adv.*

desirable /dɪˈzaɪərəb(ə)l/ *adj. & n.* ● *adj.* **1** worth having or wishing for (*it is desirable that nobody should smoke*). **2** arousing sexual desire; very attractive. ● *n.* a desirable person or thing. □ **desirably** *adv.* **desirableness** *n.* **desirability** /-ˌzaɪərəˈbɪlɪtɪ/ *n.* [ME f. OF (as DESIRE)]

desire /dɪˈzaɪə(r)/ *n. & v.* ● *n.* **1 a** an unsatisfied longing or craving. **b** an expression of this; a request (*expressed a desire to rest*). **2** sexual appetite; lust. **3** something desired (*achieved his heart's desire*). ● *v.tr.* **1 a** (often foll. by *to* + infin., or *that* + clause) long for; crave. **b** feel sexual desire for, lust after. **2** request (*desires a cup of tea*). **3** *archaic* pray, entreat, or command (*desire him to wait*). [ME f. OF *desir* f. *desirer* f. L *desiderare* DESIDERATE]

desirous /dɪˈzaɪərəs/ *predic.adj.* **1** (usu. foll. by *of*) ambitious, desiring (*desirous of stardom; desirous of doing well*). **2** (usu. foll. by *to* + infin., or *that* + clause) wishful; hoping (*desirous to do the right thing*). [ME f. AF *desirous*, OF *desireus* f. Rmc (as DESIRE)]

desist /dɪˈzɪst/ *v.intr.* (often foll. by *from*) *literary* abstain; cease (*please desist from interrupting; when requested, he desisted*). [OF *desister* f. L *desistere* (as DE-, *sistere* stop, redupl. f. *stare* stand)]

desk /desk/ *n.* **1** a piece of furniture (sometimes portable) with a flat or sloped surface for writing on, and often drawers. **2** a counter in a hotel, bank, etc., which separates the customer from the assistant. **3** a section of a newspaper office, radio station, etc., dealing with a specified topic (*the sports desk; the features desk*). **4** *Mus.* a music stand in an orchestra regarded as a unit of two players. □ **desk-bound** obliged to remain working at a desk. [ME f. med.L *desca* f. L DISCUS disc]

desktop /ˈdesktɒp/ *n.* **1** the working surface of a desk. **2** (*attrib.*) (esp. of a microcomputer) suitable for use at an ordinary desk.

desktop publishing *n.* (abbr. **DTP**) the production of printed matter by means of a printer (such as a laser printer) linked to a desktop computer, with special software. The system enables reports, advertising matter, company magazines, etc., to be produced cheaply with a layout and print quality similar to that of typeset books, for xerographic or other reproduction.

desman /ˈdesmən/ *n.* (*pl.* **desmans**) either of two aquatic flesh-eating shrewlike mammals, one occurring in Russia (*Desmana moschata*) and one in the Pyrenees (*Galemys pyrenaicus*). [F & G f. Sw. *desman-råtta* muskrat]

desmid /ˈdezmɪd/ *n.* a microscopic unicellular freshwater alga of the family Desmidiaceae. [mod.L *Desmidium* genus name, f. Gk *desmos* band, chain]

Des Moines /dɪ ˈmɔɪn/ the state capital and largest city of Iowa; pop. (1990) 193,200. It was established as a military garrison in 1843.

desolate *adj. & v.* ● *adj.* /ˈdesələt/ **1** left alone; solitary. **2** (of a building or place) uninhabited, ruined, neglected, barren, dreary, empty (*a desolate moor*). **3** forlorn; wretched; miserable (*was left desolate and weeping*). ● *v.tr.* /ˈdesəˌleɪt/ **1** depopulate or devastate; lay waste to. **2** (esp. as **desolated** *adj.*) make wretched or forlorn (*desolated by grief; inconsolable and desolated*). □ **desolately** /-lətlɪ/ *adv.* **desolateness** /-lətnɪs/ *n.* **desolator** /-ˌleɪtə(r)/ *n.* [ME f. L *desolatus* past part. of *desolare* (as DE-, *solare* f. *solus* alone)]

desolation /ˌdesəˈleɪʃ(ə)n/ *n.* **1 a** the act of desolating. **b** the process of being desolated. **2** loneliness, grief, or wretchedness, esp. caused by desertion. **3** a neglected, ruined, barren, or empty state. [ME f. LL *desolatio* (as DESOLATE)]

desorb /dɪˈsɔːb, -ˈzɔːb/ *v. Chem.* **1** *tr.* cause the release of (an adsorbed substance) from a surface. **2** *intr.* (of an adsorbed substance) become released. □ **desorbent** *adj. & n.* **desorption** /-ˈsɔːpʃ(ə)n, -ˈzɔːpʃ(ə)n/ *n.* [DE-, after ADSORB]

despair /dɪˈspeə(r)/ *n. & v.* ● *n.* the complete loss or absence of hope; a feeling of hopelessness. ● *v.intr.* **1** (often foll. by *of*) lose or be without hope (*despaired of ever seeing her again*). **2** (foll. by *of*) lose hope about (*his life is despaired of*). □ **be the despair of** be the cause of despair by badness or unapproachable excellence (*he's the despair of his parents*). □ **despairingly** *adv.* [ME f. OF *desespeir, desperer* f. L *desperare* (as DE-, *sperare* hope)]

despatch var. of DISPATCH.

desperado /ˌdespəˈrɑːdəʊ/ *n.* (*pl.* **-oes** or *US* **-os**) a desperate or reckless person, esp. a criminal. [after DESPERATE (obs. n.) & words in -ADO]

desperate /ˈdespərət/ *adj.* **1** reckless from despair; violent and lawless. **2 a** extremely dangerous or serious (*a desperate situation*). **b** staking all on a small chance (*a desperate remedy*). **3** very bad (*a desperate night; desperate poverty*). **4** (usu. foll. by *for*) needing or desiring very much (*desperate for recognition*). □ **desperately** *adv.* **desperateness** *n.* **desperation** /ˌdespəˈreɪʃ(ə)n/ *n.* [ME f. L *desperatus* past part. of *desperare* (as DE-, *sperare* hope)]

despicable /ˈdespɪkəb(ə)l, dɪˈspɪk-/ *adj.* vile; contemptible, esp. morally. □ **despicably** *adv.* [LL *despicabilis* f. *despicari* (as DE-, *specere* look at)]

de Spinoza, Baruch, see SPINOZA.

despise /dɪˈspaɪz/ *v.tr.* look down on as inferior, worthless, or contemptible. □ **despiser** *n.* [ME f. *despis-* pres. stem of OF *despire* f. L *despicere* (as DE-, *specere* look at)]

despite /dɪˈspaɪt/ *prep. & n.* ● *prep.* in spite of. ● *n.* *archaic* or *literary* **1** outrage, injury. **2** malice, hatred (*died of mere despite*). □ **despite** (or **in despite**) **of** *archaic* in spite of. □ **despiteful** *adj.* [ME f. OF *despit* f. L *despectus* noun f. *despicere* (as DESPISE)]

despoil /dɪˈspɔɪl/ *v.tr.* *literary* (often foll. by *of*) plunder; rob; deprive (*despoiled the roof of its lead*). □ **despoiler** *n.* **despoliation** /-ˌspəʊlɪˈeɪʃ(ə)n/ *n.* [ME f. OF *despoill(i)er* f. L *despoliare* (as DE-, *spoliare* SPOIL)]

despond /dɪˈspɒnd/ *v. & n.* ● *v.intr.* lose heart or hope; be dejected. ● *n.* *archaic* despondency. [L *despondere* give up, abandon (as DE-, *spondere* promise)]

despondent /dɪˈspɒndənt/ *adj.* in low spirits, dejected. □ **despondence** *n.* **despondency** *n.* **despondently** *adv.*

despot /ˈdespɒt/ *n.* **1** an absolute ruler. **2** a tyrant or oppressor. □ **despotic** /deˈspɒtɪk/ *adj.* **despotically** *adv.* [F *despote* f. med.L *despota* f. Gk *despotēs* master, lord]

despotism /ˈdespəˌtɪz(ə)m/ *n.* **1 a** a rule by a despot. **b** a country ruled by a despot. **2** absolute power or control; tyranny.

des Prez /deɪ ˈpreɪ/, Josquin (also **des Prés, Deprez**) (*c.*1440–1521), Flemish composer. Regarded as one of the leading composers of the Renaissance, he wrote eighteen complete masses, 112 motets, and some seventy songs, many of them typical examples of polyphonic song. He is perhaps best known for his Italian song 'El Grillo', with its parody of the chirrup of the cricket.

desquamate /ˈdeskwəˌmeɪt/ *v.intr.* *Med.* (esp. of the skin) come off in scales (as in some diseases). □ **desquamation** /ˌdeskwəˈmeɪʃ(ə)n/ *n.* **desquamative** /ˈdeskwəˈmeɪʃ(ə)n/ *adj.* **desquamatory** *adj.* [L *desquamare* (as DE-, *squama* scale)]

des res /dez ˈrez/ *n.* *colloq.* a desirable residence. [abbr.]

Dessau /ˈdesaʊ/ an industrial city in Germany, on the River Mulde, in Anhalt about 112 km (70 miles) south-west of Berlin; pop. (1991) 95,100.

dessert /dɪˈzɜːt/ *n.* **1** the sweet course of a meal, served at or near the end. **2** *Brit.* a course of fruit, nuts, etc., served after a meal. □ **dessert wine** usu. sweet wine drunk with or following dessert. [F, past part. of *desservir* clear the table (as DIS-, *servir* SERVE)]

dessertspoon /dɪˈzɜːtspuːn/ *n.* **1** a spoon used for dessert, smaller than a tablespoon and larger than a teaspoon. **2** the amount held by this. □ **dessertspoonful** *n.* (*pl.* **-fuls**).

destabilize /diːˈsteɪbɪˌlaɪz/ *v.tr.* (also **-ise**) **1** render unstable. **2** subvert (esp. a foreign government). □ **destabilization** /-ˌsteɪbɪlaɪˈzeɪʃ(ə)n/ *n.*

de Staël /də ˈstɑːl/ Mme (née Anne Louise Germaine Necker) (1766–1817), French novelist and critic. A major precursor of the French romantics, she wrote two semi-autobiographical novels, *Delphine* (1802) and *Corinne* (1807). Her best-known critical work *De l'Allemagne* (1810) introduced late 18th-century German writers and thinkers to France; it was banned on publication by Napoleon.

De Stijl /də ˈstaɪl/ *n.* a 20th-century Dutch art movement which took its name from the periodical *De Stijl* ('Style') founded in 1917 by Theo van Doesburg (1883–1931) and Piet Mondrian. The adherents of the movement were concerned to find principles of equilibrium and harmony that would be applicable to life and society as well as art; the abstract, economical style is typified by the geometric simplicity of Mondrian's paintings and is closely related to neo-plasticism. Architects in the group included J. P. Oud (1890–1963) and Gerrit Rietveld (1888–1965). In the 1920s and 1930s De Stijl was influential on the Bauhaus and constructivist movements and on Purism.

destination /ˌdestɪˈneɪʃ(ə)n/ *n.* a place to which a person or thing is going. [OF *destination* or L *destinatio* (as DESTINE)]

destine /ˈdestɪn/ *v.tr.* (often foll. by *to, for,* or *to* + infin.) set apart; appoint; preordain; intend (*destined him for the navy*). □ **be destined to** be fated or preordained to (*was destined to become a great man*). [ME f. F *destiner* f. L *destinare* (as DE-, *stanare* (unrecorded) settle f. *stare* stand)]

destiny /ˈdestɪnɪ/ *n.* (*pl.* **-ies**) **1 a** the predetermined course of events; fate. **b** this regarded as a power. **2** what is destined to happen to a particular person etc. (*it was their destiny to be rejected*). [ME f. OF *destinée* f. Rmc, past part. of *destinare*: see DESTINE]

destitute /ˈdestɪˌtjuːt/ *adj.* **1** without food, shelter, etc.; completely impoverished. **2** (usu. foll. by *of*) lacking (*destitute of friends*). □ **destitution** /ˌdestɪˈtjuːʃ(ə)n/ *n.* [ME f. L *destitutus* past part. of *destituere* forsake (as DE-, *statuere* place)]

destrier /ˈdestrɪə(r)/ *n.* *hist.* a war-horse. [ME f. AF *destrer*, OF *destrier* ult.

f. L DEXTER[1] right (as the knight's horse was led by the squire with the right hand)]

destroy /dɪˈstrɔɪ/ v.tr. **1** pull or break down; demolish (*destroyed the bridge*). **2** put an end to, do away with (*the accident destroyed her confidence*). **3 a** wipe out, kill, utterly defeat (*destroyed the enemy*). **b** humanely kill (esp. a sick or savage animal). **4** make useless; spoil utterly. **5** ruin financially, professionally,. or in reputation. □ **destroying angel** a poisonous all-white toadstool, *Amanita virosa*. [ME f. OF *destruire* ult. f. L *destruere* (as DE-, *struere struct-* build)]

destroyer /dɪˈstrɔɪə(r)/ n. a small fast warship heavily armed with guns, torpedoes, and depth charges. The type was developed in the late 19th century.

destruct /dɪˈstrʌkt/ v. & n. ● v. **1** tr. destroy (esp. one's own rocket) deliberately, esp. for safety reasons. **2** intr. be destroyed in this way. ● n. an act of destructing a rocket etc. [L *destruere* (as DESTROY) or as back-form. f. DESTRUCTION]

destructible /dɪˈstrʌktɪb(ə)l/ adj. able to be destroyed. □ **destructibility** /-ˌstrʌktɪˈbɪlɪtɪ/ n. [F *destructible* or LL *destructibilis* (as DESTROY)]

destruction /dɪˈstrʌkʃ(ə)n/ n. **1** the act or an instance of destroying; the process of being destroyed. **2** a cause of ruin; something that destroys (*greed was their destruction*). [ME f. OF f. L *destructio -onis* (as DESTROY)]

destructive /dɪˈstrʌktɪv/ adj. **1** (often foll. by *to, of*) destroying or tending to destroy (*destructive of her peace of mind; is destructive to organisms; a destructive child*). **2** negative in attitude or criticism; refuting without suggesting, helping, amending, etc. (opp. CONSTRUCTIVE) (*has only destructive criticism to offer*). □ **destructively** adv. **destructiveness** n. [ME f. OF *destructif -ive* f. LL *destructivus* (as DESTROY)]

destructor /dɪˈstrʌktə(r)/ n. Brit. a refuse-burning furnace.

desuetude /dɪˈsjuːɪˌtjuːd, ˈdeswɪ-/ n. a state of disuse (*the custom fell into desuetude*). [F *désuétude* or L *desuetudo* (as DE-, *suescere suet-* be accustomed)]

desulphurize /diːˈsʌlfjʊˌraɪz/ v.tr. (also **-ise**, US **desulfurize**) remove sulphur or sulphur compounds from. □ **desulphurization** /-ˌsʌlfjʊraɪˈzeɪʃ(ə)n/ n.

desultory /ˈdesəltərɪ, ˈdezəl-/ adj. **1** going constantly from one subject to another, esp. in a half-hearted way. **2** disconnected; unmethodical; superficial. □ **desultorily** adv. **desultoriness** n. [L *desultorius* superficial f. *desultor* vaulter f. *desult-* (as DE-, *salt-* past part. stem of *salire* leap)]

detach /dɪˈtætʃ/ v.tr. **1** (often foll. by *from*) unfasten or disengage and remove (*detached the buttons; detached himself from the group*). **2** Mil. send (a ship, regiment, officer, messenger, etc.) on a separate mission. **3** (as **detached** adj.) **a** impartial; unemotional (*a detached viewpoint*). **b** (esp. of a house) not joined to another or others; separate. □ **detached retina** a retina that has become detached from the inner surface of the eye, causing partial or total blindness. □ **detachable** adj. **detachedly** /-tʃɪdlɪ/ adv. [F *détacher* (as DE-, ATTACH)]

detachment /dɪˈtætʃmənt/ n. **1 a** a state of aloofness from or indifference to other people, one's surroundings, public opinion, etc. **b** disinterested independence of judgement. **2 a** the act or process of detaching or being detached. **b** an instance of this. **3** Mil. a separate group or unit of an army etc. used for a specific purpose. [F *détachement* (as DETACH)]

detail /ˈdiːteɪl/ n. & v. ● n. **1 a** a small or subordinate particular; an item. **b** such a particular, considered (ironically) to be unimportant (*the truth of the statement is just a detail*). **2 a** small items or particulars (esp. in an artistic work) regarded collectively (*has an eye for detail*). **b** the treatment of them (*the detail was insufficient and unconvincing*). **3** (often in pl.) a number of particulars; an aggregate of small items (*filled in the details on the form*). **4 a** a minor decoration on a building, in a picture, etc. **b** a small part of a picture etc. shown alone. **5** Mil. **a** the distribution of orders for the day. **b** a small detachment of soldiers etc. for special duty. ● v.tr. **1** give particulars of (*detailed the plans*). **2** relate circumstantially (*detailed the anecdote*). **3** Mil. assign for special duty. **4** (as **detailed** adj.) (of a picture, story, etc.) having many details. **b** itemized (*a detailed list*). □ **go into detail** give all the items or particulars. **in detail** item by item, minutely. [F *détail*, *détailler* (as DE-, *tailler* cut, formed as TAIL[2])]

detain /dɪˈteɪn/ v.tr. **1** keep in confinement or under restraint. **2** keep waiting; delay. □ **detainment** n. [ME f. OF *detenir* ult. f. L *detinere detent-* (as DE-, *tenere* hold)]

detainee /ˌdiːteɪˈniː/ n. a person detained in custody, esp. for political reasons.

detainer /dɪˈteɪnə(r)/ n. Law **1** the wrongful detaining of goods taken from the owner for distraint etc. **2** the detention of a person in prison etc. [AF *detener* f. OF *detenir* (as DETAIN)]

detect /dɪˈtekt/ v.tr. **1 a** (often foll. by *in*) reveal the guilt of; discover (*detected him in his crime*). **b** discover (a crime). **2** discover or perceive the existence or presence of (*detected a smell of burning; do I detect a note of sarcasm?*). **3** Physics use an instrument to observe (a signal, radiation, etc.). □ **detectable** adj. **detectably** adv. [L *detegere detect-* (as DE-, *tegere* cover)]

detection /dɪˈtekʃ(ə)n/ n. **1 a** the act or an instance of detecting; the process of being detected. **b** an instance of this. **2** the work of a detective. **3** Physics the extraction of a desired signal; a demodulation. [LL *detectio* (as DETECT)]

detective /dɪˈtektɪv/ n. & adj. ● n. (often *attrib.*) a person, esp. a member of a police force, employed to investigate crimes. ● adj. serving to detect. □ **private detective** a usu. freelance detective carrying out investigations for a private employer. [DETECT]

detective story n. (also **detective novel**) a story etc. describing a crime (esp. a murder) and the detection of criminals. The first detective stories in English literature are generally reckoned to be Poe's 'The Murders in the Rue Morgue' (1841) and Wilkie Collins's *The Moonstone* (1868). Sir Arthur Conan Doyle's Sherlock Holmes stories achieved worldwide popularity in the late 19th century with the ideal of the gentleman detective; in the early 20th century British writers such as G. K. Chesterton, Agatha Christie, and Dorothy Sayers built on this tradition, emphasizing the triumph of law and order, often in genteel or aristocratic settings. This so-called golden age was challenged by the American hard-boiled detective writers such as Dashiell Hammett and Raymond Chandler from the 1930s onwards; their tough, realistic style reflects a world permeated by corruption and deceit. In more recent detective fiction, both classic and hard-boiled styles are well represented, testifying to the genre's popularity as well as its versatility.

detector /dɪˈtektə(r)/ n. **1** a person or thing that detects. **2** Physics a device for the detection or demodulation of signals.

detent /dɪˈtent/ n. **1** a catch by the removal of which machinery is allowed to move. **2** (in a clock etc.) a catch that regulates striking. [F *détente* f. OF *destente* f. *destendre* slacken (as DE-, L *tendere*)]

détente /deɪˈtɒnt/ n. an easing of strained relations esp. between states. [F, = relaxation]

detention /dɪˈtenʃ(ə)n/ n. **1** detaining or being detained. **2 a** being kept in school after hours as a punishment. **b** an instance of this. **3** custody; confinement. □ **detention centre** Brit. an institution in which young offenders are detained for short periods. [F *détention* or LL *detentio* (as DETAIN)]

deter /dɪˈtɜː(r)/ v.tr. (**deterred**, **deterring**) **1** (often foll. by *from*) discourage or prevent (a person) through fear or dislike of the consequences. **2** discourage, check, or prevent (a thing, process, etc.). [L *deterrere* (as DE-, *terrere* frighten)]

detergent /dɪˈtɜːdʒ(ə)nt/ n. & adj. ● n. **1** a usu. synthetic water-soluble cleansing agent which combines with impurities and dirt to make them more soluble or dispersable in water, and differs from soap in not forming a scum with the salts present in hard water. **2** any additive with a similar action, e.g. a substance which holds dirt in suspension in a lubricating oil. ● adj. cleansing, esp. in the manner of a detergent. [L *detergere* (as DE-, *tergere ters-* wipe)]

deteriorate /dɪˈtɪərɪəˌreɪt/ v.tr. & intr. make or become bad or worse (*food deteriorates in hot weather; his condition deteriorated after the operation*). □ **deteriorative** /-rətɪv/ adj. **deterioration** /-ˌtɪərɪəˈreɪʃ(ə)n/ n. [LL *deteriorare deteriorat-* f. L *deterior* worse]

determinant /dɪˈtɜːmɪnənt/ adj. & n. ● adj. serving to determine or define. ● n. **1** a determining factor, element, word, etc. **2** Math. a quantity obtained by the addition of products of the elements of a square matrix according to a given rule. [L *determinare* (as DETERMINE)]

determinate /dɪˈtɜːmɪnət/ adj. **1** limited in time, space, or character. **2** of definite scope or nature. □ **determinacy** n. **determinately** adv. **determinateness** n. [ME f. L *determinatus* past part. (as DETERMINE)]

determination /dɪˌtɜːmɪˈneɪʃ(ə)n/ n. **1** firmness of purpose; resoluteness. **2** the process of deciding, determining, or calculating. **3 a** the conclusion of a dispute by the decision of an arbitrator. **b** the decision reached. **4** Law the cessation of an estate or interest. **5** Law a judicial decision or sentence. **6** *archaic* a tendency to move in a fixed direction. [ME (in sense 4) f. OF f. L *determinatio -onis* (as DETERMINE)]

determinative /dɪ'tɜːmɪnətɪv/ *adj. & n.* ● *adj.* serving to define, qualify, or direct. ● *n.* a determinative thing or circumstance. □ **determinatively** *adv.* [F *déterminatif -ive* (as DETERMINE)]

determine /dɪ'tɜːmɪn/ *v.* **1** *tr.* find out or establish precisely (*have to determine the extent of the problem*). **2** *tr.* decide or settle (*determined who should go*). **3** *tr.* be a decisive factor in regard to (*demand determines supply*). **4** *intr. & tr.* make or cause (a person) to make a decision (*we determined to go at once; what determined you to do it?*). **5** *tr. & intr. esp.* *Law* bring or come to an end. **6** *tr.* *Geom.* fix or define the position of. □ **be determined** be resolved (*was determined not to give up*). □ **determinable** *adj.* [ME f. OF *determiner* f. L *determinare* (as DE-, *terminus* end)]

determined /dɪ'tɜːmɪnd/ *adj.* **1** showing determination; resolute, unflinching. **2** fixed in scope or character; settled; determinate. □ **determinedly** *adv.* **determinedness** *n.*

determiner /dɪ'tɜːmɪnə(r)/ *n.* **1** a person or thing that determines. **2** *Gram.* any of a class of words (e.g. *a, the, every*) that determine the kind of reference a noun or noun-substitute has.

determinism /dɪ'tɜːmɪˌnɪz(ə)m/ *n.* the philosophical doctrine that all events are determined by causes external to the will. Logical determinism could therefore be taken as implying that individual human beings have no free will and cannot be held morally responsible for their actions, but philosophers such as Locke and Hume argued that a determinist view did not exclude the possibility of free will. □ **determinist** *n. & adj.* **deterministic** /-ˌtɜːmɪ'nɪstɪk/ *adj.* **deterministically** *adv.*

deterrent /dɪ'terənt/ *adj. & n.* ● *adj.* that deters. ● *n.* a deterrent thing or factor, esp. a nuclear weapon regarded as deterring an enemy from attack. □ **deterrence** *n.*

detest /dɪ'test/ *v.tr.* hate, loathe. □ **detester** *n.* [L *detestari* (as DE-, *testari* call to witness f. *testis* witness)]

detestable /dɪ'testəb(ə)l/ *adj.* intensely disliked; very unpleasant, hateful. □ **detestably** *adv.*

detestation /ˌdiːte'steɪʃ(ə)n/ *n.* **1** intense dislike, hatred. **2** a detested person or thing. [ME f. OF f. L *detestatio -onis* (as DETEST)]

dethrone /diː'θrəʊn/ *v.tr.* **1** remove from the throne, depose. **2** remove from a position of authority or influence. □ **dethronement** *n.*

detonate /'detəˌneɪt/ *v.* **1** *tr.* set off (an explosive charge). **2** *intr.* (of an explosive charge) be set off, explode esp. loudly. □ **detonative** *adj.* [L *detonare detonat-* (as DE-, *tonare* thunder)]

detonation /ˌdetə'neɪʃ(ə)n/ *n.* **1 a** the act or process of detonating. **b** a loud explosion. **2** the premature combustion of fuel in an internal-combustion engine, causing it to pink. [F *détonation* f. *détoner* (as DETONATE)]

detonator /'detəˌneɪtə(r)/ *n.* **1** a device or a small sensitive charge used to detonate an explosive. **2** a warning signal on a railway line which is set off when a train passes over it.

detour /'diːtʊə(r)/ *n. & v.* ● *n.* a divergence from a direct or intended route; a roundabout course. ● *v.intr. & tr.* make or cause to make a detour. [F *détour* change of direction f. *détourner* turn away (as DE-, TURN)]

detoxicate /diː'tɒksɪˌkeɪt/ *v.tr.* = DETOXIFY. □ **detoxication** /-ˌtɒksɪ'keɪʃ(ə)n/ *n.* [DE- + L *toxicum* poison, after *intoxicate*]

detoxify /diː'tɒksɪˌfaɪ/ *v.tr.* (**-ies, -ied**) remove the poison from. □ **detoxification** /-ˌtɒksɪfɪ'keɪʃ(ə)n/ *n.* [DE- + L *toxicum* poison]

detract /dɪ'trækt/ *v.* **1** *intr.* (foll. by *from*) **a** reduce. **b** diminish, belittle. **2** *tr.* (usu. foll. by *from*) take away (a part of something). □ **detractive** *adj.* **detractor** *n.* **detraction** /-'trækʃ(ə)n/ *n.* [L *detrahere detract-* (as DE-, *trahere* draw)]

detrain /diː'treɪn/ *v.intr. & tr.* alight or cause to alight from a train. □ **detrainment** *n.*

detribalize /diː'traɪbəˌlaɪz/ *v.tr.* (also **-ise**) **1** make (a person) no longer a member of a tribe. **2** destroy the tribal habits of. □ **detribalization** /-ˌtraɪbəlaɪ'zeɪʃ(ə)n/ *n.*

detriment /'detrɪmənt/ *n.* **1** harm, damage. **2** something causing this. [ME f. OF *détriment* or L *detrimentum* (as DE-, *terere trit-* rub, wear)]

detrimental /ˌdetrɪ'ment(ə)l/ *adj.* harmful; causing loss. □ **detrimentally** *adv.*

detrition /dɪ'trɪʃ(ə)n/ *n.* wearing away by friction. [med.L *detritio* (as DETRIMENT)]

detritus /dɪ'traɪtəs/ *n.* **1** matter produced by erosion, such as gravel, sand, silt, rock-debris, etc. **2** debris of any kind; rubbish, waste.

detrital /-'traɪt(ə)l/ *adj.* [after F *détritus* f. L *detritus* (n.) = wearing down (as DETRIMENT)]

Detroit /dɪ'trɔɪt/ a major industrial city and Great Lakes shipping centre in NE Michigan; pop. (1990) 1,028,000. It is the centre of the US automobile industry, containing the headquarters of Ford, Chrysler, and General Motors — whence its nickname 'Motown' (short for 'motor town'). In the 1960s it was also an important centre for rock and soul music.

de trop /də 'trəʊ/ *predic.adj.* not wanted, unwelcome, in the way. [F, = excessive]

de Troyes, Chrétien, see CHRÉTIEN DE TROYES.

detumescence /ˌdiːtjuː'mes(ə)ns/ *n.* subsidence from a swollen state. [L *detumescere* (as DE-, *tumescere* swell)]

Deucalion /djuː'keɪlɪən/ *Gk Mythol.* son of Prometheus, and the Greek equivalent of Noah. When Zeus flooded the earth in wrath at the impiety of humankind, Deucalion and his wife Pyrrha took refuge on the top of Mount Parnassus (in some stories they built an ark in which they were carried there). After the flood had subsided, to repopulate the world they were advised to throw stones over their shoulders; those thrown by Deucalion became men, and those thrown by Pyrrha women.

deuce[1] /djuːs/ *n.* **1** the two on dice or playing cards. **2** *Tennis* the score of 40 all, at which two consecutive points are needed to win. [OF *deus* f. L *duo* (accus. *duos*) two]

deuce[2] /djuːs/ *n.* misfortune, the Devil, used as an exclamation of surprise or annoyance (*who the deuce are you?*). □ **a** (or **the**) **deuce of a** a very bad or remarkable (*a deuce of a problem; a deuce of a fellow*). **the deuce to pay** trouble to be expected. [LG *duus*, formed as DEUCE[1], two aces at dice being the worst throw]

deuced /'djuːsɪd, djuːst/ *adj. & adv. archaic* damned, confounded (*a deuced liar*). □ **deucedly** /-sɪdlɪ/ *adv.*

deus ex machina /ˌdeɪʊs eks 'mækɪnə, ˌdiːəs/ *n.* an unexpected power or event saving a seemingly hopeless situation, esp. in a play or novel. [mod.L transl. Gk *theos ek mēkhanēs*, = god from the machinery (by which in the Greek theatre the gods were suspended above the stage)]

Deut. *abbr.* (in the Bible) Deuteronomy.

deuteragonist /ˌdjuːtə'rægənɪst/ *n.* the person second in importance to the protagonist in a drama. [Gk *deuteragōnistēs* (as DEUTERO-, *agōnistēs* actor)]

deuterate /'djuːtəˌreɪt/ *v.tr.* replace the usual isotope of hydrogen in (a substance) by deuterium. □ **deuteration** /ˌdjuːtə'reɪʃ(ə)n/ *n.*

deuterium /djuː'tɪərɪəm/ *n. Chem.* a stable isotope of hydrogen with a mass of about 2 (symbol: **D**); also called *heavy hydrogen*. Discovered in 1932 by Harold Urey, deuterium differs from the normal isotope (protium) in having a neutron as well as a proton in the nucleus. It is present to about 1 part in 6,000 in naturally occurring hydrogen, and is used as a fuel in thermonuclear bombs. Heavy water, D_2O, is used as a moderator in nuclear reactors. [f. DEUTERO-]

deutero- /'djuːtərəʊ/ *comb. form* second. [Gk *deuteros* second]

Deutero-Isaiah /ˌdjuːtərəʊaɪ'zaɪə/ the supposed later author of Isaiah 40–55, who wrote in the later years of the Babylonian Captivity (586–539 BC).

deuteron /'djuːtəˌrɒn/ *n. Physics* the nucleus of a deuterium atom, consisting of a proton and a neutron. [DEUTERIUM + -ON]

Deuteronomy /ˌdjuːtə'rɒnəmɪ/ *n.* the fifth book of the Bible, written in the form of a farewell address by Moses to the Israelites before entering Canaan and containing a repetition of the Ten Commandments and most of the laws in Exod. 21–4. [LL *Deuteronomium* f. Gk (*deuteros nomos* second law), from a mistranslation of Hebrew words (Deut. 17:18) meaning 'a copy or duplicate of this law']

Deutschland /'dɔɪtʃlant/ the German name for GERMANY.

Deutschmark /'dɔɪtʃmɑːk/ *n.* (also **Deutsche Mark** /'dɔɪtʃə ˌmɑːk/) the chief monetary unit of Germany. [G, = German mark (see MARK[2])]

deutzia /'djuːtsɪə, 'dɔɪt-/ *n.* an ornamental shrub of the genus *Deutzia*, with usu. white flowers. [Johann van der *Deutz* 18th-c. Dutch patron of botany]

deva /'deɪvə/ *n.* a member of a class of divine beings in the Vedic period, which in Indian mythology are benevolent (opposed to the asuras) and in Zoroastrianism are evil. (Cf. ASURA.) [Skr., = god]

de Valera /də və'leərə/, Eamon (1882–1975), American-born Irish statesman. A fervent Irish nationalist, de Valera was one of the leaders of the Easter Rising in 1916 and was sentenced to death by the British, but was released a year later. He served as leader of Sinn Fein (1917–26) and President of the Irish provisional government (1919–22), and as an opponent of the Anglo-Irish Treaty headed the militant republicans in the ensuing civil war. In 1926 he founded the Fianna Fáil Party, which he led in the Dáil. In 1932 de Valera became President of the Irish Free State, and was largely responsible for the new constitution of 1937, which created the sovereign state of Eire. He served as Taoiseach 1937–48, 1951–4, and 1957–9, and President 1959–73.

de Valois /də ˈvælwɑː/, Dame Ninette (born Edris Stannus) (b.1898), Irish choreographer, ballet-dancer, and teacher. A former soloist with Diaghilev's Ballets Russes, she turned to choreography in the 1920s. The success of her ballet *Job* in 1931 led to the formation of the Vic-Wells Ballet and the Sadler's Wells ballet school. De Valois was director of the company, which eventually became the Royal Ballet, from 1931 until 1963. Among the ballets she choreographed for the company were *The Rake's Progress* (1935) and *Checkmate* (1937).

devalue /diːˈvæljuː/ *v.tr.* (**devalues, devalued, devaluing**) **1** reduce the value of. **2** *Econ.* reduce the value of (a currency) in relation to other currencies or to gold (opp. REVALUE). □ **devaluation** /-ˌvæljuːˈeɪʃ(ə)n/ *n.*

Devanagari /ˌdeɪvəˈnɑːgərɪ/ *n.* the alphabet used for Sanskrit, Hindi, and other Indian languages. [Skr., = divine town script]

devastate /ˈdevəˌsteɪt/ *v.tr.* **1** lay waste; cause great destruction to. **2** (often in *passive*) overwhelm with shock or grief; upset deeply. □ **devastator** *n.* **devastation** /ˌdevəˈsteɪʃ(ə)n/ *n.* [L *devastare devastat-* (as DE-, *vastare* lay waste)]

devastating /ˈdevəˌsteɪtɪŋ/ *adj.* **1** crushingly effective; overwhelming. **2** *colloq.* **a** incisive, savage (*devastating accuracy, devastating wit*). **b** extremely impressive or attractive (*she wore a devastating black silk dress*). □ **devastatingly** *adv.*

develop /dɪˈveləp/ *v.* (**developed, developing**) **1** *tr.* & *intr.* **a** make or become bigger or fuller or more elaborate or systematic (*the new town developed rapidly*). **b** bring or come to an active or visible state or to maturity (*developed a plan of action*). **2 a** *tr.* begin to exhibit or suffer from (*developed a rattle*). **b** *intr.* come into existence; originate, emerge (*a fault developed in the engine*). **3** *tr.* **a** construct new buildings on (land). **b** convert (land) to a new purpose so as to use its resources more fully. **4** *tr.* treat (photographic film etc.) to make the latent image visible. **5** *tr. Mus.* elaborate (a theme) by modification of the melody, harmony, rhythm, etc. **6** *tr. Chess* bring (a piece) into position for effective use. □ **developer** *n.* [F *développer* f. Rmc (as DIS-, orig. of second element unknown)]

developable /dɪˈveləpəb(ə)l/ *adj.* that can be developed. □ **developable surface** *Geom.* a surface that can be flattened into a plane without overlap or separation, e.g. a cylinder.

developing country *n.* (also called *less developed country*) any of the poorer countries of the world, typically those that have not developed an industrial base. It is an imprecise term, applied to many countries in Africa and Latin America and to much of Asia; in 1988 the World Bank defined a developing country as one having a GNP per capita of less than US$6,000. Developing countries generally have a colonial past and now lag behind richer countries in terms of economic development, technology, infrastructure, and social and political institutions. However, the term is sometimes criticized as inadequate in that it classes together newly industrializing countries like South Korea and Singapore, which are becoming richer, with states such as Somalia and Bangladesh, which have stagnant or even falling incomes. Developing countries are sometimes known collectively as the Third World.

development /dɪˈveləpmənt/ *n.* **1** the act or an instance of developing; the process of being developed. **2 a** a stage of growth or advancement. **b** a thing that has developed, esp. an event or circumstance (*the latest developments*). **3** a full-grown state. **4** the process of developing a photograph. **5** a developed area of land. **6** *Mus.* the elaboration of a theme or themes, esp. in the middle section of a sonata movement. **7** industrialization or economic advancement of a country or area. **8** *Chess* the developing of pieces from their original position. □ **development area** *Brit.* one where new industries are encouraged in order to counteract unemployment.

developmental /dɪˌveləpˈment(ə)l/ *adj.* **1** incidental to growth (*developmental diseases*). **2** evolutionary. □ **developmentally** *adv.*

Devi /ˈdeɪvɪ/ *Hinduism* the supreme goddess, often identified with Parvati and Sakti. [Skr., = goddess]

deviant /ˈdiːvɪənt/ *adj.* & *n.* ● *adj.* that deviates from the normal, esp. in social or sexual behaviour. ● *n.* a deviant person or thing. □ **deviance** *n.* **deviancy** *n.* [ME (as DEVIATE)]

deviate *v., n.,* & *adj.* ● *v.intr.* /ˈdiːvɪˌeɪt/ (often foll. by *from*) turn aside or diverge (from a course of action, rule, truth, etc.); digress. ● *n.* /ˈdiːvɪət/ = DEVIANT *n.* ● *adj.* = DEVIANT *adj.* □ **deviator** /-vɪˌeɪtə(r)/ *n.* **deviatory** /-vɪətərɪ/ *adj.* [LL *deviare deviat-* (as DE-, *via* way)]

deviation /ˌdiːvɪˈeɪʃ(ə)n/ *n.* **1 a** deviating, digressing. **b** an instance of this. **2** *Polit.* a departure from accepted (esp. Communist) party doctrine. **3** *Statistics* the amount by which a single measurement differs from the mean. **4** *Naut.* the deflection of a ship's compass-needle caused by iron in the ship etc. □ **standard deviation** *Statistics* a quantity calculated to indicate the extent of deviation for a group as a whole. □ **deviational** *adj.* **deviationism** *n.* **deviationist** *n.* [F *déviation* f. med.L *deviatio -onis* (as DEVIATE)]

device /dɪˈvaɪs/ *n.* **1 a** a thing made or adapted for a particular purpose, esp. a mechanical contrivance. **b** an explosive contrivance, a bomb. **2** a plan, scheme, or trick. **3 a** an emblematic or heraldic design. **b** a drawing or design. **4** *archaic* make, look (*things of rare device*). □ **leave a person to his** or **her own devices** leave a person to do as he or she wishes. [ME f. OF *devis* ult. f. L (as DIVIDE)]

devil /ˈdev(ə)l/ *n.* & *v.* ● *n.* **1** (usu. **the Devil**) (in Christian and Jewish belief) the supreme spirit of evil; Satan. (*See note below.*) **2 a** an evil spirit; a demon; a superhuman malignant being. **b** a personified evil force or attribute. **3 a** a wicked or cruel person. **b** a mischievously energetic, clever, or self-willed person. **4** *colloq.* a person, a fellow (*lucky devil*). **5** fighting spirit, mischievousness (*the devil is in him tonight*). **6** *colloq.* something difficult or awkward (*this door is a devil to open*). **7** (**the devil** or **the Devil**) *colloq.* used as an exclamation of surprise or annoyance (*who the devil are you?*). **8** a literary hack exploited by an employer. **9** *Brit.* a junior legal counsel. **10** = *Tasmanian devil*. **11** an instrument or machine, used for cutting, tearing, or other destructive work. **12** *S. Afr.* = *dust devil*. ● *v.* (**devilled, devilling**; *US* **deviled, deviling**) **1** *tr.* cook (food) with hot seasoning. **2** *intr.* act as a devil for an author or barrister. **3** *tr. US* harass, worry. □ **between the devil and the deep blue sea** in a dilemma. **devil-may-care** cheerful and reckless. **a devil of a** … *colloq.* a considerable, difficult, or remarkable …. **devil a one** not even one. **devil on two sticks** see DIABOLO. **devil ray** = MANTA. **devil's advocate** a person who argues against a proposition to test it or provoke discussion. **2** *RC Ch. hist.* a person appointed to put the case against the beatification or canonization of someone. **devil's bit** a plant whose root looks bitten off; esp. a kind of scabious, *Succisa pratensis*. **devil's coach-horse** *Brit.* a large rove-beetle, *Staphylinus olens*. **devil's darning-needle** a dragonfly or damselfly. **devil's dozen** thirteen. **devils-on-horseback** a savoury of prunes or plums wrapped in slices of bacon. **the devil's own** … an exceptional or extreme instance of …. **devil take the hindmost** a motto of selfish competition. **the devil to pay** trouble to be expected. **go to the devil 1** be damned. **2** (in *imper.*) depart at once. **like the devil** with great energy. **play the devil with** cause severe damage to. **printer's devil** *hist.* an errand-boy in a printing office. **speak** (or **talk**) **of the devil** said when a person appears just after being mentioned. **the very devil** (*predic.*) *colloq.* a great difficulty or nuisance. [OE *dēofol* f. LL *diabolus* f. Gk *diabolos* accuser, slanderer f. *dia* across + *ballō* to throw]

▪ The Devil (also called *Satan* or *Lucifer*) is the supreme spirit of evil in Jewish and Christian theology, enemy of God and tempter of humankind. In theological tradition he was regarded as the chief of the fallen angels, cast out of heaven for rebellion against God, but there was no fixed teaching on the exact nature of his sin; he was held to preside over those condemned to eternal fire (see HELL). In the narrative of the Fall, the serpent which tempted Eve has traditionally been regarded as his embodiment, and in Rev. 12:7–9 the Devil is identified with the dragon cast out by Michael and his angels. Popularly, the Devil is often represented as a man with horns, cloven hooves, and a forked tail.

devilfish /ˈdev(ə)lˌfɪʃ/ *n.* **1** a manta ray. **2** any other fish of sinister appearance or bad reputation, esp. the stonefish. **3** *hist.* an octopus.

devilish /ˈdevəlɪʃ/ *adj.* & *adv.* ● *adj.* **1** of or like a devil; evil, wicked. **2** mischievous. ● *adv. colloq.* very, extremely. □ **devilishly** *adv.* **devilishness** *n.*

devilment /ˈdev(ə)lmənt/ *n.* mischief, wild spirits.

devilry /ˈdev(ə)lrɪ/ *n.* (also **deviltry** /-trɪ/) (*pl.* **-ies**) **1 a** wickedness; reckless mischief. **b** an instance of this. **2 a** black magic. **b** the Devil and his works. [OF *diablerie*: *-try* wrongly after *harlotry* etc.]

Devil's Island a rocky island off the coast of French Guiana. From

1852 it was part of a penal settlement, originally for prisoners suffering from contagious diseases, especially leprosy; later it was used largely for political prisoners, of whom the most famous was Alfred Dreyfus, and became notorious for its harsh conditions. No prisoners were sent there after 1938, and the last one was released in 1953. The island is now chiefly a tourist attraction.

devious /ˈdiːvɪəs/ adj. **1** (of a person etc.) not straightforward, underhand. **2** winding, circuitous. **3** erring, straying. □ **deviously** adv. **deviousness** n. [L devius f. DE- + via way]

devise /dɪˈvaɪz/ v. & n. ● v.tr. **1** plan or invent by careful thought. **2** Law leave (real estate) by the terms of a will (cf. BEQUEATH). ● n. **1** the act or an instance of devising. **2** Law a devising clause in a will. □ **devisable** adj. **deviser** n. **devisor** n. (in sense 2 of v). **devisee** /ˌdɪvaɪˈziː/ n. (in sense 2 of v). [ME f. OF deviser ult. f. L dividere divis- DIVIDE: (n.) f. OF devise f. med.L divisa fem. past part. of dividere]

devitalize /diːˈvaɪtəˌlaɪz/ v.tr. (also **-ise**) take away strength and vigour from. □ **devitalization** /-ˌvaɪtəlaɪˈzeɪʃ(ə)n/ n.

devitrify /diːˈvɪtrɪˌfaɪ/ v.tr. (**-ies**, **-ied**) deprive of vitreous qualities; make (glass or vitreous rock) opaque and crystalline. □ **devitrification** /-ˌvɪtrɪfɪˈkeɪʃ(ə)n/ n.

devoid /dɪˈvɔɪd/ predic.adj. (foll. by of) quite lacking or free from (a book devoid of all interest). [ME, past part. of obs. devoid f. OF devoidier (as DE-, VOID)]

devoir /deˈvwɑː(r)/ n. archaic **1** duty, one's best (do one's devoir). **2** (in pl.) courteous or formal attentions; respects (pay one's devoirs to). [ME f. AF dever = OF deveir f. L debere owe]

devolute /ˈdiːvəˌluːt, -ˌljuːt/ v.tr. transfer by devolution. [as DEVOLVE]

devolution /ˌdiːvəˈluːʃ(ə)n, -ˈljuːʃ(ə)n/ n. **1** the delegation of power, esp. by central government to local or regional administration. **2 a** descent or passing on through a series of stages. **b** descent by natural or due succession from one to another of property or qualities. **3** the lapse of an unexercised right to an ultimate owner. **4** Biol. degeneration. □ **devolutionary** adj. **devolutionist** n. [LL devolutio (as DEVOLVE)]

devolve /dɪˈvɒlv/ v. **1** (foll. by on, upon, etc.) **a** tr. pass (work or duties) to (a deputy etc.). **b** intr. (of work or duties) pass to (a deputy etc.). **2** intr. (foll. by on, to, upon) Law (of property etc.) descend or fall by succession to. □ **devolvement** n. [ME f. L devolvere devolut- (as DE-, volvere roll)]

Devon /ˈdev(ə)n/ (also **Devonshire** /ˈdev(ə)nˌʃɪə(r)/) a county of SW England; county town, Exeter.

Devonian /dɪˈvəʊnɪən/ adj. & n. ● adj. **1** of or relating to Devon. **2** Geol. of or relating to the fourth period of the Palaeozoic era, from about 408 to 360 million years ago, between the Silurian and Carboniferous periods. During this period fish became abundant, the first amphibians evolved, and the first forests appeared. ● n. **1** a native of Devon. **2** Geol. the Devonian period or geological system. [med.L Devonia Devonshire]

dévot /deɪˈvəʊ/ n. (fem. **dévote** /-ˈvəʊt/) a devotee. [F f. OF (as DEVOUT)]

devote /dɪˈvəʊt/ v.tr. & refl. **1** (foll. by to) apply or give over (resources etc. or oneself) to (a particular activity or purpose or person) (devoted their time to reading; devoted himself to his guests). **2** archaic doom to destruction. [L devovere devot- (as DE-, vovere vow)]

devoted /dɪˈvəʊtɪd/ adj. very loving or loyal (a devoted husband). □ **devotedly** adv. **devotedness** n.

devotee /ˌdevəˈtiː/ n. **1** (usu. foll. by of) a zealous enthusiast or supporter. **2** a zealously pious or fanatical person.

devotion /dɪˈvəʊʃ(ə)n/ n. **1** (usu. foll. by to) enthusiastic attachment or loyalty (to a person or cause); great love. **2 a** religious worship. **b** (in pl.) prayers. **c** devoutness, religious fervour. □ **devotional** adj. [ME f. OF devotion or L devotio (as DEVOTE)]

devour /dɪˈvaʊə(r)/ v.tr. **1** eat hungrily or greedily. **2** (of fire etc.) engulf, destroy. **3** take in greedily with the eyes or ears (devoured book after book). **4** absorb the attention of (devoured by anxiety). □ **devourer** n. **devouringly** adv. [ME f. OF devorer f. L devorare (as DE-, vorare swallow)]

devout /dɪˈvaʊt/ adj. **1** earnestly religious. **2** earnestly sincere (devout hope). □ **devoutly** adv. **devoutness** n. [ME f. OF devot f. L devotus past part. (as DEVOTE)]

Devoy /dəˈvɔɪ/, Susan (b.1954), New Zealand squash player. She was ranked first in the world when aged 20, the youngest player to achieve this distinction. She won the British Open Championship seven consecutive times (1984–90), and again in 1992, and was five times world champion between 1985 and 1992.

de Vries /də ˈvriːs/, Hugo (1848–1935), Dutch plant physiologist and geneticist. Until about 1890 he worked largely on osmosis and water

relations in plants, coining the term plasmolysis. He then switched abruptly to work on heredity and variation, carrying out plant-breeding experiments which gave similar results to those of Mendel (of which he was unaware). His extensive experiments with the evening primrose, though wrongly interpreted at the time, contributed substantially to the chromosome theory of heredity.

DEW abbr. distant early warning.

dew /djuː/ n. & v. ● n. **1** atmospheric vapour condensing in small drops on cool surfaces at night. **2** beaded or glistening moisture resembling this, e.g. tears. **3** freshness, refreshing quality. ● v.tr. wet with or as with dew. □ **dew-claw 1** a rudimentary inner toe found on some dogs. **2** a false hoof on a deer etc. **dew-fall 1** the time when dew begins to form. **2** evening. **dew-point** the temperature at which dew forms. **dew-pond** a shallow usu. artificial pond once supposed to have been fed by atmospheric condensation. [OE dēaw f. Gmc]

dewan /dɪˈwɑːn/ n. the Prime Minister or Finance Minister of an Indian state. [Arab. & Pers. diwān fiscal register]

Dewar /ˈdjuːə(r)/, Sir James (1842–1923), Scottish chemist and physicist. He is chiefly remembered for his work in cryogenics. He devised the vacuum flask, achieved temperatures close to absolute zero, and was the first to produce liquid oxygen and hydrogen in quantity. Dewar also worked on structural organic chemistry, spectroscopic analysis, high temperature reactions, thin films, and infrared radiation.

dewar /ˈdjuːə(r)/ n. a double-walled flask of steel or silvered glass with a vacuum between the walls to reduce the transfer of heat. [DEWAR]

dewberry /ˈdjuːbərɪ/ n. (pl. **-ies**) **1** a bluish fruit like the blackberry. **2** the bramble bearing this, esp. the European Rubus caesius.

dewdrop /ˈdjuːdrɒp/ n. a drop of dew.

Dewey[1] /ˈdjuːɪ/, John (1859–1952), American philosopher and educationist. Working in the pragmatic tradition of William James and C. S. Pierce, he defined knowledge as successful practice, and evolved the educational theory that children would learn best by doing. He published his ideas in The School and Society (1899) and convinced many American educationists that it was necessary to create less structured, more pupil-centred, practical schools.

Dewey[2] /ˈdjuːɪ/, Melvil (1851–1931), American librarian. He devised a decimal system of classifying books, using ten main subject categories. The system was first invented for Amherst College Library in 1876.

Dewey decimal classification (abbr. **DDC**) a decimal system of library classification invented by Melvil Dewey and applied internationally in libraries in an expanded and constantly revised form. [DEWEY[2]]

Dewi see DAVID, ST.

dewlap /ˈdjuːlæp/ n. **1** a loose fold of skin hanging from the throat of cattle, dogs, etc. **2** similar loose skin round the throat of an elderly person. [ME f. DEW + LAP[1], perh. after ON (unrecorded) döggleppr]

deworm /diːˈwɜːm/ v.tr. rid (a dog, cat, etc.) of worms.

Dewsbury /ˈdjuːzbərɪ/ a textile-manufacturing town in West Yorkshire; pop. (1980) 50,000.

dewy /ˈdjuːɪ/ adj. (**dewier**, **dewiest**) **1 a** wet with dew. **b** moist as if with dew. **2** of or like dew. □ **dewy-eyed** innocently trusting; naively sentimental. □ **dewily** adv. **dewiness** n. [OE dēawig (as DEW, -Y[1])]

Dexedrine /ˈdeksəˌdriːn, -drɪn/ n. propr. a form of amphetamine. [prob. f. DEXTROROTATORY after BENZEDRINE]

dexter[1] /ˈdekstə(r)/ adj. esp. Heraldry on or of the right-hand side (the observer's left) of a shield etc. [L, = on the right]

dexter[2] /ˈdekstə(r)/ n. a small hardy breed of Irish cattle. [19th c.: perh. f. the name of a breeder]

dexterity /dekˈsterɪtɪ/ n. **1** skill in handling. **2** manual or mental adroitness. **3** right-handedness, using the right hand. [F dextérité f. L dexteritas (as DEXTER[1])]

dexterous /ˈdekstərəs/ adj. (also **dextrous** /ˈdekstrəs/) having or showing dexterity. □ **dexterously** adv. **dexterousness** n. [L DEXTER[1] + -OUS]

dextral /ˈdekstrəl/ adj. & n. ● adj. **1** (of a person) right-handed. **2** of or on the right. **3** Zool. (of a spiral shell) with whorls rising to the right and coiling in an anticlockwise direction. **4** Zool. (of a flatfish) with the right side uppermost. □ **dextrally** adv. **dextrality** /dekˈstralɪtɪ/ n. [med.L dextralis f. L dextra right hand]

dextran /ˈdekstræn/ n. Chem. & Pharm. **1** a polymer of glucose formed by the action of certain bacteria on sucrose etc. **2** a degraded form of this used as a substitute for blood plasma. [G (as DEXTRO- + -*an* as in chem. names)]

dextrin /'dekstrɪn/ *n. Chem.* a soluble gummy substance obtained by boiling starch and used as an adhesive. [F *dextrine* f. L *dextra*: see DEXTRO-, -IN]

dextro- /'dekstrəʊ/ *comb. form* on or to the right (*dextrorotatory*; *dextrose*). [L *dexter, dextra* on or to the right]

dextrorotatory /ˌdekstrəʊrəʊ'teɪtərɪ/ *adj. Chem.* having the property of rotating the plane of a polarized light ray to the right (cf. LAEVOROTATORY). □ **dextrorotation** /-'teɪʃ(ə)n/ *n.*

dextrorse /'dekstrɔːs/ *adj.* rising towards the right, esp. of a spiral stem. [L *dextrorsus* (as DEXTRO-)]

dextrose /'dekstrəʊz, -strəʊs/ *n. Chem.* the dextrorotatory and predominant naturally occurring form of glucose. [formed as DEXTRO- + -OSE²]

dextrous var. of DEXTEROUS.

DF *abbr.* **1** see DEFENDER OF THE FAITH. **2** direction-finder.

DFC see DISTINGUISHED FLYING CROSS.

Dfl *abbr.* Dutch florin.

DFM *abbr.* (in the UK) Distinguished Flying Medal.

DG *abbr.* **1** *Dei gratia.* **2** *Deo gratias.* **3** director-general.

Dhaka /'dækə/ (also **Dacca**) the capital of Bangladesh, on the Ganges delta; pop. (1991) 3,637,890.

dhal /dɑːl/ *n.* (also **dal**) **1** split pulses, a common foodstuff in India. **2** a dish made with these. [Hindi]

Dhanbad /'dɑːnbæd/ a city in Bihar, NE India; pop. (1991) 818,000.

dharma /'dɑːmə, 'dɜːmə/ *n. Hinduism, Buddhism, & Jainism* in Hinduism, the eternal law of the cosmos, inherent in the very nature of things, upheld by the gods; in the context of individual action, the term denotes social or caste rules, right behaviour. In Buddhism, however, dharma is universal truth or law, esp. as proclaimed by the Buddha; in Jainism, it is both virtue and a fundamental substance, the medium of motion. [Skr., = decree, custom]

Dharuk /'dʌrʊk/ *n.* an Aboriginal language of the area around Sydney, Australia, now extinct.

Dhaulagiri /ˌdaʊlə'ɡɪrɪ/ a mountain massif in the Himalayas, in Nepal, with six peaks, rising to 8,172 m (26,810 ft) at its highest point.

Dhelfoí see DELPHI.

Dhílos see DELOS.

dhobi /'dəʊbɪ/ *n.* (*pl.* **dhobis**) (in the Indian subcontinent) a washerman or washerwoman. □ **dhobi** (or **dhobi's**) **itch** a tropical skin disease; an allergic dermatitis. [Hindi *dhobī* f. *dhob* washing]

Dhofar /dəʊ'fɑː(r)/ the fertile southern province of Oman.

dhole /dəʊl/ *n.* (*pl.* same or **dholes**) a wild dog, *Cuon alpinus*, native to Asia. [orig. unkn.]

dhoti /'dəʊtɪ/ *n.* (*pl.* **dhotis**) the loincloth worn by male Hindus. [Hindi *dhotī*]

dhow /daʊ/ *n.* a lateen-rigged ship used on the Arabian Sea. [19th c.: orig. unkn.]

DHSS *abbr. hist.* (in the UK) Department of Health and Social Security (cf. DoH, DSS).

dhurra var. of DURRA.

DI *abbr.* (in the UK) Defence Intelligence.

di-¹ /daɪ/ *comb. form* **1** twice, two-, double. **2** *Chem.* containing two atoms, molecules, or groups of a specified kind (*dichromate*; *dioxide*). [Gk f. *dis* twice]

di-² /daɪ, dɪ/ *prefix* form of DIS- occurring before *l, m, n, r, s* (foll. by a consonant), *v,* usu. *g,* and sometimes *j.* [L var. of *dis-*]

di-³ /daɪ/ *prefix* form of DIA- before a vowel.

dia. *abbr.* diameter.

dia- /'daɪə/ *prefix* (also **di-** before a vowel) **1** through (*diaphanous*). **2** apart (*diacritical*). **3** across (*diameter*). [Gk f. *dia* through]

diabase /'daɪəˌbeɪs/ *n. Geol.* dolerite, esp. (in British use) altered dolerite. [F, as if f. DI-¹ + BASE¹, but assoc. with Gk *diabasis* transition]

diabetes /ˌdaɪə'biːtiːz/ *n.* a disease causing excessive thirst and the production of large amounts of urine. The commonest form of diabetes is known as *diabetes mellitus* (Latin, = sweet diabetes), in which the body is unable to metabolize sugar and starch properly; there are abnormally high levels of sugar in the blood, and the urine contains glucose. In 1922, as a result of the work of F. G. Banting, C. H. Best, and others, diabetes mellitus was found to be caused by a deficiency of the

hormone insulin, which is produced in the pancreas. Diabetes mellitus can be controlled by a regulated diet and injections of insulin. If uncontrolled the result can be a build-up of ketones in the blood, leading to convulsions and coma. There is also a rare form of diabetes called *diabetes insipidus* (Latin, = insipid diabetes), which is caused by a deficiency of a pituitary hormone regulating kidney function, and does not involve sugar metabolism. [orig. = siphon: L f. Gk f. *diabainō* go through]

diabetic /ˌdaɪə'betɪk/ *adj. & n.* ● *adj.* **1** of or relating to or having diabetes. **2** for use by diabetics. ● *n.* a person suffering from diabetes.

diablerie /dɪɑː'bləri/ *n.* **1** the Devil's work; sorcery. **2** wild recklessness. **3** the realm of devils; devil-lore. [F f. *diable* f. L *diabolus* DEVIL]

diabolic /ˌdaɪə'bɒlɪk/ *adj.* **1** of the Devil. **2** devilish; inhumanly cruel or wicked. **3** fiendishly clever or cunning or annoying. [ME f. OF *diabolique* or LL *diabolicus* f. L *diabolus* (as DEVIL)]

diabolical /ˌdaɪə'bɒlɪk(ə)l/ *adj.* **1** *colloq.* disgracefully bad or defective; outrageous. **2** = DIABOLIC.

diabolism /daɪ'æbəˌlɪz(ə)m/ *n.* **1 a** a belief in or worship of the Devil. **b** sorcery. **2** devilish conduct or character. □ **diabolist** *n.* [Gk *diabolos* DEVIL]

diabolize /daɪ'æbəˌlaɪz/ *v.tr.* (also **-ise**) make into or represent as a devil.

diabolo /dɪ'æbələʊ, daɪ-/ *n.* (*pl.* **-os**) **1** a game in which a two-headed top is thrown up and caught with a string stretched between two sticks. **2** the top itself. [It., = DEVIL: formerly called *devil on two sticks*]

diachronic /ˌdaɪə'krɒnɪk/ *adj. Linguistics etc.* concerned with the historical development of a subject (esp. a language) (opp. SYNCHRONIC). □ **diachronically** *adv.* **diachronism** /daɪ'ækrəˌnɪz(ə)m/ *n.* **diachronous** /-'ækrənəs/ *adj.* **diachrony** *n.* **diachronistic** /daɪˌækrə'nɪstɪk/ *adj.* [F *diachronique* (as DIA-, CHRONIC)]

diaconal /daɪ'ækən(ə)l/ *adj.* of a deacon. [eccl.L *diaconalis* f. *diaconus* DEACON]

diaconate /daɪ'ækəˌneɪt, -nət/ *n.* **1 a** the office of deacon. **b** a person's time as deacon. **2** a body of deacons. [eccl.L *diaconatus* (as DIACONAL)]

diacritic /ˌdaɪə'krɪtɪk/ *n. & adj.* ● *n.* a sign (e.g. an accent, diaeresis, cedilla) used to indicate different sounds or values of a letter. ● *adj.* = DIACRITICAL. [Gk *diakritikos* (as DIA-, CRITIC)]

diacritical /ˌdaɪə'krɪtɪk(ə)l/ *adj. & n.* ● *adj.* distinguishing, distinctive. ● *n.* (in full **diacritical mark** or **sign**) = DIACRITIC.

diadelphous /ˌdaɪə'delfəs/ *adj. Bot.* with the stamens united in two bundles (cf. MONADELPHOUS, POLYADELPHOUS). [DI-¹ + Gk *adelphos* brother]

diadem /'daɪəˌdem/ *n. & v.* ● *n.* **1** a crown or headband worn as a sign of sovereignty. **2** a wreath of leaves or flowers worn round the head. **3** sovereignty. **4** a crowning distinction or glory. ● *v.tr.* (esp. as **diademed** *adj.*) adorn with or as with a diadem. [ME f. OF *diademe* f. L *diadema* f. Gk *diadēma* (as DIA-, *deō* bind)]

Diadochi /daɪ'ædəkɪ/ the six Macedonian generals of Alexander the Great (Antigonus, Antipater, Cassander, Lysimachus, Ptolemy, and Seleucus), among whom his empire was eventually divided after his death in 323 BC. [Gk *diadokhoi* successors]

diaeresis /daɪ'ɪərəsɪs/ *n.* (*US* **dieresis**) (*pl.* **-ses** /-ˌsiːz/) **1** a mark (as in *naïve, Zaïre*) over a vowel to indicate that it is sounded separately. **2** *Prosody* a break where a foot ends at the end of a word. [L f. Gk, = separation]

diagenesis /ˌdaɪə'dʒenɪsɪs/ *n. Geol.* the transformation occurring during the conversion of sedimentation to sedimentary rock.

Diaghilev /dɪ'ægɪˌlef/, Sergei (Pavlovich) (1872–1929), Russian ballet impresario. After the closure of his magazine *The World of Art* (1899–1904), he organized opera and ballet productions in Paris and in 1909 formed the Ballets Russes, which he directed until his death. Initially with Nijinsky as his star performer, and later with Massine, he transformed the European ballet scene into a creative centre for a large and varied array of artists, pooling the talents of leading choreographers, painters, and composers of his day.

diagnose /'daɪəɡˌnəʊz/ *v.tr.* make a diagnosis of (a disease, a mechanical fault, etc.) from its symptoms. □ **diagnosable** *adj.*

diagnosis /ˌdaɪəɡ'nəʊsɪs/ *n.* (*pl.* **diagnoses** /-siːz/) **1 a** the identification of a disease by means of a patient's symptoms. **b** an instance or formal statement of this. **2 a** the identification of the cause of a mechanical fault etc. **b** an instance of this. **3 a** the distinctive characterization in precise terms of a genus, species, etc. **b** an instance of this. [mod.L f. Gk (as DIA-, *gignōskō* recognize)]

diagnostic /ˌdaɪəg'nɒstɪk/ *adj. & n.* ● *adj.* of or assisting diagnosis. ● *n.* a symptom. □ **diagnostically** *adv.* **diagnostician** /-nɒ'stɪʃ(ə)n/ *n.* [Gk *diagnōstikos* (as DIAGNOSIS)]

diagnostics /ˌdaɪəg'nɒstɪks/ *n.* **1** (treated as *pl.*) *Computing* programs and other mechanisms used to detect and identify faults in hardware or software. **2** (treated as *sing.*) the science or study of diagnosing disease.

diagonal /daɪ'ægən(ə)l/ *adj. & n.* ● *adj.* **1** crossing a straight-sided figure from corner to corner. **2** slanting, oblique. ● *n.* a straight line joining two non-adjacent corners. □ **diagonally** *adv.* [L *diagonalis* f. Gk *diagōnios* (as DIA-, *gōnia* angle)]

diagram /'daɪəˌgræm/ *n. & v.* ● *n.* **1** a drawing showing the general scheme or outline of an object and its parts. **2** a graphic representation of the course or results of an action or process. **3** *Geom.* a figure made of lines used in proving a theorem etc. ● *v.tr.* (**diagrammed**, **diagramming**; *US* **diagramed**, **diagraming**) represent by means of a diagram. □ **diagrammatic** /ˌdaɪəgrə'mætɪk/ *adj.* **diagrammatically** *adv.* [L *diagramma* f. Gk (as DIA-, -GRAM)]

diagrid /'daɪəˌgrɪd/ *n. Archit.* a supporting structure of diagonally intersecting ribs of metal etc. [DIAGONAL + GRID]

diakinesis /ˌdaɪəkɪ'niːsɪs, -kaɪ'niːsɪs/ *n.* (*pl.* **diakineses** /-siːz/) *Biol.* a stage during the prophase of meiosis when the separation of homologous chromosomes is complete and crossing over has occurred. [mod.L f. G *Diakinese* (as DIA-, Gk *kinēsis* motion)]

dial /'daɪəl/ *n. & v.* ● *n.* **1** the face of a clock or watch, marked to show the hours etc. **2** a similar flat plate marked with a scale for measuring weight, volume, pressure, consumption, etc., indicated by a pointer. **3** a movable disc on a telephone, with finger-holes and numbers for making a connection. **4 a** a plate or disc etc. on a radio or television set for selecting wavelength or channel. **b** a similar selecting device on other equipment, e.g. a washing machine. **5** *Brit. sl.* a person's face. ● *v.* (**dialled**, **dialling**; *US* **dialed**, **dialing**) select (a telephone number) by means of a dial or set of buttons (*dialled 999*). **2** *tr.* measure, indicate, or regulate by means of a dial. □ **dialling code** a sequence of numbers dialled to connect a telephone with the exchange of the telephone being called. **dialling tone** (*N. Amer.* **dial tone**) a sound indicating that a caller may start to dial. □ **dialler** *n.* [ME, = sundial, f. med.L *diale* clock-dial ult. f. L *dies* day]

dialect /'daɪəˌlekt/ *n.* **1** a form of speech peculiar to a particular region. **2** a subordinate variety of a language with non-standard vocabulary, pronunciation, or grammar. □ **dialectal** /ˌdaɪə'lektəl/ *adj.* **dialectology** /ˌdaɪəlek'tɒlədʒɪ/ *n.* **dialectologist** *n.* [F *dialecte* or L *dialectus* f. Gk *dialektos* discourse f. *dialegomai* converse]

dialectic /ˌdaɪə'lektɪk/ *n. & adj.* ● *n.* (often in *pl.*, treated as *sing.* or *pl.*) **1 a** the art of investigating the truth of opinions. **b** logical disputation. **2** *Philos.* **a** inquiry into metaphysical contradictions and their solutions, esp. in the thought of Kant and Hegel. **b** the existence or action of opposing forces or tendencies in society etc.; the world seen as a continuing unification of opposites. ● *adj.* **1** of or relating to logical disputation. **2** fond of or skilled in logical disputation. [ME f. OF *dialectique* or L *dialectica* f. Gk *dialektikē* (*tekhnē*) (art) of debate (as DIALECT)]

dialectical /ˌdaɪə'lektɪk(ə)l/ *adj.* of dialectic or dialectics. □ **dialectically** *adv.*

dialectical materialism *n.* (in Marxist thought) the theory that physical matter is the only true reality, that every aspect of it (including human history) is subject to constant change resulting from the pressure of conflicting forces, and that mental processes are dependent on material ones. The theory includes the notion that transformations in quantity give rise to qualitative (hence revolutionary) changes over time. Dialectical materialism was adopted as the official philosophy of the Soviet Communists, but was not in fact ever used as a term by Marx or Engels. □ **dialectical materialist** *n. & adj.*

dialectician /ˌdaɪəlek'tɪʃ(ə)n/ *n.* a person skilled in dialectic. [F *dialecticien* f. L *dialecticus*]

dialectics /ˌdaɪə'lektɪks/ *n.* see DIALECTIC *n.*

dialogic /ˌdaɪə'lɒdʒɪk/ *adj.* of or in dialogue. [LL *dialogicus* f. Gk *dialogikos* (as DIALOGUE)]

dialogist /daɪ'æləˌdʒɪst/ *n.* a speaker in or writer of dialogue. [LL *dialogista* f. Gk *dialogistēs* (as DIALOGUE)]

dialogue /'daɪəˌlɒg/ *n.* (*US* **dialog**) **1 a** a conversation. **b** conversation in written form; this as a form of composition. **2 a** a discussion, esp. one between representatives of two political groups. **b** a conversation, a talk (*long dialogues between the two main characters*). [ME f. OF *dialoge* f. L *dialogus* f. Gk *dialogos* f. *dialegomai* converse]

dialyse /'daɪəˌlaɪz/ *v.tr.* (*US* **dialyze**) separate by means of dialysis.

dialysis /daɪ'ælɪsɪs/ *n.* (*pl.* **dialyses** /-ˌsiːz/) **1** *Chem.* a process in which dissolved substances are selectively removed from a liquid by virtue of their different abilities to pass through a semi-permeable membrane into another liquid. **2** (in full **kidney** or **renal dialysis**) *Med.* the use of such a process to purify the blood. Also called *haemodialysis*. (*See note below.*) □ **dialytic** /ˌdaɪə'lɪtɪk/ *adj.* [L f. Gk *dialusis* (as DIA-, *luō* set free)]

▪ Dialysis is used to substitute for normal kidney function in cases of kidney failure, disease, and poisoning. The patient's blood flows past a semi-permeable membrane on the other side of which is another liquid; impurities and waste products in the blood pass through the membrane and are removed, and the blood is cleansed. The dialysis may take place outside the body in an artificial kidney or inside it using a natural membrane such as the peritoneum.

diamagnetic /ˌdaɪəmæg'netɪk/ *adj. & n.* ● *adj.* tending to become magnetized in a direction at right angles to the applied magnetic field. ● *n.* a diamagnetic body or substance. □ **diamagnetically** *adv.* **diamagnetism** /-'mægnɪˌtɪz(ə)m/ *n.*

diamanté /dɪə'mɒnteɪ/ *adj. & n.* ● *adj.* decorated with powdered crystal or another sparkling substance. ● *n.* fabric or costume jewellery so decorated. [F, past part. of *diamanter* set with diamonds f. *diamant* DIAMOND]

diamantiferous /ˌdaɪəmæn'tɪfərəs/ *adj.* diamond-yielding. [F *diamantifère* f. *diamant* DIAMOND]

diamantine /ˌdaɪə'mæntaɪn/ *adj.* of or like diamonds. [F *diamantin* (as DIAMANTIFEROUS)]

diameter /daɪ'æmɪtə(r)/ *n.* **1 a** a straight line passing from side to side through the centre of a body or figure, esp. a circle or sphere. **b** the length of this line. **2** a transverse measurement; width, thickness. **3** a unit of linear measurement of magnifying power (*a lens magnifying 2,000 diameters*). □ **diametral** *adj.* [ME f. OF *diametre* f. L *diametrus* f. Gk *diametros* (*grammē*) (line) measuring across f. *metron* measure]

diametrical /ˌdaɪə'metrɪk(ə)l/ *adj.* (also **diametric**) **1** of or along a diameter. **2** (of opposition, difference, etc.) complete, like that between opposite ends of a diameter. □ **diametrically** *adv.* [Gk *diametrikos* (as DIAMETER)]

diamond /'daɪəmənd/ *n., adj., & v.* ● *n.* **1** a precious stone of pure carbon crystallized in octahedrons etc., the hardest naturally occurring substance. **2** a figure shaped like the cross-section of a diamond; a rhombus. **3 a** a playing card of a suit denoted by a red rhombus. **b** (in *pl.*) this suit. **4** a glittering particle or point (of frost etc.). **5** a tool with a small diamond for glass-cutting. **6** *Baseball* **a** the space delimited by the bases. **b** the entire field. ● *adj.* **1** made of or set with diamonds or a diamond. **2** rhombus-shaped. ● *v.tr.* adorn with or as with diamonds. □ **diamond-bird** *Austral.* = PARDALOTE. **diamond cut diamond** wit or cunning meets its match. **diamond jubilee** the 60th (or 75th) anniversary of an event, esp. a sovereign's accession. **diamond wedding** a 60th (or 75th) wedding anniversary. □ **diamondiferous** /ˌdaɪəmən'dɪfərəs/ *adj.* [ME f. OF *diamant* f. med.L *diamas diamant-* var. of L *adamas* ADAMANT f. Gk]

diamondback /'daɪəməndˌbæk/ *n.* **1** an edible freshwater terrapin, *Malaclemys terrapin*, native to North America, with lozenge-shaped markings on its shell. **2** a rattlesnake of the genus *Crotalus*, native to North America, with diamond-shaped markings.

Diamond Head a volcanic crater overlooking the port of Honolulu on the Hawaiian island of Oahu.

Diana /daɪ'ænə/ *Rom. Mythol.* an early Italian goddess anciently identified with Artemis and associated with hunting, virginity, and, in later literature, with the moon. [prob. = bright one]

Diana, Princess (formally called Diana, Princess of Wales; born Lady Diana Frances Spencer) (b.1961), Princess of Wales and wife of Prince Charles. The daughter of the 8th Earl Spencer, she married Charles, Prince of Wales, in 1981. The couple publicly announced their separation in 1993.

diandrous /daɪ'ændrəs/ *adj.* having two stamens. [DI-¹ + Gk *anēr andr-* man]

Dianetics /ˌdaɪə'netɪks/ *n.* a system developed by the American founder of the Church of Scientology L. Ron Hubbard (1911–86) which aims to relieve psychosomatic disorder by a process of cleansing the mind of harmful mental images. [f. Gk *dianoētikos* f. *dianoeisthai* think (as DIA- + *noein* think, suppose) + -ICS]

dianthus /daɪˈænθəs/ *n.* a flowering plant of the genus *Dianthus*, e.g. a carnation or pink. [Gk *Dios* of Zeus + *anthos* flower]

diapason /ˌdaɪəˈpeɪz(ə)n, -ˈpeɪs(ə)n/ *n.* **1** *Mus.* the compass of a voice or musical instrument. **2** *Mus.* a fixed standard of pitch. **3** (in full **open** or **stopped diapason**) *Mus.* either of two main organ-stops extending through the organ's whole compass. **4** *Mus.* **a** a combination of notes or parts in a harmonious whole. **b** a melodious succession of notes, esp. a grand swelling burst of harmony. **5** an entire compass, range, or scope. [ME in sense 'octave' f. L *diapason* f. Gk *dia pasōn* (*khordōn*) through all (notes)]

diapause /ˈdaɪəˌpɔːz/ *n. Zool.* (in some insects) a period of suspended development which can only be ended by an appropriate environmental stimulus.

diaper /ˈdaɪəpə(r)/ *n. & v.* ● *n.* **1** *N. Amer.* a baby's nappy. **2 a** a linen or cotton fabric with a small diamond pattern. **b** this pattern. **3** a similar ornamental design of diamonds etc. for panels, walls, etc. ● *v.tr.* **1** decorate with a diaper pattern. **2** *N. Amer.* change the nappy of (a baby). [ME f. OF *diapre* f. med.L *diasprum* f. med.Gk *diaspros* (adj.) (as DIA-, *aspros* white)]

diaphanous /daɪˈæfənəs/ *adj.* (of fabric etc.) light and delicate, and almost transparent. [med.L *diaphanus* f. Gk *diaphanes* (as DIA-, *phainō* show)]

diaphoresis /ˌdaɪəfəˈriːsɪs/ *n. Med.* sweating, esp. artificially induced. [LL f. Gk f. *diaphoreō* carry through]

diaphoretic /ˌdaɪəfəˈretɪk/ *adj. & n. Med.* ● *adj.* inducing perspiration. ● *n.* an agent inducing perspiration. [LL *diaphoreticus* f. Gk *diaphorētikos* (formed as DIAPHORESIS)]

diaphragm /ˈdaɪəˌfræm/ *n.* **1** a muscular partition separating the thorax from the abdomen in mammals. **2** a partition in animal and plant tissues. **3** a disc pierced by one or more holes in optical and acoustic systems etc. **4** a device for varying the effective aperture of the lens in a camera etc. **5** a thin contraceptive cap fitting over the cervix. **6** a thin sheet of material used as a partition etc. □ **diaphragm pump** a pump using a flexible diaphragm in place of a piston. □ **diaphragmatic** /ˌdaɪəfræɡˈmætɪk/ *adj.* [ME f. LL *diaphragma* f. Gk (as DIA-, *phragma* *-atos* f. *phrassō* fence in)]

diapir /ˈdaɪəˌpɪə(r)/ *n. Geol.* an anticline in which the upper strata are pierced by a rock core from below. [Gk *diapeirainein* pierce through]

diapositive /ˌdaɪəˈpɒzɪtɪv/ *n.* a positive photographic slide or transparency.

diarchy /ˈdaɪɑːkɪ/ *n.* (also **dyarchy**) (*pl.* **-ies**) **1** government by two independent authorities (esp. in India 1919–35). **2** an instance of this. □ **diarchal** /daɪˈɑːk(ə)l/ *adj.* **diarchic** *adj.* [DI-[1] + Gk *-arkhia* rule, after *monarchy*]

diarist /ˈdaɪərɪst/ *n.* a person who keeps a diary. □ **diaristic** /ˌdaɪəˈrɪstɪk/ *adj.*

diarize /ˈdaɪəˌraɪz/ *v.* (also **-ise**) **1** *intr.* keep a diary. **2** *tr.* enter in a diary.

diarrhoea /ˌdaɪəˈrɪə/ *n.* (esp. *US* **diarrhea**) a condition of excessively frequent and loose bowel movements. □ **diarrhoeal** *adj.* **diarrhoeic** *adj.* [ME f. LL f. Gk *diarrhoia* (as DIA-, *rheō* flow)]

diary /ˈdaɪərɪ/ *n.* (*pl.* **-ies**) **1** a daily record of events or thoughts. **2** a book for this or for noting future engagements, usu. printed and with a calendar and other information. [L *diarium* f. *dies* day]

Dias /ˈdiːæs/, Bartolomeu (also **Diaz**) (*c.*1450–1500), Portuguese navigator and explorer. He was the first European to round the Cape of Good Hope (1488), thereby establishing a sea route from the Atlantic to Asia via the southernmost point of Africa; he later accompanied Vasco da Gama on the first European expedition to Asia by this route.

diascope /ˈdaɪəˌskəʊp/ *n.* an optical projector giving images of transparent objects.

Diaspora /daɪˈæspərə/ *n.* **1** (prec. by *the*) **a** the dispersion of the Jews beyond Israel. (*See note below.*) **b** Jews dispersed in this way; Jews living outside Israel. **2** (also **diaspora**) **a** any group of people similarly dispersed. **b** their dispersion. [Gk f. *diaspeirō* (as DIA-, *speirō* scatter)]

▪ The main Diaspora began in the 8th–6th centuries BC, and even before the sack of Jerusalem in AD 70 the number of Jews dispersed by the Diaspora was greater than that living in Israel. Thereafter, Jews were dispersed even more widely throughout the Roman world and beyond (for example, into India). The term embraces concerns about cultural assimilation and loss of Jewish identity which are at the centre of the movement of Zionism.

diastase /ˈdaɪəˌsteɪz/ *n. Biochem.* = AMYLASE. □ **diastasic** /ˌdaɪəˈsteɪzɪk/ *adj.* **diastatic** /-ˈstætɪk/ *adj.* [F f. Gk *diastasis* separation (as DIA-, *stasis* placing)]

diastema /ˌdaɪəˈstiːmə/ *n.* a space between adjacent teeth, esp. in front of the premolars in rodents and ungulates. [late L f. Gk, = space between]

diastole /daɪˈæstəlɪ/ *n. Physiol.* the period between two contractions of the heart when the heart muscle relaxes and allows the chambers to fill with blood (cf. SYSTOLE). □ **diastolic** /ˌdaɪəˈstɒlɪk/ *adj.* [LL f. Gk *diastellō* (as DIA-, *stellō* place)]

diathermancy /ˌdaɪəˈθɜːmənsɪ/ *n.* the quality of transmitting radiant heat. □ **diathermic** *adj.* **diathermous** *adj.* [F *diathermansie* f. Gk *dia* through + *thermansis* heating: assim. to -ANCY]

diathermy /ˈdaɪəˌθɜːmɪ/ *n.* the application of high-frequency electric currents to produce heat in the deeper tissues of the body. In medical use the tissues are warmed but not sufficiently to change their nature; in surgical use there is sufficient heating to produce a local change such as destruction of tissue or coagulation of bleeding vessels. [G *Diathermie* f. Gk *dia* through + *thermon* heat]

diathesis /daɪˈæθɪsɪs/ *n. Med.* a constitutional predisposition to a certain state, esp. a diseased one. [mod.L f. Gk f. *diatithēmi* arrange]

diatom /ˈdaɪətəm/ *n.* a microscopic unicellular alga with a siliceous cell-wall, found as plankton and forming fossil deposits. [mod.L *Diatoma* (genus name) f. Gk *diatomos* (as DIA-, *temnō* cut)]

diatomaceous earth /ˌdaɪətəˈmeɪʃəs/ *n.* = KIESELGUHR. [DIATOM + -ACEOUS]

diatomic /ˌdaɪəˈtɒmɪk/ *adj.* consisting of two atoms. [DI-[1] + ATOM]

diatomite /daɪˈætəˌmaɪt/ *n. Mineral.* a deposit composed of the siliceous skeletons of diatoms (cf. KIESELGUHR).

diatonic /ˌdaɪəˈtɒnɪk/ *adj. Mus.* **1** (of a scale, interval, etc.) involving only notes proper to the prevailing key without chromatic alteration. **2** (of a melody or harmony) constructed from such a scale. [F *diatonique* or LL *diatonicus* f. Gk *diatonikos* at intervals of a tone (as DIA-, TONIC)]

diatribe /ˈdaɪəˌtraɪb/ *n.* a forceful verbal attack; a piece of bitter criticism. [F f. L *diatriba* f. Gk *diatribē* spending of time, discourse f. *diatribō* (as DIA-, *tribō* rub)]

Díaz /ˈdiːæs/, Porfirio (1830–1915), Mexican general and statesman, President 1877–80 and 1884–1911. He led a military coup in 1876 and was elected President the following year. During his second term of office he introduced a highly centralized government, backed by loyal mestizos and landowners, which removed powers from rural workers and American Indians. Díaz promoted the development of Mexico's infrastructure and industry, using foreign capital and engineers to build railways, bridges, and mines. Eventually the poor performance of Mexico's economy and the rise of a democratic movement under Francisco Madero (1873–1913) contributed to Díaz's forced resignation and exile in 1911.

diazepam /daɪˈæzɪˌpæm/ *n.* a tranquillizing muscle-relaxant drug with anticonvulsant properties used to relieve anxiety, tension, etc., e.g. Valium. [benzodiazepine + *am*]

diazo /daɪˈeɪzəʊ/ *n.* (in full **diazotype** /-ˈeɪzəʊˌtaɪp/) a copying or colouring process using a diazo compound decomposed by ultraviolet light. □ **diazo compound** *Chem.* a chemical compound containing two usu. multiply bonded nitrogen atoms, often highly coloured and used as dyes. [DI-[1] + AZO-]

dib /dɪb/ *v.intr.* (**dibbed, dibbing**) = DAP. [var. of DAB[1]]

dibasic /daɪˈbeɪsɪk/ *adj. Chem.* having two replaceable protons. [DI-[1] + BASE[1] 6]

dibber /ˈdɪbə(r)/ *n.* = DIBBLE.

dibble /ˈdɪb(ə)l/ *n. & v.* ● *n.* a hand-tool for making holes in the ground for seeds or young plants. ● *v.* **1** *tr.* sow or plant with a dibble. **2** *tr.* prepare (soil) with a dibble. **3** *intr.* use a dibble. [ME: perh. rel. to DIB]

dibs /dɪbz/ *n.pl. sl.* money. [earlier sense 'pebbles for game', also *dib-stones*, perh. f. DIB]

dice /daɪs/ *n. & v.* ● *n.pl.* **1 a** small cubes with faces bearing one to six spots used in games of chance. **b** (treated as *sing.*) one of these cubes (see DIE[2]). **2** a game played with one or more such cubes. **3** food cut into small cubes for cooking. ● *v.* **1 a** *intr.* play dice. **b** *intr.* take great risks, gamble (*dicing with death*). **c** *tr.* (foll. by *away*) gamble away. **2** *tr.* cut (food) into small cubes. **3** *tr. Austral. sl.* reject; leave alone. **4** *tr.* chequer, mark with squares. □ **no dice** *colloq.* (there is) no chance of success, cooperation, etc. □ **dicer** *n.* (in sense 1 of *v.*). [pl. of DIE[2]]

dicey /ˈdaɪsɪ/ *adj.* (**dicier, diciest**) *colloq.* risky, unreliable. [DICE + -Y[1]]

dichotomy /daɪˈkɒtəmɪ/ n. (pl. **-ies**) **1 a** a division into two, esp. a sharply defined one. **b** the result of such a division. **2** binary classification. **3** Bot. & Zool. repeated bifurcation. □ **dichotomize** v. **dichotomous** adj. [mod.L dichotomia f. Gk dikhotomia f. dikho- apart + -TOMY]

dichroic /daɪˈkrəʊɪk/ adj. (esp. of doubly refracting crystals) showing two colours. □ **dichroism** /ˈdaɪkrəʊˌɪz(ə)m/ n. [Gk dikhroos (as DI-[1], khrōs colour)]

dichromatic /ˌdaɪkrəʊˈmætɪk/ adj. **1** two-coloured. **2 a** (of animal species) having individuals that show different colorations. **b** having vision sensitive to only two of the three primary colours. □ **dichromatism** /daɪˈkrəʊməˌtɪz(ə)m/ n. [DI-[1] + Gk khrōmatikos f. khrōma -atos colour]

dick[1] /dɪk/ n. **1** Brit. colloq. (in certain set phrases) fellow; person (clever dick). **2** coarse sl. the penis. [pet-form of the name Richard]

dick[2] /dɪk/ n. sl. a detective. [perh. abbr.]

dick[3] /dɪk/ n. □ **take one's dick** (often foll. by that + clause) sl. swear, affirm. [abbr. of declaration]

dicken /ˈdɪkɪn/ int. Austral. sl. an expression of disgust or disbelief. [usu. assoc. with DICKENS or the surname Dickens]

Dickens /ˈdɪkɪnz/, Charles (John Huffam) (1812–70), English novelist. His early work consisted mainly of humorous sketches and short pieces for periodical publication; much of his fiction also first appeared in instalments in magazines. Dickens drew on his own childhood experiences of hardship and deprivation in his fiction, and many of his works are set in his native London. His novels are broad in scope and deal with all social classes, but they are particularly notable for their treatment of contemporary social problems, including the plight of the urban poor, corruption and inefficiency within the legal system, and general social injustices. Some of his most famous novels include *Oliver Twist* (1837–8), *Nicholas Nickleby* (1838–9), *Bleak House* (1852–3), and *Great Expectations* (1860–1). His satirical humour and varied characterizations, including such familiar caricatures as Scrooge (*A Christmas Carol*, 1843) and Mr Micawber (*David Copperfield*, 1850), contributed to the great popular appeal of his work.

dickens /ˈdɪkɪnz/ n. (usu. prec. by how, what, why, etc., the) colloq. (esp. in exclamations) deuce; the Devil (what the dickens are you doing here?). [16th c.: prob. a use of the surname Dickens]

Dickensian /dɪˈkenzɪən/ adj. & n. ● adj. **1** of or relating to Charles Dickens, or his work. **2** resembling or reminiscent of the situations, poor social conditions, or comically repulsive characters described in Dickens's work. ● n. an admirer or student of Dickens or his work.

dicker /ˈdɪkə(r)/ v. & n. esp. US ● v. **1 a** intr. bargain, haggle. **b** tr. barter, exchange. **2** intr. dither, hesitate. ● n. a deal, a barter. □ **dickerer** n. [perh. f. dicker set of ten (hides), as a unit of trade]

dickhead /ˈdɪkhed/ n. coarse sl. a stupid or obnoxious person.

Dickinson /ˈdɪkɪns(ə)n/, Emily (Elizabeth) (1830–86), American poet. From the age of 24 she led the life of a recluse in Amherst, Massachusetts. Her withdrawal and inner struggle are reflected in her mystical poems, expressed in her own elliptical language, with a greater emphasis on assonance and alliteration than rhyme. Although she wrote nearly 2,000 poems, only seven were published in her lifetime; the first selection appeared in 1890.

dicky[1] /ˈdɪkɪ/ n. (also **dickey**) (pl. **-ies** or **-eys**) colloq. **1** a false shirt-front. **2** (in full **dicky-bird**) a child's word for a little bird. **3** Brit. a driver's seat in a carriage. **4** Brit. an extra folding seat at the back of a vehicle. **5** (in full **dicky bow**) Brit. a bow-tie. [some senses f. Dicky (as DICK[1])]

dicky[2] /ˈdɪkɪ/ adj. (**dickier, dickiest**) Brit. sl. unsound, likely to collapse or fail. [19th c.: perh. f. 'as queer as Dick's hatband']

dicot /ˈdaɪkɒt/ n. Bot. = DICOTYLEDON. [abbr.]

dicotyledon /ˌdaɪkɒtɪˈliːd(ə)n/ n. Bot. a flowering plant with an embryo that bears two cotyledons. The dicotyledons form one of the two great divisions of flowering plants, Dicotyledoneae (or Dicotyledones), and typically have broad leaves with a network of veins (cf. MONOCOTYLEDON). □ **dicotyledonous** adj. [mod.L f. DI-[1] + COTYLEDON]

dicrotic /daɪˈkrɒtɪk/ adj. Physiol. (of the pulse) having a double beat. [Gk dikrotos]

dicta pl. of DICTUM.

Dictaphone /ˈdɪktəˌfəʊn/ n. propr. a machine for recording and playing back dictated words. [DICTATE + PHONE[1]]

dictate v. & n. ● v. /dɪkˈteɪt/ **1** tr. say or read aloud (words to be written down or recorded). **2 a** tr. prescribe or lay down authoritatively (terms, things to be done). **b** intr. lay down the law; give orders. ● n. /ˈdɪkteɪt/ (usu. in pl.) an authoritative instruction (dictates of conscience). [L dictare dictat- frequent. of dicere dict- say]

dictation /dɪkˈteɪʃ(ə)n/ n. **1 a** the saying of words to be written down or recorded. **b** an instance of this, esp. as a school exercise. **c** the material that is dictated. **2 a** authoritative prescription. **b** an instance of this. **c** a command. □ **dictation speed** a slow rate of speech suitable for dictation.

dictator /dɪkˈteɪtə(r)/ n. **1** a ruler with (often usurped) unrestricted authority. **2** a person with supreme authority in any sphere. **3** a domineering person. **4** a person who dictates for transcription. **5** Rom. Hist. a chief magistrate with absolute power, appointed in an emergency. [ME f. L (as DICTATE)]

dictatorial /ˌdɪktəˈtɔːrɪəl/ adj. **1** of or like a dictator. **2** imperious, overbearing. □ **dictatorially** adv. [L dictatorius (as DICTATOR)]

dictatorship /dɪkˈteɪtəˌʃɪp/ n. **1** a state ruled by a dictator. **2 a** the position, rule, or period of rule of a dictator. **b** rule by a dictator. **3** absolute authority in any sphere.

diction /ˈdɪkʃ(ə)n/ n. **1** the manner of enunciation in speaking or singing. **2** the choice of words or phrases in speech or writing. [F diction or L dictio f. dicere dict- say]

dictionary /ˈdɪkʃənrɪ, -nərɪ/ n. (pl. **-ies**) **1** a book that lists (usu. in alphabetical order) and explains the words of a language or gives equivalent words in another language. (See note below.) **2** a reference book on any subject, the items of which are arranged in alphabetical order (dictionary of architecture). [med.L dictionarium (manuale manual) and dictionarius (liber book) f. L dictio (as DICTION)]

▪ Dictionaries can be either bilingual or monolingual: early examples include Greek glossaries (1st–5th centuries AD) which explained the language used in literary works. In 16th-century England the demand was for bilingual dictionaries, especially Latin–English. In 1604 schoolmaster Robert Cawdrey produced *A Table Alphabetical of English Words*, regarded as the first monolingual English dictionary; during this period dictionaries contained only words regarded as difficult, but John Kersey broke with this tradition in 1702 by including the common words of the language and discarding some of the more fantastic formations. Nathan Bailey's dictionary (1721) of about 40,000 words, giving great attention to etymology, was extremely popular in the 18th century. Samuel Johnson's *Dictionary* (1755), based on a collection of citations from literature, replaced Bailey and remained the authoritative English dictionary for over a century (see JOHNSON[6]). The 19th century saw the publication of Noah Webster's *American Dictionary of the English Language* (1828), containing about 70,000 words, as well as the initiation of the largest and most comprehensive English language dictionary ever produced, the *Oxford English Dictionary* (see MURRAY[2], OXFORD ENGLISH DICTIONARY). Similarly ambitious projects were also carried out for other languages (see GRIMM, LITTRÉ). Many dictionaries are now also published in electronic form.

dictum /ˈdɪktəm/ n. (pl. **dicta** /-tə/ or **dictums**) **1** a formal utterance or pronouncement. **2** a saying or maxim. **3** Law = OBITER DICTUM. [L, = neut. past part. of dicere say]

dicty /ˈdɪktɪ/ adj. US sl. **1** conceited, snobbish. **2** elegant, stylish. [20th c.: orig. unkn.]

Dictyoptera /ˌdɪktɪˈɒptərə/ n.pl. Zool. an order of insects comprising the cockroaches and mantids. □ **dictyopteran** n. & adj. **dictyopterous** adj. [mod.L f. Gk diktuon net + pteron wing]

did past of DO[1].

didactic /daɪˈdæktɪk, dɪ-/ adj. **1** meant to instruct. **2** (of a person) tediously pedantic. □ **didactically** adv. **didacticism** /-tɪˌsɪz(ə)m/ n. [Gk didaktikos f. didaskō teach]

didakai var. of DIDICOI.

diddicoy var. of DIDICOI.

diddle /ˈdɪd(ə)l/ v. colloq. **1** tr. cheat, swindle. **2** intr. US waste time. □ **diddler** n. [prob. back-form. f. Jeremy Diddler in James Kenney's farce 'Raising the Wind' (1803)]

diddly-squat /ˈdɪdlɪˌskwɒt/ n. (also **doodly-squat** /ˈduːd-/) US sl. **1** (foll. by neg.) anything, the least bit (doesn't mean diddly-squat to me). **2** nothing at all. [20th c.: orig. uncert.]

diddums /ˈdɪdəmz/ int. expressing commiseration esp. to a child. [= did 'em, i.e. did they (tease you etc.)?]

Diderot /ˈdiːdəˌrəʊ/, Denis (1713–84), French philosopher, writer, and critic. He was a leading figure of the Enlightenment in France and principal editor of the *Encyclopédie* (1751–76), through which he disseminated and popularized philosophy and scientific knowledge. His major philosophical writings include *Pensées sur l'interprétation de la nature* (1754), which in some ways anticipated evolutionary ideas on the nature and origin of life. His *Salons* (1759–71; 1755; 1781) are the first examples of modern art criticism in France.

didgeridoo /ˌdɪdʒərɪˈduː/ *n.* (also **didjeridoo**) an Australian Aboriginal musical wind instrument of long tubular shape. [imit.]

didicoi /ˈdɪdɪˌkɔɪ/ *n.* (also **didakai** /ˈdɪdəˌkaɪ/, **diddicoy**) *sl.* a gypsy; an itinerant tinker. [Romany]

didn't /ˈdɪd(ə)nt/ *contr.* did not.

Dido /ˈdaɪdəʊ/ (in the *Aeneid*) the queen and founder of Carthage, who fell in love with the shipwrecked Aeneas and killed herself when he deserted her. According to ancient tradition, she was the great-niece of Jezebel and daughter of the king of Tyre, in which city she was called Elissa.

dido /ˈdaɪdəʊ/ *n.* (*pl.* **-oes** or **-os**) *US colloq.* an antic, a caper, a prank. □ **cut** (or **cut up**) **didoes** play pranks. [19th c.: orig. unkn.]

didst /dɪdst/ *archaic 2nd sing. past of* DO¹.

Didyma /ˈdɪdɪmə/ an ancient sanctuary of Apollo, site of one of the most famous oracles of the Aegean region, close to the west coast of Asia Minor.

didymium /dɪˈdɪmɪəm/ *n. Chem.* a mixture of praseodymium and neodymium, orig. regarded as an element. [mod.L f. Gk *didumos* twin (from being closely associated with lanthanum)]

die¹ /daɪ/ *v.* (**dies**, **dying** /ˈdaɪɪŋ/, **died**) **1** *intr.* (often foll. by *of*) (of a person, animal, or plant) cease to live; expire, lose vital force (*died of hunger*). **2** *intr.* **a** come to an end, cease to exist, fade away (*the project died within six months*). **b** cease to function; break down (*the engine died*). **c** (of a flame) go out. **3** *intr.* (foll. by *on*) die or cease to function while in the presence or charge of (a person). **4** *intr.* (usu. foll. by *of*, *from*, *with*) be exhausted or tormented (*nearly died of boredom*; *was dying from the heat*). **5** *tr.* suffer (a specified death) (*died a natural death*). □ **be dying** (foll. by *for*, or *to* + infin.) wish for longingly or intently (*was dying for a drink*; *am dying to see you*). **die away** become weaker or fainter to the point of extinction. **die-away** *adj.* having a languishing or affectedly feeble manner. **die back** (of a plant) decay from the tip towards the root. **die down** become less loud or strong. **die hard** die reluctantly, not without a struggle (*old habits die hard*). **die-hard** *n.* a conservative or stubborn person. **die off** die one after another until few or none are left. **die out** become extinct, cease to exist. **never say die** keep up courage, not give in. **to die for** *adj. colloq.* extremely good or desirable (*chocolate to die for*). [ME, prob. f. ON *deyja* f. Gmc]

die² /daɪ/ *n.* **1** *sing.* of DICE 1a. ¶ *Dice* is now standard in general British use in this sense. **2** (*pl.* **dies**) **a** an engraved device for stamping a design on coins, medals, etc. **b** a device for stamping, cutting, or moulding material into a particular shape. **3** (*pl.* **dice** /daɪs/) *Archit.* the cubical part of a pedestal between the base and the cornice; a dado or plinth. □ **as straight** (or **true**) **as a die 1** quite straight. **2** entirely honest or loyal. **die-cast** cast (hot metal) in a die or mould. **die-casting** the process or product of casting from metal moulds. **the die is cast** an irrevocable step has been taken. **die-sinker** an engraver of dies. **die-stamping** embossing paper etc. with a die. [ME f. OF *de* f. L *datum* neut. past part. of *dare* give, play]

dieffenbachia /ˌdiːfənˈbækɪə/ *n.* a tropical American plant of the genus *Dieffenbachia*, of the arum family, often grown as a house plant and having acrid toxic sap. [Ernst *Dieffenbach* (1811–55), Ger. horticulturalist]

Diego Garcia /dɪˌeɪgəʊ gɑːˈsiːə/ the largest island of the Chagos Archipelago in the middle of the Indian Ocean, site of a strategic Anglo-American naval base established in 1973.

dieldrin /ˈdiːldrɪn/ *n.* an insecticide produced by the oxidation of aldrin, now little used because of its persistence in the environment. [Otto *Diels*, Ger. chemist (1876–1954) + ALDRIN]

dielectric /ˌdaɪɪˈlektrɪk/ *adj. & n. Electr.* ● *adj.* insulating. ● *n.* an insulating medium or substance. □ **dielectric constant** permittivity. □ **dielectrically** *adv.* [DI-³ + ELECTRIC = through which electricity is transmitted (without conduction)]

Dien Bien Phu /ˌdjen bjen ˈfuː/ a village in NW Vietnam, in 1954 the site of a French military post which was captured by the Vietminh after a 55-day siege.

diene /ˈdaɪiːn/ *n. Chem.* any organic compound possessing two double bonds between carbon atoms. [DI-¹ + -ENE]

Dieppe /dɪˈep/ a channel port in northern France, from which ferries run to Newhaven and elsewhere; pop. (1990) 36,600. In August 1942,

during the Second World War, it was the scene of an amphibious raid by a joint force of 1,000 British and 5,000 Canadian troops to destroy the German-held port and airfield. The raid ended disastrously, with two-thirds of the Allied troops being killed.

dieresis *US var. of* DIAERESIS.

diesel /ˈdiːz(ə)l/ *n.* **1** (in full **diesel engine**) an internal-combustion engine in which the heat produced by the compression of air in the cylinder ignites the fuel. (*See note below.*) **2** a vehicle driven by a diesel engine. **3** fuel for a diesel engine. □ **diesel-electric** *n.* a vehicle driven by the electric current produced by a diesel-engined generator. ● *adj.* of or powered by this means. **diesel oil** a heavy petroleum fraction used as fuel in diesel engines. □ **dieselize** *v.tr.* (also **-ise**).

▪ The diesel engine was patented by the German engineer Dr Rudolf Diesel (1858–1913) in 1892. Air is compressed alone, inside the engine, causing its temperature to rise; fuel oil is sprayed into the combustion chamber when the air is hot enough to ignite the fuel, raising the temperature and pressure of the mixture to very high values at the start of the expansion process. A high compression ratio can be employed with no danger of the explosive combustion of a fuel/air mixture. The consequent high expansion ratio gives the diesel engine a higher thermal efficiency than petrol engines, over 40 per cent in large engines. Diesel engines have been mainly used to power ships, railway locomotives, commercial road vehicles, and electrical generators, although use in smaller vehicles such as private cars is increasing.

Dies irae /ˌdiːeɪz ˈɪəraɪ/ *n.* a Latin hymn sung in a mass for the dead. [L (the first words of the hymn), = day of wrath]

dies non /ˌdaɪiːz ˈnɒn/ *n. Law* **1** a day on which no legal business can be done. **2** a day that does not count for legal purposes. [L, short for *dies non juridicus* non-judicial day]

diet¹ /ˈdaɪət/ *n. & v.* ● *n.* **1** the kinds of food that a person or animal habitually eats. **2** a special course of food to which a person is restricted, esp. for medical reasons or to control weight. **3** a regular occupation or series of activities to which one is restricted or which form one's main concern, usu. for a purpose (*a diet of light reading and fresh air*). ● *v.* (**dieted**, **dieting**) **1** *intr.* restrict oneself to small amounts or special kinds of food, esp. to control one's weight. **2** *tr.* restrict (a person or animal) to a special diet. □ **dieter** *n.* [ME f. OF *diete* (n.), *dieter* (v.) f. L *diaeta* f. Gk *diaita* a way of life]

diet² /ˈdaɪət/ *n.* **1** a legislative assembly in certain countries. **2** *hist.* a national or international conference, esp. of a federal state or confederation. **3** *Sc. Law* a meeting or session of a court. [ME f. med.L *dieta* day's work, wages, etc.]

dietary /ˈdaɪətərɪ/ *adj. & n.* ● *adj.* of or relating to a diet. ● *n.* (*pl.* **-ies**) a regulated or restricted diet. [ME f. med.L *dietarium* (as DIET¹)]

dietetic /ˌdaɪəˈtetɪk/ *adj.* of or relating to diet. □ **dietetically** *adv.* [L *dieteticus* f. Gk *diaitētikos* (as DIET¹)]

dietetics /ˌdaɪəˈtetɪks/ *n.pl.* (usu. treated as *sing.*) the scientific study of diet and nutrition.

dietitian /ˌdaɪəˈtɪʃ(ə)n/ *n.* (also **dietician**) an expert in dietetics.

Diet of Worms a meeting of the Holy Roman emperor Charles V's imperial diet at Worms in Germany in 1521, at which Martin Luther was summoned to appear. Luther committed himself there to the cause of Protestant reform, and on the last day of the diet his teaching was formally condemned in the Edict of Worms.

Dietrich /ˈdiːtrɪx/, Marlene (born Maria Magdelene von Losch) (1901–92), German-born American actress and singer. She became famous for her part as Lola in *Der Blaue Engel* (1930; *The Blue Angel*), directed by Josef von Sternberg. It was her last film before going to the US, where she and von Sternberg made a series of films together, such as *Blonde Venus* (1932) and *The Devil is a Woman* (1935). From the 1950s she was also successful as an international cabaret star. She became increasingly reclusive towards the end of her life.

dif- /dɪf/ *prefix* assim. form of DIS- before *f.* [L var. of DIS-]

differ /ˈdɪfə(r)/ *v.intr.* **1** (often foll. by *from*) be unlike or distinguishable. **2** (often foll. by *with*) disagree; be at variance (with a person). [ME f. OF *differer* f. L *differre*, differ, DEFER¹, (as DIS-, *ferre* bear, tend)]

difference /ˈdɪfrəns/ *n. & v.* ● *n.* **1** the state or condition of being different or unlike. **2** a point in which things differ; a distinction. **3** a degree of unlikeness. **4 a** the quantity by which amounts differ; a deficit (*will have to make up the difference*). **b** the remainder left after subtraction. **5 a** a disagreement, quarrel, or dispute. **b** the grounds of disagreement (*put aside their differences*). **6** a notable change (*the difference in his behaviour is remarkable*). **7** *Heraldry* an alteration in a coat of arms distinguishing members of a family. ● *v.tr. Heraldry* alter (a coat of arms) to distinguish members of a family. □ **make a** (or **all the** etc.)

difference (often foll. by *to*) have a significant effect or influence (on a person, situation, etc.). **make no difference** (often foll. by *to*) have no effect (on a person, situation, etc.). **with a difference** having a new or unusual feature. [ME f. OF f. L *differentia* (as DIFFERENT)]

different /ˈdɪfrənt/ *adj.* **1** (often foll. by *from*, *to*, *than*) unlike, distinguishable in nature, form, or quality (from another). ¶ *Different from* is generally regarded as the most acceptable collocation; *to* is common in less formal use; *than* is established in US use and also found in British use, esp. when followed by a clause, e.g. *I am a different person than I was a year ago.* **2** distinct, separate; not the same one (as another). **3** *colloq.* unusual (*wanted to do something different*). **4** of various kinds; assorted, several, miscellaneous (*available in different colours*). □ **differently** *adv.* **differentness** *n.* [ME f. OF *different* f. L *different-* (as DIFFER)]

differentia /ˌdɪfəˈrenʃɪə/ *n.* (*pl.* **differentiae** /-ʃɪˌiː/) a distinguishing mark, esp. between species within a genus. [L: see DIFFERENCE]

differential /ˌdɪfəˈrenʃ(ə)l/ *adj. & n.* ● *adj.* **1 a** of, exhibiting, or depending on a difference. **b** varying according to circumstances. **2** constituting a specific difference; distinctive; relating to specific differences (*differential diagnosis*). **3** *Math.* relating to infinitesimal differences. **4** *Physics & Mech.* concerning the difference of two or more motions, pressures, etc. ● *n.* **1** a difference between individuals or examples of the same kind. **2** *Brit.* a difference in wage or salary between industries or categories of employees in the same industry. **3** a difference between rates of interest etc. **4** *Math.* **a** an infinitesimal difference between successive values of a variable. **b** a function expressing this as a rate of change with respect to another variable. **5** (in full **differential gear**) a gear allowing a vehicle's driven wheels to revolve at different speeds in cornering. □ **differential calculus** *Math.* the part of calculus concerned with derivatives and differentiation (cf. *integral calculus*). **differential coefficient** *Math.* = DERIVATIVE. **differential equation** *Math.* an equation involving differentials among its quantities. □ **differentially** *adv.* [med. & mod.L *differentialis* (as DIFFERENCE)]

differentiate /ˌdɪfəˈrenʃɪˌeɪt/ *v.* **1** *tr.* constitute a difference between or in. **2** *tr. &* (often foll. by *between*) *intr.* find differences (between); discriminate. **3** *tr. & intr.* make or become different in the process of growth or development (species, word-forms, etc.). **4** *tr. Math.* transform (a function) into its derivative. □ **differentiator** *n.* **differentiation** /-ˌrenʃɪˈeɪ(ə)n/ *n.* [med.L *differentiare differentiat-* (as DIFFERENCE)]

difficult /ˈdɪfɪkəlt/ *adj.* **1 a** needing much effort or skill. **b** troublesome, perplexing. **2** (of a person): **a** not easy to please or satisfy. **b** uncooperative, troublesome. **3** characterized by hardships or problems (*a difficult period in his life*). □ **difficultly** *adv.* **difficultness** *n.* [ME, back-form. f. DIFFICULTY]

difficulty /ˈdɪfɪkəltɪ/ *n.* (*pl.* **-ies**) **1** the state or condition of being difficult. **2 a** a difficult thing; a problem or hindrance. **b** (often in *pl.*) a cause of distress or hardship (*in financial difficulties; there was someone in difficulties in the water*). □ **make difficulties** be intransigent or unaccommodating. **with difficulty** not easily. [ME f. L *difficultas* (as DIS-, *facultas* FACULTY)]

diffident /ˈdɪfɪd(ə)nt/ *adj.* **1** shy, lacking self-confidence. **2** excessively modest and reticent. □ **diffidence** *n.* **diffidently** *adv.* [L *diffidere* (as DIS-, *fidere* trust)]

diffract /dɪˈfrækt/ *v.tr. Physics* cause (light etc.) to undergo diffraction. □ **diffractive** *adj.* **diffractively** *adv.* [L *diffringere diffract-* (as DIS-, *frangere* break)]

diffraction /dɪˈfrækʃ(ə)n/ *n. Physics* **1** the breaking up of a beam of light into a series of dark or light bands or coloured spectra by the edge of an opaque body or a narrow slit. **2** the similar breaking up of a beam of radiation or particles into a series of alternating high and low densities. □ **diffraction grating** a plate of glass or metal ruled with very close parallel lines, producing a spectrum by diffraction and interference of light.

diffractometer /ˌdɪfrækˈtɒmɪtə(r)/ *n. Physics* an instrument for measuring diffraction, esp. in crystallographic work.

diffuse *adj. & v.* ● *adj.* /dɪˈfjuːs/ **1** (of light, inflammation, etc.) spread out, not concentrated or localized. **2** (of prose, speech, etc.) not concise, long-winded, verbose. ● *v.tr. & intr.* /dɪˈfjuːz/ **1** disperse or be dispersed from a centre. **2** spread or be spread widely; reach a large area. **3** *Physics* intermingle by diffusion. □ **diffusely** /-ˈfjuːslɪ/ *adv.* **diffuseness** /-ˈfjuːsnɪs/ *n.* **diffusible** /-ˈfjuːzɪb(ə)l/ *adj.* **diffusive** /-ˈfjuːsɪv/ *adj.* [ME f. F *diffus* or L *diffusus* extensive (as DIS-, *fusus* past part. of *fundere* pour)]

diffuser /dɪˈfjuːzə(r)/ *n.* (also **diffusor**) **1** a person or thing that diffuses, esp. a device for diffusing light. **2** *Engin.* a duct for broadening an airflow and reducing its speed.

diffusion /dɪˈfjuːʒ(ə)n/ *n.* **1** the act or an instance of diffusing; the process of being diffused. **2** *Physics & Chem.* the interpenetration of substances by the natural movement of their particles. **3** *Anthropol.* the spread of elements of culture etc. to another region or people. □ **diffusionist** *n.* [ME f. L *diffusio* (as DIFFUSE)]

dig /dɪɡ/ *v. & n.* ● *v.* (**digging**; *past* and *past part.* **dug** /dʌɡ/) **1** *intr.* break up and remove or turn over soil, ground, etc., with a tool, one's hands, (of an animal) claws, etc. **2** *tr.* **a** break up and displace (the ground etc.) in this way. **b** (foll. by *up*) break up the soil of (fallow land). **3** *tr.* make (a hole, grave, tunnel, etc.) by digging. **4** *tr.* (often foll. by *up*, *out*) **a** obtain or remove by digging. **b** find or discover after searching. **5** *tr.* (also *absol.*) excavate (an archaeological site). **6** *tr. sl.* like, appreciate, or understand. **7** *tr. & intr.* (foll. by *in*, *into*) thrust or poke into or down into. **8** *intr.* make one's way by digging (*dug through the mountainside*). **9** *intr.* (usu. foll. by *into*) investigate or study closely; probe. ● *n.* **1** a piece of digging. **2** a thrust or poke (*a dig in the ribs*). **3** (often foll. by *at*) *colloq.* a pointed or critical remark. **4** an archaeological excavation. **5** (in *pl.*) *Brit. colloq.* lodgings. □ **dig one's feet** (or **heels** or **toes**) **in** be obstinate. **dig in** *colloq.* begin eating. **dig oneself in 1** prepare a defensive trench or pit. **2** establish one's position. [ME *digge*, of uncert. orig.: cf. OE *dīc* ditch]

Digambara /dɪˈɡʌmbərə/ *n.* a member of one of two principal sects of Jainism (the other is that of the Svetambaras), which was formed as a result of doctrinal schism in about AD 80 and continues today in parts of southern India. The sect's adherents reject property ownership and usually do not wear clothes. [Skr., = sky-clad]

digamma /daɪˈɡæmə/ *n.* the sixth letter (*F*, *ϝ*) of the early Greek alphabet (prob. pronounced w), later disused. [L f. Gk (as DI-¹, GAMMA)]

digastric /daɪˈɡæstrɪk/ *adj. & n. Anat.* ● *adj.* (of a muscle) having two wide parts with a tendon between. ● *n.* the muscle that opens the jaw. [mod.L *digastricus* (as DI-¹, Gk *gastēr* belly)]

digest *v. & n.* ● *v.tr.* /daɪˈdʒest, dɪ-/ **1** break down (food) in the stomach and bowels into simpler molecules that can be assimilated by the body. **2** understand and assimilate mentally. **3** *Chem.* treat (a substance) with heat, enzymes, or a solvent in order to decompose it, extract the essence, etc. **4 a** reduce to a systematic or convenient form; classify; summarize. **b** think over; arrange in the mind. **5** bear without resistance; brook, endure. ● *n.* /ˈdaɪdʒest/ **1 a** a methodical summary esp. of a body of laws. **b** (**the Digest**) the compendium of Roman law compiled (*c.*533 AD) during the reign of Justinian. **2 a** a compendium or summary of information; a résumé. **b** a regular or occasional synopsis of current literature or news. □ **digester** /daɪˈdʒestə(r), dɪ-/ *n.* **digestible** /daɪˈdʒestɪb(ə)l, dɪ-/ *adj.* **digestibility** /daɪˌdʒestɪˈbɪlɪtɪ, dɪ-/ *n.* [ME f. L *digerere digest-* distribute, dissolve, digest (as DI-², *gerere* carry)]

digestion /daɪˈdʒestʃ(ə)n, dɪ-/ *n.* **1** the process of digesting. **2** the capacity to digest food (*has a weak digestion*). **3** digesting a substance by means of heat, enzymes, or a solvent. [ME f. OF f. L *digestio -onis* (as DIGEST)]

digestive /dɪˈdʒestɪv, daɪ-/ *adj. & n.* ● *adj.* **1** of or relating to digestion. **2** aiding or promoting digestion. ● *n.* **1** a substance that aids digestion. **2** (in full **digestive biscuit**) *Brit.* a usu. round semi-sweet wholemeal biscuit. □ **digestively** *adv.* [ME f. OF *digestif -ive* or L *digestivus* (as DIGEST)]

Digger /ˈdɪɡə(r)/ *n.* a member of a group of radical dissenters formed in England in 1649 as an offshoot of the Levellers and led by Gerrard Winstanley (1609–72). They believed in a form of agrarian communism in which common land would be made available to the poor. The movement spread to a number of counties in southern England and the Midlands, but was suppressed by local landowners within a year.

digger /ˈdɪɡə(r)/ *n.* **1** a person or machine that digs, esp. a mechanical excavator. **2** a miner, esp. a gold-digger. **3** *colloq.* an Australian or New Zealander, esp. a private soldier. **4** *Austral. & NZ colloq.* (as a form of address) mate, fellow. □ **digger wasp** a solitary wasp that burrows in the soil.

diggings /ˈdɪɡɪŋz/ *n.pl.* **1 a** a mine or gold-field. **b** material dug out of a mine etc. **2** *Brit. colloq.* lodgings, accommodation.

dight /daɪt/ *adj. archaic* clothed, arrayed. [past part. of *dight* (v.) f. OE *dihtan* f. L *dictare* DICTATE]

digit /ˈdɪdʒɪt/ *n.* **1** any numeral from 0 to 9, esp. when forming part of a number. **2** *Anat. & Zool.* a finger, thumb, or toe. [ME f. L *digitus*]

digital /ˈdɪdʒɪt(ə)l/ adj. **1** designating, relating to, operating with, or created using, signals or information represented by digits (a digital recording). **2** (of a clock, watch, etc.) giving a reading by means of displayed digits as opposed to hands. **3** (of a computer) operating on data represented as a series of usu. binary digits or in similar discrete form. □ **digital audio tape** magnetic tape on which sound is recorded digitally. **digital to analog converter** Computing a device for converting digital values to analog form. □ **digitalize** v.tr. (also **-ise**). **digitally** adv. [L digitalis (as DIGIT)]

digitalin /ˌdɪdʒɪˈteɪlɪn/ n. the pharmacologically active constituent(s) of the foxglove. [DIGITALIS + -IN]

digitalis /ˌdɪdʒɪˈteɪlɪs/ n. a drug prepared from the dried leaves of foxgloves and containing substances that stimulate the heart muscle. [mod.L, genus name of foxglove after G Fingerhut thimble: see DIGITAL]

digitate /ˈdɪdʒɪˌteɪt/ adj. **1** Zool. having separate fingers or toes. **2** Bot. having deep radiating divisions. □ **digitately** adv. **digitation** /ˌdɪdʒɪˈteɪʃ(ə)n/ n. [L digitatus (as DIGIT)]

digitigrade /ˈdɪdʒɪtɪˌɡreɪd/ adj. & n. Zool. ● adj. (of an animal) walking on its toes and not touching the ground with its heels, e.g. dogs, cats, and rodents. ● n. a digitigrade animal (cf. PLANTIGRADE). [F f. L digitus + -gradus -walking]

digitize /ˈdɪdʒɪˌtaɪz/ v.tr. (also **-ise**) convert (data etc.) into digital form, esp. for processing by a computer. □ **digitization** /ˌdɪdʒɪtaɪˈzeɪʃ(ə)n/ n.

dignified /ˈdɪɡnɪˌfaɪd/ adj. having or expressing dignity; noble or stately in appearance or manner. □ **dignifiedly** adv.

dignify /ˈdɪɡnɪˌfaɪ/ v.tr. (**-ies, -ied**) **1** give dignity or distinction to. **2** ennoble; make worthy or illustrious. **3** give the form or appearance of dignity to (dignified the house with the name of mansion). [obs. F dignifier f. OF dignefier f. LL dignificare f. dignus worthy]

dignitary /ˈdɪɡnɪtərɪ/ n. (pl. **-ies**) a person holding high rank or office. [DIGNITY + -ARY¹, after PROPRIETARY]

dignity /ˈdɪɡnɪtɪ/ n. (pl. **-ies**) **1** a composed and serious manner or style. **2** the state of being worthy of honour or respect. **3** worthiness, excellence (the dignity of work). **4** a high or honourable rank or position. **5** high regard or estimation. **6** self-respect. □ **beneath one's dignity** not considered worthy enough for one to do. **stand on one's dignity** insist (esp. by one's manner) on being treated with due respect. [ME f. OF digneté, dignité f. L dignitas -tatis f. dignus worthy]

digraph /ˈdaɪɡrɑːf/ n. a group of two letters representing one sound, as in ph and ey. □ **digraphic** /daɪˈɡræfɪk/ adj.

digress /daɪˈɡres/ v.intr. depart from the main subject temporarily in speech or writing. □ **digresser** n. **digressive** adj. **digressively** adv. **digressiveness** n. **digression** /-ˈɡreʃ(ə)n/ n. [L digredi digress- (as DI-², gradi walk)]

digs see DIG n. 5.

dihedral /daɪˈhiːdrəl/ adj. & n. Math. ● adj. having or contained by two plane faces. ● n. = dihedral angle. □ **dihedral angle** an angle formed by two plane surfaces, esp. by an aircraft wing with the horizontal. [dihedron f. DI-¹ + -HEDRON]

dihybrid /daɪˈhaɪbrɪd/ n. & adj. Biol. ● n. a hybrid that is heterozygous for alleles of two different genes. ● adj. relating to inheritance of alleles of two genes.

dihydric /daɪˈhaɪdrɪk/ adj. Chem. (of an alcohol) containing two hydroxyl groups. [DI-¹ + HYDROGEN + -IC]

Dijon /ˈdiːʒɒn/ an industrial city in east central France, the former capital of Burgundy; pop. (1990) 151,640.

dik-dik /ˈdɪkdɪk/ n. a dwarf antelope of the genus Madoqua, native to Africa. [name in E. Africa and in Afrik.]

dike¹ var. of DYKE¹.

dike² var. of DYKE².

diktat /ˈdɪktæt/ n. a categorical statement or decree, esp. terms imposed after a war by a victor. [G, = DICTATE]

dilapidate /dɪˈlæpɪˌdeɪt/ v.intr. & tr. fall or cause to fall into disrepair or ruin. [L dilapidare demolish, squander (as DI-², lapis lapid- stone)]

dilapidated /dɪˈlæpɪˌdeɪtɪd/ adj. in a state of disrepair or ruin, esp. as a result of age or neglect.

dilapidation /dɪˌlæpɪˈdeɪʃ(ə)n/ n. **1 a** the process of dilapidating. **b** a state of disrepair. **2** (in pl.) repairs required at the end of a tenancy or lease. **3** Eccl. a sum charged against an incumbent for wear and tear during a tenancy. [ME f. LL dilapidatio (as DILAPIDATE)]

dilatation /ˌdaɪləˈteɪʃ(ə)n/ n. **1** the widening or expansion of a hollow organ or cavity. **2** the process of dilating. □ **dilatation and**

curettage an operation in which the cervix is expanded and the womb-lining scraped off with a curette.

dilate /daɪˈleɪt/ v. **1** tr. & intr. make or become wider or larger (esp. of an opening in the body) . **2** intr. (often foll. by on, upon) speak or write at length. □ **dilatable** adj. **dilation** /-ˈleɪʃ(ə)n/ n. [ME f. OF dilater f. L dilatare spread out (as DI-², latus wide)]

dilator /daɪˈleɪtə(r)/ n. **1** Anat. a muscle that dilates an organ or aperture. **2** Surgery an instrument for dilating a tube or cavity in the body.

dilatory /ˈdɪlətərɪ/ adj. given to or causing delay. □ **dilatorily** adv. **dilatoriness** n. [LL dilatorius (as DI-², dilat- past part. stem of differre DEFER¹)]

dildo /ˈdɪldəʊ/ n. (pl. **-os**) an object shaped like an erect penis and used, esp. by women, for sexual stimulation. [17th c.: orig. unkn.]

dilemma /daɪˈlemə, dɪ-/ n. **1** a situation in which a choice has to be made between two equally undesirable alternatives. **2** a state of indecision between two alternatives. **3** disp. a difficult situation. **4** an argument forcing an opponent to choose either of two unfavourable alternatives. [L f. Gk (as DI-¹, lēmma premiss)]

dilettante /ˌdɪlɪˈtæntɪ/ n. & adj. ● n. (pl. **dilettanti** pronunc. same or **dilettantes**) **1** a person who studies a subject or area of knowledge superficially. **2** a person who enjoys the arts. ● adj. trifling, not thorough; amateurish. □ **dilettantish** adj. **dilettantism** n. [It. f. pres. part. of dilettare delight f. L delectare]

Dili /ˈdiːliː/ a seaport on the Indonesian island of Timor, which was (until 1975) the capital of the former Portuguese colony of East Timor; pop. (1980) 60,150.

diligence¹ /ˈdɪlɪdʒəns/ n. **1** careful and persistent application or effort. **2** (as a characteristic) industriousness. [ME f. OF f. L diligentia (as DILIGENT)]

diligence² /ˈdɪlɪdʒəns/ n. hist. a public stagecoach, esp. in France. [F, for carrosse de diligence coach of speed]

diligent /ˈdɪlɪdʒ(ə)nt/ adj. **1** careful and steady in application to one's work or duties. **2** showing care and effort. □ **diligently** adv. [ME f. OF f. L diligens assiduous, part. of diligere love, take delight in (as DI-², legere choose)]

dill¹ /dɪl/ n. **1** an umbelliferous herb, Anethum graveolens, with yellow flowers and aromatic seeds. **2** the leaves or seeds of this plant used for flavouring and medicinal purposes. □ **dill pickle** pickled cucumber etc. flavoured with dill. **dill-water** a distillate of dill used as a carminative. [OE dile]

dill² /dɪl/ n. Austral. sl. **1** a fool or simpleton. **2** the victim of a trickster. [app. back-form. f. DILLY²]

dilly¹ /ˈdɪlɪ/ n. (pl. **-ies**) esp. US sl. a remarkable or excellent person or thing. [dilly (adj.) f. DELIGHTFUL or DELICIOUS]

dilly² /ˈdɪlɪ/ adj. Austral. sl. **1** odd or eccentric. **2** foolish, stupid, mad. [perh. f. DAFT, SILLY]

dillybag /ˈdɪlɪˌbæɡ/ n. Austral. a small bag or basket. [Aboriginal dili + BAG]

dilly-dally /ˌdɪlɪˈdælɪ/ v.intr. (**-ies, -ied**) colloq. **1** dawdle, loiter. **2** vacillate. [redupl. of DALLY]

dilophosaurus /daɪˌləʊfəˈsɔːrəs/ n. (also **dilophosaur** /-ˈləʊfəˌsɔː(r)/) a medium-sized bipedal carnivorous dinosaur of the genus Dilophosaurus, of the Jurassic period, with two flat ridges on the skull. [mod.L f. DI-¹ + LOPHO- + Gk sauros lizard]

diluent /ˈdɪljʊənt/ adj. & n. Chem. & Biochem. ● adj. that serves to dilute. ● n. a diluting agent. [L diluere diluent- DILUTE]

dilute /daɪˈluːt, -ˈljuːt/ v. & adj. ● v.tr. **1** reduce the strength of (a fluid) by adding water or another solvent. **2** weaken or reduce the strength or forcefulness of, esp. by adding something. ● adj. (also /ˈdaɪluːt, -ljuːt/) **1** (esp. of a fluid) diluted, weakened. **2** (of a colour) washed out; low in saturation. **3** Chem. **a** (of a solution) having relatively low concentration of solute. **b** (of a substance) in solution (dilute sulphuric acid). □ **diluter** n. **dilution** /-ˈluːʃ(ə)n, -ˈljuːʃ(ə)n/ n. [L diluere dilut- (as DI-², luere wash)]

diluvial /daɪˈluːvɪəl, dɪ-, -ˈljuːvɪəl/ adj. **1** of a flood, esp. of the Flood in Genesis. **2** Geol. of or consisting of diluvium. [LL diluvialis f. diluvium DELUGE]

diluvium /daɪˈluːvɪəm, dɪ-, -ˈljuːvɪəm/ n. (pl. **diluvia** /-vɪə/) Geol. = DRIFT n. 8. [L: see DILUVIAL]

dim /dɪm/ adj. & v. ● adj. (**dimmer, dimmest**) **1 a** only faintly luminous or visible; not bright. **b** obscure; ill-defined. **2** not clearly perceived or remembered. **3** colloq. stupid; slow to understand. **4** (of the eyes) not seeing clearly. ● v. (**dimmed, dimming**) **1** tr. & intr. make or become dim or less bright. **2** tr. US dip (headlights). □ **dim-wit** colloq. a stupid person. **dim-witted** *colloq.* stupid, unintelligent. **take a dim view of** *colloq.* **1** disapprove of. **2** feel gloomy about. □ **dimly** *adv.* **dimmish** *adj.* **dimness** *n.* [OE *dim, dimm,* of unkn. orig.]

dim. *abbr.* diminuendo.

DiMaggio /dɪˈmædʒɪˌəʊ/, Joseph Paul ('Joe') (b.1914), American baseball player. Star of the New York Yankees team 1936–51, he was renowned for his outstanding batting ability and for his outfield play. He was briefly married to Marilyn Monroe in 1954.

Dimbleby /ˈdɪmb(ə)lbɪ/, (Frederick) Richard (1913–65), English broadcaster. He was the BBC's first news correspondent (1936) and the first broadcaster to be commemorated in Westminster Abbey. He distinguished himself in his radio and television commentaries on royal, national, and international events and in his reports on current affairs. His sons David (b.1938) and Jonathan (b.1944) have both followed their father into careers in news broadcasting.

dime /daɪm/ *n.* N. Amer. **1** a ten-cent coin. **2** *colloq.* a small amount of money. □ **a dime a dozen** very cheap or commonplace. **dime novel** a cheap popular novel. **turn on a dime** *US colloq.* make a sharp turn in a vehicle. [ME (orig. = tithe) f. OF *disme* f. L *decima pars* tenth part]

dimension /daɪˈmɛnʃn, dɪ-/ *n. & v.* ● *n.* **1** a measurable extent of any kind, as length, breadth, depth, area, and volume. **2** (in *pl.*) size, scope, extent. **3** an aspect or facet of a situation, problem, etc. **4** *Algebra* one of a number of unknown or variable quantities contained as factors in a product (x^3, x^2y, xyz, are all of three dimensions). **5** *Physics* the product of mass, length, time, etc., raised to the appropriate power, in a derived physical quantity. ● *v.tr.* (usu. as **dimensioned** *adj.*) mark the dimensions on (a diagram etc.). □ **dimensional** *adj.* (also in *comb.*). **dimensionless** *adj.* [ME f. OF f. L *dimensio -onis* (as DI-[2], *metiri mensus* measure)]

dimer /ˈdaɪmə(r)/ *n. Chem.* a compound consisting of two identical molecules linked together (cf. MONOMER). □ **dimeric** /daɪˈmɛrɪk/ *adj.* [DI-[1] + *-mer* after POLYMER]

dimerize /ˈdaɪməˌraɪz/ *v.tr. & intr.* (also **-ise**) *Chem.* convert or be converted to a dimer. □ **dimerization** /ˌdaɪmərəˈraɪzeɪʃ(ə)n/ *n.*

dimerous /ˈdaɪmərəs, ˈdɪm-/ *adj.* (of a plant) having two parts in a whorl etc. [mod.L *dimerus* f. Gk *dimerēs* bipartite]

dimeter /ˈdɪmɪtə(r)/ *n. Prosody* a line of verse consisting of two metrical feet. [LL *dimetrus* f. Gk *dimetros* (as DI-[1], *metron* measure)]

dimethylsulphoxide /daɪˌmiːθaɪlsʌlˈfɒksaɪd/ *n.* (*US* **-sulfoxide**) (abbr. **DMSO**) *Chem.* a colourless liquid (chem. formula: $(CH_3)_2SO$) used as a solvent and in medicine, esp. to treat inflammation. [DI-[1] + METHYL + SULPHUR + OXIDE]

dimetrodon /daɪˈmiːtrədən/ *n.* a large carnivorous quadrupedal reptile of the genus *Dimetrodon*, of the Permian period, with long spines on its back forming a sail-like crest. [mod.L f. DI-[1] + Gk *metron* measure + *odous odontos* tooth, in sense of 'two long teeth']

diminish /dɪˈmɪnɪʃ/ *v.* **1** *tr. & intr.* make or become smaller or less. **2** *tr.* lessen the reputation or influence of (a person). □ **law of diminishing returns** *Econ.* the fact that the increase of expenditure, investment, taxation, etc., beyond a certain point ceases to produce a proportionate yield. □ **diminishable** *adj.* [ME, blending of earlier *minish* f. OF *menuisier* (formed as MINCE) and *diminue* f. OF *diminuer* f. L *diminuere diminut-* break up small]

diminished /dɪˈmɪnɪʃt/ *adj.* **1** reduced; made smaller or less. **2** *Mus.* (of an interval, usu. a seventh or fifth) less by a semitone than the corresponding minor or perfect interval. □ **diminished responsibility** *Law* the limitation of criminal responsibility on the ground of mental weakness or abnormality.

diminuendo /dɪˌmɪnjʊˈɛndəʊ/ *n., adv., adj., & v. Mus.* ● *n.* (*pl.* **-os**) **1** a gradual decrease in loudness. **2** a passage to be performed with such a decrease. ● *adv. & adj.* with a gradual decrease in loudness. ● *v.intr.* (**-oes, -oed**) decrease gradually in loudness or intensity. [It., part. of *diminuire* DIMINISH]

diminution /ˌdɪmɪˈnjuːʃ(ə)n/ *n.* **1 a** the act or an instance of diminishing. **b** the amount by which something diminishes. **2** *Mus.* the repetition of a passage in notes shorter than those originally used. [ME f. OF f. L *diminutio -onis* (as DIMINISH)]

diminutive /dɪˈmɪnjʊtɪv/ *adj. & n.* ● *adj.* **1** remarkably small; tiny. **2** *Gram.* (of a word or suffix) implying smallness, either actual or imputed in token of affection, scorn, etc. (e.g. *-let, -kins*). ● *n. Gram.* a diminutive word or suffix. □ **diminutively** *adv.* **diminutiveness** *n.* **diminutival** /-ˌmɪnjʊˈtaɪv(ə)l/ *adj.* [ME f. OF *diminutif, -ive* f. LL *diminutivus* (as DIMINISH)]

dimissory /dɪˈmɪsərɪ/ *adj.* **1** ordering or permitting to depart. **2** *Eccl.* granting permission for a candidate to be ordained outside the bishop's own see (*dimissory letters*). [ME f. LL *dimissorius* f. *dimittere dimiss-* send away (as DI-[2], *mittere* send)]

dimity /ˈdɪmɪtɪ/ *n.* (*pl.* **-ies**) a cotton fabric woven with stripes or checks. [ME f. It. *dimito* or med.L *dimitum* f. Gk *dimitos* (as DI-[1], *mitos* warp-thread)]

dimmer /ˈdɪmə(r)/ *n.* **1** (in full **dimmer switch**) a device for varying the brightness of an electric light. **2** *US* **a** (in *pl.*) small parking lights on a motor vehicle. **b** a headlight on low beam.

dimorphic /daɪˈmɔːfɪk/ *adj.* (also **dimorphous** /-ˈmɔːfəs/) *Biol., Chem.,* & *Mineral.* exhibiting, or occurring in, two distinct forms. □ **dimorphism** *n.* [Gk *dimorphos* (as DI-[1], *morphē* form)]

dimple /ˈdɪmp(ə)l/ *n. & v.* ● *n.* a small hollow or dent in the flesh, esp. in the cheeks or chin. ● *v.* **1** *intr.* produce or show dimples. **2** *tr.* produce dimples in (a cheek etc.). □ **dimply** *adj.* [ME prob. f. OE *dympel* (unrecorded) f. a Gmc root *dump-*, perh. a nasalized form rel. to DEEP]

dim sum /dɪm ˈsʌm/ *n.* (also **dim sim** /ˈsɪm/) a Chinese dish of small steamed or fried savoury dumplings with various fillings, eaten as a snack or as part of a meal. [Cantonese *tim sam*, lit. 'dot of the heart']

DIN /dɪn/ *n.* any of a series of technical standards originating in Germany and used internationally, esp. to designate electrical connections, film speeds, and paper sizes. [G, f. *Deutsche Industrie-Norm*]

din /dɪn/ *n. & v.* ● *n.* a prolonged loud and distracting noise. ● *v.* (**dinned, dinning**) **1** *tr.* (foll. by *into*) instil (something to be learned) by constant repetition. **2** *intr.* make a din. [OE *dyne, dynn, dynian* f. Gmc]

dinar /ˈdiːnɑː(r)/ *n.* **1** the chief monetary unit of the states of the former Yugoslavia. **2** a monetary unit of certain countries of the Middle East and North Africa. [Arab. & Pers. *dīnār* f. Gk *dēnarion* f. L *denarius*: see DENIER]

Dinaric Alps /dɪˈnærɪk/ a mountain range in the Balkans, running parallel to the Adriatic coast from Slovenia in the north-west, through Croatia, Bosnia, and Montenegro, to Albania in the south-east.

dine /daɪn/ *v.* **1** *intr.* **a** eat dinner. **b** (foll. by *on, upon*) eat for dinner. **2** *tr.* give dinner to. □ **dine out 1** dine away from home. **2** (foll. by *on*) be entertained to dinner etc. on account of (one's ability to relate an interesting event, story, etc.). **dining-car** a railway carriage equipped as a restaurant. **dining-room** a room in which meals are eaten. [ME f. OF *diner, disner,* ult. f. DIS- + LL *jejunare* f. *jejunus* fasting]

diner /ˈdaɪnə(r)/ *n.* **1** a person who dines, esp. in a restaurant. **2** a railway dining-car. **3** *N. Amer.* a small restaurant. **4** a small dining-room.

dinette /daɪˈnɛt/ *n.* a small room or part of a room used for eating meals.

ding[1] /dɪŋ/ *v. & n.* ● *v.intr.* make a ringing sound. ● *n.* a ringing sound, as of a bell. [imit.: infl. by DIN]

ding[2] /dɪŋ/ *n. Austral. sl.* a party or celebration, esp. a wild one. [perh. f. DING-DONG or WINGDING]

Ding an sich /ˌdɪŋ æn ˈzɪx/ *n. Philos.* a thing in itself. [G]

dingbat /ˈdɪŋbæt/ *n. sl.* **1** *N. Amer. & Austral.* a stupid or eccentric person. **2** (in *pl.*) *Austral. & NZ* **a** madness. **b** discomfort, unease (*gives me the dingbats*). [19th c.: perh. f. *ding* to beat + BAT[1]]

ding-dong /ˈdɪŋdɒŋ/ *n., adj., & adv.* ● *n.* **1** the sound of alternate chimes, as of two bells. **2** *colloq.* an intense argument or fight. **3** *colloq.* a riotous party. ● *adj.* (of a contest etc.) evenly matched and intensely waged; thoroughgoing. ● *adv.* with vigour and energy (*hammer away at it ding-dong*). [16th c.: imit.]

dinge /dɪndʒ/ *n. & v.* ● *n.* a dent or hollow caused by a blow. ● *v.tr.* make such a dent in. [17th c.: orig. unkn.]

dinghy /ˈdɪŋɪ, ˈdɪŋgɪ/ *n.* (*pl.* **-ies**) **1** a small boat carried by a ship. **2** a small pleasure-boat. **3** a small inflatable rubber boat (esp. for emergency use). [orig. a rowing-boat used on Indian rivers, f. Hindi *ḍīṅgī, ḍeṅgī*]

dingle /ˈdɪŋg(ə)l/ *n.* a deep wooded valley or dell. [ME: orig. unkn.]

dingo /ˈdɪŋgəʊ/ *n.* (*pl.* **-oes** or **-os**) **1** a wild or half-domesticated yellowish-brown Australian dog, *Canis dingo*, originally introduced by early human settlers. **2** *Austral. sl.* a coward or scoundrel. [Dharuk *din-gu* or *dayn-gu* domesticated dingo]

dingy /ˈdɪndʒɪ/ *adj.* (**dingier, dingiest**) dirty-looking, drab, dull-coloured. □ **dingily** *adv.* **dinginess** *n.* [perh. ult. f. OE *dynge* DUNG]

dink /dɪŋk/ *n. & v. Austral. sl.* ● *n.* a ride on the handlebar of a bicycle etc. ● *v.tr.* carry (a person) in this way. [20th c.: orig. unkn.]

Dinka /ˈdɪŋkə/ *n. & adj.* ● *n.* **1** a member of a Sudanese people of the

Nile basin. **2** the language of this people. ● *adj.* of or relating to this people or their language. [Dinka *Jieng* people]

dinkum /'dɪŋkəm/ *adj. Austral. & NZ colloq.* genuine, right. □ **dinkum oil** the honest truth. **fair dinkum 1** fair play. **2** genuine(ly), honest(ly), true, truly. [19th c.: orig. unkn.]

dinky[1] /'dɪŋkɪ/ *adj.* (**dinkier, dinkiest**) *colloq.* **1** *Brit.* (esp. of a thing) neat and attractive; small, dainty. **2** *N. Amer.* trifling, insignificant. [Sc. *dink* neat, trim, of unkn. orig.]

dinky[2] /'dɪŋkɪ/ *n.* (*pl.* **-ies**) *colloq.* **1** a well-off young working couple with no children. **2** either partner of this. [contr. of *double income no kids* + -Y[2]]

dinner /'dɪnə(r)/ *n.* **1** the main meal of the day, taken either at midday or in the evening. **2** a formal evening meal, often in honour of a person or event. □ **dinner-dance** a formal dinner followed by dancing. **dinner-jacket** a man's short usu. black formal jacket for evening wear. **dinner lady** a woman who supervises children's lunch in a school. **dinner service** a set of usu. matching crockery for serving a meal. [ME f. OF *diner, disner*: see DINE]

dinoflagellate /ˌdaɪnəʊ'flædʒə,leɪt/ *n.* a unicellular aquatic organism with two flagella, of a group variously classed as algae or as protozoa. [mod.L *Dinoflagellata*, f. Gk *dinos* + L FLAGELLUM]

dinosaur /'daɪnə,sɔ:(r)/ *n.* **1** an extinct reptile of the Mesozoic era, often reaching enormous size. (*See note below.*) **2** something that has not adapted to changing circumstances, a clumsy survival from earlier times. □ **dinosaurian** /ˌdaɪnə'sɔ:rɪən/ *adj. & n.* [mod.L *dinosaurus* f. Gk *deinos* terrible + *sauros* lizard]

▪ The dinosaurs belong to two major groups, the lizard-hipped dinosaurs (order Saurischia) and the bird-hipped dinosaurs (order Ornithischia). The first of these are further divided into the theropods, which were mostly bipedal flesh-eaters, and the sauropods, which were mostly huge plant-eaters. The bird-hipped dinosaurs were all herbivores, and include the iguanodon, duck-billed dinosaurs, and armoured dinosaurs such as the stegosaurus and triceratops. Dinosaurs were clearly more active animals than modern reptiles, which suggests that they may have been warm-blooded, and their closest living relatives are the birds. Along with many other Mesozoic animals, they were all extinct by the end of the Cretaceous period (65 million years ago), though the cause is still debated.

dint /dɪnt/ *n. & v.* ● *n.* **1** a dent. **2** *archaic* a blow or stroke. ● *v.tr.* mark with dints. □ **by dint of** by force or means of. [ME f. OE *dynt*, and partly f. cognate ON *dyntr*: ult. orig. unkn.]

diocesan /daɪ'ɒsɪs(ə)n/ *adj. & n.* ● *adj.* of or concerning a diocese. ● *n.* the bishop of a diocese. [ME f. F *diocésain* f. LL *diocesanus* (as DIOCESE)]

diocese /'daɪəsɪs/ *n.* a district under the pastoral care of a bishop. [ME f. OF *diocise* f. LL *diocesis* f. L *dioecesis* f. Gk *dioikēsis* administration (as DI-[3], *oikeō* inhabit)]

Diocletian /ˌdaɪə'kli:ʃ(ə)n/ (full name Gaius Aurelius Valerius Diocletianus) (245–313), Roman emperor 284–305. Faced with military problems on many frontiers and insurrection in the provinces, in 286 he divided the empire between himself in the east and Maximian (d.310) in the west. In 293 he further divided the empire, giving Galerius (d.311) control of Illyricum and the valley of the River Danube, with Constantius Chlorus (d.306) ruling Gaul, Spain, and Britain. An enthusiast for the old Roman religion, tradition, and discipline, Diocletius insisted on the maintenance of Roman law in the provinces and launched the final harsh persecution of the Christians (303). He abdicated in 305.

diode /'daɪəʊd/ *n. Electronics* **1** a semiconductor allowing the flow of current in one direction only and having two terminals. **2** a thermionic valve having two electrodes. [DI-[1] + ELECTRODE]

dioecious /daɪ'i:ʃəs/ *adj.* **1** *Bot.* having male and female organs on separate plants. **2** *Zool.* having the two sexes in separate individuals (cf. MONOECIOUS). [DI-[1] + Gk *-oîkos* -housed]

dioestrus /daɪ'i:strəs/ *n.* (*US* **diestrus**) an interval between periods of oestrus. [DI-[1] + OESTRUS]

Diogenes /daɪ'ɒdʒɪ,ni:z/ (*c*.400–*c*.325 BC), Greek philosopher. He was the most famous of the Cynics and the pupil of Antisthenes (*c*.445–*c*.365 BC). He lived in Athens in extreme poverty and asceticism (according to legend, he lived in a tub) and was accordingly nicknamed *Kuōn* ('the dog'), from which the Cynics derived their name. He emphasized self-sufficiency and the need for natural, uninhibited behaviour, regardless of social conventions. Among the many stories told of him is that he took a lantern in daylight, saying that he was seeking an honest man.

diol /'daɪɒl/ *n. Chem.* any alcohol containing two hydroxyl groups in each molecule. [DI-[1] + -OL]

Dione /daɪ'əʊni/ **1** *Gk Mythol.* a Titan, the mother of Aphrodite. **2** *Astron.* satellite IV of Saturn, the twelfth closest to the planet, discovered by Giovanni Cassini in 1684 (diameter 1,120 km). It is probably composed mainly of ice, and the surface varies between heavily cratered areas and smooth plains, with some bright wispy markings.

Dionysiac /ˌdaɪə'nɪzɪ,æk/ *adj.* (also **Dionysian** /-zɪən/) **1** wildly sensual; unrestrained. **2** *Gk Mythol.* of or relating to Dionysus or his worship. [LL *Dionysiacus* f. L DIONYSUS]

Dionysius /ˌdaɪə'nɪsɪəs/ the name of two rulers of Syracuse:

Dionysius I (known as Dionysius the Elder) (*c*.430–367 BC), ruled 405–367. After establishing himself as a tyrannical ruler in 405, he waged three wars against the Carthaginians for control of Sicily, the third of which (383–*c*.375) resulted in his defeat at Cronium. Nevertheless, his reign made him the principal power in Greek Italy, after the capture of Rhegium (386) and other Greek cities in southern Italy.

Dionysius II (known as Dionysius the Younger) (*c*.397–*c*.344 BC), son of Dionysius I, ruled 367–357 and 346–344. He lacked his father's military ambitions and signed a peace treaty with Carthage in 367. Despite his patronage of philosophers, he resisted the attempt by Plato to turn him into a philosopher-king, in 366 banishing the wealthy Syracusan Dion (*c*.408–*c*.354), the proponent of the scheme. He was subsequently overthrown by Dion in 357.

Dionysius Exiguus /ɪg'zɪgjʊəs/ (died *c*.556), Scythian monk and scholar. He is famous for introducing the system of dates BC and AD that is still in use today, accepting 753 AUC as the year of the Incarnation; this has since been shown to be mistaken. He is said to have given himself the nickname *Exiguus* ('little'), as a sign of humility.

Dionysius of Halicarnassus (1st century BC), Greek historian, literary critic, and rhetorician. He lived in Rome from 30 BC. He is best known for his detailed history of Rome, written in Greek; this covers the period from the earliest times until the outbreak of the first Punic War (264 BC).

Dionysius the Areopagite /ˌærɪ'ɒpə,gaɪt/ (1st century AD), Greek churchman. His conversion by St Paul is recorded in Acts 17:34 and according to tradition he went on to become the first bishop of Athens. He was later confused with St Denis and with a mystical theologian, Pseudo-Dionysius the Areopagite (fl. 500), whose Neoplatonic Christian writings exercised a profound influence on medieval theology.

Dionysus /ˌdaɪə'naɪsəs/ *Gk Mythol.* a Greek god, son of Zeus and Semele, also called Bacchus; his worship entered Greece from Thrace *c*.1000 BC. Originally a god of the fertility of nature, associated with wild and ecstatic religious rites, in later traditions he is a god of wine who loosens inhibitions and inspires creativity in music and poetry.

Diophantine equation /ˌdaɪə'fæntɪn, -taɪn/ *n. Math.* an equation with integral coefficients for which integral solutions are required. [DIOPHANTUS]

Diophantus /ˌdaɪə'fæntəs/ (fl. prob. *c*.250 AD), Greek mathematician, of Alexandria. Diophantus was the first to attempt an algebraical notation. In his *Arithmetica* he showed how to solve simple and quadratic equations. His work led Pierre de Fermat to take up the theory of numbers, in which he made his famous discoveries.

dioptre /daɪ'ɒptə(r)/ *n.* (*US* **diopter**) *Optics* a unit of refractive power of a lens, equal to the reciprocal of its focal length in metres. [F *dioptre* f. L *dioptra* f. Gk *dioptra*: see DIOPTRIC]

dioptric /daɪ'ɒptrɪk/ *adj. Optics* **1** serving as a medium for sight; assisting sight by refraction (*dioptric glass; dioptric lens*). **2** of refraction; refractive. [Gk *dioptrikos* f. *dioptra* a kind of theodolite]

dioptrics /daɪ'ɒptrɪks/ *n.pl.* (treated as *sing.*) *Optics* the part of optics that deals with refraction.

Dior /'di:ɔ:(r)/, Christian (1905–57), French couturier. In 1947 he showed his first collection, featuring narrow-waisted tightly fitted bodices and full pleated skirts; this became known as the New Look and initially shocked the fashion world by its extravagance. He remained an influential figure in fashion, for example creating the first A-line garments, and built up a range of quality accessories. He discovered and trained Yves Saint Laurent.

diorama /ˌdaɪə'rɑ:mə/ *n.* **1** a scenic painting in which changes in colour and direction of illumination simulate a sunrise etc. **2** a small representation of a scene with three-dimensional figures, viewed through a window etc. **3** a small-scale model or film-set. □ **dioramic** /-'ræmɪk/ *adj.* [DI-[3] + Gk *horama -atos* f. *horaō* see]

diorite /'daɪə,raɪt/ *n. Geol.* a coarse-grained plutonic igneous rock containing quartz. □ **dioritic** /ˌdaɪəˈrɪtɪk/ adj. [F f. Gk diorizō distinguish]

Dioscuri /ˌdaɪəˈskʊərɪ, daɪˈɒskjʊrɪ/ Gk & Rom. Mythol. the twins Castor and Pollux (also called Polydeuces), born to Leda, wife of Tyndareus, after her seduction by Zeus. Castor was mortal, the son of Tyndareus; his twin, Pollux, was immortal, the son of Zeus; at Pollux's request they shared his immortality between them, spending half their time below the earth with Hades and the other half on Olympus. They are often identified with the constellation Gemini, and were the patrons of mariners. [Gk, = sons of Zeus]

dioxan /daɪˈɒks(ə)n/ n. (also **dioxane** /-seɪn/) Chem. a colourless toxic liquid (chem. formula: $C_4H_8O_2$) used as a solvent.

dioxide /daɪˈɒksaɪd/ n. Chem. an oxide containing two atoms of oxygen which are not linked together (carbon dioxide).

dioxin /daɪˈɒksɪn/ n. Chem. a toxic compound produced in some manufacturing processes. Dioxins are a group of compounds whose molecules contain two benzene rings joined by a pair of oxygen atoms, but the term is applied especially to a compound known fully as 2,3,7,8-tetrachlorodibenzo-para-dioxin (chem. formula: $C_{12}H_4O_2Cl_4$). This compound (also called TCDD) is a contaminant of some commercial insecticides, notably 2,4,5-T. It resists degradation in the environment and is a persistent pollutant. It is extremely poisonous and is known to cause nervous disorders, skin diseases, cancer, and birth defects. Several incidents involving dioxin have led to evacuation of contaminated areas, e.g. at Seveso near Milan in 1976. [f. DI-¹ + OX- + -IN]

DIP /dɪp/ n. Computing a form of integrated circuit consisting of a small plastic or ceramic slab with two parallel rows of pins. □ **DIP-switch** an arrangement of switches on a printer for selecting a printing mode. [abbr. of dual in-line package]

Dip. abbr. Diploma.

dip /dɪp/ v. & n. ● v. (**dipped, dipping**) **1** tr. put or let down briefly into liquid etc.; immerse. **2** intr. **a** go below a surface or level (the sun dipped below the horizon). **b** (of a level of income, activity, etc.) decline slightly, esp. briefly (profits dipped in May). **3** intr. extend downwards; take or have a downward slope (the road dips after the bend). **4** intr. go under water and emerge quickly. **5** intr. (foll. by into) **a** read briefly from (a book etc.). **b** take a cursory interest in (a subject). **6** (foll. by into) **a** intr. put a hand, ladle, etc., into a container to take something out. **b** tr. put (a hand etc.) into a container to do this. **c** intr. spend from or make use of one's resources (dipped into our savings). **7** tr. & intr. lower or be lowered, esp. in salute. **8** tr. Brit. lower the beam of (a vehicle's headlights) to reduce dazzle. **9** tr. colour (a fabric) by immersing it in dye. **10** tr. wash (sheep) by immersion in a vermin-killing liquid. **11** tr. make (a candle) by immersing a wick briefly in hot tallow. **12** tr. baptize by immersion. **13** tr. (often foll. by up, out of) remove or scoop up (liquid, grain, etc., or something from liquid). ● n. **1** an act of dipping or being dipped. **2** a liquid into which something is dipped. **3** a brief bathe in the sea, river, etc. **4** a brief downward slope, followed by an upward one, in a road etc. **5** a sauce or dressing into which food is dipped before eating. **6** a depression in the skyline. **7** Astron. & Surveying the apparent depression of the horizon from the line of observation, due to the curvature of the earth. **8** Physics the angle made with the horizontal at any point by the earth's magnetic field. **9** Geol. the angle a stratum makes with the horizon. **10** sl. a pickpocket. **11** a quantity dipped up. **12** a candle made by dipping. □ **dip-switch** a switch for dipping a vehicle's headlight beams. [OE dyppan f. Gmc: rel. to DEEP]

Dip. A.D. abbr. Diploma in Art and Design.

Dip. Ed. abbr. Diploma in Education.

dipeptide /daɪˈpeptaɪd/ n. Biochem. a peptide formed by the combination of two amino acids.

Dip. H.E. abbr. Diploma of Higher Education.

diphtheria /dɪfˈθɪərɪə, disp. dɪpˈθɪərɪə/ n. Med. an acute infectious bacterial disease with inflammation of a mucous membrane esp. of the throat, resulting in the formation of a false membrane causing difficulty in breathing and swallowing. □ **diphtherial** adj. **diphtheric** /-ˈθerɪk/ adj. **diphtheritic** /-θəˈrɪtɪk/ adj. **diphtheroid** /ˈdɪfθəˌrɔɪd, disp. ˈdɪpθə-/ adj. [mod.L f. F diphthérie, earlier diphthérite f. Gk diphthera skin, hide]

diphthong /ˈdɪfθɒŋ, disp. ˈdɪpθɒŋ/ n. **1** a speech sound in one syllable in which the articulation begins as for one vowel and moves towards another (as in coin, loud, and side). **2 a** a digraph representing the sound of a diphthong or single vowel (as in feat). **b** a compound vowel character; a ligature (as æ). □ **diphthongal** /dɪfˈθɒŋg(ə)l, disp. dɪpˈθɒŋ-/

adj. [F diphthongue f. LL diphthongus f. Gk diphthoggos (as DI-¹, phthoggos voice)]

diphthongize /ˈdɪfθɒŋˌaɪz, disp. ˈdɪpθɒŋ-/ v.tr. (also **-ise**) pronounce as a diphthong. □ **diphthongization** /ˌdɪfθɒŋaɪˈzeɪʃ(ə)n, disp. ˌdɪpθɒŋ-/ n.

diplo- /ˈdɪpləʊ/ comb. form double. [Gk diplous double]

diplococcus /ˌdɪpləʊˈkɒkəs/ n. (pl. **diplococci** /-ˈkɒksaɪ, -ˈkɒkaɪ/) Biol. any coccus that occurs mainly in pairs.

diplodocus /dɪˈplɒdəkəs, ˌdɪpləʊˈdəʊkəs/ n. a huge plant-eating dinosaur of the genus Diplodocus, of the Jurassic period, with a long neck and long slender tail. [mod.L f. DIPLO- + Gk dokos wooden beam]

diploid /ˈdɪplɔɪd/ adj. & n. Biol. ● adj. (of an organism or cell) having two complete sets of chromosomes per cell. ● n. a diploid cell or organism. [G (as DIPLO-, -OID)]

diploidy /ˈdɪplɔɪdɪ/ n. Biol. the condition of being diploid.

diploma /dɪˈpləʊmə/ n. **1** a certificate of qualification awarded by a college etc. **2** a document conferring an honour or privilege. **3** a state paper; an official document; a charter. □ **diplomaed** /-məd/ adj. (also **diploma'd**). [L f. Gk diplōma -atos folded paper f. diploō to fold f. diplous double]

diplomacy /dɪˈpləʊməsɪ/ n. **1 a** the management of international relations. **b** expertise in this. **2** adroitness in personal relations; tact. [F diplomatie f. diplomatique DIPLOMATIC after aristocratic]

diplomat /ˈdɪpləˌmæt/ n. **1** an official representing a country abroad; a member of a diplomatic service. **2** a tactful person. [F diplomate, back-form. f. diplomatique: see DIPLOMATIC]

diplomate /ˈdɪpləˌmeɪt/ n. esp. US a person who holds a diploma, esp. in medicine.

diplomatic /ˌdɪpləˈmætɪk/ adj. **1 a** of or involved in diplomacy. **b** skilled in diplomacy. **2** tactful; adroit in personal relations. **3** (of an edition etc.) exactly reproducing the original. □ **diplomatic bag** a container in which official mail etc. is dispatched to or from an embassy, not usu. subject to customs inspection. **diplomatic corps** the body of diplomats representing other countries at a seat of government. **diplomatic immunity** the exemption of diplomatic staff abroad from arrest, taxation, etc. **diplomatic service** Brit. the branch of public service concerned with the representation of a country abroad. □ **diplomatically** adv. [mod.L diplomaticus and F diplomatique f. L DIPLOMA]

diplomatist /dɪˈpləʊmətɪst/ n. = DIPLOMAT.

diplont /ˈdɪplənt/ n. Biol. an animal or plant which has a diploid number of chromosomes in all its cells except for the gametes. [DIPLO- + Gk ont- stem of ōn being]

diplopia /dɪˈpləʊpɪə/ n. Med. double vision.

diplotene /ˈdɪpləʊˌtiːn/ n. Biol. a stage during the prophase of meiosis where paired chromosomes begin to separate. [DIPLO- + Gk tainia band]

dipolar /daɪˈpəʊlə(r)/ adj. having two poles, as in a magnet.

dipole /ˈdaɪpəʊl/ n. **1** Physics two equal and oppositely charged or magnetized poles separated by a distance. **2** Chem. a molecule in which a concentration of positive charges is separated from a concentration of negative charges. **3** an aerial consisting of a horizontal metal rod with a connecting wire at its centre. □ **dipole moment** Physics the product of the separation of the charges etc. of a dipole and their magnitudes.

dipper /ˈdɪpə(r)/ n. **1** a short-tailed songbird, Cinclus cinclus, that swims or walks under water. Also called water ouzel. **2** a ladle. **3** colloq. an Anabaptist or Baptist.

dippy /ˈdɪpɪ/ adj. (**dippier, dippiest**) sl. crazy, silly. [20th c.: orig. uncert.]

dipshit /ˈdɪpʃɪt/ n. esp. US sl. a contemptible or inept person. [perh. f. DIPPY + SHIT]

dipso /ˈdɪpsəʊ/ n. (pl. **-os**) colloq. a dipsomaniac. [abbr.]

dipsomania /ˌdɪpsəˈmeɪnɪə/ n. an abnormal craving for alcohol. □ **dipsomaniac** /-nɪˌæk/ n. [Gk dipso- f. dipsa thirst + -MANIA]

dipstick /ˈdɪpstɪk/ n. **1** a graduated rod for measuring the depth of a liquid, esp. in a vehicle's engine. **2** coarse sl. the penis. **3** sl. a stupid or contemptible person.

Diptera /ˈdɪptərə/ n.pl. Zool. an order of insects comprising the two-winged or 'true' flies, which have the hindwings reduced to form halteres or balancing organs. It includes many biting forms such as mosquitoes and tsetse flies that are of great importance as vectors of disease. □ **dipteran** n. & adj. **dipterist** n. [mod.L f. Gk diptera neut. pl. of dipterous two-winged (as DI-¹, pteron wing)]

dipteral /'dɪptərəl/ adj. Archit. having a double peristyle. [L dipteros f. Gk (as DI-[1], pteron wing)]

dipterous /'dɪptərəs/ adj. **1** (of an insect) of the order Diptera. **2** Bot. having two winglike appendages. [mod.L dipterus f. Gk dipteros: see DIPTERA]

diptych /'dɪptɪk/ n. **1** a painting, esp. an altarpiece, on two hinged usu. wooden panels which may be closed like a book. **2** an ancient writing-tablet consisting of two hinged leaves with waxed inner sides. [LL diptycha f. Gk diptukha (as DI-[1], ptukhē fold)]

Dirac /dɪ'ræk/, Paul Adrian Maurice (1902–84), English theoretical physicist. He applied Einstein's theory of relativity to quantum mechanics in order to describe the behaviour of the electron, including its spin, and later predicted the existence of the positron. He also developed a quantum theory of radiation, and was the co-inventor of Fermi–Dirac statistics, which describe the behaviour of the particles later called fermions. Dirac shared the 1933 Nobel Prize for physics.

dire /'daɪə(r)/ adj. **1 a** calamitous, dreadful (in dire straits). **b** ominous (dire warnings). **c** (predic.) colloq. very bad. **2** urgent (in dire need). □ **direly** adv. **direness** n. [L dirus]

direct /daɪ'rekt, dɪ-/ adj., adv., & v. ● adj. **1** extending or moving in a straight line or by the shortest route; not crooked or circuitous. **2 a** straightforward; going straight to the point. **b** frank; not ambiguous. **3** without intermediaries or the intervention of other factors (direct rule; the direct result; made a direct approach). **4** (of descent) lineal, not collateral. **5** exact, complete, greatest possible (esp. where contrast is implied) (the direct opposite). **6** Mus. (of an interval or chord) not inverted. **7** Astron. (of planetary etc. motion) proceeding from East to West; not retrograde. ● adv. **1** in a direct way or manner; without an intermediary or intervening factor (dealt with them direct). **2** frankly; without evasion. **3** by a direct route (send it direct to London). ● v.tr. **1** control, guide; govern the movements of. **2** (foll. by to + infin., or that + clause) give a formal order or command to. **3** (foll. by to) **a** address or give indications for the delivery of (a letter etc.). **b** tell or show (a person) the way to a destination. **4** (foll. by at, to, towards) **a** point, aim, or cause (a blow or missile) to move in a certain direction. **b** point or address (one's attention, a remark, etc.). **5** guide as an adviser, as a principle, etc. (I do as duty directs me). **6 a** (also absol.) supervise the performing, staging, etc., of (a film, play, etc.). **b** supervise the performance of (an actor etc.). **7** (also absol.) guide the performance of (a group of musicians), esp. as a participant. □ **direct access** Computing the facility of retrieving data immediately from any part of a computer file. **direct action** action such as a strike or sabotage directly affecting the community and meant to reinforce demands on a government, employer, etc. **direct address** Computing an address which specifies the location of data to be used in an operation. **direct current** (abbr. **DC, dc**) an electric current flowing in one direction only. **direct debit** an arrangement for the regular debiting of a bank account at the request of the payee. **direct-grant school** hist. (in the UK) a school receiving funds from the government and not from a local authority. **direct mail** advertising sent unsolicited through the post to prospective customers. **direct mailing** the sending of direct mail. **direct method** a system of teaching a foreign language using only that language and without the study of formal grammar. **direct object** Gram. the primary object of the action of a transitive verb. **direct proportion** a relation between quantities whose ratio is constant. **direct speech** (or **oration**) words actually spoken, not reported in the third person (opp. reported speech (see REPORT)). **direct tax** a tax levied on the person who ultimately bears the burden of it, esp. on income. □ **directness** n. [ME f. L directus past part. of dirigere direct- (as DI-[2], regere put straight)]

direction /daɪ'rekʃ(ə)n, dɪ-/ n. **1** the act or process of directing; supervision. **2** (usu. in pl.) an order or instruction, esp. each of a set guiding use of equipment etc. **3 a** the course or line along which a person or thing moves or looks, or which must be taken to reach a destination (sailed in an easterly direction). **b** (in pl.) guidance on how to reach a destination. **c** the point to or from which a person or thing moves or looks. **4** the tendency or scope of a theme, subject, or inquiry. □ **direction-finder** a device for determining the source of radio waves, esp. as an aid in navigation. □ **directionless** adj. [ME f. F direction or L directio (as DIRECT)]

directional /daɪ'rekʃ(ə)l, dɪ-/ adj. **1** of or indicating direction. **2** Electronics **a** concerned with the transmission of radio or sound waves in a particular direction. **b** (of equipment) designed to receive radio or sound waves most effectively from a particular direction or directions and not others. □ **directionally** adv. **directionality** /-ˌrekʃə'nælɪtɪ/ n.

directive /daɪ'rektɪv, dɪ-/ n. & adj. ● n. a general instruction from one in authority. ● adj. serving to direct. [ME f. med.L directivus (as DIRECT)]

directly /daɪ'rektlɪ, dɪ-/ adv. & conj. ● adv. **1 a** at once; without delay. **b** presently, shortly. **2** exactly, immediately (directly opposite; directly after lunch). **3** in a direct manner. ● conj. colloq. as soon as (will tell you directly they come).

Directoire /ˌdɪrek'twa:(r)/ adj. relating to or imitating the ornate, neoclassical style of decoration and design which prevailed in France around the time of the Directory (1795–9) and preceded the Empire style. □ **Directoire drawers** (or **knickers**) knickers which are straight, full, and knee-length. [F (as DIRECTORY)]

director /daɪ'rektə(r), dɪ-/ n. **1** a person who directs or controls something. **2** a member of the managing board of a commercial company. **3** a person who directs a film, play, etc., esp. professionally. **4** a person acting as spiritual adviser. **5** esp. US = CONDUCTOR 1. □ **director-general** the chief executive of a large (esp. public) organization. **director of public prosecutions** Brit. = public prosecutor. □ **directorship** n. (esp. in sense 2). **directorial** /ˌdaɪrek'tɔ:rɪəl, ˌdɪ-/ adj. [AF directour f. LL director governor (as DIRECT)]

directorate /daɪ'rektərət, dɪ-/ n. **1** a board of directors. **2** the office of director.

Directory /daɪ'rektərɪ, dɪ-/ the French revolutionary government constituted in 1795, comprising two councils and an executive. The executive, composed of five members, represented an attempt to avoid the one-man dictatorship previously achieved by Robespierre during the Reign of Terror. It maintained an aggressive foreign policy, but proved too weak to control events at home and was overthrown by Napoleon Bonaparte in 1799. [transl. of F Directoire]

directory /daɪ'rektərɪ, dɪ-/ n. (pl. **-ies**) **1** a book listing alphabetically or thematically a particular group of individuals (e.g. telephone subscribers) or organizations with various details. **2** a computer file listing other files or programs etc. **3** a book of rules, esp. for the order of private or public worship. [LL directorium (as DIRECT)]

directress /daɪ'rektrɪs, dɪ-/ n. (also **directrice** /ˌdi:rek'tri:s/) a woman director. [DIRECTOR, F directrice (as DIRECTRIX)]

directrix /daɪ'rektrɪks, dɪ-/ n. (pl. **directrices** /-trɪˌsi:z/) Geom. a fixed line used in describing a curve or surface. [med.L f. LL director: see DIRECTOR, -TRIX]

direful /'daɪəˌfʊl/ adj. literary terrible, dreadful. □ **direfully** adv. [DIRE + -FUL]

dirge /dɜ:dʒ/ n. **1** a lament for the dead, esp. forming part of a funeral service. **2** any mournful song or lament. □ **dirgeful** adj. [ME f. L dirige (imper.) direct, the first word in the Latin antiphon (from Ps. 5:8) in the Matins part of the Office for the Dead]

dirham /'dɪəræm/ n. **1** the basic monetary unit of Morocco, equal to 100 centimes. **2** the basic monetary unit of the United Arab Emirates, equal to 100 fils. **3** a monetary unit of Libya and Qatar. [Arab. f. L DRACHMA]

dirigible /'dɪrɪdʒɪb(ə)l, dɪ'rɪdʒ-/ adj. & n. ● adj. capable of being guided. ● n. a dirigible balloon or airship. [L dirigere arrange, direct: see DIRECT]

diriment /'dɪrɪmənt/ adj. Law nullifying. □ **diriment impediment** a factor (e.g. the existence of a prior marriage) rendering a marriage null and void from the beginning. [L dirimere f. dir- = DIS- + emere take]

dirk /dɜ:k/ n. a short dagger, esp. as formerly worn by Scottish Highlanders. [17th-c. durk, of unkn. orig.]

dirndl /'dɜ:nd(ə)l/ n. **1** a woman's dress styled in imitation of Alpine peasant costume, with close-fitting bodice, tight waistband, and full skirt. **2** (in full **dirndl skirt**) a full skirt of this kind. [G dial., dimin. of Dirne girl]

dirt /dɜ:t/ n. **1** unclean matter that soils. **2 a** earth, soil. **b** earth, cinders, etc., used to make a surface for a road etc. (usu. attrib.: dirt track; dirt road). **3 a** foul or malicious words or talk. **b** scurrilous information, scandal; gossip. **4** excrement. **5** a dirty condition. **6** a person or thing considered worthless. □ **dirt bike** a motorcycle designed for use on unmade roads and tracks, esp. in scrambling. **dirt cheap** colloq. extremely cheap. **dirt-track** a course made of rolled cinders, soil, etc., for motorcycle racing or flat racing. **do a person dirt** sl. harm or injure a person's reputation maliciously. **eat dirt 1** suffer insults etc. without retaliating. **2** US make a humiliating confession. **treat like dirt** treat (a person) contemptuously; abuse. [ME f. ON drit excrement]

dirty /'dɜ:tɪ/ adj., adv., & v. ● adj. (**dirtier, dirtiest**) **1** soiled, unclean. **2** causing one to become dirty (a dirty job). **3** sordid, lewd; morally illicit or questionable (dirty joke). **4** unpleasant, nasty. **5** dishonest, dishonourable, unfair (dirty play). **6** (of weather) rough, squally. **7** (of a colour) not pure or clear, dingy. **8** colloq. (of a nuclear weapon) producing considerable radioactive fallout. ● adv. sl. (with adjectives expressing magnitude) very (*a dirty great diamond*). ● *v.tr. & intr.* (**-ies, -ied**) make or become dirty. □ **dirty dog** *colloq.* a scoundrel; a despicable person. **the dirty end of the stick** *colloq.* the difficult or unpleasant part of an undertaking, situation, etc. **dirty linen** (or **washing**) *colloq.* intimate secrets, esp. of a scandalous nature. **dirty look** *colloq.* a look of disapproval, anger, or disgust. **dirty money** extra money paid to those who handle dirty materials. **dirty trick 1** a dishonourable and deceitful act. **2** (in *pl.*) underhand political activity, esp. to discredit an opponent. **dirty weekend** *colloq.* a weekend spent clandestinely with a lover. **dirty word 1** an offensive or indecent word. **2** a word for something which is disapproved of (*profit is a dirty word*). **dirty work** dishonourable or illegal activity, esp. done clandestinely. **do the dirty on** *colloq.* play a mean trick on. □ **dirtily** *adv.* **dirtiness** *n.*

dis- /dɪs/ *prefix* forming nouns, adjectives, and verbs: **1** expressing negation (*dishonest*). **2** indicating reversal or absence of an action or state (*disengage; disbelieve*). **3** indicating removal of a thing or quality (*dismember; disable*). **4** indicating separation (*distinguish; dispose*). **5** indicating completeness or intensification of the action (*disembowel; disgruntled*). **6** indicating expulsion from (*disbar*). [L *dis-*, sometimes through OF *des-*]

disability /ˌdɪsəˈbɪlɪtɪ/ *n.* (*pl.* **-ies**) **1** physical incapacity, either congenital or caused by injury, disease, etc. **2** a lack of some asset, quality, or attribute, that prevents one's doing something. **3** a legal disqualification.

disable /dɪsˈeɪb(ə)l/ *v.tr.* **1** render unable to function. **2** (often as **disabled** *adj.*) deprive of physical or mental ability, esp. through injury or disease. □ **disablement** *n.*

disablist /dɪsˈeɪblɪst/ *adj.* discriminating or prejudiced against disabled people.

disabuse /ˌdɪsəˈbjuːz/ *v.tr.* **1** (foll. by *of*) free from a mistaken idea. **2** disillusion, undeceive.

disaccharide /daɪˈsækəˌraɪd/ *n. Chem.* a sugar whose molecule contains two linked monosaccharides.

disaccord /ˌdɪsəˈkɔːd/ *n. & v.* ● *n.* disagreement, disharmony. ● *v.intr.* (usu. foll. by *with*) disagree; be at odds. [ME f. F *désaccorder* (as ACCORD)]

disadvantage /ˌdɪsədˈvɑːntɪdʒ/ *n. & v.* ● *n.* **1** an unfavourable circumstance or condition. **2** damage to one's interest or reputation. ● *v.tr.* cause disadvantage to. □ **at a disadvantage** in an unfavourable position or aspect. [ME f. OF *desavantage*: see ADVANTAGE]

disadvantaged /ˌdɪsədˈvɑːntɪdʒd/ *adj.* placed in unfavourable circumstances (esp. of a person lacking the normal social opportunities).

disadvantageous /dɪsˌædvənˈteɪdʒəs/ *adj.* **1** involving disadvantage or discredit. **2** tending to the discredit of a person or thing; derogatory. □ **disadvantageously** *adv.*

disaffected /ˌdɪsəˈfektɪd/ *adj.* **1** disloyal, esp. to one's superiors. **2** estranged; no longer friendly; discontented. □ **disaffectedly** *adv.* [past part. of *disaffect* (v.), orig. = dislike, disorder (as DIS-, AFFECT²)]

disaffection /ˌdɪsəˈfekʃ(ə)n/ *n.* **1** disloyalty. **2** political discontent.

disaffiliate /ˌdɪsəˈfɪlɪˌeɪt/ *v.* **1** *tr.* end the affiliation of. **2** *intr.* end one's affiliation. **3** *tr. & intr.* detach. □ **disaffiliation** /-ˌfɪlɪˈeɪʃ(ə)n/ *n.*

disaffirm /ˌdɪsəˈfɜːm/ *v.tr. Law* **1** reverse (a previous decision). **2** repudiate (a settlement). □ **disaffirmation** /dɪsˌæfəˈmeɪʃ(ə)n/ *n.*

disafforest /ˌdɪsəˈfɒrɪst/ *v.tr. Brit.* **1** clear of forests or trees. **2** reduce from the legal status of forest to that of ordinary land. □ **disafforestation** /-ˌfɒrɪˈsteɪʃ(ə)n/ *n.* [ME f. AL *disafforestare* (as DIS-, AFFOREST)]

disagree /ˌdɪsəˈɡriː/ *v.intr.* (**-agrees, -agreed, -agreeing**) (often foll. by *with*) **1** hold a different opinion. **2** quarrel. **3** (of factors or circumstances) not correspond. **4** have an adverse effect on (a person's health, digestion, etc.). □ **disagreement** *n.* [ME f. OF *desagreer* (as DIS-, AGREE)]

disagreeable /ˌdɪsəˈɡriːəb(ə)l/ *adj.* **1** unpleasant, not to one's liking. **2** quarrelsome; rude or bad-tempered. □ **disagreeableness** *n.* **disagreeably** *adv.* [ME f. OF *desagreable* (as DIS-, AGREEABLE)]

disallow /ˌdɪsəˈlaʊ/ *v.tr.* refuse to allow or accept as valid; prohibit. □ **disallowance** *n.* [ME f. OF *desalouer* (as DIS-, ALLOW)]

disambiguate /ˌdɪsæmˈbɪɡjʊˌeɪt/ *v.tr.* remove ambiguity from. □ **disambiguation** /-ˌbɪɡjʊˈeɪʃ(ə)n/ *n.*

disamenity /ˌdɪsəˈmiːnɪtɪ, -ˈmenɪtɪ/ *n.* (*pl.* **-ies**) an unpleasant feature (of a place etc.); a disadvantage.

disappear /ˌdɪsəˈpɪə(r)/ *v.* **1** *intr.* cease to be visible; pass from sight.

2 *intr.* cease to exist or be in circulation or use (*trams had all but disappeared*). **3** *intr.* (of a person or thing) go missing. **4** *tr.* cause to disappear, esp. (in some Central and South American countries) secretly abduct and kill (a political enemy). □ **disappearance** *n.*

disappoint /ˌdɪsəˈpɔɪnt/ *v.tr.* **1** (also *absol.*) fail to fulfil a desire or expectation of (a person). **2** frustrate (hopes etc.); cause the failure of (a plan etc.). □ **be disappointed** (foll. by *with*, *at*, *in*, or *to* + infin., or *that* + clause) fail to have one's expectation etc. fulfilled in some regard (*was disappointed with you; disappointed at the result; am disappointed to be last*). □ **disappointed** *adj.* **disappointedly** *adv.* **disappointing** *adj.* **disappointingly** *adv.* [ME f. F *désappointer* (as DIS-, APPOINT)]

disappointment /ˌdɪsəˈpɔɪntmənt/ *n.* **1** an event, thing, or person that disappoints. **2** a feeling of distress, vexation, etc., resulting from this (*I cannot hide my disappointment*).

disapprobation /dɪsˌæprəˈbeɪʃ(ə)n/ *n.* strong (esp. moral) disapproval.

disapprove /ˌdɪsəˈpruːv/ *v.* **1** *intr.* (usu. foll. by *of*) have or express an unfavourable opinion. **2** *tr.* be displeased with. □ **disapproval** *n.* **disapprover** *n.* **disapproving** *adj.* **disapprovingly** *adv.*

disarm /dɪsˈɑːm/ *v.* **1** *tr.* **a** take weapons away from (a person, state, etc.) (often foll. by *of*: *were disarmed of their rifles*). **b** *Fencing* etc. deprive of a weapon. **2** *tr.* deprive (a ship etc.) of its means of defence. **3** *intr.* (of a state etc.) disband or reduce its armed forces. **4** *tr.* remove the fuse from (a bomb etc.). **5** *tr.* deprive of the power to injure. **6** *tr.* pacify or allay the hostility or suspicions of; mollify; placate. □ **disarmer** *n.* **disarming** *adj.* (esp. in sense 6). **disarmingly** *adv.* [ME f. OF *desarmer* (as DIS-, ARM²)]

disarmament /dɪsˈɑːməmənt/ *n.* the reduction by a state of its military forces and weapons.

disarrange /ˌdɪsəˈreɪndʒ/ *v.tr.* make disordered. □ **disarrangement** *n.*

disarray /ˌdɪsəˈreɪ/ *n. & v.* ● *n.* (often prec. by *in*, *into*) disorder, confusion (esp. among people). ● *v.tr.* throw into disorder.

disarticulate /ˌdɪsɑːˈtɪkjʊˌleɪt/ *v.tr. & intr.* separate at the joints. □ **disarticulation** /-ˌtɪkjʊˈleɪʃ(ə)n/ *n.*

disassemble /ˌdɪsəˈsemb(ə)l/ *v.tr.* take (a machine etc.) to pieces. □ **disassembly** *n.*

disassembler /ˌdɪsəˈsemblə(r)/ *n. Computing* a program for converting machine code into assembly language.

disassociate /ˌdɪsəˈsəʊʃɪˌeɪt, ˌdɪsəˈsəʊsɪˌeɪt/ *v.tr. & intr.* = DISSOCIATE. □ **disassociation** /-ˌsəʊsɪˈeɪʃ(ə)n/ *n.*

disaster /dɪˈzɑːstə(r)/ *n.* **1** a great or sudden misfortune. **2 a** a complete failure. **b** a person or enterprise ending in failure. □ **disaster area 1** an area in which a major disaster has recently occurred. **2** *colloq.* a person who tends to cause or be prone to accidents or misfortunes. **disaster movie** *colloq.* a film whose plot centres on a major disaster such as a fire, aircrash, etc. □ **disastrous** *adj.* **disastrously** *adv.* [orig. 'unfavourable aspect of a star', f. F *désastre* or It. *disastro* (as DIS-, *astro* f. L *astrum* star)]

disavow /ˌdɪsəˈvaʊ/ *v.tr.* disclaim knowledge of, responsibility for, or belief in. □ **disavowal** *n.* [ME f. OF *desavouer* (as DIS-, AVOW)]

disband /dɪsˈbænd/ *v.* **1** *intr.* (of an organized group etc.) cease to work or act together; disperse. **2** *tr.* cause (such a group) to disband. □ **disbandment** *n.* [obs. F *desbander* (as DIS-, BAND¹ 6)]

disbar /dɪsˈbɑː(r)/ *v.tr.* (**disbarred, disbarring**) *Law* deprive (a barrister) of legal status and privileges; expel from the Bar. □ **disbarment** *n.*

disbelieve /ˌdɪsbɪˈliːv/ *v.* **1** *tr.* be unable or unwilling to believe (a person or statement). **2** *intr.* have no faith. □ **disbelief** *n.* **disbeliever** *n.* **disbelievingly** *adv.*

disbound /dɪsˈbaʊnd/ *adj.* (of a pamphlet etc.) removed from a bound volume.

disbud /dɪsˈbʌd/ *v.tr.* (**disbudded, disbudding**) remove (esp. superfluous) buds from.

disburden /dɪsˈbɜːd(ə)n/ *v.tr.* **1** (often foll. by *of*) relieve (a person, one's mind, etc.) of a burden. **2** get rid of, discharge (a duty, anxiety, etc.).

disburse /dɪsˈbɜːs/ *v.* **1** *tr.* expend (money). **2** *tr.* defray (a cost). **3** *intr.* pay money. □ **disbursal** *n.* **disbursement** *n.* **disburser** *n.* [OF *desbourser* (as DIS-, BOURSE)]

disc /dɪsk/ *n.* (also **disk** esp. *US* and in sense 4) **1 a** a flat thin circular object. **b** a round flat or apparently flat surface (*the sun's disc*). **c** a mark of this shape. **d** *Bot.* the close-packed cluster of tubular florets in the centre of a composite flower. **2** *Anat.* a layer of cartilage between vertebrae. **3** a gramophone record. **4** (usu. **disk**) *Computing* **a** (in full

magnetic disk) a computer storage device consisting of a rotatable magnetically coated disc or discs. (See FLOPPY n., hard disk.) **b** (in full **optical disk**) a smooth non-magnetic disc with large storage capacity for data recorded and read by laser, esp. a CD-ROM. **5** a device with a pointer or rotating disc indicating time of arrival or latest permitted time of departure, for display in a parked motor vehicle. □ **disc brake** a brake employing the friction of pads against a disc. **disk drive** *Computing* a mechanism for rotating a disk and reading or writing data from or to it. **disc harrow** a harrow with cutting edges consisting of a row of concave discs set at an oblique angle. **disc jockey** a person who introduces and plays recorded music on a radio programme etc. [F *disque* or L *discus*: see DISCUS]

discalced /dɪs'kælst/ *adj.* (of a friar or a nun) barefoot or wearing only sandals. [var. of *discalceated* (after F *déchaux*) f. L *discalceatus* (as DIS-, *calceatus* f. *calceus* shoe)]

discard v. & n. ● *v.tr.* /dɪs'kɑːd/ **1** reject or get rid of as unwanted or superfluous. **2** (also *absol.*) *Cards* remove or put aside (a card) from one's hand. ● *n.* /'dɪskɑːd/ (often in *pl.*) a discarded item, esp. a card in a card-game. □ **discardable** /dɪs'kɑːdəb(ə)l/ *adj.* [DIS- + CARD¹]

discarnate /dɪs'kɑːnət/ *adj.* having no physical body; separated from the flesh. [DIS-, L *caro carnis* flesh]

discern /dɪ's3ːn/ *v.tr.* **1** perceive clearly with the mind or the senses. **2** make out by thought or by gazing, listening, etc. □ **discerner** n. **discernible** *adj.* **discernibly** *adv.* [ME f. OF *discerner* f. L (as DIS-, *cernere cret-* separate)]

discerning /dɪ's3ːnɪŋ/ *adj.* having or showing good judgement or insight. □ **discerningly** *adv.*

discernment /dɪ's3ːnmənt/ n. good judgement or insight.

discerptible /dɪ's3ːptɪb(ə)l/ *adj. literary* able to be plucked apart; divisible. □ **discerptibility** /-,s3ːptɪ'bɪlɪtɪ/ n. [L *discerpere discerpt-* (as DIS-, *carpere* pluck)]

discerption /dɪ's3ːpʃ(ə)n/ n. *archaic* **1 a** pulling apart; severance. **b** an instance of this. **2** a severed piece. [LL *discerptio* (as DISCERPTIBLE)]

discharge v. & n. ● v. /dɪs'tʃɑːdʒ/ **1** *tr.* **a** let go, release, esp. from a duty, commitment, or period of confinement. **b** relieve (a bankrupt) of residual liability. **2** *tr.* dismiss from office, employment, army commission, etc. **3** *tr.* **a** fire (a gun etc.). **b** (of a gun etc.) fire (a bullet etc.). **4** *tr.* (also *absol.*) pour out or cause to pour out (pus, liquid, etc.) (*the wound was discharging*). **b** *tr.* throw; eject. **c** *tr.* utter (abuse etc.). **d** *intr.* (foll. by *into*) (of a river etc.) flow into (esp. the sea). **5** *tr.* **a** carry out, perform (a duty or obligation). **b** relieve oneself of (a financial commitment) (*discharged his debt*). **6** *tr. Law* cancel (an order of court). **7** *tr. Physics* release an electrical charge from. **8** *tr.* **a** relieve (a ship etc.) of its cargo. **b** unload (a cargo) from a ship. ● n. /'dɪstʃɑːdʒ, dɪs'tʃɑːdʒ/ **1** the act or an instance of discharging; the process of being discharged. **2** a dismissal, esp. from the armed services. **3 a** a release, exemption, acquittal, etc. **b** a written certificate of release etc. **4** an act of firing a gun etc. **5 a** an emission (of pus, liquid, etc.). **b** the liquid or matter so discharged. **6** (usu. foll. by *of*) **a** the payment (of a debt). **b** the performance (of a duty etc.). **7** *Physics* **a** the release of a quantity of electric charge from an object. **b** a flow of electricity through the air or other gas, esp. when accompanied by the emission of light. **c** the conversion of chemical energy in a cell into electrical energy. **8** the unloading (of a ship or a cargo). □ **discharge tube** a tube containing gas at low pressure which gives off light when an electric current passes through it. □ **dischargeable** /dɪs'tʃɑːdʒəb(ə)l/ *adj.* **discharger** /dɪs'tʃɑːdʒə(r)/ n. (in sense 7 of v.). [ME f. OF *descharger* (as DIS-, CHARGE)]

disciple /dɪ'saɪp(ə)l/ n. **1** a follower or pupil of a leader, teacher, philosophy, etc. (*a disciple of Zen Buddhism*). **2** any early believer in Christ, esp. one of the twelve Apostles. □ **discipleship** n. **discipular** /-'sɪpjʊlə(r)/ *adj.* [OE *discipul* f. L *discipulus* f. *discere* learn]

disciplinarian /,dɪsɪplɪ'neərɪən/ n. a person who upholds or practises firm discipline (*a strict disciplinarian*).

disciplinary /'dɪsɪ,plɪnərɪ/ *adj.* of, promoting, or enforcing discipline. [med.L *disciplinarius* (as DISCIPLINE)]

discipline /'dɪsɪplɪn/ n. & v. ● n. **1 a** control or order exercised over people or animals, esp. children, prisoners, military personnel, church members, etc. **b** the system of rules used to maintain this control. **c** the behaviour of groups subjected to such rules (*poor discipline in the ranks*). **2 a** mental, moral, or physical training. **b** adversity as used to bring about such training (*left the course because he couldn't take the discipline*). **3** a branch of instruction or learning (*philosophy is a hard discipline*). **4** punishment. **5** *Eccl.* mortification by physical self-punishment, esp. scourging. ● *v.tr.* **1** punish, chastise. **2** bring under control by training in obedience; drill. □ **disciplinable** *adj.*

disciplinal /,dɪsɪ'plaɪn(ə)l, 'dɪsɪplɪn(ə)l/ *adj.* [ME f. OF *discipliner* or LL & med.L *disciplinare, disciplina* f. *discipulus* DISCIPLE]

disclaim /dɪs'kleɪm/ *v.tr.* **1** deny or disown (*disclaim all responsibility*). **2** (often *absol.*) *Law* renounce a legal claim to (property etc.). [ME f. AF *desclaim-* stressed stem of *desclamer* (as DIS-, CLAIM)]

disclaimer /dɪs'kleɪmə(r)/ n. **1** a renunciation or disavowal, esp. of responsibility. **2** *Law* an act of repudiating another's claim or renouncing one's own. [ME f. AF (= DISCLAIM as noun)]

disclose /dɪs'kləʊz/ *v.tr.* **1** make known; reveal (*disclosed the truth*). **2** remove the cover from; expose to view. □ **discloser** n. [ME f. OF *desclos-* stem of *desclore* f. Gallo-Roman (as DIS-, CLOSE²)]

disclosure /dɪs'kləʊʒə(r)/ n. **1** the act or an instance of disclosing; the process of being disclosed. **2** something disclosed; a revelation. [DISCLOSE + -URE after *closure*]

disco /'dɪskəʊ/ n. & v. *colloq.* ● n. (pl. **-os**) **1** = DISCOTHÈQUE. **2** = *disco music.* ● *v.intr.* (**-oes, -oed**) **1** attend a discothèque. **2** dance to disco music (*discoed the night away*). □ **disco music** a form of popular dance music characterized by a heavy bass rhythm, popular esp. in the 1970s. [abbr.]

discobolus /dɪs'kɒbələs/ n. (pl. **discoboli** /-,laɪ/) **1** a discus-thrower in ancient Greece. **2** a statue of a discobolus. [L f. Gk *diskobolos* f. *diskos* DISCUS + *-bolos* -throwing f. *ballō* to throw]

discography /dɪs'kɒgrəfɪ/ n. (pl. **-ies**) **1** a descriptive catalogue of gramophone records, esp. of a particular performer or composer. **2** the study of gramophone records. □ **discographer** n. [DISC + -GRAPHY after *biography*]

discoid /'dɪskɔɪd/ *adj.* disc-shaped. [Gk *diskoeidēs* (as DISCUS, -OID)]

discolour /dɪs'kʌlə(r)/ *v.tr. & intr.* (*US* **discolor**) spoil or cause to spoil the colour of; stain; tarnish. □ **discoloration** /-,kʌlə'reɪʃ(ə)n/ n. (also **discolouration**). [ME f. OF *descolorer* or med.L *discolorare* (as DIS-, COLOUR)]

discombobulate /,dɪskəm'bɒbjʊ,leɪt/ *v.tr. N. Amer. sl.* disturb; disconcert. [prob. based on *discompose* or *discomfit*]

discomfit /dɪs'kʌmfɪt/ *v.tr.* (**discomfited, discomfiting**) **1 a** disconcert or baffle. **b** thwart. **2** *archaic* defeat in battle. □ **discomfiture** n. [ME f. OF *desconfit* f. OF past part. of *desconfire* f. Rmc (as DIS-, L *conficere* put together: see CONFECTION)]

discomfort /dɪs'kʌmfət/ n. & v. ● n. **1 a** a lack of ease; slight pain (*tight collar caused discomfort*). **b** mental uneasiness (*his presence caused her discomfort*). **2** a lack of comfort. ● *v.tr.* make uneasy. [ME f. OF *desconfort(er)* (as DIS-, COMFORT)]

discommode /,dɪskə'məʊd/ *v.tr.* inconvenience (a person etc.). □ **discommodious** *adj.* [obs. F *discommoder* var. of *incommoder* (as DIS-, INCOMMODE)]

discompose /,dɪskəm'pəʊz/ *v.tr.* disturb the composure of; agitate; disturb. □ **discomposure** /-ʒə(r)/ n.

disconcert /,dɪskən's3ːt/ *v.tr.* **1** (often as **disconcerted** *adj.*) disturb the composure of; agitate; fluster (*disconcerted by his expression*). **2** spoil or upset (plans etc.). □ **disconcertedly** *adv.* **disconcerting** *adj.* **disconcertingly** *adv.* **disconcertment** n. **disconcertion** /-'s3ːʃ(ə)n/ n. [obs. F *desconcerter* (as DIS-, CONCERT)]

disconfirm /,dɪskən'f3ːm/ *v.tr. formal* disprove or tend to disprove (a hypothesis etc.). □ **disconfirmation** /dɪs,kɒnfə'meɪʃ(ə)n/ n.

disconformity /,dɪskən'fɔːmɪtɪ/ n. (pl. **-ies**) **1 a** lack of conformity. **b** an instance of this. **2** *Geol.* a difference of plane between two parallel, approximately horizontal sets of strata.

disconnect /,dɪskə'nekt/ *v.tr.* **1** (often foll. by *from*) break the connection of (things, ideas, etc.). **2** put (an electrical device) out of action by disconnecting the parts, esp. by pulling out the plug.

disconnected /,dɪskə'nektɪd/ *adj.* **1** not connected; detached; separated. **2** (of speech, writing, argument, etc.) incoherent and illogical. □ **disconnectedly** *adv.* **disconnectedness** n.

disconnection /,dɪskə'nekʃ(ə)n/ n. (also **disconnexion**) the act or an instance of disconnecting; the state of being disconnected.

disconsolate /dɪs'kɒnsələt/ *adj.* **1** forlorn or inconsolable. **2** unhappy or disappointed. □ **disconsolately** *adv.* **disconsolateness** n. **disconsolation** /-,kɒnsə'leɪʃ(ə)n/ n. [ME f. med.L *disconsolatus* (as DIS-, *consolatus* past part. of L *consolari* console)]

discontent /,dɪskən'tent/ n., adj., & v. ● n. lack of contentment; restlessness, dissatisfaction. ● *adj.* dissatisfied (*was discontent with his lot*). ● *v.tr.* (esp. as **discontented** *adj.*) make dissatisfied. □ **discontentedly** *adv.* **discontentedness** n. **discontentment** n.

discontinue /,dɪskən'tɪnjuː/ v. (**-continues, -continued,**

-continuing) 1 *intr.* & *tr.* cease or cause to cease to exist or be made (*a discontinued line*). **2** *tr.* give up, cease from (*discontinued his visits*). **3** *tr.* cease taking or paying (a newspaper, a subscription, etc.). □ **discontinuance** *n.* **discontinuation** /-ˌtɪnjʊˈeɪʃ(ə)n/ *n.* [ME f. OF *discontinuer* f. med.L *discontinuare* (as DIS-, CONTINUE)]

discontinuous /ˌdɪskənˈtɪnjʊəs/ *adj.* lacking continuity in space or time; intermittent. □ **discontinuously** *adv.* **discontinuity** /dɪsˌkɒntɪˈnjuːɪtɪ/ *n.* [med.L *discontinuus* (as DIS-, CONTINUOUS)]

discord *n.* & *v.* ● *n.* /ˈdɪskɔːd/ **1** disagreement; strife. **2** harsh clashing noise; clangour. **3** *Mus.* **a** a lack of harmony between notes sounding together. **b** an unpleasing or unfinished chord needing to be completed by another. **c** any interval except unison, an octave, a perfect fifth and fourth, a major and minor third and sixth, and their octaves. **d** a single note dissonant with another. ● *v.intr.* /dɪsˈkɔːd/ **1** (usu. foll. by *with*) **a** disagree or quarrel. **b** be different or inconsistent. **2** jar, clash, be dissonant. [ME f. OF *descord*, (n.), *descorder* (v.) f. L *discordare* f. *discors discordant* (as DIS-, *cor cord-* heart)]

discordant /dɪsˈkɔːd(ə)nt/ *adj.* (usu. foll. by *to, from, with*) **1** disagreeing; at variance. **2** (of sounds) not in harmony; dissonant. □ **discordance** *n.* **discordancy** *n.* **discordantly** *adv.* [ME f. OF, part. of *discorder*: see DISCORD]

discothèque /ˈdɪskəˌtek/ *n.* **1** a club etc. for dancing to recorded popular music. **2 a** the professional lighting and sound equipment used at a discothèque. **b** a business that provides this. **3** a party with dancing to popular music, esp. using such equipment. [F, = record-library]

discount *n.* & *v.* ● *n.* /ˈdɪskaʊnt/ **1** a deduction from a bill or amount due given esp. in consideration of prompt or advance payment or to a special class of buyers. **2** a deduction from the amount of a bill of exchange etc. by a person who gives value for it before it is due. **3** the act or an instance of discounting. ● *v.tr.* /dɪsˈkaʊnt/ **1** disregard as being unreliable or unimportant (*discounted his story*). **2** reduce the effect of (an event etc.) by previous action. **3** detract from; lessen; deduct (esp. an amount from a bill etc.). **4** give or get the present worth of (a bill not yet due). □ **at a discount 1** below the nominal or usual price (cf. PREMIUM). **2** not in demand; depreciated. **discount house 1** *Brit.* a firm that discounts bills. **2** *US* = discount store. **discount rate** *US* the minimum lending rate. **discount store** a shop etc. that sells goods at less than the normal retail price. □ **discountable** /dɪsˈkaʊntəb(ə)l/ *adj.* **discounter** *n.* [obs. F *descompte*, *-conte*, *descompter* or It. *(di)scontare* (as DIS-, COUNT¹)]

discountenance /dɪsˈkaʊntɪnəns/ *v.tr.* **1** (esp. in *passive*) disconcert (*was discountenanced by his abruptness*). **2** refuse to countenance; show disapproval of.

discourage /dɪsˈkʌrɪdʒ/ *v.tr.* **1** deprive of courage, confidence, or energy. **2** (usu. foll. by *from*) dissuade (*discouraged him from going*). **3** inhibit or seek to prevent (an action, etc.) by showing disapproval; oppose (*smoking is discouraged*). □ **discouragement** *n.* **discouragingly** *adv.* [ME f. OF *descouragier* (as DIS-, COURAGE)]

discourse *n.* & *v.* ● *n.* /ˈdɪskɔːs, dɪsˈkɔːs/ **1** *literary* a conversation; talk. **b** a dissertation or treatise on an academic subject. **c** a lecture or sermon. **2** *Linguistics* a connected series of utterances; a text. ● *v.* /dɪsˈkɔːs/ **1** *intr.* talk; converse. **2** *intr.* (usu. foll. by *of, on, upon*) speak or write learnedly or at length (on a subject). **3** *tr.* *archaic* give forth (music etc.). [ME f. L *discursus* (as DIS-, COURSE): (v.) partly after F *discourir*]

discourteous /dɪsˈkɜːtɪəs/ *adj.* impolite; rude. □ **discourteously** *adv.* **discourteousness** *n.*

discourtesy /dɪsˈkɜːtəsɪ/ *n.* (*pl.* **-ies**) **1** bad manners; rudeness. **2** an impolite act or remark.

discover /dɪsˈkʌvə(r)/ *v.tr.* **1** (often foll. by *that* + clause) **a** find out or become aware of, whether by research or searching or by chance (*discovered a new entrance*; *discovered that they had been overpaid*). **b** be the first to find or find out (*who discovered America?*). **c** devise or pioneer (*discover new techniques*). **2** *Chess* give (check) by removing one's own obstructing piece. **3** (in show business) find and promote as a new singer, actor, etc. **4** *archaic* **a** make known. **b** exhibit; manifest. **c** disclose; betray. **d** catch sight of; espy. □ **discoverable** *adj.* **discoverer** *n.* [ME f. OF *descovrir* f. LL *discooperire* (as DIS-, COVER)]

discovery /dɪsˈkʌvərɪ/ *n.* (*pl.* **-ies**) **1 a** the act or process of discovering or being discovered. **b** an instance of this (*the discovery of a new planet*). **2** a person or thing discovered. **3** *Law* the compulsory disclosure, by a party to an action, of facts or documents on which the other party wishes to rely. [DISCOVER after *recover, recovery*]

discredit /dɪsˈkredɪt/ *n.* & *v.* ● *n.* **1** harm to reputation (*brought discredit on the enterprise*). **2** a person or thing causing this (*he is a discredit to his*

family). **3** lack of credibility; doubt; disbelief (*throws discredit on her story*). **4** the loss of commercial credit. ● *v.tr.* (**-credited, -crediting**) **1** harm the good reputation of. **2** cause to be disbelieved. **3** refuse to believe.

discreditable /dɪsˈkredɪtəb(ə)l/ *adj.* bringing discredit; shameful. □ **discreditably** *adv.*

discreet /dɪsˈkriːt/ *adj.* (**discreeter, discreetest**) **1 a** circumspect in speech or action, esp. to avoid social disgrace or embarrassment. **b** tactful; trustworthy. **2** unobtrusive (*a discreet touch of rouge*). □ **discreetly** *adv.* **discreetness** *n.* [ME f. OF *discret -ete* f. L *discretus* separate (as DIS-, *cretus* past part. of *cernere* sift), with LL sense f. its derivative *discretio* discernment]

discrepancy /dɪsˈkrepənsɪ/ *n.* (*pl.* **-ies**) **1** difference; failure to correspond; inconsistency. **2** an instance of this. □ **discrepant** *adj.* [L *discrepare* be discordant (as DIS-, *crepare* creak)]

discrete /dɪsˈkriːt/ *adj.* individually distinct; separate, discontinuous. □ **discretely** *adv.* **discreteness** *n.* [ME f. L *discretus*: see DISCREET]

discretion /dɪsˈkreʃ(ə)n/ *n.* **1** being discreet; discreet behaviour (*treats confidences with discretion*). **2** prudence; self-preservation. **3** the freedom to act and think as one wishes, usu. within legal limits (*it is within his discretion to leave*). **4** *Law* a court's freedom to decide a sentence etc. □ **age** (or **years**) **of discretion** the esp. legal age at which a person is able to manage his or her own affairs. **at discretion** as one pleases. **at the discretion of** to be settled or disposed of according to the judgement or choice of. **discretion is the better part of valour** reckless courage is often self-defeating. **use one's discretion** act according to one's own judgement. □ **discretionary** *adj.* [ME f. OF f. L *discretio -onis* (as DISCREET)]

discriminate /dɪsˈkrɪmɪˌneɪt/ *v.* **1** *intr.* (often foll. by *between*) make or see a distinction; differentiate (*cannot discriminate between right and wrong*). **2** *intr.* make a distinction, esp. unjustly and on the basis of race, colour, age, or sex. **3** *intr.* (foll. by *against*) select for unfavourable treatment. **4** *tr.* (usu. foll. by *from*) make or see or constitute a difference in or between (*many things discriminate one person from another*). **5** *intr.* (esp. as **discriminating** *adj.*) observe distinctions carefully; have good judgement. **6** *tr.* mark as distinctive; be a distinguishing feature of. □ **discriminator** *n.* **discriminately** /-nətlɪ/ *adv.* **discriminative** /-nətɪv/ *adj.* **discriminatory** /-nətərɪ/ *adj.* [L *discriminare* f. *discrimen -minis* distinction f. *discernere* DISCERN]

discriminating /dɪsˈkrɪmɪˌneɪtɪŋ/ *adj.* **1** able to discern, esp. distinctions. **2** having good taste. □ **discriminatingly** *adv.*

discrimination /dɪsˌkrɪmɪˈneɪʃ(ə)n/ *n.* **1** unfavourable treatment based on prejudice, esp. regarding race, colour, age, or sex. **2** good taste or judgement in artistic matters etc. **3** the power of discriminating or observing differences. **4** a distinction made with the mind or in action.

discursive /dɪsˈkɜːsɪv/ *adj.* **1** rambling or digressive. **2** *Philos.* proceeding by argument or reasoning (opp. INTUITIVE). □ **discursively** *adv.* **discursiveness** *n.* [med.L *discursivus* f. L *discurrere discurs-* (as DIS-, *currere* run)]

discus /ˈdɪskəs/ *n.* (*pl.* **discuses**) **1** a heavy thick-centred disc thrown in ancient Greek games. **2** a similar disc thrown in modern athletic field events. [L f. Gk *diskos*]

discuss /dɪsˈkʌs/ *v.tr.* **1** hold a conversation about. **2** examine by argument, esp. written; debate. □ **discussable** *adj.* **discussant** *n.* **discusser** *n.* [ME f. L *discutere discuss-* disperse (as DIS-, *quatere* shake)]

discussion /dɪsˈkʌʃ(ə)n/ *n.* **1** a conversation, esp. on specific subjects; a debate (*had a discussion about what they should do*). **2** an examination by argument, written or spoken. [ME f. OF f. LL *discussio -onis* (as DISCUSS)]

disdain /dɪsˈdeɪn/ *n.* & *v.* ● *n.* scorn; contempt. ● *v.tr.* **1** regard with disdain. **2** think oneself superior to; reject (*disdained his offer*; *disdained to enter*; *disdained answering*). [ME f. OF *desdeign(ier)* ult. f. L *dedignari* (as DE-, *dignari* f. *dignus* worthy)]

disdainful /dɪsˈdeɪnfʊl/ *adj.* showing disdain or contempt. □ **disdainfully** *adv.* **disdainfulness** *n.*

disease /dɪˈziːz/ *n.* **1** an unhealthy condition of the body (or a part of it) or the mind; illness, sickness. **2** a corresponding physical condition of plants. **3** a particular kind of disease with special symptoms or location. [ME f. OF *desaise*]

diseased /dɪˈziːzd/ *adj.* **1** affected with disease. **2** abnormal, disordered. [ME, past part. of *disease* (v.) f. OF *desaisier* (as DISEASE)]

diseconomy /ˌdɪsɪˈkɒnəmɪ/ *n.* *Econ.* the absence or reverse of economy, esp. the increase of costs in a large-scale operation.

disembark /ˌdɪsɪmˈbɑːk/ *v.tr.* & *intr.* put or go ashore or land from a ship; remove from or leave an aircraft, train, etc. □ **disembarkation** /dɪs,embɑːˈkeɪʃ(ə)n/ n. [F *désembarquer* (as DIS-, EMBARK)]

disembarrass /ˌdɪsɪmˈbærəs/ v.tr. **1** (usu. foll. by *of*) relieve (of a load etc.). **2** free from embarrassment. □ **disembarrassment** n.

disembody /ˌdɪsɪmˈbɒdɪ/ v.tr. (**-ies**, **-ied**) **1** (esp. as **disembodied** adj.) separate or free (esp. the soul) from the body or a concrete form (*disembodied spirit*; *disembodied voice*). **2** archaic disband (troops). □ **disembodiment** n.

disembogue /ˌdɪsɪmˈbəʊg/ v.tr. & intr. (**disembogues**, **disembogued**, **disemboguing**) (of a river etc.) discharge (waters) into the sea etc. [Sp. *desembocar* (as DIS-, *en* in, *boca* mouth)]

disembowel /ˌdɪsɪmˈbaʊəl/ v.tr. (**-embowelled**, **-embowelling**; US **-emboweled**, **-emboweling**) remove the bowels or entrails of. □ **disembowelment** n.

disembroil /ˌdɪsɪmˈbrɔɪl/ v.tr. extricate from confusion or entanglement.

disenchant /ˌdɪsɪnˈtʃɑːnt/ v.tr. free from enchantment; disillusion. □ **disenchantingly** adv. **disenchantment** n. [F *désenchanter* (as DIS-, ENCHANT)]

disencumber /ˌdɪsɪnˈkʌmbə(r)/ v.tr. free from encumbrance.

disendow /ˌdɪsɪnˈdaʊ/ v.tr. strip (esp. the Church) of endowments. □ **disendowment** n.

disenfranchise /ˌdɪsɪnˈfræntʃaɪz/ v.tr. (also **disfranchise** /dɪsˈfræn-/) **1 a** deprive (a person) of the right to vote. **b** deprive (a place) of the right to send a representative to Parliament. **2** deprive (a person) of rights as a citizen or of a franchise held. □ **disenfranchisement** n.

disengage /ˌdɪsɪnˈgeɪdʒ/ v. & n. ● v. **1 a** tr. detach, free, loosen, or separate (parts etc.) (*disengaged the clutch*). **b** refl. detach oneself; get loose (*disengaged ourselves from their company*). **2** tr. Mil. remove (troops) from a battle or a battle area. **3** intr. become detached. **4** intr. Fencing pass the point of one's sword to the other side of one's opponent's. **5** intr. (as **disengaged** adj.) **a** unoccupied; free; vacant. **b** uncommitted, esp. politically. ● n. Fencing a disengaging movement.

disengagement /ˌdɪsɪnˈgeɪdʒmənt/ n. **1 a** the act of disengaging. **b** an instance of this. **2** freedom from ties; detachment. **3** the dissolution of an engagement to marry. **4** ease of manner or behaviour. **5** Fencing = DISENGAGE.

disentail /ˌdɪsɪnˈteɪl/ v.tr. Law free (property) from entail; break the entail of.

disentangle /ˌdɪsɪnˈtæŋg(ə)l/ v. **1** tr. **a** unravel, untwist. **b** free from complications; extricate (*disentangled her from the difficulty*). **2** intr. become disentangled. □ **disentanglement** n.

disenthral /ˌdɪsɪnˈθrɔːl/ v.tr. (US **disenthrall**) (**-enthralled**, **-enthralling**) literary free from enthralment. □ **disenthralment** n.

disentitle /ˌdɪsɪnˈtaɪt(ə)l/ v.tr. (usu. foll. by *to*) deprive of any rightful claim.

disentomb /ˌdɪsɪnˈtuːm/ v.tr. literary **1** remove from a tomb; disinter. **2** unearth. □ **disentombment** /-ˈtuːmmənt/ n.

disequilibrium /ˌdɪsiːkwɪˈlɪbrɪəm, ˌdɪsek-/ n. a lack or loss of equilibrium; instability.

disestablish /ˌdɪsɪˈstæblɪʃ/ v.tr. **1** deprive (a Church) of state support. **2** depose from an official position. **3** terminate the establishment of. □ **disestablishment** n.

disesteem /ˌdɪsɪˈstiːm/ v. & n. ● v.tr. have a low opinion of; despise. ● n. low esteem or regard.

diseuse /diːˈzɜːz/ n. (masc. **diseur** /-ˈzɜː(r)/) a female artiste entertaining with spoken monologues. [F, = talker f. *dire* dis- say]

disfavour /dɪsˈfeɪvə(r)/ n. & v. (US **disfavor**) ● n. **1** disapproval or dislike. **2** the state of being disliked (*fell into disfavour*). ● v.tr. regard or treat with disfavour.

disfigure /dɪsˈfɪgə(r)/ v.tr. spoil the beauty of; deform; deface. □ **disfigurement** n. [ME f. OF *desfigurer* f. Rmc (as DIS-, FIGURE)]

disforest /dɪsˈfɒrɪst/ v.tr. Brit. = DISAFFOREST. □ **disforestation** /-ˌfɒrɪˈsteɪʃ(ə)n/ n.

disfranchise var. of DISENFRANCHISE.

disfrock /dɪsˈfrɒk/ v.tr. unfrock.

disgorge /dɪsˈgɔːdʒ/ v.tr. **1** eject from the throat or stomach. **2** pour forth, discharge (contents, ill-gotten gains, etc.). □ **disgorgement** n. [ME f. OF *desgorger* (as DIS-, GORGE)]

disgrace /dɪsˈgreɪs/ n. & v. ● n. **1** the loss of reputation; shame; ignominy (*brought disgrace on his family*). **2** a dishonourable, inefficient, or shameful person, thing, state of affairs, etc. (*the bus service is a disgrace*). ● v.tr. **1** bring shame or discredit on; be a disgrace to. **2** degrade from a position of honour; dismiss from favour. □ **in disgrace** having lost respect or reputation; out of favour. [F *disgrâce*, *disgracier* f. It. *disgrazia*, *disgraziare* (as DIS-, GRACE)]

disgraceful /dɪsˈgreɪsfʊl/ adj. shameful; dishonourable; degrading. □ **disgracefully** adv.

disgruntled /dɪsˈgrʌnt(ə)ld/ adj. discontented; moody; sulky. □ **disgruntlement** n. [DIS- + gruntle obs. frequent. of GRUNT]

disguise /dɪsˈgaɪz/ v. & n. ● v.tr. **1** (often foll. by *as*) alter the appearance, sound, smell, etc., of so as to conceal the identity; make unrecognizable (*disguised herself as a policewoman*; *disguised the taste by adding sugar*). **2** misrepresent or cover up (*disguised the truth*; *disguised their intentions*). ● n. **1 a** a costume, false beard, make-up, etc., used to alter the appearance so as to conceal or deceive. **b** any action, manner, etc., used for deception. **2 a** the act or practice of disguising; the concealment of reality. **b** an instance of this. □ **in disguise 1** wearing a concealing costume etc. **2** appearing to be the opposite (*a blessing in disguise*). [ME f. OF *desgui(i)er* (as DIS-, GUISE)]

disgust /dɪsˈgʌst/ n. & v. ● n. (usu. foll. by *at*, *for*) **1** strong aversion, repugnance; profound indignation. **2** strong distaste for (some item of) food, drink, medicine, etc.; nausea. ● v.tr. cause disgust in (*their behaviour disgusts me*; *was disgusted to find a slug*). □ **in disgust** as a result of disgust (*left in disgust*). □ **disgustedly** adv. [OF *degout*, *desgouster*, or It. *disgusto*, *disgustare* (as DIS-, GUSTO)]

disgustful /dɪsˈgʌstfʊl/ adj. **1** disgusting; repulsive. **2** (of curiosity etc.) caused by disgust.

disgusting /dɪsˈgʌstɪŋ/ adj. arousing aversion or indignation (*disgusting behaviour*). □ **disgustingly** adv. **disgustingness** n.

dish /dɪʃ/ n. & v. ● n. **1 a** a shallow, usu. flat-bottomed container for cooking or serving food, made of glass, ceramics, metal, etc. **b** the food served in a dish (*all the dishes were delicious*). **c** a particular kind of food (*a meat dish*). **2** (in pl.) dirty plates, cutlery, cooking pots, etc. after a meal. **3 a** a dish-shaped receptacle, object, or cavity. **b** = *satellite dish*. **4** sl. a sexually attractive person. ● v.tr. **1** put (food) into a dish ready for serving. **2** colloq. **a** outmanoeuvre. **b** Brit. destroy (one's hopes, chances, etc.). **3** make concave or dish-shaped. □ **dish** (**up** or **out**) **the dirt** (often foll. by *on*) colloq. spread scandal or gossip. **dish out** sl. distribute, deal out. esp. carelessly or indiscriminately. **dish up 1** serve or prepare to serve (food). **2** colloq. seek to present (facts, an argument, etc.) attractively. □ **dishful** n. (pl. **-fuls**) [OE *disc* plate, bowl (with Gmc and ON cognates) f. L *discus* DISC]

dishabille /ˌdɪsæˈbiːl/ n. (also **déshabille**, **deshabille** /ˌdeɪzæ-/, or **déshabillé** /ˌdeɪzæˈbiːeɪ/) a state of being only partly or carelessly clothed. [F *déshabillé* undressed]

disharmony /dɪsˈhɑːmənɪ/ n. a lack of harmony; discord. □ **disharmonize** v.tr. (also **-ise**). **disharmonious** /ˌdɪshɑːˈməʊnɪəs/ adj. **disharmoniously** adv.

dishcloth /ˈdɪʃklɒθ/ n. a usu. open-weave cloth for washing dishes. □ **dishcloth gourd** a loofah.

dishearten /dɪsˈhɑːt(ə)n/ v.tr. cause to lose courage or confidence; make despondent. □ **dishearteningly** adv. **disheartenment** n.

dishevelled /dɪˈʃev(ə)ld/ adj. (US **disheveled**) (of the hair, a person, etc.) untidy; ruffled; disordered. □ **dishevel** v.tr. (**dishevelled**, **dishevelling**; US **disheveled**, **disheveling**). **dishevelment** n. [ME *dischevelee* f. OF *deschevelé* past part. (as DIS-, *chevel* hair f. L *capillus*)]

dishonest /dɪsˈɒnɪst/ adj. (of a person, act, or statement) fraudulent or insincere. □ **dishonestly** adv. [ME f. OF *deshoneste* (as DIS-, HONEST)]

dishonesty /dɪsˈɒnɪstɪ/ n. (pl. **-ies**) **1 a** a lack of honesty. **b** deceitfulness, fraud. **2** a dishonest or fraudulent act. [ME f. OF *deshon(n)esté* (as DIS-, HONEST)]

dishonour /dɪsˈɒnə(r)/ n. & v. (US **dishonor**) ● n. **1** a state of shame or disgrace; discredit. **2** something that causes dishonour (*a dishonour to his profession*). ● v.tr. **1** treat without honour or respect. **2** disgrace (*dishonoured his name*). **3** refuse to accept or pay (a cheque or a bill of exchange). **4** archaic violate the chastity of; rape. [ME f. OF *deshonor*, *deshonorer* f. med.L *dishonorare* (as DIS-, HONOUR)]

dishonourable /dɪsˈɒnərəb(ə)l/ adj. (US **dishonorable**) **1** causing disgrace; ignominious. **2** unprincipled. □ **dishonourableness** n. **dishonourably** adv.

dishrag /ˈdɪʃræg/ n. = DISHCLOTH.

dishwasher /ˈdɪʃˌwɒʃə(r)/ n. **1** a machine for automatically washing dishes. **2** a person employed to wash dishes.

dishwater /ˈdɪʃˌwɔːtə(r)/ n. water in which dishes have been washed.

dishy /ˈdɪʃɪ/ *adj.* (**dishier, dishiest**) *colloq.* sexually attractive. [DISH *n.* 4 + -Y[1]]

disillusion /ˌdɪsɪˈluːʒ(ə)n, -ˈljuːʒ(ə)n/ *n. & v.* ● *n.* freedom from illusions; disenchantment. ● *v.tr.* rid of illusions; disenchant. □ **disillusionize** *v.tr.* (also **-ise**). **disillusionment** *n.*

disincentive /ˌdɪsɪnˈsentɪv/ *n. & adj.* ● *n.* **1** something that tends to discourage a particular action etc. **2** *Econ.* a source of discouragement to productivity or progress. ● *adj.* tending to discourage.

disinclination /dɪsˌɪŋklɪˈneɪʃ(ə)n/ *n.* (usu. foll. by *for,* or *to* + infin.) the absence of willingness; a reluctance (*a disinclination for work; disinclination to go*).

disincline /ˌdɪsɪnˈklaɪn/ *v.tr.* (usu. foll. by *to* + infin. or *for*) **1** make unwilling or reluctant. **2** (as **disinclined** *adj.*) unwilling, averse.

disincorporate /ˌdɪsɪnˈkɔːpəˌreɪt/ *v.tr.* dissolve (a corporate body).

disinfect /ˌdɪsɪnˈfekt/ *v.tr.* cleanse (a wound, a room, clothes, etc.) of infection, esp. with a disinfectant. □ **disinfection** /-ˈfekʃ(ə)n/ *n.* [F *désinfecter* (as DIS-, INFECT)]

disinfectant /ˌdɪsɪnˈfektənt/ *n. & adj.* ● *n.* a usu. commercially produced chemical liquid that destroys germs etc. ● *adj.* causing disinfection.

disinfest /ˌdɪsɪnˈfest/ *v.tr.* rid (a person, a building, etc.) of vermin, infesting insects, etc. □ **disinfestation** /dɪsˌɪnfeˈsteɪʃ(ə)n/ *n.*

disinflation /ˌdɪsɪnˈfleɪʃ(ə)n/ *n. Econ.* a policy designed to counteract inflation without causing deflation. □ **disinflationary** *adj.*

disinformation /dɪsˌɪnfəˈmeɪʃ(ə)n/ *n.* false information, intended to mislead.

disingenuous /ˌdɪsɪnˈdʒenjʊəs/ *adj.* having secret motives; dishonest, insincere. □ **disingenuously** *adv.* **disingenuousness** *n.*

disinherit /ˌdɪsɪnˈherɪt/ *v.tr.* (**disinherited, disinheriting**) reject as one's heir; deprive of the right of inheritance. □ **disinheritance** *n.* [ME f. DIS- + INHERIT in obs. sense 'make heir']

disintegrate /dɪsˈɪntɪˌɡreɪt/ *v.* **1** *tr. & intr.* **a** separate into component parts or fragments; crumble; decay. **b** lose or cause to lose cohesion. **2** *intr. colloq.* deteriorate mentally or physically; decay. **3** *intr. & tr. Physics* undergo or cause to undergo disintegration. □ **disintegrator** *n.*

disintegration /dɪsˌɪntɪˈɡreɪʃ(ə)n/ *n.* **1** the act or an instance of disintegrating. **2** *Physics* any process in which a nucleus emits a particle or particles or divides into smaller nuclei.

disinter /ˌdɪsɪnˈtɜː(r)/ *v.tr.* (**disinterred, disinterring**) **1** remove (esp. a corpse) from the ground; unearth; exhume. **2** find after a protracted search (*disinterred the letter from the back of the drawer*). □ **disinterment** *n.* [F *désenterrer* (as DIS-, INTER)]

disinterest /dɪsˈɪntrəst, -trɪst/ *n.* **1** impartiality. **2** *disp.* lack of interest; unconcern.

disinterested /dɪsˈɪntrəstɪd, -trɪstɪd/ *adj.* **1** not influenced by one's own advantage; impartial. **2** *disp.* uninterested. □ **disinterestedly** *adv.* **disinterestedness** *n.* [past part. of *disinterest* (v.) divest of interest]

disinvest /ˌdɪsɪnˈvest/ *v.intr.* (foll. by *from,* or *absol.*) reduce or dispose of one's investment (in a place, company, etc.). □ **disinvestment** *n.*

disjecta membra /dɪsˌdʒektə ˈmembrə/ *n.pl.* scattered remains; fragments, esp. of written work. [L, alt. of *disjecti membra poetae* (Horace) limbs of a dismembered poet]

disjoin /dɪsˈdʒɔɪn/ *v.tr.* separate or disunite; part. [ME f. OF *desjoindre* f. L *disjungere junct-* join)]

disjoint /dɪsˈdʒɔɪnt/ *v. & adj.* ● *v.tr.* **1** take apart at the joints. **2** (as **disjointed** *adj.*) (esp. of conversation) incoherent; desultory. **3** disturb the working or connection of; dislocate. ● *adj.* (of two or more sets) having no elements in common. □ **disjointedly** *adv.* **disjointedness** *n.* [ME f. obs. *disjoint* (adj.) f. past part. of OF *desjoindre* (as DISJOIN)]

disjunct *n. & adj.* ● *n.* /ˈdɪsdʒʌŋkt/ **1** *Logic* each of the terms of a disjunctive proposition. **2** *Gram.* an adverb or adverbial phrase that expresses a writer's or speaker's attitude to the content of the sentence in which it occurs. ● *adj.* /dɪsˈdʒʌŋkt/ **1** disjoined; separate; distinct. **2** *Logic* designating each term of a disjunctive proposition. [ME f. L *disjunctus* past part. of *disjungere* DISJOIN]

disjunction /dɪsˈdʒʌŋkʃ(ə)n/ *n.* **1 a** the process of disjoining; separation. **b** an instance of this. **2** *Logic* the relation of two mutually incompatible alternatives; a statement expressing this (esp. using the word *or*). [ME f. OF *disjunction* or L *disjunctio* (as DISJOIN)]

disjunctive /dɪsˈdʒʌŋktɪv/ *adj. & n.* ● *adj.* **1** involving separation; disjoining. **2** *Gram.* (esp. of a conjunction) expressing a choice between two words etc., e.g. *or* in *asked if he was going or staying* (cf. COPULATIVE 2a). **3** *Logic* (of a proposition) expressing alternatives. ● *n.* **1** *Gram.* a

disjunctive conjunction or other word. **2** *Logic* a disjunctive proposition. □ **disjunctively** *adv.* [ME f. L *disjunctivus* (as DISJOIN)]

disk var. of DISC (esp. *US & Computing*).

diskette /dɪsˈket/ *n. Computing* a floppy disk (see FLOPPY *n.*).

Disko /ˈdɪskəʊ/ an island with extensive coal resources on the west coast of Greenland. Its chief settlement is Godhavn.

dislike /dɪsˈlaɪk/ *v. & n.* ● *v.tr.* have an aversion or objection to; not like. ● *n.* **1** a feeling of repugnance or not liking. **2** an object of dislike. □ **dislikable** *adj.* (also **dislikeable**).

dislocate /ˈdɪsləˌkeɪt/ *v.tr.* **1** disturb the normal connection of (esp. a joint in the body). **2** disrupt; put out of order. **3** displace. [prob. backform. f. DISLOCATION]

dislocation /ˌdɪsləˈkeɪʃ(ə)n/ *n.* **1** the act or result of dislocating. **2** *Crystallog.* the displacement of part of a crystal lattice structure. [ME f. OF *dislocation* or med.L *dislocatio* f. *dislocare* (as DIS-, *locare* place)]

dislodge /dɪsˈlɒdʒ/ *v.tr.* remove from an established or fixed position (*was dislodged from his directorship*). □ **dislodgement** *n.* (also **dislodgment**). [ME f. OF *deslog(i)er* (as DIS-, LODGE)]

disloyal /dɪsˈlɔɪəl/ *adj.* (often foll. by *to*) **1** not loyal; unfaithful. **2** untrue to one's allegiance; treacherous to one's government etc. □ **disloyalist** *n.* **disloyally** *adv.* **disloyalty** *n.* [ME f. OF *desloial* (as DIS-, LOYAL)]

dismal /ˈdɪzm(ə)l/ *adj.* **1** causing or showing gloom; miserable. **2** dreary or sombre (*dismal brown walls*). **3** *colloq.* feeble or inept (*a dismal performance*). □ **the dismals** *colloq.* melancholy. **the dismal science** *joc.* economics. □ **dismally** *adv.* **dismalness** *n.* [orig. noun = unlucky days: ME f. AF *dis mal* f. med.L *dies mali* two days in each month held to be unpropitious]

Dismal Swamp see GREAT DISMAL SWAMP.

dismantle /dɪsˈmænt(ə)l/ *v.tr.* **1** take to pieces; pull down. **2** deprive of defences or equipment. **3** (often foll. by *of*) strip of covering or protection. □ **dismantlement** *n.* **dismantler** *n.* [OF *desmanteler* (as DIS-, MANTLE)]

dismast /dɪsˈmɑːst/ *v.tr.* deprive (a ship) of masts; break down the mast or masts of.

dismay /dɪsˈmeɪ/ *v. & n.* ● *v.tr.* fill with consternation or anxiety; discourage or depress; reduce to despair. ● *n.* **1** consternation or anxiety. **2** depression or despair. [ME f. OF *desmaiier* (unrecorded) ult. f. a Gmc root = deprive of power (as DIS-, MAY)]

dismember /dɪsˈmembə(r)/ *v.tr.* **1** tear or cut the limbs from. **2** partition or divide up (an empire, country, etc.). □ **dismemberment** *n.* [ME f. OF *desmembrer* f. Rmc (as DIS-, L *membrum* limb)]

dismiss /dɪsˈmɪs/ *v.* **1 a** *tr.* send away, cause to leave one's presence, disperse; disband (an assembly or army). **b** *intr.* (of an assembly etc.) disperse; break ranks. **2** *tr.* discharge from employment, office, etc., esp. dishonourably. **3** *tr.* put out of one's thoughts; cease to feel or discuss (*dismissed him from memory*). **4** *tr.* treat (a subject) summarily (*dismissed his application*). **5** *tr. Law* refuse further hearing to (a case); send out of court. **6** *tr. Cricket* put (a batsman or a side) out (*was dismissed for 75 runs*). **7** *intr.* (in *imper.*) *Mil.* a word of command at the end of drilling. □ **dismissal** *n.* **dismissible** *adj.* [ME, orig. as past part. after OF *desmis* f. med.L *dismissus* (as DIS-, L *mittere miss-* send)]

dismissive /dɪsˈmɪsɪv/ *adj.* tending to dismiss from consideration; disdainful. □ **dismissively** *adv.* **dismissiveness** *n.*

dismount /dɪsˈmaʊnt/ *v.* **1 a** *intr.* alight from a horse, bicycle, etc. **b** *tr.* (usu. in *passive*) throw from a horse, unseat. **2** *tr.* remove (a thing) from its mounting (esp. a gun from its carriage).

Disney /ˈdɪznɪ/, Walter Elias ('Walt') (1901–66), American animator and film producer. He made his name with the creation of Mortimer (later Mickey) Mouse in 1927; many other familiar cartoon characters, including Minnie Mouse, Goofy, Pluto, and Donald Duck, were invented by Disney. He produced the first full-length cartoon feature film with sound and colour, *Snow White and the Seven Dwarfs* (1937); this was followed by *Pinocchio* (1940), *Bambi* (1942), and many others. Disney also made many animal and adventure films for children; after his death the tradition of animation and film-making was continued under the Disney name. He was also immortalized by the creation of Disneyland, an amusement park in California, which was opened in 1955; the first European Disneyland was established just outside Paris in 1992.

disobedient /ˌdɪsəˈbiːdɪənt/ *adj.* disobeying; rebellious, rule-breaking. □ **disobedience** *n.* **disobediently** *adv.* [ME f. OF *desobedient* (as DIS-, OBEDIENT)]

disobey /ˌdɪsəˈbeɪ/ v.tr. (also absol.) fail or refuse to obey; disregard (orders); break (rules) (disobeyed his mother; how dare you disobey!). □ **disobeyer** n. [ME f. OF desobeir f. Rmc (as DIS-, OBEY)]

disoblige /ˌdɪsəˈblaɪdʒ/ v.tr. **1** refuse to consider the convenience or wishes of. **2** (as **disobliging** adj.) uncooperative. [F désobliger f. Rmc (as DIS-, OBLIGE)]

disorder /dɪsˈɔːdə(r)/ n. & v. ● n. **1** a lack of order; confusion. **2** a riot; a commotion. **3** Med. a usu. minor ailment or disease. ● v.tr. **1** throw into confusion; disarrange. **2** Med. put out of good health; upset. □ **disordered** adj. [ME, alt. after ORDER v. of earlier disordain f. OF desordener (as DIS-, ORDAIN)]

disorderly /dɪsˈɔːdəlɪ/ adj. **1** untidy; confused. **2** irregular; unruly; riotous. **3** Law contrary to public order or morality. □ **disorderly house** Law a brothel. □ **disorderliness** n.

disorganize /dɪsˈɔːgəˌnaɪz/ v.tr. (also **-ise**) **1** destroy the system or order of; throw into confusion. **2** (as **disorganized** adj.) lacking organization or system. □ **disorganization** /-ˌɔːgənaɪˈzeɪʃ(ə)n/ n. [F désorganiser (as DIS-, ORGANIZE)]

disorient /dɪsˈɔːrɪent, -ˈɒrɪent/ v.tr. = DISORIENTATE. [F désorienter (as DIS-, ORIENT v.)]

disorientate /dɪsˈɔːrɪenˌteɪt, -ˈɒrɪenˌteɪt/ v.tr. **1** confuse (a person) as to his or her whereabouts or bearings. **2** (often as **disorientated** adj.) confuse (a person) (disorientated by his unexpected behaviour). □ **disorientation** /-ˌɔːrɪenˈteɪʃ(ə)n, -ˈɒrɪen-/ n.

disown /dɪsˈəʊn/ v.tr. **1** refuse to recognize; repudiate; disclaim. **2** renounce one's connection with or allegiance to. □ **disowner** n.

disparage /dɪˈspærɪdʒ/ v.tr. **1** speak slightingly of; depreciate. **2** bring discredit on. □ **disparagement** n. **disparagingly** adv. [ME f. OF desparagier marry unequally (as DIS-, parage equality of rank ult. f. L par equal)]

disparate /ˈdɪspərət/ adj. & n. ● adj. essentially different in kind; without comparison or relation. ● n. (in pl.) things so unlike that there is no basis for their comparison. □ **disparately** adv. **disparateness** n. [L disparatus separated (as DIS-, paratus past part. of parare prepare), infl. in sense by L dispar unequal]

disparity /dɪˈspærɪtɪ/ n. (pl. **-ies**) **1** inequality; difference; incongruity. **2** an instance of this. [F disparité f. LL disparitas -tatis (as DIS-, PARITY[1])]

dispassionate /dɪsˈpæʃənət/ adj. free from passion; calm; impartial. □ **dispassionately** adv. **dispassionateness** n.

dispatch /dɪˈspætʃ/ v. & n. (also **despatch**) ● v.tr. **1** send off to a destination or for a purpose (dispatched him with the message; dispatched the letter yesterday). **2** perform (business, a task, etc.) promptly; finish off. **3** kill, execute (dispatched him with the revolver). **4** colloq. eat (food, a meal, etc.) quickly. ● n. **1** the act or an instance of sending (a messenger, letter, etc.). **2** the act or an instance of killing; execution. **3 a** an official written message on state or esp. military affairs. **b** a report sent in by a newspaper's correspondent, usu. from a foreign country. **c** any written message requiring fast delivery. **4** promptness, efficiency (done with dispatch). □ **dispatch-box** (or **-case**) a container for esp. official state or military documents or dispatches. **dispatch-rider** a motorcyclist or rider on horseback carrying (esp. military) dispatches. □ **dispatcher** n. [It. dispacciare or Sp. despachar expedite (as DIS-, It. impacciare and Sp. empachar hinder, of uncert. orig.)]

dispel /dɪˈspel/ v.tr. (**dispelled**, **dispelling**) dissipate; disperse; scatter (the dawn dispelled their fears). □ **dispeller** n. [L dispellere (as DIS-, pellere drive)]

dispensable /dɪˈspensəb(ə)l/ adj. **1** able to be done without; unnecessary. **2** (of a law etc.) able to be relaxed in special cases. □ **dispensability** /-ˌspensəˈbɪlɪtɪ/ n. [med.L dispensabilis (as DISPENSE)]

dispensary /dɪˈspensərɪ/ n. (pl. **-ies**) **1** a place where medicines etc. are dispensed. **2** a public or charitable institution for medical advice and the dispensing of medicines. [med.L dispensarius (as DISPENSE)]

dispensation /ˌdɪspenˈseɪʃ(ə)n/ n. **1 a** the act or an instance of dispensing or distributing. **b** (foll. by with) the state of doing without (a thing). **c** something distributed. **2** (usu. foll. by from) **a** exemption from a penalty or duty; an instance of this. **b** permission to be exempted from a religious observance; an instance of this. **3** a religious or political system obtaining in a nation etc. (the Christian dispensation). **4 a** the ordering or management of the world by Providence. **b** a specific example of such ordering (of a community, a person, etc.). □ **dispensational** adj. [ME f. OF dispensation or L dispensatio (as DISPENSE)]

dispense /dɪˈspens/ v. **1** tr. distribute; deal out. **2** tr. administer (a sacrament, justice, etc.). **3** tr. make up and give out (medicine etc.)

according to a doctor's prescription. **4** tr. (usu. foll. by from) grant a dispensation to (a person) from an obligation, esp. a religious observance. **5** intr. (foll. by with) **a** do without; render needless. **b** give exemption from (a rule). □ **dispensing chemist** a chemist qualified to make up and give out medicine etc. **dispensing optician** see OPTICIAN 2. [ME f. OF despenser f. L dispensare frequent. of dispendere weigh or pay out (as DIS-, pendere pens- weigh)]

dispenser /dɪˈspensə(r)/ n. **1** a person or thing that dispenses something, e.g. medicine, good advice. **2** an automatic machine that dispenses an item or a specific amount of something (e.g. cash).

dispersant /dɪˈspɜːs(ə)nt/ n. Chem. an agent used to disperse small particles in a medium.

disperse /dɪˈspɜːs/ v. **1** intr. & tr. go, send, drive, or distribute in different directions or over a wide area. **2 a** intr. (of people at a meeting etc.) leave and go their various ways. **b** tr. cause to do this. **3** tr. send to or station at separate points. **4** tr. put in circulation; disseminate. **5** tr. Chem. distribute (small particles) uniformly in a medium. **6** tr. Physics divide (white light) into its coloured constituents. □ **dispersable** adj. **dispersal** n. **disperser** n. **dispersible** adj. **dispersive** adj. [ME f. L dispergere dispers- (as DIS-, spargere scatter)]

dispersion /dɪˈspɜːʃ(ə)n/ n. **1** the act or an instance of dispersing; the process of being dispersed. **2** Chem. a mixture of one substance dispersed in another. **3** Physics the separation of white light into colours or of any radiation according to wavelength. **4** Statistics the extent to which values of a variable differ from the mean. **5** Ecol. the pattern of distribution of individual animals or plants within a habitat. **6** (**the Dispersion**) = DIASPORA 1. [ME f. LL dispersio (as DISPERSE), transl. Gk diaspora: see DIASPORA]

dispirit /dɪˈspɪrɪt/ v.tr. **1** (esp. as **dispiriting** adj.) make despondent; discourage. **2** (as **dispirited** adj.) dejected; discouraged. □ **dispiritedly** adv. **dispiritedness** n. **dispiritingly** adv.

displace /dɪsˈpleɪs/ v.tr. **1** shift from its accustomed place. **2** remove from office. **3** take the place of; oust. □ **displaced person** a person who is forced to leave his or her home country because of war, persecution, etc.; a refugee.

displacement /dɪsˈpleɪsmənt/ n. **1 a** the act or an instance of displacing; the process of being displaced. **b** an instance of this. **2** Physics the amount of a fluid displaced by a solid floating or immersed in it (a ship with a displacement of 11,000 tons). **3** Psychol. **a** the substitution of one idea or impulse for another. **b** the unconscious transfer of strong unacceptable emotions from one object to another. **4** the amount by which a thing is moved from a position. □ **displacement activity** Psychol. an animal or human activity which seems irrelevant to the situation (e.g head-scratching when confused).

display /dɪˈspleɪ/ v. & n. ● v.tr. **1** expose to view; exhibit; show. **2** show ostentatiously. **3** allow to appear; reveal; betray (displayed his ignorance). ● n. **1** the act or an instance of displaying. **2** an exhibition or show. **b** a thing or things intended to be looked at. **3** ostentation; flashiness. **4** the distinct behaviour of some birds and fish, esp. used to attract a mate. **5 a** the presentation of signals or data on a visual display unit etc. **b** the information so presented. **6** Printing the arrangement and choice of type in order to attract attention. □ **displayer** n. [ME f. OF despleier f. L displicare (as DIS-, plicare fold): cf. DEPLOY]

displease /dɪsˈpliːz/ v.tr. make indignant or angry; offend; annoy. □ **be displeased** (often foll. by at, with) be indignant or dissatisfied; disapprove. □ **displeasing** adj. **displeasingly** adv. [ME f. OF desplaisir (as DIS-, L placere please)]

displeasure /dɪsˈpleʒə(r)/ n. & v. ● n. disapproval; anger; dissatisfaction. ● v.tr. archaic cause displeasure to; annoy. [ME f. OF (as DISPLEASE): assim. to PLEASURE]

disport /dɪˈspɔːt/ v. & n. ● v.intr. & refl. frolic; gambol; enjoy oneself (disported on the sand; disported themselves in the sea). ● n. archaic **1** relaxation. **2** a pastime. [ME f. AF & OF desporter (as DIS-, porter carry f. L portare)]

disposable /dɪˈspəʊzəb(ə)l/ adj. & n. ● adj. **1** intended to be used once and then thrown away (disposable nappies). **2** that can be got rid of, made over, or used. **3** (esp. of financial assets) at the owner's disposal. ● n. a thing designed to be thrown away after one use. □ **disposable income** income after tax etc. □ **disposability** /-ˌspəʊzəˈbɪlɪtɪ/ n.

disposal /dɪˈspəʊz(ə)l/ n. (usu. foll. by of) **1** the act or an instance of disposing of something. **2** the arrangement, disposition, or placing of something. **3** control or management (of a person, business, etc.). **4** (esp. as **waste disposal**) the disposing of rubbish. **5** N. Amer. colloq. a waste disposal unit. □ **at one's disposal 1** available for one's use. **2** subject to one's orders or decisions.

dispose /dɪˈspəʊz/ v. **1** tr. (usu. foll. by to, or to + infin.) **a** make willing; incline (disposed him to the idea; was disposed to release them). **b** have a tendency to (the wheel was disposed to buckle). **2** tr. place suitably or in order (disposed the pictures in sequence). **3** tr. (as **disposed** adj.) having a specified mental inclination (usu. in comb.: ill-disposed). **4** intr. determine the course of events (man proposes, God disposes). □ **dispose of 1 a** deal with. **b** get rid of. **c** finish. **d** kill. **e** distribute, dispense; bestow. **2** sell. **3** prove (a claim, an argument, an opponent, etc.) to be incorrect. **4** consume (food). □ **disposer** n. [ME f. OF disposer (as DIS-, POSE¹) after L disponere disposit-]

disposition /ˌdɪspəˈzɪʃ(ə)n/ n. **1 a** (often foll. by to) a natural tendency; an inclination (a disposition to overeat). **b** a person's temperament or attitude, esp. as displayed in dealings with others (a happy disposition). **2 a** setting in order; arranging. **b** the relative position of parts; an arrangement. **3** (usu. in pl.) **a** Mil. the stationing of troops ready for attack or defence. **b** preparations; plans. **4 a** a bestowal by deed or will. **b** control; the power of disposing. **5** ordinance, dispensation. [ME f. OF f. L dispositio (as DIS-, ponere posit- place)]

dispossess /ˌdɪspəˈzes/ v.tr. **1** dislodge; oust (a person). **2** (usu. foll. by of) deprive. □ **dispossession** /-ˈzeʃ(ə)n/ n. [OF despossesser (as DIS-, POSSESS)]

dispraise /dɪsˈpreɪz/ v. & n. ● v.tr. express disapproval or censure of. ● n. disapproval, censure. [ME f. OF despreisier ult. f. LL depretiare DEPRECIATE]

disproof /dɪsˈpruːf/ n. **1** something that disproves. **2 a** refutation. **b** an instance of this.

disproportion /ˌdɪsprəˈpɔːʃ(ə)n/ n. **1** a lack of proportion. **2** an instance of this. □ **disproportional** adj. **disproportionally** adv.

disproportionate /ˌdɪsprəˈpɔːʃənət/ adj. **1** lacking proportion. **2** relatively too large or small, long or short, etc. □ **disproportionately** adv. **disproportionateness** n.

disprove /dɪsˈpruːv/ v.tr. prove false; refute. □ **disprovable** adj. [ME f. OF desprover (as DIS-, PROVE)]

Dispur /dɪsˈpʊə(r)/ a city in NE India, capital of the state of Assam.

disputable /dɪˈspjuːtəb(ə)l/ adj. open to question; uncertain. □ **disputably** adv. [F or f. L disputabilis (as DISPUTE)]

disputation /ˌdɪspjʊˈteɪʃ(ə)n/ n. **1 a** disputing, debating. **b** an argument; a controversy. **2** a formal debate. [ME f. F disputation or L disputatio (as DISPUTE)]

disputatious /ˌdɪspjʊˈteɪʃəs/ adj. fond of or inclined to argument. □ **disputatiously** adv. **disputatiousness** n.

dispute v. & n. ● v. /dɪˈspjuːt/ **1** intr. (usu. foll. by with, against) **a** debate, argue (was disputing with them about the meaning of life). **b** quarrel. **2** tr. discuss, esp. heatedly (disputed whether it was true). **3** tr. question the truth or correctness or validity of (a statement, alleged fact, etc.) (I dispute that number). **4** tr. contend for; strive to win (disputed the crown; disputed the field). **5** tr. resist (a landing, advance, etc.). ● n. /dɪˈspjuːt, disp. ˈdɪspjuːt/ **1** a controversy; a debate. **2** a quarrel. **3** a disagreement between management and employees, esp. one leading to industrial action. **4** archaic a fight or altercation; a struggle. □ **beyond** (or **past** or **without**) **dispute 1** certainly; indisputably. **2** certain, indisputable. **in dispute 1** being argued about. **2** (of a workforce) involved in industrial action. □ **disputant** /dɪˈspjuːt(ə)nt/ n. **disputer** /dɪˈspjuːtə(r)/ n. [ME f. OF desputer f. L disputare estimate (as DIS-, putare reckon)]

disqualification /dɪsˌkwɒlɪfɪˈkeɪʃ(ə)n/ n. **1** the act or an instance of disqualifying; the state of being disqualified. **2** something that disqualifies.

disqualify /dɪsˈkwɒlɪˌfaɪ/ v.tr. (-ies, -ied) **1** (often foll. by from) debar from a competition or pronounce ineligible as a winner because of an infringement of the rules etc. (disqualified from the race for taking drugs). **2** (often foll. by for, from) make or pronounce ineligible or unsuitable (his age disqualifies him for the job; a criminal record disqualified him from applying). **3** (often foll. by from) incapacitate legally; pronounce unqualified (disqualified from practising as a doctor).

disquiet /dɪsˈkwaɪət/ v. & n. ● v.tr. deprive of peace; worry. ● n. anxiety; unrest. □ **disquieting** adj. **disquietingly** adv.

disquietude /dɪsˈkwaɪəˌtjuːd/ n. a state of uneasiness; anxiety.

disquisition /ˌdɪskwɪˈzɪʃ(ə)n/ n. a long or elaborate treatise or discourse on a subject. □ **disquisitional** adj. [F f. L disquisitio (as DIS-, quaerere quaesit- seek)]

Disraeli /dɪzˈreɪlɪ/, Benjamin, 1st Earl of Beaconsfield (1804–81), British Tory statesman, of Italian–Jewish descent; Prime Minister 1868 and 1874–80. He played a dominant role in the reconstruction of the Tory Party after Sir Robert Peel, guiding it away from protectionism and generating enthusiasm for the British Empire. He was largely responsible for the introduction of the second Reform Act (1867), which doubled the electorate. In his second term as Prime Minister he ensured that Britain bought a controlling interest in the Suez Canal (1875) and also made Queen Victoria Empress of India. At home his government passed much useful social legislation, including measures to improve public health and working conditions in factories. He wrote a number of novels, including Coningsby (1844) and Sybil (1845), which drew on his experience of political life.

disrate /dɪsˈreɪt/ v.tr. Naut. reduce (a sailor) to a lower rating or rank.

disregard /ˌdɪsrɪˈgɑːd/ v. & n. ● v.tr. **1** pay no attention to; ignore. **2** treat as of no importance. **3** archaic neglect contemptuously; slight, snub. ● n. (often foll. by of, for) indifference; neglect. □ **disregardful** adj. **disregardfully** adv.

disrelish /dɪsˈrelɪʃ/ n. & v. ● n. dislike; distaste. ● v.tr. regard with dislike or distaste.

disremember /ˌdɪsrɪˈmembə(r)/ v.tr. & intr. esp. US or dial. fail to remember; forget.

disrepair /ˌdɪsrɪˈpeə(r)/ n. poor condition due to neglect (in disrepair; in a state of disrepair).

disreputable /dɪsˈrepjʊtəb(ə)l/ adj. **1** of bad reputation; discreditable. **2** not respectable in appearance; dirty, untidy. □ **disreputableness** n. **disreputably** adv.

disrepute /ˌdɪsrɪˈpjuːt/ n. a lack of good reputation or respectability; discredit (esp. fall into disrepute).

disrespect /ˌdɪsrɪˈspekt/ n. a lack of respect; discourtesy. □ **disrespectful** adj. **disrespectfully** adv.

disrobe /dɪsˈrəʊb/ v.tr. & refl. (also absol.) **1** divest (oneself or another) of a robe or a garment; undress. **2** divest (oneself or another) of office, authority, etc.

disrupt /dɪsˈrʌpt/ v.tr. **1** interrupt the flow or continuity of (a meeting, speech, etc.); bring disorder to. **2** separate forcibly; shatter. □ **disrupter** n. (also **disruptor**). **disruptive** adj. **disruptively** adv. **disruptiveness** n. **disruption** /-ˈrʌpʃ(ə)n/ n. [L disrumpere disrupt- (as DIS-, rumpere break)]

diss /dɪs/ v.tr. US sl. put (a person) down verbally; bad-mouth. [shortened f. DISRESPECT]

dissatisfy /dɪˈsætɪsˌfaɪ/ v.tr. (-ies, -ied) (often as **dissatisfied** adj.) make discontented; fail to satisfy (dissatisfied with the accommodation; dissatisfied to find him gone). □ **dissatisfiedly** adv. **dissatisfaction** /-ˌsætɪsˈfækʃ(ə)n/ n. **dissatisfactory** /-ˈfæktərɪ/ adj.

dissect /dɪˈsekt/ v.tr. **1** cut into pieces. **2** cut up (a plant or animal) to examine its parts, structure, etc., or (a corpse) for a post mortem. **3** analyse; criticize or examine in detail. □ **dissector** n. **dissection** /-ˈsekʃ(ə)n/ n. [L dissecare dissect- (as DIS-, secare cut)]

dissemble /dɪˈsemb(ə)l/ v. **1** intr. conceal one's motives; talk or act hypocritically. **2** tr. a disguise or conceal (a feeling, intention, act, etc.). **b** simulate (dissembled grief in public). □ **dissemblance** n. **dissembler** n. **dissemblingly** adv. [ME, alt. after semblance of obs. dissimule f. OF dissimuler f. L dissimulare (as DIS-, SIMULATE)]

disseminate /dɪˈsemɪˌneɪt/ v.tr. scatter about, spread (esp. ideas) widely. □ **disseminated sclerosis** = SCLEROSIS 2. □ **disseminator** n. **dissemination** /-ˌsemɪˈneɪʃ(ə)n/ n. [L disseminare (as DIS-, semen -inis seed)]

dissension /dɪˈsenʃ(ə)n/ n. disagreement giving rise to discord. [ME f. OF f. L dissensio (as DIS-, sentire sens- feel)]

dissent /dɪˈsent/ v. & n. ● v.intr. (often foll. by from) **1** think differently, disagree; express disagreement. **2** differ in religious opinion, esp. from the doctrine of an established or orthodox church. ● n. **1 a** a difference of opinion. **b** an expression of this. **2** the refusal to accept the doctrines of an established or orthodox church; nonconformity. □ **dissenting** adj. [ME f. L dissentire (as DIS-, sentire feel)]

dissenter /dɪˈsentə(r)/ n. **1** a person who dissents. **2** (**Dissenter**) Brit. a member of a non-established church; a Nonconformist.

dissentient /dɪˈsenʃ(ə)nt/ adj. & n. ● adj. disagreeing with a majority or official view. ● n. a person who dissents. [L dissentire (as DIS-, sentire feel)]

dissertation /ˌdɪsəˈteɪʃ(ə)n/ n. a detailed discourse on a subject, esp. one submitted in partial fulfilment of the requirements of a degree or diploma. □ **dissertational** adj. [L dissertatio f. dissertare discuss, frequent. of disserere dissert- examine (as DIS-, serere join)]

disservice /dɪsˈsɜːvɪs/ n. an ill turn; an injury, esp. done when trying to help.

dissever /dɪˈsevə(r)/ v.tr. & intr. sever; divide into parts. □ **disseverance** n. **disseverment** n. [ME f. AF dis(c)everer, OF desevrer f. LL disseparare (as DIS-, SEPARATE)]

dissidence /ˈdɪsɪd(ə)ns/ n. disagreement; dissent. [F dissidence or L dissidentia (as DISSIDENT)]

dissident /ˈdɪsɪd(ə)nt/ adj. & n. ● adj. disagreeing, esp. with an established government, system, etc. ● n. a dissident person. [F or f. L dissidere disagree (as DIS-, sedere sit)]

dissimilar /dɪˈsɪmɪlə(r)/ adj. (often foll. by to) unlike, not similar. □ **dissimilarly** adv. **dissimilarity** /-ˌsɪmɪˈlærɪtɪ/ n. (pl. **-ies**).

dissimilate /dɪˈsɪmɪˌleɪt/ v. (often foll. by to) Phonet. **1** tr. change (a sound or sounds in a word) to another when the word originally had the same sound repeated, as in cinnamon, orig. cinnamom. **2** intr. (of a sound) be changed in this way. □ **dissimilatory** /-lətərɪ/ adj. **dissimilation** /-ˌsɪmɪˈleɪʃ(ə)n/ n. [L dissimilis (as DIS-, similis like), after assimilate]

dissimilitude /ˌdɪsɪˈmɪlɪˌtjuːd/ n. unlikeness, dissimilarity. [L dissimilitudo (as DISSIMILATE)]

dissimulate /dɪˈsɪmjʊˌleɪt/ v.tr. & intr. dissemble. □ **dissimulator** n. **dissimulation** /-ˌsɪmjʊˈleɪʃ(ə)n/ n. [L dissimulare (as DIS-, SIMULATE)]

dissipate /ˈdɪsɪˌpeɪt/ v. **1** a tr. cause (a cloud, vapour, fear, darkness, etc.) to disappear or disperse. **b** intr. disperse, scatter, disappear. **2** intr. & tr. break up; bring or come to nothing. **3** tr. squander or fritter away (money, energy, etc.). **4** intr. (as **dissipated** adj.) given to dissipation, dissolute. □ **dissipater** n. **dissipative** adj. **dissipator** n. [L dissipare dissipat- (as DIS-, sipare (unrecorded) throw)]

dissipation /ˌdɪsɪˈpeɪʃ(ə)n/ n. **1** intemperate, dissolute, or debauched living. **2** (usu. foll. by of) wasteful expenditure (dissipation of resources). **3** scattering, dispersion, or disintegration. **4** a frivolous amusement. [F dissipation or L dissipatio (as DISSIPATE)]

dissociate /dɪˈsəʊʃɪˌeɪt, dɪˈsəʊsɪ-/ v. **1** tr. & intr. (usu. foll. by from) disconnect or become disconnected; separate (dissociated her from their guilt). **2** tr. Chem. decompose, esp. reversibly. **3** tr. Psychol. cause (a person's mind) to develop more than one centre of consciousness. □ **dissociated personality** Psychol. the pathological coexistence of two or more distinct personalities in the same person. **dissociate oneself from 1** declare oneself unconnected with. **2** decline to support or agree with (a proposal etc.). □ **dissociative** /-ˈsəʊʃɪətɪv, -ˈsəʊsɪətɪv/ adj. [L dissociare (as DIS-, socius companion)]

dissociation /dɪˌsəʊsɪˈeɪʃ(ə)n/ n. **1** the act or an instance of dissociating. **2** Psychol. the state of suffering from dissociated personality.

dissoluble /dɪˈsɒljʊb(ə)l/ adj. able to be disintegrated, loosened, or disconnected; soluble. □ **dissolubly** adv. **dissolubility** /-ˌsɒljʊˈbɪlɪtɪ/ n. [F dissoluble or L dissolubilis (as DIS-, SOLUBLE)]

dissolute /ˈdɪsəˌluːt, -ˌljuːt/ adj. lax in morals; licentious. □ **dissolutely** adv. **dissoluteness** n. [ME f. L dissolutus past part. of dissolvere DISSOLVE]

dissolution /ˌdɪsəˈluːʃ(ə)n, -ˈljuːʃ(ə)n/ n. **1** disintegration; decomposition. **2** (usu. foll. by of) the undoing or relaxing of a bond, esp. a marriage, a partnership, or an alliance. **3** the dismissal or dispersal of an assembly, esp. of a parliament at the end of its term. **4** death. **5** bringing or coming to an end; fading away; disappearance. **6** dissipation; debauchery. [ME f. OF dissolution or L dissolutio (as DISSOLVE)]

dissolution of the monasteries the abolition of monasteries in England and Wales by Henry VIII under two Acts (1536, 1539) by which they were suppressed and their assets vested in the Crown. Though monasteries were much criticized in the later Middle Ages for their wealth, moral laxity, and stress on the contemplative life, Henry's motives were personal: to replenish his treasury and to establish royal supremacy in ecclesiastical affairs.

dissolve /dɪˈzɒlv/ v. & n. ● v. **1** tr. & intr. incorporate or become incorporated into a liquid so as to form a solution. **2** intr. & tr. disappear or cause to disappear gradually. **3 a** tr. dismiss or disperse (an assembly, esp. Parliament). **b** intr. (of an assembly) be dissolved (cf. DISSOLUTION 3). **4** tr. annul or put an end to (a partnership, marriage, etc.). **5** intr. (of a person) become enfeebled or emotionally overcome (completely dissolved when he saw her; dissolved into tears). **6** intr. (often foll. by into) Cinematog. change gradually (from one picture into another). ● n. Cinematog. the act or process of dissolving a picture. □ **dissolvable** adj. [ME f. L dissolvere dissolut- (as DIS-, solvere loosen)]

dissolvent /dɪˈzɒlv(ə)nt/ adj. & n. ● adj. tending to dissolve or dissipate. ● n. a dissolvent substance. [L dissolvere (as DISSOLVE)]

dissonant /ˈdɪsənənt/ adj. **1** Mus. harsh-toned; inharmonious. **2** incongruous; clashing. □ **dissonance** n. **dissonantly** adv. [ME f. OF dissonant or L dissonare (as DIS-, sonare sound)]

dissuade /dɪˈsweɪd/ v.tr. (often foll. by from) discourage (a person); persuade against (dissuaded him from continuing; was dissuaded from his belief). □ **dissuader** n. **dissuasion** /-ˈsweɪʒ(ə)n/ n. **dissuasive** /-ˈsweɪsɪv/ adj. [L dissuadere (as DIS-, suadere suas- persuade)]

dissyllable var. of DISYLLABLE.

dissymmetry /dɪˈsɪmɪtrɪ/ n. (pl. **-ies**) **1 a** lack of symmetry. **b** an instance of this. **2** symmetry as of mirror images or the left and right hands (esp. of crystals with two corresponding forms). □ **dissymmetrical** /ˌdɪsɪˈmetrɪk(ə)l/ adj.

distaff /ˈdɪstɑːf/ n. **1 a** a cleft stick holding wool or flax wound for spinning by hand. **b** the corresponding part of a spinning-wheel. **2** women's work. □ **distaff side** the female line of descent; the female side of a family. [OE distæf (as STAFF[1]), the first element being app. rel. to LG diesse, MLG dise(ne) bunch of flax]

distal /ˈdɪst(ə)l/ adj. Anat. situated away from the centre of the body or point of attachment (opp. PROXIMAL). □ **distally** adv. [DISTANT + -AL[1]]

distance /ˈdɪstəns/ n. & v. ● n. **1** the condition of being far off; remoteness. **2 a** a space or interval between two things. **b** the length of this (a distance of twenty miles). **3** a distant point or place (came from a distance). **4** the avoidance of familiarity; aloofness; reserve (there was a certain distance between them). **5** a remoter field of vision (saw him in the distance). **6** an interval of time (can't remember what happened at this distance). **7 a** the full length of a race etc. **b** Brit. Horse-racing a length of 240 yards from the winning-post on a racecourse. **c** Boxing the scheduled length of a fight. ● v.tr. (often refl.) **1** place far off (distanced herself from them; distanced the painful memory). **2** leave far behind in a race or competition. □ **at a distance** far off. **distance-post** Horse-racing a post at the distance on a racecourse, used to disqualify runners who have not reached it by the time the winner reaches the winning-post. **distance runner** an athlete who competes in long- or middle-distance races. **go the distance 1** Boxing complete a fight without being knocked out. **2** complete a hard task; endure an ordeal. **keep one's distance** maintain one's reserve. **middle distance** see MIDDLE. **within hailing** (or **walking**) **distance** near enough to reach by hailing or walking. [ME f. OF distance, destance f. L distantia f. distare stand apart (as DI-[2], stare stand)]

distant /ˈdɪstənt/ adj. **1 a** far away in space or time. **b** (usu. predic.; often foll. by from) at a specified distance (three miles distant from them). **2** remote or far apart in position, time, resemblance, etc. (a distant prospect; a distant relation; a distant likeness). **3** not intimate; reserved; cool (a distant nod). **4** remote; abstracted (a distant stare). **5** faint, vague (he was a distant memory to her). □ **distant early warning** a radar system in North America for the early detection of a missile attack. **distant signal** Railways a railway signal preceding a home signal to give warning. □ **distantly** adv. [ME f. OF distant or L distant- part. stem of distare: see DISTANCE]

distaste /dɪsˈteɪst/ n. (usu. foll. by for) dislike; repugnance; aversion, esp. slight (a distaste for prunes; a distaste for polite company). □ **distasteful** adj. **distastefully** adv. **distastefulness** n.

distemper[1] /dɪˈstempə(r)/ n. & v. ● n. **1** a kind of paint using glue or size instead of an oil-base, for use on walls or for scene-painting. **2** a method of mural and poster painting using this. ● v.tr. paint (walls etc.) with distemper. [earlier as verb, f. OF destremper or LL distemperare soak, macerate: see DISTEMPER[2]]

distemper[2] /dɪˈstempə(r)/ n. **1** a viral disease of some animals, esp. dogs, causing fever, coughing, and catarrh. **2** archaic political disorder. [earlier as verb, = upset, derange: ME f. LL distemperare (as DIS-, temperare mingle correctly)]

distend /dɪˈstend/ v.tr. & intr. swell out by pressure from within (distended stomach). □ **distensible** /-ˈstensɪb(ə)l/ adj. **distensibility** /-ˌstensɪˈbɪlɪtɪ/ n. **distension** /-ˈstenʃ(ə)n/ n. [ME f. L distendere (as DIS-, tendere tens- stretch)]

distich /ˈdɪstɪk/ n. Prosody a pair of verse lines; a couplet. [L distichon f. Gk distikhon (as DI-[1], stikhos line)]

distichous /ˈdɪstɪkəs/ adj. Bot. arranged in two opposite vertical rows. [L distichus (as DISTICH)]

distil /dɪˈstɪl/ v. (US **distill**) (**distilled**, **distilling**) **1** tr. Chem. purify (a liquid) by vaporizing it with heat, then condensing it with cold and collecting the resulting liquid. **2** tr. **a** Chem. extract the essence of (a plant etc.) usu. by heating it in a solvent. **b** extract the essential meaning or implications of (an idea etc.). **3** *tr.* make (whisky, essence, etc.) by distilling raw materials. **4** *tr.* (foll. by *off, out*) Chem. drive (a volatile constituent) off or out by heat. **5** *tr. & intr.* come as or give forth in drops; exude. **6** *intr.* undergo distillation. □ **distillatory** /-lətərɪ/ *adj.* [ME f. L *distillare* f. *destillare* (as DE-, *stilla* drop)]

distillate /ˈdɪstɪˌleɪt/ *n.* a product of distillation.

distillation /ˌdɪstɪˈleɪʃ(ə)n/ *n.* **1** the process of distilling or being distilled (in various senses). **2** something distilled.

distiller /dɪˈstɪlə(r)/ *n.* a person who distils, esp. a manufacturer of alcoholic liquor.

distillery /dɪˈstɪlərɪ/ *n.* (*pl.* **-ies**) a place where alcoholic liquor is distilled.

distinct /dɪˈstɪŋkt/ *adj.* **1** (often foll. by *from*) **a** not identical; separate; individual. **b** different in kind or quality; unlike. **2 a** clearly perceptible; plain. **b** clearly understandable; definite. **3** unmistakable, decided (*had a distinct impression of being watched*). □ **distinctly** *adv.* **distinctness** *n.* [ME f. L *distinctus* past part. of *distinguere* DISTINGUISH]

distinction /dɪˈstɪŋkʃ(ə)n/ *n.* **1 a** the act or an instance of discriminating or distinguishing. **b** an instance of this. **c** the difference made by distinguishing. **2 a** something that differentiates, e.g. a mark, name, or title. **b** the fact of being different. **3** special consideration or honour. **4** distinguished character; excellence; eminence (*a film of distinction; shows distinction in his bearing*). **5** a grade in an examination denoting great excellence (*passed with distinction*). □ **distinction without a difference** a merely nominal or artificial distinction. [ME f. OF f. L *distinctio -onis* (as DISTINGUISH)]

distinctive /dɪˈstɪŋktɪv/ *adj.* distinguishing, characteristic. □ **distinctively** *adv.* **distinctiveness** *n.*

distingué /dɪˈstæŋɡeɪ/ *adj.* (*fem.* **distinguée** *pronunc.* same) having a distinguished air, features, manner, etc. [F, past part. of *distinguer*: see DISTINGUISH]

distinguish /dɪˈstɪŋɡwɪʃ/ *v.* **1** *tr.* (often foll. by *from*) **a** see or point out the difference of; draw distinctions between (*cannot distinguish one from the other*). **b** constitute a difference between (*the mole distinguishes him from his twin*). **c** treat as different; differentiate. **2** *tr.* be a mark or property of; characterize (*distinguished by his greed*). **3** *tr.* discover by listening, looking, etc. (*could distinguish two voices*). **4** *tr.* (usu. *refl.*; often foll. by *by*) make prominent or noteworthy (*distinguished himself by winning first prize*). **5** *tr.* (often foll. by *into*) divide; classify. **6** *intr.* (foll. by *between*) make or point out a difference between. □ **distinguishable** *adj.* [F *distinguer* or L *distinguere* (as DIS-, *stinguere stinct-* extinguish): cf. EXTINGUISH]

distinguished /dɪˈstɪŋɡwɪʃt/ *adj.* **1** (often foll. by *for, by*) of high standing; eminent; famous. **2** = DISTINGUÉ.

Distinguished Flying Cross *n.* (abbr. **DFC**) (in the UK) a decoration for distinguished active service awarded to officers of the RAF, instituted in 1918.

Distinguished Service Order *n.* (abbr. **DSO**) (in the UK) a decoration for distinguished service awarded to officers of the army and navy, instituted in 1886.

distort /dɪˈstɔːt/ *v.tr.* **1 a** put out of shape; make crooked or unshapely. **b** distort the appearance of, esp. by curved mirrors etc. **2** misrepresent (motives, facts, statements, etc.). **3** change the form of (an electrical signal) during transmission, amplification, etc. □ **distortedly** *adv.* **distortedness** *n.* [L *distorquere distort-* (as DIS-, *torquere* twist)]

distortion /dɪˈstɔːʃ(ə)n/ *n.* **1** the act or an instance of distorting; the process of being distorted. **2** Electronics a change in the form of a signal during transmission etc. usu. with some impairment of quality. □ **distortional** *adj.* **distortionless** *adj.* [L *distortio* (as DISTORT)]

distract /dɪˈstrækt/ *v.tr.* **1** (often foll. by *from*) draw away the attention of (a person, the mind, etc.). **2** bewilder, perplex. **3** (as **distracted** *adj.*) troubled or distraught (*distracted by grief; distracted with worry*). **4** amuse, esp. in order to take the attention from pain or worry. □ **distractedly** *adv.* [ME f. L *distrahere distract-* (as DIS-, *trahere* draw)]

distraction /dɪˈstrækʃ(ə)n/ *n.* **1 a** the act of distracting, esp. the mind. **b** something that distracts; an interruption. **2** a relaxation from work; an amusement. **3** a lack of concentration. **4** confusion; perplexity. **5** frenzy; madness. □ **to distraction** almost to a state of madness. [ME f. OF *distraction* or L *distractio* (as DISTRACT)]

distrain /dɪˈstreɪn/ *v.intr.* (usu. foll. by *upon*) Law impose distraint (on a person, goods, etc.). □ **distrainer** *n.* **distrainment** *n.* **distrainor** *n.* [ME f. OF *destreindre* f. L *distringere* (as DIS-, *stringere strict-* draw tight)]

distraint /dɪˈstreɪnt/ *n.* Law the seizure of chattels to make a person

pay rent etc. or meet an obligation, or to obtain satisfaction by their sale. [DISTRAIN, after *constraint*]

distrait /dɪˈstreɪ/ *adj.* (*fem.* **distraite** /-ˈstreɪt/) not paying attention; absent-minded; distraught. [ME f. OF *destrait* past part. of *destraire* (as DISTRACT)]

distraught /dɪˈstrɔːt/ *adj.* distracted with worry, fear, etc.; extremely agitated. [ME, alt. of obs. *distract* (adj.) (as DISTRACT), after *straught* obs. past part. of STRETCH]

distress /dɪˈstres/ *n. & v.* ● *n.* **1** severe pain, sorrow, anguish, etc. **2** the lack of money or comforts. **3** Law = DISTRAINT. **4** breathlessness; exhaustion. ● *v.tr.* **1** subject to distress; exhaust, afflict. **2** cause anxiety to; make unhappy; vex. □ **distress-signal** a signal from a ship in danger. **distress-warrant** Law a warrant authorizing distraint. **in distress 1** suffering or in danger. **2** (of a ship, aircraft, etc.) in danger or damaged. □ **distressful** *adj.* **distressingly** *adv.* [ME f. OF *destresse* etc., AF *destresser*, OF *-ecier* f. Gallo-Roman (as DISTRAIN)]

distressed /dɪˈstrest/ *adj.* **1** suffering from distress. **2** impoverished (*distressed gentlefolk; in distressed circumstances*). **3** (of furniture, leather, etc.) having simulated marks of age and wear. □ **distressed area** Brit. a region of high unemployment and poverty.

distributary /dɪˈstrɪbjʊtərɪ/ *n.* (*pl.* **-ies**) a branch of a river or glacier that does not return to the main stream after leaving it (as in a delta).

distribute /dɪˈstrɪbjuːt, *disp.* ˈdɪstrɪˌbjuːt/ *v.tr.* **1** give shares of; deal out. **2** spread about; scatter (*distributed the seeds evenly over the garden*). **3** divide into parts; arrange; classify. **4** Printing separate (type that has been set up) and return the characters to their separate boxes. **5** Logic use (a term) to include every individual of the class to which it refers. □ **distributed system** Computing a number of independent computers linked by a network. □ **distributable** *adj.* [ME f. L *distribuere distribut-* (as DIS-, *tribuere* assign)]

distribution /ˌdɪstrɪˈbjuːʃ(ə)n/ *n.* **1** the act or an instance of distributing; the process of being distributed. **2** Econ. **a** the dispersal of goods etc. among consumers, brought about by commerce. **b** the extent to which different groups, classes, or individuals share in the total production or wealth of a community. **3** Statistics the way in which a characteristic is spread over members of a class. □ **distributional** *adj.* [ME f. OF *distribution* or L *distributio* (as DISTRIBUTE)]

distributive /dɪˈstrɪbjʊtɪv/ *adj. & n.* ● *adj.* **1** of, concerned with, or produced by distribution. **2** Logic & Gram. (of a pronoun etc.) referring to each individual of a class, not to the class collectively (e.g. *each, either*). ● *n.* Gram. a distributive word. □ **distributively** *adv.* [ME f. F *distributif -ive* or LL *distributivus* (as DISTRIBUTE)]

distributor /dɪˈstrɪbjʊtə(r)/ *n.* **1** a person or thing that distributes. **2** an agent who supplies goods. **3** Electr. a device in an internal-combustion engine for passing current to each spark-plug in turn.

district /ˈdɪstrɪkt/ *n. & v.* ● *n.* **1 a** (often *attrib.*) a territory marked off for special administrative purposes. **b** Brit. a division of a county or region electing its own councillors. **2** an area which has common characteristics; a region (*the wine-growing district*). ● *v.tr.* US divide into districts. □ **district attorney** (in the US) the prosecuting officer of a district. **district court** (in the US) the Federal court of first instance. **district heating** a supply of heat or hot water from one source to a district or a group of buildings. **district nurse** Brit. a peripatetic nurse serving a rural or urban area. **district visitor** Brit. a person working for a member of the clergy in a section of a parish. [F f. med.L *districtus* (territory of) jurisdiction (as DISTRAIN)]

District of Columbia /kəˈlʌmbɪə/ (abbr. **DC**) a Federal district of the US, coextensive with the city of Washington, situated on the Potomac river with boundaries on the states of Virginia and Maryland.

distrust /dɪsˈtrʌst/ *n. & v.* ● *n.* a lack of trust; doubt; suspicion. ● *v.tr.* have no trust or confidence in; doubt. □ **distruster** *n.* **distrustful** *adj.* **distrustfully** *adv.*

disturb /dɪˈstɜːb/ *v.tr.* **1** break the rest, calm, or quiet of; interrupt. **2 a** agitate; worry (*your story disturbs me*). **b** irritate. **3** move from a settled position, disarrange (*the papers had been disturbed*). **4** (as **disturbed** *adj.*) Psychol. emotionally or mentally unstable or abnormal. □ **disturber** *n.* **disturbing** *adj.* **disturbingly** *adv.* [ME f. OF *desto(u)rber* f. L *disturbare* (as DIS-, *turbare* f. *turba* tumult)]

disturbance /dɪˈstɜːb(ə)ns/ *n.* **1** the act or an instance of disturbing; the process of being disturbed. **2** a tumult; an uproar. **3** agitation; worry. **4** an interruption. **5** Law interference with rights or property; molestation. [ME f. OF *desto(u)rbance* (as DISTURB)]

disulphide /daɪˈsʌlfaɪd/ *n.* (US **disulfide**) Chem. a binary chemical containing two atoms of sulphur in each molecule.

disunion /dɪsˈjuːnɪən/ n. a lack of union; separation; dissension. □ **disunity** n. **disunite** /ˌdɪsjʊˈnaɪt/ v.tr. & intr.

disuse n. & v. ● n. /dɪsˈjuːs/ **1** lack of use or practice; discontinuance. **2** a disused state. ● v.tr. /dɪsˈjuːz/ (esp. as **disused** adj.) cease to use. □ **fall into disuse** cease to be used. [ME f. OF desuser (as DIS-, USE)]

disutility /ˌdɪsjuːˈtɪlɪtɪ/ n. (pl. **-ies**) **1** harmfulness, injuriousness. **2** a factor tending to nullify the utility of something; a drawback.

disyllable /dɪˈsɪləb(ə)l, ˈdaɪsɪlə-/ n. (also **dissyllable**) Prosody a word or metrical foot of two syllables. □ **disyllabic** /ˌdɪsɪˈlæbɪk, ˌdaɪ-/ adj. [F disyllabe f. L disyllabus f. Gk disullabos (as DI-[1], SYLLABLE)]

dit /dɪt/ n. Telegraphy (in the Morse system) = DOT[1] n. 3 (cf. DAH). [imit.]

ditch /dɪtʃ/ n. & v. ● n. **1** a long narrow excavated channel esp. for drainage or to mark a boundary. **2** a watercourse, stream, etc. ● v. **1** intr. make or repair ditches (hedging and ditching). **2** tr. provide with ditches; drain. **3** tr. colloq. leave in the lurch; abandon. **4** tr. colloq. **a** bring (an aircraft) down on the sea in an emergency. **b** drive (a vehicle) into a ditch. **5** intr. colloq. (of an aircraft) make a forced landing on the sea. **6** tr. US derail (a train). □ **ditch-water** stagnant water in a ditch. **dull as ditch-water** extremely dull. **last ditch** see LAST[1]. □ **ditcher** n. [OE dīc, of unkn. orig.: cf. DIKE[1]]

ditheism /ˈdaɪθiːˌɪz(ə)m/ n. Theol. **1** a belief in two gods; dualism. **2** a belief in equal independent ruling principles of good and evil. □ **ditheist** n.

dither /ˈdɪðə(r)/ v. & n. ● v.intr. **1** hesitate; be indecisive. **2** dial. tremble; quiver. ● n. colloq. **1** a state of agitation or apprehension. **2** a state of hesitation; indecisiveness. □ **all of a dither** colloq. in a state of extreme agitation or vacillation. □ **ditherer** n. **dithery** adj. [var. of didder DODDER[1]]

dithyramb /ˈdɪθɪˌræm, -ˌræmb/ n. **1 a** a wild choral hymn in ancient Greece, esp. to Dionysus. **b** a Bacchanalian song. **2** a passionate or inflated poem, speech, etc. □ **dithyrambic** /ˌdɪθɪˈræmbɪk/ adj. [L dithyrambus f. Gk dithurambos, of unkn. orig.]

ditsy /ˈdɪtsɪ/ adj. (also **ditzy**) US sl. **1** (of a person) stupid; conceited; cute. **2** (of a thing) fussy, intricate. [20th c.: orig. uncert.]

dittany /ˈdɪtənɪ/ n. (pl. **-ies**) **1** a (in full **dittany of Crete**) a dwarf labiate shrub, Origanum dictamnus, with white woolly leaves. **b** US a similar labiate herb, Cunila origanoides. **2** = FRAXINELLA. [ME f. OF dita(i)n f. med.L dictamus f. L dictamnus f. Gk diktamnon perh. f. Diktē, a mountain in Crete]

ditto /ˈdɪtəʊ/ n. & v. ● n. (pl. **-os**) **1** (in accounts, inventories, lists, etc.) the aforesaid, the same. ¶ Often represented by „ under the word or sum to be repeated. **2** colloq. (replacing a word or phrase to avoid repetition) the same (came in late last night and ditto the night before). **3** a similar thing; a duplicate. ● v.tr. (**-oes, -oed**) repeat (another's action or words). □ **ditto marks** inverted commas etc. representing 'ditto'. **say ditto to** colloq. agree with; endorse. [It. dial. f. L dictus past part. of dicere say]

dittography /dɪˈtɒgrəfɪ/ n. (pl. **-ies**) **1** a copyist's mistaken repetition of a letter, word, or phrase. **2** an example of this. □ **dittographic** /ˌdɪtəˈgræfɪk/ adj. [Gk dittos double + -GRAPHY]

ditty /ˈdɪtɪ/ n. (pl. **-ies**) a short simple song. [ME f. OF dité composition f. L dictatum neut. past part. of dictare DICTATE]

ditty-bag /ˈdɪtɪˌbæg/ n. (also **ditty-box** /-ˌbɒks/) a sailor's or fisherman's receptacle for odds and ends. [19th c.: orig. unkn.]

diuresis /ˌdaɪjʊˈriːsɪs/ n. Med. the increased excretion of urine, esp. in excess. [mod.L f. Gk (as DI-[3], ourēsis urination)]

diuretic /ˌdaɪjʊˈrɛtɪk/ adj. & n. ● adj. causing increased output of urine. ● n. a diuretic drug. [ME f. OF diuretique or LL diureticus f. Gk diourētikos f. dioureō urinate]

diurnal /daɪˈɜːn(ə)l/ adj. **1** of or during the day; not nocturnal. **2** daily; of each day. **3** Astron. occupying one day. **4** Zool. (of animals) active in the daytime. **5** Bot. (of plants) open only during the day. □ **diurnally** adv. [ME f. LL diurnalis f. L diurnus f. dies day]

Div. abbr. Division.

diva /ˈdiːvə/ n. (pl. **divas**) a great or famous woman singer; a prima donna. [It. f. L, = goddess]

divagate /ˈdaɪvəˌgeɪt/ v.intr. literary stray; digress. □ **divagation** /ˌdaɪvəˈgeɪʃ(ə)n/ n. [L divagari (as DI-[2], vagari wander)]

divalent /daɪˈveɪlənt/ adj. Chem. having a valency of two; bivalent.

divan /dɪˈvæn, daɪ-/ n. **1 a** a long, low, backless sofa. **b** a bed consisting of a base and mattress, usu. with no board at either end. **2** an oriental state legislative body, council-chamber, or court of justice. **3** archaic **a** a cigar-shop. **b** a smoking-room attached to such a shop. [F divan or It.

divano f. Turk. dīvān f. Arab. dīwān f. Pers. dīvān anthology, register, court, bench]

divaricate /daɪˈværɪˌkeɪt, dɪ-/ v.intr. diverge, branch; separate widely. □ **divaricate** /-kət/ adj. **divarication** /-ˌværɪˈkeɪʃ(ə)n/ n. [L divaricare (as DI-[2], varicus straddling)]

dive /daɪv/ v. & n. ● v. (past **dived** or US **dove** /dəʊv/) **1** intr. plunge head first into water, esp. as a sport. **2** intr. **a** (of an aircraft) plunge steeply downwards at speed. **b** Naut. (of a submarine) submerge. **c** (of a person) plunge downwards. **3** intr. (foll. by into) colloq. **a** put one's hand into (a pocket, handbag, container, etc.) quickly and deeply. **b** occupy oneself suddenly and enthusiastically with (a subject, meal, etc.). **4** tr. (foll. by into) plunge (a hand etc.) into. ● n. **1** an act of diving; a plunge. **2 a** the submerging of a submarine. **b** the steep descent of an aircraft. **3** a sudden darting movement. **4** colloq. a disreputable nightclub etc.; a drinking-den (found themselves in a low dive). **5** Boxing sl. a pretended knockout (took a dive in the second round). □ **dive-bomb** bomb (a target) while diving in an aircraft. **dive-bomber** an aircraft designed to dive-bomb. **dive in** colloq. help oneself (to food). **diving beetle** a predatory water beetle of the family Dytiscidae, which stores air under the elytra while diving. **diving-bell** an open-bottomed box or bell, supplied with air, in which a person can descend into deep water. **diving-board** an elevated board used for diving from. **diving duck** a duck that habitually dives for food; esp. one of the tribe Aythyini, which includes the pochard, scaup, etc. **diving-suit** a watertight suit usu. with a helmet and an air-supply, worn for working under water. [OE dūfan (v.intr.) dive, sink, and dȳfan (v.tr.) immerse, f. Gmc: rel. to DEEP, DIP]

diver /ˈdaɪvə(r)/ n. **1** a person who dives. **2 a** a person who wears a diving-suit to work under water for long periods. **b** a pearl-diver etc. **3** a large waterbird of the family Gaviidae, with a straight sharply pointed bill. Also called (N. Amer.) loon.

diverge /daɪˈvɜːdʒ/ v. **1** intr. **a** proceed in a different direction or in different directions from a point (diverging rays; the path diverges here). **b** take a different course or different courses (their interests diverged). **2** intr. **a** (often foll. by from) depart from a set course (diverged from the track; diverged from his parents' wishes). **b** differ markedly (they diverged as to the best course). **3** tr. cause to diverge; deflect. **4** intr. Math. (of a series) increase indefinitely as more of its terms are added. [med.L divergere (as DI-[2], L vergere incline)]

divergent /daɪˈvɜːdʒənt/ adj. **1** diverging. **2** Psychol. (of thought) tending to reach a variety of possible solutions when analysing a problem. **3** Math. (of a series) increasing indefinitely as more of its terms are added; not convergent. □ **divergence** n. **divergency** n. **divergently** adv.

divers /ˈdaɪvəz/ adj. archaic or literary more than one; sundry; several. [ME f. OF f. L diversus DIVERSE (as DI-[2], versus past part. of vertere turn)]

diverse /daɪˈvɜːs, dɪ-/ adj. unlike in nature or qualities; varied. □ **diversely** adv. [ME (as DIVERS)]

diversify /daɪˈvɜːsɪˌfaɪ, dɪ-/ v. (**-ies, -ied**) **1** tr. make diverse; vary; modify. **2** tr. Commerce **a** spread (investment) over several enterprises or products, esp. to reduce the risk of loss. **b** introduce a spread of investment in (an enterprise etc.). **3** intr. (often foll. by into) esp. Commerce (of a firm etc.) expand the range of products handled. □ **diversification** /-ˌvɜːsɪfɪˈkeɪʃ(ə)n/ n. [ME f. OF diversifier f. med.L diversificare (as DIVERS)]

diversion /daɪˈvɜːʃ(ə)n, dɪ-/ n. **1 a** the act of diverting; deviation. **b** an instance of this. **2 a** the diverting of attention deliberately. **b** a stratagem for this purpose (created a diversion to secure their escape). **3** a recreation or pastime. **4** Brit. an alternative route when a road is temporarily closed to traffic. □ **diversional** adj. **diversionary** adj. [LL diversio (as DIVERT)]

diversionist /daɪˈvɜːʃənɪst, dɪ-/ n. **1** a person who engages in disruptive or subversive activities. **2** Polit. (esp. used by communists) a conspirator against the state; a saboteur.

diversity /daɪˈvɜːsɪtɪ, dɪ-/ n. (pl. **-ies**) **1** being diverse; variety. **2** a different kind; a variety. [ME f. OF diversité f. L diversitas -tatis (as DIVERS)]

divert /daɪˈvɜːt, dɪ-/ v.tr. **1** (often foll. by from, to) **a** turn aside; deflect. **b** draw the attention of; distract. **2** (often as **diverting** adj.) entertain; amuse. □ **divertingly** adv. [ME f. F divertir f. L divertere (as DI-[2], vertere turn)]

diverticular /ˌdaɪvɜːˈtɪkjʊlə(r)/ adj. Med. of or relating to a diverticulum. □ **diverticular disease** a condition with abdominal pain as a result of muscle spasms in the presence of diverticula.

diverticulitis /ˌdaɪvɜːˌtɪkjʊˈlaɪtɪs/ n. Med. inflammation of a diverticulum.

diverticulum /ˌdaɪvɜːˈtɪkjʊləm/ n. (pl. **diverticula** /-lə/) Anat. a blind tube forming a side-branch in a cavity or passage, esp. (Med.) an abnormal one in the alimentary tract. □ **diverticulosis** /-ˌtɪkjʊˈləʊsɪs/ n. [med.L, var. of L deverticulum byway f. devertere (as DE-, vertere turn)]

divertimento /dɪˌvɜːtɪˈmɛntəʊ, dɪˌveɑtɪ-/ n. (pl. **divertimenti** /-tɪ/ or **-os**) Mus. a light and entertaining composition, often in the form of a suite for chamber orchestra. [It., = diversion]

divertissement /dɪˈvɜːtɪsmənt, French divɛrˈtismɑ̃/ n. **1** a diversion; an entertainment. **2** a short ballet etc. between acts or longer pieces. [F, f. divertiss- stem of divertir DIVERT]

Dives /ˈdaɪviːz/ n. a rich man. [L dives rich, from Vulgate transl. of Luke 16]

divest /daɪˈvɛst/ v.tr. **1** (usu. foll. by of; often refl.) unclothe; strip (divested himself of his jacket). **2** deprive, dispossess; free, rid (cannot divest himself of the idea). □ **divestment** n. **divesture** /-ˈvɛstʃə(r)/ n. [earlier devest f. OF desvestir etc. (as DIS-, L vestire f. vestis garment)]

divi var. of DIVVY¹.

divide /dɪˈvaɪd/ v. & n. ● v. **1** tr. & intr. (often foll. by in, into) separate or be separated into parts; break up; split (the river divides into two; the road divides; divided them into three groups). **2** tr. & intr. (often foll. by out) distribute; deal; share (divided it out between them). **3** tr. a cut off; separate; part (divide the sheep from the goats). **b** mark out into parts (a ruler divided into inches). **c** specify different kinds of, classify (people can be divided into two types). **4** tr. cause to disagree; set at variance (religion divided them). **5** Math. **a** tr. find how many times (a number) contains another (divide 20 by 4). **b** intr. (of a number) be contained in (a number) without a remainder (4 divides into 20). **c** intr. be susceptible of division (10 divides by 2 and 5). **d** tr. find how many times (a number) is contained in another (divide 4 into 20). **6** intr. Math. do division. **7** Parl. **a** intr. (of a legislative assembly etc.) part into two groups for voting (the House divided). **b** tr. so divide (a Parliament etc.) for voting. ● n. **1** a dividing or boundary line (the divide between rich and poor). **2** a watershed. □ **divide and rule** maintain supremacy by preventing one's opponents from joining forces. **divided against itself** formed into factions. **divided highway** N. Amer. a dual carriageway. **divided skirt** culottes. [ME f. L dividere divis- (as DI-², vid- separate)]

dividend /ˈdɪvɪˌdɛnd/ n. **1 a** a sum of money to be divided among a number of persons, esp. that paid by a company to shareholders. **b** a similar sum payable to winners in a football pool, to members of a cooperative, or to creditors of an insolvent estate. **c** an individual's share of a dividend. **2** Math. a number to be divided by a divisor. **3** a benefit from any action (their long training paid dividends). □ **dividend stripping** the evasion of tax on dividends by arrangement between the company liable to pay tax and another able to claim repayment of tax. **dividend warrant** Brit. the documentary authority for a shareholder to receive a dividend. **dividend yield** a dividend expressed as a percentage of a current share price. [AF dividende f. L dividendum (as DIVIDE)]

divider /dɪˈvaɪdə(r)/ n. **1** a screen, piece of furniture, etc., dividing a room into two parts. **2** (in pl.) a measuring-compass, esp. with a screw for setting small intervals.

divi-divi /ˌdɪvɪˈdɪvɪ/ n. (pl. **divi-divis**) **1** a small tree, Caesalpinia coriaria, native to tropical America, bearing curved pods. **2** these pods used as a source of tannin. [Carib]

divination /ˌdɪvɪˈneɪʃ(ə)n/ n. **1** supposed insight into the future or the unknown gained by supernatural means. **2 a** a skilful and accurate forecast. **b** a good guess. □ **divinatory** /dɪˈvɪnətərɪ/ adj. [ME f. OF divination or L divinatio (as DIVINE)]

divine /dɪˈvaɪn/ adj., v., & n. ● adj. (**diviner, divinest**) **1 a** of, from, or like God or a god. **b** devoted to God; sacred (divine service). **2 a** more than humanly excellent, gifted, or beautiful. **b** colloq. excellent; delightful. ● v. **1** tr. discover by guessing, intuition, inspiration, or magic. **2** tr. foresee, predict, conjecture. **3** intr. practise divination (divining for water). ● n. **1** a cleric, usu. an expert in theology. **2** (**the Divine**) providence or God. □ **divine office** see OFFICE 10b. **diviningrod** = dowsing-rod (see DOWSE¹). □ **divinely** adv. **divineness** n. **diviner** n. **divinize** /ˈdɪvɪˌnaɪz/ v.tr. (also **-ise**). [ME f. OF devin -ine f. L divinus f. divus godlike]

divine right of kings the doctrine that a monarch in the hereditary line of succession has authority derived directly from God, independently of the subjects' will, and that rebellion is the worst of political crimes. It was enunciated in Britain in the 16th century under

the Stuarts and is also often associated with the absolutism of Louis XIV of France.

divinity /dɪˈvɪnɪtɪ/ n. (pl. **-ies**) **1** the state or quality of being divine. **2 a** a god; a divine being. **b** (as **the Divinity**) God. **3** the study of religion; theology. [ME f. OF divinité f. L divinitas -tatis (as DIVINE)]

divisible /dɪˈvɪzɪb(ə)l/ adj. **1** capable of being divided, physically or mentally. **2** (foll. by by) Math. containing (a number) a number of times without a remainder (15 is divisible by 3 and 5). □ **divisibility** /-ˌvɪzɪˈbɪlɪtɪ/ n. [F divisible or LL divisibilis (as DIVIDE)]

division /dɪˈvɪʒ(ə)n/ n. **1** the act or an instance of dividing; the process of being divided. **2** Math. the process of dividing one number by another (see also long division (see LONG¹), short division). **3** disagreement or discord (division of opinion). **4** Parl. the separation of members of a legislative body into two sets for counting votes for and against. **5 a** one of two or more parts into which a thing is divided. **b** the point at which a thing is divided. **6** a major unit of administration or organization, esp.: **a** a group of army brigades or regiments. **b** Sport a grouping of teams within a league, usu. by ability. **7 a** a district defined for administrative purposes. **b** Brit. a part of a county or borough returning a member of parliament. **8 a** Bot. a major taxonomic grouping. **b** Zool. a subsidiary category between major levels of classification. **9** Logic a classification of kinds, parts, or senses. □ **division of labour** the improvement of efficiency by giving different parts of a manufacturing process etc. to different people. **division sign** the sign (÷) indicating that one quantity is to be divided by another. □ **divisional** adj. **divisionally** adv. **divisionary** adj. [ME f. OF divisiun f. L divisio -onis (as DIVIDE)]

divisionism /dɪˈvɪʒəˌnɪz(ə)m/ n. see POINTILLISM.

divisive /dɪˈvaɪsɪv/ adj. tending to divide, esp. in opinion; causing disagreement. □ **divisively** adv. **divisiveness** n. [LL divisivus (as DIVIDE)]

divisor /dɪˈvaɪzə(r)/ n. Math. **1** a number by which another is to be divided. **2** a number that divides another without a remainder. [ME f. F diviseur or L divisor (as DIVIDE)]

divorce /dɪˈvɔːs/ n. & v. ● n. **1 a** the legal dissolution of a marriage. **b** a legal decree of this. **2** a severance or separation (a divorce between thought and feeling). ● v. **1** tr. (usu. as **divorced** adj.) (often foll. by from) legally dissolve the marriage of (a divorced couple; he wants to get divorced from her). **b** intr. separate by divorce (they divorced last year). **c** tr. end one's marriage with (divorced him for neglect). **2** tr. (often foll. by from) detach, separate (divorced from reality). **3** tr. archaic dissolve (a union). □ **divorcement** n. [ME f. OF divorce (n.), divorcer (v.) f. LL divortiare f. L divortium f. divortere (as DI-², vertere turn)]

divorcee /ˌdɪvɔːˈsiː, dɪˈvɔːseɪ/ n. (also masc. **divorcé**, fem. **divorcée**) a divorced person. [DIVORCE + -EE or F divorcé(e)]

divot /ˈdɪvət/ n. **1** a piece of turf cut out by a golf club in making a stroke. **2** esp. Sc. a piece of turf; a sod. [16th c.: orig. unkn.]

divulge /daɪˈvʌldʒ, dɪ-/ v.tr. disclose; reveal (a secret etc.). □ **divulgement** n. **divulgence** /daɪˈvʌlʤ(ə)ns, dɪ-/ n. [L divulgare (as DI-², vulgare publish f. vulgus common people)]

divvy¹ /ˈdɪvɪ/ n. & v. (also **divi**) colloq. ● n. (pl. **-ies**) **1** Brit. a dividend; a share, esp. of profits earned by a cooperative. **2** a distribution. ● v.tr. (**-ies, -ied**) (often foll. by up) share out; divide. [abbr. of DIVIDEND]

divvy² /ˈdɪvɪ/ adj. & n. (also **divy**) dial. & sl. ● adj. foolish, idiotic, daft. ● n. a foolish or half-witted person. [20th c.: orig. uncert.]

Diwali /diːˈwɑːlɪ/ a Hindu festival with lights, held over three nights in the period October to November, to celebrate the new season at the end of the monsoon. It is particularly associated with Lakshmi, the goddess of prosperity. [Hind. dīwālī f. Skr. dīpāvalī row of lights f. dīpa lamp]

Dixie /ˈdɪksɪ/ an informal name for the southern states of the US. It was used in the song 'Dixie' (1859) by Daniel Decatur Emmett (1815–1904), a popular marching-song sung by Confederate soldiers in the American Civil War. [19th c.: orig. uncert.]

dixie /ˈdɪksɪ/ n. a large iron cooking pot used by campers etc. [Hind. degchī cooking pot f. Pers. degcha dimin. of deg pot]

Dixieland /ˈdɪksɪˌlænd/ n. **1** = DIXIE. **2** a kind of early jazz with a strong two-beat rhythm and collective improvisation. [DIXIE]

DIY abbr. Brit. do-it-yourself.

Diyarbakir /dɪˈjɑːbəˌkɪə(r)/ a city in SE Turkey, capital of a province of the same name; pop. (1990) 381,100.

dizzy /ˈdɪzɪ/ adj. & v. ● adj. (**dizzier, dizziest**) **1 a** giddy, unsteady. **b** lacking mental stability; confused. **2** causing giddiness (dizzy heights;

dizzy speed). ● *v.tr.* (**-ies, -ied**) **1** make dizzy. **2** bewilder. □ **dizzily** *adv.* **dizziness** *n.* [OE *dysig* f. WG]

DJ *abbr.* **1** *Brit.* dinner-jacket. **2** disc jockey.

Djakarta /dʒə'kɑːtə/ (also **Jakarta**) the capital of Indonesia, situated in NW Java; pop. (1980) 6,503,000. Until 1949 it was called Batavia.

djellaba /'dʒeləbə/ *n.* (also **djellabah, jellaba**) a loose hooded woollen cloak worn or as worn by Arab men. [Arab. *jallaba, jallabīya*]

Djerba /'dʒɜːbə/ (also **Jerba**) a resort island in the Gulf of Gabès off the coast of Tunisia.

djibba (also **djibbah**) var. of JIBBA.

Djibouti /dʒɪ'buːtɪ/ (also **Jibuti**) **1** a country on the NE coast of Africa; pop. (est. 1991) 441,000; languages, Arabic (official), French (official), Somali and other Cushitic languages. The territory became a French protectorate under the name of French Somaliland in 1897. It was renamed the French Territory of the Afars and Issas in 1946, the Afars (of Hamitic descent) and the Issas (Somali) forming the two main ethnic groups. In 1977, the country achieved independence as the Republic of Djibouti. **2** the capital of Djibouti, a port at the western end of the Gulf of Aden; pop. (est. 1988) 290,000. □ **Djiboutian** *adj. & n.*

djinn var. of JINNEE.

DL *abbr.* Deputy Lieutenant.

dl *abbr.* decilitre(s).

D-layer /'diː‚leɪə(r)/ *n.* the lowest layer of the ionosphere able to reflect low-frequency radio waves. [*D* (arbitrary)]

D.Litt. *abbr.* Doctor of Letters. [L *Doctor Litterarum*]

DM *abbr.* (also **D-mark** /'deɪmɑːk/) Deutschmark.

dm *abbr.* decimetre(s).

DMSO *abbr.* dimethylsulphoxide.

D.Mus. *abbr.* Doctor of Music.

DMZ *abbr.* US demilitarized zone.

DNA /‚diːen'eɪ/ *abbr. Biol.* deoxyribonucleic acid, a substance present in nearly all living organisms, esp. as a constituent of chromosomes. DNA molecules carry the genetic information necessary for the organization and functioning of most living cells and control the inheritance of characteristics (see GENE). The structure of DNA was first proposed by J. D. Watson and F. H. C. Crick in 1953: each molecule consists of two strands coiled round each other to form a double helix, a structure like a spiral ladder. Each 'rung' of the ladder consists of a pair of chemical groups called bases (of which there are four types). The genetic information is carried in the form of particular base sequences which control the manufacture of proteins needed by the cell; genes are 'copied' in the nucleus into the similar RNA, which then passes out into the cytoplasm and forms a template for the synthesis of enzymes and other proteins. The DNA molecule also has the special property of self-replication: the strands separate and each provides a template for the synthesis of a new complementary strand with which it recombines, thus producing two identical copies of the original double helix.

DNA fingerprinting *n.* (also **DNA profiling**) see GENETIC FINGERPRINTING.

DNase /‚diːen'eɪz/ *n. Biochem.* an enzyme which breaks DNA into smaller molecules. [DNA + -ASE]

DNB *abbr.* Dictionary of National Biography.

Dnieper /'dniːpə(r)/ (Russian **Dnepr** /dnɛpr/, Ukrainian **Dnipro** /dniː'pro/) a river of eastern Europe, rising in Russia west of Moscow and flowing southwards some 2,200 km (1,370 miles) through Belarus and Ukraine to the Black Sea. The cities of Kiev and Dnipropetrovsk are situated on it. Dams have been built at a number of points to provide hydroelectric power and water for Ukraine's industries.

Dniester /'dniːstə(r)/ (Russian **Dnestr** /dnɛstr/) a river of eastern Europe, rising in the Carpathian Mountains in western Ukraine and flowing 1,410 km (876 miles) through Ukraine and Moldova to the Black Sea near Odessa.

Dniprodzerzhinsk /‚dniːprədzəˈʒɪnsk/ (Russian **Dneprodzerzhinsk** /‚dniprədzɪrˈʒɪnsk/) an industrial city and river port in Ukraine, on the River Dnieper; pop. (1990) 283,600. Until 1936 it was known as Kamenskoe.

Dnipropetrovsk /‚dniːprəpeˈtrɒfsk/ (Russian **Dnepropetrovsk** /‚dniprəpɪˈtrɒfsk/) an industrial city and river port in Ukraine, on the River Dnieper; pop. (1990) 1,187,000. It was founded in 1787 and known as Yekaterinoslav (Ekaterinoslav) until 1926.

D-notice /'diː‚nəʊtɪs/ *n. Brit.* a government notice to news editors not to publish items on specified subjects, for reasons of security. [defence + NOTICE]

do[1] /duː, də/ *v. & n.* ● *v.* (3rd sing. present **does** /dʌz/; past **did** /dɪd/; past part. **done**) **1** *tr.* perform, carry out, achieve, complete (work etc.) (*did his homework; there's a lot to do; he can do anything*). **2** *tr.* **a** produce, make (*she was doing a painting; I did a translation; decided to do a casserole*). **b** provide (*do you do lunches?*). **3** *tr.* bestow, grant; have a specified effect on (*a walk would do you good; do me a favour*). **4** *intr.* act, behave, proceed (*do as I do; she would do well to accept the offer*). **5** *tr.* **a** work at for a living, be occupied with (*what does your father do?*). **b** work at, study; take as one's subject (*he did chemistry at university; we're doing Chaucer next term*). **6 a** *intr.* be suitable or acceptable; suffice (*this dress won't do for a wedding; a sandwich will do until we get home; that will never do*). **b** *tr.* satisfy; be suitable for (*that hotel will do me nicely*). **7** *tr.* deal with; put in order (*the garden needs doing; the barber will do you next; I must do my hair before we go*). **8** *intr.* a fare; get on (*the patients were doing excellently; he did badly in the test*). **b** perform, work (*could do better*). **9** *tr.* **a** solve; work out (*we did the puzzle*). **b** (prec. by *can* or *be able to*) be competent at (*can you do cartwheels?; I never could do maths*). **10** *tr.* **a** traverse (a certain distance) (*we did fifty miles today*). **b** travel at a specified speed (*he overtook us doing about eighty*). **11** *tr. colloq.* **a** act or behave like (*did a Houdini*). **b** play the part of (*she was asked to do hostess*). **12** *intr. colloq.* **a** finish (*have you done annoying me?; I've done in the bathroom*). **b** (as **done** *adj.*) be over (*the day is done*). **13** *tr.* produce or give a performance of (*the school does many plays and concerts; we've never done 'Hamlet'*). **14** *tr.* cook, esp. to the right degree (*do it in the oven; the potatoes aren't done yet*). **15** *intr.* be in progress (*what's doing?*). **16** *tr. colloq.* visit; see the sights of (*we did all the art galleries*). **17** *tr. colloq.* **a** (often as **done** *adj.*; often foll. by *up*) exhaust; tire out (*the climb has completely done me*). **b** beat up, defeat, kill. **c** ruin (*now you've done it*). **18** *tr.* (foll. by *into*) translate or transform (*the book was done into French*). **19** *tr.* (with qualifying adverb) *colloq.* provide food etc. for in a specified way (*they do one very well here*). **20** *tr. sl.* **a** rob (*they did a shop in Soho*). **b** swindle (*I was done at the market*). **21** *tr. sl.* prosecute, convict (*they were done for shoplifting*). **22** *tr. sl.* undergo (a specified term of imprisonment) (*he did two years for fraud*). **23** *tr. coarse sl.* have sexual intercourse with. **24** *tr. sl.* take (a drug). ● *v.aux.* **1 a** (except with *be, can, may, ought, shall, will*) in questions and negative statements (*do you understand?; I don't smoke*). **b** (except with *can, may, ought, shall, will*) in negative commands (*don't be silly; do not come tomorrow*). **2** ellipt. or in place of verb or verb and object (*you know her better than I do; I wanted to go and I did so; tell me, do!*). **3** forming present and past tenses (*I do want to; do tell me; they did go but she was out*). **4** in inversion for emphasis (*rarely does it happen; did he but know it*). ● *n.* (pl. **dos** or **do's**) **1** *colloq.* an elaborate event, party, or operation. **2** *Brit. sl.* a swindle or hoax. □ **be done with** see DONE. **be nothing to do with 1** be no business or concern of (*his financial situation is nothing to do with me*). **2** be unconnected with (*his depression is nothing to do with his father's death*). **be to do with** be concerned or connected with (*the argument was to do with money*). **do about** see ABOUT prep. 1d. **do away with** *colloq.* **1** abolish. **2** kill. **do battle** enter into combat. **do one's best** see BEST. **do one's bit** see BIT[1]. **do by** treat or deal with in a specified way (*do as you would be done by*). **do credit to** see CREDIT. **do down** *colloq.* **1** cheat, swindle. **2** get the better of; overcome. **do for 1** be satisfactory or sufficient for. **2** *colloq.* (esp. as **done for** *adj.*) destroy, ruin, kill (*he knew he was done for*). **3** *colloq.* act as housekeeper for. **do one's head** (or **nut**) *sl.* be extremely angry or agitated. **do the honours** see HONOUR. **do in 1** *sl.* **a** kill. **b** ruin, do injury to. **2** *colloq.* exhaust, tire out. **do-it-yourself** *adj.* (of work, esp. building, painting, decorating, etc.) done or to be done by an amateur at home. ● *n.* such work. **do justice to** see JUSTICE. **do nothing for** (or **to**) *colloq.* detract from the appearance or quality of (*such behaviour does nothing for our reputation*). **do or die** persist regardless of danger. **do out** *colloq.* clean or redecorate (a room). **do a person out of** *colloq.* unjustly deprive a person of; swindle out of (*he was done out of his holiday*). **do over 1** *sl.* attack; beat up. **2** *colloq.* redecorate, refurbish. **3** *N. Amer. colloq.* do again. **do proud** see PROUD. **dos and don'ts** rules of behaviour. **do something for** (or **to**) *colloq.* enhance the appearance or quality of (*that carpet does something for the room*). **do one's stuff** see STUFF. **do to** (archaic **unto**) = *do by.* **do to death** see DEATH. **do the trick** see TRICK. **do up 1** fasten, secure. **2** *colloq.* a refurbish, renovate. **b** adorn, dress up. **3** *sl.* a ruin, get the better of. **b** beat up. **do well for oneself** prosper. **do well out of** profit by. **do with** (prec. by *could*) would be glad to have; would profit by (*I could do with a rest; you could do with a wash*). **do without** manage without; forgo (also *absol.: we shall just have to do without*). **have nothing to do with 1** have no connection or dealings with (*our problem has nothing to do with the latest news; after the disagreement he had nothing to do with his father*). **2** be no business or concern of (*the decision has nothing to do with him*). **have to do** (or **something to do**) **with** be connected with (*his limp has to do with a car accident*). [OE *dōn* f. Gmc: rel. to Skr *dádhāmi* put, Gk *tithemi* place, L *facere* do]

do[2] var. of DOH.

do. *abbr.* ditto.

DOA *abbr.* dead on arrival (at hospital etc.).

doable /'du:əb(ə)l/ *adj.* that can be done.

dob /dɒb/ *v.tr.* (**dobbed, dobbing**) (foll. by *in*) *Austral. sl.* inform against; implicate; betray. [var. of DAB[1]]

dobbin /'dɒbɪn/ *n.* a draught-horse; a farm horse. [pet-form of the name *Robert*]

dobe /'dəʊbɪ/ *n. US colloq.* adobe. [abbr.]

Dobell /dəʊ'bel/, Sir William (1899–1970), Australian painter. Noted for his portraits, he won the 1943 Archibald Prize, awarded by the Art Gallery of New South Wales, for his portrait of fellow artist Joshua Smith. The award was contested in court by two of the unsuccessful competitors on the grounds that it was not a portrait but a caricature, and created a *cause célèbre* for modernism in Australia.

Dobermann /'dəʊbəmən/ *n.* (in full **Dobermann pinscher** /'pɪnʃə(r), 'pɪntʃə(r)/) a breed of large dog with a smooth coat, originating in Germany. [Ludwig *Dobermann*, 19th-c. Ger. dog-breeder + G *Pinscher* terrier]

Dobrich /'dɒbrɪtʃ/ a city in NE Bulgaria, the centre for an agricultural region; pop. (1990) 115,800. It was named Tolbukhin (1949–91) after the Soviet marshal Fyodor Ivanovich Tolbukhin.

Dobruja /'dɒbrʊjə/ a district of eastern Romania and NE Bulgaria on the Black Sea coast, bounded on the north and west by the River Danube.

dobsonfly /'dɒbs(ə)n,flaɪ/ *n.* (pl. **-flies**) a large neuropterous insect of the family Corydalidae; esp. *Corydalis cornutus* of North America, whose predatory aquatic larvae are used as bait. [19th c.: orig. unkn.]

doc /dɒk/ *n. colloq.* **1** doctor. **2** documentary (*video-doc*). **3** *Computing* document. [abbr.]

Docetae /də'si:ti:/ *n.pl.* a group of early Christian heretics believing in Docetism.

Docetism /də'si:tɪz(ə)m, 'dəʊsɪ,tɪz(ə)m/ *n.* the doctrine or belief that Christ's body was not human but either a phantasm or of real but celestial substance and that therefore his sufferings were only apparent. This is an important tenet of Gnosticism. The followers of this belief include the Docetae, a group of 2nd-century heretics. □ **Docetist** *n.* [med. L *Docetae* f. Gk *Dokētai* f. *dokeō* seem, appear]

doch an dorris var. of DEOCH AN DORIS.

docile /'dəʊsaɪl/ *adj.* **1** submissive, easily managed. **2** *archaic* teachable. □ **docilely** *adv.* **docility** /dəʊ'sɪlɪtɪ/ *n.* [ME f. L *docilis* f. *docere* teach]

dock[1] /dɒk/ *n. & v.* ● *n.* **1** an artificially enclosed body of water for the loading, unloading, and repair of ships. **2** (in *pl.*) a range of docks with wharves and offices; a dockyard. **3** *US* a ship's berth, a wharf. **4** = dry dock. **5** *Theatr.* = scene-dock. ● *v.* **1** *tr. & intr.* bring or come into a dock. **2 a** *tr.* join (spacecraft) together in space. **b** *intr.* (of spacecraft) be joined. **3** *tr.* provide with a dock or docks. □ **dock-glass** a large glass for wine-tasting. **in dock** *Brit. colloq.* in hospital or (of a vehicle) laid up for repairs. [MDu. *docke*, of unkn. orig.]

dock[2] /dɒk/ *n.* the enclosure in a criminal court for the accused. □ **dock brief** a brief handed direct to a barrister selected by a prisoner in the dock. **in the dock** on trial. [16th c.: prob. orig. cant = Flem. *dok* cage, of unkn. orig.]

dock[3] /dɒk/ *n.* a coarse weed of the genus *Rumex*, with broad leaves. [OE *docce*]

dock[4] /dɒk/ *v. & n.* ● *v.tr.* **1 a** cut short (an animal's tail). **b** cut short the tail of (an animal). **2 a** (often foll. by *from*) deduct (a part) from wages, supplies, etc. **b** reduce (wages etc.) in this way. ● *n.* **1** the solid bony part of an animal's tail. **2** the crupper of a saddle or harness. □ **dock-tailed** having a docked tail. [ME, of uncert. orig.]

dockage /'dɒkɪdʒ/ *n.* **1** the charge made for using docks. **2** dock accommodation. **3** the berthing of ships in docks.

docker /'dɒkə(r)/ *n.* a person employed to load and unload ships.

docket /'dɒkɪt/ *n. & v.* ● *n.* **1** *Brit.* **a** a document or label listing goods delivered or the contents of a package, or recording payment of customs dues etc. **b** a voucher; an order form. **2** *US* a list of causes for trial or persons having causes pending. **3** *US* a list of things to be done. ● *v.tr.* (**docketed, docketing**) label with a docket. [15th c.: orig. unkn.]

dockland /'dɒkland/ *n.* a district near docks.

dockyard /'dɒkjɑ:d/ *n.* an area with docks and equipment for building and repairing ships, esp. for naval use.

doctor /'dɒktə(r)/ *n. & v.* ● *n.* **1 a** a qualified practitioner of medicine; a physician. **b** *N. Amer.* a qualified dentist or veterinary surgeon. **2** a person who holds a doctorate (*Doctor of Civil Law*). **3** *colloq.* a person who carries out repairs. **4** *archaic* a teacher or learned person. **5** *sl.* a cook on board a ship or in a camp. **6** (in full **doctor-blade**) *Printing* a blade for removing surplus ink etc. **7** an artificial fishing-fly. ● *v.* *colloq.* **1 a** *tr.* treat medically. **b** *intr.* (esp. as **doctoring** *n.*) practise as a physician. **2** *tr.* castrate or spay. **3** *tr.* patch up (machinery etc.); mend. **4** *tr.* adulterate. **5** *tr.* tamper with, falsify. **6** *tr.* confer a degree of doctor on. □ **go for the doctor** *Austral. sl.* **1** make an all-out effort. **2** bet all one has. **what the doctor ordered** *colloq.* something beneficial or desirable. □ **doctorly** *adj.* **doctorship** *n.* **doctorial** /dɒk'tɔ:rɪəl/ *adj.* [ME f. OF *doctour* f. L *doctor* f. *docere* doct- teach]

doctoral /'dɒktərəl/ *adj.* of or for a degree of doctor.

doctorate /'dɒktərət/ *n.* the highest university degree in any faculty, often honorary.

Doctor Martens /'mɑ:tɪnz/ *n. propr.* (also **Doc Martens, Dr Martens**) a type of heavy (esp. laced) boot or shoe with a cushioned sole. [Dr Klaus *Maertens* 20th-c. Ger. inventor of the sole]

Doctor of Philosophy *n.* **1** a doctorate in any faculty except law, medicine, or sometimes theology. **2** a person holding such a degree.

Doctor of the Church *n.* a title conferred since the Middle Ages on certain early Christian (and later Roman Catholic) theologians judged to be of outstanding merit and acknowledged saintliness. Originally there were four Doctors of the Church (Gregory the Great, Ambrose, Augustine of Hippo, and Jerome) but in later times the list was gradually extended, making a total of more than thirty.

doctrinaire /,dɒktrɪ'neə(r)/ *adj. & n.* ● *adj.* seeking to apply a theory or doctrine in all circumstances without regard to practical considerations; theoretical and impractical. ● *n.* a doctrinaire person; a pedantic theorist. □ **doctrinairism** *n.* **doctrinarian** *n.* [F f. *doctrine* DOCTRINE + *-aire* -ARY[1]]

doctrinal /dɒk'traɪn(ə)l, 'dɒktrɪn(ə)l/ *adj.* of or inculcating a doctrine or doctrines. □ **doctrinally** *adv.* [LL *doctrinalis* (as DOCTRINE)]

doctrine /'dɒktrɪn/ *n.* **1** what is taught; a body of instruction. **2 a** a principle of religious or political etc. belief. **b** a set of such principles; dogma. □ **doctrinism** *n.* **doctrinist** *n.* [ME f. OF f. L *doctrina* teaching (as DOCTOR)]

docudrama /'dɒkjʊ,drɑ:mə/ *n.* a dramatized television film based on real events. [DOCUMENTARY + DRAMA]

document *n. & v.* ● *n.* /'dɒkjʊmənt/ a piece of written or printed matter that provides a record or evidence of events, an agreement, ownership, identification, etc. ● *v.tr.* /'dɒkjʊ,ment/ **1** prove by or provide with documents or evidence. **2** record in a document. □ **documental** /,dɒkjʊ'ment(ə)l/ *adj.* [ME f. OF f. L *documentum* proof f. *docere* teach]

documentalist /,dɒkjʊ'mentəlɪst/ *n.* a person engaged in documentation.

documentary /,dɒkjʊ'mentərɪ/ *adj. & n.* ● *adj.* **1** consisting of documents (*documentary evidence*). **2** providing a factual record or report. ● *n.* (pl. **-ies**) a documentary film or programme. (*See note below.*) □ **documentarily** *adv.*

■ Documentaries are films or TV or radio programmes concerned with fact: they depict real people, events, or landscapes. The documentary film as social comment emerged most strongly in the 1930s, notably in Britain with the work of John Grierson. Such films were used as propaganda by both sides in the Second World War, but after the war their production sharply declined until the growth of television provided a new outlet. In recent years the genre has been extended; drama-documentaries, for example, represent a combination of fact and fiction in which real events are dramatized or used as the basis for a fictionalized account.

documentation /,dɒkjʊmen'teɪʃ(ə)n/ *n.* **1** the accumulation, classification, and dissemination of information. **2** the material collected or disseminated. **3** the collection of documents relating to a process or event, esp. the written specification and instructions accompanying a computer program.

DOD *abbr. US* Department of Defense.

dodder[1] /'dɒdə(r)/ *v.intr.* tremble or totter, esp. from age. □ **dodder-grass** quaking-grass. □ **dodderer** *n.* **doddering** *adj.* [17th c.: var. of obs. dial. *dadder*]

dodder[2] /ˈdɒdə(r)/ n. a climbing parasitic plant of the genus *Cuscuta*, with slender leafless threadlike stems. [ME f. Gmc]

doddered /ˈdɒdəd/ adj. (of a tree, esp. an oak) having lost its top or branches. [prob. f. obs. *dod* poll, lop]

doddery /ˈdɒdərɪ/ adj. tending to tremble or totter, esp. from age. □ **dodderiness** n. [DODDER[1] + -Y[1]]

doddle /ˈdɒd(ə)l/ n. Brit. colloq. an easy task. [perh. f. *doddle* = TODDLE]

dodeca- /ˈdəʊdɪkə/ comb. form twelve. [Gk *dōdeka* twelve]

dodecagon /dəʊˈdekəgən/ n. a plane figure with twelve sides.

dodecahedron /ˌdəʊdɪkəˈhiːdrən/ n. Geom. a solid figure with twelve faces. □ **dodecahedral** adj.

Dodecanese /ˌdəʊdɪkəˈniːz/ a group of twelve islands in the SE Aegean, of which the largest is Rhodes. They were occupied by Italy in 1912 during the war with Turkey and ceded to Greece in 1947. [Gk (*dōdeka* twelve, *nēsos* island)]

dodecaphonic /ˌdəʊdɪkəˈfɒnɪk/ adj. Mus. = TWELVE-NOTE.

dodge /dɒdʒ/ v. & n. ● v. **1** intr. (often foll. by *about*, *behind*, *round*) move quickly to one side or quickly change position, to elude a pursuer, blow, etc. (*dodged behind the chair*). **2** tr. **a** evade by cunning or trickery (*dodged paying the fare*). **b** elude (a pursuer, opponent, blow, etc.) by a sideward movement etc. **3** tr. Austral. sl. acquire dishonestly. **4** intr. (of a bell in change-ringing) move one place contrary to the normal sequence. ● n. **1** a quick movement to avoid or evade something. **2** a clever trick or expedient. **3** the dodging of a bell in change-ringing. □ **dodge the column** see COLUMN. [16th c.: orig. unkn.]

Dodge City a city in SW Kansas; pop. (1990) 21,130. Established in 1872 as a railhead on the Santa Fe Trail, it rapidly gained a reputation as a rowdy frontier town.

dodgem /ˈdɒdʒəm/ n. a small electrically driven car in an enclosure at a funfair, driven round and bumped into other dodgems. [DODGE + 'EM]

dodger /ˈdɒdʒə(r)/ n. **1** a person who dodges, esp. an artful or elusive person. **2** a screen on a ship's bridge etc. as protection from spray. **3** US a small handbill. **4** US a maize-flour cake. **5** sl. a sandwich; bread; food.

Dodgson /ˈdɒdʒsən/, Charles Lutwidge, see CARROLL.

dodgy /ˈdɒdʒɪ/ adj. (**dodgier**, **dodgiest**) **1** colloq. awkward, unreliable, tricky. **2** Brit. cunning, artful.

dodo /ˈdəʊdəʊ/ n. (pl. **-os** or **-oes**) **1** a large extinct flightless bird, *Raphus cucullatus*, formerly native to Mauritius. (*See note below.*) **2** an old-fashioned, stupid, or inactive person. □ **as dead as the** (or **a**) **dodo 1** completely or unmistakably dead. **2** entirely obsolete. [Port. *doudo* simpleton]

▪ The dodo and the solitaires are large flightless relatives of the pigeons, and were confined to the Mascarene Islands. They had large heads, a heavy hooked bill, short stocky legs, and stumpy wings. The dodo was exterminated in the late 17th century by hunting and by pigs and other introduced animals.

Dodoma /dəʊˈdəʊmə/ the capital of Tanzania, in the centre of the country; pop. (1988) 203,830.

DoE abbr. (in the UK) Department of the Environment.

doe /dəʊ/ n. a female fallow deer, reindeer, hare, rabbit, etc. [OE *dā*]

doek /dʊk/ n. S. Afr. a cloth, esp. a head-cloth. [Afrik.]

doer /ˈduːə(r)/ n. **1** a person who does something. **2** a person who acts rather than merely talking or thinking. **3** (in full **hard doer**) Austral. an eccentric or amusing person.

does 3rd sing. present of DO[1].

doeskin /ˈdəʊskɪn/ n. **1 a** the skin of a doe fallow deer. **b** leather made from this. **2** a fine cloth resembling it.

doesn't /ˈdʌz(ə)nt/ contr. does not.

doest /ˈduːɪst/ archaic 2nd sing. present of DO[1].

doeth /ˈduːɪθ/ archaic = DOTH.

doff /dɒf/ v.tr. literary take off (one's hat, clothing). [ME, = *do off*]

dog /dɒg/ n. & v. ● n. **1 a** a domesticated flesh-eating mammal, *Canis familiaris*, usu. having a long snout and non-retractile claws, and occurring in many different breeds kept as pets or for work or sport. (*See note below.*) **b** a wild member of the genus *Canis*, which includes wolves, jackals, and coyotes, or of the family Canidae, which also includes foxes. **2** the male of the dog, or of the fox (also **dog-fox**) or wolf (also **dog-wolf**). **3 a** colloq. a despicable person. **b** colloq. a person of a specified kind (*a lucky dog*). **c** US & Austral. sl. an informer; a traitor. **d** sl. a horse that is difficult to handle. **e** sl. derog. an unattractive or slovenly woman. **4** a mechanical device for gripping. **5** N. Amer. sl.

something poor; a failure. **6** = FIREDOG. **7** (in pl.; prec. by *the*) Brit. colloq. greyhound-racing. ● v.tr. (**dogged**, **dogging**) **1** follow closely and persistently; pursue, track. **2** Mech. grip with a dog. □ **die like a dog** die miserably or shamefully. **dog-biscuit** a hard thick biscuit for feeding dogs. **dog-box** Austral. sl. a compartment in a railway carriage without a corridor. **dog-clutch** Mech. a device for coupling two shafts in the transmission of power, one member having teeth which engage with slots in another. **dog-collar 1** a collar for a dog. **2 a** colloq. a clerical collar. **b** a straight high collar. **dog days** the hottest period of the year (reckoned in antiquity from the heliacal rising of the dog-star). **dog-eared** (of a book etc.) with the corners worn or battered with use. **dog-eat-dog** colloq. ruthlessly competitive. **dog-end** sl. a cigarette-end. **dog-fall** a fall in which wrestlers touch the ground together. **dog in the manger** a person who prevents others from using something, although that person has no use for it. **dog it** esp. US colloq. act lazily; idle, shirk. **dog-leg** (or **-legged**) bent like a dog's hind leg. **dog-leg hole** Golf a hole at which a player cannot aim directly at the green from the tee. **dog-paddle** (or **doggy-paddle**) n. an elementary swimming-stroke like that of a dog. ● v.intr. swim using this stroke. **dog rose** a wild rose with usu. pale pink flowers, esp. *Rosa canina* (also called *brier-rose*, *bramble*). **dog's breakfast** (or **dinner**) colloq. a mess. **dog's disease** Austral. sl. influenza. **dog's life** a life of misery or harassment. **dog's meat** horse's or other flesh as food for dogs; carrion. **dog's mercury** a woodland plant, *Mercurialis perennis*, of the spurge family, with small green flowers. **dogs of war** poet. the havoc accompanying war. **dog's-tail** (or **dog-tail**) a grass of the genus *Cynosurus*; esp. *C. cristatus*, a common pasture grass. **dog-star** = SIRIUS. **dog's tooth 1** (in full **dog's tooth violet**) a liliaceous plant of the genus *Erythronium*; esp. *E. dens-canis* with speckled leaves, purple flowers, and a toothed perianth. **2** = dog-tooth 2. **dog-tired** tired out. **dog-tooth 1** a small pointed ornament or moulding esp. in Norman and Early English architecture. **2** a broken check pattern used esp. in cloth for suits. **dog trials** Austral. & NZ a public competitive display of the skills of sheepdogs. **dog-violet** a scentless wild violet, esp. *Viola riviniana*. **go to the dogs** sl. deteriorate, be ruined. **hair of the dog** further alcoholic drink to cure the effects of alcoholic drink. **like a dog's dinner** colloq. smartly or flashily (dressed, arranged, etc.). **not a dog's chance** no chance at all. **put on the dog** N. Amer. behave pretentiously. □ **doglike** adj. [OE *docga*, of unkn. orig.]

▪ The dog family is characterized by limbs adapted for running, great stamina, and an acute sense of smell, suitable for catching prey by chasing it until it tires. The dog was domesticated early in the neolithic period: its chief wild ancestor was the wolf, *Canis lupus*, but the golden jackal, *C. aureus*, may also have been involved. The social habits of the wolf made it particularly suitable for domestication, and the dog usually regards its owner as the 'pack leader'. Such wild dogs as the dingo are actually feral domestic dogs, but the African hunting dog and the Asiatic wild dog (or dhole) are separate and more distantly related species.

dogberry /ˈdɒgbərɪ/ n. (pl. **-ies**) the fruit of the dogwood.

dogcart /ˈdɒgkɑːt/ n. a two-wheeled cart for driving in with cross seats back to back.

doge /dəʊdʒ/ n. hist. the chief magistrate of Venice or Genoa. [F f. It. f. Venetian *doze* f. L *dux ducis* leader]

dogfight /ˈdɒgfaɪt/ n. & v. ● n. **1** a close combat between fighter aircraft. **2** uproar; a fight like that between dogs. ● v.intr. take part in a dogfight.

dogfish /ˈdɒgfɪʃ/ n. a small shark of the family Scyliorhinidae or Squalidae, esp. the genus *Scyliorhinus* or *Squalus*.

dogged /ˈdɒgɪd/ adj. tenacious; grimly persistent. □ **it's dogged as does it** colloq. persistence succeeds. □ **doggedly** adv. **doggedness** n. [ME f. DOG + -ED[1]]

dogger[1] /ˈdɒgə(r)/ n. a two-masted bluff-bowed Dutch fishing-boat. [ME f. MDu., = fishing-boat]

dogger[2] /ˈdɒgə(r)/ n. Geol. a large spherical concretion occurring in sedimentary rock. [dial., = kind of iron-stone, perh. f. DOG]

Dogger Bank a submerged sandbank in the North Sea, about 115 km (70 miles) off the NE coast of England. This part of the central North Sea is covered by the shipping forecast area *Dogger*.

doggerel /ˈdɒgərəl/ n. poor or trivial verse. [ME, app. f. DOG: cf. -REL]

Doggett's Coat and Badge /ˈdɒgɪts/ an orange livery with a silver badge offered as a trophy for victory in a rowing contest held each year among Thames watermen. It was instituted in 1715 by an Irish comic actor, Thomas Doggett (1620–1721), and is now the oldest sculling race in the world.

doggie var. of DOGGY n.

doggish /'dɒgɪʃ/ adj. **1** of or like a dog. **2** currish, malicious, snappish.

doggo /'dɒgəʊ/ adv. □ **lie doggo** sl. lie motionless or hidden, making no sign. [prob. f. DOG: cf. -O]

doggone /'dɒgɒn/ adj., adv., & int. esp. N. Amer. sl. ● adj. & adv. damned. ● int. expressing annoyance. [prob. f. dog on it = God damn it]

doggy /'dɒgɪ/ adj. & n. ● adj. **1** of or like a dog. **2** devoted to dogs. ● n. (also **doggie**) (pl. **-ies**) a little dog; a pet name for a dog. □ **doggy bag** a bag given to a customer in a restaurant or to a guest at a party etc. for putting leftovers in to take home. **doggy-paddle** = dog-paddle. □ **dogginess** n.

doghouse /'dɒghaʊs/ n. N. Amer. a dog's kennel. □ **in the doghouse** sl. in disgrace or disfavour.

dogie /'dəʊgɪ/ n. N. Amer. a motherless or neglected calf. [19th c.: orig. unkn.]

dogma /'dɒgmə/ n. **1 a** a principle, tenet, or system of these, esp. as laid down by the authority of a Church. **b** such principles collectively. **2** an arrogant declaration of opinion. [L f. Gk dogma -matos opinion f. dokeō seem]

dogman /'dɒgmən/ n. (pl. **-men**) Austral. a person giving directional signals to the operator of a crane, often while sitting on the crane's load.

dogmatic /dɒg'mætɪk/ adj. **1 a** (of a person) given to asserting or imposing personal opinions; arrogant. **b** intolerantly authoritative. **2 a** of or in the nature of dogma; doctrinal. **b** based on a priori principles, not on induction. □ **dogmatically** adv. [LL dogmaticus f. Gk dogmatikos (as DOGMA)]

dogmatics /dɒg'mætɪks/ n. **1** the study of religious dogmas; dogmatic theology. **2** a system of dogma. [DOGMATIC]

dogmatism /'dɒgmə,tɪz(ə)m/ n. a tendency to be dogmatic. □ **dogmatist** n. [F dogmatisme f. med.L dogmatismus (as DOGMA)]

dogmatize /'dɒgmə,taɪz/ v. (also **-ise**) **1** intr. make positive unsupported assertions; speak dogmatically. **2** tr. express (a principle etc.) as a dogma. [F dogmatiser or f. LL dogmatizare f. Gk (as DOGMA)]

do-gooder /du:'gʊdə(r)/ n. a well-meaning but unrealistic philanthropist or reformer. □ **do-goodery** n. **do-goodism** n. **do-good** /du:gʊd/ adj. & n.

dog-racing /'dɒg,reɪsɪŋ/ n. see GREYHOUND RACING.

dogsbody /'dɒgz,bɒdɪ/ n. (pl. **-ies**) **1** colloq. a drudge. **2** Naut. sl. a junior officer.

dogshore /'dɒgʃɔː(r)/ n. a temporary wooden support for a ship just before launching.

dogskin /'dɒgskɪn/ n. leather made of or imitating dog's skin, used for gloves.

dogtrot /'dɒgtrɒt/ n. a gentle easy trot.

dogwatch /'dɒgwɒtʃ/ n. Naut. either of two short watches (4–6 or 6–8 p.m.).

dogwood /'dɒgwʊd/ n. **1** a shrub of the genus Cornus; esp. the wild cornel, C. sanguinea, with dark red branches, greenish-white flowers, and purple berries, found in woods and hedgerows. **2** any similar tree. **3** the wood of the dogwood.

DoH abbr. (in the UK) Department of Health.

doh /dəʊ/ n. (also **do**) Mus. **1** (in tonic sol-fa) the first and eighth note of a major scale. **2** the note C in the fixed-doh system. [18th c.: f. It. do]

Doha /'dəʊhɑː/ the capital of Qatar; pop. (est. 1990) 300,000.

doily /'dɔɪlɪ/ n. (also **doyley**) (pl. **-ies** or **-eys**) a small ornamental mat of paper, lace, etc., on a plate for cakes etc. [orig. the name of a fabric: f. Doiley, the surname of a 17th-c. London draper]

doing /'du:ɪŋ/ n. **1 a** (usu. in pl.) an action; the performance of a deed (famous for his doings; it was my doing). **b** activity, effort (it takes a lot of doing). **2** colloq. a scolding; a beating. **3** (in pl.) sl. things needed; adjuncts; things whose names are not known (have we got all the doings?).

Doisneau /dwʌnəʊ/, Robert (1912–94), French photographer. He is best known for his photos portraying the city and inhabitants of Paris, which he began taking in the 1930s; one of his most famous images is 'The Kiss at the Hôtel de Ville' (1950). His photojournalism includes pictures taken during the liberation of Paris in 1944.

doit /dɔɪt/ n. archaic a very small amount of money. [MLG doyt, MDu. duit, of unkn. orig.]

dojo /'dəʊdʒəʊ/ n. (pl. **-os**) **1** a room or hall in which judo and other martial arts are practised. **2** a mat on which judo etc. is practised. [Jap.]

dol. abbr. dollar(s).

Dolby /'dɒlbɪ/ n. propr. an electronic noise-reduction system used esp. in tape-recording to reduce hiss. [Ray Milton Dolby, Amer. inventor (b.1933)]

dolce /'dɒltʃeɪ/ adj. & adv. Mus. to be performed sweetly and softly. [It., = sweet]

dolce far niente /,dɒltʃeɪ ,fɑː nɪ'entɪ/ n. pleasant idleness. [It., = sweet doing nothing]

Dolcelatte /,dɒltʃeɪ'lætɪ/ n. a kind of soft creamy blue-veined cheese from Italy. [It., = sweet milk]

dolce vita /,dɒltʃeɪ 'viːtə/ n. a life of pleasure and luxury. [It., = sweet life]

doldrums /'dɒldrəmz/ n.pl. (usu. prec. by the) **1** low spirits; a feeling of boredom or depression. **2** a period of inactivity or state of stagnation. **3** an equatorial ocean region of calms, sudden storms, and light unpredictable winds. [prob. after dull and tantrum]

dole[1] /dəʊl/ n. & v. ● n. **1** (usu. prec. by the) Brit. colloq. benefit claimable by the unemployed from the state. **2 a** charitable distribution. **b** a charitable (esp. sparing, niggardly) gift of food, clothes, or money. **3** archaic one's lot or destiny. ● v.tr. (usu. foll. by out) deal out sparingly. □ **dole-bludger** Austral. sl. a person who allegedly prefers the dole to work. **on the dole** Brit. colloq. receiving state benefit for the unemployed. [OE dāl f. Gmc]

dole[2] /dəʊl/ n. poet. grief, woe; lamentation. [ME f. OF do(e)l etc. f. pop.L dolus f. L dolere grieve]

doleful /'dəʊlfʊl/ adj. **1** mournful, sad. **2** dreary, dismal. □ **dolefully** adv. **dolefulness** n. [ME f. DOLE[2] + -FUL]

dolerite /'dɒlə,raɪt/ n. a coarse basaltic rock. [F dolérite f. Gk doleros deceptive (because it is difficult to distinguish from diorite)]

dolichocephalic /,dɒlɪ,kəʊsɪ'fælɪk, -kɪ'fælɪk/ adj. Anthropol. having a skull that is longer than it is wide, with a cephalic index of 75.9 or less. □ **dolichocephalous** /-'sefələs, -'kefələs/ adj. **dolichocephaly** n. [Gk dolikhos long + -CEPHALIC, -CEPHALOUS]

Dolin /'dəʊlɪn, 'dɒl-/, Anton (born Sydney Francis Patrick Chippendall Healey-Kay) (1904–83), English ballet-dancer and choreographer. From 1923 until 1926, and again from 1928 to 1929, he was a principal with the Ballets Russes. With Alicia Markova he founded the Markova–Dolin Ballet in 1935; the company lasted until 1937. After a period spent abroad, Dolin returned to Britain in 1948, and in 1950 became artistic director and first soloist of the newly founded London Festival Ballet, a post he held until 1961.

dolina /də'liːnə/ n. (also **doline** /-'liːn/) Geol. an extensive depression or basin. [Russ. dolina valley]

Doll /dɒl/, Sir (William) Richard (Shaboe) (b.1912), English physician. He investigated the aetiology of lung cancer, leukaemia, and other cancers, and (with Sir A. Bradford Hill, 1897–1991) was the first to provide a statistical link between smoking and lung cancer. They later showed that stopping smoking is immediately effective in reducing the risk of lung cancer. Doll also studied peptic ulcers, the effects of ionizing radiation, and oral contraceptives.

doll /dɒl/ n. & v. ● n. **1** a small model of a human figure, esp. a baby or a child, as a child's toy. **2 a** colloq. a pretty but silly young woman. **b** sl. a young woman, esp. an attractive one. **3** a ventriloquist's dummy. ● v.tr. & intr. (foll. by up; often refl.) dress up smartly. □ **doll's house 1** a miniature toy house for dolls. **2** a very small house. [pet-form of the name Dorothy]

dollar /'dɒlə(r)/ n. **1** the chief monetary unit of the US, Canada, and Australia. (See note below.) **2** the chief monetary unit of certain countries in the Pacific, West Indies, SE Asia, Africa, and South America. □ **dollar area** the area in which currency is linked to the US dollar. **dollar diplomacy** diplomatic activity aimed at advancing a country's international influence by furthering its financial and commercial interests abroad. **dollar gap** the excess of a country's import trade with the dollar area over the corresponding export trade. **dollar mark** (or **sign**) the sign $, representing a dollar. **dollar spot 1** a fungal disease of lawns etc. **2** a discoloured patch caused by this. [LG daler f. G Taler, short for Joachimstaler, a coin from the silver-mine of Joachimstal, now Jáchymov in the Czech Republic]

▪ The dollar's ancestors were the Spanish peso (widely used as currency in North and South America in the 17th–18th centuries) and the Bohemian Joachimsthaler of the early 16th century. The Spanish dollar was formally retained as the standard unit of currency (with decimal subdivisions) in 1785. The first true American dollar was minted in 1794, but Spanish dollars continued in use and remained legal tender until 1857. The origin of the dollar sign $ is uncertain; it may be a modification of the figure 8 (representing 8 reals, the value of the peso), with upright strokes symbolizing the two architectural columns (representing the Pillars of Hercules), which for several centuries were conspicuous on the obverse of the Spanish peso.

Dollfuss /ˈdɒlfʊs/, Engelbert (1892–1934), Austrian statesman, Chancellor of Austria 1932–4. He was elected leader of the Christian Socialist Party and Chancellor in 1932. From 1933 Dollfuss attempted to govern without Parliament in order better to oppose Austrian Nazi attempts to force the *Anschluss*. Five months after promulgating a new Fascist constitution in 1934, he was assassinated by Austrian Nazis in an abortive coup.

dollhouse /ˈdɒlhaʊs/ *n. US* = doll's house (see DOLL).

dollop /ˈdɒləp/ *n. & v.* ● *n.* a shapeless lump of food etc. ● *v.tr.* (**dolloped**, **dolloping**) (usu. foll. by *out*) serve out in large shapeless quantities. [perh. f. Scand.]

dolly /ˈdɒlɪ/ *n., v., & adj.* ● *n.* (*pl.* **-ies**) **1** a child's name for a doll. **2** a movable platform for a cine-camera. **3** *Cricket colloq.* an easy catch or hit. **4** a stick for stirring in clothes-washing. **5** = corn dolly (see CORN[1]). **6** *colloq.* = dolly-bird. ● *v.* (**-ies**, **-ied**) **1** *tr.* (foll. by *up*) dress up smartly. **2** *intr.* (foll. by *in*, *up*) move a cine-camera in or up to a subject, or out from it. ● *adj.* (**dollier**, **dolliest**) **1** *Brit. colloq.* (esp. of a girl) attractive, stylish. **2** *Cricket colloq.* easily hit or caught. □ **dolly-bird** *Brit. colloq.* an attractive and stylish young woman. **dolly mixture** any of a mixture of small variously shaped and coloured sweets.

Dolly Varden /ˈvɑːd(ə)n/ *n.* **1** a woman's large hat with one side drooping and with a floral trimming. **2** a brightly spotted char, *Salvelinus malma*, a sporting fish of western North America. [a character in Dickens's *Barnaby Rudge*]

dolma /ˈdɒlmə/ *n.* (*pl.* **dolmas** or **dolmades** /dɒlˈmɑːðez/) a SE European delicacy of spiced rice or meat etc. wrapped in vine or cabbage leaves. [Turk. f. *dolmak* fill, be filled: *dolmades* f. mod.Gk]

dolman /ˈdɒlmən/ *n.* **1** a long Turkish robe open in front. **2** a hussar's jacket worn with the sleeves hanging loose. **3** a woman's mantle with capelike or dolman sleeves. □ **dolman sleeve** a loose sleeve cut in one piece with the body of the coat etc. [ult. f. Turk. *dolama*]

dolmen /ˈdɒlmən/ *n.* a megalithic tomb with a large flat stone laid on upright ones. [F, perh. f. Cornish *tolmên* hole of stone]

dolomite /ˈdɒləmaɪt/ *n.* a mineral or rock of calcium magnesium carbonate. □ **dolomitic** /ˌdɒləˈmɪtɪk/ *adj.* [F f. Déodat de Gratet de *Dolomieu*, Fr. geologist (1750–1801)]

Dolomite Mountains (also **Dolomites**) a range of the Alps in northern Italy, so named because the characteristic rock of the region is dolomitic limestone.

dolorous /ˈdɒlərəs/ *adj. literary* or *joc.* **1** distressing, painful; doleful, dismal. **2** distressed, sad. □ **dolorously** *adv.* [ME f. OF *doleros* f. LL *dolorosus* (as DOLOUR)]

dolour /ˈdɒlə(r)/ *n.* (*US* **dolor**) *literary* sorrow, distress. [ME f. OF f. L *dolor -oris* pain, grief]

dolphin /ˈdɒlfɪn/ *n.* **1** a porpoise-like sea mammal of the family Delphinidae, having a slender beaklike snout. **2** = DORADO 1. **3** a bollard, pile, or buoy for mooring. **4** a structure for protecting the pier of a bridge. **5** a curved fish in heraldry, sculpture, etc. [ME, also *delphin* f. L *delphinus* f. Gk *delphis -inos*]

dolphinarium /ˌdɒlfɪˈneərɪəm/ *n.* (*pl.* **dolphinariums**) an aquarium for dolphins, esp. one open to the public.

dolt /dəʊlt/ *n.* a stupid person. □ **doltish** *adj.* **doltishly** *adv.* **doltishness** *n.* [app. related to *dol, dold*, obs. var. of DULL]

Dom /dɒm/ *n.* **1** a title prefixed to the names of some Roman Catholic dignitaries, and Benedictine and Carthusian monks. **2** the Portuguese equivalent of Don (see DON[1] 2). [L *dominus* master: sense 2 through Port.]

-dom /dəm/ *suffix* forming nouns denoting: **1** state or condition (*freedom*). **2** rank or status (*earldom*). **3** domain (*kingdom*). **4** a class of people (or the attitudes etc. associated with them) regarded collectively (*officialdom*). [OE *-dōm*, orig. = DOOM]

domain /dəˈmeɪn/ *n.* **1** an area under one rule; a realm. **2** an estate or lands under one control. **3** a sphere of control or influence. **4** *Math.* the set of possible values of an independent variable. **5** *Physics* a discrete region of magnetism in ferromagnetic material. □ **domanial** *adj.* [ME f. F *domaine*, OF *demeine* DEMESNE, assoc. with L *dominus* lord]

domaine /dɒˈmeɪn/ *n.* a vineyard. [F: see DOMAIN]

dome /dəʊm/ *n. & v.* ● *n.* **1 a** a rounded vault as a roof, with a circular, elliptical, or polygonal base; a large cupola. **b** the revolving openable

hemispherical roof of an observatory. **2 a** a natural vault or canopy (of the sky, trees, etc.). **b** the rounded summit of a hill etc. **3** *Geol.* a dome-shaped structure. **4** *sl.* the head. **5** *poet.* a stately building. ● *v.tr.* (usu. as **domed** *adj.*) cover with or shape as a dome. □ **dome tent** a dome-shaped tent supported by flexible hoops. □ **domelike** *adj.* [F *dôme* f. It. DUOMO f. L *domus* house]

Dome of the Rock an Islamic shrine in Jerusalem, surrounding the sacred rock on which, according to tradition, Abraham prepared to sacrifice his son Isaac (Gen. 22:9) and from which the prophet Muhammad made his miraculous midnight ascent into heaven (see NIGHT JOURNEY). Built in the area of Solomon's temple and dating from the end of the 7th century, for Muslims it is the third most holy place, after Mecca and Medina. The expanse of rough irregular rock that forms its centre contrasts starkly with the strict geometry and ornate Byzantine-style decoration of the surrounding structure, while the exterior is of rich mosaic work capped by a golden dome.

Domesday Book /ˈduːmzdeɪ/ an extensive survey of the lands of England, excluding London, Winchester, and the four northern counties, compiled on the orders of William the Conqueror in 1086 in order to provide an informed basis for taxation. The most comprehensive survey of property carried out in medieval times, it caused considerable popular discontent at the time of its compilation, and was given its name because, like the Day of Judgement, there could be no appeal against it. [ME var. of DOOMSDAY]

domestic /dəˈmestɪk/ *adj. & n.* ● *adj.* **1** of the home, household, or family affairs. **2 a** of one's own country, not foreign or international. **b** home-grown or home-made. **3** (of an animal) kept by or living with humans. **4** fond of home life. ● *n.* a household servant. □ **domestic science** the study of household management. □ **domestically** *adv.* [F *domestique* f. L *domesticus* f. *domus* home]

domesticate /dəˈmestɪˌkeɪt/ *v.tr.* **1** tame (an animal) to live with humans. **2** accustom to home life and management. **3** naturalize (a plant or animal). □ **domesticable** /-kəb(ə)l/ *adj.* **domestication** /-ˌmestɪˈkeɪʃ(ə)n/ *n.* [med.L *domesticare* (as DOMESTIC)]

domesticity /ˌdɒmeˈstɪsɪtɪ, ˌdəʊm-/ *n.* **1** the state of being domestic. **2** domestic or home life.

domicile /ˈdɒmɪˌsaɪl, -sɪl/ *n. & v.* (also **domicil** /-sɪl/) ● *n.* **1** a dwelling-place; one's home. **2** *Law* **a** a place of permanent residence. **b** the fact of residing. **3** the place at which a bill of exchange is made payable. ● *v.tr.* **1** (usu. as **domiciled** *adj.*) (usu. foll. by *at*, *in*) establish or settle in a place. **2** (usu. foll. by *at*) make (a bill of exchange) payable at a certain place. [ME f. OF f. L *domicilium* f. *domus* home]

domiciliary /ˌdɒmɪˈsɪlɪərɪ/ *adj.* of a dwelling-place (esp. of a doctor's, official's, etc., visit to a person's home). [F *domiciliaire* f. med.L *domiciliarius* (as DOMICILE)]

dominance /ˈdɒmɪnəns/ *n.* **1** the state of being dominant. **2** control, authority.

dominant /ˈdɒmɪnənt/ *adj. & n.* ● *adj.* **1** dominating, prevailing, most influential. **2** (of a high place) prominent, overlooking others. **3** *Biol.* **a** (of an allele) expressed even when inherited from only one parent. **b** (of an inherited characteristic) appearing in an individual even when its allelic counterpart is also inherited (cf. RECESSIVE). ● *n. Mus.* the fifth note of the diatonic scale of any key. □ **dominantly** *adv.* [F f. L *dominari* (as DOMINATE)]

dominate /ˈdɒmɪˌneɪt/ *v.* **1** *tr.* & (foll. by *over*) *intr.* have a commanding influence on; exercise control over (*fear dominated them for years*; *dominates over his friends*). **2** *intr.* (of a person, sound, feature of a scene, etc.) be the most influential or conspicuous. **3** *tr.* & (foll. by *over*) *intr.* (of a building etc.) have a commanding position over; overlook. □ **dominator** *n.* [L *dominari dominat-* f. *dominus* lord]

domination /ˌdɒmɪˈneɪʃ(ə)n/ *n.* **1 a** command, control. **b** oppression, tyranny. **2** the act or an instance of dominating; the process of being dominated. **3** (in *pl.*) angelic beings of the fourth order of the celestial hierarchy (see ORDER *n.* 19). [ME f. OF f. L *dominatio -onis* (as DOMINATE)]

dominatrix /ˌdɒmɪˈneɪtrɪks/ *n.* **1** a female dominator or ruler. **2** a female who plays the dominant role in a sado-masochistic relationship. [L f. *dominator* (as DOMINATE) + -TRIX]

domineer /ˌdɒmɪˈnɪə(r)/ *v.intr.* (often as **domineering** *adj.*) behave in an arrogant and overbearing way. □ **domineeringly** *adv.* [Du. *domineren* f. F *dominer*]

Domingo /dəˈmɪŋɡəʊ/, Placido (b.1941), Spanish-born tenor. He moved to Mexico with his family in 1950, and made his operatic début in 1957. He established his reputation as one of the world's leading operatic tenors in the 1970s; his performances in operas by Verdi and Puccini have met with particular acclaim.

Dominic, St /ˈdɒmɪnɪk/ (c.1170–1221), (Spanish name Domingo de Guzmán) Spanish priest and friar. In 1203 he began a mission to reconcile the Albigenses to the Church. Although largely unsuccessful in this aim he established a number of religious communities and undertook the training of preachers; in 1216 this led to the foundation of the Order of Friars Preachers (see DOMINICAN¹) at Toulouse in France. At one time St Dominic was believed to have originated the practice of saying the rosary, but this belief has now been discredited. Feast day, 8 Aug.

Dominica /ˌdɒmɪˈniːkə, dəˈmɪnɪkə/ a mountainous island in the West Indies, the loftiest of the Lesser Antilles and the northernmost and largest of the Windward Islands; pop. (est. 1991) 72,000; languages, English (official), Creole; capital, Roseau. The island, originally inhabited by Caribs, was named by Columbus, who encountered it on a Sunday (L *dies domenica* the Lord's day) in 1493. After much Anglo-French rivalry it came into British possession at the end of the 18th century, becoming an independent republic within the Commonwealth in 1978.

dominical /dəˈmɪnɪk(ə)l/ adj. **1** of the Lord's day, of Sunday. **2** of the Lord (Jesus Christ). □ **dominical letter** the one of the seven letters A–G indicating the dates of Sundays in a year. [F *dominical* or L *dominicalis* f. L *dominicus* f. *dominus* lord]

Dominican¹ /dəˈmɪnɪkən/ adj. & n. ● adj. **1** of or relating to St Dominic or the order of preaching friars which he founded. (*See note below.*) **2** of or relating to either of the two orders of nuns founded on Dominican principles. ● n. a Dominican friar, nun, or sister. [med.L *Dominicanus* f. *Dominicus* L name of *Domingo de Guzmán* (DOMINIC, ST)]

- The Dominican order (whose members are also called *Black Friars*, after the colour of their cloak) was founded by St Dominic in 1215–16, and followed the established Rule of St Augustine. Dominicans are specially devoted to preaching and study and their chief motivation was, and is, educational and missionary. During the Middle Ages they supplied many leaders of European thought, including St Albertus Magnus and St Thomas Aquinas; they were also involved in the Inquisition. Although they became less prominent with the rise of new orders at the time of the Reformation (especially the Jesuits, often their rivals), they have retained considerable influence as champions of Roman Catholic orthodoxy.

Dominican² /dəˈmɪnɪkən/ adj. & n. ● adj. of or relating to the Dominican Republic or its people. ● n. a native or national of the Dominican Republic.

Dominican³ /ˌdɒmɪˈniːkən, dəˈmɪnɪkən/ adj. & n. ● adj. of or relating to the island of Dominica or its people. ● n. a native or national of the island of Dominica.

Dominican Republic /dəˈmɪnɪkən/ a country in the Caribbean occupying the eastern part of the island of Hispaniola; pop. (est. 1991) 7,197,000; official language, Spanish; capital, Santo Domingo. The Republic is the former Spanish colony of Santo Domingo, the part of Hispaniola which Spain retained when it ceded the western portion (now Haiti) to France in 1697. The republic, proclaimed in 1844, has had a turbulent history, culminating in the ruthless dictatorship (1930–61) of Rafael Trujillo. An unsettled period followed this, with civil war and US military intervention; a new constitution was introduced in 1966.

dominie /ˈdɒmɪnɪ/ n. Sc. a schoolmaster. [later spelling of *domine* sir, vocative of L *dominus* lord]

dominion /dəˈmɪnjən/ n. **1** sovereignty, control. **2** the territory of a sovereign or government; a domain. **3** hist. the title of each of the self-governing territories of the British Commonwealth. [ME f. OF f. med.L *dominio -onis* f. L *dominium* f. *dominus* lord]

Domino /ˈdɒmɪˌnəʊ/, Fats (born Antoine Domino) (b.1928), American pianist, singer, and songwriter. His music represents the transition from rhythm and blues to rock and roll and shows the influence of jazz, boogie-woogie, and gospel music. He made most of his recordings in the 1950s and early 1960s; his many songs include 'Ain't That a Shame' (1955) and 'Blueberry Hill' (1956).

domino /ˈdɒmɪˌnəʊ/ n. (pl. **-oes**) **1 a** any of twenty-eight small oblong pieces marked on one side with nought to six pips in each half. **b** (in pl., usu. treated as *sing.*) a game played with these. (*See note below.*) **2** a loose cloak with a mask for the upper part of the face, worn at masquerades. [F, prob. f. L *dominus* lord, but unexpl.]

- In its basic form the game involves forming a line of dominoes, each player laying down a domino to match the value of the one at either

end of the line already formed. It was played in ancient times in China but was unknown in Europe (where it may have arisen independently) before the 18th century. It was introduced to England possibly by French prisoners in about 1800.

domino theory n. the theory that a political event in one country will cause similar events in neighbouring countries, like a row of falling dominoes. The term was coined by President Eisenhower in 1954 to describe the fear of the spread of Communist rule and was later used to justify American involvement in Vietnam and Central America.

Domitian /dəˈmɪʃ(ə)n/ (full name Titus Flavius Domitianus) (AD 51–96), son of Vespasian, Roman emperor 81–96. An energetic but autocratic ruler, on succeeding his brother Titus he embarked on a large building programme, including monumental palaces on the Palatine Hill in Rome. His wife was implicated in his assassination, which ended a period of terror that had lasted a number of years.

Don /dɒn/ **1** a river in Russia which rises near Tula, south-east of Moscow, and flows for a distance of 1,958 km (1,224 miles) to the Sea of Azov. **2** a river in Scotland which rises in the Grampians and flows 131 km (82 miles) eastwards to the North Sea at Aberdeen. **3** a river in South Yorkshire which rises in the Pennines and flows 112 km (70 miles) eastwards to join the Ouse shortly before it, in turn, joins the Humber.

don¹ /dɒn/ n. **1** a university teacher, esp. a senior member of a college at Oxford or Cambridge. **2** (**Don**) **a** a Spanish title prefixed to a male forename. **b** a Spanish gentleman; a Spaniard. [Sp. f. L *dominus* lord]

don² /dɒn/ v.tr. (**donned, donning**) put on (clothing). [= *do on*]

dona /ˈdəʊnə/ n. (also **donah**) Brit. sl. a woman; a sweetheart. [Sp. *doña* or Port. *dona* f. L (as DONNA)]

Donald Duck /ˈdɒn(ə)ld/ a Walt Disney cartoon character, who first appeared in 1934. Clarence Nash (1904–85) spoke the soundtrack for his voice.

donate /dəʊˈneɪt/ v.tr. give or contribute (money etc.), esp. voluntarily to a fund or institution. □ **donator** n. [back-form. f. DONATION]

Donatello /ˌdɒnəˈtɛləʊ/ (born Donato di Betto Bardi) (1386–1466), Italian sculptor. He was one of the pioneers of scientific perspective, and is especially famous for his lifelike sculptures, including the bronze *David* (probably created between 1430 and 1460), his most classical work. He was in Padua from 1443 to 1453, where he made the *Gattamelata*, the first equestrian statue to be created since antiquity. On his return to Florence he reacted somewhat against classical principles, evolving a late style in which distortion is used to convey dramatic and emotional intensity, as in his carved wooden statue *St Mary Magdalene* (c.1455).

donation /dəʊˈneɪʃ(ə)n/ n. **1** the act or an instance of donating. **2** something, esp. an amount of money, donated. [ME f. OF f. L *donatio -onis* f. *donare* give f. *donum* gift]

Donatist /ˈdəʊnətɪst/ n. a member of a schismatic Christian group in North Africa, whose leader was Donatus (died c.355) and which was formed in 311 in opposition to the election of Caecilian (died c.345) as bishop of Carthage. The Donatists held that only those living a blameless life belonged to the Church. They survived until the 7th century. □ **Donatism** n. [LL *Donatista* follower of Donatus]

donative /ˈdəʊnətɪv, ˈdɒn-/ n. & adj. ● n. a gift or donation, esp. one given formally or officially as a largesse. ● adj. **1** given as a donation or bounty. **2** hist. (of a benefice) given directly, not presentative. [ME f. L *donativum* gift, largesse f. *donare*: see DONATION]

Donatus /dəˈneɪtəs/, Aelius (4th century), Roman grammarian. His treatises on Latin grammar were collected in the *Ars Grammatica* (undated); it was the sole textbook used in schools in the Middle Ages.

Donau see DANUBE.

Donbas /dɒnˈbɑːs/ the Ukrainian name for the DONETS BASIN.

Donbass /danˈbas/ the Russian name for the DONETS BASIN.

Doncaster /ˈdɒŋkəstə(r)/ an industrial town in South Yorkshire; pop. (1991) 284,300.

done /dʌn/ past part. of DO¹. ● adj. **1** colloq. socially acceptable (*the done thing; it isn't done*). **2** (often with *in, up*) colloq. tired out. **3** (esp. as *int.* in reply to an offer etc.) accepted. □ **be done with** have finished with, be finished with. **done for** colloq. in serious trouble. **have done** have ceased or finished. **have done with** be rid of; have finished dealing with.

donee /dəʊˈniː/ n. the recipient of a gift. [DONOR + -EE]

Donegal /ˌdɒnɪˈɡɔːl/ a county in the extreme north-west of the

Republic of Ireland, part of the old province of Ulster; capital, Lifford.

doner /'dɒnə(r), 'dəʊn-/ n. (in full **doner kebab**) a Turkish dish consisting of spiced lamb roasted on a vertical spit and sliced thinly. [Turk. *döner kebap* rotating kebab]

Donets /dɒ'njets/ a river in eastern Europe, rising near Belgorod in southern Russia and flowing south-eastwards for some 1,000 km (630 miles) through Ukraine before re-entering Russia and joining the Don near Rostov.

Donets Basin (called in Ukrainian *Donbas*, Russian *Donbass*) a coal-mining and industrial region of SE Ukraine, stretching between the valley of the Donets and lower Dnieper rivers.

Donetsk /dɒ'njetsk/ the leading city of the Donets Basin in Ukraine; pop. (1990) 1,117,000. The city was called Yuzovka from 1872 until 1924, in honour of John Hughes (1814–89), a Welshman who established its first ironworks, and Stalin or Stalino from 1924 until 1961.

dong[1] /dɒŋ/ v. & n. ● v. **1** intr. make the deep sound of a large bell. **2** tr. Austral. & NZ colloq. hit, punch. ● n. **1** the deep sound of a large bell. **2** Austral. & NZ colloq. a heavy blow. **3** coarse sl. the penis. [imit.]

dong[2] /dɒŋ/ n. the chief monetary unit of Vietnam. [Vietnamese]

donga /'dɒŋgə/ n. S. Afr. & Austral. **1** a dry watercourse. **2** a ravine caused by erosion. [Zulu]

dongle /'dɒŋg(ə)l/ n. Computing a security attachment required by a computer to enable protected software to be used. [arbitrary form.]

Donizetti /ˌdɒnɪ'tsetɪ/, Gaetano (1797–1848), Italian composer. He is generally regarded as the leading Italian composer of the 1830s, bridging the gap between Giuseppe Verdi and Vincenzo Bellini. He wrote seventy-five operas, including tragedies such as *Anna Bolena* (1830) and *Lucia di Lammermoor* (1835) and the comedies *L'Elisir d'amore* (1832) and *Don Pasquale* (1843).

donjon /'dɒndʒən, 'dʌn-/ n. the great tower or innermost keep of a castle. [archaic spelling of DUNGEON]

Don Juan /dɒn 'dʒuːən, 'hwɑːn/ a legendary Spanish nobleman of dissolute life, famous for seducing women. According to a Spanish story first dramatized by Gabriel Téllez (1584–1641), he was Don Juan Tenorio of Seville. Molière, Byron, and Mozart (in *Don Giovanni*) have all based works on the legend.

donkey /'dɒŋkɪ/ n. (pl. **-eys**) **1** a domestic ass. **2** colloq. a stupid or foolish person. □ **donkey engine** a small auxiliary engine. **donkey jacket** a thick weatherproof jacket worn by workers and as a fashion garment. **donkey's years** colloq. a very long time. **donkey-work** the laborious part of a job; drudgery. [earlier with pronunc. as *monkey*: perh. f. DUN[1], or the name *Duncan*]

Donkin /'dɒŋkɪn/, Bryan (1768–1855), English engineer. He made pioneering contributions in several fields, including paper-making and printing, patenting (with Richard Mackenzie Bacon, an English writer and printer, 1775–1844) the first rotary press. In the 1830s, he successfully developed a method of food preservation by heat sterilization, sealing the food inside a container made of sheet steel, so producing the first tin can (see CAN[2]).

donna /'dɒnə/ n. **1** an Italian, Spanish, or Portuguese lady. **2** (**Donna**) an Italian, Spanish, or Portuguese title prefixed to a female forename. [It. f. L *domina* mistress fem. of *dominus*: cf. DON[1]]

Donne /dʌn/, John (1572–1631), English poet and preacher. Generally regarded as the first of the metaphysical poets, he is most famous for his *Satires* (c.1590–9), *Elegies* (c.1590–9), and love poems, which appeared in the collection *Songs and Sonnets* (undated). He also wrote religious poems and, as dean of St Paul's from 1621, was one of the most celebrated preachers of his age.

donnée /'dɒneɪ/ n. (also **donné**) **1** the subject or theme of a story etc. **2** a basic fact or assumption. [F, fem. or masc. past part. of *donner* give]

donnish /'dɒnɪʃ/ adj. like or resembling a college don, esp. in supposed pedantry. □ **donnishly** adv. **donnishness** n.

Donnybrook /'dɒnɪˌbrʊk/ n. a scene of uproar; a free fight. [*Donnybrook* near Dublin, Republic of Ireland, formerly site of annual fair]

donor /'dəʊnə(r)/ n. **1** a person who gives or donates something (e.g. to a charity). **2** a person who provides blood for a transfusion, semen for insemination, or an organ or tissue for transplantation. **3** Chem. an atom or molecule that provides a pair of electrons in forming a coordinate bond. **4** Physics an impurity atom in a semiconductor which contributes a conducting electron to the material. □ **donor card** an official card authorizing use of organs for transplant, carried by the donor. [ME f. AF *donour*, OF *deneur* f. L *donator -oris* f. *donare* give]

Don Quixote /dɒn ˈkwɪksəʊt, kiːˈhəʊtɪ/ the hero of a romance (1605–15) by Cervantes, a satirical account of chivalric beliefs and conduct. The character of Don Quixote is typified by a romantic vision and naive idealism. [Sp., DON[1] + *quixote* thigh armour]

don't /dəʊnt/ contr. do not. ● n. a prohibition (*dos and don'ts*).

donut US var. of DOUGHNUT.

doodad /'duːdæd/ n. N. Amer. = DOODAH. [20th c.: orig. unkn.]

doodah /'duːdɑː/ n. colloq. **1** a fancy article; a trivial ornament. **2** a gadget or 'thingummy'. □ **all of a doodah** excited, dithering. [from the refrain of the song 'Camptown Races']

doodle /'duːd(ə)l/ v. & n. ● v.intr. scribble or draw, esp. absent-mindedly. ● n. a scrawl or drawing so made. □ **doodler** n. [orig. = foolish person; cf. LG *dudelkopf*]

doodlebug /'duːd(ə)lˌbʌg/ n. **1** US the larva of an ant-lion, or a similar insect. **2** US an unscientific device for locating minerals. **3** colloq. a flying bomb of the Second World War.

doohickey /'duːˌhɪkɪ/ n. (pl. **-eys**) N. Amer. colloq. a small object, esp. mechanical. [DOODAD + HICKEY]

Doolittle /'duːˌlɪt(ə)l/, Hilda (1886–1961), American poet. From 1911 she lived in London, where she met Ezra Pound and other imagist poets, whose style and concerns are reflected in her own work. Her many volumes of poetry (published under the pseudonym H.D.) also show the influence of classical mythology; they include *Sea Garden* (1916).

doom /duːm/ n. & v. ● n. **1 a** a grim fate or destiny. **b** death or ruin. **2 a** a condemnation; a judgement or sentence. **b** the Last Judgement (*the crack of doom*). **3** hist. a statute, law, or decree. ● v.tr. **1** (usu. foll. by *to*) condemn or destine (*a city doomed to destruction*). **2** (esp. as **doomed** adj.) consign to misfortune or destruction. [OE *dōm* statute, judgement f. Gmc: rel. to DO[1]]

doomsday /'duːmzdeɪ/ n. the day of the Last Judgement (cf. DOMESDAY BOOK). □ **till doomsday** for ever. [OE *dōmes dæg*: see DOOM]

doomwatch /'duːmwɒtʃ/ n. organized vigilance or observation to avert danger, esp. from environmental pollution. □ **doomwatcher** n.

door /dɔː(r)/ n. **1 a** a hinged, sliding, or revolving barrier for closing and opening an entrance to a building, room, cupboard, etc. **b** this as representing a house etc. (*lives two doors away*). **2 a** an entrance or exit; a doorway. **b** a means of access or approach. □ **close the door to** exclude the opportunity for. **door-case** (or **-frame**) the structure into which a door is fitted. **door-head** the upper part of a door-case. **door-keeper** = DOORMAN. **door-plate** a plate on the door of a house or room bearing the name of the occupant. **door-to-door** (of selling etc.) done at each house in turn. **lay** (or **lie**) **at the door of** impute (or be imputable) to. **leave the door open** ensure that an option remains available. **next door** in or to the next house or room. **next door to 1** in the next house to. **2** nearly, almost, near to. **open the door to** create an opportunity for. **out of doors** in or into the open air. □ **doored** adj. (also in comb.). [OE *duru, dor* f. Gmc]

doorbell /'dɔːbel/ n. a bell in a house etc. rung by visitors outside to signal their arrival.

doorknob /'dɔːnɒb/ n. a knob for turning to release the latch of a door.

doorman /'dɔːmən/ n. (pl. **-men**) a person on duty at the door to a large building; a janitor or porter.

doormat /'dɔːmæt/ n. **1** a mat at an entrance for wiping mud etc. from the shoes. **2** a feebly submissive person.

doornail /'dɔːneɪl/ n. a nail with which doors were studded for strength or ornament. □ **dead as a doornail** completely or unmistakably dead.

Doornik see TOURNAI.

doorpost /'dɔːpəʊst/ n. each of the uprights of a door-frame, on one of which the door is hung.

Doors, the an American rock group formed in 1965 and associated with the drug culture and psychedelia of the late 1960s. The group revolved around the flamboyant personality of lead singer Jim Morrison (1943–71) and is remembered for dramatic songs such as 'Light My Fire' (1967) and 'Riders on the Storm' (1971), featuring the characteristic playing of organist Ray Manzarek (b.1935).

doorstep /'dɔːstep/ n. & v. ● n. **1** a step leading up to the outer door of a house etc. **2** sl. a thick slice of bread. ● v. (**-stepped, -stepping**) **1** intr. go from door to door selling, canvassing, etc. **2** tr. (of a reporter) wait on the doorstep for (a person), to obtain an interview, photograph, etc. **3** *tr.* leave (a child) to the care of someone else. □ **on one's** (or **the**) **doorstep** very close.

doorstop /'dɔːstɒp/ *n.* a device for keeping a door open or to prevent it from striking a wall etc. when opened.

doorway /'dɔːweɪ/ *n.* an opening filled by a door.

dooryard /'dɔːjɑːd/ *n. N. Amer.* a yard or garden near a house-door.

doozy /'duːzɪ/ *adj. & n. esp. N. Amer. & Austral. sl.* ● *adj.* excellent, stunning. ● *n.* a very large or good thing; a stunner. [20th c.: orig. uncert., perh. rel. to DOUSE]

dop /dɒp/ *n. S. Afr.* **1** a cheap kind of brandy. **2** a tot of liquor. [Afrik.]

dopa /'dəʊpə/ *n. Biochem.* an amino acid which is a metabolic precursor of dopamine in the body. Dopamine deficiency, believed to cause the weakness and tremor of Parkinsonism, can be corrected by treatment with dopa (specifically, with the laevorotatory form of dopa, called *levodopa* or *L-dopa*). [G f. *Dihydroxyphenylalanine*, former name of the compound]

dopamine /'dəʊpə,miːn/ *n. Biochem.* a compound found in nervous tissue, acting as a neurotransmitter and a precursor of other substances including adrenalin. (See also DOPA, PARKINSON'S DISEASE.) [DOPA + AMINE]

dopant /'dəʊp(ə)nt/ *n. Electronics* a substance used in doping a semiconductor.

dope /dəʊp/ *n. & v.* ● *n.* **1** a varnish applied to the cloth surface of aeroplane parts to strengthen them, keep them airtight, etc. **2** a thick liquid used as a lubricant etc. **3** a substance added to petrol etc. to increase its effectiveness. **4** *sl.* **a** a narcotic; a stupefying drug. **b** a drug etc. given to a horse or greyhound, or taken by an athlete, to affect performance. **5** *sl.* a stupid person. **6** *sl.* **a** information about a subject, esp. if not generally known. **b** misleading information. ● *v.* **1** *tr.* administer dope to, drug. **2** *tr. Electronics* add an impurity to (a semiconductor) to produce a desired electrical characteristic. **3** *tr.* smear, daub; apply dope to. **4** *intr.* take addictive drugs. □ **dope out** *sl.* discover. □ **doper** *n.* [Du. *doop* sauce f. *doopen* to dip]

dopey /'dəʊpɪ/ *adj.* (also **dopy**) (**dopier**, **dopiest**) *colloq.* **1 a** half asleep. **b** stupefied by or as though by a drug. **2** stupid, silly. □ **dopily** *adv.* **dopiness** *n.*

doppelgänger /'dɒp(ə)l,geŋə(r)/ *n.* an apparition or double of a living person. [G, = double-goer]

Dopper /'dɒpə(r)/ *n. S. Afr.* a member of the Gereformeerde Kerk, a strictly orthodox Calvinistic denomination of the Dutch Reformed Church, usu. regarded as old-fashioned in ideas etc. [Afrik.: orig. unkn.]

Doppler effect /'dɒplə(r)/ *n.* (also **Doppler shift**) *Physics* an increase (or decrease) in the frequency of sound, light, or other waves as the source and observer move towards (or away) from each other. [Christian Johann *Doppler*, Austrian physicist (1803–53)]

dopy var. of DOPEY.

dor /dɔː(r)/ *n.* (in full **dor-beetle**) a large beetle that makes a buzzing sound in flight, esp. a dung-beetle of the genus *Geotrupes*. [OE *dora*]

Dorado /də'rɑːdəʊ/ *Astron.* a southern constellation (the Goldfish), containing most of the Large Magellanic Cloud. It is sometimes known as Xiphias (the Swordfish). [f. DORADO 1]

dorado /də'rɑːdəʊ/ *n.* (*pl.* **-os**) **1** a blue and silver marine fish, *Coryphaena hippurus*, showing brilliant colours when alive. Also called *dolphin*. **2** a golden-coloured freshwater fish, *Salminus maxillosus*, native to South America. [Sp. f. LL *deauratus* gilt f. *aurum* gold]

Dorchester /'dɔːtʃɪstə(r)/ a town in southern England, the county town of Dorset; pop. (est. 1985) 14,000.

Dordogne /dɔː'dɔɪn, French dɔrdɔɲ/ **1** a river of western France which rises in the Auvergne and flows 472 km (297 miles) westwards to meet the Garonne and form the Gironde estuary. **2** an inland department of SW France. It contains numerous caves and rock-shelters that have yielded abundant remains of early humans and their artefacts and art, for example the cave at Lascaux.

Dordrecht /'dɔːdrext/ (also **Dort** /dɔːt/) an industrial city and river port in the Netherlands, near the mouth of the Rhine (there called the Waal), 20 km (12 miles) south-east of Rotterdam; pop. (1991) 110,500. Situated on one of the busiest river junctions in the world, it was the wealthiest town in the Netherlands until surpassed by Rotterdam in the 18th century.

Doré /'dɔːreɪ/, Gustave (1832–83), French book illustrator. He was widely known for his dark, detailed woodcut illustrations of books such as Dante's *Inferno* (1861), Cervantes' *Don Quixote* (1863), and the Bible (1865–6); he produced so many of these that at one time he employed more than forty block-cutters.

Dorian /'dɔːrɪən/ *n. & adj.* ● *n.* a member of an ancient Hellenic people speaking the Doric dialect of Greek, inhabiting the Peloponnese and elsewhere. (*See note below.*) (Cf. ACHAEAN, IONIAN.) ● *adj.* of or relating to the Dorians. □ **Dorian mode** *n. Mus.* the mode represented by the natural diatonic scale D–D. [L *Dorius* f. Gk *Dōrios* f. *Dōros*, the mythical ancestor of this people]

▪ The Dorians had settled in most of the Peloponnese (including Sparta and Corinth), the southernmost Aegean Islands, and the southwestern corner of Asia Minor by the 8th century BC. They are traditionally held to have entered Greece from the north in the 11th or 12th century BC, displacing the Achaeans and the Ionians and bringing on a dark age. The Dorians later founded colonies in Sicily (including Syracuse) and southern Italy.

Doric /'dɒrɪk/ *adj. & n.* ● *adj.* **1** (of a dialect) broad, rustic. **2** *Archit.* of the oldest, sturdiest, and simplest of the Greek orders. **3** of or relating to the ancient Greek dialect of the Dorians. ● *n.* **1** rustic English or esp. Scots. **2** *Archit.* the Doric order. **3** the dialect of the Dorians in ancient Greece. [L *Doricus* f. Gk *Dōrikos* (as DORIAN)]

dork /dɔːk/ *n. coarse sl.* **1** the penis. **2** a stupid or contemptible person. [20th c.: orig. uncert., perh. f. DIRK, infl. by DICK[1] 2]

dorm /dɔːm/ *n. colloq.* dormitory. [abbr.]

dormant /'dɔːmənt/ *adj.* **1** lying inactive as in sleep; sleeping. **2 a** (of a volcano etc.) temporarily inactive. **b** (of potential faculties etc.) in abeyance. **3** (of plants) alive but not actively growing. **4** *Heraldry* (of a beast) lying with its head on its paws. □ **dormancy** *n.* [ME f. OF, pres. part. of *dormir* f. L *dormire* sleep]

dormer /'dɔːmə(r)/ *n.* (in full **dormer window**) a projecting upright window in a sloping roof. [OF *dormeor* (as DORMANT)]

dormition /dɔː'mɪʃ(ə)n/ *n.* (also **Dormition**) (in the Orthodox Church) **1** the passing of the Virgin Mary from earthly life. **2** the feast held in honour of this on 15 August, corresponding to the Assumption in the Western Church. [F f. L *dormitio* f. *dormire dormit-* sleep]

dormitory /'dɔːmɪtərɪ/ *n.* (*pl.* **-ies**) **1** a sleeping-room with several beds, esp. in a school or institution. **2** (in full **dormitory town** etc.) a small town or suburb from which people travel to work in a city etc. **3** *US* a university or college hall of residence or hostel. [ME f. L *dormitorium* f. *dormire dormit-* sleep]

Dormobile /'dɔːmə,biːl/ *n. propr.* a type of motor caravan with a rear compartment convertible for sleeping and eating in. [blend of DORMITORY, AUTOMOBILE]

dormouse /'dɔːmaʊs/ *n.* (*pl.* **dormice**) an arboreal mouselike rodent of the family Gliridae, often having a bushy tail; esp. *Muscardinus avellanarius*, noted for its long hibernation. [ME: orig. unkn., but assoc. with F *dormir*, L *dormire*: see DORMANT]

dormy /'dɔːmɪ/ *adj. Golf* (of a player or side) ahead by as many holes as there are holes left to play (*dormy five*). [19th c.: orig. unkn.]

doronicum /də'rɒnɪkəm/ *n.* = leopard's bane (see LEOPARD). [mod.L (Linnaeus) ult. f. Arab. *darānaj*]

dorp /dɔːp/ *n. S. Afr.* a village or small township. [Du. (as THORP)]

dorsal /'dɔːs(ə)l/ *adj. Anat., Zool., & Bot.* **1** of, on, or near the back (cf. VENTRAL). **2** ridge-shaped. □ **dorsally** *adv.* [F *dorsal* or LL *dorsalis* f. L *dorsum* back]

Dorset /'dɔːsɪt/ a county of SW England; county town, Dorchester.

dorsum /'dɔːsəm/ *n. Anat. & Zool.* the dorsal part of an organism or structure. [L., = back]

Dort see DORDRECHT.

Dortmund /'dɔːtmʊnd/ an industrial city in NW Germany, in North Rhine-Westphalia; pop. (1991) 601,000. It is the southern terminus of the Dortmund-Ems Canal, which links the Ruhr industrial area with the North Sea.

dory[1] /'dɔːrɪ/ *n.* (*pl.* **-ies**) a marine fish of the family Zeidae, having a thin deep body, esp. the John Dory. [ME f. F *dorée* fem. past part. of *dorer* gild (as DORADO)]

dory[2] /'dɔːrɪ/ *n.* (*pl.* **-ies**) *N. Amer. & West Indies* a flat-bottomed fishing-boat with high sides. [Miskito *dóri* dugout]

DOS /dɒs/ *n. Computing* a program for manipulating information on a disk. [abbr. of disk operating system]

dos-à-dos /,dəʊzæ'dəʊ/ *adj. & n.* ● *adj.* (of two books) bound together with a shared central board and facing in opposite directions. ● *n.* (*pl.* same) a seat, carriage, etc., in which the occupants sit back to back (cf. DO-SE-DO). [F, = back to back]

dosage /ˈdəʊsɪdʒ/ n. **1** the giving of medicine in doses. **2** the size of a dose.

dose /dəʊs/ n. & v. ● n. **1** an amount of a medicine or drug taken or recommended to be taken at one time. **2** a quantity of something administered or allocated (e.g. work, praise, punishment, etc.). **3** the amount of ionizing radiation received by a person or thing. **4** sl. a venereal infection. ● v.tr. **1** treat (a person or animal) with doses of medicine. **2** give a dose or doses to. **3** adulterate or blend (esp. wine with spirit). □ **like a dose of salts** colloq. very fast and efficiently. [F f. LL dosis f. Gk dosis gift f. didōmi give]

do-se-do /ˌdəʊziˈdəʊ, ˌdəʊsɪ-/ n. (also **do-si-do**) (pl. **-os**) a figure in square dancing in which two dancers pass round each other back to back and return to their original positions. [corrupt. of DOS-À-DOS]

dosh /dɒʃ/ n. sl. money. [20th c.: orig. unkn.]

dosimeter /dəʊˈsɪmɪtə(r)/ n. (also **dosemeter**) a device used to measure an absorbed dose of ionizing radiation. □ **dosimetry** n. **dosimetric** /ˌdəʊsɪˈmetrɪk/ adj.

Dos Passos /dɒs ˈpæsɒs/, John (Roderigo) (1896–1970), American novelist. He is chiefly remembered for his collage-like portrayal of the energy and diversity of American life in the first decades of the 20th century in such novels as *Manhattan Transfer* (1925) and *U.S.A.* (1938).

doss /dɒs/ v. & n. Brit. sl. ● v.intr. **1** (often foll. by *down*) sleep, esp. roughly or in cheap lodgings. **2** (often foll. by *about, around*) spend time idly. ● n. **1** a bed, esp. in cheap lodgings. **2** an easy task; an opportunity for, or spell of, idling. □ **doss-house** a cheap lodging-house, esp. for vagrants. [prob. = *doss* ornamental covering for a seat-back etc. f. OF *dos* ult. f. L *dorsum* back]

dossal /ˈdɒs(ə)l/ n. Eccl. a hanging cloth behind an altar or round a chancel. [med.L *dossale* f. LL *dorsalis* DORSAL]

dosser /ˈdɒsə(r)/ n. Brit. sl. **1** a person who dosses. **2** = doss-house.

dossier /ˈdɒsɪə(r), -sɪˌeɪ/ n. a set of documents, esp. a collection of information about a person, event, or subject. [F, so called from the label on the back, f. *dos* back f. L *dorsum*]

dost /dʌst/ archaic 2nd sing. present of DO[1].

Dostoevsky /ˌdɒstɔɪˈefskɪ/, Fyodor Mikhailovich (also **Dostoyevsky**) (1821–81), Russian novelist. His early socialist activism led to his being sentenced to death, but a last-minute reprieve led instead to exile in Siberia (1849–54). During this time he suffered periods of great mental and physical pain and recurring bouts of epilepsy; his experiences are recounted in *Notes from the House of the Dead* (1860–1). From the 1860s he wrote the novels on which his reputation is based, including *Crime and Punishment* (1866), *The Idiot* (1868), and *The Brothers Karamazov* (1880). These dark works reveal Dostoevsky's keen psychological insight, savage humour, and his concern with profound religious, political, and moral problems, especially that of human suffering.

DoT abbr. (in the UK) Department of Transport.

dot[1] /dɒt/ n. & v. ● n. **1 a** a small spot, speck, or mark. **b** such a mark written or printed as part of an *i* or *j*, as a diacritical mark, as one of a series of marks to signify omission, or as a full stop. **c** a decimal point. **2** Mus. a dot used to denote the lengthening of a note or rest, or to indicate staccato. **3** the shorter signal of the two used in Morse code (cf. DASH n. 6). **4** a tiny or apparently tiny object (*a dot on the horizon*). ● v.tr. (**dotted**, **dotting**) **1 a** mark with a dot or dots. **b** place a dot over (a letter). **2** Mus. mark (a note or rest) to show that the time value is increased by half. **3** (often foll. by *about*) scatter like dots. **4** partly cover as with dots (*a sea dotted with ships*). **5** sl. hit (*dotted him one in the eye*). □ **dot the i's and cross the t's** colloq. **1** be minutely accurate, emphasize details. **2** add the final touches to a task, exercise, etc. **dot matrix printer** Computing a printer with characters formed from dots printed by configurations of the tips of small wires. **dotted line** a line of dots on a document, esp. to show a place left for a signature. **on the dot** exactly on time. **the year dot** Brit. colloq. far in the past. □ **dotter** n. [OE *dott* head of a boil, perh. infl. by Du. *dot* knot]

dot[2] /dɒt/ n. a woman's dowry. [F f. L *dos dotis*]

dotage /ˈdəʊtɪdʒ/ n. feeble-minded senility (*in his dotage*). [ME f. DOTE + -AGE]

dotard /ˈdəʊtəd/ n. a person who is feeble-minded, esp. from senility. [ME f. DOTE + -ARD]

dote /dəʊt/ v.intr. **1** (foll. by *on, upon*) be foolishly or excessively fond of. **2** be silly or feeble-minded, esp. from old age. □ **doter** n. **dotingly** adv. [ME, corresp. to MDu. *doten* be silly]

doth /dʌθ/ archaic 3rd sing. present of DO[1].

dotterel /ˈdɒtərəl/ n. a small migrant plover, *Eudromias morinellus*. [ME f. DOTE + -REL, named from the ease with which it is caught, taken to indicate stupidity]

dottle /ˈdɒt(ə)l/ n. a remnant of unburnt tobacco in a pipe. [DOT[1] + -LE[1]]

dotty /ˈdɒtɪ/ adj. (**dottier**, **dottiest**) colloq. **1** feeble-minded, silly. **2** eccentric. **3** absurd. **4** (foll. by *about, on*) infatuated with; obsessed by. □ **dottily** adv. **dottiness** n. [19th c.: orig. uncert.]

Douala /duːˈɑːlə/ the chief port and largest city of Cameroon; pop. (1991) 884,000.

douane /duːˈɑːn/ n. a foreign custom-house. [F f. It. *do(g)ana* f. Turk. *duwan*, Arab. *dīwān*: cf. DIVAN]

Douay Bible /ˈduːeɪ, ˈdaʊeɪ/ n. (also **Douay version**) an English translation of the Bible formerly used in the Roman Catholic Church, completed at Douai in France early in the seventeenth century.

double /ˈdʌb(ə)l/ adj., adv., n., & v. ● adj. **1 a** consisting of two usu. equal parts or things; twofold. **b** consisting of two identical parts. **2** twice as much or many (*double the amount; double the number; double thickness*). **3** having twice the usual size, quantity, strength, etc. (*double whisky*). **4** designed for two people (*double bed*). **5 a** having some part double. **b** (of a flower) having more than one circle of petals. **c** (of a domino) having the same number of pips on each half. **6** having two different roles or interpretations, esp. implying confusion or deceit (*double meaning; leads a double life*). **7** Mus. lower in pitch by an octave (*double bassoon*). ● adv. **1** at or to twice the amount etc. (*counts double*). **2** two together (*sleep double*). ● n. **1 a** a double quantity or thing; twice as much or many. **b** a double measure of spirits. **2 a** a counterpart of a person or thing; a person who looks exactly like another. **b** an understudy. **c** a wraith. **3** (in *pl.*) (in tennis, badminton, etc.) a game between two pairs of players. **4** Sport a pair of victories over the same team, a pair of championships at the same game, etc. **5** a system of betting in which the winnings and stake from the first bet are transferred to a second. **6** Bridge the doubling of an opponent's bid. **7** Darts a hit on the narrow ring enclosed by the two outer circles of a dartboard, scoring double. **8** a sharp turn, esp. of the tracks of a hunted animal, or the course of a river. ● v. **1** tr. & intr. make or become twice as much or many; increase twofold; multiply by two. **2** tr. amount to twice as much as. **3 a** tr. fold or bend (paper, cloth, etc.) over on itself. **b** intr. become folded. **4 a** tr. (of an actor) play (two parts) in the same piece. **b** intr. (often foll. by *for*) be understudy etc. **5** intr. (usu. foll. by *as*) play a twofold role. **6** intr. turn sharply in flight or pursuit; take a tortuous course. **7** tr. Naut. sail round (a headland). **8** tr. Bridge make a call increasing the value of the points to be won or lost on (an opponent's bid). **9** Mus. **a** intr. (often foll. by *on*) play two or more musical instruments (*the clarinettist doubles on tenor sax*). **b** tr. add the same note in a higher or lower octave to (a note). **10** tr. clench (a fist). **11** intr. move at twice the usual speed; run. **12** Billiards **a** intr. rebound. **b** tr. cause to rebound. □ **at the double** running, hurrying. **bent double** folded, stooping. **double acrostic** see ACROSTIC. **double-acting** (of an engine etc.) in which pistons are pushed from both sides alternately. **double agent** a person who purports to spy for one country while actually working for a rival one. **double axe** an axe with two blades, especially as a characteristic Minoan and Mycenaean tool (also one of the most common Minoan religious symbols). **double back** take a new direction opposite to the previous one. **double-banking 1** double-parking. **2** Austral. & NZ riding two on a horse etc. **double-barrelled 1** (of a gun) having two barrels. **2** Brit. (of a surname) having two parts joined by a hyphen. **3** twofold. **double bill** a programme with two principal items. **double bind** a dilemma. **double-blind** adj. (of a test or experiment) in which neither the tester nor the subject has knowledge of identities etc. that might lead to bias. ● n. such a test or experiment. **double bluff** an action or statement intended to appear as a bluff, but in fact genuine. **double boiler** a saucepan with a detachable upper compartment heated by boiling water in the lower one. **double bond** Chem. a pair of bonds between two atoms in a molecule. **double-book** accept two reservations simultaneously for (the same seat, room, etc.). **double-breasted** (of a coat etc.) having two fronts overlapping across the body. **double-check** verify twice or in two ways. **double chin** a chin with a fold of loose flesh below it. **double-chinned** having a double chin. **double concerto** a concerto for two solo instruments. **double cream** thick cream with a high fat-content. **double-cross** v.tr. deceive or betray (a person one is supposedly helping). ● n. an act of doing this. **double-crosser** a person who double-crosses. **double dagger** Printing = double obelus. **double-dealer** a deceiver. **double-dealing** n. deceit, esp. in business. ● adj. deceitful; practising deceit. **double-decker 1** esp. Brit. a bus having an upper and lower deck. **2** colloq. anything consisting of two layers. **double-declutch** see DECLUTCH. **double decomposition** Chem. a reaction in which two compounds exchange radicals or ions (also called metathesis). **double density** Computing designating a storage device, esp. a disk, having twice the basic capacity. **double dummy** Bridge play with two hands exposed, allowing every card to be located. **double Dutch** Brit. colloq. incomprehensible talk. **double-dyed** deeply affected with guilt. **double eagle 1** a figure of a two-headed eagle. **2** US Golf = ALBATROSS 2. **3** a former US coin worth twenty dollars. **double-edged 1** having two functions or (often contradictory) applications. **2** (of a knife etc.) having two cutting-edges. **double entry** a system of bookkeeping in which each transaction is entered as a debit in one account and a credit in another. **double exposure** Photog. the accidental or deliberate repeated exposure of a plate, film, etc. **double-faced 1** insincere. **2** (of a fabric or material) finished on both sides so that either may be used as the right side. **double fault** Tennis n. & v. ● n. two consecutive faults in serving. ● v.intr. (**double-fault**) serve a double fault. **double feature** a cinema programme with two full-length films. **double figures** the numbers from 10 to 99. **double first** Brit. **1** first-class honours in two subjects or examinations at a university. **2** a person achieving this. **double-fronted** (of a house) with principal windows on either side of the front door. **double-ganger** = DOPPELGÄNGER. **double glazing 1** a window consisting of two layers of glass with a space between them, designed to reduce loss of heat and exclude noise. **2** the provision of this. **double Gloucester** a kind of hard cheese orig. made in Gloucestershire. **double header 1** a train pulled by two locomotives coupled together. **2** N. Amer. two games etc. in succession between the same opponents. **3** Austral. colloq. a coin with a head on both sides. **double helix** a pair of parallel helices with a common axis, esp. in the structure of the DNA molecule. **double-jointed** having joints that allow unusual bending of the fingers, limbs, etc. **double knitting 1** yarn or knitting that is twice the usual thickness. **2** tubular knitting for ties, belts, etc. **double-lock** lock by a double turn of the key. **double negative** Gram. a negative statement containing two negative elements (e.g. didn't say nothing). ¶ Considered ungrammatical in standard English. **double obelus** (or **obelisk**) Printing a sign (‡) used to introduce a reference. **double or quits** a gamble to decide whether a player's loss or debt be doubled or cancelled. **double-park** park (a vehicle) alongside one that is already parked at the roadside. **double play** Baseball putting out two runners. **double pneumonia** pneumonia affecting both lungs. **double-quick** very quick or quickly. **double refraction** Optics refraction forming two separate rays from a single incident ray. **double rhyme** a rhyme including two syllables. **double salt** Chem. a salt composed of two simple salts and having different crystal properties from either. **double saucepan** Brit. = double boiler. **double shuffle** Dancing a shuffle executed twice with one foot and then twice with the other. **double standard 1** a rule or principle applied more strictly to some people than to others (or to oneself). **2** bimetallism. **double star** two stars actually or apparently very close together. **double-stopping** Mus. the sounding of two strings at once on a violin etc. **double take** a delayed reaction to a situation etc. immediately after one's first reaction. **double-talk** verbal expression that is (usu. deliberately) ambiguous or misleading. **double-think** the mental capacity to accept contrary opinions or beliefs at the same time esp. as a result of political indoctrination. **double time 1** payment of an employee at twice the normal rate. **2** Mil. the regulation running-pace. **double-tonguing** Mus. a method of tonguing using two alternate tongue movements, usu. made by sounding t and k alternately, in order to facilitate rapid playing of a wind instrument. **double top** Darts a score of double twenty. **double up 1 a** bend or curl up. **b** cause to do this, esp. by a blow. **2** be overcome with pain or laughter. **3** share or assign to a room, quarters, etc., with another or others. **4** fold or become folded. **5** use winnings from a bet as stake for another. **double whammy** colloq. a twofold blow or setback. □ **doubler** n. **doubly** adv. [ME f. OF doble, duble (n.), dobler, dubler (v.) f. L duplus DUPLE]

double bass /beɪs/ n. the largest and lowest-pitched instrument of the violin family. In orchestral and chamber music the instrument is bowed while being held upright between the legs of the seated musician, whereas players of jazz and rock and roll generally pluck the strings with their fingers while standing.

double entendre /ˌduːblə ɒnˈtɒndrə/ n. (pl. **double entendres** pronunc. same) **1** a word or phrase open to two interpretations, one usu. risqué or indecent. **2** humour using such words or phrases. [obs. F, = double understanding]

doublespeak /ˈdʌb(ə)lˌspiːk/ n. language or talk that is (usu. deliberately) ambiguous or obscure.

doublet /ˈdʌblɪt/ n. **1** either of a pair of similar things, esp. either of two words of the same derivation but different sense (e.g. fashion and faction, cloak and clock). **2** hist. a man's short close-fitting jacket, with or without sleeves. **3** a historical or biblical account occurring twice in differing contexts, usu. traceable to different sources. **4** (in pl.) the same number on two dice thrown at once. **5** a pair of associated lines close together in a spectrum. **6** a combination of two simple lenses. [ME f. OF f. double: see DOUBLE]

doubloon /dʌbˈluːn/ n. **1** hist. a Spanish gold coin. **2** (in pl.) sl. money. [F doublon or Sp. doblón (as DOUBLE)]

doublure /duːˈbljʊə(r)/ n. an ornamental lining, usu. leather, inside a book-cover. [F, = lining (doubler to line)]

doubt /daʊt/ n. & v. ● n. **1** a feeling of uncertainty; an undecided state of mind (be in no doubt about; have no doubt that). **2** (often foll. by of, about) an inclination to disbelieve (have one's doubts about). **3** an uncertain state of things. **4** a lack of full proof or clear indication (benefit of the doubt). ● v. **1** tr. (often foll. by whether, if, that + clause; also foll. (after neg. or interrog.) by but, but that) feel uncertain or undecided about (I doubt that you are right; I do not doubt but that you are wrong). **2** tr. hesitate to believe or trust. **3** intr. (often foll. by of) feel uncertain or undecided; have doubts (never doubted of success). **4** tr. call in question. **5** tr. Brit. archaic or dial. rather think that; suspect or fear that (I doubt we are late). □ **beyond doubt** certainly. **doubting Thomas** an incredulous or sceptical person (see THOMAS, St). **in doubt** uncertain; open to question. **no doubt** certainly; probably; admittedly. **without doubt** (or **a doubt**) certainly. □ **doubtable** adj. **doubter** n. **doubtingly** adv. [ME doute f. OF doute (n.), douter (v.) f. L dubitare hesitate; mod. spelling after L]

doubtful /ˈdaʊtfʊl/ adj. **1** feeling doubt or misgivings; unsure or guarded in one's opinion. **2** causing doubt; ambiguous; uncertain in meaning etc. **3** unreliable (a doubtful ally). □ **doubtfully** adv. **doubtfulness** n.

doubtless /ˈdaʊtlɪs/ adv. (often qualifying a sentence) **1** certainly; no doubt. **2** probably. □ **doubtlessly** adv.

douce /duːs/ adj. Sc. sober, gentle, sedate. [ME f. OF dous douce f. L dulcis sweet]

douche /duːʃ/ n. & v. ● n. **1** a jet of liquid applied to part of the body for cleansing or medicinal purposes. **2** a device for producing such a jet. ● v. **1** tr. treat with a douche. **2** intr. use a douche. [F f. It. doccia pipe f. docciare pour by drops ult. f. L ductus: see DUCT]

dough /dəʊ/ n. **1** a thick mixture of flour etc. and liquid (usu. water), for baking into bread, pastry, etc. **2** sl. money. [OE dāg f. Gmc]

doughboy /ˈdəʊbɔɪ/ n. **1** a boiled dumpling. **2** US colloq. a United States infantryman, esp. in the war of 1914–18.

doughnut /ˈdəʊnʌt/ n. (US **donut**) **1** a small fried cake of sweetened dough, usu. in the shape of a ball or ring. **2** a ring-shaped object, esp. Physics a vacuum chamber for acceleration of particles in a betatron or synchrotron.

doughty /ˈdaʊtɪ/ adj. (**doughtier**, **doughtiest**) archaic or joc. valiant, stout-hearted. □ **doughtily** adv. **doughtiness** n. [OE dohtig var. of dyhtig f. Gmc]

doughy /ˈdəʊɪ/ adj. (**doughier**, **doughiest**) **1** having the form or consistency of dough. **2** pale and sickly in colour. □ **doughiness** n.

Douglas /ˈdʌɡləs/ the capital of the Isle of Man; pop. (1991) 22,210.

Douglas fir n. (also **Douglas pine** or **spruce**) a large conifer of the genus Pseudotsuga, of western North America. [David Douglas, Sc. botanist (1798–1834)]

Douglas-Home /ˌdʌɡləsˈhjuːm/, Sir Alec, Baron Home of the Hirsel of Coldstream (born Alexander Frederick Douglas-Home) (1903–95), British Conservative statesman, Prime Minister 1963–4. He served as private secretary to Neville Chamberlain in the negotiations with Hitler from 1937 to 1940. Various ministerial offices followed before his appointment as Foreign Secretary under Harold Macmillan in 1960. When Macmillan resigned in 1963, Douglas-Home became Prime Minister, relinquishing his hereditary peerage as 14th Earl of Home (to which he had succeeded in 1951), and for a time being neither a member of the House of Commons nor of the House of Lords, in an unprecedented situation for a British Prime Minister. His government was defeated by the Labour Party in the 1964 elections. Douglas-Home later served as Foreign Secretary under Edward Heath (1970–4).

doum /duːm/ n. (in full **doum-palm**) a palm tree, Hyphaene thebaica, native to Egypt, with large edible fruit. [Arab. dawm, dūm]

dour /dʊə(r)/ adj. severe, stern, or sullenly obstinate in manner or appearance. □ **dourly** *adv.* **dourness** *n.* [ME (orig. Sc.), prob. f. Gael. *dúr* dull, obstinate, perh. f. L *durus* hard]

Douro /ˈdʊəruː/ (Spanish **Duero** /ˈdwero/) a river of the Iberian peninsula, rising in central Spain and flowing west for 900 km (556 miles) through Portugal to the Atlantic Ocean near Oporto. In its valley in Portugal grow the grapes from which port wine is made.

douroucouli /ˌdʊərʊˈkuːlɪ/ *n.* (pl. **douroucoulis**) a nocturnal monkey of the genus *Aotus*, native to South America, having large staring eyes. Also called *night-monkey*, *owl-monkey*. [Amer. Indian name]

douse /daʊs/ *v.tr.* (also **dowse**) **1 a** throw water over. **b** plunge into water. **2** extinguish (a light). **3** *Naut.* **a** lower (a sail). **b** close (a porthole). [16th c.: perh. rel. to MDu., LG *dossen* strike]

dove[1] /dʌv/ *n.* **1** a bird of the family Columbidae, with short legs, small head, and large breast; a small pigeon. **2** a gentle or innocent person. **3** *Polit.* an advocate of peace or peaceful policies, esp. in foreign affairs (opp. HAWK[1] *n.* 2). **4** (**Dove**) *Relig.* a representation of the Holy Spirit (John 1:32). **5** a soft grey colour. □ **dove's-foot** a cranesbill, *Geranium molle*. **dove-tree** a tree with dovelike flowers, *Davidia involucrata*, native to China. □ **dovelike** *adj.* [ME f. ON *dúfa* f. Gmc]

dove[2] *US past* and *past part.* of DIVE.

dovecote /ˈdʌvkɒt/ *n.* (also **dovecot**) a shelter with nesting-holes for domesticated pigeons.

dovekie /ˈdʌvkɪ/ *n. N. Amer. & Sc.* = *little auk* (see AUK). [Sc., dimin. of DOVE[1]]

Dover /ˈdəʊvə(r)/ **1** a ferry port in Kent, the largest of the Cinque Ports, on the coast of the English Channel; pop. (1981) 34,300. It is mainland Britain's nearest point to the Continent, being only 35 km (22 miles) from Calais. **2** a shipping forecast area covering the Strait of Dover. **3** the state capital of Delaware; pop. (1990) 23,500.

Dover, Strait of a sea passage between England and France, connecting the English Channel with the North Sea. At its narrowest it is 35 km (22 miles) wide.

dovetail /ˈdʌvteɪl/ *n. & v.* ● *n.* **1** a joint formed by a mortise with a tenon shaped like a dove's spread tail or a reversed wedge. **2** such a tenon. ● *v.* **1** *tr.* join together by means of a dovetail. **2** *tr. & intr.* (often foll. by *into*, *with*) fit readily together; combine neatly or compactly.

dowager /ˈdaʊədʒə(r)/ *n.* **1** a widow with a title or property derived from her late husband (*Queen dowager*; *dowager duchess*). **2** *colloq.* a dignified elderly woman. [OF *douag(i)ere* f. *douage* (as DOWER)]

dowdy /ˈdaʊdɪ/ *adj. & n.* ● *adj.* (**dowdier**, **dowdiest**) **1** (of clothes) unattractively dull; unfashionable. **2** (of a person, esp. a woman) dressed in dowdy clothes. ● *n.* (pl. **-ies**) a dowdy woman. □ **dowdily** *adv.* **dowdiness** *n.* [ME *dowd* slut, of unkn. orig.]

dowel /ˈdaʊəl/ *n. & v.* ● *n.* a headless peg of wood, metal, or plastic for holding together components of a structure. ● *v.tr.* (**dowelled**, **dowelling**; *US* **doweled**, **doweling**) fasten with a dowel or dowels. [ME f. MLG *dovel*: cf. THOLE[1]]

dowelling /ˈdaʊəlɪŋ/ *n.* (*US* **doweling**) cylindrical rods for cutting into dowels.

dower /ˈdaʊə(r)/ *n. & v.* ● *n.* **1** a widow's share for life of her husband's estate. **2** *archaic* a dowry. **3** a natural gift or talent. ● *v.tr.* **1** *archaic* give a dowry to. **2** (foll. by *with*) endow with talent etc. □ **dower house** *Brit.* a smaller house near a big one, forming part of a widow's dower. □ **dowerless** *adj.* [ME f. OF *douaire* f. med.L *dotarium* f. L *dos dotis*]

dowitcher /ˈdaʊɪtʃə(r)/ *n.* a wading bird of the genus *Limnodromus*, breeding in northern North America and related to the sandpipers. [Iroquoian]

Dow–Jones index /daʊˈdʒəʊnz/ *n.* (also **Dow–Jones average**) a figure based on the average price of selected stocks, indicating the relative price of shares on the New York Stock Exchange. Dow Jones & Co. compiled the first average of US stock prices in 1884, and this developed in 1897 into computation of a daily average of (originally twelve) industrial stocks; the system was reorganized in 1928. [Charles Henry Dow (1851–1902) & Edward D. Jones (1856–1920), Amer. economists]

Down /daʊn/ one of the Six Counties of Northern Ireland, formerly an administrative area; chief town, Downpatrick.

down[1] /daʊn/ *adv., prep., adj., v., & n.* ● *adv.* (*superl.* **downmost**) **1** into or towards a lower place, esp. to the ground (*fall down*; *knelt down*). **2** in a lower place or position (*blinds were down*). **3** to or in a place regarded as lower, esp.: **a** southwards. **b** *Brit.* away from a major city or a university. **4 a** in or into a low or weaker position, mood, or condition (*hit a man when he's down*; *many down with colds*). **b** *Brit.* in a position of lagging behind or loss (*our team was three goals down*; *£5 down on the transaction*). **c** (of a computer system) out of action or unavailable for use (esp. temporarily). **5** from an earlier to a later time (*customs handed down*; *down to 1600*). **6** to a finer or thinner consistency or a smaller amount or size (*grind down*; *water down*; *boil down*). **7** cheaper; lower in price or value (*bread is down*; *shares are down*). **8** into a more settled state (*calm down*). **9** in writing; in or into recorded or listed form (*copy it down*; *I got it down on tape*; *you are down to speak next*). **10** (of part of a larger whole) paid, dealt with (*£5 down, £20 to pay*; *three down, six to go*). **11** *Naut.* **a** with the current or wind. **b** (of a ship's helm) with the rudder to windward. **12** inclusively of the lower limit in a series (*read down to the third paragraph*). **13** (as *int.*) lie down, put (something) down, etc. **14** (of a crossword clue or answer) read vertically (*cannot do five down*). **15** downstairs, esp. after rising (*is not down yet*). **16** swallowed (*could not get the pill down*). **17** *Amer. Football* (of the ball) out of play. ● *prep.* **1** downwards along, through, or into. **2** from top to bottom of. **3** along (*walk down the road*; *cut down the middle*). **4** at or in a lower part of (*situated down the river*). ● *adj.* (*superl.* **downmost**) **1** directed downwards. **2** *Brit.* of travel away from a capital or centre (*the down train*; *the down platform*). **3** *colloq.* unhappy, depressed. ● *v.tr. colloq.* **1** knock or bring down. **2** swallow (a drink). ● *n.* **1** an act of putting down (esp. an opponent in wrestling, or the ball in American football). **2** a reverse of fortune (*ups and downs*). **3** *colloq.* a period of depression. **4** the play of the first piece in dominoes. □ **be** (or **have a**) **down on** *colloq.* disapprove of; show animosity towards. **be down to 1** be attributable to. **2** be the responsibility of. **3** have used up everything except (*down to their last tin of rations*). **down and out 1** penniless, destitute. **2** *Boxing* unable to resume the fight. **down-and-out** *n.* a destitute person. **down at heel 1** (of a shoe) with the heel worn down. **2** (of a person) shabby, slovenly; impoverished. **down draught** a downward draught, esp. one down a chimney into a room. **down grade 1** a descending slope of a road or railway. **2** a deterioration (see also DOWNGRADE). **down in the mouth** *colloq.* looking unhappy. **down-market** *adj. & adv. colloq.* towards or relating to the cheaper or less affluent sector of the market. **down on one's luck** *colloq.* **1** temporarily unfortunate. **2** dispirited by misfortune. **down payment** a partial payment made at the time of purchase. **down stage** *Theatr.* at or to the front of the stage. **down-stroke** a stroke made or written downwards. **down time** time during which a machine, esp. a computer, is out of action or unavailable for use. **down-to-earth** practical, realistic. **down to the ground** *colloq.* completely. **down tools** *colloq.* cease work, esp. to go on strike. **down town 1** into a town from a higher or outlying part. **2** *esp. N. Amer.* to or in the business part of a city (see also DOWNTOWN). **down under** *colloq.* in the antipodes, esp. Australia. **down with** *int.* expressing strong disapproval or rejection of a specified person or thing. [OE *dūn* f. *adūne* downward]

down[2] /daʊn/ *n.* **1 a** the first covering of young birds. **b** a bird's underplumage, used in cushions etc. **c** a layer of fine soft feathers. **2** fine soft hair esp. on the face. **3** short soft hairs on some leaves, fruit, seeds, etc. **4** a fluffy substance, e.g. thistledown. [ME f. ON *dúnn*]

down[3] /daʊn/ *n.* **1** an area of open rolling land. **2** (in *pl.*; usu. prec. by *the*) **a** undulating chalk and limestone uplands esp. in southern England, with few trees and used mainly for pasture. **b** (**Downs**) see DOWNS, THE. □ **downy** *adj.* [OE *dūn* perh. f. OCelt.]

downbeat /ˈdaʊnbiːt/ *n. & adj.* ● *n. Mus.* an accented beat, usu. the first of the bar. ● *adj.* **1** pessimistic, gloomy. **2** relaxed.

downcast /ˈdaʊnkɑːst/ *adj. & n.* ● *adj.* **1** (of eyes) looking downwards. **2** (of a person) dejected. ● *n.* a shaft dug in a mine for extra ventilation.

downcomer /ˈdaʊnkʌmə(r)/ *n.* a pipe for downward transport of water or gas.

downer /ˈdaʊnə(r)/ *n. sl.* **1** a depressant or tranquillizing drug, esp. a barbiturate. **2** a depressing person or experience; a failure. **3** = DOWNTURN.

downfall /ˈdaʊnfɔːl/ *n.* **1 a** a fall from prosperity or power. **b** the cause of this. **2** a sudden heavy fall of rain etc.

downfold /ˈdaʊnfəʊld/ *n. Geol.* a syncline.

downgrade *v. & n.* ● *v.tr.* **1** make lower in rank or status. **2** speak disparagingly of. ● *n.* /ˈdaʊngreɪd/ *US* a downward gradient or course. □ **on the downgrade** *US* in decline.

downhearted /daʊnˈhɑːtɪd/ *adj.* dejected; in low spirits. □ **downheartedly** *adv.* **downheartedness** *n.*

downhill *adv., adj., & n.* ● *adv.* /daʊnˈhɪl/ in a descending direction, esp. towards the bottom of an incline. ● *adj.* /ˈdaʊnhɪl/ **1** sloping down, descending. **2** declining; deteriorating. ● *n.* /ˈdaʊnhɪl/ **1** *Skiing* a downhill race. **2** a downward slope. **3** a decline. □ **go downhill** *colloq.* decline, deteriorate (in health, state of repair, moral state, etc.). □ **downhiller** /daʊn'hɪlə(r)/ *n.* Skiing.

Downing Street /'daʊnɪŋ/ a street in Westminster, London, between Whitehall and St James's Park, frequently used allusively for the British government or the Prime Minister. No. 10 has been the official town residence of the Prime Minister since the time of Sir Robert Walpole; No. 11 is the home of the Chancellor of the Exchequer, and the Foreign and Commonwealth Office is also situated in the street. It is named after the original developer of the site, Sir George Downing (*c.*1624–84), a diplomat under both Oliver Cromwell and Charles II.

downland /'daʊnlənd/ *n.* = DOWN³.

download /daʊn'ləʊd/ *v.tr.* Computing transfer (data) from one storage device or system to another (esp. smaller remote one).

downmost /'daʊnməʊst/ *adj. & adv.* the furthest down.

downpipe /'daʊnpaɪp/ *n.* Brit. a pipe to carry rainwater from a roof to a drain or to ground level.

downplay /daʊn'pleɪ/ *v.tr.* play down; minimize the importance of.

downpour /'daʊnpɔː(r)/ *n.* a heavy fall of rain.

downright /'daʊnraɪt/ *adj. & adv.* ● *adj.* **1** plain, definite, straightforward, blunt. **2** utter, complete (*a downright lie; downright nonsense*). ● *adv.* thoroughly, completely, positively (*downright rude*). □ **downrightness** *n.*

downriver *adv. & adj.* ● *adv.* /daʊn'rɪvə(r)/ at or to a point away from the source of a river. ● *adj.* /'daʊn,rɪvə(r)/ situated or occurring downriver.

Downs, the /daʊnz/ **1** a region of chalk hills in southern England. It consists of two ranges, the *North Downs* in Surrey and Kent and the *South Downs* in West and East Sussex. **2** a part of the sea (opposite the North Downs) off east Kent.

downscale *v. & adj.* US ● *v.tr.* /daʊn'skeɪl/ reduce or restrict in size, scale, or extent. ● *adj.* /'daʊnskeɪl/ at the lower end of a scale, esp. a social scale; inferior.

downshift /'daʊnʃɪft/ *n. & v.* ● *n.* a movement downwards, esp. a change to a lower gear in a motor vehicle. ● *v.intr.* make a downshift.

downside /'daʊnsaɪd/ *n.* **1** a downward movement of share prices etc. **2** a negative aspect of something; a drawback or disadvantage.

downsize /daʊn'saɪz/ *v.tr. & intr.* **1** reduce (something) in size or scale. **2** reduce the size of (a company etc.) by making staff redundant.

downspout /'daʊnspaʊt/ *n.* US = DOWNPIPE.

Down's syndrome /daʊnz/ *n.* Med. a congenital disorder marked by diminished intelligence, short stature, and a broad flattened facial profile. The usual cause of Down's syndrome is the presence of an extra chromosome. It is also called *mongolism*, but this is now considered offensive. [John Langdon *Down*, Engl. physician (1828–96)]

downstairs *adv., adj., & n.* ● *adv.* /daʊn'steəz/ **1** down a flight of stairs. **2** to or on a lower floor. ● *adj.* /'daʊn'steəz/ (also **downstair**) situated downstairs. ● *n.* /'daʊnsteəz/ the lower floor.

downstate *adj., n., & adv.* US ● *adj.* /'daʊnsteɪt/ of or in a part of a state remote from large cities, esp. the southern part. ● *n.* /'daʊnsteɪt/ a downstate area. ● *adv.* /daʊn'steɪt/ in a downstate area.

downstream *adv. & adj.* ● *adv.* /daʊn'striːm/ in the direction of the flow of a stream etc. ● *adj.* /'daʊnstriːm/ moving downstream.

downthrow /'daʊnθrəʊ/ *n.* Geol. a downward dislocation of strata.

downtown *adj., n., & adv.* N. Amer. ● *adj.* /'daʊntaʊn/ of or in the lower or more central part, or the business part, of a town or city. ● *n.* /'daʊntaʊn/ a downtown area. ● *adv.* /daʊn'taʊn/ in or into a downtown area.

downtrodden /'daʊn,trɒd(ə)n/ *adj.* oppressed; badly treated; kept under.

downturn /'daʊntɜːn/ *n.* a decline, esp. in economic or business activity.

downward /'daʊnwəd/ *adv. & adj.* ● *adv.* (also **downwards** /-wədz/) towards what is lower, inferior, less important, or later. ● *adj.* moving, extending, pointing, or leading downward. □ **downwardly** *adv.*

downwarp /'daʊnwɔːp/ *n.* Geol. a broad surface depression; a syncline.

downwind *adj.* /'daʊnwɪnd/ & *adv.* /daʊn'wɪnd/ in the direction in which the wind is blowing.

downy /'daʊnɪ/ *adj.* (**downier, downiest**) **1 a** of, like, or covered with down. **b** soft and fluffy. **2** Brit. sl. aware, knowing. □ **downily** *adv.* **downiness** *n.*

dowry /'daʊərɪ/ *n.* (pl. **-ies**) **1** property or money brought to a bride to

her husband. **2** a talent, a natural gift. [ME f. AF *dowarie*, OF *douaire* DOWER]

dowse¹ /daʊz/ *v.intr.* search for underground water or minerals by holding a Y-shaped stick or rod which dips abruptly when over the right spot. □ **dowsing-rod** such a stick or rod. □ **dowser** *n.* [17th c.: orig. unkn.]

dowse² var. of DOUSE.

doxology /dɒk'sɒlədʒɪ/ *n.* (pl. **-ies**) a liturgical formula of praise to God. □ **doxological** /,dɒksə'lɒdʒɪk(ə)l/ *adj.* [med.L *doxologia* f. Gk *doxologia* f. *doxa* glory + -LOGY]

doxy /'dɒksɪ/ *n.* (pl. **-ies**) *literary* **1** a lover or mistress. **2** a prostitute. [16th-c. cant: orig. unkn.]

doyen /'dɔɪən/ *n.* (fem. **doyenne** /dɔɪ'en/) the senior or most prominent of a particular category or body of people. [F (as DEAN¹)]

Doyle /dɔɪl/, Sir Arthur Conan (1859–1930), Scottish novelist. He is chiefly remembered for establishing the detective story as a major fictional genre with his creation of the private detective Sherlock Holmes. Holmes and his friend Dr Watson (the narrator of the stories) first appeared in *A Study in Scarlet* (1887) and continued to demonstrate their ingenuity at crime-solving in a long line of stories contained in such collections as *The Adventures of Sherlock Holmes* (1892) and *The Hound of the Baskervilles* (1902). Doyle also wrote historical and other romances.

doyley var. of DOILY.

D'Oyly Carte /,dɔɪlɪ 'kɑːt/, Richard (1844–1901), English impresario and producer. He brought together the librettist Sir W. S. Gilbert and the composer Sir Arthur Sullivan, producing many of their operettas in London's Savoy Theatre, which he had established in 1881.

doz. abbr. dozen.

doze /dəʊz/ *v. & n.* ● *v.intr.* sleep lightly; be half asleep. ● *n.* a short light sleep. □ **doze off** fall lightly asleep. [17th c.: cf. Danish *døse* make drowsy]

dozen /'dʌz(ə)n/ *n.* **1** (prec. by *a* or a number) (pl. **dozen**) twelve, regarded collectively (*a dozen eggs; two dozen packets; ordered three dozen*). **2** a set or group of twelve (*packed in dozens*). **3** *colloq.* about twelve, a fairly large indefinite number. **4** (in *pl.*; usu. foll. by *of*) *colloq.* very many (*made dozens of mistakes*). **5** (**the dozens**) a black American game or ritualized exchange of verbal insults. □ **by the dozen** in large quantities. **talk nineteen to the dozen** Brit. talk incessantly. □ **dozenth** *adj. & n.* [ME f. OF *dozeine*, ult. f. L *duodecim* twelve]

dozer /'dəʊzə(r)/ *n. colloq.* = BULLDOZER. [abbr.]

dozy /'dəʊzɪ/ *adj.* (**dozier, doziest**) **1** drowsy; tending to doze. **2** Brit. *colloq.* stupid or lazy. □ **dozily** *adv.* **doziness** *n.*

DP abbr. **1** data processing. **2** displaced person.

D.Phil. abbr. Doctor of Philosophy.

DPP abbr. (in the UK) Director of Public Prosecutions.

Dr abbr. **1** Doctor. **2** Drive. **3** debtor.

dr. abbr. **1** drachm(s). **2** drachma(s). **3** dram(s).

drab¹ /dræb/ *adj. & n.* ● *adj.* (**drabber, drabbest**) **1** dull, uninteresting. **2** of a dull brownish colour. ● *n.* **1** drab colour. **2** monotony. □ **drably** *adv.* **drabness** *n.* [prob. f. obs. *drap* cloth f. OF f. LL *drappus*, perh. of Celt. orig.]

drab² see DRIBS AND DRABS.

drab³ /dræb/ *n.* **1** a slut; a slattern. **2** a prostitute. [perh. rel. to LG *drabbe* mire, Du. *drab* dregs]

Drabble /'dræb(ə)l/, Margaret (b.1939), English novelist. Her early works, such as *The Garrick Year* (1964) and *The Millstone* (1966), deal mainly with the concerns of the individual and are characterized by their atmosphere of reflection and introspection; her later novels, including *The Ice Age* (1977) and *The Radiant Way* (1987), have a more documentary approach to social change. Drabble is also a literary critic and biographer. She is the younger sister of the novelist A. S. Byatt.

drabble /'dræb(ə)l/ *v.intr. & tr.* become or make dirty and wet with water or mud. [ME f. LG *drabbelen* paddle in water or mire: cf. DRAB³]

dracaena /drə'siːnə/ *n.* a shrub or tree of the genus *Dracaena* or *Cordyline*, of the agave family, grown for its foliage. (See also dragon-tree.) [mod.L f. Gk *drakaina* fem. of *drakōn* dragon]

drachm /dræm/ *n.* Brit. a weight or measure formerly used by apothecaries, equivalent to 60 grains or one eighth of an ounce, or (in full **fluid drachm**) 60 minims, one eighth of a fluid ounce. [ME *dragme* f. OF *dragme* or LL *dragma* f. L *drachma* f. Gk *drakhmē* Attic weight and coin]

drachma /'drækmə/ *n.* (pl. **drachmas** or **drachmae** /-miː/) **1** the chief monetary unit of Greece. **2** a silver coin of ancient Greece. [L f. Gk *drakhmē*]

drack /dræk/ *adj. Austral. sl.* **1** (esp. of a woman) unattractive. **2** dismal, dull. [20th c.: orig. unkn.]

Draco[1] /'dreɪkəʊ/ *Astron.* a large constellation (the Dragon) stretching around the north celestial pole, said to represent the dragon killed by Hercules. It has no bright stars. [L]

Draco[2] /'dreɪkəʊ/ (7th century BC), Athenian legislator. His codification of Athenian law was notorious for its severity in that the death penalty was imposed for both serious and trivial crimes; this gave rise to the adjective *draconian* in English.

dracone /'drækəʊn/ *n.* a large flexible container for liquids, towed on the surface of the sea. [L *draco -onis* (as DRAGON)]

draconian /drə'kəʊnɪən, dreɪ'kəʊ-/ *adj.* (also **draconic** /-'kɒnɪk/) very harsh or severe (esp. of laws and their application). [see DRACO[2]]

Dracula /'drækjʊlə/ the Transylvanian vampire in Bram Stoker's novel *Dracula* (1897). The name comes from Vlad Ţepeş (Vlad the Impaler), a 15th-century prince of Wallachia, also known as Dracula, who was renowned for his cruelty. The figure of Count Dracula has become the representative type of vampire on stage and screen.

draff /dræf, drɑːf/ *n.* **1** dregs, lees. **2** refuse. [ME, perh. repr. OE *dræf* (unrecorded)]

draft /drɑːft/ *n. & v.* ● *n.* **1 a** a preliminary written version of a speech, document, etc. **b** a rough preliminary outline of a scheme. **c** a sketch of work to be carried out. **2 a** a written order for payment of money by a bank. **b** the drawing of money by means of this. **3** (foll. by *on*) a demand made on a person's confidence, friendship, etc. **4 a** a party detached from a larger group for a special duty or purpose. **b** the selection of this. **5** *US* compulsory military service. **6** a reinforcement. **7** *US* = DRAUGHT. ● *v.tr.* **1** prepare a draft of (a document, scheme, etc.). **2** select for a special duty or purpose. **3** *US* conscript for military service. □ **draft-dodge** *US* avoid military conscription. **draft-dodger** *US* a person who avoids military conscription. □ **drafter** *n.* **draftee** /drɑːf'tiː/ *n.* [phonetic spelling of DRAUGHT]

draftsman /'drɑːftsmən/ *n.* (*pl.* **-men**) **1** a person who drafts documents. **2** = DRAUGHTSMAN 1. [phonetic spelling of DRAUGHTSMAN]

drafty *US* var. of DRAUGHTY.

drag /dræg/ *v. & n.* ● *v.* (**dragged, dragging**) **1** *tr.* pull along with effort or difficulty. **2 a** *tr.* allow (one's feet, tail, etc.) to trail along the ground. **b** *intr.* trail along the ground. **c** *intr.* (of time etc.) go or pass heavily or slowly or tediously. **3 a** *intr.* (usu. foll. by *for*) use a grapnel or drag (to find a drowned person or lost object). **b** *tr.* search the bottom of (a river etc.) with grapnels, nets, or drags. **4** *tr.* (often foll. by *to*) *colloq.* take (a person to a place etc., esp. against his or her will). **5** *intr.* (foll. by *on, at*) *colloq.* draw on (a cigarette etc.). **6** *intr.* (often foll. by *on*) continue at tedious length. ● *n.* **1 a** an obstruction to progress. **b** *Aeron.* the longitudinal retarding force exerted by air. **c** slow motion; impeded progress. **d** an iron shoe for retarding a horse-drawn vehicle downhill. **2** *colloq.* a boring or dreary person, duty, performance, etc. **3 a** a strong-smelling lure drawn before hounds as a substitute for a fox. **b** a hunt using this. **4** an apparatus for dredging or recovering drowned persons etc. from under water. **5** = drag-net 1. **6** *sl.* a draw on a cigarette etc. **7** *sl.* **a** women's clothes worn by men. **b** a party at which these are worn. **c** clothes in general. **8** an act of dragging. **9 a** *sl.* a motor car. **b** (in full **drag race**) an acceleration race between cars usu. for a quarter of a mile. **10** *US sl.* influence, pull. **11** *sl.* a street or road (*the main drag*). **12** *hist.* a private vehicle like a stagecoach, drawn by four horses. □ **drag anchor** (of a ship) move from a moored position when the anchor fails to hold. **drag-anchor** *n.* = sea anchor. **drag one's feet** (or **heels**) be deliberately slow or reluctant to act. **drag-hound** a hound used to hunt with a drag. **drag in** introduce (a subject) irrelevantly. **drag-line** an excavator with a bucket pulled in by a wire rope. **drag-net 1** a net drawn through a river or across ground to trap fish or game. **2** a systematic hunt for criminals etc. **drag out** protract. **drag queen** *sl.* a male homosexual transvestite. **drag through the mud** see MUD. **drag up** *colloq.* **1** deliberately mention (an unwelcome subject). **2** rear (a child) roughly and without proper training. [ME f. OE *dragan* or ON *draga* DRAW]

dragée /'dræʒeɪ/ *n.* **1** a sugar-coated almond etc. **2** a small silver ball for decorating a cake. **3** a chocolate-coated sweet. [F: see DREDGE[2]]

draggle /'dræg(ə)l/ *v.* **1** *tr.* make dirty or wet or limp by trailing. **2** *intr.* hang trailing. **3** *intr.* lag; straggle in the rear. □ **draggle-tailed** (of a woman) with untidily trailing skirts. [DRAG + -LE[4]]

draggy /'drægɪ/ *adj. colloq.* **1** inclined to drag; heavy, slow. **2** *colloq.* tedious, boring.

dragoman /'drægəmən/ *n.* (*pl.* **dragomans** or **dragomen**) an interpreter or guide, esp. in countries speaking Arabic, Turkish, or Persian. [F f. It. *dragomano* f. med.Gk *dragomanos* f. Arab. *tarjumān* f. *tarjama* interpret, f. Aram. *targēm* f. Assyr. *targumânu* interpreter]

dragon /'drægən/ *n.* **1** a mythical monster like a reptile, usu. with wings and claws and able to breathe out fire. (*See note below.*) **2** a fierce person, esp. a woman. **3** (in full **flying dragon**) a flying lizard, *Draco volans.* □ **dragon's blood** a red gum that exudes from the fruit of some palms and the dragon-tree. **dragon's teeth** *Mil. colloq.* obstacles resembling teeth pointed upwards, used esp. against tanks. **dragon-tree** a palmlike tree, *Dracaena draco*, of the agave family, native to the Canary Islands (see also DRACAENA). [ME f. OF f. L *draco -onis* f. Gk *drakōn* serpent]

▪ The dragon is probably the commonest emblem in the art of the Far East, and the most ancient. A form with five claws on each foot (oriental dragons normally have four) was adopted as the chief imperial emblem in China. The dragon represented fertilizing power and cosmic energy and tended to live in water; in springtime it moved in heaven among the clouds. While the dragon of the Far East is largely beneficent, in the Christian art of the West it has often been a symbol of threat and destruction, as well as paganism, as illustrated by the legend of St George slaying the dragon.

dragonet /'drægənɪt/ *n.* a spiny marine fish of the family Callionymidae, the male of which is brightly coloured. [ME f. F, dimin. of DRAGON]

dragonfish /'drægənˌfɪʃ/ *n.* a deep-water marine fish of the order Stomiiformes, having a long slender body and a barbel on the chin with luminous tissue, serving to attract prey.

dragonfly /'drægənˌflaɪ/ *n.* (*pl.* **-flies**) a predatory insect of the order Odonata, having a long slender body, two pairs of large transparent wings that are spread while resting, and aquatic larvae.

dragonnade /ˌdrægə'neɪd/ *n. & v.* ● *n.* a persecution by use of troops, esp. (in *pl.*) of French Protestants under Louis XIV by quartering dragoons on them. ● *v.tr.* subject to a dragonnade. [F f. *dragon*: see DRAGOON]

dragoon /drə'guːn/ *n. & v.* ● *n.* **1** a cavalryman (orig. a mounted infantryman armed with a carbine). **2** a rough fierce man. **3** a variety of pigeon. ● *v.tr.* **1** (foll. by *into*) coerce into doing something, esp. by use of rigorous or harassing methods. **2** persecute, esp. with troops. [orig. = carbine (thought of as breathing fire) f. F *dragon* DRAGON]

dragster /'drægstə(r)/ *n.* a car built or modified to take part in drag races.

drail /dreɪl/ *n.* a fish-hook and line weighted with lead for dragging below the surface of the water. [app. var. of TRAIL]

drain /dreɪn/ *v. & n.* ● *v.* **1** *tr.* draw off liquid from, esp.: **a** make (land etc.) dry by providing an outflow for moisture. **b** (of a river) carry off the superfluous water of (a district). **c** remove purulent matter from (an abscess). **2** *tr.* (foll. by *off, away*) draw off (liquid) esp. by a pipe. **3** *intr.* (foll. by *away, off, through*) flow or trickle away. **4** *intr.* (of a wet cloth, a container, etc.) become dry as liquid flows away (*put it there to drain*). **5** *tr.* (often foll. by *of*) exhaust or deprive (a person or thing) of strength, resources, property, etc. **6** *tr.* **a** drink (liquid) to the dregs. **b** empty (a container) by drinking the contents. ● *n.* **1 a** a channel, conduit, or pipe carrying off liquid, esp. an artificial conduit for water or sewage. **b** a tube for drawing off the discharge from an abscess etc. **2** a constant outflow, withdrawal, or expenditure (*a great drain on my resources*). □ **down the drain** *colloq.* lost, wasted. **laugh like a drain** laugh copiously; guffaw. [OE *drē(a)hnian* f. Gmc]

drainage /'dreɪnɪdʒ/ *n.* **1** the process or means of draining (*the land has poor drainage*). **2** a system of drains, artificial or natural. **3** what is drained off, esp. sewage.

drainboard /'dreɪnbɔːd/ *n. N. Amer.* = DRAINING-BOARD.

drainer /'dreɪnə(r)/ *n.* **1** a device for draining; anything on which things are put to drain, e.g. a draining-board. **2** a person who drains.

draining-board /'dreɪnɪŋˌbɔːd/ *n.* a sloping usu. grooved surface beside a sink, on which washed dishes etc. are left to drain.

drainpipe /'dreɪnpaɪp/ *n.* **1** a pipe for carrying off water, sewage, etc., from a building. **2** (*attrib.*) (of trousers etc.) very narrow. **3** (in *pl.*) very narrow trousers.

Drake /dreɪk/, Sir Francis (c.1540–96), English sailor and explorer. He spent his early career privateering in Spanish seas. He was the first Englishman to circumnavigate the globe; he set off in 1577 with five ships under the sponsorship of Elizabeth I to investigate the Strait of Magellan, tried unsuccessfully to find the North-west Passage, and finally returned to England via the Cape of Good Hope with only his own ship, the *Golden Hind*, in 1580. He was knighted the following year. Drake's raid on Cadiz in 1587 delayed the sailing of the Armada for a year by destroying its supply-ships, and the next year he played an important part in its defeat in the English Channel.

drake /dreɪk/ *n.* a male duck. [ME prob. f. Gmc]

Drakensberg Mountains /ˈdrɑːkənzˌbɜːɡ/ a range of mountains in southern Africa, stretching in a NE–SW direction for a distance of 1,126 km (700 miles) through Lesotho and the South African provinces of KwaZulu/Natal, Orange Free State, and Eastern and Northern Transvaal. The highest peak is Thabana Ntlenyana (3,482 m, 11,425 ft).

Drake Passage an area of ocean connecting the South Atlantic and the South Pacific and separating the southern tip of South America (Cape Horn) from the Antarctic Peninsula. It is named after Sir Francis Drake.

Dralon /ˈdreɪlɒn/ *n. propr.* **1** a synthetic acrylic fibre used in textiles. **2** a fabric made from this. [invented word, after NYLON]

dram /dræm/ *n.* **1** a small drink of spirits. **2** = DRACHM. [ME f. OF *drame* or med.L *drama, dragma*: cf. DRACHM]

drama /ˈdrɑːmə/ *n.* **1** a play for acting on stage or for broadcasting. **2** (often prec. by *the*) the art of writing and presenting plays. **3** an exciting or emotional event, set of circumstances, etc. **4** dramatic quality (*the drama of the situation*). □ **drama-documentary** a film (esp. for television) dramatizing or based on real events. [LL f. Gk *drama -atos* f. *draō* do]

dramatic /drəˈmætɪk/ *adj.* **1** of drama or the study of drama. **2** (of an event, circumstance, etc.) sudden and exciting or unexpected. **3** vividly striking. **4** (of a gesture etc.) theatrical, overdone, absurd. □ **dramatic irony** = *tragic irony*. □ **dramatically** *adv.* [LL *dramaticus* f. Gk *dramatikos* (as DRAMA)]

dramatics /drəˈmætɪks/ *n.pl.* (often treated as *sing.*) **1** the production and performance of plays. **2** exaggerated or showy behaviour.

dramatis personae /ˌdræmətɪs pɜːˈsəʊnaɪ, -niː/ *n.pl.* (often treated as *sing.*) **1** the characters in a play. **2** a list of these. [L, = persons of the drama]

dramatist /ˈdræmətɪst/ *n.* a writer of dramas.

dramatize /ˈdræməˌtaɪz/ *v.* (also **-ise**) **1 a** *tr.* adapt (a novel etc.) to form a stage play. **b** *intr.* admit of such adaptation. **2** *tr.* make a drama or dramatic scene of. **3** *tr.* (also *absol.*) express or react to in a dramatic way. □ **dramatization** /ˌdræmətaɪˈzeɪʃ(ə)n/ *n.*

dramaturge /ˈdræməˌtɜːdʒ/ *n.* **1** a specialist in theatrical production. **2** a dramatist. [F f. Gk *dramatourgos* (as DRAMA, *-ergos* worker)]

dramaturgy /ˈdræməˌtɜːdʒɪ/ *n.* **1** the art of theatrical production; the theory of dramatics. **2** the application of this. □ **dramaturgic** /ˌdræməˈtɜːdʒɪk/ *adj.* **dramaturgical** *adj.*

Drambuie /dræmˈbjuːɪ, -ˈbuːɪ/ *n. propr.* a Scotch whisky liqueur. [Gael. *dram buidheach* satisfying drink]

Drammen /ˈdrɑːmən/ a seaport in SE Norway, on an inlet of Oslofjord; pop. (1991) 51,900.

drank *past* of DRINK.

drape /dreɪp/ *v. & n.* ● *v.tr.* **1** hang, cover loosely, or adorn with cloth etc. **2** arrange (clothes or hangings) carefully in folds. ● *n.* **1** (often in *pl.*) a curtain or drapery. **2** a piece of drapery. **3** the way in which a garment or fabric hangs. [ME f. OF *draper* f. *drap* f. LL *drappus* cloth]

draper /ˈdreɪpə(r)/ *n. Brit.* a retailer of textile fabrics. [ME f. AF, OF *drapier* (as DRAPE)]

drapery /ˈdreɪpərɪ/ *n.* (*pl.* **-ies**) **1** clothing or hangings arranged in folds. **2** (often in *pl.*) a curtain or hanging. **3** *Brit.* cloth; textile fabrics. **4** *Brit.* the trade of a draper. **5** the arrangement of clothing in sculpture or painting. [ME f. OF *draperie* f. *drap* cloth]

drastic /ˈdræstɪk, ˈdrɑːs-/ *adj.* having a strong or far-reaching effect; severe. □ **drastically** *adv.* [Gk *drastikos* f. *draō* do]

drat /dræt/ *v. & int. colloq.* ● *v.tr.* (**dratted, dratting** (usu. as an exclamation)) curse, confound (*drat the thing!*). ● *int.* expressing anger or annoyance. □ **dratted** *adj.* [for *'od* (= God) *rot*]

draught /drɑːft/ *n. & v.* (*US* **draft**) ● *n.* **1** a current of air in a confined space (e.g. a room or chimney). **2** pulling, traction. **3** *Naut.* the depth of water needed to float a ship. **4** the drawing of liquor from a cask etc. **5 a** a single act of drinking. **b** the amount drunk in this. **c** a dose of liquid medicine. **6** (in *pl.*; usu. treated as *sing.*) *Brit.* a game for two played with twelve pieces each on a draughtboard. (See DRAUGHTS.) **7 a** the

drawing in of a fishing-net. **b** the fish taken at one drawing. **8** = DRAFT. ● *v.tr.* = DRAFT. □ **draught beer** beer drawn from a cask, not bottled. **draught-horse** a horse used for pulling heavy loads, esp. a cart or plough. **feel the draught** *colloq.* suffer from adverse (usu. financial) conditions. [ME *draht*, perh. f. ON *drahtr, dráttr* f. Gmc, rel. to DRAW]

draughtboard /ˈdrɑːftbɔːd/ *n.* a chequered board, identical to a chessboard, used in draughts.

draughts /drɑːfts/ *n.* a simple board game for two people, played on a chequered board divided into sixty-four squares. Each player starts with twelve disc-shaped pieces in three rows on one side of the board, moving them diagonally with the aim of capturing all of the opponent's pieces. Games similar to draughts were played in ancient Egypt; the game appears to have reached its modern form at the beginning of the 17th century.

draughtsman /ˈdrɑːftsmən/ *n.* (*pl.* **-men**) **1** a person who makes drawings, plans, or sketches. **2** /ˈdrɑːftsmæn/ a piece in the game of draughts. **3** = DRAFTSMAN. □ **draughtsmanship** *n.* [*draught's* + MAN]

draughty /ˈdrɑːftɪ/ *adj.* (*US* **drafty**) (**-ier, -iest**) (of a room etc.) letting in sharp currents of air. □ **draughtily** *adv.* **draughtiness** *n.*

Dravidian /drəˈvɪdɪən/ *n. & adj.* ● *n.* **1** a member of a dark-skinned aboriginal people of southern India and Sri Lanka (including the Tamils and Kanarese). **2** the family of languages spoken by this people, including Tamil, Telugu, Malayalam, and Kannada. (*See note below.*) ● *adj.* of or relating to this people or this family of languages. [Skr. *drāviḍa* relating to the Tamils f. *Draviḍa* Tamil]

▪ It is generally agreed that Dravidian languages were once spoken throughout the Indian subcontinent and that they were pushed south following the arrival of speakers of Indic languages *c.*1000 BC. Today approximately seventy Dravidian languages are still in use, with over 160 million speakers. Generally they are agglutinative, with many clitics, and the verb is normally at the end of its clause.

draw /drɔː/ *v. & n.* ● *v.* (*past* **drew** /druː/; *past part.* **drawn**) **1** *tr.* pull or cause to move towards or after one. **2** *tr.* pull (a thing) up, over, or across. **3** *tr.* pull (curtains etc.) open or shut. **4** *tr.* take (a person) aside, esp. to talk to. **5** *tr.* attract; bring to oneself or to something; take in (*drew a deep breath; I felt drawn to her; drew my attention to the matter; draw him into conversation; the match drew large crowds*). **6** *intr.* (foll. by *at, on*) suck smoke from (a cigarette, pipe, etc.). **7** *tr.* **a** (also *absol.*) take out; remove (e.g. a tooth, a gun from a holster, etc.). **b** select by taking out (e.g. a card from a pack). **8** *tr.* obtain or take from a source (*draw a salary; draw inspiration; drew £100 from my account*). **9** *tr.* trace (a line, mark, furrow, or figure). **10 a** *tr.* produce (a picture) by tracing lines and marks. **b** *tr.* represent (a thing) by this means. **c** *absol.* make a drawing. **11** *tr.* (also *absol.*) finish (a contest or game) with neither side winning. **12** *intr.* make one's or its way, proceed, move, come (*drew near the bridge; draw to a close; the second horse drew level; drew ahead of the field; the time draws near*). **13** *tr.* infer, deduce (a conclusion). **14 a** elicit, evoke (*draw criticism; draw ruin upon oneself*). **b** bring about, entail. **c** induce (a person) to reveal facts, feelings, or talent (*refused to be drawn*). **d** (foll. by *to* + infin.) induce (a person) to do something. **e** *Cards* cause to be played (*drew all the trumps*). **15** *tr.* haul up (water) from a well. **16** *tr.* bring out (liquid from a container or blood from a wound). **17** *tr.* extract a liquid essence from. **18** *intr.* (of a chimney or pipe) promote or allow a draught. **19** *intr.* (of tea) infuse. **20 a** *tr.* distribute by lot (*drew the winner*). **b** *absol.* draw lots. **21** *intr.* (foll. by *on*) make a demand on a person, a person's skill, memory, imagination, etc. **22** *tr.* write out (a bill, cheque, or draft) (*drew a cheque on the bank*). **23** *tr.* frame (a document) in due form, compose. **24** *tr.* formulate or perceive (a comparison or distinction). **25** *tr.* (of a ship) require (a specified depth of water) to float in. **26** *tr.* disembowel (*hang, draw, and quarter; draw the fowl before cooking it*). **27** *tr. Hunting* search (cover) for game. **28** *tr.* drag (a badger or fox) from a hole. **29** *tr.* **a** protract, stretch, elongate (*long-drawn agony*). **b** make (wire) by pulling a piece of metal through successively smaller holes. **30** *tr.* **a** *Golf* drive (the ball) to the left (or, of a left-handed player, the right) esp. purposely. **b** *Bowls* cause (a bowl) to travel in a curve to the desired point. **31** *intr.* (of a sail) swell tightly in the wind. ● *n.* **1** an act of drawing. **2 a** a person or thing that draws custom, attention, etc. **b** the power to attract attention. **3** the drawing of lots, esp. a raffle. **4** a drawn game. **5** a suck on a cigarette etc. **6** the act of removing a gun from its holster in order to shoot (*quick on the draw*). **7** strain, pull. **8** *US* the movable part of a drawbridge. □ **draw back** withdraw from an undertaking. **draw a bead on** see BEAD. **draw bit** = *draw rein* (see REIN). **draw a blank** see BLANK. **draw bridle** = *draw rein* (see REIN). **draw a person's fire** attract hostility, criticism, etc., away from a more important target. **draw in 1 a** (of successive days) become shorter because of the changing seasons. **b** (of a day) approach its end. **c** (of successive evenings or nights) start earlier because of the changing seasons. **2** persuade to join, entice. **3** (of a train etc.) arrive at a station. **draw in one's horns** become less assertive or ambitious; draw back. **draw the line at** set a limit (of tolerance etc.) at. **draw lots** see LOT. **draw off 1** withdraw (troops). **2** drain off (a liquid), esp. without disturbing sediment. **draw on 1** approach, come near. **2** lead to, bring about. **3** allure. **4** put (gloves, boots, etc.) on. **draw out 1** prolong. **2** elicit. **3** induce to talk. **4** (of successive days) become longer because of the changing seasons. **5** (of a train etc.) leave a station etc. **6** write out in proper form. **7** lead out, detach, or array (troops). **draw rein** see REIN. **draw-sheet** a sheet that can be taken from under a patient without remaking the bed. **draw-string** a string that can be pulled to tighten the mouth of a bag, the waist of a garment, etc. **draw stumps** Cricket take the stumps out of the ground at the close of play. **draw one's sword against** attack. **draw up 1** compose or draft (a document etc.). **2** bring or come into regular order. **3** come to a halt. **4** make (oneself) stiffly erect. **5** (foll. by with, to) gain on or overtake. **draw-well** a deep well with a rope and a bucket. **quick on the draw** quick to act or react. [OE dragan f. Gmc]

drawback /ˈdrɔːbæk/ n. **1** a thing that impairs satisfaction; a disadvantage. **2** (foll. by from) a deduction. **3** an amount of excise or import duty paid back or remitted on goods exported. □ **drawback lock** a lock with a spring bolt that can be drawn back by an inside knob.

drawbridge /ˈdrɔːbrɪdʒ/ n. a bridge, esp. over water, hinged at one end so that it may be raised to prevent passage or to allow ships etc. to pass.

drawee /drɔːˈiː/ n. the person on whom a draft or bill is drawn.

drawer n. **1** /ˈdrɔːə(r)/ a person or thing that draws, esp. a person who draws a cheque etc. **2** /drɔː(r), ˈdrɔːə(r)/ a boxlike storage compartment without a lid, sliding in and out of a frame, table, etc. (chest of drawers). **3** (in pl.) /drɔːz/ knickers, underpants. □ **drawerful** n. /ˈdrɔːfʊl, ˈdrɔːəˌfʊl/ (pl. **-fuls**).

drawing /ˈdrɔːɪŋ/ n. **1 a** the art of representing by line. **b** delineation without colour or with a single colour. **c** the art of representing with pencils, pens, crayons, etc., rather than paint. **2** a picture produced in this way. □ **back to the drawing-board** colloq. back to begin afresh (after earlier failure). **drawing-board** a board for spreading drawing-paper on. **drawing-paper** paper for drawing pictures etc. on. **drawing-pin** Brit. a flat-headed pin for fastening paper etc. (orig. drawing-paper) to a surface. **out of drawing** incorrectly depicted.

drawing-room /ˈdrɔːɪŋruːm, -ˌrʊm/ n. **1** a room for comfortable sitting or entertaining in a private house. **2** (attrib.) restrained; observing social proprieties (drawing-room conversation). **3** US a private compartment in a train. **4** hist. a levee, a formal reception esp. at court. [earlier withdrawing-room, because orig. used for women to withdraw to after dinner]

drawl /drɔːl/ v. & n. ● v. **1** intr. speak with drawn-out vowel sounds. **2** tr. utter in this way. ● n. a drawling utterance or way of speaking. □ **drawler** n. [16th c.: prob. orig. cant, f. LG, Du. dralen delay, linger]

drawn /drɔːn/ past part. of DRAW. ● adj. **1** looking strained from fear, anxiety, or pain. **2** (of butter) melted. **3** (of a position in chess etc.) that will result in a draw if both players make the best moves available. □ **drawn-work** (or **drawn-thread-work**) ornamental work on linen etc., done by drawing out threads, usu. with additional needlework.

dray[1] /dreɪ/ n. **1** a low cart without sides for heavy loads, esp. beer-barrels. **2** Austral. & NZ a two-wheeled cart. □ **dray-horse** a large, powerful horse. [ME f. OE dræge drag-net, dragan DRAW]

dray[2] var. of DREY.

drayman /ˈdreɪmən/ n. (pl. **-men**) a brewer's driver.

dread /dred/ v., n., & adj. ● v.tr. **1** (foll. by that, or to + infin.) fear greatly. **2** shrink from; look forward to with great apprehension. **3** be in great fear of. ● n. **1** great fear, apprehension, awe. **2** an object of fear or awe. ● adj. **1** dreaded. **2** archaic awe-inspiring, revered. [OE ādrǣdan, ondrǣdan]

dreadful /ˈdredfʊl/ adj. & adv. ● adj. **1** terrible; inspiring fear or awe. **2** colloq. troublesome, disagreeable; very bad. ● adv. US colloq. dreadfully, very. □ **dreadfully** adv. **dreadfulness** n.

dreadlocks /ˈdredlɒks/ n.pl. **1** a Rastafarian hairstyle in which the hair is twisted into tight braids or ringlets hanging down on all sides. **2** hair dressed in this way. □ **dreadlocked** adj.

dreadnought /ˈdrednɔːt/ n. **1** hist. a type of battleship of the early 20th century, with main armament consisting entirely of big guns of

one calibre. (See note below.) **2** archaic a fearless person. **3** archaic **a** a thick coat for stormy weather. **b** the cloth used for such coats.

▪ Dreadnoughts, of which Britain's HMS Dreadnought was the first to be completed (1906), made all earlier battleship designs obsolete and were regarded as the supreme naval weapon of the early 20th century. Their exclusively large-calibre guns outclassed the mixed-calibre armament of their predecessors and they were also faster (with a speed of 21 knots or more), at first through the adoption of turbines. All battleships built after 1906 were dreadnoughts.

dream /driːm/ n. & v. ● n. **1 a** a series of pictures or events in the mind of a sleeping person. **b** the act or time of seeing this. **c** (in full **waking dream**) a similar experience of one awake. **2** a day-dream or fantasy. **3** an ideal, aspiration, or ambition, esp. of a nation. **4** a beautiful or ideal person or thing. **5** a state of mind without proper perception of reality (goes about in a dream). ● v. (past and past part. **dreamt** /dremt/ or **dreamed**) **1** intr. experience a dream. **2** tr. imagine in or as if in a dream. **3** (usu. with neg.) **a** intr. (foll. by of) contemplate the possibility of, have any conception or intention of (would not dream of upsetting them). **b** tr. (often foll. by that + clause) think of as a possibility (never dreamt that he would come). **4** tr. (foll. by away) spend (time) unprofitably. **5** intr. be inactive or unpractical. **6** intr. fall into a reverie. □ **dreamtime** Austral. (in the mythology of some Australian Aboriginals) the alcheringa, or 'golden age', when the first ancestors were created. **dream up** imagine, invent. **like a dream** colloq. easily, effortlessly. □ **dreamful** adj. **dreamless** adj. **dreamlike** adj. [ME f. OE drēam joy, music]

dreamboat /ˈdriːmbəʊt/ n. colloq. **1** a very attractive or ideal person, esp. of the opposite sex. **2** a very desirable or ideal thing.

dreamer /ˈdriːmə(r)/ n. **1** a person who dreams. **2** a romantic or unpractical person.

dreamland /ˈdriːmlænd/ n. an ideal or imaginary land.

dreamy /ˈdriːmɪ/ adj. (**dreamier, dreamiest**) **1** given to day-dreaming; fanciful; unpractical. **2** dreamlike; vague; misty. **3** colloq. delightful; marvellous. **4** poet. full of dreams. □ **dreamily** adv. **dreaminess** n.

drear /drɪə(r)/ adj. poet. = DREARY. [abbr.]

dreary /ˈdrɪərɪ/ adj. (**drearier, dreariest**) dismal, dull, gloomy. □ **drearily** adv. **dreariness** n. [OE drēorig f. drēor gore: rel. to drēosan to drop f. Gmc]

dreck /drek/ n. esp. US sl. rubbish, trash. [Yiddish drek f. G Dreck filth, dregs]

dredge[1] /dredʒ/ v. & n. ● v. **1** tr. **a** (often foll. by up) bring up (lost or hidden material) as if with a dredge (don't dredge all that up again). **b** (often foll. by away, up, out) bring up or clear (mud etc.) from a river, harbour, etc. with a dredge. **2** tr. clean (a harbour, river, etc.) with a dredge. **3** intr. use a dredge. ● n. an apparatus used to scoop up objects or to clear mud etc. from a river or seabed. [15th-c. Sc. dreg, perh. rel. to MDu. dregghe]

dredge[2] /dredʒ/ v.tr. **1** sprinkle with flour, sugar, etc. **2** (often foll. by over) sprinkle (flour, sugar, etc.) on. [obs. dredge sweetmeat f. OF dragie, dragee, perh. f. L tragemata f. Gk tragēmata spices]

dredger[1] /ˈdredʒə(r)/ n. **1** a machine used for dredging rivers etc.; a dredge. **2** a boat containing this.

dredger[2] /ˈdredʒə(r)/ n. a container with a perforated lid used for sprinkling flour, sugar, etc.

dree /driː/ v.tr. (**drees, dreed, dreeing**) Sc. or archaic endure. □ **dree one's weird** submit to one's destiny. [OE drēogan f. Gmc]

dreg /dreg/ n. **1** (usu. in pl.) **a** a sediment; grounds, lees, etc. **b** a worthless part; refuse (the dregs of humanity). **2** a small remnant (not a dreg). □ **drain** (or **drink**) **to the dregs** consume leaving nothing (drained life to the dregs). □ **dreggy** adj. colloq. [ME prob. f. ON dreggjar]

Dreiser /ˈdraɪsə(r)/, Theodore (Herman Albert) (1871–1945), American novelist. His first novel, Sister Carrie (1900), an account of a young working girl's rise to success, caused controversy for its frank treatment of the heroine's sexuality and ambition. His later works, such as America is Worth Saving (1941), express a growing faith in socialism that replaces the pessimism of his earlier writings.

drench /drentʃ/ v. & n. ● v.tr. **1 a** wet thoroughly (was drenched by the rain). **b** saturate; soak (in liquid). **2** force (an animal) to take medicine. **3** archaic cause to drink. ● n. **1** a soaking; a downpour. **2** medicine administered to an animal. **3** archaic a medicinal or poisonous draught. [OE drencan, drenc f. Gmc: rel. to DRINK]

Drenthe /ˈdrentə/ a sparsely populated agricultural province in the NE Netherlands; capital, Assen.

Dresden[1] /'drezdən/ a city in eastern Germany, the capital of Saxony, on the River Elbe; pop. (1991) 485,130. Famous for its baroque architecture from the 18th century until the Second World War, it was almost totally destroyed by Allied bombing on the night of 13 Feb. 1945.

Dresden[2] /'drezdən/ n. **1** (in full **Dresden china**) a type of china with elaborate decoration and delicate colourings, made originally at Dresden and (after 1710) at Meissen. **2** (attrib.) delicately pretty.

dress /dres/ v. & n. ● v. **1** tr. clothe; array (dressed in rags; dressed her quickly). **b** intr. wear clothes of a specified kind or in a specified way (dresses well). **2** intr. **a** put on clothes. **b** put on formal or evening clothes, esp. for dinner. **3** tr. decorate or adorn. **4** tr. Med. **a** treat (a wound) with ointment etc. **b** apply a dressing to (a wound). **5** tr. trim, comb, brush, or smooth (the hair). **6** tr. **a** clean and prepare (poultry, a crab, etc.) for cooking or eating. **b** add a dressing to (a salad etc.). **7** tr. apply manure etc. to a field, garden, etc. **8** tr. finish the surface of (fabric, building-stone, etc.). **9** tr. groom (a horse). **10** tr. curry (leather etc.). **11** Mil. **a** tr. correct the alignment of (troops etc.). **b** intr. (of troops) come into alignment. **12** tr. make (an artificial fly) for use in fishing. ● n. **1** a one-piece woman's garment consisting of a bodice and skirt. **2** clothing, esp. a whole outfit etc. (fussy about his dress; wore the dress of a highlander). **3** formal or ceremonial costume (evening dress; morning dress). **4** an external covering; the outward form (birds in their winter dress). □ **dress circle** the first gallery in a theatre, in which evening dress was formerly required. **dress coat** a man's swallow-tailed evening coat. **dress down** colloq. **1** reprimand or scold. **2** dress informally. **dress length** a piece of material sufficient to make a dress. **dress out** attire conspicuously. **dress parade 1** Mil. a military parade in full dress uniform. **2** a display of clothes worn by models. **dress rehearsal** the final rehearsal of a play etc., wearing costume. **dress-shield** (or **-preserver**) a piece of waterproof material fastened in the armpit of a dress to protect it from sweat. **dress-shirt** a man's usu. starched white shirt worn with evening dress. **dress up 1** dress (oneself or another) elaborately for a special occasion. **2** dress in fancy dress. **3** disguise (unwelcome facts) by embellishment. [ME f. OF dresser ult. f. L directus DIRECT]

dressage /'dresɑːʒ, -sɑːdʒ/ n. the training of a horse in obedience and deportment, esp. for competition. [F f. dresser to train]

dresser[1] /'dresə(r)/ n. **1** a kitchen sideboard with shelves above for displaying plates etc. **2** N. Amer. a dressing-table or chest of drawers. [ME f. OF dresseur f. dresser prepare: cf. med.L directorium]

dresser[2] /'dresə(r)/ n. **1** a person who assists actors to dress, takes care of their costumes, etc. **2** Med. a surgeon's assistant in operations. **3** a person who dresses elegantly or in a specified way (a snappy dresser).

dressing /'dresɪŋ/ n. **1** in senses of DRESS v. **2 a** a sauce for salads, esp. a mixture of oil, vinegar, etc. (French dressing). **b** N. Amer. stuffing. **3 a** a bandage for a wound. **b** ointment etc. used to dress a wound. **4** size or stiffening used to finish fabrics. **5** compost etc. spread over land (a top dressing of peat). □ **dressing-case** a case containing toiletries etc. **dressing-down** colloq. a scolding; a severe reprimand. **dressing-gown** a loose usu. belted robe worn over nightwear or while resting. **dressing-room 1** a room for changing the clothes etc. in a theatre, sports-ground, etc. **2** a small room attached to a bedroom, containing clothes. **dressing-station** esp. Mil. a place for giving emergency treatment to wounded people. **dressing-table** a table with a mirror, drawers, etc., used while applying make-up etc.

dressmaker /'dres,meɪkə(r)/ n. a person, esp. a woman, who makes clothes professionally. □ **dressmaking** n.

dressy /'dresɪ/ adj. (**dressier**, **dressiest**) **1 a** fond of smart clothes. **b** overdressed. **c** (of clothes) stylish or elaborate. **2** over-elaborate (the design is rather dressy). □ **dressiness** n.

drew past of DRAW.

drey /dreɪ/ n. (also **dray**) a squirrel's nest. [17th c.: orig. unkn.]

Dreyfus /'dreɪfəs/, Alfred (1859–1935), French army officer, of Jewish descent. In 1894 he was falsely accused of providing military secrets to the Germans; his trial, imprisonment, and eventual rehabilitation in 1906 caused a major political crisis in France, polarizing deep-set anti-militarist and anti-Semitic trends in a society still coming to terms with defeat and revolution in 1870–1. Notable among his supporters was the novelist Émile Zola, whose J'accuse, published in 1898, accused the judges at the trial of having convicted Dreyfus at the behest of the War Office.

dribble /'drɪb(ə)l/ v. & n. ● v. **1** intr. allow saliva to flow from the mouth. **2** intr. & tr. flow or allow to flow in drops or a trickling stream. **3** tr. (also absol.) esp. Football & Hockey move (the ball) forward with slight touches of the feet, the stick, etc. ● n. **1** the act or an instance of dribbling. **2** a small trickling stream. □ **dribbler** n. **dribbly** adj. [frequent. of obs. drib, var. of DRIP]

driblet /'drɪblɪt/ n. **1 a** a small quantity. **b** a petty sum. **2** a thin stream; a dribble. [drib (see DRIBBLE) + -LET]

dribs and drabs /drɪbz, dræbz/ n.pl. colloq. small scattered amounts (did the work in dribs and drabs). [as DRIBBLE + drab redupl.]

dried past and past part. of DRY.

drier[1] compar. of DRY.

drier[2] /'draɪə(r)/ n. (also **dryer**) **1** a machine for drying the hair, laundry, etc. **2** a substance mixed with oil-paint or ink to promote drying.

driest superl. of DRY.

drift /drɪft/ n. & v. ● n. **1 a** slow movement or variation. **b** such movement caused by a slow current. **2** the intention, meaning, scope, etc. of what is said etc. (didn't understand his drift). **3** a large mass of snow, sand, etc., accumulated by the wind. **4** esp. derog. a state of inaction. **5 a** Naut. a ship's deviation from its course, due to currents. **b** Aeron. an aircraft's deviation due to side winds. **c** a projectile's deviation due to its rotation. **d** a controlled slide of a racing car etc. **6** Mining a horizontal passage following a mineral vein. **7** a large mass of esp. flowering plants (a drift of bluebells). **8** Geol. **a** material deposited by the wind, a current of water, etc. **b** (**Drift**) Pleistocene ice detritus, e.g. boulder clay. **9** hist. the movement of cattle, esp. a gathering in a forest on an appointed day to determine ownership etc. **10** a tool for enlarging or shaping a hole in metal. **11** S. Afr. a ford. ● v. **1** intr. be carried by or as if by a current of air or water. **2** intr. move or progress passively, casually, or aimlessly (drifted into teaching). **3** tr. & intr. pile or be piled by the wind into drifts. **b** tr. cover (a field, a road, etc.) with drifts. **4** tr. form or enlarge (a hole) with a drift. **5** tr. (of a current) carry. □ **drift-ice** ice driven or deposited by water. **drift-net** a large net for herrings etc., allowed to drift with the tide. □ **driftage** n. [ME f. ON & MDu., MHG trift movement of cattle: rel. to DRIVE]

drifter /'drɪftə(r)/ n. **1** an aimless or rootless person. **2** a boat used for drift-net fishing.

driftwood /'drɪftwʊd/ n. wood etc. driven or deposited by water or wind.

drill[1] /drɪl/ n. & v. ● n. **1 a** a pointed, esp. revolving, steel tool or machine used for boring cylindrical holes, sinking wells, etc. **b** a dentist's rotary tool for cutting away part of a tooth etc. **2 a** esp. Mil. instruction or training in military exercises. **b** rigorous discipline or methodical instruction, esp. when learning or performing tasks. **c** routine procedure to be followed in an emergency (fire-drill). **d** a routine or exercise (drills in irregular verb patterns). **3** colloq. a recognized procedure (I expect you know the drill). **4** a marine gastropod mollusc, Urosalpinx cinera, that bores into the shells of young oysters. ● v. **1** tr. (also absol.) **a** (of a person or a tool) make a hole with a drill through or into (wood, metal, etc.). **b** make (a hole) with a drill. **2** tr. & intr. esp. Mil. subject to or undergo discipline by drill. **3** tr. impart (knowledge etc.) by a strict method. **4** tr. sl. shoot with a gun (drilled him full of holes). □ **drill-sergeant 1** Mil. a non-commissioned officer who trains soldiers in drill. **2** a strict disciplinarian. **drilling rig** a structure above an oil well etc. containing the machinery needed to drill it. □ **driller** n. [earlier as verb, f. MDu. drillen bore, of unkn. orig.]

drill[2] /drɪl/ n. & v. ● n. **1** a machine used for making furrows, sowing, and covering seed. **2** a small furrow for sowing seed in. **3** a ridge with such furrows on top. **4** a row of plants so sown. ● v.tr. **1** sow (seed) with a drill. **2** plant (the ground) in drills. [perh. f. obs. drill rill (17th c., of unkn. orig.)]

drill[3] /drɪl/ n. a West African baboon, Mandrillus leucophaeus, related to the mandrill. [prob. of Afr. orig.: cf. MANDRILL]

drill[4] /drɪl/ n. a coarse twilled cotton or linen fabric. [earlier drilling f. G Drillich f. L trilix -licis f. tri- three + licium thread]

drily /'draɪlɪ/ adv. (also **dryly**) **1** (said) in a dry manner; humorously. **2** in a dry way or condition.

drink /drɪŋk/ v. & n. ● v. (past **drank** /dræŋk/; past part. **drunk** /drʌŋk/) **1 a** tr. swallow (a liquid). **b** tr. swallow the liquid contents of (a container). **c** intr. swallow liquid, take draughts (drank from the stream). **2** intr. take alcohol, esp. to excess (I have heard that he drinks). **3** tr. (of a plant, porous material, etc.) absorb (moisture). **4** refl. bring (oneself etc.) to a specified condition by drinking (drank himself into a stupor). **5** tr. (usu. foll. by away) spend (wages etc.) on drink (drank away the money). **6** tr. wish (a person's good health, luck, etc.) by drinking (drank his health). ● n. **1 a** a liquid for drinking (milk is a sustaining drink). **b** a draught or specified amount of this (*had a drink of milk*). **2 a** alcoholic liquor (*got the drink in for Christmas*). **b** a portion, glass, etc. of this (*have a drink*). **c** excessive indulgence in alcohol (*drink is his vice*). **3** (as **the drink**) *colloq.* the sea. □ **drink deep** take a large draught or draughts. **drink-driver** a person who drives a vehicle with an excess of alcohol in the blood. **drink-driving** the act or an instance of this. **drink in** listen to closely or eagerly (*drank in his every word*). **drinking-song** a song sung while drinking, usu. concerning drink. **drinking-up time** *Brit.* a short period legally allowed for finishing drinks bought before closing time in a public house. **drinking-water** water pure enough for drinking. **drink off** drink the whole (contents) of at once. **drink to** toast; wish success to. **drink a person under the table** remain sober longer than one's drinking companion. **drink up** drink the whole of; empty. **in drink** drunk. **strong drink** alcohol, esp. spirits. □ **drinkable** *adj.* [OE *drincan* (v.), *drinc*(a) (n.) f. Gmc]

drinker /ˈdrɪŋkə(r)/ *n.* **1 a** a person who drinks (something). **b** a person who takes alcohol, esp. to excess. **2** (in full **drinker moth**) a large brownish moth, *Euthrix potatoria*, whose caterpillar is noted for drinking dew.

drip /drɪp/ *v. & n.* ● *v.* (**dripped**, **dripping**) **1** *intr. & tr.* fall or let fall in drops. **2** *intr.* (often foll. by *with*) be so wet as to shed drops (*dripped with blood*). ● *n.* **1 a** the act or an instance of dripping (*the steady drip of rain*). **b** a drop of liquid (*a drip of paint*). **c** a sound of dripping. **2** *colloq.* a stupid, dull, or ineffective person. **3** (*Med.* **drip-feed**) the drip-by-drip intravenous administration of a solution of salt, sugar, etc. **4** *Archit.* a projection, esp. from a window-sill, keeping the rain off the walls. □ **drip-dry** *v.* (**-dries**, **-dried**) **1** *intr.* (of fabric etc.) dry crease-free when hung up to drip. **2** *tr.* leave (a garment etc.) hanging up to dry. ● *adj.* able to be drip-dried. **drip-mat** a small mat under a glass. **drip-moulding** *Archit.* a stone or other projection that deflects rain etc. from walls (also called *dripstone*). **be dripping with** be full of or covered with (*a duchess dripping with pearls*). **dripping wet** very wet. [Middle Danish *drippe* f. Gmc (cf. DROP)]

dripping /ˈdrɪpɪŋ/ *n.* **1** fat melted from roasted meat and used for cooking or as a spread. **2** (in *pl.*) water, grease, etc., dripping from anything.

drippy /ˈdrɪpɪ/ *adj.* (**drippier**, **drippiest**) **1** tending to drip. **2** *colloq.* ineffectual; stupid; sloppily sentimental. □ **drippily** *adv.* **drippiness** *n.*

dripstone /ˈdrɪpstəʊn/ *n.* **1** *Archit.* = *drip-moulding*. **2** a stone formation made by dripping water, e.g. a stalactite.

drive /draɪv/ *v. & n.* ● *v.* (*past* **drove** /drəʊv/; *past part.* **driven** /ˈdrɪv(ə)n/) **1** *tr.* (usu. foll. by *away, back, in, out,* etc.) urge in some direction, esp. forcibly (*drove back the wolves*). **2** *tr.* **a** (usu. foll. by *to* + infin., or *to* + verbal noun) compel or constrain forcibly (*was driven to complain; drove her to stealing*). **b** (often foll. by *to*) force into a specified state (*drove him mad; driven to despair*). **c** (often *refl.*) urge to overwork (*drives himself too hard*). **3 a** *tr.* (also *absol.*) operate and direct the course of (a vehicle, a locomotive, etc.) (*drove a sports car; drives well*). **b** *tr. & intr.* convey or be conveyed in a vehicle, esp. a private car (*drove them to the station; he drives to work*) (cf. RIDE). **c** *tr.* (also *absol.*) be licensed or competent to drive (a vehicle) (*does he drive?*). **d** *tr.* (also *absol.*) urge and direct the course of (an animal drawing a vehicle or plough). **4** *tr.* (of wind, water, etc.) carry along, propel, send, or cause to go in some direction. **5** *tr.* **a** (often foll. by *into*) force (a stake, nail, etc.) into place by blows (*drove the nail home*). **b** *Mining* bore (a tunnel, horizontal cavity, etc.). **6** *tr.* effect or conclude forcibly (*drove a hard bargain; drove his point home*). **7** *tr.* (of steam or other power) set or keep (machinery) going. **8** *intr.* (usu. foll. by *at*) work hard; dash, rush, or hasten. **9** *tr. Cricket & Tennis* hit (the ball) hard from a freely swung bat or racket. **10** *tr.* (often *absol.*) *Golf* strike (a ball) with a driver from the tee. **11** *tr.* chase or frighten (game, wild beasts, an enemy in warfare, etc.) from a large area to a smaller, to kill or capture; corner. **12** *tr. hist.* hold a drift in (a forest etc.) (see DRIFT *n.* 9). ● *n.* **1** an act of driving in a motor vehicle; a journey or excursion in such a vehicle (*went for a pleasant drive; lives an hour's drive from us*). **2 a** the capacity for achievement; motivation and energy (*lacks the drive needed to succeed*). **b** *Psychol.* an inner urge to attain a goal or satisfy a need (*unconscious emotional drives*). **3 a** a usu. landscaped street or road. **b** a usu. private road through a garden or park to a house. **4** *Cricket, Golf, & Tennis* a driving stroke of the bat etc. **5** an organized effort to achieve a usu. charitable purpose (*a famine-relief drive*). **6 a** the transmission of power to machinery, the wheels of a motor vehicle, etc. (*belt drive; front-wheel drive*). **b** the position of a steering-wheel in a motor vehicle (*left-hand drive*). **c** *Computing* = disk drive (see DISC). **7** *Brit.* an organized competition, for many players, of whist, bingo, etc. **8** an act of driving game or an enemy. **9** *Austral. & NZ* a line of partly cut trees

on a hillside felled when the top one topples on the others. □ **drive at** seek, intend, or mean (*what is he driving at?*). **drive-by** (of a crime etc.) carried out from a moving vehicle. **drive-in** *attrib.adj.* (of a bank, cinema, etc.) able to be used while sitting in one's car. ● *n.* such a bank, cinema, etc. **drive-on** (or **drive-on/drive-off**) (*attrib.*) (of a ship) on to and from which motor vehicles may be driven. **drive out** take the place of; oust; exorcize, cast out (evil spirits etc.). **driving-licence** a licence permitting a person to drive a motor vehicle. **driving rain** an excessive windblown downpour. **driving-range** *Golf* an area for practising drives. **driving test** an official test of a motorist's competence which must be passed to obtain a driving licence. **driving-wheel 1** the large wheel of a locomotive. **2** a wheel transmitting motive power in machinery. **let drive** aim a blow or missile. □ **drivable** *adj.* [OE *drīfan* f. Gmc]

drivel /ˈdrɪv(ə)l/ *n. & v.* ● *n.* silly nonsense; twaddle. ● *v.* (**drivelled**, **drivelling**; *US* **driveled**, **driveling**) **1** *intr.* run at the mouth or nose; dribble. **2** *intr.* talk childishly or idiotically. **3** *tr.* (foll. by *away*) fritter; squander away. □ **driveller** *n.* (*US* **driveler**). [OE *dreflian* (v.)]

driven *past part.* of DRIVE.

driver /ˈdraɪvə(r)/ *n.* **1** (often in *comb.*) a person who drives a vehicle (*bus-driver; engine-driver*). **2** *Golf* a club with a flat face and wooden head, used for driving from the tee. **3** *Electr.* a device or part of a circuit providing power for output. **4** *Mech.* a wheel etc. receiving power directly and transmitting motion to other parts. **5** *Computing* a program that controls the operation of a device. □ **in the driver's seat** in charge. □ **driverless** *adj.*

driveway /ˈdraɪvweɪ/ *n.* = DRIVE *n.* 3b.

drizzle /ˈdrɪz(ə)l/ *n. & v.* ● *n.* **1** very fine rain. **2** esp. *Cookery* fine drops; a fine trickle. ● *v.intr.* (esp. of rain) fall in very fine drops (*it's drizzling again*). □ **drizzly** *adj.* [prob. f. ME *drēse*, OE *drēosan* fall]

Drogheda /ˈdrɒɪdə/ a port in the NE Republic of Ireland; pop. (1991) 23,000. The Battle of the Boyne was fought near there in 1690.

drogue /drəʊg/ *n.* **1** *Naut.* **a** a buoy at the end of a harpoon line. **b** a sea anchor. **2** *Aeron.* a truncated cone of fabric used as a brake, a target for gunnery, a wind-sock, etc. [18th c.: orig. unkn.]

droit /drɔɪt/ *n. Law* a right or due. [ME f. OF f. L *directum* (n.) f. *directus* DIRECT]

droit de seigneur /ˌdrwʌ də seɪˈnjɜː(r)/ *n. hist.* the alleged right of a feudal lord to have sexual intercourse with a vassal's bride on her wedding night. [F, = lord's right]

droll /drəʊl/ *adj. & n.* ● *adj.* **1** quaintly amusing. **2** strange; odd; surprising. ● *n. archaic* **1** a jester; an entertainer. **2** a quaintly amusing person. □ **drollery** *n.* (*pl.* **-ies**). **drolly** /ˈdrəʊllɪ/ *adv.* **drollness** *n.* [F *drôle*, perh. f. MDu. *drolle* little man]

drome /drəʊm/ *n. colloq. archaic* aerodrome. [abbr.]

-drome /drəʊm/ *comb. form* forming nouns denoting: **1** a place for running, racing, or other forms of movement (*aerodrome; hippodrome*). **2** a thing that runs or proceeds in a certain way (*palindrome; syndrome*). [Gk *dromos* course, running]

dromedary /ˈdrɒmɪdərɪ, ˈdrʌm-/ *n.* (*pl.* **-ies**) = Arabian camel. [ME f. OF *dromedaire* or LL *dromedarius* ult. f. Gk *dromas -ados* runner]

dromond /ˈdrɒmənd, ˈdrʌm-/ *n. hist.* a large medieval ship used for war or commerce. [ME f. OF *dromon*(t) f. LL *dromo -onis* f. late Gk *dromōn* light vessel]

drone /drəʊn/ *n. & v.* ● *n.* **1** a non-working male of the honey bee, whose sole function is to mate with fertile females. **2** an idler. **3** a deep humming sound. **4** a monotonous speech or speaker. **5** *Mus.* **a** a pipe, esp. of a bagpipe, sounding a continuous note of fixed low pitch. **b** the note emitted by this. **6** a remote-controlled pilotless aircraft or missile. ● *v.* **1** *intr.* make a deep humming sound. **2** *intr. & tr.* speak or utter monotonously. **3 a** *intr.* be idle. **b** *tr.* (often foll. by *away*) idle away (one's time etc.). [OE *drān, drǣn* prob. f. WG]

drongo /ˈdrɒŋgəʊ/ *n.* (*pl.* **-os** or **-oes**) **1** a black bird of the family Dicruridae, native to India, Africa, and Australia, having a long forked tail. **2** *Austral. & NZ sl. derog.* a simpleton. [Malagasy]

droob /druːb/ *n. Austral. sl.* a hopeless-looking ineffectual person. [perh. f. DROOP]

drool /druːl/ *v. & n.* ● *v.intr.* **1** drivel; slobber. **2** (often foll. by *over*) show much pleasure or infatuation. ● *n.* slobbering; drivelling. [contr. of drivel]

droop /druːp/ *v. & n.* ● *v.* **1** *intr. & tr.* hang or allow to hang down; languish, decline, or sag, esp. from weariness. **2** *intr.* **a** (of the eyes) look downwards. **b** *poet.* (of the sun) sink. **3** *intr.* lose heart; be dejected; flag.

● *n.* **1** a drooping attitude. **2** a loss of spirit or enthusiasm. □ **droop-snoot** *colloq.* ● *attrib.adj.* (of an aircraft) having an adjustable nose or leading-edge flap. ● *n.* **1** such an aircraft. **2** the nose or leading-edge flap of such an aircraft. [ME f. ON *drúpa* hang the head f. Gmc: cf. DROP]

droopy /ˈdruːpɪ/ *adj.* (**droopier, droopiest**) **1** drooping. **2** dejected, gloomy. □ **droopily** *adv.* **droopiness** *n.*

drop /drɒp/ *n. & v.* ● *n.* **1 a** a small round or pear-shaped portion of liquid that hangs or falls or adheres to a surface (*drops of dew; tears fell in large drops*). **b** a very small amount of usu. drinkable liquid (*just a drop left in the glass*). **c** a glass etc. of alcoholic liquor (*take a drop with us*). **2 a** an abrupt fall or slope. **b** the amount of this (*a drop of fifteen feet*). **c** an act of falling or dropping (*had a nasty drop*). **d** a reduction in prices, temperature, etc. **e** a deterioration or worsening (*a drop in status*). **3** something resembling a drop of liquid, esp.: **a** a pendant or earring. **b** a crystal ornament on a chandelier etc. **c** (often in *comb.*) a sweet or lozenge (*pear-drop; cough drop*). **4** something that drops or is dropped, esp.: **a** *Theatr.* a painted curtain or scenery let down on to the stage. **b** a platform or trapdoor on a gallows, the opening of which causes the victim to fall. **5** *Med.* **a** the smallest separable quantity of a liquid. **b** (in *pl.*) liquid medicine to be measured in drops (*eye drops*). **6** a minute quantity (*not a drop of pity*). **7** *sl.* **a** a hiding-place for stolen or illicit goods. **b** a secret place where documents etc. may be left or passed on in espionage. **8** *sl.* a bribe. **9** *US* a letter box. ● *v.* (**dropped, dropping**) **1** *intr. & tr.* fall or let fall in drops (*tears dropped on to the book; dropped the soup down his shirt*). **2** *intr. & tr.* fall or allow to fall; relinquish; let go (*dropped the box; the egg dropped from my hand*). **3 a** *intr.* sink or cause to sink or fall to the ground from exhaustion, a blow, a wound, etc. **b** *intr.* die. **4 a** *intr. & tr.* cease or cause to cease; lapse or let lapse; abandon (*the connection dropped; dropped the friendship; drop everything and come at once*). **b** *tr. colloq.* cease to associate with. **5** *tr.* set down (a passenger etc.) (*drop me at the station*). **6** *tr. & intr.* utter or be uttered casually (*dropped a hint; the remark dropped into the conversation*). **7** *tr.* send casually (*drop me a postcard*). **8 a** *intr. & tr.* fall or allow to fall in direction, amount, condition, degree, pitch, etc. (*his voice dropped; the wind dropped; we dropped the price by £20; the road dropped southwards*). **b** *intr.* (of a person) jump down lightly; let oneself fall. **c** *tr.* remove (clothes, esp. trousers) rapidly, allowing them to fall to the ground. **9** *tr. colloq.* lose (money, esp. in gambling). **10** *tr.* omit (*drop this article*). **b** omit (a letter, esp. aitch, a syllable etc.) in speech. **11** *tr.* (as **dropped** *adj.*) in a lower position than usual (*dropped handlebars; dropped waist*). **12** *tr.* give birth to (esp. a lamb, a kitten, etc.). **13** *a intr.* (of a card) be played in the same trick as a higher card. **b** *tr.* play or cause (a card) to be played in this way. **14** *tr. Sport* lose (a game, a point, a contest, a match, etc.). **15** *tr. Aeron.* deliver (supplies etc.) by parachute. **16** *tr. Rugby* **a** send (a ball) by a drop-kick. **b** score (a goal) by a drop-kick. **17** *tr. colloq.* dismiss or exclude (*was dropped from the team*). □ **at the drop of a hat** given the slightest excuse. **drop anchor** anchor ship. **drop asleep** fall gently asleep. **drop away** decrease or depart gradually. **drop back** (or **behind** or **to the rear**) fall back; get left behind. **drop back into** return to (a habit etc.). **drop a brick** *colloq.* make an indiscreet or embarrassing remark. **drop-curtain** (or **-scene**) *Theatr.* a painted curtain or scenery (cf. sense 4 of *n.*). **drop a curtsy** make a curtsy. **drop dead!** *sl.* an exclamation of intense scorn. **drop down** descend a hill etc. **drop-forging** a method of forcing white-hot metal through an open-ended die by a heavy weight. **drop-hammer** a heavy weight raised mechanically and allowed to drop, as used in drop-forging and pile-driving. **drop-head** *Brit.* the adjustable fabric roof of a car. **drop in** (or **by**) *colloq.* call casually as a visitor. **drop-in centre** a meeting-place where people may call casually for advice, conversation, etc. **a drop in the ocean** (or **a bucket**) a very small amount, esp. compared with what is needed or expected. **drop into** *colloq.* **1** call casually at (a place). **2** fall into (a habit etc.). **drop it!** *sl.* stop that! **drop-kick** (in rugby, football, etc.) a kick made by dropping the ball and kicking it on the bounce. **drop-leaf** (of a table etc.) having a hinged flap. **drop off 1** decline gradually. **2** *colloq.* fall asleep. **3** = sense 5 of *v.* **drop on** reprimand or punish. **drop out** *colloq.* cease to participate, esp. in a race, a course of study, or in conventional society. **drop-out** *n.* **1** *colloq.* a person who has dropped out. **2** *Rugby* the restarting of a game by a drop-kick. **drop scone** *Brit.* a small thick pancake made by dropping batter into a frying pan etc. **drop-shot** *Tennis* a shot dropping abruptly over the net. **drop a stitch** let a stitch fall off the end of a knitting needle. **drop-test** *Engin. n.* a test done by dropping under standard conditions. ● *v.tr.* carry out a drop-test on. **drop to** *sl.* become aware of. **fit** (or **ready**) **to drop** extremely tired. **have the drop on** *colloq.* have the advantage over. **have had a drop too much** *colloq.* be slightly drunk. □ **droplet** *n.* [OE *dropa, drop(p)ian* ult. f. Gmc: cf. DRIP, DROOP]

dropper /ˈdrɒpə(r)/ *n.* **1** a device for administering liquid, esp. medicine, in drops. **2** *Austral., NZ, & S. Afr.* a light vertical stave in a fence.

droppings /ˈdrɒpɪŋz/ *n.pl.* **1** the dung of animals or birds. **2** something that falls or has fallen in drops, e.g. wax from candles.

dropsy /ˈdrɒpsɪ/ *n.* (*pl.* **-ies**) **1** = OEDEMA. **2** *sl.* a tip or bribe. □ **dropsical** *adj.* (in sense 1). [ME f. *idrop(e)sie* f. OF *idropesie* ult. f. L *hydropisis* f. Gk *hudrōps* dropsy (as HYDRO-)]

dropwort /ˈdrɒpwɜːt/ *n.* a rosaceous grassland plant, *Filipendula vulgaris*, with tuberous root fibres.

droshky /ˈdrɒʃkɪ/ *n.* (*pl.* **-ies**) a Russian low four-wheeled open carriage. [Russ. *drozhki* dimin. of *drogi* wagon f. *droga* shaft]

drosophila /drəˈsɒfɪlə/ *n.* a small fruit fly of the genus *Drosophila*, used extensively in genetic research because of its large chromosomes, numerous varieties, and rapid rate of reproduction. [mod.L f. Gk *drosos* dew, moisture + *philos* loving]

dross /drɒs/ *n.* **1** rubbish, refuse. **2 a** the scum separated from metals in melting. **b** foreign matter mixed with anything; impurities. □ **drossy** *adj.* [OE *drōs*: cf. MLG *drōsem*, OHG *truosana*]

Drottningholm /ˈdrɒtnɪŋˌhɒlm/ the winter palace of the Swedish royal family, on an island to the west of Stockholm. It was built in 1662 for Queen Eleonora of Sweden. [Sw., lit. 'queen's island']

drought /draʊt/ *n.* **1** the continuous absence of rain; dry weather. **2** the prolonged lack of something. **3** *archaic* a lack of moisture; thirst; dryness. □ **droughty** *adj.* [OE *drūgath* f. *drȳge* DRY]

drouth /draʊθ/ *n.* *Sc., Ir., US, & poet.* var. of DROUGHT.

Drouzhba /ˈdruːʒbə/ (also **Druzba**) a resort town on the Black Sea coast of Bulgaria.

drove[1] *past of* DRIVE.

drove[2] /drəʊv/ *n.* **1 a** a large number (of people etc.) moving together; a crowd; a multitude; a shoal. **b** (in *pl.*) *colloq.* a great number (*people arrived in droves*). **2** a herd or flock being driven or moving together. □ **drove-road** an ancient cattle track. [OE *drāf* f. *drīfan* DRIVE]

drover /ˈdrəʊvə(r)/ *n.* a person who drives herds to market; a cattle-dealer. □ **drove** *v.tr.* **droving** *n.*

drown /draʊn/ *v.* **1** *tr. & intr.* kill or be killed by submersion in liquid. **2** *tr.* submerge; flood; drench (*drowned the fields in six feet of water*). **3** *tr.* deaden (grief etc.) with drink (*drowned his sorrows*). **4** *tr.* (often foll. by *out*) make (a sound) inaudible by means of a louder sound. □ **drowned valley** a valley partly or wholly submerged by a change in land-levels. **drown out** drive out by flood. **like a drowned rat** *colloq.* extremely wet and bedraggled. [ME (orig. north.) *drun(e), droun(e)*, perh. f. OE *drūnian* (unrecorded), rel. to DRINK]

drowse /draʊz/ *v. & n.* ● *v.* **1** *intr.* be dull and sleepy or half asleep. **2** *tr.* **a** (often foll. by *away*) pass (the time) in drowsing. **b** make drowsy. **3** *intr.* *archaic* be sluggish. ● *n.* a condition of sleepiness. [back-form. f. DROWSY]

drowsy /ˈdraʊzɪ/ *adj.* (**drowsier, drowsiest**) **1** half asleep; dozing. **2** soporific; lulling. **3** sluggish. □ **drowsily** *adv.* **drowsiness** *n.* [prob. rel. to OE *drūsian* be languid or slow, *drēosan* fall: cf. DREARY]

drub /drʌb/ *v.tr.* (**drubbed, drubbing**) **1** thump; belabour. **2** beat in a fight. **3** (usu. foll. by *into, out of*) beat (an idea, attitude, etc.) into or out of a person. □ **drubbing** *n.* [ult. f. Arab. *ḍaraba* beat]

drudge /drʌdʒ/ *n. & v.* ● *n.* a servile worker, esp. at menial tasks; a hack. ● *v.intr.* (often foll. by *at*) work slavishly (at menial, hard, or dull work). □ **drudgery** /ˈdrʌdʒərɪ/ *n.* [15th c.: perh. rel. to DRAG]

drug /drʌɡ/ *n. & v.* ● *n.* **1** a medicinal substance. **2** a narcotic, hallucinogen, or stimulant, esp. one causing addiction. ● *v.* (**drugged, drugging**) **1** *tr.* add (a drug) to (food or drink). **2** *tr.* administer a drug to. **b** stupefy with a drug. **3** *intr.* take drugs as an addict. □ **drug addict** a person who is addicted to a narcotic drug. **drug peddler** (*colloq.* **pusher**) a person who sells esp. addictive drugs illegally. **drug squad** a division of a police force investigating crimes involving illegal drugs. [ME *drogges, drouges* f. OF *drogue*, of unkn. orig.]

drugget /ˈdrʌɡɪt/ *n.* **1** a coarse woven fabric used as a floor or table covering. **2** such a covering. [F *droguet*, of unkn. orig.]

druggist /ˈdrʌɡɪst/ *n.* esp. *N. Amer.* a pharmacist. [F *droguiste* (as DRUG)]

druggy /ˈdrʌɡɪ/ *n. & adj. colloq.* ● *n.* (also **druggie**) (*pl.* **-ies**) a drug addict. ● *adj.* of or associated with narcotic drugs.

drugstore /ˈdrʌɡstɔː(r)/ *n.* *N. Amer.* a chemist's shop also selling light refreshments and other articles.

Druid /ˈdruːɪd/ *n.* (also **druid**) **1** an ancient Celtic priest, magician, or soothsayer of Gaul, Britain, or Ireland. (*See note below.*) **2** a member of a Welsh Druidic order, esp. the Gorsedd. **3** a member of any of various groups that are held to be present-day representatives of ancient

Druidism. □ **Druidism** *n.* **Druidic** /druːˈɪdɪk/ *adj.* **Druidical** *adj.* [F *druide* or L pl. *druidae, -des,* Gk *druidai* f. Gaulish *druides*]

▪ Knowledge of the Druids is based chiefly on the hostile accounts of them in the writings of Julius Caesar and Tacitus. Caesar reports that they had judicial and priestly functions, and were proficient in physical science; they also worshipped in groves (clearings in the forest), cut mistletoe from the sacred oak with a golden sickle, and were believed to offer human sacrifices. The religion was stamped out by the Romans with unrelenting ferocity. Druidism of the Roman period may well have contained elements of older faiths and, although its association with Stonehenge is now generally rejected, the modern Druidical order seeks to make ceremonial use of this and other sites.

drum[1] /drʌm/ *n. & v.* ● *n.* **1 a** a percussion instrument or toy made of a hollow cylinder or hemisphere covered at one or both ends with stretched skin or parchment and sounded by striking (*bass drum; kettledrum*). (*See note below.*) **b** (often in *pl.*) a drummer or a percussion section (*the drums are playing too loud*). **c** a sound made by or resembling that of a drum. **2** something resembling a drum in shape, esp.: **a** a cylindrical container or receptacle for oil, dried fruit, etc. **b** a cylinder or barrel in machinery on which something is wound etc. **c** *Archit.* the solid part of a Corinthian or composite capital. **d** *Archit.* a stone block forming a section of a shaft. **e** *Austral. & NZ* swag, a bundle. **3** *Zool. & Anat.* the membrane of the middle ear; the eardrum. **4** *sl.* **a** a house. **b** a nightclub. **c** a brothel. **5** (in full **drum-fish**) a marine fish of the family Sciaenidae, having a swim-bladder that produces a drumming sound. **6** *hist.* an evening or afternoon tea party. **7** *Austral. sl.* a piece of reliable information, esp. a racing tip. ● *v.* (**drummed, drumming**) **1** *intr. & tr.* play on a drum. **2** *tr. & intr.* beat, tap, or thump (knuckles, feet, etc.) continuously (on something) (*drummed on the table; drummed his feet; drumming at the window*). **3** *intr.* (of a bird or an insect) make a loud, hollow noise with quivering wings. **4** *tr. Austral. sl.* provide with reliable information. □ **drum brake** a brake in which shoes on a vehicle press against the drum on a wheel. **drum into** drive (a lesson) into (a person) by persistence. **drum machine** an electronic device that imitates the sound of a drum-kit. **drum major 1** the leader of a marching band. **2** *archaic* an NCO commanding the drummers of a regiment. **drum majorette** esp. *US* a member of a female baton-twirling parading group. **drum out** *Mil.* cashier (a soldier) by the beat of a drum; dismiss with ignominy. **drum up** summon, gather, or call up (*needs to drum up more support*). [obs. *drombslade, drombyllsclad,* f. LG *trommelslag* drum-beat f. *trommel* drum + *slag* beat]

▪ Drums have been known since neolithic times, being played, often with the hands, in religious, social, and military contexts. The first use in the orchestra appears to have been in the mid-17th century; orchestral drums are largely timpani, played with a variety of beaters. (See also DRUM-KIT.)

drum[2] /drʌm/ *n. Sc. & Ir.* a long narrow hill often separating two parallel valleys; a drumlin. [Gael. & Ir. *druim* back, ridge]

drumbeat /ˈdrʌmbiːt/ *n.* a stroke or the sound of a stroke on a drum.

drumfire /ˈdrʌmˌfaɪə(r)/ *n.* **1** *Mil.* heavy continuous rapid artillery fire, usu. heralding an infantry attack. **2** a barrage of criticism etc.

drumhead /ˈdrʌmhed/ *n.* **1** the skin or membrane of a drum. **2** an eardrum. **3** the circular top of a capstan. **4** (*attrib.*) improvised (*drumhead court martial*).

drum-kit /ˈdrʌmkɪt/ *n.* a set of drums, cymbals, and other percussion instruments used with drumsticks in jazz or popular music. Its basic constituents are a foot-operated bass drum, a snare drum, and one or more tom-toms.

drumlin /ˈdrʌmlɪn/ *n.* a low oval mound of compacted boulder clay moulded by past glacial action. [app. f. DRUM[2] + *-lin* repr. *-LING*[1]]

drummer /ˈdrʌmə(r)/ *n.* **1** a person who plays a drum or drums. **2** esp. *US colloq.* a commercial traveller. **3** *sl.* a thief.

drumstick /ˈdrʌmstɪk/ *n.* **1** a stick used for beating a drum. **2** the lower joint of the leg of a cooked chicken, turkey, etc.

drunk /drʌŋk/ *adj. & n.* ● *adj.* **1** rendered incapable by alcohol (*blind drunk; dead drunk; drunk as a lord*). **2** (often foll. by *with*) overcome or elated with joy, success, power, etc. ● *n.* **1** a habitually drunk person. **2** *sl.* a drinking-bout; a period of drunkenness. [past part. of DRINK]

drunkard /ˈdrʌŋkəd/ *n.* a person who is drunk, esp. habitually.

drunken /ˈdrʌŋkən/ *adj.* (usu. *attrib.*) **1** = DRUNK. **2** caused by or exhibiting drunkenness (*a drunken brawl*). **3** fond of drinking; often drunk. □ **drunkenly** *adv.* **drunkenness** *n.*

drupe /druːp/ *n.* any fleshy or pulpy fruit enclosing a stone containing one or a few seeds, e.g. an olive, plum, or peach. □ **drupaceous** /druːˈpeɪʃəs/ *adj.* [L *drupa* f. Gk *druppa* olive]

drupel /ˈdruːp(ə)l/ *n.* (also **drupelet** /-plɪt/) a small drupe usu. in an aggregate fruit, e.g. a blackberry or raspberry.

Drury Lane /ˈdruəri/ the Theatre Royal in Drury Lane, one of London's most famous theatres. The first theatre on the site opened in 1663 (see *patent theatre*), the second (with which David Garrick, Richard Sheridan, and Sir Richard Steele were all associated) in 1674, and the third in 1794. The present theatre, dating from 1812, was not particularly successful until the 1880s, when it became famous for its melodramas and spectacles. Since the 1920s it has staged musicals, including *The Desert Song* in 1927, *Oklahoma!* in 1947, and *My Fair Lady* in 1958.

druse /druːz/ *n. Geol.* **1** a crust of crystals lining a rock-cavity. **2** a cavity lined with this. [F f. G, = weathered ore]

Druzba see DROUZHBA.

Druze /druːz/ *n.* (also **Druse**) a member of a political and religious sect, originally an offshoot of the Ismaili Shiite Muslims, living chiefly in Lebanon, with smaller groups in Syria and Israel. The Druzes broke away from Islam in the 11th century over a disagreement about leadership and followed the sixth caliph of the Fatimid dynasty, al-Hakim b'illah (996–1021). They regard al-Hakim as a deity, and thus are considered heretics by the Muslim community at large. In recent years their militia have been involved in various Arab conflicts, esp. in Lebanon. [F f. Arab. *durūz* (pl.), prob. f. one of their founders al-Darazī (d.1019)]

Dr Watson see WATSON, DR.

dry /draɪ/ *adj., v., & n.* ● *adj.* (**drier** /ˈdraɪə(r)/; **driest** /ˈdraɪɪst/) **1** free from moisture, not wet, esp.: **a** with any moisture having evaporated, drained, or been wiped away (*the clothes are not dry yet*). **b** (of the eyes) free from tears. **c** (of a climate etc.) with insufficient rainfall; not rainy (*a dry spell*); (of land etc.) receiving little rain. **d** (of a river, well, etc.) dried up; not yielding water. **e** (of a liquid) having disappeared by evaporation etc. **f** not connected with or for use without moisture (*dry shampoo*). **g** (of a shave) with an electric razor. **2** (of wine etc.) not sweet (*dry sherry*). **3 a** meagre, plain, or bare (*dry facts*). **b** uninteresting; dull (*dry as dust*). **4** (of a sense of humour, a joke, etc.) subtle, ironic, and quietly expressed; not obvious. **5 a** (of a country, of legislation, etc.) prohibiting the sale of alcoholic drink. **b** (of a person) abstaining from alcohol or drugs. **6** (of toast, bread, etc.) without butter, margarine, etc. **7** (of provisions, groceries, etc.) solid, not liquid (*dry goods*). **8** impassive, unsympathetic; hard; cold. **9** (of a cow etc.) not yielding milk. **10** *colloq.* thirsty or thirst-making (*feel dry; this is dry work*). **11** *Brit. colloq.* (of a Conservative politician) uncompromisingly opposed to high government spending (see DRY *n.* 3). ● *v.* (**dries, dried**) **1** *tr. & intr.* make or become dry by wiping, evaporation, draining, etc. **2** *tr.* (usu. as **dried** *adj.*) preserve (food etc.) by removing the moisture (*dried egg; dried fruit; dried flowers*). **3** *intr.* (often foll. by *up*) *Theatr. colloq.* forget one's lines. **4** *tr. & intr.* (often foll. by *off*) cease or cause (a cow etc.) to cease yielding milk. ● *n.* (pl. **dries**) **1** the process or an instance of drying. **2** (prec. by *the*) a dry place (*come into the dry*). **3** *Brit. colloq.* a Conservative politician who advocates individual responsibility, free trade, and economic stringency, and opposes high government spending. **4 a** (prec. by *the*) esp. *Austral. colloq.* the dry season. **b** *Austral.* a desert area, waterless country. **5 a** dry ginger ale. **b** dry wine, sherry, etc. □ **dry battery** *Electr.* an electric battery consisting of dry cells. **dry cell** *Electr.* a cell in which the electrolyte is absorbed in a solid and cannot be spilled. **dry-clean** clean (clothes etc.) with organic solvents without using water. **dry-cleaner** a firm that specializes in dry-cleaning. **dry cough** a cough not producing phlegm. **dry-cure** cure (meat etc.) without pickling in liquid. **dry dock** an enclosure for the building or repairing of ships, from which water can be pumped out. **dry-fly** *adj.* (of fishing) with an artificial fly floating on the surface. ● *v.intr.* (**-flies, -flied**) fish by such a method. **dry ice** solid carbon dioxide. **dry land 1** land as opposed to the sea, a river, etc. **2** (**dryland**) (usu. in *pl.*) esp. *N. Amer.* an area or land where rainfall is low (also *attrib.: dryland farming*). **dry measure** a measure of capacity for dry goods. **dry milk** *US* dried milk. **dry-nurse** a nurse for young children, not required to breast-feed. **dry out 1** become fully dry. **2** (of a drug addict, alcoholic, etc.) undergo treatment to cure addiction. **dry painting** see SAND PAINTING. **dry-plate** *Photog.* a photographic plate with sensitized film hard and dry for convenience of keeping, developing at leisure, etc. **dry-point 1** a needle for engraving on a bare copper plate without acid. **2** an engraving produced with this. **dry rot 1** a type of decay in wood in poorly ventilated conditions, caused by certain fungi. **2** these fungi. **dry run** *colloq.* a rehearsal. **dry-salt** = dry-cure. **dry-salter** *hist.*

a dealer in dyes, gums, drugs, oils, pickles, tinned meats, etc. **dry-shod** without wetting the shoes. **dry up 1** make utterly dry. **2** dry dishes. **3** (of moisture) disappear utterly. **4** (of a well etc.) cease to yield water. **5** (esp. in *imper.*) *colloq.* cease talking. **go dry** enact legislation for the prohibition of alcohol. □ **dryish** *adj.* **dryness** *n.* [OE *drȳge, drygan,* rel. to MLG *dröge,* MDu. *dröghe,* f. Gmc]

dryad /ˈdraɪæd, ˈdraɪəd/ *n. Mythol.* a nymph inhabiting a tree; a wood nymph. [ME f. OF *dryade* f. L f. Gk *druas -ados* f. *drus* tree]

Dryden /ˈdraɪd(ə)n/, John (1631–1700), English poet, dramatist, and critic. One of the principal exponents of Augustan literature in England, he is remembered for his codification of verse metres and the establishment of the heroic couplet as the favoured verse form. He wrote many plays, including comedies (*Marriage à la mode;* 1673), tragedies such as the blank verse drama *All for Love* (1678), and satires, of which the best known is *Absalom and Achitophel* (1681). His prose writing style is often considered the model for modern English literature. Many of his critical works appear as prefaces to his plays; they include *Of Dramatic Poesie* (1668).

dryer var. of DRIER[2].

dryly var. of DRILY.

Dryopithecus /ˌdraɪəˈpɪθɪkəs/ *n.* a genus of fossil anthropoid apes that existed in Europe, Asia, and Africa during the middle Miocene to early Pliocene period (*c.*12–8 million years ago). It is believed that the line leading to humans diverged from these through an as yet unspecified species of *Dryopithecus.* □ **dryopithecine** /-ˈpɪθɪˌsiːn, -ˌsaɪn/ *adj.* [mod.L f. Gk *drus* tree + *pithēkos* ape]

Drysdale /ˈdraɪzdeɪl/, Sir Russell (1912–81), British-born Australian painter. His subject-matter is the harsh life of the Australian bush, as in *The Rabbiter and Family* (1938); he also deals with the plight of Aboriginals in contact with white settlement, as in *Mullaloonah Tank* (1953).

drystone /ˈdraɪstəʊn/ *adj.* (of a wall etc.) built without mortar.

DS *abbr.* **1** dal segno. **2** disseminated sclerosis.

DSC *abbr.* Distinguished Service Cross.

D.Sc. *abbr.* Doctor of Science.

DSM *abbr.* Distinguished Service Medal.

DSO see DISTINGUISHED SERVICE ORDER.

DSS *abbr.* (in the UK) Department of Social Security (formerly DHSS).

DT *abbr.* (also **DT's** /ˌdiːˈtiːz/) delirium tremens.

DTI *abbr.* (in the UK) Department of Trade and Industry.

DTP *abbr.* desktop publishing.

dual /ˈdjuːəl/ *adj., n.,* & *v.* ● *adj.* **1** of two; twofold. **2** divided in two; double (*dual ownership*). **3** *Gram.* (in some languages) denoting two persons or things (additional to singular and plural). ● *n.* (also **dual number**) *Gram.* a dual form of a noun, verb, etc. ● *v.tr.* (**dualled, dualling**) *Brit.* convert (a road) into a dual carriageway. □ **dual carriageway** *Brit.* a road with a dividing strip between the traffic in opposite directions. **dual control** (of a vehicle or an aircraft) having two sets of controls, one of which is used by the instructor. **dual in-line package** *Computing* see DIP. **dual-purpose** (of a vehicle) usable for passengers or goods. □ **dualize** *v.tr.* (also **-ise**). **dually** *adv.* **duality** /djuːˈælɪtɪ/ *n.* [L *dualis* f. *duo* two]

dualism /ˈdjuːəˌlɪz(ə)m/ *n.* **1** being twofold; duality. **2** *Philos.* the theory that in any domain of reality there are two independent underlying principles, e.g. mind and matter, form and content (opp. MONISM 1). **3** *Theol.* **a** the theory that the forces of good and evil are equally balanced in the universe. Dualism was a central doctrine in the religious system of Manichaeism, which influenced medieval heretics like the Cathars. **b** the theory that Christ had both a human and a divine personality. □ **dualist** *n.* **dualistic** /ˌdjuːəˈlɪstɪk/ *adj.* **dualistically** *adv.*

dub[1] /dʌb/ *v.tr.* (**dubbed, dubbing**) **1** make (a person) a knight by touching the shoulder with a sword. **2** give (a person) a name, nickname, or title (*dubbed him a crank*). **3** *Brit.* dress (an artificial fishing-fly). **4** smear (leather) with grease. [OE f. AF *duber, aduber,* OF *adober* equip with armour, repair, of unkn. orig.]

dub[2] /dʌb/ *v.tr.* (**dubbed, dubbing**) **1** provide (a film etc.) with an alternative soundtrack, esp. in a different language. **2** add (sound effects or music) to a film or a broadcast. **3** combine (soundtracks) into one. **4** transfer or make a copy of (recorded sound or images). [abbr. of DOUBLE]

dub[3] /dʌb/ *n.* esp. *US sl.* an inexperienced or unskilful person. [perh. f. DUB[1] in sense 'beat flat']

dub[4] /dʌb/ *v.intr.* (**dubbed, dubbing**) *sl.* (foll. by *in, up*) pay up; contribute money. [19th c.: orig. uncert.]

Dubai /duːˈbaɪ/ (also **Dubayy**) **1** a member state of the United Arab Emirates; pop. (1985) 419,100. It is the second largest of the seven states in area and population. **2** its capital city, a port on the Persian Gulf; pop. (1980) 265,700.

dubbin /ˈdʌbɪn/ *n. & v.* ● *n.* (also **dubbing** /ˈdʌbɪŋ/) prepared grease for softening and waterproofing leather. ● *v.tr.* (**dubbined, dubbining**) apply dubbin to (boots etc.). [see DUB[1] 4]

dubbing /ˈdʌbɪŋ/ *n.* an alternative soundtrack to a film etc.

Dubček /ˈdʊbtʃek/, Alexander (1921–92), Czechoslovak statesman, First Secretary of the Czechoslovak Communist Party 1968–9. He is generally regarded as the driving force behind the attempted democratization of Czech political life in 1968 that became known as the Prague Spring. At this time he and other liberal members of the government made plans for a new constitution as well as legislation for civil liberties and began to pursue a foreign policy independent of the Soviet Union. In response, Warsaw Pact forces invaded Czechoslovakia in August 1968 and Dubček was removed from office the following year. After the abandonment of Communism at the end of 1989 he returned to public life and was elected speaker of the Federal Assembly in a new democratic regime.

dubiety /djuːˈbaɪətɪ/ *n.* (*pl.* **-ies**) *literary* **1** a feeling of doubt. **2** a doubtful matter. [LL *dubietas* f. *dubium* doubt]

dubious /ˈdjuːbɪəs/ *adj.* **1** hesitating or doubting (*dubious about going*). **2** of questionable value or truth (*a dubious claim*). **3** unreliable; suspicious (*dubious company*). **4** of doubtful result. □ **dubiously** *adv.* **dubiousness** *n.* [L *dubiosus* f. *dubium* doubt]

dubitation /ˌdjuːbɪˈteɪʃ(ə)n/ *n. literary* doubt, hesitation. [ME f. OF *dubitation* or L *dubitatio* f. *dubitare* DOUBT]

dubitative /ˈdjuːbɪtətɪv/ *adj. literary* of, expressing, or inclined to doubt or hesitation. [F *dubitatif -ive* or LL *dubitativus* (as DUBITATION)]

Dublin /ˈdʌblɪn/ **1** the capital city of the Republic of Ireland, situated on the Irish Sea at the mouth of the River Liffey; pop. (1991) 477,700. It is built on the site of a Viking settlement and is noted for its 18th-century architecture. It was the birthplace of many writers, including Jonathan Swift, Edmund Burke, Richard Sheridan, Oscar Wilde, J. M. Synge, Sean O'Casey, James Joyce, and Brendan Behan. **2** a county of the Republic of Ireland, in the province of Leinster; county town, Dublin.

Dublin Bay prawn *n.* **1** the Norway lobster. **2** (in *pl.*) scampi.

Du Bois /duː ˈbɔɪs/, W(illiam) E(dward) B(urghardt) (1868–1963), American writer and political activist. While professor of economics and sociology at Atlanta University (1897–1910), he researched the social conditions of blacks in the US and wrote his influential works *The Philadelphia Negro; A Social Study* (1899) and *The Souls of Black Folk* (1903). In the latter Du Bois launched a fierce attack on Booker T. Washington's policy of appeasement towards whites, arguing that racial equality could only be achieved by political organization and struggle. In 1905 he formed the pressure group known as the Niagara Movement to campaign against the Washington lobby. Four years later Du Bois co-founded the National Association for the Advancement of Colored People, working as the editor of its magazine *Crisis* from 1910 to 1934.

Dubonnet /djuːˈbɒneɪ/ *n. propr.* **1** a sweet French aperitif. **2** a glass of this. [name of a family of French wine-merchants]

Dubrovnik /dʊˈbrɒvnɪk/ (called in Italian *Ragusa*) a port and resort on the Adriatic coast of Croatia; pop. (1981) 66,100. It was founded as Ragusium in the 7th century AD and is situated on a promontory jutting into the Adriatic, with a beautiful natural harbour.

ducal /ˈdjuːk(ə)l/ *adj.* of, like, or bearing the title of a duke. [F f. *duc* DUKE]

ducat /ˈdʌkət/ *n.* **1** *hist.* a gold coin, formerly current in most European countries. **2 a** a coin. **b** (in *pl.*) money. [ME f. It. *ducato* or med.L *ducatus* DUCHY]

Duccio /ˈduːtʃɪˌəʊ/ (full name Duccio di Buoninsegna) (*c.*1255–*c.*1320), Italian painter. The founder of the Sienese school of painting, he built on elements of the Byzantine tradition. The only fully documented surviving work by him is the *Maestà* for the high altar of Siena cathedral (completed 1311). His work conveys emotion through facial expression, sequence of colour, and arrangement of scenery, while keeping the composition within Byzantine conventions.

Duce /ˈduːtʃeɪ/ *n.* a leader, esp. (**Il Duce**) the title assumed by Mussolini (1883–1945). [It., = leader]

Duchamp /dju:'ʃɒm/, Marcel (1887–1968), French-born painter, sculptor, and art theorist. His main influence on 20th-century art has been in the area of anti-art movements: the origins of conceptual art can be traced to him, he was a leader of the Dada movement, and in 1912 he originated the ready-made art form (see also READY-MADE *n.*). His most famous provocative gesture was adding a moustache and goatee beard to a reproduction of the *Mona Lisa* in 1920, in a bid to destroy the mystique of good taste and aesthetic beauty. He became an American citizen in 1955.

Duchenne muscular dystrophy /du:'ʃen/ *n. Med.* a severe form of muscular dystrophy, usually affecting boys. [Guillaume-Benjamin-Armand *Duchenne* (1806–75), Fr. neurologist]

duchess /'dʌtʃɪs/ *n.* (as a title usu. **Duchess**) **1** a duke's wife or widow. **2** a woman holding the rank of duke in her own right. **3** *Brit. sl.* a girl or woman, esp. a man's wife or mother. [ME f. OF *duchesse* f. med.L *ducissa* (as DUKE)]

duchesse /du:'ʃes, 'dʌtʃɪs/ *n.* **1** a soft heavy kind of satin. **2** a dressing-table with a pivoting mirror. □ **duchesse lace** a kind of Brussels pillow-lace. **duchesse potatoes** mashed potatoes mixed with egg, shaped or piped into small portions, and baked. **duchesse set** a cover or a set of covers for a dressing-table. [F, = DUCHESS]

duchy /'dʌtʃɪ/ *n.* (*pl.* **-ies**) **1** the territory of a duke or duchess; a dukedom. **2** (often as **the Duchy**) the royal dukedom of Cornwall or Lancaster, each with certain estates, revenues, and jurisdiction of its own. (*See note below.*) [ME f. OF *duché(e)* f. med.L *ducatus* f. L *dux ducis* leader]

■ Since 1503 the heir apparent has succeeded to the Duchy of Cornwall by inheritance. The Duchy of Lancaster has been attached to the Crown since 1399 and its revenues are paid direct to the privy purse. It is controlled by the Chancellor of the Duchy of Lancaster, a post usually held by a member of the Cabinet.

duck[1] /dʌk/ *n.* (*pl.* same or **ducks**) **1 a** a waterbird of the family Anatidae, with a broad flat bill and large webbed feet. (*See note below.*) **b** the female of this (opp. DRAKE). **c** the flesh of a duck as food. **2** *Cricket* the score of a batsman dismissed for nought. **3** (also **ducks**) *Brit. colloq.* (esp. as a form of address) dear, darling. □ **duck-hawk 1** *dial.* a marsh harrier. **2** *N. Amer.* a peregrine. **ducks and drakes** a game of making a flat stone skim along the surface of water. **duck's arse** *sl.* a haircut with the hair on the back of the head shaped like a duck's tail. **duck soup** *US sl.* an easy task. **like a duck to water** adapting very readily. **like water off a duck's back** *colloq.* (of remonstrances etc.) producing no effect. **play ducks and drakes with** *colloq.* squander. [OE *duce, dūce*: rel. to DUCK[2]]

■ The ducks are generally smaller than geese and swans, and most kinds have shorter legs and necks. They are divided into dabbling ducks and diving ducks, according to their method of feeding. The name 'duck' is often used specifically to mean one of the domesticated varieties of the mallard or wild duck.

duck[2] /dʌk/ *v. & n.* ● *v.* **1** *intr. & tr.* plunge, dive, or dip under water and emerge (*ducked him in the pond*). **2** *intr. & tr.* bend (the head or the body) quickly to avoid a blow or being seen, or as a bow or curtsy; bob (*ducked out of sight; ducked his head under the beam*). **3** *tr. & intr. colloq.* avoid or dodge; withdraw (from) (*ducked out of the engagement; ducked the meeting*). **4** *intr. Bridge* lose a trick deliberately by playing a low card. ● *n.* **1** a quick dip or swim. **2** a quick lowering of the head etc. □ **ducking-stool** *hist.* a chair fastened to the end of a pole, which could be plunged into a pond, used formerly for punishing minor offenders by ducking. □ **ducker** *n.* [OE *dūcan* (unrecorded) f. Gmc]

duck[3] /dʌk/ *n.* **1** a strong untwilled linen or cotton fabric used for small sails and the outer clothing of sailors. **2** (in *pl.*) trousers made of this (*white ducks*). [MDu. *doek*, of unkn. orig.]

duck[4] /dʌk/ *n. colloq.* an amphibious landing-craft. [DUKW, its official designation]

duckbill /'dʌkbɪl/ *n.* (also **duck-billed platypus**) = PLATYPUS.

duckboard /'dʌkbɔːd/ *n.* (usu. in *pl.*) a path of wooden slats placed over muddy ground or in a trench.

duckling /'dʌklɪŋ/ *n.* **1** a young duck. **2** its flesh as food.

duckweed /'dʌkwiːd/ *n.* a tiny aquatic plant of the family Lemnaceae, esp. the genus *Lemna*, which frequently carpets the surface of still water.

ducky /'dʌkɪ/ *n. & adj. Brit. colloq.* ● *n.* (*pl.* **-ies**) darling, dear. ● *adj.* sweet, pretty; splendid.

duct /dʌkt/ *n. & v.* ● *n.* **1** a channel or tube for conveying fluid, cable,

etc. **2 a** a tube in the body conveying lymph or glandular secretions such as tears etc. **b** *Bot.* a tube formed by cells that have lost their intervening end walls, holding air, water, etc. ● *v.tr.* convey through a duct. [L *ductus* leading, aqueduct f. *ducere duct-* lead]

ductile /'dʌktaɪl/ *adj.* **1** (of a metal) capable of being drawn into wire; pliable, not brittle. **2** (of a substance) easily moulded. **3** (of a person) docile, gullible. □ **ductility** /dʌk'tɪlɪtɪ/ *n.* [ME f. OF *ductile* or L *ductilis* f. *ducere duct-* lead]

ducting /'dʌktɪŋ/ *n.* **1** a system of ducts. **2** material in the form of a duct or ducts.

ductless /'dʌktlɪs/ *adj.* lacking or not using a duct or ducts. □ **ductless gland** *Anat.* a gland secreting directly into the bloodstream (also called *endocrine gland*).

dud /dʌd/ *n. & adj. sl.* ● *n.* **1** a futile or ineffectual person or thing (*a dud at the job*). **2** a counterfeit article. **3** a shell etc. that fails to explode. **4** (in *pl.*) clothes. ● *adj.* **1** useless, worthless, unsatisfactory or futile. **2** counterfeit. [ME: orig. unkn.]

dude /dju:d, du:d/ *n. sl.* **1** a fastidious aesthetic person, usu. male; a dandy. **2** a holiday-maker on a ranch in the western US. **3** a fellow; a guy. □ **dude ranch** a cattle ranch converted to a holiday centre for tourists etc. □ **dudish** *adj.* [19th c.: prob. f. G dial. *dude* fool]

dudgeon /'dʌdʒən/ *n.* a feeling of offence; resentment. □ **in high dudgeon** very angry or angrily. [16th c.: orig. unkn.]

Dudley /'dʌdlɪ/, Robert, Earl of Leicester (*c*.1532–88), English nobleman. He became a favourite of Elizabeth I soon after her accession in 1558. Following the mysterious death of Dudley's wife, Amy Robsart, in 1560 it was rumoured that he would marry the queen; this did not happen, although Dudley remained in favour with Elizabeth throughout his life and in 1564 was created Earl of Leicester. He was later given the command of the military campaign in the Netherlands (1585–7) and of the forces preparing to resist the Armada (1588).

due /dju:/ *adj., n., & adv.* ● *adj.* **1** (*predic.*) owing or payable as a debt or an obligation (*our thanks are due to him; £500 was due on the 15th*). **2** (often foll. by *to*) merited; appropriate; fitting (*his due reward; received the applause due to a hero*). **3** (*attrib.*) rightful; proper; adequate (*after due consideration*). **4** (*predic.*; foll. by *to*) to be ascribed to (a cause, an agent, etc.) (*the discovery was due to Newton*). **5** (*predic.*) intended to arrive at a certain time (*a train is due at 7.30*). **6** (foll. by *to* + infin.) under an obligation or agreement to do something (*due to speak tonight*). ● *n.* **1** a person's right; what is owed to a person (*a fair hearing is my due*). **2** (in *pl.*) **a** what one owes (*pays his dues*). **b** an obligatory payment; a fee; a legal charge (*harbour dues; university dues*). ● *adv.* (of a point of the compass) exactly, directly (*went due east; a due north wind*). □ **due date** the date on which payment, delivery, etc. falls due. **due to** *disp.* because of, owing to (*was late due to an accident*) (cf. sense 4 of *adj.*). **fall** (or **become**) **due** (of a bill etc.) be immediately payable. **in due course 1** at about the appropriate time. **2** in the natural order. [ME f. OF *deü* ult. f. L *debitus* past part. of *debere* owe]

duel /'dju:əl/ *n. & v.* ● *n.* **1** *hist.* a contest with deadly weapons between two people, in the presence of two seconds, to settle a point of honour. **2** any contest between two people, parties, causes, animals, etc. ● *v.intr.* (**duelled, duelling**; *US* **dueled, dueling**) fight a duel or duels. □ **dueller** *n.* (*US* **dueler**). **duellist** *n.* (*US* **duelist**). [It. *duello* or L *duellum* (archaic form of *bellum* war), in med.L = single combat]

duende /du'endɪ/ *n.* **1** an evil spirit. **2** inspiration. [Sp.]

duenna /dju:'enə/ *n.* an older woman acting as a governess and companion in charge of girls, esp. in a Spanish family; a chaperon. [Sp. *dueña* f. L *domina* mistress]

Duero see DOURO.

duet /dju:'et/ *n. & v.* ● *n.* **1** *Mus.* **a** a performance by two voices, instrumentalists, etc. **b** a composition for two performers. **2** a dialogue. ● *v.intr.* (**duetted, duetting**) perform a duet. □ **duettist** *n.* [G *Duett* or It. *duetto* dimin. of *duo* duet f. L *duo* two]

Dufay /dju:'feɪ/, Guillaume (*c*.1400–74), French composer. He was a noted teacher and made a significant contribution to the development of Renaissance polyphony. Of his works, almost 200 are extant; they include much church music (his own Requiem Mass, now lost, was sung at his funeral), motets, and eighty-four songs.

duff[1] /dʌf/ *n.* a boiled pudding. [N.Engl. form of DOUGH]

duff[2] /dʌf/ *adj. Brit. sl.* **1** worthless, counterfeit. **2** useless, broken. [perh. = DUFF[1]]

duff[3] /dʌf/ *v.tr. sl.* **1** *Brit. Golf* mishit (a shot, a ball); bungle. **2** *Austral.* steal and alter brands on (cattle). □ **duff up** *sl.* beat; thrash. [perh. back-form. f. DUFFER]

duffer /'dʌfə(r)/ *n. sl.* **1** an inefficient, useless, or stupid person. **2** *Austral.* a person who duffs cattle. **3** *Austral. & NZ* an unproductive mine. [perh. f. obs. Sc. *dowfart* stupid person f. *douf* spiritless]

duffle /'dʌf(ə)l/ *n.* (also **duffel**) **1** a coarse woollen cloth with a thick nap. **2** a sportsman's or camper's equipment. □ **duffle bag** a cylindrical canvas bag closed by a draw-string and carried over the shoulder. **duffle-coat** a warm hooded coat made of duffle, usu. fastened with toggles. [the town of *Duffel* in Belgium]

Dufy /dju:'fi:/, Raoul (1877–1953), French painter and textile designer. In his early work he was much influenced by fauvism but he later developed his own characteristic style using bright colours, with calligraphic outlines sketched on brilliant background washes. His chief subjects were the racecourse, boating scenes, and society life on the French Riviera and in London.

dug[1] past and past part. of DIG.

dug[2] /dʌg/ *n.* **1** the udder, breast, teat, or nipple of a female animal. **2** *derog.* the breast of a woman. [16th c.: orig. unkn.]

dugong /'du:gɒŋ/ *n.* (*pl.* same or **dugongs**) a coastal marine sirenian mammal, *Dugong dugon*, of the Indian Ocean and SW Pacific, with paddle-like forelimbs and a notched tail fluke. [ult. f. Malay *dūyong*]

dugout /'dʌgaʊt/ *n.* **1 a** a roofed shelter esp. for troops in trenches. **b** an underground air-raid or nuclear shelter. **c** *N. Amer. Sport* = BENCH *n.* 4. **2** a canoe made from a hollowed tree-trunk. **3** *sl.* a retired officer etc. recalled to service.

duiker /'daɪkə(r)/ *n.* **1** (also **duyker**) a small African antelope of the genus *Cephalophus* or *Silvicapra*; esp. *S. grimmia*, having a crest of long hair between its horns. **2** *S. Afr.* a cormorant. [Du. *duiker* diver: in sense 1, from plunging through bushes when pursued]

Duisburg /'dju:sbɜ:g/ an industrial city in NW Germany, in North Rhine-Westphalia; pop. (1991) 537,440. It is the largest inland port in Europe, situated at the junction of the Rhine and Ruhr rivers.

duke /dju:k/ *n.* (as a title usu. **Duke**) **1 a** a person holding the highest hereditary title of the nobility. The title was introduced into England by Edward III, who in 1337 created his eldest son Edward (the Black Prince) Duke of Cornwall. **b** a sovereign prince ruling a duchy or small state. **2** (usu. in *pl.*) *sl.* the hand; the fist (*put up your dukes!*). **3** (in full **duke cherry**) a hybrid between the sweet cherry, *Prunus avium*, and sour cherry, *P. cerasus*. □ **royal duke** a duke who is also a royal prince. [ME f. OF *duc* f. L *dux ducis* leader]

dukedom /'dju:kdəm/ *n.* **1** a territory ruled by a duke. **2** the rank of duke.

dulcet /'dʌlsɪt/ *adj.* (esp. of sound) sweet and soothing. [ME, earlier *doucet* f. OF dimin. of *doux* f. L *dulcis* sweet]

dulcify /'dʌlsɪfaɪ/ *v.tr.* (**-ies, -ied**) *literary* **1** make gentle. **2** sweeten. □ **dulcification** /ˌdʌlsɪfɪ'keɪʃ(ə)n/ *n.* [L *dulcificare* f. *dulcis* sweet]

dulcimer /'dʌlsɪmə(r)/ *n.* **1** a musical instrument consisting of a shallow closed box over which metal strings of graduated length are stretched to be struck by wood, cane, or wire hammers, the prototype of the piano. **2** a plucked musical instrument of the zither type, fretted and with steel strings. It is played esp. in Kentucky and Alabama as an accompaniment to songs and dances. [OF *doulcemer*, said to repr. L *dulce* sweet, *melos* song]

dulcitone /'dʌlsɪˌtəʊn/ *n. Mus.* a keyboard instrument with steel tuning-forks which are struck by hammers. [L *dulcis* sweet + TONE]

dulia /'dju:lɪə/ *n. RC Ch.* the reverence accorded to saints and angels. [med.L f. Gk *douleia* servitude f. *doulos* slave]

dull /dʌl/ *adj. & v.* ● *adj.* **1** slow to understand; stupid. **2** tedious; boring. **3** (of the weather) overcast; gloomy. **4 a** (esp. of a knife edge etc.) blunt. **b** (of colour, light, sound, or taste) not bright, shining, vivid, or keen. **5** (of a pain etc.) usu. prolonged and indistinct; not acute (*a dull ache*). **6 a** (of a person, an animal, trade, etc.) sluggish, slow-moving, or stagnant. **b** (of a person) listless; depressed. **7** (of the ears, eyes, etc.) without keen perception. ● *v.tr. & intr.* make or become dull. □ **dull the edge of** make less sensitive, interesting, effective, amusing, etc.; blunt. **dull-witted** = DULL *adj.* 1. □ **dullish** *adj.* **dullness** *n.* (also **dulness**). **dully** /'dʌllɪ/ *adv.* [ME f. MLG, MDu. *dul*, corresp. to OE *dol* stupid]

dullard /'dʌləd/ *n.* a stupid person.

Dulles /'dʌlɪs/, John Foster (1888–1959), American Republican statesman and international lawyer. He was adviser to the US delegation at the conference which set up the United Nations in 1945

and negotiated the Peace Treaty with Japan in 1951. As Secretary of State under Eisenhower (1953–9) he strove to improve the position of the US in the cold war, to which end he strengthened NATO and urged that the US should stockpile nuclear arms as a deterrence against Soviet aggression.

dulse /dʌls/ *n.* an edible seaweed, *Rhodymenia palmata*, with red wedge-shaped fronds. [Ir. & Gael. *duileasg*]

duly /'dju:lɪ/ *adv.* **1** in due time or manner. **2** rightly, properly, fitly.

Duma /'du:mə/ **1** *hist.* any of four elected legislative bodies existing in Russia between 1906 and 1917, which were introduced by Tsar Nicholas II in response to popular unrest. In practice they had little effective power. **2** *hist.* any similar pre-19th century advisory municipal council in Russia. **3** a legislative body in the ruling assembly of Russia and of some other republics of the former USSR. [Russ.]

Dumas /'dju:mɑ:/, Alexandre (known as Dumas *père*) (1802–70), French novelist and dramatist. A pioneer of the romantic theatre in France, he first achieved fame with his historical dramas, such as *Henry III et sa cour* (1829). His reputation now rests on his historical adventure novels, including *The Three Musketeers* (1844–5) and *The Count of Monte Cristo* (1844–5). His son Alexandre Dumas (1824–95) (known as Dumas *fils*) wrote the novel (and play) *La Dame aux camélias* (1848), which formed the basis of Verdi's opera *La Traviata* (1853).

Du Maurier[1] /dju: 'mɒrɪˌeɪ/, Dame Daphne (1907–89), English novelist, granddaughter of George du Maurier. Many of her popular novels and period romances are set in the West Country of England, where she spent most of her life. Her works include *Jamaica Inn* (1936) and *Rebecca* (1938).

Du Maurier[2] /dju: 'mɒrɪˌeɪ/, George (Louis Palmella Busson) (1834–96), French-born cartoonist, illustrator, and novelist. He is chiefly remembered for his novel *Trilby* (1894), which included the character Svengali and gave rise to the word *Svengali* for a person who has a hypnotic influence.

dumb /dʌm/ *adj.* **1 a** (of a person) unable to speak, usu. because of a congenital defect or deafness. **b** (of an animal) naturally unable to speak (*our dumb friends*). **2** silenced by surprise, fear, etc. (*struck dumb by this revelation*). **3** taciturn or reticent, esp. insultingly (*dumb insolence*). **4** (of an action etc.) performed without speech. **5** (often in *comb.*) giving no sound; without voice or some other property normally belonging to things of the name (*a dumb piano*). **6** esp. *US colloq.* often *offens.* stupid; ignorant. **7** (usu. of a class, population, etc.) having no voice in government; inarticulate (*the dumb masses*). **8** (esp. of a computer terminal) able only to transmit data to or receive data from a computer; not programmable (opp. INTELLIGENT 3b). □ **dumb animals** animals, esp. as objects of pity. **dumb-bell 1** a short bar with a weight at each end, used for exercise, muscle-building, etc. **2** *sl.* a stupid person, esp. a woman. **dumb blonde** a pretty but stupid blonde woman. **dumb cluck** *sl.* a stupid person. **dumb crambo** a game in which one side has to guess a word chosen by the other side, after being given a word which rhymes with it, by acting in mime various words until they find it. **dumb-iron** the curved side-piece of a motor-vehicle chassis, joining it to the front springs. **dumb piano** a set of piano keys for exercising the fingers. **dumb show 1** significant gestures or mime, used when words are inappropriate. **2** a part of a play in early drama, acted in mime. **dumb waiter 1** a small lift for carrying food, plates, etc., between floors. **2** a movable table, esp. with revolving shelves, used in a dining-room. □ **dumbly** *adv.* **dumbness** *n.* [OE: orig. unkn.: sense 6 f. G *dumm*]

dumbfound /dʌm'faʊnd/ *v.tr.* (also **dumfound**; esp. as **dumbfounded** *adj.*) strike dumb; confound; nonplus. [DUMB, CONFOUND]

dumbhead /'dʌmhed/ *n.* esp. *US sl.* often *offens.* a stupid person.

dumbo /'dʌmbəʊ/ *n.* (*pl.* **-os**) *sl.* a stupid person; a fool. [DUMB + -O]

dumbstruck /'dʌmstrʌk/ *adj.* greatly shocked or surprised and so lost for words.

dumdum /'dʌmdʌm/ *n.* (in full **dumdum bullet**) a kind of soft-nosed bullet that expands on impact and inflicts laceration. [*Dum-Dum* in India, where it was first produced]

Dumfries /dʌm'fri:s/ a market town in SW Scotland, administrative centre of Dumfries and Galloway region; pop. (1981) 32,100.

Dumfries and Galloway a local government region in SW Scotland; administrative centre, Dumfries.

Dumfriesshire /dʌm'fri:sʃɪə(r)/ a former county of SW Scotland. It became part of Dumfries and Galloway region in 1975.

dummy /'dʌmɪ/ *n., adj., & v.* ● *n.* (*pl.* **-ies**) **1** a model of a human being, esp.: **a** a ventriloquist's doll. **b** a figure used to model clothes in a shop window etc. **c** a target used for firearms practice. **2** (often *attrib.*) **a** a counterfeit object used to replace or resemble a real or normal one. **b** a prototype, esp. in publishing. **3** *colloq.* a stupid person. **4** a person taking no significant part; a figurehead. **5** *Brit.* a rubber or plastic teat for a baby to suck on. **6** an imaginary fourth player at whist, whose hand is turned up and played by a partner. **7** *Bridge* **a** the partner of the declarer, whose cards are exposed after the first lead. **b** this player's hand. **8** *Mil.* a blank round of ammunition. **9** *Rugby & Football* a pretended pass. ● *adj.* sham; counterfeit. ● *v.tr.* (**-ies, -ied**) (usu. *absol.*) *Rugby & Football* pretend to pass (the ball). □ **dummy run 1** a practice attack, etc.; a trial run. **2** a rehearsal. **dummy up** *US sl.* keep quiet; give no information. **sell the** (or **a**) **dummy** *Rugby & Football colloq.* deceive (an opponent) by pretending to pass the ball. [DUMB + -Y²]

dump /dʌmp/ *n. & v.* ● *n.* **1 a** a place for depositing rubbish. **b** a heap of rubbish. **2** *colloq.* an unpleasant or dreary place. **3** *Mil.* a temporary store of ammunition, provisions, etc. **4** an accumulated pile of ore, earth, etc. **5** *Computing* **a** a printout of stored data. **b** the process or result of dumping data. ● *v.tr.* **1** put down firmly or clumsily (*dumped the shopping on the table*). **2** deposit or dispose of (rubbish etc.). **3** *colloq.* abandon, desert. **4** *Mil.* leave (ammunition etc.) in a dump. **5** *Econ.* send (goods unsaleable at a high price in the home market) to a foreign market for sale at a low price, to keep up the price at home, and to capture a new market. **6** *Computing* **a** copy (stored data) to a different location. **b** reproduce the contents of (a store) externally. □ **dump on** esp. *US sl.* criticize or abuse; get the better of. **dump truck** *N. Amer.* = DUMPER 1b. □ **dumping** *n.* [ME perh. f. Norse; cf. Danish *dumpe*, Norw. *dumpa* fall suddenly]

dumper /dʌmpə(r)/ *n.* **1 a** a person or thing that dumps. **b** (in full **dumper truck**) a truck with a body that tilts or opens at the back for unloading. **2** *Austral. & NZ* a large wave that breaks and hurls the swimmer or surfer on to the beach.

dumpling /dʌmplɪŋ/ *n.* **1 a** a small ball of usu. suet, flour, and water, boiled in stew or water. **b** a pudding consisting of apple or other fruit enclosed in dough and baked. **2** a small fat person. [17th c.: app. dimin. of *dump* small round object, but recorded much earlier]

dumps /dʌmps/ *n.pl. colloq.* depression; melancholy (esp. *down in the dumps*). [prob. f. LG or Du., figurative use of MDu. *domp* exhalation, haze, mist: rel. to DAMP]

dumpy /dʌmpɪ/ *adj.* (**dumpier, dumpiest**) short and stout. □ **dumpily** *adv.* **dumpiness** *n.* [*dump* (cf. DUMPLING) + -Y¹]

Dumyât see DAMIETTA.

dun¹ /dʌn/ *adj. & n.* ● *adj.* **1** dull greyish-brown. **2** *poet.* dark, dusky. ● *n.* **1** a dun colour. **2** a dun horse. **3** a dark fishing-fly. □ **dun-bird** a pochard. **dun-diver** a female or young male goosander. [OE *dun, dunn*]

dun² /dʌn/ *n. & v.* ● *n.* **1** a debt-collector; an importunate creditor. **2** a demand for payment. ● *v.tr.* (**dunned, dunning**) importune for payment of a debt; pester. [abbr. of obs. *dunkirk* privateer, f. DUNKIRK]

Dunbar /dʌnˈbɑː(r)/, William (*c.*1456–*c.*1513), Scottish poet. His first major poem, 'The Thrissill and the Rois' ('The Thistle and the Rose', 1503), is a political allegory on the marriage of James IV to Margaret Tudor (the daughter of Henry VII) and the first of his many satires. Dunbar is also remembered for his elegies, such as 'Lament for the Makaris' (on Chaucer and other fellow poets).

Dunbartonshire /dʌnˈbɑːt(ə)n͵ʃɪə(r)/ a former county of west central Scotland, on the Clyde, which became part of Strathclyde region in 1975.

Duncan /dʌŋkən/, Isadora (1878–1927), American dancer and teacher. A pioneer of modern dance, she developed a form of 'free' barefoot dancing based on instinctive movements and inspired by classical Greek art. She was much admired in Europe, where she settled and founded several schools of dance; her informal style deeply influenced Diaghilev. She died through being accidentally strangled when her scarf became entangled in the wheels of a car.

dunce /dʌns/ *n.* a person slow at learning; a dullard. □ **dunce's cap** a paper cone formerly put on the head of a dunce at school as a mark of disgrace. [DUNS SCOTUS, whose followers were ridiculed by 16th-c. humanists and reformers as enemies of learning]

Dundalk /dʌnˈdɔːk/ the county town of Louth, in the Republic of Ireland, a port on the east coast; pop. (1991) 25,800.

Dundee /dʌnˈdiː/ a city in eastern Scotland, the administrative centre of Tayside region, on the north side of the Firth of Tay; pop. (1991) 165,500.

Dundee cake *n.* esp. *Brit.* a rich fruit cake usu. decorated with almonds. [DUNDEE]

dunderhead /dʌndəˌhed/ *n.* a stupid person. □ **dunderheaded** *adj.* [17th c.: perh. rel. to dial. *dunner* resounding noise]

dune /djuːn/ *n.* a mound or ridge of loose sand etc. formed by the wind, esp. beside the sea or in a desert. □ **dune buggy** = *beach buggy*. [F f. MDu. *dūne*: cf. DOWN³]

Dunedin /dʌˈniːdɪn/ a city and port in South Island, New Zealand; pop. (est. 1990) 113,900. It was founded in 1848 by Scottish settlers; its name is based on the Gaelic word (Duneideann) for Edinburgh.

Dunfermline /dʌnˈfɜːmlɪn/ an industrial city in Fife, Scotland, near the Firth of Forth; pop. (1981) 52,000. A number of Scottish kings, including Robert the Bruce, are buried in its Benedictine abbey.

dung /dʌŋ/ *n. & v.* ● *n.* the excrement of animals; manure. ● *v.tr.* apply dung to; manure (land). □ **dung-beetle** a beetle whose larvae develop in dung, esp. one of the family Scarabaeidae. **dung-fly** a fly of the family Scatophagidae, laying its eggs on dung. **dung-worm** an earthworm found in cow-dung, esp. one used as bait. [OE, rel. to OHG *tunga*, Icel. *dyngja*, of unkn. orig.]

dungaree /͵dʌŋgəˈriː/ *n.* **1** a coarse Indian calico. **2** (in *pl.*) **a** overalls etc. made of dungaree or similar material. **b** trousers with a bib worn by children or as a fashion garment. [Hindi *dungrī*]

Dungarvan /dʌnˈgɑːv(ə)n/ a town on the south coast of the Republic of Ireland, the administrative centre of Waterford; pop. (1990) 6,920.

dungeon /dʌndʒən/ *n. & v.* ● *n.* **1** a strong underground cell for prisoners. **2** *archaic* a donjon. ● *v.tr.* (usu. foll. by *up*) *archaic* imprison in a dungeon. [orig. = *donjon*: ME f. OF *donjon* ult. f. L *dominus* lord]

dunghill /dʌŋhɪl/ *n.* a heap of dung or refuse, esp. in a farmyard.

dunk /dʌŋk/ *v.tr.* **1** dip (bread, a biscuit, etc.) into soup, coffee, etc. while eating. **2** immerse, dip (*was dunked in the river*). [Pennsylvanian G *dunke* to dip f. G *tunken*]

Dunkirk /dʌnˈkɜːk/ (French **Dunkerque** /dœ̃kɛrk/) a French port on the English Channel; pop. (1990) 71,070. It was the scene of the evacuation of the British Expeditionary Force in 1940. Forced to retreat to the Channel by the German breakthrough at Sedan, 225,000 British troops, as well as 110,000 of their French allies, were evacuated from Dunkirk between 27 May and 2 June by warships, requisitioned civilian ships, and a host of small boats, under constant attack from the air.

Dun Laoghaire /dʌn ˈliəri, ˈleərə/ a ferry port and resort town in the Republic of Ireland, near Dublin; pop. (1986) 54,715.

dunlin /dʌnlɪn/ *n.* a long-billed sandpiper, *Calidris alpina*. [prob. f. DUN¹ + -LING¹]

Dunlop /dʌnlɒp/, John Boyd (1840–1921), Scottish inventor. He worked in Edinburgh and Belfast as a veterinary surgeon but is best known for having invented the first successful pneumatic bicycle tyre (1888), which was manufactured by the company named after him. Though his invention was crucial to the development of the motor car, Dunlop himself received little profit from it.

Dunmow flitch /dʌnməʊ/ *n.* a side of bacon awarded on Whit Monday to any married couple who will swear that they have not quarrelled or repented of their marriage vows for at least a year and a day. The custom was instituted in Great Dunmow in Essex; the earliest evidence for it is 1244, though its origin may be earlier.

dunnage /dʌnɪdʒ/ *n. Naut.* **1** mats, brushwood, etc., stowed under or among cargo to prevent wetting or chafing. **2** *colloq.* miscellaneous baggage. [AL *dennagium*, of unkn. orig.]

Dunne /dʌn/, John William (1875–1949), English philosopher. His work is especially concerned with time and includes *An Experiment with Time* (1927) and *The Serial Universe* (1934), both of which influenced the plays of J. B. Priestley.

Dunnet Head /dʌnɪt/ a headland on the north coast of Scotland, between Thurso and John o'Groats. It is the most northerly point on the British mainland.

dunno /dəˈnəʊ/ *contr. colloq.* (I) do not know. [corrupt.]

dunnock /dʌnək/ *n.* a small European songbird, *Prunella modularis*, with a brown back and dark grey head and breast. Also called *hedge sparrow*. [app. f. DUN¹ + -OCK, from its plumage]

dunny /dʌnɪ/ *n.* (*pl.* **-ies**) **1** *Sc.* an underground passage or cellar, esp. in a tenement. **2** esp. *Austral. & NZ sl.* an earth-closet; an outdoor privy. [20th c.: orig. uncert.]

Duns Scotus /dʌnz ˈskəʊtəs/, John (known as 'the Subtle Doctor') (*c.*1265–1308), Scottish theologian and scholar. In opposition to the teaching of St Thomas Aquinas he argued that faith was a matter of will, not dependent on logical proofs. He was also the first major theologian to defend the theory of the Immaculate Conception. His system was accepted by the Franciscans as their doctrinal basis and exercised a profound influence in the Middle Ages. In the Renaissance his followers were ridiculed for their conservatism and abused as enemies of learning, which gave rise to the word *dunce*.

Dunstable /'dʌnstəb(ə)l/, John (c.1390–1453), English composer. He was a significant early exponent of counterpoint; his works include secular songs, masses, and motets.

Dunstan, St /'dʌnstən/ (c.909–88), Anglo-Saxon prelate. During his tenure as abbot at Glastonbury the monastery became a centre of religious teaching. He was appointed Archbishop of Canterbury by King Edgar in 960, and together they carried through a reform of Church and state. He introduced the strict Benedictine rule into England and succeeded in restoring monastic life; a zealous supporter of education, he also achieved fame as a musician, illuminator, and metalworker. Feast day, 19 May.

duo /'dju:əʊ/ n. (pl. **-os**) **1** a pair of actors, entertainers, singers, etc. (a comedy duo). **2** Mus. a duet. [It. f. L, = two]

duodecimal /ˌdju:əʊ'desɪm(ə)l/ adj. & n. ● adj. relating to or using a system of numerical notation that has twelve as a base. ● n. **1** the duodecimal system. **2** duodecimal notation. □ **duodecimally** adv. [L duodecimus twelfth f. duodecim twelve]

duodecimo /ˌdju:əʊ'desɪ,məʊ/ n. (pl. **-os**) Printing **1** a book-size in which each leaf is one-twelfth of the size of the printing-sheet. **2** a book of this size. [L (in) duodecimo in a twelfth (as DUODECIMAL)]

duodenary /ˌdju:əʊ'di:nərɪ/ adj. proceeding by twelves or in sets of twelve. [L duodenarius f. duodeni distributive of duodecim twelve]

duodenum /ˌdju:əʊ'di:nəm/ n. Anat. the first part of the small intestine immediately beyond the stomach. □ **duodenal** adj. **duodenitis** /-dɪ'naɪtɪs/ n. [ME f. med.L f. duodeni (see DUODENARY) from its length of about twelve fingers' breadth]

duologue /'dju:ə,lɒg/ n. **1** a conversation between two people. **2** a play or part of a play for two actors. [irreg. f. L duo or Gk duo two, after monologue]

duomo /'dwəʊməʊ/ n. (pl. **-os**) an Italian cathedral. [It., = DOME]

duopoly /dju:'ɒpəlɪ/ n. (pl. **-ies**) Econ. the possession of trade in a commodity etc. by only two sellers. [Gk duo two + pōleō sell, after MONOPOLY]

duotone /'dju:ə,təʊn/ n. & adj. Printing ● n. **1** a half-tone illustration in two colours from the same original with different screen angles. **2** the process of making a duotone. ● adj. in two colours. [L duo two + TONE]

dupe /dju:p/ n. & v. ● n. a victim of deception. ● v.tr. make a fool of; cheat; gull. □ **dupable** adj. **duper** n. **dupery** n. [F f. dial. F dupe hoopoe, from the bird's supposedly stupid appearance]

dupion /'dju:pɪən/ n. **1** a rough silk fabric woven from the threads of double cocoons. **2** an imitation of this with other fibres. [F doupion f. It. doppione f. doppio double]

duple /'dju:p(ə)l/ adj. of two parts. □ **duple ratio** Math. a ratio of two to one. **duple time** Mus. that with two beats to the bar. [L duplus f. duo two]

duplet /'dju:plɪt/ n. a set of two things, esp. of two equal musical notes played in the time of three. [L duplus DUPLE, after DOUBLET]

duplex /'dju:pleks/ n. & adj. ● n. esp. N. Amer. **1** a flat or maisonette on two levels. **2** a house subdivided for two families. ● adj. **1** having two elements; twofold. **2** esp. US **a** (of a flat) two-storeyed. **b** (of a house) for two families. **3** (of a communication system, computer circuit, etc.) allowing the transmission of signals in both directions simultaneously (opp. SIMPLEX). □ **half-duplex** (of a communication system, computer circuit, etc.) allowing the transmission of signals in both directions but not simultaneously. [L duplex duplicis f. duo two + plic- fold]

duplicate adj., n., & v. ● adj. /'dju:plɪkət/ **1** exactly like something already existing; copied (esp. in large numbers). **2 a** having two corresponding parts. **b** existing in two examples; paired. **c** twice as large or many; doubled. ● n. /'dju:plɪkət/ **1 a** one of two identical things, esp. a copy of an original. **b** one of two or more specimens of a thing exactly or almost identical. **2** Law a second copy of a letter or document. **3** (in full **duplicate bridge** or **whist**) a form of bridge or whist in which the same hands are played successively by different players. **4** archaic a pawnbroker's ticket. ● v.tr. /'dju:plɪ,keɪt/ **1** multiply by two; double. **2 a** make or be an exact copy of. **b** make or supply copies of (duplicated the leaflet for distribution). **3** repeat (an action etc.), esp. unnecessarily. □ **duplicate ratio** Math. the proportion of the squares of two numbers. **in duplicate** consisting of two exact copies.

□ **duplicable** /-kəb(ə)l/ adj. **duplication** /ˌdju:plɪ'keɪʃ(ə)n/ n. [L duplicatus past part. of duplicare (as DUPLEX)]

duplicator /'dju:plɪ,keɪtə(r)/ n. **1** a machine for making copies of a document, leaflet, etc. **2** a person or thing that duplicates.

duplicity /dju:'plɪsɪtɪ/ n. **1** double-dealing; deceitfulness. **2** archaic doubleness. □ **duplicitous** adj. [ME f. OF duplicité or LL duplicitas (as DUPLEX)]

duppy /'dʌpɪ/ n. (pl. **-ies**) West Indies a malevolent spirit or ghost. [perh. f. Afr. orig.]

du Pré /dju: 'preɪ/, Jacqueline (1945–87), English cellist. She made her solo début in London at the age of 16. She became famous for her interpretations of the cello concertos, especially that of Elgar. When her performing career was halted in 1972 by multiple sclerosis, she gave a notable series of master classes. She was married to the pianist and conductor Daniel Barenboim (b.1942).

dura var. of DURRA.

durable /'djʊərəb(ə)l/ adj. & n. ● adj. **1** capable of lasting; hard-wearing. **2** (of goods) not for immediate consumption; able to be kept. ● n. (in pl.) durable goods. □ **durably** adv. **durableness** n. **durability** /ˌdjʊərə'bɪlɪtɪ/ n. [ME f. OF f. L durabilis f. durare endure f. durus hard]

Duralumin /djʊə'ræljʊmɪn/ n. propr. a light hard alloy of aluminium with copper etc. [perh. f. Düren in the Rhineland or L durus hard + ALUMINIUM]

dura mater /ˌdjʊərə 'meɪtə(r)/ n. Anat. the tough outermost membrane enveloping the brain and spinal cord (see MENINX). [med.L = hard mother, transl. Arab. al-'umm al-jāfiya ('mother' in Arab. indicating the relationship of things)]

duramen /djʊə'reɪmen/ n. = HEARTWOOD. [L f. durare harden]

durance /'djʊərəns/ n. archaic imprisonment (in durance vile). [ME f. F f. durer last f. L durare: see DURABLE]

Durango /dʊə'ræŋgəʊ/ **1** a state of north central Mexico. **2** (in full **Victoria de Durango**) its capital city; pop. (1990) 1,352,160.

Duras /'djʊərɑ:s/, Marguerite (pseudonym of Marguerite Donnadieu) (b.1914), French novelist, film-maker, and dramatist. She established her reputation as a novelist in the 1950s and won the Prix Goncourt with *L'Amant* (1984). As well as directing a number of her own films, she has also written screenplays, for example, *Hiroshima mon amour* (1959).

duration /djʊə'reɪʃ(ə)n/ n. **1** the length of time for which something continues. **2** a specified length of time (for the duration of your visit). □ **for the duration 1** for a very long time. **2** until the end of the war. □ **durational** adj. [ME f. OF f. med.L duratio -onis (as DURANCE)]

durative /'djʊərətɪv/ adj. Gram. denoting continuing action.

Durazzo see DURRËS.

Durban /'dɜ:b(ə)n/ a seaport and resort in South Africa, on the coast of KwaZulu/Natal; pop. (1985) 634,300. Formerly known as Port Natal, it was renamed in 1835 after Sir Benjamin d'Urban, then governor of Cape Colony.

durbar /'dɜ:bɑ:(r)/ n. hist. **1** the court of an Indian ruler. **2** a public levee of an Indian prince or an Anglo-Indian governor or viceroy. [Urdu f. Pers. darbār court]

durchkomponiert /ˌdʊəx,kɒmpɒ'nɪət/ adj. Mus. (of a song) having different music for each verse. [G f. durch through + komponiert composed]

Dürer /'djʊərə(r)/, Albrecht (1471–1528), German painter and engraver. He is generally regarded as the leading German artist of the Renaissance. He was responsible for developing techniques and raising standards of craftsmanship in his favoured media of woodcut and copper engraving; examples of his work include a series of ninety-two woodcut blocks in honour of the Emperor Maximilian. He also painted detailed watercolour studies of plants and animals.

duress /djʊə'res/ n. **1** compulsion, esp. imprisonment, threats, or violence, illegally used to force a person to act against his or her will (under duress). **2** forcible restraint or imprisonment. [ME f. OF duresse f. L duritia f. durus hard]

Durex /'djʊəreks/ n. propr. a contraceptive sheath; a condom. [20th c.: invented word]

Durey /djʊə'reɪ/, Louis (1888–1979), French composer. Until 1921 he belonged to the group known as Les Six, writing mainly chamber music and songs. After 1945 he was one of a group of French composers who wrote music of a deliberate mass appeal, in accordance with Communist doctrines on art; his works from this period include the cantata *La Longue marche* (1949).

Durga /ˈdʊəgə/ *Hinduism* a fierce goddess, wife of Siva (see PARVATI), often identified with Kali. She is usually depicted riding a tiger or lion and slaying the buffalo demon, and with eight or ten arms. [Skr., = inaccessible]

Durham /ˈdʌrəm/ **1** a county of NE England. **2** a city in NE England on the River Wear, with a magnificent Norman cathedral. It is the county town of Durham; pop. (1991) 85,800.

durian /ˈdʊəriən/ *n.* **1** a large tree, *Durio zibethinus*, native to SE Asia, bearing oval spiny fruits containing a creamy pulp with a fetid smell and an agreeable taste. **2** the fruit of this tree. [Malay *durian* f. *dūri* thorn]

duricrust /ˈdjʊərɪˌkrʌst/ *n. Geol.* a hard mineral crust formed near the surface of soil in semi-arid regions by evaporation of groundwater. [L *durus* hard + CRUST]

during /ˈdjʊərɪŋ/ *prep.* **1** throughout the course or duration of (*read during the meal*). **2** at some point in the duration of (*came in during the evening*). [ME f. OF *durant* ult. f. L *durare* last, continue]

Durkheim /ˈdɜːkhaɪm/, Émile (1858–1917), French sociologist. Now regarded as one of the founders of modern sociology, he wrote about the influence of social structures on the behaviour of individuals (*The Division of Labour in Society*, 1893), formalized a methodology for sociological investigation, and examined the social causes of suicide (*Suicide*, 1897).

durmast /ˈdɜːmɑːst/ *n.* (in full **durmast oak**) an oak tree, *Quercus petraea*, having sessile acorns. This is the usual form of oak found on acid soils, as in northern and western Britain. [*dur*- (perh. erron. for DUN) + MAST²]

durn *US* var. of DARN².

durned *US dial.* var. of DARNED.

durra /ˈdʌrə/ *n.* (also **dura, dhurra**) a sorghum, a variety of *Sorghum bicolor*, native to Asia, Africa, and the US. [Arab. *ḏura, durra*]

Durrell¹ /ˈdʌrəl/, Gerald (Malcolm) (1925–95), English zoologist and writer. He acquired an interest in animals as a child in Corfu. From 1947 he organized a number of animal-collecting expeditions, which resulted in a popular series of broadcasts and books, including the autobiographical *My Family and Other Animals* (1956). Durrell became increasingly concerned with conservation and captive breeding, and in 1958 he founded a zoo in Jersey, Channel Islands, which later became the Jersey Wildlife Preservation Trust. He is the younger brother of the novelist Lawrence Durrell.

Durrell² /ˈdʌrəl/, Lawrence (George) (1912–90), English novelist, poet, and travel writer. He spent his childhood in India and his adolescence in Corfu, moving to France in the late 1950s. He first achieved fame with the *Alexandria Quartet* (1957–60), a series of novels set in Alexandria before the Second World War and written in an ornate, poetic style. He published several collections of poetry and a number of travel books, including *Bitter Lemons* (1957), about Cyprus. He was the elder brother of the zoologist and writer Gerald Durrell.

Durrës /ˈdʊrəs/ (Italian **Durazzo** /duˈrattso/) a port and resort in Albania, on the Adriatic coast; pop. (1990) 72,000.

durst /dɜːst/ *archaic* or *dial.* past of DARE.

durum /ˈdjʊərəm/ *n.* a kind of wheat, *Triticum durum*, having hard seeds and yielding a flour used in the manufacture of spaghetti etc. [L, neut. of *durus* hard]

durzi /ˈdɜːzɪ/ *n.* (*pl.* **durzis**) an Indian tailor. [Hindi f. Pers. *darzī* f. *darz* sewing]

Dushanbe /duːˈʃænbeɪ/ the capital of Tajikistan; pop. (est. 1990) 602,000. It was known as Stalinabad from 1929 to 1961.

dusk /dʌsk/ *n., adj.,* & *v.* ● *n.* **1** the darker stage of twilight. **2** shade; gloom. ● *adj. poet.* shadowy; dim; dark-coloured. ● *v.tr.* & *intr. poet.* make or become shadowy or dim. [ME *dosk, dusk* f. OE *dox* dark, swarthy, *doxian* darken in colour]

dusky /ˈdʌskɪ/ *adj.* (**duskier, duskiest**) **1** shadowy; dim. **2** dark-coloured, darkish. □ **duskily** *adv.* **duskiness** *n.*

Düsseldorf /ˈdʊs(ə)lˌdɔːf/ an industrial city of NW Germany, on the Rhine; pop. (1991) 577,560. It is the capital of North Rhine-Westphalia.

dust /dʌst/ *n.* & *v.* ● *n.* **1 a** a finely powdered earth, dirt, etc., lying on the ground or on surfaces, and blown about by the wind. **b** fine powder of any material (*pollen dust; gold-dust*). **c** a cloud of dust. **2** a dead person's remains (*honoured dust*). **3** confusion or turmoil (*raised quite a dust*). **4** *archaic* or *poet.* the mortal human body (*we are all dust*). **5** the ground; the earth (*kissed the dust*). **6** an act of dusting (*give the table a dust*). ● *v.* **1** *tr.* (also *absol.*) clear (furniture etc.) of dust etc. by wiping, brushing,

etc. **2** *tr.* **a** sprinkle (esp. a cake) with powder, dust, sugar, etc. **b** sprinkle or strew (sugar, powder, etc.). **3** *tr.* make dusty. **4** *intr. archaic* (of a bird) take a dust-bath. □ **dust and ashes** something very disappointing. **dust-bath** a bird's rolling in dust to freshen its feathers. **dust cover 1** = *dust-sheet*. **2** = *dust-jacket*. **dust devil** a whirlwind visible as a column of dust. **dust down 1** wipe or brush the dust from. **2** *colloq.* reprimand. **dusting-powder 1** talcum powder. **2** any dusting or drying powder. **dust-jacket** a usu. decorated paper cover used to protect a book from dirt etc. **dust off 1** remove the dust from (an object on which it has long been allowed to settle). **2** use and enjoy again after a long period of neglect. **dust-sheet** *Brit.* a cloth put over furniture to protect it from dust. **dust-shot** the smallest size of shot. **dust-storm** a storm with clouds of dust carried in the air. **dust-trap** something on, in, or under which dust gathers. **dust-up** *colloq.* a fight. **dust-wrapper** = *dust-jacket*. **in the dust 1** humiliated. **2** dead. **not see a person for dust** find a person hastily departed. **when the dust settles** when things quieten down. □ **dustless** *adj.* [OE *dūst*: cf. LG *dunst* vapour]

dustbin /ˈdʌstbɪn/ *n. Brit.* a container for household refuse, esp. one kept outside.

dust bowl *n.* an arid or unproductive dry region, specifically an area in the US prairie states, esp. Oklahoma, that suffered from land erosion in the 1930s after a period of drought combined with overgrazing and poor land management. A large part of the topsoil blew away, and great hardship resulted as thousands were forced to leave the area. Increased rainfall, re-grassing, and measures to prevent erosion such as contour ploughing have since reduced the problem.

dustcart /ˈdʌstkɑːt/ *n. Brit.* a vehicle used for collecting household refuse.

duster /ˈdʌstə(r)/ *n.* **1 a** a cloth for dusting furniture etc. **b** a person or contrivance that dusts. **2** a woman's light, loose, full-length coat.

dustman /ˈdʌstmən/ *n.* (*pl.* **-men**) *Brit.* **1** a man employed to clear household refuse. **2** a sandman.

dustpan /ˈdʌstpæn/ *n.* a small pan into which dust etc. is brushed from the floor.

dusty /ˈdʌstɪ/ *adj.* (**dustier, dustiest**) **1** full of, covered with, or resembling dust. **2** (esp. of a document or book) uninteresting, dull. **3** (of a colour) dull or muted. □ **dusty answer** a curt rejection of a request. **dusty miller 1** a plant with greyish or whitish leaves; esp. *Senecio cineraria, Cerastium tomentosum*, or *Artemisia stelleriana*. **2** an artificial fishing-fly. **not so dusty** *Brit. sl.* fairly good. □ **dustily** *adv.* **dustiness** *n.* [OE *dūstig* (as DUST)]

Dutch /dʌtʃ/ *adj.* & *n.* ● *adj.* **1** of, relating to, or associated with the Netherlands or its people or language. **2** *US sl.* German. **3** *S. Afr.* of Dutch descent. **4** *archaic* of Germany including the Netherlands. ● *n.* **1 a** the language of the Netherlands. (*See note below.*) **b** *S. Afr.* usu. *derog.* Afrikaans. **2** (prec. by *the*; treated as *pl.*) **a** the people of the Netherlands. **b** *S. Afr.* Afrikaans-speakers. **3** *archaic* the language of Germany including the Netherlands. □ **Dutch auction** see AUCTION. **Dutch bargain** a bargain concluded by drinking together. **Dutch barn** a barn roof over hay etc., set on poles and having no walls. **Dutch cap 1** a contraceptive diaphragm. **2** a woman's lace cap with triangular flaps on each side. **Dutch courage** false courage gained from alcohol. **Dutch doll** a jointed wooden doll. **Dutch door** a door divided into two parts horizontally allowing one part to be shut and the other open. **Dutch hoe** a hoe used with a pushing action. **Dutch interior** a painting of Dutch domestic life, esp. by Pieter de Hooch (c.1629–83). **Dutch metal** a copper-zinc alloy imitating gold leaf. **Dutch oven 1** a metal box the open side of which is turned towards a fire. **2** a covered cooking pot for braising etc. **Dutch treat** a party, outing, etc. to which each person makes a contribution. **Dutch uncle** a person giving advice with benevolent firmness. **Dutch wife** a framework of cane etc., or a bolster, used for resting the legs in bed. **go Dutch** share expenses equally. [MDu. *dutsch* Netherlandish, German, OHG *diutisc* national]

▪ Dutch belongs to the Germanic language group and is most closely related to German and English; as well as being used by the 15 million inhabitants of the Netherlands, it is also the official language of Suriname and the Netherlands Antilles. In addition, it is spoken in northern Belgium, where it is called Flemish. The English word *Dutch* originally denoted speakers of both High and Low German and is cognate with the German word *Deutsch*. After the United Provinces became an independent state in 1579, using the Low German of Holland as the national language, the English term was gradually restricted to the language and people of the Netherlands.

dutch /dʌtʃ/ *n. Brit. sl.* = DUCHESS 3. [abbr.]

Dutch East India Company a Dutch trading company founded in 1602 to protect Dutch trading interests in the Indian Ocean. It was dissolved in 1799.

Dutch East Indies the former name (until 1949) of INDONESIA.

Dutch elm disease a disease of elm trees, often fatal, causing withering of foliage and death of branches. It is caused by the fungus *Ceratocystis ulmi* and spread by bark beetles. It was first recognized in the Netherlands but probably originated in Asia. A virulent strain of the disease destroyed many British elms in the 1960s and succeeding decades.

Dutch Guiana the former name (until 1948) of SURINAME.

Dutchman /ˈdʌtʃmən/ n. (pl. **-men**; fem. **Dutchwoman**, pl. **-women**) **1 a** a native or national of the Netherlands. **b** a person of Dutch descent. **2** a Dutch ship. (See also FLYING DUTCHMAN.) **3** US sl. a German. □ **Dutchman's breeches** a plant of the fumitory family, *Dicentra cucullaria*, which is native to eastern North America, with white spurred flowers and finely divided leaves. **Dutchman's pipe** a climbing vine, *Aristolochia durior*, of eastern North America, with hooked tubular flowers. **I'm a Dutchman** expression of disbelief or refusal.

Dutch New Guinea a former name (until 1963) for IRIAN JAYA.

Dutch Reformed Church 1 a branch of the Protestant Church in the Netherlands, formed during the Reformation. It was disestablished in 1798 and replaced in 1816 by the Netherlands Reformed Church. **2** the dominant branch of the Protestant Church among Afrikaners in South Africa, which, after heavy criticism in the early 1980s, denounced (in 1986) its former theological support of apartheid.

Dutch West India Company a Dutch trading company founded in 1621 to develop Dutch trading interests in competition with Spain and Portugal and their colonies in western India, South America, and West Africa. It was dissolved in 1794.

duteous /ˈdjuːtɪəs/ adj. literary (of a person or conduct) dutiful; obedient. □ **duteously** adv. **duteousness** n. [DUTY + -OUS: cf. *beauteous*]

dutiable /ˈdjuːtɪəb(ə)l/ adj. liable to customs or other duties.

dutiful /ˈdjuːtɪˌfʊl/ adj. doing or observant of one's duty; obedient. □ **dutifully** adv. **dutifulness** n.

duty /ˈdjuːtɪ/ n. (pl. **-ies**) **1 a** a moral or legal obligation; a responsibility (*his duty to report it*). **b** the binding force of what is right (*strong sense of duty*). **c** what is required of one (*do one's duty*). **2** payment to the public revenue, esp.: **a** that levied on the import, export, manufacture, or sale of goods (*customs duty*). **b** that levied on the transfer of property, licences, the legal recognition of documents, etc. (*death duty*; *probate duty*). **3** (often in pl.) a job or function (*his duties as caretaker*). **4** the behaviour due to a superior; deference; respect. **5** the measure of an engine's effectiveness in units of work done per unit of fuel. **6** Eccl. the performance of church services. □ **do duty for** serve as or pass for (something else). **duty-bound** obliged by duty. **duty-free** (of goods) on which duty is not leviable. **duty-free shop** a shop at an airport etc. at which duty-free goods can be bought. **duty-officer** the officer currently on duty. **duty-paid** (of goods) on which duty has been paid. **duty visit** a visit paid from obligation, not from pleasure. **on** (or **off**) **duty** engaged (or not engaged) in one's work. [AF *deweté*, *dueté* (as DUE)]

duumvir /djuːˈʌmvə(r), ˈdjuːəmˌvɪə(r)/ n. (pl. **duumviri** /djuːˈʌmvɪˌraɪ/) Rom. Hist. one of two coequal magistrates or officials. □ **duumvirate** /djuːˈʌmvɪrət/ n. [L f. *duum virum* of the two men]

Duvalier /djuːˈvælɪˌeɪ/, François (known as 'Papa Doc') (1907–71), Haitian statesman, President 1957–71. His regime was noted for being authoritarian and oppressive; many of his opponents were either assassinated or forced into exile by his security force, known as the Tontons Macoutes. He proclaimed himself President for life in 1964 and was succeeded on his death by his son Jean-Claude (known as 'Baby Doc', b.1951); the Duvalier regime ended in 1986 when a mass uprising forced Jean-Claude to flee the country.

duvet /ˈduːveɪ/ n. a thick soft quilt with a detachable cover, used instead of other beclothes. [F]

Du Vigneaud /duː ˈviːnjəʊ/, Vincent (1901–78), American biochemist. He specialized in the study of vitamins and hormones that contain sulphur, beginning with insulin. He went on to study the function of methionine, to isolate and determine the structure of biotin, and to contribute to the synthesis of penicillin G. For isolating and synthesizing the pituitary hormones oxytocin and vasopressin, Du Vigneaud was awarded the Nobel Prize for chemistry in 1955.

dux /dʌks/ n. Sc., Austral., NZ, & S. Afr. the top pupil in a class or in a school. [L, = leader]

duyker var. of DUIKER 1.

DV abbr. Deo volente.

Dvořák /ˈdvɔːʒæk, -ʒɑːk/, Antonín (1841–1904), Czech composer. Living at a time of strong national consciousness, he combined ethnic folk elements with the Viennese musical tradition from Haydn to Brahms. He is probably best known for his ninth symphony ('From the New World', 1892–5), which he wrote while working in the US as director of the New York Conservatoire; it contains motifs from negro spirituals as well as Bohemian melodies. Dvořák also wrote chamber music, operas, and songs.

Dvr. abbr. Driver.

dwale /dweɪl/ n. deadly nightshade. [prob. f. Scand.]

dwarf /dwɔːf/ n. & v. ● n. (pl. **dwarfs** or **dwarves** /dwɔːvz/) **1 a** a person of abnormally small stature, esp. one with a normal-sized head and body but short limbs. ¶ Alternative terms such as *person of restricted growth* are now sometimes preferred. **b** an animal or plant much below the ordinary size for the species. **2** a small mythological being with supernatural powers. **3** (in full **dwarf star**) Astron. a small usu. dense star. **4** (attrib.) **a** of a kind very small in size (*dwarf bean*). **b** puny, stunted. ● v.tr. **1** stunt in growth. **2** cause (something similar or comparable) to seem small or insignificant (*dwarfed by their rivals' achievements*). □ **dwarfish** adj. [OE *dweorg* f. Gmc]

dwarfism /ˈdwɔːfɪz(ə)m/ n. the condition of being a dwarf.

dweeb /dwiːb/ n. US sl. a studious or tedious person. [20th c.: orig. unkn.]

dwell /dwel/ v. & n. ● v.intr. (past and past part. **dwelt** /dwelt/ or **dwelled**) **1** (usu. foll. by *in*, *at*, *near*, *on*, etc.) literary live, reside (*dwelt in the forest*). **2** (of a horse) be slow in raising its feet; pause before taking a fence. ● n. a slight, regular pause in the motion of a machine. □ **dwell on** (or **upon**) **1** spend time on, linger over; write, brood, or speak at length on (a specified subject) (*always dwells on his grievances*). **2** prolong (a note, a syllable, etc.). □ **dweller** n. [OE *dwellan* lead astray, later 'continue in a place', f. Gmc]

dwelling /ˈdwelɪŋ/ n. (also **dwelling-place**) formal a house; a residence; an abode. □ **dwelling-house** a house used as a residence, not as an office etc.

dwindle /ˈdwɪnd(ə)l/ v.intr. **1** become gradually smaller; shrink; waste away. **2** lose importance; decline; degenerate. [*dwine* fade away f. OE *dwīnan*, ON *dvina*]

dwt. abbr. hist. pennyweight.

d.w.t. abbr. dead-weight tonnage.

Dy symb. Chem. the element dysprosium.

dyad /ˈdaɪæd/ n. Math. an operator which is a combination of two vectors. □ **dyadic** /daɪˈædɪk/ adj. [LL *dyas dyad*- f. Gk *duas duados* f. *duo* two]

Dyak /ˈdaɪæk/ n. & adj. (also **Dayak**) ● n. (pl. same or **Dyaks**) **1** a member of the aboriginal peoples inhabiting parts of Borneo. **2** the group of Malayo-Polynesian languages spoken by these peoples. ● adj. of or relating to these peoples or their languages. [Malay *dayak* up-country]

▪ The Dyak peoples are subdivided into several groups including the *Land Dyak* of SW Borneo and the *Sea Dyak* or *Iban* of NW Borneo (Sarawak). They are thought to be early inhabitants of the island, forced inland by subsequent migrations of Malays to the coasts; they live in traditional longhouse communities and their economy is based on rice cultivation, hunting, and fishing. The term Sea Dyak is a misnomer, since these people live primarily in the hilly inland areas; they were so called because of their involvement in coastal raids on their more passive coastal neighbours the Land Dyak.

dyarchy var. of DIARCHY.

dybbuk /ˈdɪbʊk/ n. (pl. **dybbukim** /-kɪm/ or **dybbuks**) (in Jewish folklore) the malevolent spirit of a dead person that enters and controls the body of a living person until exorcized. [Heb. *dibbūk* f. *dābaḳ* cling]

dye /daɪ/ n. & v. ● n. **1 a** a substance used to change the colour of hair, fabric, wood, etc. **b** a colour produced by this. **2** (in full **dyestuff**) a substance yielding a dye, esp. when in solution. ● v.tr. (**dyeing**) **1** impregnate with dye. **2** make (a thing) a specified colour with dye (*dyed it yellow*). □ **dyed in the wool** (or **grain**) **1** out and out; unchangeable, inveterate. **2** (of a fabric) made of yarn dyed in its raw state. **dye-line** a print made by the diazo process. □ **dyeable** adj. [OE *deag*, *deagian*]

dyer /ˈdaɪə(r)/ n. a person who dyes cloth etc. □ **dyer's greenweed** (or **broom**) a bushy yellow-flowered leguminous plant, *Genista tinctoria*, formerly used to make a green dye. **dyer's oak** an eastern

Mediterranean oak, *Quercus infectoria*, yielding galls that were formerly used to make a yellow dye.

Dyfed /'dʌvɪd/ a county of SW Wales, comprising the former counties of Cardiganshire, Carmarthenshire, and Pembroke; administrative centre, Carmarthen.

dying /'daɪɪŋ/ *adj. & n.* ● *attrib.adj.* **1** about to die, mortally ill. **2** connected with, or at the time of, death (*his dying words*). ● *n.* (prec. by *the*; treated as *pl.*) those who are dying. □ **dying oath** an oath made at, or with the solemnity proper to, death. **to one's dying day** for the rest of one's life. [pres. part. of DIE[1]]

dyke[1] /daɪk/ *n. & v.* (also **dike**) ● *n.* **1** a long wall or embankment built to prevent flooding, esp. from the sea. **2 a** a ditch or artificial watercourse. **b** *Brit.* a natural watercourse. **3 a** a low wall, esp. of turf. **b** a causeway. **4** a barrier or obstacle; a defence. **5** *Geol.* an intrusion of igneous rock across sedimentary strata. **6** esp. *Austral. sl.* a lavatory. ● *v.tr.* provide or defend with a dyke or dykes. [ME f. ON *dík* or MLG *dík* dam, MDu. *dijc* ditch, dam: cf. DITCH]

dyke[2] /daɪk/ *n.* (also **dike**) *sl.* a lesbian. [20th c.: orig. unkn.]

Dylan /'dɪlən/, Bob (born Robert Allen Zimmerman) (b.1941), American singer and songwriter. The leader of the urban folk-music revival in the 1960s, he became known for his political and protest songs, including 'A Hard Rain's A-Gonna Fall', 'Blowin' in the Wind' (both 1963), and 'The Times They Are A-Changin' (1964). When on tour in 1966 he caused controversy and aroused severe criticism for using an amplified backing group. His albums include *Highway 61 Revisited* (1965) and *Blood on the Tracks* (1975).

dyn *abbr.* dyne.

dynamic /daɪ'næmɪk/ *adj. & n.* ● *adj.* (also **dynamical**) **1** energetic; active; potent. **2** *Physics* **a** concerning motive force (opp. STATIC *adj.* 2a). **b** concerning force in actual operation. **3** of or concerning dynamics. **4** *Mus.* relating to the volume of sound. **5** *Philos.* relating to dynamism. **6** (as **dynamical**) *Theol.* (of inspiration) endowing with divine power, not impelling mechanically. ● *n.* **1** an energizing or motive force. **2** *Mus.* = DYNAMICS 3. □ **dynamic equilibrium** a state of balance between continuing processes. **dynamic viscosity** see VISCOSITY. □ **dynamically** *adv.* [F *dynamique* f. Gk *dunamikos* f. *dunamis* power]

dynamics /daɪ'næmɪks/ *n.pl.* **1** (usu. treated as *sing.*) **a** the branch of mechanics concerned with the motion of bodies under the action of forces (cf. STATICS 1). **b** the branch of any science in which forces or changes are considered (*aerodynamics; population dynamics*). **2** the motive forces, physical or moral, affecting behaviour and change in any sphere (*group dynamics*). **3** *Mus.* the varying degree of volume of sound in musical performance. □ **dynamicist** /-'næmɪsɪst/ *n.* (in sense 1)

dynamism /'daɪnəˌmɪz(ə)m/ *n.* **1** energizing or dynamic action or power. **2** *Philos.* the theory that phenomena of matter or mind are due to the action of forces (rather than to motion or matter). □ **dynamist** *n.* [Gk *dunamis* power + -ISM]

dynamite /'daɪnəˌmaɪt/ *n. & v.* ● *n.* **1** a plastic high explosive consisting of nitroglycerine mixed with an absorbent solid. (*See note below.*) **2** a potentially dangerous person, thing, or situation. **3** *sl.* a narcotic, esp. heroin. **4** (often *attrib.*) *sl.* a powerful or impressive person or thing. ● *v.tr.* charge or shatter with dynamite. □ **dynamiter** *n.* [formed as DYNAMISM + -ITE[1]]

▪ Dynamite was invented by Alfred B. Nobel in 1866, and is chiefly used for commercial blasting. The purpose of the absorbent is to improve handling quality, and the explosive is usually moulded into sticks. The solid may be inert (e.g. kieselguhr), or active (e.g. a mixture of sodium nitrate and wood pulp).

dynamo /'daɪnəˌməʊ/ *n.* (*pl.* **-os**) **1** a machine converting mechanical into electrical energy, esp. by rotating coils of copper wire in a magnetic field. **2** *colloq.* an energetic person. [abbr. of *dynamo-electric machine* f. Gk *dunamis* power, force]

dynamometer /ˌdaɪnə'mɒmɪtə(r)/ *n.* an instrument measuring energy expended by a mechanism, vehicle, etc. [F *dynamomètre* f. Gk *dunamis* power, force]

dynast /'dɪnæst, 'daɪ-/ *n.* **1** a ruler. **2** a member of a dynasty. [L f. Gk *dunastēs* f. *dunamai* be able]

dynasty /'dɪnəstɪ/ *n.* (*pl.* **-ies**) **1** a line of hereditary rulers. **2** a succession of leaders in any field. □ **dynastic** /dɪ'næstɪk/ *adj.* **dynastically** *adv.* [F *dynastie* or LL *dynastia* f. Gk *dunasteia* lordship (as DYNAST)]

dynatron /'daɪnəˌtrɒn/ *n. Electronics* a thermionic valve, used to generate continuous oscillations. [Gk *dunamis* power + -TRON]

dyne /daɪn/ *n. Physics* a unit of force (symbol: **dyn**), equal to 10^{-5} newtons. One dyne is that force which would give a mass of one gram an acceleration of one centimetre per second per second in the direction of the force. [F f. Gk *dunamis* force, power]

dys- /dɪs/ *comb. form* esp. *Med.* bad, difficult. [Gk *dus-* bad]

dysentery /'dɪsəntrɪ/ *n.* a disease with inflammation of the intestines, causing severe diarrhoea with blood and mucus. □ **dysenteric** /ˌdɪsən'terɪk/ *adj.* [OF *dissenterie* or L *dysenteria* f. Gk *dusenteria* (as DYS-, *enteria* f. *entera* bowels)]

dysfunction /dɪs'fʌŋkʃ(ə)n/ *n.* an abnormality or impairment of function. □ **dysfunctional** *adj.*

dysgraphia /dɪs'græfɪə/ *n.* an inability to write coherently. □ **dysgraphic** *adj.* [DYS- + Gk *graphia* writing]

dyslexia /dɪs'leksɪə/ *n.* a developmental disorder marked by severe difficulty in reading and spelling (cf. ALEXIA). □ **dyslexic** *adj. & n.* **dyslectic** /-'lektɪk/ *adj. & n.* [G *Dyslexie* (as DYS-, Gk *lexis* speech)]

dysmenorrhoea /ˌdɪsmenə'rɪə/ *n.* painful or difficult menstruation.

dyspepsia /dɪs'pepsɪə/ *n.* indigestion. [L *dyspepsia* f. Gk *duspepsia* (as DYS-, *peptos* cooked, digested)]

dyspeptic /dɪs'peptɪk/ *adj. & n.* ● *adj.* of or relating to dyspepsia or the resulting depression. ● *n.* a person suffering from dyspepsia.

dysphasia /dɪs'feɪzɪə/ *n. Med.* lack of coordination in speech, owing to brain damage. □ **dysphasic** *adj.* [Gk *dusphatos* hard to utter (as DYS-, PHATIC)]

dysphoria /dɪs'fɔːrɪə/ *n.* a state of unease or mental discomfort. □ **dysphoric** /-'fɒrɪk/ *adj.* [Gk *dusphoria* f. *dusphoros* hard to bear (as DYS-, *pherō* bear)]

dysplasia /dɪs'pleɪzɪə/ *n. Med.* abnormal growth of tissues etc. □ **dysplastic** /-'plæstɪk/ *adj.* [mod.L, formed as DYS- + Gk *plasis* formation]

dyspnoea /dɪsp'nɪə/ *n.* (US **dyspnea**) *Med.* difficult or laboured breathing. □ **dyspnoeic** *adj.* [L f. Gk *duspnoia* (as DYS-, *pneō* breathe)]

dysprosium /dɪs'prəʊzɪəm/ *n.* a soft silvery-white metallic chemical element (atomic number 66; symbol **Dy**). A member of the lanthanide series, dysprosium was discovered by the French chemist Paul-Émile Lecoq de Boisbaudran (1838–1912) in 1886. Dysprosium has various specialized uses, including as a component in certain magnetic alloys. [f. Gk *dusprositos* hard to get at]

dystocia /dɪs'təʊʃə/ *n. Med.* difficult or prolonged childbirth. [DYS- + Gk *tokos* childbirth]

dystopia /dɪs'təʊpɪə/ *n.* a nightmare vision of society (opp. UTOPIA), often as one dominated by a totalitarian or technological state. Now common in science fiction, two of the best-known examples are Orwell's *Nineteen Eighty-four* (1949) and Aldous Huxley's *Brave New World* (1932). □ **dystopian** *adj. & n.* [DYS- + UTOPIA]

dystrophic /dɪs'trəʊfɪk, -'trɒfɪk/ *adj.* **1** *Med.* relating to or affected by dystrophy. **2** *Ecol.* (of a lake) containing much dissolved organic matter and little oxygen.

dystrophy /'dɪstrəfɪ/ *n.* defective nutrition. □ **muscular dystrophy** any of a group of hereditary diseases marked by progressive weakening and wasting of the muscles. [mod.L *dystrophia* formed as DYS- + Gk *-trophia* nourishment]

dysuria /dɪs'jʊərɪə/ *n.* painful or difficult urination. [LL f. Gk *dusouria* (as DYS-, *ouron* urine)]

Dzaoudzi /'dzaʊdzɪ/ the former capital of the French island of Mayotte and of the Comoros; pop. (1985) 5,675.

Dzaudzhikau /ˌdzaʊdʒɪ'kaʊ/ a former name (1944–54) for VLADIKAVKAZ.

Dzerzhinsk /dzɜː'ʒɪnsk/ a city in west central Russia, west of Nizhni Novgorod; pop. (1990) 286,000. It was formerly known as Chernoreche (until 1919) and Rastyapino (1919–29).

Dzerzhinsky /dzɜː'ʒɪnskɪ/, Feliks (Edmundovich) (1877–1926), Russian Bolshevik leader, of Polish descent. He was the organizer and first head of the post-revolutionary Soviet security police (see CHEKA, OGPU, NKVD).

dzo /zəʊ/ *n.* (also **dzho, zho**) (*pl.* same or **-os**) a hybrid of a cow and a yak. [Tibetan *mdso*]

Dzongkha /'zɒŋkə/ *n.* a Tibetan dialect that is the official language of Bhutan. [Tibetan, = language of the fortress]

Ee

E[1] /iː/ *n.* (also **e**) (*pl.* **Es** or **E's**) **1** the fifth letter of the alphabet. **2** *Mus.* the third note of the diatonic scale of C major.

E[2] *abbr.* (also **E.**) **1** East, Eastern. **2** Egyptian (*£E*). **3** Engineering (*M.I.Mech.E.* etc.). **4** see E-NUMBER. **5** *sl.* **a** the drug Ecstasy (see ECSTASY 3). **b** a dose of this drug.

E[3] *symb. Physics* energy ($E = mc^2$).

e *symb.* **1** *Math.* the base of natural logarithms, equal to approx. 2.71828. **2** used on packaging (in conjunction with specification of weight, size, etc.) to indicate compliance with EC regulations.

e- /ɪ, e/ *prefix* form of EX-[1] 1 before some consonants.

ea. *abbr.* each.

each /iːtʃ/ *adj. & pron.* ● *adj.* every one of two or more persons or things, regarded separately (*each person; five in each class*). ● *pron.* each person or thing (*each of us; have two books each; cost a penny each*). □ **each and every** every single. **each other** one another (used as a compound reciprocal pron.: *they hate each other; they wore each other's hats*). **each way** *adj. & adv. Brit.* (of a bet) backing a horse etc. for either a win or a place. [OE *ǽlc* f. WG *phr.* = ever alike, f. as AYE[2], ALIKE]

Eadwig see EDWY.

eager /ˈiːgə(r)/ *adj.* **1 a** (of a person) full of keen desire, enthusiastic. **b** (of passions etc.) keen, impatient. **2** (foll. by *to* + infin., *for*) keen, impatient, strongly desirous (*eager to learn; eager for news*). □ **eager beaver** *colloq.* an overzealous or extremely diligent person. □ **eagerly** *adv.* **eagerness** *n.* [ME f. AF *egre*, OF *aigre* keen, ult. f. L *acer acris*]

eagle /ˈiːg(ə)l/ *n. & v.* ● *n.* **1 a** a large bird of prey of the family Accipitridae, esp. of the genus *Aquila*, with keen vision and powerful flight. **b** a figure of an eagle, esp. as a symbol of the US (the *bald eagle*), or formerly as a Roman or French ensign. **2** *Golf* a score of two strokes under par at any hole. **3** *US* a coin worth ten dollars. ● *v.tr. Golf* play (a hole) in two strokes less than par. □ **eagle eye** keen sight, watchfulness. **eagle-eyed** keen-sighted, watchful. **eagle owl** a large owl of the genus *Bubo*, with long ear tufts. **eagle ray** a large ray (fish) of the family Myliobatidae, with long pointed pectoral fins. [ME f. AF *egle*, OF *aigle* f. L *aquila*]

eaglet /ˈiːglɪt/ *n.* a young eagle.

eagre /ˈeɪgə(r), ˈiːg-/ *n.* = BORE[3]. [17th c.: orig. unkn.]

Eakins /ˈiːkɪnz/, Thomas (1844–1916), American painter and photographer. Influenced by Diego Velázquez and José Ribera, he was a dominant figure in American realist painting of the 19th century. He was noted for his portraits and genre pictures of the life of his native city, Philadelphia; boating and bathing were favourite themes. His most famous picture, *The Gross Clinic* (1875), aroused controversy because of its explicit depiction of surgery.

Ealing Studios /ˈiːlɪŋ/ a film studio in Ealing, West London, active 1929–55 and directed from 1937 by Michael Balcon. The studio is remembered chiefly for the comedies it made in the postwar decade, including *Passport to Pimlico, Kind Hearts and Coronets,* and *Whisky Galore.*

-ean /ˈiːən, ɪən/ *suffix* var. of -AN.

E. & O. E. *abbr.* errors and omissions excepted.

ear[1] /ɪə(r)/ *n.* **1 a** the organ of hearing and balance in man and vertebrates, esp. the external part of this. **b** an organ sensitive to sound in other animals. **2** the faculty for discriminating sounds (*an ear for music*). **3** an ear-shaped thing, esp. the handle of a jug. **4** listening, attention. □ **all ears** listening attentively. **bring about one's ears** bring down upon oneself. **ear-drops 1** medicinal drops for the ear. **2** hanging earrings. **ear lobe** the lower soft pendulous external part of the ear. **ear-piercing** *adj.* loud and shrill. ● *n.* the piercing of a hole in the lobe of the ear so as to facilitate the wearing of earrings. **ear-splitting** excessively loud. **ear-stud** see STUD[1] *n.* 1b. **ear-trumpet** a trumpet-shaped device formerly used as a hearing-aid. **give ear to** listen to. **have a person's ear** receive a favourable hearing. **have (or keep) an ear to the ground** be alert to rumours or the trend of opinion. **in one ear and out the other** heard but disregarded or quickly forgotten. **out on one's ear** dismissed ignominiously. **up to one's ears** (often foll. by *in*) *colloq.* deeply involved or occupied. **wet behind the ears** immature, inexperienced. □ **eared** *adj.* (also in *comb.*). **earless** *adj.* [OE *ēare* f. Gmc: rel. to L *auris*, Gk *ous*]

▪ The ear of a mammal is composed of three parts, the outer, middle, and inner ears. The outer ear is the external part consisting of a fleshy flap or pinna that concentrates the sound and is often movable, and a tube leading to the eardrum or tympanum. The middle ear is an air-filled cavity connected to the throat, containing three small linked bones (the incus, malleus, and stapes) that transmit vibrations from the eardrum to the inner ear. This is a complex fluid-filled labyrinth that includes the spiral cochlea (where vibrations are converted to nerve impulses) and the three semicircular canals (constituting the organ of balance). The ears of other vertebrates are broadly similar, though lacking some of the components, and many insects also have organs for detecting sound.

ear[2] /ɪə(r)/ *n.* the seed-bearing head of a cereal plant. [OE *ēar* f. Gmc]

earache /ˈɪəreɪk/ *n.* a (usu. prolonged) pain in the ear.

earbash /ˈɪəbæʃ/ *v.tr. sl.* talk inordinately to; harangue. □ **earbasher** *n.* **earbashing** *n.*

eardrum /ˈɪədrʌm/ *n.* the membrane of the middle ear (= *tympanic membrane*).

earful /ˈɪəfʊl/ *n.* (*pl.* **-fuls**) *colloq.* **1** a copious or prolonged amount of talking. **2** a strong reprimand.

Earhart /ˈeəhɑːt/, Amelia (1898–1937), American aviator. In 1932 she became the first woman to fly across the Atlantic solo, completing the journey from Newfoundland to Londonderry in a time of 13¼ hours. The aircraft carrying Earhart and her navigator disappeared over the Pacific Ocean during a subsequent round-the-world flight.

earl /ɜːl/ *n.* a British nobleman ranking between a marquess and a viscount. The title is the oldest in the British nobility, having first been used in England under the Danish king Canute, and was the highest until 1337, when that of duke was created. It corresponds to the title of count in the nobility of other European nations. □ **earldom** *n.* [OE *eorl*, of unkn. orig.]

Earl Grey *n.* (in full **Earl Grey tea**) a type of tea flavoured with bergamot. [prob. f. the 2nd *Earl Grey* (1764–1845), supposed to have been given the recipe by a Chinese mandarin]

Earl Marshal *n.* (in the UK) the officer presiding over the College of Heralds, with ceremonial duties on various royal occasions.

Earl Palatine see COUNT PALATINE.

early /ˈɜːlɪ/ *adj., adv., & n.* ● *adj. & adv.* (**earlier, earliest**) **1** before the due, usual, or expected time (*was early for my appointment; the train arrived early*). **2 a** not far on in the day or night, or in time (*early evening; at the earliest opportunity*). **b** prompt; promptly (*early payment appreciated; reply early for a reduction*). **3 a** not far on in a period, development, or process of evolution; being the first stage (*the early Christians; early Spring*). **b** of the distant past (*early man*). **c** not far on in a sequence or serial order (*the early chapters; appears early in the list*). **4 a** of childhood, esp. the preschool years (*early learning*). **b** (of a piece of writing, music, etc.) immature, youthful (*an early work*). **5** flowering, ripening, etc., before other varieties (*early peaches*). ● *n.* (*pl.* **-ies**) (usu. in *pl.*) an early fruit or vegetable, esp. potatoes. □ **at the earliest** (often placed after a specified time) not before (*will arrive on Monday at the earliest*). **early bird** *colloq.* a person who arrives, gets up, etc. early. **early closing** *Brit.* the shutting of business premises on the afternoon of one particular day of the week. **early days** early in time for something to happen etc. **early grave** an untimely or premature death. **early hours** the very early morning, usu. before dawn. **early music** medieval, Renaissance, and baroque music, esp. as revived and played on instruments of the period. **early musician** a student or performer of early music. **early night** a night on which a person goes to bed early; an instance of retiring to bed early. **early** (or **earlier**) **on** at an early (or earlier) stage. **early retirement** retirement from one's occupation before the statutory retirement age, esp. on advantageous financial terms. **early warning** advance warning of an imminent (esp. nuclear) attack (esp. *attrib.: early warning system*). □ **earliness** *n.* [orig. as adv., f. OE *ǣrlīce, ārlīce* (*ǣr* ERE)]

Early English *Archit.* the first stage of English Gothic (late 12th and 13th centuries), marked by the use of pointed arches and simple lancet windows without tracery.

earmark /ˈɪəmɑːk/ *n.* & *v.* ● *n.* **1** an identifying mark. **2** an owner's mark on the ear of an animal. ● *v.tr.* **1** (usu. foll. by *for*) set aside (money etc.) for a special purpose. **2** mark (sheep etc.) with such a mark.

earmuff /ˈɪəmʌf/ *n.* a wrap or cover for the ears, protecting them from cold, noise, etc.

earn /ɜːn/ *v.tr.* **1** (also *absol.*) **a** (of a person) obtain (income) in the form of money in return for labour or services (*earn a weekly wage; happy to be earning at last*). **b** (of capital invested) bring in as interest or profit. **2 a** deserve; be entitled to; obtain as the reward for hard work or merit (*have earned a holiday; earned our respect; earn one's keep*). **b** incur (a reproach, reputation, etc.). □ **earned income** income derived from wages etc. (opp. *unearned income*). [OE *earnian* f. WG, rel. to Gmc roots assoc. with reaping]

earner /ˈɜːnə(r)/ *n.* **1** a person or thing that earns (often in *comb.: wage-earner*). **2** *colloq.* a lucrative job or enterprise.

earnest¹ /ˈɜːnɪst/ *adj.* & *n.* ● *adj.* ardently or intensely serious; zealous; not trifling or joking. ● *n.* seriousness. □ **in** (or in **real**) **earnest** serious(ly), not joking(ly); with determination. □ **earnestly** *adv.* **earnestness** *n.* [OE *eornust, eornost* (with Gmc cognates): cf. ON *ern* vigorous]

earnest² /ˈɜːnɪst/ *n.* **1** money paid as an instalment, esp. to confirm a contract etc. **2** a token or foretaste (*in earnest of what is to come*). [ME *ernes*, prob. var. of *erles, arles* prob. f. med.L *arrhula* (unrecorded) f. *arr(h)a* pledge]

earnings /ˈɜːnɪŋz/ *n.pl.* money earned. □ **earnings-related** (of benefit, a pension, etc.) calculated on the basis of past or present income.

EAROM /ˈɪərɒm/ *n.* *Computing* electrically alterable read-only memory. [acronym]

earphone /ˈɪəfəʊn/ *n.* a device applied to the ear to aid hearing or receive radio or telephone communications.

earpiece /ˈɪəpiːs/ *n.* the part of a telephone etc. applied to the ear during use.

earplug /ˈɪəplʌg/ *n.* a piece of wax etc. placed in the ear to protect against cold air, water, or noise.

earring /ˈɪərɪŋ/ *n.* a piece of jewellery worn in or on (esp. the lobe of) the ear.

earshot /ˈɪəʃɒt/ *n.* the distance over which something can be heard (esp. *within* or *out of earshot*).

earth /ɜːθ/ *n.* & *v.* ● *n.* **1 a** (also **Earth**; often prec. by *the*) the planet of the solar system on which human beings live. (*See note below.*) **b** (prec. by *the*) land and sea, as distinct from sky. **2 a** dry land; the ground (*fell to earth*). **b** soil, clay, mould. **c** bodily matter (*earth to earth*). **3** *Relig.* the present abode of humankind, as distinct from heaven or hell; the world. **4** *Electr.* a connection to the ground as representing zero potential in an electrical circuit. **5** the hole of a badger, fox, etc. **6** (prec. by *the*) *colloq.* a huge amount; everything (*cost the earth; want the earth*). ● *v.* **1** *tr.* (foll. by *up*) cover (the roots and lower stems of plants) with heaped-up earth. **2 a** *tr.* drive (a fox) to its earth. **b** *intr.* (of a fox etc.) run to its earth. **3** *tr. Electr.* connect to earth. □ **come back** (or **down**) **to earth** return to realities. **earth-closet** a lavatory with dry earth

used to cover excreta. **earth-hog** (or **-pig**) = AARDVARK. **earth mother 1** *Mythol.* a female spirit or deity symbolizing the earth. **2** a sensual and maternal woman. **earth-nut 1 a** a plant with an edible roundish tuber; esp. an umbelliferous woodland plant, *Conopodium majus*. **b** the tuber of this plant. **2** the peanut. **earth sciences** the sciences concerned with the earth or part of it, or its atmosphere (e.g. geology, oceanography, meteorology). **earth-shattering** (or **-shaking**) *colloq.* having a traumatic or devastating effect. **earth-shatteringly** *colloq.* devastatingly, remarkably. **earth tremor** see TREMOR *n.* 3. **gone to earth** in hiding. **on earth** *colloq.* **1** existing anywhere (*the happiest man on earth; looked like nothing on earth*). **2** as an intensifier (*what on earth?*). □ **earthward** *adj.* & *adv.* **earthwards** *adv.* [OE *eorthe* f. Gmc]

▪ The earth is the third planet from the sun in the solar system, orbiting between Venus and Mars at an average distance of 149.6 million km from the sun, and has one natural satellite, the moon. It has an equatorial diameter of 12,756 km, an average density 5.5 times that of water, and is believed to have formed about 4,600 million years ago. The earth, which is three-quarters covered by oceans and has a dense atmosphere of nitrogen and oxygen, is the only planet known to support life. The interior contains three concentric regions: core, mantle, and crust. The rocky crust, which forms the surface of the solid earth, is only a few kilometres thick, although thicker beneath the continents than beneath the oceans. The mantle, extending to a depth of roughly 3,000 km, consists mainly of silicate rocks, and constitutes the bulk of the earth's volume. Beneath it is the core, which extends to the centre of the earth and consists largely of iron and nickel; its outer part is liquid and is thought to be the seat of the earth's magnetic field.

earthbound /ˈɜːθbaʊnd/ *adj.* **1** attached to the earth or earthly things. **2** moving towards the earth.

earthen /ˈɜːθən/ *adj.* **1** made of earth. **2** made of baked clay.

earthenware /ˈɜːθənˌweə(r)/ *n.* & *adj.* ● *n.* pottery, vessels, etc., made of clay fired to a porous state which can be made impervious to liquids by the use of a glaze. (See also POTTERY.) ● *adj.* made of fired clay. [EARTHEN + WARE¹]

earthling /ˈɜːθlɪŋ/ *n.* an inhabitant of the earth, esp. as regarded in science fiction by aliens.

earthly /ˈɜːθlɪ/ *adj.* **1 a** of the earth; terrestrial. **b** of human life on earth; worldly, material; carnal (opp. SPIRITUAL, HEAVENLY). **2** (usu. with *neg.*) *colloq.* remotely possible or conceivable (*is no earthly use; there wasn't an earthly reason*). □ **not an earthly** *colloq.* no chance whatever. □ **earthliness** *n.*

earthquake /ˈɜːθkweɪk/ *n.* **1** a convulsive shaking or movement of the earth's surface. (*See note below.*) **2** a social etc. disturbance.

▪ While gentle earth tremors can occur in any region of the globe, major earthquakes are generally confined to particular active regions of the earth's crust. These correspond to the edges of the crustal plates, and most earthquakes are due to the release of strain energy associated with the relative motions of the plates. The focus of such an earthquake can be as deep as about 700 km; earthquakes associated with volcanic activity are much shallower. The intensity of earthquakes is expressed by the Richter scale, destructive earthquakes generally measuring between about 7 and 9.

earthshine /ˈɜːθʃaɪn/ *n.* *Astron.* the glow caused by sunlight reflected from the earth, esp. on the darker portion of a crescent moon.

earthstar /ˈɜːθstɑː(r)/ *n.* a woodland fungus of the genus *Geastrum*, with a spherical fruiting body surrounded by a fleshy star-shaped structure; esp. *G. triplex*, found in woodland.

earthwork /ˈɜːθwɜːk/ *n.* **1** an embankment, fortification, etc. made of earth. **2** the process of excavating soil in civil engineering work.

earthworm /ˈɜːθwɜːm/ *n.* an annelid worm, esp. of the genus *Lumbricus* or *Allolobophora*, living and burrowing in the soil.

earthy /ˈɜːθɪ/ *adj.* (**earthier, earthiest**) **1** of or like earth or soil. **2** somewhat coarse or crude; unrefined (*earthy humour*). □ **earthily** *adv.* **earthiness** *n.*

earwax /ˈɪəwæks/ *n.* a yellow waxy secretion produced by the ear. Also called *cerumen*.

earwig /ˈɪəwɪg/ *n.* & *v.* ● *n.* **1** a small elongate insect of the order Dermaptera, with a pair of terminal appendages in the shape of forceps. **2** *US* a small centipede. ● *v.tr.* (**earwigged, earwigging**) *archaic* influence (a person) by secret communication. [OE *ēarwicga* f. *ēare* EAR¹ + *wicga* earwig, prob. rel. to *wiggle*: once thought to enter the head through the ear]

ease /iːz/ *n. & v.* ● *n.* **1** absence of difficulty; facility, effortlessness (*did it with ease*). **2 a** freedom or relief from pain, anxiety, or trouble. **b** freedom from embarrassment or awkwardness. **c** freedom or relief from constraint or formality. **d** freedom from poverty. ● *v.* **1** *tr.* **a** relievè from pain or anxiety etc. (often foll. by *of: eased my mind; eased me of the burden*). **b** make easy or easier; help, facilitate. **2** *intr.* (often foll. by *off, up*) **a** become less painful or burdensome. **b** relax; begin to take it easy. **c** slow down; moderate one's behaviour, habits, etc. **3** *tr. joc.* rob or extract money etc. from (*let me ease you of your loose change*). **4** *intr. Meteorol.* become less severe (*the wind will ease tonight*). **5 a** *tr.* relax; slacken; make a less tight fit. **b** *tr. & intr.* (foll. by *through, into,* etc.) move or be moved carefully into place (*eased it into the hole*). **6** *intr.* (often foll. by *off*) *Stock Exch.* (of shares etc.) descend in price or value. □ **at ease 1** free from anxiety or constraint. **2** *Mil.* **a** in a relaxed attitude while on parade, with the feet apart. **b** the order to stand in this way. **at one's ease** free from embarrassment, awkwardness, or undue formality. **ease away** (or **down** or **off**) *Naut.* slacken (a rope, sail, etc.). □ **easer** *n.* [ME f. AF *ese*, OF *eise*, ult. f. L *adjacens* ADJACENT]

easeful /ˈiːzfʊl/ *adj.* that gives ease, comfort, or relief; soothing.

easel /ˈiːz(ə)l/ *n.* **1** a standing frame, usu. of wood, for supporting an artist's work, a blackboard, etc. **2** an artist's work collectively. [Du. *ezel* = G *Esel* ASS[1]]

easement /ˈiːzmənt/ *n. Law* a right of way or a similar right over another's land. [ME f. OF *aisement*]

easily /ˈiːzɪlɪ/ *adv.* **1** without difficulty. **2** by far (*easily the best*). **3** very probably (*it could easily snow*).

east /iːst/ *n., adj., & adv.* ● *n.* **1 a** the point of the horizon where the sun rises at the equinoxes (cardinal point 90° to the right of north). **b** the compass point corresponding to this. **c** the direction in which this lies. **2** (usu. **the East**) **a** the regions or countries lying to the east of the Mediterranean, esp. China, Japan, and other countries of eastern Asia. **b** *hist.* the Communist states of eastern Europe. **3** the eastern part of a country, town, etc. **4** (**East**) *Bridge* a player occupying the position designated 'east'. ● *adj.* **1** towards, at, near, or facing east. **2** coming from the east (*east wind*). ● *adv.* **1** towards, at, or near the east. **2** (foll. by *of*) further east than. □ **east-north-** (or **south-**) **east** the direction or compass point midway between east and north-east (or south-east). **to the east** (often foll. by *of*) in an easterly direction. [OE *ēast-* f. Gmc]

East Africa the eastern part of the African continent, especially the countries of Kenya, Uganda, and Tanzania.

East Anglia /ˈæŋɡlɪə/ a region of eastern England consisting of the counties of Norfolk, Suffolk, and parts of Essex and Cambridgeshire.

East Bengal the part of the former Indian province of Bengal that was ceded to Pakistan in 1947, forming the greater part of the province of East Pakistan. It gained independence as Bangladesh in 1971.

eastbound /ˈiːstbaʊnd/ *adj.* travelling or leading eastwards.

Eastbourne /ˈiːstbɔːn/ a town on the south coast of England, in East Sussex; pop. (1981) 78,000.

East Cape a peninsular region of North Island, New Zealand. Its tip forms the most easterly point of the island.

East China Sea see CHINA SEA.

East End the eastern part of London to the north of the Thames, noted for its converted docklands and its cockney culture. □ **East Ender** *n.*

Easter /ˈiːstə(r)/ *n.* **1** (also **Easter Day** or **Sunday**) the festival of the Resurrection of Christ, the greatest and oldest festival of the Christian Church, celebrated in the Western Church on the first Sunday after the first full moon following the northern vernal equinox (21 Mar.). It is a day of rejoicing after the fasting and penance of Lent. The Christian feast superseded an old pagan festival. **2** the season in which this occurs, esp. the weekend from Good Friday to Easter Monday. □ **Easter week** the week beginning on Easter Sunday. [OE *ēastre* app. f. *Ēostre*, a goddess associated with spring, f. Gmc]

Easter egg *n.* an edible artificial (usually chocolate) egg given as a gift at Easter. The egg is an ancient symbol of renewed life; Easter eggs were known as early as the 13th century. They were originally ordinary hard-boiled eggs, with the shells dyed different colours.

Easter Island an island in the SE Pacific west of Chile; pop. (est. 1988) 2,000. It was named by the Dutch navigator Roggeveen, who visited it on Easter Day, 1722. It has been administered by Chile since 1888. The island, first settled by Polynesians in about AD 400, is famous for its large monolithic statues of human heads, believed to date from the period 1000–1600.

easterly /ˈiːstəlɪ/ *adj., adv., & n.* ● *adj. & adv.* **1** in an eastern position or

direction. **2** (of a wind) blowing from the east. ● *n.* (*pl.* **-ies**) a wind blowing from the east.

Easter Monday the day after Easter Sunday.

eastern /ˈiːstən/ *adj.* **1** of or in the east; inhabiting the east. **2** lying or directed towards the east. **3** (**Eastern**) of or in the Far, Middle, or Near East. □ **easternmost** *adj.* [OE *ēasterne* (as EAST, -ERN)]

Eastern bloc the countries of eastern and central Europe which were under Soviet domination from the end of the Second World War until the collapse of the Soviet Communist system in 1989–91.

Eastern Cape a province of south-eastern South Africa, formerly part of Cape Province; capital, Bisho.

Eastern Church ● *n.* **1** any of the Christian Churches originating in eastern Europe and the Middle East. **2** (in full **Eastern Orthodox Church**) the Orthodox Church.

Eastern Desert an alternative name for the ARABIAN DESERT.

Eastern Empire the eastern part of the Roman Empire, after its division in AD 395. (See also BYZANTINE EMPIRE.)

easterner /ˈiːstənə(r)/ *n.* a native or inhabitant of the east.

Eastern Ghats see GHATS, THE.

eastern hemisphere the half of the earth containing Europe, Asia, and Africa.

Eastern Standard Time (also **Eastern Time**) (abbr. **EST**) **1** the standard time in a zone including most of eastern Canada and the US, five hours behind GMT. **2** the standard time in a zone including eastern Australia, ten hours ahead of GMT.

Eastern Transvaal a province of north-eastern South Africa, formerly part of Transvaal; capital, Nelspruit.

Eastern Zhou see ZHOU.

Easter Rising the uprising in Dublin and other cities in Ireland against British rule, Easter 1916. It ended with the surrender of the protesters, some of whose leaders were subsequently executed, but was a contributory factor in the establishment of the Irish Free State (1921).

Eastertide /ˈiːstəˌtaɪd/ *n.* the period including Easter.

East Flanders a province of northern Belgium; capital, Ghent. (See also FLANDERS.)

East Frisian Islands see FRISIAN ISLANDS.

East Germany *hist.* the German Democratic Republic (see GERMANY).

East India Company a trading company formed in 1600 to develop commerce in the newly colonized East Indies. In 1765, after a series of victories over local princes, the company took over the administration of Bengal and later other parts of India, maintaining its own army and political service. Government remained in the hands of the company until 1858, when the British Crown took over in the wake of the Indian Mutiny.

East Indiaman *n. hist.* a large ship engaged in trade with the East Indies.

East Indies 1 the islands of SE Asia, esp. the Malay Archipelago. **2** *hist.* (also **East India**) the whole of SE Asia to the east of and including India.

easting /ˈiːstɪŋ/ *n. Naut.* etc. the distance travelled or the angle of longitude measured eastward from either a defined north–south grid line or a meridian.

East Kilbride /kɪlˈbraɪd/ a town in west central Scotland, in Strathclyde region, south-east of Glasgow; pop. (1991) 81,400.

East London a port and resort in South Africa, on the coast of the province of Eastern Cape; pop. (1985) 193,800.

East Lothian a former county of east central Scotland. It became a district of Lothian region in 1975.

Eastman /ˈiːstmən/ George (1854–1932), American inventor and manufacturer of photographic equipment. In 1884 he established a company which in 1892 became the Eastman Kodak Company ('Kodak' was a name Eastman invented). He invented flexible roll-film coated with light-sensitive emulsion, and the Kodak camera (1888) to use it. These did much to popularize amateur photography, as did his subsequent development of colour photography.

East Prussia the north-eastern part of the former kingdom of Prussia, on the Baltic coast, later part of Germany. Designated a province of Prussia in 1815, it extended from the Vistula in the west to an eastern boundary beyond the town of Klaipeda, now in Lithuania. Its capital was Königsberg (now Kaliningrad). Reduced in size after the First World War, when it was separated from the rest of Germany by the Polish Corridor, it was divided after the Second World War between the Soviet Union and Poland.

East River an arm of the Hudson River in New York City, separating Manhattan and the Bronx from Brooklyn and Queens.

East Siberian Sea a part of the Arctic Ocean lying between the New Siberian Islands and Wrangel Island, to the north of eastern Siberia.

East Side a part of Manhattan in New York City, lying between the East River and Fifth Avenue.

East Sussex a county of SE England; county town, Lewes.

East Timor the eastern part of the island of Timor in the southern Malay Archipelago. Formerly a Portuguese colony, the region declared itself independent in 1975. In 1976 it was invaded by Indonesia, who annexed and claimed it as the 27th state of Indonesia, a claim which has never been recognized by the United Nations. Since then the region has been the scene of bitter fighting and of alleged mass killings by the Indonesian government and military forces. The first in a series of talks aiming to reach a settlement of the conflict was held in December 1992. (See also TIMOR.) □ **East Timorese** n. & adj.

eastward /ˈiːstwəd/ adj., adv., & n. ● adj. & adv. (also **eastwards** /-wədz/) towards the east. ● n. an eastward direction or region. □ **eastwardly** adj. & adv.

East–West /iːstˈwest/ attrib.adj. of or relating to countries of the East (to countries of the former Eastern bloc) and the West (*East–West relations*).

Eastwood /ˈiːstwʊd/, Clint (b.1930), American film actor and director. He became a star with his role in *A Fistful of Dollars* (1964), the first cult spaghetti western. His performance as the 'dirty cop' in *Dirty Harry* (1971) proved as much of a box-office success as his many western performances. He started directing in 1971, receiving acclaim for his portrait of the saxophonist Charlie Parker in *Bird* (1988) and for his uncompromising western *Unforgiven* (1992).

easy /ˈiːzɪ/ adj., adv., & int. ● adj. (**easier**, **easiest**) **1** not difficult; achieved without great effort. **2 a** free from pain, discomfort, anxiety, etc. **b** comfortably off, affluent (*easy circumstances*). **3** free from embarrassment, awkwardness, constraint, or pressure; relaxed and pleasant (*an easy manner*). **4 a** not strict; tolerant. **b** compliant, obliging; easily persuaded (*an easy touch*). **5** *Stock Exch.* (of goods, money on loan, etc.) not much in demand. ● adv. with ease; in an effortless or relaxed manner. ● int. go carefully; move gently. □ **easy as pie** see PIE¹. **easy chair** a large comfortable chair, usu. an armchair. **easy come easy go** *colloq.* what is easily got is soon lost or spent. **easy does it** *colloq.* go carefully. **easy money** money got without effort (esp. of dubious legality). **easy of access** easily entered or approached. **easy on the eye** (or **ear** etc.) *colloq.* pleasant to look at (or listen to etc.). **easy-peasy** *sl.* very simple. **easy terms** payment by instalments. **go easy** (usu. foll. by *with*, *on*) be sparing or cautious. **I'm easy** *colloq.* I have no preference. **of easy virtue** (of a woman) sexually promiscuous. **stand easy!** *Brit. Mil.* permission to a squad standing at ease to relax their attitude further. **take it easy 1** proceed gently or carefully. **2** relax; avoid overwork. □ **easiness** n. [ME f. AF *aisé*, OF *aisié* past part. of *aisier* EASE]

easygoing /ˌiːzɪˈɡəʊɪŋ/ adj. **1** placid and tolerant; relaxed in manner; accepting things as they are. **2** (of a horse) having an easy gait.

Easy Street n. *colloq.* comfortable circumstances, affluence.

eat /iːt/ v. (past **ate** /et, eɪt/; past part. **eaten** /ˈiːt(ə)n/) **1 a** tr. take into the mouth, chew, and swallow (food). **b** intr. consume food; take a meal. **c** tr. devour (*eaten by a lion*). **2** intr. (foll. by (away) *at*, *into*) **a** destroy gradually, esp. by corrosion, erosion, disease, etc. **b** begin to consume or diminish (resources etc.). **3** tr. *colloq.* trouble, vex (*what's eating you?*). □ **eat dirt** see DIRT. **eat one's hat** *colloq.* admit one's surprise in being wrong (only as a proposition unlikely to be fulfilled: *said he would eat his hat*). **eat one's heart out** suffer from excessive longing or envy. **eat humble pie** see HUMBLE. **eat out** have a meal away from home, esp. in a restaurant. **eat out of a person's hand** be entirely submissive to a person. **eat salt with** see SALT. **eat up 1** (also *absol.*) eat or consume completely. **2** use or deal with rapidly or wastefully (*eats up petrol*; *eats up the miles*). **3** encroach upon or annex (*eating up the neighbouring states*). **4** absorb, preoccupy (*eaten up with pride*). **eat one's words** admit that one was wrong. [OE *etan* f. Gmc]

eatable /ˈiːtəb(ə)l/ adj. & n. ● adj. that is in a condition to be eaten (cf. EDIBLE). ● n. (usu. in *pl.*) food.

eater /ˈiːtə(r)/ n. **1** a person who eats (*a big eater*). **2** *Brit.* an eating apple etc.

eatery /ˈiːtərɪ/ n. esp. N. Amer. (pl. **-ies**) *colloq.* a restaurant or eating-place.

eating /ˈiːtɪŋ/ adj. **1** suitable for eating (*eating apple*). **2** used for eating (*eating-house*).

eats /iːts/ n.pl. *colloq.* food.

eau-de-Cologne /ˌəʊdəkəˈləʊn/ n. an alcohol-based perfume of a kind made orig. at Cologne. [F, lit. 'water of Cologne']

eau-de-Nil /ˌəʊdəˈniːl/ n. & adj. ● n. a pale greenish colour. ● adj. of this colour. [F, lit. 'water of the Nile' (from the supposed resemblance)]

eau-de-vie /ˌəʊdəˈviː/ n. spirits, esp. brandy. [F, lit. 'water of life']

eaves /iːvz/ n.pl. the underside of a projecting roof. [orig. sing., f. OE *efes*: prob. rel. to OVER]

eavesdrop /ˈiːvzdrɒp/ v.intr. (**-dropped**, **-dropping**) listen secretly to a private conversation. □ **eavesdropper** n. [*eavesdropper* orig. 'one who listens under walls' prob. f. ON *upsardropi* (cf. OE *yfæsdrypæ*): *eavesdrop* by back-form.]

ebb /eb/ n. & v. ● n. **1** the movement of the tide out to sea (also *attrib.*: *ebb tide*). **2** the process of draining away of flood-water etc. **3** the process of declining or diminishing; the state of being in decline. ● v.intr. (often foll. by *away*) **1** (of tidewater) flow out to sea; recede; drain away. **2** decline; run low (*his life was ebbing away*). □ **at a low ebb** in a poor condition or state of decline. **ebb and flow** a continuing process of decline and upturn in circumstances. **on the ebb** in decline. [OE *ebba*, *ebbian*]

Ebla /ˈeblə/ a city in ancient Syria, situated to the south-west of Aleppo. It became very powerful in the mid-third millennium BC, when it dominated a region corresponding to modern Lebanon, northern Syria, and SE Turkey. It was a thriving trading city and centre of scholarship, as testified in some 15,000 cuneiform tablets discovered among the city's ruins in 1975.

E-boat /ˈiːbəʊt/ n. *hist.* a German torpedo-boat used in the Second World War. [enemy + BOAT]

ebonite /ˈebənaɪt/ n. = VULCANITE. [EBONY + -ITE¹]

ebony /ˈebənɪ/ n. & adj. ● n. (pl. **-ies**) **1** a heavy hard very dark wood used for furniture. **2** a tropical tree of the genus *Diospyros* producing this. ● adj. **1** made of ebony. **2** black like ebony. [earlier *hebeny* f. (h)*eben(e)* = *ebon*, perh. after *ivory*]

Eboracum /ɪˈbɒrəkəm, ˌiːbəˈrɑːkəm/ the Roman name for YORK.

Ebro /ˈiːbrəʊ, ˈeb-/ the principal river of NE Spain, rising in the mountains of Cantabria and flowing 910 km (570 miles) south-eastwards into the Mediterranean Sea.

ebullient /ɪˈbʌlɪənt, disp. ɪˈbʊl-/ adj. **1** exuberant, high-spirited. **2** *Chem.* boiling. □ **ebullience** n. **ebulliency** n. **ebulliently** adv. [L *ebullire* *ebullient-* bubble out (as E-, *bullire* boil)]

EC abbr. **1** East Central (London postal district). **2** executive committee. **3 a** see EUROPEAN COMMUNITY. **b** see EUROPEAN COMMISSION.

ecad /ˈiːkæd/ n. *Ecol.* an organism modified by its environment. [Gk *oikos* house + -AD¹]

écarté /eɪˈkɑːteɪ/ n. **1** a card-game for two persons in which cards from a player's hand may be exchanged for others from the pack. **2** a position in classical ballet with one arm and leg extended. [F, past part. of *écarter* discard]

Ecce Homo /ˌekeɪ ˈhəʊməʊ/ n. (in Renaissance painting) a depiction of Christ wearing the crown of thorns. [L, = 'behold the man', the words of Pilate to the Jews after the crowning with thorns (John 19:5)]

eccentric /ɪkˈsentrɪk/ adj. & n. ● adj. **1** odd or capricious in behaviour or appearance; whimsical. **2 a** not placed, not having its axis etc. placed centrally. **b** (often foll. by *to*) (of a circle) not concentric (to another). **c** (of an orbit) not circular. ● n. **1** an eccentric person. **2** *Mech.* an eccentric contrivance for changing rotatory into backward-and-forward motion, e.g. the cam used in an internal-combustion engine. □ **eccentrically** adv. **eccentricity** /ˌeksenˈtrɪsɪtɪ/ n. (pl. **-ies**). [LL *eccentricus* f. Gk *ekkentros* f. *ek* out of + *kentron* CENTRE]

ecchymosis /ˌekɪˈməʊsɪs/ n. (pl. **ecchymoses** /-siːz/) *Med.* an area of discoloration due to bleeding under the skin. [mod.L f. Gk *ekkhumōsis* escape of blood]

Eccles /ˈek(ə)lz/, Sir John Carew (b.1903), Australian physiologist. He demonstrated the means by which nerve impulses are conducted, showing that a chemical neurotransmitter is released to initiate propagation across the synapses, followed by another that inhibits the propagation. This discovery greatly influenced the medical treatment of nervous diseases and physiological research. Eccles was awarded a Nobel Prize in 1963.

Eccles. abbr. (in the Bible) Ecclesiastes.

Eccles cake /ˈek(ə)lz/ n. a round flat cake made of pastry filled with currants etc. [the town of *Eccles* in northern England]

ecclesial /ɪˈkliːzɪəl/ adj. = ECCLESIASTICAL. [Gk *ekklesia* assembly, church f. *ekklētos* summoned out f. *ek* out + *kaleō* call]

Ecclesiastes /ɪˌkliːzɪˈæstiːz/ a book of the Bible traditionally attributed to Solomon, consisting largely of reflections on the vanity of human life.

ecclesiastic /ɪˌkliːzɪˈæstɪk/ n. & adj. ● n. a priest or clergyman. ● adj. = ECCLESIASTICAL. □ **ecclesiasticism** /-tɪˌsɪz(ə)m/ n. [F *ecclésiastique* or LL *ecclesiasticus* f. Gk *ekklēsiastikos* f. *ekklēsia* assembly, church: see ECCLESIAL]

ecclesiastical /ɪˌkliːzɪˈæstɪk(ə)l/ adj. of the Church or the clergy. □ **ecclesiastically** adv.

Ecclesiasticus /ɪˌkliːzɪˈæstɪkəs/ a book of the Apocrypha containing moral and practical maxims, probably composed or compiled in the early 2nd century BC.

ecclesiology /ɪˌkliːzɪˈɒlədʒɪ/ n. **1** the study of churches, esp. church building and decoration. **2** theology as applied to the nature and structure of the Christian Church. □ **ecclesiologist** n. **ecclesiological** /-zɪəˈlɒdʒɪk(ə)l/ adj. [Gk *ekklēsia* assembly, church (see ECCLESIAL) + -LOGY]

Ecclus. abbr. Ecclesiasticus (Apocrypha).

eccrine /ˈekriːn, -rɪn/ adj. (of a gland, e.g. a sweat gland) secreting without loss of cell material. [Gk *ek* out of + *krinō* sift]

ecdysis /ˈekdaɪsɪs/ n. (pl. **ecdyses** /-siːz/) the action of casting off skin or shedding an exoskeleton etc. [mod.L f. Gk *ekdusis* f. *ekduō* put off]

ECG abbr. electrocardiogram.

echelon /ˈeʃəˌlɒn/ n. & v. ● n. **1** a level or rank in an organization, in society, etc.; those occupying it (often in *pl.*: *the upper echelons*). **2** *Mil.* a formation of troops, ships, aircraft, etc., in parallel rows with the end of each row projecting further than the one in front (*in echelon*). ● v.tr. arrange in an echelon. [F *échelon* f. *échelle* ladder f. L *scala*]

echeveria /ˌetʃəˈvɪərɪə/ n. a succulent plant of the genus *Echeveria*, native to Central and South America. [mod.L f. M. *Echeveri*, 19th-c. Mexican botanical draughtsman]

echidna /ɪˈkɪdnə/ n. an egg-laying insectivorous mammal of the genus *Tachyglossus* or *Zaglossus*, native to Australia and New Guinea, with a covering of spines, a long snout, and long claws. Also called *spiny anteater*. [mod.L f. Gk *ekhidna* viper]

echinoderm /ɪˈkaɪnəˌdɜːm/ n. *Zool.* a marine invertebrate animal of the phylum Echinodermata, which comprises the starfishes, sea urchins, brittle-stars, crinoids, and sea cucumbers. Their bodies typically show fivefold radial symmetry, most have an internal calcareous skeleton, and they possess a unique system of hydraulic tubes which operate numerous tiny tube feet. Echinoderms are well represented as fossils, and are first found from the Cambrian period. Their embryological development indicates that they are distantly related to the vertebrates. [ECHINUS + Gk *derma -atos* skin]

echinoid /ˈekɪnɔɪd, ɪˈkaɪn-/ n. a sea urchin.

echinus /ɪˈkaɪnəs/ n. **1** a sea urchin of the genus *Echinus*; esp. *E. esculentus*, the common European edible urchin. **2** *Archit.* a rounded moulding below an abacus on a Doric or Ionic capital. [ME f. L f. Gk *ekhinos* hedgehog, sea urchin]

Echo /ˈekəʊ/ *Gk Mythol.* a nymph deprived of speech by Hera in order to stop her chatter, and left able only to repeat what others had said. On being repulsed by Narcissus she wasted away with grief until there was nothing left of her but her voice. In another account she was vainly loved by the god Pan, who finally caused some shepherds to go mad and tear her to pieces; Earth hid the fragments, which could still imitate other sounds.

echo /ˈekəʊ/ n. & v. ● n. (pl. **-oes**) **1 a** the repetition of a sound by the reflection of sound waves. **b** the secondary sound produced. **2** a reflected radio or radar beam. **3** a close imitation or repetition of something already done. **4** a person who slavishly repeats the words or opinions of another. **5** (often in *pl.*) circumstances or events reminiscent of or remotely connected with earlier ones. **6** *Bridge* etc. a conventional mode of play to show the number of cards held in the suit led etc. ● v. (**-oes, -oed**) **1** *intr.* **a** (of a place) resound with an echo. **b** (of a sound) be repeated; resound. **2** *tr.* repeat (a sound) by an echo. **3** *tr.* **a** repeat (another's words). **b** imitate the words, opinions, or actions of (a person). □ **echo chamber** an enclosure with sound-reflecting walls. **echo-sounder** a piece of sounding apparatus for determining the depth of the sea beneath a ship by measuring the time taken for an echo to be received (see also SONAR). **echo-sounding** the use of an echo-sounder. **echo verse** a verse form in which a line repeats the last syllables of the previous line. □ **echoer** n. **echoless** adj. [ME f. OF f. L f. Gk *ēkhō*, rel. to *ēkhē* a sound: see ECHO]

echocardiogram /ˌekəʊˈkɑːdɪəˌgræm/ n. *Med.* a record produced by echocardiography.

echocardiography /ˌekəʊˌkɑːdɪˈɒgrəfɪ/ n. *Med.* the use of ultrasound waves to investigate the action of the heart. □ **echocardiographer** n. **echocardiograph** /-ˈkɑːdɪəˌgrɑːf/ n.

echoencephalogram /ˌekəʊenˈsefələˌgræm, -ˈkefələˌgræm/ n. *Med.* a record produced by echoencephalography.

echoencephalography /ˌekəʊenˌsefəˈlɒgrəfɪ, -ˌkefəˈlɒgrəfɪ/ n. *Med.* the use of ultrasound waves to investigate intracranial structures.

echogram /ˈekəʊˌgræm/ n. a record made by an echo-sounder.

echograph /ˈekəʊˌgrɑːf/ n. a device for automatically recording echograms.

echoic /eˈkəʊɪk/ adj. (of a word) imitating the sound it represents; onomatopoeic. □ **echoically** adv.

echoism /ˈekəʊˌɪz(ə)m/ n. = ONOMATOPOEIA.

echolalia /ˌekəʊˈleɪlɪə/ n. **1** the meaningless repetition of another person's spoken words. **2** the repetition of speech by a child learning to talk. [mod.L f. Gk *ēkhō* echo + *lalia* talk]

echolocation /ˌekəʊləʊˈkeɪʃ(ə)n/ n. the location of objects by reflected sound, esp. ultrasound. It is used by bats and many marine mammals. (See also SONAR.)

echovirus /ˈekəʊˌvaɪərəs/ n. (also **ECHO virus**) any of a group of enteroviruses sometimes causing mild meningitis, encephalitis, etc. [f. *enteric cytopathogenic human orphan* (because not originally assignable to any known disease) + VIRUS]

echt /ext/ adj. authentic, genuine, typical. [G]

éclair /eɪˈkleə(r), ɪˈkleə(r)/ n. a small elongated cake of choux pastry filled with cream and iced with chocolate or coffee icing. [F, lit. 'lightning']

éclaircissement /ˌeɪkleəˈsiːsmɒn/ n. *archaic* an enlightening explanation of something hitherto inexplicable (e.g. conduct etc.). [F f. *éclaircir* clear up]

eclampsia /ɪˈklæmpsɪə/ n. a condition involving convulsions leading to coma, occurring esp. in pregnant women. □ **eclamptic** /ɪˈklæmptɪk/ adj. [mod.L f. F *eclampsie* f. Gk *eklampsis* sudden development f. *eklampō* shine forth]

éclat /eɪˈklɑː/ n. **1** brilliant display; dazzling effect. **2** social distinction; conspicuous success; universal approbation (*with great éclat*). [F f. *éclater* burst out]

eclectic /ɪˈklektɪk/ adj. & n. ● adj. **1** deriving ideas, tastes, style, etc., from various sources. **2** *Philos.* & *Art* selecting one's beliefs etc. from various sources; attached to no particular school of philosophy. ● n. **1** an eclectic person. **2** a person who subscribes to an eclectic school of thought. □ **eclectically** adv. **eclecticism** /-tɪˌsɪz(ə)m/ n. [Gk *eklektikos* f. *eklegō* pick out]

Eclipse /ɪˈklɪps/ the most famous racehorse of the 18th century, one of the ancestors in the direct male line of all thoroughbred racehorses throughout the world. The Eclipse Stakes, a horse-race run annually at Sandown Park near London since 1886, is named in the horse's honour.

eclipse /ɪˈklɪps/ n. & v. ● n. **1** *Astron.* the obscuring of the light from one celestial body by the presence of another. (*See note below.*) **2 a** a deprivation of light or the period of this. **b** obscuration or concealment; a period of this. **3** a rapid or sudden loss of importance or prominence, esp. in relation to another or a newly arrived person or thing. **4** a phase during which a bird's distinctive markings are obscured by moulting of the breeding plumage. ● v.tr. **1** (of a celestial body) obscure the light from or to (another). **2** intercept (light, esp. of a lighthouse). **3** deprive of prominence or importance; outshine, surpass. □ **eclipsing binary** *Astron.* a binary star whose brightness varies periodically as one of the two components passes in front of the other. **in eclipse 1** surpassed; in decline. **2** (of a bird) having lost its breeding plumage. [ME f. OF f. L f. Gk *ekleipsis* f. *ekleipō* fail to appear, be eclipsed f. *leipō* leave]

▪ An eclipse can be caused by the moon passing across the sun's disc (a solar eclipse, or eclipse of the sun), by the moon passing into the earth's shadow (a lunar eclipse, or eclipse of the moon), or by one star of a binary system passing behind the other. Eclipses can be partial, total, or annular.

ecliptic /ɪˈklɪptɪk/ n. & adj. Astron. ● n. the sun's apparent path among the stars during the year. (See note below.) ● adj. of an eclipse or the ecliptic. [ME f. L f. Gk ekleiptikos (as ECLIPSE)]

▪ The ecliptic is the average plane of the earth's orbit around the sun, and, since it also marks the approximate plane of the solar system, the planets are always seen close to it. It is shown as a great circle on the celestial sphere, and is so called because lunar and solar eclipses can only occur when the moon crosses it.

eclogue /ˈeklɒg/ n. a short poem, esp. a pastoral dialogue. [L ecloga f. Gk eklogē selection f. eklegō pick out]

eclosion /ɪˈkləʊʒ(ə)n/ n. Zool. the emergence of an insect from a pupa-case or of a larva from an egg. [F éclosion f. éclore hatch (as EX-[1], L claudere to close)]

Eco /ˈekəʊ/, Umberto (b.1932), Italian novelist and semiotician. Professor of semiotics at the University of Bologna since 1971, he is known both for his extensive writings on his subject, such as Travels in Hyperreality (1986), and as a novelist. His best-known fictional work is The Name of the Rose (1981), a complex detective novel set in a medieval monastery.

eco- /ˈiːkəʊ/ comb. form ecology, ecological.

ecocide /ˈiːkəʊˌsaɪd/ n. destruction of the natural environment. [ECO- + -CIDE]

ecoclimate /ˈiːkəʊˌklaɪmɪt/ n. climate considered as an ecological factor.

eco-friendly /ˌiːkəʊˈfrendlɪ/ adj. not harmful to the environment.

eco-label /ˈiːkəʊˌleɪb(ə)l/ n. a label used to identify products that satisfy certain environmental considerations regarding their manufacture, consumption, or disposal. □ **eco-labelling** n.

ecology /ɪˈkɒlədʒɪ/ n. 1 the branch of biology dealing with the relations of organisms to one another and to their physical surroundings. (See note below.) 2 (in full **human ecology**) the study of the interaction of people with their environment. □ **ecologist** n. **ecological** /ˌiːkəˈlɒdʒɪk(ə)l/ adj. **ecologically** adv. [G Ökologie f. Gk oikos house]

▪ Ecology is a relatively new science; the word is said to have been coined by E. H. Haeckel in 1869. It may be concerned with the study of a single species in its environment, or of a complete ecosystem. Aspects that are investigated by ecologists include the dynamics and distribution of populations, the interactions between species, the effects of the physical environment, and the energy flow through the system. The study of human environmental impact is forming an increasingly important part of ecology, especially as applied to conservation.

Econ. abbr. Economics.

econometrics /ɪˌkɒnəˈmetrɪks/ n.pl. (usu. treated as sing.) a branch of economics concerned with the application of mathematical measures to economic data by the use of statistics. □ **econometric** adj. **econometrical** adj. **econometrist** n. **econometrician** /-məˈtrɪʃ(ə)n/ n. [ECONOMY + -METRIC]

economic /ˌiːkəˈnɒmɪk, ˌek-/ adj. 1 of or relating to economics. 2 maintained for profit; on a business footing. 3 adequate to repay or recoup expenditure with some profit (not economic to run buses on Sunday; an economic rent). 4 practical; considered or studied with regard to human needs (economic geography). [ME f. OF economique or L oeconomicus f. Gk oikonomikos (as ECONOMY)]

economical /ˌiːkəˈnɒmɪk(ə)l, ˌek-/ adj. sparing in the use of resources; avoiding waste. □ **economically** adv.

Economic and Social Committee (abbr. **ESC**) a consultative body of the European Community, set up in 1957 to represent the interests of the Community's citizens and composed of representatives of employers, workers, consumers, farmers, professional people, etc. of the member states. It meets in Brussels.

economics /ˌiːkəˈnɒmɪks, ˌek-/ n.pl. (often treated as sing.) the science of the production and distribution of wealth. Early schools of economics, principally mercantilism, were concerned chiefly with a nation's balance of trade and with the amount of precious metal held; they often resulted simply in protectionism. Criticism of mercantilism came from the philosopher David Hume and the French physiocrats, who advocated free trade and believed that government should be limited to the preservation of the natural order. Such laissez-faire or free-market ideas were taken up by Adam Smith, often regarded as the founder of modern economics. His work formed the basis of the

classical economics of the 19th century, in which the central tenet was competition and in which the law of supply and demand was expected to regulate the prices of commodities satisfactorily. These ideas held sway until the 1930s, when John Maynard Keynes attempted to address the problems of the Depression by opposing the classical ideal of the balanced budget, arguing that government spending, financed by borrowing, was necessary to provide full employment. Keynes's theories were enormously influential for forty years, but were later criticized as tending to lead to inflation. In the 1970s and 1980s monetarism became influential in the West and was utilized by governments (e.g. under Margaret Thatcher in the UK) intent on fighting inflation. With the collapse of the centrally planned economies of the former Soviet bloc, the free market has held increasing sway, although monetarist ideas have attracted criticism as leading to unemployment.

economist /ɪˈkɒnəmɪst/ n. 1 an expert in or student of economics. 2 a person who manages financial or economic matters. [Gk oikonomos (as ECONOMY) + -IST]

economize /ɪˈkɒnəˌmaɪz/ v.intr. (also **-ise**) 1 be economical; make economies; reduce expenditure. 2 (foll. by on) use sparingly; spend less on. □ **economizer** n. **economization** /ɪˌkɒnəmaɪˈzeɪʃ(ə)n/ n.

economy /ɪˈkɒnəmɪ/ n. (pl. **-ies**) 1 a the wealth and resources of a community, esp. in terms of the production and consumption of goods and services. b a particular kind of this (a capitalist economy). c the administration or condition of an economy. 2 a the careful management of (esp. financial) resources; frugality. b (often in pl.) an instance of this (made many economies). 3 sparing or careful use (economy of language). 4 (also **economy class**) the cheapest class of air travel. 5 (attrib.) (also **economy-size**) (of goods) consisting of a large quantity for a proportionally lower cost. [F économie or L oeconomia f. Gk oikonomia household management f. oikos house + nemō manage]

ecosphere /ˈiːkəʊˌsfɪə(r)/ n. 1 the region of space including planets where conditions are such that living things can exist. 2 = BIOSPHERE.

écossaise /ˌeɪkɒˈseɪz/ n. 1 an energetic dance in duple time. 2 the music for this. [F, fem. of écossais Scottish]

ecosystem /ˈiːkəʊˌsɪstəm/ n. a biological community of interacting organisms and their physical environment.

eco-terrorism /ˌiːkəʊˈterəˌrɪz(ə)m/ n. terrorist activity carried out to further environmentalist ends. □ **eco-terrorist** n.

ecotype /ˈiːkəʊˌtaɪp/ n. Biol. a distinct form of a species occupying a particular habitat.

ecru /ˈeɪkruː/ n. the colour of unbleached linen; light fawn. [F écru unbleached]

ECSC see EUROPEAN COAL AND STEEL COMMUNITY.

ecstasize /ˈekstəˌsaɪz/ v.tr. & intr. (also **-ise**) throw or go into ecstasies.

ecstasy /ˈekstəsɪ/ n. (pl. **-ies**) 1 an overwhelming feeling of joy or rapture. 2 Psychol. an emotional or religious frenzy or trance-like state. 3 (usu. **Ecstasy**) sl. methylenedioxymethamphetamine, a powerful stimulant and hallucinatory drug (see also MDMA). [ME f. OF extasie f. LL extasis f. Gk ekstasis standing outside oneself f. ek out + histēmi to place]

ecstatic /ɪkˈstætɪk/ adj. & n. ● adj. 1 in a state of ecstasy. 2 very enthusiastic or excited (was ecstatic about his new job). 3 producing ecstasy; sublime (an ecstatic embrace). ● n. a person subject to (usu. religious) ecstasy. □ **ecstatically** adv. [F extatique f. Gk ekstatikos (as ECSTASY)]

ECT abbr. electroconvulsive therapy.

ecto- /ˈektəʊ/ comb. form outside. [Gk ekto- stem of ektos outside]

ectoderm /ˈektəʊˌdɜːm/ n. Zool. the outermost layer of an embryo in early development. □ **ectodermal** /ˌektəʊˈdɜːm(ə)l/ adj.

ectogenesis /ˌektəʊˈdʒenɪsɪs/ n. Biol. the production of structures outside the organism. □ **ectogenetic** /-dʒɪˈnetɪk/ adj. **ectogenic** /-ˈdʒenɪk/ adj. **ectogenous** /ekˈtɒdʒɪnəs/ adj. [mod.L (as ECTO-, GENESIS)]

ectomorph /ˈektəʊˌmɔːf/ n. a person with a lean and delicate build of body and large skin surface in comparison with weight (cf. ENDOMORPH, MESOMORPH). □ **ectomorphy** n. **ectomorphic** /ˌektəʊˈmɔːfɪk/ adj. [ECTO- + Gk morphē form]

-ectomy /ˈektəmɪ/ comb. form denoting a surgical operation in which a part of the body is removed (appendectomy). [Gk ektomē excision f. ek out + temnō cut]

ectoparasite /ˌektəʊˈpærəˌsaɪt/ n. a parasite that lives on the body surface of its host (cf. ENDOPARASITE).

ectopic /'ɛktɒpɪk/ *adj. Med.* in an abnormal place or position. □ **ectopic pregnancy** a pregnancy occurring outside the womb. [mod.L *ectopia* f. Gk *ektopos* out of place]

ectoplasm /'ɛktəʊˌplæz(ə)m/ *n.* **1** *Biol.* the clear viscous outer layer of the cytoplasm in unicellular animals such as amoeba (cf. ENDOPLASM). **2** a viscous substance supposedly exuding from the body of a medium during a spiritualistic trance. □ **ectoplasmic** /ˌɛktəʊˈplæzmɪk/ *adj.*

ecu /'eɪkjuː/ *n.* (also **Ecu, ECU**) European Currency Unit, the official monetary unit of the European Community, used to evaluate on a common basis the exchange rates and reserves of members of the European Monetary System and in trading Eurobonds. There are plans to make the ecu eventually the common European unit of currency, but these have been fiercely resisted, especially by Britain. It was introduced in 1979.

Ecuador /'ɛkwəˌdɔː(r)/ an equatorial republic in South America, on the Pacific coast; pop. (est. 1991) 10,503,000; languages, Spanish (official), Quechua; capital, Quito. Ranges and plateaux of the Andes separate the coastal plain from the tropical forests of the Amazon basin. Formerly part of the Inca empire, Ecuador was conquered by the Spanish in 1534 and remained part of Spain's American empire until, after the first uprising against Spanish rule in 1809, independence was gained in 1822. □ **Ecuadorean** /ˌɛkwəˈdɔːrɪən/ *adj. & n.*

ecumenical /ˌiːkjuːˈmɛnɪk(ə)l, ˌɛk-/ *adj.* **1** of or representing the whole Christian world. **2** seeking or promoting worldwide Christian unity. □ **Ecumenical Patriarch** a title of the Orthodox Patriarch of Constantinople. □ **ecumenically** *adv.* [LL *oecumenicus* f. Gk *oikoumenikos* of the inhabited earth (*oikoumenē*)]

ecumenical movement a movement for the reunification of the various branches of the Christian Church worldwide. Its beginnings can be traced back to the evangelicalism of the 18th and 19th centuries, and to the missionary expansion of the 19th century, which benefited from interdenominational cooperation. Interest aroused by the Edinburgh Missionary Conference in 1910 contributed to the eventual formation of the World Council of Churches in 1948.

ecumenism /iːˈkjuːməˌnɪz(ə)m/ *n.* the principle or aim of the unity of Christians worldwide transcending differences of doctrines (see ECUMENICAL MOVEMENT).

eczema /'ɛksɪmə/ *n.* inflammation of the skin, with itching and discharge from blisters. □ **eczematous** /ɛkˈsiːmətəs, ɛkˈsɛm-/ *adj.* [mod.L f. Gk *ekzema -atos* f. *ek* out + *zeō* boil]

ed. *abbr.* **1** edited by. **2** edition. **3** editor. **4** educated; education.

-ed¹ /əd, ɪd/ *suffix* forming adjectives: **1** from nouns, meaning 'having, wearing, affected by, etc.' (*talented; trousered; diseased*). **2** from phrases of adjective and noun (*good-humoured; three-cornered*). [OE -*ede*]

-ed² /əd, ɪd/ *suffix* forming: **1** the past tense and past participle of weak verbs (*needed; risked*). **2** participial adjectives (*escaped prisoner; a pained look*). [OE -*ed*, -*ad*, -*od*]

edacious /ɪˈdeɪʃəs/ *adj. literary* or *joc.* **1** greedy. **2** of or relating to eating. □ **edacity** *n.* /ɪˈdæsɪtɪ/ [L *edax -acis* f. *edere* eat]

Edam¹ /'iːdæm/ a town in the Netherlands, to the north-east of Amsterdam; pop. (1991) 24,840 (with Volendam). It is noted for its cheese.

Edam² /'iːdæm/ *n.* a round Dutch cheese, usu. pale yellow with a red rind.

edaphic /ɪˈdæfɪk/ *adj.* **1** *Bot.* of or relating to the soil. **2** *Ecol.* produced or influenced by the soil. [G *edaphisch* f. Gk *edaphos* floor]

Edda /'ɛdə/ either of two 13th-century Icelandic books, the *Elder* or *Poetic Edda*, a collection of Old Norse poems on mythical or traditional subjects, and the *Younger* or *Prose Edda* (by the Icelandic historian Snorri Sturluson), a handbook to Icelandic poetry, with prosodic and grammatical treatises, quotations, and prose paraphrases from old poems. The Eddas are the chief source of knowledge of Scandinavian mythology. [perh. a name in a Norse poem or f. ON *ōthr* poetry]

Eddington /'ɛdɪŋtən/, Sir Arthur Stanley (1882–1944), English astronomer, founder of the science of astrophysics. He established the fundamental principles of stellar structure, discovered the relationship between stellar mass and luminosity, and suggested possible sources of the energy within stars. He wrote one of the finest presentations of Einstein's theory of relativity, and provided some of the best evidence in support of it when his observations of star positions during the solar eclipse of 1919 demonstrated the bending of light by gravity.

eddo /'ɛdəʊ/ *n.* (*pl.* **-oes**) = TARO. [W. Afr. word]

Eddy /'ɛdɪ/, Mary Baker (1821–1910), American religious leader and founder of the Christian Science movement. Long a victim to various ailments, she believed herself cured by a faith-healer, Phineas Quimby (1802–66). After his death she evolved her own system of spiritual healing, set out in her book *Science and Health* (1875), and established the Church of Christ, Scientist, in Boston in 1879. (See also CHRISTIAN SCIENCE.)

eddy /'ɛdɪ/ *n. & v.* ● *n.* (*pl.* **-ies**) **1** a circular movement of water causing a small whirlpool. **2** a movement of wind, fog, or smoke resembling this. ● *v.tr. & intr.* (**-ies, -ied**) whirl round in eddies. □ **eddy current** *Electr.* a localized current induced in a conductor by a varying magnetic field. [prob. OE *ed-* again, back, perh. of Scand. orig.]

Eddystone Rocks /'ɛdɪstən/ a rocky reef off the coast of Cornwall, 22 km (14 miles) SW of Plymouth. The reef was the site of the earliest lighthouse built on rocks fully exposed to the sea, which was a timber structure completed in 1699 and destroyed in a storm four years later. The third Eddystone Lighthouse (1759) was a revolutionary structure of interlocking stone, which was replaced by the present one in 1882.

edelweiss /'eɪd(ə)lˌvaɪs/ *n.* a composite alpine plant, *Leontopodium alpinum*, with woolly white bracts around the flower-heads, growing in rocky places. [G f. *edel* noble + *weiss* white]

edema US var. of OEDEMA.

Eden¹ /'iːd(ə)n/ (also **Garden of Eden**) the abode of Adam and Eve in the biblical account of the Creation, from which they were expelled for their disobedience in eating the forbidden fruit of the tree of knowledge. [ME f. LL f. Gk *Ēdēn* f. Heb. *'ēḏen*, orig. = delight]

Eden² /'iːd(ə)n/, (Robert) Anthony, 1st Earl of Avon (1897–1977), British Conservative statesman, Prime Minister 1955–7. He served as War Secretary under Churchill in 1940, in addition to three terms as Foreign Secretary (1935–8; 1940–5; 1951–5). His premiership was dominated by the Suez crisis of 1956. Widespread opposition to Britain's role in this, together with his own failing health, led to his resignation.

edentate /ɪˈdɛnteɪt/ *adj. & n.* ● *adj.* **1** having no or few teeth. **2** *Zool.* belonging to the order Edentata (or Xenarthra), which comprises mammals without incisor or canine teeth, and includes the anteaters, sloths, and armadillos. ● *n. Zool.* an edentate mammal. [L *edentatus* (as E-, *dens dentis* tooth)]

Edgar /'ɛdɡə(r)/ (944–75), king of England 959–75. He became king of Northumbria and Mercia in 957 when these regions renounced their allegiance to his elder brother Edwy, succeeding to the throne of England on Edwy's death. Edgar worked closely with St Dunstan during his reign and was renowned for his support of organized religion; he played an important role in the growth of monasticism.

edge /ɛdʒ/ *n. & v.* ● *n.* **1** a boundary line or margin of an area or surface. **2** a narrow surface of a thin object. **3** the meeting-line of two surfaces of a solid. **4 a** the sharpened side of the blade of a cutting instrument or weapon. **b** the sharpness of this (*the knife has lost its edge*). **5** the area close to a steep drop (*along the edge of the cliff*). **6** anything compared to an edge, esp. the crest of a ridge. **7 a** effectiveness, force; incisiveness. **b** keenness, excitement (esp. as an element in an otherwise routine situation). **8** an advantage, superiority. ● *v. tr. & intr.* (often foll. by *in, into, out*, etc.) move gradually or furtively towards an objective (*edged it into the corner; they edged towards the door*). **2** *tr.* **a** provide with an edge or border. **b** form a border to. **c** trim the edge of. **3** *tr.* sharpen (a knife, tool, etc.). **4** *tr. Cricket* strike (the ball) with the edge of the bat. □ **edge-tool** **1** a hand-worked or machine-operated cutting tool. **2** any implement with a sharp cutting edge. **have the edge on** (or **over**) have a slight advantage over. **on edge 1** tense and restless or irritable. **2** eager, excited. **on the edge of** almost involved in or affected by. **set a person's teeth on edge** (of a taste or sound) cause an unpleasant nervous sensation. **take the edge off** dull, weaken; make less effective or intense. □ **edgeless** *adj.* **edger** *n.* [OE *ecg* f. Gmc]

Edgehill, Battle of /ɛdʒˈhɪl/ the first pitched battle of the English Civil War (1642), fought at the village of Edgehill in the West Midlands. The Parliamentary army attempted to halt the Royalist army's march on London. Despite an early advantage by the Royalist cavalry, the battle ended with no clear winner and with heavy losses on both sides. Charles I and his army resumed their march but were still prevented from reaching London.

edgeways /'ɛdʒweɪz/ *adv.* (also **edgewise** /-waɪz/) **1** with the edge uppermost or towards the viewer. **2** edge to edge. □ **get a word in edgeways** contribute to a conversation when the dominant speaker pauses briefly.

Edgeworth /ˈedʒwəθ/, Maria (1767–1849), Irish novelist, born in England. She is best known for works such as *Castle Rackrent* (1800), a novel of Irish life, and she is considered an important figure in the development of the historical and regional novel. Other significant works include *Belinda* (1801), a portrait of contemporary English society.

edging /ˈedʒɪŋ/ n. **1** something forming an edge or border, e.g. a fringe of lace. **2** the process of making an edge. □ **edging-shears** shears for trimming the edges of a lawn.

edgy /ˈedʒɪ/ adj. (**edgier**, **edgiest**) **1** irritable; nervously anxious. **2** disjointed (*edgy rhythms*). □ **edgily** adv. **edginess** n.

edh var. of ETH.

edible /ˈedɪb(ə)l/ adj. & n. ● adj. fit or suitable to be eaten (cf. EATABLE). ● n. (in pl.) food. □ **edibility** /ˌedɪˈbɪlɪtɪ/ n. [LL *edibilis* f. *edere* eat]

edict /ˈiːdɪkt/ n. an order proclaimed by authority. □ **edictal** /ɪˈdɪktəl/ adj. [ME f. L *edictum* f. *edicere* proclaim]

Edict of Nantes see NANTES, EDICT OF.

edifice /ˈedɪfɪs/ n. **1** a building, esp. a large imposing one. **2** a complex organizational or conceptual structure. [ME f. OF f. L *aedificium* f. *aedis* dwelling + *-ficium* f. *facere* make]

edify /ˈedɪˌfaɪ/ v.tr. (**-ies**, **-ied**) (of a circumstance, experience, etc.) instruct and improve morally or intellectually. □ **edifying** adj. **edifyingly** adv. **edification** /ˌedɪfɪˈkeɪʃ(ə)n/ n. [ME f. OF *edifier* f. L *aedificare* (as EDIFICE)]

Edinburgh /ˈedɪnbərə/ the capital of Scotland, lying on the southern shore of the Firth of Forth; pop. (1991) 421,200. The city grew up round the 11th-century castle built by Malcolm III on a rocky ridge which dominates the landscape. Since 1947 it has hosted the Edinburgh Festival.

Edinburgh, Duke of see PHILIP, PRINCE.

Edinburgh Festival (in full **Edinburgh International Festival**) an international festival of the arts (esp. music and drama) held annually in Edinburgh since 1947, usually for the last three weeks of August. In addition to the main festival programme a flourishing fringe festival (the Fringe) has developed, in which plays and revues are staged by student, amateur, and experimental companies.

Edison /ˈedɪs(ə)n/, Thomas (Alva) (1847–1931), American inventor. He was employed by the age of 15 as a telegraph operator, from which he developed an interest in electricity and its applications. He took out the first of more than a thousand patents at the age of 21. His chief inventions include automatic telegraph systems, the mimeograph, the carbon microphone for telephones, the phonograph, the carbon filament lamp, and the nickel-iron accumulator. He created the precursor of the thermionic valve, and devised systems for generating and distributing electricity. Edison also established the practice of installing industrial laboratories in commercial organizations.

edit /ˈedɪt/ v. & n. ● v.tr. (**edited**, **editing**) **1 a** assemble, prepare, modify, or condense (written material, esp. the work of another or others) for publication. **b** prepare an edition of (an author's work). **2** be in overall charge of the content and arrangement of (a newspaper etc.). **3** take extracts from and collate (films, tape-recordings, etc.) to form a unified sequence. **4 a** prepare (data) for processing by a computer. **b** alter (a text entered in a word processor etc.). **5 a** reword to correct, or to alter the emphasis. **b** (foll. by *out*) remove (part) from a text etc. ● n. **1 a** a piece of editing. **b** an edited item. **2** a facility for editing. [F *éditer* (as EDITION): partly a back-form. f. EDITOR]

edition /ɪˈdɪʃ(ə)n/ n. **1 a** one of the particular forms in which a text is published (*paperback edition*; *pocket edition*). **b** a copy of a book in a particular form (*a first edition*). **2** a whole number of copies of a book, newspaper, etc., issued at one time. **3** a particular version or instance of a broadcast, esp. of a regular programme or feature. **4** a person or thing similar to or resembling another (*a miniature edition of her mother*). [F *édition* f. L *editio -onis* f. *edere* edit- put out (as E-, *dare* give)]

editio princeps /ɪˌdɪʃɪəʊ ˈprɪnseps/ n. (pl. ***editiones principes*** /ɪˌdɪʃɪˌəʊniːz ˈprɪnsɪˌpiːz/) the first printed edition of a book, text, etc. [L]

editor /ˈedɪtə(r)/ n. **1** a person who edits material for publication or broadcasting. **2** a person who directs the preparation of a newspaper or periodical, or a particular section of one (*sports editor*). **3** a person who selects or commissions material for publication. **4** a person who edits film, soundtracks, etc. **5** a computer program enabling the user to alter or rearrange text held in a computer. □ **editorship** n. [LL, = producer (of games), publisher (as EDIT)]

editorial /ˌedɪˈtɔːrɪəl/ adj. & n. ● adj. **1** of or concerned with editing or editors. **2** written or approved by an editor. ● n. a newspaper article written by or on behalf of an editor, esp. one giving an opinion on a topical issue. □ **editorialist** n. **editorialize** v.intr. (also **-ise**). **editorially** adv.

-edly /ɪdlɪ/ suffix forming adverbs from verbs, meaning 'in a manner characterized by performance of or undergoing of the verbal action' (*allegedly*; *disgustedly*; *hurriedly*).

Edmonton /ˈedməntən/ the capital of Alberta, Canada; pop. (1991) 703,070.

Edmund /ˈedmənd/ the name of two kings of England:

Edmund I (921–46), reigned 939–46. Soon after Edmund succeeded Athelstan, a Norse army took control of York and its dependent territories. From 941 Edmund set about recovering these northern territories, but after his death Northumbria fell again under Norse control.

Edmund II (known as Edmund Ironside) (c.980–1016), son of Ethelred the Unready, reigned 1016. Edmund led the resistance to Canute's forces in 1015 and on his father's death was proclaimed king. After some initial success he was defeated at Ashingdon in Essex (1016) and was forced to divide the kingdom with Canute, retaining only Wessex. On Edmund's death Canute became king of all England.

Edmund, St (born Edmund Rich) (c.1175–1240), English churchman and teacher. Archbishop of Canterbury 1234–40, he was the last Primate of all England and the first Oxford University teacher to be canonized. The Oxford college St Edmund Hall takes its name from him. Feast day, 16 Nov.

Edmund Campion, St see CAMPION, ST EDMUND.

Edmund Ironside /ˈaɪənˌsaɪd/ Edmund II of England (see EDMUND).

Edmund the Martyr, St (c.841–70), king of East Anglia 855–70. After the defeat of his army by the invading Danes in 870, tradition holds that he was captured and shot with arrows for refusing to reject the Christian faith or to share power with his pagan conqueror. His body was interred at Bury St Edmunds, Suffolk. Feast day, 20 Nov.

Edomite /ˈiːdəˌmaɪt/ adj. & n. ● adj. of Edom, an ancient region south of the Dead Sea, or its people. ● n. a member of an ancient people traditionally descended from Esau, living in Edom.

EDP abbr. electronic data processing.

EDT abbr. Eastern Daylight Time, one hour ahead of Eastern Standard Time.

EDTA abbr. Chem. ethylenediamine tetra-acetic acid, a common chelating agent.

educate /ˈedjʊˌkeɪt/ v.tr. (also absol.) **1** give intellectual, moral, and social instruction to (a pupil, esp. a child), esp. as a formal and prolonged process. **2** provide education for. **3** (often foll. by *in*, or to + infin.) train or instruct for a particular purpose. **4** advise; give information to. □ **educatable** adj. **educator** n. **educative** /-kətɪv/ adj. **educable** /-kəb(ə)l/ adj. **educability** /ˌedjʊkəˈbɪlɪtɪ/ n. [L *educare educat-*, rel. to *educere* EDUCE]

educated /ˈedjʊˌkeɪtɪd/ adj. **1** having had an education, esp. to a higher level than average. **2** resulting from a (good) education (*an educated accent*). **3** based on experience or study (*an educated guess*).

education /ˌedjʊˈkeɪʃ(ə)n/ n. **1 a** the act or process of educating or being educated; systematic instruction. **b** the knowledge gained from this. **2** a particular kind of or stage in education (*further education*; *a classical education*). **3 a** development of character or mental powers. **b** a stage in or aspect of this (*travel will be an education for you*). □ **educational** adj. **educationalist** n. **educationally** adv. **educationist** n. [F *éducation* or L *educatio* (as EDUCATE)]

educe /ɪˈdjuːs/ v.tr. **1** bring out or develop from latent or potential existence; elicit. **2** infer; elicit a principle, number, etc., from data. □ **educible** adj. **eduction** /ɪˈdʌkʃ(ə)n/ n. **eductive** /ɪˈdʌktɪv/ adj. [ME f. L *educere educt-* lead out (as E-, *ducere* lead)]

edutainment /ˌedjʊˈteɪnmənt/ n. entertainment with an educational aspect; infotainment. [EDUCATION + ENTERTAINMENT]

Edw. abbr. Edward.

Edward /ˈedwəd/ the name of six kings of England since the Norman Conquest and also one of Great Britain and Ireland and one of the United Kingdom:

Edward I (known as 'the Hammer of the Scots') (1239–1307), son of Henry III, reigned 1272–1307. After coming to the throne Edward did much to improve the ineffectual central administration he had inherited; this included summoning the Model Parliament (1295). His campaign against the Welsh Prince Llewelyn ended with the annexation of Wales in 1284. He failed to conquer Scotland, though he had a successful first campaign there in 1296, deposing the Scottish king, John de Baliol (c.1250–1313), who had made an alliance with the French against him. From 1297 to 1305 the Scots were in a state of armed insurrection, initially under the leadership of Sir William Wallace. Edward died on his way north to begin a new campaign against the Scots, who were now led by Robert the Bruce.

Edward II (1284–1327), son of Edward I, reigned 1307–27. The first English Prince of Wales, he soon proved unequal to the problems left him by his more military father, and early trouble with his barons led to civil war. In 1314 he invaded Scotland, only to be defeated by Robert the Bruce at Bannockburn in the same year. In 1326 Edward's wife, Isabella of France, allied herself with the exiled Roger de Mortimer to invade England. Edward was deposed in favour of his son and was murdered at Berkeley Castle, Gloucestershire.

Edward III (1312–77), son of Edward II, reigned 1327–77. In 1330 Edward ended the four-year regency of his mother Isabella and her lover Roger de Mortimer by taking control of his kingdom, banishing Isabella, and executing Mortimer. He supported Edward de Baliol (c.1283–1364), the pretender to the Scottish throne, and started the Hundred Years War with France by claiming the French throne in right of his mother (1337). Towards the end of his reign, effective government fell into the hands of his fourth son, John of Gaunt.

Edward IV (1442–83), son of Richard, Duke of York, reigned 1461–83. He became king after defeating the Lancastrian king Henry VI in battle in 1461. The early years of his reign were troubled by Lancastrian plots, but the most serious threat arose in 1470–1 as a result of an alliance between his old Lancastrian enemies and his disaffected former lieutenant, the Earl of Warwick. Edward was briefly forced into exile, but returned to crush his opponents at Tewkesbury (1471), thereafter ruling in relative peace until his death.

Edward V (1470–c.1483), son of Edward IV, reigned 1483, but was not crowned. Following his father's death, he was illegitimized on debatable evidence of the illegality of Edward IV's marriage; his throne was taken by his uncle, Richard III. (See PRINCES IN THE TOWER.)

Edward VI (1537–53), son of Henry VIII, reigned 1547–53. During his brief reign as a minor, England was effectively ruled by two protectors, the Duke of Somerset and the Duke of Northumberland. Nevertheless, the king's Protestant beliefs contributed significantly to the establishment of Protestantism as the state religion, especially with the publication of the *Book of Common Prayer* (1549). He was succeeded by his elder sister, Mary I.

Edward VII (1841–1910), son of Queen Victoria, reigned 1901–10. Edward was kept away from the conduct of royal affairs during the long reign of his mother. Although he played little part in government on finally coming to the throne, his popularity and willingness to make public appearances, both at home and abroad, helped revitalize the monarchy.

Edward VIII (1894–1972), son of George V, reigned 1936, but was not crowned. A popular Prince of Wales, Edward abdicated eleven months after coming to the throne in order to marry the American divorcee Mrs Wallis Simpson. Created Duke of Windsor, he served as Governor-General of the Bahamas during the Second World War before spending the rest of his life in France.

Edward, Lake a lake on the border between Uganda and Zaire. It is linked to Lake Albert by the Semliki river. Sir Henry Morton Stanley named the lake as Albert Edward Nyanza after Albert Edward, Prince of Wales, in 1889.

Edward, Prince, Edward Antony Richard Louis (b.1964), third son of Elizabeth II. He served in the Royal Marines from 1986 to 1987 after graduating from Cambridge, and has more recently worked in the theatre.

Edward, Prince of Wales see BLACK PRINCE.

Edwardian /ed'wɔːdɪən/ adj. & n. ● adj. of, characteristic of, or associated with the reign (1901–10) of King Edward VII. ● n. a person belonging to this period.

Edwards /'edwədz/, Gareth (Owen) (b.1947), Welsh Rugby Union player. His international career, during which he played chiefly at half-back, lasted from 1967 to 1978. He was appointed captain of the Welsh team in 1968, the youngest person ever to hold that position.

Edward the Confessor, St (c.1003–66), son of Ethelred the Unready, king of England 1042–66. Famed for his piety, Edward founded Westminster Abbey, where he was eventually buried. He was dominated through much of his reign by his wife's father, Earl Godwin (d.1053). In later years Edward took less interest in affairs of state,

letting effective control fall to Godwin's son, who eventually succeeded him as Harold II. He was canonized in 1161. Feast day, 13 Oct.

Edward the Elder (c.870–924), son of Alfred the Great, king of Wessex 899–924. During his reign he conquered lands previously held by the Danes, including East Anglia and the Midlands; on the death of his sister, the ruler of Mercia, in 918 he merged the kingdoms of Wessex and Mercia. His conquests made it possible for his son Athelstan to become the first king of all England in 925.

Edward the Martyr, St (c.963–78), son of Edgar, king of England 975–8. Shortly after his accession, the youthful Edward was faced by a challenge for the throne from supporters of his half-brother Ethelred, who eventually had him murdered at Corfe Castle, Dorset (978). Canonized in 1001, he became the subject of an important medieval cult. Feast day, 18 Mar.

Edwy /'edwɪ/ (also **Eadwig** /'edwɪg/) (d.959), king of England 955–7. He was probably only 15 years old when he became king. Edwy alienated a large part of his kingdom during his short reign; after Mercia and Northumbria had renounced him in favour of his brother Edgar in 957, he ruled only over the lands south of the Thames.

-ee /iː/ suffix forming nouns denoting: **1** the person affected by the verbal action (*addressee; employee; lessee*). **2** a person concerned with or described as (*absentee; bargee; refugee*). **3** an object of smaller size (*bootee*). [from or after AF past part. in -é f. L -atus]

EEC see EUROPEAN ECONOMIC COMMUNITY.

EEG abbr. electroencephalogram.

eejit /'iːdʒɪt/ n. (also **eegit**) esp. Ir. & Sc. dial. an idiot. [repr. dial. pronunc. of IDIOT]

eel /iːl/ n. **1** a snakelike fish with a slender body and poorly developed fins; esp. one of the genus *Anguilla*, living mainly in fresh water but breeding in the sea. **2** a slippery or evasive person or thing. □ **eel-grass 1** a marine plant of the genus *Zostera*, with long ribbon-like leaves. **2** a submerged freshwater plant of the genus *Vallisneria*. □ **eel-like** adj. **eely** adj. [OE ǣl f. Gmc]

Eelam /'iːləm/ the proposed homeland of the Tamil people of Sri Lanka. Since the early 1980s Tamil separatists (see TAMIL TIGER) have been fighting a virtual civil war for control of provinces in the north and east of the country, where there is a Tamil majority.

eelpout /'iːlpaʊt/ n. a small thick-lipped slender-bodied fish of the family Zoarcidae, with the dorsal and anal fins meeting to fuse with the tail. Also called *pout*. [OE ǣleputa (as EEL, POUT²)]

eelworm /'iːlwɜːm/ n. a small nematode worm, esp. one infesting plant roots.

e'en¹ /iːn/ archaic or poet. var. of EVEN¹.

e'en² /iːn/ Sc. var. of EVEN².

-een /iːn/ suffix Ir. forming diminutive nouns (*colleen*). [Ir. -ín dimin. suffix]

e'er /eə(r)/ poet. var. of EVER.

-eer /ɪə(r)/ suffix forming: **1** nouns meaning 'person concerned with or engaged in' (*auctioneer; mountaineer; profiteer*). **2** verbs meaning 'be concerned with' (*electioneer*). [from or after F -ier f. L -arius: cf. -IER, -ARY¹]

eerie /'ɪərɪ/ adj. (**eerier, eeriest**) gloomy and strange; weird, frightening (*an eerie silence*). □ **eerily** adv. **eeriness** n. [orig. N.Engl. and Sc. eri, of obscure orig.: cf. OE earg cowardly]

EETPU abbr. (in the UK) Electrical, Electronic, Telecommunications, and Plumbing Union.

EEU abbr. (in the UK) Electricians' and Engineers' Union.

ef- /ef, ɪf/ prefix assim. form of EX-¹ before f.

eff /ef/ v. sl. euphem. **1** tr. & intr. (often foll. by off) = FUCK (in expletive use). **2** intr. say fuck or similar coarse slang words. □ **effing and blinding** using coarse slang. [name of the letter F, as a euphemistic abbr.]

efface /ɪ'feɪs/ v. **1** tr. rub or wipe out (a mark etc.). **2** tr. (in abstract senses) obliterate; wipe out (*effaced it from his memory*). **3** tr. utterly surpass; eclipse (*success has effaced all previous attempts*). **4** refl. treat or regard oneself as unimportant. □ **effacement** n. [F effacer (as EX-¹, FACE)]

effect /ɪ'fekt/ n. & v. ● n. **1** the result or consequence of an action etc.; the significance or implication of this. **2** efficacy (*had little effect*). **3** an impression produced on a spectator, hearer, etc. (*the bright lights created an amazing effect*). **4** (in pl.) property, luggage. **5** (in pl.) the lighting, sound, etc., used to accompany a play, film, broadcast, etc. **6** Physics a physical phenomenon, usually named after its discoverer (*Doppler effect*). **7** the state of being operative. ● v.tr. bring about; accomplish. □ **bring** (or **carry**) **into effect** accomplish. **for effect** to create an impression. **give effect to** make operative. **in effect** for practical purposes; in reality. **take effect** become operative. **to the effect that** the general substance or gist being. **to that effect** having that result or implication. **with effect from** coming into operation at or on (a stated time or day). [ME f. OF *effect* or L *effectus* (as EX-[1], *facere* make)]

effective /ɪˈfektɪv/ *adj. & n.* ● *adj.* **1** having a definite or desired effect. **2** powerful in effect; efficient; impressive. **3 a** actual; existing in fact rather than officially or theoretically (*took effective control in their absence*). **b** actually usable; realizable; equivalent in its effect (*effective money; effective demand*). **4** coming into operation (*effective as from 1 May*). **5** (of manpower) fit for work or service. ● *n.* a soldier available for service. □ **effectively** *adv.* **effectiveness** *n.* **effectivity** /ˌefekˈtɪvɪtɪ/ *n.* [ME f. L *effectivus* (as EFFECT)]

effector /ɪˈfektə(r)/ *adj. & n. Biol.* ● *adj.* acting in response to a stimulus. ● *n.* an effector organ.

effectual /ɪˈfektjʊəl, -tjʊəl/ *adj.* **1** capable of producing the required result or effect; answering its purpose. **2** valid. □ **effectually** *adv.* **effectualness** *n.* **effectuality** /ɪˌfektjʊˈælɪtɪ, ɪˌfektjʊ-/ *n.* [ME f. med.L *effectualis* (as EFFECT)]

effectuate /ɪˈfektjʊˌeɪt, ɪˈfektjʊ-/ *v.tr.* cause to happen; accomplish. □ **effectuation** /ɪˌfektjʊˈeɪʃ(ə)n/ *n.* [med.L *effectuare* (as EFFECT)]

effeminate /ɪˈfemɪnət/ *adj.* (of a man) feminine in appearance or manner; unmasculine. □ **effeminacy** *n.* **effeminately** *adv.* [ME f. L *effeminatus* past part. of *effeminare* (as EX-[1], *femina* woman)]

effendi /eˈfendɪ/ *n.* (*pl.* **effendis**) **1** a man of education or standing in eastern Mediterranean or Arab countries. **2** a former title of respect or courtesy in Turkey. [f. Turk. *efendi* f. mod.Gk *afentēs* f. Gk *authentēs* lord, master: see AUTHENTIC]

efferent /ˈefərənt/ *adj. Physiol.* conducting outwards (*efferent nerves*; *efferent vessels*) (opp. AFFERENT). [L *efferre* (as EX-[1], *ferre* carry)]

effervesce /ˌefəˈves/ *v.intr.* **1** give off bubbles of gas; bubble. **2** (of a person) be lively or energetic. □ **effervescence** *n.* **effervescency** *n.* **effervescent** *adj.* [L *effervescere* (as EX-[1], *fervere* be hot)]

effete /ɪˈfiːt/ *adj.* **1 a** feeble and incapable. **b** effeminate. **2** worn out; exhausted of its essential quality or vitality. □ **effeteness** *n.* [L *effetus* worn out by bearing young (as EX-[1], FOETUS)]

efficacious /ˌefɪˈkeɪʃəs/ *adj.* (of a thing) producing or sure to produce the desired effect. □ **efficaciously** *adv.* **efficaciousness** *n.* **efficacy** /ˈefɪkəsɪ/ *n.* [L *efficax* (as EFFICIENT)]

efficiency /ɪˈfɪʃənsɪ/ *n.* (*pl.* **-ies**) **1** the state or quality of being efficient. **2** *Mech. & Physics* the ratio of useful work performed to the total energy expended or heat taken in. □ **efficiency bar** a point on a salary scale requiring evidence of efficiency for further promotion. [L *efficientia* (as EFFICIENT)]

efficient /ɪˈfɪʃ(ə)nt/ *adj.* **1** productive with minimum waste or effort. **2** (of a person) capable; acting effectively. □ **efficient cause** *Philos.* an agent that brings a thing into being or initiates a change. □ **efficiently** *adv.* [ME f. L *efficere* (as EX-[1], *facere* make, accomplish)]

effigy /ˈefɪdʒɪ/ *n.* (*pl.* **-ies**) a sculpture or model of a person. □ **in effigy** in the form of a (usu. crude) representation of a person. [L *effigies* f. *effingere* to fashion]

effleurage /ˌefluˈrɑːʒ/ *n. & v.* ● *n.* a form of massage involving a circular inward stroking movement made with the palm of the hand, used esp. during childbirth. ● *v.tr.* massage with a circular stroking movement. [F f. *effleurer* to skim]

effloresce /ˌefloˈres/ *v.intr.* **1** burst out into flower. **2** *Chem.* **a** (of a substance) turn to a fine powder on exposure to air. **b** (of salts) come to the surface and crystallize on it. **c** (of a surface) become covered with salt particles. □ **efflorescence** *n.* **efflorescent** *adj.* [L *efflorescere* (as EX-[1], *florere* to bloom f. *flos floris* flower)]

effluence /ˈefluəns/ *n.* **1** a flowing out (of light, electricity, etc.). **2** that which flows out. [F *effluence* or med.L *effluentia* f. L *effluere efflux-* flow out (as EX-[1], *fluere* flow)]

effluent /ˈefluənt/ *adj. & n.* ● *adj.* flowing forth or out. ● *n.* **1** sewage or industrial waste discharged into a river, the sea, etc. **2** a stream or lake flowing from a larger body of water. [L (as EFFLUENCE)]

effluvium /ɪˈfluːvɪəm/ *n.* (*pl.* **effluvia** /-vɪə/) an unpleasant or noxious odour or exhaled substance affecting the lungs or the sense of smell etc. [L, f. *effluere* flow out (see EFFLUENCE)]

efflux /ˈeflʌks/ *n.* = EFFLUENCE. □ **effluxion** /eˈflʌkʃ(ə)n/ *n.* [med.L *effluxus* (as EFFLUENCE)]

effort /ˈefət/ *n.* **1** strenuous physical or mental exertion. **2** a vigorous or determined attempt. **3** *Mech.* a force exerted. **4** *colloq.* the result of an attempt; something accomplished (*not bad for a first effort*). □ **effortful** *adj.* **effortfully** *adv.* [F f. OF *esforcier* ult. f. L *fortis* strong]

effortless /ˈefətlɪs/ *adj.* requiring or seeming to require no effort; natural, easy (*effortless progress; effortless grace*). □ **effortlessly** *adv.* **effortlessness** *n.*

effrontery /ɪˈfrʌntərɪ/ *n.* (*pl.* **-ies**) **1** shameless insolence; impudent audacity (esp. *have the effrontery to*). **2** an instance of this. [F *effronterie* f. *effronté* ult. f. LL *effrons -ontis* shameless (as EX-[1], *frons* forehead)]

effulgent /ɪˈfʌldʒ(ə)nt/ *adj. literary* radiant; shining brilliantly. □ **effulgence** *n.* **effulgently** *adv.* [L *effulgere* shine forth (as EX-[1], *fulgere* shine)]

effuse *adj. & v.* ● *adj.* /ɪˈfjuːs/ *Bot.* (of an inflorescence etc.) spreading loosely. ● *v.tr.* /ɪˈfjuːz/ **1** pour forth (liquid, light, etc.). **2** give out (ideas etc.). [ME f. L *effusus* past part. of *effundere effus-* pour out (as EX-[1], *fundere* pour)]

effusion /ɪˈfjuːʒ(ə)n/ *n.* **1** a copious outpouring. **2** *usu. derog.* an unrestrained flow of speech or writing. [ME f. OF *effusion* or L *effusio* (as EFFUSE)]

effusive /ɪˈfjuːsɪv/ *adj.* **1** gushing, demonstrative, exuberant (*effusive praise*). **2** *Geol.* (of igneous rock) poured out when molten and later solidified; volcanic. □ **effusively** *adv.* **effusiveness** *n.*

Efik /ˈefɪk/ *n. & adj.* ● *n.* (*pl.* same) **1** a member of a people of southern Nigeria. **2** the Niger-Congo language of this people. ● *adj.* of or relating to the Efik or their language.

EFL *abbr.* English as a foreign language.

eft /eft/ *n.* a newt. [OE *efeta*, of unkn. orig.]

EFTA /ˈeftə/ (also **Efta**) see EUROPEAN FREE TRADE ASSOCIATION.

EFTPOS /ˈeftpɒz, -pɒs/ *abbr.* electronic funds transfer at point-of-sale.

e.g. *abbr.* for example. [L *exempli gratia*]

egad /iːˈgæd/ *int. archaic or joc.* by God. [prob. orig. *a* ah + *God*]

egalitarian /ɪˌgælɪˈteərɪən/ *adj. & n.* ● *adj.* **1** of or relating to the principle of equal rights and opportunities for all (*an egalitarian society*). **2** advocating this principle. ● *n.* a person who advocates or supports egalitarian principles. □ **egalitarianism** *n.* [F *égalitaire* f. *égal* EQUAL]

Egas Moniz /ˌiːgæʒ məˈniːz/, Antonio Caetano de Abreu Freire (1874–1955), Portuguese neurologist. He developed cerebral angiography as a diagnostic technique, and pioneered the treatment of certain psychotic disorders by the use of prefrontal leucotomy. He shared a Nobel Prize for this in 1949. Egas Moniz was also an active politician and diplomat.

Egbert /ˈegbət/ (d.839), king of Wessex 802–39. In 825 he won a decisive victory near Swindon, bringing Mercian supremacy to an end, and annexed Kent, Essex, Surrey, and Sussex. In 829 Mercia itself fell to Egbert and Northumbria acknowledged his rule. By his death, Mercia had become independent again, but his reign foreshadowed the supremacy that Wessex later secured over all England.

Eger /ˈegə(r)/ a spa town in the north of Hungary, noted for the 'Bull's Blood' red wine produced in the surrounding region; pop. (1993) 63,365.

egg[1] /eg/ *n.* **1 a** the spheroidal reproductive body produced by females of animals such as birds, reptiles, fish, etc., enclosed in a protective layer and capable of developing into a new individual. **b** the egg of the domestic hen, used for food. **2** *Biol.* the female reproductive cell in animals and plants. **3** *colloq.* a person or thing qualified in some way (*a tough egg*). **4** anything resembling or imitating an egg, esp. in shape or appearance. □ **as sure as eggs is** (or **are**) **eggs** *colloq.* without any doubt. **bad egg 1** an egg that is rotten. **2** a thoroughly despicable person. **egg-beater 1** a device for beating eggs. **2** *US sl.* a helicopter. **egg-custard** = CUSTARD 1. **egg-flip** (or **-nog**) a drink of alcoholic spirit with beaten egg, milk, etc. **eggs** (or **egg**) **and bacon** a plant with orange-marked yellow flowers, esp. bird's-foot trefoil or yellow toadflax. **egg-spoon** a small spoon for eating a boiled egg. **egg-timer** a device for timing the cooking of an egg. **egg-tooth** a projection of an embryo bird or reptile used for breaking out of the shell. **egg-white** the white of an egg. **have** (or **put**) **all one's eggs in one basket** *colloq.* risk everything on a single venture. **with egg on one's face** *colloq.* made to look foolish. □ **eggless** *adj.* **eggy** *adj.* (**eggier**, **eggiest**). [ME f. ON, rel. to OE *æg*]

egg[2] /eg/ *v.tr.* (foll. by *on*) urge (*egged us on; egged them on to do it*). [ME f. ON *eggja* = EDGE]

eggar /ˈegə(r)/ *n.* (also **egger**) a large brown moth of the family

Lasiocampidae; esp. the *oak eggar* (*Lasiocampa quercus*), the larva of which spins an egg-shaped cocoon. [prob. f. EGG[1] + -ER[1]]

eggcup /ˈegkʌp/ n. a cup for holding a boiled egg.

egghead /ˈeghed/ n. *colloq.* an intellectual; an expert.

eggplant /ˈegplɑːnt/ n. esp. *N. Amer.* = AUBERGINE.

eggshell /ˈegʃel/ n. & adj. ● n. **1** the shell of an egg. **2** anything very fragile. ● adj. **1** (of china) thin and fragile. **2** (of paint) with a slight gloss finish.

eglantine /ˈeglənˌtaɪn/ n. sweet-brier. [ME f. F *églantine* f. OF *aiglent* ult. f. L *acus* needle]

Egmont, Mount /ˈegmənt/ (called in Maori *Taranaki*) a volcanic peak in North Island, New Zealand, rising to a height of 2,518 m (8,260 ft). It was named by Captain James Cook in honour of the British politician John Perceval, 2nd Earl of Egmont (1711–70).

ego /ˈiːgəʊ/ n. (pl. **-os**) **1** *Metaphysics* a conscious thinking subject. **2** *Psychol.* the part of the mind that reacts to reality and has a sense of individuality. **3** self-esteem, self-importance; a person's sense of this (*has a big ego*). □ **ego-ideal 1** *Psychol.* the part of the mind developed from the ego by an awareness of social standards. **2** (in general use) idealization of oneself. **ego-trip** *colloq.* activity etc. devoted entirely to one's own interests or feelings. [L, = I]

egocentric /ˌiːgəʊˈsentrɪk/ adj. **1** centred in the ego. **2** self-centred, egoistic. □ **egocentrically** adv. **egocentrism** n. **egocentricity** /-senˈtrɪsɪtɪ/ n. [EGO + -CENTRIC after *geocentric* etc.]

egoism /ˈiːgəʊˌɪz(ə)m/ n. **1** an ethical theory that treats self-interest as the foundation of morality. **2** systematic selfishness. **3** self-opinionatedness. **4** = EGOTISM 1. □ **egoist** n. **egoistic** /ˌiːgəʊˈɪstɪk/ adj. **egoistical** adj. [F *égoïsme* ult. f. mod.L *egoismus* (as EGO)]

egomania /ˌiːgəʊˈmeɪnɪə/ n. morbid egotism. □ **egomaniac** /-ˈmeɪnɪˌæk/ n. **egomaniacal** /-məˈnaɪək(ə)l/ adj.

egotism /ˈiːgəˌtɪz(ə)m/ n. **1** excessive use of 'I' and 'me'. **2** the practice of continually talking about oneself. **3** an exaggerated opinion of oneself. **4** selfishness. □ **egotize** v.intr. (also **-ise**). **egotist** n. **egotistic** /ˌiːgəˈtɪstɪk/ adj. **egotistical** adj. **egotistically** adv. [EGO + -ISM with intrusive -t-]

egregious /ɪˈɡriːdʒəs/ adj. **1** outstandingly bad; shocking (*egregious folly; an egregious ass*). **2** *archaic* or *joc.* remarkable. □ **egregiously** adv. **egregiousness** n. [L *egregius* illustrious, lit. 'standing out from the flock' f. *grex gregis* flock]

egress /ˈiːgres/ n. **1 a** a going out. **b** the right of going out. **2** an exit; a way out. **3** *Astron.* the end of an eclipse or transit. [L *egressus* f. *egredi egress-* (as E-, *gradi* to step)]

egret /ˈiːgrɪt/ n. a white heron of the genus *Egretta* or *Bulbulcus*, usu. having long white feathers in the breeding season. [ME, var. of AIGRETTE]

Egypt /ˈiːdʒɪpt/ a country in NE Africa bordering on the Mediterranean Sea; pop. (est. 1991) 53,087,000; official language, Arabic; capital, Cairo. The population of Egypt is concentrated chiefly along the fertile valley of the River Nile, the rest of the country being largely desert. Egypt's history spans 5,000 years, dating back to the neolithic period. The ancient kingdoms of Upper and Lower Egypt became united, according to tradition, under Menes (*c*.3100 BC), founder of the first of the thirty-one dynasties which successively ruled ancient Egypt. The period of the Old Kingdom (*c*.2575–2134 BC, 4th–8th dynasty), the 'Pyramid Age', was characterized by strong central government, based at Memphis. The Middle Kingdom (*c*.2040–1640 BC, 11th–14th dynasty) is considered to be the classical age of ancient Egyptian culture; the period ended with the Hyksos usurpation. The New Kingdom (*c*.1550–1070 BC, 18th–20th dynasty), with its capital at Thebes, began with the expulsion of the Hyksos (who had formed the 15th and 16th dynasties) and was marked by imperial expansion in Syria, Palestine, and Nubia. A period of decline followed, and indigenous rule was finally ended by Alexander the Great, who took Egypt in 332 BC. For the next three centuries the country, and particularly Alexandria, was a major centre of Hellenistic culture. On the death of Cleopatra (30 BC), Egypt became a Roman province. It fell under Islamic rule in 642, from 1517 forming part of the Ottoman Empire except for a brief period (1798–1801) under French rule following Napoleon's invasion. The opening of the Suez Canal in 1869 made Egypt strategically important, and when the Turks became allies of Germany in the First World War the British (who had installed themselves following a nationalist revolt in 1882) declared the country a British protectorate. Independence was declared in 1922 and a kingdom was established, becoming a republic after the overthrow of the monarchy in 1953. From 1958 to 1961 Egypt was united with Syria as the United Arab Republic, a title it retained until 1971. Wars with Israel were fought in 1967 (the Six Day War) and 1973 (the Yom Kippur or October War); the countries signed a peace treaty in 1979.

Egyptian /ɪˈdʒɪpʃ(ə)n/ adj. & n. ● adj. **1** of or relating to Egypt. **2** of or for Egyptian antiquities (e.g. in a museum) (*Egyptian room*). ● n. **1** a native of ancient or modern Egypt; a national of the Arab Republic of Egypt. **2** the Afro-Asiatic language used in ancient Egypt, attested from *c*.3000 BC. It is represented in its oldest stages by hieroglyphic inscriptions and in its latest form by Coptic. □ **Egyptianize** v.tr. (also **-ise**). **Egyptianization** /ɪˌdʒɪpʃənaɪˈzeɪʃ(ə)n/ n.

Egyptology /ˌiːdʒɪpˈtɒlədʒɪ/ n. the study of the language, history, and culture of ancient Egypt. □ **Egyptologist** n.

eh /eɪ/ int. *colloq.* **1** expressing enquiry or surprise. **2** inviting assent. **3** asking for something to be repeated or explained. [ME *ey*, instinctive exclam.]

Ehrenburg /ˈeərənˌbɜːg/, Ilya (Grigorevich) (1891–1967), Russian novelist and journalist. As a journalist, he became famous during the Second World War for his anti-German propaganda in *Pravda* and *Red Star*. His novels include *The Thaw* (1954), a work containing open criticism of Stalinism and dealing with the temporary period of liberalization following Stalin's death.

Ehrlich /ˈeəlɪx/, Paul (1854–1915), German medical scientist, one of the founders of modern immunology. He developed techniques for staining specific tissues, from which he became convinced that a disease organism could be destroyed by an appropriate magic bullet, thus pioneering the study of chemotherapy. Success in this came in 1911 when a synthetic compound of arsenic proved effective against syphilis.

-eian /ɪən/ suffix corresp. to *-ey* (or *-y*) + *-an* (*Bodleian*).

Eichmann /ˈaɪxmən/, (Karl) Adolf (1906–62), German Nazi administrator. He was responsible for carrying out Hitler's final solution and for administering the concentration camps, to which 6 million Jews were shipped from all over Europe. After the war he went into hiding in Argentina, but in 1960 he was traced by Israeli agents, abducted, and executed after trial in Israel.

Eid /iːd/ (also **Id**) a Muslim festival, esp.: **1** (in full **Eid ul-Adha** /ʊlˈɑːdə/ = 'feast of sacrifice') a festival marking the culmination of the annual pilgrimage (*Hajj*) to Mecca. **2** (in full **Eid ul-Fitr** /ʊlˈfiːtrə/ = 'feast of breaking fast') a festival marking the end of Ramadan. [Arab. *ʿīd* feast]

eider /ˈaɪdə(r)/ n. **1** (in full **eider duck**) a large northern sea-duck of the genus *Somateria* or *Polystica*; esp. *S. mollissima*, which is the source of eider-down. **2** (in full **eider-down**) small soft feathers from the breast of the eider duck. [Icel. *aethr*]

eiderdown /ˈaɪdəˌdaʊn/ n. a quilt stuffed with down (orig. from the eider) or some other soft material, esp. as the upper layer of bedclothes.

eidetic /aɪˈdetɪk/ adj. & n. ● adj. *Psychol.* (of a mental image) having unusual vividness and detail, as if actually visible. ● n. a person able to see eidetic images. □ **eidetically** adv. [G *eidetisch* f. Gk *eidētikos* f. *eidos* form]

eidolon /aɪˈdəʊlɒn/ n. (pl. **eidolons** or **eidola** /-lə/) **1** a spectre; a phantom. **2** an idealized figure. [Gk *eidōlon*: see IDOL]

Eiffel /ˈaɪf(ə)l, French ɛfɛl/, Alexandre Gustave (1832–1923), French engineer. He is best known as the designer and builder of the Eiffel Tower and as the architect of the inner structure of the Statue of Liberty.

Eiffel Tower a wrought-iron structure erected in Paris for the World Exhibition of 1889 and still a famous landmark, though at first greatly disliked. With a height of 300 metres (984 ft), it remained the tallest man-made structure for many years.

eigen- /ˈaɪgən/ comb. form *Math. & Physics* proper, characteristic. [G *eigen* OWN]

eigenfrequency /ˈaɪgənˌfriːkwənsɪ/ n. (pl. **-ies**) *Math. & Physics* one of the natural resonant frequencies of a system.

eigenfunction /ˈaɪgənˌfʌŋkʃ(ə)n/ n. *Math. & Physics* that function which under a given operation generates some multiple of itself.

eigenvalue /ˈaɪgənˌvæljuː/ n. *Math. & Physics* that value by which an eigenfunction of an operation is multiplied after the eigenfunction has been subjected to that operation.

Eiger /ˈaɪgə(r)/ a mountain peak in the Bernese Alps in central Switzerland, which rises to 3,970 m (13,101 ft).

Eigg /eg/ an island of the Inner Hebrides, off the west coast of Scotland to the south of Skye.

eight /eɪt/ n. & adj. • n. **1** one more than seven, or two less than ten; the product of two units and four units. **2** a symbol for this (8, viii, VIII). **3** (in full **figure of eight**) a figure resembling the form of 8, esp. in ice-skating and country dancing. **4** a size etc. denoted by eight. **5** an eight-oared rowing-boat or its crew. **6** eight o'clock. **7** a card with eight pips. • adj. that amount to eight. □ **have one over the eight** sl. get slightly drunk. **number eight** (in Rugby Union) the forward at the back of the scrum. [OE *ehta, eahta*]

Eight, the a group of American realist painters who exhibited together in 1908, united by opposition to institutions such as the National Academy of Design and by a concern to bring painting closer to contemporary life. They painted mainly urban scenes but were not unified stylistically; the dominant member of the group was Robert Henri.

eighteen /eɪˈtiːn/ n. & adj. • n. **1** one more than seventeen, or eight more than ten; the product of two units and nine units. **2** a symbol for this (18, xviii, XVIII). **3** a size etc. denoted by eighteen. **4** a set or team of eighteen individuals. **5** (**18**) Brit. (of films) classified as suitable for persons of 18 years and over. • adj. that amount to eighteen. □ **eighteenth** adj. & n. [OE *ehtatēne, eaht-*]

eighteenmo /eɪˈtiːnməʊ/ n. = OCTODECIMO.

eightfold /ˈeɪtfəʊld/ adj. & adv. **1** eight times as much or as many. **2** consisting of eight parts. **3** amounting to eight.

eightfold path n. the fourth of the four noble truths of Buddhism, prescribing the way to nirvana. (See BUDDHISM.)

eighth /eɪtθ/ n. & adj. • n. **1** the position in a sequence corresponding to the number 8 in the sequence 1–8. **2** something occupying this position. **3** one of eight equal parts of a thing. • adj. that is the eighth. □ **eighth note** esp. N. Amer. Mus. = QUAVER. □ **eighthly** adv.

eightsome /ˈeɪtsəm/ n. **1** (in full **eightsome reel**) a lively Scottish reel for eight dancers. **2** the music for this.

8vo abbr. octavo.

eighty /ˈeɪtɪ/ n. & adj. • n. (pl. **-ies**) **1** the product of eight and ten. **2** a symbol for this (80, lxxx, LXXX). **3** (in pl.) the numbers from 80 to 89, esp. the years of a century or of a person's life. • adj. that amount to eighty. □ **eighty-first, -second**, etc. the ordinal numbers between eightieth and ninetieth. **eighty-one, -two**, etc. the cardinal numbers between eighty and ninety. □ **eightieth** adj. & n. **eightyfold** adj. & adv. [OE *-eahtatig* (as EIGHT, -TY²)]

Eijkman /ˈeɪkmən/, Christiaan (1858–1930), Dutch physician. Working in Indonesia, Eijkman discovered the cause of beriberi to be dietary rather than bacteriological. Although he did not correctly recognize the reason for this, his work resulted in a simple cure for the disease. It also led later to the discovery of the vitamin thiamine, a deficiency of which causes beriberi. He shared a Nobel Prize in 1929.

Eilat /eɪˈlæt/ (also **Elat**) the southernmost town in Israel, a port and resort at the head of the Gulf of Aqaba; pop. (est. 1982) 19,500. Founded in 1949 near the ruins of biblical Elath, it is Israel's only outlet to the Red Sea.

Eindhoven /ˈaɪndˌhəʊv(ə)n, Dutch ˈeɪntˌhoːvə/ a city in the south of the Netherlands; pop. (1991) 192,900. The city is a major producer of electrical and electronic goods.

einkorn /ˈaɪnkɔːn/ n. a kind of wheat, *Triticum monococcum*, now grown only for fodder. [G f. *ein* one + *Korn* seed]

Einstein /ˈaɪnstaɪn/, Albert (1879–1955), German-born American theoretical physicist, founder of the theory of relativity, often regarded as the greatest scientist of the 20th century. In 1905 he published outstanding papers dealing with the photoelectric effect, Brownian motion, and his special theory of relativity (see RELATIVITY). In 1915, he succeeded in incorporating gravitation in his general theory of relativity, which was vindicated when one of its predictions was observed during the solar eclipse of 1919. He was in America when Hitler came to power in 1933 and decided to stay there, spending the remainder of his life searching without success for a unified field theory embracing electromagnetism, gravitation, relativity, and quantum mechanics. In 1939 he wrote to President Roosevelt about the military potential of nuclear energy, greatly influencing the decision to build an atom bomb. After the war he spoke out passionately against nuclear weapons.

einsteinium /aɪnˈstaɪnɪəm/ n. an artificial radioactive metallic chemical element (atomic number 99; symbol **Es**). A member of the actinide series, einsteinium was discovered in 1953 in debris from the first hydrogen bomb explosion. [EINSTEIN]

Einthoven /ˈaɪntˌhəʊv(ə)n, Dutch ˈeɪntˌhoːvə/, Willem (1860–1927), Dutch physiologist. He devised the first electrocardiograph, using a string galvanometer with an optical system to amplify the deflection of a fine wire. He was subsequently able to link the resulting electrocardiograms with specific muscular contractions in the heart, and thus begin to diagnose various heart diseases.

Eire /ˈeərə/ the Gaelic name for Ireland. It was the official name of the former Irish Free State from 1937 to 1949, when the country became a republic, and is often used loosely to refer to the Republic of Ireland.

eirenic var. of IRENIC.

eirenicon /aɪˈriːnɪˌkɒn/ n. (also **irenicon**) a proposal made as a means of achieving peace. [Gk, neut. of *eirēnikos* (adj.) f. *eirēnē* peace]

Eisenhower /ˈaɪz(ə)nˌhaʊə(r)/, Dwight David ('Ike') (1890–1969), American general and Republican statesman, 34th President of the US 1953–61. In the Second World War he was Commander-in-Chief of Allied forces in North Africa and Italy 1942–3 and Supreme Commander of Allied Expeditionary Forces in western Europe 1943–5. As President, he adopted a hard line towards Communism both in his domestic and foreign policy; in the US an extreme version of this was reflected in McCarthyism.

Eisenstadt /ˈaɪz(ə)nˌʃtæt/ a city in eastern Austria, capital of the state of Burgenland; pop. (1991) 10,500. Formerly Hungarian, the city was ceded to Austria in 1920.

Eisenstein /ˈaɪz(ə)nˌstaɪn/, Sergei (Mikhailovich) (1898–1948), Soviet film director, born in Latvia. He made his name with *The Battleship Potemkin* (1925), a film commemorating the Russian Revolution of 1905; its innovative use of montage received international acclaim. At odds with the prevailing style of socialist realism, Eisenstein fell into disfavour from 1932 and had to wait for the release of *Alexander Nevsky* (1938) to regain his reputation. His final film was *Ivan the Terrible*; although the first part (1944) was well received, the second (1946) earned Stalin's disapproval and was not released until a decade after Eisenstein's death.

eisteddfod /aɪˈstedfəd, aɪˈsteðvɒd/ n. (pl. **eisteddfods** or **eisteddfodau** /ˌaɪsteðˈvɒdaɪ/) a congress of Welsh bards; a national or local festival for musical and poetic competitions. The annual National Eisteddfod dates from the 12th century, and was revived in the late 19th century. □ **eisteddfodic** /ˌaɪstedˈfɒdɪk/ adj. [Welsh, lit. 'session', f. *eistedd* sit]

either /ˈaɪðə(r), ˈiːðə(r)/ adj., pron., adv., & conj. • adj. & pron. **1** one or the other of two (*either of you can go; you may have either book*). **2** each of two (*houses on either side of the road; either will do*). • adv. & conj. **1** as one possibility (*is either black or white*). **2** as one choice or alternative; which way you will (*either come in or go out*). **3** (with neg.) any more than the other (*I didn't like it either; if you do not go, I shall not either*). **b** moreover (*there is no time to lose, either*). □ **either-or** n. an unavoidable choice between alternatives. • adj. involving such a choice. **either way** in either case or event. [OE *ægther* f. Gmc]

eiusdem generis /eɪˌʊsdem ˈdʒenərɪs/ adj. Law of the same kind. [L]

ejaculate v. & n. • v.tr. /ɪˈdʒækjʊˌleɪt/ (also absol.) **1** utter suddenly (words esp. of prayer or other emotion). **2** eject (fluid etc., esp. semen) from the body. • n. /ɪˈdʒækjʊlət/ semen etc. that has been ejaculated from the body. □ **ejaculator** /-ˌleɪtə(r)/ n. **ejaculatory** /-ˌleɪtərɪ/ adj. **ejaculation** /ɪˌdʒækjʊˈleɪʃ(ə)n/ n. [L *ejaculari* to dart (as E-, *jaculum* javelin)]

eject /ɪˈdʒekt/ v.tr. **1 a** send or drive out precipitately or by force, esp. from a building or other property; compel to leave. **b** dismiss from employment or office. **2 a** cause (the pilot etc.) to be propelled from an aircraft or spacecraft in an emergency. **b** (absol.) (of the pilot etc.) be ejected in this way (*they both ejected at 1,000 feet*). **3** cause to be removed or drop out (e.g. a spent cartridge from a gun). **4** dispossess (a tenant etc.) by legal process. **5** dart forth; emit. □ **ejective** adj. **ejectment** n. [L *ejicere eject-* (as E-, *jacere* throw)]

ejecta /ɪˈdʒektə/ n.pl. (also treated as sing.) material that is thrown out, esp. from a volcano or a star. [L, f.*ejicere eject-* EJECT]

ejection /ɪˈdʒekʃ(ə)n/ n. the act or an instance of ejecting; the process of being ejected. □ **ejection seat** = *ejector seat*.

ejector /ɪˈdʒektə(r)/ n. a device for ejecting. □ **ejector seat** a device for the automatic ejection of the pilot etc. of an aircraft or spacecraft in an emergency.

Ekaterinburg see YEKATERINBURG.

Ekaterinodar /jəˌkætəˈriːnəˌdɑː(r)/ (also **Yekaterinodar**) a former name (until 1922) for KRASNODAR.

Ekaterinoslav /jəˌkætəˈriːnəˌslɑːf/ (also **Yekaterinoslav**) a former name (1787–1926) for DNIPROPETROVSK.

eke /iːk/ v.tr. □ **eke out 1** (foll. by *with, by*) supplement; make the best use of (defective means etc.). **2** contrive to make (a livelihood) or support (an existence). [OE *ēacan*, rel. to L *augere* increase]

ekka /ˈekə/ n. Ind. a small one-horse vehicle. [Hindi *ekkā* unit]

Ekman /ˈekmən/, Vagn Walfrid (1874–1954), Swedish oceanographer. He recognized the importance of the Coriolis effect on ocean currents, showing that it can be responsible for surface water moving at an angle to the prevailing wind direction. He also explained why water flow at different depths can vary in both velocity and direction, and devised various instruments including a type of current meter that is still in use.

-el var. of -LE².

El Aaiún see LAʼYOUN.

elaborate adj. & v. ● adj. /ɪˈlæbərət/ **1** carefully or minutely worked out. **2** highly developed or complicated. ● v. /ɪˈlæbəˌreɪt/ **1 a** tr. work out or explain in detail. **b** tr. make more intricate or ornate. **c** intr. (often foll. by *on*) go into details (*I need not elaborate*). **2** tr. produce by labour. **3** tr. (of a natural agency) produce (a substance etc.) from its elements or sources. □ **elaborately** /-rətlɪ/ adv. **elaborateness** /-rətnɪs/ n. **elaborative** /-rətɪv/ adj. **elaborator** /-ˌreɪtə(r)/ n. **elaboration** /ɪˌlæbəˈreɪʃ(ə)n/ n. [L *elaboratus* past part. of *elaborare* (as E-, *labor* work)]

Elagabalus see HELIOGABALUS.

El Alamein, Battle of /el ˈæləˌmeɪn/ a battle of the Second World War fought in 1942 at El Alamein in Egypt, 90 km (60 miles) west of Alexandria. The German Afrika Korps under Rommel was halted in its advance towards the Nile by the British 8th Army under Montgomery, giving a decisive British victory.

Elam /ˈiːləm/ an ancient kingdom east of the Tigris, established in the 4th millennium BC. Its capital was at Susa. □ **Elamite** adj. & n.

élan /eɪˈlæn, eɪˈlɒn/ n. vivacity, dash. [F f. *élancer* launch]

eland /ˈiːlənd/ n. a large African antelope of the genus *Tragelaphus*, with spirally twisted horns. The giant eland, *T. derbianus*, is the largest living antelope. [Afrik. f. Du., = elk]

elapse /ɪˈlæps/ v.intr. (of time) pass by. [L *elabor elaps-* slip away]

elasmobranch /ɪˈlæzməˌbræŋk/ n. Zool. a cartilaginous fish of the subclass Elasmobranchii, comprising sharks, skates, and rays. [mod.L f. Gk *elasmos* beaten metal + *bragkhia* gills]

elasmosaurus /ɪˌlæzməˈsɔːrəs/ n. (also **elasmosaur** /ɪˈlæzməˌsɔː(r)/) a large plesiosaur with a very long flexible neck. [mod.L f. Gk *elasmos* beaten metal + *sauros* lizard]

elastane /ɪˈlæsteɪn/ n. an elastic polyurethane fibre used esp. for hosiery, underwear, and close-fitting clothing. [f. ELASTIC + -ANE²]

elastic /ɪˈlæstɪk/ adj. & n. ● adj. **1** able to resume its normal bulk or shape spontaneously after contraction, dilatation, or distortion. **2** springy. **3** (of a person or feelings) buoyant. **4** flexible, adaptable (*elastic conscience*). **5** Econ. (of demand) variable according to price. **6** Physics (of a collision) involving no decrease of kinetic energy. ● n. elastic cord or fabric, usu. woven with strips of rubber. □ **elastic band** = *rubber band* (see RUBBER¹). □ **elastically** adv. **elasticity** /ˌiːlæˈstɪsɪtɪ/ n.

elasticize /ɪˈlæstɪˌsaɪz/ v.tr. (also **-ise**) [mod.L *elasticus* f. Gk *elastikos* propulsive f. *elaunō* drive]

elasticated /ɪˈlæstɪˌkeɪtɪd/ adj. (of a fabric) made elastic by weaving with rubber thread.

elastin /ɪˈlæstɪn/ n. an elastic fibrous glycoprotein found in connective tissue. [ELASTIC + -IN]

elastomer /ɪˈlæstəmə(r)/ n. a natural or synthetic rubber or rubber-like plastic. □ **elastomeric** /ɪˌlæstəˈmerɪk/ adj. [ELASTIC, after *isomer*]

Elastoplast /ɪˈlæstəˌplɑːst/ n. propr. a type of sticking plaster. [ELASTIC + -O- + PLASTER]

Elat see EILAT.

elate /ɪˈleɪt/ v. & adj. ● v.tr. **1** (esp. as **elated** adj.) inspirit, stimulate. **2** make proud. ● adj. archaic in high spirits; exultant, proud. □ **elatedly** adv. **elatedness** n. **elation** /ɪˈleɪʃ(ə)n/ n. [ME f. L *efferre elat-* raise]

elater /ˈeɪlətə(r)/ n. a click beetle. [mod.L f. Gk *elatēr* driver f. *elaunō* drive]

E-layer /ˈiːˌleɪə(r)/ (also **E-region**) n. a layer of the ionosphere able to reflect medium-frequency radio waves. [E (arbitrary) + LAYER]

Elba /ˈelbə/ a small island off the west coast of Italy, famous as the place of Napoleon's first exile (1814–15).

Elbasan /ˌelbəˈsɑːn/ an industrial town in central Albania; pop. (1990) 70,000.

Elbe /elb/ a river of central Europe, flowing 1,159 km (720 miles) from the Czech Republic through Dresden, Magdeburg, and Hamburg, to the North Sea.

El Beqaʼa see BEKAA.

Elbert, Mount /ˈelbət/ a mountain in Colorado, to the east of the resort town of Aspen. Rising to 4,399 m (14,431 ft), it is the highest peak in the Rocky Mountains.

elbow /ˈelbəʊ/ n. & v. ● n. **1 a** the joint between the forearm and the upper arm. **b** the part of the sleeve of a garment covering the elbow. **2** an elbow-shaped bend or corner; a short piece of piping bent through a right angle. ● v.tr. (foll. by *in, out, aside*, etc.) **1** thrust or jostle (a person or oneself). **2** make (one's) way by thrusting or jostling. **3** nudge or poke with the elbow. □ **at one's elbow** close at hand. **elbow-grease** colloq. vigorous polishing; hard work. **elbow-room** colloq. adequate space to move or work in. **give a person the elbow** colloq. send a person away; dismiss or reject a person. **out at** (or **out at the**) **elbows 1** (of a garment) worn out. **2** (of a person) scruffy, poor. [OE *elboga, elnboga*, f. Gmc (as ELL, BOW¹)]

Elbrus /elˈbruːs/ a peak in the Caucasus mountains, on the border between Russia and Georgia. Rising to 5,642 m (18,481 ft), it is the highest mountain in Europe.

Elburz Mountains /elˈbʊəz/ a mountain range in NW Iran, close to the southern shore of the Caspian Sea. Damavand is the highest peak, rising to 5,604 m (18,386 ft).

Elche /ˈeltʃɪ/ a town in the province of Alicante in SE Spain; pop. (1991) 181,200. A 13th-century mystery play is performed every August in the church of Santa María.

El Cid, see CID, EL.

eld /eld/ n. archaic or poet. **1** old age. **2** olden time. [OE *(i)eldu* f. Gmc: cf. OLD]

elder¹ /ˈeldə(r)/ adj. & n. ● attrib.adj. (of two indicated persons, esp. when related) senior; of a greater age (*my elder brother*). ● n. (often prec. by *the*) **1 a** the older or more senior of two indicated (esp. related) persons (*which is the elder?*; *is my elder by ten years*). **b** (**Elder**) as a title denoting the elder holder of the same name (*Pitt the Elder*). **2** (in pl.) **a** persons of greater age or seniority (*respect your elders*). **b** persons venerable because of age. **3** a person advanced in life. **4** hist. a member of a senate or governing body. **5** an official in the early Christian, Presbyterian, or Mormon Churches. □ **elder brother** (pl. **elder brethren**) each of thirteen senior members of Trinity House. **elder hand** Cards the first player. **elder statesman** an influential experienced person, esp. a politician, of advanced age. □ **eldership** n. [OE *eldra*, rel. to OLD]

elder² /ˈeldə(r)/ n. a shrub or tree of the genus *Sambucus*, of the honeysuckle family; esp. *S. nigra*, with white flowers and black berries. [OE *ellærn*]

elderberry /ˈeldəbərɪ/ n. (pl. **-ies**) the berry of the elder, *Sambucus nigra*, used for making jelly, wine, etc.

elderflower /ˈeldəˌflaʊə(r)/ n. the flower of the elder, used mainly for making wine etc.

elderly /ˈeldəlɪ/ adj. & n. ● adj. **1** somewhat old. **2** (of a person) past middle age. ● n. (prec. by *the*; treated as pl.) elderly people. □ **elderliness** n.

eldest /ˈeldɪst/ adj. & n. ● adj. first-born or oldest surviving (member of a family, son, daughter, etc.). ● n. (often prec. by *the*) the eldest of three or more indicated (*who is the eldest?*). □ **eldest hand** Cards the first player. [OE (as ELDER¹)]

El Djem /el ˈdʒem/ a town in eastern Tunisia, noted for its well-preserved Roman amphitheatre.

El Dorado /ˌel dəˈrɑːdəʊ/ the name of a fictitious country (or city) abounding in gold, believed to exist somewhere in the region of the Orinoco and Amazon rivers. The belief, which led Spanish conquistadors to converge on the area in search of treasure and Sir Walter Raleigh to lead his second expedition up the Orinoco, appears to have originated in rumours of an Indian ruler who ritually coated his body with gold dust and then plunged into a sacred lake while his subjects threw in gold and jewels. [Sp., = the gilded]

eldorado /ˌeldəˈrɑːdəʊ/ n. (pl. **-os**) **1** any imaginary country or city abounding in gold. **2** a place of great abundance. [EL DORADO]

eldritch /ˈeldrɪtʃ/ adj. Sc. **1** supernatural. **2** hideous. [16th c.: perh. f. OE *elfrīce* (unrecorded) 'fairy realm']

Eleanor Cross /ˈelɪnə(r)/ any of the stone crosses erected by Edward I to mark the stopping-places of the cortège that brought the body of his queen, Eleanor of Castile (1246–90), from Nottinghamshire to London in 1290. Three of the twelve crosses survive.

Eleanor of Aquitaine /'elɪnə(r)/ (c.1122–1204), daughter of the Duke of Aquitaine, queen of France 1137–52 and of England 1154–89. She was married to Louis VII of France (c.1120–80) from 1137; in 1152, with the annulment of their marriage, she married the future Henry II of England. Her ten children included the monarchs Richard I (Richard the Lion-heart) and John, whose accession she strove to secure. She acted as regent (1190–4), while Richard was away on the Crusades.

Eleatic /,elɪ'ætɪk/ adj. & n. ● adj. of Elea, an ancient Greek city in SW Italy, or the school of philosophers which flourished there in about the 5th century BC, which included Xenophanes, Parmenides, and Zeno. ● n. an Eleatic philosopher.

elecampane /,elɪkæm'peɪn/ n. a sunflower-like composite plant, Inula helenium, with bitter aromatic leaves and roots, used in herbal medicine and cookery. [corrupt. of med.L enula (for L inula f. Gk helenion) campana (prob. = of the fields)]

elect /ɪ'lekt/ v., adj., & n. ● v.tr. (usu. foll. by to + infin.) **1** choose (the principles they elected to follow). **2** choose (a person) by vote (elected a new chairman). **3** (in Christian theology) (of God) choose (persons) in preference to others for salvation. ● adj. **1** chosen. **2** select, choice. **3** (in Christian theology) chosen by God. **4** (in comb., after a noun designating office) chosen but not yet in office (president-elect). ● n. (prec. by the) those chosen by God for salvation. □ **electable** adj. [ME f. L electus past part. of eligere elect- (as E-, legere pick)]

election /ɪ'lekʃ(ə)n/ n. **1** the process of electing or being elected, esp. of members of a political body. **2** the act or an instance of electing. **3** (in Christian theology) God's choice of some persons in preference to others for salvation. In medieval times the nature and conditions of election were much disputed. They later formed a fundamental issue in the teaching of Calvin, who held that God's choice was wholly without relation to faith or good works. [ME f. OF f. L electio -onis (as ELECT)]

electioneer /ɪ,lekʃə'nɪə(r)/ v. & n. ● v.intr. **1** take part in an election campaign. **2** derog. act in a way calculated to appeal to voters. ● n. a person who electioneers. □ **electioneering** n.

elective /ɪ'lektɪv/ adj. & n. ● adj. **1 a** (of an office or its holder) filled or appointed by election. **b** (of authority) derived from election. **2** (of a body) having the power to elect. **3** having a tendency to act on or be concerned with some things rather than others (elective affinity). **4** (of a course of study) chosen by the student; optional. **5** (of a surgical operation etc.) optional; not urgently necessary. ● n. US an elective course of study. □ **electively** adv. [F électif -ive f. LL electivus (as ELECT)]

elector /ɪ'lektə(r)/ n. **1** a person who has the right of voting to elect an MP etc. **2** (**Elector**) hist. a German prince entitled to take part in the election of the emperor. **3** US a member of an electoral college. □ **electorship** n. [ME f. F électeur f. L elector (as ELECT)]

electoral /ɪ'lektərəl/ adj. relating to or ranking as electors. □ **electoral college 1** a body of persons representing the states of the US, who cast votes for the election of the President. **2** a body of electors. **electoral roll** (or **register**) an official list of those in a district, parish, etc., entitled to vote. □ **electorally** adv.

electorate /ɪ'lektərət/ n. **1** a body of electors. **2** Austral. & NZ an area represented by one member of parliament. **3** hist. the office or territories of a German Elector.

Electra /ɪ'lektrə/ Gk Mythol. the daughter of Agamemnon and Clytemnestra. She persuaded her brother Orestes to kill Clytemnestra and Aegisthus (their mother's lover) in revenge for the murder of Agamemnon. □ **Electra complex** Psychol. a daughter's subconscious sexual attraction to her father and hostility towards her mother, corresponding to the Oedipus complex in a son.

electret /ɪ'lektrɪt/ n. Physics a permanently polarized piece of dielectric material, analogous to a permanent magnet. [ELECTRICITY + MAGNET]

electric /ɪ'lektrɪk/ adj. & n. ● adj. **1** of, worked by, or charged with electricity; producing or capable of generating electricity. **2** causing or charged with sudden and dramatic excitement (the news had an electric effect; the atmosphere was electric). ● n. **1** an electric light, vehicle, etc. **2** (in pl.) electrical equipment. □ **electric blanket** a blanket that can be heated electrically by an internal element. **electric blue** a steely or brilliant light blue. **electric chair** an electrified chair used for capital punishment. **electric eel** an eel-like freshwater fish, Electrophorus electricus, native to South America, that can give a severe electric shock. **electric eye** colloq. a photoelectric cell operating a relay when the beam of light illuminating it is obscured. **electric fence** a fence charged with electricity, often consisting of one strand. **electric field** a region of electrical influence. **electric fire** an electrically operated incandescent or convector heater, usu. portable

and for domestic use. **electric guitar** a guitar with a built-in electrical sound pick-up rather than a soundbox. **electric organ 1** Biol. an organ in some fishes able to produce an electrical discharge for stunning prey or sensing the surroundings, or as a defence. **2** Mus. an electrically operated organ. **electric ray** a ray of the family Torpedinidae, that can give an electric shock. **electric shaver** (or **razor**) an electrical device for shaving, with oscillating or rotating blades behind a metal guard. **electric shock** the effect of a sudden discharge of electricity on a person or animal, usually with stimulation of the nerves and contraction of the muscles. **electric storm** a violent disturbance of the electrical condition of the atmosphere. □ **electrically** adv. [mod.L electricus f. L electrum f. Gk elektron amber, the rubbing of which causes electrostatic phenomena]

electrical /ɪ'lektrɪk(ə)l/ adj. **1** of or concerned with or of the nature of electricity. **2** operating by electricity. **3** = ELECTRIC 2.

electrician /,ɪlek'trɪʃ(ə)n, ,el-/ n. a person who installs or maintains electrical equipment, esp. professionally.

electricity /,ɪlek'trɪsɪtɪ, ,el-/ n. **1** a form of energy resulting from the existence of charged particles (electrons, protons, etc.), either statically as an accumulation of charge or dynamically as a current. (See note below.) **2** the branch of physics dealing with electricity. **3** a supply of electric current for heating, lighting, etc. (See note below.) **4** a state of heightened emotion; excitement, tension.

▪ It was known in ancient times that a piece of dry amber, rubbed with a cloth, developed the power of attracting small pieces of paper. In the late 16th century William Gilbert showed that other substances could be electrified by friction, and seems to have been the first to use the Latin word electrum (amber) to refer to the phenomenon. In the 18th century it was discovered that there are two opposite kinds of electricity, which Benjamin Franklin arbitrarily named positive and negative. Like charges repel each other and unlike charges attract. However the nature of electricity still remained a mystery until the late 19th century when J. J. Thomson located negative electricity in subatomic particles called electrons; positive electricity was found to reside in larger particles, protons, in the atomic nucleus. Electric currents were shown to be flows of electrons. Electric forces are responsible for binding protons and electrons together in atoms, binding atoms together to form matter, and for many other natural phenomena, from lightning to the transmission of nerve impulses.

▪ The town of Godalming in Surrey, England, was the first in the world to have electricity supplied for public and private use, in a three-year experiment from 1881; the electricity supply was generated at first by water-power, then by a steam engine. The first successful electric lamps were developed almost simultaneously by Joseph Swan in England and Thomas Edison in America in 1879. Use of electrical power in homes and factories did not become common until the years between the two world wars.

electrify /ɪ'lektrɪ,faɪ/ v.tr. (**-ies, -ied**) **1** charge (a body) with electricity. **2** convert (machinery or the place or system employing it) to the use of electric power. **3** cause dramatic or sudden excitement in. □ **electrifier** n. **electrification** /ɪ,lektrɪfɪ'keɪʃ(ə)n/ n.

electro /ɪ'lektrəʊ/ n. & v. ● n. (pl. **-os**) **1** = ELECTROTYPE n. **2** = ELECTROPLATE n. ● v.tr. (**-oes, -oed**) colloq. **1** = ELECTROTYPE v. **2** = ELECTROPLATE v. [abbr.]

electro- /ɪ'lektrəʊ/ comb. form of, relating to, or caused by electricity (electrocute; electromagnet). [Gk elektron amber: see ELECTRIC]

electro-acoustic /ɪ,lektrəʊə'kuːstɪk/ adj. **1** involving the direct conversion of electrical into acoustic energy or vice versa. **2** (of music) performed or composed with the creative use of electronic equipment.

electrobiology /ɪ,lektrəʊbaɪ'ɒlədʒɪ/ n. the study of the electrical phenomena of living things.

electrocardiogram /ɪ,lektrəʊ'kɑːdɪə,græm/ n. a record of the heartbeat traced by an electrocardiograph. [G Elektrocardiogramm (as ELECTRO-, CARDIO-, -GRAM)]

electrocardiograph /ɪ,lektrəʊ'kɑːdɪə,grɑːf/ n. an instrument recording the electric currents generated by the heart. When the heart muscle contracts and relaxes, changes in potential are produced, which can be picked up from the skin surface by sensitive electrodes. The electrical activity of the heart was first recorded by William Einthoven in 1903. □ **electrocardiographic** /-,kɑːdɪə'græfɪk/ adj. **electrocardiography** /-,kɑːdɪ'ɒgrəfɪ/ n.

electrochemistry /ɪ,lektrəʊ'kemɪstrɪ/ n. the branch of chemistry concerned with the relations between electrical and chemical phenomena. □ **electrochemical** *adj.* **electrochemically** *adv.* **electrochemist** *n.*

electroconvulsive /ɪˌlektrəʊkən'vʌlsɪv/ *adj.* (of a therapy) employing the application of electric shocks to the brain of a patient so as to produce a convulsion, used in some treatments for mental illness.

electrocute /ɪ'lektrəˌkjuːt/ *v.tr.* **1** kill by electricity (as a form of capital punishment). **2** cause the death of by electric shock. □ **electrocution** /ɪˌlektrə'kjuːʃ(ə)n/ *n.* [ELECTRO-, after EXECUTE]

electrode /ɪ'lektrəʊd/ *n.* a conductor through which electricity enters or leaves an electrolyte, gas, vacuum, etc. [ELECTRIC + Gk *hodos* way]

electrodialysis /ɪˌlektrəʊdaɪ'ælɪsɪs/ *n.* dialysis in which the movement of ions is aided by an electric field applied across the semi-permeable membrane.

electrodynamics /ɪˌlektrəʊdaɪ'næmɪks/ *n.pl.* (usu. treated as *sing.*) the branch of mechanics concerned with electric current applied to motive forces. □ **electrodynamic** *adj.*

electroencephalogram /ɪˌlektrəʊen'sefələˌgræm, -'kefələˌgræm/ *n.* a record of the brain's activity traced by an electroencephalograph. [G *Elektrenkephalogramm* (as ELECTRO-, ENCEPHALO-, -GRAM)]

electroencephalograph /ɪˌlektrəʊen'sefələˌgrɑːf, -'kefələˌgrɑːf/ *n.* an instrument recording the electrical activity of the brain. Its use was pioneered in 1929 by Hans Berger. □ **electroencephalography** /-ˌsefə'lɒgrəfɪ, -ˌkefə-/ *n.*

electroluminescence /ɪˌlektrəʊˌluːmɪ'nes(ə)ns/ *n.* luminescence produced electrically, esp. by the application of a voltage. □ **electroluminescent** *adj.*

electrolyse /ɪ'lektrəˌlaɪz/ *v.tr.* (US **-yze**) subject to or treat by electrolysis. □ **electrolyser** *n.* [ELECTROLYSIS after *analyse*]

electrolysis /ˌɪlek'trɒlɪsɪs, ˌel-/ *n.* **1** *Chem.* the decomposition of a substance by the application of an electric current. **2** this process applied to the destruction of tumours, hair-roots, etc. □ **electrolytic** /ɪˌlektrəʊ'lɪtɪk/ *adj.* **electrolytical** *adj.* **electrolytically** *adv.* [ELECTRO- + -LYSIS]

electrolyte /ɪ'lektrəˌlaɪt/ *n.* **1** a liquid which contains ions and can be decomposed by electrolysis, esp. that present in a battery. **2** (usu. in *pl.*) *Physiol.* the ionized or ionizable constituents of a living cell, blood, or other tissue. [ELECTRO- + Gk *lutos* released f. *luō* loosen]

electromagnet /ɪˌlektrəʊ'mægnɪt/ *n.* a soft metal core made into a magnet by the passage of electric current through a coil surrounding it. The first simple electromagnet capable of supporting its own weight was made by William Sturgeon (1783–1850) of Manchester, a self-taught electrical engineer, in 1825.

electromagnetic /ɪˌlektrəʊmæg'netɪk/ *adj.* having both an electrical and a magnetic character or properties. □ **electromagnetic radiation** a kind of radiation including visible light, radio waves, gamma rays, X-rays, etc., in which electric and magnetic fields vary simultaneously. **electromagnetic spectrum** the range of wavelengths over which electromagnetic radiation extends. **electromagnetic units** a largely disused system of units derived primarily from the magnetic properties of electric currents. □ **electromagnetically** *adv.*

electromagnetism /ɪˌlektrəʊ'mægnɪˌtɪz(ə)m/ *n.* **1** the magnetic forces produced by electricity. **2** the study of this.
  ▪ Hans Christian Oersted in 1820 was the first to show that an electric current creates a magnetic field and affects a magnetic compass exactly as an ordinary magnet does. This discovery led to the development of the electromagnet. Electromagnetic phenomenona are the basis of many devices, including electric motors, generators, relays, solenoids, microphones, and loudspeakers.

electromechanical /ɪˌlektrəʊmɪ'kænɪk(ə)l/ *adj.* relating to the application of electricity to mechanical processes, devices, etc.

electrometer /ˌɪlek'trɒmɪtə(r), ˌel-/ *n.* an instrument for measuring electrical potential without drawing any current from the circuit. □ **electrometry** *n.* **electrometric** /ɪˌlektrəʊ'metrɪk/ *adj.*

electromotive /ɪˌlektrəʊ'məʊtɪv/ *adj.* producing or tending to produce an electric current. □ **electromotive force** a difference in potential that tends to give rise to an electric current.

electron /ɪ'lektrɒn/ *n.* a subatomic particle with a unit charge of negative electricity, which is found in all atoms and acts as the primary carrier of electricity in solids. (*See note below.*) □ **electron beam** a stream of electrons in a gas or vacuum. **electron diffraction** the diffraction of a beam of electrons by atoms or molecules, used for determining crystal structures etc. **electron gun** a device for producing a narrow stream of electrons from a heated cathode.

electron lens a device for focusing a stream of electrons by means of electric or magnetic fields. **electron pair 1** *Chem.* two electrons in the same orbital in an atom or molecule. **2** *Physics* an electron and a positron. **electron spin resonance** (abbr. **ESR**) a spectroscopic method of studying electrons within the molecules of a paramagnetic substance. [ELECTRIC + -ON]
  ▪ The electron was discovered by J. J. Thomson in his study of cathode rays (1897); the word *electron* had been coined in 1891 by the Irish mathematical physicist George Johnstone Stoney (1826–1911), though he used it for the unit of charge on a hydrogen ion. The electron's mass is about 9×10^{-28}g, 1,836 times less than that of the proton. Electrons orbit the positively charged nuclei of atoms and are responsible for binding atoms together in molecules, as well as for the electrical, thermal, optical, and magnetic properties of solids. Electric currents in metals and in semiconductors consist of a flow of electrons, and light, radio waves, X-rays, and much heat radiation are all produced by accelerating and decelerating electrons.

electronegative /ɪˌlektrəʊ'negətɪv/ *adj.* **1** electrically negative. **2** *Chem.* (of an element) tending to acquire electrons in chemical reactions.

electronic /ˌɪlek'trɒnɪk, ˌel-/ *adj.* **1 a** produced by or involving the flow of electrons. **b** of or relating to electrons or electronics. **2** (of a device) using electronic components. **3 a** (of music) produced by electronic means and usu. recorded on tape. **b** (of a musical instrument) producing sounds by electronic means. □ **electronic flash** a flash from a gas-discharge tube, used in high-speed photography. **electronic mail** messages distributed by electronic means esp. from one computer system to one or more recipients. **electronic publishing** the publication of books etc. in machine-readable form rather than on paper. **electronic tagging** the attaching of electronic markers to people, goods, etc., enabling them to be traced. □ **electronically** *adv.*

electronics /ˌɪlek'trɒnɪks, ˌel-/ *n.pl.* (usu. treated as *sing.*) **1** the branch of physics and technology concerned with the behaviour and movement of electrons in a vacuum, gas, semiconductor, etc. **2** the circuits used in this.

electron microscope *n.* a type of microscope employing a beam of electrons rather than light. First developed in the early 1930s, the electron microscope employs magnetic or electrostatic fields to focus the beam. Because the de Broglie wavelength of a high-speed electron is very much shorter than that of light, a correspondingly greater resolving power is possible, so that magnifications of up to one million times can be achieved. In one type of instrument the electron beam is focused on to a very thin specimen and the transmitted electrons create an image on a fluorescent screen; other types use reflection from the specimen or a scanning beam. In all cases the image can be stored as a photograph. □ **electron microscopy** *n.*

electronvolt /ɪ'lektrɒnˌvəʊlt/ *n. Physics* a unit of energy (symbol: **eV**) equal to the work done on an electron in accelerating it through a potential difference of one volt. One electronvolt is equal to 1.602×10^{-19} joule.

electrophilic /ɪˌlektrəʊ'fɪlɪk/ *adj. Chem.* having an affinity for electrons. □ **electrophile** /ɪ'lektrəʊˌfaɪl/ *n.*

electrophoresis /ɪˌlektrəʊfɔː'riːsɪs/ *n. Physics & Chem.* the movement of charged particles in a fluid under the influence of an electric field. □ **electrophoretic** /-fə'retɪk/ *adj.* [ELECTRO- + Gk *phorēsis* being carried]

electrophorus /ˌɪlek'trɒfərəs, ˌel-/ *n.* a device for repeatedly generating static electricity by induction. [mod.L f. ELECTRO- + Gk *-phoros* bearing]

electrophysiology /ɪˌlektrəʊˌfɪzɪ'ɒlədʒɪ/ *n.* the branch of physiology that deals with the electrical phenomena associated with nervous and other bodily activity. □ **electrophysiological** /-zɪə'lɒdʒɪk(ə)l/ *adj.*

electroplate /ɪ'lektrəˌpleɪt/ *v. & n.* ● *v.tr.* coat (a utensil etc.) by electrolytic deposition with chromium, silver, etc. ● *n.* electroplated articles. □ **electroplater** *n.*

electroplexy /ɪ'lektrəˌpleksɪ/ *n. Brit.* electroconvulsive therapy. [ELECTRO- + APOPLEXY]

electroporation /ɪˌlektrəʊpə'reɪʃ(ə)n/ *n. Biol.* the action or process of introducing DNA or chromosomes into the cells of bacteria etc., using a pulse of electricity to open the pores in the cell membrane briefly. [ELECTRO- + PORE¹ + -ATION]

electropositive /ɪˌlektrəʊ'pɒzɪtɪv/ *adj.* **1** electrically positive. **2** *Chem.* (of an element) tending to lose electrons in chemical reactions.

electroscope /ɪˈlektrəˌskəʊp/ n. an instrument for detecting and measuring electricity, esp. as an indication of the ionization of air by radioactivity. □ **electroscopic** /ɪˌlektrəˈskɒpɪk/ adj.

electro-shock /ɪˈlektrəʊˌʃɒk/ attrib.adj. (of medical treatment) by means of electric shocks; electroconvulsive.

electrostatic /ɪˌlektrəʊˈstætɪk/ adj. of stationary electric charges or electrostatics. □ **electrostatic units** a system of units based primarily on the forces between electric charges. [ELECTRO- + STATIC after hydrostatic]

electrostatics /ɪˌlektrəʊˈstætɪks/ n.pl. (treated as sing.) the study of stationary electric charges or fields, as opposed to electric currents.

electrotechnology /ɪˌlektrəʊtekˈnɒlədʒɪ/ n. the science of the application of electricity in technology. □ **electrotechnic** /-ˈteknɪk/ adj. **electrotechnical** adj. **electrotechnics** n.

electrotherapy /ɪˌlektrəʊˈθerəpɪ/ n. the treatment of diseases by the use of electricity. □ **electrotherapist** n. **electrotherapeutic** /-ˌθerəˈpjuːtɪk/ adj. **electrotherapeutical** adj.

electrothermal /ɪˌlektrəʊˈθɜːm(ə)l/ adj. relating to heat electrically derived.

electrotype /ɪˈlektrəʊˌtaɪp/ v. & n. ● v.tr. copy by the electrolytic deposition of copper on a mould, esp. for printing. ● n. a copy so formed. □ **electrotyper** n.

electrovalent /ɪˌlektrəʊˈveɪlənt/ adj. Chem. of bonding resulting from electrostatic attraction between ions. □ **electrovalence** n. **electrovalency** n.

electroweak /ɪˌlektrəʊˈwiːk/ adj. Physics relating to or denoting electromagnetic and weak interactions regarded as manifestations of the same interaction.

electrum /ɪˈlektrəm/ n. 1 an alloy of silver and gold used in ancient times. 2 native argentiferous gold ore. [ME f. L f. Gk ēlektron amber, electrum]

electuary /ɪˈlektʃʊərɪ, -tjʊərɪ/ n. (pl. **-ies**) medicinal powder etc. mixed with honey or other sweet substance. [ME f. LL electuarium, prob. f. Gk ekleikton f. ekleikhō lick up]

eleemosynary /ˌeliːˈmɒsɪnərɪ, -ˈmɒzɪnərɪ/ adj. 1 of or dependent on alms. 2 charitable. 3 gratuitous. [med.L eleemosynarius f. LL eleemosyna: see ALMS]

elegant /ˈelɪɡənt/ adj. 1 graceful in appearance or manner. 2 tasteful, refined. 3 (of a mode of life etc.) of refined luxury. 4 ingeniously simple and pleasing. 5 US excellent. □ **elegance** n. **elegantly** adv. [F élégant or L elegant-, rel. to eligere: see ELECT]

elegiac /ˌelɪˈdʒaɪək/ adj. & n. ● adj. 1 (of a metre) used for elegies. 2 mournful. ● n. (in pl.) verses in an elegiac metre. □ **elegiac couplet** a pair of lines consisting of a dactylic hexameter and a pentameter, esp. in Greek and Latin verse. □ **elegiacally** adv. [F élégiaque or f. LL elegiacus f. Gk elegeiakos: see ELEGY]

elegize /ˈelɪˌdʒaɪz/ v. (also **-ise**) 1 intr. (often foll. by upon) write an elegy. 2 intr. write in a mournful strain. 3 tr. write an elegy upon. □ **elegist** n.

elegy /ˈelɪdʒɪ/ n. (in Greek and Roman poetry) a poem written in elegiac couplets, as notably by Catullus and Propertius. In modern literature an elegy is a poem of serious reflection, frequently a lament for the dead: examples include Milton's *Lycidas*, Shelley's *Adonais*, and *Elegy Written in a Country Church-Yard* by Thomas Gray. [F élégie or L elegia f. Gk elegeia f. elegos mournful poem]

element /ˈelɪmənt/ n. 1 a component part or group; a contributing factor or thing. 2 Chem. & Physics a substance that cannot be resolved by chemical means into simpler substances. (*See note below.*) 3 a any of the four substances (earth, water, air, and fire) in ancient and medieval philosophy. b any of these as a being's natural abode or environment. c a person's appropriate or preferred sphere of operation. 4 a resistance wire that heats up in an electric heater, cooker, etc.; an electrode. 5 (in pl.) atmospheric agencies, esp. wind and storm. 6 (in pl.) the rudiments of learning or of a branch of knowledge. 7 (in pl.) the bread and wine of the Eucharist. 8 Math. & Logic an entity that is a single member of a set. □ **in** (or **out of**) **one's element** in (or out of) one's accustomed or preferred surroundings. **reduced to its elements** analysed. [ME f. OF f. L elementum]

▪ All stable matter consists of elements either individually or in combination. More than 100 elements are known although not all of them occur naturally, those heavier than uranium having been synthesized in nuclear reactions. The atoms of a given element all have the same atomic number, which represents the number of protons in the nucleus; the atomic number of each element is different. (See also ATOM, PERIODIC TABLE.)

elemental /ˌelɪˈment(ə)l/ adj. & n. ● adj. 1 of the four elements. 2 of the powers of nature (*elemental worship*). 3 comparable to a force of nature (*elemental grandeur*; *elemental tumult*). 4 uncompounded (*elemental oxygen*). 5 essential. ● n. an entity or force thought to be physically manifested by occult means. □ **elementalism** n. (in senses 1, 2). [med.L elementalis (as ELEMENT)]

elementary /ˌelɪˈmentərɪ/ adj. 1 a dealing with or arising from the simplest facts of a subject; rudimentary, introductory. b simple. 2 Chem. not decomposable. □ **elementary particle** Physics a subatomic particle, esp. one not known to be decomposable into simpler particles. **elementary school** a school in which elementary subjects are taught to young children. □ **elementarily** adv. **elementariness** n. [ME f. L elementarius (as ELEMENT)]

elenchus /ɪˈleŋkəs/ n. (pl. **elenchi** /-kaɪ/) Logic logical refutation. □ **Socratic elenchus** an attempted refutation of an opponent's position by short question and answer. □ **elenctic** adj. [L f. Gk elegkhos]

elephant /ˈelɪfənt/ n. (pl. same or **elephants**) 1 the largest living land animal, with a prehensile trunk and long curved ivory tusks. (*See note below.*) 2 a size of paper (711 x 584 mm). □ **elephant-bird** = AEPYORNIS. **elephant grass** a tall African grass, esp. *Pennisetum purpureum*. **elephant seal** a very large seal of the genus *Mirounga*, the male of which has an inflatable proboscis (also called *sea elephant*). **elephant shrew** a small insectivorous mammal of the order Macroscelidea, native to Africa, having a long snout and long hind limbs. □ **elephantoid** /ˌelɪˈfæntɔɪd/ adj. [ME olifaunt etc. f. OF oli-, elefant ult. f. L elephantus, elephans f. Gk elephas -antos ivory, elephant (ivory was known to the Greeks long before the elephant)]

▪ There are two species of elephant today, of which the African elephant, *Loxodonta africana*, is the larger (though a small variety exists). It has larger ears, a single-domed head, and two projections at the tip of the trunk. The Indian elephant, *Elephas maximus*, has smaller ears, a two-domed head, and a single-tipped trunk, and is used as a beast of burden in India and some countries of SE Asia. The tusks are actually modified upper incisors; the only other teeth are massive molars, of which only two pairs are in use at a time. Many elephant populations are threatened by the illegal ivory trade.

elephantiasis /ˌelɪfənˈtaɪəsɪs/ n. gross enlargement of the body, esp. the limbs, due to lymphatic obstruction by a nematode parasite that is transmitted by mosquitoes. [L f. Gk (as ELEPHANT)]

elephantine /ˌelɪˈfæntaɪn/ adj. 1 of elephants. 2 a huge. b clumsy, unwieldy (*elephantine movements*; *elephantine humour*). [L elephantinus f. Gk elephantinos (as ELEPHANT)]

Elephant Pass a narrow strip of land at the north end of Sri Lanka, linking the Jaffna peninsula with the rest of the island.

Eleusinian mysteries /ˌeljuːˈsɪnɪən/ Gk Hist. the annual rites held at ancient Eleusis near Athens in honour of Demeter and Persephone. (See also MYSTERY[1].) [L Eleusinius f. Gk Eleusinios]

elevate /ˈelɪˌveɪt/ v.tr. 1 bring to a higher position. 2 Eccl. hold up (the Host or the chalice) for adoration. 3 raise, lift (one's eyes etc.). 4 raise the axis of (a gun). 5 raise (a railway etc.) above ground level. 6 exalt in rank etc. 7 (usu. as **elevated** adj.) a raise the spirits of, elate. b raise morally or intellectually (*elevated style*). 8 (as **elevated** adj.) colloq. slightly drunk. □ **elevatory** adj. [L elevare raise (as E-, levis light)]

elevation /ˌelɪˈveɪʃ(ə)n/ n. 1 a the process of elevating or being elevated. b the angle with the horizontal, esp. of a gun or of the direction of a celestial body. c the height above a given level, esp. sea level. d a raised area; a swelling on the skin. 2 loftiness, grandeur, dignity. 3 a a drawing or diagram made by projection on a vertical plane (cf. PLAN n. 2). b a flat drawing of the front, side, or back of a house etc. 4 Ballet a the capacity of a dancer to attain height in springing movements. b the action of tightening the muscles and uplifting the body. □ **elevational** adj. (in sense 2). [ME f. OF elevation or L elevatio: see ELEVATE]

elevator /ˈelɪˌveɪtə(r)/ n. 1 a hoisting machine. 2 Aeron. the movable part of a tailplane for changing the pitch of an aircraft. 3 esp. US a a platform or compartment housed in a shaft for raising and lowering persons or things to different floors of a building or different levels of a mine etc. (*See note below.*) b a place for lifting and storing quantities of grain. 4 that which elevates, esp. a muscle that raises a limb. [mod.L (as ELEVATE)]

▪ The first elevator with a safety device was devised by E. G. Otis in 1852 and the first passenger service was installed by him in a New York department store five years later, although simple arrangements for taking miners up and down mineshafts had been in use much earlier. The first elevator in Europe was built at the Paris

Exhibition of 1867; it was operated hydraulically. The development of electric elevators near the end of the 19th century made skyscrapers feasible, although the height of such buildings is constrained by the increasing proportion of floor-space occupied by elevators.

eleven /ɪˈlev(ə)n/ n. & adj. ● n. **1** one more than ten; the sum of six units and five units. **2** a symbol for this (11, xi, XI). **3** a size etc. denoted by eleven. **4** a set or team of eleven individuals, esp. a cricket or football team (was picked for the second eleven). **5** eleven o'clock. ● adj. that amount to eleven. □ **eleven-plus** esp. hist. (in the UK) an examination taken at the age of 11–12 to determine the type of secondary school a child should enter. [OE endleofon f. Gmc]

elevenfold /ɪˈlev(ə)nˌfəʊld/ adj. & adv. **1** eleven times as much or as many. **2** consisting of eleven parts.

elevenses /ɪˈlev(ə)nzɪz/ n. (usu. in pl.) Brit. colloq. light refreshment, usu. with tea or coffee, taken about 11 a.m.

eleventh /ɪˈlev(ə)nθ/ n. & adj. ● n. **1** the position in a sequence corresponding to the number 11 in the sequence 1–11. **2** something occupying this position. **3** one of eleven equal parts of a thing. **4** Mus. **a** an interval or chord spanning an octave and a fourth in the diatonic scale. **b** a note separated from another by this interval. ● adj. that is the eleventh. □ **the eleventh hour** the last possible moment.

elevon /ˈelɪˌvɒn/ n. Aeron. the movable part of the trailing edge of a delta wing. [ELEVATOR + AILERON]

elf /elf/ n. (pl. **elves** /elvz/) **1** a mythological being, esp. one that is small and mischievous. **2** a sprite or little creature. □ **elf-lock** a tangled mass of hair. □ **elfish** adj. **elvish** adj. [OE f. Gmc]

elfin /ˈelfɪn/ adj. & n. ● adj. of elves; elflike; tiny, dainty. ● n. archaic a dwarf; a child. [ELF, perh. infl. by ME elvene genitive pl. of elf, and by Elphin in Arthurian romance]

Elgar /ˈelɡɑː(r)/, Sir Edward (William) (1857–1934), British composer. A self-taught musician from Worcester, he made his mark with the Enigma Variations (1899), a set of fourteen orchestral variations (on an undisclosed theme), thirteen of which are titled by the initials of his friends. He gained an international reputation with the oratorio The Dream of Gerontius (1900), the violin concerto (1910), and the cello concerto (1919). In Britain he is perhaps most famous for patriotic pieces such as the five Pomp and Circumstance marches (1901–30).

Elgin /ˈelɡɪn/, 8th Earl of (title of James Bruce) (1811–63), British colonial statesman. After serving as governor of Jamaica (1842–6), he became Governor-General of Canada (1847–54). He commissioned Louis Hippolyte Lafontaine (1807–64) to form Canada's first Cabinet government in 1848. He maintained good relationships with subsequent administrations and successfully negotiated a reciprocity treaty between Canada and the US in 1854.

Elgin Marbles a collection of ancient Greek marble sculptures and architectural fragments, chiefly from the frieze and pediment of the Parthenon in Athens, which the 7th Earl of Elgin (Thomas Bruce, 1766–1841) acquired from the occupying Turkish authorities in 1801–3 and sold to the British nation in 1816 for £35,000. The sculptures were executed by Phidias in the 5th century BC. Their original exhibition in London had an enormous impact, it being the first time authentic classical Greek sculpture had been on public display. They are currently housed in the British Museum, but are the subject of a repatriation request from the Greek government, who do not accept the legality of the Turkish sale.

El Giza see GIZA.

Elgon, Mount /ˈelɡɒn/ an extinct volcano on the border between Kenya and Uganda, rising to 4,321 m (14,178 ft).

El Greco /el ˈɡrekəʊ/ (Spanish, = the Greek; born Domenikos Theotokopoulos) (1541–1614), Cretan-born Spanish painter. After studying in Venice and working in Rome, he settled in Toledo in 1577. His portraits and religious works are characterized by distorted perspective, elongated figures, and strident use of colour. Famous works include the altarpiece The Assumption of the Virgin (1577–9) and the painting The Burial of Count Orgaz (1586).

Eli /ˈiːlaɪ/ (in the Bible) a priest who acted as a teacher to the prophet Samuel (1 Sam. 1–3).

Elia /ˈiːlɪə/ the pseudonym adopted by Charles Lamb in his Essays of Elia (1823) and Last Essays of Elia (1833).

elicit /ɪˈlɪsɪt/ v.tr. (**elicited, eliciting**) **1** draw out, evoke (an admission, response, etc.). **2** draw forth (what is latent). □ **elicitor** n. **elicitation** /ɪˌlɪsɪˈteɪʃ(ə)n/ n. [L elicere elicit- (as E-, lacere entice)]

elide /ɪˈlaɪd/ v.tr. omit (a vowel or syllable) by elision. [L elidere elis- crush out (as E-, laedere knock)]

eligible /ˈelɪdʒɪb(ə)l/ adj. **1** (often foll. by for) fit or entitled to be chosen (eligible for a rebate). **2** desirable or suitable, esp. as a partner in marriage. □ **eligibly** adv. **eligibility** /ˌelɪdʒəˈbɪlɪtɪ/ n. [F éligible f. LL eligibilis (as ELECT)]

Elijah /ɪˈlaɪdʒə/ (9th century BC) a Hebrew prophet in the time of Jezebel, who maintained the worship of Jehovah against that of Baal and other pagan gods. Legends about him include that he was miraculously fed by ravens, raised a widow's son from the dead, and was carried to heaven in a chariot of fire (1 Kings 17–2 Kings 2).

eliminate /ɪˈlɪmɪˌneɪt/ v.tr. **1 a** remove, get rid of. **b** kill, murder. **2** exclude from consideration; ignore as irrelevant. **3** exclude from further participation in a competition etc. on defeat. **4** Physiol. discharge (waste matter). **5** Chem. remove (a simpler substance) from a compound. **6** Algebra remove (a quantity) by combining equations. □ **eliminator** n. **eliminable** /-nəb(ə)l/ adj. **eliminatory** /-nətərɪ/ adj. **elimination** /ɪˌlɪmɪˈneɪʃ(ə)n/ n. [L eliminare (as E-, limen liminis threshold)]

ELINT /ˈelɪnt/ n. (also **Elint**) covert intelligence-gathering by electronic means. [abbr. of electronic intelligence]

Eliot[1] /ˈelɪət/, George (pseudonym of Mary Ann Evans) (1819–80), English novelist. She is best known for her novels of provincial life, including Adam Bede (1859), The Mill on the Floss (1860), and Middlemarch (1871–2). Famed for her intellect, scholarly style, and moral sensibility, she is regarded as one of the great English novelists. Early influenced by evangelicalism, she later adopted agnostic views, although religious themes continued to feature prominently in her novels.

Eliot[2] /ˈelɪət/, T(homas) S(tearns) (1888–1965), American-born British poet, critic, and dramatist. Associated with the rise of literary modernism, he struck a new note in modern poetry with his verse collection Prufrock and other Observations (1917), which combined satire with allusion, cosmopolitanism, and lyricism. In his newly founded literary quarterly The Criterion, he published The Waste Land (1922), which established him as the voice of a disillusioned generation. In 1927 he became an Anglo-Catholic and subsequent works, such as Four Quartets (1943), reveal his increasing involvement with Christianity. He was awarded the Nobel Prize for literature in 1948.

Elisabethville /ɪˈlɪzəbəˌvɪl/ the former name (until 1966) for Lubumbashi in SE Zaire.

Elisha /ɪˈlaɪʃə/ (9th century BC) a Hebrew prophet, disciple and successor of Elijah.

elision /ɪˈlɪʒ(ə)n/ n. **1** the omission of a vowel or syllable in pronouncing (as in I'm, let's, e'en). **2** the omission of a passage in a book etc. [LL elisio (as ELIDE)]

Elista /eˈlɪstə/ a city in SW Russia, capital of the autonomous republic of Kalmykia; pop. (1990) 85,000.

élite /eɪˈliːt/ adj. & n. ● n. **1** (prec. by the) the best or choice part of a larger body or group. **2** a select group or class. **3** a size of letter in typewriting (12 per inch). ● adj. of or belonging to an élite; exclusive. [F f. past part. of élire f. Rmc: rel. to ELECT]

élitism /eɪˈliːtɪz(ə)m/ n. **1** advocacy of or reliance on leadership or dominance by a select group. **2** a sense of belonging to an élite. □ **élitist** n. & adj.

elixir /ɪˈlɪksɪə(r)/ n. **1** Alchemy **a** a preparation supposedly able to change metals into gold. **b** (in full **elixir of life**) a preparation supposedly able to prolong life indefinitely. **c** a supposed remedy for all ills. **2** Pharm. an aromatic solution used as a medicine or flavouring. **3** the quintessence or kernel of a thing. [ME f. med.L f. Arab. al-iksīr f. al the + iksīr prob. f. Gk xērion powder for drying wounds f. xēros dry]

Elizabeth I /ɪˈlɪzəbəθ/ (1533–1603), daughter of Henry VIII, queen of England and Ireland 1558–1603. Succeeding her Catholic sister Mary I, Elizabeth re-established a moderate form of Protestantism as the religion of the state. None the less, her reign was dominated by the threat of a Catholic restoration (eventually leading to the execution of Mary, Queen of Scots) and by war with Spain, during which the country was saved from invasion by the defeat of the Armada in 1588. Her reign was characterized by a flowering of national culture, particularly in the field of literature, in which Shakespeare, Marlowe, and Spenser were all active. Although frequently courted, she never married.

Elizabeth II /ɪˈlɪzəbəθ/ (born Princess Elizabeth Alexandra Mary) (b.1926), daughter of George VI, queen of the United Kingdom since 1952. She has always shown a strong personal commitment to the Commonwealth, and is one of the most travelled 20th-century monarchs, having made extensive overseas tours and many public appearances at home.

Elizabeth, the Queen Mother (born Lady Elizabeth Angela Marguerite Bowes-Lyon) (b.1900), wife of George VI. She married George VI in 1923, when he was Duke of York; they had two daughters, Elizabeth II and Princess Margaret.

Elizabethan /ɪˌlɪzə'biːθən/ adj. & n. ● adj. of or characteristic of the time of Queen Elizabeth I or (occasionally) Queen Elizabeth II. (See also TUDOR¹.) ● n. a person, esp. a writer, of the time of Queen Elizabeth I or (occasionally) Queen Elizabeth II.

Elizavetpol /jəˌliːzə'vjetpɒl/ (also **Yelizavetpol**) the former Russian name (1804–1918) for GÄNCÄ.

elk /elk/ n. (pl. same or **elks**) **1** the largest living deer, *Alces alces*, native to northern Eurasia and North America, with palmate antlers and a growth of skin hanging from the neck. Also called (esp. N. Amer.) moose. **2** N. Amer. a wapiti. □ **elk-hound** a large Scandinavian breed of hunting dog with a shaggy coat. [ME, prob. repr. OE *elh, eolh*]

ell /el/ n. hist. a former measure of length, about 45 inches. [OE *eln*, rel. to L *ulna*: see ULNA]

Ellesmere Island /'elzmɪə(r)/ the northernmost island of the Canadian Arctic. Discovered in 1616 by William Baffin, the island was eventually named after the British statesman Francis Egerton, Earl of Ellesmere (1800–57).

Ellesmere Port /'elzmɪə(r)/ a port in NW England, in Cheshire, on the estuary of the River Mersey; pop. (1981) 65,800.

Ellice Islands /'elɪs/ the former name for TUVALU.

Ellington /'elɪŋtən/, Duke (born Edward Kennedy Ellington) (1899–1974), American jazz pianist, composer, and band-leader. His band established its fame in the early 1930s and some of its members remained with him for over thirty years. Ellington wrote over 900 compositions and was one of the first popular musicians to write extended pieces; his first worldwide success was *Mood Indigo* (1930).

ellipse /ɪ'lɪps/ n. a regular oval, traced by a point moving in a plane so that the sum of its distances from two other points is constant, or resulting when a cone is cut by a plane which does not intersect (and is not parallel to) the base (cf. HYPERBOLA). [F f. L *ellipsus* f. Gk *elleipsis* f. *elleipō* come short f. *en* in + *leipō* leave]

ellipsis /ɪ'lɪpsɪs/ n. (also **ellipse**) (pl. **ellipses** /-siːz/) **1** the omission from a sentence of words needed to complete the construction or sense. **2** the omission of a sentence at the end of a paragraph. **3** a set of three dots etc. indicating an omission.

ellipsoid /ɪ'lɪpsɔɪd/ n. a solid of which all the plane sections normal to one axis are circles and all the other plane sections are ellipses. □ **ellipsoidal** /ˌɪlɪp'sɔɪd(ə)l, ˌel-/ adj.

elliptic /ɪ'lɪptɪk/ adj. (also **elliptical**) of, relating to, or having the form of an ellipse or ellipsis. □ **elliptically** adv. **ellipticity** /ˌɪlɪp'tɪsɪtɪ, ˌel-/ n. [Gk *elleiptikos* deriche f. *elleipō* (as ELLIPSE)]

Ellis Island /'elɪs/ an island in the bay of New York. Formerly the site of a fort, from 1892 until 1943 the island served as an entry point for immigrants to the US, and later (until 1954) as a detention centre for people awaiting deportation. In 1965 it became part of the Statue of Liberty National Monument. The island is named after Samuel Ellis, a Manhattan merchant who owned it in the 1770s.

Ellsworth /'elzwəθ/, Lincoln (1880–1951), American explorer. He participated in a number of polar expeditions and was the first person to fly over both the North (1926) and South (1935) Poles. During his Antarctic explorations of 1935 and 1939 he discovered new mountain ranges and named Ellsworth Land after his father.

Ellsworth Land a plateau region of Antarctica between the Walgreen Coast and Palmer Land. It rises at the Vinson Massif, the highest point in Antarctica, to 5,140 m (16,863 ft). Parts of the region are the subject of rival claims by the UK and Chile.

elm /elm/ n. **1** a tree of the genus *Ulmus*, with rough serrated leaves; esp. (in full **English elm**) *U. procera*, which was common as a tall hedgerow tree in England before it was devastated by Dutch elm disease. **2** (in full **elmwood**) the wood of the elm. □ **elmy** adj. [OE, rel. to L *ulmus*]

El Niño /el 'niːnjəʊ/ n. (pl. **El Niños**) an irregularly occurring and complex series of climatic changes affecting the equatorial Pacific region. The name was originally applied to a warm ocean current which affected the waters off northern Peru and Ecuador annually, beginning usually in late December. Every few years, this warming is very marked, and the nutrient-poor warmer water has a disastrous effect on fisheries and the breeding success of seabirds. This warm current is now recognized to be one manifestation of a larger cycle of abnormal phenomena, sometimes lasting for more than a year, in which cause and effect have not been fully distinguished. The effects of El Niño, which may reach beyond the Pacific region, include reversal of wind patterns across the Pacific, drought in Australasia, and unseasonal heavy rain in South America. [Sp., = (Christ) child, as commonly occurring around Christmas.]

elocution /ˌelə'kjuːʃ(ə)n/ n. **1** the art of clear and expressive speech, esp. of distinct pronunciation and articulation. **2** a particular style of speaking. □ **elocutionary** adj. **elocutionist** n. [L *elocutio* f. *eloqui* elocut-speak out (as E-, *loqui* speak)]

Elohim /e'ləʊhɪm, 'eləʊˌhiːm/ in the Bible, a name used frequently for God. Cf. JEHOVAH, YAHWEH. (See also ELOHIST.) [Heb. *'ĕlōhīm* god(s)]

Elohist /e'ləʊhɪst/ the postulated author or authors of parts of the Hexateuch in which God is regularly named Elohim (cf. YAHWIST). [Heb. *'ĕlōhīm* god(s) + -IST]

elongate /'iːlɒŋˌgeɪt/ v. & adj. ● v. **1** tr. lengthen, prolong. **2** intr. be of slender or tapering form. ● adj. Bot. & Zool. = ELONGATED 1. [LL *elongare* (as E-, L *longus* long)]

elongated /'iːlɒŋˌgeɪtɪd/ adj. **1** long in relation to its width. **2** that has been made longer.

elongation /ˌiːlɒŋ'geɪʃ(ə)n/ n. **1** the act or an instance of lengthening; the process of being lengthened. **2** a part of a line etc. formed by lengthening. **3** Mech. the amount of extension under stress. **4** Astron. the angular separation of a planet from the sun or of a satellite from a planet. [ME f. LL *elongatio* (as ELONGATE)]

elope /ɪ'ləʊp/ v.intr. **1** run away to marry secretly, esp. without parental consent. **2** run away with a lover. □ **elopement** n. **eloper** n. [AF *aloper* perh. f. a ME form *alope*, rel. to LEAP]

eloquence /'eləkwəns/ n. **1** fluent and effective use of language. **2** rhetoric. [ME f. OF f. L *eloquentia* f. *eloqui* speak out (as E-, *loqui* speak)]

eloquent /'eləkwənt/ adj. **1** possessing or showing eloquence. **2** (often foll. by *of*) clearly expressive or indicative. □ **eloquently** adv. [ME f. OF f. L *eloqui* (as ELOQUENCE)]

El Paso /el 'pæsəʊ/ a city in western Texas on the Rio Grande, on the border with Mexico; pop. (1990) 515,300.

El Qahira see CAIRO.

El Salvador /el 'sælvəˌdɔː(r)/ a country in Central America, on the Pacific coast; pop. (est. 1991) 5,308,000; official language, Spanish; capital, San Salvador. The territory, part of a native kingdom, was conquered by the Spanish in 1524. El Salvador gained its independence in 1821, finally emerging as an independent republic in 1839. Between 1979 and 1992 the country was devastated by a civil conflict characterized by guerrilla fighting and harsh repression; several senior army officers were removed following a UN investigation of human rights abuses during the war. □ **Salvadorean** /ˌsælvə'dɔːrɪən/ adj. & n.

Elsan /'elsæn/ n. Brit. propr. a type of transportable chemical lavatory. [app. f. E. L. Jackson (its manufacturer) + SANITATION]

else /els/ adv. **1** (prec. by indefinite or interrog. pron.) besides; in addition (*someone else; nowhere else; who else*). **2** instead; other, different (*what else could I say?; he did not love her, but someone else*). **3** otherwise; if not (*run, (or) else you will be late*). [OE *elles*, rel. to L *alius*, Gk *allos*]

elsewhere /'elsweə(r)/ adv. in or to some other place. [OE *elles hwǣr* (as ELSE, WHERE)]

Elsinore /ˈelsɪˌnɔː(r)/ (Danish **Helsingør** /ˌhɛlseŋ'øːr/) a port on the NE coast of the island of Zealand, Denmark; pop. (1990) 56,750. It is the site of the 16th-century Kronborg Castle, which is the setting for Shakespeare's *Hamlet*.

eluant var. of ELUENT.

Éluard /'elʊˌɑː(r)/, Paul (pseudonym of Eugène Grindel) (1895–1952), French poet. Following the First World War he became a leading figure in the surrealist movement; his poetry from this period includes the collection *Capitale de la douleur* (1926). He broke with the surrealists in 1938 and joined the Communist Party in 1942. He then became an active figure in the French resistance movement, secretly circulating his poetry denouncing the German occupation (such as the collection *Poésie et vérité*, 1942).

eluate /'eljʊˌeɪt/ n. Chem. a solution or gas stream obtained by elution. [formed as ELUENT]

elucidate /ɪ'luːsɪˌdeɪt, ɪ'ljuː-/ v.tr. throw light on; explain. □ **elucidative** adj. **elucidator** n. **elucidatory** adj. **elucidation** /ɪˌluːsɪ'deɪʃ(ə)n, ɪˌljuː-/ n. [LL *elucidare* (as E-, LUCID)]

elude /ɪ'luːd, ɪ'ljuːd/ v.tr. **1** escape adroitly from (a danger, difficulty, pursuer, etc.); dodge. **2** avoid compliance with (a law, request, etc.) or fulfilment of (an obligation). **3** (of a fact, solution, etc.) escape from or baffle (a person's memory or understanding). □ **elusion** /ɪ'luːʒ(ə)n, ɪˈljuː-/ *n.* **elusory** /ɪˈluːsərɪ, ɪˈljuː-/ *adj.* [L *eludere elus-* (as E-, *ludere* play)]

eluent /ˈeljʊənt/ *n.* (also **eluant**) *Chem.* a fluid used for elution. [L *eluere* wash out (as E-, *luere lut-* wash)]

Elul /ˈiːlʌl, ˈelʌl/ *n.* (in the Jewish calendar) the twelfth month of the civil and sixth of the religious year, usually coinciding with parts of August and September. [Heb. *ʾelūl*]

El Uqsur see LUXOR.

elusive /ɪˈluːsɪv, ɪˈljuː-/ *adj.* **1** difficult to find or catch; tending to elude. **2** difficult to remember or recall. **3** (of a reply etc.) avoiding the point raised; seeking to elude. □ **elusively** *adv.* **elusiveness** *n.*

elute /ɪˈluːt, ɪˈljuːt/ *v.tr. Chem.* remove (an adsorbed substance) by washing with a solvent, esp. in chromatography. □ **elution** /ɪˈluːʃ(ə)n, ɪˈljuː-/ *n.* [G *eluieren* (as ELUENT)]

elutriate /ɪˈluːtrɪˌeɪt, ɪˈljuː-/ *v.tr. Chem.* separate (lighter and heavier particles in a mixture) by suspension in an upward flow of liquid or gas. □ **elutriation** /ɪˌluːtrɪˈeɪʃ(ə)n, ɪˌljuː-/ *n.* [L *elutriare elutriat-* (as E-, *lutriare* wash)]

elver /ˈelvə(r)/ *n.* a young eel. [var. of *eel-fare* (see FARE) = a brood of young eels]

elves *pl.* of ELF.

elvish see ELF.

Ely /ˈiːlɪ/ a cathedral city in the fenland of Cambridgeshire, on the River Ouse; pop. (1981) 9,100.

Ely, Isle of a former county of England extending over the northern part of present-day Cambridgeshire. Before widespread drainage it formed a fertile 'island' in the surrounding fenland.

Elysée Palace /eɪˈliːzeɪ/ (called in French *Palais de l'Elysée*) a building in Paris which has been the official residence of the French President since 1870. It was built in 1718 for the Comte d'Evreux and was occupied by Madame de Pompadour, Napoleon I, and Napoleon III.

Elysium /ɪˈlɪzɪəm/ *n.* **1** *Gk Mythol.* the fields at the ends of the earth to which certain favoured heroes, exempted from death, were conveyed by the gods. This concept appears to be a survival from Minoan religion. Also called *Elysian Fields*. **2** a place or state of ideal happiness. □ **Elysian** *adj.* [L f. Gk *Elusion* (*pedion* plain)]

elytron /ˈelɪˌtrɒn/ *n.* (pl. **elytra** /-trə/) *Zool.* the usu. hard wing-case of a beetle, a pair of which protects the hindwings and abdomen. [Gk *elutron* sheath]

Elzevir /ˈelzəˌvɪə(r)/ a family of Dutch printers. Fifteen members were active 1581–1712. Louis (*c.*1542–1617) founded the business at Leiden *c.*1580. His son Bonaventure (1583–1652) and grandson Abraham (1592–1652) managed the firm in its prime, when it published elegant editions of the works of classical authors (1634–6) and a series on countries called *Petites Républiques* (1625–49).

em /em/ *n. Printing* **1** a unit for measuring the amount of printed matter in a line, usually equal to the nominal width of capital M. **2** a unit of measurement equal to 12 points. □ **em rule** (or **dash**) a long dash used in punctuation. [name of the letter M]

em- /em, ɪm/ *prefix* assim. form of EN-¹, EN-² before *b, p*.

'em /əm/ *pron. colloq.* them (*let 'em all come*). [orig. a form of ME *hem*, dative and accus. 3rd pers. pl. pron.: now regarded as an abbr. of THEM]

emaciate /ɪˈmeɪsɪˌeɪt, ɪˈmeɪʃɪ-/ *v.tr.* (esp. as **emaciated** *adj.*) make abnormally thin or feeble. □ **emaciation** /ɪˌmeɪsɪˈeɪʃ(ə)n/ *n.* [L *emaciare emaciat-* (as E-, *macies* leanness)]

email /ˈiːmeɪl/ *n. & v.* (also **e-mail**) = *electronic mail.* ● *v.tr.* **1** send email to (a person). **2** send by email.

emalangeni *pl.* of LILANGENI.

emanate /ˈeməˌneɪt/ *v.* **1** *intr.* (usu. foll. by *from*) (of an idea, rumour, etc.) issue, originate (from a source). **2** *intr.* (usu. foll. by *from*) (of gas, light, etc.) proceed, issue. **3** *tr.* emit; send forth. [L *emanare* flow out]

emanation /ˌeməˈneɪʃ(ə)n/ *n.* **1** the act or process of emanating. **2** something that emanates from a source (esp. of virtues, qualities, etc.). **3** *Chem.* a radioactive gas formed by radioactive decay. □ **emanative** /ˈemənətɪv/ *adj.* [LL *emanatio* (as EMANATE)]

emancipate /ɪˈmænsɪˌpeɪt/ *v.tr.* **1** free from restraint, esp. legal, social, or political. **2** (usu. as **emancipated** *adj.*) cause to be less inhibited by moral or social convention. **3** free from slavery. □ **emancipator** *n.* **emancipatory** /-pətərɪ/ *adj.* **emancipation** /ɪˌmænsɪˈpeɪʃ(ə)n/ *n.* [L *emancipare* transfer property (as E-, *manus* hand + *capere* take)]

Emancipation Proclamation (in the American Civil War) the announcement made by President Lincoln on 22 Sept. 1862 emancipating all black slaves in states still engaged in rebellion against the Federal Union with effect from the beginning of 1863. Although implementation was strictly beyond Lincoln's powers, the declaration was important in that it turned the war into a crusade against slavery and led to the recruitment of many thousands of black soldiers.

emasculate *v. & adj.* ● *v.tr.* /ɪˈmæskjʊˌleɪt/ **1** deprive of force or vigour; make feeble or ineffective. **2** castrate. ● *adj.* /ɪˈmæskjʊlət/ **1** deprived of force or vigour. **2** castrated. **3** effeminate. □ **emasculator** /-ˌleɪtə(r)/ *n.* **emasculatory** /-lətərɪ/ *adj.* **emasculation** /ɪˌmæskjʊˈleɪʃ(ə)n/ *n.* [L *emasculatus* past part. of *emasculare* (as E-, *masculus* dimin. of *mas* male)]

embalm /ɪmˈbɑːm/ *v.tr.* **1** preserve (a corpse) from decay orig. with spices, now by means of arterial injection. **2** preserve from oblivion. **3** endow with balmy fragrance. □ **embalmer** *n.* **embalmment** *n.* [ME f. OF *embaumer* (as EN-¹, BALM)]

embank /ɪmˈbæŋk/ *v.tr.* shut in or confine (a river etc.) with an artificial bank.

embankment /ɪmˈbæŋkmənt/ *n.* an earth or stone bank for keeping back water, or for carrying a road or railway.

embargo /emˈbɑːɡəʊ/ *n. & v.* ● *n.* (pl. **-oes**) **1** an order of a state forbidding foreign ships to enter, or any ships to leave, its ports. **2** an official suspension of commerce or other activity (*be under an embargo*). **3** an impediment. ● *v.tr.* (**-oes, -oed**) **1** place (ships, trade, etc.) under embargo. **2** seize (a ship, goods) for state service. [Sp. f. *embargar* arrest f. Rmc (as IN-², BAR¹)]

embark /ɪmˈbɑːk/ *v.* **1** *tr. & intr.* (often foll. by *for*) put or go on board a ship or aircraft (to a destination). **2** *intr.* (foll. by *on, upon*) engage in an activity or undertaking. □ **embarkation** /ˌembɑːˈkeɪʃ(ə)n/ *n.* (in sense 1). [F *embarquer* (as IN-², BARK³)]

embarras de choix /ɒmbæˌrɑː də ˈʃwʌ/ *n.* (also **embarras de richesses** /riːˈʃes/) more choices than one needs or can deal with. [F, = embarrassment of choice, riches]

embarrass /ɪmˈbærəs/ *v.tr.* **1 a** cause (a person) to feel awkward or self-conscious or ashamed. **b** (as **embarrassed** *adj.*) having or expressing a feeling of awkwardness or self-consciousness. **2** (as **embarrassed** *adj.*) encumbered with debts. **3** hamper, impede. **4** complicate (a question etc.). **5** perplex. □ **embarrassedly** *adv.* **embarrassing** *adj.* **embarrassingly** *adv.* **embarrassment** *n.* [F *embarrasser* (orig. = hamper) f. Sp. *embarazar* f. It. *imbarrare* bar in (as IN-², BAR¹)]

embassy /ˈembəsɪ/ *n.* (pl. **-ies**) **1 a** the residence or offices of an ambassador. **b** the ambassador and staff attached to an embassy. **2** a deputation or mission to a foreign country. [earlier *ambassy* f. OF *ambassée* etc. f. med.L *ambasciata* f. Rmc (as AMBASSADOR)]

embattle /ɪmˈbæt(ə)l/ *v.tr.* **1** set (an army etc.) in battle array. **b** fortify against attack. **2** provide (a building or wall) with battlements. **3** (as **embattled** *adj.*) **a** prepared or arrayed for battle. **b** involved in a conflict or difficult undertaking. **c** *Heraldry* like battlements in form. [ME f. OF *embataillier* (as EN-¹, BATTLE): see BATTLEMENT]

embay /ɪmˈbeɪ/ *v.tr.* (usu. in *passive*) **1** enclose in or as in a bay; shut in. **2** form (a coast) into bays. □ **embayment** *n.*

embed /ɪmˈbed/ *v.tr.* (also **imbed**) (**-bedded, -bedding**) **1** (esp. as **embedded** *adj.*) fix firmly in a surrounding mass (*embedded in concrete*). **2** (of a mass) surround so as to fix firmly. **3** place in or as in a bed. □ **embedment** *n.*

embellish /ɪmˈbelɪʃ/ *v.tr.* **1** beautify, adorn. **2** add interest to (a narrative) with fictitious additions. □ **embellisher** *n.* **embellishment** *n.* [ME f. OF *embellir* (as EN-¹, *bel* handsome f. L *bellus*)]

ember /ˈembə(r)/ *n.* **1** (usu. in *pl.*) a small piece of glowing coal or wood in a dying fire. **2** an almost extinct residue of a past activity, feeling, etc. [OE *æmyrge* f. Gmc]

Ember days *n.pl.* a group of three days in each season, observed as days of fasting and prayer in some Christian Churches. Their early history and original purpose is obscure; at first there were apparently only three groups, perhaps taken over from pagan religious observances connected with seed-time, harvest, and autumn vintage. They are now associated almost entirely with the ordination of ministers. [OE *ymbren* (n.), perh. f. *ymbryne* period f. *ymb* about + *ryne* course]

embezzle /ɪmˈbez(ə)l/ *v.tr.* (also *absol.*) fraudulently divert (money etc.) from a company, fund, etc., to one's own use. □ **embezzlement** *n.* **embezzler** *n.* [AF *embesiler* (as EN-¹, OF *besillier* maltreat, ravage, of unkn. orig.)]

embitter /ɪmˈbɪtə(r)/ *v.tr.* **1** arouse bitter feelings in (a person). **2** make more bitter or painful. **3** render (a person or feelings) hostile. □ **embitterment** *n.*

emblazon /ɪmˈbleɪz(ə)n/ *v.tr.* **1 a** portray conspicuously, as on a heraldic shield. **b** adorn (a shield) with heraldic devices. **2** adorn brightly and conspicuously. **3** celebrate, extol. □ **emblazonment** *n.*

emblem /'embləm/ *n.* **1** a symbol or representation typifying or identifying an institution, quality, etc. **2** (foll. by *of*) (of a person) the type (*the very emblem of courage*). **3** a heraldic device or symbolic object as a distinctive badge. □ **emblematic** /ˌemblə'mætɪk/ *adj.* **emblematical** *adj.* **emblematically** *adv.* [ME f. L *emblema* f. Gk *emblēma* -*matos* insertion f. *emballō* throw in (as EN-[1], *ballō* throw)]

emblematize /ɪm'blemə,taɪz/ *v.tr.* (also **-ise**) **1** serve as an emblem of. **2** represent by an emblem.

emblements /'emblɪmənts/ *n.pl.* Law crops normally harvested annually, regarded as personal property. [ME f. OF *emblaement* f. *emblaier* (as EN-[1], *blé* corn)]

embody /ɪm'bɒdɪ/ *v.tr.* (**-ies**, **-ied**) **1** give a concrete or discernible form to (an idea, concept, etc.). **2** (of a thing or person) be an expression of (an idea etc.). **3** express tangibly (*courage embodied in heroic actions*). **4** form into a body. **5** include, comprise. **6** provide (a spirit) with bodily form. □ **embodiment** *n.*

embolden /ɪm'bəʊld(ə)n/ *v.tr.* (often foll. by *to* + infin.) make bold; encourage.

embolism /'embə,lɪz(ə)m/ *n.* Med. an obstruction of an artery by a clot of blood, air-bubble, etc. [ME, = 'intercalation' f. LL *embolismus* f. Gk *embolismos* f. *emballō* (as EMBLEM)]

embolus /'embələs/ *n.* (pl. **emboli** /-ˌlaɪ/) Med. an object causing an embolism. [L, = piston, f. Gk *embolos* peg, stopper]

embonpoint /ˌɒmbɒn'pwʌn/ *n.* plumpness (of a person). [F *en bon point* in good condition]

embosom /ɪm'bʊz(ə)m/ *v.tr. literary* **1** embrace. **2** enclose, surround.

emboss /ɪm'bɒs/ *v.tr.* **1** carve or mould in relief. **2** form figures etc. so that they stand out on (a surface). **3** make protuberant. □ **embosser** *n.* **embossment** *n.* [ME, f. OF (as EN-[1], BOSS²)]

embouchure /'ɒmbʊ,ʃʊə(r)/ *n.* **1** Mus. **a** the mode of applying the mouth to the mouthpiece of a brass or wind instrument. **b** the mouthpiece of some instruments. **2** the mouth of a river. **3** the opening of a valley. [F f. *s'emboucher* discharge itself by the mouth (as EN-[1], *bouche* mouth)]

embowel /ɪm'baʊəl/ *v.tr.* (**embowelled**, **embowelling**; US **emboweled**, **emboweling**) *archaic* = DISEMBOWEL. [OF *emboweler* f. *esboueler* (as EX-[1], BOWEL)]

embower /ɪm'baʊə(r)/ *v.tr. literary* enclose as in a bower.

embrace /ɪm'breɪs/ *v. & n.* ● *v.tr.* **1** **a** hold (a person) closely in the arms, esp. as a sign of affection. **b** (*absol.*, of two people) hold each other closely. **2** clasp, enclose. **3** accept eagerly (an offer, opportunity, etc.). **4** adopt (a course of action, doctrine, cause, etc.). **5** include, comprise. **6** take in with the eye or mind. ● *n.* an act of embracing; holding in the arms. □ **embraceable** *adj.* **embracement** *n.* **embracer** *n.* [ME f. OF *embracer*, ult. f. L *in-* IN-[1] + *bracchium* arm]

embranchment /ɪm'brɑːntʃmənt/ *n.* a branching-out (of the arm of a river etc.). [F *embranchement* BRANCH (as EN-[1], BRANCH)]

embrasure /ɪm'breɪʒə(r)/ *n.* **1** the bevelling of a wall at the sides of a door or window; splaying. **2** a small opening in a parapet of a fortified building, splayed on the inside. □ **embrasured** *adj.* [F f. *embraser* splay, of unkn. orig.]

embrittle /ɪm'brɪt(ə)l/ *v.tr.* make brittle. □ **embrittlement** *n.*

embrocation /ˌembrəʊ'keɪʃ(ə)n/ *n.* a liquid used for rubbing on the body to relieve muscular pain etc. [F *embrocation* or med.L *embrocatio* ult. f. Gk *embrokhē* lotion]

embroider /ɪm'brɔɪdə(r)/ *v.tr.* **1** (also *absol.*) **a** decorate (cloth etc.) with needlework. **b** create (a design) in this way. **2** add interest to (a narrative) with fictitious additions. □ **embroiderer** *n.* [ME f. AF *enbrouder* (as EN-[1], OF *brouder*, *broisder* f. Gmc)]

embroidery /ɪm'brɔɪdərɪ/ *n.* (pl. **-ies**) **1** the art of embroidering. **2** embroidered work; a piece of this. **3** unnecessary or extravagant ornament. **4** fictitious additions to a story etc. [ME f. AF *enbrouderie* (as EMBROIDER)]

embroil /ɪm'brɔɪl/ *v.tr.* **1** (often foll. by *with*) involve (a person) in conflict or difficulties. **2** bring (affairs) into a state of confusion. □ **embroilment** *n.* [F *embrouiller* (as EN-[1], BROIL²)]

embryo /'embrɪəʊ/ *n.* (pl. **-os**) **1** **a** an unborn or unhatched offspring. **b** a human offspring in the first eight weeks from conception. **2** a rudimentary plant contained in a seed. **3** a thing in a rudimentary stage. **4** (*attrib.*) undeveloped, immature. □ **in embryo** undeveloped. □ **embryoid** *adj.* **embryonal** /'embrɪən(ə)l/ *adj.* **embryonic**

/ˌembrɪ'ɒnɪk/ *adj.* **embryonically** *adv.* [LL *embryo* -*onis* f. Gk *embruon* foetus (as EN-², *bruō* swell, grow)]

embryo- /'embrɪəʊ/ *comb. form* embryo.

embryogenesis /ˌembrɪəʊ'dʒenɪsɪs/ *n.* the formation of an embryo.

embryology /ˌembrɪ'ɒlədʒɪ/ *n.* the study of embryos. □ **embryologist** *n.* **embryologic** /ˌembrɪə'lɒdʒɪk/ *adj.* **embryological** *adj.* **embryologically** *adv.*

embus /ɪm'bʌs/ *v.* (**embused**, **embusing** or **embussed**, **embussing**) Mil. **1** tr. put (troops etc. or equipment) into a motor vehicle. **2** intr. board a motor vehicle.

emcee /em'siː/ *n. & v. colloq.* ● *n.* a master of ceremonies or compère. ● *v.tr. & intr.* (**emcees**, **emceed**) compère. [the letters *MC*]

-eme /iːm/ *suffix* Linguistics forming nouns denoting units of structure etc. (*grapheme; morpheme*). [F *-ème* unit f. Gk *-ēma*]

emend /ɪ'mend/ *v.tr.* edit (a text etc.) to remove errors and corruptions. □ **emendatory** /-dətərɪ/ *adj.* **emendation** /ˌiːmen'deɪʃ(ə)n/ *n.* **emendator** /'iːmen,deɪtə(r)/ *n.* [ME f. L *emendare* (as E-, *menda* fault)]

emerald /'emərəld/ *n.* **1** a bright-green precious stone, a variety of beryl. **2** (also **emerald green**) the colour of this. **3** (also **emerald moth**) a green moth of the family Geometridae, e.g. the *large emerald*, *Geometra papilionaria*. □ **emeraldine** /-ˌdaɪn, -ˌdɪn/ *adj.* [ME f. OF *emeraude*, *esm-*, ult. f. Gk *smaragdos*]

Emerald Isle *literary* Ireland.

emerge /ɪ'mɜːdʒ/ *v.intr.* (often foll. by *from*) **1** come up or out into view, esp. when formerly concealed. **2** come up out of a liquid. **3** (of facts, circumstances, etc.) come to light, become known, esp. as a result of inquiry etc. (*one day the truth will emerge; it emerged that he had been drinking*). **4** become recognized or prominent (*emerged as a leading contender*). **5** (of a question, difficulty, etc.) become apparent. **6** survive (an ordeal etc.) with a specified result (*emerged unscathed*). □ **emergence** *n.* [L *emergere emers-* (as E-, *mergere* dip)]

emergency /ɪ'mɜːdʒənsɪ/ *n.* (pl. **-ies**) **1** a sudden state of danger, conflict, etc., requiring immediate action. **2** **a** a medical condition requiring immediate treatment. **b** a patient with such a condition. **3** (*attrib.*) characterized by or for use in an emergency (*emergency exit*). **4** Austral. Sport a reserve player. □ **state of emergency** a condition of danger or disaster affecting a country, esp. with normal constitutional procedures suspended. [med.L *emergentia* (as EMERGE)]

emergent /ɪ'mɜːdʒənt/ *adj.* **1** becoming apparent; emerging. **2** (of a nation) newly formed or made independent.

emeritus /ɪ'merɪtəs/ *adj.* **1** retired and retaining one's title as an honour (*emeritus professor; professor emeritus*). **2** honourably discharged from service. [L, past part. of *emereri* (as E-, *mereri* earn)]

emersion /ɪ'mɜːʃ(ə)n/ *n.* **1** the act or an instance of emerging, esp. from water. **2** Astron. the reappearance of a celestial body after its eclipse or occultation. [LL *emersio* (as EMERGE)]

Emerson /'eməs(ə)n/, Ralph Waldo (1803–82), American philosopher and poet. While visiting England in 1832, he met Coleridge, Wordsworth, and Carlyle, through whom he became associated with German idealism. On his return to the US he evolved the concept of Transcendentalism, which found expression in his essay *Nature* (1836). He and Thoreau are regarded as the central figures of New England Transcendentalism, which, in its reverence for nature, foreshadowed the ecological movement of the 20th century.

emery /'emərɪ/ *n.* **1** a coarse rock of corundum and magnetite or haematite used for polishing metal or other hard materials. **2** (*attrib.*) covered with emery. □ **emery-board** a strip of thin wood or board coated with emery or another abrasive, used as a nail-file. **emery paper** (or **cloth**) paper or cloth covered with emery, used for polishing or cleaning metals etc. [F *émeri(l)* f. It. *smeriglio* ult. f. Gk *smuris*, *smēris* polishing powder]

Emesa /'eməsə/ a city in ancient Syria, on the River Orontes on the site of present-day Homs. It was famous for its temple to the sun-god Elah-Gabal. Ruled by priest-kings throughout the period of the Roman Empire, Emesa fell to the Muslims in AD 636.

emetic /ɪ'metɪk/ *adj. & n.* ● *adj.* that causes vomiting. ● *n.* an emetic medicine. [Gk *emetikos* f. *emeō* vomit]

EMF *abbr.* **1** (usu. **emf**) electromotive force. **2** electromagnetic field(s). **3** European Monetary Fund.

-emia US var. of -AEMIA.

emigrant /'emɪgrənt/ *n. & adj.* ● *n.* a person who emigrates. ● *adj.* emigrating.

emigrate /'emɪ,greɪt/ *v.* **1** *intr.* (often foll. by *to*) leave one's own country to settle in another (*decided to emigrate; thousands emigrated to America*).

2 *tr.* assist (a person) to emigrate. □ **emigratory** *adj.* **emigration** /ˌemɪˈɡreɪʃ(ə)n/ *n.* [L *emigrare emigrat-* (as E-, *migrare* depart)]

émigré /ˈemɪˌɡreɪ/ *n.* (also **emigre**) an emigrant, esp. a political exile. [F, past part. of *émigrer* EMIGRATE]

Emi Koussi /ˌemɪ ˈkuːsə/ a volcanic mountain in the Sahara, in northern Chad. Rising to 3,415 m (11,202 ft), it is the highest peak in the Tibesti Mountains.

Emilia-Romagna /eˌmiːljərəʊˈmɑːnjə/ a region of northern Italy; capital, Bologna.

eminence /ˈemɪnəns/ *n.* **1** distinction; recognized superiority. **2** a piece of rising ground. **3** (**Eminence**) a title used in addressing or referring to a cardinal (*Your Eminence*; *His Eminence*). **4** an important person. [L *eminentia* (as EMINENT)]

éminence grise /ˌeɪmɪnɒns ˈɡriːz/ *n.* (*pl.* **éminences grises** pronunc. same) **1** a person who exercises power or influence without holding office. **2** a confidential agent. [F, = grey cardinal (see EMINENCE): orig. applied to Cardinal Richelieu's grey-cloaked private secretary, Père Joseph du Tremblay (1577–1638)]

eminent /ˈemɪnənt/ *adj.* **1** distinguished, notable. **2** (of qualities) remarkable in degree. □ **eminent domain** sovereign control over all property in a state, with the right of expropriation. □ **eminently** *adv.* [ME f. L *eminere eminent-* jut]

emir /eˈmɪə(r)/ *n.* **1** a title of various Muslim rulers. **2** *archaic* a male descendant of Muhammad. [F *émir* f. Arab. *ʾamīr*: cf. AMIR]

emirate /ˈemɪərət/ *n.* the rank, domain, or reign of an emir.

emissary /ˈemɪsəri/ *n.* (*pl.* **-ies**) a person sent on a special mission (usu. diplomatic, formerly usu. odious or underhand). [L *emissarius* scout, spy (as EMIT)]

emission /ɪˈmɪʃ(ə)n/ *n.* **1** (often foll. by *of*) the process or an act of emitting. **2** a thing emitted. □ **emission nebula** *Astron.* a nebula that emits its own light. **emission spectrum** *Physics* a spectrum of the electromagnetic radiation emitted by a source. [L *emissio* (as EMIT)]

emissive /ɪˈmɪsɪv/ *adj.* having the power to radiate light, heat, etc. □ **emissivity** /ˌiːmɪˈsɪvɪti/ *n.*

emit /ɪˈmɪt/ *v.tr.* (**emitted**, **emitting**) **1 a** send out (heat, light, vapour, etc.). **b** discharge from the body. **2** utter (a cry etc.). [L *emittere emiss-* (as E-, *mittere* send)]

emitter /ɪˈmɪtə(r)/ *n.* **1** that which emits something. **2** *Electronics* the region in a transistor producing carriers of current.

Emmental /ˈemənˌtɑːl/ *n.* (also **Emmenthal**) a kind of hard Swiss cheese with many holes in it, similar to Gruyère. [G *Emmentaler* f. *Emmental* in Switzerland]

emmer /ˈemə(r)/ *n.* a kind of wheat, *Triticum dicoccum*, grown mainly for fodder. [G dial.]

emmet /ˈemɪt/ *n.* **1** *archaic* or *dial.* an ant. **2** *sl.* (in Cornwall) a holiday-maker, a tourist. [OE *æmete*: see ANT]

Emmy /ˈemɪ/ *n.* (*pl.* **Emmys**) (in the US) a statuette awarded by the American Academy of Television Arts and Sciences for an outstanding television programme, production, or performance. A series of such awards has been made annually since 1949. [perh. f. *Immy* = *image orthicon* (a form of television camera)]

emollient /ɪˈmɒlɪənt/ *adj. & n.* ● *adj.* that softens or soothes the skin. ● *n.* an emollient agent. □ **emollience** *n.* [L *emollire* (as E-, *mollis* soft)]

emolument /ɪˈmɒljʊmənt/ *n.* a salary, fee, or profit from employment or office. [ME f. OF *emolument* or L *emolumentum*, orig. prob. 'payment for corn-grinding', f. *emolere* (as E-, *molere* grind)]

Emona /ɪˈməʊnə/ the Roman name for LJUBLJANA.

emote /ɪˈməʊt/ *v.intr. colloq.* show excessive emotion. □ **emoter** *n.* [back-form. f. EMOTION]

emotion /ɪˈməʊʃ(ə)n/ *n.* **1** a strong mental or instinctive feeling such as love or fear. **2** emotional intensity or sensibility (*he spoke with emotion*). □ **emotionless** *adj.* [earlier = agitation, disturbance of the mind, f. F *émotion* f. *émouvoir* excite]

emotional /ɪˈməʊʃən(ə)l/ *adj.* **1** of or relating to the emotions. **2** (of a person) liable to or displaying excessive emotion (*couldn't discuss it without getting emotional*). **3** expressing or based on emotion (*an emotional appeal*). **4** likely to excite emotion (*an emotional issue*). □ **emotionalism** *n.* **emotionalist** *n.* **emotionalize** *v.tr.* (also **-ise**) **emotionally** *adv.* **emotionality** /ɪˌməʊʃəˈnælɪti/ *n.*

emotive /ɪˈməʊtɪv/ *adj.* **1** of or characterized by emotion. **2** tending to excite emotion. **3** arousing feeling; not purely descriptive. □ **emotively** *adv.* **emotiveness** *n.* **emotivity** /ˌiːməʊˈtɪvɪti/ *n.* [L *emovere emot-* (as E-, *movere* move)]

empanel /ɪmˈpæn(ə)l/ *v.tr.* (also **impanel**) (**-panelled, -panelling**; *US* **-paneled, -paneling**) enrol or enter on a panel (those eligible for jury service). □ **empanelment** *n.* [AF *empaneller* (as EN-[1], PANEL)]

empathize /ˈempəˌθaɪz/ *v.* (also **-ise**) *Psychol.* **1** *intr.* (usu. foll. by *with*) exercise empathy. **2** *tr.* treat with empathy.

empathy /ˈempəθi/ *n. Psychol.* the power of identifying oneself mentally with (and so fully comprehending) a person or object of contemplation. □ **empathist** *n.* **empathetic** /ˌempəˈθetɪk/ *adj.* **empathetically** *adv.* **empathic** /emˈpæθɪk/ *adj.* **empathically** *adv.* [transl. G *Einfühlung* f. *ein* in + *Fühlung* feeling, after Gk *empatheia*: see SYMPATHY]

Empedocles /emˈpedəˌkliːz/ (*c.*493–*c.*433 BC), Greek philosopher, born in Sicily. His hexametric poem *On Nature* taught that the universe is composed of the four imperishable elements of fire, air, water, and earth, which mingle and separate under the influence of the opposing principles of Love and Strife. According to legend, he leapt into the crater of Mount Etna in order that he might be thought a god.

empennage /emˈpenɪdʒ/ *n. Aeron.* an arrangement of stabilizing surfaces at the tail of an aircraft. [F f. *empenner* to feather (an arrow)]

emperor /ˈempərə(r)/ *n.* **1** the sovereign of an empire. **2** a sovereign of higher rank than a king. □ **emperor moth** a large European moth, *Saturnia pavonia*, related to the silk moths, with eye-spots on all four wings. **emperor penguin** the largest kind of penguin, *Aptenodytes forsteri*, of the Antarctic. □ **emperorship** *n.* [ME f. OF *emperere, empereor* f. L *imperator -oris* f. *imperare* command]

emphasis /ˈemfəsɪs/ *n.* (*pl.* **emphases** /-ˌsiːz/) **1** special importance or prominence attached to a thing, fact, idea, etc. (*emphasis on economy*). **2** stress laid on a word or words to indicate special meaning or importance. **3** vigour or intensity of expression, feeling, action, etc. **4** prominence, sharpness of contour. [L f. Gk f. *emphainō* exhibit (as EN-[2], *phainō* show)]

emphasize /ˈemfəˌsaɪz/ *v.tr.* (also **-ise**) **1** bring (a thing, fact, etc.) into special prominence. **2** lay stress on (a word in speaking).

emphatic /ɪmˈfætɪk/ *adj.* **1** (of language, tone, or gesture) forcibly expressive. **2** of words: bearing the stress. **3** used to give emphasis. **3** expressing oneself with emphasis. **4** (of an action or process) forcible, significant. □ **emphatically** *adv.* [LL *emphaticus* f. Gk *emphatikos* (as EMPHASIS)]

emphysema /ˌemfɪˈsiːmə/ *n.* **1** enlargement of the air sacs of the lungs causing breathlessness. **2** a swelling caused by the presence of air in the connective tissues of the body. [LL f. Gk *emphusēma* f. *emphusaō* puff up]

empire /ˈempaɪə(r)/ *n.* **1** an extensive group of states or countries under a single supreme authority, esp. an emperor or empress. **2 a** supreme dominion. **b** (often foll. by *over*) *archaic* absolute control. **3** a large commercial organization etc. owned or directed by one person or group. **4** (**the Empire**) *hist.* **a** the British Empire. **b** the Holy Roman Empire. **c** the period of the reign of Napoleon I as emperor of the French (1804–15), also called *First Empire* to distinguish it from the *Second Empire* of Napoleon III (1852–70). **5** a type or period of government in which the sovereign is called emperor or empress. □ **empire-builder** a person who deliberately acquires extra territory, authority, etc., esp. just for the sake of power. [ME f. OF f. L *imperium* rel. to *imperare*: see EMPEROR]

Empire Day the former name of Commonwealth Day, orig. 24 May.

Empire State Building a skyscraper on Fifth Avenue, New York City, which was for several years the tallest building in the world. When first erected, in 1930–1, it measured 381 m (1,250 ft); the addition of a television mast in 1951 brought its height to 449 m (1,472 ft). It is named after New York, the Empire State.

Empire style *n.* a style of furniture, interior decoration, and dress which started in Paris around the time that Napoleon became emperor (1804) and spread through Continental Europe, corresponding to the Regency style in England. The style was neoclassical but marked by an interest in ancient (particularly Egyptian) motifs that was probably inspired by Napoleon's Egyptian campaigns. In women's dress there was a distinctive high-waisted fashion embellished with elaborate embroidery.

empiric /ɪmˈpɪrɪk/ *adj. & n.* ● *adj.* = EMPIRICAL. ● *n. archaic* **1** a person relying solely on experiment. **2** a quack doctor. [L *empiricus* f. Gk *empeirikos* f. *empeiria* experience f. *empeiros* skilled]

empirical /ɪmˈpɪrɪk(ə)l/ *adj.* **1** based or acting on observation or experiment, not on theory. **2** *Philos.* regarding sense-data as valid information. **3** deriving knowledge from experience alone.

◻ **empirical formula** *Chem.* a formula giving the proportions of the elements present in a compound but not the actual numbers or arrangement. ◻ **empirically** *adv.*

empiricism /ɪmˈpɪrɪˌsɪz(ə)m/ *n. Philos.* the theory that all knowledge is derived from sense-experience. Empiricism, which denies a priori concepts, developed in the 17th and 18th centuries, stimulated by the rise of experimental science; Locke, Berkeley, and Hume were its principal exponents. (opp. RATIONALISM 1; cf. SENSATIONALISM 2). ◻ **empiricist** *n. & adj.*

emplacement /ɪmˈpleɪsmənt/ *n.* **1** the act or an instance of putting something in position. **2** a platform or defended position where a gun is placed for firing. **3** situation, position. [F as EN-¹, PLACE)]

emplane /ɪmˈpleɪn/ *v.intr. & tr.* (also **enplane** /ɪnˈpleɪn/) go or put on board an aeroplane.

employ /ɪmˈplɔɪ/ *v. & n.* ● *v.tr.* **1** use the services of (a person) in return for payment; keep (a person) in one's service. **2** (often foll. by *for, in, on*) use (a thing, time, energy, etc.) esp. to good effect. **3** (often foll. by *in*) keep (a person) occupied. ● *n.* the state of being employed, esp. for wages. ◻ **in the employ of** employed by. ◻ **employer** *n.* **employable** *adj.* **employability** /-ˌplɔɪəˈbɪlɪti/ *n.* [ME f. OF *employer* ult. f. L *implicari* be involved f. *implicare* enfold: see IMPLICATE]

employee /ˌemplɔɪˈiː, emˈplɔɪ/ *n.* (*US* **employe**) a person employed for wages or salary, esp. at non-executive level.

employment /ɪmˈplɔɪmənt/ *n.* **1** the act of employing or the state of being employed. **2** a person's regular trade or profession. ◻ **employment agency** a business that finds employers or employees for those seeking them. **employment office** (formerly **employment exchange**) *Brit.* any of a number of government offices concerned with advising and finding work for the unemployed.

empolder var. of IMPOLDER.

emporium /emˈpɔːrɪəm/ *n.* (*pl.* **emporia** /-rɪə/ or **-ums**) **1** a large retail store selling a wide variety of goods. **2** a centre of commerce, a market. [L f. Gk *emporion* f. *emporos* merchant]

empower /ɪmˈpaʊə(r)/ *v.tr.* (foll. by *to* + infin.) **1** authorize, license. **2** give power to; make able. ◻ **empowerment** *n.*

empress /ˈemprɪs/ *n.* **1** the wife or widow of an emperor. **2** a woman emperor. [ME f. OF *emperesse* fem. of *emperere* EMPEROR]

Empson /ˈemps(ə)n/, Sir William (1906–84), English poet and literary critic. His intricate, closely reasoned poems, published in the *Collected Poems* (1955), reflect his training as a mathematician. His influential literary criticism includes *Seven Types of Ambiguity* (1930).

empty /ˈempti/ *adj., v., & n.* ● *adj.* (**emptier, emptiest**) **1** containing nothing. **2** (of a space, place, house etc.) unoccupied, uninhabited, deserted; unfurnished. **3** (of a transport vehicle etc.) without a load, passengers, etc. **4 a** meaningless, hollow, insincere (*empty threats; an empty gesture*). **b** without substance or purpose (*an empty existence*). **c** (of a person) lacking sense or knowledge; vacant, foolish. **5** *colloq.* hungry. **6** (foll. by *of*) devoid, lacking. ● *v.* (**-ies, -ied**) **1** *tr.* a make empty; remove the contents of. **b** (foll. by *of*) deprive of certain contents (*emptied the room of its chairs*). **c** remove (contents) from a container etc. **2** *tr.* (often foll. by *into*) transfer (the contents of a container). **3** *intr.* become empty. **4** *intr.* (usu. foll. by *into*) (of a river) discharge itself (into the sea etc.). ● *n.* (*pl.* **-ies**) *colloq.* a container (esp. a bottle) left empty of its contents. ◻ **empty-handed 1** bringing or taking nothing. **2** having achieved or obtained nothing. **empty-headed** foolish; lacking common sense. **empty-nester** esp. *N. Amer.* either of a couple whose children have grown up and left home. **on an empty stomach** see STOMACH. ◻ **emptily** *adv.* **emptiness** *n.* [OE *æmtig, æmetig* f. *æmetta* leisure]

Empty Quarter an alternative name for the RUB' AL KHALI.

empurple /ɪmˈpɜːp(ə)l/ *v.tr.* **1** make purple or red. **2** make angry.

empyema /ˌempaɪˈiːmə, ˌempɪ-/ *n. Med.* a collection of pus in a cavity, esp. in the pleura. [LL f. Gk *empuēma* f. *empueō* suppurate (as EN-², *puon* pus)]

empyrean /ˌempaɪˈriːən, emˈpɪrɪən/ *n. & adj.* ● *n.* **1** the highest heaven, as the sphere of fire in ancient cosmology or as the abode of God in early Christianity. **2** *poet.* the visible heavens. ● *adj.* of the empyrean. ◻ **empyreal** /ˌempaɪˈriːəl, emˈpɪrɪəl/ *adj.* [med.L *empyreus* f. Gk *empurios* (as EN-², *pur* fire)]

EMS see EUROPEAN MONETARY SYSTEM.

EMU see EUROPEAN MONETARY UNION.

emu /ˈiːmjuː/ *n.* a large flightless bird, *Dromaius novaehollandiae*, native to Australia, and capable of running at high speed. [earlier *emia, eme* f. Port. *ema*]

e.m.u. *abbr.* electromagnetic unit(s).

emulate /ˈemjʊˌleɪt/ *v.tr.* **1** try to equal or excel. **2** imitate zealously. **3** rival. ◻ **emulator** *n.* **emulative** /-lətɪv/ *adj.* **emulation** /ˌemjʊˈleɪʃ(ə)n/ *n.* [L *aemulari* (as EMULOUS)]

emulous /ˈemjʊləs/ *adj.* **1** (usu. foll. by *of*) seeking to emulate. **2** actuated by a spirit of rivalry. ◻ **emulously** *adv.* [ME f. L *aemulus* rival]

emulsifier /ɪˈmʌlsɪˌfaɪə(r)/ *n.* **1** a substance that stabilizes an emulsion, esp. a food additive used to stabilize processed foods. **2** an apparatus used for producing an emulsion.

emulsify /ɪˈmʌlsɪˌfaɪ/ *v.tr.* (**-ies, -ied**) convert into an emulsion. ◻ **emulsifiable** *adj.* **emulsification** /ɪˌmʌlsɪfɪˈkeɪʃ(ə)n/ *n.*

emulsion /ɪˈmʌlʃ(ə)n/ *n. & v.* ● *n.* **1** a fine dispersion of one liquid in another, esp. as paint, medicine, etc. **2** a mixture of a silver compound suspended in gelatin etc. for coating photographic plates or films. ● *v.tr.* paint with emulsion paint. ◻ **emulsion paint** a paint consisting of pigment bound in a synthetic resin which forms an emulsion with water. ◻ **emulsionize** *v.tr.* (also **-ise**). **emulsive** /ɪˈmʌlsɪv/ *adj.* [F *émulsion* or mod.L *emulsio* f. *emulgere* (as E-, *mulgere muls-* to milk)]

en /en/ *n. Printing* a unit of measurement equal to half an em. ◻ **en rule** (or **dash**) a short dash used in punctuation. [name of the letter *N*]

en-¹ /en, ɪn/ *prefix* (also **em-** before *b, p*) forming verbs, = IN-¹: **1** from nouns, meaning 'put into or on' (*engulf; entrust; embed*). **2** from nouns or adjectives, meaning 'bring into the condition of' (*enslave*); often with the suffix *-en* (*enlighten*). **3** from verbs: **a** in the sense 'in, into, on' (*enfold*). **b** as an intensive (*entangle*). [from or after F *en-* f. L *in-*]

en-² /en, ɪn/ *prefix* (also **em-** before *b, p*) in, inside (*energy; enthusiasm*). [Gk]

-en¹ /ən/ *suffix* forming verbs: **1** from adjectives, usu. meaning 'make or become so or more so' (*deepen; fasten; moisten*). **2** from nouns (*happen; strengthen*). [OE *-nian* f. Gmc]

-en² /ən/ *suffix* (also **-n**) forming adjectives from nouns, meaning: **1** made or consisting of (often with extended and figurative senses) (*wooden*). **2** resembling; of the nature of (*golden; silvern*). [OE f. Gmc]

-en³ /ən/ *suffix* (also **-n**) forming past participles of strong verbs: **1** as a regular inflection (*spoken; sworn*). **2** with restricted sense (*drunken*). [OE f. Gmc]

-en⁴ /ən/ *suffix* forming the plural of a few nouns (*children; brethren; oxen*). [ME reduction of OE *-an*]

-en⁵ /ən, ɪn/ *suffix* forming diminutives of nouns (*chicken; maiden*). [OE f. Gmc]

-en⁶ /ən/ *suffix* **1** forming feminine nouns (*vixen*). **2** forming abstract nouns (*burden*). [OE f. Gmc]

enable /ɪnˈeɪb(ə)l/ *v.tr.* **1** (foll. by *to* + infin.) give (a person etc.) the means or authority to do something. **2** make possible. **3** esp. *Computing* make (a device) operational; switch on. ◻ **enabling act 1** a statute empowering a person or body to take certain action. **2** *US* a statute legalizing something which is otherwise unlawful. ◻ **enablement** *n.* **enabler** *n.*

enact /ɪnˈækt/ *v.tr.* **1 a** (often foll. by *that* + clause) ordain, decree. **b** make (a bill etc.) law. **2** play (a part or scene on stage or in life). ◻ **enactable** *adj.* **enactive** *adj.* **enactor** *n.* **enaction** /-ˈækʃ(ə)n/ *n.*

enactment /ɪnˈæktmənt/ *n.* **1** a law enacted. **2** the process of enacting.

enamel /ɪˈnæm(ə)l/ *n. & v.* ● *n.* **1** a glasslike opaque or semi-transparent coating on metallic or other hard surfaces for ornament or as a preservative lining. **2 a** a smooth hard coating. **b** a cosmetic simulating this. **3** the hard glossy natural coating over the crown of a tooth. **4** a picture or other work of art done in enamel. **5** *poet.* a smooth bright surface colouring, verdure, etc. ● *v.tr.* (**enamelled, enamelling**; *US* **enameled, enameling**) **1** inlay or encrust (a metal etc.) with enamel. **2** portray (figures etc.) with enamel. **3** *archaic* adorn with varied colours. ◻ **enamel paint** a paint that dries to give a smooth hard coat. ◻ **enameller** *n.* **enamelwork** *n.* [ME f. AF *enameler, enamailler* (as EN-¹, OF *esmail* f. Gmc)]

enamelling /ɪˈnæməlɪŋ/ *n.* the use of enamel as a decorative surface on metal, ceramics, etc. Enamel was used in making jewellery by the ancient Egyptians, Greeks, and Romans, and extensively in the ecclesiastical art of the Byzantine world. In the later Middle Ages Limoges in France became an important centre for enamelling, mass-producing religious artefacts such as crucifixes as well as rings, caskets, and other secular articles; enamel was also used in England from the 17th century for miniature portraits. Techniques in enamelling include *champlevé* and *cloisonné*.

turn a cog-wheel. □ **endlessly** adv. **endlessness** n. [OE endelēas (as END, -LESS)]

endmost /ˈendməʊst/ adj. nearest the end.

endnote /ˈendnəʊt/ n. a note printed at the end of a book or section of a book.

endo- /ˈendəʊ/ comb. form internal. [Gk endon within]

endocarditis /ˌendəʊkɑːˈdaɪtɪs/ n. Med. inflammation of the endocardium. □ **endocarditic** /-ˈdɪtɪk/ adj.

endocardium /ˌendəʊˈkɑːdɪəm/ n. (pl. **endocardia** /-dɪə/) Anat. the lining membrane of the heart. [ENDO- + Gk kardia heart]

endocarp /ˈendəʊˌkɑːp/ n. Bot. the innermost layer of the pericarp. □ **endocarpic** /ˌendəʊˈkɑːpɪk/ adj. [ENDO- + PERICARP]

endocrine /ˈendəʊˌkraɪn, -ˌkrɪn/ adj. (of a gland) secreting directly into the blood; ductless (cf. EXOCRINE). [ENDO- + Gk krinō sift]

endocrinology /ˌendəʊkrɪˈnɒlədʒɪ/ n. the study of the structure and physiology of endocrine glands. □ **endocrinologist** n. **endocrinological** /-ˌkrɪnəˈlɒdʒɪk(ə)l/ adj.

endoderm /ˈendəʊˌdɜːm/ n. Zool. the innermost layer of cells or tissue of an embryo in early development. □ **endodermal** /ˌendəʊˈdɜːm(ə)l/ adj. [ENDO- + Gk derma skin]

endogamy /enˈdɒgəmɪ/ n. **1** Anthropol. marrying within the same tribe. **2** Bot. pollination from the same plant. □ **endogamous** adj. [ENDO- + Gk gamos marriage]

endogenous /enˈdɒdʒɪnəs/ adj. growing or originating from within. □ **endogeny** n. **endogenesis** /ˌendəʊˈdʒenɪsɪs/ n.

endolymph /ˈendəʊˌlɪmf/ n. Physiol. the fluid in the membranous labyrinth of the ear.

endometriosis /ˌendəʊˌmiːtrɪˈəʊsɪs/ n. Med. the appearance of endometrial tissue outside the womb.

endometrium /ˌendəʊˈmiːtrɪəm/ n. Anat. the membrane lining the womb. □ **endometrial** adj. [ENDO- + Gk mētra womb]

endomorph /ˈendəʊˌmɔːf/ n. **1** a person with a soft round build of body and a high proportion of fat tissue (cf. ECTOMORPH, MESOMORPH). **2** Mineral. a mineral enclosed within another. □ **endomorphy** n. **endomorphic** /ˌendəʊˈmɔːfɪk/ adj. [ENDO- + Gk morphē form]

endoparasite /ˌendəʊˈpærəˌsaɪt/ n. Biol. a parasite that lives inside its host (cf. ECTOPARASITE).

endoplasm /ˈendəʊˌplæz(ə)m/ n. Biol. the more fluid granular inner layer of the cytoplasm in unicellular animals such as amoeba (cf. ECTOPLASM 1).

endoplasmic reticulum /ˌendəʊˈplæzmɪk/ n. Biol. a network of membranous tubules within the cytoplasm of a eukaryotic cell, continuous with the nuclear membrane and usu. having ribosomes attached.

endorphin /enˈdɔːfɪn/ n. Biochem. any of a group of peptide neurotransmitters occurring naturally in the brain and having pain-relieving properties. [F endorphine f. endogène endogenous + MORPHINE]

endorse /ɪnˈdɔːs/ v.tr. (also **indorse**) **1 a** confirm (a statement or opinion). **b** declare one's approval of. **2** sign or write on the back of (a document), esp. the back of (a bill, cheque, etc.) as the payee or to specify another as payee. **3** write (an explanation or comment) on the back of a document. **4** Brit. enter details of a conviction for a motoring offence on (a driving licence). □ **endorsable** adj. **endorser** n. **endorsee** /ˌendɔːˈsiː/ n. [med.L indorsare (as IN-², L dorsum back)]

endorsement /ɪnˈdɔːsmənt/ n. (also **indorsement**) **1** the act or an instance of endorsing. **2** something with which a document etc. is endorsed, esp. a signature. **3** a record in a driving licence of a conviction for a motoring offence. **4** a recommendation of a product which can be cited in advertising material.

endoscope /ˈendəʊˌskəʊp/ n. Surgery an instrument for viewing the internal parts of the body. □ **endoscopic** /ˌendəʊˈskɒpɪk/ adj. **endoscopically** adv. **endoscopist** /enˈdɒskəpɪst/ n. **endoscopy** n.

endoskeleton /ˈendəʊˌskelɪt(ə)n/ n. Zool. an internal skeleton, as found in vertebrates and most echinoderms.

endosperm /ˈendəʊˌspɜːm/ n. albumen enclosed with the germ in seeds.

endospore /ˈendəʊˌspɔː(r)/ n. **1** a spore formed by certain bacteria. **2** the inner coat of a spore.

endothelium /ˌendəʊˈθiːlɪəm/ n. (pl. **endothelia** /-lɪə/) Anat. a layer of cells lining the blood vessels, heart, and lymphatic vessels. □ **endothelial** adj. [ENDO- + Gk thēlē teat]

endothermic /ˌendəʊˈθɜːmɪk/ adj. Chem. occurring or formed with the absorption of heat.

endotoxin /ˈendəʊˌtɒksɪn/ n. Biochem. a toxin present inside a bacterial cell, released only when the cell disintegrates (cf. EXOTOXIN).

endow /ɪnˈdaʊ/ v.tr. **1** bequeath or give a permanent income to (a person, institution, etc.). **2** (esp. as **endowed** adj.) (usu. foll. by with) give a quality or attribute to, esp. provide (a person) with talent, ability, etc. □ **endower** n. [ME f. AF endouer (as EN-¹, OF douer f. L dotare f. dos dotis DOWER)]

endowment /ɪnˈdaʊmənt/ n. **1** the act or an instance of endowing. **2** assets, esp. property or income with which a person or body is endowed. **3** (usu. in pl.) skill, talent, etc., with which a person is endowed. **4** (attrib.) denoting forms of life insurance involving payment by the insurer of a sum on a specified date, or on the death of the insured person if earlier. □ **endowment mortgage** a mortgage linked to endowment insurance of the mortgagor's life, the capital being paid from the sum insured.

endpaper /ˈendˌpeɪpə(r)/ n. a usu. blank leaf of paper at the beginning and end of a book, fixed to the inside of the cover.

endue /ɪnˈdjuː/ v.tr. (also **indue**) (**-dues**, **-dued**, **-duing**) (foll. by with) invest or provide (a person) with qualities, powers, etc. [earlier = induct, put on clothes: ME f. OF enduire f. L inducere lead in, assoc. in sense with L induere put on (clothes)]

endurance /ɪnˈdjʊərəns/ n. **1** the power or habit of enduring (beyond endurance). **2** the ability to withstand prolonged strain (endurance test). **3** the act of enduring. **4** ability to last; enduring quality. [OF f. endurer: see ENDURE]

endure /ɪnˈdjʊə(r)/ v. **1** tr. undergo (a difficulty, hardship, etc.). **2** tr. **a** tolerate (a person) (cannot endure him). **b** (esp. with neg.; foll. by to + infin.) bear. **3** intr. (often as **enduring** adj.) remain in existence; last. **4** tr. submit to. □ **enduringly** adv. **endurable** adj. **endurability** /-ˌdjʊərəˈbɪlɪtɪ/ n. [ME f. OF endurer f. L indurare harden (as IN-², durus hard)]

enduro /ɪnˈdjʊərəʊ/ n. (pl. **-os**) a long-distance race for motor vehicles, designed to test endurance.

endways /ˈendweɪz/ adv. **1** with its end uppermost or foremost or turned towards the viewer. **2** end to end.

endwise /ˈendwaɪz/ adv. = ENDWAYS.

Endymion /enˈdɪmɪən/ Gk Mythol. a remarkably beautiful young man, loved by the Moon (Selene). A well-known tradition claims that he had fifty daughters by Selene; in some versions, he was visited nightly by her as he lay in an eternal sleep. According to one story, Endymion was put to sleep for ever by Zeus for having dared to fall in love with the latter's wife, Hera.

ENE abbr. east-north-east.

-ene /iːn/ suffix **1** forming names of inhabitants of places (Nazarene). **2** Chem. forming names of unsaturated hydrocarbons containing a double bond (benzene; ethylene). [from or after Gk -ēnos]

enema /ˈenɪmə/ n. (pl. **enemas** or **enemata** /ɪˈnemətə/) Med. **1** the injection of liquid or gas into the rectum, esp. to expel its contents. **2** a fluid or syringe used for this. [LL f. Gk enema f. eniēmi inject (as EN-², hiēmi send)]

enemy /ˈenəmɪ/ n. (pl. **-ies**) **1** a person or group actively opposing or hostile to another, or to a cause etc. **2 a** a hostile nation or army, esp. in war. **b** a member of this. **c** a hostile ship or aircraft. **3** (usu. foll. by of, to) an adversary or opponent. **4** a thing that harms or injures. **5** (attrib.) of or belonging to an enemy (destroyed by enemy action). [ME f. OF enemi f. L inimicus (as IN-¹, amicus friend)]

energetic /ˌenəˈdʒetɪk/ adj. **1** strenuously active. **2** forcible, vigorous. **3** powerfully operative. □ **energetically** adv. [Gk energētikos f. energeō (as EN-², ergon work)]

energetics /ˌenəˈdʒetɪks/ n.pl. the science of energy.

energize /ˈenəˌdʒaɪz/ v.tr. (also **-ise**) **1** infuse energy into (a person or work). **2** provide energy for the operation of (a device). □ **energizer** n.

energumen /ˌenəːˈgjuːmen/ n. an enthusiast or fanatic. [LL energumenus f. Gk energoumenos passive part. of energeō: see ENERGETIC]

energy /ˈenədʒɪ/ n. (pl. **-ies**) **1** force, vigour; capacity for activity. **2** (in pl.) individual powers in use (devote your energies to this). **3** Physics the capacity of matter or radiation to do work. (See note below.) **4** the means of doing work provided by the utilization of physical or chemical resources (nuclear energy). [F énergie or LL energia f. Gk energeia f. ergon work]

▪ The modern scientific concept of energy was formulated in the 19th century mainly through the study of the generation of mechanical power by means of heat engines and electrical machines. Energy can be neither created nor destroyed, but only changed from one form into another. These forms include kinetic energy (energy of motion), potential energy (the energy an object has by virtue of its position), heat energy, electrical energy, and so on. Einstein (among others) showed that matter can also be regarded as a form of energy, as the two are interconvertible. The standard unit of energy is the joule.

enervate v. & adj. ● v.tr. /'enə,veɪt/ deprive of vigour or vitality. ● adj. /ɪ'nɜ:vət/ enervated. □ **enervation** /,enə'veɪʃ(ə)n/ n. [L enervatus past part. of enervare (as E-, nervus sinew)]

Enewetak see ENIWETOK.

en famille /,ɒn fæ'mi:/ adv. **1** in or with one's family. **2** at home. [F, = in family]

enfant gâté /,ɒnfɒn 'gæteɪ/ n. (pl. **enfants gâtés** pronunc. same) a person given undue flattery or indulgence. [F, = spoilt child]

enfant terrible /,ɒnfɒn te'ri:blə/ n. (pl. **enfants terribles** pronunc. same) a person who causes embarrassment by indiscreet or unruly behaviour. [F, = terrible child]

enfeeble /ɪn'fi:b(ə)l/ v.tr. make feeble. □ **enfeeblement** n. [ME f. OF enfeblir (as EN-[1], FEEBLE)]

en fête /ɒn 'fet/ adv. & predic.adj. holding or ready for a holiday or celebration. [F, = in festival]

enfetter /ɪn'fetə(r)/ v.tr. literary **1** bind in or as in fetters. **2** (foll. by to) enslave.

enfilade /,enfɪ'leɪd/ n. & v. ● n. gunfire directed along a line from end to end. ● v.tr. direct an enfilade at (troops, a road, etc.). [F f. enfiler (as EN-[1], fil thread)]

enfold /ɪn'fəʊld/ v.tr. (also **infold**) **1** (usu. foll. by in, with) wrap up; envelop. **2** clasp, embrace.

enforce /ɪn'fɔ:s/ v.tr. **1** compel observance of (a law etc.). **2** (foll. by on, upon) impose (an action, conduct, one's will). **3** persist in (a demand or argument). □ **enforceable** adj. **enforcer** n. **enforcedly** /-sɪdlɪ/ adv. **enforceability** /-,fɔ:sə'bɪlɪtɪ/ n. [ME f. OF enforcir, -ier ult. f. L fortis strong]

enforcement /ɪn'fɔ:smənt/ n. the act or an instance of enforcing. □ **enforcement notice** Brit. an official notification to remedy a breach of planning legislation. [ME f. OF, as ENFORCE + -MENT]

enfranchise /ɪn'fræntʃaɪz/ v.tr. **1** give (a person) the right to vote. **2** give (a town) municipal rights, esp. that of representation in Parliament. **3** hist. free (a slave, villein, etc.). □ **enfranchisement** /-tʃɪzmənt/ n. [OF enfranchir (as EN-[1], franc franche FRANK)]

ENG abbr. electronic news gathering.

engage /ɪn'geɪdʒ/ v. **1** tr. employ or hire (a person). **2** tr. **a** (usu. in passive) employ busily; occupy (are you engaged tomorrow?). **b** hold fast (a person's attention). **3** tr. (usu. in passive) bind by a promise, esp. of marriage. **4** tr. (usu. foll. by to + infin.) bind by a contract. **5** tr. arrange beforehand to occupy (a room, seat, etc.). **6** (usu. foll. by with) Mech. **a** tr. interlock (parts of a gear etc.); cause (a part) to interlock. **b** intr. (of a part, gear, etc.) interlock. **7 a** intr. (usu. foll. by with) (of troops etc.) come into battle. **b** tr. bring (troops) into battle. **c** tr. come into battle with (an enemy etc.). **8** intr. take part (engage in politics). **9** intr. (foll. by that + clause or to + infin.) pledge oneself. **10** tr. (usu. as **engaged** adj.) Archit. attach (a column) to a wall. **11** tr. (of fencers etc.) interlock (weapons). [F engager, rel. to GAGE[1]]

engagé /ɒn'gæʒeɪ/ adj. (of a writer etc.) morally committed. [F, past part. of engager: see ENGAGE]

engaged /ɪn'geɪdʒd/ adj. **1** under a promise to marry. **2 a** occupied, busy. **b** reserved, booked. **3** Brit. (of a telephone line) unavailable because already in use. □ **engaged signal** (or **tone**) Brit. a sound indicating that a telephone line is engaged.

engagement /ɪn'geɪdʒmənt/ n. **1** the act or state of engaging or being engaged. **2** an appointment with another person. **3** a betrothal. **4** an encounter between hostile forces. **5** a moral commitment. **6** a period of paid employment, a job. □ **engagement ring** a finger-ring given by a man to a woman when they promise to marry. [F f. engager: see ENGAGE]

engaging /ɪn'geɪdʒɪŋ/ adj. attractive, charming. □ **engagingly** adv. **engagingness** n.

Engels /'eŋ(ə)lz/, Friedrich (1820–95), German socialist and political philosopher, resident chiefly in England from 1842. The founder of modern communism with Karl Marx, he collaborated with him in the

writing of the Communist Manifesto (1848). Engels also completed the second and third volumes of Marx's Das Kapital (1885; 1894). Engels's own writings include The Condition of the Working Classes in England in 1844 (1845).

engender /ɪn'dʒendə(r)/ v.tr. **1** give rise to; bring about (a feeling etc.). **2** archaic beget. [ME f. OF engendrer f. L ingenerare (as IN-[2], generare GENERATE)]

engine /'endʒɪn/ n. **1** a machine for producing energy of motion from another form of energy, esp. heat that the machine itself generates. **2 a** a railway locomotive. **b** = fire-engine. **c** = STEAM ENGINE. **3** archaic a machine or instrument, esp. a contrivance used in warfare. □ **engine-driver** the driver of a railway locomotive. **engine-room** a room containing engines (esp. in a ship). □ **engined** adj. (also in comb.). **engineless** adj. [OF engin f. L ingenium talent, device: cf. INGENIOUS]

engineer /,endʒɪ'nɪə(r)/ n. & v. ● n. **1** a person qualified in a branch of engineering, esp. as a professional. **2** = civil engineer. **3 a** a person who makes or is in charge of engines. **b** a person who maintains machines; a mechanic; a technician. **4** N. Amer. an engine-driver. **5** a person who designs and constructs military works; a soldier trained for this purpose. **6** (foll. by of) a skilful or artful contriver. ● v. **1** tr. arrange, contrive, or bring about, esp. artfully. **2** intr. act as an engineer. **3** tr. construct or manage as an engineer. [ME f. OF engigneor f. med.L ingeniator -oris f. ingeniare (as ENGINE)]

engineering /,endʒɪ'nɪərɪŋ/ n. the application of science to the design, building, and use of machines, constructions, etc. □ **engineering science** engineering as a field of study.

enginery /'endʒɪnrɪ/ n. engines and machinery generally.

engird /ɪn'gɜ:d/ v.tr. surround with or as with a girdle.

engirdle /ɪn'gɜ:d(ə)l/ v.tr. engird.

England /'ɪŋglənd/ a part of Great Britain and the United Kingdom, largely made up of the area south of the River Tweed, and containing the capital, London; pop. (1991) 46,170,300. There were settlements in England from at least palaeolithic times, and considerable remains exist of neolithic and Bronze Age cultures. These were followed by the arrival of the Celtic peoples, whose civilization spread over the whole country. The Romans under Julius Caesar raided the south of Britain in 55 and 54 BC, but full-scale invasion did not take place until a century later; the country was then administered as a Roman province until the withdrawal of the last Roman garrison in the early 5th century. In the 3rd–7th centuries Germanic-speaking tribes, traditionally known as Angles, Saxons, and Jutes, raided and then settled, establishing independent kingdoms. When the kingdom of Wessex became dominant in the 9th century England emerged as a distinct political entity, before being conquered by William, Duke of Normandy, in 1066. The neighbouring principality of Wales was gradually conquered during the Middle Ages and politically incorporated in the 16th century. During the period of Tudor rule (1485–1603) England emerged as a Protestant state with a strong stable monarchy and as a naval power. Scotland and England have been ruled by one monarch since 1603, and their two Parliaments were formally united in 1707. (See also GREAT BRITAIN.)

English /'ɪŋglɪʃ/ adj. & n. ● adj. of or relating to England or its people or language. (See note below.) ● n. **1** the language of England; its literary or standard form. (See note below.) **2** (prec. by the; treated as pl.) the people of England. **3** US Billiards = SIDE n. 10. □ **Queen's** (or **King's**) **English** the English language as correctly written or spoken in Britain. □ **Englishness** n. [OE englisc, ænglisc (as ANGLE, -ISH[1])]

▪ English is the principal language of Great Britain, the US, Ireland, Canada, Australia, New Zealand, and many other countries. There are some 400 million native speakers, and it is the medium of communication for many millions more; it is the most widely used second language in the world. Its history can be divided into three stages: Old English (up to 1150), Middle English (1150–1500), and modern English (1500 onwards). Old English began with the settlement of Germanic-speaking tribes (Angles, Saxons, and Jutes) in Britain in the mid-5th century. It was an inflected language that slowly evolved a written literary form. Extension of vocabulary was brought by the spread of Christian culture, with some words adopted or translated from Latin, and by Scandinavian invaders in the 9th–10th centuries. After the Norman Conquest, Anglo-Norman was the language of the ruling classes, but from the 14th century English again became the standard, although over the Middle English period about half of its vocabulary was replaced by Norman French and Latin words. The modern English period has seen the language taken and used abroad, leading to new regional varieties such as American,

West Indian, Indian, Australian, and Nigerian English, while the remarkable number of foreign words which were adopted in return have given English an incomparably large and international vocabulary. Despite the foreign influences, the Germanic nature of English has been maintained in its syntax and morphology, although most of its inflected endings had disappeared by the 15th century. While the pronunciation of English kept changing, the spelling (which was largely standardized by the publication of Dr Johnson's dictionary in 1755) in many cases remained unaltered, which explains why it is often at variance with pronunciation.

English bond *n.* *Building* a bond of brickwork arranged in alternate courses of stretchers and headers.

English Channel the sea channel separating southern England from northern France. It is 35 km (22 miles) wide at its narrowest point. A railway tunnel beneath it linking England and France was opened in 1994 (see CHANNEL TUNNEL).

English Civil War the war between Charles I and his Parliamentary opponents, 1642–9. Conflict between the parties had led the king to dissolve Parliament and rule alone 1629–40. Although he continued to levy taxes, he was forced to recall Parliament by the need for money to pursue war against the Scots; civil war broke out after Charles refused a series of demands made by the new Parliament. The king's forces (the Royalists, or Cavaliers) were dominated by nobles, Anglicans, and Catholics, while the Parliamentary forces (or Roundheads) contained many Puritans, Scottish Covenanters, merchants, and artisans. After several years of warfare, the better-organized Parliamentary forces under Oliver Cromwell gained the upper hand in 1644–5. Royalist resistance collapsed in 1646 and an attempt by Charles to regain power in alliance with the Scots was defeated at Preston in 1648, Charles himself being tried and executed by Parliament in 1649. Fought against a background of confused political and religious issues, the English Civil War dramatically changed the nature of English society and government, even though the period of republican government (the Commonwealth), at first under Cromwell, eventually ended with the restoration of Charles II (1660).

English Heritage (in the UK) a body responsible since 1983 for England's ancient monuments, listed buildings, and conservation areas.

English horn *n.* = COR ANGLAIS.

Englishman /ˈɪŋglɪʃmən/ *n.* (*pl.* **-men**) a man who is English by birth or descent.

English Pale 1 a small area round Calais, the only part of France remaining in English hands after the Hundred Years War. It was recaptured by France in 1558. **2** (also **the Pale**) that part of Ireland over which England exercised jurisdiction before the whole was conquered. Centred on Dublin, it varied in extent at different times from the reign of Henry II until the full conquest under Elizabeth I. [PALE²]

Englishwoman /ˈɪŋglɪʃˌwʊmən/ *n.* (*pl.* **-women**) a woman who is English by birth or descent.

engorge /ɪnˈgɔːdʒ/ *v.tr.* **1** (in passive) **a** be crammed. **b** *Med.* be congested with blood. **2** devour greedily. □ **engorgement** *n.* [F *engorger* (as EN-¹, GORGE)]

engraft /ɪnˈgrɑːft/ *v.tr.* (also **ingraft**) **1** (usu. foll. by *into, upon*) *Bot.* insert (a scion of one tree) into another tree. **2** (usu. foll. by *in*) implant (principles etc.) in a person's mind. **3** (usu. foll. by *into*) incorporate permanently. □ **engraftment** *n.*

engrail /ɪnˈgreɪl/ *v.tr.* (usu. as **engrailed** *adj.*) esp. *Heraldry* indent the edge of; give a serrated appearance to. [ME f. OF *engresler* (as EN-¹, *gresle* hail)]

engrain /ɪnˈgreɪn/ *v.tr.* **1** implant (a habit, belief, or attitude) ineradicably in a person (see also INGRAINED). **2** cause (dye etc.) to sink deeply into a thing. [ME f. OF *engrainer* dye in grain (*en graine*): see GRAIN]

engrained /ɪnˈgreɪnd/ *adj.* inveterate (= INGRAINED).

engram /ˈengræm/ *n.* a memory trace, a supposed permanent change in the brain accounting for the existence of memory. □ **engrammatic** /ˌengrəˈmætɪk/ *adj.* [G *Engramm* f. Gk *en* in + *gramma* letter of the alphabet]

engrave /ɪnˈgreɪv/ *v.tr.* **1** (often foll. by *on*) inscribe, cut, or carve (a text or design) on a hard surface. **2** (often foll. by *with*) inscribe or ornament (a surface) in this way. **3** cut (a design) as lines on a metal plate, block, etc., for printing. **4** (often foll. by *on*) impress deeply on a person's memory etc. □ **engraver** *n.* [EN-¹ + GRAVE³]

engraving /ɪnˈgreɪvɪŋ/ *n.* **1** any of various techniques of making prints by cutting a design into a plate, block, or other surface. (*See note below.*) **2** a print made from such a plate etc.

▪ The term engraving most commonly refers to line engraving, in which a design is cut into a smooth metal (usually copper) plate with a tool called a burin, and the plate is inked and pressed against paper. Until the invention of photography, engraving was the chief method of reproducing works of art; it was also used as a medium by artists and illustrators such as Dürer, Hogarth, Thomas Bewick, and William Blake. (See also MEZZOTINT.)

engross /ɪnˈgrəʊs/ *v.tr.* **1** absorb the attention of; occupy fully (*engrossed in studying*). **2** make a fair copy of a legal document. **3** reproduce (a document etc.) in larger letters or larger format. **4** *archaic* monopolize (a conversation etc.). □ **engrossing** *adj.* (in sense 1). **engrossment** *n.* [ME f. AF *engrosser*: senses 2 and 3 f. *en* in + *grosse* large writing: senses 1 and 4 f. *en gros* wholesale]

engulf /ɪnˈgʌlf/ *v.tr.* (also **ingulf**) **1** flow over and swamp; overwhelm. **2** swallow or plunge into a gulf. □ **engulfment** *n.*

enhance /ɪnˈhɑːns/ *v.tr.* heighten or intensify (qualities, powers, value, etc.); improve (something already of good quality). □ **enhancement** *n.* **enhancer** *n.* [ME f. AF *enhauncer*, prob. alt. f. OF *enhaucier* ult. f. L *altus* high]

enharmonic /ˌenhɑːˈmɒnɪk/ *adj.* *Mus.* of or having intervals smaller than a semitone (esp. such intervals as that between G sharp and A flat, these notes being made the same in a scale of equal temperament). □ **enharmonically** *adv.* [LL *enharmonicus* f. Gk *enarmonikos* (as EN-², *harmonia* HARMONY)]

enigma /ɪˈnɪgmə/ *n.* **1** a puzzling thing or person. **2** a riddle or paradox. □ **enigmatize** *v.tr.* (also **-ise**). **enigmatic** /ˌenɪgˈmætɪk/ *adj.* **enigmatical** *adj.* **enigmatically** *adv.* [L *aenigma* f. Gk *ainigma -matos* f. *ainissomai* speak allusively f. *ainos* fable]

Eniwetok /eˈniːwəˌtɔːk, ˌenɪˈwiːtɔːk/ (also **Enewetak**) an uninhabited island in the North Pacific, one of the Marshall Islands. Cleared of its native population, it was used by the US as a testing ground for atom bombs from 1948 to 1954.

enjambment /enˈdʒæmmənt/ *n.* (also **enjambement**) *Prosody* the continuation of a sentence without a pause beyond the end of a line, couplet, or stanza. [F *enjambement* f. *enjamber* (as EN-¹, *jambe* leg)]

enjoin /ɪnˈdʒɔɪn/ *v.tr.* **1 a** (foll. by *to* + infin.) command or order (a person). **b** (foll. by *that* + clause) issue instructions. **2** (often foll. by *on*) impose or prescribe (an action or conduct). **3** (usu. foll. by *from*) *Law* prohibit (a person) by order. □ **enjoinment** *n.* [ME f. OF *enjoindre* f. L *injungere* (as IN-², *jungere* join)]

enjoy /ɪnˈdʒɔɪ/ *v.tr.* **1** take delight or pleasure in. **2** have the use or benefit of. **3** experience (*enjoy poor health*). □ **enjoy oneself** experience pleasure. □ **enjoyer** *n.* **enjoyment** *n.* [ME f. OF *enjoier* give joy to or *enjoïr* enjoy, ult. f. L *gaudere* rejoice]

enjoyable /ɪnˈdʒɔɪəb(ə)l/ *adj.* pleasant; giving enjoyment. □ **enjoyably** *adv.* **enjoyableness** *n.* **enjoyability** /-ˌdʒɔɪəˈbɪlɪtɪ/ *n.*

enkephalin /enˈkefəlɪn/ *n.* (also **encephalin** /enˈsef-, enˈkef-/) *Biochem.* either of two morphine-like endorphins occurring naturally in the brain and thought to control levels of pain. [Gk *egkephalos* brain]

enkindle /enˈkɪnd(ə)l/ *v.tr.* *literary* **1 a** cause (flames) to flare up. **b** stimulate (feeling, passion, etc.). **2** inflame with passion.

enlace /ɪnˈleɪs/ *v.tr.* **1** encircle tightly. **2** entwine. **3** enfold. □ **enlacement** *n.* [ME f. OF *enlacier* ult. f. L *laqueus* noose]

enlarge /ɪnˈlɑːdʒ/ *v.* **1** *tr. & intr.* make or become larger or wider. **2** *intr.* (usu. foll. by *upon*) expatiate on. **3** *tr. Photog.* produce an enlargement of (a negative). [ME f. OF *enlarger* (as EN-¹, LARGE)]

enlargement /ɪnˈlɑːdʒmənt/ *n.* **1** the act or an instance of enlarging; the state of being enlarged. **2** *Photog.* a print that is larger than the negative from which it is produced.

enlarger /ɪnˈlɑːdʒə(r)/ *n.* *Photog.* an apparatus for enlarging or reducing negatives or positives.

enlighten /ɪnˈlaɪt(ə)n/ *v.tr.* **1 a** (often foll. by *on*) instruct or inform (a person) about a subject. **b** (as **enlightened** *adj.*) well-informed, knowledgeable. **2** (esp. as **enlightened** *adj.*) free from prejudice or superstition. **3** *rhet.* or *poet.* **a** shed light on (an object). **b** give spiritual insight to (a person). □ **enlightener** *n.*

enlightenment /ɪnˈlaɪt(ə)nmənt/ *n.* the act or an instance of enlightening; the state of being enlightened.

Enlightenment, the *n.* (also **Age of Enlightenment**) a European intellectual movement of the late 17th and 18th centuries, heavily influenced by the thinking of 17th-century philosophers and scientists such as Descartes, Locke, and Newton, and having at its core a belief in reason as the key to human knowledge and progress, a sense of religious tolerance, and a distrust of superstition. In Germany, it is associated with writers and philosophers such as Leibniz, Kant, Goethe, and Schiller. In France, Voltaire, Jean-Jacques Rousseau, and *L'Encyclopédie* (editors Diderot and d'Alembert) figure prominently, while in Britain, the movement is particularly represented in Scotland by the *Encyclopaedia Britannica* and by figures such as Adam Smith and David Hume. The ideas of the Enlightenment and its criticism of government and Church fuelled revolutionary thinking in France and the US and were influential on writers and politicians such as Jefferson and Paine.

enlist /ɪnˈlɪst/ v. **1** intr. & tr. enrol in the armed services. **2** tr. secure as a means of help or support. □ **enlisted man** US a soldier or sailor below the rank of officer. □ **enlister** n. **enlistment** n.

enliven /ɪnˈlaɪv(ə)n/ v.tr. **1** give life or spirit to. **2** make cheerful, brighten (a picture or scene). □ **enlivener** n. **enlivenment** n.

en masse /ɒn ˈmæs/ adv. **1** all together. **2** in a mass. [F]

enmesh /ɪnˈmeʃ/ v.tr. entangle in or as in a net. □ **enmeshment** n.

enmity /ˈenmɪtɪ/ n. (pl. **-ies**) **1** the state of being an enemy. **2** a feeling of hostility. [ME f. OF *enemitié* f. Rmc (as ENEMY)]

ennead /ˈenɪˌæd/ n. a group of nine. [Gk *enneas enneados* f. *ennea* nine]

Ennis /ˈenɪs/ the county town of Clare, in the Republic of Ireland; pop. (1991) 13,700.

Enniskillen /ˌenɪsˈkɪlɪn/ a town in Northern Ireland; pop. (1981) 10,400. The old spelling *Inniskilling* is preserved as a regimental name in the British army, commemorating the defence of Enniskillen by its townsmen against the supporters of the deposed King James II in 1689.

Ennius /ˈenɪəs/, Quintus (239–169 BC), Roman poet and dramatist. He was largely responsible for the creation of a native Roman literature based on Greek models. Of his many works (surviving only in fragments) the most important was the *Annals* (undated), a hexametric epic on the history of Rome, which was a major influence on Virgil.

ennoble /ɪˈnəʊb(ə)l/ v.tr. **1** make (a person) a noble. **2** make noble; elevate. □ **ennoblement** n. [F *ennoblir* (as EN-[1], NOBLE)]

ennui /ɒnˈwiː/ n. mental weariness from lack of occupation or interest; boredom. [F f. L in *odio*: cf. ODIUM]

Enoch /ˈiːnɒk/ (in the Bible) **1** the eldest son of Cain. **2** the first city, built by Cain and named after Enoch (Gen. 4:17). **3** a Hebrew patriarch, father of Methuselah. Two works ascribed to him, the *Book of Enoch* and the *Book of the Secrets of Enoch*, date from the 2nd–1st centuries BC and 1st century AD respectively. A third treatise likewise dates from the Christian era.

enology US var. of OENOLOGY.

enormity /ɪˈnɔːmɪtɪ/ n. (pl. **-ies**) **1** extreme wickedness. **2** an act of extreme wickedness. **3** a serious error. **4** disp. great size; enormousness. [ME f. F *énormité* f. L *enormitas -tatis* f. *enormis* (as ENORMOUS)]

enormous /ɪˈnɔːməs/ adj. very large; huge (*enormous animals*; *an enormous difference*). □ **enormously** adv. **enormousness** n. [L *enormis* (as E-, *norma* pattern, standard)]

enosis /ˈenəʊsɪs/ n. the political union of Cyprus and Greece, as an ideal or proposal. [mod.Gk *enōsis* f. *ena* one]

enough /ɪˈnʌf/ adj., n., adv., & int. ● adj. as much or as many as required (*we have enough apples*; *we do not have enough sugar*; *earned enough money to buy a house*). ● n. an amount or quantity that is enough (*we have enough of everything now*; *enough is as good as a feast*). ● adv. **1** to the required degree, adequately (*are you warm enough?*). **2** fairly (*she sings well enough*). **3** very, quite (*you know well enough what I mean*; *oddly enough*). ● int. that is enough (in various senses, esp. to put an end to an action, thing said, etc.). □ **have had enough of** want no more of; be satiated with or tired of. [OE *genog* f. Gmc]

en passant /ɒn ˈpæsɒn/ adv. **1** by the way. **2** Chess used with reference to the permitted capture of an opponent's pawn that has just advanced two squares in its first move with a pawn that could have taken it if it had advanced only one square. [F, = in passing]

en pension /ɒn ˈpɒnsjɒn/ adv. as a boarder or resident. [F: see PENSION[2]]

enplane var. of EMPLANE.

enprint /ˈenprɪnt/ n. a standard-sized photographic print. [*en*larged *print*]

enquire /ɪnˈkwaɪə(r)/ v. **1** intr. (often foll. by *of*) seek information; ask a question (of a person). **2** intr. = INQUIRE 1. **3** intr. (foll. by *about, after, for*)

ask about a person, a person's health, etc. **4** intr. (foll. by *for*) ask about the availability of. **5** tr. ask for information as to (*enquired my name*; *enquired whether we were coming*). **6** tr. (foll. by *into*) investigate, look into. □ **enquirer** n. **enquiring** adj. **enquiringly** adv. [ME *enquere* f. OF *enquerre* ult. f. L *inquirere* (as IN-[2], *quaerere quaesit-* seek)]

enquiry /ɪnˈkwaɪərɪ/ n. (pl. **-ies**) **1** the act or an instance of asking or seeking information. **2** = INQUIRY 1.

enrage /ɪnˈreɪdʒ/ v.tr. (often foll. by *at, by, with*) make furious. □ **enragement** n. [F *enrager* (as EN-[1], RAGE)]

en rapport /ˌɒn ræˈpɔː(r)/ adv. (usu. foll. by *with*) in harmony or rapport. [F: see RAPPORT]

enrapture /ɪnˈræptʃə(r)/ v.tr. give intense delight to.

enrich /ɪnˈrɪtʃ/ v.tr. **1** make rich or richer. **2** make richer in quality, flavour, nutritive value, etc. **3** add to the contents of (a collection, museum, or book). **4** increase the content of an isotope in (material) esp. enrich (uranium) with isotope U-235. □ **enrichment** n. [ME f. OF *enrichir* (as EN-[1], RICH)]

enrobe /ɪnˈrəʊb/ v.intr. put on a robe, vestment, etc.

enrol /ɪnˈrəʊl/ v. (US **enroll**) (**enrolled, enrolling**) **1** intr. enter one's name on a list, esp. as a commitment to membership. **2** tr. **a** write the name of (a person) on a list. **b** (usu. foll. by *in*) incorporate (a person) as a member of a society etc. **3** tr. hist. enter (a deed etc.) among the rolls of a court of justice. **4** tr. record. □ **enrollee** /ˌenrəʊˈliː/ n. [ME f. OF *enroller* (as EN-[1], *rolle* ROLL)]

enrolment /ɪnˈrəʊlmənt/ n. (US **enrollment**) **1** the act or an instance of enrolling; the state of being enrolled. **2** US the number of persons enrolled, esp. at a school or college.

en route /ɒn ˈruːt/ adv. (often foll. by *to, for*) on the way. [F]

Enschede /ˈenskəˌdeɪ/ a city in the Netherlands; pop. (1991) 146,500.

ensconce /ɪnˈskɒns/ v.tr. (usu. refl. or in *passive*) establish or settle comfortably, safely, or secretly.

ensemble /ɒnˈsɒmb(ə)l/ n. **1 a** a thing viewed as the sum of its parts. **b** the general effect of this. **2** a set of clothes worn together; an outfit. **3** a group of actors, dancers, musicians, etc., performing together, esp. subsidiary dancers in ballet etc. **4** Mus. **a** a concerted passage for an ensemble. **b** the manner in which this is performed (*good ensemble*). **5** Math. a group of systems with the same constitution but possibly in different states. [F, ult. f. L *insimul* (as IN-[2], *simul* at the same time)]

enshrine /ɪnˈʃraɪn/ v.tr. **1** enclose in or as in a shrine. **2** serve as a shrine for. **3** preserve or cherish. □ **enshrinement** n.

enshroud /ɪnˈʃraʊd/ v.tr. literary **1** cover with or as with a shroud. **2** cover completely; hide from view.

ensign /ˈensaɪn, -s(ə)n/ n. **1 a** a banner or flag, esp. the military or naval flag of a nation. **b** Brit. a flag with the union in the corner. **2** a standard-bearer. **3 a** hist. the lowest commissioned infantry officer. **b** US the lowest commissioned officer in the navy. □ **blue ensign** the ensign of government departments and formerly of the naval reserve etc. **red ensign** the ensign of the merchant service. **white ensign** the ensign of the Royal Navy and the Royal Yacht Squadron. □ **ensigncy** n. [ME f. OF *enseigne* f. L *insignia*: see INSIGNIA]

ensilage /ˈensɪlɪdʒ/ n. & v. ● n. the process of making silage. ● v.tr. treat (fodder) by ensilage. [F (as ENSILE)]

ensile /ɪnˈsaɪl/ v.tr. **1** put (fodder) into a silo. **2** preserve (fodder) in a silo. [F *ensiler* f. Sp. *ensilar* (as EN-[1], SILO)]

enslave /ɪnˈsleɪv/ v.tr. make (a person) a slave. □ **enslavement** n. **enslaver** n.

ensnare /ɪnˈsneə(r)/ v.tr. catch in or as in a snare; entrap. □ **ensnarement** n.

Ensor /ˈensɔː(r)/, James (Sydney), Baron (1860–1949), Belgian painter and engraver. An artist whose work is significant both for symbolism and for the development of 20th-century expressionism, he began to use the characteristic elements of fantasy and the macabre in his paintings in the late 1880s. Works such as *The Entry of Christ into Brussels* (1888) typically depict brightly coloured and bizarre carnival scenes crowded with skeletons or other grotesque or masked figures.

ensue /ɪnˈsjuː/ v.intr. **1** happen afterwards. **2** (often foll. by *from, on*) occur as a result. [ME f. OF *ensuivre* ult. f. L *sequi* follow]

en suite /ɒn ˈswiːt/ adv. & adj. ● adv. forming a single unit (*bedroom with bathroom en suite*). ● adj. **1** forming a single unit (*en suite bathroom*). **2** with a bathroom attached (*seven en suite bedrooms*). [F, = in sequence]

ensure /ɪnˈʃʊə(r)/ v.tr. **1** (often foll. by *that* + clause) make certain. **2** (usu. foll. by *to, for*) secure (a thing for a person etc.). **3** (usu. foll. by *against*) make safe. [ME f. AF *enseürer* f. OF *aseürer* ASSURE]

enswathe /ɪnˈsweɪð/ v.tr. bind or wrap in or as in a bandage. □ **enswathement** n.

ENT abbr. ear, nose, and throat.

-ent /ənt, ent/ suffix **1** forming adjectives denoting attribution of an action (consequent) or state (existent). **2** forming nouns denoting an agent (coefficient; president). [from or after F -ent or L -ent- pres. part. stem of verbs (cf. -ANT)]

entablature /ɪnˈtæblətʃə(r)/ n. Archit. the upper part of a classical building supported by columns or a colonnade, comprising architrave, frieze, and cornice. [It. intavolatura f. intavolare board up (as IN-², tavola table)]

entablement /ɪnˈteɪb(ə)lmənt/ n. a platform supporting a statue, above the dado and base. [F, f. entabler (as EN-¹, TABLE)]

entail /ɪnˈteɪl/ v. & n. ● v.tr. **1 a** necessitate or involve unavoidably (the work entails much effort). **b** give rise to, involve. **2** Law bequeath (property etc.) so that it remains within a family. **3** (usu. foll. by on) bestow (a thing) inalienably. ● n. Law **1** an entailed estate. **2** the succession to such an estate. □ **entailment** n. [ME, f. EN-¹ + AF taile TAIL²]

entangle /ɪnˈtæŋg(ə)l/ v.tr. **1** cause to get caught in a snare or among obstacles. **2** cause to become tangled. **3** involve in difficulties or illicit activities. **4** make (a thing) tangled or intricate; complicate.

entanglement /ɪnˈtæŋg(ə)lmənt/ n. **1** the act or condition of entangling or being entangled. **2 a** a thing that entangles. **b** Mil. an extensive barrier erected to obstruct an enemy's movements (esp. one made of stakes and interlaced barbed wire). **3** a compromising (esp. amorous) relationship.

entasis /ˈentəsɪs/ n. (pl. **entases** /-ˌsiːz/) Archit. a slight convex curve in a column shaft to correct the visual illusion that straight sides give of curving inwards. [mod.L f. Gk f. enteinō to stretch]

Entebbe /enˈtebɪ/ a town in southern Uganda, on the north shore of Lake Victoria; pop. (1980) 20,500. It was the capital of Uganda during the period of British rule, from 1894 to 1962.

entelechy /enˈtelɪkɪ/ n. **1** the realization of potential. **2** the supposed essential nature or guiding principle of a living thing. [Late L entelechia f. Gk entelekheia, f. as EN-² + telos end, perfection + ekhein be in a (certain) state]

entellus /ɪnˈteləs/ n. = HANUMAN 1. [name of a Trojan in Virgil's Aeneid]

entente /ɒnˈtɒnt/ n. **1** = ENTENTE CORDIALE. **2** a group of states in such a relation. [F, = understanding (as INTENT)]

entente cordiale /ˌɒntɒnt ˌkɔːdɪˈɑːl/ n. (pl. **ententes cordiales** pronunc. same) a friendly understanding between states, especially that reached in 1904 between Britain and France. Although not an alliance in its own right, it drew Britain away from a position of isolation in relation to the rest of Europe, and formed the basis of Anglo-French cooperation in the First World War. (See also TRIPLE ENTENTE.) [F, = cordial understanding: see ENTENTE]

enter /ˈentə(r)/ v. **1 a** intr. (often foll. by into) go or come in. **b** tr. go or come into. **2** tr. come on stage (as a direction: enter Macbeth). **2** tr. penetrate; go through; spread through (a bullet entered his chest; a smell of toast entered the room). **3** tr. (often foll. by up) write (a name, details, etc.) in a list, book, etc. **4 a** intr. register or announce oneself as a competitor (entered for the long jump). **b** tr. become a competitor in (an event). **c** tr. record the name of (a person etc.) as a competitor (entered two horses for the Derby). **5 a** tr. become a member of (a society etc.). **b** enrol as a member or prospective member of a society, school, etc.; admit or obtain admission for. **6** tr. make known; present for consideration (entered a protest). **7** tr. put into an official record. **8** intr. (foll. by into) **a** engage in (conversation, relations, an undertaking, etc.). **b** subscribe to; bind oneself by (an agreement etc.). **c** form part of (one's calculations, plans, etc.). **d** sympathize with (feelings etc.). **9** intr. (foll. by on, upon) **a** begin, undertake; begin to deal with (a subject). **b** assume the functions of (an office). **c** assume possession of (property). **10** intr. (foll. by up) complete a series of entries in (account-books etc.). [ME f. OF entrer f. L intrare]

enteric /enˈterɪk/ adj. & n. ● adj. of the intestines. ● n. (in full **enteric fever**) typhoid. [Gk enterikos (as ENTERO-)]

enteritis /ˌentəˈraɪtɪs/ n. Med. inflammation of the small intestine, often causing diarrhoea.

entero- /ˈentərəʊ/ comb. form intestine. [Gk enteron intestine]

enterostomy /ˌentəˈrɒstəmɪ/ n. (pl. **-ies**) Surgery a surgical operation in which the small intestine is brought through the abdominal wall and opened, in order to bypass the stomach or the colon.

enterotomy /ˌentəˈrɒtəmɪ/ n. (pl. **-ies**) a surgical operation to cut open the intestine.

enterovirus /ˌentərəʊˈvaɪərəs/ n. a virus infecting the intestines and sometimes spreading to other parts of the body, esp. the central nervous system.

enterprise /ˈentəˌpraɪz/ n. **1** an undertaking, esp. a bold or difficult one. **2** (as a personal attribute) readiness to engage in such undertakings (has no enterprise). **3** a business firm. □ **enterprise zone** Brit. a depressed (usu. urban) area where state incentives such as tax concessions are designed to encourage investment. □ **enterpriser** n. [ME f. OF entreprise fem. past part. of entreprendre var. of emprendre ult. f. L prendere, prehendere take]

enterprising /ˈentəˌpraɪzɪŋ/ adj. **1** ready to engage in enterprises. **2** resourceful, imaginative, energetic. □ **enterprisingly** adv.

entertain /ˌentəˈteɪn/ v.tr. **1** amuse; occupy agreeably. **2 a** receive or treat as a guest. **b** (absol.) receive guests (they entertain a great deal). **3** give attention or consideration to (an idea, feeling, or proposal). [ME f. F entretenir ult. f. L tenere hold]

entertainer /ˌentəˈteɪnə(r)/ n. a person who entertains, esp. professionally on stage etc.

entertaining /ˌentəˈteɪnɪŋ/ adj. amusing, diverting (a most entertaining book). □ **entertainingly** adv.

entertainment /ˌentəˈteɪnmənt/ n. **1** the act or an instance of entertaining; the process of being entertained. **2** a public performance or show. **3** diversions or amusements for guests etc. **4** amusement (much to my entertainment). **5** hospitality.

enthalpy /ˈenθəlpɪ, enˈθæl-/ n. Physics the total thermodynamic heat content of a system. [Gk enthalpō warm in (as EN-¹, thalpō to heat)]

enthral /ɪnˈθrɔːl/ v.tr. (US **enthrall**, **inthrall**) (**-thralled**, **-thralling**) **1** (often as **enthralling** adj.) captivate, please greatly. **2** enslave. □ **enthralment** n. (US **enthrallment**). [EN-¹ + THRALL]

enthrone /ɪnˈθrəʊn/ v.tr. **1** install (a king, bishop, etc.) on a throne, esp. ceremonially. **2** exalt. □ **enthronement** n.

enthuse /ɪnˈθjuːz, -ˈθuːz/ v.intr. & tr. colloq. be or make enthusiastic. [back-form. f. ENTHUSIASM]

enthusiasm /ɪnˈθjuːzɪˌæz(ə)m, ɪnˈθuː-/ n. **1** (often foll. by for, about) **a** strong interest or admiration. **b** great eagerness. **2** an object of enthusiasm. **3** archaic extravagant religious emotion. [F enthousiasme or LL enthusiasmus f. Gk enthousiasmos f. entheos possessed by a god, inspired (as EN-², theos god)]

enthusiast /ɪnˈθjuːzɪˌæst, ɪnˈθuː-/ n. **1** (often foll. by for) a person who is full of enthusiasm. **2** a visionary; a self-deluded person. [F enthousiaste or eccl.L enthousiastes f. Gk (as ENTHUSIASM)]

enthusiastic /ɪnˌθjuːzɪˈæstɪk, ɪnˌθuː-/ adj. having or showing enthusiasm. □ **enthusiastically** adv. [Gk enthousiastikos (as ENTHUSIASM)]

enthymeme /ˈenθɪˌmiːm/ n. Logic a syllogism in which one premiss is not explicitly stated. [L enthymema f. Gk enthumēma f. enthumeomai consider (as EN-², thumos mind)]

entice /ɪnˈtaɪs/ v.tr. (often foll. by from, into, or to + infin.) persuade by the offer of pleasure or reward. □ **enticement** n. **enticer** n. **enticing** adj. **enticingly** adv. [ME f. OF enticier prob. f. Rmc]

entire /ɪnˈtaɪə(r)/ adj. & n. ● adj. **1** whole, complete. **2** not broken or decayed. **3** unqualified, absolute (an entire success). **4** in one piece; continuous. **5** not castrated. **6** Bot. without indentation. ● n. an uncastrated animal. [ME f. AF enter, OF entier f. L integer (as IN-², tangere touch)]

entirely /ɪnˈtaɪəlɪ/ adv. **1** wholly, completely (the stock is entirely exhausted). **2** solely, exclusively (did it entirely for my benefit).

entirety /ɪnˈtaɪərətɪ/ n. (pl. **-ies**) **1** completeness. **2** (usu. foll. by of) the sum total. □ **in its entirety** in its complete form; completely. [ME f. OF entiereté f. L integritas -tatis f. integer: see ENTIRE]

entitle /ɪnˈtaɪt(ə)l/ v.tr. **1 a** (usu. foll. by to) give (a person etc.) a just claim. **b** (foll. by to + infin.) give (a person etc.) a right. **2 a** give (a book etc.) the title of. **b** archaic give (a person) the title of (entitled him sultan). □ **entitlement** n. [ME f. AF entitler, OF entiteler f. LL intitulare (as IN-², TITLE)]

entity /ˈentɪtɪ/ n. (pl. **-ies**) **1** a thing with distinct existence, as opposed to a quality or relation. **2** a thing's existence regarded distinctly; a thing's essential nature. □ **entitative** /-tətɪv/ adj. [F entité or med.L entitas f. LL ens being]

ento- /ˈentəʊ/ comb. form within. [Gk entos within]

entomb /ɪn'tuːm/ v.tr. **1** place in or as in a tomb. **2** serve as a tomb for. □ **entombment** /-'tuːmmənt/ n. [OF entomber (as EN-¹, TOMB)]

entomo- /'entəməʊ/ comb. form insect. [Gk entomos cut up (in neut. = INSECT) f. EN-² + temnō cut]

entomology /,entə'mɒlədʒɪ/ n. the branch of zoology concerned with the study of insects. □ **entomologist** n. **entomological** /-mə'lɒdʒɪk(ə)l/ adj. [F entomologie or mod.L entomologia (as ENTOMO-, -LOGY)]

entomophagous /,entə'mɒfəgəs/ adj. Zool. insect-eating.

entomophilous /,entə'mɒfɪləs/ adj. Biol. pollinated by insects.

entoparasite /,entəʊ'pærəˌsaɪt/ n. Biol. = ENDOPARASITE.

entophyte /'entəʊˌfaɪt/ n. Bot. a plant growing inside a plant or animal.

entourage /'ɒntʊəˌrɑːʒ/ n. **1** the group of people attending an important person. **2** surroundings. [F f. entourer surround]

entr'acte /'ɒntrækt/ n. **1** an interval between two acts of a play. **2** a piece of music or a dance performed during this. [F f. entre between + acte act]

entrails /'entreɪlz/ n.pl. **1** the intestines of a person or animal. **2** the innermost parts (entrails of the earth). [ME f. OF entrailles f. med.L intralia alt. f. L interaneus internal f. inter among]

entrain¹ /ɪn'treɪn/ v.intr. & tr. go or put on board a train.

entrain² /ɪn'treɪn/ v.tr. **1** (of a fluid) carry (particles etc.) along in its flow. **2** drag along. □ **entrainment** n. [F entraîner (as EN-¹, traîner drag, formed as TRAIN)]

entrain³ /ɒn'træn/ n. enthusiasm, animation. [F]

entrammel /ɪn'træm(ə)l/ v.tr. (**entrammelled, entrammelling**; US **entrammeled, entrammeling**) entangle, hamper.

entrance¹ /'entrəns/ n. **1** the act or an instance of going or coming in. **2** a door, passage, etc., by which one enters. **3** right of admission. **4** the coming of an actor on stage. **5** Mus. = ENTRY 8. **6** (foll. by into, upon) entering into office etc. **7** (in full **entrance fee**) a fee paid for admission to a society, club, exhibition, etc. [OF (as ENTER, -ANCE)]

entrance² /ɪn'trɑːns/ v.tr. **1** enchant, delight. **2** put into a trance. **3** (often foll. by with) overwhelm with strong feeling. □ **entrancement** n. **entrancing** adj. **entrancingly** adv.

entrant /'entrənt/ n. a person who enters (esp. an examination, profession, etc.). [F, part. of entrer: see ENTER]

entrap /ɪn'træp/ v.tr. (**entrapped, entrapping**) **1** catch in or as in a trap. **2** (often foll. by into + verbal noun) beguile or trick (a person). [OF entraper (as EN-¹, TRAP¹)]

entrapment /ɪn'træpmənt/ n. **1** the act or an instance of entrapping; the process of being entrapped. **2** Law inducement to commit a crime, esp. by the authorities to secure a prosecution.

entreat /ɪn'triːt/ v.tr. **1 a** (foll. by to + infin. or that + clause) ask (a person) earnestly. **b** ask earnestly for (a thing). **2** archaic treat; act towards (a person). □ **entreatingly** adv. [ME f. OF entraiter (as EN-¹, traiter TREAT)]

entreaty /ɪn'triːtɪ/ n. (pl. **-ies**) an earnest request; a supplication. [ENTREAT, after TREATY]

entrechat /'ɒntrəˌʃʌ/ n. a leap in ballet, with one or more crossings of the legs while in the air. [F f. It. (capriola) intrecciata complicated (caper)]

entrecôte /'ɒntrəˌkəʊt/ n. a boned steak cut off the sirloin. [F f. entre between + côte rib]

entrée /'ɒntreɪ/ n. **1 a** Brit. a dish served between the fish and meat courses. **b** esp. N. Amer. the main dish of a meal. **2** the right or privilege of admission, esp. at court. [F, = ENTRY]

entremets /'ɒntrəˌmeɪ/ n. **1** a sweet dish. **2** a light dish served between two courses. [F f. entre between + mets dish]

entrench /ɪn'trentʃ/ v. (also **intrench**) **1** tr. **a** establish firmly (in a defensible position, in office, etc.). **b** (as **entrenched** adj.) (of an attitude etc.) not easily modified. **2** tr. surround (a post, army, town, etc.) with a trench as a fortification. **3** tr. apply extra safeguards to (rights etc. guaranteed by legislation). **4** intr. entrench oneself. **5** intr. (foll. by upon) encroach, trespass. □ **entrench oneself** adopt a well-defended position. □ **entrenchment** n.

entre nous /ˌɒntrə 'nuː/ adv. **1** between you and me. **2** in private. [F, = between ourselves]

entrepôt /'ɒntrəˌpəʊ/ n. **1** a warehouse for temporary storage of goods in transit. **2** a commercial centre for import and export, and for collection and distribution. [F f. entreposer store f. entre- INTER- + poser place]

entrepreneur /ˌɒntrəprə'nɜː(r)/ n. **1** a person who undertakes an enterprise or business, with the chance of profit or loss. **2** a contractor acting as an intermediary. **3** the person in effective control of a commercial undertaking. **4** a person who organizes entertainments, esp. musical performances. □ **entrepreneurial** /-'nɜːrɪəl, -'njʊərɪəl/ adj. **entrepreneurialism** n. (also **entrepreneurism**). **entrepreneurship** n. [F f. entreprendre undertake: see ENTERPRISE]

entresol /'ɒntrəˌsɒl/ n. a low storey between the first and the ground floor; a mezzanine floor. [F f. entre between + sol ground]

entrism var. of ENTRYISM.

entropy /'entrəpɪ/ n. **1** Physics a measure of the unavailability of a system's thermal energy for conversion into mechanical work. (See note below.) **2** a measure of the rate of transfer of information in a message etc. □ **entropic** /en'trɒpɪk/ adj. **entropically** adv. [G Entropie (as EN-², Gk tropē transformation)]

■ The concept of entropy first arose in the 19th century as a mathematical quantity in thermodynamics. It was later given a physical interpretation as representing the degree of disorder or randomness of the constituents of any physical system, expressed as the probability of occurrence of its particular arrangement of particles. Thermodynamic theory indicates that the entropy of an isolated system can increase but will never decrease. Considered as a closed system, therefore, the universe must always increase in entropy, the eventual consequence being a heat death — a state in which structure is absent and temperature is uniform.

entrust /ɪn'trʌst/ v.tr. (also **intrust**) **1** (foll. by to) give responsibility for (a person or a thing) to a person in whom one has confidence. **2** (foll. by with) assign responsibility for a person or a thing to (a person). □ **entrustment** n.

entry /'entrɪ/ n. (pl. **-ies**) **1 a** the act or an instance of going or coming in. **b** the coming of an actor on stage. **c** ceremonial entrance. **2** liberty to go or come in. **3 a** a place of entrance; a door, gate, etc. **b** a lobby. **4** Brit. a passage between buildings. **5** the mouth of a river. **6 a** an item entered (in a diary, list, account-book, etc.). **b** the recording of this. **7 a** a person or thing competing in a race, contest, etc. **b** a list of competitors. **8** the start or resumption of music for a particular instrument in an ensemble. **9** Law the act of taking possession. **10** Bridge **a** an opportunity to transfer the lead to one's partner's hand. **b** a card providing this. □ **entry form** an application form for a competition. **entry permit** an authorization to enter a particular country etc. [ME f. OF entree ult. f. L intrare ENTER]

entryism /'entrɪˌɪz(ə)m/ n. (also **entrism** /'entrɪz(ə)m/) infiltration into a political organization to change or subvert its policies or objectives. □ **entrist** n. **entryist** n.

Entryphone /'entrɪˌfəʊn/ n. propr. an intercom device at an entrance to a building by which callers may identify themselves to gain admission.

entwine /ɪn'twaɪn/ v. (also **intwine**) **1** tr. & intr. (foll. by with, about, round) twine together (a thing with or round another). **2** tr. (as **entwined** adj.) entangled. **3** tr. interweave. □ **entwinement** n.

enucleate /ɪ'njuːklɪˌeɪt/ v.tr. Surgery extract (a tumour etc.). □ **enucleation** /ɪ,njuːklɪ'eɪʃ(ə)n/ n. [L enucleare (as E-, NUCLEUS)]

E-number /'iːˌnʌmbə(r)/ n. the letter E followed by a code number, designating food additives according to EC directives.

enumerate /ɪ'njuːməˌreɪt/ v.tr. **1** specify (items); mention one by one. **2** count; establish the number of. □ **enumerable** /-rəb(ə)l/ adj. **enumerative** /-rətɪv/ adj. **enumeration** /ɪ,njuːmə'reɪʃ(ə)n/ n. [L enumerare (as E-, NUMBER)]

enumerator /ɪ'njuːməˌreɪtə(r)/ n. **1** a person who enumerates. **2** a person employed in census-taking.

enunciate /ɪ'nʌnsɪˌeɪt/ v.tr. **1** pronounce (words) clearly. **2** express (a proposition or theory) in definite terms. **3** proclaim. □ **enunciator** n. **enunciative** /-sɪətɪv/ adj. **enunciation** /ɪ,nʌnsɪ'eɪʃ(ə)n/ n. [L enuntiare (as E-, nuntiare announce f. nuntius messenger)]

enure /ɪ'njʊə(r)/ v.intr. Law take effect. [var. of INURE]

enuresis /ˌenjʊə'riːsɪs/ n. Med. involuntary urination. □ **enuretic** /-'retɪk/ adj. & n. [mod.L f. Gk enoureō urinate in (as EN-², ouron urine)]

envelop /ɪn'veləp/ v.tr. (**enveloped, enveloping**) **1** (often foll. by in) **a** wrap up or cover completely. **b** make obscure; conceal (was enveloped in mystery). **2** Mil. completely surround (an enemy). □ **envelopment** n. [ME f. OF envoluper (as EN-¹: cf. DEVELOP)]

envelope /'envəˌləʊp, 'ɒn-/ n. **1** a folded paper container, usu. with a sealable flap, for a letter etc. **2** a wrapper or covering. **3** the structure within a balloon or airship containing the gas. **4** the outer metal or glass housing of a vacuum tube, electric light, etc. **5** Electr. a curve joining the successive peaks of a modulated wave. **6** *Bot.* an enveloping structure, esp. the calyx or corolla (or both). **7** *Math.* a line or curve tangent to each line or curve of a given family. [F *enveloppe* (as ENVELOP)]

envenom /ɪnˈvɛnəm/ *v.tr.* **1** put poison on or into; make poisonous. **2** infuse venom or bitterness into (feelings, words, or actions). [ME f. OF *envenimer* (as EN-¹, *venim* VENOM)]

Enver Pasha /ˌɛnvə ˈpɑːʃə/ (1881–1922), Turkish political and military leader. A leader of the Young Turks in the Revolution of 1908, he came to power as part of a ruling triumvirate following a coup d'état in 1913. He played a significant role in creating Turkey's alliance with Germany during the First World War, and served as Minister of War (1914–18).

enviable /ˈɛnvɪəb(ə)l/ *adj.* (of a person or thing) exciting or likely to excite envy. □ **enviably** *adv.*

envious /ˈɛnvɪəs/ *adj.* (often foll. by *of*) feeling or showing envy. □ **enviously** *adv.* [ME f. AF *envious*, OF *envieus* f. *envie* ENVY]

environ /ɪnˈvaɪərən/ *v.tr.* encircle, surround (esp. hostilely or protectively). [ME f. OF *environer* f. *environ* surroundings f. *en* in + *viron* circuit f. *virer* turn, VEER¹]

environment /ɪnˈvaɪərənmənt/ *n.* **1** surroundings and conditions, esp. as affecting people's lives (*a stimulating environment; would benefit from a change of environment*). **2** the region surrounding a place. **3** external conditions affecting the existence of plants and animals. **4** a structure designed to be experienced from inside as a work of art. **5** *Computing* the overall structure within which a user, computer, or program operates. □ **the environment** the totality of the physical conditions of the earth or a part of it, esp. as affected by human activity. **environment-friendly** not harmful to the environment. □ **environmental** /-ˌvaɪərənˈment(ə)l/ *adj.* **environmentally** *adv.*

environmentalism /ɪnˌvaɪərənˈmentəˌlɪz(ə)m/ *n.* **1** concern with or advocacy of the protection of the environment. (*See note below.*) **2** the view that environment has the primary influence on the development of a person or group. □ **environmentalist** *n.*

▪ Since the Second World War concern for the human and natural environment has been growing. Interest in town planning and the conservation of buildings was followed by increasing awareness of the damage to wildlife, habitats, and planetary systems resulting from human activities. Pollution, deforestation, and the burning of fossil fuels were discovered to be leading to previously unsuspected phenomena such as global warming, acid rain, and ozone depletion. Campaigning groups such as Greenpeace and Friends of the Earth were set up, while Green political parties began to gain a voice, particularly in Europe. Environmentalists stress that the earth's resources are finite and that environmental damage cannot be halted without movement away from policies aimed at continual economic growth.

Environmentally Sensitive Area *n.* (abbr. **ESA**) *Brit.* an area officially designated as containing landscapes or wildlife threatened by unrestricted development.

environs /ɪnˈvaɪərənz, ˈɛnvɪrənz/ *n.pl.* a surrounding district, esp. round an urban area.

envisage /ɪnˈvɪzɪdʒ/ *v.tr.* **1** have a mental picture of (a thing or conditions not yet existing). **2** contemplate or conceive, esp. as a possibility or desirable future event. **3** *archaic* **a** face (danger, facts, etc.). **b** look in the face of. □ **envisagement** *n.* [F *envisager* (as EN-¹, VISAGE)]

envision /ɪnˈvɪʒ(ə)n/ *v.tr.* envisage, visualize.

envoy¹ /ˈɛnvɔɪ/ *n.* **1** a messenger or representative, esp. on a diplomatic mission. **2** (in full **envoy extraordinary**) a minister plenipotentiary, ranking below ambassador and above chargé d'affaires. □ **envoyship** *n.* [F *envoyé*, past part. of *envoyer* send f. *en voie* on the way f. L *via*]

envoy² /ˈɛnvɔɪ/ *n.* (also **envoi**) **1** a short stanza concluding a ballade etc. **2** *archaic* an author's concluding words. [ME f. OF *envoi* f. *envoyer* (as ENVOY¹)]

envy /ˈɛnvɪ/ *n. & v.* ● *n.* (*pl.* **-ies**) **1** a feeling of discontented or resentful longing aroused by another's better fortune etc. **2** the object or ground of this feeling (*their house is the envy of the neighbourhood*). ● *v.tr.* (**-ies**, **-ied**) feel envy of (a person, circumstances, etc.) (*I envy you your position*). □ **envier** *n.* [ME f. OF *envie* f. L *invidia* f. *invidere* envy (as IN-¹, *videre* see)]

enweave var. of INWEAVE.

enwrap /ɪnˈræp/ *v.tr.* (also **inwrap**) (**-wrapped**, **-wrapping**) (often foll. by *in*) *literary* wrap or enfold.

enwreathe /ɪnˈriːð/ *v.tr.* (also **inwreathe**) *literary* surround with or as with a wreath.

Enzed /ɛnˈzɛd/ *n. Austral. & NZ colloq.* **1** New Zealand. **2** a New Zealander. □ **Enzedder** *n.* [pronunc. of NZ]

enzootic /ˌɛnzəʊˈɒtɪk/ *adj. & n.* ● *adj.* (of a disease) regularly affecting animals in a particular district or at a particular season (cf. ENDEMIC, EPIZOOTIC). ● *n.* an enzootic disease. [Gk *en* in + *zōion* animal]

enzyme /ˈɛnzaɪm/ *n. Biochem.* a substance which is produced by an organism and serves to control and promote a specific biochemical reaction. (*See note below.*) □ **enzymatic** /ˌɛnzaɪˈmætɪk/ *adj.* **enzymology** /-ˈmɒlədʒɪ/ *n.* **enzymic** /ɛnˈzaɪmɪk/ *adj.* [G *Enzym* f. med. Gk *enzumos* leavened f. Gk *en* in + *zumē* leaven]

▪ Enzymes are biochemical catalysts which occur in all living cells and are involved in most metabolic processes. They are typically proteins with large complex molecules. Most are very specific in their action, the nature of which depends on their particular molecular shape. Some enzymes control reactions within cells and some, e.g. the digestive enzymes, outside them.

EOC *abbr.* Equal Opportunities Commission.

Eocene /ˈiːəˌsiːn/ *adj. & n. Geol.* of, relating to, or denoting the second epoch of the Tertiary period, between the Palaeocene and the Oligocene. The Eocene lasted from about 54.9 to 38 million years ago, and was a time of rising temperatures. There is evidence of an abundance of mammals, including horses, bats, and whales. [Gk *ēōs* dawn + *kainos* new]

EOF *abbr. Computing* end of file.

EOKA /eɪˈəʊkə/ (also **Eoka**) a Greek-Cypriot liberation movement active in Cyprus in the 1950s and in the early 1970s, which fought for the independence of Cyprus from Britain and for its eventual union (enosis) with Greece. [Gk *Ethnikē Organōsis Kupriakou Agōnos* = National Organization of Cypriot Struggle]

eolian *US* var. of AEOLIAN.

eolith /ˈiːəlɪθ/ *n. Archaeol.* a roughly chipped stone found in Tertiary strata, originally thought to be an early artefact but probably of natural origin. [Gk *ēōs* dawn + *lithos* stone]

eolithic /ˌiːəˈlɪθɪk/ *adj. Archaeol.* of the period preceding the palaeolithic age, thought to include the earliest use of flint tools. [F *éolithique* (as EOLITH)]

eon var. of AEON.

Eos /ˈiːɒs/ *Gk Mythol.* the Greek goddess of the dawn, corresponding to the Roman Aurora. [Gk, = dawn]

eosin /ˈiːəsɪn/ *n.* a red fluorescent dye used esp. as a stain in optical microscopy. [Gk *ēōs* dawn + -IN]

eosinophil /ˌiːəˈsɪnəfɪl/ *n. Physiol.* a white blood cell containing large granules that are readily stained by eosin.

EOT *abbr.* **1** *Computing* end of tape. **2** *Telegraphy* end of transmission.

-eous /ɪəs/ *suffix* forming adjectives meaning 'of the nature of' (*erroneous; gaseous*).

EP *abbr.* **1** electroplate. **2** extended-play (gramophone record etc.). **3** extreme pressure (used in grading lubricants).

Ep. *abbr.* Epistle.

e.p. *abbr. Chess* en passant.

ep- /ɛp, ɪp, iːp/ *prefix* form of EPI- before a vowel or *h*.

EPA *abbr.* (in the US) Environmental Protection Agency.

epact /ˈiːpækt/ *n.* the number of days by which the solar year exceeds the lunar year. [F *épacte* f. LL *epactae* f. Gk *epaktai* (*hēmerai*) intercalated (days) f. *epagō* intercalate (as EPI-, *agō* bring)]

eparch /ˈɛpɑːk/ *n.* the chief bishop of an eparchy. [Gk *eparkhos* (as EPI-, *arkhos* ruler)]

eparchy /ˈɛpɑːkɪ/ *n.* (*pl.* **-ies**) a province of the Orthodox Church. [Gk *eparkhia* (as EPARCH)]

epaulette /ˈɛpəlɛt, ˈɛpɔːˌlɛt/ *n.* (*US* **epaulet**) an ornamental shoulder-piece on a coat, dress, etc., esp. on a uniform. [F *épaulette* dimin. of *épaule* shoulder f. L *spatula* SPATULA (in late L) shoulder-blade + -ETTE]

épée /ˈeɪpeɪ, ˈepeɪ/ *n.* (*pl.* same) a sharp-pointed duelling-sword, used (with the end blunted) in fencing. □ **épéeist** *n.* [F, = sword, f. OF *espee*: see SPAY]

epeirogenesis /eˌpaɪərəʊˈdʒɛnɪsɪs/ *n.* (also **epeirogeny** /ˌɛpaɪˈrɒdʒənɪ/) *Geol.* the regional uplift of extensive areas of the earth's crust. □ **epeirogenic** /-ˈdʒɛnɪk/ *adj.* [Gk *ēpeiros* mainland + GENESIS, -GENY]

epenthesis /eˈpɛnθɪsɪs/ *n.* (*pl.* **epentheses** /-ˌsiːz/) the insertion of a letter or sound within a word, e.g. *b* in *thimble*. □ **epenthetic**

/ˌepen'θetɪk/ adj. [LL f. Gk f. epentithēmi insert (as EPI- + EN-² + tithēmi place)]

epergne /ɪ'pɜːn/ n. an ornament (esp. in branched form) for the centre of a dinner-table, holding flowers or fruit. [18th c.: perh. a corrupt. of F épargne saving, economy]

epexegesis /eˌpeksɪ'dʒiːsɪs/ n. (pl. **epexegeses** /-siːz/) **1** the addition of words to clarify meaning (e.g. to do in difficult to do). **2** the words added. □ **epexegetic** /-'dʒetɪk/ adj. **epexegetical** adj. **epexegetically** adv. [Gk epexēgēsis (as EPI-, EXEGESIS)]

Eph. abbr. Epistle to the Ephesians (New Testament).

ephebe /'efiːb/ n. Gk Hist. a young man of 18–20 undergoing military training. □ **ephebic** /e'fiːbɪk/ adj. [L ephebus f. Gk ephēbos (as EPI-, hēbē early manhood)]

ephedra /ɪ'fedrə/ n. an evergreen gymnospermous shrub of the genus Ephedra, with trailing stems and scalelike leaves. [mod.L f. Gk ephedra sitting upon]

ephedrine /ɪ'fedrɪn, 'efɪˌdriːn, -drɪn/ n. Biochem. an alkaloid drug found in some ephedras, which causes constriction of the blood vessels and widening of the bronchial passages, and is used to relieve asthma etc. [EPHEDRA + -INE⁴]

ephemera¹ /ɪ'femərə, ɪ'fiːm-/ n. (pl. **ephemeras** or **ephemerae** /-ˌriː/) **1 a** an insect living only a day or a few days. **b** an insect of the order Ephemeroptera; a mayfly. **2** = EPHEMERON. [mod.L f. Gk ephēmeros lasting only a day (as EPI-, hēmera day)]

ephemera² pl. of EPHEMERON 1.

ephemeral /ɪ'femərəl, ɪ'fiːm-/ adj. **1** lasting or of use for only a short time; transitory. **2** lasting only a day. **3** (of an insect, flower, etc.) lasting a day or a few days. □ **ephemerally** adv. **ephemeralness** n. **ephemerality** /ɪˌfemə'rælɪtɪ, ɪˌfiːm-/ n. [Gk ephēmeros: see EPHEMERA¹]

ephemeris /ɪ'femərɪs, ɪ'fiːm-/ n. (pl. **ephemerides** /ˌefɪ'merɪˌdiːz/) Astron. **1** a table of the predicted positions of a celestial body. **2** a book of such tables, an almanac. □ **ephemeris time** time on a scale defined by the earth's orbital period rather than its axial rotation. [L f. Gk ephēmeris diary (as EPHEMERAL)]

ephemerist /ɪ'femərɪst, ɪ'fiːm-/ n. a collector of ephemera.

ephemeron /ɪ'femərən, ɪ'fiːm-/ n. **1** (pl. **ephemera** /-rə/) (usu. in pl.) **a** a thing (esp. a printed item) of short-lived interest or usefulness. **b** a short-lived thing. **2** (pl. **ephemerons**) = EPHEMERA¹ 1. [as EPHEMERA¹]

Ephemeroptera /ɪˌfemə'rɒptərə, ɪˌfiːm-/ n.pl. Zool. an order of insects comprising the mayflies. □ **ephemeropteran** n. & adj. **ephemeropterous** adj. [mod.L f. Gk ephēmeros lasting only a day + pteron wing]

Ephesians, Epistle to the /ɪ'fiːʒ(ə)nz/ a book of the New Testament ascribed to St Paul, an epistle to the Church at Ephesus.

Ephesus /'efɪsəs/ an ancient Greek city on the west coast of Asia Minor, in present-day Turkey, site of the temple of Diana, one of the Seven Wonders of the World. It was an important centre of early Christianity; St Paul preached there and St John is traditionally said to have lived there. Because of silting, the remains of the city are now more than 5 km (3 miles) inland.

ephod /'iːfɒd, 'ef-/ n. a Jewish priestly vestment. [ME f. Heb. 'ēphôd]

ephor /'efɔː(r)/ n. (pl. **ephors** or **ephori** /'efəˌraɪ/) Gk Hist. each of five senior magistrates in ancient Sparta. □ **ephorate** /'efərət/ n. [Gk ephoros overseer (as EPI-, horaō see)]

epi- /'epɪ-/ prefix (usu. **ep-** before a vowel or h) **1** upon (epicycle). **2** above (epicotyl). **3** in addition (epiphenomenon). [Gk epi (prep.)]

epiblast /'epɪˌblæst/ n. Biol. the outermost layer of a gastrula etc.; the ectoderm. [EPI- + -BLAST]

epic /'epɪk/ n. & adj. ● n. **1** a long poem narrating the adventures or deeds of one or more heroic or legendary figures. (See note below.) **2** an imaginative work of any form, embodying a nation's conception of its past history. **3** a book or film based on an epic narrative or heroic in type or scale. **4** a subject fit for recital in an epic. ● adj. **1** of or like an epic. **2** grand, heroic. □ **epical** adj. **epically** adv. [L epicus f. Gk epikos f. epos word, song]

▪ The early or primary epics, of which the most important Western examples are the Iliad and Odyssey of Homer, were derived from long-established, orally transmitted traditional tales and were probably intended to be recited in sections at feasts. Other primary epics include the Old English Beowulf, the Sumerian epic of Gilgamesh, the Indian Mahabharata, and the Germanic Nibelungenlied. Secondary epics such as Virgil's Aeneid and Milton's Paradise Lost were composed

later as works of literature by single authors in imitation of the primary models. The epic form was highly valued in the Renaissance, and was employed by Italians such as Ariosto (Orlando Furioso); in modern times the term has been used of ambitious, large-scale books and films such as Tolstoy's War and Peace and D. W. Griffith's Birth of a Nation.

epicarp /'epɪˌkɑːp/ n. Bot. the outermost layer of the pericarp. [EPI- + Gk karpos fruit]

epicedium /ˌepɪ'siːdɪəm/ n. (pl. **epicedia** /-dɪə/) a funeral ode. □ **epicedian** adj. [L f. Gk epikēdeion (as EPI-, kēdos care)]

epicene /'epɪˌsiːn/ adj. & n. ● adj. **1** Gram. denoting either sex without change of gender. **2** of, for, or used by both sexes. **3** having characteristics of both sexes. **4** having no characteristics of either sex. **5** effete, effeminate. ● n. an epicene person. [ME f. LL epicoenus f. Gk epikoinos (as EPI-, koinos common)]

epicentre /'epɪˌsentə(r)/ n. (US **epicenter**) **1** Geol. the point on the earth's surface directly above the focus of an earthquake. **2** the central point of a difficulty. □ **epicentral** /ˌepɪ'sentrəl/ adj. [Gk epikentros (adj.) (as EPI-, CENTRE)]

epicontinental /ˌepɪˌkɒntɪ'nent(ə)l/ adj. (of the sea) over the continental shelf.

epicotyl /'epɪˌkɒtɪl/ n. Bot. the region of an embryo or seedling stem above the cotyledon(s).

Epictetus /ˌepɪk'tiːtəs/ (c.55–c.135 AD), Greek philosopher. Originally a slave, he preached the common brotherhood of man and advocated a Stoic philosophy. His teachings were published posthumously in the Enchiridion.

epicure /'epɪˌkjʊə(r)/ n. a person with refined tastes, esp. in food and drink. □ **epicurism** n. [med.L epicurus one preferring sensual enjoyment: see EPICUREAN]

Epicurean /ˌepɪkjʊə'riːən/ n. & adj. ● n. **1** a disciple or student of the Greek philosopher Epicurus or of Epicureanism. ● adj. **1** of or concerning Epicurus or his ideas. [F épicurien or L epicureus f. Gk epikoureios f. Epikouros EPICURUS]

epicurean /ˌepɪkjʊə'riːən/ adj. & n. ● adj. fond of refined sensuous pleasure and luxury. ● n. a person with epicurean tastes. [EPICUREAN]

Epicureanism /ˌepɪkjʊə'riːəˌnɪz(ə)m/ n. an ancient school of philosophy, founded in Athens by Epicurus. The school rejected determinism and advocated hedonism, but of a restrained kind: mental pleasure was regarded more highly than physical and the ultimate pleasure was held to be freedom from anxiety and mental pain, especially that arising from needless fear of death and of gods.

Epicurus /ˌepɪ'kjʊərəs/ (341–270 BC), Greek philosopher. His physics (later expounded by the Roman writer Lucretius) is based on the theory of a materialist universe, unregulated by divine providence, composed of indestructible atoms moving in a void. From this follows his philosophy, a type of hedonism (see EPICUREANISM).

epicycle /'epɪˌsaɪk(ə)l/ n. **1** Geom. a small circle whose centre moves round the circumference of a larger one. **2** hist. any of such circles used to describe planetary orbits in the Ptolemaic system. □ **epicyclic** /ˌepɪ'saɪklɪk, -'sɪklɪk/ adj. [ME f. OF or LL epicyclus f. Gk epikuklos (as EPI-, kuklos circle)]

epicycloid /ˌepɪ'saɪklɔɪd/ n. Math. a curve traced by a point on the circumference of a circle rolling on the exterior of another circle. □ **epicycloidal** /-saɪ'klɔɪd(ə)l/ adj.

Epidaurus /ˌepɪ'dɔːrəs/ (Greek **Epídhavros** /ɛ'piðavrɔs/) an ancient Greek city and port on the NE coast of the Peloponnese, site of a temple dedicated to Asclepius and a well-preserved Greek theatre dating from the 4th century BC.

epideictic /ˌepɪ'daɪktɪk/ adj. meant for effect or display, esp. in speaking. [Gk epideiktikos (as EPI-, deiknumi show)]

epidemic /ˌepɪ'demɪk/ n. & adj. ● n. **1** a widespread occurrence of a disease in a community at a particular time. **2** such a disease. **3** (foll. by of) a wide prevalence of something usu. undesirable. ● adj. in the nature of an epidemic (cf. ENDEMIC). □ **epidemically** adv. [F épidémique f. épidémie f. LL epidemia f. Gk epidēmia prevalence of disease f. epidēmios (adj.) (as EPI-, dēmos the people)]

epidemiology /ˌepɪˌdiːmɪ'ɒlədʒɪ/ n. the study of the incidence and distribution of diseases, and of their control and prevention. □ **epidemiologist** n. **epidemiological** /-mɪə'lɒdʒɪk(ə)l/ adj.

epidermis /ˌepɪ'dɜːmɪs/ n. **1** the outer cellular layer of the skin. **2** Bot. the outer layer of cells of leaves, stems, roots, etc. □ **epidermal** adj. **epidermic** adj. **epidermoid** adj. [LL f. Gk (as EPI- + derma skin)]

Epídhavros see EPIDAURUS.

epidiascope /ˌɛpɪˈdaɪəˌskəʊp/ n. an optical projector capable of giving images of both opaque and transparent objects. [EPI- + DIA- + -SCOPE]

epididymis /ˌɛpɪˈdɪdɪmɪs/ n. (pl. **epididymides** /-dɪˈdɪmɪˌdiːz/) Anat. a convoluted duct behind the testis, along which sperm passes to the vas deferens. [Gk epididumis (as EPI-, didumoi testicles)]

epidural /ˌɛpɪˈdjʊərəl/ adj. & n. ● adj. **1** Anat. on or around the dura mater. **2** (of an anaesthetic) introduced into the space around the dura mater of the spinal cord. ● n. an epidural anaesthetic, used esp. in childbirth to produce loss of sensation below the waist. [EPI- + DURA MATER]

epifauna /ˈɛpɪˌfɔːnə/ n. Ecol. the animal life which lives on the surface of the ocean floor, a river bed, etc., or attached to submerged objects or to aquatic plants or animals (cf. INFAUNA). [Danish (as EPI-, FAUNA)]

epigastrium /ˌɛpɪˈɡæstrɪəm/ n. (pl. **epigastria** /-rɪə/) Anat. the part of the abdomen immediately over the stomach. □ **epigastric** adj. [LL f. Gk epigastrion (neut. adj.) (as EPI-, gastēr belly)]

epigeal /ˌɛpɪˈdʒiːəl/ adj. Bot. **1** having one or more cotyledons above the ground. **2** growing above the ground. [Gk epigeios (as EPI-, gē earth)]

epigene /ˈɛpɪˌdʒiːn/ adj. Geol. produced on the surface of the earth. [F épigène f. Gk epigenēs (as EPI-, genēs born)]

epigenesis /ˌɛpɪˈdʒɛnɪsɪs/ n. Biol. the theory (now generally held) that an embryo develops from an undifferentiated egg cell. [EPI- + GENESIS]

epigenetic /ˌɛpɪdʒɪˈnɛtɪk/ adj. **1** Biol. **a** relating to epigenesis. **b** due to external rather than genetic influences. **2** Geol. formed later than the surrounding rock.

epiglottis /ˌɛpɪˈɡlɒtɪs/ n. Anat. a flap of cartilage at the root of the tongue, which is depressed during swallowing to cover the windpipe. □ **epiglottal** adj. **epiglottic** adj. [Gk epiglōttis (as EPI-, glōtta tongue)]

epigone /ˈɛpɪˌɡəʊn/ n. (pl. **epigones** or **epigoni** /ɪˈpɪɡəˌnaɪ/) a member of a later (and less distinguished) generation. [pl. f. F épigones f. L epigoni f. Gk epigonoi those born afterwards (as EPI-, root of gignomai be born)]

epigram /ˈɛpɪˌɡræm/ n. **1** a short witty poem. **2** a saying or maxim, esp a proverbial one. **3 a** a pointed remark or expression, esp. a witty one. **b** the use of concise witty remarks. □ **epigrammatic** /ˌɛpɪɡrəˈmætɪk/ adj. **epigrammatically** adv. **epigrammatist** /ˌɛpɪˈɡræmətɪst/ n. **epigrammatize** v.tr. & intr. (also **-ise**). [F épigramme or L epigramma f. Gk epigramma -atos (as EPI-, -GRAM)]

epigraph /ˈɛpɪˌɡrɑːf/ n. an inscription on a statue or coin, at the head of a chapter, etc. [Gk epigraphē f. epigraphō (as EPI-, graphō write)]

epigraphy /ɪˈpɪɡrəfɪ/ n. the study of (esp. ancient) inscriptions. □ **epigraphist** n. **epigraphic** /ˌɛpɪˈɡræfɪk/ adj. **epigraphical** adj. **epigraphically** adv.

epilate /ˈɛpɪˌleɪt/ v.tr. Med. remove hair by the roots from. □ **epilation** /ˌɛpɪˈleɪʃ(ə)n/ n. [F épiler (cf. DEPILATE)]

epilepsy /ˈɛpɪˌlɛpsɪ/ n. a neurological disorder marked by episodes of sensory disturbance, loss of consciousness, or convulsions. Seizures are accompanied by changes in the rhythm of the electrical currents of the brain. There are two main forms: grand mal involves loss of consciousness and convulsions; petit mal lasts for only a few seconds, with partial loss of consciousness. [F épilepsie or LL epilepsia f. Gk epilēpsia f. epilambanō attack (as EPI-, lambanō take)]

epileptic /ˌɛpɪˈlɛptɪk/ adj. & n. ● adj. of or relating to epilepsy. ● n. a person with epilepsy. [F épileptique f. LL epilepticus f. Gk epilēptikos (as EPILEPSY)]

epilimnion /ˌɛpɪˈlɪmnɪən/ n. (pl. **epilimnia** /-nɪə/) the upper, warmer layer of water in a stratified lake. [EPI- + Gk limnion dimin. of limnē lake]

epilogist /ɪˈpɪlədʒɪst/ n. the writer or speaker of an epilogue.

epilogue /ˈɛpɪˌlɒɡ/ n. **1 a** the concluding part of a literary work. **b** an appendix. **2** a speech or short poem addressed to the audience by an actor at the end of a play. **3** Brit. a short piece at the end of a day's broadcasting (cf. PROLOGUE). [ME f. F épilogue f. L epilogus f. Gk epilogos (as EPI-, logos speech)]

epimer /ˈɛpɪmə(r)/ n. Chem. either of two isomers with different configurations of atoms about one of several asymmetric carbon atoms present. □ **epimeric** /ˌɛpɪˈmɛrɪk/ adj. **epimerism** /ɪˈpɪməˌrɪz(ə)m/ n. [G (as EPI-, -MER)]

epimerize /ɪˈpɪməˌraɪz/ v.tr. (also **-ise**) Chem. convert from one epimeric form into the other.

epinasty /ˈɛpɪˌnæstɪ/ n. Bot. a tendency in the organs of plants to grow more rapidly on the upper side. [EPI- + Gk nastos pressed]

epinephrine /ˌɛpɪˈnɛfrɪn/ n. Biochem. = ADRENALIN. [EPI- + nephros kidney]

epiphany /ɪˈpɪfənɪ/ n. (pl. **-ies**) **1** (**Epiphany**) **a** the manifestation of Christ to the Magi according to the biblical account. **b** a festival of the Christian Church held on 6 Jan., celebrating the coming of the Magi. (See note below.) **2** any manifestation of a god or demigod. **3** any sudden and important manifestation or realization. □ **epiphanic** /ˌɛpɪˈfænɪk/ adj. [ME f. Gk epiphaneia manifestation f. epiphainō reveal (as EPI-, phainō show): sense 1 through OF epiphanie and eccl.L epiphania]

▪ The festival of Epiphany originated in the East, where it has been celebrated from the 3rd century in honour of Christ's baptism. It was introduced into the Western Church in the 4th century, becoming associated with the manifestation of Christ to the Gentiles in the persons of the Magi.

epiphenomenon /ˌɛpɪfɪˈnɒmɪnən/ n. (pl. **epiphenomena** /-nə/) **1** a secondary symptom, which may occur simultaneously with a disease etc. but is not regarded as its cause or result. **2** Psychol. consciousness regarded as a by-product of brain activity. □ **epiphenomenal** adj.

epiphysis /ɪˈpɪfɪsɪs/ n. (pl. **epiphyses** /-ˌsiːz/) Anat. **1** the end of a long bone, initially growing separately from the shaft. **2** = pineal body. [mod.L f. Gk epiphusis (as EPI-, phusis growth)]

epiphyte /ˈɛpɪˌfaɪt/ n. Bot. a plant growing but not parasitic on another, e.g. a moss. □ **epiphytal** /ˌɛpɪˈfaɪt(ə)l/ adj. **epiphytic** /-ˈfɪtɪk/ adj. [EPI- + Gk phuton plant]

Epirus /ɪˈpaɪərəs/ (Greek **Ípiros** /ˈipirɒs/) **1** a coastal region of NW Greece; capital, Ioánnina. **2** an ancient country of which the modern region of Epirus corresponds to the south-western part, extending northwards to Illyria and eastwards to Macedonia and Thessaly.

episcopacy /ɪˈpɪskəpəsɪ/ n. (pl. **-ies**) **1** government of a Church by bishops. **2** (prec. by the) the bishops.

episcopal /ɪˈpɪskəp(ə)l/ adj. **1** of a bishop or bishops. **2** (of a Church) constituted on the principle of government by bishops. □ **episcopalism** n. **episcopally** adv. [ME f. F épiscopal or eccl.L episcopalis f. episcopus BISHOP]

Episcopal Church n. the Anglican Church in Scotland and the US, with elected bishops.

episcopalian /ɪˌpɪskəˈpeɪlɪən/ adj. & n. ● adj. **1** of or advocating government of a Church by bishops. **2** of or belonging to an episcopal Church or (**Episcopalian**) the Episcopal Church. ● n. **1** an adherent of episcopacy. **2** (**Episcopalian**) a member of the Episcopal Church. □ **episcopalianism** n.

episcopate /ɪˈpɪskəpət/ n. **1** the office or tenure of a bishop. **2** (prec. by the) the bishops collectively. [eccl.L episcopatus f. episcopus BISHOP]

episcope /ˈɛpɪˌskəʊp/ n. an optical projector giving images of opaque objects.

episematic /ˌɛpɪsɪˈmætɪk/ adj. Zool. (of coloration, markings, etc.) serving to help recognition by animals of the same species. [EPI- + Gk sēma sēmatos sign]

episiotomy /ɪˌpiːzɪˈɒtəmɪ/ n. (pl. **-ies**) Med. a surgical cut made at the opening of the vagina during childbirth, to aid delivery. [Gk epision pubic region]

episode /ˈɛpɪˌsəʊd/ n. **1** one event or a group of events as part of a sequence. **2** each of the parts of a serial story or broadcast. **3** an incident or set of incidents in a narrative. **4** an incident that is distinct but contributes to a whole (a romantic episode in her life). **5** Mus. a passage containing distinct material or introducing a new subject. **6** the part between two choric songs in Greek tragedy. [Gk epeisodion (as EPI- + eisodos entry f. eis into + hodos way)]

episodic /ˌɛpɪˈsɒdɪk/ adj. (also **episodical** /-ˈsɒdɪk(ə)l/) **1** occurring as separate episodes. **2** sporadic; occurring at irregular intervals. □ **episodically** adv.

epistaxis /ˌɛpɪˈstæksɪs/ n. Med. bleeding from the nose. [mod.L f. Gk (as EPI-, stazō drip)]

epistemic /ˌɛpɪˈstiːmɪk, -ˈstɛmɪk/ adj. Philos. relating to knowledge or to the degree of its validation. □ **epistemically** adv. [Gk epistēmē knowledge]

epistemology /ɪˌpɪstɪˈmɒlədʒɪ/ n. Philos. the theory of knowledge, esp. with regard to its methods and validation. Epistemology is the investigation of what distinguishes justified belief from opinion. The traditional debate has concerned the foundation of knowledge: opposing theories within epistemology are rationalism and empiricism, while scepticism holds that justification, and hence knowledge with any kind of certainty, is not possible. □ **epistemologist** *n.* **epistemological** /-mə'lɒdʒɪk(ə)l/ *adj.* **epistemologically** *adv.*

epistle /ɪ'pɪs(ə)l/ *n.* **1** *formal or joc.* a letter, esp. a long one on a serious subject. **2** (also **Epistle**) **a** any of the letters of the Apostles in the New Testament. **b** an extract from an Epistle read in a church service. **3** a poem or other literary work in the form of a letter or series of letters. [ME f. OF f. L *epistola* f. Gk *epistolē* f. *epistellō* send news (as EPI-, *stellō* send)]

Epistle to the Colossians, Epistle to the Ephesians, etc. see COLOSSIANS, EPISTLE TO THE; EPHESIANS, EPISTLE TO THE, etc.

epistolary /ɪ'pɪstələrɪ/ *adj.* **1** in the style or form of a letter or letters. **2** of, carried by, or suited to letters. [F *épistolaire* or L *epistolaris* (as EPISTLE)]

epistolary novel *n.* a novel consisting wholly or chiefly of letters written by one or more of the characters. The form was developed particularly by Samuel Richardson (in *Pamela* (1740–1) and *Clarissa Harlowe* (1747–8)) and was also used by Smollett and Goethe. The epistolary novel was largely an 18th-century form, although more recent novels have often used letters in combination with narrative.

epistrophe /ɪ'pɪstrəfɪ/ *n.* the repetition of a word at the end of successive clauses. [Gk (as EPI-, *strophē* turning)]

epistyle /'epɪˌstaɪl/ *n. Archit.* = ARCHITRAVE. [F *épistyle* or L *epistylium* f. Gk *epistulion* (as EPI-, *stulos* pillar)]

epitaph /'epɪˌtɑːf/ *n.* words written in memory of a person who has died, esp. as a tomb inscription. [ME f. OF *epitaphe* f. L *epitaphium* f. Gk *epitaphion* funeral oration (as EPI-, *taphos* tomb)]

epitaxy /'epɪˌtæksɪ/ *n. Crystallog.* the growth of crystals on a crystalline substrate that determines their orientation. □ **epitaxial** /ˌepɪ'tæksɪəl/ *adj.* [F *épitaxie* (as EPI-, Gk *taxis* arrangement)]

epithalamium /ˌepɪθə'leɪmɪəm/ *n.* (pl. **epithalamiums** or **epithalamia** /-mɪə/) a song or poem celebrating a marriage. □ **epithalamial** *adj.* **epithalamic** /-'læmɪk/ *adj.* [L f. Gk *epithalamion* (as EPI-, *thalamos* bridal chamber)]

epithelium /ˌepɪ'θiːlɪəm/ *n.* (pl. **epitheliums** or **epithelia** /-lɪə/) *Anat.* the tissue forming the outer layer of the body surface and lining the digestive tract and other hollow structures. □ **epithelial** *adj.* [mod.L f. EPI- + Gk *thēlē* teat]

epithet /'epɪˌθet/ *n.* **1** an adjective or other descriptive word expressing a quality or attribute, esp. used with or as a name. **2** such a word as a term of abuse. □ **epithetic** /ˌepɪ'θetɪk/ *adj.* **epithetical** *adj.* **epithetically** *adv.* [F *épithète* or L *epitheton* f. Gk *epitheton* f. *epitithēmi* add (as EPI-, *tithēmi* place)]

epitome /ɪ'pɪtəmɪ/ *n.* **1** a person or thing embodying a quality, class, etc. **2** a thing representing another in miniature. **3** a summary of a written work; an abstract. □ **epitomist** *n.* [L f. Gk *epitomē* f. *epitemnō* abridge (as EPI-, *temnō* cut)]

epitomize /ɪ'pɪtəˌmaɪz/ *v.tr.* (also **-ise**) **1** be a perfect example of (a quality etc.); typify. **2** make an epitome of (a work). □ **epitomization** /ɪˌpɪtəmaɪ'zeɪʃ(ə)n/ *n.*

epizoon /ˌepɪ'zəʊɒn/ *n.* (pl. **epizoa** /-'zəʊə/) *Zool.* an animal living on another animal. [mod.L (as EPI-, Gk *zōion* animal)]

epizootic /ˌepɪzəʊ'ɒtɪk/ *adj. & n.* ● *adj.* (of a disease) temporarily prevalent and widespread among an animal population (cf. ENZOOTIC). ● *n.* an outbreak of such a disease. [F *épizootique* f. *épizootie* (as EPIZOON)]

EPLF *abbr.* Eritrean People's Liberation Front.

EPNS *abbr.* electroplated nickel silver.

epoch /'iːpɒk/ *n.* **1** a period of history or of a person's life marked by notable events. **2** the beginning of an era. **3** *Geol.* a division of a period, corresponding to a set of strata. □ **epoch-making** remarkable, historic; of major importance. □ **epochal** /'epək(ə)l/ *adj.* [mod.L *epocha* f. Gk *epokhē* stoppage]

epode /'epəʊd/ *n.* **1** a form of lyric poem written in couplets each of a long line followed by a shorter one. **2** the third section of an ancient Greek choral ode or of one division of it. [F *épode* or L *epodos* f. Gk *epōidos* (as EPI-, ODE)]

eponym /'epənɪm/ *n.* **1** a person (real or imaginary) after whom a discovery, invention, place, institution, etc., is named or thought to be named. **2** the name given. □ **eponymous** /ɪ'pɒnɪməs/ *adj.* [Gk *epōnumos* (as EPI-, *-ōnumos* f. *onoma* name)]

EPOS /'iːpɒz, 'iːpɒs/ *abbr.* electronic point-of-sale (of retail outlets recording information electronically).

epoxide /ɪ'pɒksaɪd/ *n. Chem.* a compound containing an oxygen atom bonded in a triangular arrangement to two carbon atoms. [EPI- + OXIDE]

epoxy /ɪ'pɒksɪ/ *adj. Chem.* relating to or derived from an epoxide. □ **epoxy resin** a synthetic thermosetting resin containing epoxy groups. [EPI- + OXY-²]

EPROM /'iːprɒm/ *n. Computing* a read-only memory whose contents can be erased and replaced by a special process. [*e*rasable *p*rogrammable *ROM*]

epsilon /'epsɪˌlɒn/ *n.* the fifth letter of the Greek alphabet (E, ϵ). [ME f. Gk, = bare E f. *psilos* bare]

Epsom /'epsəm/ a town in Surrey, SE England; pop. (1981) 68,500. Its natural mineral waters were used in the production of the purgative known as Epsom salts. The annual Derby and Oaks horse-races are held at its racecourse on Epsom Downs.

Epsom salts *n.* a preparation of magnesium sulphate used as a purgative etc. [EPSOM]

EPSRC *abbr.* (in the UK) Engineering and Physical Sciences Research Council.

Epstein /'epstaɪn/, Sir Jacob (1880–1959), American-born British sculptor. His first important commission was a group of eighteen figures for the British Medical Association in the Strand (1907–8); it was the first of many works to arouse violent criticism for the use of distortion and alleged obscenity. A founder member of the vorticist group, he moved towards abstract sculpture and later scored great success in his modelled portraits of the famous, in particular his *Einstein* (1933).

Epstein–Barr virus /ˌepstaɪn'bɑː(r)/ *n.* a herpesvirus causing glandular fever and associated with certain cancers, e.g. Burkitt's lymphoma. [M. A. *Epstein*, Br. virologist (b.1921), and Y. M. *Barr*, Ir.-born virologist (b.1932)]

epyllion /ɪ'pɪlɪən/ *n.* (pl. **epyllia** /-lɪə/) a miniature epic poem. [Gk *epullion* dimin. of *epos* word, song]

equable /'ekwəb(ə)l/ *adj.* **1** even; not varying. **2** uniform and moderate (*an equable climate*). **3** (of a person) not easily disturbed or angered. □ **equably** *adv.* **equability** /ˌekwə'bɪlɪtɪ/ *n.* [L *aequabilis* (as EQUATE)]

equal /'iːkwəl/ *adj., n., & v.* ● *adj.* **1** (often foll. by *to, with*) the same in quantity, quality, size, degree, rank, level, etc. **2** evenly balanced (*an equal contest*). **3** having the same rights or status (*human beings are essentially equal*). **4** uniform in application or effect. ● *n.* a person or thing equal to another, esp. in rank, status, or characteristic quality (*their treatment of the subject has no equal*; *is the equal of any man*). ● *v.tr.* (**equalled, equalling**; US **equaled, equaling**) **1** be equal to in number, quality, etc. **2** achieve something that is equal to (an achievement) or to the achievement of (a person). □ **be equal to** have the ability or resources for. **equal opportunity** (often in *pl.*) the opportunity or right to be employed, paid, etc., without discrimination on grounds of sex, race, etc. **equal** (or **equals**) **sign** the symbol =. [ME f. L *aequalis* f. *aequus* even]

equalitarian /ɪˌkwɒlɪ'teərɪən/ *n.* = EGALITARIAN. □ **equalitarianism** *n.* [EQUALITY, after *humanitarian* etc.]

equality /ɪ'kwɒlɪtɪ/ *n.* the state of being equal. [ME f. OF *equalité* f. L *aequalitas -tatis* (as EQUAL)]

equalize /'iːkwəˌlaɪz/ *v.* (also **-ise**) **1** *tr. & intr.* make or become equal. **2** *intr.* score a goal etc. to make the scores equal in a game (*Grimsby equalized in the last minute*). □ **equalization** /ˌiːkwəlaɪ'zeɪʃ(ə)n/ *n.*

equalizer /'iːkwəˌlaɪzə(r)/ *n.* (also **-iser**) **1** an equalizing score or goal etc. in a game. **2** *sl.* a weapon, esp. a gun. **3** *Electr.* a connection in a system which compensates for any undesirable frequency or phase response with the system.

equally /'iːkwəlɪ/ *adv.* **1** in an equal manner (*treated them all equally*). **2** to an equal degree (*is equally important*). ¶ In sense 2 construction with *as* (*equally as important*) is often found, but is *disp.*).

equanimity /ˌekwə'nɪmɪtɪ, ˌiːk-/ *n.* mental composure, evenness of temper, esp. in misfortune. □ **equanimous** /ɪ'kwænɪməs/ *adj.* [L *aequanimitas* f. *aequanimis* f. *aequus* even + *animus* mind]

equate /ɪ'kweɪt/ *v.* **1** *tr.* (usu. foll. by *to, with*) regard as equal or equivalent. **2** *intr.* (foll. by *with*) **a** be equal or equivalent to. **b** agree or correspond. □ **equatable** *adj.* [ME f. L *aequare aequat-* f. *aequus* equal]

equation /ɪ'kweɪʒ(ə)n/ *n.* **1** the process of equating or making equal; the state of being equal. **2** *Math.* a statement that two mathematical expressions are equal (indicated by the sign =). **3** *Chem.* a formulaic representation of a chemical reaction expressed in terms of the molecules etc. taking part. □ **equation of the first order, second order**, etc. an equation involving only the first derivative, second derivative, etc. □ **equational** adj. [ME f. OF equation or L aequatio (as EQUATE)]

equator /ɪˈkweɪtə(r)/ n. **1** an imaginary line round the earth or other body, equidistant from the poles; the parallel of latitude 0°. **2** Astron. = celestial equator. [ME f. OF equateur or med.L aequator (as EQUATION)]

equatorial /ˌekwəˈtɔːrɪəl, ˌiːk-/ adj. of or near the equator. □ **equatorial mount** (or **mounting**) a mount for an astronomical telescope that rotates about the polar and declination axes, usu. with a drive on the polar axis so that celestial objects remain in view as the earth rotates (cf. altazimuth mount). **equatorial telescope** an astronomical telescope on an equatorial mount. □ **equatorially** adv.

Equatorial Guinea a small country of West Africa on the Gulf of Guinea, comprising several offshore islands and a coastal settlement between Cameroon and Gabon; pop. (est. 1991) 426,000; languages, Spanish (official), local Niger-Congo languages, pidgin; capital, Malabo (on the island of Bioko). Formerly a Spanish colony, the country became fully independent in 1968. It is the only independent Spanish-speaking state in the continent of Africa. □ **Equatorial Guinean** adj. & n.

equerry /ˈekwərɪ, ɪˈkwerɪ/ n. (pl. **-ies**) **1** an officer of the British royal household attending members of the royal family. **2** hist. an officer of a prince's or noble's household having charge over the horses. [earlier esquiry f. OF esquierie company of squires, prince's stables, f. OF esquier ESQUIRE: perh. assoc. with L equus horse]

equestrian /ɪˈkwestrɪən/ adj. & n. ● adj. **1** of or relating to horses and horse-riding. **2** on horseback. ● n. (fem. **equestrienne** /ɪˌkwestrɪˈen/) a rider or performer on horseback. [L equestris f. eques horseman, knight, f. equus horse]

equestrianism /ɪˈkwestrɪəˌnɪz(ə)m/ n. the skill or sport of horse-riding. As an Olympic sport it is divided into three categories: showjumping, dressage, and three-day eventing (combining showjumping, dressage, and cross-country riding). It is governed by the International Equestrian Federation. Carriage driving is a separate form of equestrianism.

equi- /ˈiːkwɪ, ˈek-/ comb. form equal. [L aequi- f. aequus equal]

equiangular /ˌiːkwɪˈæŋɡjʊlə(r), ˌek-/ adj. having equal angles.

equidistant /ˌiːkwɪˈdɪstənt, ˌek-/ adj. at equal distances. □ **equidistantly** adv.

equilateral /ˌiːkwɪˈlætərəl, ˌek-/ adj. having all its sides equal in length.

equilibrate /ɪˈkwɪlɪˌbreɪt, ˌiːkwɪˈlaɪbreɪt/ v. **1** tr. cause (two things) to balance. **2** intr. be in equilibrium; balance. □ **equilibration** /ɪˌkwɪlɪˈbreɪʃ(ə)n, ˌiːkwɪlaɪ-/ n. **equilibrator** /ɪˈkwɪlɪˌbreɪtə(r)/ n. [LL aequilibrare aequilibrat- (as EQUI-, libra balance)]

equilibrist /ɪˈkwɪlɪbrɪst/ n. an acrobat, esp. on a high rope.

equilibrium /ˌiːkwɪˈlɪbrɪəm, ˌek-/ n. (pl. **equilibria** /-rɪə/ or **equilibriums**) **1** a state of physical balance. **2** a state of mental or emotional equanimity. **3** a state in which the energy in a system is evenly distributed and forces, influences, etc., balance each other. [L (as EQUI-, libra balance)]

equine /ˈekwaɪn/ adj. of or like a horse. [L equinus f. equus horse]

equinoctial /ˌiːkwɪˈnɒkʃ(ə)l, ˌek-/ adj. & n. ● adj. **1** happening at or near the time of an equinox (equinoctial gales). **2** of or relating to equal day and night. **3** at or near the (terrestrial) equator. ● n. (in full **equinoctial line**) = celestial equator. □ **equinoctial point** the point at which the ecliptic cuts the celestial equator (twice each year at an equinox) (see also PRECESSION). **equinoctial year** see YEAR. [ME f. OF equinoctial or L aequinoctialis (as EQUINOX)]

equinox /ˈiːkwɪˌnɒks, ˈek-/ n. **1** the time or date (twice each year) at which the sun crosses the celestial equator, when day and night are of equal length. **2** = equinoctial point. □ **autumn** (or **autumnal**) **equinox 1** the equinox occurring at the start of autumn, about 22 Sept. in the northern hemisphere and about 20 Mar. in the southern hemisphere. **2** Astron. the equinox occurring in September. **spring** (or **vernal**) **equinox 1** the equinox occurring at the start of spring, about 20 Mar. in the northern hemisphere and about 22 Sept. in the southern hemisphere. **2** Astron. the equinox occurring in March (see also ARIES, FIRST POINT OF). [ME f. OF equinoxe or med.L equinoxium for L aequinoctium (as EQUI-, nox noctis night)]

equip /ɪˈkwɪp/ v.tr. (**equipped**, **equipping**) supply with what is needed, fit out. □ **equipper** n. [F équiper, prob. f. ON skipa to man (a ship) f. skip SHIP]

equipage /ˈekwɪpɪdʒ/ n. **1 a** requisites for an undertaking. **b** an outfit for a special purpose. **2** a carriage and horses with attendants. [F équipage (as EQUIP)]

equipment /ɪˈkwɪpmənt/ n. **1** the necessary articles, clothing, etc., for a purpose. **2** the process of equipping or being equipped. [F équipement (as EQUIP)]

equipoise /ˈekwɪˌpɔɪz/ n. & v. ● n. **1** equilibrium; a balanced state. **2** a counterbalancing thing. ● v.tr. counterbalance.

equipollent /ˌiːkwɪˈpɒlənt, ˌek-/ adj. & n. ● adj. **1** equal in power, force, etc. **2** practically equivalent. ● n. an equipollent thing. □ **equipollence** n. **equipollency** n. [ME f. OF equipolent f. L aequipollens -entis of equal value (as EQUI-, pollere be strong)]

equipotential /ˌiːkwɪpəˈtenʃ(ə)l, ˌek-/ adj. & n. Physics ● adj. (of a surface or line) having the potential of a force the same or constant at all its points. ● n. an equipotential line or surface.

equiprobable /ˌiːkwɪˈprɒbəb(ə)l, ˌek-/ adj. Logic equally probable. □ **equiprobability** /-ˌprɒbəˈbɪlɪtɪ/ n.

equisetum /ˌekwɪˈsiːtəm/ n. (pl. **equiseta** /-tə/, **-tums**) Bot. a horsetail. [mod.L f. L equus horse + saeta bristle]

equitable /ˈekwɪtəb(ə)l/ adj. **1** fair, just. **2** Law valid in equity as distinct from law. □ **equitableness** n. **equitably** adv. [F équitable (as EQUITY)]

equitation /ˌekwɪˈteɪʃ(ə)n/ n. the art of horsemanship and horse-riding. [F équitation or L equitatio f. equitare ride a horse f. eques equitis horseman f. equus horse]

equity /ˈekwɪtɪ/ n. (pl. **-ies**) **1** fairness. **2** the application of general principles of justice to correct or supplement the law. **3 a** the value of the shares issued by a company. **b** (in pl.) stocks and shares not bearing fixed interest. **4** the net value of a mortgaged property after the deduction of charges. **5** (**Equity**) (in the UK) the actors' trade union. [ME f. OF equité f. L aequitas -tatis f. aequus fair]

equivalent /ɪˈkwɪvələnt/ adj. & n. ● adj. **1** (often foll. by to) equal in value, amount, importance, etc. **2** corresponding. **3** (of words) having the same meaning. **4** having the same result. **5** Chem. (of a substance) equal in combining or displacing capacity. ● n. **1** an equivalent thing, amount, word, etc. **2** (in full **equivalent weight**) Chem. the weight of a substance that can combine with or displace one gram of hydrogen or eight grams of oxygen. □ **equivalence** n. **equivalency** n. **equivalently** adv. [ME f. OF f. LL aequivalere (as EQUI-, valere be worth)]

equivocal /ɪˈkwɪvək(ə)l/ adj. **1** of double or doubtful meaning; ambiguous. **2** of uncertain nature. **3** (of a person, character, etc.) questionable, suspect. □ **equivocally** adv. **equivocalness** n. **equivocality** /ɪˌkwɪvəˈkælɪtɪ/ n. [LL aequivocus (as EQUI-, vocare call)]

equivocate /ɪˈkwɪvəˌkeɪt/ v.intr. use ambiguity to conceal the truth, prevaricate. □ **equivocator** n. **equivocatory** adj. **equivocacy** /-kəsɪ/ n. **equivocation** /ɪˌkwɪvəˈkeɪʃ(ə)n/ n. [ME f. LL aequivocare (as EQUIVOCAL)]

equivoque /ˈekwɪˌvəʊk/ n. (also **equivoke**) **1** a pun, wordplay. **2** ambiguity. [ME in the sense 'equivocal' f. OF equivoque or LL aequivocus EQUIVOCAL]

Equuleus /eˈkwʊlɪəs/ Astron. a small constellation (the Foal or Little Horse), perhaps representing the brother of Pegasus. It has no bright stars. [L, = little horse, f. equus horse]

ER abbr. **1** Queen Elizabeth. **2** King Edward. [L Elizabetha Regina, Edwardus Rex]

Er symb. Chem. the element erbium.

er /ɜː(r)/ int. expressing hesitation or a pause in speech. [imit.]

-er[1] /ə(r)/ suffix forming nouns from nouns, adjectives, and many verbs, denoting: **1** a person, animal, or thing that performs a specified action or activity (cobbler; lover; executioner; poker; computer; eye-opener). **2** a person or thing that has a specified attribute or form (foreigner; four-wheeler; second-rater). **3** a person concerned with a specified thing or subject (hatter; geographer). **4** a person belonging to a specified place or group (villager; New Zealander; sixth-former). [orig. 'one who has to do with': OE -ere f. Gmc]

-er[2] /ə(r)/ suffix forming the comparative of adjectives (wider; hotter) and adverbs (faster). [OE -ra (adj.), -or (adv.) f. Gmc]

-er[3] /ə(r)/ suffix used in slang formations usu. distorting the root word (rugger; soccer). [prob. an extension of -ER[1]]

-er[4] /ə(r)/ suffix forming iterative and frequentative verbs (blunder; glimmer; twitter). [OE -erian, -rian f. Gmc]

-er[5] /ə(r)/ suffix **1** forming nouns and adjectives through OF or AF, corresponding to: **a** L -aris (sampler) (cf. -AR[1]). **b** L -arius, -arium (butler; carpenter; danger). **c** (through OF -eüre) L -atura or (through OF -eör) L -atorium (see COUNTER[1]). **2** = -OR[1].

-er[6] /ə(r)/ *suffix* esp. *Law* forming nouns denoting verbal action or a document effecting this (*cesser; disclaimer; misnomer*). ¶ The same ending occurs in *dinner* and *supper*. [AF infin. ending of verbs]

era /'ɪərə/ *n.* **1** a system of chronology reckoning from a noteworthy event (*the Christian era*). **2** a large distinct period of time, esp. regarded historically (*the pre-Roman era*). **3** a date at which an era begins. **4** *Geol.* a major division of time. [LL *aera* number expressed in figures (pl. of *aes aeris* money, treated as fem. sing.)]

eradicate /ɪ'rædɪˌkeɪt/ *v.tr.* root out; destroy completely; get rid of. □ **eradicator** *n.* **eradicable** /-kəb(ə)l/ *adj.* **eradication** /ɪˌrædɪ'keɪʃ(ə)n/ *n.* [ME f. L *eradicare* tear up by the roots (as E-, *radix -icis* root)]

erase /ɪ'reɪz/ *v.tr.* **1** rub out; obliterate. **2** remove all traces of (*erased it from my memory*). **3** remove recorded material from (a magnetic tape or medium). □ **erasable** *adj.* **erasure** /ɪ'reɪʒə(r)/ *n.* [L *eradere eras-* (as E-, *radere* scrape)]

eraser /ɪ'reɪzə(r)/ *n.* a thing that erases, esp. a piece of rubber or plastic used for removing pencil and ink marks.

Erasmus /ɪ'ræzməs/, Desiderius (Dutch name Gerhard Gerhards) (c.1469–1536), Dutch humanist and scholar. During his lifetime he was the most famous scholar in Europe and the first there to achieve renown through the printed word. He published his own Greek edition of the New Testament (1516), followed by a Latin translation, and paved the way for the Reformation with his satires on the Church, including the *Colloquia Familiaria* (1518). However, his abhorrence of violence prevented him from joining the Protestants and he condemned Luther in *De Libero Arbitrio* (1523).

Erastianism /ɪ'ræstɪəˌnɪz(ə)m/ *n.* the doctrine that the state should have supremacy over the Church in ecclesiastical matters (wrongly attributed to Erastus). The ascendancy of the secular power was defended by the Anglican churchman Richard Hooker in his *Ecclesiastical Polity* (1594), and the principle remains in a modified form in Britain, where Parliament, though its members may profess any or no religion, retains the right to legislate on religious matters concerning the established Church. □ **Erastian** *n. & adj.*

Erastus /ɪ'ræstəs/ (Swiss name Thomas Lieber; also called Liebler or Lüber) (1524–83), Swiss theologian and physician. He was professor of medicine at Heidelberg University from 1558. A follower of Zwingli, he opposed the imposition of a Calvinistic system of church government in Heidelberg because of the Calvinists' excessive use of excommunication. The doctrine of Erastianism was later attributed to him, although his views were less extreme.

Erato /'erəˌtəʊ/ *Gk & Rom. Mythol.* the Muse of lyric poetry and hymns. [Gk, = lovely]

Eratosthenes /ˌerə'tɒsθəˌniːz/ (c.275–194 BC), Greek scholar, geographer, and astronomer. He was a pupil of Callimachus and head of the library at Alexandria. Active in the fields of literary criticism and chronology, he was also the first systematic geographer of antiquity. He accurately calculated the circumference of the earth by measuring the angle of the sun's rays at different places at the same time, and he attempted (less successfully) to determine the magnitude and distance of the sun and of the moon.

erbium /'ɜːbɪəm/ *n.* a soft silvery-white metallic chemical element (atomic number 68; symbol **Er**). A member of the lanthanide series, erbium was discovered in 1843 by the Swedish chemist C. G. Mosander, who separated it from yttrium and terbium. It has few commercial uses. [f. *Ytterby* in Sweden: see YTTERBIUM]

ere /eə(r)/ *prep. & conj. poet.* or *archaic* before (of time) (*ere noon; ere they come*). [OE *ǣr* f. Gmc]

Erebus /'erəbəs/ *Gk Mythol.* the primeval god of darkness, son of Chaos.

Erebus, Mount a volcanic peak on Ross Island, Antarctica. Rising to 3,794 m (12,452 ft), it is the world's most southerly active volcano. It is named after the *Erebus*, the ship of Sir James Ross's expedition to the Antarctic.

Erech see URUK.

Erechtheum /ɪ'rekθɪəm/ a marble temple built on the Acropolis in Athens c.421–406 BC, with shrines to Athene, Poseidon, and Erechtheus, a legendary king of Athens. A masterpiece of the Ionic order, it is most famous for its southern portico, in which the entablature is supported by six caryatids.

erect /ɪ'rekt/ *adj. & v.* ● *adj.* **1** upright, vertical. **2** (of the penis, clitoris, or nipples) enlarged and rigid, esp. in sexual excitement. **3** (of hair) bristling, standing up from the skin. ● *v.tr.* **1** raise; set upright. **2** build.

3 establish (*erect a theory*). □ **erectable** *adj.* **erectly** *adv.* **erectness** *n.* **erector** *n.* [ME f. L *erigere erect-* set up (as E-, *regere* direct)]

erectile /ɪ'rektaɪl/ *adj.* that can be erected or become erect. □ **erectile tissue** *Physiol.* animal tissue that is capable of becoming erect, esp. with sexual excitement. [F *érectile* (as ERECT)]

erection /ɪ'rekʃ(ə)n/ *n.* **1** the act or an instance of erecting; the state of being erected. **2** a building or structure. **3 a** an enlarged and erect state of erectile tissue, esp. of the penis. **b** an occurrence of this. [F *érection* or L *erectio* (as ERECTILE)]

E-region var. of E-LAYER.

eremite /'erɪˌmaɪt/ *n.* a hermit or recluse (esp. Christian). □ **eremitic** /ˌerɪ'mɪtɪk/ *adj.* **eremitical** *adj.* [ME f. OF, var. of *hermite, ermite* HERMIT]

erethism /'erɪˌθɪz(ə)m/ *n.* **1** an excessive sensitivity to stimulation of any part of the body, esp. the sexual organs. **2** a state of abnormal mental excitement or irritation. [F *éréthisme* f. Gk *erethismos* f. *erethizō* irritate]

Erfurt /'eəfʊət/ an industrial city in central Germany, capital of Thuringia; pop. (1991) 204,910.

erg[1] /ɜːg/ *n. Physics* a unit of work or energy, equal to the work done by a force of one dyne when its point of application moves one centimetre in the direction of action of the force. One erg is equal to 10^{-7} joule. [Gk *ergon* work]

erg[2] /ɜːg/ *n.* (pl. **ergs** or **areg** /'ɑːreg/) an area of shifting sand-dunes in the Sahara. [F f. Arab. *'irj*]

ergative /'ɜːgətɪv/ *n. & adj. Gram.* ● *n.* **1** (in some languages) a case of nouns that identifies the doer of an action as the object rather than the subject of a verb. **2** a noun in this case. ● *adj.* **1** of or in this case. **2** denoting a language in which the object of a verb is typically the doer of an action and the subject is typically the recipient of the action. Linguists distinguish ergative languages such as Basque from accusative languages such as Latin. □ **ergativity** /ˌɜːgə'tɪvɪtɪ/ *n.* [Gk *ergatēs* worker + -IVE]

ergo /'ɜːgəʊ/ *adv.* therefore. [L]

ergocalciferol /ˌɜːgəʊkæl'sɪfəˌrɒl/ *n.* = CALCIFEROL. [ERGOSTEROL + CALCIFEROL]

ergonomics /ˌɜːgə'nɒmɪks/ *n.* the study of the efficiency of persons in their working environment. □ **ergonomic** *adj.* **ergonomist** /ɜː'gɒnəmɪst/ *n.* [Gk *ergon* work: cf. ECONOMICS]

ergosterol /ɜː'gɒstəˌrɒl/ *n. Biochem.* a plant sterol that is found in ergot and many other fungi and produces vitamin D_2 when irradiated with ultraviolet light. [ERGOT, after CHOLESTEROL]

ergot /'ɜːgət/ *n.* **1** a disease of rye and other cereals caused by the fungus *Claviceps purpurea*, the presence of which can cause food poisoning. **2 a** this fungus. **b** the dried spore-containing structures of this, used as a source of medicine to aid childbirth. **3** a small horny protuberance on the back of a horse's fetlock. [F f. OF *argot* cock's spur, from the appearance produced]

ergotism /'ɜːgəˌtɪz(ə)m/ *n.* poisoning produced by eating food affected by ergot.

erica /'erɪkə/ *n.* a hardy shrub of the genus *Erica*, with small leathery leaves and bell-like flowers; a heath or heather. □ **ericaceous** /ˌerɪ'keɪʃəs/ *adj.* [L f. Gk *ereikē* heath]

Ericsson[1] /'erɪks(ə)n/, John (1803–89), Swedish engineer. His inventions included a steam railway locomotive for the 1829 competition which was won by Stephenson's *Rocket*, and the marine screw propeller (1836). He then moved to the US, where he built the ironclad *Monitor*, which was the first ship to have a revolving armoured turret and was used in a battle on the Union side in the American Civil War. Ericsson was also a pioneer of solar energy, constructing a steam pump supplied from a boiler heated by a concentrating mirror.

Ericsson[2] /'erɪks(ə)n/, Leif (also **Ericson, Eriksson**), Norse explorer, son of Eric the Red. He sailed westward from Greenland (c.1000) and reputedly discovered land (variously identified as Labrador, Newfoundland, or New England), which he named Vinland because of the vines he claimed to have found growing there.

Eric the Red /'erɪk/ (c.940–c.1010), Norse explorer. He left Iceland in 982 in search of land to the west, exploring Greenland and establishing a Norse settlement there in 986.

Eridanus /e'rɪdənəs/ *Astron.* a long straggling constellation (the River), said to represent the river into which Phaethon fell when struck by Zeus' thunderbolt. Its most southerly point is marked by the bright star Achernar.

Erie, Lake /ˈɪərɪ/ one of the five Great Lakes of North America, situated on the border between Canada and the US. It is linked to Lake Huron by the Detroit River and to Lake Ontario by the Welland Ship Canal and the Niagara River, which is its only natural outlet. Cleveland and Buffalo lie on its shores.

erigeron /ɪˈrɪgəˌrɒn/ n. a hardy composite herbaceous plant of the genus *Erigeron*, with daisy-like flowers. [Gk *ērigerōn* f. *ēri* early + *gerōn* old man, because some species bear grey down]

Eriksson, Leif, see ERICSSON².

Erin /ˈɛrɪn, ˈɪərɪn/ n. archaic or poet. Ireland. [Ir.]

Erinys /eˈrɪnɪs/ n. (pl. **Erinyes** /-nɪˌiːz/) Gk Mythol. a Fury. [Gk]

eristic /eˈrɪstɪk/ adj. & n. ● adj. **1** of or characterized by disputation. **2** (of an argument or arguer) aiming at winning rather than at reaching the truth. ● n. **1** the practice of disputation. **2** an exponent of disputation. □ **eristically** adv. [Gk *eristikos* f. *erizō* wrangle f. *eris* strife]

Eritrea /ˌɛrɪˈtreɪə/ an independent state in NE Africa, on the Red Sea; pop. (est. 1991) 3,500,000; language, Tigre and other Cushitic languages; capital, Asmara. It was an Italian colony from 1890 to 1952, when it became part of Ethiopia. After a long guerrilla war it became internally self-governing in 1991 and fully independent in 1993. □ **Eritrean** adj. & n.

erk /ɜːk/ n. Brit. sl. **1** a naval rating. **2** an aircraftman. **3** a disliked person. [20th c.: orig. unkn.]

Erlanger /ˈɜːlæŋə(r)/, Joseph (1874–1965), American physiologist. He worked mainly in cardiac physiology, designing a sphygmomanometer to study the components of the pulse wave and examining the conduction of impulses in the heart. Using an oscilloscope with H. Gasser, Erlanger found that the velocity of a nerve impulse is proportional to the diameter of the fibre. Gasser and Erlanger shared a Nobel Prize for this in 1944.

erl-king /ˈɜːlkɪŋ/ n. Germanic Mythol. a bearded giant or goblin who lures little children to the land of death. [G *Erlkönig* alder-king, mistransl. Danish *ellerkonge* king of the elves]

ERM see EXCHANGE RATE MECHANISM.

ermine /ˈɜːmɪn/ n. (pl. same or **ermines**) **1** the stoat, esp. when in its white winter fur. **2** its white fur, used as trimming for the robes of judges, peers, etc. **3** Heraldry a white fur marked with black spots. □ **ermined** adj. [ME f. OF (h)*ermine* prob. f. med.L (*mus*) *Armenius* Armenian (mouse)]

ern US var. of ERNE.

-ern /ən/ suffix forming adjectives (*northern*). [OE -*erne* f. Gmc]

erne /ɜːn/ n. (US **ern**) poet. a sea eagle. [OE *earn* f. Gmc]

Ernie /ˈɜːnɪ/ (in the UK) a device used (from 1956) for drawing prize-winning numbers of Premium Bonds. [initial letters of *electronic random number indicator equipment*]

Ernst /ɜːnst/, Max (1891–1976), German artist. In 1919 he became leader of the Cologne Dada group, and was responsible for adapting the techniques of collage and photomontage to surrealist uses. He is best known, however, for the surrealist paintings he did after moving to Paris, including *L'Eléphant de Célèbes* (1921). In 1925 he developed the technique of frottage, using such surfaces as leaves and wood grain, as in *Habit of Leaves* (1925). He spent much of the 1940s in the US but later returned to France, adopting French citizenship in 1958.

erode /ɪˈrəʊd/ v. tr. & intr. wear away, destroy or be destroyed gradually. **2** tr. Med. (of ulcers etc.) destroy (tissue) little by little. □ **erodible** adj. [F *éroder* or L *erodere eros-* (as E-, *rodere ros-* gnaw)]

erogenous /ɪˈrɒdʒɪnəs/ adj. **1** (esp. of a part of the body) sensitive to sexual stimulation. **2** giving rise to sexual desire or excitement. [as EROTIC +-GENOUS]

Eros /ˈɪərɒs/ **1** Gk Mythol. the god of love (see CUPID). **2** (in Freudian psychology) the urge for self-preservation and sexual pleasure. **3** Astron. asteroid 433, discovered in 1898, which comes at times nearer to earth than any celestial body except the moon. **4** a winged statue of an archer over the fountain in Piccadilly Circus, London, made by Sir Alfred Gilbert (1854–1934), erected as a memorial to the philanthropist the Earl of Shaftesbury and unveiled in 1893. [Gk *erōs*, = sexual love]

erosion /ɪˈrəʊʒ(ə)n/ n. **1** Geol. the wearing away of the earth's surface by the action of water, wind, etc. **2** the act or an instance of eroding; the process of being eroded. □ **erosional** adj. **erosive** /ɪˈrəʊsɪv/ adj. [F *érosion* f. L *erosio* (as ERODE)]

erotic /ɪˈrɒtɪk/ adj. of or causing sexual love, esp. tending to arouse sexual desire or excitement. □ **erotically** adv. [F *érotique* f. Gk *erōtikos* f. *erōs erōtos* sexual love]

erotica /ɪˈrɒtɪkə/ n.pl. erotic literature or art.

eroticism /ɪˈrɒtɪˌsɪz(ə)m/ n. **1** erotic nature or character. **2** the use of or response to erotic images or stimulation.

erotism /ˈerəˌtɪz(ə)m/ n. sexual desire or excitement; eroticism.

eroto- /ɪˈrɒtəʊ/ comb. form erotic, eroticism. [Gk *erōs erōtos* sexual love]

erotogenic /ɪˌrɒtəˈdʒenɪk/ adj. (also **erotogenous** /ˌerəˈtɒdʒɪnəs/) = EROGENOUS.

erotology /ˌerəˈtɒlədʒɪ/ n. the study of sexual love.

erotomania /ɪˌrɒtəʊˈmeɪnɪə/ n. **1** excessive or abnormal erotic desire. **2** a preoccupation with sexual passion. □ **erotomaniac** /-nɪˌæk/ n.

err /ɜː(r)/ v.intr. **1** be mistaken or incorrect. **2** do wrong; sin. □ **err on the right side** act so that the least harmful of possible errors is the most likely to occur. **err on the side of** act with a specified bias (*errs on the side of generosity*). [ME f. OF *errer* f. L *errare* stray: rel. to Goth. *airzei* error, *airzjan* lead astray]

errand /ˈerənd/ n. **1** a short journey, esp. on another's behalf, to take a message, collect goods, etc. **2** the object of such a journey. □ **errand of mercy** a journey to relieve suffering etc. [OE *ærende* f. Gmc]

errant /ˈerənt/ adj. **1** erring; deviating from an accepted standard. **2** literary or archaic travelling in search of adventure (*knight errant*). □ **errancy** n. (in sense 1). **errantry** n. (in sense 2). [ME: sense 1 formed as ERR: sense 2 f. OF *errer* ult. f. LL *itinerare* f. *iter* journey]

erratic /ɪˈrætɪk/ adj. **1** inconsistently variable in conduct, opinions, etc.; unpredictable, eccentric. **2** uncertain in movement. □ **erratic block** Geol. a large rock carried from a distance by glacial action. □ **erratically** adv. [ME f. OF *erratique* f. L *erraticus* (as ERR)]

erratum /ɪˈrɑːtəm/ n. (pl. **errata** /-tə/) an error in printing or writing, esp. (in pl.) a list of corrected errors attached to a book etc. [L, neut. past part. (as ERR)]

Er Rif see RIF MOUNTAINS.

erroneous /ɪˈrəʊnɪəs/ adj. incorrect; arising from error. □ **erroneously** adv. **erroneousness** n. [ME f. OF *erroneus* or L *erroneus* f. *erro -onis* vagabond (as ERR)]

error /ˈerə(r)/ n. **1** a mistake. **2** the condition of being wrong in conduct or judgement (*led into error*). **3** a wrong opinion or judgement. **4** the amount by which something is incorrect or inaccurate in a calculation or measurement. □ **errorless** adj. [ME f. OF *errour* f. L *error -oris* (as ERR)]

ersatz /ˈɜːzæts, ˈeəz-/ adj. & n. ● adj. substitute, imitation (esp. of inferior quality). ● n. an ersatz thing. [G, = replacement]

Erse /ɜːs/ n. & adj. archaic ● n. the Gaelic language, esp. Irish Gaelic. ● adj. of or relating to (esp. Irish) Gaelic. [early Sc. form of IRISH]

erst /ɜːst/ adv. archaic formerly; of old. [OE *ærest* superl. of *ær:* see ERE]

erstwhile /ˈɜːstwaɪl/ adj. & adv. ● adj. former, previous. ● adv. archaic = ERST.

Erté /ˈeəteɪ/ (born Romain de Tirtoff) (1892–1990), Russian-born French fashion designer and illustrator. From 1912 he worked in Paris as a fashion designer, and during the First World War his garments became internationally famous through his decorative magazine illustrations. In the 1920s he became a noted art deco designer, moving into the design of household items and fabrics and creating elaborate *tableaux vivants* for Broadway shows such as the Ziegfeld Follies.

Ertebølle /ˈeətəˌbɜːlə/ adj. (also **Ertebolle**) Archaeol. of or relating to a late mesolithic culture in the western Baltic, the final phases of which, *c.*5000 BC, show neolithic influence (permanent coastal fishing and collecting sites, use of skin boats). [*Ertebølle* in Jutland, Denmark]

erubescent /ˌeruˈbes(ə)nt/ adj. reddening, blushing. [L *erubescere* (as E-, *rubescere* f. *rubere* be red)]

eructation /ˌiːrʌkˈteɪʃ(ə)n/ n. the act or an instance of belching. [L *eructatio* f. *eructare* (as E-, *ructare* belch)]

erudite /ˈeruˌdaɪt/ adj. **1** (of a person) learned. **2** (of writing etc.) showing great learning. □ **eruditely** adv. **erudition** /ˌeruˈdɪʃ(ə)n/ n. [ME f. L *eruditus* past part. of *erudire* instruct, train (as E-, *rudis* untrained)]

erupt /ɪˈrʌpt/ v.intr. **1** break out suddenly or dramatically. **2** (of a volcano) become active and eject lava etc. **3 a** (of a rash, boil, etc.) appear on the skin. **b** (of the skin) produce a rash etc. **4** (of the teeth) break through the gums in normal development. □ **eruptive** adj. **eruption** /ɪˈrʌpʃ(ə)n/ n. [L *erumpere erupt-* (as E-, *rumpere* break)]

-ery /ərɪ/ suffix (also **-ry** /rɪ/) forming nouns denoting: **1** a class or kind (*greenery; machinery; citizenry*). **2** an occupation or employment; state, condition, or behaviour (*archery; dentistry; slavery; bravery; mimicry*). **3** a place set aside for an activity or characterized by a gathering together of things, animals, etc. (*brewery*; *orangery*). **4** often *derog.* all that has to do with (*knavery*; *popery*; *tomfoolery*). [ME, from or after F -*erie*, -*ere* ult. f. L -*ario*-, -*ator*]

erysipelas /ˌerɪˈsɪpɪləs/ *n.* *Med.* a streptococcal infection producing inflammation and a deep red colour on the skin, esp. of the face and scalp. [ME f. L f. Gk *erusipelas*, perh. rel. to *eruthros* red + a root *pel*- skin]

erythema /ˌerɪˈθiːmə/ *n.* a superficial reddening of the skin, usu. in patches. □ **erythemal** *adj.* **erythematic** /ˌerɪθɪˈmætɪk/ *adj.* [mod.L f. Gk *eruthēma* f. *eruthainō* be red f. *eruthros* red]

erythrism /ˈerɪˌθrɪz(ə)m/ *n.* abnormal red coloration, esp. in a bird or animal. [ERYTHRO- + -ISM]

erythro- /ɪˈrɪθrəʊ/ *comb. form* red. [Gk *eruthros* red]

erythroblast /ɪˈrɪθrəʊˌblæst/ *n.* an immature erythrocyte. [G]

erythrocyte /ɪˈrɪθrəʊˌsaɪt/ *n.* the predominant cell in the blood of vertebrates, containing the red pigment haemoglobin and transporting oxygen and carbon dioxide to and from the tissues. The erythrocytes of mammals are usually small biconcave discs without nuclei or other organelles. Also called *red (blood) cell* or *corpuscle*. (See also BLOOD.) □ **erythrocytic** /ɪˌrɪθrəʊˈsɪtɪk/ *adj.*

erythroid /ˈerɪˌθrɔɪd/ *adj.* of or relating to erythrocytes.

Erzgebirge /ˈeətsɡəˌbɪəɡə/ (also called the *Ore Mountains*) a range of mountains on the border between Germany and the Czech Republic.

Erzurum /ˈeəzʊˌrʊm/ a city in NE Turkey, capital of a mountainous province of the same name; pop. (1990) 242,400.

Es *symb.* *Chem.* the element einsteinium.

-es[1] /ɪz/ *suffix* forming plurals of nouns ending in sibilant sounds (such words in -*e* dropping the *e*) (*kisses*; *cases*; *boxes*; *churches*). [var. of -s[1]]

-es[2] /ɪz, z/ *suffix* forming the 3rd person sing. present of verbs ending in sibilant sounds (such words in -*e* dropping the *e*) and ending in -*o* (but not -*oo*) (*goes*; *places*; *pushes*). [var. of -s[2]]

ESA *abbr.* **1** European Space Agency. **2** Environmentally Sensitive Area.

Esaki /eˈzɑːkɪ/, Leo (b.1925), Japanese physicist. He investigated and pioneered the development of quantum-mechanical tunnelling of electrons in semiconductor devices, and designed the tunnel diode. These (also known as *Esaki diodes*), small and fast in operation, are now widespread in electronic devices. Esaki shared the Nobel Prize for physics in 1973.

Esau /ˈiːsɔː/ (in the Bible) the elder of the twin sons of Isaac and Rebecca, who sold his birthright to his brother Jacob and was tricked out of his father's blessing by his brother (Gen. 25, 27). He is the traditional ancestor of the Edomites.

Esbjerg /ˈesbjɜːɡ/ a port in Denmark, on the west coast of Jutland; pop. (1990) 81,500. It has ferry links with Britain and the Faeroe Islands.

ESC see ECONOMIC AND SOCIAL COMMITTEE.

escadrille /ˌeskəˈdrɪl/ *n.* a French squadron of aeroplanes. [F]

escalade /ˌeskəˈleɪd/ *n.* the scaling of fortified walls with ladders, as a military attack. [F f. Sp. *escalada*, -*ado* f. med.L *scalare* f. *scala* ladder]

escalate /ˈeskəˌleɪt/ *v.* **1** *intr.* & *tr.* increase or develop (usu. rapidly) by stages. **2** *tr.* cause (an action, activity, or process) to become more intense. □ **escalation** /ˌeskəˈleɪʃ(ə)n/ *n.* [back-form. f. ESCALATOR]

escalator /ˈeskəˌleɪtə(r)/ *n.* a moving staircase consisting of a circulating belt forming steps. The name was first applied (originally as a proprietary term) to a moving stairway at the Paris Exposition of 1900, although a similar device had been invented earlier in the US. By 1896-8 such staircases were being installed in department stores. [f. the stem of *escalade* 'climb a wall by ladder' + -ATOR]

escallonia /ˌeskəˈləʊnɪə/ *n.* an evergreen shrub of the genus *Escallonia*, native to South America, with pink or white flowers. [*Escallon*, 18th-c. Sp. traveller]

escallop /ɪˈskæləp/ *n.* **1** = SCALLOP 1, 2. **2** = ESCALOPE. **3** (in *pl.*) = SCALLOP 3. **4** *Heraldry* a scallop shell as a device. [formed as ESCALOPE]

escalope /ˈeskəˌlɒp/ *n.* a thin slice of meat without any bone, esp. from a leg of veal. [F (in OF = shell): see SCALLOP]

escapade /ˈeskəˌpeɪd/ *n.* a piece of daring or reckless behaviour. [F f. Prov. *escapada* (as ESCAPE)]

escape /ɪˈskeɪp/ *v.* & *n.* ● *v.* **1** *intr.* (often foll. by *from*) get free of the restriction or control of a place, person, etc. **2** *intr.* (of a gas, liquid, etc.) leak from a container or pipe etc. **3** *intr.* succeed in avoiding danger, punishment, etc.; get off safely. **4** *tr.* get completely free of (a person, grasp, etc.). **5** *tr.* avoid or elude (a commitment, danger, etc.). **6** *tr.* elude the notice or memory of (*nothing escapes you*; *the name escaped me*). **7** *tr.*

(of words etc.) issue unawares from (a person, a person's lips). ● *n.* **1** the act or an instance of escaping; avoidance of danger, injury, etc. **2** the state of having escaped (*was a narrow escape*). **3** a means of escaping (often *attrib.*: *escape hatch*). **4** a leakage of gas etc. **5** a temporary relief from reality or worry. **6** a garden plant running wild. **7** (in full **escape key**) *Computing* a key which either ends the current operation, or converts subsequent characters to a control sequence (*escape routine*, *escape sequence*). □ **escape clause** *Law* a clause specifying the conditions under which a contracting party is free from an obligation. **escape road** a road for a vehicle to turn into if unable to negotiate a bend, descent, etc., safely (esp. on a racetrack). **escape velocity** the minimum velocity needed to escape from the gravitational field of a body. **escape wheel** a toothed wheel in the escapement of a watch or clock. □ **escapable** *adj.* **escaper** *n.* [ME f. AF, ONF *escaper* ult. f. med.L (as EX-[1], *cappa* cloak)]

escapee /ˌɪskeɪˈpiː/ *n.* a person, esp. a prisoner, who has escaped.

escapement /ɪˈskeɪpmənt/ *n.* **1** the part of a clock or watch that connects and regulates the motive power. **2** the part of the mechanism in a piano that enables the hammer to fall back immediately it has struck the string. **3** the mechanism in a typewriter that controls the leftward motion of the carriage. **4** *archaic* a means of escape. [F *échappement* f. *échapper* ESCAPE]

escapism /ɪˈskeɪpɪz(ə)m/ *n.* the tendency to seek distraction and relief from reality, esp. in the arts or through fantasy. □ **escapist** *n.* & *adj.*

escapology /ˌeskəˈpɒlədʒɪ/ *n.* the methods and techniques of escaping from confinement, esp. as a form of entertainment. □ **escapologist** *n.*

escargot /eˈskɑːɡəʊ/ *n.* an edible snail. [F]

escarpment /ɪˈskɑːpmənt/ *n.* (also **escarp**) *Geol.* a long steep slope at the edge of a plateau etc. [F *escarpement* f. *escarpe* SCARP]

Escaut /ɛsko/ the French name for the SCHELDT.

-esce /es/ *suffix* forming verbs, usu. initiating action (*effervesce*; *fluoresce*). [from or after L -*escere*]

-escent /ˈes(ə)nt/ *suffix* forming adjectives denoting the beginning of a state or action (*effervescent*; *fluorescent*). □ **-escence** *suffix* forming nouns. [from or after F -*escent* or L -*escent*-, pres. part. stem of verbs in -*escere*]

eschatology /ˌeskəˈtɒlədʒɪ/ *n.* **1** the part of theology concerned with death and final destiny. **2 a** beliefs about the destiny of humankind and of the world. **b** the study of these. □ **eschatologist** *n.* **eschatological** /-təˈlɒdʒɪk(ə)l/ *adj.* [Gk *eskhatos* last + -LOGY]

escheat /ɪsˈtʃiːt/ *n.* & *v. hist.* ● *n.* **1** the reversion of property to the state, or (in feudal law) to a lord, on the owner's dying without legal heirs. **2** property affected by this. ● *v.* **1** *tr.* hand over (property) as an escheat. **2** *tr.* confiscate. **3** *intr.* revert by escheat. [ME f. OF *eschete*, ult. f. L *excidere* (as EX-[1], *cadere* fall)]

eschew /ɪsˈtʃuː/ *v.tr. literary* avoid; abstain from. □ **eschewal** *n.* [ME f. OF *eschiver*, ult. f. Gmc: rel. to SHY[1]]

eschscholtzia /ɪsˈkɒlʃə, eˈʃɒltsɪə/ *n.* a yellow- or orange-flowered poppy of the genus *Eschscholtzia*, esp. the Californian poppy. [J. F. von *Eschscholtz*, Ger. botanist (1793–1831)]

Escoffier /eˈskɒfɪeɪ/, Georges-Auguste (1846–1935), French chef. He gained an international reputation while working in London at the Savoy Hotel (1890–9) and later at the Carlton (1899–1919). His many culinary inventions include peach Melba, first made in 1893 in honour of the singer Dame Nellie Melba when she was staying at the Savoy.

Escorial /ˌeskɒrɪˈɑːl/ a monastery and palace in central Spain, near Madrid, built in the late 16th century by Philip II. It encompasses a mausoleum, library, college, and art collection.

escort *n.* & *v.* ● *n.* /ˈeskɔːt/ **1** one or more persons, vehicles, ships, etc., accompanying a person, vehicle, etc., esp. for protection or security or as a mark of rank or status. **2** a person accompanying a person of the opposite sex socially. **3** a person or group acting as a guide or leader, esp. on a journey. ● *v.tr.* /ɪˈskɔːt/ act as an escort to. [F *escorte*, *escorter* f. It. *scorta* fem. past part. of *scorgere* conduct]

escritoire /ˌeskrɪˈtwɑː(r)/ *n.* a writing-desk with drawers etc. [F f. L *scriptorium* writing-room: see SCRIPTORIUM]

escrow /ˈeskrəʊ, eˈskrəʊ/ *n.* & *v. Law* ● *n.* **1** money, property, or a written bond, kept in the custody of a third party until a specified condition has been fulfilled. **2** the status of this (*in escrow*). ● *v.tr.* place in escrow. [AF *escrowe*, OF *escroe* scrap, scroll, f. med.L *scroda* f. Gmc]

escudo /eˈskjuːdəʊ/ *n.* (*pl.* -**os**) the chief monetary unit of Portugal. [Sp. & Port. f. L *scutum* shield]

esculent /'eskjʊlənt/ adj. & n. ● adj. fit to eat; edible. ● n. an edible substance. [L esculentus f. esca food]

escutcheon /ɪ'skʌtʃən/ n. **1** Heraldry a shield or emblem bearing a coat of arms. **2** reputation (a blot on one's escutcheon). **3** the protective plate around a keyhole or door-handle. **4** the middle part of a ship's stern, where the name was formerly placed. □ **escutcheoned** adj. [AF & ONF escuchon ult. f. L scutum shield]

Esd. abbr. Esdras (Apocrypha).

Esdras /'ezdrəs/ **1** either of two books of the Apocrypha, of which the first is mainly a compilation from Chronicles, Nehemiah, and Ezra, and the second is a record of angelic revelation. **2** (in the Vulgate) the books of Ezra and Nehemiah.

ESE abbr. east-south-east.

-ese /iːz/ suffix forming adjectives and nouns denoting: **1** an inhabitant or language of a country or city (Japanese; Milanese; Viennese). ¶ Plural forms are the same. **2** often derog. character or style, esp. of language (officialese). [OF -eis ult. f. L -ensis]

Esfahan see ISFAHAN.

esker /'eskə(r)/ n. (also **eskar**) Geol. a long ridge of post-glacial gravel in river valleys. [Ir. eiscir]

Eskimo /'eskɪˌməʊ/ n. & adj. ● n. (pl. same or **-os**) **1** a member of a people inhabiting northern Canada, Alaska, Greenland, Denmark, and eastern Siberia. (See note below.) **2** the languages of this people. ● adj. of or relating to the Eskimos or their languages. ¶ In Canada and, increasingly, elsewhere the term Inuit is used to refer to Canadian Eskimos and also to Eskimos generally. The term Eskimo is regarded as offensive by some people. [Danish f. F Esquimaux (pl.) f. Algonquian, lit. 'eaters of raw flesh']

  ■ Originally a semi-nomadic hunting-and-gathering people, Eskimos are noted for their adaptation to a harsh environment and for the self-sufficiency of the family units (although during the winter months families gather in groups which disperse in summer). They are associated with the igloo or ice shelter, although this is not as common traditionally as huts made of turf or driftwood. Their way of life has changed considerably in recent years, and many now live in permanent settlements. The Eskimo languages, which are polysynthetic and ergative, belong to the Eskimo-Aleut family and are divided into two main branches: the Inupiaq or Inuit (spoken in Greenland, Labrador, the Arctic coast of Canada, and northern Alaska) and the Yupik (spoken in southern Alaska and Siberia). There are approximately 40,000 speakers in Greenland, 7,000 in Denmark, 25,000 in Alaska, 18,000 in Canada, and several hundred in Siberia.

Esky /'eskɪ/ n. (pl. **-ies**) Austral. propr. a portable insulated container for keeping food or drink cool. [prob. f. ESKIMO, with ref. to their cold climate]

ESN abbr. educationally subnormal.

ESOL /'esɒl/ abbr. English for speakers of other languages.

esophagus US var. of OESOPHAGUS.

esoteric /ˌiːsə'terɪk, ˌes-/ adj. **1** intelligible only to those with special knowledge. **2** (of a belief etc.) intended only for the initiated. □ **esoterical** adj. **esoterically** adv. **esotericism** /-'terɪˌsɪz(ə)m/ n. **esotericist** n. [Gk esōterikos f. esōterō compar. of esō within]

ESP abbr. extrasensory perception.

espadrille /ˌespə'drɪl/ n. a light canvas shoe with a plaited fibre sole. [F f. Prov. espardillo f. espart ESPARTO]

espalier /ɪ'spælɪə(r)/ n. **1** a lattice or framework along which the branches of a tree or shrub are trained to grow flat against a wall etc. **2** a tree or shrub trained in this way. [F f. It. spalliera f. spalla shoulder]

España see SPAIN.

esparto /e'spɑːtəʊ/ n. (pl. **-os**) (in full **esparto grass**) a coarse grass, Stipa tenacissima, native to Spain and North Africa, with tough narrow leaves, used to make ropes, wickerwork, and good-quality paper. [Sp. f. L spartum f. Gk sparton rope]

especial /ɪ'speʃ(ə)l/ adj. **1** notable, exceptional. **2** attributed or belonging chiefly to one person or thing (your especial charm). [ME f. OF f. L specialis special]

especially /ɪ'speʃəlɪ/ adv. chiefly; much more than in other cases.

Esperanto /ˌespə'ræntəʊ/ n. an artificial language devised in 1887 by the Polish physician L. L. Zamenhof (1859–1917) as a medium of communication for persons of all languages. Its words are based mainly on roots commonly found in Romance and other European

languages, and while it has the advantage of grammatical regularity and ease of pronunciation it retains the structure of these languages, which makes Esperanto little easier than any other European language for a speaker whose native tongue falls outside this group. □ **Esperantist** n. [the pen-name (f. L sperare hope) of its inventor]

espial /ɪ'spaɪəl/ n. **1** the act or an instance of catching sight of or of being seen. **2** archaic spying. [ME f. OF espiaille f. espier: see ESPY]

espionage /'espɪəˌnɑːʒ/ n. the practice of spying or of using spies, esp. by governments. [F espionnage f. espionner f. espion SPY]

Espírito Santo /eˌspɪrɪˌtuː 'sæntuː/ a state of eastern Brazil, on the Atlantic coast; capital, Vitória.

esplanade /ˌesplə'neɪd/ n. **1** a long open level area for walking on, esp. beside the sea. **2** a level space separating a fortress from a town. [F f. Sp. esplanada f. esplanar make level f. L explanare (as EX-¹, planus level)]

espousal /ɪ'spaʊz(ə)l/ n. **1** (foll. by of) the espousing of a cause etc. **2** archaic a marriage or betrothal. [ME f. OF espousailles f. L sponsalia neut. pl. of sponsalis (as ESPOUSE)]

espouse /ɪ'spaʊz/ v.tr. **1** adopt or support (a cause, doctrine, etc.). **2** archaic **a** (usu. of a man) marry. **b** (usu. foll. by to) give (a woman) in marriage. □ **espouser** n. [ME f. OF espouser f. L sponsare f. sponsus past part. of spondere betroth]

espressivo /espre'siːvəʊ/ adv. & adj. Mus. with expression of feeling. [It.]

espresso /e'spresəʊ/ n. (also **expresso** /ek'spresəʊ/) (pl. **-os**) **1** strong concentrated black coffee made under steam pressure. **2** a machine for making this. [It., = pressed out]

esprit /e'spriː/ n. vivacious wit. □ **esprit de corps** /də 'kɔː(r)/ a feeling of devotion to and pride in the group one belongs to. **esprit de l'escalier** /də le'skælɪˌeɪ/ an apt retort or clever remark that comes to mind after the chance to make it is gone. [F f. L spiritus SPIRIT (+ corps body, escalier stairs)]

espy /ɪ'spaɪ/ v.tr. (**-ies**, **-ied**) literary catch sight of; perceive. [ME f. OF espier: see SPY]

Esq. abbr. Esquire.

-esque /esk/ suffix forming adjectives meaning 'in the style of' or 'resembling' (romanesque; Schumannesque; statuesque). [F f. It. -esco f. med.L -iscus]

Esquimau /'eskɪˌməʊ/ n. & adj. (pl. **Esquimaux** /-məʊz/) = ESKIMO. [F]

Esquipulas /ˌeskiː'puːlæs/ a town in SE Guatemala, near the border with Honduras; pop. (1981) 18,840. Noted for the image of the 'Black Christ of Esquipulas' in its church, the town is a centre of pilgrimage.

esquire /ɪ'skwaɪə(r)/ n. **1** (usu. as abbr. **Esq.**) Brit. a title appended to a man's surname when no other form of address is used, esp. as a form of address for letters. **2** archaic = SQUIRE. [ME f. OF esquier f. L scutarius shield-bearer f. scutum shield]

ESR abbr. Physics electron spin resonance.

ESRC abbr. (in the UK) Economic and Social Research Council.

-ess¹ /ɪs/ suffix forming nouns denoting females (actress; lioness; mayoress). [from or after F -esse f. LL -issa f. Gk -issa]

-ess² /es/ suffix forming abstract nouns from adjectives (duress). [ME f. F -esse f. L -itia; cf. -ICE]

essay n. & v. ● n. /'eseɪ/ **1** a composition, usu. short and in prose, on any subject. **2** /'eseɪ, e'seɪ/ (often foll. by at, in) formal an attempt. ● v.tr. /e'seɪ/ formal attempt, try. □ **essayist** /'eseɪɪst/ n. [ME f. ASSAY, assim. to F essayer ult. f. LL exagium weighing f. exigere weigh: see EXACT]

Essen /'es(ə)n/ an industrial city in the Ruhr valley, in NW Germany; pop. (1991) 626,990.

essence /'es(ə)ns/ n. **1** the indispensable quality or element identifying a thing or determining its character; fundamental nature or inherent characteristics. **2 a** an extract obtained by distillation etc., esp. a volatile oil. **b** a perfume or scent, esp. made from a plant or animal substance. **3** the constituent of a plant that determines its chemical properties. **4** an abstract entity; the reality underlying a phenomenon or all phenomena. □ **in essence** fundamentally. **of the essence** indispensable, vital. [ME f. OF f. L essentia f. esse be]

Essene /'esiːn, e'siːn/ n. a member of an ancient Jewish ascetic sect of the 2nd century BC–2nd century AD in Palestine, who lived in highly organized groups and held property in common. The Dead Sea scrolls are thought to have belonged to the Essenes. [L pl. Esseni f. Gk pl. Essēnoi]

essential /ɪ'senʃ(ə)l/ adj. & n. ● adj. **1** absolutely necessary; indispensable. **2** fundamental, basic. **3** of or constituting the essence of a person or thing. **4** (of a disease) with no known external stimulus or cause; idiopathic. ● *n.* (esp. in *pl.*) a basic or indispensable element or thing. □ **essential element** a chemical element required by living organisms for normal growth. **essential oil** an oil present in and having the characteristic odour of a plant etc., from which it can be obtained by distillation (see also OIL). □ **essentially** *adv.* **essentialness** *n.* **essentiality** /ɪˌsɛnʃɪˈælɪtɪ/ *n.* [ME f. LL *essentialis* (as ESSENCE)]

Essequibo /ˌɛsɪˈkiːbəʊ/ a river in Guyana, rising in the Guiana Highlands and flowing about 965 km (600 miles) northwards to the Atlantic.

Essex /ˈɛsɪks/ a county of eastern England; county town, Chelmsford.

EST *abbr.* **1** Eastern Standard Time. **2** electro-shock treatment.

-est[1] /ɪst/ *suffix* forming the superlative of adjectives (*widest*; *nicest*; *happiest*) and adverbs (*soonest*). [OE -*ost-*, -*ust-*, -*ast-*]

-est[2] /ɪst/ *suffix* (also **-st**) *archaic* forming the 2nd person sing. of verbs (*canst*; *findest*; *gavest*). [OE -*est*, -*ast*, -*st*]

establish /ɪˈstæblɪʃ/ *v.tr.* **1** set up or consolidate (a business, system, etc.) on a permanent basis. **2** (foll. by *in*) settle (a person or oneself) in some capacity. **3** (esp. as **established** *adj.*) achieve permanent acceptance for (a custom, belief, practice, institution, etc.). **4 a** validate; place beyond dispute (a fact etc.). **b** find out, ascertain. □ **established Church** a Church recognized by the state as the national Church. □ **establisher** *n.* [ME f. OF *establir* (stem *establiss-*) f. L *stabilire* f. *stabilis* STABLE[1]]

establishment /ɪˈstæblɪʃmənt/ *n.* **1** the act or an instance of establishing; the process of being established. **2 a** a business organization or public institution. **b** a place of business. **c** a residence. **3 a** the staff or equipment of an organization. **b** a household. **4** any organized body permanently maintained for a purpose. **5** a Church system organized by law. **6 a** (**the Establishment**) the group in a society exercising authority or influence, and seen as resisting change. **b** any influential or controlling group (*the literary Establishment*).

establishmentarian /ɪˌstæblɪʃmənˈtɛərɪən/ *adj.* & *n.* ● *adj.* adhering to or advocating the principle of an established Church. ● *n.* a person adhering to or advocating this. □ **establishmentarianism** *n.*

estaminet /eˈstæmɪˌneɪ/ *n.* a small French café etc. selling alcoholic drinks. [F f. Walloon *staminé* byre f. *stamo* a pole for tethering a cow, prob. f. G *Stamm* stem]

estate /ɪˈsteɪt/ *n.* **1** a property consisting of an extensive area of land usu. with a large house. **2** *Brit.* a modern residential or industrial area with integrated design or purpose. **3** all of a person's assets and liabilities, esp. at death. **4** a property where rubber, tea, grapes, etc., are cultivated. **5** (in full **estate of the realm**) an order or class forming (or regarded as) a part of the body politic. **6** *archaic* or *literary* a state or position in life (*the estate of holy matrimony*; *poor man's estate*). **7** *colloq.* = estate car. □ **estate agent** *Brit.* **1** a person whose business is the sale or lease of buildings and land on behalf of others. **2** the steward of an estate. **estate car** *Brit.* a car with the passenger area extended and combined with space for luggage, usu. with an extra door at the rear. **estate duty** *Brit. hist.* death duty levied on property. ¶ Replaced in 1975 by *capital transfer tax* and in 1986 by *inheritance tax*. [ME f. OF *estat* (as STATUS)]

Estates General see STATES GENERAL.

esteem /ɪˈstiːm/ *v.* & *n.* ● *v.tr.* **1** (usu. in *passive*) have a high regard for; greatly respect; think favourably of. **2** *formal* consider, deem (*esteemed it an honour*). ● *n.* high regard; respect; favour (*held them in esteem*). [ME f. OF *estimer* f. L *aestimare* fix the price of]

ester /ˈɛstə(r)/ *n. Chem.* an organic compound produced by replacing the hydrogen of an acid by an alkyl, aryl, etc., radical, many examples of which occur naturally as oils and fats. □ **esterify** /eˈstɛrɪˌfaɪ/ *v.tr.* (**-ies, -ied**). [G, prob. f. *Essig* vinegar + *Äther* ether]

Esth. *abbr.* (in the Bible & Apocrypha) Esther.

Esther /ˈɛstə(r)/ **1** (in the Bible) a woman who was chosen on account of her beauty by the Persian king Ahasuerus (generally supposed to be Xerxes I) to be his queen and who used her influence with him to save the Israelites in captivity from persecution. **2** the book of the Bible containing an account of these events; a part survives only in Greek and is included in the Apocrypha.

esthete *US* var. of AESTHETE.

esthetic *US* var. of AESTHETIC.

estimable /ˈɛstɪməb(ə)l/ *adj.* worthy of esteem. □ **estimably** *adv.* [F f. L *aestimabilis* (as ESTEEM)]

estimate *n.* & *v.* ● *n.* /ˈɛstɪmət/ **1** an approximate judgement, esp. of cost, value, size, etc. **2** a price specified as that likely to be charged for work to be undertaken. **3** opinion, judgement, estimation. ● *v.tr.* (also *absol.*) /ˈɛstɪˌmeɪt/ **1** form an estimate or opinion of. **2** (foll. by *that* + clause) make a rough calculation. **3** (often foll. by *at*) value or measure by estimation; adjudge. □ **estimative** /-mətɪv/ *adj.* **estimator** /-ˌmeɪtə(r)/ *n.* [L *aestimare aestimat-* fix the price of]

estimation /ˌɛstɪˈmeɪʃ(ə)n/ *n.* **1** the process or result of estimating. **2** judgement or opinion of worth (*in my estimation*). **3** *archaic* esteem (*hold in estimation*). [ME f. OF *estimation* or L *aestimatio* (as ESTIMATE)]

estival *US* var. of AESTIVAL.

estivate *US* var. of AESTIVATE.

Estonia /ɪˈstəʊnɪə/ a Baltic country on the south coast of the Gulf of Finland; pop. (est. 1991) 1,591,000; languages, Estonian (official), Russian; capital, Tallinn. Estonia is a flat, lowland country with marshland, lakes, and forest. Previously ruled by the Teutonic Knights and then by Sweden, Estonia was ceded to Russia in 1721. It was proclaimed an independent republic in 1918 but was annexed by the USSR in 1940 as a constituent republic, the Estonian SSR. With the breakup of the Soviet Union Estonia regained its independence in 1991.

Estonian /ɪˈstəʊnɪən/ *n.* & *adj.* ● *n.* **1 a** a native of Estonia. **b** a person of Estonian descent. **2** the Finno-Ugric language of Estonia, most closely related to Finnish and spoken by about a million people. ● *adj.* of or relating to Estonia or its people or language.

estop /ɪˈstɒp/ *v.tr.* (**estopped, estopping**) (foll. by *from*) *Law* bar or preclude, esp. by estoppel. □ **estoppage** *n.* [ME f. AF, OF *estoper* f. LL *stuppare* stop up f. L *stuppa* tow: cf. STOP, STUFF]

estoppel /ɪˈstɒp(ə)l/ *n. Law* the principle which precludes a person from asserting something contrary to what is implied by a previous action or statement of that person or by a previous pertinent judicial determination. [OF *estouppail* bung f. *estoper* (as ESTOP)]

Estoril /ˌɛʃtəˈrɪl/ a resort on the Atlantic coast of Portugal; pop. (1991) 24,850.

estovers /ɪˈstəʊvəz/ *n.pl. hist.* necessaries allowed by law to a tenant (esp. fuel, or wood for repairs). [AF *estover*, OF *estoveir* be necessary, f. L *est opus*]

estrange /ɪˈstreɪndʒ/ *v.tr.* (usu. in *passive*; often foll. by *from*) **1** cause (a person or group) to turn away in feeling or affection; alienate. **2** (as **estranged** *adj.*) (of a husband or wife) no longer living with his or her spouse. □ **estrangement** *n.* [ME f. AF *estraunger*, OF *estranger* f. L *extraneare* treat as a stranger f. *extraneus* stranger]

estreat /ɪˈstriːt/ *n.* & *v. Law* **1** a copy of a court record of a fine etc. for use in prosecution. **2** the enforcement of a fine or forfeiture of a recognizance. ● *v.tr.* enforce the forfeit of (a fine etc., esp. surety for bail). [ME f. AF *estrete*, OF *estraite* f. *estraire* f. L *extrahere* EXTRACT]

Estremadura /ˌɛʃtrəməˈdʊərə/ a coastal region and former province of west central Portugal.

estrogen *US* var. of OESTROGEN.

estrus etc. *US* var. of OESTRUS etc.

estuary /ˈɛstjʊərɪ/ *n.* (*pl.* **-ies**) a wide tidal mouth of a river. □ **estuarine** /-ˌraɪn/ *adj.* [L *aestuarium* tidal channel f. *aestus* tide]

e.s.u. *abbr.* electrostatic unit(s).

esurient /ɪˈsjʊərɪənt/ *adj. archaic* or *joc.* **1** hungry. **2** impecunious and greedy. □ **esuriently** *adv.* [L *esurire* (v.) hunger f. *edere es-* eat]

Esztergom /ˈɛstəˌɡɒm/ a town and river port on the Danube in Hungary; pop. (est. 1984) 31,000.

ET *abbr.* extraterrestrial.

-et[1] /ɪt/ *suffix* forming nouns (orig. diminutives) (*baronet*; *bullet*; *sonnet*). [OF -*et* -*ete*]

-et[2] /ɪt/ *suffix* (also **-ete** /iːt/) forming nouns usu. denoting persons (*comet*; *poet*; *athlete*). [Gk -*ētēs*]

ETA[1] *abbr.* estimated time of arrival.

ETA[2] /ˈɛtə/ a Basque separatist movement in Spain which has waged a terrorist campaign since its foundation in 1959 for an independent Basque state. [Basque acronym, f. *Euzkadi ta Azkatasuna* Basque homeland and liberty]

eta /ˈiːtə/ *n.* the seventh letter of the Greek alphabet (H, η). [Gk]

et al. /ɛt ˈæl/ *abbr.* and others. [L *et alii, et alia,* etc.]

etalon /ˈɛtəˌlɒn/ *n. Physics* a device consisting of two reflecting plates, for producing interfering light-beams. [F *étalon* standard]

etc. *abbr.* = ET CETERA.

et cetera /et ˈsetərə, ˈsetrə/ adv. & n. (also **etcetera**) ● adv. **1 a** and the rest; and similar things or people. **b** or similar things or people. **2** and so on. ● n. (in pl.) the usual sundries or extras. [ME f. L]

etch /etʃ/ v. & n. ● v. **1** tr. reproduce (a picture etc.) by engraving a design on a metal plate with acid (esp. to print copies). **b** tr. engrave (a plate) in this way. **2** intr. practise this craft. **3** tr. (foll. by on, upon) impress deeply (esp. on the mind). ● n. the action or process of etching. □ **etcher** n. [Du. etsen f. G ätzen etch f. OHG azzen cause to eat or to be eaten f. Gmc]

etchant /ˈetʃənt/ n. a corrosive used in etching.

etching /ˈetʃɪŋ/ n. **1** a print made from an etched plate. **2** the art of producing such plates. The first etchings date from the early 16th century, though the basic principle, that of corroding a design into a metal plate, had been used earlier for the decoration of armour.

-ete suffix var. of -ET².

eternal /ɪˈtɜːn(ə)l/ adj. **1** existing always; without an end or (usu.) beginning in time. **2** essentially unchanging (eternal truths). **3** colloq. constant; seeming not to cease (your eternal nagging). □ **eternal triangle** a relationship of three people involving sexual rivalry. □ **eternalize** v.tr. (also **-ise**). **eternally** adv. **eternalness** n. **eternality** /ˌiːtɜːˈnælɪtɪ/ n. [ME f. OF f. LL aeternalis f. L aeternus f. aevum age]

Eternal, the God.

Eternal City, the Rome.

eternity /ɪˈtɜːnɪtɪ/ n. (pl. **-ies**) **1** infinite or unending (esp. future) time. **2** (in Christian theology) endless life after death. **3** the state of being eternal. **4** (often prec. by an) colloq. a very long time. **5** (in pl.) eternal truths. □ **eternity ring** a finger-ring set with gems all round, usu. given as a token of lasting affection. [ME f. OF eternité f. L aeternitas -tatis f. aeternus: see ETERNAL]

Etesian winds /ɪˈtiːʒ(ə)n/ n.pl. north-west winds blowing each summer in the eastern Mediterranean. [L etesius f. Gk etēsios annual f. etos year]

eth /eθ/ n. (also **edh** /eð/) the name of an Old English and Icelandic letter, ð, capital Ð (= th). [Icel.]

-eth¹ var. of -TH¹.

-eth² /ɪθ/ suffix (also **-th**) archaic forming the 3rd person sing. present of verbs (doeth; saith). [OE -eth, -ath, -th]

ethanal /ˈeθəˌnæl, ˈiːθ-/ n. Chem. = ACETALDEHYDE. [ETHANE + ALDEHYDE]

ethane /ˈiːθeɪn, ˈeθ-/ n. Chem. a gaseous hydrocarbon of the alkane series (chem. formula: C_2H_6), occurring in natural gas. [ETHER + -ANE²]

ethanediol /ˈiːθeɪnˌdaɪɒl, ˈeθ-/ n. Chem. = ethylene glycol. [ETHANE + DIOL]

ethanoic acid /ˌeθəˈnəʊɪk, ˌiːθ-/ n. Chem. = acetic acid. □ **ethanoate** /-ˈnəʊeɪt/ n. [ETHANE + -OIC]

ethanol /ˈeθəˌnɒl, ˈiːθ-/ n. Chem. = ALCOHOL 1. [ETHANE + -OL]

Ethelred /ˈeθəlˌred/ the name of two kings of England, notably:

Ethelred II (known as Ethelred the Unready) (OE unrǣd = lacking good advice; rash) (c.969–1016), king of England 978–1016. Ethelred's inability to confront the Danes after he succeeded his murdered half-brother St Edward the Martyr led to his payment of tribute to prevent their attacks. In 1013 he briefly lost his throne to the Danish king Sweyn I.

ethene /ˈeθiːn, ˈiːθ-/ n. Chem. = ETHYLENE. [ETHER + -ENE]

ether /ˈiːθə(r)/ n. **1** Chem. **a** (in full **diethyl ether**) a colourless volatile organic liquid (chem. formula: $C_2H_5OC_2H_5$) used as an anaesthetic or solvent. Also called ethoxyethane. **b** any of a class of organic compounds with a similar structure to this, having an oxygen joined to two alkyl etc. groups. **2** (also **aether**) a clear sky; the upper regions of air beyond the clouds. **3** (also **aether**) hist. **a** a medium formerly assumed to permeate space and fill the interstices between particles of matter. **b** a medium through which electromagnetic waves were formerly thought to be transmitted. □ **etheric** /iːˈθerɪk/ adj. [ME f. OF ether or L aether f. Gk aithēr f. root of aithō burn, shine]

ethereal /ɪˈθɪərɪəl/ adj. (also **etherial**) **1** light, airy. **2** highly delicate, esp. in appearance. **3** heavenly, celestial. **4** Chem. of or relating to ether. □ **ethereally** adv. **ethereality** /ɪˌθɪərɪˈælɪtɪ/ n. [L aethereus, -ius f. Gk aitherios (as ETHER)]

etherial var. of ETHEREAL.

etherize /ˈiːθəˌraɪz/ v.tr. (also **-ise**) hist. treat or anaesthetize with ether. □ **etherization** /ˌiːθəraɪˈzeɪʃ(ə)n/ n.

Ethernet /ˈiːθəˌnet/ n. Computing a communication system for local area networks using coaxial cable. [ETHER + NETWORK]

ethic /ˈeθɪk/ n. & adj. ● n. a set of moral principles (the Quaker ethic). ● adj. = ETHICAL. [ME f. OF éthique or L ethicus f. Gk ēthikos (as ETHOS)]

ethical /ˈeθɪk(ə)l/ adj. **1** relating to morals, esp. as concerning human conduct. **2** morally correct; honourable. **3** (of a medicine or drug) not advertised to the general public, and usu. available only on a doctor's prescription. □ **ethical investment** investment in companies that meet ethical and moral criteria specified by the investor. □ **ethically** adv. **ethicality** /ˌeθɪˈkælɪtɪ/ n.

ethics /ˈeθɪks/ n.pl. (also treated as sing.) **1** the study of morals in human conduct; moral philosophy. (See note below.) **2** the rules of conduct recognized as appropriate to a particular profession or area of life. □ **ethicist** /ˈeθɪsɪst/ n.

▪ There has long been debate as to whether moral action can be justified or demonstrated to be rational; discussion has also centred on whether such a thing as objective moral truth exists. Schools of ethics in Western philosophy can be divided, very roughly, into three sorts. The first, drawing on the work of Aristotle, holds that the virtues (such as justice, charity, and generosity) are dispositions to act in ways that benefit both the person possessing them and that person's society. The second, defended particularly by Kant, makes the concept of duty central to morality: humans are bound, from a knowledge of their duty as rational beings, to obey the categorical imperative to respect other rational beings. Thirdly, utilitarianism asserts that the guiding principle of conduct should be the greatest happiness or benefit of the greatest number.

Ethiopia /ˌiːθɪˈəʊpɪə/ a country in NE Africa, on the Red Sea; pop. (est. 1993) 45,892,000; languages, Amharic (official), several other Afro-Asiatic languages; capital, Addis Ababa. Formerly known as Abyssinia, Ethiopia is the oldest independent country in Africa. Its earliest recorded civilization, known to the ancient Egyptians as Punt, dates from the 2nd millennium BC. Christianized in the 4th century, Ethiopia was isolated by Muslim conquests to the north three centuries later, remaining little known to Europeans until the late 19th century. It successfully resisted Italian attempts at colonization in the 1890s but was invaded and conquered by Italy in 1935. The emperor Haile Selassie was restored by the British in 1941 and ruled until overthrown in a Marxist coup in 1974. The subsequent period was marked by civil war, fighting against separatist guerrillas in Eritrea (now a separate country) and Tigray, and by repeated famines due to drought; after the fall of the military government in 1991 a multi-party system was adopted.

Ethiopian /ˌiːθɪˈəʊpɪən/ n. & adj. ● n. **1 a** a native or national of Ethiopia. **b** a person of Ethiopian descent. **2** archaic a black person. ● adj. **1** of or relating to Ethiopia. **2** Zool. of or relating to sub-Saharan Africa as a zoogeographical region. [Ethiopia f. L Aethiops f. Gk Aithiops f. aithō burn + ōps face]

Ethiopic /ˌiːθɪˈɒpɪk, -ˈəʊpɪk/ n. & adj. ● n. the liturgical language of the Coptic Church of Ethiopia. (See GE'EZ.) ● adj. of or in this language. [L aethiopicus f. Gk aithiopikos: see ETHIOPIAN]

ethmoid /ˈeθmɔɪd/ adj. Anat. sievelike. □ **ethmoid bone** a square bone at the root of the nose, with many perforations through which the olfactory nerves pass to the nose. □ **ethmoidal** /eθˈmɔɪd(ə)l/ adj. [Gk ēthmoeidēs f. ēthmos sieve]

ethnic /ˈeθnɪk/ adj. & n. ● adj. **1 a** (of a social group) having a common national or cultural tradition. **b** (of clothes, music, etc.) influenced by the traditions of a particular people or culture, esp. a non-European one regarded as exotic. **2** denoting origin by birth or descent rather than nationality (ethnic Turks). **3** relating to race or culture (ethnic group; ethnic origins). **4** archaic pagan, heathen. ● n. **1** N. Amer. & Austral. a member of an (esp. minority) ethnic group. **2** (in pl., usu. treated as sing.) = ETHNOLOGY. □ **ethnic cleansing** euphem. the practice of mass expulsion or killing of people from opposing ethnic or religious groups within a certain area. **ethnic minority** a (usu. identifiable) group differentiated from the main population of a community by racial origin or cultural background. □ **ethnically** adv. **ethnicity** /eθˈnɪsɪtɪ/ n. [ME f. eccl.L ethnicus f. Gk ethnikos heathen f. ethnos nation]

ethnical /ˈeθnɪk(ə)l/ adj. relating to ethnology.

ethno- /ˈeθnəʊ/ comb. form ethnic, ethnological. [Gk ethnos nation]

ethnoarchaeology /ˌeθnəʊˌɑːkɪˈɒlədʒɪ/ n. the study of a society's institutions based on examination of its material attributes. □ **ethnoarchaeologist** n. **ethnoarchaeological** /-ˌɑːkɪəˈlɒdʒɪk(ə)l/ adj.

ethnobotany /ˌeθnəʊˈbɒtənɪ/ n. **1** the traditional knowledge of a people concerning plants and their uses. **2** the study of such knowledge. [ETHNO- + BOTANY]

ethnocentric /ˌeθnəʊˈsentrɪk/ adj. evaluating other races and cultures by criteria specific to one's own. □ **ethnocentrically** *adv.* **ethnocentrism** *n.* **ethnocentricity** /-sen'trɪsɪtɪ/ *n.*

ethnography /eθ'nɒgrəfɪ/ *n.* the scientific description of races and cultures of humankind. □ **ethnographer** *n.* **ethnographic** /ˌeθnə'græfɪk/ *adj.* **ethnographical** *adj.* **ethnographically** *adv.*

ethnology /eθ'nɒlədʒɪ/ *n.* the comparative scientific study of human peoples. □ **ethnologist** *n.* **ethnologic** /ˌeθnə'lɒdʒɪk/ *adj.* **ethnological** *adj.* **ethnologically** *adv.*

ethnomusicology /ˌeθnəʊˌmjuːzɪ'kɒlədʒɪ/ *n.* the study of the music of one or more (esp. non-European) cultures. □ **ethnomusicologist** *n.* **ethnomusicological** /-kə'lɒdʒɪk(ə)l/ *adj.*

ethogram /'iːθəˌgræm/ *adj.* *Zool.* a list of the kinds of behaviour or activity observed in an animal. [Gk *ĕtho-* (see ETHOS) + -GRAM]

ethology /iː'θɒlədʒɪ/ *n.* **1** the science of animal behaviour. **2** the science of character-formation in human behaviour. □ **ethologist** *n.* **ethological** /ˌiːθə'lɒdʒɪk(ə)l/ *adj.* [L *ethologia* f. Gk *ēthologia* (as ETHOS)]

ethos /'iːθɒs/ *n.* the characteristic spirit or attitudes of a community, people, or system, or of a literary work etc. [mod.L f. Gk *ēthos* nature, disposition]

ethoxyethane /iːˌθɒksɪ'iːθeɪn, -'eθeɪn/ *n.* *Chem.* = ETHER 1a. [ETHER + OXY-[2] + ETHANE]

ethyl /'iːθaɪl, 'eθɪl/ *n.* (*attrib.*) *Chem.* the monovalent radical $-C_2H_5$ derived from ethane by removal of a hydrogen atom. □ **ethyl alcohol** = ALCOHOL 1. [ETHER, -YL]

ethylene /'eθɪliːn/ *n.* *Chem.* a gaseous hydrocarbon of the alkene series (chem. formula: C_2H_4), occurring in natural gas and used in the manufacture of polythene. Also called *ethene.* □ **ethylene glycol** *Chem.* a colourless viscous hygroscopic liquid (chem. formula: $C_2H_6O_2$) used as an antifreeze and in the manufacture of polyesters (also called *ethanediol*). □ **ethylenic** /ˌeθɪ'liːnɪk/ *adj.*

ethyne /'iːθaɪn, 'eθ-/ *n.* *Chem.* = ACETYLENE. [ETHANE + -YNE]

-etic /'etɪk/ *suffix* forming adjectives and nouns (*ascetic*; *emetic*; *genetic*; *synthetic*). [Gk *-etikos* or *-ētikos*: cf. -IC]

etiolate /'iːtɪəʊˌleɪt/ *v.tr.* **1** make (a plant) pale by excluding light. **2** give a sickly hue to (a person). □ **etiolation** /ˌiːtɪəʊ'leɪʃ(ə)n/ *n.* [F *étioler* f. Norman F *étieuler* make into haulm f. *éteule* ult. f. L *stipula* straw]

etiology *US* var. of AETIOLOGY.

etiquette /'etɪˌket/ *n.* **1** the conventional rules of social behaviour. **2 a** the customary behaviour of members of a profession towards each other. **b** the unwritten code governing this (*medical etiquette*). [F *étiquette* label, etiquette]

Etna, Mount /'etnə/ a volcano in eastern Sicily, rising to 3,323 m (10,902 ft). It is the highest and most active volcano in Europe.

Eton /'iːt(ə)n/ (in full **Eton College**) a boys public school near Windsor, Berkshire, perhaps the most famous in England, founded in 1440 by Henry VI to prepare scholars for King's College, Cambridge. □ **Eton collar** a broad stiff collar worn outside the coat-collar, esp. of an Eton jacket. **Eton fives** see FIVES. **Eton jacket** a short jacket reaching only to the waist, as formerly worn by pupils of Eton.

Etonian /iː'təʊnɪən/ *n.* & *adj.* ● *n.* a past or present member of Eton. ● *adj.* of or relating to Eton.

Eton wall game *n.* one of the oldest forms of football in existence, played only on a site at Eton where a red brick wall separates a playing-field from a road. The game consists chiefly of bullies or scrimmages against the wall; the player with the ball attempts to force a way through the opposition, keeping the ball against the wall. Progress is slow, and although points are scored in various ways goals are a rarity. A match has been played on St Andrew's Day (30 Nov.) at least since 1820.

Etosha Pan /ɪ'tɒʃə/ a depression in the plateau of northern Namibia, filled with salt water and having no outlets, extending over an area of 4,800 sq. km (1,854 sq. miles). The wetlands are home to a large population of flamingo.

étrier /'eɪtrɪˌeɪ/ *n.* *Mountaineering* a short rope ladder with a few rungs of wood or metal. [F, = stirrup]

Etruria /ɪ'trʊərɪə/ an ancient state of western Italy, situated between the rivers Arno and Tiber and corresponding approximately to modern Tuscany and parts of Umbria. It was the centre of the Etruscan civilization, which flourished in the middle centuries of the first millennium BC.

Etruscan /ɪ'trʌskən/ *n.* & *adj.* ● *n.* **1** a native of ancient Etruria. (*See note below.*) **2** the language of the Etruscans. (*See note below.*) ● *adj.* of or relating to ancient Etruria or its people or language. [L *Etruscus*]

▪ The Etruscans were the earliest historical inhabitants of Etruria. Their power in Italy was at its height *c.*500 BC, and their sophisticated civilization was an important influence on the Romans, who had completely subdued them by the end of the 3rd century BC. The Etruscan language was written in an alphabet derived from Greek, and is known from documentary texts; some features are understood, but not enough for Etruscan to be linked with certainty to any known family of languages.

et seq. *abbr.* (also **et seqq.**) and the following (pages etc.). [L *et sequentia*]

-ette /et/ *suffix* forming nouns meaning: **1** small (*kitchenette*; *cigarette*). **2** imitation or substitute (*leatherette*; *flannelette*). **3** female (*usherette*; *suffragette*). [from or after OF *-ette*, fem. of -ET[1]]

étude /'eɪtjuːd, eɪ'tjuːd/ *n.* a short musical composition or exercise, usu. for one instrument, designed to improve the technique of the player. [F, = study]

étui /e'twiː/ *n.* a small case for needles etc. [F *étui* f. OF *estui* prison]

-etum /'iːtəm/ *suffix* forming nouns denoting a collection of trees or other plants (*arboretum*; *pinetum*). [L]

etymologize /ˌetɪ'mɒləˌdʒaɪz/ *v.* (also **-ise**) **1** *tr.* give or trace the etymology of. **2** *intr.* study etymology. [med.L *etymologizare* f. L *etymologia* (as ETYMOLOGY)]

etymology /ˌetɪ'mɒlədʒɪ/ *n.* (*pl.* **-ies**) **1 a** the historically verifiable sources of the formation of a word and the development of its meaning. **b** an account of these. **2** the branch of linguistics concerned with etymologies. □ **etymologist** *n.* **etymological** /-mə'lɒdʒɪk(ə)l/ *adj.* **etymologically** *adv.* [OF *ethimologie* f. L *etymologia* f. Gk *etumologia* (as ETYMON, -LOGY)]

etymon /'etɪmən/ *n.* (*pl.* **etyma** /-mə/) the word that gives rise to a derivative or a borrowed or later form. [L f. Gk *etumon* (neut. of *etumos* true), the literal sense or original form of a word]

EU *abbr.* European Union (see EUROPEAN COMMUNITY).

Eu *symb.* *Chem.* the element europium.

eu- /juː/ *comb. form* well, easily. [Gk]

Euboea /juː'biːə/ (Greek **Évvoia** /'ɛvɪa/) an island of Greece in the western Aegean Sea, separated from the mainland by only a narrow channel at its capital, Chalcis.

eucalyptus /ˌjuːkə'lɪptəs/ *n.* (also **eucalypt** /'juːkəˌlɪpt/) (*pl.* **eucalyptuses**, **eucalypti** /-taɪ/, or **eucalypts**) **1** a flowering evergreen tree of the large Australian genus *Eucalyptus*, many members of which are cultivated for their timber, the aromatic oil in their leaves, various gums and resins, or for ornamental use. See also *gum-tree* (GUM[1]). **2** (in full **eucalyptus oil**) this oil used as an antiseptic etc. [mod.L f. EU- + Gk *kaluptos* covered f. *kaluptō* to cover, the unopened flower being protected by a cap]

eucaryote var. of EUKARYOTE.

eucharis /'juːkərɪs/ *n.* an evergreen bulbous plant of the genus *Eucharis*, native to South America, with white bell-shaped flowers. [Gk *eukharis* pleasing (as EU-, *kharis* grace)]

Eucharist /'juːkərɪst/ *n.* **1** the Christian sacrament commemorating the Last Supper, in which bread and wine are consecrated and consumed. (*See note below.*) **2** the consecrated elements, esp. the bread (*receive the Eucharist*). □ **Eucharistic** /ˌjuːkə'rɪstɪk/ *adj.* **Eucharistical** *adj.* [ME f. OF *eucariste*, ult. f. eccl.Gk *eukharistia* thanksgiving f. Gk *eukharistos* grateful (as EU-, *kharizomai* offer willingly)]

▪ The origins of the Eucharist are found in the Gospels of Matthew, Mark, and Luke and in 1 Corinthians 10, 11. From the first it was accepted that the Eucharist conveys to the believer the body and blood of Christ, but there have been different interpretations of the sense in which Christ is present (see CONSUBSTANTIATION, TRANSUBSTANTIATION).

euchre /'juːkə(r)/ *n.* & *v.* ● *n.* an American card-game for two, three, or four players. ● *v.tr.* **1** (in euchre) gain the advantage over (another player) when that player fails to take three tricks. **2** deceive, outwit. **3** *Austral.* exhaust, ruin. [19th c.: orig. unkn.]

Euclid /'juːklɪd/ (*c.*300 BC), Greek mathematician. He taught at Alexandria, and is famous for his great work *Elements of Geometry*, which covered plane geometry, the theory of numbers, irrationals, and solid geometry. This was the standard work until other kinds of geometry were discovered in the 19th century.

Euclidean /juːˈklɪdɪən/ adj. of or relating to Euclid. □ **Euclidean geometry** the geometry of ordinary experience, based on the postulates used by Euclid (cf. NON-EUCLIDEAN). **Euclidean space** space for which Euclidean geometry is valid. [L *Euclideus* f. Gk *Eukleideios*]

eudemonic /ˌjuːdɪˈmɒnɪk/ adj. (also **eudaemonic**) conducive to happiness. [Gk *eudaimonikos* (as EUDEMONISM)]

eudemonism /juːˈdiːməˌnɪz(ə)m/ n. (also **eudaemonism**) a system of ethics that bases moral obligation on the likelihood of actions producing happiness. □ **eudemonist** n. **eudemonistic** /-diːməˈnɪstɪk/ adj. [Gk *eudaimonismos* system of happiness f. *eudaimōn* happy (as EU-, *daimōn* guardian spirit)]

eudiometer /ˌjuːdɪˈɒmɪtə(r)/ n. Chem. a graduated glass tube in which gases may be chemically combined by an electric spark, used to measure changes in volume of gases during chemical reactions. □ **eudiometry** n. **eudiometric** /-dɪəˈmetrɪk/ adj. **eudiometrical** adj. [Gk *eudios* clear (weather): orig. used to measure the amount of oxygen, thought to be greater in clear air]

eugenics /juːˈdʒenɪks/ n.pl. (also treated as *sing.*) the science or practice of altering a population, esp. of humans, by controlled breeding for desirable inherited characteristics. The term was coined in 1883 by Francis Galton, who was an advocate of 'improving' the human race by modifying the fertility of different categories of people. Eugenics fell into disfavour after the perversion of its doctrines by the Nazis. □ **eugenic** adj. **eugenically** adv. **eugenicist** /-nɪsɪst/ n. **eugenist** /ˈjuːdʒɪnɪst/ n.

Eugénie /juːˈʒeɪnɪ/ (born Eugénia María de Montijo de Guzmán) (1826–1920), Spanish empress of France 1853–71 and wife of Napoleon III. Throughout her husband's reign she contributed much to the brilliance of his court and was an important influence on his foreign policy. She acted as regent on three occasions (1859; 1865; 1870).

eukaryote /juːˈkærɪˌɒt/ n. (also **eucaryote**) Biol. an organism consisting of a cell or cells having DNA in the form of chromosomes, a distinct nucleus with a membrane, and other specialized organelles. This includes all organisms apart from bacteria and blue-green algae (cf. PROKARYOTE). □ **eukaryotic** /-ˌkærɪˈɒtɪk/ adj. [EU- + KARYO- + -ote as in ZYGOTE]

Euler[1] /ˈɔɪlə(r)/, Leonhard (1707–83), Swiss mathematician. Euler, who worked mainly in St Petersburg and Berlin, was a prolific and original contributor to all branches of mathematics. His attempts to elucidate the nature of functions and his successful (though logically dubious) study of infinite series led his successors, notably Abel and Cauchy, to introduce ideas of convergence and rigorous argument into mathematics. One of his best-known theorems defines a connection between two of the most important constants in mathematics, expressed in the equation $e^i\pi = -1$.

Euler[2] /ˈɔɪlə(r)/, Ulf Svante von (1905–83), Swedish physiologist, the son of Hans Euler-Chelpin. He was the first to discover a prostaglandin, which he isolated from semen. He then searched for the principal chemical neurotransmitter of the sympathetic nervous system, and identified it as noradrenalin. Euler was awarded a Nobel Prize for this in 1970.

Euler-Chelpin /ˌɔɪləˈkelpɪn/, Hans Karl August Simon von (1873–1964), German-born Swedish biochemist. He worked mainly on enzymes and vitamins, and explained the role of enzymes in the alcoholic fermentation of sugar. He shared the Nobel Prize for chemistry in 1929.

eulogium /juːˈləʊdʒɪəm/ n. (pl. **eulogia** /-dʒɪə/ or **-ums**) = EULOGY. [med.L: see EULOGY]

eulogize /ˈjuːləˌdʒaɪz/ v.tr. (also **-ise**) praise in speech or writing. □ **eulogist** n. **eulogistic** /ˌjuːləˈdʒɪstɪk/ adj. **eulogistically** adv.

eulogy /ˈjuːlədʒɪ/ n. (pl. **-ies**) **1 a** speech or writing in praise of a person. **b** an expression of praise. **2** US a funeral oration in praise of a person. [med.L *eulogium* f. (app. by confusion with L *elogium* epitaph) LL *eulogia* praise f. Gk]

Eumenides /juːˈmenɪˌdiːz/ Gk Mythol. a group of goddesses or spirits, identified from an early date with the Furies. The Eumenides probably originated as well-disposed deities of fertility, whose name was given to the Furies either by confusion or euphemistically. [Gk, = kindly ones]

eunuch /ˈjuːnək/ n. **1** a castrated man, esp. one formerly employed at an oriental harem or court. **2** a person lacking effectiveness (*political eunuch*). [ME f. L *eunuchus* f. Gk *eunoukhos* lit. 'bedchamber attendant' f. *eunē* bed + second element rel. to *ekhō* hold]

euonymus /juːˈɒnɪməs/ n. a tree or shrub of the genus *Euonymus*, esp. a spindle tree. [L f. Gk *euōnumos* of lucky name (as EU-, *onoma* name)]

eupeptic /juːˈpeptɪk/ adj. of or having good digestion. [Gk *eupeptos* (as EU-, *peptō* digest)]

euphemism /ˈjuːfəˌmɪz(ə)m/ n. **1** a mild or vague expression substituted for one thought to be too harsh or direct (e.g. *pass over* for *die*). **2** the use of such expressions. □ **euphemize** v.tr. & intr. (also **-ise**). **euphemist** n. **euphemistic** /ˌjuːfəˈmɪstɪk/ adj. **euphemistically** adv. [Gk *euphēmismos* f. *euphēmos* (as EU-, *phēmē* speaking)]

euphonious /juːˈfəʊnɪəs/ adj. **1** sounding pleasant, harmonious. **2** concerning euphony. □ **euphoniously** adv.

euphonium /juːˈfəʊnɪəm/ n. a brass wind instrument of the tuba family. [mod.L f. Gk *euphōnos* (as EUPHONY)]

euphony /ˈjuːfənɪ/ n. (pl. **-ies**) **1 a** pleasantness of sound, esp. of a word or phrase; harmony. **b** a pleasant sound. **2** the tendency to make a phonetic change for ease of pronunciation. □ **euphonize** v.tr. (also **-ise**). **euphonic** /juːˈfɒnɪk/ adj. [F *euphonie* f. LL *euphonia* f. Gk *euphōnia* (as EU-, *phōnē* sound)]

euphorbia /juːˈfɔːbɪə/ n. a plant of the genus *Euphorbia*, esp. a spurge. [ME f. L *euphorbea* f. *Euphorbus*, 1st-c. Gk physician]

euphoria /juːˈfɔːrɪə/ n. a feeling of well-being, esp. one based on over-confidence or over-optimism. □ **euphoric** /-ˈfɒrɪk/ adj. **euphorically** adv. [Gk f. *euphoros* well-bearing (as EU-, *pherō* bear)]

euphoriant /juːˈfɔːrɪənt/ adj. & n. ● adj. inducing euphoria. ● n. a euphoriant drug.

Euphrates /juːˈfreɪtiːz/ a river of SW Asia which rises in the mountains of eastern Turkey and flows through Syria and Iraq to join the Tigris, forming the Shatt al-Arab waterway.

euphuism /ˈjuːfjuːˌɪz(ə)m/ n. an affected or high-flown style of writing or speaking. □ **euphuist** n. **euphuistic** /ˌjuːfjuːˈɪstɪk/ adj. **euphuistically** adv. [Gk *euphuēs* well endowed by nature: orig. of writing imitating John Lyly's *Euphues and his England* (1580)]

Eurasian /jʊəˈreɪʃ(ə)n, -ˈreɪʒ(ə)n/ adj. & n. ● adj. **1** of mixed European and Asian (formerly esp. Indian) parentage. **2** of Europe and Asia. ● n. a Eurasian person.

Euratom /jʊəˈrætəm/ see EUROPEAN ATOMIC ENERGY COMMUNITY.

eureka /jʊəˈriːkə/ int. & n. ● int. I have found it! (a cry expressing joy at a discovery etc.). ● n. an exultant cry of 'eureka'. [Gk *heurēka* 1st pers. sing. perfect of *heuriskō* find: attributed to Archimedes]

eurhythmic /jʊəˈrɪðmɪk/ adj. of or in harmonious proportion (esp. of architecture). [*eurhythmy* harmony of proportions f. L *eur(h)ythmia* f. Gk *euruthmia* (as EU-, *rhuthmos* proportion, rhythm)]

eurhythmics /jʊəˈrɪðmɪks/ n.pl. (also treated as *sing.*) (US **eurythmics**) harmony of bodily movement, esp. as developed with music and dance into a system of education.

Euripides /jʊəˈrɪpɪˌdiːz/ (480–c.406 BC), Greek dramatist. He was the last of the trio of important tragedians after Aeschylus and Sophocles. His nineteen surviving plays show important innovations in the handling of the traditional myths, such as their introduction of a low realism into grand subject-matter, their interest in feminine psychology, and their portrayal of abnormal and irrational states of mind. They include *Medea*, *Hippolytus*, *Electra*, *Trojan Women*, and *Bacchae*.

euro /ˈjʊərəʊ/ n. (pl. **-os**) Austral. a large reddish wallaby, *Macropus robustus*. [Aboriginal]

Euro- /ˈjʊərəʊ/ comb. form Europe, European. [abbr.]

Eurobond /ˈjʊərəʊˌbɒnd/ n. Commerce an international bond issued outside the country in whose currency its value is stated.

Eurocommunism /ˌjʊərəʊˈkɒmjʊˌnɪz(ə)m/ n. a form of Communism in western European countries independent of the former Soviet Communist Party. □ **Eurocommunist** adj. & n.

Eurocrat /ˈjʊərəʊˌkræt/ n. usu. derog. a bureaucrat in the administration of the European Community.

Eurodollar /ˈjʊərəʊˌdɒlə(r)/ n. a US dollar held in a bank outside the US (not necessarily in Europe).

Euro-MP /ˈjʊərəʊemˌpiː/ n. a member of the European Parliament.

Europa /jʊəˈrəʊpə/ **1** Gk Mythol. the daughter of Agenor, king of Tyre, or of Agenor's son Phoenix. Wooed by Zeus in the form of a bull, she was carried off to Crete, where she bore him three sons (Minos, Rhadamanthus, and Sarpedon). **2** Astron. satellite II of Jupiter, the sixth closest to the planet, and one of the Galilean moons (diameter

3,138 km). It has a bright and smooth icy surface, criss-crossed by an intricate network of dark lines.

Europarliament /ˈjʊərəʊˌpɑːləmənt/ *n.* = EUROPEAN PARLIAMENT.

Europarliamentarian /ˌjʊərəʊˌpɑːləmenˈteərɪən/ *n.* a member of the European Parliament.

Europarliamentary /ˌjʊərəʊˌpɑːləˈmentərɪ/ *adj.* of or relating to the European Parliament.

Europe /ˈjʊərəp/ **1** a continent of the northern hemisphere. (*See note below.*) **2** Continental Europe, Europe excluding Britain. **3** *colloq.* the European Community.

■ Europe consists of the western part of the land mass of which Asia forms the eastern (and greater) part, and includes Scandinavia and the British Isles. It contains approximately 20 per cent of the world's population. The western part of Europe was consolidated within the Roman Empire, but the subsequent barbarian invasions brought political chaos. European history has remained turbulent, with only relatively short periods of stability, some countries gaining local dominance for a time while others were seeking to establish their existence. Politically and economically pre-eminent in the 18th and 19th centuries, Europe to an extent been overshadowed in the 20th century, especially by the rise of the superpowers and latterly of Japan. After the Second World War, Europe was divided between the non-Communist nations of western Europe (which generally sought closer association with each other) and the Soviet-dominated Eastern bloc. With the collapse of Communism in the late 1980s, a new era of change began with many states in eastern Europe seeking to establish or re-establish their independence.

Europe, Council of an association of European states, independent of the European Community, founded in 1949 to safeguard the political and cultural heritage of Europe and promote economic and social cooperation. Most of its conclusions take the form of international agreements (known as European Conventions) or recommendations to governments. One of the Council's principal achievements is the European Convention on Human Rights.

European /ˌjʊərəˈpɪən/ *adj. & n.* ● *adj.* **1** of or in Europe. **2 a** descended from natives of Europe. **b** originating in or characteristic of Europe. **3 a** happening in or extending over Europe. **b** concerning Europe as a whole rather than its individual countries. **4** of or relating to the European Community. ● *n.* **1 a** a native or inhabitant of Europe. **b** a person descended from natives of Europe. **c** a white person. **2** a person concerned with European matters. □ **European plan** *N. Amer.* a system of charging for a hotel room only without meals. □ **Europeanism** *n.* **Europeanize** *v.tr. & intr.* (also **-ise**). **Europeanization** /-ˌpɪənaɪˈzeɪʃ(ə)n/ *n.* [F *européen* f. L *europaeus* f. L *Europa* f. Gk *Eurōpē* Europe]

European Atomic Energy Community (abbr. **Euratom**) an institution established in 1957 by members of the European Coal and Steel Community (ECSC) to create within a short period the technical and industrial conditions needed to exploit nuclear discoveries and especially to produce nuclear energy on a large scale. It is one of the communities (with the ECSC and EEC) that forms the European Community.

European Bank for Reconstruction and Development a bank established in London in Apr. 1991 with the aim of assisting formerly state-controlled economies in what were the Communist countries of eastern Europe, and in the Soviet Union, to make the transition to free-market economies.

European Coal and Steel Community (abbr. **ECSC**) an organization established in 1952 to regulate pricing, transport, tariffs, etc., for the coal and steel industries of the member countries, which were originally France, West Germany, Italy, Belgium, the Netherlands, and Luxembourg. It was the first of the communities which now make up the European Community (the others are Euratom and the EEC).

European Commission a group, appointed by agreement among the governments of the European Community since 1987, which acts as the initiator of Community action and the guardian of its treaties. Its members are pledged to independence of their governments and of national or other particular interests. The commission meets in Brussels.

European Commission for Human Rights an institution of the Council of Europe, set up under the European Convention on Human Rights to examine complaints of alleged breaches of the Convention. It is based in Strasbourg. (See also EUROPEAN COURT OF HUMAN RIGHTS.)

European Community (abbr. **EC**) an organization of western

European countries, committed to European economic and political integration as envisaged by the Treaty of Rome. A single market within the EC came into force on 1 Jan. 1993, under the terms of the Single European Act (1987). Monetary union is projected, but is still a distant goal; the European currency unit (see ECU) is used only for a restricted class of international transactions. The EC came into being in 1967 through the merger of the European Economic Community, Euratom, and the European Coal and Steel Community, although the combined organization was (until 1987) commonly known as the European Economic Community. It comprises also the European Commission, the European Parliament, and the European Court of Justice. Its original membership of six — Belgium, France, West Germany, Italy, Luxembourg, and the Netherlands — was augmented by Denmark, Ireland, and the UK in 1973, Greece in 1981, and Spain and Portugal in 1986; in 1994 the European Parliament approved the admission of Austria, Finland, and Sweden (Norway voted to withdraw). On 1 Nov. 1993, with the coming into force of the Maastricht Treaty, the European Community was encompassed by the newly created European Union (EU), which comprises the EC together with two intergovernmental 'pillars' for dealing with foreign affairs and with immigration and justice.

European Convention on Human Rights an international agreement set up by the Council of Europe in 1950 to protect human rights. Under the convention were established the European Commission for Human Rights and the European Court of Human Rights.

European Council a grouping of the heads of government of the European Community countries, inaugurated in 1975, which meets two or three times a year to discuss EC business.

European Court of Human Rights an institution of the Council of Europe, set up by the European Convention on Human Rights to protect human rights in conjunction with the European Commission for Human Rights. The court, based in Strasbourg, is called to give judgement in cases where the commission has failed to secure a settlement.

European Court of Justice an institution of the European Community, with thirteen judges appointed by its member governments, meeting in Luxembourg. It was established in 1958, and exists to safeguard the law in the interpretation and application of Community treaties and to determine violations of these. Cases may be brought to it by member states, Community institutions, firms, or individuals.

European currency unit see ECU.

European Economic Community (abbr. **EEC**) an institution of the European Community, an economic association of western European countries set up by the Treaty of Rome (1957). Its aims include the free movement of labour and capital between member countries, especially by the abolition of customs barriers and cartels, and the fostering of common agricultural and trading policies. Until 1987 the term was generally used also to signify the European Community. (See also EUROPEAN COMMUNITY.)

European Free Trade Association (abbr. **EFTA**) a customs union of western European countries, established in 1960 as a trade grouping without the political implications of the European Community. The original members were Austria, Denmark, Norway, Portugal, Sweden, Switzerland, and the UK, although Denmark, Portugal, and the UK have since left to join the EC. All tariffs between EFTA and EEC countries had been abolished by 1984.

European Investment Bank a bank set up in 1958 by the Treaty of Rome to finance capital investment projects promoting the balanced development of members of the European Community. It is based in Luxembourg.

European Monetary System (abbr. **EMS**) a monetary system inaugurated by the European Community in 1979 to coordinate and stabilize the exchange rates of the currencies of member countries, as a prelude to monetary union. It is based on the use of the Exchange Rate Mechanism.

European Monetary Union (abbr. **EMU**) a European Community programme intended to work towards full economic unity in Europe based on the phased introduction of a common currency (the ecu). Announced in 1989, it was delayed by difficulties with the Exchange Rate Mechanism, but under the terms of the Maastricht Treaty the second stage came into effect on 1 Jan. 1994.

European Parliament the Parliament of the European Community, originally established in 1952. From 1958 to 1979 it was composed of representatives drawn from the Parliaments of member countries, but since 1979 direct elections have taken place every five years. Through the Single European Act (1987) it assumed a degree of sovereignty over national Parliaments. The European Parliament meets (about twelve times a year) in Strasbourg, and its committee is in Brussels.

European Recovery Program see MARSHALL PLAN.

European Space Agency (abbr. **ESA**) an organization set up in 1975 and based in Paris, formed from two earlier organizations (the European Space Research Organization and the European Launcher Development Organization). Its aims are to advance space research and technology, especially by the design, development, and deployment of satellites, and to coordinate the national space programmes of the collaborating countries.

European Union (abbr. **EU**) see EUROPEAN COMMUNITY.

europium /juə'rəupɪəm/ n. a soft silvery-white metallic chemical element (atomic number 63; symbol **Eu**). A member of the lanthanide series, europium was discovered by the French chemist E. A. Demarçay in 1901. The oxide is used together with yttrium oxide as a red phosphor in colour television screens. [f. EUROPE]

Europoort /'juərəu,pɔːt/ a major European port facility in the Netherlands, near Rotterdam.

Euro-sceptic /'juərəu,skeptik/ n. a person (esp. a politician) who is sceptical about the supposed benefits of increasing British cooperation with other countries of the European Community. □ **Euro-sceptical** adj. **Euro-scepticism** n.

Eurovision /'juərəu,vɪʒ(ə)n/ n. a network of European television production administered by the European Broadcasting Union.

Eurydice /juə'rɪdɪsɪ/ Gk Mythol. the wife of Orpheus. After she was killed by a snake, Orpheus secured her release from the underworld on the condition that he did not look back at her on their way back to the world of the living. But Orpheus did look back, whereupon Eurydice disappeared.

eurythmics US var. of EURHYTHMICS.

Eusebius /juː'siːbɪəs/ (known as Eusebius of Caesaria) (c.264–c.340 AD), bishop and Church historian. His *Ecclesiastical History* is the principal source for the history of Christianity (especially in the Eastern Church) from the age of the Apostles until 324.

Eustachian tube /juː'steɪʃ(ə)n/ n. Anat. a tube leading from the pharynx to the cavity of the middle ear and equalizing the pressure on each side of the eardrum. [L *Eustachius* = Bartolomeo *Eustachio*, It. anatomist (1520–74)]

eustasy /'juːstəsɪ/ n. a change in sea level throughout the world caused by tectonic movements, melting of glaciers, etc. □ **eustatic** /juː'stætɪk/ adj. [back-form. f. G *eustatisch* (adj.) (as EU-, STATIC)]

eutectic /juː'tektɪk/ adj. & n. Chem. ● adj. (of a mixture, alloy, etc.) having the lowest freezing-point of any possible proportions of its constituents. ● n. a eutectic mixture. □ **eutectic point** (or **temperature**) the minimum freezing-point for a eutectic mixture. [Gk *eutēktos* (as EU-, *tēkō* melt)]

Euterpe /juː'tɜːpɪ/ Gk & Rom. Mythol. the Muse of flutes. [Gk, = well-pleasing]

euthanasia /ˌjuːθə'neɪzɪə/ n. the bringing about of a gentle and easy death, as for a person suffering from an incurable and painful disease or in an irreversible coma. The practice is illegal in most countries, although euthanasia in cases where the patient has given active consent is accepted in practice in the Netherlands. [Gk (as EU-, *thanatos* death)]

eutherian /juː'θɪərɪən/ n. & adj. Zool. ● n. a mammal of the infraclass Eutheria, giving nourishment to its developing young through a placenta. (*See note below.*) ● adj. of or relating to this infraclass. [mod.L f. EU- + Gk *thēr* wild beast]

▪ The infraclass Eutheria comprises all modern mammals except for the monotremes and marsupials, and probably arose during the Cretaceous period. The young are usually born at an advanced stage of development and are suckled by the mother. Except in Australasia, which they did not reach naturally, the eutherians have almost entirely replaced the more primitive mammals.

eutrophic /juː'trəufɪk, -'trɒfɪk/ adj. (of a lake etc.) rich in nutrients and therefore supporting a dense plant population, which kills animal life by depriving it of oxygen. □ **eutrophicate** v.tr. **eutrophication** /-ˌtrɒfɪ'keɪʃ(ə)n, -ˌtrəufɪ-/ n. **eutrophy** /'juːtrəfɪ/ n. [*eutrophy* f. Gk *eutrophia* (as EU-, *trephō* nourish)]

eV abbr. electronvolt.

EVA abbr. extravehicular activity.

evacuate /ɪ'vækjʊ,eɪt/ v.tr. **1 a** remove (people) from a place of danger to stay elsewhere for the duration of the danger. **b** empty or leave (a place) in this way. **2** make empty (a vessel of air etc.). **3** (of troops) withdraw from (a place). **4 a** empty (the bowels or other bodily organ). **b** discharge (faeces etc.). □ **evacuant** n. & adj. **evacuator** n. **evacuative** /-jʊətɪv/ adj. & n. **evacuation** /ɪˌvækjʊ'eɪʃ(ə)n/ n. [L *evacuare* (as E-, *vacuus* empty)]

evacuee /ɪ,vækjuː'iː/ n. a person evacuated from a place of danger.

evade /ɪ'veɪd/ v.tr. **1 a** escape from, avoid, esp. by guile or trickery. **b** avoid doing (one's duty etc.). **c** avoid answering (a question) or yielding to (an argument). **2 a** avoid paying (tax) by illegitimate presentation of one's finances. **b** defeat the intention of (a law etc.), esp. while complying with its letter. **3** (of a thing) elude or baffle (a person). □ **evadable** adj. **evader** n. [F *évader* f. L *evadere* (as E-, *vadere* vas- go)]

evaginate /ɪ'vædʒɪ,neɪt/ v.tr. Med. & Physiol. turn (a tubular organ) inside out. □ **evagination** /ɪ,vædʒɪ'neɪʃ(ə)n/ n. [L *evaginare* (as E-, *vaginare* as VAGINA)]

evaluate /ɪ'væljʊ,eɪt/ v.tr. **1** assess, appraise. **2 a** find or state the number or amount of. **b** find a numerical expression for. □ **evaluator** n. **evaluative** /-jʊətɪv/ adj. **evaluation** /ɪ,væljʊ'eɪʃ(ə)n/ n. [back-form. f. evaluation f. F *évaluation* f. *évaluer* (as E-, VALUE)]

evanesce /ˌiːvə'nes, ,evə-/ v.intr. **1** fade from sight; disappear. **2** become effaced. [L *evanescere* (as E-, *vanus* empty)]

evanescent /ˌiːvə'nes(ə)nt, ,ev-/ adj. (of an impression or appearance etc.) quickly fading. □ **evanescence** n. **evanescently** adv.

evangel /ɪ'vændʒəl/ n. **1** archaic **a** the gospel. **b** any of the four Gospels. **2** a basic doctrine or set of principles. **3** N. Amer. = EVANGELIST. [ME f. OF *evangile* f. eccl.L *evangelium* f. Gk *euaggelion* good news (as EU-, ANGEL)]

evangelic /ˌiːvæn'dʒelɪk/ adj. = EVANGELICAL adj.

evangelical /ˌiːvæn'dʒelɪk(ə)l/ adj. & n. ● adj. **1** of or denoting a school of Protestants (originating in the 18th century) that lays special stress on personal conversion and salvation by faith in the Atonement. **2** eager to share an enthusiasm or belief with others. **3** of or denoting certain Churches in Europe (esp. in Germany) which are, or were, Lutheran rather than Calvinistic. ● n. a member of the evangelical school or an evangelical Church; an adherent of evangelicalism. □ **evangelically** adv. [eccl.L *evangelicus* f. eccl.Gk *euaggelikos* (as EVANGEL)]

evangelicalism /ˌiːvæn'dʒelɪkə,lɪz(ə)m/ n. the doctrine or ethos of any of the evangelical Christian schools or Churches. The term is now applied particularly to Protestant groups which emphasize the authority of the Bible, missionary work, and personal commitment to Christ after being born again in him. Conservative evangelicalism is closely associated with Christian fundamentalism; there also exists a strand of liberal evangelicalism, which takes a critical approach to the Bible while still asserting the importance of the believer's relationship with God.

evangelism /ɪ'vændʒə,lɪz(ə)m/ n. **1** the preaching or promulgation of the gospel. **2** zealous advocacy of a cause or doctrine.

evangelist /ɪ'vændʒəlɪst/ n. **1** any of the writers of the four Gospels (Matthew, Mark, Luke, John). **2** a preacher of the gospel. **3** a lay person doing missionary work.

evangelistic /ɪ,vændʒə'lɪstɪk/ adj. **1** = EVANGELICAL. **2** of preachers of the gospel. **3** of the four evangelists.

evangelize /ɪ'vændʒə,laɪz/ v.tr. (also **-ise**) **1** (also absol.) preach the gospel to. **2** convert (a person) to Christianity. □ **evangelizer** n. **evangelization** /ɪ,vændʒəlaɪ'zeɪʃ(ə)n/ n. [ME f. eccl.L *evangelizare* f. Gk *euaggelizomai* (as EVANGEL)]

Evans[1] /'ev(ə)nz/, Sir Arthur (John) (1851–1941), English archaeologist. He is best known for his excavations at Knossos (1899–1935), which resulted in the discovery of the Bronze Age civilization of Crete; he called this civilization Minoan after the legendary Cretan king Minos.

Evans[2] /'ev(ə)nz/, Dame Edith (Mary) (1888–1976), English actress. Her stage repertoire encompassed a wide range of Shakespearian and contemporary roles; she acted in the first production of George Bernard Shaw's *Heartbreak House* (1921). She is particularly remembered as Lady Bracknell in Oscar Wilde's *The Importance of Being Earnest*, a role which she portrayed for the first time on stage in 1939 and on film in 1952.

Evans-Pritchard /,ev(ə)nz'prɪtʃəd/, Sir Edward (Evan) (1902–73), English anthropologist. He is noted for his studies of African tribal life and cultures, especially those based on his time spent living in Sudan with the Azande and Nuer peoples in the 1920s and 1930s. Among the important works of social anthropology he wrote are *Witchcraft, Oracles and Magic among the Azande* (1937) and *The Nuer* (1940).

evaporate /ɪˈvæpəˌreɪt/ v. **1** intr. turn from solid or liquid into vapour. **2** intr. & tr. lose or cause to lose moisture as vapour. **3** intr. & tr. disappear or cause to disappear (*our courage evaporated*). □ **evaporated milk** milk concentrated by partial evaporation. □ **evaporator** n. **evaporable** /-rəb(ə)l/ adj. **evaporative** /-rətɪv/ adj. **evaporation** /ɪˌvæpəˈreɪʃ(ə)n/ n. [L *evaporare* (as E-, *vaporare* as VAPOUR)]

evaporite /ɪˈvæpəˌraɪt/ n. Geol. a natural salt or mineral deposit formed by evaporation of water. [EVAPORATE + -ITE¹]

evasion /ɪˈveɪʒ(ə)n/ n. **1** a means of evading; an evasive answer or a prevaricating excuse. **2** the action of evading; prevarication. [ME f. OF f. L *evasio -onis* (as EVADE)]

evasive /ɪˈveɪsɪv/ adj. **1** seeking to evade something. **2** not direct in one's answers etc. **3** enabling or effecting evasion (*evasive action*). **4** (of a person) tending to evasion; habitually practising evasion. □ **evasively** adv. **evasiveness** n.

Eve /iːv/ (in the Bible) the first woman, wife of Adam. Among her children were Cain and Abel.

eve /iːv/ n. **1** the evening or day before a church festival or any date or event (*Christmas Eve; the eve of the funeral*). **2** the time just before anything (*the eve of the election*). **3** archaic evening. [ME, = EVEN²]

evection /ɪˈvekʃ(ə)n/ n. Astron. a perturbation of the moon's motion caused by the sun's attraction. [L *evectio* (as E-, *vehere vect-* carry)]

Evelyn /ˈiːvlɪn/, John (1620–1706), English diarist and writer. He is remembered chiefly for his *Diary* (published posthumously in 1818), which covers most of his life, describing his travels abroad, his contemporaries, and such important historical events as the Great Plague and the Great Fire of London. He was also a pioneer of English forestry and gardening, and a founder member of the Royal Society.

even¹ /ˈiːv(ə)n/ adj., adv., & v. ● adj. (**evener, evenest**) **1** level; flat and smooth. **2 a** uniform in quality; constant. **b** equal in number or amount or value etc. **c** equally balanced. **3** (usu. foll. by *with*) in the same plane or line. **4** (of a person's temper etc.) equable, calm. **5 a** (of a number such as 4, 6) divisible by two without a remainder. **b** bearing such a number (*no parking on even dates*). **c** not involving fractions; exact (*in even dozens*). ● adv. **1** used to invite comparison of the stated assertion, negation, etc., with an implied one that is less strong or remarkable (*never even opened* [let alone read] *the letter; does he even suspect* [not to say realize] *the danger?; ran even faster* [not just as fast as before]; *even if my watch is right we shall be late* [later if it is slow]). **2** used to introduce an extreme case (*even you must realize it; it might even cost £100*). **3** (sometimes foll. by *with* or *though*) in spite of, notwithstanding (*even with the delays, we arrived on time*). ● v. **1** tr. & intr. (often foll. by *up* or *out*) make or become even. **2** tr. (often foll. by *to*) archaic treat as equal or comparable. □ **even as** at the very moment that. **even break** colloq. an equal chance. **even chance** an equal chance of success or failure. **even money 1** betting odds offering the gambler the chance of winning the amount he or she staked. **2** equally likely to happen or not (*it's even money he'll fail to arrive*). **even now 1** now as well as before. **2** at this very moment. **even so 1** notwithstanding that; nevertheless. **2** quite so. **3** in that case as well as in others. **even Stephens** (or **Stevens**) colloq. even, equal, level. **even though** despite the fact that. **get** (or **be**) **even with** have one's revenge on. **of even date** Law & Commerce of the same date. **on an even keel 1** (of a ship or aircraft) not listing. **2** (of a plan or person) untroubled. □ **evenly** adv. **evenness** /ˈiːv(ə)nnɪs/ n. [OE *efen, efne*]

even² /ˈiːv(ə)n/ n. poet. evening. [OE *æfen*]

even-handed /ˌiːv(ə)nˈhændɪd/ adj. impartial, fair. □ **even-handedly** adv. **even-handedness** n.

evening /ˈiːvnɪŋ/ n. & int. ● n. **1** the end part of the day, esp. from about 6 p.m., or sunset if earlier, to bedtime (*this evening; during the evening; evening meal*). **2** this time spent in a particular way (*had a lively evening*). **3** a time compared with this, esp. the last part of a person's life. ● int. good afternoon (see GOOD adj. 14). □ **evening dress** formal dress for evening wear. **evening primrose** a plant of the genus *Oenothera*, with pale yellow flowers that open in the evening, and from whose seeds an oil is extracted for medicinal use. **evening star** a planet, esp. Venus, when visible in the west after sunset. [OE *æfnung*, rel. to EVEN²]

evens /ˈiːv(ə)nz/ n.pl. Brit. = even money.

evensong /ˈiːv(ə)nˌsɒŋ/ n. a service of evening prayer in the Church of England. [EVEN² + SONG]

event /ɪˈvent/ n. **1** a thing that happens or takes place, esp. one of importance. **2 a** the fact of a thing's occurring. **b** a result or outcome. **3** an item in a sports programme, or the programme as a whole. **4** Physics a single occurrence of a process, e.g. the ionization of one atom. **5** something on the result of which money is staked. □ **at all events** (or **in any event**) whatever happens. **event horizon** Astron. a notional boundary around a black hole beyond which no light or other radiation can escape (see BLACK HOLE). **in the event** as it turns (or turned) out. **in the event of** if (a specified thing) happens. **in the event that** if it happens that. [L *eventus* f. *evenire event-* happen (as E-, *venire* come)]

eventful /ɪˈventfʊl/ adj. marked by noteworthy events. □ **eventfully** adv. **eventfulness** n.

eventide /ˈiːv(ə)nˌtaɪd/ n. archaic or poet. = EVENING. □ **eventide home** a home for the elderly, orig. one run by the Salvation Army. [OE *æfentīd* (as EVEN², TIDE)]

eventing /ɪˈventɪŋ/ n. Brit. participation in horse trials, esp. dressage and showjumping. [EVENT 3 as in *three-day event*]

eventless /ɪˈventlɪs/ adj. without noteworthy or remarkable events.

eventual /ɪˈventʃʊəl, -tjʊəl/ adj. occurring or existing in due course or at last; ultimate. □ **eventually** adv. [as EVENT, after *actual*]

eventuality /ɪˌventʃʊˈælɪti, ɪˌventjʊ-/ n. (pl. **-ies**) a possible event or outcome.

eventuate /ɪˈventʃʊˌeɪt, ɪˈventjʊ-/ v.intr. formal **1** turn out in a specified way as the result. **2** (often foll. by *in*) result. □ **eventuation** /ɪˌventʃʊˈeɪʃ(ə)n, ɪˌventjʊ-/ n. [as EVENT, after *actuate*]

ever /ˈevə(r)/ adv. **1** at all times; always (*ever hopeful; ever after*). **2** at any time (*have you ever been to Paris?; nothing ever happens; as good as ever*). **3** as an emphatic word: **a** in any way; at all (*how ever did you do it?; when will they ever learn?*). **b** (prec. by *as*) in any manner possible (*be as quick as ever you can*). **4** (often foll. by *comb.*) constantly (*ever-present; ever-recurring*). **5** (foll. by *so*, *such*) Brit. colloq. very; very much (*is ever so easy; was ever such a nice man; thanks ever so*). **6** (foll. by *compar.*) constantly, continually; increasingly (*grew ever larger; ever more sophisticated*). □ **did you ever?** colloq. did you ever hear or see the like? **ever since** throughout the period since. **for ever 1** for all future time. **2** colloq. for a long time (cf. FOREVER). [OE *æfre*]

Everest, Mount /ˈevərɪst/ a mountain in the Himalayas, on the border between Nepal and Tibet. Rising to 8,848 m (29,028 ft), it is the highest mountain in the world. Named after Sir George Everest (1790–1866), British surveyor-general of India, it was first climbed in 1953 by Sir Edmund Hillary and Tenzing Norgay.

Everglades, the /ˈevəˌgleɪdz/ a vast area of marshland and coastal mangrove in southern Florida. A national park protects endangered species.

evergreen /ˈevəˌgriːn/ adj. & n. ● adj. **1** always green or fresh. **2** (of a plant) retaining green leaves throughout the year. ● n. an evergreen plant (cf. DECIDUOUS).

everlasting /ˌevəˈlɑːstɪŋ/ adj. & n. ● adj. **1** lasting for ever. **2** lasting for a long time, esp. so as to become unwelcome. **3** (of flowers) keeping their shape and colour when dried. ● n. **1** eternity. **2** = IMMORTELLE. □ **everlastingly** adv. **everlastingness** n.

evermore /ˌevəˈmɔː(r)/ adv. for ever; always.

Evert /ˈevət/, Christine Marie ('Chris') (b.1954), American tennis player. Her career began at an early age as a Wightman Cup player (1971) and included winning both the US and French Open championships six times and three Wimbledon titles (1974; 1976; 1981).

evert /ɪˈvɜːt/ v.tr. Physiol. turn (an organ etc.) outwards or inside out. □ **eversion** /ɪˈvɜːʃ(ə)n/ n. [L *evertere* (as E-, *vertere vers-* turn)]

every /ˈevri/ adj. **1** each single (*heard every word; watched her every movement*). **2** each at a specified interval in a series (*take every third one; comes every four days*). **3** all possible; the utmost degree of (*there is every prospect of success*). □ **every bit as** colloq. (in comparisons) quite as (*every bit as good*). **every now and again** (or **now and then**) from time to time. **every one** each one (see also EVERYONE). **every other** each second in a series (*every other day*). **every so often** at intervals; occasionally. **every time** colloq. **1** without exception. **2** certainly. **every which way** N. Amer. colloq. **1** in all directions. **2** in a disorderly manner. [OE *æfre ælc* ever each]

everybody /ˈevrɪˌbɒdi/ pron. every person.

everyday /ˈevrɪˌdeɪ/ adj. **1** occurring every day. **2** suitable for or used on ordinary days. **3** commonplace, usual. **4** mundane; mediocre, inferior.

Everyman /ˈevrɪˌmæn/ n. the ordinary or typical human being; the 'man in the street'. [the principal character in a 15th-c. morality play]

everyone /ˈevrɪˌwʌn/ pron. every person; everybody.

everything /ˈevrɪˌθɪŋ/ pron. **1** all things; all the things of a group or class. **2** colloq. **a** a great deal (he owes her everything). **b** the essential consideration (speed is everything). □ **have everything** colloq. possess every attraction, advantage, etc.

everywhere /ˈevrɪˌweə(r)/ adv. **1** in every place. **2** colloq. in many places.

evict /ɪˈvɪkt/ v.tr. expel (a tenant) from a property by legal process. □ **evictor** n. **eviction** /ɪˈvɪkʃ(ə)n/ n. [L evincere evict- (as E-, vincere conquer)]

evidence /ˈevɪd(ə)ns/ n. & v. ● n. **1** (often foll. by for, of) the available facts, circumstances, etc. supporting or otherwise a belief, proposition, etc., or indicating whether or not a thing is true or valid. **2** Law **a** information given personally or drawn from a document etc. and tending to prove a fact or proposition. **b** statements or proofs admissible as testimony in a lawcourt. **3** clearness, obviousness. ● v.tr. be evidence of; attest. □ **call in evidence** Law summon (a person) as a witness. **in evidence** noticeable, conspicuous. **Queen's** (or **King's** or **state's**) **evidence** Law evidence for the prosecution given by a participant in or accomplice to the crime at issue. [ME f. OF f. L evidentia (as EVIDENT)]

evident /ˈevɪd(ə)nt/ adj. plain or obvious (visually or intellectually); manifest. [ME f. OF evident or L evidere evident- (as E-, videre see)]

evidential /ˌevɪˈdenʃ(ə)l/ adj. of or providing evidence. □ **evidentially** adv.

evidentiary /ˌevɪˈdenʃərɪ/ adj. = EVIDENTIAL.

evidently /ˈevɪd(ə)ntlɪ/ adv. **1** plainly, obviously. **2** (used parenthetically or qualifying a whole sentence) it is plain that; it would seem that (evidently, she was too late for the train). **3** (said in reply) so it appears.

evil /ˈiːv(ə)l/ adj. & n. ● adj. **1** morally bad; wicked. **2** harmful or tending to harm, esp. intentionally or characteristically. **3** disagreeable or unpleasant (has an evil temper). **4** unlucky; causing misfortune (evil days). ● n. **1** an evil thing; an instance of something evil. **2** evil quality; wickedness, harm. □ **evil eye** a gaze or stare superstitiously believed to be able to cause material harm. **speak evil of** slander. **evil-minded** having evil intentions. □ **evilly** adv. **evilness** n. [OE yfel f. Gmc]

evildoer /ˈiːv(ə)lˌduːə(r)/ n. a person who does evil. □ **evildoing** n.

evince /ɪˈvɪns/ v.tr. **1** indicate or make evident. **2** show that one has (a quality). □ **evincible** adj. **evincive** adj. [L evincere: see EVICT]

eviscerate /ɪˈvɪsəˌreɪt/ v.tr. formal **1** disembowel. **2** empty or deprive of essential contents. □ **evisceration** /ɪˌvɪsəˈreɪʃ(ə)n/ n. [L eviscerare eviscerat- (as E-, VISCERA)]

evocative /ɪˈvɒkətɪv/ adj. tending to evoke (esp. feelings or memories). □ **evocatively** adv. **evocativeness** n.

evoke /ɪˈvəʊk/ v.tr. **1** inspire or draw forth (memories, feelings, a response, etc.). **2** = INVOKE 3. □ **evoker** n. **evocation** /ˌevəˈkeɪʃ(ə)n/ n. [L evocare (as E-, vocare call)]

evolute /ˈiːvəˌluːt, -ˌljuːt/ n. (in full **evolute curve**) Math. a curve which is the locus of the centres of curvature of another curve that is its involute. [L evolutus past part. (as EVOLVE)]

evolution /ˌiːvəˈluːʃ(ə)n, -ˈljuːʃ(ə)n/ n. **1** gradual development, esp. from a simple to a more complex form. **2** Biol. a process by which species develop from earlier forms. (See note below.) **3** the appearance or presentation of events etc. in due succession (the evolution of the plot). **4** a change in the disposition of troops or ships. **5** the giving off or evolving of gas, heat, etc. **6** an opening out. **7** the unfolding of a curve. **8** Math. the extraction of a root from any given power (cf. INVOLUTION). □ **evolutional** adj. **evolutionally** adv. **evolutionary** adj. **evolutionarily** adv. [L evolutio unrolling (as EVOLVE)]

▪ The philosophical speculation that primitive organisms may change into more complex forms and ascend some imaginary hierarchical ladder is an idea that may be traced back to the ancient Greeks. Jean Baptiste de Lamarck was one of the first to propose a theory of organic evolution, against the prevailing belief in special creation: his suggestion, not now accepted, was that species transform by the inheritance of acquired characteristics. Sir Charles Lyell's demonstration that geological deposits are the product of slow small changes over long periods of time allowed Darwin to develop a theory of evolution by gradual natural selection. In On the Origin of Species (1859) he proposed that an isolated variety within a population might give rise to a new species by survival of the form best suited to its new surroundings; this is what the term 'survival of the fittest' means. Most modern explanations of evolution are essentially modifications of Darwin's theory. Combined with recent genetical theory, and the new understanding of the role of DNA, evolution is the great unifying concept in modern biology. (See also SOCIOBIOLOGY.)

evolutionist /ˌiːvəˈluːʃənɪst, -ˈljuːʃənɪst/ n. a person who believes in evolution as explaining the origin of species. □ **evolutionism** n. **evolutionistic** /-ˌluːʃəˈnɪstɪk, -ˌljuːʃə-/ adj.

evolve /ɪˈvɒlv/ v. **1** intr. & tr. develop gradually by a natural process. **2** tr. work out or devise (a theory, plan, etc.). **3** intr. & tr. unfold; open out. **4** tr. give off (gas, heat, etc.). □ **evolvable** adj. **evolvement** n. [L evolvere evolut- (as E-, volvere roll)]

Évros /ˈɛvrɔs/ the Greek name for the MARITSA.

Évvoia see EUBOEA.

evzone /ˈevzəʊn/ n. a member of a select Greek infantry regiment. [mod.Gk euzōnos f. Gk, = dressed for exercise (as EU-, zōnē belt)]

ewe /juː/ n. a female sheep. □ **ewe lamb** a person's most cherished possession (2 Sam. 12). **ewe-necked** (of a horse) having a thin concave neck. [OE ēowu f. Gmc]

ewer /ˈjuːə(r)/ n. a large pitcher or water-jug with a wide mouth. [ME f. ONF eviere, aiguiere, ult. f. L aquarius of water f. aqua water]

ex[1] /eks/ prep. **1** (of goods) sold from (ex-works). **2** (of stocks or shares) without, excluding. [L, = out of]

ex[2] /eks/ n. colloq. a former husband or wife. [absol. use of EX-[1] 2]

ex-[1] prefix (also **e-** before some consonants, **ef-** before f) **1** /eks, egz, ɪks, ɪgz/ forming verbs meaning: **a** out, forth (exclude; exit). **b** upward (extol). **c** thoroughly (excruciate). **d** bring into a state (exasperate). **e** remove or free from (expatriate; exonerate). **2** /eks/ forming nouns from titles of office, status, etc., meaning 'formerly' (ex-convict; ex-president; ex-wife). [L f. ex out of]

ex-[2] /eks/ prefix out (exodus). [Gk f. ex out of]

exa- /ˈeksə/ comb. form denoting a factor of 10^{18}. [perh. f. HEXA-]

exacerbate /ɪɡˈzæsəˌbeɪt/ v.tr. **1** make (pain, anger, etc.) worse. **2** irritate (a person). □ **exacerbation** /-ˌzæsəˈbeɪʃ(ə)n/ n. [L exacerbare (as EX-[1], acerbus bitter)]

exact /ɪɡˈzækt/ adj. & v. ● adj. **1** accurate; correct in all details (an exact description). **2 a** precise. **b** (of a person) tending to precision. ● v.tr. (often foll. by from, of) **1** demand and enforce payment of (money, fees, etc.) from a person. **2 a** demand; insist on. **b** (of circumstances) require urgently. □ **exact science** a science admitting of absolute or quantitative precision. □ **exactable** adj. **exactitude** n. **exactness** n. **exactor** n. [L exigere exact- (as EX-[1], agere drive)]

exacting /ɪɡˈzæktɪŋ/ adj. **1** making great demands. **2** calling for much effort. □ **exactingly** adv. **exactingness** n.

exaction /ɪɡˈzækʃ(ə)n/ n. **1** the act or an instance of exacting; the process of being exacted. **2 a** an illegal or exorbitant demand; an extortion. **b** a sum or thing exacted. [ME f. L exactio (as EXACT)]

exactly /ɪɡˈzæktlɪ/ adv. **1** accurately, precisely; in an exact manner (worked it out exactly). **2** in exact terms (exactly when did it happen?). **3** (said in reply) quite so; I quite agree. **4** just; in all respects. □ **not exactly** colloq. **1** by no means. **2** not precisely.

exaggerate /ɪɡˈzædʒəˌreɪt/ v.tr. **1** (also absol.) give an impression of (a thing), esp. in speech or writing, that makes it seem larger or greater etc. than it really is. **2** enlarge or alter beyond normal or due proportions (spoke with exaggerated politeness). □ **exaggeratedly** adv. **exaggeratingly** adv. **exaggerator** n. **exaggerative** /-rətɪv/ adj. **exaggeration** /-ˌzædʒəˈreɪʃ(ə)n/ n. [L exaggerare (as EX-[1], aggerare heap up f. agger heap)]

exalt /ɪɡˈzɔːlt/ v.tr. **1** raise in rank or power etc. **2** praise highly. **3** (usu. as **exalted** adj.) **a** make lofty or noble (exalted aims; an exalted style). **b** make rapturously excited. **4** (as **exalted** adj.) elevated in rank or character; eminent, celebrated. **5** stimulate (a faculty, etc.) to greater activity; intensify, heighten. □ **exaltedly** adv. **exaltedness** n. [ME f. L exaltare (as EX-[1], altus high)]

exaltation /ˌegzɔːlˈteɪʃ(ə)n/ n. **1** the act or an instance of exalting; the state of being exalted. **2** elation; rapturous emotion. [ME f. OF exaltation or LL exaltatio (as EXALT)]

exam /ɪɡˈzæm/ n. colloq. = EXAMINATION 3.

examination /ɪɡˌzæmɪˈneɪʃ(ə)n/ n. **1** the act or an instance of examining; the state of being examined. **2** a detailed inspection. **3 a** the testing of the proficiency or knowledge of students or other candidates for a qualification by oral or written questions. **b** a test of this kind. **4** an instance of examining or being examined medically. **5** Law the formal questioning of the accused or of a witness in court.

□ **examination paper 1** the printed questions in an examination. **2** a candidate's set of answers. □ **examinational** adj. [ME f. OF f. L examinatio -onis (as EXAMINE)]

examine /ɪɡˈzæmɪn/ v. **1** tr. inquire into the nature or condition etc. of. **2** tr. look closely or analytically at. **3** tr. test the proficiency of, esp. by examination (see EXAMINATION 3a). **4** tr. check the health of (a patient) by inspection or experiment. **5** tr. Law formally question (the accused or a witness) in court. **6** intr. (foll. by into) inquire. □ **examinable** adj. **examiner** n. **examinee** /-ˌzæmɪˈniː/ n. [ME f. OF examiner f. L examinare weigh, test f. examen tongue of a balance, ult. f. exigere examine, weigh: see EXACT]

example /ɪɡˈzɑːmp(ə)l/ n. & v. ● n. **1** a thing characteristic of its kind or illustrating a general rule. **2** a person, thing, or piece of conduct, regarded in terms of its fitness to be imitated or likelihood of being imitated (must set him an example; you are a bad example). **3** a circumstance or treatment seen as a warning to others; a person so treated (shall make an example of you). **4** a problem or exercise designed to illustrate a rule. ● v.tr. (usu. in passive) serve as an example of. □ **for example** by way of illustration. [ME f. OF f. L exemplum (as EXEMPT)]

exanimate /ɪɡˈzænɪmət/ adj. **1** dead, lifeless (esp. in appearance); inanimate. **2** lacking animation or courage. [L exanimatus past part. of exanimare deprive of life, f. EX-¹ + anima breath of life]

ex ante /eks ˈænti/ adj. Econ. based on predicted or expected results; forecast, anticipated. [mod.L, = from before]

exanthema /ˌeksænˈθiːmə/ n. (pl. **exanthemata** /-mətə/) Med. a skin rash accompanying any eruptive disease or fever. [LL f. Gk exanthēma eruption f. exantheō (as EX-², anthos blossom)]

exarch /ˈeksɑːk/ n. in the Orthodox Church, a bishop lower in rank than a patriarch and having jurisdiction wider than the metropolitan of a diocese. □ **exarchate** /ˈeksɑːˌkeɪt, eksˈɑːkeɪt/ n. [eccl.L f. Gk exarkhos (as EX-², arkhos ruler)]

exasperate /ɪɡˈzɑːspəˌreɪt/ v.tr. **1** (often as **exasperated** adj. or **exasperating** adj.) irritate intensely; infuriate, enrage. **2** make (a pain, ill feeling, etc.) worse. □ **exasperatedly** adv. **exasperatingly** adv. **exasperation** /-ˌzɑːspəˈreɪʃ(ə)n/ n. [L exasperare exasperat- (as EX-¹, asper rough)]

Excalibur /eksˈkælɪbə(r)/ (in Arthurian legend) King Arthur's magic sword.

ex cathedra /ˌeks kəˈθiːdrə/ adj. & adv. with full authority (esp. of a papal pronouncement, implying infallibility as doctrinally defined). [L, = from the (teacher's) chair]

excavate /ˈekskəˌveɪt/ v.tr. **1 a** make (a hole or channel) by digging. **b** dig out material from (the ground). **2** reveal or extract by digging. **3** (also absol.) Archaeol. dig systematically into the ground to explore (a site). □ **excavator** n. **excavation** /ˌekskəˈveɪʃ(ə)n/ n. [L excavare (as EX-¹, cavus hollow)]

exceed /ɪkˈsiːd/ v.tr. **1** (often foll. by by an amount) be more or greater than (in number, extent, etc.). **2** go beyond or do more than is warranted by (a set limit, esp. of one's instructions or rights). **3** surpass, excel (a person or achievement). [ME f. OF exceder f. L excedere (as EX-¹, cedere cess- go)]

exceeding /ɪkˈsiːdɪŋ/ adj. & adv. ● adj. **1** surpassing in amount or degree. **2** pre-eminent. ● adv. archaic = EXCEEDINGLY 2.

exceedingly /ɪkˈsiːdɪŋli/ adv. **1** very; to a great extent. **2** surpassingly, pre-eminently.

excel /ɪkˈsel/ v. (**excelled, excelling**) (often foll. by in, at) **1** tr. be superior to. **2** intr. be pre-eminent or the most outstanding (excels at games). □ **excel oneself** surpass one's previous performance. [ME f. L excellere (as EX-¹, celsus lofty)]

excellence /ˈeksələns/ n. **1** the state of excelling; surpassing merit or quality. **2** the activity etc. in which a person excels. [ME f. OF excellence or L excellentia (as EXCEL)]

Excellency /ˈeksələnsi/ n. (pl. **-ies**) (usu. prec. by Your, His, Her, Their) a title used in addressing or referring to certain high officials, e.g. ambassadors and governors, and (in some countries) senior Church dignitaries. [ME f. L excellentia (as EXCEL)]

excellent /ˈeksələnt/ adj. extremely good; pre-eminent. □ **excellently** adv. [ME f. OF (as EXCEL)]

excelsior /ɪkˈselsɪˌɔː(r)/ int. & n. ● int. higher, outstanding (esp. as a motto or trademark). ● n. soft wood shavings used for stuffing, packing, etc. [L, compar. of excelsus lofty]

excentric var. of ECCENTRIC (in technical senses).

except /ɪkˈsept/ v., prep., & conj. ● v.tr. (often as **excepted** adj. placed after object) exclude from a general statement, condition, etc. (excepted him from the amnesty; present company excepted). ● prep. (often foll. by for or that) not including; other than (all failed except him; all here except for John; is all right except that it is too long). ● conj. archaic unless (except he be born again). [ME f. L excipere except- (as EX-¹, capere take)]

excepting /ɪkˈseptɪŋ/ prep. & conj. ● prep. = EXCEPT prep. ● conj. archaic = EXCEPT conj.

exception /ɪkˈsepʃ(ə)n/ n. **1** the act or an instance of excepting; the state of being excepted (made an exception in my case). **2** a thing that has been or will be excepted. **3** an instance that does not follow a rule. □ **take exception** (often foll. by to) object; be resentful (about). **with the exception of** except; not including. [ME f. OF f. L exceptio -onis (as EXCEPT)]

exceptionable /ɪkˈsepʃənəb(ə)l/ adj. open to objection. □ **exceptionably** adv.

exceptional /ɪkˈsepʃən(ə)l/ adj. **1** forming an exception. **2** unusual; not typical (exceptional circumstances). **3** unusually good; outstanding. □ **exceptionally** adv. **exceptionality** /-ˌsepʃəˈnælɪti/ n.

excerpt n. & v. ● n. /ˈeksɜːpt/ a short extract from a book, film, piece of music, etc. ● v.tr. /ɪkˈsɜːpt/ (also absol.) **1** take an excerpt or excerpts from (a book etc.). **2** take (an extract) from a book etc. □ **excerptible** /ɪkˈsɜːptɪb(ə)l/ adj. **excerption** /ɪkˈsɜːpʃ(ə)n/ n. [L excerpere excerpt- (as EX-¹, carpere pluck)]

excess /ɪkˈses, ˈekses/ n. & adj. ● n. **1** the state or an instance of exceeding. **2** the amount by which one quantity or number exceeds another. **3** exceeding of a proper or permitted limit. **4 a** the overstepping of the accepted limits of moderation, esp. intemperance in eating or drinking. **b** (in pl.) outrageous or immoderate behaviour. **5** an extreme or improper degree or extent (an excess of cruelty). **6** part of an insurance claim to be paid by the insured, esp. by prior agreement. ● attrib.adj. /usu. ˈekses/ **1** that exceeds a limited or prescribed amount (excess weight). **2** required as extra payment (excess postage). □ **excess baggage** (or **luggage**) that exceeding a weight allowance and liable to an extra charge. **in** (or **to**) **excess** exceeding the proper amount or degree. **in excess of** more than; exceeding. [ME f. OF exces f. L excessus (as EXCEED)]

excessive /ɪkˈsesɪv/ adj. **1** too much or too great. **2** more than what is normal or necessary. □ **excessively** adv. **excessiveness** n.

exchange /ɪksˈtʃeɪndʒ/ n. & v. ● n. **1** the act or an instance of giving one thing and receiving another in its place. **2 a** the giving of money for its equivalent in the money of the same or esp. another country. **b** the fee or percentage charged for this. **3** a place or installation containing the apparatus for connecting telephone calls. **4 a** a place where merchants, bankers, etc. gather to transact business. **b** (**the Exchange**) see STOCK EXCHANGE. **5 a** an office where certain information is given or a service provided, usu. involving two parties. **b** an employment office. **6** a system of settling debts between persons (esp. in different countries) without the use of money, by bills of exchange (see BILL¹). **7 a** a short conversation, esp. a disagreement or quarrel. **b** a sequence of letters between correspondents. **8** Chess the capture of an important piece (esp. a rook) by one player at the loss of a minor piece to the opposing player. **9** (attrib.) forming part of an exchange, e.g. of personnel between institutions (an exchange student). ● v. **1** tr. (often foll. by for) give or receive (one thing) in place of another. **2** tr. give and receive as equivalents (e.g. things or people, blows, information, etc.); give one and receive another of. **3** intr. (often foll. by with) make an exchange. □ **exchange rate** the value of one currency in terms of another. **in exchange** (often foll. by for) as a thing exchanged (for). □ **exchanger** n. **exchangeable** adj. **exchangeability** /-ˌtʃeɪndʒəˈbɪlɪti/ n. [ME f. OF eschangier f. Rmc (as EX-¹, CHANGE)]

exchange control n. governmental restriction on the movement of currency between countries. Exchange controls are chiefly used by developing countries, but have also been used from time to time by Western nations. Restrictions on the export of currency from the UK were lifted in 1979.

Exchange Rate Mechanism (abbr. **ERM**) in the European Community, a system for allowing the value of participating currencies to fluctuate to a defined degree in relation to each other so as to control exchange rates within the European Monetary System. The currency of each participating country is given a rate of exchange with the ecu, from which it is allowed to fluctuate by no more than a specified amount; if a currency moves outside its allotted range the government must alter its economic policies or reset the currency's rate with the ecu. Britain joined the ERM in 1990, but withdrew in

1992, as did Italy. In mid-1993 the ERM experienced a near-collapse when the Bundesbank decided not to lower its main interest rate and the French and Belgian francs and Danish krone came under pressure from currency dealers and had difficulty in remaining within their allotted ranges. As a result, it was agreed to allow most currencies (apart from those of Germany and the Netherlands) to fluctuate within a far broader band than before.

exchequer /ɪksˈtʃekə(r)/ n. **1** (also **Exchequer**) Brit. the former government department in charge of national revenue. (See note below.) **2** a royal or national treasury. **3** the money of a private individual or group. [ME f. AF escheker, OF eschequier f. med.L scaccarium chessboard (its orig. sense, with ref. to keeping accounts on a chequered cloth)]

▪ The Normans in England created two departments dealing with finance: one was the Treasury, which received and paid out money on behalf of the monarch, and the other was the Exchequer, which was itself divided into two parts, lower and upper. The lower Exchequer was an office for receiving money and was connected to the Treasury; the upper Exchequer was a court of law dealing with cases related to revenue, and was merged with the High Court of Justice in 1880. The Exchequer now denotes the account at the Bank of England into which tax receipts and other public monies are paid, the balance of which forms the Consolidated Fund. Its name survives also in the title Chancellor of the Exchequer.

excimer /ˈeksɪmə(r)/ n. Chem. & Physics a dimer existing only in an excited state, used in some lasers. [EXCITED + DIMER]

excise[1] n. & v. ● n. /ˈeksaɪz/ **1 a** a duty or tax levied on goods and commodities produced or sold within the country of origin. **b** a tax levied on certain licences. **2** Brit. a former government office collecting excise. ¶ Now the Board of Customs and Excise. ● v.tr. /ɪkˈsaɪz/ **1** charge excise on (goods). **2** force (a person) to pay excise. [MDu. excijs, accijs, perh. f. Rmc: rel. to CENSUS]

excise[2] /ɪkˈsaɪz/ v.tr. **1** remove (a passage of a book etc.). **2** cut out (an organ etc.) by surgery. □ **excision** /-ˈsɪʒ(ə)n/ n. [L excidere excis- (as EX-[1], caedere cut)]

exciseman /ˈeksaɪzˌmæn/ n. (pl. -men) Brit. hist. an officer responsible for collecting excise duty.

excitable /ɪkˈsaɪtəb(ə)l/ adj. **1** (esp. of a person) easily excited. **2** (of an organism, tissue, etc.) responding to a stimulus, or susceptible to stimulation. □ **excitably** adv. **excitability** /-ˌsaɪtəˈbɪlɪti/ n.

excitation /ˌeksɪˈteɪʃ(ə)n/ n. **1 a** the act or an instance of exciting. **b** the state of being excited; excitement. **2** the action of an organism, tissue, etc., resulting from stimulation. **3** Electr. **a** the process of applying current to the winding of an electromagnet to produce a magnetic field. **b** the process of applying a signal voltage to the control electrode of an electron tube or the base of a transistor. **4** Physics the process in which an atom etc. acquires a higher energy state.

excite /ɪkˈsaɪt/ v.tr. **1** rouse the feelings or emotions of (a person). **b** bring into play; rouse up (feelings, faculties, etc.). **c** arouse sexually. **2** provoke; bring about (an action or active condition). **3** promote the activity of (an organism, tissue, etc.) by stimulus. **4** Electr. **a** cause (a current) to flow in the winding of an electromagnet. **b** supply a signal. **5** Physics **a** cause the emission of (a spectrum). **b** cause (a substance) to emit radiation. **c** put (an atom etc.) into a state of higher energy. □ **excitement** n. **exciter** n. (esp. in senses 4, 5). **excitative** /-tətɪv/ adj. **excitatory** /-tətərɪ/ adj. **excitant** /ˈeksɪt(ə)nt, ɪkˈsaɪt(ə)nt/ adj. & n. [ME f. OF exciter or L excitare frequent. of exciere (as EX-[1], ciere set in motion)]

excited /ɪkˈsaɪtɪd/ adj. **1** that has been excited. **2** roused by strong emotion; characterized by excitement. □ **excited state** Physics a state of a quantized system with more energy than the ground state. □ **excitedly** adv. **excitedness** n.

exciting /ɪkˈsaɪtɪŋ/ adj. arousing great interest or enthusiasm; stirring. □ **excitingly** adv. **excitingness** n.

exciton /ˈeksaɪtɒn, ˈeksɪˌtɒn/ n. Physics a mobile concentration of energy in a crystalline material consisting of an excited electron and an associated hole. [EXCITATION + -ON]

exclaim /ɪkˈskleɪm/ v. **1** intr. cry out suddenly, esp. in anger, surprise, pain, etc. **2** tr. (foll. by that) utter by exclaiming. [F exclamer or L exclamare (as EX-[1]: cf. CLAIM)]

exclamation /ˌekskləˈmeɪʃ(ə)n/ n. **1** the act or an instance of exclaiming. **2** words exclaimed; a strong sudden cry. □ **exclamation mark** (US **point**) a punctuation mark (!) indicating an exclamation. [ME f. OF exclamation or L exclamatio (as EXCLAIM)]

exclamatory /ɪkˈsklæmətərɪ/ adj. of or serving as an exclamation.

exclave /ˈekskleɪv/ n. a portion of territory of one state completely surrounded by territory of another or others, as viewed by the home territory (cf. ENCLAVE). [EX-[1] + ENCLAVE]

exclosure /ɪkˈskləʊʒə(r)/ n. Forestry etc. an area from which unwanted animals are excluded. [EX-[1] + ENCLOSURE]

exclude /ɪkˈskluːd/ v.tr. **1** shut or keep out (a person or thing) from a place, group, privilege, etc. **2** expel and shut out. **3** remove from consideration (no theory can be excluded). **4** prevent the occurrence of; make impossible (excluded all doubt). □ **excluded middle** Logic the principle that of two contradictory propositions one must be true. □ **excludable** adj. **excluder** n. [ME f. L excludere exclus- (as EX-[1], claudere shut)]

exclusion /ɪkˈskluːʒ(ə)n/ n. the act or an instance of excluding; the state of being excluded. □ **exclusion order** Brit. an official order preventing a person (esp. a suspected terrorist) from entering the UK. **exclusion principle** Physics see PAULI. **to the exclusion of** so as to exclude. □ **exclusionary** adj. [L exclusio (as EXCLUDE)]

exclusionist /ɪkˈskluːʒənɪst/ adj. & n. ● adj. favouring exclusion, esp. from rights or privileges. ● n. **1** a person favouring exclusion. **2** Austral. hist. an Australian settler opposed to the emancipation of ex-convicts.

exclusive /ɪkˈskluːsɪv/ adj. & n. ● adj. **1** excluding other things. **2** (predic.; foll. by of) not including; except for. **3** tending to exclude others, esp. socially; select. **4** catering for few or select customers; high-class. **5 a** (of a commodity) not obtainable elsewhere. **b** (of a newspaper article) not published elsewhere. **6** (predic.; foll. by to) restricted or limited to; existing or available only in. **7** (of terms etc.) excluding all but what is specified. **8** employed or followed or held to the exclusion of all else (my exclusive occupation; exclusive rights). ● n. an article or story published by only one newspaper or periodical. □ **exclusively** adv. **exclusiveness** n. **exclusivity** /ˌeksklu-ˈsɪvɪti/ n. [med.L exclusivus (as EXCLUDE)]

Exclusive Brethren n.pl. one of two principal divisions of the Plymouth Brethren (the other is the Open Brethren), which was formed in 1849 as a result of doctrinal and other differences; collectively, the members of this division. The Exclusive Brethren are the more rigorous of the two; members greatly restrict their contact with those outside the Brethren and with modern technology etc.

excogitate /eksˈkɒdʒɪˌteɪt/ v.tr. think out; contrive. □ **excogitation** /-ˌkɒdʒɪˈteɪʃ(ə)n/ n. [L excogitare excogitat- (as EX-[1], cogitare COGITATE)]

excommunicate v., adj., & n. Eccl. ● v.tr. /ˌekskəˈmjuːnɪˌkeɪt/ officially exclude (a person) from participation in the sacraments, or from formal communion with the Church. ● adj. /ˌekskəˈmjuːnɪkət/ excommunicated. ● n. /ˌekskəˈmjuːnɪkət/ an excommunicated person. □ **excommunicative** /-ˈkeɪtɪv/ adj. **excommunicator** /-ˌkeɪtə(r)/ n. **excommunicatory** /-ˌkeɪtərɪ/ adj. **excommunication** /-ˌmjuːnɪˈkeɪʃ(ə)n/ n. [L excommunicare -atus (as EX-[1], communis COMMON)]

ex-con /eksˈkɒn/ n. colloq. an ex-convict; a former inmate of a prison. [abbr.]

excoriate /ɪkˈskɔːrɪˌeɪt/ v.tr. **1** remove part of the skin of (a person etc.) by abrasion. **b** strip or peel off (skin). **2** censure severely. □ **excoriation** /-ˌskɔːrɪˈeɪʃ(ə)n/ n. [L excoriare excoriat- (as EX-[1], corium hide)]

excrement /ˈekskrɪmənt/ n. (in sing. or pl.) faeces. □ **excremental** /ˌekskrɪˈment(ə)l/ adj. [F excrément or L excrementum (as EXCRETE)]

excrescence /ɪkˈskres(ə)ns/ n. **1** an abnormal or morbid outgrowth on the body or a plant. **2** an ugly addition. □ **excrescent** adj. **excrescential** /ˌekskrɪˈsenʃ(ə)l/ adj. [L excrescentia (as EX-[1], crescere grow)]

excreta /ɪkˈskriːtə/ n.pl. waste discharged from the body, esp. faeces and urine. [L neut. pl.: see EXCRETE]

excrete /ɪkˈskriːt/ v.tr. (also absol.) (of an animal or plant) separate and expel (waste matter) as a result of metabolism. □ **excreter** n. **excretive** adj. **excretory** adj. **excretion** /-ˈskriːʃ(ə)n/ n. [L excernere excret- (as EX-[1], cernere sift)]

excruciate /ɪkˈskruːʃɪˌeɪt/ v.tr. (esp. as **excruciating** adj.) torment acutely (a person's senses); torture mentally. □ **excruciatingly** adv. **excruciation** /-ˌskruːʃɪˈeɪʃ(ə)n/ n. [L excruciare excruciat- (as EX-[1], cruciare torment f. crux crucis cross)]

exculpate /ˈekskʌlˌpeɪt/ v.tr. formal **1** free from blame. **2** (foll. by from) clear (a person) of a charge. □ **exculpation** /ˌekskʌlˈpeɪʃ(ə)n/ n. **exculpatory** /eksˈkʌlpətərɪ/ adj. [med.L exculpare exculpat- (as EX-[1], culpa blame)]

excursion /ɪkˈskɜːʃ(ə)n/ n. **1** a short journey or ramble for pleasure, with return to the starting-point. **2** a digression. **3** Astron. a deviation from a regular path. **4** *archaic* a sortie (cf. *alarums and excursions*). □ **excursional** *adj.* **excursionary** *adj.* **excursionist** *n.* [L *excursio* f. *excurrere excurs-* (as EX-[1], *currere* run)]

excursive /ɪkˈskɜːsɪv/ *adj.* digressive; diverse. □ **excursively** *adv.* **excursiveness** *n.*

excursus /ekˈskɜːsəs/ *n.* (*pl.* **excursuses** or same) **1** a detailed discussion of a special point in a book, usu. in an appendix. **2** a digression in a narrative. [L, verbal noun formed as EXCURSION]

excuse *v. & n.* ● *v.tr.* /ɪkˈskjuːz/ **1** attempt to lessen the blame attaching to (a person, act, or fault). **2** (of a fact or circumstance) serve in mitigation of (a person or act). **3** obtain exemption for (a person or oneself). **4** (foll. by *from*, or with double object) release (a person) from a duty etc. (*excused from supervision duties*; *excused him the fee*). **5** overlook or forgive (a fault or offence). **6** (foll. by *for*) forgive (a person) for a fault. **7** not insist upon (what is due). **8** *refl.* apologize for leaving. ● *n.* /ɪkˈskjuːs/ **1** a reason put forward to mitigate or justify an offence, fault, etc. **2** an apology (*made my excuses*). **3** (foll. by *for*) *colloq.* a poor or inadequate example of. **4** the action of excusing; indulgence, pardon. □ **be excused** be allowed to leave a room etc., e.g. to go to the lavatory. **excuse me** a polite apology for lack of ceremony, for an interruption etc., or for disagreeing. **excuse-me** a dance in which dancers may interrupt other pairs to change partners. □ **excusable** /-ˈskjuːzəb(ə)l/ *adj.* **excusably** *adv.* **excusatory** /-ˈskjuːzətəri/ *adj.* [ME f. OF *escuser* f. L *excusare* (as EX-[1], *causa* CAUSE, accusation)]

ex-directory /ˌeksdaɪˈrektərɪ, -dɪˈrektərɪ/ *adj. Brit.* not listed in a telephone directory, at the wish of the subscriber.

ex div. *abbr.* ex dividend.

ex dividend *adj. & adv.* (of stocks or shares) not including the next dividend.

exeat /ˈeksɪˌæt/ *n. Brit.* **1** a permission for temporary absence from a college or other institution. **2** a permission granted by a bishop to a priest to move to another diocese. [L, 3rd sing. pres. subjunctive of *exire* go out (as EX-[1], *ire* go)]

exec /ɪgˈzek/ *n. colloq.* an executive. [abbr.]

execrable /ˈeksɪkrəb(ə)l/ *adj.* abominable, detestable. □ **execrably** *adv.* [ME f. OF f. L *execrabilis* (as EXECRATE)]

execrate /ˈeksɪˌkreɪt/ *v.* **1** *tr.* express or feel abhorrence for. **2** *tr.* curse (a person or thing). **3** *intr.* utter curses. □ **execrative** *adj.* **execratory** *adj.* **execration** /ˌeksɪˈkreɪʃ(ə)n/ *n.* [L *exsecrare* (as EX-[1], *sacrare* devote f. *sacer* sacred, accursed)]

executant /ɪgˈzekjʊt(ə)nt/ *n. formal* **1** a performer, esp. of music. **2** a person who carries something into effect. [F *exécutant* pres. part. (as EXECUTE)]

execute /ˈeksɪˌkjuːt/ *v.tr.* **1 a** carry out a sentence of death on (a condemned person). **b** *euphem.* kill as a political act. **2** carry into effect, perform (a plan, duty, command, operation, etc.). **3 a** carry out a design for (a product of art or skill). **b** perform (a musical composition, dance, etc.). **4** make (a legal instrument) valid by signing, sealing, etc. **5** put into effect (a judicial sentence, the terms of a will, etc.). □ **executable** *adj.* [ME f. OF *executer* f. med.L *executare* f. L *exsequi exsecut-* (as EX-[1], *sequi* follow)]

execution /ˌeksɪˈkjuːʃ(ə)n/ *n.* **1** the carrying out of a sentence of death. **2** the act or an instance of carrying out or performing something. **3** technique or style of performance in the arts, esp. music. **4 a** seizure of the property or person of a debtor in default of payment. **b** a judicial writ enforcing a judgement. □ **executionary** *adj.* [ME f. OF f. L *executio -onis* (as EXECUTE)]

executioner /ˌeksɪˈkjuːʃənə(r)/ *n.* an official who carries out a sentence of death.

executive /ɪgˈzekjʊtɪv/ *n. & adj.* ● *n.* **1** a person or body with managerial or administrative responsibility in a business organization etc.; a senior businessman or businesswoman. **2** a branch of a government or organization concerned with executing laws, agreements, etc., or with other administration or management. ● *adj.* **1** concerned with executing laws, agreements, etc., or with other administration or management. **2** relating to or having the function of executing. □ **executive session** *US* a usu. private meeting of a legislative body for executive business. □ **executively** *adv.* [med.L *executivus* (as EXECUTE)]

Executive Council *n.* the Australian constitutional body which gives legal form to Cabinet decisions.

executor /ɪgˈzekjʊtə(r)/ *n.* (*fem.* **executrix** /-trɪks/) a person appointed by a testator to carry out the terms of his or her will. □ **literary executor** a person entrusted with a writer's papers, unpublished

works, etc. □ **executorship** *n.* **executory** *adj.* **executorial** /-ˌzekjʊˈtɔːrɪəl/ *adj.* [ME f. AF *executor*, *-our* f. L *executor -oris* (as EXECUTE)]

exegesis /ˌeksɪˈdʒiːsɪs/ *n.* (*pl.* **exegeses** /-siːz/) critical explanation of a text, esp. of Scripture. □ **exegetist** /-ˈdʒiːtɪst/ *n.* **exegetic** /-ˈdʒetɪk/ *adj.* **exegetical** *adj.* **exegete** /ˈeksɪˌdʒiːt/ *n.* [Gk *exēgēsis* f. *exēgeomai* interpret (as EX-[2], *hēgeomai* lead)]

exemplar /ɪgˈzemplə(r), -plɑː(r)/ *n.* **1** a model or pattern. **2** a typical instance of a class of things. **3** a parallel instance. [ME f. OF *exemplaire* f. LL *exemplarium* (as EXAMPLE)]

exemplary /ɪgˈzemplərɪ/ *adj.* **1** fit to be imitated; outstandingly good. **2 a** serving as a warning. **b** *Law* (of damages) exceeding the amount needed for simple compensation. **3** illustrative, representative. □ **exemplarily** *adv.* **exemplariness** *n.* [LL *exemplaris* (as EXAMPLE)]

exemplify /ɪgˈzemplɪˌfaɪ/ *v.tr.* (**-ies, -ied**) **1** illustrate by example. **2** be an example of. **3** *Law* make an attested copy of (a document) under an official seal. □ **exemplification** /-ˌzemplɪfɪˈkeɪʃ(ə)n/ *n.* [ME f. med.L *exemplificare* (as EXAMPLE)]

exemplum /ɪgˈzempləm/ *n.* (*pl.* **exempla** /-plə/) an example or model, esp. a moralizing or illustrative story. [L: see EXAMPLE]

exempt /ɪgˈzempt/ *adj., n., & v.* ● *adj.* **1** free from an obligation or liability etc. imposed on others. **2** (foll. by *from*) not liable to. ● *n.* **1** a person who is exempt, esp. from payment of tax. **2** *Brit.* = EXON. ● *v.tr.* (usu. foll. by *from*) free from an obligation, esp. one imposed on others. □ **exemption** /-ˈzempʃ(ə)n/ *n.* [ME f. L *exemptus* past part. of *eximere* *exempt-* (as EX-[1], *emere* take)]

exequies /ˈeksɪkwɪz/ *n.pl. formal* funeral rites. [ME f. OF f. L *exsequiae* (as EX-[1], *sequi* follow)]

exercise /ˈeksəˌsaɪz/ *n. & v.* ● *n.* **1** activity requiring physical effort, done esp. as training or to sustain or improve health. **2** mental or spiritual activity, esp. as practice to develop a faculty. **3** (often in *pl.*) a particular task or set of tasks devised as exercise, practice in a technique, etc. **4 a** the use or application of a mental faculty, right, etc. **b** practice of an ability, quality, etc. **5** (often in *pl.*) military drill or manoeuvres. **6** (foll. by *in*) a process directed at or concerned with something specified (*was an exercise in public relations*). ● *v.* **1** *tr.* use or apply (a faculty, right, influence, restraint, etc.). **2** *tr.* perform (a function). **3 a** *intr.* take (esp. physical) exercise; do exercises. **b** *tr.* provide (an animal) with exercise. **c** *tr.* train (a person). **4** *tr.* a tax the powers of. **b** perplex, worry. □ **exercise book 1** a book containing exercises. **2** a book for writing school work, notes, etc., in. □ **exercisable** *adj.* **exerciser** *n.* [ME f. OF *exercice* f. L *exercitium* f. *exercere exercit-* keep at work (as EX-[1], *arcere* restrain)]

exergue /ekˈsɜːg/ *n.* **1** a small space usu. on the reverse of a coin or medal, below the principal device. **2** an inscription on this space. [F f. med.L *exergum* f. Gk *ex-* (as EX-[2]) + *ergon* work]

exert /ɪgˈzɜːt/ *v.tr.* **1** exercise, bring to bear (a quality, force, influence, etc.). **2** *refl.* (often foll. by *for, or to* + infin.) use one's efforts or endeavours; strive. □ **exertion** /-ˈzɜːʃ(ə)n/ *n.* [L *exserere exsert-* put forth (as EX-[1], *serere* bind)]

Exeter /ˈeksɪtə(r)/ the county town of Devon, on the River Exe; pop. (1990) 101,100. Exeter was founded by the Romans, who called it Isca. The library of the Norman cathedral contains the Exeter Book, the largest surviving collection of Anglo-Saxon poems.

exeunt /ˈeksɪˌʌnt/ *v.intr.* (as a stage direction) (actors) leave the stage. □ **exeunt omnes** /ˈɒmneɪz/ all leave the stage. [L, = they go out: 3rd pl. pres. of *exire* go out: see EXIT]

exfiltrate /ˈeksfɪlˌtreɪt/ *v.tr.* (also *absol.*) withdraw (troops, spies, etc.) surreptitiously, esp. from danger. □ **exfiltration** /ˌeksfɪlˈtreɪʃ(ə)n/ *n.*

exfoliate /eksˈfəʊlɪˌeɪt/ *v.* **1** *intr.* (of bone, the skin, a mineral, etc.) come off in scales or layers. **2** *tr.* a shed (material) in scales or layers. **b** (also *absol.*) cause (the skin etc.) to shed flakes or scales. **3** *intr.* (of a tree) throw off layers of bark. □ **exfoliative** /-lɪətɪv/ *adj.* **exfoliation** /-ˌfəʊlɪˈeɪʃ(ə)n/ *n.* [LL *exfoliare exfoliat-* (as EX-[1], *folium* leaf)]

ex gratia /eks ˈɡreɪʃə/ *adv. & adj.* ● *adv.* as a favour rather than from an (esp. legal) obligation. ● *adj.* granted on this basis. [L, = from favour]

exhalation /ˌekshəˈleɪʃ(ə)n/ *n.* **1 a** an expiration of air. **b** a puff of breath. **2** a mist, vapour. **3** an emanation or effluvium. [ME f. L *exhalatio* (as EXHALE)]

exhale /eksˈheɪl/ *v.* **1** *tr.* (also *absol.*) breathe out (esp. air or smoke) from the lungs. **2** *tr. & intr.* give off or be given off in vapour. □ **exhalable** *adj.* [ME f. OF *exhaler* f. L *exhalare* (as EX-[1], *halare* breathe)]

exhaust /ɪgˈzɔːst/ *v. & n.* ● *v.tr.* **1** consume or use up the whole of. **2** (often as **exhausted** *adj.* or **exhausting** *adj.*) use up the strength or resources of (a person); tire out. **3** study or expound on (a subject) completely. **4** (often foll. by *of*) empty (a container etc.) of its contents. **5** (often as **exhausted** *adj.*) drain of strength or resources; make (land) barren. ● *n.* **1 a** waste gases etc. expelled from an engine after combustion. **b** (also **exhaust-pipe**) the pipe or system by which these are expelled. **c** the process of expulsion of these gases. **2 a** the production of an outward current of air by the creation of a partial vacuum. **b** an apparatus for this. □ **exhauster** *n.* **exhaustible** *adj.* **exhaustibility** /-ˌzɔːstəˈbɪlɪtɪ/ *n.* [L *exhaurire* exhaust- (as EX-¹, *haurire* draw (water), drain)]

exhaustion /ɪɡˈzɔːstʃən/ *n.* **1** the action or process of draining or emptying something; the state of being depleted or emptied. **2** a total loss of strength or vitality. **3** the process of establishing a conclusion by eliminating alternatives. [LL *exhaustio* (as EXHAUST)]

exhaustive /ɪɡˈzɔːstɪv/ *adj.* **1** thorough, comprehensive. **2** tending to exhaust a subject. □ **exhaustively** *adv.* **exhaustiveness** *n.*

exhibit /ɪɡˈzɪbɪt/ *v. & n.* ● *v.tr.* (**exhibited, exhibiting**) **1** show or reveal publicly (for interest, amusement, in competition, etc.). **2 a** show, display. **b** manifest (a quality). **3** submit for consideration. ● *n.* **1** a thing or collection of things forming part or all of an exhibition. **2** a document or other item or object produced in a lawcourt as evidence. □ **exhibitory** *adj.* [L *exhibere* exhibit- (as EX-¹, *habere* hold)]

exhibition /ˌeksɪˈbɪʃ(ə)n/ *n.* **1** a display (esp. public) of works of art, industrial products, etc. (See also WORLD FAIR.) **2** the act or an instance of exhibiting; the state of being exhibited. **3** *Brit.* a scholarship, esp. from the funds of a school, college, etc. □ **make an exhibition of oneself** behave so as to appear foolish or ridiculous. [ME f. OF f. LL *exhibitio -onis* (as EXHIBIT)]

exhibitioner /ˌeksɪˈbɪʃənə(r)/ *n. Brit.* a student who has been awarded an exhibition.

exhibitionism /ˌeksɪˈbɪʃəˌnɪz(ə)m/ *n.* **1** a tendency towards display or extravagant behaviour. **2** *Psychol.* a mental condition characterized by the compulsion to display one's genitals in public. □ **exhibitionist** *n.* **exhibitionistic** /-ˌbɪʃəˈnɪstɪk/ *adj.* **exhibitionistically** *adv.*

exhibitor /ɪɡˈzɪbɪtə(r)/ *n.* a person who provides an item or items for an exhibition.

exhilarate /ɪɡˈzɪləˌreɪt/ *v.tr.* (often as **exhilarating** *adj.* or **exhilarated** *adj.*) affect with great liveliness or joy; raise the spirits of. □ **exhilarant** *adj. & n.* **exhilaratingly** *adv.* **exhilarative** /-rətɪv/ *adj.* **exhilaration** /-ˌzɪləˈreɪʃ(ə)n/ *n.* [L *exhilarare* (as EX-¹, *hilaris* cheerful)]

exhort /ɪɡˈzɔːt/ *v.tr.* (often foll. by *to* + infin.) urge or advise strongly or earnestly. □ **exhorter** *n.* **exhortative** /-tətɪv/ *adj.* **exhortatory** /-tətərɪ/ *adj.* [ME f. OF *exhorter* or L *exhortari* (as EX-¹, *hortari* exhort)]

exhortation /ˌeɡzɔːˈteɪʃ(ə)n/ *n.* **1** the act or an instance of exhorting; the state of being exhorted. **2** a formal or liturgical address. [ME f. OF *exhortation* or L *exhortatio* (as EXHORT)]

exhume /eksˈhjuːm, ɪɡˈzjuːm/ *v.tr.* dig out, unearth (esp. a buried corpse). □ **exhumation** /ˌekshʊˈmeɪʃ(ə)n, ˌɪɡzjʊ-/ *n.* [F *exhumer* f. med.L *exhumare* (as EX-¹, *humus* ground)]

ex hypothesi /ˌeks haɪˈpɒθəsɪ/ *adv.* according to the hypothesis proposed. [mod.L]

exigency /ˈeksɪdʒənsɪ, ɪɡˈzɪd-/ *n.* (also **exigence** /ˈeksɪdʒəns/) (*pl.* **-ies**) **1** an urgent need or demand. **2** an emergency. [F *exigence* & LL *exigentia* (as EXIGENT)]

exigent /ˈeksɪdʒ(ə)nt/ *adj.* **1** requiring much; exacting. **2** urgent, pressing. [ME f. L *exigere* examine, weigh: see EXACT]

exiguous /ɪɡˈzɪɡjʊəs/ *adj.* scanty, small. □ **exiguously** *adv.* **exiguousness** *n.* **exiguity** /ˌeksɪˈɡjuːɪtɪ/ *n.* [L *exiguus* scanty f. *exigere* weigh exactly: see EXACT]

exile /ˈeksaɪl, ˈeɡzaɪl/ *n. & v.* ● *n.* **1** expulsion, or the state of being expelled, from one's native land or (**internal exile**) native town etc. **2** long absence abroad, esp. enforced. **3** a person expelled or long absent from his or her native country. **4** (**the Exile**) the captivity of the Jews in Babylon in the 6th century BC. ● *v.tr.* (foll. by *from*) officially expel (a person) from his or her native country or town etc. □ **exilic** /ekˈsɪlɪk, eɡˈzɪlɪk/ *adj.* (esp. in sense 4 of *n.*). [ME f. OF *exil, exiler* f. L *exilium* banishment]

exist /ɪɡˈzɪst/ *v.intr.* **1** have a place as part of objective reality. **2 a** have being under specified conditions. **b** (foll. by *as*) exist in the form of. **3** (of circumstances etc.) occur; be found. **4** live with no pleasure under adverse conditions (*felt he was merely existing*). **5** continue in being; maintain life (*can hardly exist on this salary*). **6** be alive, live. [prob. back-form. f. EXISTENCE; cf. LL *existere*]

existence /ɪɡˈzɪst(ə)ns/ *n.* **1** the fact or condition of being or existing. **2** continued being; the manner of one's existing or living, esp. under

adverse conditions (*a wretched existence*). **3** an existing thing. **4** all that exists. [ME f. OF *existence* or LL *existentia* f. L *existere* (as EX-¹, *stare* stand)]

existent /ɪɡˈzɪstənt/ *adj.* existing, actual, current.

existential /ˌeɡzɪˈstenʃ(ə)l/ *adj.* **1** of or relating to existence. **2** *Logic* (of a proposition etc.) affirming or implying the existence of a thing. **3** *Philos.* concerned with existence, esp. with human existence as viewed by existentialism. □ **existentially** *adv.* [LL *existentialis* (as EXISTENCE)]

existentialism /ˌeɡzɪˈstenʃəˌlɪz(ə)m/ *n.* a philosophical theory emphasizing the existence of the individual person as a free and responsible agent determining his or her own development through acts of the will. The term denotes recurring themes in modern philosophy and literature rather than a single school of thought. Existentialism tends to be atheistic (although there is a strand of Christian existentialism deriving from the work of Kierkegaard), to disparage scientific knowledge, and deny the existence of objective values, stressing instead the reality and significance of human freedom and experience. Kierkegaard and Nietzsche (in the 19th century) are generally taken to be the first existentialist philosophers, although the theory was developed chiefly in Continental Europe in the 20th century, notably by Martin Heidegger, Jean-Paul Sartre, Albert Camus, and Simone de Beauvoir. □ **existentialist** *n. & adj.* [G *Existentialismus* (as EXISTENTIAL)]

exit /ˈeksɪt, ˈeɡzɪt/ *n. & v.* ● *n.* **1** a passage or door by which to leave a room, building, etc. **2 a** the act of going out. **b** the freedom or opportunity to go out. **3** a place where vehicles can leave a motorway or major road. **4** the departure of an actor from the stage. **5** *literary* a person's death. ● *v.intr.* (**exited, exiting**) **1** go out of a room, building, etc. **2** (as a stage direction) (an actor) leaves the stage (*exit Macbeth*). **3** die. □ **exit permit** (or **visa** etc.) authorization to leave a particular country. [L, 3rd sing. pres. of *exire* go out (as EX-¹, *ire* go): cf. L *exitus* going out]

ex-libris /eksˈliːbrɪs/ *n.* (*pl.* same) a usu. decorated book-plate or label bearing the owner's name, pasted into the front of a book. [L *ex libris* among the books of]

Exmoor /ˈeksmʊə(r), -mɔː(r)/ an area of moorland in north Devon and west Somerset, SW England, rising to 520 m (1,706 ft) at Dunkery Beacon. The area is designated a national park.

ex nihilo /eks ˈnaɪhɪˌləʊ/ *adv.* out of nothing (*creation ex nihilo*). [L]

exo- /ˈeksəʊ/ *comb. form* external. [Gk *exō* outside]

exobiology /ˌeksəʊbaɪˈɒlədʒɪ/ *n.* the branch of science that deals with the possibility of life outside the earth. □ **exobiologist** *n.*

Exocet /ˈeksəˌset/ *n. propr.* a French-made short-range guided missile used esp. in sea warfare. [F *exocet* flying fish]

exocrine /ˈeksəʊˌkraɪn, -ˌkrɪn/ *adj.* (of a gland) secreting through a duct (cf. ENDOCRINE). [EXO- + Gk *krinō* sift]

Exod. *abbr.* (in the Bible) Exodus.

exoderm /ˈeksəʊˌdɜːm/ *n. Biol.* = ECTODERM.

Exodus /ˈeksədəs/ the second book of the Bible, relating the departure of the Israelites under Moses from their slavery in Egypt and their journey towards the promised land of Canaan, including the crossing of the Red Sea and the giving of the Ten Commandments. This journey is ascribed by scholars to various dates within the limits *c.*1580–*c.*1200 BC. [eccl.L f. Gk *exodos* (as EX-², *hodos* way)]

exodus /ˈeksədəs/ *n.* a mass departure of people (esp. emigrants). [EXODUS]

ex officio /ˌeks əˈfɪʃɪˌəʊ/ *adv. & adj.* by virtue of one's office or status. [L]

exogamy /ekˈsɒɡəmɪ/ *n.* **1** *Anthropol.* marriage of a man outside his own tribe. **2** *Biol.* the fusion of reproductive cells from distantly related or unrelated individuals. □ **exogamous** *adj.*

exogenous /ekˈsɒdʒɪnəs/ *adj. Biol.* growing or originating from outside. □ **exogenously** *adv.*

exon /ˈeksɒn/ *n. Brit.* each of the four officers acting as commanders of the Yeomen of the Guard. [repr. F pronunc. of EXEMPT]

exonerate /ɪɡˈzɒnəˌreɪt/ *v.tr.* (often foll. by *from*) **1** free or declare free from blame etc. **2** release from a duty etc. □ **exonerative** /-rətɪv/ *adj.* **exoneration** /-ˌzɒnəˈreɪʃ(ə)n/ *n.* [L *exonerare* exonerat- (as EX-¹, *onus, oneris* burden)]

exophthalmic /ˌeksɒfˈθælmɪk/ *adj. Med.* characterized by protruding eyes. □ **exophthalmic goitre** = GRAVES' DISEASE.

exophthalmos /ˌeksɒfˈθælmɒs/ *n.* (also **exophthalmus, exophthalmia** /-mɪə/) *Med.* abnormal protrusion of the eyeball.

[mod.L f. Gk *exophthalmos* having prominent eyes (as EX-², *ophthalmos* eye)]

exoplasm /ˈeksəʊˌplæz(ə)m/ *n. Biol.* = ECTOPLASM.

exor. *abbr.* executor.

exorbitant /ɪgˈzɔːbɪt(ə)nt/ *adj.* (of a price, demand, etc.) grossly excessive. □ **exorbitance** *n.* **exorbitantly** *adv.* [LL *exorbitare* (as EX-¹, *orbita* ORBIT)]

exorcize /ˈeksɔːˌsaɪz/ *v.tr.* (also **-ise**) **1** expel (a supposed evil spirit) by invocation or by use of a holy name. **2** (often foll. by *of*) free (a person or place) of a supposed evil spirit. □ **exorcism** *n.* **exorcist** *n.* **exorcization** /ˌeksɔːsaɪˈzeɪʃ(ə)n/ *n.* [F *exorciser* or eccl.L *exorcizare* f. Gk *exorkizō* (as EX-², *horkos* oath)]

exordium /ekˈsɔːdɪəm/ *n.* (*pl.* **exordiums** or **exordia** /-dɪə/) the beginning or introductory part, esp. of a discourse or treatise. □ **exordial** *adj.* [L f. *exordiri* (as EX-¹, *ordiri* begin)]

exoskeleton /ˈeksəʊˌskelɪt(ə)n/ *n.* a rigid external covering for the body in certain animals, esp. arthropods, providing support and protection. □ **exoskeletal** *adj.*

exosphere /ˈeksəʊˌsfɪə(r)/ *n.* the layer of atmosphere furthest from the earth.

exothermic /ˌeksəʊˈθɜːmɪk/ *adj.* (also **exothermal** /-m(ə)l/) esp. *Chem.* occurring or formed with the evolution of heat. □ **exothermally** *adv.* **exothermically** *adv.*

exotic /ɪgˈzɒtɪk/ *adj. & n.* ● *adj.* **1** introduced from or originating in a foreign (esp. tropical) country (*exotic fruits*). **2** attractively or remarkably strange or unusual; bizarre. **3** (of a fuel, metal, etc.) of a kind newly brought into use. ● *n.* an exotic person or thing. □ **exotic dancer** a striptease dancer. □ **exotically** *adv.* **exoticism** /-tɪˌsɪz(ə)m/ *n.* [L *exoticus* f. Gk *exōtikos* f. *exō* outside]

exotica /ɪgˈzɒtɪkə/ *n.pl.* remarkably strange or rare objects. [L, neut. pl. of *exoticus*: see EXOTIC]

exotoxin /ˌeksəʊˈtɒksɪn/ *n. Biochem.* a toxin released by a living bacterial cell into its surroundings (cf. ENDOTOXIN).

expand /ɪkˈspænd/ *v.* **1** *tr. & intr.* increase in size or bulk or importance. **2** *intr.* (often foll. by *on*) give a fuller description or account. **3** *intr.* become more genial or effusive; discard one's reserve. **4** *tr.* set or write out in full (something condensed or abbreviated). **5** *tr. & intr.* spread out flat. □ **expanded metal** sheet metal slit and stretched into a mesh, used to reinforce concrete and other brittle materials. □ **expandable** *adj.* **expander** *n.* **expansible** /-ˈspænsɪb(ə)l/ *adj.* **expansibility** /-ˌspænsɪˈbɪlɪtɪ/ *n.* [ME f. L *expandere expans-* spread out (as EX-¹, *pandere* spread)]

expanse /ɪkˈspæns/ *n.* **1** a wide continuous area or extent of land, space, etc. **2** an amount of expansion. [mod.L *expansum* neut. past part. (as EXPAND)]

expansile /ɪkˈspænsaɪl/ *adj.* **1** of expansion. **2** capable of expansion.

expansion /ɪkˈspænʃ(ə)n/ *n.* **1** the act or an instance of expanding; the state of being expanded. **2** enlargement of the scale or scope of (esp. commercial) operations. **3** increase in the amount of a state's territory or area of control. **4** an increase in the volume of fuel etc. on combustion in the cylinder of an engine. **5** the action of making or becoming greater in area, bulk, capacity, etc.; dilatation; the degree of this (*alternate expansion and contraction of the muscle*). □ **expansion card** (or **board**) *Computing* a circuit board that can be inserted in a computer to give extra facilities. **expansion joint** a joint that allows for the thermal expansion of the parts joined. **expansion slot** *Computing* a place in a computer where an expansion card can be added. □ **expansionary** *adj.* **expansionism** *n.* **expansionist** *n.* **expansionistic** /-ˌspænʃəˈnɪstɪk/ *adj.* (all in senses 2, 3). [LL *expansio* (as EXPAND)]

expansive /ɪkˈspænsɪv/ *adj.* **1** able or tending to expand. **2** extensive, wide-ranging. **3** effusive, open. □ **expansively** *adv.* **expansiveness** *n.* **expansivity** /ˌekspænˈsɪvɪtɪ/ *n.*

ex parte /eks ˈpɑːtɪ/ *adj. & adv. Law* in the interests of one side only or of an interested outside party. [L]

expat /eksˈpæt/ *n. & adj. colloq.* = EXPATRIATE. [abbr.]

expatiate /ɪkˈspeɪʃɪˌeɪt/ *v.intr.* (usu. foll. by *on, upon*) speak or write at length or in detail. □ **expatiatory** /-ʃɪətərɪ/ *adj.* **expatiation** /-ˌspeɪʃɪˈeɪʃ(ə)n/ *n.* [L *exspatiari* digress (as EX-¹, *spatium* SPACE)]

expatriate *adj., n., & v.* ● *adj.* /eksˈpætrɪət, -ˈpeɪtrɪət/ **1** living abroad, esp. for a long period. **2** expelled from one's country; exiled. ● *n.* /eksˈpætrɪət, -ˈpeɪtrɪət/ an expatriate person. ● *v.tr.* /eksˈpætrɪˌeɪt, -ˈpeɪtrɪˌeɪt/ **1** expel or remove (a person) from his or her native country.

2 *refl.* withdraw (oneself) from one's citizenship or allegiance. □ **expatriation** /-ˌpætrɪˈeɪʃ(ə)n, -ˌpeɪtrɪ-/ *n.* [med.L *expatriare* (as EX-¹, *patria* native country)]

expect /ɪkˈspekt/ *v.tr.* **1** (often foll. by *to* + infin., or *that* + clause) **a** regard as likely; assume a future event or occurrence. **b** (often foll. by *of*) look for as appropriate or one's due (from a person) (*I expect cooperation*; *expect you to be here*; *expected better of you*). **2** (often foll. by *that* + clause) *colloq.* think, suppose (*I expect we'll be on time*). **3** be shortly to have (a baby) (*is expecting twins*). □ **be expecting** *colloq.* be pregnant (with). □ **expectable** *adj.* [L *exspectare* (as EX-¹, *spectare* look, frequent. of *specere* see)]

expectancy /ɪkˈspektənsɪ/ *n.* (*pl.* **-ies**) **1** a state of expectation. **2** a prospect, esp. of future possession. **3** (foll. by *of*) a prospective chance. [L *exspectantia, exp-* (as EXPECT)]

expectant /ɪkˈspektənt/ *adj. & n.* ● *adj.* **1** (often foll. by *of*) expecting. **2** having the expectation of possession, status, etc. **3** (*attrib.*) expecting a baby (said of the mother or father). ● *n.* **1** a person who expects. **2** a candidate for office etc. □ **expectantly** *adv.*

expectation /ˌekspekˈteɪʃ(ə)n/ *n.* **1** the act or an instance of expecting or looking forward. **2** something expected or hoped for. **3** (foll. by *of*) the probability of an event. **4** (in *pl.*) one's prospects of inheritance. [L *expectatio* (as EXPECT)]

expectorant /ekˈspektərənt/ *adj. & n.* ● *adj.* causing the coughing out of phlegm etc. ● *n.* an expectorant medicine.

expectorate /ekˈspektəˌreɪt/ *v.tr.* (also *absol.*) cough or spit out (phlegm etc.) from the throat or lungs. □ **expectorator** *n.* **expectoration** /-ˌspektəˈreɪʃ(ə)n/ *n.* [L *expectorare expectorat-* (as EX-¹, *pectus -oris* breast)]

expedient /ɪkˈspiːdɪənt/ *adj. & n.* ● *adj.* **1** advantageous; advisable on practical rather than moral grounds. **2** suitable, appropriate. ● *n.* a means of attaining an end; a resource. □ **expedience** *n.* **expediency** *n.* **expediently** *adv.* [ME f. L *expedire*: see EXPEDITE]

expedite /ˈekspɪˌdaɪt/ *v.tr.* **1** assist the progress of; hasten (an action, process, etc.). **2** accomplish (business) quickly. □ **expediter** *n.* [L *expedire expedit-* extricate, put in order (as EX-¹, *pes pedis* foot)]

expedition /ˌekspɪˈdɪʃ(ə)n/ *n.* **1** a journey or voyage for a particular purpose, esp. exploration, scientific research, or war. **2** the personnel or ships etc. undertaking this. **3** promptness, speed. □ **expeditionist** *n.* [ME f. OF f. L *expeditio -onis* (as EXPEDITE)]

expeditionary /ˌekspɪˈdɪʃənərɪ/ *adj.* of or used in an expedition, esp. military.

expeditious /ˌekspɪˈdɪʃəs/ *adj.* **1** acting or done with speed and efficiency. **2** suited for speedy performance. □ **expeditiously** *adv.* **expeditiousness** *n.* [EXPEDITION + -OUS]

expel /ɪkˈspel/ *v.tr.* (**expelled, expelling**) (often foll. by *from*) **1** deprive (a person) of the membership of or involvement in (a school, society, etc.). **2** force out or eject (a thing from its container etc.). **3** order or force to leave a building etc. □ **expellable** *adj.* **expellent** *adj.* **expeller** *n.* **expellee** /ˌekspeˈliː/ *n.* [ME f. L *expellere expuls-* (as EX-¹, *pellere* drive)]

expend /ɪkˈspend/ *v.tr.* spend or use up (money, time, etc.). [ME f. L *expendere expens-* (as EX-¹, *pendere* weigh)]

expendable /ɪkˈspendəb(ə)l/ *adj.* **1** that may be sacrificed or dispensed with, esp. to achieve a purpose. **2 a** not regarded as worth preserving or saving. **b** unimportant, insignificant. **3** not normally reused. □ **expendably** *adv.* **expendability** /-ˌspendəˈbɪlɪtɪ/ *n.*

expenditure /ɪkˈspendɪtʃə(r)/ *n.* **1** the process or an instance of spending or using up. **2** a thing (esp. a sum of money) expended. [EXPEND, after obs. *expenditor* officer in charge of expenditure, f. med.L f. *expenditus* irreg. past part. of L *expendere*]

expense /ɪkˈspens/ *n.* **1** cost incurred; payment of money. **2** (usu. in *pl.*) **a** the cost incurred in doing a particular job etc. (*will pay your expenses*). **b** an amount paid to reimburse this (*offered me £40 per day expenses*). **3** a thing that is a cause of much expense (*the house is a real expense to run*). □ **at the expense of 1** so as to cause loss, deprivation, or harm to. **2** so as to incur the cost of. **expense account** a list of an employee's expenses to be reimbursed by the employer. [ME f. AF, alt. of OF *espense* f. LL *expensa* (money) spent, past part. of L *expendere* EXPEND]

expensive /ɪkˈspensɪv/ *adj.* **1** costing much. **2** making a high charge. **3** causing much expense (*has expensive tastes*). □ **expensively** *adv.* **expensiveness** *n.*

experience /ɪkˈspɪərɪəns/ *n. & v.* ● *n.* **1** actual observation of or practical acquaintance with facts or events. **2** knowledge or skill resulting from this. **3 a** an event regarded as affecting one (*an unpleasant experience*). **b** the fact or process of being so affected (*learned by experience*). ● *v.tr.* **1** have experience of; undergo. **2** feel or be affected by (an emotion etc.). □ **experienceable** *adj.* [ME f. OF f. L *experientia* f. *experiri expert-* try]

experienced /ɪkˈspɪərɪənst/ *adj.* **1** having had much experience. **2** skilled from experience (*an experienced driver*).

experiential /ɪkˌspɪərɪˈenʃ(ə)l/ *adj.* involving or based on experience. □ **experiential philosophy** a philosophy that treats all knowledge as based on experience. □ **experientialism** *n.* **experientialist** *n.* **experientially** *adv.*

experiment *n. & v.* ● *n.* /ɪkˈsperɪmənt/ **1** a procedure undertaken to make a discovery, test a hypothesis, or demonstrate a known fact. **2** (foll. by *of*) a test or trial of. ● *v.intr.* /ɪkˈsperɪˌment/ (often foll. by *on, with*) make an experiment. □ **experimenter** /-ˌmentə(r)/ *n.* **experimentation** /-ˌsperɪmenˈteɪʃ(ə)n/ *n.* [ME f. OF *experiment* or L *experimentum* (as EXPERIENCE)]

experimental /ɪkˌsperɪˈment(ə)l/ *adj.* **1** based on or making use of experiment (*experimental psychology*). **2 a** used in experiments. **b** serving or resulting from (esp. incomplete) experiment; tentative, provisional. **3** based on experience, not on authority or conjecture. □ **experimentalism** *n.* **experimentalist** *n.* **experimentalize** *v.intr.* (also **-ise**). **experimentally** *adv.* [ME f. med.L *experimentalis* (as EXPERIMENT)]

expert /ˈekspɜːt/ *adj. & n.* ● *adj.* **1** (often foll. by *at, in*) having special skill at a task or knowledge in a subject. **2** (*attrib.*) involving or resulting from this (*expert evidence; an expert piece of work*). ● *n.* (often foll. by *at, in*) a person having special knowledge or skill. □ **expert system** *Computing* a system which can provide intelligent advice or decisions based on expert knowledge incorporated in the software. □ **expertly** *adv.* **expertness** *n.* [ME f. OF f. L *expertus* past part. of *experiri*: see EXPERIENCE]

expertise /ˌekspɜːˈtiːz/ *n.* expert skill, knowledge, or judgement. [F as EXPERT]

expertize /ˈekspɜːˌtaɪz/ *v.* (also **-ise**) **1** *intr.* give an expert opinion. **2** *tr.* give an expert opinion concerning.

expiate /ˈekspɪˌeɪt/ *v.tr.* **1** pay the penalty for (wrongdoing). **2** make amends for. □ **expiator** *n.* **expiable** /-pɪəb(ə)l/ *adj.* **expiatory** /-pɪətəri, -pɪˌeɪtəri/ *adj.* **expiation** /ˌekspɪˈeɪʃ(ə)n/ *n.* [L *expiare expiat-* (as EX-¹, *pius* devout)]

expiration /ˌekspɪˈreɪʃ(ə)n/ *n.* **1** breathing out. **2** expiry. [L *expiratio* (as EXPIRE)]

expire /ɪkˈspaɪə(r)/ *v.* **1** *intr.* (of a period of time, validity, etc.) come to an end. **2** *intr.* (of a document, authorization, etc.) cease to be valid; become void. **3** *intr.* (of a person) die. **4** *tr.* (usu. foll. by *from*; also *absol.*) exhale (air etc.) from the lungs. □ **expiratory** *adj.* (in sense 4). [ME f. OF *expirer* f. L *exspirare* (as EX-¹, *spirare* breathe)]

expiry /ɪkˈspaɪəri/ *n.* **1** the end of the validity or duration of something. **2** death.

explain /ɪkˈspleɪn/ *v.tr.* **1 a** make clear or intelligible (also *absol.*: *let me explain*). **b** make known in detail. **2** (foll. by *that* + clause) say by way of explanation. **3** account for (one's conduct etc.). □ **explain away** minimize the significance of (a difficulty or mistake) by explanation. **explain oneself 1** make one's meaning clear. **2** give an account of one's motives or conduct. □ **explainable** *adj.* **explainer** *n.* [L *explanare* (as EX-¹, *planus* flat, assim. to PLAIN¹)]

explanation /ˌekspləˈneɪʃ(ə)n/ *n.* **1** the act or an instance of explaining. **2** a statement or circumstance that explains something. **3** a declaration made with a view to mutual understanding or reconciliation. [ME f. L *explanatio* (as EXPLAIN)]

explanatory /ɪkˈsplænətəri/ *adj.* serving or intended to serve to explain. □ **explanatorily** *adv.* [LL *explanatorius* (as EXPLAIN)]

explant /eksˈplɑːnt/ *v. & n. Biol.* ● *v.tr.* transfer (living cells, tissues, or organs) from animals or plants to a nutrient medium. ● *n.* a piece of explanted tissue etc. □ **explantation** /ˌeksplɑːnˈteɪʃ(ə)n/ *n.* [mod.L *explantare* (as EX-¹, *plantare* PLANT)]

expletive /ɪkˈspliːtɪv/ *n. & adj.* ● *n.* **1** an oath, swear-word, or other expression, used in an exclamation. **2** a word used to fill out a sentence etc., esp. in verse. ● *adj.* serving to fill out (esp. a sentence, line of verse, etc.). [LL *expletivus* (as EX-¹, *plere plet-* fill)]

explicable /ɪkˈsplɪkəb(ə)l, ˈeksplɪ-/ *adj.* that can be explained.

explicate /ˈeksplɪˌkeɪt/ *v.tr.* **1** develop the meaning or implication of (an idea, principle, etc.). **2** make clear, explain (esp. a literary text). □ **explicator** *n.* **explication** /ˌeksplɪˈkeɪʃ(ə)n/ *n.* **explicative**

/ɪkˈsplɪkətɪv, ˈeksplɪˌkeɪtɪv/ *adj.* **explicatory** /ɪkˈsplɪkətəri, ˈeksplɪˌkeɪtəri/ *adj.* [L *explicare explicat-* unfold (as EX-¹, *plicare plicat-* or *plicit-* fold)]

explicit /ɪkˈsplɪsɪt/ *adj.* **1 a** expressly stated, leaving nothing merely implied; stated in detail. **b** describing or representing nudity or intimate sexual activity. **2** (of knowledge, a notion, etc.) definite, clear. **3** (of a person, book, etc.) expressing views unreservedly; outspoken. □ **explicitly** *adv.* **explicitness** *n.* [F *explicite* or L *explicitus* (as EXPLICATE)]

explode /ɪkˈspləʊd/ *v.* **1 a** *intr.* (of gas, gunpowder, a bomb, a boiler, etc.) expand suddenly with a loud noise owing to a release of internal energy. **b** *tr.* cause (a bomb etc.) to explode. **2** *intr.* give vent suddenly to emotion, esp. anger. **3** *intr.* (of a population etc.) increase suddenly or rapidly. **4** *tr.* show (a theory etc.) to be false or baseless. **5** *tr.* (as **exploded** *adj.*) (of a drawing etc.) showing the components of a mechanism as if separated by an explosion but in the normal relative positions. □ **exploder** *n.* [earliest in sense 4: L *explodere* hiss off the stage (as EX-¹, *plodere plos-* = *plaudere* clap)]

exploit *n. & v.* ● *n.* /ˈeksplɔɪt/ a bold or daring feat. ● *v.tr.* /ɪkˈsplɔɪt/ **1** make use of (a resource etc.); derive benefit from. **2** usu. *derog.* utilize or take advantage of (esp. a person) for one's own ends. □ **exploitable** /ɪkˈsplɔɪtəb(ə)l/ *adj.* **exploitation** /ˌeksplɔɪˈteɪʃ(ə)n/ *n.* **exploitative** /ɪkˈsplɔɪtətɪv/ *adj.* **exploiter** /ɪkˈsplɔɪtə(r)/ *n.* **exploitive** /ɪkˈsplɔɪtɪv/ *adj.* [ME f. OF *esploit, exploiter* ult. f. L *explicare*: see EXPLICATE]

exploration /ˌekspləˈreɪʃ(ə)n/ *n.* **1** an act or instance of exploring. **2** the process of exploring. □ **explorational** *adj.*

exploratory /ɪkˈsplɒrətəri/ *adj.* **1** (of discussion etc.) preliminary, serving to establish procedure etc. **2** of or concerning exploration or investigation (*exploratory surgery*).

explore /ɪkˈsplɔː(r)/ *v.tr.* **1** travel extensively through (a country etc.) in order to learn or discover about it. **2** inquire into; investigate thoroughly. **3** *Surgery* examine (a part of the body) in detail. □ **explorative** /-ˈsplɒrətɪv/ *adj.* [F *explorer* f. L *explorare*]

explorer /ɪkˈsplɔːrə(r)/ *n.* a traveller into undiscovered or uninvestigated territory, esp. to get scientific information.

explosion /ɪkˈspləʊʒ(ə)n/ *n.* **1** the act or an instance of exploding. **2 a** loud noise caused by something exploding. **3 a** a sudden outburst of noise. **b** a sudden outbreak of feeling, esp. anger. **4** a rapid or sudden increase, esp. of population. [L *explosio* scornful rejection (as EXPLODE)]

explosive /ɪkˈspləʊsɪv/ *adj. & n.* ● *adj.* **1** able or tending or likely to explode. **2** likely to cause a violent outburst etc.; (of a situation etc.) dangerously tense. ● *n.* an explosive substance. □ **explosively** *adv.* **explosiveness** *n.*

Expo /ˈekspəʊ/ *n.* (also **expo**) (*pl.* **-os**) a large international exhibition. (See also WORLD FAIR.) [abbr. of EXPOSITION 4]

exponent /ɪkˈspəʊnənt/ *n. & adj.* ● *n.* **1** a person who favours or promotes an idea etc. **2** a representative or practitioner of an activity, profession, etc. **3** a person who explains or interprets something. **4** an executant (of music etc.). **5** a type or representative. **6** *Math.* a raised symbol or expression beside a numeral indicating how many times it is to be multiplied by itself (e.g. $2^3 = 2 \times 2 \times 2$). ● *adj.* that sets forth or interprets. [L *exponere* (as EX-¹, *ponere posit-* put)]

exponential /ˌekspəˈnenʃ(ə)l/ *adj.* **1** *Math.* of or indicated by a mathematical exponent. **2** (of an increase etc.) more and more rapid. □ **exponential function** *Math.* a function which increases as a quantity raised to a power determined by the variable on which the function depends. **exponential growth** *Biol.* a form of population growth in which the rate of growth at any one time is related to the number of individuals present, so that the growth curve is increasingly steep. □ **exponentially** *adv.* [F *exponentiel* (as EXPONENT)]

export *v. & n.* ● *v.tr.* /ɪkˈspɔːt, ˈekspɔːt/ send out (goods or services) esp. for sale in another country. ● *n.* /ˈekspɔːt/ **1** the process of exporting. **2 a** an exported article or service. **b** (in *pl.*) the amount or value of goods exported (*exports exceeded £50m.*). **3** (*attrib.*) suitable for export, esp. of better quality. □ **export reject** an article sold in its country of manufacture, as being below the standard for export. □ **exportable** /ɪkˈspɔːtəb(ə)l, ˈekspɔːt-/ *adj.* **exportability** /ɪkˌspɔːtəˈbɪlɪtɪ, ˌekspɔːt-/ *n.* **exportation** /ˌekspɔːˈteɪʃ(ə)n/ *n.* **exporter** /ɪkˈspɔːtə(r), ˈekspɔːt-/ *n.* [L *exportare* (as EX-¹, *portare* carry)]

expose /ɪkˈspəʊz/ *v.tr.* **1** leave uncovered or unprotected, esp. from the weather. **2** (foll. by *to*) **a** cause to be liable to or in danger of (*was exposed to great danger*). **b** lay open to the action or influence of; introduce to (*exposed to bad influences; exposed to Chaucer at a young age*). **3** (as **exposed** *adj.*) **a** (foll. by *to*) open to; unprotected from (*exposed to the east*).

b vulnerable, risky. **4** *Photog.* subject (a film) to light, esp. by operation of a camera. **5** reveal the identity or fact of (esp. a person or thing disapproved of or guilty of crime etc.). **6** disclose; make public. **7** exhibit, display. **8** leave (a child) in the open to die. □ **expose oneself** display one's body, esp. the genitals, publicly and indecently. □ **exposer** *n.* [ME f. OF *exposer* after L *exponere*: see EXPONENT, POSE[1]]

exposé /ek'spəʊzeɪ/ *n.* **1** an orderly statement of facts. **2** the act or an instance of revealing something discreditable. [F, past part. of *exposer* (as EXPOSE)]

exposition /ˌekspə'zɪʃ(ə)n/ *n.* **1** an explanatory statement or account. **2** an explanation or commentary; an interpretative article or treatise. **3** *Mus.* the part of a movement, esp. in sonata form, in which the principal themes are first presented. **4** a large public exhibition. **5** *archaic* exposure. □ **expositional** *adj.* **expositive** /ɪk'spɒzɪtɪv/ *adj.* [ME f. OF *exposition*, or L *expositio* (as EXPONENT)]

expositor /ɪk'spɒzɪtə(r)/ *n.* an expounder or interpreter. □ **expository** *adj.*

ex post /eks 'pəʊst/ *adj. Econ.* based on or determined by actual results rather than forecasts or expectations; calculated retrospectively. [mod.L: perh. shortened f. EX POST FACTO]

ex post facto /ˌeks pəʊst 'fæktəʊ/ *adj.* & *adv.* with retrospective action or force. [L *ex postfacto* in the light of subsequent events]

expostulate /ɪk'spɒstjʊˌleɪt/ *v.intr.* (often foll. by *with* a person) make a protest; remonstrate earnestly. □ **expostulatory** /-lətərɪ/ *adj.* **expostulation** /-ˌspɒstjʊ'leɪʃ(ə)n/ *n.* [L *expostulare expostulat-* (as EX-[1], *postulare* demand)]

exposure /ɪk'spəʊʒə(r)/ *n.* **1** (foll. by *to*) the act or condition of exposing or being exposed (to air, cold, danger, etc.). **2** the condition of being exposed to the elements, esp. in severe conditions (*died from exposure*). **3** the revelation of an identity or fact, esp. when concealed or likely to find disapproval. **4** *Photog.* **a** the action of exposing a film etc. to the light. **b** the duration of this action. **c** the area of film etc. affected by it. **5** an aspect or outlook (*has a fine southern exposure*). **6** (often foll. by *to*) experience, esp. of a specified kind of work. □ **exposure meter** *Photog.* a device for measuring the strength of the light to determine the correct duration of exposure. [EXPOSE after *enclosure* etc.]

expound /ɪk'spaʊnd/ *v.tr.* **1** set out in detail (a doctrine etc.). **2** explain or interpret (esp. Scripture). □ **expounder** *n.* [ME f. OF *espondre* (as EXPONENT)]

express[1] /ɪk'spres/ *v.tr.* **1** represent or make known (thought, feelings, etc.) in words or by gestures, conduct, etc. **2** *refl.* say what one thinks or means. **3** esp. *Math.* represent by symbols. **4** squeeze out (liquid or air). □ **expresser** *n.* **expressible** *adj.* [ME f. OF *expresser* f. Rmc (as EX-[1], PRESS[1])]

express[2] /ɪk'spres/ *adj., adv., n.,* & *v.* ● *adj.* **1** operating at high speed. **2** (also /'ekspres/) **a** definitely stated, not merely implied. **b** *archaic* (of a likeness) exact. **3 a** done, made, or sent for a special purpose. **b** (of messages or goods) delivered by a special messenger or service. ● *adv.* **1** at high speed. **2** by express messenger or train. ● *n.* **1 a** an express train or messenger. **b** an express rifle. **2** *US* a company undertaking the transport of parcels etc. ● *v.tr.* send by express messenger or delivery. □ **express rifle** a rifle that discharges a bullet at high speed. **express train** a fast train, stopping at few intermediate stations. □ **expressly** *adv.* (in senses 2 and 3a of *adj.*) [ME f. OF *expres* f. L *expressus* distinctly shown, past part. of *exprimere* (as EX-[1], *premere* press)]

expression /ɪk'spreʃ(ə)n/ *n.* **1** the act or an instance of expressing. **2 a** a word or phrase expressed. **b** manner or means of expressing in language; wording, diction. **3** *Math.* a collection of symbols expressing a quantity. **4** a person's facial appearance or intonation of voice, esp. as indicating feeling. **5** the depiction of feeling, movement, etc., in art. **6** the conveying of feeling in the performance of a piece of music. □ **expression-mark** *Mus.* a sign or word indicating the required manner of performance. □ **expressional** *adj.* **expressionless** *adj.* **expressionlessly** *adv.* **expressionlessness** *n.* [ME f. OF *expression* or L *expressio* f. *exprimere*: see EXPRESS[1]]

expressionism /ɪk'spreʃəˌnɪz(ə)m/ *n.* a broad term used in art, literature, music, etc. to describe a general artistic approach characterized by a rejection of traditional ideas of beauty or harmony and by the use of distortion, exaggeration, and other non-naturalistic devices in order to emphasize and express the inner world of emotion rather than any external reality. The paintings of El Greco and Grünewald exemplify expressionism in this broad sense, but the term is also used of a late 19th and 20th-century European and specifically German movement tracing its origins to Van Gogh, Edvard Munch, and James Ensor, which insisted on the primacy of the artist's feelings and mood, often incorporating violence and the grotesque. Major expressionist painters include Chaim Soutine and Georges Rouault in France and Oskar Kokoschka (1886-1980) in Austria, while in the theatre the term has been associated with Strindberg, and with German dramatists such as Frank Wedekind and early Brecht. In cinema, films identified with expressionism include *Nosferatu* (1922) by F. W. Murnau (1888-1931) and Fritz Lang's *Metropolis* (1927). □ **expressionist** *n.* & *adj.* **expressionistic** /-ˌspreʃə'nɪstɪk/ *adj.* **expressionistically** *adv.*

expressive /ɪk'spresɪv/ *adj.* **1** full of expression (*an expressive look*). **2** (foll. by *of*) serving to express (*words expressive of contempt*). □ **expressively** *adv.* **expressiveness** *n.* **expressivity** /ˌekspre'sɪvɪtɪ/ *n.* [ME f. F *expressif -ive* or med.L *expressivus* (as EXPRESSION)]

expresso var. of ESPRESSO.

expressway /ɪk'spreswer/ *n. N. Amer.* & *Austral.* an urban motorway.

expropriate /eks'prəʊprɪˌeɪt/ *v.tr.* **1** (esp. of the state) take away (property) from its owner. **2** (foll. by *from*) dispossess. □ **expropriator** *n.* **expropriation** /-ˌprəʊprɪ'eɪʃ(ə)n/ *n.* [med.L *expropriare expropriat-* (as EX-[1], *proprium* property: see PROPER)]

expulsion /ɪk'spʌlʃ(ə)n/ *n.* the act or an instance of expelling; the process of being expelled. □ **expulsive** /-'spʌlsɪv/ *adj.* [ME f. L *expulsio* (as EXPEL)]

expunge /ɪk'spʌndʒ/ *v.tr.* (often foll. by *from*) erase, remove (esp. a passage from a book or a name from a list). □ **expunger** *n.* **expunction** /-'spʌŋkʃ(ə)n/ *n.* [L *expungere expunct-* (as EX-[1], *pungere* prick)]

expurgate /'ekspəˌgeɪt/ *v.tr.* **1** remove matter thought to be objectionable from (a book etc.). **2** remove (such matter). □ **expurgator** *n.* **expurgation** /ˌekspə'geɪʃ(ə)n/ *n.* **expurgatorial** /ˌekspɜːgə'tɔːrɪəl/ *adj.* **expurgatory** /ek'spɜːgətərɪ/ *adj.* [L *expurgare expurgat-* (as EX-[1], *purgare* cleanse)]

exquisite /'ekskwɪzɪt, disp. ɪk'skwɪzɪt/ *adj.* & *n.* ● *adj.* **1** extremely beautiful or delicate. **2** acute; keenly felt (*exquisite pleasure*). **3 a** keen; highly sensitive or discriminating (*exquisite taste*). **b** elaborately devised or accomplished; consummate, perfect. ● *n.* a person of refined (esp. affected) tastes. □ **exquisitely** *adv.* **exquisiteness** *n.* [ME f. L *exquirere exquisit-* (as EX-[1], *quaerere* seek)]

exsanguinate /ɪk'sæŋgwɪˌneɪt/ *Med.* *v.tr.* drain of blood. □ **exsanguination** /-ˌsæŋgwɪ'neɪʃ(ə)n/ *n.* [L *exsanguinatus* (as EX-[1], *sanguis -inis* blood)]

exsert /ɪk'sɜːt/ *v.tr. Biol.* put forth. [L *exserere*: see EXERT]

ex-service /eks'sɜːvɪs/ *adj.* **1** having formerly been a member of the armed forces. **2** relating to former servicemen and women.

ex-serviceman /eks'sɜːvɪsmən/ *n.* (pl. **-men**) a former member of the armed forces.

ex-servicewoman /eks'sɜːvɪsˌwʊmən/ *n.* (pl. **-women**) a former woman member of the armed forces.

ex silentio /ˌeks sɪ'lenʃɪˌəʊ/ *adv.* by the absence of contrary evidence. [L, = from silence]

ext. *abbr.* **1** exterior. **2** external.

extant /ek'stænt, 'ekstənt/ *adj.* (esp. of a document etc.) still existing, surviving. [L *exstare exstant-* (as EX-[1], *stare* stand)]

extemporaneous /ɪkˌstempə'reɪnɪəs/ *adj.* spoken or done without preparation. □ **extemporaneously** *adv.* **extemporaneousness** *n.*

extemporary /ɪk'stempərərɪ/ *adj.* = EXTEMPORANEOUS. □ **extemporarily** *adv.* **extemporariness** *n.*

extempore /ɪk'stempərɪ/ *adj.* & *adv.* without preparation. [L *ex tempore* on the spur of the moment, lit. 'out of the time' f. *tempus* time]

extemporize /ɪk'stempəˌraɪz/ *v.tr.* (also **-ise**) (also *absol.*) compose or produce (music, a speech, etc.) without preparation; improvise. □ **extemporization** /-ˌstempəraɪ'zeɪʃ(ə)n/ *n.*

extend /ɪk'stend/ *v.* **1** *tr.* & *intr.* lengthen or make larger in space or time. **2 a** *tr.* stretch or lay out at full length. **b** *tr.* & *intr.* (often foll. by *over*) (cause to) stretch or span over a period of time. **3** *intr.* & *tr.* (foll. by *to, over*) reach or be or make continuous over a certain area. **4** *intr.* (foll. by *to*) have a certain scope (*the permit does not extend to camping*). **5** *tr.* offer or accord (an invitation, hospitality, kindness, etc.). **6** *tr.* (usu. *refl.* or in *passive*) tax the powers of (an athlete, horse, etc.) to the utmost. □ **extended family** a family including relatives living near. **extended-play** (of a gramophone record, compact disc, etc.) playing for longer than most singles, usu. with four songs. □ **extendable** *adj.* **extendability** /-ˌstendə'bɪlɪtɪ/ *n.* **extendible** /-'stendɪb(ə)l/ *adj.* **extendibility**

/-ˌstendɪˈbɪlɪtɪ/ *n.* **extensible** /-ˈstensɪb(ə)l/ *adj.* **extensibility** /-ˌstensɪˈbɪlɪtɪ/ *n.* [ME f. L *extendere* extens- or extent- stretch out (as EX-[1], *tendere* stretch)]

extender /ɪkˈstendə(r)/ *n.* **1** a person or thing that extends. **2** a substance added to paint, ink, glue, etc., to dilute its colour or increase its bulk.

extensile /ɪkˈstensaɪl/ *adj.* capable of being stretched out or protruded.

extension /ɪkˈstenʃ(ə)n/ *n.* **1** the act or an instance of extending; the process of being extended. **2** prolongation; enlargement. **3** a part enlarging or added on to a main structure or building. **4** an additional part of anything. **5 a** a subsidiary telephone on the same line as the main one. **b** its number. **6 a** an additional period of time, esp. extending allowance for a project etc. **b** permission for the sale of alcoholic drinks until later than usual, granted to licensed premises on special occasions. **7** extramural instruction by a university or college (*extension course*). **8** extent, range. **9** *Logic* a group of things denoted by a term. □ **extensional** *adj.* [ME f. LL *extensio* (as EXTEND)]

extensive /ɪkˈstensɪv/ *adj.* **1** covering a large area in space or time. **2** having a wide scope; far-reaching (*an extensive knowledge of music*). **3** *Agriculture* involving cultivation from a large area, with a minimum of special resources (cf. INTENSIVE). □ **extensively** *adv.* **extensiveness** *n.* [F *extensif* -ive or LL *extensivus* (as EXTENSION)]

extensometer /ˌekstenˈsɒmɪtə(r)/ *n.* **1** an instrument for measuring deformation of metal under stress. **2** an instrument using such deformation to record elastic strains in other materials. [L *extensus* (as EXTEND) + -METER]

extensor /ɪkˈstensə(r)/ *n.* (in full **extensor muscle**) *Anat.* a muscle that extends or straightens out part of the body (cf. FLEXOR). [mod.L (as EXTEND)]

extent /ɪkˈstent/ *n.* **1** the space over which a thing extends. **2** the width or limits of application; scope (*to a great extent; to the full extent of their power*). [ME f. AF *extente* f. med.L *extenta* past part. of L *extendere*: see EXTEND]

extenuate /ɪkˈstenjʊˌeɪt/ *v.tr.* (often as **extenuating** *adj.*) lessen the seeming seriousness of (guilt or an offence) by reference to some mitigating factor. □ **extenuatingly** *adv.* **extenuatory** /-jʊətərɪ/ *adj.* **extenuation** /-ˌstenjʊˈeɪʃ(ə)n/ *n.* [L *extenuare* extenuat- (as EX-[1], *tenuis* thin)]

exterior /ɪkˈstɪərɪə(r)/ *adj. & n.* ● *adj.* **1 a** of or on the outer side (opp. INTERIOR *adj.* 1). **b** (foll. by *to*) situated on the outside of (a building etc.). **c** coming from outside. **2** *Cinematog.* outdoor. ● *n.* **1** the outward aspect or surface of a building etc. **2** the outward or apparent behaviour or demeanour of a person. **3** *Cinematog.* an outdoor scene. □ **exterior angle** *Geom.* the angle between the side of a rectilinear figure and the adjacent side extended outward. □ **exteriorize** *v.tr.* (also **-ise**). **exteriority** /-ˌstɪərɪˈɒrɪtɪ/ *n.* [L, compar. of *exterus* outside]

exterminate /ɪkˈstɜːmɪˌneɪt/ *v.tr.* **1** destroy utterly (esp. something living). **2** get rid of; eliminate (a pest, disease, etc.). □ **exterminator** *n.* **exterminatory** /-nətərɪ/ *adj.* **extermination** /-ˌstɜːmɪˈneɪʃ(ə)n/ *n.* [L *exterminare* exterminat- (as EX-[1], *terminus* boundary)]

external /ɪkˈstɜːn(ə)l/ *adj. & n.* ● *adj.* **1 a** of or situated on the outside or visible part (opp. INTERNAL). **b** coming or derived from the outside or an outside source. **2** relating to a country's foreign affairs. **3** outside the conscious subject (*the external world*). **4** (of medicine etc.) for use on the outside of the body. **5** for or concerning students taking the examinations of a university without attending it. ● *n.* (in *pl.*) **1** the outward features or aspect. **2** external circumstances. **3** inessentials. □ **external evidence** evidence derived from a source independent of the thing discussed. □ **externally** *adv.* **externality** /ˌekstɜːˈnælɪtɪ/ *n.* (pl. **-ies**). [med.L f. L *externus* f. *exterus* outside]

externalize /ɪkˈstɜːnəˌlaɪz/ *v.tr.* (also **-ise**) give or attribute external existence to. □ **externalization** /-ˌstɜːnəlaɪˈzeɪʃ(ə)n/ *n.*

exteroceptive /ˌekstərəʊˈseptɪv/ *adj. Biol.* relating to stimuli produced outside an organism. [irreg. f. L *externus* exterior + RECEPTIVE]

exterritorial /ˌeksterɪˈtɔːrɪəl/ *adj.* extraterritorial. □ **exterritoriality** /-ˌtɔːrɪˈælɪtɪ/ *n.*

extinct /ɪkˈstɪŋkt/ *adj.* **1** (of a family, class, species, etc.) that has died out. **2 a** (of fire etc.) no longer burning. **b** (of a volcano) that no longer erupts. **3** (of life, hope, etc.) terminated, quenched. **4** (of an office etc.) obsolete. **5** (of a title of nobility) having no qualified claimant. [ME f. L *exstinguere* exstinct- (as EX-[1], *stinguere* quench)]

extinction /ɪkˈstɪŋkʃ(ə)n/ *n.* **1** the act of making extinct; the state of being or process of becoming extinct. (*See note below.*) **2** the act of extinguishing; the state of being extinguished. **3** total destruction or

annihilation. **4** the wiping out of a debt. **5** *Physics* a reduction in the intensity of radiation by absorption, scattering, etc. □ **extinctive** /-ˈstɪŋktɪv/ *adj.* [L *extinctio* (as EXTINCT)]

▪ The idea that it was possible for a species to become extinct came to be accepted gradually in the mid-19th century. Fossils were at first seen as being victims of the biblical flood, until Darwin and others demonstrated the vast expanse of geological time and the likelihood of biological evolution. The dodo and Steller's sea cow, exterminated in the 17th–18th centuries, were probably the first examples of extinction to be recognized, and it came to be realized that the dinosaurs must have become extinct in the much more remote past. It is now known that there have been several major mass extinctions in the history of the earth, perhaps arising from catastrophic events that affected the climate, thereby making way for the dramatic diversification of new species.

extinguish /ɪkˈstɪŋgwɪʃ/ *v.tr.* **1** cause (a flame, light, etc.) to die out; put out. **2** make extinct; annihilate, destroy (*a programme to extinguish disease*). **3** put an end to; terminate; obscure utterly (a feeling, quality, etc.). **4 a** abolish; wipe out (a debt). **b** *Law* render void. **5** *archaic* surpass by superior brilliance. □ **extinguishable** *adj.* **extinguishment** *n.* [irreg. f. L *extinguere* (as EXTINCT): cf. *distinguish*]

extinguisher /ɪkˈstɪŋgwɪʃə(r)/ *n.* a person or thing that extinguishes, esp. = *fire extinguisher*.

extirpate /ˈekstəˌpeɪt/ *v.tr.* root out; destroy completely. □ **extirpator** *n.* **extirpation** /ˌekstəˈpeɪʃ(ə)n/ *n.* [L *exstirpare* exstirpat- (as EX-[1], *stirps* stem)]

extol /ɪkˈstəʊl, -ˈstɒl/ *v.tr.* (**extolled, extolling**) praise enthusiastically. □ **extoller** *n.* **extolment** *n.* [L *extollere* (as EX-[1], *tollere* raise)]

extort /ɪkˈstɔːt/ *v.tr.* obtain by force, threats, persistent demands, etc. □ **extorter** *n.* **extortive** *adj.* [L *extorquere* extort- (as EX-[1], *torquere* twist)]

extortion /ɪkˈstɔːʃ(ə)n/ *n.* **1** the act or an instance of extorting, esp. money. **2** illegal exaction. □ **extortioner** *n.* **extortionist** *n.* [ME f. LL *extortio* (as EXTORT)]

extortionate /ɪkˈstɔːʃənət/ *adj.* **1** (of a price etc.) exorbitant. **2** using or given to extortion (*extortionate methods*). □ **extortionately** *adv.*

extra /ˈekstrə/ *adj., adv., & n.* ● *adj.* additional; more than is usual or necessary or expected. ● *adv.* **1** more than usually. **2** additionally (*was charged extra*). ● *n.* **1** an extra thing. **2** a thing for which an extra charge is made; such a charge. **3** a person engaged temporarily to fill out a scene in a film or play, esp. as one of a crowd. **4** a special issue of a newspaper etc. **5** *Cricket* a run scored other than from a hit with the bat, and added to the total runs scored by the batting side. □ **extra cover** *Cricket* a fielding position on a line between cover-point and mid-off, but beyond these. **extra size** outsize. **extra time** *Sport* a further period of play at the end of a match when the scores are equal. [prob. a shortening of EXTRAORDINARY]

extra- /ˈekstrə/ *comb. form* **1** outside, beyond (*extragalactic*). **2** beyond the scope of (*extracurricular*). [med.L f. L *extra* outside]

extracellular /ˌekstrəˈseljʊlə(r)/ *adj.* situated or taking place outside a cell or cells.

extract *v. & n.* ● *v.tr.* /ɪkˈstrækt/ **1** remove or take out, esp. by effort or force (anything firmly rooted). **2** obtain (money, an admission, etc.) with difficulty or against a person's will. **3** obtain (a natural resource) from the earth. **4** select or reproduce for quotation or performance (a passage of writing, music, etc.). **5** obtain (juice etc.) by suction, pressure, distillation, etc. **6** derive (pleasure etc.). **7** *Math.* find (the root of a number). **8** *archaic* deduce (a principle etc.). ● *n.* /ˈekstrækt/ **1** a short passage taken from a book, piece of music, etc.; an excerpt. **2** a preparation containing the active principle of a substance in concentrated form (*malt extract*). □ **extractable** /ɪkˈstræktəb(ə)l/ *adj.* **extractability** /ɪkˌstræktəˈbɪlɪtɪ/ *n.* [L *extrahere* extract- (as EX-[1], *trahere* draw)]

extraction /ɪkˈstrækʃ(ə)n/ *n.* **1** the act or an instance of extracting; the process of being extracted. **2** the removal of a tooth. **3** origin, lineage, descent (*of Indian extraction*). **4** something extracted; an extract. [ME f. F f. LL *extractio* -onis (as EXTRACT)]

extractive /ɪkˈstræktɪv/ *adj.* of or involving extraction, esp. extensive extracting of natural resources without provision for their renewal.

extractor /ɪkˈstræktə(r)/ *n.* **1** a person or machine that extracts. **2** (*attrib.*) (of a device) that extracts stale air etc. or ventilates a room (*extractor fan; extractor hood*).

extracurricular /ˌekstrəkəˈrɪkjʊlə(r)/ *adj.* (of a subject of study) not included in the normal curriculum.

extraditable /ˈekstrəˌdaɪtəb(ə)l/ adj. **1** liable to extradition. **2** (of a crime) warranting extradition.

extradite /ˈekstrəˌdaɪt/ v.tr. hand over (a person accused or convicted of a crime) to the foreign state etc. in which the crime was committed.

extradition /ˌekstrəˈdɪʃ(ə)n/ n. **1** the extraditing of a person accused or convicted of a crime. **2** Psychol. the localizing of a sensation at a distance from the centre of sensation.

extrados /ekˈstreɪdɒs/ n. Archit. the upper or outer curve of an arch (opp. INTRADOS). [F, f. as EXTRA- + dos back f. L dorsum]

extragalactic /ˌekstrəgəˈlæktɪk/ adj. occurring or existing outside the Galaxy.

extrajudicial /ˌekstrədʒuːˈdɪʃ(ə)l/ adj. **1** not legally authorized. **2 a** outside the jurisdiction of a court. **b** (of a confession etc.) not made in court. □ **extrajudicially** adv.

extramarital /ˌekstrəˈmærɪt(ə)l/ adj. (esp. of sexual relations) occurring outside marriage. □ **extramaritally** adv.

extramundane /ˌekstrəˈmʌndeɪn/ adj. outside or beyond the physical world.

extramural /ˌekstrəˈmjʊərəl/ adj. **1** taught or conducted off the premises of a university, college, or school. **2** additional to normal teaching or studies, esp. for non-resident students. **3** outside the walls or boundaries of a town or city. □ **extramurally** adv. [L extra muros outside the walls]

extraneous /ɪkˈstreɪnɪəs/ adj. **1** of external origin. **2** (often foll. by to) **a** separate from the object to which it is attached etc. **b** external to; irrelevant or unrelated to. **c** inessential, superfluous. □ **extraneously** adv. **extraneousness** n. [L extraneus]

extraordinary /ɪkˈstrɔːdɪnərɪ, ˌekstrəˈɔːdɪnərɪ, -d(ə)nrɪ/ adj. **1** unusual or remarkable; out of the usual course. **2** unusually great (an extraordinary talent). **3** so exceptional as to provoke astonishment or admiration. **4 a** (of an official etc.) additional; specially employed (envoy extraordinary). **b** (of a meeting) specially convened. □ **extraordinarily** adv. **extraordinariness** n. [L extraordinarius f. extra ordinem outside the usual order]

extrapolate /ɪkˈstræpəˌleɪt/ v.tr. (also absol.) **1** Math. & Philos. **a** extend (a range of values, a curve) by inferring unknown values from trends in the known data. **b** calculate (unknown values etc.) by extension of trends in a known range of values etc. **2** infer more widely from a limited range of known facts. □ **extrapolator** n. **extrapolative** /-lətɪv/ adj. **extrapolation** /-ˌstræpəˈleɪʃ(ə)n/ n. [EXTRA- + INTERPOLATE]

extrasensory /ˌekstrəˈsensərɪ/ adj. regarded as derived by means other than the known senses, e.g. by telepathy, clairvoyance, etc. □ **extrasensory perception** a person's supposed faculty of perceiving by such means.

extraterrestrial /ˌekstrətɪˈrestrɪəl/ adj. & n. ● adj. **1** outside the earth or its atmosphere. **2** (in science fiction) from outer space. ● n. (in science fiction) a being from outer space.

extraterritorial /ˌekstrəˌterɪˈtɔːrɪəl/ adj. **1** situated or (of laws etc.) valid outside a country's territory. **2** (of an ambassador etc.) free from the jurisdiction of the territory of residence. □ **extraterritoriality** /-ˌtɔːrɪˈælɪtɪ/ n. [L extra territorium outside the territory]

extravagance /ɪkˈstrævəɡəns/ n. **1** excessive spending or use of resources; being extravagant. **2** an instance or item of this. **3** unrestrained or absurd behaviour, speech, thought, or writing. □ **extravagancy** n. (pl. **-ies**). [F (as EXTRAVAGANT)]

extravagant /ɪkˈstrævəɡənt/ adj. **1** spending (esp. money) excessively; immoderate or wasteful in use of resources. **2** exorbitant; costing much. **3** exceeding normal restraint or sense; unreasonable, absurd (extravagant claims). □ **extravagantly** adv. [ME f. med.L extravagari (as EXTRA-, vagari wander)]

extravaganza /ɪkˌstrævəˈɡænzə/ n. **1** a fanciful literary, musical, or dramatic composition. **2** a spectacular theatrical or television production, esp. of light entertainment. [It. estravaganza extravagance]

extravasate /ɪkˈstrævəˌseɪt/ v. esp. Med. **1** tr. force out (a fluid, esp. blood) from its proper vessel. **2** intr. (of blood, lava, etc.) flow out. □ **extravasation** /-ˌstrævəˈseɪʃ(ə)n/ n. [L extra outside + vas vessel]

extravehicular /ˌekstrəvɪˈhɪkjʊlə(r)/ adj. outside a vehicle, esp. a spacecraft.

extrema pl. of EXTREMUM.

Extremadura /ˌestrəməˈdʊərə/ (Spanish **Estremadura** /ˌestrəmaðura/) an autonomous region of western Spain, on the border with Portugal; capital, Mérida.

extreme /ɪkˈstriːm/ adj. & n. ● adj. **1** reaching a high or the highest degree; exceedingly great or intense; exceptional (extreme old age; in extreme danger). **2 a** severe, stringent; lacking restraint or moderation (take extreme measures; an extreme reaction). **b** (of a person, opinion, etc.) going to great lengths; advocating immoderate measures. **3** outermost; furthest from the centre; situated at either end (the extreme edge). **4** Polit. on the far left or right of a party. **5** utmost; last. ● n. **1** (often in pl.) one or other of two things as remote or as different as possible. **2** a thing at either end of anything. **3** the highest or most extreme degree of anything. **4** Math. the first or the last term of a ratio or series. **5** Logic the subject or predicate in a proposition; the major or the minor term in a syllogism. □ **extreme unction** esp. RC Ch. (former name for) the sacrament of Anointing of the Sick, esp. when administered to the dying. **go to extremes** take an extreme course of action. **go to the other extreme** take a diametrically opposite course of action. **in the extreme** to an extreme degree. □ **extremely** adv. **extremeness** n. [ME f. OF f. L extremus superl. of exterus outward]

extremist /ɪkˈstriːmɪst/ n. (also attrib.) a person who holds extreme or fanatical political or religious views and esp. resorts to or advocates extreme action. □ **extremism** n.

extremity /ɪkˈstremɪtɪ/ n. (pl. **-ies**) **1** the extreme point; the very end. **2** (in pl.) the hands and feet. **3** a condition of extreme adversity or difficulty. **4** excessiveness; extremeness. [ME f. OF extremité or L extremitas (as EXTREME)]

extremum /ɪkˈstriːməm/ n. (pl. **extremums** or **extrema** /-mə/) Math. the maximum or minimum value of a function. □ **extremal** adj. [L, neut. of extremus EXTREME]

extricate /ˈekstrɪˌkeɪt/ v.tr. (often foll. by from) free or disentangle from a constraint or difficulty. □ **extricable** /-kəb(ə)l/ adj. **extrication** /ˌekstrɪˈkeɪʃ(ə)n/ n. [L extricare extricat- (as EX-[1], tricae perplexities)]

extrinsic /ekˈstrɪnsɪk/ adj. **1** not inherent or intrinsic; not essential (opp. INTRINSIC). **2** (often foll. by to) extraneous; lying outside; not belonging (to). **3** originating or operating from without. □ **extrinsically** adv. [LL extrinsicus outward f. L extrinsecus (adv.) f. exter outside + secus beside]

extrovert /ˈekstrəˌvɜːt/ n. & adj. ● n. **1** an outgoing or sociable person. **2** Psychol. a person predominantly concerned with external things or objective considerations. ● adj. typical or characteristic of an extrovert. □ **extroverted** adj. **extroversion** /ˌekstrəˈvɜːʃ(ə)n/ n. [extro- = EXTRA- (after intro-) + L vertere turn]

extrude /ɪkˈstruːd/ v.tr. **1** (foll. by from) thrust or force out. **2** shape metal, plastic, etc., by forcing it through a die. □ **extrusion** /-ˈstruːʒ(ə)n/ n. **extrusile** /-ˈstruːsaɪl/ adj. **extrusive** /-ˈstruːsɪv/ adj. [L extrudere extrus- (as EX-[1], trudere thrust)]

exuberant /ɪɡˈzjuːbərənt/ adj. **1** lively, high-spirited. **2** (of a plant etc.) prolific; growing copiously. **3** (of feelings etc.) abounding, lavish, effusive. □ **exuberance** n. **exuberantly** adv. [F exubérant f. L exuberare (as EX-[1], uberare be fruitful f. uber fertile)]

exuberate /ɪɡˈzjuːbəˌreɪt/ v.intr. be exuberant.

exude /ɪɡˈzjuːd/ v. **1** tr. & intr. (of a liquid, moisture, etc.) escape or cause to escape gradually; ooze out; give off. **2** tr. emit (a smell). **3** tr. display (an emotion etc.) freely or abundantly (exuded displeasure). □ **exudative** /-dətɪv/ adj. **exudate** /ˈeksjʊˌdeɪt/ n. **exudation** /ˌeksjʊˈdeɪʃ(ə)n/ n. [L exsudare (as EX-[1], sudare sweat)]

exult /ɪɡˈzʌlt/ v.intr. (often foll. by at, in, over, or to + infin.) **1** be greatly joyful. **2** (often foll. by over) have a feeling of triumph (over a person). □ **exultancy** n. **exultant** adj. **exultantly** adv. **exultation** /ˌeɡzʌlˈteɪʃ(ə)n/ n. [L exsultare (as EX-[1], saltare frequent. of salire salt- leap)]

Exuma Cays /ɪkˈsuːmə/ a group of some 350 small islands in the Bahamas.

exurb /ˈeksɜːb/ n. a district outside a city or town, esp. a prosperous area beyond the suburbs. □ **exurban** /eksˈɜːb(ə)n/ adj. **exurbanite** n. [L ex out of + urbs city, or back-form. f. exurban (as EX-[1] + URBAN, after suburban)]

exurbia /eksˈɜːbɪə/ n. the exurbs collectively; the region beyond the suburbs. [EX-[1], after suburbia]

exuviae /ɪɡˈzjuːvɪˌiː/ n.pl. (also treated as sing.) an animal's cast skin or covering. □ **exuvial** adj. [L, = animal's skins, spoils of the enemy, f. exuere divest oneself of]

exuviate /ɪɡˈzjuːvɪˌeɪt/ v.tr. shed (a skin etc.). □ **exuviation** /-ˌzjuːvɪˈeɪʃ(ə)n/ n.

ex voto /eks ˈvəʊtəʊ/ n. (pl. **-os**) an offering made in pursuance of a vow. [L, = out of a vow]

-ey /ɪ/ suffix var. of -Y[2].

eyas /ˈaɪəs/ n. a young hawk, esp. one taken from the nest for training in falconry. [orig. *nyas* f. F *niais* ult. f. L *nidus* nest: for loss of *n*- cf. ADDER]

eye /aɪ/ n. & v. ● n. **1** the organ of sight in man and other animals; the light-detecting organ in some invertebrates. (*See note below.*) **2** the eye characterized by the colour of the iris (*has blue eyes*). **3** the region round the eye (*eyes red from weeping*). **4** (in *sing.* or *pl.*) sight; the faculty of sight (*demonstrate to the eye; need perfect eyes to be a pilot*). **5** a particular visual faculty or talent; visual appreciation; perspicacity (*a straight eye; cast an expert eye over*). **6 a** (in *sing.* or *pl.*) a look, gaze, or glance, esp. as indicating the disposition of the viewer (*a friendly eye*). **b** (**the eye**) a flirtatious or sexually provocative glance. **7** mental awareness; consciousness. **8** a person or animal etc. that sees on behalf of another. **9** = *electric eye*. **10** *sl.* = *private eye*. **11** a thing like an eye, esp.: **a** a spot on a peacock's tail (cf. EYELET n. 3). **b** the leaf bud of a potato. **12** the centre of something circular, e.g. a flower or target. **13** the relatively calm region at the centre of a storm or hurricane. **14** an aperture in an implement, esp. a needle, for the insertion of something, e.g. thread. **15** a ring or loop for a bolt or hook etc. to pass through. ● v.tr. (**eyes**, **eyed**, **eyeing** or **eying**) (often foll. by *up*) watch or observe closely, esp. admiringly or with curiosity or suspicion. □ **all eyes 1** watching intently. **2** general attention (*all eyes were on us*). **before one's** (or **one's very**) **eyes** right in front of one. **do a person in the eye** *colloq.* defraud or thwart a person. **eye-bolt** a bolt or bar with an eye at the end for a hook etc. **eye-catching** striking, attractive. **eye contact** looking directly into another person's eyes. **an eye for an eye** retaliation in kind (Exod. 21:24). **eye language** the process of communication by the expression of the eyes. **eye-level** the level seen by the eyes looking horizontally (*eye-level grill*). **eye-liner** a cosmetic applied as a line round the eye. **eye mask 1** a covering of soft material saturated with a lotion for refreshing the eyes. **2** a covering for the eyes. **eye-opener** *colloq.* **1** an enlightening experience; an unexpected revelation. **2** *US* an alcoholic drink taken on waking up. **eye-rhyme** a correspondence of words in spelling but not in pronunciation (e.g. *love* and *move*). **eyes front** (or **left** or **right**) *Mil.* a command to turn the head in the direction stated. **eye-shade** a device to protect the eyes, esp. from strong light. **eye-shadow** a coloured cosmetic applied to the skin round the eyes. **eye-spot 1 a** a light-sensitive area on the bodies of some invertebrate animals, e.g. flatworms, starfish, etc. (also called *ocellus*). **b** *Bot.* an area of light-sensitive pigment found in some algae etc. **2** a fungus disease of plants characterized by yellowish oval spots on the leaves and stems. **eye-stalk** *Zool.* a movable stalk carrying the eye, esp. in crabs, shrimps, etc. **eye strain** fatigue of the (internal or external) muscles of the eye. **eye-tooth** a canine tooth just under or next to the eye, esp. one in the upper jaw. **eye-worm** a nematode worm, *Loa loa*, parasitic on humans and other primates in Central and West Africa. **get** (or **keep**) **one's eye in** *Sport* accustom oneself (or keep oneself accustomed) to the conditions of play so as to judge speed, distance, etc. **have one's eye on** wish or plan to procure. **have an eye for 1** be quick to notice. **2** be partial to. **have eyes for** be interested in; wish to acquire. **hit a person in the eye** (or **between the eyes**) *colloq.* be very obvious or impressive. **keep an eye on 1** pay attention to. **2** look after; take care of. **keep an eye open** (or **out**) (often foll. by *for*) watch out carefully. **keep one's eyes open** (or **peeled** or **skinned**) watch out; be on the alert. **lower one's eyes** look modestly or sheepishly down or away. **make eyes** (or **sheep's eyes**) (foll. by *at*) look amorously or flirtatiously at. **my** (or **all my**) **eye** *int.* nonsense, rubbish. **one in the eye** (foll. by *for*) a disappointment or setback. **open a person's eyes** be enlightening or revealing to a person. **raise one's eyes** look upwards. **run** (or **pass**) **one's eye over** read cursorily. **see eye to eye** (often foll. by *with*) be in full agreement. **set eyes on** (usu. with *neg.*) see, catch sight of. **take one's eyes off** (usu. in *neg.*) stop watching; stop paying attention to. **under the eye of** under the supervision or observation of. **up to the** (or **one's**) **eyes in 1** deeply engaged or involved in; inundated with (*up to the eyes in work*). **2** to the utmost limit (*mortgaged up to the eyes*). **with one's eyes open** deliberately; with full awareness. **with one's eyes shut** (or **closed**) **1** easily; with little effort. **2** without awareness; unobservantly (*goes around with his eyes shut*). **with an eye to** with a view to; prudently considering. **with a friendly** (or **jealous** etc.) **eye** with a feeling of friendship, jealousy, etc. **with one eye on** directing one's attention partly to. **with one eye shut** easily; with little effort (*could do this with one eye shut*). □ **eyed** adj. (also in *comb.*). **eyeless** adj. [OE *ēage* f. Gmc]

▪ In the most primitive animals, from protozoa to flatworms, the eye or eyes consist merely of a few light-sensitive cells on the skin surface.

Many other invertebrates have ocelli with a simple lens, while the compound eyes of insects consist of numerous tiny facets that together form a crude mosaic image. Cephalopod molluscs have evolved a complex eye that is remarkably similar to that of vertebrates. The basic components of the vertebrate eye are a transparent cornea, an adjustable iris, a lens for focusing, and a sensitive retina lining the back of the eye. The retina contains varying proportions of two light-sensitive cells — rods for detecting black and white images, and cones for coloured light. Primates and predatory birds and mammals have forward-facing eyes that provide overlapping images and good binocular vision.

eyeball /ˈaɪbɔːl/ n. & v. ● n. the ball of the eye within the lids and socket. ● v. N. Amer. sl. **1** tr. look or stare at. **2** intr. look or stare. □ **eyeball to eyeball** confronting closely. **to** (or **up to**) **the eyeballs 1** deeply involved; very busy. **2** completely (permeated, soaked, etc.).

eyebath /ˈaɪbɑːθ/ n. (also **eyecup** /ˈaɪkʌp/) a small glass or vessel for applying lotion etc. to the eye.

eyeblack /ˈaɪblæk/ n. = MASCARA.

eyebright /ˈaɪbraɪt/ n. a small plant of the genus *Euphrasia*, traditionally used as a remedy for weak eyes.

eyebrow /ˈaɪbraʊ/ n. the line of hair growing on the ridge above the eye-socket. □ **raise one's eyebrows** (or **an eyebrow**) show surprise, disbelief, or mild disapproval.

eyeful /ˈaɪfʊl/ n. (pl. **-fuls**) *colloq.* **1** a long steady look. **2** a visually striking person or thing. **3** anything thrown or blown into the eye.

eyeglass /ˈaɪglɑːs/ n. **1 a** a lens for correcting or assisting defective sight. **b** (in *pl.*) a pair of these held in the hand or kept in position on the nose by means of a frame or a spring. **2** a small glass vessel for applying lotion etc. to the eye.

eyehole /ˈaɪhəʊl/ n. a hole to look through.

eyelash /ˈaɪlæʃ/ n. each of the hairs growing on the edges of the eyelids. □ **by an eyelash** by a very small margin.

eyelet /ˈaɪlɪt/ n. & v. ● n. **1** a small hole in paper, leather, cloth, etc., for string or rope etc. to pass through. **2** a metal ring reinforcement for this. **3** a small eye, esp. the ocellus on a butterfly's wing (cf. EYE n. 11a). **4** a form of decoration in embroidery. **5** a small hole for observation, shooting through, etc. ● v.tr. (**eyeleted**, **eyeleting**) provide with eyelets. [ME f. OF *oillet* dimin. of *oil* eye f. L *oculus*]

eyelid /ˈaɪlɪd/ n. the upper or lower fold of skin closing to cover the eye.

eyepiece /ˈaɪpiːs/ n. the lens or lenses to which the eye is applied at the end of a microscope, telescope, etc.

eyeshot /ˈaɪʃɒt/ n. seeing-distance (*out of eyeshot*).

eyesight /ˈaɪsaɪt/ n. the faculty or power of seeing.

eyesore /ˈaɪsɔː(r)/ n. a visually offensive or ugly thing, esp. a building.

eyespot /ˈaɪspɒt/ n. an eyelike marking, esp. on the wing of a butterfly.

Eyetie /ˈaɪtaɪ/ n. & adj. sl. offens. ● n. an Italian. ● adj. Italian. [joc. pronunc. of *Italian*]

eyewash /ˈaɪwɒʃ/ n. **1** lotion for the eye. **2** *colloq.* nonsense, bunkum; pretentious or insincere talk.

eyewitness /ˈaɪ͵wɪtnɪs/ n. a person who has personally seen a thing done or happen and can give evidence of it.

eyot var. of AIT.

eyra /ˈeərə/ n. Zool. a red form of jaguarundi. [Tupi (*e*)*irara*]

Eyre /eə(r)/, Edward John (1815–1901), British-born Australian explorer and colonial statesman. He undertook explorations in the interior deserts of Australia (1840–1) and discovered what came to be known as Lake Eyre. He later served as Lieutenant-Governor of New Zealand (1847–53) and Governor of Jamaica (1864–6).

Eyre, Lake a lake in South Australia, named after the explorer Edward John Eyre. It is Australia's largest salt lake.

eyrie /ˈaɪərɪ, ˈɪərɪ, ˈeərɪ/ n. (also **aerie**) **1** a nest of a bird of prey, esp. an eagle, built high up. **2** a house etc. perched high up. [med.L *aeria, aerea*, etc., prob. f. OF *aire* lair ult. f. L *agrum* piece of ground]

Eysenck /ˈaɪseŋk/, Hans Jürgen (1916–92), German-born British psychologist. Noted for his strong criticism of conventional psychotherapy, particularly Freudian psychoanalysis, he developed an alternative treatment for mental disorders in the form of behaviour therapy. Eysenck also devised methods for assessing intelligence and personality, and published his controversial ideas in *Race, Intelligence, and Education* (1971).

Ezek. *abbr.* (in the Bible) Ezekiel.

Ezekiel /ɪˈziːkɪəl/ **1** a Hebrew prophet of the 6th century BC who prophesied the forthcoming destruction of Jerusalem and the Jewish nation and inspired hope for the future well-being of a restored state. **2** a book of the Bible containing his prophecies.

Ezra /ˈezrə/ **1** a Jewish priest and scribe who played a central part in the reform of Judaism in the 5th or 4th century BC, continuing the work of Nehemiah and forbidding mixed marriages. **2** a book of the Bible dealing with Ezra, the return of the Jews from Babylon, and the rebuilding of the Temple.

Ff

F¹ /ef/ *n.* (also **f**) (*pl.* **Fs** or **F's**) **1** the sixth letter of the alphabet. **2** *Mus.* the fourth note of the diatonic scale of C major.

F² *abbr.* (also **F.**) **1** Fahrenheit. **2** female. **3** fine (pencil-lead). **4** *Biol.* filial generation (as F_1 for the first filial generation, F_2 for the second, etc.).

F³ *symb.* **1** *Chem.* the element fluorine. **2** farad(s). **3** force. **4** faraday.

f¹ *abbr.* (also **f.**) **1** female. **2** feminine. **3** following page etc. **4** (*f*) *Mus.* forte. **5** folio. **6** filly. **7** foreign.

f² *symb.* **1** focal length (cf. F-NUMBER). **2** femto-. **3** frequency.

FA *abbr.* **1** (in the UK) Football Association. **2** see FANNY ADAMS 1.

fa *var. of* FAH.

FAA *abbr.* (in the UK) Fleet Air Arm.

fab /fæb/ *adj. colloq.* fabulous, marvellous. [abbr.]

Fabergé /ˈfæbəˌʒeɪ/, Peter Carl (1846–1920), Russian goldsmith and jeweller, of French descent. He is famous for the intricate and imaginative Easter eggs and many other ornaments that he made for the family of Tsar Alexander III and royal households in other countries.

Fabian /ˈfeɪbɪən/ *n. & adj.* ● *n.* a member of the Fabian Society or a supporter of its ideals. (*See note below.*) ● *adj.* relating to or characteristic of the Fabians. □ **Fabianism** *n.* **Fabianist** *n.* [L *Fabianus* f. FABIUS, noted for cautious strategies]

▪ The Fabian Society is an organization of socialists formed in 1884, whose members have included Sidney and Beatrice Webb and George Bernard Shaw and which has close links with the British Labour Party. Fabians advocate social change through gradual reform rather than violent revolutionary action, hence their name.

Fabius /ˈfeɪbɪəs/ (full name Quintus Fabius Maximus Verrucosus, known as 'Cunctator') (d.203 BC), Roman general and statesman. After Hannibal's defeat of the Roman army at Cannae in 216 BC, Fabius successfully pursued a strategy of caution and delay in order to wear down the Carthaginian invaders. This earned him his nickname, which means 'delayer'.

fable /ˈfeɪb(ə)l/ *n. & v.* ● *n.* **1 a** a story, esp. a supernatural one, not based on fact. **b** a tale, esp. with animals as characters, conveying a moral. **2** (*collect.*) myths and legendary tales (*in fable*). **3 a** a false statement; a lie. **b** a thing only supposed to exist. ● *v.* **1** *intr.* tell fictitious tales. **2** *tr.* describe fictitiously. **3** *tr.* (as **fabled** *adj.*) celebrated in fable; famous, legendary. □ **fabler** *n.* [ME f. OF *fabler* f. L *fabulari* f. *fabula* discourse f. *fari* speak]

fabliau /ˈfæblɪˌəʊ/ *n.* (*pl.* **fabliaux** /-ˌəʊz/) a metrical tale in early French poetry, often coarsely humorous. [F f. OF dialect *fabliaux*, -*ax* pl. of *fablel* dimin. (as FABLE)]

Fabriano, Gentile da, see GENTILE DA FABRIANO.

fabric /ˈfæbrɪk/ *n.* **1 a** a woven material; a textile. **b** other material resembling woven cloth. **2** a structure or framework, esp. the walls, floor, and roof of a building. **3** (in abstract senses) the essential structure or essence of a thing (*the fabric of society*). □ **fabric conditioner** a liquid added as a conditioner when washing clothes etc. [ME f. F *fabrique* f. L *fabrica* f. *faber* metal-worker etc.]

fabricate /ˈfæbrɪˌkeɪt/ *v.tr.* **1** construct or manufacture, esp. from prepared components. **2** invent or concoct (a story, evidence, etc.). **3** forge (a document). □ **fabricator** *n.* [L *fabricare fabricat-* (as FABRIC)]

fabrication /ˌfæbrɪˈkeɪʃ(ə)n/ *n.* **1** the action or process of manufacturing or constructing something. **2** the invention of a lie, forging of a document, etc. **3** an invention or falsehood; a forgery. [L *fabricatio* (as FABRICATE)]

fabulist /ˈfæbjʊlɪst/ *n.* **1** a composer of fables. **2** a liar. [F *fabuliste* f. L *fabula*: see FABLE]

fabulous /ˈfæbjʊləs/ *adj.* **1** incredible, exaggerated, absurd (*fabulous wealth*). **2** *colloq.* excellent, marvellous (*looking fabulous*). **3 a** celebrated in fable. **b** legendary, mythical. □ **fabulously** *adv.* **fabulousness** *n.* **fabulosity** /ˌfæbjʊˈlɒsɪtɪ/ *n.* [F *fabuleux* or L *fabulosus* (as FABLE)]

façade /fəˈsɑːd/ *n.* (also **facade**) **1** the face of a building, esp. its principal front. **2** an outward appearance or front, esp. a deceptive one. [F (as FACE)]

face /feɪs/ *n. & v.* ● *n.* **1** the front of the head from the forehead to the chin. **2 a** the expression of the facial features (*had a happy face*). **b** an expression of disgust; a grimace (*make a face*). **3** composure, coolness, effrontery. **4** the surface of a thing, esp. as regarded or approached, esp.: **a** the visible part of a celestial body. **b** a side of a mountain etc. (*the north face*). **c** the (usu. vertical) surface of a coal-seam. **d** *Geom.* each surface of a solid. **e** the façade of a building. **f** the plate of a clock or watch bearing the digits, hands, etc. **5 a** the functional or working side of a tool etc. **b** the distinctive side of a playing card. **c** the obverse of a coin. **6** = TYPEFACE. **7 a** the outward appearance or aspect (*the unacceptable face of capitalism*). **b** outward show; disguise, pretence (*put on a brave face*). **8** a person, esp. conveying some quality or association (*a face from the past; some young faces for a change*). **9** credibility or respect; good reputation; dignity (*lose face*). ● *v.* **1** *tr. & intr.* look or be positioned towards or in a certain direction (*face towards the window*; *facing the window*; *the room faces north*). **2** *tr.* be opposite (*facing page 20*). **3** *tr.* **a** (often foll. by *down, out*) meet resolutely or defiantly; confront (*face one's critics*). **b** not shrink from (*face the facts*). **4** *tr.* present itself to; confront (*the problem that faces us; faces us with a problem*). **5** *tr.* **a** cover the surface of (a thing) with a coating, extra layer, etc. **b** put a facing on (a garment). **6** *intr. & tr.* turn or cause to turn in a certain direction. □ **face-ache 1** neuralgia. **2** *sl.* a mournful-looking or ugly person. **face-card** = court-card. **face-cloth 1** a cloth for washing one's face. **2** a smooth-surfaced woollen cloth. **face-cream** a cosmetic cream applied to the face to improve the complexion. **face down** (or **downwards**) with the face or surface turned towards the ground, floor, etc. (see also sense 3a of *v.*). **face a person down** overcome a person by a show of determination or by browbeating. **face facts** (or **the facts**) recognize the truth. **face-flannel** = face-cloth 1. **face-lift 1** cosmetic surgery to remove wrinkles etc. by tightening the skin of the face. **2** a procedure to improve the appearance of a thing. **face the music** put up with or stand up to unpleasant consequences, esp. criticism. **face off** *Ice Hockey* start or restart play by dropping the puck between the sticks of two opposing players. **face-off** *n. Ice Hockey* an act of facing off. **face-pack** a preparation beneficial to the complexion, spread over the face and removed when dry. **face-powder** a cosmetic powder for reducing the shine on the face. **face-saving** preserving one's reputation, credibility, etc. **face to face** (often foll. by *with*) facing; confronting each other. **face up** (or **upwards**) with the face or surface turned upwards to view. **face up to** accept bravely; confront; stand up to. **face value 1** the nominal value as printed or stamped on money. **2** the superficial appearance or implication of a thing. **face-worker** a miner who works at the coalface. **get out of a person's face** esp. *US sl.* leave a person alone; stop harassing or annoying a person. **have the face** be shameless enough. **in one's** (or **the**) **face 1** straight against one; as one approaches. **2** confronting.

in face (or **the face**) **of 1** despite. **2** confronted by. **in-your-face** esp. *US sl.* bold or aggressive; blatant, provocative. **in your face** esp. *US sl.* an exclamation of scorn or derision. **let's face it** we must be honest or realistic about it. **on the face of it** as it would appear. **put a bold** (or **brave**) **face on it** accept difficulty etc. cheerfully or with courage. **put one's face on** *colloq.* apply make-up to one's face. **put a good face on** make (a matter) look well. **put a new face on** alter the aspect of. **save face** preserve esteem; avoid humiliation. **save a person's face** enable a person to save face; forbear from humiliating a person. **show one's face** see SHOW. **set one's face against** oppose or resist with determination. **to a person's face** openly in a person's presence. □ **faced** *adj.* (also in *comb.*). **facing** *adj.* (also in *comb.*). [ME f. OF ult. f. L *facies*]

faceless /ˈfeɪslɪs/ *adj.* **1** without identity; purposely not identifiable. **2** lacking character. **3** without a face. □ **facelessly** *adv.* **facelessness** *n.*

facer /ˈfeɪsə(r)/ *n. colloq.* **1** a sudden difficulty or obstacle. **2** a blow in the face.

facet /ˈfæsɪt/ *n.* **1** a particular aspect of a thing. **2** one side of a many-sided body, esp. of a cut gem. **3** one segment of a compound eye. □ **faceted** *adj.* (also in *comb.*). [F *facette* dimin. (as FACE, -ETTE)]

facetiae /fəˈsiːʃɪˌiː/ *n.pl.* **1** pleasantries, witticisms. **2** (in bookselling) pornography. [L, pl. of *facetia* jest f. *facetus* witty]

facetious /fəˈsiːʃəs/ *adj.* **1** characterized by flippant or inopportune humour. **2** (of a person) intending to be amusing, esp. inopportunely. □ **facetiously** *adv.* **facetiousness** *n.* [F *facétieux* f. *facétie* f. L *facetia* jest]

facia var. of FASCIA.

facial /ˈfeɪʃ(ə)l/ *adj. & n.* ● *adj.* of or for the face. ● *n.* a beauty treatment for the face. □ **facially** *adv.* [med.L *facialis* (as FACE)]

-facient /ˈfeɪʃ(ə)nt/ *comb. form* forming adjectives and nouns indicating an action or state produced (*abortifacient*). [from or after L *-faciens -entis* part. of *facere* make]

facies /ˈfeɪʃɪˌiːz/ *n.* (pl. same) **1** *Med.* the appearance or facial expression of an individual. **2** *Geol.* the character of rock etc. expressed by its composition, fossil content, etc. [L, = FACE]

facile /ˈfæsaɪl/ *adj. usu. derog.* **1** easily achieved but of little value. **2** (of speech, writing, etc.) fluent, ready, glib. □ **facilely** *adv.* **facileness** *n.* [F *facile* or L *facilis* f. *facere* do]

facilitate /fəˈsɪlɪˌteɪt/ *v.tr.* make easy or less difficult or more easily achieved. □ **facilitator** *n.* **facilitative** /-tətɪv/ *adj.* [F *faciliter* f. It. *facilitare* f. *facile* easy f. L *facilis*]

facilitation /fəˌsɪlɪˈteɪʃ(ə)n/ *n.* **1** the action of facilitating something. **2** *Physiol.* the enhanced response of a neurone to a stimulus following prior stimulation.

facility /fəˈsɪlɪtɪ/ *n.* (pl. **-ies**) **1** ease; absence of difficulty. **2** fluency, dexterity, aptitude (*facility of expression*). **3** (esp. in *pl.*) an opportunity, the equipment, or the resources for doing something. **4** *US* a plant, installation, or establishment. **5** (in *pl.*) *euphem.* a (public) lavatory. [F *facilité* or L *facilitas* (as FACILE)]

facing /ˈfeɪsɪŋ/ *n.* **1 a** a layer of material covering part of a garment etc. for contrast or strength. **b** (in *pl.*) the cuffs, collar, etc., of a military jacket. **2** an outer layer covering the surface of a wall etc.

facsimile /fækˈsɪmɪlɪ/ *n. & v.* ● *n.* **1** an exact copy, esp. of writing, printing, a picture, etc. (often *attrib.*: *facsimile edition*). **2 a** production of an exact copy of a document etc. by electronic scanning and transmission of the resulting data (see also FAX). **b** a copy produced in this way. ● *v.tr.* (**facsimiled, facsimileing**) make a facsimile of. □ **in facsimile** as an exact copy. [mod.L f. L *fac* imper. of *facere* make + *simile* neut. of *similis* like]

fact /fækt/ *n.* **1** a thing that is known to have occurred, to exist, or to be true. **2** a datum of experience (often foll. by an explanatory clause or phrase: *the fact that fire burns*; *the fact of my having seen them*). **3** (usu. in *pl.*) an item of verified information; a piece of evidence. **4** truth, reality. **5** a thing assumed as the basis for argument or inference. □ **before** (or **after**) **the fact** before (or after) the committing of a crime. **a fact of life** something that must be accepted. **facts and figures** precise details. **fact-sheet** a paper setting out relevant information. **the facts of life** information about sexual functions and practices. **in** (or **in point of**) **fact 1** in reality; as a matter of fact. **2** (in summarizing) in short. [L *factum* f. *facere* do]

factice /ˈfæktɪs/ *n. Chem.* a rubber-like substance obtained by vulcanizing unsaturated vegetable oils. [G *Faktis* f. L *facticius* FACTITIOUS]

faction[1] /ˈfækʃ(ə)n/ *n.* **1** a small organized dissenting group within a larger one, esp. in politics. **2** a state of dissension within an organization. [F f. L *factio -onis* f. *facere* fact- do, make]

faction[2] /ˈfækʃ(ə)n/ *n.* a book, film, etc., using real events as a basis for a fictional narrative or dramatization. [blend of FACT and FICTION]

-faction /ˈfækʃ(ə)n/ *comb. form* forming nouns of action from verbs in *-fy* (*petrifaction*; *satisfaction*). [from or after L *-factio -factionis* f. *-facere* do, make]

factional /ˈfækʃən(ə)l/ *adj.* **1** of or characterized by dissent or faction. **2** belonging to a faction. □ **factionalism** *n.* **factionalize** *v.tr. & intr.* (also **-ise**). **factionally** *adv.* [FACTION[1]]

factious /ˈfækʃəs/ *adj.* of, characterized by, or inclined to faction or factions. □ **factiously** *adv.* **factiousness** *n.*

factitious /fækˈtɪʃəs/ *adj.* **1** specially contrived, not genuine (*factitious value*). **2** artificial, not natural (*factitious joy*). □ **factitiously** *adv.* **factitiousness** *n.* [L *facticius* f. *facere* fact- do, make]

factitive /ˈfæktɪtɪv/ *adj. Gram.* (of a verb) having a sense of regarding or designating, and taking a complement as well as an object (e.g. *appointed me captain*). [mod.L *factitivus*, irreg. f. L *factitare* frequent. of *facere* fact- do, make]

factoid /ˈfæktɔɪd/ *n. & adj.* ● *n.* an assumption or speculation that is reported and repeated so often that it becomes accepted as fact; a simulated or imagined fact. ● *adj.* being or having the character of a factoid; containing factoids.

factor /ˈfæktə(r)/ *n. & v.* ● *n.* **1** a circumstance, fact, or influence contributing to a result. **2** *Math.* a whole number etc. that when multiplied with another produces a given number or expression. **3** *Biol.* a gene etc. determining hereditary character. **4** (foll. by identifying number) *Med.* any of a group of substances in the blood contributing to coagulation (*factor eight*). **5 a** a business agent; a merchant buying and selling on commission. **b** *Sc.* a land-agent or steward. **c** an agent or deputy. **6** an agent or company that buys a manufacturer's invoices and takes responsibility for collecting the payments due on them. ● *v.tr.* **1** *Math.* resolve into factors or components. **2** *tr.* sell (one's receivable debts) to a factor. □ **factor analysis** *Statistics* a process by which the relative importance of variables in the study of a sample is assessed by mathematical techniques. □ **factorable** *adj.* [F *facteur* or L *factor* f. *facere* fact- do, make]

factorage /ˈfæktərɪdʒ/ *n.* commission or charges payable to a factor.

factor VIII *n.* (also **factor eight**) *Med.* a blood-clotting protein whose deficiency causes haemophilia.

factorial /fækˈtɔːrɪəl/ *n. & adj. Math.* ● *n.* **1** the product of a number and all the whole numbers below it (symbol: !; as in 4! (= 4 x 3 x 2 x 1)). **2** the product of a series of factors in an arithmetical progression. ● *adj.* of a factor or factorial. □ **factorially** *adv.*

factorize /ˈfæktəˌraɪz/ *v.* (also **-ise**) *Math.* **1** *tr.* resolve into factors. **2** *intr.* be capable of resolution into factors. □ **factorization** /ˌfæktəraɪˈzeɪʃ(ə)n/ *n.*

factory /ˈfæktərɪ/ *n.* (pl. **-ies**) **1** a building or buildings containing plant or equipment for manufacturing something. **2** *hist.* a merchant company's foreign trading station. □ **factory floor** workers in industry as distinct from management. **factory ship** a fishing ship with facilities for immediate processing of the catch. [Port. *feitoria* and LL *factorium*]

Factory Acts *n.pl.* (in the UK) a series of laws regulating the operation of factories, designed to improve the working conditions of employees, esp. women and children. The most important was that of 1833, which set a minimum age of 9 years and a maximum of eight hours a day for child employees and which also instituted inspectors to ensure compliance with these regulations.

factory farming *n.* a system of rearing livestock using industrial or intensive methods, by which poultry, pigs, or cattle are confined indoors under strictly controlled and often cramped conditions. It is strongly opposed by those concerned with animal welfare. Also called *battery farming*. □ **factory farm** *n.*

factotum /fækˈtəʊtəm/ *n.* (pl. **factotums**) an employee who does all kinds of work. [med.L f. L *fac* imper. of *facere* do, make + *totum* neut. of *totus* whole]

factual /ˈfæktʃʊəl, ˈfæktjʊəl/ *adj.* **1** based on or concerned with fact or facts. **2** actual, true. □ **factually** *adv.* **factualness** *n.* **factuality** /ˌfæktʃʊˈælɪtɪ, ˌfæktjʊ-/ *n.* [FACT, after *actual*]

factum /ˈfæktəm/ *n.* (pl. **factums** or **facta** /-tə/) *Law* **1** an act or deed. **2** a statement of facts. [F f. L: see FACT]

facture /ˈfæktʃə(r)/ n. the quality of execution, esp. of the surface of a painting. [ME f. OF f. L *factura* f. *facere fact-* do, make]

facula /ˈfækjʊlə/ n. (pl. **faculae** /-ˌliː/) Astron. a bright spot or streak on the sun. □ **facular** adj. **faculous** adj. [L, dimin. of *fax facis* torch]

facultative /ˈfækəltətɪv/ adj. **1** Law enabling an act to take place. **2** that may occur. **3** Biol. not restricted to a particular function, mode of life, etc. **4** of a faculty. □ **facultatively** adv. [F *facultatif -ive* (as FACULTY)]

faculty /ˈfækəltɪ/ n. (pl. **-ies**) **1** an aptitude or ability for a particular activity. **2** an inherent mental or physical power. **3 a** a department of a university etc. teaching a specific branch of learning (*faculty of modern languages*). **b** N. Amer. the staff of a university or college. **c** a branch of art or science; those qualified to teach it. **4** the members of a particular profession, esp. medicine. **5** authorization, esp. by a Church authority. [ME f. OF *faculté* f. L *facultas -tatis* f. *facilis* easy]

Faculty of Advocates Law the society or association of Scottish advocates (barristers).

FA Cup the major annual knock-out competition in English Association football, first held in 1872. The Cup final is held at Wembley Stadium.

FAD abbr. flavin adenine dinucleotide.

fad /fæd/ n. **1** a craze. **2** a peculiar notion or idiosyncrasy. □ **faddish** adj. **faddishly** adv. **faddishness** n. **faddism** n. **faddist** n. [19th c. (orig. dial.): prob. f. *fidfad* f. FIDDLE-FADDLE]

faddy /ˈfædɪ/ adj. (**faddier**, **faddiest**) having arbitrary likes and dislikes, esp. about food. □ **faddily** adv. **faddiness** n.

fade /feɪd/ v. & n. ● v. **1** intr. & tr. lose or cause to lose colour. **2** intr. lose freshness or strength; (of flowers etc.) droop, wither. **3** intr. **a** (of colour, light, etc.) disappear gradually; grow pale or dim. **b** (of sound) grow faint. **4** intr. (of a feeling etc.) diminish. **5** intr. (foll. by *away*, *out*) (of a person etc.) disappear or depart gradually. **6** tr. (foll. by *in*, *out*) Cinematog. & Broadcasting **a** cause (a picture) to come gradually in or out of view on a screen, or to merge into another shot. **b** make (the sound) more or less audible. **7** intr. (of a radio signal) vary irregularly in intensity. **8** intr. (of a brake) temporarily lose effectiveness. **9** Golf **a** intr. (of a ball) deviate from a straight course, esp. in a deliberate slice. **b** tr. cause (a ball) to fade. ● n. the action or an instance of fading. □ **do a fade** sl. depart. **fade away** languish, grow thin. **fade-in** Cinematog. & Broadcasting the action or an instance of fading in a picture or sound. **fade-out 1** sl. disappearance, death. **2** Cinematog. & Broadcasting the action or an instance of fading out a picture or sound. □ **fadeless** adj. **fader** n. (in sense 6 of v.). [ME f. OF *fader* f. *fade* dull, insipid prob. ult. f. L *fatuus* silly + *vapidus* VAPID]

fadge /fædʒ/ n. Austral. & NZ **1** a limp package of wool. **2** a loosely packed wool bale. [16th-c. Engl. dial.: orig. uncert.]

faeces /ˈfiːsiːz/ n.pl. (US **feces**) waste matter discharged from the bowels. □ **faecal** /ˈfiːk(ə)l/ adj. [L, pl. of *faex* dregs]

Faenza /faːˈentsə/ a town in Emilia-Romagna in northern Italy; pop. (1990) 54,050. The town gave its name to the type of pottery known as faience, which was originally produced there.

faerie /ˈfeɪərɪ, ˈfeə-/ n. (also **faery**) archaic **1** fairyland; the fairies, esp. as represented by Spenser (*Faerie Queene*). **2** (attrib.) visionary, fancied. [var. of FAIRY]

Faeroe Islands /ˈfeərəʊ/ (also **Faeroes**) a group of islands in the North Atlantic between Iceland and the Shetland Islands, belonging to Denmark but partly autonomous; pop. (est. 1988) 47,660; languages, Faeroese (official), Danish; capital, Tórshavn. The shipping forecast area *Faeroes* covers this area of the Atlantic.

Faeroese /ˌfeərəʊˈiːz/ adj. & n. (also **Faroese**) ● adj. of or relating to the Faeroes. ● n. (pl. same) **1** a native of the Faeroes; a person of Faeroese descent. **2** the Scandinavian language of the Faeroes.

faff /fæf/ v. & n. Brit. colloq. ● v.intr. (often foll. by *about*, *around*) fuss, dither. ● n. a fuss. [imit.]

fag¹ /fæg/ n. & v. ● n. **1** esp. Brit. colloq. a piece of drudgery; a wearisome or unwelcome task. **2** esp. Brit. colloq. a cigarette. **3** Brit. (at public schools) a junior pupil who runs errands for a senior. ● v. (**fagged**, **fagging**) **1** colloq. tr. (often foll. by *out*) tire out; exhaust. **2** intr. toil. **3** intr. Brit. (at public schools) act as a fag. **3** tr. (often foll. by *out*) Naut. fray (the end of a rope etc.). □ **fag-end** colloq. **1** Brit. a cigarette-end. **2** an inferior or useless remnant. [18th c.: orig. unkn.: cf. FLAG¹]

fag² /fæg/ n. esp. N. Amer. sl. often offens. a male homosexual. □ **fag hag** a heterosexual woman who prefers the company of homosexual men. [abbr. of FAGGOT]

faggot /ˈfægət/ n. & v. (US **fagot**) ● n. **1** (usu. in pl.) a ball or roll of seasoned chopped liver etc., baked or fried. **2** a bundle of sticks or twigs bound together as fuel. **3** a bundle of iron rods for heat treatment. **4** a bunch of herbs. **5** sl. derog. **a** an unpleasant woman. **b** N. Amer. often offens. a male homosexual. ● v.tr. (**faggoted**, **faggoting**) **1** bind in or make into faggots. **2** join by faggoting (see FAGGOTING). □ **faggoty** adj. [ME f. OF *fagot*, of uncert. orig.]

faggoting /ˈfægətɪŋ/ n. **1** embroidery in which threads are fastened together like a faggot. **2** the joining of materials in a similar manner.

fah /faː/ n. (also **fa**) Mus. **1** (in tonic sol-fa) the fourth note of a major scale. **2** the note F in the fixed-doh system. [ME *fa* f. L *famuli*: see GAMUT]

Fahr. abbr. Fahrenheit.

Fahrenheit¹ /ˈfærənˌhaɪt/, Gabriel Daniel (1686–1736), German physicist. Becoming interested in manufacturing scientific instruments, Fahrenheit set up his own business. He improved the performance of thermometers, found that liquids have their own characteristic boiling-point, developed an instrument to determine atmospheric pressure from the boiling-point of water, and designed a hydrometer. Fahrenheit is best known, however, for his thermometer scale, which he originally planned with fixed points at the human body temperature and at the coldest temperature he could achieve by mixing ice and salt.

Fahrenheit² /ˈfærənˌhaɪt/ adj. of or denoting a temperature on the Fahrenheit scale. □ **Fahrenheit scale** a scale of temperature on which water freezes at 32° and boils at 212° under standard conditions. [FAHRENHEIT¹]

faience /faɪˈɒns/ n. decorated and glazed earthenware and porcelain, e.g. delft or majolica. [F *faience* f. FAENZA]

fail /feɪl/ v. & n. ● v. **1** intr. not succeed (*failed in persuading*; *failed to qualify*; *tried but failed*). **2 a** tr. & intr. be unsuccessful in (an examination, test, interview, etc.); be rejected as a candidate. **b** tr. (of a commodity etc.) not pass (a test of quality). **c** tr. reject (a candidate etc.); adjudge unsuccessful. **3** intr. be unable to; neglect to; choose not to (*I fail to see the reason*; *he failed to appear*). **4** tr. disappoint; let down; not serve when needed. **5** intr. (of supplies, crops, etc.) be or become lacking or insufficient. **6** intr. become weaker; cease functioning; break down (*her health is failing*; *the engine has failed*). **7** intr. **a** (of an enterprise) collapse; come to nothing. **b** become bankrupt. ● n. a failure in an examination or test. □ **fail-safe** reverting to a safe condition in the event of a breakdown etc. **without fail** for certain, whatever happens. [ME f. OF *faillir* (v.), *fail(l)e* (n.) ult. f. L *fallere* deceive]

failed /feɪld/ adj. **1** unsuccessful; not good enough (*a failed actor*). **2** weak, deficient; broken down (*a failed crop*; *a failed battery*).

failing /ˈfeɪlɪŋ/ n. & prep. ● n. a fault or shortcoming; a weakness, esp. in character. ● prep. in default of; if not.

failure /ˈfeɪljə(r)/ n. **1** lack of success; failing. **2** an unsuccessful person, thing, or attempt. **3** non-performance, non-occurrence. **4** breaking down or ceasing to function (*heart failure*; *engine failure*). **5** running short of supply etc. **6** bankruptcy, collapse. [earlier *failer* f. AF, = OF *faillir* FAIL]

fain /feɪn/ adj. & adv. archaic ● predic.adj. (foll. by *to* + infin.) **1** willing under the circumstances to. **2** left with no alternative but to. ● adv. gladly (*esp. would fain*). [OE *fægen* f. Gmc]

fainéant /ˈfeɪnɪənt/ n. & adj. ● n. an idle or ineffective person. ● adj. idle, inactive. □ **fainéancy** n. [F f. *fait* does + *néant* nothing]

faint /feɪnt/ adj., v., & n. ● adj. **1** indistinct, pale, dim; quiet; not clearly perceived. **2** (of a person) weak or giddy; inclined to faint. **3** slight, remote, inadequate (*a faint chance*). **4** feeble, half-hearted (*faint praise*). **5** = FEINT². ● v.intr. **1** lose consciousness from a drop in blood pressure. **2** become faint. ● n. a sudden loss of consciousness; fainting. □ **faint-hearted** cowardly, timid. **faint-heartedly** in a faint-hearted manner. **faint-heartedness** cowardliness, timidity. **not have the faintest** colloq. have no idea. □ **faintness** n. [ME f. OF, past part. of *faindre* FEIGN]

faintly /ˈfeɪntlɪ/ adv. **1** very slightly (*faintly amused*). **2** indistinctly, feebly.

fair¹ /feə(r)/ adj., adv., n., & v. ● adj. **1** just, unbiased, equitable; in accordance with the rules. **2** blond; light or pale in colour or complexion. **3 a** of (only) moderate quality or amount; average. **b** considerable, satisfactory (*a fair chance of success*). **4** (of weather) fine and dry; (of the wind) favourable. **5** clean, clear, unblemished (*fair copy*). **6** beautiful, attractive. **7** archaic kind, gentle. **8 a** specious (*fair speeches*). **b** complimentary (*fair words*). **9** Austral. & NZ complete, unquestionable. **10** unobstructed, open. ● adv. **1** in a fair manner (*play fair*). **2** exactly, completely (*was hit fair on the jaw*). ● n. **1** a fair thing. **2** archaic a beautiful woman. ● v. **1** tr. make (the surface of a ship, aircraft, etc.) smooth and streamlined. **2** intr. dial. (of the weather) become fair.

falsies /ˈfɒlsɪz, ˈfɔːl-/ n.pl. colloq. pads of material used to increase the apparent size of the breasts.

falsify /ˈfɒlsɪˌfaɪ, ˈfɔːls-/ v.tr. (-ies, -ied) 1 fraudulently alter or make false (a document, evidence, etc.). 2 misrepresent. 3 make wrong; pervert. 4 show to be false. 5 disappoint (a hope, fear, etc.). □ **falsifiable** adj. **falsifiability** /ˌfɒlsɪˌfaɪəˈbɪlɪtɪ, ˌfɔːls-/ n. **falsification** /ˌfɒlsɪfɪˈkeɪʃ(ə)n ˌfɔːls-/ n. [ME f. F falsifier or med.L falsificare f. L falsificus making false f. falsus false]

Falster /ˈfɑːlstə(r)/ a Danish island in the Baltic Sea, south of Zealand. Its southern tip is the most southerly point of Denmark.

falter /ˈfɒltə(r), ˈfɔːl-/ v. 1 intr. stumble, stagger; go unsteadily. 2 intr. waver; lose courage. 3 tr. & intr. stammer; speak hesitatingly. □ **falterer** n. **falteringly** adv. [ME: orig. uncert.]

fame /feɪm/ n. 1 renown; the state of being famous. 2 archaic reputation. 3 archaic public report; rumour. □ **house of ill fame** archaic a brothel. **ill fame** disrepute. [ME f. OF f. L fama]

famed /feɪmd/ adj. 1 (foll. by for) famous; much spoken of (famed for its good food). 2 archaic currently reported.

familial /fəˈmɪlɪəl/ adj. of, occurring in, or characteristic of a family or its members. [F f. L familia FAMILY]

familiar /fəˈmɪlɪə(r)/ adj. & n. ● adj. 1 a (often foll. by to) well known; no longer novel. b common, usual; often encountered or experienced. 2 (foll. by with) knowing a thing well or in detail (am familiar with all the problems). 3 (often foll. by with) a well acquainted (with a person); in close friendship; intimate. b sexually intimate. 4 excessively informal; impertinent. 5 unceremonious, informal. ● n. 1 a close friend or associate. 2 RC Ch. a person rendering certain services in a pope's or bishop's household. 3 (in full **familiar spirit**) a demon supposedly attending and obeying a witch etc. □ **familiarly** adv. [ME f. OF familier f. L familiaris (as FAMILY)]

familiarity /fəˌmɪlɪˈærɪtɪ/ n. (pl. -ies) 1 the state of being well known (the familiarity of the scene). 2 (foll. by with) close acquaintance. 3 a close relationship. 4 a sexual intimacy. b (in pl.) acts of physical intimacy. 5 behaviour that is familiar or informal, esp. excessively so. [ME f. OF familiarité f. L familiaritas -tatis (as FAMILIAR)]

familiarize /fəˈmɪlɪəˌraɪz/ v.tr. (also -ise) 1 (foll. by with) make (a person) conversant or well acquainted. 2 make (a thing) well known. □ **familiarization** /-ˌmɪlɪəraɪˈzeɪʃ(ə)n/ n. [F familiariser f. familiaire (as FAMILIAR)]

famille /fæˈmiː/ n. a Chinese enamelled porcelain with a predominant colour: (**famille jaune** /ʒəʊn/) yellow, (**famille noire** /nwɑː(r)/) black, (**famille rose** /rəʊz/) red, (**famille verte** /veət/) green. [F, = family]

family /ˈfæmɪlɪ, ˈfæmlɪ/ n. (pl. -ies) 1 a set of parents and children, or of relations, living together or not. 2 a the members of a household, esp. parents and their children. b a person's children. c (attrib.) serving the needs of families (family butcher). 3 a all the descendants of a common ancestor; a house, a lineage. b a race or group of peoples from a common stock. 4 all the languages ultimately derived from a particular early language, regarded as a group. 5 a group of persons or nations united by political or religious ties. 6 a group of objects distinguished by common features. 7 Math. a group of curves etc. obtained by varying one quantity. 8 Biol. a group of related genera of organisms within an order in taxonomic classification. □ **family allowance** Brit. a former name for child benefit. **family credit** (in the UK) a regular payment by the state to a family with an income below a certain level. **family man** a man having a wife and children, esp. one fond of family life. **family name** a surname. **family planning** birth control. **family tree** a chart showing relationships and lines of descent. **in the** (or **a**) **family way** colloq. pregnant. [ME f. L familia household f. famulus servant]

Family Division (in the UK) a division of the High Court dealing with adoption, divorce, etc.

famine /ˈfæmɪn/ n. 1 a extreme scarcity of food. b a shortage of something specified (a labour famine). 2 archaic hunger, starvation. [ME f. OF f. faim f. L fames hunger]

famish /ˈfæmɪʃ/ v.tr. & intr. (usu. in passive) 1 reduce or be reduced to extreme hunger. 2 colloq. (esp as **famished** adj.) feel very hungry. [ME f. obs. fame f. OF afamer ult. f. L fames hunger]

famous /ˈfeɪməs/ adj. 1 (often foll. by for) celebrated; well known. 2 colloq. excellent. □ **famousness** n. [ME f. AF, OF fameus f. L famosus f. fama fame]

famously /ˈfeɪməslɪ/ adv. 1 colloq. excellently (got on famously). 2 notably.

famulus /ˈfæmjʊləs/ n. (pl. **famuli** /-ˌlaɪ/) hist. an attendant on a magician or scholar. [L, = servant]

Fan /fæn/ n. & adj. (also **Fang** /fæŋ/) ● n. (pl. same or **Fans**) 1 a member of a people inhabiting parts of Cameroon, Equatorial Guinea, and Gabon. 2 the Bantu language of this people. ● adj. of or relating to this people or their language. [F, prob. f. Fan]

fan¹ /fæn/ n. & v. ● n. 1 an apparatus, usu. with rotating blades, giving a current of air for ventilation etc. 2 a device, usu. folding and forming a semicircle when spread out, for agitating the air to cool oneself. 3 anything spread out like a fan, e.g. a bird's tail or kind of ornamental vaulting (fan tracery). 4 a device for winnowing grain. 5 a fan-shaped deposit of alluvium, esp. where a stream begins to descend a gentler slope. 6 a small sail for keeping the head of a windmill towards the wind. ● v. (**fanned, fanning**) 1 tr. a blow a current of air on, with or as with a fan. b agitate (the air) with a fan. 2 tr. (of a breeze) blow gently on; cool. 3 tr. a winnow (grain). b winnow away (chaff). 4 tr. sweep away by or as by the wind from a fan. 5 intr. & tr. (usu. foll. by out) spread out in the shape of a fan. □ **fan belt** a belt that drives a fan to cool the radiator in a motor vehicle. **fan dance** a dance in which the dancer is (apparently) nude and partly concealed by fans. **fan heater** an electric heater in which a fan drives air over an element. **fan-jet** = TURBOFAN. **fan palm** a palm tree with fan-shaped leaves. □ **fanlike** adj. **fanner** n. [OE fann (in sense 4 of n.) f. L vannus winnowing-fan]

fan² /fæn/ n. a devotee of a particular activity, performer, etc. (film fan; football fan). □ **fan club** an organized group of devotees. **fan mail** letters from fans. □ **fandom** n. [abbr. of FANATIC]

fanatic /fəˈnætɪk/ n. & adj. ● n. a person filled with excessive and often misguided enthusiasm for something. ● adj. excessively enthusiastic. □ **fanatical** adj. **fanatically** adv. **fanaticism** /-ˌtɪsɪz(ə)m/ n. **fanaticize** /-tɪˌsaɪz/ v.intr. & tr. (also **-ise**). [F fanatique or L fanaticus f. fanum temple (orig. in religious sense)]

fancier /ˈfænsɪə(r)/ n. a connoisseur or follower of some activity or thing (pigeon-fancier).

fanciful /ˈfænsɪˌfʊl/ adj. 1 existing only in the imagination or fancy. 2 indulging in fancies; whimsical, capricious. 3 fantastically designed, ornamented, etc.; odd-looking. □ **fancifully** adv. **fancifulness** n.

fancy /ˈfænsɪ/ n., adj., & v. ● n. (pl. **-ies**) 1 an individual taste or inclination. 2 a caprice or whim. 3 a thing favoured, e.g. a horse to win a race. 4 an arbitrary supposition. 5 a the faculty of using imagination or of inventing imagery. b a mental image. 6 delusion; unfounded belief. 7 (prec. by the) those who have a certain hobby; fanciers, esp. patrons of boxing. ● adj. (usu. attrib.) (**fancier, fanciest**) 1 ornamental; not plain. 2 capricious, whimsical, extravagant (at a fancy price). 3 based on imagination, not fact. 4 (of foods etc.) of fine quality. 5 (of flowers etc.) particoloured. 6 (of an animal) bred for particular points of beauty etc. ● v.tr. (**-ies, -ied**) 1 (foll. by that + clause) be inclined to suppose; rather think. 2 Brit. colloq. feel a desire for (do you fancy a drink?). 3 Brit. colloq. find sexually attractive. 4 colloq. have an unduly high opinion of (oneself, one's ability, etc.). 5 (in imper.) an exclamation of surprise (fancy their doing that!). 6 a picture to oneself; conceive, imagine. b (as **fancied** adj.) having no basis in fact; imaginary. □ **catch** (or **take**) **a person's fancy** please or appeal to a person. **take a fancy to** become (esp. inexplicably) fond of. **fancy dress** fanciful costume, esp. for masquerading as a different person or as an animal etc. at a party. **fancy-free** (esp. in phr. **footloose and fancy-free**) without (esp. emotional) commitments. **fancy goods** ornamental novelties etc. **fancy man** sl. derog. 1 a woman's lover. 2 a pimp. **fancy woman** sl. derog. a mistress. **fancy-work** ornamental sewing etc. □ **fanciable** adj. colloq. (in sense 3 of v.). **fancily** adv. **fanciness** n. [contr. of FANTASY]

fandangle /fænˈdæŋg(ə)l/ n. 1 a fantastic ornament. 2 nonsense, tomfoolery. [perh. f. FANDANGO after newfangle]

fandango /fænˈdæŋgəʊ/ n. (pl. **-oes** or **-os**) 1 a a lively Spanish dance for two. b the music for this. 2 nonsense, tomfoolery. [Sp.: orig. unkn.]

fane /feɪn/ n. poet. = TEMPLE¹. [ME f. L fanum]

fanfare /ˈfænfeə(r)/ n. 1 a short showy or ceremonious sounding of trumpets, bugles, etc. 2 an elaborate display; a burst of publicity. [F, imit.]

fanfaronade /ˌfænfærəˈneɪd/ n. 1 arrogant talk; brag. 2 a fanfare. [F fanfaronnade f. fanfaron braggart (as FANFARE)]

Fang var. of FAN.

fang /fæŋ/ n. 1 a large sharp tooth, esp. the canine tooth of a dog or wolf. 2 (usu. in pl.) a the tooth of a venomous snake, by which poison is injected. b the chelicera of a spider. 3 the root of a tooth or its prong. 4 Brit. colloq. a person's tooth. □ **fanged** adj. (also in comb.). **fangless** adj. [OE f. ON fang f. a Gmc root = to catch]

Fangio /ˈfændʒɪˌəʊ/, Juan Manuel (1911–95), Argentinian motor-racing driver. He first won the world championship in 1951, and then held the title from 1954 until 1957. He retired from racing in 1958.

fanlight /ˈfænlaɪt/ n. a small, orig. semicircular window over a door or another window.

fanny /ˈfænɪ/ n. (pl. **-ies**) **1** Brit. coarse sl. the female genitals. **2** N. Amer. sl. the buttocks. □ **fanny pack** N. Amer. sl. = bum-bag. [20th c.: orig. unkn.]

Fanny Adams /ˌfænɪ ˈædəmz/ n. Brit. sl. **1** (abbr. **FA**) (also **sweet Fanny Adams**) nothing at all. ¶ Sometimes understood as a euphemism for *fuck all*. **2** Naut. **a** tinned meat. **b** stew. [name of a murder victim c.1870]

fantail /ˈfænteɪl/ n. **1** a pigeon with a broad tail. **2** a monarch flycatcher of the genus *Rhipidura*, with a fan-shaped tail. **3** a fan-shaped tail or end. **4** the fan of a windmill, which turns the sails to face the wind. **5** the projecting part of a boat's stern. □ **fantailed** adj.

fan-tan /ˈfæntæn/ n. **1** a Chinese gambling game in which players try to guess the remainder after the banker has divided a number of hidden objects into four groups. **2** a card-game in which players build on sequences of sevens. [Chin., = repeated divisions]

fantasia /fænˈteɪzɪə, ˌfæntəˈzɪə/ n. **1** a musical or other composition free in form and often in improvisatory style. **2** a composition which is based on several familiar tunes. [It., = FANTASY]

fantasize /ˈfæntəˌsaɪz/ v. **1** intr. have a fantasy or fanciful vision. **2** tr. imagine; create a fantasy about. □ **fantasist** n.

fantast /ˈfæntæst/ n. (also **phantast**) a visionary; a dreamer. [med.L f. Gk *phantastēs* boaster f. *phantazomai* make a show f. *phainō* show]

fantastic /fænˈtæstɪk/ adj. (also **fantastical** /-k(ə)l/) **1** colloq. excellent, extraordinary. **2** extravagantly fanciful; capricious, eccentric. **3** grotesque or quaint in design etc. □ **fantastically** adv. **fantasticality** /-ˌtæstɪˈkælɪtɪ/ n. [ME f. OF *fantastique* f. med.L *fantasticus* f. LL *phantasticus* f. Gk *phantastikos* (as FANTAST)]

fantasticate /fænˈtæstɪˌkeɪt/ v.tr. make fantastic. □ **fantastication** /-ˌtæstɪˈkeɪʃ(ə)n/ n.

fantasy /ˈfæntəsɪ, -təzɪ/ n. & v. ● n. (pl. **-ies**) **1** the faculty of inventing images, esp. extravagant or visionary ones. **2** a fanciful mental image; a day-dream. **3** a whimsical speculation. **4** a fantastic invention or composition; a fantasia. **5** fabrication, pretence; make-believe (*his account was pure fantasy*). **6** a literary genre concerned with imaginary worlds and peoples. ● v.tr. (**-ies**, **-ied**) imagine, fancy. □ **fantasy football** (or **cricket** etc.) a competition in which participants select imaginary teams from among the players in a league etc. and score points according to the actual performance of their players. [ME f. OF *fantasie* f. L *phantasia* appearance f. Gk (as FANTAST)]

Fanti /ˈfæntɪ/ n. & adj. (also **Fante**) ● n. (pl. same or **Fantis**) **1** a member of a people inhabiting southern Ghana. **2** the language of this tribe. ● adj. of or relating to this people or their language. [Fanti]

fanzine /ˈfænziːn/ n. a magazine for fans, esp. those of sport, popular music, or science fiction. [FAN[2] + MAGAZINE]

FAO abbr. Food and Agriculture Organization (of the United Nations).

far /fɑː(r)/ adv. & adj. ● adv. **1** at or to or by a great distance (*far away*; *far off*; *far out*). **2** a long way (off) in space or time (*are you travelling far?*; *we talked far into the night*). **3** to a great extent or degree; by much (*far better*; *far too early*). ● adj. (**further**, **furthest** or **farther**, **farthest**) **1** situated at or extending over a great distance in space or time; remote (*a far country*). **2** more distant (*the far end of the hall*). **3** extreme (*far left*). □ **as far as 1** to the distance of (a place). **2** to the extent that (*travel as far as you like*). **by far 1** by a great amount. **2** (as an intensifier) without doubt. **far and away** by a very large amount. **far and near** everywhere. **far and wide** over a large area. **far-away 1** remote; long-past. **2** (of a look) dreamy. **3** (of a voice) sounding as if from a distance. **be far from me** (foll. by to + infin.) I am reluctant to (esp. express criticism etc.). **far cry** a long way. **far-fetched** (of an explanation etc.) strained, unconvincing. **far-flung 1** extending far; widely distributed. **2** remote, distant. **far from** very different from being; tending to the opposite of (*the problem is far from being solved*). **far from it** colloq. certainly not, on the contrary. **far gone 1** advanced in time. **2** colloq. in an advanced state of illness, drunkenness, etc. **3** colloq. in a dilapidated state; beyond help. **far-off** remote. **far out** sl. **1** avant-garde, unconventional. **2** excellent. **far-reaching 1** widely applicable. **2** having important consequences or implications. **far-seeing** shrewd in judgement; prescient. **go far 1** achieve much. **2** contribute greatly. **3** be adequate. **go too far** go beyond the limits of what is reasonable, polite, etc. **how far** to what extent. **in so far as** (or

that) to the extent that. **so far 1** to such an extent or distance; to this point. **2** until now. **so far as** (or **that**) to the extent that. **so far so good** progress has been satisfactory up to now. [OE *feorr*]

farad /ˈfærəd/ n. Electr. the SI unit of capacitance (symbol: **F**), equal to the capacitance of a capacitor in which one coulomb of charge causes a potential difference of one volt. [shortening of FARADAY]

faradaic /ˌfærəˈdeɪɪk/ adj. (also **faradic** /fəˈrædɪk/) Electr. inductive, induced. [FARADAY]

Faraday /ˈfærəˌdeɪ/, Michael (1791–1867), English physicist and chemist. One of the greatest experimentalists, Faraday was largely self-educated. Appointed by Sir Humphry Davy as his assistant at the Royal Institution, he initially concentrated on analytical chemistry, and discovered benzene in 1825. His most important work was in electromagnetism, in which field he demonstrated electromagnetic rotation and discovered electromagnetic induction (the key to the development of the electric dynamo and motor). Faraday's concept of magnetic lines of force formed the basis of the classical field theory of electromagnetic behaviour. He also discovered the laws of electrolysis.

faraday /ˈfærəˌdeɪ/ n. Chem. a unit of electric charge equal to Faraday's constant (symbol: **F**). [FARADAY]

Faraday cage n. Electr. an earthed metal screen used for excluding electrostatic influences.

Faraday effect n. Physics the rotation of the plane of polarization of electromagnetic waves in certain substances in a magnetic field.

Faraday's constant n. Chem. the quantity of electric charge carried by one mole of electrons (= 96.49 coulombs). (See also FARADAY.)

farandole /ˌfærənˈdɒl/ n. **1** a lively Provençal dance. **2** the music for this. [F f. mod. Prov. *farandoulo*]

farce /fɑːs/ n. **1** a comic dramatic work using buffoonery and horseplay and often including crude characterization and improbable situations; this branch of drama. (*See note below.*) **2** absurdly futile proceedings; pretence, mockery; an instance of this. [F, orig. = stuffing, f. OF *farsir* f. L *farcire* to stuff]

■ The term was first used in 15th-century France to describe a type of acrobatic entertainment which had originally existed as comic interludes improvised and inserted (or 'stuffed', hence the name) by actors into the texts of religious plays. Dramatists of the 17th century, for example Molière, used elements of farce in their plays, but the genre became especially popular during the 18th and 19th centuries, particularly in France with writers such as Feydeau, and continues to be popular today.

farceur /fɑːˈsɜː(r)/ n. **1** a joker or wag. **2** an actor or writer of farces. [F f. *farcer* act farces]

farcical /ˈfɑːsɪk(ə)l/ adj. **1** extremely ludicrous or futile. **2** of or like farce. □ **farcically** adv. **farcicality** /ˌfɑːsɪˈkælɪtɪ/ n.

farcy /ˈfɑːsɪ/ n. glanders with inflammation of the lymph vessels. □ **farcy bud** (or **button**) a small lymphatic tumour as a result of farcy. [ME f. earlier & OF *farcin* f. LL *farciminum* f. *farcire* to stuff]

farded /ˈfɑːdɪd/ adj. archaic (of a face etc.) painted with cosmetics. [past part. of obs. *fard* f. OF *farder*]

fare /feə(r)/ n. & v. ● n. **1 a** the price a passenger has to pay to be conveyed a certain distance by bus, train, etc. **b** a passenger paying to travel in a public vehicle. **2** a range of food, esp. one provided by a restaurant etc. ● v.intr. literary **1** progress; get on (*how did you fare?*). **2** happen; turn out. **3** journey, go, travel. □ **fare-stage** Brit. **1** a section of a bus etc. route for which a fixed fare is charged. **2** a stop marking this. [OE *fær*, *faru* journeying, *faran* (v.), f. Gmc]

Far East China, Japan, and other countries of eastern Asia. □ **Far Eastern** of or in the Far East.

farewell /feəˈwel/ int. & n. ● int. = GOODBYE int. ● n. **1** leave-taking, departure (also attrib.: *a farewell kiss*). **2** parting good wishes. [ME f. imper. of FARE + WELL[1]]

Farewell, Cape 1 (called in Danish *Kap Farvel*) the southernmost point of Greenland. **2** the northernmost point of South Island, New Zealand. The cape was named by Captain James Cook as the last land sighted before he left for Australia in March 1770.

Fargo, William, see WELLS, FARGO, & CO.

Faridabad /fəˈriːdəˌbæd/ an industrial city in northern India, south of Delhi, in the state of Haryana; pop. (1991) 614,000.

farina /fəˈraɪnə, -ˈriːnə/ n. **1** the flour or meal of cereal, nuts, or starchy roots. **2** a powdery substance. **3** Brit. starch. □ **farinaceous** /ˌfærɪˈneɪʃəs/ adj. [L f. *far* corn]

farl /fɑːl/ n. Sc. a thin cake, orig. quadrant-shaped, of oatmeal or flour. [obs. *fardel* quarter (as FOURTH, DEAL¹)]

farm /fɑːm/ n. & v. ● n. **1** an area of land and its buildings used under one management for growing crops, rearing animals, etc. **2** a place or establishment for breeding a particular type of animal, growing fruit, etc. (*trout-farm*; *mink-farm*). **3** = FARMHOUSE. **4** a place for the storage of oil or oil products. **5** = *sewage farm*. ● v. **1 a** tr. use (land) for growing crops, rearing animals, etc. **b** intr. be a farmer; work on a farm. **2** tr. breed (fish etc.) commercially. **3** tr. (often foll. by *out*) **a** delegate or subcontract (work) to others. **b** contract (the collection of taxes) to another for a fee. **c** arrange for (a person, esp. a child) to be looked after by another, with payment. **4** tr. let the labour of (a person) for hire. **5** tr. contract to maintain and care for (a person, esp. a child) for a fixed sum. □ **farm-hand** a worker on a farm. □ **farmable** adj. **farming** n. [ME f. OF *ferme* f. med.L *firma* fixed payment f. L *firmus* FIRM¹: orig. applied only to leased land]

farmer /'fɑːmə(r)/ n. **1** a person who runs or cultivates a farm. **2** a person to whom the collection of taxes is contracted for a fee. **3** a person who looks after children for payment. [ME f. AF *fermer*, OF *fermier* f. med.L *firmarius, firmator* f. *firma* FIRM²]

farmhouse /'fɑːmhaʊs/ n. a dwelling-place (esp. the main one) attached to a farm.

farmstead /'fɑːmsted/ n. a farm and its buildings regarded as a unit.

farmyard /'fɑːmjɑːd/ n. & adj. ● n. a yard or enclosure attached to a farmhouse. ● attrib.adj. disgusting or uncouth.

Farnborough /'fɑːnbərə/ a town in southern England, in Hampshire; pop. (1990) 48,300. Noted as a centre of aviation, it is the site of an annual air show.

Farne Islands /fɑːn/ a group of seventeen small islands off the coast of Northumberland, scene of a shipwreck in 1838 when the survivors of the *Forfarshire* were rescued by Grace Darling. The islands are noted for their wildlife.

Farnese¹ /fɑː'neɪsɪ/, Alessandro, see PAUL III.

Farnese² /fɑː'neɪsɪ/, Alessandro, Duke of Parma (1545–92), Italian general and statesman. While in the service of Philip II of Spain, he acted as Governor-General of the Netherlands (1578–92). He captured Antwerp in 1585, securing the southern Netherlands for Spain.

Faro /'fɑːruː/ a seaport on the south coast of Portugal, capital of the Algarve; pop. (1990) 31,970.

faro /'feərəʊ/ n. a gambling card-game in which bets are placed on the order of appearance of the cards. [F *pharaon* PHARAOH (said to have been the name of the king of hearts)]

Faroese var. of FAEROESE.

farouche /fə'ruːʃ/ adj. sullen, shy. [F f. OF *faroche, farouche* f. med.L *forasticus* f. L *foras* out of doors]

Farouk /fæ'ruːk/ (1920–65), king of Egypt, reigned 1936–52. On assuming power he dismissed Prime Minister Nahas Pasha, but was forced by the British government to reinstate him. Farouk's defeat in the Arab–Israeli conflict of 1948, together with the general corruption of his reign, led to a military coup in 1952, headed by General Neguib (1901–84) and masterminded by Nasser. Farouk was forced to abdicate in favour of his infant son, Fuad; he was sent into exile and eventually became a citizen of Monaco.

Farquhar /'fɑːkə(r)/, George (1678–1707), Irish dramatist. A principal figure in Restoration comedy, he is remembered for *The Recruiting Officer* (1706) and *The Beaux' Stratagem* (1707), plays marked by realism and genial merriment as well as by pungent satire.

farrago /fə'rɑːgəʊ/ n. (pl. **-os** or US **-oes**) a medley or hotchpotch, a confused mixture. □ **farraginous** /-'rædʒɪnəs/ adj. [L *farrago farraginis* mixed fodder f. *far* corn]

Farrell¹ /'færəl/, J(ames) G(ordon) (1935–79), English novelist. He is best known for his novels *The Siege of Krishnapur* (1973), dealing with events of the Indian Mutiny, and *The Singapore Grip* (1978), describing the fall of Singapore to the Japanese.

Farrell² /'færəl/, J(ames) T(homas) (1904–79), American novelist. He achieved fame with his trilogy about Studs Lonigan, a young Chicago Catholic of Irish descent: *Young Lonigan* (1932), *The Young Manhood of Studs Lonigan* (1934), and *Judgement Day* (1935).

farrier /'færɪə(r)/ n. Brit. **1** a smith who shoes horses. **2** a person who treats the diseases and injuries of horses. □ **farriery** n. [OF *ferrier* f. L *ferrarius* f. *ferrum* iron, horseshoe]

farrow /'færəʊ/ n. & v. ● n. **1** a litter of pigs. **2** the birth of a litter. ● v.tr. (also absol.) (of a sow) produce (pigs). [OE *fearh, færh* pig f. WG]

farruca /fə'ruːkə/ n. a type of flamenco dance. [Sp.]

Farsi /'fɑːsiː/ n. the modern Persian language, the official language of Iran. [Pers.: cf. PARSEE]

far-sighted /fɑː'saɪtɪd/ adj. **1** having foresight, prudent. **2** esp. N. Amer. = LONG-SIGHTED. □ **far-sightedly** adv. **far-sightedness** n.

fart /fɑːt/ v. & n. coarse sl. ● v.intr. **1** emit wind from the anus. **2** (foll. by *about, around*) behave foolishly; waste time. ● n. **1** an emission of wind from the anus. **2** an unpleasant person. [OE (recorded in *feorting* verbal noun) f. Gmc]

farther var. of FURTHER (esp. with ref. to physical distance).

farthest var. of FURTHEST (esp. with ref. to physical distance).

farthing /'fɑːðɪŋ/ n. **1** hist. (in the UK) a former coin and monetary unit equal to a quarter of a penny. ¶ Withdrawn in 1961. **2** the least possible amount (*it doesn't matter a farthing*). [OE *feorthing* f. *feortha* fourth]

farthingale /'fɑːðɪŋgeɪl/ n. hist. a hooped petticoat or a stiff curved roll to extend a woman's skirt. [earlier *vardingale, verd-* f. F *verdugale* f. Sp. *verdugado* f. *verdugo* rod]

fartlek /'fɑːtlek/ n. Athletics a method of training for middle- and long-distance running, mixing fast with slow work. [Sw. f. *fart* speed + *lek* play]

Far West 1 the regions of North America in the Rocky Mountains and along the Pacific coast. **2** hist. the area of North America west of the earliest European settlements (now called the *Middle West*).

fasces /'fæsiːz/ n.pl. **1** Rom. Hist. a bundle of rods with a projecting axe-blade, carried by a lictor as a symbol of a magistrate's power. **2** hist. (in Fascist Italy) emblems of authority. [L, pl. of *fascis* bundle]

fascia /'feɪʃə/ n. (also **facia**) (pl. **fasciae** /-ʃɪˌiː-/ or **fascias**) **1** Brit. **a** the instrument panel of a motor vehicle. **b** any similar panel or plate for operating machinery. **2** the upper part of a shopfront with the proprietor's name etc. **3** Archit. **a** a long flat surface between mouldings on the architrave in classical architecture. **b** a flat board, usu. of wood, covering the ends of rafters. **4** a stripe or band. **5** /'fæʃə/ Anat. a thin sheath of fibrous tissue, esp. that enclosing a muscle or other organ. □ **fascial** adj. [L, = band, door-frame, etc.]

fasciate /'fæʃɪˌeɪt/ adj. (also **fasciated**) **1** Bot. (of contiguous parts) compressed or growing into one. **2** striped or banded. □ **fasciation** /ˌfæʃɪ'eɪʃ(ə)n/ n. [L *fasciatus* past part. of *fasciare* swathe (as FASCIA)]

fascicle /'fæsɪk(ə)l/ n. **1** (also **fascicule** /-ˌkjuːl/) a separately published instalment of a book, usu. not complete in itself. **2** a bunch or bundle. **3** (also **fasciculus** /fæ'sɪkjʊləs/) Anat. a bundle of fibres. □ **fascicled** adj. **fascicular** /fə'sɪkjʊlə(r)/ adj. **fasciculate** /-'sɪkjʊlət/ adj. **fasciculation** /-ˌsɪkjʊ'leɪʃ(ə)n/ n. [L *fasciculus* bundle, dimin. of *fascis*: see FASCES]

fasciitis /ˌfæsɪ'aɪtɪs, ˌfæʃɪ-/ n. Med. inflammation of the fascia of a muscle etc.

fascinate /'fæsɪˌneɪt/ v.tr. **1** capture the interest of; attract irresistibly. **2** (esp. of a snake) paralyse (a victim) with fear. □ **fascinated** adj. **fascinating** adj. **fascinatingly** adv. **fascinator** n. **fascination** /ˌfæsɪ'neɪʃ(ə)n/ n. [L *fascinare* f. *fascinum* spell]

fascine /fæ'siːn/ n. a long faggot used for engineering purposes and (esp. in war) for lining trenches, filling ditches, etc. [F f. L *fascina* f. *fascis* bundle: see FASCES]

Fascism /'fæʃɪz(ə)m/ n. **1** the principles and organization of the extreme right-wing nationalist movement, prevailing in Italy under Mussolini (1922–43). (See note below.) **2** (also **fascism**) **a** any similar nationalist and authoritarian movement. **b** any system of extreme right-wing or authoritarian views. □ **Fascist** n. & adj. (also **fascist**). **Fascistic** /fə'ʃɪstɪk/ adj. (also **fascistic**). [It. *fascismo* f. *fascio* political group f. L *fascis* bundle: see FASCES]

▪ Fascism reached its peak of influence in Europe between 1930 and 1945, notably in Italy, Spain, and perhaps most importantly in Germany, where it underpinned the ideology of the Nazi Party. Although there is no coherent body of political doctrine associated with Fascism, it tends to include a belief in the supremacy of one national or ethnic group over others, a contempt for democracy, an insistence on obedience to a powerful and absolute leader, and a strong demagogic approach. After 1945 the influence of Fascism declined, although it survived in Spain (till 1975) and Portugal (till 1974) and in the style of some governments of South America; in the 1990s, however, Fascism underwent a resurgence, with neo-Fascist groups achieving electoral successes in France, Russia, and Italy.

fashion /'fæʃ(ə)n/ n. & v. ● n. **1** the current popular custom or style, esp. in dress or social conduct. **2** a manner or style of doing something (*in a peculiar fashion*). **3** (in comb.) in a specified manner (*walk crab-fashion*).

4 fashionable society (*a woman of fashion*). ● *v.tr.* (often foll. by *into*) make into a particular or the required form. □ **after** (or **in**) **a fashion** as well as is practicable, though not satisfactorily. **fashion victim** a slavish follower of trends in fashion. **in** (or **out of**) **fashion** fashionable (or not fashionable) at the time in question. □ **fashioner** *n.* [ME f. AF *fasun*, OF *façon*, f. L *factio -onis* f. *facere fact-* do, make]

fashionable /ˈfæʃnəb(ə)l/ *adj.* **1** following, suited to, or influenced by the current fashion. **2** characteristic of or favoured by those who are leaders of social fashion. □ **fashionableness** *n.* **fashionably** *adv.* **fashionability** /ˌfæʃnəˈbɪlɪtɪ/ *n.*

Fassbinder /ˈfæsˌbɪndə(r)/, Rainer Werner (1946–82), German film director. Fassbinder is remembered for films such as *The Bitter Tears of Petra von Kant* (1972) and the allegorical *The Marriage of Maria Braun* (1979). Influenced by Brecht, Marx, and Sigmund Freud, Fassbinder's films dealt largely with Germany during the Second World War and postwar West German society.

fast[1] /fɑːst/ *adj. & adv.* ● *adj.* **1** rapid, quick-moving. **2** capable of high speed (*a fast car*). **3** enabling or causing or intended for high speed (*a fast road; fast lane*). **4** (of a clock etc.) showing a time ahead of the correct time. **5** (of a pitch or ground etc. in a sport) likely to make the ball bounce or run quickly. **6 a** (of a photographic film) needing only a short exposure. **b** (of a lens) having a large aperture. **7 a** firmly fixed or attached. **b** secure; firmly established (*a fast friendship*). **8** (of a colour) not fading in light or when washed. **9** (of a person) immoral, dissipated. ● *adv.* **1** quickly; in quick succession. **2** firmly, fixedly, tightly, securely (*stand fast; eyes fast shut*). **3** soundly, completely (*fast asleep*). **4** close, immediately (*fast on their heels*). **5** in a dissipated manner; extravagantly, immorally. □ **fast breeder** (or **fast breeder reactor**) a reactor using fast neutrons to produce the same fissile material as it uses. **fast buck** see BUCK[2]. **fast food** food that can be prepared and served quickly and easily, esp. in a snack bar or restaurant. **fast neutron** a neutron with high kinetic energy, esp. not slowed by a moderator etc. **fast reactor** a nuclear reactor using mainly fast neutrons. **fast-talk** *N. Amer. colloq.* persuade by rapid or deceitful talk. **fast-wind** wind (magnetic tape) rapidly backwards or forwards. **fast worker** *colloq.* a person who achieves quick results, esp. in love affairs. **pull a fast one** (often foll. by *on*) *colloq.* try to deceive or gain an unfair advantage. [OE *fæst* f. Gmc]

fast[2] /fɑːst/ *v. & n.* ● *v.intr.* abstain from all or some kinds of food or drink, esp. as a religious observance. ● *n.* an act or period of fasting. [ON *fasta* f. Gmc]

fastback /ˈfɑːstbæk/ *n.* **1** a motor car with the rear sloping continuously down to the bumper. **2** such a rear.

fasten /ˈfɑːs(ə)n/ *v.* **1** *tr.* make or become fixed or secure. **2** *tr.* (foll. by *in, up*) lock securely; shut in. **3** *tr.* & (foll. by *on, upon*) direct (a look, thoughts, etc.) fixedly or intently. **b** focus or direct the attention fixedly upon (*fastened him with her eyes*). **4** *tr.* (foll. by *on, upon*) fix (a designation or imputation etc.). **5** *intr.* (foll. by *on, upon*) **a** take hold of. **b** single out. **6** (foll. by *off*) fix with stitches or a knot. □ **fastener** *n.* [OE *fæstnian* f. Gmc]

fastening /ˈfɑːs(ə)nɪŋ/ *n.* a device that fastens something; a fastener.

fastidious /fæˈstɪdɪəs/ *adj.* **1** very careful in matters of choice or taste; fussy. **2** easily disgusted; squeamish. □ **fastidiously** *adv.* **fastidiousness** *n.* [ME f. L *fastidiosus* f. *fastidium* loathing]

fastigiate /fæˈstɪdʒɪət/ *adj. Bot.* **1** having a conical or tapering outline. **2** having parallel upright branches. [L *fastigium* gable-top]

fastness /ˈfɑːstnɪs/ *n.* **1** a stronghold or fortress. **2** the state of being secure. [OE *fæstnes* (as FAST[1])]

Fastnet /ˈfɑːstnɛt/ **1** a rocky islet off the SW coast of Ireland. **2** a shipping forecast area covering the Celtic Sea off the south coast of Ireland as far as the latitude of the Scilly Isles.

fat /fæt/ *n., adj., & v.* ● *n.* **1** a natural oily or greasy substance occurring esp. in animal bodies. **2** the part of anything containing this. **3** excessive presence of fat in a person or animal; corpulence. **4** *Chem.* any of a group of natural esters of glycerol and various fatty acids existing as solids at room temperature. ● *adj.* (**fatter, fattest**) **1** (of a person or animal) having excessive fat; corpulent. **2** (of an animal) made plump for slaughter; fatted. **3** containing much fat. **4** greasy, oily, unctuous. **5** (of land or resources) fertile, rich; yielding abundantly. **6 a** thick, substantial in content (*a fat book*). **b** substantial as an asset or opportunity (*a fat cheque; was given a fat part in the play*). **7 a** (of coal) bituminous. **b** (of clay etc.) sticky. **8** *colloq. iron.* very little; not much (*a fat chance; a fat lot*). ● *v.tr. & intr.* (**fatted, fatting**) make or become fat. □ **fat cat** *sl.* a wealthy person, esp. a wealthy businessman, politician, or civil servant. **fat-head** *colloq.* a stupid person. **fat-**

headed *colloq.* stupid. **fat-headedness** *colloq.* stupidity. **fat hen** the white goosefoot, *Chenopodium album*. **the fat is in the fire** trouble is imminent. **kill the fatted calf** celebrate, esp. at a prodigal's return (Luke 15). **live off** (or **on**) **the fat of the land** have the best of everything. □ **fatless** *adj.* **fatly** *adv.* **fatness** *n.* **fattish** *adj.* [OE *fæt* (adj.), *fættian* (v.) f. Gmc]

Fatah, Al /ˈfætə, æl/ a Palestinian political and military organization founded in 1958 by Yasser Arafat and others to bring about the establishment of a Palestinian state. It came to dominate the Palestine Liberation Organization in the 1960s, a position which under Arafat it has continued to hold, despite challenges from more extreme groups. [Arab., = 'Palestinian national liberation movement' or 'victory']

fatal /ˈfeɪt(ə)l/ *adj.* **1** causing or ending in death (*a fatal accident*). **2** (often foll. by *to*) destructive; ruinous; ending in disaster (*was fatal to their chances; made a fatal mistake*). **3** fateful, decisive. □ **fatally** *adv.* [ME f. OF *fatal* or L *fatalis* (as FATE)]

fatalism /ˈfeɪtəˌlɪz(ə)m/ *n.* **1** the belief that all events are predetermined and therefore inevitable. **2** a submissive attitude to events as being inevitable. □ **fatalist** *n.* **fatalistic** /ˌfeɪtəˈlɪstɪk/ *adj.* **fatalistically** *adv.*

fatality /fəˈtælɪtɪ/ *n.* (pl. **-ies**) **1 a** an occurrence of death by accident or in war etc. **b** a person killed in this way. **2** a fatal influence. **3** a predestined liability to disaster. **4** subjection to or the supremacy of fate. **5** a disastrous event; a calamity. [F *fatalité* or LL *fatalitas* f. L *fatalis* FATAL]

Fata Morgana /ˌfɑːtə mɔːˈɡɑːnə/ a kind of mirage most frequently seen in the Strait of Messina between Italy and Sicily. It was attributed to fairy agency, particularly to Morgana le Fay, sister of King Arthur, whose legend and reputation as an enchantress were carried to Sicily by Norman settlers. [It., = fairy Morgan]

fate /feɪt/ *n. & v.* ● *n.* **1** a power regarded as predetermining events unalterably. **2 a** the future regarded as determined by such a power. **b** an individual's appointed lot. **c** the ultimate condition or end of a person or thing (*that sealed our fate*). **3** death, destruction. **4** (usu. **Fate**) a goddess of destiny, esp. one of the Fates (see FATES, THE) or the Norns. ● *v.tr.* **1** (usu. in *passive*) preordain (*was fated to win*). **2** (as **fated** *adj.*) **a** doomed to destruction. **b** unavoidable, preordained; fateful. □ **fate worse than death** see DEATH. [ME f. It. *fato* & L *fatum* that which is spoken, f. *fari* speak]

fateful /ˈfeɪtfʊl/ *adj.* **1** important, decisive; having far-reaching consequences. **2** controlled as if by fate. **3** causing or likely to cause disaster. **4** prophetic. □ **fatefully** *adv.*

Fates, the /feɪts/ *Gk & Rom. Mythol.* the three goddesses (also called the *Moirai* and the *Parcae*) who presided over the birth and life of humans. Each person was thought of as a spindle, around which the three Fates (Clotho, Lachesis, and Atropos) would spin the thread of human destiny.

father /ˈfɑːðə(r)/ *n. & v.* ● *n.* **1 a** a man in relation to a child or children born from his fertilization of an ovum. **b** (in full **adoptive father**) a man who has continuous care of a child, esp. by adoption. **2** any male animal in relation to its offspring. **3** (usu. in *pl.*) a progenitor or forefather. **4** an originator, designer, or early leader. **5** a person who deserves special respect (*the father of his country*). **6** (**Fathers** or **Fathers of the Church**) early Christian theologians whose writings are regarded as especially authoritative. **7** (also **Father**) **a** (often as a title or form of address) a priest, esp. of a religious order. **b** a religious leader. **8** (**the Father**) (in Christian belief) the first person of the Trinity. **9** (**Father**) a venerable person, esp. as a title in personifications (*Father Time*). **10** the oldest member or doyen (see also FATHER OF THE HOUSE). **11** (usu. in *pl.*) the leading men or elders in a city or state (*city fathers*). ● *v.tr.* **1** beget; be the father of. **2** behave as a father towards. **3** originate (a scheme etc.). **4** appear as or admit that one is the father or originator of. **5** (foll. by *on*) assign the paternity of (a child, book) to a person. □ **father-figure** an older man who is respected like a father; a trusted leader. **father-in-law** (pl. **fathers-in-law**) the father of one's husband or wife. **father of chapel** see CHAPEL. □ **fatherhood** *n.* **fatherless** *adj.* **fatherlessness** *n.* **fatherlike** *adj. & adv.* [OE *fæder* with many Gmc cognates: rel. to L *pater*, Gk *patēr*]

Father Christmas an imaginary person said to bring presents for children on Christmas Eve. Traditionally he arrives from the far north, but his origin is obscure and his conventionalized appearance as a jolly old man with a long white beard and red cloak trimmed with white fur is comparatively recent. In late medieval Europe he became identified with St Nicholas (Santa Claus). In England Father Christmas was a personification of Christmas, a genial red-robed old man who appeared in many 16th-century masques and in mummers' plays. In the 19th century there was a great revival of the celebration of Christmas, and Father Christmas acquired (from St Nicholas) the association of present-bringing.

fatherland /ˈfɑːðəˌlænd/ n. one's native country, now esp. Germany.

fatherly /ˈfɑːðəlɪ/ adj. **1** like or characteristic of a father in affection, care, etc. (*fatherly concern*). **2** of or proper to a father. □ **fatherliness** n.

Father of the House n. the member of the House of Commons with the longest continuous service.

Father's Day a day (usu. the third Sunday in June) established for a special tribute to fathers.

Father Time see TIME.

fathom /ˈfæðəm/ n. & v. ● n. (*pl.* often **fathom** when prec. by a number) **1** a measure of six feet, esp. used in taking depth soundings. **2** *Brit.* a quantity of wood six feet square in cross-section. ● *v.tr.* **1** grasp or comprehend (a problem or difficulty). **2** measure the depth of (water) with a sounding-line. □ **fathomable** adj. **fathomless** adj. [OE *fæthm* outstretched arms f. Gmc]

Fathometer /fəˈðɒmɪtə(r)/ n. a type of echo-sounder.

fatigue /fəˈtiːɡ/ n. & v. ● n. **1** extreme tiredness after exertion. **2** weakness in materials, esp. metal, caused by repeated variations of stress. **3** a reduction in the efficiency of a muscle, organ, etc., after prolonged activity. **4** an activity that causes fatigue. **5 a** a non-military duty in the army, often as a punishment. **b** (in full **fatigue-party**) a group of soldiers ordered to do such a duty. **c** (in *pl.*) clothing worn for such a duty. ● *v.tr.* (**fatigues, fatigued, fatiguing**) **1** cause fatigue in; tire, exhaust. **2** (as **fatigued** adj.) weary; listless. □ **fatiguable** adj. (also **fatigable**). **fatigueless** adj. **fatiguability** /-ˌtiːɡəˈbɪlɪtɪ/ n. (also **fatigability**). [F *fatigue, fatiguer* f. L *fatigare* tire out]

Fatiha /ˈfɑːtɪˌhɑː/ n. (also **Fatihah**) the short first sura of the Koran, used by Muslims as a prayer. [Arab. *fātiḥa* opening f. *fataḥa* to open]

Fatima /ˈfætɪmə/ (AD c.606–32), youngest daughter of the prophet Muhammad and wife of the fourth caliph, Ali (d.661). The descendants of Muhammad trace their lineage through her; she is revered especially by Shiite Muslims as the mother of the imams Hasan (624–80) and Husayn (626–80).

Fátima /ˈfætɪmə/ a village in west central Portugal, north-east of Lisbon; pop. (1991) 5,445. It became a centre of Roman Catholic pilgrimage after the reported sighting in the village in 1917 of the Virgin Mary.

Fatimid /ˈfætɪmɪd/ n. & adj. (also **Fatimite** /-ˌmaɪt/) ● n. a descendant of Fatima, the daughter of Muhammad; in particular, a member of an Arabian dynasty claiming descent from her which ruled in parts of northern Africa, Egypt, and Syria from 909 to 1171. The capital city of the Fatimids, Cairo, was founded in 969. ● adj. of or relating to the Fatimids.

fatling /ˈfætlɪŋ/ n. a young fatted animal.

fatso /ˈfætsəʊ/ n. (*pl.* **-oes**) *sl. joc.* or *offens.* a fat person. [prob. f. FAT or the designation *Fats*]

fatstock /ˈfætstɒk/ n. livestock fattened for slaughter.

fatten /ˈfæt(ə)n/ v. **1** *tr.* & *intr.* (esp. with ref. to meat-producing animals) make or become fat. **2** *tr.* enrich (soil).

fatty /ˈfætɪ/ adj. & n. ● adj. (**fattier, fattiest**) **1** like fat; oily, greasy. **2** consisting of or containing fat; adipose. **3** marked by abnormal deposition of fat, esp. in fatty degeneration. ● n. (*pl.* **-ies**) *colloq.* a fat person (esp. as a nickname). **fatty acid** *Chem.* any of a class of organic compounds consisting of a hydrocarbon chain and a terminal carboxyl group, esp. those occurring as constituents of lipids. **fatty oil** = *fixed oil.* □ **fattiness** n.

fatuous /ˈfætjʊəs/ adj. vacantly silly; purposeless, idiotic. □ **fatuously** adv. **fatuousness** n. **fatuity** /fəˈtjuːɪtɪ/ n. (*pl.* **-ies**). [L *fatuus* foolish]

fatwa /ˈfætwɑː/ n. an authoritative (usu. written) ruling on a point of Islamic law given by a mufti or Islamic leader. In 1989 a fatwa calling for the death of the novelist Salman Rushdie was issued by Ayatollah Khomeini following the publication of Rushdie's allegedly blasphemous novel *The Satanic Verses*. [Arab. *fatwā*]

faubourg /ˈfəʊbʊə(r)/ n. a suburb, esp. of Paris. [F: cf. med.L *falsus burgus* not the city proper]

fauces /ˈfɔːsiːz/ n.pl. *Anat.* a cavity at the back of the mouth. □ **faucial** /ˈfɔːʃ(ə)l/ adj. [L, = throat]

faucet /ˈfɔːsɪt/ n. *US* a tap. ¶ In British use only in special applications. [ME f. OF *fausset* vent-peg f. Prov. *falset* f. *falsar* to bore]

Faulkner /ˈfɔːknə(r)/, William (1897–1962), American novelist. His works deal with the history and legends of the American South and have a strong sense of a society in decline; in the first of his major novels, *The Sound and the Fury* (1929), he was also influenced by modernist concerns of form. Other important works include *As I Lay Dying* (1930) and *Absalom! Absalom!* (1936). He was awarded the Nobel Prize for literature in 1949.

fault /fɒlt, fɔːlt/ n. & v. ● n. **1** a defect or imperfection of character or of structure, appearance, etc. **2** a break or other defect in an electric circuit. **3** a transgression, offence, or thing wrongly done. **4 a** *Tennis* etc. a service of the ball not in accordance with the rules. **b** (in showjumping) a penalty for an error. **5** responsibility for wrongdoing, error, etc. (*it will be your own fault*). **6** a defect regarded as the cause of something wrong (*the fault lies in the teaching methods*). **7** *Geol.* an extended break in the continuity of strata or a vein. ● v. **1** *tr.* find fault with; blame. **2** *tr.* declare to be faulty. **3** *tr. Geol.* break the continuity of (strata or a vein). **4** *intr.* commit a fault. **5** *intr. Geol.* show a fault. □ **at fault** guilty; to blame. **fault-finder** a person given to continually finding fault. **fault-finding** continual criticism. **find fault** (often foll. by *with*) make an adverse criticism; complain. **to a fault** (usu. of a commendable quality etc.) excessively (*generous to a fault*). [ME *faut(e)* f. OF ult. f. L *fallere* FAIL]

faultless /ˈfɒltlɪs, ˈfɔːlt-/ adj. without fault; free from defect or error. □ **faultlessly** adv. **faultlessness** n.

faulty /ˈfɒltɪ, ˈfɔːltɪ/ (**faultier, faultiest**) adj. having faults; imperfect, defective. □ **faultily** adv. **faultiness** n.

faun /fɔːn/ n. *Rom. Mythol.* a Latin rural deity with a human face and torso and a goat's horns, legs, and tail, identified with the Greek satyrs. [ME f. OF *faune* or L FAUNUS]

fauna /ˈfɔːnə/ n. (*pl.* **faunae** /-niː/ or **faunas**) **1** the animal life of a region or geological period (cf. FLORA). **2** a treatise on or list of this. □ **faunal** adj. **faunist** n. **faunistic** /fɔːˈnɪstɪk/ adj. [mod.L f. the name of a rural goddess, sister of Faunus: see FAUN]

Fauntleroy /ˈfɔːntləˌrɔɪ/ (also (**Little**) **Lord Fauntleroy**) the gentle-mannered boy hero of Frances Hodgson Burnett's novel *Little Lord Fauntleroy* (1886). The name was also applied to the style of dress (velvet suits with lace collars and cuffs) which the book popularized.

Faunus /ˈfɔːnəs/ *Rom. Mythol.* an ancient Italian pastoral god, grandson of Saturn. His association with wooded places caused him to be identified with the Greek Pan.

Fauré /ˈfɔːreɪ/, Gabriel (Urbain) (1845–1924), French composer and organist. He composed songs throughout his career, incorporating some in cycles such as *La Bonne chanson* (1891–2). His best-known work is the *Requiem* (1887) for solo voices, choir, and orchestra; he also wrote piano pieces, chamber music, and incidental music for the theatre.

Faust /faʊst/ (also **Faustus** /-təs/) (died c.1540), German astronomer and necromancer. Reputed to have sold his soul to the Devil, he became the subject of many legends and was the subject of a drama by Goethe, an opera by Gounod, and a novel by Thomas Mann. □ **Faustian** adj.

faute de mieux /ˌfəʊt də ˈmjɜː/ adv. for want of a better alternative. [F]

fauteuil /fəʊˈtɜːɪ/ n. a kind of wooden seat in the form of an armchair with open sides and upholstered arms. [F f. OF *faudestuel, faldestoel* FALDSTOOL]

fauve /fəʊv/ n. an artist of the fauvist movement, a practitioner of fauvism. [F, = wild beast]

fauvism /ˈfəʊvɪz(ə)m/ n. a movement in painting chiefly associated with Matisse, characterized mainly by a vivid expressionistic and non-naturalistic use of colour. The name originated from a remark of the French art critic Louis Vauxcelles at the Salon of 1905; coming across a quattrocento-style statue in the midst of works by Matisse and his associates, he is reputed to have said, 'Donatello au milieu des fauves!' ('Donatello among the wild beasts'). Fauvism flourished in Paris from 1905 and, although short-lived, had an important influence on subsequent artists, especially the German expressionists. Its adherents included Dufy, Derain, and Vlaminck. □ **fauvist** n. & adj.

faux pas /fəʊ ˈpɑː/ n. (*pl.* same /ˈpɑːz/) **1** a tactless mistake; a blunder. **2** a social indiscretion. [F, = false step]

fave /feɪv/ n. & adj. *sl.* = FAVOURITE (esp. in show business). [abbr.]

favela /fəˈvelə/ n. a Brazilian shack, slum, or shanty town. [Port.]

favour /ˈfeɪvə(r)/ n. & v. (*US* **favor**) ● n. **1** an act of kindness beyond what is due or usual (*did it as a favour*). **2** esteem, liking, approval, goodwill; friendly regard (*gained their favour; look with favour on*). **3** partiality; too lenient or generous treatment. **4** aid, support (*under favour of night*). **5** a thing given or worn as a mark of favour or support, e.g. a badge or a knot of ribbons. **6** *archaic* leave, pardon (*by your favour*). **7** *Commerce archaic* a letter (*your favour of yesterday*). ● *v.tr.* **1** regard or treat with favour or partiality. **2** give support or approval to; promote, prefer. **3 a** be to the advantage of (a person). **b** facilitate (a process etc.). **4** tend to confirm (an idea or theory). **5** (foll. by *with*) oblige (*favour me with a reply*). **6** (as **favoured** *adj.*) a having special advantages. **b** preferred; favourite. **7** *colloq.* resemble in features. □ **in favour 1** meeting with approval. **2** (foll. by *of*) **a** in support of. **b** to the advantage of. **out of favour** lacking approval. □ **favourer** *n.* [ME f. OF f. L *favor -oris* f. *favere* show kindness to]

favourable /ˈfeɪvərəb(ə)l/ *adj.* (*US* **favorable**) **1 a** well disposed; propitious. **b** commendatory, approving. **2** giving consent (*a favourable answer*). **3** promising, auspicious, satisfactory (*a favourable aspect*). **4** (often foll. by *to*) helpful, suitable. □ **favourableness** *n.* **favourably** *adv.* [ME f. OF *favorable* f. L *favorabilis* (as FAVOUR)]

favourite /ˈfeɪvərɪt/ *adj. & n.* (*US* **favorite**) ● *adj.* preferred to all others (*my favourite book*). ● *n.* **1** a specially favoured person. **2** *Sport* a competitor thought most likely to win. □ **favourite son** *US* a person supported as a presidential candidate by delegates from the candidate's home state. [obs. F *favorit* f. It. *favorito* past part. of *favorire* favour]

favouritism /ˈfeɪvərɪˌtɪz(ə)m/ *n.* (*US* **favoritism**) the unfair favouring of one person or group at the expense of another.

Fawkes /fɔːks/, Guy (1570–1606), English conspirator. He was hanged for his part in the Gunpowder Plot of 5 Nov. 1605. The occasion is commemorated annually with fireworks, bonfires, and the burning of a guy (see BONFIRE NIGHT).

fawn[1] /fɔːn/ *n., adj., & v.* ● *n.* **1** a young deer in its first year. **2** a light yellowish-brown. ● *adj.* fawn-coloured. ● *v.tr.* (also *absol.*) (of a deer) bring forth (young). □ **in fawn** (of a deer) pregnant. [ME f. OF *faon* etc. ult. f. L *fetus* offspring: cf. FOETUS]

fawn[2] /fɔːn/ *v.intr.* **1** (often foll. by *on*, *upon*) (of a person) behave servilely, show cringing or abject affection. **2** (of an animal, esp. a dog) show extreme affection. □ **fawning** *adj.* **fawningly** *adv.* [OE *fagnian*, *fægnian* (var. of FAIN)]

fax /fæks/ *n. & v.* ● *n.* **1** facsimile transmission (see FACSIMILE *n.* 2). (See note below.) **2 a** a copy produced or message sent by this. **b** a machine for transmitting and receiving these. ● *v.tr.* transmit (a document) in this way. [abbr. of FACSIMILE]

▪ Most fax systems involve the scanning of the original image by a device which converts each unit area of it into a specified amount of electric current and transmits this as a signal, over a telecommunications link, to a receiver which produces an image that is a copy of the original. The basic principles were described by the Scottish inventor Alexander Bain (1810–77) in 1843 in his patent for producing copies, at distant places, of surfaces (e.g. printer's types) composed of conducting and non-conducting materials; his device used an array of electrical conductors in contact with the printing surface.

fay /feɪ/ *n. literary* a fairy. [ME f. OF *fae*, *faie* f. L *fata* (pl.); see FATES, THE]

faze /feɪz/ *v.tr.* (often as **fazed** *adj.*) *colloq.* disconcert, perturb, disorientate. [var. of *feeze* drive off, f. OE *fēsian*, of unkn. orig.]

FBA *abbr.* Fellow of the British Academy.

FBI see FEDERAL BUREAU OF INVESTIGATION.

FC *abbr.* Football Club.

FCC *abbr.* (in the US) Federal Communications Commission.

FCO *abbr.* (in the UK) Foreign and Commonwealth Office.

fcp. *abbr.* foolscap.

FD *abbr.* = DEFENDER OF THE FAITH. [L *Fidei Defensor*]

FDA *abbr.* **1** (in the US) Food and Drug Administration. **2** (in the UK) First Division (Civil Servants) Association (cf. AFDCS).

FDR the nickname of President Franklin Delano Roosevelt (see ROOSEVELT[2]).

Fe *symb. Chem.* the element iron.

fealty /ˈfiːəltɪ/ *n.* (*pl.* **-ies**) **1** *hist.* **a** a feudal tenant's or vassal's fidelity to a lord. **b** an acknowledgement of this. **2** allegiance. [ME f. OF *feaulté* f. L *fidelitas -tatis* f. *fidelis* faithful f. *fides* faith]

fear /fɪə(r)/ *n. & v.* ● *n.* **1 a** an unpleasant emotion caused by exposure to danger, expectation of pain, etc. **b** a state of alarm (*be in fear*). **2** a cause of fear (*all fears removed*). **3** (often foll. by *of*) dread or fearful respect (towards) (*had a fear of heights*). **4** anxiety for the safety of (*in fear of their lives*). **5** danger; likelihood (of something unwelcome) (*there is little fear of failure*). ● *v.* **1 a** *tr.* feel fear about or towards (a person or

thing). **b** *intr.* feel fear. **2** *intr.* (foll. by *for*) feel anxiety or apprehension about (*feared for my life*). **3** *tr.* apprehend; have uneasy expectation of (*fear the worst*). **4** *tr.* (usu. foll. by *that* + clause) apprehend with fear or regret (*I fear that you are wrong*). **5** *tr.* **a** (foll. by *to* + infin.) hesitate. **b** (foll. by verbal noun) shrink from; be apprehensive about (*he feared meeting his ex-wife*). **6** *tr.* show reverence towards. □ **for fear of** (or **that**) to avoid the risk of (or that). **never fear** there is no danger of that. **no fear** *colloq.* expressing strong denial or refusal. **without fear or favour** impartially. [OE f. Gmc]

fearful /ˈfɪəfʊl/ *adj.* **1** (usu. foll. by *of*, or *that* + clause) afraid. **2** terrible, awful. **3** *colloq.* extremely unwelcome or unpleasant (*a fearful row*). □ **fearfully** *adv.* **fearfulness** *n.*

fearless /ˈfɪəlɪs/ *adj.* **1** courageous, brave. **2** (foll. by *of*) without fear. □ **fearlessly** *adv.* **fearlessness** *n.*

fearsome /ˈfɪəsəm/ *adj.* appalling or frightening, esp. in appearance. □ **fearsomely** *adv.* **fearsomeness** *n.*

feasibility /ˌfiːzɪˈbɪlɪtɪ/ *n.* the state or degree of being feasible. □ **feasibility study** a study of the practicability of a proposed project.

feasible /ˈfiːzɪb(ə)l/ *adj.* **1** practicable, possible; easily or conveniently done. **2** *disp.* likely, probable (*it is feasible that it will rain*). □ **feasibly** *adv.* [ME f. OF *faisable, -ible* f. *fais-* stem of *faire* f. L *facere* do, make]

feast /fiːst/ *n. & v.* ● *n.* **1** a large or sumptuous meal. **2** a gratification to the senses or mind. **3 a** an annual religious celebration. **b** a day dedicated to a particular saint. **4** an annual village festival. ● *v.* **1** *intr.* partake of a feast; eat and drink sumptuously. **2** *tr.* **a** regale. **b** pass (time) in feasting. □ **feast-day** a day on which a feast (esp. in sense 3) is held. **feast one's eyes on** take pleasure in beholding. **feast of reason** intellectual talk. □ **feaster** *n.* [ME f. OF *feste*, *fester* f. L *festus* joyous]

feat /fiːt/ *n.* a noteworthy act or achievement. [ME f. OF *fait*, *fet* (as FACT)]

feather /ˈfeðə(r)/ *n. & v.* ● *n.* **1** any of the appendages growing from a bird's skin, with a horny hollow stem and fine strands. **2** one or more of these as decoration etc. **3** (*collect.*) **a** plumage. **b** game birds. ● *v.* **1** *tr.* cover or line with feathers. **2** *tr. Rowing* turn (an oar) so that it passes through the air edgeways. **3** *tr. Aeron. & Naut.* **a** cause (the propeller blades) to rotate in such a way as to lessen the air or water resistance. **b** vary the angle of incidence of (helicopter blades). **4** *intr.* float, move, or wave like feathers. □ **feather bed** a bed with a mattress stuffed with feathers. **feather-bed** *v.tr.* (**-bedded**, **-bedding**) provide with (esp. financial) advantages. **feather-bedding** making or being made comfortable by favourable economic treatment, esp. the employment of excess staff. **feather-brain** (or **-head**) a silly or absent-minded person. **feather-brained** (or **-headed**) silly, absent-minded. **feather-edge** the fine edge of a wedge-shaped board. **a feather in one's cap** an achievement to one's credit. **feather one's nest** enrich oneself. **feather-stitch** ornamental zigzag sewing. **in fine** (or **high**) **feather** in good spirits. □ **feathered** *adj.* (also in *comb.*). **featherless** *adj.* **feathery** *adj.* **featheriness** *n.* [OE *fether*, *gefithrian*, f. Gmc]

feathering /ˈfeðərɪŋ/ *n.* **1** bird's plumage. **2** the feathers of an arrow. **3** a feather-like structure in an animal's coat. **4** *Archit.* cusps in tracery.

featherweight /ˈfeðəˌweɪt/ *n.* **1 a** a weight in certain sports intermediate between bantamweight and lightweight, in the amateur boxing scale 54–7 kg but differing for professionals, wrestlers, and weightlifters. **b** a sportsman of this weight. **2** a very light person or thing. **3** (usu. *attrib.*) a trifling or unimportant thing.

feature /ˈfiːtʃə(r)/ *n. & v.* ● *n.* **1** a distinctive or characteristic part of a thing. **2** (usu. in *pl.*) a distinctive part of) the face, esp. with regard to shape and visual effect. **3 a** a distinctive or regular article in a newspaper or magazine. **b** a special attraction at an event etc. **4 a** (in full **feature film**) a full-length film intended as the main item in a cinema programme. **b** (in full **feature programme**) a broadcast devoted to a particular topic. ● *v.* **1** *tr.* make a special display or attraction of; give special prominence to. **2** *tr. & intr.* have as or be an important actor, participant, or topic in a film, broadcast, etc. **3** *intr.* be a feature. □ **featured** *adj.* (also in *comb.*). **featureless** *adj.* [ME f. OF *feture*, *faiture* form f. L *factura* formation: see FACTURE]

Feb. *abbr.* February.

febrifuge /ˈfebrɪˌfjuːdʒ/ *n.* a medicine or treatment that reduces fever; a cooling drink. □ **febrifugal** /fɪˈbrɪfjʊɡ(ə)l, ˌfebrɪˈfjuːɡ(ə)l/ *adj.* [F *fébrifuge* f. L *febris* fever + -FUGE]

febrile /ˈfiːbraɪl/ *adj.* of or relating to fever; feverish. □ **febrility** /fɪˈbrɪlɪtɪ/ *n.* [F *fébrile* or med.L *febrilis* f. L *febris* fever]

February /ˈfebrʊərɪ/ n. (pl. **-ies**) the second month of the year. [ME f. OF *feverier* ult. f. L *februarius* f. *februa* a purification feast held in this month]

February Revolution see RUSSIAN REVOLUTION.

feces US var. of FAECES.

Fechner /ˈfexnə(r)/, Gustav Theodor (1801–87), German physicist and psychologist. His early work was in electricity, but after a long illness he became interested in psychology. He sought to define the quantitative relationship between degrees of physical stimulation and the resulting sensations, the study of which he termed *psychophysics*. By associating sensations with numerical values, Fechner hoped to make psychology a truly objective science.

feckless /ˈfeklɪs/ adj. **1** feeble, ineffective. **2** unthinking, irresponsible (*feckless gaiety*). □ **fecklessly** adv. **fecklessness** n. [Sc. *feck* f. *effeck* var. of EFFECT]

feculent /ˈfekjʊlənt/ adj. **1** murky; filthy. **2** containing sediments or dregs. □ **feculence** n. [F *féculent* or L *faeculentus* (as FAECES)]

fecund /ˈfiːkənd, ˈfek-/ adj. **1** prolific, fertile. **2** fertilizing. □ **fecundability** /fɪˌkʌndəˈbɪlɪtɪ/ n. **fecundity** /fɪˈkʌndɪtɪ/ n. [ME f. F *fécond* or L *fecundus*]

fecundate /ˈfiːkənˌdeɪt, ˈfek-/ v.tr. **1** make fruitful. **2** = FERTILIZE 2. □ **fecundation** /ˌfiːkənˈdeɪʃ(ə)n, ˌfek-/ n. [L *fecundare* f. *fecundus* fruitful]

Fed /fed/ n. US sl. a Federal official, esp. a member of the FBI. [abbr. of FEDERAL]

fed /fed/ past and past part. of FEED. □ **fed up** (or **fed to death**) (often foll. by *with*) discontented or bored, esp. from a surfeit of something (*am fed up with the rain*).

fedayeen /ˌfedəˈjiːn/ n.pl. Arab guerrillas operating esp. against Israel. [colloq. Arab. *fidāʾiyīn* pl. f. Arab. *fidāʾī* adventurer]

federal /ˈfedərəl/ adj. **1** of a system of government in which several states form a unity but remain independent in internal affairs. **2** relating to or affecting such a federation. **3** (also **Federal**) of or relating to central government as distinguished from the separate units constituting it. **4** favouring centralized government. **5** comprising an association of largely independent units. **6** (**Federal**) US of the northern states in the Civil War. □ **federally** adv. **federalize** v.tr. (also **-ise**). **federalization** /ˌfedərəlaɪˈzeɪʃ(ə)n/ n. [L *foedus -eris* league, covenant]

Federal Bureau of Investigation (abbr. **FBI**) an agency of the US Federal government that deals principally with internal security and counter-intelligence and that also conducts investigations in Federal law enforcement. It was established in 1908 as a branch of the Department of Justice, but was substantially reorganized under the controversial directorship (1924–72) of J. Edgar Hoover.

federalism /ˈfedərəˌlɪz(ə)m/ n. a system of government which unites separate states while allowing each to have a substantial degree of autonomy. A federal style of government usually has a written constitution (e.g. the US) and tends to stress the importance of decentralized power and, in its democratic form, direct lines of communication between the government and its citizens. □ **federalist** n.

Federalist Party an early political party in the US, joined by George Washington during his presidency (1789–97) and in power until 1801. The party's emphasis on strong central government was extremely important in the early years after independence, but by the 1820s it had been superseded by the Democratic Republican Party.

Federal Republic of Germany hist. West Germany (see GERMANY).

Federal Reserve System the banking authority, created in 1913, that performs the functions of a central bank in the US and is used to implement the country's monetary policy. The system consists of twelve Federal Reserve Districts, each having a Federal Reserve Bank that is controlled from Washington, DC by the Federal Reserve Board.

Federal Union see UNION, THE 1.

federate v. & adj. ● v.tr. & intr. /ˈfedəˌreɪt/ organize or be organized on a federal basis. ● adj. /ˈfedərət/ having a federal organization. □ **federative** /-rətɪv/ adj. [LL *foederare foederat-* (as FEDERAL)]

Federated States of Micronesia see MICRONESIA 2.

federation /ˌfedəˈreɪʃ(ə)n/ n. **1** a federal group of states. **2** a federated society or group. **3** the act or an instance of federating. □ **federationist** n. [F *fédération* f. LL *foederatio* (as FEDERAL)]

fedora /fɪˈdɔːrə/ n. a low soft felt hat with a crown creased lengthways. [*Fédora*, a drama (1882) by Victorien Sardou (1831–1908)]

fee /fiː/ n. & v. ● n. **1** a payment made to a professional person or to a professional or public body in exchange for advice or services. **2** money paid as part of a special transaction, for a privilege, admission to a society, etc. (*enrolment fee*). **3** (in pl.) money regularly paid (esp. to a school) for continuing services. **4** Law an inherited estate, unlimited (**fee simple**) or limited (**fee tail**) as to the category of heir. **5** hist. a fief; a feudal benefice. ● v.tr. (**fee'd** or **feed**) **1** pay a fee to. **2** engage for a fee. [ME f. AF, = OF *feu, fieu,* etc. f. med.L *feodum, feudum,* perh. f. Frank.: cf. FEUD², FIEF]

feeble /ˈfiːb(ə)l/ adj. (**feebler, feeblest**) **1** weak, infirm. **2** lacking energy, force, or effectiveness. **3** dim, indistinct. **4** deficient in character or intelligence. □ **feebleness** n. **feeblish** adj. **feebly** adv. [ME f. AF & OF *feble, fieble, fleible* f. L *flebilis* lamentable f. *flere* weep]

feeble-minded /ˌfiːb(ə)lˈmaɪndɪd/ adj. **1** unintelligent. **2** mentally deficient. □ **feeble-mindedly** adv. **feeble-mindedness** n.

feed /fiːd/ v. & n. ● v. (past and past part. **fed**) **1** tr. **a** supply with food. **b** put food into the mouth of. **2** tr. **a** give as food, esp. to animals. **b** graze (cattle). **3** tr. serve as food for. **4** intr. (usu. foll. by on) (esp. of animals, or colloq. of people) take food; eat. **5** tr. nourish; make grow. **6 a** tr. maintain supply of raw material, fuel, etc., to (a fire, machine, etc.). **b** tr. (foll. by into) supply (material) to a machine etc. **c** intr. (often foll. by into) (of a river etc.) flow into another body of water. **d** tr. insert further coins into (a meter) to continue its function, validity, etc. **7** intr. (foll. by on) **a** be nourished by. **b** derive benefit from. **8** tr. use (land) as pasture. **9** tr. Theatr. sl. supply (an actor etc.) with cues. **10** tr. Sport send passes to (a player) in a ball game. **11** tr. gratify (vanity etc.). **12** tr. provide (advice, information, etc.) to. ● n. **1** an amount of food, esp. for animals or infants. **2** the act or an instance of feeding; the giving of food. **3** colloq. a meal. **4** pasturage; green crops. **5 a** a supply of raw material to a machine etc. **b** the provision of this or a device for it. **6** the charge of a gun. **7** Theatr. sl. an actor who supplies another with cues. □ **feed back** produce feedback. **feeding-bottle** a bottle with a teat for feeding infants. **feed pipe** (or **tube**) a pipe (or tube) which takes liquid from a storage tank to a pump, boiler, or other mechanism. **feed up 1** fatten. **2** satiate (cf. *fed up*). [OE *fēdan* f. Gmc]

feedback /ˈfiːdbæk/ n. **1** information about reactions to a proposal, change, person's performance, etc. **2** Electronics **a** the return of a fraction of the output signal from one stage of a circuit, amplifier, etc., to the input of the same or a preceding stage. **b** a signal so returned. **3** Biol. etc. the modification or control of a process or system by its results or effects, esp. in a biochemical pathway or behavioural response.

feeder /ˈfiːdə(r)/ n. **1** a person or thing that feeds. **2** a person who eats in a specified manner. **3** a child's feeding-bottle. **4** Brit. a bib for an infant. **5** a tributary stream. **6** a branch road, railway line, etc., linking outlying districts with a main communication system. **7** Electr. a main carrying electricity to a distribution point. **8** a hopper or feeding apparatus in a machine.

feel /fiːl/ v. & n. ● v. (past and past part. **felt** /felt/) **1** tr. **a** examine or search by touch. **b** (absol.) have the sensation of touch (*was unable to feel*). **2** tr. perceive or ascertain by touch; have a sensation of (*could feel the warmth; felt that it was cold*). **3** tr. **a** undergo, experience (*shall feel my anger*). **b** exhibit or be conscious of (an emotion, sensation, conviction, etc.). **4** intr. have a specified feeling or reaction (*felt strongly about it*). **b** tr. be emotionally affected by (*felt the rebuke deeply*). **5** tr. (foll. by that + clause) have a vague or unreasoned impression (*I feel that I am right*). **6** tr. consider, think (*I feel it useful to go*). **7** intr. seem; give an impression of being; be perceived as (*the air feels chilly*). **8** intr. be consciously; consider oneself (*I feel happy; do not feel well*). **9** intr. **a** (foll. by with) have sympathy with. **b** (foll. by for) have pity or compassion for. **10** tr. (often foll. by up) colloq. fondle clumsily for sexual pleasure. ● n. **1** the act or an instance of feeling; testing by touch. **2** the sensation characterizing a material, situation, etc. **3** the sense of touch. □ **feel free** (often foll. by to + infin.) not be reluctant or hesitant (*do feel free to criticize*). **feel like** have a wish for; be inclined towards. **feel one's oats** see OAT. **feel oneself** be fit or confident etc. **feel out** investigate cautiously. **feel strange** see STRANGE. **feel up to** be ready to face or deal with. **feel one's way** proceed carefully; act cautiously. **get the feel of** become accustomed to using. **make one's influence** (or **presence** etc.) **felt** assert one's influence; make others aware of one's presence etc. [OE *fēlan* f. WG]

feeler /ˈfiːlə(r)/ n. **1** an organ in certain animals for testing things by touch or for searching for food. **2** a tentative proposal or suggestion, esp. to elicit a response (*put out feelers*). **3** a person or thing that feels. □ **feeler gauge** a gauge equipped with blades for measuring narrow gaps etc.

feel-good /ˈfiːlgʊd/ adj. that induces or seeks to induce (esp. spurious) feelings of well-being, confidence, etc.

feeling /ˈfiːlɪŋ/ *n. & adj.* ● *n.* **1 a** the capacity to feel; a sense of touch (*lost all feeling in his arm*). **b** a physical sensation. **2 a** (often foll. by *of*) a particular emotional reaction, an atmosphere (*a feeling of despair*). **b** (in *pl.*) emotional susceptibilities or sympathies (*hurt my feelings; had strong feelings about it*). **c** intense emotion (*said it with such feeling*). **3** a particular sensitivity (*had a feeling for literature*). **4 a** an opinion or notion, esp. a vague or irrational one (*my feelings on the subject; had a feeling she would be there*). **b** vague awareness (*had a feeling of safety*). **c** sentiment (*the general feeling was against it*). **5** readiness to feel sympathy or compassion. **6 a** the general emotional response produced by a work of art, piece of music, etc. **b** emotional commitment or sensibility in artistic execution (*played with feeling*). ● *adj.* **1** sensitive, sympathetic. **2** showing emotion or sensitivity. □ **feelingless** *adj.* **feelingly** *adv.*

feet *pl.* of FOOT.

feign /feɪn/ *v.* **1** *tr.* simulate; pretend to be affected by (*feign madness*). **2** *tr. archaic* invent (an excuse etc.). **3** *intr.* indulge in pretence. [ME f. *feign-* stem of OF *feindre* f. L *fingere* mould, contrive]

feijoa /feɪˈdʒəʊə, faɪ-/ *n.* **1** an evergreen shrub or tree of the genus *Feijoa*, native to South America, bearing edible guava-like fruit. **2** this fruit. [mod.L f. J. da Silva *Feijo*, 19th-c. Brazilian naturalist]

feint[1] /feɪnt/ *n. & v.* ● *n.* **1** a sham attack or blow etc. to divert attention or fool an opponent or enemy. **2** pretence. ● *v.intr.* make a feint. [F *feinte*, fem. past part. of *feindre* FEIGN]

feint[2] /feɪnt/ *adj.* esp. *Printing* designating the pale lines ruled on paper as a guide for handwriting. [var. f. OF (as FEINT[1]): see FAINT]

feisty /ˈfaɪstɪ/ *adj.* (**feistier**, **feistiest**) *N. Amer. colloq.* **1** aggressive, spirited, exuberant. **2** touchy. □ **feistiness** *n.* [*feist* small dog]

felafel /feˈlɑːf(ə)l/ *n.* (also **falafel** /fæˈlɑː-/) a spicy Middle Eastern dish of fried rissoles made from mashed chick peas or beans. [Arab. *falāfil*]

feldspar /ˈfeldspɑː(r), ˈfelspɑː(r)/ *n.* (also **felspar** /ˈfelspɑː(r)/) *Mineral.* any of a group of aluminosilicates of potassium, sodium, or calcium, which are the most abundant rock-forming minerals in the earth's crust and frequently occur as large whitish crystals. □ **feldspathic** /feldˈspæθɪk, felˈspæθ-/ *adj.* **feldspathoid** /ˈfeldspəˌθɔɪd, ˈfelspə-/ *n.* [G *Feldspat, -spath* f. *Feld* FIELD + *Spat, Spath* SPAR[3]: *felspar* by false assoc. with G *Fels* rock]

felicitate /fəˈlɪsɪˌteɪt/ *v.tr.* (usu. foll. by *on*) congratulate. □ **felicitation** /-ˌlɪsɪˈteɪʃ(ə)n/ *n.* (usu. in *pl.*). [LL *felicitare* make happy f. L *felix -icis* happy]

felicitous /fəˈlɪsɪtəs/ *adj.* (of an expression, quotation, civilities, or a person making them) strikingly apt; pleasantly ingenious. □ **felicitously** *adv.* **felicitousness** *n.*

felicity /fəˈlɪsɪtɪ/ *n.* (*pl.* **-ies**) **1** intense happiness; being happy. **2** a cause of happiness. **3 a** a capacity for apt expression; appropriateness. **b** an appropriate or well-chosen phrase. **4** a fortunate trait. [ME f. OF *félicité* f. L *felicitas -tatis* f. *felix -icis* happy]

feline /ˈfiːlaɪn/ *adj. & n.* ● *adj.* **1** of or relating to the cat family Felidae. **2** catlike, esp. in beauty or slyness. ● *n.* an animal of the cat family. □ **felinity** /fɪˈlɪnɪtɪ/ *n.* [L *felinus* f. *feles* cat]

Felixstowe /ˈfiːlɪkˌstəʊ/ a port on the east coast of England, in Suffolk; pop. (1981) 24,460.

fell[1] *past* of FALL *v.*

fell[2] /fel/ *v. & n.* ● *v.tr.* **1** cut down (esp. a tree). **2** strike or knock down (a person or animal). **3** stitch down (the edge of a seam) to lie flat. ● *n.* an amount of timber cut. □ **feller** *n.* [OE *fellan* f. Gmc, rel. to FALL]

fell[3] /fel/ *n. N. Engl.* **1** a hill. **2** a stretch of hills or moorland. [ME f. ON *fjall, fell* hill]

fell[4] /fel/ *adj. poet.* or *rhet.* **1** fierce, ruthless. **2** terrible, destructive. □ **at (or in) one fell swoop** in a single action, in one go. [ME f. OF *fel* f. Rmc FELON[1]]

fell[5] /fel/ *n.* an animal's hide or skin with its hair. [OE *fel, fell* f. Gmc]

fellah /ˈfelə/ *n.* (*pl.* **fellahin** /ˌfeləˈhiːn/) an Egyptian peasant. [Arab. *fallāḥ* husbandman f. *falaḥa* till the soil]

fellatio /feˈleɪʃɪˌəʊ, feˈlɑːtɪˌəʊ/ *n.* oral stimulation of the penis. □ **fellate** /fɪˈleɪt/ *v.tr.* **fellator** *n.* [mod.L f. L *fellare* suck]

feller /ˈfelə(r)/ *n.* = FELLOW 1, 2. [repr. an affected or sl. pronunc.]

Fellini /fəˈliːnɪ/, Federico (1920–93), Italian film director. He rose to international fame in the 1950s with *La Strada* (1954), which won an Oscar for best foreign film. Other major films include *La Dolce vita* (1960) — a satire on Rome's high society and winner of the Grand Prix at Cannes — and the semi-autobiographical *8½* (1963).

felloe /ˈfeləʊ/ *n.* (also **felly** /ˈfelɪ/) (*pl.* **-oes** or **-ies**) the outer circle (or a section of it) of a wheel, to which the spokes are fixed. [OE *felg*, of unkn. orig.]

fellow /ˈfeləʊ/ *n.* **1** *colloq.* a man or boy (*poor fellow!; my dear fellow*). **2** *derog.* a person regarded with contempt. **3** (usu. in *pl.*) a person associated with another; a comrade (*were separated from their fellows*). **4** a counterpart or match; the other of a pair. **5** an equal; one of the same class. **6** a contemporary. **7 a** an incorporated senior member of a college. **b** an elected graduate receiving a stipend for a period of research. **c** a member of the governing body in some universities. **8** a member of a learned society. **9** (*attrib.*) belonging to the same class or activity (*fellow soldier; fellow-countryman*). □ **fellow-feeling** sympathy from common experience. **fellow-traveller 1** a person who travels with another. **2** a sympathizer with, or a secret member of, the Communist Party. [OE *féolaga* f. ON *félagi* f. *fé* cattle, property, money: see LAY[1]]

fellowship /ˈfeləʊˌʃɪp/ *n.* **1** companionship, friendliness. **2** participation, sharing; community of interest. **3** a body of associates; a company. **4** a brotherhood or fraternity. **5** a guild or corporation. **6** the status or emoluments of a fellow of a college or society.

felly var. of FELLOE.

felon[1] /ˈfelən/ *n. & adj.* ● *n.* a person who has committed a felony. ● *adj. archaic* cruel, wicked. □ **felonry** *n.* [ME f. OF f. med.L *felo -onis*, of unkn. orig.]

felon[2] /ˈfelən/ *n.* an inflammatory sore on the finger near the nail. [ME, perh. as FELON[1]: cf. med.L *felo, fello* in the same sense]

felonious /fɪˈləʊnɪəs/ *adj.* **1** criminal. **2** *Law* **a** of or involving felony. **b** who has committed felony. □ **feloniously** *adv.*

felony /ˈfelənɪ/ *n.* (*pl.* **-ies**) any of a class of usu. violent crimes formerly regarded by the law as of graver character than those called misdemeanours. In English law the class originally comprised those offences (murder, wounding, arson, rape, and robbery) for which the penalty included forfeiture of land and goods. Forfeiture was abolished in 1870, but procedural differences applied until 1967 when all distinctions between felonies and misdemeanours were removed; the distinction never existed in Scotland. In the US most jurisdictions distinguish between felonies and misdemeanours, the distinction usually depending on the penalties or consequences attaching to the crime. [ME f. OF *felonie* (as FELON[1])]

felspar var. of FELDSPAR.

felt[1] /felt/ *n. & v.* ● *n.* **1** a kind of cloth made by rolling and pressing wool etc., or by weaving and shrinking it. **2** a similar material made from other fibres. ● *v.* **1** *tr.* make into felt; mat together. **2** *tr.* cover with felt. **3** *intr.* become matted. □ **felt-tipped** (or **felt-tip**) **pen** a pen with a writing-point made of felt or fibre. □ **felty** *adj.* [OE f. WG]

felt[2] *past* and *past part.* of FEEL.

felucca /fɪˈlʌkə/ *n.* a small Mediterranean coasting vessel with oars or lateen sails or both. [It. *felucca* f. obs. Sp. *faluca* f. Arab. *fulk*, perh. f. Gk *epholkion* sloop]

felwort /ˈfelwɜːt/ *n.* a purple-flowered gentian, *Gentianella amarella*, flowering in late summer. [OE *feldwyrt* (as FIELD, WORT)]

female /ˈfiːmeɪl/ *adj. & n.* ● *adj.* **1** of the sex that can bear offspring or produce eggs. **2** (of plants or their parts) fruit-bearing; having a pistil and no stamens. **3** of or consisting of women or female animals or female plants. **4** (of a screw, socket, etc.) manufactured hollow to receive a corresponding inserted part. ● *n.* a female person, animal, or plant. □ **female condom** a contraceptive sheath worn inside the vagina. **female impersonator** a male performer impersonating a woman. □ **femaleness** *n.* [ME f. OF *femelle* (n.) f. L *femella* dimin. of *femina* a woman, assim. to *male*]

feme /fem, fiːm/ *n. Law* a woman or wife. □ **feme covert** a married woman. **feme sole** a woman without a husband (esp. if divorced). [ME f. AF & OF f. L *femina* woman]

feminal /ˈfemɪn(ə)l/ *adj. archaic* womanly. □ **feminality** /ˌfemɪˈnælɪtɪ/ *n.* [med.L *feminalis* f. L *femina* woman]

femineity /ˌfemɪˈniːɪtɪ/ *n. archaic* womanliness; womanishness. [L *femineus* womanish f. *femina* woman]

feminine /ˈfemɪnɪn/ *adj. & n.* ● *adj.* **1** of or characteristic of women. **2** having qualities associated with women. **3** womanly, effeminate. **4** *Gram.* of or denoting the gender proper to women's names. ● *n. Gram.* a feminine gender or word. □ **femininely** *adv.* **feminineness** *n.* **femininity** /ˌfemɪˈnɪnɪtɪ/ *n.* [ME f. OF *feminin -ine* or L *femininus* f. *femina* woman]

feminism /ˈfemɪˌnɪz(ə)m/ *n.* **1** the advocacy of women's rights on the ground of the equality of the sexes. (*See note below.*) **2** *Med.* the development of female characteristics in a male person. □ **feminist** *n. & adj.* (in sense 1). [L *femina* woman (in sense 1 after F *féminisme*)]

▪ Feminism is a broad term which does not refer to a single unified movement and is used to cover a great variety of political and social campaigns and ideas. The issue of rights for women became prominent during the French and American revolutions in the late 18th century; Mary Wollstonecraft's *A Vindication of the Rights of Woman* was published in 1792. In Britain, however, it was not until the emergence of the suffragette movement in the late 19th century that there was significant political change, women finally gaining the vote in 1918 (see SUFFRAGETTE). The 'second wave' of feminism that arose during the 1960s in Europe and North America formed part of the widespread social changes and political protests of that era; in general, the idea of unity or 'sisterhood' was considered more important than issues (such as ethnicity and class) which could divide women. Seminal figures included Simone de Beauvoir, Betty Friedan, and Germaine Greer. During the 1970s and 1980s the movement became more fragmented and critical of the ideals of unity expressed in the 1960s: a division arose between mainstream or socialist feminists, interested in working to improve the position of women in society as it is, and radical feminists, who felt that any interaction with men tended to result in exploitation or violence. In recent years intellectual feminism, associated with psychoanalysis and deconstruction, has come to prominence, particularly in France; there has also been something of a backlash against feminism and the hailing by some of a post-feminist age.

feminity /fe'mɪnɪtɪ/ n. = FEMININITY. [ME f. OF *feminité* f. med.L *feminitas -tatis* f. L *femina* woman]

feminize /'femɪˌnaɪz/ v.tr. & intr. (also **-ise**) make or become feminine or female. □ **feminization** /ˌfemɪnaɪˈzeɪʃ(ə)n/ n.

femme fatale /ˌfæm fæˈtɑːl/ n. (pl. **femmes fatales** pronunc. same) a seductively attractive woman. [F]

femto- /'femtəʊ/ comb. form denoting a factor of 10^{-15} (*femtometre*). [Danish or Norw. *femten* fifteen]

femur /'fiːmə(r)/ n. (pl. **femurs** or **femora** /'femərə/) **1** Anat. the thigh-bone, the thick bone between the hip and the knee. **2** Zool. the third joint of the leg in insects etc., usu. seen as the thickest and longest segment closest to the body. □ **femoral** /'femərəl/ adj. [L *femur femoris* thigh]

fen /fen/ n. a low marshy or flooded area of land. □ **fen-berry** (pl. **-berries**) a cranberry. **fen-fire** a will-o'-the-wisp. □ **fenny** adj. [OE *fenn* f. Gmc]

fence /fens/ n. & v. ● n. **1** a barrier or railing or other upright structure enclosing an area of ground, esp. to prevent or control access. **2** a large upright obstacle in steeplechasing or showjumping. **3** sl. a receiver of stolen goods. **4** a guard or guide in machinery. ● v. **1** tr. surround with or as with a fence. **2** tr. **a** (foll. by *in*, *off*) enclose or separate with or as with a fence. **b** (foll. by *up*) seal with or as with a fence. **3** tr. (foll. by *from*, *against*) screen, shield, protect. **4** tr. (foll. by *out*) exclude with or as with a fence; keep out. **5** tr. (also *absol.*) sl. deal in (stolen goods). **6** intr. practise the sport of fencing; use a sword. **7** intr. (foll. by *with*) evade answering (a person or question). **8** intr. (of a horse etc.) leap fences. □ **sit on the fence** remain neutral or undecided in a dispute etc. □ **fenceless** adj. **fencer** n. [ME f. DEFENCE]

fencible /'fensɪb(ə)l/ n. hist. a soldier liable only for home service. [ME f. DEFENSIBLE]

fencing /'fensɪŋ/ n. **1** a set or extent of fences. **2** material for making fences. **3** the activity or sport of fighting with foils or other kinds of sword. (*See note below.*)

▪ Skilful use of a sword according to established rules and movements was practised by the ancient Egyptians, Persians, Greeks, and Romans, not only in war but also as a pastime. The modern style of fencing, with its emphasis on skill and speed and its use of a light slender sword, arose in 16th-century Italy, although the organized sport did not begin until the late 19th century. Fencers wear protective jackets and masks, and score points by registering hits with the sword (a *foil*, *sabre*, or *épée*); the officials judging competitions are now assisted by electronic registering of hits.

fend /fend/ v. **1** intr. (foll. by *for*) look after (esp. oneself). **2** tr. (usu. foll. by *off*) keep away; ward off (an attack etc.). [ME f. DEFEND]

Fender /'fendə(r)/, Leo (1907–91), American guitar-maker. He pioneered the production of electric guitars, designing the Fender Broadcaster of 1948 (later called the Telecaster), which was the first solid-body electric guitar to be widely available, and the Fender Stratocaster, first marketed in 1956. These two types are still produced and sold worldwide by the Fender company. Leo Fender also pioneered a number of electric bass guitars.

fender /'fendə(r)/ n. **1** a low frame bordering a fireplace to keep in falling coals etc. **2** Naut. a piece of old cable, matting, etc., hung over a vessel's side to protect it against impact. **3** a thing used to keep something off, prevent a collision, etc. **4** N. Amer. **a** the mudguard or area around the wheel well of a bicycle or motor vehicle. **b** the bumper of a motor vehicle.

fenestella /ˌfenɪˈstelə/ n. Archit. a niche in a wall south of an altar, holding the piscina and often the credence. [L, dimin. of *fenestra* window]

fenestra /fɪˈnestrə/ n. (pl. **fenestrae** /-triː/) **1** Anat. a small hole or opening in a bone etc., esp. one of two (*fenestra ovalis*, *fenestra rotunda*) in the inner ear. **2** a perforation in a surgical instrument. **3** a hole made by surgical fenestration. [L, = window]

fenestrate /fɪˈnestreɪt/ adj. Bot. & Zool. having small window-like perforations or transparent areas. [L *fenestratus* past part. of *fenestrare* f. *fenestra* window]

fenestrated /fɪˈnestreɪtɪd, 'fenɪˌstreɪtɪd/ adj. **1** Archit. having windows. **2** perforated. **3** = FENESTRATE. **4** Surgery having fenestrae.

fenestration /ˌfenɪˈstreɪʃ(ə)n/ n. **1** Archit. the arrangement of windows in a building. **2** Bot. & Zool. being fenestrate. **3** Surgery an operation in which a new opening is formed, esp. in the bony labyrinth of the inner ear, as a form of treatment in some cases of deafness.

Fenian /'fiːnɪən/ n. & adj. ● n. a member of the Irish Republican Brotherhood, a 19th-century revolutionary nationalist organization founded among the Irish in the US. (*See note below.*) ● adj. of or relating to the Fenians. □ **Fenianism** n. [OIr. *féne* name of an ancient Irish people, confused with *fiann* guard of legendary kings]

▪ The organization was founded in 1858 by James Stephens, a veteran of the failed 1848 Irish uprising; he and his members encouraged revolutionary activity and aimed for the overthrow of the British government in Ireland. The Fenians staged an unsuccessful revolt in Ireland in 1867 and were responsible for isolated revolutionary acts against the British until the early 20th century, when they were gradually eclipsed by the IRA.

fennec /'fenɪk/ n. a small desert fox, *Vulpes zerda*, native to North Africa and Arabia, having very large ears. [Arab. *fanak*]

fennel /'fen(ə)l/ n. **1 a** a yellow-flowered umbelliferous plant, *Foeniculum vulgare*, with fragrant feathery leaves used in sauces, salad dressings, etc. **b** (in full **Florence** (or **sweet**) **fennel**) a variety of this with swollen leaf-bases, used as a vegetable. **2** the fragrant seeds of fennel used as flavouring. [OE *finugl* etc. & OF *fenoil* f. L *feniculum* f. *fenum* hay]

Fennoscandia /ˌfenəʊˈskændɪə/ the land mass in NW Europe comprising Scandinavia, Finland, and the adjacent area of NE Russia.

Fens, the /fenz/ the flat low-lying areas of Lincolnshire, Cambridgeshire, and Norfolk, in eastern England. Formerly marshland, they have been drained for agriculture since the 17th century.

fenugreek /'fenjuːˌɡriːk/ n. **1** a leguminous plant, *Trigonella foenum-graecum*, having aromatic seeds. **2** these seeds used as flavouring, esp. ground and used in curry powder. [OE *fenogrecum*, superseded in ME f. OF *fenugrec* f. L *faenugraecum* (*fenum graecum* Greek hay), used by the Romans as fodder]

feoffment /'fefmənt, 'fiːf-/ n. hist. a mode of conveying a freehold estate by a formal transfer of possession. □ **feoffor** n. **feoffee** /feˈfiː, fiːˈfiː/ n. [ME f. AF *feoffement*, rel. to FEE]

feral /'fɪərəl, 'fer-/ adj. **1** (of an animal or plant) wild, untamed, uncultivated. **2 a** (of an animal) in a wild state after escape from captivity. **b** born in the wild of such an animal. **3** brutal. [L *ferus* wild]

fer-de-lance /ˌfeədəˈlɑːns/ n. (pl. **fers-de-lance** or **fer-de-lances** pronunc. same) a large highly venomous pit viper, *Bothrops atrox*, native to Central and South America. [F, = iron (head) of a lance]

Ferdinand /'fɜːdɪnənd/ of Aragon (known as Ferdinand the Catholic) (1452–1516), king of Castile 1474–1516 and of Aragon 1479–1516. His marriage to Isabella of Castile in 1469 ensured his accession (as Ferdinand V) to the throne of Castile with her. During this time, they instituted the Spanish Inquisition (1478). Ferdinand subsequently succeeded to the throne of Aragon (as Ferdinand II) and was joined as monarch by Isabella. Together, they supported Columbus's expedition in 1492. Their capture of Granada from the Moors in the same year effectively united Spain as one country. Their daughter Catherine of Aragon became the first wife of Henry VIII of England. (See also ISABELLA I.)

feretory /'ferɪtərɪ/ n. (pl. **-ies**) **1** a shrine for a saint's relics. **2** a chapel containing such a shrine. [ME f. OF *fiertre* f. L *feretrum* f. Gk *pheretron* f. *pherō* bear]

ferial /ˈfɪərɪəl, ˈfer-/ *adj. Eccl.* **1** (of a day) ordinary; not appointed for a festival or fast. **2** (of a service etc.) for use on a ferial day. [ME f. OF *ferial* or med.L *ferialis* f. L *feriae*: see FAIR²]

Ferm. *abbr.* Fermanagh.

Fermanagh /fəˈmænə/ one of the Six Counties of Northern Ireland, formerly an administrative area; chief town, Enniskillen.

Fermat /ˈfɜːmɑː/, Pierre de (1601–65), French lawyer and mathematician. His study of the problems of finding tangents to curves, finding areas under curves, and maxima and minima, led directly to the general methods of calculus introduced by Newton and Leibniz. Fermat made many discoveries about integers, for which he is seen as the founder of the theory of numbers.

fermata /fəˈmɑːtə/ *n.* (*pl.* **fermatas**) *Mus.* **1** an unspecified prolongation of a note or rest. **2** a sign indicating this. [It.]

Fermat's last theorem *n.* (also **Fermat's theorem**) the conjecture by Fermat that if n is greater than 2 then there is no integer whose nth power can be expressed as the sum of two smaller nth powers. The conjecture (of which Fermat noted that he had 'a truly wonderful proof') has been demonstrated by calculation to be true for very many possible values of n, and in 1993 a general proof was announced by the Princeton-based British mathematician Andrew Wiles.

ferment *n. & v.* ● *n.* /ˈfɜːment/ **1** agitation, excitement, tumult. **2 a** fermenting, fermentation. **b** a fermenting-agent or leaven. ● *v.* /fəˈment/ **1** *intr. & tr.* undergo or subject to fermentation. **2** *intr. & tr.* effervesce or cause to effervesce. **3** *tr.* excite; stir up; foment. □ **fermentable** /fəˈmentəb(ə)l/ *adj.* **fermenter** *n.* [ME f. OF *ferment* or L *fermentum* f. L *fervere* boil]

fermentation /ˌfɜːmenˈteɪʃ(ə)n/ *n.* **1** the breakdown of a substance by micro-organisms, such as yeasts and bacteria, usu. in the absence of oxygen, esp. of sugar to ethyl alcohol in making beers, wines, and spirits. **2** agitation, excitement. □ **fermentative** /fəˈmentətɪv/ *adj.* [ME f. LL *fermentatio* (as FERMENT)]

Fermi /ˈfɜːmɪ/, Enrico (1901–54), Italian-born American atomic physicist. Working at first in Italy, he invented (with Paul Dirac) Fermi–Dirac statistics, a mathematical tool of great value in atomic, nuclear, and solid-state physics. He predicted the existence of the neutrino, and produced radioactive isotopes by bombarding atomic nuclei with neutrons. He was awarded the Nobel Prize for physics in 1938. Moving to the US, Fermi directed the first controlled nuclear chain reaction in 1942, and joined the Manhattan Project to work on the atom bomb. The artificial element fermium and a class of subatomic particles, the fermions, are named after him.

fermi /ˈfɜːmɪ/ *n.* (*pl.* **fermis**) a unit of length equal to 10^{-15} metre, formerly used in nuclear physics. [FERMI]

fermion /ˈfɜːmɪˌɒn/ *n. Physics* a subatomic particle, such as a nucleon, which has half-integral spin and follows the statistical description given by Fermi and Dirac (cf. BOSON). [FERMI]

fermium /ˈfɜːmɪəm/ *n.* an artificial radioactive metallic chemical element (atomic number 100; symbol **Fm**). A member of the actinide series, fermium was discovered in 1953 in the debris of the first hydrogen bomb explosion. [FERMI]

fern /fɜːn/ *n.* (*pl.* same or **ferns**) a flowerless pteridophyte plant of the order Filicopsida, reproducing by spores and usu. having feathery fronds. □ **fernery** *n.* (*pl.* **-ies**). **ferny** *adj.* [OE *fearn* f. WG]

Fernando Póo /fəˌnændəʊ ˈpəʊ/ the former name (until 1973) for BIOKO.

ferocious /fəˈrəʊʃəs/ *adj.* **1** fierce, savage; wildly cruel. **2** *esp. US colloq.* (as an intensifier) very great, extreme. □ **ferociously** *adv.* **ferociousness** *n.* [L *ferox -ocis*]

ferocity /fəˈrɒsɪtɪ/ *n.* (*pl.* **-ies**) a ferocious nature or act. [F *férocité* or L *ferocitas* (as FEROCIOUS)]

-ferous /fərəs/ *comb. form* (usu. **-iferous** /ˈɪfərəs/) forming adjectives with the sense 'bearing', 'having' (*auriferous*; *odoriferous*). □ **-ferously** *suffix* forming adverbs. **-ferousness** *suffix* forming nouns. [f. or after F *-fère* or L *-fer* producing f. *ferre* bear]

Ferranti /fəˈræntɪ/, Sebastian Ziani de (1864–1930), English electrical engineer. He was one of the pioneers of electricity generation and distribution in Britain, his chief contribution being the use of high voltages for economical transmission over a distance.

Ferrara /fəˈrɑːrə/ a city in northern Italy, capital of a province of the

same name; pop. (1990) 140,600. Ferrara grew to prominence in the 13th century under the rule of the powerful Este family.

Ferrari /fəˈrɑːrɪ/, Enzo (1898–1988), Italian car designer and manufacturer. He became a racing driver for Alfa Romeo in 1920 and proceeded to work as one of their designers. In 1929 he founded the company named after him, launching the famous Ferrari marque in 1947 and producing a range of high-quality sports and racing cars. Since the early 1950s Ferraris have won the greatest number of world championship Grands Prix of any car.

ferrate /ˈfereɪt/ *n. Chem.* a salt of (the hypothetical) ferric acid. [L *ferrum* iron]

ferrel var. of FERRULE.

ferret /ˈferɪt/ *n. & v.* ● *n.* **1** a small half-tamed domesticated polecat, *Mustela furo*, used in catching rabbits, rats, etc. (*See note below.*) **2** a person who searches assiduously. ● *v.* (**ferreted**, **ferreting**) **1** *intr.* hunt with ferrets. **2** *intr.* rummage; search about. **3** *tr.* (often foll. by *about*, *away*, *out*, etc.) **a** clear out (holes or an area of ground) with ferrets. **b** take or drive away (rabbits etc.) with ferrets. **4** *tr.* (foll. by *out*) search out (secrets, criminals, etc.). □ **ferreter** *n.* **ferrety** *adj.* [ME f. OF *fu(i)ret* alt. f. *fu(i)ron* f. LL *furo -onis* f. L *fur* thief]

■ The domestic ferret is derived mainly from the European polecat, *Mustela putorius*, though it does have some characteristics that may be derived from the steppe polecat, *M. eversmanni*. The usual form is albino, though a brown form (the 'polecat ferret') occurs, and feral animals usually revert to this more natural colouring.

ferri- /ˈferɪ/ *comb. form Chem.* containing iron, esp. in ferric compounds. [L *ferrum* iron]

ferriage /ˈferɪdʒ/ *n.* **1** conveyance by ferry. **2** a charge for using a ferry.

ferric /ˈferɪk/ *adj.* **1** of iron. **2** *Chem.* containing iron in a trivalent form (cf. FERROUS).

Ferrier /ˈferɪə(r)/, Kathleen (1912–53), English contralto. She made her operatic début as Lucretia in the first performance of Britten's *The Rape of Lucretia* (1946), and is particularly famous for her performance in 1947 of Mahler's song cycle *Das Lied von der Erde*.

ferrimagnetism /ˌferɪˈmægnɪˌtɪz(ə)m/ *n. Physics* a form of ferromagnetism with antiparallel alignment of neighbouring atoms or ions. □ **ferrimagnetic** /-mægˈnetɪk/ *adj.* [F *ferrimagnétisme* (as FERRI-, MAGNETISM)]

Ferris wheel /ˈferɪs/ *n.* a fairground ride consisting of a giant revolving vertical wheel with passenger cars suspended on its outer edge. [G. W. G. *Ferris*, Amer. engineer (1859–96)]

ferrite /ˈferaɪt/ *n. Chem.* **1** a salt of (the hypothetical) ferrous acid $H_2Fe_2O_4$, often with magnetic properties. **2** an allotrope of pure iron occurring in low-carbon steel. □ **ferritic** /feˈrɪtɪk/ *adj.* [L *ferrum* iron]

ferro- /ˈferəʊ/ *comb. form Chem.* **1** iron, esp. in ferrous compounds (*ferrocyanide*). **2** (of alloys) containing iron (*ferromanganese*). [L *ferrum* iron]

ferroconcrete /ˌferəʊˈkɒnkriːt/ *n. & adj.* ● *n.* concrete reinforced with steel. ● *adj.* made of reinforced concrete.

ferroelectric /ˌferəʊɪˈlektrɪk/ *adj. & n. Physics* ● *adj.* exhibiting permanent electric polarization which varies in strength with the applied electric field. ● *n.* a ferroelectric substance. □ **ferroelectricity** /-ˌɪlekˈtrɪsɪtɪ, -ˌelek-/ *n.* [ELECTRIC after *ferromagnetic*]

ferromagnetism /ˌferəʊˈmægnɪˌtɪz(ə)m/ *n. Physics* a phenomenon in which a material has a high susceptibility to magnetization, the strength of which varies with the applied magnetizing field, and which may persist after removal of the applied field. □ **ferromagnetic** /-mægˈnetɪk/ *adj.*

ferrous /ˈferəs/ *adj.* **1** containing iron (*ferrous and non-ferrous metals*). **2** *Chem.* containing iron in a divalent form (cf. FERRIC). [L *ferrum* iron]

ferruginous /fəˈruːdʒɪnəs/ *adj.* **1** of or containing iron-rust, or iron as a chemical constituent. **2** rust-coloured; reddish-brown. □ **ferruginous duck** a European diving duck, *Aythya nyroca*, with mainly red-brown plumage. [L *ferrugo -ginis* rust f. *ferrum* iron]

ferrule /ˈferuːl/ *n.* (also **ferrel** /ˈferəl/) **1** a ring or cap strengthening the end of a stick or tube. **2** a band strengthening or forming a joint. [earlier *verrel* etc. f. OF *virelle*, *virol(e)*, f. L *viriola* dimin. of *viriae* bracelet: assim. to L *ferrum* iron]

ferry /ˈferɪ/ *n. & v.* ● *n.* (*pl.* **-ies**) **1** a boat or aircraft etc. for conveying passengers and goods, esp. across water and as a regular service. **2** the service itself or the place where it operates. ● *v.* (**-ies**, **-ied**) **1** *tr. & intr.*

convey or go in a boat etc. across water. **2** *intr.* (of a boat etc.) pass to and fro across water. **3** *tr.* transport from one place to another, esp. as a regular service. □ **ferryman** *n.* (*pl.* **-men**). [ME f. ON *ferja* f. Gmc]

fertile /ˈfɜːtaɪl/ *adj.* **1 a** (of soil) producing abundant vegetation or crops. **b** fruitful. **2 a** (of a seed, egg, etc.) capable of becoming a new individual. **b** (of animals and plants) able to conceive young or produce fruit. **3** (of the mind) inventive. **4** (of nuclear material) able to become fissile by the capture of neutrons. □ **fertility** /fəˈtɪlɪtɪ/ *n.* [ME f. F f. L *fertilis*]

Fertile Crescent a crescent-shaped area of fertile land in the Middle East extending from the eastern Mediterranean coast through the valley of the Tigris and Euphrates rivers to the Persian Gulf. This formed the cradle of the Assyrian, Sumerian, Phoenician, and Babylonian civilizations.

fertilization /ˌfɜːtɪlaɪˈzeɪʃ(ə)n/ *n.* (also **-isation**) **1** *Biol.* the fusion of male and female gametes during sexual reproduction to form a zygote. **2 a** the act or an instance of fertilizing. **b** the process of being fertilized.

fertilize /ˈfɜːtɪˌlaɪz/ *v.tr.* (also **-ise**) **1** make (soil etc.) fertile or productive. **2** cause (an egg, female animal, or plant) to develop a new individual by introducing male reproductive material. □ **fertilizable** *adj.*

fertilizer /ˈfɜːtɪˌlaɪzə(r)/ *n.* a chemical or natural substance added to soil to make it more fertile.

Fertö Tó /ˈfɜːtø ˈtoː/ the Hungarian name for the NEUSIEDLER SEE.

ferula /ˈferjʊlə/ *n.* **1** an umbelliferous plant of the genus *Ferula*, comprising the fennels; esp. the giant fennel (*F. communis*), having a tall sticklike stem and thick roots. **2** = FERULE. [ME f. L, = giant fennel, rod]

ferule /ˈferuːl/ *n. & v.* ● *n.* a flat ruler with a widened end formerly used for beating children. ● *v.tr.* beat with a ferule. [ME (AS FERULA)]

fervent /ˈfɜːv(ə)nt/ *adj.* **1** ardent, impassioned, intense (*fervent admirer; fervent hatred*). **2** hot, glowing. □ **fervency** *n.* **fervently** *adv.* [ME f. OF f. L *fervēre* boil]

fervid /ˈfɜːvɪd/ *adj.* **1** ardent, intense. **2** *poet.* hot, glowing. □ **fervidly** *adv.* [L *fervidus* (as FERVENT)]

fervour /ˈfɜːvə(r)/ *n.* (*US* **fervor**) **1** vehemence, passion, zeal. **2** a glowing condition; intense heat. [ME f. OF f. L *fervor -oris* (as FERVENT)]

Fès see FEZ.

fescue /ˈfeskjuː/ *n.* a fine-leaved grass of the genus *Festuca*, valuable for pasture and fodder. [ME *festu(e)* f. OF *festu* ult. f. L *festuca* stalk, straw]

fess /fes/ *n.* (also **fesse**) *Heraldry* a horizontal stripe across the middle of a shield. □ **fess point** a point at the centre of a shield. **in fess** arranged horizontally. [ME f. OF f. L *fascia* band]

Fessenden /ˈfesəndən/, Reginald Aubrey (1866–1932), Canadian-born American physicist and radio engineer. He pioneered radio-telephony, devised the amplitude modulation of radio waves for carrying audio signals, and invented the heterodyne receiver. He made the first sound broadcast at Christmas 1906 in the US. Fessenden was involved in both industrial and academic research, and obtained hundreds of patents for his inventions.

festal /ˈfest(ə)l/ *adj.* **1** joyous, merry. **2** engaging in holiday activities. **3** of a feast. □ **festally** *adv.* [OF f. LL *festalis* (as FEAST)]

fester /ˈfestə(r)/ *v.* **1** *tr. & intr.* make or become septic. **2** *intr.* cause continuing annoyance. **3** *intr.* rot, stagnate. [ME f. obs. *fester* (n.) or OF *festrir*, f. OF *festre* f. L *fistula*: see FISTULA]

festival /ˈfestɪv(ə)l/ *n. & adj.* ● *n.* **1** a day or period of celebration, religious or secular. **2** a concentrated series of concerts, plays, etc., held regularly in a town etc. (*Bath Festival*). ● *attrib.adj.* of or concerning a festival. [earlier as adj.: ME f. OF f. med.L *festivalis* (as FESTIVE)]

Festival of Britain a festival celebrated with lavish exhibitions and shows throughout Britain in May 1951 to mark the centenary of the Great Exhibition of 1851.

festive /ˈfestɪv/ *adj.* **1** of or characteristic of a festival. **2** cheerful, joyous. **3** fond of feasting, jovial. □ **festively** *adv.* **festiveness** *n.* [L *festivus* f. *festum* (as FEAST)]

festivity /feˈstɪvɪtɪ/ *n.* (*pl.* **-ies**) **1** gaiety, rejoicing. **2 a** a festive celebration. **b** (in *pl.*) festive proceedings. [ME f. OF *festivité* or L *festivitas* (as FESTIVE)]

festoon /feˈstuːn/ *n. & v.* ● *n.* **1** a chain of flowers, leaves, ribbons, etc., hung in a curve as a decoration. **2** a carved or moulded ornament representing this. ● *v.tr.* (often foll. by *with*) adorn with or form into festoons; decorate elaborately. □ **festoonery** *n.* [F *feston* f. It. *festone* f. *festa* FEAST]

Festschrift /ˈfestʃrɪft/ *n.* (also **festschrift**) (*pl.* **-schriften** /-ˌʃrɪftən/ or **-schrifts**) a collection of writings published in honour of a scholar. [G f. *Fest* celebration + *Schrift* writing]

feta /ˈfetə/ *n.* (also **fetta**) a white ewe's-milk or goat's-milk cheese made esp. in Greece. [mod.Gk *pheta*]

fetch[1] /fetʃ/ *v. & n.* ● *v.tr.* **1** go for and bring back (a person or thing) (*fetch a doctor*). **2** cause to come (*the thought of food fetched him*). **3** cause (blood, tears, a sigh) to come out. **4** draw (breath). **5** *colloq.* give (a blow, slap, etc.) (usu. with recipient stated: *fetched him a slap on the face*). **6** be sold for; realize (a price) (*fetched £10*). ● *n.* **1** an act of fetching. **2** a dodge or trick. **3** *Naut.* **a** the distance travelled by wind or waves across open water. **b** the distance a vessel must sail to reach open water. □ **fetch and carry** run backwards and forwards with things, be a mere servant. **fetch up 1** arrive, come to rest. **2** vomit. □ **fetcher** *n.* [OE *fecc(e)an* var. of *fetian*, prob. rel. to a Gmc root = grasp]

fetch[2] /fetʃ/ *n.* a person's wraith or double. [18th c.: orig. unkn.]

fetching /ˈfetʃɪŋ/ *adj.* attractive. □ **fetchingly** *adv.*

fête /feɪt/ *n. & v.* ● *n.* **1** an outdoor function with the sale of goods, amusements, etc., esp. to raise funds for charity etc. **2** a great entertainment; a festival. **3** a saint's day. ● *v.tr.* honour or entertain lavishly. [F *fête* (as FEAST)]

fête champêtre /ˌfet ʃɒmˈpetrə/ (*pl.* **fêtes champêtres** *pronunc.* same) *n.* an outdoor entertainment; a rural festival. [F (as FÊTE, *champêtre* rural)]

fête galante /ˌfet gæˈlɒnt/ *n.* (*pl.* **fêtes galantes** *pronunc.* same) **1** an outdoor entertainment or rural festival, esp. as depicted in 18th-century French painting. **2** a painting in this genre.

fetid /ˈfetɪd, ˈfiːt-/ *adj.* (also **foetid**) stinking. □ **fetidly** *adv.* **fetidness** *n.* [L *fetidus* f. *fetere* stink]

fetish /ˈfetɪʃ/ *n.* **1** *Psychol.* a thing abnormally stimulating or attracting sexual desire. **2 a** an inanimate object worshipped by primitive peoples for its supposed inherent magical powers or as being inhabited by a spirit. **b** a thing evoking irrational devotion or respect. □ **fetishism** *n.* **fetishist** *n.* **fetishistic** /ˌfetɪˈʃɪstɪk/ *adj.* [F *fétiche* f. Port. *feitiço* charm: orig. adj. = made by art, f. L *factitius* FACTITIOUS]

fetlock /ˈfetlɒk/ *n.* the part of a horse's leg between the cannon bone and the pastern, forming a projection above and behind the hoof where there is a tuft of hair often grows. [ME *fetlak* etc. rel. to G *Fessel* fetlock f. Gmc]

fetor /ˈfiːtə(r)/ *n.* a stench. [L (as FETID)]

fetta var. of FETA.

fetter /ˈfetə(r)/ *n. & v.* ● *n.* **1 a** a shackle for holding a prisoner by the ankles. **b** any shackle or bond. **2** (in *pl.*) captivity. **3** a restraint or check. ● *v.tr.* **1** put into fetters. **2** restrict, restrain, impede. [OE *feter* f. Gmc]

fetterlock /ˈfetəˌlɒk/ *n.* **1** a D-shaped fetter for tethering a horse by the leg. **2** a heraldic representation of this.

fettle /ˈfet(ə)l/ *n. & v.* ● *n.* condition or trim (*in fine fettle*). ● *v.tr.* trim or clean (the rough edge of a metal casting, pottery before firing, etc.). [earlier as verb, f. dial. *fettle* (n.) = girdle, f. OE *fetel* f. Gmc]

fettler /ˈfetlə(r)/ *n.* **1** *Brit. & Austral.* a railway maintenance worker. **2** a person who fettles.

fettuccine /ˌfetuˈtʃiːnɪ/ *n.pl.* (also **fettucine**) pasta in the form of ribbons. [It., pl. of dim. of *fetta* slice, ribbon]

fetus *US* var. of FOETUS.

feu /fjuː/ *n. & v. Sc.* ● *n.* **1** a perpetual lease at a fixed rent. **2** a piece of land so held. ● *v.tr.* (**feus, feued, feuing**) grant (land) on feu. [OF: see FEE]

feud[1] /fjuːd/ *n. & v.* ● *n.* **1** prolonged mutual hostility, esp. between two families, tribes, etc., with murderous assaults in revenge for a previous injury (*be at feud with*). **2** a prolonged or bitter quarrel or dispute (*a family feud*). ● *v.intr.* conduct a feud. [ME *fede* f. OF *feide, fede* f. MDu., MLG *vēde* f. Gmc, rel. to FOE]

feud[2] /fjuːd/ *n.* a piece of land held under the feudal system or in fee; a fief. [med.L *feudum*: see FEE]

feudal /ˈfjuːd(ə)l/ *adj.* **1** of, according to, or resembling the feudal system. **2** of a feud or fief. **3** outdated (*had a feudal attitude*). □ **feudally** *adv.* **feudalize** *v.tr.* (also **-ise**). **feudalization** /ˌfjuːdəlaɪˈzeɪʃ(ə)n/ *n.* [med.L *feudalis, feodalis* f. *feudum, feodum* FEE, perh. f. Gmc]

feudalism /ˈfjuːdəˌlɪz(ə)m/ *n.* the feudal system or its principles. □ **feudalist** *n.* **feudalistic** /ˌfjuːdəˈlɪstɪk/ *adj.*

feudality /fjuːˈdælɪtɪ/ *n.* (*pl.* **-ies**) **1** the feudal system or its principles. **2** a feudal holding, a fief. [F *féodalité* f. *féodal* (as FEUDAL)]

feudal system *n.* a medieval European politico-economic system of land-holding which was based on a reciprocal arrangement between vassal, peasant, and lord. The nobility held lands from the Crown in exchange for a specified amount of military service; the knights or vassals were the tenants of the nobles; and the unfree peasantry (villeins or serfs) lived on their lord's land, gave him homage, and had to provide him with labour or a share of their produce, notionally in exchange for military protection. Having been introduced into England by William I, the feudal system began to break down there in the 13th and 14th centuries, although feudal tenures were not actually abolished by statute until 1666; the related system of serfdom survived in some countries for much longer (see SERF). In Scotland land is still technically held under a feudal tenure from the Crown.

feudatory /ˈfjuːdətərɪ/ *adj. & n.* ● *adj.* (often foll. by *to*) feudally subject, under overlordship. ● *n.* (*pl.* **-ies**) a feudal vassal. [med.L *feudatorius* f. *feudare* enfeoff (as FEUD²)]

feu de joie /ˌfɜː də ˈʒwʌ/ *n.* (*pl.* **feux** *pronunc.* same) a salute by firing rifles etc. on a ceremonial occasion. [F, = fire of joy]

feudist /ˈfjuːdɪst/ *n. US* a person who is conducting a feud.

feuilleton /ˈfɜːɪˌtɒn/ *n.* **1** a part of a newspaper etc. devoted to fiction, criticism, light literature, etc. **2** an item printed in this. [F, = leaflet]

fever /ˈfiːvə(r)/ *n. & v.* ● *n.* **1 a** an abnormally high body temperature, often with delirium etc. **b** a disease characterized by this (*scarlet fever; typhoid fever*). **2** nervous excitement; agitation. ● *v.tr.* (esp. as **fevered** *adj.*) affect with fever or excitement. □ **fever pitch** a state of extreme excitement. **fever tree 1** a yellow-flowered southern African tree, *Acacia xanthophloea*. **2** a tree yielding a febrifuge; esp. *Pickneya pubens* of the south-eastern US. [OE *fefor* & AF *fevre*, OF *fievre* f. L *febris*]

feverfew /ˈfiːvəˌfjuː/ *n.* an aromatic bushy composite plant, *Tanacetum parthenium*, with white daisy-like flowers, used to treat migraine and formerly to reduce fever. [OE *feferfuge* f. L *febrifuga* (as FEBRIFUGE)]

feverish /ˈfiːvərɪʃ/ *adj.* **1** having the symptoms of a fever. **2** excited, fitful, restless. **3** (of a place) infested by fever; feverous. □ **feverishly** *adv.* **feverishness** *n.*

feverous /ˈfiːvərəs/ *adj.* **1** infested with or apt to cause fever. **2** *archaic* feverish.

few /fjuː/ *adj. & n.* ● *adj.* **1** not many (*few doctors smoke; visitors are few*). **2** (prec. by *a*) a small number of (*everything depended on a few people*). ● *n.* (as *pl.*) **1** (prec. by *a*) some but not many (*he only kept a few; a few of his friends were there*). **2** a small number, not many (*many are called but few are chosen*). **3** (prec. by *the*) **a** a minority. **b** the elect. **4** (**the Few**) the RAF pilots who took part in the Battle of Britain. □ **every few** once in every small group of (*every few days*). **few and far between** scarce. **a good few** a fairly large number. **have a few** *colloq.* take several alcoholic drinks. **no fewer than** as many as (a specified number). **not a few** a considerable number. **some few** some but not at all many. [OE *feawe, feawa* f. Gmc]

fey /feɪ/ *adj.* **1 a** strange, other-worldly; elfin; whimsical. **b** clairvoyant. **2** *Sc.* **a** fated to die soon. **b** overexcited or elated, as formerly associated with the state of mind of a person about to die. □ **feyly** *adv.* **feyness** *n.* [OE *fæge* f. Gmc]

Feydeau /ˈfeɪdəʊ/, Georges (1862–1921), French dramatist. His name has become a byword for French bedroom farce. He wrote some forty plays, including *Hotel Paradiso* (1894) and *Le Dindon* (1896).

Feynman /ˈfeɪnmən/, Richard Phillips (1918–88), American theoretical physicist. He worked in quantum electrodynamics, and introduced important new techniques for studying the electromagnetic interactions between subatomic particles. This approach is expressed in diagrams that describe the exchange of particles (see FEYNMAN DIAGRAM). He shared the Nobel Prize for physics in 1965.

Feynman diagram *n. Physics* a diagram of interactions between subatomic particles. [FEYNMAN]

Fez /fez/ (also **Fès**) a city in northern Morocco, founded in 808; pop. (1982) 448,800.

fez /fez/ *n.* (*pl.* **fezzes**) a flat-topped conical red cap with a tassel, worn by men in some Muslim countries. □ **fezzed** *adj.* [Turk., perh. f. FEZ]

ff *abbr. Mus.* fortissimo.

ff. *abbr.* **1** following pages etc. **2** folios.

fiacre /fɪˈɑːkrə/ *n. hist.* a small four-wheeled cab. [the Hôtel de St Fiacre, Paris]

fiancé /fɪˈɒnseɪ/ *n.* (*fem.* **fiancée** *pronunc.* same) a person to whom another is engaged to be married. [F, past part. of *fiancer* betroth f. OF *fiance* a promise, ult. f. L *fidere* to trust]

fianchetto /ˌfɪənˈtʃetəʊ, -ˈketəʊ/ *n. & v. Chess* ● *n.* (*pl.* **-oes**) the development of a bishop to a long diagonal of the board. ● *v.tr.* (**-oes, -oed**) develop (a bishop) in this way. [It., dimin. of *fianco* FLANK]

Fianna Fáil /ˌfiːənə ˈfɔɪl/ one of the major political parties in the Republic of Ireland, larger and traditionally more republican than its rival Fine Gael. It was formed in 1926 in opposition to the Anglo-Irish Treaty of 1921 by Eamon de Valera together with some of the moderate members of Sinn Fein. The phrase *Fianna Fáil* was used in 15th-century poetry in the neutral sense 'people of Ireland', but the founders of the political party interpreted it to mean 'soldiers of destiny'. [Ir. *fianna* bands of hunters + *Fáil* genitive of *Fál*, an ancient name for Ireland]

fiasco /fɪˈæskəʊ/ *n.* (*pl.* **-os**) a ludicrous or humiliating failure or breakdown (orig. in a dramatic or musical performance); an ignominious result. [It., = bottle (with unexplained allusion): see FLASK]

fiat /ˈfaɪæt, ˈfaɪət/ *n.* **1** an authorization. **2** a decree or order. □ **fiat money** *US* inconvertible paper money made legal tender by a government decree. [L, = let it be done]

fib /fɪb/ *n. & v.* ● *n.* a trivial or venial lie. ● *v.intr.* (**fibbed, fibbing**) tell a fib. □ **fibber** *n.* **fibster** *n.* [perh. f. obs. *fible-fable* nonsense, redupl. of FABLE]

fiber *US* var. of FIBRE.

Fibonacci series /ˌfɪbəˈnɑːtʃɪ/ *n. Math.* a series of numbers in which each number (*Fibonacci number*) is the sum of the two preceding numbers, esp. 1, 1, 2, 3, 5, 8, etc. [Leonardo *Fibonacci*, It. mathematician *fl.* 1200]

fibre /ˈfaɪbə(r)/ *n.* (*US* **fiber**) **1** any of the threads or filaments forming animal or vegetable tissue and textile substances. **2** a piece of glass in the form of a thread. **3 a** a substance formed of fibres. **b** a substance that can be spun, woven, or felted. **4** the structure, grain, or character of something (*lacks moral fibre*). **5** dietary material that is resistant to the action of digestive enzymes; roughage. □ **fibred** *adj.* (also in *comb.*). **fibreless** *adj.* **fibriform** /ˈfaɪbrɪˌfɔːm/ *adj.* [ME f. F f. L *fibra*]

fibreboard /ˈfaɪbəˌbɔːd/ *n.* (*US* **fiberboard**) a building material made of wood or other plant fibres compressed into boards.

fibreglass /ˈfaɪbəˌglɑːs/ *n.* (*US* **fiberglass**) glass in fibrous form; material made from or containing this. Plastic reinforced by glass fibres is a strong lightweight structural material used e.g. for boat-hulls; matted fibreglass is used mainly as a thermal insulation material. Fibreglass was developed commercially in the US in the 1930s.

fibre optics *n.* the use of thin flexible fibres of glass or other transparent solids to transmit light-signals. Fibre optics was developed and named by Dr Narinder Singh Kapany in the mid-1950s. The fibres used are less than 1 mm in thickness and have a high refractive index. Applications include the internal inspection of the body (endoscopy) and communications. An optical communications system consists of a transmitter, a length of fibre, and a receiver. The transmitter converts an electrical signal into light by modulating the output of a semiconductor laser or light-emitting diode. The modulated light passes into the fibre and propagates through it by a series of total internal reflections at the fibre walls; the receiver detects the optical signal and reconstructs the original electrical signal. Because of the high frequencies employed, one fibre can carry nearly 2,000 channels simultaneously.

fibril /ˈfaɪbrɪl/ *n.* **1** a small fibre. **2** a subdivision of a fibre. □ **fibrillar** *adj.* **fibrillary** *adj.* [mod.L *fibrilla* dimin. of L *fibra* fibre]

fibrillate /ˈfɪbrɪˌleɪt, ˈfaɪ-/ *v.* **1** *intr.* **a** (of a fibre) split up into fibrils. **b** (of a muscle, esp. in the heart) undergo a quivering movement due to uncoordinated contraction of the individual fibrils. **2** *tr.* break (a fibre) into fibrils. □ **fibrillation** /ˌfɪbrɪˈleɪʃ(ə)n, ˌfaɪ-/ *n.*

fibrin /ˈfaɪbrɪn/ *n.* an insoluble protein formed during blood clotting from fibrinogen. □ **fibrinoid** *adj.* [FIBRE + -IN]

fibrinogen /faɪˈbrɪnədʒən/ *n.* a soluble blood plasma protein which produces fibrin when acted upon by the enzyme thrombin.

fibro /ˈfaɪbrəʊ/ *n.* (*pl.* **-os**) *Austral.* **1** fibro-cement. **2** a house constructed mainly of this. [abbr.]

fibro- /ˈfaɪbrəʊ/ *comb. form* fibre.

fibroblast /ˈfaɪbrəʊˌblæst/ *n. Anat.* a cell producing collagen fibres in connective tissue. [FIBRO- + -BLAST]

fibro-cement /ˌfaɪbrəʊsɪˈment/ *n.* a mixture of various fibrous materials, such as glass fibre, cellulose fibre, etc., and cement, used in sheets for building etc.

fibroid /'faɪbrɔɪd/ *adj. & n.* ● *adj.* **1** of or characterized by fibrous tissue. **2** resembling or containing fibres. ● *n.* a benign tumour of muscular and fibrous tissues, one or more of which may develop in the wall of the womb.

fibroin /'faɪbrəʊɪn/ *n.* a protein which is the chief constituent of silk. [FIBRO- + -IN]

fibroma /faɪ'brəʊmə/ *n.* (*pl.* **fibromas** or **fibromata** /-mətə/) *Med.* a fibrous tumour. [f. L *fibra* fibre]

fibrosis /faɪ'brəʊsɪs/ *n. Med.* a thickening and scarring of connective tissue, usu. as a result of injury. □ **fibrotic** /-'brɒtɪk/ *adj.* [mod.L f. L *fibra* fibre + -OSIS]

fibrositis /ˌfaɪbrə'saɪtɪs/ *n.* an inflammation of fibrous connective tissue, usu. rheumatic and painful. □ **fibrositic** /-'sɪtɪk/ *adj.* [mod.L f. L *fibrosus* fibrous + -ITIS]

fibrous /'faɪbrəs/ *adj.* consisting of or like fibres. □ **fibrously** *adv.* **fibrousness** *n.*

fibula /'fɪbjʊlə/ *n.* (*pl.* **fibulae** /-ˌliː/ or **fibulas**) **1** *Anat.* the smaller and outer of the two bones between the knee and the ankle in terrestrial vertebrates. **2** *Antiq.* a brooch or clasp. □ **fibular** *adj.* [L, perh. rel. to *figere* fix]

-fic /fɪk/ *suffix* (usu. as **-ific** /'ɪfɪk/) forming adjectives meaning 'producing', 'making' (*prolific*; *pacific*). □ **-fically** *suffix* forming adverbs. [from or after F *-fique* or L *-ficus* f. *facere* do, make]

-fication /fɪ'keɪʃ(ə)n/ *suffix* (usu. as **-ification** /ɪfɪ-/) forming nouns of action from verbs in *-fy* (*acidification*; *purification*; *simplification*). [from or after F *-fication* or L *-ficatio -onis* f. *-ficare*: see -FY]

fiche /fiːʃ/ *n.* (*pl.* same or **fiches**) a microfiche. [F, = slip of paper]

Fichte /'fɪxtə/, Johann Gottlieb (1762–1814), German philosopher. A pupil of Kant, he postulated that the ego is the only basic reality; the world around it, or the 'non-ego', is posited by the ego in defining and delimiting itself. Fichte preached moral virtues and encouraged patriotic values; his political addresses had some influence on the development of German nationalism and the overthrow of Napoleon.

fichu /'fiːʃuː, 'fiːʃuː/ *n.* a woman's small triangular shawl of lace etc. for the shoulders and neck. [F]

fickle /'fɪk(ə)l/ *adj.* inconstant, changeable, esp. in loyalty. □ **fickleness** *n.* **fickly** *adv.* [OE *ficol*; cf. *befician* deceive, *fæcne* deceitful]

fictile /'fɪktaɪl/ *adj.* **1** made of earth or clay by a potter. **2** of pottery. [L *fictilis* f. *fingere fict-* fashion]

fiction /'fɪkʃ(ə)n/ *n.* **1** an invented idea or statement or narrative; an imaginary thing. **2** literature, esp. novels, describing imaginary events and people; the genre comprising novels and stories. **3** a conventionally accepted falsehood (*legal fiction*; *polite fiction*). **4** the act or process of inventing imaginary things. □ **fictional** *adj.* **fictionally** *adv.* **fictionist** *n.* **fictionalize** *v.tr.* (also **-ise**). **fictionalization** /ˌfɪkʃənəlaɪ'zeɪʃ(ə)n/ *n.* **fictionality** /ˌfɪkʃə'nælɪtɪ/ *n.* [ME f. OF f. L *fictio -onis* (as FICTILE)]

fictitious /fɪk'tɪʃəs/ *adj.* **1** imaginary, unreal. **2** counterfeit; not genuine. **3** (of a name or character) assumed. **4** of or in novels. **5** regarded as or called such by a legal or conventional fiction. □ **fictitiously** *adv.* **fictitiousness** *n.* [L *ficticius* (as FICTILE)]

fictive /'fɪktɪv/ *adj.* **1** creating or created by imagination. **2** not genuine. □ **fictiveness** *n.* [F *fictif -ive* or med.L *fictivus* (as FICTILE)]

ficus /'fiːkəs/ *n.* a tree or shrub of the large genus *Ficus*, of the mulberry family, including the figs and the rubber plant. [L, = fig, fig tree]

fid /fɪd/ *n.* **1** a small thick piece or wedge or heap of anything. **2** *Naut.* **a** a square wooden or iron bar to support the topmast. **b** a conical wooden pin used in splicing. [17th c.: orig. unkn.]

Fid. Def. *abbr.* = DEFENDER OF THE FAITH. [L *Fidei Defensor*]

fiddle /'fɪd(ə)l/ *n. & v.* ● *n.* **1** *colloq.* or *derog.* a stringed instrument played with a bow, esp. a violin. **2** *colloq.* an instance of cheating or fraud. **3** a fiddly task. **3** *Naut.* a contrivance for stopping things from rolling or sliding off a table in bad weather. ● *v.* **1** *intr.* **a** (often foll. by *with*, *at*) play restlessly. **b** (often foll. by *about*) move aimlessly. **c** act idly or frivolously. **d** (usu. foll. by *with*) make minor adjustments; tinker (esp. in an attempt to make improvements). **2** *tr. sl.* a cheat, swindle. **b** falsify. **c** get by cheating. **3** *intr.* play the fiddle. **b** *tr.* play (a tune etc.) on the fiddle. □ **as fit as a fiddle** in very good health. **face as long as a fiddle** a dismal face. **on the fiddle** engaged in cheating or swindling. **fiddle-back** a fiddle-shaped back of a chair or front of a chasuble. **fiddle-head** a scroll-like carving at a ship's bows. **fiddle pattern** the pattern

of spoons and forks with fiddle-shaped handles. **play second** (or **first**) **fiddle** take a subordinate (or leading) role. [OE *fithele* f. Gmc f. a Rmc root rel. to VIOL]

fiddle-de-dee /ˌfɪd(ə)ldɪ'diː/ *int. & n.* nonsense.

fiddle-faddle /'fɪd(ə)lˌfæd(ə)l/ *n., v., int., & adj.* ● *n.* trivial matters. ● *v.intr.* fuss, trifle. ● *int.* nonsense! ● *adj.* (of a person or thing) petty, fussy. [redupl. of FIDDLE]

fiddler /'fɪdlə(r)/ *n.* **1** a fiddle-player. **2** *sl.* a swindler, a cheat. **3** (in full **fiddler crab**) a small crab of the genus *Uca*, the male having one of its claws larger than the other and held in position like a violinist's arm. [OE *fithelere* (as FIDDLE)]

Fiddler's Green the sailor's Elysium, traditionally a place of wine, women, and song.

fiddlestick /'fɪd(ə)lˌstɪk/ *n.* **1** (in *pl.*; as *int.*) nonsense! **2** *colloq.* a bow for a fiddle.

fiddling /'fɪdlɪŋ/ *adj.* **1 a** petty, trivial. **b** contemptible, futile. **2** = FIDDLY. **3** that fiddles.

fiddly /'fɪdlɪ/ *adj.* (**fiddlier**, **fiddliest**) intricate, awkward, or tiresome to do or use.

Fidei Defensor /ˌfaɪdiˌaɪ dɪ'fensɔː(r), ˌfiːdeɪ/ = DEFENDER OF THE FAITH. [L]

fideism /'faɪdiˌɪz(ə)m, 'fiːdeɪ-/ *n. Theol.* the doctrine that all or some knowledge depends on faith or revelation. □ **fideist** *n.* **fideistic** /ˌfaɪdɪ'ɪstɪk, ˌfiːdeɪ-/ *adj.* [L *fides* faith + -ISM]

fidelity /fɪ'delɪtɪ/ *n.* **1** (often foll. by *to*) faithfulness, loyalty. **2** strict conformity to truth or fact. **3** exact correspondence to the original. **4** precision in reproduction of sound (*high fidelity*). □ **fidelity insurance** insurance taken out by an employer against losses incurred through an employee's dishonesty etc. [F *fidélité* or L *fidelitas* (as FEALTY)]

fidget /'fɪdʒɪt/ *v. & n.* ● *v.* (**fidgeted**, **fidgeting**) **1** *intr.* move or act restlessly or nervously, usu. while maintaining basically the same posture. **2** *intr.* be uneasy, worry. **3** *tr.* make (a person) uneasy or uncomfortable. ● *n.* **1** a person who fidgets. **2** (usu. in *pl.*) **a** a bodily uneasiness seeking relief in spasmodic movements; such movements. **b** a restless mood. □ **fidgety** *adj.* **fidgetiness** *n.* [obs. or dial. *fidge* to twitch]

Fido /'faɪdəʊ/ *n.* a device enabling aircraft to land by dispersing fog by means of petrol-burners on the ground. [acronym f. Fog Intensive Dispersal Operation]

fiducial /fɪ'djuːʃ(ə)l/ *adj. Surveying, Astron.*, etc. (of a line, point, etc.) assumed as a fixed basis of comparison. [LL *fiducialis* f. *fiducia* trust f. *fidere* to trust]

fiduciary /fɪ'djuːʃərɪ/ *adj. & n.* ● *adj.* **1 a** of a trust, trustee, or trusteeship. **b** held or given in trust. **2** (of a paper currency) depending for its value on public confidence or securities. ● *n.* (*pl.* **-ies**) a trustee. [L *fiduciarius* (as FIDUCIAL)]

fidus Achates /ˌfaɪdəs ə'keɪtiːz/ *n.* a faithful friend; a devoted follower. [L, = faithful Achates (a companion of Aeneas in Virgil's *Aeneid*)]

fie /faɪ/ *int. archaic* expressing disgust, shame, or a pretence of outraged propriety. [ME f. OF f. L *fi* exclam. of disgust at a stench]

fief /fiːf/ *n.* **1** a piece of land held under the feudal system or in fee. **2** a person's sphere of operation or control. [F (as FEE)]

fiefdom /'fiːfdəm/ *n.* a fief.

Field /fiːld/, John (1782–1837), Irish composer and pianist. He is noted for the invention of the nocturne and for his twenty compositions in this form.

field /fiːld/ *n. & v.* ● *n.* **1** an area of open land, esp. one used for pasture or crops, often bounded by hedges, fences, etc. **2** an area rich in some natural product (*gas field*; *diamond field*). **3** a piece of land for a specified purpose, esp. an area marked out for games (*football field*). **4 a** the participants in a contest or sport. **b** all the competitors in a race or all except those specified. **5** *Cricket* **a** the side fielding. **b** a fielder. **6** an expanse of ice, snow, sea, sky, etc. (*star field*). **7 a** the ground on which a battle is fought; a battlefield (*left his rival in possession of the field*). **b** the scene of a campaign. **c** (*attrib.*) (of artillery etc.) light and mobile for use on campaign. **d** a battle (*Bosworth Field*). **8** an area of operation or activity; a subject of study (*each supreme in his own field*). **9** the region in which a force is effective (*gravitational field*; *magnetic field*); the force exerted in such an area. (*See note below.*) **10** a range of perception (*field of view*; *wide field of vision*; *filled the field of the telescope*). **11** *Math.* a system subject to two operations analogous to those for the multiplication and addition of real numbers. **12** (*attrib.*) **a** (of an animal or plant) found in the countryside, wild (*field mouse*). **b** carried out or working in the natural environment, not in a laboratory etc. (*field test*). **13 a** the background of a picture, coin, flag, etc. **b** *Heraldry* the surface of an escutcheon or of one of its divisions. **14** *Computing* a part of a record, representing an item of data. ● *v.* **1** *Cricket, Baseball*, etc. **a** *intr.* act as a fielder. **b** *tr.* stop (and return) (the ball). **2** *tr.* **a** select (a team or individual) to play in a game. **b** deploy (an army). **c** propose (a candidate). **3** *tr.* deal with (a succession of questions etc.). □ **field-book** a book used in the field by a surveyor for technical notes. **field-cornet** *S. Afr. hist.* a minor magistrate. **field-day 1** wide scope for action or success; a time occupied with exciting events (*when crowds form, pickpockets have a field-day*). **2** *Mil.* an exercise, esp. in manoeuvring; a review. **3** a day spent in exploration, scientific investigation, etc., in the natural environment. **field-effect transistor** a semiconductor device with current flowing through a channel controlled by a transverse electric field. **field events** athletic sports other than races (e.g. shot-putting, jumping, discus-throwing). **field-glasses** binoculars for outdoor use. **field goal** *Amer. Football & Basketball* a goal scored when the ball is in normal play. **field hockey** *N. Amer.* = HOCKEY[1] 1. **field hospital** a temporary hospital near a battlefield. **field mouse** = wood mouse. **field mushroom** the common edible mushroom, *Agaricus campestris.* **field mustard** = CHARLOCK. **field officer** an army officer of field rank. **field of honour** the place where a duel or battle is fought. **field rank** any rank in an army above captain and below general. **field sports** outdoor sports, esp. hunting, shooting, and fishing. **field telegraph** a movable telegraph for use on campaign. **field vole** a common Eurasian grassland vole, *Microtus agrestis*, with a short tail. **hold the field** not be superseded. **in the field 1** campaigning. **2** working etc. away from one's laboratory, headquarters, etc. **keep the field** continue a campaign. **play the field** *colloq.* avoid exclusive attachment to one person or activity etc. **take the field 1** begin a campaign. **2** (of a sports team) go on to a pitch to begin a game. [OE *feld* f. WG]

▪ The means by which a magnet attracts a piece of iron or another magnet some distance away, even when wood or stone is interposed, has perplexed scientists from antiquity and is still poorly understood. The explanation most widely accepted today, originally proposed by Faraday, holds that a magnet produces an intermediate condition in space, called a magnetic field, which in turn acts directly on other magnets placed in the field; this condition can exist even in the absence of any material medium and is not directly affected by the presence of any such medium. Electric fields are similarly introduced to explain the action of one stationary electric charge on another. It is now known that the magnetism of an object is due to tiny circulating electric currents, and that electric charges in motion are responsible for all magnetic fields. When electric charges accelerate, a third type of field is produced, the electromagnetic radiation field. In summary: stationary electric charges produce electric fields, steadily moving charges produce in addition a magnetic field, and accelerating charges produce electromagnetic radiation.

fielder /ˈfiːldə(r)/ *n. Cricket* etc. a member (other than the bowler) of the side that is fielding.

fieldfare /ˈfiːldfeə(r)/ *n.* a large thrush, *Turdus pilaris*, breeding in northern Europe, having a grey head and rump. [ME *feldefare*, perh. as FIELD + FARE]

Fielding /ˈfiːldɪŋ/, Henry (1707–54), English novelist. After writing several comedies and farces, he provoked the introduction of censorship with his sharp political satire *The Historical Register for 1736*; the resultant Licensing Act of 1737 effectively ended his career as a dramatist. He turned to writing picaresque novels, including *Joseph Andrews* (1742) – which begins as a parody of Samuel Richardson's *Pamela* – and *Tom Jones* (1749). Fielding became Justice of the Peace for Westminster in 1748, and was responsible for the formation of the Bow Street Runners the following year.

Field Marshal *n.* an officer of the highest rank in the British and some other armies.

Field of the Cloth of Gold the scene of a meeting between Henry VIII of England and Francis I of France near Calais in 1520, for which both monarchs erected elaborate temporary palaces and pavilions, including a sumptuous display of golden cloth. Little of importance was achieved, although the meeting functioned as a symbol of Henry's determination to play a full part in European dynastic politics.

Fields[1] /fiːldz/, Dame Gracie (born Grace Stansfield) (1898–1979), English singer and comedienne. During the 1930s she enjoyed great success with English music-hall audiences. She went on to star in a series of popular films, including *Sing as We Go* (1934).

Fields[2] /fiːldz/, W. C. (born William Claude Dukenfield) (1880–1946), American comedian. Having made his name as a comedy juggler he became a vaudeville star, appearing in the *Ziegfeld Follies* revues between 1915 and 1921. His films established him as an internationally famous comic; they include *The Bank Dick* (1940).

fieldsman /ˈfiːldzmən/ *n.* (*pl.* **-men**) *Cricket* = FIELDER.

fieldstone /ˈfiːldstəʊn/ *n.* stone used in its natural form.

fieldwork /ˈfiːldwɜːk/ *n.* **1** the practical work of a surveyor, collector of scientific data, sociologist, etc., conducted in the natural environment rather than a laboratory, office, etc. **2** a temporary fortification. □ **fieldworker** *n.*

fiend /fiːnd/ *n.* **1 a** an evil spirit, a demon. **b** (prec. by *the*) the Devil. **2 a** a very wicked or cruel person. **b** a person causing mischief or annoyance. **3** (with a qualifying word) *colloq.* a devotee or addict (*a fitness fiend*). **4** something difficult or unpleasant. □ **fiendlike** *adj.* [OE *fēond* f. Gmc]

fiendish /ˈfiːndɪʃ/ *adj.* **1** like a fiend; extremely cruel or unpleasant. **2** *colloq.* extremely difficult or exasperating. □ **fiendishly** *adv.* **fiendishness** *n.*

fierce /fɪəs/ *adj.* (**fiercer**, **fiercest**) **1** vehemently aggressive or frightening in temper or action, violent. **2** eager, intense, ardent. **3** unpleasantly strong or intense; uncontrolled (*fierce heat*). **4** (of a mechanism) not smooth or easy in action. □ **fiercely** *adv.* **fierceness** *n.* [ME f. AF *fers*, OF *fiers fier* proud f. L *ferus* savage]

fieri facias /ˌfaɪəraɪ ˈfeɪʃɪˌæs/ *n. Law* a writ to a sheriff for executing a judgement. [L, = cause to be made or done]

fiery /ˈfaɪərɪ/ *adj.* (**fierier**, **fieriest**) **1 a** consisting of or flaming with fire. **b** (of an arrow etc.) fire-bearing. **2** like fire in appearance, bright red. **3 a** hot as fire. **b** acting like fire; producing a burning sensation. **4 a** flashing, ardent (*fiery eyes*). **b** eager, pugnacious, spirited, irritable (*fiery temper*). **c** (of a horse) mettlesome. **5** (of gas, a mine, etc.) inflammable; liable to explosions. **6** *Cricket* (of a pitch) making the ball rise dangerously. □ **fiery cross** a wooden cross charred or set on fire as a symbol. □ **fierily** *adv.* **fieriness** *n.*

fiesta /fiˈestə/ *n.* **1** a holiday or festivity. **2** a religious festival in Spanish-speaking countries. [Sp., = feast]

FIFA /ˈfiːfə/ *abbr.* International Federation of Association Football. [F *Fédération Internationale de Football Association*]

fi. fa. *abbr.* fieri facias.

Fife /faɪf/ a local government region and former county (until 1975) of east central Scotland; capital, Glenrothes.

fife /faɪf/ *n. & v.* ● *n.* **1** a kind of small shrill flute used with the drum in military music. **2** a player of this instrument. ● *v.* **1** *intr.* play the fife. **2** *tr.* play (an air etc.) on the fife. □ **fifer** *n.* [G *Pfeife* PIPE, or F *fifre* f. Swiss G *Pfifre* piper]

fife-rail /ˈfaɪfreɪl/ *n. Naut.* a rail round the mainmast with belaying-pins. [18th c.: orig. unkn.]

fifteen /fɪfˈtiːn/ *n. & adj.* ● *n.* **1** one more than fourteen, or five more than ten; the product of three units and five units. **2** a symbol for this (15, xv, XV). **3** a size etc. denoted by fifteen. **4** a team of fifteen players, esp. in Rugby football. **5** (**the Fifteen**) *hist.* the Jacobite rebellion of 1715. **6** (**15**) *Brit.* (of films) classified as suitable for persons of 15 years and over. ● *adj.* that amount to fifteen. □ **fifteenth** *adj. & n.* [OE *fíftēne* (as FIVE, -TEEN)]

fifth /fɪfθ/ *n. & adj.* ● *n.* **1** the position in a sequence corresponding to that of the number 5 in the sequence 1–5. **2** something occupying this position. **3** the fifth person etc. in a race or competition. **4** any of five equal parts of a thing. **5** *Mus.* **a** an interval or chord spanning five consecutive notes in the diatonic scale (e.g. C to G). **b** a note separated from another by this interval. **6** *US colloq.* **a** a fifth of a gallon of liquor. **b** a bottle containing this. ● *adj.* that is the fifth. □ **fifth wheel 1** an extra wheel of a coach. **2** a superfluous person or thing. **3** a horizontal turntable over the front axle of a carriage as an extra support to prevent its tipping. **take the fifth** (in the US) exercise the right guaranteed by the Fifth Amendment to the Constitution of refusing to answer questions in order to avoid incriminating oneself. □ **fifthly** *adv.* [earlier and dial. *fift* f. OE *fífta* f. Gmc, assim. to FOURTH]

fifth column *n.* a group of people within a country at war who are sympathetic to or working for its enemies. The term dates from the Spanish Civil War, when General Mola, leading four columns of troops towards Madrid, declared that he had a fifth column inside the city. □ **fifth-columnist** *n.*

Fifth Monarchy the last of the five great kingdoms predicted in Daniel 2:44. □ **Fifth-monarchy-man** *hist.* a 17th-century zealot expecting the immediate second coming of Christ and repudiating all other government.

Fifth Republic 1 the republican regime established in France with de Gaulle's introduction of a new constitution in 1958. **2** this period in France.

fifty /'fɪftɪ/ *n. & adj.* ● *n.* (*pl.* **-ies**) **1** the product of five and ten. **2** a symbol for this (50, l (letter), L). **3** (in *pl.*) the numbers from 50 to 59, esp. the years of a century or of a person's life. **4** a set of fifty persons or things. **5** a large indefinite number (*have fifty things to tell you*). ● *adj.* that amount to fifty. □ **fifty-fifty** *adj.* equal, with equal shares or chances (*on a fifty-fifty basis*). ● *adv.* equally, half and half (*go fifty-fifty*). **fifty-first, -second,** etc. the ordinal numbers between fiftieth and sixtieth. **fifty-one, -two,** etc. the cardinal numbers between fifty and sixty. □ **fiftieth** *adj. & n.* **fiftyfold** *adj. & adv.* [OE *fiftig* (as FIVE, -TY²)]

fig¹ /fɪg/ *n.* **1 a** a soft pear-shaped fruit with many seeds, eaten fresh or dried. **b** (in full **fig tree**) a deciduous tree of the genus *Ficus*; esp. *F. carica*, having broad leaves and bearing figs. **2** a valueless thing (*don't care a fig for*). □ **fig-leaf 1** a leaf of a fig tree. **2** a device for concealing something, esp. the genitals (Gen. 3:7). **not care** (or **give**) **a fig** not care at all. [ME f. OF *figue* f. Prov. *fig(u)a* ult. f. L *ficus*]

fig² /fɪg/ *n. & v.* ● *n.* **1** dress or equipment (*in full fig*). **2** condition or form (*in good fig*). ● *v.tr.* (**figged, figging**) **1** (foll. by *out*) dress up (a person). **2** (foll. by *out, up*) make (a horse) lively. [var. of obs. *feague* (v.) f. G *fegen*: see FAKE¹]

fig. *abbr.* figure.

fight /faɪt/ *v. & n.* ● *v.* (*past* and *past part.* **fought** /fɔːt/) **1** *intr.* **a** (often foll. by *against, with*) contend or struggle in war, battle, single combat, etc. **b** (often foll. by *with*) argue, quarrel. **2** *tr.* contend with (an opponent) in this way. **3** *tr.* take part or engage in (a battle, war, duel, etc.). **4** *tr.* contend about (an issue, an election); maintain (a lawsuit, cause, etc.) against an opponent. **5** *intr.* campaign or strive determinedly to achieve something. **6** *tr.* strive to overcome (disease, fire, fear, etc.). **7** *tr.* make (one's way) by fighting. **8** *tr.* cause (cocks or dogs) to fight. **9** *tr.* handle (troops, a ship, etc.) in battle. ● *n.* **1 a** a combat, esp. unpremeditated, between two or more persons, animals, or parties. **b** a boxing-match. **c** a battle. **d** an argument. **2** a conflict or struggle; a vigorous effort in the face of difficulty. **3** power or inclination to fight (*has no fight left*; *showed fight*). □ **fight back 1** counter-attack. **2** suppress (one's feelings, tears, etc.). **fight down** suppress (one's feelings, tears, etc.). **fight for 1** fight on behalf of. **2** fight to secure (a thing). **fighting chair** US a fixed chair on a boat for use when catching large fish. **fighting chance** an opportunity of succeeding by great effort. **fighting fish** (in full **Siamese fighting fish**) a freshwater fish, *Betta splendens*, native to Thailand, the males of which fight vigorously. **fighting fit** fit enough to fight; at the peak of fitness. **fighting fund** money raised to support a campaign. **fighting-top** *Naut.* a circular gun-platform high on a warship's mast. **fighting words** *colloq.* words indicating a willingness to fight. **fight off** repel with effort. **fight out** (usu. **fight it out**) settle (a dispute etc.) by fighting. **fight shy of** avoid; be unwilling to approach (a person, task, etc.). **make a fight of it** (or **put up a fight**) offer resistance. [OE *feohtan, feoht(e)*, f. WG]

fighter /'faɪtə(r)/ *n.* **1** a person or animal that fights. **2** a fast military aircraft designed for attacking other aircraft. □ **fighter-bomber** an aircraft serving as both fighter and bomber.

figment /'fɪgmənt/ *n.* a thing invented or existing only in the imagination. [ME f. L *figmentum*, rel. to *fingere* fashion]

figura /fɪ'gjʊərə/ *n.* a person or thing representing or symbolizing a fact etc. [mod.L f. L, = FIGURE]

figural /'fɪgərəl, 'fɪgjʊr-/ *adj.* **1** figurative. **2** relating to figures or shapes. **3** *Mus.* florid in style. [OF *figural* or LL *figuralis* f. *figura* FIGURE]

figurant /'fɪgjʊˌrɒn/ *n.* (*fem.* **figurante** /-ˌrɒnt/) a ballet-dancer appearing only in a group. [F, pres. part. of *figurer* FIGURE]

figurante /ˌfɪgjʊ'ræntɪ/ *n.* (*pl.* **figuranti** *pronunc.* same) = FIGURANT. [It., pres. part. of *figurare* FIGURE]

figuration /ˌfɪgjʊ'reɪʃ(ə)n/ *n.* **1 a** the act of formation. **b** a mode of formation; a form. **c** a shape or outline. **2 a** ornamentation by designs. **b** *Mus.* ornamental patterns of scales, arpeggios, etc., often derived from an earlier motif. **3** allegorical representation. [ME f. F or f. L *figuratio* (as FIGURE)]

figurative /'fɪgjʊrətɪv, 'fɪgər-/ *adj.* **1 a** metaphorical, not literal. **b** metaphorically so called. **2** characterized by or addicted to figures of speech. **3** of pictorial or sculptural representation. **4** emblematic,

serving as a type. □ **figuratively** *adv.* **figurativeness** *n.* [ME f. LL *figurativus* (as FIGURE)]

figure /'fɪgə(r)/ *n. & v.* ● *n.* **1 a** the external form or shape of a thing. **b** bodily shape (*has a well-developed figure*). **2 a** a person as seen in outline but not identified (*saw a figure leaning against the door*). **b** a person as contemplated mentally (*a public figure*). **3** appearance as giving a certain impression (*cut a poor figure*). **4 a** a representation of the human form in drawing, sculpture, etc. **b** an image or likeness. **c** an emblem or type. **5** *Geom.* a two-dimensional space enclosed by a line or lines, or a three-dimensional space enclosed by a surface or surfaces; any of the classes of these, e.g. the triangle, the sphere. **6 a** a numerical symbol, esp. any of the ten in Arabic notation. **b** a number so expressed. **c** an amount of money, a value (*cannot put a figure on it*). **d** (in *pl.*) arithmetical calculations. **7** a diagram or illustrative drawing. **8** a decorative pattern. **9 a** a division of a set dance, an evolution. **b** (in skating) a prescribed pattern of movements from a stationary position. **10** *Mus.* a short succession of notes producing a single impression, a brief melodic or rhythmic formula out of which longer passages are developed. **11** (in full **figure of speech**) a recognized form of rhetorical expression giving variety, force, etc., esp. metaphor or hyperbole. **12** *Gram.* a permitted deviation from the usual rules of construction, e.g. ellipsis. **13** *Logic* the form of a syllogism, classified according to the position of the middle term. **14** a horoscope. ● *v.* **1** *intr.* appear or be mentioned, esp. prominently. **2** *tr.* represent in a diagram or picture. **3** *tr.* imagine; picture mentally. **4** *tr.* **a** embellish with a pattern (*figured satin*). **b** *Mus.* embellish with figures. **5** *tr.* mark with numbers (*figured bass*) or prices. **6 a** *tr.* calculate. **b** *intr.* do arithmetic. **7** *tr.* be a symbol of, represent typically. **8** esp. *N. Amer.* **a** *tr.* understand, ascertain, consider. **b** *intr. colloq.* be likely or understandable (*that figures*). □ **figured bass** *Mus.* = CONTINUO. **figure of eight** see EIGHT *n.* 3. **figure of fun** a ridiculous person. **figure on** US count on, expect. **figure out 1** work out by arithmetic or logic. **2** estimate. **3** understand. **figure-skater** a person who practises figure-skating. **figure-skating** skating in prescribed patterns from a stationary position. □ **figureless** *adj.* [ME f. OF *figure* (n.), *figurer* (v.) f. L *figura*, *figurare*, rel. to *fingere* fashion]

figurehead /'fɪgəˌhed/ *n.* **1** a nominal leader or head without real power. **2** a carving, usu. a bust or a full-length figure, at a ship's prow.

figurine /ˌfɪgə'riːn/ *n.* a statuette. [F f. It. *figurina* dimin. of *figura* FIGURE]

figwort /'fɪgwɜːt/ *n.* a plant of the genus *Scrophularia*, with dull purplish-brown flowers, once believed to be useful against scrofula.

Fiji /'fiːdʒiː/ a country in the South Pacific consisting of a group of some 840 islands, of which about a hundred are inhabited; pop. (est. 1990) 800,000; languages, English (official), Fijian, Hindi; capital, Suva. The population contains an almost equal mix of indigenous Pacific islanders and Indians descended from those brought in to work the sugar plantations in the 19th century. Discovered by Tasman in 1643 and visited by Captain James Cook in 1774, the Fiji Islands became a British Crown Colony in 1874 and an independent Commonwealth state in 1970. In 1987, following a coup, Fiji became a republic and withdrew from the Commonwealth.

Fijian /fɪ'dʒiːən/ *n. & adj.* ● *n.* **1** a native or national of Fiji. **2** the Malayo-Polynesian language of the Fijians. ● *adj.* of or relating to Fiji, the Fijians, or their language.

filagree var. of FILIGREE.

filament /'fɪləmənt/ *n.* **1** a slender threadlike body or fibre (esp. in animal or vegetable structures). **2** a conducting wire or thread with a high melting-point in an electric bulb or thermionic valve, heated or made incandescent by an electric current. **3** *Bot.* the part of the stamen that supports the anther. **4** *archaic* (of air, light, etc.) a notional train of particles following each other. □ **filamented** /-ˌmentɪd/ *adj.* **filamentary** /ˌfɪlə'mentərɪ/ *adj.* **filamentous** *adj.* [F *filament* or mod.L *filamentum* f. LL *filare* spin f. L *filum* thread]

filaria /fɪ'leərɪə/ *n.* (*pl.* **filariae** /-rɪˌiː/) a threadlike parasitic nematode worm of the family Filariidae, introduced into the blood by certain biting flies and mosquitoes. □ **filarial** *adj.* [mod.L f. L *filum* thread]

filariasis /ˌfɪlə'raɪəsɪs, fɪˌleərɪ'eɪsɪs/ *n.* (*pl.* **filariases** /-ˌsiːz/) *Med.* a tropical disease caused by the presence of filarial worms esp. in the lymph vessels.

filature /'fɪlətʃə(r)/ *n.* an establishment for or the action of reeling silk from cocoons. [F f. It. *filatura* f. *filare* spin]

filbert /'fɪlbət/ *n.* **1** a cultivated hazel, esp. *Corylus maxima*, bearing relatively elongated edible nuts. **2** this nut. **3** (in full **filbert brush**) an oval brush used in oil painting. [ME *philliberd* etc. f. AF *philbert*, dial. F *noix de filbert*, a nut ripe about St Philibert's day (20 Aug.)]

filch /fɪltʃ/ v.tr. pilfer, steal. □ **filcher** n. [16th-c. thieves' sl.: orig. unkn.]

file¹ /faɪl/ n. & v. ● n. **1** a folder, box, etc., for holding loose papers, esp. arranged for reference. **2** a set of papers kept in this. **3** Computing a collection of (usu. related) data stored under one name. **4** a series of issues of a newspaper etc. in order. **5** a stiff pointed wire on which documents etc. are impaled for keeping. ● v.tr. **1** place (papers) in a file or among (esp. public) records; classify or arrange (papers etc.). **2** submit (a petition for divorce, an application for a patent, etc.) to the appropriate authority. **3** (of a reporter) send (a story, information, etc.) to a newspaper. □ **file away** place in a file, or make a mental note of, for future reference. **filing cabinet** a case with drawers for storing documents. **file server** Computing a device which controls access to one or more separately stored files. **on file** in a file or filing system. □ **filer** n. [F fil f. L filum thread]

file² /faɪl/ n. & v. ● n. **1** a line of persons or things one behind another. **2** a small detachment of men (now usu. two). **3** Chess a line of squares from player to player (cf. RANK¹ n. 6). ● v.intr. walk in a file. □ **file off** (or **away**) Mil. go off by files. [F file f. LL filare spin or L filum thread]

file³ /faɪl/ n. & v. ● n. a tool with a roughened surface or surfaces, usu. of steel, for smoothing or shaping wood, fingernails, etc. ● v.tr. **1** smooth or shape with a file. **2** elaborate or improve (a thing, esp. a literary work). □ **file away** remove (roughness etc.) with a file. □ **filer** n. [OE fil f. WG]

filefish /ˈfaɪlfɪʃ/ n. a fish of the family Balistidae, with a spiny dorsal fin and rough skin.

filet /ˈfɪlɪt/ n. **1** a kind of net or lace with a square mesh. **2** a fillet of meat. □ **filet mignon** /ˌfɪleɪ ˈmiːnjɒn/ a small tender piece of beef from the end of the undercut. [F, = thread]

filial /ˈfɪlɪəl/ adj. **1** of or due from a son or daughter. **2** Biol. bearing the relation of offspring (cf. F² 4). □ **filially** adv. [ME f. OF filial or LL filialis f. filius son, filia daughter]

filiation /ˌfɪlɪˈeɪʃ(ə)n/ n. **1** being the child of one or two specified parents. **2** (often foll. by from) descent or transmission. **3** the formation of offshoots. **4** a branch of a society or language. **5** a genealogical relation or arrangement. [F f. LL filiatio -onis f. L filius son]

filibeg /ˈfɪlɪbeg/ n. (also **philabeg** /ˈfɪləbeg/) Sc. a kilt. [Gael. feileadh-beag little fold]

filibuster /ˈfɪlɪbʌstə(r)/ n. & v. ● n. **1 a** the obstruction of progress in a legislative assembly, esp. by prolonged speaking. **b** esp. N. Amer. a person who engages in a filibuster. **2** esp. hist. a person engaging in unauthorized warfare against a foreign state. ● v. **1** intr. act as a filibuster. **2** tr. act in this way against (a motion etc.). [ult. f. Du. vrijbuiter FREEBOOTER, infl. by F flibustier, Sp. filibustero]

filigree /ˈfɪlɪgriː/ n. (also **filagree** /ˈfɪləgriː/) **1** ornamental work of fine (usu. gold or silver) wire formed into delicate tracery; fine metal openwork. **2** anything delicate resembling this. □ **filigreed** adj. [earlier filigreen, filigrane f. F filigrane f. It. filigrana f. L filum thread + granum seed]

filing /ˈfaɪlɪŋ/ n. (usu. in pl.) a particle rubbed off by a file.

Filioque /ˌfiːlɪˈəʊkwiː/ the word (L, = and from the Son) inserted in the Western version of the Nicene Creed to assert the doctrine of the procession of the Holy Ghost from the Son as well as from the Father, which is not admitted by the Eastern Church. It was one of the central issues in the Great Schism of 1054.

Filipino /ˌfɪlɪˈpiːnəʊ/ n. & adj. ● n. (pl. **-os**; fem. **Filipina** /-nə/) **1** a native or national of the Philippines. **2** a language of the Philippines, a standardized form of Tagalog resembling Pilipino. ● adj. of or relating to the Philippines, their people, or the language Filipino. [Sp., = Philippine]

Filippoi see PHILIPPI.

fill /fɪl/ v. & n. ● v. **1** tr. & intr. (often foll. by with) make or become full. **2** tr. occupy completely; spread over or through; pervade. **3** tr. block up (a cavity or hole in a tooth) with cement, amalgam, gold, etc.; drill and put a filling into (a decayed tooth). **4** tr. appoint a person to hold (a vacant post). **5** tr. hold (a position); discharge the duties of (an office). **6** tr. carry out or supply (an order, commission, etc.). **7** tr. occupy (vacant time). **8** intr. (of a sail) be distended by wind. **9** tr. (usu. as **filling** adj.) (esp. of food) satisfy, satiate. **10** tr. satisfy, fulfil (a need or requirement). **11** tr. Poker etc. complete (a holding) by drawing the necessary cards. **12** tr. stock abundantly. ● n. **1** (prec. by possessive) as much as one wants or can bear (eat your fill). **2** enough to fill something (a fill of tobacco). **3** material used for filling. □ **fill the bill** be suitable or adequate. **fill in 1** add information to complete (a form, document,

blank cheque, etc.). **2 a** complete (a drawing etc.) within an outline. **b** fill (an outline) in this way. **3** fill (a hole etc.) completely. **4** (often foll. by for) act as a substitute. **5** occupy oneself during (time between other activities). **6** colloq. inform (a person) more fully. **7** sl. thrash, beat. **fill out 1** enlarge to the required size. **2** become enlarged or plump. **3** US fill in (a document etc.). **fill up 1** make or become completely full. **2** fill in (a document etc.). **3** fill the petrol tank of (a car etc.). **4** provide what is needed to occupy vacant parts or places or deal with deficiencies in. **5** do away with (a pond etc.) by filling. **fill-up** n. **1** a thing that fills something up. **2** an act of filling up a petrol tank etc. [OE fyllan f. Gmc, rel. to FULL¹]

fille de joie /ˌfiː də ˈʒwa/ n. (pl. **filles de joie** pronunc. same) a prostitute. [F, lit. 'daughter of joy']

filler¹ /ˈfɪlə(r)/ n. **1** material or an object used to fill a cavity or increase bulk. **2** an item filling space in a newspaper etc. **3** a person or thing that fills. □ **filler cap** a cap closing the pipe leading to the petrol tank of a motor vehicle.

filler² /ˈfɪlɜː(r)/ n. (pl. same) a monetary unit of Hungary, equal to one-hundredth of a forint. [Hungarian fillér]

fillet /ˈfɪlɪt/ n. & v. ● n. **1 a** a fleshy boneless piece of meat from near the loins or the ribs. **b** (in full **fillet steak**) the undercut of a sirloin. **c** a boned longitudinal section of a fish. **2 a** a headband, ribbon, string, or narrow band, for binding the hair or worn round the head. **b** a band or bandage. **3 a** a thin narrow strip of anything. **b** a raised rim or ridge on any surface. **4** Archit. **a** a narrow flat band separating two mouldings. **b** a small band between the flutes of a column. **5** Carpentry an added triangular piece of wood to round off an interior angle. **6 a** a plain line impressed on the cover of a book. **b** a roller used to impress this. **7** Heraldry a horizontal division of a shield, a quarter of the depth of a chief. ● v.tr. (**filleted**, **filleting**) **1 a** remove bones from (fish or meat). **b** divide (fish or meat) into fillets. **2** bind or provide with a fillet or fillets. **3** encircle with an ornamental band. □ **filleter** n. [ME f. OF filet f. Rmc dimin. of L filum thread]

filling /ˈfɪlɪŋ/ n. **1** material that fills or is used to fill, esp.: **a** a piece of material used to fill a cavity in a tooth. **b** the edible substance between the bread in a sandwich or between the pastry in a pie. **2** US weft. □ **filling-station** an establishment selling petrol etc. to motorists.

fillip /ˈfɪlɪp/ n. & v. ● n. **1** a stimulus or incentive. **2 a** a sudden release of a finger or thumb when it has been bent and checked by a thumb or finger. **b** a slight smart stroke given in this way. ● v. (**filliped**, **filliping**) **1** tr. stimulate (fillip one's memory). **2** tr. strike slightly and smartly. **3** tr. propel (a coin, marble, etc.) with a fillip. **4** intr. make a fillip. [imit.]

fillis /ˈfɪlɪs/ n. loosely twisted string used as a horticultural tying material. [F filasse tow]

fillister /ˈfɪlɪstə(r)/ n. a rabbet or rabbet plane for window-sashes etc. [19th c.: perh. f. F feuilleret]

Fillmore /ˈfɪlmɔː(r)/, Millard (1800–74), American Whig statesman, 13th President of the US 1850–3. He succeeded to the presidency on the death of Zachary Taylor. Fillmore was an advocate of compromise on the slavery issue. However, his unpopular enforcement of the 1850 Fugitive Slave Act hastened the end of the Whig Party.

filly /ˈfɪlɪ/ n. (pl. **-ies**) **1** a young female horse, usu. before it is four years old. **2** colloq. a girl or young woman. [ME, prob. f. ON fylja f. Gmc (as FOAL)]

film /fɪlm/ n. & v. ● n. **1** a thin coating or covering layer. **2** Photog. a strip or sheet of plastic or other flexible base coated with light-sensitive emulsion for exposure in a camera, either as individual visual representations or as a sequence which form the illusion of movement when shown in rapid succession. **3 a** a representation of a story, episode, etc., on a film, with the illusion of movement. **b** a story represented in this way. **c** (in pl.) the cinema industry. **4** a slight veil or haze etc. **5** a dimness or abnormal opacity affecting the eyes. **6** a fine thread or filament. ● v. **1 a** tr. make a photographic film of (a scene, person, etc.). **b** tr. (also absol.) make a cinema or television film of (a book etc.). **c** tr. be (well or ill) suited for reproduction on film. **2** tr. & intr. cover or become covered with or as with a film. □ **film-goer** a person who frequents the cinema. **film star** a celebrated actor or actress in films. **film-strip** a series of transparencies in a strip for projection. [OE filmen membrane f. WG f. fell, rel. to FELL⁵]

filmic /ˈfɪlmɪk/ adj. of or relating to films or cinematography.

film noir /film ˈnwɑː(r)/ n. a style or genre of cinematographic film marked by a mood of pessimism, fatalism, and menace. The term was originally applied (by a group of French critics) to American thriller or detective films made in the period 1944–54 and to the work of directors such as Orson Welles, Fritz Lang, and Billy Wilder. The films are often seen as reflecting postwar disillusion, and unease at the onset of the cold war and the McCarthy era. [F, lit. 'black film']

filmography /fɪlˈmɒɡrəfɪ/ n. (pl. **-ies**) a list of films by one director etc. or on one subject. [FILM + -GRAPHY after *bibliography*]

filmset /ˈfɪlmset/ v.tr. (**-setting**; past and past part. **-set**) Printing set (material for printing) by filmsetting. □ **filmsetter** n.

filmsetting /ˈfɪlmˌsetɪŋ/ n. Printing typesetting using characters on photographic film.

filmy /ˈfɪlmɪ/ adj. (**filmier, filmiest**) **1** thin and translucent. **2** covered with or as with a film. □ **filmily** adv. **filminess** n.

filo /ˈfiːləʊ/ n. dough that can be stretched into very thin layers; pastry made from this dough. [mod.Gk *phullo* leaf]

Filofax /ˈfaɪləʊˌfæks/ n. propr. a portable loose-leaf filing system for personal or office use. [FILE[1] + *facts* pl. of FACT]

filoplume /ˈfɪləˌpluːm, ˈfaɪl-/ n. Zool. a slender feather of a type found between and beneath the flight feathers of some birds. [mod.L *filopluma* f. L *filum* thread, *pluma* feather]

filoselle /ˈfɪləˌsel/ n. floss silk. [F]

fils[1] /fiːs/ n. (added to a surname to distinguish a son from a father of the same name) the son, junior (cf. PÈRE). [F, = son]

fils[2] /fɪls/ n. a monetary unit in certain countries of the Middle East and North Africa. [colloq. pronunc. of Arab. *fals* small copper coin]

filter /ˈfɪltə(r)/ n. & v. ● n. **1** a porous device for removing impurities or solid particles from a liquid or gas passed through it. **2** = *filter tip*. **3** a screen or attachment for absorbing or modifying light, X-rays, etc. **4** a device for suppressing electrical or sound waves of frequencies not required. **5** Brit. **a** an arrangement for filtering traffic. **b** a traffic light signalling this. ● v. **1** tr. & intr. pass or cause to pass through a filter. **2** tr. (foll. by *out*) remove (impurities etc.) by means of a filter. **3** intr. (foll. by *through, into*, etc.) make way gradually. **4** intr. (foll. by *out*) leak or cause to leak. **5** tr. & intr. Brit. allow (traffic) or (of traffic) be allowed to pass to the left or right at a junction while traffic going straight ahead is halted (esp. at traffic lights). □ **filter-bed** a tank or pond containing a layer of sand etc. for filtering large quantities of liquid. **filter-feeding** Zool. feeding by filtering out plankton or nutrients suspended in water. **filter-paper** porous paper for filtering. **filter tip** a filter attached to a cigarette for removing impurities from the inhaled smoke. **2** a cigarette with this. **filter-tipped** having a filter tip. [F *filtre* f. med.L *filtrum* felt used as a filter, f. WG]

filterable /ˈfɪltərəb(ə)l/ adj. (also **filtrable** /ˈfɪltrəb(ə)l/) **1** Med. (of a virus) able to pass through a filter that retains bacteria. **2** that can be filtered.

filth /fɪlθ/ n. **1** repugnant or extreme dirt; excrement, refuse. **2** vileness, corruption, obscenity. **3** foul or obscene language. **4** (prec. by *the*) sl. derog. the police. [OE *fylth* (as FOUL, -TH[2])]

filthy /ˈfɪlθɪ/ adj. & adv. ● adj. (**filthier, filthiest**) **1** extremely or disgustingly dirty. **2** obscene. **3** colloq. (of weather) very unpleasant. **4** vile; disgraceful. ● adv. **1** filthily (*filthy dirty*). **2** colloq. extremely (*filthy rich*). □ **filthy lucre 1** dishonourable gain (Tit. 1:11). **2** joc. money. □ **filthily** adv. **filthiness** n.

filtrable var. of FILTERABLE.

filtrate /ˈfɪltreɪt/ v. & n. ● v.tr. filter. ● n. filtered liquid. □ **filtration** /fɪlˈtreɪʃ(ə)n/ n. [mod.L *filtrare* (as FILTER)]

fimbriate /ˈfɪmbrɪˌeɪt/ adj. (also **fimbriated** /-ˌeɪtɪd/) **1** Bot. & Zool. fringed or bordered with hairs etc. **2** Heraldry having a narrow border. [L *fimbriatus* f. *fimbriae* fringe]

fin /fɪn/ n. & v. ● n. **1** a flattened appendage on various parts of the body of many aquatic vertebrates and some invertebrates, including fish and cetaceans, for propelling, steering, and balancing (*dorsal fin; anal fin*). **2** a small projecting surface or attachment on an aircraft, rocket, or motor car for ensuring aerodynamic stability. **3** an underwater swimmer's flipper. **4** a sharp lateral projection on the share or coulter of a plough. **5** a finlike projection on any device, for improving heat transfer etc. ● v. (**finned, finning**) **1** tr. provide with fins. **2** intr. swim under water. □ **fin whale** (or **fin-back**) a large rorqual, *Balaenoptera physalus*, with a prominent dorsal fin. □ **finless** adj. **finned** adj. (also in comb.). [OE *fin(n)*]

finable see FINE[2].

finagle /fɪˈneɪɡ(ə)l/ v.intr. & tr. colloq. act or obtain dishonestly. □ **finagler** n. [dial. *fainaigue* cheat]

final /ˈfaɪn(ə)l/ adj. & n. ● adj. **1** situated at the end, coming last. **2** conclusive, decisive, unalterable, putting an end to doubt.

3 concerned with the purpose or end aimed at. ● n. **1** the last or deciding heat or game in sports or in a competition (*Cup Final*). **2** the edition of a newspaper published latest in the day. **3** (usu. in pl.) the series of examinations at the end of a degree course. **4** Mus. the principal note in any mode. □ **final cause** Philos. the end towards which a thing naturally develops or at which an action aims. **final clause** Gram. a clause expressing purpose, introduced by *in order that, lest*, etc. **final drive** the last part of the transmission system in a motor vehicle. □ **finally** adv. [ME f. OF or f. L *finalis* f. *finis* end]

finale /fɪˈnɑːlɪ, -leɪ/ n. **1 a** the last movement of an instrumental composition. **b** a piece of music closing an act in an opera. **2** the close of a drama etc. **3** a conclusion. [It. (as FINAL)]

finalism /ˈfaɪnəˌlɪz(ə)m/ n. the doctrine that natural processes (e.g. evolution) are directed towards some goal. □ **finalistic** /ˌfaɪnəˈlɪstɪk/ adj.

finalist /ˈfaɪnəlɪst/ n. a competitor in the final of a competition etc.

finality /faɪˈnælɪtɪ/ n. (pl. **-ies**) **1** the quality or fact of being final. **2** the belief that something is final. **3** a final act, state, or utterance. **4** the principle of final cause viewed as operative in the universe. [F *finalité* f. LL *finalitas -tatis* (as FINAL)]

finalize /ˈfaɪnəˌlaɪz/ v.tr. (also **-ise**) **1** put into final form. **2** complete; bring to an end. **3** approve the final form or details of. □ **finalization** /ˌfaɪnəlaɪˈzeɪʃ(ə)n/ n.

final solution the policy under the German Nazi regime of exterminating European Jews. Introduced by Heinrich Himmler and administered by Adolf Eichmann, the policy resulted in the murder of 6 million Jews in concentration camps between 1941 and 1945.

finance /ˈfaɪnæns, fɪˈnæns, faɪˈnæns/ n. & v. ● n. **1** the management of (esp. public) money. **2** monetary support for an enterprise. **3** (in pl.) the money resources of a state, company, or person. ● v.tr. provide capital for (a person or enterprise). □ **finance company** (or **house**) a company concerned mainly with providing money for hire-purchase transactions. [ME f. OF *finer* settle a debt f. *fin* end: see FINE[2]]

financial /faɪˈnænʃ(ə)l, fɪ-/ adj. **1** of finance. **2** Austral. & NZ sl. possessing money. □ **financial year** a year as reckoned for taxing or accounting (e.g. the British tax year, reckoned from 6 Apr.). □ **financially** adv.

Financial Times index (abbr. **FT–SE**) (also called *FT Index*, (colloq.) *Footsie*) in Britain, the Financial Times–Stock Exchange 100 share index, an index based on the share values of Britain's one hundred largest public companies, which was set up in 1984.

financier n. & v. /faɪˈnænsɪə(r), fɪ-/ a person engaged in large-scale finance. ● v.intr. /ˌfaɪnænˈsɪə(r), ˌfɪ-/ usu. derog. conduct financial operations. [F (as FINANCE)]

finch /fɪntʃ/ n. a small seed-eating songbird, usu. having a short stout bill; esp. one of the family Fringillidae, which includes canaries, chaffinches, crossbills, etc. [OE *finc* f. WG]

find /faɪnd/ v. & n. ● v.tr. (past and past part. **found** /faʊnd/) **1 a** discover by chance or effort (*found a key*). **b** become aware of. **c** (absol.) discover game, esp. a fox. **2 a** get possession of by chance (*found a treasure*). **b** obtain, receive (*idea found acceptance*). **c** succeed in obtaining (*cannot find the money; can't find time to read*). **d** summon up (*found courage to protest*). **e** sl. steal. **3 a** seek out and provide (*will find you a book*). **b** supply, furnish (*each finds his own equipment*). **4** ascertain by study or calculation or inquiry (*could not find the answer*). **5 a** perceive or experience (*find no sense in it; find difficulty in breathing*). **b** (often in passive) recognize or discover to be present (*the word is not found in Shakespeare*). **c** regard or discover from experience (*finds England too cold; you'll find it pays; find it impossible to reply*). **6** Law (of a jury, judge, etc.) decide and declare (*found him guilty; found that he had done it; found it murder*). **7** reach by a natural or normal process (*water finds its own level*). **8 a** (of a letter) reach (a person). **b** (of an address) be adequate to enable a letter etc. to reach (a person). **9** archaic reach the conscience of. ● n. **1 a** a discovery of treasure, minerals, etc. **b** Hunting the finding of a fox. **2** a thing or person discovered, esp. when of value. □ **all found** (of an employee's wages) with board and lodging provided free. **find against** Law decide against (a person), judge to be guilty. **find fault** see FAULT. **find favour** prove acceptable. **find one's feet 1** become able to walk. **2** develop one's independent ability. **find for** Law decide in favour of (a person), judge to be innocent. **find it in one's heart** (esp. with neg.; foll. by to + infin.) prevail upon oneself, be willing. **find oneself 1** discover that one is (*woke to find myself in hospital; found herself agreeing*). **2** discover one's vocation. **3** provide for one's own needs. **find out 1** discover or detect (a wrongdoer etc.). **2** (often foll. by *about*) get information (*find out about holidays abroad*). **3** discover (*find out where we are*). **4** (often foll. by *about*) discover the truth, a fact, etc. (*he never found out*). **5** devise.

6 solve. **find-spot** *Archaeol.* the place where an object is found. **find one's way 1** (often foll. by *to*) manage to reach a place. **2** (often foll. by *into*) be brought or get. □ **findable** *adj.* [OE *findan* f. Gmc]

finder /ˈfaɪndə(r)/ *n.* **1** a person who finds. **2** a small telescope attached to a large one to locate an object for observation. **3** the viewfinder of a camera. □ **finders keepers** *colloq.* whoever finds a thing is entitled to keep it.

fin de siècle /ˌfæ̃ də ˈsjɛklə/ *adj.* **1** characteristic of the end of the nineteenth century. **2** decadent. [F, = end of century]

finding /ˈfaɪndɪŋ/ *n.* **1** (often in *pl.*) a conclusion reached by an inquiry. **2** (in *pl.*) *US* small parts or tools used by workmen.

fine[1] /faɪn/ *adj., n., adv., & v.* ● *adj.* **1** of high quality (*they sell fine fabrics*). **2 a** excellent; of notable merit (*a fine painting*). **b** good, satisfactory (*that will be fine*). **c** fortunate (*has been a fine thing for him*). **d** well conceived or expressed (*a fine saying*). **3 a** pure, refined. **b** (of gold or silver) containing a specified proportion of pure metal. **4** of handsome appearance or size; imposing, dignified (*fine buildings; a person of fine presence*). **5** in good health (*I'm fine, thank you*). **6** (of weather etc.) bright and clear with sunshine; free from rain. **7 a** thin; sharp. **b** in small particles. **c** worked in slender thread. **d** (esp. of print) small. **e** (of a pen) narrow-pointed. **8** *Cricket* behind the wicket and near the line of flight of the ball. **9** tritely complimentary; euphemistic (*say fine things about a person; call things by fine names*). **10** ornate, showy, smart. **11** fastidious, dainty, pretending refinement; (of speech or writing) affectedly ornate. **12 a** capable of delicate perception or discrimination. **b** perceptible only with difficulty (*a fine distinction*). **13 a** delicate, subtle, exquisitely fashioned. **b** (of feelings) refined, elevated. **14** (of wine or other goods) of a high standard; conforming to a specified grade. ● *n.* **1** fine weather (*in rain or fine*). **2** (in *pl.*) very small particles in mining, milling, etc. ● *adv.* **1** finely. **2** *colloq.* very well (*suits me fine*). ● *v.* **1** (often foll. by *down*) **a** *tr.* make (beer or wine) clear. **b** *intr.* (of liquid) become clear. **2** *tr.* & *intr.* (often foll. by *away, down, off*) make or become finer, thinner, or less coarse; dwindle or taper, or cause to do so. □ **cut** (or **run**) **it fine** allow very little margin of time etc. **fine arts** those appealing to the mind or to the sense of beauty, as poetry, music, and esp. painting, sculpture, and architecture. **fine chemicals** see CHEMICAL. **fine-draw** sew together (two pieces of cloth, edges of a tear, parts of a garment) so that the join is imperceptible. **fine-drawn 1** extremely thin. **2** subtle. **fine print** detailed printed information, esp. in legal documents, instructions, etc. **fine-spun 1** delicate. **2** (of a theory etc.) too subtle, unpractical. **fine-tooth comb** a comb with narrow close-set teeth. **fine-tune** make small adjustments to (a mechanism etc.) in order to obtain the best possible results. **fine up** *Austral. colloq.* (of the weather) become fine. **go over with a fine-tooth comb** check or search thoroughly. **not to put too fine a point on it** (as a parenthetic remark) to speak bluntly. □ **finely** *adv.* **fineness** *n.* [ME f. OF *fin* ult. f. L *finire* finish]

fine[2] /faɪn/ *n. & v.* ● *n.* **1** a sum of money exacted as a penalty. **2** *hist.* a sum of money paid by an incoming tenant in return for the rent's being small. ● *v.tr.* punish by a fine (*fined him £5*). □ **in fine** to sum up; in short. □ **finable** *adj.* [ME f. OF *fin* f. med.L *finis* sum paid on settling a lawsuit f. L *finis* end]

fine[3] /fiːn/ *n.* = FINE CHAMPAGNE. [abbr.]

fine champagne /ˌfiːn ʃæmˈpɑ̃ːnjə/ *n.* old liqueur brandy. [F, = fine (brandy from) Champagne (vineyards in Charente)]

Fine Gael /ˌfiːnə ˈɡeɪl/ one of the major political parties in the Republic of Ireland, the major rival of Fianna Fáil. Founded in 1923 and first known as Cumann na nGaedheal, it changed its name in 1933. It has advocated the concept of a united Ireland achieved by peaceful means. [Ir., = tribe of Gaels]

finery[1] /ˈfaɪnərɪ/ *n.* showy dress or decoration. [FINE[1] + -ERY, after BRAVERY]

finery[2] /ˈfaɪnərɪ/ *n.* (*pl.* **-ies**) *hist.* a hearth where pig-iron was converted into wrought iron. [F *finerie* f. *finer* refine, FINE[1]]

fines herbes /fiːn ˈzɛəb/ *n.pl.* mixed herbs used in cooking, esp. chopped as omelette-flavouring. [F, = fine herbs]

finesse /fɪˈnes/ *n. & v.* ● *n.* **1** refinement. **2** subtle or delicate manipulation. **3** artfulness, esp. in handling a difficulty tactfully. **4** *Cards* an attempt to win a trick with a card that is not the highest held. ● *v.* **1** *intr.* & *tr.* use or achieve by finesse. **2** *Cards* **a** *intr.* make a finesse. **b** *tr.* play (a card) by way of finesse. **3** *tr.* evade or trick by finesse. [F, rel. to FINE[1]]

Fingal /ˈfɪŋɡ(ə)l/ a character in an epic poem by the Scottish poet James Macpherson (1736–96), based on the legendary Irish hero Finn MacCool but fictionally transformed and depicted as fighting both the Norse invaders and the Romans (under Caracalla) from an invented kingdom in NW Scotland.

Fingal's Cave a cave on the island of Staffa in the Inner Hebrides, noted for the clustered basaltic pillars that form its cliffs. It is said to have been the inspiration of Mendelssohn's overture *The Hebrides* (also known as *Fingal's Cave*) but in fact he noted down the principal theme before his visit to Staffa.

finger /ˈfɪŋɡə(r)/ *n. & v.* ● *n.* **1** any of the terminal projections of the hand (including or excluding the thumb). **2** the part of a glove etc. intended to cover a finger. **3 a** a finger-like object (*fish finger*). **b** a long narrow structure. **4** *colloq.* a measure of liquor in a glass, based on the breadth of a finger. **5** *sl.* **a** an informer. **b** a pickpocket. **c** a police officer. ● *v.tr.* **1** touch, feel, or turn about with the fingers. **2** *Mus.* **a** play (a passage) with fingers used in a particular way. **b** mark (music) with signs showing which fingers are to be used. **c** play upon (an instrument) with the fingers. **3** *N. Amer. sl.* indicate (a victim, or a criminal to the police). □ **all fingers and thumbs** clumsy. **finger alphabet** a form of sign language using the fingers. **finger-board** a flat strip at the top end of a stringed instrument, against which the strings are pressed to determine tones. **finger-bowl** (or **-glass**) a small bowl for rinsing the fingers during a meal. **finger-dry** dry and style (the hair) by running one's fingers through it. **finger language** language expressed by means of the finger alphabet. **finger-mark** a mark left on a surface by a finger. **finger-paint** *n.* paint that can be applied with the fingers. ● *v.intr.* apply paint with the fingers. **finger-plate** a plate fixed to a door above the handle to prevent finger-marks. **finger-post** a signpost at a road junction. **one's fingers itch** (often foll. by *to* + infin.) one is longing or impatient. **finger-stall** a cover to protect a finger, esp. when injured. **get** (or **pull**) **one's finger out** *sl.* cease procrastinating and start to act. **give a person the finger** *sl.* make a gesture with the middle finger raised as an obscene sign of contempt. **have a finger in** (or **in the pie**) be (esp. officiously) concerned in (the matter). **lay a finger on** touch however slightly. **point the finger at** *colloq.* accuse, blame. **put one's finger on** locate or identify exactly. **put the finger on** *sl.* **1** inform against. **2** identify (an intended victim). **slip through one's fingers** escape. **twist** (or **wind**) **round one's finger** (or **little finger**) persuade (a person) without difficulty, dominate (a person) completely. **work one's fingers to the bone** see BONE. □ **fingered** *adj.* (also in *comb.*). **fingerless** *adj.* [OE f. Gmc]

fingering[1] /ˈfɪŋɡərɪŋ/ *n.* **1** a manner or technique of using the fingers, esp. to play an instrument. **2** an indication of this in a musical score.

fingering[2] /ˈfɪŋɡərɪŋ/ *n.* fine wool for knitting. [earlier *fingram*, perh. f. F *fin grain*, as GROGRAM f. *gros grain*]

fingerling /ˈfɪŋɡəlɪŋ/ *n.* a parr.

fingernail /ˈfɪŋɡəˌneɪl/ *n.* the nail at the tip of each finger.

fingerprint /ˈfɪŋɡəˌprɪnt/ *n. & v.* ● *n.* **1** an impression made on a surface by the fingertips, esp. as used for identifying individuals. (*See note below.*) **2** a distinctive characteristic. ● *v.tr.* record the fingerprints of (a person).

▪ No two persons have exactly the same pattern of ridges and marks on the fingertips, and the patterns can be classified and recorded systematically. Sir William James Herschel (1833–1917) introduced the use of fingerprints for identification purposes while a magistrate in India in the late 1870s. Fingerprint evidence was first accepted in an English court in 1902 and has been widely used for criminal identification ever since. (See also GENETIC FINGERPRINTING.)

fingertip /ˈfɪŋɡəˌtɪp/ *n.* the tip of a finger. □ **have at one's fingertips** be thoroughly familiar with (a subject etc.).

finial /ˈfɪnɪəl/ *n. Archit.* **1** an ornament finishing off the apex of a roof, pediment, gable, tower-corner, canopy, etc. **2** the topmost part of a pinnacle. [ME f. OF *fin* f. L *finis* end]

finical /ˈfɪnɪk(ə)l/ *adj.* = FINICKY. □ **finicality** /ˌfɪnɪˈkælɪtɪ/ *n.* **finically** *adv.* **finicalness** *n.* [16th c.: prob. orig. university sl. f. FINE[1] + -ICAL]

finicking /ˈfɪnɪkɪŋ/ *adj.* = FINICKY. [FINICAL + -ING[2]]

finicky /ˈfɪnɪkɪ/ *adj.* **1** over-particular, fastidious. **2** needing much attention to detail; fiddly. □ **finickiness** *n.*

finis /ˈfɪnɪs, ˈfiːn-, ˈfaɪn-/ *n.* **1** (at the end of a book) the end. **2** the end of anything, esp. of life. [L]

finish /ˈfɪnɪʃ/ *v. & n.* ● *v.* **1** *tr.* **a** (often foll. by *off*) bring to an end; come to the end of; complete. **b** (usu. foll. by *off*) *colloq.* kill; overcome completely. **c** (often foll. by *off, up*) consume or get through the whole or the remainder of (food or drink) (*finish up your dinner*). **2** *intr.* **a** come to an end, cease. **b** reach the end, esp. of a race. **c** = *finish up*. **3** *tr.*

a complete the manufacture of (cloth, woodwork, etc.) by surface treatment. **b** put the final touches to; make perfect or highly accomplished (*finished manners*). **c** prepare (a girl) for entry into fashionable society. ● *n.* **1 a** the end, the last stage. **b** the point at which a race etc. ends. **c** the death of a fox in a hunt (*be in at the finish*). **2** a method, material, or texture used for surface treatment of wood, cloth, etc. (*mahogany finish*). **3** what serves to give completeness. **4** an accomplished or completed state. □ **fight to the finish** fight till one party is completely beaten. **finishing-school** a private college where girls are prepared for entry into fashionable society. **finish off** provide with an ending. **finish up** (often foll. by *in*, *up*) end in something, end by doing something (*he finished up last in the race; the plan finished up in the waste-paper basket; finished up by apologizing*). **finish with** have no more to do with, complete one's use of or association with. [ME f. OF *fenir* f. L *finire* f. *finis* end]

finisher /ˈfɪnɪʃə(r)/ *n.* **1** a person who finishes something. **2** a worker or machine doing the last operation in manufacture. **3** *colloq.* a discomfiting thing, a crushing blow, etc.

Finisterre, Cape /ˌfɪnɪˈsteə(r)/ a promontory of NW Spain, forming the westernmost point of the mainland. The shipping forecast area *Finisterre* covers part of the Atlantic off NW Spain, west of the Bay of Biscay.

finite /ˈfaɪnaɪt/ *adj.* **1** limited, bounded; not infinite. **2** *Gram.* (of a part of a verb) having a specific number and person. **3** not infinitely small. □ **finitely** *adv.* **finiteness** *n.* **finitude** /ˈfɪnɪtjuːd/ *n.* [L *finitus* past part. of *finire* FINISH]

finitism /ˈfaɪnaɪˌtɪz(ə)m/ *n.* belief in the finiteness of the world, God, etc. □ **finitist** *n.*

fink /fɪŋk/ *n. & v. N. Amer. sl.* ● *n.* **1** an unpleasant person. **2** an informer. **3** a strikebreaker; a blackleg. ● *v.intr.* **1** (foll. by *on*) inform on. **2 a** (foll. by *out*) back out of something. **b** (foll. by *out on*) back out of (a thing); let (a person) down. [20th c.: orig. unkn.]

Finland /ˈfɪnlənd/ (called in Finnish *Suomi*) a country on the Baltic Sea, between Sweden and Russia; pop. (1990) 4,998,500; official languages, Finnish and Swedish; capital, Helsinki. Finland is mainly a lowland country, with extensive forests and many lakes; the northern third of the country lies within the Arctic Circle. Converted to Christianity by Eric IX of Sweden in the 12th century, Finland became an area of Swedish–Russian rivalry. A Grand Duchy from the 16th century, it was ceded to Russia in 1809, becoming an independent republic after the Russian Revolution. Wars with the USSR in 1939–40 and 1941–4 cost Finland Karelia and Petsamo, and the country remained neutral throughout the cold war. Finland's application to join the European Community was approved by the European Parliament in 1994.

Finland, Gulf of an arm of the Baltic Sea between Finland and Estonia, extending eastwards to St Petersburg in Russia.

Finn /fɪn/ *n.* a native or national of Finland; a person of Finnish descent. [OE *Finnas* pl.]

finnan /ˈfɪnən/ *n.* (in full **finnan haddock**) a haddock cured with the smoke of green wood, turf, or peat. [*Findhorn* or *Findon* in Scotland]

finnesko /ˈfɪnəˌskəʊ/ *n.* (*pl.* same) a boot of tanned reindeer-skin with the hair on the outside. [Norw. *finnsko* (as FINN, *sko* SHOE)]

Finnic /ˈfɪnɪk/ *adj.* **1** of the group of peoples related to the Finns. **2** of the group of languages related to Finnish, including Estonian and Lappish.

Finnish /ˈfɪnɪʃ/ *adj. & n.* ● *adj.* of the Finns or their language. ● *n.* the Finno-Ugric language spoken by the Finns. (*See note below.*)

▪ Finnish is spoken by about 4.6 million people in Finland (where it is one of the two official languages), and is also spoken in parts of Russia and Sweden. Closely related to Estonian and distantly related to Hungarian, it is noted for its morphological complexity: a Finnish noun can have thirteen different case-forms.

Finn MacCool /ˌfɪn məˈkuːl/ (also **Finn Mac Cumhaill**) *Ir. Mythol.* the warrior-hero of a cycle of legends about a band of warriors defending Ireland. Father of the legendary Irish warrior and bard Ossian, he is supposed to have lived in the 3rd century AD. (See also FINGAL.)

Finno-Ugric /ˌfɪnəʊˈuːɡrɪk, -ˈjuːɡrɪk/ *adj. & n.* (also **Finno-Ugrian** /-ˈuːɡrɪən, -ˈjuːɡrɪən/) ● *adj.* of or relating to a group of Uralic languages consisting of the Finnic and Ugric branches, and several languages of central Russia, from where they originally spread. (See also FINNIC 2, UGRIC *adj.*) ● *n.* this group of languages.

finny /ˈfɪnɪ/ *adj.* **1** having fins; like a fin. **2** *poet.* of or teeming with fish.

fino /ˈfiːnəʊ/ *n.* (*pl.* **-os**) a light-coloured dry sherry. [Sp., = fine]

fiord /fjɔːd/ *n.* (also **fjord**) a long narrow inlet of sea between high cliffs, as in Norway. [Norw. f. ON *fjörthr* f. Gmc: cf. FIRTH, FORD]

fioritura /ˌfɪˌɔːrɪˈtʊərə/ *n.* (*pl.* **fioriture** /-rɪ/) *Mus.* the usu. improvised decoration of a melody. [It., = flowering f. *fiorire* to flower]

fipple /ˈfɪp(ə)l/ *n.* a plug at the mouth-end of a wind instrument. □ **fipple flute** a flute played by blowing endwise, e.g. a recorder. [17th c.: orig. unkn.]

fir /fɜː(r)/ *n.* **1** (in full **fir-tree**) an evergreen coniferous tree, esp. of the genus *Abies*, with needles borne singly on the stems (cf. PINE¹). **2** the wood of the fir. □ **fir-cone** the fruit of the fir. □ **firry** *adj.* [ME, prob. f. ON *fyri-* f. Gmc]

fire /ˈfaɪə(r)/ *n. & v.* ● *n.* **1 a** the state or process of combustion, in which substances combine chemically with oxygen from the air and usu. give out bright light and heat. **b** the active principle operative in this. **c** flame or incandescence. **2** a conflagration, a destructive burning (*forest fire*). **3 a** burning fuel in a grate, furnace, etc. **b** = *electric fire*. **c** = *gas fire*. **4** firing of guns (*open fire*). **5 a** fervour, spirit, vivacity. **b** poetic inspiration, lively imagination. **c** vehement emotion. **6** burning heat, fever. **7** luminosity, glow (*St Elmo's fire*). ● *v.* **1 a** *tr.* discharge (a gun etc.). **b** *tr.* propel (a missile) from a gun etc. **c** *intr.* (often foll. by *at, into, on*) fire a gun or missile. **d** *tr.* produce (a broadside, salute, etc.) by discharge of guns. **e** *intr.* (of a gun etc.) be discharged. **2** *tr.* cause (explosive) to explode. **3** *tr.* deliver or utter in rapid succession (*fired insults at us*). **4** *tr. sl.* dismiss (an employee) from a job. **5** *tr.* **a** set fire to with the intention of destroying. **b** kindle (explosives). **6** *intr.* catch fire. **7** *intr.* (of an internal-combustion engine, or a cylinder in one) undergo ignition of its fuel. **8** *tr.* supply (a furnace, engine, boiler, or power station) with fuel. **9** *tr.* **a** stimulate (the imagination or emotion). **b** fill (a person) with enthusiasm. **10** *tr.* bake or dry (pottery, bricks, etc.). **b** cure (tea or tobacco) by artificial heat. **11** *intr.* become heated or excited. **12** *tr.* cause to glow or redden. □ **catch fire** begin to burn. **fire-alarm** a device for giving warning of fire. **fire and brimstone** the supposed torments of hell. **fire away** *colloq.* begin; go ahead. **fire-ball 1** a large meteor. **2** a ball of flame, esp. from a nuclear explosion. **3** an energetic person. **4** ball lightning. **5** *Mil. hist.* a ball filled with combustibles. **fire-balloon** a balloon made buoyant by the heat of a fire burning at its mouth. **fire-blight** a disease of plants, esp. hops and fruit trees, causing a scorched appearance. **fire-bomb** *n.* an incendiary bomb. ● *v.tr.* attack or destroy with a fire-bomb. **fire-break** an obstacle to the spread of fire in a forest etc., esp. an open space. **fire-brick** a fireproof brick used in a grate. **fire brigade** esp. *Brit.* an organized body of firemen trained and employed to extinguish fires. **fire-bug** *colloq.* a pyromaniac. **fire company 1** = *fire brigade*. **2** a fire-insurance company. **fire-control** a system of regulating the fire of a ship's or a fort's guns. **fire department** *US* = *fire brigade*. **fire door** a fire-resistant door to prevent the spread of fire. **fire-drill 1** a rehearsal of the procedures to be used in case of fire. **2** a primitive device for kindling fire with a stick and wood. **fire-eater 1** a conjuror who appears to swallow fire. **2** a person fond of quarrelling or fighting. **fire-engine** a vehicle carrying equipment for fighting large fires. **fire-escape** an emergency staircase or apparatus for escape from a building on fire. **fire extinguisher** an apparatus with a jet for discharging liquid chemicals, water, or foam to extinguish a fire. **fire-fighter** a person whose task is to extinguish fires. **fire-guard 1** a protective screen or grid placed in front of a fireplace. **2** *N. Amer.* a fire-watcher. **3** *N. Amer.* a fire-break. **fire-hose** a hose-pipe used in extinguishing fires. **fire-irons** tongs, poker, and shovel, for tending a domestic fire. **fire-lighter** *Brit.* a piece of inflammable material to help start a fire in a grate. **fire-office** a fire-insurance company. **fire-opal** girasol. **fire-plug** a hydrant for a fire-hose. **fire-power 1** the destructive capacity of guns etc. **2** financial, intellectual, or emotional strength. **fire-practice** a fire-drill. **fire-raiser** *Brit.* an arsonist. **fire-raising** *Brit.* arson. **fire-screen 1** a screen to keep off the direct heat of a fire. **2** a fire-guard. **3** an ornamental screen for a fireplace. **fire-ship** *hist.* a ship loaded with combustibles and set adrift to ignite an enemy's ships etc. **fire station** the headquarters of a fire brigade. **fire-step** = *firing-step*. **fire-stone** stone that resists fire, used for furnaces and kilns. **fire-storm** a high wind or storm following a fire caused by bombs. **fire-tongs** tongs for picking up pieces of coal etc. in tending a fire. **fire-trap** a building without proper provision for escape in case of fire. **fire up 1** start (an engine etc.); get ready for action. **2** fill with enthusiasm. **fire-walking** the (often ceremonial) practice of walking barefoot over red-hot stones, wood-ashes, etc. **fire warden** *N. Amer.* a person employed to prevent or extinguish fires. **fire-watcher** a person keeping watch for fires, esp. those caused by bombs. **fire-water** *colloq.* strong alcoholic liquor. **go on fire** *Sc. & Ir.* catch fire. **go through fire and water** face all perils.

on fire 1 burning. **2** excited. **set fire to** (or **set on fire**) ignite, kindle, cause to burn. **set the world** (or **Thames**) **on fire** do something remarkable or sensational. **take fire** catch fire. **under fire 1** being shot at. **2** being rigorously criticized or questioned. □ **fireless** *adj.* **firer** *n.* [OE *fӯr, fӯrian,* f. WG]

firearm /ˈfaɪərˌɑːm/ *n.* (usu. in *pl.*) a gun, esp. a pistol or rifle.

fireback /ˈfaɪəˌbæk/ *n.* **1 a** the back wall of a fireplace. **b** an iron sheet for this. **2** a SE Asian pheasant of the genus *Lophura.*

firebox /ˈfaɪəˌbɒks/ *n.* the fuel-chamber of a steam engine or boiler.

firebrand /ˈfaɪəˌbrænd/ *n.* **1** a piece of burning wood. **2** a cause of trouble, esp. a person causing unrest.

fireclay /ˈfaɪəˌkleɪ/ *n.* clay capable of withstanding high temperatures, often used to make fire-bricks.

firecracker /ˈfaɪəˌkrækə(r)/ *n.* an explosive firework; a banger.

firecrest /ˈfaɪəˌkrest/ *n.* a European woodland kinglet, *Regulus ignicapillus,* with a red and orange crest.

firedamp /ˈfaɪəˌdæmp/ *n. Mining* methane, which is explosive when mixed in certain proportions with air.

firedog /ˈfaɪəˌdɒg/ *n.* a metal support for burning wood or for a grate or fire-irons.

firefly /ˈfaɪəˌflaɪ/ *n.* (*pl.* **-flies**) a soft-bodied beetle of the glow-worm family Lampyridae, usu. emitting light in flashes.

firehouse /ˈfaɪəˌhaʊs/ *n. US* a fire station.

firelight /ˈfaɪəˌlaɪt/ *n.* light from a fire in a fireplace. [OE *fӯr-leoht* (as FIRE, LIGHT[1])]

firelock /ˈfaɪəˌlɒk/ *n. hist.* a musket in which the priming was ignited by sparks.

fireman /ˈfaɪəmən/ *n.* (*pl.* **-men**) **1** a member of a fire brigade; a person employed to extinguish fires. **2** a person who tends a furnace or the fire of a steam engine or steamship.

Firenze see FLORENCE.

Fire of London (also called *Great Fire* (*of London*)) the huge and devastating fire which destroyed some 13,000 houses over 400 acres in London between 2 and 6 Sept. 1666. The fire started in a bakery in Pudding Lane in the City of London; perhaps the best-known account of it is in Samuel Pepys's *Diary.* (See also GREAT PLAGUE.)

fireplace /ˈfaɪəˌpleɪs/ *n. Archit.* **1** a place for a domestic fire, esp. a grate or hearth at the base of a chimney. **2** a structure surrounding this. **3** the area in front of this.

fireproof /ˈfaɪəˌpruːf/ *adj. & v.* ● *adj.* able to resist fire or great heat. ● *v.tr.* make fireproof.

fireside /ˈfaɪəˌsaɪd/ *n.* **1** the area round a fireplace. **2** a person's home or home-life. □ **fireside chat** an informal talk.

firethorn /ˈfaɪəˌθɔːn/ *n.* = PYRACANTHA.

fireweed /ˈfaɪəˌwiːd/ *n.* a plant which thrives on burnt land; esp. rosebay willowherb, *Chamerion angustifolium.*

firewood /ˈfaɪəˌwʊd/ *n.* wood for use as fuel.

firework /ˈfaɪəˌwɜːk/ *n.* **1** a device containing combustible chemicals that cause explosions or spectacular effects. (*See note below.*) **2** (in *pl.*) **a** an outburst of passion, esp. anger. **b** a display of wit or brilliance.

▪ Fireworks are of ancient Chinese origin, and developed from the military use of gunpowder for rockets and missiles. Displays of fireworks on festive occasions began to be staged in western Europe in the 16th century. The brilliance and variety of displays was increased in the 19th century by the introduction of new ingredients, in particular metal salts giving bright colours to the flames. Different effects are obtained by varying the composition and particle size of the combustible mixture, and the dimensions of the container.

firing /ˈfaɪərɪŋ/ *n.* **1** the discharging of guns. **2** material for a fire, fuel. **3** the heating process which hardens clay into pottery etc. □ **firing-line 1** the front line in a battle. **2** the leading part in an activity etc. **firing-party** a group detailed to fire the salute at a military funeral. **firing-squad 1** a group detailed to shoot a condemned person. **2** a firing-party. **firing-step** a step on which soldiers in a trench stand to fire.

firkin /ˈfɜːkɪn/ *n.* **1** a small cask for liquids, butter, fish, etc. **2** *Brit.* (as a measure of beer etc.) half a kilderkin (usu. 9 imperial gallons or about 41 litres). [ME *ferdekyn,* prob. f. MDu. *vierdekijn* (unrecorded) dimin. of *vierde* fourth]

firm[1] /fɜːm/ *adj., adv., & v.* ● *adj.* **1 a** of solid or compact structure. **b** fixed, stable. **c** steady; not shaking. **2 a** resolute, determined. **b** not easily shaken (*firm belief*). **c** steadfast, constant (*a firm friend*). **3 a** (of an offer etc.) not liable to cancellation after acceptance. **b** (of a decree, law, etc.) established, immutable. **4** *Commerce* (of prices or goods) maintaining their level or value. ● *adv.* firmly (*stand firm; hold firm to*). ● *v.* **1** *tr. & intr.* (often foll. by *up*) make or become firm, secure, compact, or solid. **2** *tr.* (often foll. by *in*) fix (plants) firmly in the soil. □ **firmly** *adv.* **firmness** *n.* [ME f. OF *ferme* f. L *firmus*]

firm[2] /fɜːm/ *n.* **1 a** a business concern. **b** the partners in such a concern. **2** a group of persons working together, esp. of hospital doctors and assistants. [earlier = signature, style: Sp. & It. *firma* f. med.L, f. L *firmare* confirm f. *firmus* FIRM[1]]

firmament /ˈfɜːməmənt/ *n. literary* the sky regarded as a vault or arch. □ **firmamental** /ˌfɜːməˈment(ə)l/ *adj.* [ME f. OF f. L *firmamentum* f. *firmare* (as FIRM[2])]

firman /fɜːˈmɑːn, ˈfɜːmən/ *n.* **1** an oriental sovereign's edict. **2** a grant or permit. [Pers. *fermān,* Skr. *pramāṇam* right measure]

firmware /ˈfɜːmweə(r)/ *n. Computing* a permanent kind of software programmed into a read-only memory.

firry see FIR.

first /fɜːst/ *adj., n., & adv.* ● *adj.* **1 a** earliest in time or order. **b** coming next after a specified or implied time (*shall take the first train; the first cuckoo*). **2** foremost in position, rank, or importance (*First Lord of the Treasury; first mate*). **3** *Mus.* performing the highest or chief of two or more parts for the same instrument or voice. **4** most willing or likely (*should be the first to admit the difficulty*). **5** basic or evident (*first principles*). ● *n.* **1** (prec. by *the*) the person or thing first mentioned or occurring. **2** the first occurrence of something notable. **3 a** a place in the first class in an examination. **b** a person having this. **4** the first day of a month. **5** = *first gear* (see GEAR). **6 a** first place in a race. **b** the winner of this. **7** (in *pl.*) goods of the best quality. ● *adv.* **1** before any other person or thing (*first of all; first and foremost; first come first served*). **2** before someone or something else (*must get this done first*). **3** for the first time (*when did you first see her?*). **4** in preference; rather (*will see him damned first*). **5** first-class (*I usually travel first*). □ **at first** at the beginning. **at first hand** directly from the original source. **first aid** help given to an injured person until proper medical treatment is available. **first and last** taking one thing with another, on the whole. **first blood** see BLOOD. **first-born** *adj.* eldest. ● *n.* the eldest child of a person. **first cousin** see COUSIN. **first-day cover** an envelope with stamps postmarked on their first day of issue. **first-degree** *Med.* denoting burns that affect only the surface of the skin, causing reddening. **first finger** the finger next to the thumb. **first floor** see FLOOR. **first-foot** *Sc. n.* the first person to cross a threshold in the New Year. ● *v.intr.* be a first-foot. **first-fruit** (usu. in *pl.*) **1** the first agricultural produce of a season, esp. as offered to God. **2** the first results of work etc. **3** *hist.* a payment to a superior by the new holder of an office. **first gear** see GEAR. **first intention** see INTENTION. **first lesson** the first of two or more passages from the Bible read at a service in the Church of England. **first lieutenant** *US* an army or air force officer next below captain. **first light** the time when light first appears in the morning. **first mate** (on a merchant ship) the officer second in command to the master. **first name** a personal or Christian name. **first night** the first public performance of a play etc. **first-nighter** a habitual attender of first nights. **first off** esp. *US colloq.* at first, first of all. **first offender** a criminal against whom no previous conviction is recorded. **first officer** the mate on a merchant ship. **first or last** sooner or later. **first past the post 1** winning a race etc. by being the first to reach the finishing line. **2** (of an electoral system) selecting a candidate or party by simple majority (see also PROPORTIONAL REPRESENTATION). **first person** see PERSON. **first post** see POST[3]. **first-rate** *adj.* **1** of the highest class, excellent. **2** *colloq.* very well (*feeling first-rate*). **first reading** the occasion when a Bill is presented to a legislature to permit its introduction. **first refusal** see REFUSAL. **first school** *Brit.* a school for children from 5 to 9 years old. **first sergeant** *US* the highest-ranking non-commissioned officer in a company. **first-strike** denoting a first aggressive attack with nuclear weapons. **first thing** *colloq.* before anything else; very early in the morning (*shall do it first thing*). **the first thing** even the most elementary fact or principle (*does not know the first thing about it*). **first things first** the most important things before any others (*we must do first things first*). **first up 1** first of all. **2** *Austral.* at the first attempt. **from the first** from the beginning. **from first to last** throughout. **get to first base** *US* achieve the first step towards an objective. **in the first place** as the first consideration. **of the first water** see WATER. [OE *fyrst* f. Gmc]

First Adar see ADAR.

First Boer War see BOER WAR.

First Cause *n.* the Creator of the universe.

first class *n., adj., & adv.* ● *n.* **1** a set of persons or things grouped together as the best. **2** the best accommodation in a train, ship, etc. **3** the class of mail given priority in handling. **4 a** the highest division in an examination list. **b** a place in this. ● *adj.* (**first-class**) **1** belonging to or travelling by the first class. **2** of the highest standard or best quality; very good. **3** (of cricket) played between sides of recognized stature and with matches of two innings per side. ● *adv.* (**first-class**) by the first class (*travels first-class*).

First Empire see EMPIRE 4c.

firsthand /fɜːstˈhænd/ *attrib.adj. & adv.* from the original source; direct.

First International see INTERNATIONAL.

First Lady *n.* (in the US) the wife of the President.

firstling /ˈfɜːstlɪŋ/ *n.* (usu. in *pl.*) **1** the first result of anything, first-fruits. **2** the first offspring; the first born in a season.

firstly /ˈfɜːstlɪ/ *adv.* (in enumerating topics, arguments, etc.) in the first place, first (cf. FIRST *adv.*).

First Point of Aries see ARIES, FIRST POINT OF.

First Republic *hist.* **1** the republican regime in France from the abolition of the monarchy in 1792 until Napoleon's accession as emperor (1804). **2** this period in France.

First World War (1914–18), a war between the Central Powers (Germany and Austria–Hungary, joined later by Turkey and Bulgaria) and the Allies (Britain and its dominions, France, Russia, and others, joined later by Italy and the US). Political tensions in central Europe relating mainly to the rise of the German Empire were its principal cause, although it was set off by the assassination of Archduke Franz Ferdinand of Austria by a Bosnian Serb nationalist in Sarajevo on 28 June 1914. This event was used as a pretext by Austria for declaring war on Serbia. Most of the fighting took place on land in Europe (the only significant naval battle was at Jutland in 1916), and was generally characterized by long periods of bloody stalemate. In the east, the demands of the war were a factor in the eventual collapse of Russia in 1917. In the west, an initial German offensive through Belgium was stemmed and a long period of attritional trench warfare ensued (see also WESTERN FRONT). The balance eventually shifted in the Allies' favour in 1917 when the US joined the war, largely in response to the German policy of unrestricted submarine warfare against Atlantic shipping. The stalemate in the west was broken by Allied offensives in 1918; after the retreat of German forces, the war was ended by a series of armistices in late 1918, and peace terms were settled at Versailles in 1919. Total casualties of the war are estimated at 10 million killed. One of the consequences of the war was the collapse of the German, Austro-Hungarian, Russian, and Ottoman empires.

firth /fɜːθ/ *n.* (also **frith** /frɪθ/) **1** a narrow inlet of the sea. **2** an estuary. [ME orig. Sc.) f. ON *fjǫrthr* FIORD]

fisc /fɪsk/ *n. Rom. Hist.* the public treasury; the emperor's privy purse. [F *fisc* or L *fiscus* rush-basket, purse, treasury]

fiscal /ˈfɪsk(ə)l/ *adj. & n.* ● *adj.* of public revenue; esp. *US* of financial matters. ● *n.* **1** a legal official in some countries. **2** *Sc.* = *procurator fiscal*. □ **fiscal year** = *financial year.* □ **fiscally** *adv.* [F *fiscal* or L *fiscalis* (as FISC)]

fiscality /fɪˈskælɪtɪ/ *n.* (*pl.* **-ies**) **1** (in *pl.*) fiscal matters. **2** excessive regard for these.

Fischer¹ /ˈfɪʃə(r)/, Emil Hermann (1852–1919), German organic chemist. He studied the structure of sugars, other carbohydrates, and purines, and synthesized many of them. He also worked on peptides and proteins, and confirmed that they consist of chains of amino acids. Fischer's work was largely the basis for the German drug industry, and he was awarded the Nobel Prize for chemistry in 1902.

Fischer² /ˈfɪʃə(r)/, Hans (1881–1945), German organic chemist. His work was largely concerned with the important porphyrin group of natural pigments. He determined the complex structure of the red oxygen-carrying part of haemoglobin, the green chlorophyll pigments found in plants, and the orange bile pigment bilirubin. He also synthesized some of these, and was awarded the Nobel Prize for chemistry in 1930.

Fischer³ /ˈfɪʃə(r)/, Robert James ('Bobby') (b.1943), American chess player. He was world champion 1972–5.

Fischer-Dieskau /ˌfɪʃəˈdiːskaʊ/, Dietrich (b.1925), German baritone. He is noted for his interpretations of German lieder, in particular Schubert's song cycles. He has made more recordings of songs than any other recording artist and has the largest vocal repertoire of any contemporary singer (more than 1,000 songs).

fish¹ /fɪʃ/ *n. & v.* ● *n.* (*pl.* same or **fishes**) **1** a cold-blooded vertebrate animal with gills and fins, living wholly in water. (*See note below.*) **2** an invertebrate animal living wholly in water, e.g. cuttlefish, shellfish, jellyfish. **3** the flesh of fish as food. **4** *colloq.* a person remarkable in some way (usu. unfavourable) (*an odd fish*). **5** (**the Fish** or **Fishes**) the zodiacal sign or constellation Pisces. **6** *Naut. sl.* a torpedo; a submarine. ● *v.* **1** *intr.* try to catch fish, esp. with a line or net. **2** *tr.* fish for (a certain kind of fish) or in (a certain stretch of water). **3** *intr.* (foll. by *for*) **a** search for in water or a concealed place. **b** seek by indirect means (*fishing for compliments*). **4** *tr.* (foll. by *up, out,* etc.) retrieve with careful or awkward searching. □ **drink like a fish** drink excessively. **fish-bowl** a usu. round glass bowl for keeping pet fish in. **fish cake** a cake of shredded fish and mashed potato, usu. eaten fried. **fish eagle** an eagle, esp. of the genus *Haliaeetus*, that catches and feeds on fish. **fish-eye lens** *Photog.* a very wide-angle lens with a convex front, giving a field of vision covering up to 180° and reducing the scale of the image more at the edges than in the centre. **fish farm** a place where fish are bred for food. **fish finger** *Brit.* a small oblong piece of fish in batter or breadcrumbs. **fish-glue** isinglass. **fish-hawk** an osprey. **fish-hook** a barbed hook for catching fish. **fish-kettle** an oval pan for boiling fish. **fish-knife** a knife for eating or serving fish. **fish-meal** ground dried fish used as fertilizer or animal feed. **fish out of water** a person in an unsuitable or unwelcome environment or situation. **fish-pond** a pond or pool in which fish are kept. **fish-slice** a flat utensil for lifting fish and fried foods during and after cooking. **other fish to fry** other matters to attend to. □ **fishlike** *adj.* [OE *fisc, fiscian* f. Gmc]

▪ Fish represent a form of life rather than a single related group of animals. The most primitive forms are the jawless fishes, which first appeared about 500 million years ago and are represented today by the lampreys. True jawed fishes appeared about 100 million years later, and present-day forms are divided into two main classes: cartilaginous fish and bony fish. The former are mostly elasmobranchs (sharks, dogfish, skates, and rays), which have a skeleton of cartilage, and lack a swim-bladder to give buoyancy. Bony fish are mostly teleosts, which constitute the great majority of the 30,000 or so living fish species, usually having a swim-bladder, and possessing fins that allow great manoeuvrability.

fish² /fɪʃ/ *n. & v.* ● *n.* **1** a flat plate of iron, wood, etc., to strengthen a beam or joint. **2** *Naut.* a piece of wood, convex and concave, used to strengthen a mast etc. ● *v.tr.* **1** mend or strengthen (a spar etc.) with a fish. **2** join (rails) with a fish-plate. □ **fish-bolt** a bolt used to fasten fish-plates and rails together. **fish-plate a** a flat piece of iron etc. connecting railway rails. **b** a flat piece of metal with ends like a fish's tail, used to position masonry. [orig. as verb: f. F *ficher* fix ult. f. L *figere*]

fish³ /fɪʃ/ *n.* a piece of ivory etc. used as a counter in games. [F *fiche* (*ficher*; see FISH²)]

Fisher¹ /ˈfɪʃə(r)/ a shipping forecast area in the North Sea off northern Jutland and the mouth of the Skagerrak.

Fisher² /ˈfɪʃə(r)/, Sir Ronald Aylmer (1890–1962), English statistician and geneticist. Fisher made major contributions to the development of statistics, publishing influential books on statistical theory, the design of experiments, statistical methods for research workers, and the relationship between Mendelian genetics and evolutionary theory. He also carried out experimental work in agriculture and on the genetics of blood groups.

fisher /ˈfɪʃə(r)/ *n.* **1 a** an animal that catches fish. **b** a tree-living North American marten, *Martes pennanti*, valued for its fur. Also called *pekan*. **2** *archaic* a fisherman. [OE *fiscere* f. Gmc (as FISH¹)]

Fisher, St John /ˈfɪʃə(r)/ (1469–1535), English churchman. In 1504 he became bishop of Rochester and earned the disfavour of Henry VIII by opposing his divorce from Catherine of Aragon. When he refused to accept the king as supreme head of the Church, he was condemned to death. Feast day, 22 June.

fisherman /ˈfɪʃəmən/ *n.* (*pl.* **-men**) **1** a person who catches fish as a livelihood or for sport. **2** a fishing-boat.

fishery /ˈfɪʃərɪ/ *n.* (*pl.* **-ies**) **1** a place where fish are caught or reared. **2** the occupation or industry of catching or rearing fish.

fishing /ˈfɪʃɪŋ/ *n.* the activity of catching fish, esp. for food or as a sport. □ **fishing-line** a long thread of silk etc. with a baited hook, sinker, float, etc., used for catching fish. **fishing-rod** a long tapering usu. jointed rod to which a fishing-line is attached. **fishing story** *Brit. colloq.* an exaggerated account.

fishmonger /ˈfɪʃˌmʌŋɡə(r)/ *n.* esp. *Brit.* a dealer in fish for food.

fishnet /ˈfɪʃnet/ *n.* (often *attrib.*) an open-meshed fabric (*fishnet stockings*).

fishpot /ˈfɪʃpɒt/ *n.* a wicker trap for eels, lobsters, etc.

fishtail /ˈfɪʃteɪl/ n. & v. ● n. anything resembling a fish's tail in shape or movement. ● v.intr. move the tail of a vehicle from side to side. □ **fishtail burner** a kind of gas burner producing a broadening jet of flame.

fishwife /ˈfɪʃwaɪf/ n. (pl. **-wives**) 1 a coarse-mannered or noisy woman. 2 a woman who sells fish.

fishy /ˈfɪʃɪ/ adj. (**fishier, fishiest**) 1 of or like (a) fish. 2 joc. or poet. abounding in fish. 3 sl. arousing suspicion, questionable, mysterious (something fishy going on). □ **fishily** adv. **fishiness** n.

fisk /fɪsk/ n. Sc. the state treasury, the exchequer. [var. of FISC]

fissile /ˈfɪsaɪl/ adj. 1 capable of undergoing nuclear fission. 2 cleavable; tending to split. □ **fissility** /fɪˈsɪlɪtɪ/ n. [L fissilis (as FISSURE)]

fission /ˈfɪʃ(ə)n/ n. & v. ● n. 1 the action of dividing or splitting into two or more parts. 2 Physics = NUCLEAR FISSION. 3 Biol. the division of a cell etc. into new cells etc. as a mode of reproduction. ● v.intr. & tr. undergo or cause to undergo fission. □ **fission bomb** an atom bomb. □ **fissionable** adj. [L fissio (as FISSURE)]

fissiparous /fɪˈsɪpərəs/ adj. 1 Biol. reproducing by fission. 2 tending to split. □ **fissiparously** adv. **fissiparousness** n. **fissiparity** /ˌfɪsɪˈpærɪtɪ/ n. [L fissus past part. (as FISSURE) after viviparous]

fissure /ˈfɪʃə(r)/ n. & v. ● n. 1 an opening, usu. long and narrow, made esp. by cracking, splitting, or separation of parts. 2 Bot. & Anat. a narrow opening in an organ etc., esp. a depression between convolutions of the brain. 3 a cleavage. ● v.tr. & intr. split or crack. [ME f. OF fissure or L fissura f. findere fiss- cleave]

fist /fɪst/ n. & v. ● n. 1 a tightly closed hand. 2 sl. handwriting (writes a good fist; I know his fist). 3 sl. a hand (give us your fist). ● v.tr. 1 strike with the fist. 2 Naut. handle (a sail, an oar, etc.). □ **make a good** (or **poor** etc.) **fist** (foll. by at, of) colloq. make a good (or poor etc.) attempt at. □ **fisted** adj. (also in comb.). **fistful** n. (pl. **-fuls**). [OE fŷst f. WG]

fistic /ˈfɪstɪk/ adj. (also **fistical** /-ɪk(ə)l/) joc. pugilistic.

fisticuffs /ˈfɪstɪˌkʌfs/ n.pl. fighting with the fists. [prob. obs. fisty adj. = FISTIC, + CUFF²]

fistula /ˈfɪstjʊlə/ n. (pl. **fistulas** or **fistulae** /-ˌliː/) 1 Med. an abnormal or surgically made passage between a hollow organ and the body surface or between two hollow organs. 2 Zool. a natural pipe or spout in whales, insects, etc. □ **fistular** adj. **fistulous** adj. [L, = pipe, flute]

fit¹ /fɪt/ adj., v., n., & adv. ● adj. (**fitter, fittest**) 1 a (usu. foll. by for, or to + infin.) well adapted or suited. b (foll. by to + infin.) qualified, competent, worthy. c (foll. by for, or to + infin.) in a suitable condition, ready. d (foll. by for) good enough (a dinner fit for a king). e (foll. by to + infin.) sufficiently exhausted, troubled, or angry (fit to drop). 2 in good health or athletic condition. 3 proper, becoming, right (it is fit that). ● v. (**fitted, fitting**) 1 a tr. (also absol.) be of the right shape and size for (the dress fits her; the key doesn't fit the lock; these shoes don't fit). b tr. make, fix, or insert (a thing) so that it is of the right size or shape (fitted shelves in the alcoves). c intr. (often foll. by in, into) (of a component) be correctly positioned (that bit fits here). d tr. find room for (can't fit another person on the bench). 2 tr. (foll. by for, or to + infin.) a make suitable; adapt. b make competent (fitted him to be a priest). 3 tr. (usu. foll. by with) supply, furnish (fitted the boat with a new rudder). 4 tr. fix in place (fit a lock on the door). 5 tr. = fit on. 6 tr. be in harmony with, befit, become (it fits the occasion; the punishment fits the crime). 7 (often foll. by up) esp. Austral. secure enough (genuine or false) evidence to convict; frame. ● n. the way in which a garment, component, etc., fits (a bad fit; a tight fit). ● adv. (foll. by to + infin.) colloq. in a suitable manner, appropriately (was laughing fit to bust). □ **fit the bill** = fill the bill. **fit in** (often foll. by with) be (esp. socially) compatible or accommodating (doesn't fit in with the rest of the group; tried to fit in with their plans). 2 find space or time for (an object, engagement, etc.) (the dentist fitted me in at the last minute). **fit on** try on (a garment). **fit out** (or **up**) (often foll. by with) equip. **fit-up** Theatr. sl. 1 a temporary stage etc. 2 a travelling company. **see** (or **think**) **fit** (often foll. by to + infin.) decide or choose (a specified course of action). □ **fitly** adv. **fitness** n. [ME: orig. unkn.]

fit² /fɪt/ n. 1 a sudden seizure of epilepsy, hysteria, apoplexy, fainting, or paralysis, with unconsciousness or convulsions. 2 a sudden brief attack of an illness or of symptoms (fit of coughing). 3 a sudden short bout or burst (fit of energy; fit of giggles). 4 colloq. an attack of strong feeling (fit of rage). 5 a capricious impulse; a mood (when the fit was on him). □ **by** (or **in**) **fits and starts** spasmodically. **give a person a fit** colloq. surprise or outrage him or her. **have a fit** colloq. be greatly surprised or outraged. **in fits** laughing uncontrollably. [ME, = position of danger, perh. = OE fitt conflict (?)]

fit³ /fɪt/ n. (also **fytte**) archaic a section of a poem. [OE fitt]

fitch /fɪtʃ/ n. 1 a polecat. 2 a the hair of a polecat. b a brush made from this or similar hair. [MDu. fisse etc.: cf. FITCHEW]

fitchew /ˈfɪtʃuː/ n. a polecat. [14th c. f. OF ficheau, fissel dimin. of MDu. fisse]

fitful /ˈfɪtfʊl/ adj. active or occurring spasmodically or intermittently. □ **fitfully** adv. **fitfulness** n.

fitment /ˈfɪtmənt/ n. (usu. in pl.) a fixed item of furniture.

fitted /ˈfɪtɪd/ adj. 1 made or shaped to fill a space or cover something closely or exactly (a fitted carpet). 2 provided with appropriate equipment, fittings, etc. (a fitted kitchen). 3 built-in; filling an alcove etc. (fitted cupboards).

fitter /ˈfɪtə(r)/ n. 1 a person who supervises the cutting, fitting, altering, etc. of garments. 2 a mechanic who fits together and adjusts machinery.

fitting /ˈfɪtɪŋ/ n. & adj. ● n. 1 the process or an instance of having a garment etc. fitted (needed several fittings). 2 a (in pl.) the fixtures and fitments of a building. b a piece of apparatus or furniture. ● adj. proper, becoming, right. □ **fitting-shop** a place where machine parts are put together. □ **fittingly** adv. **fittingness** n.

Fitzgerald¹ /fɪtsˈdʒerəld/, Edward (1809–83), English scholar and poet. He is remembered for his free poetic translation of The Rubáiyát of Omar Khayyám (1859).

Fitzgerald² /fɪtsˈdʒerəld/, Ella (b.1918), American jazz singer. In the 1940s she evolved a distinctive style of scat singing. Fitzgerald joined the American impresario Norman Granz (b.1918) on his world tours in 1946, appearing with Count Basie and Duke Ellington. From the mid-1950s she made a successful series of recordings of songs by George Gershwin and Cole Porter.

Fitzgerald³ /fɪtsˈdʒerəld/, F(rancis) Scott (Key) (1896–1940), American novelist. His novels, particularly The Great Gatsby (1925), provide a vivid portrait of the US during the jazz era of the 1920s. From the mid-1920s Fitzgerald and his wife, the writer Zelda Sayre (1900–47), became part of an affluent and fashionable set living on the French Riviera; their lifestyle is reflected in the semi-autobiographical novel Tender is the Night (1934).

FitzGerald⁴ /fɪtsˈdʒerəld/, George Francis (1851–1901), Irish physicist. He suggested that length, time, and mass depend on the relative motion of the observer, while the speed of light is constant. This hypothesis, postulated independently by Lorentz, prepared the way for Einstein's special theory of relativity.

FitzGerald contraction n. (also **FitzGerald-Lorentz**) Physics the contraction or foreshortening, in a direction parallel to its line of motion, of a body moving relative to an observer. The effect is insignificant except at speeds close to that of light. [FITZGERALD⁴, LORENTZ]

Fiume /ˈfjuːme/ the Italian name for RIJEKA.

five /faɪv/ n. & adj. ● n. 1 one more than four or one half of ten; the sum of three units and two units. 2 a symbol for this (5, v, V). 3 a size etc. denoted by five. 4 a set or team of five individuals. 5 the time of five o'clock (is it five yet?). 6 a card with five pips. 7 Cricket a hit scoring five runs. ● adj. that amount to five. □ **bunch of fives** Brit. sl. 1 a hand or fist. 2 a punch. **five-a-side** (in full **five-a-side football**) (also attrib.) a form of Association football played between teams of five players. **five-corner** (or **-corners**) Austral. 1 a shrub of the genus Styphelia. 2 the pentagonal fruit of this. **five-eighth** Austral. & NZ Rugby either of two players between the scrum-half and the centre three-quarter. **five-finger exercise** 1 an exercise on the piano involving all the fingers. 2 an easy task. **five hundred** a form of euchre in which 500 points make a game. **five o'clock shadow** beard-growth visible on a man's face in the latter part of the day. **five-star** of the highest class. **five-year plan** 1 (in the former USSR) a government plan for economic development over five years, inaugurated in 1928. 2 a similar plan in another country. [OE fīf f. Gmc]

fivefold /ˈfaɪvfəʊld/ adj. & adv. 1 five times as much or as many. 2 consisting of five parts. 3 amounting to five.

Five Pillars of Islam the five duties expected of every Muslim (see ISLAM).

fiver /ˈfaɪvə(r)/ n. colloq. 1 Brit. a five-pound note. 2 N. Amer. a five-dollar bill.

fives /faɪvz/ n. a game in which a ball is hit with a gloved hand or a bat against the walls of a court with three walls (Eton fives) or four walls (Rugby fives). [pl. of FIVE used as sing.: orig. uncert.]

fivestones /ˈfaɪvstəʊnz/ n. Brit. jacks played with five pieces of metal etc. and usu. without a ball.

fix /fɪks/ *v. & n.* ● *v.* **1** *tr.* make firm or stable; fasten, secure (*fixed a picture to the wall*). **2** *tr.* decide, settle, specify (a price, date, etc.). **3** *tr.* mend, repair. **4** *tr.* implant (an idea or memory) in the mind (*couldn't get the rules fixed in his head*). **5** *tr.* **a** (foll. by *on, upon*) direct steadily, set (one's eyes, gaze, attention, or affection). **b** attract and hold (a person's attention, eyes, etc.). **c** (foll. by *with*) single out with one's eyes etc. **6** *tr.* place definitely or permanently, establish, station. **7** *tr.* determine the exact nature, position, etc., of; refer (a thing or person) to a definite place or time; identify, locate. **8 a** *tr.* make (eyes, features, etc.) rigid. **b** *intr.* (of eyes, features, etc.) become rigid. **9** *tr.* N. Amer. colloq. prepare (food or drink) (*fixed me a drink*). **10 a** *tr.* deprive of fluidity or volatility; congeal. **b** *intr.* lose fluidity or volatility, become congealed. **11** *tr.* colloq. punish, kill, silence, deal with, take revenge on (a person). **12** *tr.* colloq. **a** secure the support of (a person) fraudulently, esp. by bribery. **b** arrange the result of (a race, match, etc.) fraudulently (*the competition was fixed*). **13** *sl.* **a** *tr.* inject (a person, esp. oneself) with a narcotic. **b** *intr.* take an injection of a narcotic. **14** *tr.* **a** make (a colour, photographic image, etc.) fast or permanent. **b** *Biol.* preserve or stabilize (a specimen) prior to treatment and microscopic examination. **15** *tr.* (of a plant or micro-organism) assimilate (nitrogen or carbon dioxide) by forming a non-gaseous compound. **16** *tr.* castrate or spay (an animal). **17** *tr.* arrest changes or development in (a language or literature). **18** *tr.* determine the incidence of (liability etc.). **19** *intr.* archaic take up one's position. **20** (as **fixed** *adj.*) **a** permanently placed, stationary. **b** without moving, rigid; (of a gaze, etc.) steady or intent. **c** definite. **d** *sl.* dishonest, fraudulent. ● *n.* **1** colloq. a position hard to escape from; a dilemma or predicament. **2 a** the act of finding one's position by bearings or astronomical observations (*get a fix on that star*). **b** a position found in this way. **3** *sl.* a dose of a narcotic drug to which one is addicted. **4** *sl.* bribery; an illicit arrangement. □ **be fixed** (usu. foll. by *for*) be disposed or affected (regarding) (*how is he fixed for money?*; *how are you fixed for Friday?*). **fixed capital** machinery etc. that remains in the owner's use. **fixed-doh** *Mus.* applied to a system of sight-singing in which C is called 'doh', D is called 'ray', etc., irrespective of the key in which they occur (cf. *movable-doh*). **fixed focus** *Photog.* a camera focus that cannot be adjusted, typically used with a small aperture lens having a large depth of field. **fixed idea** = IDÉE FIXE. **fixed income** income deriving from a pension, investment at fixed interest, etc. **fixed odds** predetermined odds in racing etc. (opp. *starting price*). **fixed oil** a non-volatile oil of animal or plant origin, used in varnishes, lubricants, illuminants, soaps, etc. (see also OIL). **fixed point** *Physics* a well-defined reproducible temperature. **fixed star** *Astron.* a true star, being so far from the earth as to appear motionless except for the diurnal revolution of the heavens (opp. *planet, comet,* etc.). **fix on** (or **upon**) choose, decide on. **fix up 1** arrange, organize, prepare. **2** accommodate. **3** (often foll. by *with*) provide (a person) (*fixed me up with a job*). □ **fixable** *adj.* **fixedly** /ˈfɪksɪdlɪ/ *adv.* **fixedness** /-sɪdnɪs/ *n.* [ME, partly f. obs. *fix* fixed f. OF *fix* or L *fixus* past part. of *figere* fix, fasten, partly f. med.L *fixare* f. *fixus*]

fixate /ˈfɪkseɪt/ *v.tr.* **1** direct one's gaze on. **2** *Psychol.* **a** (usu. in *passive*; often foll. by *on, upon*) cause (a person) to acquire an abnormal attachment to persons or things (*was fixated on his son*). **b** arrest (part of the libido) at an immature stage, causing such attachment. [L *fixus* (see FIX) + -ATE³]

fixation /fɪkˈseɪʃ(ə)n/ *n.* **1** the act or an instance of being fixated. **2** an obsession, concentration on a single idea. **3** fixing or being fixed. **4** the process of rendering solid; coagulation. **5** the process of assimilating a gas to form a solid compound. [ME f. med.L *fixatio* f. *fixare*: see FIX]

fixative /ˈfɪksətɪv/ *adj. & n.* ● *adj.* tending to fix or secure. ● *n.* a substance used to fix colours, hair, biological specimens, etc.

fixer /ˈfɪksə(r)/ *n.* **1** a person or thing that fixes. **2** *Photog.* a substance used for fixing a photographic image etc. **3** colloq. a person who makes arrangements, esp. of an illicit kind.

fixing /ˈfɪksɪŋ/ *n.* (in *pl.*) US **1** a method or means of fixing. **2 a** apparatus or equipment. **b** the trimmings for a dish. **c** the trimmings of a dress etc.

fixity /ˈfɪksɪtɪ/ *n.* **1** a fixed state. **2** stability; permanence. [obs. *fix* fixed: see FIX]

fixture /ˈfɪkstʃə(r)/ *n.* **1 a** something fixed or fastened in position. **b** (usu. *predic.*) colloq. a person or thing confined to or established in one place (*he seems to be a fixture*). **2 a** a sporting event, esp. a match, race, etc. **b** the date agreed for this. **3** (in *pl.*) *Law* articles attached to a house or land and regarded as legally part of it. [alt. of obs. *fixure* f. LL *fixura* f. L *figere* fix- fix]

fizgig /ˈfɪzgɪg/ *n. & adj.* ● *n.* **1** archaic a silly or flirtatious young woman.

2 archaic a kind of small firework; a cracker. **3** *Austral. sl.* a police informer. ● *adj.* archaic flighty. [prob. f. FIZZ + obs. *gig* flighty girl]

fizz /fɪz/ *v. & n.* ● *v.intr.* **1** make a hissing or spluttering sound. **2** (of a drink) make bubbles; effervesce. ● *n.* **1** effervescence. **2** colloq. an effervescent drink, esp. champagne. [imit.]

fizzer /ˈfɪzə(r)/ *n.* **1** an excellent or first-rate thing. **2** *Cricket colloq.* a very fast ball, or one that deviates with unexpected speed. **3** *Austral. sl.* a disappointing failure or fiasco.

fizzle /ˈfɪz(ə)l/ *v. & n.* ● *v.intr.* make a feeble hissing or spluttering sound. ● *n.* such a sound. □ **fizzle out** end feebly (*the party fizzled out at 10 o'clock*). [formed as FIZZ + -LE⁴]

fizzy /ˈfɪzɪ/ *adj.* (**fizzier, fizziest**) (of a drink) effervescent. □ **fizzily** *adv.* **fizziness** *n.*

FJI *abbr.* Fellow of the Institute of Journalists.

fjord *var. of* FIORD.

FL *abbr. US* Florida (in official postal use).

fl. *abbr.* **1** floor. **2** (*fl.*) floruit. **3** fluid.

Fla. *abbr.* Florida.

flab /flæb/ *n.* colloq. fat; flabbiness. [imit., or back-form. f. FLABBY]

flabbergast /ˈflæbəˌgɑːst/ *v.tr.* (esp. as **flabbergasted** *adj.*) colloq. overwhelm with astonishment; dumbfound. [18th c.: perh. f. FLABBY + AGHAST]

flabby /ˈflæbɪ/ *adj.* (**flabbier, flabbiest**) **1** (of flesh etc.) hanging down; limp; flaccid. **2** (of language or character) feeble. □ **flabbily** *adv.* **flabbiness** *n.* [alt. of earlier *flappy* f. FLAP]

flaccid /ˈflæksɪd, ˈflæsɪd/ *adj.* **1 a** (of flesh etc.) hanging loose or wrinkled; limp, flabby. **b** (of plant tissue) soft; less rigid. **2** relaxed, drooping. **3** lacking vigour; feeble. □ **flaccidly** *adv.* **flaccidity** /flækˈsɪdɪtɪ, fləˈsɪd-/ *n.* [F *flaccide* or L *flaccidus* f. *flaccus* flabby]

flack¹ /flæk/ *n.* US sl. a publicity agent. [20th c.: orig. unkn.]

flack² *var. of* FLAK.

flag¹ /flæg/ *n. & v.* ● *n.* **1 a** a piece of cloth, usu. oblong or square, attachable by one edge to a pole or rope and used as a country's emblem or as a standard, signal, etc. **b** a small toy, device, etc., resembling a flag. **2** a device that is raised to indicate that a taxi is for hire. **3** *Naut.* a flag carried by a flagship as an emblem of an admiral's rank afloat. ● *v.* (**flagged, flagging**) **1** *intr.* **a** grow tired; lose vigour; lag (*his energy flagged after the first lap*). **b** hang down; droop; become limp. **2** *tr.* **a** place a flag on or over. **b** mark out with or as if with a flag or flags. **3** *tr.* (often foll. by *that*) **a** inform (a person) by flag-signals. **b** communicate (information) by flagging. □ **black flag 1** a pirate's ensign. **2** *hist.* a flag hoisted outside a prison to announce an execution. **flag-boat** a boat serving as a mark in sailing-matches. **flag-captain** the captain of a flagship. **flag-day** *Brit.* a day on which money is raised for a charity by the sale of small paper flags etc. in the street. **flag down** signal to (a vehicle or driver) to stop. **flag-lieutenant** *Naut.* an admiral's ADC. **flag-list** *Naut.* a roll of flag-officers. **flag of convenience** a foreign flag under which a ship is registered, usu. to avoid financial charges etc. **flag-officer** *Naut.* an admiral, vice admiral, or rear admiral, or the commodore of a yacht-club. **flag of truce** a white flag indicating a desire for a truce. **flag-pole** = FLAGSTAFF. **flag-rank** *Naut.* the rank attained by flag-officers. **flag-station** a station at which trains stop only if signalled. **flag-wagging** *sl.* **1** signalling with hand-held flags. **2** = flag-waving. **flag-waver** a populist agitator; a chauvinist. **flag-waving** populist agitation, chauvinism. **keep the flag flying** continue the fight. **put the flag out** celebrate victory, success, etc. **show the flag 1** make an official visit to a foreign port etc. **2** ensure that notice is taken of one's country, oneself, etc.; make a patriotic display. □ **flagger** *n.* [16th c.: perh. f. obs. *flag* drooping]

flag² /flæg/ *n. & v.* ● *n.* (also **flagstone**) **1** a flat usu. rectangular stone slab used for paving. **2** (in *pl.*) a pavement made of these. ● *v.tr.* (**flagged, flagging**) pave with flags. [ME, = sod: cf. Icel. *flag* spot from which a sod has been cut out, ON *flaga* slab of stone, and FLAKE¹]

flag³ /flæg/ *n.* **1** a plant with sword-shaped leaves, esp. the *yellow flag* (*Iris pseudacorus*) (cf. *sweet flag*). **2** the leaf of such a plant. [ME: cf. MDu. *flag,* Danish *flæg*]

flag⁴ /flæg/ *n.* (in full **flag-feather**) a quill-feather of a bird's wing. [perh. rel. to obs. *fag* loose flap: cf. FLAG¹ *v.*]

Flag Day *US* 14 June, the anniversary of the adoption of the Stars and Stripes in 1777.

flagellant /ˈflædʒələnt, fləˈdʒel-/ *n. & adj.* ● *n.* **1** a person who scourges himself or herself or others as a religious discipline. **2** a person who engages in flogging as a sexual stimulus. ● *adj.* of or concerning flagellation. [L *flagellare* to whip f. FLAGELLUM]

flagellate[1] /ˈflædʒəˌleɪt/ *v.tr.* scourge, flog (cf. FLAGELLANT). □ **flagellator** *n.* **flagellatory** /-lətərɪ/ *adj.*

flagellate[2] /ˈflædʒəlɪt/ *adj. & n.* ● *adj.* having flagella. ● *n. Zool.* a protozoan having one or more flagella.

flagellation /ˌflædʒəˈleɪʃ(ə)n/ *n.* the act or practice of flagellating others or (esp.) oneself, as a sexual stimulus or religious discipline.

flagellum /fləˈdʒeləm/ *n.* (*pl.* **flagella** /-lə/) **1** *Biol.* a long lashlike appendage found esp. on microscopic organisms. **2** *Bot.* a runner; a creeping shoot. □ **flagellar** *adj.* **flagelliform** *adj.* [L, = whip, dimin. of *flagrum* scourge]

flageolet[1] /ˌflædʒəˈlet/ *n.* **1** a small flute blown at the end, like a recorder but with two thumb-holes. **2** an organ stop having a similar sound. [F, dimin. of OF *flag(e)ol* f. Prov. *flajol*, of unkn. orig.]

flageolet[2] /ˈflædʒəˌleɪ, -ˌlet/ *n.* a kind of French kidney bean. [F]

flagitious /fləˈdʒɪʃəs/ *adj.* deeply criminal; utterly villainous. □ **flagitiously** *adv.* **flagitiousness** *n.* [ME f. L *flagitiosus* f. *flagitium* shameful crime]

flagman /ˈflægmən/ *n.* (*pl.* **-men**) a person who signals with or as with a flag, e.g. at races.

flagon /ˈflægən/ *n.* **1** a large bottle in which wine, cider, etc., is sold, usu. holding 1.13 litres (about 2 pints). **2 a** a large vessel usu. with a handle, spout, and lid, to hold wine etc. **b** a similar vessel used for the Eucharist. [ME *flakon* f. OF *flacon* ult. f. LL *flasco -onis* FLASK]

flagrant /ˈfleɪgrənt/ *adj.* (of an offence or an offender) glaring; notorious; scandalous. □ **flagrancy** *n.* **flagrantly** *adv.* [F *flagrant* or L *flagrant-* part. stem of *flagrare* blaze]

flagship /ˈflægʃɪp/ *n.* **1** a ship having an admiral on board. **2** something that is held to be the best or most important of its kind; a leader.

flagstaff /ˈflægstɑːf/ *n.* a pole on which a flag may be hoisted.

flagstone /ˈflægstəʊn/ *n.* = FLAG[2].

flail /fleɪl/ *n. & v.* **●** *n.* a threshing-tool consisting of a wooden staff with a short heavy stick swinging from it. ● *v.* **1** *tr.* beat or strike with or as if with a flail. **2** *intr.* wave or swing wildly or erratically (*went into the fight with arms flailing*). [OE prob. f. L FLAGELLUM]

flair /fleə(r)/ *n.* **1** an instinct for selecting or performing what is excellent, useful, etc.; a talent (*has a flair for knowing what the public wants; has a flair for languages*). **2** talent or ability, esp. artistic or stylistic. [F *flairer* to smell ult. f. L *fragrare*: see FRAGRANT]

flak /flæk/ *n.* (also **flack**) **1** anti-aircraft fire. **2** adverse criticism; abuse. □ **flak jacket** a protective jacket of heavy camouflage fabric reinforced with metal, worn by soldiers etc. [abbr. of G *Fliegerabwehrkanone*, lit. 'aviator-defence-gun']

flake[1] /fleɪk/ *n. & v.* ● *n.* **1 a** a small thin light piece of snow. **b** a similar piece of another material. **2** a thin broad piece of material peeled or split off. **3** *Archaeol.* a piece of hard stone chipped off and used as a tool. **4** a natural division of the flesh of some fish. **5** the dogfish or other shark as food. **6** esp. *N. Amer. sl.* a crazy or eccentric person. ● *v.tr. & intr.* (often foll. by *away, off*) **1** take off or come away in flakes. **2** sprinkle with or fall in snowlike flakes. □ **flake out** *colloq.* fall asleep or drop from exhaustion; faint. [ME: orig. unkn.: cf. ON *flakna* flake off]

flake[2] /fleɪk/ *n.* **1** a stage for drying fish etc. **2** a rack for storing oatcakes etc. [ME, perh. f. ON *flaki, fleki* wicker shield]

flaky /ˈfleɪkɪ/ *adj.* (**flakier, flakiest**) **1** of or like flakes; separating easily into flakes. **2** esp. *N. Amer. sl.* crazy, eccentric. □ **flaky pastry** pastry consisting of thin light layers. □ **flakily** *adv.* **flakiness** *n.*

flambé /ˈflɒmbeɪ/ *adj.* (of food) covered with alcohol and set alight briefly. [F, past part. of *flamber* singe (as FLAMBEAU)]

flambeau /ˈflæmbəʊ/ *n.* (*pl.* **flambeaus** or **flambeaux** /-bəʊz/) **1** a flaming torch, esp. composed of several thick waxed wicks. **2** a branched candlestick. [F f. *flambe* f. L *flammula* dimin. of *flamma* flame]

Flamborough Head /ˈflæmbərə/ a rocky promontory on the east coast of England, in the former East Riding of Yorkshire (now part of Humberside).

flamboyant /flæmˈbɔɪənt/ *adj.* **1** ostentatious; showy. **2** floridly decorated. **3** gorgeously coloured. **4** *Archit.* (of decoration) marked by wavy flamelike lines. □ **flamboyance** *n.* **flamboyancy** *n.* **flamboyantly** *adv.* [F (in Archit. sense), pres. part. of *flamboyer* f. *flambe*: see FLAMBEAU]

flame /fleɪm/ *n. & v.* ● *n.* **1 a** ignited gas (*the fire burnt with a steady flame*). **b** one portion of this (*the flame flickered and died*). **c** (usu. in *pl.*) visible combustion (*burst into flames*). **2 a** a bright light; brilliant colouring. **b** a brilliant orange-red colour. **3 a** strong passion, esp. love (*fan the flame*). **b** *colloq.* a boyfriend or girlfriend. ● *v.* **1** *intr. & tr.* (often foll. by *away, forth, out, up*) emit or cause to emit flames. **2** *intr.* (often foll. by *out, up*) **a** (of passion) break out. **b** (of a person) become angry. **3** *intr.* shine or glow like flame (*leaves flamed in the autumn sun*). **4** *intr. poet.* move like flame. **5** *tr.* send (a signal) by means of flame. **6** *tr.* subject to the action of flame. □ **flame gun** a device for throwing flames to destroy weeds etc. **flame out** (of a jet engine) lose power through the extinction of the flame in the combustion chamber. **flame-proof** *adj. & v.* ● *adj.* (esp. of a fabric) treated so as to be non-flammable. ● *v.tr.* make flame-proof. **flame-thrower** (or **-projector**) a weapon for throwing a spray of flame. **flame tree** a tree with brilliant red flowers, esp. a Madagascan poinciana, *Delonix regia*, or an Australian bottle tree, *Brachychiton australis*. **go up in flames** be consumed by fire. □ **flameless** *adj.* **flamelike** *adj.* **flamy** *adj.* [ME f. OF *flame, flam(m)e* f. L *flamma*]

flamen /ˈfleɪmən/ *n.* (*pl.* **flamens, flamines** /ˈflæmɪˌniːz/) *Rom. Hist.* a priest serving a particular deity. [ME f. L]

flamenco /fləˈmeŋkəʊ/ *n.* (*pl.* **-os**) **1** a type of traditional Spanish music originating in Andalusia, played (esp. on the guitar) and sung originally by gypsies. **2** a dance performed to this music, characterized by stamping and hand-clapping. [Sp., = Flemish]

flaming /ˈfleɪmɪŋ/ *adj.* **1** emitting flames. **2** very hot (*flaming June*). **3** *colloq.* **a** passionate; intense (*a flaming row*). **b** expressing annoyance, or as an intensifier (*that flaming dog*). **4** bright-coloured (*flaming red hair*).

flamingo /fləˈmɪŋgəʊ/ *n.* (*pl.* **-os** or **-oes**) a tall long-necked long-legged wading bird of the family Phoenicopteridae, with a crooked bill and pink or scarlet plumage. [Port. *flamengo* f. Prov. *flamenc* f. *flama* flame + *-enc* = -ING[3]]

flammable /ˈflæməb(ə)l/ *adj.* inflammable. ¶ Often used because *inflammable* can be mistaken for a negative (the true negative being *non-flammable*). □ **flammability** /ˌflæməˈbɪlɪtɪ/ *n.* [L *flammare* f. *flamma* flame]

Flamsteed /ˈflæmstiːd/, John (1646–1719), English astronomer. He was appointed the first Astronomer Royal at the Royal Greenwich Observatory, with the task of accurately providing the positions of stars for use in navigation. He eventually produced the first star catalogue, which gave the positions of nearly 3,000 stars. Flamsteed also worked on the motions of the sun and moon, tidal tables, and other measurements.

flan /flæn/ *n.* **1 a** a pastry case with a savoury or sweet filling. **b** a sponge base with a sweet topping. **2** a disc of metal from which a coin etc. is made. [F (orig. = round cake) f. OF *flaon* f. med.L *flado -onis* f. Frank.]

flanch /flɑːntʃ/ *v.tr. & intr.* (also **flaunch** /flɔːntʃ/) (esp. with ref. to a chimney) slope inwards or cause to slope inwards towards the top. □ **flanching** *n.* [perh. f. OF *flanchir* f. *flanche, flanc* FLANK]

Flanders /ˈflɑːndəz/ a region in the south-western part of the Low Countries, now divided between Belgium (where it forms the provinces of East and West Flanders), France, and the Netherlands. It was a powerful medieval principality. The area was the scene of prolonged fighting during the First World War, when British troops held the sector of the Western Front round the town of Ypres.

Flanders poppy *n.* **1** a red poppy, used as an emblem of the soldiers of the Allies who fell in the First World War. **2** an artificial red poppy made for wearing on Remembrance Sunday, sold in aid of needy ex-servicemen and -servicewomen.

flânerie /ˈflænərɪ/ *n.* idling, idleness. [F f. *flâner* lounge]

flâneur /flæˈnɜː(r)/ *n.* an idler; a lounger. [F (as FLÂNERIE)]

flange /flændʒ/ *n. & v.* ● *n.* a projecting flat rim, collar, or rib, used for strengthening or attachment. ● *v.tr.* provide with a flange. □ **flangeless** *n.* [17th c.: perh. f. *flange* widen out f. OF *flangir* (as FLANCH)]

flank /flæŋk/ *n. & v.* ● *n.* **1 a** the side of the body between the ribs and the hip. **b** the side of an animal carved as meat (*flank of beef*). **2** the side of a mountain, building, etc. **3** the right or left side of an army or other body of persons. ● *v.tr.* **1** (often in *passive*) be situated at both sides of (*a road flanked by mountains*). **2** *Mil.* **a** guard or strengthen on the flank. **b** menace the flank of. **c** rake with sweeping gunfire; enfilade. □ **flank forward** *Rugby* a wing forward. **in flank** at the side. [ME f. OF *flanc* f. Frank.]

flanker /ˈflæŋkə(r)/ *n.* **1** *Mil.* a fortification guarding or menacing the flank. **2** anything that flanks another thing. **3 a** (in Rugby) a wing forward. **b** (in American football) an offensive player who lines up to the outside of an end. **4** *sl.* a trick; a swindle (*pulled a flanker*).

flannel /ˈflæn(ə)l/ *n. & v.* ● *n.* **1 a** a kind of woven woollen fabric, usu. without a nap. **b** (in *pl.*) flannel garments, esp. trousers. **2** *Brit.* a small usu. towelling cloth, used for washing oneself. **3** *Brit. sl.* nonsense; flattery. ● *v.* (**flannelled, flannelling**; *US* **flanneled, flanneling**) **1** *Brit. sl.* a *tr.* flatter. **b** *intr.* use flattery. **2** *tr.* wash or clean with a flannel. □ **flannel-mouth** *US sl.* a flatterer; a braggart. [perh. f. Welsh *gwlanen* f. *gwlân* wool]

flannelboard /ˈflæn(ə)lˌbɔːd/ *n.* a piece of flannel as a base for paper or cloth cut-outs, used as a toy or a teaching aid.

flannelette /ˌflænəˈlet/ *n.* a napped cotton fabric imitating flannel. [FLANNEL]

flannelgraph /ˈflæn(ə)lˌɡrɑːf/ *n.* = FLANNELBOARD.

flannelled /ˈflæn(ə)ld/ *adj.* (*US* also **flanneled**) wearing flannel trousers. [FLANNEL]

flap /flæp/ *v. & n.* ● *v.* (**flapped, flapping**) **1 a** *tr.* move (wings, the arms, etc.) up and down when flying, or as if flying. **b** *intr.* (of wings, the arms, etc.) move up and down; beat. **2** *intr. colloq.* be agitated or panicky. **3** *intr.* (esp. of curtains, loose cloth, etc.) swing or sway about; flutter. **4** *tr.* (usu. foll. by *away, off*) strike (flies etc.) with something broad; drive. **5** *intr. colloq.* (of ears) listen intently. ● *n.* **1** a piece of cloth, wood, paper, etc. hinged or attached by one side only and often used to cover a gap, e.g. a pocket-cover, the folded part of an envelope, a table-leaf. **2** one up-and-down motion of a wing, an arm, etc. **3** *colloq.* a state of agitation; panic (*don't get into a flap*). **4** a hinged or sliding section of a wing used to control lift; an aileron. **5** a light blow with something broad. **6** an open mushroom-top. □ **flappy** *adj.* [ME, prob. imit.]

flapdoodle /flæpˈduːd(ə)l/ *n. colloq.* nonsense. [19th c.: orig. unkn.]

flapjack /ˈflæpdʒæk/ *n.* **1** a cake made from oats and golden syrup etc. **2** *sl. US* a pancake. [FLAP + JACK[1]]

flapper /ˈflæpə(r)/ *n.* **1** a person or thing that flaps. **2** an instrument that is flapped to kill flies, scare birds, etc. **3** a person who panics easily or is easily agitated. **4** *sl.* (in the 1920s) a young unconventional or lively woman. **5** a young mallard or partridge.

flare /fleə(r)/ *v. & n.* ● *v.* **1** *intr.* widen or cause to widen gradually towards the top or bottom (*flared trousers*). **2** *intr. & tr.* burn or cause to burn suddenly with a bright unsteady flame. **3** *intr.* burst into anger; burst forth. ● *n.* **1 a** a dazzling irregular flame or light, esp. in the open air. **b** a sudden outburst of flame. **2 a** a signal light used at sea. **b** a bright light used as a signal. **c** a flame dropped from an aircraft to illuminate a target etc. **3** *Astron.* a sudden burst of radiation from a star. **4 a** a gradual widening, esp. of a skirt or trousers. **b** (in *pl.*) wide-bottomed trousers. **5** an outward bulge in a ship's sides. **6** *Photog.* extraneous illumination on film caused by internal reflection in the lens etc. □ **flare-path** an area illuminated to enable an aircraft to land or take off. **flare up 1** burst into a sudden blaze. **2** become suddenly angry or active. **flare-up** *n.* an outburst of flame, anger, activity, etc. [16th c.: orig. unkn.]

flash /flæʃ/ *v., n., & adj.* ● *v.* **1** *intr. & tr.* emit or reflect or cause to emit or reflect light briefly, suddenly, or intermittently; gleam or cause to gleam. **2** *intr.* break suddenly into flame; give out flame or sparks. **3** *tr.* send or reflect like a sudden flame or blaze (*his eyes flashed fire*). **4** *intr.* **a** burst suddenly into view or perception (*the explanation flashed upon me*). **b** move swiftly (*the train flashed through the station*). **5** *tr.* **a** send (news etc.) by radio, telegraph, etc. (*flashed a message to her*). **b** signal to (a person) by shining lights or headlights briefly. **6** *tr. colloq.* show ostentatiously (*flashed her engagement ring*). **7** *intr. sl.* indecently expose oneself. ● *n.* **1** a sudden bright light or flame, e.g. of lightning. **2** a very brief time; an instant (*all over in a flash*). **3 a** a brief, sudden burst of feeling (*a flash of hope*). **b** a sudden display (of wit, understanding, etc.). **4** = NEWSFLASH. **5** *Photog.* = FLASHGUN. **6 a** a rush of water, esp. down a weir to take a boat over shallows. **b** a contrivance for producing this. **7** *Brit. Mil.* a coloured patch of cloth on a uniform etc. as a distinguishing emblem. **8** vulgar display, ostentation. **9** a bright patch of colour. **10** *Cinematog.* the momentary exposure of a scene. **11** excess plastic or metal oozing from a mould during moulding. ● *adj. colloq.* **1** gaudy; showy; vulgar (*a flash car*). **2** counterfeit (*flash notes*). **3** connected with thieves, the underworld, etc. □ **flash-board** a board used for sending more water from a mill-dam into a mill-race. **flash bulb** *Photog.* a bulb for a flashgun. **flash burn** a burn caused by sudden intense heat, esp. from a nuclear explosion. **flash card** a card containing a small amount of information, held up for pupils to see, as an aid to learning. **flash-cube** *Photog.* a set of four flash bulbs arranged as a cube and operated in

turn. **flash-flood** a sudden local flood due to heavy rain etc. **flashing-point** = FLASHPOINT. **flash in the pan** a promising start followed by failure (from the priming of old guns). **flash-lamp** a portable flashing electric lamp, esp. an electric torch. **flash out** (or **up**) show sudden passion. **flash over** *Electr.* make an electric circuit by sparking across a gap. **flash-over** *n.* an instance of this. [ME orig. with ref. to the rushing of water: cf. SPLASH]

flashback /ˈflæʃbæk/ *n. Cinematog.* a scene in a film, novel, etc. set in a time earlier than the main action.

flasher /ˈflæʃə(r)/ *n.* **1** *sl.* a man who indecently exposes himself. **2 a** an automatic device for switching lights rapidly on and off. **b** a sign or signal using this. **3** a person or thing that flashes.

flashgun /ˈflæʃɡʌn/ *n. Photog.* a device producing an intense flash of light, used for taking photographs by night, indoors, etc.

flashing /ˈflæʃɪŋ/ *n.* a usu. metallic strip used to prevent water penetration at the junction of a roof with a wall etc. [dial. *flash* seal with lead sheets or obs. *flash* flashing]

flashlight /ˈflæʃlaɪt/ *n.* **1** an electric torch. **2** a flashing light used for signals and in lighthouses.

flashpoint /ˈflæʃpɔɪnt/ *n.* **1** the temperature at which vapour from oil etc. will ignite in air. **2** the point at which anger, indignation, etc. becomes uncontrollable.

flashy /ˈflæʃɪ/ *adj.* (**flashier, flashiest**) showy; gaudy; cheaply attractive. □ **flashily** *adv.* **flashiness** *n.*

flask /flɑːsk/ *n.* **1** a narrow-necked bulbous bottle for wine etc. or as used in chemistry. **2** = *hip-flask* (see HIP[1]). **3** = VACUUM FLASK. **4** *hist.* = *powder-flask.* [F *flasque* & (prob.) It. *fiasco* f. med.L *flasca, flasco*: cf. FLAGON]

flat[1] /flæt/ *adj., adv., n., & v.* ● *adj.* (**flatter, flattest**) **1 a** horizontally level (*a flat roof*). **b** even; smooth; unbroken; without projection or indentation (*a flat stomach*). **c** with a level surface and little depth; shallow (*a flat cap; a flat heel*). **2** unqualified; plain; downright (*a flat refusal; a flat denial*). **3 a** dull; lifeless; monotonous (*spoke in a flat tone*). **b** without energy; dejected. **4** (of a fizzy drink) having lost its effervescence; stale. **5** (of an accumulator, a battery, etc.) having exhausted its charge. **6** *Mus.* below the true or normal pitch (*the violins are flat*). **b** (of a key) having a flat or flats in the signature. **c** (as **B flat, E flat**, etc.) a semitone lower than B, E, etc. **7** *Photog.* lacking contrast. **8 a** (of paint etc.) not glossy; matt. **b** (of a tint) uniform. **9** (of a tyre) punctured; deflated. **10** (of a market, prices, etc.) inactive; sluggish. **11** of or relating to flat racing. ● *adv.* **1** lying at full length; spread out, esp. on another surface (*lay flat on the floor; the ladder was flat against the wall*). **2** *colloq.* **a** completely, absolutely (*turned it down flat; flat broke*). **b** exactly (*in five minutes flat*). **3** *Mus.* below the true or normal pitch (*always sings flat*). ● *n.* **1** the flat part of anything; something flat (*the flat of the hand*). **2** level ground, esp. a plain or swamp. **3** *Mus.* **a** a note lowered a semitone below natural pitch. **b** the sign (♭) indicating this. **4** (as **the flat**) *Brit.* **a** flat racing. **b** the flat racing season. **5** *Theatr.* a flat section of scenery mounted on a frame. **6** *colloq.* a flat tyre, a puncture. **7** *sl.* a foolish person. ● *v.tr.* (**flatted, flatting**) **1** make flat, flatten (esp. in technical use). **2** *US Mus.* make (a note) flat. □ **fall flat** fail to live up to expectations; not win applause. **flat arch** *Archit.* an arch with a flat lower or inner curve. **flat** (or **flat-bottomed**) **boat** a boat with a flat bottom for transport in shallow water. **flat foot** a foot with a less than normal arch. **flat-four** (of an engine) having four cylinders all horizontal, two on each side of the crankshaft. **flat-head 1** a marine fish of the family Platycephalidae, having a flattened body with both eyes on the top side. **2** *sl.* a foolish person. **flat-iron** *hist.* an iron heated externally and used for pressing clothes etc. **flat out 1** at top speed. **2** without hesitation or delay. **3** using all one's strength, energy, or resources. **flat race** a horse-race over level ground, as opposed to a steeplechase or hurdles. **flat racing** the racing of horses in flat races (see HORSE-RACING). **flat rate** a rate that is the same in all cases, not proportional. **flat spin 1** *Aeron.* a nearly horizontal spin. **2** *colloq.* a state of agitation or panic. **flat-top 1** *US sl.* an aircraft-carrier. **2** *sl.* a man's short flat haircut. **that's flat** *colloq.* let there be no doubt about it. □ **flatly** *adv.* **flatness** *n.* **flattish** *adj.* [ME f. ON *flatr* f. Gmc]

flat[2] /flæt/ *n. & v.* ● *n.* a set of rooms, usu. on one floor, used as a residence. ● *v.intr.* (**flatted, flatting**) (often foll. by *with*) *Austral.* share a flat with. □ **flatlet** *n.* [alt. f. obs. *flet* floor, dwelling f. Gmc (as FLAT[1])]

flatcar /ˈflætkɑː(r)/ *n.* a railway wagon without raised sides or ends.

flatfish /ˈflætfɪʃ/ *n.* (*pl.* same or **flatfishes**) a flat-bodied bottom-dwelling marine fish of the order Pleuronectiformes, having an asymmetric appearance with both eyes on one side, including sole, turbot, plaice, flounders, etc.

flatfoot /ˈflætfʊt/ *n.* (*pl.* **-foots** or **-feet**) *sl.* a police officer.

flat-footed /flæt'fʊtɪd/ adj. **1** having flat feet. **2** colloq. downright, positive. **3** colloq. unprepared; off guard (was caught flat-footed). □ **flat-footedly** adv. **flat-footedness** n.

flatmate /'flætmeɪt/ n. Brit. a person in relation to one or more others living in the same flat.

flatten /'flæt(ə)n/ v. **1** tr. & intr. make or become flat. **2** tr. colloq. **a** humiliate. **b** knock down. □ **flatten out** bring an aircraft parallel to the ground. □ **flattener** n.

flatter /'flætə(r)/ v.tr. **1** compliment unduly; overpraise, esp. for gain or advantage. **2** (usu. refl.) please, congratulate, or delude (oneself etc.) (I flatter myself that I can sing). **3 a** (of a colour, a style, etc.) make (a person) appear to the best advantage (that blouse flatters you). **b** (esp. of a portrait, a painter, etc.) represent too favourably. **4** gratify the vanity of; make (a person) feel honoured. **5** inspire (a person) with hope, esp. unduly (was unduly flattered into thinking himself invulnerable). **6** please or gratify (the ear, the eye, etc.). □ **flattering unction** a salve that one administers to one's own conscience or self-esteem (Shakespeare esp. Hamlet III. iv. 136). □ **flatterer** n. **flattering** adj. **flatteringly** adv. [ME, perh. rel. to OF flater to smooth]

flattery /'flætərɪ/ n. (pl. **-ies**) **1** exaggerated or insincere praise. **2** the act or an instance of flattering.

flattie /'flætɪ/ n. (also **flatty**) (pl. **-ies**) colloq. **1** a flat-heeled shoe. **2** a flat-bottomed boat. **3** a police officer.

flatulent /'flætjʊlənt/ adj. **1 a** causing formation of gas in the alimentary canal. **b** caused by or suffering from this. **2** (of speech etc.) inflated, pretentious. □ **flatulence** n. **flatulency** n. **flatulently** adv. [F f. mod.L flatulentus (as FLATUS)]

flatus /'fleɪtəs/ n. wind in or from the stomach or bowels. [L, = blowing f. flare blow]

flatware /'flætweə(r)/ n. **1** plates, saucers, etc. (opp. HOLLOWWARE). **2** N. Amer. domestic cutlery.

flatworm /'flætwɜːm/ n. a worm of the phylum Platyhelminthes, having a flattened body with no body-cavity or blood vessels, including turbellarians, flukes, and tapeworms.

Flaubert /'fləʊbeə(r)/, Gustave (1821–80), French novelist and short-story writer. A dominant figure in the French realist school, he achieved fame with his first published novel, Madame Bovary (1857). Its portrayal of the adulteries and suicide of a provincial doctor's wife caused Flaubert to be accused of immorality, but he was tried and acquitted. His Trois contes (1877) demonstrates Flaubert's versatility with different modes of narrative and anticipates Maupassant's experiments with the short story.

flaunch var. of FLANCH.

flaunt /flɔːnt/ v. & n. • v.tr. & intr. **1** (often refl.) display ostentatiously (oneself or one's finery); show off; parade (liked to flaunt his gold cufflinks; flaunted themselves before the crowd). ¶ Often confused with flout. **2** wave or cause to wave proudly (flaunted the banner). • n. an act or instance of flaunting. □ **flaunter** n. **flaunty** adj. [16th c.: orig. unkn.]

flautist /'flɔːtɪst/ n. a flute-player. [It. flautista f. flauto FLUTE]

flavescent /flə'ves(ə)nt/ adj. turning yellow; yellowish. [L flavescere f. flavus yellow]

Flavian /'fleɪvɪən/ adj. & n. • adj. of or relating to a dynasty (69–96 AD) of Roman emperors including Vespasian and his sons Titus and Domitian. • n. a member of this dynasty. [L Flavianus f. Flavius name of family]

flavin /'fleɪvɪn/ n. (also **flavine** /-viːn/) **1** Biochem. any of a group of cyclic compounds forming the nucleus of various natural yellow pigments. **2** a yellow dye formerly obtained from dyer's oak. □ **flavin adenine dinucleotide** (abbr. **FAD**) a coenzyme derived from riboflavin, important in various biochemical reactions. [L flavus yellow + -IN]

flavine /'fleɪviːn/ n. Pharm. an antiseptic derived from acridine. [as FLAVIN + -INE⁴]

flavone /'fleɪvəʊn/ n. Biochem. any of a group of naturally occurring white or yellow pigments found in plants. [as FLAVINE + -ONE]

flavoprotein /ˌfleɪvəʊ'prəʊtiːn/ n. Biochem. any of a group of conjugated proteins containing flavin that are involved in oxidation reactions in cells. [FLAVINE + PROTEIN]

flavorous /'fleɪvərəs/ adj. having a pleasant or pungent flavour.

flavour /'fleɪvə(r)/ n. & v. (US **flavor**) • n. **1** a distinctive mingled sensation of smell and taste (has a cheesy flavour). **2** an indefinable characteristic quality (music with a romantic flavour). **3** (usu. foll. by of) slight admixture of a usu. undesirable quality (the flavour of failure hangs over the enterprise). **4** esp. US = FLAVOURING. • v.tr. give flavour to; season. □ **flavour of the month** (or **week**) a temporary trend or fashion. □ **flavourful** adj. **flavourless** adj. **flavoursome** adj. [ME f. OF flaor perh. f. L flatus blowing & foetor stench: assim. to savour]

flavouring /'fleɪvərɪŋ/ n. (US **flavoring**) a substance used to flavour food or drink.

flaw¹ /flɔː/ n. & v. • n. **1** an imperfection; a blemish (has a character without a flaw). **2** a crack or similar fault (the cup has a flaw). **3** Law an invalidating defect in a legal matter. • v.tr. & intr. crack; damage; spoil. □ **flawed** adj. **flawless** adj. **flawlessly** adv. **flawlessness** n. [ME perh. f. ON flaga slab f. Gmc: cf. FLAKE¹, FLAG²]

flaw² /flɔː/ n. a squall of wind; a short storm. [prob. f. MDu. vlāghe, MLG vlāge, perh. = stroke]

flax /flæks/ n. **1 a** a blue-flowered plant, Linum usitatissimum, cultivated for its textile fibre and its seeds (see LINSEED). **b** a plant resembling this. **2 a** dressed or undressed flax fibres. **b** archaic linen, cloth of flax. □ **flax-lily** (pl. **-ies**) a New Zealand plant, Phormium tenax, of the agave family, yielding valuable fibre; also called New Zealand flax. **flax-seed** linseed. [OE flæx f. WG]

flaxen /'flæks(ə)n/ adj. **1** of flax. **2** (of hair) coloured like dressed flax; pale yellow.

Flaxman /'flæksmən/, John (1755–1826), English sculptor and draughtsman. He worked for Josiah Wedgwood 1775–87, designing medallion portraits and plaques. In 1793 he published engraved illustrations to Homer, influenced by Greek vase-painting; these won him international fame. Flaxman also sculpted church monuments such as the memorial to the Earl of Mansfield in Westminster Abbey (1793–1801).

flay /fleɪ/ v.tr. **1** strip the skin or hide off, esp. by beating. **2** criticize severely (the play was flayed by the critics). **3** peel off (skin, bark, peel, etc.). **4** strip (a person) of wealth by extortion or exaction. □ **flayer** n. [OE flēan f. Gmc]

F-layer /'ef ˌleɪə(r)/ n. the highest and most strongly ionized region of the ionosphere. [F (arbitrary) + LAYER]

flea /fliː/ n. **1** a small wingless jumping insect of the order Siphonaptera, feeding on human and other blood. **2 a** (in full **flea beetle**) a small jumping leaf-beetle, often infesting hops, cabbages, etc. **b** (in full **water flea**) daphnia. □ **flea-bite 1** the bite of a flea. **2 a** trivial injury or inconvenience. **flea-bitten 1** bitten by or infested with fleas. **2** shabby. **flea-bug** US = FLEA 2a. **flea-circus** a show of performing fleas. **flea-collar** an insecticidal collar for pets. **a flea in one's ear** a sharp reproof. **flea market** a street market selling second-hand goods etc. **flea-pit** a dingy dirty place, esp. a run-down cinema. **flea-wort** a composite plant of the genus Tephroseris, related to ragwort, formerly thought to drive away fleas. [OE flēa, flēah f. Gmc]

fleabag /'fliːbæg/ n. sl. a shabby or unattractive person or thing.

fleabane /'fliːbeɪn/ n. a composite plant of the genus Inula or Pulicaria, formerly thought to drive away fleas.

flèche /fleɪʃ, fleʃ/ n. Archit. a slender spire, often perforated with windows, esp. at the intersection of the nave and the transept of a church. [F, orig. = arrow]

fleck /flek/ n. & v. • n. **1** a small patch of colour or light (eyes with green flecks). **2** a small particle or speck, esp. of dust. **3** a spot on the skin; a freckle. • v.tr. mark with flecks; dapple; variegate. [perh. f. ON flekkr (n.), flekka (v.), or MLG, MDu. vlecke, OHG flec, fleccho]

Flecker /'flekə(r)/, James (Herman) Elroy (1884–1915), English poet. His best-known works are the verse collection The Golden Journey to Samarkand (1913) and the poetic Eastern play Hassan (1922), for which Delius wrote incidental music.

flection US var. of FLEXION.

fled past and past part. of FLEE.

fledge /fledʒ/ v. **1** intr. (of a bird) grow feathers. **2** tr. provide (an arrow) with feathers. **3** tr. bring up (a young bird) until it can fly. **4** tr. (as **fledged** adj.) **a** able to fly. **b** independent; mature. **5** tr. deck or provide with feathers or down. [obs. fledge (adj.) 'fit to fly', f. OE flycge (recorded in unfligge) f. a Gmc root rel. to FLY¹]

fledgling /'fledʒlɪŋ/ n. (also **fledgeling**) **1** a young bird. **2** an inexperienced person. [FLEDGE + -LING¹]

flee /fliː/ v. (past and past part. **fled** /fled/) **1** intr. (often foll. by from, before) **a** run away. **b** seek safety by fleeing. **2** tr. run away from; leave abruptly; shun (fled the room; fled his attentions). **3** intr. vanish; cease; pass away. [OE flēon f. Gmc]

fleece /fliːs/ n. & v. • n. **1 a** the woolly covering of a sheep or a similar animal. **b** the amount of wool sheared from a sheep at one time. **2** something resembling a fleece, esp.: **a** a woolly or rough head of hair. **b** a soft warm fabric with a pile, used for lining coats etc. **c** a white cloud, a blanket of snow, etc. **3** *Heraldry* a representation of a fleece suspended from a ring. ● *v.tr.* **1** (often foll. by *of*) strip (a person) of money, valuables, etc.; swindle. **2** remove the fleece from (a sheep etc.); shear. **3** cover as if with a fleece (*a sky fleeced with clouds*). □ **fleece-picker** *Austral. & NZ* = FLEECY *n.* □ **fleeceable** *adj.* **fleeced** *adj.* (also in *comb.*). [OE *flēos, flēs* f. WG]

fleecy /ˈfliːsɪ/ *adj. & n.* ● *adj.* (**fleecier, fleeciest**) **1** of or like a fleece. **2** covered with a fleece. ● *n.* (also **fleecie**) (*pl.* **-ies**) *Austral. & NZ* a person whose job is to pick up fleeces in a shearing shed. □ **fleecily** *adv.* **fleeciness** *n.*

fleer /flɪə(r)/ *v. & n.* ● *v.intr.* laugh impudently or mockingly; sneer; jeer. ● *n.* a mocking look or speech. [ME, prob. f. Scand.: cf. Norw. & Sw. dial. *flira* to grin]

fleet[1] /fliːt/ *n.* **1 a** a number of warships under one commander-in-chief. **b** (prec. by *the*) all the warships and merchant-ships of a nation. **2** a number of ships, aircraft, buses, lorries, taxis, etc. operating together or owned by one proprietor. [OE *flēot* ship, shipping f. *flēotan* float, FLEET[5]]

fleet[2] /fliːt/ *adj. poet. or literary* swift; nimble. □ **fleetly** *adv.* **fleetness** *n.* [prob. f. ON *fljótr* f. Gmc: cf. FLEET[5]]

fleet[3] /fliːt/ *n. dial.* a creek; an inlet. [OE *flēot* f. Gmc: cf. FLEET[5]]

fleet[4] /fliːt/ *adj. & adv. dial.* ● *adj.* (of water) shallow. ● *adv.* at or to a small depth (*plough fleet*). [orig. uncert.: perh. f. OE *flēat* (unrecorded), rel. to FLEET[5]]

fleet[5] /fliːt/ *v.intr. archaic* **1** glide away; vanish; be transitory. **2** (usu. foll. by *away*) (of time) pass rapidly; slip away. **3** move swiftly; fly. [OE *flēotan* float, swim f. Gmc]

Fleet Admiral *n. US* = ADMIRAL OF THE FLEET.

Fleet Air Arm *n. hist.* the aviation service of the Royal Navy.

fleeting /ˈfliːtɪŋ/ *adj.* transitory; brief. □ **fleetingly** *adv.* [FLEET[5] + -ING[2]]

Fleet Street a street in central London in which the offices of leading national newspapers were located until the mid-1980s. The name is also used allusively to refer to the British press.

Fleming[1] /ˈflemɪŋ/, Sir Alexander (1881–1955), Scottish bacteriologist. He worked mainly at St Mary's Hospital, London, where he studied the body's defences against bacteriological infection. In 1928 he fortuitously discovered the effect of penicillin on bacteria, and twelve years later Florey and Chain established its therapeutic use as an antibiotic. In 1942 Fleming was officially publicized as a British scientific hero, and so achieved fame retrospectively for his work in the 1920s. He was jointly awarded a Nobel Prize in 1945.

Fleming[2] /ˈflemɪŋ/, Ian (Lancaster) (1908–64), English novelist. He is known for his spy novels whose hero is the secret agent James Bond. Many of these stories (of which Fleming completed one a year from 1953 until his death) were successfully turned into feature films, making the character of James Bond world famous.

Fleming[3] /ˈflemɪŋ/, Sir John Ambrose (1849–1945), English electrical engineer. He is chiefly remembered for his invention of the thermionic valve (1900), which was the basis for all electronic devices until the transistor began to supersede it more than fifty years later. He also worked on transformers, radio-telegraphy, and telephony.

Fleming[4] /ˈflemɪŋ/ *n.* **1** a native of medieval Flanders in the Low Countries. **2** a member of a Flemish-speaking people inhabiting northern and western Belgium (see also WALLOON). [OE f. ON *Flǣmingi* & MDu. *Vlāming* f. root of *Vlaanderen* Flanders]

Flemish /ˈflemɪʃ/ *adj. & n.* ● *adj.* of or relating to Flanders or its people or language. ● *n.* the West Germanic language of Flanders, comprising a group of Dutch dialects, now one of the two official languages of Belgium. [MDu. *Vlāmisch* (as FLEMING[4])]

Flemish bond *n. Building* a bond in which each course consists of alternate headers and stretchers.

flense /flenz/ *v.tr.* (also **flench** /flentʃ/, **flinch** /flɪntʃ/) **1** cut up (a whale or seal). **2** flay (a seal). [Danish *flense*: cf. Norw. *flinsa, flunsa* flay]

flesh /fleʃ/ *n. & v.* ● *n.* **1 a** the soft, esp. muscular, substance between the skin and bones of an animal or a human. **b** plumpness; fat (*has put on flesh*). **c** *archaic* meat, esp. excluding poultry, game, and offal. **2** the body as opposed to the mind or the soul, esp. considered as sinful. **3** the pulpy substance of a fruit or a plant. **4 a** the visible surface of the human body with ref. to its colour or appearance. **b** (also **flesh-colour**) a yellowish-pink colour. **5** animal or human life. ● *v.tr.*

1 embody in flesh. **2** incite (a hound etc.) by the taste of blood. **3** initiate, esp. by aggressive or violent means, esp.: **a** use (a sword etc.) for the first time on flesh. **b** use (wit, the pen, etc.) for the first time. **c** inflame (a person) by the foretaste of success. □ **all flesh** all human and animal creation. **flesh and blood** ● *n.* **1** the body or its substance. **2** humankind. **3** human nature, esp. as being fallible. ● *adj.* actually living, not imaginary or supernatural. **flesh-fly** (*pl.* **-flies**) a fly of the family Sarcophagidae that deposits eggs or larvae on dead flesh. **flesh out** make or become substantial. **flesh side** the side of a hide that adjoined the flesh. **flesh tints** flesh-colours as rendered by a painter. **flesh-wound** a wound not reaching a bone or a vital organ. **in the flesh** in bodily form, in person. **lose** (or **put on**) **flesh** grow thinner or fatter. **make a person's flesh creep** frighten or horrify a person, esp. with tales of the supernatural etc. **one flesh** (of two people) intimately united, esp. by virtue of marriage (Gen. 2:24). **one's own flesh and blood** near relatives; descendants. **sins of the flesh** unchastity. **the way of all flesh** experience common to all humankind. □ **fleshless** *adj.* [OE *flǣsc* f. Gmc]

flesher /ˈfleʃə(r)/ *n. Sc.* a butcher.

fleshings /ˈfleʃɪŋz/ *n.pl.* an actor's flesh-coloured tights.

fleshly /ˈfleʃlɪ/ *adj.* (**fleshlier, fleshliest**) **1** (of desire etc.) bodily; lascivious; sensual. **2** mortal, not divine. **3** worldly. [OE *flǣsclic* (as FLESH)]

fleshpots /ˈfleʃpɒts/ *n.pl.* luxurious living (Exod. 16:3).

fleshy /ˈfleʃɪ/ *adj.* (**fleshier, fleshiest**) **1** plump, fat. **2** of flesh, without bone. **3** (of plant or fruit tissue) pulpy. **4** like flesh. □ **fleshiness** *n.*

Fletcher /ˈfletʃə(r)/, John (1579–1625), English dramatist. A writer of Jacobean tragicomedies, he wrote some fifteen plays with Francis Beaumont, including *The Maid's Tragedy* (1610–11). He is also believed to have collaborated with Shakespeare on such plays as *The Two Noble Kinsmen* and *Henry VIII* (both c.1613).

fletcher /ˈfletʃə(r)/ *n. archaic* a maker or seller of arrows. [ME f. OF *flech(i)er* f. *fleche* arrow]

fleur-de-lis /ˌflɜːdəˈliː/ *n.* (also **fleur-de-lys**) (*pl.* **fleurs-** pronunc. same) **1** the iris flower. **2** *Heraldry* **a** a lily composed of three petals bound together near their bases. **b** the former royal arms of France. [ME f. OF *flour de lys* flower of lily]

fleuret /flʊəˈret/ *n.* an ornament like a small flower. [F *fleurette* f. *fleur* flower]

fleuron /ˈflʊərɒn/ *n.* a flower-shaped ornament of a building, a coin, a book, etc. [ME f. OF *floron* f. *flour* FLOWER]

fleury /ˈflʊərɪ/ *adj.* (also **flory** /ˈflɔːrɪ/) *Heraldry* decorated with fleurs-de-lis. [ME f. OF *flo(u)ré* (as FLEURON)]

Flevoland /ˈfleɪvəˌlɑːnt/ a province of the Netherlands, created in 1986, comprising an area reclaimed from the Zuider Zee during the 1950s and 1960s.

flew *past of* FLY[1].

flews /fluːz/ *n.pl.* the hanging lips of a bloodhound etc. [16th c.: orig. unkn.]

flex[1] /fleks/ *v.* **1** *tr. & intr.* bend (a joint, limb, etc.) or be bent. **2** *tr. & intr.* move (a muscle) or (of a muscle) be moved to bend a joint. **3** *tr. Geol.* bend (strata). **4** *tr. Archaeol.* place (a corpse) with the legs drawn up under the chin. [L *flectere flex-* bend]

flex[2] /fleks/ *n. Brit.* a flexible insulated cable used for carrying electric current to an appliance. [abbr. of FLEXIBLE]

flexible /ˈfleksɪb(ə)l/ *adj.* **1** able to bend without breaking; pliable; pliant. **2** easily led; manageable; docile. **3** adaptable; versatile; variable (*works flexible hours*). □ **flexibly** *adv.* **flexibility** /ˌfleksɪˈbɪlɪtɪ/ *n.* [ME f. OF *flexible* or L *flexibilis* (as FLEX[1])]

flexile /ˈfleksaɪl/ *adj. archaic* **1** supple; mobile. **2** tractable; manageable. **3** versatile. □ **flexility** /flekˈsɪlɪtɪ/ *n.* [L *flexilis* (as FLEX[1])]

flexion /ˈflekʃ(ə)n/ *n.* (US **flection**) **1 a** the act of bending or the condition of being bent, esp. of a limb or joint. **b** a bent part; a curve. **2** *Gram.* inflection. **3** *Math.* = FLEXURE. □ **flexional** *adj.* (in sense 2). **flexionless** *adj.* (in sense 2). [L *flexio* (as FLEX[1])]

flexitime /ˈfleksɪˌtaɪm/ *n. Brit.* **1** a system of working a set number of hours with the starting and finishing times chosen within agreed limits by the employee. **2** the hours worked in this way. [FLEXIBLE + TIME]

flexography /flekˈsɒɡrəfɪ/ *n. Printing* a rotary letterpress technique using rubber or plastic plates and synthetic inks or dyes for printing on fabrics, plastics, etc., as well as on paper. □ **flexographic** /ˌfleksəˈɡræfɪk/ *adj.* [L *flexus* a bending f. *flectere* bend + -GRAPHY]

flexor /ˈfleksə(r)/ n. (in full **flexor muscle**) a muscle that bends part of the body (cf. EXTENSOR). [mod.L as FLEX¹)]

flexuous /ˈfleksjʊəs/ adj. full of bends; winding. □ **flexuously** adv. **flexuosity** /ˌfleksjʊˈɒsɪtɪ/ n. [L flexuosus f. flexus bending formed as FLEX¹]

flexure /ˈflekʃə(r)/ n. **1 a** the act of bending or the condition of being bent. **b** a bend, curve, or turn. **2** Math. the curving of a line, surface, or solid, esp. from a straight line, plane, etc. **3** Geol. the bending of strata under pressure. □ **flexural** adj. [L flexura (as FLEX¹)]

flibbertigibbet /ˌflɪbətɪˈdʒɪbɪt/ n. a gossiping, frivolous, or restless person. [imit. of chatter]

flick /flɪk/ n. & v. ● n. **1 a** a light, sharp, quickly retracted blow with a whip etc. **b** the sudden release of a bent finger or thumb, esp. to propel a small object. **2** a sudden movement or jerk. **3** a quick turn of the wrist in playing games, esp. in throwing or striking a ball. **4** a slight, sharp sound. **5** Brit. colloq. **a** a cinema film. **b** (in pl.; prec. by the) the cinema. ● v. **1** tr. (often foll. by away, off) strike or move with a flick (flicked the ash off his cigar; flicked away the dust). **2** tr. give a flick with (a whip, towel, etc.). **3** intr. make a flicking movement or sound. □ **flick-knife** a weapon with a blade that springs out from the handle when a button is pressed. **flick through 1** turn over (cards, pages, etc.). **2 a** turn over the pages etc. of, by a rapid movement of the fingers. **b** look cursorily through (a book etc.). [ME, imit.]

flicker¹ /ˈflɪkə(r)/ v. & n. ● v.intr. **1** (of light) shine unsteadily or fitfully. **2** (of a flame) burn unsteadily, alternately flaring and dying down. **3 a** (of a flag, a reptile's tongue, an eyelid, etc.) move or wave to and fro; quiver; vibrate. **b** (of the wind) blow lightly and unsteadily. **4** (of hope etc.) increase and decrease unsteadily and intermittently. ● n. **1** a flickering movement or light. **2** a brief spell of hope, recognition, etc. □ **flicker out** die away after a final flicker. [OE flicorian, flycerian]

flicker² /ˈflɪkə(r)/ n. a woodpecker of the genus Colaptes, native to North America. [imit. of its call]

flier var. of FLYER.

flight¹ /flaɪt/ n. & v. ● n. **1 a** the act or manner of flying through the air (studied swallows' flight). **b** the swift movement or passage of a projectile etc. through the air (the flight of an arrow). **2 a** a journey made through the air or in space. **b** a timetabled journey made by an airline. **c** Brit. an RAF unit of about six aircraft. **3 a** a flock or large body of birds, insects, etc., esp. when migrating. **b** a migration. **4** (usu. foll. by of) a series, esp. of stairs between floors, or of hurdles across a race track (lives up six flights). **5** an extravagant soaring, a mental or verbal excursion or sally (of wit etc.) (a flight of fancy; a flight of ambition). **6** the trajectory and pace of a ball in games. **7** the distance that a bird, aircraft, or missile can fly. **8** (usu. foll. by of) a volley (a flight of arrows). **9** the tail of a dart. **10** the pursuit of game by a hawk. **11** swift passage (of time). ● v.tr. **1** vary the trajectory and pace of (a cricket-ball etc.). **2** provide (an arrow) with feathers. **3** shoot (wildfowl etc.) in flight. □ **flight bag** a small, zipped, shoulder bag carried by air travellers. **flight control** an internal or external system directing the movement of aircraft. **flight-deck 1** the deck of an aircraft-carrier used for take-off and landing. **2** the accommodation for the pilot, navigator, etc. in an aircraft. **flight-feather** a primary or secondary feather in a bird's wing, supporting it in flight. **flight lieutenant** an RAF officer next in rank below squadron leader. **flight officer** a rank in the WRAF, corresponding to flight lieutenant. **flight path** the planned course of an aircraft or spacecraft. **flight-recorder** a device in an aircraft to record technical details during a flight, that may be used in the event of an accident to discover its cause. **flight sergeant** an RAF rank next above sergeant. **flight-test** test (an aircraft, rocket, etc.) during flight. **in the first** (or **top**) **flight** taking a leading place. **take** (or **wing**) **one's flight** fly. [OE flyht f. WG: rel to FLY¹]

flight² /flaɪt/ n. **1** the act or manner of fleeing. **b** a hasty retreat. **2** Econ. the selling of currency, investments, etc. in anticipation of a fall in value (flight from sterling). □ **put to flight** cause to flee. **take** (or **take to**) **flight** flee. [OE f. Gmc: rel. to FLEE]

flightless /ˈflaɪtlɪs/ adj. (of a bird etc.) naturally unable to fly.

flighty /ˈflaɪtɪ/ adj. (**flightier**, **flightiest**) **1** (usu. of a girl) frivolous, fickle, changeable. **2** crazy. □ **flightily** adv. **flightiness** n. [FLIGHT¹ + -Y¹]

flimflam /ˈflɪmflæm/ n. & v. ● n. **1** a trifle; nonsense; idle talk. **2** humbug; deception. ● v.tr. (**flimflammed**, **flimflamming**) cheat; deceive. □ **flimflammer** n. **flimflammery** n. (pl. **-ies**). [imit. redupl.]

flimsy /ˈflɪmzɪ/ adj. & n. ● adj. (**flimsier**, **flimsiest**) **1** lightly or carelessly assembled; insubstantial, easily damaged (a flimsy structure). **2** (of an excuse etc.) unconvincing (a flimsy pretext). **3** paltry; trivial; superficial

(a flimsy play). **4** (of clothing) thin (a flimsy blouse). ● n. (pl. **-ies**) **1** a very thin paper. **b** a document, esp. a copy, made on this. **2** a flimsy thing, esp. women's underwear. □ **flimsily** adv. **flimsiness** n. [17th c.: prob. f. FLIMFLAM]

flinch¹ /flɪntʃ/ v. & n. ● v.intr. **1** draw back in pain or expectation of a blow etc.; wince. **2** (often foll. by from) give way; shrink, turn aside (flinched from his duty). ● n. an act or instance of flinching. □ **flincher** n. **flinchingly** adv. [OF flenchir, flainchir f. WG]

flinch² var. of FLENSE.

Flinders /ˈflɪndəz/, Matthew (1774–1814), English explorer. He explored the coast of New South Wales (1795–1800) before being commissioned by the Royal Navy to circumnavigate Australia (1801–3). During this voyage he charted much of the west coast of the continent for the first time.

flinders /ˈflɪndəz/ n.pl. fragments; splinters. [ME, prob. f. Scand.]

Flinders Island the largest island in the Furneaux group, situated in the Bass Strait between Tasmania and mainland Australia. It is named after Matthew Flinders.

fling /flɪŋ/ v. & n. ● v. (past and past part. **flung** /flʌŋ/) **1** tr. throw or hurl (an object) forcefully. **2** refl. **a** (usu. foll. by into) rush headlong (into a person's arms, a train, etc.). **b** (usu. foll. by into) embark wholeheartedly (on an enterprise). **c** (usu. foll. by on) throw (oneself) on a person's mercy. **3** tr. utter (words) forcefully. **4** tr. (usu. foll. by out) suddenly spread (the arms). **5** tr. (foll. by on, off) put on or take off (clothes) carelessly or rapidly. **6** intr. go angrily or violently; rush (flung out of the room). **7** tr. put or send suddenly or violently (was flung into jail). **8** tr. (foll. by away) discard or put aside thoughtlessly or rashly (flung away their reputation). **9** intr. (usu. foll. by out) (of a horse etc.) kick and plunge. **10** tr. archaic send, emit (sound, light, smell). ● n. **1** an act or instance of flinging; a throw; a plunge. **2 a** a spell of indulgence or wild behaviour (he's had his fling). **b** colloq. an attempt (give it a fling). **3** an impetuous, whirling Scottish dance, esp. the Highland fling. □ **have a fling at 1** make an attempt at. **2** jeer at. □ **flinger** n. [ME, perh. f. ON]

flint /flɪnt/ n. **1 a** a hard grey stone of nearly pure silica occurring naturally as nodules or bands in chalk. **b** a piece of this esp. as flaked or ground to form a primitive tool or weapon. **2** a piece of hard alloy of rare-earth metals used to give an igniting spark in a cigarette-lighter etc. **3** a piece of flint used with steel to produce fire, esp. in a flintlock gun. **4** anything hard and unyielding. □ **flint corn** a variety of maize having hard translucent grains. **flint glass** an optical glass of low refractive index containing lead oxide, originally made with flint. □ **flinty** adj. (**flintier**, **flintiest**). **flintily** adv. **flintiness** n. [OE]

flintlock /ˈflɪntlɒk/ n. hist. **1** an old type of gun fired by a spark from a flint. **2** the lock producing such a spark.

Flintshire /ˈflɪntʃɪə(r)/ a former county of NE Wales. It was made part of Clwyd in 1974.

flip¹ /flɪp/ v., n., & adj. ● v. (**flipped**, **flipping**) **1** tr. **a** flick or toss (a coin, pellet, etc.) with a quick movement so that it spins in the air. **b** remove (a small object) from a surface with a flick of the fingers. **2** tr. **a** strike or flick (a person's ear, cheek, etc.) lightly or smartly. **b** move (a fan, whip, etc.) with a sudden jerk. **3** tr. turn (a small object) over. **4** intr. **a** make a fillip or flicking noise with the fingers. **b** (foll. by at) strike smartly at. **5** intr. move about with sudden jerks. **6** intr. sl. become suddenly angry, excited, or enthusiastic. ● n. **1** a smart light blow; a flick. **2** colloq. **a** a short pleasure flight in an aircraft. **b** a quick tour etc. **3** an act of flipping over (gave the stone a flip). ● adj. colloq. glib; flippant. □ **flip chart** a large pad erected on a stand and bound so that one page can be turned over at the top to reveal the next. **flip one's lid** sl. **1** lose self-control. **2** go mad. **flip side** colloq. the less important side of something (orig. of a gramophone record). **flip through** = flick through. [prob. f. FILLIP]

flip² /flɪp/ n. **1** a drink of heated beer and spirit. **2** = egg-flip (see EGG¹). [perh. f. FLIP¹ in the sense whip up]

flip-flop /ˈflɪpflɒp/ n. & v. ● n. **1** a usu. plastic sandal with a thong between the big and second toe. **2** N. Amer. a backward somersault. **3** an electronic switching circuit changed from one stable state to another, or through an unstable state back to its stable state, by a triggering pulse. ● v.intr. (**-flopped**, **-flopping**) move with a sound or motion suggested by 'flip-flop'. [imit.]

flippant /ˈflɪp(ə)nt/ adj. lacking in seriousness; treating serious things lightly; disrespectful. □ **flippancy** n. **flippantly** adv. [FLIP¹ + -ANT]

flipper /ˈflɪpə(r)/ n. **1** a broadened limb of a turtle, penguin, etc., used in swimming. **2** a flat rubber etc. attachment worn on the foot for underwater swimming. **3** *sl.* a hand.

flipping /ˈflɪpɪŋ/ *adj. & adv. Brit. sl.* expressing annoyance, or as an intensifier (*where's the flipping towel?*; *he flipping beat me*). [FLIP¹ + -ING²]

flirt /flɜːt/ *v. & n.* ● *v.* **1** *intr.* (usu. foll. by *with*) behave in a frivolously amorous or sexually enticing manner. **2** *intr.* (usu. foll. by *with*) **a** superficially interest oneself (with an idea etc.). **b** trifle (with danger etc.) (*flirted with disgrace*). **3** *tr.* wave or move (a fan, a bird's tail, etc.) briskly. **4** *intr. & tr.* move or cause to move with a jerk. ● *n.* **1** a person who indulges in flirting. **2** a quick movement; a sudden jerk. □ **flirty** *adj.* (**flirtier, flirtiest**). **flirtation** /flɜːˈteɪʃ(ə)n/ *n.* **flirtatious** *adj.* **flirtatiously** *adv.* **flirtatiousness** *n.* [imit.]

flit /flɪt/ *v. & n.* ● *v.intr.* (**flitted, flitting**) **1** move lightly, softly, or rapidly (*flitted from one room to another*). **2** fly lightly; make short flights (*flitted from branch to branch*). **3** *Brit. colloq.* leave one's house etc. secretly to escape creditors or obligations. **4** *esp. Sc. & N. Engl.* change one's home; move. ● *n.* **1** an act of flitting. **2** (also **moonlight flit**) a secret change of abode in order to escape creditors etc. □ **flitter** *n.* [ME f. ON *flytja*: rel. to FLEET⁵]

flitch /flɪtʃ/ *n.* **1** a side of bacon. **2** a slab of timber from a tree-trunk, usu. from the outside. **3** (in full **flitch-plate**) a strengthening plate in a beam etc. □ **flitch-beam** a compound beam, esp. of an iron plate between two slabs of wood. [OE *flicce* f. Gmc]

flitter /ˈflɪtə(r)/ *v.intr.* flit about; flutter. □ **flitter-mouse** = BAT². [FLIT + -ER⁴]

flivver /ˈflɪvə(r)/ *n. US sl.* **1** a cheap car or aircraft. **2** a failure. [20th c.: orig. uncert.]

flixweed /ˈflɪkswiːd/ *n.* a cruciferous plant, *Descurainia sophia*, with small yellow flowers, formerly thought to cure dysentery. [earlier *fluxweed*]

FLN see FRONT DE LIBÉRATION NATIONALE.

float /fləʊt/ *v. & n.* ● *v.* **1** *intr. & tr.* **a** rest or move or cause (a buoyant object) to rest or move on the surface of a liquid without sinking. **b** get afloat or set (a stranded ship) afloat. **2** *intr.* move with a liquid or current of air; drift (*the clouds floated high up*). **3** *intr. colloq.* **a** move in a leisurely or casual way (*floated about humming quietly*). **b** (often foll. by *before*) hover before the eye or mind (*the prospect of lunch floated before them*). **4** *intr.* (often foll. by *in*) move or be suspended freely in a liquid or a gas. **5** *tr.* **a** bring (a company, scheme, etc.) into being; launch. **b** offer (stock, shares, etc.) on the stock market. **6** *Commerce* **a** *intr.* (of currency) be allowed to have a fluctuating exchange rate. **b** *tr.* cause (currency) to float. **c** *intr.* (of an acceptance) be in circulation. **7** *tr.* (of water etc.) support; bear along (a buoyant object). **8** *intr. & tr.* circulate or cause (a rumour or idea) to circulate. **9** *tr.* waft (a buoyant object) through the air. **10** *tr. archaic* cover with liquid; inundate. ● *n.* **1** a thing that floats, esp.: **a** a raft. **b** a cork or quill on a fishing-line as an indicator of a fish biting. **c** a cork supporting the edge of a fishing-net. **d** the hollow or inflated part or organ supporting an organism in the water; an air bladder. **e** a hollow structure fixed underneath an aircraft enabling it to float on water. **f** a floating device on the surface of water, petrol, etc., controlling the flow. **2** a small vehicle or cart, esp. one powered by electricity (*milk float*). **3** a platform mounted on a lorry and carrying a display in a procession etc. **4 a** a sum of money used at the beginning of a period of selling in a shop, a fête, etc. to provide change. **b** a small sum of money for minor expenditure; petty cash. **5** (in *sing.* or *pl.*) *Theatr.* footlights. **6** a tool used for smoothing plaster. **7** a soft drink with a scoop of ice cream floating in it. □ **float-board** one of the boards of a water-wheel or paddle-wheel. **float glass** a high quality plate glass with perfectly true surfaces, made by drawing the molten glass continuously on to the surface of molten metal while it hardens. **float process** the process used to make float glass. **float-stone** a light, porous stone that floats. □ **floatable** *adj.* **floatability** /ˌfləʊtəˈbɪlɪtɪ/ *n.* [OE *flot, flotian* float, OE *flota* ship, ON *flota, floti* rel. to FLEET⁵: in ME infl. by OF *floter*]

floatage /ˈfləʊtɪdʒ/ *n.* **1** the act or state of floating. **2** *Brit.* **a** floating objects or masses; flotsam. **b** the right of appropriating flotsam. **3 a** ships etc. afloat on a river. **b** the part of a ship above the water-line. **4** buoyancy; floating power.

floatation var. of FLOTATION.

floater /ˈfləʊtə(r)/ *n.* **1** a person or thing that floats. **2** a floating voter. **3** *sl.* a mistake; a gaffe. **4** a person who frequently changes occupation. **5** *Stock Exch.* a government stock certificate etc. recognized as a security.

floating /ˈfləʊtɪŋ/ *adj.* not settled in a definite place; fluctuating; variable (*the floating population*). □ **floating anchor** a sea anchor. **floating bridge 1** a bridge on pontoons etc. **2** a ferry working on

chains. **floating debt** a debt repayable on demand, or at a stated time. **floating dock** a floating structure usable as a dry dock. **floating kidney 1** an abnormal condition in which the kidneys are unusually movable. **2** such a kidney. **floating light 1** a lightship. **2** a lifebuoy with a lantern. **floating point** *Computing* a decimal etc. point that does not occupy a fixed position in the numbers processed. **floating rib** any of the lower ribs, which are not attached to the breastbone. **floating voter** a voter without allegiance to any political party. □ **floatingly** *adv.*

floaty /ˈfləʊtɪ/ *adj.* (esp. of a woman's garment or a fabric) light and airy. [FLOAT]

floc /flɒk/ *n.* a flocculent mass of fine particles. [abbr. of FLOCCULUS]

flocculate /ˈflɒkjʊˌleɪt/ *v.tr. & intr.* form into flocculent masses. □ **flocculation** /ˌflɒkjʊˈleɪʃ(ə)n/ *n.*

floccule /ˈflɒkjuːl/ *n.* a small portion of matter resembling a tuft of wool.

flocculent /ˈflɒkjʊlənt/ *adj.* **1** like tufts of wool. **2** consisting of or showing tufts, downy. **3** *Chem.* (of precipitates) loosely massed. □ **flocculence** *n.* [L *floccus* FLOCK²]

flocculus /ˈflɒkjʊləs/ *n.* (*pl.* **flocculi** /-ˌlaɪ/) **1** a floccule. **2** *Anat.* a small ovoid lobe in the undersurface of the cerebellum. **3** *Astron.* a small cloudy wisp on the sun's surface. [mod.L, dimin. of FLOCCUS]

floccus /ˈflɒkəs/ *n.* (*pl.* **flocci** /ˈflɒksaɪ/) a tuft of woolly hairs or filaments. [L, = FLOCK²]

flock¹ /flɒk/ *n. & v.* ● *n.* **1 a** a number of animals of one kind, esp. birds, feeding, resting, or travelling together. **b** a number of domestic animals, esp. sheep, goats, or geese, kept together. **2** a large crowd of people. **3 a** a Christian congregation or body of believers, esp. in relation to one minister. **b** a family of children, a number of pupils, etc. ● *v.intr.* **1** congregate; mass. **2** (usu. foll. by *to, in, out, together*) go together in a crowd; troop (*thousands flocked to Wembley*). [OE *flocc*]

flock² /flɒk/ *n.* **1** a lock or tuft of wool, cotton, etc. **2 a** (often *attrib.*) material for quilting and stuffing made of wool-refuse or torn-up cloth (*a flock pillow*). **b** powdered wool or cloth. □ **flock-paper** (or **-wallpaper**) wallpaper sized and sprinkled with powdered wool to make a raised pattern. □ **flocky** *adj.* [ME f. OF *floc* f. L *floccus*]

Flodden, Battle of /ˈflɒd(ə)n/ (in full **Flodden Field**) a decisive battle of the Anglo-Scottish war of 1513, which took place at Flodden, a hill near the Northumbrian village of Branxton. A Scottish army under James IV was defeated by a smaller but better-led English force under the Earl of Surrey (sent northwards by Henry VIII, who was on campaign in France). The Scottish side suffered heavy losses, including the king and most of his nobles.

floe /fləʊ/ *n.* a sheet of floating ice. [prob. f. Norw. *flo* f. ON *fló* layer]

flog /flɒg/ *v.* (**flogged, flogging**) **1** *tr.* **a** beat with a whip, stick, etc. (as a punishment or to urge on). **b** make work through violent effort (*flogged the engine*). **2** *tr.* (often foll. by *off*) *Brit. sl.* sell. **3** *tr.* (usu. foll. by *into, out of*) drive (a quality, knowledge, etc.) into or out of a person, esp. by physical punishment. **4** *intr. & refl. sl.* proceed by violent or painful effort. □ **flog a dead horse** waste energy on something unalterable. **flog to death** *colloq.* talk about or treat at tedious length. □ **flogger** *n.* [17th-c. cant: prob. imit. or f. L *flagellare* to whip]

flong /flɒŋ/ *n. Printing* prepared paper for making stereotype moulds. [F *flan* FLAN]

flood /flʌd/ *n. & v.* ● *n.* **1 a** an overflowing or influx of water beyond its normal confines, esp. over land; an inundation. **b** the water that overflows. **2 a** an outpouring of water; a torrent (*a flood of rain*). **b** something resembling a torrent (*a flood of tears*; *a flood of relief*). **c** an abundance or excess. **3** the inflow of the tide (also in *comb.*: *flood-tide*). **4** *colloq.* a floodlight. **5** *poet.* a river; a stream; a sea. ● *v.* **1** *tr.* cover with or overflow in a flood (*rain flooded the cellar*). **b** overflow as if with a flood (*the market was flooded with foreign goods*). **2** *tr.* irrigate (*flooded the paddy fields*). **3** *tr.* deluge (a burning house, a mine, etc.) with water. **4** *intr.* (often foll. by *in, through*) arrive in great quantities (*complaints flooded in*; *fear flooded through them*). **5** *intr.* become inundated (*the bathroom flooded*). **6** *tr.* overfill (a carburettor) with petrol. **7** *intr.* experience a uterine haemorrhage. **8** *tr.* (of rain etc.) fill (a river) to overflowing. □ **flood and field** sea and land. **flood out** drive out (of one's home etc.) with a flood. **flood-tide** the periodical exceptional rise of the tide because of lunar or solar attraction. [OE *flōd* f. Gmc]

Flood, the the flood described in Genesis, brought by God upon the earth in the time of Noah (Gen. 6 ff.) because of the wickedness of the human race. In the biblical story, all but a chosen few (Noah's family and the animals sheltered on the ark) perished in the Flood, which lasted for forty days and forty nights. Parallel stories are found in other traditions (see GILGAMESH and DEUCALION). Although archaeological evidence at Ur and Kish shows that the Tigris–Euphrates valley was subject to periodic inundations, this cannot be equated directly with the biblical narrative.

floodgate /ˈflʌdgeɪt/ n. **1** a gate opened or closed to admit or exclude water, esp. the lower gate of a lock. **2** (usu. in pl.) a last restraint holding back tears, rain, anger, etc.

floodlight /ˈflʌdlaɪt/ n. & v. ● n. **1** a large powerful light (usu. one of several) to illuminate a building, sports ground, stage, etc. **2** the illumination so provided. ● v.tr. (past and past part. **floodlit** /-lɪt/) illuminate with floodlight.

floor /flɔː(r)/ n. & v. ● n. **1 a** the lower surface of a room. **b** the boards etc. of which it is made. **2 a** the bottom of the sea, a cave, a cavity, etc. **b** any level area. **3** all the rooms etc. on the same level of a building; a storey (lives on the ground floor; walked up to the sixth floor). **4 a** (in a legislative assembly) the part of the house in which members sit and from which they speak. **b** the right to speak next in debate (gave him the floor). **5** Stock Exch. the large central hall where trading takes place. **6** the minimum of prices, wages, etc. **7** colloq. the ground. ● v.tr. **1** furnish with a floor; pave. **2** bring to the ground; knock (a person) down. **3** colloq. confound, baffle (was floored by the puzzle). **4** colloq. get the better of; overcome. **5** serve as the floor of (leopard skins floored the hall). □ **first** (US **second**) **floor** the floor above the ground floor. **floor-lamp** US a standard lamp. **floor-leader** US the leader of a party in a legislative assembly. **floor manager 1** the stage-manager of a television production. **2** a shopwalker. **floor plan** a diagram of the rooms etc. on one storey of a building. **floor-polish** a manufactured substance used for polishing floors. **floor show** an entertainment presented on the floor (as opposed to the stage) of a nightclub etc. **floor-walker** US a shopwalker. **from the floor** (of a speech etc.) given by a member of the audience, not by those on the platform etc. **take the floor 1** begin to dance on a dance-floor etc. **2** speak in a debate. □ **floorless** adj. [OE flōr f. Gmc]

floorboard /ˈflɔːbɔːd/ n. a long wooden board used for flooring.

floorcloth /ˈflɔːklɒθ/ n. a cloth for washing the floor.

flooring /ˈflɔːrɪŋ/ n. the boards etc. of which a floor is made.

floozie /ˈfluːzɪ/ n. (also **floozy**) (pl. **-ies**) colloq. a girl or a woman, esp. a disreputable one. [20th c.: cf. FLOSSY and dial. floosy fluffy]

flop /flɒp/ v., n., & adv. ● v.intr. (**flopped, flopping**) **1** sway about heavily or loosely (hair flopped over his face). **2** move in an ungainly way (flopped along the beach in flippers). **3** (often foll. by down, on, into) sit, kneel, lie, or fall awkwardly or suddenly (flopped down on to the bench). **4** sl. (esp. of a play, film, book, etc.) fail; collapse (flopped on Broadway). **5** sl. sleep. **6** make a dull sound as of a soft body landing, or of a flat thing slapping water. ● n. **1 a** a flopping movement. **b** the sound made by it. **2** sl. a failure. **3** sl. esp. US a bed. ● adv. with a flop. □ **flop-house** sl. esp. US a doss-house. [var. of FLAP]

-flop /flɒp/ comb. form Computing floating-point operations per second (20 megaflops). [acronym]

floppy /ˈflɒpɪ/ adj. & n. ● adj. (**floppier, floppiest**) tending to flop; not firm or rigid. ● n. (pl. **-ies**) (in full **floppy disk**) Computing a flexible removable magnetic disk for the storage of data. □ **floppily** adv. **floppiness** n.

flor. abbr. floruit.

Flora /ˈflɔːrə/ Rom. Mythol. the goddess of flowering plants. [f. L flos floris flower]

flora /ˈflɔːrə/ n. (pl. **floras** or **florae** /-riː/) **1** the plants of a particular region, geological period, or environment. **2** a treatise on or list of these. [mod.L f. FLORA]

floral /ˈflɔːrəl, ˈflɒr-/ adj. **1** of flowers. **2** decorated with or depicting flowers. **3** of flora or floras. □ **florally** adv. [L floralis or flos floris flower]

floreat /ˈflɒrɪˌæt/ v.intr. may (he, she, or it) flourish. [L, 3rd sing. pres. subjunctive of florere flourish]

Florence /ˈflɒrəns/ (Italian **Firenze** /fiˈrɛntse/) a city in west central Italy, the capital of Tuscany, on the River Arno; pop. (1990) 408,400. Florence was a leading centre of the Italian Renaissance from the 14th to the 16th century, especially under the rule of the Medici family during the 15th century.

Florentine /ˈflɒrənˌtaɪn/ adj. & n. ● adj. **1** of or relating to Florence in Italy. **2** (**florentine** /-ˌtiːn/) (of a dish) served on a bed of spinach. ● n. **1** a native or inhabitant of Florence. **2** a kind of biscuit consisting mainly of nuts and preserved fruit, coated on one side with chocolate. [F Florentin -ine or L Florentinus f. Florentia Florence]

Flores /ˈflɔːres/ the largest of the Lesser Sunda Islands in Indonesia.

florescence /flɔːˈres(ə)ns/ n. the process, state, or time of flowering. [mod.L florescentia f. L florescere f. florere bloom]

floret /ˈflɒrɪt/ n. **1** Bot. each of the small flowers making up a composite flower-head. **2** each of the flowering stems making up a head of cauliflower, broccoli, etc. **3** a small flower. [L flos floris flower]

Florey /ˈflɔːrɪ/, Howard Walter, Baron (1898–1968), Australian pathologist. In collaboration with Sir Ernst Chain, he isolated and purified penicillin, developed techniques for its large-scale production, and performed the first clinical trials. Florey and Chain shared a Nobel Prize in 1945 with Sir Alexander Fleming.

Florianópolis /ˌflɔːrɪəˈnɒpəlɪs/ a city in southern Brazil, on the Atlantic coast, capital of the state of Santa Catarina; pop. (1990) 293,300.

floriate /ˈflɔːrɪˌeɪt/ v.tr. decorate with flower-designs etc.

floribunda /ˌflɒrɪˈbʌndə/ n. a plant, esp. a rose, bearing dense clusters of flowers. [mod.L f. floribundus freely flowering f. L flos floris flower, infl. by L abundus copious]

floriculture /ˈflɒrɪˌkʌltʃə(r)/ n. the cultivation of flowers. □ **floricultural** /ˌflɒrɪˈkʌltʃərəl/ adj. **floriculturist** n. [L flos floris flower + CULTURE, after horticulture]

florid /ˈflɒrɪd/ adj. **1** ruddy; flushed; high-coloured (a florid complexion). **2** (of a book, a picture, music, architecture, etc.) elaborately ornate; ostentatious; showy. **3** adorned with or as with flowers; flowery. □ **floridly** adv. **floridness** n. **floridity** /flɒˈrɪdɪtɪ/ n. [F floride or L floridus f. flos floris flower]

Florida /ˈflɒrɪdə/ a state forming a peninsula of the south-eastern US; pop. (1990) 12,937,900; capital, Tallahassee. Explored by Ponce de León in 1513, it was held by Spain from then until 1819, when it was purchased by the US; it became the 27th state of the Union in 1845.

Florida Keys a chain of small islands off the tip of the Florida peninsula. Linked to each other and to the mainland by a series of causeways and bridges forming the Overseas Highway, the islands extend south-westwards over a distance of 160 km (100 miles). Key Largo, the longest island, is closest to the mainland.

floriferous /flɔːˈrɪfərəs/ adj. (of a seed or plant) producing many flowers. [L florifer f. flos floris flower]

florilegium /ˌflɔːrɪˈliːdʒəm/ n. (pl. **florilegia** /-dʒɪə/ or **florilegiums**) an anthology. [mod.L f. L flos floris flower + legere gather, transl. Gk anthologion ANTHOLOGY]

florin /ˈflɒrɪn/ n. hist. **1 a** a former British coin and monetary unit worth two shillings. **b** an English gold coin of the 14th century, worth 6s. 8d. **2** a foreign coin of gold or silver, esp. a Dutch guilder. [ME f. OF f. It. fiorino dimin. of fiore flower f. L flos floris, the orig. coin bearing a fleur-de-lis]

Florio /ˈflɔːrɪˌəʊ/, John (c.1553–1625), English lexicographer, of Italian descent. In 1598 he produced an Italian–English dictionary entitled A Worlde of Wordes. His most important work was the first translation into English of Montaigne's essays (1603), on which Shakespeare drew in The Tempest.

florist /ˈflɒrɪst/ n. a person who deals in or grows flowers. □ **floristry** n. [L flos floris flower + -IST]

floristic /flɒˈrɪstɪk/ adj. Bot. relating to the study of the distribution of plants. □ **floristically** adv. **floristics** n.

floruit /ˈflɒrʊɪt/ v. & n. ● v.intr. (he or she) was alive and working; flourished (used of a person, esp. a painter, a writer, etc., whose exact dates are unknown). ● n. the period or date at which a person lived or worked. [L, = he or she flourished]

flory var. of FLEURY.

floscular /ˈflɒskjʊlə(r)/ adj. (also **flosculous** /-ləs/) Bot. having florets or composite flowers. [L flosculus dimin. of flos flower]

floss /flɒs/ n. & v. ● n. **1** the rough silk enveloping a silkworm's cocoon. **2** untwisted silk thread used in embroidery. **3** = dental floss. ● v.tr. (also absol.) clean (the teeth) with dental floss. □ **floss silk** a rough silk used in cheap goods. [F (soie) floche floss(-silk) f. OF flosche down, nap of velvet]

flossy /ˈflɒsɪ/ adj. (**flossier, flossiest**) **1** of or like floss. **2** colloq. fancy, showy.

flotation /fləʊˈteɪʃ(ə)n/ n. (also **floatation**) **1** the process of launching or financing a commercial enterprise. **2** the separation of the components of crushed ore etc. by their different capacities to float. **3** the capacity to float. □ **centre of flotation** the centre of gravity in a floating body. [alt. of floatation f. FLOAT, after rotation etc.]

flotilla /fləˈtɪlə/ n. **1** a small fleet. **2** a fleet of boats or small ships. [Sp., dimin. of *flota* fleet, OF *flote* multitude]

flotsam /ˈflɒtsəm/ n. wreckage found floating. □ **flotsam and jetsam 1** odds and ends; rubbish. **2** vagrants etc. [AF *floteson* f. *floter* FLOAT]

flounce[1] /flaʊns/ v. & n. ● v.intr. (often foll. by *away, about, off, out*) go or move with an agitated, violent, or impatient motion (*flounced out in a huff*). ● n. a flouncing movement. [16th c.: orig. unkn.: perh. imit., as *bounce, pounce*]

flounce[2] /flaʊns/ n. & v. ● n. a wide ornamental strip of material gathered and sewn to a skirt, dress, etc.; a frill. ● v.tr. trim with a flounce or flounces. [alt. of earlier *frounce* fold, pleat, f. OF *fronce* f. *froncir* wrinkle]

flounder[1] /ˈflaʊndə(r)/ v. & n. ● v.intr. **1** struggle in mud, or as if in mud, or when wading. **2** perform a task badly or without knowledge; be out of one's depth. ● n. an act of floundering. □ **flounderer** n. [imit.: perh. assoc. with *founder, blunder*]

flounder[2] /ˈflaʊndə(r)/ n. **1** a small edible flatfish, *Pleuronectes flesus*, native to European coastal waters. **2** any small flatfish of the families Pleuronectidae or Bothidae. [ME f. AF *floundre*, OF *flondre*, prob. of Scand. orig.]

flour /ˈflaʊə(r)/ n. & v. ● n. **1** a meal or powder obtained by grinding and usu. sifting cereals, esp. wheat. **2** any fine powder. ● v.tr. **1** sprinkle with flour. **2** *US* grind into flour. □ **floury** adj. (**flourier, flouriest**). **flouriness** n. [ME, differentiated spelling of FLOWER in the sense 'finest part']

flourish /ˈflʌrɪʃ/ v. & n. ● v. **1** intr. **a** grow vigorously; thrive. **b** prosper; be successful. **c** be in one's prime. **d** be in good health. **e** (as **flourishing** adj.) successful, prosperous. **2** intr. (usu. foll. by *in, at, about*) spend one's life; be active (at a specified time) (*flourished in the Middle Ages*) (cf. FLORUIT). **3** tr. show ostentatiously (*flourished his cheque-book*). **4** tr. wave (a weapon, one's limbs, etc.) vigorously. ● n. **1** an ostentatious gesture with a weapon, a hand, etc. (*removed his hat with a flourish*). **2** an ornamental curving decoration of handwriting. **3** a florid verbal expression; a rhetorical embellishment. **4** *Mus.* **a** a fanfare played by brass instruments. **b** an ornate musical passage. **c** an extemporized addition played esp. at the beginning or end of a composition. **5** *archaic* an instance of prosperity; a flourishing. □ **flourisher** n. [ME f. OF *florir* ult. f. L *florere* f. *flos floris* flower]

flout /flaʊt/ v. & n. ● v. **1** tr. express contempt for (the law, rules, etc.) by word or action; mock; insult (*flouted convention by shaving her head*). ¶ Often confused with *flaunt*. **2** intr. (often foll. by *at*) mock or scoff. ● n. a flouting speech or act. [perh. f. Du. *fluiten* whistle, hiss: cf. FLUTE]

flow /fləʊ/ v. & n. ● v.intr. **1** glide along as a stream (*the Thames flows under London Bridge*). **2 a** (of a liquid, esp. water) gush out; spring. **b** (of blood, liquid, etc.) be spilt. **3** (of blood, money, electric current, etc.) circulate. **4** (of people or things) come or go in large numbers or smoothly (*traffic flowed down the hill*). **5** (of talk, literary style, etc.) proceed easily and smoothly. **6** (of a garment, hair, etc.) hang easily or gracefully; undulate. **7** (often foll. by *from*) result from; be caused by (*his failure flows from his diffidence*). **8** (esp. of the tide) be in flood; run full. **9** (of wine) be poured out copiously. **10** (of a rock or metal) undergo a permanent change of shape under stress. **11** (foll. by *with*) *archaic* be plentifully supplied with (*land flowing with milk and honey*). ● n. **1 a** a flowing movement in a stream. **b** the manner in which a thing flows (*a sluggish flow*). **c** a flowing liquid (*couldn't stop the flow*). **d** a copious outpouring; a stream (*a continuous flow of complaints*). **2** the rise of a tide or a river (*ebb and flow*). **3** the gradual deformation of a rock or metal under stress. **4** *Sc.* a bog or morass. □ **flow chart** (or **diagram** or **sheet**) **1** a diagram of the movement or action of things or persons engaged in a complex activity. **2** a graphical representation of a computer program in relation to its sequence of functions (as distinct from the data it processes). **flow of spirits** habitual cheerfulness. **flow-on** *Austral.* a wage or salary adjustment made as a consequence of one already made in a similar or related occupation. [OE *flōwan* f. Gmc. rel. to FLOOD]

flower /ˈflaʊə(r)/ n. & v. ● n. **1 a** *Bot.* the reproductive organ in a plant from which the fruit or seed develops, containing one or more pistils or stamens or both, and usu. a corolla and calyx. **b** such an organ when brightly coloured and conspicuous; a blossom; a bloom. **c** a blossom on a cut stem, used in bunches for decoration. **2** a plant cultivated or noted for its flowers. **3** (in *pl.*) ornamental phrases (*flowers of speech*). ● v. **1** intr. (of a plant) produce flowers; bloom or blossom. **2** intr. reach a peak. **3** tr. cause or allow (a plant) to flower. **4** tr. decorate with worked flowers or a floral design. □ **flower-bed** a garden bed in which flowers are grown. **flower-girl** a woman who sells flowers, esp. in the street. **flower-head** = HEAD n. 4d. **the flower of** the best or best part of. **flower people** hippies carrying or wearing flowers as symbols of peace and love. **flower power** the ideas of the flower people regarded as an instrument in changing the world. **flowers of sulphur** *Chem.* a fine powder produced when sulphur evaporates and condenses. **in flower** with the flowers out. □ **flowered** adj. (also in *comb.*). **flowerless** adj. **flowerlike** adj. [ME f. AF *flur*, OF *flour, flor*, f. L *flos floris*]

flowerer /ˈflaʊərə(r)/ n. a plant that flowers at a specified time (*a late flowerer*).

floweret /ˈflaʊərɪt/ n. a small flower.

flowering /ˈflaʊərɪŋ/ adj. **1** (of a plant) in bloom. **2** capable of bearing flowers, esp. compared with a similar plant in which the flowers are less conspicuous or absent.

flowering plant n. *Bot.* an angiosperm. The flowering plants are divided into the dicotyledons and the monocotyledons.

flowerpot /ˈflaʊəˌpɒt/ n. a pot in which a plant may be grown.

flowery /ˈflaʊərɪ/ adj. **1** decorated with flowers or floral designs. **2** (of literary style, manner of speech, etc.) high-flown; ornate. **3** full of flowers (*a flowery meadow*). □ **floweriness** n.

flowing /ˈfləʊɪŋ/ adj. **1** (of literary style etc.) fluent; easy. **2** (of a line, a curve, or a contour) smoothly continuous, not abrupt. **3** (of hair, a garment, a sail, etc.) unconfined. □ **flowingly** adv.

flowmeter /ˈfləʊˌmiːtə(r)/ n. an instrument for measuring the rate of flow of water, gas, fuel, etc., esp. in a pipe.

flown past part. of FLY[1].

flowstone /ˈfləʊstəʊn/ n. *Geol.* rock deposited in a thin sheet by a flow of water.

FLQ abbr. *Can.* Front de Libération du Québec.

Flt. Lt. abbr. Flight Lieutenant.

Flt. Off. abbr. Flight Officer.

Flt. Sgt. abbr. Flight Sergeant.

flu /fluː/ n. *colloq.* influenza. [abbr.]

flub /flʌb/ v. & n. *N. Amer. colloq.* ● v.tr. & intr. (**flubbed, flubbing**) botch; bungle. ● n. something badly or clumsily done. [20th c.: orig. unkn.]

fluctuate /ˈflʌktjʊˌeɪt/ v.intr. vary irregularly; be unstable, vacillate; rise and fall, move to and fro. □ **fluctuation** /ˌflʌktjʊˈeɪʃ(ə)n/ n. [L *fluctuare* f. *fluctus* flow, wave f. *fluere fluct-* flow]

flue /fluː/ n. **1** a smoke-duct in a chimney. **2** a channel for conveying heat, esp. a hot-air passage in a wall; a tube for heating water in some kinds of boiler. □ **flue-cure** cure (tobacco) by artificial heat from flues. **flue-pipe** an organ pipe into which the air enters directly, not striking a reed. [16th c.: orig. unkn.]

fluence /ˈfluːəns/ n. *colloq.* influence. □ **put the fluence on** apply hypnotic etc. power to (a person). [shortening of INFLUENCE]

fluency /ˈfluːənsɪ/ n. **1** a smooth, easy flow, esp. in speech or writing. **2** a ready command of words or of a specified foreign language.

fluent /ˈfluːənt/ adj. **1 a** (of speech or literary style) flowing naturally and readily. **b** having command of a foreign language (*is fluent in German*). **c** able to speak quickly and easily. **2** flowing easily or gracefully (*the fluent line of her arabesque*). **3** *archaic* liable to change; unsettled. □ **fluently** adv. [L *fluere* flow]

fluff /flʌf/ n. & v. ● n. **1** soft, light, feathery material coming off blankets etc. **2** soft fur or feathers. **3** *sl.* **a** a mistake in delivering theatrical lines, in playing music, etc. **b** a mistake in playing a game. **4** something insubstantial or trifling, esp. sentimental writing. ● v. **1** tr. & intr. (often foll. by *up*) shake into or become a soft mass. **2** tr. & intr. *colloq.* make a mistake in (a theatrical part, a game, playing music, a speech, etc.); blunder (*fluffed his opening line*). **3** tr. make into fluff. **4** tr. put a soft surface on (the flesh side of leather). □ **bit of fluff** *sl. offens.* a woman regarded as an object of sexual desire. [prob. dial. alt. of *flue* fluff]

fluffy /ˈflʌfɪ/ adj. (**fluffier, fluffiest**) **1** of or like fluff. **2** covered in fluff; downy. □ **fluffily** adv. **fluffiness** n.

flugelhorn /ˈfluːɡ(ə)lˌhɔːn/ n. a valved brass wind instrument like a cornet but with a broader tone. [G *Flügelhorn* f. *Flügel* wing + *Horn* horn]

fluid /ˈfluːɪd/ n. & adj. ● n. **1** a substance, esp. a gas or liquid, lacking definite shape and capable of flowing and yielding to the slightest pressure. **2** a fluid part or secretion. ● adj. **1** able to flow and alter shape freely. **2** constantly changing or fluctuating (*the situation is fluid*). **3** (of a clutch, coupling, etc.) in which liquid is used to transmit power. □ **fluid drachm** see DRACHM. **fluid mechanics** the study of forces and flow within fluids. **fluid ounce** see OUNCE[1]. □ **fluidly** adv.

fluidness n. **fluidity** /fluː'ɪdɪtɪ/ n. **fluidify** /-'ɪdɪˌfaɪ/ v.tr. (**-ies, -ied**). [F *fluide* or L *fluidus* f. *fluere* flow]

fluidics /fluː'ɪdɪks/ n.pl. (usu. treated as *sing.*) the study and technique of using small interacting flows and fluid jets for functions usu. performed by electronic devices. □ **fluidic** adj.

fluidize /'fluːɪˌdaɪz/ v.tr. (also **-ise**) cause (a finely divided solid) to acquire the characteristics of a fluid by the upward passage of a gas etc. □ **fluidized bed** a layer of a fluidized solid, used in chemical processes and in the efficient burning of coal for power generation. □ **fluidization** /ˌfluːɪdaɪ'zeɪʃ(ə)n/ n.

fluidounce /'fluːɪdˌaʊns/ n. US a fluid ounce (see OUNCE[1]).

fluidram /'fluːɪˌdræm/ n. US a fluid drachm (see DRACHM).

fluke[1] /fluːk/ n. & v. ● n. **1** a lucky accident (*won by a fluke*). **2** a chance breeze. ● v.tr. achieve by a fluke (*fluked that shot*). [19th c.: perh. f. dial. *fluke* guess]

fluke[2] /fluːk/ n. **1** a parasitic flatworm of the class Trematoda, esp. of the subclass Digenea, which includes liver flukes and blood flukes (cf. TREMATODE). **2** a flatfish, esp. a flounder. [OE *flōc*]

fluke[3] /fluːk/ n. **1** *Naut.* a broad triangular plate on the arm of an anchor. **2** the barbed head of a lance, harpoon, etc. **3** either of the lobes of a whale's tail. [16th c.: perh. f. FLUKE[2]]

fluky /'fluːkɪ/ adj. (**flukier, flukiest**) of the nature of a fluke; obtained more by chance than skill. □ **flukily** adv. **flukiness** n.

flume /fluːm/ n. & v. ● n. **1** an artificial channel conveying water etc. for industrial use. **2** a ravine with a stream. **3** a water chute or water slide at an amusement park or swimming-pool. ● v. **1** intr. build flumes. **2** tr. convey down a flume. [ME f. OF *flum, flun* f. L *flumen* river f. *fluere* flow]

flummery /'flʌmərɪ/ n. (pl. **-ies**) **1** empty compliments; trifles; nonsense. **2** a sweet dish made with beaten eggs, sugar, etc. [Welsh *llymru*, of unkn. orig.]

flummox /'flʌməks/ v.tr. colloq. bewilder, confound, disconcert. [19th c.: prob. dial., imit.]

flump /flʌmp/ v. & n. ● v. (often foll. by *down*) **1** intr. fall or move heavily. **2** tr. set or throw down with a heavy thud. ● n. the action or sound of flumping. [imit.]

flung past and past part. of FLING.

flunk /flʌŋk/ v. & n. colloq. ● v. **1** tr. **a** fail (an examination etc.). **b** fail (an examination candidate). **2** intr. (often foll. by *out*) fail utterly; give up. ● n. an instance of flunking. □ **flunk out** be dismissed from school etc. after failing an examination. [cf. FUNK[1] and obs. *flink* be a coward]

flunkey /'flʌŋkɪ/ n. (also **flunky**) (pl. **-eys** or **-ies**) usu. derog. **1** a liveried servant; a footman. **2** a toady; a snob. **3** US a cook, waiter, etc. □ **flunkeyism** n. [18th c. (orig. Sc.): perh. f. FLANK with the sense 'sidesman, flanker']

fluoresce /fluə'res/ v.intr. be or become fluorescent.

fluorescein /'fluərəˌsiːn, -sɪn/ n. Chem. an orange dye with a yellowish-green fluorescence, used in solution as an indicator in biochemistry and medicine. [FLUORESCENCE + -IN]

fluorescence /fluə'res(ə)ns/ n. **1** the visible or invisible radiation produced from certain substances as a result of incident radiation of a shorter wavelength as X-rays, ultraviolet light, etc. **2** the property of absorbing light of short (invisible) wavelength and emitting light of longer (visible) wavelength. [FLUORSPAR (which fluoresces) after *opalescence*]

fluorescent /fluə'res(ə)nt/ adj. (of a substance) having or showing fluorescence. □ **fluorescent lamp** (or **bulb**) a lamp or bulb radiating largely by fluorescence, esp. a tubular lamp in which phosphor on the inside surface of the tube is made to fluoresce by ultraviolet radiation from mercury vapour. **fluorescent screen** a screen coated with fluorescent material to show images from X-rays etc.

fluoridate /'fluərɪˌdeɪt/ v.tr. add traces of fluoride to (drinking-water etc.).

fluoridation /ˌfluərɪ'deɪʃ(ə)n/ n. (also **fluoridization** /-daɪ'zeɪʃ(ə)n/) the addition of traces of fluoride to drinking-water in order to prevent or reduce tooth-decay.

fluoride /'fluəraɪd/ n. Chem. a binary compound of fluorine.

fluorinate /'fluərɪˌneɪt/ v.tr. **1** = FLUORIDATE. **2** introduce fluorine into (a compound) (*fluorinated hydrocarbons*). □ **fluorination** /ˌfluərɪ'neɪʃ(ə)n/ n.

fluorine /'fluəriːn/ n. a pale-yellow gaseous chemical element (atomic number 9; symbol **F**). Fluorine was identified as an element by Scheele in 1771, but its very difficult and dangerous isolation was not achieved until 1886, by the French chemist Moissan. Fluorine is a halogen and the most reactive of all the elements, causing very severe burns on contact with skin. It occurs in fluorite, cryolite, and other minerals. Fluorine compounds are used to protect against dental decay, and the very unreactive fluorocarbons are used to form non-stick surfaces and as lubricants. [F (as FLUORSPAR)]

fluorite /'fluəraɪt/ n. Mineral. a mineral form of calcium fluoride. [It. (as FLUORSPAR)]

fluoro- /'fluərəʊ/ comb. form **1** fluorine (*fluorocarbon*). **2** fluorescence (*fluoroscope*). [FLUORINE, FLUORESCENCE]

fluorocarbon /ˌfluərəʊ'kɑːb(ə)n/ n. Chem. a compound formed by replacing one or more of the hydrogen atoms in a hydrocarbon with fluorine atoms.

fluoroscope /'fluərəˌskəʊp/ n. an instrument with a fluorescent screen on which X-ray images may be viewed without taking and developing X-ray photographs.

fluorosis /fluə'rəʊsɪs/ n. poisoning by fluorine or its compounds. [F *fluorose* (as FLUORO- 1)]

fluorspar /'fluəspɑː(r)/ n. = FLUORITE. [archaic *fluor* a mineral used as a flux, (later) fluorspar, f. L *fluor* f. *fluere* flow + SPAR[3]]

flurry /'flʌrɪ/ n. & v. ● n. (pl. **-ies**) **1** a gust or squall (of snow, rain, etc.). **2** a sudden burst of activity. **3** a commotion; excitement; nervous agitation (*a flurry of speculation; the flurry of the city*). ● v.tr. (**-ies, -ied**) confuse by haste or noise; agitate. [imit.: cf. obs. *flurr* ruffle, *hurry*]

flush[1] /flʌʃ/ v. & n. ● v. **1** intr. **a** blush, redden (*he flushed with embarrassment*). **b** glow with a warm colour (*sky flushed pink*). **2** tr. (usu. as **flushed** adj.) cause to glow, blush, or be elated (often foll. by *with*: *flushed with pride*). **3** tr. **a** cleanse (a drain, lavatory, etc.) by a rushing flow of water. **b** (often foll. by *away, down*) dispose of (an object) in this way (*flushed away the cigarette*). **4** intr. rush out, spurt. **5** tr. flood (*the river flushed the meadow*). **6** intr. (of a plant) throw out fresh shoots. ● n. **1** a **a** blush. **b** a glow of light or colour. **2** a a rush of water. **b** the cleansing of a drain, lavatory, etc. by flushing. **3** a a rush of emotion. **b** the elation produced by a victory etc. (*the flush of triumph*). **4** sudden abundance. **5** freshness; vigour (*in the first flush of womanhood*). **6** a (also **hot flush**) a sudden feeling of heat during the menopause. **b** a feverish temperature. **c** facial redness, esp. caused by fever, alcohol, etc. **7** a fresh growth of grass etc. □ **flusher** n. [ME, perh. = FLUSH[4] infl. by *flash* and *blush*]

flush[2] /flʌʃ/ adj. & v. ● adj. **1** (often foll. by *with*) in the same plane; level; even (*the sink is flush with the cooker; fitted it flush with the wall*). **2** (usu. predic.) colloq. **a** having plenty of money. **b** (of money) abundant, plentiful. **3** full to overflowing; in flood. ● v.tr. **1** make (surfaces) level. **2** fill in (a joint) level with a surface. □ **flushness** n. [prob. f. FLUSH[1]]

flush[3] /flʌʃ/ n. a hand of cards all of one suit, esp. in poker. □ **royal flush** a straight poker flush headed by an ace. **straight flush** a flush that is a numerical sequence. [OF *flus, flux* f. L *fluxus* FLUX]

flush[4] /flʌʃ/ v. **1** tr. cause (esp. a game bird) to fly up from cover. **2** intr. (of a bird) fly up and away. □ **flush out 1** reveal. **2** drive out. [ME, imit.: cf. *fly, rush*]

Flushing /'flʌʃɪŋ/ (Dutch **Vlissingen** /'vlɪsɪŋə/) a port in the SW Netherlands; pop. (1991) 43,800.

fluster /'flʌstə(r)/ v. & n. ● v. **1** tr. & intr. make or become nervous or confused; flurry (*was flustered by the noise; he flusters easily*). **2** tr. confuse with drink; half-intoxicate. **3** intr. bustle. ● n. a confused or agitated state. [ME: orig. unkn.: cf. Icel. *flaustr*(a) hurry, bustle]

flute /fluːt/ n. & v. ● n. **1** a a high-pitched wind instrument of metal or wood in which the air is directed against an edge, and having holes along its length. (*See note below.*) **b** an organ stop having a similar sound. **c** any of various wind instruments resembling a flute. **d** a flute-player. **2** a Archit. an ornamental vertical groove in a column. **b** a trumpet-shaped frill on a dress etc. **c** any similar cylindrical groove. **3** a tall narrow wineglass. ● v. **1** intr. play the flute. **2** intr. speak, sing, or whistle in a fluting way. **3** tr. make flutes or grooves in. **4** tr. play (a tune etc.) on a flute. □ **flutelike** adj. **fluting** n. **flutist** n. US (cf. FLAUTIST). **fluty** adj. (in sense 1a of n.). [ME f. OF *flēute, flaute, flahute*, prob. f. Prov. *flaüt*]

▪ The modern orchestral flute is usually made in silver, stainless steel, or occasionally wood, and has holes along its length stopped by fingers or (since the 19th century) keys, with a mouthpiece in the side, near one end. It has a range of three octaves. From the 14th century the flute had an association with military marching music; it became a standard orchestral instrument in the early 18th century.

flutter /'flʌtə(r)/ v. & n. ● v. **1** a intr. flap the wings in flying or trying to fly (*butterflies fluttered in the sunshine*). **b** tr. flap (the wings). **2** intr. fall with a quivering motion (*leaves fluttered to the ground*). **3** *intr. & tr.* move or cause to move irregularly or tremblingly (*the wind fluttered the flag*). **4** *intr.* go about restlessly; flit; hover. **5** *tr.* agitate, confuse. **6** *intr.* (of a pulse or heartbeat) beat feebly or irregularly. **7** *intr.* tremble with excitement or agitation. ● *n.* **1 a** the act of fluttering. **b** an instance of this. **2** a tremulous state of excitement; a sensation (*was in a flutter; caused a flutter with his behaviour*). **3** *Brit. sl.* a small bet, esp. on a horse. **4** an abnormally rapid but regular heartbeat. **5** *Aeron.* an undesired oscillation in a part of an aircraft etc. under stress. **6** *Mus.* a rapid movement of the tongue (as when rolling one's rs) in playing a wind instrument. **7** *Electronics* a rapid variation of pitch, esp. of recorded sound (cf. wow²). **8** a vibration. □ **flutter the dovecots** cause alarm among normally imperturbable people. □ **flutterer** *n.* **fluttery** *adj.* [OE *floterian, flotorian,* frequent. form rel. to FLEET⁵]

fluvial /ˈfluːvɪəl/ *adj.* of or found in a river or rivers. [ME f. L *fluvialis* f. *fluvius* river f. *fluere* flow]

fluviatile /ˈfluːvɪəˌtaɪl/ *adj.* of, found in, or produced by a river or rivers. [F f. L *fluviatilis* f. *fluviatus* moistened f. *fluvius*]

fluvio- /ˈfluːvɪəʊ/ *comb. form* river (*fluviometer*). [L *fluvius* river f. *fluere* flow]

fluvioglacial /ˌfluːvɪəʊˈɡleɪsɪəl, -ˈɡleɪʃ(ə)l/ *adj.* *Geol.* of or caused by streams from glacial ice, or the combined action of rivers and glaciers.

fluviometer /ˌfluːvɪˈɒmɪtə(r)/ *n.* an instrument for measuring the rise and fall of rivers.

flux /flʌks/ *n. & v.* ● *n.* **1** a process of flowing or flowing out. **2** an issue or discharge. **3** continuous change (*in a state of flux*). **4** a substance mixed with a metal etc. to promote fusion. **5** *Physics* **a** the rate of flow of any fluid across a given area. **b** the amount of fluid crossing an area in a given time. **6** *Physics* the amount of radiation or particles incident on an area in a given time. **7** *Electr.* the total electric or magnetic field passing through a surface. **8** *Med.* an abnormal discharge of blood or excrement from the body. ● *v.* **1** *tr. & intr.* make or become fluid. **2** *tr.* **a** fuse. **b** treat with a fusing flux. [ME f. OF *flux* or L *fluxus* f. *fluere* flux-flow]

fluxion /ˈflʌkʃ(ə)n/ *n.* *Math.* the rate at which a variable quantity changes; a derivative. [F *fluxion* or L *fluxio* f. FLUX]

fly¹ /flaɪ/ *v. & n.* ● *v.* (**flies**; *past* **flew** /fluː/; *past part.* **flown** /fləʊn/) **1** *intr.* move through the air under control, esp. with wings. **2** (of an aircraft or its occupants): **a** *intr.* travel through the air or through space. **b** *tr.* traverse (a region or distance) (*flew the Channel*). **3** *tr.* **a** control the flight of (esp. an aircraft). **b** transport in an aircraft. **4** **a** *tr.* cause to fly or remain aloft. **b** *intr.* (of a flag, hair, etc.) wave or flutter. **5** *intr.* pass or rise quickly through the air or over an obstacle. **6** *intr.* go or move quickly; pass swiftly (*time flies*). **7** *intr.* **a** flee. **b** *colloq.* depart hastily. **8** *intr.* be driven or scattered; be forced off suddenly (*sent me flying; the door flew open*). **9** *intr.* (foll. by *at, upon*) **a** hasten or spring violently. **b** attack or criticize fiercely. **10** *tr.* flee from; escape in haste. ● *n.* (pl. **-ies**) **1** (usu. in pl.) **a** a flap on a garment, esp. trousers, to contain or cover a fastening. **b** this fastening. **2** a flap at the entrance of a tent. **3** (in pl.) the space over the proscenium in a theatre. **4** the act or an instance of flying. **5** (pl. usu. **flys**) *Brit. hist.* a one-horse hackney carriage. **6** a speed-regulating device in clockwork and machinery. □ **fly-away** (of hair etc.) tending to fly out or up; streaming. **fly-by** (pl. **-bys**) a flight past a position, esp. the approach of a spacecraft to a planet for observation. **fly-by-night** *adj.* **1** unreliable. **2** short-lived. ● *n.* an unreliable person. **fly-by-wire** (often *attrib.*) a semi-automatic and usu. computer-regulated system for controlling the flight of an aircraft, spacecraft, etc. **fly-drive** *n.* (often *attrib.*) a holiday which combines the cost of a flight and of car rental. ● *v.intr.* take such a holiday. **fly-half** *Rugby* a stand-off half. **fly high 1** pursue a high ambition. **2** excel, prosper. **fly in the face of** openly disregard or disobey; conflict roundly with (probability, the evidence, etc.). **fly into a rage** (or **temper** etc.) become suddenly or violently angry. **fly a kite 1** try something out; test public opinion. **2** raise money by an accommodation bill. **fly off the handle** *colloq.* lose one's temper suddenly and unexpectedly. **fly-past** a ceremonial flight of aircraft past a person or a place. **fly-pitcher** *sl.* a street-trader. **fly-pitching** *sl.* street-trading. **fly-post** display (posters etc.) rapidly in unauthorized places. **fly-tip** illegally dump (waste). **fly-tipper** a person who engages in fly-tipping. □ **flyable** *adj.* [OE *flēogan* f. Gmc]

fly² /flaɪ/ *n.* (pl. **flies**) **1** an insect of the order Diptera, with only one pair of usu. transparent wings. **2** any other winged insect, e.g. a firefly or mayfly. **3** a disease of plants or animals caused by flies. **4** a natural or artificial fly used as bait in fishing. □ **fly agaric** a poisonous fungus, *Amanita muscaria,* forming bright red toadstools with white flecks. **fly-blow** flies' eggs contaminating food, esp. meat. **fly-blown** *adj.* tainted,

esp. by flies. **fly-fish** *v.intr.* fish with a fly. **fly in the ointment** a minor irritation that spoils enjoyment. **fly on the wall** an unnoticed observer. **fly-paper** sticky treated paper for catching flies. **fly-trap** a plant that catches flies, esp. the Venus fly-trap. **like flies** in large numbers (usu. of people dying in an epidemic etc.). **no flies on** *colloq.* nothing to diminish (a person's) astuteness. [OE *flȳge, flēoge* f. WG]

fly³ /flaɪ/ *adj.* (**flyer, flyest**) **1** *Brit. sl.* knowing, clever, alert. **2** *US sl.* stylish, good-looking. □ **flyness** *n.* [19th c.: orig. unkn.]

flycatcher /ˈflaɪˌkætʃə(r)/ *n.* a passerine bird that catches flying insects, esp. in short flights from a perch. Typical flycatchers constitute the Old World family Muscicapidae; others are found in the Old World family Monarchidae (*monarch flycatchers*), and in the New World family Tyrannidae (*tyrant flycatchers*).

flyer /ˈflaɪə(r)/ *n.* (also **flier**) **1** an airman or airwoman. **2** a thing that flies in a specified way (*a poor flyer*). **3** *colloq.* a fast-moving animal or vehicle. **4** an ambitious or outstanding person; an outstanding thing. **5** a small handbill. **6** *US* a speculative investment. **7** a flying jump.

flying /ˈflaɪɪŋ/ *adj. & n.* ● *adj.* **1** fluttering or waving in the air; hanging loose. **2** hasty, brief (*a flying visit*). **3** designed for rapid movement. **4** (of an animal) able to make very long leaps by using winglike membranes etc. ● *n.* flight, esp. in an aircraft. □ **flying boat** a large seaplane with a fuselage that resembles a boat. **flying bomb** a pilotless aircraft with an explosive warhead. **flying buttress** *Archit.* a buttress slanting from a separate column, usu. forming an arch with the wall it supports. **flying doctor** a doctor (esp. in a large sparsely populated area) who visits distant patients by aircraft. **flying dragon** see DRAGON 3. **flying fish** a tropical fish of the family Exocoetidae, with winglike pectoral fins for gliding through the air. **flying fox** a large fruit-bat, esp. of the genus *Pteropus,* with a foxlike head. **flying lemur** a lemur-like SE Asian mammal of the genus *Cynocephalus,* constituting the order Dermoptera, having a membrane between the fore and hind limbs for gliding from tree to tree (also called *colugo*). **flying lizard** a lizard of the genus *Draco,* having membranes on elongated ribs for gliding (also called *dragon, flying dragon*). **flying officer** the RAF rank next below flight lieutenant. **flying phalanger** an Australian marsupial of the genus *Petaurus* or *Petauroides,* having a membrane between the fore and hind limbs for gliding (also called *glider*). **flying picket** an industrial picket that can be moved rapidly from one site to another, esp. to reinforce local pickets. **flying saucer** an unidentified, esp. circular, flying object, popularly supposed to have come from space; a UFO. **flying squad** a police detachment or other body organized for rapid movement. **flying squirrel** a squirrel of the subfamily Pteromyinae, with skin joining the fore and hind limbs for gliding from tree to tree. **flying start 1** a start (of a race etc.) in which the starting-point is passed at full speed. **2** a vigorous start giving an initial advantage. **flying wing** an aircraft with little or no fuselage and no tailplane. **with flying colours** with distinction.

Flying Dutchman a legendary spectral ship supposed to be seen in the region of the Cape of Good Hope and presaging disaster. The legend is the basis of a music drama (1843) by Wagner.

Flying Scotsman **1** a daily express train service between London (Kings Cross) and Edinburgh. **2** a famous steam locomotive of Sir Nigel Gresley's A3 pacific design, once used on this service and now preserved. It was the first locomotive to be authentically recorded as travelling at 100 m.p.h.

flyleaf /ˈflaɪliːf/ *n.* (pl. **-leaves**) a blank leaf at the beginning or end of a book.

Flynn /flɪn/, Errol (born Leslie Thomas Flynn) (1909–59), Australian-born (or Irish-born) American actor. His usual role was the swashbuckling hero of romantic costume dramas in films such as *Captain Blood* (1935) and *The Adventures of Robin Hood* (1938).

flyover /ˈflaɪˌəʊvə(r)/ *n.* **1** *Brit.* a bridge carrying one road or railway over another. **2** *US* = fly-past (see FLY¹).

flysheet /ˈflaɪʃiːt/ *n.* **1** a tract or circular of two or four pages. **2** a fabric cover pitched outside and over a tent to give extra protection against bad weather.

flyweight /ˈflaɪweɪt/ *n.* **1** a weight in certain sports intermediate between light flyweight and bantamweight, in the amateur boxing scale 48–51 kg but differing for professionals, wrestlers, and weightlifters. **2** a sportsman of this weight. □ **light flyweight 1** a weight in amateur boxing up to 48 kg. **2** an amateur boxer of this weight.

flywheel /ˈflaɪwiːl/ *n.* a heavy wheel on a revolving shaft used to regulate machinery or accumulate power.

FM *abbr.* **1** Field Marshal. **2** frequency modulation.

Fm *symb. Chem.* the element fermium.

fm. *abbr.* (also **fm**) fathom(s).

f-number /ˈefˌnʌmbə(r)/ *n. Photog. & Astron.* the ratio of the focal length to the effective diameter of a lens or telescope mirror (e.g. *f*5, indicating that the focal length is five times the diameter), used to indicate the amount of light collected by the lens or mirror. [*f* (denoting focal length) + NUMBER]

FO *abbr.* Flying Officer.

Fo /fəʊ/, Dario (b.1926), Italian dramatist. After many years of performing and writing revues and plays for assorted theatre groups, he made his name internationally with the political satire *Accidental Death of an Anarchist* (1970). Subsequent successes include *Trumpets and Raspberries* (1980) and the farcical *Open Couple* (1983), written with his wife, the Italian dramatist Franca Rame (b.1929).

fo. *abbr.* folio.

foal /fəʊl/ *n. & v.* ● *n.* the young of a horse or related animal. ● *v.tr.* (of a mare etc.) give birth to (a foal). □ **in** (or **with**) **foal** (of a mare etc.) pregnant. [OE *fola* f. Gmc: cf. FILLY]

foam /fəʊm/ *n. & v.* ● *n.* **1** a mass of small bubbles formed on or in liquid by agitation, fermentation, etc. **2** a froth of saliva or sweat. **3** a substance resembling these, e.g. rubber (in full **foam rubber**) or plastic (in full **foam plastic**) in a cellular mass. ● *v.intr.* **1** emit foam; froth. **2** run with foam. **3** (of a vessel) be filled and overflow with foam. □ **foam at the mouth** be very angry. □ **foamless** *adj.* **foamy** *adj.* (**foamier, foamiest**). [OE *fām* f. WG]

fob[1] /fɒb/ *n. & v.* ● *n.* **1** (in full **fob-chain**) a chain attached to a watch for carrying in a waistcoat or waistband pocket. **2** a small pocket for carrying a watch. **3** a tab on a key-ring. ● *v.tr.* (**fobbed, fobbing**) put in one's fob; pocket. [orig. cant, prob. f. G]

fob[2] /fɒb/ *v.tr.* (**fobbed, fobbing**) □ **fob off 1** (often foll. by *with* a thing) deceive into accepting something inferior. **2** (often foll. by *on* to a person) palm or pass off (an inferior thing). [16th c.: cf. obs. *fop* to dupe, G *foppen* to banter]

f.o.b. *abbr.* free on board.

focaccia /fɒˈkætʃə/ *n.* a type of flat Italian bread made with yeast and olive oil and often flavoured with herbs etc. [It.]

focal /ˈfəʊk(ə)l/ *adj.* of, at, or in terms of a focus. □ **focal length** (or **distance**) **1** the distance between the centre of a lens or curved mirror and its focus. **2** the equivalent distance in a compound lens or telescope. **focal plane** the plane through the focus perpendicular to the axis of a mirror or lens. **focal point 1** = FOCUS *n.* 1. **2** = FOCUS *n.* 3. [mod.L *focalis* (as FOCUS)]

focalize /ˈfəʊkəˌlaɪz/ *v.tr.* (also **-ise**) = FOCUS *v.* □ **focalization** /ˌfəʊkəlaɪˈzeɪʃ(ə)n/ *n.*

Foch /fɒʃ/, Ferdinand (1851–1929), French general. He strongly supported the use of offensive warfare, which resulted in many of his 20th Corps being killed by German machine-guns in August 1914. He became Supreme Commander of all Allied Forces on the Western Front in early 1918, and served as the senior French representative at the Armistice negotiations.

fo'c's'le var. of FORECASTLE.

focus /ˈfəʊkəs/ *n. & v.* ● *n.* (pl. **focuses** or **foci** /ˈfəʊsaɪ/) **1** *Physics* **a** the point at which rays or waves meet after reflection or refraction. **b** the point from which diverging rays or waves appear to proceed. Also called *focal point.* **2** **a** *Optics* the point at which an object must be situated for an image of it given by a lens or mirror to be well defined (*bring into focus*). **b** the adjustment of the eye or a lens necessary to produce a clear image (*the binoculars were not in focus*). **c** a state of clear definition (*the photograph was out of focus*). **3** the centre of interest or activity (*focus of attention*). **4** *Geom.* one of the points from which the distances to any point of a given curve are connected by a linear relation. **5** *Med.* the principal site of an infection or other disease. **6** *Geol.* the point of origin of an earthquake in the earth's crust. ● *v.* (**focused, focusing** or **focussed, focussing**) **1** *tr.* bring into focus. **2** *tr.* adjust the focus of (a lens, the eye, etc.). **3** *tr. & intr.* (often foll. by *on*) concentrate or be concentrated on. **4** *intr. & tr.* converge or make converge to a focus. □ **focuser** *n.* [L, = hearth]

fodder /ˈfɒdə(r)/ *n. & v.* ● *n.* dried hay or straw etc. for cattle, horses, etc. ● *v.tr.* give fodder to. [OE *fōdor* f. Gmc, rel. to FOOD]

foe /fəʊ/ *n. esp. poet.* or *formal* an enemy or opponent. [OE *fāh* hostile, rel. to FEUD[1]]

foehn var. of FÖHN.

foetid var. of FETID.

foetus /ˈfiːtəs/ *n.* (US **fetus**) (pl. **-tuses**) an unborn or unhatched offspring of an animal; esp. an unborn human more than eight weeks after conception. □ **foetal** *adj.* **foeticide** /-tɪˌsaɪd/ *n.* [ME f. L *fetus* offspring]

fog[1] /fɒg/ *n. & v.* ● *n.* **1** **a** a thick cloud of water droplets or smoke suspended in the atmosphere at or near the earth's surface restricting or obscuring visibility. **b** obscurity in the atmosphere caused by this. **2** *Photog.* cloudiness on a developed negative etc. obscuring the image. **3** an uncertain or confused position or state. ● *v.* (**fogged, fogging**) **1** *tr.* **a** (often foll. by *up*) cover with fog or condensed vapour. **b** bewilder or confuse as if with a fog. **2** *intr.* (often foll. by *up*) become covered with fog or condensed vapour. **3** *tr. Photog.* make (a negative etc.) obscure or cloudy. □ **fog-bank** a mass of fog at sea. **fog-bound** unable to proceed because of fog. **fog-bow** a manifestation like a rainbow, produced by light on fog. **fog-lamp** a lamp used to improve visibility in fog. **fog-signal** a detonator placed on a railway line in fog to warn train drivers. **in a fog** puzzled; at a loss. [perh. back-form. f. FOGGY]

fog[2] /fɒg/ *n. & v. esp. Brit.* ● *n.* **1** a second growth of grass after cutting; aftermath. **2** long grass left standing in winter (cf. YORKSHIRE FOG). ● *v.tr.* (**fogged, fogging**) **1** leave (land) under fog. **2** feed (cattle) on fog. [ME: orig. unkn.]

fogey /ˈfəʊgɪ/ *n.* (also **fogy**) (pl. **-ies** or **-eys**) *colloq.* a dull old-fashioned person (*old fogey*). □ **young fogey** a young person with conservative tastes or ideas. □ **fogeydom** *n.* **fogeyish** *adj.* [18th c.: rel. to sl. *fogram,* of unkn. orig.]

Foggia /ˈfɒdʒə/ a town in SE Italy, in Apulia; pop. (1990) 159,540.

foggy /ˈfɒgɪ/ *adj.* (**foggier, foggiest**) **1** (of the atmosphere) thick or obscure with fog. **2** of or like fog. **3** vague, confused, unclear. □ **not have the foggiest** *colloq.* have no idea at all. □ **fogginess** *n.* [perh. f. FOG[2]]

foghorn /ˈfɒghɔːn/ *n.* **1** a deep-sounding instrument for warning ships in fog. **2** *colloq.* a loud penetrating voice.

fogy var. of FOGEY.

föhn /fɜːn/ *n.* (also **foehn**) **1** a hot southerly wind on the northern slopes of the Alps. **2** a warm dry wind on the lee side of mountains. [G, ult. f. L *Favonius* mild west wind]

foible /ˈfɔɪb(ə)l/ *n.* **1** a minor weakness or idiosyncrasy. **2** *Fencing* the part of a sword-blade from the middle to the point. [F, obs. form of *faible* (as FEEBLE)]

foie gras /fwʌ ˈgrɑː/ *n. colloq.* = *pâté de foie gras.*

foil[1] /fɔɪl/ *v. & n.* ● *v.tr.* **1** frustrate, baffle, defeat. **2** *Hunting* **a** run over or cross (ground or a scent) to confuse the hounds. **b** (*absol.*) (of an animal) spoil the scent in this way. ● *n.* **1** *Hunting* the track of a hunted animal. **2** *archaic* a repulse or defeat. [ME, = trample down, perh. f. OF *fouler* to full cloth, trample, ult. f. L *fullo* FULLER[1]]

foil[2] /fɔɪl/ *n.* **1** **a** metal hammered or rolled into a thin sheet (*tin foil*). **b** a sheet of this, or of tin amalgam, attached to mirror glass as a reflector. **c** a leaf of foil placed under a precious stone etc. to brighten or colour it. **2** a person or thing that enhances the qualities of another by contrast. **3** *Archit.* a leaf-shaped curve formed by the cusping of an arch or circle. [ME f. OF f. L *folium* leaf, and f. OF *foille* f. L *folia* (pl.)]

foil[3] /fɔɪl/ *n.* a light blunt-edged sword with a button on its point used in fencing. □ **foilist** *n.* [16th c.: orig. unkn.]

foil[4] /fɔɪl/ *n.* = HYDROFOIL. [abbr.]

foist /fɔɪst/ *v.tr.* **1** (foll. by *on, upon*) impose (an unwelcome person or thing). **2** (foll. by *on, upon*) falsely fix the authorship of (a composition). **3** (foll. by *in, into*) introduce surreptitiously or unwarrantably. [orig. of palming a false die, f. Du. dial. *vuisten* take in the hand f. *vuist* FIST]

Fokine /ˈfəʊkɪn/, Michel (born Mikhail Mikhailovich Fokin) (1880–1942), Russian-born American dancer and choreographer. He became known as a reformer of modern ballet, striving for a greater dramatic, stylistic, and directional unity. From 1909 he was Diaghilev's chief choreographer and staged the premières of Stravinsky's *The Firebird* (1910) and Ravel's *Daphnis and Chloë* (1912).

Fokker /ˈfɒkə(r)/, Anthony Herman Gerard (1890–1939), Dutch-born American pioneer aircraft designer and pilot. He built his first aircraft in 1908, the monoplane Eindecker, a type used by Germany as a fighter aircraft in the First World War. He also designed the successful Trimotor F-7 airliners, later versions of which provided the backbone of continental airlines in the 1930s.

fol. *abbr.* folio.

folacin /ˈfəʊləsɪn/ *n.* = FOLIC ACID. [*folic acid* + -IN]

fold[1] /fəʊld/ *v. & n.* ● *v.* **1** *tr.* **a** bend or close (a flexible thing) over upon itself. **b** (foll. by *back, over, down*) bend a part of (a flexible thing) in the manner specified (*fold down the flap*). **2** *intr.* become or be able to be folded. **3** *tr.* (foll. by *away, up*) make compact by folding. **4** *intr.* (often foll. by *up*) *colloq.* **a** collapse, disintegrate. **b** (of an enterprise) fail; go bankrupt. **5** *tr. poet.* embrace (esp. *fold in the arms* or *to the breast*). **6** *tr.* (foll. by *about, round*) clasp (the arms); wrap, envelop. **7** *tr.* (foll. by *in*) mix (an ingredient with others) using a gentle cutting and turning motion. ● *n.* **1** the act or an instance of folding. **2** a line made by or for folding. **3** a folded part. **4** a hollow among hills. **5** *Geol.* a curvature of strata. □ **fold one's arms** place one's arms across the chest, side by side or entwined. **fold one's hands** clasp them. **folding door** a door with jointed sections, folding on itself when opened. **folding money** esp. *US colloq.* banknotes. **fold-out** an oversize page in a book etc. to be unfolded by the reader. □ **foldable** *adj.* [OE *falden, fealden* f. Gmc]

fold² /fəʊld/ *n. & v.* ● *n.* **1** = SHEEPFOLD. **2** a body of believers or members of a Church. ● *v.tr.* enclose (sheep) in a fold. [OE *fald*]

-fold /fəʊld/ *suffix* forming adjectives and adverbs from cardinal numbers, meaning: **1** in an amount multiplied by (*repaid tenfold*). **2** consisting of so many parts (*threefold blessing*). [OE *-fald, -feald*, rel. to FOLD¹: orig. sense 'folded in so many layers']

foldaway /ˈfəʊldəˌweɪ/ *adj.* adapted or designed to be folded away.

folder /ˈfəʊldə(r)/ *n.* **1** a folding cover or holder for loose papers. **2** a folded leaflet.

folderol var. of FALDERAL.

foliaceous /ˌfəʊlɪˈeɪʃəs/ *adj.* **1** of or resembling a leaf or leaves. **2** (esp. *Geol.*) laminated. [L *foliaceus* leafy f. *folium* leaf]

foliage /ˈfəʊlɪdʒ/ *n.* **1** leaves, leafage. **2** a design in art resembling leaves. □ **foliage leaf** a leaf excluding petals and other modified leaves. [ME f. F *feuillage* f. *feuille* leaf f. OF *foille*: see FOIL²]

foliar /ˈfəʊlɪə(r)/ *adj.* of or relating to leaves. □ **foliar feed** feed supplied to leaves of plants. [mod.L *foliaris* f. L *folium* leaf]

foliate *adj. & v.* ● *adj.* /ˈfəʊlɪət/ **1** leaflike. **2** having leaves. **3** (in *comb.*) having a specified number of leaflets (*trifoliate*). ● *v.* /ˈfəʊlɪˌeɪt/ **1** *intr.* split into laminae. **2** *tr. Archit.* decorate (an arch or door-head) with foils. **3** *tr.* number leaves (not pages) of (a volume) consecutively. □ **foliation** /ˌfəʊlɪˈeɪʃ(ə)n/ *n.* [L *foliatus* leaved f. *folium* leaf]

folic acid /ˈfəʊlɪk/ *n. Biochem.* a vitamin of the B complex, found in leafy green vegetables, liver, and kidney, a deficiency of which causes pernicious anaemia. Also called *folacin, pteroylglutamic acid*, or (esp. *US*) *vitamin M*. [L *folium* leaf (because found esp. in green leaves) + -IC]

Folies-Bergère /ˌfɒlibɛəˈʒeə(r)/ a variety theatre in Paris, opened in 1869, known for its circus acts, vaudeville entertainment, and lavish productions featuring nude and semi-nude female performers.

folio /ˈfəʊlɪəʊ/ *n. & adj.* ● *n.* (*pl.* **-os**) **1** a leaf of paper etc., esp. one numbered only on the front. **2** a leaf-number of a book. **3** a sheet of paper folded once making two leaves of a book. **4** a book made of such sheets. ● *adj.* (of a book) made of folios, of the largest size. □ **in folio** made of folios. [L, ablat. of *folium* leaf, = *on leaf* (as specified)]

foliole /ˈfəʊlɪˌəʊl/ *n. Bot.* a division of a compound leaf; a leaflet. [F f. LL *foliolum* dimin. of L *folium* leaf]

foliose /ˈfəʊlɪˌəʊz, -ˌəʊs/ *adj. Bot.* (of a lichen) having a lobed, leaflike shape. [L *foliosus* f. *folium* leaf + -OSE¹]

foliot /ˈfəʊlɪət, ˈfɒl-/ *n.* a type of clock escapement consisting of a bar with adjustable weights on the ends. [OF, = watch-spring, perh. f. *folier* play the fool, dance about]

folk /fəʊk/ *n.* (*pl.* **folk** or **folks**) **1** (treated as *pl.*) people in general or of a specified class (*few folk about; townsfolk*). **2** (in *pl.*) (usu. **folks**) one's parents or relatives. **3** (treated as *sing.*) a people. **4** (treated as *sing.*) = FOLK MUSIC. **5** (*attrib.*) of popular origin; traditional (*folk art, folk hero*). □ **folk etymology** a popular modifying of the form of a word or phrase to make it seem to be derived from a more familiar word (e.g. *sparrowgrass* for *asparagus*). **folk memory** recollection of the past persisting among a people. **folk rock** folk music incorporating the stronger beat of rock music and usu. also electric stringed instruments etc. **folk-singer** a singer of folk-songs. **folk-song** a song of popular or traditional origin or style. **folk-tale** a popular or traditional story. **folk-ways** the traditional behaviour of a people. [OE *folc* f. Gmc]

folk-dance /ˈfəʊkdɑːns/ *n.* a dance of traditional origin in any human culture, now usually performed as a recreation but often originally of ritual, magical, or warlike significance. Many folk-dances developed into formal dances, such as the minuet, the waltz, and the cotillion.

Folkestone /ˈfəʊkstən/ a seaport and resort in Kent, on the SE coast of England; pop. (1981) 44,000. The English terminal of the Channel Tunnel is at Cheriton, near Folkestone.

folkish /ˈfəʊkɪʃ/ *adj.* of the common people; traditional, unsophisticated.

folklore /ˈfəʊklɔː(r)/ *n.* the traditional beliefs and stories of a people; the study of these. □ **folkloric** *adj.* **folklorist** *n.* **folkloristic** /ˌfəʊklɔːˈrɪstɪk/ *adj.*

folk music *n.* instrumental or vocal music of traditional origin, transmitted orally from generation to generation, whose authorship is often unknown. Folk music tends to have a relatively simple structure and melody, and to use portable instruments such as guitar, violin, harmonica, accordion, and bagpipes. Folk music is often monophonic, consisting of simple unaccompanied tunes, although vocal polyphony is common in southern and eastern Europe. While in some regions of Europe (e.g. Bulgaria, Romania, the Basque country, Macedonia, etc.) there has been an uninterrupted tradition of living folk music, concern began to be felt in the late 19th century in Britain and elsewhere that the folk tradition would be lost. Pioneering collectors and revivers of folk music include Cecil Sharp and Percy Grainger in Britain, Dvořák in Bohemia and Moravia, and Bartók and Kodály in Transylvania. During the Depression years in the US, Woody Guthrie revived interest in the form with his political protest songs; he had a strong influence on later figures such as Bob Dylan, who performed traditional folk material and also wrote new songs in a folk style. In Britain in the 1960s and 1970s groups such as Fairport Convention and Steeleye Span explored and expanded the form; electronic and other less traditional instruments began to be used, giving rise to a style known as *folk rock*. Recently much attention has been given to folk music from other cultures (see *world music*).

folksy /ˈfəʊksɪ/ *adj.* (**folksier, folksiest**) **1** friendly, sociable, informal. **2 a** having the characteristics of folk art, culture, etc. **b** ostensibly or artificially folkish. □ **folksiness** *n.*

folkweave /ˈfəʊkwiːv/ *n.* a rough loosely woven fabric.

folky /ˈfəʊkɪ/ *adj.* (**folkier, folkiest**) **1** = FOLKSY. **2** = FOLKISH. □ **folkiness** *n.*

follicle /ˈfɒlɪk(ə)l/ *n.* **1** *Anat.* a small secretory cavity, sac, or gland; esp.: **a** (in full **hair follicle**) the gland or cavity at the root of a hair. **b** see GRAAFIAN FOLLICLE. **2** *Bot.* a single-carpelled dry fruit opening on one side only to release its seeds. □ **follicle-stimulating hormone** (abbr. **FSH**) *Physiol.* a pituitary hormone which promotes the formation of ova or sperm. □ **follicular** /fɒˈlɪkjʊlə(r)/ *adj.* **folliculate** /-lət/ *adj.* **folliculated** /-ˌleɪtɪd/ *adj.* [L *folliculus* dimin. of *follis* bellows]

follow /ˈfɒləʊ/ *v.* **1** *tr.* or (foll. by *after*) *intr.* go or come after (a person or thing proceeding ahead). **2** *tr.* go along (a route, path, etc.). **3** *tr. & intr.* come after in order or time (*Nero followed Claudius; dessert followed; my reasons are as follows*). **4** *tr.* take as a guide or leader. **5** *tr.* conform to (*follow your example*). **6** *tr.* practise (a trade or profession). **7** *tr.* undertake (a course of study etc.). **8** *tr.* understand the meaning or tendency of (a speaker or argument). **9** *tr.* maintain awareness of the current state or progress of (events etc. in a particular sphere). **10** *tr.* (foll. by *with*) provide with a sequel or successor. **11** *intr.* happen after something else; ensue. **12** *intr.* **a** be necessarily true as a result of something else. **b** (foll. by *from*) be a result of. **13** *tr.* strive after; aim at; pursue (*followed fame and fortune*). □ **follow-my-leader** a game in which players must do as the leader does. **follow one's nose** trust to instinct. **follow on 1** continue. **2** (of a cricket team) have to bat again immediately after the first innings. **follow-on** *n.* an instance of following on, esp. by a cricket team. **follow out** carry out; adhere precisely to (instructions etc.). **follow suit 1** *Cards* play a card of the suit led. **2** conform to another person's actions. **follow through 1** continue (an action etc.) to its conclusion. **2** *Sport* continue the movement of a stroke after the ball has been struck. **follow-through** *n.* the act or an instance of following through. **follow up 1** (foll. by *with*) pursue, develop, supplement. **2** make further investigation of. **follow-up** *n.* a subsequent or continued action, measure, experience, etc. [OE *folgian* f. Gmc]

follower /ˈfɒləʊə(r)/ *n.* **1** an adherent or devotee. **2** a person or thing that follows. **3** *Austral. Rules* either of two players who, with the rover, do not have fixed positions and so follow play.

following /ˈfɒləʊɪŋ/ *prep., n., & adj.* ● *prep.* coming after in time; as a sequel to. ● *n.* **1** a body of adherents or devotees. **2 (the following)** (treated as *sing.* or *pl.*) the person(s) or thing(s) now to be mentioned. ● *adj.* that follows or comes after.

folly /ˈfɒlɪ/ *n.* (*pl.* **-ies**) **1** foolishness; lack of good sense. **2** a foolish act, behaviour, idea, etc. **3** a costly ornamental building, usu. a tower or mock Gothic ruin. **4** (in *pl.*) *Theatr.* **a** a revue with glamorous female performers, esp. scantily clad. **b** the performers in such a revue. [ME f. OF *folie* f. *fol* mad, FOOL¹]

Folsom /ˈfəʊlsəm/ *adj. & n. Archaeol.* of, relating to, or denoting a prehistoric culture and its remains first found at a site near the US village of Folsom in NE New Mexico, and referring in particular to a distinctive type of fluted lanceolate projectile point or spearhead, flaked from stone, found at the site in association with the bones of an extinct bison. The discovery in 1926 of the points, which date from *c*.11,000–10,000 to *c*.8,000 years ago, forced a radical rethinking of the date at which humans first inhabited the New World. (See also CLOVIS¹.)

Fomalhaut /ˈfɒm(ə)lˌhɔːt, -məˌləʊt/ *Astron.* the brightest star in the constellation of Piscis Austrinus, which is only briefly visible from the northern hemisphere. [Arab., = mouth of the fish]

foment /fəˈment, fəʊˈment/ *v.tr.* **1** instigate or stir up (trouble, sedition, etc.). **2 a** bathe with warm or medicated liquid. **b** apply warmth to. □ **fomenter** *n.* [ME f. F *fomenter* f. LL *fomentare* f. L *fomentum* poultice, lotion f. *fovere* heat, cherish]

fomentation /ˌfəʊmenˈteɪʃ(ə)n/ *n.* **1** the act or an instance of fomenting. **2** materials prepared for application to a wound etc. [ME f. OF or LL *fomentatio* (as FOMENT)]

Fon /fɒn/ *n. & adj.* ● *n.* (*pl.* same or **Fons**) **1** a member of a people inhabiting the southern part of Benin. **2** the Kwa language of the Fon. ● *adj.* of or relating to the Fon or their language. [Fon]

fond /fɒnd/ *adj.* **1** (*predic.*; foll. by *of*) having affection or a liking for. **2** (*attrib.*) affectionate, loving, doting. **3** (*attrib.*) (of beliefs etc.) foolishly optimistic or credulous; naive. □ **fondly** *adv.* **fondness** *n.* [ME f. obs. *fon* fool, be foolish]

Fonda /ˈfɒndə/ a family of American actors. Henry Fonda (1905–82) is noted for his roles in such films as *The Grapes of Wrath* (1939) and *Twelve Angry Men* (1957). He won his only Oscar for his role in his final film, *On Golden Pond* (1981). His daughter Jane (b.1937) was a model and stage actress before becoming a screen star. Her films include *Klute* (1971), for which she won an Oscar, and *The China Syndrome* (1979); she also acted alongside her father in *On Golden Pond*. In the 1980s she became known for her fitness routine, *Jane Fonda's Workout*. Her brother Peter (b.1939) is also an actor, as is his daughter Bridget (b.1964).

fondant /ˈfɒndənt/ *n.* a soft sweet of flavoured sugar. [F, pres. part. of *fondre* melt f. L *fundere* pour]

fondle /ˈfɒnd(ə)l/ *v. & n.* ● *v.tr.* touch or stroke lovingly; caress. ● *n.* an act of fondling. □ **fondler** *n.* [back-form. f. *fondling* fondled person (as FOND, -LING¹)]

fondue /ˈfɒndjuː, -duː/ *n.* **1** a dish of flavoured melted cheese into which pieces of bread are dipped. **2** any other dish in which small pieces of food are dipped into a hot or boiling liquid. [F, fem. past part. of *fondre* melt f. L *fundere* pour]

font¹ /fɒnt/ *n.* **1** a receptacle in a church for baptismal water. **2** the reservoir for oil in a lamp. □ **fontal** *adj.* (in sense 1). [OE *font, fant* f. OIr. *fant, font* f. L *fons fontis* fountain, baptismal water]

font² /fɒnt/ *n. Printing* a set of type of one face or size. [F *fonte* f. *fondre* FOUND³]

fontanelle /ˌfɒntəˈnel/ *n.* (*US* **fontanel**) *Anat.* a membranous space in an infant's skull at the angles of the parietal bones. [F *fontanelle* f. mod.L *fontanella* f. OF *fontenelle* dimin. of *fontaine* fountain]

Fonteyn /fɒnˈteɪn/, Dame Margot (born Margaret Hookham) (1919–91), English ballet-dancer. She had her first major role in Sir Frederick Ashton's *Le Baiser de la fée* (1935), later dancing all the classical ballerina roles and creating many new ones for the Royal Ballet. In 1962 she began a celebrated partnership with Rudolf Nureyev, dancing in *Giselle* and *Romeo and Juliet*. In 1979 she was named *prima ballerina assoluta*, a title given only three times in the history of ballet.

Foochow see FUZHOU.

food /fuːd/ *n.* **1** a nutritious substance, esp. solid in form, that can be taken into an animal or a plant to maintain life and growth. **2** ideas as a resource for or stimulus to mental work (*food for thought*). □ **food additive** a substance added to food to enhance its colour, flavour, or presentation, or for any other non-nutritional purpose. **food-chain** *Ecol.* a sequence of organisms each dependent on the next organism in the chain as its source of food. **food poisoning** illness due to bacteria or toxins in food. **food processor** a machine for chopping and mixing food materials. **food value** the relative nourishing power of a food. [OE *fōda* f. Gmc: cf. FEED]

Food and Agriculture Organization an agency of the United Nations established in 1945 to secure improvements in the production

and distribution of all food and agricultural products and to raise levels of nutrition. Its headquarters are in Rome.

foodie /ˈfuːdɪ/ *n.* (also **foody**) (*pl.* **-ies**) *colloq.* a person with a particular interest in food; a gourmet.

foodstuff /ˈfuːdstʌf/ *n.* any substance suitable as food.

fool¹ /fuːl/ *n., v., & adj.* ● *n.* **1** a person who acts unwisely or imprudently; a stupid person. **2** *hist.* a jester; a clown. **3** a dupe. ● *v.* **1** *tr.* deceive so as to cause to appear foolish. **2** *tr.* (foll. by *into* + verbal noun, or *out of*) trick; cause to do something foolish. **3** *tr.* play tricks on; dupe. **4** *intr.* act in a joking, frivolous, or teasing way. **5** *intr.* (foll. by *about, around*) behave in a playful or silly way. ● *adj. N. Amer. colloq.* foolish, silly. □ **act** (or **play**) **the fool** behave in a silly way. **fool's errand** a fruitless venture. **fool's gold** iron pyrites. **fool's paradise** happiness founded on an illusion. **fool's parsley** an umbelliferous plant resembling parsley. **make a fool of** make (a person or oneself) look foolish; trick or deceive. **no** (or **nobody's**) **fool** a shrewd or prudent person. [ME f. OF *fol* f. L *follis* bellows, empty-headed person]

fool² /fuːl/ *n.* a dessert of usu. stewed fruit crushed and mixed with cream, custard, etc. [16th c.: perh. f. FOOL¹]

foolery /ˈfuːlərɪ/ *n.* (*pl.* **-ies**) **1** foolish behaviour. **2** a foolish act.

foolhardy /ˈfuːlˌhɑːdɪ/ *adj.* (**foolhardier, foolhardiest**) rashly or foolishly bold; reckless. □ **foolhardily** *adv.* **foolhardiness** *n.* [ME f. OF *folhardi* f. *fol* foolish + *hardi* bold]

foolish /ˈfuːlɪʃ/ *adj.* (of a person, action, etc.) lacking good sense or judgement; unwise. □ **foolishly** *adv.* **foolishness** *n.*

foolproof /ˈfuːlpruːf/ *adj.* (of a procedure, mechanism, etc.) so straightforward or simple as to be incapable of misuse or mistake.

foolscap /ˈfuːlskæp/ *n. Brit.* **1** a size of paper, about 330 x 200 (or 400) mm. **2** foolscap paper. [named from the former watermark representing a fool's cap]

foot /fʊt/ *n. & v.* ● *n.* (*pl.* **feet** /fiːt/) **1 a** the lower extremity of the leg below the ankle. **b** the part of a sock etc. covering the foot. **2 a** the lower or lowest part of anything, e.g. a mountain, a page, stairs, etc. **b** the lower end of a table. **c** the end of a bed where the user's feet normally rest. **d** a part of a chair, appliance, etc. on which it rests. **3** the base, often projecting, of anything extending vertically. **4** a step, pace, or tread; a manner of walking (*fleet of foot*). **5** (*pl.* **feet** or **foot**) a unit of linear measure equal to 12 inches (30.48 cm) (Symbol: ′). **6** *Prosody* **a** a group of syllables (one usu. stressed) constituting a metrical unit. **b** a similar unit of speech etc. **7** *Brit. hist.* infantry (*a regiment of foot*). **8** *Zool.* the locomotive or adhesive organ of invertebrates. **9** *Bot.* the part by which a petal is attached. **10** a device on a sewing machine for holding the material steady as it is sewn. **11** (*pl.* **foots**) *a* dregs; oil refuse. **b** coarse sugar. ● *v.* **1** (usu. as **foot it**) **a** traverse (esp. a long distance) by foot. **b** dance. **2** pay (a bill, esp. one considered large). □ **at a person's feet** as a person's disciple or subject. **feet of clay** a fundamental weakness in a person otherwise revered. **foot-and-mouth disease** a contagious viral disease of cattle etc. **foot-fault** *n.* incorrect placement of the feet while serving in tennis etc. ● *v.* **1** *intr.* make a foot-fault. **2** *tr.* award a foot-fault against the player. **foot-passenger** a passenger who boards (a ship etc.) on foot rather than by car etc; a pedestrian. **foot-pound** the amount of energy required to raise 1 lb a distance of 1 foot. **foot-pound-second system** a system of measurement with the foot, pound, and second as basic units. **foot-rot** a bacterial disease of the feet in sheep and cattle. **foot-rule** a ruler 1 foot long. **foot-soldier** a soldier who fights on foot. **get one's feet wet** begin to participate. **have one's** (or **both**) **feet on the ground** be practical. **have a foot in the door** have a prospect of success. **have one foot in the grave** be near death or very old. **my foot!** *int. colloq.* expressing strong contradiction. **not put a foot wrong** make no mistakes. **off one's feet** so as to be unable to stand, or in a state compared with this (*was rushed off my feet*). **on one's feet** standing or walking. **on foot** walking, not riding etc. **put one's best foot forward** make every effort; proceed with determination. **put one's feet up** *colloq.* take a rest. **put one's foot down** *colloq.* **1** be firmly insistent or repressive. **2** accelerate a motor vehicle. **put one's foot in it** *colloq.* commit a blunder or indiscretion. **set foot on** (or **in**) enter; go into. **set on foot** put (an action, process, etc.) in motion. **under one's feet** in the way. **under foot** on the ground. □ **footed** *adj.* (also in comb.). **footless** *adj.* [OE *fōt* f. Gmc]

footage /ˈfʊtɪdʒ/ *n.* **1** length or distance in feet. **2** an amount of film made for showing, broadcasting, etc.

football /ˈfʊtbɔːl/ *n. & v.* ● *n.* **1** any of several esp. outdoor games between two teams played with a ball on a rectangular pitch with goals at each end, esp. (*Brit.*) Association football. (See note below.) **2** a large inflated ball of a kind used in these games. **3** a person or thing treated like a football. **4** a topical issue or problem that is the subject of continued argument or controversy (*political football*). ● *v.intr.* play football. □ **football pool** (or **pools**) a form of gambling on the results of football matches, the winners receiving sums accumulated from entry money. □ **footballer** *n.*

■ A form of football was played at an early date in China (*c.*206 BC) and in ancient Greece and Rome; the ball was handled as well as kicked. A form of the game was also known in Japan in the 7th century AD. In Europe in medieval and later times it was a rowdy game with ill-defined rules and frequently large numbers of participants. By the 19th century forms of football were played regularly in the English public schools, but it was not until the middle of the century that the establishment of common rules made it possible to hold matches between different schools, clubs, and (eventually) countries. (See also AMERICAN FOOTBALL, ASSOCIATION FOOTBALL, AUSTRALIAN RULES FOOTBALL, RUGBY.)

footboard /ˈfʊtbɔːd/ *n.* **1** a board to support the feet or a foot. **2** an upright board at the foot of a bed.

footbrake /ˈfʊtbreɪk/ *n.* a brake operated by the foot in a motor vehicle.

footbridge /ˈfʊtbrɪdʒ/ *n.* a bridge for use by pedestrians.

footer[1] /ˈfʊtə(r)/ *n.* (in *comb.*) **1** a person or thing of so many feet in length or height (*six-footer*). **2** a ball, shot, etc. struck by the specified foot (*scored a goal with a fine right-footer*).

footer[2] /ˈfʊtə(r)/ *n. Brit. colloq.* = FOOTBALL 1.

footfall /ˈfʊtfɔːl/ *n.* the sound of a footstep.

foothill /ˈfʊthɪl/ *n.* (often in *pl.*) any of the low hills around the base of a mountain.

foothold /ˈfʊthəʊld/ *n.* **1** a place, esp. in climbing, where a foot can be supported securely. **2** a secure initial position or advantage.

footie var. of FOOTY.

footing /ˈfʊtɪŋ/ *n.* **1** a foothold; a secure position (*lost his footing*). **2** the basis on which an enterprise is established or operates; the position or status of a person in relation to others (*on an equal footing*). **3** (usu. in *pl.*) the foundations of a wall, usu. with a course of brickwork wider than the base of the wall.

footle /ˈfuːt(ə)l/ *v.intr.* (usu. foll. by *about*) · *colloq.* behave foolishly or trivially. [19th c.: perh. f. dial. *footer* idle]

footlights /ˈfʊtlaɪts/ *n.pl.* a row of lights along the front of a stage at the level of the actors' feet.

footling /ˈfuːtlɪŋ/ *adj. colloq.* trivial, silly.

footloose /ˈfʊtluːs/ *adj.* (esp. in phr. **footloose and fancy-free**) free to go where or act as one pleases.

footman /ˈfʊtmən/ *n.* (*pl.* **-men**) **1** a liveried servant attending at the door, at table, or on a carriage. **2** *hist.* an infantryman.

footmark /ˈfʊtmɑːk/ *n.* a footprint.

footnote /ˈfʊtnəʊt/ *n. & v.* ● *n.* a note printed at the foot of a page. ● *v.tr.* supply with a footnote or footnotes.

footpad /ˈfʊtpæd/ *n. hist.* an unmounted highwayman.

footpath /ˈfʊtpɑːθ/ *n.* a path for pedestrians; a pavement.

footplate /ˈfʊtpleɪt/ *n.* esp. *Brit.* the platform in the cab of a locomotive for the crew.

footprint /ˈfʊtprɪnt/ *n.* **1** the impression left by a foot or shoe. **2** *Computing* the area of desk space etc. occupied by a microcomputer or other piece of hardware. **3** the ground area covered by a communications satellite or affected by noise etc. from aircraft.

footrest /ˈfʊtrest/ *n.* a support for the feet or a foot.

Footsie /ˈfʊtsɪ/ *n. colloq.* (also **Footsie index** or **footsie**) the Financial Times Index. [repr. pronunc. of FT–SE]

footsie /ˈfʊtsɪ/ *n. colloq.* amorous play with the feet. [joc. dimin. of FOOT]

footslog /ˈfʊtslɒg/ *v. & n.* ● *v.intr.* (**-slogged, -slogging**) walk or march, esp. laboriously for a long distance. ● *n.* a laborious walk or march. □ **footslogger** *n.*

footsore /ˈfʊtsɔː(r)/ *adj.* having sore feet, esp. from walking.

footstalk /ˈfʊtstɔːk/ *n.* **1** *Bot.* a stalk of a leaf or peduncle of a flower. **2** *Zool.* an attachment of a barnacle etc.

footstep /ˈfʊtstep/ *n.* **1** a step taken in walking. **2** the sound of this. □ **follow** (or **tread**) **in a person's footsteps** do as another person did before.

footstool /ˈfʊtstuːl/ *n.* a low stool for resting the feet on when sitting.

footway /ˈfʊtweɪ/ *n.* a path or way for pedestrians.

footwear /ˈfʊtweə(r)/ *n.* shoes, socks, etc.

footwork /ˈfʊtwɜːk/ *n.* the use of the feet, esp. skilfully, in sports, dancing, etc.

footy /ˈfʊtɪ/ *n.* (also **footie**) *colloq.* football.

fop /fɒp/ *n.* an affectedly elegant or fashionable man; a dandy. □ **foppery** *n.* **foppish** *adj.* **foppishly** *adv.* **foppishness** *n.* [17th c.: perh. f. earlier *fop* fool]

for /fə(r), fɔː(r)/ *prep. & conj.* ● *prep.* **1** in the interest or to the benefit of; intended to go to (*these flowers are for you; wish to see it for myself; did it all for my country; silly for you to go*). **2** in defence, support, or favour of (*fight for one's rights*). **3** suitable or appropriate to (*a dance for beginners; not for me to say*). **4** in respect of or with reference to; regarding; so far as concerns (*usual for ties to be worn; don't care for him at all; ready for bed; MP for Lincoln*). **5** representing or in place of (*here for my uncle*). **6** in exchange against (*swopped it for a bigger one*). **7 a** as the price of (*give me £5 for it*). **b** at the price of (*bought it for £5*). **c** to the amount of (*a bill for £100; all out for 45*). **8** as the penalty of (*fined them heavily for it*). **9** in requital of (*that's for upsetting my sister*). **10** as a reward for (*here's £5 for your trouble*). **11** with a view to; in the hope or quest of; in order to get (*go for a walk; run for a doctor; did it for the money*). **12** corresponding to (*word for word*). **13** to reach; in the direction of; towards (*left for Rome; ran for the end of the road*). **14** conducive or conducively to; in order to achieve (*take the pills for a sound night's sleep*). **15** so as to start promptly at (*the meeting is at seven-thirty for eight*). **16** through or over (a distance or period); during (*walked for miles; sang for two hours*). **17** in the character of; as being (*for the last time; know it for a lie; I for one refuse*). **18** because of; on account of (*could not see for tears*). **19** in spite of; notwithstanding (*for all we know; for all your fine words*). **20** considering or making due allowance in respect of (*good for a beginner*). ● *conj.* because, since, seeing that. □ **be for it** *Brit. colloq.* be in imminent danger of punishment or other trouble. **for ever** see EVER (cf. FOREVER). **o** (or **oh**) **for** I wish I had. [OE, prob. a reduction of Gmc *fora* (unrecorded) BEFORE (of place and time)]

f.o.r. *abbr.* free on rail.

for- /fɔː(r), fə(r)/ *prefix* forming verbs and their derivatives meaning: **1** away, off, apart (*forget; forgive*). **2** prohibition (*forbid*). **3** abstention or neglect (*forgo; forsake*). **4** excess or intensity (*forlorn*). [OE *for-, fær-*]

forage /ˈfɒrɪdʒ/ *n. & v.* ● *n.* **1** food for horses and cattle. **2** the act or an instance of searching for food. ● *v.* **1** *intr.* go searching; rummage (esp. for food). **2** *tr.* obtain food from; plunder. **3** *tr.* **a** get by foraging. **b** supply with food. □ **forage cap** an infantry undress cap. □ **forager** *n.* [ME f. OF *fourrage, fourrager*, rel. to FODDER]

foramen /fɒˈreɪmen/ *n.* (*pl.* **foramina** /-ˈræmɪnə/) *Anat.* an opening, hole, or passage, esp. in a bone. □ **foramen magnum** the hole in the base of the skull through which the spinal cord passes. [L *foramen -minis* f. *forare* bore a hole]

foraminifer /ˌfɒrəˈmɪnɪfə(r)/ *n.* (also **foraminiferan** /fɒˌræmɪˈnɪfərən/) *Zool.* a protozoan of the mainly marine order Foraminifera, having a perforated shell through which amoeba-like pseudopodia emerge. The shells eventually drop to the seabed to form a calcareous ooze, and fossil deposits of foraminifers form the main constituent of chalk. □ **foraminiferous** /fɒˌræmɪˈnɪfərəs/ *adj.*

forasmuch as /ˌfɒrəzˈmʌtʃ/ *conj. archaic* because, since. [= for as much]

foray /ˈfɒreɪ/ *n. & v.* ● *n.* a sudden attack; a raid or incursion. ● *v.intr.* make or go on a foray. [ME, prob. earlier as verb: back-form. f. *forayer* f. OF *forrier* forager, rel. to FODDER]

forbade (also **forbad**) *past of* FORBID.

forbear[1] /fɔːˈbeə(r)/ *v.intr. & tr.* (*past* **forbore** /-ˈbɔː(r)/; *past part.* **forborne** /-ˈbɔːn/) (often foll. by *from*, or *to* + infin.) *literary* abstain or desist (from) (*could not forbear (from) speaking out; forbore to mention it*). [OE *forberan* (as FOR-, BEAR[1])]

forbear[2] var. of FOREBEAR.

forbearance /fɔːˈbeərəns/ *n.* patient self-control; tolerance.

forbearing /fɔːˈbeərɪŋ/ *adj.* patient, long-suffering. □ **forbearingly** *adv.*

forbid /fəˈbɪd/ *v.tr.* (**forbidding**; *past* **forbade** /-ˈbæd, -ˈbeɪd/ or **forbad** /-ˈbæd/; *past part.* **forbidden** /-ˈbɪd(ə)n/) **1** (foll. by *to* + infin.) order not (*I forbid you to go*). **2** refuse to allow (a thing, or a person to have a thing) (*I forbid it; was forbidden any wine*). **3** refuse a person entry to (*the gardens are forbidden to children*). □ **forbidden degrees** see DEGREE. **forbidden fruit** something desired or enjoyed all the more because not allowed.

God forbid! may it not happen! □ **forbiddance** n. [OE forbēodan (as FOR-, BID)]

Forbidden City 1 see LHASA. **2** see BEIJING.

forbidding /fəˈbɪdɪŋ/ adj. uninviting, repellent, stern. □ **forbiddingly** adv.

forbore past of FORBEAR¹.

forborne past part. of FORBEAR¹.

forbye /fɔːˈbaɪ/ prep. & adv. archaic or Sc. ● prep. besides. ● adv. in addition.

force¹ /fɔːs/ n. & v. ● n. **1** power; exerted strength or impetus; intense effort. **2** coercion or compulsion, esp. with the use or threat of violence. **3 a** military strength. **b** (in pl.) troops; fighting resources. **c** an organized body of people, esp. soldiers, police, or workers. **4** binding power; validity. **5** effect; precise significance (the force of their words). **6 a** mental or moral strength; influence, efficacy (force of habit). **b** vividness of effect (described with much force). **7** Physics an influence tending to change the motion of a body or produce motion or stress in a stationary body; the intensity of this (calculated e.g. by multiplying the mass of the body and its acceleration). (See note below.) **8** a person or thing regarded as exerting influence (is a force for good). ● v. **1** tr. constrain (a person) by force or against his or her will. **2** tr. make a way through or into by force; break open by force. **3** tr. (usu. with prep. or adv.) drive or propel violently or against resistance (forced it into the hole; the wind forced them back). **4** tr. (foll. by on, upon) impose or press (on a person) (forced their views on us). **5** tr. **a** cause or produce by effort (forced a smile). **b** attain by strength or effort (forced an entry; must force a decision). **c** make (a way) by force. **6** tr. strain or increase to the utmost; overstrain. **7** tr. artificially hasten the development or maturity of (a plant). **8** tr. seek or demand quick results from; accelerate the process of (force the pace). **9** intr. Cards make a play that compels another particular play. □ **by force of** by means of. **force the bidding** (at an auction) make bids to raise the price rapidly. **forced labour** compulsory labour, esp. under harsh conditions. **forced landing** the unavoidable landing of an aircraft in an emergency. **forced march** a long and vigorous march esp. by troops. **force-feed** force (esp. a prisoner) to take food. **force field** (in science fiction) an invisible barrier of force. **force a person's hand** make a person act prematurely or unwillingly. **force the issue** render an immediate decision necessary. **force-land** land an aircraft in an emergency. **force-pump** a pump that forces water under pressure. **in force 1** valid, effective. **2** in great strength or numbers. **join forces** combine efforts. □ **forceable** adj. **forcer** n. [ME f. OF force, forcer ult. f. L fortis strong]

▪ Force in physics is that which causes a mass to undergo acceleration or deformation, or the quantity 'mass times acceleration' itself. Forces were first described mathematically by Newton in his laws of motion, and the present-day standard unit of force, the newton, is named after him. Newton showed that a force is not, as earlier thinkers had supposed, necessary for maintaining motion itself; its effect is rather to cause a body to tend to accelerate (or decelerate). Physical phenomena are now believed to be ultimately dependent on four fundamental forces that govern the interaction between all particles of matter: the strong nuclear force which binds protons and neutrons to one another in the atomic nucleus; the weak nuclear force responsible for some radioactive phenomena; the electromagnetic force acting between charged particles; and the gravitational force.

force² /fɔːs/ n. N. Engl. a waterfall. [ON fors]

forced /fɔːst/ adj. **1** obtained or imposed by force (forced entry). **2** (of a gesture, etc.) produced or maintained with effort; affected, unnatural (a forced smile).

forceful /ˈfɔːsfʊl/ adj. **1** vigorous, powerful. **2** (of speech) compelling, impressive. □ **forcefully** adv. **forcefulness** n.

force majeure /ˌfɔːs mæˈʒɜː(r)/ n. **1** irresistible compulsion or coercion. **2** an unforeseeable course of events excusing a person from the fulfilment of a contract. [F, = superior strength]

forcemeat /ˈfɔːsmiːt/ n. meat or vegetables etc. chopped and seasoned for use as a stuffing or a garnish. [obs. force, farce stuff f. OF farsir: see FARCE]

forceps /ˈfɔːseps/ n. (pl. same) **1** surgical pincers, used for grasping and holding. **2** Bot. & Zool. an organ or structure resembling forceps. [L forceps forcipis]

forcible /ˈfɔːsɪb(ə)l/ adj. done by or involving force; forceful. □ **forcibly** adv. [ME f. AF & OF (as FORCE¹)]

Ford¹ /fɔːd/, Ford Madox (born Ford Hermann Hueffer) (1873–1939), English novelist. He was the grandson of the Pre-Raphaelite painter Ford Madox Brown (1821–93) and is chiefly remembered as the author of the novel The Good Soldier (1915). As founder of both the English Review (1908) and the Transatlantic Review (1924), he published works by such writers as Ernest Hemingway, James Joyce, and Ezra Pound.

Ford² /fɔːd/, Gerald R(udolph) (b.1913), American Republican statesman, 38th President of the US 1974–7. He became President on the resignation of Richard Nixon in the wake of the Watergate affair. The free pardon he granted Nixon two months later aroused controversy.

Ford³ /fɔːd/, Harrison (b.1942), American actor. He made his screen début in 1966. Ford became internationally famous with his leading roles in the science-fiction film Star Wars (1977) and its two sequels, and in the adventure film Raiders of the Lost Ark (1981) and its two sequels (including Indiana Jones and the Temple of Doom, 1984). Other films include The Fugitive (1993).

Ford⁴ /fɔːd/, Henry (1863–1947), American motor manufacturer. He was a pioneer of mass production and had a profound influence on the widespread use of motor vehicles. By 1903 he had evolved a reliable car and founded his own firm, the Ford Motor Company. In 1909 Ford produced his famous Model T, of which 15 million were made over the next 19 years at gradually reducing prices due to large-scale manufacture, a succession of simple assembly tasks, and the use of a conveyor belt. He went on to produce a cheap and effective farm tractor, the Fordson, which had a great effect on agricultural mechanization. Control of the Ford Motor Company passed to his grandson, Henry Ford II (1917–1987), in 1945 and today it is a huge multinational corporation. Among the first Henry Ford's philanthropic legacies is the Ford Foundation (established 1936), a major charitable trust.

Ford⁵ /fɔːd/, John (born Sean Aloysius O'Feeney) (1895–1973), American film director. He is chiefly known for his westerns, which depict the early pioneers and celebrate the frontier spirit. His many films starring John Wayne include Stagecoach (1939) and She Wore a Yellow Ribbon (1949). Notable films in other genres include The Grapes of Wrath (1940), for which he won an Oscar.

ford /fɔːd/ n. & v. ● n. a shallow place where a river or stream may be crossed by wading or in a vehicle. ● v.tr. cross (water) at a ford. □ **fordable** adj. **fordless** adj. [OE f. WG]

fore /fɔː(r)/ adj., n., int., & prep. ● adj. situated in front. ● n. the front part, esp. of a ship; the bow. ● int. Golf a warning to a person in the path of a ball. ● prep. archaic (in oaths) in the presence of (fore God). □ **come to the fore** take a leading part. **fore and aft** Naut. at bow and stern; all over the ship. **fore-and-aft** adj. Naut. (of a sail or rigging) set lengthwise, not on the yards. **to the fore** in front; conspicuous. [OE f. Gmc.: (adj. & n.) ME f. compounds with FORE-]

fore- /fɔː(r)/ prefix forming: **1** verbs meaning: **a** in front (foreshorten). **b** beforehand; in advance (foreordain; forewarn). **2** nouns meaning: **a** situated in front of (forecourt). **b** the front part of (forehead). **c** of or near the bow of a ship (forecastle). **d** preceding (forerunner).

forearm¹ /ˈfɔːrɑːm/ n. **1** the part of the arm from the elbow to the wrist or the fingertips. **2** the corresponding part in a foreleg or wing.

forearm² /fɔːrˈɑːm/ v.tr. prepare or arm beforehand.

forebear /ˈfɔːbeə(r)/ n. (also **forbear**) (usu. in pl.) an ancestor. [FORE + obs. bear, beer (as BE, -ER¹)]

forebode /fɔːˈbəʊd/ v.tr. **1** betoken; be an advance warning of (an evil or unwelcome event). **2** have a presentiment of (usu. evil).

foreboding /fɔːˈbəʊdɪŋ/ n. an expectation of trouble or evil; a presage or omen. □ **forebodingly** adv.

forebrain /ˈfɔːbreɪn/ n. Anat. the anterior part of the brain, including the cerebrum, thalamus, hypothalamus, etc. Also called prosencephalon. (See also BRAIN.)

forecast /ˈfɔːkɑːst/ v. & n. ● v.tr. (past and past part. **-cast** or **-casted**) predict; estimate or calculate beforehand. ● n. a calculation or estimate of something future, esp. coming weather. □ **forecaster** n.

forecastle /ˈfəʊks(ə)l/ n. (also **fo'c's'le**) Naut. **1** the forward part of a ship where the crew has quarters. **2** hist. a short raised deck at the bow.

foreclose /fɔːˈkləʊz/ v.tr. **1** (also absol.; foll. by on) stop (a mortgage) from being redeemable or (a mortgager) from redeeming, esp. as a result of defaults in payment. **2** exclude, prevent. **3** shut out; bar. □ **foreclosure** /-ˈkləʊʒə(r)/ n. [ME f. OF forclos past part. of forclore f. for- out f. L foras + CLOSE²]

forecourt /ˈfɔːkɔːt/ n. **1** an enclosed space in front of a building. **2** the part of a filling-station where petrol is supplied. **3** Tennis the part of a tennis-court between the service line and the net.

foredoom /fɔːˈduːm/ v.tr. (often foll. by to) doom or condemn beforehand.

fore-edge /ˈfɔːredʒ/ n. (also **foredge**) the front or outer edge (esp. of the pages of a book).

forefather /ˈfɔːˌfɑːðə(r)/ n. (usu. in pl.) **1** an ancestor. **2** a member of a past generation of a family or people.

forefinger /ˈfɔːˌfɪŋɡə(r)/ n. the finger next to the thumb.

forefoot /ˈfɔːfʊt/ n. (pl. **-feet**) **1** either of the front feet of a four-footed animal. **2** Naut. the foremost section of a ship's keel.

forefront /ˈfɔːfrʌnt/ n. **1** the foremost part. **2** the leading position.

foregather var. of FORGATHER.

forego[1] /fɔːˈɡəʊ/ v.tr. & intr. (**-goes**; past **-went** /-ˈwent/; past part. **-gone**) precede in place or time. □ **foregoer** n. [OE foregān]

forego[2] var. of FORGO.

foregoing /fɔːˈɡəʊɪŋ/ adj. preceding; previously mentioned.

foregone /fɔːˈɡɒn/ past part. of FOREGO[1]. ● attrib.adj. previous, preceding, completed. □ **foregone conclusion** an easily foreseen or predictable result.

foreground /ˈfɔːɡraʊnd/ n. & v. ● n. **1** the part of a view, esp. in a picture, that is nearest the observer. **2** the most conspicuous position. ● v.tr. place in the foreground; make prominent. [Du. voorgrond (as FORE-, GROUND[1])]

forehand /ˈfɔːhænd/ n. **1** Tennis etc. **a** a stroke played with the palm of the hand facing the opponent. **b** (attrib.) (also **forehanded**) of or made with a forehand. **2** the part of a horse in front of the seated rider.

forehead /ˈfɒrɪd, ˈfɔːhed/ n. the part of the face above the eyebrows. [OE forhēafod (as FORE-, HEAD)]

forehock /ˈfɔːhɒk/ n. a foreleg cut of pork or bacon.

foreign /ˈfɒrɪn/ adj. **1** of or from or situated in or characteristic of a country or a language other than one's own. **2** dealing with other countries (Foreign Service) (opp. HOME adj. 2b, INTERIOR adj. 3). **3** of another district, society, etc. **4** (often foll. by to) unfamiliar, strange, uncharacteristic (his behaviour is foreign to me). **5** (of an object or particle) introduced from outside into the body, food, machinery, etc. (a foreign body lodged in my eye). □ **foreign aid** money, food, etc. given or lent by one country to another. **foreign exchange 1** the currency of other countries. **2** dealings in these. **Foreign Minister** (or **Secretary**) a government minister in charge of his or her country's relations with other countries. **Foreign Office** hist. or colloq. = FOREIGN AND COMMONWEALTH OFFICE. □ **foreignness** /-rɪnnɪs/ n. [ME f. OF forein, forain ult. f. L foras, -is outside: for -g- cf. sovereign]

Foreign and Commonwealth Office n. the UK government department dealing with foreign affairs.

foreigner /ˈfɒrɪnə(r)/ n. **1** a person born in or coming from a foreign country or place. **2** esp. dial. a person not belonging to a particular place or society. **3 a** a foreign ship. **b** an imported animal or article.

Foreign Legion a body of foreign volunteers in an army, especially a military formation of the French army founded in the 1830s to fight France's colonial wars and composed, except for the higher ranks, of non-Frenchmen. The French Foreign Legion is famed for its audacity and endurance; its most famous campaigns were in French North Africa in the late 19th and early 20th centuries. Although its original purpose has been lost, it is still in existence, in greatly reduced form.

forejudge /fɔːˈdʒʌdʒ/ v.tr. judge or determine before knowing the evidence.

foreknow /fɔːˈnəʊ/ v.tr. (past **-knew** /-ˈnjuː/; past part. **-known** /-ˈnəʊn/) know beforehand; have prescience of. □ **foreknowledge** /-ˈnɒlɪdʒ/ n.

forelady /ˈfɔːˌleɪdɪ/ n. (pl. **-ies**) US = FOREWOMAN.

foreland /ˈfɔːlænd/ n. **1** a cape or promontory. **2** a piece of land in front of something.

foreleg /ˈfɔːleɡ/ n. either of the front legs of a quadruped.

forelimb /ˈfɔːlɪm/ n. either of the front limbs of an animal.

forelock /ˈfɔːlɒk/ n. a lock of hair growing just above the forehead. □ **take time by the forelock** seize an opportunity. **touch** (or **tug**) **one's forelock** defer to a person of higher social rank.

foreman /ˈfɔːmən/ n. (pl. **-men**) **1** a worker with supervisory responsibilities. **2** the member of a jury who presides over its deliberations and speaks on its behalf.

foremast /ˈfɔːmɑːst, -məst/ n. Naut. the forward (lower) mast of a ship.

foremost /ˈfɔːməʊst/ adj. & adv. ● adj. **1** the chief or most notable. **2** the most advanced in position; the front. ● adv. before anything else in position; in the first place (first and foremost). [earlier formost, formest, superl. of OE forma first, assim. to FORE, MOST]

forename /ˈfɔːneɪm/ n. a first or Christian name.

forenoon /ˈfɔːnuːn/ n. Naut. or Law or archaic the part of the day before noon.

forensic /fəˈrensɪk, -ˈrenzɪk/ adj. & n. ● adj. **1** of or used in connection with courts of law, esp. in relation to the detection of crime. **2** of or involving forensic science (forensic examination). ● n. (usu. in pl.) colloq. forensic science. □ **forensic medicine** the application of medical knowledge to the investigation of crime, esp. in determining the cause of death. **forensic science** the application of biochemical and other scientific techniques to the investigation of crime. □ **forensically** adv. [L forensis f. FORUM]

foreordain /ˌfɔːrɔːˈdeɪn/ v.tr. predestinate; ordain beforehand. □ **foreordination** /-dɪˈneɪʃ(ə)n/ n.

forepaw /ˈfɔːpɔː/ n. either of the front paws of a quadruped.

forepeak /ˈfɔːpiːk/ n. Naut. the end of the forehold in the angle of the bows.

foreplay /ˈfɔːpleɪ/ n. stimulation preceding sexual intercourse.

forerun /fɔːˈrʌn/ v.tr. (**-running**; past **-ran** /-ˈræn/; past part. **-run**) **1** go before. **2** indicate the coming of; foreshadow.

forerunner /ˈfɔːˌrʌnə(r)/ n. **1** a predecessor. **2** an advance messenger.

foresail /ˈfɔːseɪl, -s(ə)l/ n. Naut. the principal sail on a foremast (the lowest square sail, or the fore-and-aft bent on the mast, or the triangular before the mast).

foresee /fɔːˈsiː/ v.tr. (past **-saw** /-ˈsɔː/; past part. **-seen** /-ˈsiːn/) (often foll. by that + clause) see or be aware of beforehand. □ **foreseer** /-ˈsiːə(r)/ n. **foreseeable** adj. **foreseeably** adv. **foreseeability** /ˌfɔːsiːəˈbɪlɪtɪ/ n. [OE foresēon (as FORE- + SEE[1])]

foreshadow /fɔːˈʃædəʊ/ v.tr. be a warning or indication of (a future event).

foresheets /ˈfɔːʃiːts/ n.pl. Naut. the inner part of the bows of a boat with gratings for the bowman to stand on.

foreshock /ˈfɔːʃɒk/ n. a lesser shock preceding the main shock of an earthquake.

foreshore /ˈfɔːʃɔː(r)/ n. the part of the shore between high- and low-water marks, or between the water and cultivated or developed land.

foreshorten /fɔːˈʃɔːt(ə)n/ v.tr. show or portray (an object) with the apparent shortening due to visual perspective.

foreshow /fɔːˈʃəʊ/ v.tr. (past part. **-shown** /-ˈʃəʊn/) **1** foretell. **2** foreshadow, portend, prefigure.

foresight /ˈfɔːsaɪt/ n. **1** regard or provision for the future. **2** the process of foreseeing. **3** the front sight of a gun. **4** Surveying a sight taken forwards. □ **foresighted** adj. **foresightedly** adv. **foresightedness** n. [ME, prob. after ON forsjá, forsjó (as FORE-, SIGHT)]

foreskin /ˈfɔːskɪn/ n. the fold of skin covering the end of the penis. Also called prepuce.

forest /ˈfɒrɪst/ n. & v. ● n. **1 a** (often attrib.) a large area covered chiefly with trees and undergrowth. **b** the trees growing in it. **c** a large number or dense mass of vertical objects (a forest of masts). **2** a district formerly a forest but now cultivated (Sherwood Forest). **3 a** hist. an area usu. owned by the sovereign, kept for hunting and having its own laws. **b** see deer-forest. ● v.tr. **1** plant with trees. **2** convert into a forest. □ **forest-tree** a large tree suitable for a forest. [ME f. OF f. LL forestis silva wood outside the walls of a park f. L foris outside]

forestall /fɔːˈstɔːl/ v.tr. **1** act in advance of in order to prevent. **2** anticipate (the action of another, or an event). **3** anticipate the action of. **4** deal with beforehand. **5** hist. buy up (goods) in order to profit by an enhanced price. □ **forestaller** n. **forestalment** n. [ME in sense 5: cf. AL forestallare f. OE foresteall an ambush (as FORE-, STALL[1])]

forestay /ˈfɔːsteɪ/ n. Naut. a stay from the head of the foremast to the ship's deck to support the foremast.

Forester /ˈfɒrɪstə(r)/, C(ecil) S(cott) (pseudonym of Cecil Lewis Troughton Smith) (1899–1966), English novelist. He is remembered for his seafaring novels set during the Napoleonic Wars, featuring Captain Horatio Hornblower. His other works include The African Queen (1935), later made into a celebrated film by John Huston (1951).

forester /ˈfɒrɪstə(r)/ n. **1** a person in charge of a forest or skilled in forestry. **2** a person or animal living in a forest. **3** (**Forester**) a member of the Ancient Order of Foresters (a friendly society). [ME f. OF forestier (as FOREST)]

forestry /ˈfɒrɪstrɪ/ n. **1** the science or management of forests. (*See note below*.) **2** wooded country; forests.

■ Forests are a source of many valuable commodities besides timber, such as wood pulp for papermaking, rubber, resin, dyes, gums, and other chemical or pharmaceutical products. Forestry is concerned with the management of forests as a renewable resource by control of density and composition, protection from pests and disease, planned harvesting, and appropriate replanting. These practices have been applied mainly in the coniferous forests of the north; elsewhere forests have often been exploited and destroyed rather than managed. (See also DEFORESTATION.)

Forestry Commission the government department responsible for forestry policy in the UK, established in 1919.

foretaste n. & v. ● n. /ˈfɔːteɪst/ partial enjoyment or suffering in advance; anticipation. ● v.tr. /fɔːˈteɪst/ taste beforehand; anticipate the experience of.

foretell /fɔːˈtel/ v.tr. (*past* and *past part.* **-told** /-ˈtəʊld/) tell of or presage (an event etc.) before it takes place; predict, prophesy. □ **foreteller** n.

forethought /ˈfɔːθɔːt/ n. **1** care or provision for the future. **2** previous thinking or devising. **3** deliberate intention.

foretoken n. & v. ● n. /ˈfɔːˌtəʊkən/ a sign of something to come. ● v.tr. /fɔːˈtəʊkən/ portend; indicate beforehand. [OE *foretācn* (as FORE-, TOKEN)]

foretold *past* and *past part.* of FORETELL.

foretop /ˈfɔːtɒp, -təp/ n. Naut. a platform at the top of a foremast (see TOP¹ n. 10).

fore-topgallant /ˌfɔːtɒpˈgælənt, ˌfɔːtəˈgæl-/ n. (in full **fore-topgallant mast**) Naut. a part of a mast above the fore-topmast. □ **fore-topgallant sail** the sail above the fore-topsail.

fore-topmast /fɔːˈtɒpmaːst, -məst/ n. Naut. the mast above the foremast.

fore-topsail /fɔːˈtɒpseɪl, -s(ə)l/ n. Naut. the sail above the foresail.

forever /fəˈrevə(r)/ adv. continually, persistently (*is forever complaining*) (cf. for ever (see EVER)).

forevermore /fəˌrevəˈmɔː(r)/ adv. esp. US an emphatic form of FOREVER or for ever (see EVER).

forewarn /fɔːˈwɔːn/ v.tr. warn beforehand. □ **forewarner** n.

forewent *past* of FOREGO¹, FOREGO².

forewing /ˈfɔːwɪŋ/ n. either of the two front wings of a four-winged insect.

forewoman /ˈfɔːˌwʊmən/ n. (*pl.* **-women**) **1** a female worker with supervisory responsibilities. **2** a woman who presides over a jury's deliberations and speaks on its behalf.

foreword /ˈfɔːwɜːd/ n. introductory remarks at the beginning of a book, often by a person other than the author. [FORE- + WORD after G *Vorwort*]

foreyard /ˈfɔːjaːd/ n. Naut. the lowest yard on a foremast.

Forfar /ˈfɔːfə(r)/ a town in eastern Scotland, in Tayside. It is noted for its castle, the meeting-place in 1057 of an early Scottish Parliament and the home of several Scottish kings.

Forfarshire /ˈfɔːfəˌʃɪə(r)/ the former name (from the 16th century until 1928) for ANGUS.

forfeit /ˈfɔːfɪt/ n., adj., & v. ● n. **1 a** a penalty for a breach of contract or neglect; a fine. **2 a** a trivial fine for a breach of rules in clubs etc. or in games. **b** (in pl.) a game in which forfeits are exacted. **3** something surrendered as a penalty. **4** the process of forfeiting. **5** Law property or a right or privilege lost as a legal penalty. ● adj. lost or surrendered as a penalty. ● v.tr. (**forfeited**, **forfeiting**) lose the right to, be deprived of, or have to pay as a penalty. □ **forfeitable** adj. **forfeiter** n. **forfeiture** /-fɪtʃə(r)/ n. [ME (= crime) f. OF *forfet*, *forfait* past part. of *forfaire* transgress (f. L *foris* outside) + *faire* f. L *facere* do]

forfend /fɔːˈfend/ v.tr. **1** US protect by precautions. **2** archaic avert; keep off.

forgather /fɔːˈgæðə(r)/ v.intr. (also **foregather**) assemble; meet together; associate. [16th-c. Sc. f. Du. *vergaderen*, assim. to FOR-, GATHER]

forgave *past* of FORGIVE.

forge¹ /fɔːdʒ/ v. & n. ● v.tr. **1 a** make (money etc.) in fraudulent imitation. **b** write (a document or signature) in order to pass it off as written by another. **2** fabricate, invent. **3** shape (esp. metal) by heating in a fire and hammering. ● n. **1** a blacksmith's workshop; a smithy. **2 a** a furnace or hearth for melting or refining metal. **b** a workshop containing this. □ **forgeable** adj. **forger** n. [ME f. OF *forge* (n.), *forger* (v.) f. L *fabricare* FABRICATE]

forge² /fɔːdʒ/ v.intr. move forward gradually or steadily. □ **forge ahead 1** take the lead in a race. **2** move forward or make progress rapidly. [18th c.: perh. an aberrant pronunc. of FORCE¹]

forgery /ˈfɔːdʒərɪ/ n. (*pl.* **-ies**) **1** the act or an instance of forging, counterfeiting, or falsifying a document etc. **2** a forged or spurious thing, esp. a document or signature.

forget /fəˈget/ v. (**forgetting**; *past* **forgot** /-ˈgɒt/; *past part.* **forgotten** /-ˈgɒt(ə)n/ or esp. US **forgot**) **1** tr. & (often foll. by *about*) intr. lose the remembrance of; not remember (a person or thing). **2** tr. (foll. by clause or to + infin.) not remember; neglect (*forgot to come*; *forgot how to do it*). **3** tr. inadvertently omit to bring or mention or attend to. **4** tr. (also absol.) put out of mind; cease to think of (*forgive and forget*). □ **forget-me-not** a low-growing plant of the genus *Myosotis*; esp. *M. scorpioides*, with small bright blue yellow-centred flowers. **forget oneself 1** neglect one's own interests. **2** act unbecomingly or unworthily. □ **forgettable** adj.

forgetter n. [OE *forgietan* f. WG (as FOR-, GET)]

forgetful /fəˈgetfʊl/ adj. **1** apt to forget, absent-minded. **2** (often foll. by *of*) forgetting, neglectful. □ **forgetfully** adv. **forgetfulness** n.

forgive /fəˈgɪv/ v.tr. (also absol. or with double object) (*past* **forgave** /-ˈgeɪv/; *past part.* **forgiven** /-ˈgɪv(ə)n/) **1** cease to feel angry or resentful towards; pardon (an offender or offence) (*forgive us our mistakes*). **2** remit or let off (a debt or debtor). □ **forgivable** adj. **forgivably** adv. **forgiver** n. [OE *forgiefan* (as FOR-, GIVE)]

forgiveness /fəˈgɪvnɪs/ n. **1** the act of forgiving; the state of being forgiven. **2** readiness to forgive. [OE *forgiefenes* (as FORGIVE)]

forgiving /fəˈgɪvɪŋ/ adj. inclined readily to forgive. □ **forgivingly** adv.

forgo /fɔːˈgəʊ/ v.tr. (also **forego**) (**-goes**; *past* **-went** /-ˈwent/; *past part.* **-gone** /-ˈgɒn/) **1** abstain from; go without; relinquish. **2** omit or decline to take or use (a pleasure, advantage, etc.). [OE *forgān* (as FOR-, GO¹)]

forgot *past* of FORGET.

forgotten *past part.* of FORGET.

forint /ˈfɒrɪnt/ n. the chief monetary unit of Hungary. [Hungarian f. It. *fiorino*: see FLORIN]

fork /fɔːk/ n. & v. ● n. **1** an instrument with two or more prongs used in eating or cooking. **2** a similar much larger instrument used for digging, lifting, etc. **3** any pronged device or component (*tuning-fork*). **4** a forked support for a bicycle wheel. **5 a** a divergence of anything, e.g. a stick or road, or N. Amer. a river, into two parts. **b** the place where this occurs. **c** either of the two parts (*take the left fork*). **6** a flash of forked lightning. **7** Chess a simultaneous attack on two pieces by one. ● v. **1** intr. form a fork or branch by separating into two parts. **2** intr. take one or other road etc. at a fork (*fork left for Banbury*). **3** tr. dig or lift etc. with a fork. **4** tr. Chess attack (two pieces) simultaneously with one. □ **fork-lift truck** a vehicle with a horizontal fork in front for lifting and carrying loads. **fork lunch** (or **supper** etc.) a light meal eaten with a fork at a buffet etc. **fork out** (or **up**) colloq. hand over or pay, usu. reluctantly. [OE *forca*, *force* f. L *furca*]

Forkbeard /ˈfɔːkbɪəd/, Sweyn, see SWEYN I.

forked /fɔːkt/ adj. **1** having a fork or forklike end or branches. **2** divergent, cleft. **3** (in comb.) having so many prongs (*three-forked*). □ **forked lightning** a lightning flash in the form of a zigzag or branching line.

forlorn /fɔːˈlɔːn/ adj. **1** sad and abandoned or lonely. **2** in a pitiful state; of wretched appearance. □ **forlorn hope 1** a faint remaining hope or chance. **2** a desperate enterprise. □ **forlornly** adv. **forlornness** /-ˈlɔːnnɪs/ n. [past part. of obs. *forlese* f. OE *forlēosan* (as FOR-, LOSE): *forlorn hope* f. Du. *verloren hoop* lost troop, orig. of a storming-party etc.]

form /fɔːm/ n. & v. ● n. **1 a** a shape; an arrangement of parts. **b** the outward aspect (esp. apart from colour) or shape of a body. **2** a person or animal as visible or tangible (*the familiar form of the postman*). **3** the mode in which a thing exists or manifests itself (*took the form of a book*). **4** a species, kind, or variety. **5 a** a printed document with blank spaces for information to be inserted. **b** a regularly drawn document. **6** esp. Brit. a class in a school. **7** a customary method; what is usually done (*common form*). **8** a set order of words; a formula. **9** behaviour according to a rule or custom. **10** (prec. by *the*) correct procedure (*knows the form*). **11 a** (of an athlete, horse, etc.) condition of health and training (*is in top form*). **b** Horse-racing details of previous performances. **12** general state or disposition (*was in great form*). **13** sl. a criminal record. **14** formality or mere ceremony. **15** Gram. **a** one of the ways in which a word may be spelt or pronounced or inflected. **b** the external characteristics of words apart from meaning. **16** arrangement and style in literary or musical composition. **17** *Philos.* the essential nature of a species or thing. **18** a long bench without a back. **19** *esp. US Printing* = FORME. **20** a hare's lair. **21** formwork (see SHUTTERING 1). ● *v.* **1** *tr.* make or fashion into a certain shape or form. **2** *intr.* take a certain shape; be formed. **3** *tr.* be the material of; make up or constitute (*together form a unit; forms part of the structure*). **4** *tr.* train or instruct. **5** *tr.* develop or establish as a concept, institution, or practice (*form an idea; formed an alliance; form a habit*). **6** *tr.* (foll. by *into*) embody, organize. **7** *tr.* articulate (a word). **8** *tr. & intr.* (often foll. by *up*) *esp. Mil.* bring or be brought into a certain arrangement or formation. **9** *tr.* construct (a new word) by derivation, inflection, etc. □ **bad form** an offence against current social conventions. **form class** *Linguistics* a class of linguistic forms with grammatical or syntactical features in common. **form criticism** textual analysis of the Bible etc. by tracing the history of its content by forms (e.g. proverbs, myths). **form letter** a standardized letter to deal with frequently occurring matters. **good form** what complies with current social conventions. **in** (or **in good**) **form** in a state of good health or training. **off form** not playing or performing well. **on** (or **on good**) **form 1** playing or performing well. **2** in good spirits. **out of form** not fit for racing etc. [ME f. OF *forme* f. L *forma* mould, form]

-form /fɔːm/ *comb. form* (usu. as **-iform** /ɪfɔːm/) forming adjectives meaning: **1** having the form of (*cruciform; cuneiform*). **2** having such a number of (*uniform; multiform*). [from or after F *-forme* f. L *-formis* f. *forma* FORM]

formal /ˈfɔːm(ə)l/ *adj. & n.* ● *adj.* **1** used or done or held in accordance with rules, convention, or ceremony (*formal dress; a formal occasion*). **2** ceremonial; required by convention (*a formal call*). **3** precise or symmetrical (*a formal garden*). **4** prim or stiff in manner. **5** perfunctory, having the form without the spirit. **6** valid or correctly so called because of its form; explicit and definite (*a formal agreement*). **7** in accordance with recognized forms or rules. **8** of or concerned with (outward) form or appearance, esp. as distinct from content or matter. **9** *Logic* concerned with the form and not the matter of reasoning. **10** *Philos.* of the essence of a thing; essential not material. ● *n. N. Amer.* **1** evening dress. **2** an occasion on which evening dress is worn. □ **formally** *adv.* [ME f. L *formalis* (as FORM)]

formaldehyde /fɔːˈmaldɪˌhaɪd/ *n.* a colourless pungent gas (chem. formula: CH_2O) used as a disinfectant and preservative and in the manufacture of synthetic resins. Also called *methanal*. [FORMIC ACID + ALDEHYDE]

formalin /ˈfɔːməlɪn/ *n.* a colourless solution of formaldehyde in water used as a preservative for biological specimens etc.

formalism /ˈfɔːməˌlɪz(ə)m/ *n.* **1 a** excessive adherence to prescribed forms. **b** the use of forms without regard to inner significance. **2** *derog.* an artist's concentration on form at the expense of content. **3** the treatment of mathematics as a manipulation of meaningless symbols. **4** *Theatr.* a symbolic and stylized manner of production. **5** *Physics & Math.* the mathematical description of a physical situation etc. □ **formalist** *n.* **formalistic** /ˌfɔːməˈlɪstɪk/ *adj.*

formality /fɔːˈmalɪtɪ/ *n.* (*pl.* **-ies**) **1 a** a formal or ceremonial act, requirement of etiquette, regulation, or custom (often with an implied lack of real significance). **b** a thing done simply to comply with a rule. **2** the rigid observance of rules or convention. **3** ceremony; elaborate procedure. **4** being formal; precision of manners. **5** stiffness of design. [F *formalité* or med.L *formalitas* (as FORMAL)]

formalize /ˈfɔːməˌlaɪz/ *v.tr.* (also **-ise**) **1** give definite shape or legal formality to. **2** make ceremonious, precise, or rigid; imbue with formalism. □ **formalization** /ˌfɔːməlaɪˈzeɪʃ(ə)n/ *n.*

formant /ˈfɔːmənt/ *n.* **1** *Phonet.* the characteristic pitch-constituent of a vowel. **2** *Linguistics* a morpheme occurring only in combination in a word or word-stem. [G f. L *formare* *formant-* to form]

format /ˈfɔːmat/ *n. & v.* ● *n.* **1** the shape and size of a book, periodical, etc. **2** the style or manner of an arrangement or procedure. **3** *Computing* a defined structure for holding data etc. in a record for processing or storage. ● *v.tr.* (**formatted, formatting**) **1** arrange or put into a format. **2** *Computing* prepare (a storage medium) to receive data. [F f. G f. L *formatus (liber)* shaped (book), past part. of *formare* FORM]

formate /ˈfɔːmeɪt/ *n. Chem.* a salt or ester of formic acid. [FORMIC ACID + -ATE[1]]

formation /fɔːˈmeɪʃ(ə)n/ *n.* **1** the act or an instance of forming; the process of being formed. **2** a thing formed. **3** a structure or arrangement of parts. **4** a particular arrangement, e.g. of troops, aircraft in flight, etc. (also *attrib.: formation dancing, formation flying*).

5 *Geol.* an assemblage of rocks or series of strata having some common characteristic. □ **formational** *adj.* [ME f. OF *formation* or L *formatio* (as FORM)]

formative /ˈfɔːmətɪv/ *adj. & n.* ● *adj.* **1** serving to form or fashion; of formation. **2** *Gram.* (of a flexional or derivative suffix or prefix) used in forming words. ● *n. Gram.* a formative element. □ **formatively** *adv.* [ME f. OF *formatif -ive* or med.L *formativus* (as FORM)]

Formby /ˈfɔːmbɪ/, George (born George Booth) (1904–61), English comedian. He became famous for his numerous musical films in the 1930s, in which he projected the image of a Lancashire working lad and accompanied his songs on the ukulele.

forme /fɔːm/ *n.* (*US* **form**) *Printing* **1** a body of type secured in a chase for printing. **2** a quantity of film arranged for making a plate etc. [var. of FORM]

Formentera /ˌfɔːmənˈteərə/ a small island in the Mediterranean, south of Ibiza. It is the southernmost of the Balearic Islands.

former[1] /ˈfɔːmə(r)/ *attrib.adj.* **1** of or occurring in the past or an earlier period (*in former times*). **2** having been previously (*her former husband*). **3** (*prec. by the*; often *absol.*) the first or first mentioned of two (opp. LATTER). [ME f. *forme* first, after FOREMOST]

former[2] /ˈfɔːmə(r)/ *n.* **1** a person or thing that forms. **2** *Electr.* a frame or core for winding a coil on. **3** *Aeron.* a transverse strengthening member in a wing or fuselage. **4** (in *comb.*) a pupil of a specified form in a school (*fourth-former*).

formerly /ˈfɔːməlɪ/ *adv.* in the past; in former times.

Formica /fɔːˈmaɪkə/ *n. propr.* a hard durable plastic laminate used for working surfaces, cupboard doors, etc. [20th c.: orig. uncert.]

formic acid /ˈfɔːmɪk/ *n. Chem.* a colourless irritant volatile acid (chem. formula: HCOOH) contained in the fluid emitted by some ants. Also called *methanoic acid.* [L *formica* ant]

formication /ˌfɔːmɪˈkeɪʃ(ə)n/ₙ *n.* a sensation as of ants crawling over the skin. [L *formicatio* f. *formica* ant]

formidable /ˈfɔːmɪdəb(ə)l/, *disp.* fɔːˈmɪd-/ *adj.* **1** inspiring fear or dread. **2** inspiring respect or awe. **3** likely to be hard to overcome, resist, or deal with. □ **formidableness** *n.* **formidably** *adv.* [F *formidable* or L *formidabilis* f. *formidare* fear]

formless /ˈfɔːmlɪs/ *adj.* shapeless; without determinate or regular form. □ **formlessly** *adv.* **formlessness** *n.*

Formosa /fɔːˈməʊsə/ the former name for TAIWAN. [Port. *formosa* beautiful]

formula /ˈfɔːmjʊlə/ *n.* (*pl.* **formulas** or (esp. in senses 1, 2) **formulae** /-ˌliː/) **1** *Chem.* a set of chemical symbols showing the constituents of a substance and their relative proportions. **2** *Math.* a mathematical rule expressed in symbols. **3 a** a fixed form of words, esp. one used on social or ceremonial occasions. **b** a rule unintelligently or slavishly followed; an established or conventional usage. **c** a form of words embodying or enabling agreement, resolution of a dispute, etc. **4 a** a list of ingredients; a recipe. **b** esp. *N. Amer.* an infant's food made up from a recipe. **5** a classification of racing car, esp. by the engine capacity (*formula one*). □ **formularize** *v.tr.* (also **-ise**). **formulize** *v.tr.* (also **-ise**) **formulaic** /ˌfɔːmjʊˈleɪɪk/ *adj.* [L, dimin. of *forma* form]

formulary /ˈfɔːmjʊlərɪ/ *n. & adj.* ● *n.* (*pl.* **-ies**) **1** a collection of formulas or set forms, esp. for religious use. **2** *Pharm.* a compendium of formulae used in the preparation of medicinal drugs. ● *adj.* **1** using formulae. **2** in or of formulae. [(n.) *formulaire* or f. med.L *formularius (liber* book) f. L (as FORMULA): (adj.) f. FORMULA]

formulate /ˈfɔːmjʊˌleɪt/ *v.tr.* **1** express in a formula. **2** express clearly and precisely. **3** create or devise (a plan etc.). □ **formulator** *n.* **formulation** /ˌfɔːmjʊˈleɪʃ(ə)n/ *n.*

formulism /ˈfɔːmjʊˌlɪz(ə)m/ *n.* adherence to or dependence on conventional formulas. □ **formulistic** /ˌfɔːmjʊˈlɪstɪk/ *adj.*

formwork /ˈfɔːmwɜːk/ *n.* = SHUTTERING 1.

fornicate /ˈfɔːnɪˌkeɪt/ *v.intr. archaic* or *joc.* (of people not married or not married to each other) have sexual intercourse voluntarily. □ **fornicator** *n.* **fornication** /ˌfɔːnɪˈkeɪʃ(ə)n/ *n.* [eccl.L *fornicari* f. L *fornix -icis* brothel]

forrader /ˈfɒrədə(r)/ *colloq. compar.* of FORWARD.

Forrest /ˈfɒrɪst/, John, 1st Baron (1847–1918), Australian explorer and statesman, Premier of Western Australia 1890–1901. From 1864, as colonial surveyor, he was one of the principal explorers of Western Australia. He did much to secure the colony's self-government and became its first Premier.

forsake /fəˈseɪk/ *v.tr.* (*past* **forsook** /-ˈsʊk/; *past part.* **forsaken**

/-'seɪkən/) **1** give up; break off from; renounce. **2** withdraw one's help, friendship, or companionship from; desert, abandon. □ **forsakenness** /-kənnis/ n. **forsaker** n. [OE *forsacan* deny, renounce, refuse, f. WG; cf. OE *sacan* quarrel]

forsooth /fə'su:θ/ adv. archaic or joc. truly; in truth; no doubt. [OE *forsōth* (as FOR, SOOTH)]

Forster /'fɔ:stə(r)/, E(dward) M(organ) (1879–1970), English novelist and literary critic. His novels, many of which have been made into successful films, include *A Room with a View* (1908) and *A Passage to India* (1924). Forster's novel *Maurice*, dealing with homosexual themes, was written in 1914, and appeared posthumously in 1971. He is also noted for his critical work *Aspects of the Novel* (1927).

forswear /fɔ:'sweə(r)/ v.tr. (past **forswore** /-'swɔ:(r)/; past part. **forsworn** /-'swɔ:n/) **1** abjure; renounce on oath. **2** (in refl. or passive) swear falsely; commit perjury. [OE *forswerian* (as FOR-, SWEAR)]

Forsyth /fɔ:'saɪθ/, Frederick (b.1938), English novelist. He is known for political thrillers such as *The Day of the Jackal* (1971), *The Odessa File* (1972), and *The Fourth Protocol* (1984).

forsythia /fɔ:'saɪθɪə/ n. an ornamental shrub of the genus *Forsythia*, bearing bright yellow flowers in early spring. [mod.L f. William *Forsyth*, Engl. botanist (1737–1804)]

fort /fɔ:t/ n. **1** a fortified building or position. **2** hist. a trading-station, orig. fortified. [F *fort* or It. *forte* f. L *fortis* strong]

Fortaleza /ˌfɔ:tə'leɪzə/ a port in NE Brazil, on the Atlantic coast, capital of the state of Ceará; pop. (1990) 1,708,700.

Fort-de-France /ˌfɔ:də'frɑ:ns/ the capital of Martinique; pop. (1990) 101,540.

forte¹ /'fɔ:teɪ/ n. **1** a person's strong point; a thing in which a person excels. **2** Fencing the part of a sword-blade from the hilt to the middle (cf. FOIBLE 2). [F *fort* strong f. L *fortis*]

forte² /'fɔ:tɪ/ adj., adv., & n. Mus. ● adj. performed loudly. ● adv. loudly. ● n. a passage to be performed loudly. □ **forte piano** adj. & adv. loud and then immediately soft. [It., = strong, loud]

fortepiano /ˌfɔ:tɪpɪ'ænəʊ/ n. (pl. **-os**) Mus. = PIANOFORTE, esp. with ref. to an instrument of the 18th to early 19th centuries. [FORTE² + PIANO²]

Forth /fɔ:θ/ **1** a river of central Scotland, rising on Ben Lomond and flowing eastwards through Stirling into the North Sea. **2** a shipping forecast area covering Scottish coastal waters roughly from Berwick in the south to Aberdeen in the north, including the Firth of Forth.

forth /fɔ:θ/ adv. archaic except in set phrases and after certain verbs, esp. *bring*, *come*, *go*, and *set*. **1** forward; into view. **2** onwards in time (*from this time forth*; *henceforth*). **3** forwards. **4** out from a starting-point (*set forth*). □ **and so forth** and so on; and the like. [OE f. Gmc]

Forth, Firth of the estuary of the River Forth, separating the regions of Fife and Lothian. It is spanned by a cantilever railway bridge (opened 1890) and a road suspension bridge (1964).

forthcoming /fɔ:θ'kʌmɪŋ/ attrib.adj. **1 a** about or likely to appear or become available. **b** approaching. **2** produced when wanted (*no reply was forthcoming*). **3** (of a person) informative, responsive. □ **forthcomingness** n.

forthright adj. & adv. ● adj. /'fɔ:θraɪt/ **1** direct and outspoken; straightforward. **2** decisive, unhesitating. ● adv. /fɔ:θ'raɪt/ in a direct manner; bluntly. □ **forthrightly** /'fɔ:θˌraɪtlɪ/ adv. **forthrightness** n. [OE *forthriht* (as FORTH, RIGHT)]

forthwith /fɔ:θ'wɪθ, -'wɪð/ adv. immediately; without delay. [earlier *forthwithal* (as FORTH, WITH, ALL)]

Forties /'fɔ:tɪz/ **1** (**the Forties**) the central North Sea between Scotland and southern Norway, so called from its prevailing depth of forty fathoms or more (cf. ROARING FORTIES). The area is an important centre of North Sea oil production. **2** a shipping forecast area covering the central North Sea east of Scotland.

fortification /ˌfɔ:tɪfɪ'keɪʃ(ə)n/ n. **1** the act or an instance of fortifying; the process of being fortified. **2** Mil. **a** the art or science of fortifying. **b** (usu. in pl.) defensive works fortifying a position. [ME f. F f. LL *fortificatio -onis* act of strengthening (as FORTIFY)]

fortify /'fɔ:tɪˌfaɪ/ v.tr. (**-ies**, **-ied**) **1** provide or equip with defensive works so as to strengthen against attack. **2** strengthen or invigorate physically, mentally, or morally. **3** strengthen the structure of. **4** strengthen (wine) with alcohol. **5** increase the nutritive value of (food, esp. with vitamins). □ **fortifiable** adj. **fortifier** n. [ME f. OF *fortifier* f. LL *fortificare* f. L *fortis* strong]

fortissimo /fɔ:'tɪsɪˌməʊ/ adj., adv., & n. Mus. ● adj. performed very loudly. ● adv. very loudly. ● n. (pl. **-os** or **fortissimi** /-mɪ/) a passage to be performed very loudly. [It., superl. of FORTE²]

fortitude /'fɔ:tɪˌtju:d/ n. courage in pain or adversity. [ME f. F f. L *fortitudo -dinis* f. *fortis* strong]

Fort Knox /nɒks/ a US military reservation in Kentucky, famous as the site of the depository (built in 1936) which holds the bulk of the nation's gold bullion in its vaults.

Fort Lamy /'lɑ:mɪ/ the former name (until 1973) for N'DJAMENA.

fortnight /'fɔ:tnaɪt/ n. **1** a period of two weeks. **2** (prec. by a specified day) two weeks after (that day) (*Tuesday fortnight*). [OE *fēowertīene niht* fourteen nights]

fortnightly /'fɔ:tˌnaɪtlɪ/ adj., adv., & n. ● adj. done, produced, or occurring once a fortnight. ● adv. every fortnight. ● n. (pl. **-ies**) a magazine etc. issued every fortnight.

Fortran /'fɔ:træn/ n. (also **FORTRAN**) Computing a high-level programming language used esp. for scientific calculations. [*formula translation*]

fortress /'fɔ:trɪs/ n. a military stronghold, esp. a strongly fortified town fit for a large garrison. [ME f. OF *forteresse*, ult. f. L *fortis* strong]

fortuitous /fɔ:'tju:ɪtəs/ adj. due to or characterized by chance; accidental, casual. □ **fortuitously** adv. **fortuitousness** n. [L *fortuitus* f. *forte* by chance]

fortuity /fɔ:'tju:ɪtɪ/ n. (pl. **-ies**) **1** a chance occurrence. **2** accident or chance; fortuitousness.

fortunate /'fɔ:tjʊnət, -tʃənət/ adj. **1** favoured by fortune; lucky, prosperous. **2** auspicious, favourable. [ME f. L *fortunatus* (as FORTUNE)]

fortunately /'fɔ:tjʊnətlɪ, -tʃənətlɪ/ adv. **1** luckily, successfully. **2** (qualifying a whole sentence) it is fortunate that.

fortune /'fɔ:tju:n, -tʃu:n/ n. **1 a** a chance or luck as a force in human affairs. **b** a person's destiny. **2** (**Fortune**) this force personified, often as a deity. **3** (in sing. or pl.) the good or bad luck that befalls a person or an enterprise. **4** good luck. **5** prosperity; a prosperous condition. **6** (also colloq. **small fortune**) great wealth; a huge sum of money. □ **fortune-hunter** colloq. a person seeking wealth by marriage. **Fortune's wheel** = *wheel of Fortune*. **fortune-teller** a person who claims to predict future events in a person's life. **fortune-telling** the practice of this. **make a** (or **one's**) **fortune** acquire wealth or prosperity. **tell a person's fortune** make predictions about a person's future. [ME f. OF f. L *fortuna* luck, chance]

Fort William a town in western Scotland, on Loch Linnhe near Ben Nevis; pop. (est. 1985) 4,400.

Fort Worth /wɜ:θ/ a city in northern Texas; pop. (1990) 447,600.

forty /'fɔ:tɪ/ n. & adj. ● n. (pl. **-ies**) **1** the product of four and ten. **2** a symbol for this (40, xl, XL). **3** (in pl.) the numbers from 40 to 49, esp. the years of a century or of a person's life. ● adj. that amount to forty. □ **forty-first, -second**, etc. the ordinal numbers between fortieth and fiftieth. **forty-five** a gramophone record played at 45 r.p.m. **forty-niner** a seeker for gold etc. in the Californian gold rush of 1849. **forty-one, -two**, etc. the cardinal numbers between forty and fifty. **forty winks** colloq. a short sleep. □ **fortieth** adj. & n. **fortyfold** adj. & adv. [OE *fēowertig* (as FOUR, -TY²)]

Forty-five, the /ˌfɔ:tɪ'faɪv/ the Jacobite rebellion of 1745. (See also JACOBITE².)

forty-ninth parallel /'fɔ:tɪˌnaɪnθ/ the parallel of latitude 49° north of the equator, especially as forming the boundary between Canada and the US west of the Lake of the Woods.

forum /'fɔ:rəm/ n. (pl. **forums** or **fora** /-rə/) **1** a place of or meeting for public discussion. **2** a periodical etc. giving an opportunity for discussion. **3** a court or tribunal. **4** hist. a public square or market-place in an ancient Roman city used for judicial and other business. [L, in sense 4]

forward /'fɔ:wəd/ adj., n., adv., & v. ● adj. **1** lying in one's line of motion. **2 a** onward or towards the front. **b** /'fɒrəd/ Naut. belonging to the fore part of a ship. **3** precocious; bold in manner; presumptuous. **4 a** Commerce relating to future produce, delivery, etc. (*forward contract*). **b** prospective; advanced; with a view to the future (*forward planning*). **5 a** advanced; progressing towards or approaching maturity or completion. **b** (of a plant etc.) well advanced or early. ● n. an attacking player positioned near the front of a team in football, hockey, etc. ● adv. **1** to the front; into prominence (*come forward*; *move forward*). **2** in advance; ahead (*sent them forward*). **3** onward so as to make progress (*not getting any further forward*). **4** towards the future; continuously onwards (*from this time forward*). **5** (also **forwards** /-wədz/) **a** towards a the front in the direction one is facing. **b** in the normal direction of motion or of traversal. **c** with continuous forward motion (*backwards and forwards; rushing forward*). **6** /ˈfɔrəd/ *Naut. & Aeron.* in, near, or towards the bow or nose. ● *v.tr.* **1 a** send (a letter etc.) on to a further destination. **b** dispatch (goods etc.) (*forwarding agent*). **2** help to advance; promote. □ **forward-looking** progressive; favouring change. □ **forwarder** *n.* **forwardly** *adv.* **forwardness** *n.* (esp. in sense 3 of *adj.*). [OE *forweard*, var. of *forthweard* (as FORTH, -WARD)]

forwards var. of FORWARD *adv.* 5.

forwent past of FORGO.

Fosbury /ˈfɔzbəri/, Richard (b.1947), American high jumper. He originated the style of jumping known as the 'Fosbury flop', in which the jumper clears the bar head-first and backwards. In 1968 he won the Olympic gold medal using this technique.

fossa /ˈfɔsə/ *n.* (*pl.* **fossae** /-siː/) *Anat.* a shallow depression or cavity. [L, = ditch, fem. past part. of *fodere* dig]

fosse /fɔs/ *n.* **1** a long narrow trench or excavation, esp. in a fortification. **2** *Anat.* = FOSSA. [ME f. OF f. L *fossa*: see FOSSA]

Fosse Way an ancient road in Britain, so called from the fosse or ditch that used to run along each side of it. It ran from Axminster to Lincoln, via Bath and Leicester (about 300 km, 200 miles), and marked the limit of the first stage of the Roman occupation (mid-1st century AD).

fossick /ˈfɔsɪk/ *v.intr.* esp. *Austral. & NZ colloq.* **1** (foll. by *about, around*) rummage, search. **2** search for gold etc. in abandoned workings. □ **fossicker** *n.* [19th c.: cf. dial. *fossick* bustle about]

fossil /ˈfɔs(ə)l/ *n. & adj.* ● *n.* **1** the remains or impression of a prehistoric animal or plant, usu. petrified while embedded in rock, amber, etc. (often *attrib.: fossil bones; fossil shells*). (*See note below.*) **2** *colloq.* an antiquated or unchanging person or thing. **3** a word that has become obsolete except in set phrases or forms, e.g. *hue* in *hue and cry*. ● *adj.* **1** of or like a fossil. **2** antiquated; out of date. □ **fossil fuel** a natural fuel such as coal or gas formed in the geological past from the remains of living organisms. **fossil ivory** see IVORY. □ **fossilize** *v.tr. & intr.* (also **-ise**). **fossilization** /ˌfɔsɪlaɪˈzeɪʃ(ə)n/ *n.* **fossiliferous** /ˌfɔsɪˈlɪfərəs/ *adj.* [F *fossile* f. L *fossilis* f. *fodere foss-* dig]

▪ Fossils are usually the hard parts of organisms, such as bones, teeth, shells, or wood, preserved in petrified form or as moulds or casts in rock. *Trace fossils* represent burrows, footprints, etc. Rarely, the fossils of such insubstantial organisms as jellyfish are preserved in fine-grained rocks such as shales, or the entire bodies of animals may be preserved frozen in ice or amber or tar. Fossils may be used to date accurately a particular rock stratum, and they have provided much of the information upon which the present subdivisions of the geological time-scale are based.

fossorial /fɒˈsɔːrɪəl/ *adj.* **1** (of animals) burrowing. **2** (of limbs etc.) used in burrowing. [med.L *fossorius* f. *fossor* digger (as FOSSIL)]

Foster /ˈfɔstə(r)/, Stephen (Collins) (1826–64), American composer. He wrote more than 200 songs, and, though a Northerner, he was best known for songs which captured the Southern plantation spirit, such as 'Oh! Susannah' (1848), 'Camptown Races' (1850), and 'Old Folks at Home' (1851).

foster /ˈfɔstə(r)/ *v. & adj.* ● *v.tr.* **1 a** promote the growth or development of. **b** encourage or harbour (a feeling). **2** (of circumstances) be favourable to. **3 a** bring up (a child that is not one's own by birth). **b** (often foll. by *out*) *Brit.* (of a local authority etc.) assign (a child) to be fostered. **4** cherish; have affectionate regard for (an idea, scheme, etc.). ● *adj.* **1** having a family connection by fostering and not by birth (*foster brother; foster child; foster parent*). **2** involving or concerned with fostering a child (*foster care; foster home*). □ **fosterage** *n.* (esp. in sense 3 of *v.*). **fosterer** *n.* [OE *fōstrian, fōster*, rel. to FOOD]

fosterling /ˈfɔstəlɪŋ/ *n.* a foster-child; a nursling or protégé. [OE *fōsterling* (as FOSTER)]

Foucault[1] /ˈfuːkəʊ/, Jean Bernard Léon (1819–68), French physicist. He is chiefly remembered for the huge pendulum which he hung from the roof of the Panthéon in Paris in 1851: as the pendulum swung, the path of its oscillations slowly rotated, demonstrating the rotation of the earth. He obtained the first reasonably accurate determination of the velocity of light, invented the gyroscope, introduced the technique of silvering glass for the reflecting telescope, pioneered astronomical photography, discovered eddy currents in the cores of electrical equipment, and improved a host of devices such as the arc lamp and the induction coil.

Foucault[2] /ˈfuːkəʊ/, Michel (1929–84), French philosopher. A student

of Louis Althusser, he was mainly concerned with exploring how society defines categories of abnormality such as insanity, sexuality, and criminality, and the manipulation of social attitudes towards such things by those in power. Major works include *Histoire de la folie* (1961; *Madness and Civilization*, 1967) and *L'Histoire de la sexualité* (three volumes 1976–84; *The History of Sexuality*, 1978–86). □ **Foucauldian** /fuːˈkəʊdɪən/ *adj.* **Foucaultian** /-ˈkəʊɪən/ *adj.*

fouetté /ˈfweteɪ/ *n. Ballet* a quick whipping movement of the raised leg. [F, past part. of *fouetter* whip]

fought past and past part. of FIGHT.

Fou-hsin see FUXIN.

foul /faʊl/ *adj., n., adv., & v.* ● *adj.* **1** offensive to the senses; loathsome, stinking. **2** dirty, soiled, filthy. **3** *colloq.* revolting, disgusting. **4 a** containing or charged with noxious matter (*foul air*). **b** clogged, choked. **5** morally polluted; disgustingly abusive or offensive (*foul language; foul deeds*). **6** unfair; against the rules of a game etc. (*by fair means or foul*). **7** (of the weather) wet, rough, stormy. **8** (of a rope etc.) entangled. **9** (of a ship's bottom) overgrown with weeds, barnacles, etc. ● *n.* **1** *Sport* an unfair or invalid stroke or piece of play. **2** a collision or entanglement, esp. in riding, rowing, or running. **3** a foul thing. ● *adv.* unfairly; contrary to the rules. ● *v.* **1** *tr. & intr.* make or become foul or dirty. **2** *tr.* (of an animal) make dirty with excrement. **3 a** *tr. Sport* commit a foul against (a player). **b** *intr.* commit a foul. **4 a** *tr.* (often foll. by *up*) cause (an anchor, cable, etc.) to become entangled or muddled. **b** *intr.* become entangled. **5** *tr.* jam or block (a crossing, railway line, or traffic). **6** *tr.* (usu. foll. by *up*) *colloq.* spoil or bungle. **7** *tr.* run foul of; collide with. **8** *tr.* pollute with guilt; dishonour. □ **foul brood** a fatal disease of larval bees caused by bacteria. **foul line** *Baseball* either of the straight lines extending from the home plate and marking the limit of the playing area, within which a hit is deemed to be fair. **foul mouth** a person who uses foul language. **foul play 1** unfair play in games. **2** treacherous or violent activity, esp. murder. **foul-up** a muddled or bungled situation. □ **foully** /ˈfaʊllɪ/ *adv.* **foulness** *n.* [OE *fūl* f. Gmc]

foulard /fuːˈlɑːd/ *n.* **1** a thin soft material of silk or silk and cotton. **2** an article made of this. [F]

foumart /ˈfuːmɑːt/ *n.* a polecat. [ME *fulmert* etc. (as FOUL, *mart* MARTEN)]

found[1] past and past part. of FIND.

found[2] /faʊnd/ *v.* **1** *tr.* **a** establish (esp. with an endowment). **b** originate or initiate (an institution). **2** *tr.* be the original builder or begin the building of (a town etc.). **3** *tr.* lay the base of (a building etc.). **4** (foll. by *on, upon*) **a** *tr.* construct or base (a story, theory, rule, etc.) according to a specified principle or ground. **b** *intr.* have a basis in. □ **founding father** a person associated with a founding, esp. an American statesman at the time of the Revolution. [ME f. OF *fonder* f. L *fundare* f. *fundus* bottom]

found[3] /faʊnd/ *v.tr.* **1 a** melt and mould (metal). **b** fuse (materials for glass). **2** make by founding. □ **founder** *n.* [ME f. OF *fondre* f. L *fundere fus-* pour]

foundation /faʊnˈdeɪʃ(ə)n/ *n.* **1 a** the solid ground or base, natural or artificial, on which a building rests. **b** (usu. in *pl.*) the lowest load-bearing part of a building, usu. below ground level. **2 a** body or ground on which other parts are overlaid. **3** a basis or underlying principle; groundwork (*the report has no foundation*). **4 a** the act or an instance of establishing or constituting (esp. an endowed institution) on a permanent basis. **b** such an institution, e.g. a monastery, college, or hospital. **5** (in full **foundation garment**) a woman's supporting undergarment, e.g. a corset. □ **foundation cream** a cream used as a base for applying cosmetics. **foundation-stone 1** a stone laid with ceremony to celebrate the founding of a building. **2** the main ground or basis of something. □ **foundational** *adj.* [ME f. OF *fondation* f. L *fundatio -onis* (as FOUND[2])]

founder[1] /ˈfaʊndə(r)/ *n.* a person who founds an institution.

founder[2] /ˈfaʊndə(r)/ *v. & n.* ● *v.* **1 a** *intr.* (of a ship) fill with water and sink. **b** *tr.* cause (a ship) to founder. **2** *intr.* (of a plan etc.) fail. **3** *intr.* (of earth, a building, etc.) fall down or in, give way. **4 a** *intr.* (of a horse or its rider) fall to the ground, fall from lameness, stick fast in mud etc. **b** *tr.* cause (a horse) to break down, esp. with founder. ● *n.* **1** inflammation of a horse's foot from overwork. **2** rheumatism of the chest-muscles in horses. [ME f. OF *fondrer, esfondrer* submerge, collapse, ult. f. L *fundus* bottom]

foundling /ˈfaʊndlɪŋ/ *n.* an abandoned infant of unknown parentage. [ME, perh. f. obs. *funding* (as FIND, -ING[3]), assim. to -LING[1]]

foundry /ˈfaʊndrɪ/ *n.* (*pl.* **-ies**) a workshop for or a business of casting metal.

fount[1] /faʊnt/ n. poet. a spring or fountain; a source. [back-form. f. FOUNTAIN after MOUNT[2]]

fount[2] /faʊnt, fɒnt/ n. Printing = FONT[2].

fountain /'faʊntɪn/ n. **1 a** a jet or jets of water made to spout for ornamental purposes or for drinking. **b** a structure provided for this. **2** a structure for the constant public supply of drinking-water. **3** a natural spring of water. **4** a source (in physical or abstract senses). **5** = soda fountain. **6** a reservoir for oil, ink, etc. □ **fountain-head** an original source. **fountain-pen** a pen with a reservoir or cartridge holding ink. □ **fountained** adj. (also in comb.). [ME f. OF fontaine f. LL fontana fem. of L fontanus (adj.) f. fons fontis a spring]

four /fɔː(r)/ n. & adj. ● n. **1** one more than three, or six less than ten; the product of two units and two units. **2** a symbol for this (4, iv, IV, rarely iiii, IIII). **3** a size etc. denoted by four. **4** a four-oared rowing-boat or its crew. **5** four o'clock. **6** a card with four pips. **7** a hit at cricket scoring four runs. ● adj. that amount to four. □ **four-eyes** sl. a person wearing glasses. **four-flush** US Cards a poker hand of little value, having four cards of the same suit and one of another. **four-flusher** US a bluffer or humbug. **four hundred** US the social élite of a community. **four-in-hand 1** a vehicle with four horses driven by one person. **2** US a necktie worn with a knot and two hanging ends superposed. **four-leaf** (or **-leaved**) **clover** a clover leaf with four leaflets thought to bring good luck. **four-letter word** any of several short words referring to sexual or excretory functions, regarded as coarse or offensive. **four o'clock** = marvel of Peru. **four-part** Mus. arranged for four voices to sing or instruments to play. **four-poster** a bed with a post at each corner supporting a canopy. **four-square** adj. **1** solidly based. **2** steady, resolute; forthright. **3** square-shaped. ● adv. steadily, resolutely. **four-wheel drive** drive acting on all four wheels of a vehicle. **on all fours** on hands and knees. [OE fēower f. Gmc]

Four Cantons, Lake of the an alternative name for Lake Lucerne (see LUCERNE, LAKE).

fourchette /fʊə'ʃet/ n. Anat. a thin fold of skin at the back of the vulva. [F, dimin. of fourche (as FORK)]

fourfold /'fɔːfəʊld/ adj. & adv. **1** four times as much or as many. **2** consisting of four parts. **3** amounting to four.

four freedoms the four essential human freedoms as proclaimed in a speech to Congress by Franklin D. Roosevelt in 1941: freedom of speech and expression, freedom of worship, freedom from want, and freedom from fear.

Fourier /'fʊərɪˌeɪ/, Jean Baptiste Joseph (1768–1830), French mathematician. His theory of the diffusion of heat involved him in the solution of partial differential equations by the method of separation of variables and superposition. This led him to study the series and integrals that are now known by his name. His belief that a wide class of periodic phenomena could be described by means of Fourier series was substantially vindicated by later mathematicians, and this theory now provides one of the most important methods for solving many partial differential equations that occur in physics and engineering.

Fourier analysis n. Math. the resolution of periodic data into harmonic functions using a Fourier series.

Fourier series n. Math. an infinite series of trigonometric functions representing or approximating a given periodic function.

four noble truths the four central beliefs containing the essence of Buddhist teaching. (See BUDDHISM.)

fourpence /'fɔːp(ə)ns/ n. Brit. the sum of four pence, esp. before decimalization.

fourpenny /'fɔːp(ə)nɪ/ adj. Brit. costing four pence, esp. before decimalization. □ **fourpenny one** colloq. a hit or blow.

fourscore /fɔː'skɔː(r)/ n. archaic eighty.

foursome /'fɔːsəm/ n. **1** a group of four persons. **2** a golf match between two pairs with partners playing the same ball.

four-stroke /'fɔːstrəʊk/ adj. & n. ● attrib.adj. **1** (of an internal-combustion engine) having a power cycle of four strokes (intake, compression, combustion, and exhaust). **2** (of a vehicle) having a four-stroke engine. ● n. a four-stroke engine or vehicle.

fourteen /fɔː'tiːn/ n. & adj. ● n. **1** one more than thirteen, or four more than ten; the product of two units and seven units. **2** a symbol for this (14, xiv, XIV). **3** a size etc. denoted by fourteen. ● adj. that amount to fourteen. □ **fourteenth** adj. & n. [OE fēowertīene (as FOUR, -TEEN)]

fourth /fɔːθ/ n. & adj. ● n. **1** the position in a sequence corresponding to that of the number 4 in the sequence 1–4. **2** something occupying this position. **3** the fourth person etc. in a race or competition. **4** each of four equal parts of a thing; a quarter. **5** the fourth (and often highest) in a sequence of gears. **6** Mus. **a** an interval or chord spanning four consecutive notes in the diatonic scale (e.g. C to F). **b** a note separated from another by this interval. ● adj. that is the fourth. □ **fourth dimension 1** a postulated dimension additional to those determining area and volume. **2** time regarded as equivalent to linear dimensions. **fourth estate** joc. the press; journalism. □ **fourthly** adv. [OE fēortha, fēowertha f. Gmc]

Fourth International see INTERNATIONAL.

Fourth of July (also called Independence Day) in the US, a national holiday celebrating the anniversary of the adoption of the Declaration of Independence in 1776.

Fourth Republic hist. **1** the republican regime in France between the end of the Second World War (1945) and the introduction of a new constitution by Charles de Gaulle in 1958. **2** this period in France.

4to abbr. quarto.

fovea /'fəʊvɪə/ n. (pl. **foveae** /-vɪˌiː/) Anat. a small depression or pit, esp. the pit in the retina of the eye for focusing images. □ **foveal** adj. **foveate** /-vɪˌeɪt, -vɪət/ adj. [L]

fowl /faʊl/ n. & v. ● n. (pl. same or **fowls**) **1 a** (in full **domestic fowl**) a domestic cock or hen, a gallinaceous bird kept chiefly for its eggs and flesh. (See note below.) **b** any other domesticated bird kept for eggs or flesh, e.g. turkey, duck, goose, and guinea-fowl. **2** the flesh of birds, esp. a domestic cock or hen, as food. **3** archaic except in collect. or comb. a bird (guinea-fowl; wildfowl). ● v.intr. catch or hunt wildfowl. □ **fowl cholera** see CHOLERA. **fowl pest** an infectious viral disease of fowls. □ **fowler** n. **fowling** n. [OE fugol f. Gmc]

▪ The domestic fowl is derived from the Indian red jungle fowl, Gallus gallus, which was domesticated at least 4,000 years ago. Many breeds have been developed, including bantams and others kept mainly for show, and some types used for cockfighting. It has become the basis of a massive industry, and highly intensive (and controversial) methods are used to rear and keep the birds in order to provide a cheap supply of meat and eggs.

Fowler /'faʊlə(r)/, H(enry) W(atson) (1858–1933), English lexicographer and grammarian. With his brother F(rancis) G(eorge) Fowler (1870–1918), he compiled the first Concise Oxford Dictionary (1911). He is most famous for his moderately prescriptive guide to style and idiom, Modern English Usage, first published in 1926.

Fowles /faʊlz/, John (Robert) (b.1926), English novelist. His works include the psychological thriller The Collector (1963), the magic realist novel The Magus (1966), and the semi-historical novel The French Lieutenant's Woman (1969).

Fox /fɒks/, George (1624–91), English preacher and founder of the Society of Friends (Quakers). He began preaching in 1647, teaching that truth is the inner voice of God speaking to the soul, and rejecting priesthood and ritual. Despite repeated imprisonment, he established a society called the 'Friends of the Truth' (c.1650), which later became the Society of Friends.

fox /fɒks/ n. & v. ● n. **1 a** a flesh-eating canine mammal of the genus Vulpes or a related genus, with a sharp snout and bushy tail; esp. V. vulpes, with reddish fur. **b** the fur of a fox. **2** a cunning or sly person. **3** N. Amer. sl. an attractive woman. ● v. **1 a** intr. act craftily. **b** tr. deceive, baffle, trick. **2** tr. (usu. as **foxed** adj.) discolour (the leaves of a book, an engraving, etc.) with brownish marks. □ **fox-terrier** a short-haired breed of terrier originally used for unearthing foxes. □ **foxing** n. (in sense 2 of v.). **foxlike** adj. [OE f. WG]

Foxe /fɒks/, John (1516–87), English religious writer. After fleeing to the Continent on the accession of Queen Mary I, he published his Actes and Monuments in Strasbourg in 1554; popularly known as The Book of Martyrs, it appeared in England in 1563. This passionate account of the persecution of English Protestants fuelled hostility to Catholicism for generations.

foxglove /'fɒksglʌv/ n. a tall plant, Digitalis purpurea, with erect spikes of purple or white bell-shaped flowers. It is the source of the drug digitalis.

foxhole /'fɒkshəʊl/ n. **1** Mil. a hole in the ground used as a shelter against enemy fire or as a firing-point. **2** a place of refuge or concealment.

foxhound /'fɒkshaʊnd/ n. a kind of hound bred and trained to hunt foxes.

fox-hunt /'fɒkshʌnt/ n. & v. ● n. **1** a hunt for a fox with hounds. **2** a particular group of people engaged in this. ● v.intr. engage in a fox-hunt. □ **fox-hunter** n. **fox-hunting** n. & adj.

foxtail /'fɒksteɪl/ *n.* a grass of the genus *Alopecurus*, with brushlike spikes.

Fox Talbot, William Henry, see TALBOT.

foxtrot /'fɒkstrɒt/ *n. & v.* ● *n.* **1** a ballroom dance with slow and quick steps. **2** the music for this. ● *v.intr.* (**foxtrotted, foxtrotting**) perform this dance.

foxy /'fɒksɪ/ *adj.* (**foxier, foxiest**) **1** of or like a fox. **2** sly or cunning. **3** reddish-brown. **4** (of paper) damaged, esp. by mildew. **5** *N. Amer. sl.* (of a woman) sexually attractive. □ **foxily** *adv.* **foxiness** *n.*

foyer /'fɔɪeɪ/ *n.* the entrance hall or other large area in a hotel, theatre, etc. [F, = hearth, home, ult. f. L *focus* fire]

FP *abbr.* freezing-point.

fp *abbr.* forte piano.

FPA *abbr.* (in the UK) Family Planning Association.

FPS *abbr.* Fellow of the Pharmaceutical Society of Great Britain.

fps *abbr.* (also **f.p.s.**) **1** feet per second. **2** foot-pound-second.

Fr *symb. Chem.* the element francium.

Fr. *abbr.* (also **Fr**) **1** Father. **2** French.

fr. *abbr.* franc(s).

Fra /frɑː/ *n.* a prefixed title given to an Italian monk or friar. [It., abbr. of *frate* brother]

frabjous /'fræbdʒəs/ *adj.* delightful, joyous. □ **frabjously** *adv.* [devised by Lewis Carroll, app. to suggest *fair* and *joyous*]

fracas /'frækɑː/ *n.* (*pl.* same /-kɑːz/) a noisy disturbance or quarrel. [F f. *fracasser* f. It. *fracassare* make an uproar]

fractal /'fræktəl/ *n. & adj. Math.* ● *n.* a curve or geometrical figure, each part of which has the same statistical character as the whole. (*See note below.*) ● *adj.* of or relating to a fractal. [F f. L *fract-*: see FRACTION]

- Fractals are theoretically useful in describing partly random or chaotic natural phenomena such as crystal growth, fluid turbulence, galaxy formation, etc. Non-uniform structures in which similar patterns recur at progressively smaller scales, such as snowflakes and eroded coastlines, can be realistically modelled using fractals, which have become familiar through striking computer-graphic images. The concept was introduced (and the word coined) by the Polish-born mathematician Benoit B. Mandelbrot (b.1924) in 1975.

fraction /'frækʃ(ə)n/ *n.* **1** a numerical quantity that is not a whole number (e.g. $\frac{1}{2}$, 0.5). **2** a small, esp. very small, part, piece, or amount. **3** a portion of a mixture separated by distillation etc. **4** *Polit.* any organized dissentient group, esp. a group of communists in a non-communist organization. **5** *Eccl.* the division of the Eucharistic bread. □ **fractionary** *adj.* **fractionize** *v.tr.* (also **-ise**). [ME f. OF f. LL *fractio -onis* f. L *frangere fract-* break]

fractional /'frækʃən(ə)l/ *adj.* **1** of or relating to or being a fraction. **2** very slight; incomplete. **3** *Chem.* relating to the separation of parts of a mixture by making use of their different physical properties (*fractional crystallization; fractional distillation*). □ **fractionalize** *v.tr.* (also **-ise**). **fractionally** *adv.* (esp. in sense 2).

fractionate /'frækʃəneɪt/ *v.tr.* **1** break up into parts. **2** separate (a mixture) by fractional distillation etc. □ **fractionation** /ˌfrækʃə'neɪʃ(ə)n/ *n.*

fractious /'frækʃəs/ *adj.* **1** irritable, peevish. **2** unruly. □ **fractiously** *adv.* **fractiousness** *n.* [FRACTION in obs. sense 'brawling', prob. after *factious* etc.]

fracto- /'fræktəʊ/ *comb. form Meteorol.* (of a cloud form) broken or fragmentary (*fracto-cumulus; fracto-nimbus*). [L *fractus* broken: see FRACTION]

fracture /'fræktʃə(r)/ *n. & v.* ● *n.* **1 a** breakage or breaking, esp. of a bone or cartilage. **b** the result of breaking; a crack or split. **2** the surface appearance of a freshly broken rock or mineral. **3** *Linguistics* **a** the substitution of a diphthong for a simple vowel owing to an influence esp. of a following consonant. **b** a diphthong substituted in this way. ● *v.intr. & tr.* **1** *Med.* undergo or cause to undergo a fracture. **2** break or cause to break. [ME f. F *fracture* or f. L *fractura* (as FRACTION)]

fraenulum /'friːnjʊləm/ *n.* (also **frenulum**) (*pl.* **-la** /-lə/) *Anat.* a small fraenum. [mod.L, dimin. of FRAENUM]

fraenum /'friːnəm/ *n.* (also **frenum**) (*pl.* **-na** /-nə/) *Anat.* a fold of mucous membrane or skin serving to check the motion of the part to which it is attached, esp. any of those underneath the tongue. [L, = bridle]

fragile /'frædʒaɪl, -dʒɪl/ *adj.* **1** easily broken; weak. **2** of delicate frame or constitution; not strong. □ **fragilely** *adv.* **fragility** /frə'dʒɪlɪtɪ/ *n.* [F *fragile* or L *fragilis* f. *frangere* break]

fragment *n. & v.* ● *n.* /'frægmənt/ **1** a part broken off; a detached piece. **2** an isolated or incomplete part. **3** the remains of an otherwise lost or destroyed whole, esp. the extant remains or unfinished portion of a book or work of art. ● *v.tr. & intr.* /fræg'ment/ break or separate into fragments. □ **fragmental** /fræg'ment(ə)l/ *adj.* **fragmentize** /'frægmən,taɪz/ *v.tr.* (also **-ise**). [ME f. F *fragment* or L *fragmentum* (as FRAGILE)]

fragmentary /'frægməntərɪ/ *adj.* **1** consisting of fragments. **2** disconnected. **3** *Geol.* composed of fragments of previously existing rocks. □ **fragmentarily** *adv.*

fragmentation /ˌfrægmən'teɪʃ(ə)n/ *n.* the process or an instance of breaking into fragments. □ **fragmentation bomb** a bomb designed to break up into small rapidly-moving fragments when exploded.

Fragonard /'frægə'nɑː(r)/, Jean-Honoré (1732–1806), French painter. His paintings, usually landscapes, gardens, and family scenes, embody the rococo spirit. He is most famous for erotic canvases such as *The Swing* (c.1766) and *The Progress of Love* (1771).

fragrance /'freɪgrəns/ *n.* **1** sweetness of smell. **2** a sweet scent. □ **fragranced** *adj.* [F *fragrance* or L *fragrantia* (as FRAGRANT)]

fragrancy /'freɪgrənsɪ/ *n.* (*pl.* **-ies**) = FRAGRANCE.

fragrant /'freɪgrənt/ *adj.* sweet-smelling. □ **fragrantly** *adv.* [ME f. F *fragrant* or L *fragrare* smell sweet]

frail /freɪl/ *adj. & n.* ● *adj.* **1** fragile, delicate. **2** in weak health. **3** morally weak; unable to resist temptation. **4** transient, insubstantial. ● *n.* *US sl.* a woman. □ **frailly** /'freɪllɪ/ *adv.* **frailness** *n.* [ME f. OF *fraile, frele* f. L *fragilis* FRAGILE]

frailty /'freɪltɪ/ *n.* (*pl.* **-ies**) **1** the condition of being frail. **2** liability to err or yield to temptation. **3** a fault, weakness, or foible. [ME f. OF *frailete* f. L *fragilitas -tatis* (as FRAGILE)]

Fraktur /'fræktʊə(r)/ *n. Printing* a German style of black-letter type. [G]

framboesia /fræm'biːzɪə/ *n.* (*US* **frambesia**) *Med.* = YAWS. [mod.L f. F *framboise* raspberry f. L *fraga ambrosia* ambrosial strawberry]

frame /freɪm/ *n. & v.* ● *n.* **1** a case or border enclosing a picture, window, door, etc. **2** the basic rigid supporting structure of anything, e.g. of a building, motor vehicle, or aircraft. **3** (in *pl.*) the structure of spectacles holding the lenses. **4** a human or animal body, esp. with reference to its size or structure (*his frame shook with laughter*). **5** a framed work or structure (*the frame of heaven*). **6 a** an established order, plan, or system (*the frame of society*). **b** construction, constitution, build. **7** a temporary state (esp. in **frame of mind**). **8** a single complete image or picture on a cinema film or transmitted in a series of lines by television. **9 a** a triangular structure for positioning the balls in snooker etc. **b** the balls positioned in this way. **c** a round of play in snooker etc. **10** a boxlike structure of glass etc. for protecting plants. **11** a removable box of slats for the building of a honeycomb in a beehive. **12** *N. Amer. sl.* = frame-up. ● *v.tr.* **1 a** set in or provide with a frame. **b** serve as a frame for. **2** construct by a combination of parts or in accordance with a design or plan. **3** formulate or devise the essentials of (a complex thing, idea, theory, etc.). **4** (foll. by *to, into*) adapt or fit. **5** *colloq.* concoct a false charge or evidence against; devise a plot with regard to. **6** articulate (words). □ **frame-house** a house constructed of a wooden skeleton covered with boards etc. **frame of reference 1** a set of standards or principles governing behaviour, thought, etc. **2** *Geom.* a system of geometrical axes for defining position. **frame-saw** a saw stretched in a frame to make it rigid. **frame-up** *colloq.* a conspiracy, esp. to make an innocent person appear guilty. □ **framable** *adj.* **frameless** *adj.* **framer** *n.* [OE *framian* be of service f. *fram* forward: see FROM]

framework /'freɪmwɜːk/ *n.* **1** an essential supporting structure. **2** a basic system.

framing /'freɪmɪŋ/ *n.* a framework; a system of frames.

franc /fræŋk/ *n.* the chief monetary unit of France, its dependencies and former colonies, Belgium, Switzerland, Luxembourg, and several other countries. [ME f. OF f. *Francorum Rex* king of the Franks, the legend on the earliest gold coins so called (14th c.): see FRANK[2]]

France[1] /frɑːns/ a country in western Europe; pop. (1990) 56,556,000; official language, French; capital, Paris. Julius Caesar subdued the area in the 1st century BC and it became the Roman province of Gaul. It was politically splintered by barbarian invasions from the 3rd century onwards and, although briefly united under the Merovingian and Carolingian kings, did not emerge as a unified state until the ejection of the English and Burgundians at the end of the Middle Ages. France became a major power under the Valois and Bourbon dynasties in the 16th–18th centuries, and, after the overthrow of the monarchy in the

French Revolution, briefly dominated Europe under Napoleon. France was defeated in the Franco–Prussian War (1870–1) and suffered much destruction and loss of life in the First World War. During the Second World War France was occupied by the Germans and nominally controlled by a French puppet government based in Vichy. The postwar period saw France involved in war with the Vietminh in French Indo-China (1946–54). During the 1950s France was also in conflict with Algeria over that country's independence. France was a founder member of the EEC in 1957, and under Charles de Gaulle successfully rebuilt its economy in the 1960s.

France[2] /fra:ns/, Anatole (pseudonym of Jacques-Anatole-François Thibault) (1844–1924), French writer. He achieved success as a novelist with *Le Crime de Sylvestre Bonnard* (1881). His later work was more satirical, notably *L'Île des pingouins* (1908), an ironic version of the Dreyfus case, and *Les Dieux ont soif* (1912), a study of fanaticism during the French Revolution. He was awarded the Nobel Prize for literature in 1921.

Franche-Comté /ˌfra:nʃkɒmˈteɪ/ a region of eastern France, in the northern foothills of the Jura mountains. Originally part of the kingdom of Burgundy, it was independent from 1137 to 1384. After a period of Habsburg rule, it was annexed by France in 1678.

franchise /ˈfræntʃaɪz/ n. & v. ● n. **1 a** the right to vote at state (esp. parliamentary) elections. **b** the principle of qualification for this. **2** full membership of a state or corporation; citizenship. **3** authorization granted to an individual or group by a company to sell its goods or services in a particular way. **4** *hist.* legal immunity or exemption from a burden or jurisdiction. **5** a right or privilege granted to a person or corporation. ● *v.tr.* grant a franchise to. □ **franchiser** n. (also **franchisor**). **franchisee** /ˌfræntʃaɪˈzi:/ n. [ME f. OF f. *franc, franche* free: see FRANK]

Francis /ˈfra:nsɪs/, Richard Stanley ('Dick') (b.1920), English jockey and writer. He was champion jockey 1953–4 and began writing after his retirement in 1957. He has written a series of thrillers, mostly set in the world of horse-racing.

Francis I /ˈfra:nsɪs/ (1494–1547), king of France 1515–47. He succeeded his cousin Louis XII in 1515 and soon afterwards took the duchy of Milan. The greater part of his reign (1521–44) was spent at war with the Holy Roman emperor Charles V, with the result that Francis relinquished all claims to Italy. In the early years of his reign Francis generally practised religious toleration towards supporters of the Reformation, although his policies became harsher from the mid-1530s. A noted patron of the arts, he supported Rabelais and Cellini, and commissioned many new buildings in Paris, including the Louvre.

Franciscan /frænˈsɪskən/ n. & adj. ● n. a friar, nun, or sister of a Christian order founded in 1209 by St Francis of Assisi. (*See note below.*) ● *adj.* of St Francis or his order. [F *franciscain* f. mod.L *Franciscanus* f. *Franciscus* Francis]

▪ The rule for the Franciscan order originally laid down by St Francis in 1209 (including a vow of complete poverty) was felt to be too strict; it was reformulated in 1221 and received papal approval in 1223. The debate about degrees of strictness continued, however, and eventually resulted in the division of the order of friars into three branches: the Friars Minor of the Observance (or Observants), the Friars Minor Capuchin, and the Friars Minor Conventual. The Observants and the Conventuals (the least strict of the branches) separated in 1517; the strictest branch is that of the Capuchins, whose rule was drawn up in 1529. The Franciscan nuns, of an order founded by St Clare (*c.*1212) under the guidance of St Francis, are known as Poor Clares. There is also a third order of tertiaries or lay brothers and sisters (founded 1221). The friars have been important preachers and missionaries. There is also a Franciscan order in the Anglican Church.

Francis of Assisi, St (born Giovanni di Bernardone) (*c.*1181–1226), Italian monk, founder of the Franciscan order. Born into a wealthy family, he renounced his inheritance in favour of a life of poverty after experiencing a personal call to rebuild the semi-derelict church of San Damiano of Assisi. He soon attracted followers, founding the Franciscan order in 1209 and drawing up its original rule (based on complete poverty). His generosity, simple faith, deep humility, and love of nature have made him one of the most cherished saints. Feast day, 4 Oct.

Francis of Sales, St /sa:l/ (1567–1622), French bishop. One of the leaders of the Counter-Reformation, he was bishop of Geneva (1602–22) and co-founder of the Order of the Visitation, an order of nuns

(1610). The Salesian order (founded in 1859) is named after him. Feast day, 24 Jan.

Francis Xavier, St, see XAVIER, ST FRANCIS.

francium /ˈfrænsɪəm/ n. a radioactive metallic chemical element (atomic number 87; symbol **Fr**). Francium was discovered in 1939 by the French scientist Marguerite Perey (1909–75); it is the heaviest member of the alkali-metal group. It only occurs naturally as a radioactive decay product; all its isotopes have short half-lives, the most stable about 22 minutes. [f. FRANCE[1], the country of its discoverer]

Franck[1] /frɒŋk/, César (Auguste) (1822–90), Belgian-born French composer. He was a noted organist, becoming organ professor at the Paris Conservatoire in 1872. His reputation rests on a few works composed late in life, particularly the *Symphonic Variations* for piano and orchestra (1885), the D minor Symphony (1886–8), and the String Quartet (1889).

Franck[2] /fræŋk/, James (1882–1964), German-born American physicist. He worked on the bombardment of atoms by electrons, and found that the atoms absorb and lose energy in discrete increments or quanta. He then studied the vibration and rotation of dissociated molecules. After moving to America in 1935 he eventually joined the atom bomb project, and in the Franck report, completed in 1945, he and other scientists proposed the explosion of the bomb in an uninhabited area to demonstrate its power to Japan, rather than using it directly in war.

Franco /ˈfræŋkəʊ/, Francisco (1892–1975), Spanish general and statesman, head of state 1939–75. After commanding the Spanish Foreign Legion in Morocco, Franco was among the leaders of the military uprising against the Republican government which led to the Spanish Civil War. In 1937 he became leader of the Falange (Fascist) Party and proclaimed himself 'Caudillo' (leader) of Spain. With the surrender of Madrid and the defeat of the republic in 1939, he took control of the government and established a dictatorship. Despite pressure from Germany and Italy, Franco kept Spain neutral during the Second World War. In 1969 he named Prince Juan Carlos as his successor and heir to the reconstituted Spanish throne.

Franco- /ˈfræŋkəʊ/ *comb. form* **1** French; French and (*Franco-Prussian*). **2** regarding France or the French (*Francophile*). [med.L *Francus* FRANK[2]]

francolin /ˈfræŋkəʊlɪn/ n. a partridge of the genus *Francolinus*, native to Africa and Asia. [F f. It. *francolino*]

Franconia /frænˈkəʊnɪə/ a medieval duchy of southern Germany, inhabited by the Franks. It was partitioned in 939 into Western, or Rhenish, Franconia and Eastern Franconia. The region now falls mainly within Bavaria, Hesse, and Baden-Württemberg.

Francophile /ˈfræŋkəˌfaɪl/ n. a person who is fond of France or the French.

francophone /ˈfræŋkəˌfəʊn/ n. & adj. ● n. a French-speaking person. ● adj. French-speaking. [FRANCO- + Gk *phōnē* voice]

Franco-Prussian War /ˈfræŋkəʊ ˌprʌʃən/ the war of 1870–1 between France (under Napoleon III) and Prussia, in which Prussian troops advanced into France and decisively defeated the French at Sedan. The defeat marked the end of the French Second Empire. For Prussia, the proclamation of the new German Empire at Versailles was the climax of Bismarck's ambitions to unite Germany.

frangible /ˈfrændʒɪb(ə)l/ adj. breakable, fragile. [OF *frangible* or med.L *frangibilis* f. L *frangere* to break]

frangipane /ˈfrændʒɪˌpeɪn/ n. **1 a** an almond-flavoured cream or paste. **b** a flan filled with this. **2** = FRANGIPANI. [F (as FRANGIPANI)]

frangipani /ˌfrændʒɪˈpɑːnɪ/ n. (pl. **frangipanis**) **1** a tree or shrub of the genus *Plumeria*, native to tropical America; esp. *P. rubra*, with clusters of fragrant white, pink, or yellow flowers. **2** the perfume from this plant. [Marquis Muzio *Frangipani*, 16th-c. Italian perfumer]

franglais /ˈfrɒŋgleɪ/ n. a corrupt version of French using many words and idioms borrowed from English. [F f. *français* French + *anglais* English]

Frank[1] /fræŋk/, Anne (1929–45), German Jewish girl. Her diary (1947; *The Diary of a Young Girl*, 1953) records the experiences of her family living for two years in hiding from the Nazis in occupied Amsterdam. They were eventually betrayed and sent to concentration camps; Anne died in Belsen. Her diary has been translated into over thirty languages.

Frank[2] /fræŋk/ n. **1** a member of the Germanic nation or coalition that conquered Gaul in the 6th century. **2** (in the eastern Mediterranean region) a person of Western nationality. □ **Frankish** adj. [OE *Franca*, OHG *Franko*, perh. f. the name of a weapon: cf. OE *franca* javelin]

frank /fræŋk/ adj., v., & n. ● adj. **1** candid, outspoken (*a frank opinion*).

2 undisguised, avowed (*frank admiration*). **3** ingenuous, open (a *frank face*). **4** *Med.* unmistakable. ● *v.tr.* **1** stamp (a letter) with an official mark (esp. other than a normal postage stamp) to record the payment of postage. **2** *hist.* superscribe (a letter etc.) with a signature ensuring conveyance without charge; send without charge. **3** *archaic* facilitate the coming and going of (a person). ● *n.* **1** a franking signature or mark. **2** a franked cover. □ **frankable** *adj.* **frankness** *n.* [ME f. OF *franc* f. med.L *francus* free, f. FRANK² (since only Franks had full freedom in Frankish Gaul)]

Frankenstein /'fræŋkən,staɪn/ a character in the novel *Frankenstein, or the Modern Prometheus* (1818) by Mary Shelley. Baron Frankenstein is a scientist who creates and brings to life a manlike monster which eventually turns on him and destroys him (Frankenstein is not the name of the monster itself as often assumed). The story forms the basis of many films, notably *Frankenstein* (1931), starring Boris Karloff.

Frankfort /'fræŋkfət/ the state capital of Kentucky; pop. (1990) 26,000.

Frankfurt /'fræŋkfɜːt/ (in full **Frankfurt am Main** /æm 'maɪn/) a commercial city in western Germany, in Hesse; pop. (1991) 654,080. The headquarters of the *Bundesbank* are located there.

frankfurter /'fræŋk,fɜːtə(r)/ *n.* a seasoned smoked sausage made of beef and pork. [G *Frankfurter Wurst* Frankfurt sausage]

Frankfurt School a school of philosophy associated with the Frankfurt Institute for Social Research after 1923, whose adherents were involved in a reappraisal of Marxism, particularly in terms of the cultural and aesthetic dimension of modern industrial society. The principal figures include Theodor Adorno, Max Horkheimer, and Herbert Marcuse.

frankincense /'fræŋkɪn,sens/ *n.* an aromatic gum resin obtained from trees of the genus *Boswellia*, used for burning as incense. Also called *olibanum*. [ME f. OF *franc encens* pure incense]

Franklin¹ /'fræŋklɪn/, Aretha (b.1942), American soul and gospel singer. She made her name with the album *I Never Loved a Man (the Way I Love You)* (1967), going on to record over thirty albums, including the live gospel set *Amazing Grace* (1972).

Franklin² /'fræŋklɪn/, Benjamin (1706–90), American statesman, inventor, and scientist. A wealthy printer and publisher, he was one of the signatories to the peace between the US and Great Britain after the War of American Independence. His main scientific achievements were the formulation of a theory of electricity, based on the concept of an electric fluid, which introduced (and arbitrarily defined) positive and negative electricity, and a demonstration of the electrical nature of lightning, which led to the invention of the lightning conductor. His inventions include the 'Franklin stove' (a kind of free-standing cast-iron heater) and bifocal spectacles.

Franklin³ /'fræŋklɪn/, Rosalind Elsie (1920–58), English physical chemist and molecular biologist. Her early work was on the structure of coals, and she went on to investigate the various forms of carbon by means of X-ray crystallography. She then used this technique on DNA, and contributed to the discovery of its structure (see WILKINS). Franklin was using the technique to investigate the structure of viruses at the time of her premature death from cancer.

franklin /'fræŋklɪn/ *n. hist.* a landowner of free but not noble birth in the 14th and 15th centuries in England. [ME *francoleyn* etc. f. AL *francalanus* f. *francalis* held without dues f. *francus* free: see FRANK]

frankly /'fræŋklɪ/ *adv.* **1** in a frank manner. **2** (qualifying a whole sentence) to be frank.

frantic /'fræntɪk/ *adj.* **1** wildly excited; frenzied. **2** characterized by great hurry or anxiety; desperate, violent. **3** *colloq.* extreme; very great. □ **frantically** *adv.* **franticness** *n.* [ME *frentik, frantik* f. OF *frenetique* f. L *phreneticus*: see FRENETIC]

Franz Josef /frænts 'jəʊzef/ (1830–1916), emperor of Austria 1848–1916 and king of Hungary 1867–1916. The early part of his reign was characterized by his efforts to rule as an absolutist monarch. Later, he made concessions, granting Austria a parliamentary constitution in 1861 and giving Hungary equal status with Austria (see AUSTRIA–HUNGARY). His annexation of Bosnia–Herzegovina (1908) contributed to European political tensions, and the assassination in Sarajevo of his nephew and heir apparent, Archduke Franz Ferdinand (1863–1914), prompted Austria's attack on Serbia and precipitated the First World War.

Franz Josef Land a group of islands in the Arctic Ocean, discovered in 1873 by an Austrian expedition and named after the Austrian emperor Franz Josef. The islands were annexed by the USSR in 1928.

frap /fræp/ *v.tr.* (**frapped, frapping**) *Naut.* bind tightly. [F *frapper* bind, strike]

frappé /'fræpeɪ/ *adj. & n.* ● *adj.* (esp. of wine) iced, cooled. ● *n.* **1** an iced drink. **2** a soft water-ice. [F, past part. of *frapper* strike, ice (drinks)]

frascati /fræ'skɑːtɪ/ *n.* a usu. white wine produced in the Frascati region of Italy.

Fraser¹ /'freɪzə(r)/ a river of British Columbia. It rises in the Rocky Mountains and flows in a wide curve 1,360 km (850 miles) into the Strait of Georgia, just south of Vancouver.

Fraser² /'freɪzə(r)/, Dawn (b.1937), Australian swimmer. She won the Olympic gold medal for the 100-metres freestyle in 1956, 1960, and 1964, the first competitor to win the same title at three successive Olympics. She was the first woman to swim 100 metres in under one minute (1964), and set many world records.

frass /fræs/ *n.* **1** a fine powdery refuse left by insects boring into wood etc. **2** the excrement of insect larvae. [G f. *fressen* devour (as FRET¹)]

fraternal /frə'tɜːn(ə)l/ *adj.* **1** of a brother or brothers. **2** suitable to a brother; brotherly. **3** (of twins) developed from separate ova and not necessarily closely similar. **4** *N. Amer.* of or concerning a fraternity (see FRATERNITY 3). □ **fraternalism** *n.* **fraternally** *adv.* [med.L *fraternalis* f. L *fraternus* f. *frater* brother]

fraternity /frə'tɜːnɪtɪ/ *n. (pl.* **-ies**) **1** a religious brotherhood. **2** a group or company with common interests, or of the same professional class. **3** *N. Amer.* a male students' society in a university or college. **4** being fraternal; brotherliness. [ME f. OF *fraternité* f. L *fraternitas -tatis* (as FRATERNAL)]

fraternize /'frætə,naɪz/ *v.intr.* (also **-ise**) (often foll. by *with*) **1** associate; make friends; behave as intimates. **2** (of troops) enter into friendly relations with enemy troops or the inhabitants of an occupied country. □ **fraternization** /,frætənaɪ'zeɪʃ(ə)n/ *n.* [F *fraterniser* & med.L *fraternizare* f. L *fraternus*: see FRATERNAL]

fratricide /'frætrɪ,saɪd/ *n.* **1** the killing of one's brother or sister. **2** a person who does this. □ **fratricidal** /,frætrɪ'saɪd(ə)l/ *adj.* [F *fratricide* or LL *fratricidium*, L *fratricida*, f. *frater fratris* brother]

Frau /fraʊ/ *n. (pl.* **Frauen** /'fraʊən/) (often as a title) a married or widowed German-speaking woman. [G]

fraud /frɔːd/ *n.* **1** criminal deception; the use of false representations to gain an unjust advantage. **2** a dishonest artifice or trick. **3** a person or thing not fulfilling what is claimed or expected of him, her, or it. □ **fraudster** *n.* [ME f. OF *fraude* f. L *fraus fraudis*]

fraudulent /'frɔːdjʊlənt/ *adj.* **1** characterized or achieved by fraud. **2** guilty of fraud; intending to deceive. □ **fraudulence** *n.* **fraudulently** *adv.* [ME f. OF *fraudulent* or L *fraudulentus* (as FRAUD)]

fraught /frɔːt/ *adj.* **1** (foll. by *with*) filled or attended with (*fraught with danger*). **2** *colloq.* causing or affected by great anxiety or distress. [ME, past part. of obs. *fraught* (v.) load with cargo f. MDu. *vrachten* f. *vracht* FREIGHT]

Fräulein /'frɔɪlaɪn/ *n.* (often as a title) an unmarried (esp. young) German-speaking woman. [G, dimin. of FRAU]

Fraunhofer /'fraʊn,həʊfə(r)/, Joseph von (1787–1826), German optician and pioneer in spectroscopy. He observed and mapped a large number of fine dark lines in the solar spectrum and plotted their wavelengths. These lines, now named after him, were later used to determine the chemical elements present in the spectra of the sun and stars. He became noted for his finely ruled diffraction gratings, used to determine the wavelengths of specific colours of light and of the major spectral lines.

fraxinella /,fræksɪ'nelə/ *n.* an aromatic plant, *Dictamnus albus*, of the rue family, having foliage that emits an ethereal inflammable oil. Also called *dittany, gas plant, burning bush*. [mod.L, dimin. of L *fraxinus* ash tree]

fray¹ /freɪ/ *v.* **1** *tr. & intr.* wear through or become worn, esp. (of woven material) unweave at the edges. **2** *intr.* (of nerves, temper, etc.) become strained; deteriorate. [F *frayer* f. L *fricare* rub]

fray² /freɪ/ *n.* **1** conflict, fighting (*eager for the fray*). **2** a noisy quarrel or brawl. [ME f. *fray* to quarrel f. *affray* (v.) (as AFFRAY)]

Fray Bentos /freɪ 'bentɒs/ a port and meat-packing centre in western Uruguay; pop. (1985) 20,000.

Frazer /'freɪzə(r)/, Sir James George (1854–1941), Scottish anthropologist. He is often regarded as the founder of British social anthropology and ethnology. In a series of essays, *The Golden Bough* (1890–1915), he proposed an evolutionary theory of the development of human thought, from the magical and religious to the scientific.

The first chair in anthropology was created for him at Liverpool University in 1907.

frazil /'freɪzɪl/ n. N. Amer. ice crystals that form in a stream or on its bed. [Canad. F frasil snow floating in the water; cf. F fraisil cinders]

frazzle /'fræz(ə)l/ n. & v. colloq. ● n. a worn or exhausted state (burnt to a frazzle). ● v.tr. (esp. as **frazzled** adj.) wear out; exhaust. [19th c.: orig. uncert.]

freak /friːk/ n. & v. ● n. **1** (also **freak of nature**) a monstrosity; an abnormally developed individual or thing. **2** (often attrib.) an abnormal, irregular, or bizarre occurrence (a freak storm). **3** colloq. **a** an unconventional person. **b** a person with a specified enthusiasm or interest (health freak). **c** a person who undergoes hallucinations; a drug addict (see sense 2 of v.). **4 a** a caprice or vagary. **b** capriciousness. ● v. (often foll. by out) colloq. **1** intr. & tr. become or make very angry. **2** intr. & tr. undergo or cause to undergo hallucinations or a strong emotional experience, esp. from use of narcotics. **3** intr. adopt a wildly unconventional lifestyle. □ **freak-out** colloq. an act of freaking out; a hallucinatory or strong emotional experience. [16th c.: prob. f. dial.]

freakish /'friːkɪʃ/ adj. **1** of or like a freak. **2** bizarre, unconventional. □ **freakishly** adv. **freakishness** n.

freaky /'friːkɪ/ adj. (**freakier, freakiest**) = FREAKISH. □ **freakily** adv. **freakiness** n.

freckle /'frek(ə)l/ n. & v. ● n. (often in pl.) a light brown spot on the skin, usu. caused by exposure to the sun. ● v. **1** tr. (usu. as **freckled** adj.) spot with freckles. **2** intr. be spotted with freckles. □ **freckly** adj. [ME fracel etc. f. dial. freken f. ON freknur (pl.)]

Frederick I /'fredrɪk/ (known as Frederick Barbarossa = 'Redbeard') (c.1123–90), king of Germany and Holy Roman emperor 1152–90. He made a sustained attempt to subdue Italy and the papacy, but was eventually defeated at the battle of Legnano in 1176. He was drowned in Asia Minor while on his way to the Third Crusade.

Frederick II /'fredrɪk/ (known as Frederick the Great) (1712–86), king of Prussia 1740–86. On his succession, Frederick promptly claimed Silesia, launching Europe into the War of the Austrian Succession (1740–8). During the Seven Years War (1756–63), he joined with Britain and Hanover against a coalition of France, Russia, Austria, Spain, Sweden, and Saxony, and succeeded in considerably strengthening Prussia's position. By the end of his reign, he had doubled the area of his country. He was a distinguished patron of the arts.

Frederick William (known as 'the Great Elector') (1620–88), Elector of Brandenburg 1640–88. His programme of reconstruction and reorganization following the Thirty Years War, including the strengthening of the army and the development of the civil service, brought stability to his country and laid the basis for the expansion of Prussian power in the 18th century. In his foreign policy he sought to create a balance of power by the formation of shifting strategic alliances.

Fredericton /'fredrɪktən/ the capital of New Brunswick, Canada; pop. (1991) 45,360. The city was founded in 1785 by the United Empire Loyalists, colonists who left America after the War of American Independence out of loyalty to the British Crown. They named the city after Frederick Augustus, second son of George III.

free /friː/ adj., adv., & v. ● adj. (**freer** /'friːə(r)/; **freest** /'friːɪst/) **1** not in bondage to or under the control of another; having personal rights and social and political liberty. **2** (of a state, or its citizens or institutions) subject neither to foreign domination nor to despotic government; having national and civil liberty (a free press; a free society). **3 a** unrestricted, unimpeded; not restrained or fixed. **b** at liberty; not confined or imprisoned. **c** released from ties or duties; unimpeded. **d** unrestrained as to action; independent (set free). **4** (foll. by of, from) **a** not subject to; exempt from (free of tax). **b** not containing or subject to a specified (usu. undesirable) thing (free of preservatives; free from disease). **5** (foll. by to + infin.) able or permitted to take a specified action (you are free to choose). **6** unconstrained (free gestures). **7 a** available without charge; costing nothing. **b** not subject to tax, duty, trade-restraint, or fees. **8 a** clear of engagements or obligations (are you free tomorrow?). **b** not occupied or in use (the bathroom is free now). **c** clear of obstructions. **9** spontaneous, unforced (free compliments). **10** open to all comers. **11** lavish, profuse; using or used without restraint (very free with their money). **12** frank, unreserved. **13** (of a literary style) not observing the strict laws of form. **14** (of a translation) conveying the broad sense; not literal. **15** forward, familiar, impudent. **16** (of talk, stories, etc.) slightly indecent. **17** Physics **a** not modified by an external force. **b** not bound in an atom or molecule. **18** Chem. not combined (free oxygen). **19** (of power or energy) disengaged or available. ● adv. **1** in

a free manner. **2** without cost or payment. **3** Naut. not close-hauled. ● v.tr. **1** make free; set at liberty. **2** (foll. by of, from) relieve from (something undesirable). **3** disengage, disentangle. □ **for free** colloq. free of charge, gratis. **free agent** a person with freedom of action. **free and easy** informal, unceremonious. **free association** Psychol. a method of investigating a person's unconscious by eliciting from him or her spontaneous associations with ideas proposed by the examiner. **free-born** inheriting a citizen's rights and liberty. **free enterprise** a system in which private business operates in competition and largely free of state control. **free fall** movement under the force of gravity only, esp.: **1** the part of a parachute descent before the parachute opens. **2** the movement of a spacecraft in space without thrust from the engines. **free fight** a general fight in which all present join. **free-for-all** a free fight, unrestricted discussion, etc. **free-form** (attrib.) of an irregular shape or structure. **free hand** freedom to act at one's own discretion (see also FREEHAND). **free-handed** generous. **free-handedly** generously. **free-handedness** generosity. **free house** Brit. an inn or public house not controlled by a brewery and therefore not restricted to selling particular brands of beer or liquor. **free kick** Football a set kick allowed to be taken by one side without interference from the other. **free labour** the labour of workers not in a trade union. **free-living 1** indulgence in pleasures, esp. that of eating. **2** Biol. living freely and independently; not attached to a substrate. **free love** sexual relations according to choice and unrestricted by marriage. **free market** a market in which prices are determined by unrestricted competition (usu. hyphenated when attrib.: a free-market economy). **free on board** (or **rail**) without charge for delivery to a ship or railway wagon. **free pass** an authorization of free admission, travel, etc. **free port 1** a port area where goods in transit are exempt from customs duty. **2** a port open to all traders. **free radical** Chem. an uncharged atom or group of atoms with one or more unpaired electrons. **free-range** esp. Brit. **1** (of hens etc.) kept in natural conditions with freedom of movement. **2** (of eggs) produced by such birds. **free rein** see give free rein to (see REIN). **free school 1** a school for which no fees are charged. **2** a school run on the basis of freedom from restriction for the pupils. **free speech** the right to express opinions freely. **free-spoken** speaking candidly; not concealing one's opinions. **free-standing** not supported by another structure. **free throw** Basketball an unimpeded throw at the basket from 15 ft (4.6 m) in front of the backboard, awarded to a player who is fouled and scoring one point if successful. **free trade** international trade left to its natural course without restriction on imports or exports. **free verse** = VERS LIBRE. **free vote** a Parliamentary vote not subject to party discipline. **free wheel** the driving wheel of a bicycle, able to revolve with the pedals at rest. **free-wheel** v.intr. **1** ride a bicycle with the pedals at rest, esp. downhill. **2** move or act without constraint or effort. **free will 1** the power of acting without the constraint of necessity or fate. **2** the ability to act at one's own discretion (I did it of my own free will). **free world** esp. US hist. non-Communist countries. □ **freely** adv. **freeness** n. [OE frēo, frēon f. Gmc]

-free /friː/ comb. form free of or from (duty-free; trouble-free).

freebase /'friːbeɪs/ n. & v. sl. ● n. cocaine that has been purified by heating with ether, and is taken by inhaling the fumes or smoking the residue. ● v.tr. purify (cocaine) for smoking or inhaling.

freebie /'friːbɪ/ n. colloq. a thing provided free of charge. [arbitrary formation f. FREE]

freeboard /'friːbɔːd/ n. Naut. the part of a ship's side between the water-line and the deck.

freebooter /'friːˌbuːtə(r)/ n. a pirate or lawless adventurer. □ **freeboot** v.intr. [Du. vrijbuiter (as FREE, BOOTY): cf. FILIBUSTER]

Free Church n. a Church dissenting or seceding from an established Church.

Free Church of Scotland a strict Presbyterian Church organized by dissenting members of the established Church of Scotland in 1843. In 1900 its majority amalgamated with the United Presbyterian Church to form the United Free Church. Its name was retained by the minority group, nicknamed the Wee Free Kirk (see WEE FREE).

freedman /'friːdmən/ n. (pl. **-men**) an emancipated slave.

freedom /'friːdəm/ n. **1** the condition of being free or unrestricted. **2** personal or civic liberty; absence of slave status. **3** the power of self-determination; the quality of not being controlled by fate or necessity. **4** the state of being free to act (often foll. by to + infin.: we have the freedom to leave). **5** frankness, outspokenness; undue familiarity. **6** (foll. by from) the condition of being exempt from or not subject to (a defect, burden, etc.). **7** (foll. by of) **a** full or honorary participation in (membership, privileges, etc.). **b** unrestricted use of (facilities etc.). **8** a privilege possessed by a city or corporation. **9** facility or ease in action. **10** boldness of conception. □ **freedom fighter** a person who takes part in violent resistance to an established political system. [OE *frēodōm* (as FREE, -DOM)]

Freedom Trail a historic route through Boston, Massachusetts, which begins and ends at Faneuil Hall, where Bostonians met to protest against British 'taxation without representation' (see BOSTON TEA PARTY) in the months preceding the War of American Independence.

Freefone /ˈfriːfəʊn/ n. *propr.* (also **Freephone**) a telephone service, operated in the UK by British Telecom, whereby telephone calls can be made free of charge to subscriber organizations.

Free French an organization of French troops and volunteers in exile formed under General de Gaulle in 1940. Based in London, the movement continued the war against the Axis Powers after de Gaulle appealed by radio from London for French resistance to the Franco-German armistice. Apart from organizing forces that participated in military campaigns in French Equatorial Africa, Lebanon, and elsewhere, and cooperating with the French Resistance, its French National Committee (established in 1941) eventually developed into a provisional government for liberated France. The Free French were also involved in the liberation of Paris in 1944.

freehand /ˈfriːhænd/ adj. & adv. ● adj. (of a drawing or plan etc.) done by hand without special instruments or guides. ● adv. in a freehand manner.

freehold /ˈfriːhəʊld/ n. & adj. Law ● n. **1** tenure of land or property in fee simple or fee tail or for life. **2** land or property or an office held by such tenure. ● adj. held by or having the status of freehold. □ **freeholder** n.

freelance /ˈfriːlɑːns/ n., v., & adv. ● n. **1** (also **freelancer** /-ˌlɑːnsə(r)/) a person, usu. self-employed, offering services on a temporary basis, esp. to several businesses etc. for particular assignments (often *attrib.*: a *freelance editor*). **2** (usu. **free lance**) hist. a medieval mercenary. ● v.intr. act as a freelance. ● adv. as a freelance. [19th c.: orig. in sense 2 of n.]

freeloader /ˈfriːˌləʊdə(r)/ n. esp. US sl. a person who eats or drinks at others' expense; a sponger. □ **freeload** v.intr.

freeman /ˈfriːmən/ n. (pl. **-men**) **1** a person who has the freedom of a city, company, etc. **2** a person who is not a slave or serf.

freemartin /ˈfriːˌmɑːtɪn/ n. a hermaphrodite or imperfect female calf of oppositely sexed twins. [17th c.: orig. unkn.]

Freemason /ˈfriːˌmeɪs(ə)n/ n. a member of an international fraternity for mutual help and fellowship, with an elaborate ritual and system of secret signs. The original *free masons* date from the 14th century and were emancipated skilled itinerant stonemasons who found work wherever important buildings were being erected, while the *accepted masons* were honorary members of the fraternity who began to be admitted early in the 17th century. All of them recognized their fellow craftsmen by secret signs. In 1717 four of these societies or 'lodges' in London united to form a 'grand lodge', with a new constitution and ritual and a new objective of mutual help and fellowship. The London grand lodge became the parent of other lodges in Britain and around the world; present-day bodies of Freemasons are largely composed of members of the upper-middle-class professional and business worlds.

Freemasonry /ˈfriːˌmeɪsənrɪ/ n. **1** the system and institutions of the Freemasons. **2** (**freemasonry**) a secret or tacit fellowship; instinctive sympathy.

Freepost /ˈfriːpəʊst/ n. a postal service whereby postage is paid by the addressee.

freer compar. of FREE.

freesia /ˈfriːzjə, ˈfriːʒə/ n. a bulbous plant of the genus *Freesia*, native to Africa, having fragrant coloured flowers. [mod.L f. Friedrich Heinrich Theodor *Freese*, Ger. physician (d.1876)]

freest superl. of FREE.

freestone /ˈfriːstəʊn/ n. **1** any fine-grained stone which can be cut easily, esp. sandstone or limestone. **2** a stone-fruit, esp. a peach, in which the stone is loose when the fruit is ripe (cf. CLINGSTONE).

freestyle /ˈfriːstaɪl/ adj. & n. ● adj. (of a race or contest) in which all styles are allowed, esp.: **1** *Swimming* in which any stroke may be used. **2** *Wrestling* with few restrictions on the holds permitted. ● n. freestyle swimming or wrestling. □ **freestyler** n.

freethinker /friːˈθɪŋkə(r)/ n. a person who rejects dogma or authority, esp. in religious belief. □ **freethinking** n. & adj.

Freetown /ˈfriːtaʊn/ the capital and chief port of Sierra Leone; pop. (est. 1988) 469,800.

freeware /ˈfriːweə(r)/ n. Computing software that is distributed free and without technical support to users. [FREE + SOFTWARE]

freeway /ˈfriːweɪ/ n. esp. US **1** an express highway, esp. with controlled access. **2** a toll-free highway.

freeze /friːz/ v. & n. ● v. (past **froze** /frəʊz/; past part. **frozen** /ˈfrəʊz(ə)n/) **1** tr. & intr. **a** turn or be turned into ice or another solid by cold. **b** (often foll. by *over, up*) make or become rigid or solid as a result of the cold. **2** intr. be or feel very cold. **3** tr. & intr. cover or become covered with ice. **4** intr. (foll. by *to, together*) adhere or be fastened by frost (the curtains froze to the window). **5** tr. preserve (food) by refrigeration below freezing-point. **6** tr. & intr. **a** make or become motionless or powerless through fear, surprise, etc. **b** react or cause to react with sudden aloofness or detachment. **7** tr. stiffen or harden, injure or kill, by chilling (frozen to death). **8** tr. make (credits, assets, etc.) temporarily or permanently unrealizable. **9** tr. fix or stabilize (prices, wages, etc.) at a certain level. **10** tr. arrest (an action) at a certain stage of development. **11** tr. arrest (a movement in a film) by repeating a frame or stopping the film at a frame. ● n. **1** a state of frost; a period or the coming of frost or very cold weather. **2** the fixing or stabilization of prices, wages, etc. **3** = *freeze-frame* n. □ **freeze-dry** (**-dries**, **-dried**) freeze and dry by the sublimation of ice in a high vacuum. **freeze-frame** n. & v. ● n. (also *attrib.*) the facility of stopping a film or videotape in order to view a motionless image. ● v.tr. use freeze-frame on (an image, a recording, etc.). **freeze on to** colloq. take or keep tight hold of. **freeze out** US colloq. exclude from business, society, etc. by competition or boycott etc. **freeze up** obstruct or be obstructed by the formation of ice. **freeze-up** n. a period or conditions of extreme cold. **freezing** adj. (also **freezing cold**) colloq. very cold. **freezing-mixture** salt and snow or some other mixture used to freeze liquids. **freezing-point** the temperature at which a liquid, esp. water, freezes. **freezing works** Austral. & NZ a place where animals are slaughtered and carcasses frozen for export. **frozen mitt** colloq. a cool reception. **frozen shoulder** Med. a shoulder joint which is painfully stiff. □ **freezable** adj. **frozenly** adv. [OE *frēosan* f. Gmc]

freezer /ˈfriːzə(r)/ n. a refrigerated cabinet or room for preserving food at very low temperatures; = DEEP-FREEZE n.

Frege /ˈfreɪgə/, Gottlob (1848–1925), German philosopher and mathematician, founder of modern logic. He developed a logical system for the expression of mathematics which was a great improvement on the syllogistic logic which it replaced, and he also worked on general questions of philosophical logic and semantics. His theory of meaning, based on his use of a distinction between what a linguistic term refers to and what it expresses, is still influential. Frege tried to provide a rigorous foundation for mathematics on the basis of purely logical principles, but abandoned the attempt when Bertrand Russell, on whose work he had a profound influence, pointed out that his system was inconsistent.

Freiburg /ˈfraɪbɜːg/ (in full **Freiburg im Breisgau** /ɪm ˈbraɪsgaʊ/) an industrial city in SW Germany, in Baden-Württemberg, on the edge of the Black Forest; pop. (1991) 193,775.

freight /freɪt/ n. & v. ● n. **1** the transport of goods more slowly and cheaply than by express delivery. **2** goods transported; cargo. **3** a charge for transportation of goods. **4** the hire of a ship or aircraft for transporting goods. **5** a load or burden. ● v.tr. **1** transport (goods) as freight. **2** load with freight. **3** hire or let out (a ship) for the carriage of goods and passengers. □ **freight ton** see TON¹. [MDu., MLG *vrecht* var. of *vracht*: cf. FRAUGHT]

freightage /ˈfreɪtɪdʒ/ n. **1 a** the transportation of freight. **b** the cost of this. **2** freight transported.

freighter /ˈfreɪtə(r)/ n. **1** a ship or aircraft designed to carry freight. **2** US a wagon for freight. **3** a person who loads or charters and loads a ship. **4** a person who consigns goods for carriage inland. **5** a person whose business is to receive and forward freight.

Freightliner /ˈfreɪtˌlaɪnə(r)/ n. propr. a train carrying goods in containers.

Frelimo /freˈliːməʊ/ Frente de Libertação de Moçambique, the nationalist liberation party of Mozambique, founded in 1962. After independence in 1975, Frelimo governed Mozambique as a one-party state until 1990, when a multi-party system was introduced. [Port., acronym]

Fremantle /ˈfriːˌmænt(ə)l/ the principal port of Western Australia; pop. (est. 1987) 24,000. The city is part of the Perth metropolitan area.

Frémont /ˈfriːmɒnt/, John Charles (known as 'the Pathfinder') (1813–

90), American explorer and politician. He was responsible for exploring several viable routes to the Pacific across the Rockies in the 1840s. He made an unsuccessful bid for the presidency in 1856, losing to James Buchanan.

French /frentʃ/ adj. & n. ● adj. **1** of or relating to France or its people or language. **2** having the characteristics attributed to the French people. ● n. **1** the language of France, also used in Belgium, Switzerland, Canada, and elsewhere. (*See note below.*) **2** (prec. by *the*; treated as *pl.*) the people of France. **3** colloq. bad language (*excuse my French*). **4** colloq. dry vermouth (*gin and French*). □ **French bean** Brit. **1** a bean plant, *Phaseolus vulgaris*, having many varieties cultivated for their pods and seeds. **2 a** the pod used as food. **b** the seed used as food (see also HARICOT, *kidney bean*). **French bread** white bread in a long crisp loaf. **French chalk** a kind of steatite used for marking cloth and removing grease and as a dry lubricant. **French cricket** an informal type of cricket without stumps and played with a soft ball. **French cuff** a double cuff formed by turning back a long cuff and fastening it. **French curve** a template used for drawing curved lines. **French door** = *French window*. **French dressing** a salad dressing of vinegar and oil, usu. seasoned. **French fried potatoes** (or **French fries**) potato chips. **French kiss** a kiss with one partner's tongue inserted in the other's mouth. **French knickers** wide-legged knickers. **French leave** absence without permission. **French letter** Brit. colloq. a condom. **French mustard** Brit. a mild mustard mixed with vinegar. **French polish** shellac polish for wood. **French-polish** v.tr. polish with this. **French roof** a mansard. **French seam** a seam with the raw edges enclosed. **French toast 1** Brit. bread buttered on one side and toasted on the other. **2** bread dipped in egg and milk and fried. **French vermouth** dry vermouth. **French window** (usu. in pl.) a glazed door in an outside wall, serving as a window and door. □ **Frenchness** n. [OE *frencisc* f. Gmc]

▪ French is spoken as a first language by some 110,000 million people in France and neighbouring European countries, in parts of Canada, and in a number of African states, having spread as a result of French colonization. It is a Romance language which developed from the Latin spoken in Gaul after its conquest in 58–51 BC. A number of dialects of French arose, but the northern dialects became dominant after Paris became the French capital in the 10th century. From the 11th to the 14th century France was the leading country in Europe and its influence and language spread, so that in most European countries it became customary for the upper classes to learn French, while from the 13th century until well into the 20th century it was the language of international diplomacy. In 1635 the French Academy was founded (see ACADÉMIE FRANÇAISE), determining what should be considered correct French; modern literary French remains much the same as the language of the 17th century. (See also PROVENÇAL, ANGLO-FRENCH.)

French Canadian n. & adj. ● n. **1** a Canadian whose principal language is French. **2** the language spoken by such a person. ● adj. of or relating to French Canadians or their language.

French Community a political union established by France in 1958 and lasting informally until the late 1970s (its formal operations ceased in the 1960s). Comprising metropolitan France, its overseas departments and territories, and several former French colonies, it coordinated matters of foreign and economic policy, currency, and defence.

French Congo the name (until 1910) for FRENCH EQUATORIAL AFRICA.

French Equatorial Africa a former federation of French territories in west central Africa (1910–58). Originally called French Congo, its constituent territories were Chad, Ubanghi Shari (now the Central African Republic), Gabon, and Middle Congo (now Congo).

French Guiana an overseas department of France, in northern South America; pop. (est. 1991) 96,000; capital, Cayenne.

French horn n. a coiled brass wind instrument with a wide bell. It developed from the simple hunting horn (see HORN); a Czech sonata of the 1670s is its first known musical use. The French horn is frequently played with the right hand inserted into the bell as a mute.

Frenchify /ˈfrentʃɪˌfaɪ/ v.tr. (**-ies, -ied**) (usu. as **Frenchified** adj.) make French in form, character, or manners.

Frenchman /ˈfrentʃmən/ n. (pl. **-men**) a man who is French by birth or descent.

French Polynesia an overseas territory of France in the South Pacific; pop. (est. 1991) 200,000; capital, Papeete (on Tahiti). French Polynesia comprises the Society Islands, the Gambier Islands, the Tuamotu Archipelago, the Tubuai Islands, and the Marquesas. It

became an overseas territory of France in 1946, and was granted partial autonomy in 1977.

French Revolution the overthrow of the Bourbon monarchy in France (1789–99), the profound political and social effects of which mark a turning point in French history. It was the first of a series of European political upheavals, in which various groups in French society found common cause in opposing the feudal structure of the state, with its privileged Establishment and discredited monarchy. It began with the meeting of the legislative assembly (the States General) in May 1789, when the French government was already in crisis; the Bastille was stormed in July of the same year. The Revolution became steadily more radical and ruthless with power increasingly in the hands of the Jacobins and Robespierre, who in 1793 gained control of the Committee of Public Safety, a governing body originally dominated by Danton. Louis XVI's execution in Jan. 1793 was followed by Robespierre's Reign of Terror (Sept. 1793–July 1794). The Revolution failed to produce a stable form of republican government and after several different forms of administration the last, the Directory, was overthrown by Napoleon in 1799.

French Somaliland /səˈmɑːlɪˌlænd/ the former name (until 1967) for DJIBOUTI.

French Southern and Antarctic Territories an overseas territory of France, comprising Adélie Land in Antarctica, and the Kerguelen and Crozet archipelagos and the islands of Amsterdam and St Paul in the southern Indian Ocean.

French Sudan the former name for MALI.

French Wars of Religion a series of religious and political conflicts in France (1562–98) involving the Protestant Huguenots on one side and Catholic groups on the other. The wars were complicated by interventions from Spain, Rome, England, the Netherlands, and elsewhere, and were not brought to an end until the defeat of the Holy League and the settlement of the Edict of Nantes (1598).

French West Africa a former federation of French territories in NW Africa (1895–1959). Its constituent territories were Senegal, Mauritania, French Sudan (now Mali), Upper Volta (now Burkina), Niger, French Guinea (now Guinea), the Ivory Coast, and Dahomey (now Benin).

Frenchwoman /ˈfrentʃˌwʊmən/ n. (pl. **-women**) a woman who is French by birth or descent.

frenetic /frəˈnetɪk/ adj. **1** frantic, frenzied. **2** fanatic. □ **frenetically** adv. [ME f. OF *frenetique* f. L *phreneticus* f. Gk *phrenitikos* f. *phrenitis* delirium f. *phrēn phrenos* mind]

frenulum var. of FRAENULUM.

frenum var. of FRAENUM.

frenzy /ˈfrenzɪ/ n. & v. ● n. (pl. **-ies**) **1** mental derangement; wild excitement or agitation. **2** delirious fury. ● v.tr. (**-ies, -ied**) (usu. as **frenzied** adj.) drive to frenzy; infuriate. □ **frenziedly** adv. [ME f. OF *frenesie* f. med.L *phrenesia* f. L *phrenesis* f. Gk *phrēn* mind]

Freon /ˈfriːɒn/ n. propr. any of a group of halogenated hydrocarbons containing fluorine, chlorine, and sometimes bromine, used in aerosols, refrigerants, etc. (See also CFC.)

frequency /ˈfriːkwənsɪ/ n. (pl. **-ies**) **1** commonness of occurrence. **2 a** the state of being frequent; frequent occurrence. **b** the process of being repeated at short intervals. **3** (abbr. **f**) Physics the rate of recurrence of a vibration, oscillation, cycle, etc.; the number of repetitions in a given time, esp. per second. **4** Statistics the ratio of the number of actual to possible occurrences of an event. □ **frequency band** Electronics = BAND¹ n. 3a. **frequency distribution** Statistics a measurement of the frequency of occurrence of the values of a variable. **frequency modulation** (abbr. **FM**) Electronics the modulation of a radio etc. wave by variation of its frequency, esp. as a means of carrying an audio signal. **frequency response** Electronics the dependence on signal-frequency of the output-input ratio of an amplifier etc. [L *frequentia* (as FREQUENT)]

frequent adj. & v. ● adj. /ˈfriːkwənt/ **1** occurring often or in close succession. **2** habitual, constant (*a frequent caller*). **3** found near together; numerous, abundant. **4** (of the pulse) rapid. ● v.tr. /frɪˈkwent/ attend or go to habitually. □ **frequentation** /ˌfriːkwenˈteɪʃ(ə)n/ n. **frequenter** /frɪˈkwentə(r)/ n. **frequently** /ˈfriːkwəntlɪ/ adv. [F *fréquent* or L *frequens -entis* crowded]

frequentative /frɪˈkwentətɪv/ adj. & n. Gram. ● adj. expressing frequent repetition or intensity of action. ● n. a verb or verbal form or conjugation expressing this (e.g. *chatter, twinkle*). [F *fréquentatif -ive* or L *frequentativus* (as FREQUENT)]

fresco /ˈfreskəʊ/ n. (pl. **-os** or **-oes**) **1** a method of painting in which watercolour is applied to wet, freshly laid plaster, so that the colours penetrate and become fixed as it dries. (See note below.) **2** a painting produced by this method. □ **fresco secco** = SECCO. □ **frescoed** adj. [It., = cool, fresh]

▪ The fresco painter has to work rapidly, before the plaster dries, and corrections are almost impossible; but, once completed, the fresco will last as long as the wall itself. Fresco was used in the wall-paintings at Pompeii, and by the great masters of the Italian Renaissance, Giotto, Masaccio, Piero della Francesca, Raphael, and Michelangelo. It works best in dry climates, and has been used chiefly in Italy and rarely in northern Europe, although attempts were made to revive it by the German Nazarenes and English Pre-Raphaelites in the 19th century. Its most notable revival in the 20th century, stimulated by the work of the Mexican painter Diego Rivera, was in Latin America and the US.

fresh /freʃ/ adj., adv., & n. ● adj. **1** newly made or obtained (fresh sandwiches). **2 a** other, different; not previously known or used (start a fresh page; we need fresh ideas). **b** additional (fresh supplies). **3** (foll. by from) lately arrived from (a specified place or situation). **4** not stale or musty or faded (fresh flowers; fresh memories). **5** (of food) not preserved by salting, tinning, freezing, etc. **6** not salty (fresh water). **7 a** pure, untainted, refreshing, invigorating (fresh air). **b** bright and pure in colour (a fresh complexion). **8** (of the wind) brisk; of fair strength. **9** alert, vigorous, fit (never felt fresher). **10** colloq. **a** cheeky, presumptuous. **b** amorously impudent. **11** young and inexperienced. ● adv. newly, recently (esp. in comb.: fresh-baked; fresh-cut). ● n. the fresh part of the day, year, etc. (in the fresh of the morning). □ **freshly** adv. **freshness** n. [ME f. OF freis fresche ult. f. Gmc]

freshen /ˈfreʃ(ə)n/ v. **1** tr. & intr. make or become fresh or fresher. **2** intr. & tr. (foll. by up) **a** change one's clothes, etc. **b** revive, refresh, renew.

fresher /ˈfreʃə(r)/ n. Brit. colloq. = FRESHMAN.

freshet /ˈfreʃɪt/ n. **1** a rush of fresh water flowing into the sea. **2** the flood of a river from heavy rain or melted snow. [prob. f. OF freschete f. frais FRESH]

freshman /ˈfreʃmən/ n. (pl. **-men**) a first-year student at university or N. Amer. at high school.

freshwater /ˈfreʃˌwɔːtə(r)/ adj. **1** of or found in fresh water; not of the sea. **2** US (esp. of a school or college) rustic or provincial. □ **freshwater flea** = DAPHNIA.

Fresnel /ˈfreɪn(ə)l, French frɛnɛl/, Augustin Jean (1788–1827), French physicist and civil engineer. He took up the study of polarized light and postulated that light moves in a wavelike motion, which had already been suggested by, among others, Christiaan Huygens and Thomas Young. They, however, assumed the waves to be longitudinal, while Fresnel was sure that they vibrated transversely to the direction of propagation, and he used this to explain successfully the phenomenon of double refraction. He invented the lens that is named after him.

fresnel /ˈfreɪn(ə)l/ n. (also **Fresnel**) (in full **fresnel lens**) a flat lens made of a number of concentric rings to reduce spherical aberration, used in lighthouses, searchlights, camera viewfinders, etc. [FRESNEL]

Fresno /ˈfreznəʊ/ a city in central California, in the San Joaquin valley; pop. (1990) 354,200.

fret¹ /fret/ v. & n. ● v. (**fretted, fretting**) **1** intr. **a** be greatly and visibly worried or distressed. **b** be irritated or resentful. **2** tr. **a** cause anxiety or distress to. **b** irritate, annoy. **3** tr. wear or consume by gnawing or rubbing. **4** tr. form (a channel or passage) by wearing away. **5** intr. (of running water) flow or rise in little waves. ● n. irritation, vexation, querulousness (esp. in a fret). [OE fretan f. Gmc, rel. to EAT]

fret² /fret/ n. & v. ● n. **1** an ornamental pattern made of continuous combinations of straight lines joined usu. at right angles. **2** Heraldry a device of narrow bands and a diamond interlaced. ● v.tr. (**fretted, fretting**) **1** embellish or decorate with a fret. **2** adorn (esp. a ceiling) with carved or embossed work. [ME f. OF frete trellis-work and freter (v.)]

fret³ /fret/ n. each of a sequence of bars or ridges on the finger-board of some stringed musical instruments (esp. the guitar) fixing the positions of the fingers to produce the desired notes. □ **fretless** adj. [15th c.: orig. unkn.]

fretful /ˈfretfʊl/ adj. visibly anxious, distressed, or irritated. □ **fretfully** adv. **fretfulness** n.

fretsaw /ˈfretsɔː/ n. a saw consisting of a narrow blade stretched on a frame, for cutting thin wood in patterns.

fretwork /ˈfretwɜːk/ n. ornamental work in wood, done with a fretsaw.

Freud¹ /frɔɪd/, Lucian (b.1922), German-born painter, grandson of Sigmund Freud. He came to Britain in 1931, becoming a British citizen in 1939. Since the 1950s he has established a reputation as a powerful figurative painter. His subjects, especially his portraits and nudes, are painted in a meticulously detailed style based on firm draughtsmanship, often using striking angles.

Freud² /frɔɪd/, Sigmund (1856–1939), Austrian neurologist and psychotherapist. He was the first to draw attention to the significance of unconscious processes in normal and neurotic behaviour, and was the founder of psychoanalysis as both a theory of personality and a therapeutic practice (see PSYCHOANALYSIS). He proposed the existence of an unconscious element in the mind which influences consciousness, and of conflicts in it between various sets of forces. Freud also emphasized the importance of a child's semiconsciousness of sex as a factor in mental development, while his theory of the sexual origin of neuroses aroused great controversy. His works include The Interpretation of Dreams (1899), Totem and Taboo (1913), and The Ego and the Id (1923).

Freudian /ˈfrɔɪdɪən/ adj. & n. Psychol. ● adj. of or relating to Sigmund Freud or his methods of psychoanalysis, esp. with reference to the importance of sexuality in human behaviour. ● n. a follower of Freud or his methods. □ **Freudian slip** an unintentional error regarded as revealing subconscious feelings. □ **Freudianism** n.

Frey /freɪ/ (also **Freyr** /ˈfreɪə(r)/) Scand. Mythol. the god of fertility and dispenser of rain and sunshine.

Freya /ˈfreɪə/ Scand. Mythol. the goddess of love and of the night, sister of Frey. She is often identified with Frigga.

Freyberg /ˈfraɪbɜːg/, Bernard Cyril, 1st Baron Freyberg of Wellington and of Munstead (1889–1963), British-born New Zealand general. He served in both world wars, winning the VC in France in 1917. He was Governor-General of New Zealand 1946–1952. Appointed Commander-in-Chief of the New Zealand Expeditionary Forces in 1939, Freyberg led the unsuccessful Commonwealth expedition to Greece and Crete (1941), and later took an active role in the North African and Italian campaigns.

Fri. abbr. Friday.

friable /ˈfraɪəb(ə)l/ adj. easily crumbled. □ **friableness** n. **friability** /ˌfraɪəˈbɪlɪti/ n. [F friable or L friabilis f. friare crumble]

friar /ˈfraɪə(r)/ n. a member of any of certain religious orders of men, esp. the four mendicant orders (Augustinians, Carmelites, Dominicans, and Franciscans). □ **Friar Minor** a Franciscan friar, so called because the Franciscans regarded themselves as of humbler rank than members of other orders. **friar's** (or **friars'**) **balsam** a tincture of benzoin etc. used esp. as an inhalant. □ **friarly** adj. [ME & OF frere f. L frater fratris brother]

friary /ˈfraɪəri/ n. (pl. **-ies**) a convent of friars.

fricandeau /ˈfrɪkənˌdəʊ/ n. & v. ● n. (pl. **fricandeaux** /-ˌdəʊz/) **1** a slice of meat, esp. veal, cut from the leg. **2** a dish made from this, usu. fried or stewed and served with a sauce. ● v.tr. (**fricandeaus, fricandeaued** /-ˌdəʊd/, **fricandeauing** /-ˌdəʊɪŋ/) make into fricandeaux. [F]

fricassee /ˈfrɪkəˌsiː/ n. & v. ● n. a dish of stewed or fried pieces of meat served in a thick white sauce. ● v.tr. (**fricassees, fricasseed**) make a fricassee of. [F fricassée fem. past part. of fricasser (v.)]

fricative /ˈfrɪkətɪv/ adj. & n. Phonet. ● adj. made by the friction of breath in a narrow opening. ● n. a consonant made in this way, e.g. f and th. [mod.L fricativus f. L fricare rub]

friction /ˈfrɪkʃ(ə)n/ n. **1** the action of one object rubbing against another. **2** the resistance an object encounters in moving over another. **3** a clash of wills, temperaments, or opinions; mutual animosity arising from disagreement. **4** (in comb.) of devices that transmit motion by frictional contact (friction-clutch; friction-disc). □ **friction-ball** a ball used in bearings to lessen friction. □ **frictional** adj. **frictionless** adj. [F f. L frictio -onis f. fricare frict- rub]

Friday /ˈfraɪdeɪ, -dɪ/ n. & adv. ● n. the sixth day of the week, following Thursday. ● adv. colloq. **1** on Friday. **2** (**Fridays**) on Fridays; each Friday. □ **girl** (or **man**) **Friday** a helper or follower (after Man Friday in Daniel Defoe's Robinson Crusoe). [OE frīgedæg f. Gmc (named after FRIGGA)]

fridge /frɪdʒ/ n. Brit. colloq. = REFRIGERATOR. □ **fridge-freezer** an upright unit comprising a refrigerator and a freezer, each self-contained. [abbr.]

Friedan /friːˈdæn/, Betty (b.1921), American feminist and writer. After the birth of her three children, she published The Feminine Mystique

(1963), an instant best seller which presented femininity as an artificial construct and traced the ways in which American women are socialized to become mothers and housewives. In 1966 she founded the National Organization for Women, serving as its president until 1970.

Friedman /ˈfriːdmən/, Milton (b.1912), American economist. A principal exponent of monetarism, he was awarded the Nobel Prize for economics in 1976. He acted as a policy adviser to President Reagan from 1981 to 1989.

Friedrich /ˈfriːdrɪx/, Caspar David (1774–1840), German painter. Noted for his romantic landscapes in which he saw a spiritual significance, he caused controversy with his altarpiece *The Cross in the Mountains* (1808), which lacked a specifically religious subject.

friend /frend/ *n. & v.* ● *n.* **1** a person with whom one enjoys mutual affection and regard (usu. exclusive of sexual or family bonds). **2** a sympathizer, helper, or patron (*no friend to virtue; a friend of order*). **3** a person who is not an enemy or who is on the same side (*friend or foe?*). **4 a** a person already mentioned or under discussion (*my friend at the next table then left the room*). **b** a person known by sight. **c** used as a polite or ironic form of address. **5** a regular contributor of money or other assistance to an institution. **6** (**Friend**) a member of the Society of Friends, a Quaker. **7** (in *pl.*) one's near relatives, those responsible for one. **8** a helpful thing or quality. ● *v.tr. archaic* or *poet.* befriend, help. □ **be** (or **keep**) **friends with** be friendly with. **friend at court** a friend whose influence may be made use of. **my honourable friend** *Brit.* used in the House of Commons to refer to another member of one's own party. **my learned friend** used by a lawyer in court to refer to another lawyer. **my noble friend** *Brit.* used in the House of Lords to refer to another member of one's own party. □ **friended** *adj.* **friendless** *adj.* [OE *frēond* f. Gmc]

friendly /ˈfrendlɪ/ *adj., n., & adv.* ● *adj.* (**friendlier, friendliest**) **1** acting as or like a friend, well-disposed, kindly. **2 a** (often foll. by *with*) on amicable terms. **b** not hostile. **3** characteristic of friends, showing or prompted by kindness. **4** favourably disposed, ready to approve or help. **5 a** (of a thing) serviceable, convenient, opportune. **b** = *user-friendly*. **6** (esp. in *comb.*) not harming; helping (*ozone-friendly; user-friendly*). ● *n.* (*pl.* **-ies**) = *friendly match*. ● *adv.* in a friendly manner. □ **friendly action** *Law* an action brought merely to get a point decided. **friendly fire** *Mil.* fire coming from one's own side in a conflict, esp. as the cause of accidental injury or damage to one's forces. **friendly match** a match played for enjoyment and not in competition for a cup etc. □ **friendlily** *adv.* **friendliness** *n.*

Friendly Islands an alternative name for TONGA.

Friendly Society any of a number of mutual-aid associations in Britain whose members pay regular contributions and in return receive financial aid in sickness or old age and, on death, provision for their families. Such societies arose in the 17th–18th centuries and became numerous in the 19th and early 20th centuries; many closed after 1946, when the welfare state was established, but some developed into life insurance companies. The Friendly Society was originally the name of a particular fire insurance company operating *c.*1700.

friendship /ˈfrendʃɪp/ *n.* **1** being friends, the relationship between friends. **2** a friendly disposition felt or shown. [OE *frēondscipe* (as FRIEND, -SHIP)]

Friends of the Earth an international pressure group established in 1971 to campaign for a better awareness of and response to environmental problems.

frier var. of FRYER.

Friesian /ˈfriːʒ(ə)n, ˈfriːzɪən/ *adj.* of, relating to, or denoting a breed of large, usu. black and white dairy cattle orig. from Friesland. [var. of FRISIAN]

Friesland /ˈfriːzlənd/ **1** the western part of the ancient region of Frisia. It became a province of the Netherlands in 1597. **2** a northern province of the Netherlands, bounded to the west and north by the IJsselmeer and the North Sea; capital, Leeuwarden.

frieze[1] /friːz/ *n.* **1** *Archit.* the part of an entablature between the architrave and the cornice. **2** *Archit.* a horizontal band of sculpture filling this. **3** a band of decoration elsewhere, esp. along a wall near the ceiling. [F *frise* f. med.L *frisium, frigium* f. L *Phrygium* (*opus*) (work) of Phrygia]

frieze[2] /friːz/ *n.* coarse woollen cloth with a nap, usu. on one side only. [ME f. F *frise*, prob. rel. to FRISIAN]

frig[1] /frɪg/ *v. & n. coarse sl.* ● *v.tr. & intr.* (**frigged, frigging**) **1** = FUCK *v.*

2 masturbate. ● *n.* = FUCK *n.* 1a. [perh. imit.: orig. senses 'move about, rub']

frig[2] /frɪdʒ/ *n. Brit. colloq.* = REFRIGERATOR. [abbr.]

frigate /ˈfrɪgɪt/ *n.* **1 a** *Brit.* a naval escort-vessel between a corvette and a destroyer in size. **b** *US* a similar ship between a destroyer and a cruiser in size. **2** *hist.* a warship next in size to ships of the line. □ **frigate bird** a large predatory tropical seabird of the family Fregatidae, with long wings, a deeply forked tail, and a long hooked bill (also called *man-of-war bird*). [F *frégate* f. It. *fregata*, of unkn. orig.]

Frigga /ˈfrɪgə/ *Scand. Mythol.* the wife of Odin and goddess of married love and of the hearth, often identified with Freya. Friday is named after her.

fright /fraɪt/ *n. & v.* ● *n.* **1 a** sudden or extreme fear. **b** an instance of this (*gave me a fright*). **2** a person or thing looking grotesque or ridiculous. ● *v.tr. poet.* frighten. □ **take fright** become frightened. [OE *fryhto*, metathetic form of *fyrhto*, f. Gmc]

frighten /ˈfraɪt(ə)n/ *v.* **1** *tr.* fill with fright; terrify (*was frightened at the bang; is frightened of dogs*). **2** *tr.* (foll. by *away, off, out of, into*) drive or force by fright (*frightened it out of the room; frightened them into submission; frightened me into agreeing*). **3** *intr.* become frightened (*I frighten easily*). □ **frightening** *adj.* **frighteningly** *adv.*

frightener /ˈfraɪt(ə)nə(r)/ *n.* a person or thing that frightens. □ **put the frighteners on** *sl.* intimidate.

frightful /ˈfraɪtfʊl/ *adj.* **1 a** dreadful, shocking, revolting. **b** ugly, hideous. **2** *colloq.* extremely bad (*a frightful idea*). **3** *colloq.* very great, extreme. □ **frightfully** *adv.*

frightfulness /ˈfraɪtfʊlnɪs/ *n.* **1** being frightful. **2** (transl. G *Schrecklichkeit*) the terrorizing of a civilian population as a military resource.

frigid /ˈfrɪdʒɪd/ *adj.* **1 a** lacking friendliness or enthusiasm; apathetic, formal, forced. **b** dull, flat, insipid. **c** chilling, depressing. **2** (of a woman) sexually unresponsive. **3** (esp. of climate or air) cold. □ **frigid zones** the parts of the earth north of the Arctic Circle and south of the Antarctic Circle. □ **frigidly** *adv.* **frigidness** *n.* **frigidity** /frɪˈdʒɪdɪtɪ/ *n.* [L *frigidus* f. *frigere* be cold f. *frigus* (n.) cold]

frigidarium /ˌfrɪdʒɪˈdeərɪəm/ *n.* (*pl.* **frigidariums** or **frigidaria** /-rɪə/) *Rom. Antiq.* in a Roman bath, the room containing the final, cold, bath. (See also ROMAN BATHS.) [L (as FRIGID)]

frijoles /friːˈhəʊles/ *n.pl.* beans. [Sp., pl. of *frijol* bean ult. f. L *phaseolus*]

frill /frɪl/ *n. & v.* ● *n.* **1 a** a strip of material with one side gathered or pleated and the other left loose with a fluted appearance, used as an ornamental edging. **b** a similar paper ornament on a ham-knuckle, chop, etc. **c** a natural fringe of feathers, hair, etc., on an animal (esp. a bird) or a plant. **2** (in *pl.*) **a** unnecessary embellishments or accomplishments. **b** airs, affectation (*put on frills*). ● *v.tr.* **1** decorate with a frill. **2** form into a frill. □ **frill** (or **frilled**) **lizard** a large Australian lizard, *Chlamydosaurus kingii*, with an erectile membrane round the neck. □ **frilled** *adj.* **frillery** *n.* [16th c.: orig. unkn.]

frilling /ˈfrɪlɪŋ/ *n.* **1** a set of frills. **2** material for frills.

frilly /ˈfrɪlɪ/ *adj. & n.* ● *adj.* (**frillier, frilliest**) **1** having a frill or frills. **2** resembling a frill. ● *n.* (*pl.* **-ies**) (in *pl.*) *colloq.* women's underwear. □ **frilliness** *n.*

fringe /frɪndʒ/ *n. & v.* ● *n.* **1 a** an ornamental bordering of threads left loose or formed into tassels or twists. **b** such a bordering made separately. **c** any border or edging. **2 a** a portion of the front hair hanging over the forehead. **b** a natural border of hair etc. in an animal or plant. **3 a** an outer edge or margin of something; the outer limit of an area, population, etc. or boundary of convention, acceptability, etc. **b** a thing, part, or area of secondary or minor importance. **c** (usu. **the Fringe**) a secondary festival taking place on the periphery of a mainstream event. **4** (*attrib.*) peripheral; secondary; unconventional. **5 a** a band of contrasting brightness or darkness produced by diffraction or interference of light. **b** a strip of false colour in an optical image. **6** *US* a fringe benefit. ● *v.tr.* **1** adorn or encircle with a fringe. **2** serve as a fringe to. □ **fringe benefit** an employee's benefit supplementing a money wage or salary. **fringe medicine** systems of treatment of disease etc. not regarded as orthodox by the medical profession. **fringe theatre** theatre that operates outside the mainstream because it is experimental or has a different audience from conventional theatre; alternative theatre. **fringing reef** a coral reef that fringes the shore. □ **fringeless** *adj.* **fringy** *adj.* [ME & OF *frenge* ult. f. LL *fimbria* (earlier only in pl.) fibres, fringe]

fringing /ˈfrɪndʒɪŋ/ *n.* material for a fringe or fringes.

Frink /frɪŋk/, Dame Elisabeth (1930–93), English sculptor and graphic artist. She made her name with somewhat angular bronzes, often of birds. During the 1960s her figures — typically male nudes, horses, and riders — became smoother, although she retained a feeling for the bizarre.

frippery /ˈfrɪpərɪ/ n. & adj. ● n. (pl. **-ies**) **1** showy, tawdry, or unnecessary finery or ornament, esp. in dress. **2** empty display in speech, literary style, etc. **3 a** knick-knacks, trifles. **b** a knick-knack or trifle. ● adj. **1** frivolous. **2** contemptible. [F friperie f. OF freperie f. frepe rag]

frippet /ˈfrɪpɪt/ n. sl. a frivolous or showy young woman. [20th c.: orig. unkn.]

Frisbee /ˈfrɪzbɪ/ n. propr. a concave plastic disc for skimming through the air as an outdoor game. [perh. f. Frisbie bakery (Bridgeport, Conn.), whose pie-tins could be used similarly]

Frisch¹ /frɪʃ/, Karl von (1886–1982), Austrian zoologist. He is noted for his work on the behaviour of the honey bee. He studied the vision of bees, showing that they can use polarized light for navigation and see ultraviolet. He also investigated communication between bees, and concluded that they perform an elaborate dance in the hive to show other bees the direction and distance of a source of food. Von Frisch shared a Nobel Prize in 1973 with Konrad Lorenz and Nikolaas Tinbergen.

Frisch² /frɪʃ/, Otto Robert (1904–79), Austrian-born British physicist. With his aunt, Lise Meitner, he recognized that Otto Hahn's experiments with uranium had produced a new type of nuclear reaction. Frisch named it nuclear fission, and indicated the explosive potential of its chain reaction. During the Second World War he continued his research in England, and worked on nuclear weapons in the US at Los Alamos.

Frisch³ /frɪʃ/, Ragnar (Anton Kittil) (1895–1973), Norwegian economist. A pioneer of econometrics, he shared the first Nobel Prize for economics with Jan Tinbergen (1969).

Frisia /ˈfriːʒə, ˈfrɪzɪə/ an ancient region of NW Europe. It consisted of the Frisian Islands and parts of the mainland corresponding to the modern provinces of Friesland and Groningen in the Netherlands and the regions of Ostfriesland and Nordfriesland in NW Germany.

Frisian /ˈfrɪʒ(ə)n, ˈfrɪzɪən/ adj. & n. ● adj. of Friesland or (formerly) Frisia or its people or language. ● n. **1** a native or inhabitant of Friesland or (formerly) Frisia. **2** the Germanic language of Frisia (now largely restricted in use to a small area of the NW Netherlands), most closely related to English and Dutch, with about 700,000 speakers. [L Frisii pl. f. OFris. Frīsa, Frēsa FRISIA]

Frisian Islands a chain of islands lying off the coast of NW Europe, extending from the IJsselmeer in the Netherlands to Jutland. The islands consist of three groups; the West Frisian Islands, which form part of the Netherlands, the East Frisian Islands, which form part of Germany, and the North Frisian Islands, which are divided between Germany and Denmark.

frisk /frɪsk/ v. & n. ● v. **1** intr. leap or skip playfully. **2** tr. feel over or search (a person) for a weapon etc. (usu. rapidly). ● n. **1** a playful leap or skip. **2** the frisking of a person. □ **frisker** n. [obs. frisk (adj.) f. OF frisque lively, of unkn. orig.]

frisket /ˈfrɪskɪt/ n. Printing a thin iron frame keeping the sheet in position during printing on a hand-press. [F frisquette f. Prov. frisqueto f. Sp. frasqueta]

frisky /ˈfrɪskɪ/ adj. (**friskier, friskiest**) lively, playful. □ **friskily** adv. **friskiness** n.

frisson /ˈfriːsɒn/ n. an emotional thrill. [F, = shiver]

frit /frɪt/ n. & v. ● n. **1** a calcined mixture of sand and fluxes as material for glass-making. **2** a vitreous composition from which soft porcelain, enamel, etc., are made. ● v.tr. (**fritted, fritting**) make into frit, partially fuse, calcine. [It. fritta fem. past part. of friggere FRY¹]

frit-fly /ˈfrɪtflaɪ/ n. (pl. **-flies**) a small black fly, Oscinella frit, the larvae of which are destructive to cereals. [19th c.: orig. unkn.]

Frith /frɪθ/, William Powell (1819–1909), English painter. He is remembered for his panoramic paintings of Victorian life, including Derby Day (1858) and The Railway Station (1862).

frith var. of FIRTH.

fritillary /frɪˈtɪlərɪ/ n. (pl. **-ies**) **1** a liliaceous plant of the genus Fritillaria, esp. the snake's head. **2** a butterfly, esp. of the genus Argynnis or a related genus, having red-brown wings chequered with black. [mod.L fritillaria f. L fritillus dice-box]

fritter¹ /ˈfrɪtə(r)/ v.tr. **1** (usu. foll. by away) waste (money, time, energy, etc.) triflingly, indiscriminately, or on divided aims. **2** archaic subdivide.

[obs. n. fritter(s) fragments = obs. fitters (n.pl.), perh. rel. to MHG vetze rag]

fritter² /ˈfrɪtə(r)/ n. a piece of fruit, meat, etc., coated in batter and deep-fried (apple fritter). [ME f. OF friture ult. f. L frigere frict- FRY¹]

fritto misto /ˌfriːtəʊ ˈmiːstəʊ/ n. a mixed grill. [It., = mixed fry]

fritz /frɪts/ n. □ **on the fritz** N. Amer. sl. out of order, unsatisfactory. [20th c.: orig. unkn.]

Friuli-Venezia Giulia /frɪˌuːliːveˌnetsiə ˈdʒuːliə/ a region in NE Italy, on the border with Slovenia and Austria; capital, Trieste.

frivol /ˈfrɪv(ə)l/ v. (**frivolled, frivolling**; US **frivoled, frivoling**) **1** intr. be a trifler; trifle. **2** tr. (foll. by away) spend (money or time) foolishly. [back-form. f. FRIVOLOUS]

frivolous /ˈfrɪvələs/ adj. **1** paltry, trifling, trumpery. **2** lacking seriousness; given to trifling; silly. □ **frivolously** adv. **frivolousness** n. **frivolity** /frɪˈvɒlɪtɪ/ n. (pl. **-ies**). [L frivolus silly, trifling]

frizz /frɪz/ v. & n. ● v. **1 a** tr. form (hair) into a mass of small curls. **b** intr. (of hair) form itself into small curls. **2** tr. dress (wash-leather etc.) with pumice or a scraping-knife. ● n. **1 a** frizzed hair. **b** a row of curls. **2** a frizzed state. [F friser, perh. f. the stem of frire FRY¹]

frizzle¹ /ˈfrɪz(ə)l/ v.intr. & tr. **1** fry, toast, or grill, with a sputtering noise. **2** (often foll. by up) burn or shrivel. [frizz (in the same sense) f. FRY¹, with imit. ending + -LE⁴]

frizzle² /ˈfrɪz(ə)l/ v. & n. ● v. **1** tr. form (hair) into tight curls. **2** intr. (often foll. by up) (of hair etc.) curl tightly. ● n. frizzled hair. [16th c.: orig. unkn. (earlier than FRIZZ)]

frizzly /ˈfrɪzlɪ/ adj. in tight curls.

frizzy /ˈfrɪzɪ/ adj. (**frizzier, frizziest**) in a mass of small curls. □ **frizziness** n.

Frl. abbr. Fräulein.

fro /frəʊ/ adv. back (now only in to and fro: see TO). [ME f. ON frá FROM]

Frobisher /ˈfrəʊbɪʃə(r)/, Sir Martin (c.1535–94), English explorer. In 1576 he led an unsuccessful expedition in search of the North-west Passage, discovering Frobisher Bay (in Baffin Island) and landing in Labrador. He returned to Canada in each of the following two years in a fruitless search for gold, before serving in Sir Francis Drake's West Indies expedition of 1585–6 and playing a prominent part in the defeat of the Spanish Armada (for which he was knighted). He died from wounds received in an attack on a Spanish fort in Brittany.

frock /frɒk/ n. & v. ● n. **1** a woman's or girl's dress. **2 a** a monk's or priest's long gown with loose sleeves. **b** priestly office. **3** a smock. **4 a** a frock-coat. **b** a military coat of similar shape. **5** a sailor's woollen jersey. ● v.tr. invest with priestly office (cf. DEFROCK). □ **frock-coat** a man's long-skirted coat not cut away in front. [ME f. OF froc f. Frank.]

froe /frəʊ/ n. (also **frow**) US a cleaving tool with a handle at right angles to the blade. [abbr. of frower f. FROWARD 'turned away']

Froebel /ˈfrəʊb(ə)l, ˈfrɜːb-/, Friedrich (Wilhelm August) (1782–1852), German educationist and founder of the kindergarten system. Believing that play materials, practical occupations, and songs are needed to develop a child's real nature, he opened a school for young children in 1837, later naming it the Kindergarten (= children's garden). He also established a teacher training school. □ **Froebelism** n. **Froebelian** /frəʊˈbiːlɪən, frɜːˈbiː-/ adj.

frog¹ /frɒg/ n. **1** a tailless amphibian of the order Anura, having moist smooth skin and legs developed for jumping. **2** (Frog) Brit. sl. offens. a Frenchman or Frenchwoman. **3** a hollow in the top face of a brick for holding the mortar. **4** the nut of a violin bow etc. □ **frog in the** (or **one's**) **throat** colloq. an irritation or apparent impediment in the throat; hoarseness. **frog-spawn** the spawn of a frog. [OE frogga f. Gmc]

frog² /frɒg/ n. an elastic horny substance in the sole of a horse's foot. [17th c.: orig. uncert. (perh. a use of FROG¹)]

frog³ /frɒg/ n. **1** an ornamental coat-fastening of a spindle-shaped button and loop. **2** an attachment to a waist-belt to support a sword, bayonet, etc. □ **frogged** adj. **frogging** n. [18th c.: orig. unkn.]

frog⁴ /frɒg/ n. a grooved piece of iron at a place in a railway where tracks cross. [19th c.: orig. unkn.]

frogbit /ˈfrɒgbɪt/ n. a floating plant of the family Hydrocharitaceae, found in stagnant water; esp. Hydrocharis morsus-ranae. [FROG¹ + BIT², f. mod.L]

frogfish /ˈfrɒgfɪʃ/ n. an angler fish, esp. of the family Antennariidae.

froggy /ˈfrɒgɪ/ adj. & n. ● adj. **1** of or like a frog or frogs. **2** Brit. sl. offens. French. ● n. (**Froggy**) (pl. **-ies**) sl. offens. a Frenchman or Frenchwoman.

froghopper /ˈfrɒɡˌhɒpə(r)/ n. a jumping plant-sucking bug of the family Cercopidae, having larvae that produce a protective mass of cuckoo-spit.

frogman /ˈfrɒɡmən/ n. (pl. **-men**) a person equipped with a rubber suit, flippers, and an oxygen supply for underwater swimming.

frogmarch /ˈfrɒɡmɑːtʃ/ v. & n. esp. Brit. ● v.tr. **1** hustle (a person) forward holding and pinning the arms from behind. **2** carry (a person) in a frogmarch. ● n. the carrying of a person face downwards by four others each holding a limb.

frogmouth /ˈfrɒɡmaʊθ/ n. a nightjar-like bird of the family Podargidae, native to Australia and SE Asia, having a large wide mouth.

frolic /ˈfrɒlɪk/ v., n., & adj. ● v.intr. (**frolicked**, **frolicking**) play about cheerfully, gambol. ● n. **1** cheerful play. **2** a prank. **3** a merry party. **4** an outburst of gaiety. **5** merriment. ● adj. archaic **1** full of pranks, sportive. **2** joyous, mirthful. □ **frolicker** n. [Du. vrolijk (adj.) f. vro glad + -lijk -LY¹]

frolicsome /ˈfrɒlɪksəm/ adj. merry, playful. □ **frolicsomely** adv. **frolicsomeness** n.

from /frəm, frɒm/ prep. expressing separation or origin, followed by: **1** a person, place, time, etc., that is the starting-point of motion or action, or of extent in space or time (rain comes from the clouds; repeated from mouth to mouth; dinner is served from 8; from start to finish). **2** a place, object, etc. whose distance or remoteness is reckoned or stated (ten miles from Rome; I am far from admitting it; absent from home; apart from its moral aspect). **3 a** a source (dig gravel from a pit; a man from Italy; draw a conclusion from information; quotations from Shaw). **b** a giver or sender (presents from Father Christmas; have not heard from her). **4 a** a thing or person avoided, escaped, lost, etc. (released him from prison; cannot refrain from laughing; dissuaded from folly). **b** a person or thing deprived (took his gun from him). **5** a reason, cause, or motive (died from fatigue; suffering from mumps; did it from jealousy; from his looks you might not believe it). **6** a thing distinguished or unlike (know black from white). **7** a lower limit (saw from 10 to 20 boats; tickets from £5). **8** a state changed for another (from being the victim he became the attacker; raised the penalty from a fine to imprisonment). **9** an adverb or preposition of time or place (from long ago; from abroad; from under the bed). **10** the position of a person who observes or considers (saw it from the roof; from his point of view). **11** a model (painted it from nature). □ **from a child** since childhood. **from day to day** (or **hour to hour** etc.) daily (or hourly etc.); as the days (or hours etc.) pass. **from home** out, away. **from now on** henceforward. **from time to time** occasionally. **from year to year** each year; as the years pass. [OE fram, from f. Gmc]

fromage blanc /ˌfrɒmɑːʒ ˈblɒŋk/ n. a type of soft French cheese made from cow's milk and having a creamy sour taste. [F, lit. 'white cheese']

fromage frais /ˌfrɒmɑːʒ ˈfreɪ/ n. a kind of mild low-fat soft cheese. [F, lit. 'fresh cheese']

frond /frɒnd/ n. **1** Bot. **a** a large usu. divided foliage leaf in various flowerless plants, esp. ferns and palms. **b** the leaflike thallus of some algae. **2** Zool. a leaflike expansion. □ **frondage** n. **frondose** /-dəʊz, -dəʊs/ adj. [L frons frondis leaf]

Fronde /frɒnd/ a series of civil wars in France 1648–53, in which French nobles, whose power had been weakened by the policies of Cardinal Richelieu, rose in rebellion against Mazarin and the court during the minority of Louis XIV. Although some concessions were obtained after the rebellion of the Parliament in 1648, the nobles were not successful in curbing the power of the monarchy, the Fronde marks the last serious challenge to the monarchy until the Revolution of 1789. The name derives from the French word for a type of sling used in a children's game played in the streets of Paris at the time.

frondeur /frɒnˈdɜː(r)/ n. a political rebel. [F, = slinger; see FRONDE]

front /frʌnt/ n., adj., & v. ● n. **1** the side or part normally nearer or towards the spectator or the direction of motion (the front of the car; the front of the chair; the front of the mouth). **2** any face of a building, esp. that of the main entrance. **3** Mil. **a** the foremost line or part of an army etc. **b** line of battle. **c** the part of the ground towards a real or imaginary enemy. **d** a scene of actual fighting (go to the front). **e** the direction in which a formed line faces (change front). **4 a** a sector of activity regarded as resembling a military front. **b** an organized political group. **5** a demeanour, bearing (show a bold front). **b** outward appearance. **6 a** a forward or conspicuous position (come to the front). **7 a** a bluff. **b** a pretext. **8** a person etc. serving to cover subversive or illegal activities. **9** (prec. by the) the promenade of a seaside resort. **10** Meteorol. the forward edge of an advancing mass of cold or warm air. **11** (prec. by the) the auditorium of a theatre. **12 a** a face. **b** poet. or rhet. a forehead.

13 a the breast of a person's clothes. **b** a false shirt-front. **14** impudence. ● attrib.adj. **1** of the front. **2** situated in front. **3** Phonet. (of a speech sound) formed at the front of the mouth. ● v. **1** intr. (foll. by on, to, towards, upon) have the front facing or directed. **2** intr. (foll. by for) sl. act as a front or cover for. **3** tr. furnish with a front (fronted with stone). **4** tr. lead (a band, organization, etc.). **5** tr. **a** stand opposite to, front towards. **b** have its front on the side of (a street etc.). **6** tr. archaic confront, meet, oppose. **7** tr. act as presenter or host of (a programme). □ **front bench** Brit. the foremost seats in the House of Commons, occupied by leading members of the government and opposition. **front-bencher** Brit. such a member. **front door 1** the chief entrance of a building. **2** a chief means of approach or access to a place, situation, etc. **front line** Mil. = sense 3 of n. **front-line States** hist. countries in southern Africa bordering on and opposed to South Africa's policy of apartheid. **front man 1** a person acting as a front or cover. **2** Broadcasting a programme's presenter or host. **3** the leader of a group of musicians, an organization, etc. **front matter** Printing the title-page, preface, etc. preceding the text proper. **front office** a main office, esp. police headquarters. **front page** the first page of a newspaper, esp. as containing important or remarkable news. **front passage** colloq. the vagina. **front runner 1** the contestant most likely to succeed. **2** an athlete or horse running best when in the lead. **in front 1** in an advanced position. **2** facing the spectator. **in front of 1** ahead of, in advance of. **2** in the presence of, confronting. **on the front burner** see BURNER. □ **frontless** adj. **frontward** adj. & adv. **frontwards** adv. [ME f. OF front (n.), fronter (v.) f. L frons frontis]

frontage /ˈfrʌntɪdʒ/ n. **1** the front of a building. **2 a** land abutting on a street or on water. **b** the land between the front of a building and the road. **3** extent of front (a shop with little frontage). **4 a** the way a thing faces. **b** outlook. □ **frontage road** N. Amer. a service road. □ **frontager** n.

frontal¹ /ˈfrʌnt(ə)l/ adj. **1 a** of, at, or on the front (a frontal attack). **b** of the front as seen by an onlooker (a frontal view). **2** Anat. of the forehead or front part of the skull (frontal bone). □ **frontal lobe** Anat. each of the paired lobes of the brain lying immediately behind the forehead, including areas concerned with behaviour, learning, and voluntary movement. □ **frontally** adv. [mod.L frontalis (as FRONT)]

frontal² /ˈfrʌnt(ə)l/ n. **1** a covering for the front of an altar. **2** the façade of a building. [ME f. OF frontel f. L frontale (as FRONT)]

Front de Libération Nationale /ˌfrɒn də ˌliːbəˌræsjɒn ˌnæsjɒˈnɑːl/ (abbr. **FLN**) a revolutionary political party in Algeria that supported the Algerian war of independence against France 1954–62.

frontier /ˈfrʌntɪə(r), frʌnˈtɪə(r)/ n. **1 a** the border between two countries. **b** the district on each side of this. **2** the limits of attainment or knowledge in a subject. **3** (in the US) the borders between settled and unsettled country. □ **frontierless** adj. [ME f. AF frounter, OF frontiere ult. f. L frons frontis FRONT]

frontiersman /ˈfrʌntɪəzmən, frʌnˈtɪəz-/ n. (pl. **-men**; fem. **frontierswoman**, pl. **-women**) a person living in the region of a frontier, esp. between settled and unsettled country.

frontispiece /ˈfrʌntɪsˌpiːs/ n. **1** an illustration facing the title-page of a book or of one of its divisions. **2** Archit. **a** the principal face of a building. **b** a decorated entrance. **c** a pediment over a door etc. [F frontispice or LL frontispicium façade f. L frons frontis FRONT + -spicium f. specere look: assim. to PIECE]

frontlet /ˈfrʌntlɪt/ n. **1** a piece of cloth hanging over the upper part of an altar frontal. **2** a band worn on the forehead. **3** a phylactery. **4** an animal's forehead. [OF frontelet (as FRONTAL²)]

fronton /ˈfrʌntən/ n. Archit. a pediment. [F f. It. frontone f. fronte forehead]

frore /frɔː(r)/ adj. poet. frozen, frosty. [archaic past part. of FREEZE]

Frost /frɒst/, Robert (Lee) (1874–1963), American poet. Much of his poetry reflects his affinity with New England, including the verse collections North of Boston (1914) and New Hampshire (1923). Noted for his ironic tone, conversational manner, and simple language, he won the Pulitzer Prize on three occasions (1924; 1931; 1937).

frost /frɒst/ n. & v. ● n. **1 a** (also **white frost**) a white frozen dew coating esp. the ground at night (windows covered with frost). **b** a consistent temperature below freezing-point causing frost to form. **2** a chilling dispiriting atmosphere. **3** sl. a failure. ● v. **1** intr. (usu. foll. by over, up) become covered with frost. **2** tr. **a** cover with or as if with frost, powder, etc. **b** injure (a plant etc.) with frost. **3** tr. give a roughened or finely granulated surface to (glass, metal) (frosted glass). **4** tr. US cover or decorate (a cake etc.) with icing. □ **black frost** a frost without white dew. **degrees of frost** Brit. degrees below freezing-point (ten degrees of frost tonight). **frost-work** tracery made by frost on glass etc. □ **frostless** adj. [OE f. Gmc]

frostbite /ˈfrɒstbaɪt/ n. injury to body tissues, esp. the nose, fingers, or toes, due to freezing and often resulting in gangrene.

frosting /ˈfrɒstɪŋ/ n. **1** US icing. **2** a rough surface on glass etc.

frosty /ˈfrɒstɪ/ adj. (**frostier, frostiest**) **1** cold with frost. **2** covered with or as with hoar-frost. **3** unfriendly in manner, lacking in warmth of feeling. □ **frostily** adv. **frostiness** n.

froth /frɒθ/ n. & v. ● n. **1 a** a collection of small bubbles in liquid, caused by shaking, fermenting, etc.; foam. **b** impure matter on liquid, scum. **2 a** idle talk or ideas. **b** anything unsubstantial or of little worth. ● v. **1** intr. emit or gather froth (frothing at the mouth). **2** tr. cause (beer etc.) to foam. □ **froth-blower** Brit. joc. a beer-drinker (formerly esp. as a designation of a member of a charitable organization). □ **frothily** adv. **frothiness** n. **frothy** adj. (**frothier, frothiest**). [ME f. ON frotha, frauth f. Gmc]

frottage /frɒˈtɑːʒ/ n. **1** Psychol. the practice of touching or rubbing against the clothed body of another person as a means of obtaining sexual gratification. **2** Art the technique or process of taking a rubbing from an uneven surface to form the basis of a work of art. [F, = rubbing f. frotter rub f. OF froter]

frou-frou /ˈfruːfruː/ n. a rustling, esp. of a dress. [F, imit.]

frow[1] /fraʊ/ n. **1** a Dutchwoman. **2** a housewife. [ME f. Du. vrouw woman]

frow[2] var. of FROE.

froward /ˈfrəʊəd/ adj. archaic perverse; difficult to deal with. □ **frowardly** adv. **frowardness** n. [ME f. FRO + -WARD]

frown /fraʊn/ v. & n. ● v. **1** intr. wrinkle one's brows, esp. in displeasure or deep thought. **2** intr. (foll. by at, on, upon) express disapproval. **3** intr. (of a thing) present a gloomy aspect. **4** tr. compel with a frown (frowned them into silence). **5** tr. express (defiance etc.) with a frown. ● n. **1** an action of frowning; a vertically furrowed or wrinkled state of the brow. **2** a look expressing severity, disapproval, or deep thought. □ **frowner** n. **frowningly** adv. [ME f. OF frongnier, froignier f. froigne surly look f. Celt.]

frowst /fraʊst/ n. & v. Brit. colloq. ● n. fusty warmth in a room. ● v.intr. stay in or enjoy frowst. □ **frowster** n. [back-form. f. FROWSTY]

frowsty /ˈfraʊstɪ/ adj. Brit. (**frowstier, frowstiest**) fusty, stuffy. □ **frowstiness** n. [var. of FROWZY]

frowzy /ˈfraʊzɪ/ adj. (also **frowsy**) (**-ier, -iest**) **1** fusty, musty, ill-smelling, close. **2** slatternly, unkempt, dingy. □ **frowziness** n. [17th c.: orig. unkn.: cf. earlier frowy]

froze past of FREEZE.

frozen past part. of FREEZE.

FRS abbr. (in the UK) Fellow of the Royal Society.

FRSE abbr. Fellow of the Royal Society of Edinburgh.

fructiferous /frʌkˈtɪfərəs/ adj. bearing fruit. [L fructifer f. fructus FRUIT]

fructification /ˌfrʌktɪfɪˈkeɪʃ(ə)n/ n. Bot. **1** the process of fructifying. **2** any spore-bearing structure esp. in ferns, fungi, and mosses. [LL fructificatio (as FRUCTIFY)]

fructify /ˈfrʌktɪˌfaɪ/ v. (**-ies, -ied**) **1** intr. bear fruit. **2** tr. make fruitful; impregnate. [ME f. OF fructifier f. L fructificare f. fructus FRUIT]

fructose /ˈfrʌktəʊz, -təʊs, ˈfrʊk-/ n. Chem. a simple sugar found in honey and fruits. Also called laevulose, fruit sugar. [L fructus FRUIT + -OSE[2]]

fructuous /ˈfrʌktjʊəs/ adj. full of or producing fruit. [ME f. OF fructuous or L fructuosus (as FRUIT)]

frugal /ˈfruːg(ə)l/ adj. **1** (often foll. by of) sparing or economical, esp. as regards food. **2** sparingly used or supplied, meagre, costing little. □ **frugally** adv. **frugalness** n. **frugality** /fruːˈgælɪtɪ/ n. [L frugalis f. frugi economical]

frugivorous /fruːˈdʒɪvərəs/ adj. feeding on fruit. [L frux frugis fruit + -VOROUS]

fruit /fruːt/ n. & v. ● n. **1 a** the usu. sweet and fleshy edible product of a plant or tree, containing seed. **b** (in sing.) these in quantity (eats fruit). **2** the seed of a plant or tree with its covering, e.g. an acorn, pea pod, cherry, etc. **3** (usu. in pl.) vegetables, grains, etc. used for food (fruits of the earth). **4** (usu. in pl.) the result of action etc., esp. as financial reward (fruits of his labours). **5** sl. esp. US a male homosexual. **6** Bibl. an offspring (the fruit of the womb; the fruit of his loins). ● v.intr. & tr. bear or cause to bear fruit. □ **fruit bar** a piece of dried and pressed fruit. **fruit-bat** a large tropical bat that feeds on fruit (see also BAT[2], flying fox). **fruit-** (or

fruiting-) **body** (pl. **-ies**) the spore-bearing part of a fungus. **fruit cake 1** a cake containing dried fruit. **2** sl. an eccentric or mad person. **fruit cocktail** a finely chopped usu. tinned fruit salad. **fruit fly** (pl. **flies**) a fly, esp. of the genus Drosophila, having larvae that feed on fruit. **fruit machine** Brit. a coin-operated gaming machine giving random combinations of symbols often representing fruit. **fruit salad 1** various fruits cut up and served in syrup, juice, etc. **2** sl. a display of medals etc. **fruit sugar** fructose. **fruit-tree** a tree grown for its fruit. **fruit-wood** the wood of a fruit-tree, esp. when used in furniture. □ **fruitage** n. **fruited** adj. (also in comb.). [ME f. OF f. L fructus fruit, enjoyment f. frui enjoy]

fruitarian /fruːˈteərɪən/ n. a person who eats only fruit. [FRUIT, after vegetarian]

fruiter /ˈfruːtə(r)/ n. **1** a tree producing fruit, esp. with reference to its quality (a poor fruiter). **2** Brit. a fruit grower. **3** a ship carrying fruit. [ME f. OF fruitier (as FRUIT, -ER[3]): later f. FRUIT + -ER[1]]

fruiterer /ˈfruːtərə(r)/ n. esp. Brit. a dealer in fruit.

fruitful /ˈfruːtfʊl/ adj. **1** producing much fruit; fertile; causing fertility. **2** producing good results; beneficial, remunerative. **3** producing offspring, esp. prolifically. □ **fruitfully** adv. **fruitfulness** n.

fruition /fruːˈɪʃ(ə)n/ n. **1 a** the bearing of fruit. **b** the production of results. **2** the realization of aims or hopes. **3** enjoyment. [ME f. OF f. LL fruitio -onis f. frui enjoy, erron. assoc. with FRUIT]

fruitless /ˈfruːtlɪs/ adj. **1** not bearing fruit. **2** useless, unsuccessful, unprofitable. □ **fruitlessly** adv. **fruitlessness** n.

fruitlet /ˈfruːtlɪt/ n. = DRUPEL.

fruity /ˈfruːtɪ/ adj. (**fruitier, fruitiest**) **1 a** of fruit. **b** tasting or smelling like fruit, esp. (of wine) tasting of the grape. **2** (of a voice etc.) of full rich quality. **3** colloq. full of rough humour or (usu. scandalous) interest; suggestive. □ **fruitily** adv. **fruitiness** n.

frumenty /ˈfruːməntɪ/ n. (also **furmety** /ˈfɜːmɪtɪ/) hulled wheat boiled in milk and seasoned with cinnamon, sugar, etc. [ME f. OF frumentee f. frument f. L frumentum corn]

frump /frʌmp/ n. a dowdy unattractive old-fashioned woman. □ **frumpish** adj. **frumpishly** adv. [16th c.: perh. f. dial. frumple (v.) wrinkle f. MDu. verrompelen (as FOR-, RUMPLE)]

frumpy /ˈfrʌmpɪ/ adj. (**frumpier, frumpiest**) dowdy, unattractive, and old-fashioned. □ **frumpily** adv. **frumpiness** n.

Frunze /ˈfruːnzɪ/ a former name (1926–91) for BISHKEK.

frustrate v. & adj. ● v.tr. /frʌˈstreɪt, ˈfrʌstreɪt/ **1** make (efforts) ineffective. **2** prevent (a person) from achieving a purpose. **3** (as **frustrated** adj.) **a** discontented because unable to achieve one's desire. **b** sexually unfulfilled. **4** disappoint (a hope). ● adj. /ˈfrʌstreɪt/ archaic frustrated. □ **frustratedly** adv. **frustrater** n. **frustrating** adj. **frustratingly** adv. **frustration** /frʌˈstreɪʃ(ə)n/ n. [ME f. L frustrari frustrat- f. frustra in vain]

frustule /ˈfrʌstjuːl/ n. Bot. the siliceous cell wall of a diatom. [F f. L frustulum (as FRUSTUM)]

frustum /ˈfrʌstəm/ n. (pl. **frusta** /-tə/ or **frustums**) Geom. **1** the remainder of a cone or pyramid whose upper part has been cut off by a plane parallel to its base. **2** the part of a cone or pyramid intercepted between two planes. [L, = piece cut off]

frutescent /fruːˈtes(ə)nt/ adj. Bot. of the nature of a shrub. [irreg. f. L frutex bush]

frutex /ˈfruːteks/ n. (pl. **frutices** /-tɪˌsiːz/) Bot. a woody-stemmed plant smaller than a tree; a shrub. [L frutex fruticis]

fruticose /ˈfruːtɪˌkəʊz, -ˌkəʊs/ adj. Bot. resembling a shrub. [L fruticosus (as FRUTEX)]

Fry[1] /fraɪ/, Christopher (Harris) (b.1907), English dramatist. He is chiefly remembered for his comic verse dramas, especially The Lady's not for Burning (1948) and Venus Observed (1950). He also wrote several screenplays and translated other dramatists, notably Jean Anouilh.

Fry[2] /fraɪ/, Elizabeth (1780–1845), English Quaker prison reformer. In the forefront of the early 19th-century campaign for penal reform, she concerned herself particularly with conditions in Newgate and other prisons, the plight of convicts transported to Australia, and the vagrant population in London and the south-east.

fry[1] /fraɪ/ v. & n. ● v. (**fries, fried**) **1** tr. & intr. cook or be cooked in hot fat. **2** tr. & intr. sl. electrocute or be electrocuted. **3** tr. (as **fried** adj.) sl. drunk. **4** intr. colloq. be very hot. ● n. (pl. **fries**) **1** various internal parts of animals usu. eaten fried (lamb's fry). **2** a dish of fried food, esp. meat. **3** US a social gathering to eat fried food. □ **frying-** (US **fry-**) **pan** a shallow pan used in frying. **fry up** heat or reheat (food) in a frying-pan. **fry-up** n. Brit. colloq. a dish of miscellaneous fried food. **out of the frying-pan into the fire** from a bad situation to a worse one. [ME f. OF frire f. L frigere]

fry[2] /fraɪ/ n.pl. **1** young or newly hatched fishes. **2** the young of other creatures produced in large numbers, e.g. bees or frogs. □ **small fry** people of little importance; children. [ME f. ON frjó]

Frye /fraɪ/, (Herman) Northrop (1912–91), Canadian literary critic. His first major work was Fearful Symmetry (1947), a study of William Blake and of the role of myth and symbol in various literary genres. He subsequently analysed the structure and mythology of the Bible in The Great Code: The Bible and Literature (1982).

fryer /ˈfraɪə(r)/ n. (also **frier**) **1** a person who fries. **2** a vessel for frying esp. fish. **3** US a chicken suitable for frying.

FSA abbr. Fellow of the Society of Antiquaries.

FSH abbr. follicle-stimulating hormone.

FT abbr. Financial Times.

Ft. abbr. Fort.

ft abbr. foot, feet.

FTC abbr. US Federal Trade Commission.

FT Index see FINANCIAL TIMES INDEX.

FTP abbr. Computing file-transfer protocol.

FT–SE (also **FT-SE 100**) see FINANCIAL TIMES INDEX.

Fuad /ˈfuːæd/ the name of two kings of Egypt:

Fuad I (1868–1936), reigned 1922–36. Formerly sultan of Egypt (1917–1922), he became Egypt's first king after independence.

Fuad II (b.1952), grandson of Fuad I, reigned 1952–3. Named king as an infant on the forced abdication of his father, Farouk, he was deposed when Egypt became a republic.

fubsy /ˈfʌbzɪ/ adj. (**fubsier, fubsiest**) Brit. fat or squat. [obs. fubs small fat person + -Y[1]]

Fuchs[1] /fʊks/, (Emil) Klaus (Julius) (1911–88), German-born British physicist. Fuchs was a Communist who came to England to escape Nazi persecution. During the 1940s he passed to the USSR secret information acquired while working in the US, where he was involved in the development of the atom bomb, and in Britain, where he held a senior post in the Atomic Energy Research Establishment at Harwell. He was imprisoned from 1950 to 1959, and on his release he returned to East Germany.

Fuchs[2] /fʊks/, Sir Vivian (Ernest) (b.1908), English geologist and explorer. He led the Commonwealth Trans-Antarctic Expedition (1955–8). His party met Sir Edmund Hillary's New Zealand contingent, approaching from the opposite direction, at the South Pole, and went on to complete the first overland crossing of the continent.

fuchsia /ˈfjuːʃə/ n. an ornamental shrub of the genus Fuchsia, with drooping red, purple, or white flowers. [mod.L f. Leonhard Fuchs, Ger. botanist (1501–66)]

fuchsine /ˈfuːksiːn, -sɪn/ n. a deep red aniline dye used in the pharmaceutical and textile-processing industries, rosaniline. [FUCHSIA (from its resemblance to the colour of the flower)]

fuck /fʌk/ v., int., & n. coarse sl. ● v. **1** tr. & intr. have sexual intercourse (with). **2** intr. (foll. by about, around) mess about; fool around. **3** tr. (usu. as an exclamation) curse, confound (fuck the thing!). **4** intr. (as **fucking** adj., adv.) used as an intensive to express annoyance etc. ● int. expressing anger or annoyance. ● n. **1 a** an act of sexual intercourse. **b** a partner in sexual intercourse. **2** the slightest amount (don't give a fuck). □ **fuck all** nothing. **fuck off** go away. **fuck up 1** make a mess of. **2** disturb emotionally. **3** make a blunder. **fuck-up** n. a mess or muddle. □ **fucker** n. (often as a term of abuse). [16th c.: orig. unkn.]

fucus /ˈfjuːkəs/ n. (pl. **fuci** /ˈfjuːsaɪ/) a brown seaweed of the genus Fucus, with flat leathery fronds. □ **fucoid** adj. [L, = rock-lichen, f. Gk phukos, of Semitic orig.]

fuddle /ˈfʌd(ə)l/ v. & n. ● v. **1** tr. confuse or stupefy, esp. with alcoholic liquor. **2** intr. tipple, booze. ● n. **1** confusion. **2** intoxication. **3** a spell of drinking (on the fuddle). [16th c.: orig. unkn.]

fuddy-duddy /ˈfʌdɪˌdʌdɪ/ adj. & n. sl. ● adj. old-fashioned or quaintly fussy. ● n. (pl. **-ies**) a fuddy-duddy person. [20th c.: orig. unkn.]

fudge /fʌdʒ/ n., v., & int. ● n. **1** a soft toffee-like sweet made with milk, sugar, butter, etc. **2** nonsense. **3** a piece of dishonesty or faking. **4** a piece of late news inserted in a newspaper page. ● v. **1** tr. put together in a makeshift or dishonest way; fake. **2** tr. deal with incompetently. **3** intr. practise such methods. ● int. expressing disbelief or annoyance. [perh. f. obs. fadge (v.) fit]

fuehrer var. of FÜHRER.

fuel /ˈfjuːəl/ n. & v. ● n. **1** material, esp. coal, wood, oil, etc., burnt or used as a source of heat or power. **2** food as a source of energy. **3** material used as a source of nuclear energy. **4** anything that sustains or inflames emotion or passion. ● v. (**fuelled, fuelling;** US **fueled, fueling**) **1** tr. supply with fuel. **2** tr. sustain or inflame (an argument, feeling, etc.) (drink fuelled his anger). **3** intr. take in or get fuel. □ **fuel cell** a cell producing an electric current direct from a chemical reaction. **fuel element** an element of nuclear fuel etc. for use in a reactor. **fuel-injected** having fuel injection. **fuel injection** the direct introduction of fuel under pressure into the combustion units of an internal-combustion engine. **fuel oil** oil used as fuel in an engine or furnace. **fuel rod** a rod-shaped fuel element, esp. one used in a nuclear reactor. [ME f. AF fuaille, fewaile, OF fouaille, ult. f. L focus hearth]

Fuentes /ˈfwentes/, Carlos (b.1928), Mexican novelist and writer. His first novel, Where the Air is Clear (1958), took Mexico City as its theme and was an immediate success. Other novels include Terra nostra (1975), which explores the Spanish heritage in Mexico, and The Old Gringo (1984).

fug /fʌg/ n. & v. colloq. ● n. stuffiness or fustiness of the air in a room. ● v.intr. (**fugged, fugging**) stay in or enjoy a fug. □ **fuggy** adj. [19th c.: orig. unkn.]

fugacious /fjuːˈɡeɪʃəs/ adj. literary fleeting, evanescent, hard to capture or keep. □ **fugaciously** adv. **fugaciousness** n. **fugacity** /-ˈɡæsɪtɪ/ n. [L fugax fugacis f. fugere flee]

fugal /ˈfjuːɡ(ə)l/ adj. of the nature of a fugue. □ **fugally** adv.

Fugard /ˈfuːɡɑːd/, Athol (b.1932), South African dramatist. His plays, including Blood Knot (1963) and The Road to Mecca (1985), are mostly set in contemporary South Africa and deal with social deprivation and other aspects of life under apartheid.

-fuge /fjuːdʒ/ comb. form forming adjectives and nouns denoting expelling or dispelling (febrifuge; vermifuge). [from or after mod.L -fugus f. L fugare put to flight]

fugitive /ˈfjuːdʒɪtɪv/ adj. & n. ● adj. **1** fleeing; that runs or has run away. **2** transient, fleeting; of short duration. **3** (of literature) of passing interest, ephemeral. **4** flitting, shifting. ● n. **1** (often foll. by from) a person who flees, esp. from justice, an enemy, danger, or a master. **2** an exile or refugee. □ **fugitively** adv. [ME f. OF fugitif -ive f. L fugitivus f. fugere fugit- flee]

fugle /ˈfjuːɡ(ə)l/ v.intr. act as a fugleman. [back-form. f. FUGLEMAN]

fugleman /ˈfjuːɡ(ə)lmən/ n. (pl. **-men**) **1** hist. a soldier placed in front of a regiment etc. while drilling to show the motions and time. **2** a leader, organizer, or spokesman. [G Flügelmann f. Flügel wing + Mann man]

fugue /fjuːɡ/ n. & v. ● n. **1** Mus. a piece of music in which a short melodic theme is introduced by one part or voice and successively taken up by others, which are then interwoven and developed. The form was brought to a high level by J. S. Bach. **2** Psychol. loss of awareness of one's identity, often coupled with flight from one's usual environment. ● v.intr. (**fugues, fugued, fuguing**) Mus. compose or perform a fugue. □ **fuguist** n. [F or It. f. L fuga flight]

fugued /fjuːɡd/ adj. in the form of a fugue.

führer /ˈfjʊərə(r)/ n. (also **fuehrer**) a leader, esp. a tyrannical one. [G, = leader: part of the title (Führer und Reichskanzler) assumed in 1934 by Hitler]

Fujairah /fuːˈdʒaɪərə/ (also **Al Fujayrah**) **1** one of the seven member states of the United Arab Emirates; pop. (1985) 54,400. **2** its capital city.

Fuji, Mount /ˈfuːdʒɪ/ (also **Fujiyama** /ˌfuːdʒɪˈjɑːmə/) a dormant volcano in the Chubu region of Japan. Rising to 3,776 m (12,385 ft), it is Japan's highest mountain, with a symmetrical, conical, snow-capped peak. Its last eruption was in 1707. Mount Fuji, regarded by the Japanese as sacred, has been celebrated in art and literature for centuries.

Fujian /ˌfuːdʒɪˈæn/ (also **Fukien** /fuːˈkjen/) a province of SE China, on the China Sea; capital, Fuzhou.

Fukuoka /ˌfuːkuːˈəʊkə/ an industrial city and port in southern Japan, capital of Kyushu island; pop. (1990) 1,237,100.

-ful /fʊl/ comb. form forming: **1** adjectives from nouns, meaning: **a** full of (beautiful). **b** having the qualities of (masterful). **2** adjectives from adjectives or Latin stems with little change of sense (direful; grateful). **3** adjectives from verbs, meaning 'apt to', 'able to', 'accustomed to' (forgetful; mournful; useful). **4** nouns (pl. **-fuls**) meaning 'the amount needed to fill' (handful; spoonful).

Fulani /fuːˈlɑːnɪ/ n. (pl. **Fulanis**, same) a member of an African people of northern Nigeria and adjacent territories. [Hausa]

Fulbright /ˈfʊlbraɪt/, (James) William (1905–95), American senator. His name designates grants awarded under the Fulbright Act of 1946, which authorized funds from the sale of surplus war materials overseas to be used to finance exchange programmes of students and teachers between the US and other countries. The scheme was later supported by grants from the US government.

fulcrum /ˈfʊlkrəm, ˈfʌl-/ n. (pl. **fulcra** /-rə/ or **fulcrums**) 1 the point against which a lever is placed to get a purchase or on which it turns or is supported. 2 the means by which influence etc. is brought to bear. [L, = post of a couch, f. *fulcire* to prop]

fulfil /fʊlˈfɪl/ v.tr. (US **fulfill**) (**fulfilled**, **fulfilling**) 1 bring to consummation, carry out (a prophecy or promise). 2 a satisfy (a desire or prayer). b (as **fulfilled** adj.) completely happy. 3 a execute, obey (a command or law). b perform, carry out (a task). 4 comply with (conditions). 5 answer (a purpose). 6 bring to an end, finish, complete (a period or piece of work). □ **fulfil oneself** develop one's gifts and character to the full. □ **fulfillable** adj. **fulfiller** n. **fulfilment** n. (US **fulfillment**). [OE *fullfyllan* (as FULL¹, FILL)]

fulgent /ˈfʌldʒənt/ adj. poet. or rhet. shining, brilliant. [ME f. L *fulgere* shine]

fulguration /ˌfʌlɡjʊˈreɪʃ(ə)n/ n. Surgery the destruction of tissue by means of high-voltage electric sparks. [L *fulguratio* sheet lightning f. *fulgur* lightning]

fulgurite /ˈfʌlɡjʊraɪt/ n. Geol. a rocky substance of sand fused or vitrified by lightning. [L *fulgur* lightning]

fuliginous /fjuːˈlɪdʒɪnəs/ adj. sooty, dusky. [LL *fuliginosus* f. *fuligo -ginis* soot]

full¹ /fʊl/ adj., adv., n., & v. ● adj. 1 (often foll. by of) holding all its limits will allow (the bucket is full; full of water). 2 having eaten to one's limits or satisfaction. 3 abundant, copious, satisfying, sufficient (a full programme of events; led a full life; turned it to full account; give full details; the book is very full on this point). 4 (foll. by of) having or holding an abundance of, showing marked signs of (full of vitality; full of interest; full of mistakes). 5 (foll. by of) a engrossed in thinking about (full of himself; full of his work). b unable to refrain from talking about (full of the news). 6 a complete, perfect, reaching the specified or usual or utmost limit (full membership; full daylight; waited a full hour; it was full summer; in full bloom). b Bookbinding used for the entire cover (full leather). 7 a (of tone or colour) deep and clear, mellow. b (of light) intense. c (of motion etc.) vigorous (a full pulse; at full gallop). 8 plump, rounded, protuberant (a full figure). 9 (of clothes) made of much material arranged in folds or gathers. 10 (of the heart etc.) overcharged with emotion. 11 sl. drunk. 12 (foll. by of) archaic having had plenty of (full of years and honours). ● adv. 1 very (you know full well). 2 quite, fully (full six miles; full ripe). 3 exactly (hit him full on the nose). 4 more than sufficiently (full early). ● n. 1 height, acme (season is past the full). 2 the state or time of full moon. 3 the whole (cannot tell you the full of it). ● v.intr. & tr. be or become or make (esp. clothes) full. □ **at full length** 1 lying stretched out. 2 without abridgement. **come full circle** see CIRCLE. **full age** adult status (esp. with ref. to legal rights and duties). **full and by** Naut. close-hauled but with sails filling. **full back** a defensive player, or a position near the goal, in football, hockey, etc. **full-blooded** 1 vigorous, hearty, sensual. 2 not hybrid. **full-bloodedly** forcefully, wholeheartedly. **full-bloodedness** being full-blooded. **full-blown** fully developed, complete, (of flowers) quite open. **full board** provision of accommodation and all meals at a hotel etc. **full-bodied** rich in quality, tone, etc. **full-bottomed** (of a wig) long at the back. **full brother** a brother born of the same parents. **full-cream** of or made from unskimmed milk. **full dress** formal clothes worn on great occasions. **full-dress** attrib.adj. (of a debate etc.) of major importance. **full employment** 1 the condition in which there is no idle capital or labour of any kind that is in demand. 2 the condition in which virtually all who are able and willing to work are employed. **full face** with all the face visible to the spectator. **full-fashioned** = fully-fashioned. **full-fledged** mature. **full forward** Austral. Rules any of three players who constitute the forward line, esp. the centrally positioned one. **full-frontal** 1 (of nudity or a nude figure) with full exposure at the front. 2 unrestrained, explicit, with nothing concealed. **full-grown** having reached maturity. **full hand** Poker a hand with three of a kind and a pair. **full-hearted** full of feeling; confident, zealous. **full-heartedly** in a full-hearted manner. **full-heartedness** fullness of feeling, ardour, zeal. **full house 1** a maximum or large attendance at a theatre, in Parliament, etc. **2** = full hand. **full-length 1** of normal, standard, or maximum length; not

shortened or abbreviated. 2 (of a mirror, portrait, etc.) showing the whole height of the human figure. **full lock** see LOCK¹ n. 3b. **full marks** the maximum award in an examination, in assessment of a person, etc. **full measure** not less than the professed amount. **full moon 1** the moon with its whole disc illuminated. 2 the time when this occurs. **full-mouthed 1** (of cattle or sheep) having a full set of teeth. 2 (of a dog) baying loudly. 3 (of oratory etc.) sonorous, vigorous. **full out 1** Printing flush with the margin. 2 at full power. 3 completely. **full page** an entire page of a newspaper etc. (hyphenated when attrib.: full-page spread) **full pitch** = full toss. **full point** = full stop 1. **full professor** a professor of the highest grade in a university etc. **full-scale** not reduced in size, complete. **full score** Mus. a score giving the parts for all performers on separate staves. **full service** with a church service performed by a choir without solos, or performed with music wherever possible. **full sister** a sister born of the same parents. **full speed** (or **steam**) **ahead!** an order to proceed at maximum speed or to pursue a course of action energetically. **full stop 1** a punctuation mark (.) used at the end of a sentence or an abbreviation. 2 a complete cessation. **full term** the completion of a normal pregnancy. **full tilt** see TILT. **full time 1** the total normal duration of work etc. 2 the end of a football etc. match. **full-time** adj. occupying or using the whole of the available working time. **full-timer** a person who does a full-time job. **full toss** Cricket n. a ball pitched right up to the batsman. ● adv. without the ball's having touched the ground. **full up** colloq. completely full. **in full** without abridgement. 2 to or for the full amount (paid in full). **in full swing** at the height of activity. **in full view** entirely visible. **on a full stomach** see STOMACH. **to the full** to the utmost extent. [OE f. Gmc]

full² /fʊl/ v.tr. cleanse and thicken (cloth). [ME, back-form. f. FULLER¹: cf. OF *fouler* (FOIL¹)]

Fuller¹ /ˈfʊlə(r)/, R(ichard) Buckminster (1895–1983), American designer and architect. An advocate of the use of technology to produce efficiency in many aspects of life, he is best known for his postwar invention of the geodesic dome. These domes enable large spaces to be enclosed with great efficiency — in line with Fuller's ideals of using the world's resources with maximum purpose and least waste.

Fuller² /ˈfʊlə(r)/, Thomas (1608–61), English cleric and historian. He is chiefly remembered for *The History of the Worthies of England* (1662), a description of the counties with short biographies of local personages.

fuller¹ /ˈfʊlə(r)/ n. a person who fulls cloth. [OE *fullere* f. L *fullo*]

fuller² /ˈfʊlə(r)/ n. & v. ● n. 1 a grooved or rounded tool on which iron is shaped. 2 a groove made by this esp. in a horseshoe. ● v.tr. stamp with a fuller. [19th c.: orig. unkn.]

fullerene /ˈfʊləriːn/ n. Chem. a form of carbon having a large spheroidal molecule consisting of a hollow cage of atoms, of which buckminsterfullerene was the first known example. Fullerenes are produced by the action of an arc discharge between carbon electrodes in an inert atmosphere. They are also found in some natural carbon deposits resulting from burning. [BUCKMINSTERFULLERENE]

fuller's earth n. a type of clay used for many centuries in fulling (see FULL²). It possesses the ability to absorb water, grease, oil, and colouring matter, and now has a variety of industrial uses, especially in the purification of oils and as an absorbent and a bleaching agent.

fullness /ˈfʊlnɪs/ n. (also **fulness**) 1 being full. 2 (of sound, colour, etc.) richness, volume, body. 3 all that is contained (in the world etc.). □ **the fullness of the heart** emotion, genuine feelings. **the fullness of time** the appropriate or destined time.

fully /ˈfʊlɪ/ adv. 1 completely, entirely. 2 no less or fewer than (fully 60). □ **fully-fashioned** (of women's clothing) shaped to fit the body. **fully-fledged** mature. [OE *fullice* (as FULL¹, -LY²)]

-fully /fʊlɪ/ comb. form forming adverbs corresp. to adjectives in -ful.

fulmar /ˈfʊlmə(r)/ n. a gull-sized grey and white seabird, *Fulmarus glacialis*, of the petrel family, with a stout body and robust bill. [orig. Hebridean dial.: perh. f. ON *fúll* FOUL (with ref. to its smell) + *már* gull (cf. MEW GULL)]

fulminant /ˈfʌlmɪnənt, ˈfʊl-/ adj. 1 fulminating. 2 Med. (of a disease or symptom) developing suddenly. [F *fulminant* or L *fulminant-* (as FULMINATE)]

fulminate /ˈfʌlmɪneɪt, ˈfʊl-/ v. & n. ● v.intr. 1 (often foll. by against) express censure loudly and forcefully. 2 explode violently; flash like lightning (fulminating mercury). 3 Med. (of a disease or symptom) develop suddenly. ● n. Chem. a salt or ester of fulminic acid; esp. any of a number of unstable explosive salts of metals, used in detonators. □ **fulminatory** adj. **fulmination** /ˌfʌlmɪˈneɪʃ(ə)n, ˌfʊl-/ n. [L *fulminare fulminat-* f. *fulmen -minis* lightning]

fulminic acid /fʌlˈmɪnɪk, fʊl-/ n. Chem. an isomer of cyanic acid that is stable only in solution (chem. formula: HCNO). (See also ISOCYANIC ACID.) [L fulmen: see FULMINATE]

fulness var. of FULLNESS.

fulsome /ˈfʊlsəm/ adj. **1** disgusting by excess of flattery, servility, or expressions of affection; excessive, cloying. **2** disp. copious. ¶ In fulsome praise, fulsome means 'excessive', not 'generous'. □ **fulsomely** adv. **fulsomeness** n. [ME f. FULL¹ + -SOME¹]

Fulton /ˈfʊlt(ə)n/, Robert (1765–1815), American pioneer of the steamship. During the Napoleonic Wars he spent some time in France and proposed both torpedoes and submarines, constructing a steam-propelled 'diving-boat' called Nautilus in 1800 which submerged to a depth of 7.6 m (25 ft). He returned to America in 1806 and built the first successful paddle-steamer, the Clermont. Eighteen other steamships were subsequently built, inaugurating the era of commercial steam navigation.

fulvous /ˈfʌlvəs, ˈfʊl-/ adj. reddish-yellow, tawny. □ **fulvescent** /fʌlˈves(ə)nt, fʊl-/ adj. [L fulvus]

fumarole /ˈfjuːməˌrəʊl/ n. an opening in or near a volcano, through which hot vapours emerge. □ **fumarolic** /ˌfjuːməˈrɒlɪk/ adj. [F fumarolle]

fumble /ˈfʌmb(ə)l/ v. & n. ● v. **1** intr. (often foll. by at, with, for, after) use the hands awkwardly, grope about. **2** tr. a handle or deal with clumsily or nervously. **b** Sport fail to stop (a ball) cleanly. ● n. an act of fumbling. □ **fumbler** n. **fumblingly** adv. [LG fummeln, fommeln, Du. fommelen]

fume /fjuːm/ n. & v. ● n. **1** (usu. in pl.) exuded gas or smoke or vapour, esp. when harmful or unpleasant. **2** a fit of anger (in a fume). ● v. **1** a intr. emit fumes. **b** tr. give off as fumes. **2** intr. (often foll. by at) be affected by (esp. suppressed) anger (was fuming at their inefficiency). **3** tr. a fumigate. **b** subject to fumes esp. those of ammonia (to darken tints in oak, photographic film, etc.). **4** tr. perfume with incense. □ **fume cupboard** (or **chamber** etc.) a ventilated structure in a laboratory, for storing or experimenting with noxious chemicals. □ **fumeless** adj. **fumingly** adv. **fumy** adj. (in sense 1 of n.). [ME f. OF fum f. L fumus smoke & OF fume f. fumer f. L fumare to smoke]

fumigate /ˈfjuːmɪˌgeɪt/ v.tr. **1** disinfect or purify with fumes. **2** apply fumes to. □ **fumigator** n. **fumigant** n. **fumigation** /ˌfjuːmɪˈgeɪʃ(ə)n/ n. [L fumigare fumigat- f. fumus smoke]

fumitory /ˈfjuːmɪtərɪ/ n. a small plant of the genus Fumaria, with tubular flowers; esp. F. officinalis, formerly used against scurvy. [ME f. OF fumeterre f. med.L fumus terrae earth-smoke]

fun /fʌn/ n. & adj. ● n. **1** amusement, esp. lively or playful. **2** a source of this. **3** (in full **fun and games**) exciting or amusing goings-on. ● adj. colloq. amusing, entertaining, enjoyable (a fun thing to do). □ **be great** (or **good**) **fun** be very amusing. **for fun** (or **for the fun of it**) not for a serious purpose. **fun run** colloq. an uncompetitive run, esp. for sponsored runners in support of a charity. **have fun** enjoy oneself. **in fun** as a joke, not seriously. **like fun 1** vigorously, quickly. **2** much. **3** iron. not at all. **make fun of** tease, ridicule. **what fun!** how amusing! [obs. fun (v.) var. of fon befool: cf. FOND]

Funafuti /ˌfuːnəˈfuːtɪ/ the capital of Tuvalu, situated on an island of the same name; pop. (est. 1981) 2,500.

funambulist /fjuːˈnæmbjʊlɪst/ n. a rope-walker. [F funambule or L funambulus f. funis rope + ambulare walk]

Funchal /fʊnˈʃaːl/ the capital and chief port of Madeira, on the south coast of the island; pop. (est. 1987) 44,110.

function /ˈfʌŋkʃ(ə)n/ n. & v. ● n. **1** a an activity proper to a person or institution. **b** a mode of action or activity by which a thing fulfils its purpose. **c** an official or professional duty; an employment, profession, or calling. **2** a a public ceremony or occasion. **b** a social gathering, esp. a large, formal, or important one. **3** Math. a variable quantity regarded in relation to another or others in terms of which it may be expressed or on which its value depends (x is a function of y and z). **4** Computing a part of a program that corresponds to a single value. ● v.intr. fulfil a function, operate; be in working order. □ **function key** Computing a key which is used to generate instructions. □ **functionless** adj. [F fonction f. L functio -onis f. fungi funct- perform]

functional /ˈfʌŋkʃən(ə)l/ adj. **1** of or serving a function. **2** (esp. of buildings) designed or intended to be practical rather than attractive; utilitarian. **3** Physiol. **a** (esp. of disease) of or affecting only the functions of an organ etc., not structural or organic. **b** (of mental disorder) having no discernible organic cause. **c** (of an organ) having a function, not functionless or rudimentary. **4** Math. of a function. □ **functional group** Chem. a group of atoms that determine the reactions of a compound containing the group. □ **functionally** adv. **functionality** /ˌfʌŋkʃəˈnælɪtɪ/ n.

functionalism /ˈfʌŋkʃənəˌlɪz(ə)m/ n. **1** in the arts, the doctrine that the design of an object should be determined solely by its function, rather than by aesthetic considerations, and that anything practically designed will be inherently beautiful. Although similar ideas are found in ancient Greek thought and in 18th-century aesthetic theory, it was not until the 20th century that functionalism was established as a new aesthetic, primarily in architecture, where it is associated particularly with figures such as Le Corbusier (who defined a house as 'a machine for living in'), Frank Lloyd Wright, and Mies van der Rohe. Functionalism had a strong impact on industrial design (e.g. of furniture) of the 1930s and after. **2** in the social sciences, the theory that all aspects of a society serve a function and are necessary for the survival of that society. Functionalism gained ground among 19th-century sociologists such as Émile Durkheim and was also adopted by early 20th-century anthropologists such as Bronisław Malinowski. □ **functionalist** n. & adj.

functionary /ˈfʌŋkʃənərɪ/ n. (pl. **-ies**) a person who has to perform official functions or duties; an official.

fund /fʌnd/ n. & v. ● n. **1** a permanent stock of something ready to be drawn upon (a fund of knowledge; a fund of tenderness). **2** a stock of money, esp. one set apart for a purpose. **3** (in pl.) money resources. **4** (in pl.; prec. by the) Brit. the stock of the national debt (as a mode of investment). ● v.tr. **1** provide with money. **2** convert (a floating debt) into a more or less permanent debt at fixed interest. **3** put into a fund. □ **fund-raiser** a person who seeks financial support for a cause, enterprise, etc. **fund-raising** the seeking of financial support. **in funds** colloq. having money to spend. [L fundus bottom, piece of land]

fundament /ˈfʌndəmənt/ n. joc. the buttocks. [ME f. OF fondement f. L fundamentum (as FOUND²)]

fundamental /ˌfʌndəˈment(ə)l/ adj. & n. ● adj. of, affecting, or serving as a base or foundation, essential, primary, original (a fundamental change; the fundamental rules; the fundamental form). ● n. **1** (usu. in pl.) a fundamental rule, principle, or article. **2** Mus. a fundamental note or tone. □ **fundamental note** Mus. the lowest note of a chord in its original (uninverted) form. **fundamental particle** Physics a subatomic particle. **fundamental tone** Mus. the tone produced by vibration of the whole of a sonorous body (opp. HARMONIC). □ **fundamentally** adv. **fundamentality** /-menˈtælɪtɪ/ n. [ME f. F fondamental or LL fundamentalis (as FUNDAMENT)]

fundamentalism /ˌfʌndəˈmentəˌlɪz(ə)m/ n. **1** strict maintenance of traditional orthodox religious beliefs, esp. in Protestant groups in the US after the First World War. (See note below.) **2** strict maintenance of ancient or fundamental doctrines of any religion, esp. Islam. (See ISLAMIC FUNDAMENTALISM.) □ **fundamentalist** n. & adj.

▪ Christian fundamentalism in America grew out of millenarian thinking of the 19th century, stressing a literal interpretation of the Bible and hence a belief in, for example, the physical second coming of Christ and the virgin birth. It gained momentum in the early 20th century specifically as a reaction against the growth of liberalism, and in the 1950s came to be associated with anti-Communism. In more recent years fundamentalist groups also fostered the emergence of the right-wing Moral Majority, ran a campaign to insist on creationist teaching in American schools, and generally opposed abortion and homosexual rights.

fundholder /ˈfʌndˌhəʊldə(r)/ n. a general practitioner who is provided with and controls his or her own budget. □ **fundholding** n. & adj.

fundus /ˈfʌndəs/ n. (pl. **fundi** /-daɪ/) Anat. the base of a hollow organ; the part furthest from the opening. [L, = bottom]

Fundy, Bay of /ˈfʌndɪ/ an arm of the Atlantic Ocean extending between the Canadian provinces of New Brunswick and Nova Scotia. It is subject to fast-running tides, the highest in the world, which reach 12–15 m (50–80 ft) and are now used to generate electricity.

funeral /ˈfjuːnərəl/ n. & adj. ● n. **1** a the burial or cremation of a dead person with its ceremonies. **b** a burial or cremation procession. **c** US a burial or cremation service. **2** sl. one's (usu. unpleasant) concern (that's your funeral). ● attrib.adj. of or used etc. at a funeral (funeral oration). □ **funeral director** an undertaker. **funeral parlour** (US **home**) an establishment where the dead are prepared for burial or cremation. **funeral pile** (or **pyre**) a pile of wood etc. on which a corpse is burnt. **funeral urn** an urn holding the ashes of a cremated body. [ME f. OF funeraille f. med.L funeralia neut. pl. of LL funeralis f. L funus -eris funeral: (adj.) OF f. L funeralis]

funerary /'fjuːnərəri/ adj. of or used at a funeral or funerals. [LL *funerarius* (as FUNERAL)]

funereal /fjuːˈnɪərɪəl/ adj. **1** of or appropriate to a funeral. **2** gloomy, dismal, dark. □ **funereally** adv. [L *funereus* (as FUNERAL)]

funfair /'fʌnfeə(r)/ n. Brit. a fair, or part of one, consisting of amusements and sideshows.

fungi pl. of FUNGUS.

fungible /'fʌndʒɪb(ə)l/ adj. Law (of goods etc. contracted for, when an individual specimen is not meant) that can serve for, or be replaced by, another answering to the same definition. □ **fungibility** /ˌfʌndʒɪˈbɪlɪtɪ/ n. [med.L *fungibilis* f. *fungi* (vice) serve (in place of)]

fungicide /'fʌndʒɪˌsaɪd/ n. a fungus-destroying substance. □ **fungicidal** /ˌfʌndʒɪˈsaɪd(ə)l/ adj.

fungistatic /ˌfʌndʒɪˈstætɪk/ adj. inhibiting the growth of fungi. □ **fungistatically** adv.

fungoid /'fʌŋɡɔɪd/ adj. & n. ● adj. **1** resembling a fungus in texture or in rapid growth. **2** of a fungus or fungi. ● n. a fungoid plant.

fungous /'fʌŋɡəs/ adj. **1** having the nature of a fungus. **2** springing up like a mushroom; transitory. [ME f. L *fungosus* (as FUNGUS)]

fungus /'fʌŋɡəs/ n. (pl. **fungi** /-ɡaɪ, -dʒaɪ/ or **funguses**) **1** a unicellular, multicellular, or multinucleate spore-producing organism feeding on organic matter, e.g. moulds, yeasts, mushrooms, toadstools, etc. (See note below.) **2** anything similar usu. growing suddenly and rapidly. **3** Med. a spongy morbid growth. **4** sl. a beard. □ **fungal** adj. **fungiform** /'fʌŋɡɪˌfɔːm, ˈfʌndʒɪ-/ adj. **fungivorous** /fʌnˈdʒɪvərəs/ adj. [L, perh. f. Gk *sp(h)oggos* SPONGE]

■ Fungi lack chlorophyll and are therefore incapable of photosynthesis. Many play an ecologically vital role in breaking down dead organic matter, some are an important source of antibiotics or are used in fermentation, and others cause disease. The familiar mushrooms and toadstools are merely the fruiting bodies of organisms that exist mainly as a threadlike mycelium in the soil. Some fungi form associations with other plants, growing with algae to form lichens, or in the roots of higher plants to form mycorrhizas. Although believed to be an ancient group, fungi are scarce as fossils. Their relationship with other plants is obscure, and they are now often classified as a separate kingdom distinct from the green plants.

funicular /fjuːˈnɪkjʊlə(r)/ adj. & n. ● adj. **1** (of a railway, esp. on a mountainside) operating by cable with ascending and descending cars counterbalanced. **2** of a rope or its tension. ● n. a funicular railway. [L *funiculus* f. *funis* rope]

Funk /fʌŋk/, Casimir (1884–1967), Polish-born American biochemist. He showed that a number of diseases, including scurvy, rickets, beriberi, and pellagra, were each caused by the deficiency of a particular dietary component. He coined the term *vitamins* for the chemicals concerned, from his belief (later shown to be inaccurate) that they were all amines. His work formed the basis of modern dietary studies.

funk[1] /fʌŋk/ n. & v. sl. ● n. **1** fear, panic. **2** a coward. ● v. Brit. **1** intr. flinch, shrink, show cowardice. **2** tr. try to evade (an undertaking), shirk. **3** tr. be afraid of. [18th-c. Oxford sl.: perh. f. sl. FUNK[2] = tobacco-smoke]

funk[2] /fʌŋk/ n. sl. **1** funky music, now esp. a style of popular dance music of US black origin, popularized by singers such as James Brown. **2** US a strong smell. [*funk* blow smoke on, perh. f. F dial. *funkier* f. L (as FUMIGATE)]

funkia /'fʌŋkɪə/ n. = HOSTA. [mod.L f. Heinrich Christian *Funck*, Prussian botanist (1771–1839)]

funky[1] /'fʌŋkɪ/ adj. (**funkier**, **funkiest**) sl. **1** (esp. of jazz or popular music) earthy, bluesy, with a heavy rhythmical beat. **2** fashionable. **3** US having a strong smell. □ **funkily** adv. **funkiness** n.

funky[2] /'fʌŋkɪ/ adj. (**funkier**, **funkiest**) sl. **1** terrified. **2** cowardly.

funnel /'fʌn(ə)l/ n. & v. ● n. **1** a narrow tube or pipe widening at the top, for pouring liquid, powder, etc., into a small opening. **2** a metal chimney on a steam engine or ship. **3** something resembling a funnel in shape or use. ● v.tr. & intr. (**funnelled**, **funnelling**; US **funneled**, **funneling**) guide or move through or as through a funnel. □ **funnel-like** adj. [ME f. Prov. *fonilh* f. LL *fundibulum* f. L *infundibulum* f. *infundere* (as IN-[2], *fundere* pour)]

funniosity /ˌfʌnɪˈɒsɪtɪ/ n. (pl. **-ies**) joc. **1** comicality. **2** a comical thing. [FUNNY + -OSITY]

funny /'fʌnɪ/ adj. & n. ● adj. (**funnier**, **funniest**) **1** amusing, comical. **2** strange, perplexing, hard to account for. **3** colloq. slightly unwell, eccentric, etc. ● n. (pl. **-ies**) (usu. in pl.) colloq. **1** a comic strip in a newspaper. **2** a joke. □ **funny-bone** the part of the elbow over which the ulnar nerve passes. **funny business 1** sl. misbehaviour or deception. **2** comic behaviour, comedy. **funny-face** joc. colloq. an affectionate form of address. **funny farm** sl. a mental hospital. **funny-ha-ha** colloq. = sense 1 of adj. **funny man** a clown or comedian, esp. a professional. **funny money** colloq. inflated or counterfeit currency. **funny paper** a newspaper etc. containing humorous matter. **funny-peculiar** colloq. = senses 2, 3 of adj. □ **funnily** adv. **funniness** n. [FUN + -Y[1]]

fur /fɜː(r)/ n. & v. ● n. **1 a** the short fine soft hair of certain animals, distinguished from the longer hair. **b** the skin of such an animal with the fur on it; a pelt. **2 a** the coat of certain animals as material for making, trimming, or lining clothes. **b** a trimming or lining made of the dressed coat of such animals, or of material imitating this. **c** a garment made of or trimmed or lined with fur. **3** (collect.) furred animals. **4 a** a coating formed on the tongue in sickness. **b** Brit. a coating formed on the inside surface of a pipe, kettle, etc., by hard water. **c** a crust adhering to a surface, e.g. a deposit from wine. **5** Heraldry a representation of tufts on a plain ground. ● v. (**furred**, **furring**) **1** tr. (esp. as **furred** adj.) **a** line or trim (a garment) with fur. **b** provide (an animal) with fur. **c** clothe (a person) with fur. **d** coat (a tongue, the inside of a kettle) with fur. **2** intr. (often foll. by up) (of a kettle etc.) become coated with fur. **3** tr. level (floor-timbers) by inserting strips of wood. □ **fur and feather** game animals and birds. **fur seal** an eared seal of the genus *Arctocephalus* or *Callorhinus*, with thick fur used commercially as sealskin. **the fur will fly** colloq. there will be a quarrel or disturbance. □ **furless** adj. [ME (earlier as v.) f. OF *forrer* f. *forre, fuerre* sheath f. Gmc]

fur. abbr. furlong(s).

furbelow /'fɜːbɪˌləʊ/ n. & v. ● n. **1** a gathered strip or pleated border of a skirt or petticoat. **2** (in pl.) derog. showy ornaments. ● v.tr. adorn with a furbelow or furbelows. [18th-c. var. of *falbala* flounce, trimming]

furbish /'fɜːbɪʃ/ v.tr. (often foll. by up) **1** remove rust from, polish, burnish. **2** give a new look to, renovate, revive (something antiquated). □ **furbisher** n. [ME f. OF *forbir* f. Gmc]

furcate /'fɜːkeɪt/ adj. & v. ● adj. (also /'fɜːkət/) forked, branched. ● v.intr. form a fork, divide. □ **furcation** /fɜːˈkeɪʃ(ə)n/ n. [LL *furca* fork: (adj.) f. LL *furcatus*]

furcula /'fɜːkjʊlə/ n. (pl. **furculae** /-liː/) Zool. & Anat. a forked organ or structure, e.g. the wishbone of a bird. [L, dimin. of *furca* fork]

furfuraceous /ˌfɜːfəˈreɪʃəs/ adj. **1** Med. (of skin) resembling bran or dandruff; scaly. **2** Bot. covered with branlike scales. [*furfur* scurf f. L *furfur* bran]

Furies, the /'fjʊəriːz/ Gk Mythol. the spirits of punishment, often represented as three goddesses (Alecto, Megaera, and Tisiphone) with hair composed of snakes, who executed the curses pronounced upon criminals, tortured the guilty with stings of conscience, and inflicted famines and pestilences. They were identified at an early date with the Eumenides.

furious /'fjʊəriəs/ adj. **1** extremely angry. **2** full of fury. **3** raging, violent, intense. □ **fast and furious** adv. **1** rapidly. **2** eagerly, uproariously. ● adj. (of mirth etc.) eager, uproarious. □ **furiously** adv. **furiousness** n. [ME f. OF *furieus* f. L *furiosus* (as FURY)]

furl /fɜːl/ v. **1** tr. roll up and secure (a sail, umbrella, flag, etc.). **2** intr. become furled. **3 a** close (a fan). **b** fold up (wings). **c** draw away (a curtain). **d** relinquish (hopes). □ **furlable** adj. [F *ferler* f. OF *fer(m)* FIRM[1] + *lier* bind f. L *ligare*]

furlong /'fɜːlɒŋ/ n. one-eighth of a mile, 220 yards (201.17 metres). The term originally meant the length of a furrow in a common field, regarded as a square of 10 acres. As early as the 9th century it also came to be regarded as the equivalent of the Roman *stadium*, which was one-eighth of a Roman mile, and hence furlong has become the name for the eighth part of an English mile, even though this does not coincide with the original agricultural measure. [OE *furlang* f. *furh* FURROW + *lang* LONG[1]]

furlough /'fɜːləʊ/ n. & v. ● n. leave of absence, esp. granted to a member of the services or to a missionary. ● v. US **1** tr. grant furlough to. **2** intr. spend furlough. [Du. *verlof* after G *Verlaub* (as FOR-, LEAVE[2])]

furmety var. of FRUMENTY.

furnace /'fɜːnɪs/ n. **1** an enclosed structure for intense heating by fire, esp. of metals or water. **2** a very hot place. [ME f. OF *fornais* f. L *fornax -acis* f. *fornus* oven]

Furneaux Islands /ˈfɜːnəʊ/ a group of islands off the coast of NE Tasmania, in the Bass Strait. The largest island is Flinders Island.

furnish /ˈfɜːnɪʃ/ v.tr. **1** provide (a house, room, etc.) with all necessary contents, esp. movable furniture. **2** (foll. by *with*) cause to have possession or use of. **3** provide, afford, yield. [OF *furnir* ult. f. WG]

furnished /ˈfɜːnɪʃt/ adj. (of a house, flat, etc.) let with furniture.

furnisher /ˈfɜːnɪʃə(r)/ n. **1** a person who sells furniture. **2** a person who furnishes.

furnishings /ˈfɜːnɪʃɪŋz/ n.pl. the furniture and fitments in a house, room, etc.

furniture /ˈfɜːnɪtʃə(r)/ n. **1** the movable equipment of a house, room, etc., e.g. tables, chairs, and beds. **2** Naut. a ship's equipment, esp. tackle etc. **3** accessories, e.g. the handles and lock of a door. **4** Printing pieces of wood or metal placed round or between type to make blank spaces and fasten the matter in the chase. □ **furniture beetle** a small beetle, *Anobium punctatum*, the larva of which is the woodworm. **part of the furniture** colloq. a person or thing taken for granted. [F *fourniture* f. *fournir* (as FURNISH)]

furore /fjʊəˈrɔːrɪ/ n. (US **furor** /ˈfjʊərɔːr(r)/) **1** an uproar; an outbreak of fury. **2** a wave of enthusiastic admiration, a craze. [It. f. L *furor -oris* f. *furere* be mad]

furphy /ˈfɜːfɪ/ n. (pl. **-ies**) Austral. sl. **1** a false report or rumour. **2** an absurd story. [water and sanitary *Furphy carts* of the war of 1914–18, made at a foundry set up by the Furphy family]

furrier /ˈfʌrɪə(r)/ n. a dealer in or dresser of furs. [ME *furrour* f. OF *forreor* f. *forrer* trim with fur, assim. to -IER]

furriery /ˈfʌrɪərɪ/ n. the work of a furrier.

furrow /ˈfʌrəʊ/ n. & v. ● n. **1** a narrow trench made in the ground by a plough. **2** a rut, groove, or deep wrinkle. **3** a ship's track. ● v. **1** tr. plough. **2** tr. **a** make furrows, grooves, etc. in. **b** mark with wrinkles. **3** intr. (esp. of the brow) become furrowed. □ **furrow-slice** the slice of earth turned up by the mould-board of a plough. □ **furrowless** adj. **furrowy** adj. [OE *furh* f. Gmc]

furry /ˈfɜːrɪ/ adj. (**furrier, furriest**) **1** of or like fur. **2** covered with or wearing fur. □ **furriness** n.

further /ˈfɜːðə(r)/ adv., adj., & v. ● adv. (also **farther** /ˈfɑːðə(r)/ esp. with ref. to physical distance) **1** to or at a more advanced point in space or time (*unsafe to proceed further*). **2** at a greater distance (*nothing was further from his thoughts*). **3** to a greater extent, more (*will enquire further*). **4** in addition; furthermore (*I may add further*). ● adj. (also **farther** /ˈfɑːðə(r)/) **1** more distant or advanced (*on the further side*). **2** more, additional, going beyond what exists or has been dealt with (*threats of further punishment*). ● v.tr. promote, favour, help on (a scheme, undertaking, movement, or cause). □ **further education** Brit. education for persons above school age but usu. below degree level. **further to** formal following on from (esp. an earlier letter etc.). **till further notice** (or **orders**) to continue until explicitly changed. □ **furtherer** n. **furthermost** adj. [OE *furthor* (adv.), *furthra* (adj.), *fyrthrian* (v.), formed as FORTH, -ER²]

furtherance /ˈfɜːðərəns/ n. furthering or being furthered; the advancement of a scheme etc.

furthermore /ˌfɜːðəˈmɔː(r)/ adv. in addition, besides (esp. introducing a fresh consideration in an argument).

furthest /ˈfɜːðɪst/ adj. & adv. (also **farthest** /ˈfɑːðɪst/ esp. with ref. to physical distance) ● adj. most distant. ● adv. to or at the greatest distance. □ **at the furthest** (or **at furthest**) at the greatest distance; at the latest; at most. [ME, superl. f. FURTHER]

furtive /ˈfɜːtɪv/ adj. **1** done by stealth, clandestine, meant to escape notice. **2** sly, stealthy. **3** stolen, taken secretly. □ **furtively** adv. **furtiveness** n. [F *furtif -ive* or L *furtivus* f. *furtum* theft]

Furtwängler /ˈfʊətˌvɛŋlə(r)/, Wilhelm (1886–1954), German conductor. He was chief conductor of the Berlin Philharmonic Orchestra from 1922, and worked often at Bayreuth. He is noted particularly for his interpretations of Beethoven and Wagner.

furuncle /ˈfjʊərʌŋk(ə)l/ n. Med. = BOIL². □ **furuncular** /fjʊəˈrʌŋkjʊlə(r)/ adj. **furunculous** adj. [L *furunculus* f. *fur* thief]

furunculosis /fjʊəˌrʌŋkjʊˈləʊsɪs/ n. **1** a diseased condition in which boils appear. **2** a bacterial disease of salmon and trout. [mod.L (as FURUNCLE)]

fury /ˈfjʊərɪ/ n. (pl. **-ies**) **1 a** wild and passionate anger, rage. **b** a fit of rage (*in a blind fury*). **c** impetuosity in battle etc. **2** violence of a storm, disease, etc. **3** an avenging spirit. **4** an angry or malignant woman, a

virago. □ **like fury** colloq. with great force or effect. [ME f. OF *furie* f. L *furia* f. *furere* be mad]

furze /fɜːz/ n. Brit. = GORSE. □ **furzy** /-zɪ/ adj. [OE *fyrs*, of unkn. orig.]

fuscous /ˈfʌskəs/ adj. sombre, dark-coloured. [L *fuscus* dusky]

fuse¹ /fjuːz/ v. & n. ● v. **1** tr. & intr. melt with intense heat; liquefy. **2** tr. & intr. blend or amalgamate into one whole by or as by melting. **3** tr. provide (a circuit, plug, etc.) with a fuse. **4 a** intr. (of an appliance) cease to function when a fuse blows. **b** tr. cause (an appliance) to do this. ● n. a device or component for protecting an electric circuit, containing a strip or wire of easily melted metal and placed in the circuit so as to break it by melting when an excessive current passes through. □ **blow a fuse** lose one's temper. **fuse-box** a box housing the fuses for circuits in a building. [L *fundere fus-* pour, melt]

fuse² /fjuːz/ n. & v. (also **fuze**) ● n. **1** a device for igniting a bomb or explosive charge, consisting of a tube or cord etc. filled or saturated with combustible matter. **2** a component in a shell, mine, etc., designed to detonate an explosive charge on impact, after an interval, or when subjected to a magnetic or vibratory stimulation. ● v.tr. fit a fuse to. □ **fuseless** adj. [It. *fuso* f. L *fusus* spindle]

fusee /fjuːˈziː/ n. (US **fuzee**) **1** a conical pulley or wheel esp. in a watch or clock. **2** a large-headed match for lighting a cigar or pipe in a wind. **3** US a railway signal-flare. [F *fusée* spindle ult. f. L *fusus*]

fuselage /ˈfjuːzəˌlɑːʒ, -lɪdʒ/ n. the body of an aeroplane. [F f. *fuseler* cut into a spindle f. *fuseau* spindle f. OF *fusel* ult. f. L *fusus*]

fusel oil /ˈfjuːz(ə)l/ n. a mixture of several alcohols, chiefly amyl alcohol, produced usu. in small amounts during alcoholic fermentation. [G *Fusel* bad brandy etc.: cf. *fuseln* to bungle]

Fushun /fuːˈʃʊn/ a coal-mining city in NE China, in the province of Liaoning; pop. (1990) 1,330,000.

fusible /ˈfjuːzɪb(ə)l/ adj. that can be easily fused or melted. □ **fusibility** /ˌfjuːzɪˈbɪlɪtɪ/ n.

fusiform /ˈfjuːzɪˌfɔːm/ adj. Bot. & Zool. shaped like a spindle or cigar, tapering at both ends. [L *fusus* spindle + -FORM]

fusil /ˈfjuːzɪl/ n. hist. a light musket. [F ult. f. L *focus* hearth, fire]

fusilier /ˌfjuːzɪˈlɪə(r)/ n. (US **fusileer**) **1** a member of any of several British regiments formerly armed with fusils. **2** hist. a soldier armed with a fusil. [F (as FUSIL)]

fusillade /ˌfjuːzɪˈleɪd/ n. & v. ● n. **1 a** a continuous discharge of firearms. **b** a wholesale execution by this means. **2** a sustained outburst of criticism etc. ● v.tr. **1** assault (a place) by a fusillade. **2** shoot down (persons) with a fusillade. [F f. *fusiller* shoot]

fusilli /fʊˈziːlɪ/ n.pl. pasta pieces in the form of short spirals. [It., lit. 'little spindles', dimin. of *fuso* spindle]

fusion /ˈfjuːʒ(ə)n/ n. **1** the act or an instance of fusing or melting. **2** a fused mass. **3** the blending of different things into one. **4** a coalition. **5** Physics = NUCLEAR FUSION. □ **fusion bomb** a bomb involving nuclear fusion, esp. a hydrogen bomb. □ **fusional** adj. [F *fusion* or L *fusio* (as FUSE¹)]

fuss /fʌs/ n. & v. ● n. **1** excited commotion, bustle, ostentatious or nervous activity. **2 a** excessive concern about a trivial thing. **b** abundance of petty detail. **3** a sustained protest or dispute. **4** a person who fusses. ● v. **1** intr. **a** make a fuss. **b** busy oneself restlessly with trivial things. **c** (often foll. by *about, up and down*) move fussily. **2** tr. agitate, worry. □ **make a fuss** complain vigorously. **make a fuss of** (or **over**) treat (a person or animal) with great or excessive attention. □ **fusser** n. [18th c.: perh. Anglo-Ir.]

fusspot /ˈfʌspɒt/ n. colloq. a person given to fussing.

fussy /ˈfʌsɪ/ adj. (**fussier, fussiest**) **1** inclined to fuss. **2** full of unnecessary detail or decoration. **3** fastidious. □ **fussily** adv. **fussiness** n.

fustanella /ˌfʌstəˈnelə/ n. a man's stiff white kilt worn in Albania and Greece. [It. dimin. of mod.Gk *phoustani* prob. f. It. *fustagno* FUSTIAN]

fustian /ˈfʌstɪən/ n. & adj. ● n. **1** thick twilled cotton cloth with a short nap, usu. dyed in dark colours. **2** turgid speech or writing, bombast. ● adj. **1** made of fustian. **2** bombastic. **3** worthless. [ME f. OF *fustaigne* f. med.L *fustaneus* (adj.) relating to cloth from *Fostat* a suburb of Cairo]

fustic /ˈfʌstɪk/ n. a yellow dye obtained from either of two kinds of wood, esp. old fustic. □ **old fustic 1** a tropical tree, *Chlorophora tinctoria*, native to America. **2** the wood of this tree. **young fustic 1** an ornamental sumac, *Cotinus coggyria* (see *smoke-plant*). **2** the wood of this tree. [F f. Sp. *fustoc* f. Arab. *fustuḳ* f. Gk *pistakē* pistachio]

fusty /ˈfʌstɪ/ adj. (**fustier, fustiest**) **1** stale-smelling, musty, mouldy. **2** stuffy, close. **3** antiquated, old-fashioned. □ **fustily** adv. **fustiness** n.

[ME f. OF *fusté* smelling of the cask f. *fust* cask, tree-trunk, f. L *fustis* cudgel]

futhorc /ˈfuːθɔːk/ *n.* the Scandinavian runic alphabet. [its first six letters *f*, *u*, *þ* (th), *ö*, *r*, *k*]

futile /ˈfjuːtaɪl/ *adj.* **1** useless, ineffectual, vain. **2** frivolous, trifling. □ **futilely** *adv.* **futility** /fjuːˈtɪlɪtɪ/ *n.* [L *futilis* leaky, futile, rel. to *fundere* pour]

futon /ˈfuːtɒn/ *n.* **1** a Japanese quilted mattress rolled out on the floor for use as a bed. **2** a type of low-slung wooden bed using this kind of mattress. [Jap.]

futtock /ˈfʌtək/ *n. Naut.* each of the middle timbers of a ship's frame, between the floor and the top timbers. [ME *votekes* etc. pl. f. MLG f. *fōt* FOOT + -*ken* -KIN]

future /ˈfjuːtʃə(r)/ *adj. & n.* ● *adj.* **1 a** going or expected to happen or be or become (*his future career*). **b** that will be something specified (*my future wife*). **c** that will be after death (*a future life*). **2 a** of time to come (*future years*). **b** *Gram.* (of a tense or participle) describing an event yet to happen. ● *n.* **1** time to come (*past, present, and future*). **2** what will happen in the future (*the future is uncertain*). **3** the future condition of a person, country, etc. **4** a prospect of success etc. (*there's no future in it*). **5** *Gram.* the future tense. **6** (in *pl.*) *Stock Exch.* **a** goods and stocks sold for future delivery. **b** contracts for these. □ **for the future** = *in future*. **future perfect** *Gram.* a tense giving the sense *will have done*. **future shock** inability to cope with rapid progress. **in future** from now onwards. □ **futureless** *adj.* [ME f. OF *futur* -*ure* f. L *futurus* future part. of *esse* be f. stem *fu*- be]

futurism /ˈfjuːtʃəˌrɪz(ə)m/ *n.* an early 20th-century movement in art, literature, music, etc., concerned with celebrating and incorporating into art the energy and dynamism of modern technology. It was launched by the Italian poet Filippo Marinetti in 1909 and widely publicized by a series of manifestos, demonstrations, and exhibitions. Although effectively ended by the First World War, it had an important influence throughout Europe and particularly in Russia, where it is associated with painters such as Kasimir Malevich (1878–1935) and poets such as Mayakovsky. [FUTURE + -ISM, after It. *futurismo*, F *futurisme*]

futurist /ˈfjuːtʃərɪst/ *n.* (often *attrib.*) **1** an adherent of futurism. **2** a believer in human progress. **3** a student of the future. **4** (in Christian theology) a person who believes that biblical prophecies, esp. those of the Apocalypse, are still to be fulfilled.

futuristic /ˌfjuːtʃəˈrɪstɪk/ *adj.* **1** suitable for the future; ultra-modern. **2** of or relating to futurism. **3** relating to the future. □ **futuristically** *adv.*

futurity /fjuːˈtjʊərɪtɪ/ *n.* (*pl.* -**ies**) **1** future time. **2** (in *sing.* or *pl.*) future events. **3** future condition; existence after death. □ **futurity stakes** *US* stakes raced for long after entries or nominations are made.

futurology /ˌfjuːtʃəˈrɒlədʒɪ/ *n.* systematic forecasting of the future esp. from present trends in society. □ **futurologist** *n.*

Fuxin /fuːˈʃɪn/ (also **Fou-hsin**) an industrial city in NE China, in Liaoning province; pop. (1984) 653,200.

fuze var. of FUSE².

fuzee *US* var. of FUSEE.

Fuzhou /fuːˈdʒəʊ/ (also **Foochow** /-ˈtʃəʊ/) a port in SE China, capital of Fujian province; pop. (1990) 1,270,000.

fuzz /fʌz/ *n. & v.* ● *n.* **1** fluff. **2** fluffy or frizzled hair. **3** *sl.* **a** the police. **b** a policeman. ● *v.tr. & intr.* make or become fluffy or blurred. □ **fuzz-ball** a puff-ball fungus. [17th c.: prob. f. LG or Du.: sense 3 perh. a different word]

fuzzy /ˈfʌzɪ/ *adj.* (**fuzzier, fuzziest**) **1 a** like fuzz. **b** frayed, fluffy. **c** frizzy. **2** blurred, indistinct. **3** *Computing & Logic* (of a set) of which membership is determined imprecisely according to probability functions; of or relating to such sets (*fuzzy logic*). □ **fuzzy-wuzzy** (*pl.* -**ies**) *sl. offens.* **1** *hist.* a Sudanese soldier. **2** a black person, esp. one with tightly curled hair. □ **fuzzily** *adv.* **fuzziness** *n.*

fwd *abbr.* forward.

f.w.d. *abbr.* **1** four-wheel drive. **2** front-wheel drive.

f.y. *abbr. US* fiscal year.

-fy /faɪ/ *suffix* forming: **1** verbs from nouns, meaning: **a** make, produce (*pacify; speechify*). **b** make into (*deify; petrify*). **2** verbs from adjectives, meaning 'bring or come into such a state' (*Frenchify; solidify*). **3** verbs in causative sense (*horrify; stupefy*). [from or after F -*fier* f. L -*ficare*, -*facere* f. *facere* do, make]

fylfot /ˈfɪlfət/ *n.* a swastika. [perh. f. *fill-foot*, pattern to fill the foot of a painted window]

fyrd /fɜːd, fɪəd/ *n. hist.* **1** the English militia before 1066. **2** the duty to serve in this. [OE f. Gmc (as FARE)]

fytte var. of FIT³.

Gg

G[1] /dʒiː/ n. (also **g**) (pl. **Gs** or **G's**) **1** the seventh letter of the alphabet. **2** Mus. the fifth note in the diatonic scale of C major.

G[2] abbr. (also **G.**) N. Amer. colloq. = GRAND n. 2.

G[3] symb. **1** gauss. **2** giga-. **3** gravitational constant.

g[1] abbr. (also **g.**) **1** gelding. **2** gas.

g[2] symb. **1** gram(s). **2 a** gravity. **b** acceleration due to gravity.

G7 see GROUP OF SEVEN.

GA abbr. US Georgia (in official postal use).

Ga symb. Chem. the element gallium.

Ga. abbr. Georgia (US).

gab /gæb/ n. & v. colloq. ● n. talk, chatter, twaddle. ● v.intr. talk incessantly, trivially, or indiscreetly; chatter. □ **gift of the gab** the facility of speaking eloquently or profusely. [17th-c. var. of GOB[1]]

gabardine /ˈgæbəˌdiːn/ n. (also **gaberdine**) **1** a smooth durable twill-woven cloth esp. of worsted or cotton. **2** a garment made of this, esp. a raincoat. [var. of GABERDINE]

gabble /ˈgæb(ə)l/ v. & n. ● v. **1** intr. **a** talk volubly or inarticulately. **b** read aloud too fast. **2** tr. utter too fast, esp. in reading aloud. ● n. fast unintelligible talk. □ **gabbler** n. [MDu. gabbelen (imit.)]

gabbro /ˈgæbrəʊ/ n. (pl. **-os**) Geol. a dark granular plutonic rock of crystalline texture. □ **gabbroid** adj. **gabbroic** /gæˈbrəʊɪk/ adj. [It. f. Gabbro in Tuscany]

gabby /ˈgæbɪ/ adj. (**gabbier**, **gabbiest**) colloq. talkative. [GAB + -Y[1]]

gaberdine /ˈgæbəˌdiːn/ n. **1** var. of GABARDINE. **2** hist. a loose long upper garment worn esp. by Jews and beggars. [OF gauvardine perh. f. MHG wallevart pilgrimage]

Gabès /ˈgɑːbɪs/ (also **Qabis**) an industrial seaport in eastern Tunisia; pop. (1984) 92,250.

gabion /ˈgeɪbɪən/ n. a cylindrical wicker or metal basket for filling with earth or stones, used in engineering or (formerly) in fortification. □ **gabionage** n. [F f. It. gabbione f. gabbia CAGE]

Gable /ˈgeɪb(ə)l/, (William) Clark (1901–60), American actor. He became famous through his numerous roles in Hollywood films of the 1930s; they include It Happened One Night (1934), for which he won an Oscar, and Gone with the Wind (1939), in which he starred as Rhett Butler. His last film was The Misfits (1961), in which he played opposite Marilyn Monroe.

gable /ˈgeɪb(ə)l/ n. **1 a** the triangular upper part of a wall at the end of a ridged roof. **b** (in full **gable-end**) a gable-topped wall. **2** a gable-shaped canopy over a window or door. □ **gabled** adj. (also in comb.). [ME gable f. ON gafl]

Gabo /ˈgɑːbəʊ/, Naum (born Naum Neemia Pevsner) (1890–1977), Russian-born American sculptor. With his brother, Antoine Pevsner, he was a founder of Russian constructivism, and was one of the first sculptors to use transparent materials. The brothers' Realistic Manifesto of 1920 set down their artistic principles, including the idea of introducing time and movement into sculpture; Gabo made his first kinetic sculpture, a vibrating metal rod powered by an electric motor, in 1920.

Gabon /gəˈbɒn/ an equatorial country in West Africa, on the Atlantic coast; pop. (est. 1991) 1,200,000; languages, French (official), West African languages; capital, Libreville. French traders gained a foothold in the 1830s and 1840s, and in 1888 Gabon became a French Territory. Part of French Equatorial Africa from 1910 to 1958, it became an independent republic in 1960. □ **Gabonese** /ˌgæbəˈniːz/ adj. & n.

Gabor /ˈgɑːbɔː(r), gəˈbɔː(r)/, Dennis (1900–79), Hungarian-born British electrical engineer. Having left Hungary for Britain in 1934, he became a British citizen in 1946. The following year he conceived the idea of holography, originally as a microscopic technique, greatly improving it after the invention of lasers in 1960. He was awarded the Nobel Prize for physics in 1971.

Gaborone /ˌgæbəˈrəʊnɪ/ the capital of Botswana, in the south of the country near the border with South Africa; pop. (1991) 133,470.

Gabriel /ˈgeɪbrɪəl/ (in the Bible) the archangel who foretold the birth of Jesus to the Virgin Mary (Luke 1:26–38), and who also appeared to Zacharias, father of John the Baptist, and to Daniel. In Islam, Gabriel revealed the Koran to Muhammad.

Gad /gad/ **1** a Hebrew patriarch, son of Jacob and Zilpah. **2** the tribe of Israel traditionally descended from him.

gad[1] /gæd/ v. & n. ● v.intr. (**gadded**, **gadding**) (foll. by about, abroad, around) go about idly or in search of pleasure. ● n. idle wandering or adventure (esp. in **on the gad**). [back-form. f. obs. gadling companion f. OE gædeling f. gæd fellowship]

gad[2] /gæd/ int. (also **by gad**) an expression of surprise or emphatic assertion. [= God]

gadabout /ˈgædəˌbaʊt/ n. a person who gads about; an idle pleasure-seeker.

Gadarene /ˈgædəˌriːn/ adj. involving or engaged in headlong or suicidal rush or flight. [LL Gadarenus f. Gk Gadarēnos of Gadara in ancient Palestine, with ref. to Matthew 8:28–32]

Gaddafi /gəˈdɑːfɪ/, Mu'ammer Muhammad al (also **Qaddafi**) (b.1942), Libyan colonel, head of state since 1970; President since 1977. After leading the coup which overthrew King Idris (1890–1983) in 1969, he gained power as chairman of the revolutionary council and established the Libyan Arab Republic. As self-appointed head of state, Gaddafi has pursued an anti-colonial policy at home, expelling foreigners from Libya and seeking to establish an Islamic Socialist regime. He has been accused of supporting international terrorism and has been involved in a number of conflicts with the West, as also with neighbouring Arab countries.

gadfly /ˈgædflaɪ/ n. (pl. **-flies**) **1** a cattle-biting fly, esp. a warble fly, horsefly, or bot-fly. **2** an irritating or harassing person. [f. GAD[1] or obs. gad spike, goad f. ON gaddr, rel. to YARD[1]]

gadget /ˈgædʒɪt/ n. any small and usu. ingenious mechanical device or tool. □ **gadgetry** n. **gadgety** adj. **gadgeteer** /ˌgædʒɪˈtɪə(r)/ n. [19th-c. Naut.: orig. unkn.]

gadoid /ˈgeɪdɔɪd/ n. & adj. Zool. ● n. a marine fish resembling the cod, or belonging to the cod family Gadidae (see COD[1]). ● adj. belonging to or resembling the Gadidae. [mod.L gadus f. Gk gados cod + -OID]

gadolinite /ˈgædəlɪˌnaɪt/ n. a dark crystalline mineral consisting of ferrous silicate of beryllium. [Johan Gadolin, Finnish mineralogist (1760–1852)]

gadolinium /ˌgædəˈlɪnɪəm/ n. a soft silvery-white metallic chemical element (atomic number 64; symbol **Gd**). A member of the lanthanide series, gadolinium was isolated by the Swiss chemist Jean Charles Galissard de Marignac (1817–94) in 1886. It is strongly magnetic below 17°C. The oxide is used as a phosphor in colour television sets. [f. GADOLINITE]

gadroon /gəˈdruːn/ n. a decoration on silverware etc., consisting of convex curves in a series forming an ornamental edge like inverted fluting. □ **gadrooned** adj. [F godron: cf. goder pucker]

Gadsden Purchase /ˈgædzdən/ an area in New Mexico and Arizona, near the Rio Grande. Extending over 77,700 sq. km (30,000 sq. miles), it was purchased from Mexico in 1853 by the American diplomat James Gadsden (1788–1858), with the intention of ensuring a southern railroad route to the Pacific.

gadwall /ˈgædwɔːl/ n. a brownish-grey freshwater duck, Anas strepera. [17th c.: orig. unkn.]

gadzooks /gædˈzuːks/ int. archaic an expression of asseveration etc. [GAD[2] + zooks of unkn. orig.]

Gaea var. of GAIA.

Gael /geɪl/ n. **1** a Scottish Celt. **2** a Gaelic-speaking Celt. □ **Gaeldom** n. [Gael. Gaidheal: cf. GOIDEL]

Gaelic /ˈgeɪlɪk, ˈgæl-/ n. & adj. ● n. a Celtic language spoken in Ireland and Scotland, especially the Scottish variety. (See note below.) ● adj. of or relating to the Gaels or Gaelic. □ **Gaelic coffee** coffee served with cream and esp. Scotch whisky.

■ Gaelic exists in two distinct varieties, Irish and Scottish Gaelic, which together with Manx form the Goidelic group of Celtic languages. From about the 5th century AD the language was carried to Scotland by settlers from Ireland and became the language of most of the Highlands and islands; in time the Scottish variety diverged to the point where it was clearly a different language. Scottish Gaelic is now spoken by only about 75,000 people in the west of Scotland; there is a small but flourishing literary movement. A number of English words are taken from it, e.g. bog, cairn, slogan, whisky. (See also DALRIADA, IRISH.)

Gaelic football n. a type of football played mainly in Ireland between teams of fifteen players, with a goal resembling that used in rugby but having a net attached. The object is to kick or punch the round ball into the net (scoring three points) or over the crossbar (one point).

Gaeltacht /ˈgeɪltəxt/ n. any of the regions in Ireland where the vernacular language is Irish. [Ir.]

gaff[1] /gæf/ n. & v. ● n. **1 a** a stick with an iron hook for landing large fish. **b** a barbed fishing-spear. **2** Naut. a spar to which the head of a fore-and-aft sail is bent. ● v.tr. seize (a fish) with a gaff. [ME f. Prov. gaf hook]

gaff[2] /gæf/ n. Brit. sl. □ **blow the gaff** let out a plot or secret. [19th c., = nonsense: orig. unkn.]

gaffe /gæf/ n. a blunder; an indiscreet act or remark. [F]

gaffer /ˈgæfə(r)/ n. **1** an old fellow; an elderly rustic. **2** Brit. colloq. a foreman or boss. **3** colloq. the chief electrician in a film or television production unit. [prob. contr. of GODFATHER]

Gafsa /ˈgæfsə/ (also **Qafsah**) an industrial town in central Tunisia; pop. (1984) 61,000. The town was known to the Romans as Capsa, this name being applied to the Capsian culture of the palaeolithic period, found in this part of North Africa.

gag /gæg/ n. & v. ● n. **1** a piece of cloth etc. thrust into or held over the mouth to prevent speaking or crying out, or to hold it open in surgery. **2** a joke or comic scene in a play, film, etc., or as part of a comedian's act. **3** an actor's interpolation in a dramatic dialogue. **4** a thing or circumstance restricting free speech. **5 a** a joke or hoax. **b** a humorous action or situation. **6** an imposture or deception. **7** Parl. a closure or guillotine. ● v. (**gagged**, **gagging**) **1** tr. apply a gag to. **2** tr. silence; deprive of free speech. **3** tr. apply a gag-bit to (a horse). **4 a** intr. choke or retch. **b** tr. cause to do this. **5** intr. Theatr. make gags. □ **gag-bit** a specially powerful bit for horse-breaking. **gag man** a deviser or performer of theatrical gags. [ME, orig. as verb: orig. uncert.]

gaga /ˈgɑːgɑː/ adj. sl. **1** senile. **2** fatuous; slightly crazy. [F, = senile]

Gagarin /gəˈgɑːrɪn/, Yuri (Alekseevich) (1934–68), Russian cosmonaut. In 1961 he made the first manned space flight, completing a single orbit of the earth in 108 minutes. He was killed in a crash while testing an aeroplane.

gage[1] /geɪdʒ/ n. & v. ● n. **1** a pledge; a thing deposited as security. **2 a** a challenge to fight. **b** a symbol of this, esp. a glove thrown down. ● v.tr. archaic stake, pledge; offer as a guarantee. [ME f. OF gage (n.), gager (v.) ult. f. Gmc, rel. to WED]

gage[2] US var. of GAUGE.

gage[3] /geɪdʒ/ n. = GREENGAGE. [abbr.]

gaggle /ˈgæg(ə)l/ n. & v. ● n. **1** a flock of geese. **2** colloq. a disorderly group of people. ● v.intr. (of geese) cackle. [ME, imit.: cf. gabble, cackle]

gagster /ˈgægstə(r)/ n. = gag man.

Gaia /ˈgeɪə, ˈgaɪə/ **1** (also **Gaea** /ˈdʒiːə/, **Ge** /dʒiː, giː/) Gk Mythol. the Earth personified as a goddess, daughter of Chaos. She was the mother and wife of Uranus (Heaven); their offspring included the Titans and the Cyclops. **2** the earth viewed as a vast living organism (see GAIA HYPOTHESIS). □ **Gaian** adj. & n.

Gaia hypothesis n. the theory, put forward by the British scientist James Lovelock (b.1919) in 1969, that the entire range of living matter on the earth collectively defines and regulates the material conditions necessary for the continuance of life. The earth is thus likened to a vast self-regulating organism, modifying the biosphere to suit its needs.

gaiety /ˈgeɪətɪ/ n. (US **gayety**) **1** the state of being light-hearted or merry; mirth. **2** merrymaking, amusement. **3** a bright appearance. □ **gaiety of nations** the cheerfulness or pleasure of numerous people. [F gaieté (as GAY)]

gaijin /ˈgaɪdʒɪn/ n. & adj. ● n. (pl. same) (in Japan) a foreigner, an alien. ● adj. foreign, alien. [Jap., contr. of gaikoku-jin (gaikaku foreign country, jin person)]

gaillardia /geɪˈlɑːdɪə/ n. a composite plant of the genus Gaillardia, with showy flowers. [mod.L f. Gaillard de Marentonneau, 18th-c. Fr. botanist]

gaily /ˈgeɪlɪ/ adv. **1** in a gay or light-hearted manner. **2** with a bright or colourful appearance.

gain /geɪn/ v. & n. ● v. **1** tr. obtain or secure (usu. something desired or favourable) (gain an advantage; gain recognition). **2** tr. acquire (a sum) as profits or as a result of changed conditions; earn. **3** tr. obtain as an increment or addition (gain momentum; gain weight). **4** tr. **a** win (a victory). **b** reclaim (land from the sea). **5** intr. (foll. by in) make a specified advance or improvement (gained in stature). **6 a** tr. (of a clock etc.) have the fault of becoming fast. **b** tr. become fast by a specific amount of time. **7** intr. (often foll. by on, upon) come closer to a person or thing pursued. **8** tr. **a** bring over to one's interest or views. **b** (foll. by over) win by persuasion etc. **9** tr. reach or arrive at (a desired place). ● n. **1** something gained, achieved, etc. **2** an increase of possessions etc.; a profit, advance, or improvement. **3** the acquisition of wealth. **4** (in pl.) sums of money acquired by trade etc., emoluments, winnings. **5** an increase in amount. **6** Electronics the factor by which power etc. is increased. **b** the logarithm of this. □ **gain ground** see GROUND[1]. **gain time** improve one's chances by causing or accepting delay. □ **gainable** adj. **gainer** n. **gainings** n.pl. [OF gaigner, gaaignier to till, acquire, ult. f. Gmc]

gainful /ˈgeɪnfʊl/ adj. **1** (of employment) paid. **2** lucrative, remunerative. □ **gainfully** adv. **gainfulness** n.

gainsay /geɪnˈseɪ/ v.tr. (past and past part. **gainsaid** /-ˈsed/) archaic or literary deny, contradict. □ **gainsayer** n. [ME f. obs. gain- against f. ON gegn straight f. Gmc + SAY]

Gainsborough /ˈgeɪnzbərə/, Thomas (1727–88), English painter. From 1760 he worked in Bath and from 1774 in London, where he became a society portrait painter. Although he was famous for his portraits, such as Mr and Mrs Andrews (1748) and The Blue Boy (c.1770), landscape was his preferred subject; his works reflect the influence of the naturalistic approach to landscape of 17th-century Dutch painting and include The Watering Place (1777). He was a founder member of the Royal Academy of Arts (1768).

'gainst /genst/ prep. poet. = AGAINST. [abbr.]

gait /geɪt/ n. **1** a manner of walking; one's bearing or carriage as one walks. **2** the manner of forward motion of a runner, horse, vehicle, etc. □ **go one's** (or **one's own**) **gait** pursue one's own course. [var. of GATE[2]]

gaiter /ˈgeɪtə(r)/ n. a covering of cloth, leather, etc. for the leg below the knee, for the ankle, for part of a machine, etc. □ **gaitered** adj. [F guêtre, prob. rel. to WRIST]

Gaitskell /ˈgeɪtskɪl/, Hugh (Todd Naylor) (1906–63), British Labour politician. Having served in the Attlee government in several ministerial posts, including Chancellor of the Exchequer (1950–1), he became leader of the Labour Party in opposition from 1955 until his death. Although his leadership covered a period of upheaval and reassessment within his party following successive election defeats, he eventually succeeded in restoring party unity.

Gal. abbr. Epistle to the Galatians (New Testament).

gal[1] /gæl/ n. sl. a girl. [repr. var. pronunc.]

gal[2] /gæl/ n. Physics a unit of acceleration equal to one centimetre per second per second. [GALILEO GALILEI]

gal. abbr. gallon(s).

gala /ˈgɑːlə/ n. **1** (often attrib.) a festive or special occasion (a gala performance). **2** Brit. a festive gathering for sports, esp. swimming. [F or It. f. Sp. f. OF gale rejoicing f. Gmc]

galactagogue /ɡəˈlæktəˌɡɒɡ/ adj. & n. Pharm. ● adj. inducing a flow of milk. ● n. a galactagogue substance. [Gk gala galaktos milk, + agōgos leading]

galactic /ɡəˈlæktɪk/ adj. of or relating to a galaxy or galaxies, esp. the Galaxy. [Gk galaktias, var. of galaxias: see GALAXY]

galactose /ɡəˈlæktəʊz, -təʊs/ n. Biochem. a hexose sugar present in many polysaccharides. [Gk gala galakt- milk + -OSE²]

galago /ɡəˈleɪɡəʊ/ n. (pl. **-os**) a bushbaby, esp. one of the genus Galago. [mod.L]

galah /ɡəˈlɑː/ n. **1** a small rose-breasted grey-backed cockatoo, Eulophus roseicapillus. **2** Austral. sl. a fool or simpleton. [Aboriginal]

Galahad /ˈɡæləˌhæd/ (in Arthurian legend) a knight of immaculate purity, destined to find the Holy Grail.

galangal var. of GALINGALE.

galantine /ˈɡælənˌtiːn/ n. white meat or fish boned, cooked, pressed, and served cold in aspic etc. [ME f. OF, alt. f. galatine jellied meat f. med.L galatina]

galanty show /ɡəˈlæntɪ/ n. hist. a performance of shadow theatre. [perh. f. It. galanti gallants]

Galapagos Islands /ɡəˈlæpəɡəs/ (official Spanish name Archipiélago de Colón) a Pacific archipelago on the equator, about 1,045 km (650 miles) west of Ecuador, to which it belongs; pop. (1990) 9,7500. It is noted for its abundant wildlife, including giant tortoises, flightless cormorants, and many other endemic species. It is the site of Charles Darwin's observations of 1835, which helped him to form his theory of natural selection. Fragments of pottery found there, made by the Chimu Indians, indicate that the islands were visited by people who travelled from the mainland of South America before the Spanish conquest. [Sp., = tortoises]

Galatea /ˌɡæləˈtɪə/ Gk Mythol. **1** a sea-nymph courted by the Cyclops Polyphemus, who in jealousy killed his rival Acis. **2** the name given to the statue fashioned by Pygmalion and brought to life (see PYGMALION¹).

Galaţi /ɡæˈlæts/ an industrial city in eastern Romania, a river port on the lower Danube; pop. (1989) 307,380.

Galatia /ɡəˈleɪʃə/ an ancient region in central Asia Minor, settled by invading Gauls (the Galatians) in the 3rd century BC. In 64 BC it became a protectorate of Rome and, in 25 BC, with the addition of some further territories, was made a province of the Roman Empire. □ **Galatian** adj. & n.

Galatians, Epistle to the /ɡəˈleɪʃ(ə)nz/ a book of the New Testament, an epistle of St Paul to the Church in Galatia.

galaxy /ˈɡæləksɪ/ n. (pl. **-ies**) **1** each of many independent systems of millions or billions of stars, with gas and dust, bound together by gravity. (See note below.) **2** (**the Galaxy**) **a** the galaxy of which the solar system is a part. (See note below.) **b** = MILKY WAY. **3** (foll. by of) a brilliant company or gathering. [ME f. OF galaxie f. med.L galaxia, LL galaxias f. Gk f. gala galaktos milk]

- Galaxies may be elliptical, irregular, or disc-shaped in form, and frequently have well-defined spiral arms. The Galaxy in which the earth is located is a collection of approximately 100,000 million stars. It has two main components: the halo, about 50,000 light-years across, containing a sparse population of cool dim stars and globular clusters, and the disc, 100,000 light-years across but only a few thousand light-years thick, rich in gas, dust, and young stars, with the spiral arms indicating the regions of most recent star formation. Our sun is located about two-thirds of the way out from the centre. The centre of the disc has an almost spherical nuclear bulge containing powerful radio and infrared sources, apparently associated with disturbed complexes of ionized hydrogen. These streaming gas-clouds behave as if ejected explosively from something not yet understood, perhaps a massive black hole, at the exact centre. (See also MILKY WAY.)

Galba /ˈɡælbə/ (full name Servius Sulpicius Galba) (c.3 BC–AD 69), Roman emperor AD 68–9. He was a governor in Spain when invited to succeed Nero as emperor in AD 68. Once in power, he aroused hostility by his severity and parsimony, as well as alienating the legions in Germany by removing their commander. In AD 69 he was murdered in a conspiracy organized by Otho.

galbanum /ˈɡælbənəm/ n. a bitter aromatic gum resin produced from kinds of ferula. [ME f. L f. Gk khalbanē, prob. of Semitic orig.]

Galbraith /ɡælˈbreɪθ/, John Kenneth (b.1908), Canadian-born American economist. He is well known for his criticism of consumerism, the power of large multinational corporations, and a perceived preoccupation in Western society with economic growth for its own sake. His books have a broad appeal and include The Affluent Society (1958) and The New Industrial State (1967).

gale¹ /ɡeɪl/ n. **1 a** a very strong wind. **b** Meteorol. a wind intermediate between a strong breeze and a storm, between force 7 and force 9 on the Beaufort scale (32–54 m.p.h. or 13.9–24.4 metres per second). **2** Naut. a storm. **3** an outburst, esp. of laughter. [16th c.: orig. unkn.]

gale² /ɡeɪl/ n. (in full **sweet-gale**) bog myrtle. [OE gagel(le), MDu. gaghel]

galea /ˈɡeɪlɪə/ n. (pl. **galeae** /-lɪˌiː/ or **-as**) Bot. & Zool. a structure like a helmet in shape, form, or function. □ **galeate** /-lɪət/ adj. **galeated** /-lɪˌeɪtɪd/ adj. [L, = helmet]

Galen /ˈɡeɪlən/ (129–99), Greek physician (full name Claudios Galenos; Latin name Claudius Galenus). He spent the latter part of his life in Rome, becoming court physician to the emperor Marcus Aurelius in 169. He was the author of numerous works which attempted to systematize the whole of medicine, and was especially productive as a practical anatomist and experimental physiologist. He demonstrated that the arteries carry blood not (as had been thought) air, but postulated the presence of minute pores in the wall between the ventricles of the heart, allowing blood to pass through, a theory which was accepted until medieval times. His works reached Europe in the 12th century and were widely influential.

galena /ɡəˈliːnə/ n. a grey mineral of metallic appearance, consisting of lead sulphide (chem. formula: PbS). [L, = lead ore (in a partly purified state)]

galenic /ɡəˈlenɪk/ adj. & n. (also **galenical**) ● adj. **1** of or relating to Galen or his methods. **2** made of natural as opposed to synthetic components. ● n. a drug or medicament produced directly from animal or vegetable tissues.

galia melon /ˈɡɑːlɪə/ n. a small roundish variety of melon with rough skin and orange flesh. [f. Galia, Heb. first name]

Galibi /ɡəˈliːbɪ/ n. & adj. ● n. **1** (pl. same or **Galibis**) a member of a South American Indian people inhabiting French Guiana. **2** the Carib language of this people. ● adj. of or relating to this people or their language. [Carib, lit. 'strong man']

Galicia /ɡəˈlɪsɪə, -ˈlɪʃə/ **1** an autonomous region and former kingdom of NW Spain; capital, Santiago de Compostela. **2** a region of east central Europe, north of the Carpathian Mountains. A former province of Austria (until 1918–20), it now forms part of SE Poland and western Ukraine. □ **Galician** adj. & n.

Galilean¹ /ˌɡælɪˈleɪən/ adj. of or relating to Galileo or his methods. □ **Galilean satellites** (or **moons**) the four largest moons of Jupiter (Callisto, Europa, Ganymede, and Io), discovered by Galileo in 1610 and independently by the German astronomer Simon Marius (1573–1624), visible through binoculars and small telescopes. **Galilean telescope** the first type of astronomical telescope, used by Galileo, with a biconvex objective and biconcave eyepiece.

Galilean² /ˌɡælɪˈliːən/ adj. & n. ● adj. **1** of Galilee in Palestine. **2** Christian. ● n. **1** a native of Galilee. **2** a Christian. **3** (prec. by the) derog. Christ.

Galilee /ˈɡælɪˌliː/ a northern region of ancient Palestine, west of the River Jordan, associated with the ministry of Jesus. The region is now part of Israel.

Galilee, Sea of (also called Lake Tiberias) a lake in northern Israel. The River Jordan flows through it from north to south.

Galileo /ˌɡælɪˈleɪəʊ/ an American space probe to Jupiter, launched in 1989. It obtained the first close-up photographs of two asteroids (Gaspra and Ida) while en route to the planet. Galileo went into orbit around Jupiter in 1995. [GALILEO GALILEI]

Galileo Galilei /ˌɡælɪˌleɪəʊ ˌɡælɪˈleɪɪ/ (1564–1642), Italian astronomer and physicist, one of the founders of modern science. His discoveries include the constancy of a pendulum's swing, later applied to the regulation of clocks. He formulated the law of uniform acceleration of falling bodies, and described the parabolic trajectory of projectiles. Galileo applied the telescope to astronomy and observed craters on the moon, sunspots, the stars of the Milky Way, Jupiter's satellites, and the phases of Venus. His acceptance of the Copernican system was rejected by the Catholic Church, and under threat of torture from the Inquisition he publicly recanted his heretical views.

galingale /ˈɡælɪŋˌɡeɪl/ n. (also **galangal** /ˈɡæləŋɡ(ə)l/) **1** a spice made from the aromatic rhizome of an East Asian plant of the genus Alpinia or *Kaempferia* (ginger family). **2** (in full **sweet** or **English galingale**) a sedge, *Cyperus longus*, with an aromatic root. [OF f. Arab. ḵalanjān, perh. f. Chin. *Gāoliáng* district in Canton + *jiāng* ginger]

galiot var. of GALLIOT.

galipot /ˈɡælɪˌpɒt/ *n.* a hardened deposit of resin formed on the stem of the cluster pine. [F: orig. unkn.]

gall[1] /ɡɔːl/ *n.* **1** *sl.* impudence. **2** asperity, rancour. **3** bitterness; anything bitter (*gall and wormwood*). **4** the bile of animals. **5** the gall-bladder and its contents. □ **gall-bladder** *Anat.* the organ storing bile after its secretion by the liver and before release into the intestine. [ON, corresp. to OE *gealla*, f. Gmc]

gall[2] /ɡɔːl/ *n. & v.* ● *n.* **1** a sore on the skin made by chafing. **2 a** irritation or vexation. **b** a cause of this. **3** a place rubbed bare. ● *v.tr.* **1** rub sore; injure by rubbing. **2** vex, annoy, humiliate. □ **gallingly** *adv.* [ME f. LG or Du. *galle*, corresp. to OE *gealla* sore on a horse]

gall[3] /ɡɔːl/ *n.* **1** a growth produced by insects or fungus etc. on plants and trees, esp. on oak. **2** (*attrib.*) of insects producing galls (*gall fly*). □ **gall wasp** a small gall-forming insect of the hymenopteran superfamily Cynipoidea. [ME f. OF *galle* f. L *galla*]

gall. *abbr.* gallon(s).

gallant *adj., n., & v.* ● *adj.* /ˈɡælənt/ **1** brave, chivalrous. **2 a** (of a ship, horse, etc.) grand, fine, stately. **b** *archaic* finely dressed. **3** /ˈɡælənt, ɡəˈlænt/ **a** markedly attentive to women. **b** concerned with sexual love; amatory. ● *n.* /ˈɡælənt, ɡəˈlænt/ **1** a ladies' man; a lover or paramour. **2** *archaic* a man of fashion; a fine gentleman. ● *v.* /ɡəˈlænt/ **1** *tr.* flirt with. **2** *tr.* escort; act as an escort to (a woman). **3** *intr.* **a** play the gallant. **b** (foll. by *with*) flirt. □ **gallantly** /ˈɡæləntlɪ/ *adv.* [ME f. OF *galant* part. of *galer* make merry]

gallantry /ˈɡæləntrɪ/ *n.* (*pl.* **-ies**) **1** bravery; dashing courage. **2** courtliness; devotion to women. **3** a polite act or speech. **4** the conduct of a gallant; sexual intrigue; immorality. [F *galanterie* (as GALLANT)]

Galle /ˈɡɑːlə/ a seaport on the SW coast of Sri Lanka; pop. (1981) 76,800.

galleon /ˈɡælɪən/ *n.* a type of sailing-ship that was developed from the carrack in the late 16th century, originally as a warship. The elimination of the high forecastle (which caught the wind), with which all large ships were then built, produced a ship that was more weatherly and manoeuvrable. Although the design was essentially English, the galleon was mainly developed in Spain, and formed part of the Spanish Armada of 1588. In general the galleon was used both as a warship and a trading ship, and became the principal trading ship of the 17th century. As a warship the galleon was gradually replaced during the 17th and 18th centuries by the ship of the line and an early version of the cruiser, while Spain continued to use it for trade until the late 17th century. [MDu. *galjoen* f. F *galion* f. *galie* galley, or f. Sp. *galeón*]

galleria /ˌɡæləˈrɪə/ *n.* a collection of small shops under a single roof; an arcade. [It.]

gallery /ˈɡælərɪ/ *n.* (*pl.* **-ies**) **1** a room or building for showing works of art. **2** a balcony, esp. a platform projecting from the inner wall of a church, hall, etc., providing extra room for spectators etc. or reserved for musicians etc. (*minstrels' gallery*). **3 a** the highest balcony in a theatre. **b** its occupants. **4 a** a covered space for walking in, partly open at the side; a portico or colonnade. **b** *Archit.* a long narrow passage in the thickness of a wall or supported on corbels, open towards the interior of the building. **5** a long narrow room, passage, or corridor. **6** *Mil. & Mining* a horizontal underground passage. **7** a group of spectators at a golf-match etc. □ **play to the gallery** seek to win approval by appealing to popular taste. □ **galleried** *adj.* [F *galerie* f. It. *galleria* f. med.L *galeria*]

galleryite /ˈɡælərɪˌaɪt/ *n.* a member of the audience in the gallery of a theatre; a playgoer.

galley /ˈɡælɪ/ *n.* (*pl.* **-eys**) **1** *hist.* **a** a low flat single-decked vessel using sails and oars, esp. an ancient Greek or Roman warship with one or more banks of oars. (*See note below.*) **b** a large open rowing-boat, e.g. that used by the captain of a man-of-war. **2** a ship's or aircraft's kitchen. **3** *Printing* **a** an oblong tray for set type. **b** the corresponding part of a composing-machine. **c** (in full **galley proof**) a proof in the form of long single-column strips from type in a galley, not in sheets or pages. □ **galley-slave 1** *hist.* a person condemned to row in a galley. **2** a drudge. [ME f. OF *galie* f. med.L *galea*, med.Gk *galaia*]

▪ The galley was the oared fighting-ship of the Mediterranean, dating from about 3000 BC and lasting into the 18th century. Such ships had up to three banks of oars (as in the trireme) as well as one or more sails, which were lowered or beached before action. The weapon was the ram, a pointed spur fixed to the bow of the galley on or just below the waterline; the ancient Greek technique of fighting was to ram and then depart at speed, while the Romans preferred to grapple and fight on deck, or use catapult weapons. Slaves and criminals were later sometimes used as rowers, although free men were preferred as being more reliable in battle. With its slim light-draught design the galley was an unstable vessel suitable only for calm waters.

Gallia Narbonensis /ˌɡælɪə ˌnɑːbəˈnɛnsɪs/ the southern province of Transalpine Gaul (see GAUL[1]).

galliard /ˈɡælɪˌɑːd/ *n. hist.* **1** a lively dance usu. in triple time for two persons. **2** the music for this. [ME f. OF *gaillard* valiant]

Gallic /ˈɡælɪk/ *adj.* **1** French or typically French. **2** of the Gauls; Gaulish. □ **Gallicize** /-lɪˌsaɪz/ *v.tr. & intr.* (also **-ise**) [L *Gallicus* f. *Gallus* a Gaul]

gallic acid /ˈɡælɪk/ *n. Chem.* an acid extracted from oak galls and other vegetable products, formerly used in making ink. [F *gallique* f. *galle* GALL[3]]

Gallican /ˈɡælɪkən/ *adj & n.* ● *adj.* **1** of the ancient church of Gaul or France. **2** of or holding a doctrine (reaching its peak in the 17th century) asserting the freedom of the Roman Catholic Church, especially in France, from the ecclesiastical authority of the papacy (cf. ULTRAMONTANE *adj.* 2). ● *n.* an adherent of this doctrine. □ **Gallicanism** *n.* [as GALLIC]

gallice /ˈɡælɪˌsiː/ *adv.* in French. [L, = in Gaulish]

Gallicism /ˈɡælɪˌsɪz(ə)m/ *n.* a French idiom, esp. one adopted in another language. [F *gallicisme* (as GALLIC)]

Gallic Wars Julius Caesar's campaigns 58–51 BC, which established Roman control over Gaul north of the Alps and west of the River Rhine (Transalpine Gaul). During this period Caesar twice invaded Britain (55 and 54 BC), which he regarded as a refuge for his Belgic opponents. Largely disunited, the Gauls combined in 53–52 BC under the chieftain Vercingetorix (died c.46 BC) but were eventually defeated.

galligaskins /ˌɡælɪˈɡæskɪnz/ *n.pl. hist.* or *joc.* breeches, trousers. [orig. wide hose of the 16th–17th centuries, f. obs. F *garguesque* for *greguesque* f. It. *grechesca* fem. of *grechesco* Greek]

gallimaufry /ˌɡælɪˈmɔːfrɪ/ *n.* (*pl.* **-ies**) a heterogeneous mixture; a jumble or medley. [F *galimafrée*, of unkn. orig.]

gallimimus /ˌɡælɪˈmaɪməs/ *n.* a slender bipedal dinosaur of the genus *Gallimimus*, of the Cretaceous period, adapted for swift running. [mod.L f. L *gallus* cock + *mimus* mime]

gallinaceous /ˌɡælɪˈneɪʃəs/ *adj.* of or relating to the Galliformes, an order of birds which includes domestic fowl, pheasants, grouse, etc. [L *gallinaceus* f. *gallina* hen f. *gallus* cock]

gallinule /ˈɡælɪˌnjuːl/ *n.* a moorhen or related bird, esp. of the genus *Porphyrula* or *Porphyrio*. [mod.L *gallinula*, dimin. of L *gallina* hen (see GALLINACEOUS)]

galliot /ˈɡælɪət/ *n.* (also **galiot**) **1** a Dutch single-masted cargo-boat or fishing-vessel. **2** *hist.* a small (usu. Mediterranean) galley. [ME f. OF *galiote* f. It. *galeotta* f. med.L *galea* galley]

Gallipoli /ɡəˈlɪpəlɪ/ a major campaign of the First World War which took place on the Gallipoli peninsula, on the European side of the Dardanelles. In early 1915, after a naval attempt to force the Dardanelles had failed, the Allies (with heavy involvement of troops from Australia and New Zealand) invaded the peninsula, hoping to remove Turkey from the war and open supply lines to Russia's Black Sea ports. The campaign reached stalemate and became bogged down in trench warfare. After each side had suffered a quarter of a million casualties, the Allies evacuated the peninsula without further loss in Jan. 1916.

gallipot /ˈɡælɪˌpɒt/ *n.* a small pot of earthenware, metal, etc., used for ointments etc. [prob. GALLEY + POT[1], because brought in galleys from the Mediterranean]

gallium /ˈɡælɪəm/ *n.* a soft silvery-white metallic chemical element (atomic number 31; symbol **Ga**). Gallium was discovered by the French chemist Paul-Émile Lecoq de Boisbaudran (1838–1912) in 1875, after its existence had been predicted by Dmitri Mendeleev on the basis of his periodic table. A rare element found in traces in bauxite and other ores, gallium now has a number of uses, e.g. in high-temperature thermometers and in semiconductors. Gallium melts at about 30°C, just above room temperature. [L *Gallia* France]

gallivant /ˈɡælɪˌvænt/ *v.intr. colloq.* **1** gad about. **2** flirt. [19th c.: orig. uncert.]

galliwasp /ˈgælɪˌwɒsp/ n. a West Indian lizard, *Diploglossus monotropis.* [18th c.: orig. unkn.]

gallnut /ˈgɔːlnʌt/ n. = GALL³.

Gallo- /ˈgæləʊ/ comb. form **1** French; French and. **2** Gaul (*Gallo-Roman*). [L *Gallus* a Gaul]

gallon /ˈgælən/ n. **1 a** (in full **imperial gallon**) *Brit.* a measure of capacity equal to eight pints and equivalent to 4546 cc, used for liquids and corn etc. **b** *US* a measure of capacity equivalent to 3785 cc, used for liquids. **2** (usu. in *pl.*) *colloq.* a large amount. □ **gallonage** n. [ME f. ONF *galon*, OF *jalon*, f. base of med.L *gallēta, gallētum*, perh. of Celtic orig.]

galloon /gəˈluːn/ n. a narrow closely woven braid of gold, silver, silk, cotton, nylon, etc., for binding dresses etc. [F *galon* f. *galonner* trim with braid, of unkn. orig.]

gallop /ˈgæləp/ n. & v. ● n. **1** the fastest pace of a horse or other quadruped, with all the feet off the ground together in each stride. **2** a ride at this pace. **3** a track or ground for this. ● v. (**galloped, galloping**) **1 a** *intr.* (of a horse etc. or its rider) go at the pace of a gallop. **b** *tr.* make (a horse etc.) gallop. **2** *intr.* (foll. by *through, over*) read, recite, or talk at great speed. **3** *intr.* move or progress rapidly (*galloping inflation*). □ **at a gallop** at the pace of a gallop. □ **galloper** n. [OF *galop, galoper*: see WALLOP]

Galloway /ˈgæləˌweɪ/ an area of SW Scotland consisting of the two former counties of Kirkcudbrightshire and Wigtownshire, and now part of Dumfries and Galloway region. The area is noted for its western peninsula, called the Rhinns, of which the southern tip, the Mull of Galloway, is the most southerly point of Scotland.

galloway /ˈgæləˌweɪ/ n. **1** a breed of hornless black beef cattle from Galloway. **2** (in full **belted galloway**) a breed of galloway cattle with a broad white band.

gallows /ˈgæləʊz/ n.pl. (usu. treated as *sing.*) **1** a structure, usu. of two uprights and a crosspiece, for the hanging of criminals. **2** (prec. by *the*) execution by hanging. □ **gallows humour** grim and ironical humour. [ME f. ON *gálgi*]

gallstone /ˈgɔːlstəʊn/ n. a small hard mass of bile pigments, cholesterol, and calcium salts, formed in the gall-bladder or bile ducts.

Gallup poll /ˈgæləp/ n. an assessment of public opinion by questioning a representative sample, esp. as the basis for forecasting the results of voting. The system was devised by the US statistician George Horace Gallup (1901–84).

galluses /ˈgæləsɪz/ n.pl. dial. & US trouser-braces. [pl. of *gallus* var. of GALLOWS]

Galois /ˈgælwʌ/, Évariste (1811–32), French mathematician. His memoir on the conditions for solubility of polynomial equations was highly innovative but was not published until after his death in 1846. He was imprisoned for his republican activities aged 19, and died aged 20 after a duel.

Galois theory n. Math. a method of applying group theory to the solution of algebraic equations.

galoot /gəˈluːt/ n. colloq. a person, esp. a strange or clumsy one. [19th-c. Naut. sl.: orig. unkn.]

galop /ˈgæləp/ n. & v. ● n. **1** a lively dance in duple time. **2** the music for this. ● v.intr. (**galoped, galoping**) perform this dance. [F: see GALLOP]

galore /gəˈlɔː(r)/ adv. in abundance (placed after noun: *flowers galore*). [Ir. *go leór* to sufficiency]

galosh /gəˈlɒʃ/ n. (also **golosh**) (usu. in pl.) a waterproof overshoe, usu. of rubber. [ME f. OF *galoche* f. LL *gallicula* small Gallic shoe]

Galsworthy /ˈgɔːlzˌwɜːðɪ/, John (1867–1933), English novelist and dramatist. He wrote several plays on social and moral themes, but is remembered chiefly for his sequence of novels known collectively as *The Forsyte Saga* (1906–28), tracing the declining fortunes of an affluent middle-class family in the years leading up to the First World War. These novels gained a wider audience through their adaptation for television in 1967. Galsworthy was awarded the Nobel Prize for literature in 1932.

Galtieri /ˌgæltɪˈeərɪ/, Leopoldo Fortunato (b.1926), Argentinian general and statesman, President 1981–2. He was one of the leaders of the right-wing junta that ordered the invasion of the Falkland Islands in 1982, precipitating the Falklands War. After the British victory in the war, Galtieri was court-martialled and, in 1986, sentenced to twelve years in prison.

Galton /ˈgɔːlt(ə)n/, Sir Francis (1822–1911), English scientist. A man of wide interests and a cousin of Charles Darwin, he is remembered chiefly for his founding and advocacy of eugenics. He introduced methods of measuring human mental and physical abilities, and developed statistical techniques to analyse his data. Galton also carried out important work in meteorology, especially on the theory of anticyclones, and pioneered the use of fingerprints as a means of identification.

galumph /gəˈlʌmf/ v.intr. colloq. **1** move noisily or clumsily. **2** go prancing in triumph. [coined by Lewis Carroll (in sense 2), perh. f. GALLOP + TRIUMPH]

Galvani /gælˈvɑːnɪ/, Luigi (1737–98), Italian anatomist. He studied the structure of organs and the physiology of tissues, but he is best known for his discovery of the twitching of frogs' legs in an electric field. He concluded that these convulsions were caused by 'animal electricity' found in the body. This was disputed by Alessandro Volta who, in the course of this argument, invented his electrochemical cell. The current produced by this device was for many years called *galvanic electricity.*

galvanic /gælˈvænɪk/ adj. **1 a** sudden and remarkable (*had a galvanic effect*). **b** stimulating; full of energy. **2** = VOLTAIC. □ **galvanically** adv.

galvanism /ˈgælvəˌnɪz(ə)m/ n. hist. **1** electricity produced by chemical action. **2** the therapeutic use of electricity. □ **galvanist** n. [F *galvanisme* f. GALVANI]

galvanize /ˈgælvəˌnaɪz/ v.tr. (also **-ise**) **1** (often foll. by *into*) rouse forcefully, esp. by shock or excitement (*was galvanized into action*). **2** stimulate by or as if by electricity. **3** coat (iron) with zinc (usu. without the use of electricity) as a protection against rust. □ **galvanizer** n. **galvanization** /ˌgælvənaɪˈzeɪʃ(ə)n/ n. [F *galvaniser*: see GALVANISM]

galvanometer /ˌgælvəˈnɒmɪtə(r)/ n. an instrument for detecting and measuring small electric currents. □ **galvanometric** /-nəˈmetrɪk/ adj.

Galveston /ˈgælvɪstən/ a port in Texas, south-east of Houston; pop. (1990) 59,100. It is situated on Galveston Bay, an inlet of the Gulf of Mexico.

Galway /ˈgɔːlweɪ/ **1** a county of the Republic of Ireland, on the west coast in the province of Connacht. **2** its county town, a seaport at the head of Galway Bay; pop. (1991) 50,800.

Galway Bay an inlet of the Atlantic Ocean on the west coast of Ireland.

Gama, Vasco da, see DA GAMA.

gambade /gæmˈbɑːd, -ˈbeɪd/ n. (also **gambado** /-ˈbɑːdəʊ, -ˈbeɪdəʊ/) (pl. **gambades**; **-os** or **-oes**) **1** a horse's leap or bound. **2** a fantastic movement. **3** an escapade. [F *gambade* & Sp. *gambado* f. It. & Sp. *gamba* leg]

Gambia /ˈgæmbɪə/ **1** (also **the Gambia**) a country on the coast of West Africa; pop. (est. 1991) 900,000; languages, English (official), Malinke and other indigenous languages, Creole; capital, Banjul. The Gambia consists of a narrow strip of territory on either side of the Gambia River extending upstream from its mouth, forming an enclave in Senegal. It was created a British colony in 1843, becoming an independent member of the Commonwealth in 1965, and a republic in 1970. **2** a river of West Africa, which rises near Labé in Guinea and flows 800 km (500 miles) through Senegal and the Gambia to meet the Atlantic at Banjul. □ **Gambian** adj. & n.

gambier /ˈgæmbɪə(r)/ n. an astringent extract of an Eastern plant used in tanning etc. [Malay *gambir* name of the plant]

Gambier Islands /ˈgæmbɪə(r)/ a group of coral islands in the South Pacific, forming part of French Polynesia; pop. (1986) 582.

gambit /ˈgæmbɪt/ n. **1** a chess opening in which a player sacrifices a piece or pawn to secure an advantage. **2** an opening move in a discussion etc. **3** a trick or device. [earlier *gambett* f. It. *gambetto* tripping up f. *gamba* leg]

gamble /ˈgæmb(ə)l/ v. & n. ● v. **1** intr. play games of chance for money, esp. for high stakes. **2** tr. **a** bet (a sum of money) in gambling. **b** (often foll. by *away*) lose (assets) by gambling. **3** intr. take great risks in the hope of substantial gain. **4** intr. (foll. by *on*) act in the hope or expectation of (*gambled on fine weather*). ● n. **1** a risky undertaking or attempt. **2** a spell or an act of gambling. □ **gambler** n. [obs. *gamel* to sport, *gamene* GAME¹]

gamboge /gæmˈbəʊdʒ, -ˈbuːʒ/ n. a gum resin produced by various eastern Asian trees and used as a yellow pigment and as a purgative. [mod.L *gambaugium* f. *Cambodia* in SE Asia]

gambol /ˈgæmb(ə)l/ v. & n. ● v.intr. (**gambolled, gambolling**; US **gamboled, gamboling**) skip or frolic playfully. ● n. a playful frolic. [GAMBADE]

gambrel /'gæmbrəl/ n. (in full **gambrel roof**) **1** Brit. a roof like a hipped roof but with gable-like ends. **2** US = curb roof. [ONF gamberel f. gambier forked stick f. gambe leg (from the resemblance to the shape of a horse's hind leg)]

game[1] /geɪm/ n., adj., & v. ● n. **1** a form or spell of play or sport, esp. a competitive one played according to rules and decided by skill, strength, or luck. **2** a single portion of play forming a scoring unit in some contests, e.g. bridge or tennis. **3** (in pl.) **a** athletics or sports as organized in a school etc. **b** a meeting for athletic etc. contests (Olympic Games). **4** a winning score in a game; the state of the score in a game (the game is two all). **5** the equipment for a game. **6** one's level of achievement in a game, as specified (played a good game). **7 a** a piece of fun; a jest (was only playing a game with you). **b** (in pl.) dodges, tricks (none of your games!). **8** a scheme or undertaking etc. regarded as a game (so that's your game). **9 a** a policy or line of action. **b** an occupation or profession (the fighting game). **10** (collect.) **a** wild animals or birds hunted for sport or food. **b** the flesh of these. **11** a hunted animal; a quarry or object of pursuit or attack. **12** a kept flock of swans. ● adj. **1** spirited; eager and willing. **2** (foll. by for, or to + infin.) having the spirit or energy; eagerly prepared. ● v.intr. play at games of chance for money; gamble. □ **the game is up** the scheme is revealed or foiled. **game bird 1** a bird shot for sport or food. **2** a bird of the order Galliformes, which includes the pheasants, grouse, etc. **game plan 1** a winning strategy worked out in advance for a particular match. **2** a plan of campaign, esp. in politics. **game point** Tennis etc. a point which, if won, would win the game. **game show** a television light-entertainment programme in which people compete in a game or quiz, often for prizes. **game-warden** an official locally supervising game and hunting. **gaming-house** a place frequented for gambling; a casino. **gaming-table** a table used for gambling. **make game** (or **a game**) **of** mock, taunt. **off** (or **on**) **one's game** playing badly (or well). **on the game** Brit. sl. involved in prostitution or thieving. **play the game** behave fairly or according to the rules. □ **gamely** adv. **gameness** n. **gamester** n. [OE gamen]

game[2] /geɪm/ adj. (of a leg, arm, etc.) lame, crippled. [18th-c. dial.: orig. unkn.]

gamebook /'geɪmbʊk/ n. a book for recording game killed by a sportsman or sportswoman.

Game Boy n. (also **Gameboy**) propr. a hand-held, electronic device, incorporating a small screen, which is used to play computer games loaded in the form of cartridges.

gamecock /'geɪmkɒk/ n. (also **gamefowl** /-faʊl/) a cock bred and trained for cock-fighting.

gamekeeper /'geɪm,ki:pə(r)/ n. a person employed to breed and protect game. □ **gamekeeping** n.

gamelan /'gæmə,læn/ n. the standard instrumental ensemble of Indonesia, comprising sets of tuned gongs, gong-chimes, and other percussion instruments as well as string and woodwind instruments. [Javanese]

gamesman /'geɪmzmən/ n. (pl. **-men**) an exponent of gamesmanship.

gamesmanship /'geɪmzmən,ʃɪp/ n. the art or practice of winning games or other contests by gaining a psychological advantage over an opponent.

gamesome /'geɪmsəm/ adj. merry, sportive. □ **gamesomely** adv. **gamesomeness** n.

gametangium /,gæmɪ'tændʒɪəm/ n. (pl. **gametangia** /-dʒɪə/) Bot. an organ in which gametes are formed. [as GAMETE + aggeion vessel]

gamete /'gæmi:t, gə'mi:t/ n. Biol. a mature germ cell able to unite with another in sexual reproduction. □ **gametic** /gə'metɪk/ adj. [mod.L gameta f. Gk gametē wife f. gamos marriage]

game theory n. (also **games theory**) the branch of mathematics that deals with the selection of best strategies for participants in competitive situations where the outcome of a person's choice of action depends critically on the actions of other players. The theory was established by John von Neumann in 1928 and was expanded in collaboration with the economist Oskar Morgenstern (1902–77) in their book Theory of Games and Economic Behavior (1944). As well as being influential for economics and business, game theory has found uses in military strategy, sport, biology, and other fields.

gameto- /gə'mi:təʊ/ comb. form Biol. gamete.

gametocyte /gə'mi:təʊ,saɪt/ n. Biol. any cell that is in the process of developing into one or more gametes.

gametogenesis /gə,mi:təʊ'dʒenɪsɪs/ n. Biol. the process by which cells undergo meiosis to form gametes.

gametophyte /gə'mi:təʊ,faɪt/ n. Bot. the gamete-producing form of a plant that has alternation of generations between this and the asexual form (cf. SPOROPHYTE). □ **gametophytic** /-,mi:təʊ'fɪtɪk/ adj.

gamin /'gæmɪn/ n. **1** a street urchin. **2** an impudent child. [F]

gamine /'gæmi:n, gæ'mi:n/ n. **1** a girl gamin. **2** a girl with mischievous or boyish charm. [F]

gamma /'gæmə/ n. **1** the third letter of the Greek alphabet (Γ, γ). **2** a third-class mark given for a piece of work or in an examination. **3** (**Gamma**) Astron. the third (usu. third-brightest) star in a constellation (foll. by Latin genitive: Gamma Orionis). **4** the third member of a series. □ **gamma globulin** Biochem. a mixture of blood plasma proteins, mainly immunoglobulins of relatively low electrophoretic mobility, often given to boost immunity. **gamma radiation** (or **rays**) Physics penetrating electromagnetic radiation of shorter wavelength than X-rays. [ME f. Gk]

gammer /'gæmə(r)/ n. archaic an old woman, esp. as a rustic name. [prob. contr. of GODMOTHER: cf. GAFFER]

gammon[1] /'gæmən/ n. & v. ● n. **1** the bottom piece of a flitch of bacon including a hind leg. **2** the ham of a pig cured like bacon. ● v.tr. cure (bacon). [ONF gambon f. gambe leg: cf. JAMB]

gammon[2] /'gæmən/ n. & v. ● n. a victory in backgammon (carrying a double score) in which the winner removes all his or her pieces before the loser has removed any. ● v.tr. defeat in this way. [app. = ME gamen GAME[1]]

gammon[3] /'gæmən/ n. & v. colloq. ● n. humbug, deception. ● v. **1** intr. **a** talk speciously. **b** pretend. **2** tr. hoax, deceive. [18th c.: orig. uncert.]

gammy /'gæmɪ/ adj. (**gammier**, **gammiest**) Brit. colloq. (esp. of a leg) lame; permanently injured. [dial. form of GAME[2]]

Gamow /'geɪməʊ/, George (1904–68), Russian-born American physicist. He explained (with Ralph Asher Alpher, b.1921, and Hans Albrecht Bethe, b.1906) the abundances of chemical elements in the universe, and was a proponent of the big bang theory. He also suggested the triplet code of bases in DNA, which governs the synthesis of amino acids.

gamp /gæmp/ n. Brit. colloq. an umbrella, esp. a large unwieldy one. [Mrs Gamp in Charles Dickens's Martin Chuzzlewit]

gamut /'gæmət/ n. **1** the whole series or range or scope of anything (the whole gamut of crime). **2** Mus. **a** the whole series of notes used in medieval or modern music. **b** a major diatonic scale. **c** a people's or a period's recognized scale. **d** a voice's or instrument's compass. **3** Mus. the lowest note in the medieval sequence of hexachords, = modern G on the lowest line of the bass staff. □ **run the gamut** of experience or perform the complete range of. [med.L gamma ut f. GAMMA taken as the name for a note one tone lower than A of the classical scale + ut the first of six arbitrary names of notes forming the hexachord, being syllables (ut, re, mi, fa, so, la) of the Latin hymn beginning Ut queant laxis)]

gamy /'geɪmɪ/ adj. (**gamier**, **gamiest**) **1** having the flavour or scent of game kept till it is high. **2** N. Amer. scandalous, sensational. **3** = GAME[1] adj. □ **gamily** adv. **gaminess** n.

Ganapati /,gʌnə'pʌtɪ/ Hinduism see GANESHA. [as GANESHA]

Gäncä /'gəndʒə/ (Russian **Gyandzhe** /'gjandʒə/) an industrial city in Azerbaijan; pop. (1990) 281,000. The city was formerly called Yelizavetpol (Elizavetpol) (1804–1918) and Kirovabad (1935–89).

Gance /gɒns/, Abel (1889–1991), French film director. He was a notable early pioneer of technical experimentation in film; Napoléon (1926), for example, was significant for its use of the split-screen, hand-held camera, and wide-angle photography. His other films include J'accuse (1918) and La Roue (1921).

Gand see GHENT.

Gander /'gændə(r)/ a town on the island of Newfoundland, on Lake Gander; pop. (1991) 10,100. Its airport served the first regular transatlantic flights during the Second World War.

gander /'gændə(r)/ n. & v. ● n. **1** a male goose. **2** sl. a look, a glance (take a gander). ● v.intr. look or glance. [OE gandra, rel. to GANNET]

Gandhi[1] /'gændɪ/, Mrs Indira (1917–84), Indian stateswoman, Prime Minister 1966–77 and 1980–4. The daughter of Jawaharlal Nehru, she had already served as president of the Indian National Congress (1959–60) and Minister of Information (1964) when she succeeded Lal Bahadur Shastri (1904–66) as Prime Minister. In her first term of office she sought to establish a secular state and to lead India out of poverty. However, in 1975 she introduced an unpopular state of emergency to deal with growing political unrest, and the Congress Party lost the

1977 election. Mrs Gandhi lost her seat and was unsuccessfully tried for corruption. Having formed a breakaway group from the Congress Party — known as the Indian National Congress (I) — in 1978, she was elected Prime Minister again in 1980. Her second period of office was marked by prolonged religious disturbance, during which she alienated many Sikhs by allowing troops to storm the Golden Temple at Amritsar; she was assassinated by her own Sikh bodyguards.

Gandhi[2] /'gændɪ/, Mahatma (born Mohandas Karamchand Gandhi) (1869–1948), Indian nationalist and spiritual leader. After early civil-rights activities as a lawyer in South Africa, in 1914 Gandhi returned to India, where he became prominent in the opposition to British rule, pursuing a policy of passive resistance and non-violent civil disobedience. The president of the Indian National Congress (1925–34), he never held government office, but was regarded as the country's supreme political and spiritual leader and the principal force in achieving India's independence. He was assassinated by a Hindu following his agreement to the creation of the state of Pakistan for the Muslim minority.

Gandhi[3] /'gændɪ/, Rajiv (1944–91), Indian statesman, Prime Minister 1984–9. The eldest son of Indira Gandhi, he entered politics following the accidental death of his brother Sanjay (1946–80), becoming Prime Minister after his mother's assassination. His premiership, at the head of the Indian National Congress (I) party, was marked by continuing unrest and he resigned in 1989; he was assassinated during the election campaign of 1991.

Gandhinagar /ˌgændɪˈnʌgə(r)/ a city in western India, capital of the state of Gujarat; pop. (1991) 121,750.

Ganesha /gəˈneɪʃə/ (also called *Ganapati*) *Hinduism* an elephant-headed deity, son of Siva and Parvati. Worshipped as the remover of obstacles and patron of learning, he is invoked at the beginning of literary works, rituals, or any new undertaking. He is usually depicted coloured red, with a pot belly and one broken tusk, riding a rat. [Skr., = lord of ganas (Siva's attendants)]

gang[1] /gæŋ/ *n. & v.* ● *n.* **1 a** a band of persons acting or going about together, esp. for criminal purposes. **b** *colloq.* such a band pursuing a purpose causing disapproval. **2** a set of workers, slaves, or prisoners. **3** a set of tools arranged to work simultaneously. ● *v.tr.* arrange (tools etc.) to work in coordination. □ **gang-bang** *sl.* an occasion on which several men have sexual intercourse with one woman. **gang rape** the successive rape of a person by a group of people. **gang show** a variety show performed annually by members of the Scout and Guide Association. **gang up** *colloq.* **1** (often foll. by *with*) act in concert. **2** (foll. by *on*) combine against. [orig. = going, journey, f. ON *gangr, ganga* GOING, corresp. to OE *gang*]

gang[2] /gæŋ/ *v.intr. Sc.* go. □ **gang agley** (of a plan etc.) go wrong. [OE *gangan*: cf. GANG[1]]

Ganga see GANGES.

gangboard /'gæŋbɔːd/ *n.* = GANGPLANK.

ganger /'gæŋə(r)/ *n. Brit.* the foreman of a gang of workers, esp. navvies.

Ganges /'gændʒiːz/ (Hindi **Ganga** /'gʌŋgə/) a river of northern India and Bangladesh, which rises in the Himalayas and flows some 2,700 km (1,678 miles) south-east to the Bay of Bengal, where it forms the world's largest delta. The river is regarded by Hindus as sacred.

gangland /'gæŋlænd/ *n.* the world of gangs and gangsters.

gangle /'gæŋg(ə)l/ *v.intr.* move ungracefully. [back-form. f. GANGLING]

gangling /'gæŋglɪŋ/ *adj.* (of a person) loosely built; lanky. [frequent. of GANG[2]]

ganglion /'gæŋglɪən/ *n.* (pl. **ganglia** /-lɪə/ or **ganglions**) **1 a** an enlargement or knot on a nerve etc. containing an assemblage of nerve cells. **b** a mass of grey matter in the central nervous system forming a nerve-nucleus. **2** *Med.* a cyst, esp. on a tendon sheath. **3** a centre of activity or interest. □ **gangliar** *adj.* **gangliform** *adj.* **ganglionated** *adj.* **ganglionic** /ˌgæŋglɪˈɒnɪk/ *adj.* [Gk *gagglion*]

gangly /'gæŋglɪ/ *adj.* (**ganglier, gangliest**) = GANGLING.

Gang of Four *hist.* **1** (in China) a group of four associates involved in implementing many of Mao Zedong's policies during the Cultural Revolution. The four (Wang Hongwen, Zhang Chunjao, Yao Wenyuan, and Mao's wife Jiang Qing) were among the groups competing for power on Mao's death in 1976, but were arrested and imprisoned. **2** (in the UK) a group of four Labour MPs (Shirley Williams, Roy Jenkins, David Owen, and William Rodgers) who broke away from the Labour Party in 1981 to form the Social Democratic Party.

gangplank /'gæŋplæŋk/ *n.* a movable plank usu. with cleats nailed on it for boarding or disembarking from a ship etc.

gangrene /'gæŋgriːn/ *n. & v.* ● *n.* **1** *Med.* death and decomposition of a part of the body tissue, resulting from either obstructed circulation or bacterial infection. **2** moral corruption. ● *v.tr. & intr.* affect or become affected with gangrene. □ **gangrenous** /-grɪnəs/ *adj.* [F *gangrène* f. L *gangraena* f. Gk *gangraina*]

gangsta /'gæŋstə/ *n.* **1** *sl.* = GANGSTER. **2** (in full **gangsta rap**) a type of rap music featuring aggressive, macho lyrics, often with reference to gang warfare, gun battles, etc. [repr. pronunc. of GANGSTER]

gangster /'gæŋstə(r)/ *n.* a member of a gang of violent criminals. □ **gangsterism** *n.*

Gangtok /gæŋˈtɒk/ a city in northern India, in the foothills of the Kanchenjunga mountain range, capital of the state of Sikkim; pop. (1991) 24,970.

gangue /gæŋ/ *n.* valueless earth etc. in which ore is found. [F f. G *Gang* lode, way, course = GANG[1]]

gangway /'gæŋweɪ/ *n. & int.* ● *n.* **1** *Brit.* a passage, esp. between rows of seats. **2 a** an opening in the bulwarks by which a ship is entered or left. **b** a bridge laid from ship to shore. **c** a passage on a ship, esp. a platform connecting the quarterdeck and forecastle. **3** a temporary bridge on a building site etc. ● *int.* make way!

ganister /'gænɪstə(r)/ *n.* a close-grained hard siliceous stone found in the coal measures of northern England, and used for furnace-linings. [19th c.: orig. unkn.]

ganja /'gændʒə/ *n.* marijuana. [Hindi *gānjhā*]

gannet /'gænɪt/ *n.* **1** a large mainly white seabird of the genus *Sula*, catching fish by plunge-diving; esp. *Sula bassana*, the North Atlantic (also called *solan goose*). **2** *sl.* a greedy person. □ **gannetry** *n.* (*pl.* **-ies**) [OE *ganot* f. Gmc, rel. to GANDER]

ganoid /'gænɔɪd/ *adj. & n.* ● *adj.* **1** (of fish scales) enamelled; smooth and bright. **2** having ganoid scales. ● *n.* a fish having ganoid scales. [F *ganoïde* f. Gk *ganos* brightness]

Gansu /gænˈsuː/ (also **Kansu** /kænˈ-/) a province of NW central China, between Mongolia and Tibet; capital, Lanzhou. This narrow, mountainous province, traversed by the valleys of the upper Yellow River, forms a corridor through which passed, in ancient times, the trade route to the west known as the Silk Road.

gantlet *US* var. of GAUNTLET[2].

gantry /'gæntrɪ/ *n.* (*pl.* **-ies**) **1** an overhead structure with a platform supporting a travelling crane, or railway or road signals. **2** a structure supporting a space rocket prior to launching. **3** (also **gauntry** /'gɔːntrɪ/) a wooden stand for barrels. [prob. f. *gawn*, dial. form of GALLON + TREE]

Ganymede /'gænɪˌmiːd/ **1** *Gk Mythol.* a Trojan youth who was so beautiful that he was carried off (in one version, by an eagle) to be Zeus' cup-bearer. **2** *Astron.* satellite III of Jupiter, the seventh closest to the planet, and one of the Galilean moons. With a diameter of 5,262 km it is the largest satellite in the solar system. The surface is a mixture of dark cratered areas and younger, brighter areas of parallel linear grooves.

gaol *Brit.* var. of JAIL.

gaoler *Brit.* var. of JAILER.

gap /gæp/ *n.* **1** an unfilled space or interval; a blank; a break in continuity. **2** a breach in a hedge, fence, or wall. **3** a wide (usu. undesirable) divergence in views, sympathies, development, etc. (*generation gap*). **4** a gorge or pass. □ **fill** (or **close** etc.) **a gap** make up a deficiency. **gap-toothed** having gaps between the teeth. □ **gapped** *adj.* **gappy** *adj.* [ME f. ON, = chasm, rel. to GAPE]

gape /geɪp/ *v. & n.* ● *v.intr.* **1 a** open one's mouth wide, esp. in amazement or wonder. **b** be or become wide open. **2** (foll. by *at*) gaze curiously or wondrously. **3** split; part asunder. **4** yawn. ● *n.* **1** an open-mouthed stare. **2** a yawn. **3** (in *pl.*; prec. by *the*) **a** a disease of birds with gaping as a symptom, caused by infestation with gapeworm. **b** *joc.* a fit of yawning. **4 a** an expanse of open mouth or beak. **b** the part of a beak that opens. **5** a rent or opening. □ **gapingly** *adv.* [ME f. ON *gapa*]

gaper /'geɪpə(r)/ *n.* **1** a bivalve mollusc of the genus *Mya*, with the shell open at one or both ends. **2** the comber fish, which gapes when dead. **3** a person who gapes.

gapeworm /'geɪpwɜːm/ *n.* a nematode worm of the family Syngamidae, infesting the trachea and bronchi of birds and causing the gapes.

gar /gɑː(r)/ *n.* = GARFISH 2.

garage /'gærɑːdʒ, -rɑːʒ, -rɪdʒ/ *n. & v.* ● *n.* **1** a building or shed for the storage of a motor vehicle or vehicles. **2** an establishment which sells petrol etc., or repairs and sells motor vehicles, or does both. **3** esp. *US* **a** the style of rock music played by garage bands. **b** a variety of house music influenced by soul and gospel. ● *v.tr.* put or keep (a motor vehicle) in a garage. □ **garage band** esp. *US* a rock group whose loud, energetic, but unpolished performance style is reminiscent of an amateur band rehearsing in a garage. **garage sale** *US* a sale of miscellaneous household goods, usu. for charity, held in the garage of a private house. [F f. *garer* shelter; sense 3b from Paradise *Garage*, a Manhattan dance club]

garam masala /ˈɡʌrəm/ *n.* a spice mixture used in Indian cookery. [Urdu *garam maṣālāḥ*: see MASALA]

garb /ɡɑːb/ *n. & v.* ● *n.* **1** clothing, esp. of a distinctive kind. **2** the way a person is dressed. ● *v.tr.* **1** (usu. in *passive* or *refl.*) put (esp. distinctive) clothes on (a person). **2** attire. [obs. F *garbe* f. It. *garbo* f. Gmc, rel. to GEAR]

garbage /ˈɡɑːbɪdʒ/ *n.* **1 a** refuse, filth. **b** domestic waste. **2** foul or rubbishy literature etc. **3 a** nonsense. **b** *Computing* incorrect or useless data (*garbage in, garbage out*). □ **garbage can** *N. Amer.* a dustbin. [AF: orig. unkn.]

garble /ˈɡɑːb(ə)l/ *v.tr.* **1** unintentionally distort or confuse (facts, messages, etc.). **2 a** mutilate in order to misrepresent. **b** make (usu. unfair or malicious) selections from (facts, statements, etc.). □ **garbler** *n.* [It. *garbellare* f. Arab. *ġarbala* sift, perh. f. LL *cribellare* to sieve f. L *cribrum* sieve]

Garbo /ˈɡɑːbəʊ/, Greta (born Greta Gustafsson) (1905–90), Swedish-born American actress. In 1924 her first important Swedish film led to a Hollywood contract; she gained instant recognition for her compelling screen presence and enigmatic beauty in *The Torrent* (1925). She made the transition from silent pictures to sound in *Anna Christie* (1930), later starring in *Mata Hari* (1931) and *Anna Karenina* (1935). She also starred in two comedies; the second was less successful than the first and in 1941 she retired and lived as a recluse for the rest of her life.

garboard /ˈɡɑːbəd/ *n.* (in full **garboard strake**) *Naut.* the first range of planks or plates laid on a ship's bottom next to the keel. [Du. *gaarboord*, perh. f. *garen* GATHER + *boord* BOARD]

garbologist /ɡɑːˈbɒlədʒɪst/ *n.* **1** a person who examines discarded refuse in order to gain information about a society or culture. **2** *joc.* a dustman, a refuse collector. □ **garbology** *n.* [f. GARBAGE]

García Lorca see LORCA.

García Márquez /ɡɑːˈsiːə ˈmɑːkes/, Gabriel (b.1928), Colombian novelist. His left-wing sympathies brought him into conflict with the Colombian government and he spent the 1960s and 1970s in voluntary exile in Mexico and Spain. During this time he wrote *One Hundred Years of Solitude* (1967), which has come to be regarded as a classic example of magic realism. More recent novels include *The General in His Labyrinth* (1990). He was awarded the Nobel Prize for literature in 1982 and was formally invited back to Colombia, where he has since lived.

garçon /ˈɡɑːsɒn/ *n.* a waiter in a French restaurant, hotel, etc. [F, lit. 'boy']

Garda /ˈɡɑːdə/ *n.* **1** the state police force of the Republic of Ireland. **2** (also **garda**) (*pl.* **-dai** /-diː/) a member of this. [Ir. *Garda Síochána* Civic Guard]

Garda, Lake /ˈɡɑːdə/ a lake in NE Italy, lying between Lombardy and Venetia.

garden /ˈɡɑːd(ə)n/ *n. & v.* ● *n.* **1** esp. *Brit.* a piece of ground, usu. partly grassed and adjoining a private house, used for growing flowers, fruit, or vegetables, and as a place of recreation. **2** (esp. in *pl.*) ornamental grounds laid out for public enjoyment (*botanical gardens*). **3** a similar place with the service of refreshments (*tea garden*). **4** (*attrib.*) **a** (of plants) cultivated, not wild. **b** for use in a garden (*garden seat*). **5** (usu. in *pl.* prec. by a name) *Brit.* a street, square, etc. (*Onslow Gardens*). **6** an especially fertile region. **7** *US* a large public hall. **8** (**the Garden**) the philosophy or school of Epicurus. ● *v.intr.* cultivate or work in a garden. □ **garden centre** an establishment where plants and garden equipment etc. are sold. **garden cress** a hot-tasting cruciferous plant, *Lepidium sativum*, used in salads. **garden party** a social event held on a lawn or in a garden. **garden suburb** *Brit.* a suburb laid out spaciously with open spaces, parks, etc. **garden warbler** a greyish-brown Eurasian warbler, *Sylvia borin*. □ **gardening** *n.* **gardenesque** /ˌɡɑːdəˈnesk/ *adj.* [ME f. ONF *gardin* (OF *jardin*) ult. f. Gmc: cf. YARD[2]]

garden city *n.* a town of moderate size laid out systematically with wide streets, public parks and gardens, and rural surroundings. The first garden city in the UK was Letchworth in Hertfordshire (begun 1903), followed by nearby Welwyn Garden City (1919).

gardener /ˈɡɑːdnə(r)/ *n.* a person who gardens or is employed to tend a garden. □ **gardener-bird** a bowerbird that makes a 'garden' of moss etc. in front of a bower. [ME ult. f. OF *jardinier* (as GARDEN)]

gardenia /ɡɑːˈdiːnɪə/ *n.* a tree or shrub of the genus *Gardenia*, with large white or yellow flowers and usu. a fragrant scent. [mod.L f. Dr Alexander *Garden*, Sc. naturalist (1730–91)]

Garden of Eden see EDEN[1].

Garden of Gethsemane see GETHSEMANE, GARDEN OF.

Gardner /ˈɡɑːdnə(r)/, Erle Stanley (1899–1970), American novelist and short-story writer. He practised as a defence lawyer (1922–38) and went on to become famous for his series of novels featuring the lawyer-detective Perry Mason, many of which end with a dramatic courtroom scene.

Garfield /ˈɡɑːfiːld/, James A(bram) (1831–81), American Republican statesman, 20th President of the US Mar.–Sept. 1881. A major-general who had fought for the Union side in the American Civil War, he resigned his command to enter Congress, where he served as leader of the Republican Party (1863–80). He was assassinated within months of taking presidential office.

garfish /ˈɡɑːfɪʃ/ *n.* (*pl.* same) **1** a mainly marine fish of the family Belonidae, having long beaklike jaws with sharp teeth, esp. *B. belone*. Also called *needlefish*. **2** *US* a similar freshwater fish of the genus *Lepisosteus*, with ganoid scales. Also called *gar* or *garpike*. **3** *NZ & Austral.* a half-beak. [app. f. OE *gār* spear + *fisc* FISH[1]]

garganey /ˈɡɑːɡənɪ/ *n.* (*pl.* **-eys**) a small duck, *Anas querquedula*, the drake of which has a white stripe from the eye to the neck. [It., dial. var. of *garganello*]

gargantuan /ɡɑːˈɡæntjʊən/ *adj.* enormous, gigantic. [the name of a giant in Rabelais' book *Gargantua* (1534)]

garget /ˈɡɑːɡɪt/ *n.* **1** inflammation of a cow's or ewe's udder. **2** *US* pokeweed. [perh. f. obs. *garget* throat f. OF *gargate*, *-guete*]

gargle /ˈɡɑːɡ(ə)l/ *v. & n.* ● *v.* **1** *tr.* (also *absol.*) wash (one's mouth and throat), esp. for medicinal purposes, with a liquid kept in motion by breathing through it. **2** *intr.* make a sound as when doing this. ● *n.* **1** a liquid used for gargling. **2** *sl.* an alcoholic drink. [F *gargouiller* f. *gargouille*: see GARGOYLE]

gargoyle /ˈɡɑːɡɔɪl/ *n.* a grotesque carved human or animal face or figure projecting from the gutter of (esp. a Gothic) building usu. as a spout to carry water clear of a wall. [OF *gargouille* throat, gargoyle]

gargoylism /ˈɡɑːɡɔɪˌlɪz(ə)m/ *n. Med.* = HURLER'S SYNDROME.

Garibaldi /ˌɡærɪˈbɔːldɪ/, Giuseppe (1807–82), Italian patriot and military leader. He was a hero of the Risorgimento, who began his political activity as a member of the Young Italy Movement. After involvement in the early struggles against Austrian rule in the 1830s and 1840s he commanded a volunteer force on the Sardinian side in 1859, and successfully led his 'Red Shirts' to victory in Sicily and southern Italy in 1860–1, thus playing a vital part in the establishment of a united kingdom of Italy. He was less successful in his attempts to conquer the papal territories around French-held Rome in 1862 and 1867.

garibaldi /ˌɡærɪˈbɔːldɪ/ *n.* (*pl.* **garibaldis**) **1** *hist.* a kind of loose blouse worn by women and children, originally bright red, imitating the shirts worn by Garibaldi and his followers. **2** *Brit.* a biscuit containing a layer of currants. **3** a small red Californian damselfish, *Hypsypops rubicundus*. [GARIBALDI]

garish /ˈɡeərɪʃ/ *adj.* **1** obtrusively bright; showy. **2** gaudy; over-decorated. □ **garishly** *adv.* **garishness** *n.* [16th-c. *gaurish* app. f. obs. *gaure* stare]

Garland /ˈɡɑːlənd/, Judy (born Frances Gumm) (1922–69), American singer and actress. The daughter of vaudeville entertainers, she became a child star and was under contract to MGM at the age of 13. Her most famous early film role was as Dorothy in *The Wizard of Oz* (1939), in which she sang 'Over the Rainbow'. Later successful films included *Meet Me in St Louis* (1944) and *A Star is Born* (1954). She apparently died of a drug overdose after suffering many personal problems. Among her children is the actress Liza Minelli (b.1946), her daughter from her marriage to the film director Vincente Minelli (1910–86).

garland /ˈɡɑːlənd/ *n. & v.* ● *n.* **1** a wreath of flowers, leaves, etc., worn on the head or hung as a decoration. **2** a prize or distinction. **3** a literary anthology or miscellany. ● *v.tr.* **1** adorn with garlands. **2** crown with a garland. [ME f. OF *garlande*, of unkn. orig.]

garlic /ˈɡɑːlɪk/ *n.* **1** a strong-smelling allium plant, esp. *Allium sativum*. (See also RAMSONS.) **2** the pungent-tasting bulb of *A. sativum*, used as a flavouring in cookery. □ **garlicky** adj. [OE gārleac f. gār spear + lēac LEEK]

garment /'gɑːmənt/ n. & v. ● n. **1 a** an article of dress. **b** (in pl.) clothes. **2** the outward and visible covering of anything. ● v.tr. (usu. in passive) rhet. attire. [ME f. OF garnement (as GARNISH)]

garner /'gɑːnə(r)/ v. & n. ● v.tr. **1** collect. **2** store, deposit. ● n. literary a storehouse or granary. [ME (orig. as noun) f. OF gernier f. L granarium GRANARY]

garnet /'gɑːnɪt/ n. a vitreous silicate mineral, esp. a transparent deep-red kind used as a gem. [ME f. OF grenat f. med.L granatum POMEGRANATE, from its resemblance to the pulp of the fruit]

garnish /'gɑːnɪʃ/ v. & n. ● v.tr. **1** decorate or embellish (esp. food). **2** Law **a** serve notice on (a person) for the purpose of legally seizing money belonging to a debtor or defendant. **b** summon (a person) as a party to litigation started between others. ● n. (also **garnishing**) a decoration or embellishment, esp. to food. □ **garnishment** n. (in sense 2). [ME f. OF garnir f. Gmc]

garnishee /ˌgɑːnɪˈʃiː/ n. & v. Law ● n. a person garnished. ● v.tr. (**garnishees, garnisheed**) **1** garnish (a person). **2** attach (money etc.) by way of garnishment.

garniture /'gɑːnɪtʃə(r)/ n. **1** decoration or trimmings, esp. of food. **2** accessories, appurtenances. [F (as GARNISH)]

Garonne /gæˈrɒn/ a river of SW France, which rises in the Pyrenees and flows 645 km (400 miles) north-west through Toulouse and Bordeaux to join the Dordogne at the Gironde estuary.

garotte var. of GARROTTE.

Garoua /gæˈruːə/ a river port in northern Cameroon, on the River Bénoué; pop. (1981) 77,850.

garpike /'gɑːpaɪk/ n. = GARFISH 2. [OE gār spear + PIKE[1]]

garret /'gærɪt/ n. **1** a top-floor or attic room, esp. a dismal one. **2** an attic. [ME f. OF garite watch-tower f. Gmc]

Garrick /'gærɪk/, David (1717–79), English actor, manager, and dramatist. His style of acting was characterized by an easy, natural manner of speech, and he was equally successful in tragic and comic roles in both Shakespearian and contemporary plays. In 1747 he became involved in the management of the Drury Lane Theatre and later became its sole manager.

garrison /'gærɪs(ə)n/ n. & v. ● n. **1** the troops stationed in a fortress, town, etc., to defend it. **2** the building occupied by them. ● v.tr. **1** provide (a place) with or occupy as a garrison. **2** place on garrison duty. □ **garrison town** a town having a permanent garrison. [ME f. OF garison f. garir defend, furnish f. Gmc]

garrotte /gəˈrɒt/ v. & n. (also **garotte**; US **garrote**) ● v.tr. **1** execute or kill by strangulation, esp. with an iron collar or with a length of wire etc. **2** throttle in order to rob. ● n. **1 a** a Spanish method of execution by garrotting. **b** the apparatus used for this. **2** hist. highway robbery in which the victim is throttled. [F garrotter or Sp. garrotear f. garrote a cudgel, of unkn. orig.]

garrulous /'gærʊləs/ adj. **1** talkative, esp. on trivial matters. **2** loquacious, wordy. □ **garrulously** adv. **garrulousness** n. **garrulity** /gəˈruːlɪtɪ/ n. [L garrulus f. garrire chatter]

garter /'gɑːtə(r)/ n. & v. ● n. **1** a band worn to keep a sock or stocking up. **2 (the Garter)** Brit. **a** the highest order of English knighthood. (See note below.) **b** the badge of this. **c** membership of this. **3** N. Amer. a suspender for a sock or stocking. ● v.tr. fasten (a stocking) or encircle (a leg) with a garter. □ **garter-belt** N. Amer. a suspender belt. **garter snake** a harmless semiaquatic snake of the genus Thamnophis, native to North America, frequently having lengthwise stripes. **garter stitch** a plain knitting stitch or pattern, forming ridges in alternate rows. [ME f. OF gartier f. garet bend of the knee]

■ The Order of the Garter was founded by Edward III c.1344. The traditional story of the order's founding is that the garter was that of the Countess of Salisbury, which the king placed on his own leg after it fell off while she was dancing with him. The king's comment to those present, 'Honi soit qui mal y pense' (shame be to him who thinks evil of it), was adopted as the motto of the order, which became the highest in English knighthood and the model for others founded by other late medieval kings. The badge of the order is a garter of dark blue velvet bearing the motto embroidered in gold; garters also form part of the ornament of the collar worn by members.

Garter King of Arms see KING OF ARMS.

garth /gɑːθ/ n. Brit. **1** an open space within cloisters. **2** archaic **a** a close or yard. **b** a garden or paddock. [ME f. ON garthr = OE geard YARD[2]]

Garuda /gæˈruːdə/ Hinduism an eagle-like being that serves as the mount of the god Vishnu.

Garvey /'gɑːvɪ/, Marcus (Mosiah) (1887–1940), Jamaican political activist and black nationalist leader. He was the leader of the Back to Africa Movement, which advocated the establishment of an African homeland for black Americans, and founder of the Universal Negro Improvement Association (1914). He was chiefly active in the US, attracting a large following in support of his calls for black civil rights and economic independence. He died in obscurity, however, after his movement lost support during the Depression, but his thinking was later an important influence in the growth of Rastafarianism.

Gary /'gærɪ/ an industrial city in NW Indiana, on Lake Michigan south-east of Chicago; pop. (1991) 116,600.

gas /gæs/ n. & v. ● n. (pl. **gases**) **1 a** a fluid which is compressible and expands to fill any space in which it is enclosed; a substance which normally exists in this state. (See note below.) **b** a gas which at a given temperature cannot be liquefied by pressure alone (other gases generally being known as 'vapours' in this context). **2 a** such a substance (esp. found naturally or extracted from coal) used as a domestic or industrial fuel (also attrib.: gas cooker; gas fire). (See note below.) **b** an explosive mixture of firedamp with air. **3** nitrous oxide or another gas used as an anaesthetic (esp. in dentistry). **4** a gas or vapour used as a poisonous agent to disable an enemy in warfare. **5** N. Amer. colloq. petrol, gasoline. **6** sl. pointless idle talk; boasting. **7** sl. an enjoyable, attractive, or amusing thing or person. ● v. (**gases, gassed, gassing**) **1** tr. expose to gas, esp. to kill or make unconscious. **2** intr. give off gas. **3** tr. (usu. foll. by up) N. Amer. colloq. fill (the tank of a motor vehicle) with petrol. **4** intr. colloq. talk idly or boastfully. □ **gas chamber** an airtight chamber that can be filled with poisonous gas to kill people or animals. **gas chromatography** Chem. chromatography employing gas as the eluent. **gas-cooled** (of a nuclear reactor etc.) cooled by a current of gas. **gas field** an area yielding natural gas. **gas fire** a domestic fire using gas as its fuel. **gas-fired** (of a power station etc.) using gas as its fuel. **gas gangrene** a rapidly spreading gangrene of injured tissue infected by a soil bacterium and accompanied by the evolution of foul-smelling gas. **gas giant** Astron. a large planet of relatively low density consisting predominantly of hydrogen and helium, as Jupiter, Saturn, Uranus, and Neptune. **gas–liquid chromatography** (abbr. **GLC**) Chem. gas chromatography in which the stationary phase is liquid. **gas mask** a respirator used as a defence against poison gas. **gas meter** an apparatus recording the amount of gas consumed. **gas oil** a type of fuel oil distilled from petroleum and heavier than paraffin oil. **gas-permeable** (of a contact lens) allowing the diffusion of gases into and out of the cornea. **gas plant** Bot. = FRAXINELLA. **gas ring** a hollow ring perforated with gas jets, used esp. for cooking. **gas station** N. Amer. a filling-station. **gas-tight** proof against the leakage of gas. [invented by J. B. van Helmont, after Gk khaos chaos]

■ Different 'airs', vapours, and exhalations were recognized in antiquity and classified with the element 'air', distinguishing them from the other Aristotelian elements, earth, water, and fire. It was not until the second half of the 18th century that different chemical species of gases were clearly distinguished and identified, and recognized as elements in their own right. During the same period the four Aristotelian elements were definitively replaced by three states of matter — solid, liquid, and gas. In gases, the atoms or molecules are not bound together and can move about relatively independently. (See also KINETIC THEORY.)

■ The technical founder of the modern gas industry was William Murdock (1754–1839), a Scotsman, who prepared coal gas, and in 1792 succeeded in lighting his office with it. In 1807 a street in London was lit with gas, and gas soon began to be used for cooking and heating as well. In the second half of the 20th century coal gas has tended to be replaced as a fuel by natural gas, large deposits of which are found under the North Sea and in Russia, the US, Algeria, Iran, the Netherlands, Saudi Arabia, and Kuwait.

gasbag /'gæsbæg/ n. **1** a container of gas, esp. for holding the gas for a balloon or airship. **2** sl. an idle talker.

Gascogne see GASCONY.

Gascon /'gæskən/ n. & adj. ● n. **1** a native of Gascony. **2** the dialect of Gascony. **3 (gascon)** a braggart. ● adj. of or relating to Gascony or its people or dialect. [F f. L Vasco -onis]

Gascony /'gæskənɪ/ (French **Gascogne** /gaskɔɲ/) a region and former province of SW France, in the northern foothills of the Pyrenees. Having united with Aquitaine in the 11th century, it was held by England between 1154 and 1453.

gaseous /ˈgæsɪəs, ˈgeɪs-/ *adj.* of or like gas. □ **gaseousness** *n.*

gash[1] /gæʃ/ *n. & v.* ● *n.* **1** a long and deep slash, cut, or wound. **2 a** a cleft such as might be made by a slashing cut. **b** the act of making such a cut. ● *v.tr.* make a gash in; cut. [var. of ME *garse* f. OF *garcer* scarify, perh. ult. f. Gk *kharassō*]

gash[2] /gæʃ/ *n. & adj. Brit. sl.* ● *n.* rubbish, waste. ● *adj.* spare, extra. [20th-c. Naut. sl.: orig. unkn.]

gasholder /ˈgæsˌhəʊldə(r)/ *n.* a large receptacle for storing gas; a gasometer.

gasify /ˈgæsɪˌfaɪ/ *v.tr. & intr.* (**-ies, -ied**) convert or be converted into gas. □ **gasification** /ˌgæsɪfɪˈkeɪʃ(ə)n/ *n.*

Gaskell /ˈgæsk(ə)l/, Mrs Elizabeth (Cleghorn) (1810–65), English novelist. An active humanitarian from a Unitarian background, she is famous for *Mary Barton* (1848), *Cranford* (1853), and *North and South* (1855); all of these display her interest in social concerns. She also wrote a biography (1857) of her friend Charlotte Brontë.

gasket /ˈgæskɪt/ *n.* **1** a sheet or ring of rubber etc., shaped to seal the junction of metal surfaces. **2** *Naut.* a small cord securing a furled sail to a yard. □ **blow a gasket** *sl.* lose one's temper. [perh. f. F *garcette* thin rope (orig. little girl)]

gaskin /ˈgæskɪn/ *n.* the hinder part of a horse's thigh. [perh. erron. f. GALLIGASKINS]

gaslight /ˈgæslaɪt/ *n.* **1** a jet of burning gas, usu. heating a mantle, to provide light. **2** light emanating from this. □ **gaslit** /-lɪt/ *adj.*

gasman /ˈgæsmən/ *n.* (*pl.* **-men**) a man who installs or services gas appliances, or reads gas meters.

gasohol /ˈgæsəˌhɒl/ *n.* a mixture of petrol and ethyl alcohol used as fuel. [GAS + ALCOHOL]

gasoline /ˈgæsəˌliːn/ *n.* (also **gasolene**) **1** a volatile inflammable liquid distilled from petroleum and used for heating and lighting. **2** *N. Amer.* petrol. [GAS + -OL + -INE[4], -ENE]

gasometer /gæˈsɒmɪtə(r)/ *n.* a large tank in which gas is stored for distribution by pipes to users. [F *gazomètre* f. *gaz* gas + *-mètre* -METER]

gasp /gɑːsp/ *v. & n.* ● *v.* **1** *intr.* catch one's breath with an open mouth as in exhaustion or astonishment. **2** *intr.* (foll. by *for*) strain to obtain by gasping (*gasped for air*). **3** *tr.* (often foll. by *out*) utter with gasps. ● *n.* a convulsive catching of breath. □ **at one's last gasp 1** at the point of death. **2** exhausted. [ME f. ON *geispa*: cf. *geip* idle talk]

gasper /ˈgɑːspə(r)/ *n.* **1** a person who gasps. **2** *Brit. sl.* a cigarette.

Gaspra /ˈgæsprə/ *Astron.* asteroid 951, photographed closely by the Galileo spacecraft in 1991. It is roughly triangular, 19 × 12 × 11 km in size, and has numerous small craters as well as some linear features that may result from earlier fractures. [Crimean resort and spa where Tolstoy was treated]

Gassendi /gæˈsendi/, Pierre (1592–1655), French astronomer and philosopher. He is best known for his atomic theory of matter, which was based on his interpretation of the works of Epicurus, and he was an outspoken critic of Aristotle. He observed a new comet, a lunar eclipse, and a transit of Mercury (confirming Kepler's theories), and he coined the term *aurora borealis*.

Gasser /ˈgæsə(r)/, Herbert Spencer (1888–1963), American physiologist. Collaborating with Joseph Erlanger, he used an oscilloscope to show that the velocity of a nerve impulse is proportional to the diameter of the fibre. He also demonstrated the differences between sensory and motor nerves. Gasser and Erlanger shared a Nobel Prize in 1944.

gasser /ˈgæsə(r)/ *n.* **1** *colloq.* an idle talker. **2** *sl.* a very attractive or impressive person or thing.

gassy /ˈgæsɪ/ *adj.* (**gassier, gassiest**) **1 a** of or like gas. **b** full of gas. **2** *colloq.* (of talk etc.) pointless, verbose. □ **gassiness** *n.*

Gastarbeiter /ˈgɑːstɑːˌbaɪtə(r)/ *n.* (*pl.* **Gastarbeiters** or same) a person with temporary permission to work in another country (esp. in western Europe). [G f. *Gast* GUEST + *Arbeiter* worker]

gasteropod var. of GASTROPOD.

Gasthaus /ˈgæsthaʊs/ *n.* a small inn or hotel in German-speaking countries. [G f. *Gast* GUEST + *Haus* HOUSE]

Gasthof /ˈgæsthɒf/ *n.* (*pl.* **Gasthofs** or **Gasthöfe** /-ˌhɜːfə/) a hotel in German-speaking countries, usu. larger than a *gasthaus*. [G f. *Gast* GUEST + *Hof* hotel, large house]

gastrectomy /gæˈstrektəmɪ/ *n.* (*pl.* **-ies**) the surgical removal of the whole or a part of the stomach. [GASTRO- + -ECTOMY]

gastric /ˈgæstrɪk/ *adj.* of the stomach. □ **gastric flu** a popular name

for an intestinal disorder of unknown cause. **gastric juice** a thin clear virtually colourless acid fluid secreted by the stomach glands and active in promoting digestion. [mod.L *gastricus* f. Gk *gastēr gast(e)ros* stomach]

gastritis /gæˈstraɪtɪs/ *n. Med.* inflammation of the lining of the stomach.

gastro- /ˈgæstrəʊ/ *comb. form* (also **gastr-** before a vowel) stomach. [Gk *gastēr gast(e)ros* stomach]

gastro-enteric /ˌgæstrəʊenˈterɪk/ *adj. Med.* of or relating to the stomach and intestines.

gastro-enteritis /ˌgæstrəʊˌentəˈraɪtɪs/ *n. Med.* inflammation of the stomach and intestines.

gastroenterology /ˌgæstrəʊˌentəˈrɒlədʒɪ/ *n.* the branch of medicine which deals with disorders of the stomach and intestines. □ **gastroenterologist** *n.* **gastroenterological** /-rəˈlɒdʒɪk(ə)l/ *adj.* [GASTRO- + ENTERO- + -LOGY]

gastrointestinal /ˌgæstrəʊˌɪnteˈstaɪn(ə)l, -ɪnˈtestɪn(ə)l/ *adj. Med.* of or relating to the stomach and the intestines.

gastrolith /ˈgæstrəʊˌlɪθ/ *n.* **1** *Zool.* a small stone swallowed by a bird, reptile, or fish, to aid digestion in the gizzard. **2** *Med.* a hard concretion in the stomach. [GASTRO- + -LITH]

gastronome /ˈgæstrəˌnəʊm/ *n.* a gourmet. [F f. *gastronomie* GASTRONOMY]

gastronomy /gæˈstrɒnəmɪ/ *n.* the practice, study, or art of eating and drinking well. □ **gastronomic** /ˌgæstrəˈnɒmɪk/ *adj.* **gastronomical** *adj.* **gastronomically** *adv.* [F *gastronomie* f. Gk *gastronomia* (as GASTRO-, *-nomia* f. *nomos* law)]

gastropod /ˈgæstrəˌpɒd/ *n. & adj.* (also **gasteropod** /ˈgæstərə-/) *Zool.* ● *n.* a mollusc of the class Gastropoda, moving along by means of a large muscular foot. (See note below.) ● *adj.* of or relating to this class. □ **gastropodous** /gæˈstrɒpədəs/ *adj.* [F *gastéropode* f. mod.L *gasteropoda* (as GASTRO-, Gk *pous podos* foot)]

▪ The gastropods are the only molluscs to have conquered terrestrial as well as aquatic habitats. They have a definite head with stalked eyes, and most forms have a single spirally coiled shell. The body is asymmetrical and usually twisted round inside the shell. The gastropods include snails, slugs, whelks, conches, cowries, etc. Numerous fossil forms are known only from their shells.

gastroscope /ˈgæstrəˌskəʊp/ *n.* an optical instrument used for inspecting the interior of the stomach.

gastrula /ˈgæstrʊlə/ *n.* (*pl.* **gastrulae** /-ˌliː/) *Zool.* an embryonic stage developing from the blastula. [mod.L f. Gk *gastēr gast(e)ros* belly]

gas turbine *n.* a turbine driven by a flow of expanding hot gases following combustion. Modern gas turbines date from the 1930s, when they were independently developed for electric power generation and, in the form of the jet engine, for aircraft propulsion. Gas turbines are a form of internal-combustion engine in which air is compressed, heated by means of fuel sprayed into a combustion chamber, then expanded in a turbine to produce power, some of which is used to drive the compressor. The output may be used to drive a shaft or generator; in a jet engine, thrust is obtained from the expanding exhaust gases.

gasworks /ˈgæswɜːks/ *n.pl.* a place where gas is manufactured and processed.

gat[1] /gæt/ *n. sl.* a revolver or other firearm. [abbr. of GATLING]

gat[2] /gæt/ *archaic past of* GET *v.*

gate[1] /geɪt/ *n. & v.* ● *n.* **1** a barrier, usu. hinged, used to close an opening made for entrance and exit through a wall, fence, etc. **2** such an opening, esp. in the wall of a city, enclosure, or large building. **3** a means of entrance or exit. **4** a numbered place of access to aircraft at an airport. **5** a mountain pass. **6** an arrangement of slots into which the gear lever of a motor vehicle moves to engage the required gear. **7** a device for holding the frame of a cine film momentarily in position behind the lens of a camera or projector. **8 a** an electrical signal that causes or controls the passage of other signals. **b** an electrical circuit with an output which depends on the combination of several inputs. **9** a device regulating the passage of water in a lock etc. **10 a** the number of people entering by payment at the gates of a sports ground etc. **b** (in full **gate-money**) the proceeds taken for admission. **11** *sl.* the mouth. **12** = *starting-gate.* ● *v.tr.* **1** *Brit.* confine to college or school entirely or after certain hours. **2** (as **gated** *adj.*) (of a road) having a gate or gates to control the movement of traffic or animals. □ **gate valve** a valve in which a sliding part controls the extent of the aperture. **get** (or **be given**) **the gate** *N. Amer. sl.* be dismissed. [OE *gæt, geat,* pl. *gatu,* f. Gmc]

gate[2] /geɪt/ *n.* (preceded or prefixed by a name) *Brit.* a street (*Westgate*). [ME f. ON *gata,* f. Gmc]

-gate /geɪt/ *suffix* forming nouns denoting an actual or alleged scandal comparable in some way to the Watergate scandal of 1972 (*Irangate*). [after WATERGATE]

gateau /'gætəʊ/ *n.* (pl. **gateaus** or **gateaux** /-təʊz/) any of various rich cakes, usu. containing cream or fruit. [F *gâteau* cake]

gatecrasher /'geɪtˌkræʃə(r)/ *n.* an uninvited guest at a party etc. □ **gatecrash** *v.tr. & intr.*

gatefold /'geɪtfəʊld/ *n.* a page in a book or magazine etc. that folds out to be larger than the page-format.

gatehouse /'geɪthaʊs/ *n.* **1** a house standing by a gateway, esp. to a large house or park. **2** *hist.* a room over a city gate, often used as a prison.

gatekeeper /'geɪtˌkiːpə(r)/ *n.* **1** an attendant at a gate, controlling entrance and exit. **2** a brown satyrid butterfly, *Pyronia tithonus,* frequenting hedgerows and woodland.

gateleg /'geɪtleg/ *n.* (in full **gateleg table**) a table with folding flaps supported by legs swung open like a gate. □ **gatelegged** *adj.*

gateman /'geɪtmən/ *n.* (pl. **-men**) = GATEKEEPER 1.

gatepost /'geɪtpəʊst/ *n.* a post on which a gate is hung or against which it shuts. □ **between you and me and the gatepost** in strict confidence.

Gates /geɪts/, William (Henry) ('Bill') (b.1955), American computer entrepreneur. In 1975 he co-founded Micro-Soft (later Microsoft), a private company for the manufacture and sale of computers. As its chairman and chief executive, Gates expanded the firm overseas in the early 1980s; by the end of the decade, Microsoft was a leading multinational computer company and had made Gates the youngest multi-billionaire in American history. In the early 1990s, he was the richest man in the US, with his personal assets valued at $6.5 billion.

Gateshead /'geɪtshed/ an industrial town in Tyne and Wear, NE England, on the south bank of the River Tyne opposite Newcastle; pop. (1991) 196,500.

gateway /'geɪtweɪ/ *n.* **1** an entrance with or opening for a gate. **2** a frame or structure built over a gate. **3** a means of access or entry (*gateway to Scotland; gateway to success*). **4** *Computing* a device used to connect two different networks.

gather /'gæðə(r)/ *v. & n.* ● *v.* **1** *tr. & intr.* bring or come together; assemble, accumulate. **2** *tr.* (usu. foll. by *up*) **a** bring together from scattered places or sources. **b** take up together from the ground, a surface, etc. **c** draw into a smaller compass; contract. **3** *tr.* acquire by gradually collecting; amass. **4** *tr.* **a** pick a quantity of (flowers etc.). **b** collect (grain etc.) as a harvest. **5** *tr.* (often foll. by *that* + clause) infer or understand. **6** *tr.* be subjected to or affected by the accumulation or increase of (*unread books gathering dust; gather speed; gather strength*). **7** *tr.* (often foll. by *up*) summon up (one's thoughts, energy, etc.) for a purpose. **8** *tr.* gain or recover (one's breath). **9** *tr.* **a** draw (material, or one's brow) together in folds or wrinkles. **b** pucker or draw together (part of a dress) by running a thread through. **10** *intr.* come to a head; develop a purulent swelling. ● *n.* (in *pl.*) a part of a garment that is gathered or drawn in. □ **gather way** (of a ship) begin to move. □ **gatherer** *n.* [OE *gaderian* f. WG]

gathering /'gæðərɪŋ/ *n.* **1** an assembly or meeting. **2** a purulent swelling. **3** a group of leaves taken together in bookbinding.

Gatling /'gætlɪŋ/ *n.* (also **Gatling gun**) a rapid-fire crank-driven gun with clustered barrels into which cartridges were automatically loaded. The first practical machine-gun, it was adopted by the US army in 1866, too late to be of significant use in the American Civil War. It was named after its inventor Richard Jordan Gatling (1818–1903).

GATT /gæt/ *abbr.* (also **Gatt**) General Agreement on Tariffs and Trade, a treaty to promote international trade and economic development by reducing tariffs and other restrictions. The treaty came into effect in 1948; over 100 countries eventually ratified it, while others applied its provisions de facto. The eighth round of GATT talks, begun in Uruguay in 1986, eventually led to an agreement (signed 1994) to set up the World Trade Organization to govern international trade. This was established on 1 Jan. 1995 as the successor to GATT.

Gatwick /'gætwɪk/ an international airport in SE England, to the south of London.

gauche /gəʊʃ/ *adj.* **1** lacking ease or grace; socially awkward. **2** tactless. □ **gauchely** *adv.* **gaucheness** *n.* [F, = left-handed, awkward]

gaucherie /'gəʊʃəˌriː/ *n.* **1** gauche manners. **2** a gauche action. [F]

gaucho /'gaʊtʃəʊ/ *n.* (pl. **-os**) a cowboy from the South American pampas. [Amer. Sp., prob. f. Araucanian *kauču*]

gaud /gɔːd/ *n.* **1** a gaudy thing; a showy ornament. **2** (in *pl.*) showy ceremonies. [perh. through AF f. OF *gaudir* rejoice f. L *gaudere*]

Gaudí /'gaʊdi/, Antonio (full surname Gaudí y Cornet) (1853–1926), Spanish architect. A leading but idiosyncratic exponent of art nouveau, he worked chiefly in Barcelona, designing distinctive buildings such as the Parc Güell (begun 1900) and the Casa Batlló (begun 1905), notable for their use of ceramics, wrought-iron work, flowing lines, and organic forms. He began work on his most ambitious project, the church of the Sagrada Familia, in 1884; unfinished at his death, it is still under construction.

Gaudier-Brzeska /ˌgəʊdɪˌeɪ'bʒeskə/, Henri (1891–1915), French sculptor. He settled in London in 1911 and became a leading member of the vorticist movement. His stylistically varied and advanced work, such as the faceted bust of Horace Brodzky (1912) and the semi-abstract *Bird Swallowing a Fish* (1913), only achieved wide recognition after his death.

gaudy[1] /'gɔːdɪ/ *adj.* (**gaudier, gaudiest**) tastelessly or extravagantly bright or showy. □ **gaudily** *adv.* **gaudiness** *n.* [prob. f. GAUD + -Y[1]]

gaudy[2] /'gɔːdɪ/ *n.* (pl. **-ies**) *Brit.* an annual feast or entertainment, esp. a college dinner for old members etc. [L *gaudium* joy or *gaude* imper. of *gaudere* rejoice]

gauge /geɪdʒ/ *n. & v.* (*US* **gage**: see also sense 7) ● *n.* **1** a standard measure to which certain things must conform, esp.: **a** the measure of the capacity or contents of a barrel. **b** the fineness of a textile. **c** the diameter of a bullet. **d** the thickness of sheet metal. **2** an instrument for measuring or determining this, or for measuring length, thickness, or other dimensions or properties. **3** the distance between a pair of rails or the wheels on one axle. **4** the capacity, extent, or scope of something. **5** a means of estimating; a criterion or test. **6** a graduated instrument measuring the force or quantity of rainfall, stream, tide, wind, etc. **7** (usu. **gage**) *Naut.* a relative position with respect to the wind. ● *v.tr.* **1** measure exactly (esp. objects of standard size). **2** determine the capacity or content of. **3** estimate or form a judgement of (a person, temperament, situation, etc.). **4** make uniform; bring to a standard size or shape. □ **gauge pressure** the amount by which a pressure exceeds that of the atmosphere. **gauge theory** *Physics* a form of quantum theory using mathematical functions to describe subatomic interactions in terms of particles not directly detectable. **take the gauge of** estimate. □ **gaugeable** *adj.* **gauger** *n.* [ME f. ONF *gauge, gauger,* of unkn. orig.]

Gauguin /'gəʊgæn/, (Eugène Henri) Paul (1848–1903), French painter. He left Paris in search of an environment that would bring him closer to nature, going first to Brittany, where he painted works such as *The Vision After the Sermon* (1888), and later briefly to stay with Van Gogh at Arles. In 1891 he left France for Tahiti; he spent most of the rest of his life there, painting enigmatic works such as *Faa Iheihe* (1898). He was a post-impressionist whose painting was influenced by primitive art, freeing colour from its representational function to use it in flat contrasting areas to achieve decorative or emotional effects. His work influenced both the Nabi Group and the symbolist movement.

Gauhati /gaʊ'hɑːtɪ/ an industrial city in NE India, in Assam, a river port on the Brahmaputra; pop. (1991) 578,000.

Gaul[1] /gɔːl/ an ancient region of Europe, corresponding to modern France, Belgium, the south Netherlands, SW Germany, and northern Italy. The area was settled by groups of Celts, who had begun migration across the Rhine in 900 BC, spreading further south beyond the Alps from 400 BC onwards and ousting the Etruscans. The area south of the Alps was conquered in 222 BC by the Romans, who called it *Cisalpine Gaul.* The area north of the Alps, known to the Romans as *Transalpine Gaul,* was taken by Julius Caesar between 58 and 51 BC, remaining under Roman rule until the 5th century AD. Within Transalpine Gaul the southern province, parts of which had fallen to the Romans in the previous century, became known as *Gallia Narbonensis.*

Gaul[2] /gɔːl/ *n.* a native or inhabitant of ancient Gaul. [GAUL[1]; F f. Gmc, = foreigners]

gauleiter /'gaʊˌlaɪtə(r)/ *n.* **1** *hist.* an official governing a district under Nazi rule. **2** a local or petty tyrant. [G f. *Gau* administrative district + *Leiter* leader]

Gaulish /'gɔːlɪʃ/ *adj. & n.* ● *adj.* of or relating to the ancient Gauls. ● *n.* the language of the ancient Gauls.

Gaulle, Charles de, see DE GAULLE.

Gaullism /'gəʊlɪz(ə)m/ n. **1** the principles and policies of Charles de Gaulle, characterized by their conservatism, nationalism, and advocacy of centralized government. **2** adherence to these. □ **Gaullist** n. [F *Gaullisme*]

gault /gɔːlt/ n. Geol. **1** a series of clay and marl beds between the upper and lower greensand in southern England. **2** clay obtained from these beds. [16th c.: orig. unkn.]

Gaunt[1] /gɔːnt/ a former name for GHENT.

Gaunt[2] /gɔːnt/ John of, see JOHN OF GAUNT.

gaunt /gɔːnt/ adj. **1** lean, haggard. **2** grim or desolate in appearance. □ **gauntly** adv. **gauntness** n. [ME: orig. unkn.]

gauntlet[1] /'gɔːntlɪt/ n. **1** a stout glove with a long loose wrist. **2** hist. an armoured glove. **3** the part of a glove covering the wrist. □ **pick up** (or **take up**) **the gauntlet** accept a challenge. **throw down the gauntlet** issue a challenge. [ME f. OF *gantelet* dimin. of *gant* glove f. Gmc]

gauntlet[2] /'gɔːntlɪt/ n. (US **gantlet** /'gænt-/) □ **run the gauntlet 1** be subjected to harsh criticism. **2** pass between two rows of people and receive blows from them, as a punishment or ordeal. [earlier *gantlope* f. Sw. *gatlopp* f. *gata* lane, *lopp* course, assim. to GAUNTLET[1]]

gauntry var. of GANTRY 3.

gaur /'gaʊə(r)/ n. a wild ox, *Bos gaurus*, found in forests from India to Malaysia. [Skr. *gaura*]

Gauss /gaʊs/, Karl Friedrich (1777–1855), German mathematician, astronomer, and physicist. Regarded as the 'prince of mathematics', he laid the foundations of number theory, and in 1801 he rediscovered the lost asteroid Ceres using advanced computational techniques. He contributed to many areas of mathematics, and applied rigorous mathematical analysis to such subjects as geometry, geodesy, electrostatics, and electromagnetism. He was involved in the first worldwide survey of the earth's magnetic field. Two of Gauss's most interesting discoveries, which he did not pursue, were non-Euclidean geometry and quaternions.

gauss /gaʊs/ n. (pl. same or **gausses**) Physics a unit of magnetic induction equal to 10^{-4} tesla (symbol: **G**). [GAUSS]

Gaussian distribution /'gaʊsɪən/ n. Statistics = normal distribution. [GAUSS]

Gautama /'gaʊtəmə/, Siddhartha, see BUDDHA. [Skr.]

gauze /gɔːz/ n. **1** a thin transparent fabric of silk, cotton, etc. **2** a fine mesh of wire etc. **3** a slight haze. [F *gaze* f. *Gaza* in Palestine]

gauzy /'gɔːzɪ/ adj. (**gauzier**, **gauziest**) **1** like gauze; thin and translucent. **2** flimsy, delicate. □ **gauzily** adv. **gauziness** n.

Gavaskar /gə'væskə(r)/, Sunil Manohar (b.1949), Indian cricketer. He made his Test début in the West Indies at the age of 20, making his mark by scoring an aggregate of 774 runs. He later captained India and achieved several world batting records, in 1987 becoming the first batsman to score 10,000 runs in Test cricket.

gave past of GIVE.

gavel /'gæv(ə)l/ n. & v. ● n. a small hammer used by an auctioneer, or for calling a meeting to order. ● v. (**gavelled**, **gavelling**; US **gaveled**, **gaveling**) **1** intr. use a gavel. **2** tr. (often foll. by *down*) end (a meeting) or dismiss (a speaker) by use of a gavel. [19th c.: orig. unkn.]

gavial var. of GHARIAL.

gavotte /gə'vɒt/ n. **1** a medium-paced dance popular in the 18th century. **2 a** a piece of music for this, composed in common time with each phrase beginning on the third beat of the bar. **b** a piece of music in this rhythm as a movement in a suite. [F f. Prov. *gavoto* f. *Gavot* native of a region in the Alps]

Gawain /'gɑːweɪn, gə'weɪn/ (in Arthurian legend) one of the knights of the Round Table who participated in the quest for the Holy Grail. He is the hero of the medieval poem *Sir Gawain and the Green Knight*.

Gawd /gɔːd/ n. & int. sl. God. [alt. of *God*]

gawk /gɔːk/ v. & n. ● v.intr. colloq. stare stupidly. ● n. an awkward or bashful person. □ **gawkish** adj. [rel. to obs. *gaw* gaze f. ON *gá* heed]

gawky /'gɔːkɪ/ adj. (**gawkier**, **gawkiest**) awkward or ungainly. □ **gawkily** adv. **gawkiness** n.

gawp /gɔːp/ v.intr. Brit. colloq. stare stupidly or obtrusively. □ **gawper** n. [earlier *gaup*, *galp* f. ME *galpen* yawn, rel. to YELP]

Gay /geɪ/, John (1685–1732), English poet and dramatist. He is now chiefly known for *The Beggar's Opera* (1728), a ballad opera combining burlesque and political satire and dealing with life in low society. A major success in its day, it has been revived several times this century, and was adapted by Brecht in *The Threepenny Opera* (1928).

gay /geɪ/ adj. & n. ● adj. (**gayer**, **gayest**) **1** light-hearted and carefree; mirthful. **2** characterized by cheerfulness or pleasure (*a gay life*). **3** brightly coloured; showy, brilliant (*a gay scarf*). **4** colloq. **a** homosexual. **b** intended for or used by homosexuals (*a gay bar*). ¶ Generally informal in use, but favoured by homosexuals with ref. to themselves. **5** colloq. dissolute, immoral. ● n. colloq. a homosexual, esp. male. □ **gay liberation 1** the liberation of homosexuals from social and legal discrimination. **2** (**Gay Liberation**) a movement campaigning for gay liberation. **gay plague** colloq. sometimes offens. Aids (so called because first identified amongst homosexual men). **gay rights** (often attrib.) equal rights for homosexuals. □ **gayness** n. [ME f. OF *gai*, of unkn. orig.]

Gaya /'gɑːjə/ a city in NE India, in the state of Bihar south of Patna; pop. (1991) 291,000. It is a place of Hindu pilgrimage.

gayal /gə'jæl/ n. a semi-domesticated ox, *Bos frontalis*, found in India and SE Asia. [Bengali]

gayety US var. of GAIETY.

Gay-Lussac /geɪ'luːsæk/, Joseph Louis (1778–1850), French chemist and physicist. He is best known for his work on gases, and in 1808 he formulated the law usually known by his name, that gases which combine chemically do so in volumes which are in a simple ratio to each other. He developed techniques of quantitative chemical analysis, confirmed that iodine was an element, discovered cyanogen, improved the process for manufacturing sulphuric acid, prepared potassium and boron, and made two balloon ascents to study the atmosphere and terrestrial magnetism.

gazania /gə'zeɪnɪə/ n. a herbaceous composite plant of the genus *Gazania*, with showy yellow or orange daisy-like flowers. [18th c.: f. Theodore of *Gaza*, Greek scholar (1398–1478)]

Gazankulu /ˌgæzən'kuːluː/ a former homeland established in South Africa for the Tsonga people, now part of the provinces of Northern and Eastern Transvaal.

Gaza Strip /'gɑːzə/ a strip of territory in Palestine, on the SE Mediterranean coast, including the town of Gaza; pop. (est. 1988) 564,000. Administered by Egypt from 1949, it was occupied by Israel from 1967 until the implementation of the PLO–Israeli accord in 1994. Under the terms of the agreement Israeli troops were withdrawn and the Gaza Strip became a self-governing enclave with partial autonomy, including its own police force.

gaze /geɪz/ v. & n. ● v.intr. (foll. by *at*, *into*, *on*, *upon*, etc.) look fixedly. ● n. a fixed or intent look. □ **gazer** n. [ME: orig. unkn.; cf. obs. *gaw* GAWK]

gazebo /gə'ziːbəʊ/ n. (pl. **-os** or **-oes**) a small building or structure such as a summer-house or turret, designed to give a wide view. [perh. joc. f. GAZE, in imitation of L futures in *-ēbo*: cf. LAVABO]

gazelle /gə'zel/ n. a small graceful African or Asian antelope, esp. of the genus *Gazella*. [F prob. f. Sp. *gacela* f. Arab. *ġazāl*]

gazette /gə'zet/ n. & v. ● n. **1** a newspaper, esp. the official one of an organization or institution (*University Gazette*). **2** hist. a news-sheet; a periodical publication giving current events. **3** Brit. an official journal with a list of government appointments, bankruptcies, and other public notices (*London Gazette*). ● v.tr. Brit. announce or publish in an official gazette. [F f. It. *gazzetta* f. *gazeta*, a Venetian small coin]

gazetteer /ˌgæzɪ'tɪə(r)/ n. a geographical index or dictionary. [earlier = journalist, for whom such an index was provided: f. F *gazettier* f. It. *gazzettiere* (as GAZETTE)]

Gaziantep /ˌgæzɪən'tep/ a city in southern Turkey, near the border with Syria; pop. (1990) 603,400. Until 1921 it was called Aintab.

gazpacho /gæz'pɑːtʃəʊ/ n. (pl. **-os**) a Spanish soup made with tomatoes, peppers, garlic, cucumber, etc., and served cold. [Sp.]

gazump /gə'zʌmp/ v.tr. (also absol.) Brit. colloq. **1** (of a seller) raise the price of a property after having accepted an offer by (an intending buyer). **2** swindle. □ **gazumper** n. [20th c.: orig. uncert.]

gazunder /gə'zʌndə(r)/ v.tr. (also absol.) Brit. colloq. (of a buyer) lower the amount of an offer made to (the seller) for a property, esp. just before exchange of contracts. [GAZUMP + UNDER]

GB abbr. Great Britain.

GBE abbr. (in the UK) Knight (or Dame) Grand Cross (of the Order of the British Empire.

GBH abbr. grievous bodily harm.

GC abbr. (in the UK) George Cross.

GCB *abbr.* (in the UK) Knight (or Dame) Grand Cross (of the Order) of the Bath.

GCE see GENERAL CERTIFICATE OF EDUCATION.

GCHQ *abbr.* (in the UK) Government Communications Headquarters.

GCMG *abbr.* (in the UK) Knight (or Dame) Grand Cross (of the Order) of St Michael and St George.

GCSE see GENERAL CERTIFICATE OF SECONDARY EDUCATION.

GCVO *abbr.* (in the UK) Knight (or Dame) Grand Cross of the Royal Victorian Order.

Gd *symb. Chem.* the element gadolinium.

Gdańsk /gdænsk/ (German **Danzig** /'dantsıç/) an industrial port and shipbuilding centre in northern Poland, on an inlet of the Baltic Sea; pop. (1990) 465,100. Originally a member of the Hanseatic League, it was disputed between Prussia and Poland during the 19th century. It was a free city under a League of Nations mandate from 1919 until 1939, when it was annexed by Nazi Germany, precipitating hostilities with Poland and the outbreak of the Second World War. It passed to Poland in 1945. In the 1980s the Gdańsk shipyards were the site of the activities of the Solidarity movement, which eventually led to the collapse of the Communist regime in Poland in 1989.

Gdn. *abbr.* Garden.

Gdns. *abbr.* Gardens.

GDP *abbr.* gross domestic product.

GDR *abbr. hist.* German Democratic Republic.

Gdynia /'gdɪnjə/ a port and naval base in northern Poland, on the Baltic Sea north-west of Gdańsk; pop. (1990) 251,500.

Ge[1] *symb. Chem.* the element germanium.

Ge[2] /geɪ/ *Gk Mythol.* = GAIA. [Gk, = earth]

gean /gi:n/ *n.* **1** the wild sweet cherry, *Prunus avium.* **2** the fruit of this. [OF *guine* (mod. *guigne*)]

gear /gɪə(r)/ *n. & v.* ● *n.* **1** (often in *pl.*) **a** a set of toothed wheels that work together to transmit and control motion from an engine, esp. to the road wheels of a vehicle. **b** a mechanism for doing this. **2** a particular function or state of adjustment of engaged gears (*low gear*; *second gear*). **3** a mechanism of wheels, levers, etc., usu. for a special purpose (*winding-gear*). **4** a particular apparatus or mechanism, as specified (*landing-gear*). **5** equipment or tackle for a special purpose. **6** *colloq.* **a** clothing, esp. when modern or fashionable. **b** possessions in general. **7** goods; household utensils. **8** rigging. **9** a harness for a draught animal. ● *v.* **1** *tr.* (foll. by *to*) adjust or adapt to suit a special purpose or need. **2** *tr.* (often foll. by *up*) equip with gears. **3** *tr.* (foll. by *up*) make ready or prepared. **4** *tr.* put (machinery) in gear. **5** *intr.* **a** be in gear. **b** (foll. by *with*) work smoothly with. □ **be geared** (or **all geared**) **up** (often foll. by *for*, or *to* + infin.) *colloq.* be ready or enthusiastic. **first** (or **bottom**) **gear** the lowest gear in a motor vehicle or bicycle. **gear change 1** an act of engaging a different gear in a vehicle. **2** *US* a gear lever. **gear down** (or **up**) provide with a low (or high) gear. **gear lever** (or **shift**) a lever used to engage or change gear, esp. in a motor vehicle. **high** (or **low**) **gear** a gear such that the driven end of a transmission revolves faster (or slower) than the driving end. **in gear** with a gear engaged. **out of gear 1** with no gear engaged. **2** out of order. **top gear** the highest gear in a motor vehicle or bicycle. [ME f. ON *gervi* f. Gmc]

gearbox /'gɪəbɒks/ *n.* **1** the casing that encloses a set of gears. **2** a set of gears with its casing, esp. in a motor vehicle.

gearing /'gɪərɪŋ/ *n.* **1** a set or arrangement of gears in a machine. **2** *Finance* the ratio of a company's loan capital (debt) to the value of its ordinary shares (equity).

gearstick /'gɪəstɪk/ *n. Brit.* = gear lever.

gearwheel /'gɪəwiːl/ *n.* **1** a toothed wheel in a set of gears. **2** (in a bicycle) the cog-wheel driven directly by the chain.

Geber /'dʒi:bə(r)/ (Latinized name of Jabir ibn Hayyan, *c.*721–*c.*815), Arab chemist. He was a member of the court of Harun ar-Rashid, and although many works are attributed to him, there is doubt about the authenticity of some of them, and his name was used by later writers. He was familiar with many chemicals and laboratory techniques, including distillation and sublimation.

GEC *abbr.* General Electric Company.

gecko /'gekəʊ/ *n.* (*pl.* **-os** or **-oes**) a vocal and mainly nocturnal lizard of the family Gekkonidae, found in warm climates, with adhesive feet for climbing vertical surfaces. [Malay *chichak* etc., imit. of its cry]

gee[1] /dʒi:/ *int.* (also **gee whiz** /wɪz/) *N. Amer. colloq.* a mild expression of surprise, discovery, etc. [perh. abbr. of JESUS]

gee[2] /dʒi:/ *int. & v.* ● *int.* (often foll. by *up*) a command to a horse etc., esp. to go faster. ● *v.tr.* (**geed**, **geeing**) command (a horse etc.) to go faster. [17th c.: orig. unkn.]

gee[3] /dʒi:/ *n. US sl.* (usu. in *pl.*) a thousand dollars. [the letter *G*, as initial of GRAND *n.* 2]

gee-gee /'dʒi:dʒi:/ *n. Brit. colloq.* a horse. [orig. a child's word, f. GEE[2]]

geek /giːk/ *n.* **1** esp. *US sl.* a person who is socially inept or tediously conventional. **2** *Austral. sl.* a look. [var. of dial. *geck* fool, dupe]

Geelong /dʒiː'lɒŋ/ a port and oil-refining centre on the south coast of Australia, in the state of Victoria; pop. (1991) 126,300.

geese *pl.* of GOOSE.

gee-string var. of G-STRING 2.

Ge'ez /'giːez/ *n.* the classical literary language of Ethiopia, a Semitic language thought to have been introduced from Arabia in the 1st century BC. It is the ancestor of the modern Ethiopian languages such as Amharic, and survives as the liturgical language of the Coptic Church in Ethiopia, in which context it is also called Ethiopic. [Ge'ez]

geezer /'giːzə(r)/ *n. sl.* a person, esp. an old man. [dial. pronunc. of *guiser* mummer]

Gehenna /gɪ'henə/ (in Judaism and the New Testament) a name for hell as a place of fiery torment for the wicked. The term is derived from the name of the valley of Hinnom near Jerusalem, where children were thought to have been burnt in ancient times in sacrifice to pagan gods. [eccl.L f. Gk f. Heb. *gê hinnōm*]

Gehrig /'gerɪg/, Henry Louis ('Lou') (1903–41), American baseball player. He set a record of playing 2,130 major-league games for the New York Yankees from 1925 to 1939; his stamina caused him to be known as the 'Iron Horse'. He died from a form of motor neurone disease now often called *Lou Gehrig's disease.*

Geiger /'gaɪgə(r)/, Hans (Johann) Wilhelm (1882–1945), German nuclear physicist. He worked with Sir Ernest Rutherford at Manchester on radioactivity, and in 1908 developed his prototype radiation counter for detecting alpha particles. In 1925 he was appointed professor of physics at Kiel, where he improved the sensitivity of his device with Walther Müller.

Geiger counter *n.* a device for measuring radioactivity and other radiation by detecting and counting ionizing particles. Also called *Geiger-Müller counter.* (See GEIGER.)

Geikie /'giːkɪ/, Sir Archibald (1835–1924), Scottish geologist. He carried out most of his field work in Scotland. He specialized in Pleistocene geology, especially the geomorphological effects of glaciations and the resulting deposits. Geikie wrote several important works and was a leading figure in British geology, eventually becoming director-general of the British Geological Survey. He was also president of the Geological Society, the Royal Society, and the Classical Association.

geisha /'geɪʃə/ *n.* (*pl.* same or **geishas**) **1** a Japanese hostess trained in entertaining men with dance and song. **2** a Japanese prostitute. [Jap.]

Geissler tube /'gaɪslə(r)/ *n.* a sealed tube of glass or quartz with a central constriction, filled with vapour for the production of a luminous electrical discharge. [Heinrich *Geissler*, Ger. mechanic (1814–79)]

Gejiu /ge'dʒuː/ (also **Geju**) a tin-mining city in southern China, near the border with Vietnam; pop. (1986) 349,700.

gel /dʒel/ *n. & v.* ● *n.* **1** a semi-solid colloidal suspension or jelly, of a solid dispersed in a liquid. **2** a jelly-like substance used for setting the hair. ● *v.intr.* (**gelled**, **gelling**) **1** form a gel. **2** = JELL *v.* 1b. **3** = JELL *v.* 2. □ **gelation** /dʒɪ'leɪʃ(ə)n/ *n.* [abbr. of GELATIN]

gelada /dʒə'lɑːdə/ *n.* a brownish gregarious baboon, *Theropithecus gelada*, with a bare red patch on its chest, native to Ethiopia. [Amharic *č'ällada*]

gelatin /'dʒelətɪn/ *n.* (also **gelatine** /-ˌtiːn/) a virtually colourless tasteless transparent water-soluble protein derived from collagen and used in food preparation, photography, etc. □ **gelatin paper** a paper coated with sensitized gelatin for photography. □ **gelatinize** /dʒɪ'lætɪˌnaɪz/ *v.tr. & intr.* (also **-ise**). **gelatinization** /-ˌlætɪnaɪ'zeɪʃ(ə)n/ *n.* [F *gélatine* f. It. *gelatina* f. *gelata* JELLY]

gelatinous /dʒɪ'lætɪnəs/ *adj.* **1** of or like gelatin. **2** of a jelly-like consistency. □ **gelatinously** *adv.*

gelation /dʒɪ'leɪʃ(ə)n/ *n.* solidification by freezing. [L *gelatio* f. *gelare* freeze]

gelato /gə'lɑːtəʊ/ *n. Austral.* a kind of ice-cream. [It.]

geld /geld/ *v.tr.* **1** deprive (usu. a male animal) of the ability to reproduce. **2** castrate or spay; excise the testicles or ovaries of. [ME f. ON *gelda* f. *geldr* barren f. Gmc]

Gelderland /ˈɡeldəˌlænd/ a province of the Netherlands, on the border with Germany; capital, Arnhem. Formerly a duchy, the province was variously occupied by the Spanish, the French, and the Prussians, until in 1815 it joined the newly united kingdom of the Netherlands.

gelding /ˈɡeldɪŋ/ *n.* a gelded animal, esp. a male horse. [ME f. ON *geldingr*: see GELD]

gelid /ˈdʒelɪd/ *adj.* **1** icy, ice-cold. **2** chilly, cool. [L *gelidus* f. *gelu* frost]

gelignite /ˈdʒelɪɡˌnaɪt/ *n.* a high explosive consisting of a gel of nitroglycerine and nitrocellulose in a base of wood pulp and sodium or potassium nitrate. It is a preferred explosive for rock-blasting under wet conditions, because of its good handling properties, water resistance, and low fume emission. [GELATIN + L *ignis* fire + -ITE[1]]

Gell-Mann /ɡelˈmæn/, Murray (b.1929), American theoretical physicist. He coined the word *quark*, proposed the concept of strangeness in quarks, and made major contributions on the classification and interactions of subatomic particles. He was awarded the Nobel Prize for physics in 1969.

gelly /ˈdʒeli/ *n. Brit. sl.* gelignite. [abbr.]

gelsemium /dʒelˈsiːmɪəm/ *n.* **1** the rhizome of *Gelsemium sempervirens*, a twining shrub of southern North America. **2** a preparation of this used medicinally, esp. in the treatment of neuralgia. [mod.L f. It. *gelsomino* jasmine]

Gelsenkirchen /ˈɡelz(ə)nˌkɪəx(ə)n/ an industrial city in western Germany, in North Rhine-Westphalia north-east of Essen; pop. (1991) 293,840.

gem /dʒem/ *n. & v.* ● *n.* **1** a precious stone, esp. when cut and polished or engraved. **2** an object or person of great beauty or worth. ● *v.tr.* (**gemmed**, **gemming**) adorn with or as with gems. □ **gemlike** *adj.* **gemmy** *adj.* [ME f. OF *gemme* f. L *gemma* bud, jewel]

Gemara /ɡɪˈmɑːrə/ *n.* a rabbinical commentary on the Mishnah, forming the second part of the Talmud. [Aram. *gᵉmārā* completion]

Gemayel /dʒəˈmaɪəl/, Pierre (1905–84), Lebanese political leader. A Maronite Christian, he founded the right-wing Phalange Party (1936) and served as a member of parliament 1960–84; during this time he held several government posts and led the Phalange militia forces during the civil war (1975–6). His youngest son, Bashir (1947–82), was assassinated while President-elect; his eldest son, Amin (b.1942), served as President 1982–8.

geminal /ˈdʒemɪn(ə)l/ *adj. Chem.* (of molecules) having two functional groups attached to the same atom. □ **geminally** *adv.* [as GEMINATE + -AL[1]]

geminate *adj. & v.* ● *adj.* /ˈdʒemɪnət/ combined in pairs. ● *v.tr.* /ˈdʒemɪˌneɪt/ **1** double, repeat. **2** arrange in pairs. □ **gemination** /ˌdʒemɪˈneɪʃ(ə)n/ *n.* [L *geminatus* past part. of *geminare* f. *geminus* twin]

Gemini /ˈdʒemɪˌnaɪ, -ˌniː/ *n.* **1** *Astron.* a northern constellation (the Twins), said to represent the twins Castor and Pollux (see DIOSCURI), whose names are given to its two brightest stars. **2** *Astrol.* **a** the third sign of the zodiac, which the sun enters about 21 May. **b** a person born when the sun is in this sign. **3** a series of twelve manned American orbiting spacecraft, launched in the 1960s in preparation for the Apollo programme. □ **Geminian** /ˌdʒemɪˈnaɪən/ *n. & adj.* [ME f. L, = twins]

Geminids /ˈdʒemɪnɪdz/ *n.pl. Astron.* an annual meteor shower with a radiant in the constellation of Gemini, reaching a peak about 13 Dec. [GEMINI]

gemma /ˈdʒemə/ *n.* (*pl.* **gemmae** /-miː/) *Bot.* a small cellular body in cryptogams that separates from the mother-plant and starts a new one; an asexual spore. [L: see GEM]

gemmation /dʒeˈmeɪʃ(ə)n/ *n. Bot.* reproduction by gemmae. [F f. *gemmer* to bud, *gemme* bud]

gemmiferous /dʒeˈmɪfərəs/ *adj.* **1** producing precious stones. **2** bearing buds. [L *gemmifer* (as GEMMA, -FEROUS)]

gemmiparous /dʒeˈmɪpərəs/ *adj. Bot.* of or propagating by gemmation. [mod.L *gemmiparus* f. L *gemma* bud + *parere* bring forth]

gemmology /dʒeˈmɒlədʒi/ *n.* the study of gems. □ **gemmologist** *n.* [L *gemma* gem + -LOGY]

gemmule /ˈdʒemjuːl/ *n. Zool.* a dormant tough-coated cluster of embryonic cells produced by a freshwater sponge, for development in more favourable conditions. [F *gemmule* or L *gemmula* little bud (as GEM)]

gemsbok /ˈɡemzbɒk/ *n.* a large antelope, *Oryx gazella*, of SW and East Africa. [Afrik. f. Du., = chamois]

gemstone /ˈdʒemstəʊn/ *n.* a precious stone used as a gem.

gemütlich /ɡəˈmuːtlɪx/ *adj.* **1** pleasant and comfortable. **2** genial, agreeable. [G]

Gen. *abbr.* **1** General. **2** (in the Bible) Genesis.

gen /dʒen/ *n. & v. Brit. sl.* ● *n.* information. ● *v.tr. & intr.* (**genned**, **genning**) (foll. by *up*) provide with or obtain information. [perh. f. first syllable of *general information*]

-gen /dʒən/ *comb. form* **1** *Chem.* that which produces (*hydrogen*; *antigen*). **2** *Bot.* growth (*endogen*; *exogen*). [F *-gène* f. Gk *-genēs* -born, of a specified kind f. *gen-* root of *gignomai* be born, become]

genco /ˈdʒenkəʊ/ *n.* a power-generating company, esp. a private company selling electricity. [contr. of *generating company*]

gendarme /ˈʒɒndɑːm/ *n.* **1** (in France and French-speaking countries) a police officer belonging to a military force engaged in policing. **2** *Geol.* a rock-tower on a mountain, occupying and blocking an arête. [F f. *gens d'armes* men of arms]

gendarmerie /ʒɒnˈdɑːməri/ *n.* **1** a force of gendarmes. **2** the headquarters of such a force.

gender /ˈdʒendə(r)/ *n.* **1 a** the grammatical classification of nouns and related words, roughly corresponding to the two sexes and sexlessness. **b** each of the classes of nouns (see MASCULINE *n.*, FEMININE *n.*, NEUTER *n.* 1, COMMON *adj.* 6). **2** (of nouns and related words) the property of belonging to such a class. **3 a** esp. *colloq.* or *euphem.* a person's sex. **b** sex as expressed by social or cultural distinctions. [ME f. OF *gendre* ult. f. L GENUS]

gene /dʒiːn/ *n.* a unit of heredity which is transmitted from parent to offspring, usually as part of a chromosome. (*See note below.*) □ **gene pool** the whole stock of different genes in an interbreeding population. **gene therapy** *Med.* the introduction of normal genes into cells in place of defective or missing ones in order to correct genetic disorders. [G *Gen*: see -GEN]

■ Genes were originally (and are still informally) regarded as inherited factors each controlling one feature (e.g. eye colour) of an organism, but this simple view has gradually been modified. Genes are now identified with lengths of DNA (RNA in some viruses), which determine the synthesis of particular protein molecules; a character usually results from the action of many genes. (See also DNA.)

genealogical /ˌdʒiːnɪəˈlɒdʒɪk(ə)l/ *adj.* **1** of or concerning genealogy. **2** tracing family descent. □ **genealogical tree** a chart like an inverted branching tree showing the descent of a family or of an animal species. □ **genealogically** *adv.* [F *généalogique* f. Gk *genealogikos* (as GENEALOGY)]

genealogy /ˌdʒiːnɪˈælədʒi/ *n.* (*pl.* **-ies**) **1 a** a line of descent traced continuously from an ancestor. **b** an account or exposition of this. **2** the study and investigation of lines of descent. **3** a plant's or animal's line of development from earlier forms. □ **genealogist** *n.* **genealogize** *v.tr. & intr.* (also **-ise**). [ME f. OF *genealogie* f. LL *genealogia* f. Gk *genealogia* f. *genea* race]

genera *pl.* of GENUS.

general /ˈdʒenrəl/ *adj. & n.* ● *adj.* **1 a** completely or almost universal. **b** including or affecting all or nearly all parts or cases of things. **2** prevalent, widespread, usual. **3** not partial, particular, local, or sectional. **4** not limited in application; relating to whole classes or all cases. **5** including points common to the individuals of a class and neglecting the differences (*a general term*). **6** not restricted or specialized (*general knowledge*). **7 a** roughly corresponding or adequate. **b** sufficient for practical purposes. **8** not detailed (*a general resemblance*; *a general idea*). **9** vague, indefinite (*spoke only in general terms*). **10** chief or principal; having overall authority (*general manager*; *Secretary-General*). ● *n.* **1 a** an army officer ranking next below Field Marshal or above lieutenant general. **b** *US* = lieutenant general, major-general. **2** a commander of an army. **3** a tactician or strategist of specified merit (*a great general*). **4** the head of a religious order, e.g. of the Jesuits or Dominicans or the Salvation Army. **5** (prec. by *the*) *archaic* the public. □ **as a general rule** in most cases. **general anaesthesia** see ANAESTHETIC. **general delivery** *US* the delivery of letters to callers at a post office. **general election** the election of representatives to a legislature (esp. in the UK to the House of Commons) from constituencies throughout the country. **general headquarters** the headquarters of a military commander. **general meeting** a meeting open to all the members of a society etc. **general of the army** (or **air force**) *US* the officer of the highest rank in the army or air force.

general practice the work of a general practitioner. **general practitioner** a doctor working in the community and treating cases of all kinds in the first instance, as distinct from a consultant or specialist. **general-purpose** having a range of potential uses or functions; not specialized in design. **general staff** the staff assisting a military commander in planning and administration. **general theory of relativity** see RELATIVITY. **in general 1** as a normal rule; usually. **2** for the most part. [ME f. OF f. L *generalis* (as GENUS)]

General American *n.* a form of US speech not markedly dialectal or regional.

General Certificate of Education *n.* (abbr. **GCE**) (in England, Wales, and Northern Ireland) General Certificate of Education, an examination taken (until 1988) at ordinary level (O level) at about the age of 16 and at advanced level (A level) or advanced supplementary level (S level) at about 18. The O-level examination has now been replaced by the GCSE.

General Certificate of Secondary Education *n.* (abbr. **GCSE**) (in England, Wales and Northern Ireland) General Certificate of Secondary Education, an examination taken at about age 16, which replaced the GCE O level and CSE examinations in 1988. It puts a greater emphasis on practical and course work than its predecessors did.

generalissimo /ˌdʒɛnrəˈlɪsɪˌməʊ/ *n.* (*pl.* **-os**) the commander of a combined military force consisting of army, navy, and air-force units. [It., superl. of *generale* GENERAL]

generalist /ˈdʒɛnrəlɪst/ *n.* a person competent in several different fields or activities (opp. SPECIALIST).

generality /ˌdʒɛnəˈrælɪtɪ/ *n.* (*pl.* **-ies**) **1** a statement or principle etc. having general validity or force. **2** applicability to a whole class of instances. **3** vagueness; lack of detail. **4** the state of being general. **5** (foll. by *of*) the main body or majority. [F *généralité* f. LL *generalitas -tatis* (as GENERAL)]

generalization /ˌdʒɛnrəlaɪˈzeɪʃ(ə)n/ *n.* (also **-isation**) **1** a general notion or proposition obtained by inference from (esp. limited or inadequate) particular cases. **2** the act or an instance of generalizing. [F *généralisation* (as GENERALIZE)]

generalize /ˈdʒɛnrəˌlaɪz/ *v.* (also **-ise**) **1** *intr.* **a** speak in general or indefinite terms. **b** form general principles or notions. **2** *tr.* reduce to a general statement, principle, or notion. **3** *tr.* **a** give a general character to. **b** call by a general name. **4** *tr.* infer (a law or conclusion) by induction. **5** *tr. Math. & Philos.* express in a general form; extend the application of. **6** *tr.* (in painting) render only the typical characteristics of. **7** *tr.* bring into general use. □ **generalizer** *n.* **generalizable** *adj.* **generalizability** /ˌdʒɛnrəˌlaɪzəˈbɪlɪtɪ/ *n.* [F *généraliser* (as GENERAL)]

generally /ˈdʒɛnrəlɪ/ *adv.* **1** usually; in most cases. **2** in a general sense; without regard to particulars or exceptions (*generally speaking*). **3** for the most part; extensively (*not generally known*). **4** in most respects (*they were generally well-behaved*).

General Motors (abbr. **GM**) a motor-manufacturing firm founded in the US in 1908 as an amalgamation of several other car companies and now one of the world's largest corporations.

generalship /ˈdʒɛnrəlˌʃɪp/ *n.* **1** the art or practice of exercising military command. **2** military skill; strategy. **3** skilful management; tact, diplomacy.

general strike *n.* (also **General Strike**) a strike of workers in all or the most important trades, esp. that called in the UK on 3 May 1926 by the Trades Union Congress in support of the miners, who were threatened with the imposition of wage cuts and longer hours. Although more than 2 million workers eventually came out on strike, the Conservative government of Stanley Baldwin was able, with the use of troops and volunteers, to end the strike after nine days.

General Synod the highest governing body in the Church of England consisting of three houses, of bishops, and of elected representatives of clergy and laity.

General Thanksgiving *n.* a form of thanksgiving in the Book of Common Prayer or the Alternative Service Book.

generate /ˈdʒɛnəˌreɪt/ *v.tr.* **1** bring into existence; produce, evolve. **2** produce (electricity). **3** *Math.* (of a point, line, or surface) move and so notionally form (a line, surface, or solid). **4** *Math. & Linguistics* produce (a set or sequence of items) by performing specified operations on or applying specified rules to an initial set. □ **generable** /-rəb(ə)l/ *adj.* [L *generare* beget (as GENUS)]

generation /ˌdʒɛnəˈreɪʃ(ə)n/ *n.* **1** all the people born at a particular time, regarded collectively (*my generation*; *the rising generation*). **2** a

single step in descent or pedigree (*have known them for three generations*). **3** a stage in technological or other development, esp. of computers (see also COMPUTER). **4** the average time in which children are ready to take the place of their parents (usu. reckoned at about thirty years). **5** production by natural or artificial process, esp. the production of electricity or heat. **6 a** procreation; the propagation of species. **b** the act of begetting or being begotten. □ **generation gap** differences of outlook or opinion between those of different generations. □ **generational** *adj.* [ME f. OF f. L *generatio -onis* (as GENERATE)]

generative /ˈdʒɛnərətɪv/ *adj.* **1** of or concerning procreation. **2** able to produce, productive. [ME f. OF *generatif* or LL *generativus* (as GENERATE)]

generative grammar *n.* the theory that for each language a set of rules can be formulated capable of 'generating' the infinite number of possible sentences of that language and providing them with the correct structural description; a set of rules in such a grammar. The theory, which was greatly influenced by mathematical logic, is best known in its transformational versions (see TRANSFORMATIONAL GRAMMAR).

generator /ˈdʒɛnəˌreɪtə(r)/ *n.* **1** a machine for converting mechanical into electrical energy; a dynamo. **2** an apparatus for producing gas, steam, etc. **3** a person who generates an idea etc.; an originator.

generic /dʒɪˈnɛrɪk/ *adj.* **1** characteristic of or relating to a class; general, not specific or special. **2** *Biol.* characteristic of or belonging to a genus. **3** (of goods, esp. a drug) having no brand name; not protected by a registered trademark. □ **generically** *adv.* [F *générique* f. L GENUS]

generous /ˈdʒɛnərəs/ *adj.* **1** giving or given freely. **2** magnanimous, noble-minded, unprejudiced. **3 a** ample, abundant, copious (*a generous portion*). **b** (of wine) rich and full. □ **generously** *adv.* **generousness** *n.* **generosity** /ˌdʒɛnəˈrɒsɪtɪ/ *n.* [OF *genereus* f. L *generosus* noble, magnanimous (as GENUS)]

Genesis /ˈdʒɛnɪsɪs/ the first book of the Bible, containing an account of the creation of the universe and of the early history of humankind. (See PENTATEUCH.) [as GENESIS]

genesis /ˈdʒɛnɪsɪs/ *n.* the origin, or mode of formation or generation, of a thing. [L f. Gk *gen-* be produced, root of *gignomai* become]

Genet /ʒəˈneɪ/, Jean (1910–86), French novelist, poet, and dramatist. He began to write while serving a prison sentence; much of his work portrayed life in the criminal and homosexual underworlds, of which he was a part, and his first novel *Notre-Dame des fleurs* (*Our Lady of the Flowers*) caused a sensation when it was first published in 1944. He also wrote the autobiography *Journal du voleur* (1949, *The Thief's Journal*) and a number of plays, including *Les Bonnes* (1947, *The Maids*) and *Les Nègres* (1958).

genet /ˈdʒɛnɪt/ *n.* (also **genette** /dʒɪˈnɛt/) **1** a catlike mammal of the genus *Genetta*, native to Africa and southern Europe, with spotted fur and a long ringed bushy tail. **2** the fur of the genet. [ME f. OF *genete* f. Arab. *jarnaiṭ*]

genetic /dʒɪˈnɛtɪk/ *adj.* **1** of genetics or genes; inherited. **2** of, in, or concerning origin; causal. □ **genetically** *adv.* [GENESIS after *antithetic*]

genetic code *n. Biochem.* the means by which DNA and RNA molecules carry genetic information in living cells. Each triplet of three bases codes for a particular amino acid, so that a sequence of such triplets embodies the instructions for making a specific protein. (See also DNA.)

genetic engineering *n.* the deliberate modification of the characters of an organism by the manipulation of genetic material. The purpose of genetic engineering is to introduce desirable properties of one organism into another. Two lines of approach have produced results in the last two decades. The first, the production of cultures of hybrid cells, has been used to produce monoclonal antibodies. Recombinant DNA techniques are capable of transferring genetic activity from one organism to another. DNA fragments known to confer the desired activity are isolated and then spliced into DNA from the host organism; these processes are carried out using particular enzymes. Recombinant DNA methods are used, for example, to transform bacteria into forms able to synthesize particular biochemicals (such as insulin, now manufactured in this way). Genetically engineered organisms can now be patented. The explosive growth of genetic engineering, and its potential for developments with controversial implications, is leading to the framing of codes of practice and legal controls.

genetic fingerprinting *n.* (also **genetic profiling**) the analysis of DNA from samples of bodily tissues or fluids in order to identify individuals. It is possible to locate patterns in genetic material which are believed (except in cases of identical twins) to be specific to the person or animal from which the sample came. The technique has been used in forensic analysis (the first conviction based on the technique was obtained in 1987), in the determination of parentage, and to establish whether twins are identical.

genetics /dʒɪˈnetɪks/ n.pl. (usu. treated as sing.) the study of heredity and the variation of inherited characteristics. □ **geneticist** /-tɪsɪst/ n.

genette var. of GENET.

Geneva /dʒɪˈniːvə/ (French **Genève** /ʒənɛv/) a city in SW Switzerland, on Lake Geneva; pop. (1990) 167,200. In the 16th century it was a stronghold of John Calvin, who rewrote its laws and constitution. It was the site of the conclusion of the Geneva Conventions (1846–1949) and the headquarters of the League of Nations (1920–46). It is the headquarters of international bodies such as the Red Cross, various organizations of the United Nations, and the World Health Organization.

Geneva, Lake (called in French *Lac Léman*) a lake in SW central Europe, between the Jura mountains and the Alps. Its southern shore forms part of the border between France and Switzerland.

Geneva bands n.pl. two white cloth strips attached to the collar of some Protestants' clerical dress (orig. worn by Calvinists in Geneva).

Geneva Bible an English translation of the Bible prepared by Protestant exiles at Geneva and printed there in 1560. (See also BREECHES BIBLE.)

Geneva Conventions a series of international agreements concluded at Geneva between 1864 and 1949 governing the status and treatment of captured and wounded service personnel and civilians in wartime.

Genève see GENEVA.

genever /dʒɪˈniːvə/ n. (also literary **geneva**) Dutch gin. [Du. f. OF genevre f. alt. of L juniperus juniper, with assim. of var. to GENEVA]

Genghis Khan /ˌɡɛŋɡɪs ˈkɑːn, ˌdʒɛŋ-/ (1162–1227), the founder of the Mongol empire. Originally named Temujin, he took the name Genghis Khan (= 'ruler of all') in 1206 after uniting the nomadic Mongol tribes under his command and becoming master of both eastern and western Mongolia. He then attacked China, capturing Beijing in 1215. When he died his empire extended from the shores of the Pacific to the northern shores of the Black Sea. His grandson Kublai Khan completed the conquest of China.

genial[1] /ˈdʒiːnɪəl/ adj. **1** jovial, sociable, kindly, cheerful. **2** (of the climate) mild and warm; conducive to growth. **3** cheering, enlivening. □ **genially** adv. **geniality** /ˌdʒiːnɪˈælɪtɪ/ n. [L genialis (as GENIUS)]

genial[2] /dʒɪˈniːəl/ adj. Anat. of or relating to the chin. [Gk geneion chin f. genus jaw]

genic /ˈdʒiːnɪk/ adj. of or relating to genes.

-genic /ˈdʒenɪk/ comb. form forming adjectives meaning: **1** producing (carcinogenic; pathogenic). **2** well suited to (photogenic; radiogenic). **3** produced by (iatrogenic). □ **-genically** suffix forming adverbs. [-GEN + -IC]

genie /ˈdʒiːnɪ/ n. (pl. usu. **genii** /-nɪˌaɪ/) a spirit or jinnee (in Arabian stories), esp. one trapped in a bottle, lamp, etc. and capable of granting wishes. [F génie f. L GENIUS: cf. JINNEE]

genii pl. of GENIE, GENIUS.

genista /dʒɪˈnɪstə/ n. an almost leafless leguminous shrub of the genus Genista, with a profusion of yellow pea-shaped flowers, e.g. dyer's greenweed. [L]

genital /ˈdʒenɪt(ə)l/ adj. & n. ● adj. of or relating to the organs of reproduction. ● n. (in pl.) the external organ or organs of reproduction, esp. of the male. [OF génital or L genitalis f. gignere genit- beget]

genitalia /ˌdʒenɪˈteɪlɪə/ n.pl. the genitals. [L, neut. pl. of genitalis: see GENITAL]

genitive /ˈdʒenɪtɪv/ n. & adj. Gram. ● n. the case of nouns and pronouns (and words in grammatical agreement with them) corresponding to of, from, and other prepositions and indicating possession or close association. ● adj. of or in the genitive. □ **genitival** /ˌdʒenɪˈtaɪv(ə)l/ adj. **genitivally** adv. [ME f. OF genetif, -ive or L genitivus f. gignere genit- beget]

genito- /ˈdʒenɪtəʊ/ comb. form genital.

genito-urinary /ˌdʒenɪtəʊˈjʊərɪnərɪ/ adj. Anat. & Med. of the genital and urinary organs.

genius /ˈdʒiːnɪəs/ n. (pl. **geniuses** or **genii** /-nɪˌaɪ/) **1** (pl. **geniuses**) **a** an exceptional intellectual or creative power or other natural ability or tendency. **b** a person having this. **2** the tutelary spirit of a person, place, institution, etc. **3** a person or spirit regarded as powerfully influencing a person for good or evil. **4** the prevalent feeling or

associations etc. of a nation, age, etc. [L (in sense 2) f. the root of gignere beget]

genizah /ɡeˈniːzə/ n. a room attached to a synagogue and housing damaged, discarded, or heretical texts and sacred relics. The term is often used with specific reference to the genizah of an ancient synagogue in Cairo, where vast quantities of fragments of biblical and other Jewish manuscripts were discovered in 1896–8. [Heb. genīzāh, lit. 'hiding-place' f. gānaz hide, set aside]

Genoa /ˈdʒenəʊə/ (Italian **Genova** /ˈdʒɛnova/) a seaport on the NW coast of Italy, capital of Liguria region; pop. (1990) 701,000. It was the birthplace of Christopher Columbus. □ **Genoese** /ˌdʒenəʊˈiːz/ adj. & n.

Genoa cake n. a rich fruit cake with almonds on top.

Genoa jib n. a large jib or foresail used esp. on racing yachts.

genocide /ˈdʒenəˌsaɪd/ n. the mass extermination of human beings, esp. of a particular race or nation. □ **genocidal** /ˌdʒenəˈsaɪd(ə)l/ adj. [Gk genos race + -CIDE]

genome /ˈdʒiːnəʊm/ n. Biol. **1** the haploid set of chromosomes of an organism. **2** the genetic material of an organism. [GENE + CHROMOSOME]

genotype /ˈdʒiːnəˌtaɪp/ n. Biol. the genetic constitution of an individual. □ **genotypic** /ˌdʒenəˈtɪpɪk/ adj. [G Genotypus (as GENE, TYPE)]

-genous /ˈdʒenəs/ comb. form forming adjectives meaning 'produced' (endogenous).

Genova see GENOA.

genre /ˈʒɒnrə/ n. **1** a kind or style, esp. of art or literature (e.g. tragedy, drama, satire). **2** (in full **genre painting**) a style of painting depicting scenes from ordinary life, especially domestic situations; also, a painting in this style. Genre painting is associated particularly with 17th-century Dutch and Flemish artists such as Jan Vermeer, Pieter de Hooch, and Adriaen Brouwer. [F, = a kind (as GENDER)]

gens /dʒenz/ n. (pl. **gentes** /ˈdʒentiːz/) **1** Rom. Hist. a group of families sharing a name and claiming a common origin. **2** Anthropol. a number of people sharing descent through the male line. [L, f. the root of gignere beget]

Gent see GHENT.

gent /dʒent/ n. colloq. (often joc.) **1** a gentleman. **2** (in pl.) (in shop titles) men (gents' outfitters). **3** (the Gents) Brit. a men's public lavatory. [abbr. of GENTLEMAN]

genteel /dʒenˈtiːl/ adj. **1** affectedly or ostentatiously refined or stylish. **2** often iron. of or appropriate to the upper classes. □ **genteelly** adv. **genteelness** n. [earlier gentile, readoption of F gentil GENTLE]

genteelism /dʒenˈtiːlɪz(ə)m/ n. a word used because it is thought to be less vulgar than the commoner word (e.g. perspire for sweat).

gentes pl. of GENS.

gentian /ˈdʒenʃ(ə)n/ n. **1** a plant of the genus Gentiana or Gentianella, found esp. in mountainous regions, usu. having violet or vivid blue trumpet-shaped flowers. **2** (in full **gentian bitter**) a liquor extracted from the root of the gentian. □ **gentian violet** a violet dye used as an antiseptic, esp. in the treatment of burns. [OE f. L gentiana, named, according to Pliny, after Gentius king of Illyria in the 2nd c. BC, said to have discovered the medicinal properties of the gentian root]

gentile /ˈdʒentaɪl/ adj. & n. ● adj. **1** (**Gentile**) **a** not Jewish. **b** not belonging to one's Church, esp. (formerly) a non-Mormon. **c** pagan, heathen. **2** Gram. (of a word) indicating nationality. ● n. **1** (**Gentile**) **a** a person who is not Jewish. **b** a person not belonging to one's Church, esp. a non-Mormon. **2** Gram. a word indicating nationality. [ME f. L gentilis f. gens gentis family: see GENS]

Gentile da Fabriano /ˌdʒenˌtiːleɪ dɑː ˌfæbrɪˈɑːnəʊ/ (c.1370–1427), Italian painter. He painted a number of frescos and altarpieces and worked chiefly in Florence and Rome. Most of the work on which his contemporary reputation was based has been destroyed; his major surviving work, the altarpiece *The Adoration of the Magi* (1423), is notable for its rich detail and naturalistic treatment of light.

gentility /dʒenˈtɪlɪtɪ/ n. **1** social superiority. **2** often derog. good manners; habits associated with the nobility. **3** people of noble birth. [ME f. OF gentilité (as GENTLE)]

gentle /ˈdʒent(ə)l/ adj., v, & n. ● adj. (**gentler, gentlest**) **1** (of a person, action, etc.) mild or kind; free from sternness or aggression. **2** moderate; not severe or drastic (gentle slope; gentle breeze). **3** usu. archaic honourable, or of fit for people of good social position. **4** archaic generous, courteous. ● v.tr. **1** make gentle or docile. **2** make (a horse etc.) docile or tractable. ● n. a maggot, the larva of the meat-fly or

bluebottle used as fishing-bait. □ **gentle art** the sport of angling.
□ **gentleness** n. **gently** adv. [ME f. OF gentil f. L gentilis: see GENTILE]

gentlefolk /ˈdʒent(ə)lˌfəʊk/ n.pl. literary people of good social position.

gentleman /ˈdʒent(ə)lmən/ n. (pl. **-men**) **1** a man (in polite or formal
use). **2** a chivalrous, courteous, or well-educated man. **3 a** a man of
good social position or of wealth and leisure (country gentleman). **b** hist.
a man of good birth who is entitled to bear arms but who is not a
nobleman. **4** a man of noble birth attached to a royal household
(gentleman in waiting). **5** (in pl. as a form of address) a male audience or
the male part of an audience. □ **gentleman-at-arms** one of the
bodyguards of the British monarch on ceremonial occasions.
gentleman's (or **-men's**) **agreement** one which is binding in
honour but not legally enforceable. **the gentlemen's** Brit. a men's
public lavatory. [GENTLE + MAN after OF gentilz hom]

gentlemanly /ˈdʒent(ə)lmənlɪ/ adj. like a gentleman in looks or
behaviour; befitting a gentleman. □ **gentlemanliness** n.

gentlewoman /ˈdʒent(ə)lˌwʊmən/ n. (pl. **-women**) archaic a woman of
good birth or breeding.

gentoo /ˈdʒentuː/ n. a penguin, Pygoscelis papua, found in the Falklands
and other antarctic islands. [perh. f. Anglo-Ind. Gentoo = Hindu, f. Port.
gentio GENTILE]

gentrify /ˈdʒentrɪˌfaɪ/ v.tr. (**-ies, -ied**) convert (a working-class or inner-
city district etc.) into an area of middle-class residence. □ **gentrifier**
n. **gentrification** /ˌdʒentrɪfɪˈkeɪʃ(ə)n/ n.

gentry /ˈdʒentrɪ/ n.pl. **1** the class of people next below the nobility in
position and birth. **2** derog. people (these gentry). [prob. f. obs. gentrice f.
OF genterise var. of gentelise nobility f. gentil GENTLE]

genuflect /ˈdʒenjʊˌflekt/ v.intr. bend the knee, esp. in worship or as a
sign of respect. □ **genuflector** n. **genuflection** /ˌdʒenjʊˈflekʃ(ə)n/ n.
(also **genuflexion**). [eccl.L genuflectere genuflex- f. L genu the knee +
flectere bend]

genuine /ˈdʒenjʊɪn/ adj. **1** really coming from its stated, advertised,
or reputed source. **2** properly so called; not sham. **3** pure-bred. **4** (of a
person) free from affectation or hypocrisy; honest. □ **genuinely** adv.
genuineness n. [L genuinus f. genu knee, with ref. to a father's
acknowledging a newborn child by placing it on his knee: later
associated with GENUS]

genus /ˈdʒiːnəs, ˈdʒen-/ n. (pl. **genera** /ˈdʒenərə/) **1** Biol. a taxonomic
grouping of organisms having common characteristics distinct from
those of other genera, usu. containing several or many species and
being one of a series constituting a taxonomic family. **2** a kind or
class having common characteristics. **3** Logic kinds of things including
subordinate kinds or species. [L genus -eris birth, race, stock]

-geny /dʒənɪ/ comb. form forming nouns meaning 'mode of production
or development of' (anthropogeny; ontogeny; pathogeny). [F -génie (as -GEN,
-Y³)]

Geo. abbr. George.

geo- /ˈdʒiːəʊ/ comb. form earth. [Gk geō- f. gē earth]

geobotany /ˌdʒiːəʊˈbɒtənɪ/ n. the study of the geographical
distribution of plants. □ **geobotanist** n.

geocentric /ˌdʒiːəʊˈsentrɪk/ adj. **1** considered as viewed from the
centre of the earth. **2** having or representing the earth as the centre;
not heliocentric. □ **geocentric latitude** the latitude at which a
planet would appear if viewed from the centre of the earth.
□ **geocentrically** adv.

geochemistry /ˌdʒiːəʊˈkemɪstrɪ/ n. the chemistry of the earth and
its rocks, minerals, etc. □ **geochemical** adj. **geochemist** n.

geochronology /ˌdʒiːəʊkrəˈnɒlədʒɪ/ n. **1** the study and measurement
of geological time by means of geological events. **2** the ordering of
geological events. □ **geochronologist** n. **geochronological**
/-ˌkrɒnəˈlɒdʒɪk(ə)l/ adj.

geode /ˈdʒiːəʊd/ n. **1** a small cavity lined with crystals or other mineral
matter. **2** a rock containing such a cavity. □ **geodic** /dʒiːˈɒdɪk/ adj. [L
geodes f. Gk geōdēs earthy f. gē earth]

geodesic /ˌdʒiːəʊˈdiːzɪk/ adj. **1** of or relating to geodesy (cf. GEODETIC).
2 designating, or designed according to, constructional principles
based on spheres and geodesic lines. □ **geodesic line** the shortest
possible line between two points on a curved surface.

geodesic dome n. a dome constructed of short struts along geodesic
lines, forming an open framework of triangles or polygons. The
principles of its construction were described by Buckminster Fuller,
and it combines the structural advantages of the sphere and the
tetrahedron.

geodesy /dʒɪˈɒdɪsɪ/ n. the branch of mathematics dealing with the
shape and area of the earth or large portions of it. □ **geodesist** n.
[mod.L f. Gk geōdaisia (as GEO-, daiō divide)]

geodetic /ˌdʒiːəʊˈdetɪk/ adj. of or relating to geodesy, esp. as applied
to land surveying (cf. GEODESIC). [Gk geōdaitēs land surveyor, f. geōdaisia
GEODESY]

Geoffrey of Monmouth /ˈdʒefrɪ/ (c.1100–c.1154), Welsh chronicler.
His Historia Regum Britanniae (c.1139; first printed in 1508), which
purports to give an account of the kings of Britain, is now thought to
contain little historical fact; it was, however, a major source for English
literature, including stories of King Arthur and the plots of some of
Shakespeare's plays.

geographical /ˌdʒiːəˈgræfɪk(ə)l/ adj. (also **geographic**) of or relating
to geography. □ **geographical latitude** the angle made with the
plane of the equator by a perpendicular to the earth's surface at
any point. **geographical mile** a distance equal to one minute of
longitude or latitude at the equator (about 1850 metres).
□ **geographically** adv. [geographic f. F géographique or LL geographicus
f. Gk geōgraphikos (as GEO-, -GRAPHIC)]

geography /dʒɪˈɒgrəfɪ/ n. **1** the study of the earth's physical features,
resources, natural and political divisions, climate, products,
population, etc. **2** the main physical features of an area. **3** the layout
or arrangement of rooms in a building. □ **geographer** n. [F géographie
or L geographia f. Gk geōgraphia (as GEO-, -GRAPHY)]

geoid /ˈdʒiːɔɪd/ n. **1** the shape of the earth. **2** a shape formed by the
mean sea level and its imagined extension under land areas. **3** an
oblate spheroid. [Gk geōeidēs (as GEO-, -OID)]

geology /dʒɪˈɒlədʒɪ/ n. **1** the science of the earth, including the
composition, structure, and origin of its rocks. **2** this science applied
to any other planet or celestial body. **3** the geological features of a
district. □ **geologist** n. **geologize** v.tr. & intr. (also **-ise**). **geologic**
/ˌdʒiːəˈlɒdʒik/ adj. **geological** adj. **geologically** adv. [med.L geologia
(as GEO-, -LOGY)]

geomagnetism /ˌdʒiːəʊˈmægnɪˌtɪz(ə)m/ n. the study of the magnetic
properties of the earth. □ **geomagnetic** /-mæg'netɪk/ adj.
geomagnetically adv.

geomancy /ˈdʒiːəʊˌmænsɪ/ n. **1** the art of siting buildings etc.
auspiciously. **2** divination from the configuration of a handful of earth
or random dots. □ **geomantic** /ˌdʒiːəʊˈmæntɪk/ adj.

geometer /dʒɪˈɒmɪtə(r)/ n. **1** a person skilled in geometry. **2** Zool. **a** =
LOOPER 1. **b** = GEOMETRID. [ME f. LL geometra f. L geometres f. Gk geōmetrēs
(as GEO-, metrēs measurer)]

geometric /ˌdʒiːəˈmetrɪk/ adj. (also **geometrical**) **1** of, according to,
or like geometry. **2** (of a design, architectural feature, etc.)
characterized by or decorated with regular lines and shapes.
□ **geometric mean** the central number in a geometric progression,
also calculable as the nth root of a product of n numbers (as 9 from 3
and 27). **geometric progression** a progression of numbers with a
constant ratio between each number and the one before (as 1, 3, 9, 27,
81). **geometric tracery** tracery with openings of geometric form.
□ **geometrically** adv. [F géométrique f. L geometricus f. Gk geōmetrikos (as
GEOMETER)]

geometrid /ˌdʒiːəʊˈmetrɪd/ n. & adj. Zool. ● n. a moth of the large
family Geometridae, whose twiglike caterpillars move by looping and
straightening the body as if measuring the ground. (See also LOOPER
1.) ● adj. of or relating to this family. [mod.L Geometra genus name, f. L
geometres GEOMETER]

geometry /dʒɪˈɒmɪtrɪ/ n. **1** the branch of mathematics concerned
with the properties and relations of points, lines, surfaces, and solids.
(See note below.) **2** the relative arrangement of objects or parts.
□ **geometrician** /ˌdʒiːəmɪˈtrɪʃ(ə)n/ n. [ME f. OF geometrie f. L geometria
f. Gk (as GEO-, -METRY)]

▪ Geometry was developed for surveying by the ancient Egyptians
and Mesopotamians, and until the 19th century was still essentially
based on the work of the Greek mathematician Euclid. A major
advance was the introduction of coordinates by Descartes in the 17th
century, enabling other mathematical techniques to be applied to
geometrical problems. Since the early 19th century mathematics
has embraced other forms of geometry, especially non-Euclidean and
higher dimensional geometry.

geomorphology /ˌdʒiːəʊmɔːˈfɒlədʒɪ/ n. the study of the physical
features of the surface of the earth and their relation to its geological structures. □ **geomorphologist** n. **geomorphological**
/-ˌmɔːfəˈlɒdʒɪk(ə)l/ adj.

geophagy /dʒɪˈɒfədʒɪ/ n. (also **geophagia** /ˌdʒiːəʊˈfeɪdʒə/) Med. or Anthropol. the practice of eating earth. [GEO- + Gk phagō eat]

geophysics /ˌdʒiːəʊˈfɪzɪks/ n. the physics of the earth. □ **geophysical** adj. **geophysicist** /-zɪsɪst/ n.

geopolitics /ˌdʒiːəʊˈpɒlɪtɪks/ n. **1** the politics of a country as determined by its geographical features. **2** the study of this. □ **geopolitical** /-pəˈlɪtɪk(ə)l/ adj. **geopolitically** adv. **geopolitician** /-ˌpɒlɪˈtɪʃ(ə)n/ n.

Geordie /ˈdʒɔːdɪ/ n. & adj. Brit. colloq. ● n. **1** a native of Tyneside. **2** the dialect spoken on Tyneside. ● adj. of or relating to Tyneside, its people, or its dialect. [the name George + -IE]

George[1] /dʒɔːdʒ/ the name of four kings of Great Britain and Ireland, one of Great Britain and Ireland (from 1920 of the United Kingdom), and one of the United Kingdom:

George I (1660–1727), great-grandson of James I, reigned 1714–27, Elector of Hanover 1698–1727. George succeeded to the British throne as a result of the Act of Settlement (1701). Unpopular in England as a foreigner who never learned English, he left the administration of his new kingdom to his ministers and devoted himself to diplomacy and the interests of Hanover. However, the relatively easy suppression of the Jacobite uprisings of 1715 and 1719 demonstrated that he was generally preferred to the Catholic Old Pretender (James Stuart, see STUART[3]).

George II (1683–1760), son of George I, reigned 1727–60, Elector of Hanover 1727–60. Like his father, he depended heavily on his ministers, although he took an active part in the War of the Austrian Succession (1740–8), successfully leading a British army against the French at Dettingen in 1743, the last occasion on which a British king was present on the field of battle. In the latter years of his reign, George largely withdrew from active politics, allowing advances in the development of constitutional monarchy which his successor George III was ultimately unable to reverse.

George III (1738–1820), grandson of George II, reigned 1760–1820, Elector of Hanover 1760–1815 and king of Hanover 1815–20. He took great interest in British domestic politics and attempted to exercise royal control of government to the fullest possible extent. His determination to suppress the War of American Independence dominated British foreign policy 1775–83, but his political influence declined from 1788 after a series of bouts of mental illness. In 1811 it became clear that the king's mental health made him unfit to rule and his son was made regent.

George IV (1762–1830), son of George III, reigned 1820–30. Known as a patron of the arts and *bon viveur*, he was Prince Regent during his father's final period of mental illness. His lifestyle gained him a bad reputation which was further damaged by his attempt to divorce his estranged wife Caroline of Brunswick just after coming to the throne. His only child, Charlotte, died in 1817.

George V (1865–1936), son of Edward VII, reigned 1910–36. He won respect for his punctilious attitude towards royal duties and responsibilities, especially during the First World War. He exercised restrained but none the less important influence over British politics, playing an especially significant role in the formation of the government in 1931.

George VI (1894–1952), son of George V, reigned 1936–52. He was created Duke of York in 1920 and came to the throne on the abdication of his elder brother Edward VIII. Despite a retiring disposition he became a popular monarch, gaining respect for the staunch example he and his family set during the London Blitz.

George[2] /dʒɔːdʒ/ n. Brit. sl. the automatic pilot of an aircraft. [the name George]

George, St, patron saint of England. Little is known of his life, and his historical existence was once challenged but is now generally accepted. He may have been martyred near Lydda in Palestine some time before the reign of Constantine (d.337), but his cult did not become popular until the 6th century. The slaying of the dragon (possibly derived from the legend of Perseus) was not attributed to him until the 12th century. His rank as patron saint of England (in place of St Edward the Confessor) probably dates from the reign of Edward III. The latter founded the Order of the Garter (c.1344) under the patronage of St George, who by that time was honoured as the ideal of chivalry. Feast day, 23 Apr.

George Cross n. (also **George Medal**) (in the UK) each of two (different) decorations for bravery awarded esp. to civilians, instituted in 1940 by King George VI.

Georgetown /ˈdʒɔːdʒtaʊn/ the capital of Guyana, a port at the mouth of the Demerara river; pop. (est. 1983) 188,000.

George Town 1 the capital of the Cayman Islands, on the island of Grand Cayman; pop. (est. 1988) 12,000. **2** (also called Penang) the chief port of Malaysia and capital of the state of Penang, on Penang island; pop. (1980) 250,500. It was founded in 1786 by the British East India Company.

georgette /dʒɔːˈdʒet/ n. a thin silk or crêpe dress-material. [Georgette de la Plante (fl. c.1900), Fr. dressmaker]

Georgia /ˈdʒɔːdʒə/ **1** a country of SE Europe, on the eastern shore of the Black Sea; pop. (1989) 5,478,000; languages, Georgian (official), Russian, and Armenian; capital, Tbilisi. Georgia was an independent kingdom in medieval times until divided between Persia and Turkey in 1555. It was acquired by the Russian empire in the 19th century, and after the Russian Revolution of 1917 it was absorbed into the Soviet Union, becoming first (1922) part of the Transcaucasian Soviet Federated Socialist Republic, and later, in 1936, a constituent republic of the USSR. On the breakup of the USSR in 1991, Georgia became an independent republic outside the Commonwealth of Independent States. Since then separatist movements among the Abkhazian and South Ossetian minorities have led to outbreaks of ethnic conflict. **2** a state of the south-eastern US, on the Atlantic coast; pop. (1990) 5,463,100; capital, Atlanta. Founded as an English colony in 1732 and named after George II, it became one of the original thirteen states of the Union (1788).

Georgian[1] /ˈdʒɔːdʒən/ adj. **1 a** of or characteristic of the time of kings George I–IV (1714–1830). **b** designating or resembling the style of architecture characteristic of this period. (See note below.) **2 a** of or characteristic of the time of kings George V and VI (1910–52). **b** of or relating to British literature of 1910–20, especially poetry (by writers such as Rupert Brooke) characterized by pastoral themes and strongly attacked by the early modernists.

▪ Georgian architecture encompassed a wide range of styles, such as Palladianism, neoclassicism, and chinoiserie. It is perhaps best exemplified by the simplicity and restraint of the typical red-brick house of the latter half of the period, with regularly spaced windows, white paintwork, and pedimented doorway, and by the rather grander stone-built terraces and crescents in cities such as Bath and Edinburgh.

Georgian[2] /ˈdʒɔːdʒən/ adj. & n. ● adj. **1** of or relating to Georgia in the Caucasus. **2** of or relating to Georgia in the US. ● n. **1** a native of Georgia in the Caucasus; a person of Georgian descent. **2** the language of Georgia, a Caucasian language spoken by some 4 million people. The origin of the characteristic Georgian alphabet is obscure but it is known to have been invented in the 5th century AD. **3** a native of Georgia in the US.

geosphere /ˈdʒiːəˌsfɪə(r)/ n. **1** the solid surface of the earth. **2** any of the almost spherical concentric regions of the earth and its atmosphere.

geostationary /ˌdʒiːəʊˈsteɪʃənərɪ/ adj. (of an artificial satellite of the earth) moving in such an orbit as to remain above the same point on the earth's surface (see also GEOSYNCHRONOUS).

geostrophic /ˌdʒiːəʊˈstrɒfɪk, -ˈstrəʊfɪk/ adj. Meteorol. depending upon the rotation of the earth. [GEO- + Gk strophē a turning f. strephō to turn]

geosynchronous /ˌdʒiːəʊˈsɪŋkrənəs/ adj. (of an artificial satellite of the earth) moving in an orbit equal to the earth's period of rotation (see also GEOSTATIONARY).

geothermal /ˌdʒiːəʊˈθɜːm(ə)l/ adj. relating to, originating from, or produced by the internal heat of the earth. In a number of countries, geothermal energy has been harnessed by using boreholes to extract steam that is then used to generate electricity. In volcanically active areas such as Iceland the heat is also used directly for industrial and domestic purposes.

geotropism /dʒɪˈɒtrəˌpɪz(ə)m/ n. plant growth in relation to gravity. □ **negative geotropism** the tendency of stems etc. to grow away from the centre of the earth. **positive geotropism** the tendency of roots to grow towards the centre of the earth. □ **geotropic** /ˌdʒiːəʊˈtrəʊpɪk, -ˈtrɒpɪk/ adj. [GEO- + Gk tropikos f. tropē a turning f. trepō to turn]

Ger. abbr. German.

Gera /ˈɡeərə/ an industrial city in east central Germany, in Thuringia; pop. (1991) 126,520.

Geraldton /ˈdʒerəldtən/ a seaport and resort on the west coast of Australia, to the north of Perth; pop. (1991) 24,360.

geranium /dʒəˈreɪnɪəm/ n. **1** a herbaceous plant or shrub of the genus *Geranium*, bearing fruit shaped like the bill of a crane, e.g. cranesbill. **2** (in general use) a cultivated pelargonium. **3** the colour of the scarlet pelargonium. [L f. Gk *geranion* f. *geranos* crane]

gerbera /ˈdʒɜːbərə/ n. a composite plant of the genus *Gerbera*, native to Africa and Asia, esp. the Transvaal daisy. [Traugott *Gerber*, Ger. naturalist (d.1743)]

gerbil /ˈdʒɜːbɪl/ n. a usu. nocturnal mouselike desert rodent of the subfamily Gerbillinae; esp. *Meriones unguiculatus*, kept as a pet. [F *gerbille* f. mod.L *gerbillus* dimin. of *gerbo* JERBOA]

gerenuk /ˈɡerɪˌnʊk/ n. an antelope, *Litocranius walleri*, native to East Africa, with a very long neck. [Somali]

gerfalcon var. of GYRFALCON.

geri /ˈdʒerɪ/ n. Austral. colloq. a geriatric person. [abbr.]

geriatric /ˌdʒerɪˈætrɪk/ adj. & n. ● adj. **1** of or relating to old people. **2** colloq. old, outdated. ● n. **1** an old person, esp. one receiving special care. **2** colloq. a person or thing considered as relatively old or outdated. [Gk *gēras* old age + *iatros* doctor]

geriatrics /ˌdʒerɪˈætrɪks/ n.pl. (usu. treated as sing.) a branch of medicine or social science dealing with the health and care of old people. □ **geriatrician** /-rɪəˈtrɪʃ(ə)n/ n.

Géricault /ˈʒerɪˌkəʊ/, (Jean Louis André) Théodore (1791–1824), French painter. His rejection of the prevailing classicism of his day and use of bright colours in a bold, romantic style brought him criticism from the art world. His most famous work, *The Raft of the Medusa* (1819), depicts the survivors of a famous shipwreck of 1816, with realistic treatment of the macabre. He also painted landscapes and scenes of horse-races.

germ /dʒɜːm/ n. **1** a micro-organism, esp. one which causes disease. **2 a** a portion of an organism capable of developing into a new one; the rudiment of an animal or plant. **b** an embryo of a seed (*wheat germ*). **3** an original idea etc. from which something may develop; an elementary principle. □ **germ cell** Biol. **1** a cell containing half the number of chromosomes of a somatic cell and able to unite with one from the opposite sex to form a new individual; a gamete. **2** any embryonic cell with the potential of developing into a gamete. **germ layer** Biol. each of the three layers of cells (ectoderm, mesoderm, and endoderm) that arise in the early stages of embryonic development. **germ line** Biol. a series of germ cells each descended from earlier cells in the series, regarded as continuing through successive generations of an organism. **germ plasm** Biol. **1** hist. the part of a germ cell which carries hereditary factors and is itself transmitted unchanged from generation to generation. **2** germ cells collectively; their genetic material. **germ warfare** the systematic spreading of micro-organisms to cause disease in an enemy population. **in germ** not yet developed. □ **germy** adj. [F *germe* f. L *germen germinis* sprout]

German /ˈdʒɜːmən/ n. & adj. ● n. **1** a native or national of Germany; a person of German descent. **2** the language of Germany. (See note below.) **3** hist. a member of any of the Germanic peoples of north and central Europe. ● adj. of or relating to Germany or Germans or their language. □ **German measles** a contagious viral disease, rubella, with symptoms like those of mild measles. **German shepherd** (or **shepherd dog**) an Alsatian (dog). **German silver** a white alloy of nickel, zinc, and copper. [L *Germanus* with ref. to related peoples of central and northern Europe, a name perh. given by Celts to their neighbours: cf. OIr. *gair* neighbour]

▪ German is a Germanic language spoken by some 100 million people. It is the official language of Germany and of Austria and one of the official languages of Switzerland; there are also large German-speaking communities in the US. (See also HIGH GERMAN, LOW GERMAN.)

german /ˈdʒɜːmən/ adj. (placed after *brother*, *sister*, or *cousin*) **1** having both parents the same (*brother german*). **2** having both grandparents the same on one side (*cousin german*). **3** archaic germane. [ME f. OF *germain* f. L *germanus* genuine, of the same parents]

German Bight a shipping forecast area covering the eastern North Sea off the northern Netherlands, Germany, and southern Denmark.

German Democratic Republic (abbr. **GDR, DDR**) hist. East Germany (see GERMANY).

germander /dʒɜːˈmændə(r)/ n. a labiate plant of the genus *Teucrium*, with oaklike leaves. □ **germander speedwell** a creeping plant, *Veronica chamaedrys*, with germander-like leaves and blue flowers. [ME f. med.L *germandra* ult. f. Gk *khamaidrus* f. *khamai* on the ground + *drus* oak]

germane /dʒɜːˈmeɪn/ adj. (usu. foll. by *to*) relevant (to a subject under consideration). □ **germanely** adv. **germaneness** n. [var. of GERMAN]

German East Africa a former German protectorate in East Africa (1891–1918), corresponding to present-day Tanzania, Rwanda, and Burundi.

German Empire (also called *Second Reich*) an empire in German-speaking central Europe, created by Bismarck in 1871 after the Franco-Prussian War by the union of twenty-five German states under the Hohenzollern king of Prussia. Forming an alliance with Austria-Hungary, the German Empire became the greatest industrial power in Europe and engaged in colonial expansion in Africa, China, and the Far East. Tensions arising with other colonial powers led to the First World War, after which the German Empire collapsed and the Weimar Republic was created.

Germanic /dʒɜːˈmænɪk/ adj. & n. ● adj. **1** having German characteristics. **2** hist. of the Germans. **3** of the Scandinavians, Anglo-Saxons, or Germans. **4** of the languages or language group called Germanic. ● n. **1** a branch of Indo-European languages including English, German, Dutch, and the Scandinavian languages. (See note below.) **2** the (unrecorded) early language from which other Germanic languages developed. [L *Germanicus* (as GERMAN)]

▪ Present-day Germanic languages reflect an original dialectal split into West Germanic (English, Dutch, and German), North Germanic (the Scandinavian languages, for which the oldest evidence is that of Old Norse), and East Germanic, which has died out but for which Gothic provides the oldest evidence.

germanic /dʒɜːˈmænɪk/ adj. Chem. of or containing germanium, esp. in its tetravalent state.

Germanist /ˈdʒɜːmənɪst/ n. an expert in or student of the language, literature, and civilization of Germany, or Germanic languages.

germanium /dʒɜːˈmeɪnɪəm/ n. a shiny grey semi-metallic chemical element (atomic number 32; symbol **Ge**). Germanium was discovered by the German chemist Clemens Winkler (1838–1904) in 1886, its existence having been suggested by Mendeleev on the basis of his periodic table. A rare element, germanium became very important after the Second World War in the making of transistors and other semiconductor devices, but it has been to some extent displaced by silicon for this purpose. [*Germanus* GERMAN]

Germanize /ˈdʒɜːməˌnaɪz/ v.tr. & intr. (also **-ise**) make or become German; adopt or cause to adopt German customs etc. □ **Germanizer** n. **Germanization** /ˌdʒɜːmənaɪˈzeɪʃ(ə)n/ n.

Germano- /dʒɜːˈmænəʊ/ comb. form German; German and.

germanous /dʒɜːˈmeɪnəs/ adj. Chem. containing germanium in the divalent state.

German South West Africa a former German protectorate in SW Africa (1884–1918), corresponding to present-day Namibia.

Germany /ˈdʒɜːmənɪ/ (called in German *Deutschland*) a country in central Europe; pop. (1991) 78,700,000; official language, German; capital, Berlin; seat of government, Bonn. In the early history of the region German tribes came repeatedly into conflict with the Romans, and after the collapse of the Roman Empire they overran the Rhine which had been its northern frontier. Loosely unified under the Holy Roman Empire during the Middle Ages, the multiplicity of small German states achieved real unity only with the rise of Prussia and the formation of the German Empire in the mid-19th century. After being defeated in the First World War, Germany was taken over in the 1930s by the Nazi dictatorship which led to a policy of expansionism and eventually to complete defeat in the Second World War. Germany was occupied for a time by the victorious Allies during which it was partitioned; some territory in the east, including East Prussia, was lost, mainly to Poland and the USSR. The western part (including West Berlin), which was occupied by the US, Britain, and France, became the Federal Republic of Germany (*West Germany*), with its capital at Bonn. The eastern part, occupied by the Soviet Union, became the German Democratic Republic (*East Germany*), with its capital in East Berlin. West Germany emerged as a major European industrial power within the Western defence community, and was a founder member of the EEC, while the East remained under Soviet domination. After the general collapse of Communism in eastern Europe towards the end of 1989, East and West Germany reunited on 3 Oct. 1990. Germany remains the strongest economy within the EC, although the costs of reunification have strained the country's resources.

germicide /ˈdʒɜːmɪˌsaɪd/ n. a substance destroying germs, esp. those causing disease. □ **germicidal** /ˌdʒɜːmɪˈsaɪd(ə)l/ adj.

germinal /'dʒɜːmɪn(ə)l/ adj. **1** relating to or of the nature of a germ or germs (see GERM 1). **2** in the earliest stage of development. **3** productive of new ideas. □ **germinally** adv. [L germen germin- sprout: see GERM]

germinate /'dʒɜːmɪˌneɪt/ v. **1 a** intr. sprout, bud, or put forth shoots. **b** tr. cause to sprout or shoot. **2 a** tr. cause (ideas etc.) to originate or develop. **b** intr. come into existence. □ **germinator** n. **germinative** /-nətɪv/ adj. **germination** /ˌdʒɜːmɪˈneɪʃ(ə)n/ n. [L germinare germinat- (as GERM)]

Germiston /'dʒɜːmɪstən/ a city in South Africa, in the province of Pretoria-Witwatersrand-Vereeniging, south-east of Johannesburg; pop. (1980) 155,400. It is the site of a large gold refinery, which serves the Witwatersrand gold-mining region.

germon /'dʒɜːmən/ n. = ALBACORE. [F]

Geronimo[1] /dʒəˈrɒnɪˌməʊ/ (c.1829–1909), Apache chief. He led his people in resistance to white encroachment on tribal reservations in Arizona, waging war against settlers and US troops in a series of raids, before surrendering in 1886.

Geronimo[2] /dʒəˈrɒnɪˌməʊ/ int. expressing exhilaration on performing a daring leap etc. (orig. adopted as a slogan by US paratroopers in the Second World War). [GERONIMO[1]]

gerontology /ˌdʒerɒnˈtɒlədʒɪ/ n. the scientific study of old age, the process of ageing, and the special problems of old people. □ **gerontologist** n. **gerontological** /-təˈlɒdʒɪk(ə)l/ adj. [Gk gerōn -ontos old man + -LOGY]

-gerous /dʒərəs/ comb. form forming adjectives meaning 'bearing' (lanigerous). [f. L -ger bearing + -OUS]

gerrymander /'dʒerɪˌmændə(r)/ v. & n. (also **jerrymander**) ● v.tr. **1** manipulate the boundaries of (a constituency etc.) so as to give undue influence to some party or class. (See note below.) **2** manipulate (a situation etc.) to gain advantage. ● n. this practice. □ **gerrymanderer** n.

▪ The term is derived from a combination of the name of Governor Elbridge Gerry (1744–1814) of Massachusetts with the word salamander, in reference to the supposed resemblance to a salamander of the outline of an electoral district formed by Gerry in 1812 for party purposes.

Gershwin /'gɜːʃwɪn/, George (born Jacob Gershovitz) (1898–1937), American composer and pianist, of Russian-Jewish descent. He had no formal musical training, but gained early experience of popular music from working in New York's Tin Pan Alley. He made his name in 1919 with the song 'Swanee' and went on to compose many successful songs and musicals. The lyrics for many of these were written by his brother Ira (Israel, 1896–1983), who was also the librettist for the opera Porgy and Bess (1935). In 1924 George Gershwin successfully turned to orchestral music with his jazz-influenced Rhapsody in Blue.

gerund /'dʒerənd/ n. Gram. a form of a verb functioning as a noun, orig. in Latin ending in -ndum (declinable), in English ending in -ing and used distinctly as a part of a verb (e.g. do you mind my asking you?). [LL gerundium f. gerundum var. of gerendum, the gerund of L gerere do]

gerundive /dʒɪˈrʌndɪv/ n. Gram. a form of a Latin verb, ending in -ndus (declinable) and functioning as an adjective meaning 'that should or must be done' etc. [LL gerundivus (modus mood) f. gerundium: see GERUND]

gesso /'dʒesəʊ/ n. (pl. **-oes**) plaster of Paris or gypsum as used in painting or sculpture. [It. f. L gypsum: see GYPSUM]

gestalt /gəˈstɑːlt, -ˈʃtɑːlt/ n. Psychol. an organized whole that is perceived as more than the sum of its parts. □ **gestaltism** n. **gestaltist** n. [G, = form, shape]

gestalt psychology n. a movement in psychology founded in Germany in 1912 by three psychologists, Max Wertheimer (1880–1943), Wolfgang Köhler (1887–1967), and Kurt Koffka (1886–1941). Formed in response to previous theories of perception which tended to analyse perception and experience by breaking them down into their constituent parts, gestalt psychology was an attempt to explain human perception in terms of gestalts; it aimed to show how the mind can perceive organized wholes by understanding relations between otherwise unconnected physical stimuli, a commonly given example being that of the illusion of the moving picture created from a series of 'stills'. Gestalt principles were later used in numerous other fields, including learning, aesthetics, and psychotherapy.

gestalt therapy n. a type of psychotherapy developed by the German-born psychoanalyst Fritz Perls (1893–1970), which concentrates on awareness and personal responsibility and a recognition of 'gestalts',

significant patterns or constructs in oneself and in one's relation to the world.

Gestapo /geˈstɑːpəʊ/ the German secret police under Nazi rule. Founded in 1933 by Hermann Goering, the Gestapo ruthlessly suppressed opposition to the Nazis within Germany and in occupied Europe and was responsible for rounding up Jews and other groups to be sent to concentration camps. From 1936 it was headed by Heinrich Himmler. [G, f. Geheime Staatspolizei Secret State Police]

gestate /'dʒesteɪt, dʒeˈsteɪt/ v.tr. **1** carry (a foetus) in gestation. **2** develop (an idea etc.).

gestation /dʒeˈsteɪʃ(ə)n/ n. **1 a** the process of carrying or being carried in the womb between conception and birth. **b** this period. **2** the private development of a plan, idea, etc. [L gestatio f. gestare frequent. of gerere carry]

gesticulate /dʒeˈstɪkjʊˌleɪt/ v. **1** intr. use gestures instead of or in addition to speech. **2** tr. express with gestures. □ **gesticulator** n. **gesticulative** /-lətɪv/ adj. **gesticulatory** /-lətərɪ/ adj. **gesticulation** /-ˌstɪkjʊˈleɪʃ(ə)n/ n. [L gesticulari f. gesticulus dimin. of gestus GESTURE]

gesture /'dʒestʃə(r)/ n. & v. ● n. **1** a movement of the body or limbs as an expression of thought or feeling (she silenced him with a gesture of the hand). **2** the use of such movements esp. to convey feeling or as a rhetorical device. **3** an action to evoke a response or convey intention, usu. friendly (a gesture of friendship). ● v.tr. & intr. gesticulate. □ **gestural** adj. [ME f. med.L gestura f. L gerere gest- wield]

gesundheit /gəˈzʊnthaɪt/ int. expressing a wish of good health, esp. before drinking or to a person who has sneezed. [G, = health]

get /get/ v. & n. ● v. (**getting**; past **got** /gɒt/ or archaic **gat** /gæt/; past part. **got** or N. Amer. (and in comb.) **gotten** /'gɒt(ə)n/) **1** tr. come into the possession of; receive or earn (get a job; got £200 a week; got first prize). **2** tr. **a** fetch, obtain, procure, purchase (get my book for me; got a new car). **b** capture, get hold of (a person). **3** tr. go to reach or catch (a bus, train, etc.). **4** tr. prepare (a meal etc.). **5** intr. & tr. reach or cause to reach a certain state or condition; become or cause to become (get rich; get one's feet wet; get to be famous; got them ready; got him into trouble). **6** tr. obtain as a result of calculation. **7** tr. contract (a disease etc.). **8** tr. establish or be in communication with via telephone or radio; receive (a radio signal). **9** tr. experience or suffer; have inflicted on one; receive as one's lot or penalty (got four years in prison). **10 a** tr. succeed in bringing, placing, etc. (get it round the corner; cannot get the key into the lock; get it on to the agenda; flattery will get you nowhere). **b** intr. & tr. succeed or cause to succeed in coming or going (will get you there somehow; got absolutely nowhere). **11** tr. (prec. by have) **a** possess (have not got a penny). **b** (foll. by to + infin.) be bound or obliged (have got to see you). **12** tr. (foll. by to + infin.) induce; prevail upon (got them to help me). **13** tr. colloq. understand (a person or argument) (have you got that?; I get your point; do you get me?). **14** tr. colloq. inflict punishment or retribution on, esp. in retaliation (I'll get you for that). **15** tr. colloq. **a** annoy. **b** move; affect emotionally. **c** attract, obsess. **d** amuse. **16** tr. (foll. by to + infin.) develop an inclination as specified (am getting to like it). **17** intr. (foll. by pres. part.) begin (get going). **18** tr. (esp. in past or perfect) catch in an argument; corner, puzzle. **19** tr. establish (an idea etc.) in one's mind. **20** intr. sl. be off; go away. **21** tr. archaic beget. **22** tr. archaic learn; acquire (knowledge) by study. ● n. **1 a** an act of begetting (of animals). **b** an offspring (of animals). **2** sl. a fool or idiot. □ **be getting on for** be approaching (a specified time, age, etc.). **get about** (or **around**) **1** travel extensively or fast; go from place to place. **2** manage to walk, move about, etc. (esp. after illness). **3** (of news) be circulated, esp. orally. **get across 1** manage to communicate (an idea etc.). **2** (of an idea etc.) be communicated successfully. **3** colloq. annoy, irritate. **get ahead** be or become successful. **get along** (or **on**) **1** (foll. by together, with) live harmoniously, accord. **2** (as imper.) colloq. expressing scepticism. **3** leave, depart (I must be getting along). **get at 1** reach; get hold of. **2** colloq. imply (what are you getting at?). **3** colloq. nag, criticize, bully. **get away 1** escape. **2** leave, esp. on holiday. **3** (as imper.) colloq. expressing disbelief or scepticism. **4** (foll. by with) escape blame or punishment for. **get back 1** move back or away. **2** return, arrive home. **3** recover (something lost). **4** (usu. foll. by to) contact later (I'll get back to you). **get back at** colloq. retaliate against. **get by** colloq. **1** just manage, even with difficulty. **2** be acceptable. **get down 1** alight, descend (from a vehicle, ladder, etc.). **2** record in writing. **3** esp. US colloq. be successful; participate fully, esp. dance enthusiastically. **get a person down** depress or deject him or her. **get down to 1** begin working on or discussing. **2** consider the essentials of. **get even** (often foll. by with) **1** achieve revenge; act in retaliation. **2** equalize the score. **get his** (or **hers** etc.) sl. be killed. **get hold of 1** grasp (physically). **2** grasp (intellectually); understand. **3** make contact with (a person). **4** acquire. **get in 1** gain entrance;

arrive. **2** be elected. **get into** become interested or involved in. **get it** *sl.* be punished or in trouble. **get it into one's head** (foll. by *that +* clause) firmly believe or maintain; realize. **get off 1** *colloq.* be acquitted; escape with little or no punishment. **2** leave. **3** alight; alight from (a bus etc.). **4** go, or cause to go, to sleep. **5** (foll. by *with, together*) *Brit. colloq.* form an amorous or sexual relationship, esp. abruptly or quickly. **get a person off** *colloq.* cause a person to be acquitted. **get off on** *sl.* be excited or aroused by, enjoy. **get on 1** make progress; manage. **2** enter (a bus etc.). **3** see *get along.* **4** (usu. as **getting on** *adj.*) *colloq.* grow old. **get on to** *colloq.* **1** make contact with. **2** understand; become aware of. **get out 1** leave or escape. **2** manage to go outdoors. **3** alight from a vehicle. **4** transpire; become known. **5** succeed in uttering, publishing, etc. **6** solve or finish (a puzzle etc.). **7** *Cricket* be dismissed. **get-out** *n.* a means of avoiding something. **get a person out 1** help a person to leave or escape. **2** *Cricket* dismiss (a batsman). **get out of 1** avoid or escape (a duty etc.). **2** abandon (a habit) gradually. **get a thing out of** manage to obtain something (esp. information) from (a person) esp. with difficulty. **get outside** (or **outside of**) *sl.* eat or drink. **get over 1** recover from (an illness, upset, etc.). **2** overcome (a difficulty). **3** manage to communicate (an idea etc.). **get a thing over** (or **over with**) complete (a tedious task) promptly. **get one's own back** *colloq.* have one's revenge. **get-rich-quick** *adj.* concerned or designed (esp. exclusively) to make a lot of money fast. **get rid of** see RID. **get round** (US **around**) **1** successfully coax or cajole (a person) esp. to secure a favour. **2** evade (a law etc.). **get round to** deal with (a task etc.) in due course. **get somewhere** make progress; be initially successful. **get there** *colloq.* **1** succeed. **2** understand what is meant. **get through 1** pass or assist in passing (an examination, an ordeal, etc.). **2** finish or use up (esp. resources). **3** (often foll. by *to*) make contact by telephone. **4** (foll. by *to*) succeed in making (a person) listen or understand. **get a thing through** cause it to overcome obstacles, difficulties, etc. **get to 1** reach. **2** = *get down to.* **get together** gather, assemble. **get-together** *n. colloq.* a social gathering. **get up 1** rise or cause to rise from sitting etc., or from bed after sleeping or an illness. **2** ascend or mount, e.g. on horseback. **3** (of fire, wind, or the sea) begin to be strong or agitated. **4** prepare or organize. **5** enhance or refine one's knowledge of (a subject). **6** work up (a feeling, e.g. anger). **7** produce or stimulate (*get up steam; get up speed*). **8** (often *refl.*) dress or arrange elaborately; make presentable; arrange the appearance of. **9** (foll. by *to*) *colloq.* indulge or be involved in (*always getting up to mischief*). **get-up** *n. colloq.* **1** a style or arrangement of dress etc., esp. an elaborate one. **2** a style of production or finish, esp. of a book. **get-up-and-go** *colloq.* energy, vim, enthusiasm. **get the wind up** see WIND¹. **get with child** *archaic* make pregnant. **have got it bad** (or **badly**) *sl.* be obsessed or affected emotionally. □ **gettable** *adj.* [ME f. ON *geta* obtain, beget, guess, corresp. to OE *gietan* (recorded only in compounds), f. Gmc]

get-at-able /ɡet'ætəb(ə)l/ *adj.* accessible.

getaway /'ɡetəˌweɪ/ *n.* an escape, esp. after committing a crime.

Gethsemane, Garden of /ɡeθ'semənɪ/ a garden lying in the valley between Jerusalem and the Mount of Olives, where Jesus went with his disciples after the Last Supper and which was the scene of his agony and betrayal (Matt. 26:36–46). [Heb. *gath-shemen* oil-press]

getter /'ɡetə(r)/ *n. & v.* ● *n.* **1** in senses of GET *v.* **2** *Physics* a substance used to remove residual gas from an evacuated vessel. ● *v.tr. Physics* remove (gas) or evacuate (a vessel) with a getter.

Getty /'ɡetɪ/, Jean Paul (1892–1976), American industrialist. He made a large fortune in the oil industry and was also a noted art collector. He founded the museum which bears his name at Malibu in California.

Gettysburg, Battle of /'ɡetɪzˌbɜːɡ/ a decisive battle of the American Civil War, fought near the town of Gettysburg in Pennsylvania in the first three days of July 1863. The Confederate army of northern Virginia, commanded by General Lee, was repulsed in a bloody engagement by the Union army of the Potomac, commanded by General Meade, forcing Lee to abandon his invasion of the north.

Gettysburg address a speech delivered on 18 Nov. 1863 by President Abraham Lincoln at the dedication of the national cemetery on the site of the Battle of Gettysburg.

geum /'dʒiːəm/ *n.* a rosaceous plant of the genus *Geum* (rose family), with rosettes of leaves and yellow, red, or white flowers, e.g. herb bennet. [mod.L, var. of L *gaeum*]

GeV *abbr.* gigaelectronvolt (equivalent to 10⁹ electronvolts). Also called *BeV.*

gewgaw /'ɡjuːɡɔː/ *n.* a gaudy plaything or ornament; a bauble. [ME: orig. unkn.]

Gewürztraminer /ɡə'vʊətstrɑːˌmiːnə(r)/ *n.* **1** a variety of Traminer grape grown esp. in the Rhine valley, Alsace, and Austria. **2** a mildly spicy white wine made from these grapes. [G f. *Gewürz* spice + TRAMINER]

geyser *n.* **1** /'ɡaɪzə(r), 'ɡiːz-/ an intermittently gushing hot spring that throws up a tall column of water. **2** /'ɡiːzə(r)/ *Brit.* an apparatus for heating water rapidly for domestic use. [Icel. *Geysir*, the name of a particular spring in Iceland, rel. to *geysa* to gush]

GG *abbr.* Governor-General.

Ghana /'ɡɑːnə/ a country of West Africa, with its southern coastline bordering on the Atlantic Ocean; pop. (est. 1991) 16,500,000; languages, English (official), West African languages; capital, Accra. In the 15th century the region was visited by Portuguese and other European traders, who called it the Gold Coast, and it became a centre of the slave trade until the 19th century. Britain established control over the area and it became the British colony of Gold Coast in 1874. In 1957 it gained independence as a member of the Commonwealth, under the leadership of Kwame Nkrumah. It was the first former British colony to become independent, and took the name of Ghana from a kingdom that had flourished in that region in medieval times. Ghana became a republic in 1960. □ **Ghanaian** /ɡɑː'neɪən/ *adj. & n.*

gharial /'ɡeərɪəl, 'ɡær-/ *n.* (also **gavial** /'ɡeɪvɪəl/) a large fish-eating crocodile, *Gavialis gangeticus*, native to India, having a long narrow snout widening at the nostrils. □ **false gharial** a similar crocodile, *Tomistoma schlegelii*, of Indonesia and Malaya. [Hindi]

ghastly /'ɡɑːstlɪ/ *adj. & adv.* ● *adj.* (**ghastlier, ghastliest**) **1** horrible, frightful. **2** *colloq.* objectionable, unpleasant. **3** deathlike, pallid. ● *adv.* in a ghastly or sickly way (*ghastly pale*). □ **ghastlily** *adv.* **ghastliness** *n.* [ME *gastlich* f. obs. *gast* terrify: *gh* after *ghost*]

ghat /ɡɑːt/ *n.* (also **ghaut**) in the Indian subcontinent: **1** steps leading down to a river. **2** a landing-place. **3** a defile or mountain pass. □ **burning-ghat** a level area at the top of a river ghat where Hindus burn their dead. [Hindi *ghāṭ*]

Ghats, the /ɡɑːts/ two mountain ranges in central and southern India. Known as the *Eastern Ghats* and the *Western Ghats*, they run parallel to the coast on either side of the Deccan plateau, meeting at the southern tip of India.

Ghazi /'ɡɑːzɪ/ *n.* (*pl.* **Ghazis**) (often as an honorific title) a Muslim fighter against non-Muslims. [Arab. *al-ġāzī* part. of *ġazā* raid]

Ghaziabad /'ɡɑːzɪəˌbæd/ a city in northern India, in Uttar Pradesh east of Delhi; pop. (1991) 461,000.

Ghaznavid /ɡæz'nɑːvɪd/ *n. & adj.* ● *n.* a member of a Turkish Muslim dynasty founded in Ghazna, Afghanistan, in AD 977. The dynasty extended its power into Persia and the Punjab but in the 11th century fragmented under pressure from the Seljuk Turks, and was finally destroyed in 1186. ● *adj.* of or relating to this dynasty.

ghee /ɡiː/ *n.* (also **ghi**) Indian clarified butter esp. from the milk of a buffalo or cow. [Hindi *ghī* f. Skr. *ghṛitá-* sprinkled]

Gheg /ɡeɡ/ *n. & adj.* ● *n.* (*pl.* **Ghegs** or same) **1** a member of one of the main ethnic groups of Albania, living mainly in the north of the country. **2** the dialect of Albanian spoken by this people. ● *adj.* of or relating to the Ghegs or their dialect. [Albanian *Geg*]

Ghent /ɡent/ (Flemish **Gent** /xɛnt/, French **Gand** /ɡɑ̃/) a city in Belgium, capital of the province of East Flanders; pop. (1991) 230,200. Founded in the 10th century, it became the capital of the medieval principality of Flanders. A port on the River Scheldt, it is connected by canal to the North Sea at Zeebrugge. It was formerly known in English as Gaunt (surviving in names, e.g. John of Gaunt).

gherao /ɡe'raʊ/ *n.* (*pl.* **-os**) (in India and Pakistan) coercion of employers, by which their workers prevent them from leaving the premises until certain demands are met. [Hindi *gherna* besiege]

gherkin /'ɡɜːkɪn/ *n.* a small kind of cucumber, or a young green cucumber, used for pickling. [Du. *gurkkijn* (unrecorded), dimin. of *gurk*, f. Slav., ult. f. med. Gk *aggourion*]

ghetto /'ɡetəʊ/ *n. & v.* ● *n.* (*pl.* **-os**) **1** a part of a city, esp. a slum area, occupied by a minority group or groups. **2** *hist.* the Jewish quarter in a city. **3** a segregated group or area. ● *v.tr.* (**-oes, -oed**) put or keep (people) in a ghetto. □ **ghetto-blaster** a large portable radio, esp. used to play loud pop music. □ **ghettoize** *v.tr.* (also **ghettoise**). [perh. f. It. *getto* foundry (applied to the site of the first ghetto in Venice in 1516)]

ghi var. of GHEE.

Ghibelline /'ɡɪbɪˌlaɪn/ *n.* a member of one of the two great political factions in Italian medieval politics, traditionally supporting the Holy Roman emperor against the pope and his supporters, the Guelphs, during the long struggle between the papacy and the Empire. [It. *Ghibellino* perh. f. G *Waiblingen* estate belonging to Hohenstaufen emperors]

Ghiberti /gɪ'beətɪ/, Lorenzo (1378–1455), Italian sculptor and goldsmith. His career was dominated by his work on two successive pairs of bronze doors for the baptistery in Florence. The second, more famous, pair (1425–52) depicts episodes from the Bible laid out on carefully constructed perspective stages.

ghillie var. of GILLIE.

Ghirlandaio /ˌɡɪəlæn'daɪəʊ/ (born Domenico di Tommaso Bigordi) (c.1448–94), Italian painter. Born the son of a goldsmith, he acquired the name Ghirlandaio (= garland-maker) in recognition of his father's skill in metal garland-work. He worked mainly in Florence and is noted for his religious frescos, painted in a naturalistic style and including detailed portraits of leading contemporary citizens. His major works include the fresco *Christ Calling Peter and Andrew* (1482–4) in the Sistine Chapel, Rome.

ghost /ɡəʊst/ n. & v. ● n. **1** the soul of a dead person which supposedly manifests itself to the living visibly (as a shadowy apparition), audibly etc. **2** a faint mark; a mere semblance (*not a ghost of a chance*). **3** a secondary or duplicated image produced by defective television reception or by a telescope. **4** *archaic* the spirit of God (*Holy Ghost*). ● v. **1** *intr.* (often foll. by *for*) act as ghost-writer. **2** *tr.* act as ghost-writer of (a work). □ **ghost town** a deserted town with few or no remaining inhabitants. **ghost train** (at a funfair) an open-topped miniature railway in which the rider experiences ghoulish sights, sounds, etc. **ghost-write** act as ghost-writer (of). **ghost-writer** a person who writes on behalf of the credited author of a work. □ **ghostlike** adj. [OE *gāst* f. WG: *gh-* occurs first in Caxton, prob. infl. by Flem. *gheest*]

ghostbuster /ˈɡəʊstˌbʌstə(r)/ n. colloq. a person who professes to banish ghosts, poltergeists, etc.

Ghost Dance an American Indian religious cult of the second half of the 19th century. The cult was based on the performance of a ritual dance, lasting sometimes for several days, which, it was believed, would drive away white people, bring the dead back to life, and restore the traditional lands and way of life. Advocated by the Sioux chief Sitting Bull, the cult was central to the uprising that was crushed at the Battle of Wounded Knee.

ghosting /ˈɡəʊstɪŋ/ n. the appearance of a 'ghost' (see GHOST n. 3) or secondary image in a television picture.

ghostly /ˈɡəʊstlɪ/ adj. (**ghostlier**, **ghostliest**) like a ghost; spectral. □ **ghostliness** n. [OE *gāstlic* (as GHOST)]

ghoul /ɡuːl/ n. **1** an evil spirit or phantom. **2** a spirit in Arabic folklore preying on travellers. **3** a person morbidly interested in death etc. □ **ghoulish** adj. **ghoulishly** adv. **ghoulishness** n. [Arab. *ḡūl* protean desert demon]

GHQ abbr. General Headquarters.

Ghulghuleh /ɡʊlˈɡʊlə/ an ancient city, now ruined, near Bamian in central Afghanistan. The city was destroyed by Genghis Khan c.1221.

ghyll Brit. var. of GILL³.

GI /dʒiːˈaɪ/ n. & adj. ● n. (pl. **GIs**) a private soldier in the US army. ● adj. of or for US servicemen. [abbr. of *government* (or *general*) *issue*]

Giacometti /ˌdʒækəˈmetɪ/, Alberto (1901–66), Swiss sculptor and painter. He experimented with both naturalistic sculpture and surrealism, but his most characteristic style, which he adopted after the Second World War, features human figures that are notable for their emaciated and extremely elongated forms, exemplified in such works as *Pointing Man* (1947).

giant /ˈdʒaɪənt/ n. & adj. ● n. **1** (fem. **giantess**) an imaginary or mythical being of human form but superhuman size. **2** *Gk Mythol.* a member of a race of beings of enormous size and great strength, sons of Gaia (Earth), who tried to overthrow the Olympian gods and were defeated with the help of Hercules. **3** an abnormally tall or large person, animal, or plant. **4** a person of exceptional ability, integrity, courage, etc. **5** *Astron.* a star of relatively high luminosity. (*See note below.*) ● attrib.adj. **1** of extraordinary size or force, gigantic; monstrous. **2** *colloq.* extra large (*giant packet*). **3** (of a plant or animal) of a very large kind. □ **giant-killer** a person who defeats a seemingly much more powerful opponent. **giant-killing** defeat of or victory over a seemingly much stronger opponent. **giant sequoia** see SEQUOIA. □ **giantism** n. **giant-like** adj. [ME *geant* (later infl. by L) f. OF, ult. f. L *gigas gigant-* f. Gk]

▪ Giant stars have a very high luminosity and a radius 10 to 100 times that of the sun. Most are red giants, which have greatly expanded after the nuclear reactions which fuel them have transferred to regions outside the core.

Giant's Causeway a geological formation of basalt columns, dating from the Tertiary period, on the north coast of Northern Ireland. It was once believed to be the end of a road made by a legendary giant to Staffa in the Inner Hebrides, where there is a similar formation.

giaour /ˈdʒaʊə(r)/ n. archaic derog. a non-Muslim, esp. a Christian. [Pers. *gaur*, prob. f. Arab. *kāfir* KAFFIR]

Giap /dʒæp/, Vo Nguyen (b.1912), Vietnamese military and political leader. As North Vietnamese Vice-Premier and Defence Minister, he was responsible for the strategy leading to the withdrawal of American forces from South Vietnam in 1973 and the subsequent reunification of the country in 1975. His book, *People's War, People's Army* (1961), was an influential text for revolutionaries.

giardiasis /ˌdʒiːɑːˈdaɪəsɪs/ n. Med. infection of the gut with a flagellate protozoan of the genus *Giardia*, esp. *G. lamblia*, causing diarrhoea etc. [mod.L *Giardia* genus name (f. A. *Giard*, Fr. biologist (1846–1908)) + -IASIS]

Gib. /dʒɪb/ n. colloq. Gibraltar. [abbr.]

gib /dʒɪb, ɡɪb/ n. a wood or metal bolt, wedge, or pin for holding a machine part etc. in place. [18th c.: orig. unkn.]

gibber¹ /ˈdʒɪbə(r)/ v. & n. ● v.intr. speak fast and inarticulately; chatter incoherently. ● n. such speech or sound. [imit.]

gibber² /ˈɡɪbə(r)/ n. Austral. a boulder or large stone. [Dharuk *giba* stone]

gibberellin /ˌdʒɪbəˈrelɪn/ n. any of a group of plant hormones that promote stem elongation, germination, flowering, etc. [*Gibberella* a genus of fungi, dimin. of genus name *Gibbera* f. L *gibber* hump]

gibberish /ˈdʒɪbərɪʃ/ n. unintelligible or meaningless speech; nonsense. [perh. f. GIBBER¹ (but attested earlier) + -ISH¹ as used in *Spanish*, *Swedish*, etc.]

gibbet /ˈdʒɪbɪt/ n. & v. ● n. hist. **1 a** a gallows. **b** an upright post with an arm on which the bodies of executed criminals were hung up. **2** (prec. by *the*) death by hanging. ● v.tr. (**gibbeted**, **gibbeting**) **1** put to death by hanging. **2 a** expose on a gibbet. **b** hang up as on a gibbet. **3** hold up to contempt. [ME f. OF *gibet* gallows dimin. of *gibe* club, prob. f. Gmc]

Gibbon /ˈɡɪb(ə)n/, Edward (1737–94), English historian. A Catholic convert at 16, he reconverted to Protestantism the following year while studying in Lausanne. While in Rome in 1764 he formed the plan for his book *The History of the Decline and Fall of the Roman Empire* (1776–88), chapters of which aroused controversy for their critical account of the spread of Christianity. The complete work, originally published in six volumes, covers a period from the age of Trajan to the fall of Constantinople in 1453; it is regarded as a contribution to literature as well as historical analysis.

gibbon /ˈɡɪb(ə)n/ n. a small arboreal ape of the genus *Hylobates*, native to SE Asia, having a slender body, long arms, and no tail. [F, of unkn. orig.]

Gibbons /ˈɡɪb(ə)nz/, Grinling (1648–1721), Dutch-born English sculptor. From 1671 he worked in England, where he was introduced to Charles II by the writer John Evelyn, and became Master Carver in Wood to the Crown. He is famous for his decorative carvings (chiefly in wood) of fruit and flowers, small animals, and cherubs' heads; examples of his work can be seen in the choir stalls of St Paul's Cathedral, London.

gibbous /ˈɡɪbəs/ adj. **1** convex or protuberant. **2** (of a moon or planet) having the bright part greater than a semicircle and less than a circle. **3** humped or humpbacked. □ **gibbously** adv. **gibbousness** n. **gibbosity** /ɡɪˈbɒsɪtɪ/ n. [ME f. LL *gibbosus* f. *gibbus* hump]

Gibbs¹ /ɡɪbz/, James (1682–1754), Scottish architect. An admirer of Wren, he developed Wren's ideas for London's city churches, especially in his masterpiece, St Martin's-in-the-Fields (1722–6). The latter's combination of steeple and portico was influential in subsequent church design.

Gibbs² /ɡɪbz/, Josiah Willard (1839–1903), American physical chemist. He was the founder of the study of chemical thermodynamics and statistical mechanics, though the importance of his theoretical work was not generally appreciated until after his death.

gibe /dʒaɪb/ v. & n. (also **jibe**) ● v.intr. (often foll. by *at*) jeer, mock. ● n. an instance of gibing; a taunt. □ **giber** n. [perh. f. OF *giber* handle roughly]

giblets /ˈdʒɪblɪts/ n.pl. the liver, gizzard, neck, etc., of a bird, usu.

removed and kept separate when the bird is prepared for cooking. [OF *gibelet* game stew, perh. f. *gibier* game]

Gibraltar /dʒɪˈbrɔːltə(r)/ a British dependency near the southern tip of the Iberian peninsula, at the eastern end of the Strait of Gibraltar; pop. (est. 1988) 29,140; languages, English (official), Spanish. Occupying a site of great strategic importance, Gibraltar consists of a fortified town and military base at the foot of a rocky headland (the *Rock of Gibraltar*) 426 m (1398 ft) high. It was captured by Britain during the War of the Spanish Succession in 1704 and formally ceded by the Peace of Utrecht (1713–14); Britain is responsible for defence, external affairs, and internal security. Since the Second World War Spain has forcefully urged her claim to the territory; in 1969 Spain closed the border with Gibraltar, although this blockade was subsequently lifted in 1985. Remains of what was later called Neanderthal man were first discovered in Gibraltar in 1848. (See also PILLARS OF HERCULES.) □ **Gibraltarian** /ˌdʒɪbrɔːlˈteərɪən/ adj. & n. [Arab. *gebel-al-Tarik* hill of Tarik (8th-c. Saracen commander)]

Gibraltar, Strait of a channel between the southern tip of the Iberian peninsula and North Africa, forming the only outlet of the Mediterranean Sea to the Atlantic. Varying in width from 24 km (15 miles) to 40 km (25 miles) at its western extremity, it stretches east–west for some 60 km (38 miles).

Gibson /ˈɡɪbs(ə)n/, Althea (b.1927), American tennis player. She was the first black player to compete successfully at the highest level of tennis, winning all the major world women's singles titles, including the French and Italian championships (1956) and the British and American titles (1957 and 1958).

Gibson Desert a desert region in Western Australia, to the south-east of the Great Sandy Desert. The first European to cross it was Ernest Giles; Giles named the desert after his companion Alfred Gibson, who went missing on an expedition in the 1870s.

Gibson girl n. a girl typifying the fashionable ideal of around 1900. [Charles Dana *Gibson* (1867–1944), Amer. artist]

giddy /ˈɡɪdɪ/ adj. & v. ● adj. (**giddier**, **giddiest**) **1** having a sensation of whirling and a tendency to fall, stagger, or spin round. **2 a** overexcited, mentally intoxicated. **b** excitable, frivolous. **3** tending to make one giddy. ● v.tr. & intr. (**-ies**, **-ied**) make or become giddy. □ **giddily** adv. **giddiness** n. [OE *gidig* insane, lit. 'possessed by a god']

giddy-up /ˌɡɪdɪˈʌp, ˌɡɪdɪˈʌp/ int. & v. ● int. commanding a horse to go or go faster. ● v.tr. urge to go or go faster.

Gide /ʒiːd/, André (Paul Guillaume) (1869–1951), French novelist, essayist, and critic. The first of a series of visits to North Africa in 1893 caused him to rebel against his strict Protestant upbringing and to acknowledge the importance of individual expression and sexual freedom, including his own bisexuality. He was a prolific writer, completing more than fifty books in his lifetime and coming to be regarded as the father of modern French literature. His works include *The Immoralist* (1902), *La Porte étroite* (1909, *Strait is the Gate*), *Si le grain ne meurt* (1926, *If I die…*), *The Counterfeiters* (1927), and his *Journal* (1939–50). He was awarded the Nobel Prize for literature in 1947.

Gideon /ˈɡɪdɪən/ **1** (in the Bible) an Israelite leader, described in Judges 6:11 ff. **2** a member of the Christian organization Gideons International. □ **Gideon bible** a bible purchased by this organization and placed in a hotel room etc.

Gideons International an international Christian organization of business and professional people, founded in 1899 in the US with the aim of spreading the Christian faith by placing bibles in hotel rooms, hospital wards, etc.

gie /ɡiː/ v.tr. & intr. Sc. = GIVE.

Gielgud /ˈɡiːlɡʊd/, Sir (Arthur) John (b.1904), English actor and director. A notable Shakespearian actor, he is particularly remembered for his interpretation of the role of Hamlet, which he performed for the first time in 1929; he also appeared in contemporary plays, such as those by Harold Pinter. Gielgud has also appeared on television and in numerous films, including many roles late in his life; he won an Oscar for his role as a butler in *Arthur* (1980).

GIFT /ɡɪft/ n. gamete intrafallopian transfer, a technique for assisting conception in women with healthy Fallopian tubes. Eggs are removed from the ovary, mixed with sperm, and introduced into a Fallopian tube, where fertilization takes place. [acronym]

gift /ɡɪft/ n. & v. ● n. **1** a thing given; a present. **2** a natural ability or talent. **3** the power to give (*in his gift*). **4** the act or an instance of giving. **5** colloq. an easy task. ● v.tr. **1** endow with gifts. **2 a** (foll. by *with*) give to as a gift. **b** bestow as a gift. □ **gift of tongues** see TONGUE. **gift token** (or **voucher**) a voucher used as a gift and exchangeable for goods. **gift wrap** decorative paper etc. for wrapping gifts. **gift-wrap** v.tr. (**-wrapped**, **-wrapping**) wrap attractively as a gift. **look a gift-horse in the mouth** (usu. neg.) find fault with what has been given. [ME f. ON *gipt* f. Gmc, rel. to GIVE]

gifted /ˈɡɪftɪd/ adj. exceptionally talented or intelligent. □ **giftedly** adv. **giftedness** n.

giftware /ˈɡɪftweə(r)/ n. goods sold as suitable gifts.

Gifu /ˈɡiːfuː/ a city in central Japan, on the island of Honshu; pop. (1990) 410,300.

gig[1] /ɡɪɡ/ n. **1** a light two-wheeled one-horse carriage. **2** a light ship's boat for rowing or sailing. **3** a rowing-boat esp. for racing. [ME in var. senses: prob. imit.]

gig[2] /ɡɪɡ/ n. & v. colloq. ● n. **1** an engagement of a musician or musicians to play jazz, pop music, etc., usu. for a single appearance. **2** a performance of this kind. ● v.intr. (**gigged**, **gigging**) perform a gig. [20th c.: orig. unkn.]

gig[3] /ɡɪɡ/ n. a kind of fishing-spear. [short for *fizgig, fishgig*: cf. Sp. *fisga* harpoon]

giga- /ˈɡaɪɡə, ˈɡɪɡə, ˈdʒaɪɡə, ˈdʒɪɡə/ comb. form **1** denoting a factor of 10^9 (*gigawatt*). **2** Computing denoting a factor of 2^{30} (*gigabyte*). [Gk *gigas* giant]

gigaflop /ˈɡaɪɡəˌflɒp, ˈɡɪɡə-, ˈdʒaɪɡə-, ˈdʒɪɡə-/ n. Computing a unit of computing speed equal to one thousand million floating-point operations per second.

gigametre /ˈɡaɪɡəˌmiːtə(r), ˈɡɪɡə-, ˈdʒaɪɡə-, ˈdʒɪɡə-/ n. a metric unit equal to 10^9 metres.

gigantic /dʒaɪˈɡæntɪk/ adj. **1** very large; enormous. **2** like or suited to a giant. □ **gigantically** adv. **gigantesque** /ˌdʒaɪɡænˈtesk/ adj. [L *gigas gigantis* GIANT]

gigantism /ˈdʒaɪɡənˌtɪz(ə)m/ n. abnormal largeness, esp. (Med.) excessive growth due to hormonal imbalance, or (Bot.) excessive size due to polyploidy.

Gigantopithecus /dʒaɪˌɡæntəˈpɪθɪkəs/ n. a genus of very large fossil anthropoid apes, sometimes considered hominids, known from remains found in Asia that date from the Upper Miocene to the Lower Pleistocene epochs. [mod.L f. Gk *gigas* giant + *pithēkos* ape]

giggle /ˈɡɪɡ(ə)l/ v. & n. ● v.intr. laugh in half-suppressed spasms, esp. in an affected or silly manner. ● n. **1** such a laugh. **2** colloq. an amusing person or thing; a joke. □ **giggler** n. **giggly** adj. (**gigglier**, **giggliest**). [imit.: cf. Du. *gichelen*, G *gickeln*]

Gigli /ˈdʒiːliː/, Beniamino (1890–1957), Italian operatic tenor. He made his Milan début with the conductor Toscanini in 1918, and retained his singing talents to a considerable age, touring the US in 1955. Notable roles included Rodolfo in *La Bohème*, the Duke of Mantua in *Rigoletto*, and Cavaradossi in *Tosca*.

GIGO /ˈɡaɪɡəʊ/ abbr. Computing garbage in, garbage out.

gigolo /ˈʒɪɡəˌləʊ, ˈdʒɪɡ-/ n. (pl. **-os**) **1** a young man paid by an older woman to be her escort or lover. **2** a professional male dancing partner or escort. [F, formed as masc. of *gigole* dance-hall woman]

gigot /ˈdʒɪɡət, ˈʒiːɡəʊ/ n. a leg of mutton or lamb. □ **gigot sleeve** a leg-of-mutton sleeve. [F, dimin. of dial. *gigue* leg]

gigue /ʒiːɡ/ n. **1** = JIG 1. **2** a lively piece of music in duple or triple time, usu. with dotted rhythms. [F: see JIG]

Gijón /ɡɪˈhɒn/ a port and industrial city in northern Spain, on the Bay of Biscay; pop. (1991) 260,250.

Gila monster /ˈhiːlə/ n. a large venomous lizard, *Heloderma suspectum*, found in Mexico and the south-western US. [*Gila*, river in New Mexico and Arizona]

Gilbert[1] /ˈɡɪlbət/, Sir Humphrey (c.1539–83), English explorer. After a distinguished career as a soldier, Gilbert led an unsuccessful attempt to colonize the New World (1578–9). On a second voyage in 1583 he claimed Newfoundland for Elizabeth I and established a colony at St John's, but was lost on the trip homewards when his ship foundered in a storm off Nova Scotia.

Gilbert[2] /ˈɡɪlbət/, William (1544–1603), English physician and physicist. He worked on terrestrial magnetism, discovered how to make magnets, and first used the term *magnetic pole*. His book *De Magnete* (1600) is one of the most important early works on physics published in England.

Gilbert[3] /ˈɡɪlbət/, Sir W(illiam) S(chwenck) (1836–1911), English dramatist. His early writing career was chiefly devoted to humorous verse, such as his *Bab Ballads* (1866–73). However, he is best known for his collaboration with the composer Sir Arthur Sullivan; between 1871 and 1896 he wrote the libretti for fourteen light operas, including *Trial by Jury* (1875), *HMS Pinafore* (1878), *The Pirates of Penzance* (1879), *Iolanthe* (1882), *The Mikado* (1885), and *The Gondoliers* (1889).

Gilbert and Ellice Islands a former British colony (1915–75) in the central Pacific, consisting of two groups of islands: the Gilbert Islands, now a part of Kiribati, and the Ellice Islands, now Tuvalu.

Gilbert Islands a group of islands in the central Pacific, forming part of Kiribati. The islands straddle the equator and lie immediately west of the International Date Line. They were formerly part of the British colony of the Gilbert and Ellice Islands. The islands were named by the British after Thomas Gilbert, an English adventurer who arrived there in 1788.

gild[1] /gɪld/ *v.tr.* (*past part.* **gilded** or as adj. in sense 1 **gilt**) **1** cover thinly with gold. **2** tinge with a golden colour or light. **3** give a specious or false brilliance to. □ **gilded cage** a luxurious but restrictive environment. **gilded youth** young people of wealth, fashion, and flair. **gild the lily** try to improve what is already beautiful or excellent. □ **gilder** *n.* [OE *gyldan* f. Gmc]

gild[2] var. of GUILD.

gilding /'gɪldɪŋ/ *n.* **1** the act or art of applying gilt. **2** material used in applying gilt.

gilet /dʒɪ'leɪ/ *n.* **1** a light women's garment resembling a waistcoat. **2** a sleeveless padded jacket. [F, = waistcoat]

gilgai /'gɪlgaɪ/ *n. Austral.* a saucer-like natural reservoir for rainwater. [Wiradhuri *gilgaay*]

Gilgamesh /'gɪlgəmeʃ/ a legendary king of the Sumerian city-state of Uruk in southern Mesopotamia who ruled some time during the first half of the 3rd millennium BC. He is the hero of the epic of Gilgamesh, one of the best-known works of ancient literature, an epic which recounts Gilgamesh's exploits in an ultimately unsuccessful quest for immortality. It contains an account of a flood that has close parallels with the biblical story of Noah.

Gilgit /'gɪlgɪt/ a town in a mountainous district of the same name, in the northern part of Pakistani Kashmir.

Gill /gɪl/, (Arthur) Eric (Rowton) (1882–1940), English sculptor, engraver, and typographer. His best-known sculptures are the relief carvings *Stations of the Cross* (1914–18) at Westminster cathedral and the *Prospero and Ariel* (1931) on Broadcasting House in London. He illustrated many books for the Golden Cockerell Press, and designed printing types for the Monotype Corporation. Among his famous type designs was the first sanserif type, Gill Sans.

gill[1] /gɪl/ *n. & v.* ● *n.* (usu. in *pl.*) **1** the respiratory organ in fishes and other aquatic animals. **2** the vertical radial plates on the underside of mushrooms and other fungi. **3** the flesh below a person's jaws and ears (*green about the gills*). **4** the wattles or dewlap of fowls. ● *v.tr.* **1** gut (a fish). **2** cut off the gills of (a mushroom). **3** catch in a gill-net. □ **gill-cover** a bony case protecting a fish's gills; an operculum. **gill-net** a net for entangling fishes by the gills. □ **gilled** *adj.* (also in *comb.*). [ME f. ON *gil* (unrecorded) f. Gmc]

gill[2] /dʒɪl/ *n.* **1** a unit of liquid measure, equal to a quarter of a pint. **2** *Brit. dial.* half a pint. [ME f. OF *gille*, med.L *gillo* f. LL *gello*, *gillo* water-pot]

gill[3] /gɪl/ *n.* (also **ghyll**) *Brit.* **1** a deep usu. wooded ravine. **2** a narrow mountain torrent. [ME f. ON *gil* glen]

gill[4] /dʒɪl/ *n.* (also **Gill**, **jill**, **Jill**) **1** *derog.* a young woman. **2** *colloq.* or *dial.* a female ferret. [ME, abbr. of *Gillian* f. OF *Juliane* f. L *Juliana* (*Julius*)]

Gillespie /gɪ'lespɪ/, Dizzy (born John Birks Gillespie) (1917–93), American jazz trumpet player and band-leader. He was a virtuoso trumpet player and a leading exponent of the bebop style. After working with various other groups he formed his own in New York in 1944, and thereafter toured the world almost annually.

gillie /'gɪlɪ/ *n.* (also **ghillie**) *Sc.* **1** a man or boy attending a person hunting or fishing. **2** *hist.* a Highland chief's attendant. [Gael. *gille* lad, servant]

Gillingham /'dʒɪlɪŋəm/ a town in Kent, on the Medway estuary south-east of London; pop. (1981) 93,700.

gillyflower /'dʒɪlɪ,flaʊə(r)/ *n.* **1** (in full **clove gillyflower**) a clove-scented pink (see CLOVE[1] 2). **2** a similarly scented flower, such as the wallflower or white stock. [ME *gilofre*, *gerofle* f. OF *gilofre*, *girofle*, f. med.L f. Gk *karuophullon* clove-tree f. *karuon* nut + *phullon* leaf, assim. to FLOWER; orig. the name of the spice: see CLOVE[1]]

gilt[1] /gɪlt/ *adj. & n.* ● *adj.* **1** covered thinly with gold. **2** gold-coloured. ● *n.* **1** gold or a goldlike substance applied in a thin layer to a surface.

2 (often in *pl.*) a gilt-edged security. □ **gilt-edged 1** (of securities, stocks, etc.) having a high degree of reliability as an investment. **2** having a gilded edge. [past part. of GILD[1]]

gilt[2] /gɪlt/ *n.* a young unbred sow. [ME f. ON *gyltr*]

giltwood /'gɪltwʊd/ *adj.* made of wood and gilded.

gimbals /'dʒɪmb(ə)lz/ *n.pl.* a contrivance, usu. of rings and pivots, for keeping instruments such as a compass and chronometer horizontal at sea, in the air, etc. □ **gimballed** *adj.* [var. of earlier *gimmal* f. OF *gemel* double finger-ring f. L *gemellus* dimin. of *geminus* twin]

gimcrack /'dʒɪmkræk/ *adj. & n.* ● *adj.* showy but flimsy and worthless. ● *n.* a cheap showy ornament; a knick-knack. □ **gimcrackery** *n.* **gimcracky** *adj.* [ME *gibecrake* a kind of ornament, of unkn. orig.]

gimlet /'gɪmlɪt/ *n.* **1** a small tool with a screw-tip for boring holes. **2** a cocktail usu. of gin and lime juice. □ **gimlet eye** an eye with a piercing glance. [ME f. OF *guimbelet*, dimin. of *guimble*]

gimmick /'gɪmɪk/ *n. colloq.* a trick or device, esp. to attract attention, publicity, or trade. □ **gimmickry** *n.* **gimmicky** *adj.* [20th-c. US: orig. unkn.]

gimp[1] /gɪmp/ *n.* (also **guimp**, **gymp**) **1** a twist of silk etc. with cord or wire running through it, used esp. as trimming. **2** fishing-line of silk etc. bound with wire. **3** a coarser thread outlining the design of lace. [Du.: orig. unkn.]

gimp[2] /gɪmp/ *n. & v. sl.* ● *n.* **1** a lame person or leg. **2** a stupid or contemptible person. ● *v.intr.* limp, hobble. □ **gimpy** *adj.* [orig. uncert.: perh. alt. of GAMMY]

gin[1] /dʒɪn/ *n.* an alcoholic spirit distilled from grain or malt and flavoured with juniper berries, originating in the Netherlands in the 17th century. □ **gin rummy** a form of the card-game rummy. [abbr. of GENEVER]

gin[2] /dʒɪn/ *n. & v.* ● *n.* **1** a snare or trap. **2** a machine for separating cotton from its seeds. **3** a kind of crane and windlass. ● *v.tr.* (**ginned**, **ginning**) **1** treat (cotton) in a gin. **2** trap. [ME f. OF *engin* ENGINE]

gin[3] /dʒɪn/ *n. Austral.* an Aboriginal woman. [Dharuk *diyin* woman, wife]

ging /gɪŋ/ *n. Austral. colloq.* a catapult. [20th c.: orig. uncert.]

ginger /'dʒɪndʒə(r)/ *n., adj., & v.* ● *n.* **1 a** a hot spicy root usu. powdered for use in cooking, or preserved in syrup, or candied. **b** the plant, *Zingiber officinale*, of SE Asia, having this root. **2** a light reddish-yellow colour; the sandy or reddish colour of a person's hair. **3** spirit, mettle. **4** stimulation. ● *adj.* sandy, reddish, esp. of hair. ● *v.tr.* **1** flavour with ginger. **2** (foll. by *up*) rouse or enliven. **3** *Austral. colloq.* steal from (a person). □ **black ginger** unscraped ginger. **ginger ale** an effervescent non-alcoholic clear drink flavoured with ginger extract. **ginger beer** an effervescent mildly alcoholic cloudy drink, made by fermenting a mixture of ginger and syrup. **ginger group** *Brit.* a group within a party or movement that presses for stronger or more radical policy or action. **ginger-nut** a ginger-flavoured biscuit. **ginger-pop** *colloq.* ginger beer or ginger ale. **ginger-snap** a thin brittle biscuit flavoured with ginger. **ginger wine** a drink of fermented sugar, water, and bruised ginger. □ **gingery** *adj.* [ME f. OE *gingiber* & OF *gingi(m)bre*, both f. med.L *gingiber* ult. f. Skr. *śṛṅgaveram* f. *śṛṅgam* horn + -*vera* body, with ref. to the antler-shape of the root]

gingerbread /'dʒɪndʒə,bred/ *n.* **1** a cake made with treacle or syrup and flavoured with ginger. **2** (often *attrib.*) a gaudy or tawdry decoration or ornament. □ **take the gilt off the gingerbread** strip something of its attractions.

gingerly /'dʒɪndʒəlɪ/ *adv. & adj.* ● *adv.* in a careful or cautious manner. ● *adj.* showing great care or caution. □ **gingerliness** *n.* [perh. f. OF *gensor* delicate, compar. of *gent* graceful f. L *genitus* (well-)born]

gingham /'gɪŋəm/ *n.* a plain-woven cotton cloth esp. striped or checked. [Du. *gingang* f. Malay *ginggang* (orig. adj. = striped)]

gingili /'dʒɪndʒɪlɪ/ *n.* **1** sesame. **2** sesame oil. [Hindi *jinjalī* f. Arab. *juljulān*]

gingival /dʒɪn'dʒaɪv(ə)l, 'dʒɪndʒɪ-/ *adj.* of or relating to the gums. [L *gingiva* gum]

gingivitis /,dʒɪndʒɪ'vaɪtɪs/ *n. Med.* inflammation of the gums.

gingko var. of GINKGO.

ginglymus /'dʒɪŋglɪməs/ *n.* (*pl.* **ginglymi** /-,maɪ/) *Anat.* a hingelike joint in the body with motion in one plane only, e.g. the elbow or knee. [mod.L f. Gk *gigglumos* hinge]

gink[1] /gɪŋk/ *n. sl.* often *derog.* a fellow; a man. [20th-c. US: orig. unkn.]

gink[2] /gɪŋk/ *n. Austral.* a scrutinizing look. [prob. alt. of GEEK 2]

ginkgo /'gɪŋkgəʊ/ *n.* (also **gingko** /'gɪŋkəʊ/) (*pl.* **-os** or **-oes**) an ornamental tree, *Ginkgo biloba*, native to China and Japan, with fan-shaped leaves and yellow flowers. Also called *maidenhair tree*. [Jap. *ginkyo* f. Chin. *yinxing* silver apricot]

ginormous /dʒaɪˈnɔːməs/ *adj. Brit. sl.* very large; enormous. [GIANT + ENORMOUS]

Ginsberg /ˈɡɪnzbɜːɡ/, Allen (b.1926), American poet. A leading poet of the beat generation, and later influential in the hippy movement of the 1960s, he is notable for his *Howl and Other Poems* (1956), in which he attacked American society for its materialism and complacency. He later campaigned for civil rights, gay liberation, and the peace movement.

ginseng /ˈdʒɪnseŋ/ *n.* **1** a medicinal plant of the genus *Panax*, found in East Asia and North America. **2** the root of this. [Chin. *renshen* perh. = man-image, with allusion to its forked root]

Giolitti /dʒəˈlɪti/, Giovanni (1842–1928), Italian statesman, Prime Minister five times between 1892 and 1921. A former lawyer, as Prime Minister he was responsible for the introduction of a wide range of social reforms, including national insurance (1911) and universal male suffrage (1912).

Giorgione /ˌdʒɔːdʒɪˈəʊni/ (also called Giorgio Barbarelli or Giorgio da Castelfranco) (*c.*1478–1510), Italian painter. He was an influential figure in Renaissance art, especially for his introduction of the small easel picture in oils intended for private collectors. Although the attribution of many of his works is doubtful, paintings such as *The Tempest* typically feature enigmatic figures in pastoral settings and gave a new prominence to landscape. His pupils included Titian, who is said to have completed some of his works, such as *Sleeping Venus* (*c.*1510), after his death.

Giotto[1] /ˈdʒɒtəʊ/ (full name Giotto di Bondone) (*c.*1267–1337), Italian painter. He is an important figure in the development of painting for his rejection of the flat, formulaic, and static images of Italo-Byzantine art in favour of a more naturalistic style showing human expression. Notable works include the frescos in the Arena Chapel, Padua (1305–8) and those in the church of Santa Croce in Florence (*c.*1320).

Giotto[2] /ˈdʒɒtəʊ/ a space probe of the European Space Agency, which photographed the nucleus of Halley's Comet in March 1986, and encountered Comet Grigg–Skjellerup in July 1992. [GIOTTO[1] (a comet, possibly Halley's, was depicted by Giotto (1304) as the Star of Bethlehem on a fresco in Padua)]

Giovanni de' Medici /dʒɒˈvæni/ the name of the Pope Leo X (see LEO[2]).

gippy tummy /ˈdʒɪpɪ/ *n.* (also **gyppy tummy**) *colloq.* diarrhoea affecting visitors to hot countries. [abbr. of EGYPTIAN]

gipsy var. of GYPSY.

giraffe /dʒɪˈrɑːf, -ˈræf/ *n.* (*pl.* same or **giraffes**) a ruminant mammal, *Giraffa camelopardalis* of Africa. It is the tallest living animal, reaching a height of 5.5 m (18 ft) or more, with a long neck and forelegs, and a skin of dark patches separated by lighter lines. [F *girafe*, It. *giraffa*, ult. f. Arab. *zarāfa*]

girandole /ˈdʒɪrənˌdəʊl/ *n.* **1** a revolving cluster of fireworks. **2** a branched candle-bracket or candlestick. **3** an earring or pendant with a large central stone surrounded by small ones. [F f. It. *girandola* f. *girare* GYRATE]

girasol /ˈdʒɪrəˌsɒl/ *n.* (also **girasole** /-ˌsəʊl/) a kind of opal reflecting a reddish glow; a fire-opal. [orig. = sunflower, f. F *girasol* or It. *girasole* f. *girare* (as GIRANDOLE) + *sole* sun]

gird[1] /ɡɜːd/ *v.tr.* (*past* and *past part.* **girded** or **girt**) *literary* **1** encircle, attach, or secure with a belt or band. **2** secure (clothes) on the body with a girdle or belt. **3** enclose or encircle. **4 a** (foll. by *with*) equip with a sword in a belt. **b** fasten (a sword) with a belt. **5** (foll. by *round*) place (cord etc.) round. □ **gird** (or **gird up**) **one's loins** prepare for action. [OE *gyrdan* f. Gmc (as GIRTH)]

gird[2] /ɡɜːd/ *v. & n.* ● *v.intr.* (foll. by *at*) jeer or gibe. ● *n.* a gibe or taunt. [ME, = strike etc.: orig. unkn.]

girder /ˈɡɜːdə(r)/ *n.* a large iron or steel beam or compound structure for bearing loads, esp. in bridge-building. [GIRD[1] + -ER[1]]

girdle[1] /ˈɡɜːd(ə)l/ *n. & v.* ● *n.* **1** a belt or cord worn round the waist. **2** a woman's corset extending from waist to thigh. **3** a thing that surrounds like a girdle. **4** the bony support for a limb (*pelvic girdle*). **5** the part of a cut gem dividing the crown from the base and embraced by the setting. **6** a ring round a tree made by the removal of bark. ● *v.tr.* **1** surround with a girdle. **2** remove a ring of bark from (a tree), esp. to make it more fruitful. [OE *gyrdel*: see GIRD[1]]

girdle[2] /ˈɡɜːd(ə)l/ *n. Sc. & N. Engl.* = GRIDDLE *n.* 1. [var. of GRIDDLE]

girl /ɡɜːl/ *n.* **1** a female child or youth. **2** *colloq.* a young (esp. unmarried) woman. **3** *colloq.* a girlfriend or sweetheart. **4** a female servant. □ **girl Friday** see FRIDAY. □ **girlhood** *n.* [ME *gurle*, *girle*, *gerle*, perh. rel. to LG *gör* child]

girlfriend /ˈɡɜːlfrend/ *n.* **1** a regular female companion or lover. **2** a female friend.

Girl Guide *n.* a member of the Guides Association.

Girl Guides Association see GUIDES ASSOCIATION.

girlie /ˈɡɜːlɪ/ *n. & adj.* (also **girly**) *colloq.* ● *n.* (*pl.* **-ies**) a girl (esp. as a term of endearment). ● *adj.* **1** (of a magazine etc.) depicting nude or partially nude young women in erotic poses. **2** girlish.

girlish /ˈɡɜːlɪʃ/ *adj.* of or like a girl. □ **girlishly** *adv.* **girlishness** *n.*

Girl Scout *n.* a member of an American organization corresponding to the Guides Association.

giro /ˈdʒaɪrəʊ/ *n. & v.* ● *n.* (*pl.* **-os**) **1** a system of credit transfer between banks, post offices, etc. **2** a cheque or payment by giro. ● *v.tr.* (**-oes**, **-oed**) pay by giro. [G f. It., = circulation (of money)]

Gironde /ʒɪˈrɒnd/ a river estuary in SW France, formed at the junction of the Garonne and Dordogne rivers, north of Bordeaux, and flowing north-west for 72 km (45 miles) into the Bay of Biscay.

Girondin /dʒɪˈrɒndɪn/ *n.* = GIRONDIST. [F]

Girondist /dʒɪˈrɒndɪst/ *n.* a member of the French moderate republican Party in power during the Revolution 1791–3, and so called because the party leaders were the deputies from the department of the Gironde. [F *Girondiste* (now *Girondin*)]

girt[1] *past part.* of GIRD[1].

girt[2] var. of GIRTH.

girth /ɡɜːθ/ *n. & v.* (also **girt** /ɡɜːt/) ● *n.* **1** the distance around any more or less cylindrical object, as the trunk of a tree, the human body, etc. **2** a band round the body of a horse to secure the saddle etc. ● *v.* **1** *tr.* **a** secure (a saddle etc.) with a girth. **b** put a girth on (a horse). **2** *tr.* surround, encircle. **3** *intr.* measure (an amount) in girth. [ME f. ON *gjörth*, Goth. *gairda* f. Gmc]

Gisborne /ˈɡɪzbən/ a port and resort on the east coast of North Island, New Zealand; pop. (1991) 31,480.

Giscard d'Estaing /ˌʒiːskɑː deˈstæŋ/, Valéry (b.1926), French statesman, President 1974–81. As Secretary of State for Finance (1959–62) and Finance Minister (1962–6) under President de Gaulle, he was responsible for the policies which formed the basis for France's economic growth. Dismissed following mounting opposition to his policies, he regained the finance portfolio under President Pompidou, whose death in 1974 paved the way for Giscard d'Estaing's own election to the presidency. However, French economic conditions worsened during his term of office and he was defeated by François Mitterrand. Since 1989 he has been a member of the European Parliament.

Gish /ɡɪʃ/, Lillian (1896–1993), American actress. She and her sister Dorothy (1898–1968) appeared in a number of D. W. Griffith's films, including *Hearts of the World* (1918) and *Orphans of the Storm* (1922).

gismo /ˈɡɪzməʊ/ *n.* (also **gizmo**) (*pl.* **-os**) *sl.* a gadget. [20th c.: orig. unkn.]

Gissing /ˈɡɪsɪŋ/, George (Robert) (1857–1903), English novelist. His own experiences of poverty and failure provided material for much of his fiction. His many novels include *New Grub Street* (1891), *Born in Exile* (1892), and *The Private Papers of Henry Ryecroft* (1903). He also wrote a notable biography of Charles Dickens (1898).

gist /dʒɪst/ *n.* **1** the substance or essence of a matter. **2** *Law* the real ground of an action etc. [OF, 3rd sing. pres. of *gesir* lie f. L *jacere*]

git /ɡɪt/ *n. Brit. sl.* a silly or contemptible person. [var. of GET *n.*]

gîte /ʒiːt/ *n.* a furnished holiday house in France, usu. small and in a rural district. [orig. = lodging: F f. OF *giste*, rel. to *gésir* lie]

gittern /ˈɡɪtɜːn/ *n.* a medieval stringed instrument, a forerunner of the guitar. [ME f. OF *guiterne*: cf. CITTERN, GUITAR]

give /ɡɪv/ *v. & n.* ● *v.* (*past* **gave** /ɡeɪv/; *past part.* **given**) **1** *tr.* (also *absol.*; often foll. by *to*) transfer the possession of freely; hand over as a present (*gave them her old curtains*; *gives to cancer research*). **2** *tr.* **a** transfer the ownership of with or without actual delivery; bequeath (*gave him £200 in her will*). **b** transfer, esp. temporarily or for safe keeping; hand over; provide with (*gave him the dog to hold*; *gave them a drink*). **c** administer (medicine). **d** communicate or impart (a message, compliments, etc.) (*give her my best wishes*). **3** *tr.* (usu. foll. by *for*) pay (*gave him £30 for the bicycle*). **b** sell (*gave him the bicycle for £30*) **4** *tr.* **a** confer; grant (a benefit, an honour, etc.). **b** accord; bestow (one's affections, confidence, etc.). **c** award; administer (one's approval, blame, etc.); tell, offer (esp.

something unpleasant) (*gave him a talking-to*; *gave him my blessing*; *gave him the sack*). **d** pledge, assign as a guarantee (*gave his word*). **5** *tr.* **a** effect or perform (an action etc.) (*gave him a kiss*; *gave a jump*). **b** utter (*gave a shriek*). **6** *tr.* allot; assign; grant (*was given the contract*). **7** *tr.* (in passive; foll. by *to*) be inclined to or fond of (*is given to speculation*). **8** *tr.* yield as a product or result (*the lamp gives a bad light*; *the field gives fodder for twenty cows*). **9** *intr.* **a** yield to pressure; become relaxed; lose firmness (*they pushed the door but it wouldn't give*). **b** collapse (*the roof gave under the pressure*). **10** *intr.* (usu. foll. by *of*) grant; bestow (*gave freely of his time*). **11** *tr.* **a** commit, consign, or entrust (*gave him into custody*; *give her into your care*). **b** sanction the marriage of (a daughter etc.). **12** *tr.* devote; dedicate (*gave his life to table tennis*; *shall give it my attention*). **13** *tr.* (usu. *absol.* *colloq.* tell what one knows (*What happened? Come on, give!*). **14** *tr.* present; offer; show; hold out (*gives no sign of life*; *gave her his arm*; *give him your ear*). **15** *tr. Theatr.* read, recite, perform, act, etc. (*gave them Hamlet's soliloquy*). **16** *tr.* impart; be a source of (*gave him my sore throat*; *gave its name to the battle*; *gave them to understand*; *gives him a right to complain*). **17** *tr.* allow (esp. a fixed amount of time) (*can give you five minutes*). **18** *tr.* (usu. foll. by *for*) value (something) (*gives nothing for their opinions*). **19** *tr.* concede; yield (*I give you the victory*). **20** *tr.* deliver (a judgement etc.) authoritatively (*gave his verdict*). **21** *tr. Cricket* (of an umpire) declare (a batsman) out or not out. **22** *tr.* toast (a person, cause, etc.) (*I give you our president*). **23** *tr.* provide (a party, meal, etc.) as host (*gave a banquet*). ● *n.* **1** capacity to yield or bend under pressure; elasticity (*there is no give in a stone floor*). **2** ability to adapt or comply (*no give in his attitudes*). □ **give and take** *v.tr.* exchange (words, blows, or concessions). ● *n.* an exchange of words etc.; a compromise. **give as good as one gets** retort adequately in words or blows. **give away 1** transfer as a gift. **2** hand over (a bride) ceremonially to a bridegroom. **3** betray or expose to ridicule or detection. **4** *Austral.* abandon, desist from, lose faith or interest in. **give-away** *n. colloq.* **1** an inadvertent betrayal or revelation. **2** an act of giving away. **3** a free gift; a low price. **give back** return (something) to its previous owner or in exchange. **give a person the best** see BEST. **give birth (to)** see BIRTH. **give chase** pursue a person, animal, etc.; hunt. **give down** (often *absol.*) (of a cow) let (milk) flow. **give forth** emit; publish; report. **give the game** (or **show**) **away** reveal a secret or intention. **give a hand** see HAND. **give a person** (or **the devil**) **his** (or **her**) **due** acknowledge, esp. grudgingly, a person's rights, abilities, etc. **give in 1** cease fighting or arguing; yield. **2** hand in (a document etc.) to an official etc. **give in marriage** sanction the marriage of (one's daughter etc.). **give it to a person** *colloq.* scold or punish. **give me** I prefer or admire (*give me the Greek islands any day*). **give off** emit (vapour etc.). **give oneself** (of a woman) yield sexually. **give oneself airs** act pretentiously or snobbishly. **give oneself up to 1** abandon oneself to an emotion, esp. despair. **2** addict oneself to. **give on to** (or **into**) (of a window, corridor, etc.) overlook or lead into. **give or take** *colloq.* add or subtract (a specified amount or number) in estimating. **give out 1** announce; emit; distribute. **2** cease or break down from exhaustion etc. **3** run short. **give over 1** *colloq.* cease from doing; abandon (a habit etc.); desist (*give over sniffing*). **2** hand over. **3** devote. **give rise to** cause, induce, suggest. **give tongue 1** speak one's thoughts. **2** (of hounds) bark, esp. on finding a scent. **give a person to understand** inform authoritatively. **give up 1** resign; surrender. **2** part with. **3** deliver (a wanted person etc.). **4** pronounce incurable or insoluble; renounce hope of. **5** renounce or cease (an activity). **give up the ghost** *archaic* or *colloq.* die. **give way** see WAY. **give a person what for** *colloq.* punish or scold severely. **give one's word** (or **word of honour**) promise solemnly. **not give a damn** (or **monkey's** or **toss** etc.) *colloq.* not care at all. **what gives?** *colloq.* what is the news?; what's happening? **would give the world** (or **anything, one's ears, eyes, right arm,** etc.) **for** covet or wish for desperately. □ **giveable** *adj.* **giver** *n.* [OE g(i)efan f. Gmc]

given /ˈgɪv(ə)n/ *past part.* of GIVE *v.* ● *adj.* **1** as previously stated or assumed; granted; specified (*given that he is a liar, we cannot trust him*; *a given number of people*). **2** Law (of a document) signed and dated (*given this day the 30th June*). ● *n.* a known fact or situation. □ **given name** US a name given at, or as if at, baptism; a forename.

Giza /ˈgiːzə/ (also **El Giza**; Arabic **Al Jizah** /ælˈdʒiːzə/) a city south-west of Cairo in northern Egypt, on the west bank of the Nile, site of the pyramids of Cheops (including the Great Pyramid) and of the Sphinx; pop. (est. 1986) 1,670,800.

gizmo var. of GISMO.

gizzard /ˈgɪzəd/ *n.* **1** a muscular thick-walled part of a bird's stomach, for grinding food usu. with grit. **2** a muscular stomach of some fish, insects, molluscs, and other invertebrates. □ **stick in one's gizzard**

colloq. be distasteful. [ME *giser* f. OF *giser*, *gesier* etc., ult. f. L *gigeria* cooked entrails of fowl]

glabella /gləˈbelə/ *n.* (*pl.* **glabellae** /-liː/) *Anat.* the smooth part of the forehead above and between the eyebrows. □ **glabellar** *adj.* [mod.L f. L *glabellus* (adj.) dimin. of *glaber* smooth]

glabrous /ˈgleɪbrəs/ *adj.* free from hair or down; smooth skinned. [L *glaber glabri* hairless]

glacé /ˈglæseɪ/ *adj.* **1** (of fruit, esp. cherries) preserved in sugar, usu. resulting in a glossy surface. **2** (of cloth, leather, etc.) smooth; polished. □ **glacé icing** icing made with icing sugar and water. [F, past part. of *glacer* to ice, gloss f. *glace* ice: see GLACIER]

glacial /ˈgleɪʃ(ə)l, ˈgleɪsɪəl/ *adj.* **1** of ice; icy. **2** *Geol.* characterized or produced by the presence or agency of ice. **3** *Chem.* forming icelike crystals upon freezing (*glacial acetic acid*). □ **glacially** *adv.* [F *glacial* or L *glacialis* icy f. *glacies* ice]

glacial period *n.* (also **glacial epoch**) a period in the earth's history characterized by an unusual extension of polar and mountain ice-sheets over the earth's surface. The Pliocene and Pleistocene epochs included a number of such periods, interrupted by warmer phases (interglacials), and it may be that the climate of the present day represents such a warm phase and that another ice age is to follow. During the coldest time, about 18,000 years ago, extensive ice-sheets covered much of Europe, North America, and Asia, and sea levels were as much as 200 metres lower than they are today. Also called (loosely) *ice age*.

glaciated /ˈgleɪsɪˌeɪtɪd, ˈglæs-/ *adj.* **1** marked or polished by the action of ice. **2** covered or having been covered by glaciers or ice sheets. [past part. of *glaciate* f. L *glaciare* freeze f. *glacies* ice]

glaciation /ˌgleɪsɪˈeɪʃ(ə)n, ˌglæs-/ *n.* **1** the covering of a landscape or region by ice. **2** = GLACIAL PERIOD.

glacier /ˈglæsɪə(r)/ *n.* a slowly moving mass or river of ice formed by the accumulation and compaction of snow in high mountains and at lower altitudes near the poles. Glaciers flow typically at a rate of a few centimetres per day, ending where the ice melts or reaches the sea. They are powerful agents of erosion producing very characteristic land-forms. [F f. *glace* ice ult. f. L *glacies*]

Glacier Bay National Park a national park in SE Alaska, on the Pacific coast. Extending over an area of 12,880 sq. km (4,975 sq. miles), it contains the terminus of the Grand Pacific Glacier.

glaciology /ˌgleɪsɪˈɒlədʒɪ, ˌglæs-/ *n.* the science of the internal dynamics and effects of glaciers. □ **glaciologist** *n.* **glaciological** /-sɪəˈlɒdʒɪk(ə)l/ *adj.* [L *glacies* ice + -LOGY]

glacis /ˈglæsɪs, -siː/ *n.* (*pl.* same /-sɪz, -siːz/) a bank sloping down from a fort, on which attackers are exposed to the defenders' missiles etc. [F f. OF *glacier* to slip f. *glace* ice: see GLACIER]

glad¹ /glæd/ *adj.* & *v.* ● *adj.* (**gladder, gladdest**) **1** (*predic.*; usu. foll. by *of*, *about*, or *to* + infin.) pleased; willing (*shall be glad to come*; *would be glad of a chance to talk about it*). **2 a** marked by, filled with, or expressing, joy (*a glad expression*). **b** (of news, events, etc.) giving joy (*glad tidings*). ● *v.tr.* (**gladded, gladding**) *archaic* make glad. □ **the glad eye** *colloq.* an amorous glance. **glad hand** the hand of welcome. **glad-hand** *v.tr.* greet cordially; welcome. **glad rags** *colloq.* best clothes; evening dress. □ **gladly** *adv.* **gladness** *n.* **gladsome** *adj. poet.* [OE *glæd* f. Gmc, orig. in sense 'bright, beautiful']

glad² /glæd/ *n. colloq.* a gladiolus. [abbr.]

gladden /ˈglæd(ə)n/ *v.tr.* & *intr.* make or become glad.

gladdie /ˈglædɪ/ *n. Austral. colloq.* = GLAD².

gladdon /ˈglæd(ə)n/ *n.* a purple-flowered iris, *Iris foetidissima*, of western Europe and North Africa, having an unpleasant odour when bruised. Also called *stinking iris*. [OE *glædene*]

glade /gleɪd/ *n.* an open space in a wood or forest. [16th c.: orig. unkn.]

gladiator /ˈglædɪˌeɪtə(r)/ *n.* **1** *hist.* a man trained to fight with a sword or other weapons at ancient Roman shows. **2** a person defending or opposing a cause; a controversialist. □ **gladiatorial** /ˌglædɪəˈtɔːrɪəl/ *adj.* [L f. *gladius* sword]

gladiolus /ˌglædɪˈəʊləs/ *n.* (*pl.* **gladioli** /-laɪ/ or **gladioluses**) a plant of the genus *Gladiolus* (iris family), with sword-shaped leaves and usu. brightly coloured flower-spikes. [L, dimin. of *gladius* sword]

Gladstone /ˈglædstən/, William Ewart (1809–98), British Liberal statesman, Prime Minister 1868–74, 1880–5, 1886, and 1892–4. After an early career as a Conservative minister, he joined the Liberal Party, becoming its leader in 1867. His ministries were notable for the introduction of a series of social and political reforms (including the introduction of elementary education, and the passing of the Irish Land Acts (see LAND ACTS) and the third Reform Act) and for his campaign in favour of Home Rule for Ireland, which led to the defection of the Unionists from the Liberal Party.

Gladstone bag *n.* a bag like a briefcase having two equal compartments joined by a hinge. [GLADSTONE]

Glagolitic /ˌglæɡəˈlɪtɪk/ *n.* an alphabet, based largely on Greek minuscules, formerly used in writing some Slavonic languages. It was introduced about the same time as Cyrillic (9th century), and may have been devised by St Cyril, but its origin is obscure. Although long superseded by the Cyrillic alphabet, it is still used in the liturgy of some Churches in Dalmatia, Montenegro, etc. [mod.L *glagoliticus* f. Serbo-Croat *glagóljica* Glagolitic alphabet f. Old Ch. Slav. *glagolŭ* word]

glair /gleə(r)/ *n.* (also **glaire**) **1** white of egg. **2** an adhesive preparation made from this, used in bookbinding etc. □ **glairy** *adj.* [ME f. OF *glaire*, ult. f. L *clara* fem. of *clarus* clear]

glaive /gleɪv/ *n. archaic poet.* **1** a broadsword. **2** any sword. [ME f. OF, app. f. L *gladius* sword]

Glam. *abbr.* Glamorgan.

glam /glæm/ *adj., n., & v. colloq.* ● *adj.* glamorous. ● *n.* glamour. ● *v.tr.* (**glammed, glamming**) glamorize. □ **glam rock** a style of rock music first popular in the early 1970s, characterized by male performers wearing glamorous clothes, make-up, etc., often with the suggestion of androgyny or sexual ambiguity. [abbr.]

Glamorgan /gləˈmɔːɡən/ a former county of South Wales. It was divided in 1974 into the counties of West Glamorgan, Mid Glamorgan, and South Glamorgan.

glamorize /ˈglæməˌraɪz/ *v.tr.* (also **glamourize, -ise**) make glamorous or attractive. □ **glamorization** /ˌglæməraɪˈzeɪʃ(ə)n/ *n.*

glamour /ˈglæmə(r)/ *n. & v.* (*US* **glamor**) ● *n.* **1** physical attractiveness, esp. when achieved by make-up etc. **2** alluring or exciting beauty or charm (*the glamour of New York*). ● *v.tr.* **1** *poet.* affect with glamour; bewitch; enchant. **2** *colloq.* make glamorous. □ **cast a glamour over** enchant. **glamour girl** (or **boy**) an attractive young woman (or man), esp. a model etc. □ **glamorous** *adj.* **glamorously** *adv.* [18th c.: orig. sense 'magic, enchantment', var. of GRAMMAR, with ref. to the occult practices associated with learning in the Middle Ages]

glance[1] /glɑːns/ *v. & n.* ● *v.* **1** *intr.* (often foll. by *down, up,* etc.) cast a momentary look (*glanced up at the sky*). **2** *intr.* (often foll. by *off*) (of a weapon or tool) glide or bounce (off an object). **3** *intr.* (usu. foll. by *over, off, from*) (of talk or a talker) pass quickly over a subject or subjects (*glanced over the question of payment*). **4** *intr.* (of a bright object or light) flash, dart, or gleam; reflect (*the sun glanced off the knife*). **5** *tr.* (esp. of a weapon) strike (an object) obliquely. **6** *tr. Cricket* deflect (the ball) with an oblique stroke. ● *n.* **1** (usu. foll. by *at, into, over,* etc.) a brief look (*took a glance at the paper; threw a glance over her shoulder*). **2 a** a flash or gleam (*a glance of sunlight*). **b** a sudden movement producing this. **3** *Cricket* a stroke with the bat's face turned slantwise to deflect the ball. □ **at a glance** immediately upon looking. **glance at 1** give a brief look at. **2** make a passing and usu. sarcastic allusion to. **glance one's eye** (foll. by *at, over,* etc.) look at briefly (esp. a document). **glance over** (or **through**) read cursorily. □ **glancingly** *adv.* [ME *glence* etc., prob. a nasalized form of obs. *glace* in the same sense, f. OF *glacier* to slip: see GLACIS]

glance[2] /glɑːns/ *n.* any lustrous sulphide ore (*copper glance; lead glance*). [G *Glanz* lustre]

gland[1] /glænd/ *n.* **1 a** an organ in an animal body secreting substances for use in the body or for ejection. **b** a structure resembling this, such as a lymph gland. **2** *Bot.* a secreting cell or group of cells on the surface of a plant-structure. [F *glande* f. OF *glandre* f. L *glandulae* throat-glands]

gland[2] /glænd/ *n.* a sleeve used to produce a seal round a piston rod or other shaft. [19th c.: perh. var. of *glam, glan* a vice, rel. to CLAMP[1]]

glanders /ˈglændəz/ *n.pl.* (also treated as *sing.*) **1** a contagious disease of horses, caused by a bacterium and characterized by swellings below the jaw and mucous discharge from the nostrils. **2** this disease in humans or other animals. □ **glandered** *adj.* **glanderous** *adj.* [OF *glandre*: see GLAND[1]]

glandular /ˈglændjʊlə(r)/ *adj.* of or relating to a gland or glands. □ **glandular fever** an infectious viral disease characterized by swelling of the lymph glands and prolonged lassitude; infectious mononucleosis (see MONONUCLEOSIS). [F *glandulaire* (as GLAND[1])]

glans /glænz/ *n.* (*pl.* **glandes** /ˈglændiːz/) the rounded part forming the end of the penis or clitoris. [L, = acorn]

glare[1] /gleə(r)/ *v. & n.* ● *v.* **1** *intr.* (usu. foll. by *at, upon*) look fiercely or fixedly. **2** *intr.* shine dazzlingly or disagreeably. **3** *tr.* express (hatred, defiance, etc.) by a look. **4** *intr.* be over-conspicuous or obtrusive. ● *n.* **1 a** strong fierce light, esp. sunshine. **b** oppressive public attention (*the glare of fame*). **2** a fierce or fixed look (*a glare of defiance*). **3** tawdry brilliance. □ **glary** *adj.* [ME, prob. ult. rel. to GLASS: cf. MDu. and MLG *glaren* gleam, glare]

glare[2] /gleə(r)/ *adj. N. Amer.* (esp. of ice) smooth and glassy. [perh. f. *glare* frost (16th c., of uncert. orig.)]

glaring /ˈgleərɪŋ/ *adj.* **1** obvious, conspicuous (*a glaring error*). **2** shining oppressively. **3** staring fiercely. □ **glaringly** *adv.*

Glasgow /ˈglɑːzɡəʊ/ a city in Scotland on the River Clyde; pop. (1991) 654,500. It was formerly a major shipbuilding centre, is still an important commercial and cultural centre, and is the largest city in Scotland.

glasnost /ˈglæznɒst, ˈglɑːs-/ *n.* (in the former Soviet Union) the policy or practice of more open consultative government and wider dissemination of information, initiated by Soviet leader Mikhail Gorbachev from 1985. Glasnost (often mentioned in conjunction with *perestroika*, restructuring) entailed more open reporting of events in the USSR and abroad, the rehabilitation of dissidents, and tolerance of religious worship. [Russ. *glasnost'*, lit. 'publicity, openness']

Glass /glɑːs/, Philip (b.1937), American composer. A leading minimalist, he studied with the Indian musician and composer Ravi Shankar; his work is influenced by Asian and North African music as well as reflecting his interest in jazz and rock. Major works include the opera *Einstein on the Beach* (1976), the ballet *Glass Pieces* (1982), and his first orchestral symphony, *Low Symphony* (1993).

glass /glɑːs/ *n., v., & adj.* ● *n.* **1 a** a hard, brittle, usu. transparent, translucent, or shiny substance, made by fusing sand with soda and lime and sometimes other ingredients. (*See note below.*) **b** any similar substance which has solidified from a molten state without crystallizing (*volcanic glass*). **2** (often *collect.*) an object or objects made from, or partly from, glass, esp.: **a** a drinking-vessel. **b** a mirror; a looking-glass. **c** an hourglass. **d** a window. **e** a greenhouse (*rows of lettuce under glass*). **f** glass ornaments. **g** a barometer. **h** a glass disc covering a watch-face. **i** a magnifying lens. **j** a monocle. **3** (in *pl.*) **a** spectacles. **b** binoculars; opera-glasses. **4** the amount of liquid contained in a glass; a drink (*he likes a glass*). ● *v.tr.* **1** (usu. as **glassed** *adj.*) fit with glass; glaze. **2** *poet.* reflect as in a mirror. **3** *Mil.* look at or for with field-glasses. ● *adj.* of or made from glass. □ **glass-blower** a person who blows semi-molten glass to make glassware. **glass-blowing** this occupation. **glass case** an exhibition display case made mostly from glass. **glass ceiling** an officially unacknowledged barrier to personal advancement. **glass-cloth 1** a linen cloth for drying glasses. **2** a cloth covered with powdered glass or abrasive, like glass-paper. **3** a woven fabric of fine-spun glass. **glass-cutter 1** a worker who cuts glass. **2** a tool used for cutting glass. **glass eye** a false eye made from glass. **glass fibre** = FIBREGLASS. **glass-gall** = SANDIVER. **glass-making** the manufacture of glass. **glass-paper** paper covered with powdered glass or abrasive and used for smoothing and polishing. **glass snake** a snakelike lizard of the genus *Ophisaurus*, with a very brittle tail. **glass wool** glass in the form of fine fibres used for packing and insulation. □ **glassful** *n.* (*pl.* **-fuls**). **glassless** *adj.* **glasslike** *adj.* [OE *glæs* f. Gmc: cf. GLAZE]

▪ Because of its special properties of transparency and resistance to corrosion, glass has found widespread uses for windows, food containers, chemical apparatus, and optical lenses. It has the property of being smoothly and reversibly converted to a liquid by the application of heat. Objects composed of glass paste appeared in Egypt *c.*1500 BC, made by modelling or moulding molten glass. The invention of glass-blowing in the 1st century BC (probably in Syria) enabled cheaper mass production of glass objects. Evidence for window glass made by rolling has been found at Pompeii. The addition of particular minerals produces special properties: lead oxide, for instance, is used in crystal and cut glass, and borax produces glass with low thermal expansion suitable for cookware and laboratory equipment; other minerals are used to add colour.

glasshouse /ˈglɑːshaʊs/ *n.* **1** a greenhouse. **2** *Brit. sl.* a military prison. **3** a building where glass is made.

glassie var. of GLASSY *n.*

glassine /glæˈsiːn/ *n.* a glossy transparent paper. [GLASS]

glassware /ˈglɑːsweə(r)/ *n.* articles made from glass, esp. drinking-glasses, tableware, etc.

glasswort /ˈglɑːswɜːt/ *n.* a salt marsh plant of the genus *Salicornia* or *Salsola*, formerly burnt for use in glass-making (cf. SALTWORT).

glassy /ˈglɑːsɪ/ adj. & n. ● adj. (**glassier**, **glassiest**) **1** of or resembling glass, esp. in smoothness. **2** (of the eye, the expression, etc.) abstracted; dull; fixed (*fixed her with a glassy stare*). ● n. (also **glassie**) esp. Austral. a glass marble. □ **the** (or **just the**) **glassy** Austral. colloq. the most excellent person or thing. □ **glassily** adv. **glassiness** n.

Glastonbury /ˈglæstənbərɪ/ a town in Somerset; pop. (1981) 6,770. It is the legendary burial place of King Arthur and Queen Guinevere and the site of a ruined abbey held by legend to have been founded by Joseph of Arimathea. It was identified in medieval times with the mythical Avalon.

Glaswegian /glæzˈwiːdʒən, glɑːz-/ adj. & n. ● adj. of or relating to Glasgow. ● n. a native or citizen of Glasgow. [GLASGOW after Norwegian etc.]

Glatzer Neisse /ˌglætsə ˈnaɪsə/ an alternative German name for the River Neisse in Poland (see NEISSE 2).

Glauber's salt /ˈglaʊbəz, ˈglɔːb-/ n. (also **Glauber's salts**) a crystalline hydrated form of sodium sulphate used esp. as a laxative. [Johann Rudolf Glauber, Ger. chemist (1604–68)]

glaucoma /glɔːˈkəʊmə/ n. an eye-condition with increased pressure within the eyeball, causing gradual loss of sight. □ **glaucomatous** adj. [L f. Gk glaukōma -atos, ult. f. glaukos: see GLAUCOUS]

glaucous /ˈglɔːkəs/ adj. **1** of a dull greyish-green or blue. **2** covered with a powdery bloom as of grapes. □ **glaucous gull** a large grey and white gull, Larus hyperboreus, of Arctic coasts. [L glaucus f. Gk glaukos]

glaze /gleɪz/ v. & n. ● v. **1** tr. **a** fit (a window, picture, etc.) with glass. **b** provide (a building) with glass windows. **2** tr. **a** cover (pottery etc.) with a glaze. **b** fix (paint) on pottery with a glaze. **3** tr. cover (pastry, meat, etc.) with a glaze. **4** intr. (often foll. by over) (of the eyes) become fixed or glassy (*his eyes glazed over*). **5** tr. cover (cloth, paper, leather, a painted surface, etc.) with a glaze or other similar finish. **6** tr. give a glassy surface to, e.g. by rubbing. ● n. **1** a vitreous substance, usu. a special glass, used to glaze pottery. **2** a smooth shiny coating of milk, sugar, gelatine, etc., on food. **3** a thin topcoat of transparent paint used to modify the tone of the underlying colour. **4** a smooth surface formed by glazing. **5** US a thin coating of ice. □ **glazed frost** a glassy coating of ice caused by frozen rain or a sudden thaw succeeded by a frost. **glaze in** enclose (a building, a window frame, etc.) with glass. □ **glazer** n. **glazy** adj. [ME f. an oblique form of GLASS]

glazier /ˈgleɪzɪə(r)/ n. a person whose trade is glazing windows etc. □ **glaziery** n.

glazing /ˈgleɪzɪŋ/ n. **1** the act or an instance of glazing. **2** windows (see also double glazing). **3** material used to produce a glaze.

Glazunov /ˈglæzʊˌnɒf/, Aleksandr (Konstantinovich) (1865–1936), Russian composer. He was a pupil of Rimsky-Korsakov (1880–1) and his work was influenced by Liszt and Wagner. His output includes orchestral and chamber music, songs, and the ballet The Seasons (1901).

GLC 1 see GREATER LONDON COUNCIL. **2** Chem. gas–liquid chromatography.

gleam /gliːm/ n. & v. ● n. **1** a faint or brief light (*a gleam of sunlight*). **2** a faint, sudden, intermittent, or temporary show (*not a gleam of hope*). ● v.intr. **1** emit gleams. **2** shine with a faint or intermittent brightness. **3** (of a quality) be indicated (*fear gleamed in his eyes*). □ **gleamingly** adv. **gleamy** adj. [OE glǣm: cf. GLIMMER]

glean /gliːn/ v.tr. **1** collect or scrape together (news, facts, gossip, etc.) in small quantities. **2 a** (also absol.) gather (ears of corn etc.) after the harvest. **b** strip (a field etc.) after a harvest. □ **gleaner** n. [ME f. OF glener f. LL glennare, prob. of Celt. orig.]

gleanings /ˈgliːnɪŋz/ n.pl. things gleaned, esp. facts.

glebe /gliːb/ n. **1** a piece of land serving as part of a clergyman's benefice and providing income. **2** poet. earth; land; a field. [ME f. L gl(a)eba clod, soil]

glee /gliː/ n. **1** mirth; delight, esp. expressing triumph (*watched the enemy's defeat with glee*). **2** a song for three or more, esp. adult male, voices, singing different parts simultaneously, usu. unaccompanied. □ **glee club** a society for singing part-songs. □ **gleesome** adj. [OE glío, glēo minstrelsy, jest f. Gmc]

gleeful /ˈgliːfʊl/ adj. triumphantly or exuberantly joyful. □ **gleefully** adv. **gleefulness** n.

gleet /gliːt/ n. Med. a watery discharge from the urethra caused by gonorrhoeal infection. [OF glette slime, secretion]

Gleichschaltung /ˈglaɪxˌʃæltʊŋ/ n. the standardization of political, economic, and social institutions in authoritarian states. [G]

glen /glen/ n. a narrow valley. [Gael. & Ir. gleann]

Glencoe, Massacre of /glenˈkəʊ/ a massacre in 1692 of members of the Jacobite MacDonald clan by Campbell soldiers, which took place near Glencoe in the Scottish Highlands. The massacre happened after the MacDonald clan had failed to swear allegiance to William III, making them liable to military punishment. Soldiers from the Campbell clan, who had been billeted on the MacDonalds for twelve days, murdered the MacDonald chief and about thirty of his followers, while the rest of the clan escaped. The chief of the Campbell clan, which had a long-standing feud with the MacDonalds, was held responsible, although the massacre was almost certainly instigated by the government.

Glendower /glenˈdaʊə(r)/, Owen (also **Glyndwr**) (c.1354–c.1417), Welsh chief. A legendary symbol of Welsh nationalism, he was leader first of armed resistance to English overlordship and then of a national uprising against Henry IV. He proclaimed himself Prince of Wales and allied himself with Henry's English opponents, including Henry Percy ('Hotspur'); by 1404 this policy had proved sufficiently successful for Glendower to hold his own Parliament. Though suffering subsequent defeats, he continued fighting against the English until his death.

Gleneagles /glenˈiːg(ə)lz/ a valley in eastern Scotland, south-west of Perth, site of a noted hotel and golfing centre.

glengarry /glenˈgærɪ/ n. (pl. -**ies**) a brimless hat with a cleft down the centre and usu. two ribbons hanging at the back, chiefly worn as part of Highland dress. [Glengarry in Scotland]

Glen More /mɔː(r)/ an alternative name for the GREAT GLEN.

glenoid cavity /ˈgliːnɔɪd/ n. a shallow depression on a bone, esp. the scapula and temporal bone, receiving the projection of another bone to form a joint. [F glénoïde f. Gk glēnoeidēs f. glēnē socket]

Glenrothes /glenˈrɒθɪs/ a town in eastern Scotland, capital of Fife region; pop. (1981) 32,970.

gley /gleɪ/ n. a sticky waterlogged soil grey to blue in colour. [Ukrainian, = sticky blue clay, rel. to CLAY]

glia /ˈgliːə/ n. = NEUROGLIA. □ **glial** adj. [Gk, = glue]

glib /glɪb/ adj. (**glibber**, **glibbest**) **1** (of a speaker, speech, etc.) fluent and voluble but insincere and shallow. **2** archaic smooth; unimpeded. □ **glibly** adv. **glibness** n. [rel. to obs. glibbery slippery f. Gmc: perh. imit.]

glide /glaɪd/ v. & n. ● v. **1** intr. (of a stream, bird, snake, ship, train, skater, etc.) move with a smooth continuous motion. **2** intr. **a** (of an aircraft) fly without engine-power. **b** (of a pilot) fly a glider. **3** intr. of time etc.: **a** pass gently and imperceptibly. **b** (often foll. by into) pass and change gradually and imperceptibly (*night glided into day*). **4** intr. move quietly or stealthily. **5** tr. cause to glide (*breezes glided the ship on its course*). ● n. **1 a** the act of gliding. **b** an instance of this. **2** Phonet. a gradually changing sound made in passing from one position of the speech-organs to another. **3** a gliding dance or dance-step. **4** a flight in a glider. **5** Cricket = GLANCE[1] n. 3. □ **glide path** an aircraft's line of descent to land, esp. as indicated by ground radar. [OE glīdan f. WG]

glider /ˈglaɪdə(r)/ n. **1 a** a fixed-wing aircraft that is not power-driven when in flight. (See note below.) **b** a glider pilot. **2** a person or thing that glides. **3** a small Australian marsupial that glides, esp. a flying phalanger.

▪ A glider is usually launched by being towed from an aircraft or car, or by catapult or winch; some are fitted with a small retractable engine and propeller to enable them to be self-launched. The first glider to fly in free flight was devised by Sir George Cayley in 1853.

gliding /ˈglaɪdɪŋ/ n. the sport or pastime of flying in a glider.

glim /glɪm/ n. **1** a faint light. **2** archaic sl. a candle; a lantern. [17th c.: perh. abbr. of GLIMMER or GLIMPSE]

glimmer /ˈglɪmə(r)/ v. & n. ● v.intr. shine faintly or intermittently. ● n. **1** a feeble or wavering light. **2** (usu. foll. by of) a faint gleam (of hope, understanding, etc.). **3** a glimpse. □ **glimmeringly** adv. [ME prob. f. Scand. f. WG: see GLEAM]

glimmering /ˈglɪmərɪŋ/ n. **1** = GLIMMER n. **2** an act of glimmering.

glimpse /glɪmps/ n. & v. ● n. (often foll. by of) **1** a momentary or partial view (*caught a glimpse of her*). **2** a faint and transient appearance (*glimpses of the truth*). ● v. **1** tr. see faintly or partly (*glimpsed his face in the crowd*). **2** intr. (often foll. by at) cast a passing glance. **3** intr. **a** shine faintly or intermittently. **b** poet. appear faintly; dawn. [ME glimse corresp. to MHG glimsen f. WG (as GLIMMER)]

Glinka /ˈglɪŋkə/, Mikhail (Ivanovich) (1804–57), Russian composer. Regarded as the father of the Russian national school of music, he is best known for his operas A Life for the Tsar (1836), inspired by Russian folk music, and *Russlan and Ludmilla* (1842), based on a poem by Pushkin.

glint /glɪnt/ *v. & n.* ● *v.intr. & tr.* flash or cause to flash; glitter; sparkle; reflect (*eyes glinted with amusement; the sword glinted fire*). ● *n.* a brief flash of light; a sparkle. [alt. of ME *glent*, prob. of Scand. orig.]

glissade /glɪˈsɑːd, -ˈseɪd/ *n. & v.* ● *n.* **1** an act of sliding down a steep slope of snow or ice, usu. on the feet with the support of an ice-axe etc. **2** a gliding step in ballet. ● *v.intr.* perform a glissade. [F f. *glisser* slip, slide]

glissando /glɪˈsændəʊ/ *n.* (*pl.* **glissandi** /-dɪ/ *or* **-os**) *Mus.* a continuous slide of adjacent notes upwards or downwards. [It. f. F *glissant* sliding (as GLISSADE)]

glissé /ˈgliːseɪ/ *n.* (in full **pas glissé**) (*pl.* **glissés** *pronunc.* same) *Ballet* a sliding step in which the flat of the foot is often used. [F, past part. of *glisser*: see GLISSADE]

glisten /ˈglɪs(ə)n/ *v. & n.* ● *v.intr.* shine, esp. like a wet object, snow, etc.; glitter. ● *n.* a glitter; a sparkle. [OE *glisnian* f. *glisian* shine]

glister /ˈglɪstə(r)/ *v. & n. archaic* ● *v.intr.* sparkle; glitter. ● *n.* a sparkle; a gleam. [ME f. MLG *glistern*, MDu *glisteren*, rel. to GLISTEN]

glitch /glɪtʃ/ *n. colloq.* a sudden irregularity or malfunction (of equipment etc.). [20th c.: orig. unkn.]

glitter /ˈglɪtə(r)/ *v. & n.* ● *v.intr.* **1** shine, esp. with a bright reflected light; sparkle. **2** (usu. foll. by *with*) **a** be showy or splendid (*glittered with diamonds*). **b** be ostentatious or flashily brilliant (*glittering rhetoric*). ● *n.* **1** a gleam; a sparkle. **2** showiness; splendour. **3** tiny pieces of sparkling material as on Christmas-tree decorations. □ **glitteringly** *adv.* **glittery** *adj.* [ME f. ON *glitra* f. Gmc]

glitterati /ˌglɪtəˈrɑːtɪ/ *n.pl. sl.* the fashionable set of literary or show-business people. [GLITTER + LITERATI]

Glittertind /ˈglɪtətɪn/ a mountain in Norway, in the Jotunheim range. Rising to 2,470 m (8,104 ft), it is the highest mountain in the country.

glitz /glɪts/ *n. sl.* extravagant but superficial display; show-business glamour. [back-form. f. GLITZY]

glitzy /ˈglɪtsɪ/ *adj.* (**glitzier**, **glitziest**) *sl.* extravagant, ostentatious; tawdry, gaudy. □ **glitzily** *adv.* **glitziness** *n.* [GLITTER, after RITZY: cf. G *glitzerig* glittering]

Gliwice /gliˈviːtsə/ a mining and industrial city in southern Poland, near the border with the Czech Republic; pop. (1990) 214,200.

gloaming /ˈgləʊmɪŋ/ *n. poet.* twilight; dusk. [OE *glōmung* f. *glōm* twilight, rel. to GLOW]

gloat /gləʊt/ *v. & n.* ● *v.intr.* (often foll. by *on*, *upon*, *over*) consider or contemplate with lust, greed, malice, triumph, etc. (*gloated over his collection*). ● *n.* **1** the act of gloating. **2** a look or expression of triumphant satisfaction. □ **gloater** *n.* **gloatingly** *adv.* [16th c.: orig. unkn., but perh. rel. to ON *glotta* grin, MHG *glotzen* stare]

glob /glɒb/ *n.* a mass or lump of semi-liquid substance, e.g. mud. [20th c.: perh. f. BLOB and GOB²]

global /ˈgləʊb(ə)l/ *adj.* **1** worldwide (*global conflict*). **2 a** relating to or embracing a group of items etc.; total. **b** *Computing* operating or applying through the whole of a file, program, etc. □ **globally** *adv.* [F (as GLOBE)]

globalize /ˈgləʊbəˌlaɪz/ *v.tr.* (also **-ise**) make global. □ **globalization** /ˌgləʊbəlaɪˈzeɪʃ(ə)n/ *n.*

global warming *n.* the increase in average temperatures on the earth caused by the greenhouse effect. While there is broad scientific agreement that an increase is occurring, there is no consensus regarding its likely extent. An average rise of up to 5°C by the year 2100 has been predicted, which would cause a significant raising of sea level, due to melting of polar ice, and unpredictable climatic changes, probably with far-reaching consequences. Remedial measures would include reducing the world's production of carbon dioxide and other greenhouse gases, and reversing the trend of deforestation. A UN conference in Brazil in 1992 produced an international commitment to address the problem but did not set firm targets for action.

globe /gləʊb/ *n. & v.* ● *n.* **1 a** (prec. by *the*) the planet earth. **b** a planet, star, or sun. **c** any spherical body; a ball. **2** a spherical representation of the earth or of the constellations with a map on the surface. **3** a golden sphere as an emblem of sovereignty; an orb. **4** any spherical glass vessel, esp. a fish bowl, a lamp, etc. **5** the eyeball. ● *v.tr. & intr.* make (usu. in *passive*) or become globular. □ **globe artichoke** the head of the artichoke plant, of which parts are edible. **globe-fish** a tropical fish of the family Tetraodontidae, able to inflate itself into a spherical form (also called *puffer fish*). **globe lightning** = *ball lightning* (see BALL¹).

globe-trotter a person who travels widely. **globe-trotting** such travel. □ **globelike** *adj.* **globoid** *adj. & n.* **globose** /ˈgləʊbəʊs, gləʊˈbəʊs/ *adj.* [F *globe* or L *globus*]

globeflower /ˈgləʊbˌflaʊə(r)/ *n.* a ranunculaceous plant of the genus *Trollius*, with globular usu. yellow flowers.

Globe Theatre a theatre in Southwark, London, erected in 1599, where many of Shakespeare's plays were first publicly performed by Richard Burbage and his company. Shakespeare had a share in the theatre and also acted there. It caught fire in 1613 from a discharge of stage gunfire during a play, and was destroyed, although it was rebuilt in 1614 and used until all London theatres were closed on the outbreak of the Civil War in 1642. The theatre's site was rediscovered in 1989 and a project to build a reconstruction of the original theatre was launched.

globigerina /gləʊˌbɪdʒəˈraɪnə/ *n.* (*pl.* same or **globigerinae** /-niː/) *Zool.* a marine planktonic foraminifer of the genus *Globigerina*, with a calcareous shell. □ **globigerina ooze** a deposit of pale mud covering a large part of the ocean floor, consisting of the shells of globigerinae. [mod.L f. L *globus* globe + *-ger* carrying + -INA]

globular /ˈglɒbjʊlə(r)/ *adj.* **1** globe-shaped, spherical. **2** composed of globules. □ **globular cluster** *Astron.* a large compact spherical cluster of old stars, usu. in the outer regions of a galaxy. □ **globularly** *adv.* **globularity** /ˌglɒbjʊˈlærɪtɪ/ *n.*

globule /ˈglɒbjuːl/ *n.* **1** a small globe or round particle; a drop. **2** a pill. □ **globulous** *adj.* [F *globule* or L *globulus* (as GLOBE)]

globulin /ˈglɒbjʊlɪn/ *n. Biochem.* a simple protein of a group characterized by solubility only in salt solutions and esp. forming a large fraction of blood serum protein.

glockenspiel /ˈglɒkənˌspiːl, -ˌʃpiːl/ *n.* a percussion instrument formed from a set of tuned metal bars, each supported at two points but with both ends free and struck in the centre with small hand-held hammers. It has a range of about two-and-a-half octaves, and became a fairly regular member of the orchestra only in the 19th century. [G, = bell-play]

glom /glɒm/ *v. US sl.* (**glommed**, **glomming**) **1** *tr.* steal; grab. **2** *intr.* (usu. foll. by *on to*) steal; grab. [var. of Sc. *glaum* (18th c., of unkn. orig.)]

glomerate /ˈglɒmərət/ *adj. Bot. & Anat.* compactly clustered. [L *glomeratus* past part. of *glomerare* f. *glomus -eris* ball]

glomerule /ˈglɒməˌruːl/ *n.* a clustered flower-head.

glomerulus /gləˈmerələs/ *n.* (*pl.* **glomeruli** /-ˌlaɪ/) a cluster of nerve endings, spores, or small blood vessels, esp. of capillaries at the end of a kidney tubule. □ **glomerular** *adj.* [mod.L, dimin. of L *glomus -eris* ball]

gloom /gluːm/ *n. & v.* ● *n.* **1** darkness; obscurity. **2** melancholy; despondency. **3** *poet.* a dark place. ● *v.* **1** *intr.* be gloomy or melancholy; frown. **2** *intr.* (of the sky etc.) be dull or threatening; lour. **3** *intr.* appear darkly or obscurely. **4** *tr.* cover with gloom; make dark or dismal. [ME *gloum(b)e*, of unkn. orig.: cf. GLUM]

gloomy /ˈgluːmɪ/ *adj.* (**gloomier**, **gloomiest**) **1** dark; unlighted. **2** depressed; sullen. **3** dismal; depressing. □ **gloomily** *adv.* **gloominess** *n.*

gloop /gluːp/ *n. colloq.* semifluid or sticky material. [imit.: cf. GLOP]

glop /glɒp/ *n. US sl.* a liquid or sticky mess, esp. inedible food. [imit.: cf. obs. *glop* swallow greedily]

Gloria /ˈglɔːrɪə/ *n.* **1** any of various doxologies beginning with *Gloria*, esp. the hymn beginning with *Gloria in excelsis Deo* (Glory be to God in the highest) as part of the Mass. **2** an aureole. [L, = glory]

Gloriana /ˌglɔːrɪˈɑːnə/ the nickname of Elizabeth I of England and Ireland.

glorify /ˈglɔːrɪˌfaɪ/ *v.tr.* (**-ies**, **-ied**) **1** exalt to heavenly glory; make glorious. **2** transform into something more splendid. **3** extol; praise. **4** (as **glorified** *adj.*) seeming or pretending to be more splendid than in reality (*just a glorified office boy*). □ **glorifier** *n.* **glorification** /ˌglɔːrɪfɪˈkeɪʃ(ə)n/ *n.* [ME f. OF *glorifier* f. eccl.L *glorificare* f. LL *glorificus* f. L *gloria* glory]

gloriole /ˈglɔːrɪˌəʊl/ *n.* an aureole; a halo. [F f. L *gloriola* dimin. of *gloria* glory]

glorious /ˈglɔːrɪəs/ *adj.* **1** possessing glory; illustrious. **2** conferring glory; honourable. **3** *colloq.* splendid; magnificent; delightful (*a glorious day; glorious fun*). **4** *iron.* intense; unmitigated (*a glorious muddle*). **5** *colloq.* happily intoxicated. □ **gloriously** *adv.* **gloriousness** *n.* [ME f. AF *glorious*, OF *glorios, -eus* f. L *gloriosus* (as GLORY)]

Glorious Revolution the events (1688–9) that led to the removal of

James II from the English throne and his replacement in 1689 by his daughter Mary II and her husband William of Orange (who became William III) as joint monarchs. The bloodless 'revolution' greatly enhanced the constitutional powers of Parliament, with William and Mary's acceptance of the conditions laid down in the Bill of Rights.

glory /'glɔːrɪ/ n. & v. ● n. (pl. **-ies**) **1** high renown or fame; honour. **2** adoring praise and thanksgiving (*Glory to the Lord*). **3** resplendent majesty or magnificence; great beauty (*the glory of Versailles; the glory of the rose*). **4** a thing that brings renown or praise; a distinction. **5** the bliss and splendour of heaven. **6** *colloq.* a state of exaltation, prosperity, happiness, etc. (*is in his glory playing with his trains*). **7** an aureole, a halo. **8** an anthelion. ● v.intr. (**-ies, -ied**) (often foll. by *in*, or *to* + infin.) pride oneself; exult (*glory in their skill*). □ **glory be! 1** expressing enthusiastic piety. **2** *colloq.* an exclamation of surprise or delight. **glory-box** *Austral. & NZ* a box for women's clothes etc., stored in preparation for marriage. **glory-hole 1** *colloq.* an untidy room, drawer, or receptacle. **2** *N. Amer.* an open quarry. **glory-of-the-snow** = CHIONODOXA. **go to glory** *sl.* die; be destroyed. [ME f. AF & OF *glorie* f. L *gloria*]

Glos. *abbr.* Gloucestershire.

gloss[1] /glɒs/ n. & v. ● n. **1 a** surface shine or lustre. **b** an instance of this; a smooth finish. **2 a** deceptively attractive appearance. **b** an instance of this. **3** (in full **gloss paint**) paint formulated to give a hard glossy finish (cf. MATT n. 2). ● v.tr. make glossy. □ **gloss over 1** seek to conceal beneath a false appearance. **2** conceal or evade by mentioning briefly or misleadingly. □ **glosser** n. [16th c.: orig. unkn.]

gloss[2] /glɒs/ n. & v. ● n. **1 a** an explanatory word or phrase inserted between the lines or in the margin of a text. **b** a comment, explanation, interpretation, or paraphrase. **2** a misrepresentation of another's words. **3 a** a glossary. **b** an interlinear translation or annotation. ● v. **1** tr. **a** add a gloss or glosses to (a text, word, etc.). **b** read a different sense into; explain away. **2** intr. (often foll. by *on*) make (esp. unfavourable) comments. **3** intr. write or introduce glosses. [alt. of GLOZE after med.L *glossa*]

glossal /'glɒs(ə)l/ adj. *Anat.* of the tongue; lingual. [Gk *glōssa* tongue]

glossary /'glɒsərɪ/ n. (pl. **-ies**) **1** an alphabetical list of terms or words found in or relating to a specific subject or text, esp. dialect, with explanations; a brief dictionary. **2** a collection of glosses. □ **glossarist** n. **glossarial** /glɒ'seərɪəl/ adj. [L *glossarium* f. *glossa* GLOSS[2]]

glossator /glɒ'seɪtə(r)/ n. **1** a writer of glosses. **2** *hist.* a commentator on, or interpreter of, medieval law-texts. [ME f. med.L f. *glossare* f. *glossa* GLOSS[2]]

glosseme /'glɒsiːm/ n. any meaningful feature of a language that cannot be analysed into smaller meaningful units. [Gk *glōssēma* f. *glōssa* GLOSS[2]]

glossitis /glɒ'saɪtɪs/ n. *Med.* inflammation of the tongue. [Gk *glōssa* tongue + -ITIS]

glossographer /glɒ'sɒgrəfə(r)/ n. a writer of glosses or commentaries. [GLOSS[2] + -GRAPHER]

glossolalia /ˌglɒsə'leɪlɪə/ n. = *gift of tongues* (see TONGUE). [mod.L f. Gk *glōssa* tongue + *-lalia* speaking]

glossopharyngeal /ˌglɒsəfə'rɪndʒɪəl/ adj. & n. *Anat.* ● adj. of or relating to the tongue and the pharynx. ● n. (in full **glossopharyngeal nerve**) either of the ninth pair of cranial nerves, which supply these organs.

glossy /'glɒsɪ/ adj. & n. ● adj. (**glossier, glossiest**) **1** having a shine; smooth. **2** (of paper etc.) smooth and shiny. **3** (of a magazine etc.) printed on such paper; expensively produced and attractively presented, but sometimes lacking in content or depth. ● n. (pl. **-ies**) *colloq.* **1** a glossy magazine. **2** a photograph with a glossy surface. □ **glossily** adv. **glossiness** n.

glottal /'glɒt(ə)l/ adj. of or produced by the glottis.

glottal stop n. a sound produced by the sudden opening or shutting of the glottis.

■ The glottal stop is used in some dialects of English, notably cockney, in place of the consonant 't', as in *bu'er* for *butter*. It also functions as a phoneme in some other languages, including Arabic and many American Indian languages.

glottis /'glɒtɪs/ n. the space at the upper end of the windpipe and between the vocal cords, affecting voice modulation through expansion or contraction. □ **glottic** adj. [mod.L f. Gk *glōttis* f. *glōtta* var. of *glōssa* tongue]

Gloucester[1] /'glɒstə(r)/ a city in SW England, the county town of Gloucestershire; pop. (1991) 91,800. It was founded by the Romans,

who called it Glevum, in AD 96. Its Norman cathedral contains the tomb of Edward II.

Gloucester[2] /'glɒstə(r)/ n. (usu. **double Gloucester**, orig. a richer kind) a kind of hard cheese orig. made in Gloucestershire.

Gloucestershire /'glɒstəˌʃɪə(r)/ a county of SW England; county town, Gloucester.

glove /glʌv/ n. & v. ● n. **1** a covering for the hand, of wool, leather, cotton, etc., worn esp. for protection against cold or dirt, and usu. having separate fingers. **2** a padded protective glove, esp.: **a** a boxing glove. **b** a wicket-keeper's glove. ● v.tr. cover or provide with a glove or gloves. □ **fit like a glove** fit exactly. **glove box 1** a box for gloves. **2** a closed chamber with sealed-in gloves for handling radioactive material etc. **3** = *glove compartment*. **glove compartment** a recess for small articles in the dashboard of a motor vehicle. **glove puppet** a small cloth puppet fitted on the hand and worked by the fingers. **throw down** (or **take up**) **the glove** issue (or accept) a challenge. **with the gloves off** mercilessly; unfairly; with no compunction. □ **gloveless** adj. **glover** n. [OE *glōf*, corresp. to ON *glófi*, perh. f. Gmc]

glow /gləʊ/ v. & n. ● v.intr. **1 a** throw out light and heat without flame; be incandescent. **b** shine like something heated in this way. **2** (of cheeks) redden, esp. from cold or exercise. **3** (often foll. by *with*) **a** (of the body) be heated, esp. from exertion; sweat. **b** express or experience strong emotion (*glowed with pride; glowing with indignation*). **4** show a warm colour (*the painting glows with warmth*). **5** (as **glowing** adj.) expressing pride or satisfaction (*a glowing report*). ● n. **1** a glowing state. **2** a bright warm colour, esp. the red of cheeks. **3** ardour; passion. **4** a feeling induced by good health, exercise, etc.; well-being. □ **glow discharge** a luminous sparkless electrical discharge from a pointed conductor in a gas at low pressure. **glow-worm** a soft-bodied beetle of the genus *Lampyris*, whose wingless female emits light from the abdomen. **in a glow** *colloq.* hot or flushed; sweating. □ **glowingly** adv. [OE *glōwan* f. Gmc]

glower /'glaʊə(r)/ v. & n. ● v.intr. (often foll. by *at*) stare or scowl, esp. angrily. ● n. a glowering look. □ **gloweringly** adv. [orig. uncert.: perh. Sc. var. of ME *glore* f. LG or Scand., or f. obs. (ME) *glow* stare + -ER[4]]

gloxinia /glɒk'sɪnɪə/ n. a tropical plant of the genus *Gloxinia*, native to South America, with large bell flowers of various colours. [mod.L f. Benjamin Peter *Gloxin*, 18th-c. Ger. botanist]

gloze /gləʊz/ v. **1** tr. (also **gloze over**) explain away; extenuate; palliate. **2** intr. archaic (usu. foll. by *on, upon*) comment. **b** talk speciously; fawn. [ME f. OF *gloser* f. *glose* f. med.L *glosa, gloza* f. L *glossa* tongue, GLOSS[2]]

glucagon /'gluːkəgɒn/ n. a polypeptide hormone formed in the pancreas, which aids the breakdown of glycogen to glucose. [Gk *glukus* sweet + *agōn* leading]

Gluck /glʊk/, Christoph Willibald von (1714–87), German composer. From early operas in the Italian style he went on to seek a balance of music and drama in his 'reform' operas, reducing the emphasis on the star singer and attempting a continuous musical unfolding of the narrative. He spent much of his working life in Paris as a protégé of Marie Antoinette. His most notable works include *Orfeo ed Euridice* (1762) and *Iphigénie en Aulide* (1774).

glucose /'gluːkəʊz, -kəʊs/ n. **1** a simple sugar containing six carbon atoms (chem. formula: $C_6H_{12}O_6$), which is an important energy source in living organisms and is obtained from some carbohydrates by hydrolysis. (See also DEXTROSE.) **2** a syrup containing glucose and other sugars from the incomplete hydrolysis of starch. [F f. Gk *gleukos* sweet wine, rel. to *glukus* sweet]

glucoside /'gluːkəˌsaɪd/ n. a compound giving glucose and other products upon hydrolysis. □ **glucosidic** /ˌgluːkə'sɪdɪk/ adj.

glue /gluː/ n. & v. ● n. an adhesive substance used for sticking objects or materials together. ● v.tr. (**glues, glued, gluing** or **glueing**) **1** fasten or join with glue. **2** keep or put very close (*an eye glued to the keyhole*). □ **glue ear** blocking of the Eustachian tube by mucus, esp. in children. **glue-pot 1** a pot with an outer vessel holding water to heat glue. **2** *colloq.* an area of sticky mud etc. **glue-sniffing** the inhalation of intoxicating fumes from the solvents in adhesives etc. □ **gluelike** adj. **gluey** /'gluːɪ/ adj. (**gluier, gluiest**). **glueyness** n. [ME f. OF *glu* (n.), *gluer* (v.), f. LL *glus glutis* f. L *gluten*]

glug /glʌg/ n. & v. ● n. a hollow, usu. repetitive gurgling sound. ● v.intr. make a gurgling sound as of water from a bottle. [imit.]

glühwein /'gluːvaɪn/ n. mulled wine. [G f. *glühen* mull + *Wein* wine]

glum /glʌm/ adj. (**glummer, glummest**) looking or feeling dejected; sullen; morose. □ **glumly** adv. **glumness** n. [rel. to dial. *glum* (v.) frown, var. of *gloume* GLOOM v.]

glume /gluːm/ n. **1** a membranous bract surrounding the spikelet of grasses or the florets of sedges. **2** the husk of grain. □ **glumaceous** /gluːˈmeɪʃəs/ adj. **glumose** /ˈgluːməʊs, -məʊz/ adj. [L *gluma* husk]

gluon /ˈgluːɒn/ n. Physics a hypothetical subatomic particle believed to transmit the force binding quarks together in hadrons. [GLUE]

glut /glʌt/ v. & n. ● v.tr. (**glutted, glutting**) **1** feed (a person, one's stomach, etc.) or indulge (an appetite, a desire, etc.) to the full; satiate; cloy. **2** fill to excess; choke up. **3** Econ. overstock (a market) with goods. ● n. **1** Econ. supply exceeding demand; a surfeit (*a glut in the market*). **2** full indulgence; one's fill. [ME prob. f. OF *gloutir* swallow f. L *gluttire*: cf. GLUTTON]

glutamate /ˈgluːtəˌmeɪt/ n. Chem. a salt or ester of glutamic acid. (See also MONOSODIUM GLUTAMATE.)

glutamic acid /gluːˈtæmɪk/ n. Biochem. a naturally occurring amino acid, a constituent of many proteins. [GLUTEN + AMINE + -IC]

glutamine /ˈgluːtəˌmiːn/ n. Biochem. a hydrophilic amino acid present in many proteins. [GLUTAMIC ACID + AMINE]

gluten /ˈgluːt(ə)n/ n. **1** a mixture of two proteins present in flour, esp. wheat flour. **2** archaic a sticky substance. [F f. L *gluten glutinis* glue]

gluteus /ˈgluːtɪəs/ n. (pl. **glutei** /-tɪˌaɪ/) any of the three muscles (in full *gluteus maximus, gluteus medius,* and *gluteus minimus*) in each buttock. □ **gluteal** adj. [mod.L f. Gk *gloutos* buttock]

glutinous /ˈgluːtɪnəs/ adj. sticky; like glue. □ **glutinously** adv. **glutinousness** n. [F *glutineux* or L *glutinosus* (as GLUTEN)]

glutton /ˈglʌt(ə)n/ n. **1** an excessively greedy eater. **2** (often foll. by *for*) colloq. a person insatiably eager (*a glutton for work*). **3** the wolverine. □ **a glutton for punishment** a person eager to take on hard or unpleasant tasks. □ **gluttonize** v.intr. (also **-ise**). **gluttonous** adj. **gluttonously** adv. [ME f. OF *gluton, gloton* f. L *glutto -onis* f. *gluttire* swallow, *gluttus* greedy]

gluttony /ˈglʌtənɪ/ n. habitual greed or excess in eating. [OF *glutonie* (as GLUTTON)]

glyceride /ˈglɪsəˌraɪd/ n. Chem. a fatty-acid ester of glycerol.

glycerine /ˈglɪsəˌriːn/ n. (US **glycerin** /-rɪn/) = GLYCEROL. [F *glycerin* f. Gk *glukeros* sweet]

glycerol /ˈglɪsəˌrɒl/ n. a colourless sweet viscous liquid (chem. formula: $C_3H_8O_3$), formed as a by-product in the manufacture of soap and used as an emollient and laxative, as a cryoprotectant, in explosives, etc. Also called *glycerine*. [GLYCERINE + -OL]

glycine /ˈglaɪsiːn/ n. Biochem. the simplest naturally occurring amino acid (chem. formula H_2NCH_2COOH), a general constituent of proteins. [G *Glycin* f. Gk *glukus* sweet]

glyco- /ˈglaɪkəʊ/ comb. form sugar. [Gk *glukus* sweet]

glycogen /ˈglaɪkədʒən/ n. Biochem. a polysaccharide serving as a store of carbohydrates, esp. in animal tissues, and yielding glucose on hydrolysis. □ **glycogenic** /ˌglaɪkəˈdʒenɪk/ adj.

glycogenesis /ˌglaɪkəʊˈdʒenɪsɪs/ n. Biochem. the formation of glycogen from sugar.

glycol /ˈglaɪkɒl/ n. Chem. a diol, esp. ethylene glycol. □ **glycolic** /glaɪˈkɒlɪk/ adj. [GLYCERINE + -OL, orig. as being intermediate between glycerine and alcohol]

glycolysis /glaɪˈkɒlɪsɪs/ n. Biochem. the breakdown of glucose by enzymes in most living organisms to release energy and pyruvic acid.

glycoprotein /ˌglaɪkəʊˈprəʊtiːn/ n. Biochem. any of a class of compounds consisting of a protein combined with a carbohydrate.

glycoside /ˈglaɪkəˌsaɪd/ n. Chem. any compound giving sugar and other products on hydrolysis. □ **glycosidic** /ˌglaɪkəˈsɪdɪk/ adj. [GLYCO-, after GLUCOSIDE]

glycosuria /ˌglaɪkəʊˈsjʊərɪə/ n. Med. a condition characterized by an excess of sugar in the urine, associated with diabetes, kidney disease, etc. □ **glycosuric** adj. [F *glycose* glucose + -URIA]

Glyndebourne /ˈglaɪndbɔːn/ (in full **Glyndebourne Festival**) an annual festival of opera, held at the estate of Glyndebourne near Lewes in East Sussex, England. The original opera house was built by the owner of the estate, John Christie (1882–1962), who founded the festival in 1934. A new opera house, built after the old one was pulled down in 1992, was opened in 1994.

Glyndwr see GLENDOWER.

glyph /glɪf/ n. **1** a sculptured character or symbol. **2** a vertical groove, esp. that on a Greek frieze. □ **glyphic** adj. [F *glyphe* f. Gk *gluphē* carving f. *gluphō* carve]

glyptal /ˈglɪptæl/ n. an alkyd resin, esp. one formed from glycerine and phthalic acid or anhydride. [perh. f. *glycerol* + *phthalic*]

glyptic /ˈglɪptɪk/ adj. of or concerning carving, esp. on precious stones. [F *glyptique* or Gk *gluptikos* f. *gluptēs* carver f. *gluphō* carve]

glyptodon /ˈglɪptəˌdɒn/ n. (also **glyptodont** /-ˌdɒnt/) an extinct armoured edentate mammal of the genus *Glyptodon*, native to South America. It is related to (but much larger than) the armadillo, with the body covered in a hard thick bony shell. [mod.L f. Gk *gluptos* carved + *odous odontos* tooth]

glyptography /glɪpˈtɒgrəfɪ/ n. the art or scientific study of gem-engraving. [Gk *gluptos* carved + -GRAPHY]

GM abbr. **1** (in the UK) George Medal. **2** see GENERAL MOTORS. **3** general manager.

gm abbr. gram(s).

G-man /ˈdʒiːmæn/ n. (pl. **G-men**) **1** US colloq. a federal criminal-investigation officer. **2** Ir. a political detective. [Government + MAN]

GMB abbr. (in the UK) General and Municipal Boilermakers (Union).

GMS abbr. (in the UK, with reference to schools) grant maintained status.

GMT abbr. Greenwich Mean Time.

GMWU abbr. (in the UK) General and Municipal Workers' Union.

gnamma /ˈnæmə/ n. (also **namma**) Austral. a natural hole in a rock, containing water; a water-hole. [Aboriginal]

gnarl /nɑːl/ n. a contorted knotty protuberance, esp. on a tree. [back-form. f. GNARLED]

gnarled /nɑːld/ adj. (also **gnarly** /ˈnɑːlɪ/) (of a tree, hands, etc.) knobbly, twisted, rugged. [var. of *knarled*, rel. to KNURL]

gnash /næʃ/ v. & n. ● v. **1** tr. grind (the teeth). **2** intr. (of the teeth) strike together; grind. ● n. an act of grinding the teeth. [var. of obs. *gnacche* or *gnast*, rel. to ON *gnastan* a gnashing (imit.)]

gnashers /ˈnæʃəz/ n.pl. sl. teeth, esp. false teeth.

gnat /næt/ n. **1** a small fragile dipterous insect with long thin legs; esp. a slender biting fly of the family Culicidae, which includes the common gnat, *Culex pipiens*. **2** an insignificant annoyance. **3** a tiny thing. [OE *gnætt*]

gnathic /ˈnæθɪk/ adj. of or relating to the jaws. [Gk *gnathos* jaw]

gnaw /nɔː/ v. (past part. **gnawed** or **gnawn** /nɔːn/) **1 a** tr. (usu. foll. by *away, off, in two,* etc.) bite persistently; wear away by biting. **b** intr. (often foll. by *at, into*) bite, nibble. **2 a** intr. (often foll. by *at, into*) (of a destructive agent, pain, fear, etc.) corrode; waste away; consume; torture. **b** tr. corrode, consume, torture, etc. with pain, fear, etc. (*was gnawed by doubt*). **3** tr. (as **gnawing** adj.) persistent; worrying. □ **gnawingly** adv. [OE *gnagen*, ult. imit.]

gneiss /naɪs/ n. a usu. coarse-grained metamorphic rock foliated by mineral layers, principally of feldspar, quartz, and ferromagnesian minerals. □ **gneissic** adj. **gneissoid** adj. **gneissose** /-səʊz/ adj. [G]

gnocchi /ˈnɒkɪ, ˈnjɒkɪ/ n.pl. an Italian dish of small dumplings usu. made from potato, semolina flour, etc., often flavoured with spinach or cheese. [It., pl. of *gnocco* f. *nocchio* knot in wood]

gnome[1] /nəʊm/ n. **1 a** a dwarfish legendary creature supposed to guard the earth's treasures underground; a goblin. **b** a figure of a gnome, esp. as a garden ornament. **2** (esp. in pl.) colloq. a person with sinister influence, esp. financial (*gnomes of Zurich*). □ **gnomish** adj. [F f. mod.L *gnomus* (word invented by Paracelsus)]

gnome[2] /ˈnəʊmɪ, nəʊm/ n. a maxim; an aphorism. [Gk *gnōmē* opinion f. *gignōskō* know]

gnomic /ˈnəʊmɪk/ adj. **1** of, consisting of, or using gnomes or aphorisms; sententious (see GNOME[2]). **2** Gram. (of a tense) used without the implication of time to express a general truth, e.g. *men were deceivers ever.* □ **gnomically** adv. [Gk *gnōmikos* (as GNOME[2])]

gnomon /ˈnəʊmɒn/ n. **1** the rod or pin etc. on a sundial that shows the time by the position of its shadow. **2** Geom. the part of a parallelogram left when a similar parallelogram has been taken from its corner. **3** Astron. a column etc. used in observing the sun's meridian altitude. □ **gnomonic** /nəʊˈmɒnɪk/ adj. [F or L *gnomon* f. Gk *gnōmōn* indicator etc. f. *gignōskō* know]

gnosis /ˈnəʊsɪs/ n. knowledge of spiritual mysteries. [Gk *gnōsis* knowledge (as GNOMON)]

gnostic /ˈnɒstɪk/ adj. & n. ● adj. **1** relating to knowledge, esp. esoteric mystical knowledge. **2** (**Gnostic**) concerning the Gnostics; occult; mystic. ● n. (**Gnostic**) (usu. in pl.) a Christian heretic of the 1st–3rd centuries claiming gnosis. (See GNOSTICISM.) □ **gnosticize** /-tɪˌsaɪz/ *v.tr. & intr.* (also **-ise**). [eccl.L *gnosticus* f. Gk *gnōstikos* (as GNOSIS)]

Gnosticism /ˈnɒstɪˌsɪz(ə)m/ *n.* a heretical movement prominent in the Christian Church in the 2nd century, in part at least of pre-Christian origin and with ideas drawn from Greek philosophy and other pagan sources. Gnostics emphasized the power of gnosis, the supposed revealed knowledge of God, to redeem the spiritual element in humankind; they contrasted the supreme remote divine being with the demiurge or creator god, who controlled the world and was antagonistic to all that was purely spiritual. Christ came as an emissary from the supreme god, bringing gnosis. Gnostic teaching for long was known only from anti-heretical writers, such as Irenaeus and Tertullian, until in 1945–6 a collection of Gnostic texts was found in Egypt.

GNP *abbr.* gross national product.

Gnr. *abbr. Brit.* Gunner.

gns. *abbr. Brit. hist.* guineas.

gnu /nuː, njuː, gəˈnuː/ *n.* a large grazing antelope of the genus *Connochaetes*, native to southern and East Africa, with a long shaggy head, hooked horns, and a long tail. Also called *wildebeest*. [Bushman *nqu*, prob. through Du. *gnoe*]

go[1] /gəʊ/ *v., n., & adj.* ● *v.* (*3rd sing. present* **goes** /gəʊz/; *past* **went** /went/; *past part.* **gone**) **1** *intr.* **a** start moving or be moving from one place or point in time to another; travel, proceed. **b** (foll. by *to* + infin., or *and* + verb) proceed in order to (*went to find him; go and buy some bread*). **c** (foll. by *and* + verb) *colloq.* expressing annoyance (*you went and told him; they've gone and broken it; she went and won*). **2** *intr.* (foll. by verbal noun) make a special trip for; participate in; proceed to do (*went skiing; then went shopping; often goes running*). **3** *intr.* lie or extend in a certain direction; lead to (*the road goes to London; where does that door go?*). **4** *intr.* **a** leave; depart (*they had to go*). **b** *colloq.* disappear, vanish (*my bag has gone*). **5** *intr.* move, act, work, etc. (*the clock doesn't go; his brain is going all the time*). **6** *intr.* **a** make a specified movement (*go like this with your foot*). **b** make a sound (of a specified kind) (*the gun went bang; the cow went 'moo'*). **c** (of a bell etc.) make a sound in functioning (*an ambulance with its sirens going; the door bell went*). **d** *colloq.* say (*so he goes to me 'Why didn't you like it?'*). **7** *intr.* be in a specified state (*go hungry; went in fear of his life*). **8** *intr.* **a** pass into a specified condition (*gone bad; went mad; went to sleep*). **b** *colloq.* die. **c** proceed or escape in a specified condition (*the poet went unrecognized; the crime went unnoticed*). **9** *intr.* **a** (of time or distance) pass, elapse; be traversed (*ten days to go before Easter; the last mile went quickly*). **b** be finished (*the film went quickly*). **10** *intr.* **a** (of a document, verse, song, etc.) have a specified content or wording; run (*the tune goes like this*). **b** be current or accepted (*so the story goes*). **c** be suitable; fit; match (*the shoes don't go with the hat*). **d** be regularly kept or put (*the forks go here*). **e** be accommodated; fit (*this won't go into the cupboard*). **11** *intr.* **a** turn out, proceed; take a course or view (*things went well; Liverpool went Labour*). **b** be successful (*make the party go*). **c** progress (*we've still a long way to go*). **12** *intr.* **a** be sold (*went for £1; went cheap*). **b** (of money) be spent (*£200 went on a new jacket*). **13** *intr.* **a** be relinquished, dismissed, or abolished (*the car will have to go*). **b** fail, decline; give way, collapse (*his sight is going; the bulb has gone*). **14** *intr.* be acceptable or permitted; be accepted without question (*anything goes; what I say goes*). **15** *intr.* (often foll. by *by, with, on, upon*) be guided by; judge or act on or in harmony with (*have nothing to go on; a good rule to go by*). **16** *intr.* attend or visit or travel to regularly (*goes to church; goes to school; this train goes to Bristol*). **17** *intr.* (foll. by pres. part.) *colloq.* proceed (often foolishly) to do (*went running to the police; don't go making him angry*). **18** *intr.* act or proceed to a certain point (*will go so far and no further; went as high as £100*). **19** *intr.* (of a number) be capable of being contained in another (*6 into 12 goes twice; 6 into 5 won't go*). **20** *tr. Cards* bid; declare (*go nap; has gone two spades*). **21** *intr.* (usu. foll. by *to*) be allotted or awarded; pass (*first prize went to the girl; the job went to his rival*). **22** *intr.* (foll. by *to, towards*) amount to; contribute to (*12 inches go to make a foot; this will go towards your holiday*). **23** *intr.* (in *imper.*) begin motion (a starter's order in a race) (*ready, steady, go!*). **24** *intr.* (usu. foll. by *to*) refer or appeal (*go to him for help*). **25** *intr.* (often foll. by *on*) take up a specified profession (*went on the stage; gone soldiering; went to sea*). **26** *intr.* (usu. foll. by *by, under*) be known or called (*goes by the name of Droopy*). **27** *tr. colloq.* proceed to (*go jump in the lake*). **28** *intr.* (foll. by *for*) apply to; have relevance for (*that goes for me too*). ● *n.* (*pl.* **goes**) **1** the act or an instance of going. **2** mettle; spirit; dash; animation (*she has a lot of go in her*). **3** vigorous activity (*it's all go*). **4** *colloq.* a success (*made a go of it*). **5** *colloq.* a turn; an attempt (*I'll have a go; it's my go; all in one go*). **6** *colloq.* a state of affairs (*a rum go*). **7** *colloq.* an attack of illness (*a bad go of flu*). **8** *colloq.* a quantity of liquor, food, etc. served at one time. ● *adj. colloq.* **1** functioning properly (*all systems are go*). **2** fashionable; progressive. □ **all the go** *colloq.* in fashion. **as** (or **so**) **far as it goes** an expression of caution against taking a statement too positively (*the work is good as far as it goes*). **as (a person or thing) goes** as the average is (*a good actor as actors go*). **from the word go** *colloq.* from the very beginning. **give it a go** *colloq.* make an effort to succeed. **go about 1** busy oneself with; set to work at. **2** be socially active. **3** (foll. by pres. part.) make a habit of doing (*goes about telling lies*). **4** *Naut.* change to an opposite tack. **go against 1** be contrary to (*goes against my principles*). **2** have an unfavourable result for (*the decision went against them*). **go ahead** proceed without hesitation. **go-ahead** *n.* permission to proceed. ● *adj.* enterprising. **go along with** agree to; take the same view as. **go around 1** (foll. by *with*) be regularly in the company of. **2** = *go about* 3. **3** = *go on* 4. **go-as-you-please** untrammelled; free. **go at** take in hand energetically; attack. **go away** depart, esp. from home for a holiday etc. **go back 1** return (to). **2** extend backwards in space or time (*goes back to the 18th century*). **3** (of the hour, a clock, etc.) be set to an earlier standard time (*the clocks go back in the autumn*). **go back on** fail to keep (one's word, promise, etc.). **go bail** see BAIL[1]. **go begging** see BEG. **go-between** an intermediary; a negotiator. **go by 1** pass. **2** be dependent on; be guided by. **go-by** *colloq.* a snub; a slight (*gave it the go-by*). **go by default** see *by default* (see DEFAULT). **go-cart 1** a handcart. **2** a pushchair. **3** = *go-kart*. **4** *archaic* a baby-walker. **go-devil** *US* an instrument used to clean the inside of pipes etc. **go down 1 a** (of an amount) become less (*the coffee has gone down a lot*). **b** subside (*the flood went down*). **c** decrease in price; lose value. **2 a** (of a ship) sink. **b** (of the sun) set. **3** (usu. foll. by *to*) be continued to a specified point. **4** deteriorate; fail; (of a computer network etc.) cease to function. **5** be recorded in writing. **6** be swallowed. **7** (often foll. by *with*) be received (in a specified way). **8** *Brit. colloq.* leave university. **9** *colloq.* be sent to prison (*went down for ten years*). **10** (often foll. by *before*) fall (before a conqueror). **go down with** *Brit.* begin to suffer from (a disease). **go Dutch** see DUTCH. **go far** see FAR. **go for 1** go to fetch. **2** be accounted as or achieve (*went for nothing*). **3** prefer; choose (*that's the one I go for*). **4** *colloq.* strive to attain (*go for it!*). **5** *colloq.* attack (*the dog went for him*). **go forward 1** proceed, progress (*go forward into the next round*). **2** (of the hour, a clock, etc.) be set to a later standard time. **go-getter** *colloq.* an aggressively enterprising person, esp. a businessman. **go-go** *colloq.* **1** (of a dancer, music, etc.) in modern style, lively, and rhythmic. **2** unrestrained; energetic. **3** (of investment) speculative. **go great guns** see GUN. **go halves** (or **shares**) (often foll. by *with*) share equally. **go in 1** enter a room, house, etc. **2** (usu. foll. by *for*) enter as a competitor. **3** *Cricket* take or begin an innings. **4** (of the sun etc.) become obscured by cloud. **go in for** take as one's object, style, pursuit, principle, etc. **going!, gone!** an auctioneer's announcement that bidding is closing or closed. **go into 1** enter (a place); go to stay in (hospital etc.). **2** pass into (a state or condition) (*he has gone into hiding; the company went into liquidation*). **3** investigate. **4** (of resources etc.) be invested in (*a lot of effort went into this*). **5** start a career or interest in. **6** dress oneself in (mourning etc.). **go it** *colloq.* **1** act vigorously, furiously, etc. **2** indulge in dissipation. **go it alone** see ALONE. **go it strong** *colloq.* go to great lengths; exaggerate. **go-kart** a miniature racing car with a skeleton body. **go a long way 1** (often foll. by *towards*) have a great effect; contribute or progress significantly. **2** (of food, money, etc.) last a long time, buy much. **3** = *go far* (see FAR). **go off 1** explode. **2 a** leave the stage. **b** leave, depart. **3** gradually cease to be felt. **4** (esp. of foodstuffs) deteriorate; decompose. **5** go to sleep; become unconscious. **6** be extinguished. **7** die. **8** be got rid of by sale etc. **9** *Brit. colloq.* begin to dislike (*I've gone off him*). **go-off** *colloq.* a start (*at the first go-off*). **go off at** *Austral. & NZ sl.* reprimand, scold. **go off well** (or **badly** etc.) (of an enterprise etc.) be received or accomplished well (or badly etc.). **go on 1** (often foll. by pres. part.) continue, persevere (*decided to go on with it; went on trying; unable to go on*). **2** *colloq.* **a** talk at great length. **b** (foll. by *at*) admonish (*went on and on at him*). **3** (foll. by *to* + infin.) proceed (*went on to become a star*). **4** happen. **5** conduct oneself (*shameful, the way they went on*). **6** *Theatr.* appear on stage. **7** *Cricket* begin bowling. **8** (of a garment) be large enough for its wearer. **9** take one's turn to do something. **10** (also **go upon**) *colloq.* use as evidence (*police don't have anything to go on*). **11** (esp. in *neg.*) *colloq.* **a** concern oneself about. **b** care for (*don't go much on red hair*). **12** become chargeable to (the parish etc.). **go on!** *colloq.* an expression of encouragement or disbelief. **go out 1** leave a room, house, etc. **2** be broadcast. **3** be extinguished. **4** (often foll. by *with*) be courting. **5** (of a government) leave office. **6** cease to be fashionable. **7** (usu. foll. by *to*) depart, esp. to a colony etc. **8** *colloq.* lose consciousness. **9** (of workers) strike. **10** (usu. foll. by *to*) (of the heart etc.) expand with sympathy etc. towards (*my heart goes out to them*). **11** *Golf* play the first nine holes in a round. **12** *Cards* be the first to dispose of one's hand. **13** (of a tide) ebb, recede to low tide. **14** mix socially; attend (social) events. **go over 1** inspect the details of; rehearse; retouch. **2** (often foll. by *to*) change one's allegiance or religion. **3** (of a play etc.) be received, esp. favourably (*went over well in Dundee*). **go round 1** spin, revolve. **2** be long enough to encompass. **3** (of food etc.) suffice for everybody. **4** (usu. foll. by *to*) visit informally. **5** = go around. **go slow** work slowly, as a form of industrial action. **go-slow** *Brit.* such industrial action. **go through 1** be dealt with or completed. **2** discuss in detail; scrutinize in sequence. **3** perform (a ceremony, a recitation, etc.). **4** undergo. **5** *colloq.* use up; spend (money etc.). **6** make holes in. **7** (of a book) be successively published (in so many editions). **8** *Austral. sl.* abscond. **go through with** not leave unfinished; complete. **go to!** *archaic* an exclamation of disbelief, impatience, admonition, etc. **go to the bar** become a barrister. **go to blazes** (or **hell** or **Jericho** etc.) *sl.* an exclamation of dismissal, contempt, etc. **go to the country** see COUNTRY. **go together 1** match; fit. **2** be courting. **go to it!** *colloq.* begin work! **go-to-meeting** (of a hat, clothes, etc.) suitable for going to church in. **go to show** (or **prove**) serve as evidence (or proof) (also *absol.*). **go under** sink; fail; succumb. **go up 1** increase in price. **2** *Brit. colloq.* enter university. **3** be consumed (in flames etc.); explode. **go up in the world** attain a higher social position. **go well** (or **ill** etc.) (often foll. by *with*) turn out well, (or ill etc.). **go with 1** be harmonious with; match. **2** agree to; take the same view as. **3 a** be a pair with. **b** be courting. **4** follow the drift of. **go without** manage without; forgo (also *absol.*: *we shall just have to go without*). **go with the tide** do as others do; follow the drift. **have a go at** (or **times**) **1** attack, criticize. **2** attempt, try. **on the go** *colloq.* **1** in constant motion. **2** constantly working. **to go** *N. Amer.* **1** still to be dealt with. **2** (of refreshments etc.) to be eaten or drunk off the premises. **who goes there?** a sentry's challenge. [OE *gān* f. Gmc: *went* orig. past of WEND]

go² /gəʊ/ *n.* a complex Japanese board game of territorial possession and capture, played by placing small pieces or stones on a large grid. Go has a handicapping system which allows novice and expert to play each other. It originated in China *c.*1500 BC. [Jap.]

Goa /ˈgəʊə/ a state on the west coast of India; capital, Panaji. Formerly a Portuguese territory, it was seized by India in 1961. With two other former Portuguese territories, Daman and Diu, it then formed a Union Territory in India, which was made a state in 1987. □ **Goan** *adj. & n.* **Goanese** /ˌgəʊəˈniːz/ *adj. & n.*

goa /ˈgəʊə/ *n.* a Tibetan antelope, *Procapra picticaudata*, with backward curving horns.

goad /gəʊd/ *n. & v.* ● *n.* **1** a spiked stick used for urging cattle forward. **2** anything that torments, incites, or stimulates. ● *v.tr.* **1** urge on with a goad. **2** (usu. foll. by *on*, *into*) irritate; stimulate (*goaded him into retaliating*; *goaded me on to win*). [OE *gād*, rel. to Lombard *gaida* arrowhead f. Gmc]

goal /gəʊl/ *n.* **1** the object of a person's ambition or effort; a destination; an aim (*fame is his goal*; *London was our goal*). **2 a** *Football* the space between a pair of posts with a crossbar into which the ball has to be sent to score. **b** a cage or basket used similarly in other games. **c** a successful attempt to get the ball etc. into this space; the point or points won (*scored 3 goals*). **3** a point marking the end of a race. □ **goal area** *Football* a rectangular area in front of the goal from within which goal-kicks must be taken. **goal average** *Football* the ratio of the numbers of goals scored for and against a team in a series of matches. **goal difference** *Football* the difference of goals scored for and against a team. **goal-kick 1** *Football* a kick by the defending side after attackers send the ball over the goal-line without scoring. **2** *Rugby* an attempt to kick a goal. **goal-line 1** *Football*, *Hockey*, etc. the line between each pair of goalposts, extended to form the end-boundary of a field of play (cf. *touch-line*). **2** *Amer. Football & Rugby* the line separating an end zone from the rest of the field, over which the ball must be carried or passed to score a touchdown or try. **goal-mouth** *Football* the space between or near the goalposts. **goal-tender** (or **-minder**) *N. Amer.* a goalkeeper at ice hockey. **in goal** in the position of goalkeeper. **in-goal area** *Rugby* the area at either end of the pitch, between the goal-line and the dead-ball line. □ **goalless** /ˈgəʊllɪs/ *adj.* [16th c.: orig. unkn.: perh. identical with ME *gol* boundary]

goalball /ˈgəʊlbɔːl/ *n.* a team ball game for blind and visually handicapped players.

goalie /ˈgəʊlɪ/ *n. colloq.* = GOALKEEPER.

goalkeeper /ˈgəʊlˌkiːpə(r)/ *n.* a player stationed to protect the goal in various sports. □ **goalkeeping** *n.*

goalpost /ˈgəʊlpəʊst/ *n.* either of the two upright posts of a goal. □ **move the goalposts** alter the basis or scope of a procedure during its course, so as to fit adverse circumstances encountered.

goanna /gəʊˈænə/ *n. Austral.* a monitor lizard. [corrupt. of IGUANA]

goat /gəʊt/ *n.* **1 a** hardy usu. short-haired domesticated ruminant mammal, *Capra hircus*, having horns and (in the male) a beard, and kept for its milk and meat. (*See note below.*) **2 a** any other mammal of the genus *Capra*, including the ibexes and markhor. **b** the mountain goat. **3** a lecherous man. **4** *colloq.* a foolish person. **5** (**the Goat**) the zodiacal sign Capricorn or the constellation Capricornus. **6** *US* a scapegoat. □ **get a person's goat** *colloq.* irritate a person. **goat-antelope** a goatlike ruminant of the subfamily Caprinae, esp. an agile one of the tribe Rupicaprini, which includes the chamois, goral, serow, and Rocky Mountain goat. **goat moth** a large moth of the family Cossidae; esp. *Cossus cossus*, with a wood-boring caterpillar that smells like a goat. **goat's-beard 1** a composite meadow plant, *Tragopogon pratensis*, with yellow flowers. **2** a rosaceous plant, *Aruncus dioicus*, with long plumes of white flowers. □ **goatish** *adj.* **goaty** *adj.* [OE *gāt* she-goat f. Gmc]

■ The goat was first domesticated about 9,000 years ago in SW Asia. Its main wild ancestor, *Capra aegagrus*, is still found from the Greek islands to India. The 'wild goats' found in parts of Britain etc. are actually feral domestic goats.

goatee /gəʊˈtiː, ˈgəʊtɪ/ *n.* (also **goatee beard**) a small pointed beard like that of a goat.

goatherd /ˈgəʊθɜːd/ *n.* a person who tends goats.

goatskin /ˈgəʊtskɪn/ *n.* **1** the skin of a goat. **2** a garment or bottle made out of goatskin.

goatsucker /ˈgəʊtˌsʌkə(r)/ *n.* = NIGHTJAR.

gob¹ /gɒb/ *n. esp. Brit. sl.* the mouth. □ **gob-stopper** a very large hard sweet. [perh. f. Gael. & Ir., = beak, mouth]

gob² /gɒb/ *n. & v. Brit. sl.* ● *n.* a clot of slimy matter. ● *v.intr.* (**gobbed**, **gobbing**) spit. [ME f. OF *go(u)be* mouthful]

gob³ /gɒb/ *n. sl.* a US sailor. [20th c.: cf. GOBBY]

Göbbels see GOEBBELS.

gobbet /ˈgɒbɪt/ *n.* **1 a** a piece, lump, or portion. **b** a clot of slimy matter. **2** an extract from a text, esp. one set for translation or comment in an examination. [ME f. OF *gobet* (as GOB²)]

Gobbi /ˈgɒbɪ/, Tito (1915–84), Italian operatic baritone. Particularly renowned for his interpretations of Verdi's baritone roles, he also gained notable successes with his performances in the title role of Berg's *Wozzeck* and as Scarpia in Puccini's *Tosca*.

gobble¹ /ˈgɒb(ə)l/ *v.tr. & intr.* eat hurriedly and noisily. □ **gobbler** *n.* [prob. dial. f. GOB²]

gobble² /ˈgɒb(ə)l/ *v.intr.* **1** (of a turkeycock) make a characteristic swallowing sound in the throat. **2** make such a sound when speaking, esp. when excited, angry, etc. [imit.: perh. based on GOBBLE¹]

gobbledegook /ˈgɒb(ə)ldɪˌguːk, -ˌgʊk/ *n.* (also **gobbledygook**) *colloq.* pompous or unintelligible jargon. [prob. imit. of a turkeycock]

gobbler /ˈgɒblə(r)/ *n. colloq.* a turkeycock.

gobby /ˈgɒbɪ/ *n.* (pl. **-ies**) *sl.* **1** a coastguard. **2** an American sailor. [perh. f. GOB² + -Y¹]

Gobelins /ˈgəʊbəlɪnz, French ɡɔblɛ̃/ a carpet and tapestry factory on the Left Bank in Paris. It was established by the Gobelins family *c.*1440 and was taken over by the French Crown in 1662. During the late 17th and 18th centuries it became extremely important and successful, with leading French painters providing cartoons, and with tapestry panels becoming used as alternatives to oil paintings. The factory still exists, as a state-owned concern.

gobemouche /ˈgɒbmuːʃ/ *n.* (pl. **gobemouches** *pronunc.* same) a gullible listener. [F *gobe-mouches*, = fly-catcher f. *gober* swallow + *mouches* flies]

Gobi Desert /ˈgəʊbɪ/ a barren plateau of southern Mongolia and northern China.

Gobineau /ˈgɒbɪˌnəʊ/, Joseph Arthur, Comte de (1816–82), French writer and anthropologist. Gobineau is best known for his theories of racial superiority; he claimed the existence of a hierarchy of races, at the top of which was the 'Aryan race' and noted that Aryan civilization and culture degenerated when its members interbred with the so-called black and yellow races, which he considered inferior. Gobineau cited 'Aryan' Germans as the sole remaining ideal of racial purity. His theories anticipated the 'master-class' philosophy of Nietzsche, and later influenced the ideology and policies of the Nazis.

goblet /ˈgɒblɪt/ *n.* **1** a drinking-vessel with a foot and a stem, usu. of glass. **2 a** *archaic* a metal or glass bowl-shaped drinking-cup without handles, sometimes with a foot and a cover. **b** *poet.* any drinking-cup.

3 a goblet-shaped receptacle for food forming part of a blender or liquidizer. [ME f. OF *gobelet* dimin. of *gobel* cup, of unkn. orig.]

goblin /ˈgɒblɪn/ *n.* a mischievous ugly dwarflike creature of folklore. [ME prob. f. AF *gobelin*, med.L *gobelinus*, prob. f. name dimin. of *Gobel*, rel. to G *Kobold*: see COBALT]

gobsmacked /ˈgɒbsmækt/ *adj. sl.* flabbergasted; struck dumb with awe or amazement. □ **gobsmacking** *adj.* [GOB[1] + SMACK[1]]

goby /ˈgəʊbɪ/ *n.* (*pl.* **-ies**) a small marine fish of the family Gobiidae, having ventral fins joined to form a sucker or disc. [L *gobius, cobius* f. Gk *kōbios* GUDGEON[1]]

GOC *abbr.* General Officer Commanding.

god /gɒd/ *n. & int.* ● *n.* **1 a** (in many religions) a superhuman being or spirit worshipped as having power over nature, human fortunes, etc.; a deity. **b** an image, idol, animal, or other object worshipped as divine or symbolizing a god. **2** (**God**) (in Christian and other monotheistic religions) the creator and ruler of the universe; the supreme being. **3 a** an adored, admired, or influential person. **b** something worshipped like a god (*makes a god of success*). **4** (in *pl.*) *Theatr.* **a** the gallery. **b** the people sitting in it. ● *int.* (**God!**) an exclamation of surprise, anger, etc. □ **by God!** an exclamation of surprise etc. **for God's sake!** see SAKE[1]. **God-awful** *sl.* extremely unpleasant, nasty, etc. **God bless** an expression of good wishes on parting. **God bless me** (or **my soul**) see BLESS. **God damn** (**you, him,** etc.) may (you, him, etc.) be damned. **god-damn** (or **-dam** or **-damned**) *sl.* accursed, damnable. **god-daughter** a female godchild. **God the Father, Son, and Holy Ghost** (in the Christian tradition) the persons of the Trinity. **God-fearing** earnestly religious. **God forbid** (foll. by *that* + clause, or *absol.*) may it not happen! **God-forsaken** devoid of all merit; dismal; dreary. **God-given** received as if from God; possessed from birth or by divine authority. **God grant** (foll. by *that* + clause) may it happen. **God help** (**you, him,** etc.) an expression of concern for or sympathy with a person. **God knows 1** it is beyond all knowledge (*God knows what will become of him*). **2** I call God to witness that (*God knows we tried hard enough*). **God's Acre** a churchyard. **God's book** the Bible. **God's country** (or **God's own country**) an earthly paradise, esp. with reference to the United States. **God's gift** often *iron.* a godsend. **God squad** *sl.* **1** a religious organization, esp. an evangelical Christian group. **2** its members. **God's truth** the absolute truth. **God willing** if Providence allows. **good God!** an exclamation of surprise, anger, etc. **in God's name** an appeal for help. **in the name of God** an expression of surprise or annoyance. **my** (or **oh**) **God!** an exclamation of surprise, anger, etc. **play God** assume importance or superiority. **thank God!** an exclamation of pleasure or relief. **with God** dead and in heaven. □ **godhood** *n.* **godship** *n.* **godward** *adj. & adv.* **godwards** *adv.* [OE f. Gmc]

Godard /ˈgɒdɑː(r)/, Jean-Luc (b.1930), French film director. He was one of the leading figures of the *nouvelle vague*; his films frequently deal with existentialist themes and are notable for their use of improvised dialogue, disjointed narratives, and unconventional shooting and cutting techniques. In addition to his more commercial works, such as *Breathless* (1960), *Alphaville* (1965), and *Slow Motion* (1980), he has also explored the use of film for more overtly political purposes, for example in *Wind from the East* (1969).

Godavari /gəʊˈdɑːvərɪ/ a river in central India which rises in the state of Maharashtra and flows about 1,440 km (900 miles) south-east across the Deccan plateau to the Bay of Bengal.

godchild /ˈgɒdtʃaɪld/ *n.* (*pl.* **-children**) a person in relation to a godparent.

Goddard /ˈgɒdɑːd/, Robert Hutchings (1882–1945), American physicist. He carried out pioneering work in rocketry, and designed and built the first successful liquid-fuelled rocket. NASA's Goddard Space Flight Center is named after him.

goddess /ˈgɒdɪs/ *n.* **1** a female deity. **2** a woman who is adored, esp. for her beauty.

Gödel /ˈgɜːd(ə)l/, Kurt (1906–78), Austrian-born American mathematician. He made several important contributions to mathematical logic, especially the *incompleteness theorem* (Gödel's proof): that in any sufficiently powerful, logically consistent formulation of logic or mathematics there must be true formulas which are neither provable nor disprovable. This makes such formulations essentially incomplete, and entails the corollary that the consistency of such a system cannot be proved within that system.

godet /gəʊˈdeɪ/ *n.* a triangular piece of material inserted in a dress, glove, etc. [F]

godetia /gəˈdiːʃə/ *n.* a plant of the genus *Godetia*, with showy rose-purple or reddish flowers. [mod.L f. C. H. *Godet*, Swiss botanist (1797–1879)]

godfather /ˈgɒdˌfɑːðə(r)/ *n.* **1** a male godparent. **2** esp. *US* a person directing an illegal organization, esp. the Mafia. □ **my godfathers!** *euphem.* my God!

Godhavn /ˈgɒdˌhɑːv(ə)n, Danish ˈgoðhaʊn/ a town in western Greenland, on the south coast of the island of Disko.

godhead /ˈgɒdhed/ *n.* (also **Godhead**) **1 a** the state of being God or a god. **b** divine nature. **2** a deity. **3** (**the Godhead**) God.

Godiva /gəˈdaɪvə/, Lady (d.1080), English noblewoman, wife of Leofric, Earl of Mercia (d.1057). According to a 13th-century legend, she agreed to her husband's proposition that he would reduce some particularly unpopular taxes only if she rode naked on horseback through the market-place of Coventry. Later versions of the story describe how all the townspeople stayed indoors at Lady Godiva's request, except for peeping Tom, who as a result was struck blind.

godless /ˈgɒdlɪs/ *adj.* **1** impious; wicked. **2** without a god. **3** not recognizing God. □ **godlessness** *n.*

godlike /ˈgɒdlaɪk/ *adj.* **1** resembling God or a god in some quality, esp. in physical beauty. **2** befitting or appropriate to a god.

godly /ˈgɒdlɪ/ *adj.* (**-ier, -iest**) religious, pious, devout. □ **godliness** *n.*

godmother /ˈgɒdˌmʌðə(r)/ *n.* a female godparent.

godown /ˈgəʊdaʊn/ *n.* a warehouse in parts of eastern Asia, esp. in India. [Port. *gudão* f. Malay *godong* perh. f. Telugu *gidangi* place where goods lie f. *kidu* lie]

godparent /ˈgɒdˌpeərənt/ *n.* a person who presents a child at baptism and responds on the child's behalf.

God Save the King (or **Queen**) the British national anthem. The origins of both words and tune are obscure: the phrase 'God save the king' occurs in various passages in the Old Testament, while as early as 1545 it was a watchword in the navy, with 'long to reign over us' as a countersign. Evidence points to a 17th-century origin for the complete words and tune of the anthem.

godsend /ˈgɒdsend/ *n.* an unexpected but welcome event or acquisition.

godson /ˈgɒdsʌn/ *n.* a male godchild.

Godspeed /gɒdˈspiːd/ *int.* an expression of good wishes to a person starting a journey.

Godthåb /ˈgɒdhɔːb/ the former name (until 1979) for NUUK.

Godunov /ˈgɒduˌnɒf/, Boris (1550–1605), tsar of Russia 1598–1605. He rose to prominence as a counsellor of Ivan the Terrible and eventually succeeded Ivan's son as tsar. His reign was overshadowed by famine, doubts over his involvement in the earlier death of Ivan's eldest son, and the appearance of a pretender, the so-called False Dmitri. Godunov died suddenly while his army was resisting an invasion by the pretender. He has been made famous outside Russia by Mussorgsky's opera *Boris Godunov*.

Godwin /ˈgɒdwɪn/, William (1756–1836), English social philosopher and novelist. At first a dissenting minister, he subsequently became an atheist and expounded theories of anarchic social organization based on a belief in the goodness of human reason and on his doctrine of extreme individualism. His ideological novel *Caleb Williams* (1794), which exposes the tyranny exercised by the ruling classes, was an early example of the crime and detection novel. In 1797 he married Mary Wollstonecraft; the couple's daughter was Mary Shelley.

Godwin-Austen, Mount /ˌgɒdwɪnˈɔːstɪn/ an alternative name for K2.

godwit /ˈgɒdwɪt/ *n.* a long-legged wading bird of the genus *Limosa*, with a long straight or slightly upcurved bill. [16th c.: of unkn. orig.]

Godwottery /gɒdˈwɒtərɪ/ *n. joc.* affected, archaic, or excessively elaborate speech or writing, esp. regarding gardens. [*God wot* (in a line from T. E. Brown's poem 'My Garden', 1876)]

Goebbels /ˈgɜːb(ə)lz/, (Paul) Joseph (also **Göbbels**) (1897–1945), German Nazi leader and politician. In 1933 he became Hitler's Minister of Propaganda, with control of the press, radio, and all aspects of culture, and manipulated these in order to further Nazi aims. A supporter of Hitler to the last, he committed suicide rather than surrender to the Allies.

goer /ˈgəʊə(r)/ *n.* **1** a person or thing that goes (*a slow goer*). **2** (often in *comb.*) a person who attends, esp. regularly (*a churchgoer*). **3** *colloq.* **a** a lively or persevering person. **b** a sexually promiscuous person. **4** *Austral. colloq.* a project likely to be accepted or to succeed.

Goering /ˈgɜːrɪŋ/, Hermann Wilhelm (1893–1946), German Nazi leader and politician. In 1934 he became commander of the German air force, and was responsible for the German rearmament programme. Until 1936 Goering headed the Gestapo, which he had founded; from then until 1943 he directed the German economy. In that year he fell from favour, was deprived of all authority, and was finally dismissed in 1945 after unauthorized attempts to make peace with the Allies. Sentenced to death at the Nuremberg war trials, he committed suicide in his cell.

Goes /gu:s/, Hugo van der (*fl.* *c.*1467–82), Flemish painter, born in Ghent. He worked chiefly in his birthplace, though his best-known work is the large-scale *Portinari Altarpiece* (1475), commissioned for a church in Florence.

goes 3rd *sing. present of* GO[1].

goest /'gəʊɪst/ *archaic 2nd sing. present of* GO[1].

goeth /'gəʊɪθ/ *archaic 3rd sing. present of* GO[1].

Goethe /'gɜːtə/, Johann Wolfgang von (1749–1832), German poet, dramatist, and scholar. In his early career he was involved with the *Sturm und Drang* movement. In 1775 he moved to Weimar (later becoming a close friend of Schiller) and held a number of government posts 1776–86, after which he spent two years in Italy. His writing began to move away from the energy and romanticism of *Sturm und Drang* and became more measured and classical in style, as in the 'Wilhelm Meister' novels (1796–1829), which are also the prototype of the *Bildsungsroman*. An important figure of the Enlightenment in Germany, his wide-ranging interests included philosophy, physics, and biology; he wrote drama, poetry, novels, ballads, and autobiography, and was also director of Weimar Theatre 1791–1817. His works include the epic drama *Götz von Berlichingen* (1773), the epistolary novel *The Sorrows of Young Werther* (1774), and the two-part poetic drama *Faust* (1808–32), as well as the classical dramas *Iphigenia in Tauris* (1787) and *Tasso* (1790).

Goethean /'gɜːtɪən/ *adj.* (also **Goethian**) of, relating to, or characteristic of Goethe.

gofer /'gəʊfə(r)/ *n.* esp. *N. Amer. sl.* a person who runs errands, esp. on a film set or in an office; a dogsbody. [*go for* (see GO[1])]

goffer /'gəʊfə(r)/, 'gɒf–/ *v. & n.* ● *v.tr.* **1** make wavy, flute, or crimp (a lace edge, a trimming, etc.) with heated irons. **2** (as **goffered** *adj.*) (of the edges of a book) embossed. ● *n.* **1** an iron used for goffering. **2** ornamental plaiting used for frills etc. [F *gaufrer* stamp with a patterned tool f. *gaufre* honeycomb, rel. to WAFER, WAFFLE[2]]

Gog and Magog /gɒg, 'meɪgɒg/ **1** in the Bible, the names of enemies of God's people. In Ezek. 38–9, Gog is apparently a ruler (or his people) from the land of Magog; in Rev. 20:8, Gog and Magog are nations under the dominion of Satan. **2** (in medieval legend) opponents of Alexander the Great, living north of the Caucasus. **3** two giant statues standing in Guildhall, London, from the time of Henry V (destroyed in 1666 and 1940; replaced in 1708 and 1953), representing either the last two survivors of a race of giants supposed to have inhabited Britain before Roman times, or Gogmagog, chief of the giants, and Corineus, a Roman invader.

goggle /'gɒg(ə)l/ *v., adj., & n.* ● *v.* **1** *intr.* **a** (often foll. by *at*) look with wide-open eyes. **b** (of the eyes) be rolled about; protrude. **2** *tr.* turn (the eyes) sideways or from side to side. ● *adj.* (usu. *attrib.*) (of the eyes) protuberant or rolling. ● *n.* **1** (in *pl.*) **a** spectacles for protecting the eyes from glare, dust, water, etc. **b** *colloq.* spectacles. **2** (in *pl.*) a sheep disease, the staggers. **3** a goggling expression. □ **goggle-box** *Brit. colloq.* a television set. **goggle-dive** an underwater dive in goggles. **goggle-eyed** having staring or protuberant eyes, esp. through astonishment or disbelief. [ME, prob. from a base *gog* (unrecorded) expressive of oscillating movement]

goglet /'gɒglɪt/ *n. Anglo-Ind.* a long-necked usu. porous earthenware vessel used for keeping water cool. [Port. *gorgoleta*]

Gogol /'gəʊgɒl, 'gɒg(ə)l/, Nikolai (Vasilevich) (1809–52), Russian novelist, dramatist, and short-story writer, born in Ukraine. He first became famous for his play *The Inspector General* (1836), a savagely satirical picture of life in a provincial Russian town. His St Petersburg stories (including *Notes of a Madman*, 1835, and *The Greatcoat*, 1842) also display a trenchant satirical wit. Living mainly abroad from 1836 to 1848, he wrote the comic epic novel *Dead Souls* (1842), widely regarded as the foundation of the modern Russian novel, but after a spiritual crisis he burned the manuscript of the second part.

Goiânia /gɔɪˈɑːnɪə/ a city in south central Brazil, capital of the state of Goiás; pop. (1990) 998,500. Founded as a new city in 1933, it replaced the town of Goiás as state capital in 1942.

Goiás /gɔɪˈɑːs/ a state in south central Brazil; capital, Goiânia.

Goidel /'gɔɪd(ə)l/ *n.* a member of the Gaelic people that comprises the Scottish, Irish, and Manx Celts. [OIr. *Góidel*: cf. GAEL]

Goidelic /gɔɪˈdelɪk/ *n. & adj.* ● *n.* the northern group of the Celtic languages, comprising Irish, Scottish Gaelic, and Manx. (*See note below.*) ● *adj.* of or relating to Goidelic or the Goidels. [GOIDEL]

▪ Speakers of the Celtic precursor of the Goidelic languages are thought to have invaded Ireland from Europe *c.*1000 BC, moving from about the 5th century AD to Scotland and the Isle of Man. By the 10th century the Irish, Scots, and Manx varieties had diverged into separate languages, of which Irish and Scots Gaelic remain living. Goidelic is also known as *Q-Celtic*, because it retains the Indo-European *kw* sound as *q* or *c*, whereas in Brythonic (*P-Celtic*) *kw* becomes *p*. (Cf. BRYTHONIC.)

going /'gəʊɪŋ/ *n. & adj.* ● *n.* **1 a** the act or process of going. **b** an instance of this; a departure. **2 a** the condition of the ground for walking, riding, etc. **b** progress affected by this (*found the going hard*). ● *adj.* **1** in or into action (*set the clock going*). **2** existing, available; to be had (*there's cold beef going; one of the best fellows going*). **3** current, prevalent (*the going rate*). □ **get going** start steadily talking, working, etc. (*can't stop him when he gets going*). **going away** a departure, esp. on a honeymoon. **going concern** a thriving business. **going for one** *colloq.* acting in one's favour (*he has got a lot going for him*). **going on fifteen** etc. esp. *US* approaching one's fifteenth etc. birthday. **going on for** approaching (a time, an age, etc.) (*must be going on for 6 years*). **going-over 1** *colloq.* an inspection or overhaul. **2** *sl.* a thrashing. **3** *US colloq.* a scolding. **goings-on** /ˌgəʊɪŋz'ɒn/ behaviour, esp. morally suspect. **going to** intending or intended to; about to; likely to (*it's going to sink!*). **heavy going** slow or difficult to progress with (*found Proust heavy going*). **to be going on with** to start with; for the time being. **while the going is good** while conditions are favourable. [GO[1]: in some senses f. earlier *a-going*: see A[2]]

goitre /'gɔɪtə(r)/ *n.* (*US* **goiter**) *Med.* a swelling of the neck resulting from enlargement of the thyroid gland. □ **goitred** *adj.* **goitrous** *adj.* [F, back-form. f. *goitreux* or f. Prov. *goitron*, ult. f. L *guttur* throat]

Gokhale /'gəʊkəˌleɪ/, Gopal Krishna (1866–1915), Indian political leader and social reformer. A member of the Indian National Congress, he became its president in 1905. He was a leading advocate of Indian self-government, favouring gradual reform through constitutional or moderate means; he met and influenced Mahatma Gandhi. He also founded the Servants of India Society (1905), whose members pledged themselves to assisting the underprivileged.

Golan Heights /gəʊ'lɑːn, 'gəʊlən/ a range of hills on the border between Syria and Israel, north-east of the Sea of Galilee. Formerly under Syrian control, the area was occupied by Israel in 1967 and annexed in 1981. Negotiations for the withdrawal of Israeli troops from the region began after Yitzhak Rabin became Prime Minister of Israel in 1992.

Golconda /gɒl'kɒndə/ *n.* a mine or source of wealth, advantages, etc. [ancient ruined city near Hyderabad, India, once famous for its diamonds]

gold /gəʊld/ *n. & adj.* ● *n.* **1** a precious shiny yellow metallic chemical element (atomic number 79; symbol **Au**). (*See note below.*) **2** the colour of gold. **3 a** coins or articles made of gold. **b** money in large sums, wealth. **4** something precious, beautiful, or brilliant (*all that glitters is not gold*). **5** = *gold medal*. **6** gold used for coating a surface or as a pigment, gilding. **7** the bull's-eye of an archery target (usu. gilt). ● *adj.* **1** made wholly or chiefly of gold. **2** coloured like gold. □ **age of gold** = *golden age*. **gold amalgam** an easily moulded combination of gold with mercury. **gold-beater** a person who beats gold out into gold leaf. **gold-beater's skin** a membrane used to separate leaves of gold during beating, or as a covering for slight wounds. **gold bloc** a bloc of countries having a gold standard. **gold brick** *sl.* **1** a thing with only a surface appearance of value, a sham or fraud. **2** *US* a lazy person. **goldbrick** *v.intr. US* shirk. **gold card** a kind of preferential charge card giving privileges and benefits not available to holders of the standard card. **gold-digger 1** a person who digs for gold. **2** *sl.* a woman who wheedles money out of men. **gold disc** an award given to a recording artist or group for sales of a recording exceeding a million in the US or 500,000 in Britain for a single, or 500,000 in the US or 250,000 in Britain for an album. **gold-dust 1** gold in fine particles as often found naturally. **2** a cruciferous plant, *Alyssum saxatile*, with many small yellow flowers. **gold-field** a district in which gold is found as a mineral. **gold foil** gold beaten into a thin sheet. **gold leaf** gold beaten into a very thin sheet. **gold medal** a gold-coloured medal, usu. awarded as first prize in a contest, esp. the Olympic Games. **gold-mine 1** a place where gold is mined. **2** colloq. a source of wealth. **gold of pleasure** an annual cruciferous plant, *Camelina sativa*, with yellow flowers. **gold plate 1** vessels made of gold. **2** material plated with gold. **gold-plate** v.tr. plate with gold. **gold reserve** a reserve of gold coins or bullion held by a central bank etc. **gold thread 1** a thread of silk etc. with gold wire wound round it. **2** a bitter ranunculaceous plant, *Coptis trifolia*. [OE f. Gmc]

■ Gold has been valued from ancient times for its colour and brightness, rarity, and durability: in normal conditions it does not corrode or tarnish. A transition metal, gold is quite widely distributed in nature but economical extraction is only possible from deposits of the native metal or sulphide ores, or as a by-product of copper and lead mining. In world production of gold South Africa is dominant, although Russia also has considerable resources. Its most important uses are for jewellery and other decorative purposes and to guarantee the value of currency (the use of gold itself for coinage is now limited). It is also used in electrical contacts and (in some countries) as a filling for teeth. The relative purity of gold is measured in carats. The symbol Au is from the Latin word for gold, *aurum*.

Gold Coast 1 the former name (until 1957) for GHANA. **2** a resort region on the east coast of Australia, to the south of Brisbane.

Gold Collar a classic greyhound race, inaugurated in 1933, run annually at the Catford track in south London, originally in May, now in September.

goldcrest /ˈɡəʊldkrest/ n. a European kinglet, *Regulus regulus*, with a golden crest.

golden /ˈɡəʊld(ə)n/ adj. **1 a** made or consisting of gold (*golden sovereign*). **b** yielding gold. **2** coloured or shining like gold (*golden hair*). **3** precious; valuable; excellent; important (*a golden memory*; *a golden opportunity*). □ **golden age 1** a supposed past age when people were happy and innocent. **2** the period of a nation's greatest prosperity, literary merit, etc. **golden-ager** N. Amer. an old person. **golden balls** a pawnbroker's sign. **golden boy** (or **girl**) colloq. a popular or successful person. **golden chain** the laburnum. **golden delicious** a variety of dessert apple. **golden eagle** a large eagle, *Aquila chrysaetos*, with yellow-tipped head-feathers. **goldeneye** a marine diving duck of the genus *Bucephala*, with bright yellow eyes. **golden goose** a continuing source of wealth or profit (esp. in phr. *kill the golden goose*; see GOOSE). **golden hamster** a usu. tawny hamster, *Mesocricetus auratus*, kept as a pet or laboratory animal. **golden handshake** colloq. a payment given on redundancy or early retirement. **golden hello** colloq. a payment made by an employer to a keenly sought recruit. **golden labrador** a labrador dog with a golden-coloured coat. **golden jubilee 1** the fiftieth anniversary of a sovereign's accession. **2** any other fiftieth anniversary. **golden mean 1** the principle of moderation, as opposed to excess. **2** = *golden section*. **golden number** the number of a year in the Metonic lunar cycle, used to fix the date of Easter. **golden oldie** colloq. **1** an old hit record or film etc. that is still well known and popular. **2** a person who is no longer young but is still successful in his or her field. **golden opinions** high regard. **golden oriole** a European oriole, *Oriolus oriolus*, of which the male has yellow and black plumage and the female has mainly green plumage. **golden parachute** colloq. financial compensation guaranteed to company executives dismissed as a result of a merger or takeover. **golden perch** Austral. = CALLOP. **golden retriever** a retriever with a thick golden-coloured coat. **goldenrod** a composite plant of the genus *Solidago*, with a rodlike stem and a spike of small bright-yellow flowers. **golden rule** a basic principle of action, esp. 'do as you would be done by'. **golden section** the division of a line such that the whole is to the greater part as that part is to the smaller part. **golden share** the controlling interest in a company, esp. as retained by the government after a nationalized industry is privatized. **golden syrup** Brit. a pale treacle. **golden wedding** the fiftieth anniversary of a wedding. □ **goldenly** adv. **goldenness** n.

Golden Calf n. **1** (in the Bible) an image of gold in the shape of a calf, made by Aaron in response to the Israelites' pleas for a god as they awaited Moses' return from Mount Sinai, where he was receiving the Ten Commandments (Exod. 32). **2** (**golden calf**) wealth as an object of worship.

Golden Fleece Gk Mythol. the fleece of gold taken from the ram that carried Phrixus through the air to Colchis. It was guarded by an unsleeping dragon until secured by Jason and the Argonauts with the help of Medea. (See also HELLESPONT.)

Golden Gate a deep channel connecting San Francisco Bay with the Pacific Ocean, spanned by the Golden Gate suspension bridge (completed 1937).

Golden Hind the ship in which, in 1577–80, Francis Drake circumnavigated the globe. The ship was originally called the *Pelican*, but Drake changed the name en route in honour of his patron, Sir Christopher Hatton (1540–91), whose crest was a golden hind.

Golden Horde a Mongol and Tartar army under the leadership of the descendants of Genghis Khan that overran Asia and part of eastern Europe in the 13th century, and maintained an empire of varying size until the end of the 15th century. It was so called because of the magnificence of its leader's camp.

Golden Horn (called in Turkish *Haliç*) a curved inlet of the Bosporus, forming the harbour of Istanbul.

Golden State US the state of California.

goldfinch /ˈɡəʊldfɪntʃ/ n. (pl. same or **-finches**) a finch of the genus *Carduelis*, with yellow in the plumage; esp. *C. carduelis* of Eurasia, and *C. tristis* of North America. [OE *goldfinc* (as GOLD, FINCH)]

goldfish /ˈɡəʊldfɪʃ/ n. a small reddish-golden Chinese carp, *Carassius auratus*, kept as a pet or ornamental fish. □ **goldfish bowl 1** a globular glass container for goldfish. **2** a place or situation lacking privacy.

goldilocks /ˈɡəʊldɪˌlɒks/ n. **1** a person with golden hair. 'Goldilocks' as the name of the girl in the traditional fairy story was first used in John Hassall's *Old Nursery Stories and Rhymes* (c.1904). **2 a** a buttercup, *Ranunculus auricomus*, found in hedges and woods. **b** a rare composite plant, *Aster linosyris*, that resembles goldenrod. [*goldy* f. GOLD + LOCK²]

Golding /ˈɡəʊldɪŋ/, Sir William (Gerald) (1911–93), English novelist. He achieved literary success with his first novel *Lord of the Flies* (1954) about a group of boys who revert to savagery when stranded on a desert island. The human capacity for evil and guilt is a predominant theme in Golding's work, including *The Inheritors* (1955), which describes the extermination of Neanderthal man by modern *Homo sapiens*. Later works include *The Spire* (1964) and *Fire Down Below* (1989). He was awarded the Nobel Prize for literature in 1983.

Goldman /ˈɡəʊldmən/, Emma (known as 'Red Emma') (1869–1940), Lithuanian-born American political activist. She emigrated to the US in 1885, where she later became involved in New York's anarchist movement. With Alexander Berkman (1870–1936) she founded and co-edited the anarchist monthly *Mother Earth* (1906–17). She was imprisoned in 1917 with Berkman for opposing US conscription; they were released after two years and deported to Russia. She eventually settled in France. Her disenchantment with and opposition to the Soviet system are related in *My Disillusionment in Russia* (1923).

Goldmark /ˈɡəʊldmɑːk/, Peter Carl (1906–77), Hungarian-born American inventor and engineer. He made the first colour television broadcast in 1940, invented the long-playing record in 1948, and pioneered video cassette recording.

gold rush n. the rapid movement of large numbers of people to a newly discovered gold-field. The first major gold rush, across the US to California after the discovery of gold there in 1848, was followed by similar rushes in the US and elsewhere, including those in Australia (1851–3), South Africa (1884), and the Klondike in Canada (1897–8).

Goldschmidt /ˈɡəʊldʃmɪt/, Victor Moritz (1888–1947), Swiss-born Norwegian chemist, the founder of modern geochemistry. He carried out fundamental work on crystal structure, suggesting a law relating it to chemical composition, and used X-ray crystallography to determine the structure of many compounds.

Goldsmith /ˈɡəʊldsmɪθ/, Oliver (1728–74), Irish novelist, poet, essayist, and dramatist. After studying medicine and travelling in Europe, in 1756 he settled in London, where he practised as a physician and began his literary career as a journalist and essayist. He is now best known for his novel *The Vicar of Wakefield* (1766), the poem *The Deserted Village* (1770), and the comic plays *She Stoops to Conquer* (1773) and *The Good-Natur'd Man* (1768).

goldsmith /ˈɡəʊldsmɪθ/ n. a worker in gold, a manufacturer of gold articles. [OE (as GOLD, SMITH)]

gold standard n. a system by which the value of a currency is defined in terms of gold, for which the currency may be exchanged. Most countries held to this from 1900 until it was suspended during the First World War. It was reintroduced in 1925, but collapsed in the Depression of the 1930s.

Gold Stick n. **1** (in the UK) a gilt rod carried on state occasions by the colonel of the Life Guards or the captain of the gentlemen-at-arms. **2** the officer carrying this rod.

Goldwyn /ˈɡəʊldwɪn/, Samuel (born Schmuel Gelbfisz; changed to

Goldfish then Goldwyn) (1882–1974), Polish-born American film producer. He produced his first film in 1913; his film company, Metro-Goldwyn-Mayer (MGM), which he formed with Louis B. Mayer in 1924, soon became world famous. His successful films include an adaptation of Emily Brontë's novel *Wuthering Heights* (1939), *The Little Foxes* (1941), and the musical *Guys and Dolls* (1955). Goldwyn is also famous for introducing a number of catch-phrases (such as 'Include me out') to the English language.

golem /ˈɡəʊləm/ *n*. **1** a clay figure supposedly brought to life in Jewish legend. **2** an automaton; a robot. [Yiddish *goylem* f. Heb. *gōlem* shapeless mass]

golf /ɡɒlf/ *n. & v.* ● *n.* a game played on a large, open-air course, in which a small hard ball is struck with a club from a tee into each of a series of small cylindrical holes set in the ground. (*See note below.*) ● *v.intr.* play golf. □ **golf-bag** a bag used for carrying golf clubs and balls. **golf ball 1** a ball used in golf. **2** *colloq.* a small ball used in some electric typewriters to carry the type. **golf cart 1** a trolley used for carrying clubs in golf. **2** a motorized cart for golfers and equipment. **golf club 1** a club used in golf. **2** an association for playing golf. **3** the premises used by a golf club. **golf course** (or **links**) the course on which golf is played. [15th-c. Sc.: orig. unkn.]

▪ The aim of golf is to drive the ball into all the holes of a course successively with the fewest possible strokes. A golf course now usually has eighteen (sometimes nine) holes; each hole is set in a smooth green at varying distances from the next hole and separated from it by fairways, rough ground, bunkers, and other hazards. The ball is first driven towards the green, which may be up to 450 m away, and then putted into the hole. A selection of clubs is used depending on the kind of stroke to be played. Golf is played competitively in either of two ways: in *stroke play* (now the usual arrangement in open championships and professional tournaments, in which four rounds are often played) the total number of strokes taken by each competitor decides the outcome; in *match play* each hole is competed for individually, the victor being the player or team winning the most holes. Golf is a game of considerable antiquity in Scotland, having been played at least since the 15th century (originally on the common land by the sea), and is now played all over the world.

golfer /ˈɡɒlfə(r)/ *n.* **1** a golf-player. **2** a cardigan.

Golgi /ˈɡɒldʒɪ/, Camillo (1844–1926), Italian histologist and anatomist. He devised a method of staining nerve tissue with silver salts to reveal details of the cells and nerve fibres, classified types of nerve cell, and described a complex structure in the cytoplasm of most cells (the *Golgi body* or *apparatus*) that is now known to be involved in secretion. Golgi shared a Nobel Prize with Ramón y Cajal in 1906.

Golgotha /ˈɡɒlɡəθə/ the site of the Crucifixion of Jesus; Calvary. [LL f. Gk f. Aram. f. Heb. *gulgoleth* skull; see Matt. 27:33]

Goliath /ɡəˈlaɪəθ/ (in the Bible) a Philistine giant, according to legend slain by David (1 Sam. 17), but according to another tradition slain by Elhanan (2 Sam. 21:19).

Goliath beetle *n.* a very large arboreal beetle of the genus *Goliathus*, of the scarab family, found in the tropics.

Goliath frog *n.* a giant frog, *Rana goliath*, native to Central Africa.

Gollancz /ˈɡɒlænts/, Sir Victor (1893–1967), British publisher and philanthropist. In 1928 he founded his own publishing company, Victor Gollancz Ltd., concentrating on fiction, politics, music, and philosophy. A committed socialist, he was active in campaigning against the rise of Fascism, founded the Left Book Club (1936), and also contributed to the growing influence of the Labour Party in British politics. After the Second World War he organized aid for refugees, founded the charity War on Want, and also campaigned for nuclear disarmament and the abolition of capital punishment.

golliwog /ˈɡɒlɪˌwɒɡ/ *n.* a black-faced brightly dressed soft doll with fuzzy hair. [19th c.: perh. f. GOLLY[1] + POLLIWOG]

gollop /ˈɡɒləp/ *v. & n. colloq.* ● *v.tr.* (**golloped, golloping**) swallow hastily or greedily. ● *n.* a hasty gulp. [perh. f. GULP, infl. by GOBBLE[1]]

golly[1] /ˈɡɒlɪ/ *int.* expressing surprise. [euphem. for *God*]

golly[2] /ˈɡɒlɪ/ *n.* (*pl.* **-ies**) *colloq.* = GOLLIWOG. [abbr.]

golosh var. of GALOSH.

GOM *abbr.* Grand Old Man (name orig. applied to William E. Gladstone).

gombeen /ɡɒmˈbiːn/ *n. Ir.* usury. □ **gombeen-man** a moneylender. [Ir. *gaimbín* perh. f. the same OCelt. source as med.L *cambire* CHANGE]

Gomel see HOMEL.

Gomorrah /ɡəˈmɒrə/ a town in ancient Palestine, probably south of the Dead Sea. According to Gen. 19:24, it was destroyed by fire from heaven, along with Sodom, for the wickedness of its inhabitants.

-gon /ɡən/ *comb. form* forming nouns denoting plane figures with a specified number of angles (*hexagon; polygon; n-gon*). [Gk *-gōnos* -angled]

gonad /ˈɡəʊnæd/ *n.* an animal organ producing gametes, e.g. the testis or ovary. □ **gonadal** /ɡəʊˈneɪd(ə)l/ *adj.* [mod.L *gonas gonad-* f. Gk *gonē, gonos* generation, seed]

gonadotrophic hormone /ˌɡəʊnədəʊˈtrəʊfɪk, -ˈtrɒfɪk/ *n.* (also **gonadotropic** /-ˈtrəʊpɪk, -ˈtrɒpɪk/) *Biochem.* any of a class of hormones stimulating the activity of the gonads.

gonadotrophin /ˌɡəʊnədəʊˈtrəʊfɪn/ *n.* = GONADOTROPHIC HORMONE.

Goncourt /ɡɒnˈkʊə(r)/, Edmond de (1822–96) and Jules de (1830–70), French novelists and critics. The Goncourt brothers collaborated closely in their writing, originally producing art criticism and social history. They regarded their highly detailed realist novels, such as *Germinie Lacerteux* (1864) and *Madame Gervaisais* (1869), as a form of contemporary social history. They also wrote the *Journal des Goncourt*, a detailed record of cultural life in Paris between 1851 and 1896. In his will Edmond de Goncourt provided for the establishment of the Académie Goncourt, which awards the annual Prix Goncourt for a work of French literature; this continued the tradition of the literary circle that the Goncourt brothers had gathered around them during their lifetime.

gondola /ˈɡɒndələ/ *n.* **1** a light flat-bottomed boat, usually much ornamented, with a high rising and curving stem and stern-post, worked with a single oar by one person standing near the stern, especially on the canals of Venice. The origin of the type is unknown, but gondolas are mentioned as early as 1094. **2** a car suspended from an airship or balloon. **3** a free-standing block of shelves used to display goods in a supermarket. **4** (also **gondola car**) *US* a flat-bottomed open railway goods wagon. **5** a car attached to a ski-lift. [Venetian It., f. Rhaeto-Romance *gondolà* rock, roll]

gondolier /ˌɡɒndəˈlɪə(r)/ *n.* the oarsman on a gondola. [F f. It. *gondoliere* (as GONDOLA)]

Gondwana /ɡɒnˈdwɑːnə/ (also **Gondwanaland**) a vast continental area believed to have existed in the southern hemisphere and to have resulted from the breakup of Pangaea in Mesozoic times. It comprised the present Arabia, Africa, South America, Antarctica, Australia, and the peninsula of India. (See also CONTINENTAL DRIFT.) [*Gondwana* land of the Gonds, a Dravidian people of central India]

gone /ɡɒn/ *past part.* of GO[1]. ● *adj.* **1** (of time) past (*not until gone nine*). **2a** lost; hopeless. **b** dead. **3** *colloq.* pregnant for a specified time (*already three months gone*). **4** *sl.* completely enthralled or entranced, esp. by rhythmic music, drugs, etc. □ **all gone** consumed, finished, used up. **be gone** depart (cf. BEGONE). **gone away!** a huntsman's cry, indicating that a fox has been started. **gone goose** (or **gosling**) *colloq.* a person or thing beyond hope. **gone on** *sl.* infatuated with. [past part. of GO[1]]

goner /ˈɡɒnə(r)/ *n. sl.* a person or thing that is doomed, ended, irrevocably lost, etc.; a dead person.

gonfalon /ˈɡɒnfələn/ *n.* **1** a banner, often with streamers, hung from a crossbar. **2** *hist.* such a banner as the standard of some Italian republics. □ **gonfalonier** /ˌɡɒnfələˈnɪə(r)/ *n.* [It. *gonfalone* f. Gmc (cf. VANE)]

gong /ɡɒŋ/ *n. & v.* ● *n.* **1a** a metal disc with a turned rim, giving a resonant note when struck, esp. one used as a signal for meals. **b** a percussion instrument which exists in widely varying shapes and sizes, generally comprising a large hanging bronze disc with a bossed centre which is struck in the middle with a soft-headed drumstick. The ancient gong dates from at least the 6th century. **2** a saucer-shaped bell. **3** *Brit. sl.* a medal; a decoration. ● *v.tr.* summon with a gong. [Malay *gong, gung* of imit. orig.]

goniometer /ˌɡəʊnɪˈɒmɪtə(r)/ *n.* an instrument for measuring angles. □ **goniometry** *n.* **goniometric** /-nɪəˈmetrɪk/ *adj.* **goniometrical** *adj.* [F *goniomètre* f. Gk *gōnia* angle]

gonna /ˈɡɒnə/ *contr. colloq.* going to (*we're gonna win; it's gonna be tough*). [corrupt.]

gonococcus /ˌɡɒnəˈkɒkəs/ *n.* (*pl.* **gonococci** /-ˈkɒksaɪ, -ˈkɒkaɪ/) a bacterium causing gonorrhoea. □ **gonococcal** *adj.* [Gk *gonos* generation, semen + COCCUS]

gonorrhoea /ˌɡɒnəˈrɪə/ *n.* (*US* **gonorrhea**) a venereal disease with inflammatory discharge from the urethra or vagina. □ **gonorrhoeal** *adj.* [LL f. Gk *gonorrhoia* f. *gonos* semen + *rhoia* flux]

goo /ɡuː/ *n. colloq.* **1** a sticky or slimy substance. **2** sickly sentiment. [20th c.: perh. f. *burgoo* (Naut. sl.) = porridge]

good /gʊd/ adj., n., & adv. ● adj. (**better**, **best**) **1** having the right or desired qualities; satisfactory, adequate. **2 a** (of a person) efficient, competent (*good at French; a good driver*). **b** (of a thing) reliable, efficient (*good brakes*). **c** (of health etc.) strong (*good eyesight*). **3 a** kind, benevolent (*good of you to come*). **b** morally excellent; virtuous (*a good deed*). **c** charitable (*good works*). **d** well-behaved (*a good child*). **4** enjoyable, agreeable (*a good party*). **5** thorough, considerable (*gave it a good wash*). **6 a** not less than (*waited a good hour*). **b** considerable in number, quality, etc. (*a good many people*). **7** healthy, beneficial. **8 a** valid, sound (*a good reason*). **b** financially sound (*his credit is good*). **c** (usu. foll. by *for*) *US* (of a ticket) valid. **9** in exclamations of surprise (*good heavens!*). **10** right, proper, expedient (*thought it good to have a try*). **11** fresh, eatable, untainted (*is the meat still good?*). **12** (sometimes patronizing) commendable, worthy (*good old George; your good lady wife; good men and true; my good man*). **13** attractive (*has good legs; good looks*). **14** in courteous greetings and farewells (*good afternoon*). **15** promising or favourable (*a good omen; good news*). **16** expressing approval; complimentary (*a good review*). ● n. **1** (only in *sing.*) that which is good; what is beneficial or morally right (*only good can come of it; did it for your own good; what good will it do?*). **2** (only in *sing.*) a desirable end or object; a thing worth attaining (*sacrificing the present for a future good*). **3** (in *pl.*) **a** *Law* movable property or merchandise. **b** *Brit.* things to be transported, as distinct from passengers. **c** (prec. by *the*) *colloq.* what one has undertaken to supply (esp. *deliver the goods*). **d** (prec. by *the*) *sl.* the real thing; the genuine article. **4** (as *pl.*; prec. by *the*) virtuous people. ● adv. *US colloq.* well (*doing pretty good*). □ **as good as** practically (*he as good as told me*). **as good as gold** extremely well behaved. **be so good as** (or **be good enough**) **to** (often in a request) be kind and do (a favour; *be so good as to open the window*). **be** (**a certain amount**) **to the good** have as net profit or advantage. **do good** show kindness, act philanthropically. **do a person good** be beneficial to. **for good** (**and all**) finally, permanently. **good and** *colloq.* used as an intensifier before an adj. or adv. (*raining good and hard; was good and angry*). **the good book** the Bible. **good breeding** correct or courteous manners. **good faith** see FAITH. **good for 1** beneficial to; having a good effect on. **2** able to perform; inclined for (*good for a ten-mile walk*). **3** able to be trusted to pay (*is good for £100*). **good form** see FORM. **good-for-nothing** adj. worthless. ● n. a worthless person. **good for you** (or **him, her,** etc.) *colloq.* well done. **good-hearted** kindly, well-meaning. **good humour** a genial mood. **a good job** a fortunate state of affairs (*it's a good job you came early*). **good-looker** a handsome or attractive person. **good-looking** (esp. of a person) handsome; attractive. **good luck 1** good fortune, happy chance. **2** exclamation of well-wishing. **good money 1** genuine money; money that might usefully have been spent elsewhere. **2** *colloq.* high wages. **good nature** a friendly disposition. **good oil** *Austral. sl.* reliable information. **good on you** (or **him** etc.) esp. *Austral. & NZ* good for you. **goods and chattels** see CHATTEL. **good-time** recklessly pursuing pleasure. **good-timer** a person who recklessly pursues pleasure. **good times** a period of prosperity. **good will** the intention and hope that good will result (see also GOODWILL). **a good word** (often in phr. **put in a good word for**) words in recommendation or defence of a person. **good works** charitable acts. **have a good mind** see MIND. **have the goods on a person** *sl.* have advantageous information about a person. **have a good time** enjoy oneself. **in a person's good books** see BOOK. **in** (also **on**) **good form** see FORM. **in good faith** with honest or sincere intentions. **in** (also **on**) **good form** see FORM. **in good time 1** with no risk of being late. **2** (also **all in good time**) in due course but without haste. **make good 1** make up for, compensate for, pay (an expense). **2** fulfil (a promise); effect (a purpose or an intended action). **3** demonstrate the truth of (a statement); substantiate (a charge). **4** gain and hold (a position). **5** replace or restore (a thing lost or damaged). **6** (*absol.*) accomplish what one intended. **no good 1** mischief (*is up to no good*). **2** uselessness; no advantage (*it is no good arguing*). **no-good** adj. useless. ● n. a useless thing or person. **take in good part** not be offended by. **to the good** having as profit or benefit. □ **goodish** adj. [OE gōd f. Gmc]

goodbye /gʊdˈbaɪ/ int. & n. (also *US* **goodby**) ● int. expressing good wishes on parting, ending a telephone conversation, etc., or said with reference to a thing got rid of or irrevocably lost. ● n. (pl. **goodbyes** or also *US* **goodbys**) the saying of 'goodbye'; a parting; a farewell. [contr. of *God be with you!* with *good* substituted after *good night* etc.]

Good Friday n. the Friday before Easter Sunday, commemorating the Crucifixion of Christ.

Good Hope, Cape of see CAPE OF GOOD HOPE.

good-humoured /gʊdˈhjuːməd/ adj. genial, cheerful, amiable. □ **good-humouredly** adv.

goodie var. of GOODY[1] n.

Good King Henry n. a weed of the goosefoot family, *Chenopodium bonus-henricus*, native to northern Europe.

goodly /ˈgʊdlɪ/ adj. (**goodlier**, **goodliest**) **1** archaic comely, handsome. **2** considerable in size or quantity. □ **goodliness** n. [OE gōdlic (as GOOD, -LY[1])]

Goodman /ˈgʊdmən/, Benjamin David ('Benny') (1909–86), American jazz clarinettist and band-leader. He formed his own big band in 1934, and soon gained a mass following through radio and live performances. Goodman chose players for their musicianship regardless of their colour, and his bands were the first to feature black and white musicians together. The advent of his distinctive style marked the start of a new era in the history of jazz; he soon became known as the 'King of Swing'. He also performed a wide variety of classical clarinet music; Bartók, Hindemith, and Copland all composed works for him.

goodman /ˈgʊdmən/ n. (pl. **-men**) archaic esp. *Sc.* the head of a household.

good-natured /gʊdˈneɪtʃəd/ adj. kind, patient; easygoing. □ **good-naturedly** adv.

goodness /ˈgʊdnɪs/ n. & int. ● n. **1** virtue; excellence, esp. moral. **2** kindness, generosity (*had the goodness to wait*). **3** what is good or beneficial in a thing (*vegetables with all the goodness boiled out*). ● int. (as a substitution for 'God') expressing surprise, anger, etc. (*goodness me!; goodness knows; for goodness' sake!*). [OE gōdnes (as GOOD, -NESS)]

goodo /ˈgʊdəʊ/ adj. *Austral. & NZ* good, excellent.

good-tempered /gʊdˈtempəd/ adj. having a good temper; not easily annoyed. □ **good-temperedly** adv.

goodwife /ˈgʊdwaɪf/ n. (pl. **-wives**) archaic esp. *Sc.* the female head of a household.

goodwill /gʊdˈwɪl/ n. **1** kindly feeling. **2** the established reputation of a business etc. as enhancing its value. **3** cheerful consent or acquiescence; readiness, zeal. **4** (**good will**) the intention and hope that good will result.

Goodwin Sands /ˈgʊdwɪn/ an area of sandbanks in the Strait of Dover, off the coast of SE England. Often exposed at low tide, the sandbanks are a hazard to shipping.

Goodwood /ˈgʊdwʊd/ a racecourse in West Sussex, near Chichester. Situated at the north end of Goodwood Park, it is the scene of an annual summer race meeting.

goody[1] /ˈgʊdɪ/ n. & int. ● n. (also **goodie**) (pl. **-ies**) **1** *colloq.* a good or favoured person, esp. a hero in a story, film, etc. **2** (usu. in *pl.*) something good or attractive, esp. to eat. **3** = GOODY-GOODY n. ● int. expressing childish delight.

goody[2] /ˈgʊdɪ/ n. (pl. **goodies**) archaic (often as a title prefixed to a surname) an elderly woman of humble station (*Goody Blake*). [for GOODWIFE: cf. HUSSY]

goody-goody /ˈgʊdɪˌgʊdɪ/ n. & adj. *colloq.* ● n. (pl. **-ies**) a smug or obtrusively virtuous person. ● adj. obtrusively or smugly virtuous.

gooey /ˈguːɪ/ adj. (**gooier**, **gooiest**) *colloq.* **1** viscous, sticky. **2** sickly, sentimental. □ **gooeyness** n. (also **gooiness**). [GOO + -Y[2]]

goof /guːf/ n. & v. *sl.* ● n. **1** a foolish or stupid person. **2** a mistake. ● v. **1** tr. bungle, mess up. **2** intr. blunder, make a mistake. **3** intr. (often foll. by *off*) idle. **4** tr. (as **goofed** adj.) stupefied with drugs. [var. of dial. *goff* f. F *goffe* f. It. *goffo* f. med.L *gufus* coarse]

goofy /ˈguːfɪ/ adj. (**goofier**, **goofiest**) *sl.* **1** stupid, silly, daft. **2** having or displaying protruding or crooked front teeth. □ **goofily** adv. **goofiness** n.

goog /gʊg/ n. *Austral. sl.* an egg. □ **full as a goog** very drunk. [20th c.: orig. uncert., perh. f. dial. *goggie* egg]

googly /ˈguːglɪ/ n. (pl. **-ies**) Cricket an off-break ball bowled with apparent leg-break action. [20th c.: orig. unkn.]

googol /ˈguːgɒl/ n. ten raised to the hundredth power (10^{100}). ¶ Not in formal use. [arbitrary formation]

googolplex /ˈguːgə(ʊ)ˌpleks/ n. ten raised to the power of a googol. ¶ Not in formal use. [GOOGOL + -plex as in MULTIPLEX]

gook /guːk, gʊk/ n. *US sl. offens.* a foreigner, esp. a coloured person from eastern Asia. [20th c.: orig. unkn.]

goolie /ˈguːlɪ/ n. (also **gooly**) (pl. **-ies**) **1** (usu. in *pl.*) *sl.* a testicle. **2** *Austral. sl.* a stone or pebble. [app. f. Ind. orig.; cf. Hind. *golī* bullet, ball, pill]

goombah /guːmˈbɑː/ n. *US sl.* **1** a member of a criminal gang; a

Mafioso. **2** a boss or mentor; a crony. [It., prob. dial. alt. of *compàre* godfather, friend, accomplice]

goon /guːn/ *n. sl.* **1** a stupid or playful person. **2** *N. Amer.* a person hired by racketeers etc. to terrorize political or industrial opponents. [perh. f. dial. *gooney* booby: infl. by the subhuman cartoon character 'Alice the *Goon*']

goop /guːp/ *n. sl.* a stupid or fatuous person. [20th c.: cf. GOOF]

goopy /ˈguːpɪ/ *adj. sl.* (**goopier**, **goopiest**) stupid, fatuous. □ **goopiness** *n.*

goosander /guːˈsændə(r)/ *n.* a large diving duck, *Mergus merganser*, with a narrow serrated bill. Also called (*N. Amer.*) *common merganser*. [prob. f. GOOSE + *-ander* in *bergander* sheldrake]

goose /guːs/ *n. & v.* ● *n.* (*pl.* **geese** /giːs/) **1 a** a large waterbird of the family Anatidae, with short legs, webbed feet, and a short broad bill. (*See note below.*) **b** the female of this (opp. GANDER *n.* 1). **c** the flesh of a goose as food. **2** *colloq.* a simpleton. **3** (*pl.* **gooses**) a tailor's smoothing-iron, having a handle like a goose's neck. ● *v.tr. sl.* poke (a person) in the bottom. □ **goose bumps** *N. Amer.* = goose-flesh. **goose-egg** *N. Amer.* a zero score in a game. **goose-flesh** (or **-pimples** or **-skin**) a pimply state of the skin with the hairs erect, produced by cold, fright, etc. **goose-step** *n.* a military marching step in which the knees are kept stiff. ● *v.intr.* march in this way. **kill the golden goose** or **kill the goose that lays** (or **laid**) **the golden egg** sacrifice long-term advantage to short-term gain. [OE *gōs* f. Gmc]

▪ Geese are generally larger than ducks, and usually have a longer neck and a shorter bill. Many kinds migrate in V-shaped flocks to and from their breeding grounds. The farmyard goose is a domesticated form of the greylag, though other kinds are sometimes kept.

gooseberry /ˈguzbərɪ/ *n.* (*pl.* **-ies**) **1** a round edible yellowish-green berry with a thin usu. translucent skin enclosing seeds in a juicy flesh. **2** the thorny shrub, *Ribes grossularia*, bearing this fruit. **3** (esp. in **play gooseberry**) *colloq.* an unwanted third party. [perh. f. GOOSE + BERRY]

goosefoot /ˈguːsfʊt/ *n.* (*pl.* **-foots**) a plant of the genus *Chenopodium*, with leaves shaped like the foot of a goose.

goosegog /ˈguzgɒg/ *n. Brit. colloq.* a gooseberry. [joc. corrupt.]

goosegrass /ˈguːsgrɑːs/ *n.* = CLEAVERS.

Goossens /ˈguːs(ə)nz/, Sir (Aynsley) Eugene (1893–1962), English conductor, violinist, and composer, of Belgian descent. He played the violin in a number of orchestras and chamber ensembles before deciding to concentrate on a career as a conductor and composer. After conducting in the US (1923–45), in 1947 Goossens was appointed the director of the New South Wales Conservatorium and conductor of the Sydney Symphony Orchestra. His compositions include opera, ballet, and symphonies. His brother Leon (1897–1988) was a virtuoso oboist, and his sister Marie (1894–1991) a distinguished harpist.

GOP *abbr.* US Grand Old Party (the Republican Party).

gopher[1] /ˈgəʊfə(r)/ *n.* **1** (in full **pocket gopher**) a burrowing rodent of the family Geomyidae, native to North America, having foodpouches on the cheeks. **2** *US* a ground squirrel. **3** (in full **gopher tortoise**) a tortoise, *Gopherus polyphemus*, native to the southern US, that excavates tunnels as shelter from the sun. □ **gopher snake** = *indigo snake*. [18th c.: orig. uncert.]

gopher[2] /ˈgəʊfə(r)/ *n.* **1** *Bibl.* a tree from the wood of which Noah's ark was made. **2** (in full **gopher-wood**) a tree, *Cladrastis lutea*, yielding yellowish timber. [Heb. *gōper*]

Gorakpur /ˈgɔːrəkˌpʊə(r)/ an industrial city in NE India, in Uttar Pradesh near the border with Nepal; pop. (1991) 490,000.

goral /ˈgɔːrəl/ *n.* a goat-antelope, *Nemorhaedus goral*, native to mountainous regions of northern India, having short horns curving to the rear. [local (Himalayan) name]

Gorbachev /ˌgɔːbəˈtʃɒf/, Mikhail (Sergeevich) (b.1931), Soviet statesman, General Secretary of the Communist Party of the USSR 1985–91 and President 1988–91. He was born in Russia. His foreign policy was notable for bringing about an end to the cold war, largely through arms control negotiations with the West, culminating in the signing of treaties with the US in 1987 and 1991 (see START). Within the USSR he introduced major political, economic, and cultural reforms, including the eventual removal of the Communist Party's monopoly of power and moves towards a market economy (see GLASNOST, PERESTROIKA). He retained power in spite of widespread opposition to his policies, which eventually led to an attempted coup in 1991. Ultimately his wish to retain a centrally controlled union of Soviet states conflicted with the Soviet republics' desire for autonomy and

he lost his battle with his chief political opponent, Boris Yeltsin; he resigned in Dec. 1991. He was awarded the Nobel Peace Prize in 1990.

Gorbals, the /ˈgɔːb(ə)lz/ a district of Glasgow on the south bank of the River Clyde, formerly noted for its slums and tenement buildings. Since the 1960s it has been the focus of urban regeneration schemes.

gorblimey /gɔːˈblaɪmɪ/ *int. & n. Brit. sl.* ● *int.* an expression of surprise, indignation, etc. ● *n.* (*pl.* **-eys**) a soft service cap. [corrupt. of *God blind me*]

gorcock /ˈgɔːkɒk/ *n. Sc. & N. Engl.* the male of the red grouse. [*gor-* (of unkn. orig.) + COCK[1]]

Gordian knot /ˈgɔːdɪən/ *n.* **1** an intricate knot tied by Gordius, king of Gordium, Phrygia, and cut through with a sword by Alexander the Great in response to the prophecy that whoever untied it would become the ruler of Asia. **2** an indissoluble bond; a difficult problem or task. □ **cut the Gordian knot** solve a problem with a radical solution, esp. one that eliminates the conditions that caused the problem in the first place.

Gordimer /ˈgɔːdɪmə(r)/, Nadine (b.1923), South African novelist and short-story writer. The impetus for her writing was provided by the contrast between her own wealthy background and the conditions experienced by black miners in her home town of Johannesburg. Her work avoids overt polemicism, examining the effect of apartheid on the oppressed and the privileged alike. Her novels include *The Conservationist* (1974) and *My Son's Story* (1990). She was awarded the Nobel Prize for literature in 1991.

Gordium /ˈgɔːdɪəm/ an ancient city of Asia Minor, the capital of Phrygia in the 8th and 9th centuries BC. According to legend the city was founded by Gordius, who tied the knot cut by Alexander the Great during his expedition of 334 BC (see GORDIAN KNOT). The city, now in ruins, lies beside the River Sakarya in NW Turkey, 80 km (50 miles) west of Ankara.

gordo /ˈgɔːdəʊ/ *n. Austral.* a popular variety of grape. [Sp. *gordo blanco* lit. 'fat white']

Gordon /ˈgɔːd(ə)n/, Charles George (1833–85), British general and colonial administrator. He went to China in 1860 while serving with the Royal Engineers, and became known as 'Chinese Gordon' after crushing the Taiping Rebellion (1863–4). In 1884 he was sent to rescue the Egyptian garrisons in Sudan from Mahdist forces led by Muhammad Ahmad (see MAHDI), but was trapped at Khartoum and killed before a relieving force could reach him.

Gordon Riots a series of anti-Catholic riots in London on 2–9 June 1780, in which about 300 people were killed. The riots were provoked by a petition presented to Parliament by Lord George Gordon (1751–93) against the relaxation of restrictions on the holding of landed property by Roman Catholics.

Gordon setter *n.* a black and tan breed of setter, used as a gun dog. [4th Duke of *Gordon* (1743–1827), promoter of the breed]

gore[1] /gɔː(r)/ *n.* **1** blood shed and clotted. **2** slaughter, carnage. [OE *gor* dung, dirt]

gore[2] /gɔː(r)/ *v.tr.* pierce with a horn, tusk, etc. [ME: orig. unkn.]

gore[3] /gɔː(r)/ *n. & v.* ● *n.* **1** a wedge-shaped piece in a garment. **2** a triangular or tapering piece in an umbrella etc. ● *v.tr.* shape with a gore. [OE *gāra* triangular piece of land, rel. to OE *gār* spear, a spearhead being triangular]

Górecki /gəˈretskɪ/, Henryk (Mikołaj) (b.1933), Polish composer. He studied at Katowice Conservatory and with Olivier Messiaen in Paris. His works are chiefly for orchestra and chamber ensemble; they include three symphonies and a chamber trilogy, *Genesis* (1963). In *Old Polish Music* (1969) he draws on medieval Polish religious music as his source of inspiration, while his Third Symphony (1976), also called the *Symphony of Sorrowful Songs*, is heavily influenced by church music.

Göreme /ˈgɜːrɪmɪ/ a valley in Cappadocia in central Turkey, noted for its cave-dwellings, hollowed out of the soft tufa rock. In the Byzantine era these contained hermits' cells, monasteries, and more than 400 churches.

Gore-tex /ˈgɔːteks/ *n. propr.* a synthetic waterproof fabric permeable to air and water vapour, used in outdoor and sports clothing.

gorge /gɔːdʒ/ *n. & v.* ● *n.* **1** a narrow opening between hills or a rocky ravine, often with a stream running through it. **2** an act of gorging; a feast. **3** the contents of the stomach; what has been swallowed. **4** in fortifications, the neck of a bastion or other outwork; the rear entrance to a work. **5** a mass of ice etc. blocking a narrow passage. ● *v.* **1** *intr.* feed greedily. **2** *tr.* **a** (often *refl.*) satiate, glut. **b** swallow, devour greedily. □ **cast the gorge at** reject with loathing. **one's gorge rises at**

one is sickened by. □ **gorger** *n.* [ME f. OF *gorge* throat ult. f. L *gurges* whirlpool]

gorgeous /ˈgɔːdʒəs/ *adj.* **1** richly coloured, sumptuous, magnificent. **2** *colloq.* very pleasant, splendid (*gorgeous weather*). **3** *colloq.* strikingly beautiful. □ **gorgeously** *adv.* **gorgeousness** *n.* [earlier *gorgayse, -yas* f. OF *gorgias* fine, elegant, of unkn. orig.]

gorget /ˈgɔːdʒɪt/ *n.* **1** *hist.* **a** a piece of armour for the throat. **b** a woman's wimple. **2** a patch of colour on the throat of a bird, insect, etc. [OF *gorgete* (as GORGE)]

Gorgio /ˈgɔːdʒɪˌəʊ/ *n.* (*pl.* **-os**) the gypsy name for a non-gypsy. [Romany]

gorgon /ˈgɔːgən/ *n.* **1** *Gk Mythol.* any of three sisters, Stheno, Euryale, and Medusa (the only mortal one), with snakes for hair, who had the power to turn anyone who looked at them to stone. Medusa was killed by Perseus. **2** a frightening or repulsive woman. [L *Gorgo -onis* f. Gk *Gorgō* f. *gorgos* terrible]

gorgonian /gɔːˈgəʊnɪən/ *n.* & *adj. Zool.* ● *n.* a usu. brightly coloured horny coral of the order Gorgonacea, having a treelike skeleton bearing polyps, e.g. a sea fan. ● *adj.* of or relating to the Gorgonacea. [mod.L (as GORGON), with ref. to its petrifaction]

gorgonize /ˈgɔːgəˌnaɪz/ *v.tr.* (also **-ise**) **1** stare at like a gorgon. **2** paralyse with terror etc.

Gorgonzola /ˌgɔːgənˈzəʊlə/ *n.* a type of rich, strong-flavoured cheese with bluish-green veins. [*Gorgonzola* in Italy]

gorilla /gəˈrɪlə/ *n.* **1** the largest anthropoid ape, *Gorilla gorilla*, native to central Africa, having a large head, short neck, and muscular arms. **2** *colloq.* a rough heavily built person, a thug. [adopted as the specific name in 1847 f. Gk *Gorillai* an African people noted for hairiness]

Gorky[1] /ˈgɔːkɪ/ the name from 1932 to 1991 of NIZHNI NOVGOROD.

Gorky[2] /ˈgɔːkɪ/, Arshile (1904–48), Turkish-born American painter. At first influenced by Picasso, he later gained inspiration from the surrealist techniques of artists such as Miró to develop a distinctive style of abstract expressionism. He is best known for his work of the early 1940s, which uses ambiguous biomorphic forms characterized by bright colours, fluid handling of the paint, and black sinuous outlines, and is represented by paintings such as *Waterfall* (1943).

Gorky[3] /ˈgɔːkɪ/, Maxim (pseudonym of Aleksei Maksimovich Peshkov) (1868–1936), Russian writer and revolutionary. He became famous for the short stories that he published between 1895 and 1900; he later turned to writing novels and plays. Among his best-known works are the play *The Lower Depths* (1901) and his autobiographical trilogy (1915–23). He was imprisoned for his involvement in the Russian Revolution of 1905 and exiled for his revolutionary activity. On his final return to the Soviet Union in 1931 after periods spent abroad, he was honoured as the founder of the new, officially sanctioned socialist realism.

Gorlovka /ˈgɔːləfkə/ an industrial city in SE Ukraine, in the Donets Basin; pop. (1990) 337,900.

gormandize /ˈgɔːmənˌdaɪz/ *v.* & *n.* (also **-ise**) ● *v.* **1** *intr.* & *tr.* eat or devour voraciously. **2** *intr.* indulge in good eating. ● *n.* = GOURMANDISE. □ **gormandizer** *n.* [as GOURMANDISE]

gormless /ˈgɔːmlɪs/ *adj.* esp. *Brit. colloq.* foolish, lacking sense. □ **gormlessly** *adv.* **gormlessness** *n.* [orig. *gaumless* f. dial. *gaum* understanding]

Gorno-Altai /ˌgɔːnəʊælˈtaɪ/ an autonomous republic in south central Russia, on the border with Mongolia; pop. (1989) 192,000; capital, Gorno-Altaisk.

Gorno-Altaisk /ˌgɔːnəʊælˈtaɪsk/ a city in south central Russia, capital of the republic of Gorno-Altai; pop. (1990) 39,000. It was known as Ulala until 1932 and as Oirot-Tura from 1932 until 1948.

gorse /gɔːs/ *n.* a spiny yellow-flowered leguminous shrub of the genus *Ulex*, esp. growing on European wastelands. Also called *furze*. □ **gorsy** *adj.* [OE *gors(t)* rel. to OHG *gersta*, L *hordeum*, barley]

Gorsedd /ˈgɔːseð/ *n.* a council of Welsh etc. bards and Druids (esp. as meeting daily before the eisteddfod). [Welsh, lit. 'throne']

gory /ˈgɔːrɪ/ *adj.* (**gorier, goriest**) **1** involving bloodshed; bloodthirsty (*a gory film*). **2** covered in gore. □ **gory details** *joc.* disconcertingly or unpleasantly explicit details. □ **gorily** *adv.* **goriness** *n.*

gosh /gɒʃ/ *int.* expressing surprise. [euphem. for *God*]

goshawk /ˈgɒshɔːk/ *n.* a large short-winged hawk, *Accipiter gentilis*. [OE *gōs-hafoc* (as GOOSE, HAWK[1])]

gosling /ˈgɒzlɪŋ/ *n.* a young goose. [ME, orig. *gesling* f. ON *gǽslingr*]

gospel /ˈgɒsp(ə)l/ *n.* **1** the teaching or revelation of Christ. **2** (**Gospel**) **a** the record of Christ's life and teaching in the first four books of the New Testament; each of these books. (*See note below.*) **b** a portion from these books read at a church service. **3** a thing regarded as absolutely true (*take my word as gospel*). **4** a principle one acts on or advocates. **5** = GOSPEL MUSIC. □ **Gospel side** the north side of the altar, at which the Gospel is read. **gospel truth** something regarded as absolutely true. [OE *gōdspel* (as GOOD, *spel* news, SPELL[2]), rendering eccl.L *bona annuntiatio, bonus nuntius* = *evangelium* EVANGEL: assoc. with *God*]

- The four Gospels, traditionally ascribed to St Matthew, St Mark, St Luke, and St John, all give an account of the ministry, Crucifixion, and Resurrection of Christ, but the first three (known as the Synoptic Gospels) give accounts that are similar in many ways, while the Gospel of John differs greatly. In addition to these canonical Gospels, whose authority had been established by the Church in the 2nd century, there exist several apocryphal gospels of later date.

gospeller /ˈgɒspələ(r)/ *n.* the reader of the Gospel in a Communion service. □ **hot gospeller** a zealous puritan; a rabid propagandist.

gospel music *n.* a fervent style of black American vocal music which developed in the early 20th century from the spirituals sung in Baptist and Pentecostal churches of the southern US. Gospel was a forerunner of soul music and remains popular in its own right.

gossamer /ˈgɒsəmə(r)/ *n.* & *adj.* ● *n.* **1** a filmy substance of small spiders' webs. **2** delicate filmy material. **3** a thread of gossamer. ● *adj.* light and flimsy as gossamer. □ **gossamered** *adj.* **gossamery** *adj.* [ME *gos(e)somer(e)*, app. f. GOOSE + SUMMER[1] (*goose summer* = St Martin's summer, i.e. early November when geese were eaten, gossamer being common then)]

gossip /ˈgɒsɪp/ *n.* & *v.* ● *n.* **1 a** easy or unconstrained talk or writing esp. about persons or social incidents. **b** idle talk; groundless rumour. **2** an informal chat, esp. about persons or social incidents. **3** a person who indulges in gossip. ● *v.intr.* (**gossiped, gossiping**) talk or write gossip. □ **gossip column** a section of a newspaper devoted to gossip about well-known people. **gossip columnist** a regular writer of gossip columns. **gossip-monger** a perpetrator of gossip. □ **gossiper** *n.* **gossipy** *adj.* [earlier sense 'godparent': f. OE *godsibb* person related to one in God: see GOD, SIB]

gossoon /gɒˈsuːn/ *n. Ir.* a lad. [earlier *garsoon* f. F *garçon* boy]

got past and past part. of GET.

Göteborg see GOTHENBURG.

Goth /gɒθ/ *n.* **1** a member of a Germanic tribe that invaded the Roman Empire from the east between the 3rd and 5th centuries AD. The eastern half of the tribe, the Ostrogoths, eventually founded a kingdom in Italy, while the Visigoths, their western cousins, went on to found one in Spain. **2** an uncivilized or ignorant person. [LL *Gothi* (pl.) f. Gk *Go(t)thoi* f. Goth.]

goth /gɒθ/ *n.* (also **Goth**) **1** a style of rock music with an intense or droning blend of guitars, bass, and drums, often with apocalyptic or mystical lyrics. **2** a member of a subculture favouring black clothing, white and black make-up, metal jewellery, and goth music. [cf. GOTH, but perh. contr. of GOTHIC]

Gotha /ˈgəʊtə, ˈgəʊθə/ a city in central Germany, in Thuringia; pop. (1981) 57,600. From 1640 until 1918 it was the residence of the dukes of Saxe-Gotha and Saxe-Coburg-Gotha.

Gotham 1 /ˈgəʊtəm/ a village in Nottinghamshire. It is associated with the English folk tale *The Wise Men of Gotham*, in which the inhabitants of the village demonstrated cunning by feigning stupidity. **2** /ˈgɒθəm/ a nickname for New York City, used originally by the American writer Washington Irving in *Salmagundi* (1807–8) and now associated with the Batman stories.

Gothenburg /ˈgɒθənˌbɜːg/ (Swedish **Göteborg** /ˌjœtəˈbɔrj/) a seaport in SW Sweden, on the Kattegat strait; pop. (1990) 433,000. It is the second largest city in Sweden.

Gothic /ˈgɒθɪk/ *adj.* & *n.* ● *adj.* **1** of the Goths or their language. (*See note below.*) **2** designating a style of architecture which succeeded Romanesque, originating in France in the mid-12th century and prevailing in western Europe until the 16th century. (*See note below.*) **3** barbarous, uncouth. **4** characteristic of the Middle Ages; medieval, romantic; also, reminiscent of a Gothic novel; portentously gloomy or horrifying. (*See also* GOTHIC NOVEL.) **5** *Printing* (of type) old-fashioned German, black letter, or sanserif. ● *n.* **1** the Gothic language. (*See note below.*) **2** Gothic architecture. (*See note below.*) **3** *Printing* Gothic type. □ **Gothic revival** the reintroduction in England of Gothic architecture towards the middle of the 19th century. □ **Gothically** *adv.* **Gothicism** /-θɪˌsɪz(ə)m/ *n.* **Gothicize** /-θɪˌsaɪz/ *v.tr.* & *intr.* (also **-ise**). [F *gothique* or LL *gothicus* f. *Gothi*: see GOTH]

■ The Gothic language provides the oldest manuscript evidence for the Germanic language group. It belongs to the East Germanic group and was spoken in the area to the north and west of the Black Sea. The evidence is fragmentary, the main text being part of a Gothic translation of the Greek Bible written in the 4th century by bishop Ulfilas.

■ Gothic architecture is characterized by pointed arches, rib vaults, and flying buttresses, all features which had appeared in some late Romanesque buildings, but which when used together gave a new impression of airiness and grace. This was increased by the use of windows far larger than those in Romanesque buildings. Window openings and walls were decorated with elaborate tracery so diverse and distinctive that its variations are frequently used in classifying the different stages of the style. Gothic architecture in England is commonly divided into three stages: Early English, Decorated, and Perpendicular.

Gothic novel *n.* an English genre of fiction, popular in the 18th to early 19th centuries, characterized by an atmosphere of mystery and horror and with a pseudo-medieval ('Gothic') setting. Examples include Horace Walpole's *Castle of Otranto* (1764) and Ann Radcliffe's *The Mysteries of Udolpho* (1794).

Gotland /'gɒtlənd/ an island and province of Sweden, in the Baltic Sea; pop. (1990) 57,100; capital, Visby.

gotta /'gɒtə/ *colloq.* have got a; have got to (*I gotta pain*; *we gotta go*). [corrupt.]

gotten US past part. of GET.

Götterdämmerung /ˌgɜːtə'demə ˌrʊŋ/ *n.* **1** (esp. in Germanic mythology) the twilight of the gods. **2** the complete downfall of a regime etc. [G, lit. 'twilight of the gods', esp. as the title of an opera by Wagner]

Göttingen /'gɜːtɪŋən/ a town in north central Germany, on the River Leine; pop. (1991) 124,000. It is noted for its university, which was founded in 1734 by George II of England.

gouache /gʊˈɑːʃ, gwɑːʃ/ *n.* **1** a method of painting in opaque pigments ground in water and thickened with a gluelike substance. **2** these pigments. **3** a picture painted in this way. [F f. It. *guazzo*]

Gouda¹ /'gaʊdə/ a market town in the Netherlands, just north-east of Rotterdam; pop. (1991) 65,900.

Gouda² /'gaʊdə/ *n.* a flat round usu. Dutch cheese with a yellow rind, orig. made at Gouda.

gouge /gaʊdʒ/ *n.* & *v.* ● *n.* **1** a chisel with a concave blade, used in carpentry, sculpture, and surgery. **2** an indentation or groove made with or as with this. ● *v.* **1** *tr.* cut with or as with a gouge. **2** *tr.* **a** (foll. by *out*) force out (esp. an eye with the thumb) with or as with a gouge. **b** force out the eye of (a person). **3** *tr.* *N. Amer. colloq.* overcharge, swindle. **4** *intr. Austral.* dig for opal. □ **gouger** *n.* [F f. LL *gubia*, perh. f. Celt. orig.]

Gough Island /gɒf/ an island in the South Atlantic, south of Tristan da Cunha. In 1938 it became a dependency of the British Crown Colony of St Helena.

goujons /'guːʒɒnz, -ʒɒn/ *n.pl.* deep-fried pieces of chicken, fish, etc. [F, f. as GUDGEON¹]

goulash /'guːlæʃ/ *n.* **1** a highly seasoned Hungarian dish of meat and vegetables, usu. flavoured with paprika. **2** (in contract bridge) a dealing again, several cards at a time, of the four hands (unshuffled, but with each hand arranged in suits and order of value) when no player has bid. [Hungarian *gulyás-hús* f. *gulyás* herdsman + *hús* meat]

Gould¹ /guːld/, Glenn (Herbert) (1932–82), Canadian pianist and composer. He made his début as a soloist with the Toronto Symphony Orchestra at the age of 14 and gave many performances both in Canada and abroad until 1964, when he retired from the concert platform to concentrate on recording and broadcasting. He is especially known for his performances of works by Bach, but also Beethoven, Hindemith, and Schoenberg.

Gould² /guːld/, Stephen Jay (b.1941), American palaeontologist. He has studied modifications of Darwinian evolutionary theory, proposed the concept of punctuated equilibrium, and is especially interested in the social context of scientific theory. His popular science books include *Ever Since Darwin* (1977), *Hen's Teeth and Horses' Toes* (1983), and *Bully for Brontosaurus* (1992).

Gounod /'guːnəʊ/, Charles François (1818–93), French composer, conductor, and organist. He achieved his first operatic success in 1859 with *Faust*, a grand opera in the French tradition, but with a grace and naturalism new to the genre. Gounod's later operas have not

maintained their popularity to the same extent as *Faust*, but *Roméo et Juliette* (1867) and several of his songs are still performed.

gourami /'gʊərəmɪ, gʊə'rɑːmɪ/ *n.* (*pl.* same or **gouramis**) **1** a large freshwater fish, *Osphronemus goramy*, native to SE Asia, used as food. **2** a small brightly coloured freshwater fish of the family Belontiidae, often kept in aquariums. Also called *labyrinth fish*. [Malay *gurāmī*]

gourd /gʊəd/ *n.* **1 a** a fleshy usu. large fruit with a hard skin. **b** a climbing or trailing plant of the family Cucurbitaceae bearing such a fruit. Also called *cucurbit*. **2** the hollow hard skin of a gourd, dried and used as a drinking-vessel, water container, ornament, etc. □ **gourdful** *n.* (*pl.* **-fuls**). [ME f. AF *gurde*, OF *gourde* ult. f. L *cucurbita*]

gourde /gʊəd/ *n.* the basic monetary unit of Haiti, equal to 100 centimes. [F, fem. of *gourd* heavy, dull]

gourmand /'gʊəmænd/ *n.* & *adj.* ● *n.* **1** a glutton. **2** *disp.* a gourmet. ● *adj.* gluttonous; fond of eating, esp. to excess. □ **gourmandism** *n.* [ME f. OF, of unkn. orig.]

gourmandise /ˌgʊəmən'diːz/ *n.* the habits of a gourmand; gluttony. [F (as GOURMAND)]

gourmet /'gʊəmeɪ/ *n.* **1** a connoisseur of good food. **2** (*attrib.*) of a kind or standard suitable for gourmets (*a gourmet meal*). [F, = wine-taster: sense infl. by GOURMAND]

gout /gaʊt/ *n.* **1** a disease with inflammation of the smaller joints, esp. the toe, as a result of the deposition of uric acid crystals in the joints. **2** *archaic* **a** a drop, esp. of blood. **b** a splash or spot. □ **gouty** *adj.*

goutiness *n.* [ME f. OF *goute* f. L *gutta* drop, with ref. to the medieval theory of the flowing down of humours]

Gov. *abbr.* **1** Government. **2** Governor.

govern /'gʌv(ə)n/ *v.* **1** *tr.* rule or control (a state, subject, etc.) with authority; conduct the policy and affairs of (an organization etc.). **b** *intr.* be in government. **2 a** *tr.* influence or determine (a person or a course of action). **b** *intr.* be the predominating influence. **3** *tr.* be a standard or principle for; constitute a law for; serve to decide (a case). **4** *tr.* check or control (esp. passions). **5** *tr.* *Gram.* (esp. of a verb or preposition) determine the role or case of (another component, esp. a direct object). **6** *tr.* be in military command of (a fort, town). □ **governing body** the body of managers of an institution. □ **governable** *adj.* **governability** /ˌgʌvənə'bɪlɪtɪ/ *n.* [ME f. OF *governer* f. L *gubernare* steer, rule f. Gk *kubernaō*]

governance /'gʌvənəns/ *n.* **1** the act or manner of governing. **2** the office or function of governing. **3** sway, control. [ME f. OF (as GOVERN)]

governess /'gʌvənɪs/ *n.* a woman employed to teach children in a private household. [earlier *governeress* f. OF *governeresse* (as GOVERNOR)]

governessy /'gʌvənɪsɪ/ *adj.* characteristic of a governess; prim.

government /'gʌvənmənt/ *n.* **1** the act or manner of governing. **2** the system by which a state or community is governed. **3 a** a body of persons governing a state. **b** (usu. **Government**) a particular ministry in office. **4** the state as an agent. **5** *Gram.* the relation between a governed and a governing word. □ **Government House** the official residence of a governor. **government issue** *US* (of equipment) provided by the government. **government paper** (or **securities**) bonds etc. issued by the government. **government surplus** unused equipment sold by the government. □ **governmental** /ˌgʌvən'ment(ə)l/ *adj.* **governmentally** *adv.* [ME f. OF *governement* (as GOVERN)]

governor /'gʌvənə(r)/ *n.* **1** a person who governs; a ruler. **2 a** an official governing a province, town, etc. **b** a representative of the Crown in a colony. **3** the executive head of each state of the US. **4** an officer commanding a fortress or garrison. **5** the head or a member of a governing body of an institution. **6** the official in charge of a prison. **7 a** *sl.* one's employer. **b** *sl.* one's father. **c** *colloq.* (as a form of address) sir. **8** *Mech.* an automatic regulator controlling the speed of an engine etc. □ **governorship** *n.* [ME f. AF *gouvernour*, OF *governëo(u)r* f. L *gubernator -oris* (as GOVERN)]

governorate /'gʌvənərət/ *n.* **1** the residence or office of a governor. **2** the area ruled by a governor.

Governor-General /ˌgʌvənə'dʒenrəl/ *n.* the representative of the Crown in a Commonwealth country that regards the British monarch as head of state.

Govt. *abbr.* Government.

gowan /'gaʊən/ *n.* *Sc.* & *N. Engl.* **1** a daisy. **2** any white or yellow field-flower. [prob. var. of dial. *gollan* ranunculus etc., and rel. to *gold* in *marigold*]

gowk /gaʊk/ n. esp. dial. **1** a cuckoo. **2** an awkward or halfwitted person; a fool. [ME f. ON *gaukr* f. Gmc]

gown /gaʊn/ n. & v. ● n. **1** a loose flowing garment, esp. a long dress worn by a woman. **2** the official robe of an alderman, judge, cleric, member of a university, etc. **3** a protective overall worn by a surgeon, a hospital patient, etc. **4** the members of a university as distinct from the permanent residents of the university town (cf. TOWN 3). ● v.tr. (usu. as **gowned** adj.) attire in a gown. [ME f. OF *goune*, *gon(n)e* f. LL *gunna* fur garment: cf. med. Gk *gouna* fur]

Gowon /ˈɡaʊwən/, Yakubu (b.1934), Nigerian general and statesman, head of state 1966–75. He seized power in 1966, ousting the leader of an earlier military coup. Following the Biafran civil war (1967–70), he maintained a policy of 'no victor, no vanquished' which helped to reconcile the warring factions. Gowon was himself removed in a military coup in 1975.

goy /ɡɔɪ/ n. (pl. **goyim** /ˈɡɔɪɪm/ or **goys**) sl. derog. a Jewish name for a non-Jew. □ **goyish** adj. (also **goyisch**). [Heb. *gōy* people, nation]

Goya /ˈɡɔɪə/ (full name Francisco José de Goya y Lucientes) (1746–1828), Spanish painter and etcher. Although he painted notable portraits, such as *The Family of Charles IV* (1800), he is now chiefly famous for the works which express his reaction to the French occupation of Spain (1808–14): the painting *The Shootings of May 3rd 1808* (1814), and the set of sixty-five etchings *The Disasters of War* (1810–14), depicting the cruelty and horror of war. His work influenced many artists, including Manet and Delacroix.

Gozo /ˈɡəʊzəʊ/ a Maltese island, to the north-west of the main island of Malta.

GP abbr. **1** general practitioner. **2** Grand Prix.

Gp. Capt. abbr. (in the RAF) group captain.

GPI abbr. general paralysis of the insane.

GPMU abbr. (in the UK) Graphical, Paper, and Media Union (formed by the merger of NGA and SOGAT in 1991).

GPO abbr. **1** General Post Office. **2** US Government Printing Office.

GPU abbr. a Soviet secret police agency 1922–3 (see also OGPU). [Russ., = State Political Directorate]

GR abbr. King George. [L *Georgius Rex*]

gr abbr. (also **gr.**) **1** gram(s). **2** grains. **3** gross. **4** grey.

Graafian follicle /ˈɡrɑːfɪən/ n. a follicle in the mammalian ovary in which an ovum develops prior to ovulation. [Regnier de *Graaf*, Du. anatomist (1641–73)]

grab /ɡræb/ v. & n. ● v. (**grabbed**, **grabbing**) **1** tr. **a** seize suddenly. **b** capture, arrest. **2** tr. take greedily or unfairly. **3** tr. sl. attract the attention of, impress. **4** intr. (foll. by *at*) make a sudden snatch at. **5** intr. (of the brakes of a motor vehicle) act harshly or jerkily. ● n. **1** a sudden clutch or attempt to seize. **2** a mechanical device for clutching. **3** the practice of grabbing; rapacious proceedings esp. in politics and commerce. **4** a children's card-game in which certain cards may be snatched from the table. □ **grab-bag** N. Amer. a lucky dip. **grab handle** (or **rail** etc.) a handle or rail etc. to steady passengers in a moving vehicle. **up for grabs** sl. open to offer; easily obtainable; inviting capture. □ **grabber** n. [MLG, MDu. *grabben*: cf. GRIP, GRIPE, GROPE]

grabble /ˈɡræb(ə)l/ v.intr. **1** grope about, feel for something. **2** (often foll. by *for*) sprawl on all fours, scramble (for something). [Du. & LG *grabbelen* scramble for a thing (as GRAB)]

grabby /ˈɡræbɪ/ adj. colloq. tending to grab; greedy, grasping.

graben /ˈɡrɑːb(ə)n/ n. (pl. same or **grabens**) Geol. a depression of the earth's surface between faults. [G, orig. = ditch]

Gracchus /ˈɡrækəs/, Tiberius Sempronius (c.163–133 BC), Roman tribune. He and his brother, Gaius Sempronius Gracchus (c.153–121 BC), were responsible for radical social and economic legislation, passed against the wishes of the senatorial class. Tiberius was killed by his opponents after the passing of his agrarian bill (133 BC), which aimed at a redistribution of land to the poor. Gaius continued his brother's programme and instituted other reforms to relieve poverty, but was killed in a riot.

Grace /ɡreɪs/, W(illiam) G(ilbert) (1848–1915), English cricketer. He began playing first-class cricket for Gloucestershire in 1864 and played in his last test match at the age of 50. During his career he made 126 centuries, scored 54,896 runs, and took 2,864 wickets. He twice captained England in test matches against Australia (1880 and 1882).

grace /ɡreɪs/ n. & v. ● n. **1** attractiveness, esp. in elegance of proportion or manner or movement; gracefulness. **2** courteous good will (*had the grace to apologize*). **3** an attractive feature; an accomplishment (*social* graces). **4 a** (in Christian belief) the unmerited favour of God; a divine saving and strengthening influence. The nature and conditions of divine grace were a subject of controversy between St Augustine and the adherents of Pelagianism, and between Calvin and his opponents at the Reformation. **b** the state of receiving this. **c** a divinely given talent. **5** goodwill, favour (*fall from grace*). **6** delay granted as a favour (*a year's grace*). **7** a short thanksgiving before or after a meal; often iron. **8** (**Grace**) Gk Mythol. any of the Graces (see GRACES). **9** (**Grace**) (prec. by His, Her, Your) forms of description or address for a duke, duchess, or archbishop. ● v.tr. (often foll. by *with*) lend or add grace to; enhance or embellish; confer honour or dignity on (*a vase graced the table*; often iron.: *graced us with his presence*). □ **days of grace** the time allowed by law for payment of a sum due. **grace and favour house** etc. Brit. a house etc. occupied by permission of the British monarch or government. **grace-note** Mus. an extra note as an embellishment not essential to the harmony or melody. **in a person's good** (or **bad**) **graces** regarded by a person with favour (or disfavour). **with good** (or **bad**) **grace** as if willingly (or reluctantly). [ME f. OF f. L *gratia* f. *gratus* pleasing: cf. GRATEFUL]

graceful /ˈɡreɪsfʊl/ adj. having or showing grace or elegance. □ **gracefully** adv. **gracefulness** n.

graceless /ˈɡreɪslɪs/ adj. lacking grace or elegance or charm. □ **gracelessly** adv. **gracelessness** n.

Graces, the /ˈɡreɪsɪz/ n.pl. Gk Mythol. beautiful goddesses, usually three (Aglaia, Thalia, Euphrosyne), daughters of Zeus, personifying charm, grace, and beauty, which they bestowed as physical, intellectual, artistic, and moral qualities.

Gracias a Dios, Cape /ˌɡrɑːsɪˌɑːs ɑː ˈdiːɒs/ a cape forming the easternmost extremity of the Mosquito Coast in Central America, on the border between Nicaragua and Honduras. The cape was named by Columbus, who, in 1502, had been becalmed off the coast but was able to continue his voyage with the arrival of a following wind. [Sp., = thanks (be) to God]

gracile /ˈɡræsaɪl, -sɪl/ adj. esp. Zool. & Anthropol. of slender build. [L *gracilis* slender]

gracility /ɡrəˈsɪlɪtɪ/ n. **1** slenderness. **2** (of literary style) unornamented simplicity.

gracious /ˈɡreɪʃəs/ adj. & int. ● adj. **1** (often poet.) elegant, charming, courteous. **2** (often joc. or iron.) kind; indulgent and beneficent to inferiors. **3** (of God) merciful, benign. **4** a polite epithet used of royal persons or their acts (*gracious majesty*; *the gracious speech from the throne*). ● int. expressing surprise. □ **gracious living** an elegant way of life. □ **graciously** adv. **graciousness** n. **graciosity** /ˌɡreɪsɪˈɒsɪtɪ/ n. [ME f. OF f. L *gratiosus* (as GRACE)]

grackle /ˈɡræk(ə)l/ n. **1** an American oriole, esp. of the genus *Quiscalus*, the male of which is shiny black with a blue-green sheen. Also called *blackbird*. **2** an Asian mynah of the genus *Gracula*. [mod.L *Gracula* f. L *graculus* jackdaw]

grad /ɡræd/ n. colloq. = GRADUATE n. 1. [abbr.]

gradate /ɡrəˈdeɪt/ v. **1** v.intr. & tr. pass or cause to pass by gradations from one shade to another. **2** tr. arrange in steps or grades of size etc. [back-form. f. GRADATION]

gradation /ɡrəˈdeɪʃ(ə)n/ n. (usu. in pl.) **1** a stage of transition or advance. **2 a** a certain degree in rank, intensity, merit, divergence, etc. **b** such a degree; an arrangement in such degrees. **3** (of paint etc.) the gradual passing from one shade, tone, etc., to another. **4** Philol. ablaut. □ **gradational** adj. **gradationally** adv. [L *gradatio* f. *gradus* step]

Grade /ɡreɪd/, Lew, Baron Grade of Elstree (born Louis Winogradsky) (b.1906), British television producer and executive, born in Russia. After emigrating to Britain with his brother (see DELFONT) in 1912, he established a reputation as one of the pioneers of British commercial television in the 1950s; he was long associated with the television company ATV (Associated Television), later serving as its president from 1977 to 1982. His nephew Michael Grade (b.1943) became chief executive of Channel Four in 1988.

grade /ɡreɪd/ n. & v. ● n. **1 a** a certain degree in rank, merit, proficiency, quality, etc. **b** a class of persons or things of the same grade. **2 a** a mark indicating the quality of a student's work. **b** an examination, esp. in music. **3** N. Amer. a class in school, concerned with a particular year's work and usu. numbered from the first upwards. **4 a** a gradient or slope. **b** the rate of ascent or descent. **5 a** a variety of cattle produced by crossing native stock with a superior breed. **b** a group of animals at a similar level of development. **6** Philol. a relative position in a series of forms involving ablaut. ● v. **1** tr. arrange in or allocate to grades; class, sort. **2** intr. (foll. by up, down, off, into, etc.) pass gradually between grades, or into a grade. **3** *tr.* give a grade to (a student). **4** *tr.* blend so as to affect the grade of colour with tints passing into each other. **5** *tr.* reduce (a road etc.) to easy gradients. **6** *tr.* (often foll. by *up*) cross (livestock) with a better breed. □ **at grade** *US* on the same level. **grade crossing** *US* = *level crossing*. **grade school** *US* elementary school. **make the grade** *colloq.* succeed; reach the desired standard. [F *grade* or L *gradus* step]

grader /ˈɡreɪdə(r)/ *n.* **1** a person or thing that grades. **2** a wheeled machine for levelling the ground, esp. in making roads. **3** (in *comb.*) N. Amer. a pupil of a specified grade in a school (*ninth-grader*).

gradient /ˈɡreɪdɪənt/ *n.* **1 a** a stretch of road, railway, etc., that slopes from the horizontal. **b** the amount of such a slope. **2** the rate of rise or fall of temperature, pressure, concentration, etc., in passing from one region to another. [prob. formed on GRADE after *salient*]

gradine /ˈɡreɪdiːn/ *n.* (also **gradin** /-dɪn/) **1** each of a series of low steps or a tier of seats. **2** a ledge at the back of an altar. [It. *gradino* dimin. of *grado* GRADE]

gradual /ˈɡrædjʊəl/ *adj. & n.* ● *adj.* **1** taking place or progressing slowly or by degrees. **2** not rapid or steep or abrupt. ● *n. Eccl.* **1** a response sung or recited between the Epistle and Gospel in the Mass. **2** a book of music for the sung Mass service. □ **gradually** *adv.* **gradualness** *n.* [med.L *gradualis, -ale* f. L *gradus* step, the noun referring to the altar-steps on which the response is sung]

gradualism /ˈɡrædjʊəˌlɪz(ə)m/ *n.* a policy of gradual reform rather than sudden change or revolution. □ **gradualist** *n.* **gradualistic** /ˌɡrædjʊəˈlɪstɪk/ *adj.*

graduand /ˈɡrædjʊˌænd/ *n. Brit.* a person about to receive an academic degree. [med.L *graduandus* gerundive of *graduare* GRADUATE]

graduate *n. & v.* ● *n.* /ˈɡrædjʊət/ **1** a person who has been awarded an academic degree (also *attrib.*: *graduate student*). **2** *N. Amer.* a person who has completed a school course. ● *v.* /ˈɡrædjʊˌeɪt/ **1** *intr.* take an academic degree. **2** *intr.* **a** (foll. by *from*) be a graduate of a specified university. **b** (foll. by *in*) be a graduate in a specified subject. **3** *tr. US* confer an academic qualification on. **4** *intr.* **a** (foll. by *to*) move up to (a higher grade of activity etc.). **b** (foll. by *as, in*) gain specified qualifications. **5** *tr.* mark out in degrees or parts. **6** *tr.* arrange in gradations; apportion (e.g. tax) according to a scale. **7** *intr.* (foll. by *into, away*) pass by degrees. □ **graduated pension** (in the UK) a system of pension contributions by employees in proportion to their wages or salary. **graduate school** a department of a university for advanced work by graduates. □ **graduator** /-jʊˌeɪtə(r)/ *n.* [med.L *graduari* take a degree f. L *gradus* step]

graduation /ˌɡrædjʊˈeɪʃ(ə)n/ *n.* **1** the act or an instance of graduating or being graduated. **2** a ceremony at which degrees are conferred. **3** each or all of the marks on a vessel or instrument indicating degrees of quantity etc.

Graecism /ˈɡriːsɪz(ə)m, ˈɡriːkɪz(ə)m/ *n.* (also **Grecism**) **1** a Greek idiom, esp. as imitated in another language. **2 a** the Greek spirit, style, mode of expression, etc. **b** the imitation of these. [F *grécisme* or med.L *Graecismus* f. *Graecus* GREEK]

Graecize /ˈɡriːsaɪz, ˈɡriːkaɪz/ *v.tr.* (also **Grecize, -ise**) give a Greek character or form to. [L *Graecizare* (as GRAECISM)]

Graeco- /ˈɡriːkəʊ/ *comb. form* (also **Greco-**) Greek; Greek and. [L *Graecus* GREEK]

Graeco-Roman /ˌɡriːkəʊˈrəʊmən/ *adj.* **1** of or relating to the Greeks and Romans. **2** denoting a style of wrestling like that of the Greeks and Romans, attacking only the upper part of the body.

Graf /ɡrɑːf/, Stephanie ('Steffi') (b.1969), German tennis player. She was ranked top women's player at the age of 16; in 1988 she won the Australian, French, and US Open championships, as well as the Wimbledon trophy and an Olympic gold medal. She won her fifth Wimbledon singles title in 1993.

graffiti /ɡrəˈfiːtɪ/ *n. & v.* ● *n.pl.* (*sing.* **graffito** /-təʊ/) inscriptions or drawings scribbled, scratched, or sprayed on a surface, orig. as inscribed on ancient walls. ¶ *Graffiti* is, in formal terms, a plural noun and as such should be used with a plural verb. However, in its most common use as a collective or mass noun, the word *graffiti* is usually found with a singular verb, e.g. *Graffiti is an increasing problem*; most people find this use natural and acceptable, as they do the collective use of the word *data.* ● *v.tr.* (**graffitied**) **1** cover (a surface) with graffiti. **2** write as graffiti (*graffitied initials*). □ **graffitist** *n.* [It. f. *graffio* a scratch]

graft¹ /ɡrɑːft/ *n. & v.* ● *n.* **1** *Bot.* **a** a shoot or scion inserted into a slit of stock, from which it receives sap. **b** the place where a graft is inserted. **2** *Surgery* a piece of living tissue, organ, etc., transplanted surgically. **3** *sl.* work, esp. hard work. ● *v.* **1** *tr.* **a** (often foll. by *into, on, together,* etc.)

insert (a scion) as a graft. **b** insert a graft on (a stock). **2** *intr.* insert a graft. **3** *tr. Surgery* transplant (living tissue). **4** *tr.* (foll. by *in, on*) insert or fix (a thing) permanently to another. **5** *intr. sl.* work hard. □ **grafting-clay** (or **-wax**) a substance for covering the united parts of a graft and stock. □ **grafter** *n.* [ME (earlier *graff*) f. OF *grafe, grefe* f. L *graphium* f. Gk *graphion* stylus f. *graphō* write]

graft² /ɡrɑːft/ *n. & v. colloq.* ● *n.* **1** practices, esp. bribery, used to secure illicit gains in politics or business. **2** such gains. ● *v.intr.* seek or make such gains. □ **grafter** *n.* [19th c.: orig. unkn.]

Grafton /ˈɡrɑːftən/, Augustus Henry Fitzroy, 3rd Duke of (1735–1811), British Whig statesman, Prime Minister 1768–70.

Graham¹ /ˈɡreɪəm/, Martha (1893–1991), American dancer, teacher, and choreographer. An influential teacher of modern dance, she evolved a new dance language using more flexible movements intended to express psychological complexities and emotional power. She established her own studio in 1927, and during the 1930s produced works on the theme of American roots and values, such as *Appalachian Spring* (1931). Her later works include *Care of the Heart* (1946).

Graham² /ˈɡreɪəm/, Thomas (1805–69), Scottish physical chemist. He studied the diffusion of gases, and suggested a law which relates a gas's rate of diffusion to its density. He also investigated the passage of dissolved substances through porous membranes, and coined the word *osmose* (an earlier form of *osmosis*); he was also the first (in 1861) to use the word *colloid* in its modern (chemical) sense.

Graham³ /ˈɡreɪəm/, William Franklin ('Billy') (b.1918), American evangelical preacher. A minister of the Southern Baptist Church, he became world famous as a mass evangelist; he has conducted large, theatrically staged religious meetings throughout the world, including several in Britain, as well as others in South Korea and the former Soviet Union.

Grahame /ˈɡreɪəm/, Kenneth (1859–1932), Scottish-born writer of children's stories, resident in England from 1864. He is most famous for *The Wind in the Willows* (1908), a collection of stories about animals of the river-bank, now regarded as a children's classic. The main characters (Rat, Mole, Badger, and Toad) became even more familiar to British children through A. A. Milne's musical version of the story (*Toad of Toad Hall*, 1930) and various television adaptations of Grahame's work.

Graham Land the northern part of the Antarctic Peninsula. It is the only part of Antarctica lying outside the Antarctic Circle. Discovered in 1831–2 by the English navigator John Biscoe (1794–1843), it now forms part of British Antarctic Territory, but is claimed also by Chile and Argentina.

Grail /ɡreɪl/ (in full **Holy Grail**) an object of quest in medieval legend, conferring mystical benefits. Its original nature and significance are unknown. From the late 12th century (the date of the first known reference in literature to the Grail, in Chrétien de Troyes's romance *Perceval*) it was given a Christian significance, and was supposed to be the vessel from which Christ drank at the Last Supper and in which Joseph of Arimathea, who was said to have acquired it and brought it to England, had caught some of the blood of the crucified Christ; in other legends it was the dish used at the Last Supper. By the early 13th century it had become attached to the Arthurian cycle of legends as a symbol of perfection sought by the knights of the Round Table. [ME f. OF *graal* etc. f. med.L *gradalis* dish, of unkn. orig.]

grain /ɡreɪn/ *n. & v.* ● *n.* **1** a fruit or seed of a cereal. **2 a** (*collect.*) wheat or any allied grass used as food, corn. **b** (*collect.*) their fruit. **c** any particular species of corn. **3 a** a small hard particle of salt, sand, etc. **b** a discrete particle or crystal, usu. small, in a rock or metal. **c** a piece of solid propellant for use in a rocket engine. **4** the smallest unit of weight in the troy system (equivalent to $1/480$ of an ounce), and in the avoirdupois system (equivalent to $1/437.5$ of an ounce). **5** the smallest possible quantity (*not a grain of truth in it*). **6 a** a roughness of surface. **b** *Photog.* a granular appearance on a photograph or negative. **7** the texture of skin, wood, stone, textile, etc.; the arrangement and size of constituent particles. **8 a** a pattern of lines of fibre in wood or paper. **b** lamination or planes of cleavage in stone, coal, etc. **9** nature, temper, tendency. **10 a** *hist.* kermes or cochineal, or dye made from either of these. **b** *poet.* dye; colour. ● *v.* **1** *tr.* paint in imitation of the grain of wood or marble. **2** *tr.* give a granular surface to. **3** *tr.* dye in grain. **4** *tr. & intr.* form into grains. **5** *tr.* remove hair from (hides). □ **against the grain** (often in phr. **go against the grain**) contrary to one's natural inclination or feeling. **grain-leather** leather dressed with grain-side out. **grain-side** the side of a hide on which the hair was. **grains of Paradise** capsules of a West African plant, *Aframomum melegueta*, used as a spice and a drug. **in grain** thorough, genuine, by nature, downright, indelible. □ **grained** *adj.* (also in *comb.*). **grainer** *n.* **grainless** *adj.* [ME f. OF f. L *granum*]

Grainger /'greɪndʒə(r)/, (George) Percy (Aldridge) (1882–1961), Australian-born American composer and pianist. From 1901 he lived in London, where he gained fame as a concert pianist; he later settled in the US. He joined the English Folk Song Society in 1905 and collected, edited, and arranged English folk-songs. As a composer he is best known for his light music incorporating traditional melodies, such as *Shepherd's Hey, Country Gardens*, and *Handel in the Strand*.

grainy /'greɪnɪ/ *adj.* (**grainier, grainiest**) **1** granular. **2** resembling the grain of wood. **3** *Photog.* having a granular appearance. □ **graininess** *n.*

grallatorial /ˌɡrælə'tɔːrɪəl/ *adj.* *Zool.* of or relating to long-legged wading birds, e.g. storks, flamingos, etc. [mod.L *grallatorius* f. L *grallator* stilt-walker f. *grallae* stilts]

gram¹ /ɡræm/ *n.* (also **gramme**) a metric unit of mass equal to one-thousandth of a kilogram. □ **gram-equivalent** *Chem.* the quantity of a substance equal to its equivalent weight in grams. [F *gramme* f. Gk *gramma* small weight]

gram² /ɡræm/ *n.* a pulse used as food, esp. chick pea or mung bean. [Port. *grão* f. L *granum* grain]

-gram /ɡræm/ *comb. form* forming nouns denoting a thing written or recorded (often in a certain way) (*anagram; epigram; monogram; telegram*). □ **-grammatic** /ɡrə'mætɪk/ *comb. form* forming adjectives. [from or after Gk *gramma -atos* thing written, letter of the alphabet, f. *graphō* write]

gramineous /ˌɡræmɪ'neɪʃəs/ *adj.* of or like grass; grassy. [L *gramen -inis* grass]

gramineous /ɡrə'mɪnɪəs/ *adj.* = GRAMINACEOUS. [L *gramineus* f. *gramen -inis* grass]

graminivorous /ˌɡræmɪ'nɪvərəs/ *adj.* feeding on grass, cereals, etc. [L *gramen -inis* grass + -VOROUS]

grammalogue /'ɡræməˌlɒɡ/ *n.* **1** a word represented by a single shorthand sign. **2** a logogram. [irreg. f. Gk *gramma* letter of the alphabet + *logos* word]

grammar /'ɡræmə(r)/ *n.* **1 a** the branch of language study or linguistics which deals with the means of showing the relation between words as used in speech or writing, traditionally divided into the study of inflections (or morphology) and of the structure of sentences (syntax), and often also including phonology (see also LINGUISTICS). **b** a body of form and usages in a specified language (*Latin grammar*). **2** a person's manner or quality of observance or application of the rules of standard grammar (*bad grammar*). **3** a book on grammar. **4** the elements or rudiments of an art or science. **5** *Brit. colloq.* = GRAMMAR SCHOOL. □ **grammarless** *adj.* [ME f. AF *gramere*, OF *gramaire* f. L *grammatica* f. Gk *grammatikē* (*tekhnē*) (art) of letters f. *gramma -atos* letter of the alphabet]

grammarian /ɡrə'meərɪən/ *n.* an expert in grammar or linguistics; a philologist. [ME f. OF *gramarien*]

grammar school *n.* **1** (in the UK) any of a class of (usually endowed) schools. (*See note below.*) **2** *US* a school intermediate between primary and high school.

▪ The first grammar schools were founded in the 16th century or earlier for teaching Latin, with some instruction also in literature and elementary mathematics. By the end of the 19th century they had become secondary schools with a full curriculum. In the early 20th century many new grammar schools were created and brought into the state system along with some of the established schools; entrance was generally by examination. Since 1965 many have become comprehensive schools.

grammatical /ɡrə'mætɪk(ə)l/ *adj.* **1 a** of or relating to grammar. **b** determined by grammar, esp. by form or inflection (*grammatical gender*). **2** conforming to the rules of grammar, or to the formal principles of an art, science, etc. □ **grammatically** *adv.* **grammaticalness** *n.* [F *grammatical* or LL *grammaticalis* f. L *grammaticus* f. Gk *grammatikos* (as GRAMMAR)]

gramme var. of GRAM¹.

Gram-negative /ɡræm'neɡətɪv/ see GRAM STAIN.

gramophone /'ɡræməˌfəʊn/ *n.* *archaic* an apparatus for reproducing sound recorded on a gramophone record (see RECORD *n.* 3a); now usually called *record-player*. (See also SOUND RECORDING.) □ **gramophone record** = RECORD *n.* 3a. □ **gramophonic** /ˌɡræmə'fɒnɪk/ *adj.* [formed by inversion of PHONOGRAM in sense 'recording made by a phonograph']

Grampian /'ɡræmpɪən/ a local government region in NE Scotland; capital, Aberdeen.

Grampian Mountains (also **Grampians**) **1** a mountain range in north central Scotland. Its southern edge forms a natural boundary between the Highlands and the Lowlands. **2** a mountain range in SE Australia, in Victoria. It forms a spur of the Great Dividing Range at its western extremity.

Gram-positive /ɡræm'pɒzɪtɪv/ see GRAM STAIN.

grampus /'ɡræmpəs/ *n.* (*pl.* **grampuses**) **1 a** a dolphin, *Grampus griseus*, with a blunt snout and long pointed black flippers. **b** the killer whale. **2** a person breathing heavily and loudly. [earlier *graundepose*, *grapeys* f. OF *grapois* etc. f. med.L *craspiscis* f. L *crassus piscis* fat fish]

Gramsci /'ɡræmʃɪ/, Antonio (1891–1937), Italian political theorist and activist. A founder of the Italian Communist Party, he became its leader and was elected to the Chamber of Deputies in 1924. He was imprisoned when the Communist Party was banned by the Fascists in 1926, and died shortly after his release. Gramsci's most notable writings date from his imprisonment, and include *Letters from Prison* (1947). His work on the dictatorship of the proletariat remains an important influence on left-wing thought and cultural theory.

Gram stain /ɡræm/ *n.* (also **Gram's stain**) *Med.* a staining technique for the preliminary identification of bacteria, in which a violet dye is applied, followed by a decolorizing agent and then a red dye. The cell walls of certain bacteria (denoted *Gram-positive*) retain the first dye and appear violet, while those that lose it (denoted *Gram-negative*) appear red. [Hans Christian Joachim *Gram*, Danish physician (1853–1938)]

gran /ɡræn/ *n.* *colloq.* grandmother (cf. GRANNY). [abbr.]

Granada /ɡrə'nɑːdə/ **1** a city in Andalusia in southern Spain; pop. (1991) 286,700. Founded in the 8th century, it became the capital of the Moorish kingdom of Granada in 1238. It is the site of the Alhambra palace. **2** a city in Nicaragua, on the NW shore of Lake Nicaragua; pop. (1985) 88,600. Founded by the Spanish in 1523, it is the oldest city in the country.

granadilla /ˌɡrænə'dɪlə/ *n.* (also **grenadilla** /ˌɡren-/) a passion-fruit. [Sp., dimin. of *granada* pomegranate]

granary /'ɡrænərɪ/ *n.* (*pl.* **-ies**) **1** a storehouse for threshed grain. **2** a region producing, and esp. exporting, much corn. **3** *propr.* a type of brown bread or flour containing whole grains of wheat (*granary bread*). [L *granarium* f. *granum* grain]

Gran Canaria /ˌɡræn kə'nɑːrɪə/ a volcanic island in the Atlantic Ocean off the NW coast of Africa, one of the Canary Islands. Its chief town, Las Palmas, a port on the north coast, is the capital of the Canary Islands.

Gran Chaco /ɡræn 'tʃækəʊ/ (also **Chaco**) a lowland plain in central South America, extending from southern Bolivia through Paraguay to northern Argentina. [Sp., = great hunting-ground or riches]

grand /ɡrænd/ *adj.* & *n.* ● *adj.* **1 a** splendid, magnificent, imposing, dignified. **b** solemn or lofty in conception, execution, or expression; noble. **2** main; of chief importance (*grand staircase; grand entrance*). **3** (usu. **Grand**) of the highest rank, esp. in official titles (*Grand Cross; grand vizier; Grand Inquisitor*). **4** *colloq.* excellent, enjoyable (*had a grand time; in grand condition*). **5** belonging to high society; displaying opulence (*the grand folk at the big house*). **6** (in *comb.*) in names of family relationships, denoting the second degree of ascent or descent (*granddaughter*). **7** (**Grand**) (in French phrases or imitations) great (*Grand Monarch; Grand Hotel*). **8** *Law* serious, important (*grand larceny*) (cf. COMMON *adj.* 9, PETTY 4). ● *n.* **1** = grand piano. **2** (*pl.* same) (usu. in *pl.*) *sl.* a thousand dollars or pounds. □ **grand aunt** a great-aunt (see GREAT *adj.* 11). **grand duchy** a state ruled by a grand duke or duchess. **grand duke** (or **duchess**) **1** a prince (or princess) or noble person ruling over a territory. **2** (**Grand Duke**) *hist.* the son or grandson of a Russian tsar. **grand jury** esp. *US Law* a jury selected to examine the validity of an accusation prior to trial. **grand master 1** a chess player of the highest class. **2** the head of a military order of knighthood, of Freemasons, etc. **grand nephew** (or **niece**) a great-nephew (or great-niece) (see GREAT *adj.* 11). **grand opera** opera on a serious theme, or in which the entire libretto (including dialogue) is sung. **grand piano** a large full-toned piano standing on three legs, with the body, strings, and soundboard arranged horizontally and in line with the keys. **grand total** the final amount after everything is added up; the sum of other totals. **grand tour** *hist.* a cultural tour of Europe, esp. in the 18th century for educational purposes. **grand unified theory** *Physics* a theory that attempts to give a single explanation of the electromagnetic, strong, and weak interactions between subatomic

[]

particles. □ **grandly** *adv.* **grandness** *n.* [ME f. AF *graunt*, OF *grant* f. L *grandis* full-grown]

grandad /ˈgrændæd/ *n.* (also **grand-dad**) *colloq.* **1** grandfather. **2** an elderly man.

grandam /ˈgrændæm/ *n.* **1** (also **grandame** /-deim, -dæm/) *archaic* grandmother. **2** an old woman. **3** an ancestress. [ME f. AF *graund dame* (as GRANDAM, DAME)]

Grand Banks a submarine plateau of the continental shelf off the SE coast of Newfoundland, Canada. It is a meeting-place of the warm Gulf Stream and the cold Labrador Current; this promotes the growth of plankton, making the waters an important feeding area for fish.

Grand Canal 1 a series of waterways in eastern China, extending from Beijing southwards to Hangzhou, a distance of 1,700 km (1,060 miles). First constructed in 486 BC as a link between the Yangtze and the Yellow River, it was extended over the centuries until its present length was reached in AD 1327. Its original purpose was to transport rice from the river valleys to the cities. **2** the main waterway of Venice in Italy. It is lined on each side by fine palaces and spanned by the Rialto Bridge.

Grand Canyon a deep gorge in Arizona, formed by the Colorado river. It is about 440 km (277 miles) long, 8 to 24 km (5 to 15 miles) wide, and, in places, 1,800 m (6,000 ft) deep. The area was designated a national park in 1919.

grandchild /ˈgræntʃaild, ˈgrænd-/ *n.* (*pl.* **-children**) a child of one's son or daughter.

granddaughter /ˈgrænˌdɔːtə(r), ˈgrænd-/ *n.* a female grandchild.

Grande Comore /ˌgrɒnd kəˈmɔː(r)/ the largest of the islands of the Comoros, off the NW coast of Madagascar; pop. (1980) 189,000; chief town (and capital of the Comoros), Moroni.

grande dame /grɒnd ˈdæm/ *n.* (*pl.* ***grandes dames*** *pronunc.* same) a dignified lady of high rank. [F]

grandee /grænˈdiː/ *n.* **1** a Spanish or Portuguese nobleman of the highest rank. **2** a person of high rank or eminence. [Sp. & Port. *grande*, assim. to -EE]

grandeur /ˈgrændjə(r), ˈgrændʒə(r)/ *n.* **1** majesty, splendour; dignity of appearance or bearing. **2** high rank, eminence. **3** nobility of character. [F f. *grand* great, GRAND]

grandfather /ˈgrænˌfɑːðə(r), ˈgrænd-/ *n.* a male grandparent. □ **grandfather clock** a clock in a tall wooden case, driven by weights. □ **grandfatherly** *adj.*

Grand Fleet *Brit. hist.* **1** (in the 18th century) the British fleet based at Spithead. **2** the British fleet operating in the North Sea during the First World War.

Grand Guignol /ˌgrɒn giːˈnjɒl/ *n.* a dramatic entertainment of a sensational or horrific nature, originally a sequence of short pieces as performed at the Grand Guignol theatre in Paris. (See also GUIGNOL.)

grandiflora /ˌgrændɪˈflɔːrə/ *adj.* bearing large flowers. [mod.L (often used in specific names of large-flowered plants) f. L *grandis* great + FLORA]

grandiloquent /grænˈdɪləkwənt/ *adj.* **1** pompous or inflated in language. **2** given to boastful talk. □ **grandiloquence** *n.* **grandiloquently** *adv.* [L *grandiloquus* (as GRAND, -*loquus* -speaking f. *loqui* speak), after *eloquent* etc.]

grandiose /ˈgrændɪˌəʊs/ *adj.* **1** producing or meant to produce an imposing effect. **2** planned on an ambitious or magnificent scale. □ **grandiosely** *adv.* **grandiosity** /ˌgrændɪˈɒsɪtɪ/ *n.* [F f. It. *grandioso* (as GRAND, -OSE¹)]

grandma /ˈgrænmɑː, ˈgrænd-/ *n. colloq.* grandmother.

grand mal /grɒn ˈmæl/ *n.* a serious form of epilepsy with loss of consciousness; an epileptic fit of this kind (cf. PETIT MAL). [F, = great sickness]

grandmama /ˈgrænməˌmɑː, ˈgrænd-/ *n. archaic colloq.* = GRANDMA.

grand manner *n.* **1** the lofty and rhetorical manner of history painting formerly considered appropriate for serious and elevated subjects. Nicolas Poussin and especially Raphael were regarded as the masters of the style. **2** (often in phr. **in the grand manner**) time-honoured magnificence or lavishness.

grandmother /ˈgrænˌmʌðə(r), ˈgrænd-/ *n.* a female grandparent. □ **grandmother clock** a clock like a grandfather clock but in a smaller case. **teach one's grandmother to suck eggs** presume to advise a more experienced person. □ **grandmotherly** *adj.*

Grand National an annual horse-race established in 1839, a

steeplechase run over a course of 4 miles 856 yards (about 7,200 metres) with thirty jumps, at Aintree, Liverpool, in late March or early April.

grandpa /ˈgrænpɑː, ˈgrænd-/ *n. colloq.* grandfather.

grandpapa /ˈgrænpəˌpɑː, ˈgrænd-/ *n. archaic colloq.* = GRANDPA.

grandparent /ˈgrænˌpeərənt, ˈgrænd-/ *n.* a parent of one's father or mother.

Grand Penitentiary *n.* a cardinal presiding over the penitentiary in the papal court.

Grand Pensionary *n. hist.* the first minister of Holland and Zeeland (1619–1794).

Grand Prix /grɒn ˈpriː/ (*pl.* **Grands Prix** *pronunc.* same) **1** any of various motor-racing or motorcycling contests forming part of a world championship series, held in various countries under international rules. (See MOTOR-RACING.) **2** (in full **Grand Prix de Paris**) an international horse-race for three-year-olds, founded in 1863 and run annually in June at Longchamps, Paris. [F, = great or chief prize]

grand siècle /ˌgrɒn ˈsjeklə/ *n.* the classical or golden age, esp. the reign of Louis XIV (1643–1715) in France. [F, = great century or age]

grandsire /ˈgrænˌsaɪə(r), ˈgrænd-/ *n. archaic* **1** grandfather, old man, ancestor. **2** (in bell-ringing) a method of change-ringing.

grand slam *n.* **1** *Sport* **a** a group of matches, championships, etc. in a particular sport. (See note below.) **b** the winning of all these matches etc. **2** *Bridge* the winning of all thirteen tricks.

■ Perhaps the best-known grand slams are those in tennis (the Australian Open, the French Open, Wimbledon, and the US Open), golf (the US Open, the British Open, the Masters, and the PGA), and Rugby Union (victories against all opposition in the competition between England, Wales, France, Ireland, and Scotland).

grandson /ˈgrænsʌn, ˈgrænd-/ *n.* a male grandchild.

grandstand /ˈgrænstænd, ˈgrænd-/ *n.* the main stand, usu. roofed, for spectators at a racecourse etc. □ **grandstand finish** a close and exciting finish to a race etc.

grange /greɪndʒ/ *n.* **1** a country house with farm-buildings. **2** *archaic* a barn. [ME f. AF *graunge*, OF *grange* f. med.L *granica* (*villa*) ult. f. L *granum* GRAIN]

graniferous /grəˈnɪfərəs/ *adj.* producing grain or a grainlike seed. □ **graniform** /ˈgrænɪˌfɔːm/ *adj.* [L *granum* GRAIN]

granite /ˈgrænɪt/ *n.* **1** a granular crystalline igneous rock of quartz, mica, feldspar, etc., used for building. **2** a determined or resolute quality, attitude, etc. □ **granitoid** *adj. & n.* **granitic** /grəˈnɪtɪk/ *adj.* [It. *granito*, lit. 'grained' f. *grano* f. L *granum* GRAIN]

graniteware /ˈgrænɪtˌweə(r)/ *n.* **1** a speckled form of earthenware imitating the appearance of granite. **2** a kind of enamelled ironware.

granivorous /grəˈnɪvərəs/ *adj.* feeding on grain. □ **granivore** /ˈgrænɪˌvɔː(r)/ *n.* [L *granum* GRAIN]

granny /ˈgrænɪ/ *n.* (also **grannie**) (*pl.* **-ies**) *colloq.* grandmother. □ **granny bond** *Brit. colloq.* a form of National Savings certificate orig. available only to pensioners. **granny flat** (or **annexe**) *Brit.* part of a house made into self-contained accommodation for an elderly relative. **granny knot** a reef-knot crossed the wrong way and therefore insecure. [obs. *grannam* for GRANDAM + -Y²]

Granny Smith *n.* a green Australian variety of apple. [Maria Ann ('Granny') Smith (c.1801–70)]

granodiorite /ˌgrænəʊˈdaɪəˌraɪt/ *n. Geol.* a coarse-grained plutonic rock containing quartz and plagioclase, between granite and diorite in composition. [GRANITE + DIORITE]

Grant¹ /grɑːnt/, Cary (born Alexander Archibald Leach) (1904–86), British-born American actor. He made his Hollywood screen début in *This is the Night* (1932) after appearing in Broadway musicals. He acted in more than seventy films, including *Holiday* (1938) and *The Philadelphia Story* (1940).

Grant² /grɑːnt/, Ulysses S(impson) (born Hiram Ulysses Grant) (1822–85), American general and 18th President of the US 1869–77. Having made his reputation through a series of victories on the Union side in the American Civil War (most notably the capture of Vicksburg in 1863), Grant was made supreme commander of the Unionist armies. His policy of attrition against the Confederate army eventually proved successful in ending the war. He became President in 1869, but was unable to check widespread political corruption and inefficiency.

grant /grɑːnt/ *v. & n.* ● *v.tr.* **1 a** consent to fulfil (a request, wish, etc.) (*granted all he asked*). **b** allow (a person) to have (a thing) (*granted me my freedom*). **c** (as **granted** *adj.*) *colloq.* apology accepted; pardon given. **2** give (rights, property, etc.) formally; transfer legally. **3** (often foll. by *that* + clause) admit as true; concede, esp. as a basis for argument. ● *n.* **1** the process of granting or a thing granted. **2** a sum of money given by the state for any of various purposes, esp. to finance education. **3** *Law* **a** a legal conveyance by written instrument. **b** formal conferment. □ **grant aid** = *grant-in-aid.* **grant-aid** *v.tr.* give financial assistance to. **grant-in-aid** (*pl.* **grants-in-aid**) a grant by central government to local government or an institution. **grant-maintained** (of a school) funded by central rather than local government. **take for granted 1** assume something to be true or valid. **2** cease to appreciate through familiarity. □ **grantable** *adj.* **granter** *n.* **grantor** /grɑːnˈtɔː(r)/ *n.* **grantee** /-ˈtiː/ *n.* (esp. in sense 2 of *v.*). [ME f. OF *gr(e)anter* var. of *creanter* ult. f. part. of L *credere* entrust]

Granth /grʌnt/ *n.* (more fully **Granth Sahib**) = ADI GRANTH. [f. Skr.; see GRANTHA, SAHIB]

Grantha /ˈgrʌntə/ *n.* a southern Indian alphabet dating from the 5th century AD, used by Tamil brahmans for the Sanskrit transcriptions of their sacred books. [Skr., = tying, literary composition]

gran turismo /ˌgræn tʊəˈrɪzməʊ/ *n.* (*pl.* **-os**) (abbr. **GT**) a comfortable high-performance model of motor car. [It., = great touring]

granular /ˈgrænjʊlə(r)/ *adj.* **1** of or like grains or granules. **2** having a granulated surface or structure. □ **granularly** *adv.* **granularity** /ˌgrænjʊˈlærɪti/ *n.* [LL *granulum* GRANULE]

granulate /ˈgrænjʊˌleɪt/ *v.* **1** *tr.* & *intr.* form into grains (*granulated sugar*). **2** *tr.* roughen the surface of. **3** *intr.* (of a wound etc.) form small prominences as the beginning of healing; heal, join. □ **granulator** *n.* **granulation** /ˌgrænjʊˈleɪʃ(ə)n/ *n.*

granule /ˈgrænjuːl/ *n.* a small grain. [LL *granulum,* dimin. of L *granum* grain]

granulocyte /ˈgrænjʊˌsaɪt/ *n.* *Physiol.* a white blood cell with granules in the cytoplasm. □ **granulocytic** /ˌgrænjʊˈsɪtɪk/ *adj.*

granulometric /ˌgrænjʊləˈmetrɪk/ *adj.* relating to the distribution of grain sizes in sand etc. [F *granulométrique* (as GRANULE, METRIC)]

Granville-Barker /ˌgrænvɪlˈbɑːkə(r)/, Harley (1877–1946), English dramatist, critic, theatre director, and actor. As co-manager of the Royal Court Theatre, London (1904–7), he presented plays incorporating social comment and realism, including works by Ibsen and Shaw. His own plays, such as *The Voysey Inheritance* (1905), also dealt naturalistically with social and moral issues. His Shakespearian productions and his *Prefaces to Shakespeare* (1927–46) were influential for subsequent interpretation and staging of Shakespeare's work.

grape /greɪp/ *n.* **1** a berry (usu. green, purple, or black) growing in clusters on a vine, used as fruit and in making wine. **2** (prec. by *the*) *colloq.* wine. **3** = GRAPESHOT. **4** (in *pl.*) a diseased growth like a bunch of grapes on the pastern of a horse etc., or on a pleura in cattle. □ **grape hyacinth** a liliaceous plant of the genus *Muscari,* with clusters of small ball-like blue flowers. **grape-sugar** dextrose. □ **grapey** *adj.* (also **grapy**). [ME f. OF *grape* bunch of grapes prob. f. *graper* gather (grapes) f. *grap(p)e* hook, ult. f. Gmc]

grapefruit /ˈgreɪpfruːt/ *n.* (*pl.* same) **1** a large round yellow citrus fruit with an acid juicy pulp. **2** the tree, *Citrus paradisi,* bearing this fruit.

grapeseed oil /ˈgreɪpsiːd/ *n.* oil extracted from the seeds of grapes.

grapeshot /ˈgreɪpʃɒt/ *n.* *hist.* small balls used as charge in a cannon and scattering when fired.

grapevine /ˈgreɪpvaɪn/ *n.* **1** a grape-yielding vine of the genus *Vitis,* esp. *Vitis vinifera.* **2** *colloq.* the means of transmission of unofficial information or rumour (esp. as *heard it through the grapevine*).

graph[1] /grɑːf, græf/ *n.* & *v.* ● *n.* **1** a diagram showing the relation between variable quantities, usu. of two variables, each measured along one of a pair of axes at right angles. **2** *Math.* a collection of points whose coordinates satisfy a given relation. ● *v.tr.* plot or trace on a graph. □ **graph paper** paper printed with a network of lines as a basis for drawing graphs. [abbr. of *graphic formula*]

graph[2] /grɑːf, græf/ *n.* *Linguistics* a visual symbol, esp. a letter or letters, representing a unit of sound or other feature of speech. [Gk *graphē* writing]

-graph /grɑːf, græf/ *comb. form* forming nouns and verbs meaning: **1** a thing written or drawn etc. in a specified way (*autograph; photograph*). **2** an instrument that records (*heliograph; seismograph; telegraph*).

grapheme /ˈgræfiːm/ *n.* *Linguistics* **1** a class of letters etc. representing a unit of sound. **2** a feature of a written expression that cannot be analysed into smaller meaningful units. □ **graphematic** /ˌgræfɪˈmætɪk/ *adj.* **graphemic** /grəˈfiːmɪk/ *adj.* **graphemically** *adv.* [GRAPH[2] + -EME]

-grapher /grəfə(r)/ *comb. form* forming nouns denoting a person concerned with a subject etc. with a name in -graphy (see GRAPHY) (*geographer; radiographer*). [from or after Gk -*graphos* writer + -ER[1]]

graphic /ˈgræfɪk/ *adj.* & *n.* ● *adj.* **1** of or relating to the visual or descriptive arts, esp. writing and drawing. **2** vividly descriptive; conveying all (esp. unwelcome or unpleasant) details; unequivocal. **3** (of minerals) showing marks like writing on the surface or in a fracture. **4** = GRAPHICAL. ● *n.* a product of the graphic arts (cf. GRAPHICS). □ **graphic arts** the visual and technical arts involving design, writing, drawing, printing, etc. **graphic equalizer** a device for the separate control of the strength and quality of selected frequency bands. **graphic novel** an adult novel in comic-strip format. □ **graphically** *adv.* **graphicness** *n.* [L *graphicus* f. Gk *graphikos* f. *graphē* writing]

-graphic /ˈgræfɪk/ *comb. form* (also **-graphical**) forming adjectives corresponding to nouns in -graphy (see -GRAPHY). □ **-graphically** *comb. form* forming adverbs. [from or after Gk -*graphikos* (as GRAPHIC)]

graphicacy /ˈgræfɪkəsɪ/ *n.* the ability to read a map, graph, etc., or to present information by means of diagrams. [GRAPHIC, after *literacy, numeracy*]

graphical /ˈgræfɪk(ə)l/ *adj.* **1** of or in the form of graphs (see GRAPH[1]). **2** graphic. □ **graphically** *adv.*

graphics /ˈgræfɪks/ *n.pl.* (usu. treated as *sing.*) **1** the products of the graphic arts, esp. commercial design or illustration. **2** the use of diagrams in calculation and design. **3** (in full **computer graphics**) **a** the use of computers linked to VDUs to generate and manipulate visual images. **b** visual images produced by means of computer processing.

graphite /ˈgræfaɪt/ *n.* a dark grey crystalline electrically conducting allotropic form of carbon used as a solid lubricant, in pencils, and as a moderator in nuclear reactors etc. Also called *plumbago, black lead.* □ **graphitize** /ˈgræfɪˌtaɪz/ *v.tr.* & *intr.* (also **-ise**). **graphitic** /grəˈfɪtɪk/ *adj.* [G *Graphit* f. Gk *graphō* write]

graphology /grəˈfɒlədʒɪ/ *n.* **1** the study of handwriting esp. as a supposed guide to a person's character. **2** a system of graphic formulae; notation for graphs (see GRAPH[1]). **3** *Linguistics* the study of systems of writing. □ **graphologist** *n.* **graphological** /ˌgræfəˈlɒdʒɪk(ə)l/ *adj.* [Gk *graphē* writing]

-graphy /grəfɪ/ *comb. form* forming nouns denoting: **1** a descriptive science (*bibliography; geography*). **2** a technique of producing images (*photography; radiography*). **3** a style or method of writing, drawing, etc. (*calligraphy*). [from or after F or G -*graphie* f. L -*graphia* f. Gk -*graphia* writing]

grapnel /ˈgræpn(ə)l/ *n.* **1** a device with iron claws, attached to a rope and used for dragging or grasping. **2** a small anchor with several flukes. [ME f. AF f. OF *grapon* f. Gmc: cf. GRAPE]

grappa /ˈgræpə/ *n.* a brandy distilled from the fermented residue of grapes after they have been pressed in wine-making. [It.]

Grappelli /grəˈpelɪ/, Stephane (b.1908), French jazz violinist. After early classical training, he turned to jazz in the late 1920s; in 1934, together with the guitarist Django Reinhardt, he formed the Quintette du Hot Club de France, becoming famous for his improvisational style of swing, and making many recordings with the group until it split up in 1939. Grappelli went on to lead a successful international career, both as a soloist and with groups.

grapple /ˈgræp(ə)l/ *v.* & *n.* ● *v.* **1** *intr.* (often foll. by *with*) fight at close quarters or in close combat. **2** *intr.* (foll. by *with*) try to manage or overcome a difficult problem etc. **3** *tr.* **a** grip with the hands; come to close quarters with. **b** seize with or as with a grapnel; grasp. ● *n.* **1 a** a hold or grip in or as in wrestling. **b** a contest at close quarters. **2** a clutching-instrument; a grapnel. □ **grappling-iron** (or **-hook**) = GRAPNEL. □ **grappler** *n.* [OF *grapil* (n.) f. Prov., dimin. of *grapa* hook (as GRAPNEL)]

graptolite /ˈgræptəˌlaɪt/ *n.* an extinct marine invertebrate animal found as a fossil in lower Palaeozoic rocks. [Gk *graptos* marked with letters + -LITE]

Grasmere /ˈgrɑːsmɪə(r)/ a village in Cumbria, beside a small lake of the same name; pop. (1981) 1,100. It is associated with William and Dorothy Wordsworth, who settled there in 1799.

grasp /grɑːsp/ *v.* & *n.* ● *v.* **1** *tr.* **a** clutch at; seize greedily. **b** hold firmly; grip. **2** *intr.* (foll. by *at*) try to seize; accept avidly. **3** *tr.* understand or realize (a fact or meaning). ● *n.* **1** a firm hold; a grip. **2** (foll. by *of*) **a** mastery or control (*a grasp of the situation*). **b** a mental hold or understanding (*a grasp of the facts*). **3** mental agility (*a quick grasp*).

□ **grasp the nettle** tackle a difficulty boldly. **within one's grasp** capable of being grasped or comprehended by one. □ **graspable** adj. **grasper** n. [ME graspe, grapse perh. f. OE græpsan (unrecorded) f. Gmc, rel. to GROPE: cf. LG grapsen]

grasping /ˈgrɑːspɪŋ/ adj. avaricious, greedy. □ **graspingly** adv. **graspingness** n.

Grass /grɑːs/, Günter (Wilhelm) (b.1927), German novelist, poet, and dramatist. He won international acclaim with his first novel, The Tin Drum (1959), a picaresque tale drawing on Grass's own experiences as a youth in Nazi Germany. He is known for his outspoken socialist views, which are reflected in the play The Plebians Rehearse the Uprising (1966). His other well-known novels include Dog Years (1965).

grass /grɑːs/ n. & v. ● n. **1 a** a plant of the family Gramineae, having long narrow leaves, jointed stems, and spikes of usu. inconspicuous wind-pollinated flowers. (See note below.) **b** vegetation composed of usu. short grasses, grazed by cattle, horses, sheep, etc. **2** pasture land. **3** grass-covered ground, a lawn (keep off the grass). **4** sl. marijuana. **5** Brit. sl. an informer, esp. a police informer. **6** Mining the earth's surface above a mine; the pit-head. **7** sl. asparagus. ● v. **1** tr. cover with turf. **2** tr. US provide with pasture. **3** Brit. sl. **a** tr. betray, esp. to the police. **b** intr. (foll. by on) inform the police about someone. **4** tr. knock down; fell (an opponent). **5** tr. **a** bring (a fish) to the bank. **b** bring down (a bird) by a shot. □ **at grass** at pasture. **grass bird 1** an Australasian warbler, esp. of the genus Megalurus, living among reeds. **2** a southern African warbler, Sphenoeacus afer. **grass-box** a receptacle for cut grass on a lawnmower. **grass-cloth** a linen-like cloth woven from ramie etc. **grass court** a grass-covered lawn-tennis court. **grass of Parnassus** a marsh plant, Parnassia palustris, with solitary white flowers. **grass parakeet** a small Australian parakeet, esp. of the genus Neophema, frequenting grassland. **grass roots 1** a fundamental level or source. **2** ordinary people, esp. as voters; the rank and file of an organization, esp. a political party. **grass ski** each of a pair of short skis with small wheels or rollers for skiing down grass-covered slopes. **grass skirt** a skirt made of long grass and leaves fastened to a waistband. **grass snake 1** a common Eurasian snake, Natrix natrix, greenish-brown or -grey with a yellow band around its neck. **2** N. Amer. the common green snake, Opheodrys vernalis. **grass tree** = BLACKBOY. **grass widow** (or **widower**) a person whose husband (or wife) is away for a prolonged period. **grass-wrack** eel-grass. **not let the grass grow under one's feet** be quick to act or to seize an opportunity. **out to grass 1** to pasture. **2** redundant, in retirement, on holiday. □ **grassless** adj. **grasslike** adj. [OE græs f. Gmc, rel. to GREEN, GROW]

▪ The grasses form a very large and important family of mainly herbaceous plants, and are the dominant vegetation of many areas of the world. They are able to withstand repeated grazing or cutting because of their mode of growth with the growing point mainly at ground level. This makes them suitable as the food of many grazing animals, and also ideal for lawns and playing-fields. The family includes some larger plants, such as the reeds and the bamboos, and several kinds have been cultivated as important cereal crops.

Grasse /grɑːs/ a town near Cannes in SE France, centre of the French perfume industry; pop. (1990) 42,080.

grasshopper /ˈgrɑːsˌhɒpə(r)/ n. an orthopterous insect, usu. winged and feeding on plants, that makes a chirping call and has long hind legs adapted for jumping. Those of the family Acrididae (short-horned grasshoppers) call by rubbing a hind leg against a wing, while those of the family Tettigoniidae (long-horned grasshoppers) call by rubbing two wings together. □ **grasshopper warbler** a warbler of the genus Locustella, esp. L. naevia, with a prolonged buzzing call.

grassland /ˈgrɑːslænd/ n. open country covered with grass, esp. used for grazing.

grassy /ˈgrɑːsɪ/ adj. (**grassier, grassiest**) **1** covered with or having much grass. **2** resembling grass. **3** of grass. □ **grassiness** n.

grate[1] /greɪt/ v. **1** tr. reduce to small particles by rubbing on a serrated surface. **2** intr. (often foll. by against, on) rub with a harsh scraping sound. **3** tr. utter in a harsh tone. **4** intr. **a** sound harshly or discordantly. **b** (of a hinge etc.) creak. **5** tr. grind (one's teeth). **6** intr. (often foll. by on) have an irritating effect. [ME f. OF grater ult. f. WG]

grate[2] /greɪt/ n. **1** the recess of a fireplace or furnace. **2** a metal frame confining fuel in a fireplace etc. [ME, = grating f. OF ult. f. L cratis hurdle]

grateful /ˈgreɪtfʊl/ adj. **1** thankful; feeling or showing gratitude (am grateful to you for helping). **2** pleasant, acceptable. □ **gratefully** adv. **gratefulness** n. [obs. grate (adj.) f. L gratus + -FUL]

grater /ˈgreɪtə(r)/ n. a device for reducing cheese or other food to small particles.

graticule /ˈgrætɪˌkjuːl/ n. **1** a series of fine lines or fibres incorporated in a telescope or other optical instrument as a measuring scale or as an aid in locating objects. **2** Surveying a network of lines on paper representing meridians and parallels. [F f. med.L graticula for craticula gridiron f. L cratis hurdle]

gratify /ˈgrætɪˌfaɪ/ v.tr. (**-fies, -fied**) **1 a** please, delight. **b** please by compliance; assent to the wish of. **2** indulge in or yield to (a feeling or desire). □ **gratifier** n. **gratifying** adj. **gratifyingly** adv. **gratification** /ˌgrætɪfɪˈkeɪʃ(ə)n/ n. [F gratifier or L gratificari do a favour to, make a present of, f. gratus pleasing]

gratin /ˈgrætæn/ n. Cookery a light, browned crust usu. of breadcrumbs or melted cheese; a dish cooked with this (cf. AU GRATIN).

gratiné /ˈgrætɪˌneɪ/ adj. & n. (fem. **gratinée** pronunc. same) ● adj. = AU GRATIN. ● n. = GRATIN.

grating[1] /ˈgreɪtɪŋ/ adj. **1** sounding harsh or discordant (a grating laugh). **2** having an irritating effect. □ **gratingly** adv.

grating[2] /ˈgreɪtɪŋ/ n. **1** a framework of parallel or crossed metal bars. **2** Optics a set of parallel wires, lines ruled on glass, etc., for producing spectra by diffraction.

gratis /ˈgrɑːtɪs, ˈgreɪtɪs, ˈgrætɪs/ adv. & adj. free; without charge. [L, contracted abl. pl. of gratia favour]

gratitude /ˈgrætɪˌtjuːd/ n. being thankful; readiness to show appreciation for and to return kindness. [F gratitude or med.L gratitudo f. gratus thankful]

gratuitous /grəˈtjuːɪtəs/ adj. **1** given or done free of charge. **2** uncalled-for; unwarranted; lacking good reason (a gratuitous insult). □ **gratuitously** adv. **gratuitousness** n. [L gratuitus spontaneous: cf. fortuitous]

gratuity /grəˈtjuːɪtɪ/ n. (pl. **-ies**) money given in recognition of services; a tip. [OF gratuité or med.L gratuitas gift f. L gratus grateful]

gratulatory /ˈgrætjʊlətərɪ/ adj. expressing congratulation. [LL gratulatorius f. L gratus grateful]

graunch /grɔːntʃ/ v.intr. & tr. colloq. make or cause to make a crunching or grinding sound. [imit.]

gravamen /grəˈveɪmen/ n. (pl. **gravamens** or **gravamina** /-ˈvæmɪnə/) **1** the essence or most serious part of an argument. **2** a grievance. [LL, = inconvenience, f. L gravare to load f. gravis heavy]

grave[1] /greɪv/ n. **1 a** a trench dug in the ground to receive a coffin on burial. **b** the place where someone is buried, often marked by a mound or stone. **2** (prec. by the) death, esp. as indicating mortal finality. **3** something compared to or regarded as a grave. □ **turn in one's grave** (of a dead person) be thought of in certain circumstances as likely to have been shocked or angry if still alive. [OE græf f. WG]

grave[2] /greɪv/ adj. & n. ● adj. **1 a** serious, weighty, important (a grave matter). **b** dignified, solemn, sombre (a grave look). **2** extremely serious or threatening (grave danger). **3** /grɑːv/ (of sound) low-pitched, not acute. ● n. /grɑːv/ = grave accent. □ **grave accent** /grɑːv/ a mark (`) placed over a vowel in some languages to denote pronunciation, length, etc., orig. indicating low or falling pitch. □ **gravely** adv. **graveness** n. [F grave or L gravis heavy, serious]

grave[3] /greɪv/ v.tr. (past part. **graven** /ˈgreɪv(ə)n/ or **graved**) **1** (foll. by in, on) fix indelibly (on one's memory). **2** archaic engrave, carve. □ **graven image** an idol. [OE grafan dig, engrave f. Gmc: cf. GROOVE]

grave[4] /greɪv/ v.tr. clean (a ship's bottom) by burning off accretions and by tarring. □ **graving dock** = dry dock. [perh. F dial. grave = OF greve shore]

gravedigger /ˈgreɪvˌdɪgə(r)/ n. **1** a person who digs graves. **2** (in full **gravedigger beetle**) a sexton beetle.

gravel /ˈgræv(ə)l/ n. & v. ● n. **1 a** a mixture of coarse sand and small water-worn or pounded stones, used for paths and roads and as an aggregate. **b** Geol. a stratum of this. **2** Med. aggregations of crystals formed in the urinary tract. ● v.tr. (**gravelled, gravelling;** US **graveled, graveling**) **1** lay or strew with gravel. **2** perplex, puzzle, nonplus (from an obs. sense 'run (a ship) aground'). □ **gravel-blind** literary almost completely blind ('more than sand-blind', in Shakespeare's Merchant of Venice II. ii. 33). [ME f. OF gravel(e) dimin. of grave (as GRAVE[4])]

gravelly /ˈgrævəlɪ/ adj. **1** of or like gravel. **2** having or containing gravel. **3** (of a voice) deep and rough-sounding.

graven past part. of GRAVE[3].

graver /ˈgreɪvə(r)/ *n.* **1** an engraving tool; a burin. **2** *archaic* an engraver; a carver.

Graves[1] /greɪvz/, Robert (Ranke) (1895–1985), English poet, novelist, and critic. His early poetry was written during the First World War, but he is better known for his later work, which is individualistic and cannot be associated with any school or movement. He was professor of poetry at Oxford University 1961–6. His prose includes autobiography (*Good-bye to All That*, 1929; *Occupation Writer*, 1950), historical fiction (*I, Claudius*, 1934; *Claudius the God*, 1934), and non-fiction (*The White Goddess*, 1948). All of his writing reflects his keen interest in classics and mythology.

Graves[2] /ɡrɑːv/ *n.* a red or white wine from the district of Graves, to the south of Bordeaux in France.

Graves' disease /greɪvz/ *n.* a disease with characteristic swelling of the neck and protrusion of the eyes, resulting from an overactive thyroid gland; exophthalmic goitre. [Robert J. *Graves*, Ir. physician (1796–1853)]

graveside /ˈgreɪvsaɪd/ *n.* the ground at the edge of a grave.

gravestone /ˈgreɪvstəʊn/ *n.* a stone (usu. inscribed) marking a grave.

Gravettian /grəˈvetɪən/ *adj. & n. Archaeol.* of, relating to, or denoting an upper palaeolithic culture in Europe, following the Aurignacian. [f. *La Gravette*, the type-site in the Dordogne, France]

graveyard /ˈgreɪvjɑːd/ *n.* a burial-ground, esp. by a church.

gravid /ˈgrævɪd/ *adj. literary* or *Zool.* pregnant; carrying eggs or young. [L *gravidus* f. *gravis* heavy]

gravimeter /grəˈvɪmɪtə(r)/ *n.* an instrument for measuring the difference in the force of gravity from one place to another. [F *gravimètre* f. L *gravis* heavy]

gravimetric /ˌɡrævɪˈmetrɪk/ *adj.* **1** of or relating to the measurement of weight. **2** denoting chemical analysis based on weighing reagents and products.

gravimetry /grəˈvɪmɪtrɪ/ *n.* the measurement of weight.

gravitas /ˈgrævɪˌtæs, -ˌtɑːs/ *n.* solemn demeanour; seriousness. [L f. *gravis* serious]

gravitate /ˈgrævɪˌteɪt/ *v.* **1** *intr.* (foll. by *to, towards*) move or be attracted to some source of influence. **2** *tr. & intr.* **a** move or tend by force of gravity towards. **b** sink by or as if by gravity. [mod.L *gravitare* GRAVITAS]

gravitation /ˌgrævɪˈteɪʃ(ə)n/ *n. Physics* **1** a force of attraction between any particle of matter in the universe and any other. **2** the effect of this, esp. the falling of bodies to the earth (see GRAVITY). [mod.L *gravitatio* (as GRAVITY)]

gravitational /ˌgrævɪˈteɪʃən(ə)l/ *adj.* of or relating to gravitation. □ **gravitational constant** the constant in Newton's law of gravitation relating gravity to the masses and separation of particles (symbol: **G**), equal to $6.67 \times 10^{-11} \mathrm{N\ m^2\ kg^{-2}}$. **gravitational field** the region of space surrounding a body in which another body experiences a force of attraction. **gravitational lens** *Astron.* a region of space containing a massive object whose gravitational field distorts electromagnetic radiation passing through it in a similar way to a lens. □ **gravitationally** *adv.*

graviton /ˈgrævɪˌtɒn/ *n. Physics* a hypothetical quantum of gravitational energy. [GRAVITY + -ON]

gravity /ˈgrævɪtɪ/ *n.* **1 a** the force that attracts a body to the centre of the earth or other physical body having mass. (*See note below.*) **b** the degree of intensity of this measured by acceleration. **c** gravitational force. **2** the property of having weight. **3 a** importance, seriousness; the quality of being grave. **b** solemnity, sobriety; serious demeanour. □ **gravity feed** the supply of material by its fall under gravity. [F *gravité* or L *gravitas* f. *gravis* heavy]

■ In Plato's cosmology, gravity is a tendency of like bodies to cluster together. In Aristotle's theory, it is a self-propelling force that draws a body towards the centre of the universe, coinciding with the centre of the earth. In Newton's theory, all bodies in the universe attract each other with a force which depends on the mass of each body and is inversely proportional to the square of the distance between both bodies. After 1916 the Newtonian theory gave way to Einstein's general theory of relativity (see RELATIVITY), but for most terrestrial events the predictions of the two theories do not differ significantly.

gravlax /ˈgrævlæks/ *n.* (also **gravadlax** /ˈgrævədˌlæks/) a Scandinavian dish of dry-cured salmon marinated in herbs. [Sw. *gravlax* from *grav* trench + *lax* salmon, from the former practice of marinating the salmon in a hole in the ground]

gravure /grəˈvjʊə(r)/ *n.* = PHOTOGRAVURE. [abbr.]

gravy /ˈgreɪvɪ/ *n.* (*pl.* **-ies**) **1 a** the juices exuding from meat during and after cooking. **b** a dressing or sauce for food, made from these or from other materials, e.g. stock. **2** *sl.* unearned or unexpected money. □ **gravy-boat** a boat-shaped vessel for serving gravy. **gravy train** *sl.* a source of easy financial benefit. [ME, perh. from a misreading as *gravé* of OF *grané*, prob. f. *grain* spice: see GRAIN]

Gray[1] /greɪ/, Asa (1810–88), American botanist. He was the author of many textbooks which greatly popularized botany. Finding no conflict between evolution and his view of divine design in nature, he supported Darwin's theories at a time when they were anathema to many.

Gray[2] /greɪ/, Thomas (1716–71), English poet. He first gained recognition with the poem 'Elegy Written in a Country Church-Yard' (1751) and this remains his best-known work. His other poems include two Pindaric odes, 'The Bard' (1757) and 'The Progress of Poesy' (1757); these mark a clear transition from neoclassical lucidity towards the obscure and the sublime and are regarded as precursors of romanticism.

gray[1] /greɪ/ *n. Physics* the SI unit of the absorbed dose of ionizing radiation (symbol: **Gy**), corresponding to one joule per kilogram. [Louis H. *Gray*, Engl. radiobiologist (1905–65)]

gray[2] *US var. of* GREY.

grayling /ˈgreɪlɪŋ/ *n.* **1** a silver-grey freshwater fish of the genus *Thymallus*, with a long high dorsal fin. **2** a butterfly, *Hipparchia semele*, having wings with grey undersides and eye-spots on the brown upper side. [*gray* var. of GREY + -LING[1]]

graywacke *US var. of* GREYWACKE[1].

Graz /grɑːts/ a city in southern Austria, on the River Mur, capital of the state of Styria; pop. (1991) 232,155. It is the second largest city in Austria.

graze[1] /greɪz/ *v.* **1** *intr.* (of cattle, sheep, etc.) eat growing grass. **2** *tr.* **a** feed (cattle etc.) on growing grass. **b** feed on (grass). **3** *intr.* pasture cattle. **4** *intr.* esp. *US colloq.* **a** eat snacks or small meals throughout the day. **b** flick rapidly between television channels. **c** casually sample something. □ **grazer** *n.* [OE *grasian* f. *græs* GRASS]

graze[2] /greɪz/ *v. & n.* ● *v.* **1** *tr.* rub or scrape (a part of the body, esp. the skin) so as to break the surface without causing bleeding. **2 a** *tr.* touch lightly in passing. **b** *intr.* (foll. by *against, along,* etc.) move with a light passing contact. ● *n.* an act or instance of grazing; the result of grazing. [perh. a specific use of GRAZE[1], as if 'take off the grass close to the ground' (of a shot etc.)]

grazier /ˈgreɪzɪə(r)/ *n.* **1** a person who feeds cattle for market. **2** *Austral.* a large-scale sheep-farmer or cattle-farmer. □ **graziery** *n.* [GRASS + -IER]

grazing /ˈgreɪzɪŋ/ *n.* grassland suitable for pasturage.

grease /griːs/ *n. & v.* ● *n.* **1** oily or fatty matter esp. as a lubricant. **2** the melted fat of a dead animal. **3** oily matter in unprocessed wool. ● *v.tr.* (also /griːz/) smear or lubricate with grease. □ **grease-gun** a device for pumping grease under pressure to a particular point. **grease the palm of** *colloq.* bribe. **like greased lightning** *colloq.* very fast. □ **greaseless** *adj.* [ME f. AF *grece, gresse*, OF *graisse* ult. f. L *crassus* (adj.) fat]

greasepaint /ˈgriːspeɪnt/ *n.* a waxy composition used as make-up for actors.

greaseproof /ˈgriːspruːf/ *adj.* (esp. of paper) impervious to the penetration of grease.

greaser /ˈgriːsə(r)/ *n.* **1** a person or thing that greases machinery etc. **2** *sl.* a member of a gang of youths with long hair and riding motorcycles. **3** *US sl. offens.* a Mexican or Spanish-American. **4** *sl.* a gentle landing of an aircraft.

greasewood /ˈgriːswʊd/ *n.* a resinous dwarf shrub, esp. *Sarcobatus vermiculatus*, of the western US.

greasy /ˈgriːsɪ, ˈgriːzɪ/ *adj.* (**greasier, greasiest**) **1 a** of or like grease. **b** smeared or covered with grease. **c** containing or having too much grease. **2 a** slippery. **b** (of a person or manner) unpleasantly unctuous, smooth. **c** objectionable. □ **greasy pole a** a pole smeared with grease and difficult to climb or walk on. **b** the path of advancement in a sphere in which status or success is precarious. □ **greasily** *adv.* **greasiness** *n.*

great /greɪt/ *adj., n., & adv.* ● *adj.* **1 a** of a size, amount, extent, or intensity considerably above the normal or average; big (*made a great hole; take great care; lived to a great age*). **b** also with implied surprise, admiration, contempt, etc., esp. in exclamations (*you great idiot!; great stuff!; look at that great wasp*). **c** reinforcing other words denoting size, quantity, etc. (*a great big hole; a great many*). **2** important, pre-eminent; worthy or most worthy of consideration (*the great thing is not to get caught*). **3** grand, imposing (*a great occasion; the great hall*). **4 a** (esp. of a public or historic figure) distinguished; prominent. **b** (**the Great**) as a title denoting the most important of the name (*Alfred the Great*). **5 a** (of a person) remarkable in ability, character, achievement, etc. (*great men; a great thinker*). **b** (of a thing) outstanding of its kind (*the Great Fire*). **6** (foll. by *at, on*) competent, skilled, well-informed. **7** fully deserving the name of; doing a thing habitually or extensively (*a great reader; a great believer in tolerance; not a great one for travelling*). **8** (also **greater**) the larger of the name, species, etc. (*great auk; greater celandine*). **9** (**Greater**) (of a city etc.) including adjacent urban areas (*Greater Manchester*). **10** *colloq.* **a** very enjoyable or satisfactory; attractive, fine (*had a great time; it would be great if we won*). **b** (as an exclamation) fine, very good. **11** (in *comb.*) (in names of family relationships) denoting one degree further removed upwards or downwards (*great-uncle; great-great-grandmother*). ● *n.* **1** a great or outstanding person or thing. **2** (in *pl.*) (**Greats**) *colloq.* (at Oxford University) an honours course or final examinations in classics and philosophy (cf. LITERAE HUMANIORES). ● *adv. colloq.* excellently, well, successfully. □ **great and small** all classes or types. **the great and the good** often *iron.* distinguished and worthy people. **great ape** a large ape of the family Pongidae, closely related to humans, e.g. the gorilla, orang-utan, and chimpanzee. **great circle** a circle on a sphere with the plane of the circle passing through the centre of the sphere. **great crested grebe** a large Old World grebe, *Podiceps cristatus*, with a crest and ear tufts. **great deal** see DEAL[1] *n.* 1. **great-hearted** magnanimous; having a noble or generous mind. **great-heartedness** magnanimity. **the great majority** by far the most. **great northern diver** a large diver, *Gavia immer*, breeding in North America (also called (*N. Amer.*) **common loon**). **great organ** the chief manual in a large organ, with its related pipes and mechanism. **great tit** a Eurasian songbird, *Parus major*, with black and white head markings. **great toe** the big toe. **to a great extent** largely. □ **greatness** *n.* [OE *grēat* f. WG]

Great Australian Bight a wide bay on the south coast of Australia, part of the southern Indian Ocean.

Great Barrier Reef a coral reef in the western Pacific, off the coast of Queensland, Australia. It extends for about 2,000 km (1,250 miles), roughly parallel to the coast, and is the largest coral reef in the world.

Great Basin an arid region of the western US between the Sierra Nevada and the Rocky Mountains, including most of Nevada and parts of the adjacent states.

Great Bear Lake a large lake in the Northwest Territories, Canada. It drains into the Mackenzie River via the Great Bear river.

Great Bible the edition of the English Bible which Thomas Cromwell ordered in 1538 to be set up in every parish church. It was the work of Miles Coverdale, and was first issued in 1539.

Great Britain England, Wales, and Scotland considered as a unit (see also BRITAIN); the name is also often used loosely to refer to the United Kingdom. Wales was politically incorporated with England in the 16th century, and the Act of Union formally united Scotland with England in 1707. In 1801, Great Britain was united with Ireland (from 1921, only Northern Ireland) as the United Kingdom. (See also UNITED KINGDOM.)

Great Charter = MAGNA CARTA.

greatcoat /'greɪtkəʊt/ *n.* a long heavy overcoat.

Great Dane *n.* a breed of large, powerful short-haired dog.

Great Depression see DEPRESSION, THE.

Great Dismal Swamp (also **Dismal Swamp**) an area of swampland in SE Virginia and NE Carolina.

Great Divide[1] **1** see CONTINENTAL DIVIDE. **2** see GREAT DIVIDING RANGE.

Great Divide[2] *n.* (prec. by *the*) the boundary between life and death.

Great Dividing Range (also **Great Divide**) a mountain system in eastern Australia. Curving roughly parallel to the coast, it extends from eastern Victoria to northern Queensland.

Greater Antilles see ANTILLES.

Greater London a metropolitan area comprising central London and the surrounding regions. It is divided administratively into the City of London, thirteen inner London boroughs, and nineteen outer London boroughs.

Greater London Council (abbr. **GLC**) a local government body for the whole of Greater London, established in 1963 as the successor to the London County Council and abolished in 1986.

Greater Manchester a metropolitan county of NW England including the city of Manchester and adjacent areas.

Greater Sunda Islands see SUNDA ISLANDS.

Great Exhibition the first international exhibition of the products of industry, promoted by Prince Albert and held in the Crystal Palace in London in 1851. (See also WORLD FAIR.)

Great Fire (of London) see FIRE OF LONDON.

Great Glen (also called *Glen More*) a valley in Scotland, extending from the Moray Firth south-west for 37 km (60 miles) to Loch Linnhe, and containing Loch Ness. It is traversed by the Caledonian Canal.

Great Grimsby see GRIMSBY.

Great Indian Desert an alternative name for the THAR DESERT.

Great Lakes a group of five large interconnected lakes in central North America, consisting of Lakes Superior, Michigan, Huron, Erie, and Ontario. With the exception of Lake Michigan, which is wholly within the US, they lie on the Canada–US border. They constitute the largest area of fresh water in the world. Connected to the Atlantic Ocean by the St Lawrence Seaway, the Great Lakes form an important commercial waterway.

Great Leap Forward an unsuccessful attempt made in China 1958–60 to hasten the process of industrialization and improve agricultural production. Advocated chiefly by Mao Zedong, it entailed reorganizing the population into large rural collectives and adopting labour-intensive industrial methods to obviate the need for expensive heavy machinery.

greatly /'greɪtlɪ/ *adv.* by a considerable amount; much (*greatly admired; greatly superior*).

Great Mother see MOTHER GODDESS.

Great Nebula *n.* (in full **Great Nebula in Orion**) *Astron.* a bright emission nebula in Orion, visible to the naked eye. It surrounds the multiple star Theta Orionis.

Great Northern War a conflict 1700–21 in which Russia, Denmark, Poland, and Saxony opposed Sweden. The war resulted in Sweden losing her imperial possessions in central Europe, and Russia under Peter the Great becoming a major power in the Baltic.

Great Ouse see OUSE 1.

Great Plague a serious outbreak of bubonic plague in England in 1665–6, in which about one-fifth of the population of London died. The Fire of London (1666) destroyed much of the poor housing in which the disease had flourished, and this outbreak proved to be the last in Britain. (See also BUBONIC PLAGUE.)

Great Plains a vast area of plains to the east of the Rocky Mountains in North America, extending from the valleys of the Mackenzie River in Canada to southern Texas.

Great Rebellion *hist.* the Royalist name for the English Civil War of 1642–51.

Great Rift Valley a large system of rift valleys in eastern Africa and the Middle East. It is the most extensive such system in the world, running for some 4,285 km (3,000 miles) from the Jordan valley in Syria, along the Red Sea into Ethiopia, and through Kenya, Tanzania, and Malawi into Mozambique. It is marked by a chain of lakes and a series of volcanoes, including Mount Kilimanjaro.

Great Russian *n. & adj. hist.* ● *n.* a Russian; the language of Russia. ● *adj.* Russian.

Great Salt Lake a salt lake in northern Utah, near Salt Lake City. With an area of some 2,590 sq km (1,000 sq miles) approx., it is the largest salt lake in North America.

Great Sand Sea an area of desert in NE Africa, on the border between Libya and Egypt.

Great Sandy Desert 1 a large tract of desert in north central Western Australia. **2** an alternative name for the RUB' AL KHALI.

Great Schism 1 the breach between the Eastern and the Western Churches, traditionally dated to 1054, when, in an attempt to assert the primacy of the papacy, Cardinal Humbert excommunicated the patriarch of Constantinople and the latter excommunicated the Western legates. Negotiations to restore unity continued over a long period but the breach became final in 1472. The excommunications of 1054 were mutually abolished as an ecumenical gesture in 1965. (See also FILIOQUE.) **2** the period 1378–1417, when the Western Church was divided by the creation of antipopes. The Council of Constance ended the schism by the election of Martin V in 1417.

Great Seal *n.* a seal used to authenticate important documents issued in the name of the sovereign or highest executive authority. In Britain the Great Seal is held by the Lord Chancellor, who was formerly sometimes referred to as the Lord Keeper (of the Seal); the analogous Great Seal of the US is held by the Secretary of State.

Great Slave Lake a large lake in the Northwest Territories in Canada. The deepest lake in North America, it reaches a depth of 615 m (2,015 ft). The Mackenzie River flows out of it.

Great St Bernard Pass see ST BERNARD PASS.

Great Trek the northward migration 1835–7 of large numbers of Boers, discontented with British rule in the Cape, to the areas where they eventually founded the Transvaal Republic and Orange Free State.

Great Victoria Desert a desert region of Australia, which straddles the boundary between Western Australia and South Australia.

Great Wall of China a fortified wall in northern China, extending some 2,400 km (1,500 miles) from Kansu province to the Yellow Sea north of Beijing. It dates from c.210 BC, when the country was unified under one ruler and the northern walls of existing rival states were linked to form a continuous protection against nomad invaders. It was rebuilt in medieval times largely against the Mongols; the present wall dates from the Ming dynasty. Although principally a defensive wall, it served also as a means of communication: for most of its length it was wide enough to allow five horses to travel abreast.

Great War the First World War.

Great White Way a nickname for BROADWAY.

Great Zimbabwe a complex of stone ruins in a fertile valley in Zimbabwe, about 270 km (175 miles) south of Harare, discovered by Europeans in 1868. Optimistic beliefs that they were King Solomon's Mines or the unidentified biblical city of Ophir led to the staking of many mining claims in the area. The buildings consist of an acropolis, a stone enclosure, and other scattered remains covering an area of 24 hectares (60 acres). The city probably grew up as a focus of trade routes, and in the 14th–15th centuries was the centre of a flourishing civilization. The circumstances of its eventual decline and abandonment are unknown. [Shona *dzimbabwe* walled grave]

greave /gri:v/ n. (usu. in *pl.*) a piece of armour for the shin. [ME f. OF *greve* shin, greave, of unkn. orig.]

grebe /gri:b/ n. a diving waterbird of the family Podicipedidae, with a long neck, lobed toes, and a very short tail. [F *grèbe*, of unkn. orig.]

grebo /'gri:bəʊ/ n. (pl. **-os**) **1** a British urban youth cult favouring heavy metal and punk rock music, and characterized by long hair and an antisocial manner. **2** a member of this. [perh. f. GREASER + *-bo* (after DUMBO etc.)]

Grecian /'gri:ʃ(ə)n/ adj. (of architecture or facial outline) following Greek models or ideals. □ **Grecian nose** a straight nose that continues the line of the forehead without a dip. [OF *grecien* or med.L *graecianus* (unrecorded) f. L *Graecia* Greece]

Grecism var. of GRAECISM.

Grecize var. of GRAECIZE.

Greco- var. of GRAECO-.

Greco, El see EL GRECO.

Greece /gri:s/ a country in SE Europe; pop. (1991) 10,269,000; official language, Greek; capital, Athens. Greece consists of a mountainous peninsula bounded by the Ionian, Mediterranean, and Aegean Seas, and many outlying islands, of which the largest are Crete, Euboea, Lesbos, Chios, and Rhodes. The region was invaded by Greek-speaking peoples c.2000–1700 BC, and thereafter enjoyed settled conditions which allowed the Mycenaean civilization to develop and flourish. The arrival in the 11th or 12th century of the Dorians ushered in a dark age (see DARK AGES). This was followed by the period of the city-states, of which the most prominent were Athens and Sparta. Although the city-states could combine in a crisis, as against the Persian invaders whom they unexpectedly defeated in 480 BC, they were weakened by rivalry and internal conflict. In 338 BC they fell to Philip II of Macedon and formed part of the empire of his successor Alexander the Great. In 146 BC Greece was made a Roman province, and, on the division of the Roman Empire in AD 395, became part of the Eastern Empire, centred on Constantinople. It was conquered by the Ottoman Turks (1466) and remained under Turkish rule until the war of independence (1821–30), after which it became a kingdom. In the Second World War Greece was occupied by Axis forces (1941–4); their withdrawal was followed by civil war between Communists and Royalists. The monarchy was restored with British assistance in 1946 and the Communists were finally defeated in 1949. The monarchy was overthrown in a military coup in 1967; a civilian republic was established in 1974. Greece joined the EEC in 1981.

greed /gri:d/ n. intense or excessive desire, esp. for food or wealth. [back-form. f. GREEDY]

greedy /'gri:dɪ/ adj. (**greedier, greediest**) **1** having or showing an excessive appetite for food or drink. **2** wanting wealth or pleasure to excess. **3** (foll. by *for*, or *to* + infin.) very keen or eager; needing intensely (*greedy for affection; greedy to learn*). □ **greedy-guts** *colloq.* a glutton. □ **greedily** adv. **greediness** n. [OE *grǣdig* f. Gmc]

Greek /gri:k/ n. & adj. ● n. **1 a** a native or national of modern Greece; a person of Greek descent. **b** a native or citizen of any of the ancient states of Greece. **2** the ancient or modern language of Greece, the only certain representative of the Greek or Hellenic branch of the Indo-European family. (*See note below.*) ● adj. of Greece or its people or language; Hellenic. □ **Greek cross** a cross with four equal arms. **Greek to me** *colloq.* incomprehensible to me. □ **Greekness** n. [OE *Grēcas* (pl.) f. Gmc f. L *Graecus* Greek f. Gk *Graikoi*, the prehistoric name of the Hellenes (in Aristotle)]

▪ The ancient form of Greek was spoken in the Balkan peninsula from the 2nd millennium BC; the earliest evidence is to be found in the Linear B tablets dating from c.1400–1200 BC. Like Latin, it was a highly inflected language with rather complicated grammar. The Greek alphabet, used from the first millennium BC onwards, was adapted from the Phoenician alphabet. There were a number of dialects, but with the rise of Athens in classical times the dialect of that city (Attic) predominated and formed the basis of what became the standard dialect (*koinē*) from the 3rd century BC onwards. During the period of the Byzantine Empire and in the four centuries when Greece was under Turkish rule (when Greek was the official language) the *koinē* continued to be used as a literary language, while the colloquial language evolved separately. Today Greek is spoken by some 11.5 million people in Greece, Cyprus, the US, and elsewhere, and is the official language of Greece. Modern Greek has changed in various ways: the spelling is very conservative but vowels and consonants are now very different from those of ancient Greek, and the morphology has been simplified. Next to demotic, the colloquial modern language which is the basis of the newly emerging standard, there is a puristic archaizing form of the language, katharevousa, which at various periods in the past had considerable importance.

Greek fire n. *hist.* a combustible composition for setting fire to an enemy's ships, works, etc., emitted by a flame-throwing weapon, so called from being first used by the Greeks besieged in Constantinople (673–8). It ignited on contact with water, and was probably based on naphtha and quicklime.

Greek Orthodox Church (also **Greek Church**) the national Church of Greece (see ORTHODOX CHURCH).

green /gri:n/ adj., n., & v. ● adj. **1** of the colour between blue and yellow in the spectrum; coloured like grass, emeralds, etc. **2 a** covered with leaves or grass. **b** mild and without snow (*a green Christmas*). **3** (of fruit etc. or wood) unripe or unseasoned. **4** not dried, smoked, or tanned. **5** inexperienced, naive, gullible. **6 a** (of the complexion) pale, sickly-hued. **b** jealous, envious. **7** young, flourishing. **8** not withered or worn out (*a green old age*). **9** (also **Green**) concerned with or supporting protection of the environment as a political principle. **10** *archaic* fresh; not healed (*a green wound*). ● n. **1** a green colour or pigment. **2** green clothes or material (*dressed in green*). **3 a** a piece of public or common grassy land (*village green*). **b** a grassy area used for a special purpose (*putting-green; bowling-green*). **c** Golf a putting-green. **d** Golf a fairway. **4** (in *pl.*) green vegetables. **5** vigour, youth, virility (*in the green*). **6 a** a green light. **7** a green ball, piece, etc., in a game or sport. **8** (also **Green**) a member or supporter of an environmentalist group or party. **9** (in *pl.*) *sl.* sexual intercourse. **10** *sl.* low-grade marijuana. **11** *sl.* money. **12** green foliage or growing plants. ● v. **1** *tr.* & *intr.* make or become green. **2** *tr. sl.* hoax; take in. □ **green belt** an area of open land round a city, on which building is restricted. **green card 1** an international insurance document for motorists. **2** *US* a permit enabling a foreign national to live and work permanently in the US. **green cheese 1** cheese coloured green with sage. **2** whey cheese. **3** unripened cheese. **green crop** a crop used as fodder in a green state rather than as hay etc. **green drake** the common mayfly. **green earth** a hydrous silicate of potassium, iron, and other metals. **green-eyed** jealous. **the green-eyed monster** jealousy. **green-fee** *Golf* a charge for playing one round on a course. **green-fingered** skilled in growing plants. **green fingers** skill in growing plants. **green goose** a goose killed under four months old and eaten without stuffing. **green in a person's eye** a sign of gullibility (*do you see any green in my eye?*). **green leek** *Austral.* a green parakeet, esp. *Polytelis swainsoni*. **green light 1** a green light used as a signal to proceed on a road, railway, etc. **2** *colloq.* permission to go ahead with a project. **green linnet** the greenfinch. **green man** *colloq.* a symbol of a walking figure illuminated green on a pedestrian crossing to indicate a time to cross. **green manure** growing plants ploughed into the soil as fertilizer. **green plover** the lapwing. **green pound** the exchange rate for the pound for payments for agricultural produce in the EC. **green revolution** greatly increased crop production in the developing countries by improvement of soil fertility, pest control, increased mechanization, etc. **green-room** a room in a theatre or television studio for actors and actresses who are off-stage. **green shoots** signs of growth or renewal, esp of economic recovery. **green-stick fracture** a bone-fracture, esp. in children, in which one side of the bone is broken and one only bent. **green tea** tea made from steam-dried, not fermented, leaves. **green thumb** = *green fingers.* **green turtle** a green-shelled sea turtle, *Chelonia mydas*, highly regarded as food. **green vitriol** ferrous sulphate crystals. **green woodpecker** a large green and yellow European woodpecker, *Picus viridis*, with a red crown and a laughing call. □ **greenish** *adj.* **greenly** *adv.* **greenness** /ˈgriːnnɪs/ *n.* [OE *grēne* (adj. & n.), *grēnian* (v.), f. Gmc, rel. to GROW]

Greenaway /ˈgriːnəˌweɪ/, Catherine ('Kate') (1846–1901), English artist. She is known especially for her illustrations of children's books such as *Under the Windows* (1879) and *Mother Goose* (1881). An annual award for the best children's book illustration in Britain is named after her.

greenback /ˈgriːnbæk/ *n.* **1** *US* a US legal-tender note; the US dollar. **2** a green-backed animal.

Green Beret *n. colloq.* a British or American commando.

greenbottle /ˈgriːnˌbɒt(ə)l/ *n.* a green-bodied blowfly of the genus *Lucilia*; esp. *L. sericata*, which lays eggs in the flesh of sheep.

Green Cloth (in full **Board of Green Cloth**) (in the UK) the Lord Steward's department of the Royal Household.

Greene /griːn/, (Henry) Graham (1904–91), English novelist. He became a Roman Catholic in 1926; the moral paradoxes of his faith underlie much of his work. Well-known works which explore religious themes include *Brighton Rock* (1938), *The Power and the Glory* (1940), and *Travels with My Aunt* (1969). Among his other novels are thrillers, which Greene classed as 'entertainments', such as his first successful novel, *Stamboul Train* (1932), and *The Third Man* (1950); the latter was originally written as a screenplay and filmed in 1949.

greenery /ˈgriːnərɪ/ *n.* green foliage or growing plants.

greenfeed /ˈgriːnfiːd/ *n. Austral. & NZ* forage grown to be fed fresh to livestock.

greenfield /ˈgriːnfiːld/ *n.* (*attrib.*) (of a site, in terms of its potential development) having no previous building development on it.

greenfinch /ˈgriːnfɪntʃ/ *n.* a finch, *Carduelis chloris*, native to Eurasia and North Africa, with green and yellow plumage.

greenfly /ˈgriːnflaɪ/ *n.* (*pl.* same or **-flies**) *Brit.* a green aphid.

greengage /ˈgriːngeɪdʒ/ *n.* a roundish green fine-flavoured variety of plum. [Sir William *Gage* (1657–1727), who introduced it to England]

greengrocer /ˈgriːnˌgrəʊsə(r)/ *n. Brit.* a retailer of fruit and vegetables.

greengrocery /ˈgriːnˌgrəʊsərɪ/ *n.* (*pl.* **-ies**) *Brit.* **1** the business of a greengrocer. **2** goods sold by a greengrocer.

greenhead /ˈgriːnhed/ *n.* **1** a green-eyed biting fly of the family Tabanidae. **2** an Australian ant, *Ectatomma metallicum*, with a painful sting.

greenheart /ˈgriːnhɑːt/ *n.* **1** a tropical American evergreen tree, *Ocotea rodiaei*, of the laurel family. **2** the hard greenish wood of this tree.

greenhide /ˈgriːnhaɪd/ *n. Austral.* the untanned hide of an animal.

greenhorn /ˈgriːnhɔːn/ *n.* an inexperienced or foolish person; a new recruit.

greenhouse /ˈgriːnhaʊs/ *n.* a light structure with the sides and roof mainly of glass, for rearing delicate plants or hastening the growth of plants. □ **greenhouse gas** a gas that contributes to the greenhouse effect, e.g. carbon dioxide.

greenhouse effect *n.* the trapping of solar heat in a planet's lower atmosphere due to the greater transparency of the atmosphere to visible radiation from the sun than to infrared radiation from the planet's surface. Such an effect is responsible for the surprisingly high surface temperature (460°C) of the planet Venus, whose dense carbon dioxide (CO_2) atmosphere is opaque to infrared radiation. On earth, small amounts of CO_2 and water vapour naturally present in air act in a similar but limited way. However, the increasing quantity of atmospheric CO_2 from the burning of fossil fuels, together with the release of other greenhouse gases (notably chlorofluorocarbons) and the effects of deforestation, are expected to cause a significant rise in world temperatures with unforeseen climatic and environmental effects. (See also GLOBAL WARMING.)

greenie /ˈgriːnɪ/ *n. colloq.* a person concerned about environmental issues. [f. GREEN]

greening[1] /ˈgriːnɪŋ/ *n.* **1** the process or result of making something green. **2** the planting of trees etc. in urban or desert areas. **3** the process of becoming or making aware of ecological issues. [f. GREEN]

greening[2] /ˈgriːnɪŋ/ *n.* a variety of apple that is green when ripe. [prob. f. MDu. *groeninc* (as GREEN)]

greenkeeper /ˈgriːnˌkiːpə(r)/ *n.* the keeper of a golf course.

Greenland /ˈgriːnlənd/ (Danish **Grønland** /ˈgrœnlan/; called in Inupiaq *Kalaallit Nunaat*) a large island lying to the north-east of North America and mostly within the Arctic Circle; pop. (est. 1988) 54,800; capital, Nuuk (Godthåb). Only five per cent of Greenland is habitable; the population is largely Inupiaq. Visited in 986 by the Norse explorer Eric the Red, Greenland became successively a Norse and (from 1721) a Danish settlement. It became a dependency of Denmark in 1953, with internal autonomy from 1979. Greenland withdrew from the EEC in 1985. □ **Greenlander** *n.*

Greenland Sea a sea which lies between the east coast of Greenland and the islands of Svalbard, forming part of the Arctic Ocean.

greenlet /ˈgriːnlɪt/ *n.* = VIREO.

greenmail /ˈgriːnmeɪl/ *n. Stock Exch.* the practice of buying enough shares in a company to threaten a takeover, thereby forcing the other shareholders to buy them back at a higher price in order to retain control of the business. □ **greenmailer** *n.* [GREEN + BLACKMAIL]

Greenock /ˈgriːnək/ a port in west central Scotland, on the Firth of Clyde; pop. (1981) 57,300.

Green Paper *n.* (in the UK) a preliminary report of government proposals, for discussion.

Green Party *n.* an environmentalist political party. Green Parties arose in Europe in the early 1970s, since when they have achieved a certain amount of electoral success, particularly in Germany. They are frequently organized without a recognized leader, reflecting their commitment to decentralization. The Green Party in Britain was founded in 1973 as the Ecology Party, changing its name in 1985; it has not won any parliamentary seats.

Greenpeace /ˈgriːnpiːs/ an international organization that campaigns actively but non-violently for conservation of the environment and the preservation of endangered species. It was first active in British Columbia in 1971, opposing US testing of nuclear weapons in Alaska; in 1985 its ship *Rainbow Warrior*, which had been campaigning against French nuclear tests in the Pacific, was sunk by the French secret service.

greensand /ˈgriːnsænd/ *n.* **1** a greenish kind of sandstone, often imperfectly consolidated. **2** a stratum largely formed of this sandstone.

greenshank /ˈgriːnʃæŋk/ *n.* a large sandpiper, *Tringa nebularia*, with greenish legs, breeding in northern Eurasia.

greenstone /ˈgriːnstəʊn/ *n.* **1** a greenish igneous rock containing feldspar and hornblende. **2** a variety of jade found in New Zealand, used for tools, ornaments, etc.

greenstuff /ˈgriːnstʌf/ *n.* vegetation; green vegetables.

greensward /ˈgriːnswɔːd/ *n. archaic* or *literary* **1** grassy turf. **2** an expanse of this.

greenweed /ˈgriːnwiːd/ *n.* = *dyer's greenweed.*

Greenwich /ˈgrenɪtʃ, ˈgrɪnɪdʒ/ a London borough on the south bank of the Thames. Greenwich was the original site of the Royal Greenwich Observatory, which was founded in 1675 but moved elsewhere in 1948 (see ROYAL GREENWICH OBSERVATORY). The buildings at Greenwich, together with many of the old instruments, now form part of the National Maritime Museum.

Greenwich Mean Time (abbr. **GMT**) the mean solar time at the Greenwich meridian, adopted as the standard time in a zone that includes the British Isles. It is now known formally as Universal Time. (See TIME.)

Greenwich meridian the prime meridian, which passes through the former Royal Observatory at Greenwich. It was adopted internationally as the zero of longitude in 1884.

Greenwich Village /ˈgrenɪtʃ/ a district of New York City on the lower west side of Manhattan. Formerly a separate village, it is a residential area traditionally associated with writers, artists, and musicians.

greenwood /ˈɡriːnwʊd/ *n.* woodland in summer, esp. as the scene of outlaw life.

greeny /ˈɡriːnɪ/ *adj.* greenish (*greeny-yellow*).

greenyard /ˈɡriːnjɑːd/ *n. Brit.* an enclosure for stray animals, a pound.

Greer /ɡrɪə(r)/, Germaine (b.1939), Australian feminist and writer. She first achieved recognition with her influential book *The Female Eunuch* (1970), an analysis of women's subordination in a male-dominated society. She has since become a high-profile figure in the women's movement; other books include *The Change* (1991), about social attitudes to female ageing.

greet[1] /ɡriːt/ *v.tr.* **1** address politely or welcomingly on meeting or arrival. **2** receive or acknowledge in a specified way (*was greeted with derision*). **3** (of a sight, sound, etc.) become apparent to or noticed by. □ **greeter** *n.* [OE *grētan* handle, attack, salute f. WG]

greet[2] /ɡriːt/ *v.intr. Sc.* weep. [OE *grētan, grēotan,* of uncert. orig.]

greeting /ˈɡriːtɪŋ/ *n.* **1** the act or an instance of welcoming or addressing politely. **2** words, gestures, etc., used to greet a person. **3** (often in *pl.*) an expression of goodwill. □ **greetings card** a decorative card sent to convey greetings.

gregarious /ɡrɪˈɡeərɪəs/ *adj.* **1** fond of company. **2** living in flocks or communities. **3** growing in clusters. □ **gregariously** *adv.* **gregariousness** *n.* [L *gregarius* f. *grex gregis* flock]

Gregorian calendar /ɡrɪˈɡɔːrɪən/ *n.* the modified calendar introduced by Pope Gregory XIII in 1582, adopted in Scotland in 1600 and in England and Wales in 1752, and now in use throughout most of the Western world. It is a modification of the Julian calendar, necessary to bring it into closer conformity with astronomical data. To correct errors which had accumulated because the average Julian year of 365¼ days was 11 min. 10 sec. longer than the solar year, 10 days were suppressed in 1582 (in England and Wales, 11 in 1752). As a further refinement Gregory provided that of the centenary years (1600, 1700, etc.) only those exactly divisible by 400 should be counted as leap years. Dates reckoned by the Gregorian calendar are sometimes designated 'New Style'. [med.L *Gregorianus* f. LL *Gregorius* f. Gk *Grēgorios* Gregory]

Gregorian chant /ɡrɪˈɡɔːrɪən/ *n.* plainsong ritual music or chant, in the Roman Catholic Church generally known as 'Gregorian' after Pope Gregory the Great (see GREGORY, ST), although it is likely that his contribution was one of organization and standardization rather than composition.

Gregorian telescope /ɡrɪˈɡɔːrɪən/ *n. hist.* a reflecting telescope in which light reflected from a concave elliptical secondary mirror is brought to a focus just behind a hole in the primary mirror. It was soon rendered obsolete by the introduction of Newtonian and Cassegrain telescopes. [James *Gregory,* Sc. mathematician (1638–75), who devised it]

Gregory, St /ˈɡreɡərɪ/, (known as St Gregory the Great) (*c.*540–604), pope (as Gregory I) 590–604 and Doctor of the Church. He made peace with the Lombards after their invasions of Italy and appointed governors to the Italian cities, thus establishing the temporal power of the papacy. He sent St Augustine to England to lead the country's conversion to Christianity. He is also credited with the introduction of Gregorian chant. Feast day, 12 Mar.

Gregory of Nazianzus, St /ˈɡreɡərɪ/ (329–89), Doctor of the Church, bishop of Constantinople. With St Basil and St Gregory of Nyssa he was an upholder of Orthodoxy against the Arian and Apollinarian heresies, and influential in restoring adherence to the Nicene Creed. Feast day, (in the Eastern Church) 25 and 30 Jan.; (in the Western Church) 2 Jan. (formerly 9 May).

Gregory of Nyssa, St /ˈɡreɡərɪ, ˈnɪsə/ (*c.*330–*c.*395), Doctor of the Eastern Church, bishop of Nyssa in Cappadocia. The brother of St Basil, he was an Orthodox follower of Origen and joined with St Basil and St Gregory of Nazianzus in opposing Arianism. Feast day, 9 Mar.

Gregory of Tours, St /ˈɡreɡərɪ, tʊə(r)/ (*c.*540–94), Frankish bishop and historian. He was elected bishop of Tours in 573; his writings provide the chief authority for the early Merovingian period of French history. Feast day, 17 Nov.

gregory-powder /ˈɡreɡərɪˌpaʊdə(r)/ *n. hist.* a compound powder of rhubarb, magnesia, and ginger, used as a laxative. [James *Gregory,* Sc. physician (1758–1822)]

Gregory the Great see GREGORY, ST.

gremlin /ˈɡremlɪn/ *n. colloq.* **1** an imaginary mischievous sprite regarded as responsible for mechanical faults, esp. in aircraft. **2** any similar cause of trouble. [20th c.: orig. unkn., but prob. after *goblin*]

Grenada /ɡrəˈneɪdə/ a country in the West Indies, consisting of the island of Grenada (the southernmost of the Windward Islands) and the southern Grenadines; pop. (est. 1990) 94,000; languages, English (official), English Creole; capital, St George's. The island of Grenada was sighted in 1498 by Columbus; it was inhabited by Caribs, who had wiped out the earlier Arawak people. Colonized by the French, it was ceded to Britain in 1763, recaptured by the French, and restored to Britain in 1783. It became an independent Commonwealth state in 1974. Seizure of power by a left-wing military group in 1983 prompted an invasion by the US and some Caribbean countries; they withdrew in 1985. □ **Grenadian** *adj. & n.*

grenade /ɡrɪˈneɪd/ *n.* **1** a small bomb thrown by hand (in full **hand-grenade**) or launched mechanically. **2** a glass receptacle containing chemicals which disperse on impact, for testing drains, extinguishing fires, etc. [F f. OF *grenate* and Sp. *granada* POMEGRANATE]

grenadier /ˌɡrenəˈdɪə(r)/ *n.* **1 a** *Brit.* (**Grenadiers** or **Grenadier Guards**) the first regiment of the royal household infantry. **b** *hist.* a soldier armed with grenades. **2** a deep-sea fish of the family Macrouridae, with a long tapering body and pointed tail, and becoming luminescent when disturbed. [F (as GRENADE)]

grenadilla var. of GRANADILLA.

grenadine[1] /ˈɡrenəˌdiːn/ *n.* a French cordial syrup of pomegranates etc. [F f. *grenade:* see GRENADE]

grenadine[2] /ˈɡrenəˌdiːn/ *n.* a dress-fabric of loosely woven silk or silk and wool. [F, earlier *grenade* grained silk f. *grenu* grained]

Grenadine Islands (also **Grenadines**) a chain of small islands in the West Indies, part of the Windward Islands. They are divided administratively between St Vincent and Grenada.

Grendel /ˈɡrend(ə)l/ a water-monster killed by Beowulf in the Old English epic poem *Beowulf.*

Grenoble /ɡrəˈnəʊb(ə)l/ a city in SE France; pop. (1991) 153,970. It is an important industrial city and a winter-sports centre.

Grenville /ˈɡrenvɪl/, George (1712–70), British Whig statesman, Prime Minister 1763–5. The American Stamp Act (1765), which aroused great opposition in the North American colonies, was passed during his term of office.

Gresham /ˈɡreʃəm/, Sir Thomas (*c.*1519–79), English financier. He founded the Royal Exchange in 1566 and served as the chief financial adviser to the Elizabethan government.

Gresham's law *n. Econ.* the tendency for money of lower intrinsic value to circulate more freely than money of higher intrinsic and equal nominal value (often expressed as 'Bad money drives out good'). Formulation of this principle is attributed to Sir Thomas Gresham.

Gresley /ˈɡrezlɪ/, Sir (Herbert) Nigel (1876–1941), British railway engineer. He became locomotive engineer of the Great Northern Railway in 1911, continuing as chief mechanical engineer on the newly formed London and North Eastern Railway from 1923. He is most famous for designing express steam locomotives, such as the A3 class exemplified by *Flying Scotsman.* His A4 pacifics hauled the first British streamlined train service in 1935, and in 1938 one of these engines, *Mallard,* achieved a world speed record of 126 m.p.h., never surpassed by a steam locomotive.

gressorial /ɡreˈsɔːrɪəl/ *adj. Zool.* that walks; adapted for walking. [mod.L *gressorius* f. L *gradi gress-* walk]

Gretna Green /ˈɡretnə/ a village in Scotland just north of the English border near Carlisle, formerly a popular place for runaway couples from England to be married according to Scots law without the parental consent required in England for people under a certain age. A valid marriage could be contracted in Scotland merely by a declaration of consent by the two parties before a witness. The practice of eloping to Gretna Green began after English law made it impossible to conduct a marriage clandestinely (in 1753), but became less common when a series of laws passed from 1857 diminished the differences in Scottish and English marital law; the age of consent to marriage was made uniform (at 18) throughout the UK in 1969.

Gretzky /ˈɡretskɪ/, Wayne (b.1961), Canadian ice-hockey player. He made his professional début in 1978 with the Indianapolis Racers, soon moving to the Edmonton Oilers. A prolific scorer, from 1980 to 1987 he was voted Most Valuable Player (MVP) in the National Hockey League.

Greuze /ɡrɜːz/, Jean-Baptiste (1725–1805), French painter. He first gained recognition with his narrative genre painting *A Father Reading* *the Bible to his Children* (1755). Much of his later work, such as *The Broken Pitcher* (*c.*1773), consisted of pictures of young women, often in *décolleté* dress.

grew *past* of GROW.

Grey[1] /greɪ/, Charles, 2nd Earl (1764–1845), British statesman, Prime Minister 1830–4. He was an advocate of electoral reform and his government passed the first Reform Act (1832) as well as important factory legislation and the Act abolishing slavery throughout the British Empire.

Grey[2] /greɪ/, Sir George (1812–98), British statesman and colonial administrator, Prime Minister of New Zealand 1877–9. He was appointed as governor in South Australia (1840), New Zealand (1845; 1861), and Cape Colony (1854), in each case at a time of conflict between the native peoples and European settlers. As Prime Minister of New Zealand he brought peace to the country and established good relations with the Maoris, learning their language and studying their mythology and culture.

Grey[3] /greɪ/, Lady Jane (1537–54), queen of England 9–19 July 1553. In 1553 Jane, the daughter of Henry VIII's sister, was forced by the Duke of Northumberland to marry his son to ensure a Protestant succession. Northumberland then persuaded the dying Edward VI to name Jane as his successor, but she was deposed after nine days on the throne by forces loyal to Edward's elder sister Mary. Jane was executed in the following year because she was seen as a potential focal point for Protestant opposition to Mary's Catholic regime.

grey /greɪ/ *adj., n., & v.* (*US* **gray**) ● *adj.* **1** of a colour intermediate between black and white, as of ashes or lead. **2 a** (of the weather etc.) dull, dismal; heavily overcast. **b** bleak, depressing; (of a person) depressed. **3 a** (of hair) turning white with age etc. **b** (of a person) having grey hair. **4** anonymous, nondescript, unidentifiable. ● *n.* **1 a** a grey colour or pigment. **b** grey clothes or material (*dressed in grey*). **2** a cold sunless light. **3** a grey or white horse. ● *v.tr. & intr.* make or become grey. □ **grey area 1** an ill-defined intermediate category or situation. **2** *S. Afr. hist.* an area where black and coloured people lived illicitly alongside white. **3** *Brit.* an area in economic decline. **grey economy** the part of the economy not accounted for in official statistics. **grey eminence** = ÉMINENCE GRISE. **grey goose** a goose of the genus *Anser*, with mainly grey plumage, esp. the greylag. **grey-hen** the female of the black grouse (cf. BLACKCOCK). **grey market** unofficial trade in controlled or scarce goods, or esp. in unissued shares. **grey matter 1** the darker tissues of the brain and spinal cord consisting of nerve cell bodies and branching dendrites. **2** *colloq.* intelligence. **grey mullet** a mullet of the family Mugilidae, usu. found near coasts and having a thick body and blunt head, and often used as food. **grey seal** a common large seal, *Halichoerus grypus*, found in the North Atlantic. **grey squirrel** an American squirrel, *Sciurus carolinensis*, introduced in Europe in the 19th century. **grey suit** see SUIT *n.* 1c. □ **greyish** *adj.* **greyly** *adv.* **greyness** *n.* [OE *græg* f. Gmc]

greybeard /ˈgreɪbɪəd/ *n.* **1** an old man. **2** a large stoneware jug for spirits. **3** = *traveller's joy.*

Grey Friar *n.* a Franciscan friar.

greyhound /ˈgreɪhaʊnd/ *n.* a tall slender breed of dog having keen sight and capable of high speed, used since ancient times for hunting small game and now chiefly in racing and coursing. [OE *grīghund* f. *grīeg* bitch (unrecorded: cf. ON *grey*) + *hund* dog, rel. to HOUND]

greyhound racing *n.* a spectator sport in which greyhounds race around a circular or oval track, in pursuit of a mechanical hare. Bets are laid on the outcome of races. Greyhound racing developed from hare-coursing in the early 20th century, in the US, but is now popular chiefly in the UK and Australia. Also called *dog-racing*.

greylag /ˈgreɪlæg/ *n.* (in full **greylag goose**) a wild grey goose, *Anser anser*, native to Europe. It is the ancestor of the farmyard goose. [18th c., prob. f. GREY + obs. dial. *lag* goose]

greywacke /ˈgreɪˌwækə, -wæk/ *n.* (*US* **graywacke**) *Geol.* a dark coarse-grained sandstone, usu. with an admixture of clay. [anglicized f. G *Grauwacke* f. *grau* grey: see WACKE]

grid /grɪd/ *n.* **1** a framework of spaced parallel bars; a grating. **2** a system of numbered squares printed on a map and forming the basis of map references. **3** a network of lines, electric-power connections, gas-supply lines, etc. **4** a pattern of lines marking the starting-places on a motor-racing track. **5** the wire network between the filament and the anode of a thermionic valve etc. **6** an arrangement of town streets in a rectangular pattern. □ **grid bias** *Electr.* a fixed voltage applied between the cathode and the control grid of a thermionic valve which

determines its operating conditions. □ **gridded** *adj.* [back-form. f. GRIDIRON]

griddle /ˈgrɪd(ə)l/ *n. & v.* ● *n.* **1** a circular metal plate placed over a fire or otherwise heated for baking, toasting, etc. **2** a miner's wire-bottomed sieve. ● *v.tr.* **1** cook with a griddle; grill. **2** sieve with a griddle. [ME f. OF *gredil, gridil* gridiron ult. f. L *craticula* dimin. of *cratis* hurdle; cf. GRATE[2], GRILL[1]]

gridiron /ˈgrɪdˌaɪən/ *n.* **1** a cooking utensil of metal bars for broiling or grilling. **2** a frame of parallel beams for supporting a ship in dock. **3** an American football field (with parallel lines marking out the area of play). **4** *Theatr.* a plank structure over a stage supporting the mechanism for drop curtains etc. **5** = GRID 6. [ME *gredire*, var. of *gredil* GRIDDLE, later assoc. with IRON]

gridlock /ˈgrɪdlɒk/ *n.* **1** a traffic jam affecting a network of streets, caused by continuous queues of intersecting traffic. **2** = DEADLOCK *n.* 1. □ **gridlocked** *adj.*

grief /griːf/ *n.* **1** deep or intense sorrow or mourning. **2** the cause of this. **3** *colloq.* trouble, annoyance. □ **come to grief** meet with disaster; fail. **good grief!** an exclamation of surprise, alarm, etc. [ME f. AF *gref*, OF *grief* f. *grever* GRIEVE[1]]

Grieg /griːg/, Edvard (1843–1907), Norwegian composer, conductor, and violinist. He took much of his inspiration from Norwegian folk music, as in many of his songs and the incidental music to Ibsen's play *Peer Gynt* (1876). He avoided the larger forms of opera and symphony in favour of songs, orchestral suites, and violin sonatas. Other famous works include the Piano Concerto in A minor (1869).

Grierson /ˈgrɪəs(ə)n/, John (1898–1972), Scottish film director and producer. His pioneering work in British documentary film-making is represented by films for the Empire Marketing Board (1928–33) and the GPO Film Unit (1933–6); his work for the latter includes *Night Mail* (1936), with a verse commentary by W. H. Auden. Grierson is also notable for establishing the National Film Board of Canada (1937) and for his television series *This Wonderful World* (1957–68).

grievance /ˈgriːv(ə)ns/ *n.* a real or fancied cause for complaint. [ME, = injury, f. OF *grevance* (as GRIEVE[1])]

grieve[1] /griːv/ *v.* **1** *tr.* cause grief or great distress to. **2** *intr.* suffer grief, esp. at another's death. □ **griever** *n.* [ME f. OF *grever* ult. f. L *gravare* f. *gravis* heavy]

grieve[2] /griːv/ *n. Sc.* a farm-bailiff; an overseer. [OE *græfa*: cf. REEVE[1]]

grievous /ˈgriːvəs/ *adj.* **1** (of pain etc.) severe. **2** causing grief or suffering. **3** injurious. **4** flagrant, heinous. □ **grievous bodily harm** *Law* serious injury inflicted intentionally on a person. □ **grievously** *adv.* **grievousness** *n.* [ME f. OF *grevos* (as GRIEVE[1])]

griffin /ˈgrɪfɪn/ *n.* (also **gryphon** /-f(ə)n/) a fabulous creature with an eagle's head and wings and a lion's body. [ME f. OF *grifoun* ult. f. LL *gryphus* f. L *gryps* f. Gk *grups*]

Griffith[1] /ˈgrɪfɪθ/, Arthur (1872–1922), Irish nationalist leader and statesman, President of the Irish Free State 1922. In 1905 he founded and became president of Sinn Fein. Griffith was among those who established the unofficial Irish Parliament in 1919, becoming Vice-President of the newly declared republic in the same year. He led the Irish delegation during negotiations for the Anglo-Irish Treaty (1921), and was elected President of the Irish Free State in 1922, only to die in office several months later.

Griffith[2] /ˈgrɪfɪθ/, D(avid) W(ark) (1875–1948), American film director. A significant figure in the history of film, he began to discover the elements of cinematic expression in his early one-reel films, and is responsible for introducing the techniques of flashback and fade-out. Notable films include his epic of the American Civil War *The Birth of a Nation* (1915), *Intolerance* (1916), and *Broken Blossoms* (1919). He made only two sound films, in 1930 and 1931, and then retired.

griffon /ˈgrɪf(ə)n/ *n.* **1** a small terrier-like breed of dog with coarse or smooth hair. **2** (in full **griffon vulture**) a large vulture, *Gyps fulvus*, found in Eurasia and North Africa. [F (in sense 1) or var. of GRIFFIN]

grig /grɪg/ *n.* **1** a small eel. **2** a grasshopper or cricket. □ **merry** (or **lively**) **as a grig** full of fun; extravagantly lively. [ME, orig. = dwarf: orig. unkn.]

Grignard reagent /ˈgriːnjɑː(r)/ *Chem.* any of a class of organometallic magnesium compounds, used in organic syntheses. [Victor *Grignard*, Fr. organic chemist (1871–1935)]

grike /graɪk/ *n.* (also **gryke**) *Geol.* a fissure between clints in limestone, enlarged through erosion by rainwater. [18th c. north. dial., of unkn. orig.]

grill[1] /grɪl/ *n. & v.* ● *n.* **1 a** a device on a cooker for radiating heat downwards. **b** = GRIDIRON 1. **2** a dish of food cooked on a grill (*mixed grill*). **3** (in full **grill room**) a restaurant serving grilled food. ● *v.* **1** *tr.* & *intr.* cook or be cooked under a grill or on a gridiron. **2** *tr.* & *intr.* subject or be subjected to extreme heat, esp. from the sun. **3** *tr.* subject to severe questioning or interrogation. □ **griller** *n.* **grilling** *n.* (in sense 3 of *v.*). [F *gril* (n.), *griller* (v.), f. OF forms of GRILLE]

grill² var. of GRILLE.

grillage /ˈɡrɪlɪdʒ/ *n.* a heavy framework of cross-timbering or metal beams forming a foundation for building on difficult ground. [F (as GRILLE)]

grille /ɡrɪl/ *n.* (also **grill**) **1** a grating or latticed screen, used as a partition or to allow discreet vision. **2** a metal grid protecting the radiator of a motor vehicle etc. **3** (in real tennis) a square opening in the wall at the end of the hazard side (a ball entering it scores a point). [F f. OF *graille* f. med.L *graticula*, *craticula*: see GRIDDLE]

grilse /ɡrɪls/ *n.* a young salmon that has returned to fresh water from the sea for the first time. [ME: orig. unkn.]

grim /ɡrɪm/ *adj.* (**grimmer**, **grimmest**) **1** of a stern or forbidding appearance. **2 a** harsh, merciless, severe. **b** resolute, uncompromising (*grim determination*). **3** ghastly, joyless, sinister (*has a grim truth in it*). **4** unpleasant, unattractive. □ **like grim death** with great determination. □ **grimly** *adv.* **grimness** *n.* [OE f. Gmc]

grimace /ˈɡrɪməs, ɡrɪˈmeɪs/ *n.* & *v.* ● *n.* a distortion of the face made in disgust etc. or to amuse. ● *v.intr.* make a grimace. □ **grimacer** *n.* [F f. Sp. *grimazo* f. *grima* fright]

Grimaldi¹ /ɡrɪˈmældɪ/, Francesco Maria (1618–63), Italian Jesuit physicist and astronomer, discoverer of the diffraction of light. He verified Galileo's law of the uniform acceleration of falling bodies, drew a detailed map of the moon, and began the practice of naming lunar features after astronomers and physicists.

Grimaldi² /ɡrɪˈmældɪ/, Joseph (1779–1837), English circus entertainer. He created the role of the clown in the circus; it was in his honour that later clowns were nicknamed Joey. From 1806 until his retirement in 1823 he performed at Covent Garden, where he became famous for his acrobatic skills.

grimalkin /ɡrɪˈmælkɪn, -ˈmɔːlkɪn/ *n. archaic* **1** an old she-cat. **2** a spiteful old woman. [GREY + *Malkin* dimin. of the name *Matilda*]

grime /ɡraɪm/ *n.* & *v.* ● *n.* soot or dirt ingrained in a surface, esp. of buildings or the skin. ● *v.tr.* blacken with grime; befoul. [orig. as verb: f. MLG & MDu.]

Grimm /ɡrɪm/, Jacob (Ludwig Carl) (1785–1863) and Wilhelm (Carl) (1786–1859), German philologists and folklorists. Jacob produced a historical German grammar (1819, 1822) and in 1852 the brothers jointly inaugurated a dictionary of German on historical principles; it was continued by other scholars and completed in 1960. The brothers are also remembered for the anthology of German fairy tales which they compiled; this appeared in three volumes between 1812 and 1822.

Grimm's law *n. Linguistics* the statement, first fully formulated by Jacob Grimm in the second edition of his German grammar (1822), that certain Indo-European consonants (mainly the stops) undergo regular changes in the Germanic languages (e.g. *p* becoming *f*, *k* becoming *h*), while other languages, such as Greek and Latin, are more conservative.

Grimond /ˈɡrɪmənd/, Joseph ('Jo'), Baron (1913–93), British Liberal politician. As leader of the Liberal Party (1956–67), he advocated British membership of the European Economic Community and sought to make the Liberal Party the only radical alternative to Conservatism.

Grimsby /ˈɡrɪmzbɪ/ (official name **Great Grimsby**) a port in Humberside, on the south shore of the Humber estuary; pop. (1991) 88,900. It is a major fishing port and a centre of the frozen food industry.

grimy /ˈɡraɪmɪ/ *adj.* (**grimier**, **grimiest**) covered with grime; dirty. □ **grimily** *adv.* **griminess** *n.*

grin /ɡrɪn/ *v.* & *n.* ● *v.* (**grinned**, **grinning**) **1** *intr.* smile broadly, esp. in an unrestrained, forced, or stupid manner. **2** *tr.* express by grinning (*grinned his satisfaction*). ● *n.* the act or action of grinning. □ **grin and bear it** take pain or misfortune stoically. □ **grinner** *n.* **grinningly** *adv.* [OE *grennian* f. Gmc]

grind /ɡraɪnd/ *v.* & *n.* ● *v.* (*past* and *past part.* **ground** /ɡraʊnd/) **1 a** *tr.* (esp. as **ground** *adj.*) reduce to small particles or powder by crushing esp. by passing through a mill. **b** *intr.* (of a mill, machine, etc.) move with a crushing action. **2 a** *tr.* reduce, sharpen, or smooth by friction. **b** *tr.* & *intr.* rub or rub together gratingly (*grind one's teeth*). **3** *tr.* (often foll. by *down*) oppress; harass with exactions (*grinding poverty*). **4** *intr.* **a** (often foll. by *away*) work or study hard. **b** (foll. by *out*) produce with effort

(*grinding out verses*). **c** (foll. by *on*) (of a sound) continue gratingly or monotonously. **5** *tr.* turn the handle of (a barrel-organ etc.). **6** *intr. sl.* (of a dancer) rotate the hips. **7** *intr. coarse sl.* have sexual intercourse. ● *n.* **1** the act or an instance of grinding. **2** *colloq.* hard dull work; a laborious task (*the daily grind*). **3** the size of ground particles. **4** a dancer's rotary movement of the hips. **5** *coarse sl.* an act of sexual intercourse. □ **grind to a halt** stop laboriously. **ground down** exhausted, worn down. **ground glass 1** glass made non-transparent by grinding etc. **2** glass ground to a powder. □ **grindingly** *adv.* [OE *grindan*, of unkn. orig.]

grinder /ˈɡraɪndə(r)/ *n.* **1** a person or thing that grinds, esp. a machine (often in *comb.*: *coffee-grinder*; *organ-grinder*). **2** a molar tooth.

grindstone /ˈɡraɪndstəʊn/ *n.* **1** a thick revolving disc used for grinding, sharpening, and polishing. **2** a kind of stone used for this. □ **keep one's nose to the grindstone** work hard and continuously.

gringo /ˈɡrɪŋɡəʊ/ *n.* (pl. **-os**) *colloq.* a foreigner, esp. a British or North American person, in a Spanish-speaking country. [Sp., = gibberish]

grip /ɡrɪp/ *v.* & *n.* ● *v.* (**gripped**, **gripping**) **1 a** *tr.* grasp tightly; take a firm hold of. **b** *intr.* take a firm hold, esp. by friction. **2** *tr.* (of a feeling or emotion) deeply affect (a person) (*was gripped by fear*). **3** *tr.* compel the attention or interest of (*a gripping story*). ● *n.* **1 a** a firm hold; a tight grasp or clasp. **b** a manner of grasping or holding. **2** the power of holding attention. **3 a** mental or intellectual understanding or mastery. **b** effective control of a situation or one's behaviour etc. (*lose one's grip*). **4 a** a part of a machine that grips or holds something. **b** a part or attachment by which a tool, implement, weapon, etc., is held in the hand. **5** = HAIRGRIP. **6** a travelling bag. **7** an assistant in a theatre, film studio, etc. **8** *Austral. sl.* a job or occupation. □ **come** (or **get**) **to grips with** approach (a task, an opponent, etc.) purposefully; begin to deal with. **get a grip** (**on oneself**) keep or recover one's self-control. **in the grip of** dominated or affected by (esp. an adverse circumstance or unpleasant sensation). **lose one's grip** lose control. □ **gripper** *n.* **grippingly** *adv.* **grippy** *adj.* [OE *gripe*, *gripa* handful (as GRIPE)]

gripe /ɡraɪp/ *v.* & *n.* ● *v.* **1** *intr. colloq.* complain, esp. peevishly. **2** *tr.* affect with gastric or intestinal pain. **3** *tr. archaic* clutch, grip. **4** *Naut.* **a** *tr.* secure with gripes. **b** *intr.* turn to face the wind in spite of the helm. ● *n.* **1** (usu. in *pl.*) gastric or intestinal pain; colic. **2** *colloq.* **a** a complaint. **b** the act of griping. **3** a grip or clutch. **4** (in *pl.*) *Naut.* lashings securing a boat in its place. □ **Gripe Water** *propr.* a carminative solution to relieve colic and stomach ailments in infants. □ **griper** *n.* **gripingly** *adv.* [OE *grīpan* f. Gmc: cf. GROPE]

grippe /ɡrɪp/ *n. archaic* or *colloq.* influenza. [F f. *gripper* seize]

Gris /ɡriː/, Juan (born José Victoriano Gonzales) (1887–1927), Spanish painter. His association with Braque and Picasso in Paris during their early cubist period influenced his approach, although his chief concern was not the analytical treatment of form. His main contribution was to the development of the later phase of synthetic cubism. His work, such as *The Sunblind* (1914), uses collage and paint in simple fragmented shapes which are arranged to create interplays and contrasts of textures, colours, and forms.

grisaille /ɡrɪˈzeɪl, -ˈzaɪl/ *n.* **1** a method of painting in grey monochrome, often to imitate sculpture. **2** a painting or stained-glass window of this kind. [F f. *gris* grey]

griseofulvin /ˌɡrɪzɪəˈfʊlvɪn/ *n.* an antibiotic used against fungal infections of the hair and skin. [mod.L (*Penicillium*) *griseofulvum* (mould from which obtained) f. med.L *griseus* grey + L *fulvus* reddish-yellow]

grisette /ɡriːˈzet/ *n.* a young working-class Frenchwoman. [F, orig. a grey dress-material, f. *gris* grey]

grisly /ˈɡrɪzlɪ/ *adj.* (**grislier**, **grisliest**) causing horror, disgust, or fear. □ **grisliness** *n.* [OE *grislic* terrifying]

grison /ˈɡraɪs(ə)n, ˈɡrɪz(ə)n/ *n.* a weasel-like mammal of the genus *Galictis*, native to Central and South America, with dark fur and a white stripe across the forehead. [F, app. f. *grison* grey]

grist /ɡrɪst/ *n.* **1** corn to grind. **2** malt crushed for brewing. □ **grist to the** (or **a person's**) **mill** a source of profit or advantage. [OE f. Gmc, rel. to GRIND]

gristle /ˈɡrɪs(ə)l/ *n.* tough cartilaginous tissue esp. as occurring in meat. □ **gristly** /-slɪ/ *adj.* [OE *gristle*]

grit /ɡrɪt/ *n.* & *v.* ● *n.* **1** particles of stone or sand, esp. as causing discomfort, clogging machinery, etc. **2** coarse sandstone. **3** *colloq.* pluck, endurance; strength of character. ● *v.* (**gritted**, **gritting**) **1** *tr.* spread grit on (icy roads etc.). **2** *tr.* clench (the teeth). **3** *intr.* make or move with a grating sound. □ **gritter** n. **gritty** adj. (**grittier, grittiest**). **grittily** adv. **grittiness** n. [OE grēot f. Gmc: cf. GRITS, GROATS]

grits /grɪts/ n.pl. **1** esp. US coarsely ground grain, esp. hominy. **2** oats that have been husked but not ground. [OE grytt(e): cf. GRIT, GROATS]

Grivas /ˈgriːvəs/, George (Theodorou) (1898–1974), Greek-Cypriot patriot and soldier. A lifelong supporter of the union of Cyprus with Greece, he led the guerrilla campaign against British rule in Cyprus during the 1950s, which culminated in the country's independence in 1959. Grivas was rewarded for his role in this by promotion to general in the Greek army. He returned to Cyprus in 1971 to organize guerrilla opposition to President Makarios and died a fugitive.

grizzle /ˈgrɪz(ə)l/ v.intr. Brit. colloq. **1** (esp. of a child) cry fretfully. **2** complain whiningly. □ **grizzler** n. **grizzly** adj. [19th c.: orig. unkn.]

grizzled /ˈgrɪz(ə)ld/ adj. having, or streaked with, grey hair. [grizzle grey f. OF grisel f. gris grey]

grizzly /ˈgrɪzlɪ/ adj. & n. ● adj. (**grizzlier, grizzliest**) grey, greyish, grey-haired. ● n. (pl. **-ies**) (in full **grizzly bear**) a large variety of brown bear found in North America.

groan /grəʊn/ v. & n. ● v. **1 a** intr. make a deep sound expressing pain, grief, or disapproval. **b** tr. utter with groans. **2** intr. complain inarticulately. **3** intr. (usu. foll. by under, beneath, with) be loaded or oppressed. ● n. the sound made in groaning. □ **groan inwardly** be distressed. □ **groaner** n. **groaningly** adv. [OE grānian f. Gmc, rel. to GRIN]

groat /grəʊt/ n. hist. **1** a silver coin worth four old pence. **2** archaic a small sum (don't care a groat). [ME f. MDu. groot, orig. = great, i.e. thick (penny): cf. GROSCHEN]

groats /grəʊts/ n.pl. hulled or crushed grain, esp. oats. [OE grotan (pl.): cf. grot fragment, grēot GRIT, grytt bran]

grocer /ˈgrəʊsə(r)/ n. a dealer in food and household provisions. [ME & AF grosser, orig. one who sells in the gross, f. OF grossier f. med.L grossarius (as GROSS)]

grocery /ˈgrəʊsərɪ/ n. (pl. **-ies**) **1** a grocer's trade or shop. **2** (in pl.) provisions, esp. food, sold by a grocer.

grockle /ˈgrɒk(ə)l/ n. dial. & sl. a visitor or holiday-maker, esp. from the North or Midlands to SW England. [invented word, orig. a character in a children's comic.]

Grodno see HRODNA.

grog /grɒg/ n. **1** a drink of spirit (orig. rum) and water. **2** Austral. & NZ colloq. alcoholic liquor, esp. beer. [said to be from 'Old Grog', the reputed nickname (f. his GROGRAM cloak) of Admiral Sir Edward Vernon (1684–1757), who in 1740 first had diluted instead of neat rum served out to sailors]

groggy /ˈgrɒgɪ/ adj. (**groggier, groggiest**) incapable or unsteady from being dazed or semi-conscious. □ **groggily** adv. **grogginess** n.

grogram /ˈgrɒgrəm/ n. a coarse fabric of silk, or of mohair and wool, or of a mixture of all these, often stiffened with gum. [F gros grain coarse grain (as GROSS, GRAIN)]

groin[1] /grɔɪn/ n. & v. ● n. **1** the depression between the belly and the thigh. **2** Archit. **a** an edge formed by intersecting vaults. **b** an arch supporting a vault. ● v.tr. Archit. build with groins. [ME grynde, perh. f. OE grynde depression]

groin[2] US var. of GROYNE.

grommet /ˈgrɒmɪt/ n. (also **grummet** /ˈgrʌm-/) **1** a metal, plastic, or rubber eyelet placed in a hole to protect or insulate a rope or cable etc. passed through it. **2** Med. a small plastic tube surgically implanted in the eardrum to make a communication with the middle ear. [obs. F grommette f. gourmer to curb, of unkn. orig.]

gromwell /ˈgrɒmwəl/ n. a plant of the genus Lithospermum, of the borage family, with hard seeds formerly used in medicine. [ME f. OF gromil, prob. f. med.L gruinum milium (unrecorded) crane's millet]

Gromyko /grəˈmiːkəʊ/, Andrei (Andreevich) (1909–89), Soviet statesman, President of the USSR 1985–8. Born in Russia, he pursued a career in the Soviet diplomatic service. He was appointed Foreign Minister in 1957, a post which he held until becoming President in 1985. As Foreign Minister he represented the Soviet Union abroad throughout most of the cold war. His appointment to the presidency (at that time largely a formal position) by Gorbachev was widely interpreted as a manoeuvre to reduce Gromyko's influence and make possible an ending of the cold war; he eventually retired in 1988.

Groningen /ˈgrəʊnɪŋən/ a city in the northern Netherlands, capital of a province of the same name; pop. (1991) 168,700.

Grønland see GREENLAND.

groom /gruːm/ n. & v. ● n. **1** a person employed to take care of horses. **2** = BRIDEGROOM. **3** Brit. Mil. any of certain officers of the Royal Household. ● v.tr. **1 a** tend to, esp. brush the coat of (a horse, dog, etc.). **b** (of an animal) clean and comb the fur of (another) (also refl.). **2 a** give a neat or tidy appearance to (a person etc.). **b** carefully attend to (a lawn, a ski slope, etc.). **3** prepare or train (a person) for a particular purpose or activity (was groomed for the top job). [ME, orig. = boy: orig. unkn.]

groove /gruːv/ n. & v. ● n. **1 a** a channel or hollow, esp. one made to guide motion or receive a corresponding ridge. **b** a spiral track cut in a gramophone record. **2 a** an established routine or habit. **b** a monotonous routine, a rut. **3** sl. an established rhythmic pattern (got a groove going). ● v. **1** tr. make a groove or grooves in. **2** intr. sl. **a** play music (esp. jazz or dance music) rhythmically. **b** dance or move rhythmically to music. **c** enjoy oneself, relax. □ **in the groove** sl. **1** doing or performing well. **2** fashionable. [ME, = mine-shaft, f. obs. Du. groeve furrow f. Gmc]

groovy /ˈgruːvɪ/ adj. (**groovier, grooviest**) **1** sl. fashionable and exciting; enjoyable, excellent. **2** of or like a groove. □ **groovily** adv. **grooviness** n.

grope /grəʊp/ v. & n. ● v. **1** intr. (usu. foll. by for) feel about or search blindly or uncertainly with the hands. **2** intr. (foll. by for, after) search mentally (was groping for the answer). **3** tr. feel (one's way) towards something. **4** tr. sl. fondle clumsily for sexual pleasure. ● n. the process or an instance of groping. □ **groper** n. **gropingly** adv. [OE grāpian f. Gmc]

groper /ˈgrəʊpə(r)/ n. esp. Austral. & NZ = GROUPER. [var. of GROUPER]

Gropius /ˈgrəʊpɪəs/, Walter (1883–1969), German-born American architect. He was the first director of the Bauhaus School of Design (1919–28) and a pioneer of the international style; his intention was to relate architecture more closely to social needs and to the industrial techniques and modern construction materials on which it increasingly relied. He left Germany in 1934 and eventually settled in the US in 1938; he was professor of architecture at Harvard University until 1952 and designed the Harvard Graduate Center (1949).

grosbeak /ˈgrəʊsbiːk/ n. a finch or cardinal with a stout conical bill and usu. brightly coloured plumage. [orig. = hawfinch, f. F grosbec (as GROSS)]

groschen /ˈgrɒʃ(ə)n/ n. **1** an Austrian coin and monetary unit equal to one-hundredth of a schilling. **2** colloq. a German 10-pfennig piece. **3** hist. a small German silver coin. [G f. MHG gros, grosse f. med.L (denarius) grossus thick (penny): cf. GROAT]

grosgrain /ˈgrəʊgreɪn/ n. a corded fabric of silk etc. [F (see GROGRAM)]

gros point /grəʊ ˈpɔɪnt/ n. cross-stitch embroidery on canvas. [F (as GROSS, POINT)]

gross /grəʊs/ adj., v., & n. ● adj. **1** overfed, bloated; repulsively fat. **2** (of a person, manners, or morals) noticeably coarse, unrefined, or indecent. **3** sl. very unpleasant, repulsive, disgusting. **4** flagrant; conspicuously wrong (gross negligence). **5** total; without deductions; not net (gross tonnage; gross income). **6 a** luxuriant, rank. **b** thick, solid, dense. **7** (of the senses etc.) dull; lacking sensitivity. ● v.tr. produce or earn as gross profit or income. ● n. (pl. same) an amount equal to twelve dozen (144). □ **by the gross** in large quantities; wholesale. **gross domestic product** the total value of goods produced and services provided in a country in one year. **gross national product** the gross domestic product plus the total of net income from abroad. **gross out** N. Amer. sl. disgust, esp. by repulsive or obscene behaviour. **gross up** increase (a net amount) to its value before deductions. □ **grossly** adv. **grossness** n. [ME f. OF gros grosse large f. LL grossus: (n.) f. F grosse douzaine large dozen]

Grosseteste /ˈgrəʊstest/, Robert (c.1175–1253), English churchman, philosopher, and scholar. He taught theology at Oxford before becoming bishop of Lincoln in 1235. His interests were wide-ranging and his experimental approach to science, especially in optics and mathematics, inspired his pupil Roger Bacon. His writings include translations of Aristotle, philosophical treatises, and devotional works.

Grossglockner /ˈgrəʊsˌglɒknə(r)/ the highest mountain in Austria, in the eastern Tyrolean Alps, rising to a height of 3,797 m (12,457 ft).

Grosz /grəʊs/, George (1893–1959), German painter and draughtsman. His revulsion at contemporary bourgeois society was strengthened by his experiences during the First World War. He became prominent in the Neue Sachlichkeit (New Objectivity) movement of the 1920s; his satirical drawings and paintings characteristically depicted a decaying society in which gluttony and depraved sensuality are juxtaposed with poverty and disease. Major works include the paintings Metropolis (1917) and *Suicide* (1916). In 1932 he emigrated to the United States and became an American citizen in 1938.

grot /grɒt/ n. & adj. Brit. sl. ● n. rubbish, junk. ● adj. dirty. [back-form. f. GROTTY]

grotesque /grəʊˈtesk/ adj. & n. ● adj. **1** comically or repulsively distorted; monstrous, unnatural. **2** incongruous, ludicrous, absurd. ● n. **1** a decorative form interweaving human and animal features. **2** a comically distorted figure or design. **3** Printing a family of sanserif typefaces. □ **grotesquely** adv. **grotesqueness** n. **grotesquerie** /-ˈteskəri/ n. [earlier crotesque f. F crotesque f. It. grottesca grotto-like (painting etc.) fem. of grottesco (as GROTTO, -ESQUE)]

Grotius /ˈgrəʊtɪəs/, Hugo (Latinized name of Huig de Groot) (1583–1645), Dutch jurist and diplomat. His fame rests on the legal treatise *De Jure Belli et Pacis*, written in exile in Paris and published in 1625, which established the basis of modern international law.

grotto /ˈgrɒtəʊ/ n. (pl. **-oes** or **-os**) **1** a small picturesque cave. **2** an artificial ornamental cave, e.g. in a park or large garden. □ **grottoed** adj. [It. grotta ult. f. L crypta f. Gk kruptē CRYPT]

grotty /ˈgrɒtɪ/ adj. (**grottier, grottiest**) Brit. sl. unpleasant, dirty, shabby, unattractive. □ **grottiness** n. [shortening of GROTESQUE + -Y¹]

grouch /graʊtʃ/ v. & n. colloq. ● v.intr. grumble. ● n. **1** a habitually discontented or grumpy person. **2** a fit of grumbling or the sulks. **3** a cause of discontent. [var. of grutch: see GRUDGE]

grouchy /ˈgraʊtʃɪ/ adj. (**grouchier, grouchiest**) colloq. discontented, grumpy. □ **grouchily** adv. **grouchiness** n.

ground¹ /graʊnd/ n. & v. ● n. **1 a** the surface of the earth, esp. as contrasted with the air around it. **b** a part of this specified in some way (low ground). **2** the substance of the earth's surface; soil, earth (stony ground; dug deep into the ground). **3 a** area or distance on the earth's surface. **b** the extent of activity etc. achieved or of a subject dealt with (the book covers a lot of ground). **4** (often in pl.) a foundation, motive, or reason (there is ground for concern; there are grounds for believing; excused on the grounds of ill health). **5** an area of a special kind or designated for special use (often in comb.: cricket ground; fishing-grounds). **6** (in pl.) an area of usu. enclosed land attached to a house etc. **7** an area or basis for consideration, agreement, etc. (common ground; on firm ground). **8 a** (in painting) the prepared surface giving the predominant colour or tone. **b** (in embroidery, ceramics, etc.) the undecorated surface. **9** (in full **ground bass**) Mus. a short theme in the bass constantly repeated with the upper parts of the music varied. **10** (in pl.) solid particles, esp. of coffee, forming a residue. **11** Electr. = EARTH n. 4. **12** the bottom of the sea (the ship touched ground). **13** Brit. the floor of a room etc. **14** a piece of wood fixed to a wall as a base for boards, plaster, or joinery. **15** (attrib.) **a** (of animals) living on or in the ground; (of fish) living at the bottom of water; (of plants) dwarfish or trailing. **b** relating to or concerned with the ground (ground staff). ● v. **1** tr. **a** refuse authority for (a pilot or an aircraft) to fly. **b** sl. curtail the activities of (a person, esp. a teenage child), esp. as a punishment. **2 a** tr. run (a ship) aground; strand. **b** intr. (of a ship) run aground. **3** tr. (foll. by in) instruct thoroughly (in a subject). **4** tr. (often as **grounded** adj.) (foll. by on) base (a principle, conclusion, etc.) on. **5** tr. Electr. = EARTH v. 3. **6** intr. alight on the ground. **7** tr. place or lay (esp. weapons) on the ground. □ **break new** (or **fresh**) **ground** treat a subject previously not dealt with. **cut the ground from under a person's feet** anticipate and pre-empt a person's arguments, plans, etc. **down to the ground** Brit. colloq. thoroughly; in every respect. **fall to the ground** (of a plan etc.) fail. **gain** (or **make**) **ground 1** advance steadily; make progress. **2** (foll. by on) catch (a person) up. **get in on the ground floor** become part of an enterprise in its early stages. **get off the ground** colloq. make a successful start. **give** (or **lose**) **ground 1** retreat, decline. **2** lose the advantage or one's position in an argument, contest, etc. **go to ground 1** (of a fox etc.) enter its earth or burrow etc. **2** (of a person) become inaccessible for a prolonged period. **ground-bait** bait thrown to the bottom of a fishing-ground. **ground cherry** a physalis native to America, often with edible berries. **ground control** the personnel or establishment on the ground monitoring and directing the flight of aircraft or spacecraft. **ground cover** plants covering the surface of the earth, esp. low-growing spreading plants that inhibit the growth of weeds. **ground crew** the people who maintain and service an aircraft on the ground. **ground elder** an umbelliferous plant, *Aegopodium podagraria*, spreading by means of underground stems, and common as a weed. **ground floor** the floor of a building at ground level. **ground frost** frost on the surface of the ground or in the top layer of soil. **ground ivy** a common Eurasian labiate hedge-plant, *Glechoma hederacea*, with bluish-purple flowers. **ground level 1** the level of the ground; the ground floor. **2** Physics = ground state. **ground-**

plan 1 the plan of a building at ground level. **2** the general outline of a scheme. **ground-rent** rent for land leased for building. **ground rule** a basic principle. **ground speed** an aircraft's speed relative to the ground. **ground squirrel** a ground-dwelling rodent, esp. of the genus *Spermophilus*, resembling a squirrel. **ground staff 1** the non-flying personnel of an airport or airbase. **2** Brit. a paid staff of esp. young players kept by a cricket club. **ground state** Physics the lowest energy state of an atom etc. **ground stroke** Tennis a stroke played near the ground after the ball has bounced. **ground swell 1** a heavy sea caused by a distant or past storm or an earthquake. **2** an increasingly forceful presence (esp. of public opinion). **ground zero** the point on the ground under an exploding (usu. nuclear) bomb. **hold** (or **stand**) **one's ground** not retreat or give way. **on the ground** at the point of production or operation; in practical conditions. **on one's own ground** on one's own territory or subject; on one's own terms. **thin on the ground** not numerous. **work** (or **run** etc.) **oneself into the ground** colloq. work etc. to the point of exhaustion. □ **grounder** n. [OE grund f. Gmc]

ground² past and past part. of GRIND.

groundage /ˈgraʊndɪdʒ/ n. Brit. duty levied on a ship entering a port or lying on a shore.

grounder /ˈgraʊndə(r)/ n. (esp. in baseball) a ball that is hit or passed along the ground.

groundhog /ˈgraʊndhɒg/ n. N. Amer. = WOODCHUCK.

Groundhog Day (in the US) 2 Feb., when the groundhog is said to come out of its hole at the end of hibernation. If the animal sees its shadow — i.e. if the weather is sunny — it is said to portend six weeks more of winter weather.

grounding /ˈgraʊndɪŋ/ n. basic training or instruction in a subject.

groundless /ˈgraʊndlɪs/ adj. without motive or foundation. □ **groundlessly** adv. **groundlessness** n. [OE grundlēas (as GROUND¹, -LESS)]

groundling /ˈgraʊndlɪŋ/ n. **1 a** a creeping or dwarf plant. **b** a fish that lives at the bottom of lakes etc., esp. a gudgeon or loach. **2** sl. a person on the ground as opposed to one in an aircraft. **3** a spectator or reader of inferior taste (with ref. to Shakespeare's *Hamlet* III. ii. 11).

groundnut /ˈgraʊndnʌt/ n. **1** Brit. = PEANUT 1. **2 a** a North American wild bean, *Apios americana*. **b** its edible tuber.

groundsel /ˈgraʊns(ə)l/ n. a small composite plant of the genus *Senecio*, with yellow rayless flowers; esp. *S. vulgaris*, common as a weed. [OE grundeswylige, gundæswelgiæ (perh. = pus-absorber f. gund pus, with ref. to use for poultices)]

groundsheet /ˈgraʊndʃiːt/ n. a waterproof sheet for spreading on the ground, esp. in a tent.

groundsman /ˈgraʊndzmən/ n. (pl. **-men**) a person who maintains a sports ground.

groundwater /ˈgraʊndˌwɔːtə(r)/ n. water found in soil or in pores, crevices, etc., in rock.

groundwork /ˈgraʊndwɜːk/ n. **1** preliminary or basic work. **2** a foundation or basis.

group /gruːp/ n. & v. ● n. **1** a number of persons or things located close together, or considered or classed together. **2** (attrib.) concerning or done by a group (a group photograph; group sex). **3** a number of people working together or sharing beliefs, e.g. part of a political party. **4** a number of commercial companies under common ownership. **5** an ensemble playing popular music (heavy metal group). **6** a division of an air force or air-fleet. **7** Math. a set of elements, together with an associative binary operation, which contains an inverse for each element and an identity element. **8** Chem. **a** a set of ions or radicals giving a characteristic qualitative reaction. **b** a set of elements occupying a column of the periodic table, with broadly similar properties. **c** a combination of atoms having a recognizable identity in a number of compounds. ● v. **1** tr. & intr. form or be formed into a group. **2** tr. (often foll. by with) place in a group or groups. **3** tr. form (colours, figures, etc.) into a well-arranged and harmonious whole. **4** tr. classify. □ **group captain** an RAF officer next below air commodore. **group dynamics** Psychol. the interaction of people in groups. **group practice** a medical practice in which several doctors are associated. **group therapy** therapy in which patients with a similar condition are brought together to assist one another psychologically. **group-think** the prevailing ideas and attitudes of a group; conformity to the views of a group. **group velocity** the speed of travel of the energy of a wave or wave-group. **group work** work done by a group working in close association. □ **groupage** n. [F *groupe* f. It. *gruppo* f. Gmc, rel. to CROP]

grouper /ˈgruːpə(r)/ n. a large marine fish of the family Serranidae, with a heavy body, big head, and wide mouth. [Port. *garoupa*, prob. f. a S. Amer. name]

groupie /ˈgruːpɪ/ n. sl. **1** a young woman following a pop group etc. and providing sexual favours. **2** derog. an enthusiastic fan or follower; an uncritical hanger-on (*cinema groupie*).

grouping /ˈgruːpɪŋ/ n. **1** a process or system of allocation to groups. **2** a formation or arrangement in a group or groups.

Group of Five the five countries of France, Japan, the UK, the US, and (West) Germany, whose financial representatives, at a meeting in New York in September 1985, agreed to take measures to establish greater stability of international currency.

Group of Seven 1 (abbr. **G7**) the seven leading industrial nations outside the former Communist bloc, i.e. the US, Japan, (West) Germany, France, the UK, Italy, and Canada. Since 1975 their heads of government have met regularly to discuss matters of economic and political importance. **2** a group of Canadian landscape painters, officially established in 1920, who formed the first major national movement in Canadian art. Their work exhibited a bold and colourful expressionistic style.

Group of Seventy-Seven a term sometimes used for the developing countries of the world.

Group of Ten the ten relatively prosperous industrial countries (Belgium, Canada, France, Italy, Japan, the Netherlands, Sweden, (West) Germany, the UK, and the US) who agreed in 1962 to lend money to the IMF to be used to give financial assistance to other members of the fund.

Group of Three the three largest industrialized economies (the US, Germany, and Japan).

groupware /ˈgruːpweə(r)/ n. Computing software designed to facilitate collective working by a number of different users.

grouse¹ /graʊs/ n. (pl. same) **1** a game bird of the family Tetraonidae, with a plump body and feathered legs. **2** the flesh of a grouse used as food. □ **grouse moor** Brit. an area of moorland managed for the shooting of red grouse. [16th c.: orig. uncert.]

grouse² /graʊs/ v. & n. colloq. ● v.intr. grumble or complain pettily. ● n. a complaint. □ **grouser** n. [19th c.: orig. unkn.]

grouse³ /graʊs/ adj. Austral. sl. very good or excellent (*extra grouse*). [20th c.: orig. unkn.]

grout¹ /graʊt/ n. & v. ● n. a thin fluid mortar for filling gaps in tiling etc. ● v.tr. provide or fill with grout. □ **grouter** n. [perh. f. GROUT², but cf. F dial. *grouter* grout a wall]

grout² /graʊt/ n. sediment, dregs. [OE *grūt*, rel. to GRITS, GROATS]

grouter /ˈgraʊtə(r)/ n. Austral. sl. an unfair advantage. [20th c.: orig. uncert.]

Grove /graʊv/, Sir George (1820–1900), English musicologist. He is chiefly remembered as the founder and first editor of the multi-volume *Dictionary of Music and Musicians* (1879–89). He was also instrumental in establishing the Royal College of Music (1883–95) and served as its first director (1883–94).

grove /graʊv/ n. a small wood or group of trees, esp. deliberately planted; an orchard (*olive grove*). □ **grovy** adj. [OE *grāf*, rel. to *grāfa* brushwood]

grovel /ˈgrɒv(ə)l/ v.intr. (**grovelled**, **grovelling**; US **groveled**, **groveling**) **1** behave obsequiously in seeking favour or forgiveness. **2** lie prone in abject humility. □ **groveller** n. **grovelling** adj. **grovellingly** adv. [back-form. f. obs. *grovelling* (adv.) f. *gruf* face down f. *on grufe* f. ON *á grúfu*, later taken as pres. part.]

grow /grəʊ/ v. (past **grew** /gruː/; past part. **grown** /grəʊn/) **1** intr. increase in size, height, quantity, degree, or in any way regarded as measurable (e.g. authority or reputation) (often foll. by in: *grew in stature*). **2** intr. **a** develop or exist as a living plant or natural product. **b** develop in a specific way or direction (*began to grow sideways*). **c** germinate, sprout; spring up. **3** intr. be produced; come naturally into existence; arise. **4** intr. (as **grown** adj.) fully matured; adult (*grown man*). **5** intr. **a** become gradually (*grow rich*; *grow less*). **b** (foll. by to + infin.) come by degrees (*grew to like it*). **6** intr. (foll. by into) **a** become, having grown or developed (*the acorn has grown into a tall oak*; *will grow into a fine athlete*). **b** become large enough for or suited to (*will grow into the coat*; *grew into her new job*). **7** intr. (foll. by on) become gradually more appealing to. **8** tr. **a** produce (plants, fruit, wood, etc.) by cultivation. **b** bring forth.

c allow (a beard etc.) to develop or increase in length. **9** tr. (in passive; foll. by over, up) be covered with a growth. □ **growing bag** a bag containing peat-based potting compost in which plants may be grown. **growing pains 1** early difficulties in the development of an enterprise etc. **2** neuralgic pain in children's legs due to fatigue etc. **grown-up** adj. adult. ● n. an adult person. **grow out of 1** become too large to wear (a garment). **2** become too mature to retain (a childish habit etc.). **3** be the result or development of. **grow together** coalesce. **grow up 1 a** advance to maturity. **b** (esp. in imper.) begin to behave sensibly. **2** (of a custom) arise, become common. □ **growable** adj. [OE *grōwan* f. Gmc, rel. to GRASS, GREEN]

growbag /ˈgrəʊbæg/ n. = growing bag (see GROW). [f. propr. name *Gro-bag*]

grower /ˈgrəʊə(r)/ n. **1** (often in comb.) a person growing produce (*fruit-grower*). **2** a plant that grows in a specified way (*a fast grower*).

growl /graʊl/ v. & n. ● v. **1** intr. **a** (often foll. by at) (esp. of a dog) make a low guttural sound, usu. of anger. **b** murmur angrily. **2** intr. rumble. **3** tr. (often foll. by out) utter with a growl. ● n. **1** a growling sound, esp. made by a dog. **2** an angry murmur; complaint. **3** a rumble. □ **growlingly** adv. [prob. imit.]

growler /ˈgraʊlə(r)/ n. **1** a person or thing that growls, esp. sl. a dog. **2** a small iceberg.

grown past part. of GROW.

growth /grəʊθ/ n. **1** the act or process of growing. **2** an increase in size or value. **3** something that has grown or is growing. **4** Med. a morbid formation, a tumour. **5** the cultivation of produce. **6** a crop or yield of grapes. □ **full growth** the size ultimately attained; maturity. **growth hormone** Biol. a substance which stimulates the growth of a plant or animal. **growth industry** an industry that is developing rapidly. **growth ring** Biol. a concentric layer of wood, shell, etc., developed during an annual or other regular period of growth. **growth stock** etc. stock etc. that tends to increase in capital value rather than yield high income.

groyne /grɔɪn/ n. (US **groin**) a timber framework or low broad wall built out from a shore to check erosion of a beach. [dial. *groin* snout f. OF *groign* f. LL *grunium* pig's snout]

Grozny /ˈgrɒznɪ/ a city in SW Russia, near the border with Georgia, capital of the Chechen Republic; pop. (est. 1990) 401,000. It is a major centre of the oil industry.

grub /grʌb/ n. & v. ● n. **1** the larva of an insect, esp. of a beetle. **2** colloq. food. ● v. (**grubbed**, **grubbing**) **1** tr. & intr. dig superficially. **2** tr. **a** clear (the ground) of roots and stumps. **b** clear away (roots etc.). **3** tr. (foll. by up, out) **a** fetch by digging (*grubbing up weeds*). **b** extract (information etc.) by searching in books etc. **4** intr. search, rummage. **5** intr. (foll. by on, along, away) toil, plod. □ **grub-screw** a small headless screw, esp. used to attach a handle etc. to a spindle. [ME, (v.) perh. corresp. to OE *grybban* (unrecorded) f. Gmc]

grubber /ˈgrʌbə(r)/ n. **1** an implement for digging up weeds etc. **2** (usu. in comb.) derog. a person devoted to amassing something (*money-grubber*). **3** Cricket, Rugby, etc. a ball bowled or kicked along the ground. **4** a person or animal that grubs.

grubby /ˈgrʌbɪ/ adj. (**grubbier**, **grubbiest**) **1** dirty, grimy, slovenly. **2** of or infested with grubs. □ **grubbily** adv. **grubbiness** n.

grubstake /ˈgrʌbsteɪk/ n. & v. N. Amer. colloq. ● n. material or provisions supplied to an enterprise in return for a share in the resulting profits (orig. in prospecting for ore). ● v.tr. provide with a grubstake. □ **grubstaker** n.

Grub Street n. (often attrib.) the world or class of literary hacks and impoverished authors. [name of a street (later Milton St.) in Moorgate, London, inhabited by these in the 17th c.]

grudge /grʌdʒ/ n. & v. ● n. a persistent feeling of ill will or resentment, esp. one due to an insult or injury (*bears a grudge against me*). ● v.tr. **1** be resentfully unwilling to give, grant, or allow (a thing) (*grudged him his success*). **2** (foll. by verbal noun or to + infin.) be reluctant to do (a thing) (*grudged paying so much*). □ **grudger** n. [ME *grutch* f. OF *grouchier* murmur, of unkn. orig.]

grudging /ˈgrʌdʒɪŋ/ adj. reluctant; not willing (*grudging praise*). □ **grudgingly** adv. **grudgingness** n.

gruel /ˈgruːəl/ n. a liquid food of oatmeal etc. boiled in milk or water, esp. as part of an invalid's diet. [ME f. OF, f. Frank.]

gruelling /ˈgruːəlɪŋ/ adj. & n. (US **grueling**) ● adj. extremely demanding, severe, or tiring. ● n. a harsh or exhausting experience; punishing treatment. □ **gruellingly** adv. [GRUEL as verb, = exhaust, punish]

gruesome /ˈgruːsəm/ *adj.* horrible, grisly, disgusting. □ **gruesomely** *adv.* **gruesomeness** *n.* [Sc. *grue* to shudder f. Scand. + -SOME[1]]

gruff /grʌf/ *adj.* **1 a** (of a voice) low and harsh. **b** (of a person) having a gruff voice. **2** surly, laconic, rough-mannered. □ **gruffly** *adv.* **gruffness** *n.* [Du., MLG *grof* coarse f. WG (rel. to ROUGH)]

grumble /ˈgrʌmb(ə)l/ *v.* & *n.* ● *v.* **1** *intr.* **a** (often foll. by *at, about, over*) complain peevishly. **b** be discontented. **2** *intr.* **a** utter a dull inarticulate sound; murmur, growl faintly. **b** rumble. **3** *tr.* (often foll. by *out*) utter complainingly. **4** *intr.* (as **grumbling** *adj.*) *colloq.* giving intermittent discomfort without causing illness (*a grumbling appendix*). ● *n.* **1 a** complaint. **2 a** a dull inarticulate sound; a murmur. **b** a rumble. □ **grumbler** *n.* **grumbling** *adj.* **grumblingly** *adv.* **grumbly** *adj.* [obs. *grumme*: cf. MDu. *grommen*, MLG *grommelen*, f. Gmc]

grummet var. of GROMMET.

grump /grʌmp/ *n. colloq.* **1** a grumpy person. **2** (in *pl.*) a fit of sulks. □ **grumpish** *adj.* **grumpishly** *adv.* [imit.]

grumpy /ˈgrʌmpɪ/ *adj.* (**grumpier, grumpiest**) morosely irritable; surly. □ **grumpily** *adv.* **grumpiness** *n.*

Grundy /ˈgrʌndɪ/ *n.* (*pl.* **-ies**) (in full **Mrs Grundy**) someone embodying conventional propriety and prudery. Mrs Grundy was a person repeatedly mentioned ('What will Mrs Grundy say?') in the comedy *Speed the Plough* (1798) by the English dramatist Thomas Morton (*c.*1764–1838). □ **Grundyism** *n.*

Grünewald /ˈgruːnəˌvælt/, Mathias (born Mathis Nithardt; also called Mathis Gothardt) (*c.*1460–1528), German painter. His most famous work, the nine-panel *Isenheim Altar* (completed 1516), exemplifies his style: figures with twisted limbs, contorted postures, and expressive faces, painted in glowing colour against a dark background.

grunge /grʌndʒ/ *n. esp. US sl.* **1** grime, dirt. **2** a style of rock music characterized by a raucous guitar sound and lazy vocal delivery. □ **grungy** *adj.* [app. after GRUBBY, DINGY, etc.]

grunion /ˈgrʌnjən/ *n.* a slender Californian silverside fish, *Leuresthes tenuis*, that spawns on beaches. [prob. f. Sp. *gruñón* grunter]

grunt /grʌnt/ *n.* & *v.* ● *n.* **1** a low guttural sound made by a pig. **2** a sound resembling this. **3** a tropical marine fish of the family Haemulidae, able to produce a grunting sound by grinding its pharyngeal teeth together. ● *v.* **1** *intr.* make a grunt or grunts. **2** *intr.* (of a person) make a low inarticulate sound resembling this, esp. to express discontent, dissent, fatigue, etc. **3** *tr.* utter with a grunt. [OE *grunnettan*, prob. orig. imit.]

grunter /ˈgrʌntə(r)/ *n.* **1** a person or animal that grunts, esp. a pig. **2** a fish that grunts when caught, esp. = GRUNT *n.* 3.

Gruyère /ˈgruːjeə(r)/ *n.* a firm pale cheese made from cow's milk. [*Gruyère*, a district in Switzerland where it was first made]

gryke var. of GRIKE.

gryphon var. of GRIFFIN.

grysbok /ˈgraɪsbɒk/ *n.* a small straight-horned antelope of the genus *Raphicerus*, native to central and southern Africa. [S. Afr. Du. f. Du. *grijs* grey + *bok* BUCK[1]]

Grytviken /ˈgrɪtˌviːkən/ the chief settlement on the island of South Georgia, in the South Atlantic. It was the site of a whaling station between 1904 and 1966.

gs. *abbr. Brit. hist.* guineas.

Gstaad /gəˈʃtɑːt/ a winter-sports resort in western Switzerland.

G-string /ˈdʒiːstrɪŋ/ *n.* **1** *Mus.* a string sounding the note G. **2** (also **gee-string**) a garment consisting of a string worn round the waist with a narrow strip of material covering the pubic region, worn e.g. by a striptease artist.

G-suit /ˈdʒiːsuːt, -sjuːt/ *n.* a garment with inflatable pressurized pouches, worn by pilots and astronauts to enable them to withstand high acceleration. [g = gravity + SUIT]

GT *abbr.* gran turismo.

Gt. *abbr.* Great.

GTi /ˌdʒiːtiːˈaɪ/ *adj.* designating a high performance car with a fuel-injected engine. [GT + injection]

guacamole /ˌgwɑːkəˈməʊlɪ/ *n.* a dish of puréed avocado pears mixed with chopped onion, tomatoes, chilli peppers, and seasoning. [Amer. Sp. f. Nahuatl *ahuacamolli* f. *ahuacatl* avocado + *molli* sauce]

guacharo /ˈgwɑːtʃəˌrəʊ/ *n.* (*pl.* **-os**) = oil-bird. [S. Amer. Sp.]

Guadalajara /ˌgwɑːdələˈhɑːrə/ **1** a city in central Spain, to the north-east of Madrid; pop. (1991) 67,200. **2** a city in west central Mexico, capital of the state of Jalisco; pop. (1990) 2,846,720.

Guadalcanal /ˌgwɑːdəlkəˈnæl/ an island in the western Pacific, the largest of the Solomon Islands; pop. (est. 1987) 71,300. During the Second World War it was the scene of the first major US offensive against the Japanese (Aug. 1942), bitter fighting on land and at sea continuing for several months before victory was secured.

Guadalquivir /ˌgwɑːdəlkɪˈvɪə(r)/ a river of Andalusia in southern Spain. It flows for 657 km (410 miles) through Cordoba and Seville to reach the Atlantic north-west of Cadiz.

Guadeloupe /ˌgwɑːdəˈluːp/ a group of islands in the Lesser Antilles, forming an overseas department of France; pop. (1991) 386,990; languages, French (official), French Creole; capital, Basse-Terre. □ **Guadeloupian** *adj.* & *n.*

Guadiana /ˌgwɑːdɪˈɑːnə/ a river of Spain and Portugal. Rising in a plateau region south-east of Madrid, it flows south-westwards for some 580 km (360 miles), entering the Atlantic at the Gulf of Cadiz. For the last part of its course it forms the border between Spain and Portugal.

guaiacum /ˈgwaɪəkəm/ *n.* **1** a tree or shrub of the genus *Guaiacum*, native to tropical America. **2** (also **guaiac** /ˈgwaɪæk/) **a** the hard dense oily timber of this, esp. of *G. officinale*. Also called *lignum vitae*. **b** the resin from this used medicinally. [mod.L f. Sp. *guayaco* of Haitian orig.]

Guam /gwɑːm/ the largest and southernmost of the Mariana Islands, administered as an unincorporated territory of the US; pop. (1990) 132,000; languages, English (official), Malayo-Polynesian languages; capital, Agaña. Guam was ceded to the US by Spain in 1898. □ **Guamanian** /gwɑːˈmeɪnɪən/ *adj.* & *n.*

guan /gwɑːn/ *n.* a pheasant-like bird of the family Cracidae, esp. of the genus *Penelope*, found in the rainforests of tropical America. [Amer. Sp. f. Miskito *kwamu*]

guanaco /gwəˈnɑːkəʊ/ *n.* (*pl.* **-os**) a South American mammal, *Lama guanicoe*, related to the llama, with a coat of soft pale-brown hair used for wool. [Sp. f. Quechua *huanacu*]

Guanajuato /ˌgwɑːnəˈhwɑːtəʊ/ **1** a state of central Mexico. **2** its capital city; pop. (est. 1983) 45,000. The city developed as a silver-mining centre after a rich vein of silver was discovered there in 1558.

Guangdong /gwæŋˈdʊŋ/ (also **Kwangtung** /kwæŋˈtʊŋ/) a province of southern China, on the South China Sea; capital, Guangzhou (Canton).

Guangxi Zhuang /ˌgwæŋʃiː ˈʒwæŋ/ (also **Kwangsi Chuang** /ˌkwæŋsiː ˈtʃwæŋ/) an autonomous region of southern China, on the Gulf of Tonkin; capital, Nanning.

Guangzhou /gwæŋˈdʒəʊ/ (also **Canton** /kænˈtɒn/, **Kwangchow** /kwæŋˈtʃaʊ/) a city in southern China, the capital of Guangdong province; pop. (est. 1986) 3,290,000. It is the leading industrial and commercial centre of southern China.

guanine /ˈgwɑːniːn/ *n. Biochem.* a purine found in all living organisms as a component base of DNA and RNA. [GUANO + -INE[4]]

guano /ˈgwɑːnəʊ/ *n.* & *v.* ● *n.* **1** the excrement of seabirds, found esp. on islands off South America and used as manure. **2** artificial manure, esp. that made from fish. ● *v.tr.* (**-oes, -oed**) fertilize with guano. [Sp. f. Quechua *huanu* dung]

Guantánamo Bay /gwɑːnˈtɑːnəˌmaʊ/ a bay on the SE coast of Cuba. It is the site of a US naval base established in 1903.

Guarani /ˌgwɑːrənɪ/ *n.* & *adj.* ● *n.* (*pl.* same or **Guaranis**) **1 a** a member of a native people of South America. **b** the language of this people. **2** (**guarani**) the chief monetary unit of Paraguay. ● *adj.* of or relating to the Guarani or their language. [Sp.]

guarantee /ˌgærənˈtiː/ *n.* & *v.* ● *n.* **1 a** a formal promise or assurance, esp. that an obligation will be fulfilled or that something is of a specified quality and durability. **b** a document giving such an undertaking. **2** = GUARANTY. **3** a person making a guaranty or giving a security. ● *v.tr.* (**guarantees, guaranteed**) **1 a** give or serve as a guarantee for; answer for the due fulfilment of (a contract etc.) or the genuineness of (an article). **b** assure the permanence etc. of. **c** provide with a guarantee. **2** (foll. by *that* + clause, or *to* + infin.) give a promise or assurance. **3 a** (foll. by *to*) secure the possession of (a thing) for a person. **b** make (a person) secure against a risk or in possession of a thing. □ **guarantee fund** a sum pledged as a contingent indemnity for loss. [earlier *garante*, perh. f. Sp. *garante* = F *garant* WARRANT: later infl. by F *garantie* guaranty]

guarantor /ˌgærənˈtɔː(r), ˈgærəntə(r)/ *n.* a person who gives a guarantee or guaranty.

guaranty /ˈgærəntɪ/ *n.* (*pl.* **-ies**) **1** a written or other undertaking to answer for the payment of a debt or for the performance of an obligation by another person liable in the first instance. **2** a thing serving as security for a guaranty. [AF *guarantie*, var. of *warantie* WARRANTY]

guard /gɑːd/ *v. & n.* ● *v.* **1** *tr.* (often foll. by *from*, *against*) watch over and defend or protect from harm. **2** *tr.* keep watch by (a door etc.) so as to control entry or exit. **3** *tr.* supervise (prisoners etc.) and prevent from escaping. **4** *tr.* provide (machinery) with a protective device. **5** *tr.* keep (thoughts or speech) in check. **6** *tr.* provide with safeguards. **7** *intr.* (foll. by *against*) take precautions. **8** *tr.* (in various games) protect (a piece, card, etc.) with set moves. ● *n.* **1** a state of vigilance or watchfulness. **2** a person who protects or keeps watch. serving to protect a place or person; an escort. **4** *Brit.* an official who rides with and is in general charge of a train. **5** a part of an army detached for some purpose (*advance guard*). **6** (in *pl.*) (usu. **Guards**) a body of troops nominally employed to guard a monarch (*Coldstream Guards*). **7** a thing that protects or defends. **8** (often in *comb.*) a device fitted to a machine, vehicle, weapon, etc., to prevent injury or accident to the user (*fire-guard*). **9** *N. Amer.* a prison warder. **10** in some sports: **a** a protective or defensive player. **b** a defensive posture or motion. □ **be on** (or **keep** or **stand**) **guard** act as a sentry; keep watch. **guard cell** *Bot.* either of a pair of cells surrounding the stomata in plants. **guard hair** *Zool.* each of the long coarse hairs forming an animal's outer fur. **guard-rail** a rail, e.g. a handrail, fitted as a support or to prevent an accident. **guard ring** *Electronics* a ring-shaped electrode used to limit the extent of an electric field, esp. in a capacitor. **guard's van** *Brit.* a railway vehicle or compartment occupied by a guard. **lower one's guard** reduce vigilance against attack. **off** (or **off one's**) **guard** unprepared for some surprise or difficulty. **on** (or **on one's**) **guard** prepared for all contingencies; vigilant, cautious. **raise one's guard** become vigilant against attack. [ME f. OF *garde, garder* ult. f. WG, rel. to WARD *n.*]

guardant /ˈgɑːd(ə)nt/ *adj. Heraldry* depicted with the body sideways and the face towards the viewer.

guarded /ˈgɑːdɪd/ *adj.* (of a remark etc.) cautious, avoiding commitment. □ **guardedly** *adv.* **guardedness** *n.*

guardhouse /ˈgɑːdhaʊs/ *n.* a building used to accommodate a military guard or to detain prisoners.

Guardi /ˈgwɑːdɪ/, Francesco (1712–93), Italian painter. He came from a family of artists working in Venice and was a pupil of Canaletto. His paintings of Venice are notable for their free handling of light and atmosphere. His works include *View of S. Giorgio Maggiore* (1775–80).

guardian /ˈgɑːdɪən/ *n.* **1** a defender, protector, or keeper. **2** a person having legal custody of another person and his or her property when that person is incapable of managing his or her own affairs. **3** the superior of a Franciscan convent. □ **guardian angel** a spirit conceived as watching over a person or place. □ **guardianship** *n.* [ME f. AF *gardein*, OF *garden* f. Frank., rel. to WARD, WARDEN]

guardroom /ˈgɑːdruːm, -rʊm/ *n.* a room with the same purpose as a guardhouse.

guardsman /ˈgɑːdzmən/ *n.* (*pl.* **-men**) **1** a soldier belonging to a body of guards. **2** (in the UK) a soldier of a regiment of Guards.

Guarneri /gwɑːˈneərɪ/ Giuseppe ('del Gesù') (1687–1744), Italian violin-maker. The most famous of a family of three generations of violin-makers based in Cremona, he is noted for the attention he gave to the tone quality of his instruments, which do not conform to any standard shape or dimensions.

Guatemala /ˌgwɑːtɪˈmɑːlə/ a country in Central America, bordering on the Pacific Ocean and with a short coastline on the Caribbean Sea; pop. (est. 1990) 9,200,000; official language, Spanish; capital, Guatemala City. A former centre of Mayan civilization, Guatemala was conquered by the Spanish in 1523–4. After independence it formed the core of the short-lived United Provinces of Central America (1828–38) before becoming an independent republic in its own right. □ **Guatemalan** *adj. & n.*

Guatemala City the capital of Guatemala; pop. (est. 1990) 1,900,000. Situated at an altitude of 1,500 m (4,920 ft) in the central highlands, Guatemala City was founded in 1776 to replace the former capital, Antigua Guatemala, destroyed by an earthquake in 1773.

guava /ˈgwɑːvə/ *n.* **1** a small tropical American tree, *Psidium guajava*, bearing an edible pale orange fruit with pink juicy flesh. **2** this fruit. [Sp. *guayaba* prob. f. a S. Amer. name]

Guayaquil /ˌgwaɪəˈkiːl/ a seaport in Ecuador, the country's principal port and second largest city; pop. (1982) 1,300,900.

guayule /gwaɪˈjuːlɪ/ *n.* **1** a silver-leaved shrub, *Parthenium argentatum*,

native to Mexico. **2** a rubber substitute made from the sap of this plant. [Amer. Sp. f. Nahuatl *cuauhuli*]

gubbins /ˈgʌbɪnz/ *n. Brit. colloq.* **1** equipment, paraphernalia; bits and pieces; rubbish. **2** a gadget; an oddment. **3** = MUGGINS 1b. [orig. = fragments, f. obs. *gobbon*: perh. rel. to GOBBET]

gubernatorial /ˌgjuːbənəˈtɔːrɪəl, ˌguː-/ *adj.* esp. *US* of or relating to a governor. [L *gubernator* governor]

gudgeon[1] /ˈgʌdʒən/ *n.* **1** a small European freshwater fish, *Gobio gobio*, often used as bait. **2** a credulous or easily fooled person. [ME f. OF *goujon* f. L *gobio -onis* GOBY]

gudgeon[2] /ˈgʌdʒən/ *n.* **1** any of various kinds of pivot working a wheel, bell, etc. **2** the tubular part of a hinge into which the pin fits to effect the joint. **3** a socket at the stern of a boat, into which a rudder is fitted. **4** a pin holding two blocks of stone etc. together. □ **gudgeon-pin** a pin holding a piston-rod and a connecting-rod together. [ME f. OF *goujon* dimin. of *gouge* GOUGE]

Gudrun /ˈgʊdrʊn/ (in Norse legend) the Norse equivalent of Kriemhild, wife of Sigurd and later of Atli (Attila the Hun).

guelder rose /ˈgeldə(r)/ *n.* a deciduous shrub, *Viburnum opulus*, with round bunches of creamy-white flowers. Also called *snowball tree*. [Du. *geldersch* f. GELDERLAND]

Guelph /gwelf/ *n.* **1** a member of one of the two great factions in Italian medieval politics, traditionally supporting the pope against the Holy Roman emperor (cf. GHIBELLINE). **2** a member of a princely family of Swabian origin from which the British royal house is descended through George I. [It. *Guelfo* f. MHG *Welf* name of a German noble family]

guenon /ɡəˈnɒn/ *n.* an African monkey of the genus *Cercopithecus*, having a characteristic long tail, e.g. the vervet. [F: orig. unkn.]

guerdon /ˈgɜːd(ə)n/ *n. & v. poet.* ● *n.* a reward or recompense. ● *v.tr.* give a reward to. [ME f. OF *guerdon* f. med.L *widerdonum* f. WG *widarlōn* (as WITH, LOAN[1]), assim. to L *donum* gift]

Guericke /ˈgeɪrɪkə/, Otto von (1602–86), German engineer and physicist. He invented an air pump, using it to produce a partial vacuum. He was the first to investigate the properties of a vacuum, and devised the Magdeburg hemispheres to demonstrate atmospheric pressure. Guericke also built the first known electrostatic machine.

Guernica /ɡɜːˈniːkə, ˈɡɜːnɪkə/ (in full **Guernica y Luno** /iː ˈluːnəʊ/) a town in the Basque Provinces of northern Spain, to the east of Bilbao; pop. (1981) 17,840. Formerly the seat of a Basque parliament, it was bombed in 1937, during the Spanish Civil War, by German planes in support of Franco, an event depicted in a famous painting by Picasso.

Guernsey[1] /ˈgɜːnzɪ/ an island in the English Channel, to the north-west of Jersey; pop. (1991) 58,870; capital, St Peter Port. It is the second largest of the Channel Islands.

Guernsey[2] /ˈgɜːnzɪ/ *n.* (*pl.* **-eys**) **1** a breed of dairy cattle from Guernsey. **2** (**guernsey**) **a** a thick oiled sweater of dark blue wool. **b** *Austral.* a football shirt. □ **get a guernsey** *Austral. colloq.* **1** be selected for a football team. **2** gain recognition. **Guernsey lily** a kind of nerine orig. from southern Africa, with large pink lily-like flowers.

Guerrero /ɡeˈreərəʊ/ a state of SW central Mexico, on the Pacific coast; capital, Chilpancingo.

guerrilla /ɡəˈrɪlə/ *n.* (also **guerilla**) a member of a small independently acting (usu. political) group taking part in irregular fighting, esp. against larger regular forces. □ **guerrilla war** (or **warfare**) fighting by or with guerrillas. [Sp. dimin. of *guerra* war]

guess /ges/ *v. & n.* ● *v.* **1** *tr.* (often *absol.*) estimate without calculation or measurement, or on the basis of inadequate data. **2** *tr.* (often foll. by *that* etc. + clause, or *to* + infin.) form a hypothesis or opinion about; conjecture; think likely (*cannot guess how you did it; guess them to be Italian*). **3** *tr.* conjecture or estimate correctly by guessing (*you have to guess the weight*). **4** *intr.* (foll. by *at*) make a conjecture about. ● *n.* an estimate or conjecture reached by guessing. □ **anybody's** (or **anyone's**) **guess** something very vague or difficult to determine. **I guess** *colloq.* I think it likely; I suppose. **keep a person guessing** *colloq.* withhold information. □ **guessable** *adj.* **guesser** *n.* [ME *gesse*, of uncert. orig.: cf. OSw. *gissa*, MLG, MDu. *gissen*: f. the root of GET *v.*]

guess-rope var. of GUEST-ROPE.

guesswork /ˈgeswɜːk/ *n.* **1** the process of guessing. **2** conclusions reached by guessing.

guest /gest/ *n. & v.* ● *n.* **1** a person invited to visit another's house or have a meal etc. at the expense of the inviter. **2** a person lodging at a hotel, boarding-house, etc. **3 a** an outside performer invited to take part with a regular body of performers. **b** a person who takes part by invitation in a radio or television programme (often *attrib.*: *guest artist*). **4** (*attrib.*) **a** a serving or set aside for guests (*guest-room*; *guest-night*). **b** acting as a guest (*guest speaker*). **5** an organism living in close association with another, esp. in an ant nest. ● *v.intr.* be a guest on a radio or television show or in a theatrical performance etc. □ **be my guest** *colloq.* make what use you wish of the available facilities. **guest beer** *Brit.* **1** (in a tied public house) a beer offered in addition to those produced by the brewery. **2** (in a free house) a beer available only temporarily. **guest-house** a private house offering paid accommodation. **guest of honour** the most important guest at an occasion. **guest worker** = GASTARBEITER. □ **guestship** *n.* [ME f. ON *gestr* f. Gmc]

guestimate /ˈgɛstɪmət/ *n.* (also **guesstimate**) *colloq.* an estimate based on a mixture of guesswork and calculation. [GUESS + ESTIMATE]

guest-rope /ˈgɛstrəʊp/ *n.* (also **guess-rope** /ˈgɛsrəʊp/) **1** a second rope fastened to a boat in tow to steady it. **2** a rope slung outside a ship to give a hold for boats coming alongside. [17th c.: orig. uncert.]

Guevara /gəˈvɑːrə/, 'Che' (born Ernesto Guevara de la Serna) (1928–67), Argentinian revolutionary and guerrilla leader. He played a significant part in the Cuban revolution (1956–9) and as a government minister under Castro was instrumental in the transfer of Cuba's traditional economic ties from the US to the Communist bloc. In 1967 he was captured and executed while training guerrillas for a planned uprising against the Bolivian government. He became a hero figure among radical students in the West during the 1960s and early 1970s.

guff /gʌf/ *n. sl.* empty talk; nonsense. [19th c., orig. = 'puff': imit.]

guffaw /gʌˈfɔː/ *n. & v.* ● *n.* a coarse or boisterous laugh. ● *v.* **1** *intr.* utter a guffaw. **2** *tr.* say with a guffaw. [orig. Sc.: imit.]

Guggenheim /ˈgʊgənˌhaɪm/, Meyer (1828–1905), Swiss-born American industrialist. With his seven sons he established large mining and metal-processing companies. His children established several foundations providing support for the arts, including the Guggenheim Foundation, established in 1925 to provide financial support for scholars, artists, and writers. In 1937 Guggenheim's son Solomon (1861–1949) established a foundation for the advancement of art; it now operates the Guggenheim Museum in New York and directs the Guggenheim Collection in Venice.

Guiana /gɪˈɑːnə, gaɪˈænə/ a region in northern South America, bounded by the Orinoco, Negro, and Amazon rivers and the Atlantic Ocean. It now comprises Guyana, Suriname, French Guiana, and the Guiana Highlands of SE Venezuela and northern Brazil.

Guiana Highlands a mountainous plateau region of northern South America, lying between the Orinoco and Amazon river basins, largely in SE Venezuela and northern Brazil. Its highest peak is Roraima (2,774 m; 9,094 ft).

guidance /ˈgaɪd(ə)ns/ *n.* **1 a** advice or information aimed at resolving a problem, difficulty, etc. **b** leadership or direction. **2** the process of guiding or being guided; navigation, esp. by instruments (*inertial guidance*).

guide /gaɪd/ *n. & v.* ● *n.* **1** a person who leads or shows the way, or directs the movements of a person or group. **2** a person who conducts visitors or tourists on tours etc. **3** a professional mountain-climber in charge of a group. **4** an adviser. **5** a directing principle or standard (*one's feelings are a bad guide*). **6** a book with essential information on a subject, esp. = GUIDEBOOK. **7** a thing marking a position or guiding the eye. **8** a soldier, vehicle, or ship whose position determines the movements of others. **9** *Mech.* **a** a bar, rod, etc., directing the motion of something. **b** a gauge etc. controlling a tool. **10** (**Guide**) a member of the Guides Association. ● *v.tr.* **1 a** act as guide to; lead or direct. **b** arrange the course of (events). **2** be the principle, motive, or ground of (an action, judgement, etc.). **3** direct the affairs of (a state etc.). □ **guided missile** a missile directed to its target by remote control or by equipment within itself. **guide-dog** a dog trained to guide a blind person. **guided tour** a tour led by a guide or with a guide in charge. **guide-rope** a rope guiding the movement of a crane, airship, etc. **Queen's** (or **King's**) **Guide** a Guide (sense 10) who has reached the highest rank of proficiency. □ **guidable** *adj.* **guider** *n.* [ME f. OF *guide* (n.), *guider* (v.), earlier *guier* ult. f. Gmc, rel. to WIT²]

guidebook /ˈgaɪdbʊk/ *n.* a book of information about a place for visitors, tourists, etc.

guideline /ˈgaɪdlaɪn/ *n.* a principle or criterion guiding or directing action.

guidepost /ˈgaɪdpəʊst/ *n.* = SIGNPOST.

Guider /ˈgaɪdə(r)/ *n.* an adult leader of Guides (see GUIDE *n.* 10).

Guides Association (in the UK) an organization for girls, corresponding to the Scout Association, established in 1910 by Lord Baden-Powell with his wife Olave and sister Agnes. The three sections into which it is divided, originally Brownies, Guides, and Rangers, are now called Brownie Guides (7–11 years), Guides (10–16 years), and Ranger Guides (14–19 years). The organization was known until 1992 as the Girl Guides Association. Similar organizations exist in many countries worldwide under the aegis of the World Association of Girl Guides and Girl Scouts, formed in 1928.

guideway /ˈgaɪdweɪ/ *n.* a groove or track that guides movement.

guidon /ˈgaɪd(ə)n/ *n.* a pennant narrowing to a point or fork at the free end, esp. one used as the standard of a regiment of dragoons. [F f. It. *guidone* f. *guida* GUIDE]

Guienne see GUYENNE.

Guignol /giːˈnjɒl/ the bloodthirsty chief character in a French puppet-show of that name, similar to Punch and Judy. The term is also used for the theatre where the show is performed. (See also GRAND GUIGNOL.)

guild /gɪld/ *n.* (also **gild**) an association of people for mutual aid or the pursuit of a common goal. Merchant guilds were formed in England from the 11th century (earlier on the Continent), and many became powerful in local government. Trade guilds (associations of persons exercising the same craft) came into prominence in England in the 14th century and became powerful there and elsewhere in medieval Europe, effectively controlling trade and commerce in many countries, particularly in large cities. Their monopolistic policies generally brought them into conflict with strengthening central governments and their powers were gradually curtailed. [ME prob. f. MLG, MDu. *gilde* f. Gmc: rel. to OE *gild* payment, sacrifice]

guilder /ˈgɪldə(r)/ *n.* **1** the chief monetary unit of the Netherlands. **2** *hist.* a gold coin of the Netherlands and Germany. [ME, alt. of Du. *gulden*: see GULDEN]

Guildford /ˈgɪlfəd/ a city in southern England, the county town of Surrey; pop. (1981) 63,090.

guildhall /gɪldˈhɔːl/ *n.* **1** the meeting-place of a guild or corporation; a town hall. **2** (**Guildhall**) the hall of the Corporation of the City of London, used for ceremonial occasions.

guildsman /ˈgɪldzmən/ *n.* (pl. **-men**; *fem.* **guildswoman**, pl. **-women**) a member of a guild.

guile /gaɪl/ *n.* treachery, deceit; cunning or sly behaviour. □ **guileful** *adj.* **guilefully** *adv.* **guileless** *adj.* **guilelessly** *adv.* **guilelessness** *n.* [ME f. OF, prob. f. Gmc]

Guilin /gweɪˈlɪn/ (also **Kweilin** /kweɪ-/) a city in southern China, on the Li river, in the autonomous region of Guangxi Zhuang; pop. (est. 1984) 446,900.

guillemot /ˈgɪlɪˌmɒt/ *n.* a narrow-billed auk of the genus *Uria* or *Cepphus*, nesting on cliffs or islands. [F f. *Guillaume* William]

guilloche /gɪˈlɒʃ/ *n.* an architectural or metalwork ornament imitating braided ribbons. [F *guillochis* (or *guilloche* the tool used)]

guillotine /ˈgɪləˌtiːn/ *n. & v.* ● *n.* **1** a machine with a heavy knife-blade sliding vertically in grooves, used for executing people by beheading (esp. in France during the Revolution). **2** a device for cutting paper, metal, etc. by means of a descending or sliding blade. **3** a surgical instrument with a sliding blade for excising the tonsils etc. **4** *Parl.* a method of preventing delay in the discussion of a legislative bill by fixing times at which various parts of it must be voted on. ● *v.tr.* **1** use a guillotine on. **2** *Parl.* end discussion of (a bill) by applying a guillotine. □ **guillotiner** *n.* [F f. Joseph-Ignace *Guillotin*, Fr. physician (1738–1814), who recommended its use for executions in 1789]

guilt /gɪlt/ *n.* **1** the fact of having committed a specified or implied offence. **2 a** culpability. **b** the feeling of this. □ **guilt complex** *Psychol.* a mental obsession with the idea of having done wrong. [OE *gylt*, of unkn. orig.]

guiltless /ˈgɪltlɪs/ *adj.* **1** (often foll. by *of* an offence) innocent. **2** (foll. by *of*) not having knowledge or possession of. □ **guiltlessly** *adv.* **guiltlessness** *n.* [OE *gyltlēas* (as GUILT, -LESS)]

guilty /ˈgɪltɪ/ *adj.* (**guiltier**, **guiltiest**) **1** culpable of or responsible for a wrong. **2** conscious of or affected by guilt (*a guilty conscience*; *a guilty look*). **3** concerning guilt (*a guilty secret*). **4** (often foll. by *of*) **a** having committed a (specified) offence. **b** *Law* adjudged to have committed a specified offence, esp. by a verdict in a trial (*was found guilty of manslaughter*). □ **guiltily** *adv.* **guiltiness** *n.* [OE *gyltig* (as GUILT, -Y¹)]

guimp var. of GIMP¹.

Guinea /ˈgɪnɪ/ a country on the west coast of Africa; pop. (est. 1990)

6,909,300; languages, French (official), Fulani, Malinke, and other languages; capital, Conakry. Part of a feudal Fulani empire from the 16th century, Guinea was colonized by France, becoming part of French West Africa. Guinea became an independent republic in 1958. The country is estimated to contain one-third of the world's reserves of bauxite. □ **Guinean** adj. & n.

guinea /ˈgɪnɪ/ n. **1** Brit. esp. hist. the sum of 21 shillings (£1.05), still sometimes used in determining professional fees, auction prices, etc. **2** hist. a former British gold coin worth 21 shillings, first coined for the African trade. □ **guinea-fowl** an African game bird of the family Numididae, with slate-coloured white-spotted plumage and loud cackling calls; esp. *Numida meleagris*, which is widely domesticated. **guinea pig 1** a domesticated South American cavy, *Cavia porcellus*, kept as a pet or for research in biology etc. **2** a person or thing used as a subject for experiment (*human guinea pig*). [*Guinea* in West Africa]

Guinea, Gulf of a large inlet of the Atlantic Ocean bordering on the southern coast of West Africa.

Guinea-Bissau /ˌgɪnɪbɪˈsaʊ/ a country on the west coast of Africa, between Senegal and Guinea; pop. (est. 1990) 1,000,000; languages, Portuguese (official), West African languages, Creoles; capital, Bissau. The area was explored by the Portuguese in the 15th century and was a centre of the slave trade. Formerly called Portuguese Guinea, it became a colony in 1879, and the independent republic of Guinea-Bissau in 1974.

Guinea worm n. a very long parasitic nematode worm, *Dracunculus medinensis*, which lives under the skin of infected people in rural tropical Africa and Asia.

Guinevere /ˈgwɪnɪˌvɪə(r)/ (in Arthurian legend) the wife of King Arthur and mistress of Lancelot.

Guinness /ˈgɪnɪs/, Sir Alec (b.1914), English actor. His stage career ranges from Shakespeare to contemporary drama and includes a notable interpretation of Hamlet. He made his film début in *Great Expectations* (1946); other films include *Kind Hearts and Coronets* (1949), *Bridge on the River Kwai* (1957), and *Star Wars* (1977). He is also known for his portrayal of the espionage chief George Smiley in television versions of John Le Carré's *Tinker Tailor Soldier Spy* (1979) and *Smiley's People* (1981–2).

guipure /gɪˈpjʊə(r), ˈgiːpjʊə(r)/ n. a heavy lace of linen pieces joined by embroidery. [F f. *guiper* cover with silk etc. f. Gmc]

guise /gaɪz/ n. **1** an assumed appearance; a pretence (*in the guise of*; *under the guise of*). **2** external appearance. **3** archaic style of attire, garb. [ME f. OF ult. f. Gmc]

guitar /gɪˈtɑː(r)/ n. a usu. six-stringed musical instrument with a fretted finger-board, played by plucking or strumming with the fingers or a plectrum or finger-pick. The characteristic form of the instrument, probably of Spanish origin, appeared at the end of the 15th century, although stringed instruments played with a plectrum have existed from pre-Christian times. Earlier forms of the guitar had four strings, the present six strings dating from the 1790s. Acoustic guitars include the Spanish, or classical, with gut or nylon strings; the steel-strung guitar, used especially in country and folk music; and the Hawaiian guitar, which is played by stopping the strings with a small metal bar. The electric guitar, especially important in popular music, is usually solid-bodied and uses a magnetic pick-up to convert the string vibrations into electrical signals which are then amplified by a separate amplifier and speaker. It dates from the 1930s. □ **guitarist** n. [Sp. *guitarra* (partly through F *guitare*) f. Gk *kithara*: see CITTERN, GITTERN]

guiver /ˈgaɪvə(r)/ n. (also **gyver**) Austral. & NZ sl. **1** plausible talk. **2** affectation of speech or manner. [19th c.: orig. unkn.]

Guiyang /ˈgweɪˈjæŋ/ (also **Kweiyang** /ˈkweɪ-/) an industrial city in southern China, capital of Guizhou province; pop. (est. 1989) 1,490,000.

Guizhou /ˈgweɪˈdʒəʊ/ (also **Kweichow** /ˈkweɪˈtʃaʊ/) a province of southern China; capital, Guiyang.

Gujarat /ˌgʊdʒəˈrɑːt/ a state in western India, with an extensive coastline on the Arabian Sea; capital, Gandhinagar. It was formed in 1960 from the northern and western parts of the former state of Bombay, and is one of the most industrialized parts of the country.

Gujarati /ˌgʊdʒəˈrɑːtɪ/ n. & adj. (also **Gujerati**) ● n. (pl. **Gujaratis**) **1** a language belonging to the Indic language group, descended from Sanskrit and spoken by some 45 million people, particularly in the state of Gujarat in India. It is written in a form of the Devanagari script. **2** a native of Gujarat. ● adj. of or relating to Gujarat or Gujarati. [Hind.: see -I¹]

Gujranwala /ˌgʊdʒrənˈwɑːlə/ a city in Pakistan, in Punjab province,

north-west of Lahore; pop. (1981) 597,000. It was the birthplace of the Sikh ruler Ranjit Singh, and was an important centre of Sikh influence in the early 19th century.

Gujrat /gʊdʒˈrɑːt/ a city in Pakistan, in Punjab province, north of Lahore; pop. (1981) 154,000.

Gulag /ˈguːlæg/ n. the system of forced-labour camps in the Soviet Union, specifically in the period 1930–55, in which hundreds of thousands, perhaps millions, died. Besides ordinary criminals, inmates included dissident intellectuals, members of ethnic groups suspected of disloyalty, and members of political factions who had lost power. A system of labour camps had been introduced in 1919, after the Revolution, and was much used by Stalin during the period of mass arrests. Although the Gulag was officially disbanded in 1955, a system of labour colonies remained. [Russ. acronym, f. *Glavnoe Upravlenie Ispravitel'no-Trudovykh Lagerei* Chief Administration of Corrective Labour Camps]

Gulbarga /gʊlˈbɑːgə/ a city in south central India, in the state of Karnataka; pop. (1991) 303,000. It was part of the Muslim sultanate of Delhi in the early 14th century, becoming the seat of the Bahmani kings of the Deccan in 1347 (until *c.*1424). It is a centre of the cotton trade.

Gulbenkian /gʊlˈbeŋkɪən/, Calouste Sarkis (1869–1955), Turkish-born British oil magnate and philanthropist, of Armenian descent. He was a pioneer in the development of the oil industry in the Middle East and his company (established in 1911) was the first to exploit the Iraqi oilfields. He bequeathed his large fortune and valuable art collection to the Calouste Gulbenkian Foundation (based in Lisbon); this disburses funds for the advancement of social and cultural projects in Portugal and elsewhere.

gulch /gʌltʃ/ n. N. Amer. a ravine, esp. one in which a torrent flows. [perh. dial. *gulch* to swallow]

gulden /ˈgʊld(ə)n/ n. = GUILDER. [Du. & G, = GOLDEN]

gules /gjuːlz/ n. & adj. (usu. placed after noun) Heraldry red. [ME f. OF *goules* red-dyed fur neck ornaments f. *gole* throat]

gulf /gʌlf/ n. **1** a stretch of sea consisting of a deep inlet with a narrow mouth. **2** a deep hollow; a chasm or abyss. **3** a wide difference of feelings, opinion, etc. [ME, f. OF *golfe* f. It. *golfo* ult. f. Gk *kolpos* bosom, gulf]

Gulf, the an informal name for the PERSIAN GULF.

Gulf of Aden, Gulf of Boothia, etc. see ADEN, GULF OF; BOOTHIA, GULF OF.

Gulf States 1 the states bordering on the Persian Gulf (Iran, Iraq, Kuwait, Saudi Arabia, Bahrain, Qatar, the United Arab Emirates, and Oman). **2** the states of the US bordering on the Gulf of Mexico (Florida, Alabama, Mississippi, Louisiana, and Texas).

Gulf Stream a warm ocean current which flows from the Gulf of Mexico parallel with the American coast towards Newfoundland, continuing across the Atlantic Ocean towards NW Europe as the North Atlantic Drift.

Gulf War 1 see IRAN–IRAQ WAR. **2** the war of January and February 1991 between Saddam Hussein's Iraq and an international coalition of forces determined to compel Iraqi withdrawal from Kuwait, which Iraq had invaded and occupied in August 1990. The force had assembled in Saudi Arabia under the auspices of the United Nations, but with US leadership and an American Commander-in-Chief, General Norman Schwarzkopf. The Iraqis were quickly driven out, with heavy military and civilian casualties on the Iraqi side. As they withdrew, the Iraqis sabotaged Kuwait's oil wells, starting hundreds of fires which were not all extinguished for several months.

gulfweed /ˈgʌlfwiːd/ n. = SARGASSO.

gull¹ /gʌl/ n. a long-winged web-footed seabird of the family Laridae, typically having white plumage with a grey or black mantle, and a bright bill. □ **gullery** n. (pl. **-ies**). [ME ult. f. OCelt.]

gull² /gʌl/ v.tr. (usu. in passive; foll. by into) dupe, fool. [perh. f. obs. *gull* yellow f. ON *gulr*]

Gullah /ˈgʌlə/ n. & adj. ● n. **1** a member of a black people living on the coast of South Carolina or the nearby sea islands. **2** the Creole language spoken by them, having an English base with elements from various West African languages. ● adj. of or relating to the Gullahs or their language. [perh. a shortening of ANGOLA, or f. a tribal name *Golas*]

gullet /ˈgʌlɪt/ n. **1** the food-passage extending from the mouth to the stomach; the oesophagus. **2** the throat. [ME f. OF dimin. of *go(u)le* throat f. L *gula*]

gullible /ˈɡʌlɪb(ə)l/ adj. easily persuaded or deceived, credulous. □ **gullibly** adv. **gullibility** /ˌɡʌlɪˈbɪlɪtɪ/ n. [GULL² + -IBLE]

gully /ˈɡʌlɪ/ n. & v. ● n. (also **gulley**) (pl. -ies or -eys) **1** a water-worn ravine. **2** a deep artificial channel; a gutter or drain. **3** Austral. & NZ a river valley. **4** Cricket a fielding position between point and slips. ● v.tr. (-ies, -ied) **1** form (channels) by water action. **2** make gullies in. □ **gully-hole** an opening in a street to a drain or sewer. [F goulet bottle-neck (as GULLET)]

gulp /ɡʌlp/ v. & n. ● v. **1** tr. (often foll. by down) swallow hastily, greedily, or with effort. **2** intr. swallow gaspingly or with difficulty; choke. **3** tr. (foll. by down, back) stifle, suppress (esp. tears). ● n. **1** an act of gulping (drained it in one gulp). **2** an effort to swallow. **3** a large mouthful of a drink. □ **gulper** n. **gulpy** adj. [ME prob. f. MDu. gulpen (imit.)]

GUM abbr. genito-urinary medicine.

gum¹ /ɡʌm/ n. & v. ● n. **1 a** a viscous secretion of some trees and shrubs that hardens on drying but is soluble in water (cf. RESIN n. 1). **b** an adhesive substance made from this. **2** N. Amer. chewing gum. **3** = GUMDROP. **4** = gum arabic. **5** = gum-tree. **6** a secretion collecting in the corner of the eye. **7** US = GUMBOOT. ● v. (**gummed**, **gumming**) **1** tr. smear or cover with gum. **2** tr. (usu. foll. by down, together, etc.) fasten with gum. **3** intr. exude gum. □ **gum arabic** a gum exuded by some kinds of acacia and used as glue and in incense. **gum benjamin** = BENZOIN. **gum dragon** tragacanth. **gum juniper** sandarac. **gum resin** a vegetable secretion of resin mixed with gum, e.g. gamboge. **gum-tree** a tree exuding gum, esp. a eucalyptus. **gum up 1** (of a mechanism etc.) become clogged or obstructed with stickiness. **2** colloq. interfere with the smooth running of (gum up the works). **up a gum-tree** colloq. in great difficulties. [ME f. OF gomme ult. f. L gummi, cummi f. Gk kommi f. Egypt. kemai]

gum² /ɡʌm/ n. (usu. in pl.) the firm flesh around the roots of the teeth. □ **gum-shield** a pad protecting a boxer's teeth and gums. [OE gōma rel. to OHG guomo, ON gómr roof or floor of the mouth]

gum³ /ɡʌm/ n. colloq. □ **by gum** a mild oath. [corrupt. of God]

gumbo /ˈɡʌmbəʊ/ n. (pl. -os) N. Amer. **1** okra. **2** (in Cajun cooking) a spicy chicken or seafood stew thickened with okra, rice, etc. **3** (**Gumbo**) a patois of blacks and Creoles spoken esp. in Louisiana. [of Afr. orig.]

gumboil /ˈɡʌmbɔɪl/ n. a small abscess on the gums.

gumboot /ˈɡʌmbuːt/ n. a rubber boot; a wellington.

gumdrop /ˈɡʌmdrɒp/ n. a soft coloured sweet made with gelatin or gum arabic.

gumma /ˈɡʌmə/ n. (pl. **gummas** or **gummata** /-mətə/) Med. a small soft swelling occurring in the connective tissue of the liver, brain, testes, and heart, and characteristic of the late stages of syphilis. □ **gummatous** adj. [mod.L f. L gummi GUM¹]

gummy¹ /ˈɡʌmɪ/ adj. (**gummier**, **gummiest**) **1** viscous, sticky. **2** abounding with or exuding gum. □ **gumminess** n. [ME f. GUM¹ + -Y¹]

gummy² /ˈɡʌmɪ/ adj. & n. ● adj. (**gummier**, **gummiest**) toothless. ● n. (pl. -ies) **1** (also **gummy shark**) a small shark of Australasian coasts, Mustelus antarcticus, having rounded teeth with which it crushes hard-shelled prey. **2** Austral. & NZ a toothless sheep. □ **gummily** adv. [GUM² + -Y¹]

gumption /ˈɡʌmpʃ(ə)n/ n. colloq. **1** resourcefulness, initiative; enterprising spirit. **2** common sense. [18th-c. Sc.: orig. unkn.]

gumshoe /ˈɡʌmʃuː/ n. **1** a galosh. **2** N. Amer. sl. a detective.

gun /ɡʌn/ n. & v. ● n. **1** a weapon (of any size from a massive mounted piece of artillery to a hand-sized pistol) consisting of a metal tube from which bullets or other missiles are propelled by the force of a controlled explosion. **2** any device imitative of this, e.g. a starting pistol. **3** a device for discharging insecticide, grease, electrons, etc., in a required direction (often in comb.: grease-gun). **4** a member of a shooting-party. **5** N. Amer. a gunman. **6** the firing of a gun. **7** (in pl.) Naut. sl. a gunnery officer. ● v. (**gunned**, **gunning**) **1** tr. **a** (usu. foll. by down) shoot (a person) with a gun. **b** shoot at with a gun. **2** tr. colloq. accelerate (an engine or vehicle). **3** intr. go shooting. **4** intr. (foll. by for) seek out determinedly to attack or rebuke. □ **go great guns** colloq. proceed forcefully or vigorously or successfully. **gun-carriage** a wheeled support for a gun. **gun-cotton** an explosive used for blasting, made by steeping cotton in nitric and sulphuric acids. **gun crew** a team manning a gun. **gun dog** a dog trained to follow sportsmen using guns. **gun-shy** (esp. of a sporting dog) alarmed at the sound of a gun. **gun-site** a (usu. fortified) emplacement for a gun. **jump the gun** colloq. start before a signal is given, or before an agreed time. **stick to one's guns** colloq. maintain one's position under attack. □ **gunless**

adj. **gunned** adj. [ME gunne, gonne, perh. f. the Scand. name Gunnhildr]

gunboat /ˈɡʌnbəʊt/ n. a small vessel of shallow draught and with relatively heavy guns. □ **gunboat diplomacy** uncompromising diplomacy supported by the use or threat of military force.

gundy /ˈɡʌndɪ/ n. Austral. colloq. □ **no good to gundy** no good at all. [20th c.: orig. unkn.]

gunfight /ˈɡʌnfaɪt/ n. a fight with firearms. □ **gunfighter** n.

gunfire /ˈɡʌnˌfaɪə(r)/ n. **1** the firing of a gun or guns, esp. repeatedly. **2** the noise from this.

gunge /ɡʌndʒ/ n. & v. Brit. colloq. ● n. sticky or viscous matter, esp. when messy or indeterminate. ● v.tr. (usu. foll. by up) clog or obstruct with gunge. □ **gungy** adj. [20th c.: orig. uncert.: cf. GOO, GUNK]

gung-ho /ɡʌŋˈhəʊ/ adj. eager for action, trigger-happy; uninhibited. [Chin. gonghe work together, slogan adopted by US Marines in 1942]

gunk /ɡʌŋk/ n. sl. viscous or liquid material. [20th c.: orig. the name of a detergent (propr.)]

gunlock /ˈɡʌnlɒk/ n. a mechanism by which the charge of a gun is exploded.

gunmaker /ˈɡʌnˌmeɪkə(r)/ n. a manufacturer of guns.

gunman /ˈɡʌnmən/ n. (pl. -**men**) a man armed with a gun, esp. in committing a crime.

gun-metal /ˈɡʌnˌmet(ə)l/ n. **1** a dull bluish-grey colour. **2** an alloy of copper and tin or zinc (formerly used for guns).

Gunn /ɡʌn/, Thomson William ('Thom') (b.1929), English poet. His first volume of poems, Fighting Terms, was published in 1954; shortly afterwards he moved to California, where he has lived ever since. Subsequent volumes of verse include The Sense of Movement (1959) and My Sad Captains (1961). His poetry is written in a predominantly low-key, laconic, and colloquial style.

gunnel¹ /ˈɡʌn(ə)l/ n. a small eel-shaped marine fish of the family Pholidae, esp. Pholis gunnellus. Also called butterfish. [17th c.: orig. unkn.]

gunnel² var. of GUNWALE.

gunner /ˈɡʌnə(r)/ n. **1** an artillery soldier (esp. as an official term for a private). **2** Naut. a warrant-officer in charge of a battery, magazine, etc. **3** a member of an aircraft crew who operates a gun. **4** a person who hunts game with a gun.

gunnera /ˈɡʌnərə/ n. a plant of the genus Gunnera, native to South America and New Zealand, having large leaves and often grown for ornament. [mod.L f. J. Ernst Gunnerus, Norw. botanist (1718–73)]

gunnery /ˈɡʌnərɪ/ n. **1** the construction and management of large guns. **2** the firing of guns.

gunny /ˈɡʌnɪ/ n. (pl. -ies) **1** coarse sacking, usu. of jute fibre. **2** a sack made of this. [Hindi & Marathi gōnī f. Skr. gōṇi sack]

gunplay /ˈɡʌnpleɪ/ n. the use of guns.

gunpoint /ˈɡʌnpɔɪnt/ n. □ **at gunpoint** at the point of a gun, threatened with a gun.

gunpowder /ˈɡʌnˌpaʊdə(r)/ n. **1** an explosive consisting of a powdered mixture of potassium nitrate, charcoal, and sulphur. (See note below.) **2** a fine green tea of granular appearance.

■ The earliest known propellant explosive, gunpowder was known to the Chinese, and possibly also to the Arabs, by the 10th century. Roger Bacon described its manufacture c.1250, and by the mid-14th century its military uses were well established in Europe. It remained the foremost military explosive for five centuries, having a profound effect on weapons and warfare. Gunpowder has now largely been superseded in military applications by high explosives, although it is still widely used for quarry blasting and in fuses and fireworks.

Gunpowder Plot a conspiracy by a small group of Catholic extremists to blow up James I and his Parliament on 5 Nov. 1605. The plot was uncovered when a Catholic MP was sent an anonymous letter telling him to stay away from the House on the appointed day. Guy Fawkes was arrested in the cellars of the Houses of Parliament the day before the scheduled attack and betrayed his colleagues under torture. The leader of the plot, Robert Catesby, was killed resisting arrest and the rest of the conspirators were captured and executed. The plot is commemorated by the traditional searching of the vaults before the opening of each session of Parliament, and by bonfires and fireworks, with the burning of an effigy of Guy Fawkes, annually on 5 Nov.

gunpower /ˈɡʌnˌpaʊə(r)/ n. the strength or quantity of available guns.

gunroom /ˈɡʌnruːm, -rʊm/ n. Brit. **1** a room in a house for storing sporting-guns. **2** quarters for junior officers (orig. for gunners) in a warship.

gunrunner /ˈgʌnˌrʌnə(r)/ n. a person engaged in the illegal sale or importing of firearms. □ **gunrunning** n.

gunsel /ˈgʌns(ə)l/ n. US sl. a criminal, esp. a gunman. [Yiddish *gendzel* = G *Gänslein* gosling; infl. by GUN]

gunship /ˈgʌnʃɪp/ n. a heavily armed helicopter or other aircraft.

gunshot /ˈgʌnʃɒt/ n. **1** a shot fired from a gun. **2** the range of a gun (*within gunshot*).

gunsight /ˈgʌnsaɪt/ n. a sight on a gun (see SIGHT n. 6a).

gunslinger /ˈgʌnˌslɪŋə(r)/ n. colloq. **1** a gunman, esp. in the Wild West. **2** a forceful (esp. young) participant in any sphere. □ **gunslinging** n.

gunsmith /ˈgʌnsmɪθ/ n. a person who makes, sells, and repairs small firearms.

gunstock /ˈgʌnstɒk/ n. the wooden mounting of the barrel of a gun.

Gunter's chain /ˈgʌntəz/ n. Surveying a measuring instrument 66 ft (20.1 m) long, subdivided into 100 links (1 rod or perch = 25 links), each link being a short section of wire connected to the next link by a loop. It was long used for land surveying and became a unit of length (80 chains = 1 mile), but has now been superseded by the steel tape and electronic equipment. [Edmund *Gunter*, Engl. mathematician (1581–1626)]

Gunther /ˈgʊntə(r)/ (in the Nibelungenlied) the husband of Brunhild and brother of Kriemhild, by whom he was beheaded in revenge for Siegfried's murder.

Guntur /ɡʊnˈtʊə(r)/ a city in eastern India, in Andhra Pradesh; pop. (1991) 471,000.

gunwale /ˈgʌn(ə)l/ n. (also **gunnel**) the upper edge of the side of a boat or ship. [GUN + WALE n. 3 (because formerly used to support guns)]

gunyah /ˈgʌnjə/ n. Austral. an Aboriginal bush hut. [Dharuk *ganʸi* house, hut]

Guomindang var. of KUOMINTANG.

guppy /ˈgʌpɪ/ n. (pl. **-ies**) a small freshwater fish, *Poecilia reticulata*, native to the West Indies and South America. Guppies give birth to live young, and are frequently kept in aquariums. [R. J. L. *Guppy*, Trinidadian clergyman (1836–1916) who sent the first specimen to the British Museum]

Gupta /ˈgʊptə/ a Hindu dynasty established in AD 320 by Chandragupta I in Bihar. The Gupta empire eventually stretched across most of the north of the Indian subcontinent, but began to disintegrate towards the end of the 5th century, only northern Bengal being left by the middle of the 6th century. □ **Guptan** adj.

Gurdjieff /ˈgɜːdjef/, George (Ivanovich) (1877–1949), Russian spiritual leader and occultist. After travels in India, Tibet, and the Middle East, he founded the Institute for the Harmonious Development of Man in Paris (1922). Those who attended the Institute were taught to attain a higher level of consciousness through a programme of lectures, dance, and physical labour. His ideas were published posthumously in *All and Everything* (1950) and *Meetings with Remarkable Men* (1963).

gurdwara /gɜːˈdwɑːrə, gʊəˈdwɑːrə/ n. a Sikh temple. [Punjabi *gurduārā* f. Skr. *guru* teacher + *dvāra* door]

gurgle /ˈgɜːg(ə)l/ v. & n. ● v. **1** intr. make a bubbling sound as of water from a bottle. **2** tr. utter with such a sound. ● n. a gurgling sound. [imit., or f. Du. *gorgelen*, G *gurgeln*, or med.L *gurgulare* f. L *gurgulio* gullet]

Gurkha /ˈgɜːkə, ˈgʊəkə/ n. **1** a member of a people of Hindu descent, speaking an Indic language, who have been settled in the province of Gurkha, Nepal, since the 18th century. They are renowned for their military prowess. **2** a member of one of the Gurkha regiments in the British army, established specifically for Nepalese recruits in the mid-19th century. [name of locality, f. Skr. *goraksha* cowherd (epithet of patron deity) f. *go* cow + *raksh* protect]

gurn /gɜːn/ v.intr. (also **girn**) esp. dial. grimace, pull a face (esp. for a competition). □ **gurner** n. **gurning** n. [dial. var. of GRIN]

gurnard /ˈgɜːnəd/ n. (also **gurnet** /-nɪt/) a marine fish of the family Triglidae, having a large spiny head with armoured sides, and three finger-like pectoral rays used for walking on the seabed. [ME f. OF *gornart* f. *grondir* to grunt f. L *grunnire*]

Gurney /ˈgɜːnɪ/, Ivor (Bertie) (1890–1937), English poet and composer. He fought on the Western Front during the First World War, and wrote the verse collections *Severn and Somme* (1917) and *War's Embers* (1919). Gurney also wrote nearly 300 songs, many of which were influenced by Elizabethan music. He suffered a breakdown and spent the last fifteen years of his life in a mental hospital.

guru /ˈgʊruː, ˈguːruː/ n. **1** (in Hinduism) a spiritual teacher. **2** (in Sikhism) any of the first ten leaders of the Sikh religion. **3** an influential teacher, a revered mentor (*management guru*). [Hindi *gurū* teacher f. Skr. *gurús* grave, dignified]

gush /gʌʃ/ v. & n. ● v. **1** tr. & intr. emit or flow in a sudden and copious stream. **2** intr. speak or behave with effusiveness or sentimental affectation. ● n. **1** a sudden or copious stream. **2** an effusive or sentimental manner. □ **gushing** adj. **gushingly** adv. [ME *gosshe*, *gusche*, prob. imit.]

gusher /ˈgʌʃə(r)/ n. **1** an oil well from which oil flows without being pumped. **2** an effusive person.

gushy /ˈgʌʃɪ/ adj. (**gushier**, **gushiest**) excessively effusive or sentimental. □ **gushily** adv. **gushiness** n.

gusset /ˈgʌsɪt/ n. **1** a piece let into a garment etc. to strengthen or enlarge a part. **2** a bracket strengthening an angle of a structure. □ **gusseted** adj. [ME f. OF *gousset* flexible piece filling up a joint in armour f. *gousse* pod, shell]

gussy /ˈgʌsɪ/ v.tr. (**-ies**, **-ied**) (esp. in passive; foll. by up) sl. smarten up, dress up. [perh. f. *Gussie* pet-form of the name *Augustus*]

gust /gʌst/ n. & v. ● n. **1** a sudden strong rush of wind. **2** a burst of rain, fire, smoke, or sound. **3** a passionate or emotional outburst. ● v.intr. blow in gusts. [ON *gustr*, rel. to *gjósa* to gush]

gustation /gʌˈsteɪʃ(ə)n/ n. the act or capacity of tasting. □ **gustative** /ˈgʌstətɪv/ adj. **gustatory** /-tərɪ/ adj. [F *gustation* or L *gustatio* f. *gustare* f. *gustus* taste]

Gustavus Adolphus /gʊˌstɑːvəs əˈdɒlfəs/ (1594–1632), king of Sweden 1611–32. He raised Sweden to the status of a European power by his victories against Denmark, Poland, and Russia in the first part of his reign. In 1630 he intervened on the Protestant side in the Thirty Years War, revitalizing the anti-imperialist cause with several victories and earning himself the title of 'Lion of the North'. At home he instituted reforms in administration, economic development, and education, laying the foundation of the modern state.

gusto /ˈgʌstəʊ/ n. **1** zest; enjoyment or vigour in doing something. **2** (foll. by for) archaic relish or liking. **3** archaic a style of artistic execution. [It. f. L *gustus* taste]

gusty /ˈgʌstɪ/ adj. (**gustier**, **gustiest**) **1** characterized by or blowing in gusts (*gusty wind*). **2** characterized by gusto. □ **gustily** adv. **gustiness** n.

gut /gʌt/ n. & v. ● n. **1** the lower alimentary canal or a part of this; the intestine. **2** (in pl.) the bowel or entrails, esp. of animals. **3** (in pl.) colloq. personal courage and determination; vigorous application and perseverance. **4** colloq. **a** (in pl.) the belly as the source of appetite. **b** belly or abdomen. **5** (in pl.) **a** the contents of anything, esp. representing substantiality. **b** the essence of a thing, e.g. of an issue or problem. **6 a** material for violin or racket strings or surgical use made from the intestines of animals. **b** material for fishing-lines made from the silk-glands of silkworms. **7 a** narrow water-passage; a sound, straits. **b** a defile or narrow passage. **8** (attrib.) **a** instinctive (*a gut reaction*). **b** fundamental (*a gut issue*). ● v.tr. (**gutted**, **gutting**) **1** remove or destroy (esp. by fire) the internal fittings of (a house etc.). **2** take out the guts of (a fish). **3** (as **gutted** adj.) sl. bitterly disappointed; deeply upset. **4** extract the essence of (a book etc.). □ **gut flora** = *intestinal flora*. **gut-rot** sl. **1** = *rot-gut*. **2** a stomach upset. **hate a person's guts** colloq. dislike a person intensely. **sweat** (or **work**) **one's guts out** colloq. work extremely hard. [OE *guttas* (pl.), prob. rel. to *gēotan* pour]

Gutenberg /ˈguːt(ə)nˌbɜːg/, Johannes (c.1400–68), German printer. He is remembered as the first in the West to print using movable type; he introduced typecasting using a matrix, and was the first to use a press. By c.1455 he had produced what later became known as the Gutenberg Bible.

Gutenberg Bible the edition of the Bible (Vulgate version) completed by Johannes Gutenberg in about 1455 in Mainz, Germany. Printed in Latin in three volumes in 42-line columns, it is the first complete book extant in the West and is also the earliest to be printed from movable type. There are about forty copies still in existence.

Guthrie /ˈgʌθrɪ/, Woody (born Woodrow Wilson Guthrie) (1912–1967), American folk-singer and songwriter. His radical political stance and commitment to causes such as the hardships of the rural poor during the Depression inspired many of his best-known songs, such as 'This Land is Your Land' (1944). His work was responsible for a revival of interest in folk music and influenced later singers and songwriters such as Bob Dylan.

Gutiérrez /ˌgʊtɪˈeərəz/, Gustavo (b.1928), Peruvian theologian. He was an important figure in the emergence of liberation theology in Latin America, outlining its principles in *A Theology of Liberation* (1971). Gutiérrez argued that the theologian should be concerned with liberating the poor and oppressed, and that this entailed responding to local needs rather than applying alien ideas and solutions.

gutless /'gʌtlɪs/ adj. colloq. lacking courage or determination; feeble. □ **gutless wonder** Austral. colloq. a person who lacks courage or determination. □ **gutlessly** adv. **gutlessness** n.

gutsy /'gʌtsɪ/ adj. (**gutsier**, **gutsiest**) colloq. **1** courageous. **2** greedy. □ **gutsily** adv. **gutsiness** n.

gutta-percha /ˌgʌtə'pɜːtʃə/ n. a tough plastic substance obtained from the latex of various Malaysian trees. [Malay getah gum + percha name of a tree]

guttate /'gʌteɪt/ adj. Biol. having droplike markings. [L guttatus speckled f. gutta drop]

gutter /'gʌtə(r)/ n. & v. ● n. **1** a shallow trough below the eaves of a house, or a channel at the side of a street, to carry off rainwater. **2** (prec. by the) a poor or degraded background or environment. **3** an open conduit along which liquid flows out. **4** a groove. **5** a track made by the flow of water. ● v. **1** intr. flow in streams. **2** tr. furrow, channel. **3** intr. (of a candle) burn unsteadily and melt away rapidly forming channels in the sides. □ **gutter press** sensational journalism concerned esp. with the private lives of public figures. [ME f. AF gotere, OF gotiere ult. f. L gutta drop]

guttering /'gʌtərɪŋ/ n. **1** the gutters of a building etc. **2** material for gutters.

guttersnipe /'gʌtəˌsnaɪp/ n. a street urchin.

guttural /'gʌtərəl/ adj. & n. ● adj. **1** throaty, harsh-sounding. **2 a** Phonet. (of a consonant) produced in the throat or by the back of the tongue and palate. **b** (of a sound) coming from the throat. **c** of the throat. ● n. Phonet. a guttural consonant (e.g. k, g). □ **gutturally** adv. [F guttural or med.L gutturalis f. L guttur throat]

gutzer /'gʌtsə(r)/ n. (also **gutser**) Austral. colloq. **1** a heavy fall. **2** a failure. [20th c.: orig. uncert.]

guv /gʌv/ n. Brit. sl. = GOVERNOR 7. [abbr.]

guy[1] /gaɪ/ n. & v. ● n. **1** colloq. a man; a fellow. **2** (usu. in pl.) N. Amer. a person of either sex. **3** Brit. an effigy of Guy Fawkes in ragged clothing, burnt on a bonfire on 5 Nov. **4** Brit. a grotesquely dressed person. ● v.tr. **1** ridicule. **2** exhibit in effigy. [Guy Fawkes (see FAWKES)]

guy[2] /gaɪ/ n. & v. ● n. a rope or chain to secure a tent or steady a crane-load etc. (often attrib.: guy-rope). ● v.tr. secure with a guy or guys. [prob. of LG orig.: cf. LG & Du. gei brail etc.]

Guyana /gaɪ'ænə/ a country on the NE coast of South America; pop. (est. 1991) 800,000; languages, English (official), English Creole, Hindi; capital, Georgetown. The Spaniards explored the area in 1499, and the Dutch settled there in the 17th century. It was occupied by the British from 1796 and established, with adjacent areas, as the colony of British Guiana in 1831; many indentured workers were brought in, especially from India and China, to work on sugar plantations. In 1966 it became an independent Commonwealth state, adopting the official designation of Co-operative Republic in 1970. The name Guyana is an American Indian word meaning 'land of waters'. □ **Guyanese** /ˌgaɪə'niːz/ adj. & n.

Guyenne /gi:'en/ (also **Guienne**) a region and former province of southern France, stretching from the Bay of Biscay to the SW edge of the Massif Central.

Guy Fawkes Night see BONFIRE NIGHT.

guzzle /'gʌz(ə)l/ v.tr. & intr. eat, drink, or consume excessively or greedily. □ **guzzler** n. [perh. f. OF gosiller chatter, vomit f. gosier throat]

Gvozdena Vrata /ˌgvɒzdenə 'vraːtə/ the Serbo-Croat name for the IRON GATE.

Gwalior /'gwaːlɪˌɔː(r)/ a city in a district of the same name in Madhya Pradesh, central India; pop. (1991) 693,000. It is noted for its fortress, dating from before AD 525, which includes several palaces, temples, and shrines. The district of Gwalior corresponds to a former princely state.

Gwent /gwent/ a county of SE Wales, formed in 1974 from most of Monmouthshire, part of Breconshire, and Newport; administrative centre, Cwmbran.

GWR abbr. hist. Great Western Railway.

Gwynedd /'gwɪneð/ **1** a county of NW Wales, formed in 1974 from Anglesey, Caernarvonshire, part of Denbighshire, and most of Merionethshire; administrative centre, Caernarfon. **2** a former principality of North Wales. Powerful in the mid-13th century under Llewelyn, it was finally subjugated by the English forces of Edward I in 1282, following Llewelyn's death.

Gwynn /gwɪn/, Eleanor ('Nell') (1650–87), English actress. Originally an orange-seller, she became famous as a comedienne at the Theatre Royal, Drury Lane, London. She was a mistress of Charles II; one of her sons was later created Duke of St Albans.

Gy abbr. Physics = GRAY[1].

Gyandzhe see GÄNCÄ.

gybe /dʒaɪb/ v. & n. (US **jibe**) ● v. **1** intr. (of a fore-and-aft sail or boom) swing across in wearing or running before the wind. **2** tr. cause (a sail) to do this. **3** intr. (of a ship or its crew) change course so that this happens. ● n. a change of course causing gybing. [obs. Du. gijben]

gym /dʒɪm/ n. colloq. **1** a gymnasium. **2** gymnastics. [abbr.]

gymkhana /dʒɪm'kaːnə/ n. **1** a meeting for competition or display in sport, esp. horse-riding. **2** a public place with facilities for athletics. [Hind. gendkhāna ball-house, racket-court, assim. to GYMNASIUM]

gymnasium /dʒɪm'neɪzɪəm/ n. (pl. **gymnasiums** or **gymnasia** /-zɪə/) **1** a room or building equipped for gymnastics. **2** a school in Germany or Scandinavia that prepares pupils for university entrance. □ **gymnasial** adj. [L f. Gk gumnasion f. gumnazō exercise f. gumnos naked]

gymnast /'dʒɪmnæst/ n. an expert in gymnastics. [F gymnaste or Gk gumnastēs athlete-trainer f. gumnazō: see GYMNASIUM]

gymnastic /dʒɪm'næstɪk/ adj. of or involving gymnastics. □ **gymnastically** adv. [L gymnasticus f. Gk gumnastikos (as GYMNASIUM)]

gymnastics /dʒɪm'næstɪks/ n.pl. (also treated as sing.) **1** exercises developing or displaying physical agility and coordination, usu. in competition. (See note below.) **2** other forms of physical or mental agility (verbal gymnastics).

■ Programmes of physical exercises were carried out as fitness training by the ancient Greeks and Romans, but modern techniques originated in Germany at the end of the 18th century. Organized competitions began in the late 19th century, and gymnastics was included in the first modern Olympic Games in 1896. Competitors are judged for a series of exercises on bars, beam, floor, and vaulting-horse.

gymno- /'dʒɪmnəʊ/ comb. form Biol. bare, naked. [Gk gumnos naked]

gymnosophist /dʒɪm'nɒsəfɪst/ n. a member of an ancient Hindu sect wearing little clothing and devoted to contemplation. □ **gymnosophy** n. [ME f. F gymnosophiste f. L gymnosophistae (pl.) f. Gk gumnosophistai: see GYMNO-, SOPHIST]

gymnosperm /'dʒɪmnəʊˌspɜːm/ n. Bot. a plant of the subdivision Gymnospermae, having seeds unprotected by an ovary, comprising the conifers, cycads, and ginkgo (cf. ANGIOSPERM). □ **gymnospermous** /ˌdʒɪmnəʊ'spɜːməs/ adj. [mod.L f. Gk gumnos naked + sperma seed]

gymp var. of GIMP[1].

gymslip /'dʒɪmslɪp/ n. a sleeveless tunic, usu. belted, worn by schoolgirls.

gynaeceum var. of GYNOECIUM.

gynaeco- /'gaɪnɪkəʊ/ comb. form (US **gyneco-**) woman, women; female. [Gk gunē gunaikos woman]

gynaecology /ˌgaɪnɪ'kɒlədʒɪ/ n. (US **gynecology**) the science of the physiological functions and diseases of women and girls, esp. those affecting the reproductive system. □ **gynaecologist** n. **gynaecological** /-kə'lɒdʒɪk(ə)l/ adj. **gynaecologically** adv. **gynecologic** adj. US.

gynaecomastia /ˌgaɪnɪkəʊ'mæstɪə/ n. (US **gynecomastia**) Med. enlargement of a man's breasts, usu. due to hormone imbalance or hormone therapy.

gynandromorph /gaɪ'nændrəˌmɔːf/ n. Biol. an individual, esp. an insect, having male and female characteristics. □ **gynandromorphic** /-ˌnændrə'mɔːfɪk/ adj. **gynandromorphism** n. [formed as GYNANDROUS + Gk morphē form]

gynandrous /gaɪ'nændrəs/ adj. Bot. with stamens and pistil united in one column, as in orchids. [Gk gunandros of doubtful gender, f. gunē woman + anēr andros man]

gyneco- comb. form US var. of GYNAECO-.

gynoecium /gaɪ'niːsɪəm/ n. (also **gynaecium**) (pl. **-cia** /-sɪə/) Bot. the carpels of a flower taken collectively (as GYNAECO-, Gk oīkos house)]

-gynous /gɪnəs, dʒɪnəs/ comb. form Bot. forming adjectives meaning 'having specified female organs or pistils' (monogynous). [Gk -gunos f. gunē woman]

gyp[1] /dʒɪp/ n. Brit. colloq. **1** pain or severe discomfort. **2** a scolding (*gave them gyp*). [19th c.: perh. f. *gee-up* (see GEE[2])]

gyp[2] /dʒɪp/ n. Brit. a college servant at Cambridge and Durham. [perh. f. obs. *gippo* scullion, orig. a man's short tunic, f. obs. F *jupeau*]

gyp[3] /dʒɪp/ v. & n. sl. ● v.tr. (**gypped**, **gypping**) cheat, swindle. ● n. an act of cheating; a swindle. [19th c.: perh. f. GYP[2]]

gyppy tummy var. of GIPPY TUMMY.

gypsophila /dʒɪp'sɒfɪlə/ n. a composite plant of the genus *Gypsophila*, with a profusion of small usu. white flowers. [mod.L f. Gk *gupsos* chalk + *philos* loving]

gypsum /'dʒɪpsəm/ n. hydrated calcium sulphate, occurring naturally and used to make plaster of Paris, blackboard chalk, fertilizer, etc. □ **gypseous** /-sɪəs/ adj. **gypsiferous** /dʒɪp'sɪfərəs/ adj. [L f. Gk *gupsos*]

gypsy /'dʒɪpsɪ/ n. (also **gipsy**) (*pl.* **-ies**) **1** (also **Gypsy**) a member of a travelling people with dark skin and hair, speaking a language (Romany) related to Hindi and traditionally living by seasonal work, itinerant trade, and fortune-telling. (*See note below.*) **2** a person resembling or living like a gypsy. □ **gypsyish** adj. [earlier *gipcyan*, *gipsen* f. EGYPTIAN, from the supposed origin of gypsies when they appeared in England in the early 16th c.]

▪ Gypsies are now found mostly in Europe, parts of North Africa, and North America. Their primary identifying cultural characteristic is their language and it is generally agreed that their original homeland was the Indian subcontinent, which they probably left in three separate migrations beginning before AD 1000; they reached western Europe by the 15th century. Gypsies have at various times and in various countries been the object of persecution and policies of forced migration; at least 250,000 were systematically killed in Nazi Germany.

gypsy moth n. a tussock moth, *Lymantria dispar*, having larvae that are very destructive to foliage.

gyrate v. & adj. ● v.intr. /ˌdʒaɪə'reɪt/ go in a circle or spiral; revolve, whirl. ● adj. /'dʒaɪərət/ Bot. arranged in rings or convolutions. □ **gyration** /ˌdʒaɪə'reɪʃ(ə)n/ n. **gyrator** /ˌdʒaɪə'reɪtə(r)/ n. **gyratory** /'dʒaɪərətərɪ, ˌdʒaɪə'reɪtərɪ/ adj. [L *gyrare gyrat*- revolve f. *gyrus* ring f. Gk *guros*]

gyre /'dʒaɪə(r)/ v. & n. v.intr. whirl, gyrate. ● n. **1** a whirling; a vortex. **2** a circulatory ocean current. [L *gyrus* ring f. Gk *guros*]

gyrfalcon /'dʒɜː.fɔːlkən, -.fɒlkən/ n. (also **gerfalcon**) a large falcon, *Falco rusticolus*, of cold northern regions, frequently whitish in colour. [ME f. OF *gerfaucon* f. Frank. *gĕrfalco* f. ON *geirfálki*: see FALCON]

gyro /'dʒaɪərəʊ/ n. (*pl.* **-os**) colloq. **1** = GYROSCOPE. **2** = GYROCOMPASS. [abbr.]

gyro- /'dʒaɪərəʊ/ comb. form **1** rotation. **2** gyroscope. [Gk *guros* ring]

gyrocompass /'dʒaɪərəʊˌkʌmpəs/ n. a non-magnetic compass in which the direction of true north is maintained by means of a continuously driven gyroscope whose axis is parallel to earth's rotation. The gyrocompass was independently invented (*c.* 1908) by Hermann Anschutz-Kaempfe in Germany and Elmer A. Sperry (1860–1930) in the US. It has largely replaced the magnetic compass in ships and aircraft.

gyromagnetic /ˌdʒaɪərəʊmæg'netɪk/ adj. **1** Physics of the magnetic and mechanical properties of a rotating charged particle. **2** (of a compass) combining a gyroscope and a normal magnetic compass.

gyropilot /'dʒaɪərəʊˌpaɪlət/ n. a gyrocompass used for automatic steering.

gyroplane /'dʒaɪərəʊˌpleɪn/ n. = AUTOGIRO.

gyroscope /'dʒaɪərəˌskəʊp/ n. a wheel or disc mounted so that it can spin rapidly about an axis which is itself free to change its direction. As in a spinning-top, the spin axis of a gyroscope tends to maintain the same direction in space, regardless of how the mounting is tilted. (A torque acting on the spin axis does, however, cause precession.) This behaviour makes gyroscopes suitable for providing stability or a reference direction. Common applications of gyroscopes are in navigation systems, automatic pilots, and stabilizers, as well as in the measurement of angular velocity and acceleration. □ **gyroscopic** /ˌdʒaɪərə'skɒpɪk/ adj. **gyroscopically** adv. [F (as GYRO-, -SCOPE)]

gyrostabilizer /ˌdʒaɪərəʊ'steɪbɪˌlaɪzə(r)/ n. a gyroscopic device for maintaining the equilibrium of a ship, aircraft, platform, etc.

gyrus /'dʒaɪərəs/ n. (*pl.* **gyri** /-raɪ/) a fold or convolution, esp. of the brain. [L f. Gk *guros* ring]

gyttja /'jɪtʃə/ n. Geol. a lake deposit of a usu. black organic sediment. [Sw., = mud, ooze]

Gyumri /'gjʊmrɪ/ (Russian **Kumayri** /ku'majrɪ/) an industrial city in NW Armenia, close to the border with Turkey; pop. (1989) 120,000. Founded as a fortress in 1837, the city was destroyed by an earthquake in 1926 and again in 1988. It was formerly called Alexandropol (1840–1924) and Leninakan (1924–91).

gyver var. of GUIVER.

Hh

H[1] /eɪtʃ/ n. (also **h**) (pl. **Hs** or **H's**) **1** the eighth letter of the alphabet (see AITCH). **2** anything having the form of an H (esp. in comb.: H-girder).

H[2] abbr. (also **H.**) **1** (of a pencil-lead) hard. **2** (water) hydrant. **3** sl. heroin.

H[3] symb. **1** Chem. the element hydrogen. **2** henry(s). **3** magnetic field strength.

h[1] abbr. (also **h.**) **1** height. **2** horse. **3** hot. **4** hour(s). **5** husband.

h[2] symb. **1** hecto-. **2** Planck's constant.

Ha symb. Chem. the element hahnium.

ha[1] /hɑː/ int. & v. (also **hah**) ● int. expressing surprise, suspicion, triumph, etc. (cf. HA HA). ● v.intr. (in **hum and ha**: see HUM[1]) [ME]

ha[2] abbr. hectare(s).

haar /hɑː(r)/ n. a cold sea-fog on the east coast of England or Scotland. [perh. f. ON hárr hoar, hoary]

Haarlem /'hɑːləm/ a city in the Netherlands, near Amsterdam; pop. (1991) 148,470. It is the capital of the province of North Holland and commercial centre of the Dutch bulb industry.

Hab. abbr. (in the Bible) Habakkuk.

Habakkuk /'hæbəkək, hə'bæk-/ **1** a Hebrew minor prophet probably of the 7th century BC. **2** a book of the Bible bearing his name.

habanera /ˌhæbə'neərə/ n. **1** a Cuban dance in slow duple time. **2** the music for this. [Sp., fem. of habanero of Havana in Cuba]

habeas corpus /ˌheɪbɪəs 'kɔːpəs/ n. Law a writ requiring a person under arrest to be brought before a judge or into court, esp. to investigate the lawfulness of his or her detention, thus ensuring that imprisonment cannot take place without a legal hearing. The right to sue for such a writ was an old common-law right; formulations of it look back to a phrase in Magna Carta, but the right may predate this. It was not actually passed as an Act of Parliament in England until 1679 and before that date was frequently disregarded. A similar right is recognized in the legal system of the US. [L, = you must have the body]

Haber-Bosch process /ˌhɑːbə'bɒʃ/ n. (also **Haber process**) an industrial process for producing ammonia from atmospheric nitrogen, using hydrogen and a catalyst at high temperature and pressure. [Fritz Haber (1868-1934), and Carl Bosch (1874-1940), German chemists]

haberdasher /'hæbəˌdæʃə(r)/ n. **1** Brit. a dealer in dress accessories and sewing-goods. **2** N. Amer. a dealer in men's clothing. □ **haberdashery** n. (pl. **-ies**). [ME prob. ult. f. AF hapertas perh. the name of a fabric]

habergeon /'hæbədʒən/ n. hist. a sleeveless coat of mail. [ME f. OF haubergeon (as HAUBERK)]

habiliment /hə'bɪlɪmənt/ n. (usu. in pl.) **1** clothes suited to a particular purpose. **2** joc. ordinary clothes. [ME f. OF habillement f. habiller fit out f. habile ABLE]

habilitate /hə'bɪlɪˌteɪt/ v.intr. qualify for office (esp. as a teacher in a German university). □ **habilitation** /-ˌbɪlɪ'teɪʃ(ə)n/ n. [med.L habilitare (as ABILITY)]

habit /'hæbɪt/ n. & v. ● n. **1** a settled or regular tendency or practice (often foll. by of + verbal noun: has a habit of ignoring me). **2** a practice that is hard to give up. **3** a mental constitution or attitude. **4** Psychol. an automatic reaction to a specific situation. **5** colloq. an addictive practice, esp. of taking drugs. **6 a** the dress of a particular class, esp. of a religious order. **b** (in full **riding-habit**) a woman's riding-dress. **c** archaic dress, attire. **7** a bodily constitution. **8** Biol. & Crystallog. a mode of growth. ● v.tr. (usu. as **habited** adj.) clothe. □ **habit-forming** (of a

drug etc.) causing addiction. **make a habit of** do regularly. [ME f. OF abit f. L habitus f. habere habit- have, be constituted]

habitable /'hæbɪtəb(ə)l/ adj. that can be inhabited. □ **habitability** /ˌhæbɪtə'bɪlɪtɪ/ n. [ME f. OF f. L habitabilis (as HABITANT)]

habitant /'hæbɪt(ə)nt/ n. **1** an inhabitant. **2** (also French /abitɑ̃/) **a** an early French settler in Canada or Louisiana. **b** a descendant of these settlers. [F f. OF habiter f. L habitare inhabit (as HABIT)]

habitat /'hæbɪˌtæt/ n. **1** the natural home of an organism. **2** a habitation. [L, = it dwells: see HABITANT]

habitation /ˌhæbɪ'teɪʃ(ə)n/ n. **1** the process of inhabiting (fit for human habitation). **2** a house or home. [ME f. OF f. L habitatio -onis (as HABITANT)]

habitual /hə'bɪtjʊəl/ adj. **1** done constantly or as a habit. **2** regular, usual. **3** given to a (specified) habit (a habitual smoker). □ **habitually** adv. **habitualness** n. [med.L habitualis (as HABIT)]

habituate /hə'bɪtjʊˌeɪt/ v.tr. (often foll. by to) accustom; make used to something. □ **habituation** /-ˌbɪtjʊ'eɪʃ(ə)n/ n. [LL habituare (as HABIT)]

habitude /'hæbɪˌtjuːd/ n. **1** a mental or bodily disposition. **2** a custom or tendency. [ME f. OF f. L habitudo -dinis f. habere habit- have]

habitué /hə'bɪtjʊˌeɪ/ n. a habitual visitor or resident. [F, past part. of habituer (as HABITUATE)]

Habsburg /'hæpsbɜːg/ (also **Hapsburg**) one of the principal dynasties of central Europe from medieval to modern times. They take their name from a castle in Aargau, Switzerland. The family established a hereditary monarchy in Austria in 1282, gradually extending its power, more often by marriage than by conquest. It secured the title of Holy Roman emperor from 1452, reaching its peak of power under Emperor Charles V. Austrian and Spanish branches were created when Charles divided the territories between his son Philip II and his brother Ferdinand; the Habsburgs ruled Spain 1504–1700, while Habsburg rule in Austria ended with the collapse of Austria–Hungary in 1918.

háček /'hætʃek/ n. a diacritic mark (ˇ) placed over letters in some Slavonic and Baltic languages, esp. to indicate various types of palatalization. [Czech, dimin. of hák hook]

hachures /hæ'ʃjʊə(r)/ n.pl. parallel lines used in hill-shading on maps, their closeness indicating the steepness of gradient. [F f. hacher HATCH[3]]

hacienda /ˌhæsɪ'endə/ n. in Spanish-speaking countries: **1** an estate or plantation with a dwelling-house. **2** a factory. [Sp. f. L facienda things to be done]

hack[1] /hæk/ v. & n. ● v. **1** tr. cut or chop roughly; mangle. **2** tr. kick with the toe of a boot in football etc. **3** intr. (often foll. by at) deliver cutting blows. **4** tr. cut (one's way) through thick foliage etc. **5 a** intr. (often with into) use a computer in order to explore systems and gain unauthorized access to data. **b** tr. make (one's way) into a computer thus; gain access to by hacking. **6** tr. sl. **a** manage, cope with (left because he couldn't hack it). **b** tolerate. **c** (often foll. by off or as **hacked off** adj.) annoy, disconcert. ● n. **1** a kick with the toe of a boot. **2** a gash or wound, esp. from a kick. **3 a** a mattock. **b** a miner's pick. □ **hacking cough** a short dry frequent cough. [OE haccian cut in pieces f. WG]

hack[2] /hæk/ n., adj., & v. ● n. **1 a** a horse for ordinary riding. **b** a horse let out for hire. **c** = JADE[2] 1. **2** a writer of mediocre literary work or journalism; colloq. usu. derog. a journalist. **3** a person hired to do dull routine work. **4** N. Amer. a taxi. ● attrib.adj. **1** used as a hack. **2** typical of a hack; commonplace (hack work). ● v. **1 a** intr. ride on horseback on a road at an ordinary pace. **b** *tr.* ride (a horse) in this way. **2** *tr.* make common or trite. [abbr. of HACKNEY]

hack[3] /hæk/ *n.* **1** a board on which a hawk's meat is laid. **2** a rack holding fodder for cattle. □ **at hack** (of a young hawk) not yet allowed to prey for itself. [var. of HATCH[1]]

hackberry /'hækbərɪ/ *n.* (*pl.* **-ies**) *US* **1** a tree of the genus *Celtis*, native to North America, bearing purple edible berries. **2** the berry of this tree. [var. of *hagberry*, of Norse orig.]

hacker /'hækə(r)/ *n.* **1** a person or thing that hacks or cuts roughly. **2** an enthusiastic computer programmer or user. **3** a person who tries to gain unauthorized access to a computer or to data held in one.

hackette /hæ'ket/ *n. colloq.* usu. *derog.* a female journalist.

hackle /'hæk(ə)l/ *n. & v.* ● *n.* **1** a long feather or series of feathers on the neck or saddle of a domestic cock and other birds. **2** an artificial fishing-fly dressed with a hackle. **3** a feather in a Highland soldier's bonnet. **4** (in *pl.*) the erectile hairs along the back of a dog, which rise when it is angry or alarmed. **5** a steel comb for dressing flax. ● *v.tr.* dress or comb with a hackle. □ **make a person's hackles rise** cause a person to be angry or indignant. [ME *hechele, hakele*, prob. f. OE f. WG]

hackney /'hæknɪ/ *n.* (*pl.* **-eys**) **1** a light horse with a high-stepping trot, used in harness. **2** (*attrib.*) designating any of various vehicles kept for hire. (*See note below.*) [ME, perh. f. *Hackney* in London, where horses were pastured]

▪ Hackney coaches began to operate in London in 1625 and ten years later there were so many that Charles I issued an order restricting them. By the reign of George III the number had been increased to 1,000, operating by licence and paying a duty of five shillings a week to the king. The term *hackney carriage* is now used as an official name for a taxi.

hackneyed /'hæknɪd/ *adj.* (of a phrase etc.) made commonplace or trite by overuse.

hacksaw /'hæksɔː/ *n. & v.* ● *n.* a saw with a narrow blade set in a frame, for cutting metal. ● *v.tr.* (*past part.* **-sawn** /-sɔːn/ or **sawed**) cut using a hacksaw.

had past and past part. of HAVE.

haddock /'hædək/ *n.* (*pl.* same) an edible marine fish, *Melanogrammus aeglefinus*, of the North Atlantic, allied to cod, but smaller. [ME, prob. f. AF *hadoc*, OF *(h)adot*, of unkn. orig.]

hade /heɪd/ *n. & v. Geol.* ● *n.* an incline from the vertical. ● *v.intr.* incline from the vertical. [17th c., perh. dial. form of *head*]

Hades /'heɪdiːz/ *Gk Mythol.* one of the sons of Cronus, lord of the lower world, the abode of the spirits of the dead. In classical Greek the name is always that of the person (known also as Pluto) rather than his kingdom, but later it came to refer to the place itself. Greek mythology has no Satan, and Hades is represented as grim and unpitying rather than evil. □ **Hadean** /heɪ'diːən, 'heɪdɪən/ *adj.* [Gk *haidēs*]

Hadhramaut /ˌhɑːdrə'maʊt, -'mɔːt/ a narrow region on the southern coast of Yemen, separating the Gulf of Aden from the desert land of the southern Arabian peninsula.

Hadith /'hædɪθ, hæ'diːθ/ a collection of traditions containing sayings of the prophet Muhammad which, with accounts of his daily practice (see SUNNA), constitute the major source of guidance for Muslims after the Koran. [Arab. *ḥadīt* tradition]

Hadlee /'hædlɪ/, Sir Richard (John) (b.1951), New Zealand cricketer. An all-round cricketer, he made his test début in 1973; in 1989 he became the first bowler to take more than 400 test wickets, finishing his career with a total of 431. He was knighted in 1990 and was the first cricketer to receive this honour while still playing in test matches.

hadn't /'hæd(ə)nt/ *contr.* had not.

Hadrian /'heɪdrɪən/ (full name Publius Aelius Hadrianus) (AD 76–138), Roman emperor 117–138. He became emperor as the adopted successor of Trajan, and spent much of his reign touring the provinces of the Empire, promoting good government and loyalty to Rome, and securing the frontiers. The building of Hadrian's Wall was begun after his visit to Britain in 122.

Hadrian's Wall a Roman defensive wall across northern England, stretching from the Solway Firth in the west to the mouth of the River Tyne in the east (about 120 km, 74 miles). It was begun in AD 122, after the emperor Hadrian's visit, to defend the province of Britain against invasions of tribes from the north. The wall was built of stone and was 2.5–3 m thick, with forts and fortified posts at intervals along its length. After Hadrian's death the frontier was advanced to the Antonine Wall, which the Romans proved unable to hold; after being overrun and restored several times Hadrian's Wall was abandoned AD c.410.

hadron /'hædrɒn/ *n. Physics* a subatomic particle, such as a baryon or meson, which takes part in the strong interaction (cf. LEPTON[2]). □ **hadronic** /hə'drɒnɪk/ *adj.* [Gk *hadros* bulky]

hadrosaur /'hædrəˌsɔː(r)/ *n.* a large herbivorous mainly bipedal dinosaur of the family Hadrosauridae, of the late Cretaceous period, with jaws flattened like the bill of a duck. [mod.L *Hadrosaurus* genus name, f. Gk *hadros* thick, stout + *sauros* lizard]

hadst /hædst/ *archaic* 2nd sing. past of HAVE.

haecceity /hek'siːɪtɪ/ *n. Philos.* **1** the quality of a thing that makes it unique or describable as 'this (one)'. **2** individuality. [med.L *haecceitas* f. *haec* fem. of *hic* this]

Haeckel /'hek(ə)l/, Ernst Heinrich (1834–1919), German biologist and philosopher. His popularization of Darwin's theories introduced them to many readers for the first time. Haeckel saw evolution as providing a framework for describing the world; he rejected religion as superstition and believed that the German Empire represented the highest evolved form of civilized nation. He upheld the essential unity of mind, organic life, and inorganic matter, and developed the recapitulation theory of ontogenesis, now discredited. He is said to have coined the word *ecology* in 1869.

haem /hiːm/ *n.* (*US* **heme**) *Biochem.* a porphyrin molecule containing iron, responsible for the red colour of haemoglobin. [Gk *haima* blood or f. HAEMOGLOBIN]

haemal /'hiːm(ə)l/ *adj.* (*US* **hemal**) *Anat.* **1** of or concerning the blood. **2** situated on the same side of the body as the heart and major blood vessels (i.e. in chordates, ventral). [Gk *haima* blood]

haematic /hiː'mætɪk/ *adj.* (*US* **hematic** /hiː-, he-/) *Med.* of or containing blood. [Gk *haimatikos* (as HAEMATIN)]

haematin /'hiːmətɪn/ *n.* (*US* **hematin** /'hiːm-, 'hem-/) *Biochem.* a bluish-black derivative of haemoglobin, formed by removal of the protein part and oxidation of the iron atom. [Gk *haima -matos* blood]

haematite /'hiːmətaɪt/ *n.* (*US* **hematite** /'hiːm-, 'hem-/) a ferric oxide ore forming dark red or reddish-black masses. [L *haematites* f. Gk *haimatitēs (lithos)* bloodlike (stone) (as HAEMATIN)]

haemato- /'hiːmətəʊ/ *comb. form* (*US* **hemato-** /'hiːm-, 'hem-/) blood. [Gk *haima haimat-* blood]

haematocele /'hiːmətəʊˌsiːl/ *n.* (*US* **hematocele** /'hiːm-, 'hem-/) *Med.* a swelling caused by blood collecting in a body cavity.

haematocrit /'hiːmətəʊˌkrɪt/ *n.* (*US* **hematocrit** /'hiːm-, 'hem-/) *Physiol.* **1** the ratio of the volume of red blood cells to the total volume of blood. **2** an instrument for measuring this. [HAEMATO- + Gk *kritēs* judge]

haematology /ˌhiːmə'tɒlədʒɪ/ *n.* (*US* **hematology** /ˌhiːm-, ˌhem-/) the study of the physiology of the blood. □ **haematologist** *n.* **haematologic** /-təˌlɒdʒɪk/ *adj.* **haematological** *adj.*

haematoma /ˌhiːmə'təʊmə/ *n.* (*US* **hematoma** /ˌhiːm-, ˌhem-/) (*pl.* **haematomas** or **haematomata** /-mətə/) *Med.* a solid swelling of clotted blood within the tissues.

haematophagous /ˌhiːmə'tɒfəgəs/ *adj.* (*US* **hematophagous** /ˌhiːm-, ˌhem-/) feeding on blood. [HAEMATO- + -PHAGOUS]

haematuria /ˌhiːmə'tjʊərɪə/ *n.* (*US* **hematuria** /ˌhiːm-, ˌhem-/) *Med.* the presence of blood in urine.

-haemia var. of -AEMIA.

haemo- /'hiːməʊ/ *comb. form* (*US* **hemo-** /'hiːm-, 'hem-/) = HAEMATO-. [abbr.]

haemocoel /'hiːməˌsiːl/ *n.* (*US* **hemocoel** /'hiːm-, 'hem-/) *Zool.* the primary body cavity of most invertebrates, containing circulatory fluid. [HAEMO- + Gk *koilos* hollow, cavity]

haemocyanin /ˌhiːmə'saɪənɪn/ *n.* (*US* **hemocyanin** /ˌhiːm-, ˌhem-/) *Biochem.* an oxygen-carrying protein containing copper, present in the blood plasma of arthropods and molluscs. [HAEMO- + *cyanin* blue pigment (as CYAN)]

haemodialysis /ˌhiːmə'maʊdaɪˌælɪsɪs/ *n.* (*US* **hemodialysis** /ˌhiːm-, ˌhem-/) *Med.* = DIALYSIS 2.

haemoglobin /ˌhiːmə'gləʊbɪn/ *n.* (*US* **hemoglobin** /ˌhiːm-, ˌhem-/) *Biochem.* a red oxygen-carrying protein containing iron, present in the red blood cells of vertebrates. [shortened f. HAEMATO- + GLOBULIN]

haemolysis /hiː'mɒlɪsɪs/ *n.* (*US* **hemolysis** /hiː-, he-/) *Med.* the rupture of red blood cells leading to the loss of haemoglobin. □ **haemolytic** /ˌhiːmə'lɪtɪk/ *adj.*

haemophilia /ˌhiːməˈfɪlɪə/ n. (US **hemophilia** /ˌhiːm-, ˌhem-/) Med. a usu. hereditary disorder with a tendency to bleed severely from even a slight injury, through the failure of the blood to clot normally. □ **haemophilic** adj. [mod.L (as HAEMO-, -PHILIA)]

haemophiliac /ˌhiːməˈfɪlɪˌæk/ n. (US **hemophiliac** /ˌhiːm-, ˌhem-/) a person suffering from haemophilia.

haemorrhage /ˈhemərɪdʒ/ n. & v. (US **hemorrhage**) ● n. **1** an escape of blood from a ruptured blood vessel, esp. when profuse. **2** an extensive damaging loss suffered by a state, organization, etc., esp. of people or assets. ● v. **1** intr. suffer a haemorrhage; bleed heavily. **2** intr. spread uncontrollably. **3** tr. expend (money etc.) in large amounts; lose or dissipate, esp. wastefully. □ **haemorrhagic** /ˌheməˈrædʒɪk/ adj. [earlier *haemorrhagy* f. F *hémorr(h)agie* f. L *haemorrhagia* f. Gk *haimorrhagia* f. *haima* blood + stem of *rhēgnumi* burst]

haemorrhoid /ˈheməˌrɔɪd/ n. (US **hemorrhoid**) (usu. in pl.) swollen veins at or near the anus; piles. □ **haemorrhoidal** /ˌheməˈrɔɪd(ə)l/ adj. [ME *emeroudis* (Bibl. *emerods*) f. OF *emeroyde* f. L f. Gk *haimorrhoides* (*phlebes*) bleeding (veins) f. *haima* blood, *-rhoos* -flowing]

haemostasis /ˌhiːməˈsteɪsɪs/ n. (US **hemostasis** /ˌhiːm-, ˌhem-/) Med. the stopping of the flow of blood from a wound etc. □ **haemostatic** /ˌhiːməˈstætɪk/ adj.

haere mai /ˈhaɪərə ˌmaɪ/ int. NZ welcome. [Maori, lit. 'come hither']

hafiz /ˈhɑːfɪz/ n. a Muslim who knows the Koran by heart. [Pers. f. Arab. *ḥāfiẓ* guardian]

hafnium /ˈhæfnɪəm/ n. a hard silver-grey metallic chemical element (atomic number 72; symbol **Hf**). A transition metal, hafnium is a rare element resembling zirconium and usually found associated with it. It was discovered in 1923 by George Hevesy of Hungary and Dirk Coster (1889–1950) of the Netherlands. The metal is used in control rods of nuclear reactors because of its capacity to absorb neutrons, and its strength and corrosion resistance, and also in tungsten alloys for filaments and electrodes. [mod.L *Hafnia* Copenhagen, where it was discovered]

haft /hɑːft/ n. & v. ● n. the handle of a dagger, knife, etc. ● v.tr. provide with a haft. [OE *hæft* f. Gmc]

Hag. abbr. (in the Bible) Haggai.

hag[1] /hæg/ n. **1** an ugly old woman. **2** a witch. **3** = HAGFISH. □ **haggish** adj. [ME *hegge*, *hagge*, perh. f. OE *hægtesse*, OHG *hagazissa*, of unkn. orig.]

hag[2] /hæg/ n. Sc. & N. Engl. **1** a soft place on a moor. **2** a firm place in a bog. [ON *högg* gap, orig. 'cutting blow', rel. to HEW]

Hagar /ˈheɪgɑː(r)/ (in the Bible) the Egyptian maid of Abraham's wife Sarah, who bore Ishmael to Abraham (Gen. 16, 21). She and Ishmael were driven away after the birth of Isaac.

Hagen /ˈhɑːgən/ an industrial city in NW Germany, in North Rhine-Westphalia; pop. (1991) 214,085.

hagfish /ˈhægfɪʃ/ n. a marine jawless fish of the family Myxinidae, with an eel-like, slimy body, a slitlike mouth surrounded by barbels, and a rasp-like tongue used for feeding on dead or dying fish. [HAG[1]]

Haggadah /həˈgɑːdə, ˌhɑːgɑːˈdɑː/ n. (pl. **Haggadoth** /ˌhɑːgɑːˈdəʊt/) **1** a legend or story used to illustrate a point of the Law in the Talmud; the legendary element of the Talmud. **2** a book recited at the Passover Seder service. □ **Haggadic** /həˈgædɪk, -ˈgɑːdɪk/ adj. [Heb., = tale, f. *higgīd* tell]

Haggai /ˈhægeɪˌaɪ/ **1** a Hebrew minor prophet of the 6th century BC. **2** a book of the Bible containing his prophecies of a glorious future in the Messianic age.

Haggard /ˈhægəd/, Sir H(enry) Rider (1856–1925), English novelist. He is famous for his adventure novels, many of which have an African setting and are based on the time he spent in South Africa in the mid-1870s. Among the best known of his books are *King Solomon's Mines* (1885) and *She* (1889).

haggard /ˈhægəd/ adj. & n. ● adj. **1** looking exhausted and distraught, esp. from fatigue, worry, privation, etc. **2** (of a hawk) caught and trained as an adult. ● n. a haggard hawk. □ **haggardly** adv. **haggardness** n. [F *hagard*, of uncert. orig.: later infl. by HAG[1]]

haggis /ˈhægɪs/ n. a Scottish dish consisting of a sheep's or calf's offal mixed with suet, oatmeal, etc., and boiled in a bag made from the animal's stomach or in an artificial bag. [ME: orig. unkn.]

haggle /ˈhæg(ə)l/ v. & n. ● v.intr. (often foll. by *about*, *over*) dispute or bargain persistently. ● n. a dispute or wrangle. □ **haggler** n. [earlier sense 'hack' f. ON *höggva* HEW]

Hagia Sophia /ˌhægɪə səˈfiːə/ = ST SOPHIA. [Gk, = holy wisdom]

hagio- /ˈhægɪəʊ/ comb. form of saints or holiness. [Gk *hagios* holy]

Hagiographa /ˌhægɪˈɒgrəfə/ n.pl. the twelve books comprising the last of the three major divisions of the Hebrew Bible, additional to the Law and the Prophets.

hagiographer /ˌhægɪˈɒgrəfə(r)/ n. **1** a writer of the lives of saints. **2** a writer of any of the Hagiographa.

hagiography /ˌhægɪˈɒgrəfɪ/ n. **1** the writing of the lives of saints; hagiology. **2** uncritical or idealized biography. □ **hagiographic** /-grəˈgræfɪk/ adj. **hagiographical** adj.

hagiolatry /ˌhægɪˈɒlətrɪ/ n. the worship of saints.

hagiology /ˌhægɪˈɒlədʒɪ/ n. literature dealing with the lives and legends of saints. □ **hagiologist** n. **hagiological** /-grəˈlɒdʒɪk(ə)l/ adj.

hagridden /ˈhægˌrɪd(ə)n/ adj. afflicted by nightmares or anxieties.

Hague, The /heɪg/ (Dutch **Den Haag** /dɛn ˈhɑːx/; also called *'s-Gravenhage*) the seat of government and administrative centre of the Netherlands, on the North Sea coast, capital of the province of South Holland; pop. (1991) 444,240. The International Court of Justice is based at The Hague. The city's two names, *'s-Gravenhage* 'the count's park', and *Den Haag* 'the hedge', refer to a wooded former hunting preserve, site of a royal palace since 1250.

hah var. of HA[1].

ha-ha /ˈhɑːhɑː/ n. a ditch with a wall on its inner side below ground level, forming a boundary to a park or garden without interrupting the view. [F, perh. from the cry of surprise on encountering it]

ha ha /hɑː ˈhɑː/ int. repr. laughter. [OE: cf. HA[1]]

Hahn /hɑːn/, Otto (1879–1968), German chemist, co-discoverer of nuclear fission. He pioneered the study of radiochemistry in England, first with Ramsay in London and then with Rutherford in Manchester, identifying various radioactive isotopes of thorium. His fruitful partnership with Lise Meitner began shortly after his return to Germany and ended when she fled from the Nazis in 1938. They discovered the new element protactinium in 1917, but the culmination of their collaboration occurred in 1938 when, with Fritz Strassmann (1902–80), they discovered nuclear fission. Hahn was awarded the Nobel Prize for chemistry in 1944.

hahnium /ˈhɑːnɪəm/ n. a name proposed in the US for the chemical element unnilpentium (cf. NIELSBOHRIUM). The corresponding symbol is **Ha**. [HAHN]

Haida /ˈhaɪdə/ n. & adj. ● n. (pl. same or **Haidas**) **1** a member of an American Indian people living on the west coast of Canada. **2** the language of this people. ● adj. of or relating to this people or their language. [Haida]

Haifa /ˈhaɪfə/ the chief port of Israel, in the north-west of the country on the Mediterranean coast; pop. (1988) 222,600. It is the site of the Baha'i shrine.

Haig /heɪg/, Douglas, 1st Earl Haig of Bemersyde (1861–1928), British Field Marshal. During the First World War he served as Commander-in-Chief of British forces in France (1915–18). He believed that the war could be won only by defeating the German army on the Western Front and maintained a strategy of attrition throughout his period of command. The strength of the main German army was eventually broken by this means, albeit with a very high cost in lives. In 1921 Haig helped to establish the Royal British Legion to improve the welfare of ex-servicemen.

haik /haɪk, heɪk/ n. (also **haick**) an outer covering for head and body worn by Arabs. [Moroccan Arab. *ḥā'ik*]

Haikou /haɪˈkəʊ/ the capital of Hainan autonomous region, a port on the NE coast of Hainan island; pop. (est. 1984) 278,600.

haiku /ˈhaɪkuː/ n. (pl. same) a Japanese three-part poem of usu. 17 syllables in lines of 5, 7, 5 syllables. The form emerged in the 16th century and flourished from the 17th century to the 19th century, and traditionally dealt with images of the natural world. In the early 20th century it was admired and imitated by modernist, particularly imagist, poets as a model of clarity and precision of expression. [Jap.]

hail[1] /heɪl/ n. & v. ● n. **1** pellets of frozen rain falling in showers from cumulonimbus clouds. **2** (foll. by *of*) a barrage or onslaught (of missiles, curses, questions, etc.). ● v. **1** intr. (prec. by *it* as subject) hail falls (*it is hailing*; *if it hails*). **2 a** tr. pour down (blows, words, etc.). **b** intr. come down forcefully. [OE *hagol*, *hægl*, *hagalian* f. Gmc]

hail[2] /heɪl/ v., int., & n. ● v. **1** tr. greet enthusiastically. **2** tr. signal to or attract the attention of (*hailed a taxi*). **3** tr. (often foll. by *as*) acclaim (*hailed him king*; *was hailed as a prodigy*). **4** intr. (foll. by *from*) have one's home or origins in (a place) (*hails from Mauritius*). ● int. archaic or rhet. expressing greeting. ● n. **1** a greeting or act of hailing. **2** distance as affecting the possibility of hailing (*was within hail*). □ **hail-fellow-well-met** intimate, on familiar terms, esp. overly so. □ **hailer** *n.* [ellipt. use of obs. *hail* (adj.) f. ON *heill* sound, WHOLE]

Haile Selassie /ˌhaɪlɪ səˈlæsɪ/ (born Tafari Makonnen) (1892–1975), emperor of Ethiopia 1930–74. He lived in exile in Britain during the Italian occupation of Ethiopia (1936–41), but was restored to the throne by the Allies and ruled until deposed in a Communist military coup. As a statesman, he made his country a prominent force in Africa and helped establish the Organization of African Unity in the early 1960s. He is revered by the Rastafarian religious sect, which is named after him.

Hail Mary *n.* the Ave Maria (see AVE).

hailstone /ˈheɪlstəʊn/ *n.* a pellet of hail.

hailstorm /ˈheɪlstɔːm/ *n.* a period of heavy hail.

Hailwood /ˈheɪlwʊd/, Mike (full name Stanley Michael Bailey Hailwood) (1940–81), English racing motorcyclist. He achieved a record fourteen wins in the Isle of Man TT races between 1961 and 1979, and was world champion nine times in three different classes between 1961 and 1967.

Hainan /haɪˈnæn/ an island in the South China Sea, forming an autonomous region of China; pop. (est. 1986) 6,000,000; capital, Haikou.

Hainaut /eɪˈnəʊ/ a province of southern Belgium; capital, Mons.

Haiphong /haɪˈfɒŋ/ a port in northern Vietnam, on the delta of the Red River in the Gulf of Tonkin; pop. (1989) 456,050.

hair /heə(r)/ *n.* **1 a** any of the fine threadlike strands growing from the skin of mammals, esp. from the human head. **b** these collectively (*his hair is falling out*). **c** a hairstyle or way of wearing the hair (*I like your hair today*). **2 a** an artificially produced hairlike strand, e.g. in a brush. **b** a mass of such hairs. **3** anything resembling a hair. **4** an elongated cell growing from the epidermis of a plant. **5** a very small quantity or extent (also *attrib.*: *a hair crack*). □ **get in a person's hair** *colloq.* encumber or annoy a person. **hair-drier** (or **-dryer**) an electrical device for drying the hair by blowing warm air over it. **hair-grass** a slender grass, esp. of the genus *Deschampsia* or *Aira*. **hair of the dog** see DOG. **hair-raising** extremely alarming; terrifying. **hair's breadth** a very small amount or margin. **hair shirt** a shirt of haircloth, worn formerly by penitents and ascetics. **hair-shirt** *adj.* (*attrib.*) austere, harsh, self-sacrificing. **hair-slide** *Brit.* a (usu. ornamental) clip for keeping the hair in position. **hair-splitter** a quibbler. **hair-splitting** *adj. & n.* making overfine distinctions; quibbling. **hair-trigger** a trigger of a firearm set for release at the slightest pressure. **keep one's hair on** *Brit. colloq.* remain calm; not get angry. **let one's hair down** *colloq.* abandon restraint; behave freely or wildly. **make a person's hair stand on end** alarm or horrify a person. **not turn a hair** remain apparently unmoved or unaffected. □ **haired** *adj.* (also in *comb.*). **hairless** *adj.* **hairlike** *adj.* [OE *hǣr* f. Gmc]

hairbreadth /ˈheəbredθ/ *n.* = *hair's breadth*; (esp. *attrib.*: *a hairbreadth escape*).

hairbrush /ˈheəbrʌʃ/ *n.* a brush for arranging or smoothing the hair.

haircloth /ˈheəklɒθ/ *n.* stiff cloth woven from hair, used e.g. in upholstery.

haircut /ˈheəkʌt/ *n.* **1** a cutting of the hair. **2** the style in which the hair is cut.

hairdo /ˈheəduː/ *n.* (*pl.* **-dos**) *colloq.* the style or an act of styling a woman's hair.

hairdresser /ˈheəˌdresə(r)/ *n.* **1** a person who cuts and styles hair, esp. professionally. **2** the business or establishment of a hairdresser. □ **hairdressing** *n.*

hairgrip /ˈheəgrɪp/ *n. Brit.* a flat hairpin with the ends close together.

hairline /ˈheəlaɪn/ *n.* **1** the edge of a person's hair, esp. on the forehead. **2** a very thin line or crack etc. (esp. *attrib.*: *hairline fracture*).

hairnet /ˈheənet/ *n.* a piece of fine mesh-work for confining the hair.

hairpiece /ˈheəpiːs/ *n.* a quantity or switch of detached hair used to augment a person's natural hair.

hairpin /ˈheəpɪn/ *n.* a U-shaped pin for fastening the hair. □ **hairpin bend** a sharp U-shaped bend in a road.

hairspray /ˈheəspreɪ/ *n.* a solution sprayed on to the hair to keep it in place.

hairspring /ˈheəsprɪŋ/ *n.* a fine spring regulating the balance-wheel in a watch.

hairstreak /ˈheəstriːk/ *n.* a butterfly of the genera *Callophrys*, *Strymonidia*, etc., with fine streaks or rows of spots on its wings.

hairstyle /ˈheəstaɪl/ *n.* a particular way of arranging or dressing the hair. □ **hairstyling** *n.* **hairstylist** *n.*

hairy /ˈheərɪ/ *adj.* (**hairier**, **hairiest**) **1** made of or covered with hair. **2** having the feel of hair. **3** *sl.* **a** alarmingly unpleasant or difficult. **b** crude, clumsy. □ **hairily** *adv.* **hairiness** *n.*

Haiti /ˈheɪtɪ/ a country in the Caribbean, occupying the western part of the island of Hispaniola; pop. (est. 1990) 6,500,000; official languages, Haitian Creole, French; capital, Port-au-Prince. The western third of Hispaniola was ceded to France by Spain in 1697. The French already had settlements there, established by pirates, and after 1697 many slaves were imported from West Africa to work on sugar plantations. In 1791 the slaves rose in rebellion under Toussaint L'Ouverture and other leaders, and in 1804 the colony was proclaimed an independent state under the name of Haiti. It was the second country in the Americas (after the US) to achieve freedom from colonial rule. Haiti was administered by the US (1915–34) after a succession of corrupt dictatorships. From 1957 to 1986 the country was under the oppressive dictatorship of the Duvalier family. The overthrow of Jean-Claude Duvalier in 1986 was followed, in 1990, by the election of Jean-Bertrand Aristide (b. 1953) as Haiti's first democratically chosen President, but in the following year the military seized power. In 1994 the US sent troops to Haiti in an attempt to ensure the restoration of democracy.

Haitian /ˈheɪʃən, -ʃ(ə)n/ *n. & adj.* ● *n.* **1** a native or inhabitant of Haiti. **2** (also **Haitian Creole**) the French-based Creole language spoken in Haiti. ● *adj.* of or relating to the Haitians or their language.

hajj /hædʒ/ *n.* (also **haj**) the annual Muslim pilgrimage to Mecca. [Arab. *ḥajj* pilgrimage]

hajji /ˈhædʒɪ/ *n.* (also **haji**) (*pl.* **-is**) **1** a Muslim who has been to Mecca as a pilgrim. **2** (**Hajji**) used as a title. [Pers. *hājī* (partly through Turk. *hacı*) f. Arab. *ḥajj*: see HAJJ]

haka /ˈhɑːkə/ *n.* **1** a Maori ceremonial war-dance accompanied by chanting. **2** an imitation of this by members of a New Zealand sports team before a match. [Maori]

hake /heɪk/ *n.* an edible marine fish of the genus *Merluccius*; esp. *M. merluccius*, with an elongated body and large head. [ME perh. ult. f. dial. *hake* hook]

hakenkreuz /ˈhɑːkənˌkrɔɪts/ *n.* a swastika, esp. as a Nazi symbol. [G f. *Haken* hook + *Kreuz* CROSS]

hakim[1] /hʌˈkiːm/ *n.* (in India and Muslim countries) a physician. [Arab. *ḥakīm* wise man, physician]

hakim[2] /ˈhʌkɪm, -kiːm/ *n.* (in India and Muslim countries) a judge, ruler, or governor. [Arab. *ḥākim* governor]

Hakluyt /ˈhækluːt/, Richard (c.1552–1616), English geographer and historian. He compiled *Principal Navigations, Voyages, and Discoveries of the English Nation* (1589), a collection of accounts of famous voyages of discovery, which brought to light the hitherto obscure achievements of English navigators and gave great impetus to discovery and colonization.

Hakodate /ˌhɑːkəʊˈdɑːteɪ/ a port in northern Japan, on the southern tip of the island of Hokkaido; pop. (1990) 307,000.

Halab see ALEPPO.

Halacha /həˈlɑːxə/ *n.* (also **Halakah**) Jewish law and jurisprudence, based on the Talmud. □ **Halachic** *adj.* [Aram. *hᵃlākāh* law]

Halafian /həˈlɑːfɪən/ *adj & n. Archaeol.* of, relating to, or denoting a prehistoric culture (c.5000–4500 BC) identified primarily by the use of polychrome pottery (called *Halaf ware*) which was first noted during excavations at Tell Halaf in NE Syria. Its distribution extends to the Mediterranean coast and the region of Lake Van in eastern Turkey.

halal /hɑːˈlɑːl/ *v. & n.* ● *v.tr.* (**halalled**, **halalling**) kill (an animal) as prescribed by Muslim law. ● *n.* (often *attrib.*) meat prepared in this way; lawful food. [Arab. *ḥalāl* lawful]

halation /hæˈleɪʃ(ə)n/ *n. Photog.* the spreading of light beyond its proper extent in a developed image, caused by internal reflection in the support of the emulsion. [irreg. f. HALO + -ATION]

halberd /ˈhælbəd/ *n.* (also **halbert** /-bət/) *hist.* a combined spear and battleaxe. [ME f. F *hallebarde* f. It *alabarda* f. MHG *helmbarde* f. *helm* handle + *barde* hatchet]

halberdier /ˌhælbəˈdɪə(r)/ *n. hist.* a man armed with a halberd. [F *hallebardier* (as HALBERD)]

halcyon /ˈhælsɪən/ *adj. & n.* ● *adj.* **1** calm, peaceful (*halcyon days*). **2** (of a period) happy, prosperous. ● *n.* **1** a tropical kingfisher of the genus *Halcyon*, with brightly coloured plumage. **2** *Mythol.* a bird thought in antiquity to breed in a nest floating at sea at the winter solstice, charming the wind and waves into calm. [ME f. L *(h)alcyon* f. Gk *(h)alkuōn* kingfisher]

Haldane /ˈhɔːldeɪn/, J(ohn) B(urdon) S(anderson) (1892–1964), Scottish mathematical biologist. Haldane helped to lay the foundations of population genetics. He also worked in biochemistry, and on the effects of diving on human physiology. Haldane became well known as a popularizer of science and for his outspoken Marxist views.

Hale /heɪl/, George Ellery (1868–1938), American astronomer. He discovered that sunspots are associated with strong magnetic fields, and invented the spectroheliograph. He initiated the construction of several large telescopes, culminating in the 5-metre (200-inch) reflector at Mount Palomar in California, named the Hale reflector in his honour.

hale[1] /heɪl/ *adj.* (esp. of an old person) strong and healthy (esp. in **hale and hearty**). □ **haleness** *n.* [OE *hāl* WHOLE]

hale[2] /heɪl/ *v.tr.* drag or draw forcibly. [ME f. OF *haler* f. ON *hala*]

haler /ˈhɑːlə(r)/ *n.* (*pl.* same or **haleru** /-ˌruː/) a monetary unit of Bohemia, Moravia, and Slovakia, equal to one-hundredth of a koruna. [Czech *haléř* f. MHG *haller*]

Halesowen /ˌheɪlzˈəʊɪn/ an engineering town in the West Midlands; pop. (1981) 57,530.

Haley /ˈheɪlɪ/, William John Clifton ('Bill') (1925–81), American rock and roll singer. He was the first to popularize rock and roll with the release of his song 'Rock Around the Clock' (1954), recorded with his group The Comets.

half /hɑːf/ *n., adj., & adv.* ● *n.* (*pl.* **halves** /hɑːvz/) **1** either of two equal or corresponding parts or groups into which a thing is or might be divided. **2** *colloq.* = *half-back*. **3** *colloq.* half a pint, esp. of beer etc. **4** either of two equal periods of play in sports. **5** *colloq.* a half-price fare or ticket, esp. for a child. **6** *Golf* a score esp. for a hole that is the same as one's opponent's. ● *adj.* **1** of an amount or quantity equal to a half, or loosely to a part thought of as roughly a half (*take half the men; spent half the time reading; half a pint; a half-pint; half-price*). **2** forming a half (*a half share*). ● *adv.* **1** (often in *comb.*) to the extent of half; partly (*only half cooked; half-frozen; half-laughing*). **2** to a certain extent; somewhat (esp. in idiomatic phrases: *half dead; am half inclined to agree*). **3** (in reckoning time) by the amount of half (an hour etc.) (*half past two*). □ **at half cock** see COCK[1]. **by half** (prec. by *too* + adj.) excessively (*too clever by half*). **by halves** imperfectly or incompletely (*never does things by halves*). **half-and-half** being half one thing and half another. **half-back** (in some sports) a player between the forwards and full backs. **half-baked 1** incompletely considered or planned. **2** (of enthusiasm etc.) only partly committed. **3** foolish. **half the battle** see BATTLE. **half-beak** a usu. marine fish of the family Hemirhamphidae, related to flying fish and with the lower jaw projecting beyond the upper. **half-binding** a type of bookbinding in which the spine and corners are bound in one material (usu. leather) and the sides in another. **half-blood 1** a person having one parent in common with another. **2** this relationship. **3** = *half-breed*. **half-blooded** born from parents of different races. **half-blue** *Brit.* **1** a person who has represented a university, esp. Oxford or Cambridge in a minor sport or as a second choice (cf. BLUE[1] *n.* 3), in a sport but who has not received a full blue. **2** this distinction. **half board** provision of bed, breakfast, and one main meal at a hotel etc. **half-boot** a boot reaching up to the calf. **half-bottle 1** a bottle that is half the standard size. **2** its contents or the amount that will fill it. **half-breed** *offens.* a person of mixed race. **half-brother** a brother with only one parent in common. **half-caste** *offens. n.* a person whose parents are of different races, esp. the offspring of a European father and an Indian mother. ● *adj.* of or relating to such a person. **half-century 1** a period of fifty years. **2** a score etc. of fifty in a sporting event, esp. cricket. **half a chance** *colloq.* the slightest opportunity (esp. *given half a chance*). **half-crown** (or **half a crown**) *hist.* (in the UK) a former coin and monetary unit equal to 2s. 6d. (12½p). **half-cut** *Brit. sl.* fairly drunk. **half-deck** the quarters of cadets and apprentices on a merchant vessel. **half-dozen** (or **half a dozen**) *colloq.* six, or about six. **half-duplex** see DUPLEX. **half an eye** the slightest degree of perceptiveness. **half-hardy** (of a plant) able to grow in the open air at all times except in severe frost. **half hitch** a noose or knot formed by passing the end of a rope round its standing part and then through the loop. **half holiday** a day of which half (usu. the afternoon) is taken as a holiday. **half-hour 1** (also **half an hour**) a period of 30 minutes. **2** a point of time 30 minutes after any hour o'clock. **half-hourly** at intervals of 30 minutes. **half-hunter** a watch with a hinged cover in which a small opening allows identification of the approximate position of the hands. **half-inch** *n.* a unit of length half as large as an inch. ● *v.tr. rhyming sl.* steal (= *pinch*). **half-integral** equal

to half an odd integer. **half-landing** a landing part of the way up a flight of stairs, whose length is twice the width of the flight plus the width of the well. **half-lap** the joining of rails, shafts, etc., by halving the thickness of each at one end and fitting them together. **half-length** a portrait of a person's upper half. **half-light** a dim imperfect light. **half-litre** a unit of capacity half as large as a litre. **half-marathon** a long-distance running-race, usu. of 13 miles 352 yards (21.243 km). **half-mast 1** the position of a flag halfway down the mast, as a mark of respect for a person who has died. **2** *colloq.* the position of a garment halfway to that normal (*trousers at half-mast*). **half measures** an unsatisfactory compromise or inadequate policy. **half a mind** (see *have a good mind to* under MIND). **half moon 1** the moon when only half its illuminated surface is visible from earth. **2** the time when this occurs. **3** a semicircular object. **half nelson** *Wrestling* see NELSON. **half-note** esp. *N. Amer. Mus.* = MINIM 1. **the half of it** *colloq.* the rest or more important part of something (usu. after *neg.*: *you don't know the half of it*). **half pay** reduced income, esp. on retirement. **half-pie** *NZ sl.* imperfect, mediocre. **half-plate 1** a photographic plate measuring 16.5 by 10.8 cm. **2** a photograph reproduced from this. **half-seas-over** *Brit. sl.* fairly drunk. **half-sister** a sister with only one parent in common. **half-sole** the sole of a boot or shoe from the shank to the toe. **half-sovereign** *hist.* (in the UK) a former gold coin and monetary unit equal to ten shillings (50p). **half-starved** poorly or insufficiently fed; malnourished. **half-step** *Mus.* a semitone. **half-term** *Brit.* a period about halfway through a school term, when a short holiday is usually taken. **half-timbered** *Archit.* having walls with a timber frame and a brick or plaster filling, a structural style common in England in the 15th–16th centuries. **half-time 1** the time at which half of a game or contest is completed. **2** a short interval occurring at this time. **half the time** see TIME. **half-title 1** the title or short title of a book, printed on the recto of the leaf preceding the title-page. **2** the title of a section of a book printed on the recto of the leaf preceding it. **half-tone 1** a reproduction printed from a block (produced by photographic means) in which the various tones of grey are produced from small and large black dots. **2** esp. *US Mus.* a semitone. **half-track 1** a propulsion system for land vehicles with wheels at the front and an endless driven belt at the back. **2** a vehicle equipped with this. **half-truth** a statement that (esp. deliberately) conveys only part of the truth. **half-volley** (*pl.* **-eys**) (in ball games) the playing of a ball as soon as it bounces off the ground. **half-yearly** at intervals of six months. **not half 1** not nearly (*not half long enough*). **2** *colloq.* not at all (*not half bad*). **3** *Brit. sl.* to an extreme degree (*he didn't half get angry*). [OE *half*, *healf* f. Gmc, orig. = 'side']

half-hearted /hɑːfˈhɑːtɪd/ *adj.* lacking enthusiasm; feeble. □ **half-heartedly** *adv.* **half-heartedness** *n.*

half-life /ˈhɑːflaɪf/ *n.* **1** the time taken for half of a sample of a particular radioactive isotope to decay into other materials. (See note below.) **2** the time taken for half of a dose of drug etc. to disappear in the body after administration.

▪ The half-life is a characteristic property of any process which occurs with a constant probability, such as radioactive decay. The half-life of a particular isotope is independent of physical and chemical environment; values range from minute fractions of a second for extremely unstable nuclei to very much longer periods. Radioactive isotopes with moderately long half-lives are potentially more dangerous because their radioactivity persists longer. Knowledge of the half-lives of radiocarbon and other naturally occurring isotopes forms the basis of their use for dating purposes.

halfpenny /ˈheɪpnɪ/ *n.* (also **ha'penny**) (*pl.* **-pennies** or **-pence** /-p(ə)ns/) (in the UK) a former bronze coin worth half a penny. ¶ Withdrawn in 1984 (cf. FARTHING).

halfpennyworth /ˈheɪpnəθ/ *n.* (also **ha'p'orth**) **1** as much as could be bought for a halfpenny. **2** *colloq.* a negligible amount (esp. after *neg.*: *doesn't make a halfpennyworth of difference*).

halfway /hɑːfˈweɪ/ *adv. & adj.* ● *adv.* **1** at a point equidistant between two others (*we were halfway to Rome*). **2** to some extent; more or less (*is halfway decent*). ● *adj.* situated halfway (*reached a halfway point*). □ **halfway house 1** a compromise. **2** the halfway point in a progression. **3** a centre for rehabilitating ex-prisoners, mental patients, or others unused to normal life. **4** an inn midway between two towns. **halfway line** a line midway between the ends of a pitch, esp. in football.

halfwit /ˈhɑːfwɪt/ *n.* **1** *colloq.* an extremely foolish or stupid person. **2** a person who is mentally deficient. □ **halfwitted** /hɑːfˈwɪtɪd/ *adj.* **halfwittedly** *adv.* **halfwittedness** *n.*

halibut /ˈhælɪbət/ *n.* (also **holibut** /ˈhɒl-/) (*pl.* same) a large marine flatfish, esp. *Hippoglossus hippoglossus* of the North Atlantic, used as food. [ME f. *haly* HOLY + BUTT⁵ flatfish, perh. because eaten on holy days]

Haliç /ha'liːtʃ/ the Turkish name for the GOLDEN HORN.

Halicarnassus /ˌhælɪkɑːˈnæsəs/ an ancient Greek city on the SW coast of Asia Minor, at what is now the Turkish city of Bodrum. It was the birthplace of the historian Herodotus and is the site of the Mausoleum of Halicarnassus, one of the Seven Wonders of the World. (See also MAUSOLEUM.)

halide /ˈhælaɪd, ˈheɪl-/ *n. Chem.* a binary compound of a halogen with another element or group.

halieutic /ˌhælɪˈjuːtɪk/ *adj. formal* of or concerning fishing. [L *halieuticus* f. Gk *halieutikos* f. *halieutēs* fisherman]

Halifax /ˈhælɪˌfæks/ **1** the capital of Nova Scotia, Canada; pop. (1991) 67,800; metropolitan area pop. 320,500. Originally a French fishing station, it was settled in 1749 by the English, who named it after the second earl of Halifax, president of the Board of Trade and Plantations. It is Canada's principal ice-free port on the Atlantic coast. **2** a town in West Yorkshire, on the River Calder; pop. (1981) 77,350.

haliotis /ˌhælɪˈəʊtɪs/ *n.* an edible gastropod mollusc of the genus *Haliotis*, with an ear-shaped shell lined with mother-of-pearl. [Gk *hals* hali- sea + *ous* ōt- ear]

halite /ˈhælaɪt/ *n.* rock-salt. [mod.L *halites* f. Gk *hals* salt]

halitosis /ˌhælɪˈtəʊsɪs/ *n. = bad breath.* [mod.L f. L *halitus* breath]

Hall¹ /hɔːl/, Charles Martin (1863–1914), American industrial chemist. He investigated different processes for producing aluminium from bauxite, and settled on electrolysis, obtaining the best results with alumina dissolved in molten cryolite. This remains the usual commercial method.

Hall² /hɔːl/, (Marguerite) Radclyffe (1883–1943), English novelist and poet. Her novels attracted both acclaim and outrage; while *Adam's Breed* (1926) was awarded the James Tait Black Memorial Prize, *The Well of Loneliness* (1928), with its exploration of a lesbian relationship, was banned for obscenity (in Britain though not in the US), despite the support of writers such as Virginia Woolf, E. M. Forster, and Arnold Bennett. The ban was overturned after Hall's death.

hall /hɔːl/ *n.* **1 a** a space or passage into which the front entrance of a house etc. opens. **b** *N. Amer.* a corridor or passage in a building. **2 a** a large room or building for meetings, meals, concerts, etc. **b** (in *pl.*) music-halls. **3** a large country house, esp. with a landed estate. **4** (in full **hall of residence**) a university residence for students. **5 a** (in a college etc.) a common dining-room. **b** dinner in this. **6** the building of a guild (*Fishmongers' Hall*). **7 a** a large public room in a palace etc. **b** the principal living-room of a medieval house. □ **hall porter** *Brit.* a porter who carries baggage etc. in a hotel. **hall-stand** a stand in the hall of a house, with a mirror, pegs, etc. [OE f. Gmc, rel. to HELL]

Halle /ˈhælə/ a city in east central Germany, on the River Saale, in Saxony-Anhalt; pop. (1991) 303,000.

Hallé /ˈhæleɪ/, Sir Charles (German name Karl Halle) (1819–95), German-born pianist and conductor. He came to Manchester from Paris in 1848 to escape the revolution and remained there for the rest of his life. He founded his own orchestra (still known as the Hallé Orchestra and based in Manchester) and inaugurated a series of orchestral concerts (the Hallé Concerts) in 1858.

hallelujah var. of ALLELUIA.

Haller /ˈhælə(r)/, Albrecht von (1708–77), Swiss anatomist and physiologist. He pioneered the study of neurology and experimental physiology, and wrote the first textbook of physiology.

Halley /ˈhælɪ, ˈhɔːlɪ/, Edmond (1656–1742), English astronomer and mathematician. Halley became an influential Fellow of the Royal Society and friend of Newton, the publication of whose *Principia* was due largely to him. He became professor of geometry at Oxford and was later appointed Astronomer Royal. He realized that nebulae were clouds of luminous gas among the stars, and that the aurora was a phenomenon connected with the earth's magnetism. Halley is best known for recognizing that a bright comet (later named after him) had appeared several times, and for successfully predicting its return.

Halley's Comet a bright comet whose reappearance in 1758–9 was predicted by Edmond Halley. Its orbital period is about 76 years, and earlier appearances can be traced in historical documents. It is first recorded in 240 BC and last appeared, though rather faintly, in 1985–6. A representation of its appearance in 1066 is visible on the Bayeux Tapestry. The European space probe Giotto passed close to Halley's Comet in 1986, obtaining a large quantity of data.

hallmark /ˈhɔːlmɑːk/ *n. & v. ● n.* **1** a mark used at Goldsmiths' Hall (and by the UK assay offices) for marking the standard of gold, silver, and platinum. (*See note below.*) **2** any distinctive feature esp. of excellence. *● v.tr.* **1** stamp with a hallmark. **2** designate as excellent.

▪ In the UK hallmarking dates from a statute of 1300, in the reign of Edward I, and the Worshipful Company of Goldsmiths has been responsible for the assaying and marking of plate in London since then. With certain exceptions, all gold, silver, and platinum articles are required by law to be hallmarked before they are offered for sale. The marks impressed include a number of symbols indicating the maker, standard, date, and office. Many countries outside the UK have a system of plate marks. In the US there is no hallmarking; local regulations existed in the 18th and 19th centuries, but there is no consistent system of symbols. Signatory countries of an International Convention — the UK, Austria, Finland, Ireland, Portugal, Norway, Sweden, and Switzerland — recognize equivalents to their own hallmarks applied in the others.

hallo var. of HELLO.

Hall of Fame *n.* esp. *N. Amer.* **1** a building with memorials of celebrated people. **2** a group of people considered most famous in a particular sphere.

halloo /həˈluː/ *int., n., & v. ● int.* **1** inciting dogs to the chase. **2** calling attention. **3** expressing surprise. ● *n.* the cry 'halloo'. *● v.* (**halloos, hallooed**) **1** *intr.* cry 'halloo', esp. to dogs. **2** *intr.* shout to attract attention. **3** *tr.* urge on (dogs etc.) with shouts. [perh. f. *hallow* pursue with shouts f. OF *halloer* (imit.)]

hallow /ˈhæləʊ/ *v. & n. ● v.tr.* **1** make holy, consecrate. **2** honour as holy. ● *n. archaic* a saint or holy person. [OE *hālgian, hālga* f. Gmc]

Hallowe'en /ˌhæləʊˈiːn/ the evening of 31 October, the eve of All Saints' Day. Hallowe'en is of pre-Christian origin, being associated with Samhain, the Celtic festival marking the end of the year and the beginning of winter, when ghosts and spirits were thought to be abroad. It was adopted as a Christian festival but gradually became a secular rather than a Christian observance, involving the dressing up and wearing of masks, and was particularly strong in Scotland. These secular customs were popularized in the US in the late 19th century by newly arrived immigrants (esp. from Ireland), and later developed into the custom of children playing trick or treat. [HALLOW + EVEN²]

Hallowes /ˈhæləʊz/, Odette (born Marie Céline) (1912–95), French heroine of the Second World War. She entered occupied France in 1942 and worked secretly as a British agent until captured by the Gestapo in 1943. Imprisoned until 1945, she refused to betray her associates in spite of torture. For her work and her courage she was awarded the George Cross (1946).

Hallstatt /ˈhælʃtæt/ *adj. & n. Archaeol.* of, relating to, or denoting a cultural phase of the late Bronze and early Iron Age (*c.*700–500 BC), preceding the La Tène period. [a village in Austria, site of an ancient necropolis where remains of the period were found]

halluces pl. of HALLUX.

hallucinate /həˈluːsɪˌneɪt/ *v.* **1** *tr.* produce illusions in the mind of (a person). **2** *intr.* experience hallucinations. □ **hallucinant** *adj. & n.* **hallucinator** *n.* [L (h)*allucinari* wander in mind f. Gk *alussō* be uneasy]

hallucination /həˌluːsɪˈneɪʃ(ə)n/ *n.* **1** the apparent or alleged perception of an object or stimulus not actually present. **2** something so perceived, a visual illusion. □ **hallucinatory** /-ˈluːsɪnətərɪ/ *adj.* [L *hallucinatio* (as HALLUCINATE)]

hallucinogen /həˈluːsɪnədʒən/ *n.* a drug causing hallucinations. □ **hallucinogenic** /-ˌluːsɪnəˈdʒenɪk/ *adj.*

hallux /ˈhælʌks/ *n.* (*pl.* **halluces** /ˈhæljʊˌsiːz/) **1** *Anat.* the big toe. **2** *Zool.* the innermost digit of the hind foot of vertebrates. [mod.L f. L *allex*]

hallway /ˈhɔːlweɪ/ *n.* an entrance-hall or corridor.

halm var. of HAULM.

halma /ˈhælmə/ *n.* a game played by two or four persons using a board of 256 squares, with pieces advancing from one corner to the opposite corner by being moved over other pieces into vacant squares. [Gk, = leap]

Halmahera /ˌhælməˈhɪərə/ the largest of the Molucca Islands.

halo /ˈheɪləʊ/ *n. & v. ● n.* (*pl.* **-oes** or **-os**) **1** a disc or circle of light shown surrounding the head of a sacred person. **2** the glory associated with an idealized person etc. **3** a circle of white or coloured light round a luminous body, esp. the sun or moon. **4** a circle or ring. *● v.tr.* (**-oes, -oed**) surround with a halo. [med.L f. L f. Gk *halōs* threshing-floor, disc of the sun or moon]

halogen /ˈhælədʒən/ *n.* **1** *Chem.* any of the five elements fluorine, chlorine, bromine, iodine, and astatine, occupying group VIIA (17) of the periodic table. (*See note below.*) **2** (*attrib.*) (of lamps and radiant heat sources) using a filament surrounded by a halogen, usu. iodine vapour. □ **halogenic** /ˌhæləˈdʒɛnɪk/ *adj.* [Gk *hals* salt]

■ The halogens are reactive, electronegative non-metals, with a tendency to be monovalent; with hydrogen they form gaseous compounds which are very soluble in water, giving strongly acid solutions from which simple salts (such as common salt, sodium chloride) can be prepared.

halogenation /həˌlɒdʒɪˈneɪʃ(ə)n, ˌhælədʒɪ-/ *n.* the introduction of a halogen atom into a molecule. □ **halogenated** /həˈlɒdʒɪˌneɪtɪd, ˈhælədʒɪ-/ *adj.*

halon /ˈheɪlɒn/ *n. Chem.* any of a class of gaseous compounds of carbon, bromine, and other halogens, used to extinguish fires. [as HALOGEN + -ON]

haloperidol /ˌhæləʊˈpɛrɪˌdɒl, ˌheɪləʊ-/ *n. Pharm.* a drug used to treat psychotic disorders, esp. mania. [HALOGEN + PIPERIDINE + -OL]

halophyte /ˈhæləʊˌfaɪt, ˈheɪləʊ-/ *n. Bot.* a plant adapted to saline conditions. [Gk *hals halos* salt + -PHYTE]

halothane /ˈhæləʊˌθeɪm, ˈheɪləʊ-/ *n. Pharm.* a volatile liquid used as a general anaesthetic, a halogenated derivative of ethane. [HALOGEN + ETHANE]

Hals /hæls/, Frans (*c.*1580–1666), Dutch portrait and genre painter. His use of bold brushwork to capture the character, mood, and facial expressions of his subjects gave a vitality to his portraits and represented a departure from conventional portraiture. His best-known portraits include groups, such as *The Banquet of the Officers of the St George Militia Company* (1616), and single figures, such as *The Laughing Cavalier* (1624). His genre pictures, painted during the 1620s, reflect the influence of the Dutch followers of Caravaggio.

Hälsingborg see HELSINGBORG.

halt[1] /hɒlt, hɔːlt/ *n. & v.* ● *n.* **1** a stop (usu. temporary); an interruption of progress (*come to a halt*). **2** a temporary stoppage on a march or journey. **3** *Brit.* a minor stopping-place on a local railway line, usu. without permanent buildings. ● *v.intr. & tr.* stop; come or bring to a halt. □ **call a halt** (**to**) decide to stop. [orig. in phr. *make halt* f. G *Halt machen* f. *halten* hold, stop]

halt[2] /hɒlt, hɔːlt/ *v. & adj.* ● *v.intr.* **1** (esp. as **halting** *adj.*) fail to make smooth progress. **2** hesitate (*halt between two opinions*). **3** walk hesitatingly. **4** *archaic* be lame. ● *adj. archaic* lame or crippled. □ **haltingly** *adv.* [OE *halt, healt, healtian* f. Gmc]

halter /ˈhɒltə(r), ˈhɔːl-/ *n. & v.* ● *n.* **1** a rope or strap with a noose or headstall for horses or cattle. **2 a** a strap round the back of a woman's neck holding her dress-top and leaving her shoulders and back bare. **b** a dress-top held by this. **3 a** a rope with a noose for hanging a person. **b** death by hanging. ● *v.tr.* **1** put a halter on (a horse etc.). **2** hang (a person) with a halter. □ **halter-break** accustom (a horse) to a halter. **halter-neck** (of a woman's garment) held up by a strap around the neck. [OE *hælftre*: cf. HELVE]

haltere /ˈhæltɪə(r)/ *n.* (*pl.* **halteres** /ˈhæltɪriːz/) (usu. in *pl.*) *Zool.* either of the modified and reduced hindwings forming the balancing organs of a dipterous insect. [Gk *haltēres* = weights used to aid leaping f. *hallomai* to leap]

halva /ˈhælvɑː, -və/ *n.* (also **halvah**) a sweet confection of sesame flour and honey. [Yiddish f. Turk. *helva* f. Arab. *ḥalwa*]

halve /hɑːv/ *v.tr.* **1** divide into two halves or parts. **2** reduce by half. **3** share equally (with another person etc.). **4** *Golf* use the same number of strokes as one's opponent in (a hole or match). **5** fit (crossing timbers) together by cutting out half the thickness of each. [ME *halfen* f. HALF]

halves *pl.* of HALF.

halyard /ˈhæljəd/ *n. Naut.* a rope or tackle for raising or lowering a sail or yard etc. [ME *halier* f. HALE[2] + -IER, assoc. with YARD[1]]

Ham /hæm/ (in the Bible) a son of Noah (Gen. 10:1), traditional ancestor of the Hamites.

ham /hæm/ *n. & v.* ● *n.* **1 a** the upper part of a pig's leg salted and dried or smoked for food. **b** the meat from this. **2** the back of the thigh; the thigh and buttock. **3** (often *attrib.*) *sl.* an inexpert or unsubtle actor or piece of acting; overacting. **4** (in full **radio ham**) *colloq.* the operator of an amateur radio station. ● *v.intr. & (often foll. by up) tr.* (**hammed**, **hamming**) *sl.* overact; act or treat emotionally or sentimentally. [OE *ham, hom* f. a Gmc root meaning 'be crooked']

Hama /ˈhɑːmɑː/ (also **Hamah**) an industrial city in western Syria, on the River Orontes; pop. (1981) 176,640. Originally a Hittite city, it became the Kingdom of Hamath under the Aramaeans in the 11th century BC and was renamed Epiphania in the 2nd century BC by the Seleucids. It is noted for its water-wheels, which have been used for irrigation since the Middle Ages.

Hamada /ˈhæmədə/, Shoji (1894–1978), Japanese potter. He visited England in 1920 and collaborated with his friend Bernard Leach before returning to Japan in 1923 to set up his own kiln at Mashiko. He worked mainly in stoneware, producing utilitarian items of unpretentious simplicity; he was a firm believer in the beauty and individuality of the handmade object as opposed to the uniformity of the products of industrialization.

hamadryad /ˌhæməˈdraɪæd/ *n.* **1** (in Greek and Roman mythology) a nymph who lives in a tree and dies when it dies. **2** = *king cobra.* [ME f. L *hamadryas* f. Gk *hamadruas* f. *hama* with + *drus* tree]

hamadryas /ˌhæməˈdraɪəs/ *n.* a large baboon, *Papio hamadryas*, native to NE Africa and Arabia, with a silvery-grey cape of hair over the shoulders. It was held sacred in ancient Egypt.

Hamamatsu /ˌhæməˈmætsuː/ an industrial city on the southern coast of the island of Honshu, Japan; pop. (1990) 535,000.

hamamelis /ˌhæməˈmiːlɪs/ *n.* = WITCH-HAZEL 1. [mod.L f. Gk *hamamēlis* medlar]

hamartia /həˈmɑːtɪə/ *n.* (in Greek tragedy) the error or fatal flaw leading to the destruction of the tragic hero or heroine. [Gk = fault, failure, guilt]

Hamas /ˈhæmæs/ a Palestinian Islamic fundamentalist movement, which developed as the Palestinian wing of the Muslim Brotherhood. Hamas has become a focus for Arab resistance in the Israeli-occupied territories; it opposes peace with Israel and has come into conflict with the more moderate Palestine Liberation Organization.

hamba /ˈhæmbə/ *int. S. Afr.* be off; go away. [Nguni *-hambe* go]

hambone /ˈhæmbəʊn/ *n. Austral. colloq.* a male striptease show. [20th c.: orig. uncert.]

Hamburg /ˈhæmbɜːg/ a city-state and port in northern Germany, on the River Elbe; pop. (1991) 1,661,760. It was founded by Charlemagne in the 9th century, was a founder member of the Hanseatic League, and became a state of the German Empire in 1871. It is the largest port in Germany, with extensive shipyards.

hamburger /ˈhæmbɜːgə(r)/ *n.* a cake of minced beef usu. fried or grilled and eaten in a soft bread roll. [G, = of Hamburg]

Hameln /ˈhɑːməln/ (also **Hamelin** /ˈhæmlɪn/) a town in NW Germany, in Lower Saxony, on the River Weser; pop. (1983) 57,000. It was a medieval market town, the setting of the legend of the Pied Piper of Hamelin, which may be based on actual events which occurred in 1284.

hames /heɪmz/ *n.pl.* two curved pieces of iron or wood forming the collar or part of the collar of a draught-horse, to which the traces are attached. [ME f. MDu. *hame*]

ham-fisted /hæmˈfɪstɪd/ *adj. colloq.* clumsy, heavy-handed, bungling. □ **ham-fistedly** *adv.* **ham-fistedness** *n.*

ham-handed /hæmˈhændɪd/ *adj. colloq.* = HAM-FISTED. □ **ham-handedly** *adv.* **ham-handedness** *n.*

Hamhung /hæmˈhʌŋ/ an industrial city in eastern North Korea; pop. (est. 1984) 775,000. It was the centre of government of NE Korea during the Yi dynasty of 1392–1910.

Hamilcar /hæˈmɪlkɑː(r), ˈhæmɪlˌkɑː(r)/ (*c.*270–229 BC), Carthaginian general and father of Hannibal. He fought Rome in the first Punic War and negotiated terms of peace after the Carthaginian defeat of 241, which led to the loss of Sicily to the Romans. From 237 he and Hannibal were engaged in the conquest of Spain.

Hamilton[1] /ˈhæmɪlt(ə)n/ **1** a town in Strathclyde region, southern Scotland, near Glasgow; pop. (1981) 51,720. **2** a port and industrial city in southern Canada, at the western end of Lake Ontario; pop. (1991) 318,500; metropolitan area pop. 599,760. The city was founded in 1813 by George Hamilton, a local landowner. **3** a city on North Island, New Zealand; pop. (1991) 148,625. **4** the capital of Bermuda; pop. (1980) 1,617.

Hamilton[2] /ˈhæmɪlt(ə)n/, Alexander (*c.*1757–1804), American Federalist politician. As First Secretary of the Treasury under Washington (1789–95), he established the US central banking system. Hamilton was a prime mover behind the Federalist Party's commitment to strong central government in the aftermath of American independence. He died from a gunshot wound after a duel with Aaron Burr.

Hamilton[3] /ˈhæmɪlt(ə)n/, Sir Charles (1900–78), New Zealand inventor and motor-racing driver. He is best known for his development of the jet boat, being knighted in 1974 for his services to manufacturing. Hamilton was also a successful motor-racing driver and was the first New Zealander to exceed 100 m.p.h.

Hamilton[4] /'hæmɪlt(ə)n/, Lady Emma (born Amy Lyon) (c.1765–1815), English beauty and mistress of Lord Nelson. In 1791 she married Sir William Hamilton (1730–1803), the British ambassador to Naples, after living with him there for five years. She first met Lord Nelson in Naples in 1793; they became lovers after his second visit in 1799. She had a daughter by him in 1801 and lived with him after her husband's death.

Hamilton[5] /'hæmɪlt(ə)n/, Sir William Rowan (1806–65), Irish mathematician and theoretical physicist. Hamilton made influential contributions to optics and in the foundations of algebra. He invented quaternions while investigating the subject of complex numbers. Hamilton's formulation of mechanics was incorporated into the equations of quantum mechanics.

Hamite /'hæmaɪt/ n. a member of a group of North African peoples, including the ancient Egyptians and Berbers, supposedly descended from Ham, son of Noah.

Hamitic /hə'mɪtɪk/ adj. **1** of or relating to certain Afro-Asiatic languages, including Egyptian, Berber, Chadic, and Cushitic, formerly classified together as a subfamily. **2** of or relating to the Hamites.

Hamito-Semitic /ˌhæmɪtəʊsɪ'mɪtɪk/ adj. = AFRO-ASIATIC.

Hamlet /'hæmlɪt/ a legendary prince of Denmark, hero of a tragedy by Shakespeare. □ **Hamlet without the Prince** a performance or event taking place without the principal actor etc.

hamlet /'hæmlɪt/ n. a small village, esp. one without a church. [ME f. AF hamelet(t)e, OF hamelet dimin. of hamel dimin. of ham f. MLG hamm]

Hamm /hæm/ an industrial city in NW Germany, in North Rhine-Westphalia, on the Lippe river; pop. (1991) 180,320. It was founded in 1226 and was a member of the Hanseatic League.

Hammer /'hæmə(r)/, Hammer Film Productions, a British film company founded in 1948. From the mid-1950s it acquired a reputation for its horror films, many of the best known starring Peter Cushing or Christopher Lee. Noted for their gory scenes and often garish colour, the films include The Curse of Frankenstein (1957), Dracula (1958), and The Vampire Lovers (1970).

hammer /'hæmə(r)/ n. & v. ● n. **1 a** a tool with a heavy metal head mounted at right angles on a handle, used for driving nails, breaking objects, etc. **b** a machine with a metal block serving the same purpose. **c** a similar contrivance, as for exploding the charge in a gun, striking the strings of a piano, etc. **2** an auctioneer's mallet, indicating by a sharp tap that an article is sold. **3 a** a metal ball of about 7 kg, attached to a wire for throwing in an athletic contest. **b** the sport of throwing the hammer. **4** Anat. = MALLEUS. ● v. **1 a** tr. & intr. hit or beat with or as with a hammer. **b** intr. strike loudly; knock violently (esp. on a door). **2** tr. **a** drive in (nails) with a hammer. **b** fasten or secure by hammering (hammered the lid down). **3** tr. (often foll. by in) inculcate (ideas, knowledge, etc.) forcefully or repeatedly. **4** tr. colloq. utterly defeat; inflict heavy damage on. **5** intr. (foll. by at, away at) work hard or persistently at. **6** tr. Stock Exch. declare (a person or a firm) a defaulter on the International Stock Exchange in London (until 1970 announced by three blows of a hammer followed by the broker's name). □ **come under the hammer** be sold at an auction. **hammer and sickle** the symbols of the industrial worker and the peasant used as the emblem of the former USSR and of international communism. **hammer and tongs** colloq. with great vigour and commotion. **hammer drill** a drill with a bit that moves backwards and forwards while rotating. **hammer out 1** make flat or smooth by hammering. **2** work out the details of (a plan, agreement, etc.) laboriously. **3** play (a tune, esp. on the piano) loudly or clumsily. **hammer-toe** a deformity in which the toe is bent permanently downwards. □ **hammerless** adj. [OE hamor, hamer]

hammerbeam /'hæmə̩biːm/ n. a wooden beam (often carved) projecting from a wall to support the principal rafter or the end of an arch.

Hammerfest /'hæmə̩fest/ a port in northern Norway, on North Kvaløy island; pop. (1991) 6,900. It is the northernmost town in Europe.

hammerhead /'hæmə̩hed/ n. **1** (also **hammerhead shark**) a shark of the family Sphyrnidae, with a flattened head and eyes in lateral extensions of it. **2** a long-legged African marsh bird, Scopus umbretta, with a thick bill and a crest.

hammerlock /'hæmə̩lɒk/ n. Wrestling a hold in which the arm is twisted and bent behind the back.

Hammerstein /'hæmə̩staɪn/, Oscar (full name Oscar Hammerstein II) (1895–1960), American librettist. He collaborated with the

composers Jerome Kern (e.g. in Showboat, 1927), Sigmund Romberg, and most notably with Richard Rodgers. He also wrote the libretto for the musical Carmen Jones (1943), an adaptation of Bizet's opera Carmen.

Hammett /'hæmɪt/, (Samuel) Dashiell (1894–1961), American novelist. His detective fiction, based in part on his own experiences as a detective, is characterized by a hard-boiled style and influenced Raymond Chandler and other writers in the genre. Many of Hammett's stories, including The Maltese Falcon (1930) and The Thin Man (1932), were made into successful films. He lived for many years with the dramatist Lillian Hellman; they were both persecuted for their left-wing views during the McCarthy era.

hammock /'hæmək/ n. a bed of canvas or rope network, suspended by cords at the ends, used esp. on board ship. [earlier hamaca f. Sp., of Carib orig.]

Hammond /'hæmənd/, Dame Joan (b.1912), Australian operatic soprano, born in New Zealand. She made her operatic début in Sydney in 1929 and went on to an international career with a wide repertoire. Her operatic successes include Puccini's Madame Butterfly and Turandot; she has also performed in choral works and oratorios, such as Handel's Messiah.

Hammurabi /ˌhæmʊ'rɑːbɪ/ (d.1750 BC), the sixth king of the first dynasty of Babylon, reigned 1792–1750 BC. He made Babylon the capital of Babylonia and extended the Babylonian empire. He instituted one of the earliest known legal codes, which took the form of 282 case laws dealing with the economy and with family, criminal, and civil law.

hammy /'hæmɪ/ adj. (**hammier, hammiest**) **1** of or like ham. **2** colloq. (of an actor or acting) over-theatrical.

Hamnett /'hæmnɪt/, Katharine (b.1952), English fashion designer. After working as a freelance designer she established her own company in 1979. Her designs are characterized by their loose, simple lines and use of utilitarian fabrics. She also made a name for herself as a feminist and supporter of CND.

hamper[1] /'hæmpə(r)/ n. **1** a large basket usu. with a hinged lid and containing food (picnic hamper). **2** Brit. a selection of food, drink, etc., for an occasion. [ME f. obs. hanaper, AF f. OF hanapier case for a goblet f. hanap goblet]

hamper[2] /'hæmpə(r)/ v. & n. ● v.tr. **1** prevent the free movement or activity of. **2** impede, hinder. ● n. Naut. necessary but cumbersome equipment on a ship. [ME: orig. unkn.]

Hampshire /'hæmpʃɪə(r)/ a county on the coast of southern England; county town, Winchester. Its largest cities are Southampton, from which it takes its name, and Portsmouth.

Hampstead /'hæmpstɪd/ a residential suburb of NW London. It contains Hampstead Heath, a large tract of open common land within the city, popular for recreation.

Hampton /'hæmptən/ a city in SE Virginia, on the harbour of Hampton Roads, on Chesapeake Bay; pop. (1990) 133,790.

Hampton Court a palace on the north bank of the Thames in the borough of Richmond-upon-Thames, London. It was built by Cardinal Wolsey as his private residence but later presented by him to Henry VIII, and was a favourite royal residence until the reign of George II. William III had part of it rebuilt by Sir Christopher Wren and the gardens laid out in formal Dutch style.

Hampton Roads a deep-water estuary 6 km (4 miles) long, formed by the James river where it joins Chesapeake Bay, on the Atlantic coast in SE Virginia. The ports of Newport News and Hampton are situated on it.

hamsin var. of KHAMSIN.

hamster /'hæmstə(r)/ n. a Eurasian rodent of the subfamily Cricetinae, having a short tail and large cheek pouches for storing food; esp. the common hamster (Cricetus cricetus), and the golden hamster (Mesocricetus auratus). [G f. OHG hamustro corn-weevil]

hamstring /'hæmstrɪŋ/ n. & v. ● n. **1** each of five tendons at the back of the knee in humans. **2** the great tendon at the back of the hock in quadrupeds. ● v.tr. (past and past part. **hamstrung** /-strʌŋ/ or **hamstringed**) **1** cripple by cutting the hamstrings of (a person or animal). **2** prevent the activity or efficiency of (a person or enterprise).

Hamsun /'hæmsʊn/, Knut (pseudonym of Knut Pedersen) (1859–1952), Norwegian novelist. He worked in a variety of manual jobs before the publication of his first novel, Hunger (1890), a semi-autobiographical account of the mental and physical hardships he experienced during this period. This successful novel was followed by further works exploring the human psyche and written in a similar fragmentary, vivid style, including *Growth of the Soil* (1917). He was awarded the Nobel Prize for literature in 1920.

hamulus /'hæmjʊləs/ *n.* (*pl.* **hamuli** /-ˌlaɪ/) *Anat., Zool., & Bot.* a hooklike projection. [L, dimin. of *hamus* hook]

Han /hæn/ the Chinese dynasty that ruled from 206 BC until AD 220 with only a brief interruption. During this period Chinese rule was extended over Mongolia, administration was in the hands of an organized civil service, Confucianism was recognized as the state philosophy, and detailed historical records were kept. The arts flourished, and technological advances included the invention of paper. It was to this era that later dynasties looked for their model. The term is now also used to describe the dominant ethnic group in China (as distinct from the many minority groups).

Hancock /'hæŋkɒk/, Tony (full name Anthony John Hancock) (1924–68), English comedian. He made his name in 1954 with the radio series *Hancock's Half Hour*, in which he played a materialistic, lonely misfit and was noted for his sardonic wit. The series readily adapted to television (1956–61); Hancock later turned to writing his own material and starred in other comedy shows, as well as appearing in several films, but failed to repeat his earlier success. He committed suicide in 1968.

hand /hænd/ *n. & v.* ● *n.* **1 a** the end part of the human arm beyond the wrist, including the fingers and thumb. **b** in other primates, the end part of a forelimb, also used as a foot. **2 a** (often in *pl.*) control, management, custody, disposal (*is in good hands*). **b** agency or influence (*suffered at their hands*). **c** a share in an action; active support. **3** a thing compared with a hand or its functions, esp. the pointer of a clock or watch. **4** the right or left side or direction relative to a person or thing. **5 a** a skill, esp. in something practical (*a hand for making pastry*). **b** a person skilful in some respect. **6** a person who does or makes something, esp. distinctively (*a picture by the same hand*). **7** an individual's writing or the style of this; a signature (*a legible hand*; *in one's own hand*; *witness the hand of …*). **8** a person etc. as the source of information etc. (*at first hand*). **9** a pledge of marriage. **10** a person as a source of manual labour esp. in a factory, on a farm, or on board ship. **11 a** the playing cards dealt to a player. **b** the player holding these. **c** a round of play. **12** *colloq.* applause (*got a big hand*). **13** the unit of measure of a horse's height, equal to 4 inches (10.2 cm). **14** a forehock of pork. **15** a bunch of bananas. **16** (*attrib.*) **a** operated or held in the hand (*hand-drill*; *hand-luggage*). **b** done by hand and not by machine (*hand-knitted*). ● *v.tr.* **1** (foll. by *in, to, over*, etc.) deliver; transfer by hand or otherwise. **2** convey verbally (*handed me a lot of abuse*). **3** *colloq.* give away too readily (*handed them the advantage*). □ **all hands 1** the entire crew of a ship. **2** the entire workforce. **at hand 1** close by. **2** about to happen. **by hand 1** by a person and not a machine. **2** delivered privately and not by the public post. **from hand to mouth** satisfying only one's immediate needs (also *attrib.*: *a hand-to-mouth existence*). **get** (or **have** or **keep**) **one's hand in** become (or be or remain) practised in something. **give** (or **lend**) **a hand** assist in an action or enterprise. **hand and foot** completely; satisfying all demands (*waited on them hand and foot*). **hand-axe** a prehistoric stone implement, normally oval or pear-shaped and worked on both sides, used for chopping, cutting, and scraping things. **hand cream** an emollient for the hands. **hand down 1** pass the ownership or use of to another, esp. a successor or descendant. **2 a** transmit (a decision) from a higher court etc. **b** *US* express (an opinion or verdict). **hand-grenade** see GRENADE. **hand-held** *adj.* designed to be held in the hand; compact, portable. ● *n.* a hand-held device, esp. a microcomputer. **hand in glove** in collusion or association. **hand in hand** in close association. **hand it to** *colloq.* acknowledge the merit of (a person). **hand-me-down** an article of clothing etc. passed on from another person. **hand off** *Rugby* push off (a tackling opponent) with the hand. **hand on** pass (a thing) to the next in a series or succession. **hand out 1** serve, distribute. **2** award, allocate (*the judges handed out stiff sentences*). **hand-out** *n.* **1** something given free to a needy person. **2** a statement given to the press etc. **hand over** deliver; surrender possession of. **hand-over** *n.* the act or an instance of handing over. **hand-over-fist** *colloq.* with rapid progress. **hand-pick** choose carefully or personally. **hand-picked** carefully or personally chosen. **hand round** distribute. **hands down** (esp. of winning) with no difficulty. **hands off** a warning not to touch or interfere with something. **hands-off** not involving direct control or intervention. **hands-on 1** *Computing* of or requiring personal operation at a keyboard. **2** involving or offering active participation rather than theory; direct, practical. **hands up!** an instruction to raise one's hands in surrender or to signify assent or participation. **hand-to-hand** (of fighting) at close quarters. **hand-tool** a tool operated by hand. **have** (or **take**) **a hand** (often foll. by *in*) share or take part. **have one's hands full** be fully occupied. **have one's hands tied** *colloq.* be unable to act. **hold one's hand** = *stay one's hand* below. **in hand 1** receiving attention. **2** in reserve; at one's disposal. **3** under one's control. **lay** (or **put**) **one's hands on** see LAY[1]. **off one's hands** no longer one's responsibility. **on every hand** (or **all hands**) to or from all directions. **on hand** in attendance, available. **on one's hands** resting on one as a responsibility. **on the one** (or **the other**) **hand** from one (or another) point of view. **out of hand 1** out of control. **2** peremptorily (*refused out of hand*). **put** (or **set**) **one's hand to** start work on; engage in. **stay one's hand** *archaic* or *literary* refrain from action. **to hand 1** within easy reach. **2** (of a letter) received. **turn one's hand to** undertake (as a new activity). □ **handless** *adj.* [OE *hand*, *hond*]

handbag /'hændbæg/ *n. & v.* ● *n.* a small bag for a purse etc., carried esp. by a woman. ● *v.tr. colloq.* (of a woman) treat (a person, idea, etc.) ruthlessly or insensitively. □ **handbagging** *n.*

handball *n.* **1** /'hændbɔːl/ a game resembling fives, in which a ball is hit with the hand in a walled court. The modern form of the game was first played in Germany in the 1890s. **2** /'hændbɔːl/ a game similar to football, in which the ball is thrown rather than kicked in an attempt to get it into the opposing team's goal. **3** /hænd'bɔːl/ *Football* intentional touching of the ball with the hand or arm by a player (other than the goalkeeper in the penalty area), constituting a foul.

handbell /'hændbel/ *n.* a small bell, usu. tuned to a particular note and rung by hand, esp. one of a set giving a range of notes.

handbill /'hændbɪl/ *n.* a printed notice distributed by hand.

handbook /'hændbʊk/ *n.* a short manual or guidebook.

handbrake /'hændbreɪk/ *n.* a brake operated by hand.

h. & c. *abbr.* hot and cold (water).

handcart /'hændkɑːt/ *n.* a small cart pushed or drawn by hand.

handclap /'hændklæp/ *n.* a clapping of the hands.

handcraft /'hændkrɑːft/ *n. & v.* ● *n.* = HANDICRAFT. ● *v.tr.* make by handicraft.

handcuff /'hændkʌf/ *n. & v.* ● *n.* (in *pl.*) a pair of lockable linked metal rings for securing a prisoner's wrists. ● *v.tr.* put handcuffs on.

-handed /'hændɪd/ *adj.* (in *comb.*) **1** for or involving a specified number of hands (in various senses) (*two-handed*). **2** using chiefly the hand specified (*left-handed*). □ **-handedly** *adv.* **-handedness** *n.*

Handel /'hænd(ə)l/, George Frederick (born Georg Friedrich Händel) (1685–1759), German-born composer, resident in England from 1712. He was a major baroque composer whose prolific output included choral works, chamber music, operas, concerti grossi, and orchestral pieces. He is now chiefly remembered for his oratorios, the most famous of which is *Messiah* (1742); other choral works include *Samson* (1743) and *Judas Maccabaeus* (1747). Of his many other works, perhaps the best known is the *Water Music* suite for orchestra (c.1717), written for George I's procession down the River Thames. Handel was also a noted organist and invented the organ concerto, which he intended to be performed between the acts of his own oratorios.

handful /'hændfʊl/ *n.* (*pl.* **-fuls**) **1** a quantity that fills the hand. **2** a small number or amount. **3** *colloq.* a troublesome person or task (*found the child quite a handful*).

handglass /'hændglɑːs/ *n.* **1** a magnifying glass held in the hand. **2** a small mirror with a handle.

handgrip /'hændgrɪp/ *n.* **1** a grasp with the hand. **2** a handle designed for easy holding.

handgun /'hændgʌn/ *n.* a small firearm held in and fired with one hand.

handhold /'hændhəʊld/ *n.* something for the hands to grip on (in climbing, sailing, etc.).

handicap /'hændɪˌkæp/ *n. & v.* ● *n.* **1 a** a disadvantage imposed on a superior competitor in order to make the chances more equal. **b** a race or contest in which this is imposed. **2** the number of strokes by which a golfer normally exceeds par for the course. **3** a thing that makes progress or success difficult. **4** a physical or mental disability. ● *v.tr.* (**handicapped**, **handicapping**) **1** impose a handicap on. **2** place (a person) at a disadvantage. □ **handicapper** *n.* [prob. from the phrase *hand i' (= in) cap* describing a kind of sporting lottery]

handicapped /'hændɪˌkæpt/ *adj.* suffering from a physical or mental disability.

handicraft /ˈhændɪˌkrɑːft/ n. work that requires both manual and artistic skill. [ME, alt. of earlier HANDCRAFT after HANDIWORK]

handiwork /ˈhændɪˌwɜːk/ n. work done or a thing made by hand, or by a particular person. [OE handgeweorc]

handkerchief /ˈhæŋkəˌtʃɪf, -ˌtʃiːf/ n. (pl. **handkerchiefs** or **-chieves** /-ˌtʃɪvz, -ˌtʃiːvz/) a square of cotton, linen, silk, etc., usu. carried in the pocket for wiping one's nose, etc.

handle /ˈhænd(ə)l/ n. & v. ● n. **1** the part by which a thing is held, carried, or controlled. **2** a fact that may be taken advantage of (gave a handle to his critics). **3** colloq. a personal title. **4** the feel of goods, esp. textiles, when handled. ● v.tr. **1** touch, feel, operate, or move with the hands. **2** manage or deal with; treat in a particular or correct way (knows how to handle people; unable to handle the situation). **3** deal in (goods). **4** discuss or write about (a subject). □ **get a handle on** colloq. understand the basis of or reason for (a situation, circumstance, etc.). □ **handled** adj. **handleable** adj. **handleability** /ˌhænd(ə)ləˈbɪlɪtɪ/ n. (also in comb.). [OE handle, handlian (as HAND)]

handlebar /ˈhænd(ə)lˌbɑː(r)/ n. (often in pl.) the steering bar of a bicycle etc., with a handgrip at each end. □ **handlebar moustache** a thick moustache extended sideways usu. with curved ends.

handler /ˈhændlə(r)/ n. **1** a person who handles or deals in certain commodities. **2** a person who trains and looks after an animal (esp. a police dog).

Handley Page, Frederick, see PAGE.

handlist /ˈhændlɪst/ n. a short list of essential reading, reference books, etc.

handmade /hændˈmeɪd/ adj. made by hand and not by machine, esp. as designating superior quality.

handmaid /ˈhændmeɪd/ n. (also **handmaiden** /-ˌmeɪd(ə)n/) archaic a female servant or helper.

handpump /ˈhændpʌmp/ n. a pump operated by hand.

handrail /ˈhændreɪl/ n. a narrow rail for holding as a support on stairs etc.

handsaw /ˈhændsɔː/ n. a saw worked by one hand.

handsel /ˈhæns(ə)l/ n. & v. (also **hansel**) archaic ● n. **1** a gift at the beginning of the new year, or on coming into new circumstances. **2** = EARNEST[2]. ● v.tr. (**handselled, handselling**; US **handseled, handseling**) **1** give a handsel to. **2** inaugurate; be the first to try. [ME, corresp. to OE handselen giving into a person's hands, ON handsal giving of the hand (esp. in promise), formed as HAND + OE sellan SELL]

handset /ˈhændset/ n. a telephone mouthpiece and earpiece forming one unit.

handshake /ˈhændʃeɪk/ n. **1** the shaking of a person's hand with one's own as a greeting etc. **2** a gratuity given as compensation (cf. golden handshake).

handsome /ˈhænsəm/ adj. (**handsomer, handsomest**) **1** (of a person, esp. a man) of pleasing appearance, good-looking. **2** (of a building etc.) imposing, attractive. **3 a** generous, liberal (a handsome present; handsome treatment). **b** (of a price, fortune, etc., as assets gained) considerable. □ **handsomeness** n. [ME, = easily handled, f. HAND + -SOME[1]]

handsomely /ˈhænsəmlɪ/ adv. **1** generously, liberally. **2** finely, beautifully. **3** Naut. carefully.

handspike /ˈhændspaɪk/ n. a wooden rod tipped with metal, used as a lever.

handspring /ˈhændsprɪŋ/ n. a somersault in which one lands first on the hands and then on the feet.

handstand /ˈhændstænd/ n. an act of balancing on one's hands with the feet in the air or against a wall.

handwork /ˈhændwɜːk/ n. work done with the hands, esp. as opposed to machinery. □ **handworked** adj.

handwriting /ˈhændˌraɪtɪŋ/ n. **1** writing with a pen, pencil, etc. **2** a person's particular style of writing. □ **handwritten** /-ˌrɪt(ə)n/ adj.

handy /ˈhændɪ/ adj. (**handier, handiest**) **1** convenient to handle or use; useful. **2** ready to hand; placed or occurring conveniently. **3** clever with the hands. □ **handily** adv. **handiness** n.

handyman /ˈhændɪˌmæn/ n. (pl. **-men**) a person able or employed to do occasional domestic repairs and minor renovations.

hang /hæŋ/ v. & n. ● v. (past and past part. **hung** /hʌŋ/ except in sense 7) **1** tr. **a** secure or cause to be supported from above, esp. with the lower part free. **b** (foll. by up, on, on to, etc.) attach loosely by suspending from the top. **2** tr. set up (a door, gate, etc.) on its hinges so that it moves freely. **3** tr. place (a picture) on a wall or in an exhibition. **4** tr. attach (wallpaper) in vertical strips to a wall. **5** tr. (foll. by on) colloq. attach the blame for (a thing) to (a person) (you can't hang that on me). **6** tr. (foll. by with) decorate by hanging pictures or decorations etc. (a hall hung with tapestries). **7** tr. & intr. (past and past part. **hanged**) **a** suspend or be suspended by the neck with a noosed rope until dead, esp. as a form of capital punishment. **b** as a mild oath (hang the expense; let everything go hang). **8** tr. let droop (hang one's head). **9** tr. suspend (meat or game) from a hook and leave it until dry or tender or high. **10** intr. be or remain hung (in various senses). **11** intr. remain static in the air. **12** intr. (often foll. by over) be present or imminent, esp. oppressively or threateningly (a hush hung over the room). **13** intr. (foll. by on) **a** be contingent or dependent on (everything hangs on the discussions). **b** listen closely to (hangs on their every word). ● n. **1** the way a thing hangs or falls. **2** a downward droop or bend. □ **be hung up on** sl. have a psychological or emotional obsession or problem about (is really hung up on her father). **get the hang of** colloq. understand the technique or meaning of. **hang about** (or **around**) **1** loiter or dally; not move away. **b** linger near (a person or place). **2** (foll. by with) associate with (a person etc.). **hang back 1** show reluctance to act or move. **2** remain behind. **hang fire** be slow in taking action or in progressing. **hang heavily** (or **heavy**) (of time) pass slowly. **hang in** esp. N. Amer. colloq. **1** persist, persevere. **2** linger. **hang loose** colloq. relax, stay calm. **hang on 1** colloq. continue or persevere, esp. with difficulty. **2** (often foll. by to) cling, retain one's grip. **3** (foll. by to) retain; fail to give back. **4 a** colloq. wait for a short time. **b** (in telephoning) continue to listen during a pause in the conversation. **hang out 1** hang from a window, clothes-line, etc. **2** protrude or cause to protrude downwards. **3** (foll. by of) lean out of (a window etc.). **4** sl. reside or be often present. **5** (foll. by with) sl. accompany, be friends with. **hang-out** n. sl. a place one lives in or frequently visits. **hang together 1** make sense. **2** remain associated. **hang up 1** hang from a hook, peg, etc. **2** end a telephone conversation, esp. (foll. by on) abruptly (then he hung up on me). **3** cause delay or difficulty to. **hang-up** n. sl. an emotional problem or inhibition. **hungover** colloq. suffering from a hangover. **hung parliament** a parliament in which no party has a clear majority. **let it all hang out** sl. be uninhibited or relaxed. **not care** (or **give**) **a hang** colloq. not care at all. [ON hanga (tr.) = OE hōn, & f. OE hangian (intr.), f. Gmc]

hangar /ˈhæŋə(r)/ n. a building with extensive floor area, for housing aircraft etc. □ **hangarage** n. [F, of unkn. orig.]

Hangchow see HANGZHOU.

hangdog /ˈhæŋdɒg/ adj. having a dejected or guilty appearance; shamefaced.

hanger[1] /ˈhæŋə(r)/ n. **1** a person or thing that hangs. **2** (in full **coat-hanger**) a shaped piece of wood or plastic etc. from which clothes may be hung. □ **hanger-on** (pl. **hangers-on**) a follower or dependant, esp. an unwelcome one.

hanger[2] /ˈhæŋə(r)/ n. Brit. a wood on the side of a steep hill. [OE hangra f. hangian HANG]

hang-glider /ˈhæŋˌglaɪdə(r)/ n. a form of glider consisting of a frame with a fabric aerofoil stretched over it, from which the operator is suspended and controls flight by body movement. □ **hang-glide** v.intr. **hang-gliding** n.

hanging /ˈhæŋɪŋ/ n. & adj. ● n. **1 a** the practice or an act of executing by hanging a person. **b** (attrib.) meriting or causing this (a hanging offence). **2** (usu. in pl.) draperies hung on a wall etc. ● adj. **1** that hangs or is hung; suspended. **2** (of gardens or woodland) on a steep slope. □ **hanging valley** a valley which is cut across by a deeper valley or ends in a precipice.

Hanging Gardens of Babylon terraced gardens at Babylon, watered by pumps from the Euphrates, whose construction was ascribed to Nebuchadnezzar (c.600 BC). They were one of the Seven Wonders of the World.

hangman /ˈhæŋmən/ n. (pl. **-men**) **1** an executioner who hangs condemned persons. **2** a word-game for two players, in which the tally of failed guesses at the letters of an unknown word is kept by drawing a representation of a gallows.

hangnail /ˈhæŋneɪl/ n. = AGNAIL. [alt. of AGNAIL, infl. by HANG and taking nail as = NAIL n. 2a]

hangover /ˈhæŋˌəʊvə(r)/ n. **1** a severe headache or other after-effects caused by drinking an excess of alcohol (woke up with a hangover). **2** a survival from the past.

Hang Seng index /hæŋ ˈseŋ/ an index based on the average movement in the price of selected securities on the Hong Kong Stock Exchange. [name of a bank in Hong Kong]

Hangzhou /hæŋ'dʒəʊ/ (also **Hangchow** /-'tʃaʊ/) the capital of Zhejiang province in eastern China, situated on Hangzhou Bay, an inlet of the Yellow Sea, at the southern end of the Grand Canal; pop. (1989) 1,330,000.

hank /hæŋk/ n. **1** a coil or skein of wool or thread etc. **2** any of several measures of length of cloth or yarn, e.g. 840 yds for cotton yarn and 560 yds for worsted. **3** Naut. a ring of rope, iron, etc., for securing the staysails to the stays. [ME f. ON hönk: cf. Sw. hank string, Danish hank handle]

hanker /'hæŋkə(r)/ v.intr. (foll. by to + infin., for, or after) long for; crave. □ **hankerer** n. **hankering** n. [obs. hank, prob. rel. to HANG]

hanky /'hæŋkɪ/ n. (also **hankie**) (pl. **-ies**) colloq. a handkerchief. [abbr.]

hanky-panky /ˌhæŋkɪ'pæŋkɪ/ n. colloq. **1** naughtiness, esp. sexual misbehaviour. **2** dishonest dealing; trickery. [19th c.: perh. based on hocus-pocus]

Hannibal /'hænɪb(ə)l/ (247–182 BC), Carthaginian general. He precipitated the second Punic War by attacking the town of Saguntum in Spain, an ally of Rome. In 218, in a pre-emptive move, he led an army of about 30,000 over the Alps into Italy. There he inflicted a series of defeats on the Romans, campaigning for sixteen years undefeated but failing to take Rome itself. After being recalled to Africa he was defeated at Zama by Scipio Africanus in 202.

Hannover see HANOVER.

Hanoi /hæˈnɔɪ/ the capital of Vietnam, situated on the Red River in the north of the country; pop. (est. 1984) 925,000. It was the capital of French Indo-China from 1887 to 1946 and of North Vietnam before the reunification of North and South Vietnam.

Hanover /'hænəvə(r)/ (German **Hannover** /haˈnoːfər/) **1** an industrial city in NW Germany, on the Mittelland Canal; pop. (1991) 517,480. It is the capital of Lower Saxony. **2** a former state and province in northern Germany. It was an electorate of the Holy Roman Empire from 1692 until 1806, ruled by the Guelph dynasty, and from 1866 until 1945 was a province of Prussia. In 1714 the Elector of Hanover succeeded to the British throne as George I, and from then until 1837 the same monarch ruled both Britain and Hanover. With the accession of Victoria (1837) to the British throne, however, Hanover passed to her uncle, Ernest, Duke of Cumberland (1771–1851), the Hanoverian succession being denied to a woman as long as a male member of the Guelph family survived. **3** The British royal house from 1714 to the death of Queen Victoria in 1901. □ **Hanoverian** /ˌhænəˈvɪərɪən/ adj. & n.

Hansa /'hænsə/ n. (also **Hanse**) **1 a** a medieval merchant guild. **b** the entrance fee to a guild. **2** the Hanseatic League. [MHG hanse, OHG, Goth. hansa company]

Hansard /'hænsɑːd/ the official verbatim record of the proceedings of the Houses of Parliament, colloquially so called because for most of the 19th century it was published by Messrs Hansard. Since 1909 publication has been by HMSO.

Hanseatic League /ˌhænsɪˈætɪk/ a medieval association of north German cities, formed in 1241 as a commercial alliance for trade between the eastern and western parts of northern Europe. In the later Middle Ages the League, with about 100 member towns, functioned as an independent political power, with its own army and navy. It began to collapse in the early 17th century and only three major cities (Hamburg, Bremen, and Lübeck) remained when it finally disbanded in the 19th century. [HANSA]

hansel var. of HANDSEL.

Hansen's disease /'hæns(ə)nz/ n. Med. = LEPROSY 1. [Gerhard H. A. Hansen, Norw. physician (1841–1912)]

hansom /'hænsəm/ n. (in full **hansom cab**) hist. a two-wheeled horse-drawn cab accommodating two inside, with the driver seated behind. [Joseph A. Hansom, Engl. architect (1803–82), who designed it]

Hants abbr. Hampshire. [OE Hantescire]

Hanukkah /'hɑːnəkə, 'xɑːn-/ (also **Chanukkah** /'xɑːn-/) the eight-day Jewish festival of lights, beginning in December, commemorating the rededication of the Temple in 165 BC after its desecration by the Syrians. [Heb. ḥanukkāh consecration]

hanuman /ˌhʌnʊˈmɑːn/ n. **1** a langur monkey, Presbytis entellus, native to India and Sri Lanka and venerated by Hindus. **2** (**Hanuman**) Hinduism a semi-divine monkey-like creature credited with extraordinary powers, whose exploits are described in the Ramayana. [Skr.]

Haora see HOWRAH.

hap /hæp/ n. & v. archaic ● n. **1** chance, luck. **2** a chance occurrence.

● v.intr. (**happed, happing**) **1** come about by chance. **2** (foll. by to + infin.) happen to. [ME f. ON happ]

hapax legomenon /ˌhæpæks lɪˈɡɒmɪˌnɒn/ n. (pl. **hapax legomena** /-mɪnə/) a word of which only one instance of use is recorded. [Gk, = a thing said once]

ha'penny var. of HALFPENNY.

haphazard /hæpˈhæzəd/ adj. & adv. ● adj. done etc. by chance; random. ● adv. at random. □ **haphazardly** adv. **haphazardness** n. [HAP + HAZARD]

hapless /'hæplɪs/ adj. (esp. of a person) unfortunate (hapless victims). □ **haplessly** adv. **haplessness** n. [HAP + -LESS]

haplography /hæpˈlɒɡrəfɪ/ n. the accidental omission of letters when these are repeated in a word (e.g. philogy for philology). [Gk haplous single + -GRAPHY]

haploid /'hæplɔɪd/ adj. & n. Biol. ● adj. **1** (of a cell) containing a single set of unpaired chromosomes. **2** (of an organism) composed of haploid cells. ● n. a haploid organism or cell. [G f. Gk haplous single + eidos form]

haplology /hæpˈlɒlədʒɪ/ n. the omission of a sound when this is repeated within a word (e.g. February pronounced /'febrɪ/). [Gk haplous single + -LOGY]

ha'p'orth Brit. var. of HALFPENNYWORTH.

happen /'hæp(ə)n/ v. & adv. ● v.intr. **1** occur (by chance or otherwise). **2** (foll. by to + infin.) have the (good or bad) fortune to (I happened to meet her). **3** (foll. by to) be the (esp. unwelcome) fate or experience of (what happened to you?; I hope nothing happens to them). **4** (foll. by on) encounter or discover by chance. ● adv. N. Engl. dial. perhaps, maybe (happen it'll rain). □ **as it happens** in fact; in reality (as it happens, it turned out well). [ME f. HAP + -EN[1]]

happening /'hæp(ə)nɪŋ/ n. & adj. ● n. **1** an event or occurrence. **2** an improvised or spontaneous theatrical etc. performance. ● adj. sl. exciting, fashionable, trendy.

happenstance /'hæpənstəns/ n. esp. N. Amer. a thing that happens by chance. [HAPPEN + CIRCUMSTANCE]

happi /'hæpɪ/ n. (also **happi-coat**) (pl. **happis**) a loose informal Japanese coat. [Jap.]

happy /'hæpɪ/ adj. (**happier, happiest**) **1** feeling or showing pleasure or contentment. **2 a** fortunate; characterized by happiness. **b** (of words, behaviour, etc.) apt, pleasing. **3** colloq. slightly drunk. **4** (in comb.) colloq. inclined to use excessively or at random (trigger-happy). □ **happy event** colloq. the birth of a child. **happy families** a card-game the object of which is to acquire four members of the same 'family'. **happy-go-lucky** cheerfully casual. **happy hour** a period of the day when drinks are sold at reduced prices in bars, hotels, etc. **happy hunting-ground** a place where success or enjoyment is obtained. **happy medium** a compromise; the avoidance of extremes. □ **happily** adv. **happiness** n. [ME f. HAP + -Y[1]]

Hapsburg var. of HABSBURG.

haptic /'hæptɪk/ adj. relating to the sense of touch. [Gk haptikos able to touch f. haptō fasten]

hara-kiri /ˌhærəˈkɪrɪ/ n. ritual suicide by disembowelment with a sword, formerly practised by Samurai to avoid dishonour. [colloq. Jap. f. hara belly + kiri cutting]

harangue /həˈræŋ/ n. & v. ● n. a lengthy and earnest speech. ● v.tr. lecture or make a harangue to. □ **haranguer** n. [ME f. F f. OF arenge f. med.L harenga, perh. f. Gmc]

Harappa /həˈræpə/ an ancient city of the Indus valley civilization (c.2600–1700 BC), in northern Pakistan. The site of the ruins was discovered in 1920.

Harare /həˈrɑːrɪ/ the capital of Zimbabwe; pop. (est. 1982) 656,000. It was known as Salisbury until 1982.

harass /'hærəs, disp. həˈræs/ v.tr. **1** trouble and annoy continually or repeatedly. **2** make repeated attacks on (an enemy or opponent). □ **harasser** n. **harassingly** adv. **harassment** n. [F harasser f. OF harer set a dog on]

Harbin /hɑːˈbiːn, -ˈbɪn/ the capital of Heilongjiang province in NE China, on the Songhua river; pop. (est. 1986) 2,630,000.

harbinger /'hɑːbɪndʒə(r)/ n. & v. ● n. **1** a person or thing that announces or signals the approach of another. **2** a forerunner. ● v.tr. announce the approach of. [earlier = 'a person who provides lodging': ME herbergere f. OF f. herberge lodging f. Gmc]

harbour /'hɑːbə(r)/ n. & v. (US **harbor**) ● n. **1** a place of shelter for ships. **2** a shelter; a place of refuge or protection. ● v. **1** tr. give shelter to (esp.

a criminal or wanted person). **2** *tr.* keep in one's mind, esp. resentfully (*harbour a grudge*). **3** *intr.* come to anchor in a harbour. □ **harbourmaster** an official in charge of a harbour. **harbour seal** *N. Amer.* = *common seal*. □ **harbourless** *adj.* [OE *herebeorg* perh. f. ON, rel. to HARBINGER]

harbourage /ˈhɑːbərɪdʒ/ *n.* (US **harborage**) a shelter or place of shelter, esp. for ships.

hard /hɑːd/ *adj., adv., & n.* ● *adj.* **1** (of a substance, material, etc.) firm and solid; unyielding to pressure; not easily cut. **2 a** difficult to understand or explain (*a hard problem*). **b** difficult to accomplish (*a hard decision*). **c** (foll. by *to* + infin.) not easy (*hard to believe; hard to please*). **3** difficult to bear; entailing suffering (*a hard life*). **4** (of a person) unfeeling; severely critical. **5** (of a season or the weather) severe, harsh (*a hard winter; a hard frost*). **6** harsh or unpleasant to the senses (*a hard voice; hard colours*). **7 a** strenuous, enthusiastic, intense (*a hard worker; a hard fight*). **b** severe, uncompromising (*a hard blow; a hard bargain; hard words*). **c** politically extreme; most radical (*the hard right*). **8 a** (of liquor) strongly alcoholic. **b** (of drugs) potent and addictive. **c** (of radiation) highly penetrating. **d** (of pornography) highly obscene and explicit. **9** (of water) containing mineral salts that make lathering difficult. **10** established; not disputable; reliable (*hard facts; hard data*). **11** *Stock Exch.* (of currency, prices, etc.) high; not likely to fall in value. **12** *Phonet.* (of a consonant) guttural (as *c* in *cat*, *g* in *go*). **13** (of a shape, boundary, etc.) clearly defined, unambiguous. ● *adv.* **1** strenuously, intensely, copiously; with one's full effort (*try hard; look hard at; is raining hard; hard-working*). **2** with difficulty or effort (*hard-earned*). **3** so as to be hard or firm (*hard-baked; the jelly set hard*). ● *n. Brit.* **1** a sloping roadway across a foreshore. **2** *sl.* = *hard labour* (*got two years hard*). □ **be hard on 1** be difficult for. **2** be severe in one's treatment or criticism of. **3** be unpleasant to (the senses). **be hard put to it** (usu. foll. by *to* + infin.) find it difficult. **go hard with** turn out to (a person's) disadvantage. **hard and fast** (of a rule or a distinction made) definite, unalterable, strict. **hard at it** *colloq.* busily working or occupied. **hard by** near; close by. **a hard case 1** *colloq.* **a** an intractable person. **b** *Austral. & NZ* an amusing or eccentric person. **2** a case of hardship. **hard cash** negotiable coins and banknotes. **hard coal** anthracite. **hard copy** printed material produced by computer, usu. on paper, suitable for ordinary reading. **hard disk** *Computing* a stack of several rigid magnetic disks used for large-capacity data storage. **hard-done-by** harshly or unfairly treated. **hard-earned** that has taken a great deal of effort to earn or acquire. **hard error** *Computing* a permanent error. **hard feelings** feelings of resentment. **hard hat 1** a protective helmet worn on building sites etc. **2** *colloq.* a reactionary person. **hard hit** badly affected. **hard-hitting** aggressively critical. **hard labour** heavy manual work as a punishment, esp. in a prison. **hard landing 1** a clumsy or rough landing of an aircraft. **2** an uncontrolled landing in which a spacecraft is destroyed. **hard line** unyielding adherence to a firm policy. **hard-liner** a person who adheres rigidly to a policy. **hard lines** *Brit. colloq.* = *hard luck*. **hard luck** worse fortune than one deserves. **hard-nosed** realistic, uncompromising. **a hard nut** *sl.* a tough, aggressive person. **a hard nut to crack 1** a difficult problem. **2** a person or thing not easily understood or influenced. **hard of hearing** somewhat deaf. **hard on** (or **upon**) close to in pursuit etc. **hard-on** *n.* *coarse sl.* an erection of the penis. **hard pad** a form of distemper in dogs etc. **hard palate** the front part of the palate. **hard-paste** denoting a Chinese or 'true' porcelain made of fusible and infusible materials (usu. clay and stone) and fired at a high temperature. **hard-pressed 1** closely pursued. **2** burdened with urgent business. **hard rock** rock music with a heavy beat. **hard roe** see ROE[1] **1**. **hard sauce** a sauce of butter and sugar, often with brandy etc. added. **hard sell** aggressive salesmanship or advertising. **hard shoulder** *Brit.* a hardened strip alongside a motorway for stopping on in an emergency. **hard stuff** *colloq.* strong alcoholic drink, esp. whisky. **hard tack** a ship's biscuit. **hard up 1** short of money. **2** (foll. by *for*) at a loss for; lacking. **hard-wearing** able to stand much wear. **hard wheat** wheat with a hard grain rich in gluten. **hard-wired** involving or achieved by permanently connected circuits designed to perform a specific function. **hard-working** diligent. **put the hard word on** *Austral. & NZ sl.* ask a favour (esp. sexual or financial) of. □ **hardish** *adj.* **hardness** *n.* [OE *hard, heard* f. Gmc]

hardback /ˈhɑːdbæk/ *adj. & n.* ● *adj.* (of a book) bound in stiff covers. ● *n.* a hardback book.

hardball /ˈhɑːdbɔːl/ *n. & v. N. Amer.* ● *n.* **1** = BASEBALL. **2** *sl.* uncompromising methods or dealings, esp. in politics (*play hardball*). ● *v.tr. sl.* pressure or coerce politically.

hardbitten /hɑːdˈbɪt(ə)n/ *adj.* tough and cynical.

hardboard /ˈhɑːdbɔːd/ *n.* stiff board made of compressed and treated wood pulp.

hard-boiled /hɑːdˈbɔɪld/ *adj.* **1** (of an egg) boiled until the white and the yolk are solid. **2** (of a person) tough, shrewd. **3** relating to or denoting a tough, realistic style of detective fiction produced in the US from the 1930s by writers such as Dashiell Hammett, James M. Cain, and Raymond Chandler.

hardcore /ˈhɑːdkɔː(r)/ *adj.* **1** forming a nucleus or centre. **2** blatant, uncompromising. **3** (of pornography) = HARD *adj.* 8d. **4** (of drug addiction) relating to 'hard' drugs, esp. heroin.

hard core *n.* **1** an irreducible nucleus. **2 a** the most active or committed members of a society etc. **b** a conservative or reactionary minority. **3** (usu. **hardcore**) *Brit.* solid material, esp. rubble, forming the foundation of a road etc. **4** (usu. **hardcore**) popular music that is experimental in nature and usu. characterized by high volume and aggressive presentation.

harden /ˈhɑːd(ə)n/ *v.* **1** *tr. & intr.* make or become hard or harder. **2** *intr. & tr.* become, or make (one's attitude etc.), obdurate, uncompromising, or less sympathetic. **3** *intr.* (of prices etc.) cease to fall or fluctuate. □ **harden off** inure (a plant) to cold by gradual increase of its exposure. □ **hardener** *n.*

hardening /ˈhɑːd(ə)nɪŋ/ *n.* **1** the process or an instance of becoming hard. **2** (in full **hardening of the arteries**) *Med.* = ARTERIOSCLEROSIS.

hard-headed /hɑːdˈhedɪd/ *adj.* practical, realistic; not sentimental. □ **hard-headedly** *adv.* **hard-headedness** *n.*

hardheads /ˈhɑːdhedz/ *n.* (as *sing.* or *pl.*) a tough-stemmed knapweed, *Centaurea nigra*, with dense purple flower-heads, native to Europe and widely introduced elsewhere.

hard-hearted /hɑːdˈhɑːtɪd/ *adj.* unfeeling, unsympathetic. □ **hard-heartedly** *adv.* **hard-heartedness** *n.*

Hardie /ˈhɑːdɪ/, (James) Keir (1856–1915), Scottish Labour politician. He worked as a miner before entering Parliament in 1892, becoming the first leader of the Independent Labour Party the next year. In 1906 he became a co-founder and first leader of the Labour Party. His pacifist views prompted his withdrawal from Labour politics when the majority of his party's MPs declared their support for British participation in the First World War, although he remained an MP until his death.

hardihood /ˈhɑːdɪhʊd/ *n.* boldness, daring.

Harding /ˈhɑːdɪŋ/, Warren (Gamaliel) (1865–1923), American Republican statesman, 29th President of the US 1921–3.

hardly /ˈhɑːdlɪ/ *adv.* **1** scarcely; only just (*we hardly knew them*). **2** only with difficulty (*could hardly speak*). **3** harshly. **4** (as *int.*) certainly not, far from it. □ **hardly any** almost no; almost none. **hardly ever** very rarely.

hardpan /ˈhɑːdpæn/ *n. Geol.* a hardened layer of clay occurring in or below the soil profile.

hardshell /ˈhɑːdʃel/ *adj.* **1** having a hard shell. **2** esp. *US* rigid, orthodox, uncompromising.

hardship /ˈhɑːdʃɪp/ *n.* **1** severe suffering or privation. **2** the circumstance causing this.

hardstanding /ˈhɑːdˌstændɪŋ/ *n.* an area of hard material for a vehicle to stand on when not in use.

hardtop /ˈhɑːdtɒp/ *n.* a motor car with a rigid (usu. detachable) roof.

Hardwar /hɑːˈdwɑː(r)/ a city in Uttar Pradesh, northern India, on the River Ganges; pop. (1991) 188,960. It is a place of Hindu pilgrimage.

hardware /ˈhɑːdweə(r)/ *n.* **1** tools and household articles of metal etc. **2** heavy machinery or armaments. **3** the mechanical and electronic components of a computer etc. (cf. SOFTWARE).

hardwood /ˈhɑːdwʊd/ *n.* the wood from a deciduous broad-leaved tree as distinguished from that of conifers.

Hardy[1], Oliver, see LAUREL AND HARDY.

Hardy[2] /ˈhɑːdɪ/, Thomas (1840–1928), English novelist and poet. He spent most of his life in his native Dorset (the 'Wessex' of his novels). A recurrent theme in Hardy's work is the struggle of human beings against the indifferent force that inflicts the sufferings and ironies of life. Major novels include *The Mayor of Casterbridge* (1886), *Tess of the D'Urbervilles* (1891), and *Jude the Obscure* (1896). He turned to writing poetry in the late 1890s and published eight volumes of poems, as well as a drama in blank verse, *The Dynasts* (1904–8).

hardy /ˈhɑːdɪ/ *adj.* (**hardier, hardiest**) **1** robust; capable of enduring difficult conditions. **2** (of a plant) able to grow in the open air all the year. □ **hardy annual 1** an annual plant that may be sown in the open. **2** *joc.* a subject that comes up at regular intervals. □ **hardily** *adv.* **hardiness** *n.* [ME f. OF *hardi* past part. of *hardir* become bold, f. Gmc, rel. to HARD]

Hare /heə(r)/, William, see BURKE[4].

hare /heə(r)/ *n. & v.* ● *n.* **1** a mammal, esp. of the genus *Lepus*, resembling a large rabbit, with tawny fur, long ears, short tail, and very long hind legs, inhabiting open country; esp. *L. europaeus*, of Eurasia. **2** (in full **electric hare**) a dummy hare propelled by electricity, used in greyhound racing. ● *v.intr.* run with great speed. □ **hare and hounds** a paper-chase. **hare-brained** rash, foolish, wild. **hare's-foot** (in full **hare's-foot clover**) a clover, *Trifolium arvense*, with soft hair around the flowers. **run with the hare and hunt with the hounds** try to remain on good terms with both sides. **start a hare** raise a topic of conversation. [OE *hara* f. Gmc]

harebell /ˈheəbel/ *n.* a small bellflower, *Campanula rotundifolia*, with slender stems and pale blue flowers. Also called *bluebell* in Scotland.

Harefoot /ˈheəfʊt/, Harold, Harold I of England (see HAROLD).

Hare Krishna /ˌhɑːrɪ ˈkrɪʃnə/ *n.* **1** a religious sect in the US and elsewhere, the International Society for Krishna Consciousness, founded in New York in 1966. (*See note below.*) **2** a member of this sect. [Skr., = O Vishnu Krishna (a devotional chant)]

▪ The society takes its name from a devotional chant or mantra based on the name of the Hindu god Krishna. Its principles include strict vegetarianism, celibacy outside marriage, and a complete ban on gambling and drugs. Members dress in saffron robes (the men with shaven heads), and daily ritual includes dance-marching through the streets to the accompaniment of oriental instruments and chanting of the Hare Krishna mantra.

harelip /ˈheəlɪp/ *n.* often *offens.* a cleft lip. □ **harelipped** *adj.*

harem /ˈhɑːriːm, hɑːˈriːm/ *n.* **1 a** the women of a Muslim household, living in a separate part of the house. **b** their quarters. **2** a group of female animals sharing a mate. [Arab. *ḥarām, ḥarīm*, orig. = prohibited, prohibited place, f. *ḥarama* prohibit]

harewood /ˈheəwʊd/ *n.* stained sycamore-wood used for making furniture. [G dial. *Ehre* f. L *acer* maple + WOOD]

Hargeisa /hɑːˈgeɪsə/ (also **Hargeysa**) a city in NW Somalia; pop. (est.) 400,000. It was the capital of British Somaliland between 1941 and 1960.

Hargreaves /ˈhɑːgriːvz/, James (1720–78), English inventor. A pioneer of the Lancashire cotton industry, he invented *c.*1760 an improved carding-machine, which used a roller with multiple pins to comb out the cotton fibres. In 1770 he patented his most famous invention, the spinning-jenny. His success in speeding up the spinning process caused opposition; spinners on the old-fashioned wheels were alarmed at the threat to their employment and in 1768 his house and machinery were destroyed by a mob.

haricot /ˈhærɪkəʊ/ *n.* **1** (in full **haricot bean**) a variety of French bean with small white seeds. **2** the dried seed of this used as a vegetable. [F]

Harijan /ˈhærɪdʒən, ˈhʌrɪ-/ *n.* a member of the class of untouchables in the Indian subcontinent. The name was introduced by Mahatma Gandhi in the interests of providing a positive image for the class. [Skr., = a person dedicated to Vishnu, f. *Hari* Vishnu, *jana* person]

hark /hɑːk/ *v.intr.* (usu. in *imper.*) *archaic* listen attentively. □ **hark back** revert to a topic discussed earlier. [ME *herkien* f. OE *heorcian* (unrecorded): cf. HEARKEN: *hark back* was orig. a hunting call to retrace steps]

harken var. of HEARKEN.

harl /hɑːl/ *n.* (also **harle, herl** /hɜːl/) fibre of flax or hemp. [MLG *herle, harle* fibre of flax or hemp]

Harlech /ˈhɑːlek, -ləx/ a village in Gwynedd, on the west coast of Wales; pop. (1989) 1,313. It is noted for the ruins of a castle, built in the 13th century by Edward I.

Harlem /ˈhɑːləm/ a district of New York City, situated to the north of 96th Street in NE Manhattan. It has a large black population and in the 1920s and 1930s was noted for its nightclubs and jazz bands.

Harlem Globetrotters a US professional basketball team who tour the world giving exhibition matches combining ball skills with comedy routines. They were formed in 1927.

Harlem Renaissance *n.* a movement in US literature in the 1920s which centred on Harlem in New York, an early manifestation of black consciousness in the US. The movement, stimulated by W. E. B. Du Bois's magazine *Crisis*, included writers such as Langston Hughes (1902–67) and Zora Neale Hurston.

Harlequin /ˈhɑːlɪkwɪn/ a clownlike servant, originally a stock character in Italian *commedia dell'arte*. In English pantomime he is a mute character usually wearing a mask and multi-coloured tights, and is supposed to be invisible to the clown and Pantaloon. [F f. earlier *Herlequin* leader of a legendary nocturnal band of demon horsemen]

harlequin /ˈhɑːlɪkwɪn/ *adj.* brightly coloured; (of a bird, animal, etc.) having distinctively variegated markings. □ **harlequin duck** an Icelandic duck, *Histrionicus histrionicus*, the male of which has grey-blue plumage with chestnut and white markings.

harlequinade /ˌhɑːlɪkwɪˈneɪd/ *n.* **1** a play or section of a pantomime in which Harlequin plays the leading role. (*See note below.*) **2** a piece of buffoonery. [F *arlequinade* (as HARLEQUIN)]

▪ Harlequinade was an important element in the development of the English pantomime. It originated in Italian *commedia dell'arte* and consisted of story-telling dances in which Harlequin played the leading role. In the 19th century Joseph Grimaldi made the English character of the clown into the chief personage, and the harlequinade began to be preceded by a fairy tale. As Harlequin's importance lessened and the fairy tales became longer and more popular, the harlequinade dwindled into a short epilogue to what became the present English pantomime and was finally abandoned completely.

Harley Street /ˈhɑːlɪ/ a street in central London long associated with the premises of eminent physicians and surgeons. The name is also used allusively to refer to medical specialists.

harlot /ˈhɑːlət/ *n. archaic* a prostitute. □ **harlotry** *n.* [ME f. OF *harlot, herlot* lad, knave, vagabond]

Harlow /ˈhɑːləʊ/ a town in west Essex, north of London; pop. (1981) 79,520. It was designated as a new town in 1947.

harm /hɑːm/ *n. & v.* ● *n.* hurt, damage. ● *v.tr.* cause harm to. □ **out of harm's way** in safety. [OE *hearm, hearmian* f. Gmc]

harmattan /hɑːˈmæt(ə)n/ *n.* a parching dusty land-wind of the West African coast occurring from December to February. [Twi *haramata*]

harmful /ˈhɑːmfʊl/ *adj.* causing or likely to cause harm. □ **harmfully** *adv.* **harmfulness** *n.*

harmless /ˈhɑːmlɪs/ *adj.* **1** not able or likely to cause harm. **2** inoffensive. □ **harmlessly** *adv.* **harmlessness** *n.*

harmonic /hɑːˈmɒnɪk/ *adj. & n.* ● *adj.* **1** of or characterized by harmony; harmonious. **2** *Mus.* **a** of or relating to harmony. **b** (of a tone) produced by vibration of a string etc. in an exact fraction of its length (cf. FUNDAMENTAL). **3** *Math.* of or relating to quantities whose reciprocals are in arithmetical progression. ● *n.* **1** *Mus.* an overtone accompanying at a fixed interval (and forming a note with) a fundamental. **2** *Physics* a component frequency of wave motion. □ **harmonic motion** (in full **simple harmonic motion**) oscillatory motion under a retarding force proportional to the amount of displacement from an equilibrium position. **harmonic progression** (or **series**) *Math.* a series of quantities whose reciprocals are in arithmetical progression. □ **harmonically** *adv.* [L *harmonicus* f. Gk *harmonikos* (as HARMONY)]

harmonica /hɑːˈmɒnɪkə/ *n.* a small rectangular wind instrument with a row of metal reeds along its length, held against the lips and moved from side to side to produce different notes by blowing or sucking; a mouth-organ. [L, fem. sing. or neut. pl. of *harmonicus*: see HARMONIC]

harmonious /hɑːˈməʊnɪəs/ *adj.* **1** sweet-sounding, tuneful. **2** forming a pleasing or consistent whole; concordant. **3** free from disagreement or dissent. □ **harmoniously** *adv.* **harmoniousness** *n.*

harmonist /ˈhɑːmənɪst/ *n.* a person skilled in musical harmony, a harmonizer. □ **harmonistic** /ˌhɑːməˈnɪstɪk/ *adj.*

harmonium /hɑːˈməʊnɪəm/ *n.* a keyboard instrument in which the notes are produced by air driven through metal reeds by bellows operated by the feet. [F f. L (as HARMONY)]

harmonize /ˈhɑːmənaɪz/ *v.* (also **-ise**) **1** *tr.* add notes to (a melody) to produce harmony. **2** *tr. & intr.* (often foll. by *with*) bring into or be in harmony. **3** *intr.* make or form a pleasing or consistent whole. □ **harmonization** /ˌhɑːmənaɪˈzeɪʃ(ə)n/ *n.* [f. F *harmoniser* (as HARMONY)]

harmony /ˈhɑːmənɪ/ *n.* (pl. **-ies**) **1 a** a combination of simultaneously sounded musical notes to produce chords and chord progressions, esp. as having a pleasing effect. **b** the study of this. **2 a** an apt or aesthetic arrangement of parts. **b** the pleasing effect of this. **3** agreement, concord. **4** a collation of parallel narratives, esp. of the Gospels. □ **in harmony 1** (of singing etc.) producing chords; not discordant. **2** (often foll. by *with*) in agreement. **harmony of the spheres** see SPHERE. [ME

f. OF *harmonie* f. L *harmonia* f. Gk *harmonia* joining, concord, f. *harmos* joint]

Harmsworth /'hɑːmzwɜːθ/, Alfred Charles William, see NORTHCLIFFE.

harness /'hɑːnɪs/ *n. & v.* ● *n.* **1** the equipment of straps and fittings by which a horse or other draught animal may be linked to a cart etc. and controlled. **2** a similar arrangement for fastening a thing to a person's body, for restraining a young child, etc. ● *v.tr.* **1 a** put a harness on (esp. a horse). **b** (foll. by *to*) attach by a harness. **2** make use of (natural resources), esp. to produce energy. □ **harness racing** see TROTTING. **in harness** in the routine of daily work. □ **harnesser** *n.* [ME f. OF *harneis* military equipment f. ON *hernest* (unrecorded) f. *herr* army + *nest* provisions]

Harold /'hærəld/ the name of two kings of England:

Harold I (known as Harold Harefoot) (d.1040), reigned 1035–40. Harold was an illegitimate son of Canute and first came to the throne when his half-brother Hardecanute (Canute's legitimate heir) was king of Denmark and thus absent at the time of his father's death. When a third royal claimant was murdered a year later, Harold was formally recognized as king, although Hardecanute returned to the kingdom when Harold himself died.

Harold II (c.1019–66), reigned 1066, the last Anglo-Saxon king of England. He succeeded Edward the Confessor, having dominated the latter's court in the last years of his reign, but was faced with two invasions within months of his accession. He resisted his half-brother Tostig and the Norse king Harald Hardrada at Stamford Bridge, but was killed and his army defeated at the Battle of Hastings; the victor, William of Normandy, took the throne as William I.

Haroun-al-Raschid see HARUN AR-RASHID.

harp /hɑːp/ *n. & v.* ● *n.* **1** a large upright roughly triangular musical instrument consisting of a set of strings placed over an open frame so that they can be plucked or swept with the fingers from both sides. (*See note below.*) **2** (in full **mouth-harp**) a mouth-organ, a harmonica. ● *v.intr.* **1** (foll. by *on, on about*) talk repeatedly and tediously about. **2** play on a harp. □ **harp-seal** a seal of the NW Atlantic and Arctic Ocean, *Phoca groenlandica*, with a dark harp-shaped mark on its back. □ **harper** *n.* **harpist** *n.* [OE *hearpe* f. Gmc]

▪ The standard orchestral harp has a range of about six-and-a-half octaves, but has only seven strings to each octave, further notes being obtained by use of pedals. The harp may be the oldest stringed instrument: it is depicted in various forms in pre-Christian times, is mentioned in the Bible, and was played in Ireland, Scotland, Wales, and France from the 9th and 10th centuries onwards.

Harpers Ferry /'hɑːpəz/ a small town in Jefferson County, West Virginia, at the junction of the Potomac and Shenandoah rivers. It is famous for a raid in Oct. 1859 in which John Brown and a group of abolitionists captured a Federal arsenal located there.

Harpocrates /hɑː'pɒkrəˌtiːz/ see HORUS.

harpoon /hɑː'puːn/ *n. & v.* ● *n.* a barbed spearlike missile with a rope attached, for catching whales etc. ● *v.tr.* spear with a harpoon. □ **harpoon-gun** a gun for firing a harpoon. □ **harpooner** *n.* [F *harpon* f. *harpe* clamp f. L *harpa* f. Gk *harpē* sickle]

harpsichord /'hɑːpsɪˌkɔːd/ *n.* a wing-shaped keyboard instrument having strings plucked by a small leather or quill plectrum. The first detailed descriptions are found in the mid-15th century, but the heyday of the harpsichord was from the late 16th to the mid-18th century. Its tone is crisp and clear, but the instrument is unable to make a gradual progression between loud and soft sounds by pressure on the keys. It was overtaken in popularity by the piano, which can do this, at the end of the 18th century, although it enjoyed something of a revival in the 20th century. □ **harpsichordist** *n.* [obs. F *harpechorde* f. LL *harpa* harp, + *chorda* string, the *-s-* being unexplained]

harpy /'hɑːpɪ/ *n.* (pl. **-ies**) **1** Gk & Rom. Mythol. a rapacious monster having a woman's face and body and a bird's wings and claws. (*See note below.*) **2** a grasping unscrupulous person. □ **harpy eagle** a very large crested eagle, *Harpia harpyja*, found in South America. [F *harpie* or L *harpyia* f. Gk *harpuiai* snatchers (cf. *harpazō* snatch)]

▪ The harpies (whose names are Aello, Ocypete, and Celoeno) were originally storm winds in Greek mythology, portrayed in ancient art as winged women. Virgil, however, describes them as repellent birds with women's faces, and they were later portrayed in this form.

harquebus /'hɑːkwɪbəs/ *n.* (also **arquebus** /'ɑːk-/) *hist.* an early type of portable gun supported on a tripod or on a forked rest. [F (h)*arquebuse* ult. f. MLG *hakebusse* or MHG *hakenbühse*, f. *haken* hook + *busse* gun]

harridan /'hærɪd(ə)n/ *n.* a bad-tempered old woman. [17th-c. slang, perhaps f. F *haridelle* old horse]

harrier[1] /'hærɪə(r)/ *n.* a person who harries or lays waste.

harrier[2] /'hærɪə(r)/ *n.* **1 a** a hound used for hunting hares. **b** (in pl.) a pack of these with huntsmen. **2** (in pl.) (usu. **Harriers** as part of a club's name) cross-country runners as a group or club. [HARE + -IER, assim. to HARRIER[1]]

harrier[3] /'hærɪə(r)/ *n.* a slender bird of prey of the genus *Circus*, with long wings swooping over the ground. [*harrower* f. *harrow* harry, rob, assim. to HARRIER[1]]

Harris /'hærɪs/ the southern part of the island of Lewis with Harris, in the Outer Hebrides.

Harrisburg /'hærɪsˌbɜːg/ the state capital of Pennsylvania, on the Susquehanna river; pop. (1990) 52,370. The nearby nuclear power station at Three Mile Island suffered a serious accident in 1979 (see THREE MILE ISLAND).

Harrison[1] /'hærɪs(ə)n/, Benjamin (1833–1901), American Republican statesman, 23rd President of the US 1889–93. He was the grandson of William Henry Harrison.

Harrison[2] /'hærɪs(ə)n/, Sir Rex (born Reginald Carey Harrison) (1908–90), English actor. He first appeared on the stage in 1924 and made his film début in 1930 with *The Great Game*. His most famous role was as Professor Higgins in the stage and film musical *My Fair Lady* (1956, 1964). His other films include *Blithe Spirit* (1944) and *Cleopatra* (1962).

Harrison[3] /'hærɪs(ə)n/, William Henry (1773–1841), American Whig statesman, 9th President of the US, 1841. He died of pneumonia one month after his inauguration. He was the grandfather of Benjamin Harrison.

Harris tweed *n.* a hand-woven tweed made in the Outer Hebrides, esp. on the island of Lewis with Harris.

Harrod /'hærəd/, Charles Henry (1800–85), English grocer and tea merchant. In 1853 he took over a shop in Knightsbridge, London, which, after expansion by his son Charles Digby Harrod (1841–1905), became a prestigious department store.

harrow /'hærəʊ/ *n. & v.* ● *n.* a heavy frame with iron teeth dragged over ploughed land to break up clods, remove weeds, cover seed, etc. ● *v.tr.* **1** draw a harrow over (land). **2** (usu. as **harrowing** adj.) distress greatly. □ **harrower** *n.* **harrowingly** adv. [ME f. ON *hervi*]

Harrowing of Hell (in medieval Christian theology) the defeat of the powers of evil and the release of its victims, by the descent of Christ into hell after his death. This event is mentioned in the Epistle to the Ephesians 4:9 and included in certain early creeds (but not the Nicene Creed). It is a frequent subject of medieval mystery plays and of Orthodox icons.

Harrow School /'hærəʊ/ a boys' public school in NW London. It was founded and endowed under a charter granted by Queen Elizabeth I in 1571.

harrumph /hə'rʌmf/ *v.intr.* clear the throat or make a similar sound, esp. ostentatiously. [imit.]

harry /'hærɪ/ *v.tr.* (**-ies, -ied**) **1** ravage or despoil. **2** harass, worry. [OE *herian, hergian* f. Gmc, rel. to OE *here* army]

harsh /hɑːʃ/ *adj.* **1** unpleasantly rough or sharp, esp. to the senses. **2** severe, cruel. □ **harshen** *v.tr. & intr.* **harshly** adv. **harshness** *n.* [MLG *harsch* rough, lit. 'hairy', f. *haer* HAIR]

harslet var. of HASLET.

hart /hɑːt/ *n.* the male of a deer, esp. the red deer, usu. over five years old. □ **hart's tongue** a fern, *Phyllitis scolopendrium*, with narrow undivided fronds. [OE *heoro(t)* f. Gmc]

hartal /hɑː'tɑːl/ *n.* the closing of shops and offices in the Indian subcontinent as a mark of protest or sorrow. [Hind. *hartāl, hattāl* f. Skr. *hatta* shop + *tālaka* lock]

Harte /hɑːt/, (Francis) Bret (1836–1902), American short-story writer and poet. He is chiefly remembered for his stories about life in a gold-mining settlement, which were inspired by his own brief experience of mining and collected in works such as *The Luck of Roaring Camp* (1870).

hartebeest /'hɑːtɪˌbiːst/ *n.* a large African antelope, esp. of the genus *Alcelaphus*, with ringed horns bent back at the tips. [Afrik. f. Du. *hert* HART + *beest* BEAST]

Hartford /'hɑːtfəd/ the state capital of Connecticut, situated in the centre of the state on the Connecticut river; pop. (1990) 139,740.

Hartlepool /'hɑːtlɪˌpuːl/ a port on the North Sea coast of NE England, in the county of Cleveland; pop. (1981) 92,130.

Hartley /'hɑːtlɪ/, L(eslie) P(oles) (1895–1972), English novelist. Much of his work deals with memory and the effects of childhood experience on adult life and character, as in his trilogy *The Shrimp and the Anemone* (1944), *The Sixth Heaven* (1946), and *Eustace and Hilda* (1947), as well as the novel *The Go-Between* (1953).

Hartnell /'hɑːtn(ə)l/, Sir Norman (1901–79), English couturier. He is remembered as the dressmaker to Queen Elizabeth II (whose coronation gown he designed) and the Queen Mother.

hartshorn /'hɑːtshɔːn/ *n.* **1** a substance obtained from the horns of a hart, formerly a source of ammonia. **2** (in full **spirit of hartshorn**) *archaic* an aqueous solution of ammonia. [OE (as HART, HORN)]

harum-scarum /ˌheərəm'skeərəm/ *adj. & n.* ● *adj.* wild and reckless. ● *n.* such a person. [rhyming formation on HARE, SCARE]

Harun ar-Rashid /hæˌruːn ɑːræ'ʃiːd/ (also **Haroun-al-Raschid** /-ælræ'ʃiːd/) (763–809), fifth Abbasid caliph of Baghdad 786–809. He was the most powerful and vigorous of the Abbasid caliphs; he and his court were made famous by their portrayal in the *Arabian Nights*.

haruspex /hə'rʌspeks/ *n.* (*pl.* **haruspices** /-spɪˌsiːz/) *Rom. Antiq.* a religious official who interpreted omens from the inspection of animals' entrails. [L]

Harvard classification /'hɑːvəd/ *n. Astron.* a system of classifying stars according to their spectral types, prepared at the Harvard College Observatory and first published in 1890. The range of classes (O, B, A, F, G, K, and M) represents a series of falling surface temperatures, from bluish-white (very hot) to dull red (cool).

Harvard University /'hɑːvəd/ the oldest American university, founded in 1636 at Cambridge, Massachusetts. It is named after John Harvard (1607–38), an English settler who bequeathed to it his library and half his estate.

harvest /'hɑːvɪst/ *n. & v.* ● *n.* **1 a** the process of gathering in crops etc. **b** the season when this takes place. **2** the season's yield or crop. **3** the product or result of any action. ● *v.tr.* **1 a** gather as a harvest, reap. **b** earn, obtain as a result of harvesting. **2** experience (consequences). □ **harvest festival** a thanksgiving festival in church for the harvest. **harvest home** the close of harvesting or the festival to mark this. **harvest mite** the biting larva of a mite of the genus *Trombicula*, common at harvest-time; a jigger. **harvest moon** the full moon nearest to the autumn equinox. **harvest mouse** a small mouse, *Micromys minutus*, that frequently nests in the stalks of growing grain and has a prehensile tail. □ **harvestable** *adj.* [OE *hærfest* f. Gmc]

harvester /'hɑːvɪstə(r)/ *n.* **1** a reaper. **2** a reaping-machine, esp. with sheaf-binding.

harvestman /'hɑːvɪstmən/ *n.* (*pl.* **-men**) an arachnid of the order Opiliones, with long thin legs and a small rounded body.

Harvey /'hɑːvɪ/, William (1578–1657), English discoverer of the circulation of the blood and physician to James I and Charles I. Harvey set out to provide a satisfactory account of the motion of the heart, and in *De Motu Cordis* (1628) concluded that it forcibly expelled blood in contraction. He drew attention to the quantity of blood emerging from the heart into the arteries, and deduced that it must pass through the flesh and enter the veins, returning once more to the heart. Harvey also studied embryology and animal locomotion.

Harwich /'hærɪtʃ/ a port in Essex, on the North Sea coast of SE England; pop. (1981) 17,330. It has extensive freight terminal and ferry links with northern Europe.

Haryana /ˌhʌrɪ'ɑːnə/ a state of northern India; capital, Chandigarh. It was formed in 1966, largely from Hindi-speaking parts of the former state of Punjab.

Harz Mountains /hɑːts/ a range of mountains in central Germany, the highest of which is the Brocken. The region is the source of many legends about witchcraft and sorcery.

has 3rd sing. present of HAVE.

has-been /'hæzbiːn/ *n. colloq.* a person or thing that has lost a former popularity, importance, or usefulness.

Hasdrubal¹ /'hæzdrʊb(ə)l/ (d.221 BC), Carthaginian general. He was the son-in-law of Hamilcar, whom he accompanied to Spain in 237. Hasdrubal advanced to the Ebro, which became recognized as the boundary between Carthaginian and Roman spheres of influence.

Hasdrubal² /'hæzdrʊb(ə)l/ (d.207 BC), Carthaginian general. He was the son of Hamilcar and younger brother of Hannibal. At the start of the second Punic War in 218 he was left in command of Carthaginian forces in Spain after Hannibal had departed for Italy. After a defeat, Hasdrubal campaigned with only moderate success before crossing

the Alps with the aim of joining Hannibal, but was intercepted and killed in battle.

Hašek /'hæʃek/, Jaroslav (1883–1923), Czech novelist and short-story writer. He is chiefly known as the author of an unfinished four-volume work published in Czechoslovakia between 1921 and 1923; it first appeared in Britain in a bowdlerized form as *The Good Soldier Schweik* (1930). The book is a comic novel satirizing military life and bureaucracy; its central character is the archetypal 'little man' fighting against the system.

hash¹ /hæʃ/ *n. & v.* ● *n.* **1** a dish of cooked meat cut into small pieces and recooked. **2 a** a mixture; a jumble. **b** a mess. **3** re-used or recycled material. ● *v.tr.* (often foll. by *up*) **1** make (meat etc.) into a hash. **2** recycle (old material). □ **make a hash of** make a mess of; bungle. **settle a person's hash** *colloq.* deal with and subdue a person. [F *hacher* f. *hache* HATCHET]

hash² /hæʃ/ *n. colloq.* hashish. [abbr.]

hash³ /hæʃ/ *n.* (also **hash sign**) the symbol #. □ **hash mark** *Amer. Football* either of two broken lines running the length of the field and enclosing a central zone within which all plays must start.

Hashemite /'hæʃɪˌmaɪt/ *n. & adj.* ● *n.* a member of an Arab princely family claiming descent from Hashim, great-grandfather of Muhammad. (See also JORDAN 2.) ● *adj.* of or relating to the Hashemites.

hashish /'hæʃiːʃ/ *n.* a resinous product of the top leaves and tender parts of hemp, smoked or chewed for its narcotic effects. [f. Arab. *ḥašīš* dry herb, powdered hemp leaves]

Hasid /'hæsɪd/ *n.* (also **Chasid**) (*pl.* **Hasidim** /-sɪˌdiːm/) **1** a member of a strictly orthodox Jewish sect in Palestine in the 3rd and 2nd centuries BC which opposed Hellenizing influences on their faith and supported the Maccabean revolt. **2** a member of a Jewish sect founded in the 18th century. (See HASIDISM.) □ **Hasidic** /hə'sɪdɪk/ *adj.* [Heb. *ḥasîd* pious (person)]

Hasidism /'hæsɪˌdɪz(ə)m/ *n.* (also **Chasidism**) an influential mystical Jewish movement founded in Poland in the 18th century in reaction to the rigid academicism of rabbinical Judaism. The movement, which emphasized the importance of religious enthusiasm rather than learning, had a strong popular following; it was inspired by Israel ben Eliezer (c.1700–60), called Baal-Shem-Tov (Heb., = master of the good name) because of his reputation as a miraculous healer. Denounced in 1781 as heretical, the movement declined sharply in the 19th century, but fundamentalist communities developed from it and Hasidism is still a force in Jewish life, particularly in Israel and New York.

haslet /'hæzlɪt/ *n.* (also **harslet** /'hɑːz-/) pieces of (esp. pig's) offal cooked together and usu. compressed into a meat loaf. [ME f. OF *hastelet* dimin. of *haste* roast meat, spit, f. OLG, OHG *harst* roast]

Hasmonean /ˌhæzmə'nɪən/ *adj. & n.* ● *adj.* of or relating to the Jewish dynasty established by the Maccabees. ● *n.* a member of this dynasty. [Gk *Asmōnaios* f. Heb. *ḥašmônây* Hasmon, name of reputed ancestor]

hasn't /'hæz(ə)nt/ *contr.* has not.

hasp /hɑːsp/ *n. & v.* ● *n.* a hinged metal clasp that fits over a staple and can be secured by a padlock. ● *v.tr.* fasten with a hasp. [OE *hæpse, hæsp*]

Hasselt /'hæselt/ a city in NE Belgium, on the River Demer, capital of the province of Limburg; pop. (1991) 66,610.

hassle /'hæs(ə)l/ *n. & v. colloq.* ● *n.* **1** a prolonged trouble or inconvenience. **2** an argument or involved struggle. ● *v.* **1** *tr.* harass, annoy; cause trouble to. **2** *intr.* argue, quarrel. [20th c.: orig. dial.]

hassock /'hæsək/ *n.* **1** a thick firm cushion for kneeling on, esp. in church. **2** a tuft of matted grass etc. [OE *hassuc*]

hast /hæst/ *archaic* 2nd sing. present of HAVE.

hastate /'hæsteɪt/ *adj. Bot.* triangular like the head of a spear. [L *hastatus* f. *hasta* spear]

haste /heɪst/ *n. & v.* ● *n.* **1** urgency of movement or action. **2** excessive hurry. ● *v.intr. archaic* = HASTEN. □ **in haste** quickly, hurriedly. **make haste** hurry; be quick. [ME f. OF *haste, haster* f. WG]

hasten /'heɪs(ə)n/ *v.* **1** *intr.* (often foll. by *to* + infin.) make haste; hurry. **2** *tr.* cause to occur or be ready or be done sooner.

Hastings /'heɪstɪŋz/, Warren (1732–1818), British colonial administrator. In 1774 he became India's first Governor-General and during his term of office introduced many of the administrative reforms vital to the successful maintenance of British rule in India. On his return to England in 1785 he was impeached for corruption; he was eventually acquitted in 1795 after a seven-year trial before the House of Lords.

Hastings, Battle of a decisive battle which took place in 1066 just north of the town of Hastings, East Sussex. William of Normandy (William the Conqueror) defeated the forces of the Anglo-Saxon king Harold II; Harold died in the battle, leaving the way open for William to seize London and the vacant throne and leading to the subsequent Norman Conquest of England.

hasty /ˈheɪstɪ/ *adj.* (**hastier, hastiest**) **1** hurried; acting quickly or hurriedly. **2** said, made, or done too quickly or too soon; rash, unconsidered. **3** quick-tempered. □ **hastily** *adv.* **hastiness** *n.* [ME f. OF *hasti, hastif* (as HASTE, -IVE)]

hat /hæt/ *n. & v.* ● *n.* **1** a covering for the head, often with a brim and worn out of doors. **2** *colloq.* a person's occupation or capacity, esp. one of several (*wearing his managerial hat*). ● *v.tr.* (**hatted, hatting**) cover or provide with a hat. □ **hat trick 1** *Football* the scoring of three goals in a game by one player. **2** *Cricket* the taking of three wickets by the same bowler with three successive balls. **3** the scoring of three points, successes, etc., in sport etc. **keep it under one's hat** keep it secret. **out of a hat** by random selection. **pass the hat round** collect contributions of money. **take off one's hat to** acknowledge admiration for. **throw one's hat in the ring** take up a challenge. □ **hatful** *n.* (*pl.* **-fuls**). **hatless** *adj.* [OE *hætt* f. Gmc]

hatband /ˈhætbænd/ *n.* a band of ribbon etc. round a hat above the brim.

hatbox /ˈhætbɒks/ *n.* a box to hold a hat, esp. for travelling.

hatch[1] /hætʃ/ *n.* **1** an opening between two rooms, e.g. between a kitchen and a dining-room for serving food. **2** an opening or door in an aircraft, spacecraft, etc. **3** *Naut.* **a** = HATCHWAY. **b** a trapdoor or cover for this (often in *pl.*: *batten the hatches*). **4** a floodgate. □ **down the hatch** (as a drinking toast) drink up! cheers! **under hatches 1** below deck. **2 a** down out of sight. **b** brought low; dead. [OE *hæcc* f. Gmc]

hatch[2] /hætʃ/ *v. & n.* ● *v.* **1** *intr.* **a** (often foll. by *out*) (of a young bird or fish etc.) emerge from the egg. **b** (of an egg) produce a young animal. **2** *tr.* incubate (an egg). **3** *tr.* (also foll. by *up*) devise (a plot etc.). ● *n.* **1** the act or an instance of hatching. **2** a brood hatched. [ME *hacche*, of unkn. orig.]

hatch[3] /hætʃ/ *v.tr.* mark (a surface, e.g. a map or drawing) with close parallel lines. [ME f. F *hacher* f. *hache* HATCHET]

hatchback /ˈhætʃbæk/ *n.* **1** a car with a sloping back hinged at the top to form a door. **2** such a sloping back.

hatchery /ˈhætʃərɪ/ *n.* (*pl.* **-ies**) a place for hatching eggs, esp. of fish or poultry.

hatchet /ˈhætʃɪt/ *n.* a light short-handled axe. □ **hatchet-faced** *colloq.* sharp-featured or grim-looking. **hatchet fish 1** a tropical American freshwater fish of the family Gasteropelecidae, kept in aquariums. **2** a deep-sea fish of the family Sternoptychidae. **hatchet job** *colloq.* a fierce verbal attack on a person, esp. in print. **hatchet man** *colloq.* **1** a hired killer. **2** a person employed to carry out a hatchet job. [ME f. OF *hachette* dimin. of *hache* axe f. med.L *hapia* f. Gmc]

hatching /ˈhætʃɪŋ/ *n.* close parallel lines forming shading, esp. on a map or an architectural drawing.

hatchling /ˈhætʃlɪŋ/ *n.* a bird or fish that has just hatched.

hatchment /ˈhætʃmənt/ *n.* a large usu. diamond-shaped tablet with a deceased person's armorial bearings, affixed to that person's house, tomb, etc. [contr. of ACHIEVEMENT]

hatchway /ˈhætʃweɪ/ *n.* an opening in a ship's deck for lowering cargo into the hold.

hate /heɪt/ *v. & n.* ● *v.tr.* **1** dislike intensely; feel hatred towards. **2 a** dislike. **b** (foll. by verbal noun or *to* + infin.) be reluctant (to do something) (*I hate to disturb you*). ● *n.* **1** hatred. **2** a hated person or thing. □ **hatable** *adj.* (also **hateable**). **hater** *n.* [OE *hatian* f. Gmc]

hateful /ˈheɪtfʊl/ *adj.* arousing hatred, odious. □ **hatefully** *adv.* **hatefulness** *n.*

hath /hæθ/ *archaic* 3rd sing. present of HAVE.

Hathaway /ˈhæθəˌweɪ/, Anne (*c*.1557–1623), the wife of Shakespeare, whom she married in 1582. They had three children, a daughter (Susannah) and a twin daughter and son (Judith and Hamnet).

hatha yoga /ˈhæθə/ *n.* a system of physical exercises and breathing control used in yoga. (See YOGA.) [Skr. *haṭha* force + YOGA]

Hathor /ˈhæθɔː(r)/ *Egyptian Mythol.* a sky-goddess, the patron of love and joy, represented variously as a cow, with a cow's head or ears, or with a solar disc between a cow's horns. Her name means 'House of Horus'.

hatpin /ˈhætpɪn/ *n.* a long pin, often decorative, for securing a hat to the hair.

hatred /ˈheɪtrɪd/ *n.* intense dislike or ill will. [ME f. HATE + *-red* f. OE *rǣden* condition]

Hatshepsut /hætˈʃepsʊt/ (d.1482 BC), Egyptian queen of the 18th dynasty, reigned *c*.1503–1482 BC. She was the wife of her half-brother Tuthmosis II, on whose death she became regent for her nephew Tuthmosis III. She then proclaimed herself co-ruler and dominated the partnership until her death. Her reign was predominantly peaceful and she promoted Egypt's cultural life.

hatstand /ˈhætstænd/ *n.* a stand with hooks on which to hang hats.

hatter /ˈhætə(r)/ *n.* a maker or seller of hats. □ **as mad as a hatter** wildly eccentric.

Hattusa /ˈhætuːˌsæ/ the capital of the ancient Hittite empire, situated in central Turkey about 35 km (22 miles) east of Ankara. It is the site of the modern Turkish village of Bogazköy.

hauberk /ˈhɔːbɜːk/ *n. hist.* a coat of mail. [ME f. OF *hau(s)berc* f. Frank., = neck protection, f. *hals* neck + *berg-* f. *beorg* protection]

haughty /ˈhɔːtɪ/ *adj.* (**haughtier, haughtiest**) arrogantly self-admiring and disdainful; imperious, overbearing. □ **haughtily** *adv.* **haughtiness** *n.* [extension of *haught* (adj.), earlier *haut* f. OF *haut* f. L *altus* high]

haul /hɔːl/ *v. & n.* ● *v.* **1** *tr.* pull or drag forcibly. **2** *tr.* transport by lorry, cart, etc. **3** *intr.* turn a ship's course. **4** *tr.* (usu. foll. by *up*) *colloq.* bring for reprimand or trial. ● *n.* **1** the act or an instance of hauling. **2** an amount gained or acquired. **3** a distance to be traversed (*a short haul*). □ **haul over the coals** see COAL. [var. of HALE[2]]

haulage /ˈhɔːlɪdʒ/ *n.* **1** the commercial transport of goods. **2** a charge for this.

hauler /ˈhɔːlə(r)/ *n.* **1** a person or thing that hauls. **2** a miner who takes coal from the face to the bottom of the shaft. **3** a person or firm engaged in the transport of goods.

haulier /ˈhɔːlɪə(r)/ *n. Brit.* = HAULER.

haulm /hɔːm, hɑːm/ *n.* (also **halm**) **1** a stalk or stem. **2** the stalks or stems collectively of peas, beans, potatoes, etc., without the pods etc. [OE *h(e)alm* f. Gmc]

haunch /hɔːntʃ/ *n.* **1** (often in *pl.*) the fleshy part of the buttock with the thigh, esp. in animals. **2** the leg and loin of a deer etc. as food. **3** the side of an arch between the crown and the pier. [ME f. OF *hanche*, of Gmc orig.: cf. LG *hanke* hind leg of a horse]

haunt /hɔːnt/ *v. & n.* ● *v.* **1** *tr.* **a** (of a ghost) visit (a place) regularly, usu. reputedly giving signs of its presence. **b** (as **haunted** *adj.*) (of a place) frequented by a ghost. **2** *tr.* (of a person or animal) frequent or be persistently in (a place). **3** *tr.* (of a memory etc.) be persistently in the mind of. **4** *intr.* (foll. by *with, in*) stay habitually. ● *n.* **1** (often in *pl.*) a place frequented by a person. **2** a place frequented by animals, esp. for food and drink. □ **haunter** *n.* [ME f. OF *hanter* f. Gmc]

haunting /ˈhɔːntɪŋ/ *adj.* (of a memory, melody, etc.) poignant, wistful, evocative. □ **hauntingly** *adv.*

Hauptmann /ˈhaʊptmən/, Gerhart (1862–1946), German dramatist. He was an early pioneer of naturalism in the German theatre; his plays, such as *Before Sunrise* (1889) and *The Weavers* (1892), treat social and moral issues with directness and realism. He also wrote plays which combined naturalism with symbolism, such as *The Ascension of Joan* (1893). He was awarded the Nobel Prize for literature in 1912.

Hausa /ˈhaʊzə, ˈhaʊsə/ *n. & adj.* ● *n.* (*pl.* same or **Hausas**) **1** a member of a people of northern Nigeria and adjacent regions. **2** the Chadic language of this people, one of the most important in West Africa. It is spoken by some 25–30 million people, mainly in Nigeria and Niger. ● *adj.* of or relating to the Hausa or their language. [Hausa]

hausfrau /ˈhaʊsfraʊ/ *n.* a German housewife. [G f. *Haus* house + *Frau* woman]

hautboy /ˈəʊbɔɪ/ *archaic* = OBOE 1. [F: see OBOE]

haute couture /ˌəʊt kuːˈtjʊə(r)/ *n.* high fashion; the leading fashion houses or their products. [F, lit. 'high dressmaking']

haute cuisine /ˌəʊt kwɪˈziːn/ *n.* cookery of a high standard, esp. of the French traditional school. [F, lit. 'high cookery']

haute école /ˌəʊt eɪˈkɒl/ *n.* the art or practice of advanced classical dressage. [F, = high school]

Haute-Normandie /ˌəʊtnɔːmɒnˈdɪ/ a region of northern France, on the coast of the English Channel, including the city of Rouen. It was formed from part of the former province of Normandy.

hauteur /əʊˈtɜː(r)/ *n.* haughtiness of manner. [F f. *haut* high]

haut monde /ˌəʊ ˈmɒnd/ *n.* fashionable society. [F, lit. 'high world']

Havana[1] /həˈvænə/ the capital of Cuba, situated on the north coast; pop. (est. 1986) 2,014,800. It was founded in 1515 by the explorer Diego Velázquez de Cuéllar and is noted for its cigars.

Havana[2] /həˈvænə/ n. (also **Havana cigar**) a cigar made at Havana or elsewhere in Cuba.

Havant /ˈhæv(ə)nt/ a town in SE Hampshire; pop. (1981) 50,220.

have /hæv, həv/ v. & n. ● v. (3rd sing. present **has** /hæz, hæs, həz, həs/; past and past part. **had** /hæd, həd/) ● v.tr. **1** hold in possession as one's property or at one's disposal; be provided with (has a car; had no time to read; has nothing to wear). **2** hold in a certain relationship (has a sister; had no equals). **3** contain as a part or quality (house has two floors; has green eyes). **4 a** undergo, experience, enjoy, suffer (had a good time; had a shock; has a headache). **b** be subjected to a specified state (had my car stolen; the book has a page missing). **c** cause, instruct, or invite (a person or thing) to be in a particular state or take a particular action (had him dismissed; had us worried; had my hair cut; had a copy made; had them to stay). **5 a** engage in (an activity) (had an argument). **b** hold (a meeting, party, etc.). **6** eat or drink (had a beer). **7** (usu. in neg.) accept or tolerate; permit to (I won't have it; will not have you say such things). **8 a** let (a feeling etc.) be present (have no doubt; has a lot of sympathy for me; have nothing against them). **b** show or feel (mercy, pity, etc.) towards another person (have pity on him; have mercy!). **c** (foll. by to + infin.) show by action that one is influenced by (a feeling, quality, etc.) (have the goodness to leave now). **9 a** give birth to or beget (offspring). **b** conceive mentally (an idea etc.). **10** receive, obtain (had a letter from him; not a ticket to be had). **11** be burdened with or committed to (has a job to do; have my garden to attend to). **12 a** have obtained (a qualification) (has six O levels). **b** know (a language) (has no Latin). **13** colloq. **a** get the better of (I had him there). **b** (usu. in passive) Brit. cheat, deceive (you were had). **14** coarse sl. have sexual intercourse with. ● v.aux. (with past part. or ellipt., to form the perfect, pluperfect, and future perfect tenses, and the conditional mood) (have worked; had seen; will have been; had I known, I would have gone; have you met her? yes, I have). ● n. **1** (usu. in pl.) colloq. a person who has wealth or resources. **2** sl. a swindle. □ **had best** see BEST. **had better** would find it prudent to. **had rather** see RATHER. **have a care** see CARE. **have done, have done with** see DONE. **have an eye for, have eyes for, have an eye to** see EYE. **have a good mind to** see MIND. **have got to** = have to. **have had it** colloq. **1** have missed one's chance. **2** (of a person) have passed one's prime; (of a thing) be worn out or broken. **3** have been killed, defeated, etc. **have it 1** (foll. by that + clause) express the view that. **2** win a decision in a vote etc. **3** have found the answer etc. **have it away** (or **off**) Brit. coarse sl. have sexual intercourse. **have it both ways** see BOTH. **have it in for** colloq. be hostile or ill-disposed towards. **have it out** (often foll. by with) colloq. attempt to settle a dispute by discussion or argument. **have it one's own way** see WAY. **have-not** (usu. in pl.) colloq. a person lacking wealth or resources. **have nothing to do with** see DO[1]. **have on 1** be wearing (clothes). **2** be committed to (an engagement). **3** colloq. tease, play a trick on. **have out** get (a tooth etc.) extracted (had her tonsils out). **have sex** (often foll. by with) have sexual intercourse. **have something** (or **nothing**) **on a person 1** know something (or nothing) discreditable or incriminating about a person. **2** have an (or no) advantage or superiority over a person. **have to** be obliged to, must. **have to do with** see DO[1]. **have up** Brit. colloq. bring (a person) before a court of justice, interviewer, etc. [OE habban f. Gmc, prob. rel. to HEAVE]

Havel /ˈhɑːv(ə)l/, Václav (b.1936), Czech dramatist and statesman, President of Czechoslovakia 1989–92 and of the Czech Republic since 1993. Having written plays, such as The Garden Party (1963), which were critical of totalitarianism, in the 1970s he became the leading spokesman for Charter 77 and other human rights groups and was twice imprisoned as a dissident. Shortly after his release in 1989 he founded the opposition group Civic Forum and led a renewed campaign for political change; in December of that year he was elected President following the peaceful overthrow of Communism (the velvet revolution). He remained as President of the Czech Republic after the partition of Czechoslovakia on 1 Jan. 1993.

haven /ˈheɪv(ə)n/ n. **1** a harbour or port. **2** a place of refuge. [OE hæfen f. ON höfn]

haven't /ˈhæv(ə)nt/ contr. have not.

haver /ˈheɪvə(r)/ v. & n. ● v.intr. Brit. **1** talk foolishly; babble. **2** vacillate, hesitate. ● n. (usu. in pl.) Sc. foolish talk; nonsense. [18th c.: orig. unkn.]

haversack /ˈhævəˌsæk/ n. a stout bag for provisions etc., carried on the back or over the shoulder. [F havresac f. G Habersack f. Haber oats + Sack SACK[1]]

haversine /ˈhævəˌsaɪn/ n. (also **haversin** /-sɪn/) Math. half of a versed sine. [contr.]

havildar /ˈhævɪlˌdɑː(r)/ n. an Indian NCO corresponding to an army sergeant. [Hind. ḥavildār f. Pers. ḥawāldār trust-holder]

havoc /ˈhævək/ n. & v. ● n. widespread destruction; great confusion or disorder. ● v.tr. (**havocked, havocking**) devastate. □ **play havoc with** cause great confusion or difficulty to. [ME f. AF havok f. OF havo(t), of unkn. orig.]

haw[1] /hɔː/ n. the hawthorn or its fruit. [OE haga f. Gmc, rel. to HEDGE]

haw[2] /hɔː/ n. the nictitating membrane of a horse, dog, etc., esp. when inflamed. [16th c.: orig. unkn.]

haw[3] /hɔː/ int. & v. ● int. expressing hesitation. ● v.intr. (in **hum and haw**: see HUM[1]). [imit.: cf. HA[1]]

Hawaii /həˈwaɪi/ **1** a state of the US comprising a group of over twenty islands in the North Pacific; capital, Honolulu (on Oahu); pop. (1990) 1,108,230. First settled by Polynesians, Hawaii was discovered by Captain James Cook in 1778, who named it the Sandwich Islands. It came to be known as Hawaii, after the largest of the islands, in the 19th century. Annexed by the US in 1898, it became the 50th state in 1959. See also PEARL HARBOR. **2** the largest island in the state of Hawaii. □ **Hawaiian** /-ˈwaɪən/ adj. & n.

hawfinch /ˈhɔːfɪntʃ/ n. a large stout finch of the genus Coccothraustes, with a heavy beak for cracking seeds. [HAW[1] + FINCH]

hawk[1] /hɔːk/ n. & v. ● n. **1** a diurnal bird of prey of the family Accipitridae, esp. the genus Accipiter, having a characteristic curved beak, rounded short wings, and a long tail. **2** Polit. a person who advocates an aggressive or warlike policy, esp. in foreign affairs (opp. DOVE[1] 3). **3** a rapacious person. ● v. **1** intr. hunt game with a hawk. **2** intr. (often foll. by at) & tr. attack, as a hawk does. **3** intr. (of a bird) hunt on the wing for food. □ **hawk-eyed** keen-sighted. **hawk-nosed** having an aquiline nose. □□ **hawkish** adj. **hawkishness** n. **hawklike** adj. [OE h(e)afoc, hæbuc f. Gmc]

hawk[2] /hɔːk/ v.tr. **1** carry about or offer around (goods) for sale. **2** (often foll. by about) relate (news, gossip, etc.) freely. [back-form. f. HAWKER[1]]

hawk[3] /hɔːk/ v. **1** intr. clear the throat noisily. **2** tr. (foll. by up) bring (phlegm etc.) up from the throat. [prob. imit.]

hawk[4] /hɔːk/ n. a plasterer's square board with a handle underneath for carrying plaster or mortar. [17th c.: orig. unkn.]

hawkbit /ˈhɔːkbɪt/ n. a composite plant of the genus Leontodon, with yellow florets and rosettes of leaves. [HAWK[1] + BIT[2]: cf. devil's bit]

Hawke /hɔːk/, Robert James Lee ('Bob') (b.1929), Australian Labor statesman, Prime Minister 1983–91. He was elected leader of the Australian Labor Party in 1983, becoming Prime Minister a month later following his party's election victory over the Liberal government. During his premiership he pursued an economic programme based on free-market policies and tax reform. In 1990 he won a record fourth election victory but lost a leadership challenge the following year to Paul Keating.

Hawke Bay a bay on the east coast of North Island, New Zealand. The port of Napier lies on its southern shore. It was visited in 1769 by Captain James Cook, who named it after Edward Hawke, First Lord of the Admiralty 1766–71.

hawker[1] /ˈhɔːkə(r)/ n. a person who travels about selling goods. [16th c.: prob. f. LG or Du.; cf. HUCKSTER]

hawker[2] /ˈhɔːkə(r)/ n. a falconer. [OE hafocere]

Hawke's Bay an administrative region on the eastern coast of North Island, New Zealand.

Hawking /ˈhɔːkɪŋ/, Stephen William (b.1942), English theoretical physicist. His main work has been on space-time, quantum mechanics, and black holes, which he deduced can emit thermal radiation at a steady rate. While still a student he developed a progressive disabling neuromuscular disease: confined to a wheelchair, unable to write, and with severely impaired speech, he carries out his mathematical calculations mentally and communicates them in a developed form. Hawking's life and work are a triumph over severe physical disability, and his book A Brief History of Time (1988) has proved a popular best seller.

Hawkins[1] /ˈhɔːkɪnz/, Coleman Randolph (1904–69), American jazz saxophonist. During the 1920s and 1930s he was influential in making the tenor saxophone popular as a jazz instrument; playing with the Fletcher Henderson band (1923–34), he used a stiff reed which enabled him to be heard as a soloist over the band. His playing was characterized by its deep and rich expressive tone.

Hawkins[2] /ˈhɔːkɪnz/, Sir John (also **Hawkyns**) (1532–95), English sailor. In the 1560s and early 1570s he became involved in the slave trade and participated in early privateering raids in the Spanish West Indies. He was appointed treasurer of the Elizabethan navy in 1573 and played an important part in building up the fleet which defeated the Spanish Armada in 1588. He died at sea during an unsuccessful expedition to the West Indies.

hawkmoth /ˈhɔːkmɒθ/ n. a large darting and hovering moth of the family Sphingidae, having narrow forewings and a stout body.

Hawks /hɔːks/, Howard (1896–1977), American film director, producer, and screenwriter. He entered the film industry in 1922, directing and writing the screenplay for his first film in 1926. Over the next forty years he directed some of the most famous stars in comedies, westerns, musicals, and gangster films. His best-known films include *Scarface* (1931), *The Big Sleep* (1946), *Gentlemen Prefer Blondes* (1953), and *Rio Bravo* (1959).

hawksbill /ˈhɔːksbɪl/ n. a small tropical turtle, *Eretmochelys imbricata*, with hooked jaws.

Hawksmoor /ˈhɔːksmʊə(r), -mɔː(r)/, Nicholas (1661–1736), English architect. He began his career at the age of 18 as a clerk to Sir Christopher Wren, and from 1690 worked with Vanbrugh at Castle Howard and Blenheim Palace. In 1711 he was commissioned to design six London churches; notable examples include St Mary Woolnoth (1716–24) and St George's, Bloomsbury (1716–30).

hawkweed /ˈhɔːkwiːd/ n. a composite plant, esp. of the genus *Hieracium*, with yellow flowers.

Hawkyns see HAWKINS[2].

Haworth /ˈhaʊwəθ/, Sir Walter Norman (1883–1950), English organic chemist. He was a pioneer in carbohydrate chemistry, making major contributions to understanding the structure and classification of sugars and polysaccharides. His book *The Constitution of the Sugars* (1929) became a standard work. Haworth also determined the structure of vitamin C and later synthesized it, the first person to make a vitamin artificially. He shared the Nobel Prize for chemistry in 1937.

hawse /hɔːz/ n. **1** the part of a ship's bows in which hawse-holes or hawse-pipes are placed. **2** the space between the head of an anchored vessel and the anchors. **3** the arrangement of cables when a ship is moored with port and starboard forward anchors. □ **hawse-hole** a hole in the side of a ship through which a cable or anchor-rope passes.

hawse-pipe a metal pipe lining a hawse-hole. [ME *halse*, prob. f. ON *háls* neck, ship's bow]

hawser /ˈhɔːzə(r)/ n. Naut. a thick rope or cable for mooring or towing a ship. [ME f. AF *haucer, hauceour* f. OF *haucier* hoist ult. f. L *altus* high]

hawthorn /ˈhɔːθɔːn/ n. a thorny rosaceous shrub or tree of the genus *Crataegus*, with white, red, or pink blossom and small dark red fruit or haws; esp. the common *C. monogyna*, of Eurasia, much used for hedging. Also called *may, quickthorn, whitethorn*. [OE *hagathorn* (as HAW[1], THORN)]

Hawthorne /ˈhɔːθɔːn/, Nathaniel (1804–64), American novelist and short-story writer. Hawthorne's New England Puritan background is evident in much of his fiction, which uses allegory and symbolism to explore themes of hereditary guilt, sin, and morality. His works include collections of short stories, such as *Twice-Told Tales* (1837), and the novels *The Scarlet Letter* (1850) and *The House of Seven Gables* (1851). He also wrote a number of books for children, including *Tanglewood Tales* (1853).

hay[1] /heɪ/ n. & v. ● n. grass mown and dried for fodder. ● v. **1** intr. make hay. **2** tr. put (land) under grass for hay. **3** tr. make into hay. □ **hay fever** an allergy with catarrhal and other asthmatic symptoms, caused by pollen or dust. **make hay of** throw into confusion. **make hay (while the sun shines)** seize opportunities for profit or enjoyment. [OE *hēg, hīeg, hīg* f. Gmc]

hay[2] /heɪ/ n. (also **hey**) **1** a country dance with interweaving steps. **2** a figure in this. [obs. F *haie*]

haybox /ˈheɪbɒks/ n. a box stuffed with hay, in which heated food is left to continue cooking.

haycock /ˈheɪkɒk/ n. a conical heap of hay in a field.

Haydn /ˈhaɪd(ə)n/, Franz Joseph (1732–1809), Austrian composer. He was a major exponent of the classical style and a teacher of Mozart and Beethoven. In 1761 he joined the household of the Hungarian Prince Esterházy as musical director, a post which he held for nearly thirty years and which was conducive to his prolific output and his development of musical forms. His work comprises more than 100 symphonies and many string quartets and keyboard sonatas, and he

played a significant role in the development of the symphony and the string quartet in their classical four-movement forms. His choral music includes twelve masses and the oratorios *The Creation* (1796–8) and *The Seasons* (1799–1801).

Hayek /ˈhaɪjek/, Friedrich August von (1899–1992), Austrian-born economist. He is known as a leading advocate of the free market and a critic of government intervention and Keynesian economics. His book *The Road to Serfdom* (1944) linked state economic control with the loss of individual liberty, a topic he became increasingly concerned with in his later writings. He held various academic posts in Austria, the UK, and the US, and became a British citizen in 1938. He shared the Nobel Prize for economics in 1974.

Hayes /heɪz/, Rutherford B(irchard) (1822–93), American Republican statesman, 19th President of the US 1877–81. His administration brought the Reconstruction era in the South to an end.

hayfield /ˈheɪfiːld/ n. a field where hay is being or is to be made.

hayloft /ˈheɪlɒft/ n. = LOFT n. 2.

haymaker /ˈheɪˌmeɪkə(r)/ n. **1** a person who tosses and spreads hay to dry after mowing. **2** an apparatus for shaking and drying hay. **3** sl. a forceful blow or punch. □ **haymaking** n.

haymow /ˈheɪməʊ/ n. hay stored in a stack or barn.

hayrick /ˈheɪrɪk/ n. = HAYSTACK.

hayseed /ˈheɪsiːd/ n. **1** grass seed obtained from hay. **2** N. Amer., Austral., & NZ colloq. a rustic or yokel.

haystack /ˈheɪstæk/ n. a packed pile of hay with a pointed or ridged top.

haywire /ˈheɪˌwaɪə(r)/ adj. colloq. **1** badly disorganized, out of control. **2** (of a person) emotionally disturbed; erratic. [HAY[1] + WIRE, from the use of hay-baling wire in makeshift repairs]

Hayworth /ˈheɪwəθ/, Rita (born Margarita Carmen Cansino) (1918–87), American actress and dancer. She began her career as a dancer at the age of 12 and made her screen début in 1935. Hayworth achieved stardom with a succession of film musicals including *Cover Girl* (1944). She also played leading roles in several films of the *film noir* genre, notably *Gilda* (1946) and the *The Lady from Shanghai* (1948), in which she co-starred with her second husband Orson Welles.

hazard /ˈhæzəd/ n. & v. ● n. **1** a danger or risk. **2** a source of this. **3** chance. **4** a dice game with a complicated arrangement of chances. **5** Golf an obstruction in playing a shot, e.g. a bunker, water, etc. **6** each of the winning openings in a real-tennis court. ● v.tr. **1** venture on (*hazard a guess*). **2** run the risk of. **3** expose to hazard. [ME f. OF *hasard* f. Sp. *azar* f. Arab. *az-zahr* chance, luck]

hazardous /ˈhæzədəs/ adj. **1** risky, dangerous. **2** dependent on chance. □ **hazardously** adv. **hazardousness** n. [F *hasardeux* (as HAZARD)]

haze[1] /heɪz/ n. **1** obscuration of the atmosphere near the earth by fine particles of water, smoke, or dust. **2** mental obscurity or confusion. [prob. back-form. f. HAZY]

haze[2] /heɪz/ v.tr. **1** Naut. harass with overwork. **2** US bully; seek to disconcert. [orig. uncert.: cf. obs. F *haser* tease, insult]

hazel /ˈheɪz(ə)l/ n. **1** a shrub or small tree of the genus *Corylus*; esp. *C. avellana*, bearing round brown edible nuts. **2 a** wood from the hazel. **b** a stick made of this. **3** a reddish-brown or greenish-brown colour (esp. of the eyes). □ **hazel-grouse** a woodland grouse, *Tetrastes bonasia*, native to Europe. [OE *hæsel* f. Gmc]

hazelnut /ˈheɪz(ə)lˌnʌt/ n. the fruit of the hazel, a round brown hard-shelled nut.

Hazlitt /ˈhæzlɪt, ˈheɪz-/, William (1778–1830), English essayist and critic. From about 1812 he wrote many articles on diverse subjects for several periodicals, including the *Edinburgh Review* and the *Morning Chronicle*; his essays were collected in *Table Talk* (1821–2) and *The Plain Speaker* (1826). Among his critical works are *Lectures on the English Poets* (1818) and *The Spirit of the Age* (1825). His style, marked by clarity and conviction, brought a new vigour to English prose writing.

hazy /ˈheɪzɪ/ adj. (**hazier, haziest**) **1** misty. **2** vague, indistinct. **3** confused, uncertain. □ **hazily** adv. **haziness** n. [17th c. in Naut. use: orig. unkn.]

HB abbr. hard black (pencil-lead).

Hb symb. haemoglobin.

HBM abbr. Her or His Britannic Majesty (or Majesty's).

H-bomb /ˈeɪtʃbɒm/ n. = HYDROGEN BOMB. [H[3] + BOMB]

HC abbr. **1** Holy Communion. **2** (in the UK) House of Commons.

h.c. abbr. honoris causa.

HCF *abbr.* **1** highest common factor. **2** (in the UK) Honorary Chaplain to the Forces.

HCFC *abbr.* hydrochlorofluorocarbon (similar to CFC but thought to be less harmful to the ozone layer).

H.D. see DOOLITTLE.

HDTV *abbr.* high-definition television.

HE *abbr.* **1** His or Her Excellency. **2** His Eminence. **3** high explosive.

He *symb. Chem.* the element helium.

he /hiː, hɪ/ *pron. & n.* ● *pron.* (*obj.* **him**; *poss.* **his**; *pl.* **they**) **1** the man or boy or male animal previously named or in question. **2** a person etc. of unspecified sex, esp. referring to one already named or identified (*if anyone comes he will have to wait*). ● *n.* **1** a male; a man. **2** (in *comb.*) male (*he-goat*). **3** a children's chasing game, with the chaser designated 'he'. □ **he-man** (*pl.* **-men**) a masterful or virile man, esp. ostentatiously so. [OE f. Gmc]

Head /hed/, Edith (b.1907), American costume designer. She joined Paramount studios in 1923, and worked on films ranging from westerns to musicals and comedies. She was awarded Oscars for costume design in several films, including *All About Eve* (1950). Head later worked for Universal, where she won a further Oscar for the costumes in *The Sting* (1973).

head /hed/ *n., adj., & v.* ● *n.* **1** the upper part of the human body, or the foremost or upper part of an animal's body, containing the brain, mouth, and sense-organs. **2 a** the head regarded as the seat of intellect or repository of comprehended information. **b** intelligence; imagination (*use your head*). **c** mental aptitude or tolerance (usu. foll. by *for*: *a good head for business; no head for heights*). **3** *colloq.* a headache, esp. resulting from a blow or from intoxication. **4** a thing like a head in form or position, esp.: **a** the operative part of a tool. **b** the flattened top of a nail. **c** the ornamented top of a pillar. **d** a mass of leaves or flowers at the top of a stem. **e** the flat end of a drum. **f** the foam on top of a glass of beer etc. **g** the upper horizontal part of a window frame, door frame, etc. **5** life when regarded as vulnerable (*it cost him his head*). **6 a** a person in charge; a director or leader (esp. the principal teacher at a school or college). **b** a position of leadership or command. **7** the front or forward part of something, e.g. a queue. **8** the upper end of something, e.g. a table or bed. **9** the top or highest part of something, e.g. a page, stairs, etc. **10** a person or individual regarded as a numerical unit (*£10 per head*). **11** (*pl.* same) **a** an individual animal as a unit. **b** (as *pl.*) a number of cattle or game as specified (*20 head*). **12 a** the side of a coin bearing the image of a head. **b** (usu. in *pl.*) this side as a choice when tossing a coin. **13 a** the source of a river or stream etc. **b** the end of a lake, bay, etc., at which a river enters. **14** the height or length of a head as a measure. **15** the component of a machine that is in contact with or very close to what is being processed or worked on, esp.: **a** the component on a tape recorder that touches the moving tape in play and converts the signals. **b** the part of a record-player that holds the playing cartridge and stylus. **c** = PRINTHEAD. **16 a** a confined body of water or steam in an engine etc. **b** the pressure exerted by this. **17** a promontory (esp. in place-names) (*Beachy Head*). **18** *Naut.* **a** the bows of a ship. **b** (often in *pl.*) a ship's latrine. **19** a main topic or category for consideration or discussion. **20** *Journalism* = HEADLINE *n.* **21** a culmination, climax, or crisis. **22** the fully developed top of a boil etc. **23** *sl.* a habitual taker of drugs; a drug addict. ● *attrib.adj.* chief or principal (*head gardener; head office*). ● *v.* **1** *tr.* be at the head or front of. **2** *tr.* be in charge of (*headed a small team*). **3** *tr.* **a** provide with a head or heading. **b** (of an inscription, title, etc.) be at the top of, serve as a heading for. **4 a** *intr.* face or move in a specified direction or towards a specified result (often foll. by *for*: *is heading for trouble*). **b** *tr.* direct in a specified direction. **5** *tr. Football* strike (the ball) with the head. **6 a** *tr.* (often foll. by *down*) cut the head off (a plant etc.). **b** *intr.* (of a plant etc.) form a head. □ **above** (or **over**) **one's head** beyond one's ability to understand. **come to a head** reach a crisis. **enter** (or **come into**) **one's head** occur to one. **from head to toe** (or **foot**) all over a person's body. **get something into one's head** come, or cause, to realize or understand something. **get one's head down 1** go to bed. **2** concentrate on the task in hand. **give a person his** (or **her**) **head** allow a person to act freely. **go out of one's head** go mad. **go to one's head 1** (of liquor) make one dizzy or slightly drunk. **2** (of success) make one conceited. **head and shoulders** *colloq.* by a considerable amount. **head back 1** get ahead of so as to intercept and turn back. **2** return home etc. **headbanger** *colloq.* **1** an enthusiast for heavy metal music. **2** a crazy or eccentric person. **headbanging** vigorous head-shaking in time to heavy metal music. **head-butt** *n.* a forceful thrust with the top of the head into the chin or body of another person. ● *v.tr.* attack (another person) with a head-butt. **head**

case *colloq.* a mentally ill or unstable person. **head-dress** an ornamental covering or band for the head. **head first 1** with the head foremost. **2** precipitately. **head louse** a variety of the louse *Pediculus humanus* infesting the hair of the human head. **head off 1** get ahead of so as to intercept and turn aside. **2** forestall. **a head of hair** the hair on a person's head, esp. as a distinctive feature. **head of state** the title of the head of a state, usu. the leader of the ruling party or a monarch. **head-on 1** with the front foremost (*a head-on crash*). **2** in direct confrontation. **head over heels 1** turning over completely in forward motion as in a somersault etc. **2** topsy-turvy. **3** utterly, completely (*head over heels in love*). **head-shrinker** *sl.* a psychiatrist. **head start** an advantage granted or gained at an early stage. **heads will roll** people will be disgraced or dismissed. **head teacher** the teacher in charge of a school. **head up** take charge of (a group of people). **head-up** *adj.* (of instrument readings in an aircraft, vehicle, etc.) shown so as to be visible without lowering the eyes. **head-voice** the high register of the voice in speaking or singing. **head wind** a wind blowing from directly in front. **hold up one's head** be confident or unashamed. **in one's head 1** in one's thoughts or imagination. **2** by mental process without use of physical aids. **keep one's head** remain calm. **keep one's head above water 1** keep out of debt. **2** avoid succumbing to difficulties. **keep one's head down** remain inconspicuous in difficult or dangerous times. **lose one's head** lose self-control; panic. **make head or tail of** (usu. with *neg.* or *interrog.*) understand at all. **off one's head** *colloq.* crazy. **off the top of one's head** *colloq.* impromptu; without careful thought or investigation. **on one's** (or **one's own**) **head** as one's sole responsibility. **out of one's head 1** *colloq.* crazy. **2** from one's imagination or memory. **over one's head 1** beyond one's ability to understand. **2** without one's knowledge or involvement, esp. when one has a right to this. **3** with disregard for one's own (stronger) claim (*was promoted over their heads*). **put heads together** consult together. **put into a person's head** suggest to a person. **take** (or **get**) **it into one's head** (foll. by *that* + clause or *to* + infin.) form a definite idea or plan. **turn a person's head** make a person conceited. **with one's head in the clouds** see CLOUD. □ **headed** *adj.* (also in *comb.*). **headless** *adj.* **headward** *adj. & adv.* [OE *hēafod* f. Gmc]

-head /hed/ *suffix* = -HOOD (*godhead; maidenhead*). [ME *-hed, -hede* = -HOOD]

headache /ˈhedeɪk/ *n.* **1** a continuous pain in the head. **2** *colloq.* **a** a worrying problem. **b** a troublesome person. □ **headachy** *adj.*

headage /ˈhedɪdʒ/ *n.* the number of animals on a farm etc.

headband /ˈhedbænd/ *n.* a band worn round the head as decoration or to keep the hair off the face.

headboard /ˈhedbɔːd/ *n.* an upright panel forming the head of a bed.

headcount /ˈhedkaʊnt/ *n.* **1** a counting of individual people. **2** a total number of people, esp. the number of people employed in a particular organization.

header /ˈhedə(r)/ *n.* **1** *Football* a shot or pass made with the head. **2** *colloq.* a headlong fall or dive. **3** a brick or stone laid at right angles to the face of a wall (cf. STRETCHER *n.* 2). **4** (in full **header-tank**) a tank of water etc. maintaining pressure in a plumbing system.

headgear /ˈhedɡɪə(r)/ *n.* a hat or head-dress.

head-hunting /ˈhedˌhʌntɪŋ/ *n.* **1** the practice among some peoples of collecting the heads of dead enemies as trophies. **2** the practice of filling a (usu. senior) business position by approaching a suitable person employed elsewhere. □ **head-hunt** *v.tr.* (also *absol.*). **head-hunter** *n.*

heading /ˈhedɪŋ/ *n.* **1 a** a title at the head of a page or section of a book etc. **b** a division or section of a subject of discourse etc. **2 a** a horizontal passage made in preparation for building a tunnel. **b** *Mining* = DRIFT *n.* 6. **3** material for making cask-heads. **4** the extension of the top of a curtain above the tape that carries the hooks or the pocket for a wire. **5** the course of an aircraft, ship, etc.

headlamp /ˈhedlæmp/ *n.* = HEADLIGHT.

headland *n.* **1** /ˈhedlənd/ a promontory. **2** /ˈhedlænd/ a strip left unploughed at the end of a field, for machinery to pass along.

headlight /ˈhedlaɪt/ *n.* **1** a strong light at the front of a motor vehicle or railway engine. **2** the beam from this.

headline /ˈhedlaɪn/ *n. & v.* ● *n.* **1** a heading at the top of an article or page, esp. in a newspaper. **2** (in *pl.*) the most important items of news in a newspaper or broadcast news bulletin. ● *v.* **1** *tr.* give a headline to. **2** *intr. & tr.* appear as the chief performer (at). □ **hit** (or **make**) **the headlines** be given prominent attention as news.

headliner /ˈhedˌlaɪnə(r)/ n. a headlining performer, a star.

headlock /ˈhedlɒk/ n. *Wrestling* a hold with an arm round the opponent's head.

headlong /ˈhedlɒŋ/ adv. & adj. **1** with head foremost. **2** in a rush. [ME *headling* (as HEAD, -LING²), assim. to -LONG]

headman /ˈhedmən/ n. (pl. **-men**) the chief man of a tribe etc.

headmaster /hedˈmɑːstə(r)/ n. (fem. **headmistress** /-ˈmɪstrɪs/) the principal teacher in charge of a school. □ **headmasterly** adj.

headmost /ˈhedməʊst/ adj. (esp. of a ship) foremost.

headnote /ˈhednəʊt/ n. **1** a note or comment at the head of a document, page, etc. **2** *Law* a summary giving the principle of a decision and an outline of the facts, prefixed to the report of a decided case.

headphone /ˈhedfəʊn/ n. (usu. in pl.) a pair of earphones joined by a band placed over the head, for listening to audio equipment etc.

headpiece /ˈhedpiːs/ n. **1** an ornamental engraving at the head of a chapter etc. **2** a helmet. **3** *archaic* intellect.

headquarters /hedˈkwɔːtəz/ n. (as *sing.* or *pl.*) **1** the administrative centre of an organization. **2** the premises occupied by a military commander and the commander's staff.

headrest /ˈhedrest/ n. a support for the head, esp. on a seat or chair.

headroom /ˈhedruːm, -rʊm/ n. **1** the space or clearance between the top of a vehicle and the underside of a bridge etc. which it passes under. **2** the space above a driver's or passenger's head in a vehicle.

headsail /ˈhedseɪl/ n. *Naut.* -s(ə)l/ n. a sail on a ship's foremast or bowsprit.

headscarf /ˈhedskɑːf/ n. (pl. **-scarves** /-skɑːvz/) a scarf worn round the head and tied under the chin.

headset /ˈhedset/ n. a set of headphones, often with a microphone attached, used esp. in telephony and radio communications.

headship /ˈhedʃɪp/ n. the position of chief or leader, esp. of a headmaster or headmistress.

headsman /ˈhedzmən/ n. (pl. **-men**) **1** *hist.* an executioner who beheads. **2** a person in command of a whaling boat.

headspring /ˈhedsprɪŋ/ n. **1** the main source of a stream. **2** a principal source of ideas etc.

headsquare /ˈhedskweə(r)/ n. a rectangular scarf for wearing on the head.

headstall /ˈhedstɔːl/ n. the part of a halter or bridle that fits round a horse's head.

headstock /ˈhedstɒk/ n. a set of bearings in a machine, supporting a revolving part.

headstone /ˈhedstəʊn/ n. a (usu. inscribed) stone set up at the head of a grave.

headstrong /ˈhedstrɒŋ/ adj. self-willed and obstinate.

head-to-head /ˌhedtəˈhed/ n., adj., & adv. ● n. a conversation, confrontation, or contest between two parties. ● attrib.adj. involving two parties confronting each other. ● adv. confronting another party. [transl. F TÊTE-À-TÊTE]

headwater /ˈhedˌwɔːtə(r)/ n. (in *sing.* or *pl.*) streams flowing from the sources of a river.

headway /ˈhedweɪ/ n. **1** progress. **2** the rate of progress of a ship. **3** = HEADROOM 1. □ **make headway** see MAKE.

headword /ˈhedwɜːd/ n. a word forming a heading, e.g. of an entry in a dictionary or encyclopedia.

headwork /ˈhedwɜːk/ n. mental work or effort.

heady /ˈhedɪ/ adj. (**headier**, **headiest**) **1** (of liquor) potent, intoxicating. **2** (of success etc.) likely to cause conceit. **3** (of a person, thing, or action) impetuous, violent. **4** headachy. □ **headily** adv. **headiness** n.

heal /hiːl/ v. **1** intr. (often foll. by *up*) (of a wound or injury) become sound or healthy again. **2** tr. cause (a wound, disease, or person) to heal or be healed. **3** tr. put right (differences etc.). **4** tr. alleviate (sorrow etc.). □ **heal-all** **1** a universal remedy, a panacea. **2** a popular name of various medicinal plants. □ **healable** adj. [OE *hǣlan* f. Gmc, rel. to WHOLE]

heald /hiːld/ n. = HEDDLE. [app. f. OE *hefel*, *hefeld*, f. Gmc]

healer /ˈhiːlə(r)/ n. **1** a person who heals, esp. a faith-healer. **2** a thing which heals or assists in healing.

health /helθ/ n. **1** the state of being well in body or mind. **2** a person's mental or physical condition (*has poor health*). **3** soundness, esp. financial or moral (*the health of the nation*). **4** a toast drunk in someone's honour. □ **health centre** the headquarters of a group of local medical services. **health certificate** a certificate stating a person's fitness for work etc. **health farm** a residential establishment where people seek improved health by a regimen of dieting, exercise, etc. **health food** natural food thought to have health-giving qualities. **health physics** the branch of radiology which deals with the health of people working with radioactive materials. **health service** a public service providing medical care. **health visitor** *Brit.* a trained nurse who visits those in need of medical attention in their homes. [OE *hǣlth* f. Gmc]

healthful /ˈhelθfʊl/ adj. conducive to good health; beneficial. □ **healthfully** adv. **healthfulness** n.

healthy /ˈhelθɪ/ adj. (**healthier**, **healthiest**) **1** having, showing, or promoting good health. **2** beneficial, helpful (*a healthy respect for experience*). **3** ample, sizeable, considerable (*a healthy portion*). □ **healthily** adv. **healthiness** n.

Heaney /ˈhiːnɪ/, Seamus (Justin) (b.1939), Irish poet. He was born in Northern Ireland and his early poetry, such as *Death of a Naturalist* (1966), reflects the rural life of his youth. In 1972 he took Irish citizenship; from this time his poetry began to deal with wider social and cultural themes. Later collections include *North* (1975), which deals with the conflict in Northern Ireland, and *The Haw Lantern* (1987). He was awarded the Nobel Prize for literature in 1995.

heap /hiːp/ n. & v. ● n. **1** a collection of things lying haphazardly one on another; a pile. **2** (esp. in pl.) colloq. a large number or amount (*there's heaps of time; is heaps better*). **3** colloq. an old or dilapidated thing, esp. a motor vehicle or building. ● v. **1** tr. & intr. (foll. by *up*, *together*, etc.) collect or be collected in a heap. **2** tr. (foll. by *with*) load copiously or to excess. **3** tr. (foll. by *on*, *upon*) accord or offer copiously to (*heaped insults on them*). **4** tr. (as **heaped** adj.) (of a spoonful etc.) with the contents piled above the brim. □ **heap coals of fire on a person's head** cause a person remorse by returning good for evil. [OE *hēap*, *hēapian* f. Gmc]

hear /hɪə(r)/ v. (past and past part. **heard** /hɜːd/) **1** tr. (also absol.) perceive (sound etc.) with the ear. **2** tr. listen to (*heard them on the radio*). **3** tr. listen judicially to and judge (a case, plaintiff, etc.). **4** intr. (foll. by *about*, *of*, or *that* + clause) be told or informed. **5** intr. (foll. by *from*) be contacted by, esp. by letter or telephone. **6** tr. be ready to obey (an order). **7** tr. grant (a prayer). □ **have heard of** be aware of; know of the existence of. **hear! hear!** int. expressing agreement (esp. with something said in a speech). **hear a person out** listen to all that a person says. **hear say** (or **tell**) (usu. foll. by *of*, or *that* + clause) be informed. **will not hear of** will not allow or agree to. □ **hearable** adj. **hearer** n. [OE *hīeran* f. Gmc]

Heard and McDonald Islands /hɜːd, məkˈdɒn(ə)ld/ a group of uninhabited islands in the southern Indian Ocean, administered by Australia since 1947 as an external territory.

hearing /ˈhɪərɪŋ/ n. **1** the faculty of perceiving sounds. **2** the range within which sounds may be heard; earshot (*within hearing; in my hearing*). **3** an opportunity to state one's case (*give them a fair hearing*). **4** the listening to evidence and pleadings in a lawcourt. □ **hearing-aid** a small device to amplify sound, worn by a partially deaf person.

hearken /ˈhɑːkən/ v.intr. (also **harken**) archaic or literary (often foll. by *to*) listen. [OE *heorcnian* (as HARK)]

hearsay /ˈhɪəseɪ/ n. rumour, gossip. □ **hearsay evidence** *Law* evidence given by a witness based on information received from others rather than personal knowledge.

hearse /hɜːs/ n. a vehicle for conveying the coffin at a funeral. [ME f. OF *herse* harrow f. med.L *herpica* ult. f. L *hirpex -icis* large rake]

Hearst /hɜːst/, William Randolph (1863–1951), American newspaper publisher and tycoon. He is noted for his introduction of large headlines, sensational crime reporting, and other features designed to increase circulation; his innovations revolutionized American journalism. At the peak of his fortunes in the mid-1930s he had acquired a number of newspapers and magazines, radio stations, and two film companies. He was the model for the central character of Orson Welles's film *Citizen Kane* (1941).

heart /hɑːt/ n. **1** a hollow muscular organ maintaining the circulation of blood by rhythmic contraction and dilation. (*See note below.*) **2** the region of the heart; the breast. **3 a** the heart regarded as the centre of thought, feeling, and emotion (esp. love). **b** a person's capacity for feeling emotion (*has no heart*). **4 a** courage or enthusiasm (*take heart; lose heart*). **b** one's mood or feeling (*change of heart*). **5 a** the central or innermost part of something. **b** the vital part or essence (*the heart of the matter*). **6** the close compact head of a cabbage, lettuce, etc. **7 a** a heart-shaped thing. **b** a conventional representation of a heart with two equal curves meeting at a point at the bottom and a cusp at the top. **8 a** a playing card of a suit denoted by a red figure of a heart. **b** (in pl.) this suit. **c** (in pl.) a card-game in which players avoid taking tricks containing a card of this suit. **9** condition of land as regards fertility (*in good heart*). □ **after one's own heart** such as one likes or desires. **at heart 1** in one's inmost feelings. **2** basically, essentially. **break a person's heart** overwhelm a person with sorrow. **by heart** in or from memory. **close to** (or **near**) **one's heart 1** dear to one. **2** affecting one deeply. **from the heart** (or **the bottom of one's heart**) sincerely, profoundly. **give** (or **lose**) **one's heart** (often foll. by *to*) fall in love (with). **have a heart** be merciful. **have the heart** (usu. with *neg.*; foll. by *to* + infin.) be insensitive or hard-hearted enough (*didn't have the heart to ask him*). **have one's heart in** (or **put one's heart into**) be keenly involved in or committed to (an enterprise etc.). **have one's heart in one's mouth** be greatly alarmed or apprehensive. **have one's heart in the right place** be sincere or well-intentioned. **heart attack** a sudden occurrence of coronary thrombosis usu. resulting in the death of part of a heart muscle. **heart failure** severe failure of the heart to function properly, esp. as a cause of death. **heart-lung machine** a machine that temporarily takes over the functions of the heart and lungs, esp. in surgery. **heart of gold** a generous nature. **heart of oak** a courageous nature. **heart of stone** a stern or cruel nature. **heart-rending** very distressing. **heart-rendingly** in a heart-rending way. **heart's-blood** lifeblood, life. **heart-searching** the thorough examination of one's own feelings and motives. **heart to heart** candidly, intimately. **heart-to-heart** *adj.* (of a conversation etc.) candid, intimate. ● *n.* a candid or personal conversation. **heart-warming** emotionally rewarding or uplifting. **in heart** in good spirits. **in one's heart of hearts** in one's inmost feelings. **out of heart** in low spirits. **take to heart** be much affected or distressed by. **to one's heart's content** see CONTENT[1]. **wear one's heart on one's sleeve** make one's feelings apparent. **with all one's heart** sincerely; with all goodwill. **with one's whole heart** with enthusiasm; without doubts or reservations. □ **-hearted** *adj.* [OE *heorte* f. Gmc]

▪ The heart of mammals and birds is a four-chambered muscular pump whose function is to drive arterial blood through the body and venous blood through the lungs. Other vertebrates and some invertebrates possess a somewhat simpler organ. In humans the two sides of the heart each consist of a receiving chamber or atrium and a pumping chamber or ventricle. Venous blood from the body, laden with carbon dioxide, flows into the right atrium, is drawn into the right ventricle, and is ejected by the contraction or beat of the ventricle into the pulmonary artery and so to the lungs. Oxygenated blood from the lungs flows into the left atrium, passes to the left ventricle, and is pumped into the aorta and so to all arteries of the body.

heartache /ˈhɑːteɪk/ *n.* mental anguish or grief.

heartbeat /ˈhɑːtbiːt/ *n.* a pulsation of the heart.

heartbreak /ˈhɑːtbreɪk/ *n.* overwhelming distress. □ **heartbreaker** *n.* **heartbreaking** *adj.* **heartbroken** /-ˌbrəʊkən/ *adj.*

heartburn /ˈhɑːtbɜːn/ *n.* a burning sensation in the chest resulting from indigestion; pyrosis.

hearten /ˈhɑːt(ə)n/ *v.tr.* & *intr.* make or become more cheerful. □ **heartening** *adj.* **hearteningly** *adv.*

heartfelt /ˈhɑːtfelt/ *adj.* sincere; deeply felt.

hearth /hɑːθ/ *n.* **1 a** the floor of a fireplace. **b** the area in front of a fireplace. **2** this symbolizing the home. **3** the bottom of a blast-furnace where molten metal collects. [OE *heorth* f. WG]

hearthrug /ˈhɑːθrʌɡ/ *n.* a rug laid before a fireplace.

hearthstone /ˈhɑːθstəʊn/ *n.* **1** a flat stone forming a hearth. **2** a soft stone used to whiten hearths, doorsteps, etc.

heartily /ˈhɑːtɪli/ *adv.* **1** in a hearty manner; with sincerity, vehemence, or appetite. **2** very; to a great degree (esp. with ref. to personal feelings) (*am heartily sick of it; disliked him heartily*).

heartland /ˈhɑːtlənd/ *n.* the central or most important part of an area.

heartless /ˈhɑːtlɪs/ *adj.* unfeeling, pitiless. □ **heartlessly** *adv.* **heartlessness** *n.*

heartsease /ˈhɑːtsiːz/ *n.* (also **heart's-ease**) a pansy, esp. the wild pansy, *Viola tricolor.*

heartsick /ˈhɑːtsɪk/ *adj.* very despondent. □ **heartsickness** *n.*

heartsore /ˈhɑːtsɔː(r)/ *adj.* archaic or literary grieving, heartsick.

heartstrings /ˈhɑːtstrɪŋz/ *n.pl.* one's deepest feelings or emotions.

heartthrob /ˈhɑːtθrɒb/ *n.* **1** beating of the heart. **2** *colloq.* an extremely attractive (usu. male) person, esp. an actor or other celebrity.

heartwood /ˈhɑːtwʊd/ *n.* the dense inner part of a tree-trunk yielding the hardest timber.

hearty /ˈhɑːtɪ/ *adj.* & *n.* ● *adj.* (**heartier, heartiest**) **1** strong, vigorous, vehement. **2** spirited, energetic, unrestrainedly enthusiastic. **3** (of a meal or appetite) large. **4** warm, friendly. **5** genuine, sincere. ● *n.* **1** a hearty person, esp. one ostentatiously so. **2** (usu. in *pl.*) (as a form of address) fellows, esp. fellow sailors. □ **heartiness** *n.*

heat /hiːt/ *n.* & *v.* ● *n.* **1 a** the condition of being hot. **b** the sensation or perception of this. **c** high temperature of the body. **2** *Physics* **a** a form of energy arising from the random motion of the molecules of bodies, which may be transferred by conduction, convection, or radiation. (*See note below.*) **b** the amount of this needed to cause a specific process, or evolved in a process (*heat of formation; heat of solution*). **3** hot weather (*succumbed to the heat*). **4 a** warmth of feeling. **b** anger or excitement (*the heat of the argument*). **5** (foll. by *of*) the most intense part or period of an activity (*in the heat of the battle*). **6 a** (usu. preliminary or trial) round in a race or contest. **7** the receptive period of the sexual cycle, esp. in female mammals. **8** redness of the skin with a sensation of heat (*prickly heat*). **9** pungency of flavour. **10** *sl.* intensive pursuit, e.g. by the police. ● *v.* **1** *tr.* & *intr.* make or become hot or warm. **2** *tr.* inflame; excite or intensify. □ **heat barrier** the limitation of the speed of an aircraft etc. by heat resulting from air friction. **heat capacity** thermal capacity. **heat death** *Physics* a state of uniform distribution of energy, esp. viewed as a possible fate of the universe (see ENTROPY). **heat engine** a device for producing motive power from heat. **heat-exchanger** a device for the transfer of heat from one medium to another. **heat lamp** a lamp used for its heat as well as its light. **heat pump** a device in which mechanical energy is used to force the transfer of heat from a colder area to a hotter area, as in a refrigerator. **heat-resistant** = HEATPROOF. **heat-seeking** (of a missile etc.) able to detect infrared radiation to guide it to its target. **heat shield** a device for protection from excessive heat, esp. fitted to a spacecraft. **heat sink** a device or substance for absorbing excessive or unwanted heat. **heat-treat** subject to heat treatment. **heat treatment** the use of heat to modify the properties of a metal, ease muscular pains, etc. **in the heat of the moment** during or resulting from intense activity, without pause for thought. **on heat** (of mammals, esp. females) sexually receptive. **turn the heat on** *colloq.* concentrate an attack or criticism on (a person). [OE *hætu* f. Gmc]

▪ In the late 18th and early 19th centuries heat was believed to be a kind of fluid substance, called caloric. The view of heat as a form of energy emerged with the development of thermodynamics, and the kinetic theory related it to the random motions of the individual molecules that make up matter. Addition of heat to a body does not always lead to an increase in temperature, because the body may change state (from solid to liquid, or liquid to gas).

heated /ˈhiːtɪd/ *adj.* **1** (of a person, discussions, etc.) angry; inflamed with passion or excitement. **2** made hot. □ **heatedly** *adv.*

heater /ˈhiːtə(r)/ *n.* **1** a device for warming the air in a room, car, etc. **2** a container with an element etc. for heating the contents (*water-heater*). **3** *sl.* a gun.

Heath /hiːθ/, Sir Edward (Richard George) (b.1916), British Conservative statesman, Prime Minister 1970–4. In 1973 his long-standing commitment to European unity was realized when Britain joined the European Economic Community. His premiership was marked by problems of inflation and balance of payments (exacerbated by a marked increase in oil prices in 1973); attempts to restrain wage rises led to widespread strikes. After a second national coal strike Heath called an election to strengthen his position, but was defeated.

heath /hiːθ/ *n.* **1 a** esp. *Brit.* an area of open uncultivated land, usu. on acid sandy soil and covered with heather, coarse grasses, etc. (cf. MOOR[1] 1a). **b** *Ecol.* an area dominated by dwarf ericaceous shrubs. **2** a dwarf ericaceous shrub of the genus *Erica* or *Calluna*, growing on heaths and moors, e.g. heather. □ **heathy** *adj.* [OE *hæth* f. Gmc]

heathen /ˈhiːðən/ *n.* & *adj.* ● *n.* **1** a person who does not belong to a widely held religion (esp. who is not Christian, Jew, or Muslim) as regarded by those that do. **2** an unenlightened person; a person regarded as lacking culture or moral principles. **3** (**the heathen**) heathen people collectively. **4** *Bibl.* a Gentile. ● *adj.* **1** of or relating to heathens. **2** having no religion. □ **heathendom** *n.* **heathenism** *n.* [OE *hæthen* f. Gmc]

heather /ˈheðə(r)/ *n.* **1** a dwarf ericaceous shrub, *Calluna vulgaris*, with purple bell-shaped flowers. **2** a dwarf ericaceous shrub of the genus *Erica* or *Daboecia*, growing esp. on moors and heaths. □ **heather mixture 1** a fabric of mixed hues supposed to resemble heather. **2** the colour of this. □ **heathery** *adj.* [ME, Sc., & N.Engl. *hathir* etc., of unkn. orig.: assim. to *heath*]

heathland /'hi:θlænd/ n. an extensive area of heath.

Heath Robinson /hi:θ 'rɒbɪns(ə)n/ adj. absurdly ingenious and impracticable in design or construction. [William *Heath Robinson*, Engl. cartoonist (1872–1944) who drew such contrivances]

Heathrow /'hi:θrəʊ, hi:θ'rəʊ/ an international airport situated 25 km (15 miles) west of the centre of London. It is also known as London Airport.

heating /'hi:tɪŋ/ n. **1** the imparting or generation of heat. **2** equipment or devices used to provide heat, esp. to a building.

heatproof /'hi:tpru:f/ adj. & v. ● adj. able to resist great heat. ● v.tr. make heatproof.

heatstroke /'hi:tstrəʊk/ n. a feverish condition caused by excessive exposure to high temperature.

heatwave /'hi:tweɪv/ n. a prolonged period of abnormally hot weather.

heave /hi:v/ v. & n. ● v. (past and past part. **heaved** or esp. Naut. **hove** /həʊv/) **1** tr. lift or haul (a heavy thing) with great effort. **2** tr. utter with effort or resignation (*heaved a sigh*). **3** tr. colloq. throw. **4** intr. rise and fall rhythmically or spasmodically. **5** tr. Naut. haul by rope. **6** intr. retch. ● n. **1** an instance of heaving. **2** Geol. a sideways displacement in a fault. **3** (in pl.) a disease of horses, with laboured breathing. □ **heave-ho** **1** a sailors' cry, esp. on raising the anchor. **2** (usu. prec. by *the* or *the old*) sl. a dismissal or rejection. **heave in sight** Naut. or colloq. come into view. **heave to** esp. Naut. bring or be brought to a standstill. □ **heaver** n. [OE *hebban* f. Gmc, rel. to L *capere* take]

heaven /'hev(ə)n/ n. **1** (also **Heaven**) a place regarded in some religions as the abode of God and the angels, and of the good after death, often characterized as above the sky. (*See note below.*) **2** a place or state of supreme bliss. **3** colloq. something delightful. **4** (also **Heaven**) **a** God, Providence. **b** (in sing. or pl.) an exclamation or mild oath (*by Heaven!*). **5** (**the heavens**) esp. poet. the expanse in which the sun, moon, and stars are seen, having the appearance of a vast vault arched over the earth. □ **heaven-sent** providential; wonderfully opportune. **in heaven's name** used as an exclamation of surprise or annoyance. **move heaven and earth** (foll. by *to* + infin.) make extraordinary efforts. **seventh heaven** see SEVENTH. □ **heavenward** adj. & adv. **heavenwards** adv. [OE *heofon*]

▪ In Jewish and Christian belief heaven has traditionally been regarded as a place in which the virtuous see God face to face; in addition, Christians have believed that the body will be reunited with the soul in heaven after Judgement Day. Faith in Christ is believed to allow every Christian the opportunity to reach it, whereas in Judaism only exceptional human beings are raised to heaven. Theologians today see heaven less as a place than as a state in which the full reality of God is experienced. In Islamic belief heaven is depicted as a blissful paradise of material delights, although this description is generally supposed to be more allegorical than literal.

heavenly /'hev(ə)nlɪ/ adj. **1** of heaven; divine. **2** of the heavens or sky. **3** colloq. very pleasing; wonderful. □ **heavenly body** a natural object in outer space, e.g. the sun, a star, a planet, etc.; a celestial object. □ **heavenliness** n. [OE *heofonlic* (as HEAVEN)]

Heaviside /'hevɪsaɪd/, Oliver (1850–1925), English physicist and electrical engineer. His theoretical contributions improved long-distance telephone communication and had significance to both cable and wireless telegraphy. He studied inductance, introduced the concept of impedance, and pioneered the use of calculus for dealing with the properties of electrical networks. In 1902 he suggested (independently of A. E. Kennelly) the existence of a layer in the atmosphere responsible for reflecting radio waves back to earth. It is now known as the Heaviside or Kennelly–Heaviside layer, or the E region of the ionosphere.

heavy /'hevɪ/ adj., n., adv., & v. ● adj. (**heavier**, **heaviest**) **1** a of great or exceptionally high weight; difficult to lift. **b** (of a person) fat, overweight. **2** a of great density. **b** Physics having a greater than the usual mass (esp. of isotopes and compounds containing them). **3** abundant, considerable (*a heavy crop*). **4** severe, intense, extensive, excessive (*heavy fighting; a heavy sleep*). **5** doing something to excess (*a heavy drinker*). **6** a striking or falling with force (*heavy blows; heavy rain*). **b** (of the sea) having large powerful waves. **7** (of machinery, artillery, etc.) very large of its kind; large in calibre etc. **8** causing a strong impact (*a heavy fall*). **9** needing much physical effort (*heavy work*). **10** (foll. by *with*) laden. **11** carrying heavy weapons (*the heavy brigade*). **12** a (of a person, writing, music, etc.) serious or sombre in tone or attitude; dull, tedious. **b** (of an issue etc.) grave; important, weighty. **13** a (of food) hard to digest. **b** (of a literary work etc.) hard to read or

understand. **14** a (of temperament) dignified, stern. **b** intellectually slow. **15** (of bread etc.) too dense from not having risen. **16** (of ground) difficult to traverse or work. **17** a oppressive; hard to endure (*a heavy fate; heavy demands*). **b** (of the atmosphere, weather, etc.) overcast; oppressive, sultry. **18** a coarse, ungraceful (*heavy features*). **b** unwieldy. **19** (of rock etc. music) having a strong beat and a loud distorted sound. ● n. (pl. **-ies**) **1** colloq. a large violent person; a thug. **2** a villainous or tragic role or actor in a play etc. **3** (usu. in pl.) colloq. a serious newspaper. **4** anything large or heavy of its kind, e.g. a vehicle. **5** esp. Sc. a usu. strong, dark brown beer. ● adv. heavily (esp. in comb.: *heavy-laden*). ● v.tr. colloq. harass or pressurize (a person). □ **heavier-than-air** (of an aircraft) weighing more than the air it displaces. **heavy chemicals** see CHEMICAL. **heavy-duty** adj. intended to withstand hard use. **heavy-footed** awkward, ponderous. **heavy going** slow or difficult progress. **heavy-hearted** sad, doleful. **heavy hydrogen** = DEUTERIUM. **heavy industry** industry producing metal, machinery, etc. **heavy-lidded** sleepy. **heavy metal 1** heavy guns. **2** metal of high density. **3** (often *attrib.*) a type of highly amplified harsh-sounding rock music with a strong beat and often theatrical performance. **heavy on** using a lot of (*heavy on petrol*). **heavy petting** erotic fondling between two people, stopping short of intercourse. **heavy-set** stocky, thickset. **heavy sleeper** a person who sleeps deeply. **heavy water** a form of water consisting of deuterium oxide (chem. formula: D_2O), used as a moderator in nuclear reactors. **make heavy weather of** see WEATHER. □ **heavily** adv. **heaviness** n. **heavyish** adj. [OE *hefig* f. Gmc, rel. to HEAVE]

heavy-handed /ˌhevɪ'hændɪd/ adj. **1** clumsy. **2** overbearing, oppressive. □ **heavy-handedly** adv. **heavy-handedness** n.

heavyweight /'hevɪweɪt/ n. **1** a a weight in certain sports, in the amateur boxing scale over 81 kg but differing for professional boxers, wrestlers, and weightlifters. **b** a sportsman of this weight. **2** a person, animal, or thing of above average weight. **3** a person of influence or importance. □ **light heavyweight 1** the weight in some sports between middleweight and heavyweight, in the amateur boxing scale 75–81 kg (also called *cruiserweight*). **2** a sportsman of this weight.

Heb. abbr. **1** Hebrew. **2** Epistle to the Hebrews (New Testament).

hebdomadal /heb'dɒməd(ə)l/ adj. formal weekly, esp. meeting weekly. [LL *hebdomadalis* f. Gk *hebdomas*, *-ados* f. *hepta* seven]

Hebe /'hi:bɪ/ **1** Gk Mythol. daughter of Hera and Zeus, and cup-bearer of the gods. **2** Astron. asteroid 6, discovered in 1847 (diameter 192 km). [Gk *hēbē* = youthful beauty]

hebe /'hi:bɪ/ n. a flowering evergreen shrub of the genus *Hebe*, native to New Zealand, with alternate overlapping leaves. [mod.L (HEBE)]

Hebei /hə'beɪ/ (also **Hopeh** /həʊ'peɪ/) a province of NE central China; capital, Shijiazhuang.

hebetude /'hebɪtju:d/ n. literary dullness. [LL *hebetudo* f. *hebes*, *-etis* blunt]

Hebraic /hi:'breɪɪk/ adj. of Hebrew or the Hebrews. □ **Hebraically** adv. [LL f. Gk *Hebraikos* (as HEBREW)]

Hebraism /'hi:breɪˌɪz(ə)m/ n. **1** a Hebrew idiom or expression, esp. in the Greek of the Bible. **2** an attribute of the Hebrews. **3** the Hebrew system of thought or religion. □ **Hebraize** v.tr. & intr. (also **-ise**). [F *hébraïsme* or mod.L *Hebraismus* f. late Gk *Hebraïsmos* (as HEBREW)]

Hebraistic /ˌhi:breɪ'ɪstɪk/ adj.

Hebraist /'hi:breɪɪst/ n. an expert in Hebrew.

Hebrew /'hi:bru:/ n. & adj. ● n. **1** a member of a Semitic people orig. centred in ancient Palestine and having a descent traditionally traced from Abraham, Isaac, and Jacob; an Israelite, a Jew. **2** a the language of this people. (*See note below.*) **b** a modern form of this used esp. in the state of Israel. ● adj. **1** of or in Hebrew. **2** of the Hebrews or the Jews. [ME f. OF *Ebreu* f. med.L *Ebreus* f. L *hebraeus* f. Gk *Hebraios* f. Aram. *'ibray* f. Heb. *'ibrī* one from the other side (of the river)]

▪ Hebrew was spoken and written in ancient Palestine for more than a thousand years. It is written from right to left, originally in an alphabet consisting of twenty-two letters, all of them consonants (vowel signs were added to the text of the Hebrew Bible in the 6th century AD to facilitate reading). By c.500 BC it had come greatly under the influence of Aramaic, which had largely replaced it as a spoken language by AD c.200, but it continued as the religious language of the Jewish people. It was revived as a spoken language in the 19th century, the modern form having its roots in the ancient language but drawing words from the vocabularies of European languages, and is now the official language of the state of Israel.

Hebrew Bible the sacred writings of Judaism, called by Christians the Old Testament. The Hebrew Bible consists of the Torah (the Law or Pentateuch, ascribed to Moses), the Prophets, and the Hagiographa or Writings. (See OLD TESTAMENT.)

Hebrews, Epistle to the a book of the New Testament, traditionally included among the letters of St Paul but now generally held to be non-Pauline.

Hebrides, the /ˈhebrɪˌdiːz/ (also called *Western Isles*) a group of about 500 islands off the NW coast of Scotland. The *Inner Hebrides* are divided between Highland and Strathclyde regions and include the islands of Skye, Mull, Jura, Islay, Iona, Coll, Eigg, Rhum, Staffa, and Tiree. The Little Minch separates this group from the *Outer Hebrides*, which form part of the Western Isles region and include the islands of Lewis with Harris, Barra, North and South Uist, and the island group of St Kilda. Until the end of the 13th century the Hebrides group was also regarded as including the islands in the Firth of Clyde, the peninsula of Kintyre in SW Scotland, the Isle of Man, and the (Irish) isle of Rathlin. They formed part of the kingdom of Scotland from 1266, when they were ceded to the Scottish king Alexander III by Magnus of Norway. Norse occupation has influenced the language, customs, and place-names, although most of the present-day population have Celtic affinities and Gaelic is still spoken on several of the islands. The shipping forecast area *Hebrides* covers an area of the Atlantic off the NW coast of Scotland. □ **Hebridean** /ˌhebrɪˈdiːən/ *adj. & n.*

Hebron /ˈhebrɒn/ a Palestinian city on the West Bank of the Jordan; pop. (est. 1984) 75,000. It is one of most ancient cities in the Middle East, probably founded in the 18th century BC. As the home of Abraham it is a holy city of both Judaism and Islam.

Hebros /ˈhiːbrəs/ (also **Hebrus**) the ancient Greek name for the MARITSA.

Hecate /ˈhekətɪ/ *Gk Mythol.* a goddess of dark places, often associated with ghosts and sorcery and worshipped with offerings at crossroads. She is frequently identified with Artemis and Selene; her name means 'the distant one'.

hecatomb /ˈhekəˌtuːm/ *n.* **1** (in ancient Greece or Rome) a great public sacrifice, orig. of 100 oxen. **2** any extensive sacrifice. [L *hecatombe* f. Gk *hekatombē* f. *hekaton* hundred + *bous* ox]

heck /hek/ *int. colloq.* a mild exclamation of surprise or dismay. [alt. f. HELL]

heckelphone /ˈhekəlˌfəʊn/ *n. Mus.* a bass oboe. [G *Heckelphon* f. Wilhelm *Heckel*, Ger. instrument-maker (1856–1909)]

heckle /ˈhekəl/ *v. & n.* ● *v.tr.* **1** interrupt and harass (a public speaker). **2** dress (flax or hemp). ● *n.* an act of heckling. □ **heckler** *n.* [ME, northern and eastern form of HACKLE]

hectare /ˈhekteə(r), -tɑː(r)/ *n.* (abbr. **ha**) a metric unit of square measure, equal to 100 ares (2.471 acres or 10,000 square metres). □ **hectarage** /ˈhektərɪdʒ/ *n.* [F (as HECTO-, ARE²)]

hectic /ˈhektɪk/ *adj. & n.* ● *adj.* **1** busy and confused; characterized by feverish excitement or haste. **2** *hist.* having a hectic fever; morbidly flushed. ● *n. hist.* **1** a hectic fever or flush. **2** a patient suffering from this. □ **hectic fever** (or **flush**) *hist.* a fever which accompanies consumption and similar diseases, with flushed cheeks and hot dry skin. □ **hectically** *adv.* [ME *etik* f. OF *etique* f. LL *hecticus* f. Gk *hektikos* habitual f. *hexis* habit, assim. to F *hectique* or LL]

hecto- /ˈhektəʊ/ *comb. form* (abbr. **h**) a hundred, esp. of a unit in the metric system. [F, irreg. f. Gk *hekaton* hundred]

hectogram /ˈhektəˌɡræm/ *n.* (also **hectogramme**) a metric unit of mass, equal to one hundred grams.

hectograph /ˈhektəˌɡrɑːf/ *n.* an apparatus for copying documents by the use of a gelatin plate which receives an impression of the master copy.

hectolitre /ˈhektəˌliːtə(r)/ *n.* (US **hectoliter**) a metric unit of capacity, equal to one hundred litres.

hectometre /ˈhektəˌmiːtə(r)/ *n.* (US **hectometer**) a metric unit of length, equal to one hundred metres.

Hector /ˈhektə(r)/ *Gk Mythol.* a Trojan warrior, son of Priam and Hecuba and husband of Andromache. He was killed by Achilles, who dragged his body at the wheels of his chariot three times round the walls of Troy.

hector /ˈhektə(r)/ *v. & n.* ● *v.tr.* bully, intimidate. ● *n.* a bully. □ **hectoringly** *adv.* [*Hector*, L f. Gk *Hektōr* HECTOR, f. its earlier use to mean 'swaggering fellow']

Hecuba /ˈhekjʊbə/ *Gk Mythol.* a Trojan woman, the wife of Priam and

mother of numerous children, including Hector, Paris, Cassandra, and Troilus.

he'd /hiːd, hɪd/ *contr.* **1** he had. **2** he would.

heddle /ˈhedəl/ *n.* one of the sets of small cords or wires between which the warp is passed in a loom before going through the reed. [app. f. OE *hefeld*]

hedge /hedʒ/ *n. & v.* ● *n.* **1** a fence or boundary formed by closely growing bushes or shrubs. **2** a protection against possible loss or diminution. ● *v.* **1** *tr.* surround or bound with a hedge. **2** *tr.* enclose. **3 a** *tr.* reduce one's risk of loss on (a bet or speculation) by compensating transactions on the other side. **b** *intr.* avoid a definite decision or commitment. **4** *intr.* make or trim hedges. □ **hedge-hop** fly at a very low altitude. **hedge sparrow** = DUNNOCK. **hedge trimmer** an electric etc. device for trimming hedges. □ **hedger** *n.* [OE *hegg* f. Gmc]

hedgehog /ˈhedʒhɒɡ/ *n.* **1** a small nocturnal mammal of the family Erinaceidae, having a piglike snout and a coat of spines, eating small invertebrates, and rolling itself up into a ball for defence; esp. *Erinaceus europaeus*, of Europe. **2** a porcupine or other animal similarly covered with spines. [ME f. HEDGE (from its habitat) + HOG (from its snout)]

hedgerow /ˈhedʒrəʊ/ *n.* a row of bushes etc. forming a hedge.

hedonic /hiːˈdɒnɪk, heˈdɒnɪk/ *adj.* **1** of or characterized by pleasure. **2** *Psychol.* of pleasant or unpleasant sensations. [Gk *hēdonikos* f. *hēdonē* pleasure]

hedonism /ˈhiːdəˌnɪz(ə)m, ˈhed-/ *n.* **1** belief in pleasure as the highest good and humankind's proper or sole aim. (*See note below.*) **2** behaviour based on this; devotion to or pursuit of pleasure. □ **hedonist** *n.* **hedonistic** /ˌhiːdəˈnɪstɪk, ˌhed-/ *adj.* [Gk *hēdonē* pleasure]

▪ Hedonism as an ethical doctrine constitutes a central thesis in Cyrenaic, Epicurean, and utilitarian philosophy. In its philosophical sense it is commonly distinguished by its definition of pleasure as the satisfaction of desires or wishes rather than as immediate physical enjoyment. While the ancient Cyrenaics equated pleasure with the satisfaction of sensual desires, the Epicureans valued the mental serenity brought about by an absence of anxiety and by the rational control of desire. Utilitarians like Bentham and Mill examine the pleasure given to society in general rather than to the individual by an act.

-hedron /ˈhiːdrən, ˈhed-/ *comb. form* (*pl.* **-hedra** /-rə/) forming nouns denoting geometrical solids with various numbers or shapes of faces (*dodecahedron; rhombohedron*). □ **-hedral** *comb. form* forming adjectives. [Gk *hedra* seat]

heebie-jeebies /ˌhiːbɪˈdʒiːbɪz/ *n.pl.* (prec. by *the*) *sl.* a state of nervous depression or anxiety. [20th c.: orig. unkn.]

heed /hiːd/ *v. & n.* ● *v.tr.* attend to; take notice of. ● *n.* careful attention. □ **heedful** *adj.* **heedfully** *adv.* **heedfulness** *n.* **heedless** *adj.* **heedlessly** *adv.* **heedlessness** *n.* [OE *hēdan* f. WG]

hee-haw /ˈhiːhɔː/ *n. & v.* ● *n.* the bray of a donkey. ● *v.intr.* emit a braying sound. [imit.]

heel¹ /hiːl/ *n. & v.* ● *n.* **1** the back part of the foot below the ankle. **2** the corresponding part in vertebrate animals. **3 a** the part of a sock etc. covering the heel. **b** the part of a shoe or boot supporting the heel. **4** a thing like a heel in form or position, e.g. the part of the palm next to the wrist, the end of a violin bow at which it is held, or the part of a golf club near where the head joins the shaft. **5** the crust end of a cut loaf of bread. **6** *colloq.* a person regarded with contempt or disapproval. **7** (as *int.*) a command to a dog to walk close to its owner's heel. ● *v.* **1** *tr.* fit or renew a heel on (a shoe or boot). **2** *intr.* touch the ground with the heel as in dancing. **3** *intr.* (foll. by *out*) *Rugby* pass the ball with the heel. **4** *tr. Golf* strike (the ball) with the heel of the club. □ **at heel 1** (of a dog) close behind. **2** (of a person etc.) under control. **at** (or **on**) **the heels of** following closely after (a person or event). **cool** (or **kick**) **one's heels** be kept waiting. **down at heel** see DOWN¹. **take to one's heels** run away. **to heel 1** (of a dog) close behind. **2** (of a person etc.) under control. **turn on one's heel** turn sharply round. **well-heeled** *colloq.* wealthy. □ **heelless** /ˈhiːllɪs/ *adj.* [OE *hēla, hǣla* f. Gmc]

heel² /hiːl/ *v. & n.* ● *v.* **1** *intr.* (of a ship etc.) lean over owing to the pressure of wind or an uneven load (cf. LIST²). **2** *tr.* cause (a ship etc.) to do this. ● *n.* the act or amount of heeling. [prob. f. obs. *heeld, hield* incline, f. OE *hieldan*, OS *-heldian* f. Gmc]

heel³ /hiːl/ *v.tr.* (foll. by *in*) set (a plant) in the ground and cover its roots. [OE *helian* f. Gmc]

heelball /ˈhiːlbɔːl/ *n.* **1** a mixture of hard wax and lampblack used by shoemakers for polishing. **2** this or a similar mixture used in brass-rubbing.

heeltap /ˈhiːltæp/ *n.* **1** a layer of leather in a shoe heel. **2** liquor left at the bottom of a glass after drinking.

Hefei /heˈfeɪ/ (also **Hofei** /həʊˈfeɪ/) an industrial city in eastern China, capital of Anhui province; pop. (1990) 1,541,000.

heft /heft/ *v. & n.* ● *v.tr.* lift (something heavy), esp. to judge its weight. ● *n.* dial. or *N. Amer.* weight, heaviness. [prob. f. HEAVE after *cleft, weft*]

hefty /ˈheftɪ/ *adj.* (**heftier, heftiest**) **1** (of a person) big and strong. **2** (of a thing) large, heavy, powerful; sizeable, considerable. □ **heftily** *adv.* **heftiness** *n.*

Hegel /ˈheɪɡ(ə)l/, Georg Wilhelm Friedrich (1770–1831), German philosopher. His philosophy represents a complex system of thought with far-reaching influences and diverse applications. He is especially known for his three-stage process of dialectical reasoning (set out in his *Science of Logic*, 1812–16), which underlies his idealist concepts of historical development and the evolution of ideas; Marx based his theory of dialectical materialism on this aspect of Hegel's work. Other major works include *The Phenomenology of Mind* (1807), which describes the progression of the human mind from consciousness through self-consciousness, reason, spirit, and religion to absolute knowledge. □ **Hegelian** /hɪˈɡeɪlɪən/ *adj. & n.* **Hegelianism** *n.*

hegemonic /ˌhedʒɪˈmɒnɪk, ˌheɡɪ-/ *adj.* ruling, supreme. [Gk *hēgemonikos* (as HEGEMONY)]

hegemony /hɪˈdʒeməni, hɪˈɡem-/ *n.* leadership or dominance, esp. by one state of a confederacy. [Gk *hēgemonia* f. *hēgemōn* leader f. *hēgeomai* lead]

Hegira /ˈhedʒɪrə/ *n.* (also **Hejira, Hijra** /ˈhɪdʒrə/) Muhammad's departure from Mecca to Medina in AD 622, which marks the consolidation of the first Muslim community. The departure was made necessary by the opposition of the merchant community in Mecca to both Muhammad's preaching and his attempt to achieve a social solidarity that was not tribally based. Its significance is indicated by the fact that the Islamic calendar (which is based on lunar months) is dated from AD 622 (= 1 AH). [med.L *hegira* f. Arab. *hijra* departure from one's country f. *hajara* separate]

Heidegger /ˈhaɪˌdeɡə(r)/, Martin (1889–1976), German philosopher. In *Being and Time* (1927) he examines the setting of human existence in the world. He regards *Angst* (dread) as a fundamental part of human consciousness, a symptom of the gravity of the human situation with its radical freedom of choice and awareness of death. Consequently human beings are continually attempting to escape their destiny, either by disguising it or by distracting their attention from its inevitability. Although he did not consider himself an existentialist, his work had a significant influence on existentialist philosophers such as Sartre.

Heidelberg /ˈhaɪd(ə)l.bɜːɡ/ a city in SW Germany, on the River Neckar, in Baden-Württemberg; pop. (1991) 139,390. It is noted for its university, which received its charter in 1386 and is the oldest in Germany, and for its medieval castle.

heifer /ˈhefə(r)/ *n.* **1 a** a young cow, esp. one that has not had more than one calf. **b** a female calf. **2** *sl. derog.* a woman. [OE *heahfore*]

heigh /heɪ/ *int.* expressing encouragement or enquiry. □ **heigh-ho** expressing boredom, resignation, etc. [imit.]

height /haɪt/ *n.* **1** the measurement from base to top or (of a standing person) from head to foot. **2** the elevation above ground or a recognized level (usu. sea level). **3** any considerable elevation (*situated at a height*). **4 a** high place or area. **b** rising ground. **5** the top of something. **6** *Printing* the distance from the foot to the face of type. **7 a** the most intense part or period of anything (*the battle was at its height*). **b** an extreme instance or example (*the height of fashion*). □ **height of land** *N. Amer.* a watershed. [OE *hēhthu* f. Gmc]

heighten /ˈhaɪt(ə)n/ *v.tr. & intr.* make or become higher or more intense.

Heilbronn /ˈhaɪlbrɒn/ a city in Baden-Württemberg, southern Germany, on the River Neckar; pop. (1991) 117,430.

Heilong /heɪˈlʊŋ/ the Chinese name for the AMUR.

Heilongjiang /ˌheɪlʊŋdʒɪˈæŋ/ (also **Heilungkiang** /-lʊŋkɪˈæŋ/) a province of NE China, on the Russian frontier; capital, Harbin.

Heine /ˈhaɪnə/, (Christian Johann) Heinrich (1797–1856), German poet. His reputation rests on his lyric poetry, particularly that in *Das Buch der Lieder* (1827), much of which was set to music by Schumann and Schubert. In 1830 Heine emigrated to Paris, where his works became more political; they include *Zur Geschichte der Religion und Philosophie in Deutschland* (1834), a witty and savage attack on German thought and literature, and his two verse satires *Atta Troll* (1843) and *Deutschland* (1844).

heinous /ˈheɪnəs, ˈhiːn-/ *adj.* (of a crime or criminal) utterly odious or wicked. □ **heinously** *adv.* **heinousness** *n.* [ME f. OF *haïneus* ult. f. *hair* to hate f. Frank.]

Heinz /haɪnz/, Henry John (1844–1919), American food manufacturer. In 1869 he established a family firm for the manufacture and sale of processed foods. Heinz devised the marketing slogan '57 Varieties' in 1896 and erected New York's first electric sign to promote his company's pickles in 1900. By the turn of the century, his firm was one of the largest in the US; since his death, it has become a major multinational company.

heir /eə(r)/ *n.* **1** a person entitled to property or rank as the legal successor of its former owner (often foll. by *to*: *heir to the throne*). **2** a person deriving or morally entitled to some thing, quality, etc., from a predecessor. □ **heir apparent** (*pl.* **heirs apparent**) an heir whose claim cannot be set aside by the birth of another heir. **heir-at-law** (*pl.* **heirs-at-law**) an heir by right of blood, esp. to the real property of an intestate. **heir presumptive** (*pl.* **heirs presumptive**) an heir whose claim may be set aside by the birth of another heir. □ **heirdom** *n.* **heirless** *adj.* **heirship** *n.* [ME f. OF *eir* f. LL *herem* f. L *heres -edis*]

heiress /ˈeərɪs/ *n.* a female heir, esp. to wealth or high title.

heirloom /ˈeəluːm/ *n.* **1** a piece of personal property that has been in a family for several generations. **2** a piece of property received as part of an inheritance. [HEIR + LOOM[1] in the sense 'tool']

Heisenberg /ˈhaɪz(ə)n.bɜːɡ/, Werner Karl (1901–76), German mathematical physicist and philosopher, who developed a system of quantum mechanics based on matrix algebra. For this and his discovery of the allotropic forms of hydrogen he was awarded the 1932 Nobel Prize for physics. He stated his famous uncertainty principle in 1927 (see UNCERTAINTY PRINCIPLE).

heist /haɪst/ *n. & v. N. Amer. sl.* ● *n.* a robbery. ● *v.tr.* rob. [repr. a local pronunc. of HOIST]

hei-tiki /heɪˈtɪkɪ/ *n. NZ* a greenstone neck-ornament worn by Maoris. [Maori f. *hei* hang, TIKI]

Hejaz /hɪˈdʒæz/ (also **Hijaz**) a coastal region of western Saudi Arabia, extending along the Red Sea.

Hejira var. of HEGIRA.

Hekla /ˈheklə/ an active volcano in SW Iceland, rising to a height of 1,491 m (4,840 ft).

HeLa cell /ˈhiːlə/ *n.* (usu. in *pl.*) a human epithelial cell of a strain continuously maintained in tissue culture since 1951, widely used esp. in virology. [Henrietta Lacks, whose cervical carcinoma provided the original cells]

held *past* and *past part.* of HOLD[1].

Heldentenor /ˈheldənteˌnɔː(r)/ *n.* **1** a powerful tenor voice suitable for heroic roles in opera. **2** a singer with this voice. [G f. *Held* hero]

hele var. of HEEL[3].

Helen /ˈhelɪn/ *Gk Mythol.* the daughter of Zeus and Leda, born from an egg. In the Homeric poems she is the surpassingly beautiful wife of Menelaus, and her abduction by Paris (to whom she had been promised, as a bribe, by Aphrodite; see PARIS[2]) led to the Trojan War. Helen has a non-Greek name and is probably in origin an ancient pre-Hellenic goddess connected with vegetation and fertility.

Helena /ˈhelɪnə/ the state capital of Montana; pop. (1990) 24,570.

Helena, St /ˈhelɪnə/ (AD *c*.255–*c*.330), Roman empress and mother of Constantine the Great. She was a convert to Christianity and in 326 visited the Holy Land, where she founded basilicas on the Mount of Olives and at Bethlehem. Later tradition ascribes to her the finding of the cross on which Christ was crucified. Feast day (in the Eastern Church) 21 May; (in the Western Church) 18 Aug.

helenium /heˈliːnɪəm/ *n.* a composite plant of the genus *Helenium*, with daisy-like flowers having prominent central discs. [mod.L f. Gk *helenion*, possibly commemorating HELEN]

Helgoland see HELIGOLAND.

heli- /ˈhelɪ/ *comb. form* helicopter (*heliport*).

heliacal /hɪˈlaɪək(ə)l/ *adj. Astron.* relating to or near the sun. □ **heliacal rising** (or **setting**) the first rising (or setting) of a star after (or before) a period of invisibility due to conjunction with the sun. [LL *heliacus* f. Gk *hēliakos* f. *hēlios* sun]

helianthemum /ˌhiːlɪˈænθəməm/ *n.* an evergreen shrub of the genus *Helianthemum*, with saucer-shaped flowers. See also *rock rose* (ROCK[1]). [mod.L f. Gk *hēlios* sun + *anthemon* flower]

helianthus /ˌhiːlɪˈænθəs/ n. a composite plant of the genus *Helianthus*, including the sunflower, Jerusalem artichoke, and related plants. [mod.L f. Gk *hēlios* sun + *anthos* flower]

helical /ˈhelɪk(ə)l/ adj. having the form of a helix. □ **helically** adv. **helicoid** adj. & n.

helices pl. of HELIX.

helichrysum /ˌhelɪˈkraɪz(ə)m/ n. a composite plant of the genus *Helichrysum*, with flowers that retain their appearance when dried. [L f. Gk *helikhrusos* f. *helix* spiral + *khrusos* gold]

helicity /hiːˈlɪsɪtɪ/ n. **1** esp. *Biochem.* helical character. **2** *Physics* a combination of the spin and the linear motion of a subatomic particle. [HELICAL + -ITY]

helicon /ˈhelɪkən/ n. a large spiral bass tuba played encircling the player's head and resting on the shoulder. [L f. Gk *Helikōn* Mount Helicon: later assoc. with HELIX]

Helicon, Mount /ˈhelɪkən/ a mountain in Boeotia, central Greece, to the north of the Gulf of Corinth, rising to 1,750 m (5741 ft). It was believed by the ancient Greeks to be the home of the Muses.

helicopter /ˈhelɪˌkɒptə(r)/ n. & v. ● n. an aircraft which derives both lift and propulsive power from horizontally revolving rotors. (*See note below.*) ● v.tr. & intr. transport or fly by helicopter. [F *hélicoptère* f. Gk *helix* (see HELIX) + *pteron* wing]

■ A helicopter has one or two sets of rotor blades powered by either piston or jet engines. Control of motion is provided by varying the pitch of the blades, i.e. the angle at which they meet the airstream. Lift at take-off, for example, is obtained by steeply pitching the blades; forward thrust comes by increasing pitch at the rear of the blades' circuit. Additional directional control may sometimes be provided by a tail-rotor. The helicopter differs from the autogiro, which, since its rotors are not powered, cannot take off vertically. The first practicable helicopters were developed in the 1930s, principally by Igor Sikorsky. Experimental models had flown earlier, and the principle of their operation dates back at least to Leonardo da Vinci, who drew and described a helicopter-like flying machine in the 15th century.

Heligoland /ˈhelɪɡəʊˌlænd/ (German **Helgoland** /ˈhɛlɡoˌlantʃ/) a small island in the North Sea, one of the North Frisian Islands, off the coast of Germany. Originally the home of Frisian seamen, it was Danish from 1714 until seized by the British navy in 1807 and later ceded officially to Britain. In 1890 it was returned to Germany in exchange for Zanzibar and Pemba, and became an important naval base. The naval installations were demolished after the Second World War, and the island became part of the Federal Republic of Germany in 1952.

helio- /ˈhiːlɪəʊ/ comb. form the sun. [Gk *hēlios* sun]

heliocentric /ˌhiːlɪəˈsentrɪk/ adj. **1** regarding the sun as centre. **2** considered as viewed from the sun's centre. □ **heliocentrically** adv.

Heliogabalus /ˌhiːlɪəˈɡæbələs/ (also **Elagabalus** /ˌelə-/) (born Varius Avitus Bassianus) (AD 204–22), Roman emperor 218–22. He took his name from the Syro-Phoenician sun-god Elah-Gabal, whose hereditary priest he was. During his reign he became notorious for his dissipated lifestyle and neglect of state affairs; he and his mother were both murdered.

heliogram /ˈhiːlɪəˌɡræm/ n. a message sent by heliograph.

heliograph /ˈhiːlɪəˌɡrɑːf/ n. & v. ● n. **1 a** a signalling apparatus reflecting sunlight in flashes from a movable mirror. **b** a message sent by means of this; a heliogram. **2** an apparatus for photographing the sun. **3** an engraving obtained chemically by exposure to light. ● v.tr. send (a message) by heliograph. □ **heliography** /ˌhiːlɪˈɒɡrəfɪ/ n.

heliogravure /ˌhiːlɪəʊɡrəˈvjʊə(r)/ n. = PHOTOGRAVURE.

heliolithic /ˌhiːlɪəˈlɪθɪk/ adj. (of a civilization) characterized by sun-worship and megaliths.

heliometer /ˌhiːlɪˈɒmɪtə(r)/ n. an instrument used for finding the angular distance between two stars (orig. used for measuring the diameter of the sun).

Heliopolis /ˌhiːlɪˈɒpəlɪs/ **1** an ancient Egyptian city situated near the apex of the Nile delta at what is now Cairo. It was an important religious centre and the centre of sun worship (its name means 'city of the sun'), and was the original site of the obelisks known as Cleopatra's Needles. **2** the ancient Greek name for BAALBEK.

Helios /ˈhiːlɪəs/ *Gk Mythol.* the sun personified as a god, father of Phaethon. He is generally represented as a charioteer driving daily from east to west across the sky. In Rhodes he was the chief national god (see COLOSSUS OF RHODES). [Gk *hēlios* sun]

heliostat /ˈhiːlɪəˌstæt/ n. an apparatus with a mirror driven by clockwork to reflect sunlight in a fixed direction. □ **heliostatic** /ˌhiːlɪəˈstætɪk/ adj.

heliotherapy /ˌhiːlɪəˈθerəpɪ/ n. the use of sunlight as a therapeutic treatment.

heliotrope /ˈhiːlɪəˌtrəʊp, ˈhel-/ n. **1 a** a plant of the genus *Heliotropium* (borage family), with cymes of small flowers; esp. *H. arborescens*, a shrub with fragrant purple flowers. **b** the scent of these. **2** (in full **winter heliotrope**) a composite plant, *Petasites fragrans*, related to butterbur and producing fragrant lilac flowers in winter. **3** a light purple colour. **4** bloodstone. [L *heliotropium* f. Gk *hēliotropion* plant turning its flowers to the sun, f. *hēlios* sun + -*tropos* f. *trepō* turn]

heliotropism /ˌhiːlɪˈɒtrəˌpɪz(ə)m/ n. the directional growth of a plant in response to sunlight (cf. PHOTOTROPISM). □ **heliotropic** /ˌhiːlɪəˈtrəʊpɪk, -ˈtrɒpɪk/ adj.

heliotype /ˈhiːlɪəˌtaɪp/ n. a picture obtained from a sensitized gelatin film exposed to light.

helipad /ˈhelɪˌpæd/ n. a landing pad for helicopters.

heliport /ˈhelɪˌpɔːt/ n. a place where helicopters take off and land. [HELI-, after *airport*]

heli-skiing /ˈhelɪˌskiːɪŋ/ n. skiing in which transport up the mountain is by helicopter.

helium /ˈhiːlɪəm/ n. an inert gaseous chemical element (atomic number 2; symbol **He**). Before it was discovered on earth, helium was detected by Lockyer in 1868 in the spectrum of the sun. On earth helium occurs in traces in air, and more abundantly in natural gas deposits. The lightest of the noble gases, helium is a particularly suitable lifting gas for balloons and airships; liquid helium (boiling-point: 4.2 kelvins, −268.9°C) is used as a coolant. Helium is produced in stars as the main product of the thermonuclear fusion of hydrogen, and is the second most abundant element in the universe after hydrogen. Helium nuclei (alpha particles) are emitted in many forms of radioactive decay. It has no known chemical compounds. [Gk *hēlios* sun]

helix /ˈhiːlɪks/ n. (pl. **helices** /ˈhiːlɪˌsiːz, ˈhel-/) **1** a spiral curve (like a corkscrew) or a coiled curve (like a watch spring). **2** *Geom.* a curve that cuts a line on a solid cone or cylinder, at a constant angle with the axis. **3** *Archit.* a spiral ornament. **4** *Anat.* the rim of the external ear. [L *helix -icis* f. Gk *helix -ikos*]

hell /hel/ n. & int. ● n. **1** the abode of the dead; in Christian, Jewish, and Islamic belief, the place of punishment or torment where the souls of the damned are confined after death. (*See note below.*) **2** a place or state of misery or wickedness. **3** fun; high spirits. ● int. an exclamation of surprise or annoyance. □ **beat** (or **knock** etc.) **the hell out of** colloq. beat etc. without restraint. **come hell or high water** no matter what the difficulties. **for the hell of it** colloq. for fun; on impulse. **get** (or **catch**) **hell** colloq. be severely scolded or punished. **give a person hell** colloq. scold or punish or make things difficult for a person. **the hell** (usu. prec. by *what, where, who*, etc.) expressing anger, disbelief, etc. (*who the hell is this?; the hell you are!*). **hell-bent** (foll. by *on*) recklessly determined. **hell-cat** a spiteful violent woman. **hell-fire** the fire or fires regarded as existing in hell. **hell for leather** at full speed. **a** (or **one**) **hell of a** colloq. an outstanding example of (*a hell of a mess; one hell of a party*). **hell-hole** an oppressive or unbearable place. **hell-hound** a fiend. **hell-raiser** a person who causes trouble or creates chaos. **like hell** colloq. **1** not at all. **2** recklessly, exceedingly. **not a hope in hell** colloq. no chance at all. **play hell** (or **merry hell**) **with** colloq. be upsetting or disruptive to. **what the hell** colloq. it is of no importance. □ **hell-like** adj. **hellward** adv. & adj. [OE *hel, hell* f. Gmc]

■ The concept of hell as a place where the souls of sinners receive punishment is widespread in Christian, Jewish, and Islamic tradition. It derives from the notion of the underworld as a place where departed souls continue to exist after death, which is found in classical mythology, Judaism (see SHEOL), and other ancient religions. The physical descriptions of the underworld vary: commonly found attributes include darkness, coldness, and location underground. The notion of hell as a place of eternal punishment developed from the idea that the virtuous would be rewarded in heaven: the souls of those who were not selected for heavenly reward would remain in the underworld, where they would be punished for their misdeeds. The popular Christian idea of hell as a dark and fiery pit is derived from texts such as Matt. 25:41 and Rev. 19:20; it was powerfully elaborated in medieval Christian art and literature.

he'll /hiːl, hɪl/ contr. he will; he shall.

hellacious /heˈleɪʃəs/ adj. US sl. terrific, tremendous; remarkable, enormous. [HELL + -ACIOUS]

Helladic /heˈlædɪk/ adj. Archaeol. of, relating to, or denoting the Bronze Age cultures of mainland Greece (c.3000–1050 BC), of which the latest period is equivalent to the Mycenaean age. [Gk Helladikos f. Hellas -ados Greece]

Hellas /ˈhɛlas/ the Greek name for GREECE.

hellbender /ˈhelˌbendə(r)/ n. a large North American salamander, Cryptobranchus alleganiensis. [HELL + BEND¹ + -ER¹]

hellebore /ˈhelɪˌbɔː(r)/ n. 1 an evergreen ranunculaceous plant of the genus Helleborus, with large white, green, or purplish flowers, e.g. the Christmas rose. 2 a liliaceous plant, Veratrum album, resembling a helleborine. 3 hist. any plant supposed to cure madness. [ME f. OF ellebre, elebore or med.L eleborus f. L elleborus f. Gk (h)elleboros]

helleborine /ˈhelɪbəˌriːn, ˌhelɪˈbɔːriːn/ n. an orchid of the genus Epipactis or Cephalanthera, with a loose spike of spurless flowers. [F or L helleborine or L f. Gk helleborinē plant like hellebore (as HELLEBORE)]

Hellen /ˈhelɪn/ Gk Mythol. the son or brother of Deucalion and ancestor of all the Hellenes or Greeks.

Hellene /ˈheliːn/ n. 1 a native of modern Greece. 2 an ancient Greek of genuine Greek descent. [Gk Hellēn HELLEN]

Hellenic /heˈlenɪk, -ˈliːnɪk/ n. & adj. ● n. 1 a branch of the Indo-European language family from which is derived classical and modern Greek. 2 the Greek language. ● adj. of or relating to Greece, the Greeks, or Hellenic; Greek. [Gk Hellēnikos, f. as HELLENE]

Hellenism /ˈhelɪˌnɪz(ə)m/ n. 1 Greek character or culture (esp. of ancient Greece). 2 the study or imitation of Greek culture. □ **Hellenize** v.tr. & intr. (also **-ise**). **Hellenization** /ˌhelɪnaɪˈzeɪʃ(ə)n/ n. [Gk hellēnismos f. hellēnizō speak Greek, make Greek (as HELLENE)]

Hellenist /ˈhelɪnɪst/ n. an expert on or admirer of Greek language or culture. [Gk Hellēnistēs (as HELLENISM)]

Hellenistic /ˌhelɪˈnɪstɪk/ adj. of or relating to the period of Greek history, language, and culture from the death of Alexander the Great in 323 BC to the defeat of Cleopatra and Mark Antony by Octavian in 31 BC. The Hellenistic period saw Greek culture spread through the Mediterranean and into the Near East and Asia, centring on Alexandria in Egypt and Pergamum in Turkey; it is sometimes known as the Alexandrian period. It was a time of wealth and culture, which produced writers such as Callimachus, Theocritus, and Menander.

Heller /ˈhelə(r)/, Joseph (b.1923), American novelist. His experiences in the US air force during the Second World War inspired his best-known novel Catch-22 (1961), an absurdist black comedy satirizing war. The dilemma in which the book's hero finds himself when he tries to avoid combat duty forms the novel's title and has passed into the language (see CATCH-22).

Hellespont /ˈhelɪsˌpɒnt/ the ancient name for the Dardanelles, named after the legendary Helle, who fell into the strait and was drowned while escaping with her brother Phrixus from their stepmother, Ino, on a golden-fleeced ram. (See also GOLDEN FLEECE.) [Gk Hellēspontos sea of Helle]

hellgrammite /ˈhelɡrəˌmaɪt/ n. US an aquatic larva of an American fly, Corydalus cornutus, often used as fishing bait. [19th c.: orig. unkn.]

hellion /ˈheljən/ n. N. Amer. colloq. a mischievous or troublesome person, esp. a child. [perh. f. dial. hallion a worthless fellow, assim. to HELL]

hellish /ˈhelɪʃ/ adj. & adv. ● adj. 1 of or like hell. 2 colloq. extremely difficult or unpleasant. ● adv. Brit. colloq. (as an intensifier) extremely (hellish expensive). □ **hellishly** adv. **hellishness** n.

Hellman /ˈhelmən/, Lillian (Florence) (1907–84), American dramatist. She gained her first success with The Children's Hour (1934), which was followed by plays such as The Little Foxes (1939) and the anti-Fascist Watch on the Rhine (1941). Hellman was a socialist and a feminist, and her plays frequently reflect her political concerns. For more than thirty years she lived with the detective-story writer Dashiell Hammett; both were blacklisted during the McCarthy era.

hello /həˈləʊ/ int., n., & v. (also **hallo**, **hullo**) ● int. 1 a an expression of informal greeting, or of surprise. b used to begin a telephone conversation. 2 a cry used to call attention. ● n. (pl. **-os**) a cry of 'hello'. ● v.intr. (**-oes**, **-oed**) cry 'hello'. [var. of earlier HOLLO]

Hell's Angel n. a member of any of a number of motorcycle gangs, originating in California in the 1950s though now found widely elsewhere, originally notorious for disturbances of civil order. Their symbol is the winged death's-head.

Hell's Canyon a chasm in Idaho, cut by the Snake River and forming the deepest gorge in the US. Flanked by the Seven Devils Mountains, the canyon drops to a depth of 2,433 m (7,900 ft).

helm¹ /helm/ n. & v. ● n. 1 a tiller or wheel by which a ship's rudder is controlled. 2 the amount by which this is turned (more helm needed). ● v.tr. steer or guide as if with a helm. □ **at the helm** in control; at the head (of an organization etc.). [OE helma, prob. related to HELVE]

helm² /helm/ n. archaic helmet. □ **helmed** adj. [OE f. Gmc]

Helmand /ˈhelmənd/ the longest river in Afghanistan. Rising in the Hindu Kush, it flows 1,125 km (700 miles), generally south-west, before emptying into marshland near the Iran–Afghanistan frontier.

helmet /ˈhelmɪt/ n. 1 any of various protective head-coverings worn by soldiers, police officers, firefighters, divers, motorcyclists, etc. 2 Bot. the arched upper part of the corolla in some flowers. 3 the large thick shell of a gastropod mollusc of the genus Cassis. □ **helmeted** adj. [ME f. OF, dimin. of helme f. WG (as HELM²)]

Helmholtz /ˈhelmholts/, Hermann Ludwig Ferdinand von (1821–94), German physiologist and physicist. His investigation of animal heat led to his formulation of the principle of the conservation of energy in 1847. He produced two studies on sense perception: first with physiological optics (in the course of which he invented the ophthalmoscope), and then with physiological acoustics. Other achievements include his attempts to measure the speed of nerve impulses, and his studies of vortex motion in hydrodynamics and the properties of oscillating electric currents. He also contributed to non-Euclidean geometry.

helminth /ˈhelmɪnθ/ n. a parasitic worm, e.g. a fluke, tapeworm, or nematode. □ **helminthic** /helˈmɪnθɪk/ adj. **helminthoid** adj. **helminthology** /ˌhelmɪnˈθɒlədʒɪ/ n. [Gk helmins -inthos intestinal worm]

helminthiasis /ˌhelmɪnˈθaɪəsɪs/ n. a disease characterized by the presence of parasitic worms in the body.

Helmont /ˈhelmɒnt/, Joannes Baptista van (1577–1644), Belgian chemist and physician. He made early studies on the conservation of matter, was the first to distinguish gases, and coined the word gas. Having failed to realize that green plants take in carbon dioxide, he concluded that they are composed entirely of water.

helmsman /ˈhelmzmən/ n. (pl. **-men**) a steersman.

Héloïse /ˌeləʊˈiːz/ (1098–1164), French abbess. She is chiefly remembered for her passionate but tragic love affair with Abelard, which began after she became his pupil (at the instigation of her uncle, Fulbert) c.1118. She gave birth to a son, after which the two were secretly married; when the affair came to light relatives of Héloïse castrated Abelard, while Héloïse was forced to enter a convent. She later became abbess of Paraclete. (See also ABELARD.)

helot /ˈhelət/ n. 1 (also **Helot**) a member of a class of serfs in ancient Sparta, intermediate in status between slaves and citizens. 2 a serf or slave. □ **helotism** n. **helotry** n. [L helotes pl. f. Gk heilōtes, -ōtai, usu. derived from Helos, a town in Laconia whose inhabitants were enslaved]

help /help/ v. & n. ● v.tr. 1 provide (a person etc.) with the means towards what is needed or sought (helped me with my work; helped me (to) pay my debts). 2 (foll. by up, down, etc.) assist or give support to (a person) in moving etc. as specified (helped her into the chair; helped him on with his coat). 3 (often absol.) be of use or service to (a person) (does that help?). 4 contribute to alleviating (a pain or difficulty). 5 prevent or remedy (it can't be helped). 6 (usu. with neg.) a refrain from (can't help it; could not help laughing). b refl. refrain from acting (couldn't help himself). 7 (often foll. by to) serve (a person with food) (shall I help you to potatoes?). ● n. 1 the act of helping or being helped (we need your help; came to our help). 2 a person or thing that helps. 3 a domestic servant or employee, or several collectively. 4 a remedy or escape (there is no help for it). □ **helping hand** assistance. **help oneself** (often foll. by to) 1 serve oneself (with food etc.). 2 take without seeking help; take without permission, steal. **help a person out** give a person help, esp. in difficulty. **so help me** (or **help me God**) (as an invocation or oath) I am speaking the truth. □ **helper** n. [OE helpan f. Gmc]

helpful /ˈhelpfʊl/ adj. (of a person or thing) giving help; obliging, useful. □ **helpfully** adv. **helpfulness** n.

helping /ˈhelpɪŋ/ n. a portion of food, esp. at a meal.

helpless /ˈhelplɪs/ adj. 1 lacking help or protection; defenceless. 2 unable to act without help. □ **helplessly** adv. **helplessness** n.

helpline /ˈhelplaɪn/ n. a telephone service providing help with problems.

Helpmann /ˈhelpmən/, Sir Robert (Murray) (1909–86), Australian ballet-dancer, choreographer, director, and actor. Coming to England in 1933, he joined the Vic-Wells Ballet and in 1935 began a long partnership with Margot Fonteyn. Helpmann was noted for his dramatic ability and for his strongly theatrical choreography; his own ballets include *Comus* and *Hamlet* (both 1942).

helpmate /'helpmeɪt/ *n.* a helpful companion or partner (usu. a husband or wife).

Helsingborg /'helsɪŋˌbɔːg/ (Swedish **Hälsingborg** /ˌhɛlsɪŋ'bɔrj/) a port in southern Sweden, situated on the Øresund opposite Helsinør in Denmark; pop. (1990) 109,270.

Helsingfors /ˌhɛlsɪŋ'fɔrs/ the Swedish name for HELSINKI.

Helsingør see ELSINORE.

Helsinki /'helˌsɪŋkɪ, hel'sɪŋkɪ/ (called in Swedish *Helsingfors*) the capital of Finland, a port in the south on the Gulf of Finland; pop. (1990) 492,400.

helter-skelter /ˌheltə'skeltə(r)/ *adv., adj., & n.* ● *adv.* in disorderly haste; confusedly. ● *adj.* characterized by disorderly haste or confusion. ● *n. Brit.* a tall spiral slide round a tower, at a fairground or funfair. [imit., orig. in a rhyming jingle, perh. f. ME *skelte* hasten]

helve /helv/ *n.* the handle of a weapon or a tool. [OE *helfe* f. WG]

Helvetia /hel'viːʃə/ the Latin name for SWITZERLAND.

Helvetian /hel'viːʃ(ə)n/ *adj. & n.* ● *adj.* Swiss. ● *n.* a native of Switzerland. [L HELVETIA]

hem[1] /hem/ *n. & v.* ● *n.* the border of a piece of cloth, esp. a cut edge turned under and sewn down. ● *v.tr.* (**hemmed, hemming**) turn down and sew in the edge of (a piece of cloth etc.). □ **hem in** confine; restrict the movement of. [OE, perh. rel. to dial. *ham* enclosure]

hem[2] /hem, həm/ *int., n., & v.* ● *int.* expressing hesitation or attempting to attract attention by a slight cough or clearing of the throat. ● *n.* an utterance of this. ● *v.intr.* (**hemmed, hemming**) say *hem*; hesitate in speech. □ **hem and haw** = *hum and haw* (see HUM[1]). [imit.]

hemal etc. *US* var. of HAEMAL etc.

hemato- etc. *US* var. of HAEMATO- etc.

heme *US* var. of HAEM.

Hemel Hempstead /ˌhem(ə)l 'hempstɪd/ a town in Hertfordshire, in SE England; pop. (1981) 80,340. It was designated as a new town in 1947.

hemerocallis /ˌheməraʊ'kælɪs/ *n.* = *day lily.* [L *hemerocalles* f. Gk *hēmerokalles* a kind of lily f. *hēmera* day + *kallos* beauty]

hemi- /'hemɪ/ *comb. form* half. [Gk *hēmi-* = L *semi-*: see SEMI-]

-hemia *comb. form US* var. of -AEMIA.

hemianopsia /ˌhemɪə'nɒpsɪə/ *n.* (also **hemianopia** /-'nəʊpɪə/) blindness over half the field of vision.

hemicellulose /ˌhemɪ'seljʊˌləʊz, -ˌləʊs/ *n.* any of a class of polysaccharides forming the matrix of plant cell walls in which cellulose is embedded. [G (as HEMI-, CELLULOSE)]

hemichordate /ˌhemɪ'kɔːdeɪt/ *n. & adj. Zool.* ● *n.* a wormlike marine invertebrate of the phylum Hemichordata, which comprises the acorn worms, possessing a notochord in the larval stage. ● *adj.* of or relating to this phylum. [HEMI- + CHORDATE]

hemicycle /'hemɪˌsaɪk(ə)l/ *n.* a semicircular figure.

hemidemisemiquaver /ˌhemɪˌdemɪ'semɪˌkweɪvə(r)/ *n. Mus.* a note having the time value of half a demisemiquaver and represented by a large dot with a four-hooked stem. Also called *sixty-fourth note.*

hemihedral /ˌhemɪ'hiːdrəl/ *adj. Crystallog.* having half the number of planes required for symmetry of the holohedral form.

Hemingway /'hemɪŋˌweɪ/, Ernest (Miller) (1899–1961), American novelist and journalist. After the First World War he lived in Paris, where he came into contact with writers such as Ezra Pound and worked as a journalist before publishing short stories and then novels. His early novels reflect the disillusionment of the postwar 'lost generation'; they include *The Sun Also Rises* (1926) and *A Farewell to Arms* (1929). In his later works there is a developing theme of the strength and dignity of the human spirit; the most famous are *For Whom the Bell Tolls* (1940) and *The Old Man and the Sea* (1952). He was awarded the Nobel Prize for literature in 1954.

hemiplegia /ˌhemɪ'pliːdʒɪə/ *n. Med.* paralysis of one side of the body. □ **hemiplegic** *n. & adj.* [mod.L f. Gk *hēmiplēgia* paralysis (as HEMI-, *plēgē* stroke)]

Hemiptera /he'mɪptərə/ *n.pl. Zool.* an order of insects comprising the 'true' bugs, which have the mouthparts adapted for piercing and sucking, and undergo incomplete metamorphosis. See HETEROPTERA,

HOMOPTERA. □ **hemipteran** *n. & adj.* **hemipterous** *adj.* [mod.L f. HEMI- + Gk *pteron* wing]

hemisphere /'hemɪˌsfɪə(r)/ *n.* **1** half of a sphere. **2** a half of the earth, esp. as divided by the equator (into *northern* and *southern hemisphere*) or by an imaginary line passing through the poles (into *eastern* and *western hemisphere*). □ **hemispheric** /ˌhemɪ'sferɪk/ *adj.* **hemispherical** *adj.* [OF *emisphere* & L *hemisphaerium* f. Gk *hēmisphaira* (as HEMI-, SPHERE)]

hemistich /'hemɪˌstɪk/ *n.* half of a line of verse. [LL *hemistichium* f. Gk *hēmistichion* (as HEMI-, *stikhion* f. *stikhos* line)]

Hemkund, Lake /hem'kʊnd/ a lake in northern India, in the Himalayan foothills of Uttar Pradesh. It is regarded as holy by the Sikhs.

hemline /'hemlaɪn/ *n.* the line or level of the lower edge of a skirt, dress, or coat.

hemlock /'hemlɒk/ *n.* **1 a** a poisonous umbelliferous plant, *Conium maculatum*, with a spotted stem, fernlike leaves, and small white flowers. **b** a poisonous potion obtained from this. **2** (in full **hemlock fir** or **spruce**) **a** a coniferous tree of the genus *Tsuga*, with foliage that smells like hemlock when crushed. **b** the timber or pitch of these trees. [OE *hymlic(e)*]

hemo- *comb. form US* var. of HAEMO-.

hemp /hemp/ *n.* **1** (in full **Indian hemp**) a herbaceous plant, *Cannabis sativa*, native to Asia, grown for its fibre or for the drug marijuana or cannabis. **2** its fibre extracted from the stem and used to make rope and stout fabrics. **3** any narcotic drug made from the hemp plant, esp. marijuana or cannabis. **4** any other plant yielding fibre (*Manila hemp; sunn hemp*). □ **hemp agrimony** a composite plant, *Eupatorium cannabinum*, with pale purple flowers and hairy leaves. **hemp-nettle** a nettle-like labiate plant of the genus *Galeopsis*. [OE *henep, hænep* f. Gmc, rel. to Gk *kannabis*]

hempen /'hempən/ *adj.* made from hemp.

hemstitch /'hemstɪtʃ/ *n. & v.* ● *n.* a decorative stitch used in sewing hems. ● *v.tr.* hem with this stitch.

hen /hen/ *n.* **1 a** a female bird, esp. of a domestic fowl. **b** (in *pl.*) domestic fowls of either sex. **2** a female lobster, crab, or salmon. □ **hen and chickens** a plant that produces additional small flower-heads or offshoots, esp. the houseleek. **hen-coop** a coop for keeping fowls in. **hen harrier** a widely distributed harrier, *Circus cyaneus*, found in open country and moorland; also called (*N. Amer.*) *northern harrier, marsh hawk.* **hen-house** a small shed for fowls to roost in. **hen-party** *colloq.* often *derog.* a social gathering of women. **hen-roost** a place where fowls roost at night. **hen-run** an enclosure for fowls. [OE *henn* f. WG]

Henan /hə'næn/ (also **Honan** /həʊ'næn/) a province of NE central China; capital, Zhengzhou.

henbane /'henbeɪn/ *n.* **1** a poisonous herbaceous solanaceous plant, *Hyoscyamus niger*, with sticky hairy leaves and an unpleasant smell. **2** a narcotic drug obtained from this.

hence /hens/ *adv.* **1** from this time (*two years hence*). **2** for this reason; as a result of inference (*hence we seem to be wrong*). **3** *archaic* from here; from this place. [ME *hens, hennes, henne* f. OE *heonan* f. the root of HE]

henceforth /hens'fɔːθ/ *adv.* (also **henceforward** /-'fɔːwəd/) from this time onwards.

henchman /'hentʃmən/ *n.* (*pl.* **-men**) **1** often *derog.* **a** a trusted supporter or attendant. **b** a political supporter; a partisan. **2** *hist.* a squire; a page of honour. **3** the principal attendant of a Highland chief. [ME *henxman, hengestman* f. OE *hengst* male horse]

hendeca- /hen'dekə/ *comb. form* eleven. [Gk *hendeka* eleven]

hendecagon /hen'dekəˌgɒn/ *n.* a plane figure with eleven sides and angles.

hendiadys /hen'daɪəˌdɪs/ *n.* the expression of an idea by two words connected with 'and', instead of one modifying the other, e.g. *nice and warm* for *nicely warm*. [med.L f. Gk *hen dia duoin* one thing by two]

Hendrix /'hendrɪks/, Jimi (born James Marshall Hendrix) (1942–70), American rock guitarist and singer. Remembered for the flamboyance and originality of his improvisations, he greatly widened the scope of the electric guitar. He gave notable live performances with his groups, playing with the Jimi Hendrix Experience at the Monterey pop festival (1967) and with the Band of Gypsies at the Woodstock festival (1969). His best-known singles include 'Purple Haze' (1967) and 'All along the Watchtower' (1968).

henequen /'henɪˌken/ *n.* **1** a Mexican agave, *Agave fourcroydes*, grown for its fibre. **2** the sisal-like fibre obtained from this. [Sp. *jeniquen*]

henge /hendʒ/ *n. Archaeol.* a large neolithic monument consisting of a circular bank and ditch, and frequently containing a circle of standing stones. [back-form. f. STONEHENGE]

Hengist /ˈhɛŋɡɪst/ (d.488), semi-mythological Jutish leader. He and his brother Horsa (d.455) are said by early chroniclers to have been invited to Britain by the British king Vortigern in 449 to assist in defeating the Picts, and later to have established an independent Anglo-Saxon kingdom in Kent. The historicity of the brothers has been questioned, and they may have been mythological figures (their names mean 'gelding' and 'horse').

Henley /ˈhɛnlɪ/ (in full **Henley Royal Regatta**) the oldest rowing regatta in Europe, inaugurated in 1839 at Henley-on-Thames, Oxfordshire, as a result of the interest aroused locally by the first Oxford and Cambridge Boat Race, which took place at Henley in 1829. The regatta is held annually in the first week in July.

henna /ˈhɛnə/ n. & v. ● n. **1** a tropical shrub, *Lawsonia inermis*, having small pink, red, or white flowers. **2** the reddish dye from its shoots and leaves, used esp. to colour hair. ● v.tr. (**hennaed**, **hennaing**) dye (hair) with henna. [Arab. *ḥinnā*]

henotheism /ˈhɛnəˌθiːɪz(ə)m/ n. belief in or adoption of a particular god in a polytheistic system as the god of a tribe, class, etc. [Gk *heis henos* one + *theos* god]

henpeck /ˈhɛnpɛk/ v.tr. (usu. as **henpecked** adj.) (of a woman) constantly harass or nag (a man, esp. her husband).

Henri /ˈhɛnrɪ/, Robert (1865–1929), American painter. He was an advocate of realism and believed that the artist must be a social force; as a teacher at the New York School of Art (1898–1928), he encouraged his students to turn away from academicism and towards a realistic depiction of everyday life. The Ashcan School of painters was formed largely as a result of his influence.

Henrietta Maria /ˌhɛnrɪˌɛtə məˈriːə/ (1609–69), daughter of Henry IV of France, queen consort of Charles I of England 1625–49. Her Roman Catholicism heightened public anxieties about the court's religious sympathies and was a contributory cause of the English Civil War. From 1644 she lived mainly in France.

Henry[1] /ˈhɛnrɪ/ the name of eight kings of England:

Henry I (1068–1135), youngest son of William I, reigned 1100–35. On the death of his brother, William II, Henry seized the throne in the absence of his other brother, Robert of Normandy; Henry conquered Normandy in 1105. After his only son was drowned in 1120 there were problems with the succession, and although Henry extracted an oath of loyalty to his daughter Matilda from the barons in 1127, his death was followed almost immediately by the outbreak of civil war.

Henry II (1133–89), son of Matilda, reigned 1154–89. The first Plantagenet king, Henry restored order after the reigns of Stephen and Matilda, added Anjou and Aquitaine to the English possessions in France, established his rule in Ireland, and forced the king of Scotland to acknowledge him as overlord of that kingdom. Henry was less successful in reducing the power of the Church; opposition to his policies was led by Thomas à Becket, who was eventually murdered by four of Henry's knights.

Henry III (1207–72), son of John, reigned 1216–72. Until Henry declared himself of age to rule personally in 1227, his regent the Earl of Pembroke kept the rebellious barons in check, but afterwards the king's ineffectual government caused widespread discontent, ending in Simon de Montfort's defeat and capture of the king in 1264. Although Henry was restored after the defeat of the rebels a year later, real power resided with his son, who eventually succeeded him as Edward I.

Henry IV (known as Henry Bolingbroke) (1367–1413), son of John of Gaunt, reigned 1399–1413. He returned from exile in 1399 to overthrow Richard II and establish the Lancastrian dynasty. His reign was scarred by rebellion, both in Wales and in the north, where the Percy family raised several uprisings. Although Henry defeated Sir Henry Percy ('Hotspur') in 1403, the Percy threat did not abate until the head of the family was killed in 1408.

Henry V (1387–1422), son of Henry IV, reigned 1413–22. He renewed the Hundred Years War soon after coming to the throne and defeated the French at Agincourt in 1415. By the Treaty of Troyes (1420) Henry was named successor to Charles VI of France and betrothed to his daughter Catherine of Valois. When the Dauphin repudiated the treaty, Henry returned to France but fell ill and died, leaving his infant son to inherit the throne.

Henry VI (1421–71), son of Henry V, reigned 1422–61 and 1470–1. Succeeding his father while still an infant, Henry VI proved to have a recurrent mental illness which made him unfit to rule effectively on his own. During his reign the Hundred Years War with France was finally lost, and government by the monarchy, in the hands of a series of regents and noble favourites, became increasingly unpopular. In the 1450s opposition coalesced round the House of York, and, after intermittent civil war (see WARS OF THE ROSES), Henry was deposed in 1461 by Edward IV. In 1470 Henry briefly regained his throne following a Lancastrian uprising, but was deposed again and murdered soon after.

Henry VII (known as Henry Tudor) (1457–1509), the first Tudor king, son of Edmund Tudor, Earl of Richmond, reigned 1485–1509. Although he was the grandson of Owen Tudor, it was through his mother, a great-granddaughter of John of Gaunt, that he inherited the Lancastrian claim to the throne. Having grown up in exile in France, Henry returned to England in 1485 and ascended the throne after defeating Richard III at Bosworth Field. Threatened in the early years of his reign by a series of Yorkist plots, Henry eventually established an unchallenged Tudor dynasty, dealing ruthlessly with other claimants to the throne. As king he continued the strengthening of royal government commenced by his Yorkist predecessors.

Henry VIII (1491–1547), son of Henry VII, reigned 1509–47. Henry had six wives (Catherine of Aragon, Anne Boleyn, Jane Seymour, Anne of Cleves, Catherine Howard, Catherine Parr), two of whom he had executed and two of whom he divorced. His efforts to divorce his first wife, Catherine of Aragon, which were opposed by the pope, led to England's break with the Roman Catholic Church and indirectly to the establishment of Protestantism. Henry's ensuing dissolution of the monasteries not only destroyed most of the remaining vestiges of the old religious Establishment but also changed the pattern of land ownership. The final years of Henry's reign were marked by wars and rebellion.

Henry[2] /ˈhɛnrɪ/ (known as Henry the Navigator) (1394–1460), Portuguese prince. The third son of John I of Portugal, he was a leading patron of voyages of exploration, from which he earned his title. He established a school of navigation at Cape St Vincent and organized and funded many voyages of discovery, most notably south along the African coast. The efforts of his captains, who reached as far south as Cape Verde and the Azores, laid the groundwork for later Portuguese imperial expansion south-east round Africa to the Far East.

Henry[3] /ˈhɛnrɪ/ the name of seven kings of the Germans, six of whom were also Holy Roman emperors, notably:

Henry I (known as Henry the Fowler) (c.876–936), reigned 919–36. As duke of Saxony, he was elected king by the nobles of Saxony and Franconia following the death of Conrad I (903–18). He waged war successfully against the Slavs in Brandenburg, the Magyars, and the Danes, from whom he gained the territory of Schleswig in 934.

Henry III (1017–56), reigned 1039–56, Holy Roman emperor 1046–56. He brought stability and prosperity to the empire, defeating the Czechs in 1041 and fixing the frontier between Austria and Hungary in 1043. A devout Christian, he introduced religious reforms, attacked corruption in the Church, and strengthened the papacy, securing the appointment of four successive German popes.

Henry IV (1050–1106), son of Henry III, reigned 1056–1105, Holy Roman emperor 1084–1105. Following the end of his regency in 1065 he came into increasing conflict with Pope Gregory VII (c.1020–85), which culminated in 1076 when Henry called a council to depose the Pope. Gregory retaliated by excommunicating Henry and absolving his subjects from their oaths of loyalty to him. In 1077 Henry obtained absolution by doing penance before Gregory at Canossa in Italy, and spent the next three years waging war on his rebellious subjects. He finally managed to depose Gregory in 1084, being crowned emperor by his successor in the same year.

henry /ˈhɛnrɪ/ n. (pl. **-ies** or **henrys**) Electr. the SI unit of inductance (symbol: **H**), equal to the inductance of a circuit in which an electromotive force of one volt is produced by a current changing at the rate of one ampere per second. [Joseph *Henry*, Amer. physicist (1797–1878)]

Henry Bolingbroke, Henry IV of England (see HENRY[1]).

Henry IV /ˈhɛnrɪ, French ɑ̃ri/ (known as Henry of Navarre) (1553–1610), king of France 1589–1610. As king of Navarre, Henry was the leader of Huguenot forces in the latter stages of the French Wars of Religion, but on succeeding the Catholic Henry III he became Catholic himself in order to guarantee peace. He founded the Bourbon dynasty, established religious freedom with the Edict of Nantes (1598), and restored order after prolonged civil war. He was assassinated by a Catholic fanatic.

Henry's law n. Chem. a law stating that the amount of a gas dissolved in a given volume of solvent is proportional to the pressure of gas with which the solvent is in equilibrium. [William *Henry*, Engl. chemist (1774–1836)]

Henry the Fowler, Henry I, king of the Germans (see HENRY³).

Henry Tudor, Henry VII of England (see HENRY¹).

Henze /'hentsə/, Hans Werner (b.1926), German composer and conductor. His musical style is diverse, displaying a respect for classical forms such as the sonata and a lyricism that reveals his knowledge of Italian opera. His many works include *The Raft of the Medusa* (1968) (a requiem for Che Guevara), the opera *The English Cat* (1982), seven symphonies, and ballet and chamber music.

heortology /ˌhiːɔːˈtɒlədʒɪ/ n. the study of Church festivals. [G *Heortologie*, F *héortologie* f. Gk *heortē* feast]

hep¹ var. of HIP³.

hep² var. of HIP².

heparin /'hepərɪn/ n. Biochem. a sulphur-containing polysaccharide found in various organs and tissues which inhibits blood coagulation and is used as an anticoagulant in the treatment of thrombosis. □ **heparinize** v.tr. (also **-ise**). [L f. Gk *hēpar* liver]

hepatic /hɪˈpætɪk/ adj. **1** of or relating to the liver. **2** dark brownish-red; liver-coloured. [ME f. L *hepaticus* f. Gk *hēpatikos* f. *hēpar -atos* liver]

hepatica /hɪˈpætɪkə/ n. a ranunculaceous plant of the genus *Hepatica*, with reddish-brown lobed leaves resembling the liver. [med.L fem. of *hepaticus*: see HEPATIC]

hepatitis /ˌhepəˈtaɪtɪs/ n. inflammation of the liver; a disease in which this occurs. □ **hepatitis A** a viral form of hepatitis transmitted in food, causing fever and jaundice. **hepatitis B** a severe viral form of hepatitis transmitted in infected blood, causing fever, debility, and jaundice. [mod.L: see HEPATIC]

Hepburn¹ /'hepbɜːn/, Audrey (1929–93), British actress, born in Belgium. After pursuing a career as a stage and film actress in England, she moved to Hollywood, where she starred in such films as *Roman Holiday* (1953), for which she won an Oscar, and *War and Peace* (1956). She is perhaps best known for her performance as Eliza Doolittle in the film musical *My Fair Lady* (1964).

Hepburn² /'hepbɜːn/, Katharine (b.1909), American actress. After her screen début in 1932 she went on to star in a wide range of films, including many in which she formed a partnership with Spencer Tracy. Her films include *The Philadelphia Story* (1940), *Woman of the Year* (1942), *The African Queen* (1951), and *On Golden Pond* (1981), for which she won her fourth Oscar.

Hephaestus /hɪˈfiːstəs/ Gk Mythol. the god of fire (especially the smithy fire), son of Zeus and Hera, identified with Vulcan by the Romans. He was the god of craftsmen, and was himself a divine craftsman who was lame as the result of having interfered in a quarrel between his parents.

Hepplewhite /'hep(ə)lˌwaɪt/ n. a late 18th-century style of furniture, originally as made by the English cabinet-maker George Hepplewhite (d.1786), characterized by lightness, delicacy, and graceful curves.

hepta- /'heptə/ comb. form seven. [Gk *hepta* seven]

heptad /'heptæd/ n. a group of seven. [Gk *heptas -ados* set of seven (*hepta*)]

heptagon /'heptəgən/ n. a plane figure with seven sides and angles. □ **heptagonal** /hepˈtæg(ə)l/ adj. [F *heptagone* or med.L *heptagonum* f. Gk (as HEPTA-, -GON)]

heptahedron /ˌheptəˈhiːdrən/ n. a solid figure with seven faces. □ **heptahedral** adj. [HEPTA- + -HEDRON after POLYHEDRON]

heptameter /hepˈtæmɪtə(r)/ n. a line or verse of seven metrical feet. [L *heptametrum* f. Gk (as HEPTA-, -METER)]

heptane /'hepteɪn/ n. Chem. a liquid hydrocarbon of the alkane series (chem. formula: C_7H_{16}), obtained from petroleum. [HEPTA- + -ANE²]

heptarchy /'heptɑːkɪ/ n. (pl. **-ies**) **1 a** a government by seven rulers. **b** an instance of this. **2** hist. the supposed seven kingdoms of the Angles and the Saxons in Britain in the 7th–8th centuries. □ **heptarchic** /hepˈtɑːkɪk/ adj. **heptarchical** adj. [HEPTA- after *tetrarchy*]

Heptateuch /'heptəˌtjuːk/ n. the first seven books of the Old Testament. [L f. Gk f. *hepta* seven + *teukhos* book, volume]

heptathlon /hepˈtæθlən/ n. an athletic contest, usu. for women, in which each competitor takes part in seven events. □ **heptathlete** /-liːt/ n. [HEPTA-, after DECATHLON]

heptavalent /ˌheptəˈveɪlənt/ adj. Chem. having a valency of seven; septivalent.

Hepworth /'hepwəθ/, Dame (Jocelyn) Barbara (1903–75), English sculptor. A pioneer of abstraction in British sculpture, in the 1930s she worked chiefly in wood and stone, using forms suggested by the inherent qualities of these materials and assimilating them to organic forms or to the human figure. From the 1950s onwards she also worked in bronze, producing simple monumental works for landscape and architectural settings. Her works include the nine-piece group *The Family of Man* (1972).

her /hɜː(r), hə(r)/ pron. & poss.pron. ● pron. **1** objective case of SHE (*I like her*). **2** she (*it's her all right; am older than her*). ¶ Strictly, the pronoun following the verb *to be* should be in the subjective rather than the objective case, i.e. *am older than she* rather than *am older than her*. **3** archaic & US dial. herself (*she fell and hurt her*). ● poss.pron. (attrib.) **1** of or belonging to her or herself (*her house; her own business*). **2** (**Her**) (in titles) that she is (*Her Majesty*). □ **her indoors** Brit. colloq. or joc. one's wife. [OE *hi(e)re* dative & genitive of *hio*, *hēo* fem. of HE]

Hera /'hɪərə/ Gk Mythol. a powerful goddess, the wife and sister of Zeus and the daughter of Cronus and Rhea. She was worshipped as the queen of heaven and as a marriage-goddess and was identified by the Romans with Juno. [Gk *Hēra* (perh. as a title, = lady, fem. of *hērōs* hero)]

Heracles /'herəˌkliːz/ the Greek form of HERCULES.

Heraclitus /ˌherəˈklaɪtəs/ (c.500 BC), Greek philosopher. He regarded the universe as a ceaselessly changing conflict of opposites, all things being in a harmonious process of constant change, and held that fire, the type of this constant change, is their origin. He believed that the mind derives a false idea of the permanence of the external world from the passing impressions of experience.

Heraklion /hɪˈræklɪən/ (Greek **Iráklion** /iˈraklɪɒn/) a port in and the capital of Crete, on the north coast of the island; pop. (1991) 117,000.

herald /'herəld/ n. & v. ● n. **1** an official messenger bringing news. **2** a forerunner (*spring is the herald of summer*). **3 a** hist. an officer who made state proclamations and bore messages between princes, officiated in the tournament, arranged various state ceremonials, regulated the use of armorial bearings, settled questions of precedence, or recorded the names and pedigrees of those entitled to armorial bearings. **b** (in the UK) an official of the College of Arms. ● v.tr. proclaim the approach of; usher in (*the storm heralded trouble*). [ME f. OF *herau(l)t*, *herauder* f. Gmc]

heraldic /heˈrældɪk/ adj. of or concerning heraldry. □ **heraldically** adv. [HERALD]

heraldist /'herəldɪst/ n. an expert in heraldry. [HERALD]

heraldry /'herəldrɪ/ n. the science or art of a herald, especially in blazoning armorial bearings and settling the right of persons to bear arms. Bearings probably originated as designs on shields at a time when chain-mail was worn and a bold, simple identification device was needed. Such designs are found in western Europe from the 12th century; similar designs from Japan are found in the same period. By the 13th century heraldry in Europe had so developed that it had its own terminology, based on Old French. Armorial bearings are hereditary; strictly they belong to the head of a family. The use of heraldic devices is now widespread and they are sought by corporate bodies for reasons of prestige and identification.

Heralds' College see COLLEGE OF ARMS.

Herat /həˈræt/ a city in western Afghanistan; pop. (est. 1984) 160,000.

herb /hɜːb/ n. **1** Bot. a non-woody seed-bearing plant which dies down to the ground after flowering. **2** a plant with leaves, seeds, or flowers used for flavouring, food, medicine, scent, etc. □ **give it the herbs** Austral. colloq. accelerate. **herb bennet** a common yellow-flowered rosaceous plant, *Geum urbanum*. **herb Christopher** a white-flowered baneberry, *Actaea spicata*. **herb Gerard** ground elder. **herb Paris** a liliaceous plant, *Paris quadrifolia*, with a single flower and four leaves in a cross shape on an unbranched stem. **herb Robert** a common cranesbill, *Geranium robertianum*, with red-stemmed leaves and pink flowers. **herb tea** an infusion of herbs. **herb tobacco** a mixture of herbs smoked as a substitute for tobacco. □ **herbiferous** /-ˈbɪfərəs/ adj. **herblike** adj. [ME f. OF *erbe* f. L *herba* grass, green crops, herb; *herb bennet* prob. f. med.L *herba benedicta* blessed herb (thought of as expelling the Devil)]

herbaceous /hɜːˈbeɪʃəs/ adj. of or like herbs (see HERB 1). □ **herbaceous border** a garden border containing esp. perennial flowering plants. **herbaceous perennial** a plant whose growth dies down annually but whose roots etc. survive. [L *herbaceus* grassy (as HERB)]

herbage /'hɜːbɪdʒ/ n. **1** herbaceous vegetation. **2** the succulent part of this, esp. as pasture. **3** *Law* the right of pasture on another person's land. [ME f. OF *erbage* f. med.L *herbaticum, herbagium* right of pasture, f. L *herba* herb]

herbal /ˈhɜːb(ə)l/ *adj. & n.* ● *adj.* of herbs in medicinal and culinary use. ● *n.* a book with descriptions and accounts of the properties of these. [med.L *herbalis* (as HERB)]

herbalism /ˈhɜːbəˌlɪz(ə)m/ *n.* the use of plants and plant products in the treatment of disease. Until quite recently, all drugs were essentially plant extracts, but the pharmaceutical industry now synthesizes their active components chemically, as well as developing artificial drugs. Modern herbalism, a form of alternative medicine, is a reversion to the use of pure plant extracts to stimulate the body's own defences, based on the principle of harmony with the environment and the belief that herbal remedies represent a neglected resource.

herbalist /ˈhɜːbəlɪst/ *n.* **1** a practitioner of herbalism; a dealer in medicinal herbs. **2** a collector of or writer on plants, esp. an early botanical writer.

herbarium /hɜːˈbeərɪəm/ *n. (pl.* **herbaria** /-rɪə/) **1** a systematically arranged collection of dried plants. **2** a book, room, or building for these. [LL (as HERB)]

Herbert[1] /ˈhɜːbət/, Sir A(lan) P(atrick) (1890–1970), English writer and politician. He was a writer of versatility and humour who contributed to the magazine *Punch* for many years, wrote libretti for a number of comic operas, and published a number of novels. He also campaigned for several causes, most notably the reform of the divorce laws; as Independent MP for Oxford University (1935–50) he was responsible for introducing the Matrimonial Causes Act (1937), which radically amended the legislation.

Herbert[2] /ˈhɜːbət/, George (1593–1633), English metaphysical poet. In 1630 he became the vicar of Bemerton, near Salisbury, after a brief time spent as MP for Montgomery (1624–5). His devout religious verse is pervaded by simple piety and reflects the spiritual conflicts he experienced before submitting his will to God. His poems are marked by metrical versatility and homely imagery; most were published just after his death in *The Temple: Sacred Poems and Private Ejaculations*.

herbicide /ˈhɜːbɪˌsaɪd/ *n.* a substance toxic to plants and used to destroy unwanted vegetation.

herbivore /ˈhɜːbɪˌvɔː(r)/ *n.* an animal that feeds on plants. □ **herbivorous** /hɜːˈbɪvərəs/ *adj.* [L *herba* herb + -VORE (see -VOROUS)]

herby /ˈhɜːbɪ/ *adj.* (**herbier, herbiest**) **1** abounding in herbs. **2** of the nature of a culinary or medicinal herb.

Hercegovina see HERZEGOVINA.

Herculaneum /ˌhɜːkjʊˈleɪnɪəm/ an ancient Roman town, near Naples, on the lower slopes of Vesuvius. The volcano's eruption in AD 79 buried it deeply under volcanic ash, along with Pompeii, and thus largely preserved it until its accidental rediscovery by a well-digger in 1709. The first excavations were begun in 1738.

Herculean /ˌhɜːkjʊˈliːən, hɜːˈkjuːlɪən/ *adj.* having or requiring great strength or effort. [L *Herculeus* (as HERCULES)]

Hercules /ˈhɜːkjʊˌliːz/ **1** (called by the Greeks *Heracles*) *Gk & Rom. Mythol.* a hero of superhuman strength and courage who performed twelve immense tasks or 'labours' imposed on him by Eurystheus, king of Argos, and who after death was ranked among the gods. He is usually depicted with a lion-skin, club, and bow. **2** *Astron.* a large northern constellation, said to represent the figure of Hercules kneeling on his right knee. It contains the brightest globular cluster in the northern hemisphere, but no bright stars. [ME f. L f. Gk *Hēraklēs* (= Hera's glory; *kleos* fame)]

Hercules beetle *n.* a very large South American beetle, *Dynastes hercules*, of the scarab family, with two horns extending from the head and thorax.

Hercynian /hɜːˈsɪnɪən/ *adj. Geol.* of or relating to a mountain-forming time in the eastern hemisphere during the late Palaeozoic era. [L *Hercynia silva* forested mountains of central Germany]

herd /hɜːd/ *n. & v.* ● *n.* **1** a large number of animals, esp. cattle, feeding or travelling or kept together. **2** (often prec. by *the*) *derog.* a large number of people; a crowd, a mob (*prefers not to follow the herd*). **3** (esp. in *comb.*) a keeper of herds; a herdsman (*cowherd*). ● *v.* **1** *intr. & tr.* go or cause to go in a herd or crowd (*herded together for warmth; herded the cattle into the field*). **2** *tr.* tend (sheep, cattle, etc.) (*he herds the goats*). □ **herd-book** a book recording the pedigrees of cattle or pigs. **the herd instinct** the tendency to associate or conform with one's own kind for support etc. **ride herd on** *US* keep watch on. □ **herder** *n.* [OE *heord*, (in sense 3) *hirdi*, f. Gmc]

herdsman /ˈhɜːdzmən/ *n. (pl.* **-men**) the owner or keeper of a herd of domesticated animals.

Herdwick /ˈhɜːdwɪk/ *n.* a hardy breed of mountain sheep from northern England. [obs. *herdwick* pasture-ground (as HERD, WICK[2]), perh. because this breed originated in Furness Abbey pastures]

here /hɪə(r)/ *adv., n., & int.* ● *adv.* **1** in or at or to this place or position (*put it here; has lived here for many years; comes here every day*). **2** indicating a person's presence or a thing offered (*here is your coat; my son here will show you*). **3** at this point in the argument, situation, etc. (*here I have a question*). ● *n.* this place (*get out of here; lives near here; fill it up to here*). ● *int.* **1** calling attention: short for *come here, look here*, etc. (*here, where are you going with that?*). **2** indicating one's presence in a roll-call: short for *I am here*. □ **here and now** at this very moment; immediately. **here and there** in various places. **here goes!** an expression indicating the start of a bold act. **here's to** I drink to the health of. **here we are** said on arrival at one's destination. **here we go again** the same, usu. undesirable, events are recurring. **here you are** said on handing something to somebody. **neither here nor there** of no importance or relevance. [OE *hēr* f. Gmc]

hereabouts /ˌhɪərəˈbaʊts/ *adv.* (also **hereabout**) near this place.

hereafter /hɪərˈɑːftə(r)/ *adv. & n.* ● *adv.* **1** from now on; in the future. **2** in the world to come (after death). ● *n.* **1** the future. **2** life after death.

hereat /hɪərˈæt/ *adv. archaic* as a result of this.

hereby /hɪəˈbaɪ/ *adv.* by this means; as a result of this.

hereditable /hɪˈredɪtəb(ə)l/ *adj.* that can be inherited. [obs. F *héréditable* or med.L *hereditabilis* f. eccl.L *hereditare* f. L *heres -edis* heir]

hereditament /ˌherɪˈdɪtəmənt, hɪˈredɪ-/ *n. Law* **1** any property that can be inherited. **2** inheritance. [med.L *hereditamentum* (as HEREDITABLE)]

hereditary /hɪˈredɪtərɪ/ *adj.* **1** (of disease, instinct, etc.) able to be passed down from one generation to another. **2 a** descending by inheritance. **b** holding a position by inheritance. **3** the same as or resembling what one's parents had (*a hereditary hatred*). **4** of or relating to inheritance. □ **hereditarily** *adv.* **hereditariness** *n.* [L *hereditarius* (as HEREDITY)]

heredity /hɪˈredɪtɪ/ *n.* **1 a** the passing on of physical or mental characteristics genetically from one generation to another. (*See note below.*) **b** these characteristics. **2** the genetic constitution of an individual. [F *hérédité* or L *hereditas* heirship (as HEIR)]

▪ That 'like begets like' has been recognized since ancient times, but a real understanding of the nature of heredity had to await the discovery of the sex cells (gametes), the phenomenon of fertilization, and the existence of genes. Lamarck proposed that characteristics acquired by parents during life could be passed on to their offspring, but this view is no longer accepted. It was also widely believed in the 19th century that sexual reproduction involved 'blending' of the characteristics of the two parents in the offspring. It was Mendel who first demonstrated that the hereditary material consists of discrete hereditary factors now known as genes (see MENDELISM).

Hereford[1] /ˈherɪfəd/ a city in west central England, in the county of Hereford and Worcester, on the River Wye; pop. (1991) 49,800.

Hereford[2] /ˈherɪfəd/ *n.* a breed of red and white beef cattle. [HEREFORD[1]]

Hereford and Worcester a county of west central England, on the border with Wales; administrative centre, Worcester. It was formed in 1974 from the former counties of Herefordshire and Worcestershire.

Herefordshire /ˈherɪfədˌʃɪə(r)/ a former county of west central England. Since 1974 it has been part of the county of Hereford and Worcester.

herein /hɪərˈɪn/ *adv. formal* in this matter, book, etc.

hereinafter /ˌhɪərɪnˈɑːftə(r)/ *adv. formal esp. Law* **1** from this point on. **2** in a later part of this document etc.

hereinbefore /ˌhɪərɪnbɪˈfɔː(r)/ *adv. formal esp. Law* in a preceding part of this document etc.

hereof /hɪərˈɒv/ *adv. formal* of this.

Herero /həˈreərəʊ, -ˈrɪərəʊ/ *n. & adj.* ● *n. (pl.* same or **Hereros**) **1** a member of any of several peoples, speaking a Bantu language, of Namibia, Angola, and Botswana. **2** the language of these peoples. ● *adj.* of or relating to the Herero or their language.

heresiarch /heˈriːzɪˌɑːk/ *n.* the leader or founder of a heresy. [eccl.L *haeresiarcha* f. Gk *hairesiarkhēs* (as HERESY + *arkhēs* ruler)]

heresy /ˈherɪsɪ/ *n. (pl.* **-ies**) **1** esp. *RC Ch.* **a** a belief or practice contrary to orthodox doctrine. **b** an instance of this. **2 a** opinion contrary to what is normally accepted or maintained (*it's heresy to suggest that instant coffee is as good as the real thing*). **b** an instance of this. □ **heresiology** /ˌherɪsɪˈɒlədʒɪ/ *n.* [ME f. OF (h)eresie, f. eccl.L haeresis, in L = school of thought, f. Gk hairesis choice, sect f. haireomai choose]

heretic /ˈherɪtɪk/ *n.* **1** the holder of an unorthodox opinion. **2** a person believing in or practising religious heresy. □ **heretical** /hɒˈretɪk(ə)l/ *adj.* **heretically** *adv.* [ME f. OF heretique f. eccl.L haereticus f. Gk hairetikos able to choose (as HERESY)]

hereto /hɪɒˈtuː/ *adv. formal* to this matter.

heretofore /ˌhɪɒtʊˈfɔː(r)/ *adv. formal* before this time.

hereunder /hɪɒrˈʌndə(r)/ *adv. formal* below (in a book, legal document, etc.).

hereunto /ˌhɪɒrʌnˈtuː/ *adv. archaic* to this.

hereupon /ˌhɪɒrəˈpɒn/ *adv.* after this; in consequence of this.

Hereward the Wake /ˈherɪwəd, weɪk/ (11th century), semi-legendary Anglo-Saxon rebel leader. Although little is known of Hereward's life beyond what can be found in literary accounts of dubious reliability, he is remembered as a leader of Anglo-Saxon resistance to William I's new Norman regime, and was apparently responsible for an uprising centred on the Isle of Ely in 1070.

herewith /hɪɒˈwɪð, -ˈwɪθ/ *adv.* with this (esp. of an enclosure in a letter etc.).

heriot /ˈherɪɒt/ *n. hist.* a tribute paid to a lord on the death of a tenant, consisting of a live animal, a chattel, or, originally, the return of borrowed equipment. [OE heregeatwa f. here army + geatwa trappings]

heritable /ˈherɪtəb(ə)l/ *adj.* **1** *Law* **a** (of property) capable of being inherited by heirs-at-law (cf. MOVABLE). **b** (of a person) capable of inheriting. **2** *Biol.* (of a characteristic) transmissible from parent to offspring. □ **heritably** *adv.* **heritability** /ˌherɪtəˈbɪlɪtɪ/ *n.* [ME f. OF f. heriter f. eccl.L hereditare: see HEREDITABLE]

heritage /ˈherɪtɪdʒ/ *n.* **1** anything that is or may be inherited. **2** inherited circumstances, benefits, etc. (*a heritage of confusion*). **3** a nation's historic buildings, monuments, countryside, etc., esp. when regarded as worthy of preservation. **4** *Bibl.* **a** the ancient Israelites. **b** the Church. □ **heritage centre** a museum focusing on the cultural heritage of the surrounding area. [ME f. OF (as HERITABLE)]

heritor /ˈherɪtə(r)/ *n.* esp. *Sc. Law* a person who inherits. [ME f. AF heriter, OF heritier (as HEREDITARY), assim. to words in -OR[1]]

herl var. of HARL.

herm /hɜːm/ *n. Gk Antiq.* a squared stone pillar with a head (esp. of Hermes) on top, used as a boundary-marker etc. (cf. TERMINUS 6). [L Herma f. Gk HERMES]

hermaphrodite /hɜːˈmæfrəˌdaɪt/ *n. & adj.* ● *n.* **1 a** *Zool.* an animal having both male and female sexual organs. **b** *Bot.* a plant having stamens and pistils in the same flower. **2** a human being in which both male and female sex organs are present, or in which the sex organs contain both ovarian and testicular tissue. **3** a person or thing combining opposite qualities or characteristics. ● *adj.* **1** combining both sexes. **2** combining opposite qualities or characteristics. □ **hermaphrodite brig** *hist.* a two-masted sailing-ship rigged on the foremast as a brig and on the mainmast as a schooner. □ **hermaphroditism** *n.* **hermaphroditic** /-ˌmæfrəˈdɪtɪk/ *adj.* **hermaphroditical** *adj.* [L hermaphroditus f. Gk hermaphroditos HERMAPHRODITUS]

Hermaphroditus /hɜːˌmæfrəˈdaɪtəs/ *Gk Mythol.* a son of Hermes and Aphrodite, with whom the nymph Salmacis fell in love and prayed to be forever united. As a result Hermaphroditus and Salmacis became joined in a single body which retained characteristics of either sex.

hermeneutic /ˌhɜːmɪˈnjuːtɪk/ *adj.* concerning interpretation, esp. of Scripture or literary texts. □ **hermeneutical** *adj.* **hermeneutically** *adv.* [Gk hermēneutikos f. hermēneuō interpret]

hermeneutics /ˌhɜːmɪˈnjuːtɪks/ *n.pl.* (also treated as *sing.*) the branch of knowledge dealing with interpretation, esp. of Scripture or literary texts.

Hermes /ˈhɜːmiːz/ *Gk Mythol.* the son of Zeus and Maia, the messenger of the gods and god of merchants, thieves, and oratory. Identified by the Romans with Mercury, he was pictured as a herald equipped for travelling, with broad-brimmed hat, winged shoes, and a winged rod. He was also associated with fertility, however, and from early times was represented by a stock or stone, generally having a human head carved at the top and a phallus halfway up it. [prob. f. Gk herma heap of stones]

Hermes Trismegistus /ˌtrɪsməˈdʒɪstəs/ a legendary figure regarded by Neoplatonists and others as the author of certain works on astrology, magic, and alchemy. [L, = thrice-greatest Hermes, irreg. transl. of 'Thoth the very great' (see THOTH)]

hermetic /hɜːˈmetɪk/ *adj.* (also **hermetical**) **1** with an airtight closure. **2** protected from outside agencies. **3 a** of alchemy or other occult sciences (*hermetic art*). **b** esoteric. □ **hermetic seal** an airtight seal (orig. as used by alchemists). □ **hermetically** *adv.* **hermetism** /ˈhɜːmɪˌtɪz(ə)m/ *n.* [mod.L hermeticus irreg. f. HERMES TRISMEGISTUS]

hermit /ˈhɜːmɪt/ *n.* **1** an early Christian recluse. **2** any person living in solitude. □ **hermit-crab** a crab of the family Paguridae, living in a cast-off mollusc shell for protection. **hermit thrush** a migratory North American thrush, *Catharus guttatus*. □ **hermitic** /hɜːˈmɪtɪk/ *adj.* [ME f. OF (h)ermite or f. LL eremita f. Gk erēmitēs f. erēmia desert f. erēmos solitary]

hermitage /ˈhɜːmɪtɪdʒ/ *n.* **1** a hermit's dwelling. **2** a monastery. **3** a solitary dwelling. [ME f. OF (h)ermitage (as HERMIT)]

Hermitage, the a major art museum in St Petersburg, Russia, containing among its collections those begun by Catherine the Great. It derives its name from the 'retreat' in which she displayed her treasures to her friends.

Hermitian /hɜːˈmɪtɪən/ *adj. Math.* designating or relating to a matrix in which pairs of elements symmetrically placed with respect to the principal diagonal are complex conjugates. [Charles *Hermite*, Fr. mathematician (1822–1901)]

Hermosillo /ˌeəməˈsiːjəʊ/ a city in NW Mexico, capital of the state of Sonora; pop. (1990) 449,470.

hernia /ˈhɜːnɪə/ *n.* (pl. **hernias** or **herniae** /-nɪˌiː/) the displacement and protrusion of part of an organ through the wall of the cavity containing it, esp. of the abdomen. □ **hernial** *adj.* **herniated** /-nɪˌeɪtɪd/ *adj.* [L]

Herning /ˈhɜːnɪŋ/ a city in central Jutland, Denmark; pop. (1991) 56,690.

Hero[1] /ˈhɪɒrəʊ/ *Gk Mythol.* a priestess of Aphrodite at Sestos on the European shore of the Hellespont, whose lover Leander, a youth of Abydos on the opposite shore, swam the strait nightly to visit her until one stormy night he was drowned and Hero in grief threw herself into the sea.

Hero[2] /ˈhɪɒrəʊ/ (known as Hero of Alexandria) (1st century), Greek mathematician and inventor. His surviving works are important as a source for ancient practical mathematics and mechanics. He described a number of hydraulic, pneumatic, and other mechanical devices designed both for utility and amusement, including elementary applications of the power of steam.

hero /ˈhɪɒrəʊ/ *n.* (pl. **-oes**) **1 a** a person noted or admired for courage, outstanding achievements, nobility, etc. (*Newton, a hero of science*). **b** a great warrior. **2** the chief male character in a poem, play, story, etc. **3** *Gk Antiq.* a man of superhuman qualities, favoured by the gods; a demigod. □ **hero's welcome** a rapturous welcome, like that given to a successful warrior. **hero-worship** *n.* **1** idealization of an admired man. **2** *Gk Antiq.* worship of the superhuman heroes. ● *v.tr.* (**-worshipped**, **-worshipping**; US **-worshiped**, **-worshiping**) worship as a hero; idolize. **hero-worshipper** a person engaging in hero-worship. [ME f. L heros f. Gk hērōs]

Herod /ˈherəd/ the name of four rulers of ancient Palestine:

Herod the Great (*c.*74–4 BC), ruled 37–4 BC. He built the palace of Masada and rebuilt the Temple in Jerusalem. Jesus was born during his reign; according to the New Testament (Matt. 2:16), he ordered the massacre of the innocents.

Herod Antipas (22 BC–AD *c.*40), son of Herod the Great, tetrarch of Galilee and Peraea 4 BC–AD 40. He married Herodias and was responsible for the beheading of John the Baptist. According to the New Testament (Luke 23:7), Pilate sent Jesus to be questioned by him before the Crucifixion.

Herod Agrippa I (10 BC–AD 44), grandson of Herod the Great, king of Judaea AD 41–4. He imprisoned St Peter and put St James the Great to death.

Herod Agrippa II (AD 27–*c.*93), son of Herod Agrippa I, king of various territories in northern Palestine 50–*c.*93. He presided over the trial of St Paul (Acts 25:13 ff.).

Herodotus /hɪˈrɒdətəs/ (known as 'the Father of History') (5th century BC), Greek historian. His *History* tells of the Persian Wars of the early 5th century BC, with an account of the earlier history of the Persian empire and its relations with the Greeks to explain the origins of the conflict. He was the first historian to collect his materials systematically, test their accuracy to a certain extent, and arrange them in a well-constructed and vivid narrative.

heroic /hɪˈrəʊɪk/ *adj. & n.* ● *adj.* **1 a** (of an act, quality, etc.) of or fit for a hero. **b** (of a person) like a hero. **2 a** (of language) grand, high-flown, dramatic. **b** (of a work of art) ambitious in scale or subject; unusually large or impressive. **3** of the heroes of Greek antiquity; (of poetry) dealing with the ancient heroes. ● *n.* (in *pl.*) **1 a** high-flown language or sentiments. **b** unduly bold behaviour. **2** = *heroic verse*. □ **the heroic age** the period in Greek history before the return from the Trojan War. **heroic couplet** two lines of rhyming iambic pentameters. **heroic verse** a type of verse used for heroic poetry, esp. the hexameter, the iambic pentameter, or the alexandrine. □ **heroically** *adv.* [F *héroïque* or L *heroicus* f. Gk *hērōikos* (as HERO)]

heroi-comic /ˌhɪrəʊɪˈkɒmɪk/ *adj.* (also **heroi-comical** /-k(ə)l/) combining the heroic with the comic. [F *héroï-comique* (as HERO, COMIC)]

heroin /ˈherəʊɪn/ *n.* a highly addictive crystalline analgesic drug derived from morphine, often used as a narcotic. [G (as HERO, from its effects on the user's self-esteem)]

heroine /ˈherəʊɪn/ *n.* **1** a woman noted or admired for courage, outstanding achievements, nobility, etc. **2** the chief female character in a poem, play, story, etc. **3** *Gk Antiq.* a demigoddess. [F *héroïne* or L *heroina* f. Gk *hērōïnē*, fem. of *hērōs* HERO]

heroism /ˈherəʊˌɪz(ə)m/ *n.* heroic conduct or qualities. [F *héroïsme* f. *héros* HERO]

heroize /ˈherəʊˌaɪz/ *v.* (also **-ise**) **1** *tr.* **a** make a hero of. **b** make heroic. **2** *intr.* play the hero.

heron /ˈherən/ *n.* a tall wading bird of the family Ardeidae, with long legs and a long S-shaped neck; esp. the European *grey heron* (*Ardea cinerea*). □ **heronry** *n.* (*pl.* **-ies**). [ME f. OF *hairon* f. Gmc]

Herophilus /hɪəˈrɒfɪləs/ (4th–3rd centuries BC), Greek anatomist, regarded as the father of human anatomy. Based in Alexandria, he made fundamental discoveries concerning the anatomy of the brain, eye, and reproductive organs, and some of his terms are still in use. Herophilus distinguished nerves from tendons, recognizing that nerves are connected to the brain and that they can be either sensory or motor in function. He also distinguished between veins and arteries, showing that they contain blood, and carried out the first systematic study of the pulse. None of his works survive.

herpes /ˈhɜːpiːz/ *n.* a viral disease with outbreaks of blisters on the skin etc. □ **herpes simplex** /ˈsɪmpleks/ a viral infection which may produce cold sores, genital inflammation, or conjunctivitis. **herpes zoster** /ˈzɒstə(r)/ = SHINGLES. □ **herpetic** /hɜːˈpetɪk/ *adj.* [ME f. L f. Gk *herpēs -ētos* shingles f. *herpō* creep: *zoster* f. Gk *zōstēr* belt, girdle]

herpesvirus /ˈhɜːpiːzˌvaɪərəs/ *n.* any of a group of viruses causing herpes.

herpetology /ˌhɜːpɪˈtɒlədʒɪ/ *n.* the study of reptiles. □ **herpetologist** *n.* **herpetological** /-təˈlɒdʒɪk(ə)l/ *adj.* [Gk *herpeton* reptile f. *herpō* creep]

herptile /ˈhɜːptaɪl/ *n. & adj.* ● *n.* a reptile or amphibian. ● *adj.* of or relating to reptiles and amphibians. [HERPETOLOGY + REPTILE]

Herr /heə(r)/ *n.* (*pl.* **Herren** /ˈherən/) **1** the title of a German man; Mr. **2** a German man. [G f. OHG *hērro* compar. of *hēr* exalted]

Herrenvolk /ˈherənˌfɒlk, -ˌfəʊk/ *n.* **1** the German nation characterized by the Nazis as born to mastery. **2** a group regarding itself as naturally superior. [G, = master-race (as HERR, FOLK)]

Herrick /ˈherɪk/, Robert (1591–1674), English poet. He is best known for *Hesperides*, a collection of poems published in 1648, which contained a section of religious poems, *Noble Numbers*. His secular poems, which deal chiefly with country rituals, folklore, and love, show a clear debt to the classical poets, particularly Horace and Catullus; notable examples include 'To the Virgins, To Make Much of Time' and 'Cherry Ripe'.

herring /ˈherɪŋ/ *n.* a North Atlantic food fish, *Clupea harengus*, that comes near the coast in large shoals to spawn. □ **herring-gull** a large gull, *Larus argentatus*, with a grey mantle and dark wing-tips. [OE *hæring, hēring* f. WG]

herring-bone /ˈherɪŋˌbəʊn/ *n. & v.* ● *n.* **1** a stitch with a zigzag pattern, resembling the pattern of a herring's bones. **2** this pattern, or cloth woven in it. **3** any zigzag pattern, e.g. in building. **4** *Skiing* a method of ascending a slope with the skis pointing outwards. ● *v.* **1** *tr.* **a** work with a herring-bone stitch. **b** mark with a herring-bone pattern. **2** *intr. Skiing* ascend a slope using the herring-bone technique.

Herriot /ˈherɪət/, James (pseudonym of James Alfred Wight) (b.1916), English short-story writer and veterinary surgeon. His experiences at work as a vet in North Yorkshire inspired a series of stories, collected in *If Only They Could Talk* (1970), *All Creatures Great and Small* (1972), and *The Lord God Made Them All* (1981), which were made into a British television series as well as a number of films.

Herrnhuter /ˈheənˌhuːtə(r), ˈherən-/ *n.* a member of a Christian Moravian sect (see MORAVIAN). [G f. *Herrnhut* (= the Lord's keeping), name of their first German settlement]

hers /hɜːz/ *poss.pron.* the one or ones belonging to or associated with her (*it is hers; hers are over there*). □ **of hers** of or belonging to her (*a friend of hers*).

Herschel[1] /ˈhɜːʃ(ə)l/, Sir (Frederick) William (1738–1822), German-born British astronomer, the father of stellar astronomy. He was a skilful telescope maker whose painstaking cataloguing of the skies resulted in the discovery of the planet Uranus. His unsuccessful attempts to measure the distances of the stars convinced him of their remoteness, while his mapping of stellar distributions suggested to him that the sun was a member of a great star system forming the disc of the Milky Way. He was elected first president of the Royal Astronomical Society in 1820.

Herschel[2] /ˈhɜːʃ(ə)l/, Sir John (Frederick William) (1792–1871), British astronomer and physicist, son of William Herschel. He extended the sky survey to the southern hemisphere, cataloguing many new clusters and nebulae. He carried out pioneering work in photography, to which he introduced the words *positive* and *negative*, and also made contributions to meteorology and geophysics.

herself /həˈself/ *pron.* **1 a** *emphat. form of* SHE or HER (*she herself will do it*). **b** *refl. form of* HER (*she has hurt herself*). **2** in her normal state of body or mind (*does not feel quite herself today*). □ **be herself** act in her normal unconstrained manner. **by herself** see *by oneself*. [OE *hire self* (as HER, SELF)]

Hertford /ˈhɑːtfəd/ the county town of Hertfordshire; pop. (1981) 21,400.

Hertfordshire /ˈhɑːtfədˌʃɪə(r)/ a county of SE England, one of the Home Counties; county town, Hertford.

Herts. *abbr.* Hertfordshire.

Hertz /hɜːts/, Heinrich Rudolf (1857–94), German physicist and pioneer of radio communication. He worked for a time as Helmholtz's assistant in Berlin, and in 1886 began studying the electromagnetic waves that Maxwell had predicted. He demonstrated them experimentally, and also showed that they behaved like light and radiant heat, thus proving that these phenomena, too, were electromagnetic. In 1889 he was appointed professor of physics at Bonn, but he died of blood-poisoning at the early age of 37.

hertz /hɜːts/ *n.* (*pl.* same) the SI unit of frequency (symbol **Hz**), equal to one cycle per second. [HERTZ]

Hertzian wave /ˈhɜːtsɪən/ *n.* an electromagnetic wave of a length suitable for use in radio.

Hertzsprung–Russell diagram /ˌhɜːtssprʌŋˈrʌs(ə)l/ *n. Astron.* a two-dimensional graph, devised independently by Ejnar Hertzsprung (1873–1967) and H. N. Russell, in which the absolute magnitudes of stars are plotted against their spectral types. Stars are found to occupy only certain regions of this diagram, depending on their mass and the stage of their life cycle. Most stars fall on a line or *main sequence*, which represents a relationship between the properties of a star that is still relying on hydrogen as its energy source; when a star has exhausted its hydrogen and begun to consume other elements, it will occupy a different position on the diagram.

Herut /heˈruːt/ *hist.* a right-wing Israeli political party founded by Menachem Begin in 1948, from the remains of the Irgun group. Herut was one of the parties that combined to form the Likud coalition in 1973. [Heb., = freedom]

Herzegovina /ˌheətsəˈɡɒvɪnə, -ɡəˈviːnə/ (also **Hercegovina**) a region in the Balkans forming the southern part of Bosnia–Herzegovina and separated from the Adriatic by part of Croatia. Its chief town is Mostar. □ **Herzegovinian** /-ɡəˈvɪnɪən/ *adj. & n.*

Herzl /ˈhɜːts(ə)l/, Theodor (1860–1904), Hungarian-born journalist, dramatist, and Zionist leader. He worked for most of his life as a writer and journalist in Vienna, advocating the establishment of a Jewish state in Palestine; in 1897 he founded the Zionist movement, of which he was the most influential statesman.

he's /hiːz, hɪz/ *contr.* **1** he is. **2** he has.

Heshvan var. of HESVAN.

Hesiod /ˈhiːsɪəd/ (*c.*700 BC), Greek poet. He is one of the earliest known Greek poets and is often linked or contrasted with Homer as the other leading writer of early epic verse. His hexametric poem the *Theogony* deals with the origin and genealogies of the gods; his *Works and Days* contains moral and practical advice for living an honest life of (chiefly agricultural) work, and was the chief model for later ancient didactic poetry.

hesitant /ˈhɛzɪt(ə)nt/ adj. **1** hesitating; irresolute. **2** (of speech) stammering, faltering. □ **hesitance** n. **hesitancy** n. **hesitantly** adv.

hesitate /ˈhɛzɪˌteɪt/ v.intr. **1** (often foll. by about, over) show or feel indecision or uncertainty; pause in doubt (*hesitated over her choice*). **2** (often foll. by to + infin.) be deterred by scruples; be reluctant (*I hesitate to inform against him*). **3** stammer or falter in speech. □ **hesitater** n. **hesitatingly** adv. **hesitative** /-tətɪv, -ˌteɪtɪv/ adj. **hesitation** /ˌhɛzɪˈteɪʃ(ə)n/ n. [L *haesitare* frequent. of *haerere haes-* stick fast]

Hesperian /hɛˈspɪərɪən/ adj. poet. **1** western. **2** of or concerning the Hesperides. [L *Hesperius* f. Gk *Hesperios* (as HESPERUS)]

Hesperides /hɛˈspɛrɪˌdiːz/ Gk Mythol. the nymphs who were the daughters of Hesperus (or, in earlier versions, of Night and Hades). They were guardians, with the aid of a watchful dragon, of a tree of golden apples (given to Hera by Gaia) in a garden located beyond the Atlas Mountains at the western border of Oceanus, the river encircling the world. One of the labours of Hercules was to fetch the golden apples.

hesperidium /ˌhɛspəˈrɪdɪəm/ n. (pl. **hesperidia** /-dɪə/) Bot. a fruit with sectioned pulp inside a separable rind, e.g. an orange or grapefruit. [Gk HESPERIDES]

Hesperus /ˈhɛspərəs/ n. poet. the evening star, Venus. [ME f. L f. Gk *hesperos* (adj. & n.) western, evening (star)]

Hess[1] /hɛs/, Dame Myra (1890–1965), English pianist. She was noted for her performances of the music of Schumann, Beethoven, Mozart, and Bach. Her many piano transcriptions of baroque music include 'Jesu, Joy of Man's Desiring' from Bach's Cantata No. 147.

Hess[2] /hɛs/, Victor Francis (born Victor Franz Hess) (1883–1964), Austrian-born American physicist. He worked on atmospheric electricity and radioactivity. After making several balloon flights he was able to show that some ionizing radiation was extraterrestrial in origin, but did not come from the sun. The high-energy particles responsible were later termed *cosmic rays*, and their study led to the discovery of the positron by C. D. Anderson. They shared the Nobel Prize for physics in 1936.

Hess[3] /hɛs/, (Walther Richard) Rudolf (1894–1987), German Nazi politician. He was deputy leader of the Nazi Party (1934–41) and a close friend of Hitler. In 1941, secretly and on his own initiative, he parachuted into Scotland to negotiate peace with Britain. He was imprisoned for the duration of the war, and after his conviction at the Nuremberg war trials was sentenced to life imprisonment in Spandau prison, where he died.

Hesse[1] /hɛs/ (German **Hessen** /ˈhɛs(ə)n/) a state of western Germany; capital, Wiesbaden.

Hesse[2] /hɛs, ˈhɛsə/, Hermann (1877–1962), German-born Swiss novelist and poet. In 1911 he visited India, where he became interested in Indian mysticism; this experience, together with his involvement with Jungian analysis in 1916–17, had a marked effect on his work, which emphasizes spiritual values as expressed in Eastern religion. His novels — titles include *Siddhartha* (1922), *Der Steppenwolf* (1927), and *The Glass Bead Game* (1943) — met with renewed interest in the 1960s and 1970s. Hesse was awarded the Nobel Prize for literature in 1946.

Hessian /ˈhɛsɪən/ n. & adj. ● n. a native or inhabitant of Hesse in Germany. ● adj. of or relating to Hesse. □ **Hessian boot** a tasselled high boot first worn by Hessian troops. **Hessian fly** a midge, *Mayetiola destructor*, whose larva destroys growing wheat (thought to have been taken to America by Hessian troops). [HESSE[1]]

hessian /ˈhɛsɪən/ n. & adj. ● n. a strong coarse sacking made of hemp or jute. ● adj. made of hessian. [HESSE[1]]

hest /hɛst/ n. archaic behest. [OE *hǣs* (see HIGHT), assim. to ME nouns ending in -t]

Hesvan /ˈhɛsv(ə)n/ n. (also **Chesvan** /ˈxɛsv(ə)n/, **Heshvan** /ˈhɛsv(ə)n/) (in the Jewish calendar) the second month of the civil and eighth of the religious year, usually coinciding with parts of October and November. [Heb. *ḥ̌ešwān*]

hetaera /hɪˈtɪərə/ n. (also **hetaira** /-ˈtaɪərə/) (pl. **-as**, **hetaerae** /-ˈtɪəriː/, or **hetairai** /-ˈtaɪraɪ/) a courtesan or mistress, esp. in ancient Greece. [Gk *hetaira*, fem. of *hetairos* companion]

hetaerism /hɪˈtɪərɪz(ə)m/ n. (also **hetairism** /-ˈtaɪrɪz(ə)m/) **1** a recognized system of concubinage. **2** communal marriage in a tribe. [Gk *hetairismos* prostitution (as HETAERA)]

hetero /ˈhɛtərəʊ/ n. (pl. **-os**) colloq. a heterosexual. [abbr.]

hetero- /ˈhɛtərəʊ/ comb. form other, different (often opp. HOMO-). [Gk *heteros* other]

heterochromatic /ˌhɛtərəʊkrəˈmætɪk/ adj. of several colours.

heteroclite /ˈhɛtərəˌklaɪt/ adj. & n. ● adj. **1** abnormal. **2** Gram. (esp. of a noun) irregularly declined. ● n. **1** an abnormal thing or person. **2** Gram. an irregularly declined word, esp. a noun. [LL *heteroclitus* f. Gk (as HETERO-, *klitos* f. *klinō* bend, inflect)]

heterocyclic /ˌhɛtərəʊˈsaɪklɪk, -ˈsɪklɪk/ adj. Chem. (of a compound) with a bonded ring of atoms of more than one kind.

heterodox /ˈhɛtərəˌdɒks/ adj. (of a person, opinion, etc.) not orthodox. □ **heterodoxy** n. [LL *heterodoxus* f. Gk (as HETERO-, *doxos* f. *doxa* opinion)]

heterodyne /ˈhɛtərəʊˌdaɪn/ adj. & v. Radio ● adj. relating to the production of a lower frequency from the combination of two almost equal high frequencies. ● v.intr. produce a lower frequency in this way. [HETERO- + Gk *dunamis* power]

heterogamous /ˌhɛtəˈrɒɡəməs/ adj. **1** Bot. irregular as regards stamens and pistils. **2** Biol. characterized by heterogamy or heterogony.

heterogamy /ˌhɛtəˈrɒɡəmɪ/ n. **1** Biol. the alternation of generations, esp. of a sexual and parthenogenic generation. **2** Biol. sexual reproduction by fusion of unlike gametes. **3** Bot. a state in which the flowers of a plant are of two types.

heterogeneous /ˌhɛtərəˈdʒiːnɪəs/ adj. (also disp. **heterogenous** /ˌhɛtəˈrɒdʒɪnəs/) **1** diverse in character. **2** varied in content. **3** Math. incommensurable through being of different kinds or degrees. □ **heterogeneously** adv. **heterogeneousness** n. **heterogeneity** /-dʒɪˈniːɪtɪ/ n. [med.L *heterogeneus* f. Gk *heterogenēs* (as HETERO-, *genos* kind)]

heterogenesis /ˌhɛtərəˈdʒɛnɪsɪs/ n. **1** the birth of a living being otherwise than from parents of the same kind. **2** spontaneous generation from inorganic matter. □ **heterogenetic** /-dʒɪˈnɛtɪk/ adj.

heterogony /ˌhɛtəˈrɒɡənɪ/ n. the alternation of generations, esp. of a sexual and hermaphroditic generation. □ **heterogonous** adj.

heterograft /ˈhɛtərəʊˌɡrɑːft/ n. = XENOGRAFT.

heterologous /ˌhɛtəˈrɒləɡəs/ adj. not homologous. □ **heterology** /-ˈrɒlədʒɪ/ n.

heteromerous /ˌhɛtəˈrɒmərəs/ adj. not isomerous.

heteromorphic /ˌhɛtərəʊˈmɔːfɪk/ adj. (also **heteromorphous** /-ˈmɔːfəs/) Biol. **1** of dissimilar forms. **2** (of insects) existing in different forms at different stages in their life cycle.

heteromorphism /ˌhɛtərəʊˈmɔːfɪz(ə)m/ n. existing in various forms.

heteronomous /ˌhɛtəˈrɒnəməs/ adj. **1** subject to an external law (cf. AUTONOMOUS). **2** Biol. subject to different laws (of growth etc.).

heteronomy /ˌhɛtəˈrɒnəmɪ/ n. **1** the presence of a different law. **2** subjection to an external law.

heteropathic /ˌhɛtərəʊˈpæθɪk/ adj. **1** allopathic. **2** differing in effect.

heterophyllous /ˌhɛtərəʊˈfɪləs/ adj. bearing leaves of different forms on the same plant. □ **heterophylly** /ˈhɛtərəʊˌfɪlɪ/ n. [HETERO- + Gk *phullon* leaf]

heteropolar /ˌhɛtərəʊˈpəʊlə(r)/ adj. having dissimilar poles, esp. Electr. with an armature passing north and south magnetic poles alternately.

Heteroptera /ˌhɛtəˈrɒptərə/ n.pl. Zool. a suborder of hemipterous insects comprising predatory and plant bugs, which have forewings with a thickened base and membranous tip (cf. HOMOPTERA). □ **heteropteran** n. & adj. **heteropterous** adj. [mod.L f. HETERO- + Gk *pteron* wing]

heterosexism /ˌhɛtərəˈsɛksɪz(ə)m/ n. discrimination in favour of heterosexuals; prejudice against homosexuals. □ **heterosexist** adj. & n.

heterosexual /ˌhɛtərəˈsɛksjʊəl, -ˈsɛkʃʊəl/ adj. & n. ● adj. **1** feeling or involving sexual attraction to persons of the opposite sex. **2** concerning heterosexual relations or people. **3** relating to the opposite sex. ● n. a heterosexual person. □ **heterosexually** adv. **heterosexuality** /-ˌsɛksjʊˈælɪtɪ, -ˌsɛkʃʊ-/ n.

heterosis /ˌhɛtəˈrəʊsɪs/ n. the tendency of a cross-bred individual to show qualities superior to those of both parents. [Gk f. *heteros* different]

heterotaxy /ˈhɛtərəʊˌtæksɪ/ n. the abnormal disposition of organs or parts. [HETERO- + Gk *taxis* arrangement]

heterotransplant /ˌhɛtərəʊˈtrænsplɑːnt, -ˈtrɑːnsplɑːnt/ n. = XENOGRAFT.

heterotrophic /ˌhɛtərəʊˈtrəʊfɪk, -ˈtrɒfɪk/ adj. Biol. deriving its nourishment and carbon requirements from organic substances; not autotrophic. [HETERO- + Gk *trophos* feeder]

heterozygote /ˌhetərəʊˈzaɪgəʊt/ *n. Biol.* an individual having two different alleles of a particular gene or genes, and so giving rise to varying offspring (cf. HOMOZYGOTE).

heterozygous /ˌhetərəʊˈzaɪgəs/ *adj. Biol.* (often foll. by *for* a gene or allele) having different alleles at one or more genetic loci.

hetman /ˈhetmən/ *n.* (*pl.* **-men**) a Polish or Cossack military commander. [Pol., prob. f. G *Hauptmann* captain]

het up /het ˈʌp/ *adj. colloq.* excited, overwrought. [*het* dial. past part. of HEAT]

heuchera /ˈhjuːkərə, ˈhɔɪk-/ *n.* a North American herbaceous plant of the genus *Heuchera*, with dark green round or heart-shaped leaves and tiny flowers. [mod.L f. J. H. von *Heucher*, Ger. botanist (1677–1747)]

heuristic /hjʊəˈrɪstɪk/ *adj. & n.* ● *adj.* **1** allowing or assisting to discover. **2** *Computing* proceeding to a solution by trial and error. ● *n.* **1** the science of heuristic procedure. **2** a heuristic process or method. **3** (in *pl.*, usu. treated as *sing.*) *Computing* the study and use of heuristic techniques in data processing. □ **heuristic method** a system of education under which pupils are trained to find out things for themselves. □ **heuristically** *adv.* [irreg. f. Gk *heuriskō* find]

hevea /ˈhiːvɪə/ *n.* a South American tree of the genus *Hevea*, yielding a milky sap used for making rubber. [mod.L f. Quechua *hevé*]

Hevesy /ˈhevəʃɪ/, George Charles de (1885–1966), Hungarian-born radiochemist. He worked in seven different European countries, and made fundamental contributions to the study of radioisotopes, an interest he acquired while working with Rutherford in Manchester. He invented the technique of labelling with isotopic tracers, for which he was awarded the Nobel Prize for chemistry in 1943. Hevesy was also co-discoverer of the element hafnium (1923).

HEW *abbr.* (in the US) Department of Health, Education, and Welfare.

hew /hjuː/ *v.* (*past part.* **hewn** /hjuːn/ or **hewed**) **1** *tr.* **a** (often foll. by *down, away, off*) chop or cut (a thing) with an axe, a sword, etc. **b** cut (a block of wood etc.) into shape. **2** *intr.* (often foll. by *at, among*, etc.) strike cutting blows. **3** *intr. N. Amer.* (usu. foll. by *to*) conform. □ **hew one's way** make a way for oneself by hewing. [OE *hēawan* f. Gmc]

hewer /ˈhjuːə(r)/ *n.* **1** a person who hews. **2** a person who cuts coal from a seam. □ **hewers of wood and drawers of water** menial drudges; labourers (Josh. 9:21).

hex[1] /heks/ *v. & n.* ● *v.* **1** *intr.* practise witchcraft. **2** *tr.* cast a spell on; bewitch. ● *n.* **1** a magic spell; a curse. **2** a witch. [Pennsylvanian G *hexe* (v.), *Hex* (n.), f. G *hexen, Hexe*]

hex[2] /heks/ *adj. & n.* esp. *Computing* = HEXADECIMAL. [abbr.]

hexa- /ˈheksə/ *comb. form* six. [Gk *hex* six]

hexachord /ˈheksəˌkɔːd/ *n. Mus.* a diatonic series of six notes with a semitone between the third and fourth, used at three different pitches in medieval music. [HEXA- + CHORD[1]]

hexad /ˈheksæd/ *n.* a group of six. [Gk *hexas -ados* f. *hex* six]

hexadecimal /ˌheksəˈdesɪm(ə)l/ *adj. & n.* esp. *Computing.* ● *adj.* relating to or using a system of numerical notation that has 16 rather than 10 as a base. ● *n.* the hexadecimal system; hexadecimal notation. □ **hexadecimally** *adv.*

hexagon /ˈheksəgən/ *n.* a plane figure with six sides and angles. □ **hexagonal** /hekˈsægən(ə)l/ *adj.* [LL *hexagonum* f. Gk (as HEXA-, -GON)]

hexagram /ˈheksəˌgræm/ *n.* **1** a figure formed by two intersecting equilateral triangles. **2** a figure of six lines. [HEXA- + Gk *gramma* line]

hexahedron /ˌheksəˈhiːdrən/ *n.* a solid figure with six faces. □ **hexahedral** *adj.* [Gk (as HEXA-, -HEDRON)]

hexameter /hekˈsæmɪtə(r)/ *n.* a line or verse of six metrical feet. □ **dactylic hexameter** a hexameter having five dactyls and a spondee or trochee, any of the first four feet, and sometimes the fifth, being replaceable by a spondee. □ **hexametrist** *n.* **hexametric** /ˌheksəˈmetrɪk/ *adj.* [ME f. L f. Gk *hexametros* (as HEXA-, *metron* measure)]

hexane /ˈhekseɪn/ *n. Chem.* a liquid hydrocarbon of the alkane series (chem. formula: C_6H_{14}). [HEXA- + -ANE[2]]

hexapla /ˈheksəplə/ *n.* a sixfold text, esp. of the Old Testament, in parallel columns. [Gk neut. pl. of *hexaploos* (as HEXA-, *ploos* -fold), orig. of Origen's OT text]

hexapod /ˈheksəˌpɒd/ *n. & adj. Zool.* ● *n.* an arthropod with six legs; an insect. ● *adj.* having six legs. [Gk *hexapous, hexapod-* (as HEXA-, *pous* pod-foot)]

hexastyle /ˈheksəˌstaɪl/ *n. & adj.* ● *n.* a six-columned portico. ● *adj.* having six columns. [Gk *hexastulos* (as HEXA-, *stulos* column)]

Hexateuch /ˈheksəˌtjuːk/ *n.* the first six books of the Old Testament. [Gk *hex* six + *teukhos* book]

hexavalent /ˌheksəˈveɪlənt/ *adj. Chem.* having a valency of six; sexivalent.

hexose /ˈheksəʊz, -səʊs/ *n. Biochem.* a monosaccharide with six carbon atoms in each molecule, e.g. glucose or fructose. [HEXA- + -OSE[2]]

hey[1] /heɪ/ *int.* calling attention or expressing joy, surprise, inquiry, enthusiasm, etc. □ **hey presto!** a phrase announcing the successful completion of a trick or other surprising achievement. [ME: cf. OF *hay*, Du., G *hei*]

hey[2] var. of HAY[2].

heyday /ˈheɪdeɪ/ *n.* the flush or full bloom of youth, vigour, prosperity, etc. [archaic *heyday* expression of joy, surprise, etc.: cf. LG *heidi, heida*, excl. denoting gaiety]

Heyer /ˈheɪə(r)/, Georgette (1902–74), English novelist. She is noted especially for her historical novels, which include numerous Regency romances, such as *Regency Buck* (1935) and *Faro's Daughter* (1941). She also wrote detective stories, including *Envious Casca* (1941) and *Detection Unlimited* (1953).

Heyerdahl /ˈheɪəˌdɑːl/, Thor (b.1914), Norwegian anthropologist. He is noted for his ocean voyages in primitive craft to demonstrate his theories of cultural diffusion. His first such voyage, in 1947, was an attempt to show that Polynesian peoples could originally have been migrants from South America. His raft (the *Kon-Tiki*) successfully made the journey from Peru to the islands east of Tahiti. Later journeys included a transatlantic crossing in 1969 from Morocco towards Central America in the *Ra*, a raft made according to an ancient Egyptian design.

Heyhoe-Flint /ˌheɪhəʊˈflɪnt/, Rachel (b.1939), English cricketer. Between 1966 and 1977 she was the captain of the England women's cricket team, leading England to victory in the first Women's World Cup in 1972. She combined her long playing career for England (1960–83) with journalism, and in 1972 became the first woman sports reporter on Independent Television.

Hezbollah /ˌhezbəˈlɑː, ˈhezbʊlə/ an extremist Shiite Muslim group which has close links with Iran, created after the Iranian revolution of 1979 and active especially in Lebanon. [Arab. *ḥizbullah* Party of God, f. *ḥezb* party + *allāh* ALLAH]

HF *abbr.* high frequency.

Hf *symb. Chem.* the element hafnium.

hf. *abbr.* half.

HG *abbr.* **1** Her or His Grace. **2** Home Guard.

Hg *symb. Chem.* the element mercury. [mod.L *hydrargyrum*]

hg *abbr.* hectogram(s).

HGV *abbr. Brit.* heavy goods vehicle.

HH *abbr.* **1** Her or His Highness. **2** His Holiness. **3** double-hard (pencil-lead).

hh. *abbr.* hands (see HAND *n.* 13).

hhd. *abbr.* hogshead(s).

H-hour /ˈeɪtʃˌaʊə(r)/ *n.* the hour at which an operation is scheduled to begin. [*H* for *hour* + HOUR]

HI *abbr. US* **1** Hawaii (also in official postal use). **2** the Hawaiian Islands.

hi[1] /haɪ/ *int.* calling attention or as a greeting. [parallel form to HEY[1]]

hi[2] /haɪ/ *adj. colloq.* (esp. in special collocations) high (*hi-fi; hi-hat*). [repr. pronunc.]

hiatus /haɪˈeɪtəs/ *n.* (*pl.* **hiatuses**) **1** a break or gap, esp. in a series, account, or chain of proof. **2** *Prosody & Gram.* a break between two vowels coming together but not in the same syllable, as in *though oft the ear*. □ **hiatus hernia** *Med.* the protrusion of an organ, esp. the stomach, through the oesophageal opening in the diaphragm. □ **hiatal** *adj.* [L, = gaping f. *hiare* gape]

Hiawatha /ˌhaɪəˈwɒθə/ a legendary 16th-century North American Indian teacher and chieftain, hero of a narrative poem by Henry Wadsworth Longfellow called the *Song of Hiawatha* (1855).

Hib /hɪb/ *n. Med.* a bacterium, *Haemophilus influenzae* type B, causing infant meningitis (usu. *attrib.: Hib vaccine*). [acronym]

hibernate /ˈhaɪbəˌneɪt/ *v.intr.* **1** (of some animals) spend the winter in a dormant state. **2** remain inactive. □ **hibernator** *n.* **hibernation** /ˌhaɪbəˈneɪʃ(ə)n/ *n.* [L *hibernare* f. *hibernus* wintry]

Hibernian /haɪˈbɜːnɪən/ *adj. & n. archaic poet.* ● *adj.* of or concerning

Ireland. ● *n.* a native of Ireland. [L *Hibernia, Iverna* f. Gk *Iernē* f. OCelt.]

Hibernicism /haɪˈbɜːnɪˌsɪz(ə)m/ *n.* **1** an Irish idiom or expression. **2** = BULL³ 1. [as HIBERNIAN after *Anglicism* etc.]

Hiberno- /haɪˈbɜːnəʊ/ *comb. form* Irish (*Hiberno-British*). [med.L *hibernus* Irish (as HIBERNIAN)]

hibiscus /hɪˈbɪskəs/ *n.* (*pl.* **hibiscuses**) a tree or shrub of the genus *Hibiscus* (mallow family), cultivated for its large brightly coloured flowers. Also called *rose-mallow*. [L f. Gk *hibiskos* marsh mallow]

hic /hɪk/ *int.* expressing the sound of a hiccup, esp. a drunken hiccup. [imit.]

hiccup /ˈhɪkʌp/ *n. & v.* (also **hiccough**) ● *n.* **1 a** an involuntary spasm of the diaphragm and respiratory organs, with sudden closure of the glottis and characteristic coughlike sound. **b** (in *pl.*) an attack of such spasms. **2** a temporary or minor fault or setback. ● *v.* (**hiccuped**, **hiccuping**) **1** *intr.* make a hiccup or series of hiccups. **2** *tr.* utter with a hiccup. □ **hiccupy** *adj.* [imit.]

hic jacet /hɪk ˈjækɛt/ *n.* an epitaph. [L, = here lies]

hick /hɪk/ *n. colloq.* a country dweller; a provincial. [pet-form of the name *Richard*: cf. DICK¹]

hickey /ˈhɪkɪ/ *n.* (*pl.* **-eys**) N. Amer. *colloq.* **1** a gadget (cf. DOOHICKEY). **2** a mark on the skin, esp. caused by a lovebite. [20th c.: orig. unkn.]

hickory /ˈhɪkərɪ/ *n.* (*pl.* **-ies**) **1** a North American tree of the genus *Carya*, yielding tough heavy wood and bearing nutlike edible fruits (see also PECAN 2). **2 a** the wood of these trees. **b** a stick made of this. [native Virginian *pohickery*]

Hicks /hɪks/, Sir John Richard (1904–89), English economist. He is chiefly remembered for his pioneering work on general economic equilibrium (the theory that economic forces tend to balance one another rather than simply reflect cyclical trends), for which he shared a Nobel Prize with K. J. Arrow in 1972.

hid *past* of HIDE¹.

Hidalgo /hɪˈdælɡəʊ/ a state of southern Mexico; capital, Pachuca de Soto.

hidalgo /hɪˈdælɡəʊ/ *n.* (*pl.* **-os**) a Spanish gentleman. [Sp. f. *hijo dalgo* son of something]

hidden *past part.* of HIDE¹. □ **hiddenness** /ˈhɪd(ə)nnɪs/ *n.*

hide¹ /haɪd/ *v. & n.* ● *v.* (*past* **hid** /hɪd/; *past part.* **hidden** /ˈhɪd(ə)n/ or *archaic* **hid**) **1** *tr.* put or keep out of sight (*hid it under the cushion*; *hid her in the cupboard*). **2** *intr.* conceal oneself. **3** *tr.* (usu. foll. by *from*) keep (a fact) secret (*hid his real motive from her*). **4** *tr.* conceal (a thing) from sight intentionally or not (*trees hid the house*). ● *n.* Brit. a camouflaged shelter used for observing wildlife or hunting animals. □ **hidden agenda** a secret motivation behind a policy, statement, etc.; an ulterior motive. **hidden reserves** extra profits, resources, etc., kept concealed in reserve. **hide-and-seek 1** a children's game in which one or more players seek a child or children hiding. **2** a process of attempting to find an evasive person or thing. **hide one's head** keep out of sight, esp. from shame. **hide one's light under a bushel** conceal one's merits (Matt. 5:15). **hide out** (or **up**) remain in concealment. **hide-out** a hiding-place. □ **hidden** *adj.* **hider** *n.* [OE *hȳdan* f. WG]

hide² /haɪd/ *n. & v.* ● *n.* **1** the skin of an animal, esp. of a large animal and when tanned or dressed (cf. SKIN *n.* 2b). **2** *joc.* the human skin, esp. on the buttocks (*I'll tan your hide*). **3** esp. Austral. & NZ *colloq.* impudence, effrontery, nerve. ● *v.tr. colloq.* flog. □ **hided** *adj.* (also in comb.). [OE *hȳd* f. Gmc]

hide³ /haɪd/ *n. hist.* a former measure of land large enough to support a family and its dependants, usu. between 60 and 120 acres. [OE *hī(gi)d* f. *hīw-, hīg-* household]

hideaway /ˈhaɪdəˌweɪ/ *n.* a hiding-place or place of retreat.

hidebound /ˈhaɪdbaʊnd/ *adj.* **1 a** narrow-minded; bigoted. **b** (of the law, rules, etc.) constricted by tradition. **2** (of cattle) with the skin clinging close as a result of bad feeding. [HIDE² + BOUND⁴]

hideosity /ˌhɪdɪˈɒsɪtɪ/ *n.* (*pl.* **-ies**) **1** a hideous object. **2** hideousness.

hideous /ˈhɪdɪəs/ *adj.* **1** frightful, repulsive, or revolting, to the senses or the mind (*a hideous monster*; *a hideous pattern*). **2** *colloq.* unpleasant. □ **hideously** *adv.* **hideousness** *n.* [ME *hidous* f. AF *hidous*, OF *hidos*, *-eus*, f. OF *hide*, *hisde* fear, of unkn. orig.]

hidey-hole /ˈhaɪdɪˌhəʊl/ *n. colloq.* a hiding-place.

hiding¹ /ˈhaɪdɪŋ/ *n. colloq.* a thrashing. □ **on a hiding to nothing** in a position from which there can be no successful outcome. [HIDE² + -ING¹]

hiding² /ˈhaɪdɪŋ/ *n.* **1** the act or an instance of hiding. **2** the state of

remaining hidden (*go into hiding*). □ **hiding-place** a place of concealment. [ME, f. HIDE¹ + -ING¹]

hidrosis /hɪˈdrəʊsɪs, haɪ-/ *n. Med.* perspiration. □ **hidrotic** /-ˈdrɒtɪk/ *adj.* [mod.L f. Gk f. *hidrōs* sweat]

hie /haɪ/ *v.intr. & refl.* (**hies**, **hied**, **hieing** or **hying**) *archaic* or *poet.* go quickly (*hie to your chamber*; *hied him to the chase*). [OE *hīgan* strive, pant, of unkn. orig.]

hierarch /ˈhaɪəˌrɑːk/ *n.* **1** a chief priest. **2** an archbishop. [med.L f. Gk *hierarkhēs* f. *hieros* sacred + *-arkhēs* ruler]

hierarchy /ˈhaɪəˌrɑːkɪ/ *n.* (*pl.* **-ies**) **1 a** a system in which grades or classes of status or authority are ranked one above the other (*ranks third in the hierarchy*). **b** a hierarchical system (of government, management, etc.). **2 a** priestly government. **b** a priesthood organized in grades. **3** (in Christian theology) **a** each of the three divisions of angels. **b** the angels. □ **hierarchism** *n.* **hierarchize** *v.tr.* (also **-ise**). **hierarchic** /ˌhaɪəˈrɑːkɪk/ *adj.* **hierarchical** *adj.* [ME f. OF *ierarchie* f. med.L *(h)ierarchia* f. Gk *hierarkhia* (as HIERARCH)]

hieratic /ˌhaɪəˈrætɪk/ *adj. & n.* ● *adj.* **1** of or concerning priests; sacerdotal. **2** of or relating to a style of ancient Egyptian writing consisting of abridged forms of hieroglyphics, used originally for religious texts and eventually superseded by demotic. **3** of or concerning Egyptian or Greek traditional styles of art. ● *n.* hieratic script. □ **hieratically** *adv.* [L f. Gk *hieratikos* f. *hieraomai* be a priest f. *hiereus* priest]

hiero- /ˈhaɪərəʊ/ *comb. form* sacred, holy. [Gk *hieros* sacred + -O-]

hierocracy /ˌhaɪəˈrɒkrəsɪ/ *n.* (*pl.* **-ies**) **1** priestly rule. **2** a body of ruling priests. [HIERO- + -CRACY]

hieroglyph /ˈhaɪərəˌɡlɪf/ *n.* **1 a** a picture of an object representing a word, sound, or syllable in any of the pictorial systems of writing, especially the ancient Egyptian. (*See note below.*) **b** a writing consisting of characters of this kind. **2** a secret or enigmatic symbol. **3** (in *pl.*) *joc.* writing difficult to read. [back-form. f. HIEROGLYPHIC]

▪ Hieroglyphs were used in ancient Egypt for inscriptions on monuments from the end of the 4th millennium BC until the 4th century AD. These inscriptions were often more decorative than informative: there was, for example, no separation between words or phrases and the direction of the writing varied (vertical, right to left, or left to right). In the classic period there were about 700 signs, but signs were continually invented and modified and became innumerable. Hieroglyphs remained undeciphered until the discovery of the Rosetta Stone.

hieroglyphic /ˌhaɪərəˈɡlɪfɪk/ *adj. & n.* ● *adj.* **1** of or written in hieroglyphs. **2** symbolical. ● *n.* (in *pl.*) hieroglyphs; hieroglyphic writing. □ **hieroglyphical** *adj.* **hieroglyphically** *adv.* [F *hiéroglyphique* or LL *hieroglyphicus* f. Gk *hieroglyphikos* (as HIERO-, *gluphikos* f. *gluphē* carving)]

hierogram /ˈhaɪərəʊˌɡræm/ *n.* a sacred inscription or symbol.

hierograph /ˈhaɪərəʊˌɡrɑːf/ *n.* = HIEROGRAM.

hierolatry /ˌhaɪəˈrɒlətrɪ/ *n.* the worship of saints or sacred things.

hierology /ˌhaɪəˈrɒlədʒɪ/ *n.* sacred literature or lore.

hierophant /ˈhaɪərəˌfænt/ *n.* **1** Gk Antiq. an initiating or presiding priest; an official interpreter of sacred mysteries. **2** an interpreter of sacred mysteries or any esoteric principle. □ **hierophantic** /ˌhaɪərəˈfæntɪk/ *adj.* [LL *hierophantes* f. Gk *hierophantēs* (as HIERO-, *phantēs* f. *phainō* show)]

hi-fi /ˈhaɪfaɪ/ *adj. & n. colloq.* ● *adj.* of high fidelity. ● *n.* (*pl.* **hi-fis**) a set of equipment for high-fidelity sound reproduction. [abbr.]

higgle /ˈhɪɡ(ə)l/ *v.intr.* dispute about terms; haggle. [var. of HAGGLE]

higgledy-piggledy /ˌhɪɡ(ə)ldɪˈpɪɡ(ə)ldɪ/ *adv., adj., & n.* ● *adv. & adj.* in confusion or disorder. ● *n.* a state of disordered confusion. [rhyming jingle, prob. with ref. to the irregular herding together of pigs]

high /haɪ/ *adj., n., & adv.* ● *adj.* **1 a** of great vertical extent (*a high building*). **b** (*predic.*; often in comb.) of a specified height (*one inch high*; *water was waist-high*). **2 a** far above ground or sea level etc. (*a high altitude*). **b** inland, esp. when raised (*High Asia*). **3** extending above the normal or average level (*high boots*; *jersey with a high neck*). **4** of exalted, esp. spiritual, quality (*high minds*; *high principles*; *high art*). **5 a** of exalted rank (*in high society*; *is high in the government*). **b** important, serious, grave. **6 a** great; intense; extreme; powerful (*high praise*; *high temperature*). **b** greater than normal (*high prices*). **c** extreme in religious or political opinion (*high Tory*). **7** (of physical action, esp. athletics) performed at, to, or from a considerable height (*high diving*; *high flying*). **8 a** elated, merry. **b** (often foll. by *on*) *colloq.* intoxicated by alcohol or esp. drugs. **9** (of a sound or note) of high frequency; shrill; at the top end of the scale. **10** (of a period, an age, a time, etc.) at its peak (*high noon*; *high summer*; *High Renaissance*). **11 a** (of meat etc.) beginning to go bad; off. **b** (of game) well-hung and slightly decomposed. **12** *Geog.* (of latitude) near the North or South Pole. **13** *Phonet.* (of a vowel) close (see CLOSE[1] *adj.* 14). ● *n.* **1** a high, or the highest, level or figure. **2** an area of high barometric pressure; an anticyclone. **3** *sl.* a euphoric state, esp. drug-induced (*I'm on a high at the moment*). **4** top gear in a motor vehicle. **5** *US colloq.* high school. **6** (**the High**) *Brit. colloq.* a High Street, esp. that in Oxford. ● *adv.* **1** far up; aloft (*flew the flag high*). **2** in or to a high degree. **3** at a high price. **4** (of a sound) at or to a high pitch (*sang high*). □ **ace** (or **King** or **Queen** etc.) **high** (in card-games) having the ace etc. as the highest-ranking card. **from on high** from heaven or a high place. **high altar** the chief altar of a church. **high and dry 1** out of the current of events; stranded. **2** (of a ship) out of the water. **high and low 1** everywhere (*searched high and low*). **2** (people) of all conditions. **high and mighty 1** *colloq.* arrogant. **2** *archaic* of exalted rank. **high-born** of noble birth. **high camp** sophisticated camp (cf. CAMP[2]). **high card** a card that outranks others, esp. the ace or a court-card. **high chair** an infant's chair with long legs and a tray, for use at meals. **high-class 1** of high quality. **2** characteristic of the upper class. **high colour** a flushed complexion. **high command** an army commander-in-chief and associated staff. **high court** a supreme court of justice for civil cases (see also HIGH COURT). **high day** a festal day. **high enema** an enema delivered into the colon. **higher animal** (or **plant**) an animal or plant showing more advanced characteristics, e.g. a placental mammal or a flowering plant. **higher court** *Law* a court that can overrule the decision of another. **the higher criticism** see CRITICISM. **higher education** education at university etc., esp. to degree level. **higher mathematics** advanced mathematics as taught at university etc. **higher-up** *colloq.* a person of higher rank. **highest common factor** *Math.* the highest number that can be divided exactly into each of two or more numbers. **high explosive** an extremely explosive substance used in shells, bombs, etc. **high fashion** = HAUTE COUTURE. **high finance** financial transactions involving large sums. **high-five** *n.* *N. Amer. sl.* a gesture of celebration or greeting in which two people slap each other's palms with their arms extended over their heads. ● *v.tr.* *N. Amer. sl.* greet with a high-five. **high-flown** (of language etc.) extravagant, bombastic. **high-flyer** (or **-flier**) an ambitious person; a person or thing with great potential for achievement. **high-flying** reaching a great height; ambitious. **high frequency** a frequency, esp. in radio, of 3 to 30 megahertz. **high gear** see GEAR. **high-grade** of high quality. **high ground 1** ground that is naturally elevated and therefore strategically advantageous. **2** (usu. prec. by *the*) a position of superiority in a debate etc. (*the moral high ground*). **high hat 1** a tall hat; a top hat. **2** (also **hi-hat**) a pair of cymbals operated by a foot pedal. **3** a snobbish or overbearing person. **high-hat** *adj.* supercilious; snobbish. ● *v.* (**-hatted, -hatting**) *US* **1** *tr.* treat superciliously. **2** *intr.* assume a superior attitude. **high-heeled** having high heels. **high heels** women's shoes with high heels. **high holiday** the Jewish New Year or Yom Kippur. **high jinks** boisterous joking or merrymaking. **high jump 1** an athletic event consisting of jumping as high as possible over a bar of adjustable height. **2** *colloq.* a drastic punishment (*he's for the high jump*). **high jumper** an athlete who competes in a high-jump event. **high-key** *Photog.* consisting of light tones only. **high kick** a dancer's kick high in the air. **high-level 1** (of negotiations etc.) conducted by high-ranking people. **2** *Computing* (of a programming language) that is not machine-dependent and is usu. at a level of abstraction close to natural language. **high life** (or **living**) a luxurious existence ascribed to the upper classes. **high-lows** *archaic* boots reaching over the ankles. **high mass** see MASS[2]. **high-octane** (of petrol etc.) having good antiknock properties. **high old** *colloq.* most enjoyable (*had a high old time*). **high opinion of** a favourable opinion of. **high-pitched 1** (of a sound) high. **2** (of a roof) steep. **3** (of style etc.) lofty. **high places** the upper ranks of an organization etc. **high point** the maximum or best state reached. **high polymer** a polymer having a high molecular weight. **high-powered 1** having great power or energy. **2** important or influential. **high pressure 1** a high degree of activity or exertion. **2** a condition of the atmosphere with the pressure above average. **high priest** (*fem.* **high priestess**) **1** a chief priest, esp. Jewish. **2** the head of any cult. **high profile** exposure to attention or publicity. **high-profile** *adj.* having a high profile. **high-ranking** of high rank, senior. **high relief** see RELIEF 6a. **high-rise** *adj.* (of a building) having many storeys. ● *n.* such a building. **high-risk** (usu. *attrib.*) involving or exposed to danger (*high-risk sports*). **high road 1** a main road. **2** (usu. foll. by *to*) a direct route (*on the high road to success*). **high roller** *N. Amer. sl.* a person who gambles large sums or spends freely. **high school 1** *Brit.* a grammar school. **2** *N. Amer. & Sc.* a secondary school. **high sea** (or **seas**) open seas not within any country's jurisdiction. **high season** the period of the greatest number of visitors at a resort etc. **high sign** *US colloq.* a surreptitious gesture indicating that all is well or that the coast is clear. **high-sounding** pretentious, bombastic. **high-speed 1** operating at great speed. **2** (of steel) suitable for the blades of cutting tools even when red-hot. **high-spirited** vivacious; cheerful. **high-spiritedness** = *high spirits*. **high spirits** vivacity; energy; cheerfulness. **high spot** *colloq.* the most enjoyable feature, moment, or experience. **high-stepper 1** a horse that lifts its feet high when walking or trotting. **2** a stately person. **high street** *Brit.* a main road, esp. the principal shopping street of a town. **high-strung** = *highly strung*. **high table** a table on a platform at a public dinner or for the fellows of a college. **high tea** *Brit.* a main evening meal usu. consisting of a cooked dish, bread and butter, tea, etc. **high tech** *n.* = *high technology*. ● *adj.* **1** (of architecture, interior design, etc.) imitating styles more usual in industry etc., esp. using steel, glass, or plastic in a functional way. **2** employing, requiring, or involved in high technology. **high technology** advanced technological development, esp. in electronics. **high-tensile** (of metal) having great tensile strength. **high tension** = *high voltage*. **high tide** the time or level of the tide at its flow. **high time** a time that is late or overdue (*it is high time they arrived*). **high-toned** stylish; dignified; superior. **high treason** see TREASON 1. **high-up** *colloq.* a person of high rank. **high voltage** electrical potential large enough to cause injury or damage if diverted, as in power transmission cables. **high water** the tide at its fullest. **2** the time of this. **high-water mark 1** the level reached at high water. **2** the maximum recorded value or highest point of excellence. **high, wide, and handsome** *colloq.* in a carefree or stylish manner. **high wire** a high tightrope. **high words** angry talk. **high yellow** *US offens.* a light-skinned person of mixed black and white parentage. **in high feather** see FEATHER. **on high** in or to heaven or a high place. **on one's high horse** *colloq.* behaving superciliously or arrogantly. **play high 1** play for high stakes. **2** play a card of high value. **run high 1** (of the sea) have a strong current with high tide. **2** (of feelings) be strong. [OE *hēah* f. Gmc]

High Admiral *n. Naut.* a chief officer of admiral's rank.

highball /ˈhaɪbɔːl/ *n. N. Amer.* **1** a drink of spirits and soda etc., served with ice in a tall glass. **2** a railway signal to proceed.

highbinder /ˈhaɪˌbaɪndə(r)/ *n. US sl.* a ruffian; a swindler; an assassin.

highboy /ˈhaɪbɔɪ/ *n. N. Amer.* a tall chest of drawers on legs.

highbrow /ˈhaɪbraʊ/ *adj. & n. colloq.* ● *adj.* intellectual; cultural. ● *n.* an intellectual or cultured person.

High Church *n.* the section of the Church of England which stresses historical continuity with Catholic Christianity and emphasizes ritual, priestly authority, and the sacraments. In its modern sense the term dates back to the Oxford Movement.

High Commission *n.* an embassy from one Commonwealth country to another.

High Commissioner *n.* the head of a High Commission.

High Court (in full **High Court of Justice**) (in the UK) a court of justice established in 1873 primarily to deal with civil cases. It is composed of three divisions, Queen's Bench, Chancery, and the Family Division.

Higher /ˈhaɪə(r)/ *n.* (in Scotland) **1** a school examination leading to the Scottish Certificate of Education, Higher Grade. **2** a pass in this examination.

highfalutin /ˌhaɪfəˈluːtɪn/ *adj. & n.* (also **highfaluting** /-tɪŋ/) *colloq.* ● *adj.* absurdly pompous or pretentious. ● *n.* speech or writing of this kind. [HIGH + *-falutin*, of unkn. orig.]

high fidelity *n.* the reproduction of sound with little distortion, giving a result as close as possible to the original. In order to achieve this the system must be able to record all frequencies and their respective intensities within the audible frequency and volume range, and the reverberation and spatial sound-pattern of the original must be reproduced too.

High German *n.* the variety of German originally confined to 'high' or southern Germany, now accepted as the literary language of the whole country. The spread of this form as a standard language owes much to the biblical translations of Martin Luther in the 16th century.

high-handed /haɪˈhændɪd/ *adj.* disregarding others' feelings; overbearing. □ **high-handedly** *adv.* **high-handedness** *n.*

highland /ˈhaɪlənd/ *n. & adj.* ● *n.* (in *pl.*) **1** an area of high land, a mountainous region. **2** (**the Highlands**) (in full **the Scottish Highlands**) the mountainous part of Scotland, to the north of

Glasgow and Stirling. ● *adj.* **1** of or relating to high land or a mountainous region. **2** (**Highland**) of or in the Highlands of Scotland. □ **Highland cattle** a shaggy-haired breed of cattle with long curved widely spaced horns, developed in the Highlands of Scotland. **Highland dress** the kilt, plaid, sporran, etc. **Highland fling** see FLING *n.* 3. □ **highlander** *n.* (also **Highlander**). **Highlandman** *n.* (*pl.* **-men**). [OE *hēahlond* promontory (as HIGH, LAND)]

Highland clearances the forced removal of crofters from their land in the Highlands of Scotland in the late 18th and early 19th centuries. The clearances, carried out by absentee landlords wanting to install more profitable sheep and, later, deer on their estates, led to widespread emigration to North America and elsewhere.

Highland games *n.pl.* a sports meeting traditional to the Highlands of Scotland and typically consisting of athletic events such as tossing the caber and putting the shot, as well as other activities such as dancing and bagpipe-playing. The most famous gathering is that held annually at Braemar.

Highland Region a local government region of northern Scotland; administrative centre, Inverness.

highlight /ˈhaɪlaɪt/ *n.* & *v.* ● *n.* **1** (in a painting etc.) a light area, or one seeming to reflect light. **2** a moment or detail of vivid interest; an outstanding feature. **3** (usu. in *pl.*) a bright tint in the hair, esp. as produced by bleaching or dyeing. ● *v.tr.* **1 a** bring into prominence; draw attention to. **b** mark with a highlighter. **2** create highlights in (the hair).

highlighter /ˈhaɪˌlaɪtə(r)/ *n.* a marker pen which overlays colour on a printed word etc., leaving it legible and emphasized.

highly /ˈhaɪlɪ/ *adv.* **1** in a high degree (*highly amusing*; *highly probable*; *commend it highly*). **2** honourably; favourably (*think highly of him*). **3** in a high position or rank (*highly placed*). □ **highly strung** very sensitive or nervous. [OE *hēalīce* (as HIGH)]

high-minded /haɪˈmaɪndɪd/ *adj.* **1** having high moral principles. **2** *archaic* proud. □ **high-mindedly** *adv.* **high-mindedness** *n.*

high-muck-a-muck /ˈhaɪˌmʌkəˌmʌk/ *n.* N. Amer. a person of great self-importance. [perh. f. Chinook *hiu* plenty + *muckamuck* food]

highness /ˈhaɪnɪs/ *n.* **1** the state of being high (*highness of taxation*) (cf. HEIGHT). **2** (**Highness**) a title used in addressing and referring to a prince or princess (*Her Highness*; *Your Royal Highness*). [OE *hēanes* (as HIGH)]

High Renaissance see RENAISSANCE.

High Sheriff see SHERIFF 1a.

High Steward *n.* **1** a high officer in some English municipalities, usu. a nobleman. **2** = LORD HIGH STEWARD OF ENGLAND.

hight /haɪt/ *adj. archaic poet.*, or *joc.* called; named. [past part. (from 14th c.) of OE *hātan* command, call]

hightail /ˈhaɪteɪl/ *v.intr. N. Amer. colloq.* move at high speed.

highway /ˈhaɪweɪ/ *n.* **1 a** a public road. **b** a main route (by land or water). **2** a direct course of action (*on the highway to success*). □ **Highway Code** (in the UK) the official booklet of guidance for road-users. **Queen's** (or **King's**) **highway** a public road, regarded as being under the sovereign's protection.

highwayman /ˈhaɪweɪmən/ *n.* (*pl.* **-men**) *hist.* a robber of passengers, travellers, etc., usu. mounted.

HIH *abbr.* Her or His Imperial Highness.

hijack /ˈhaɪdʒæk/ *v.* & *n.* ● *v.tr.* **1** seize control of (a loaded lorry, an aircraft in flight, etc.), esp. to force it to a different destination. **2** seize (goods) in transit. **3** take over (an organization etc.) by force or subterfuge in order to redirect it. ● *n.* an instance of hijacking. □ **hijacker** *n.* [20th c.: orig. unkn.]

Hijaz see HEJAZ.

Hijra var. of HEGIRA.

hike /haɪk/ *n.* & *v.* ● *n.* **1** a long walk, esp. in the country for pleasure or exercise with boots, rucksack, etc. **2** esp. US an increase (of prices etc.). ● *v.* **1** *intr.* walk, esp. across country, for a long distance; go on a hike. **2** (usu. foll. by *up*) **a** *tr.* hitch up (clothing etc.); hoist; shove. **b** *intr.* work upwards out of place, become hitched up. **3** *tr.* increase (prices etc.). □ **hiker** *n.* [19th-c. dial.: orig. unkn.]

hila *pl.* of HILUM.

hilarious /hɪˈleərɪəs/ *adj.* **1** exceedingly funny. **2** boisterously merry. □ **hilariously** *adv.* **hilariousness** *n.* **hilarity** /-ˈlærɪtɪ/ *n.* [L *hilaris* f. Gk *hilaros* cheerful]

Hilary, St /ˈhɪlərɪ/ (*c.*315–*c.*367), French bishop. In *c.*350 he was appointed bishop of Poitiers, in which position he became a leading opponent of Arianism. Feast day, 13 Jan.

Hilary term *n.* Brit. a term in some universities and the law beginning in January, near the feast day of St Hilary.

Hilbert /ˈhɪlbət/, David (1862–1943), German mathematician. He proved fundamental theorems about rings, collected, systematized, and extended all that was then known about algebraic numbers, and reorganized the axiomatic foundations of geometry. He set potential theory and the theory of integral equations on its modern course with his invention of *Hilbert space* (an infinite-dimensional analogue of Euclidean space), and formulated the formalist philosophy of mathematics and mathematical logic. In 1900 Hilbert proposed twenty-three problems which crystallized mathematical thinking for the next few decades.

Hildegard of Bingen, St /ˈhɪldəˌɡɑːd, ˈbɪŋən/ (1098–1179), German abbess, scholar, composer, and mystic. A nun of the Benedictine order, she became Abbess of Diessem in 1136, later moving her community to Bingen. She described and illustrated her mystical experiences in *Scivias*, and also wrote poetry and composed sacred music. Her scientific writings covered a range of subjects, including the circulation of the blood and aspects of natural history.

Hildesheim /ˈhɪldəsˌhaɪm/ an industrial city in Lower Saxony, NW Germany; pop. (1991) 106,000.

Hill[1] /hɪl/, Benny (born Alfred Hawthorne) (1925–92), English comedian. After working in clubs and seaside shows he made an early and successful transition to television in 1949, being named TV Personality of the Year in 1954. His risqué humour, as seen in the series of programmes *The Benny Hill Show*, had an international appeal; television rights for the show were sold worldwide.

Hill[2] /hɪl/, Octavia (1838–1912), English housing reformer. An active campaigner for the improvement of housing for the poor, she met John Ruskin while working for a Christian Socialist association. In 1864 Ruskin provided financial assistance for Hill to fund the first of several housing projects, the purchase and refurbishment of three London slum houses. She was also a co-founder of the National Trust (1895).

Hill[3] /hɪl/, Sir Rowland (1795–1879), British educationist, administrator, and inventor. He was initially a teacher who introduced a system of self-government at his school in Birmingham and wrote on the challenges of mass education. In the 1830s he invented a rotary printing-press. Hill is chiefly remembered for his introduction of the penny postage-stamp system (see STAMP); he later became Secretary to the Post Office (1854–64).

hill /hɪl/ *n.* & *v.* ● *n.* **1** a naturally raised area of land, not as high as a mountain. **2** (often in *comb.*) a heap; a mound (*anthill; dunghill*). **3** a sloping piece of road. ● *v.tr.* **1** form into a hill. **2** (usu. foll. by *up*) bank up (plants) with soil. □ **hillbilly** (*pl.* **-ies**) US **1** *colloq.*, often *derog.* a person from a remote rural area in a southern state (cf. HICK). **2** folk music of or like that of the southern US. **hill climb** a race for vehicles up a steep hill. **hill-station** Anglo-Ind. a government settlement, esp. for holidays etc. during the hot season, in the low mountains of the northern Indian subcontinent. **old as the hills** very ancient. **over the hill** *colloq.* **1** past the prime of life; declining. **2** past the crisis. **the hills** Anglo-Ind. the low mountains of the northern Indian subcontinent, esp. as the location for a hill-station. **up hill and down dale** see UP. [OE *hyll*]

Hillary /ˈhɪlərɪ/, Sir Edmund (Percival) (b.1919), New Zealand mountaineer and explorer. In 1953 Hillary and Tenzing Norgay were the first people to reach the summit of Mount Everest, as members of a British expedition. Hillary later led the New Zealand contingent of the Commonwealth Trans-Antarctic Expedition (1955–8) organized by Sir Vivian Fuchs.

hill figure *n.* a design (usually either a horse or a human figure) cut into the chalk or limestone of hills, especially in southern England, and standing out white against the green turf. The oldest of these (the White Horse at Uffington, Oxfordshire) dates back to the Iron Age (1st century BC) and was probably a tribal emblem.

hill-fort /ˈhɪlfɔːt/ *n.* a fortified place built on a hilltop, with ramparts and ditches, found in western Europe and dating from between the late Bronze Age and the Roman period.

hillock /ˈhɪlək/ *n.* a small hill or mound. □ **hillocky** *adj.*

Hillsborough /ˈhɪlzbərə/ a football stadium in Sheffield, England, home of Sheffield Wednesday Football Club. It was the scene of Britain's worst sports disaster when on 15 Apr. 1989 ninety-five

Liverpool fans died in a crush at an FA Cup semi-final match between Liverpool and Nottingham Forest.

hillside /'hɪlsaɪd/ n. the sloping side of a hill.

hilltop /'hɪltɒp/ n. the summit of a hill.

hillwalking /'hɪl‚wɔːkɪŋ/ n. the pastime of hiking in hilly country. □ **hillwalker** n.

hilly /'hɪlɪ/ adj. (**hillier, hilliest**) having many hills. □ **hilliness** n.

hilt /hɪlt/ n. & v. ● n. **1** the handle of a sword, dagger, etc. **2** the handle of a tool. ● v.tr. provide with a hilt. □ **up to the hilt** completely. [OE hilt(e) f. Gmc]

hilum /'haɪləm/ n. (pl. **hila** -lə) **1** Bot. the point of attachment of a seed to its seed-vessel. **2** Anat. a notch or indentation where a vessel enters an organ. [L, = little thing, trifle]

Hilversum /'hɪlvəsəm/ a town in the Netherlands, in North Holland province, near Amsterdam; pop. (1991) 84,600. It is the centre of the Dutch radio and television network.

HIM abbr. Her or His Imperial Majesty.

him /hɪm/ pron. **1** objective case of HE (I saw him). **2** he (it's him again; is taller than him). ¶ Strictly, the pronoun following the verb to be should be in the subjective rather than the objective case, i.e. is taller than he rather than is taller than him. **3** archaic himself (fell and hurt him). [OE, masc. and neut. dative sing. of HE, IT[1]]

Himachal Pradesh /hɪ‚mɑːtʃəl prə'deʃ/ a mountainous state in northern India; capital, Simla.

Himalayas, the /‚hɪmə'leɪəz, hɪ'mɑːljəz/ a vast mountain system in southern Asia, extending 2,400 km (1,500 miles) from Kashmir eastwards to Assam. It covers most of Nepal, Sikkim, and Bhutan and forms part of the northern boundary of the Indian subcontinent. The Himalayas consist of a series of parallel ranges rising up from the Ganges basin to the Tibetan plateau, at over 3,000 m above sea level, and include the Karakoram, Zaskar, and Ladakh ranges. The backbone is the Great Himalayan Range, the highest mountain range in the world, with several peaks rising to over 7,700 m (25,000 ft), the highest being Mount Everest. □ **Himalayan** adj. [Skr. f. hima snow + ālaya abode]

himation /hɪ'mætɪən/ n. Gk Antiq. the outer garment worn by the ancient Greeks over the left shoulder and under the right. [Gk]

Himmler /'hɪmlə(r)/, Heinrich (1900–45), German Nazi leader, chief of the SS (1929–45) and of the Gestapo (1936–45). He established and oversaw the programme of systematic genocide of over 6 million Jews and other disfavoured groups between 1941 and 1945. He was captured by British forces in 1945, but committed suicide by swallowing a cyanide capsule.

Hims see HOMS.

himself /hɪm'self/ pron. **1 a** emphat. form of HE or HIM (he himself will do it). **b** refl. form of HIM (he has hurt himself). **2** in his normal state of body or mind (does not feel quite himself today). **3** esp. Ir. a third party of some importance; the master of the house. □ **be himself** act in his normal unconstrained manner. **by himself** see by oneself. [OE (as HIM, SELF)]

Hinault /i:'nəʊ/, Bernard (b.1954), French racing cyclist. During his professional career he won the Tour de France five times between 1978 and 1985; he also achieved three wins in the Tour of Italy between 1980 and 1985.

Hinayana /‚hiːnə'jɑːnə/ n. (also **Hinayana Buddhism**) a name given by the followers of Mahayana Buddhism to the more orthodox and, as they thought, less central, schools of early Buddhism. The Hinayana tradition died out in India by the 7th century AD but survived in Ceylon (Sri Lanka) as the Theravada school and was taken from there to Burma, Thailand, and other regions of SE Asia. [Skr. f. hīna lesser + yāna vehicle]

hind[1] /haɪnd/ adj. (esp. of parts of the body) situated at the back, posterior (hind leg) (opp. FORE). □ **on one's hind legs** see LEG. [ME, perh. shortened f. OE bihindan BEHIND]

hind[2] /haɪnd/ n. a female red deer (usu. a red deer or sika), esp. in and after the third year. [OE f. Gmc]

hind[3] /haɪnd/ n. hist. **1** esp. Sc. a skilled farm-worker, usu. married and with a tied cottage, and formerly having charge of two horses. **2** a steward on a farm. **3** a rustic, a boor. [ME hine f. OE hīne (pl.) app. f. hī(g)na genitive pl. of hīgan, hīwan 'members of a family' (cf. HIDE[3]): for -d cf. SOUND[1]]

hindbrain /'haɪndbreɪn/ n. Anat. the lower part of the brainstem, comprising the cerebellum, pons, and medulla oblongata. Also called rhombencephalon. (See also BRAIN.)

Hindemith /'hɪndə‚mɪt/, Paul (1895–1963), German composer. His music forms part of the neoclassical trend which began in the 1920s. Hindemith believed that music should have a social purpose and that audiences should participate as well as listen; his Gebrauchsmusik ('utility music') compositions are intended for performance by amateurs. A prolific composer, he wrote operas (such as Mathis der Maler, 1938), concertos, and orchestral and chamber music. He left Germany in the 1930s after Nazi hostility to his work, eventually settling in the US.

Hindenburg[1] /'hɪnd(ə)n‚bʊrk/ the former German name (1915–45) for ZABRZE.

Hindenburg[2] /'hɪndən‚bɜːg/, Paul Ludwig von Beneckendorff und von (1847–1934), German Field Marshal and statesman, President of the Weimar Republic 1925–34. He was recalled from retirement at the outbreak of the First World War and appointed Commander-in-Chief of German forces from 1916, directing the war effort in partnership with Erich Ludendorff. He was elected President in 1925 and re-elected in 1932, and was reluctantly persuaded to appoint Hitler as Chancellor in 1933.

Hindenburg Line (also called the Siegfried Line) in the First World War, a German fortified line of defence on the Western Front to which Hindenburg directed retreat and which was not breached until near the end of the war.

hinder[1] /'hɪndə(r)/ v.tr. (also absol.) impede; delay; prevent (you will hinder him; hindered me from working). [OE hindrian f. Gmc]

hinder[2] /'haɪndə(r)/ adj. rear, hind (the hinder part). [ME, perh. f. OE hinderweard backward: cf. HIND[1]]

Hindi /'hɪndɪ/ n. & adj. ● n. an Indic language of northern India. It is derived from Sanskrit and written in the Devanagari script and has a rich literature. It is one of the official languages of India. Hindi is the most widely spoken language in India, with over 200 million people using it as a first or second language. ● adj. of or relating to Hindi. [Urdu f. Hind India]

hindlimb /'haɪndlɪm/ n. either of the back limbs of an animal.

hindmost /'haɪndməʊst/ adj. furthest behind; most remote.

Hindoo archaic var. of HINDU.

hindquarters /'haɪnd‚kwɔːtəz/ n.pl. the hind legs and adjoining parts of a quadruped.

hindrance /'hɪndrəns/ n. **1** the act or an instance of hindering; the state of being hindered. **2** a thing that hinders; an obstacle.

hindsight /'haɪndsaɪt/ n. **1** wisdom after the event (realized with hindsight that they were wrong) (opp. FORESIGHT). **2** the backsight of a gun.

Hindu /'hɪnduː, hɪn'duː/ n. & adj. (also archaic **Hindoo**) ● n. (pl. **Hindus**) **1** a follower of Hinduism. **2** archaic an Indian. ● adj. **1** of or concerning Hindus or Hinduism. **2** archaic Indian. [Urdu f. Pers. f. Hind India]

Hinduism /'hɪnduː‚ɪz(ə)m/ n. a religious and social tradition which has developed over thousands of years and is found primarily in India, Bangladesh, Sri Lanka, and Nepal. Hinduism is extremely heterogeneous, with no one sacred text or body of doctrine; unlike most religions, it implies no single belief concerning the nature of god: it embraces polytheism, monotheism, and monism. Hinduism developed from the Vedic tradition, but perhaps more important than the Veda as a Hindu text today is the Bhagavadgita, which helped to establish cults relating to the deities Vishnu (Krishna) and Siva. It also outlines the central idea of moksha, or liberation of the soul from the cycle of perpetual death and rebirth; the state of the soul in this cycle is established by the idea of karma. Fundamental to Hinduism are the beliefs concerning the nature of the universe and the structure of society. Brahman is the ultimate reality underlying all being and is also the name given to the highest (priestly) caste. The caste system (see CASTE, VARNA) underpins Hindu society and is described partly by the key concept of dharma, both the eternal law underlying the whole of existence and the social and religious duty of the individual. □ **Hinduize** v.tr. (also **-ise**).

Hindu Kush /‚hɪnduː 'kuːʃ, 'kʊʃ/ a range of high mountains in northern Pakistan and Afghanistan, forming a westward continuation of the Himalayas. Several peaks exceed 6,150 m (20,000 ft), the highest being Tirich Mir.

Hindustan /‚hɪndʊ'stɑːn/ hist. the Indian subcontinent in general, more specifically that part of India north of the Deccan, especially the plains of the Ganges and Jumna rivers.

Hindustani /‚hɪndʊ'stɑːnɪ/ n. & adj. ● n. **1** hist. a group of mutually intelligible languages and dialects spoken in north-west India, principally Urdu and Hindi. **2** the Delhi dialect of Hindi, used as a lingua franca throughout India. ¶ *Hindustani* was the usual term in the 18th and 19th centuries for the native language of north-west India. The usual modern term is *Hindi*, although *Hindustani* is still sometimes used to refer to the lingua franca. ● *adj.* of or relating to Hindustan or its people, or Hindustani. [Urdu f. Pers. *hindūstānī*, f. *hindū* HINDU + *-stān* country]

hindwing /ˈhaɪndwɪŋ/ *n.* either of the two back wings of a four-winged insect.

hinge /hɪndʒ/ *n. & v.* ● *n.* **1 a** a movable, usu. metal, joint or mechanism such as that by which a door is hung on a side post. **b** *Biol.* a natural joint performing a similar function, e.g. that of a bivalve shell. **2** a central point or principle on which everything depends. ● *v.* (**hingeing** or **hinging**) **1** *intr.* (foll. by *on*) **a** depend (on a principle, an event, etc.) (*all hinges on his acceptance*). **b** (of a door etc.) hang and turn (on a post etc.). **2** *tr.* attach with or as if with a hinge. □ **stamp-hinge** a small piece of gummed transparent paper used for fixing postage stamps in an album etc. □ **hinged** *adj.* **hingeless** *adj.* **hingewise** *adv.* [ME *heng* etc., rel. to HANG]

hinny[1] /ˈhɪnɪ/ *n.* (*pl.* **-ies**) the offspring of a female donkey and a male horse. [L *hinnus* f. Gk *hinnos*]

hinny[2] /ˈhɪnɪ/ *n.* (also **hinnie**) (*pl.* **-ies**) *Sc. & N. Engl.* (esp. as a form of address) darling, sweetheart. □ **singing hinny** a currant cake baked on a griddle. [var. of HONEY]

Hinshelwood /ˈhɪnʃ(ə)l‚wʊd/, Sir Cyril Norman (1897–1967), English physical chemist. He made fundamental contributions to reaction kinetics in gases and liquids. He later applied the laws of kinetics to bacterial growth, and suggested the role of nucleic acids in protein synthesis. Hinshelwood was simultaneously president of both the Royal Society and the Classical Association. He shared the Nobel Prize for chemistry in 1956.

hint /hɪnt/ *n. & v.* ● *n.* **1** a slight or indirect indication or suggestion (*took the hint and left*). **2** a small piece of practical information (*handy hints on cooking*). **3** a very small trace; a suggestion (*a hint of perfume*). ● *v.tr.* (often foll. by *that* + clause) suggest slightly (*hinted the contrary*; *hinted that they were wrong*). □ **hint at** give a hint of; refer indirectly to. [app. f. obs. *hent* grasp, lay hold of, f. OE *hentan*, f. Gmc, rel. to HUNT]

hinterland /ˈhɪntə‚lænd/ *n.* **1** the often deserted or uncharted areas beyond a coastal district or a river's banks. **2** an area served by a port or other centre. **3** a remote or fringe area. [G f. *hinter* behind + *Land* LAND]

hip[1] /hɪp/ *n.* **1** a projection of the pelvis and upper thigh-bone on each side of the body in human beings and quadrupeds. **2** (often in *pl.*) the circumference of the body at the buttocks. **3** *Archit.* the sharp edge of a roof from ridge to eaves where two sides meet. □ **hip-bath** a portable bath in which a person sits. **hip-bone** a bone forming the hip, esp. the ilium. **hip-flask** a flask for spirits etc., carried in a hip-pocket. **hip-joint** the articulation of the head of the thigh-bone with the ilium. **hip-length** (of a garment) reaching down to the hips. **hip-pocket** a trouser-pocket just behind the hip. **hip-** (or **hipped-**)**roof** a roof with the sides and the ends inclined. **on the hip** *archaic* at a disadvantage. □ **hipless** *adj.* **hipped** *adj.* (also in comb.). [OE *hype* f. Gmc, rel. to HOP[1]]

hip[2] /hɪp/ *n.* (also **hep** /hep/) the fruit of a rose, esp. a wild kind. [OE *hēope*, *hīope* f. WG]

hip[3] /hɪp/ *adj.* (also **hep** /hep/) (**hipper**, **hippest** or **hepper**, **heppest**) *sl.* **1** following the latest fashion in esp. jazz or popular music, clothes, etc.; stylish, fashionable. **2** (often foll. by *to*) understanding, aware. □ **hip-cat** a hip person. □ **hipness** *n.* [20th c.: orig. unkn.]

hip[4] /hɪp/ *int.* introducing a united cheer (*hip, hip, hooray*). [19th c.: orig. unkn.]

hip-hop /ˈhɪphɒp/ *n.* (also **hip hop**) **1** a style of popular music of US black and Hispanic origin, featuring rap over a spare electronic backing. **2** the subculture associated with this music, including graffiti art, break-dancing, etc. [20th c.: perhaps from HIP[1]]

Hipparchus /hɪˈpɑːkəs/ (*c.*170–after 126 BC) Greek astronomer and geographer, working in Rhodes. His major works are lost, but his astronomical observations were developed by Ptolemy. He constructed the celestial coordinates of 800 stars, indicating their relative brightness, but rejected Aristarchus' hypothesis that the sun is the centre of the planetary system. Hipparchus is best known for his discovery of the precession of the equinoxes. He suggested improved methods of determining latitude and longitude, and is credited with the invention of trigonometry.

hippeastrum /‚hɪpɪˈæstrəm/ *n.* a South American bulbous plant of the genus *Hippeastrum*, with showy white or red flowers. [mod.L f. Gk *hippeus* horseman (the leaves appearing to ride on one another) + *astron* star (from the flower-shape)]

hipped /hɪpt/ *adj.* (usu. foll. by *on*) esp. *US sl.* obsessed, infatuated. [past part. of *hip* (v.) = make hip (HIP[3])]

hipper /ˈhɪpə(r)/ *n. Austral.* a soft pad used to protect the hip when sleeping on hard ground.

hippie var. of HIPPY[1].

hippo /ˈhɪpəʊ/ *n.* (*pl.* **-os**) *colloq.* a hippopotamus. [abbr.]

hippocampus /‚hɪpəˈkæmpəs/ *n.* (*pl.* **hippocampi** /-paɪ/) **1** a sea horse of the genus *Hippocampus*. **2** *Anat.* the elongated ridges on the floor of each lateral ventricle of the brain, thought to be the centre of emotion and the autonomic nervous system. [L f. Gk *hippokampos* f. *hippos* horse + *kampos* sea monster]

hippocras /ˈhɪpə‚kræs/ *n. hist.* wine flavoured with spices. [ME f. OF *ipocras* HIPPOCRATES, prob. because strained through a filter called 'Hippocrates' sleeve']

Hippocrates /hɪˈpɒkrə‚tiːz/ (*c.*460–377 BC), the most famous of all physicians, of whom, paradoxically, almost nothing is known. Referred to briefly by Plato, his name was later attached to a body of ancient Greek medical writings of which probably none was written by him. This collection is so varied that all subsequent physicians have been able to find within it notions that agreed with their own ideas of what medicine and doctors should be. If there are common features of an agreed Hippocratic philosophy, they might be that nature has an innate power of healing, and that diseases are closely linked to the physical environment.

Hippocratic oath /‚hɪpəˈkrætɪk/ *n.* an oath stating the obligations and proper conduct of physicians, formerly taken by those beginning medical practice. Parts of the oath are still used in some medical schools. (See HIPPOCRATES.) [med.L *Hippocraticus* f. HIPPOCRATES]

Hippocrene /ˈhɪpə‚kriːn/ a spring on Mount Helicon, Greece, according to classical mythology the inspiration of poets. It was fabled to have been produced by a stroke from the hoof of Pegasus. [L f. Gk f. *hippos* horse + *krēnē* fountain]

hippodrome /ˈhɪpə‚drəʊm/ *n.* **1** a theatre or dancehall. **2** *Antiq.* a course for chariot races etc. **3** a circus. [F *hippodrome* or L *hippodromus* f. Gk *hippodromos* f. *hippos* horse + *dromos* race, course]

hippogriff /ˈhɪpə‚grɪf/ *n.* (also **hippogryph**) a mythical griffin-like creature with the body of a horse. [F *hippogriffe* f. It. *ippogrifo* f. Gk *hippos* horse + It. *grifo* GRIFFIN]

Hippolytus /hɪˈpɒlɪtəs/ *Gk Mythol.* the son of Theseus, banished and cursed by his father after being accused by Phaedra of rape. He was killed when a sea monster, sent by Poseidon in response to the curse, frightened his horses as he drove his chariot along a seashore.

hippopotamus /‚hɪpəˈpɒtəməs/ *n.* (*pl.* **hippopotamuses** or **hippopotami** /-‚maɪ/) **1** a very large thick-skinned amphibious mammal, *Hippopotamus amphibius*, native to Africa, inhabiting rivers and lakes. **2** (in full **pigmy hippopotamus**) a smaller related mammal, *Choeropsis liberiensis*, native to Africa, inhabiting forests and swamps. [ME f. L f. Gk *hippopotamos* f. *hippos* horse + *potamos* river]

Hippo Regius /‚hɪpəʊ ˈriːdʒɪəs/ see ANNABA.

hippy[1] /ˈhɪpɪ/ *n.* (also **hippie**) (*pl.* **-ies**) (esp. in the 1960s) a person of unconventional appearance, typically with long hair, jeans, beads, etc., often associated with hallucinogenic drugs and a rejection of conventional values. [HIP[3]]

hippy[2] /ˈhɪpɪ/ *adj.* having large hips.

hipster[1] /ˈhɪpstə(r)/ *adj. & n. Brit.* ● *adj.* (of a garment) hanging from the hips rather than the waist. ● *n.* (in *pl.*) trousers hanging from the hips.

hipster[2] /ˈhɪpstə(r)/ *n. sl.* a stylish or hip person. □ **hipsterism** *n.*

hiragana /‚hɪərəˈɡɑːnə/ *n.* the cursive form of Japanese syllabic writing or kana (cf. KATAKANA). [Jap., = plain kana]

hircine /ˈhɜːsaɪn/ *adj.* goatlike. [L *hircinus* f. *hircus* he-goat]

hire /ˈhaɪə(r)/ *v. & n.* ● *v.tr.* **1** (often foll. by *from*) procure the temporary use of (a thing) for an agreed payment (*hired a van from them*). **2** esp. *US* employ (a person) for wages or a fee. **3** *US* borrow (money). ● *n.* **1** hiring or being hired. **2** payment for this. □ **for** (or **on**) **hire** ready to be hired. **hire-car** a car available for hire. **hired girl** (or **man**) *N. Amer.* a domestic servant, esp. on a farm. **hire out** grant the temporary use of (a thing) for an agreed payment. **hire purchase** *Brit.* a system by which a person may purchase a thing by regular payments while having the use of it. □ **hireable** *adj.* (*US* **hirable**). **hirer** *n.* [OE *hȳrian*, *hȳr* f. WG]

hireling /ˈhaɪəlɪŋ/ n. usu. derog. a person who works for remuneration; an employee, esp. a mercenary one. [OE hўrling (as HIRE, -LING¹)]

Hirohito /ˌhɪrəˈhiːtəʊ/ (born Michinomiya Hirohito) (1901–89), emperor of Japan 1926–89. Regarded as the 124th direct descendant of Jimmu, he ruled as a divinity and generally refrained from involvement in politics. In 1945, however, he was instrumental in obtaining his government's agreement to the unconditional surrender which ended the Second World War. He was obliged to renounce his divinity and become a constitutional monarch by the terms of the constitution established in 1946.

Hiroshima /hɪˈrɒʃmə, ˌhɪrəˈʃiːmə/ a city on the south coast of the island of Honshu, western Japan, capital of Chugoku region; pop. (1990) 1,086,000. It was the target of the first atom bomb, which was dropped by the United States on 6 Aug. 1945 and resulted in the deaths of about one-third of the city's population of 300,000. Together with a second attack, on Nagasaki three days later, this led to Japan's surrender and the end of the Second World War.

hirsute /ˈhɜːsjuːt/ adj. **1** hairy, shaggy. **2** untrimmed. □ **hirsuteness** n. [L hirsutus]

hirsutism /ˈhɜːsjuːˌtɪz(ə)m/ n. the excessive growth of hair on the face and body.

hirundine /ˈhɪrənˌdaɪn, hɪˈrʌndaɪn/ n. & adj. ● n. a bird of the swallow family Hirundinidae. ● adj. of or relating to swallows. [L hirundo swallow + -INE¹]

his /hɪz/ poss.pron. **1** (attrib.) of or belonging to him or himself (his house; his own business). **2** (**His**) (attrib.) (in titles) that he is (His Majesty). **3** the one or ones belonging to or associated with him (it is his; his are over there). □ **his and hers** joc. (of matching items) for husband and wife, or men and women. **of his** of or belonging to him (a friend of his). [OE, genitive of HE, IT¹]

Hispanic /hɪˈspænɪk/ adj. & n. ● adj. **1** of or relating to Spain or to Spain and Portugal. **2** of Spain and other Spanish-speaking countries. ● n. a Spanish-speaking person, esp. one of Latin American descent, living in the US. □ **Hispanicize** /-nɪˌsaɪz/ v.tr. (also **-ise**). [L Hispanicus f. Hispania Spain]

Hispaniola /ˌhɪspæˈnjəʊlə/ an island of the Greater Antilles in the West Indies, divided into the states of Haiti and the Dominican Republic. After its European discovery by Columbus in 1492, Hispaniola was colonized by the Spaniards, who soon enslaved or killed the native inhabitants. The western part (now Haiti) was ceded to France in 1697. [f. Sp. La Isla Española (the Spanish island), so named by Columbus]

Hispanist /ˈhɪspənɪst/ n. (also **Hispanicist** /hɪˈspænɪsɪst/) an expert in or student of the language, literature, and civilization of Spain.

Hispano- /hɪˈspænəʊ/ comb. form Spanish. [L Hispanus Spanish]

hispid /ˈhɪspɪd/ adj. Bot. & Zool. **1** rough with bristles; bristly. **2** shaggy. [L hispidus]

hiss /hɪs/ v. & n. ● v. **1** intr. (of a person, snake, goose, etc.) make a sharp sibilant sound, esp. as a sign of disapproval or derision (audience booed and hissed; the water hissed on the hotplate). **2** tr. express disapproval of (a person etc.) by hisses. **3** tr. whisper (a threat etc.) urgently or angrily ('Shut up!' he hissed). ● n. **1** a sharp sibilant sound as of the letter s, esp. as an expression of disapproval or derision. **2** Electronics unwanted interference at audio frequencies. □ **hiss away** (or **down**) drive off etc. by hisses. **hiss off** drive (an actor etc.) off the stage by hissing. [ME: imit.]

hist /hɪst/ int. archaic used to call attention, enjoin silence, incite a dog, etc. [16th c.: natural excl.]

histamine /ˈhɪstəmɪn, -ˌmiːn/ n. Biochem. an amine that causes the contraction of smooth muscle and the dilation of blood capillaries, released by most cells in response to wounding and in allergic and inflammatory reactions. □ **histaminic** /ˌhɪstəˈmɪnɪk/ adj. [HISTO- + AMINE]

histidine /ˈhɪstɪˌdiːn/ n. Biochem. an amino acid from which histamine is derived. [Gk histos web, tissue]

histo- /ˈhɪstəʊ/ comb. form (before a vowel also **hist-**) Biol. tissue. [Gk histos web]

histochemistry /ˌhɪstəʊˈkemɪstrɪ/ n. the study of the identification and distribution of the chemical constituents of tissues by means of stains, indicators, and microscopy. □ **histochemical** adj.

histogenesis /ˌhɪstəʊˈdʒenɪsɪs/ n. the formation of tissues. □ **histogenetic** /-dʒɪˈnetɪk/ adj.

histogeny /hɪˈstɒdʒɪnɪ/ n. histogenesis. □ **histogenic** /ˌhɪstəʊˈdʒenɪk/ adj.

histogram /ˈhɪstəˌɡræm/ n. Statistics a chart consisting of rectangles (usu. drawn vertically from a base line) whose areas and positions are proportional to the value or range of a number of variables. [Gk histos web, mast + -GRAM]

histology /hɪˈstɒlədʒɪ/ n. the study of the microscopic structure of tissues. □ **histologist** n. **histological** /ˌhɪstəˈlɒdʒɪk(ə)l/ adj.

histolysis /hɪˈstɒlɪsɪs/ n. the breaking down of tissues. □ **histolytic** /ˌhɪstəˈlɪtɪk/ adj.

histone /ˈhɪstəʊn/ n. Biochem. any of a group of basic proteins found in chromatin. [G Histon perh. f. Gk histamai arrest, or as HISTO-]

histopathology /ˌhɪstəʊpəˈθɒlədʒɪ/ n. **1** changes in tissues caused by disease. **2** the study of these. □ **histopathologist** n. **histopathological** /-ˌpæθəˈlɒdʒɪk(ə)l/ adj.

historian /hɪˈstɔːrɪən/ n. **1** a writer of history, esp. a critical analyst, rather than a compiler. **2** a person learned in or studying history (English historian; ancient historian). [F historien f. L (as HISTORY)]

historiated /hɪˈstɔːrɪˌeɪtɪd/ adj. = STORIATED. [med.L historiare (as HISTORY)]

historic /hɪˈstɒrɪk/ adj. **1** famous or important in history or potentially so (a historic moment). **2** Gram. (of a tense) normally used in the narration of past events (esp. Latin & Greek imperfect and pluperfect) (cf. PRIMARY adj. 9). **3** archaic or disp. = HISTORICAL. □ **historic infinitive** the infinitive when used instead of the indicative. **historic present** the present tense used instead of the past in vivid narration. [L historicus f. Gk historikos (as HISTORY)]

historical /hɪˈstɒrɪk(ə)l/ adj. **1** of or concerning history (historical evidence). **2** belonging to history, not to prehistory or legend. **3** (of the study of a subject) based on an analysis of its development over a period. **4** belonging to the past, not the present. **5** (of a novel, a film, etc.) dealing or professing to deal with historical events. **6** in connection with history, from the historian's point of view (of purely historical interest). □ **historically** adv.

historical novel n. a novel which is set in the past (usually a generation or more before the time of writing) and which attempts some historical accuracy in its treatment of period and characters. Sir Walter Scott is often regarded as the first writer of historical novels, followed by writers such as Alessandro Manzoni, Maria Edgeworth, Victor Hugo, Dumas (père), and Charles Dickens (with his A Tale of Two Cities (1859), for example). In the 20th century the genre has been attempted successfully by Robert Graves and Mary Renault, among others, while stylized sub-genres have arisen, as exemplified by the Regency romances of Georgette Heyer.

historicism /hɪˈstɒrɪˌsɪz(ə)m/ n. **1 a** the theory that social and cultural phenomena are determined by history. **b** the belief that historical events are governed by laws. **2** the tendency to regard historical development as the most basic aspect of human existence. **3** an excessive regard for past styles etc. □ **historicist** n. [HISTORIC after G Historismus]

historicity /ˌhɪstəˈrɪsɪtɪ/ n. the historical genuineness of an event etc.

historiographer /hɪˌstɔːrɪˈɒɡrəfə(r)/ n. **1** an expert in or student of historiography. **2** a writer of history, esp. an official historian. [ME f. F historiographe or f. LL historiographus f. Gk historiographos (as HISTORY, -GRAPHER)]

historiography /hɪˌstɔːrɪˈɒɡrəfɪ/ n. **1** the writing of history. **2** the study of history-writing. □ **historiographic** /-rɪəˈɡræfɪk/ adj. **historiographical** adj. [med.L historiographia f. Gk historiographia (as HISTORY, -GRAPHY)]

history /ˈhɪstərɪ, -trɪ/ n. (pl. **-ies**) **1** a continuous, usu. chronological, record of important or public events. **2 a** the study of past events, esp. human affairs. **b** the total accumulation of past events, esp. relating to human affairs or to the accumulation of developments connected with a particular nation, person, thing, etc. (our island history; the history of astronomy; he has a history of illness). **c** the past in general; antiquity. **3** an eventful past (this house has a history). **4 a** a systematic or critical account of or research into a past event or events etc. **b** a similar record or account of natural phenomena. **5** a historical play. □ **history painting** Art pictorial representation of actual historical events or of scenes (esp. of a morally edifying kind) from legend and literature. **make history 1** influence the course of history. **2** do something memorable. [ME f. L historia f. Gk historia finding out, narrative, history f. histōr learned, wise man, rel. to WIT²]

histrionic /ˌhɪstrɪˈɒnɪk/ adj. & n. ● adj. **1** of or concerning actors or acting. **2** (of behaviour) theatrical, dramatic. ● *n.* **1** (in *pl.*) **a** insincere and dramatic behaviour designed to impress. **b** theatricals; theatrical art. **2** *archaic* an actor. □ **histrionically** *adv.* [LL *histrionicus* f. L *histrio -onis* actor]

hit /hɪt/ *v. & n.* ● *v.* (**hitting**; *past* and *past part.* **hit**) **1** *tr.* **a** strike with a blow or a missile. **b** (of a moving body) strike (*the plane hit the ground*). **c** reach (a target, a person, etc.) with a directed missile (*hit the window with the ball*). **2** *tr.* cause to suffer or affect adversely; wound (*the loss hit him hard*). **3** *intr.* (often foll. by *at*, *against*, *upon*) direct a blow. **4** *tr.* (often foll. by *against*, *on*) knock (a part of the body) (*hit his head on the door-frame*). **5** *tr.* light upon; get at (a thing aimed at) (*he's hit the truth at last*; *tried to hit the right tone in his apology*) (see *hit on*). **6** *tr. colloq.* **a** encounter (*hit a snag*). **b** arrive at (*hit an all-time low*; *hit the town*). **c** indulge in, esp. liquor etc. (*hit the bottle*). **7** *tr. esp. US sl.* rob or kill. **8** *tr.* occur forcefully to (*the seriousness of the situation only hit him later*). **9** *tr. Sport* **a** propel (a ball etc.) with a bat etc. to score runs or points. **b** score (runs etc.) in this way. **c** (usu. foll. by *for*) strike (a ball or a bowler) for so many runs (*hit him for six*). **10** *tr.* represent exactly. **11** *tr. US, Austral., & NZ* (often foll. by *up*) *sl.* ask (a person) for; beg. ● *n.* **1 a** a blow; a stroke. **b** a collision. **2** a shot etc. that hits its target. **3** *colloq.* a popular success in entertainment, esp. a successful pop record. **4** a stroke of sarcasm, wit, etc. **5** a stroke of good luck. **6** *esp. US sl.* **a** a murder or other violent crime. **b** a drug injection etc. **7** a successful attempt. □ **hit-and-miss** aimed or done carelessly or at random. **hit and run** cause (accidental or wilful) damage and escape or leave the scene before being discovered. **hit-and-run** *attrib.adj.* relating to or (of a person) committing an act of this kind. **hit back** retaliate. **hit below the belt 1** *esp. Boxing* give a foul blow. **2** treat or behave unfairly. **hit for six** *Brit.* defeat in argument. **hit the hay** (or **sack**) *colloq.* go to bed. **hit the headlines** see HEADLINE. **hit the high spots** *colloq.* **1** visit the most important places. **2** go to excess or extremes. **hit home** make a salutary impression. **hit it off** (often foll. by *with*, *together*) agree or be congenial. **hit list** *sl.* a list of prospective victims. **hit man** (*pl.* **hit men**) *sl.* a hired assassin. **hit the nail on the head** state the truth exactly. **hit on** (or **upon**) find (what is sought), esp. by chance. **hit-or-miss** = *hit-and-miss*. **hit out** deal vigorous physical or verbal blows (*hit out at her enemies*). **hit-out** *n. Austral. sl.* a brisk gallop. **hit parade** *colloq.* a list of the current best-selling records of popular music. **hit the road** (*US* **trail**) *sl.* depart. **hit the roof** see ROOF. **hit up** *Cricket* score (runs) energetically. **hit wicket** *Cricket* (of a batsman) out through striking the wicket with the bat etc. **make a hit** (usu. foll. by *with*) be successful or popular. □ **hitter** *n.* [ME f. OE *hittan* f. ON *hitta* meet with, of unkn. orig.]

hitch /hɪtʃ/ *v. & n.* ● *v.* **1 a** *tr.* fasten with a loop, hook, etc.; tether (*hitched the horse to the cart*). **b** *intr.* (often foll. by *in*, *on to*, etc.) become fastened in this way (*the rod hitched in to the bracket*). **2** *tr.* move (a thing) with a jerk; shift slightly (*hitched the pillow to a comfortable position*). **3** *colloq.* **a** *intr.* = HITCHHIKE. **b** *tr.* obtain (a lift) by hitchhiking. ● *n.* **1** an impediment; a temporary obstacle. **2** an abrupt pull or push; a jerk. **3** a noose or knot of various kinds. **4** *colloq.* a free ride in a vehicle. **5** *N. Amer. sl.* a period of service. □ **get hitched** *colloq.* marry. **half hitch** a knot formed by passing the end of a rope round its standing part and then through the bight. **hitch up** lift (esp. clothing) with a jerk. **hitch one's wagon to a star** make use of powers higher than one's own. □ **hitcher** *n.* [ME: orig. uncert.]

Hitchcock /ˈhɪtʃkɒk/, Sir Alfred (Joseph) (1899–1980), English film director. Having established his reputation in Britain in the 1930s with films such as *The Thirty-Nine Steps* (1935) and *The Lady Vanishes* (1938), in 1939 he moved to Hollywood, where his first film was *Rebecca* (1940). Outstanding among his numerous later works are the thrillers *Strangers on a Train* (1951), *Psycho* (1960), with its famous shower murder, and *The Birds* (1963). His films are notable for their ability to generate suspense and for their technical ingenuity.

Hitchens /ˈhɪtʃɪnz/, Ivon (1893–1979), English painter. His main interest was in landscape, which he represented in an almost abstract style using broad fluid areas of vibrant colour, usually on a long rectangular canvas, as in *Winter Stage* (1936).

hitchhike /ˈhɪtʃhaɪk/ *v. & n.* ● *v.intr.* travel by seeking free lifts in passing vehicles. ● *n.* a journey made by hitchhiking. □ **hitchhiker** *n.*

hi-tech /ˈhaɪtek/ *adj.* = *high tech*. [abbr.]

hither /ˈhɪðə(r)/ *adv. & adj.* usu. *formal* or *literary* ● *adv.* to or towards this place. ● *adj.* situated on this side; the nearer (of two). □ **hither and thither** (or **yon**) in various directions; to and fro. [OE *hider*: cf. THITHER]

hitherto /ˌhɪðəˈtuː/ *adv.* until this time, up to now.

hitherward /ˈhɪðəwəd/ *adv. archaic* in this direction.

Hitler /ˈhɪtlə(r)/, Adolf (1889–1945), Austrian-born Nazi leader, Chancellor of Germany 1933–45. A co-founder of the National Socialist German Workers' Party (which later became known as the Nazi Party) after the First World War, Hitler was imprisoned in 1923 as one of the organizers of an unsuccessful putsch in Munich. While in prison he wrote *Mein Kampf*, an exposition of his political ideas. After his release his powers as an orator soon won prominence for him and the Nazi Party; following his appointment as Chancellor in 1933 he was able to overthrow the Weimar Republic and establish the totalitarian Third Reich, proclaiming himself *Führer*. His expansionist foreign policy precipitated the Second World War, while his fanatical anti-Semitism and desire to create an Aryan German state led to the deaths of millions of Jews. Hitler committed suicide in his Berlin headquarters as Soviet Allied forces were attacking the city. □ **Hitlerite** *n. & adj.*

Hitlerism /ˈhɪtləˌrɪz(ə)m/ *n.* the political principles or policy of the Nazi Party in Germany. [HITLER]

Hittite /ˈhɪtaɪt/ *n. & adj.* ● *n.* **1 a** a member of a powerful and widespread ancient non-Semitic people whose history can be traced *c.*1900–700 BC in Asia Minor and Syria. (*See note below.*) **b** a subject of the Hittite empire. **2** the Indo-European language of the Hittites, written in cuneiform and deciphered in the early 20th century. **3** (in the Bible) a member of a Canaanite or Syrian tribe, perhaps an offshoot of the peoples described above. ● *adj.* of or relating to the Hittites or their language. [Heb. *Ḥittīm*]

▪ The Hittites gained political control of central Anatolia *c.*1800–1200 BC and reached the zenith of their power under the rule of Suppiluliuma I (*c.*1340 BC), whose political influence extended from the capital, Hattusa, west to the Mediterranean coast and south-east into northern Syria. The decline of Hittite power may have resulted from an outbreak of famine.

HIV *abbr.* human immunodeficiency virus, a retrovirus which causes Aids.

hive /haɪv/ *n. & v.* ● *n.* **1 a** a beehive. **b** the bees in a hive. **2** a busy swarming place. **3** a swarming multitude. **4** a thing shaped like a hive in being domed. ● *v.* **1** *tr.* **a** place (bees) in a hive. **b** house (people etc.) snugly. **2** *intr.* **a** enter a hive. **b** live together like bees. □ **hive off 1** separate from a larger group. **2 a** form into or assign (work) to a subsidiary department or company. **b** denationalize or privatize (an industry etc.). **hive up** hoard. [OE *hȳf* f. Gmc]

hives /haɪvz/ *n.pl.* **1** a skin eruption, esp. nettle-rash. **2** inflammation of the larynx etc. [16th c. (orig. Sc.): orig. unkn.]

hiya /ˈhaɪjə/ *int. colloq.* a word used in greeting. [corrupt. of *how are you?*]

HK *abbr.* Hong Kong.

HL *abbr.* (in the UK) House of Lords.

hl *abbr.* hectolitre(s).

HM *abbr.* **1** Her (or His) Majesty('s). **2 a** headmaster. **b** headmistress.

hm *abbr.* hectometre(s).

h'm /hm/ *int. & n.* (also **hmm**) = HEM², HUM².

HMG *abbr.* (in the UK) Her or His Majesty's Government.

HMI *abbr.* (in the UK) Her or His Majesty's Inspector (of Schools).

HMS *abbr.* (in the UK) Her or His Majesty's Ship.

HMSO *abbr.* (in the UK) Her or His Majesty's Stationery Office.

HMV *abbr.* (in the UK) His Master's Voice.

HNC *abbr.* (in the UK) Higher National Certificate.

HND *abbr.* (in the UK) Higher National Diploma.

Ho *symb. Chem.* the element holmium.

ho /həʊ/ *int.* **1 a** an expression of admiration or (often repeated as **ho! ho!** etc.) derision, surprise, or triumph. **b** (in *comb.*) the second element of various exclamations (*heigh-ho*; *what ho*). **2** a call for attention. **3** (in *comb.*) *Naut.* an addition to the name of a destination etc. (*westward ho*). [ME, imit.: cf. ON *hó*]

ho. *abbr.* house.

hoar /hɔː(r)/ *adj. & n. literary* ● *adj.* **1** grey-haired with age. **2** greyish-white. **3** (of a thing) grey with age. ● *n.* **1** = *hoar-frost*. **2** hoariness. □ **hoar-frost** frozen water vapour deposited in clear still weather on vegetation etc. [OE *hār* f. Gmc]

hoard /hɔːd/ *n. & v.* ● *n.* **1** a stock or store (esp. of money) laid by. **2** an amassed store of facts etc. **3** *Archaeol.* an ancient store of treasure etc. ● *v.* **1** *tr.* (often *absol.*; often foll. by *up*) amass (money etc.) and put away; store. **2** *intr.* accumulate more than one's current requirements of food etc. in a time of scarcity. **3** *tr.* store in the mind. □ **hoarder** *n.* [OE *hord* f. Gmc]

hoarding /ˈhɔːdɪŋ/ n. **1** Brit. a large, usu. wooden, structure used to carry advertisements etc. **2** a board fence erected round a building site etc., often used for displaying posters etc. [obs. hoard f. AF h(o)urdis f. OF hourd, hort, rel. to HURDLE]

hoarhound var. of HOREHOUND.

hoarse /hɔːs/ adj. **1** (of the voice) rough and deep; husky; croaking. **2** having such a voice. □ **hoarsely** adv. **hoarsen** v.tr. & intr. **hoarseness** n. [ME f. ON hārs (unrecorded) f. Gmc]

hoarstone /ˈhɔːstəʊn/ n. Brit. an ancient boundary stone.

hoary /ˈhɔːrɪ/ adj. (**hoarier, hoariest**) **1 a** (of hair) grey or white with age. **b** having such hair; aged. **2** old and trite (a hoary joke). **3** Bot. & Zool. covered with short white hairs. □ **hoarily** adv. **hoariness** n.

hoatzin /hwætˈsiːn, həʊˈætsɪn/ n. a tropical South American bird, Opisthocomus hoatzin, whose young climb by means of hooked claws on their wings. [Amer. Sp. f. Nahuatl uatzin, imit.]

hoax /həʊks/ n. & v. ● n. a humorous or malicious deception; a practical joke. ● v.tr. deceive (a person) with a hoax. □ **hoaxer** n. [18th c.: prob. contr. f. HOCUS]

hob[1] /hɒb/ n. **1 a** a cooking appliance or the flat top part of a cooker, with hotplates or burners. **b** a flat metal shelf at the side of a fireplace, having its surface level with the top of the grate, used esp. for heating a pan etc. **2** a tool used for cutting gears etc. **3** a peg or pin used as a mark in quoits etc. **4** = HOBNAIL. [perh. var. of HUB, orig. = lump]

hob[2] /hɒb/ n. **1** a male ferret. **2** a hobgoblin. □ **play** (or **raise**) **hob** US cause mischief. [ME, familiar form of Rob, short for Robin or Robert]

Hobart /ˈhəʊbɑːt/ the capital and chief port of Tasmania; pop. (1991) 127,130. A penal colony named after Lord Hobart (1760–1816), Secretary of State for the Colonies, was moved to the city's present site in 1804; it became the island's capital in 1812.

Hobbema /ˈhɒbɪmə/, Meindert (1638–1709), Dutch landscape painter. He was a pupil of Jacob van Ruisdael and was one of the last 17th-century Dutch landscape painters, since demand for such work was diminishing in the late 1660s. His work features a narrow range of favourite subject-matter, often including a water-mill and trees round a pool. Among his best-known paintings is Avenue at Middelharnis (1689).

Hobbes /hɒbz/, Thomas (1588–1679), English philosopher. There were two key components in Hobbes's conception of humankind: he was a materialist, claiming that there was no more to the mind than the physical motions discovered by science, and a cynic, holding that human action was motivated entirely by selfish concerns, notably fear of death. His view of society was expressed in his most famous work, Leviathan (1651), in which he argued, by means of a version of a social contract theory, that simple rationality made social institutions and even absolute monarchy inevitable.

hobbit /ˈhɒbɪt/ n. a member of an imaginary race similar to humans, characterized by their small size and hairy feet, in stories by J. R. R. Tolkien. □ **hobbitry** n. [invented by Tolkien and said by him to mean 'hole-dweller']

hobble /ˈhɒb(ə)l/ v. & n. ● v. **1** intr. **a** walk lamely; limp. **b** proceed haltingly in action or speech (hobbled lamely to his conclusion). **2** tr. **a** tie together the legs of (a horse etc.) to prevent it from straying. **b** tie (the legs of a horse etc.). **3** tr. cause (a person etc.) to limp. ● n. **1** an uneven or infirm gait. **2** a rope, clog, etc. used for hobbling a horse etc. □ **hobble skirt** a skirt so narrow at the hem as to impede walking. □ **hobbler** n. [ME, prob. f. LG: cf. HOPPLE and Du. hobbelen rock from side to side]

hobbledehoy /ˈhɒb(ə)ldɪˌhɔɪ/ n. colloq. **1** a clumsy or awkward youth. **2** a hooligan. [16th c.: orig. unkn.]

Hobbs /hɒbz/, Sir John Berry ('Jack') (1882–1963), English cricketer. His career as a batsman in first-class cricket extended from 1905 to 1934, during which time he scored a total of 61,237 runs and 197 centuries. He first played for England in 1907 and went on to make 61 test appearances.

hobby[1] /ˈhɒbɪ/ n. (pl. **-ies**) **1** a favourite leisure-time activity or occupation. **2** archaic a small horse. **3** hist. an early type of velocipede. □ **hobbyist** n. [ME hobyn, hoby, f. pet-forms of Robin: cf. DOBBIN]

hobby[2] /ˈhɒbɪ/ n. (pl. **-ies**) a small long-winged falcon, esp. Falco subbuteo, that catches small birds, dragonflies, etc., on the wing. [ME f. OF hobé, hobet dimin. of hobe small bird of prey]

hobby-horse /ˈhɒbɪˌhɔːs/ n. **1** a child's toy consisting of a stick with a horse's head. **2** a preoccupation; a favourite topic of conversation. **3** a model of a horse, esp. of wicker, used in morris dancing etc. **4** a rocking horse. **5** a horse on a merry-go-round.

hobday /ˈhɒbdeɪ/ v.tr. operate on (a horse) to improve its breathing. [F. T. Hobday, Brit. veterinary surgeon (1869–1939)]

hobgoblin /ˈhɒbˌɡɒblɪn/ n. a mischievous imp; a bogey; a bugbear. [HOB[2] + GOBLIN]

hobnail /ˈhɒbneɪl/ n. a heavy-headed nail used for boot-soles. □ **hobnail** (or **hobnailed**) **liver** a liver having many small knobbly projections due to cirrhosis. □ **hobnailed** adj. [HOB[1] + NAIL]

hobnob /ˈhɒbnɒb/ v.intr. (**hobnobbed, hobnobbing**) **1** (usu. foll. by with) mix socially or informally. **2** drink together. [hob or nob = give or take, of alternate drinking; earlier hab nab, = have or not have]

hobo /ˈhəʊbəʊ/ n. (pl. **-oes** or **-os**) N. Amer. a wandering worker; a tramp. [19th c.: orig. unkn.]

Hobson's choice /ˈhɒbs(ə)nz/ n. a choice of taking the thing offered or nothing. [Thomas Hobson, Cambridge carrier (1554–1631), who let out horses on the basis that customers must take the one nearest the door]

Ho Chi Minh /ˌhəʊ tʃiː ˈmɪn/ (born Nguyen That Thanh) (1890–1969), Vietnamese Communist statesman, President of North Vietnam 1954–69. He was a committed nationalist who was instrumental in gaining his country's independence from French rule. He founded the Indo-Chinese Communist Party in 1930, and led the Vietminh in guerrilla warfare against the Japanese during the Second World War. He then fought the French for eight years until their defeat in 1954, when Vietnam was divided into North Vietnam, of which he became President, and South Vietnam. Committed to the creation of a united Communist country, Ho Chi Minh then deployed his forces in the guerrilla struggle that became the Vietnam War.

Ho Chi Minh City the official name (since 1975) for SAIGON.

hock[1] /hɒk/ n. **1** the joint of a quadruped's hind leg between the knee and the fetlock. **2** a knuckle of pork; the lower joint of a ham. [obs. hockshin f. OE hōhsinu: see HOUGH]

hock[2] /hɒk/ n. Brit. a German white wine from the Rhineland (properly that of Hochheim on the River Main). [abbr. of obs. hockamore f. G Hochheimer]

hock[3] /hɒk/ v. & n. esp. N. Amer. colloq. ● v.tr. pawn; pledge. ● n. a pawnbroker's pledge. □ **in hock 1** in pawn. **2** in debt. **3** in prison. [Du. hok hutch, prison, debt]

hockey[1] /ˈhɒkɪ/ n. **1** a team game played between two teams of eleven players each, using hooked sticks with which the players try to drive a small ball towards goals at opposite ends of the field. (See note below.) **2** N. Amer. ice hockey. □ **hockeyist** n. (in sense 2). [16th c.: orig. unkn.; the name probably belonged originally to the hooked stick]

▪ Games similar to hockey were played in ancient Greece, Persia, and Arabia, while the related games hurling and shinty have long been established in Ireland and Scotland. Hockey in its modern form, however, dates from the late 19th century in England, when rules were codified and men's and women's hockey associations set up. Play begins with the bully or bully off and the ball is propelled using the flat side of the stick. The pitch is similar to that used in football, but the goals are smaller, and scoring shots must be struck from within a semicircular line painted around the goal. The game is played particularly in the UK, Commonwealth countries, and Europe; in North America it is known as field hockey.

hockey[2] var. of OCHE.

Hockney /ˈhɒknɪ/, David (b.1937), English painter and draughtsman. His work of the early 1960s was associated with the pop art movement; deliberately naive and characterized by an ironic humour, it reflected the influence of graffiti and children's art. While in California in the mid-1960s he produced perhaps his best-known work: a series of paintings, such as A Bigger Splash (1967), which depict flat, almost shadowless architecture, lawns, and swimming-pools.

Hocktide /ˈhɒktaɪd/ n. hist. the second Monday and Tuesday after Easter Day, on which in pre-Reformation times money was collected for church and parish purposes and various sports and amusements took place. The merrymaking at this season survived in some places until the 19th century. [ME: orig. unkn.]

hocus /ˈhəʊkəs/ v.tr. (**hocussed, hocussing** or **hocused, hocusing**) **1** take in; hoax. **2** stupefy (a person) with drugs. **3** drug (liquor). [obs. noun hocus = HOCUS-POCUS]

hocus-pocus /ˌhəʊkəsˈpəʊkəs/ n. & v. ● n. **1** deception; trickery. **2 a** a typical verbal formula used in conjuring. **b** language intended to mystify; mumbo-jumbo. **3** conjuring, sleight of hand. ● v. (**-pocussed, -pocussing** or **-pocused, -pocusing**) **1** intr. (often foll. by with) play tricks. **2** tr. play tricks on, deceive. [17th-c. sham L]

hod /hɒd/ n. **1** a V-shaped open trough on a pole used for carrying bricks, mortar, etc. **2** a portable receptacle for coal. [prob. = dial. *hot* f. OF *hotte* pannier, f. Gmc]

hodden /'hɒd(ə)n/ n. *Sc.* a coarse woollen cloth. □ **hodden grey** grey hodden; typical rustic clothing. [16th c.: orig. unkn.]

hoddie /'hɒdi/ n. *Austral.* a bricklayer's labourer; a hodman. [HOD + -IE]

Hodeida /həʊ'deɪdə/ (Arabic **Al-Hudayda** /ˌælhuː'deɪdə/) the chief port of Yemen, on the Red Sea; pop. (1986) 155,100.

Hodge /hɒdʒ/ n. *Brit.* a typical English agricultural labourer. [pet-form of the name *Roger*]

hodgepodge /'hɒdʒpɒdʒ/ n. = HOTCHPOTCH 1, 2. [ME, assim. to HODGE]

Hodgkin /'hɒdʒkɪn/, Dorothy (Crowfoot) (1910–94), British chemist. She worked mainly at Oxford, where she developed Sir Lawrence Bragg's X-ray diffraction technique for investigating the structure of crystals and applied it to complex organic compounds. Using this method Hodgkin determined the structures of penicillin (1945), vitamin B_{12} (1956), and (after many years work) the large insulin molecule (1969). She was awarded the Nobel Prize for chemistry in 1964.

Hodgkin's disease /'hɒdʒkɪnz/ n. a malignant disease of lymphatic tissues usu. characterized by enlargement of the lymph nodes. [Thomas *Hodgkin*, Engl. physician (1798–1866)]

hodiernal /ˌhɒdɪ'ɜːn(ə)l, ˌhəʊd-/ adj. *formal* of the present day. [L *hodiernus* f. *hodie* today]

hodman /'hɒdmæn/ n. (pl. **-men**) **1** a labourer who carries a hod. **2** a literary hack. **3** a person who works mechanically.

hodograph /'hɒdəˌgrɑːf/ n. a curve in which the radius vector represents the velocity of a moving particle. [Gk *hodos* way + -GRAPH]

hodometer var. of ODOMETER.

Hoe /həʊ/, Richard March (1812–86), American inventor and industrialist. In 1846 he became the first printer to develop a successful rotary press. This greatly increased the speed of printing compared with the use of a flat plate; by 1857 *The Times* had a Hoe press printing 20,000 impressions an hour. This machine had still to be fed with individual cut sheets, but by 1871 Hoe had developed a machine fed from a continuous roll.

hoe /həʊ/ n. & v. ● n. a long-handled tool with a thin metal blade, used for weeding etc. ● v. (**hoes, hoed, hoeing**) **1** tr. weed (crops); loosen (earth); dig up or cut down with a hoe. **2** intr. use a hoe. □ **hoe-cake** *US* a coarse cake of maize flour orig. baked on the blade of a hoe. **hoe in** *Austral.* & *NZ sl.* eat eagerly. **hoe into** *Austral.* & *NZ sl.* attack (food, a person, a task). □ **hoer** n. [ME *howe* f. OF *houe* f. Gmc]

hoedown /'həʊdaʊn/ n. *N. Amer.* **1 a** a lively folk dance. **b** the music for this. **2** a party at which such dancing takes place.

Hoek van Holland see HOOK OF HOLLAND.

Hofei see HEFEI.

Hoffman /'hɒfmən/, Dustin (Lee) (b.1937), American actor. His first major film was *The Graduate* (1967); he has since appeared in a wide variety of roles, including that of a man pretending to be a woman in the comedy *Tootsie* (1983). Other films include *Midnight Cowboy* (1969) and *Rain Man* (1989), for which he received his second Oscar.

Hoffmann /'hɒfmən/, E(rnst) T(heodor) A(madeus) (1776–1822), German novelist, short-story writer, and music critic. He is best known for his extravagantly fantastic stories; his shorter tales appear in collections such as *Phantasiestücke* (1814–15), while longer works include *Elixire des Teufels* (1815–16). His stories provided the inspiration for Offenbach's opera *Tales of Hoffmann* (1881).

Hofmannsthal /'hɒfmənzˌtɑːl/, Hugo von (1874–1929), Austrian poet and dramatist. He wrote the libretti for Richard Strauss's operas *Elektra* (1909), *Der Rosenkavalier* (1911), *Ariadne auf Naxos* (1912), and *Arabella* (1933). With Strauss and Max Reinhardt he helped found the Salzburg Festival. His *Jedermann* (1912), a modernized form of a morality play, was first performed at the opening of the festival in 1920.

hog /hɒg/ n. & v. ● n. **1 a** a domesticated pig, esp. a castrated male reared for slaughter. **b** any pig of the family Suidae (*warthog*). **2** *colloq.* a greedy person. **3** (also **hogg**) *Brit. dial.* a young sheep before the first shearing. ● v. (**hogged, hogging**) **1** tr. *colloq.* take greedily; hoard selfishly; monopolize. **2** tr. & intr. raise (the back), or rise in an arch in the centre. □ **go the whole hog** *colloq.* do something completely or thoroughly. **hog-tie** *N. Amer.* **1** secure by fastening the hands and feet or all four feet together. **2** restrain, impede. □ **hogger** n. **hoggery** n. **hoglike** adj. [OE *hogg, hocg*, perh. of Celt. orig.]

hogan /'həʊgən/ n. a North American Indian hut of logs etc. [Navajo]

Hogarth /'həʊgɑːθ/, William (1697–1764), English painter and engraver. He contributed to the development of an English school of painting, both by his criticism of the contemporary taste for foreign artists and by encouraging the establishment of art institutions in England, a process which later culminated in the foundation of the Royal Academy (1768). Notable works include his series of engravings on 'modern moral subjects', such as *A Rake's Progress* (1735) and *Marriage à la Mode* (1743–5), which satirized the vices of both high and low life in 18th-century England.

hogback /'hɒgbæk/ n. (also **hog's back**) a steep-sided ridge of a hill.

Hogg /hɒg/, James (1770–1835), Scottish poet. He was a shepherd in the Ettrick Forest whose poetic talent was discovered by Sir Walter Scott; he gained the nickname 'the Ettrick Shepherd'. He made his reputation as a poet with *The Queen's Wake* (1813), but is better known today for his prose work *The Confessions of a Justified Sinner* (1824).

hogg var. of HOG n. 3.

Hoggar Mountains /'hɒgə(r)/ (also **Ahaggar Mountains** /ɑː'hɑːgɑː(r)/) a mountain range in the Saharan desert of southern Algeria, rising to a height of 2,918 m (9,573 ft) at Tahat.

hogget /'hɒgɪt/ n. *Brit.* a yearling sheep. [HOG]

hoggin /'hɒgɪn/ n. **1** a mixture of sand and gravel. **2** sifted gravel. [19th c.: orig. unkn.]

hoggish /'hɒgɪʃ/ adj. **1** of or like a hog. **2** filthy; gluttonous; selfish. □ **hoggishly** adv. **hoggishness** n.

hogmanay /'hɒgməˌneɪ/ n. *Sc.* **1** New Year's Eve. **2** a celebration on this day. **3** a gift of cake etc. demanded by children at hogmanay. [17th c.: perh. f. Norman F *hoguinané*. OF *aguillanneuf* (also = new year's gift)]

hog's back var. of HOGBACK.

hogshead /'hɒgzhed/ n. **1** a large cask. **2** a liquid or dry measure, usu. about 50 imperial gallons. [ME f. HOG, HEAD: reason for the name unkn.]

hogwash /'hɒgwɒʃ/ n. *colloq.* **1** nonsense, rubbish. **2** kitchen swill etc. for pigs.

hogweed /'hɒgwiːd/ n. a tall umbelliferous plant of the genus *Heracleum*; esp. *H. sphondylium*, that grows as a weed.

Hohenstaufen /'həʊən,ʃtaʊf(ə)n/ a German dynastic family, some of whom ruled as Holy Roman emperors between 1138 and 1254, among them Frederick I (Barbarossa).

Hohenzollern /'həʊən,zɒlən/ a German dynastic family from which came the kings of Prussia from 1701 to 1918 and German emperors from 1871 to 1918.

Hohhot /həʊ'hɒt/ (also **Huehot** /ˌhuː'heɪ'hɒt/) the capital of Inner Mongolia autonomous region, NE China; pop. (1990) 1,206,000. The original Mongol city, named Kukukhoto, was founded in the 16th century and was an important religious centre for Tibetan Buddhism. It was renamed Kwesui by the Chinese in the 18th century and Hohhot in 1954, when it became provincial capital.

ho-ho /həʊ'həʊ/ int. **1** representing deep jolly laughter. **2** expressing surprise, triumph, or derision. [redupl. of HO]

ho-hum /'həʊhʌm/ int. expressing boredom. [imit. of yawn]

hoick[1] /hɔɪk/ v. & n. *colloq.* ● v.tr. (often foll. by *out*) lift or pull, esp. with a jerk. ● n. a jerky pull; a jerk. [perh. var. of HIKE]

hoick[2] /hɔɪk/ v.intr. *sl.* spit. [perh. var. of HAWK[3]]

hoicks var. of YOICKS.

hoi polloi /ˌhɔɪ pə'lɔɪ/ n. (often prec. by *the*: see note below) **1** the masses; the common people. **2** the majority. ¶ Use with *the* is strictly unnecessary, since *hoi* = 'the', but this construction is very common. [Gk, = the many]

hoist /hɔɪst/ v. & n. ● v.tr. **1** raise or haul up. **2** raise by means of ropes and pulleys etc. ● n. **1** an act of hoisting, a lift. **2** an apparatus for hoisting. **3 a** the part of a flag nearest the staff. **b** a group of flags raised as a signal. □ **hoist the flag** stake one's claim to discovered territory by displaying a flag. **hoist one's flag** signify that one takes command. **hoist with one's own petard** see PETARD. □ **hoister** n. [16th c.: alt. of *hoise* f. (15th-c.) *hysse*, prob. of LG orig.: cf. LG *hissen*]

hoity-toity /ˌhɔɪtɪ'tɔɪtɪ/ adj., int., & n. ● adj. **1** haughty; petulant; snobbish. **2** *archaic* frolicsome. ● int. expressing surprised protest at presumption etc. ● n. *archaic* riotous or giddy conduct. [obs. *hoit* indulge in riotous mirth, of unkn. orig.]

hokey /'həʊkɪ/ adj. (also **hoky**) *N. Amer. colloq.* sentimental, melodramatic, artificial. □ **hokeyness** n. (also **hokiness**). [HOKUM + -Y²]

hokey-cokey /ˌhəʊkɪˈkəʊkɪ/ n. a communal dance performed in a circle with synchronized shaking of the limbs in turn. [perh. f. HOCUS-POCUS]

hokey-pokey /ˌhəʊkɪˈpəʊkɪ/ n. colloq. **1** = HOCUS-POCUS 1. **2** ice-cream formerly sold esp. by Italian street vendors. [HOCUS-POCUS: sense 2 of unkn. orig.]

hoki /ˈhəʊkɪ/ n. an edible marine fish, *Macruronus novaezelandiae*, related to the hake, native to the southern coasts of New Zealand. [Maori]

Hokkaido /hɒˈkaɪdəʊ/ the most northerly of the four main islands of Japan, constituting an administrative region; pop. (1990) 5,644,000; capital, Sapporo.

hokku /ˈhɒkuː/ n. (pl. same) = HAIKU. [Jap.]

hokum /ˈhəʊkəm/ n. esp. *US* colloq. **1** sentimental, popular, sensational, or unreal situations, dialogue, etc., in a film or play etc. **2** bunkum; rubbish. [20th c.: orig. unkn.]

Hokusai /ˈhəʊkʊˌsaɪ, ˌhəʊkʊˈsaɪ/, Katsushika (1760–1849), Japanese painter and wood-engraver. He was a leading artist of the ukiyo-e school who vividly represented many aspects of Japanese everyday life in his woodcuts. His best-known pictures are contained in the ten-volume *Mangwa* (1814–19) and the *Hundred Views of Mount Fuji* (1835). His prints were a significant stylistic influence on the work of impressionist and post-impressionist artists such as Van Gogh.

hoky var. of HOKEY.

Holarctic /həˈlɑːktɪk/ adj. & n. (also **holarctic**) *Zool.* ● adj. **1** of or relating to the Palaearctic and Nearctic regions considered together as a single zoogeographical region. **2** found throughout this region, circumpolar. ● n. the Holarctic region. [HOLO- + ARCTIC]

Holbein /ˈhɒlbaɪn/, Hans (known as Holbein the Younger) (1497–1543), German painter. He worked in Basle, where he produced the series of woodcuts the *Dance of Death* (c.1523–6), and in England, which he first visited in 1526. In England he became a well-known portraitist, depicting Sir Thomas Cromwell and other prominent courtiers, and painting group portraits such as *The Ambassadors* (1533). In 1536 he was appointed painter to Henry VIII; his commissions included portraits of the king's prospective brides, such as the miniatures *Christina, Duchess of Milan* (1538) and *Anne of Cleves* (1539).

hold¹ /həʊld/ v. & n. ● v. (past and past part. **held** /held/) **1** tr. **a** keep fast; grasp (esp. in the hands or arms). **b** (also refl.) keep or sustain (a thing, oneself, one's head, etc.) in a particular position (hold it to the light; held himself erect). **c** grasp so as to control (hold the reins). **2** tr. (of a vessel etc.) contain or be capable of containing (the jug holds two pints; the hall holds 900). **3** tr. possess, gain, or have, esp.: **a** be the owner or tenant of (land, property, stocks, etc.) (holds the farm from the trust). **b** gain or have gained (a degree, record, etc.) (holds the long-jump record). **c** have the position of (a job or office). **d** have (a specified playing card) in one's hand. **e** keep possession of (a place, a person's thoughts, etc.), esp. against attack (held the fort against the enemy; held his place in her estimation). **4** intr. remain unbroken; not give way (the roof held under the storm). **5** tr. observe; celebrate; conduct (a meeting, festival, conversation, etc.). **6** tr. **a** keep (a person etc.) in a specified condition, place, etc. (held him prisoner; held him at arm's length). **b** detain, esp. in custody (hold him until I arrive). **7** tr. **a** engross (a person or a person's attention) (the book held him for hours). **b** dominate (held the stage). **8** tr. (foll. by to) make (a person etc.) adhere to (terms, a promise, etc.). **9** intr. (of weather) continue fine. **10** tr. (often foll. by to + infin., or that + clause) think; believe (held it to be self-evident; held that the earth was flat). **11** tr. regard with a specified feeling (held him in contempt). **12** tr. **a** cease; restrain (hold your fire). **b** *US* colloq. withhold; not use (a burger please, and hold the onions!). **13** tr. keep or reserve (will you hold our seats please?). **14** tr. be able to drink (liquor) without effect (can't hold his drink). **15** tr. (usu. foll. by that + clause) (of a judge, a court, etc.) lay down; decide. **16** intr. keep going (held on his way). **17** tr. *Mus.* sustain (a note). **18** intr. = hold the line 2. **19** intr. archaic restrain oneself. ● n. **1** a grasp (catch hold of him; keep a hold on him). **2** (often in comb.) a thing to hold by (seized the handhold). **3** (foll. by on, over) influence over (has a strange hold over them). **4** a manner of holding in wrestling etc. **5** archaic a fortress. □ **hold (a thing) against (a person)** resent or regard it as discreditable to (a person). **hold aloof** avoid communication with people etc. **hold back 1** impede the progress of; restrain. **2** keep (a thing) to or for oneself. **3** (often foll. by from) hesitate; refrain. **hold-back** n. **1** something serving to hold a thing in place. **2** a hindrance. **hold one's breath** see BREATH. **hold by** (or **to**) adhere to (a choice, purpose, etc.). **hold cheap** not value highly;

despise. **hold the clock on** time (a sporting event etc.). **hold court** preside over one's admirers etc., like a sovereign. **hold dear** regard with affection. **hold down 1** repress. **2** be competent enough to keep (one's job etc.). **hold everything!** (or **it!**) cease action or movement. **hold the fort 1** act as a temporary substitute. **2** cope in an emergency. **hold forth 1** offer (an inducement etc.). **2** usu. derog. speak at length or tediously. **hold good** (or **true**) be valid; apply. **hold one's ground** see GROUND¹. **hold one's hand** see HAND. **hold a person's hand** give a person guidance or moral support. **hold hands** grasp one another by the hand as a sign of affection or for support or guidance. **hold hard!** stop!; wait! **hold harmless** *Law* indemnify. **hold one's head high** behave proudly and confidently. **hold one's horses** colloq. stop; slow down. **hold in** keep in check, confine. **holding pattern 1** *Aeron.* the (usu. circular) flight path maintained by an aircraft awaiting permission to land. **2** a state or period of no progress or change. **hold it good** think it advisable. **hold the line 1** not yield. **2** maintain a telephone connection. **hold one's nose** compress the nostrils to avoid a bad smell. **hold off 1** delay; not begin. **2** keep one's distance. **3** keep at a distance, fend off. **hold on 1** keep one's grasp on something. **2** wait a moment. **3** (when telephoning) not ring off. **hold out 1** stretch forth (a hand etc.). **2** offer (an inducement etc.). **3** maintain resistance. **4** persist or last. **hold out for** continue to demand. **hold out on** colloq. refuse something to (a person). **hold over 1** postpone. **2** retain. **hold-over** n. esp. *N. Amer.* a relic. **hold something over** threaten (a person) constantly with something. **hold one's own** see OWN. **hold to bail** *Law* bind by bail. **hold to a draw** *Sport* manage to achieve a draw against (an opponent thought likely to win). **hold together 1** cohere. **2** cause to cohere. **hold one's tongue** colloq. be silent. **hold to ransom 1** keep (a person) prisoner until a ransom is paid. **2** demand concessions from by threats of esp. damaging action. **hold up 1 a** support; sustain. **b** maintain (the head etc.) erect. **c** last; endure. **2** exhibit; display. **3** arrest the progress of; obstruct. **4** stop and rob by violence or threats. **hold-up** n. **1** a stoppage or delay by traffic, fog, etc. **2** a robbery, esp. by the use of threats or violence. **hold water** (of reasoning) be sound; bear examination. **hold with** (usu. with neg.) colloq. approve of (don't hold with motor bikes). **left holding the baby** left with unwelcome responsibility. **on hold 1** in abeyance; temporarily deferred. **2** (of a telephone call or caller) holding on (see hold on 3 above). **take hold** (of a custom or habit) become established. **there is no holding him** (or **her** etc.) he (or she etc.) is restive, high-spirited, determined, etc. **with no holds barred** with no restrictions, all methods being permitted. □ **holdable** adj. [OE h(e)aldan, heald]

hold² /həʊld/ n. a cavity in the lower part of a ship or aircraft in which the cargo is stowed. [obs. holl f. OE hol (orig. adj. = hollow), rel. to HOLE, assim. to HOLD¹]

holdall /ˈhəʊldɔːl/ n. a portable case or bag for miscellaneous articles.

holder /ˈhəʊldə(r)/ n. **1** (often in comb.) a device or implement for holding something (cigarette-holder). **2 a** the possessor of a title etc. **b** the occupant of an office etc. **3** = SMALLHOLDER.

Hölderlin /ˈhɜːldəˌlɪn/, (Johann Christian) Friedrich (1770–1843), German poet. His early poetry was full of political idealism fostered by the French Revolution, but most of his poems express a romantic yearning for ancient Greek harmony with nature and beauty. While working as a tutor he fell in love with his employer's wife, who is portrayed in his novel *Hyperion* (1797–9). Her death in 1802 exacerbated his already advanced schizophrenic condition.

holdfast /ˈhəʊldfɑːst/ n. **1** a firm grasp. **2** a staple or clamp securing an object to a wall etc. **3** the attachment-organ of an alga etc.

holding /ˈhəʊldɪŋ/ n. **1 a** land held by lease (cf. SMALLHOLDING). **b** the tenure of land. **2** stocks, property, etc., held. □ **holding company** a company created to hold the shares of other companies, which it then controls. **holding operation** a manoeuvre designed to maintain the status quo.

hole /həʊl/ n. & v. ● n. **1 a** an empty space in a solid body. **b** an aperture in or through something. **2** an animal's burrow. **3** a cavity or receptacle into which the ball must be got in various sports or games, esp. golf. **4 a** colloq. a small, mean, or dingy abode. **b** a dungeon, a prison cell. **5** colloq. an awkward situation. **6** *Golf* **a** a point scored by a player who gets the ball from tee to hole with the fewest strokes. **b** each of the divisions of a course consisting of the terrain or distance from tee to hole. **7** a position from which an electron is absent, esp. acting as a mobile positive particle in a semiconductor. ● v. **1** tr. make a hole or holes in. **2** tr. pierce the side of (a ship). **3** tr. put into a hole. **4 a** tr. *Golf* send (the ball) into a hole. **b** intr. (usu. foll. by out) *Golf* send the ball into a hole. **c** intr. (foll. by out) *Cricket* (of a batsman) be caught. □ **hole-and-**

corner secret; underhand. **hole in the heart** a congenital defect in the heart septum. **hole in one** *Golf* a shot that enters the hole from the tee. **hole in the wall 1** a small dingy place (esp. of business). **2** *colloq.* an automatic cash dispenser installed in the outside wall of a bank etc. **hole-proof** (of materials etc.) treated so as to be resistant to wear. **hole up** *N. Amer. colloq.* hide oneself. **in holes** worn so much that holes have formed. **make a hole in** use a large amount of. **a square peg in a round hole** see PEG. □ **holey** *adj.* [OE *hol, holian* (as HOLD²)]

Holi /ˈhəʊliː/ a Hindu spring festival celebrated in February or March in honour of Krishna. [Hindi f. Skr.]

holibut var. of HALIBUT.

Holiday /ˈhɒlɪˌdeɪ/, Billie (born Eleanora Fagan) (1915–59), American jazz singer. Following an early life of poverty and work in a brothel, she became a singer in the clubs of Harlem; in 1933 she began her recording career with Benny Goodman's band and went on to perform with many small jazz groups. Her style was characterized by dramatic intensity and vocal agility, as of a jazz musician playing a solo. Her autobiography *Lady Sings the Blues* (1956) was made into a film in 1972.

holiday /ˈhɒlɪˌdeɪ, -dɪ/ *n. & v.* ● *n.* **1** esp. *Brit.* (often in *pl.*) an extended period of recreation, esp. away from home or in travelling; a break from work (cf. VACATION). **2** a day of festivity or recreation when no work is done, esp. a religious festival etc. **3** (*attrib.*) (of clothes etc.) festive. ● *v.intr.* esp. *Brit.* spend a holiday. □ **holiday camp** *Brit.* a camp for holiday-makers with accommodation, entertainment, and facilities on site. **holiday centre** a place with many tourist attractions. **holiday-maker** esp. *Brit.* a person on holiday. **on holiday** (or **one's holidays**) in the course of one's holiday, having a break from work. **take a** (or archaic **make**) **holiday** have a break from work. [OE *hāligdæg* (HOLY, DAY)]

holily /ˈhəʊlɪlɪ/ *adv.* in a holy manner. [OE *hāliglīce* (as HOLY)]

holiness /ˈhəʊlɪnɪs/ *n.* **1** sanctity; the state of being holy. **2** (**Holiness**) a title used when referring to or addressing the pope. [OE *hālignes* (as HOLY)]

Holinshed /ˈhɒlɪnˌʃed/, Raphael (died *c.*1580), English chronicler. Although the named compiler of *The Chronicles of England, Scotland and Ireland* (1577), Holinshed in fact wrote only the *Historie of England* and had help with the remainder. In 1587 the work was revised and reissued, and this edition was widely used by Shakespeare and other dramatists.

holism /ˈhɒlɪz(ə)m, ˈhəʊl-/ *n.* (also **wholism**) **1** *Philos.* the theory that certain wholes are to be regarded as greater than the sum of their parts (cf. REDUCTIONISM). **2** *Med.* the treating of the whole person including mental and social factors rather than just the symptoms of a disease. □ **holist** *adj. & n.* **holistic** /hɒˈlɪstɪk/ *adj.* **holistically** *adv.* [as HOLO- + -ISM]

holla /ˈhɒlə/ *int., n., & v.* ● *int.* calling attention. ● *n.* a cry of 'holla'. ● *v.* (**hollas, hollaed** /-ləd/ or **holla'd, hollaing** /-ləɪŋ/) **1** *intr.* shout. **2** *tr.* call to (hounds). [F *holà* (as HO, *là* there)]

Holland /ˈhɒlənd/ **1** the Netherlands. **2** a former province of the Netherlands, comprising the coastal parts of the country. It is now divided into *North Holland* and *South Holland*. **3** one of three administrative divisions of Lincolnshire before the local government reorganization of 1974 (the others were Kesteven and Lindsey), comprising the fenland region in the south of the county, around the Wash. [Du., earlier *Holtlant* f. *Holt* wood + *-lant* land, describing Dordrecht district]

holland /ˈhɒlənd/ *n.* a smooth hard-wearing linen fabric. □ **brown holland** unbleached holland. [HOLLAND]

hollandaise sauce /ˌhɒlənˈdeɪz/ *n.* a creamy sauce of melted butter, egg-yolks, vinegar, etc., served esp. with fish. [F, fem. of *hollandais* Dutch f. *Hollande* Holland]

Hollander /ˈhɒləndə(r)/ *n.* **1** a native of Holland (the Netherlands). **2** a Dutch ship.

Hollands /ˈhɒləndz/ *n.* archaic Dutch gin, genever. [obs. Du. *hollandsch genever* Dutch gin]

holler /ˈhɒlə(r)/ *v. & n.* dial. or *N. Amer. colloq.* ● *v.* **1** *intr.* make a loud cry or noise. **2** *tr.* express with a loud cry or shout. ● *n.* a loud cry, noise, or shout. [var. of HOLLO]

Hollerith /ˈhɒlərɪθ/, Herman (1860–1929), American engineer. He invented a tabulating machine using punched cards for computation, an important precursor of the electronic computer. He founded a company in 1896 that later expanded to become the IBM Corporation.

hollo /ˈhɒləʊ/ *int., n., & v.* ● *int.* = HOLLA. ● *n.* (*pl.* **-os**) = HOLLA. ● *v.* (also **hollow** pronunc. same) (**-oes, -oed**) = HOLLA. [rel. to HOLLA]

hollow /ˈhɒləʊ/ *adj., n., v., & adv.* ● *adj.* **1 a** having a hole or cavity inside; not solid throughout. **b** having a depression; sunken (*hollow cheeks*). **2** (of a sound) echoing, as though made in or on a hollow container. **3** empty; hungry. **4** without significance; meaningless (*a hollow triumph*). **5** insincere; cynical; false (*a hollow laugh; hollow promises*). ● *n.* **1** a hollow place; a hole. **2** a valley; a basin. ● *v.tr.* (often foll. by *out*) make hollow; excavate. ● *adv. colloq.* completely (*beaten hollow*). □ **hollow-eyed** with eyes deep sunk. **hollow-hearted** insincere. **hollow square** *Mil. hist.* a body of infantry drawn up in a square with a space in the middle. **in the hollow of one's hand** entirely subservient to one. □ **hollowly** *adv.* **hollowness** *n.* [ME *holg, holu, hol(e)we* f. OE *holh* cave, rel. to HOLE]

hollowware /ˈhɒləʊˌweə(r)/ *n.* hollow articles of metal, china, etc., such as pots, kettles, jugs, etc. (opp. FLATWARE).

Holly /ˈhɒlɪ/, Buddy (born Charles Hardin Holley) (1936–59), American rock and roll singer, guitarist, and songwriter. Initially a hillbilly singer, Holly went on to become an important figure in early rock and roll, helping to shape rock guitar styling and being among the first to use a line-up of two guitars, bass, and drums. In 1955 Holly and some friends formed the band known as the Crickets, recording such hits as 'That'll be the Day' and 'Oh Boy' in 1957. The group toured Britain in 1958 (inspiring the name of the Beatles), before Holly left to go solo later the same year. He was killed in a plane crash.

holly /ˈhɒlɪ/ *n.* (*pl.* **-ies**) **1** an evergreen shrub, *Ilex aquifolium*, with prickly usu. dark green leaves, small white flowers, and red berries. **2** its branches and foliage used as decorations at Christmas. □ **holly oak** a holm-oak. [OE *hole(g)n*]

hollyhock /ˈhɒlɪˌhɒk/ *n.* a tall plant of the mallow family, *Alcea rosea*, with large showy flowers of various colours. [ME (orig. = marsh mallow) f. HOLY + obs. *hock* mallow, OE *hoc*, of unkn. orig.]

Hollywood /ˈhɒlɪˌwʊd/ a district of Los Angeles, the principal centre of the US film industry. The variety of the local landscape, the sunny weather, and the availability of cheap land go some way towards explaining the appeal of Hollywood for the early film industry. The first studio was established in Hollywood in 1911 and fifteen others followed in the same year; by 1920 Hollywood was producing 90 per cent of US films. Its heyday continued throughout the 1930s and 1940s, but the district experienced a relative decline after the 1940s; from the 1950s many films and programmes were made for television.

holm¹ /həʊm/ *n.* (also **holme**) *Brit.* **1** an islet, esp. in a river or near a mainland. **2** a piece of flat ground by a river, which is submerged in time of flood. [ON *holmr*]

holm² /həʊm/ *n.* (in full **holm-oak**) an evergreen oak, *Quercus ilex*, with holly-like young leaves. [ME, alt. of obs. *holin* (as HOLLY)]

Holmes¹ /həʊmz/, Arthur (1890–1965), English geologist and geophysicist. He pioneered the dating of rocks using isotopic decay, and he was the first to use it to provide absolute dates to the geological time-scale. He was one of the first supporters of the theory of continental drift. His book *Principles of Physical Geology* (1944) became a standard text.

Holmes² /həʊmz/, Oliver Wendell (1809–94), American physician, poet, and essayist. His main contribution to medicine was an essay (1843) on contagion as one cause of puerperal fever. While professor and dean of the medical school at Harvard (1847–82) he began to devote more time to his literary interests. His best-known works are the essays known as 'table talks', which began with *The Autocrat of the Breakfast Table* (1857–8); they take the form of humorous and erudite discourses by the author to listeners who interpose occasional remarks.

Holmes³ /həʊmz/, Sherlock, an extremely perceptive and intense private detective in stories by Sir Arthur Conan Doyle. The character was in part based on an eminent Edinburgh surgeon, Dr Joseph Bell, under whom Doyle studied medicine.

holmium /ˈhəʊlmɪəm/ *n.* a soft silvery-white metallic chemical element (atomic number 67; symbol **Ho**). A member of the lanthanide series, holmium was discovered by the Swedish chemist P. T. Cleve in 1878. It has few commercial uses. [mod.L *Holmia* Stockholm]

holo- /ˈhɒləʊ/ *comb. form* whole (*Holocene*; *holocaust*). [Gk *holos* whole]

holocaust /ˈhɒləˌkɔːst/ *n.* **1** a case of large-scale destruction or slaughter, esp. by fire or nuclear war. (See also HOLOCAUST, THE.) **2** a sacrifice wholly consumed by fire. [ME f. OF *holocauste* f. LL *holocaustum* f. Gk *holokauston* (as HOLO-, *kaustos* burnt f. *kaiō* burn)]

Holocaust, the the mass murder of Jews (and other persecuted groups such as gypsies and homosexuals) under the German Nazi regime. After the Nazis came to power in 1933, Jews were systematically deprived of civil rights, confined to ghettoes, and persecuted, but it was not until the introduction of Adolf Eichmann's 'final solution' in 1941 that the programme of extermination at concentration camps such as Auschwitz, Dachau, and Treblinka began. More than 6 million European Jews (about two-thirds of the total number) were murdered in the period 1941–5.

Holocene /ˈhɒləˌsiːn/ *adj. & n. Geol.* (also called *Recent*) of, relating to, or denoting the second epoch of the Quaternary period, following the Pleistocene and lasting from about 10,000 years ago to the present. The Holocene, which may merely represent a warmer interval between glaciations, coincides with the development of human agricultural settlement and civilization. [HOLO- + Gk *kainos* new]

holoenzyme /ˌhɒləʊˈenzaɪm/ *n. Biochem.* the active complex of an enzyme with a coenzyme.

Holofernes /ˌhɒləˈfɜːniːz, həˈlɒfəˌniːz/ (in the Bible) the Assyrian general of Nebuchadnezzar's forces who was killed by Judith (Judith 4:1 ff.).

hologram /ˈhɒləˌgræm/ *n. Physics* **1** a three-dimensional image formed by the interference of light beams from a coherent light source. (See also HOLOGRAPHY.) **2** a photograph of the interference pattern, which when suitably illuminated produces a three-dimensional image.

holograph /ˈhɒləˌgrɑːf/ *adj. & n.* ● *adj.* wholly written by hand by the person named as the author. ● *n.* a holograph document. [F *holographe* or LL *holographus* f. Gk *holographos* (as HOLO-, -GRAPH)]

holography /həˈlɒgrəfɪ/ *n. Physics* the study or production of holograms. Holography was proposed by Dennis Gabor in 1948, but only became practicable when the laser was invented in 1960. To make a hologram, laser light is split into two beams, of which one is reflected from an object. The two beams are then recombined. The reflected light is no longer coherent, and as a result the two beams interfere to produce a pattern of dark and bright areas which can be photographically recorded as a hologram. A three-dimensional image of the object can be constructed from the interference pattern by illuminating it with a laser beam. Some holograms, like those placed on credit cards to prevent fraud, can be viewed with white light, and holograms can also be made with coherent radiation other than visible light. □ **holographic** /ˌhɒləˈgræfɪk/ *adj.* **holographically** *adv.*

holohedral /ˌhɒləˈhiːdrəl/ *adj. Crystallog.* having the full number of planes required by the symmetry of a crystal system.

holophyte /ˈhɒləˌfaɪt/ *n.* an organism that synthesizes complex organic compounds by photosynthesis. □ **holophytic** /ˌhɒləˈfɪtɪk/ *adj.*

holothurian /ˌhɒləˈθjʊərɪən/ *n. & adj. Zool.* ● *n.* an echinoderm of the class Holothuroidea, with a wormlike body, e.g. a sea cucumber. ● *adj.* of or relating to this class. [mod.L *Holothuria* (n.pl.) f. Gk *holothourion*, a zoophyte]

holotype /ˈhɒləˌtaɪp/ *n.* the specimen used for naming and describing a species.

hols /hɒlz/ *n.pl. Brit. colloq.* holidays. [abbr.]

Holst /həʊlst/, Gustav (Theodore) (1874–1934), English composer, of Swedish and Russian descent. He made his reputation with the orchestral suite *The Planets* (1914–16), which was an instant success when first performed in 1919. He took inspiration for his music from a range of sources; the *St Paul's Suite* for strings (1913) reflects his interest in English folk-song, while his enthusiasm for Sanskrit literature resulted in works such as the four sets of *Choral Hymns from the Rig Veda* (1908–12).

Holstein[1] /ˈhɒlstaɪn/ a former duchy of the German kingdom of Saxony, situated in the southern part of the Jutland peninsula. It became a duchy of Denmark in 1474. Taken by Prussia in 1866, it was incorporated with the neighbouring duchy of Schleswig as the province of Schleswig-Holstein.

Holstein[2] /ˈhɒlstiːn, -staɪn/ *n. & adj. N. Amer.* = FRIESIAN. [HOLSTEIN[1]]

holster /ˈhəʊlstə(r)/ *n.* a leather case for a pistol or revolver, worn on a belt or under an arm or fixed to a saddle. [17th c., synonymous with Du. *holster*: orig. unkn.]

holt[1] /həʊlt/ *n.* **1** an animal's (esp. an otter's) lair. **2** *colloq.* or *dial.* grip, hold. [var. of HOLD[1]]

holt[2] /həʊlt/ *n. archaic* or *dial.* **1** a wood or copse. **2** a wooded hill. [OE f. Gmc]

holus-bolus /ˌhəʊləsˈbəʊləs/ *adv.* all in a lump, altogether. [app. sham L]

holy /ˈhəʊlɪ/ *adj.* (**holier, holiest**) **1** morally and spiritually excellent or perfect, and to be revered. **2** belonging to, devoted to, or empowered by God. **3** consecrated, sacred. **4** used in trivial exclamations (*holy cow!*; *holy mackerel!*; *holy Moses!*; *holy smoke!*). □ **holier-than-thou** *colloq.* self-righteous. **holy day** a religious festival. **holy Joe** *orig. Naut. sl.* **1** a clergyman. **2** a pious person. **holy land** a region sacred to the followers of a religion (see also HOLY LAND). **holy of holies 1** the inner chamber of the sanctuary in the Temple in Jerusalem, separated by a veil from the outer chamber. **2** an innermost shrine. **3** a thing regarded as most sacred. **holy orders** see ORDER. **holy place 1** (in *pl.*) places to which religious pilgrimage is made. **2** the outer chamber of the sanctuary in the Temple in Jerusalem. **holy roller** *sl.* a member of a religious group characterized by frenzied excitement or trances. **holy scripture** the Bible. **holy terror** see TERROR 2b. **holy war** a war waged in support of a religious cause. **holy water** water dedicated to holy uses, or blessed by a priest. **holy writ** holy writings collectively, esp. the Bible. [OE *hālig* f. Gmc, rel. to WHOLE]

Holy Alliance a loose alliance of European powers pledged to uphold the principles of the Christian religion. It was proclaimed at the Congress of Vienna (1814–15) by the emperors of Austria and Russia and the king of Prussia and was joined by most other European monarchs. Its influence was not great, but it is often seen as a symbol of conservatism and autocratic principles in Europe.

Holy Ark see ARK 2a.

Holy City *n.* **1** a city held sacred by the adherents of a religion, esp. Jerusalem. **2** heaven.

Holy Communion see COMMUNION 3.

Holy Cross Day the festival of the Exaltation of the Cross, 14 Sept.

Holy Family the young Jesus with his mother and St Joseph (often with St John the Baptist, St Anne, etc.) as grouped in pictures etc.

Holy Father (often as a form of address) the pope.

Holy Ghost see HOLY SPIRIT.

Holy Grail see GRAIL.

Holyhead /ˈhɒlɪˌhed/ a port on Holy Island, off Anglesey, in the Welsh county of Gwynedd; pop. (1981) 12,652. It is the chief port for ferries between the British mainland and Ireland.

Holy Island 1 an alternative name for LINDISFARNE. **2** a small island off the western coast of Anglesey in North Wales. It contains the ferry port of Holyhead.

Holy Land a region on the eastern shores of the Mediterranean, in what is now Israel and Palestine. It has religious significance for Judaism, Christianity, and Islam. In the Christian religion, the name has been applied since the Middle Ages with reference both to its having been the scene of the Incarnation and also to the existing sacred sites there, especially the Holy Sepulchre at Jerusalem.

Holy League any of various European alliances sponsored by the papacy during the 15th, 16th, and 17th centuries. They include the League of 1511–13, formed by Pope Julius II to expel Louis XII of France from Italy; the French Holy League (also called the *Catholic League*) of 1576 and 1584, a Catholic extremist league formed during the French Wars of Religion; and the Holy (or Catholic) League of 1609, a military alliance of the German Catholic princes.

Holy Name *RC Ch.* the name of Jesus as an object of formal devotion.

Holyoake /ˈhəʊlɪˌəʊk/, Sir Keith (Jacka) (1904–83), New Zealand statesman, Prime Minister 1957 and 1960–72. One of New Zealand's longest-serving statesmen, Holyoake first entered politics in 1932 as the National Party as the youngest member in the House of Representatives; after two terms as Prime Minister he went on to serve as Governor-General 1977–80.

Holy Office an ecclesiastical court of the Roman Catholic Church established in 1542 as part of the Inquisition, as the final court of appeal in trials of heresy. In 1965 it was renamed the Sacred Congregation for the Doctrine of the Faith; its function is to promote as well as to safeguard what it regards as sound doctrine in the Roman Catholic Church.

Holy Roman Empire the empire set up in western Europe following the coronation of Charlemagne as emperor in the year 800, which lasted until the title was abolished in 1806. The office was created by the medieval papacy in an attempt to provide a secular deputy to rule Christendom; in fact the emperor proved to be more often a rival than an ally. The emperor never ruled the whole of Christendom and from Otto I's coronation (962) the title was largely the preserve of German dynasties, although at times the territory of the empire was extensive and included Germany, Austria, Switzerland, and parts of Italy and the Netherlands. The ideal behind the empire of a united Christendom was shaken by the Reformation, and after the Thirty Years War (1618–48) the institution was again seriously weakened.

Holy Rood Day 1 the festival of the Invention of the Cross, 3 May. **2** = Holy Cross Day.

Holy Sacrament see SACRAMENT 3.

Holy Saturday the Saturday of Holy Week.

Holy See (also called *See of Rome*) the papacy or papal court; those associated with the pope in the government of the Roman Catholic Church at the Vatican.

Holy Sepulchre the tomb in which Christ was laid.

Holy Spirit (also called *Holy Ghost*) (in Christian theology) the third person of the Trinity. Regarded as God spiritually active in the world, the Holy Spirit is held to have guided the ministry and mission of the Church since its descent on the Apostles at Pentecost (Acts 2). It is associated also with the gifts of prophecy and of speaking in tongues, especially by Pentecostalists, and is usually symbolized by a white dove.

holystone /ˈhəʊliˌstəʊn/ *n. & v. Naut.* ● *n.* a piece of soft sandstone used for scouring decks. ● *v.tr.* scour with this. [19th c.: prob. f. HOLY + STONE: the stones were called *bibles* etc., perh. because used while kneeling]

Holy Thursday 1 (in the Anglican Church) Ascension Day. **2** *RC Ch.* Maundy Thursday.

Holy Trinity see TRINITY 3.

Holy Week the week before Easter.

Holy Year *n. RC Ch.* a period of remission from the penal consequences of sin, granted under certain conditions for a year usu. at intervals of twenty-five years.

hom /həʊm/ *n.* (also **homa** /ˈhəʊmə/) **1** the soma plant. **2** the juice of this plant as a sacred drink of the Parsees. [Pers. *hōm, hūm,* Avestan *haoma*]

homage /ˈhɒmɪdʒ/ *n.* **1** acknowledgement of superiority; respect, dutiful reverence (*pay homage to; do homage to*). **2** *hist.* formal public acknowledgement of feudal allegiance. [ME f. OF (*h*)*omage* f. med.L *hominaticum* f. L *homo -minis* man]

hombre /ˈɒmbreɪ/ *n. sl.* a man. [Sp.]

Homburg /ˈhɒmbɜːg/ *n.* a man's felt hat with a narrow curled brim and a lengthwise dent in the crown. [*Homburg* in Germany, where first worn]

home /həʊm/ *n., adj., adv., & v.* ● *n.* **1 a** the place where one lives; the fixed residence of a family or household. **b** a dwelling-house. **2** the members of a family collectively; one's family background (*comes from a good home*). **3** the native land of a person or of a person's ancestors. **4** an institution for persons needing care, rest, or refuge (*nursing home*). **5** the place where a thing originates or is native or most common. **6 a** the finishing-point in a race. **b** (in games) the place where one is free from attack; the goal. **c** *Lacrosse* a player in an attacking position near the opponents' goal. **7** *Sport* a home match or win. ● *attrib.adj.* **1 a** of or connected with one's home. **b** carried on, done, or made, at home. **2 a** proceeding from home. **2 a** carried on or produced in one's own country (*home industries; the home market*). **b** dealing with the domestic affairs of a country (cf. INTERIOR *adj.* 3; opp. FOREIGN 2). **3** *Sport* played on one's own ground etc. (*home match; home win*). **4** in the neighbourhood of home. ● *adv.* **1 a** to one's home or country (*go home*). **b** arrived at home (*is he home yet?*). **c** *N. Amer.* at home (*stay home*). **2 a** to the point aimed at (*the thrust went home*). **b** as far as possible (*drove the nail home; pressed his advantage home*). ● *v.* **1** *intr.* (esp. of a trained pigeon) return home (cf. HOMING 1). **2** *intr.* (often foll. by *on, in on*) (of a vessel, missile, etc.) be guided towards a destination or target by a landmark, radio beam, etc. **3** *tr.* send or guide homewards. **4** *tr.* provide with a home. □ **at home 1** in one's own house or native land. **2** at ease as if in one's own home (*make yourself at home*). **3** (usu. foll. by *in, on, with*) familiar or well-informed. **4** available to callers. **at-home** *n.* a social reception in a person's home. **come home to** become fully realized by. **come home to roost** see ROOST[1]. **home and dry** having achieved one's purpose. **home away from home** = *home from home*. **home-bird** *colloq.* a person who likes to stay at home. **home-brew** beer or other alcoholic drink brewed at home. **home-brewed** (of beer etc.) brewed at home. **home buyer** a person who buys a house, flat, etc. **home-coming** arrival at home. **home economics** the study of household management. **home farm** *Brit.* a farm (one of several on an estate) set aside to provide produce for the owner. **home from home** a place other than one's home where one feels at home; a place providing homelike amenities. **home-grown** grown or produced at home. **home help** *Brit.* a person employed to help in another's home,

esp. one provided by a local authority. **home, James!** *joc.* drive home quickly! **home loan** a loan advanced to a person to assist in buying a house, flat, etc. **home-made** made at home. **home-maker** a person, esp. a housewife, who creates a (pleasant) home. **home-making** the creation of a (pleasant) home. **home movie** a film made at home or of one's own activities. **home of lost causes** Oxford University. **home-owner** a person who owns his or her own home. **home perm** a permanent wave made with domestic equipment. **home plate** *Baseball* a plate beside which the batter stands. **home port** the port from which a ship originates. **home run** *Baseball* a hit that allows the batter to make a complete circuit of the bases. **home signal** a signal indicating whether a train may proceed into a station or to the next section of the line. **home straight** (*US* **stretch**) the concluding stretch of a racecourse. **home town** the town of one's birth or early life or present fixed residence. **home trade** trade carried on within a country. **home truth** basic but unwelcome information concerning oneself. **home unit** *Austral.* a private residence, usu. occupied by the owner, as one of several in a building. **near home** affecting one closely. □ **homelike** *adj.* [OE *hām* f. Gmc]

homebody /ˈhəʊmˌbɒdɪ/ *n.* (*pl.* **-ies**) a person who likes to stay at home.

homeboy /ˈhəʊmbɔɪ/ *n.* esp. *US black sl.* a person from one's own town or neighbourhood.

Home Counties the counties surrounding London, into which London has extended, comprising chiefly Essex, Kent, Surrey, and Hertfordshire.

Home Guard *n. hist.* **1** a British citizen army organized in 1940 to defend the UK against invasion, and disbanded in 1957. **2** a member of this.

Homel /ˈhɒmɪl/ (Russian **Gomel** /ˈgɒmɪlj/) an industrial city in SE Belarus; pop. (1990) 506,100.

homeland /ˈhəʊmlænd/ *n.* **1** one's native land. **2** *hist.* a partly autonomous area in South Africa reserved for a particular African people. (*See note below.*) **3** any similar semi-autonomous region.

■ In 1970, as part of the apartheid system, South Africa designated ten territories as African homelands. All people classified as black were made citizens of a homeland, regardless of where they were living. Subsequently four of the homelands — Bophuthatswana, Ciskei, Transkei, and Venda — were made republics and granted a degree of independence. Their independence was never internationally recognized and they were abolished after democratic elections in 1994.

homeless /ˈhəʊmlɪs/ *adj. & n.* ● *adj.* lacking a home. ● *n.* (prec. by *the*) homeless people. □ **homelessness** *n.*

homely /ˈhəʊmlɪ/ *adj.* (**homelier, homeliest**) **1 a** simple, plain. **b** unpretentious. **c** primitive. **2** comfortable in the manner of a home, cosy. **3** *N. Amer.* (of people or their features) not attractive in appearance, ugly. **4** skilled at housekeeping. □ **homeliness** *n.*

homeobox /ˈhəʊmɪəʊˌbɒks, ˈhɒm-/ *n.* (also **homoeobox**) *Biol.* any of a class of similar DNA sequences occurring in various genes, involved in regulating embryonic development in a wide range of species. [Gk *homoios* like + BOX[1]]

Home Office *n.* the British government department dealing with law and order, immigration, etc., in England and Wales.

Home of the Hirsel of Coldstream /ˈhɜːs(ə)l, ˈkəʊldstriːm/, Baron, see DOUGLAS-HOME.

homeopath /ˈhəʊmɪəʊˌpæθ, ˈhɒm-/ *n.* (also **homoeopath**) a person who practises homeopathy. [G *Homöopath* (as HOMEOPATHY)]

homeopathy /ˌhəʊmɪˈɒpəθɪ, ˌhɒm-/ *n.* (also **homoeopathy**) the treatment of disease by minute doses of drugs that in a healthy person would produce symptoms like those of the disease (cf. ALLOPATHY). A form of alternative medicine, homeopathy is based on the theory that 'like cures like'. This doctrine, known to the Greeks and Romans, was first given definite expression by Samuel Hahnemann (1755–1843), a German physician and chemist who had become dissatisfied with orthodox medical teaching. Homeopathy is widely practised, although both the effectiveness and the mechanism of any therapeutic action are uncertain. Hahnemann claimed that a homeopathic drug induced a condition which displaced the disease; others maintained that the drug stimulated the body's protective responses. □ **homeopathist** *n.* **homeopathic** /-mɪəˈpæθɪk/ *adj.* **homeopathically** *adv.* [G *Homöopathie* f. Gk *homoios* like + *patheia* -PATHY]

homeostasis /ˌhəʊmɪəʊˈsteɪsɪs, ˌhɒm-/ *n.* (also **homoeostasis**) (*pl.* **-stases** /-siːz/) esp. *Biol.* the tendency towards a relatively stable equilibrium between interdependent elements, esp. as maintained by physiological processes. □ **homeostatic** /-mɪəʊˈstætɪk/ adj. [mod.L f. Gk homoios like + -STASIS]

homeotherm /ˈhəʊmɪəʊˌθɜːm, ˈhɒm-/ n. (also **homoeotherm, homoiotherm**) Zool. an organism that maintains its body's internal temperature at a constant level, usu. above that of the environment, by its metabolic activity; a warm-blooded organism (cf. POIKILOTHERM). □ **homeothermy** n. **homeothermal** /ˌhəʊmɪəʊˈθɜːm(ə)l, ˌhɒm-/ adj. **homeothermic** adj. [mod.L f. Gk homoios like + thermē heat]

Homer /ˈhəʊmə(r)/ (8th century BC), Greek epic poet. He is traditionally held to be the author of the *Iliad* and the *Odyssey*. Various cities in Ionia claim to be his birthplace, and he is said to have been blind. Modern scholarship has revealed the place of the Homeric poems in a pre-literate oral tradition, in which a succession of bards elaborated the traditional stories of the heroic age; questions of authorship are thus very difficult to answer. In later antiquity Homer was regarded as the greatest and unsurpassable poet, and his poems were constantly used as a model and source by others.

homer /ˈhəʊmə(r)/ n. **1** a homing pigeon. **2** Baseball a home run.

Homeric /həʊˈmerɪk/ adj. **1** of, or in the style of, Homer or the epic poems ascribed to him. **2** of Bronze Age Greece as described in these poems. **3** epic, large-scale, titanic (*Homeric conflict*). [L Homericus f. Gk Homērikos f. Homēros HOMER]

home rule n. (also **Home Rule**) a movement for the government of a colony, dependent country, etc., by its own citizens, specifically the movement advocating devolved government for Ireland, 1870–1914. The campaign was one of the dominant forces in British politics in the late 19th and early 20th centuries, particularly in that Irish nationalists, under Charles Parnell and later John Redmond, frequently held the balance of power in the House of Commons. After 1885 the Liberal Party supported Home Rule, but the Conservatives, joined by the Liberal Unionists, continued to oppose it. The situation was complicated by the opposition to Home Rule of the Ulster Unionists on one side and Sinn Fein on the other. A Home Rule Act was finally passed in 1914 but was suspended until after the First World War. After the Easter Rising of 1916 and Sinn Fein's successes in the general election of 1918, and following three years of unrest and guerrilla warfare (1918–21), southern Ireland became the Irish Free State (1921), Northern Ireland remaining part of the UK.

Home Secretary n. (in the UK) the Secretary of State in charge of the Home Office.

home-shopping /həʊmˈʃɒpɪŋ/ n. shopping from home using satellite TV channels, catalogues, etc. □ **home-shopper** n.

homesick /ˈhəʊmsɪk/ adj. depressed by longing for one's home during absence from it. □ **homesickness** n.

homespun /ˈhəʊmspʌn/ adj. & n. ● adj. **1 a** (of cloth) made of yarn spun at home. **b** (of yarn) spun at home. **2** plain, simple, unsophisticated, homely. ● n. **1** homespun cloth. **2** anything plain or homely.

homestead /ˈhəʊmsted, -stɪd/ n. **1** a house, esp. a farmhouse, and outbuildings. **2** Austral. & NZ the owner's residence on a sheep or cattle station. **3** N. Amer. an area of land (usu. 160 acres) granted to a settler as a home. □ **homesteader** /-ˌstedə(r)/ n. **homesteading** n. [OE hāmstede (as HOME, STEAD)]

homestyle /ˈhəʊmstaɪl/ adj. N. Amer. (esp. of food) of a kind made or done at home, homely.

homeward /ˈhəʊmwəd/ adv. & adj. ● adv. (also **homewards** /-wədz/) towards home. ● adj. going or leading towards home. □ **homeward-bound** (esp. of a ship) preparing to go, or on the way, home. [OE hāmweard(es) (as HOME, -WARD)]

homework /ˈhəʊmwɜːk/ n. **1** work to be done at home, esp. by a school pupil. **2** preparatory work or study.

homeworker /ˈhəʊmˌwɜːkə(r)/ n. a person who works from home, esp. doing low-paid piecework.

homey /ˈhəʊmɪ/ adj. (also **homy**) (**homier, homiest**) suggesting home; cosy. □ **homeyness** n. (also **hominess**).

homicide /ˈhɒmɪˌsaɪd/ n. **1** the killing of a human being by another. **2** a person who kills a human being. □ **homicidal** /ˌhɒmɪˈsaɪd(ə)l/ adj. [ME f. OF f. L homicidium (sense 1), homicida (sense 2) (HOMO man)]

homiletic /ˌhɒmɪˈletɪk/ adj. & n. ● adj. of homilies. ● n. (usu. in pl.) the art of preaching. [LL homileticus f. Gk homilētikos f. homileō hold converse, consort (as HOMILY)]

homiliary /hɒˈmɪlɪərɪ/ n. (pl. **-ies**) a book of homilies. [med.L homeliarius (as HOMILY)]

homily /ˈhɒmɪlɪ/ n. (pl. **-ies**) **1** a sermon. **2** a tedious moralizing discourse. □ **homilist** n. [ME f. OF omelie f. eccl.L homilia f. Gk homilia f. homilos crowd]

homing /ˈhəʊmɪŋ/ attrib.adj. **1** (of a pigeon) trained to fly home, bred for long-distance racing. **2** (of a device) for guiding to a target etc. **3** that goes home. □ **homing instinct** the instinct of certain animals to return to the territory which they have left or from which they have been moved.

hominid /ˈhɒmɪnɪd/ n. & adj. ● n. a member of the primate family Hominidae, including humans and their fossil ancestors. The great apes are now also usually placed in this family. ● adj. of or relating to this family. [mod.L Hominidae f. L homo hominis man]

hominoid /ˈhɒmɪˌnɔɪd/ adj. & n. ● adj. **1** like a human. **2** hominid or pongid. ● n. an animal resembling a human.

hominy /ˈhɒmɪnɪ/ n. (also **hominy grits**) (esp. in the US) coarsely ground maize kernels boiled with water or milk. [Algonquian]

Homo /ˈhəʊməʊ, ˈhɒm-/ n. a genus of primates of which humans (*Homo sapiens*) are the modern representatives. The genus has existed for about 2 million years, of which *H. sapiens* has occupied perhaps the last 400,000 years. The genus appears to have evolved in East Africa with the advent of *H. habilis* ('handy man'), from Olduvai Gorge, Tanzania, and an as yet undifferentiated species from Koobi Fora, Kenya. It is believed that *Homo* diverged from *Australopithecus africanus*, being distinguished not only physically but also by social and cultural traits such as the making of tools and the use of fire. *H. erectus* ('upright man'), first appearing about 1.5 million years BP, is thought to have radiated out from East Africa to the more varied environments of the remainder of the Old World (see JAVA MAN, PEKING MAN). Archaic *H. sapiens* may have appeared as early as c.400,000 BP, at least in Europe, and survived as late as c.100,000 BP in Asia. Neanderthal man probably evolved in Europe c.120,000 BP. Modern humans, who first appeared in the upper palaeolithic (see CRO-MAGNON MAN), are classified as *H. sapiens sapiens*. The principal trends within the genus *Homo* through time are: an increase in brain size and complexity, an increase in the complexity of culture and social relationships, and the increasing ability to modify the environment. [L, = man]

homo /ˈhəʊməʊ/ n. (pl. **-os**) colloq. a homosexual. [abbr.]

homo- /ˈhəʊməʊ, ˈhɒm-/ comb. form same (often opp. HETERO-). [Gk homos same]

homocentric /ˌhəʊməʊˈsentrɪk, ˌhɒm-/ adj. having the same centre.

homoeobox var. of HOMEOBOX.

homoeopath var. of HOMEOPATH.

homoeopathy var. of HOMEOPATHY.

homoeostasis var. of HOMEOSTASIS.

homoeotherm var. of HOMEOTHERM.

homoerotic /ˌhəʊməʊɪˈrɒtɪk, ˌhɒm-/ adj. homosexual.

homogametic /ˌhəʊməʊgəˈmiːtɪk, ˌhɒm-/ adj. Biol. (of a sex or individuals of a sex) producing gametes that carry the same sex chromosome.

homogamy /həˈmɒgəmɪ/ n. Bot. **1** a state in which the flowers of a plant are hermaphrodite or of the same sex. **2** the simultaneous ripening of the stamens and pistils of a flower. □ **homogamous** adj. [Gk homogamos (as HOMO-, gamos marriage)]

homogenate /həˈmɒdʒɪnət, ˌneɪt/ n. a suspension produced by homogenizing.

homogeneous /ˌhəʊməˈdʒiːnɪəs, ˌhɒm-/ adj. (also disp. **homogenous** /həˈmɒdʒɪnəs/) **1** of the same kind. **2** consisting of parts all of the same kind; uniform. **3** Math. containing terms all of the same degree. □ **homogeneously** adv. **homogeneousness** n. **homogeneity** /-dʒɪˈniːɪtɪ/ n. [med.L homogeneus f. Gk homogenēs (as HOMO-, genēs f. genos kind)]

homogenize /həˈmɒdʒɪˌnaɪz/ v. (also **-ise**) **1** tr. & intr. make or become homogeneous. **2** tr. treat (milk) so that the fat droplets are emulsified and the cream does not separate. □ **homogenizer** n. **homogenization** /-ˌmɒdʒɪnaɪˈzeɪʃ(ə)n/ n.

homogeny /həˈmɒdʒɪnɪ/ n. Biol. similarity due to common descent. □ **homogenetic** /ˌhəʊmədʒɪˈnetɪk, ˌhɒm-/ adj.

homograft /ˈhɒməˌɡrɑːft/ n. = ALLOGRAFT.

homograph /ˈhɒməˌɡrɑːf/ n. a word spelt like another but of different meaning or origin (e.g. POLE[1], POLE[2]).

homoiotherm var. of HOMEOTHERM.

homoiousian /ˌhɒmɔɪˈuːsɪən, -ˈaʊsɪən/ n. hist. a person who held that God the Father and God the Son are of like but not identical substance (cf. HOMOOUSIAN). [eccl.L f. Gk *homoiousios* f. *homoios* like + *ousia* essence]

homolog *US* var. of HOMOLOGUE.

homologate /həˈmɒləˌgeɪt/ *v.tr.* **1** acknowledge, admit. **2** confirm, accept. **3** approve (a car, boat, engine, etc.) for sale in a particular market or use in a particular class of racing. □ **homologation** /-ˌmɒləˈgeɪʃ(ə)n/ *n.* [med.L *homologare* agree f. Gk *homologeō* (as HOMO-, *logos* word)]

homologize /həˈmɒləˌdʒaɪz/ *v.* (also **-ise**) **1** *intr.* be homologous; correspond. **2** *tr.* make homologous.

homologous /həˈmɒləgəs/ *adj.* **1 a** having the same relation, relative position, etc. **b** corresponding. **2** *Biol.* (of organs etc.) similar in position, structure, and evolutionary origin but not necessarily in function (cf. ANALOGOUS 2). **3** *Biol.* (of chromosomes) pairing at meiosis and having the same structural features and pattern of genes. **4** *Chem.* (of a series of chemical compounds) having the same functional group but differing in composition by a fixed group of atoms. [med.L *homologus* f. Gk (as HOMO-, *logos* ratio, proportion)]

homologue /ˈhɒməˌlɒg/ *n.* (*US* **homolog**) a homologous thing. [F f. Gk *homologon* (neut. adj.) (as HOMOLOGOUS)]

homology /həˈmɒlədʒɪ/ *n.* a homologous state or relation; correspondence. □ **homological** /ˌhɒməˈlɒdʒɪk(ə)l/ *adj.*

homomorphic /ˌhɒʊməʊˈmɔːfɪk, ˌhɒm-/ *adj.* (also **homomorphous** /-ˈmɔːfəs/) of the same or similar form. □ **homomorphically** *adv.* **homomorphism** *n.* **homomorphy** /ˈhɒʊməʊˌmɔːfɪ, ˈhɒm-/ *n.*

homonym /ˈhɒmənɪm/ *n.* **1** a word of the same spelling or sound as another but of different meaning; a homograph or homophone. **2** a namesake. □ **homonymic** /ˌhɒməˈnɪmɪk/ *adj.* **homonymous** /həˈmɒnɪməs/ *adj.* [L *homonymum* f. Gk *homōnumon* (neut. adj.) (as HOMO-, *onoma* name)]

homoousian /ˌhɒməʊˈuːsɪən, -ˈaʊsɪən/ *n.* (also **homousian** /hɒmˈuːsɪən, -ˈaʊsɪən/) *hist.* a person who held that God the Father and God the Son are of the same substance (cf. HOMOIOUSIAN). [eccl.L *homoousianus* f. LL *homousius* f. Gk *homoousios* (as HOMO-, *ousia* essence)]

homophobia /ˌhɒʊməˈfəʊbɪə, ˌhɒm-/ *n.* a hatred or fear of homosexuals. □ **homophobic** *adj.* **homophobe** /ˈhɒʊməˌfəʊb, ˈhɒm-/ *n.*

homophone /ˈhɒməˌfəʊn/ *n.* **1** a word having the same sound as another but of different meaning, origin, or spelling (e.g. *pair*, *pear*). **2** a symbol denoting the same sound as another.

homophonic /ˌhɒməˈfɒnɪk/ *adj. Mus.* in unison; characterized by movement of all parts to the same melody. □ **homophonically** *adv.*

homophonous /həˈmɒfənəs/ *adj.* **1** (of music) homophonic. **2** (of a word or symbol) that is a homophone. □ **homophony** *n.*

homopolar /ˌhɒʊməʊˈpəʊlə(r), ˌhɒm-/ *adj.* **1** electrically symmetrical. **2** *Electr.* (of a generator) producing direct current without the use of commutators. **3** *Chem.* (of a covalent bond) in which one atom supplies both electrons.

Homoptera /həˈmɒptərə/ *n.pl. Zool.* a suborder of hemipterous insects comprising aphids, cicadas, leafhoppers, water-boatmen, etc., which have wings of uniform texture (cf. HETEROPTERA). □ **homopteran** *n.* & *adj.* **homopterous** *adj.* [mod.L f. HOMO- + Gk *pteron* wing]

Homo sapiens /ˌhɒʊməʊ ˈsæpɪˌenz, ˌhɒm-/ *n.* the primate species to which humans belong; humans regarded as an animal species. (See HOMO.) [L, = wise man]

homosexual /ˌhɒʊməˈseksjʊəl, ˌhɒm-, -ˈsekʃʊəl/ *adj.* & *n.* ● *adj.* **1** feeling or involving sexual attraction only to persons of the same sex. **2** concerning homosexual relations or people. **3** relating to the same sex. ● *n.* a homosexual person. □ **homosexually** *adv.* **homosexuality** /-ˌseksjʊˈælɪtɪ, -ˌsekʃʊ-/ *n.*

homotransplant /ˌhɒʊməˈtrænsplɑːnt, ˌhɒm-, -ˈtrɑːnsplɑːnt/ *n.* = ALLOGRAFT.

homousian var. of HOMOOUSIAN.

homozygote /ˌhɒʊməʊˈzaɪgəʊt, ˌhɒm-/ *n. Biol.* an individual having two identical alleles of a particular gene or genes, and so breeding true for the corresponding characteristics (cf. HETEROZYGOTE). □ **homozygous** *adj.*

Homs /hɒms, hɒmz/ (also **Hims** /hɪms, hɪmz/) an industrial city in western Syria, on the River Orontes; pop. (1981) 354,500. It was named in 636 by the Muslims and occupies the site of ancient Emesa. It lies at the junction of north–south and east–west trade routes.

homunculus /həˈmʌŋkjʊləs/ *n.* (also **homuncule** /-kjuːl/) (*pl.* **homunculi** /-ˌlaɪ/ or **homuncules**) a little man, a manikin. It was formerly believed that the foetus develops from a microscopic but fully formed human being or homunculus in the spermatozoon or egg. [L *homunculus* f. *homo* -*minis* man]

homy var. of HOMEY.

Hon. *abbr.* **1** Honorary. **2** Honourable.

hon /hʌn/ *n. colloq.* = HONEY 5. [abbr.]

Honan /həˈnæn/ **1** see HENAN. **2** a former name for LUOYANG.

honcho /ˈhɒntʃəʊ/ *n.* & *v. N. Amer. sl.* ● *n.* (*pl.* **-os**) **1** a leader or manager, the person in charge. **2** an admirable man. ● *v.tr.* (**-oes**, **-oed**) be in charge of, oversee. [Jap. *han'chō* group leader]

Honda /ˈhɒndə/, Soichiro (1906–92), Japanese motor manufacturer. In 1928 he opened his own garage; his first factory, producing piston-rings, was established in 1934. He began motorcycle manufacture in 1948, becoming the world's largest producer. During the 1960s he successfully expanded his operations into car production; the Honda Corporation has since become involved in joint ventures with firms such as the British car manufacturer Rover.

Honduras /hɒnˈdjʊərəs/ a country of Central America, bordering on the Caribbean Sea and with a short coastline on the Pacific Ocean; pop. (est. 1990) 5,100,000; official language, Spanish; capital, Tegucigalpa. Honduras was at the southern limit of the Mayan empire. It was encountered by Columbus in 1502, and became a Spanish colony. In 1821 Honduras became an independent republic, and was part of the United Provinces of Central America between 1823 and 1838. (See also BRITISH HONDURAS.) □ **Honduran** *adj.* & *n.*

hone /həʊn/ *n.* & *v.* ● *n.* **1** a whetstone, esp. for razors. **2** any of various stones used as material for this. ● *v.tr.* sharpen on or as on a hone. [OE *hān* stone f. Gmc]

Honecker /ˈhɒnɪkə(r)/, Erich (1912–94), East German Communist statesman, head of state 1976–89. He was appointed First Secretary of the Socialist Unity Party in 1971, becoming effective leader of East Germany in 1973, and head of state (Chairman of the Council of State) three years later. His repressive regime was marked by a close allegiance to the Soviet Union. Honecker was ousted in 1989 after a series of pro-democracy demonstrations. In 1992 he was arrested but proceedings against him for manslaughter and embezzlement were later dropped because of his ill health.

Honegger /ˈhɒnɪgə(r)/, Arthur (1892–1955), French composer, of Swiss descent. He lived and worked chiefly in Paris, where he became a member of the anti-romantic group Les Six. His orchestral work *Pacific 231* (1924), a musical representation of a steam locomotive, brought him his first major success; his work also includes five symphonies (1930–51) and the dramatic oratorio *Joan of Arc at the Stake* (1935).

honest /ˈɒnɪst/ *adj.* & *adv.* ● *adj.* **1** fair and just in character or behaviour, not cheating or stealing. **2** free of deceit and untruthfulness, sincere. **3** fairly earned (*an honest living*). **4** (of an act or feeling) showing fairness. **5** (with patronizing effect) blameless but undistinguished (cf. WORTHY *adj.* 2). **6** (of a thing) unadulterated, unsophisticated. ● *adv. colloq.* genuinely, really. □ **earn** (or **turn**) **an honest penny** earn money fairly. **honest broker** a mediator in international, industrial, etc., disputes (orig. of Bismarck). **honest Injun** *colloq.* genuinely, really. **honest-to-God** (or **-goodness**) *colloq. adj.* genuine, real. ● *adv.* genuinely, really. **make an honest woman of** *colloq.* or *joc.* marry (esp. a pregnant woman). [ME f. OF (h)*oneste* f. L *honestus* f. *honos* HONOUR]

honestly /ˈɒnɪstlɪ/ *adv.* **1** in an honest way. **2** really (*I don't honestly know; honestly, the cheek of it!*).

honesty /ˈɒnɪstɪ/ *n.* **1** being honest. **2** truthfulness. **3** a plant of the genus *Lunaria*, with purple or white flowers, so called from its flat round semi-transparent seed-pods. [ME f. OF (h)*oneste* f. L *honestas* -*tatis* (as HONEST)]

honey /ˈhʌnɪ/ *n.* (*pl.* **-eys**) **1** a sweet sticky yellowish fluid made by bees and other insects from nectar collected from flowers. **2** the colour of this. **3 a** sweetness. **b** a sweet thing. **4** a person or thing excellent of its kind. **5** esp. *N. Amer.* (usu. as a form of address) darling, sweetheart. □ **honey-badger** a ratel. **honey bee** see BEE 1a. **honey-bun** (or **-bunch**) (esp. as a form of address) darling. **honey-buzzard** a bird of prey of the genus *Pernis*, feeding on the larvae of bees and wasps. **honey-fungus** a parasitic fungus, *Armillaria mellea*, with honey-coloured edible toadstools. **honey-guide 1** a small bird of the family Indicatoridae, feeding on beeswax and insects. **2** a marking on the corolla of a flower thought to guide bees to nectar. **honey-parrot** a lorikeet. **honey-pot 1** a pot for honey. **2** a posture with the hands clasped under the hams. **3** something very attractive or tempting. **honey sac** an enlarged part of a bee's gullet where honey is formed. **honey-sweet** sweet as honey. [OE *hunig* f. Gmc]

honeycomb /ˈhʌnɪˌkəʊm/ n. & v. ● n. **1** a structure of hexagonal cells of wax, made by bees to store honey and eggs. **2 a** a pattern arranged hexagonally. **b** fabric made with a pattern of raised hexagons etc. **3** tripe from the second stomach of a ruminant. **4** a cavernous flaw in metalwork, esp. in guns. ● v.tr. **1** fill with cavities or tunnels, undermine. **2** mark with a honeycomb pattern. [OE hunigcamb (as HONEY, COMB)]

honeycreeper /ˈhʌnɪˌkriːpə(r)/ n. **1** a Hawaiian bird of the family Drepanididae. **2** a tanager of tropical America, feeding on nectar.

honeydew /ˈhʌnɪˌdjuː/ n. **1** a sweet sticky substance found on leaves and stems, excreted by aphids. **2** a variety of melon with smooth pale skin and sweet green flesh. **3** an ideally sweet substance. **4** tobacco sweetened with molasses.

honeyeater /ˈhʌnɪˌiːtə(r)/ n. an Australasian bird of the family Meliphagidae, with a long tongue used for taking nectar from flowers.

honeyed /ˈhʌnɪd/ adj. (also **honied**) **1** of or containing honey. **2** (of words, flattery, etc.) sweet; sweet-sounding.

honeymoon /ˈhʌnɪˌmuːn/ n. & v. ● n. **1** a holiday spent together by a newly married couple. **2** an initial period of enthusiasm or goodwill. ● v.intr. (usu. foll. by in, at) spend a honeymoon. □ **honeymooner** n. [HONEY + MOON, orig. with ref. to waning affection, not to a period of a month]

honeysuckle /ˈhʌnɪˌsʌk(ə)l/ n. a climbing shrub of the genus Lonicera, with fragrant yellow and pink flowers. [ME hunisuccle, -soukel, extension of hunisuce, -souke, f. OE hunigsūce, -sūge (as HONEY, SUCK)]

Hong Kong /hɒŋ ˈkɒŋ/ a British dependency (until 1997) on the SE coast of China; pop. (est. 1990) 5,900,000; official languages, English and Cantonese; capital, Victoria. The colony comprises Hong Kong Island, ceded by China in 1841, the Kowloon peninsula, ceded in 1860, and the New Territories, additional areas of the mainland, leased for 99 years in 1898. Hong Kong has become one of the world's major financial and manufacturing centres, with the third largest container port in the world. An agreement signed between the British and Chinese governments in 1984 provided for China to resume sovereignty over the whole territory in 1997. Hong Kong would then become a Special Administrative Region, with basic laws guaranteeing present systems and lifestyles for a period of fifty years.

Honiara /ˌhəʊnɪˈɑːrə/ a port and the capital of the Solomon Islands, situated on the NW coast of the island of Guadalcanal; pop. (est. 1985) 26,000.

honied var. of HONEYED.

honk /hɒŋk/ n. & v. ● n. **1** the cry of a wild goose. **2** the harsh sound of a car horn. ● v. **1** intr. emit or give a honk. **2** tr. cause to do this. [imit.]

honky /ˈhɒŋkɪ/ n. (pl. **-ies**) US black sl. derog. **1** a white person. **2** white people collectively. [20th c.: orig. unkn.]

honky-tonk /ˈhɒŋkɪˌtɒŋk/ n. colloq. **1** ragtime piano music. **2** a cheap or disreputable nightclub, dancehall, etc. [20th c.: orig. unkn.]

honnête homme /ˌɒnet ˈɒm/ n. a decent, cultivated man of the world. [F]

Honolulu /ˌhɒnəˈluːluː/ the state capital and principal port of Hawaii, situated on the SE coast of the island of Oahu; pop. (1990) 836,230.

honor US var. of HONOUR.

honorable US var. of HONOURABLE.

honorand /ˈɒnəˌrænd/ n. a person to be honoured, esp. with an honorary degree. [L honorandus (as HONOUR)]

honorarium /ˌɒnəˈreərɪəm/ n. (pl. **honorariums** or **honoraria** /-rɪə/) a fee, esp. a voluntary payment for professional services rendered without the normal fee. [L, neut. of honorarius: see HONORARY]

honorary /ˈɒnərərɪ/ adj. **1 a** conferred as an honour, without the usual requirements, functions, etc. (honorary degree). **b** holding such a title or position (honorary colonel). **2** (of an office or its holder) unpaid (honorary secretaryship; honorary treasurer). **3** (of an obligation) depending on honour, not legally enforceable. [L honorarius (as HONOUR)]

honorific /ˌɒnəˈrɪfɪk/ adj. & n. ● adj. **1** conferring honour. **2** (esp. of oriental forms of speech) implying respect. ● n. an honorific form of words. □ **honorifically** adv. [L honorificus (as HONOUR)]

honoris causa /ɒˌnɔːrɪs ˈkaʊzə/ adv. (esp. of a degree awarded without examination) as a mark of esteem. [L, = for the sake of honour]

honour /ˈɒnə(r)/ n. & v. (US **honor**) ● n. **1** high respect; glory; credit, reputation, good name. **2** adherence to what is right or to a conventional standard of conduct. **3** nobleness of mind, magnanimity (honour among thieves). **4** a thing conferred as a distinction, esp. an official award for bravery or achievement. **5** (foll. by of + verbal noun, or to + infin.) privilege, special right (had the honour of being invited). **6 a** exalted position. **b** (**Honour**) (prec. by your, his, etc.) a title of a circuit judge, US a mayor, and Ir. or in rustic speech any person of rank. **7** (foll. by to) a person or thing that brings honour (she is an honour to her profession). **8 a** (of a woman) chastity. **b** the reputation for this. **9** (in pl.) **a** special distinction for proficiency in an examination. **b** a course of degree studies more specialized than for an ordinary pass. **10 a** Bridge the ace, king, queen, jack, and ten, esp. of trumps, or the four aces at no trumps. **b** Whist the ace, king, queen, and jack, esp. of trumps. **11** Golf the right of driving off first as having won the last hole (it is my honour). ● v.tr. **1** respect highly. **2** confer honour on. **3** accept (a bill) or pay (a cheque) when due. **4** acknowledge. □ **do the honours** perform the duties of a host to guests etc. **honour bright** colloq. = on my honour. **honour point** Heraldry the point halfway between the top of a shield and the fesse point. **honours are even** there is equality in the contest. **honours list** a list of persons awarded honours. **honours of war** privileges granted to a capitulating force, e.g. that of marching out with colours flying. **honour system** a system of examinations etc. without supervision, relying on the honour of those concerned. **honour-trick** = quick trick. **in honour bound** = on one's honour. **in honour of** as a celebration of. **on one's honour** (usu. foll. by to + infin.) under a moral obligation. **on** (or **upon**) **my honour** an expression of sincerity. [ME f. OF (h)onor (n.), onorer (v.) f. L honor, honorare]

honourable /ˈɒnərəb(ə)l/ adj. (US **honorable**) **1 a** worthy of honour. **b** bringing honour to its possessor. **c** showing honour, not base. **d** consistent with honour. **e** colloq. or joc. (of the intentions of a man courting a woman) directed towards marriage. **2** (**Honourable**) a title indicating eminence or distinction, given to certain high officials, the children of certain ranks of the nobility, and MPs. □ **honourable mention** an award of merit to a candidate in an examination, a work of art, etc., not awarded a prize. □ **honourableness** n. **honourably** adv. [ME f. OF honorable f. L honorabilis (as HONOUR)]

Hon. Sec. abbr. Honorary Secretary.

Honshu /ˈhɒnʃuː/ the largest of the four main islands of Japan; pop. (1990) 99,254,000.

Hooch, Pieter de, see DE HOOCH.

hooch /huːtʃ/ n. (also **hootch**) N. Amer. colloq. alcoholic liquor, esp. inferior or illicit whisky. [abbr. of Alaskan hoochinoo, name of a liquor-making tribe]

Hood /hʊd/, Thomas (1799–1845), English poet and humorist. He was the editor of a number of literary magazines and a friend of Charles Lamb, William Hazlitt, and Thomas De Quincey. He wrote much humorous verse but is now chiefly remembered for serious poems such as 'The Song of the Shirt' (1843) and 'The Bridge of Sighs' (1844).

hood[1] /hʊd/ n. & v. ● n. **1 a** a covering for the head and neck, whether part of a cloak etc. or separate. **b** a separate hoodlike garment worn over a university gown or a surplice to indicate the wearer's degree. **2** Brit. a folding waterproof top of a motor car, pram, etc. **3** N. Amer. the bonnet of a motor vehicle. **4** a canopy to protect users of machinery or to remove fumes etc. **5** a hoodlike flap or marking on the head or neck of a cobra, seal, etc. **6** a leather covering for a hawk's head. ● v.tr. cover with a hood. □ **hood-mould** (or **-moulding**) Archit. a dripstone. □ **hoodless** adj. **hoodlike** adj. [OE hōd f. WG, rel. to HAT]

hood[2] /hʊd, huːd/ n. esp. US sl. a gangster or gunman. [abbr. of HOODLUM]

-hood /hʊd/ suffix forming nouns: **1** of condition or state (childhood; falsehood). **2** indicating a collection or group (sisterhood; neighbourhood). [OE -hād, orig. an independent noun, = person, condition, quality]

hooded /ˈhʊdɪd/ adj. having a hood; covered with a hood. □ **hooded crow** a piebald grey and black crow that is a northern race of the carrion crow.

hoodie /ˈhʊdɪ/ n. = hooded crow.

hoodlum /ˈhuːdləm/ n. **1** a street hooligan, a young thug. **2** a gangster. [19th c.: orig. unkn.]

hoodoo /ˈhuːduː/ n. & v. esp. US ● n. **1 a** bad luck. **b** a thing or person that brings or causes this. **2** voodoo. **3** a strangely shaped rock pinnacle or column of rock formed by erosion etc. ● v.tr. (**hoodoos, hoodooed**) **1** make unlucky. **2** bewitch. [alt. of VOODOO]

hoodwink /ˈhʊdwɪŋk/ v.tr. deceive, delude. [orig. 'blindfold', f. HOOD[1] n. + WINK]

hooey /ˈhuːɪ/ n. & int. colloq. nonsense, humbug. [20th c.: orig. unkn.]

hoof /huːf/ n. & v. ● n. (pl. **hooves** /huːvz/ or **hoofs**) the horny part of the foot of a horse, antelope, and other ungulates. ● v. **1** tr. strike with a hoof. **2** *tr. sl.* kick or shove. □ **hoof it** *sl.* **1** go on foot. **2** dance. **on the hoof** (of cattle) not yet slaughtered. □ **hoofed** *adj.* (also in *comb.*). [OE *hōf* f. Gmc]

hoofer /ˈhuːfə(r)/ *n. sl.* a professional dancer.

Hooghly /ˈhuːglɪ/ (also **Hugli**) the most westerly of the rivers of the Ganges delta, in West Bengal, India. It flows for 192 km (120 miles) into the Bay of Bengal and is navigable to Calcutta.

hoo-ha /ˈhuːhɑː/ *n. colloq.* a commotion, a row; uproar, trouble. [20th c.: orig. unkn.]

hook /hʊk/ *n. & v.* ● *n.* **1 a** a piece of metal or other material bent back at an angle with a round bend, for catching hold or for hanging things on. **b** (in full **fish-hook**) a bent piece of wire, usu. barbed, attached to a fishing-line to carry bait etc. **2** a curved cutting instrument (*reaping-hook*). **3 a** a sharp bend, e.g. in a river. **b** a projecting point of land (*Hook of Holland*). **c** a sand-spit with a curved end. **4 a** *Cricket & Golf* a hooking stroke (see sense 5 of *v.*). **b** *Boxing* a short swinging blow with the elbow bent and rigid. **5** a trap, a snare. **6 a** a curved stroke in handwriting, esp. as made in learning to write. **b** *Mus.* an added stroke transverse to the stem in the symbol for a quaver etc. **7** (in *pl.*) *sl.* fingers. ● *v.* **1** *tr.* **a** grasp with a hook. **b** secure with a hook or hooks. **2** (often foll. by *on, up*) **a** *tr.* attach with or as with a hook. **b** *intr.* be or become attached with a hook. **3** *tr.* catch with or as with a hook (*he hooked a fish; she hooked a husband*). **4** *tr. sl.* steal. **5** *tr.* **a** *Cricket* play (the ball) round from the off to the on side with an upward stroke. **b** (also *absol.*) *Golf* strike (the ball) so that it deviates towards the striker. **6** *tr. Rugby* secure (the ball) and pass it backward with the foot in the scrum. **7** *tr. Boxing* strike (one's opponent) with the elbow bent and rigid. □ **by hook or by crook** by one means or another, by fair means or foul. **hook and eye** a small metal hook and loop as a fastener on a garment. **hook it** *sl.* make off, run away. **hook, line, and sinker** entirely. **hook-nose** an aquiline nose. **hook-nosed** having an aquiline or hooked nose. **hook-up** a connection, esp. an interconnection of broadcasting equipment for special transmissions. **off the hook 1** *colloq.* no longer in difficulty or trouble. **2** (of a telephone receiver) not on its rest, and so preventing incoming calls. **off the hooks** *sl.* dead. **on one's own hook** *sl.* on one's own account. **sling** (or **take**) **one's hook** *colloq.* make off, go away. □ **hookless** *adj.* **hooklet** /ˈhʊklɪt/ *n.* **hooklike** *adj.* [OE *hōc*: sense 3 of *n.* prob. influenced by Du. *hoek* corner]

hookah /ˈhʊkə/ *n.* an oriental tobacco-pipe with a long tube passing through water for cooling the smoke as it is drawn through. [Urdu f. Arab. *ḥuḳḳah* casket]

Hooke /hʊk/, Robert (1635–1703), English scientist. He began as Boyle's assistant, and soon became curator of experiments for the new Royal Society. After the Fire of London he was made a surveyor to the City, and designed several of London's prominent buildings. His scientific achievements were many and varied: he proposed an undulating theory of light, formulated the law of elasticity (*Hooke's law*), introduced the term *cell* to biology, postulated elliptical orbits for the earth and moon, and proposed the inverse square law of gravitational attraction. He improved the compound microscope and reflecting telescope, and invented many scientific instruments and mechanical devices.

hooked /hʊkt/ *adj.* **1** hook-shaped (*hooked nose*). **2** furnished with a hook or hooks. **3** in senses of HOOK *v.* **4** (foll. by *on*) *colloq.* addicted to; captivated by. **5** (of a rug or mat) made by pulling woollen yarn through canvas with a hook.

Hooker /ˈhʊkə(r)/, Sir Joseph Dalton (1817–1911), English botanist and pioneer in plant geography. Following a voyage to the Antarctic he proposed an ancient joining of land between Australia and South America, and from NE India he sent rhododendrons home and later introduced their cultivation. Hooker firmly supported Darwin's theories and applied them to plants. His many works include *Genera Plantarum* (1862–83), a classification of plants devised jointly with George Bentham (1800–84). He became director of Kew Gardens on the death of his father Sir William Jackson Hooker (1785–1865), who had greatly extended the royal gardens there, opened them to the public, and founded a museum.

hooker[1] /ˈhʊkə(r)/ *n.* **1** *Rugby* the player in the middle of the front row of the scrum who tries to hook the ball. **2** *sl.* a prostitute. **3** a person or thing that hooks.

hooker[2] /ˈhʊkə(r)/ *n.* **1** a small Dutch or Irish fishing-vessel. **2** *derog.* any ship. [Du. *hoeker* f. *hoek* HOOK]

Hooke's law *n. Physics* the law that the strain in an elastic solid is directly proportional to the applied stress. [HOOKE]

hookey /ˈhʊkɪ/ *n.* (also **hooky**) *N. Amer.* □ **blind hookey** a gambling guessing-game at cards. **play hookey** *colloq.* play truant. [19th c.: orig. unkn.]

Hook of Holland (Dutch **Hoek van Holland** /ˌhuːk fɑn ˈhɔlɑnt/) a cape and port of the Netherlands, near The Hague, linked by ferry to Harwich, Hull, and Dublin.

hookworm /ˈhʊkwɜːm/ *n.* **1** a parasitic nematode worm with hooklike mouthparts for attachment and feeding, infesting the gut of humans and animals. **2** a disease caused by such an infestation, often resulting in severe anaemia.

hooligan /ˈhuːlɪgən/ *n.* a young ruffian, esp. a member of a gang. □ **hooliganism** *n.* [19th c.: orig. unkn.]

hoon /huːn/ *n. & v. Austral. sl.* ● *n.* a lout or idiot. ● *v.intr.* behave like a hoon. [20th c.: orig. unkn.]

hoop[1] /huːp/ *n. & v.* ● *n.* **1** a circular band of metal, wood, etc., esp. for binding the staves of casks etc. or for forming part of a framework. **2 a** a ring bowled along by a child. **b** a large ring, usu. with paper stretched over it, for circus performers to jump through. **3** an arch of iron etc. through which the balls are hit in croquet. **4** *hist.* **a** a circle of flexible material for expanding a woman's petticoat or skirt. **b** (in full **hoop petticoat**) a petticoat expanded with this. **5 a** a band in contrasting colour on a footballer's shirt or a jockey's blouse, sleeves, or cap. **b** *Austral. colloq.* a jockey. ● *v.tr.* **1** bind with a hoop or hoops. **2** encircle with or as with a hoop. □ **be put** (or **go**) **through the hoop** (or **hoops**) undergo an ordeal. **hoop-iron** in long thin strips for binding casks etc. **hoop-la 1** *Brit.* a game in which rings are thrown in an attempt to encircle one of various prizes. **2** *colloq.* commotion. **3** *colloq.* pretentious nonsense. [OE *hōp* f. WG]

hoop[2] var. of WHOOP.

hoopoe /ˈhuːpuː/ *n.* a salmon-pink bird, *Upupa epops*, with black and white wings and tail, a large erectile crest, and a long decurved bill. [alt. of ME *hoop* f. OF *huppe* f. L *upupa*, imit. of its cry]

hooray /hʊˈreɪ/ *int.* **1** = HURRAH. **2** *Austral. & NZ* goodbye. □ **Hooray Henry** /ˈhʊəreɪ/ *Brit. colloq.* a rich ineffectual young man, esp. one who is fashionable, extroverted, and conventional. [var. of HURRAH]

hooroo /hʌˈruː/ *int. & n.* (also **hurroo**) *Austral. colloq.* = HURRAH. [alt. of HOORAY, HURRAH]

hoosegow /ˈhuːsgaʊ/ *n. US sl.* a prison. [Amer. Sp. *juzgao*, Sp. *juzgado* tribunal f. L *judicatum* neut. past part. of *judicare* JUDGE]

hoot /huːt/ *n. & v.* ● *n.* **1** an owl's cry. **2** the sound made by a motor horn or a steam whistle. **3** a shout expressing scorn or disapproval; an inarticulate shout. **4 a** a shout of laughter. **b** *colloq.* a cause of laughter or merriment. **5** (also **two hoots**) *colloq.* anything at all (*don't care a hoot; don't give a hoot; doesn't matter two hoots*). ● *v.* **1** *intr.* **a** (of an owl) utter its cry. **b** (of a motor horn or steam whistle) make a hoot. **c** (often foll. by *at*) make loud sounds, esp. of scorn, disapproval, or merriment (*hooted with laughter*). **2** *tr.* **a** assail with scornful shouts. **b** (often foll. by *out, away*) drive away by hooting. **3** *tr.* sound (a motor horn or steam whistle). [ME *hūten*, perh. imit.]

hootch var. of HOOCH.

hootenanny /ˈhuːtˌnænɪ/ *n.* (*pl.* **-ies**) esp. *US* an informal gathering with folk music. [orig. dial., = 'gadget']

hooter /ˈhuːtə(r)/ *n.* **1** *Brit.* a siren or steam whistle, esp. as a signal for work to begin or cease. **2** *Brit.* the horn of a motor vehicle. **3** *sl.* a nose. **4** a person or animal that hoots.

hoots /huːts/ *int. Sc. & N. Engl.* expressing dissatisfaction or impatience. [natural exclam.: cf. Sw. *hut* begone, Welsh *hwt* away, Ir. *ut* out, all in similar sense]

Hoover[1] /ˈhuːvə(r)/, Herbert C(lark) (1874–1964), American Republican statesman, 31st President of the US 1929–33. He first gained prominence for his work in organizing food production and distribution in the US and Europe during and after the First World War. As President he was faced with the long-term problems of the Depression which followed the stock market crash of 1929. He returned to relief work after the Second World War as coordinator of food supplies to avert the threat of postwar famine.

Hoover[2] /ˈhuːvə(r)/, J(ohn) Edgar (1895–1972), American lawyer and director of the FBI 1924–72. Beginning his term of office with the fight against organized crime in the 1920s and 1930s, he went on to be instrumental in reorganizing the FBI into an efficient, scientific law-enforcement agency. However, he came under criticism for the organization's role during the McCarthy era and for its reactionary political stance in the 1960s.

Hoover[3] /ˈhuːvə(r)/ *n. & v.* ● *n. propr.* a vacuum cleaner (properly one made by the Hoover company, first patented in 1927). ● *v.* (**hoover**) **1** *tr.* (also *absol.*) clean (a carpet etc.) with a vacuum cleaner. **2** (foll. by *up*) **a** *tr.* suck up with or as with a vacuum cleaner (*hoovered up the crumbs*). **b** *absol.* clean a room etc. with a vacuum cleaner (*decided to hoover up before they arrived*). [W. H. *Hoover*, Amer. manufacturer (1849–1932)]

Hooverville /ˈhuːvəˌvɪl/ *n.* (in the US) any of the shanty towns built by unemployed and destitute people during the Depression of the early 1930s. They were named Hoovervilles after the President of the day, Herbert C. Hoover.

hooves *pl.* of HOOF.

hop[1] /hɒp/ *v.* & *n.* ● *v.* (**hopped, hopping**) **1** *intr.* (of a bird, frog, etc.) spring with two or all feet at once. **2** *intr.* (of a person) jump on one foot. **3** *intr.* move or go quickly (*hopped over the fence*). **4** *tr.* cross (a ditch etc.) by hopping. **5** *intr. colloq.* **a** make a quick trip. **b** make a quick change of position or location. **6** *tr. colloq.* **a** jump into (a vehicle). **b** obtain (a ride) in this way. **7** *tr.* (usu. as **hopping** *n.*) (esp. of aircraft) pass quickly from one (place of a specified type) to another (*cloud-hopping*; *hedge-hopping*). ● *n.* **1** a hopping movement. **2** *colloq.* an informal dance. **3** a short flight in an aircraft; the distance travelled by air without landing; a stage of a flight or journey. □ **hop in** (or **out**) *colloq.* get into (or out of) a car etc. **hop it** *Brit. colloq.* go away. **hop, skip** (or **step**), **and jump** = *triple jump.* **hop the twig** (or **stick**) *sl.* **1** depart suddenly. **2** die. **on the hop** *colloq.* **1** unprepared (*caught on the hop*). **2** bustling about. [OE *hoppian*]

hop[2] /hɒp/ *n.* & *v.* ● *n.* **1** a climbing plant, *Humulus lupulus*, cultivated for the cones borne by the female. **2** (in *pl.*) **a** the ripe cones of this, used to give a bitter flavour to beer. **b** *Austral.* & *NZ colloq.* beer. **3** *US sl.* opium or any other narcotic. ● *v.* (**hopped, hopping**) **1** *tr.* flavour with hops. **2** *intr.* produce or pick hops. **3** *tr. US sl.* (foll. by *up*) (esp. as **hopped up**) stimulate or intoxicate with alcohol or a drug. □ **hop-bind** (or **-bine**) the climbing stem of the hop. **hop-sack** (or **-sacking**) **1 a** a coarse material made from hemp etc. **b** sacking for hops made from this. **2** a coarse clothing fabric of a loose plain weave. [ME *hoppe* f. MLG, MDu. *hoppe*]

Hope /həʊp/, Bob (born Leslie Townes Hope) (b.1903), British-born American comedian. His dry allusive style gave him the character of a humorously cowardly incompetent, always cheerfully failing in his attempts to become a romantic hero, particularly in *Road to Singapore* (1940) and the rest of the series of *Road* films (1940–62), in which he starred with Bing Crosby and Dorothy Lamour (b.1914).

hope /həʊp/ *n.* & *v.* ● *n.* **1** (in *sing.* or *pl.*; often foll. by *of, that*) expectation and desire combined, e.g. for a certain thing to occur (*hope of getting the job*). **2 a** a person, thing, or circumstance that gives cause for hope. **b** ground of hope, promise. **3** what is hoped for. **4** *archaic* a feeling of trust. ● *v.* **1** *intr.* (often foll. by *for*) feel hope. **2** *tr.* expect and desire. **3** *tr.* feel fairly confident. □ **hope against hope** cling to a mere possibility. **hope chest** *N. Amer.* = *bottom drawer.* **not a** (or **some**) **hope!** *colloq.* no chance at all. □ **hoper** *n.* [OE *hopa*]

hopeful /ˈhəʊpfʊl/ *adj.* & *n.* ● *adj.* **1** feeling hope. **2** causing or inspiring hope. **3** likely to succeed, promising. ● *n.* (in full **young hopeful**) **1** a person likely to succeed. **2** *iron.* a person likely to be disappointed. □ **hopefulness** *n.*

hopefully /ˈhəʊpfʊli/ *adv.* **1** in a hopeful manner. **2** (qualifying a whole sentence) *disp.* it is to be hoped (*hopefully, the car will be ready by then*).

Hopeh see HEBEI.

hopeless /ˈhəʊplɪs/ *adj.* **1** feeling no hope. **2** admitting no hope (*a hopeless case*). **3** inadequate, incompetent (*am hopeless at tennis*). **4** without hope of success; futile. □ **hopelessly** *adv.* **hopelessness** *n.*

hophead /ˈhɒphed/ *n. sl.* **1** *US* a drug addict. **2** *Austral.* & *NZ* a drunkard.

Hopi /ˈhəʊpi/ *n.* & *adj.* ● *n.* (*pl.* same or **Hopis**) **1** a member of an American Indian people living chiefly in NE Arizona. (*See note below.*) **2** the language of the Hopi. ● *adj.* of or relating to the Hopi or their language.

■ The Hopi are Pueblo Indians, traditionally living a settled life in houses of stone and adobe in pueblos (villages) and surviving through growing crops and herding. The Hopi are a small group (fewer than 10,000), but retain some of their rich cultural life, exemplified by ceremonies such as the snake dance, in which live rattlesnakes are handled.

Hopkins[1] /ˈhɒpkɪnz/, Sir Anthony (Philip) (b.1937), Welsh actor. His acting career began on the stage in 1961, with his screen début following in 1967. His films include *The Elephant Man* (1980), *The Bounty* (1984), and *The Remains of the Day* (1993). He won an Oscar for his performance in *The Silence of the Lambs* (1991).

Hopkins[2] /ˈhɒpkɪnz/ Sir Frederick Gowland (1861–1947), English biochemist, considered the father of British biochemistry. He carried out pioneering work on 'accessory food factors' essential to the diet, later called vitamins. Hopkins shared a Nobel Prize in 1929.

Hopkins[3] /ˈhɒpkɪnz/, Gerard Manley (1844–89), English poet. Influenced at Oxford by John Henry Newman, Hopkins converted to Roman Catholicism in 1866 and became a Jesuit two years later. He wrote little poetry until 1876, when the shipwreck of a vessel carrying nuns and other emigrants to America the previous year inspired him to write 'The Wreck of the Deutschland'. The poem makes bold use of Hopkins's 'sprung rhythm' technique, as do his best-known poems 'Windhover' and 'Pied Beauty', both written in 1877. His work, collected in *Poems* (1918), was published posthumously by his friend Robert Bridges.

hoplite /ˈhɒplaɪt/ *n.* a heavily armed foot-soldier of ancient Greece. [Gk *hoplitēs* f. *hoplon* weapon]

Hopper /ˈhɒpə(r)/, Edward (1882–1967), American realist painter. He supported himself as a commercial illustrator before gaining recognition for his paintings in the 1920s. He is best known for his mature works, such as *Early Sunday Morning* (1930) and *Nighthawks* (1942), depicting scenes from everyday American urban life in which still figures appear in introspective isolation, often in bleak or shabby settings.

hopper[1] /ˈhɒpə(r)/ *n.* **1** a person who hops. **2** a hopping arthropod, esp. a flea or cheese-maggot or young locust. **3 a** a container tapering downward (orig. having a hopping motion) through which grain passes into a mill. **b** a similar contrivance in various machines. **4 a** a barge carrying away mud etc. from a dredging-machine and discharging it. **b** a railway truck able to discharge coal etc. through its floor.

hopper[2] /ˈhɒpə(r)/ *n.* a hop-picker.

hopping /ˈhɒpɪŋ/ *adj.* **1** in senses of HOP[1] *v.* **2** esp. *N. Amer. colloq.* very active, lively. □ **hopping mad** *colloq.* very angry.

hopple /ˈhɒp(ə)l/ *v.* & *n.* ● *v.tr.* fasten together the legs of (a horse etc.) to prevent it from straying etc. ● *n.* an apparatus for this. [prob. LG: cf. HOBBLE and early Flem. *hoppelen* = MDu. *hobelen* jump, dance]

hopscotch /ˈhɒpskɒtʃ/ *n.* a children's game of hopping over squares or oblongs marked on the ground to retrieve a flat stone etc. [HOP[1] + SCOTCH[1]]

Horace /ˈhɒrɪs/ (full name Quintus Horatius Flaccus) (65–8 BC), Roman poet of the Augustan period. His two books of *Satires* departed from earlier convention with their realism and irony directed at both the satirist and his targets. His *Odes*, displaying a mastery of poetic form in the style of earlier Greek lyric poets, celebrate friendship, love, good wine, and the contentment of a peaceful rural life in contrast to the turmoil of politics and civil war. (Horace had fought with Brutus and Cassius at Philippi in 42.) They were much imitated by later ages, especially by the poets of 17th-century England. Horace was also a notable literary critic; his *Ars Poetica* influenced John Dryden and the 18th-century Augustans in their critical writing.

horary /ˈhɒrəri/ *adj. archaic* **1** of the hours. **2** occurring every hour, hourly. [med.L *horarius* f. L *hora* HOUR]

horde /hɔːd/ *n.* **1 a** usu. *derog.* a large group, a gang. **b** a moving swarm or pack (of insects, wolves, etc.). **2** a troop of Tartar or other nomads. [Pol. *horda* f. Turkic *ordī, ordū* camp: cf. URDU]

horehound /ˈhɔːhaʊnd/ *n.* (also **hoarhound**) **1 a** (in full **white horehound**) a herbaceous labiate plant, *Marrubium vulgare*, with a white cottony covering on its stem and leaves. **b** its bitter aromatic juice used against coughs etc. **2** (in full **black horehound**) a herbaceous labiate plant, *Ballota nigra*, with an unpleasant aroma. [OE *hāre hūne* f. *hār* HOAR + *hūne* a plant]

horizon /həˈraɪz(ə)n/ *n.* **1 a** the line at which the earth and sky appear to meet. **b** (in full **apparent** or **sensible** or **visible horizon**) the line at which the earth and sky would appear to meet but for irregularities and obstructions; a circle where the earth's surface touches a cone whose vertex is at the observer's eye. **c** (in full **celestial** or **rational** or **true horizon**) a great circle of the celestial sphere, the plane of which passes through the centre of the earth and is parallel to that of the apparent horizon of a place. **2** limit of mental perception, experience, interest, etc. **3** a geological stratum or set of strata, or layer of soil, with particular characteristics. **4** *Archaeol.* the level at which a particular set of remains is found. □ **on the horizon**

(of an event) just imminent or becoming apparent. [ME f. OF *orizon(te)* f. LL *horizon -ontis* f. Gk *horizōn (kuklos)* limiting (circle)]

horizontal /ˌhɒrɪˈzɒnt(ə)l/ *adj. & n.* ● *adj.* **1 a** parallel to the plane of the horizon, at right angles to the vertical (*horizontal plane*). **b** (of machinery etc.) having its parts working in a horizontal direction. **2 a** combining firms engaged in the same stage of production (*horizontal integration*). **b** involving social groups of equal status etc. **3** of or at the horizon. ● *n.* a horizontal line, plane, etc. □ **horizontally** *adv.* **horizontality** /-zɒnˈtælɪtɪ/ *n.* [F *horizontal* or mod.L *horizontalis* (as HORIZON)]

Horkheimer /ˈhɔːkˌhaɪmə(r)/, Max (1895–1973), German philosopher and sociologist. He was director of the Frankfurt Institute for Social Research from 1930 to 1958 and a principal figure in the Frankfurt School of philosophy. His reputation is based on a series of articles written in the 1930s expounding the school's Marxist analysis of modern industrial society and culture; these were collected into the two-volume *Critical Theory* (1968). Other works include *Dialectic of the Enlightenment* (1947), written with his colleague Theodor Adorno.

hormone /ˈhɔːməʊn/ *n.* **1** *Physiol.* a regulatory substance produced in an organism and transported in tissue fluids such as blood or sap to stimulate specific cells or tissues into action. **2** a synthetic substance with a similar effect. □ **hormone replacement therapy** (abbr. **HRT**) treatment with oestrogens to alleviate menopausal symptoms. □ **hormonal** /hɔːˈməʊn(ə)l/ *adj.* [Gk *hormōn* part. of *hormaō* impel]

Hormuz /ˈhɔːmʊz, hɔːˈmuːz/ (also **Ormuz** /ˈɔːmʊz, ɔːˈmuːz/) an Iranian island at the mouth of the Persian Gulf, in the Strait of Hormuz. It is the site of an ancient city, which was an important centre of commerce in the Middle Ages.

Hormuz, Strait of a strait linking the Persian Gulf with the Gulf of Oman, which leads to the Arabian Sea, and separating Iran from the Arabian peninsula. It is of strategic and economic importance as a waterway through which sea traffic to and from the oil-rich states of the Gulf must pass.

horn /hɔːn/ *n. & v.* ● *n.* **1 a** a hard permanent outgrowth, often curved and pointed, on the head of cattle, rhinoceroses, giraffes, and other esp. hoofed mammals, found singly, in pairs, or one in front of another. **b** the structure of a horn, consisting of a core of bone encased in keratinized skin. **2** either of two deciduous branched appendages on the head of (esp. male) deer. **3** a hornlike projection on the head of other animals, e.g. a snail's tentacle, the crest of a horned owl, etc. **4** the substance of which horns are composed. **5** anything resembling or compared to a horn in shape. **6** *Mus.* **a** = FRENCH HORN. **b** a wind instrument played by lip vibration, orig. made of horn, now usu. of brass. (*See note below.*) **c** a horn player. **7** an instrument sounding a warning or other signal (*car horn; foghorn*). **8** a receptacle or instrument made of horn, e.g. a drinking-vessel or powder-flask etc. **9** a horn-shaped projection. **10** the extremity of the moon or other crescent. **11** an arm or branch of a river, bay, etc. **12** a pyramidal peak formed by glacial action. **13** *coarse sl.* an erect penis. **14** the hornlike emblem of a cuckold. ● *v.tr.* **1** (esp. as **horned** *adj.*) provide with horns. **2** gore with the horns. □ **horn in** *sl.* **1** (usu. foll. by *on*) intrude. **2** interfere. **horn of plenty** a cornucopia. **horn-rimmed** (esp. of spectacles) having rims made of horn or a substance resembling it. **on the horns of a dilemma** faced with a decision involving equally unfavourable alternatives. □ **hornist** *n.* (in sense 6 of *n.*). **hornless** *adj.* **hornlike** *adj.* [OE f. Gmc, rel. to L *cornu*]

■ The horn of domesticated cattle, with the tip cut off (or, in Africa, with a central orifice cut) to leave a hole for blowing, has been used by herdsmen, watchmen, and huntsmen from time immemorial. Instruments of this kind were also made of bronze, ivory, and brass. Metal hunting horns with a loop in the tube have existed since the 15th century, and tightly wound multiple-coil horns since the 16th century. In written music and among musicians the term *horn* generally denotes the French horn; particularly in popular music, it can now also mean any brass wind instrument.

Horn, Cape (also **the Horn**) the southernmost point of South America, on a Chilean island south of Tierra del Fuego. It was discovered by the Dutch navigator W. C. Schouten in 1616 and named after Hoorn, his birthplace. The region is notorious for its storms, and until the opening of the Panama Canal in 1914 constituted the only sea route between the Atlantic and Pacific Oceans.

hornbeam /ˈhɔːnbiːm/ *n.* a tree of the genus *Carpinus*, with a smooth bark and a hard tough wood.

hornbill /ˈhɔːnbɪl/ *n.* a tropical bird of the family Bucerotidae, with a hornlike excrescence on its large red or yellow curved bill.

hornblende /ˈhɔːnblend/ *n.* a dark brown, black, or green mineral occurring in many igneous and metamorphic rocks, and composed of calcium, magnesium, and iron silicates. [G (as HORN, BLENDE)]

hornbook /ˈhɔːnbʊk/ *n. hist.* a leaf of paper containing the alphabet, the Lord's Prayer, etc., mounted on a wooden tablet with a handle, and protected by a thin plate of horn.

horned /hɔːnd/ *adj.* **1** having a horn or horns. **2** crescent-shaped (*horned moon*). □ **horned-owl** a large North American owl, *Bubo virginianus*, with hornlike feathers over the ears. **horned toad 1** an American lizard, *Phrynosoma cornutum*, covered with spiny scales. **2** a SE Asian toad of the family Pelobatidae, with horn-shaped extensions over the eyes.

hornet /ˈhɔːnɪt/ *n.* a large wasp, esp. *Vespa crabro*, with a brown and yellow striped body, and capable of inflicting a severe sting. □ **stir up a hornets' nest** provoke or cause trouble or opposition. [prob. f. MLG, MDu. *horn(e)te*, corresp. to OE *hyrnet*, perh. rel. to HORN]

hornfels /ˈhɔːnfelz/ *n. Geol.* a dark fine-grained metamorphic rock composed mainly of quartz, mica, and feldspars. [G, = horn rock]

Horn of Africa (also called *Somali Peninsula*) a peninsula of NE Africa, comprising Somalia and parts of Ethiopia. It lies between the Gulf of Aden and the Indian Ocean.

hornpipe /ˈhɔːnpaɪp/ *n.* **1** a lively dance, usu. for one person, orig. to the accompaniment of a wind instrument, and esp. associated with the merrymaking of sailors. **2** the music for this. [name of an obs. wind instrument partly of horn: ME, f. HORN + PIPE]

hornswoggle /ˈhɔːnˌswɒg(ə)l/ *v.tr. sl.* cheat, hoax. [19th c.: orig. unkn.]

Hornung /ˈhɔːnəŋ/, Ernest William (1866–1921), English novelist. He is remembered as the creator of the gentleman burglar Raffles, who first featured in *The Amateur Cracksman* (1899). Hornung was the brother-in-law of Sir Arthur Conan Doyle.

hornwort /ˈhɔːnwɜːt/ *n.* an aquatic rootless plant of the genus *Ceratophyllum*, with forked leaves.

horny /ˈhɔːnɪ/ *adj.* (**hornier, horniest**) **1** of or like horn. **2** hard like horn, callous (*horny-handed*). **3** *sl.* sexually excited. □ **horniness** *n.*

horologe /ˈhɒrəˌlɒdʒ/ *n. archaic* a timepiece. [ME f. OF *orloge* f. L *horologium* f. Gk *hōrologion* f. *hōra* time + *-logos* -telling]

horology /həˈrɒlədʒɪ/ *n.* the art of measuring time or making clocks, watches, etc.; the study of this. □ **horologer** *n.* **horologist** *n.* **horologic** /ˌhɒrəˈlɒdʒɪk/ *adj.* **horological** *adj.* [Gk *hōra* time + -LOGY]

horoscope /ˈhɒrəˌskəʊp/ *n. Astrol.* **1** a forecast of a person's future based on a diagram showing the relative positions of the stars and planets at that person's birth. **2** such a diagram (*cast a horoscope*). **3** observation of the sky and planets at a particular moment, esp. at a person's birth. □ **horoscopic** /ˌhɒrəˈskɒpɪk/ *adj.* **horoscopical** *adj.* **horoscopy** /həˈrɒskəpɪ/ *n.* [F f. L *horoscopus* f. Gk *hōroskopos* f. *hōra* time + *skopos* observer]

Horowitz /ˈhɒrəvɪts/, Vladimir (1904–89), Russian pianist. He first toured the US in 1928, and settled there soon afterwards. He was a leading international virtuoso and was best known for his performances of Scarlatti, Liszt, Scriabin, and Prokofiev. His concert career was interrupted by periodic bouts of illness.

horrendous /həˈrendəs/ *adj. colloq.* horrifying; awful. □ **horrendously** *adv.* **horrendousness** *n.* [L *horrendus* gerundive of *horrere*: see HORRID]

horrent /ˈhɒrənt/ *adj. poet.* **1** bristling. **2** shuddering. [L *horrere*: see HORRID]

horrible /ˈhɒrɪb(ə)l/ *adj.* **1** causing or likely to cause horror; hideous, shocking. **2** *colloq.* unpleasant, excessive (*horrible weather; horrible noise*). □ **horribleness** *n.* **horribly** *adv.* [ME f. OF (h)*orrible* f. L *horribilis* f. *horrere*: see HORRID]

horrid /ˈhɒrɪd/ *adj.* **1** horrible, revolting. **2** *colloq.* unpleasant, disagreeable (*horrid weather; horrid children*). **3** *poet.* rough, bristling. □ **horridly** *adv.* **horridness** *n.* [L *horridus* f. *horrere* bristle, shudder]

horrific /həˈrɪfɪk/ *adj.* horrifying. □ **horrifically** *adv.* [F *horrifique* or L *horrificus* f. *horrere*: see HORRID]

horrify /ˈhɒrɪˌfaɪ/ *v.tr.* (**-ies, -ied**) arouse horror in; shock, scandalize. □ **horrifying** *adj.* **horrifyingly** *adv.* **horrification** /ˌhɒrɪfɪˈkeɪʃ(ə)n/ *n.* [L *horrificare* (as HORRIFIC)]

horripilation /ˌhɒrɪpɪˈleɪʃ(ə)n/ *n. literary* = goose-flesh. [LL *horripilatio* f. L *horrere* to bristle + *pilus* hair]

horror /ˈhɒrə(r)/ *n. & adj.* ● *n.* **1** an intense feeling of loathing and fear. **2 a** (often foll. by *of*) intense dislike. **b** (often foll. by *at*) *colloq.* intense dismay. **3 a** a person or thing causing horror. **b** *colloq.* a bad or mischievous person etc. **4** (in *pl.*; prec. by *the*) a fit of horror, depression, or nervousness, esp. as in delirium tremens. **5** a terrified and revolted shuddering. **6** (in *pl.*) an exclamation of dismay. ● *attrib.adj.* (of literature, films, etc.) designed to attract by arousing pleasurable feelings of horror. □ **chamber of horrors** a place full of horrors (orig. a room of waxworks of criminals etc. in Madame Tussaud's; see TUSSAUD). **horror-struck** (or **-stricken**) horrified, shocked. [ME f. OF (h)orrour f. L *horror -oris* (as HORRID)]

Horsa /ˈhɔːsə/ see HENGIST.

hors concours /ˌɔː kɒnˈkʊə(r)/ *adj.* **1** unrivalled, unequalled. **2** (of an exhibit or exhibitor) not competing for a prize. [F, lit. 'outside competition']

hors de combat /ˌɔː də ˈkɒmbʌ/ *adj.* out of the fight or the running, disabled. [F]

hors-d'œuvre /ɔːˈdɜːvrə, -ˈdɜːv/ *n.* an appetizer served at the beginning of a meal or (occasionally) during a meal. [F, lit. 'outside the work']

horse /hɔːs/ *n. & v.* ● *n.* **1 a** a solid-hoofed plant-eating domesticated quadruped, *Equus caballus*, with flowing mane and tail, used for riding and to carry and pull loads. (*See note below.*) **b** an adult male horse; a stallion or gelding. **c** any other four-legged mammal of the genus *Equus*, including asses and zebras. **d** (*collect.*; as *sing.*) cavalry. **e** a representation of a horse. **2** a vaulting-block. **3** a supporting frame esp. with legs (*clothes-horse*). **4** *sl.* heroin. **5** *colloq.* a unit of horsepower. **6** *Naut.* any of various ropes and bars. **7** *Mining* an obstruction in a vein. ● *v.* **1** *intr.* (foll. by *around*) fool about. **2** *tr.* provide (a person or vehicle) with a horse or horses. **3** *intr.* mount or go on horseback. □ **from the horse's mouth** (of information etc.) from the person directly concerned or another authoritative source. **horse-and-buggy** *N. Amer.* old-fashioned, bygone. **horse-block** a small platform of stone or wood for mounting a horse. **horse-brass** see BRASS *n.* 5. **horse-breaker** a person who breaks in horses. **horse chestnut 1** a large ornamental tree of the genus *Aesculus*, with upright conical clusters of white, pink, or red flowers. **2** the dark brown fruit of this, resembling an edible chestnut, but with a coarse bitter taste; a conker. **horse-cloth** a cloth used to cover a horse, or as part of its trappings. **horse-coper** a horse-dealer. **horse-doctor** a veterinary surgeon attending horses. **horse-drawn** (of a vehicle) pulled by a horse or horses. **horse mackerel** a large fish of the mackerel type, e.g. the scad or the tuna. **horse-mushroom** a large edible mushroom, *Agaricus arvensis*. **horse opera** *N. Amer. sl.* a western film. **horse-pistol** a pistol for use by a horseman. **horse-pond** a pond for watering and washing horses, proverbial as a place for ducking obnoxious persons. **horse-race** a race between horses with riders. **horse sense** *colloq.* plain common sense. **horses for courses** the matching of tasks and talents. **horse's neck** *sl.* a drink of flavoured ginger ale usu. with spirits. **horse-soldier** a soldier mounted on a horse. **horse-trading 1** *N. Amer.* dealing in horses. **2** shrewd bargaining. **to horse!** (as a command) mount your horses. □ **horseless** *adj.* **horselike** *adj.* [OE *hors* f. Gmc]

■ The only surviving undomesticated horse is Przewalski's horse (*Equus ferus*) of Mongolia, which is extinct in the wild. Other 'wild horses' are actually feral domesticated animals, the surviving wild members of the horse family being the asses and zebras. The horse evolved from multi-toed ancestors in North America, but it was extinct there by the time that it was domesticated in the Old World in the Bronze Age. Horses were used from around the 19th century BC to draw war chariots, while the use of cavalry dates from the Iron Age. Heavy horses were developed in medieval times for use in battle, finding a use also in traction and gradually replacing the ox. The appearance of horses in Britain was altered in the 17th century by interbreeding with Arab and Barbary stock for racehorses and hunters.

horseback /ˈhɔːsbæk/ *n.* the back of a horse, esp. as sat on in riding. □ **on horseback** mounted on a horse.

horsebean /ˈhɔːsbiːn/ *n.* a broad bean used as fodder.

horsebox /ˈhɔːsbɒks/ *n. Brit.* a closed vehicle for transporting a horse or horses.

horseflesh /ˈhɔːsfleʃ/ *n.* **1** the flesh of a horse, esp. as food. **2** horses collectively.

horsefly /ˈhɔːsflaɪ/ *n.* (*pl.* **-flies**) a biting dipterous fly of the family Tabanidae, troublesome esp. to horses.

Horse Guards *n.* (in the UK) **1** a cavalry brigade of the household troops, now an armoured-car regiment providing a mounted squadron for ceremonial purposes, and part of the Household Cavalry regiment. **2** the headquarters of this brigade in Whitehall, London.

horsehair /ˈhɔːsheə(r)/ *n.* hair from the mane or tail of a horse, used for padding etc.

Horsehead Nebula /ˈhɔːshed/ *n. Astron.* a dust nebula in the shape of a horse's head, forming a dark silhouette against a bright emission nebula in Orion.

horse latitudes *n.* a belt of calms in each hemisphere, at the latitudes 30–35° north and south of the equator, between the trade winds and the westerlies. [18th c.: orig. uncert.]

horseleech /ˈhɔːsliːtʃ/ *n.* **1** a large freshwater leech of the genus *Haemopis*, feeding by swallowing rather than sucking. **2** an insatiable person (cf. Prov. 30:15).

horseless /ˈhɔːslɪs/ *adj.* without a horse. □ **horseless carriage** *archaic* a motor car.

horseman /ˈhɔːsmən/ *n.* (*pl.* **-men**) **1** a rider on horseback. **2** a skilled rider.

horsemanship /ˈhɔːsmənʃɪp/ *n.* the art of riding on horseback; skill in doing this.

Horsens /ˈhɔːs(ə)nz/ a port on the east coast of Denmark, situated at the head of Horsens Fjord; pop. (1990) 55,210.

horseplay /ˈhɔːspleɪ/ *n.* boisterous play.

horsepower /ˈhɔːsˌpaʊə(r)/ *n.* (*pl.* same) an imperial unit of power (symbol **hp**) equal to 550 foot-pounds per second (about 750 watts); the power of an engine etc. measured in terms of this. The term was introduced by James Watt, who modified Thomas Savery's method of estimating engine-power in terms of equivalent work done by horses.

horse-racing /ˈhɔːsˌreɪsɪŋ/ *n.* the sport of conducting races between horses with riders. Organized racing, with Newmarket as the headquarters, became popular in Britain under the patronage of Charles II. In modern racing there are two main types: flat racing, in which young thoroughbred horses run around a circuit without jumps (in races such as the Derby, the Oaks, and the St Leger), and steeplechasing, with obstacles to jump, which spread from Ireland in the 1830s and in which the horses tend to be older and not always thoroughbred (the best-known steeplechase in Britain is the Grand National). A form related to steeplechasing is hurdling, in which the horses jump less severe obstacles. Racing today is a big business, with millions of pounds changing hands annually in on-course and off-course betting.

horseradish /ˈhɔːsˌrædɪʃ/ *n.* **1** a cruciferous plant, *Armoracia rusticana*, with long lobed leaves. **2** the pungent root of this scraped or grated as a condiment, often made into a sauce.

horseshoe /ˈhɔːsʃuː/ *n.* **1** an iron shoe for a horse shaped like the outline of the hard part of the hoof. **2** a thing of this shape; an object shaped like C or U (e.g. a magnet, a table, a Spanish or Islamic arch). □ **horseshoe bat** a bat of the Old World family Rhinolophidae, usu. with a horseshoe-shaped ridge on the nose. **horseshoe crab** a large marine arthropod, *Xiphosura polyphemus*, with a horseshoe-shaped shell and a long tail-spine (also called *king-crab*).

horsetail /ˈhɔːsteɪl/ *n.* **1 a** the tail of a horse. **b** this as formerly used in Turkey as a standard, or as an ensign denoting the rank of a pasha. **2** a pteridophyte plant of the genus *Equisetum*, resembling a horse's tail, with a hollow jointed stem and scale-like leaves. **3** = *pony-tail*.

horsewhip /ˈhɔːswɪp/ *n. & v.* ● *n.* a whip for driving horses. ● *v.tr.* (**-whipped, -whipping**) beat with a horsewhip.

horsewoman /ˈhɔːsˌwʊmən/ *n.* (*pl.* **-women**) **1** a woman who rides on horseback. **2** a skilled woman rider.

horst /hɔːst/ *n. Geol.* a raised elongated block of land bounded by faults on both sides. [G, = heap]

Horst Wessel Song /hɔːst ˈves(ə)l/ the official song of the Nazi Party in Germany. The tune was that of a music-hall song popular with the German army in the First World War; the words were written by Horst Wessel (1907–30), a member of Hitler's Storm Troops killed by political enemies and regarded as a Nazi martyr.

horsy /ˈhɔːsɪ/ *adj.* (also **horsey**) (**horsier, horsiest**) **1** of or like a horse. **2** concerned with or devoted to horses or horse-racing. **3** affectedly using the dress and language of a groom or jockey. □ **horsily** *adv.* **horsiness** *n.*

Horta /ˈɔːtə/, Victor (1861–1947), Belgian architect. He was a leading figure in art nouveau architecture and worked chiefly in Brussels. His buildings include the Hôtel Tassel (1892), notable for its decorative iron staircase with a slender exposed iron support, and the Maison du Peuple (1896–9), with an innovative curved façade of iron and glass.

hortatory /ˈhɔːtətərɪ/ *adj.* (also **hortative** /-tɪv/) tending to serve to exhort. □ **hortation** /hɔːˈteɪʃ(ə)n/ *n.* [L *hortativus* f. *hortari* exhort]

hortensia /hɔːˈtensɪə/ *n.* a kind of hydrangea, *Hydrangea macrophylla*, with large rounded infertile flower-heads. [mod.L f. *Hortense* Lepaute, 18th-c. Frenchwoman]

horticulture /ˈhɔːtɪˌkʌltʃə(r)/ n. the art of garden cultivation. □ **horticultural** /ˌhɔːtɪˈkʌltʃərəl/ adj. **horticulturalist** n. **horticulturist** n. [L *hortus* garden, after AGRICULTURE]

hortus siccus /ˌhɔːtəs ˈsɪkəs/ n. **1** an arranged collection of dried plants. **2** a collection of uninteresting facts etc. [L, = dry garden]

Horus /ˈhɔːrəs/ *Egyptian Mythol.* originally a sky-god whose symbol was the hawk, usually depicted as a falcon-headed man. From early dynastic times he was regarded as the protector of the monarchy and his name was often added to the royal titles. Horus assumed various aspects: in the myth of Isis and Osiris he was the posthumous son of the latter, whose murder he avenged, and in this aspect he was known to the Greeks as Harpocrates (= 'Horus the Child'), most often represented as a chubby infant with a finger held to his mouth.

Hos. abbr. (in the Bible) Hosea.

hosanna /həʊˈzænə/ n. & int. a shout of adoration (Matt. 21:9, 15, etc.). [ME f. LL f. Gk *hōsanna* f. Heb. *hôšaʿnā* for *hôšîʿa-nnā* save now!]

hose /həʊz/ n. & v. ● n. **1** (also **hose-pipe**) a flexible tube conveying water for watering plants etc., putting out fires, etc. **2 a** (collect.; as pl.) stockings and socks (esp. in trade use). **b** *hist.* breeches (doublet and hose). ● v.tr. **1** (often foll. by down) water or spray or drench with a hose. **2** provide with hose. □ **half-hose** socks. [OE f. Gmc]

Hosea /həʊˈzɪə/ **1** a Hebrew minor prophet of the 8th century BC. **2** a book of the Bible containing his prophecies.

hosier /ˈhəʊzɪə(r), ˈhəʊʒə(r)/ n. a dealer in hosiery.

hosiery /ˈhəʊzɪərɪ, ˈhəʊʒərɪ/ n. **1** stockings and socks. **2** *Brit.* knitted or woven underwear.

hospice /ˈhɒspɪs/ n. **1** *Brit.* a home for people who are ill (esp. terminally) or destitute. **2** a lodging for travellers, esp. one kept by a religious order. [F f. L *hospitium* (as HOST²)]

hospitable /hɒˈspɪtəb(ə)l, ˈhɒsp-/ adj. **1** giving or disposed to give welcome and entertainment to strangers or guests. **2** disposed to welcome something readily, receptive. □ **hospitably** adv. [F f. *hospiter* f. med.L *hospitare* entertain (as HOST²)]

hospital /ˈhɒspɪt(ə)l/ n. **1** an institution providing medical and surgical treatment and nursing care for ill or injured people. **2** *hist.* **a** a hospice. **b** an establishment of the Knights Hospitallers. **3** *Law* a charitable institution (also in proper names, e.g. *Christ's Hospital*). □ **hospital corners** a way of tucking in sheets, used by nurses. **hospital fever** a kind of typhus formerly prevalent in crowded hospitals. **hospital ship** a ship to receive sick and wounded sailors, or to take sick and wounded soldiers home. **hospital train** a train taking wounded soldiers from a battlefield. **hospital trust** (in the UK) a trust consisting of a National Health Service hospital or hospitals no longer under local authority control. [ME f. OF f. med.L *hospitale* neut. of L *hospitalis* (adj.) (as HOST²)]

hospitaler *US* var. of HOSPITALLER.

hospitalism /ˈhɒspɪtəˌlɪz(ə)m/ n. the adverse effects of a prolonged stay in hospital.

hospitality /ˌhɒspɪˈtælɪtɪ/ n. the friendly and generous reception and entertainment of guests or strangers. [ME f. OF *hospitalité* f. L *hospitalitas -tatis* (as HOSPITAL)]

hospitalize /ˈhɒspɪtəˌlaɪz/ v.tr. (also **-ise**) send or admit (a patient) to hospital. □ **hospitalization** /ˌhɒspɪtəlaɪˈzeɪʃ(ə)n/ n.

hospitaller /ˈhɒspɪtələ(r)/ n. (*US* **hospitaler**) **1** (also **Hospitaller**) a member of a charitable religious order, orig. the Knights Hospitallers. **2** (in some London hospitals) a chaplain. [ME f. OF *hospitalier* f. med.L *hospitalarius* (as HOSPITAL)]

host¹ /həʊst/ n. **1** (usu. foll. by of) a large number of people or things. **2** *archaic* an army. **3** (in full **heavenly host**) *Bibl.* **a** the sun, moon, and stars. **b** the angels. □ **host** (or **hosts**) **of heaven** = sense 3. **is a host in himself** (or **herself**) can do as much as several ordinary people. **Lord** (or **Lord God**) **of hosts** God as Lord over earthly or heavenly armies. [ME f. OF f. L *hostis* stranger, enemy, in med.L 'army']

host² /həʊst/ n. & v. ● n. **1** a person who receives or entertains another as a guest. **2** the landlord or landlady of an inn (mine host). **3** *Biol.* an animal or plant having a parasite or commensal. **4** an animal or person that has received a transplanted organ etc. **5** the compère of a show, esp. of a television or radio programme. ● v.tr. act as host to (a person) or at (an event). [ME f. OF *oste* f. L *hospes -pitis* host, guest]

host³ /həʊst/ n. *Eccl.* the bread consecrated in the Eucharist. [ME f. OF (h)*oiste* f. L *hostia* victim]

hosta /ˈhɒstə/ n. a perennial garden plant of the genus *Hosta* (formerly *Funkia*), with green or variegated ornamental leaves, and loose clusters of tubular mauve or white flowers. [mod.L, f. Nicolaus Thomas *Host*, Austrian physician (1761–1834)]

hostage /ˈhɒstɪdʒ/ n. **1** a person seized or held as security for the fulfilment of a condition. **2** a pledge or security. □ **a hostage to fortune** an acquisition, commitment, etc., regarded as endangered by unforeseen circumstances. □ **hostageship** n. [ME f. OF (h)*ostage* ult. f. LL *obsidatus* hostageship f. L *obses obsidis* hostage]

hostel /ˈhɒst(ə)l/ n. **1** *Brit.* **a** a house of residence or lodging for students, nurses, etc. **b** a place providing temporary accommodation for the homeless etc. **2** = youth hostel. **3** *archaic* an inn. [ME f. OF (h)*ostel* f. med.L (as HOSPITAL)]

hostelling /ˈhɒstəlɪŋ/ n. (*US* **hosteling**) the practice of staying in youth hostels, esp. while travelling. □ **hosteller** n.

hostelry /ˈhɒst(ə)lrɪ/ n. (pl. **-ies**) *archaic* or *literary* an inn. [ME f. OF (h)*ostelerie* f. (h)*ostelier* (as HOSTEL)]

hostess /ˈhəʊstɪs/ n. **1** a woman who receives or entertains a guest. **2** a woman employed to welcome and entertain customers at a nightclub etc. **3** a stewardess on an aircraft, train, etc. (air hostess). [ME f. OF (h)*ostesse* (as HOST²)]

hostile /ˈhɒstaɪl/ adj. **1** of an enemy. **2** (often foll. by to) unfriendly, opposed. □ **hostile witness** *Law* a witness who appears hostile to the party calling him or her and therefore untrustworthy. □ **hostilely** adv. [F *hostile* or L *hostilis* (as HOST¹)]

hostility /hɒˈstɪlɪtɪ/ n. (pl. **-ies**) **1** being hostile, enmity. **2** a state of warfare. **3** (in pl.) acts of warfare. **4** opposition (in thought etc.). [F *hostilité* or LL *hostilitas* (as HOSTILE)]

hostler /ˈɒslə(r)/ n. **1** *hist.* = OSTLER. **2** *US* a person in charge of vehicles or machines, esp. railway engines, when they are not in use. [ME f. *hosteler* (as OSTLER)]

hot /hɒt/ adj., v., & adv. ● adj. (**hotter, hottest**) **1 a** having a relatively or noticeably high temperature. **b** (of food or drink) prepared by heating and served without cooling. **2** producing the sensation of heat (hot fever; hot flush). **3** (of pepper, spices, etc.) pungent. **4** (of a person) feeling heat. **5 a** (often foll. by for, on) eager, keen (in hot pursuit). **b** (foll. by on) colloq. strict with. **c** (foll. by on) colloq. knowledgeable about. **6 a** ardent, passionate, excited. **b** angry or upset. **c** lustful. **d** exciting. **7 a** (of news etc.) fresh, recent. **b** *Brit.* colloq. (of Treasury bills) newly issued. **8** *Hunting* (of the scent) fresh and strong, indicating that the quarry has passed recently. **9 a** (of a player) very skilful. **b** (of a competitor in a race or other sporting event) strongly fancied to win (a hot favourite). **c** (of a hit, return, etc., in ball games) difficult for an opponent to deal with. **d** colloq. currently popular or in demand. **10** (of music, esp. jazz) strongly rhythmical and emotional. **11 a** difficult or awkward to deal with. **b** sl. (of goods) stolen, esp. easily identifiable and hence difficult to dispose of. **c** sl. (of a person) wanted by the police. **12 a** live, at a high voltage. **b** sl. radioactive. **13** colloq. (of information) unusually reliable (hot tip). **14** (of a colour, shade, etc.) suggestive of heat; intense, bright. ● v. (**hotted, hotting**) (usu. foll. by up) *Brit.* colloq. **1** tr. & intr. make or become hot. **2** tr. & intr. make or become active, lively, exciting, or dangerous. ● adv. **1** angrily, severely (give it him hot). **2** eagerly. □ **go hot and cold** feel alternately hot and cold owing to fear etc. **have the hots for** sl. be sexually attracted to. **hot air** colloq. empty, boastful, or excited talk. **hot-air balloon** a balloon (see BALLOON n. 2) consisting of a bag in which air is heated by burners located below it, causing it to rise. **hot blast** a blast of heated air forced into a furnace. **hot-blooded** ardent, passionate. **hot cathode** a cathode heated to emit electrons. **hot cross bun** see BUN. **hot dog** n. **1** a hot sausage sandwiched in a soft roll. **2** *N. Amer.* sl. a person who performs stunts, esp. when skiing or surfing. ● int. *N. Amer.* sl. expressing approval. **hotdog** v.intr. (**-dogged, -dogging**) *N. Amer.* sl. perform stunts, esp. when skiing or surfing. **hot flush** see FLUSH¹ n. 6a. **hot gospeller** see GOSPELLER. **hot line** a direct exclusive line of communication, esp. for emergencies. **hot metal** *Printing* using type made from molten metal. **hot money** capital transferred at frequent intervals. **hot potato** colloq. a controversial or awkward matter or situation. **hot-press** n. a press of glazed boards and hot metal plates for smoothing paper or cloth or making plywood. ● v.tr. press (paper etc.) in this. **hot rod** a motor vehicle modified to have extra power and speed. **hot seat** sl. **1** a position of difficult responsibility. **2** the electric chair. **hot shoe** *Photog.* a socket on a camera with electrical contacts for a flash gun etc. **hot-short** (of metal) brittle in its hot state (cf. COLD-SHORT). **hot spot 1** a small region that is relatively hot. **2** a lively or dangerous place. **hot spring** a spring of naturally hot water. **hot stuff** colloq. **1** a formidably capable person. **2** an important person or thing. **3** a sexually attractive person. **4** a spirited, strong-willed, or passionate person. **5** a book, film, etc. with a strongly erotic content. **hot-tempered** impulsively angry. **hot under the collar** angry, resentful, or embarrassed. **hot war** an open war, with active hostilities. **hot water** *colloq.* difficulty, trouble, or disgrace (*be in hot water; get into hot water*). **hot-water bottle** (*US* **bag**) a container, usu. made of rubber, filled with hot water, esp. to warm a bed. **hot well 1** = *hot spring*. **2** a reservoir in a condensing steam engine. **hot-wire** *attrib.adj.* operated by the expansion of heated wire. ● *v.tr. N. Amer. sl.* start the engine of (a car etc.) by bypassing the ignition system. **make it** (or **things**) **hot for a person** persecute a person. **not so hot** *colloq.* only mediocre. **sell** (or **go**) **like hot cakes** see CAKE. □ **hotly** *adv.* **hotness** *n.* **hottish** *adj.* [OE *hāt* f. Gmc: cf. HEAT]

hotbed /ˈhɒtbed/ *n.* **1** a bed of earth heated by fermenting manure for raising or forcing plants. **2** (foll. by *of*) an environment promoting the growth of something, esp. something unwelcome (*hotbed of vice*).

Hotchkiss /ˈhɒtʃkɪs/ *n.* (in full **Hotchkiss gun**) any of various types of gun, especially a form of machine-gun, designed by the American inventor Benjamin Berkeley Hotchkiss (1826–85) or produced by the company he founded.

hotchpotch /ˈhɒtʃpɒtʃ/ *n.* (also (esp. in sense 3) **hotchpot** /-pɒt/) **1** a confused mixture, a jumble. **2** a dish of many mixed ingredients, esp. a mutton broth or stew with vegetables. **3** *Law* the reunion and blending of properties for the purpose of securing equal division (esp. of the property of an intestate parent). [ME f. AF & OF *hochepot* f. OF *hocher* shake + POT[1]: *-potch* by assim.]

hotel /həʊˈtel/ *n.* **1** an establishment providing accommodation and meals for payment. **2** *Austral. & NZ* a public house. [F *hôtel*, later form of HOSTEL]

hotelier /həʊˈtelɪə(r)/ *n.* a hotel-keeper. [F *hôtelier* f. OF *hostelier*: see HOSTELRY]

hotfoot /ˈhɒtfʊt/ *adv., v., & adj.* ● *adv.* in eager haste. ● *v.tr.* hurry eagerly (esp. *hotfoot it*). ● *adj.* acting quickly.

hothead /ˈhɒthed/ *n.* an impetuous person.

hotheaded /hɒtˈhedɪd/ *adj.* impetuous, excitable. □ **hotheadedly** *adv.* **hotheadedness** *n.*

hothouse /ˈhɒthaʊs/ *n. & adj.* ● *n.* **1** a heated building, usu. largely of glass, for rearing plants out of season or in a climate colder than is natural for them. **2** an environment that encourages the rapid growth or development of something. ● *adj.* (*attrib.*) characteristic of something reared in a hothouse; sheltered, sensitive.

hotplate /ˈhɒtpleɪt/ *n.* a heated metal plate etc. (or a set of these) for cooking food or keeping it hot.

hotpot /ˈhɒtpɒt/ *n.* a casserole of meat and vegetables, usu. with a layer of potato on top.

hotshot /ˈhɒtʃɒt/ *n. & adj.* esp. *US colloq.* ● *n.* **1** an important or exceptionally able person. **2** a player of football, basketball, etc. with exceptionally good aim. ● *attrib.adj.* important, able, expert, suddenly prominent.

Hotspur /ˈhɒtspɜː(r)/ the nickname of Sir Henry Percy (see PERCY).

hotspur /ˈhɒtspɜː(r)/ *n.* a rash person. [sobriquet of PERCY]

Hottentot /ˈhɒt(ə)n,tɒt/ *n.* (*pl.* same or **Hottentots**) a former name for the Nama or Khoikhoi of southern Africa (see NAMA). [Du., prob. orig. a repetitive formula in a Nama dancing-song, transferred by Dutch sailors to the people themselves]

hotter /ˈhɒtə(r)/ *n. Brit. sl.* a person, esp. a youth, who engages in hotting; a joyrider. [cf. HOTTING]

hottie /ˈhɒtɪ/ *n.* (also **hotty**) (*pl.* **-ies**) *colloq.* a hot-water bottle.

hotting /ˈhɒtɪŋ/ *n. Brit. sl.* joyriding in stolen, high-performance cars, esp. dangerously and for display. [f. HOT *adj.*, but perh. infl. by *hot-wire v.*]

Houdini /huːˈdiːnɪ/, Harry (born Erik Weisz) (1874–1926), Hungarian-born American magician and escape artist. In the early 1900s he became famous for his ability to escape from all kinds of bonds and containers, from prison cells to aerially suspended straitjackets.

hough /hɒk/ *n. & v. Brit.* ● *n.* **1** = HOCK[1]. **2** a cut of beef etc. from the hock and the leg above it. ● *v.tr.* hamstring. [ME *ho(u)gh* = OE *hōh* (heel) in *hōhsinu* hamstring]

hoummos var. of HUMMUS.

hound /haʊnd/ *n. & v.* ● *n.* **1 a** a dog used for hunting, esp. one able to track by scent. **b** (**the hounds**) *Brit.* a pack of foxhounds. **2** *colloq.* a despicable man. **3** a runner who follows a trail in hare and hounds.

4 a person keen in pursuit of something (usu. in *comb.*: *news-hound*). ● *v.tr.* **1** harass or pursue relentlessly. **2** chase or pursue with a hound. **3** (foll. by *at*) set (a dog or person) on (a quarry). **4** urge on or nag (a person). □ **hound's tongue** a tall downy plant, *Cynoglossum officinale*, with tongue-shaped leaves and a pungent smell. **hound's-tooth** a check pattern with notched corners suggestive of a canine tooth. **ride to hounds** go fox-hunting on horseback. [OE *hund* f. Gmc]

hour /ˈaʊə(r)/ *n.* **1** a twenty-fourth part of a day and night, 60 minutes. **2** a time of day, a point in time (*a late hour; what is the hour?*). **3** (in *pl.* with preceding numerals in form 18.00, 20.30, etc.) this number of hours and minutes past midnight on the 24-hour clock (*will assemble at 20.00 hours*). **4 a** a period set aside for some purpose (*lunch hour; keep regular hours*). **b** (in *pl.*) a fixed period of time for work, use of a building, etc. (*office hours; opening hours*). **5** a short indefinite period of time (*an idle hour*). **6** the present time (*question of the hour*). **7** a time for action etc. (*the hour has come*). **8** the distance travelled in one hour (*we are an hour from London*). **9** *RC Ch.* **a** prayers to be said at one of seven fixed times of day (*book of hours*). **b** any of these times. **10** (prec. by *the*) each time o'clock of a whole number of hours (*buses leave on the hour; on the half hour; at quarter past the hour*). **11** *Astron.* 15° of longitude or right ascension. □ **after hours** after closing-time. **hour-hand** the hand on a clock or watch which shows the hour. **hour-long** *adj.* lasting for one hour. ● *adv.* for one hour. **till all hours** till very late. [ME *ure* etc. f. AF *ure*, OF *ore*, *eure* f. L *hora* f. Gk *hōra* season, hour]

hourglass /ˈaʊəˌɡlɑːs/ *n. & adj.* ● *n.* a reversible device with two connected glass bulbs containing sand that takes an hour to pass from the upper to the lower bulb. ● *attrib.adj.* shaped like an hourglass, constricted or narrowed in the middle (*an hourglass figure*).

houri /ˈhʊərɪ/ *n.* (*pl.* **houris**) a beautiful young woman, esp. in the Muslim paradise. [F f. Pers. *ḥūrī* f. Arab. *ḥūr* pl. of *ḥawrāʾ* gazelle-like (in the eyes)]

hourly /ˈaʊəlɪ/ *adj. & adv.* ● *adj.* **1** done or occurring every hour. **2** frequent, continual. **3** reckoned hour by hour (*hourly wage*). ● *adv.* **1** every hour. **2** frequently, continually.

house *n. & v.* ● *n.* /haʊs/ (*pl.* /ˈhaʊzɪz/) **1 a** a building for human habitation. **b** (*attrib.*) (of an animal) kept in, frequenting, or infesting houses (*house-cat; housefly*). **2** a building for a special purpose (*opera-house; summer-house*). **3** a building for keeping animals or goods (*hen-house*). **4 a** a religious community. **b** the buildings occupied by it. **5 a** a body of pupils living in the same building at a boarding-school. **b** such a building. **c** a division of a day-school for games, competitions, etc. **6** a college of a university. **7** (usu **House**) a family, esp. a royal family; a dynasty (*House of York*). **8 a** a firm or institution. **b** its place of business. **c** (**the House**) *Brit. colloq.* the Stock Exchange. **9 a** a legislative or deliberative assembly. **b** the building where it meets. **c** (**the House**) (in the UK) the House of Commons or Lords; (in the US) the House of Representatives. **10 a** an audience in a theatre, cinema, etc. **b** a performance in a theatre or cinema (*second house starts at 9 o'clock*). **c** a theatre. **11** *Astrol.* a twelfth part of the heavens. **12** (*attrib.*) living in a hospital as a member of staff (*house officer; house surgeon*). **13 a** a place of public refreshment, a restaurant or inn (*coffee-house; public house*). **b** (*attrib.*) (of wine) selected by the management of a restaurant, hotel, etc. to be offered at a special price. **14** (*US*) a brothel. **15** *Sc.* a dwelling that is one of several in a building. **16** *Brit. sl.* = HOUSEY-HOUSEY. **17** an animal's den, shell, etc. **18** (**the House**) *Brit. hist. euphem.* the workhouse. **19** (also **House**) = house music. ● *v.tr.* /haʊz/ **1** provide (a person, a population, etc.) with a house or houses or other accommodation. **2** store (goods etc.). **3** serve as accommodation for, contain. **b** enclose or encase (a part or fitting). **4** fix in a socket, mortise, etc. □ **as safe as houses** thoroughly or completely safe. **house-agent** *Brit.* an agent for the sale and letting of houses. **house and home** (as an emphatic) home. **house arrest** detention in one's own house etc., not in prison. **house-broken** = house-trained. **house church 1** a charismatic church independent of traditional denominations. **2** a group meeting in a house as part of the activities of a church. **house cricket** see CRICKET[2]. **house-dog** a dog kept to guard a house. **house-father** a man in charge of a house, esp. of a home for children. **house finch** a red-breasted finch, *Carpodacus mexicanus*, common in western North America. **house-flag** a flag indicating to what firm a ship belongs. **house guest** a guest staying for some days in a private house. **house-hunting** the process of seeking a house to live in. **house-husband** a husband who carries out the household duties traditionally carried out by a housewife. **house lights** the lights in the auditorium of a theatre. **house magazine** a magazine published by a firm and dealing mainly with its own activities. **house-martin** a black and white swallow-like bird, *Delichon urbica*, which builds a mud nest on house walls etc. **house-mother** a woman in charge of a house, esp. of a home for children. **house mouse** a usu. grey mouse, *Mus musculus*, very common as a scavenger around human dwellings, and bred as a pet and experimental animal. **house music** a style of popular dance music typically using drum machines, sampled sound effects, and synthesized bass lines with sparse repetitive vocals and a fast beat. **house of cards 1** an insecure scheme etc. **2** a structure built (usu. by a child) out of playing cards. **house of God** a church, a place of worship. **house of ill fame** *archaic* a brothel. **house-parent** a house-mother or house-father. **house party** a group of guests staying at a country house etc. **house plant** a plant grown indoors. **house-proud** attentive to, or unduly preoccupied with, the care and appearance of the home. **house sparrow** a common brown and grey sparrow, *Passer domesticus*, which nests in the eaves and roofs of houses. **house style** a particular printer's or publisher's etc. preferred way of presentation. **house-to-house** performed at or carried to each house in turn. **house-trained** *Brit.* **1** (of animals) trained to be clean in the house. **2** *colloq.* well-mannered. **house-warming** a party celebrating a move to a new home. **keep house** provide for or manage a household. **keep** (or **make**) **a House** secure the presence of enough members for a quorum in the House of Commons. **keep open house** see OPEN. **keep to the house** (or **keep the house**) stay indoors. **like a house on fire 1** vigorously, fast. **2** successfully, excellently. **on the house** at the management's expense, free. **play house** play at being a family in its home. **put** (or **set**) **one's house in order** make necessary reforms. **set up house** begin to live in a separate dwelling. □ **houseful** *n.* (*pl.* **-fuls**). **houseless** *adj.* [OE *hūs, hūsian*, f. Gmc]

houseboat /ˈhaʊsbəʊt/ *n.* a boat fitted up for living in.

housebound /ˈhaʊsbaʊnd/ *adj.* unable to leave one's house through illness etc.

houseboy /ˈhaʊsbɔɪ/ *n.* a boy or man as a servant in a house, esp. in Africa.

housebreaker /ˈhaʊsˌbreɪkə(r)/ *n.* **1** a person guilty of house-breaking; a burglar. **2** *Brit.* a person who is employed to demolish houses.

housebreaking /ˈhaʊsˌbreɪkɪŋ/ *n.* the act of breaking into a building, esp. in daytime, to commit a crime. ¶ In 1968 replaced as a statutory crime in English law by *burglary*.

housecarl /ˈhaʊskɑːl/ *n.* (also **housecarle**) *hist.* a member of the bodyguard of a Danish or English king or noble. [OE *hūscarl* f. ON *húskarl* f. *hús* HOUSE + *karl* man: cf. CARL]

housecoat /ˈhaʊskəʊt/ *n.* a woman's garment for informal wear in the house, usu. a long dresslike coat.

housecraft /ˈhaʊskrɑːft/ *n.* *Brit.* skill in household management.

housefly /ˈhaʊsflaɪ/ *n.* (*pl.* **-flies**) a dipterous fly of the family Muscidae, esp. *Musca domestica*, breeding in decaying organic matter and often entering houses.

household /ˈhaʊshəʊld/ *n.* **1** the occupants of a house regarded as a unit. **2** a house and its affairs. **3** (prec. by *the*) (in the UK) the royal household. □ **household gods 1** gods presiding over a household, esp. *Rom. Hist.* the lares and penates. **2** the essentials of home life. **household troops** (in the UK) troops nominally employed to guard the sovereign. **household word** (or **name**) **1** a familiar name or saying. **2** a familiar person or thing.

householder /ˈhaʊsˌhəʊldə(r)/ *n.* **1** a person who owns or rents a house. **2** the head of a household.

housekeep /ˈhaʊskiːp/ *v.intr.* (*past* and *past part.* **-kept** /-kept/) *colloq.* keep house.

housekeeper /ˈhaʊsˌkiːpə(r)/ *n.* **1** a person, esp. a woman, employed to manage a household. **2** a person in charge of a house, office, etc.

housekeeping /ˈhaʊsˌkiːpɪŋ/ *n.* **1** the management of household affairs. **2** money allowed for this. **3** operations of maintenance, record-keeping, etc., in an organization.

houseleek /ˈhaʊsliːk/ *n.* a succulent plant, *Sempervivum tectorum*, with pink flowers, growing on walls and roofs.

housemaid /ˈhaʊsmeɪd/ *n.* a female servant in a house, esp. in charge of reception rooms and bedrooms. □ **housemaid's knee** inflammation of the kneecap, often due to excessive kneeling.

houseman /ˈhaʊsmən/ *n.* (*pl.* **-men**) **1** *Brit.* a resident doctor at a hospital etc. **2** = HOUSEBOY.

housemaster /ˈhaʊsˌmɑːstə(r)/ *n.* (*fem.* **housemistress** /-ˌmɪstrɪs/) the teacher in charge of a house at a boarding-school.

House of Commons (in the UK) the elected chamber of Parliament. (See PARLIAMENT.)

House of Keys (in the Isle of Man) the elected chamber of Tynwald.

House of Lords 1 (in the UK) the chamber of Parliament composed of peers and bishops. (See PARLIAMENT.) **2** a committee of specially qualified members of this appointed as the ultimate judicial appeal court.

House of Representatives the lower house of the US Congress and other legislatures. (See CONGRESS.)

houseroom /ˈhaʊsruːm, -rʊm/ *n.* space or accommodation in one's house. □ **not give houseroom to** not have in any circumstances.

Houses of Parliament 1 the Houses of Lords and Commons regarded together. (See PARLIAMENT.) **2** the buildings where they meet. (See WESTMINSTER, PALACE OF.)

housetop /ˈhaʊstɒp/ *n.* the roof of a house. □ **proclaim** (or **shout** etc.) **from the housetops** announce publicly.

House Un-American Activities Committee (abbr. **HUAC**) *hist.* a committee of the US House of Representatives established in 1938 to investigate subversives. It became notorious for its zealous investigations of alleged Communists, particularly in the late 1940s, although it was originally intended to pursue Fascists also. The committee's name was changed to the House Committee on Internal Security in 1969. (See also MCCARTHYISM.)

housewife *n.* (*pl.* **-wives**) **1** /ˈhaʊswaɪf/ a woman (usu. married) managing a household. **2** /ˈhʌzɪf/ a case for needles, thread, etc. □ **housewifely** /ˈhaʊsˌwaɪflɪ/ *adj.* [ME *hus(e)wif* f. HOUSE + WIFE]

housewifery /ˈhaʊsˌwɪfrɪ/ *n.* **1** housekeeping. **2** skill in this, housecraft.

housework /ˈhaʊswɜːk/ *n.* regular work done in housekeeping, e.g. cleaning and cooking.

housey-housey /ˌhaʊsɪˈhaʊsɪ, ˌhaʊzɪˈhaʊzɪ/ *n.* (also **housie-housie**) *Brit.* bingo, lotto, or tombola, esp. as played for money.

housing[1] /ˈhaʊzɪŋ/ *n.* **1 a** dwelling-houses collectively. **b** the provision of these. **2** shelter, lodging. **3** a rigid casing, esp. for moving or sensitive parts of a machine. **4** the hole or niche cut in one piece of wood to receive some part of another in order to join them. □ **housing estate** a residential area planned as a unit.

housing[2] /ˈhaʊzɪŋ/ *n.* a cloth covering put on a horse for protection or ornament. [ME = covering, f. obs. *house* f. OF *houce* f. med.L *hultia* f. Gmc]

Housman /ˈhaʊsmən/, A(lfred) E(dward) (1859–1936), English poet and classical scholar. Having failed his final examinations at Oxford University, he worked as a clerk and studied Greek and Latin in his spare time. In 1892 he was appointed professor of Latin at University College, London, on the strength of his classical publications. He is now chiefly remembered for the poems collected in *A Shropshire Lad* (1896), a series of nostalgic verses largely based on ballad forms.

Houston /ˈhjuːstən/ an inland port of Texas, linked to the Gulf of Mexico by the Houston Ship Canal; pop. (1990) 1,630,550. Since 1961 it has been a centre for space research and manned space-flight; it is the site of the NASA Space Centre. The city is named after Samuel Houston (1793–1863), American politician and military leader who led the struggle to win control of Texas (1834–6) and make it part of the US.

houting /ˈhaʊtɪŋ/ *n.* a freshwater whitefish, *Coregonus lavaretus*, of northern Eurasia. [Du., f. MDu. *houtic*]

Hove /həʊv/ a resort town on the southern coast of England in East Sussex, adjacent to Brighton; pop. (1981) 67,140.

hove *past* of HEAVE.

hovel /ˈhɒv(ə)l/ *n.* **1** a small miserable dwelling. **2** a conical building enclosing a kiln. **3** an open shed or outhouse. [ME: orig. unkn.]

hover /ˈhɒvə(r)/ *v. & n.* ● *v.intr.* **1** (of a bird, helicopter, etc.) remain in one place in the air. **2** (often foll. by *about, round*) wait close at hand, linger. **3** remain undecided. ● *n.* **1** hovering. **2** a state of suspense. □ **hoverer** *n.* [ME f. obs. *hove* hover, linger]

hovercraft /ˈhɒvəˌkrɑːft/ *n.* (*pl.* same) a vehicle or craft that travels over land or water on a cushion of air provided by a downward blast. A design was first patented by Christopher Cockerell in 1955, and a hovercraft crossing of the English Channel was made on 25 July 1959, the 50th anniversary of Louis Blériot's historic flight.

hoverfly /ˈhɒvəˌflaɪ/ *n.* (*pl.* **-flies**) a dipterous fly of the family Syrphidae, which hovers with rapidly beating wings.

hoverport /ˈhɒvəˌpɔːt/ *n.* a terminal for hovercraft.

hovertrain /ˈhɒvəˌtreɪn/ n. a train that travels on a cushion of air like a hovercraft.

how[1] /haʊ/ adv., conj., & n. ● *interrog.adv.* **1** by what means, in what way (*how do you do it?*; *tell me how you do it*; *how could you behave so disgracefully?*; *but how to bridge the gap?*). **2** in what condition, esp. of health (*how is the patient?*; *how do things stand?*). **3 a** to what extent (*how far is it?*; *how would you like to take my place?*; *how we laughed!*). **b** to what extent good or well, what … like (*how was the film?*; *how did they play?*). ● *rel.adv.* in whatever way, as (*do it how you can*). ● *conj. colloq.* that (*told us how he'd been in India*). ● n. the way a thing is done (*the how and why of it*). □ **and how!** *colloq.* very much so (chiefly used ironically or intensively). **here's how!** I drink to your good health. **how about 1** would you like (*how about a game of chess?*). **2** what is to be done about. **3** what is the news about. **how are you? 1** what is your state of health? **2** = *how do you do?* **how come?** see COME. **how do?** an informal greeting on being introduced to a stranger. **how do you do?** a formal greeting. **how-do-you-do** (or **how-d'ye-do**) n. (*pl.* **-dos**) an awkward situation. **how many** what number. **how much 1** what amount (*how much do I owe you?*; *did not know how much to take*). **2** what price (*how much is it?*). **3** (as *interrog.*) *joc.* what? ('*She is a hedonist.*' '*A how much?*'). **how now?** *archaic* what is the meaning of this? **how so?** how can you show that that is so? **how's that? 1** what is your opinion or explanation of that? **2** *Cricket* (said to an umpire) is the batsman out or not? [OE *hū* f. WG]

how[2] /haʊ/ int. a greeting attributed to North American Indians. [perh. f. Sioux *háo*, Omaha *hau*]

Howard[1] /ˈhaʊəd/, Catherine (c.1521–1542), fifth wife of Henry VIII. She married Henry soon after his divorce from Anne of Cleves in 1540, probably at the instigation of her ambitious Howard relatives. She was accused of infidelity, confessed, and was beheaded in 1542.

Howard[2] /ˈhaʊəd/, John (1726–90), English philanthropist and prison reformer. In 1773 his sense of horror at conditions in Bedford jail led him to undertake a tour of British prisons; this culminated the following year in two Acts of Parliament setting down sanitary standards. His work *The State of Prisons in England and Wales* (1777) gave further impetus to the movement for improvements in the design and management of prisons.

howbeit /haʊˈbiːɪt/ adv. *archaic* nevertheless.

howdah /ˈhaʊdə/ n. a seat for two or more, usu. with a canopy, for riding on the back of an elephant or camel. [Urdu *hawda* f. Arab. *hawdaj* litter]

howdy /ˈhaʊdɪ/ int. US = *how do you do?* [corrupt.]

Howe /haʊ/, Elias (1819–67), American inventor. In 1846 he patented a sewing machine with an eyed needle to carry the upper thread and a holder resembling a shuttle for the lower thread. The machine's principles were adapted by Isaac Merrit Singer and others, in violation of Howe's patent rights, and it took a seven-year litigation battle to secure the royalties. (See also SEWING MACHINE.)

however /haʊˈevə(r)/ adv. **1 a** in whatever way (*do it however you want*). **b** to whatever extent, no matter how (*must go however inconvenient*). **2** nevertheless. **3** *colloq.* (as an emphatic) in what way, by what means (*however did that happen?*).

howitzer /ˈhaʊɪtsə(r)/ n. a short gun for high-angle firing of shells at low velocities. [Du. *houwitser* f. G *Haubitze* f. Czech *houfnice* catapult]

howl /haʊl/ n. & v. ● n. **1** a long loud doleful cry uttered by a dog, wolf, etc. **2** a prolonged wailing noise, e.g. as made by a strong wind. **3** a loud cry of pain or rage. **4** a yell of derision or merriment. **5** *Electronics* a howling noise in a loudspeaker due to electrical or acoustic feedback. ● v. **1** *intr.* make a howl. **2** *intr.* weep loudly. **3** *tr.* utter (words) with a howl. □ **howl down** prevent (a speaker) from being heard by howls of derision. [ME *houle* (v.), prob. imit.: cf. OWL]

howler /ˈhaʊlə(r)/ n. **1** *colloq.* a glaring mistake. **2** (in full **howler monkey**) a South American monkey of the genus *Alouatta*, with loud howling calls. **3** a person or animal that howls.

howling /ˈhaʊlɪŋ/ adj. **1** that howls. **2** *colloq.* extreme (*a howling shame*). **3** *archaic* dreary (*howling wilderness*). □ **howling dervish** see DERVISH.

Howrah /ˈhaʊrə/ (also **Haora**) a city in eastern India; pop. (1991) 947,000. It is situated on the Hooghly river opposite Calcutta.

howsoever /ˌhaʊsəʊˈevə(r)/ adv. (also *poet.* **howsoe'er** /-ˈeə(r)/) **1** in whatsoever way. **2** to whatsoever extent.

Hoxha /ˈhɒdʒə/, Enver (1908–85), Albanian statesman, Prime Minister 1944–54 and First Secretary of the Albanian Communist Party 1954–85. In 1941 he founded the Albanian Communist Party and led the fight for national independence. As Prime Minister and thereafter First Secretary of the Communist Party's Central Committee, he rigorously isolated Albania from Western influences and implemented a Stalinist programme of nationalization and collectivization.

hoy[1] /hɔɪ/ int. & n. ● int. used to call attention, drive animals, or *Naut.* hail or call aloft. ● n. *Austral.* a game of chance resembling bingo, using playing cards. [ME: natural cry]

hoy[2] /hɔɪ/ n. *hist.* a small vessel, usu. rigged as a sloop, carrying passengers and goods esp. for short distances. [MDu. *hoei*, *hoede*, of unkn. orig.]

hoy[3] /hɔɪ/ v.tr. *Austral. sl.* throw. [Brit. dial.: orig. unkn.]

hoya /ˈhɔɪə/ n. a climbing shrub of the genus *Hoya*, with pink, white, or yellow waxy flowers. [mod.L f. Thomas *Hoy*, Engl. gardener c.1750–c.1821]

hoyden /ˈhɔɪd(ə)n/ n. a boisterous girl. □ **hoydenish** adj. [orig. = rude fellow, prob. f. MDu. *heiden* (= HEATHEN)]

Hoyle[1] /hɔɪl/, Sir Fred (b.1915), English astrophysicist, one of the proponents of the steady-state theory of cosmology. He also formulated theories about the origins of stars, and of the processes by which atoms of the heavier chemical elements are built up within the stars, writing a seminal paper on the subject in 1956 with the American physicist William A. Fowler (1911–95). His later theories have been controversial, including the suggestions that life on earth has an extra-terrestrial origin, and that viruses arrived from space. His publications include works of popular science and science fiction.

Hoyle[2] /hɔɪl/ n. □ **according to Hoyle** adv. correctly, exactly. ● adj. correct, exact. [Edmund *Hoyle*, Engl. writer on card-games (1672–1769)]

h.p. abbr. (also **HP**) **1** horsepower. **2** hire purchase. **3** high pressure.

HQ abbr. headquarters.

HR abbr. House of Representatives.

hr. abbr. hour.

Hradec Králové /ˌhrɑːdets ˈkrɑːləˌveɪ/ (called in German *Königgrätz*) a town in the northern Czech Republic, on the River Elbe; pop. (1991) 161,960. Founded in the 13th century, it is now the capital of East Bohemia region.

HRH abbr. Her or His Royal Highness.

Hrodna /ˈhrɒdnə/ (Russian **Grodno** /ˈgrɒdnə/) a city in western Belarus, on the Neman river near the borders with Poland and Lithuania; pop. (1990) 277,000.

hrs. abbr. hours.

HRT abbr. hormone replacement therapy.

Hrvatska see CROATIA.

HSH abbr. Her or His Serene Highness.

Hsia-men see XIAMEN.

Hsian see XIAN.

Hsining see XINING.

Hsu-chou see XUZHOU.

HT abbr. high tension.

HUAC see HOUSE UN-AMERICAN ACTIVITIES COMMITTEE.

Huainan /hwaɪˈnæn/ a city in the province of Anhui, in east central China; pop. (1986) 1,519,000.

Huallaga /hwɑːˈjɑːɡə/ a river in central Peru, one of the headwaters of the Amazon. Rising in the central Andes, it flows, generally north-eastwards, for 1,100 km (700 miles) and emerges into the Amazon Basin at Lagunas. The remote upper river valley is one of the world's chief coca-growing regions.

Huambo /ˈhwæmbəʊ/ a city in the mountains in western Angola; pop. (est. 1983) 203,000. Founded in 1912, it was known by its Portuguese name of Nova Lisboa until 1978.

Huang Hai /hwæŋ ˈhaɪ/ the Chinese name for the YELLOW SEA.

Huang Ho /hwæŋ ˈhəʊ/ (also **Huang He** /ˈhiː/) the Chinese name for the YELLOW RIVER.

Huascarán /ˌhwæskəˈrɑːn/ an extinct volcano in the Peruvian Andes, west central Peru, rising to 6,768 m (22,205 ft). It is the highest peak in Peru.

hub /hʌb/ n. **1** the central part of a wheel, rotating on or with the axle, and from which the spokes radiate. **2** a central point of interest, activity, etc. □ **hub-cap** a cover for the hub of a vehicle's wheel. [16th c.: perh. = HOB[1]]

Hubble /ˈhʌb(ə)l/, Edwin Powell (1889–1953), American astronomer. In 1929 he demonstrated that the distance of a galaxy is directly proportional to its observed velocity of recession from us (*Hubble's law*), a natural consequence of a uniformly expanding universe. It implies that the age of the universe is inversely proportional to a constant of proportionality (*Hubble's constant*) in the mathematical expression of the law. Current estimates of this constant are still uncertain to a factor of at least two, but suggest an age for the universe of between ten and twenty thousand million years.

hubble-bubble /ˈhʌb(ə)lˌbʌb(ə)l/ *n.* **1** a rudimentary form of hookah. **2** a bubbling sound. **3** confused talk. [redupl. of BUBBLE]

Hubble classification *n. Astron.* a simple method of describing the shapes of galaxies, using subdivisions of each of four basic types (elliptical, spiral, barred spiral, and irregular). Hubble's suggestion that they form an evolutionary sequence is no longer accepted.

Hubble Space Telescope an orbiting astronomical observatory launched in 1990, built and operated jointly by NASA and the European Space Agency. Despite focusing problems due to a faulty mirror, the telescope has produced images of a much higher resolution than is possible for a ground-based telescope. The faults were successfully rectified by astronauts on a special shuttle mission in 1993.

hubbub /ˈhʌbʌb/ *n.* **1** a confused din, esp. from a crowd of people. **2** a disturbance or riot. [perh. of Ir. orig.: cf. Gael. *ubub* int. of contempt, Ir. *abú*, used in battle-cries]

hubby /ˈhʌbɪ/ *n.* (*pl.* **-ies**) *colloq.* a husband. [abbr.]

Hubei /huːˈbeɪ/ (also **Hupeh** /-ˈpeɪ/) a province of eastern China; capital, Wuhan.

Hubli /ˈhuːblɪ/ (also **Hubli-Dharwad** /ˌhuːblɪdɑːˈwɑːd/, **Hubli-Dharwar** /-ˈwɑː(r)/) a city in SW India; pop. (1991) 648,000. It was united with the adjacent city of Dharwad in 1961.

hubris /ˈhjuːbrɪs/ *n.* **1** arrogant pride or presumption. **2** (in Greek tragedy) excessive pride towards or defiance of the gods, leading to nemesis. □ **hubristic** /hjuːˈbrɪstɪk/ *adj.* [Gk]

huckaback /ˈhʌkəˌbæk/ *n.* a stout linen or cotton fabric with a rough surface, used for towelling. [17th c.: orig. unkn.]

huckleberry /ˈhʌk(ə)lbərɪ/ *n.* (*pl.* **-ies**) **1** a dwarf ericaceous shrub of the genus *Gaylussacia*, native to North America. **2** the blue or black soft fruit of this plant. [prob. alt. of *hurtleberry*, WHORTLEBERRY]

huckster /ˈhʌkstə(r)/ *n. & v.* ● *n.* **1** a mercenary person. **2** *US* a publicity agent, esp. for broadcast material. **3** a pedlar or hawker. ● *v.* **1** *intr.* bargain, haggle. **2** *tr.* carry on a petty traffic in. **3** *tr.* adulterate. □ **hucksterism** *n.* [ME prob. f. LG: cf. dial. *huck* to bargain, HAWKER¹]

Huddersfield /ˈhʌdəzˌfiːld/ a town in West Yorkshire; pop. (1981) 148,540.

huddle /ˈhʌd(ə)l/ *v. & n.* ● *v.* **1** *tr. & intr.* (often foll. by *up*) crowd together; nestle closely. **2** *intr. & refl.* (often foll. by *up*) coil one's body into a small space. **3** *tr. Brit.* heap together in a muddle. ● *n.* **1** a confused or crowded mass of people or things. **2** *colloq.* a close or secret conference (esp. in **go into a huddle**). **3** confusion, bustle. [16th c.: perh. f. LG and ult. rel. to HIDE³]

Hudson /ˈhʌds(ə)n/, Henry (*c.*1565–1611), English explorer. He discovered the North American bay, river, and strait which bear his name. In 1607 and 1608 he conducted two voyages in search of the North-east Passage to Asia, reaching Greenland and Spitzbergen on the first and Novaya Zemlya on the second. In 1609 he explored the NE coast of America, sailing up the Hudson River to Albany. During his final voyage in 1610 he attempted to winter in Hudson Bay, but his crew mutinied and set Hudson and a few companions adrift, never to be seen again.

Hudson Bay a large inland sea in NE Canada. It is the largest inland sea in the world and is connected to the North Atlantic Ocean via the Hudson Strait. It was named after the explorer Henry Hudson, who discovered it in 1610.

Hudson River a river of eastern North America, which rises in the Adirondack Mountains and flows southwards for 560 km (350 miles) into the Atlantic at New York. It was named after Henry Hudson, who, in 1609 sailed 240 km (150 miles) up the river as far as Albany.

Hudson's Bay Company formerly, a British colonial trading company set up by Royal Charter in 1670 and granted all lands draining into Hudson Bay in northern Canada for purposes of commercial exploitation, principally trade in fur. The company amalgamated with the rival North-West Company in 1821 and continued to operate in the area until finally handing over control to the new Canadian government in 1870. It is now a Canadian retail and wholesale operation.

Hué /hweɪ/ a city in central Vietnam; pop. (est.) 209,000.

hue /hjuː/ *n.* **1 a** a colour or tint. **b** a variety or shade of colour caused by the admixture of another. **2** the attribute of a colour by virtue of which it is discernible as red, green, etc. □ **-hued** *adj.* **hueless** *adj.* [OE *hīew*, *hēw* form, beauty f. Gmc: cf. ON *hȳ* down on plants]

hue and cry /hjuː/ *n.* a loud clamour, orig. calling for the pursuit and capture of a criminal. In former English law, the cry had to be raised by the inhabitants of a hundred in which a robbery had been committed, if they were not to become liable for the damages suffered by the victim. [AF *hu e cri* f. OF *hu* outcry (f. *huer* shout) + *e* and + *cri* cry.]

huff /hʌf/ *v. & n.* ● *v.* **1** *intr.* give out loud puffs of air, steam, etc. **2** *intr.* bluster loudly or threateningly (*huffing and puffing*). **3** *intr. & tr.* take or cause to take offence. **4** *tr. Draughts* remove (an opponent's man that could have made a capture) from the board as a forfeit (orig. after blowing on the piece). ● *n.* a fit of petty annoyance. □ **in a huff** annoyed and offended. □ **huffish** *adj.* [imit. of the sound of blowing]

huffy /ˈhʌfɪ/ *adj.* (**huffier**, **huffiest**) **1** apt to take offence. **2** offended. □ **huffily** *adv.* **huffiness** *n.*

hug /hʌg/ *v. & n.* ● *v.tr.* (**hugged**, **hugging**) **1** squeeze tightly in one's arms, esp. with affection. **2** (of a bear) squeeze (a person) between its forelegs. **3** keep close to (the shore, kerb, etc.). **4** cherish or cling to (prejudices etc.). **5** *refl.* congratulate or be pleased with (oneself). ● *n.* **1** a strong clasp with the arms. **2** a squeezing grip in wrestling. □ **huggable** *adj.* [16th c.: prob. f. Scand.: cf. ON *hugga* console]

huge /hjuːdʒ/ *adj.* **1** extremely large; enormous. **2** (of immaterial things) very great (*a huge success*). □ **hugeness** *n.* [ME *huge* f. OF *ahuge*, *ahoge*, of unkn. orig.]

hugely /ˈhjuːdʒlɪ/ *adv.* **1** enormously (*hugely successful*). **2** very much (*enjoyed it hugely*).

hugger-mugger /ˈhʌgəˌmʌgə(r)/ *adj., adv., n., & v.* ● *adj. & adv.* **1** in secret. **2** confused; in confusion. ● *n.* **1** secrecy. **2** confusion. ● *v.intr.* proceed in a secret or muddled fashion. [prob. rel. to ME *hoder* huddle, *mokere* conceal: cf. 15th-c. *hoder moder*, 16th-c. *hucker mucker* in the same sense]

Huggins /ˈhʌgɪnz/, Sir William (1824–1910), British astronomer. He pioneered spectroscopic analysis in astronomy, showing that nebulae are composed of luminous gas, and that some comets contain hydrocarbon molecules. He discovered the red shift in stellar spectra, correctly interpreting it as being due to the Doppler effect and using it to measure recessional velocities.

Hughes¹ /hjuːz/, Edward James ('Ted') (b.1930), English poet. His work is pervaded by his vision of the natural world as a place of violence, terror, and beauty, as can be seen in his first volume of poetry, *The Hawk in the Rain* (1957). This vision is continued in later works; *Crow* (1970) explores the legends surrounding creation and birth through the character of the sinister and mocking crow. Hughes was appointed Poet Laureate in 1984. From 1956 to 1963 he was married to the American poet Sylvia Plath.

Hughes² /hjuːz/, Howard (Robard) (1905–76), American industrialist, film producer, and aviator. When his father died in 1924 he took control of the Hughes Tool Company; this formed the basis of his large fortune. He made his début as a film director in 1926; notable titles include *Hell's Angels* (1930) and *The Outlaw* (1941). From 1935 to 1938 he broke many world aviation records, sometimes while flying an aircraft of his own design. For the last twenty-five years of his life he lived as a recluse.

Hughie /ˈhjuːɪ/ *n. Austral. & NZ sl.* the imaginary being responsible for the weather (esp. *send her down, Hughie!*). [dimin. of male forename *Hugh* + -IE]

Hugli see HOOGHLY.

Hugo /ˈhjuːgəʊ/ Victor(-Marie) (1802–85), French poet, novelist, and dramatist. He was a leading figure of French romanticism, and brought a new freedom of diction, subject, and versification to French poetry; his many collections include *Les Feuilles d'automne* (1831). He set out his ideas on drama in the preface to his play *Cromwell* (1827); this included the view that the theatre should express both the grotesque and the sublime of human existence and became a manifesto of the romantic movement. The success of his drama *Hernani* (1830) signalled the triumph of romanticism over the conventions which had prevailed in French theatre since the time of Racine and Corneille. *Notre Dame de Paris* (1831) and *Les Misérables* (1862) are among his best-known novels and demonstrate Hugo's concern for social and political issues. Between 1851 and 1870 he lived in exile in Guernsey, where he wrote his satire against Napoleon III (*Les Châtiments*, 1853).

Huguenot /ˈhjuːgəˌnəʊ, -ˌnɒt/ *n.* a French Protestant of the 16th–17th centuries. Largely Calvinist, the Huguenots suffered severe persecution, most notably the Massacre of St Bartholomew in 1572. As a political faction they were involved in the French Wars of Religion against the Catholic majority, a conflict that was ended temporarily by the Edict of Nantes (1598), which guaranteed the Huguenots freedom of worship. However, civil wars occurred again in the 1620s, after which the military and political power of the Huguenots was destroyed by Cardinal Richelieu's government. The Edict of Nantes was revoked by Louis XIV in 1685 and Protestantism was not legalized again in France until after the Revolution; many thousands of Huguenots emigrated from France. [F, assim. of *eiguenot* (f. Du. *eedgenot* f. Swiss G *Eidgenoss* confederate) to the name of a Geneva burgomaster *Hugues*]

huh /hə/ *int.* expressing disgust, surprise, etc. [imit.]

Huhehot see HOHHOT.

hula /ˈhuːlə/ *n.* (also **hula-hula**) a Hawaiian dance with six basic steps and gestures. □ **hula hoop** a large hoop for spinning round the body by movement of the waist and hips. **hula skirt** a long grass skirt. [Hawaiian]

hulk /hʌlk/ *n.* **1 a** the body of a dismantled ship, used as a store vessel etc. **b** (in *pl.*) *hist.* this used as a prison. **2** an unwieldy vessel. **3** *colloq.* a large clumsy-looking person or thing. [OE *hulc* & MLG, MDu. *hulk*: cf. Gk *holkas* cargo ship]

hulking /ˈhʌlkɪŋ/ *adj. colloq.* bulky; large and clumsy.

Hull /hʌl/ (official name *Kingston-upon-Hull*) a city and port in Humberside, situated at the junction of the Hull and Humber rivers; pop. (1991) 252,200. It is linked to the south bank of the Humber estuary by a suspension bridge, completed in 1981.

hull[1] /hʌl/ *n. & v.* ● *n.* the body or frame of a ship, airship, flying boat, etc. ● *v.tr.* pierce the hull of (a ship) with gunshot etc. [ME, perh. rel. to HOLD[2]]

hull[2] /hʌl/ *n. & v.* ● *n.* **1** the outer covering of a fruit, esp. the pod of peas and beans, the husk of grain, or the green calyx of a strawberry. **2** a covering. ● *v.tr.* remove the hulls from (fruit etc.). [OE *hulu* ult. f. *helan* cover: cf. HEEL[3]]

hullabaloo /ˌhʌləbəˈluː/ *n.* (*pl.* **hullabaloos**) an uproar or clamour. [18th c.: redupl. of *hallo, hullo,* etc.]

hullo var. of HELLO.

hum[1] /hʌm/ *v. & n.* ● *v.* (**hummed, humming**) **1** *intr.* make a low steady continuous sound like that of a bee. **2** *tr.* (also *absol.*) sing (a wordless tune) with closed lips. **3** *intr.* utter a slight inarticulate sound. **4** *intr. colloq.* be in an active state (*really made things hum*). **5** *intr. Brit. colloq.* smell unpleasantly. ● *n.* **1** a humming sound. **2** an unwanted low-frequency noise caused by variation of electric current, usu. the alternating frequency of the mains, in an amplifier etc. **3** *Brit. colloq.* a bad smell. □ **hum and haw** (or **ha**) hesitate, esp. in speaking. □ **hummable** *adj.* **hummer** *n.* [ME, imit.]

hum[2] /həm/ *int.* expressing hesitation or dissent. [imit.]

human /ˈhjuːmən/ *adj. & n.* ● *adj.* **1** of or belonging to the genus *Homo*. **2** consisting of human beings (*the human race*). **3** of or characteristic of humankind as opposed to God or animals or machines, esp. susceptible to the weaknesses of humankind (*is only human*). **4** showing (esp. the better) qualities of man (*proved to be very human*). ● *n.* a human being. □ **human being** a man, woman, or child of the species *Homo sapiens*, distinguished from other animals by superior mental development, power of articulate speech, and upright stance. **human chain** a line of people formed for passing things along, e.g. buckets of water to the site of a fire. **human engineering 1** the management of industrial labour, esp. as regards man–machine relationships. **2** the study of this. **human equation** = *personal equation* 2. **human geography** the branch of geography that deals with the activities of humankind as they affect or are influenced by the earth's surface. **human interest** (in a newspaper story etc.) reference to personal experience and emotions etc. **human nature** the general characteristics and feelings of humankind. **human relations** relations with or between people or individuals. **human rights** rights held to be justifiably belonging to any person. **human shield** a person or persons placed in the line of fire in order to discourage attack. □ **humanness** /-mənnɪs/ *n.* [ME *humain(e)* f. OF f. L *humanus* f. *homo* human being]

humane /hjuːˈmeɪn/ *adj.* **1** benevolent, compassionate. **2** inflicting the minimum of pain. **3** (of a branch of learning) tending to civilize or confer refinement. □ **humane killer** an instrument for the painless slaughter of animals. □ **humanely** *adv.* **humaneness** *n.* [var. of HUMAN, differentiated in sense in the 18th c.]

humanism /ˈhjuːməˌnɪz(ə)m/ *n.* an outlook or system of thought attaching prime importance to human rather than divine or supernatural matters. The term humanism does not refer to a unified theory. Historically, it was first applied to studies promoting human culture, and especially to the cultural movement of the Renaissance, which turned away from medieval scholasticism (with its theological bias) to value the human achievement of ancient Greece and Rome. In philosophy, the term has encompassed systems of thought stressing rational enquiry and human experience over abstract theorizing or orthodox religion. More broadly, humanist beliefs stress the potential value and goodness of human beings, emphasize common human needs, and seek solely rational ways of solving human problems.

humanist /ˈhjuːmənɪst/ *n.* **1** an adherent of humanism. **2** a humanitarian. **3** a student (esp. in the 14th–16th centuries) of Roman and Greek literature and antiquities. □ **humanistic** /ˌhjuːməˈnɪstɪk/ *adj.* **humanistically** *adv.* [F *humaniste* f. It. *umanista* (as HUMAN)]

humanitarian /hjuːˌmænɪˈteərɪən/ *n. & adj.* ● *n.* **1** a person who seeks to promote human welfare. **2** a person who advocates or practises humane action; a philanthropist. ● *adj.* relating to or holding the views of humanitarians. □ **humanitarianism** *n.*

humanity /hjuːˈmænɪtɪ/ *n.* (*pl.* **-ies**) **1 a** the human race. **b** human beings collectively. **c** the fact or condition of being human. **2** humaneness, benevolence. **3** (in *pl.*) human attributes. **4** (in *pl.*) learning or literature concerned with human culture, esp. the study of Roman and Greek literature and philosophy. [ME f. OF *humanité* f. L *humanitas -tatis* (as HUMAN)]

humanize /ˈhjuːməˌnaɪz/ *v.tr.* (also **-ise**) **1** make human; give a human character to. **2** make humane. □ **humanization** /ˌhjuːmənaɪˈzeɪʃ(ə)n/ *n.* [F *humaniser* (as HUMAN)]

humankind /ˈhjuːmənˌkaɪnd/ *n.* human beings collectively.

humanly /ˈhjuːmənlɪ/ *adv.* **1** by human means (*I will do it if it is humanly possible*). **2** in a human manner. **3** from a human point of view. **4** with human feelings.

Humber /ˈhʌmbə(r)/ **1** an estuary in NE England. It is formed at the junction of the rivers Ouse and Trent, near Goole, and flows 60 km (38 miles) eastwards to enter the North Sea at Spurn Head. The major port of Hull is situated on its north bank. The estuary is spanned by the world's largest suspension bridge, opened in 1981, and with a span of 1,410 m (4,626 ft). **2** a shipping forecast area covering an area of the North Sea off eastern England, extending roughly from the latitude of north Norfolk to that of Flamborough Head.

Humberside /ˈhʌmbəˌsaɪd/ a county of NE England; administrative centre, Beverley. It was formed in 1974 from parts of the East and West Ridings of Yorkshire and the northern part of Lincolnshire.

humble /ˈhʌmb(ə)l/ *adj. & v.* ● *adj.* (**humbler, humblest**) **1 a** having or showing a low estimate of one's own importance. **b** offered with or affected by such an estimate (*if you want my humble opinion*). **2** of low social or political rank (*humble origins*). **3** (of a thing) of modest pretensions, dimensions, etc. ● *v.tr.* **1** make humble; bring low; abase. **2** lower the rank or status of. □ **eat humble pie** make a humble apology; accept humiliation. □ **humbleness** *n.* **humbly** *adv.* [ME *umble, humble* f. OF *umble* f. L *humilis* lowly f. *humus* ground: *humble pie* f. UMBLES]

humble-bee /ˈhʌmb(ə)lˌbiː/ *n.* = BUMBLE-BEE. [ME prob. f. MLG *hummelbē*, MDu. *hommel*, OHG *humbal*]

Humboldt /ˈhʌmbəʊlt/, Friedrich Heinrich Alexander, Baron von (1769–1859), German explorer and scientist. Humboldt travelled in Central and South America (1799–1804) and wrote extensively on natural history, meteorology, and physical geography. He proved that the Amazon and Orinoco river systems are connected, and ascended to 5,877 m (19,280 ft) in the Andes, the highest ascent ever made at that time, researching the relation of temperature and altitude. He spent the next twenty years in Paris writing up his results, returning later to Berlin, where he served at the Prussian court. He wrote a popular work in several volumes, *Kosmos* (1845–62), describing the structure of the universe as it was then known.

humbug /ˈhʌmbʌg/ *n. & v.* ● *n.* **1** deceptive or false talk or behaviour. **2** an impostor. **3** *Brit.* a hard boiled sweet usu. flavoured with peppermint. ● *v.* (**humbugged, humbugging**) **1** *intr.* be or behave like an impostor. **2** *tr.* deceive, hoax. □ **humbuggery** *n.* [18th c.: orig. unkn.]

humdinger /ˈhʌmˌdɪŋə(r)/ *n. sl.* an excellent or remarkable person or thing. [20th c.: orig. unkn.]

humdrum /ˈhʌmdrʌm/ *adj. & n.* ● *adj.* **1** commonplace, dull. **2** monotonous. ● *n.* **1** commonplaceness, dullness. **2** a monotonous routine etc. [16th c.: prob. f. HUM[1] by redupl.]

Hume /hjuːm/, David (1711–76), Scottish philosopher, economist, and historian. His philosophy rejected the possibility of certainty in knowledge, and he agreed with John Locke that there are no innate ideas, only a series of subjective sensations, and that all the data of reason stem from experience. His philosophical legacy is particularly evident in the work of 20th-century empiricist philosophers. In economics, he attacked mercantilism and anticipated the views of economists such as Adam Smith. Among his chief works are *A Treatise of Human Nature* (1739–40) and a five-volume *History of England* (1754–62).

humectant /hjuːˈmektənt/ adj. & n. ● adj. retaining or preserving moisture. ● n. a substance, esp. a food additive, used to reduce loss of moisture. [L (h)*umectant*- part. stem of (h)*umectare* moisten f. *umere* be moist]

humeral /ˈhjuːmərəl/ adj. **1** of the humerus or shoulder. **2** worn on the shoulder. [F *huméral* & LL *humeralis* (as HUMERUS)]

humerus /ˈhjuːmərəs/ n. (pl. **humeri** /-ˌraɪ/) **1** the bone of the upper arm in humans. **2** the corresponding bone in other vertebrates. [L, = shoulder]

humic /ˈhjuːmɪk/ adj. of or consisting of humus.

humid /ˈhjuːmɪd/ adj. (of the air or climate) warm and damp. □ **humidly** adv. [F *humide* or L *humidus* f. *umere* be moist]

humidifier /hjuːˈmɪdɪˌfaɪə(r)/ n. a device for keeping the atmosphere moist in a room etc.

humidify /hjuːˈmɪdɪˌfaɪ/ v.tr. (**-ies, -ied**) make (air etc.) humid or damp. □ **humidification** /-ˌmɪdɪfɪˈkeɪʃ(ə)n/ n.

humidity /hjuːˈmɪdɪtɪ/ n. (pl. **-ies**) **1** a humid state. **2** moisture. **3** the degree of moisture esp. in the atmosphere. □ **relative humidity** the proportion of moisture to the value for saturation at the same temperature. [ME f. OF *humidité* or L *humiditas* (as HUMID)]

humidor /ˈhjuːmɪˌdɔː(r)/ n. a room or container for keeping cigars or tobacco moist. [HUMID after CUSPIDOR]

humify /ˈhjuːmɪˌfaɪ/ v.tr. & intr. (**-ies, -ied**) make or be made into humus. □ **humification** /ˌhjuːmɪfɪˈkeɪʃ(ə)n/ n.

humiliate /hjuːˈmɪlɪˌeɪt/ v.tr. make humble; injure the dignity or self-respect of. □ **humiliating** adj. **humiliatingly** adv. **humiliator** n. **humiliation** /-ˌmɪlɪˈeɪʃ(ə)n/ n. [LL *humiliare* (as HUMBLE)]

humility /hjuːˈmɪlɪtɪ/ n. **1** humbleness, meekness. **2** a humble condition. [ME f. OF *humilité* f. L *humilitas -tatis* (as HUMBLE)]

hummingbird /ˈhʌmɪŋˌbɜːd/ n. a very small nectar-feeding tropical American bird of the family Trochilidae, that makes a humming sound by the vibration of its wings when it hovers.

humming-top /ˈhʌmɪŋˌtɒp/ n. a child's top which hums as it spins.

hummock /ˈhʌmək/ n. **1** a hillock or knoll. **2** US a piece of forested ground rising above a marsh. **3** a hump or ridge in an ice-field. □ **hummocky** adj. [16th c.: orig. unkn.]

hummus /ˈhʊməs/ n. (also **hoummos**) a thick dip or spread made from ground chick peas and sesame oil flavoured with lemon and garlic. [Turk. *humus* mashed chick peas]

humongous /hjuːˈmʌŋɡəs/ adj. (also **humungous**) sl. extremely large or massive. [20th c.: orig. uncert.]

humor US var. of HUMOUR.

humoral /ˈhjuːmərəl/ adj. **1** hist. of the four bodily humours. **2** Med. (of an immune response) of or relating to antibodies in the body fluids as distinct from cells. [F *humoral* or med.L *humoralis* (as HUMOUR)]

humoresque /ˌhjuːməˈresk/ n. a short lively piece of music. [G *Humoreske* f. *Humor* HUMOUR]

humorist /ˈhjuːmərɪst/ n. **1** a facetious person. **2** a humorous talker, actor, or writer. □ **humoristic** /ˌhjuːməˈrɪstɪk/ adj.

humorous /ˈhjuːmərəs/ adj. **1** showing humour or a sense of humour. **2** facetious, comic. □ **humorously** adv. **humorousness** n.

humour /ˈhjuːmə(r)/ n. & v. (US **humor**) ● n. **1 a** the quality of being amusing or comic. **b** the expression of humour in literature, speech, etc. **2** (in full **sense of humour**) the ability to perceive or express humour or take a joke. **3** a mood or state of mind (*bad humour*). **4** an inclination or whim (*in the humour for fighting*). **5** (in full **cardinal humour**) hist. each of the four chief fluids of the body (blood, phlegm, choler, melancholy), held (in Galen's theory) to determine a person's physical and mental qualities. ● v.tr. **1** gratify or indulge (a person or taste etc.). **2** adapt oneself to; make concessions to. □ **aqueous humour** Anat. the clear fluid in the eye between the lens and the cornea. **out of humour** displeased. **vitreous humour** (or **body**) Anat. the transparent jelly-like tissue filling the eyeball behind the lens. □ **-humoured** adj. **humourless** adj. **humourlessly** adv.

humourlessness n. [ME f. AF *umour*, *humour*, OF *umor*, *humor* f. L *humor* moisture (as HUMID)]

humous /ˈhjuːməs/ adj. like or consisting of humus.

hump /hʌmp/ n. & v. ● n. **1** a rounded protuberance on the back of a camel etc., or as an abnormality on a person's back. **2** a rounded raised mass of earth etc. **3** a mound over which railway vehicles are pushed so as to run by gravity to the required place in a marshalling yard. **4** a critical point in an undertaking, ordeal, etc. **5** (prec. by *the*) Brit. sl. a fit of depression or vexation (*it gives me the hump*). **6** coarse sl. an act of sexual intercourse; a sexual partner. ● v.tr. **1 a** (often foll. by *about*) colloq. lift or carry (heavy objects etc.) with difficulty. **b** esp. Austral. hoist up, shoulder (one's pack etc.). **2** make hump-shaped. **3** annoy, depress. **4** coarse sl. have sexual intercourse with. □ **hump bridge** = *humpback bridge*. **live on one's hump** colloq. be self-sufficient. **over the hump** over the worst; well begun. □ **humped** adj. **humpless** adj. [17th c.: perh. rel. to LG *humpel* hump, LG *humpe*, Du. *homp* lump, hunk (of bread)]

humpback /ˈhʌmpbæk/ n. **1 a** a back with a hump. **b** a person having this. **2** (in full **humpback whale**) a baleen whale, *Megaptera novaeangliae*, with a dorsal fin forming a hump, long fins, and a remarkable 'song'. □ **humpback bridge** Brit. a small bridge with a steep ascent and descent. □ **humpbacked** adj.

Humperdinck /ˈhʌmpəˌdɪŋk, ˈhʊm-/, Engelbert (1854–1921), German composer. He was influenced by Wagner, whose opera *Parsifal* he helped to prepare for performance. He is chiefly remembered as the composer of the opera *Hänsel und Gretel* (1893).

humph /həmf/ int. & n. an inarticulate sound expressing doubt or dissatisfaction. [imit.]

Humphries /ˈhʌmfrɪz/, (John) Barry (b.1934), Australian comedian. He became widely known with his creation and impersonation of the female celebrity 'Dame Edna Everage', which was first shown on British television in the late 1970s.

humpty-dumpty /ˌhʌmptɪˈdʌmptɪ/ n. (pl. **-ies**) **1** a short dumpy person. **2** a person or thing that once overthrown cannot be restored. [the nursery rhyme *Humpty-Dumpty*, perh. ult. f. HUMPY[1], DUMPY]

humpy[1] /ˈhʌmpɪ/ adj. (**humpier, humpiest**) **1** having a hump or humps. **2** humplike.

humpy[2] /ˈhʌmpɪ/ n. (pl. **-ies**) Austral. a primitive hut. [Aboriginal *oompi*, infl. by HUMP]

humungous var. of HUMONGOUS.

humus /ˈhjuːməs/ n. the organic constituent of soil, usu. formed by the decomposition of plants and leaves by soil bacteria. □ **humusify** v.tr. & intr. (**-ies, -ied**) [L, = soil]

Hun /hʌn/ n. **1** a member of a warlike Asiatic nomadic people who invaded and ravaged Europe in the 4th–5th centuries. **2** offens. a German (esp. in military contexts). **3** an uncivilized devastator; a vandal. □ **Hunnish** adj. [OE *Hūne* pl. f. LL *Hunni* f. Gk *Hounnoi* f. Turkic *Hun-yü*]

Hunan /huːˈnæn/ a province of east central China; capital, Changsha.

hunch /hʌntʃ/ v. & n. ● v. **1** tr. bend or arch into a hump. **2** tr. thrust out or up to form a hump. **3** intr. (usu. foll. by *up*) esp. N. Amer. sit with the body hunched. ● n. **1** colloq. an intuitive feeling or conjecture. **2** US colloq. a hint. **3** a hump. **4** a thick piece. [16th c.: orig. unkn.]

hunchback /ˈhʌntʃbæk/ n. = HUMPBACK. □ **hunchbacked** adj.

hundred /ˈhʌndrəd/ n. & adj. ● n. (pl. **hundreds** or (in sense 1) **hundred**) (in sing., prec. by *a* or *one*) **1** the product of ten and ten. **2** a symbol for this (100, c, C). **3** a set of a hundred things. **4** (in sing. or pl.) colloq. a large number. **5** (in pl.) the years of a specified century (*the seventeen hundreds*). **6** Brit. hist. a subdivision of a county or shire, having its own court. ● adj. **1** that amount to a hundred. **2** used to express whole hours in the 24-hour system (*thirteen hundred hours*). □ **a** (or **one**) **hundred per cent** adv. entirely, completely. ● adj. **1** entire, complete. **2** (usu. with *neg.*) fully recovered. **hundreds and thousands** tiny coloured sweets used chiefly for decorating cakes etc. □ **hundredfold** adj. & adv. **hundredth** adj. & n. [OE f. Gmc]

Hundred Flowers a period of debate in China 1956–7, when, under the slogan 'Let a hundred flowers bloom and a hundred schools of thought contend', citizens were invited to voice their opinions of the Communist regime. The period was initiated by Mao Zedong and others after Khrushchev's denunciation of Stalin, but was forcibly ended after social unrest and fierce criticism of the government and led to the prosecution of those who had exercised their right to voice their opinions.

hundredweight /ˈhʌndrədˌweɪt/ n. (pl. same or **-weights**) **1** (in full

long hundredweight) *Brit.* a unit of weight equal to 112 lb avoirdupois (about 50.8 kg). **2** (in full **metric hundredweight**) a unit of weight equal to 50 kg. **3** (in full **short hundredweight**) *US* a unit of weight equal to 100 lb (about 45.4 kg).

Hundred Years War a war between France and England, conventionally dated 1337–1453. The war consisted of a series of conflicts in which successive English kings attempted to dominate France; it began when Edward III claimed the throne of France following the death of the last Capetian king. Despite an early string of English military successes, most notably Crécy and Poitiers, the House of Valois retained its position, and in the reign of Edward's son, Richard II, hostilities ceased almost completely. The English claim was revived by Henry V, who renewed hostilities in 1415 with a crushing victory at Agincourt and occupied much of northern France. England once again proved unable to consolidate the advantage, however, and under the regency of Henry VI gradually lost control of conquered territory to French forces, revitalized in the first instance by Joan of Arc. With the exception of Calais, all English conquests had been lost by 1453.

hung past and past part. of HANG.

Hungarian /hʌŋˈɡeərɪən/ *n. & adj.* ● *n.* **1 a** a native or national of Hungary. **b** a person of Hungarian descent. **2** the official language of Hungary, spoken also by some 2.5 million people in Romania. Hungarian is a Finno-Ugric language that is the only major language of the Ugric branch. ● *adj.* of or relating to Hungary or its people or language. [med.L *Hungaria* f. *Hungari* Magyar nation]

Hungary /ˈhʌŋɡərɪ/ (called in Hungarian *Magyarország*) a country in central Europe; pop. (est. 1990) 10,600,000; official language, Hungarian; capital, Budapest. Settled by the Magyars in the 9th century, Hungary was the centre of a strong Magyar kingdom in the late Middle Ages, but was conquered first by the Turks in the 16th century and then by the Habsburgs in the 17th century. In the 19th century, nationalist pressure resulted in increased Hungarian power and autonomy within the empire, which led to the establishment of the Dual Monarchy in 1867 (see AUSTRIA–HUNGARY). Following the collapse of the empire in 1918, Hungary became an independent kingdom, although it lost territory to neighbouring states. After participation in the Second World War on the Axis side, Hungary was occupied by the USSR, and became a Communist state. A liberal reform movement was crushed by Soviet troops in 1956. The Communist system was abandoned towards the end of 1989.

hunger /ˈhʌŋɡə(r)/ *n. & v.* ● *n.* **1** a feeling of pain or discomfort, or (in extremes) an exhausted condition, caused by lack of food. **2** (often foll. by *for, after*) a strong desire. ● *v.intr.* **1** (often foll. by *for, after*) have a craving or strong desire. **2** feel hunger. □ **hunger strike** the refusal of food as a form of protest, esp. by prisoners. **hunger striker** a person who takes part in a hunger strike. [OE *hungor, hyngran* f. Gmc]

hunger march *n.* a march undertaken by a group of people in protest against unemployment or poverty, esp. those organized by unemployed workers in Britain during the 1920s and 1930s. The demonstrations started from various cities, and their destination was usually London; the first took place in 1922 from Glasgow, the most famous from Jarrow in 1936.

hungry /ˈhʌŋɡrɪ/ *adj.* (**hungrier, hungriest**) **1** feeling hunger; needing food. **2** indicating hunger. **3 a** (often foll. by *for*) eager, greedy, craving. **b** *Austral.* mean, stingy. **4** (of soil) poor, barren. □ **hungrily** *adv.* **hungriness** *n.* [OE *hungrig* (as HUNGER)]

hunk /hʌŋk/ *n.* **1 a** a large piece cut off (*a hunk of bread*). **b** a thick or clumsy piece. **2** *colloq.* **a** a very large person. **b** a sexually attractive, ruggedly handsome man. □ **hunky** *adj.* (**hunkier, hunkiest**). [19th c.: prob. f. Flem. *hunke*]

hunker /ˈhʌŋkə(r)/ *v.intr.* esp. *Sc. & US* **1** (often foll. by *down*) squat or crouch so that the haunches nearly touch the heels, esp. for shelter or concealment. **2** (foll. by *down*) apply oneself, knuckle down. [rel. to MDu. *hucken*, MLG *hūken*, ON *húka*; cf. G *hocken* squat]

hunkers /ˈhʌŋkəz/ *n.pl.* the haunches. [orig. Sc., f. HUNKER]

hunky-dory /ˌhʌŋkɪˈdɔːrɪ/ *adj. colloq.* excellent. [19th c.: orig. unkn.]

Hunt /hʌnt/, (William) Holman (1827–1910), English painter. In 1848 he co-founded the Pre-Raphaelite Brotherhood and was the only member of the group to remain true to its aims. He made several visits to Egypt and the Holy Land to ensure that his biblical scenes accurately reflected local settings, as in *The Scapegoat* (1855). Much of his painting has a didactic or moral purpose which is reinforced by his extensive use of symbolism, as in *The Light of the World* (1854).

hunt /hʌnt/ *v. & n.* ● *v.* **1** *tr.* (also *absol.*) **a** pursue and kill (wild animals

or game) for sport or food. **b** (esp. in England) pursue (esp. foxes) with hounds, usu. on horseback. **c** use (hounds or a horse) for hunting. **d** (of an animal) chase (its prey). **2** *intr.* (foll. by *after, for*) seek, search (*hunting for a pen*). **3** *intr.* **a** oscillate. **b** (of an engine etc.) run alternately too fast and too slow. **4** *tr.* **a** (foll. by *away* etc.) drive off by pursuit. **b** pursue with hostility. **5** *tr.* scour (a district) in pursuit of game. **6** *tr.* (as **hunted** *adj.*) (of a look etc.) expressing alarm or terror as of one being hunted. **7** *tr.* (foll. by *down, up*) move the place of (a bell) in ringing the changes. ● *n.* **1** the practice of hunting or an instance of this. **2 a** an association of people engaged in hunting with hounds. **b** an area where hunting takes place. **3** an oscillating motion. □ **hunt down** pursue and capture. **hunt out** find by searching; track down. [OE *huntian*, weak grade of *hentan* seize]

huntaway /ˈhʌntəˌweɪ/ *n. Austral. & NZ* a dog trained to drive sheep forward.

Hunter /ˈhʌntə(r)/, John (1728–93), Scottish anatomist. Hunter is regarded as a founder of scientific surgery and made valuable investigations in pathology, physiology, dentistry, and biology. His large museum collection of comparative anatomy was eventually passed to the Royal College of Surgeons in London.

hunter /ˈhʌntə(r)/ *n.* **1 a** (*fem.* **huntress** /-trɪs/) a person or animal that hunts. **b** a horse used in hunting. **2** a person who seeks something. **3** a watch with a hinged cover protecting the glass. □ **hunter's moon** the next full moon after the harvest moon.

hunter-gatherer /ˌhʌntəˈɡæðərə(r)/ *n.* a member of a people who live chiefly by hunting and fishing, and harvesting wild food. Such peoples are typically organized into small nomadic groups in which the women collect food while the men hunt; at one time all humans probably lived in this way, but now only a few groups (including some pygmies and Australian Aboriginals) remain as hunter-gatherers.

hunting /ˈhʌntɪŋ/ *n.* the practice of pursuing and killing wild animals, esp. for sport. □ **hunting crop** see CROP *n.* 3. **hunting dog 1** a dog used for hunting game. **2** a wild dog, *Lycaon pictus*, native to southern Africa, hunting its prey in packs. **hunting-ground 1** a place suitable for hunting. **2** a source of information or object of exploitation likely to be fruitful. **hunting horn** a straight horn used in hunting. **hunting pink** see PINK¹ *n.* 4. [OE *huntung* (as HUNT)]

Huntingdon¹ /ˈhʌntɪŋdən/ a town in Cambridgeshire, on the River Ouse; pop. (1981) 17,600.

Huntingdon² /ˈhʌntɪŋdən/, Selina, Countess of (title of Selina Hastings, née Shirley) (1707–91), English religious leader. On her husband's death in 1746 she devoted herself to religious and social work and was instrumental in introducing Methodism to the upper classes. Following the expulsion of six theological students from Oxford University on allegations of Methodism, she established Trevecca House in mid-Wales as a college for the training of evangelical clergymen in 1768. She was a follower of the English evangelical preacher George Whitefield (1714–70) and made him her chaplain. The Calvinistic Methodist chapels which she helped to establish are still known as 'Countess of Huntingdon chapels'.

Huntingdonshire /ˈhʌntɪŋdənˌʃɪə(r)/ a former county of SE England. It became part of Cambridgeshire in 1974.

Huntington /ˈhʌntɪŋtən/ a city in West Virginia, on the Ohio river; pop. (1990) 54,840. It was founded in 1871 by Collis P. Huntington (1821–1900), president of the Chesapeake and Ohio Railroad.

Huntington Beach a city on the Pacific coast, to the south of Long Beach, in southern California; pop. (1990) 181,520. It is noted as a surfing locality.

Huntington's chorea *n. Med.* see CHOREA. [George *Huntington*, Amer. neurologist (1851–1916)]

huntsman /ˈhʌntsmən/ *n.* (*pl.* **-men**) **1** a hunter. **2** a hunt official in charge of hounds.

Huntsville /ˈhʌntsvɪl/ a city in northern Alabama; pop. (1990) 159,790. It is a centre for space exploration and solar energy research.

Hupeh see HUBEI.

hurdle /ˈhɜːd(ə)l/ *n. & v.* ● *n.* **1 a** each of a series of light frames to be cleared by athletes in a race. **b** (in *pl.*) a hurdle-race. **2** an obstacle or difficulty. **3** a portable rectangular frame strengthened with withes or wooden bars, used as a temporary fence etc. **4** *hist.* a frame on which traitors were dragged to execution. ● *v.* **1** *intr.* run in a hurdle-race. **b** *tr.* clear (a hurdle). **2** *tr.* fence off etc. with hurdles. **3** *tr.* overcome (a difficulty). [OE *hyrdel* f. Gmc]

hurdler /ˈhɜːdlə(r)/ *n.* **1** an athlete who runs in hurdle-races. **2** a person who makes hurdles.

hurdy-gurdy /ˈhɜːdɪˌɡɜːdɪ/ n. (pl. **-ies**) a musical instrument producing a droning sound, played by turning a handle, esp. one with a rosined wheel turned by the right hand to sound drone-strings, and with keys played by the left hand. The instrument was known in Europe from the early 12th century, used at first in choir schools; today it is mainly heard as a folk instrument, esp. in France. The term is also used loosely to refer to a barrel-organ. [prob. imit.]

hurl /hɜːl/ v. & n. ● v. **1** tr. throw with great force. **2** tr. utter (abuse etc.) vehemently. **3** intr. play hurling. ● n. **1** a forceful throw. **2** the act of hurling. [ME, prob. imit., but corresp. in form and partly in sense with LG *hurreln*]

Hurler's syndrome /ˈhɜːləz/ n. Med. a defect in metabolism resulting in mental retardation, a protruding abdomen, and deformities of the bones, including an abnormally large head. Also called *gargoylism*. [Gertrud *Hurler*, Ger. paediatrician, *fl.* 1920]

hurling /ˈhɜːlɪŋ/ n. (also **hurley** /-lɪ/) an Irish game somewhat resembling hockey, played with broad sticks. It is the national game of Ireland, mentioned in folk-tales and possibly dating back to the second millennium BC, although its rules were not standardized until 1884.

hurly-burly /ˈhɜːlɪˌbɜːlɪ/ n. boisterous activity; commotion. [redupl. f. HURL]

Huron /ˈhjʊərən/ n. & adj. ● n. (pl. same or **Hurons**) **1** a member of a confederation of five Iroquoian peoples formerly inhabiting a region adjacent to Lake Huron. **2** the Iroquoian language of these peoples. ● adj. of or relating to the Hurons or their language.

Huron, Lake the second largest of the five Great Lakes of North America, on the border between Canada and the US.

hurrah /hʊˈrɑː/ int., n., & v. = HURRAY.

hurray /hʊˈreɪ/ int., n., & v. ● int. & n. an exclamation of joy or approval. ● v.intr. cry or shout 'hurray'. [alt. of earlier *huzza*, perh. orig. a sailor's cry when hauling]

Hurri /ˈhʊrɪ/ n. (pl. same or **Hurris**) a member of a people, originally from Armenia, who settled in northern Mesopotamia and Syria during the 3rd–2nd millennium BC and were later aborbed by the Hittites and Assyrians. (See also MITANNI.) [Hittite & Assyrian *Harri*, *Hurri*]

Hurrian /ˈhʊrɪən/ n. & adj. **1** of, relating to, or denoting the Hurri. **2** the language of the Hurri, written in cuneiform, of unknown derivation (it is neither Semitic nor Indo-European).

Hurricane /ˈhʌrɪkən, -ˌkeɪn/ a British single-seat fighter aircraft of the Second World War, powered by a single piston engine. The Hawker Hurricane is remembered in particular for its role in the Battle of Britain along with the Spitfire.

hurricane /ˈhʌrɪkən, -ˌkeɪn/ n. **1** a storm with a violent wind, esp. a tropical cyclone in the West Indies or America. **2** Meteorol. a wind of force 12 on the Beaufort scale (75 m.p.h. or 32.7 metres per second or more). **3** a violent commotion. □ **hurricane-bird** a frigate bird. **hurricane-deck** a light upper deck on a ship etc. **hurricane-lamp** an oil lamp designed to resist a high wind. [Sp. *huracan* & Port. *furacão* of Carib orig.]

hurry /ˈhʌrɪ/ n. & v. ● n. **1 a** great haste. **b** (with neg. or interrog.) a need for haste (*there is no hurry*; *what's the hurry?*). **2** (often foll. by *for*, or *to* + infin.) eagerness to get a thing done quickly. ● v. (**-ies, -ied**) **1** intr. move or act with great or undue haste. **2** tr. (often foll. by *away*, *along*) cause to move or proceed in this way. **3** tr. (as **hurried** adj.) hasty; done rapidly owing to lack of time. □ **hurry along** (or **up**) make or cause to make haste. **in a hurry** hurrying; rushed; in a rushed manner. **2** colloq. easily or readily (*you will not beat that in a hurry*; *shall not ask again in a hurry*). □ **hurriedly** adv. **hurriedness** n. [16th c.: imit.]

hurry-scurry /ˌhʌrɪˈskʌrɪ/ n., adj., & adv. ● n. disorderly haste. ● adj. & adv. in confusion. [jingling redupl. of HURRY]

hurst /hɜːst/ n. **1** a hillock. **2** a sandbank in the sea or a river. **3** a wood or wooded eminence. [OE *hyrst*, rel. to OS, OHG *hurst*, *horst*]

Hurston /ˈhɜːstən/, Zora Neale (1901–60), American novelist. In 1928 she graduated in cultural anthropology and continued her studies into the folklore of the Deep South until 1932. Her novels, which include *Jonah's Gourd Vine* (1934), *Moses, Man of the Mountain* (1938), and *Seraph on the Suwanee* (1948), reflect her continuing interest in folklore. Her work was largely ignored after her death until the novelist Alice Walker instigated a revival of interest in her writings, many of which, including her autobiography *Dust Tracks on a Road* (1942), have since been reprinted.

hurt /hɜːt/ v., n., & adj. ● v. (past and past part. **hurt**) **1** tr. (also absol.) cause pain or injury to. **2** tr. cause mental pain or distress to (a person, feelings, etc.). **3** intr. (of a part of the body) suffer pain or harm; be a source of pain (*my arm hurts*). **4** tr. cause damage to, be detrimental to. **5** intr. cause hurt or pain (*my shoes hurt*) **6** intr. US (of a person) suffer pain or distress (*she lost the job and now the whole family's hurting*) **7** intr. (foll. by *for*) US have a pressing need for. ● n. **1** bodily or material injury. **2** harm, wrong. **3** mental pain or distress. ● adj. expressing emotional pain; distressed, aggrieved. [ME f. OF *hurter*, *hurt* ult. perh. f. Gmc]

hurtful /ˈhɜːtfʊl/ adj. causing (esp. mental) hurt; causing damage or harm. □ **hurtfully** adv. **hurtfulness** n.

hurtle /ˈhɜːt(ə)l/ v. **1** intr. & tr. move or hurl rapidly or with a clattering sound. **2** intr. come with a crash. [HURT in obs. sense 'strike forcibly']

Husain var. of HUSSEIN², HUSSEIN³.

Husák /ˈhuːsæk/, Gustáv (1913–91), Czechoslovak statesman, leader of the Communist Party of Czechoslovakia 1969–87 and President 1975–89. He succeeded Alexander Dubček as leader of the Communist Party, following the latter's removal in the wake of the Soviet military invasion of 1968. Husák's objectives were to re-establish order after the Prague Spring, to purge the party of its reformist element, and to implement a new federalist constitution. He was ousted during the velvet revolution of 1989.

husband /ˈhʌzbənd/ n. & v. ● n. a married man esp. in relation to his wife. ● v.tr. manage thriftily; use (resources) economically. □ **husbander** n. **husbandhood** n. **husbandless** adj. **husbandly** adj. [OE *hūsbonda* house-dweller f. ON *húsbóndi* (as HOUSE, *bóndi* one who has a household)]

husbandry /ˈhʌzbəndrɪ/ n. **1** farming. **2 a** management of resources. **b** careful management.

hush /hʌʃ/ v., int., & n. ● v. **1** tr. & intr. (often as **hushed** adj.) make or become silent, quiet, or muted. **2** tr. calm (disturbance, disquiet, etc.); soothe, allay. ● int. calling for silence. ● n. an expectant stillness or silence. □ **hush money** money paid to prevent the disclosure of a discreditable matter. **hush puppy** US quickly fried maize bread. **hush up** suppress public mention of (an affair). [back-form. f. obs. *husht* int., = quiet!, taken as a past part.]

hushaby /ˈhʌʃəˌbaɪ/ int. (also **hushabye**) used to lull a child.

hush-hush /ˈhʌʃˈhʌʃ/ adj. colloq. (esp. of an official plan or enterprise etc.) highly secret or confidential.

husk /hʌsk/ n. & v. ● n. **1** the dry outer covering of some fruits or seeds, esp. of a nut or N. Amer. maize. **2** the worthless outside part of a thing. ● v.tr. remove a husk or husks from. [ME, prob. f. LG *hüske* sheath, dimin. of *hūs* HOUSE]

husky¹ /ˈhʌskɪ/ adj. (**huskier, huskiest**) **1** (of a person or voice) dry in the throat; hoarse. **2** of or full of husks. **3** dry as a husk. **4** tough, strong, hefty. □ **huskily** adv. **huskiness** n.

husky² /ˈhʌskɪ/ n. (pl. **-ies**) a powerful breed of dog used in the Arctic for pulling sledges. [perh. contr. f. ESKIMO]

Huss /hʌs/, John (Czech name Jan Hus) (c.1372–1415), Bohemian religious reformer. He was a preacher in Prague and a rector of Prague University; his support for the views of Wyclif and his attacks on ecclesiastical abuses aroused the hostility of the Church and he was excommunicated in 1411. He was later tried (1414) and burnt at the stake. (See also HUSSITE.)

huss /hʌs/ n. dogfish as food. [ME *husk*, of unkn. orig.]

hussar /hʊˈzɑː(r)/ n. **1** a soldier of a light cavalry regiment. **2** a Hungarian light horseman of the 15th century. [Hungarian *huszár* f. OSerb. *husar* f. It. *corsaro* CORSAIR]

Hussein¹, Abdullah ibn, see ABDULLAH IBN HUSSEIN.

Hussein² /hʊˈseɪn/, ibn Talal (also **Husain**) (b.1935), king of Jordan since 1953. Throughout his reign Hussein has steered a middle course in his policies, seeking to maintain good relations both with the West and with other Arab nations. His moderate policies led to conflict with the Palestinians who had entered Jordan after the Six Day War of 1967, and after a short civil war in 1970 the Palestinians were expelled. In 1990 he acted as a mediator between the opposing sides following the Iraqi invasion of Kuwait, but in the subsequent Gulf War Jordan was the only Middle Eastern country to give open support to Iraq. In 1994 he signed a treaty normalizing relations with Israel.

Hussein³ /hʊˈseɪn/, Saddam (also **Husain**) (full name Saddam bin Hussein at-Takriti) (b.1937), Iraqi President, Prime Minister, and head of the armed forces since 1979. In 1968 he played a leading role in the coup which returned the Baath Socialist Party to power. As President he suppressed opposing parties, built up the army and its weaponry, and made himself the object of an extensive personality cult. During his presidency Iraq fought a war with Iran (1980–8) and invaded Kuwait

(1990), from which Iraqi forces were expelled in the Gulf War of 1991. He also ordered punitive attacks on Kurdish rebels in the north of Iraq and on the Marsh Arabs in the south.

Husserl /ˈhʊsɜːl/, Edmund (Gustav Albrecht) (1859–1938), German philosopher. His work forms the basis of the school of phenomenology; having originally trained as a mathematician he turned to philosophy, seeking in his work the clarity and certainty he found in mathematics and science. He rejected metaphysical assumptions about what actually exists, and explanations of why it exists, in favour of pure subjective consciousness as the condition for all experience, with the world as the object of this consciousness. He taught at the University of Freiberg, where Martin Heidegger was among his pupils.

Hussite /ˈhʌsaɪt/ n. a member or follower of a religious and nationalist movement begun by John Huss. After Huss's execution the Hussites, with the support of many Bohemian nobles, took up arms against the Holy Roman Empire and demanded a set of reforms that anticipated the Reformation. The movement was split by a schism in which the more extreme faction was defeated, but most of the demands of the moderate Hussites were then granted (1436), and a Church was established that remained independent of the Roman Catholic Church until 1620. An early Protestant group that arose among the Hussites, the Bohemian Brethren, is thought to have formed the basis of the Moravian Church. □ **Hussitism** n.

hussy /ˈhʌsɪ, ˈhʌzɪ/ n. (pl. **-ies**) derog. an impudent or immoral girl or woman. [phonetic reduction of HOUSEWIFE (the orig. sense)]

hustings /ˈhʌstɪŋz/ n. **1** parliamentary election proceedings. **2** Brit. hist. a platform from which (before 1872) candidates for Parliament were nominated and addressed electors. [late OE husting f. ON hústhing house of assembly]

hustle /ˈhʌs(ə)l/ v. & n. ● v. **1** tr. push roughly; jostle. **2** tr. **a** (foll. by into, out of, etc.) force, coerce, or deal with hurriedly or unceremoniously (hustled them out of the room). **b** (foll. by into) coerce hurriedly (was hustled into agreeing). **3** intr. push one's way; hurry, bustle. **4** tr. colloq. **a** obtain by forceful action. **b** swindle. **5** intr. sl. engage in prostitution. ● n. **1 a** an act or instance of hustling. **b** forceful or strenuous activity. **2** colloq. a fraud or swindle. [MDu. husselen shake, toss, frequent. of hutsen, orig. imit.]

hustler /ˈhʌslə(r)/ n. **1** an active, enterprising, or unscrupulous individual. **2** sl. a prostitute.

Huston /ˈhjuːstən/, John (1906–87), American-born film director. After a varied background as a boxer, cavalryman, journalist, and actor, he made his début as a film director in 1941 with The Maltese Falcon. A number of successful adventure films followed, including The Asphalt Jungle (1950), The African Queen (1951), and Moby Dick (1956); more recent successes include Prizzi's Honour (1985). He became an Irish citizen in 1964.

hut /hʌt/ n. & v. ● n. **1** a small simple or crude house or shelter. **2** Mil. a temporary wooden etc. house for troops. ● v. (**hutted**, **hutting**) **1** tr. provide with huts. **2** tr. Mil. place (troops etc.) in huts. **3** intr. lodge in a hut. □ **hutlike** adj. [F hutte f. MHG hütte]

hutch /hʌtʃ/ n. **1** a box or cage, usu. with a wire mesh front, for keeping small pet animals. **2** derog. a small house. [ME, = coffer, f. OF huche f. med.L hutica, of unkn. orig.]

hutment /ˈhʌtmənt/ n. Mil. an encampment of huts.

Hutton[1] /ˈhʌt(ə)n/, James (1726–97), Scottish geologist. Hutton's views, controversial at the time, became accepted tenets of modern geology. In opposition to Abraham Werner's Neptunian theory, he emphasized heat as the principal agent in the formation of land masses, and held that rocks such as granite were igneous in origin. He described the processes of deposition and denudation and proposed that such phenomena, operating over millions of years, would account for the present configuration of the earth's surface; it therefore followed that the earth was very much older than was believed. Hutton's views were not widely known until a concise account was published in 1802.

Hutton[2] /ˈhʌt(ə)n/ Sir Leonard ('Len') (1916–90), English cricketer. In his long career he played for Yorkshire (1934–55) and for England (1937–55). He scored a record 364 in the 1938 test against Australia and became the first professional captain of the England team in 1953.

Hutu /ˈhuːtuː/ n. & adj. ● n. (pl. same, **Hutus**, or **Bahutu** /bəˈhuː-/) a member of a Bantu-speaking people, traditionally farmers, who form the majority of the population in Rwanda and Burundi but who were formerly dominated by the Tutsi minority. ● adj. of or relating to this people. [Bantu]

Huxley[1] /ˈhʌkslɪ/, Aldous (Leonard) (1894–1963), English novelist and essayist. During the 1920s and 1930s he lived in Italy and France; his fiction during this period included Antic Hay (1923) and the futuristic Brave New World (1932), probably his best-known work. In 1937 he left for California, where he remained for the rest of his life and pursued his interests in Eastern mysticism and parapsychology. In 1953 he experimented with psychedelic drugs, writing of his experiences with mescalin in The Doors of Perception (1954).

Huxley[2] /ˈhʌkslɪ/, Sir Julian (1887–1975), English biologist. He contributed to the early development of the study of animal behaviour, was a notable interpreter of science to the public through writing and broadcasting, and became the first director-general of UNESCO (1946–8). He was the grandson of Thomas Henry Huxley.

Huxley[3] /ˈhʌkslɪ/, Thomas Henry (1825–95), English biologist. A qualified surgeon, Huxley made his reputation as a marine biologist during service as a ship's surgeon off the coast of northern Australia. Later he turned to the study of fossils, especially of fishes and reptiles, and became a leading supporter of Darwinism in opposition to Richard Owen. On the basis of a detailed study in anthropology he wrote Man's Place in Nature (1863), and coined the word agnostic to describe his own beliefs. Huxley was a supporter of education for the less privileged and argued for the inclusion of science in the school curriculum. His grandson was Sir Julian Huxley.

Huygens /ˈhaɪɡənz/, Christiaan (1629–95), Dutch physicist, mathematician, and astronomer. Although probably best known for his pendulum-regulated clock, which he patented in 1657 (see CLOCK[1]), his work covered a number of fields. He improved the lenses of his telescope, discovered a satellite of Saturn, and also that planet's rings, whose nature had eluded Galileo. In dynamics he studied centrifugal force and the problem of colliding bodies, but his greatest contribution was his wave theory of light. He formulated the principle that every point on a wave front is the centre of a new wave, and this allowed him to explain reflection and refraction, including the double refraction of light in some minerals.

Huygens eyepiece n. a simple eyepiece consisting of two separate planoconvex lenses, used chiefly in refracting telescopes of long focal length.

Hwange /ˈhwæŋɡ/ a town in western Zimbabwe; pop. (1982) 39,000. It was known as Wankie until 1982 and is the centre of the country's coal-mining industry. Nearby is the Hwange National Park, established as a game reserve in 1928 and as a park in 1949.

HWM abbr. high-water mark.

hwyl /ˈhuːɪl/ n. an emotional quality inspiring impassioned eloquence. [Welsh]

Hy. abbr. Henry.

hyacinth /ˈhaɪəˌsɪnθ/ n. **1** a bulbous plant of the genus Hyacinthus, with racemes of usu. purplish-blue, pink, or white bell-shaped fragrant flowers. **2** = grape hyacinth. **3** the purplish-blue colour of hyacinth flowers. **4** an orange variety of zircon used as a precious stone. **5** poet. hair or locks like the hyacinth flower (as a Homeric epithet of doubtful sense). □ **wild** (or **wood**) **hyacinth** = BLUEBELL 1. □ **hyacinthine** /ˌhaɪəˈsɪnθaɪn/ adj. [F hyacinthe f. L hyacinthus f. Gk huakinthos, flower and gem, also the name of a youth loved by Apollo (see HYACINTHUS)]

Hyacinthus /ˌhaɪəˈsɪnθəs/ Gk Mythol. a pre-Hellenic god, said to have been a beautiful boy whom the god Apollo loved but killed accidentally with a discus. From his blood Apollo caused the flower that bears his name to spring up.

Hyades /ˈhaɪəˌdiːz/ n.pl. Astron. an open star cluster in the constellation of Taurus, appearing to surround the bright star Aldebaran. Its heliacal rising was once thought to foretell rain. [ME f. Gk Huades (by popular etymology f. huō rain, but perh. f. hus pig)]

hyaena var. of HYENA.

hyalin /ˈhaɪəlɪn/ n. (also **hyaline**) Anat. & Zool. a translucent substance, esp. produced as a result of the degeneration of certain body tissues. [Gk hualos glass + -IN]

hyaline /ˈhaɪəlɪn, -ˌlaɪn, -ˌliːn/ adj. & n. ● adj. **1** glasslike, vitreous, transparent. **2** Anat. & Zool. characterized by the formation of hyaline material. ● n. **1** literary a smooth sea, clear sky, etc. **2** Anat. & Zool. = HYALIN. □ **hyaline cartilage** n. Anat. a translucent bluish-white cartilage present in the joints, respiratory tract, and immature skeleton. [L hyalinus f. Gk hualinos f. hualos glass]

hyalite /ˈhaɪəˌlaɪt/ n. a colourless variety of opal. [Gk hualos glass]

hyaloid /ˈhaɪəˌlɔɪd/ adj. Anat. glassy. □ **hyaloid membrane** a thin transparent membrane enveloping the vitreous humour of the eye. [F hyaloïde f. LL hyaloides f. Gk hualoeidēs (as HYALITE)]

hyaluronic acid /ˌhaɪəljʊˈrɒnɪk/ n. Biochem. a viscous fluid carbohydrate found in synovial fluid, the vitreous humour of the eye, etc. [HYALOID + -uronic chem. suffix]

hybrid /ˈhaɪbrɪd/ n. & adj. ● n. **1** Biol. the offspring of two plants or animals of different species or varieties. **2** often offens. a person of mixed racial or cultural origin. **3** a thing composed of incongruous elements, e.g. a word with parts taken from different languages. ● adj. **1** bred as a hybrid from different species or varieties. **2** Biol. heterogeneous. **3** of mixed character; derived from incongruous elements or unlike sources. □ **hybrid vigour** Biol. heterosis. □ **hybridism** n. **hybridity** /haɪˈbrɪdɪtɪ/ n. [L hybrida, (h)ibrida offspring of a tame sow and wild boar, child of a freeman and slave, etc.]

hybridize /ˈhaɪbrɪˌdaɪz/ v. (also **-ise**) **1** tr. subject (a species etc.) to cross-breeding. **2** intr. **a** produce hybrids. **b** (of an animal or plant) interbreed. □ **hybridizable** adj. **hybridization** /ˌhaɪbrɪdaɪˈzeɪʃ(ə)n/ n.

hydatid /ˈhaɪdətɪd/ n. Med. **1** a cyst containing watery fluid (esp. one formed by, and containing, a tapeworm larva). **2** a tapeworm larva. □ **hydatidiform** /ˌhaɪdəˈtɪdɪˌfɔːm/ adj. [mod.L hydatis f. Gk hudatis -idos watery vesicle f. hudōr hudatos water]

Hyde[1] /haɪd/, Edward, see CLARENDON.

Hyde[2] /haɪd/, Mr, see JEKYLL.

Hyde Park the largest of the royal parks, in west central London, between Bayswater Road and Kensington Road. The Serpentine, Marble Arch, the Albert Memorial, and Speakers' Corner are all situated in the park.

Hyderabad /ˈhaɪdərəˌbæd/ **1** a city in central India, capital of the state of Andhra Pradesh; pop. (1991) 3,005,000. **2** hist. a former large princely state of south central India, divided in 1956 between Maharashtra, Mysore, and Andhra Pradesh. **3** a city in SE Pakistan, in the province of Sind, on the River Indus; pop. (est. 1991) 1,000,000.

Hydra /ˈhaɪdrə/ **1** Gk Mythol. a many-headed snake of the marshes of Lerna in the Peloponnese, whose heads grew again as they were cut off, killed by Hercules. **2** Astron. the largest constellation (the Water Snake or Sea Monster), said to represent the beast slain by Hercules. It contains few bright stars.

hydra /ˈhaɪdrə/ n. **1** a freshwater polyp of the genus Hydra, with tubular body and tentacles around the mouth. **2** a water snake. **3** something which is hard to destroy. [ME f. L f. Gk hudra water snake, esp. HYDRA]

hydrangea /haɪˈdreɪndʒə/ n. a shrub of the genus Hydrangea, with large white, pink, or blue flowers. [mod.L f. Gk hudōr water + aggos vessel (from the cup-shape of its seed-capsule)]

hydrant /ˈhaɪdrənt/ n. a pipe (esp. in a street) with a nozzle to which a hose can be attached for drawing water from the main. [irreg. f. HYDRO- + -ANT]

hydrate /ˈhaɪdreɪt/ n. & v. Chem. ● n. a compound of water combined with another compound or with an element. ● v.tr. **1 a** combine chemically with water. **b** (as **hydrated** adj.) chemically bonded to water. **2** cause to absorb water. □ **hydratable** adj. **hydrator** n. **hydration** /haɪˈdreɪʃ(ə)n/ n. [F f. Gk hudōr water]

hydraulic /haɪˈdrɔːlɪk, -ˈdrɒlɪk/ adj. **1** (of water, oil, etc.) conveyed through pipes or channels usu. by pressure. **2** (of a mechanism etc.) operated by liquid moving in this manner (hydraulic brakes; hydraulic lift). **3** of or concerned with hydraulics (hydraulic engineer). **4** hardening under water (hydraulic cement). □ **hydraulic press** a device in which the force applied to a fluid creates a pressure which when transmitted to a larger volume of fluid gives rise to a greater force. **hydraulic ram** an automatic pump in which the kinetic energy of a descending column of water raises some of the water above its original level. □ **hydraulically** adv. **hydraulicity** /ˌhaɪdrɔːˈlɪsɪtɪ/ n. [L hydraulicus f. Gk hudraulikos f. hudōr water + aulos pipe]

hydraulics /haɪˈdrɔːlɪks, -ˈdrɒlɪks/ n.pl. (usu. treated as sing.) the science of the conveyance of liquids through pipes etc. esp. as motive power.

hydrazine /ˈhaɪdrəˌziːn/ n. Chem. a colourless alkaline liquid (chem. formula: N_2H_4), which is a powerful reducing agent and is used as a rocket propellant. [HYDROGEN + AZO- + -INE[4]]

hydride /ˈhaɪdraɪd/ n. Chem. a binary compound of hydrogen with an element, esp. with a metal.

hydriodic acid /ˌhaɪdrɪˈɒdɪk, -draɪˈɒdɪk/ n. Chem. a strong acid and reducing agent consisting of the gas hydrogen iodide (chem. formula: HI) dissolved in water.

hydro /ˈhaɪdrəʊ/ n. (pl. **-os**) colloq. **1** a hotel or clinic etc. orig. providing hydropathic treatment. **2** a hydroelectric power plant. [abbr.]

hydro- /ˈhaɪdrəʊ/ comb. form (also **hydr-** before a vowel) **1** having to do

with water (hydroelectric). **2** Med. affected with an accumulation of serous fluid (hydrocele). **3** Chem. combined with hydrogen (hydrochloric). [Gk hudro- f. hudōr water]

hydrobromic acid /ˌhaɪdrəʊˈbrəʊmɪk/ n. Chem. a strong acid consisting of the gas hydrogen bromide (chem. formula: HBr) dissolved in water.

hydrocarbon /ˌhaɪdrəʊˈkɑːb(ə)n/ n. Chem. a compound of hydrogen and carbon.

hydrocele /ˈhaɪdrəˌsiːl/ n. Med. the accumulation of serous fluid in a body sac.

hydrocephalus /ˌhaɪdrəˈsefələs, -ˈkefələs/ n. Med. an accumulation of fluid in the brain, esp. in young children, which makes the head enlarge and can cause mental handicap. □ **hydrocephalic** /-sɪˈfælɪk, -kɪˈfælɪk/ adj.

hydrochloric acid /ˌhaɪdrəˈklɒrɪk, -ˈklɔːrɪk/ n. Chem. a strong acid consisting of the gas hydrogen chloride (chem. formula: HCl) dissolved in water.

hydrochloride /ˌhaɪdrəˈklɔːraɪd/ n. Chem. a compound of an organic base with hydrochloric acid.

hydrocortisone /ˌhaɪdrəˈkɔːtɪˌzəʊn/ n. Biochem. a steroid hormone produced by the adrenal cortex, used medicinally to treat inflammation and rheumatism. Also called cortisol.

hydrocyanic acid /ˌhaɪdrəsaɪˈænɪk/ n. Chem. a highly poisonous volatile liquid (chem. formula: HCN) with a characteristic odour of bitter almonds. Also called hydrogen cyanide, prussic acid.

hydrodynamics /ˌhaɪdrəʊdaɪˈnæmɪks/ n. the science of forces acting on or exerted by fluids (esp. liquids). □ **hydrodynamic** adj. **hydrodynamical** adj. **hydrodynamicist** /-mɪsɪst/ n. [mod.L hydrodynamicus (as HYDRO-, DYNAMIC)]

hydroelectric /ˌhaɪdrəʊɪˈlektrɪk/ adj. **1** generating electricity by utilization of water-power. (See note below.) **2** (of electricity) generated in this way. □ **hydroelectricity** /-ˌɪlekˈtrɪsɪtɪ, -ˌel-/ n.

▪ Hydroelectric power is the most widely used form of renewable energy, accounting for nearly a fifth of the world's electric power. It is particularly suitable for hilly country with substantial rainfall, and tens of gigawatts are generated by the largest power stations. The first was commissioned in northern England in 1880 and the principle has remained the same since then; a dam or barrage is used to hold back a head of water, which generates electricity as it passes out past turbines that drive generators.

hydrofining /ˈhaɪdrəʊˌfaɪnɪŋ/ n. a catalytic process in which a petroleum product is stabilized and its sulphur content reduced by treatment with gaseous hydrogen under relatively mild conditions, so that unsaturated hydrocarbons and sulphur compounds undergo selective hydrogenation. □ **hydrofined** adj. [HYDRO- + refining (see REFINE)]

hydrofluoric acid /ˌhaɪdrəʊˈfluərɪk/ n. Chem. a fuming liquid, sufficiently corrosive to attack glass, consisting of the gas hydrogen fluoride (chem. formula: HF) dissolved in water.

hydrofoil /ˈhaɪdrəˌfɔɪl/ n. a boat equipped with a device consisting of planes which work in a similar way to those of an aerofoil, serving to increase speed by lifting the hull out of the water. The first true hydrofoil is thought to have been built by Enrico Forlanini, an Italian inventor, in 1898–1905, but such craft were not widely used until the 1950s. [HYDRO-, after AEROFOIL]

hydrogel /ˈhaɪdrəʊˌdʒel/ n. a gel in which the liquid component is water.

hydrogen /ˈhaɪdrədʒən/ n. a colourless, odourless, highly inflammable, gaseous chemical element (atomic number 1; symbol H). (See note below.) □ **hydrogen bond** a weak electrostatic interaction between an electronegative atom and a hydrogen atom bonded to a different electronegative atom. **hydrogen cyanide** hydrocyanic acid. **hydrogen peroxide** a colourless viscous unstable liquid (chem. formula: H_2O_2) with strong oxidizing properties, used as a mild antiseptic and bleach. **hydrogen sulphide** a colourless poisonous gas (chem. formula: H_2S) with a disagreeable smell, formed by rotting animal matter. □ **hydrogenous** /haɪˈdrɒdʒɪnəs/ adj. [F hydrogène (as HYDRO-, -GEN)]

▪ Hydrogen was first recognized by Henry Cavendish in 1766 and named by Antoine Lavoisier. It is the lightest of the elements and has the simplest atomic structure, a single electron orbiting the nucleus which (in the commonest isotope) consists of a single proton (see also DEUTERIUM, TRITIUM). Hydrogen is by far the commonest element in the universe, although not on the earth, where it occurs chiefly combined with oxygen as water. It is also a constituent of nearly all organic compounds.

hydrogenase /haɪˈdrɒdʒɪˌneɪz, -ˌneɪs/ n. Biochem. an enzyme that catalyses the reduction of a substrate by hydrogen, as in some micro-organisms.

hydrogenate /haɪˈdrɒdʒɪˌneɪt, ˈhaɪdrədʒəˌneɪt/ v.tr. charge with or cause to combine with hydrogen. □ **hydrogenation** /-ˌdrɒdʒɪˈneɪʃ(ə)n, ˌhaɪdrədʒə-/ n.

hydrogen bomb n. a bomb whose destructive power comes from the rapid release of nuclear energy by fusion of hydrogen nuclei. In a hydrogen bomb there is a thermonuclear fusion reaction involving nuclei of two hydrogen isotopes, deuterium and tritium; the reaction produces helium nuclei and releases energy in relatively greater quantities than does nuclear fission. A temperature of about 10 million °C is needed to sustain fusion and this is provided by using an atom bomb as a trigger; various kinds of weapons combining fission and fusion have been developed. The first hydrogen bomb, more than a thousand times more powerful than the first atom bomb, was tested by the US in November 1952, the USSR following suit a few months later. Also called H-bomb.

hydrogeology /ˌhaɪdrəʊdʒɪˈɒlədʒɪ/ n. the branch of geology dealing with underground and surface water. □ **hydrogeologist** n. **hydrogeological** /-dʒɪəˈlɒdʒɪk(ə)l/ adj.

hydrography /haɪˈdrɒɡrəfɪ/ n. the science of surveying and charting seas, lakes, rivers, etc. □ **hydrographer** n. **hydrographic** /ˌhaɪdrəˈɡræfɪk/ adj. **hydrographical** adj. **hydrographically** adv.

hydroid /ˈhaɪdrɔɪd/ n. & adj. Zool. ● n. a hydrozoan coelenterate of the order Hydroida, in which the polyp phase is predominant, e.g. hydra, sea-anemones, corals, etc. ● adj. of or relating to this order. [mod.L Hydroida f. HYDRA]

hydrolase /ˈhaɪdrəʊˌleɪz, -ˌleɪs/ n. Biochem. an enzyme that catalyses the hydrolysis of a substrate.

hydrology /haɪˈdrɒlədʒɪ/ n. the science of the properties of the earth's water, esp. of its movement in relation to land. □ **hydrologist** n. **hydrologic** /ˌhaɪdrəˈlɒdʒɪk/ adj. **hydrological** adj. **hydrologically** adv.

hydrolyse /ˈhaɪdrəˌlaɪz/ v.tr. & intr. (also **hydrolyze**) subject to or undergo the chemical action of water.

hydrolysis /haɪˈdrɒlɪsɪs/ n. the chemical reaction of a substance with water, usu. resulting in decomposition. □ **hydrolytic** /ˌhaɪdrəˈlɪtɪk/ adj.

hydromagnetics /ˌhaɪdrəmæɡˈnetɪks/ n. pl. (usu. treated as sing.) = MAGNETOHYDRODYNAMICS. □ **hydromagnetic** adj.

hydromania /ˌhaɪdrəˈmeɪnɪə/ n. a craving for water.

hydromechanics /ˌhaɪdrəʊmɪˈkænɪks/ n. the mechanics of liquids; hydrodynamics.

hydrometer /haɪˈdrɒmɪtə(r)/ n. an instrument for measuring the density of liquids. □ **hydrometry** n. **hydrometric** /ˌhaɪdrəˈmetrɪk/ adj.

hydronium ion /haɪˈdrəʊnɪəm/ n. Chem. = HYDROXONIUM ION. [contr.]

hydropathy /haɪˈdrɒpəθɪ/ n. the treatment of disorders by external and internal application of water. □ **hydropathist** n. **hydropathic** /ˌhaɪdrəˈpæθɪk/ adj. [HYDRO-, after HOMEOPATHY etc.]

hydrophil /ˈhaɪdrəfɪl/ adj. (also **hydrophile** /-ˌfaɪl/) = HYDROPHILIC. [as HYDROPHILIC]

hydrophilic /ˌhaɪdrəˈfɪlɪk/ adj. Chem. 1 having an affinity for water. 2 readily mixing with or wetted by water. [HYDRO- + Gk philos loving]

hydrophobia /ˌhaɪdrəˈfəʊbɪə/ n. 1 a morbid aversion to water, esp. as a symptom of rabies in humans. 2 rabies, esp. in humans. [LL f. Gk hudrophobia (as HYDRO-, -PHOBIA)]

hydrophobic /ˌhaɪdrəˈfəʊbɪk/ adj. 1 of or suffering from hydrophobia. 2 a lacking an affinity for water. b not readily wettable.

hydrophone /ˈhaɪdrəˌfəʊn/ n. a microphone for the detection of sound-waves in water.

hydrophyte /ˈhaɪdrəˌfaɪt/ n. an aquatic plant, or a plant which needs much moisture.

hydroplane /ˈhaɪdrəˌpleɪn/ n. & v. ● n. 1 a light fast motor boat designed to skim over the surface of water. 2 a finlike attachment which enables a submarine to rise and fall in water. ● v.intr. 1 (of a boat) skim over the surface of water with its hull lifted. 2 = AQUAPLANE v. 2.

hydroponics /ˌhaɪdrəˈpɒnɪks/ n. the process of growing plants in sand, gravel, or liquid, without soil and with added nutrients. □ **hydroponic** adj. **hydroponically** adv. [HYDRO- + Gk ponos labour]

hydroquinone /ˌhaɪdrəˈkwɪnəʊn, -kwɪˈnəʊn/ n. Photog. a substance formed by the reduction of benzoquinone, used as a photographic developer.

hydrosphere /ˈhaɪdrəˌsfɪə(r)/ n. the waters of the earth's surface.

hydrostatic /ˌhaɪdrəˈstætɪk/ adj. of the equilibrium of liquids and the pressure exerted by liquid at rest. □ **hydrostatic press** = hydraulic press. □ **hydrostatical** adj. **hydrostatically** adv. [prob. f. Gk hudrostatēs hydrostatic balance (as HYDRO-, STATIC)]

hydrostatics /ˌhaɪdrəˈstætɪks/ n.pl. (usu. treated as sing.) the branch of mechanics concerned with the hydrostatic properties of liquids.

hydrotherapy /ˌhaɪdrəˈθerəpɪ/ n. the use of water in the treatment of disorders, usu. exercises in swimming-pools for arthritic or partially paralysed patients. □ **hydrotherapist** n.

hydrothermal /ˌhaɪdrəˈθɜːm(ə)l/ adj. of the action of heated water on the earth's crust. □ **hydrothermally** adv.

hydrothorax /ˌhaɪdrəˈθɔːræks/ n. Med. the condition of having fluid in the pleural cavity.

hydrotropism /haɪˈdrɒtrəˌpɪz(ə)m/ adj. a tendency of plant roots etc. to turn to or from moisture.

hydrous /ˈhaɪdrəs/ adj. Chem. & Mineral. containing water. [Gk hudōr hudro- water]

hydroxide /haɪˈdrɒksaɪd/ n. Chem. a metallic compound containing oxygen and hydrogen either in the form of the hydroxide ion (OH^-) or the hydroxyl group (–OH).

hydroxonium ion /ˌhaɪdrɒkˈsəʊnɪəm/ n. Chem. the hydrated hydrogen ion, H_3O^+. [HYDRO- + OXY-[2] + -onium]

hydroxy- /haɪˈdrɒksɪ/ comb. form Chem. having a hydroxide ion (or ions) or a hydroxyl group (or groups) (hydroxybenzoic acid). [HYDROGEN + OXYGEN]

hydroxyl /haɪˈdrɒksaɪl, -sɪl/ n. Chem. the monovalent group containing hydrogen and oxygen, as –OH. [HYDROGEN + OXYGEN + -YL]

hydrozoan /ˌhaɪdrəˈzəʊən/ n. & adj. Zool. ● n. an aquatic coelenterate of the class Hydrozoa, of mainly marine forms that are frequently colonial and have both polyp and medusa phases, e.g. hydra, Portuguese man-of-war, etc. ● adj. of or relating to this class. [mod.L Hydrozoa (as HYDRA, Gk zōion animal)]

hyena /haɪˈiːnə/ n. (also **hyaena**) a flesh-eating mammal of the family Hyaenidae, resembling a dog but with hind limbs shorter than forelimbs. □ **laughing hyena** a hyena, Crocuta crocuta, whose call is compared to a fiendish laugh. [ME f. OF hyene & L hyaena f. Gk huaina fem. of hus pig]

hygiene /ˈhaɪdʒiːn/ n. 1 a a study, or set of principles, of maintaining health. b conditions or practices conducive to maintaining health. 2 sanitary science. [F hygiène f. mod.L hygieina f. Gk hugieinē (tekhnē) (art) of health f. hugiēs healthy]

hygienic /haɪˈdʒiːnɪk/ adj. conducive to hygiene; clean and sanitary. □ **hygienically** adv.

hygienics /haɪˈdʒiːnɪks/ n.pl. (usu. treated as sing.) = HYGIENE 1a.

hygienist /ˈhaɪdʒiːnɪst/ n. a specialist in the promotion and practice of cleanliness for the preservation of health.

hygro- /ˈhaɪɡrəʊ/ comb. form moisture. [Gk hugro- f. hugros wet, moist]

hygrology /haɪˈɡrɒlədʒɪ/ n. the study of the humidity of the atmosphere etc.

hygrometer /haɪˈɡrɒmɪtə(r)/ n. an instrument for measuring the humidity of the air or a gas. □ **hygrometry** n. **hygrometric** /ˌhaɪɡrəˈmetrɪk/ adj.

hygrophilous /haɪˈɡrɒfɪləs/ adj. (of a plant) growing in a moist environment.

hygrophyte /ˈhaɪɡrəˌfaɪt/ n. = HYDROPHYTE.

hygroscope /ˈhaɪɡrəˌskəʊp/ n. an instrument which indicates but does not measure the humidity of the air.

hygroscopic /ˌhaɪɡrəˈskɒpɪk/ adj. 1 of the hygroscope. 2 (of a substance) tending to absorb moisture from the air. □ **hygroscopically** adv.

hying pres. part. of HIE.

Hyksos /ˈhɪksɒs/ n.pl. a people of mixed Semitic and Asian descent who invaded Egypt and settled in the Nile delta c.1640 BC. They formed the 15th and 16th dynasties of Egypt and ruled a large part of the country until driven out c.1532 BC. [Gk Huksōs (interpreted by Manetho as 'shepherd kings' or 'captive shepherds') f. Egyptian heqa khoswe foreign rulers]

hylic /'haɪlɪk/ adj. of matter; material. [LL hylicus f. Gk hulikos f. hulē matter]

hylo- /'haɪləʊ/ comb. form matter. [Gk hulo- f. hulē matter]

hylomorphism /ˌhaɪlə'mɔːfɪz(ə)m/ n. Philos. the theory that physical objects are composed of matter and form. [HYLO- + Gk morphē form]

hylozoism /ˌhaɪlə'zəʊɪz(ə)m/ n. Philos. the doctrine that all matter has life. [HYLO- + Gk zōē life]

Hymen /'haɪmen/ originally a cry (Hymen Hymeniae) used at ancient Greek weddings, and understood (rightly or wrongly) as an invocation of a youth of that name, a handsome young man who had been happily married.

hymen /'haɪmen/ n. Anat. a membrane which partially closes the opening of the vagina and is usu. broken at the first occurrence of sexual intercourse. □ **hymenal** adj. [LL f. Gk humēn membrane]

hymeneal /ˌhaɪmɪ'niːəl/ adj. literary of or concerning marriage. [HYMEN]

hymenium /haɪ'miːnɪəm/ n. (pl. **hymenia** /-nɪə/) the spore-bearing surface of certain fungi. [mod.L f. Gk humenion dimin. of humēn membrane]

Hymenoptera /ˌhaɪmə'nɒptərə/ n.pl. Zool. an order of insects comprising the ants, bees, wasps, ichneumons, and sawflies, which have two pairs of transparent wings and frequently a sting. Some of them form social colonies, but many are parasitic. □ **hymenopteran** n. & adj. **hymenopterous** adj. [mod.L f. Gk humenopteros membrane-winged (as HYMENIUM, pteron wing)]

hymn /hɪm/ n. & v. ● n. 1 a song of praise, esp. to God in Christian worship, usu. a metrical composition sung in a religious service. 2 a song of praise in honour of a god or other exalted being or thing. ● v. 1 tr. praise or celebrate in hymns. 2 intr. sing hymns. □ **hymn-book** a book of hymns. □ **hymnic** /'hɪmnɪk/ adj. [ME ymne etc. f. OF ymne f. L hymnus f. Gk humnos]

hymnal /'hɪmn(ə)l/ n. & adj. ● n. a hymn-book. ● adj. of hymns. [ME f. med.L hymnale (as HYMN)]

hymnary /'hɪmnərɪ/ n. (pl. **-ies**) a hymn-book.

hymnody /'hɪmnədɪ/ n. (pl. **-ies**) 1 a the singing of hymns. b the composition of hymns. 2 hymns collectively. □ **hymnodist** n. [med.L hymnodia f. Gk humnōidia f. humnos hymn: cf. PSALMODY]

hymnographer /hɪm'nɒɡrəfə(r)/ n. a writer of hymns. □ **hymnography** n. [Gk humnographos f. humnos hymn]

hymnology /hɪm'nɒlədʒɪ/ n. (pl. **-ies**) 1 the composition or study of hymns. 2 hymns collectively. □ **hymnologist** n.

hyoid /'haɪɔɪd/ n. & adj. Anat. ● n. (in full **hyoid bone**) a U-shaped bone in the neck which supports the tongue. ● adj. of or relating to this. [F hyoïde f. mod.L hyoïdes f. Gk huoeidēs shaped like the letter upsilon (hu)]

hyoscine /'haɪəˌsiːn/ n. a poisonous alkaloid found in plants of the nightshade family, esp. of the genus Scopolia, used as an antiemetic in motion sickness and a preoperative medication for examination of the eye. Also called scopolamine. [f. HYOSCYAMINE]

hyoscyamine /ˌhaɪə'saɪəˌmiːn/ n. a poisonous alkaloid obtained from henbane, having similar properties to hyoscine. [mod.L hyoscyamus f. Gk huoskuamos henbane f. hus huos pig + kuamos bean]

hypaesthesia /ˌhaɪpɪs'θiːzɪə/ n. (US **hypesthesia**) Med. a diminished capacity for sensation, esp. of the skin. □ **hypaesthetic** /-'θetɪk/ adj. [mod.L (as HYPO-, Gk -aisthēsia f. aisthanomai perceive)]

hypaethral /haɪ'piːθrəl/ adj. (also **hypethral**) 1 Archit. open to the sky; roofless. 2 literary open-air. [L hypaethrus f. Gk hupaithros (as HYPO-, aithēr air)]

hypallage /haɪ'pælədʒɪ/ n. Rhet. the transposition of the natural relations of two elements in a proposition (e.g. Melissa shook her doubtful curls). [LL f. Gk hupallagē (as HYPO-, allassō exchange)]

Hypatia /haɪ'peɪʃɪə/ (c.370–415), Greek philosopher, astronomer, and mathematician. She taught geometry, algebra, and astronomy at Alexandria, and was head of the Neoplatonist school there. Hypatia wrote several learned treatises as well as devising inventions such as an astrolabe. She was murdered by a Christian mob opposed to the scientific rationalism advocated by her Neoplatonist philosophy.

hype[1] /haɪp/ n. & v. colloq. ● n. 1 extravagant or intensive publicity promotion. 2 cheating; a trick. ● v.tr. 1 promote (a product) with extravagant publicity. 2 cheat, trick. [20th c.: orig. unkn.]

hype[2] /haɪp/ n. sl. 1 a drug addict. 2 a hypodermic needle or injection. □ **hyped up** stimulated by or as if by a hypodermic injection. [abbr. of HYPODERMIC]

hyper /'haɪpə(r)/ adj. colloq. hyperactive, highly strung; extraordinarily energetic.

hyper- /'haɪpə(r)/ prefix meaning: 1 over, beyond, above (hyperphysical). 2 exceeding (hypersonic). 3 excessively; above normal (hyperbole; hypersensitive; hyperinflation). [Gk huper over, beyond]

hyperactive /ˌhaɪpər'æktɪv/ adj. (of a person, esp. a child) abnormally active. □ **hyperactivity** /-æk'tɪvɪtɪ/ n.

hyperaemia /ˌhaɪpər'iːmɪə/ n. (US **hyperemia**) Med. an excessive quantity of blood in the vessels supplying an organ or other part of the body. □ **hyperaemic** adj. [mod.L (as HYPER-, -AEMIA)]

hyperaesthesia /ˌhaɪpəriːs'θiːzɪə/ n. (US **hyperesthesia**) Med. an excessive physical sensibility, esp. of the skin. □ **hyperaesthetic** /-'θetɪk/ adj. [mod.L (as HYPER-, Gk -aisthēsia f. aisthanomai perceive)]

hyperbaric /ˌhaɪpə'bærɪk/ adj. (of a gas) at a pressure greater than normal. [HYPER- + Gk barus heavy]

hyperbaton /haɪ'pɜːbəˌtɒn/ n. Rhet. the inversion of the normal order of words, esp. for the sake of emphasis (e.g. this I must see). [L f. Gk huperbaton (as HYPER-, bainō go)]

hyperbola /haɪ'pɜːbələ/ n. (pl. **hyperbolas** or **hyperbolae** /-ˌliː/) Geom. the plane curve of two equal branches, produced when a cone is cut by a plane that makes a larger angle with the base than the side of the cone (cf. ELLIPSE). [mod.L f. Gk huperbolē excess (as HYPER-, ballō to throw)]

hyperbole /haɪ'pɜːbəlɪ/ n. Rhet. an exaggerated statement not meant to be taken literally. □ **hyperbolism** n. **hyperbolical** /ˌhaɪpə'bɒlɪk(ə)l/ adj. **hyperbolically** adv. [L (as HYPERBOLA)]

hyperbolic /ˌhaɪpə'bɒlɪk/ adj. Geom. of or relating to a hyperbola. □ **hyperbolic function** a function related to a rectangular hyperbola, e.g. a hyperbolic cosine.

hyperboloid /haɪ'pɜːbəˌlɔɪd/ n. Geom. a solid or surface having plane sections that are hyperbolas, ellipses, or circles. □ **hyperboloidal** /ˌhaɪpəbə'lɔɪd(ə)l/ adj.

Hyperborean /ˌhaɪpəbɔː'riːən, -'bɔːrɪən/ n. Gk Mythol. a member of a fabled race worshipping Apollo and living in a land of sunshine and plenty 'beyond the north wind'.

hyperborean /ˌhaɪpəbɔː'riːən, -'bɔːrɪən/ n. & adj. ● n. an inhabitant of the extreme north of the earth. ● adj. of the extreme north of the earth. [LL hyperboreanus f. L hyperboreus f. Gk huperboreos (as HYPER-, Boreas god of the north wind)]

HyperCard /'haɪpəˌkɑːd/ n. Computing propr. a programming system that uses symbols resembling index cards onscreen, permitting the creation of hypertext links. [HYPERTEXT + CARD[1]]

hypercholesterolaemia /ˌhaɪpəkəˌlestərɒ'liːmɪə/ n. Med. an excess of cholesterol in the bloodstream. [HYPER- + CHOLESTEROL + -AEMIA]

hyperconscious /ˌhaɪpə'kɒnʃəs/ adj. (foll. by of) acutely or excessively aware.

hypercritical /ˌhaɪpə'krɪtɪk(ə)l/ adj. excessively critical, esp. of small faults. □ **hypercritically** adv.

hyperemia US var. of HYPERAEMIA.

hyperesthesia US var. of HYPERAESTHESIA.

hyperfocal distance /ˌhaɪpə'fəʊk(ə)l/ n. the distance on which a camera lens can be focused to bring the maximum range of object-distances into focus.

hypergamy /haɪ'pɜːɡəmɪ/ n. marriage to a person of equal or superior caste or class. [HYPER- + Gk gamos marriage]

hyperglycaemia /ˌhaɪpəɡlaɪ'siːmɪə/ n. (US **hyperglycemia**) Med. an excess of glucose in the bloodstream, often associated with diabetes mellitus. □ **hyperglycaemic** adj. [HYPER- + GLYCO- + -AEMIA]

hypergolic /ˌhaɪpə'ɡɒlɪk/ adj. (of a rocket propellant) igniting spontaneously on contact with an oxidant etc. [G Hypergol (perh. as HYPO-, ERG[1], -OL)]

hypericum /haɪ'perɪkəm/ n. a shrub or herbaceous plant of the genus Hypericum, with five-petalled yellow flowers. (See also ST JOHN'S WORT.) [L f. Gk hupereikon (as HYPER-, ereikē heath)]

Hyperion /haɪ'pɪərɪən/ 1 Gk Mythol. a Titan, son of Uranus (Heaven) and Gaia (Earth). 2 Astron. satellite VII of Saturn, the sixteenth closest to the planet, discovered in 1848. Its irregular shape suggests that it is a remnant of a larger body.

hyperkinesis /ˌhaɪpəkɪ'niːsɪs, -kaɪ'niːsɪs/ n. (also **hyperkinesia** /-'niːzɪə/) Med. 1 muscle spasm. 2 a disorder of children marked by hyperactivity and inability to attend. □ **hyperkinetic** /-'netɪk/ adj. [HYPER- + Gk kinēsis motion]

hypermarket /'haɪpəˌmɑːkɪt/ n. Brit. a very large self-service store with a wide range of goods and extensive car-parking facilities, usu. outside a town. [transl. F *hypermarché* (as HYPER-, MARKET)]

hypermedia /ˈhaɪpəˌmiːdɪə/ *n. Computing* an extension of hypertext allowing the provision of audio and video material cross-referenced to a computer text (often *attrib.: hypermedia database*). [HYPER- + MEDIA[1]]

hypermetropia /ˌhaɪpəmɪˈtrəʊpɪə/ *n. Med.* the condition of having long sight. □ **hypermetropic** /-ˈtrəʊpɪk, -ˈtrɒpɪk/ *adj.* [mod.L f. HYPER- + Gk *metron* measure, *ōps* eye]

hypernym /ˈhaɪpənɪm/ *n.* (in semantics) a word of more general meaning than, and therefore implied by or able to replace, other more specific terms; a superordinate (e.g. *animal* is a hypernym of *lion* and *elephant*) (cf. HYPONYM). [f. HYPER- after HYPONYM]

hyperon /ˈhaɪpəˌrɒn/ *n. Physics* an unstable baryon which is heavier than a nucleon. □ **hyperonic** /ˌhaɪpəˈrɒnɪk/ *adj.* [HYPER-]

hyperopia /ˌhaɪpəˈrəʊpɪə/ *n. Med.* = HYPERMETROPIA. □ **hyperopic** /-ˈrɒpɪk/ *adj.* [mod.L f. HYPER- + Gk *ōps* eye]

hyperphysical /ˌhaɪpəˈfɪzɪk(ə)l/ *adj.* supernatural, incorporeal. □ **hyperphysically** *adv.*

hyperplasia /ˌhaɪpəˈpleɪzɪə/ *n. Med.* the enlargement of an organ or tissue from the increased production of cells. [HYPER- + Gk *plasis* formation]

hypersensitive /ˌhaɪpəˈsensɪtɪv/ *adj.* **1** abnormally or excessively sensitive. **2** *Med.* having an abnormally strong, esp. allergic, reaction to a substance. □ **hypersensitiveness** *n.* **hypersensitivity** /-ˌsensɪˈtɪvɪtɪ/ *n.*

hypersonic /ˌhaɪpəˈsɒnɪk/ *adj.* **1** relating to speeds of more than five times the speed of sound (Mach 5). **2** relating to sound-frequencies above about 1,000 million hertz. □ **hypersonically** *adv.* [HYPER-, after SUPERSONIC, ULTRASONIC]

hyperspace /ˈhaɪpəˌspeɪs/ *n.* space of more than three dimensions, esp. (in science fiction) a notional space–time continuum in which motion and communication at speeds greater than that of light are supposedly possible.

hypersthene /ˈhaɪpəˌsθiːn/ *n.* a rock-forming mineral, magnesium iron silicate, of greenish colour. [F *hypersthène* (as HYPER-, Gk *sthenos* strength, from its being harder than hornblende]

hypertension /ˌhaɪpəˈtenʃ(ə)n/ *n.* **1** abnormally high blood pressure. **2** a state of great emotional tension. □ **hypertensive** /-ˈtensɪv/ *adj.*

hypertext /ˈhaɪpəˌtekst/ *n. Computing* a software system allowing extensive cross-referencing between related sections of text and associated graphic material.

hyperthermia /ˌhaɪpəˈθɜːmɪə/ *n. Med.* the condition of having a body-temperature greatly above normal. □ **hyperthermic** *adj.* [HYPER- + Gk *thermē* heat]

hyperthyroidism /ˌhaɪpəˈθaɪrɔɪˌdɪz(ə)m/ *n. Med.* overactivity of the thyroid gland, resulting in rapid heartbeat and an increased rate of metabolism. □ **hyperthyroid** *adj.* **hyperthyroidic** /-ˌθaɪəˈrɔɪdɪk/ *adj.*

hypertonic /ˌhaɪpəˈtɒnɪk/ *adj.* **1** *Med.* (of muscles) having high tension. **2** *Physiol.* (of a solution) having a greater osmotic pressure than another solution. □ **hypertonia** /-ˈtəʊnɪə/ *n.* (in sense 1). **hypertonicity** /-təˈnɪsɪtɪ/ *n.*

hypertrophy /haɪˈpɜːtrəfɪ/ *n. Med.* the enlargement of an organ or tissue from the increase in size of its cells. □ **hypertrophied** *adj.* **hypertrophic** /ˌhaɪpəˈtrɒfɪk/ *adj.* [mod.L *hypertrophia* (as HYPER-, Gk *-trophia* nourishment)]

hyperventilation /ˌhaɪpəˌventɪˈleɪʃ(ə)n/ *n. Med.* breathing at an abnormally rapid rate, resulting in an increased loss of carbon dioxide. □ **hyperventilate** /ˌhaɪpəˈventɪleɪt/ *v.intr.*

hypethral var. of HYPAETHRAL.

hypha /ˈhaɪfə/ *n.* (*pl.* **hyphae** /-fiː/) a filament in the mycelium of a fungus. □ **hyphal** *adj.* [mod.L f. Gk *huphē* web]

Hyphasis /ˈhaɪfəsɪs/ the ancient Greek name for the BEAS.

hyphen /ˈhaɪf(ə)n/ *n. & v.* ● *n.* the sign (-) used to join words semantically or syntactically (as in *fruit-tree*, *pick-me-up*, *rock-forming*), to indicate the division of a word at the end of a line, or to indicate a missing or implied element (as in *man-* and *womankind*). ● *v.tr.* **1** write (a compound word) with a hyphen. **2** join (words) with a hyphen. [LL f. Gk *huphen* together f. *hupo* under + *hen* one]

hyphenate /ˈhaɪfəˌneɪt/ *v.tr.* = HYPHEN *v.* □ **hyphenation** /ˌhaɪfəˈneɪʃ(ə)n/ *n.*

hypno- /ˈhɪpnəʊ/ *comb. form* sleep, hypnosis. [Gk *hupnos* sleep]

hypnogenesis /ˌhɪpnəʊˈdʒenɪsɪs/ *n.* the induction of a hypnotic state.

hypnology /hɪpˈnɒlədʒɪ/ *n.* the science of the phenomena of sleep. □ **hypnologist** *n.*

hypnopaedia /ˌhɪpnəʊˈpiːdɪə/ *n.* (*US* **hypnopedia**) learning by hearing while asleep.

Hypnos /ˈhɪpnɒs/ *Gk Mythol.* the god of sleep, son of Nyx (Night). [Gk *hupnos* sleep]

hypnosis /hɪpˈnəʊsɪs/ *n.* **1** a state of consciousness in which a person, usually under the influence of another, appears to lose the power of voluntary action and to become highly responsive to external suggestion. (*See note below.*) **2** artificially produced sleep. [mod.L f. Gk *hupnos* sleep + -OSIS]

▪ Hypnotic techniques were known in ancient times, but it was Franz Mesmer in the 18th century who popularized hypnosis in Europe, and, under the name of *mesmerism*, it enjoyed a vogue as a therapeutic technique. Some subjects under hypnosis seem able to recall memories (e.g. of childhood) that have apparently been forgotten, and the study of hypnosis seems to have influenced Sigmund Freud in his theories of the unconscious mind. Hypnotic suggestion has been used in psychotherapy, e.g. to treat addiction, but it is often distrusted because of its theatrical, mysterious, and exploitative aspects.

hypnotherapy /ˌhɪpnəʊˈθerəpɪ/ *n.* the treatment of disease by hypnosis. □ **hypnotherapist** *n.*

hypnotic /hɪpˈnɒtɪk/ *adj. & n.* ● *adj.* **1** of or producing hypnosis. **2** (of a drug) soporific. ● *n.* **1** a thing, esp. a drug, that produces sleep. **2** a person under or open to the influence of hypnotism. □ **hypnotically** *adv.* [F *hypnotique* f. LL *hypnoticus* f. Gk *hupnōtikos* f. *hupnoō* put to sleep]

hypnotism /ˈhɪpnəˌtɪz(ə)m/ *n.* the study or practice of hypnosis. □ **hypnotist** *n.*

hypnotize /ˈhɪpnəˌtaɪz/ *v.tr.* (also **-ise**) **1** produce hypnosis in. **2** fascinate; capture the mind of (a person). □ **hypnotizable** *adj.*

hypo[1] /ˈhaɪpəʊ/ *n. Photog.* the chemical sodium thiosulphate (incorrectly called hyposulphite) used as a photographic fixer. [abbr.]

hypo[2] /ˈhaɪpəʊ/ *n.* (*pl.* **-os**) *colloq.* = HYPODERMIC *n.* [abbr.]

hypo- /ˈhaɪpəʊ/ *prefix* (usu. **hyp-** before a vowel or *h*) **1** under (*hypodermic*). **2** below normal (*hypoxia*). **3** slightly (*hypomania*). **4** *Chem.* containing an element combined in low valence (*hypochlorous*). [Gk f. *hupo* under]

hypo-allergenic /ˌhaɪpəʊˌæləˈdʒenɪk/ *adj. Med.* having little tendency, or a specially reduced tendency, to cause an allergic reaction.

hypoblast /ˈhaɪpəˌblæst/ *n. Biol.* = ENDODERM. [mod.L *hypoblastus* (as HYPO-, -BLAST)]

hypocaust /ˈhaɪpəˌkɔːst/ *n.* a hollow space under the floor in ancient Roman houses, into which hot air was sent for heating a room or bath. [L *hypocaustum* f. Gk *hupokauston* place heated from below (as HYPO-, *kaiō*, *kau-* burn)]

hypochlorite /ˌhaɪpəˈklɔːraɪt/ *n. Chem.* a salt of hypochlorous acid.

hypochlorous acid /ˌhaɪpəˈklɔːrəs/ *n. Chem.* an unstable acid (chem. formula: HOCl), existing only in dilute solution and used in bleaching and water treatment. [HYPO- + CHLORINE + -OUS]

hypochondria /ˌhaɪpəˈkɒndrɪə/ *n.* **1** abnormal anxiety about one's health. **2** morbid depression without real cause. [LL f. Gk *hupokhondria* soft parts of the body below the ribs, where melancholy was thought to arise (as HYPO-, *khondros* sternal cartilage)]

hypochondriac /ˌhaɪpəˈkɒndrɪˌæk/ *n. & adj.* ● *n.* a person suffering from hypochondria. ● *adj.* (also **hypochondriacal** /-kɒnˈdraɪək(ə)l/) of or affected by hypochondria. [F *hypocondriaque* f. Gk *hupokhondriakos* (as HYPOCHONDRIA)]

hypocoristic /ˌhaɪpəkəˈrɪstɪk/ *adj. Gram.* of the nature of a pet name. [Gk *hupokoristikos* f. *hupokorizomai* call by pet names]

hypocotyl /ˌhaɪpəˈkɒtɪl/ *n. Bot.* the part of the stem of an embryo plant beneath the stalks of the seed leaves or cotyledons and directly above the root.

hypocrisy /hɪˈpɒkrɪsɪ/ *n.* (*pl.* **-ies**) **1** the assumption or postulation of moral standards to which one's own behaviour does not conform; dissimulation, pretence. **2** an instance of this. [ME f. OF *ypocrisie* f. eccl.L *hypocrisis* f. Gk *hupokrisis* acting of a part, pretence (as HYPO-, *krinō* decide, judge)]

hypocrite /ˈhɪpəˌkrɪt/ *n.* a person given to hypocrisy. □ **hypocritical** /ˌhɪpəˈkrɪtɪk(ə)l/ *adj.* **hypocritically** *adv.* [ME f. OF *ypocrite* f. eccl.L f. Gk *hupokritēs* actor (as HYPOCRISY)]

hypocycloid /ˌhaɪpəˈsaɪklɔɪd/ *n. Math.* the curve traced by a point on the circumference of a circle rolling on the interior of another circle. □ **hypocycloidal** /-saɪˈklɔɪd(ə)l/ *adj.*

hypodermic /ˌhaɪpəˈdɜːmɪk/ *adj. & n.* ● *adj. Med.* **1** of or relating to the area beneath the skin. **2 a** (of a drug etc. or its application) injected beneath the skin. **b** (of a needle, syringe, etc.) used to do this. ● *n.* a hypodermic injection or syringe. □ **hypodermically** *adv.* [HYPO- + Gk *derma* skin]

hypogastrium /ˌhaɪpəˈgæstrɪəm/ *n.* (*pl.* **hypogastria** /-trɪə/) *Anat.* the part of the central abdomen which is situated below the region of the stomach. □ **hypogastric** *adj.* [mod.L f. Gk *hupogastrion* (as HYPO-, *gastēr* belly)]

hypogeal /ˌhaɪpəˈdʒiːəl/ *adj.* **1** esp. *Biol.* (existing or growing) underground. **2** *Bot.* (of seed germination) with the seed leaves remaining below the ground. [LL *hypogeus* f. Gk *hupogeios* (as HYPO-, *gē* earth)]

hypogene /ˈhaɪpəˌdʒiːn/ *adj. Geol.* produced under the surface of the earth. [HYPO- + Gk *gen-* produce]

hypogeum /ˌhaɪpəˈdʒiːəm/ *n.* (*pl.* **hypogea** /-ˈdʒiːə/) an underground chamber. [L f. Gk *hupogeion* neut. of *hupogeios*: see HYPOGEAL]

hypoglycaemia /ˌhaɪpəʊglaɪˈsiːmɪə/ *n.* (US **hypoglycemia**) *Med.* a deficiency of glucose in the bloodstream. □ **hypoglycaemic** *adj.* [HYPO- + GLYCO- + -AEMIA]

hypoid /ˈhaɪpɔɪd/ *n.* a gear with the pinion offset from the centre-line of the wheel, to connect non-intersecting shafts. [perh. f. HYPERBOLOID]

hypolimnion /ˌhaɪpəˈlɪmnɪən/ *n.* (*pl.* **hypolimnia** /-nɪə/) the lower, cooler layer of water in stratified lakes. [HYPO- + Gk *limnion* dimin. of *limnē* lake]

hypomania /ˌhaɪpəˈmeɪnɪə/ *n.* a minor form of mania. □ **hypomanic** /-ˈmænɪk/ *adj.* [mod.L f. G *Hypomanie* (as HYPO-, MANIA)]

hyponasty /ˈhaɪpəˌnæstɪ/ *n. Bot.* the tendency in plant-organs for growth to be more rapid on the under-side. □ **hyponastic** /ˌhaɪpəˈnæstɪk/ *adj.* [HYPO- + Gk *nastos* pressed]

hyponym /ˈhaɪpəʊnɪm/ *n.* **1** (in semantics) a word of more specific meaning than, and therefore implying or able to be replaced by, another more general or superordinate term (e.g. *scarlet* is a hyponym of *red*) (cf. HYPERNYM). **2** (in taxonomy) a name made invalid by the lack of adequate contemporary description of the taxon it was intended to designate. [f. HYPO- + Gk *onoma* name]

hyponymy /haɪˈpɒnɪmɪ/ *n.* (in semantics) the relation of a word to another word of which the former is a hyponym. [f. HYPO-, after SYNONYMY]

hypophysis /haɪˈpɒfɪsɪs/ *n.* (*pl.* **hypophyses** /-ˌsiːz/) *Anat.* the pituitary. □ **hypophyseal** /ˌhaɪpəˈfɪzɪəl/ *adj.* (also **hypophysial**). [mod.L f. Gk *hupophusis* offshoot (as HYPO-, *phusis* growth)]

hypostasis /haɪˈpɒstəsɪs/ *n.* (*pl.* **hypostases** /-ˌsiːz/) **1** *Med.* an accumulation of fluid or blood in the lower parts of the body or organs under the influence of gravity, in cases of poor circulation. **2** *Metaphysics* an underlying substance, as opposed to attributes or to that which is unsubstantial. **3** (in Christian theology) **a** the person of Christ, combining human and divine natures. **b** each of the three persons of the Trinity. □ **hypostasize** *v.tr.* (also **-ise**) (in senses 1, 2). [eccl.L f. Gk *hupostasis* (as HYPO-, STASIS standing, state)]

hypostatic /ˌhaɪpəˈstætɪk/ *adj.* (also **hypostatical**) (in Christian theology) relating to the three persons of the Trinity. □ **hypostatic union** the union of divine and human natures in Christ, a doctrine formally accepted by the Church in 451.

hypostyle /ˈhaɪpəˌstaɪl/ *adj. Archit.* having a roof supported by pillars. [Gk *hupostulos* (as HYPO-, STYLE)]

hypotaxis /ˌhaɪpəˈtæksɪs/ *n. Gram.* the subordination of one clause to another. □ **hypotactic** /-ˈtæktɪk/ *adj.* [Gk *hupotaxis* (as HYPO-, *taxis* arrangement)]

hypotension /ˌhaɪpəˈtenʃ(ə)n/ *n. Med.* abnormally low blood pressure. □ **hypotensive** /-ˈtensɪv/ *adj.*

hypotenuse /haɪˈpɒtəˌnjuːz/ *n.* the side opposite the right angle of a right-angled triangle. [L *hypotenusa* f. Gk *hupoteinousa* (*grammē*) subtending (line) fem. part. of *hupoteinō* (as HYPO-, *teinō* stretch)]

hypothalamus /ˌhaɪpəˈθæləməs/ *n.* (*pl.* **-mi** /-ˌmaɪ/) *Anat.* the region of the brain which controls body-temperature, thirst, hunger, etc. □ **hypothalamic** /-θəˈlæmɪk/ *adj.* [mod.L formed as HYPO-, THALAMUS]

hypothec /haɪˈpɒθɪk/ *n.* (in Roman and Scottish law) a right established by law over property belonging to a debtor. □ **hypothecary** *adj.* [F *hypothèque* f. LL *hypotheca* f. Gk *hupothēkē* deposit (as HYPO-, *tithēmi* place)]

hypothecate /haɪˈpɒθɪˌkeɪt/ *v.tr.* **1** pledge, mortgage. **2** hypothesize. □ **hypothecator** *n.* **hypothecation** /-ˌpɒθɪˈkeɪʃ(ə)n/ *n.* [med.L *hypothecare* (as HYPOTHEC)]

hypothermia /ˌhaɪpəʊˈθɜːmɪə/ *n. Med.* the condition of having an abnormally low body-temperature. [HYPO- + Gk *thermē* heat]

hypothesis /haɪˈpɒθɪsɪs/ *n.* (*pl.* **hypotheses** /-ˌsiːz/) **1** a proposition made as a basis for reasoning, without the assumption of its truth. **2** a supposition made as a starting-point for further investigation from known facts (cf. THEORY 1). **3** a groundless assumption. [LL f. Gk *hupothesis* foundation (as HYPO-, THESIS)]

hypothesize /haɪˈpɒθɪˌsaɪz/ *v.* (also **-ise**) **1** *intr.* frame a hypothesis. **2** *tr.* assume as a hypothesis. □ **hypothesist** *n.* **hypothesizer** *n.*

hypothetical /ˌhaɪpəˈθetɪk(ə)l/ *adj.* **1** of or based on or serving as a hypothesis. **2** supposed but not necessarily real or true. □ **hypothetically** *adv.*

hypothyroidism /ˌhaɪpəʊˈθaɪrɔɪˌdɪz(ə)m/ *n. Med.* subnormal activity of the thyroid gland, resulting in cretinism in children, and mental and physical slowing in adults. □ **hypothyroid** *n. & adj.* **hypothyroidic** /-ˌθaɪəˈrɔɪdɪk/ *adj.*

hypoventilation /ˌhaɪpəʊˌventɪˈleɪʃ(ə)n/ *n. Med.* breathing at an abnormally slow rate, resulting in an increased amount of carbon dioxide in the blood.

hypoxaemia /ˌhaɪpɒkˈsiːmɪə/ *n.* (US **hypoxemia**) *Med.* an abnormally low concentration of oxygen in the blood. [mod.L (as HYPO-, OXYGEN, -AEMIA)]

hypoxia /haɪˈpɒksɪə/ *n. Med.* a deficiency of oxygen reaching the tissues. □ **hypoxic** *adj.* [HYPO- + OX- + -IA¹]

hypsilophodon /ˌhɪpsɪˈlɒfədɒn/ *n.* a small bipedal dinosaur of the genus *Hypsilophodon*, of the Cretaceous period, adapted for swift running. [mod.L f. Gk *hupsi-* high + LOPHO- + *odous odont* tooth]

hypso- /ˈhɪpsəʊ/ *comb. form* height. [Gk *hupsos* height]

hypsography /hɪpˈsɒgrəfɪ/ *n.* a description or mapping of the contours of the earth's surface. □ **hypsographic** /ˌhɪpsəˈgræfɪk/ *adj.* **hypsographical** *adj.*

hypsometer /hɪpˈsɒmɪtə(r)/ *n.* **1** a device for calibrating thermometers at the boiling-point of water. **2** this instrument when used to estimate height above sea level. □ **hypsometric** /ˌhɪpsəˈmetrɪk/ *adj.*

hyrax /ˈhaɪəræks/ *n.* a small mammal of the order Hyracoidea, resembling a short-eared rabbit (but actually allied to the hoofed mammals), e.g. rock-rabbit and dassie. [mod.L f. Gk *hurax* shrew-mouse]

hyson /ˈhaɪs(ə)n/ *n.* a kind of green China tea. [Chin. *xichun*, lit. 'bright spring']

hyssop /ˈhɪsəp/ *n.* **1** a small bushy aromatic herb of the genus *Hyssopus*; esp. *H. officinalis*, formerly used medicinally. **2** *Bibl.* **a** a plant whose twigs were used for sprinkling in Jewish rites. **b** a bunch of this used in purification. [OE (*h*)*ysope* (reinforced in ME by OF *ysope*) f. L *hyssopus* f. Gk *hyssōpos*, of Semitic orig.]

hysterectomy /ˌhɪstəˈrektəmɪ/ *n.* (*pl.* **-ies**) the surgical removal of the womb. □ **hysterectomize** *v.tr.* (also **-ise**). [Gk *hustera* womb + -ECTOMY]

hysteresis /ˌhɪstəˈriːsɪs/ *n. Physics* the lagging behind of an effect when its cause varies in amount etc., esp. of magnetic induction behind the magnetizing force. [Gk *husterēsis* f. *hustereō* be behind f. *husteros* coming after]

hysteria /hɪˈstɪərɪə/ *n.* **1** a wild uncontrollable emotion or excitement. **2** a functional disturbance of the nervous system, of psychoneurotic origin. [mod.L (as HYSTERIC)]

hysteric /hɪˈsterɪk/ *n. & adj.* ● *n.* **1** (in *pl.*) **a** a fit of hysteria. **b** *colloq.* overwhelming mirth or laughter (*we were in hysterics*). **2** a hysterical person. ● *adj.* = HYSTERICAL. [L f. Gk *husterikos* of the womb (*hustera*), hysteria being thought to occur more frequently in women than in men and to be associated with the womb]

hysterical /hɪˈsterɪk(ə)l/ *adj.* **1** of or affected with hysteria. **2** morbidly or uncontrolledly emotional. **3** *colloq.* extremely funny or amusing. □ **hysterically** *adv.*

hysteron proteron /ˌhɪstəˌrɒn ˈprɒtəˌrɒn/ *n. Rhet.* a figure of speech in which what should come last is put first; an inversion of the natural order (e.g. *I die! I faint! I fail!*). [LL f. Gk *husteron proteron* the latter (put in place of) the former]

Hz *abbr.* hertz.

Ii

I¹ /aɪ/ n. (also **i**) (pl. **Is** or **I's**) **1** the ninth letter of the alphabet. **2** (as a Roman numeral) one. □ **I-beam** a girder of I-shaped section.

I² /aɪ/ pron. & n. ● pron. (obj. **me**; poss. **my**, **mine**; pl. **we**) used by a speaker or writer to refer to himself or herself. ● n. (**the I**) Metaphysics the ego; the subject or object of self-consciousness. [OE f. Gmc]

I³ abbr. (also **I.**) Island(s), Isle(s).

I⁴ symb. **1** Chem. the element iodine. **2** electric current.

i symb. Math. the imaginary square root of minus one.

-i¹ /ɪ, aɪ/ suffix forming the plural of nouns from Latin in -us or from Italian in -e or -o (foci; dilettanti; timpani). ¶ Plural in -s or -es is often also possible.

-i² /ɪ/ suffix forming adjectives from names of countries or regions in the Near or Middle East (Israeli; Pakistani). [adj. suffix in Semitic and Indo-Iranian languages]

-i- /ɪ/ a connecting vowel esp. forming words in -ana, -ferous, -fic, -form, -fy, -gerous, -vorous (cf. -O-). [from or after F f. L]

IA abbr. US Iowa (in official postal use).

Ia. abbr. Iowa.

-ia¹ /ɪə/ suffix **1** forming abstract nouns (mania; utopia), often in Med. (anaemia; pneumonia). **2** Bot. forming names of classes and genera (dahlia; fuchsia). **3** forming names of countries (Australia; India). [from or after L & Gk]

-ia² /ɪə/ suffix forming plural nouns or the plural of nouns: **1** from Greek in -ion or Latin in -ium (paraphernalia; regalia; amnia; labia). **2** Zool. the names of groups (Mammalia).

IAA abbr. indoleacetic acid.

IAEA abbr. International Atomic Energy Agency.

-ial /ɪəl/ suffix forming adjectives (celestial; dictatorial; trivial). [from or after F -iel or L -ialis: cf. -AL¹]

iamb /ˈaɪæm, ˈaɪæmb/ n. Prosody an iambus. [anglicized f. IAMBUS]

iambic /aɪˈæmbɪk/ adj. & n. Prosody ● adj. of or using iambuses. ● n. (usu. in pl.) iambic verse. [F iambique f. LL iambicus f. Gk iambikos (as IAMBUS)]

iambus /aɪˈæmbəs/ n. (pl. **iambuses** or **-bi** /-baɪ/) Prosody a foot consisting of one short (or unstressed) followed by one long (or stressed) syllable. [L f. Gk iambos iambus, lampoon, f. iaptō assail in words, from its use by Gk satirists]

-ian /ɪən/ suffix var. of -AN. [from or after F -ien or L -ianus]

Iapetus /aɪˈæpɪtəs/ **1** Gk Mythol. a Titan, son of Uranus (Heaven) and Gaia (Earth). **2** Astron. satellite VIII of Saturn, the seventeenth closest to the planet, discovered by Giovanni Cassini in 1671 (diameter 1,440 km). It is unusual in having one side bright, icy, and cratered, and the other side covered with very dark material.

Iaşi /ˈjæʃɪ/ (German **Jassy** /ˈjasi/) a city in eastern Romania; pop. (1989) 330,195. Between 1565 and 1859 it was the capital of the principality of Moldavia.

-iasis /ˈaɪəsɪs/ suffix the usual form of -ASIS.

IATA /aɪˈɑːtə/ abbr. International Air Transport Association.

iatrogenic /aɪˌætrəˈdʒenɪk/ adj. (of a disease etc.) caused by medical examination or treatment. [Gk iatros physician + -GENIC]

ib. var. of IBID.

IBA see INDEPENDENT BROADCASTING AUTHORITY.

Ibadan /ɪˈbæd(ə)n/ the second largest city of Nigeria, situated 160 km (100 miles) north-east of Lagos; pop. (1983) 1,060,000.

Iban /ˈiːbæn/ n. & adj. ● n. (pl. same) **1** (also called Sea Dyak) a member of a group of non-Muslim indigenous peoples of Sarawak (see DYAK). **2** the language of the Iban, spoken by about 380,000 people, belonging to the Indonesian branch of the Malayo-Polynesian group of languages. ● adj. of or relating to these peoples or their language. [Iban]

Ibarruri Gomez /ɪˌbaːrʊrɪ ˈgəʊmez/, Dolores (known as 'La Pasionaria') (1895–1989), Spanish Communist politician. A founder of the Spanish Communist Party (1920), she was elected to Parliament in 1936. During the Spanish Civil War (1936–9) she became famous as an inspirational leader of the Republicans. She left Spain after the Nationalist victory and did not return until 1977, when she won re-election to the National Assembly at the age of 81.

Iberia /aɪˈbɪərɪə/ the ancient name for what is now Spain and Portugal; the Iberian peninsula. □ **Iberian** adj. & n. [L f. Gk Ibēr Spaniard]

Iberian peninsula /aɪˈbɪərɪən/ the extreme SW peninsula of Europe, containing present-day Spain and Portugal. In ancient times it was a centre of Carthaginian colonization until the third Punic War (149–146 BC), after which it came increasingly under Roman influence. It was invaded by the Visigoths in the 4th–5th centuries AD and by the Moors in the 8th century.

Ibero- /ˈɪbɪərəʊ/ comb. form Iberian; Iberian and (Ibero-American).

ibex /ˈaɪbeks/ n. (pl. **ibexes**) a bearded wild goat of the genus Capra, found in mountainous areas, with thick curved ridged horns. One kind, C. ibex, is found in the Alps, central Asia, and NE Africa, and another, C. pyrenaica (the Spanish ibex or Spanish goat), in the Pyrenees. [L]

ibid. /ˈɪbɪd/ abbr. (also **ib.**) in the same book or passage etc. [L ibidem in the same place]

-ibility /ɪˈbɪlɪtɪ/ suffix forming nouns from, or corresponding to, adjectives in -ible (possibility; credibility). [F -ibilité or L -ibilitas]

ibis /ˈaɪbɪs/ n. (pl. **ibises**) a wading bird of the family Threskiornithidae, with a long down-curved bill, long neck, and long legs, and nesting in colonies. □ **sacred ibis** a mainly white ibis, Threskiornis aethiopica, native to Africa and Madagascar, venerated by the ancient Egyptians. [ME f. L f. Gk]

Ibiza /ɪˈbiːθə/ **1** the westernmost of the Balearic Islands. **2** its capital city and port; pop. (1981) 25,490.

-ible /ɪb(ə)l/ suffix forming adjectives meaning 'that may or may be' (see -ABLE) (terrible; forcible; possible). [F -ible or L -ibilis]

-ibly /ɪblɪ/ suffix forming adverbs corresponding to adjectives in -ible.

IBM abbr. International Business Machines Corporation, a leading American computer manufacturer with an important share of the international market. The company was instituted in 1911.

Ibn Batuta /ˌɪb(ə)n baːˈtuːtaː/ (c.1304–68), Arab explorer. From 1325 to 1354, he journeyed through North and West Africa, India, and China, writing a vivid account of his travels in the Rihlah (undated).

ibn Hussein, Abdullah, see ABDULLAH IBN HUSSEIN.

Ibo /ˈiːbəʊ/ n. & adj. (also **Igbo**) ● n. (pl. same or **Ibos**) **1** a member of a people of SE Nigeria. **2** the Kwa language of this people, one of the official languages of Nigeria, spoken by some 15 million people. ● adj. of or relating to the Ibo or their language. [African name]

IBRD see INTERNATIONAL BANK FOR RECONSTRUCTION AND DEVELOPMENT.

Ibsen /ˈɪbs(ə)n/, Henrik (1828–1906), Norwegian dramatist. After the success of his verse drama Peer Gynt (1867), he turned to writing prose plays on social issues, including A Doll's House (1879) and Ghosts (1881).

He is credited with being the first major dramatist to write tragedy about ordinary people in prose, and was an important influence on George Bernard Shaw. Ibsen's later works, such as *The Master Builder* (1892), deal increasingly with the forces of the unconscious and were admired by Sigmund Freud.

ibuprofen /ˌaɪbjuːˈprəʊfen/ *n. Pharm.* an analgesic and anti-inflammatory drug used esp. as a more powerful alternative to aspirin. [ISO- + BUTYL + PROPIONIC ACID + -*fen* (representing PHENYL)]

IC *abbr.* integrated circuit.

i/c *abbr.* **1** in charge. **2** in command. **3** internal combustion.

-ic /ɪk/ *suffix* **1** forming adjectives (*Arabic*; *classic*; *public*) and nouns (*critic*; *epic*; *mechanic*; *music*). **2** *Chem.* denoting a higher valency or oxidation state than the corresponding word in -*ous* (*ferric*; *sulphuric*). **3** denoting a particular form or instance of a noun in -*ics* (*aesthetic*; *tactic*). [from or after F -*ique* or L -*icus* or Gk -*ikos*: cf. -ATIC, -ETIC, -FIC, -OTIC]

-ical /ɪk(ə)l/ *suffix* **1** forming adjectives corresponding to nouns or adjectives, usu. in -*ic* (*classical*; *comical*; *farcical*; *musical*). **2** forming adjectives corresponding to nouns in -*y* (*pathological*).

-ically /ɪkəlɪ/ *suffix* forming adverbs corresponding to adjectives in -*ic* or -*ical* (*comically*; *musically*; *tragically*).

ICAO *abbr.* International Civil Aviation Organization.

Icarus /ˈɪkərəs/ *Gk Mythol.* the son of Daedalus, who escaped from Crete on wings made by his father but was killed when he flew too near the sun and the wax attaching his wings melted.

ICBM *abbr.* intercontinental ballistic missile.

ICE *abbr.* **1** (in the UK) Institution of Civil Engineers. **2** internal-combustion engine.

ice /aɪs/ *n. & v.* ● *n.* **1 a** frozen water, a brittle transparent crystalline solid. **b** a sheet of this on the surface of water (*fell through the ice*). **2** *Brit.* a portion of ice-cream or water-ice (*would you like an ice?*). **3** *sl.* diamonds. ● *v.* **1** *tr.* mix with or cool in ice (*iced drinks*). **2** *tr. & intr.* (often foll. by *over*, *up*) **a** cover or become covered with ice. **b** freeze. **3** *tr.* cover (a cake etc.) with icing. **4** *US sl.* kill. □ **ice age** a glacial period, esp. in the Pleistocene epoch (see GLACIAL PERIOD). **ice-axe** a tool used by mountain-climbers for cutting footholds. **ice-bag** an ice-filled rubber bag for medical use. **ice-blue** a very pale blue. **ice-boat 1** a boat mounted on runners for travelling on ice. **2** a boat used for breaking ice on a river etc. **ice-bound** confined by ice. **ice-breaker 1** = *ice-boat* 2. **2** something that serves to relieve inhibitions, start a conversation, etc. **ice bucket** a bucket-like container with chunks of ice, used to keep a bottle of wine chilled. **ice-cap** a permanent covering of ice, e.g. in polar regions. **ice-cold** as cold as ice. **ice-cream** a sweet creamy frozen food, usu. flavoured. **ice-cube** a small block of ice made in a refrigerator. **ice-dancing** skating to choreographed dance moves, esp. competitively and in pairs. **ice-fall** a steep part of a glacier like a frozen waterfall. **ice-field** an expanse of ice, esp. in polar regions. **ice-fish 1** = CAPELIN. **2** a semi-transparent Antarctic fish of the family Chaenichthyidae. **ice floe** = FLOE. **ice house** a building often partly or wholly underground for storing ice. **ice** (or **iced**) **lolly** *Brit.* a piece of flavoured ice, often with chocolate or ice-cream, on a stick. **ice-pack 1** a quantity of ice applied to the body for medical etc. purposes. **2** see PACK[1] *n.* 9. **ice-pick 1** a needle-like implement with a handle for splitting up small pieces of ice. **2** a mountaineer's pick. **ice-plant** a mesembryanthemum having leaves covered with watery vesicles that look like ice crystals, esp. *Mesembryanthemum crystallinum*. **ice-rink** = RINK 1. **ice sheet** *Geog.* a thick permanent layer of ice covering a large area of land. **ice shelf** *Geog.* a floating sheet of ice permanently attached to a landmass. **ice-skate** *n.* a skate consisting of a boot with a blade beneath, for skating on ice. ● *v.intr.* skate on ice. (See ICE-SKATING.) **ice-skater** a person who skates on ice. **ice station** a meteorological research centre in polar regions. **on ice 1** (of an entertainment, sport, etc.) performed by skaters. **2** *colloq.* held in reserve; awaiting further attention. **on thin ice** in a risky situation. [OE *īs* f. Gmc]

-ice /ɪs/ *suffix* forming (esp. abstract) nouns (*avarice*; *justice*; *service*) (cf. -ISE[2]).

iceberg /ˈaɪsbɜːg/ *n.* **1** a large floating mass of ice detached from a glacier or ice-sheet and carried out to sea. **2** an unemotional or cold-blooded person. □ **iceberg lettuce** a crisp lettuce with a freely blanching head. **the tip of the iceberg** a small perceptible part of something (esp. a difficulty) the greater part of which is hidden. [prob. f. Du. *ijsberg* f. *ijs* ice + *berg* hill]

iceblink /ˈaɪsblɪŋk/ *n.* a luminous appearance on the horizon, caused by a reflection from ice.

iceblock /ˈaɪsblɒk/ *n. Austral. & NZ* = *ice lolly*.

icebox /ˈaɪsbɒks/ *n.* **1** a compartment in a refrigerator for making or storing ice. **2** *US* a refrigerator.

ice hockey *n.* a game played on ice between two teams of six skaters, who attempt to drive a small rubber disc or puck into the opposing goal with hooked or angled sticks. It is a fast game with much physical contact, and the players wear masks and protective clothing. Developed from hockey in Canada in the second half of the 19th century, ice hockey is today played particularly in Canada, the US, and Russia.

Iceland /ˈaɪslənd/ (Icelandic **Island** /ˈiːsland/) an island country in the North Atlantic; pop. (est. 1990) 300,000; official language, Icelandic; capital, Reykjavik. Iceland lies just south of the Arctic Circle, and only about 21% of the land area is habitable. Situated at the north end of the Mid-Atlantic Ridge, it is volcanically active. First settled by Norse colonists in the 9th century, Iceland was under Norwegian rule from 1262 to 1380, when it passed to Denmark. Granted internal self-government in 1874, it became a fully fledged independent republic in 1944. □ **Icelander** *n.*

Icelandic /aɪsˈlændɪk/ *adj & n.* ● *adj.* of or relating to Iceland or its language. ● *n.* the official language of Iceland, spoken by its 300,000 inhabitants, a Scandinavian language which is the purest descendant of Old Norse. Its purity is due partly to the geographical position of Iceland but also to a policy of avoiding loan-words.

Iceland lichen *n.* (or **moss**) a mountain and moorland lichen, *Cetraria islandica*, with edible branching fronds.

Iceland poppy *n.* an Arctic poppy; esp. *Papaver nudicaule*, with white or yellow flowers.

Iceland spar *n.* a transparent variety of calcite with the optical property of strong double refraction.

iceman /ˈaɪsmən/ *n.* (pl. **-men**) **1** a person skilled in crossing ice, esp. in Alpine or polar regions. **2** esp. *N. Amer.* a person who sells or delivers ice.

Iceni /aɪˈsiːnaɪ/ *n.pl.* a tribe of ancient Britons inhabiting an area of SE England in present-day Norfolk and Suffolk. Their queen, Boudicca, led an unsuccessful rebellion against the Romans in AD 60. [L]

ice-skating /ˈaɪsˌskeɪtɪŋ/ *n.* skating on ice as a sport or pastime. Ice-skating became a recognized sport after the opening of the first artificial ice rink in London in 1876. The first international competition was held in 1896; ice-skating has been included in the Winter Olympics since their beginning in 1924. Contests, frequently televised, are now popular: skaters are marked for technical and artistic excellence in performing a series of prescribed patterns (*figure-skating*) esp. to music, with improvised passages, or a choreographed series of dance moves (*ice-dancing*).

ICFTU see INTERNATIONAL CONFEDERATION OF FREE TRADE UNIONS.

I.Chem.E. *abbr.* (in the UK) Institution of Chemical Engineers.

I Ching /iː ˈtʃɪŋ/ an ancient Chinese manual based on a system of symbols known as the eight trigrams and sixty-four hexagrams that are symbolically interpreted in terms of the principles of yin and yang. Originally used for divination, it was later included as one of the 'five classics' of Confucianism. Also called *Book of Changes*. [Chin. *yijing* book of changes]

ichneumon /ɪkˈnjuːmən/ *n.* **1** (in full **ichneumon wasp**) a small parasitic hymenopterous insect of the family Ichneumonidae, depositing eggs in or on the larva of another insect as food for its own larva. **2** a mongoose of North Africa, *Herpestes ichneumon*, noted for destroying crocodile eggs. [L f. Gk *ikhneumōn* spider-hunting wasp f. *ikhneuō* trace f. *ikhnos* footstep]

ichnography /ɪkˈnɒɡrəfɪ/ *n.* (pl. **-ies**) **1** the ground-plan of a building, map of a region, etc. **2** a drawing of this. [F *ichnographie* or L *ichnographia* f. Gk *ikhnographia* f. *ikhnos* track: see -GRAPHY]

ichor /ˈaɪkɔː(r)/ *n.* **1** *Gk Mythol.* fluid flowing like blood in the veins of the gods. **2** *poet.* bloodlike fluid. **3** *archaic* a watery fetid discharge from a wound etc. □ **ichorous** /ˈaɪkərəs/ *adj.* [Gk *ikhōr*]

ichthyo- /ˈɪkθɪəʊ/ *comb. form* fish. [Gk *ikhthus* fish]

ichthyoid /ˈɪkθɪ,ɔɪd/ *adj. & n.* ● *adj.* fishlike. ● *n.* any fishlike vertebrate.

ichthyolite /ˈɪkθɪə,laɪt/ *n.* a fossil fish.

ichthyology /ˌɪkθɪˈɒlədʒɪ/ *n.* the study of fishes. □ **ichthyologist** *n.* **ichthyological** /-θɪəˈlɒdʒɪk(ə)l/ *adj.*

ichthyophagous /ˌɪkθɪˈɒfəɡəs/ *adj.* fish-eating. □ **ichthyophagy** /-ˈɒfədʒɪ/ *n.*

ichthyosaur /ˈɪkθɪə,sɔː(r)/ *n.* (also **ichthyosaurus** /ˌɪkθɪəˈsɔːrəs/) an extinct marine reptile of the order Ichthyosauria, of the Jurassic and Cretaceous periods, resembling a dolphin. It had a streamlined body with a long, toothed jaw, four flippers, and a tail fin. [mod.L f. ICHTHYO- + Gk *sauros* lizard]

ichthyosis /ˌɪkθɪˈəʊsɪs/ *n. Med.* a skin disease which causes the epidermis to become dry and horny like fish scales. □ **ichthyotic** /-ˈɒtɪk/ *adj.* [Gk *ikhthus* fish + -OSIS]

I-chun see YICHUN.

ICI *abbr.* Imperial Chemical Industries Limited, a British corporation manufacturing chemicals, plastics, pharmaceutical products, and synthetic fibres, founded in 1926. In 1993 the company underwent a demerger, with Zeneca, its new pharmaceuticals, seeds, and speciality chemicals group, separating from the main ICI organization.

-ician /ˈɪʃ(ə)n/ *suffix* forming nouns denoting persons skilled in or concerned with subjects having nouns (usu.) in -ic or -ics (*magician*; *politician*). [from or after F -*icien* (as -IC, -IAN)]

icicle /ˈaɪsɪk(ə)l/ *n.* a hanging tapering piece of ice, formed by the freezing of dripping water. [ME f. ICE + *ickle* (now dial.) icicle]

icing /ˈaɪsɪŋ/ *n.* **1** a coating of sugar paste on a cake or biscuit. **2** the formation of ice on a ship or aircraft. □ **icing on the cake** an attractive though inessential addition or enhancement. **icing sugar** *Brit.* finely powdered sugar for making icing for cakes etc.

-icist /ɪsɪst/ *suffix* = -ICIAN (*classicist*). [-IC + -IST]

-icity /ˈɪsɪtɪ/ *suffix* forming abstract nouns esp. from adjectives in -ic (*authenticity*; *publicity*). [-IC + -ITY]

-ick /ɪk/ *suffix* archaic var. of -IC.

Icknield Way /ˈɪkniːld/ an ancient pre-Roman track, which crosses England in a wide curve from Wiltshire to Norfolk.

icky /ˈɪkɪ/ *adj.* (also **ikky**) *colloq.* **1** sweet, sticky, sickly. **2** (as a general term of disapproval) nasty, repulsive. [20th c.: orig. unkn.]

-icle /ɪk(ə)l/ *suffix* forming (orig. diminutive) nouns (*article*; *particle*). [formed as -CULE]

icon /ˈaɪkɒn/ *n.* (also **ikon**) **1** a devotional painting or carving, usu. on wood, of Christ or another holy figure, esp. in the Eastern Church. **2** an image or statue. **3** *Computing* a symbol or graphic representation on a VDU screen of a program, option, or window, esp. one of several for selection. **4** *Linguistics* a sign which has a characteristic in common with the thing it signifies. [L f. Gk *eikōn* image]

iconic /aɪˈkɒnɪk/ *adj.* **1** of or having the nature of an image or portrait. **2** (of a statue) following a conventional type. **3** *Linguistics* that is an icon. □ **iconicity** /ˌaɪkəˈnɪsɪtɪ/ *n.* (esp. in sense 3). [L *iconicus* f. Gk *eikonikos* (as ICON)]

icono- /aɪˈkɒnəʊ/ *comb. form* an image or likeness. [Gk *eikōn*]

iconoclasm /aɪˈkɒnəˌklæz(ə)m/ *n.* **1** the breaking of images. **2** the assailing of cherished beliefs. [ICONOCLAST after *enthusiasm* etc.]

iconoclast /aɪˈkɒnəˌklæst/ *n.* **1** a person who attacks cherished beliefs. **2** a person who destroys images used in religious worship, esp. one who took part in a movement in the 8th–9th centuries against the use of images in religious worship in churches in the Eastern Roman Empire, or a Puritan of the 16th–17th centuries. □ **iconoclastic** /aɪˌkɒnəˈklæstɪk/ *adj.* **iconoclastically** *adv.* [med.L *iconoclastes* f. eccl.Gk *eikonoklastēs* (as ICONO-, *klaō* break)]

iconography /ˌaɪkəˈnɒɡrəfɪ/ *n.* (*pl.* -ies) **1** the illustration of a subject by drawings or figures. **2 a** the study of portraits, esp. of an individual. **b** the study of artistic images or symbols. **3** a treatise on pictures or statuary. **4** a book of illustrations. □ **iconographer** *n.* **iconographic** /aɪˌkɒnəˈɡræfɪk/ *adj.* **iconographical** *adj.* **iconographically** *adv.* [Gk *eikonographia* sketch (as ICONO- + -GRAPHY)]

iconolatry /ˌaɪkəˈnɒlətrɪ/ *n.* the worship of images. [eccl.Gk *eikonolatreia* (as ICONO-, -LATRY)]

iconology /ˌaɪkəˈnɒlədʒɪ/ *n.* **1** the study of visual imagery and its symbolism and interpretation. **2** symbolism.

iconostasis /ˌaɪkəˈnɒstəsɪs, aɪˌkɒnəˈstæsɪs/ *n.* (*pl.* **iconostases** /-ˌsiːz/) (in the Eastern Church) a screen bearing icons and separating the sanctuary from the nave. [mod.Gk *eikonostasis* (as ICONO-, STASIS)]

icosahedron /ˌaɪkəsəˈhiːdrən/ *n.* (*pl.* **icosahedra** /-drə/) a solid figure with twenty faces. □ **icosahedral** *adj.* [LL *icosahedrum* f. Gk *eikosaedron* f. *eikosi* twenty + -HEDRON]

-ics /ɪks/ *suffix* (treated as *sing.* or *pl.*) forming nouns denoting arts or sciences or branches of study or action (*athletics*; *politics*) (cf. -IC 3). [from or after F pl. -*iques* or L pl. -*ica* or Gk pl. -*ika*]

icterus /ˈɪktərəs/ *n. Med.* = JAUNDICE. □ **icteric** /ɪkˈterɪk/ *adj.* [L f. Gk *ikteros*]

Ictinus /ɪkˈtaɪnəs/ (5th century BC), Greek architect. His most famous building was the Parthenon in Athens, which he is said to have designed with the architect Callicrates and the sculptor Phidias between 448 and 437 BC.

ictus /ˈɪktəs/ *n.* (*pl.* same or **ictuses**) **1** *Prosody* rhythmical or metrical stress. **2** *Med.* a stroke or seizure; a fit. [L, = blow f. *icere* strike]

icy /ˈaɪsɪ/ *adj.* (**icier**, **iciest**) **1** very cold. **2** covered with or having much ice. **3** (of a tone or manner) unfriendly, hostile (*an icy stare*). □ **icily** *adv.* **iciness** *n.*

ID *abbr.* **1** identification, identity (*ID card*). **2** US Idaho (in official postal use).

Id var. of EID.

I'd /aɪd/ *contr.* **1** I had. **2** I should; I would.

id /ɪd/ *n. Psychol.* the inherited instinctive impulses of the individual as part of the unconscious. [L, = that, transl. G *es*]

id. *abbr.* = IDEM.

i.d. *abbr.* inner diameter.

-id[1] /ɪd/ *suffix* forming adjectives (*arid*; *rapid*). [F -*ide* f. L -*idus*]

-id[2] /ɪd/ *suffix* forming nouns: **1** general (*pyramid*). **2** *Biol.* of structural constituents (*plastid*). **3** *Bot.* of a plant belonging to a family with a name in -*idaceae* (*orchid*). [from or after F -*ide* f. L -*is* -*idis* f. Gk -*is* -*ida* or -*idos*]

-id[3] /ɪd/ *suffix* forming nouns denoting: **1** *Zool.* an animal belonging to a family with a name in -*idae* or a class with a name in -*ida* (*canid*; *arachnid*). **2** a member of a person's family (*Seleucid* from Seleucus). **3** *Astron.* a meteor in a group radiating from a specified constellation (*Leonid* from Leo). **b** a star of a class like one in a specified constellation (*cepheid*). [from or after L -*ides*, pl. -*idae* or -*ida*]

-id[4] /ɪd/ *suffix* esp. *US* var. of -IDE.

IDA see INTERNATIONAL DEVELOPMENT ASSOCIATION.

Ida /ˈaɪdə/ **1** a mountain in central Crete, associated in classical times with the god Zeus. Rising to 2,456 m (8,058 ft), it is the highest peak in the island. **2** *Astron.* asteroid 243, photographed closely in 1993 by the Galileo spacecraft. It is irregular in shape, 52 km long, and has many craters, some of which are large and degraded. It also has a tiny moon, about 1.5 km across.

Idaho /ˈaɪdəˌhəʊ/ a state of the north-western US, bordering on British Columbia to the north and containing part of the Rocky Mountains; pop. (1990) 1,007,750; capital, Boise. It became the 43rd state of the US in 1890.

ide /aɪd/ *n.* = ORFE. [mod.L *idus* f. Sw. *id*]

-ide /aɪd/ *suffix* (also esp. *US* **-id**) *Chem.* forming nouns denoting: **1** binary compounds of an element (the suffix -*ide* being added to the abbreviated name of the more electronegative element etc.) (*sodium chloride*; *lead sulphide*; *calcium carbide*). **2** various other compounds (*amide*; *anhydride*; *peptide*; *saccharide*). **3** elements of a series in the periodic table (*actinide*; *lanthanide*). [orig. in OXIDE]

idea /aɪˈdɪə/ *n.* **1** a conception or plan formed by mental effort (*have you any ideas?; had the idea of writing a book*). **2 a** a mental impression or notion; a concept. **b** a vague belief or fancy (*had an idea you were married; had no idea where you were*). **c** an opinion, or outlook or point of view (*had some funny ideas about marriage*). **3** an intention, purpose, or essential feature (*the idea is to make money*). **4** an archetype or pattern as distinguished from its realization in individual cases. **5** *Philos.* **a** (in Platonism) an eternally existing pattern of which individual things in any class are imperfect copies. **b** a concept of pure reason which transcends experience. □ **get** (or **have**) **ideas** *colloq.* be ambitious, rebellious, etc. **have no idea** *colloq.* **1** not know at all. **2** be completely incompetent. **not one's idea of** *colloq.* not what one regards as (*not my idea of a pleasant evening*). **put ideas into a person's head** suggest ambitions etc. he or she would not otherwise have had. **that's an idea** *colloq.* that proposal etc. is worth considering. **the very idea!** *colloq.* an exclamation of disapproval or disagreement. □ **idealess** *adj.* [Gk *idea* form, pattern f. stem *id-* see]

ideal /aɪˈdiːl/ *adj. & n.* ● *adj.* **1 a** answering to one's highest conception. **b** perfect or supremely excellent. **2 a** existing only as an idea. **b** visionary. **3** embodying an idea. **4** relating to or consisting of ideas; dependent on the mind. ● *n.* **1** a perfect type, or a conception of this. **2 a** an actual thing as a standard for imitation. **b** (often in *pl.*) a moral principle or standard of behaviour. □ **ideal gas** a hypothetical gas consisting of molecules occupying negligible space and without attraction for each other, thereby obeying simple laws. □ **ideally** /aɪˈdɪəlɪ/ *adv.* [ME f. F *idéal* f. LL *idealis* (as IDEA)]

idealism /ˈaɪdɪəlɪz(ə)m/ n. **1** the practice of forming or pursuing ideals, esp. unrealistically (cf. REALISM). **2** the representation of things in ideal or idealized form. **3** imaginative treatment. **4** Philos. any of various systems of thought in which the object of perception is held to consist of ideas not resulting from any perceived material substance (cf. REALISM). (See note below.) □ **idealist** n. **idealistic** /ˌaɪdɪəˈlɪstɪk/ adj. **idealistically** adv. [F idéalisme or G Idealismus (as IDEAL)]

▪ Idealism in philosophy derives from Plato, who held that 'ideas' are the only objects of knowledge, of which the world of ordinary perceived experience is only a shadow. In George Berkeley's view of idealism, no material things exist independently of minds, but the mind of God, being infinite, gives objects a continuous existence. While Berkeley believed that other minds had an independent existence, Hegel rejected the reality of separate objects, and of minds in space and time, and believed instead in an all-embracing unity (the absolute).

ideality /ˌaɪdɪˈælɪtɪ/ n. (pl. **-ies**) **1** the quality of being ideal. **2** an ideal thing.

idealize /ˈaɪdɪəlaɪz/ v.tr. (also **-ise**) **1** regard or represent (a thing or person) in ideal form or character. **2** exalt in thought to ideal perfection or excellence. □ **idealizer** n. **idealization** /ˌaɪˌdɪəlaɪˈzeɪʃ(ə)n/ n.

ideate /ˈaɪdɪˌeɪt/ v. Psychol. **1** tr. imagine, conceive. **2** intr. form ideas. □ **ideation** /ˌaɪdɪˈeɪʃ(ə)n/ n. **ideational** adj. **ideationally** adv. [med.L ideare form an idea (as IDEA)]

idée fixe /ˌiːdeɪ ˈfiːks/ n. (pl. **idées fixes** pronunc. same) an idea that dominates the mind; an obsession. [F, lit. 'fixed idea']

idée reçue /ˌiːdeɪ rəˈsjuː/ n. (pl. **idées reçues** pronunc. same) a generally accepted notion or opinion. [F, lit. 'received idea']

idem /ˈɪdem/ adv. & n. ● adv. in the same author. ● n. the same word or author. [ME f. L]

identical /aɪˈdentɪk(ə)l/ adj. **1** (often foll. by with) (of different things) agreeing in every detail. **2** (of one thing viewed at different times) one and the same. **3** (of twins) developed from a single fertilized ovum, therefore of the same sex and usu. very similar in appearance. **4** Logic & Math. expressing an identity. □ **identically** adv. [med.L identicus (as IDENTITY)]

identification /aɪˌdentɪfɪˈkeɪʃ(ə)n/ n. **1 a** the act or an instance of identifying; recognition, pinpointing. **b** association of oneself with the feelings, situation, characteristics, etc. of another person or group of people. **2** a means of identifying a person. **3** (attrib.) serving to identify (esp. the bearer) (identification card). □ **identification parade** an assembly of persons from whom a suspect is to be identified.

identifier /aɪˈdentɪˌfaɪə(r)/ n. **1** a person or thing that identifies. **2** Computing a sequence of characters used to identify or refer to a set of data.

identify /aɪˈdentɪˌfaɪ/ v. (**-ies, -ied**) **1** tr. establish the identity of; recognize. **2** tr. establish or select by consideration or analysis of the circumstances (identify the best method of solving the problem). **3** tr. (foll. by with) associate (a person or oneself) inseparably or very closely (with a party, policy, etc.). **4** tr. (often foll. by with) treat (a thing) as identical. **5** intr. (foll. by with) **a** regard oneself as sharing characteristics of (another person). **b** associate oneself. □ **identifiable** adj. **identifiably** adv. [med.L identificare (as IDENTITY)]

Identikit /aɪˈdentɪˌkɪt/ n. (often attrib.) propr. a reconstructed picture of a person (esp. one sought by the police) assembled from transparent strips showing typical facial features according to witnesses' descriptions. [IDENTITY + KIT[1]]

identity /aɪˈdentɪtɪ/ n. (pl. **-ies**) **1 a** the quality or condition of being a specified person or thing. **b** individuality, personality (felt he had lost his identity). **2** identification or the result of it (a case of mistaken identity; identity card). **3** the state of being the same in substance, nature, qualities, etc.; absolute sameness (no identity of interests between them). **4** Algebra **a** the equality of two expressions for all values of the quantities expressed by letters. **b** an equation expressing this, e.g. $(x + 1)^2 = x^2 + 2x + 1$. **5** Math. **a** (in full **identity element**) an element in a set, left unchanged by any operation to it. **b** a transformation that leaves an object unchanged. □ **identity crisis** a temporary period during which an individual experiences a feeling of loss or breakdown of identity. **identity parade** = identification parade. [LL identitas f. L idem same]

ideogram /ˈɪdɪəˌgræm/ n. = IDEOGRAPH. [Gk idea form + -GRAM]

ideograph /ˈɪdɪəˌɡrɑːf/ n. a character symbolizing the idea of a thing without indicating the sequence of sounds in its name (e.g. a numeral, and many Chinese characters). □ **ideographic** /ˌɪdɪəˈɡræfɪk/ adj. **ideography** /ˌɪdɪˈɒɡrəfɪ/ n. [Gk idea form + -GRAPH]

ideologue /ˈaɪdɪəˌlɒɡ/ n. **1** a theorist; a visionary. **2** an adherent of an ideology. [F idéologue f. Gk idea (see IDEA) + -LOGUE]

ideology /ˌaɪdɪˈɒlədʒɪ/ n. (pl. **-ies**) **1** the system of ideas at the basis of an economic or political theory (Marxist ideology). **2** the manner of thinking characteristic of a class or individual (bourgeois ideology). **3** visionary speculation. **4** archaic the science of ideas. □ **ideologist** n. **ideological** /ˌaɪdɪəˈlɒdʒɪk(ə)l/ adj. **ideologically** adv. [F idéologie (as IDEOLOGUE)]

ides /aɪdz/ n.pl. the eighth day after the nones in the ancient Roman calendar (the 15th day of March, May, July, October, the 13th of other months). [ME f. OF f. L idus (pl.), perh. f. Etruscan]

idiocy /ˈɪdɪəsɪ/ n. (pl. **-ies**) **1** utter foolishness; idiotic behaviour or an idiotic action. **2** extremely low intelligence. [ME f. IDIOT, prob. after lunacy]

idiolect /ˈɪdɪəˌlekt/ n. the form of language used by an individual person. [Gk idios own + -lect in DIALECT]

idiom /ˈɪdɪəm/ n. **1** a group of words established by usage and having a meaning not deducible from those of the individual words (as in over the moon, see the light). **2** a form of expression peculiar to a language, person, or group of people. **3 a** the language of a people or country. **b** the specific character of this. **4** a characteristic mode of expression in music, art, etc. [F idiome or LL idioma f. Gk idiōma -matos private property f. idios own, private]

idiomatic /ˌɪdɪəˈmætɪk/ adj. **1** relating to or conforming to idiom. **2** characteristic of a particular language. □ **idiomatically** adv. [Gk idiōmatikos peculiar (as IDIOM)]

idiopathy /ˌɪdɪˈɒpəθɪ/ n. Med. any disease or condition of unknown cause or that arises spontaneously. □ **idiopathic** /ˌɪdɪəˈpæθɪk/ adj. [mod.L idiopathia f. Gk idiopatheia f. idios own + -PATHY]

idiosyncrasy /ˌɪdɪəʊˈsɪŋkrəsɪ/ n. (pl. **-ies**) **1** a mental constitution, view or feeling, or mode of behaviour, peculiar to a person. **2** anything highly individualized or eccentric. **3** a mode of expression peculiar to an author. **4** Med. a physical constitution peculiar to a person. □ **idiosyncratic** /-sɪŋˈkrætɪk/ adj. **idiosyncratically** adv. [Gk idiosugkrasia f. idios own + sun together + krasis mingling]

idiot /ˈɪdɪət/ n. **1** colloq. a stupid person; an utter fool. **2** a person of extremely low intelligence. □ **idiot board** (or **card**) colloq. a board displaying a television script to a speaker as an aid to memory. **idiot box** colloq. a television set. □ **idiotic** /ˌɪdɪˈɒtɪk/ adj. **idiotically** adv. [ME f. OF f. L idiota ignorant person f. Gk idiōtēs private person, layman, ignorant person f. idios own, private]

idle /ˈaɪd(ə)l/ adj. & v. ● adj. (**idler, idlest**) **1** lazy, indolent. **2** not in use; not working; unemployed. **3** (of time etc.) unoccupied. **4** having no special basis or purpose (idle rumour; idle curiosity). **5** useless. **6** (of an action, thought, or word) ineffective, worthless, vain. ● v. **1 a** intr. (of an engine) run slowly without doing any work. **b** tr. cause (an engine) to idle. **2** intr. be idle. **3** tr. (foll. by away) pass (time etc.) in idleness. □ **idle wheel** an intermediate wheel between two geared wheels, esp. to allow them to rotate in the same direction. □ **idleness** n. **idly** adv. [OE īdel empty, useless]

idler /ˈaɪdlə(r)/ n. **1** a habitually lazy person. **2** = idle wheel.

Ido /ˈiːdəʊ/ n. an artificial universal language based on Esperanto. [Ido, = offspring]

idol /ˈaɪd(ə)l/ n. **1** an image of a deity etc. used as an object of worship. **2** Bibl. a false god. **3** a person or thing that is the object of excessive or supreme adulation (cinema idol). **4** archaic a phantom. [ME f. OF idole f. L idolum f. Gk eidōlon phantom f. eidos form]

idolater /aɪˈdɒlətə(r)/ n. (fem. **idolatress** /-trɪs/) **1** a worshipper of idols. **2** (often foll. by of) a devoted admirer. □ **idolatrous** adj. [ME idolatrer f. OF or f. idolatry or f. OF idolâtre, ult. f. Gk eidōlolatrēs (as IDOL, -LATER)]

idolatry /aɪˈdɒlətrɪ/ n. **1** the worship of idols. **2** great adulation. [OF idolatrie (as IDOLATER)]

idolize /ˈaɪdəˌlaɪz/ v. (also **-ise**) **1** tr. venerate or love extremely or excessively. **2** tr. make an idol of. **3** intr. practise idolatry. □ **idolizer** n. **idolization** /ˌaɪdəlaɪˈzeɪʃ(ə)n/ n.

Idomeneus /aɪˈdɒmɪˌniːəs/ Gk Mythol. king of Crete, son of Deucalion and descendant of Minos. He was forced to sacrifice his son in consequence of a vow, made on his return from fighting in the Trojan War, to sacrifice the first living thing that met him on his return.

Id ul-Adha see EID 1.

Id ul-Fitr see EID 2.

idyll /'ɪdɪl/ n. (also **idyl**) **1** a short description in verse or prose of a picturesque scene or incident, esp. in rustic life. **2** an episode suitable for such treatment; a picturesque, blissful, or romantic event or period. □ **idyllist** n. **idyllize** v.tr. (also **-ise**). [L idyllium f. Gk eidullion, dimin. of eidos form]

idyllic /ɪ'dɪlɪk/ adj. **1** blissfully peaceful and happy. **2** of or like an idyll. □ **idyllically** adv.

i.e. abbr. that is to say. [L id est]

-ie /ɪ/ suffix **1** var. of -Y² (dearie; nightie). **2** archaic var. of -Y¹, -Y³ (litanie; prettie). [earlier form of -Y]

IEA see INTERNATIONAL ENERGY AGENCY.

IEE abbr. (in the UK) Institution of Electrical Engineers.

Ieper see YPRES.

-ier /ɪə(r)/ suffix forming personal nouns denoting an occupation or interest: **1** with stress on the preceding element (grazier). **2** with stress on the suffix (cashier; brigadier). [sense 1 ME of various orig.; sense 2 F -ier f. L -arius]

IF abbr. intermediate frequency.

if /ɪf/ conj. & n. ● conj. **1** introducing a conditional clause: **a** on the condition or supposition that; in the event that (if he comes I will tell him; if you are tired we will rest). **b** (with past tense) implying that the condition is not fulfilled (if I were you; if I knew I would say). **2** even though (I'll finish it, if it takes me all day). **3** whenever (if I am not sure I ask). **4** whether (see if you can find it). **5 a** expressing wish or surprise (if I could just try!; if it isn't my old hat!). **b** expressing a request (if you wouldn't mind opening the door?). **6** with implied reservation, = and perhaps not (very rarely if at all). **7** (with reduction of the protasis to its significant word) if there is or it is etc. (took little if any). **8** despite being (a useful if cumbersome device). ● n. a condition or supposition (too many ifs about it). □ **if anything** if any degree, perhaps even (if anything, it's too large; if anything, he finds maths easier). **if only 1** even if for no other reason than (I'll come if only to see her). **2** (often ellipt.) an expression of regret (if only I had thought of it; if only I could swim!). **if so** if that is the case. [OE gif]

IFAD see INTERNATIONAL FUND FOR AGRICULTURAL DEVELOPMENT.

IFC abbr. International Finance Corporation.

Ife /'iːfeɪ/ an industrial city in SW Nigeria; pop. (1981) 240,600. It was a major centre of the Yoruba kingdom from the 14th to the 17th centuries and is noted for its bronze art work, which dates back to the 12th century.

-iferous /'ɪfərəs/ comb. form see -FEROUS.

iff /ɪf/ conj. Logic & Math. = if and only if. [arbitrary extension of if]

iffy /'ɪfɪ/ adj. (**iffier, iffiest**) colloq. uncertain, doubtful.

-ific /'ɪfɪk/ comb. form see -FIC.

-ification /ɪfɪ'keɪʃ(ə)n/ comb. form see -FICATION.

Ifni /'ɪfnɪ/ a former overseas province of Spain, on the SW coast of Morocco. It was settled by Spain in the late 15th century, then abandoned until reclaimed in 1860. It was formally ceded to Morocco in 1969.

-iform /ɪ'fɔːm/ comb. form see -FORM.

IFP abbr. Inkatha Freedom Party (see INKATHA).

Ig abbr. immunoglobulin.

Igbo var. of IBO.

Iglesias /ɪ'ɡleɪzɪˌæs/, Julio (b.1943), Spanish singer. He has recorded more than sixty albums since the start of his singing career in 1970 and is famous for love songs and ballads; his many hits include 'Begin the Beguine' (1981) and 'Yours' (1982).

igloo /'ɪɡluː/ n. an Eskimo dome-shaped dwelling, esp. one built of blocks of snow. [Eskimo, = house]

Ignatius Loyola, St /ɪɡˌneɪʃəs 'lɔɪələ, lɔɪ'əʊlə/ (1491–1556), Spanish theologian and founder of the Jesuits. After sustaining a leg wound as a soldier, he renounced military life and turned to prayer and mortification. In 1534 he founded the Society of Jesus and became its first general. His Spiritual Exercises (1548), an ordered scheme of meditations on the life of Christ and the truths of the Christian faith, is still used in the training of Jesuits. Feast day, 31 July.

igneous /'ɪɡnɪəs/ adj. **1** of fire; fiery. **2** Geol. (esp. of rocks) produced by volcanic or magmatic action. (See note below.) [L igneus f. ignis fire]

▪ Igneous rocks are commonly divided into two categories: volcanic rocks, which solidified at the earth's surface (such as basalt and other lavas); and plutonic rocks, which solidified from magma within

the earth's crust (such as granite). They constitute 95 per cent of the known crust of the earth. (See METAMORPHIC, SEDIMENTARY.)

ignimbrite /'ɪɡnɪmˌbraɪt/ n. Geol. a volcanic rock, esp. a tuff, formed by the consolidation of material from a pyroclastic flow. [L ignis fire + imbr- imber shower of rain, stormcloud + -ITE¹]

ignis fatuus /ˌɪɡnɪs 'fætjʊəs/ n. (pl. **ignes fatui** /ˌɪɡniːz 'fætjʊˌaɪ/) a will-o'-the-wisp. [mod.L, = foolish fire, because of its erratic movement]

ignite /ɪɡ'naɪt/ v. **1** tr. set fire to; cause to burn. **2** intr. catch fire. **3** tr. Chem. heat to the point of combustion or chemical change. **4** tr. provoke or excite (feelings etc.). □ **ignitable** adj. **ignitability** /-ˌnaɪtə'bɪlɪtɪ/ n. [L ignire ignit- f. ignis fire]

igniter /ɪɡ'naɪtə(r)/ n. **1** a device for igniting a fuel mixture in an engine. **2** a device for causing an electric arc.

ignition /ɪɡ'nɪʃ(ə)n/ n. **1** a mechanism for, or the action of, starting the combustion of mixture in the cylinder of an internal-combustion engine. **2** the act or an instance of igniting or being ignited. □ **ignition key** a key to operate the ignition of a motor vehicle. [F ignition or med.L ignitio (as IGNITE)]

ignitron /ɪɡ'naɪtrən/ n. Electr. a kind of rectifier with a mercury cathode, able to carry large currents. [IGNITE + -TRON]

ignoble /ɪɡ'nəʊb(ə)l/ adj. **1** dishonourable, mean, base. **2** of low birth, position, or reputation. □ **ignobly** adv. **ignobility** /ˌɪɡnə'bɪlɪtɪ/ n. [F ignoble or L ignobilis (as IN-¹, nobilis noble)]

ignominious /ˌɪɡnə'mɪnɪəs/ adj. **1** causing or deserving ignominy. **2** humiliating. □ **ignominiously** adv. **ignominiousness** n. [ME f. F ignominieux or L ignominiosus]

ignominy /'ɪɡnəˌmɪnɪ/ n. **1** dishonour, infamy. **2** archaic infamous conduct. [F ignominie or L ignominia (as IN-¹, nomen name)]

ignoramus /ˌɪɡnə'reɪməs/ n. (pl. **ignoramuses**) an ignorant person. [L, = we do not know: in legal use (formerly of a grand jury rejecting a bill) we take no notice of it; mod. sense perh. from a character in George Ruggle's comedy Ignoramus (1615) exposing lawyers' ignorance]

ignorance /'ɪɡnərəns/ n. (often foll. by of) lack of knowledge (about a thing). [ME f. OF f. L ignorantia (as IGNORANT)]

ignorant /'ɪɡnərənt/ adj. **1 a** lacking knowledge or experience. **b** (foll. by of, in) uninformed (about a fact or subject). **2** colloq. ill-mannered, uncouth. □ **ignorantly** adv. [ME f. OF f. L ignorare ignorant- (as IGNORE)]

ignore /ɪɡ'nɔː(r)/ v.tr. **1** refuse to take notice of or accept. **2** intentionally disregard. □ **ignorer** n. [F ignorer or L ignorare not know, ignore (as IN-¹, gno- know)]

Iguaçu /ˌɪɡwə'suː/ (Spanish **Iguazú** /ˌɪɣwa'θu/) a river of southern Brazil. It rises in the Serra do Mar in SE Brazil and flows westwards for 1,300 km (800 miles) to the Paraná river, which it joins at the point where the frontiers between Brazil, Paraguay, and Argentina meet, shortly below the Iguaçu Falls, a spectacular series of waterfalls.

iguana /ɪ'ɡwɑːnə/ n. a large lizard of the family Iguanidae, native to tropical America, the West Indies, and some Pacific islands; esp. Iguana iguana, having a spiny crest along the back. [Sp. f. Carib iwana]

iguanodon /ɪ'ɡwɑːnəˌdɒn/ n. a large mainly bipedal plant-eating dinosaur of the genus Iguanodon, of the Cretaceous period, with a broad stiff tail and a spike on each thumb. [mod.L f. IGUANA + Gk odous odontos tooth, from the resemblance of its teeth to those of the iguana]

i.h.p. abbr. indicated horsepower.

IHS abbr. Jesus. [ME f. LL, repr. Gk IHΣ = Iēs(ous) Jesus: often taken as an abbr. of various Latin words]

IJssel /'aɪs(ə)l/ a river in the Netherlands. In part it is a distributary of the Rhine, which it leaves at Arnhem, joining the Oude IJssel ('Old IJssel') a few kilometres downstream, and flows 115 km (72 miles) northwards through the eastern Netherlands to the IJsselmeer.

IJsselmeer /ˌaɪs(ə)l'meə(r)/ a shallow lake in the NW Netherlands, created in 1932 by the building of a dam across the entrance to the old Zuider Zee. The salt water of the Zuider Zee was gradually replaced by fresh water flowing in from the River IJssel and large areas of land that had been under water were reclaimed as polders.

ikebana /ˌɪkɪ'bɑːnə/ n. the art of Japanese flower arrangement, with formal display according to strict rules. [Jap., = living flowers]

Ikhnaton see AKHENATEN.

ikky var. of ICKY.

ikon var. of ICON.

IL abbr. US Illinois (in official postal use).

il- /ɪl/ prefix assim. form of IN-¹, IN-² before l.

-il /ɪl/ *suffix* (also **-ile** /aɪl/) forming adjectives or nouns denoting relation (*civil*; *utensil*) or capability (*agile*; *sessile*). [OF f. L *-ilis*]

ilang-ilang var. of YLANG-YLANG.

ILEA /ˈɪliə/ *abbr. hist.* Inner London Education Authority.

ilea pl. of ILEUM.

Île-de-France /ˌiːldəˈfrɑːns/ a region of north central France, incorporating the city of Paris.

ileitis /ˌɪlɪˈaɪtɪs/ *n. Med.* inflammation of the ileum. [ILEUM + -ITIS]

ileostomy /ˌɪlɪˈɒstəmɪ/ *n.* (pl. **-ies**) a surgical operation in which the ileum is shortened to remove a damaged part and its cut end directed to an opening in the abdominal wall. [ILEUM + Gk *stoma* mouth]

Ilesha /ɪˈleɪʃə/ a city in SW Nigeria; pop. (1981) 306,200.

ileum /ˈɪlɪəm/ *n.* (pl. **ilea** /ˈɪlɪə/) *Anat.* the third and last portion of the small intestine. □ **ileac** /ˈɪlɪˌæk/ *adj.* [var. of ILIUM]

ileus /ˈɪlɪəs/ *n. Med.* any painful obstruction of the intestine, esp. of the ileum. [L f. Gk (e)*ileos* colic]

ilex /ˈaɪleks/ *n.* **1** a tree or shrub of the genus *Ilex*, esp. the common holly. **2** the holm-oak. [ME f. L]

ilia pl. of ILIUM.

iliac /ˈɪlɪˌæk/ *adj. Anat.* of the lower body or ilium (*iliac artery*). [LL *iliacus* (as ILIUM)]

Iliad /ˈɪlɪəd/ a Greek hexameter epic poem in twenty-four books, traditionally ascribed to Homer. It tells of the climax of the Trojan War between Greeks and Trojans. The greatest of the Greek heroes, Achilles, retires to his tent enraged at Agamemnon's abduction of his mistress, the captive Briseis. In his absence the Trojan forces under Hector push the Greeks back to their ships, and Achilles' close companion Patroclus is killed in combat by Hector; at this the grief-stricken Achilles takes the field and kills Hector under the walls of Troy.

Ilium /ˈɪlɪəm/ the alternative name for TROY, denoting especially the Greek city built there in the 7th century BC.

ilium /ˈɪlɪəm/ *n.* (pl. **ilia** /ˈɪlɪə/) **1** *Anat.* the bone forming the upper part of each half of the human pelvis. **2** the corresponding bone in animals. [ME f. L]

ilk /ɪlk/ *n.* **1** *colloq.* a family, class, or set (*not of the same ilk as you*). ¶ Usu. *derog.* and therefore best avoided. **2** (in **of that ilk**) *Sc.* of the same (name) (*Guthrie of that ilk* = of Guthrie). [OE *ilca* same]

Ill. *abbr.* Illinois.

I'll /aɪl/ *contr.* I shall; I will.

ill /ɪl/ *adj., adv., & n.* ● *adj.* (attrib. except in sense 1) **1** (usu. *predic.*; often foll. by *with*) out of health; sick (*is ill*; *was taken ill with pneumonia*; *mentally ill people*). **2** wretched, unfavourable (*ill fortune*; *ill luck*). **3** harmful (*ill effects*). **4** hostile, unkind (*ill feeling*). **5** *archaic* morally bad. **6** faulty, unskilful (*ill taste*; *ill management*). **7** (of manners or conduct) improper. ● *adv.* **1** badly, wrongly (*ill-matched*). **2 a** imperfectly (*ill-provided*). **b** scarcely (*can ill afford to do it*). **3** unfavourably (*it would have gone ill with them*). ● *n.* **1** injury, harm. **2** evil; the opposite of good. □ **do an ill turn to** harm (a person or a person's interests). **ill-advised 1** (of a person) foolish or imprudent. **2** (of a plan etc.) not well formed or considered. **ill-advisedly** /-ədˈvaɪzɪdlɪ/ in a foolish or badly considered manner. **ill-affected** (foll. by *towards*) not well disposed. **ill-assorted** not well matched. **ill at ease** embarrassed, uneasy. **ill-behaved** see BEHAVE. **ill blood** bad feeling; animosity. **ill-bred** badly brought up; rude. **ill breeding** bad manners. **ill-considered** = *ill-advised*. **ill-defined** not clearly defined. **ill-disposed 1** (often foll. by *towards*) unfavourably disposed. **2** disposed to evil; malevolent. **ill-equipped** (often foll. by *to* + infin.) not adequately equipped or qualified. **ill fame** see FAME. **ill-fated** destined to or bringing bad fortune. **ill-favoured** (US **-favored**) unattractive, displeasing, objectionable. **ill feeling** bad feeling; animosity. **ill-founded** (of an idea etc.) not well founded; baseless. **ill-gotten** gained by wicked or unlawful means. **ill humour** moroseness, irritability. **ill-humoured** bad-tempered. **ill-judged** unwise; badly considered. **ill-mannered** having bad manners; rude. **ill nature** churlishness, unkindness. **ill-natured** churlish, unkind. **ill-naturedly** churlishly. **ill-omened** attended by bad omens. **ill-starred** unlucky; destined to failure. **ill success** partial or complete failure. **ill temper** moroseness. **ill-tempered** morose, irritable. **ill-timed** done or occurring at an inappropriate time. **ill-treat** (or **-use**) treat badly; abuse. **ill treatment** (or **ill use**) abuse; bad treatment. **ill will** bad feeling; animosity. **an ill wind** an unfavourable or untoward circumstance (with ref. to the proverb *it's an ill wind that blows nobody good*). **speak**

ill of say something unfavourable about. [ME f. ON *illr*, of unkn. orig.]

illation /ɪˈleɪʃ(ə)n/ *n.* **1** a deduction or conclusion. **2** a thing deduced. [L *illatio* f. *illatus* past part. of *inferre* INFER]

illative /ɪˈleɪtɪv, ˈɪlətɪv/ *adj.* **1 a** (of a word) stating or introducing an inference. **b** inferential. **2** *Gram.* (of a case) denoting motion into. □ **illatively** *adv.* [L *illativus* (as ILLATION)]

illegal /ɪˈliːg(ə)l/ *adj.* **1** not legal. **2** contrary to law. □ **illegally** *adv.* **illegality** /ˌɪlɪˈgælɪtɪ/ *n.* (pl. **-ies**). [F *illégal* or med.L *illegalis* (as IN-¹, LEGAL)]

illegible /ɪˈledʒɪb(ə)l/ *adj.* not legible. □ **illegibly** *adv.* **illegibility** /ɪˌledʒɪˈbɪlɪtɪ/ *n.*

illegitimate *adj., n., & v.* ● *adj.* /ˌɪlɪˈdʒɪtɪmət/ **1** (of a child) born of parents not married to each other. **2** not authorized by law; unlawful. **3** improper. **4** wrongly inferred. **5** physiologically abnormal. ● *n.* /ˌɪlɪˈdʒɪtɪmət/ a person whose position is illegitimate, esp. by birth. ● *v.tr.* /ˌɪlɪˈdʒɪtɪˌmeɪt/ declare or pronounce illegitimate. □ **illegitimacy** /-məsɪ/ *n.* **illegitimately** /-mətlɪ/ *adv.* [LL *illegitimus*, after LEGITIMATE]

illiberal /ɪˈlɪbərəl/ *adj.* **1** intolerant, narrow-minded. **2** without liberal culture. **3** not generous; stingy. **4** vulgar, sordid. □ **illiberally** *adv.* **illiberality** /ɪˌlɪbəˈrælɪtɪ/ *n.* (pl. **-ies**). [F *illibéral* f. L *illiberalis* mean, sordid (as IN-¹, LIBERAL)]

Illich /ˈɪlɪtʃ/, Ivan (b.1926), Austrian-born American educationist and writer. He is chiefly known as a critic of the centralized nature of Western industrial society and as an advocate of the deinstitutionalization of education, religion, and medicine. His books include *Deschooling Society* (1971) and *Limits to Medicine* (1978).

illicit /ɪˈlɪsɪt/ *adj.* **1** unlawful, forbidden (*illicit dealings*). **2** secret, furtive (*an illicit cigarette*). □ **illicitly** *adv.* **illicitness** *n.*

illimitable /ɪˈlɪmɪtəb(ə)l/ *adj.* limitless. □ **illimitably** *adv.* **illimitability** /ɪˌlɪmɪtəˈbɪlɪtɪ/ *n.* [LL *illimitatus* f. L *limitatus* (as IN-¹, L *limitatus* past part. of *limitare* LIMIT)]

Illinois /ˌɪlɪˈnɔɪ/ a state in the Middle West of the US; pop. (1990) 11,430,600; capital, Springfield. It was colonized by the French but was ceded to Britain in 1763. It was acquired by the US in 1783 and became the 21st state in 1818.

illiquid /ɪˈlɪkwɪd/ *adj.* (of assets) not easily converted into cash. □ **illiquidity** /ˌɪlɪˈkwɪdɪtɪ/ *n.*

illiterate /ɪˈlɪtərət/ *adj. & n.* ● *adj.* **1** unable to read. **2** uneducated. ● *n.* an illiterate person. □ **illiteracy** *n.* **illiterately** *adv.* **illiterateness** *n.* [L *illitteratus* (as IN-¹, *litteratus* LITERATE)]

illness /ˈɪlnɪs/ *n.* **1** a disease, ailment, or malady. **2** the state of being ill.

illogical /ɪˈlɒdʒɪk(ə)l/ *adj.* devoid of or contrary to logic. □ **illogically** *adv.* **illogicality** /ɪˌlɒdʒɪˈkælɪtɪ/ *n.* (pl. **-ies**).

illude /ɪˈluːd, ɪˈljuːd/ *v.tr. literary* trick or deceive. [ME, = mock, f. L *illudere* (as ILLUSION)]

illume /ɪˈluːm, ɪˈljuːm/ *v.tr. poet.* light up; make bright. [shortening of ILLUMINE]

illuminance /ɪˈluːmɪnəns, ɪˈljuː-/ *n. Physics* the amount of luminous flux per unit area.

illuminant /ɪˈluːmɪnənt, ɪˈljuː-/ *n. & adj.* ● *n.* a means of illumination. ● *adj.* serving to illuminate. [L *illuminant-* part. stem of *illuminare* ILLUMINATE]

illuminate /ɪˈluːmɪˌneɪt, ɪˈljuː-/ *v.tr.* **1** light up; make bright. **2** decorate (buildings etc.) with lights as a sign of festivity. **3** decorate (an initial letter, a manuscript, etc.) with gold, silver, or brilliant colours. **4** help to explain (a subject etc.). **5** enlighten spiritually or intellectually. **6** shed lustre on. □ **illuminating** *adj.* **illuminatingly** *adv.* **illuminator** *n.* **illuminative** /-nətɪv/ *adj.* [L *illuminare* (as IN-², *lumen luminis* light)]

illuminati /ɪˌluːmɪˈnɑːtɪ, ɪˌljuː-/ *n.pl.* **1** persons claiming to possess special knowledge or enlightenment. **2** (**Illuminati**) *hist.* any of various intellectual movements or societies of illuminati. □ **illuminism** /ɪˈluːmɪˌnɪz(ə)m, ɪˈljuː-/ *n.* **illuminist** *n.* [pl. of L *illuminatus* or It. *illuminato* past part. (as ILLUMINATE)]

illumination /ɪˌluːmɪˈneɪʃ(ə)n, ɪˌljuː-/ *n.* **1** the act or an instance of shedding light on something or lighting something up. **2** (usu. in *pl.*) coloured lights arranged in designs to decorate a building, street, town, etc., or as an attraction. **3** intellectual enlightenment; information, learning. **4 a** the decoration of a medieval manuscript with elaborate tracery or designs in gold, silver, and brilliant colours. **b** a design or illustration used in such decoration, an illuminated page. [ME f. OF f. L *illuminatio -onis* (as ILLUMINATE)]

illumine /ɪˈluːmɪn, ɪˈljuː-/ v.tr. *literary* **1** light up; make bright. **2** enlighten spiritually. [ME f. OF *illuminer* f. L (as ILLUMINATE)]

illusion /ɪˈluːʒ(ə)n, ɪˈljuː-/ n. **1** deception, delusion. **2** a misapprehension of the true state of affairs. **3 a** the faulty perception of an external object. **b** an instance of this. **4** a figment of the imagination. **5** = *optical illusion*. □ **be under the illusion** (foll. by *that* + clause) believe mistakenly. □ **illusional** adj. [ME f. F f. L *illusio -onis* f. *illudere* mock (as IN-², *ludere lus-* play)]

illusionist /ɪˈluːʒənɪst, ɪˈljuː-/ n. a person who produces illusions; a conjuror. □ **illusionism** n. **illusionistic** /ɪˌluːʒəˈnɪstɪk, ɪˌljuː-/ adj.

illusive /ɪˈluːsɪv, ɪˈljuː-/ adj. = ILLUSORY. [med.L *illusivus* (as ILLUSION)]

illusory /ɪˈluːsərɪ, ɪˈljuː-/ adj. **1** deceptive (esp. as regards value or content). **2** having the character of an illusion. □ **illusorily** adv. **illusoriness** n. [eccl.L *illusorius* (as ILLUSION)]

illustrate /ˈɪləˌstreɪt/ v.tr. **1 a** provide (a book, newspaper, etc.) with pictures. **b** elucidate (a description etc.) by drawings or pictures. **2** serve as an example of. **3** explain or make clear, esp. by examples. [L *illustrare* (as IN-², *lustrare* light up)]

illustration /ˌɪləˈstreɪʃ(ə)n/ n. **1** a drawing or picture illustrating a book, magazine article, etc. **2** an example serving to elucidate. **3** the act or an instance of illustrating. □ **illustrational** adj. [ME f. OF f. L *illustratio -onis* (as ILLUSTRATE)]

illustrative /ˈɪləstrətɪv/ adj. (often foll. by *of*) serving as an explanation or example. □ **illustratively** adv.

illustrator /ˈɪləˌstreɪtə(r)/ n. a person who makes illustrations, esp. for magazines, books, advertising copy, etc.

illustrious /ɪˈlʌstrɪəs/ adj. distinguished, renowned. □ **illustriously** adv. **illustriousness** n. [L *illustris* (as ILLUSTRATE)]

Illyria /ɪˈlɪrɪə/ an ancient region along the east coast of the Adriatic Sea, including Dalmatia and what is now Montenegro and northern Albania. It was subsequently the Roman province of Illyricum and was later divided into the provinces of Dalmatia and Pannonia. It was overrun by the Huns and the Visigoths between the 3rd and 5th centuries AD. The name was revived, as the Illyrian Provinces, in 1809 after Napoleon's defeat of the Austrians and the annexation of the region to France. The region was reclaimed by Austria in 1814, retaining its identity as the kingdom of Illyria until 1849. □ **Illyrian** adj. & n.

illywhacker /ˈɪlɪˌwækə(r)/ n. *Austral. sl.* a professional trickster. [20th c.: orig. unkn.]

ilmenite /ˈɪlmənaɪt/ n. a black ore of titanium. [*Ilmen* mountains in the Urals]

ILO see INTERNATIONAL LABOUR ORGANIZATION.

Iloilo /ˌiːləʊˈiːləʊ/ a port on the south coast of the island of Panay in the Philippines; pop. (1990) 309,500.

Ilorin /ɪˈlɒrɪn/ a city in western Nigeria; pop. (1986) 390,000. In the 18th century it was the capital of a Yoruba kingdom that was eventually absorbed into a Fulani state in the early 19th century.

ILP see INDEPENDENT LABOUR PARTY.

ILR *abbr.* Independent Local Radio.

-ily /ɪlɪ/ *suffix* forming adverbs corresponding to adjectives in *-y* (see -Y¹, -LY²).

I'm /aɪm/ *contr.* I am.

im- /ɪm/ *prefix* assim. form of IN-¹, IN-² before *b*, *m*, *p*.

image /ˈɪmɪdʒ/ n. & v. ● n. **1** a representation of the external form of a person or thing in sculpture, painting, etc. **2** the character or reputation of a person or thing as generally perceived. **3** an optical appearance or counterpart produced by light or other radiation from an object reflected in a mirror, refracted through a lens, etc. **4** semblance, likeness (*God created man in His own image*). **5** a person or thing that closely resembles another (*is the image of his father*). **6** a typical example. **7** a simile or metaphor. **8 a** a mental representation. **b** an idea or conception. **9** *Math.* a set formed by mapping from another set. ● v.tr. **1** make an image of; portray. **2** reflect, mirror. **3** describe or imagine vividly. **4** typify. □ **image intensifier** a device used to make a brighter version of an image on a photoelectric screen. **image processing** the analysis and manipulation of a (usu. digitized) image, esp. to improve its quality. **image processor** a device or system that performs image processing. □ **imageable** adj. **imageless** adj. [ME f. OF f. L *imago -ginis*, rel. to IMITATE]

imagery /ˈɪmɪdʒərɪ/ n. **1** figurative illustration, esp. as used by an author for particular effects. **2** images collectively. **3** statuary, carving. **4** mental images collectively. [ME f. OF *imagerie* (as IMAGE)]

imaginable /ɪˈmædʒɪnəb(ə)l/ adj. that can be imagined (*the greatest difficulty imaginable*). □ **imaginably** adv. [ME f. LL *imaginabilis* (as IMAGINE)]

imaginal /ɪˈmædʒɪn(ə)l/ adj. **1** of an image or images. **2** *Zool.* of or relating to an adult insect or imago. [L *imago imagin-*: see IMAGE]

imaginary /ɪˈmædʒɪnərɪ/ adj. **1** existing only in the imagination. **2** *Math.* being the square root of a negative quantity, and plotted graphically in a direction usu. perpendicular to the axis of real quantities (see REAL¹). □ **imaginarily** adv. [ME f. L *imaginarius* (as IMAGE)]

imagination /ɪˌmædʒɪˈneɪʃ(ə)n/ n. **1** a mental faculty forming images or concepts of external objects not present to the senses. **2** the ability of the mind to be creative or resourceful. **3** the process of imagining. [ME f. OF f. L *imaginatio -onis* (as IMAGINE)]

imaginative /ɪˈmædʒɪnətɪv/ adj. **1** having or showing in a high degree the faculty of imagination. **2** given to using the imagination. □ **imaginatively** adv. **imaginativeness** n. [ME f. OF *imaginatif -ive* f. med.L *imaginativus* (as IMAGINE)]

imagine /ɪˈmædʒɪn/ v.tr. **1 a** form a mental image or concept of. **b** picture to oneself (something non-existent or not present to the senses). **2** (often foll. by *to* + infin.) think or conceive (*imagined them to be soldiers*). **3** guess (*cannot imagine what they are doing*). **4** (often foll. by *that* + clause) suppose; be of the opinion (*I imagine you will need help*). **5** (in *imper.*) as an exclamation of surprise (*just imagine!*). □ **imaginer** n. [ME f. OF *imaginer* f. L *imaginari* (as IMAGE)]

imagines *pl.* of IMAGO.

imaginings /ɪˈmædʒɪnɪŋz/ n.pl. fancies, fantasies.

imagism /ˈɪmɪˌdʒɪz(ə)m/ n. a movement in English and American poetry *c.*1910–17 which, in a revolt against romanticism, sought clarity of expression through the use of precise images. The movement derived in part from the aesthetic philosophy of T. E. Hulme and involved Ezra Pound, James Joyce, Amy Lowell, and others. □ **imagist** n. **imagistic** /ˌɪmɪˈdʒɪstɪk/ adj.

imago /ɪˈmeɪɡəʊ/ n. (pl. **-os** or **imagines** /ɪˈmædʒɪˌniːz/) **1** *Zool.* the final and fully developed stage of an insect after all metamorphoses, e.g. a butterfly or beetle. **2** *Psychol.* an idealized mental picture of oneself or others, esp. a parent. [mod.L sense of *imago* IMAGE]

imam /ɪˈmɑːm/ n. **1** a leader of prayers in a mosque. **2** a title of various Muslim leaders, esp. of one succeeding Muhammad as leader of Shiite Islam. (See also SHIA.) □ **imamate** /-meɪt/ n. [Arab. *'imām* leader f. *'amma* precede]

Imax /ˈaɪmæks/ n. (also **IMAX**) *Cinematog. propr.* a technique of wide-screen cinematography which produces an image approx. ten times larger than that from standard 35-mm film. [f. *i-* (prob. repr. pronunc. EYE) + MAXIMUM]

imbalance /ɪmˈbæləns/ n. **1** lack of balance. **2** disproportion.

imbecile /ˈɪmbɪˌsiːl/ n. & adj. ● n. **1** *Psychol.* a person of abnormally weak intellect, esp. an adult with a mental age of about 5. **2** *colloq.* a stupid person. ● adj. mentally weak; stupid, idiotic. □ **imbecilic** /ˌɪmbɪˈsɪlɪk/ adj. **imbecility** n. (pl. **-ies**). [F *imbécil(l)e* f. L *imbecillus* (as IN-¹, *baculum* stick) orig. in sense 'without supporting staff']

imbed var. of EMBED.

imbibe /ɪmˈbaɪb/ v.tr. **1** (also *absol.*) drink (esp. alcoholic liquor). **2 a** absorb or assimilate (ideas etc.). **b** absorb (moisture etc.). **3** inhale (air etc.). □ **imbiber** n. **imbibition** /ˌɪmbɪˈbɪʃ(ə)n/ n. [ME f. L *imbibere* (as IN-², *bibere* drink)]

imbricate v. & adj. ● v.tr. & intr. /ˈɪmbrɪˌkeɪt/ arrange (leaves, the scales of a fish, etc.), or be arranged, so as to overlap like roof-tiles. ● adj. /ˈɪmbrɪkət/ having scales etc. arranged in this way. □ **imbrication** /ˌɪmbrɪˈkeɪʃ(ə)n/ n. [L *imbricare imbricat-* cover with rain-tiles f. *imbrex -icis* rain-tile f. *imber* shower]

imbroglio /ɪmˈbrəʊlɪˌəʊ/ n. (pl. **-os**) **1** a confused or complicated situation. **2** a confused heap. [It. *imbrogliare* confuse (as EMBROIL)]

Imbros /ˈɪmbrɒs/ (Turkish **Imroz** /ɪmˈrɔz/) a Turkish island in the north-east Aegean Sea, near the entrance to the Dardanelles.

imbrue /ɪmˈbruː/ v.tr. (**imbrues, imbrued, imbruing**) (foll. by *in, with*) *literary* stain (one's hand, sword, etc.). [OF *embruer* bedabble (as IN-², *breu* ult. f. Gmc, rel. to BROTH)]

imbue /ɪmˈbjuː/ v.tr. (**imbues, imbued, imbuing**) (often foll. by *with*) **1** inspire or permeate (with feelings, opinions, or qualities). **2** saturate. **3** dye. [orig. as past part., f. F *imbu* or L *imbutus* f. *imbuere* moisten]

I.Mech.E. *abbr.* (in the UK) Institution of Mechanical Engineers.

IMF see INTERNATIONAL MONETARY FUND.

Imhotep /ˈɪmˈhəʊtep/ (*fl.* 27th century BC), Egyptian architect and scholar. He is usually credited with designing the step pyramid built at Saqqara for the 3rd-dynasty pharaoh Djoser (*c.*2686–*c.*2613 BC) and, through this, with pioneering the use of hewn stone in building. He was later deified; in Egypt, he was worshipped as the patron of architects, scribes, and doctors, while in Greece he was identified with the god Asclepius.

imide /ˈɪmaɪd/ *n. Chem.* an organic compound containing the group (–CO.NH.CO.–) formed by replacing two of the hydrogen atoms in ammonia by carbonyl groups. [orig. F: arbitrary alt. of AMIDE]

I.Min.E. *abbr.* (in the UK) Institution of Mining Engineers.

imine /ˈɪmiːn/ *n. Chem.* a compound containing the group (–NH–) formed by replacing two of the hydrogen atoms in ammonia by other groups. [G *Imin* arbitrary alt. of *Amin* AMINE]

imitate /ˈɪmɪteɪt/ *v.tr.* **1** follow the example of; copy the action(s) of. **2** mimic. **3** make a copy of; reproduce. **4** be (consciously or not) like. □ **imitator** *n.* **imitable** /-təb(ə)l/ *adj.* [L *imitari imitat-*, rel. to *imago* IMAGE]

imitation /ˌɪmɪˈteɪʃ(ə)n/ *n. & adj.* ● *n.* **1** the act or an instance of imitating or being imitated. **2** a copy. **3** *Mus.* the repetition of a phrase etc., usu. at a different pitch, in another part or voice. ● *adj.* made in imitation of something genuine; counterfeit, fake (*imitation leather*). [F *imitation* or L *imitatio* (as IMITATE)]

imitative /ˈɪmɪtətɪv/ *adj.* **1** (often foll. by *of*) imitating; following a model or example. **2** counterfeit. **3** of a word: **a** that reproduces a natural sound (e.g. *fizz*). **b** whose sound is thought to correspond to the appearance etc. of the object or action described (e.g. *blob*). □ **imitative arts** painting and sculpture. □ **imitatively** *adv.* **imitativeness** *n.* [LL *imitativus* (as IMITATE)]

immaculate /ɪˈmækjʊlət/ *adj.* **1** pure, spotless; perfectly clean or neat and tidy. **2** perfectly or extremely well executed (*an immaculate performance*). **3** free from fault; innocent. **4** *Biol.* not spotted. □ **immaculacy** *n.* **immaculately** *adv.* **immaculateness** *n.* [ME f. L *immaculatus* (as IN-[1], *maculatus* f. *macula* spot)]

Immaculate Conception (in Christian theology) the doctrine that the Virgin Mary was conceived, and remained, free from all original sin. The belief, which sought biblical support from Gen. 3:15 and Luke 1:28, was much disputed in the Middle Ages, but was generally accepted by Roman Catholics from the 16th century; it was defined as a dogma of the Roman Catholic Church in 1854. The feast is kept on 8 December.

immanent /ˈɪmənənt/ *adj.* **1** (often foll. by *in*) indwelling, inherent. **2** (of the supreme being) permanently pervading the universe (opp. TRANSCENDENT). □ **immanence** *n.* **immanency** *n.* **immanentism** *n.* **immanentist** *n.* [LL *immanere* (as IN-[2], *manere* remain)]

Immanuel /ɪˈmænjʊəl/ the name given to Christ as the deliverer of Judah prophesied by Isaiah (Isa. 7:14, 8:8; Matt. 1:23). [Heb., = God with us]

immaterial /ˌɪməˈtɪərɪəl/ *adj.* **1** of no essential consequence; unimportant. **2** not material; incorporeal. □ **immaterialize** *v.tr.* (also **-ise**). **immaterially** *adv.* **immateriality** /-ˌtɪərɪˈælɪtɪ/ *n.* [ME f. LL *immaterialis* (as IN-[1], MATERIAL)]

immaterialism /ˌɪməˈtɪərɪəˌlɪz(ə)m/ *n. Philos.* the doctrine that matter has no objective existence. □ **immaterialist** *n.*

immature /ˌɪməˈtjʊə(r)/ *adj.* **1** not mature or fully developed. **2** lacking emotional or intellectual development. **3** unripe. □ **immaturely** *adv.* **immaturity** *n.* [L *immaturus* (as IN-[1], MATURE)]

immeasurable /ɪˈmeʒərəb(ə)l/ *adj.* not measurable; immense. □ **immeasurably** *adv.* **immeasurability** /ɪˌmeʒərəˈbɪlɪtɪ/ *n.*

immediate /ɪˈmiːdɪət/ *adj.* **1** occurring or done at once or without delay (*an immediate reply*). **2** nearest, next; not separated by others (*the immediate vicinity; the immediate future; my immediate neighbour*). **3** most pressing or urgent; of current concern (*our immediate concern was to get him to hospital*). **4** (of a relation or action) having direct effect; without an intervening medium or agency (*the immediate cause of death*). **5** (of knowledge, reactions, etc.) intuitive, gained or exhibited without reasoning. □ **immediacy** *n.* **immediateness** *n.* [ME f. F *immédiat* or LL *immediatus* (as IN-[1], MEDIATE)]

immediately /ɪˈmiːdɪətlɪ/ *adv. & conj.* ● *adv.* **1** without pause or delay. **2** without intermediary. ● *conj.* as soon as.

immedicable /ɪˈmedɪkəb(ə)l/ *adj.* that cannot be healed or cured. [L *immedicabilis* (as IN-[1], MEDICABLE)]

immemorial /ˌɪmɪˈmɔːrɪəl/ *adj.* **1** ancient beyond memory or record.

2 very old. □ **immemorially** *adv.* [med.L *immemorialis* (as IN-[1], MEMORIAL)]

immense /ɪˈmens/ *adj.* **1** immeasurably large or great; huge. **2** very great; considerable (*made an immense difference*). **3** *colloq.* very good. □ **immenseness** *n.* **immensity** *n.* [ME f. F f. L *immensus* immeasurable (as IN-[1], *mensus* past part. of *metiri* measure)]

immensely /ɪˈmenslɪ/ *adv.* **1** very much (*enjoyed myself immensely*). **2** to an immense degree.

immerse /ɪˈmɜːs/ *v.tr.* **1 a** (often foll. by *in*) dip, plunge. **b** cause (a person) to be completely under water. **2** (often *refl.* or in *passive*; often foll. by *in*) absorb or involve deeply. **3** (often foll. by *in*) bury, embed. [L *immergere* (as IN-[2], *mergere mers-* dip)]

immersion /ɪˈmɜːʃ(ə)n/ *n.* **1** the act or an instance of immersing; the process of being immersed. **2** baptism by immersing the whole person in water. **3** mental absorption. **4** *Astron.* the disappearance of a celestial body behind another or into its shadow. □ **immersion heater** an electric heater designed for direct immersion in a liquid to be heated, esp. as a fixture in a hot-water tank. [ME f. LL *immersio* (as IMMERSE)]

immigrant /ˈɪmɪɡrənt/ *n. & adj.* ● *n.* a person who immigrates. ● *adj.* **1** immigrating. **2** of or concerning immigrants.

immigrate /ˈɪmɪɡreɪt/ *v.* **1** *intr.* come as a permanent resident to a country other than one's native land. **2** *tr.* bring in (a person) as an immigrant. □ **immigratory** *adj.* **immigration** /ˌɪmɪˈɡreɪʃ(ə)n/ *n.* [L *immigrare* (as IN-[2], MIGRATE)]

imminent /ˈɪmɪnənt/ *adj.* **1** (of an event, esp. danger) impending; about to happen. **2** *archaic* overhanging. □ **imminence** *n.* **imminently** *adv.* [L *imminere* imminent- overhang, project]

Immingham /ˈɪmɪŋəm/ a port on the east coast of England, on the Humber estuary north-west of Grimsby; pop. (1981) 11,500.

immiscible /ɪˈmɪsɪb(ə)l/ *adj.* (often foll. by *with*) that cannot be mixed. □ **immiscibly** *adv.* **immiscibility** /ɪˌmɪsɪˈbɪlɪtɪ/ *n.* [LL *immiscibilis* (as IN-[1], MISCIBLE)]

immitigable /ɪˈmɪtɪɡəb(ə)l/ *adj.* that cannot be mitigated. □ **immitigably** *adv.* [LL *immitigabilis* (as IN-[1], MITIGATE)]

immittance /ɪˈmɪt(ə)ns/ *n. Electr.* admittance or impedance (when not distinguished). [impedance + admittance]

immixture /ɪˈmɪkstʃə(r)/ *n.* **1** the process of mixing up. **2** (often foll. by *in*) being involved.

immobile /ɪˈməʊbaɪl/ *adj.* **1** not moving. **2** not able to move or be moved. □ **immobility** /ˌɪməˈbɪlɪtɪ/ *n.* [ME f. OF f. L *immobilis* (as IN-[1], MOBILE)]

immobilize /ɪˈməʊbɪˌlaɪz/ *v.tr.* (also **-ise**) **1** make or keep immobile. **2** make (esp. a vehicle or troops) incapable of being moved. **3** keep (a limb or patient) restricted in movement for healing purposes. **4** restrict the free movement of. **5** withdraw (coins) from circulation to support banknotes. □ **immobilizer** *n.* **immobilization** /ɪˌməʊbɪlaɪˈzeɪʃ(ə)n/ *n.* [F *immobiliser* (as IMMOBILE)]

immoderate /ɪˈmɒdərət/ *adj.* excessive; lacking moderation. □ **immoderately** *adv.* **immoderateness** *n.* **immoderation** /ɪˌmɒdəˈreɪʃ(ə)n/ *n.* [ME f. L *immoderatus* (as IN-[1], MODERATE)]

immodest /ɪˈmɒdɪst/ *adj.* **1** lacking modesty; forward, impudent. **2** lacking due decency. □ **immodestly** *adv.* **immodesty** *n.* [F *immodeste* or L *immodestus* (as IN-[1], MODEST)]

immolate /ˈɪməˌleɪt/ *v.tr.* **1** kill or offer as a sacrifice. **2** *literary* sacrifice (a valued thing). □ **immolator** *n.* **immolation** /ˌɪməˈleɪʃ(ə)n/ *n.* [L *immolare* sprinkle with sacrificial meal (as IN-[2], *mola* MEAL[2])]

immoral /ɪˈmɒrəl/ *adj.* **1** not conforming to accepted standards of morality (cf. AMORAL). **2** morally wrong (esp. in sexual matters). **3** depraved, dissolute. □ **immorally** *adv.* **immorality** /ˌɪməˈrælɪtɪ/ *n.* (*pl.* **-ies**).

immortal /ɪˈmɔːt(ə)l/ *adj. & n.* ● *adj.* **1 a** living for ever; not mortal. **b** divine. **2** unfading, incorruptible. **3** likely or worthy to be famous for all time. ● *n.* **1 a** an immortal being. **b** (in *pl.*) the gods of antiquity. **2** a person (esp. an author) of enduring fame. **3** (**Immortal**) a member of the French Academy. □ **immortalize** *v.tr.* (also **-ise**). **immortally** *adv.* **immortality** /ˌɪmɔːˈtælɪtɪ/ *n.* **immortalization** /ɪˌmɔːtəlaɪˈzeɪʃ(ə)n/ *n.* [ME f. L *immortalis* (as IN-[1], MORTAL)]

immortelle /ˌɪmɔːˈtel/ *n.* a composite flower of papery texture retaining its shape and colour after being dried, esp. a helichrysum. [F, fem. of *immortel* IMMORTAL]

immovable /ɪˈmuːvəb(ə)l/ *adj. & n.* (also **immoveable**) ● *adj.* **1** that cannot be moved. **2** steadfast, unyielding. **3** emotionless. **4** not subject to change (*immovable law*). **5** motionless. **6** *Law* (of property) consisting of land, houses, etc. ● *n.* (in *pl.*) *Law* immovable property. □ **immovable feast** a religious feast-day that occurs on the same date each year. □ **immovably** *adv.* **immovability** /ɪˌmuːvəˈbɪlɪtɪ/ *n.*

immune /ɪˈmjuːn/ *adj.* **1** *Biol.* **a** (often foll. by *against, from, to*) resistant to a particular infection, toxin, etc., owing to the presence of specific antibodies or sensitized white blood cells, acquired actively by injection, inoculation, etc., or passively from injected antiserum or from the mother. **b** relating to immunity (*immune mechanism*). **2** (foll. by *from, to*) free or exempt from or not subject to (some undesirable factor or circumstance). □ **immune response** the reaction of the body to the introduction into it of an antigen. [ME f. L *immunis* exempt from public service or charge (as IN-¹, *munis* ready for service): sense 1 f. F *immun*]

immunity /ɪˈmjuːnɪtɪ/ *n.* (*pl.* **-ies**) **1** *Biol.* the ability of an organism to resist a specific infection, toxin, etc., owing to the presence of phagocytic cells or (in vertebrates) antibodies and sensitized white blood cells. **2** freedom or exemption from an obligation, penalty, or unfavourable circumstance. [ME f. L *immunitas* (as IMMUNE): sense 1 f. F *immunité*]

immunize /ˈɪmjʊnaɪz/ *v.tr.* (also **-ise**) make immune, esp. to infection, usu. by vaccination. (See also VACCINATE.) □ **immunizer** *n.* **immunization** /ˌɪmjʊnaɪˈzeɪʃ(ə)n/ *n.*

immuno- /ˈɪmjʊnəʊ/ *comb. form* immunity to infection.

immunoassay /ˌɪmjʊnəʊˈæseɪ/ *n.* *Biochem.* the determination of the presence or quantity of a substance, esp. a protein, through its properties as an antigen or antibody.

immunochemistry /ˌɪmjʊnəʊˈkemɪstrɪ/ *n.* *Biochem.* **1** the chemical study of immune systems. **2** the use of specific immune reactions in the investigation of biological molecules.

immunocompetent /ˌɪmjʊnəʊˈkɒmpɪt(ə)nt/ *adj.* *Med.* having a normal immune response.

immunocompromised /ˌɪmjʊnəʊˈkɒmprəˌmaɪzd/ *adj.* *Med.* having an impaired immune system.

immunodeficiency /ˌɪmjʊnəʊdɪˈfɪʃənsɪ/ *n.* a reduction in a person's normal immune defences.

immunogenic /ˌɪmjʊnəʊˈdʒenɪk/ *adj.* *Biochem.* of, relating to, or possessing the ability to elicit an immune response.

immunoglobulin /ˌɪmjʊnəʊˈɡlɒbjʊlɪn/ *n.* (abbr. **Ig**) *Biochem.* any of a group of structurally related blood proteins which function as antibodies.

immunology /ˌɪmjʊˈnɒlədʒɪ/ *n.* the scientific study of immunity. □ **immunologist** *n.* **immunologic** /-nəˈlɒdʒɪk/ *adj.* **immunological** *adj.* **immunologically** *adv.*

immunosuppressed /ˌɪmjʊˌnəʊsəˈprest/ *adj.* (of an individual) rendered partially or completely unable to react immunologically.

immunosuppression /ˌɪmjʊˌnəʊsəˈpreʃ(ə)n/ *n.* *Biochem.* the partial or complete suppression of the immune response of an individual, esp. to maintain the survival of an organ after a transplant operation. □ **immunosuppressant** /-ˈpres(ə)nt/ *n.*

immunosuppressive /ˌɪmjʊˌnəʊsəˈpresɪv/ *adj. & n.* ● *adj.* partially or completely suppressing the immune response of an individual. ● *n.* an immunosuppressive drug.

immunotherapy /ˌɪmjʊnəʊˈθerəpɪ/ *n.* *Med.* the prevention or treatment of disease with substances that stimulate the immune response.

immure /ɪˈmjʊə(r)/ *v.tr.* **1** enclose within walls; imprison. **2** *refl.* shut oneself away. □ **immurement** *n.* [F *emmurer* or med.L *immurare* (as IN-², *murus* wall)]

immutable /ɪˈmjuːtəb(ə)l/ *adj.* **1** unchangeable. **2** not subject to variation in different cases. □ **immutably** *adv.* **immutability** /ɪˌmjuːtəˈbɪlɪtɪ/ *n.* [ME f. L *immutabilis* (as IN-¹, MUTABLE)]

IMO *abbr.* International Maritime Organization.

imp /ɪmp/ *n. & v.* ● *n.* **1** a mischievous child. **2** a small mischievous devil or sprite. ● *v.tr.* **1** add feathers to (the wing of a falcon) to restore or improve its flight. **2** *archaic* enlarge; add by grafting. [OE *impa, impe* young shoot, scion, *impian* graft: ult. f. Gk *emphutos* implanted, past part. of *emphuō*]

impact *n. & v.* ● *n.* /ˈɪmpækt/ **1** (often foll. by *on, against*) the action of one body coming forcibly into contact with another. **2** an effect or influence, esp. when strong. ● *v.* /ɪmˈpækt/ **1** *tr.* (often foll. by *in, into*) press or fix firmly. **2** *tr.* (as **impacted** *adj.*) **a** (of a tooth) wedged between another tooth and the jaw. **b** (of a fractured bone) with the parts crushed together. **c** (of faeces) lodged in the intestine. **3** *intr.* **a** (foll. by *against, on*) come forcibly into contact with a (larger) body or surface. **b** (foll. by *on*) have a pronounced effect. □ **impaction** /ɪmˈpækʃ(ə)n/ *n.* [L *impact-* part. stem of *impingere* IMPINGE]

impair /ɪmˈpeə(r)/ *v.tr.* damage or weaken. □ **impairment** *n.* [ME *empeire* f. OF *empeirier* (as IN-¹, LL *pejorare* f. L *pejor* worse)]

impala /ɪmˈpɑːlə, -ˈpælə/ *n.* (*pl.* same) a medium-sized antelope, *Aepyceros melampus*, of southern and East Africa, capable of long high jumps. [Zulu]

impale /ɪmˈpeɪl/ *v.tr.* **1** (foll. by *on, upon, with*) transfix or pierce with a sharp instrument. **2** *Heraldry* combine (two coats of arms) by placing them side by side on one shield separated by a vertical line down the middle. □ **impalement** *n.* [F *empaler* or med.L *impalare* (as IN-², *palus* stake)]

impalpable /ɪmˈpælpəb(ə)l/ *adj.* **1** not easily grasped by the mind; intangible. **2** imperceptible to the touch. **3** (of powder) very fine; not containing grains that can be felt. □ **impalpably** *adv.* **impalpability** /-ˌpælpəˈbɪlɪtɪ/ *n.* [F *impalpable* or LL *impalpabilis* (as IN-¹, PALPABLE)]

impanel var. of EMPANEL.

impark /ɪmˈpɑːk/ *v.tr. hist.* enclose (land) for a park. [ME f. AF *enparker*, OF *emparquer* (as IN-², *parc* PARK)]

impart /ɪmˈpɑːt/ *v.tr.* (often foll. by *to*) **1** communicate (news etc.). **2** give a share of (a thing). □ **impartation** /ˌɪmpɑːˈteɪʃ(ə)n/ *n.* [ME f. OF *impartir* f. L *impartire* (as IN-², *pars* part)]

impartial /ɪmˈpɑːʃ(ə)l/ *adj.* treating all sides in a dispute etc. equally; unprejudiced, fair. □ **impartially** *adv.* **impartiality** /-ˌpɑːʃɪˈælɪtɪ/ *n.*

impassable /ɪmˈpɑːsəb(ə)l/ *adj.* that cannot be traversed. □ **impassably** *adv.* **impassableness** *n.* **impassability** /-ˌpɑːsəˈbɪlɪtɪ/ *n.*

impasse /ˈæmpæs, ˈɪm-/ *n.* a position from which progress is impossible; deadlock. [F (as IN-¹, *passer* PASS¹)]

impassible /ɪmˈpæsɪb(ə)l/ *adj.* **1** impassive. **2** incapable of feeling or emotion. **3** incapable of suffering injury. **4** (in Christian theology) not subject to suffering. □ **impassibly** *adv.* **impassibility** /-ˌpæsɪˈbɪlɪtɪ/ *n.* [ME f. OF f. eccl.L *impassibilis* (as IN-¹, PASSIBLE)]

impassion /ɪmˈpæʃ(ə)n/ *v.tr.* fill with passion; arouse emotionally. [It. *impassionare* (as IN-², PASSION)]

impassioned /ɪmˈpæʃ(ə)nd/ *adj.* deeply felt; ardent (*an impassioned plea*).

impassive /ɪmˈpæsɪv/ *adj.* **1 a** deficient in or incapable of feeling emotion. **b** undisturbed by passion; serene. **2** without sensation. **3** not subject to suffering. □ **impassively** *adv.* **impassiveness** *n.* **impassivity** /ˌɪmpæˈsɪvɪtɪ/ *n.*

impasto /ɪmˈpæstəʊ/ *n.* *Art* **1** the process of laying on paint thickly. **2** this technique of painting. [It. *impastare* (as IN-², *pastare* paste)]

impatiens /ɪmˈpætɪˌenz, ɪmˈpeɪʃɪ-/ *n.* a plant of the genus *Impatiens*, including busy Lizzie, balsam, and touch-me-not. [mod.L f. IMPATIENT]

impatient /ɪmˈpeɪʃ(ə)nt/ *adj.* **1 a** (often foll. by *at, with*) lacking patience or tolerance. **b** (of an action) showing a lack of patience. **2** (often foll. by *for*, or *to* + infin.) restlessly eager. **3** (foll. by *of*) intolerant. □ **impatience** *n.* **impatiently** *adv.* [ME f. OF f. L *impatiens* (as IN-¹, PATIENT)]

impeach /ɪmˈpiːtʃ/ *v.tr.* **1** *Brit.* charge with a crime against the state, esp. treason. (*See note below.*) **2** esp. *US* charge (the holder of a public office) with misconduct. (*See note below.*) **3** call in question, disparage (a person's integrity etc.). □ **impeachable** *adj.* **impeachment** *n.* [ME f. OF *empecher* impede f. LL *impedicare* entangle (as IN-², *pedica* fetter f. *pes pedis* foot)]

■ In Britain impeachment was first used in the 14th century; it became important in the 17th century as a weapon with which to remove unpopular ministers but thereafter declined; there have been no impeachment proceedings since 1806. In the US, where the House of Representatives institutes proceedings, impeachment results in dismissal and disqualification from office. Andrew Johnson is the only US President to have been impeached; Richard Nixon resigned in 1974 before impeachment proceedings against him could begin.

impeccable /ɪmˈpekəb(ə)l/ *adj.* **1** (of behaviour, performance, etc.) faultless, exemplary. **2** not liable to sin. □ **impeccably** *adv.* **impeccability** /-ˌpekəˈbɪlɪtɪ/ *n.* [L *impeccabilis* (as IN-¹, *peccare* sin)]

impecunious /ˌɪmpɪˈkjuːnɪəs/ *adj.* having little or no money. □ **impecuniousness** *n.* **impecuniosity** /-ˌkjuːnɪˈɒsɪtɪ/ *n.* [IN-¹ + obs. *pecunious* having money f. L *pecuniosus* f. *pecunia* money f. *pecu* cattle]

impedance /ɪmˈpiːd(ə)ns/ *n.* **1** *Electr.* the total effective resistance of an electric circuit etc. to alternating current, arising from ohmic resistance and reactance. **2** an analogous mechanical property. [IMPEDE + -ANCE]

impede /ɪmˈpiːd/ v.tr. retard by obstructing; hinder. [L *impedire* shackle the feet of (as IN-², *pes* foot)]

impediment /ɪmˈpedɪmənt/ n. **1** a hindrance or obstruction. **2** a defect in speech, e.g. a lisp or stammer. □ **impedimental** /-ˌpedɪˈment(ə)l/ adj. [ME f. L *impedimentum* (as IMPEDE)]

impedimenta /ɪmˌpedɪˈmentə/ n.pl. **1** encumbrances. **2** travelling equipment, esp. of an army. [L, pl. of *impedimentum*: see IMPEDIMENT]

impel /ɪmˈpel/ v.tr. (**impelled, impelling**) **1** drive, force, or urge into action. **2** drive forward; propel. □ **impeller** n. [ME f. L *impellere* (as IN-², *pellere puls-* drive)]

impend /ɪmˈpend/ v.intr. **1** be about to happen. **2** (often foll. by *over*) **a** (of a danger) be threatening. **b** hang; be suspended. □ **impending** adj. [L *impendere* (as IN-², *pendere* hang)]

impenetrable /ɪmˈpenɪtrəb(ə)l/ adj. **1** that cannot be penetrated. **2** inscrutable, unfathomable. **3** inaccessible to ideas, influences, etc. **4** *Physics* (of matter) having the property such that a body is incapable of occupying the same place as another body at the same time. □ **impenetrably** adv. **impenetrableness** n. **impenetrability** /-ˌpenɪtrəˈbɪlɪtɪ/ n. [ME f. F *impénétrable* f. L *impenetrabilis* (as IN-¹, PENETRATE)]

impenitent /ɪmˈpenɪt(ə)nt/ adj. not repentant or penitent. □ **impenitence** n. **impenitency** n. **impenitently** adv. [eccl.L *impaenitens* (as IN-¹, PENITENT)]

imperative /ɪmˈperətɪv/ adj. & n. ● adj. **1** urgent. **2** obligatory. **3** commanding, peremptory. **4** *Gram.* (of a mood) expressing a command (e.g. *come here!*). ● n. **1** *Gram.* the imperative mood. **2** a command. **3** an essential or urgent thing. □ **imperatively** adv. **imperativeness** n. **imperatival** /-ˌperəˈtaɪv(ə)l/ adj. [LL *imperativus* f. *imperare* command (as IN-², *parare* make ready)]

imperator /ˌɪmpəˈrɑːtɔː(r)/ n. Rom. Hist. commander (a title conferred under the Republic on a victorious general and under the Empire on the emperor). □ **imperatorial** /ɪmˌperəˈtɔːrɪəl/ adj. [L (as IMPERATIVE)]

imperceptible /ˌɪmpəˈseptɪb(ə)l/ adj. **1** that cannot be perceived. **2** very slight, gradual, or subtle. □ **imperceptibly** adv. **imperceptibility** /-ˌseptɪˈbɪlɪtɪ/ n. [F *imperceptible* or med.L *imperceptibilis* (as IN-¹, PERCEPTIBLE)]

impercipient /ˌɪmpəˈsɪpɪənt/ adj. lacking in perception. □ **impercipience** n.

imperfect /ɪmˈpɜːfɪkt/ adj. & n. ● adj. **1** not fully formed or done; faulty, incomplete. **2** *Gram.* (of a tense) denoting a (usu. past) action in progress but not completed at the time in question (e.g. *they were singing*). **3** *Mus.* (of a cadence) ending on the dominant chord. ● n. *Gram.* the imperfect tense. □ **imperfect rhyme** *Prosody* a rhyme that only partly satisfies the usual criteria (e.g. *love* and *move*). **imperfectly** adv. [ME *imparfit* etc. f. OF *imparfait* f. L *imperfectus* (as IN-¹, PERFECT)]

imperfection /ˌɪmpəˈfek(ʃ)(ə)n/ n. **1** incompleteness. **2 a** faultiness. **b** a fault or blemish. [ME f. OF *imperfection* or LL *imperfectio* (as IMPERFECT)]

imperfective /ˌɪmpəˈfektɪv/ adj. & n. *Gram.* ● adj. (of a verb aspect etc.) expressing an action without reference to its completion (opp. PERFECTIVE). ● n. an imperfective aspect or form of a verb.

imperforate /ɪmˈpɜːfərət/ adj. **1** not perforated. **2** *Anat.* lacking the normal opening. **3** (of a postage stamp) lacking perforations.

imperial /ɪmˈpɪərɪəl/ adj. & n. ● adj. **1** of or characteristic of an empire or comparable sovereign state. **2 a** of or characteristic of an emperor or empress. **b** supreme in authority. **c** majestic, august. **d** magnificent. **3** (of non-metric weights and measures) used or formerly used by statute in the UK (*imperial gallon*). (*See note below.*) ● n. a former size of paper, 30 x 22 inches (762 x 559 mm). □ **imperially** adv. [ME f. OF f. L *imperialis* f. *imperium* command, authority]

▪ Many of the imperial measures are of great antiquity, but until defined by Parliament they varied from place to place. Units of length include the mile and the yard as well as feet and inches; area of land is measured in acres. The basic unit of capacity is the gallon, divided into 8 pints, while 8 gallons make up a bushel (these units are larger than their counterparts in US measures). The basic unit of weight is the pound, divided into 16 ounces; the hundredweight is 112 pounds, and the ton is 20 hundredweight. A unit peculiar to Britain is the stone (14 pounds). Most of the imperial measures are being or have been replaced by the metric system.

imperialism /ɪmˈpɪərɪəˌlɪz(ə)m/ n. usu. derog. a policy of acquiring dependent territories or of extending a country's influence over less powerful or less developed countries through trade, diplomacy, etc.

□ **imperialize** v.tr. (also **-ise**). **imperialistic** /-ˌpɪərɪəˈlɪstɪk/ adj. **imperialistically** adv.

imperialist /ɪmˈpɪərɪəlɪst/ n. & adj. ● n. usu. derog. an advocate or agent of imperial rule or of imperialism. ● adj. of or relating to imperialism or imperialists.

imperil /ɪmˈperɪl/ v.tr. (**imperilled, imperilling;** US **imperiled, imperiling**) bring or put into danger.

imperious /ɪmˈpɪərɪəs/ adj. **1** overbearing, domineering. **2** urgent, imperative. □ **imperiously** adv. **imperiousness** n. [L *imperiosus* f. *imperium* command, authority]

imperishable /ɪmˈperɪʃəb(ə)l/ adj. that cannot perish. □ **imperishably** adv. **imperishableness** n. **imperishability** /-ˌperɪʃəˈbɪlɪtɪ/ n.

imperium /ɪmˈpɪərɪəm, -ˈperɪəm/ n. absolute power or authority. [L, = command, authority]

impermanent /ɪmˈpɜːmənənt/ adj. not permanent; transient. □ **impermanence** n. **impermanency** n. **impermanently** adv.

impermeable /ɪmˈpɜːmɪəb(ə)l/ adj. **1** that cannot be penetrated. **2** that does not permit the passage of fluids. □ **impermeability** /-ˌpɜːmɪəˈbɪlɪtɪ/ n. [F *imperméable* or LL *impermeabilis* (as IN-¹, PERMEABLE)]

impermissible /ˌɪmpəˈmɪsɪb(ə)l/ adj. not to be allowed. □ **impermissibility** /-ˌmɪsɪˈbɪlɪtɪ/ n.

impersonal /ɪmˈpɜːsən(ə)l/ adj. **1** having no personal feeling or reference. **2** having no personality. **3** *Gram.* **a** (of a verb) used only with a formal subject (usu. *it*) and expressing an action not attributable to a definite subject (e.g. *it is snowing*). **b** (of a pronoun) = INDEFINITE. □ **impersonally** adv. **impersonality** /-ˌpɜːsəˈnælɪtɪ/ n. [LL *impersonalis* (as IN-¹, PERSONAL)]

impersonate /ɪmˈpɜːsəˌneɪt/ v.tr. **1** pretend to be (another person) for the purpose of entertainment or fraud. **2** act (a character). □ **impersonator** n. **impersonation** /-ˌpɜːsəˈneɪʃ(ə)n/ n. [IN-² + L *persona* PERSON]

impertinent /ɪmˈpɜːtɪnənt/ adj. **1** rude or insolent; lacking proper respect. **2** out of place; absurd. **3** esp. *Law* irrelevant, intrusive. □ **impertinence** n. **impertinently** adv. [ME f. OF or LL *impertinens* (as IN-¹, PERTINENT)]

imperturbable /ˌɪmpəˈtɜːbəb(ə)l/ adj. not excitable; calm. □ **imperturbably** adv. **imperturbableness** n. **imperturbability** /-ˌtɜːbəˈbɪlɪtɪ/ n. [ME f. LL *imperturbabilis* (as IN-¹, PERTURB)]

impervious /ɪmˈpɜːvɪəs/ adj. (usu. foll. by *to*) **1** not responsive to an argument etc. **2** not affording passage to a fluid etc. □ **imperviously** adv. **imperviousness** n. [L *impervius* (as IN-¹, PERVIOUS)]

impetigo /ˌɪmpɪˈtaɪɡəʊ/ n. Med. a contagious bacterial skin infection forming pustules and yellow crusty sores. □ **impetiginous** /-ˈtɪdʒɪnəs/ adj. [ME f. L *impetigo -ginis* f. *impetere* assail]

impetuous /ɪmˈpetjʊəs/ adj. **1** acting or done rashly or with sudden energy. **2** moving forcefully or rapidly. □ **impetuously** adv. **impetuousness** n. **impetuosity** /-ˌpetjʊˈɒsɪtɪ/ n. [ME f. OF *impetueux* f. LL *impetuosus* (as IMPETUS)]

impetus /ˈɪmpɪtəs/ n. **1** the force or energy with which a body moves. **2** a driving force or impulse. [L, = assault, force, f. *impetere* assail (as IN-², *petere* seek)]

Imphal /ˈɪmfəl, ɪmˈfɑːl/ the capital of the state of Manipur in the far north-east of India, lying close to the border with Burma (Myanmar); pop. (1991) 156,620. It was the scene of an important victory in 1944 by Anglo-Indian forces over the Japanese.

impi /ˈɪmpɪ/ n. (pl. **impis**) esp. hist. a body of Zulu warriors or armed tribesmen. [Zulu, = regiment, armed band]

impiety /ɪmˈpaɪətɪ/ n. (pl. **-ies**) **1** a lack of piety or reverence. **2** an act etc. showing this. [ME f. OF *impieté* or L *impietas* (as IN-¹, PIETY)]

impinge /ɪmˈpɪndʒ/ v.tr. (**impinging**) (usu. foll. by *on, upon*) **1** make an impact; have an effect. **2** encroach. □ **impingement** n. **impinger** n. [L *impingere* drive (a thing) at (as IN-², *pangere* fix, drive)]

impious /ˈɪmpɪəs/ adj. **1** not pious. **2** wicked, profane. □ **impiously** adv. **impiousness** n. [L *impius* (as IN-¹, PIOUS)]

impish /ˈɪmpɪʃ/ adj. of or like an imp; mischievous. □ **impishly** adv. **impishness** n.

implacable /ɪmˈplækəb(ə)l/ adj. that cannot be appeased; inexorable. □ **implacably** adv. **implacability** /-ˌplækəˈbɪlɪtɪ/ n. [ME f. F *implacable* or L *implacabilis* (as IN-¹, PLACABLE)]

implant v. & n. ● v.tr. /ɪmˈplɑːnt/ **1** (often foll. by *in*) insert or fix. **2** (often foll. by *in*) instil (a principle, idea, etc.) in a person's mind. **3** plant. **4** *Med.* **a** insert (tissue, a substance, or an artificial object) into the body. **b** (in *passive*) (of a fertilized ovum) become attached to the wall of the womb. ● *n.* /ˈɪmplɑːnt/ **1** a thing implanted. **2** a thing implanted in the body, e.g. a piece of tissue or a capsule containing material for radium therapy. □ **implantation** /ˌɪmplɑːnˈteɪʃ(ə)n/ *n.* [F *implanter* or LL *implantare* engraft (as IN-², PLANT)]

implausible /ɪmˈplɔːzɪb(ə)l/ *adj.* not plausible. □ **implausibly** *adv.* **implausibility** /-ˌplɔːzɪˈbɪlɪtɪ/ *n.*

implead /ɪmˈpliːd/ *v.tr. Law* **1** prosecute or take proceedings against (a person). **2** involve (a person etc.) in a suit. [ME f. AF *empleder*, OF *empleidier* (as EM-, IM- + PLEAD)]

implement *n. & v.* ● *n.* /ˈɪmplɪmənt/ **1** a tool, instrument, or utensil. **2** (in *pl.*) equipment; articles of furniture, dress, etc. **3** *Law* performance of an obligation. ● *v.tr.* /ˈɪmplɪˌment/ **1 a** put (a decision, plan, etc.) into effect. **b** fulfil (an undertaking). **2** complete (a contract etc.). **3** fill up; supplement. □ **implementer** /ˈɪmplɪˌmentə(r)/ *n.* **implementation** /ˌɪmplɪmenˈteɪʃ(ə)n/ *n.* [ME f. med.L *implementa* (pl.) f. *implere* employ (as IN-², L *plere plet-* fill)]

implicate *v. & n.* ● *v.tr.* /ˈɪmplɪˌkeɪt/ **1** (often foll. by *in*) show (a person) to be concerned or involved (in a charge, crime, etc.). **2** (in *passive*; often foll. by *in*) be affected or involved. **3** lead to as a consequence or inference. ● *n.* /ˈɪmplɪkət/ a thing implied. □ **implicative** /ɪmˈplɪkətɪv/ *adj.* **implicatively** *adv.* [L *implicatus* past part. of *implicare* (as IN-², *plicare*, *plicat-* or *plicit-* fold)]

implication /ˌɪmplɪˈkeɪʃ(ə)n/ *n.* **1** what is involved in or implied by something else. **2** the act of implicating or implying. □ **by implication** by what is implied or suggested rather than by formal expression. [ME f. L *implicatio* (as IMPLICATE)]

implicit /ɪmˈplɪsɪt/ *adj.* **1** implied though not plainly expressed. **2** (often foll. by *in*) virtually contained. **3** absolute, unquestioning, unreserved (*implicit obedience*). **4** *Math.* (of a function) not expressed directly in terms of independent variables. □ **implicitly** *adv.* **implicitness** *n.* [F *implicite* or L *implicitus* (as IMPLICATE)]

implode /ɪmˈpləʊd/ *v.intr. & tr.* burst or cause to burst inwards. □ **implosion** /-ˈpləʊʒ(ə)n/ *n.* **implosive** /-ˈpləʊsɪv, -ˈpləʊzɪv/ *adj.* [IN-² + L *-plodere*, after EXPLODE]

implore /ɪmˈplɔː(r)/ *v.tr.* **1** (often foll. by *to* + infin.) entreat (a person). **2** beg earnestly for. □ **imploring** *adj.* **imploringly** *adv.* [F *implorer* or L *implorare* invoke with tears (as IN-², *plorare* weep)]

imply /ɪmˈplaɪ/ *v.tr.* (**-ies**, **-ied**) **1** (often foll. by *that* + clause) strongly suggest the truth or existence of (a thing not expressly asserted). **2** insinuate, hint (*what are you implying?*). **3** signify. □ **implied** *adj.* **impliedly** /-ˈplaɪdlɪ/ *adv.* [ME f. OF *emplier* f. L *implicare* (as IMPLICATE)]

impolder /ɪmˈpəʊldə(r)/ *v.tr.* (also **empolder**) *Brit.* **1** make a polder of. **2** reclaim from the sea. [Du. *inpolderen* (as IN-², POLDER)]

impolite /ˌɪmpəˈlaɪt/ *adj.* ill-mannered, uncivil, rude. □ **impolitely** *adv.* **impoliteness** *n.* [L *impolitus* (as IN-¹, POLITE)]

impolitic /ɪmˈpɒlɪtɪk/ *adj.* **1** inexpedient, unwise. **2** not politic. □ **impoliticly** *adv.*

imponderable /ɪmˈpɒndərəb(ə)l/ *adj. & n.* ● *adj.* **1** that cannot be estimated or assessed in any definite way. **2** very light. **3** *Physics* having no weight. ● *n.* (usu. in *pl.*) something difficult or impossible to assess. □ **imponderably** *adv.* **imponderability** /-ˌpɒndərəˈbɪlɪtɪ/ *n.*

import *v. & n.* ● *v.tr.* /ɪmˈpɔːt, ˈɪmpɔːt/ **1** bring in (esp. foreign goods or services) to a country. **2** (often foll. by *that* + clause) **a** imply, indicate, signify. **b** express, make known. ● *n.* /ˈɪmpɔːt/ **1** the process of importing. **2 a** an imported article or service. **b** (in *pl.*) an amount imported (*imports exceeded £50m.*). **3** what is implied; meaning. **4** importance. □ **importable** /ɪmˈpɔːtəb(ə)l/ *adj.* **importation** /ˌɪmpɔːˈteɪʃ(ə)n/ *n.* **importer** /ɪmˈpɔːtə(r)/ *n.* (all in sense 1 of *v.*). [ME f. L *importare* bring in, in med.L = imply, be of consequence (as IN-², *portare* carry)]

importance /ɪmˈpɔːt(ə)ns/ *n.* **1** the state of being important. **2** import, significance. **3** personal consequence; dignity. [F f. med.L *importantia* (as IMPORT)]

important /ɪmˈpɔːt(ə)nt/ *adj.* **1** (often foll. by *to*) of great effect or consequence; momentous. **2** (of a person) having high rank or status, or great authority. **3** pretentious, pompous. **4** (*absol.* in parenthetic construction prec. by *more* or *most*) what is a more, or most, significant point or matter (*they are willing and, more important, able*). ¶ Use of *importantly* here is *disp.* □ **importantly** *adv.* (see note above). [F f. med.L (as IMPORT)]

importunate /ɪmˈpɔːtjʊnət/ *adj.* **1** making persistent or pressing requests. **2** (of affairs) urgent. □ **importunately** *adv.* **importunity**

/ˌɪmpɔːˈtjuːnɪtɪ/ *n.* [L *importunus* inconvenient (as IN-¹, *portunus* f. *portus* harbour)]

importune /ɪmˈpɔːtjuːn, ˌɪmpɔːˈtjuːn/ *v.tr.* **1** solicit (a person) pressingly. **2** solicit for an immoral purpose. [F *importuner* or med.L *importunari* (as IMPORTUNATE)]

impose /ɪmˈpəʊz/ *v.* **1** *tr.* (often foll. by *on*, *upon*) require (a tax, duty, charge, or obligation) to be paid or undertaken (by a person etc.). **2** *tr.* enforce compliance with. **3** *intr. & refl.* (foll. by *on*, *upon*, or *absol.*) demand the attention or commitment of (a person); take advantage of (*I do not want to impose on you any longer; I did not want to impose*). **4** *tr.* (often foll. by *on*, *upon*) palm (a thing) off on (a person). **5** *tr. Printing* lay (pages of type) in the proper order ready for printing. **6** *intr.* (foll. by *on*, *upon*) exert influence by an impressive character or appearance. **7** *intr.* (often foll. by *on*, *upon*) practise deception. **8** *tr.* (foll. by *upon*) *archaic* place (a thing). [ME f. F *imposer* f. L *imponere* inflict, deceive (as IN-², *ponere* put)]

imposing /ɪmˈpəʊzɪŋ/ *adj.* impressive, formidable, esp. in appearance. □ **imposingly** *adv.* **imposingness** *n.*

imposition /ˌɪmpəˈzɪʃ(ə)n/ *n.* **1** the act or an instance of imposing; the process of being imposed. **2** an unfair or resented demand or burden. **3** a tax or duty. **4** *Brit.* work set as a punishment at school. [ME f. OF *imposition* or L *impositio* f. *imponere*: see IMPOSE]

impossibility /ɪmˌpɒsɪˈbɪlɪtɪ/ *n.* (*pl.* **-ies**) **1** the fact or condition of being impossible. **2** an impossible thing or circumstance. [F *impossibilité* or L *impossibilitas* (as IMPOSSIBLE)]

impossible /ɪmˈpɒsɪb(ə)l/ *adj.* **1** not possible; that cannot be done, occur, or exist (*it is impossible to alter them; such a thing is impossible*). **2** (loosely) not easy; not convenient; not easily believable. **3** *colloq.* (of a person or thing) outrageous, intolerable. □ **impossibly** *adv.* [ME f. OF *impossible* or L *impossibilis* (as IN-¹, POSSIBLE)]

impost¹ /ˈɪmpəʊst/ *n.* **1** a tax, duty, or tribute. **2** a weight carried by a horse in a handicap race. [F f. med.L *impost-* part. stem of L *imponere*: see IMPOSE]

impost² /ˈɪmpəʊst/ *n. Archit.* the upper course of a pillar, carrying an arch. [F *imposte* or It. *imposta* fem. past part. of *imporre* f. L *imponere*: see IMPOSE]

impostor /ɪmˈpɒstə(r)/ *n.* (also **imposter**) **1** a person who assumes a false character or pretends to be someone else. **2** a swindler. [F *imposteur* f. LL *impostor* (as IMPOST¹)]

imposture /ɪmˈpɒstʃə(r)/ *n.* the act or an instance of fraudulent deception. [F f. LL *impostura* (as IMPOST¹)]

impotent /ˈɪmpət(ə)nt/ *adj.* **1 a** powerless; lacking all strength. **b** helpless. **c** ineffective. **2 a** (esp. of a male) unable, esp. for a prolonged period, to achieve a sexual erection or orgasm. **b** *colloq.* unable to procreate; infertile. □ **impotence** *n.* **impotency** *n.* **impotently** *adv.* [ME f. OF f. L *impotens* (as IN-¹, POTENT¹)]

impound /ɪmˈpaʊnd/ *v.tr.* **1** confiscate. **2** take possession of. **3** shut up (animals) in a pound. **4** shut up (a person or thing) as in a pound. **5** (of a dam etc.) collect or confine (water). □ **impoundable** *adj.* **impounder** *n.* **impoundment** *n.*

impoverish /ɪmˈpɒvərɪʃ/ *v.tr.* (often as **impoverished** *adj.*) **1** make poor. **2** exhaust the strength or natural fertility of. □ **impoverishment** *n.* [ME f. OF *empoverir* (as EN-¹, *povre* POOR)]

impracticable /ɪmˈpræktɪkəb(ə)l/ *adj.* **1** impossible in practice. **2** (of a road etc.) impassable. **3** (of a person or thing) unmanageable. □ **impracticably** *adv.* **impracticableness** *n.* **impracticability** /-ˌpræktɪkəˈbɪlɪtɪ/ *n.*

impractical /ɪmˈpræktɪk(ə)l/ *adj.* **1** not practical. **2** esp. *US* not practicable. □ **impractically** *adv.* **impracticality** /-ˌpræktɪˈkælɪtɪ/ *n.*

imprecation /ˌɪmprɪˈkeɪʃ(ə)n/ *n.* **1** a spoken curse; a malediction. **2** the act of uttering an imprecation; cursing. [L *imprecatio* f. *imprecari* (as IN-², *precari* pray)]

imprecatory /ˈɪmprɪˌkeɪtərɪ/ *adj.* expressing or involving imprecation.

imprecise /ˌɪmprɪˈsaɪs/ *adj.* not precise. □ **imprecisely** *adv.* **impreciseness** *n.* **imprecision** /-ˈsɪʒ(ə)n/ *n.*

impregnable¹ /ɪmˈpregnəb(ə)l/ *adj.* **1** (of a fortified position) that cannot be taken by force. **2** resistant to attack or criticism. □ **impregnably** *adv.* **impregnability** /-ˌpregnəˈbɪlɪtɪ/ *n.* [ME f. OF *imprenable* (as IN-¹, *prendre* take)]

impregnable² /ɪmˈpregnəb(ə)l/ *adj.* that can be impregnated.

impregnate *v. & adj.* ● *v.tr.* /ˈɪmpregˌneɪt/ **1** (often foll. by *with*) fill or saturate. **2** (often foll. by *with*) imbue, fill (with feelings, moral qualities, etc.). **3 a** make (a female) pregnant. **b** *Biol.* fertilize (a female reproductive cell or ovum). ● *adj.* /ɪmˈpregnət/ **1** pregnant. **2** (often foll. by *with*) permeated. □ **impregnation** /ˌɪmpreɡˈneɪʃ(ə)n/ *n.* [LL *impregnare impregnat-* (as IN-², *pregnare* be pregnant)]

impresario /ˌɪmprɪˈsɑːrɪəʊ/ *n.* (*pl.* **-os**) an organizer of public entertainments, esp. the manager of an operatic, theatrical, or concert company. [It. f. *impresa* undertaking]

imprescriptible /ˌɪmprɪˈskrɪptɪb(ə)l/ *adj. Law* (of rights) that cannot be taken away by prescription or lapse of time. [med.L *imprescriptibilis* (as IN-¹, PRESCRIBE)]

impress¹ *v. & n.* ● *v.tr.* /ɪmˈpres/ **1** (often foll. by *with*) **a** affect or influence deeply. **b** (also *absol.*) evoke a favourable opinion or reaction from (a person) (*was most impressed with your efforts*). **2** (often foll. by *on*, *upon*) emphasize (an idea etc.) (*must impress on you the need to be prompt*). **3** (often foll. by *on*) **a** imprint or stamp. **b** apply (a mark etc.) with pressure. **4** make a mark or design on (a thing) with a stamp, seal, etc. **5** *Electr.* apply (voltage etc.) from outside. ● *n.* /ˈɪmpres/ **1** the act or an instance of impressing. **2** a mark, stamp, etc. **3** a characteristic mark or quality. **4** = IMPRESSION 1. □ **impressible** /ɪmˈpresɪb(ə)l/ *adj.* [ME f. OF *empresser* (as EN-¹, PRESS¹)]

impress² /ɪmˈpres/ *v.tr. hist.* **1** force (men) to serve in the army or navy. **2** seize (goods etc.) for public service. □ **impressment** *n.* [IN-² + PRESS²]

impression /ɪmˈpreʃ(ə)n/ *n.* **1** an effect produced (esp. on the mind or feelings). **2** a notion or belief (esp. a vague or mistaken one) (*my impression is they are afraid*). **3** an imitation of a person or sound, esp. done to entertain. **4 a** the impressing of a mark. **b** a mark impressed. **5** an unaltered reprint of a book etc. from standing type or plates (esp. as distinct from *edition*). **6 a** the number of copies of a book, newspaper, etc., issued at one time. **b** the printing of these. **7** a print taken from a wood engraving. **8** *Dentistry* a negative copy of the teeth or gums made by pressing them into a soft substance. □ **impressional** *adj.* [ME f. OF f. L *impressio -onis* f. *imprimere impress-* (as IN-², PRESS¹)]

impressionable /ɪmˈpreʃənəb(ə)l/ *adj.* easily influenced; susceptible to impressions. □ **impressionably** *adv.* **impressionability** /-ˌpreʃənəˈbɪlɪtɪ/ *n.* [F *impressionnable* f. *impressionner* (as IMPRESSION)]

impressionism /ɪmˈpreʃəˌnɪz(ə)m/ *n.* **1** an artistic style or movement originating in France in the late 19th century, characterized by a concern with depicting the visual impression of the moment, esp. in terms of the shifting effect of light and colour. (*See note below.*) **2** a style of music or writing that seeks to describe a feeling or experience rather than achieve accurate depiction or systematic structure. □ **impressionist** *n.*

▪ The name and reputation of impressionism derive from eight exhibitions held in Paris in 1874–86. The term *impressionist* was coined by a critic as a derisory comment on Monet's *Impression: soleil levant* in 1874, and in the early years the movement was the object of much critical hostility. The artists were trying to move away from precise academic draughtsmanship and also away from the romantic notion that art should convey personal emotion; instead, they were interested in capturing momentary visual perception in an objective way, and in this they were strongly influenced by contemporary developments in photography. Their work is characterized by free, loose brushwork, a sense of the immediacy of the resultant image — the picture's 'unfinished' quality — as well as a complex interplay of light and colour; the subject-matter is typically landscape, although not exclusively so. Among the principal members of the group were Monet, Auguste Renoir, Camille Pissarro, Cézanne, Degas, and Alfred Sisley. The influence of impressionism was enormous, and it is often regarded as the most important artistic development of the 19th century.

impressionistic /ɪmˌpreʃəˈnɪstɪk/ *adj.* **1** in the style of impressionism. **2** subjective, unsystematic. □ **impressionistically** *adv.*

impressive /ɪmˈpresɪv/ *adj.* **1** impressing the mind or senses, esp. so as to cause approval or admiration. **2** (of language, a scene, etc.) tending to excite deep feeling. □ **impressively** *adv.* **impressiveness** *n.*

imprest /ˈɪmprest/ *n.* money advanced to a person for use in state business. [orig. *in prest* f. OF *prest* loan, advance pay: see PRESS²]

imprimatur /ˌɪmprɪˈmeɪtə(r), -ˈmɑːtə(r), -tʊə(r)/ *n.* **1** *RC Ch.* an official licence to print (an ecclesiastical or religious book etc.). **2** official approval. [L, = let it be printed]

imprimatura /ɪmˌpriːməˈtʊərə/ *n.* (in painting) a coloured transparent glaze as a primer. [It. *imprimitura* f. *imprimere* IMPRESS¹]

imprint *v. & n.* ● *v.tr.* /ɪmˈprɪnt/ **1** (often foll. by *on*) impress or establish firmly, esp. on the mind. **2 a** (often foll. by *on*) make a stamp or impression of (a figure etc.) on a thing. **b** make an impression on (a thing) with a stamp etc. **3** (usu. in *passive*; often foll. by *on*, *to*) *Biol.* cause

(esp. a young animal) to recognize another as a parent or other object of habitual trust. ● *n.* /ˈɪmprɪnt/ **1** an impression or stamp. **2** the printer's or publisher's name and other details printed in a book. [ME f. OF *empreinter empreint* f. L *imprimere*: see IMPRESSION]

imprison /ɪmˈprɪz(ə)n/ *v.tr.* **1** put into prison. **2** confine; shut up. □ **imprisonment** *n.* [ME f. OF *emprisoner* (as EN-¹, PRISON)]

impro /ˈɪmprəʊ/ *n.* (*pl.* **-os**) (also *attrib.*) esp. *Theatr. colloq.* **1** improvisation, esp. in performance or as a theatrical technique. **2** an instance of this. [abbr.: cf. IMPROV]

improbable /ɪmˈprɒbəb(ə)l/ *adj.* **1** not likely to be true or to happen. **2** difficult to believe. □ **improbably** *adv.* **improbability** /-ˌprɒbəˈbɪlɪtɪ/ *n.* [F *improbable* or L *improbabilis* (as IN-¹, PROBABLE)]

improbity /ɪmˈprəʊbɪtɪ/ *n.* (*pl.* **-ies**) **1** wickedness; lack of moral integrity. **2** dishonesty. **3** a wicked or dishonest act. [L *improbitas* (as IN-¹, PROBITY)]

impromptu /ɪmˈprɒmptjuː/ *adj., adv., & n.* ● *adj. & adv.* extempore, unrehearsed. ● *n.* (*pl.* **impromptus**) **1** an extempore performance or speech. **2** a short piece of usu. solo instrumental music, often songlike. [F f. L *in promptu* in readiness: see PROMPT]

improper /ɪmˈprɒpə(r)/ *adj.* **1 a** unseemly; indecent. **b** not in accordance with accepted rules of behaviour. **2** inaccurate, wrong. **3** not properly so called. □ **improper fraction** *Math.* a fraction in which the numerator is greater than or equal to the denominator. □ **improperly** *adv.* [F *impropre* or L *improprius* (as IN-¹, PROPER)]

impropriate /ɪmˈprəʊprɪˌeɪt/ *v.tr. Brit.* **1** annex (an ecclesiastical benefice) to a corporation or person as property. **2** place (tithes or ecclesiastical property) in lay hands. □ **impropriation** /-ˌprəʊprɪˈeɪʃ(ə)n/ *n.* [AL *impropriare* (as IN-², *proprius* own)]

impropriator /ɪmˈprəʊprɪˌeɪtə(r)/ *n. Brit.* a person to whom a benefice is impropriated.

impropriety /ˌɪmprəˈpraɪətɪ/ *n.* (*pl.* **-ies**) **1** lack of propriety; indecency. **2** an instance of improper conduct etc. **3** incorrectness. **4** unfitness. [F *impropriété* or L *improprietas* (as IN-¹, *proprius* proper)]

improv /ˈɪmprɒv/ *n.* (also *attrib.*) esp. *Theatr. colloq.* = IMPRO. [abbr.]

improvable /ɪmˈpruːvəb(ə)l/ *adj.* **1** that can be improved. **2** suitable for cultivation. □ **improvability** /-ˌpruːvəˈbɪlɪtɪ/ *n.*

improve /ɪmˈpruːv/ *v.* **1 a** *tr. & intr.* make or become better. **b** *intr.* (foll. by *on*, *upon*) produce something better than. **2** *absol.* (as **improving** *adj.*) giving moral benefit (*improving literature*). [orig. *emprove*, *improwe* f. AF *emprower* f. OF *emprou* f. *prou* profit, infl. by PROVE]

improvement /ɪmˈpruːvmənt/ *n.* **1** the act or an instance of improving or being improved. **2** something that improves, esp. an addition or alteration that adds to value. **3** something that has been improved. [ME f. AF *emprowement* (as IMPROVE)]

improver /ɪmˈpruːvə(r)/ *n.* **1** a person who improves. **2** *Brit. hist.* a person who works for low wages while acquiring skill and experience in a trade.

improvident /ɪmˈprɒvɪd(ə)nt/ *adj.* **1** lacking foresight or care for the future. **2** not frugal; thriftless. **3** heedless, incautious. □ **improvidence** *n.* **improvidently** *adv.*

improvisation /ˌɪmprəvaɪˈzeɪʃ(ə)n/ *n.* **1** the act of improvising; something done on the spur of the moment. **2** a piece of improvised music or verse. □ **improvisational** *adj.* **improvisatorial** /-zəˈtɔːrɪəl/ *adj.* **improvisatory** /-ˈzeɪtərɪ/ *adj.*

improvise /ˈɪmprəˌvaɪz/ *v.tr.* (also *absol.*) **1** compose or perform (music, verse, etc.) extempore. **2** provide or construct (a thing) extempore. □ **improviser** *n.* [F *improviser* or It. *improvvisare* f. *improvviso* extempore, f. L *improvisus* past part. (as IN-¹, PROVIDE)]

imprudent /ɪmˈpruːd(ə)nt/ *adj.* rash, indiscreet. □ **imprudence** *n.* **imprudently** *adv.* [ME f. L *imprudens* (as IN-¹, PRUDENT)]

impudent /ˈɪmpjʊd(ə)nt/ *adj.* **1** insolently disrespectful; impertinent. **2** shamelessly presumptuous. **3** unblushing. □ **impudence** *n.* **impudently** *adv.* [ME f. L *impudens* (as IN-¹, *pudere* be ashamed)]

impudicity /ˌɪmpjʊˈdɪsɪtɪ/ *n.* shamelessness, immodesty. [F *impudicité* f. L *impudicus* (as IMPUDENT)]

impugn /ɪmˈpjuːn/ *v.tr.* challenge or call in question (a statement, action, etc.). □ **impugnable** *adj.* **impugnment** *n.* [ME f. L *impugnare* assail (as IN-², *pugnare* fight)]

impuissant /ɪmˈpjuːɪs(ə)nt, -ˈpwiːs(ə)nt/ *adj.* impotent, weak. □ **impuissance** *n.* [F (as IN-¹, PUISSANT)]

impulse /ˈɪmpʌls/ *n.* **1** the act or an instance of impelling; a push. **2** an impetus. **3** *Physics* **a** an indefinitely large force acting for a very short time but producing a finite change of momentum (e.g. the blow of a hammer). **b** the change of momentum produced by this or any force. **4** *Physiol.* a wave of excitation in a nerve. **5** mental incitement. **6** a sudden desire or tendency to act without reflection (*did it on impulse*). □ **impulse buying** the unpremeditated buying of goods as a result of a whim or impulse. [L *impulsus* (as IMPEL)]

impulsion /ɪmˈpʌlʃ(ə)n/ *n.* **1** the act or an instance of impelling. **2** a mental impulse. **3** impetus. [ME f. OF f. L *impulsio -onis* (as IMPEL)]

impulsive /ɪmˈpʌlsɪv/ *adj.* **1** (of a person or conduct etc.) apt to be affected or determined by sudden impulse. **2** tending to impel. **3** *Physics* acting as an impulse. □ **impulsively** *adv.* **impulsiveness** *n.* [ME f. F *impulsif -ive* or LL *impulsivus* (as IMPULSION)]

impunity /ɪmˈpjuːnɪtɪ/ *n.* exemption from punishment or from the injurious consequences of an action. □ **with impunity** without having to suffer the normal injurious consequences (of an action). [L *impunitas* f. *impunis* (as IN-[1], *poena* penalty)]

impure /ɪmˈpjʊə(r)/ *adj.* **1** mixed with foreign matter; adulterated. **2 a** dirty. **b** ceremonially unclean. **3** unchaste. **4** (of a colour) mixed with another colour. □ **impurely** *adv.* **impureness** *n.* [ME f. L *impurus* (as IN-[1], *purus* pure)]

impurity /ɪmˈpjʊərɪtɪ/ *n.* (*pl.* **-ies**) **1** the quality or condition of being impure. **2 a** a thing or constituent which impairs the purity of something. **b** *Electronics* a trace element deliberately added to a semiconductor, a dopant. [F *impurité* or L *impuritas* (as IMPURE)]

impute /ɪmˈpjuːt/ *v.tr.* (foll. by *to*) **1** regard (esp. something undesirable) as being done or caused or possessed by. **2** (in Christian theology) ascribe (righteousness, guilt, etc.) to (a person) by virtue of a similar quality in another. □ **imputable** *adj.* **imputative** /-ˈtətɪv/ *adj.* **imputation** /ˌɪmpjʊˈteɪʃ(ə)n/ *n.* [ME f. OF *imputer* f. L *imputare* enter in the account (as IN-[2], *putare* reckon)]

Imroz see IMBROS.

imshi /ˈɪmʃiː/ *int. Austral. colloq.* be off! [colloq. (Egyptian) Arabic]

I.Mun.E. *abbr.* (in the UK) Institution of Municipal Engineers.

IN *abbr. US* Indiana (in official postal use).

In *symb. Chem.* the element indium.

in /ɪn/ *prep., adv., & adj.* ● *prep.* **1** expressing inclusion or position within limits of space, time, circumstance, etc. (*in England; in bed; in the rain*). **2** during the time of (*in the night; in 1989*). **3** within the time of (*will be back in two hours*). **4 a** with respect to (*blind in one eye; good in parts*). **b** as a kind of (*the latest thing in luxury*). **5** as a proportionate part of (*one in three failed; a gradient of one in six*). **6** with the form or arrangement of (*packed in tens; falling in folds*). **7** as a member of (*in the army*). **8** concerned with (*is in politics*). **9** as or regarding the content of (*there is something in what you say*). **10** within the ability of (*does he have it in him?*). **11** having the condition of; affected by (*in bad health; in danger*). **12** having as a purpose (*in search of; in reply to*). **13** by means of or using as material (*drawn in pencil; modelled in bronze*). **14 a** using as the language of expression (*written in French*). **b** (of music) having as its key (*symphony in C*). **15** (of a word) having a beginning or ending (*words in un-*). **16** wearing as dress (*in blue; in a suit*). **17** with the identity of (*found a friend in Mary*). **18** (of an animal) pregnant with (*in calf*). **19** into (with a verb of motion or change: *put it in the box; cut it in two*). **20** introducing an indirect object after a verb (*believe in; engage in; share in*). **21** forming adverbial phrases (*in any case; in reality; in short*). ● *adv.* expressing position within limits, or motion to such a position: **1** into a room, house, etc. (*come in*). **2** at home, in one's office, etc. (*is not in*). **3** so as to be enclosed or confined (*locked in*). **4** in a publication (*is the advertisement in?*). **5** in or to the inward side (*rub it in*). **6** in a fashion, season, or office (*long skirts are in; strawberries are not yet in*). **b** elected (*the Democrat got in*). **7** exerting favourable action or influence (*their luck was in*). **8** *Cricket* (of a player or side) batting. **9** (of transport) at the platform etc. (*the train is in*). **10** (of a season, harvest, order, etc.) having arrived or been received. **11** *Brit.* (of a fire) continuing to burn. **12** denoting effective action (*join in*). **13** (of the tide) at the highest point. **14** (in *comb.*) *colloq.* denoting prolonged or concerted action, esp. by large numbers (*sit-in; teach-in*). ● *adj.* **1** internal; living in; inside (*in-patient*). **2** fashionable, esoteric (*the in thing to do*). **3** confined to or shared by a group of people (*in-joke*). □ **in all** see ALL. **in at** present at; contributing to (*in at the kill*). **in between** see BETWEEN *adv.* **in-between** *attrib.adj. colloq.* intermediate (*at an in-between stage*). **in for 1** about to undergo (esp. something unpleasant). **2** competing in or for. **3** involved in; committed to. **in on** sharing in; privy to (a secret etc.). **ins and outs** (often foll. by *of*) all the details (of a procedure etc.). **in so far as** see *so far as* (see FAR). **in that** because; in so far as. **in with** on good terms with. [OE *in, inn,* orig. as *adv.* with verbs of motion]

in. *abbr.* inch(es).

in-[1] /ɪn/ *prefix* (also **il-, im-, ir-**) added to: **1** adjectives, meaning 'not' (*inedible; insane*). **2** nouns, meaning 'without, lacking' (*inaction*). [L]

in-[2] /ɪn/ *prefix* (also **il-** before *l*, **im-** before *b, m, p,* **ir-** before *r*) in, on, into, towards, within (*induce; influx; insight; intrude*). [IN, or from or after L *in* in *prep.*]

-in /ɪn/ *suffix Chem.* forming names of organic compounds, pharmaceutical products, etc. (*gelatin, penicillin, dioxin*). [-INE[4]]

-ina /ˈiːnə/ *suffix* denoting: **1** feminine names and titles (*Georgina; tsarina*). **2** names of musical instruments (*concertina*). **3** names of zoological classification categories (*globigerina*). [It. or Sp. or L]

inability /ˌɪnəˈbɪlɪtɪ/ *n.* **1** the state of being unable. **2** a lack of power or means.

in absentia /ˌɪn æbˈsentɪə/ *adv.* in (his, her, or their) absence. [L]

inaccessible /ˌɪnækˈsesɪb(ə)l/ *adj.* **1** not accessible; that cannot be reached. **2** (of a person) not open to advances or influence; unapproachable. □ **inaccessibly** *adv.* **inaccessibility** /-ˌsesɪˈbɪlɪtɪ/ *n.* [ME f. F *inaccessible* or LL *inaccessibilis* (as IN-[1], ACCESSIBLE)]

inaccurate /ɪnˈækjʊrət/ *adj.* not accurate. □ **inaccuracy** *n.* (*pl.* **-ies**). **inaccurately** *adv.*

inaction /ɪnˈækʃ(ə)n/ *n.* **1** lack of action. **2** sluggishness, inertness.

inactivate /ɪnˈæktɪˌveɪt/ *v.tr.* make inactive or inoperative. □ **inactivation** /-ˌæktɪˈveɪʃ(ə)n/ *n.*

inactive /ɪnˈæktɪv/ *adj.* **1** not active or inclined to act. **2** passive. **3** indolent. □ **inactively** *adv.* **inactivity** /ˌɪnækˈtɪvɪtɪ/ *n.*

inadequate /ɪnˈædɪkwət/ *adj.* (often foll. by *to*) **1** not adequate; insufficient. **2** (of a person) incompetent; unable to deal with a situation. □ **inadequacy** *n.* (*pl.* **-ies**). **inadequately** *adv.*

inadmissible /ˌɪnədˈmɪsɪb(ə)l/ *adj.* that cannot be admitted or allowed. □ **inadmissibly** *adv.* **inadmissibility** /-ˌmɪsɪˈbɪlɪtɪ/ *n.*

inadvertent /ˌɪnədˈvɜːt(ə)nt/ *adj.* **1** (of an action) unintentional. **2 a** not properly attentive. **b** negligent. □ **inadvertence** *n.* **inadvertency** *n.* **inadvertently** *adv.* [IN-[1] + obs. *advertent* attentive (as ADVERT[2])]

inadvisable /ˌɪnədˈvaɪzəb(ə)l/ *adj.* not advisable. □ **inadvisability** /-ˌvaɪzəˈbɪlɪtɪ/ *n.* [ADVISABLE]

inalienable /ɪnˈeɪlɪənəb(ə)l/ *adj.* that cannot be transferred to another; not alienable. □ **inalienably** *adv.* **inalienability** /-ˌeɪlɪənəˈbɪlɪtɪ/ *n.*

inalterable /ɪnˈɔːltərəb(ə)l, ɪnˈɒl-/ *adj.* not alterable; that cannot be changed. □ **inalterably** *adv.* **inalterability** /-ˌɔːltərəˈbɪlɪtɪ, -ˌɒl-/ *n.* [med.L *inalterabilis* (as IN-[1], *alterabilis* alterable)]

inamorato /ɪnˌæməˈrɑːtəʊ/ *n.* (*pl.* **-os**; *fem.* **inamorata** /-tə/) a lover. [It., past part. of *inamorare* enamour (as IN-[2], *amore* f. L *amor* love)]

inane /ɪˈneɪn/ *adj.* **1** silly, senseless. **2** empty, void. □ **inanely** *adv.* **inaneness** *n.* **inanity** /ɪˈnænɪtɪ/ *n.* (*pl.* **-ies**). [L *inanis* empty, vain]

inanimate /ɪnˈænɪmət/ *adj.* **1** not animate; not endowed with (esp. animal) life. **2** lifeless; showing no sign of life. **3** spiritless, dull. □ **inanimate nature** everything other than the animal world. □ **inanimately** *adv.* **inanimation** /-ˌænɪˈmeɪʃ(ə)n/ *n.* [LL *inanimatus* (as IN-[1], ANIMATE)]

inanition /ˌɪnəˈnɪʃ(ə)n/ *n.* emptiness, esp. exhaustion from lack of nourishment. [ME f. LL *inanitio* f. L *inanire* make empty (as INANE)]

inappellable /ˌɪnəˈpeləb(ə)l/ *adj.* that cannot be appealed against. [obs.F *inappelable* (as IN-[1], *appeler* APPEAL)]

inapplicable /ɪnˈæplɪkəb(ə)l, ˌɪnəˈplɪk-/ *adj.* (often foll. by *to*) not applicable; unsuitable. □ **inapplicably** *adv.* **inapplicability** /-ˌæplɪkəˈbɪlɪtɪ, -əˌplɪkə-/ *n.*

inapposite /ɪnˈæpəzɪt/ *adj.* not apposite; out of place. □ **inappositely** *adv.* **inappositeness** *n.*

inappreciable /ˌɪnəˈpriːʃəb(ə)l/ *adj.* **1** imperceptible; not worth reckoning. **2** that cannot be appreciated. □ **inappreciably** *adv.*

inappreciation /ˌɪnəˌpriːʃɪˈeɪʃ(ə)n, ˌɪnəˌpriːsɪ-/ *n.* failure to appreciate. □ **inappreciative** /-ˈpriːʃətɪv/ *adj.*

inappropriate /ˌɪnəˈprəʊprɪət/ *adj.* not appropriate; unsuitable. □ **inappropriately** *adv.* **inappropriateness** *n.*

inapt /ɪnˈæpt/ *adj.* **1** not apt or suitable. **2** unskilful. □ **inaptitude** *n.* **inaptly** *adv.*

inarch /ɪnˈɑːtʃ/ *v.tr.* graft (a plant) by connecting a growing branch without separation from the parent stock. [IN-[2] + ARCH[1] *v.*]

inarguable /ɪnˈɑːgjʊəb(ə)l/ *adj.* that cannot be argued about or disputed. □ **inarguably** *adv.*

inarticulate /ˌɪnɑːˈtɪkjʊlət/ *adj.* **1** unable to speak distinctly or express oneself clearly. **2** (of speech) not articulate; indistinctly pronounced. **3** dumb. **4** esp. *Anat.* not jointed. □ **inarticulately** *adv.* **inarticulateness** *n.* **inarticulacy** *n.* [LL *inarticulatus* (as IN-[1], ARTICULATE)]

inartistic /ˌɪnɑːˈtɪstɪk/ *adj.* **1** not following the principles of art. **2** lacking skill or talent in art; not appreciating art. □ **inartistically** *adv.*

inasmuch /ˌɪnəzˈmʌtʃ/ *adv.* (foll. by *as*) **1** since, because. **2** to the extent that. [ME, orig. *in as much*]

inattentive /ˌɪnəˈtentɪv/ *adj.* **1** not paying due attention; heedless. **2** neglecting to show courtesy. □ **inattentively** *adv.* **inattentiveness** *n.* **inattention** /-ˈtenʃ(ə)n/ *n.*

inaudible /ɪnˈɔːdɪb(ə)l/ *adj.* that cannot be heard. □ **inaudibly** *adv.* **inaudibility** /-ˌɔːdɪˈbɪlɪtɪ/ *n.*

inaugural /ɪnˈɔːɡjʊrəl/ *adj. & n.* ● *adj.* **1** of inauguration. **2** (of a lecture etc.) given by a person being inaugurated. ● *n.* an inaugural speech etc. [F f. *inaugurer* (as INAUGURATE)]

inaugurate /ɪnˈɔːɡjʊˌreɪt/ *v.tr.* **1** admit (a person) formally to office. **2** initiate the public use of (a building etc.). **3** begin, introduce. **4** enter with ceremony upon (an undertaking etc.). □ **inaugurator** *n.* **inauguratory** /-rətərɪ/ *adj.* **inauguration** /-ˌɔːɡjʊˈreɪʃ(ə)n/ *n.* [L *inaugurare* (as IN-[2], *augurare* take omens: see AUGUR)]

inauspicious /ˌɪnɔːˈspɪʃəs/ *adj.* **1** ill-omened, unpropitious. **2** unlucky. □ **inauspiciously** *adv.* **inauspiciousness** *n.*

inboard /ˈɪnbɔːd/ *adv. & adj.* ● *adv.* within the sides of or towards the centre of a ship, aircraft, or vehicle. ● *adj.* situated inboard.

inborn /ɪnˈbɔːn/ *adj.* existing from birth; implanted by nature.

inbreathe /ɪnˈbriːð/ *v.tr.* **1** breathe in or absorb. **2** inspire (a person).

inbred /ɪnˈbred/ *adj.* **1** inborn. **2** produced by inbreeding.

inbreeding /ɪnˈbriːdɪŋ/ *n.* breeding from closely related animals or persons. □ **inbreed** *v.tr. & intr.* (*past* and *past part.* **inbred**).

inbuilt /ɪnˈbɪlt/ *adj.* incorporated as part of a structure.

Inc. *abbr.* N. Amer. Incorporated.

Inca /ˈɪŋkə/ *n. & adj.* ● *n.* (*pl.* same or **Incas**) a member of a native people of the central Andes. (*See note below.*) ● *adj.* of or relating to the Incas. □ **Incan** *adj.* **Incaic** /ɪŋˈkeɪɪk/ *adj.* [Quechua, = lord, royal person]

▪ The Incas arrived in the valley of Cuzco *c.*1200. Their origin and early history are uncertain. In the first part of the 15th century they began a series of rapid conquests and a century later their power extended over most of modern Ecuador and Peru, large areas of Bolivia, and parts of Argentina and Chile. Their empire was highly centralized and governed by a despotic monarchy supported by an aristocratic bureaucracy, with Cuzco as its capital city and religious centre. They were skilled engineers, and built a network of roads; technology and architecture were highly developed despite the absence of wheeled vehicles and a system of writing, and many of their palaces, temples, fortifications, and irrigation systems survive. The empire was weakened by civil war in the early 16th century and fell to the invading Spaniards in the early 1530s. Quechua-speaking descendants of the Incas make up about half of Peru's population.

incalculable /ɪnˈkælkjʊləb(ə)l/ *adj.* **1** too great for calculation. **2** that cannot be reckoned beforehand. **3** (of a person, character, etc.) uncertain. □ **incalculably** *adv.* **incalculability** /-ˌkælkjʊləˈbɪlɪtɪ/ *n.*

in camera see CAMERA.

incandesce /ˌɪnkænˈdes/ *v.intr. & tr.* glow or cause to glow with heat. [back-form. f. INCANDESCENT]

incandescent /ˌɪnkænˈdes(ə)nt/ *adj.* **1** glowing with heat. **2** shining brightly. **3** (of an electric or other light) produced by a glowing white-hot filament. □ **incandescence** *n.* **incandescently** *adv.* [F f. L *incandescere* (as IN-[2], *candescere* inceptive of *candere* be white)]

incantation /ˌɪnkænˈteɪʃ(ə)n/ *n.* **1 a** a magical formula. **b** the use of this. **2** a spell or charm. □ **incantational** *adj.* **incantatory** /ɪnˈkæntətərɪ/ *adj.* [ME f. OF f. LL *incantatio -onis* f. *incantare* chant, bewitch (as IN-[2], *cantare* sing)]

incapable /ɪnˈkeɪpəb(ə)l/ *adj.* **1** (often foll. by *of*) **a** not capable. **b** lacking the required quality or characteristic (favourable or adverse) (*incapable of hurting anyone*). **2** not capable of rational conduct or of managing one's own affairs (*drunk and incapable*). □ **incapably** *adv.* **incapability** /-ˌkeɪpəˈbɪlɪtɪ/ *n.* [F *incapable* or LL *incapabilis* (as IN-[1], *capabilis* CAPABLE)]

incapacitate /ˌɪnkəˈpæsɪˌteɪt/ *v.tr.* **1** render incapable or unfit. **2** disqualify. □ **incapacitant** *n.* **incapacitation** /-ˌpæsɪˈteɪʃ(ə)n/ *n.*

incapacity /ˌɪnkəˈpæsɪtɪ/ *n.* (*pl.* -ies) **1** inability; lack of the necessary power or resources. **2** legal disqualification. **3** an instance of incapacity. [F *incapacité* or LL *incapacitas* (as IN-[1], CAPACITY)]

incarcerate /ɪnˈkɑːsəˌreɪt/ *v.tr.* imprison or confine. □ **incarcerator** *n.* **incarceration** /-ˌkɑːsəˈreɪʃ(ə)n/ *n.* [med.L *incarcerare* (as IN-[2], L *carcer* prison)]

incarnadine /ɪnˈkɑːnəˌdaɪn/ *adj. & v. poet.* ● *adj.* flesh-coloured or crimson. ● *v.tr.* dye this colour. [F *incarnadin -ine* f. It. *incarnadino* (for *-tino*) f. *incarnato* INCARNATE *adj.*]

incarnate *adj. & v.* ● *adj.* /ɪnˈkɑːnət/ **1** (of a person, spirit, quality, etc.) embodied in flesh, esp. in human form (*is the devil incarnate*). **2** represented in a recognizable or typical form (*folly incarnate*). ● *v.tr.* /ˈɪnkɑːˌneɪt, ɪnˈkɑːneɪt/ **1** embody in flesh. **2** put (an idea etc.) into concrete form; realize. **3** (of a person etc.) be the living embodiment of (a quality). [ME f. eccl.L *incarnare incarnat-* make flesh (as IN-[2], L *caro carnis* flesh)]

incarnation /ˌɪnkɑːˈneɪʃ(ə)n/ *n.* **1 a** embodiment in (esp. human) flesh. **b** (**the Incarnation**) (in Christian theology) the embodiment of God the Son in human flesh as Jesus Christ. **2** (often foll. by *of*) a living type (of a quality etc.). **3** *Med.* the process of forming new flesh. [ME f. OF f. eccl.L *incarnatio -onis* (as INCARNATE)]

incase var. of ENCASE.

incautious /ɪnˈkɔːʃəs/ *adj.* heedless, rash. □ **incaution** *n.* **incautiously** *adv.* **incautiousness** *n.*

incendiary /ɪnˈsendɪərɪ/ *adj. & n.* ● *adj.* **1** (of a substance or device, esp. a bomb) designed to cause fires. **2 a** of or relating to the malicious setting on fire of property. **b** guilty of this. **3** tending to stir up strife; inflammatory. ● *n.* (*pl.* -ies) **1** an incendiary bomb or device. **2** an incendiary person. □ **incendiarism** *n.* [ME f. L *incendiarius* f. *incendium* conflagration f. *incendere incens-* set fire to]

incense[1] /ˈɪnsens/ *n. & v.* ● *n.* **1** a gum or spice producing a sweet smell when burned. **2** the smoke of this, esp. in religious ceremonial. ● *v.tr.* **1** treat or perfume (a person or thing) with incense. **2** burn incense to (a deity etc.). **3** suffuse with fragrance. □ **incensation** /ˌɪnsenˈseɪʃ(ə)n/ *n.* [ME f. OF *encens, encenser* f. eccl.L *incensum* a thing burnt, incense: see INCENDIARY]

incense[2] /ɪnˈsens/ *v.tr.* (often foll. by *at, with, against*) enrage; make angry. [ME f. OF *incenser* (as INCENDIARY)]

incensory /ˈɪnsensərɪ/ *n.* (*pl.* -ies) = CENSER. [med.L *incensorium* (as INCENSE[1])]

incentive /ɪnˈsentɪv/ *n. & adj.* ● *n.* **1** (often foll. by *to*) a motive or incitement, esp. to action. **2** a payment or concession to stimulate greater output by workers. ● *adj.* serving to motivate or incite. [ME f. L *incentivus* setting the tune f. *incinere incent-* sing to (as IN-[2], *canere* sing)]

incept /ɪnˈsept/ *v.* **1** *tr. Biol.* (of an organism) digest (food). **2** *intr. Brit. hist.* take a master's or doctor's degree at a university. □ **inceptor** *n.* (in sense 2). [L *incipere incept-* begin (as IN-[2], *capere* take)]

inception /ɪnˈsepʃ(ə)n/ *n.* a beginning. [ME f. OF *inception* or L *inceptio* (as INCEPT)]

inceptive /ɪnˈseptɪv/ *adj. & n.* ● *adj.* **1 a** beginning. **b** initial. **2** *Gram.* (of a verb) that denotes the beginning of an action. ● *n.* an inceptive verb. [LL *inceptivus* (as INCEPT)]

incertitude /ɪnˈsɜːtɪˌtjuːd/ *n.* uncertainty, doubt. [F *incertitude* or LL *incertitudo* (as IN-[1], CERTITUDE)]

incessant /ɪnˈses(ə)nt/ *adj.* unceasing, continual, repeated. □ **incessancy** *n.* **incessantly** *adv.* **incessantness** *n.* [F *incessant* or LL *incessans* (as IN-[1], *cessans* pres. part. of L *cessare* CEASE)]

incest /ˈɪnsest/ *n.* sexual intercourse between persons regarded as too closely related to marry each other. [ME f. L *incestus* (as IN-[1], *castus* CHASTE)]

incestuous /ɪnˈsestjʊəs/ *adj.* **1** involving or guilty of incest. **2** (of human relations generally) excessively restricted or resistant to wider influence. □ **incestuously** *adv.* **incestuousness** *n.* [LL *incestuosus* (as INCEST)]

inch[1] /ɪntʃ/ *n. & v.* ● *n.* **1** a unit of linear measure equal to one-twelfth of a foot (2.54 cm) (symbol: ″). **2 a** (as a unit of rainfall) a quantity that would cover a horizontal surface to a depth of 1 inch. **b** (of atmospheric or other pressure) an amount that balances the weight of a column of mercury 1 inch high. **3** (as a unit of map-scale) so many inches representing 1 mile on the ground (*a 4-inch map*). **4** a small amount (usu. with *neg.*: *would not yield an inch*). ● *v.tr. & intr.* move gradually in a specified way (*inched forward*). □ **every inch 1** entirely (*looked every inch a queen*). **2** the whole distance or area (*combed every inch of the garden*). **give a person an inch and he** or **she will take a mile** (or orig.

an ell) a person once conceded to will demand much. **inch by inch** gradually; bit by bit. **within an inch of** almost to the point of. [OE *ynce* f. L *uncia* twelfth part: cf. OUNCE[1]]

inch[2] /ɪntʃ/ n. esp. *Sc.* a small island (esp. in place-names). [ME f. Gael. *innis*]

Inchcape Rock /ˈɪntʃkeɪp/ a sandstone reef in the North Sea, off the mouth of the River Tay in Scotland. The Scottish civil engineers Robert Stevenson (1772–1850) and John Rennie designed a lighthouse that was built there in 1807–*c*.1811.

inchmeal /ˈɪntʃmiːl/ adv. by inches; little by little; gradually. [f. INCH[1] + MEAL[1]]

inchoate /ɪnˈkəʊeɪt/ adj. & v. ● adj. **1** just begun. **2** undeveloped, rudimentary, unformed. ● v.tr. begin; originate. □ **inchoately** adv. **inchoateness** n. **inchoative** /-ˈkəʊətɪv/ adj. [L *inchoatus* past part. of *inchoare* (as IN-², *choare* begin)]

Inchon /ɪnˈtʃɒn/ a port on the west coast of South Korea, on the Yellow Sea near Seoul; pop. (1990) 1,818,300.

inchworm /ˈɪntʃwɜːm/ n. = LOOPER 1.

incidence /ˈɪnsɪd(ə)ns/ n. **1** (often foll. by *of*) the fact, manner, or rate, of occurrence or action. **2** the range, scope, or extent of influence of a thing. **3** *Physics* the falling of a line, or of a thing moving in a line, upon a surface. **4** the act or an instance of coming into contact with a thing. □ **angle of incidence** *Physics* the angle which an incident line, ray, etc., makes with the perpendicular to the surface at the point of incidence. [ME f. OF *incidence* or med.L *incidentia* (as INCIDENT)]

incident /ˈɪnsɪd(ə)nt/ n. & adj. ● n. **1 a** an event or occurrence. **b** a minor or detached event attracting general attention or noteworthy in some way. **2** a hostile clash, esp. of troops of countries at war (*a frontier incident*). **3** a distinct piece of action in a play or a poem. **4** *Law* a privilege, burden, etc., attaching to an obligation or right. ● adj. **1 a** (often foll. by *to*) apt or liable to happen; naturally attaching or dependent. **b** (foll. by *to*) *Law* attaching to. **2** (often foll. by *on, upon*) (of light etc.) falling or striking. [ME f. F *incident* or L *incidere* (as IN-², *cadere* fall)]

incidental /ˌɪnsɪˈdent(ə)l/ adj. & n. ● adj. **1** (often foll. by *to*) **a** having a minor role in relation to a more important thing, event, etc. **b** not essential. **c** casual, happening by chance. **2** (foll. by *to*) liable to happen. **3** (foll. by *on, upon*) following as a subordinate event. ● n. (usu. in *pl.*) a minor detail, expense, event, etc. □ **incidental music** music used as a background to the action of a film, broadcast, etc.

incidentally /ˌɪnsɪˈdentəlɪ/ adv. **1** by the way; as an unconnected remark. **2** in an incidental way.

incinerate /ɪnˈsɪnəˌreɪt/ v.tr. **1** consume (a body etc.) by fire. **2** reduce to ashes. □ **incineration** /-ˌsɪnəˈreɪʃ(ə)n/ n. [med.L *incinerare* (as IN-², *cinis -eris* ashes)]

incinerator /ɪnˈsɪnəˌreɪtə(r)/ n. a furnace or apparatus for burning esp. refuse to ashes.

incipient /ɪnˈsɪpɪənt/ adj. **1** beginning. **2** in an initial stage. □ **incipience** n. **incipiency** n. **incipiently** adv. [L *incipere incipient-* (as INCEPT)]

incise /ɪnˈsaɪz/ v.tr. **1** make a cut in. **2** engrave. [F *inciser* f. L *incidere incis-* (as IN-², *caedere* cut)]

incision /ɪnˈsɪʒ(ə)n/ n. **1** a cut; a division produced by cutting; a notch. **2** the act of cutting into a thing. [ME f. OF *incision* or LL *incisio* (as INCISE)]

incisive /ɪnˈsaɪsɪv/ adj. **1** mentally sharp; acute. **2** clear and effective. **3** cutting, penetrating. □ **incisively** adv. **incisiveness** n. [med.L *incisivus* (as INCISE)]

incisor /ɪnˈsaɪzə(r)/ n. a narrow-edged tooth at the front of the mouth, adapted for cutting. [med.L, = cutter (as INCISE)]

incite /ɪnˈsaɪt/ v.tr. (often foll. by *to*) urge or stir up. □ **inciter** n. **incitement** n. **incitation** /ˌɪnsaɪˈteɪʃ(ə)n/ n. [ME f. F *inciter* f. L *incitare* (as IN-², *citare* rouse)]

incivility /ˌɪnsɪˈvɪlɪtɪ/ n. (pl. **-ies**) **1** rudeness, discourtesy. **2** a rude or discourteous act. [F *incivilité* or LL *incivilitas* (as IN-¹, CIVILITY)]

inclement /ɪnˈklemənt/ adj. (of the weather or climate) severe, esp. cold or stormy. □ **inclemency** n. (pl. **-ies**). **inclemently** adv. [F *inclément* or L *inclemens* (as IN-¹, CLEMENT)]

inclination /ˌɪnklɪˈneɪʃ(ə)n/ n. **1** (often foll. by *to*) a disposition or propensity. **2** (often foll. by *for*) a liking or affection. **3 a** a leaning, slope, or slant. **b** a bending of the body or head in a bow. **4** the difference of direction of two lines or planes, esp. as measured by the

angle between them. **5** the dip of a magnetic needle. [ME f. OF *inclination* or L *inclinatio* (as INCLINE)]

incline v. & n. ● v. /ɪnˈklaɪn/ **1** tr. (usu. in *passive*; often foll. by *to, for,* or *to* + infin.) **a** dispose (a person, feelings, etc.) willing or favourably disposed (*am inclined to think so; does not incline me to agree*). **b** give a specified tendency to (a thing) (*the door is inclined to bang*). **2** intr. **a** be disposed (*I incline to think so*). **b** (often foll. by *to, towards*) tend. **3** intr. & tr. lean or turn away from a given direction, esp. the vertical. **4** tr. bend (the head, body, or oneself) forward or downward. ● n. /ˈɪnklaɪn/ **1** a slope. **2** an inclined plane. □ **inclined plane** a sloping plane (esp. as a means of reducing the force needed to raise a load). **incline one's ear** (often foll. by *to*) listen favourably. □ **incliner** /ɪnˈklaɪnə(r)/ n. [ME *encline* f. OF *encliner* f. L *inclinare* (as IN-², *clinare* bend)]

inclinometer /ˌɪnklɪˈnɒmɪtə(r)/ n. **1** an instrument for measuring the angle between the direction of the earth's magnetic field and the horizontal. **2** an instrument for measuring the inclination of an aircraft or ship to the horizontal. **3** an instrument for measuring a slope. [L *inclinare* INCLINE v. + -METER]

inclose var. of ENCLOSE.

inclosure var. of ENCLOSURE.

include /ɪnˈkluːd/ v.tr. **1** comprise or reckon in as part of a whole; place in a class or category. **2** (as **including** prep.) if one takes into account (*six members, including the chairman*). **3** treat or regard as so included. **4** (as **included** adj.) shut in; enclosed. □ **include out** *colloq.* or *joc.* specifically exclude. □ **includable** adj. **includible** adj. [ME f. L *includere inclus-* (as IN-², *claudere* shut)]

inclusion /ɪnˈkluːʒ(ə)n/ n. **1** the act of including something. **2 a** the condition of being included. **b** an instance of this. **3** a thing which is included. **4** a body or particle distinct from the substance in which it is embedded.

inclusive /ɪnˈkluːsɪv/ adj. **1** (foll. by *of*) including, comprising. **2** with the inclusion of the extreme limits stated (*pages 7 to 26 inclusive*). **3** including all the normal services etc. (*a hotel offering inclusive terms*). **4 a** not excluding any section of society. **b** (of language) deliberately non-sexist, esp. avoiding the use of masculine pronouns to cover both men and women. □ **inclusively** adv. **inclusiveness** n. [med.L *inclusivus* (as INCLUDE)]

incog /ɪnˈkɒg/ adj., adv., & n. colloq. = INCOGNITO. [abbr.]

incognito /ˌɪnkɒgˈniːtəʊ/ adj., adv., & n. ● adj. & adv. with one's name or identity kept secret (*was travelling incognito*). ● n. (pl. **-os**) **1** a person who is incognito. **2** the pretended identity or anonymous character of such a person. [It., = unknown, f. L *incognitus* (as IN-¹, *cognitus* past part. of *cognoscere* know)]

incognizant /ɪnˈkɒgnɪz(ə)nt/ adj. (also **incognisant**) (foll. by *of*) unaware; not knowing. □ **incognizance** n.

incoherent /ˌɪnkəʊˈhɪərənt/ adj. **1** (of a person) unable to speak intelligibly. **2** (of speech etc.) lacking logic or consistency. **3** *Physics* (of waves) having no definite or stable phase relationship. □ **incoherence** n. **incoherency** n. (pl. **-ies**). **incoherently** adv.

incombustible /ˌɪnkəmˈbʌstɪb(ə)l/ adj. that cannot be burnt or consumed by fire. □ **incombustibility** /-ˌbʌstɪˈbɪlɪtɪ/ n. [ME f. med.L *incombustibilis* (as IN-¹, COMBUSTIBLE)]

income /ˈɪnkʌm/ n. the money or other assets received, esp. periodically or in a year, from one's business, lands, work, investments, etc. □ **income group** a section of the population determined by income. **income support** (in the UK) a system by which people on low incomes can, according to their circumstances, claim a payment from the state. [ME (orig. = arrival), prob. f. ON *innkoma* in later use f. *come in*]

incomer /ˈɪnˌkʌmə(r)/ n. **1** a person who comes in. **2** a person who arrives to settle in a place; an immigrant. **3** an intruder. **4** a successor.

-incomer /ˈɪnˌkʌmə(r)/ comb. form earning a specified kind or level of income (*middle-incomer*).

income tax n. a tax levied on income. Such a tax was first introduced in Britain in 1799 to help pay for the war against revolutionary France. Through most of the 19th century income tax was regarded as a temporary measure to meet extraordinary expenses, and the rate was generally kept below two shillings (10p) in the pound. By the end of the century, however, it had become the major source of government revenue. The rate of taxation rose dramatically during the two world wars, and during the Second (1944) the pay-as-you-earn system, deducting tax at the source of income, was introduced.

incoming /ˈɪnˌkʌmɪŋ/ adj. & n. ● adj. **1** coming in (*the incoming tide; incoming telephone calls*). **2** succeeding another person or persons (*the* *incoming tenant).* **3** immigrant. **4** (of profit) accruing. ● *n.* **1** (usu. in *pl.*) revenue, income. **2** the act of arriving or entering.

incommensurable /ˌɪnkəˈmenʃərəb(ə)l, -ˈmensjərəb(ə)l/ *adj. & n.* ● *adj.* **1** having no common standard of measurement; not comparable in respect of magnitude or value. **2** (foll. by *with*) not worthy of being compared with; utterly disproportionate to. **3** *Math.* **a** (often foll. by *with*) (of a magnitude or magnitudes) having no common factor, integral or fractional. **b** irrational. ● *n.* (usu. in *pl.*) an incommensurable quantity. □ **incommensurably** *adv.* **incommensurability** /-ˌmenʃərəˈbɪlɪtɪ, -ˌmensjərə-/ *n.* [LL *incommensurabilis* (as IN-[1], COMMENSURABLE)]

incommensurate /ˌɪnkəˈmenʃərət, -ˈmensjərət/ *adj.* **1** (often foll. by *with*, *to*) out of proportion; inadequate. **2** = INCOMMENSURABLE *adj.* □ **incommensurately** *adv.* **incommensurateness** *n.*

incommode /ˌɪnkəˈməʊd/ *v.tr.* **1** hinder, inconvenience. **2** trouble, annoy. [F *incommoder* or L *incommodare* (as IN-[1], *commodus* convenient)]

incommodious /ˌɪnkəˈməʊdɪəs/ *adj.* not affording good accommodation; uncomfortable. □ **incommodiously** *adv.* **incommodiousness** *n.*

incommunicable /ˌɪnkəˈmjuːnɪkəb(ə)l/ *adj.* **1** that cannot be communicated or shared. **2** that cannot be uttered or told. **3** that does not communicate; uncommunicative. □ **incommunicably** *adv.* **incommunicableness** *n.* **incommunicability** /-ˌmjuːnɪkəˈbɪlɪtɪ/ *n.* [LL *incommunicabilis* (as IN-[1], COMMUNICABLE)]

incommunicado /ˌɪnkəˌmjuːnɪˈkɑːdəʊ/ *adj.* **1** without or deprived of the means of communication with others. **2** (of a prisoner) held in solitary confinement. [Sp. *incomunicado* past part. of *incomunicar* deprive of communication]

incommunicative /ˌɪnkəˈmjuːnɪkətɪv/ *adj.* not communicative; taciturn. □ **incommunicatively** *adv.* **incommunicativeness** *n.*

incommutable /ˌɪnkəˈmjuːtəb(ə)l/ *adj.* **1** not changeable. **2** not commutable. □ **incommutably** *adv.* [ME f. L *incommutabilis* (as IN-[1], COMMUTABLE)]

incomparable /ɪnˈkɒmpərəb(ə)l/ *adj.* **1** without an equal; matchless. **2** (often foll. by *with*, *to*) not to be compared. □ **incomparably** *adv.* **incomparability** /-ˌkɒmpərəˈbɪlɪtɪ/ *n.* [ME f. OF f. L *incomparabilis* (as IN-[1], COMPARABLE)]

incompatible /ˌɪnkəmˈpætɪb(ə)l/ *adj.* **1** opposed in character; discordant. **2** (often foll. by *with*) inconsistent. **3** (of persons) unable to live, work, etc., together in harmony. **4** (of drugs) not suitable for taking at the same time. **5** (of equipment, machinery, etc.) not capable of being used in combination. □ **incompatibly** *adv.* **incompatibleness** *n.* **incompatibility** /-ˌpætɪˈbɪlɪtɪ/ *n.* [med.L *incompatibilis* (as IN-[1], COMPATIBLE)]

incompetent /ɪnˈkɒmpɪt(ə)nt/ *adj. & n.* ● *adj.* **1** (often foll. by *to* + infin.) not qualified or able to perform a particular task or function (*an incompetent builder*). **2** showing a lack of skill (*an incompetent performance*). **3** *Med.* (esp. of a valve or sphincter) not able to perform its function. ● *n.* an incompetent person. □ **incompetence** *n.* **incompetency** *n.* **incompetently** *adv.* [F *incompétent* or LL *incompetens* (as IN-[1], COMPETENT)]

incomplete /ˌɪnkəmˈpliːt/ *adj.* not complete. □ **incompletely** *adv.* **incompleteness** *n.* [ME f. LL *incompletus* (as IN-[1], COMPLETE)]

incompleteness theorem *n. Math.* see GÖDEL.

incomprehensible /ɪnˌkɒmprɪˈhensɪb(ə)l/ *adj.* (often foll. by *to*) that cannot be understood. □ **incomprehensibly** *adv.* **incomprehensibleness** *n.* **incomprehensibility** /-ˌhensɪˈbɪlɪtɪ/ *n.* [ME f. L *incomprehensibilis* (as IN-[1], COMPREHENSIBLE)]

incomprehension /ɪnˌkɒmprɪˈhenʃ(ə)n/ *n.* failure to understand.

incompressible /ˌɪnkəmˈpresɪb(ə)l/ *adj.* that cannot be compressed. □ **incompressibility** /-ˌpresɪˈbɪlɪtɪ/ *n.*

inconceivable /ˌɪnkənˈsiːvəb(ə)l/ *adj.* **1** that cannot be imagined. **2** *colloq.* very remarkable. □ **inconceivably** *adv.* **inconceivableness** *n.* **inconceivability** /-ˌsiːvəˈbɪlɪtɪ/ *n.*

inconclusive /ˌɪnkənˈkluːsɪv/ *adj.* (of an argument, evidence, or action) not decisive or convincing. □ **inconclusively** *adv.* **inconclusiveness** *n.*

incondensable /ˌɪnkənˈdensəb(ə)l/ *adj.* that cannot be condensed, esp. that cannot be reduced to a liquid or solid condition.

incongruous /ɪnˈkɒŋɡrʊəs/ *adj.* **1** out of place; absurd. **2** (often foll. by *with*) disagreeing; out of keeping. □ **incongruously** *adv.* **incongruousness** *n.* **incongruity** /ˌɪnkɒnˈɡruːɪtɪ/ *n.* (*pl.* **-ies**). [L *incongruus* (as IN-[1], CONGRUOUS)]

inconsecutive /ˌɪnkənˈsekjʊtɪv/ *adj.* lacking sequence; inconsequent. □ **inconsecutively** *adv.*

inconsequent /ɪnˈkɒnsɪkwənt/ *adj.* **1** not following naturally; irrelevant. **2** lacking logical sequence. **3** disconnected. □ **inconsequence** *n.* **inconsequently** *adv.* [L *inconsequens* (as IN-[1], CONSEQUENT)]

inconsequential /ɪnˌkɒnsɪˈkwenʃ(ə)l/ *adj.* **1** unimportant. **2** = INCONSEQUENT. □ **inconsequentially** *adv.* **inconsequentialness** *n.* **inconsequentiality** /-ˌkwenʃɪˈælɪtɪ/ *n.* (*pl.* **-ies**).

inconsiderable /ˌɪnkənˈsɪdərəb(ə)l/ *adj.* **1** of small size, value, etc. **2** not worth considering. [obs. F *inconsidérable* or LL *inconsiderabilis* (as IN-[1], CONSIDERABLE)]

inconsiderate /ˌɪnkənˈsɪdərət/ *adj.* **1** (of a person or action) thoughtless, rash. **2** lacking in regard for the feelings of others. □ **inconsiderately** *adv.* **inconsiderateness** *n.* **inconsideration** /-ˌsɪdəˈreɪʃ(ə)n/ *n.* [L *inconsideratus* (as IN-[1], CONSIDERATE)]

inconsistent /ˌɪnkənˈsɪst(ə)nt/ *adj.* **1** acting at variance with one's own principles or former conduct. **2** (often foll. by *with*) not in keeping; discordant, incompatible. **3** (of a single thing) incompatible or discordant; having self-contradictory parts. □ **inconsistency** *n.* (*pl.* **-ies**). **inconsistently** *adv.*

inconsolable /ˌɪnkənˈsəʊləb(ə)l/ *adj.* (of a person, grief, etc.) that cannot be consoled or comforted. □ **inconsolably** *adv.* **inconsolability** /-ˌsəʊləˈbɪlɪtɪ/ *n.* [F *inconsolable* or L *inconsolabilis* (as IN-[1], *consolabilis* f. *consolari* CONSOLE[1])]

inconsonant /ɪnˈkɒnsənənt/ *adj.* (often foll. by *with*, *to*) not harmonious; not compatible. □ **inconsonance** *n.* **inconsonantly** *adv.*

inconspicuous /ˌɪnkənˈspɪkjʊəs/ *adj.* **1** not conspicuous; not easily noticed. **2** *Bot.* (of flowers) small, pale, or green. □ **inconspicuously** *adv.* **inconspicuousness** *n.* [L *inconspicuus* (as IN-[1], CONSPICUOUS)]

inconstant /ɪnˈkɒnst(ə)nt/ *adj.* **1** (of a person) fickle, changeable. **2** frequently changing; variable, irregular. □ **inconstancy** *n.* (*pl.* **-ies**). **inconstantly** *adv.* [ME f. OF f. L *inconstans -antis* (as IN-[1], CONSTANT)]

incontestable /ˌɪnkənˈtestəb(ə)l/ *adj.* that cannot be disputed. □ **incontestably** *adv.* **incontestability** /-ˌtestəˈbɪlɪtɪ/ *n.* [F *incontestable* or med.L *incontestabilis* (as IN-[1], *contestabilis* f. L *contestari* CONTEST)]

incontinent /ɪnˈkɒntɪnənt/ *adj.* **1** unable to control movements of the bowels or bladder or both. **2** lacking self-restraint (esp. in regard to sexual desire). **3** (foll. by *of*) unable to control. □ **incontinence** *n.* **incontinently** *adv.* [ME f. OF or L *incontinens* (as IN-[1], CONTINENT[2])]

incontrovertible /ˌɪnkɒntrəˈvɜːtɪb(ə)l/ *adj.* indisputable, indubitable. □ **incontrovertibly** *adv.* **incontrovertibility** /-ˌvɜːtɪˈbɪlɪtɪ/ *n.*

inconvenience /ˌɪnkənˈviːnɪəns/ *n. & v.* ● *n.* **1** lack of suitability to personal requirements or ease. **2** cause or instance of this. ● *v.tr.* cause inconvenience to. [ME f. OF f. LL *inconvenientia* (as INCONVENIENT)]

inconvenient /ˌɪnkənˈviːnɪənt/ *adj.* **1** unfavourable to ease or comfort; not convenient. **2** awkward, troublesome. □ **inconveniently** *adv.* [ME f. OF f. L *inconveniens -entis* (as IN-[1], CONVENIENT)]

inconvertible /ˌɪnkənˈvɜːtɪb(ə)l/ *adj.* **1** not convertible. **2** (esp. of currency) not convertible into another form on demand. □ **inconvertibly** *adv.* **inconvertibility** /-ˌvɜːtɪˈbɪlɪtɪ/ *n.* [F *inconvertible* or LL *inconvertibilis* (as IN-[1], CONVERTIBLE)]

incoordination /ˌɪnkəʊˌɔːdɪˈneɪʃ(ə)n/ *n.* lack of coordination, esp. of muscular action.

incorporate *v. & adj.* ● *v.* /ɪnˈkɔːpəˌreɪt/ **1** *tr.* (often foll. by *in*, *with*) unite; form into one body or whole. **2** *intr.* become incorporated. **3** *tr.* combine (ingredients) into one substance. **4** *tr.* admit as a member of a company etc. **5** *tr.* **a** constitute as a legal corporation. **b** (as **incorporated** *adj.*) forming a legal corporation. ● *adj.* /ɪnˈkɔːpərət/ **1** (of a company etc.) formed into a legal corporation. **2** embodied. □ **incorporation** /-ˌkɔːpəˈreɪʃ(ə)n/ *n.* **incorporator** /-ˈkɔːpəˌreɪtə(r)/ *n.* [ME f. LL *incorporare* (as IN-[2], L *corpus -oris* body)]

incorporeal /ˌɪnkɔːˈpɔːrɪəl/ *adj.* **1** not composed of matter. **2** of immaterial beings. **3** *Law* having no physical existence. □ **incorporeally** *adv.* **incorporeality** /-ˌpɔːrɪˈælɪtɪ/ *n.* **incorporeity** /-pəˈriːɪtɪ/ *n.* [L *incorporeus* (as INCORPORATE)]

incorrect /ˌɪnkəˈrekt/ *adj.* **1** not in accordance with fact; wrong. **2** (of style etc.) improper, faulty. □ **incorrectly** *adv.* **incorrectness** *n.* [ME f. OF or L *incorrectus* (as IN-[1], CORRECT)]

incorrigible /ɪnˈkɒrɪdʒɪb(ə)l/ *adj.* **1** (of a person or habit) incurably bad or depraved. **2** not readily improved. □ **incorrigibly** *adv.*

incorrigibleness *n.* **incorrigibility** /-ˌkɒrɪdʒɪˈbɪlɪtɪ/ *n.* [ME f. OF *incorrigible* or L *incorrigibilis* (as IN-¹, CORRIGIBLE)]

incorruptible /ˌɪnkəˈrʌptɪb(ə)l/ *adj.* **1** that cannot be corrupted, esp. by bribery. **2** that cannot decay; everlasting. □ **incorruptibly** *adv.* **incorruptibility** /-ˌrʌptɪˈbɪlɪtɪ/ *n.* [ME f. OF *incorruptible* or eccl.L *incorruptibilis* (as IN-¹, CORRUPT)]

increase *v. & n.* ● *v.* /ɪnˈkriːs/ **1** *tr. & intr.* make or become greater in size, amount, etc., or more numerous. **2** *intr.* advance (in quality, attainment, etc.). **3** *tr.* intensify (a quality). ● *n.* /ˈɪnkriːs/ **1** the act or process of becoming greater or more numerous; growth; enlargement. **2** (of people, animals, or plants) growth in numbers; multiplication. **3** the amount or extent of an increase. □ **on the increase** increasing, esp. in frequency. □ **increasable** /ɪnˈkriːsəb(ə)l/ *adj.* **increasingly** *adv.* [ME f. OF *encreiss-* stem of *encreistre* f. L *increscere* (as IN-², *crescere* grow)]

incredible /ɪnˈkredɪb(ə)l/ *adj.* **1** that cannot be believed. **2** *colloq.* hard to believe; amazing. □ **incredibly** *adv.* **incredibility** /-ˌkredɪˈbɪlɪtɪ/ *n.* [ME f. L *incredibilis* (as IN-¹, CREDIBLE)]

incredulous /ɪnˈkredjʊləs/ *adj.* (often foll. by *of*) unwilling to believe. □ **incredulously** *adv.* **incredulousness** *n.* **incredulity** /ˌɪnkrɪˈdjuːlɪtɪ/ *n.* [L *incredulus* (as IN-¹, CREDULOUS)]

increment /ˈɪnkrɪmənt/ *n.* **1 a** an increase or addition, esp. one of a series on a fixed scale. **b** the amount of this. **2** *Math.* a small amount by which a variable quantity increases. □ **incremental** /ˌɪnkrɪˈment(ə)l/ *adj.* **incrementally** *adv.* [ME f. L *incrementum* f. *increscere* INCREASE]

incriminate /ɪnˈkrɪmɪˌneɪt/ *v.tr.* **1** (often as **incriminating** *adj.*) tend to prove the guilt of (*incriminating evidence*). **2** involve in an accusation. **3** charge with a crime. □ **incriminatory** /-nətərɪ/ *adj.* **incrimination** /-ˌkrɪmɪˈneɪʃ(ə)n/ *n.* [LL *incriminare* (as IN-², L *crimen* offence)]

incrust var. of ENCRUST.

incrustation /ˌɪnkrʌˈsteɪʃ(ə)n/ *n.* **1** the process of encrusting or state of being encrusted. **2** a crust or hard coating, esp. of fine material. **3** a concretion or deposit on a surface. **4** a facing of marble etc. on a building. [F *incrustation* or LL *incrustatio* (as ENCRUST)]

incubate /ˈɪŋkjʊˌbeɪt/ *v.* **1** *tr.* sit on or artificially heat (eggs) in order to bring forth young birds etc. **2** *tr.* cause the development of (bacteria etc.) by creating suitable conditions. **3** *intr.* sit on eggs; brood. **4** *tr. & intr.* develop slowly. [L *incubare* (as IN-², *cubare* cubit- or cubat- lie)]

incubation /ˌɪŋkjʊˈbeɪʃ(ə)n/ *n.* **1 a** the act of incubating. **b** brooding. **2** *Med.* **a** (in full **incubation period**) the period between exposure to an infection and the appearance of the first symptoms. **b** the processes occurring during this. □ **incubative** /ˈɪŋkjʊˌbeɪtɪv/ *adj.* **incubatory** *adj.* [L *incubatio* (as INCUBATE)]

incubator /ˈɪŋkjʊˌbeɪtə(r)/ *n.* **1** an apparatus used to provide a suitable temperature and environment for a premature baby or one of low birth-weight. **2** an apparatus used to hatch eggs or grow micro-organisms etc.

incubus /ˈɪŋkjʊbəs/ *n.* (*pl.* **incubi** /-ˌbaɪ/) **1** a male demon believed to have sexual intercourse with sleeping women. **2** a nightmare. **3** a person or thing that oppresses like a nightmare. [ME f. LL, = L *incubo* nightmare (as INCUBATE)]

incudes *pl.* of INCUS.

inculcate /ˈɪnkʌlˌkeɪt/ *v.tr.* (often foll. by *upon*, *in*) urge or impress (a fact, habit, or idea) persistently. □ **inculcator** *n.* **inculcation** /ˌɪnkʌlˈkeɪʃ(ə)n/ *n.* [L *inculcare* (as IN-², *calcare* tread f. *calx calcis* heel)]

inculpate /ˈɪnkʌlˌpeɪt/ *v.tr.* **1** involve in a charge. **2** accuse, blame. □ **inculpation** /ˌɪnkʌlˈpeɪʃ(ə)n/ *n.* **inculpative** /ɪnˈkʌlpətɪv/ *adj.* **inculpatory** *adj.* [LL *inculpare* (as IN-², *culpare* blame f. *culpa* fault)]

incumbency /ɪnˈkʌmbənsɪ/ *n.* (*pl.* **-ies**) the office, tenure, or sphere of an incumbent.

incumbent /ɪnˈkʌmb(ə)nt/ *adj. & n.* ● *adj.* **1** (foll. by *on*, *upon*) resting as a duty (*it is incumbent on you to warn them*). **2** (often foll. by *on*) lying, pressing. **3** in occupation or having the tenure of a post or position. ● *n.* the holder of an office or post, esp. an ecclesiastical benefice. [ME f. AL *incumbens* pres. part. of L *incumbere* lie upon (as IN-², *cubare* lie)]

incunable /ɪnˈkjuːnəb(ə)l/ *n.* = INCUNABULUM 1. [F, formed as INCUNABULUM]

incunabulum /ˌɪnkjʊˈnæbjʊləm/ *n.* (*pl.* **incunabula** /-lə/) **1** a book printed at an early date, esp. before 1501. **2** (in *pl.*) the early stages of the development of a thing. [L *incunabula* swaddling-clothes, cradle (as IN-², *cunae* cradle)]

incur /ɪnˈkɜː(r)/ *v.tr.* (**incurred**, **incurring**) suffer, experience, or become subject to (something unpleasant) as a result of one's own behaviour etc. (*incurred huge debts*). □ **incurrable** *adj.* [ME f. L *incurrere incurs-* (as IN-², *currere* run)]

incurable /ɪnˈkjʊərəb(ə)l/ *adj. & n.* ● *adj.* that cannot be cured. ● *n.* a person who cannot be cured. □ **incurably** *adv.* **incurableness** *n.* **incurability** /-ˌkjʊərəˈbɪlɪtɪ/ *n.* [ME f. OF *incurable* or LL *incurabilis* (as IN-¹, CURABLE)]

incurious /ɪnˈkjʊərɪəs/ *adj.* **1** lacking curiosity. **2** heedless, careless. □ **incuriously** *adv.* **incuriousness** *n.* **incuriosity** /-ˌkjʊərɪˈɒsɪtɪ/ *n.* [L *incuriosus* (as IN-¹, CURIOUS)]

incursion /ɪnˈkɜːʃ(ə)n, -ˈkɜːʒ(ə)n/ *n.* an invasion or attack, esp. when sudden or brief. □ **incursive** /-ˈkɜːsɪv/ *adj.* [ME f. L *incursio* (as INCUR)]

incurve /ɪnˈkɜːv/ *v.tr.* **1** bend into a curve. **2** (as **incurved** *adj.*) curved inwards. □ **incurvation** /ˌɪnkɜːˈveɪʃ(ə)n/ *n.* [L *incurvare* (as IN-², CURVE)]

incus /ˈɪŋkəs/ *n.* (*pl.* **incudes** /ɪnˈkjuːdiːz/) *Anat.* the small anvil-shaped bone in the middle ear, in contact with the malleus and stapes. Also called *anvil*. [L, = anvil]

incuse /ɪnˈkjuːz/ *n., v., & adj.* ● *n.* an impression hammered or stamped on a coin. ● *v.tr.* **1** mark (a coin) with a figure by stamping. **2** impress (a figure) on a coin by stamping. ● *adj.* hammered or stamped on a coin. [L *incusus* past part. of *incudere* (as IN-², *cudere* forge)]

Ind. *abbr.* **1** Independent. **2 a** India. **b** Indian. **3** Indiana.

indaba /ɪnˈdɑːbə/ *n.* *S. Afr.* **1** a conference between or with members of southern African native peoples. **2** *colloq.* one's problem or concern. [Zulu, = business]

Indebele *pl.* of NDEBELE.

indebted /ɪnˈdetɪd/ *adj.* (usu. foll. by *to*) **1** owing gratitude or obligation. **2** owing money. □ **indebtedness** *n.* [ME f. OF *endetté* past part. of *endetter* involve in debt (as EN-¹, *detter* f. *dette* DEBT)]

indecent /ɪnˈdiːs(ə)nt/ *adj.* **1** offending against recognized standards of decency. **2** unbecoming; highly unsuitable (*with indecent haste*). □ **indecent assault** a sexual attack not involving rape. **indecent exposure** the intentional act of publicly and indecently exposing one's body, esp. the genitals. □ **indecency** *n.* (*pl.* **-ies**). **indecently** *adv.* [F *indécent* or L *indecens* (as IN-¹, DECENT)]

indecipherable /ˌɪndɪˈsaɪfərəb(ə)l/ *adj.* that cannot be deciphered.

indecision /ˌɪndɪˈsɪʒ(ə)n/ *n.* lack of decision; hesitation. [F *indécision* (as IN-¹, DECISION)]

indecisive /ˌɪndɪˈsaɪsɪv/ *adj.* **1** not decisive. **2** undecided, hesitating. □ **indecisively** *adv.* **indecisiveness** *n.*

indeclinable /ˌɪndɪˈklaɪnəb(ə)l/ *adj.* *Gram.* **1** that cannot be declined. **2** having no inflections. [ME f. F *indéclinable* f. L *indeclinabilis* (as IN-¹, DECLINE)]

indecorous /ɪnˈdekərəs/ *adj.* **1** improper. **2** in bad taste. □ **indecorously** *adv.* **indecorousness** *n.* [L *indecorus* (as IN-¹, *decorus* seemly)]

indecorum /ˌɪndɪˈkɔːrəm/ *n.* **1** lack of decorum. **2** improper behaviour. [L, neut. of *indecorus*: see INDECOROUS]

indeed /ɪnˈdiːd/ *adv. & int.* ● *adv.* **1** in truth; really; yes, that is so (*they are, indeed, a remarkable family*). **2** expressing emphasis or intensification (*I shall be very glad indeed; indeed it is; very, indeed inordinately, proud of it*). **3** admittedly (*there are indeed exceptions*). **4** in point of fact (*if indeed such a thing is possible*). **5** expressing an approving or ironic echo (*who is this Mr Smith? — who is he indeed?*). ● *int.* expressing irony, contempt, incredulity, etc.

indefatigable /ˌɪndɪˈfætɪɡəb(ə)l/ *adj.* (of a person, quality, etc.) that cannot be tired out; unwearying, unremitting. □ **indefatigably** *adv.* **indefatigability** /-ˌfætɪɡəˈbɪlɪtɪ/ *n.* [obs. F *indéfatigable* or L *indefatigabilis* (as IN-¹, *defatigare* wear out)]

indefeasible /ˌɪndɪˈfiːzɪb(ə)l/ *adj.* *literary* (esp. of a claim, rights, etc.) unable to be forfeited or annulled. □ **indefeasibly** *adv.* **indefeasibility** /-ˌfiːzɪˈbɪlɪtɪ/ *n.*

indefectible /ˌɪndɪˈfektɪb(ə)l/ *adj.* **1** unfailing; not liable to defect or decay. **2** faultless. [IN-¹ + *defectible* f. LL *defectibilis* (as DEFECT)]

indefensible /ˌɪndɪˈfensɪb(ə)l/ *adj.* that cannot be defended or justified. □ **indefensibly** *adv.* **indefensibility** /-ˌfensɪˈbɪlɪtɪ/ *n.*

indefinable /ˌɪndɪˈfaɪnəb(ə)l/ *adj.* that cannot be defined or exactly described. □ **indefinably** *adv.*

indefinite /ɪnˈdefɪnɪt/ *adj.* **1** vague, undefined. **2** unlimited. **3** *Gram.* not determining the person, thing, time, etc., referred to. □ **indefinite article** see ARTICLE. **indefinite integral** see INTEGRAL. **indefinite pronoun** *Gram.* a pronoun indicating a person, amount, etc., without being definite or particular, e.g. *any, some, anyone.* □ **indefiniteness** *n.* [L *indefinitus* (as IN-¹, DEFINITE)]

indefinitely /ɪnˈdefɪnɪtlɪ/ adv. **1** for an unlimited time (*was postponed indefinitely*). **2** in an indefinite manner.

indehiscent /ˌɪndɪˈhɪs(ə)nt/ adj. Bot. (of fruit) not splitting open when ripe. □ **indehiscence** n.

indelible /ɪnˈdelɪb(ə)l/ adj. **1** that cannot be rubbed out or (in abstract senses) removed. **2** (of ink etc.) that makes indelible marks. □ **indelibly** adv. **indelibility** /-ˌdelɪˈbɪlɪtɪ/ n. [F *indélébile* or L *indelebilis* (as IN-¹, *delebilis* f. *delere* efface)]

indelicate /ɪnˈdelɪkət/ adj. **1** coarse, unrefined. **2** tactless. **3** tending to indecency. □ **indelicacy** n. (pl. **-ies**). **indelicately** adv.

indemnify /ɪnˈdemnɪˌfaɪ/ v.tr. (**-ies, -ied**) **1** (often foll. by *from*, *against*) protect or secure (a person) in respect of harm, a loss, etc. **2** (often foll. by *for*) secure (a person) against legal responsibility for actions. **3** (often foll. by *for*) compensate (a person) for a loss, expenses, etc. □ **indemnifier** n. **indemnification** /-ˌdemnɪfɪˈkeɪʃ(ə)n/ n. [L *indemnis* unhurt (as IN-¹, *damnum* loss, damage)]

indemnity /ɪnˈdemnɪtɪ/ n. (pl. **-ies**) **1 a** compensation for loss incurred. **b** a sum paid for this, esp. a sum exacted by a victor in war etc. as one condition of peace. **2** security against loss. **3** legal exemption from penalties etc. incurred. [ME f. F *indemnité* or LL *indemnitas -tatis* (as INDEMNIFY)]

indemonstrable /ɪnˈdemənstrəb(ə)l, ˌɪndɪˈmɒn-/ adj. that cannot be proved (esp. of primary or axiomatic truths).

indene /ˈɪndiːn/ n. Chem. a colourless flammable liquid hydrocarbon obtained from coal tar and used in making synthetic resins. [INDOLE + -ENE]

indent¹ v. & n. ● v. /ɪnˈdent/ **1** tr. start (a line of print or writing) further from the margin than other lines, e.g. to mark a new paragraph. **2** tr. **a** divide (a document drawn up in duplicate) into its two copies with a zigzag line dividing them and ensuring identification. **b** draw up (usu. a legal document) in exact duplicate. **3** Brit. **a** intr. (often foll. by *on*, *upon* a person, *for* a thing) make a requisition (orig. a written order with a duplicate). **b** tr. order (goods) by requisition. **4** tr. make toothlike notches in. **5** tr. form deep recesses in (a coastline etc.). ● n. /ˈɪndent/ **1** Brit. **a** an order (esp. from abroad) for goods. **b** an official requisition for stores. **2** an indented line. **3** indentation. **4** an indenture. □ **indenter** /ɪnˈdentə(r)/ n. **indentor** n. [ME f. AF *endenter* f. AL *indentare* (as IN-², L *dens dentis* tooth)]

indent² /ɪnˈdent/ v.tr. **1** make a dent in. **2** impress (a mark etc.). [ME f. IN-² + DENT]

indentation /ˌɪndenˈteɪʃ(ə)n/ n. **1** the act or an instance of indenting; the process of being indented. **2** a cut or notch. **3** a zigzag. **4** a deep recess in a coastline etc.

indention /ɪnˈdenʃ(ə)n/ n. **1** the indenting of a line in printing or writing. **2** = INDENTATION.

indenture /ɪnˈdentʃə(r)/ n. & v. ● n. **1** an indented document (see INDENT¹ v. 2). **2** (usu. in pl.) a sealed agreement or contract. **3** a formal list, certificate, etc. ● v.tr. hist. bind (a person) by indentures, esp. as an apprentice. □ **indentureship** n. [ME (orig. Sc.) f. AF *endenture* (as INDENT¹)]

independence /ˌɪndɪˈpendəns/ n. **1 a** (often foll. by *of*, *from*) the state of being independent. **b** the act or process of becoming independent (*a new constitution was introduced after independence*). **2** independent income.

Independence Day n. a day celebrating the anniversary of national independence, esp. 4 July in the US.

Independence Hall a building in Philadelphia where the US Declaration of Independence was proclaimed and outside which the Liberty Bell is kept.

independency /ˌɪndɪˈpendənsɪ/ n. (pl. **-ies**) **1** an independent state. **2** = INDEPENDENCE.

independent /ˌɪndɪˈpendənt/ adj. & n. ● adj. **1 a** (often foll. by *of*) not depending on authority or control. **b** self-governing. **2 a** not depending on another person for one's opinion or livelihood. **b** (of income or resources) making it unnecessary to earn one's living. **3** unwilling to be under an obligation to others. **4** Polit. not belonging to or supported by a party. **5** not depending on something else for its validity, efficiency, value, etc. (*independent proof*). **6** (of broadcasting, a school, etc.) not supported by public funds. **7** (**Independent**) hist. Congregational. ● n. **1** a person who is politically independent. **2** (**Independent**) hist. a Congregationalist. □ **independent school** (in the UK) a school that receives no grant from the government and is not subject to the control of a local authority. □ **independently** adv.

Independent Broadcasting Authority (abbr. **IBA**) hist. (in the UK) the body responsible for regulating commercial television and radio, until its replacement in 1991 by the Independent Television Commission and the Radio Authority. It had taken the place of the Independent Television Authority in 1972.

Independent Labour Party (abbr. **ILP**) a British socialist political party formed in 1893 under the leadership of Keir Hardie. In 1900 it became one of the constituent groups of the Labour Representation Committee and was instrumental in the formation of the Labour Party in 1906, but tension between the two parties grew in the 1930s over the questions of pacifism and support for Communism, and by the early 1950s the Independent Labour Party had lost all its parliamentary representation.

Independent Television Authority (abbr. **ITA**) hist. (in the UK) the predecessor of the Independent Broadcasting Authority, founded in 1954, replaced in 1972.

Independent Television Commission (abbr. **ITC**) (in the UK) an organization responsible for licensing and regulating commercial television. It replaced the IBA in 1991.

in-depth see DEPTH.

indescribable /ˌɪndɪˈskraɪbəb(ə)l/ adj. **1** too unusual or extreme to be described. **2** vague, indefinite. □ **indescribably** adv. **indescribability** /-ˌskraɪbəˈbɪlɪtɪ/ n.

indestructible /ˌɪndɪˈstrʌktɪb(ə)l/ adj. that cannot be destroyed. □ **indestructibly** adv. **indestructibility** /-ˌstrʌktɪˈbɪlɪtɪ/ n.

indeterminable /ˌɪndɪˈtɜːmɪnəb(ə)l/ adj. **1** that cannot be ascertained. **2** (of a dispute etc.) that cannot be settled. □ **indeterminably** adv. [ME f. LL *indeterminabilis* (as IN-¹, L *determinare* DETERMINE)]

indeterminate /ˌɪndɪˈtɜːmɪnət/ adj. **1** not fixed in extent, character, etc. **2** left doubtful; vague. **3** Math. (of a quantity) not limited to a fixed value by the value of another quantity. **4** (of a judicial sentence) such that the convicted person's conduct determines the date of release. □ **indeterminate vowel** the obscure vowel /ə/ heard in 'a moment ago'; a schwa. □ **indeterminacy** n. **indeterminately** adv. **indeterminateness** n. [ME f. LL *indeterminatus* (as IN-¹, DETERMINATE)]

indetermination /ˌɪndɪˌtɜːmɪˈneɪʃ(ə)n/ n. **1** lack of determination. **2** the state of being indeterminate.

indeterminism /ˌɪndɪˈtɜːmɪˌnɪz(ə)m/ n. the belief that human action is not wholly determined by motives. □ **indeterminist** n. **indeterministic** /-ˌtɜːmɪˈnɪstɪk/ adj.

index /ˈɪndeks/ n. & v. ● n. (pl. **indexes** or esp. in technical use **indices** /ˈɪndɪˌsiːz/) **1** an alphabetical list of names, subjects, etc., with references, usu. at the end of a book. **2** = *card index*. **3** (in full **index number**) a number showing the variation of prices or wages as compared with a chosen base period (*retail price index*; *Dow-Jones index*). **4** Math. **a** the exponent of a number. **b** the power to which it is raised. **5 a** a pointer, esp. on an instrument, showing a quantity, a position on a scale, etc. **b** an indicator of a trend, direction, tendency, etc. **c** (usu. foll. by *of*) a sign, token, or indication of something. **6** Physics a number expressing a physical property etc. in terms of a standard (*refractive index*). **7** Computing a set of items each of which specifies one of the records of a file and contains information about its address. **8** (**Index**) see INDEX LIBRORUM PROHIBITORUM. **9** Printing a symbol shaped like a pointing hand, used to draw attention to a note etc. ● v.tr. **1** provide (a book etc.) with an index. **2** enter in an index. **3** relate (wages etc.) to the value of a price index. □ **index finger** the forefinger.

index-linked related to the value of a retail price index. □ **indexer** n. **indexless** adj. **indexible** /ˈɪndeksɪb(ə)l/, ɪnˈdeks-/ adj. **indexical** /ɪnˈdeksɪk(ə)l/ adj. **indexation** /ˌɪndekˈseɪʃ(ə)n/ n. [ME f. L *index indicis* forefinger, informer, sign]

Index Librorum Prohibitorum /lɪˌbrɔːrʊm prəʊˌhɪbɪˈtɔːrʊm/ an official list of books which Roman Catholics were forbidden to read or which were to be read only in expurgated editions, as contrary to Catholic faith or morals. The first Index was issued in 1557; it was revised at intervals until abolished in 1966. [L, = index of forbidden books]

India /ˈɪndɪə/ (called in Hindi *Bharat*) a country in southern Asia occupying the greater part of the Indian subcontinent; pop. (est. 1991) 859,200,000; official languages, Hindi, English; capital, New Delhi. Of the many other languages spoken in India, fourteen are recognized as official in certain regions; of these, Bengali, Gujarati, Marathi, Tamil, Telugu, and Urdu have most first-language speakers. The history of the subcontinent began in the 3rd millennium BC, when the Indus valley was the site of a fully developed civilization. This collapsed c.1760 BC, when Aryan invaders spread from the west through the northern part of the country. Consolidated first within the Buddhist empire of Asoka and then the Hindu empire of the Gupta dynasty, much of India was united under a Muslim sultanate based on Delhi from the 12th century until incorporated in the Mogul empire by Babur and Akbar the Great in the 16th century. The decline of Mogul power in the late 17th and early 18th centuries coincided with increasing European penetration, with Britain eventually triumphing over her colonial rivals. British interest had begun in the 17th century with the formation of the East India Company, which in 1765 acquired the right to administer Bengal and afterwards other parts (see BRITISH INDIA). In 1858, after the Indian Mutiny, the Crown took over the Company's authority, and in 1876 Queen Victoria was proclaimed Empress of India. Limited autonomy was granted under the provisions of the Government of India Act (1919); a system of diarchy was established whereby British governors headed provincial governments consisting of British-appointed councillors and Indian ministers selected by governors from locally elected officials. This system met with opposition, led by Mahatma Gandhi, and was replaced in 1935 by full provincial autonomy. However, rising nationalism eventually resulted in independence in 1947, at which time India was partitioned, Pakistan being created from mainly Muslim territories in the north-east (now Bangladesh) and the north-west. A member of the Commonwealth, India is the second most populous country in the world.

India ink n. esp. N. Amer. = Indian ink. [INDIA: see INDIAN]

Indiaman /ˈɪndɪəmən/ n. (pl. **-men**) Naut. hist. a ship engaged in trade with India or the East Indies.

Indian /ˈɪndɪən/ n. & adj. ● n. **1 a** a native or national of India. **b** a person of Indian descent. **2** an American Indian or Native American. ● adj. **1** of or relating to India, or to the subcontinent comprising India, Pakistan, and Bangladesh. **2** of or relating to American Indians or Native Americans. □ **Indian clubs** a pair of bottle-shaped clubs swung to exercise the arms in gymnastics. **Indian corn** maize. **Indian elephant**, Elephas maximus, of India (see ELEPHANT). **Indian file** = single file. **Indian hemp** see HEMP 1. **Indian ink** Brit. **1** a black pigment made orig. in China and Japan. **2** a dark ink made from this, used esp. in drawing and technical graphics. **Indian rope-trick** the supposed Indian feat of climbing an upright unsupported length of rope. **Indian summer 1** a period of unusually dry warm weather sometimes occurring in late autumn. **2** a late period of life characterized by comparative calm. [ME f. INDIA ult. f. Gk Indos the River Indus f. Pers. Hind: cf. HINDU]

Indiana /ˌɪndɪˈænə/ a state in the Middle West of the US; pop. (1990) 5,544,160; capital, Indianapolis. It was colonized by the French in the 18th century and ceded to Britain in 1763. It passed to the US in 1783 and became the 19th state in 1816.

Indianapolis /ˌɪndɪəˈnæpəlɪs/ the state capital of Indiana; pop. (1990) 741,950. The city hosts an annual 500-mile (804.5-km) motor-race, known as the Indy 500.

Indian Mutiny (also called Sepoy Mutiny) a revolt of Indians against British rule, 1857–8. At a time when the number of British troops in India had reached a low point and the ruling East India Company was almost totally dependent on Indian soldiers (sepoys), discontent with British administration resulted in widespread mutinies in British garrison towns, with accompanying massacres of white soldiers and inhabitants. After a series of sieges (most notably that of Lucknow) and battles, the revolt was put down; it was followed by the institution of direct rule by the British Crown in place of the East India Company administration.

Indian National Congress a broad-based political party in India, founded in 1885. It developed a powerful central organization under Mahatma Gandhi in the 1920s and dominated the independence movement in the following decade; it has been the principal party in government since independence in 1947. Following splits in the party the Indian National Congress (I), formed by Indira Gandhi as a breakaway group, was confirmed in 1981 as the official Congress party.

Indian Ocean the ocean to the south of India, extending from the east coast of Africa to the East Indies and Australia.

Indian subcontinent the part of Asia south of the Himalayas which forms a peninsula extending into the Indian Ocean, between the Arabian Sea and the Bay of Bengal. Historically forming the whole territory of greater India, the region is now divided between India, Pakistan, and Bangladesh. Geologically, the Indian subcontinent is a distinct unit, formerly part of the ancient supercontinent of Gondwana. As a result of continental drift it became joined to the rest

of Asia, perhaps as recently as 40 million years ago, in a collision which created the Himalayas.

India paper n. **1** a soft absorbent kind of paper orig. imported from China, used for proofs of engravings. **2** a very thin tough opaque printing-paper.

indiarubber /ˌɪndɪəˈrʌbə(r)/ n. = RUBBER[1] 2.

Indic /ˈɪndɪk/ adj. & n. (also called Indo-Aryan) ● adj. of or relating to a group of Indo-European languages comprising Sanskrit and the modern Indian languages which are its descendants. ● n. this group of languages. [L Indicus f. Gk Indikós INDIAN]

indicate /ˈɪndɪˌkeɪt/ v. (often foll. by that + clause) **1** tr. point out; make known; show. **2** tr. be a sign or symptom of; express the presence of. **3** tr. (often in passive) suggest; call for; require or show to be necessary (stronger measures are indicated). **4** tr. admit to or state briefly (indicated his disapproval). **5** tr. (of a gauge etc.) give as a reading. **6** intr. signal one's intention to turn etc. using an indicator. [L indicare (as IN-[2], dicare make known)]

indication /ˌɪndɪˈkeɪʃ(ə)n/ n. **1 a** the act or an instance of indicating. **b** something that suggests or indicates; a sign or symptom. **2** something indicated or suggested; esp. in Med., a remedy or treatment that is suggested by the symptoms. **3** a reading given by a gauge or instrument. [F f. L indicatio (as INDICATE)]

indicative /ɪnˈdɪkətɪv/ adj. & n. ● adj. **1** (foll. by of) suggestive; serving as an indication. **2** Gram. (of a mood) denoting simple statement of a fact. ● n. Gram. **1** the indicative mood. **2** a verb in this mood. □ **indicatively** adv. [ME f. F indicatif -ive f. LL indicativus (as INDICATE)]

indicator /ˈɪndɪˌkeɪtə(r)/ n. **1** a person or thing that indicates. **2** a device indicating the condition of a machine etc. **3** a recording instrument attached to an apparatus etc. **4** a board in a railway station etc. giving current information. **5** a device (esp. a flashing light) on a vehicle to show that it is about to change direction. **6** esp. Chem. a substance which changes to a characteristic colour in the presence of a particular concentration of an ion, e.g. to show the degree of acidity. **7** Physics & Med. a radioactive tracer. **8** Ecol. a species or group which acts as a sign of particular environmental conditions.

indicatory /ɪnˈdɪkətərɪ/ adj. = INDICATIVE adj. 1.

indices pl. of INDEX.

indicia /ɪnˈdɪʃɪə/ n.pl. **1** distinguishing marks. **2** signs, indications. [pl. of L indicium (as INDEX)]

indicial /ɪnˈdɪʃ(ə)l/ adj. **1** of the nature or form of an index. **2** of the nature of indicia; indicative.

indict /ɪnˈdaɪt/ v.tr. accuse (a person) formally by legal process. □ **indicter** n. **indictee** /ˌɪndaɪˈtiː/ n. [ME f. AF enditer indict f. OF enditier declare f. Rmc indictare (unrecorded: as IN-[2], DICTATE)]

indictable /ɪnˈdaɪtəb(ə)l/ adj. **1** (of an offence) rendering the person who commits it liable to be charged with a crime. **2** (of a person) so liable.

indictment /ɪnˈdaɪtmənt/ n. **1** the act of indicting. **2 a** a formal accusation. **b** a legal process in which this is made. **c** a document containing a charge. **3** something that serves to condemn or censure. [ME f. AF enditement (as INDICT)]

indie /ˈɪndɪ/ n. & adj. colloq. ● n. **1** an independent record or film company. **2** a musician or band whose music is recorded by an independent company. ● adj. **1** (of a pop group or record label) independent, not belonging to one of the major recording companies. **2** characteristic of the deliberately unpolished or uncommercialized style of indie bands. [abbr. of INDEPENDENT]

Indies /ˈɪndɪz/ n.pl. (prec. by the) archaic India and adjacent regions. □ **East Indies**, **West Indies** see separate entries. [pl. of obs. Indy India]

indifference /ɪnˈdɪfrəns/ n. **1** lack of interest or attention. **2** unimportance (a matter of indifference). **3** neutrality. [L indifferentia (as INDIFFERENT)]

indifferent /ɪnˈdɪfrənt/ adj. **1** neither good nor bad; average, mediocre. **2 a** not especially good. **b** fairly bad. **3** (often prec. by very) decidedly inferior. **4** (foll. by to) having no partiality for or against; having no interest in or sympathy for. **5** chemically, magnetically, etc., neutral. □ **indifferently** adv. [ME f. OF indifferent or L indifferens (as IN-[1], DIFFERENT)]

indifferentism /ɪnˈdɪfrənˌtɪz(ə)m/ n. an attitude of indifference, esp. in religious matters. □ **indifferentist** n.

indigenize /ɪnˈdɪdʒɪˌnaɪz/ v.tr. (also **-ise**) **1** make indigenous; subject to native influence. **2** subject to increased use of indigenous people in government etc. □ **indigenization** /-ˌdɪdʒɪnaɪˈzeɪʃ(ə)n/ n.

indigenous /ɪnˈdɪdʒɪnəs/ adj. **1 a** (esp. of flora or fauna) originating naturally in a region. **b** (of people) born in a region. **2** (foll. by to) belonging naturally to a place. □ **indigenously** adv. **indigenousness** n. [L indigena f. indi- = IN-² + gen- be born]

indigent /ˈɪndɪdʒənt/ adj. needy, poor. □ **indigence** n. [ME f. OF f. LL indigere f. indi- = IN-² + egere need]

indigested /ˌɪndaɪˈdʒestɪd, ˌɪndɪ-/ adj. **1** shapeless. **2** ill-considered. **3** not digested.

indigestible /ˌɪndɪˈdʒestɪb(ə)l/ adj. **1** difficult or impossible to digest. **2** too complex or awkward to read or comprehend easily. □ **indigestibly** adv. **indigestibility** /-ˌdʒestɪˈbɪlɪtɪ/ n. [F indigestible or LL indigestibilis (as IN-¹, DIGEST)]

indigestion /ˌɪndɪˈdʒestʃən/ n. **1** difficulty in digesting food. **2** pain or discomfort caused by this. □ **indigestive** /-ˈdʒestɪv/ adj. [ME f. OF indigestion or LL indigestio (as IN-¹, DIGESTION)]

Indigirka /ˌɪndɪˈɡɪəkə/ a river of far eastern Siberia, which flows northwards for 1,779 km (1,112 miles) to the Arctic Ocean, where it forms a wide delta.

indignant /ɪnˈdɪɡnənt/ adj. feeling or showing scornful anger or a sense of injured innocence. □ **indignantly** adv. [L indignari indignant-regard as unworthy (as IN-¹, dignus worthy)]

indignation /ˌɪndɪɡˈneɪʃ(ə)n/ n. scornful anger at supposed unjust or unfair conduct or treatment. [ME f. OF indignation or L indignatio (as INDIGNANT)]

indignity /ɪnˈdɪɡnɪtɪ/ n. (pl. **-ies**) **1** unworthy treatment. **2** a slight or insult. **3** the humiliating quality of something (the indignity of my position). [F indignité or L indignitas (as INDIGNANT)]

indigo /ˈɪndɪˌɡəʊ/ n. (pl. **-os**) **1 a** a natural blue dye obtained from the indigo plant. **b** a synthetic form of this dye. **2** a tropical leguminous plant of the genus Indigofera. **3** (in full **indigo blue**) a colour between blue and violet in the spectrum. □ **indigo snake** a large non-venomous blue-black, brown, or particoloured snake, Drymarchon corais, native to America (also called cribo, gopher snake). □ **indigotic** /ˌɪndɪˈɡɒtɪk/ adj. [16th-c. indico (f. Sp.), indigo (f. Port.) f. L indicum f. Gk indikon INDIAN (dye)]

Indira Gandhi Canal /ɪnˌdɪərə ˈɡændɪ, ˌɪndərə/ (formerly called Rajasthan Canal) a massive canal in NW India, bringing water to the Thar Desert of Rajasthan from the Harike Barrage on the Sutlej river. The canal, which is 650 km (406 miles) long, was completed in 1986.

indirect /ˌɪndaɪˈrekt, ˌɪndɪ-/ adj. **1** not going straight to the point. **2** (of a route etc.) not straight. **3** not directly sought or aimed at (an indirect result). **4** (of lighting) from a concealed source and diffusely reflected. □ **indirect object** Gram. a person or thing affected by a verbal action but not primarily acted on (e.g. him in give him the book). **indirect question** Gram. a question in reported speech (e.g. they asked who I was). **indirect speech** (or **oration**) = reported speech (see REPORT). **indirect tax** a tax levied on goods and services and not on income or profits. □ **indirectly** adv. **indirectness** n. [ME f. OF indirect or med.L indirectus (as IN-¹, DIRECT)]

indiscernible /ˌɪndɪˈsɜːnɪb(ə)l/ adj. that cannot be discerned or distinguished from another. □ **indiscernibly** adv. **indiscernibility** /-ˌsɜːnɪˈbɪlɪtɪ/ n.

indiscipline /ɪnˈdɪsɪplɪn/ n. lack of discipline.

indiscreet /ˌɪndɪˈskriːt/ adj. **1** not discreet; revealing secrets. **2** injudicious, unwary. □ **indiscreetly** adv. **indiscreetness** n. [ME f. LL indiscretus (as IN-¹, DISCREET)]

indiscrete /ˌɪndɪˈskriːt/ adj. not divided into distinct parts. [L indiscretus (as IN-¹, DISCRETE)]

indiscretion /ˌɪndɪˈskreʃ(ə)n/ n. **1** lack of discretion; indiscreet conduct. **2** an indiscreet action, remark, etc. [ME f. OF indiscretion or LL indiscretio (as IN-¹, DISCRETION)]

indiscriminate /ˌɪndɪˈskrɪmɪnət/ adj. **1** making no distinctions. **2** confused, promiscuous. □ **indiscriminately** adv. **indiscriminateness** n. **indiscriminative** adj. **indiscrimination** /-ˌskrɪmɪˈneɪʃ(ə)n/ n. [IN-¹ + discriminate (adj.) f. L discriminatus past part. (as DISCRIMINATE)]

indispensable /ˌɪndɪˈspensəb(ə)l/ adj. **1** (often foll. by to, for) that cannot be dispensed with; necessary. **2** (of a law, duty, etc.) that is not to be set aside. □ **indispensably** adv. **indispensableness** n. **indispensability** /-ˌspensəˈbɪlɪtɪ/ n. [med.L indispensabilis (as IN-¹, DISPENSABLE)]

indispose /ˌɪndɪˈspəʊz/ v.tr. **1** (often foll. by for, or to + infin.) make unfit or unable. **2** (often foll. by towards, from, or to + infin.) make averse.

indisposed /ˌɪndɪˈspəʊzd/ adj. **1** slightly unwell. **2** averse or unwilling.

indisposition /ˌɪndɪspəˈzɪʃ(ə)n/ n. **1** ill health, a slight or temporary ailment. **2** disinclination. **3** aversion. [F indisposition or IN-¹ + DISPOSITION]

indisputable /ˌɪndɪˈspjuːtəb(ə)l/ adj. **1** that cannot be disputed. **2** unquestionable. □ **indisputably** adv. **indisputability** /-ˌspjuːtəˈbɪlɪtɪ/ n. [LL indisputabilis (as IN-¹, DISPUTABLE)]

indissolubilist /ˌɪndɪˈsɒljʊbɪlɪst/ n. & adj. ● n. (in the Church of England) a person who believes that the Church should not remarry divorcees. ● adj. of or holding this belief.

indissoluble /ˌɪndɪˈsɒljʊb(ə)l/ adj. **1** that cannot be dissolved or decomposed. **2** lasting, stable (an indissoluble bond). □ **indissolubly** adv. **indissolubility** /-ˌsɒljʊˈbɪlɪtɪ/ n. [L indissolubilis (as IN-¹, DISSOLUBLE)]

indistinct /ˌɪndɪˈstɪŋkt/ adj. **1** not distinct. **2** confused, obscure. □ **indistinctly** adv. **indistinctness** n. [ME f. L indistinctus (as IN-¹, DISTINCT)]

indistinctive /ˌɪndɪˈstɪŋktɪv/ adj. not having distinctive features. □ **indistinctively** adv. **indistinctiveness** n.

indistinguishable /ˌɪndɪˈstɪŋgwɪʃəb(ə)l/ adj. (often foll. by from) not distinguishable. □ **indistinguishableness** n. **indistinguishably** adv.

indite /ɪnˈdaɪt/ v.tr. formal or joc. **1** put (a speech etc.) into words. **2** write (a letter etc.). [ME f. OF enditier: see INDICT]

indium /ˈɪndɪəm/ n. a soft silvery-white metallic chemical element (atomic number 49; symbol **In**). Indium was discovered spectroscopically by the German physicists Ferdinand Reich (1799–1882) and (Hieronymus) Theodor Richter (1824–98) in 1863. A rare element, it mainly occurs in association with zinc and other metals. It is used in semiconductor devices and in alloys of low melting-point. [L indicum indigo with ref. to its characteristic spectral lines]

indivertible /ˌɪndɪˈvɜːtɪb(ə)l/ adj. that cannot be turned aside. □ **indivertibly** adv.

individual /ˌɪndɪˈvɪdjʊəl/ adj. & n. ● adj. **1** single. **2** particular, special; not general. **3** having a distinct character. **4** characteristic of a particular person. **5** designed for use by one person. ● n. **1** a single member of a class. **2** a single human being as distinct from a family or group. **3** colloq. a person (a most unpleasant individual). **4** a distinctive person. [ME, = indivisible, f. med.L individualis (as IN-¹, dividuus f. dividere DIVIDE)]

individualism /ˌɪndɪˈvɪdjʊəˌlɪz(ə)m/ n. **1** the habit or principle of being independent and self-reliant. **2** a social theory favouring the free action of individuals. **3** self-centred feeling or conduct; egoism. □ **individualist** n. **individualistic** /-ˌvɪdjʊəˈlɪstɪk/ adj. **individualistically** adv.

individuality /ˌɪndɪˌvɪdjʊˈælɪtɪ/ n. (pl. **-ies**) **1** individual character, esp. when strongly marked. **2** (in pl.) individual tastes etc. **3** separate existence.

individualize /ˌɪndɪˈvɪdjʊəˌlaɪz/ v.tr. (also **-ise**) **1** give an individual character to. **2** specify. (esp. as **individualized** adj.) personalize or tailor to suit the individual (individualized notepaper; individualized training course). □ **individualization** /-ˌvɪdjʊəlaɪˈzeɪʃ(ə)n/ n.

individually /ˌɪndɪˈvɪdjʊəlɪ/ adv. **1** personally; in an individual capacity. **2** in a distinctive manner. **3** one by one; not collectively.

individuate /ˌɪndɪˈvɪdjʊˌeɪt/ v.tr. individualize; form into an individual. □ **individuation** /-ˌvɪdjʊˈeɪʃ(ə)n/ n. [med.L individuare (as INDIVIDUAL)]

indivisible /ˌɪndɪˈvɪzɪb(ə)l/ adj. **1** not divisible. **2** not distributable among a number. □ **indivisibly** adv. **indivisibility** /-ˌvɪzɪˈbɪlɪtɪ/ n. [ME f. LL indivisibilis (as IN-¹, DIVISIBLE)]

Indo- /ˈɪndəʊ/ comb. form Indian; Indian and. [L Indus f. Gk Indos]

Indo-Aryan /ˌɪndəʊˈeərɪən/ n. & adj. ● n. **1** a member of any of the Aryan peoples of India. **2** the Indic group of languages. ● adj. of or relating to the Indo-Aryans or Indo-Aryan.

Indo-China /ˌɪndəʊˈtʃaɪnə/ the peninsula of SE Asia containing Burma (Myanmar), Thailand, Malaya, Laos, Cambodia, and Vietnam; especially, the part of this area consisting of Laos, Cambodia, and Vietnam, which was a French dependency (French Indo-China) from 1862 to 1954. □ **Indo-Chinese** /-tʃaɪˈniːz/ adj. & n.

indocile /ɪnˈdəʊsaɪl/ adj. not docile. □ **indocility** /ˌɪndəʊˈsɪlɪtɪ/ n. [F indocile or L indocilis (as IN-¹, DOCILE)]

indoctrinate /ɪnˈdɒktrɪˌneɪt/ v.tr. **1** teach (a person or group) systematically or for a long period to accept (esp. partisan or tendentious) ideas uncritically. **2** teach, instruct. □ **indoctrinator** n.

indoctrination /-ˌdɒktrɪ'neɪʃ(ə)n/ n. [IN-² + DOCTRINE + -ATE³]

Indo-European /ˌɪndəʊˌjʊərə'pɪən/ adj. & n. ● adj. of or relating to the family of languages spoken for at least the last 3,000 years over the greater part of Europe and extending into Asia as far as northern India. Also (now less usually) called *Indo-Germanic* or *Aryan*. (*See note below.*) ● n. **1** the Indo-European family of languages. **2** (usu. in pl.) a speaker of an Indo-European language, esp. the ancestral form.

▪ The Indo-European languages are thought to have originated in a language which has been called *Proto-Indo-European*, spoken at some distant time in the past, considerably before 2000 BC. No records of this language survive, but its existence can be inferred from a comparison of the Indo-European languages and most of its phonology and morphology and some of its vocabulary can be reconstructed at a theoretical level. The main divisions into which it eventually split are the Indo-Iranian branch, the Hellenic branch or Greek, the Italic branch (including Latin, from which came French and the other Romance languages), the Germanic languages (to which English belongs), the Celtic branch, the Baltic languages, and the Slavonic languages. In addition to these, Albanian forms a distinct member of the family, as does Armenian. Two important discoveries of the 20th century have added to the family the ancient Anatolian languages, such as Hittite (the oldest attested Indo-European language, dating from the 2nd millennium BC), and Tocharian. Recognition of the breadth of this language family came about when Sanskrit was first studied by Europeans in the late 18th and early 19th centuries and the strong affinity that it bore to Greek and Latin was noticed. Most of the research on which the language groupings and the reconstruction of the parent language are based was the work of German scholars in the 19th century.

Indo-Germanic /ˌɪndəʊdʒɜː'mænɪk/ adj. & n. = INDO-EUROPEAN.

Indo-Iranian /ˌɪndəʊɪ'reɪnɪən/ adj. & n. ● adj. of or relating to a branch of Indo-European languages spoken in northern India and Iran. The group can be divided into the Indic (or Indo-Aryan) group and the Iranian. Also (now less usually) called *Aryan*. ● n. the Indo-Iranian group of languages.

indole /'ɪndəʊl/ n. *Chem.* an organic compound with a characteristic odour formed on the reduction of indigo. [INDIGO + L oleum oil]

indoleacetic acid /ˌɪndəʊlə'siːtɪk/ n. (abbr. **IAA**) *Biochem.* any of a group of isomeric acetic acid derivatives of indole, esp. one found as a natural growth hormone (auxin) in plants. [INDOLE + ACETIC]

indolent /'ɪndələnt/ adj. **1** lazy; wishing to avoid activity or exertion. **2** *Med.* causing no pain (*an indolent tumour*). □ **indolence** n. **indolently** adv. [LL *indolens* (as IN-¹, *dolere* suffer pain)]

Indology /ɪn'dɒlədʒɪ/ n. the study of Indian history, literature, etc. □ **Indologist** n.

indomitable /ɪn'dɒmɪtəb(ə)l/ adj. **1** that cannot be subdued; unyielding. **2** stubbornly persistent. □ **indomitably** adv. **indomitableness** n. **indomitability** /-ˌdɒmɪtə'bɪlɪtɪ/ n. [LL *indomitabilis* (as IN-¹, L *domitare* tame)]

Indonesia /ˌɪndə'niːzɪə, -'niːʒə, -'niːʃə/ a SE Asian country consisting of many islands in the Malay Archipelago; pop. (est. 1990) 184,300,000; languages, Indonesian (official), Malay, Balinese, Chinese, Javanese, and others; capital, Djakarta (on Java). Indonesia consists of the territories of the former Dutch East Indies, of which the largest are Java, Sumatra, southern Borneo, western New Guinea, the Moluccas, and Sulawesi. Originally settled by Malayo-Polynesian peoples, Indonesia came under Hindu and Buddhist, and later Islamic, influence from India, especially in the west. By the 16th century, it was divided into small, often competing, territories under the control of local rulers. The European colonial powers rapidly became influential, with the Dutch establishing control by the late 17th century. The region was occupied by the Japanese in 1942. On liberation, Indonesia was proclaimed a republic by nationalists. After trying unsuccessfully to regain control, the Dutch agreed to Indonesian independence in 1949, although western New Guinea (Irian Jaya) was not handed over until 1963. In 1965 an attempted Communist coup (in which President Sukarno was said to have participated) was crushed by the army. Indonesia invaded East Timor in 1976, annexing it and claiming it as Indonesia's 27th state; there have since been allegations of mass killings in the region. The majority of Indonesia's population is Javanese, with Java the most densely populated of the country's islands. Since the mid-1980s the Indonesian government has attempted to relieve overcrowding in Java by encouraging transmigration to Irian Jaya and other less populated islands, a policy which has encountered much opposition from the islands' native inhabitants.

Indonesian /ˌɪndə'niːzɪən, -'niːʒ(ə)n, -'niːʃ(ə)n/ n. & adj. ● adj. **1 a** a native or national of Indonesia. **b** a person of Indonesian descent. **2** a member of the chief pre-Malay population of Indonesia and neighbouring islands. **3** the group of languages spoken in Indonesia and neighbouring islands, esp. the official language (also called *Bahasa Indonesia*) of Indonesia. (*See note below.*) ● adj. of or relating to Indonesia or its people or language(s).

▪ Indonesian is virtually the same language as Malay; the apparent differences are mainly due to the different spelling systems, the Indonesian one having been developed by the Dutch and the Malay by the British.

indoor /'ɪndɔː(r)/ adj. situated, carried on, or used within a building or under cover (*indoor aerial; indoor games*). [earlier *within-door*: cf. INDOORS]

indoors /ɪn'dɔːz/ adv. into or within a building. [earlier *within doors*]

Indo-Pacific /ˌɪndəʊpə'sɪfɪk/ adj. & n. ● adj. **1** of or relating to the Indian Ocean and the adjacent parts of the Pacific. **2** of or relating to the group of languages spoken in the islands of this region; Malayo-Polynesian. ● n. the Indo-Pacific seas or ocean.

Indore /ɪn'dɔː(r)/ a manufacturing city of Madhya Pradesh in central India; pop. (1991) 1,087,000.

indorse var. of ENDORSE.

indorsement var. of ENDORSEMENT.

Indra /'ɪndrə/ *Hinduism* the warrior-king of the heavens, god of war and storm, to whom many of the prayers in the Rig-veda are addressed. His weapons are the thunderbolt and lightning, his helpers are the Maruts. His role in later Hinduism is small. [Skr., = lord]

indraught /'ɪndrɑːft/ n. (US **indraft**) **1** the drawing in of something. **2** an inward flow or current.

indrawn /ɪn'drɔːn/ adj. **1** (of breath etc.) drawn in. **2** aloof.

indri /'ɪndrɪ/ n. (pl. **indris**) a large woolly lemur, *Indri indri*, of Madagascar. [Malagasy *indry* behold, mistaken for its name]

indubitable /ɪn'djuːbɪtəb(ə)l/ adj. that cannot be doubted. □ **indubitably** adv. [F *indubitable* or L *indubitabilis* (as IN-¹, *dubitare* to doubt)]

induce /ɪn'djuːs/ v.tr. **1** (often foll. by to + infin.) prevail on; persuade. **2** bring about; give rise to. **3** *Med.* bring on (labour) artificially, esp. by use of drugs. **4** *Electr.* produce (a current) by induction. **5** *Physics* cause (radioactivity) by bombardment. **6** infer; derive as a deduction. □ **inducer** n. **inducible** adj. [ME f. L *inducere induct-* (as IN-², *ducere* lead)]

inducement /ɪn'djuːsmənt/ n. **1** (often foll. by to) an attraction that leads one on. **2** a thing that induces.

induct /ɪn'dʌkt/ v.tr. (often foll. by to, into) **1 a** introduce formally into office. **b** introduce (a member of the clergy) formally into possession of a benefice. **2** introduce, initiate. **3** *US* enlist (a person) for military service. **4** *archaic* lead (to a seat, into a room, etc.); install. □ **inductee** /ˌɪndʌk'tiː/ n. [ME (as INDUCE)]

inductance /ɪn'dʌktəns/ n. *Electr.* the property of an electric circuit that causes an electromotive force to be generated by a change in the current flowing.

induction /ɪn'dʌkʃ(ə)n/ n. **1** the act or an instance of inducting or inducing. **2** *Med.* the process of bringing on (esp. labour) by artificial means. **3** *Logic* **a** the inference of a general law from particular instances (cf. DEDUCTION). **b** *Math.* a means of proving a theorem by showing that if it is true of any particular case it is true of the next case in a series, and then showing that it is indeed true in one particular case. **c** (foll. by of) the production of (facts) to prove a general statement. **4** (often *attrib.*) a formal introduction to a new job, position, etc. (*attended an induction course*). **5** *Electr.* **a** the production of an electric or magnetic state by the proximity (without contact) of an electrified or magnetized body. **b** the production of an electric current in a conductor by a change of magnetic field. **6** the drawing of a fuel mixture into the cylinders of an internal-combustion engine. **7** *US* enlistment for military service. □ **induction-coil** *Electr.* a coil for generating intermittent high voltage from a direct current. **induction heating** heating by an induced electric current. **induction loop** a sound system in which a loop of wire around an area in a building etc. produces an electromagnetic signal received directly by hearing aids for the partially deaf. [ME f. OF *induction* or L *inductio* (as INDUCE)]

inductive /ɪn'dʌktɪv/ adj. **1** (of reasoning etc.) of or based on induction.

2 of electric or magnetic induction. □ **inductively** *adv.* **inductiveness** *n.* [LL *inductivus* (as INDUCE)]

inductor /ɪnˈdʌktə(r)/ *n.* **1** *Electr.* a component (in a circuit) which possesses inductance. **2** a person who inducts a member of the clergy to a benefice. [L (as INDUCE)]

indue var. of ENDUE.

indulge /ɪnˈdʌldʒ/ *v.* **1** *intr.* (often foll. by *in*) take pleasure freely. **2** *tr.* yield freely to (a desire etc.). **3** *tr.* gratify the wishes of; favour (*indulged them with money*). **4** *intr. colloq.* take alcoholic liquor. □ **indulger** *n.* [L *indulgere indult-* give free rein to]

indulgence /ɪnˈdʌldʒəns/ *n.* **1 a** the act of indulging. **b** the state of being indulgent. **2** something indulged in. **3** *RC Ch.* the remission of punishment in purgatory, still due for sins even after sacramental absolution. (*See note below.*) **4** a privilege granted. [ME f. OF f. L *indulgentia* (as INDULGENT)]

- The later Middle Ages saw the growth of considerable abuses within the Christian Church, such as the unrestricted sale of indulgences by professional pardoners, a growth which was attacked by Martin Luther and became an immediate occasion of the Reformation. In the Roman Catholic Church the granting of indulgences is now ordinarily confined to the pope.

indulgent /ɪnˈdʌldʒənt/ *adj.* **1** ready or too ready to overlook faults etc. **2** indulging or tending to indulge. □ **indulgently** *adv.* [F *indulgent* or L *indulgere indulgent-* (as INDULGE)]

indumentum /ˌɪndjʊˈmentəm/ *n.* (pl. **indumenta** /-tə/) *Bot.* the covering of hairs on part of a plant, esp. when dense. [L, = garment]

induna /ɪnˈduːnə/ *n. S. Afr.* **1** a tribal councillor or headman. **2 a** an African foreman. **b** a person in authority. [Nguni *inDuna* captain, councillor]

indurate /ˈɪndjʊəˌreɪt/ *v.* **1** *tr.* & *intr.* make or become hard. **2** *tr.* make callous or unfeeling. **3** *intr.* become inveterate. □ **indurative** *adj.* **induration** /ˌɪndjʊəˈreɪʃ(ə)n/ *n.* [L *indurare* (as IN-[2], *durus* hard)]

Indus /ˈɪndəs/ a river of southern Asia, about 2,900 km (1,800 miles) in length, flowing from Tibet through Kashmir and Pakistan to the Arabian Sea. Along its valley an early civilization flourished from *c.*2600 to 1760 BC, with important centres at Mohenjo-Daro and Harappa, characterized by towns built on a grid-like plan with granaries, drainage systems, and public buildings, copper–bronze technology, a standard system of weights and measures, and steatite seals with hieroglyphic inscriptions, which remain undeciphered. This civilization's economic wealth was derived from well-attested sea and land trade with the rest of the Indian subcontinent, Afghanistan, the Persian Gulf, and Mesopotamia. In the early 2nd millennium its power declined, probably because of incursions by the Aryans.

indusium /ɪnˈdjuːzɪəm/ *n.* (pl. **indusia** /-zɪə/) **1** *Bot.* a membranous shield covering the fruit-cluster of a fern. **2** *Bot.* a collection of hairs enclosing the stigma of some flowers. **3** *Zool.* the case of a larva. □ **indusial** *adj.* [L, = tunic, f. *induere* put on (a garment)]

industrial /ɪnˈdʌstrɪəl/ *adj.* & *n.* ● *adj.* **1** of or relating to industry or industries. **2** designed or suitable for use in industry (*industrial alcohol*). **3** characterized by highly developed industries (*the industrial nations*). ● *n.* (in *pl.*) shares in industrial companies. □ **industrial action** *Brit.* any action, esp. a strike or work to rule, taken by employees as a protest. **industrial archaeology** the study of machines, factories, bridges, etc., formerly used in industry. **industrial estate** *Brit.* an area of land developed for the siting of industrial enterprises. **industrial relations** the relations between management and workers in industries. □ **industrially** *adv.* [INDUSTRY + -AL[1]: in 19th c. partly f. F *industriel*]

industrialism /ɪnˈdʌstrɪəˌlɪz(ə)m/ *n.* a social or economic system in which manufacturing industries are prevalent.

industrialist /ɪnˈdʌstrɪəlɪst/ *n.* a person engaged in the management of industry.

industrialize /ɪnˈdʌstrɪəˌlaɪz/ *v.* (also **-ise**) **1** *tr.* introduce industries to (a country or region etc.). **2** *intr.* become industrialized. □ **industrialization** /-ˌdʌstrɪəlaɪˈzeɪʃ(ə)n/ *n.*

Industrial Revolution the rapid development of a nation's industry, especially that which occurred in Britain in the second half of the 18th century and the first half of the 19th century, in which the bulk of the working population turned from agriculture to industry. Preceded by major changes in agricultural methods which freed workers for the factories, it was made possible by the rise of modern industrial methods, with steam power replacing the use of muscle, wind, and water, the growth of factories, and the mass production of manufactured goods. The textile industry was the prime example of industrialization, and created a demand for machines, and for tools for their manufacture, which stimulated further mechanization. Improved transport was needed, provided by canals, roads, railways, and steamships; construction of these required a large labour force, and the skills acquired were exported to other countries. It made Britain the most powerful industrial country in the world but radically changed the face of British society, throwing up large cities (particularly in the Midlands) as the population shifted from the countryside, and causing or exacerbating a series of profound social and economic problems.

Industrial Workers of the World (abbr. **IWW**) a radical US labour movement, popularly known as the *Wobblies*, founded in 1905 and, as part of the syndicalist movement, dedicated to the overthrow of capitalism. Its popularity declined after the First World War and by 1925 its membership was insignificant.

industrious /ɪnˈdʌstrɪəs/ *adj.* diligent, hard-working. □ **industriously** *adv.* **industriousness** *n.* [F *industrieux* or LL *industriosus* (as INDUSTRY)]

industry /ˈɪndəstrɪ/ *n.* (pl. **-ies**) **1 a** a branch of trade or manufacture. **b** trade and manufacture collectively (*incentives to industry*). **2** concerted or copious activity (*the building was a hive of industry*). **3 a** diligence. **b** *colloq.* the diligent study of a particular topic (*the Shakespeare industry*). **4** habitual employment in useful work. [ME, = skill, f. F *industrie* or L *industria* diligence]

indwell /ɪnˈdwel/ *v.* (*past and past part.* **indwelt** /-ˈdwelt/) *literary* **1** *intr.* (often foll. by *in*) be permanently present as a spirit, principle, etc. **2** *tr.* inhabit spiritually. □ **indweller** *n.*

Indy /ˈɪndɪ/ *n.* **1** (in full **Indy 500**) the annual Indianapolis 500-mile motor-race. **2** (usu. *attrib.*) any of a series of similar competitive circuit races. □ **Indy car** a rear-engine, turbocharged racing car designed to compete in an Indy race. [abbr. of INDIANAPOLIS]

Ine /ˈiːnə/ king of Wessex 688–726. He extended the prestige and power of the throne, developing an extensive legal code.

-ine[1] /aɪn, ɪn/ *suffix* forming adjectives, meaning 'belonging to, of the nature of' (*Alpine*; *asinine*). [from or after F *-in -ine*, or f. L *-inus*]

-ine[2] /aɪn/ *suffix* forming adjectives esp. from names of minerals, plants, etc. (*crystalline*). [L *-inus* from or after Gk *-inos*]

-ine[3] /ɪn, iːn/ *suffix* forming feminine nouns (*heroine*; *margravine*). [F f. L *-ina* f. Gk *-inē*, or f. G *-in*]

-ine[4] *suffix* **1** /ɪn/ forming (esp. abstract) nouns (*discipline*; *medicine*). **2** /iːn, ɪn/ *Chem.* forming nouns denoting derived substances, esp. alkaloids, halogens, amines, and amino acids. [F f. L *-ina* (fem.) = -INE[1]]

inebriate *v., adj.,* & *n.* ● *v.tr.* /ɪˈniːbrɪˌeɪt/ **1** make drunk; intoxicate. **2** excite. ● *adj.* /ɪˈniːbrɪət/ drunken. ● *n.* /ɪˈniːbrɪət/ a drunken person, esp. a habitual drunkard. □ **inebriation** /ɪˌniːbrɪˈeɪʃ(ə)n/ *n.* **inebriety** /ˌɪnɪˈbraɪətɪ/ *n.* [ME f. L *inebriatus* past part. of *inebriare* (as IN-[2], *ebrius* drunk)]

inedible /ɪnˈedɪb(ə)l/ *adj.* not edible, esp. not suitable for eating (cf. UNEATABLE). □ **inedibility** /-ˌedɪˈbɪlɪtɪ/ *n.*

ineducable /ɪnˈedjʊkəb(ə)l/ *adj.* incapable of being educated, esp. through mental handicap. □ **ineducability** /-ˌedjʊkəˈbɪlɪtɪ/ *n.*

ineffable /ɪnˈefəb(ə)l/ *adj.* **1** unutterable; too great for description in words. **2** that must not be uttered. □ **ineffably** *adv.* **ineffability** /-ˌefəˈbɪlɪtɪ/ *n.* [ME f. OF *ineffable* or L *ineffabilis* (as IN-[1], *effari* speak out, utter)]

ineffaceable /ˌɪnɪˈfeɪsəb(ə)l/ *adj.* that cannot be effaced. □ **ineffaceably** *adv.* **ineffaceability** /-ˌfeɪsəˈbɪlɪtɪ/ *n.*

ineffective /ˌɪnɪˈfektɪv/ *adj.* **1** not producing any effect or the desired effect. **2** (of a person) inefficient; not achieving results. **3** lacking artistic effect. □ **ineffectively** *adv.* **ineffectiveness** *n.*

ineffectual /ˌɪnɪˈfektʃʊəl, -tjʊəl/ *adj.* **1 a** without effect. **b** not producing the desired or expected effect. **2** (of a person) lacking the ability to achieve results (*an ineffectual leader*). □ **ineffectually** *adv.* **ineffectualness** *n.* **ineffectuality** /-ˌfektʃʊˈælɪtɪ, -ˌfektjʊ-/ *n.* [ME f. med.L *ineffectualis* (as IN-[1], EFFECTUAL)]

inefficacious /ˌɪnefɪˈkeɪʃəs/ *adj.* (of a remedy etc.) not producing the desired effect. □ **inefficaciously** *adv.* **inefficaciousness** *n.* **inefficacy** /ɪnˈefɪkəsɪ/ *n.*

inefficient /ˌɪnɪˈfɪʃ(ə)nt/ *adj.* **1** not efficient. **2** (of a person) not fully capable; not well qualified. □ **inefficiency** *n.* **inefficiently** *adv.*

inelastic /ˌɪnɪˈlæstɪk/ *adj.* **1** not elastic. **2** unadaptable, inflexible, unyielding. **3** *Physics* (of a collision etc.) involving an overall loss of translational kinetic energy. □ **inelastically** adv. **inelasticity** /ˌɪnɪlæˈstɪsɪtɪ/ n.

inelegant /ɪnˈelɪgənt/ adj. **1** ungraceful. **2 a** unrefined. **b** (of a style) unpolished. □ **inelegance** n. **inelegantly** adv. [F inélégant f. L inelegans (as IN-[1], ELEGANT)]

ineligible /ɪnˈelɪdʒɪb(ə)l/ adj. **1** not eligible. **2** undesirable. □ **ineligibly** adv. **ineligibility** /-ˌelɪdʒɪˈbɪlɪtɪ/ n.

ineluctable /ˌɪnɪˈlʌktəb(ə)l/ adj. **1** against which it is useless to struggle. **2** that cannot be escaped from. □ **ineluctably** adv. **ineluctability** /-ˌlʌktəˈbɪlɪtɪ/ n. [L ineluctabilis (as IN-[1], eluctari struggle out)]

inept /ɪnˈept/ adj. **1** unskilful. **2** absurd, silly. **3** out of place. □ **ineptitude** n. **ineptly** adv. **ineptness** n. [L ineptus (as IN-[1], APT)]

inequable /ɪnˈekwəb(ə)l/ adj. **1** not fairly distributed. **2** not uniform. [L inaequabilis uneven (as IN-[1], EQUABLE)]

inequality /ˌɪnɪˈkwɒlɪtɪ/ n. (pl. **-ies**) **1 a** lack of equality in any respect. **b** an instance of this. **2** the state of being variable. **3** (of a surface) irregularity. **4** Math. a formula affirming that two expressions are not equal. [ME f. OF inequalité or L inaequalitas (as IN-[1], EQUALITY)]

inequitable /ɪnˈekwɪtəb(ə)l/ adj. unfair, unjust. □ **inequitably** adv.

inequity /ɪnˈekwɪtɪ/ n. (pl. **-ies**) unfairness, bias.

ineradicable /ˌɪnɪˈrædɪkəb(ə)l/ adj. that cannot be eradicated or rooted out. □ **ineradicably** adv.

inerrant /ɪnˈerənt/ adj. not liable to err. □ **inerrancy** n. [L inerrans (as IN-[1], ERR)]

inert /ɪnˈɜːt/ adj. **1** without inherent power of action, motion, or resistance. **2** without active chemical or other properties. **3** sluggish, slow. □ **inert gas** = NOBLE GAS. □ **inertly** adv. **inertness** n. [L iners inert- (as IN-[1], ars ART[1])]

inertia /ɪnˈɜːʃə, -ʃɪə/ n. **1** Physics a property of matter by which it continues in its existing state of rest or uniform motion in a straight line, unless that state is changed by an external force. (See also MASS[1].) **2 a** inertness, sloth. **b** a tendency to remain unchanged. □ **inertia reel** a reel which allows a vehicle seat-belt to unwind freely but which locks under force of impact or rapid deceleration. **inertia selling** the sending of unsolicited goods in the hope of making a sale. □ **inertialess** adj. [L (as INERT)]

inertial /ɪˈnɜːʃ(ə)l, -ʃɪəl/ adj. of, relating to, or involving inertia. □ **inertial guidance** guidance of a missile by internal instruments which measure its acceleration and compare the calculated position with stored data.

inescapable /ˌɪnɪˈskeɪpəb(ə)l/ adj. that cannot be escaped or avoided. □ **inescapably** adv. **inescapability** /-ˌskeɪpəˈbɪlɪtɪ/ n.

-iness /ɪnɪs/ suffix forming nouns corresponding to adjectives in -y (see -Y[1], -LY[2]).

inessential /ˌɪnɪˈsenʃ(ə)l/ adj. & n. ● adj. **1** not necessary. **2** dispensable. ● n. an inessential thing.

inestimable /ɪnˈestɪməb(ə)l/ adj. too great, intense, precious, etc., to be estimated. □ **inestimably** adv. [ME f. OF f. L inaestimabilis (as IN-[1], ESTIMABLE)]

inevitable /ɪnˈevɪtəb(ə)l/ adj. & n. ● adj. **1 a** unavoidable; sure to happen. **b** that is bound to occur or appear. **2** colloq. that is tiresomely familiar (dinner was the inevitable curry). ● n. (prec. by the) an inevitable fact, truth, etc. □ **inevitably** adv. **inevitableness** n. **inevitability** /-ˌevɪtəˈbɪlɪtɪ/ n. [L inevitabilis (as IN-[1], evitare avoid)]

inexact /ˌɪnɪɡˈzækt/ adj. not exact. □ **inexactitude** n. **inexactly** adv. **inexactness** n.

inexcusable /ˌɪnɪkˈskjuːzəb(ə)l/ adj. (of a person, action, etc.) that cannot be excused or justified. □ **inexcusably** adv. [ME f. L inexcusabilis (as IN-[1], EXCUSE)]

inexhaustible /ˌɪnɪɡˈzɔːstɪb(ə)l/ adj. **1** that cannot be exhausted or used up. **2** that cannot be worn out. □ **inexhaustibly** adv. **inexhaustibility** /-ˌzɔːstɪˈbɪlɪtɪ/ n.

inexorable /ɪnˈeksərəb(ə)l/ adj. **1** relentless. **2** (of a person or attribute) that cannot be persuaded by request or entreaty. □ **inexorably** adv. **inexorability** /-ˌeksərəˈbɪlɪtɪ/ n. [F inexorable or L inexorabilis (as IN-[1], exorare entreat)]

inexpedient /ˌɪnɪkˈspiːdɪənt/ adj. not expedient. □ **inexpediency** n.

inexpensive /ˌɪnɪkˈspensɪv/ adj. not expensive, cheap. □ **inexpensively** adv. **inexpensiveness** n.

inexperience /ˌɪnɪkˈspɪərɪəns/ n. lack of experience, or of the resulting knowledge or skill. □ **inexperienced** adj. [F inexpérience f. LL inexperientia (as IN-[1], EXPERIENCE)]

inexpert /ɪnˈekspɜːt/ adj. unskilful; lacking expertise. □ **inexpertly** adv. **inexpertness** n. [OF f. L inexpertus (as IN-[1], EXPERT)]

inexpiable /ɪnˈekspɪəb(ə)l/ adj. (of an act or feeling) that cannot be expiated or appeased. □ **inexpiably** adv. [L inexpiabilis (as IN-[1], EXPIATE)]

inexplicable /ˌɪnɪkˈsplɪkəb(ə)l, ɪnˈeks-/ adj. that cannot be explained or accounted for. □ **inexplicably** adv. **inexplicability** /-ɪkˌsplɪkəˈbɪlɪtɪ, -ˌeksplɪkə-/ n. [F inexplicable or L inexplicabilis that cannot be unfolded (as IN-[1], EXPLICABLE)]

inexplicit /ˌɪnɪkˈsplɪsɪt/ adj. not definitely or clearly expressed. □ **inexplicitness** n.

inexpressible /ˌɪnɪkˈspresɪb(ə)l/ adj. that cannot be expressed. □ **inexpressibly** adv.

inexpressive /ˌɪnɪkˈspresɪv/ adj. not expressive. □ **inexpressively** adv. **inexpressiveness** n.

inexpungible /ˌɪnɪkˈspʌndʒɪb(ə)l/ adj. that cannot be expunged or obliterated.

in extenso /ˌɪn ɪkˈstensəʊ/ adv. in full; at length. [L]

inextinguishable /ˌɪnɪkˈstɪŋgwɪʃəb(ə)l/ adj. **1** not quenchable; indestructible. **2** (of laughter etc.) irrepressible.

in extremis /ˌɪn ɪkˈstriːmɪs/ adj. **1** at the point of death. **2** in great difficulties. [L]

inextricable /ɪnˈekstrɪkəb(ə)l, ˌɪnɪkˈstrɪk-/ adj. **1** (of a circumstance) that cannot be escaped from. **2** (of a knot, problem, etc.) that cannot be unravelled or solved. **3** intricately confused. □ **inextricably** adv. **inextricability** /-ˌekstrɪkəˈbɪlɪtɪ, -ɪkˌstrɪkə-/ n. [ME f. L inextricabilis (as IN-[1], EXTRICATE)]

INF abbr. intermediate-range nuclear force(s).

infallible /ɪnˈfælɪb(ə)l/ adj. **1** incapable of error. **2** (of a method, test, proof, etc.) unfailing; sure to succeed. **3** RC Ch. (of the pope) unable to err in pronouncing dogma as doctrinally defined. □ **infallibly** adv. **infallibility** /-ˌfælɪˈbɪlɪtɪ/ n. [ME f. F infaillible or LL infallibilis (as IN-[1], FALLIBLE)]

infamous /ˈɪnfəməs/ adj. **1** notoriously bad; having a bad reputation. **2** abominable. **3** (in ancient law) deprived of all or some citizen's rights on account of serious crime. □ **infamously** adv. **infamy** n. (pl. **-ies**). [ME f. med.L infamosus f. L infamis (as IN-[1], FAME)]

infancy /ˈɪnfənsɪ/ n. (pl. **-ies**) **1** early childhood; babyhood. **2** an early state in the development of an idea, undertaking, etc. **3** Law the state of being a minor. [L infantia (as INFANT)]

infant /ˈɪnf(ə)nt/ n. **1 a** a child during the earliest period of its life. **b** Brit. a schoolchild below the age of 7 years. **2** (esp. attrib.) a thing in an early stage of its development. **3** Law a minor; a person under 18. □ **infant mortality** death before the age of 1. [ME f. OF enfant f. L infans unable to speak (as IN-[1], fans fantis pres. part. of fari speak)]

infanta /ɪnˈfæntə/ n. hist. a daughter of the ruling monarch of Spain or Portugal (usu. the eldest daughter who is not heir to the throne). [Sp. & Port., fem. of INFANTE]

infante /ɪnˈfæntɪ/ n. hist. the second son of the ruling monarch of Spain or Portugal. [Sp. & Port. f. L (as INFANT)]

infanticide /ɪnˈfæntɪˌsaɪd/ n. **1** the killing of an infant soon after birth. **2** the practice of killing newborn infants. **3** a person who kills an infant. □ **infanticidal** /-ˌfæntɪˈsaɪd(ə)l/ adj. [F f. LL infanticidium, -cida (as INFANT)]

infantile /ˈɪnfənˌtaɪl/ adj. **1 a** like or characteristic of a child. **b** childish, immature (infantile humour). **2** in its infancy. □ **infantile paralysis** Med. poliomyelitis. □ **infantility** /ˌɪnfənˈtɪlɪtɪ/ n. (pl. **-ies**). [F infantile or L infantilis (as INFANT)]

infantilism /ɪnˈfæntɪˌlɪz(ə)m/ n. **1** childish behaviour. **2** Psychol. the persistence of infantile characteristics or behaviour in adult life.

infantry /ˈɪnfəntrɪ/ n. (pl. **-ies**) a body of soldiers who march and fight on foot; foot-soldiers collectively. [F infanterie f. It. infanteria f. infante youth, infantryman (as INFANT)]

infantryman /ˈɪnfəntrɪmən/ n. (pl. **-men**) a soldier of an infantry regiment.

infarct /ɪnˈfɑːkt/ n. Med. a small localized area of dead tissue caused by an inadequate blood supply. □ **infarction** /-ˈfɑːkʃ(ə)n/ n. [mod.L infarctus (as IN-[2], L farcire farct- stuff)]

infatuate /ɪnˈfætjʊˌeɪt/ v.tr. **1** inspire with intense usu. transitory fondness or admiration. **2** affect with extreme folly. □ **infatuation** /-ˌfætjʊˈeɪʃ(ə)n/ n. [L infatuare (as IN-[2], fatuus foolish)]

infatuated /ɪnˈfætjʊˌeɪtɪd/ adj. (often foll. by with) affected by an intense fondness or admiration.

infauna /ˈɪnˌfɔːnə/ n. Ecol. the animal life which lives within the sediments of the ocean floor, a river bed, etc. (cf. EPIFAUNA). [Danish ifauna (as IN-², FAUNA)]

infeasible /ɪnˈfiːzɪb(ə)l/ adj. not feasible; that cannot easily be done. □ **infeasibility** /-ˌfiːzɪˈbɪlɪtɪ/ n.

infect /ɪnˈfekt/ v.tr. **1** contaminate (air, water, etc.) with harmful organisms or noxious matter. **2 a** affect (a person) with disease etc. **b** affect (a computer) with a computer virus. **3** instil bad feeling or opinion into. □ **infector** n. [ME f. L inficere infect- taint (as IN-², facere make)]

infection /ɪnˈfekʃ(ə)n/ n. **1 a** the process of infecting or state of being infected. **b** an instance of this; an infectious disease. **c** the entry of a virus into, or the presence of a virus in, a computer. **2** communication of disease, esp. by the agency of air or water etc. **3 a** moral contamination. **b** the diffusive influence of example, sympathy, etc. [ME f. OF infection or LL infectio (as INFECT)]

infectious /ɪnˈfekʃəs/ adj. **1** infecting with disease. **2** (of a disease) liable to be transmitted by air, water, etc. **3** (of emotions etc.) apt to spread; quickly affecting others (infectious laughter). □ **infectiously** adv. **infectiousness** n.

infective /ɪnˈfektɪv/ adj. **1** capable of infecting with disease. **2** infectious. □ **infectiveness** n. [L infectivus (as INFECT)]

infelicitous /ˌɪnfɪˈlɪsɪtəs/ adj. not felicitous; unfortunate. □ **infelicitously** adv.

infelicity /ˌɪnfɪˈlɪsɪtɪ/ n. (pl. **-ies**) **1 a** inaptness of expression etc. **b** an instance of this. **2 a** unhappiness. **b** a misfortune. [ME f. L infelicitas (as IN-¹, FELICITY)]

infer /ɪnˈfɜː(r)/ v.tr. (**inferred**, **inferring**) (often foll. by that + clause) **1** deduce or conclude from facts and reasoning. **2** disp. imply, suggest. □ **inferable** adj. (also **inferrable**). [L inferre (as IN-², ferre bring)]

inference /ˈɪnfərəns/ n. **1** the act or an instance of inferring. **2** Logic **a** the forming of a conclusion from premisses. **b** a thing inferred. □ **inferential** /ˌɪnfəˈrenʃ(ə)l/ adj. **inferentially** adv. [med.L inferentia (as INFER)]

inferior /ɪnˈfɪərɪə(r)/ adj. & n. ● adj. **1** (often foll. by to) **a** lower; in a lower position. **b** of lower rank, quality, etc. **2** poor in quality. **3** Astron. (of a planet) having an orbit closer to the sun than the earth's, i.e. Mercury and Venus. **4** Bot. situated below an ovary or calyx. **5** (of figures or letters) written or printed below the line. ● n. **1** a person inferior to another, esp. in rank. **2** an inferior letter or figure. □ **inferiorly** adv. [ME f. L, compar. of inferus that is below]

inferiority /ɪnˌfɪərɪˈɒrɪtɪ/ n. the state of being inferior. □ **inferiority complex** Psychol. an unrealistic feeling of general inadequacy caused by actual or supposed inferiority in one sphere, sometimes marked by aggressive behaviour in compensation.

infernal /ɪnˈfɜːn(ə)l/ adj. **1 a** of hell or the underworld. **b** hellish, fiendish. **2** colloq. detestable, tiresome. □ **infernally** adv. [ME f. OF f. LL infernalis f. L infernus situated below]

inferno /ɪnˈfɜːnəʊ/ n. (pl. **-os**) **1** a raging fire. **2** a scene of horror or distress. **3** hell, esp. with ref. to Dante's Divine Comedy. [It. f. LL infernus (as INFERNAL)]

infertile /ɪnˈfɜːtaɪl/ adj. not fertile. □ **infertility** /ˌɪnfəˈtɪlɪtɪ/ n. [F infertile or LL infertilis (as IN-¹, FERTILE)]

infest /ɪnˈfest/ v.tr. (of harmful persons or things, esp. vermin or disease) overrun (a place) in large numbers. □ **infestation** /ˌɪnfeˈsteɪʃ(ə)n/ n. [ME f. F infester or L infestare assail f. infestus hostile]

infibulate /ɪnˈfɪbjʊˌleɪt/ v.tr. (esp. as **infibulated** adj.) perform infibulation on (a woman). [L infibulat- past part. stem of infibulare (as IN-² + FIBULA)]

infibulation /ɪnˌfɪbjʊˈleɪʃ(ə)n/ n. the partial stitching together of the labia, often after excision of the clitoris, to prevent copulation.

infidel /ˈɪnfɪd(ə)l/ n. & adj. ● n. **1** a person who does not believe in religion or in a particular religion; an unbeliever. **2** usu. hist. an adherent of a religion other than Christianity, esp. a Muslim. ● adj. **1** that is an infidel. **2** of unbelievers. [ME f. F infidèle or L infidelis (as IN-¹, fidelis faithful)]

infidelity /ˌɪnfɪˈdelɪtɪ/ n. (pl. **-ies**) **1 a** disloyalty or unfaithfulness, esp. lack of sexual faithfulness to a partner. **b** an instance of this. **2** disbelief in Christianity or another religion. [ME f. F infidélité or L infidelitas (as INFIDEL)]

infield /ˈɪnfiːld/ n. **1** Cricket **a** the part of the ground near the wicket.

b the fielders stationed there. **2** Baseball **a** the area between the four bases. **b** the four fielders stationed on its boundaries. **3** farm land around or near a homestead. **4 a** arable land. **b** land regularly manured and cropped. □ **infielder** n. (in sense 2).

infighting /ˈɪnˌfaɪtɪŋ/ n. **1** hidden conflict or competitiveness within an organization. **2** boxing at closer quarters than arm's length. □ **infighter** n.

infill /ˈɪnfɪl/ n. & v. ● n. **1** material used to fill a hole, gap, etc. **2** the placing of buildings to occupy the space between existing ones. ● v.tr. fill in (a cavity etc.).

infilling /ˈɪnˌfɪlɪŋ/ n. = INFILL n.

infiltrate /ˈɪnfɪlˌtreɪt/ v. **1** tr. **a** gain entrance or access to surreptitiously and by degrees (as spies etc.). **b** cause to do this. **2** tr. permeate by filtration. **3** tr. (often foll. by into, through) introduce (fluid) by filtration. □ **infiltrator** n. **infiltration** /ˌɪnfɪlˈtreɪʃ(ə)n/ n. [IN-² + FILTRATE]

infinite /ˈɪnfɪnɪt/ adj. & n. ● adj. **1** boundless, endless. **2** very great. **3** (usu. with pl.) innumerable; very many (infinite resources). **4** Math. **a** greater than any assignable quantity or countable number. **b** (of a series) that may be continued indefinitely. **5** Gram. (of a verb part) not limited by person or number, e.g. infinitive, gerund, and participle. ● n. **1** (the Infinite) God. **2** (the infinite) infinite space. □ **infinitely** adv. **infiniteness** n. [ME f. L infinitus (as IN-¹, FINITE)]

infinitesimal /ˌɪnfɪnɪˈtesɪm(ə)l/ adj. & n. ● adj. infinitely or very small. ● n. an infinitesimal amount. □ **infinitesimal calculus** Math. the differential and integral calculuses regarded as one subject. □ **infinitesimally** adv. [mod.L infinitesimus f. INFINITE: cf. CENTESIMAL]

infinitive /ɪnˈfɪnɪtɪv/ n. & adj. Gram. ● n. a form of a verb expressing the verbal notion without reference to a particular subject, tense, etc. (e.g. see in we came to see, let him see). ● adj. having this form. □ **infinitival** /ˌɪnfɪnɪˈtaɪv(ə)l/ adj. **infinitivally** adv. [L infinitivus (as IN-¹, finitivus definite f. finire finit- define)]

infinitude /ɪnˈfɪnɪˌtjuːd/ n. **1** the state of being infinite; boundlessness. **2** (often foll. by of) a boundless number or extent. [L infinitus: see INFINITE, -TUDE]

infinity /ɪnˈfɪnɪtɪ/ n. (pl. **-ies**) **1** the state of being infinite. **2** an infinite number or extent. **3** infinite distance. **4** Math. infinite quantity (symbol: ∞). [ME f. OF infinité or L infinitas (as INFINITE)]

infirm /ɪnˈfɜːm/ adj. **1** physically weak, esp. through age. **2** (of a person, mind, judgement, etc.) weak, irresolute. □ **infirmity** n. (pl. **-ies**). **infirmly** adv. [ME f. L infirmus (as IN-¹, FIRM¹)]

infirmary /ɪnˈfɜːmərɪ/ n. (pl. **-ies**) **1** a hospital. **2** a place for those who are ill in a monastery, school, etc. [med.L infirmaria (as INFIRM)]

infix v. & n. ● v.tr. **1** (often foll. by in) **a** fix (a thing in another). **b** impress (a fact etc. in the mind). **2** Gram. insert (a formative element) into the body of a word. ● n. /ˈɪnfɪks/ Gram. a formative element inserted in a word. □ **infixation** /ˌɪnfɪkˈseɪʃ(ə)n/ n. [IN-² + FIX] (n.) after prefix, suffix]

in flagrante delicto /ɪn fləˌgræntɪ dɪˈlɪktəʊ/ adv. in the very act of committing an offence. [L, = in blazing crime.]

inflame /ɪnˈfleɪm/ v. **1** tr. & intr. (often foll. by with, by) provoke or become provoked to strong feeling, esp. anger. **2** Med. **a** intr. become hot, reddened, and sore. **b** tr. (esp. as **inflamed** adj.) cause inflammation or fever in (a body etc.); make hot. **3** tr. aggravate. **4** intr. & tr. catch or set on fire. **5** tr. light up with or as if with flames. □ **inflamer** n. [ME f. OF enflammer f. L inflammare (as IN-², flamma flame)]

inflammable /ɪnˈflæməb(ə)l/ adj. & n. ● adj. **1** easily set on fire; flammable. **2** easily excited. ● n. (usu. in pl.) an inflammable substance. □ **inflammably** adv. **inflammableness** n. **inflammability** /-ˌflæməˈbɪlɪtɪ/ n. [INFLAME after F inflammable]

inflammation /ˌɪnfləˈmeɪʃ(ə)n/ n. **1** the act or an instance of inflaming. **2** Med. a localized physical condition with heat, swelling, redness, and usu. pain, esp. as a reaction to injury or infection. [L inflammatio (as INFLAME)]

inflammatory /ɪnˈflæmətərɪ/ adj. **1** (esp. of speeches, leaflets, etc.) tending to cause anger etc. **2** of or tending to inflammation of the body.

inflatable /ɪnˈfleɪtəb(ə)l/ adj. & n. ● adj. that can be inflated. ● n. an inflatable plastic or rubber object.

inflate /ɪnˈfleɪt/ v.tr. **1** distend (a balloon etc.) with air. **2** (usu. foll. by with; usu. in passive) puff up (a person with pride etc.). **3 a** (often absol.) bring about inflation of (a currency). **b** raise (prices) artificially. **4** exaggerate or embellish. **5** (as **inflated** adj.) (esp. of language, sentiments, etc.) bombastic. □ **inflatedly** adv. **inflatedness** n. **inflater** n. **inflator** n. [L inflare inflat- (as IN-², flare blow)]

inflation /ɪnˈfleɪʃ(ə)n/ n. **1 a** the act or condition of inflating or being inflated. **b** an instance of this. **2** Econ. **a** a general increase in prices and fall in the purchasing value of money. **b** an increase in available currency regarded as causing this. □ **inflationary** adj. **inflationism** n. **inflationist** n. & adj. [ME f. L inflatio (as INFLATE)]

inflect /ɪnˈflekt/ v. **1** tr. change the pitch of (the voice, a musical note, etc.). **2** Gram. **a** tr. change the form of (a word) to express tense, gender, number, mood, etc. **b** intr. (of a word, language, etc.) undergo such change. **3** tr. bend inwards; curve. □ **inflective** adj. [ME f. L inflectere inflex- (as IN-², flectere bend)]

inflection /ɪnˈflekʃ(ə)n/ n. (also **inflexion**) **1 a** the act or condition of inflecting or being inflected. **b** an instance of this. **2** Gram. **a** the process or practice of inflecting words. **b** an inflected form of a word. **c** a suffix etc. used to inflect, e.g. -ed. **3** a modulation of the voice. **4** Geom. a change of curvature from convex to concave at a particular point on a curve. □ **inflectional** adj. **inflectionally** adv. **inflectionless** adj. [F inflection or L inflexio (as INFLECT)]

inflexible /ɪnˈfleksɪb(ə)l/ adj. **1** unbendable. **2** stiff; immovable; obstinate (old and inflexible in his attitudes). **3** unchangeable; inexorable. □ **inflexibly** adv. **inflexibility** /-ˌfleksɪˈbɪlɪtɪ/ n. [L inflexibilis (as IN-¹, FLEXIBLE)]

inflict /ɪnˈflɪkt/ v.tr. (usu. foll. by on, upon) **1** administer, deal (a stroke, wound, defeat, etc.). **2** (also refl.) often joc. impose (suffering, a penalty, oneself, one's company, etc.) on (shall not inflict myself on you any longer). □ **inflictable** adj. **inflicter** n. **inflictor** n. [L infligere inflict- (as IN-², fligere strike)]

infliction /ɪnˈflɪkʃ(ə)n/ n. **1** the act or an instance of inflicting. **2** something inflicted, esp. a troublesome or boring experience. [LL inflictio (as INFLICT)]

inflight /ˈɪnflaɪt/ attrib.adj. occurring or provided during an aircraft flight.

inflorescence /ˌɪnfləˈres(ə)ns/ n. **1** Bot. **a** the complete flower-head of a plant including stems, stalks, bracts, and flowers. **b** the arrangement of this. **2** the process of flowering. [mod.L inflorescentia f. LL inflorescere (as IN-², FLORESCENCE)]

inflow /ˈɪnfləʊ/ n. **1** a flowing in. **2** something that flows in. □ **inflowing** n. & adj.

influence /ˈɪnfluəns/ n. & v. ● n. **1 a** (usu. foll. by on, upon) the effect a person or thing has on another. **b** (usu. foll. by over, with) moral ascendancy or power. **c** a thing or person exercising such power (is a good influence on them). **2** Astrol. an ethereal fluid supposedly flowing from the stars and affecting character and destiny. **3** Electr. archaic = INDUCTION 5. ● v.tr. exert influence on; have an effect on. □ **under the influence** colloq. affected by alcoholic drink. □ **influenceable** adj. **influencer** n. [ME f. OF influence or med.L influentia inflow f. L influere flow in (as IN-², fluere flow)]

influent /ˈɪnfluənt/ adj. & n. ● adj. flowing in. ● n. a tributary stream. [ME f. L (as INFLUENCE)]

influential /ˌɪnfluˈenʃ(ə)l/ adj. having a great influence or power (influential in the financial world). □ **influentially** adv. [med.L influentia INFLUENCE]

influenza /ˌɪnfluˈenzə/ n. Med. a highly contagious virus infection causing fever, severe aching, and catarrh. The strain responsible for epidemics constantly changes by mutation, so that there is no lasting immunity, and secondary lung infection is a common and serious complication. (See also Spanish influenza.) □ **influenzal** adj. [It. f. med.L influentia INFLUENCE]

influx /ˈɪnflʌks/ n. **1** a continual stream of people or things (an influx of complaints). **2** (usu. foll. by into) a flowing in, esp. of a stream etc. [F influx or LL influxus (as IN-², FLUX)]

info /ˈɪnfəʊ/ n. colloq. information. [abbr.]

infold var. of ENFOLD.

infomercial /ˌɪnfəˈmɜːʃ(ə)l/ n. esp. US (also **informercial**) an advertising film, esp. on television, promoting a product in an informative and purportedly objective style. [blend of INFORMATION + COMMERCIAL]

inform /ɪnˈfɔːm/ v. **1** tr. (usu. foll. by of, about, on, or that, how + clause) tell (informed them of their rights; informed us that the train was late). **2** intr. (usu. foll. by against, on) give incriminating information about a person to the authorities. **3** tr. (usu. foll. by with) literary inspire or imbue (a person, heart, or thing) with a feeling, principle, quality, etc. **4** tr. impart its

quality to; permeate. □ **informant** n. [ME f. OF enfo(u)rmer f. L informare give shape to, fashion, describe (as IN-², forma form)]

informal /ɪnˈfɔːm(ə)l/ adj. **1** without ceremony or formality (just an informal chat). **2** (of language, clothing, etc.) everyday; normal. □ **informal vote** NZ & Austral. an invalid vote or voting paper. □ **informally** adv. **informality** /ˌɪnfɔːˈmælɪtɪ/ n. (pl. **-ies**)

informatics /ˌɪnfəˈmætɪks/ n.pl. (usu. treated as sing.) the science of processing data for storage and retrieval; information science. [transl. Russ. informatika (as INFORMATION, -ICS)]

information /ˌɪnfəˈmeɪʃ(ə)n/ n. **1 a** something told; knowledge. **b** (usu. foll. by on, about) items of knowledge; news (the latest information on the crisis). **2** (usu. foll. by against) Law a charge or complaint lodged with a court or magistrate. **3 a** the act of informing or telling. **b** an instance of this. **4 a** that which is inherent in or can be conveyed by a particular arrangement or sequence of things. **b** (in information theory) a mathematical quantity expressing the probability of occurrence of a particular sequence of symbols etc. as against a number of possible alternatives. □ **information retrieval** the tracing of information stored in books, computers, etc. **information science** the study of the processes and principles of the collection, storage, and retrieval of (esp. scientific or technical) information. **information technology** the technology involved in the recording, storage, and dissemination of information, esp. using computers, telecommunications, etc. □ **informational** adj. **informationally** adv. [ME f. OF f. L informatio -onis (as INFORM)]

information theory n. the mathematical study of the coding and transmission of information in the form of sequences of symbols, impulses, etc. Information is defined statistically (see INFORMATION 4b), and information theory seeks to establish how much information (measured in bits, or bits per second) can be transmitted in any given situation, e.g. by a telecommunications channel. Information theory was pioneered by C. E. Shannon in the 1940s. Besides its great contribution to the development of computers and communications systems, it has assisted the understanding of other processes, including thermodynamics and the use of language.

informative /ɪnˈfɔːmətɪv/ adj. (also **informatory** /-tərɪ/) giving information; instructive. □ **informatively** adv. **informativeness** n. [med.L informativus (as INFORM)]

informed /ɪnˈfɔːmd/ adj. knowing the facts; educated, knowledgeable (his answers show that he is badly informed; informed opinion). □ **informedly** /-ˈfɔːmɪdlɪ/ adv. **informedness** n.

informer /ɪnˈfɔːmə(r)/ n. **1** a person who informs against another. **2** a person who informs or advises.

infotainment /ˌɪnfəˈteɪnmənt/ n. broadcast matter that seeks both to inform and to entertain. [INFORMATION + ENTERTAINMENT]

infra /ˈɪnfrə/ adv. below, further on (in a book or writing). [L, = below]

infra- /ˈɪnfrə/ comb. form **1** below (opp. SUPRA-). **2** Anat. below or under a part of the body. [from or after L infra below, beneath]

infraclass /ˈɪnfrəˌklɑːs/ n. Biol. a taxonomic category below a subclass.

infraction /ɪnˈfrækʃ(ə)n/ n. esp. Law a violation or infringement. □ **infract** /-ˈfrækt/ v.tr. **infractor** n. [L infractio (as INFRINGE)]

infra dig /ˌɪnfrə ˈdɪg/ predic.adj. colloq. beneath one's dignity; unbecoming. [abbr. of L infra dignitatem]

infrangible /ɪnˈfrændʒɪb(ə)l/ adj. **1** unbreakable. **2** inviolable. □ **infrangibly** adv. **infrangibility** /-ˌfrændʒɪˈbɪlɪtɪ/ n. [obs.F infrangible or med.L infrangibilis (as IN-¹, FRANGIBLE)]

infrared /ˌɪnfrəˈred/ adj. & n. (abbr. **IR**) ● adj. **1** (of electromagnetic radiation) having a wavelength just greater than that of the red end of the visible light spectrum but less than that of microwaves. (See note below.) **2** of or using such radiation. ● n. the infrared part of the spectrum.

■ Infrared radiation spans wavelengths from about 800 nm to 1 mm, and is particularly emitted by heated bodies. It has numerous uses in electronic devices such as remote controls, rangefinders, and lasers, and infrared sensors are used in security systems and heat-seeking missiles. Infrared is also valuable for its ability to form photographic and electronic images in the dark (e.g. in snooperscopes) and through mist and haze (e.g. in remote sensing). There are a number of uses in science ranging from spectroscopy to thermography.

infrasonic /ˌɪnfrəˈsɒnɪk/ adj. of or relating to sound waves with a frequency below the lower limit of human audibility. □ **infrasonically** adv.

infrasound /ˈɪnfrəˌsaʊnd/ n. sound waves with frequencies below the lower limit of human audibility.

infrastructure /ˈɪnfrəˌstrʌktʃə(r)/ n. **1 a** the basic structural foundations of a society or enterprise; a substructure or foundation. **b** roads, bridges, sewers, etc., regarded as a country's economic foundation. **2** permanent installations as a basis for military etc. operations. □ **infrastructural** adj. [F (as INFRA-, STRUCTURE)]

infrequent /ɪnˈfriːkwənt/ adj. not frequent. □ **infrequency** n. **infrequently** adv. [L infrequens (as IN-[1], FREQUENT)]

infringe /ɪnˈfrɪndʒ/ v. **1** tr. **a** act contrary to; violate (a law, an oath, etc.). **b** act in defiance of (another's rights etc.). **2** intr. (usu. foll. by on, upon) encroach; trespass. □ **infringement** n. **infringer** n. [L infringere infract- (as IN-[2], frangere break)]

infula /ˈɪnfjʊlə/ n. (pl. **infulae** /-ˌliː/) Eccl. either of the two ribbons on a bishop's mitre. [L, = woollen fillet worn by priest etc.]

infundibular /ˌɪnfʌnˈdɪbjʊlə(r)/ adj. funnel-shaped. [L infundibulum funnel f. infundere pour in (as IN-[2], fundere pour)]

infuriate v. & adj. ● v.tr. /ɪnˈfjʊərɪˌeɪt/ fill with fury; enrage. ● adj. /ɪnˈfjʊərɪət/ literary excited to fury; frantic. □ **infuriating** /-rɪˌeɪtɪŋ/ adj. **infuriatingly** adv. [med.L infuriare infuriat- (as IN-[2], L furia FURY)]

infuse /ɪnˈfjuːz/ v. **1** tr. (usu. foll. by with) imbue; pervade (anger infused with resentment). **2** tr. steep (herbs, tea, etc.) in liquid to extract the content. **3** tr. (usu. foll. by into) instil (grace, spirit, life, etc.). **4** intr. undergo infusion (let it infuse for five minutes). **5** tr. (usu. foll. by into) pour (a thing). □ **infusable** adj. **infuser** n. [ME f. L infundere infus- (as IN-[2], fundere pour)]

infusible /ɪnˈfjuːzɪb(ə)l/ adj. not able to be fused or melted. □ **infusibility** /-ˌfjuːzɪˈbɪlɪtɪ/ n.

infusion /ɪnˈfjuːʒ(ə)n/ n. **1** a liquid obtained by infusing. **2** an infused element; an admixture. **3** Med. a slow injection of a substance into a vein or tissue. **4 a** the act of infusing. **b** an instance of this. [ME f. F infusion or L infusio (as INFUSE)]

infusoria /ˌɪnfjʊˈsɔːrɪə/ n.pl. Zool. hist. unicellular organisms of the former group Infusoria, which includes mainly the ciliate protozoans. □ **infusorial** adj. [mod.L Infusoria (as INFUSE), because orig. found in infusions of decaying organic matter]

-ing[1] /ɪŋ/ suffix forming gerunds and nouns from verbs (or sometimes from nouns), denoting: **1 a** the verbal action or its result (asking; carving; fighting; learning). **b** the verbal action as described or classified in some way (tough going). **2** material used for or associated with a process etc. (piping; washing). **3** an occupation or event (banking; wedding). **4** a set or arrangement of (colouring; feathering). [OE -ung, -ing f. Gmc]

-ing[2] /ɪŋ/ suffix **1** forming the present participle of verbs (asking; fighting), often as adjectives (charming; strapping). **2** forming adjectives from nouns (hulking). [ME, alt. of OE -ende, later -inde]

-ing[3] /ɪŋ/ suffix forming nouns meaning 'one belonging to or of the kind of', used to form diminutives, fractional parts, patronymics, etc. (atheling; farthing; gelding; riding). [OE f. Gmc]

ingather /ɪnˈgæðə(r)/ v.tr. gather in; assemble.

ingathering /ɪnˈgæðərɪŋ/ n. the act or an instance of gathering in, esp. of a harvest.

ingeminate /ɪnˈdʒemɪˌneɪt/ v.tr. literary repeat; reiterate. □ **ingeminate peace** constantly urge peace. [L ingeminare ingeminat- (as IN-[2], GEMINATE)]

Ingenhousz /ˈɪŋənˌhuːs/, Jan (1730–99), Dutch scientist. His early work was in medicine, and he popularized the inoculation of live smallpox vaccine as a protection against the disease. He is best known, however, for his work on photosynthesis, discovering that sunlit green plants take in carbon dioxide, fix the carbon, and 'restore' the air (oxygen) required by animals for respiration. Ingenhousz discovered Brownian motion, and introduced the use of cover slips for microscopy. He also worked in physics.

ingenious /ɪnˈdʒiːnɪəs/ adj. **1** clever at inventing, constructing, organizing, etc.; skilful; resourceful. **2** (of a machine, theory, etc.) cleverly contrived. □ **ingeniously** adv. **ingeniousness** n. [ME, = talented, f. F ingénieux or L ingeniosus f. ingenium cleverness: cf. ENGINE]

ingénue /ˈænʒeɪˌnjuː/ n. **1** an innocent or unsophisticated young woman. **2** Theatr. **a** such a part in a play. **b** the actress who plays this part. [F, fem. of ingénu INGENUOUS]

ingenuity /ˌɪndʒɪˈnjuːɪtɪ/ n. skill in devising or contriving; ingeniousness. [L ingenuitas ingenuousness (as INGENUOUS): Engl. meaning by confusion of INGENIOUS with INGENUOUS]

ingenuous /ɪnˈdʒenjʊəs/ adj. **1** innocent; artless. **2** open; frank. □ **ingenuously** adv. **ingenuousness** n. [L ingenuus free-born, frank (as IN-[2], root of gignere beget)]

ingest /ɪnˈdʒest/ v.tr. **1** take in (food etc.); eat. **2** absorb (facts, knowledge, etc.). □ **ingestive** adj. **ingestion** /-ˈdʒestʃən/ n. [L ingerere ingest- (as IN-[2], gerere carry)]

inglenook /ˈɪŋg(ə)lˌnʊk/ n. a space within the opening on either side of a large fireplace. [dial. (orig. Sc.) ingle fire burning on a hearth, perh. f. Gael. aingeal fire, light + NOOK]

inglorious /ɪnˈglɔːrɪəs/ adj. **1** shameful; ignominious. **2** not famous. □ **ingloriously** adv. **ingloriousness** n.

-ingly /ɪŋlɪ/ suffix forming adverbs esp. denoting manner of action or nature or condition (dotingly; charmingly; slantingly).

ingoing /ˈɪnˌgəʊɪŋ/ adj. **1** going in; entering. **2** penetrating; thorough.

ingot /ˈɪŋgɒt, -gət/ n. a usu. oblong piece of cast metal, esp. of gold, silver, or steel. [ME: perh. f. IN-[1] + goten past part. of OE geotan cast]

ingraft var. of ENGRAFT.

ingrain adj. & v. ● adj. /ˈɪngreɪn/ **1** inherent; ingrained. **2** (of textiles) dyed in the fibre, before being woven. ● v.tr. /ɪnˈgreɪn/ cause (a dye) to sink deeply into the texture of a fabric; cause to become embedded. □ **ingrain carpet** a reversible carpet, with different colours interwoven.

ingrained /ɪnˈgreɪnd/ attrib.adj. **1** deeply rooted; inveterate. **2** thorough. **3** (of dirt etc.) deeply embedded. □ **ingrainedly** /-ˈgreɪnɪdlɪ/ adv. [var. of engrained: see ENGRAIN]

ingrate /ˈɪngreɪt/ n. & adj. formal or literary ● n. an ungrateful person. ● adj. ungrateful. [ME f. L ingratus (as IN-[1], gratus grateful)]

ingratiate /ɪnˈgreɪʃɪˌeɪt/ v.refl. (usu. foll. by with) bring oneself into favour. □ **ingratiating** adj. **ingratiatingly** adv. **ingratiation** /-ˌgreɪʃɪˈeɪʃ(ə)n/ n. [L in gratiam into favour]

ingratitude /ɪnˈgrætɪˌtjuːd/ n. a lack of due gratitude. [ME f. OF ingratitude or LL ingratitudo (as INGRATE)]

ingravescent /ˌɪngrəˈves(ə)nt/ adj. Med. (of a disease etc.) growing worse. □ **ingravescence** n. [L ingravescere (as IN-[2], gravescere grow heavy f. gravis heavy)]

ingredient /ɪnˈgriːdɪənt/ n. a component part or element in a recipe, mixture, or combination. [ME f. L ingredi ingress- enter (as IN-[2], gradi step)]

Ingres /ˈæŋgrə/, Jean Auguste Dominique (1780–1867), French painter. He was a pupil of David and a vigorous opponent of Delacroix's romanticism, upholding neoclassicism in paintings such as Ambassadors of Agamemnon (1801). Ingres's many nudes, including the Bather (1808), reflect his skills as a draughtsman. In his feeling for pure form, he was admired by Degas.

ingress /ˈɪngres/ n. **1 a** the act or right of going in or entering. **b** an entrance. **2** Astron. the start of an eclipse or transit. □ **ingression** /ɪnˈgreʃ(ə)n/ n. [ME f. L ingressus (as INGREDIENT)]

in-group /ˈɪngruːp/ n. a small exclusive group of people with a common interest.

ingrowing /ˈɪnˌgrəʊɪŋ/ adj. growing inwards, esp. (of a toenail) growing into the flesh. □ **ingrown** adj. **ingrowth** n.

inguinal /ˈɪŋgwɪnəl/ adj. of the groin. □ **inguinally** adv. [L inguinalis f. inguen -inis groin]

ingulf var. of ENGULF.

ingurgitate /ɪnˈgɜːdʒɪˌteɪt/ v.tr. **1** swallow greedily. **2** engulf. □ **ingurgitation** /-ˌgɜːdʒɪˈteɪʃ(ə)n/ n. [L ingurgitare ingurgitat- (as IN-[2], gurges gurgitis whirlpool)]

inhabit /ɪnˈhæbɪt/ v.tr. (**inhabited, inhabiting**) (of a person or animal) dwell in; occupy (a region, town, house, etc.). □ **inhabitable** adj. **inhabitant** n. **inhabitability** /-ˌhæbɪtəˈbɪlɪtɪ/ n. **inhabitation** /-ˈteɪʃ(ə)n/ n. [ME inhabite, enhabite f. OF enhabiter or L inhabitare (as IN-[2], habitare dwell): see HABIT]

inhabitancy /ɪnˈhæbɪtənsɪ/ n. (also **inhabitance** /-t(ə)ns/) residence as an inhabitant, esp. during a specified period so as to acquire rights etc.

inhalant /ɪnˈheɪlənt/ n. & adj. ● n. **1** a medicinal preparation for inhaling. **2** a substance inhaled by drug-abusers. ● adj. of or relating to inhalation or inhalants.

inhale /ɪnˈheɪl/ v.tr. (often absol.) breathe in (air, gas, tobacco-smoke, etc.). □ **inhalation** /ˌɪnhəˈleɪʃ(ə)n/ n. [L inhalare breathe in (as IN-[2], halare breathe)]

inhaler /ɪnˈheɪlə(r)/ *n.* a portable device used for relieving nasal or bronchial congestion, esp. asthma, by inhaling.

inharmonic /ˌɪnhɑːˈmɒnɪk/ *adj.* esp. *Mus.* not harmonic.

inharmonious /ˌɪnhɑːˈməʊnɪəs/ *adj.* esp. *Mus.* not harmonious. □ **inharmoniously** *adv.*

inhere /ɪnˈhɪə(r)/ *v.intr.* (often foll. by *in*) **1** exist essentially or permanently in (*goodness inheres in that child*). **2** (of rights etc.) be vested in (a person etc.). [L *inhaerere inhaes-* (as IN-², *haerere* to stick)]

inherent /ɪnˈhɪərənt, -ˈherənt/ *adj.* (often foll. by *in*) **1** existing in something, esp. as a permanent or characteristic attribute. **2** vested in (a person etc.) as a right or privilege. □ **inherence** *n.* **inherently** *adv.* [L *inhaerere inhaerent-* (as INHERE)]

inherit /ɪnˈherɪt/ *v.* (**inherited, inheriting**) **1** *tr.* receive (property, rank, title, etc.) by legal descent or succession. **2** *tr.* derive (a quality or characteristic) from one's parents or ancestors. **3** *absol.* succeed as an heir (*a younger son rarely inherits*). □ **inheritor** *n.* (*fem.* **inheritress** /-trɪs/ or **inheritrix** /-trɪks/). [ME f. OF *enheriter* f. LL *inhereditare* (as IN-², L *heres heredis* heir)]

inheritable /ɪnˈherɪtəb(ə)l/ *adj.* **1** capable of being inherited. **2** capable of inheriting. □ **inheritability** /-ˌherɪtəˈbɪlɪtɪ/ *n.* [ME f. AF (as INHERIT)]

inheritance /ɪnˈherɪt(ə)ns/ *n.* **1** something that is inherited. **2 a** the act of inheriting. **b** an instance of this. □ **inheritance tax** a tax levied on property etc. acquired by gift or inheritance. ¶ Introduced in the UK in 1986 to replace *capital transfer tax*. [ME f. AF *inheritaunce* f. OF *enheriter*: see INHERIT]

inhesion /ɪnˈhiːʒ(ə)n/ *n. formal* the act or fact of inhering. [LL *inhaesio* (as INHERE)]

inhibit /ɪnˈhɪbɪt/ *v.tr.* (**inhibited, inhibiting**) **1** hinder, restrain, or prevent (an action or progress). **2** (as **inhibited** *adj.*) subjected to inhibition. **3 a** (usu. foll. by *from* + verbal noun) forbid or prohibit (a person etc.). **b** (esp. in ecclesiastical law) forbid (an ecclesiastic) to exercise clerical functions. □ **inhibitive** *adj.* **inhibitor** *n.* **inhibitory** *adj.* [L *inhibere* (as IN-², *habere* hold)]

inhibition /ˌɪnɪˈbɪʃ(ə)n, ˌɪnhɪ-/ *n.* **1** *Psychol.* a restraint on the direct expression of an instinct. **2** *colloq.* an emotional resistance to a thought, an action, etc. (*has no inhibitions about singing in public*). **3** *Law* an order forbidding alteration to property rights. **4 a** the act of inhibiting. **b** the process of being inhibited. [ME f. OF *inhibition* or L *inhibitio* (as INHIBIT)]

inhomogeneous /ɪnˌhəʊməʊˈdʒiːnɪəs, ɪnˌhɒm-/ *adj.* not homogeneous. □ **inhomogeneity** /-dʒiːˈniːɪtɪ/ *n.*

inhospitable /ˌɪnhɒˈspɪtəb(ə)l, ɪnˈhɒsp-/ *adj.* **1** not hospitable. **2** (of a region, coast, etc.) not affording shelter etc. □ **inhospitableness** *n.* **inhospitably** *adv.* [obs. F (as IN-¹, HOSPITABLE)]

inhospitality /ɪnˌhɒspɪˈtælɪtɪ/ *n.* the act or process of being inhospitable. [L *inhospitalitas* (as IN-¹, HOSPITALITY)]

in-house *adj. & adv.* ● *adj.* /ˈɪnhaʊs/ done or existing within an institution, company, etc. (*an in-house project*). ● *adv.* /ɪnˈhaʊs/ internally, without outside assistance.

inhuman /ɪnˈhjuːmən/ *adj.* **1** (of a person, conduct, etc.) brutal; unfeeling; barbarous. **2** not of a human type. □ **inhumanly** *adv.* [L *inhumanus* (as IN-¹, HUMAN)]

inhumane /ˌɪnhjuːˈmeɪn/ *adj.* not humane. □ **inhumanely** *adv.* [L *inhumanus* (see INHUMAN) & f. IN-¹ + HUMANE, orig. = INHUMAN]

inhumanity /ˌɪnhjuːˈmænɪtɪ/ *n.* (*pl.* **-ies**) **1** brutality; barbarousness; callousness. **2** an inhumane act.

inhume /ɪnˈhjuːm/ *v.tr. literary* bury. □ **inhumation** /ˌɪnhjuːˈmeɪʃ(ə)n/ *n.* [L *inhumare* (as IN-², *humus* ground)]

inimical /ɪˈnɪmɪk(ə)l/ *adj.* (usu. foll. by *to*) **1** hostile. **2** harmful. □ **inimically** *adv.* [LL *inimicalis* f. L *inimicus* (as IN-¹, *amicus* friend)]

inimitable /ɪˈnɪmɪtəb(ə)l/ *adj.* impossible to imitate. □ **inimitably** *adv.* **inimitability** /ɪˌnɪmɪtəˈbɪlɪtɪ/ *n.* [F *inimitable* or L *inimitabilis* (as IN-¹, *imitabilis* imitable)]

iniquity /ɪˈnɪkwɪtɪ/ *n.* (*pl.* **-ies**) **1** wickedness; unrighteousness. **2** a gross injustice. □ **iniquitous** *adj.* **iniquitously** *adv.* **iniquitousness** *n.* [ME f. OF *iniquité* f. L *iniquitas -tatis* f. *iniquus* (as IN-¹, *aequus* just)]

initial /ɪˈnɪʃ(ə)l/ *adj., n., & v.* ● *adj.* of, existing, or occurring at the beginning (*initial stage; initial expenses*). ● *n.* **1** = *initial letter*. **2** (usu. in *pl.*) the first letter or letters of the words of a (esp. a person's) name or names. ● *v.tr.* (**initialled, initialling**; *US* **initialed, initialing**) mark or sign with one's initials. □ **initial letter** (or **consonant**) a letter or consonant at the beginning of a word. **initial teaching alphabet** a 44-letter phonetic alphabet used to help those beginning to read and write English. □ **initially** *adv.* [L *initialis* f. *initium* beginning f. *inire* init-go in]

initialism /ɪˈnɪʃəˌlɪz(ə)m/ *n.* a group of initial letters used as an abbreviation for a name or expression, each letter being pronounced separately (e.g. *BBC*) (cf. ACRONYM).

initialize /ɪˈnɪʃəˌlaɪz/ *v.tr.* (also **-ise**) (often foll. by *to*) *Computing* set to the value or put in the condition appropriate to the start of an operation. □ **initialization** /ɪˌnɪʃəlaɪˈzeɪʃ(ə)n/ *n.*

initiate *v., n., & adj.* ● *v.tr.* /ɪˈnɪʃɪˌeɪt/ **1** begin; set going; originate. **2 a** (usu. foll. by *into*) admit (a person) into a society, an office, a secret, etc., esp. with a ritual. **b** (usu. foll. by *in, into*) instruct (a person) in science, art, etc. ● *n.* /ɪˈnɪʃɪət/ a person who has been newly initiated. ● *adj.* /ɪˈnɪʃɪət/ (of a person) newly initiated (*an initiate member*). □ **initiation** /ɪˌnɪʃɪˈeɪʃ(ə)n/ *n.* **initiator** /ɪˈnɪʃɪˌeɪtə(r)/ *n.* **initiatory** /ɪˈnɪʃɪətərɪ, ɪˈnɪʃətərɪ/ *adj.* [L *initiare* f. *initium*: see INITIAL]

initiative /ɪˈnɪʃətɪv, ɪˈnɪʃɪə-/ *n. & adj.* ● *n.* **1** the ability to initiate things; enterprise (*I'm afraid he lacks initiative*). **2** a first step; origination (*a peace initiative*). **3** the power or right to begin something. **4** *Polit.* (esp. in Switzerland and some US states) the right of citizens outside the legislature to originate legislation. ● *adj.* beginning; originating. □ **have the initiative** esp. *Mil.* be able to control the enemy's movements. **on one's own initiative** without being prompted by others. **take the initiative** (often foll. by *in* + verbal noun) be the first to take action. [F (as INITIATE)]

inject /ɪnˈdʒekt/ *v.tr.* **1** *Med.* **a** (usu. foll. by *into*) drive or force (a solution, medicine, etc.) by or as if by a syringe. **b** (usu. foll. by *with*) fill (a cavity etc.) by injecting. **c** administer medicine etc. to (a person) by injection. **2** introduce by way of interruption; insert; suggest (*may I inject a note of realism?*). □ **injectable** *adj. & n.* **injector** *n.* [L *injicere* (as IN-², *jacere* throw)]

injection /ɪnˈdʒekʃ(ə)n/ *n.* **1 a** the act of injecting. **b** an instance of this. **2** a liquid or solution (to be) injected (*prepare a morphine injection*). □ **injection moulding** the shaping of rubber or plastic articles by injecting heated material into a mould. [F *injection* or L *injectio* (as INJECT)]

injudicious /ˌɪndʒuːˈdɪʃəs/ *adj.* unwise; ill-judged. □ **injudiciously** *adv.* **injudiciousness** *n.*

Injun /ˈɪndʒən/ *n. US colloq.* or *dial.* a North American Indian. [corrupt.]

injunction /ɪnˈdʒʌŋkʃ(ə)n/ *n.* **1** an authoritative warning or order. **2** *Law* a judicial order restraining a person from an act or compelling redress to an injured party. □ **injunctive** /-ˈdʒʌŋktɪv/ *adj.* [LL *injunctio* f. L *injungere* ENJOIN]

injure /ˈɪndʒə(r)/ *v.tr.* **1** do physical harm or damage to; hurt (*was injured in a road accident*). **2** harm or impair (*illness might injure her chances*). **3** do wrong to. □ **injurer** *n.* [back-form. f. INJURY]

injured /ˈɪndʒəd/ *adj.* **1** harmed or hurt (*the injured passengers*). **2** offended; wronged (*in an injured tone*).

injurious /ɪnˈdʒʊərɪəs/ *adj.* **1** hurtful. **2** (of language) insulting; libellous. **3** wrongful. □ **injuriously** *adv.* **injuriousness** *n.* [ME f. F *injurieux* or L *injuriosus* (as INJURY)]

injury /ˈɪndʒərɪ/ *n.* (*pl.* **-ies**) **1 a** physical harm or damage. **b** an instance of this (*suffered head injuries*). **2** esp. *Law* **a** a wrongful action or treatment. **b** an instance of this. **3** damage to one's good name etc. □ **injury time** *Football* extra playing-time allowed by a referee to compensate for time lost in dealing with injuries. [ME f. AF *injurie* f. L *injuria* a wrong (as IN-¹, *jus juris* right)]

injustice /ɪnˈdʒʌstɪs/ *n.* **1** a lack of fairness or justice. **2** an unjust act. □ **do a person an injustice** judge a person unfairly. [ME f. OF f. L *injustitia* (as IN-¹, JUSTICE)]

ink /ɪŋk/ *n. & v.* ● *n.* **1** a coloured fluid used for writing with a pen, marking with a rubber stamp, etc.; a coloured paste used in printing, duplicating, in ball-point pens, etc. (*See note below.*) **2** *Zool.* a black liquid ejected by a cuttlefish, octopus, etc. to confuse a predator. ● *v.tr.* **1** (usu. foll. by *in, over*) mark with ink. **2** cover (type etc.) with ink before printing. **3** apply ink to. **4** (as **inked** *adj.*) *Austral. sl.* drunk. □ **ink-blot test** = RORSCHACH TEST. **ink-cap** a fungus of the genus *Coprinus*, turning into black liquid when it decays. **ink-horn** *hist.* a small portable horn container for ink. **ink out** obliterate with ink. **ink-pad** an ink-soaked pad, usu. in a box, used for inking a rubber stamp etc. **ink-well** a pot for ink usu. housed in a hole in a desk. □ **inker** *n.* [ME *enke, inke* f. OF *enque* f. LL *encau(s)tum* f. Gk *egkauston* purple ink used by Roman emperors for signature (as EN-², CAUSTIC)]

▪ The ancient Egyptians and Chinese made ink from lampblack mixed with gum or glue; such inks continued in use in medieval Europe.

Plant juices and other substances were also used as colouring-matter, especially an extract of tannin with a soluble iron salt. Oil-based printing inks were developed in the mid-15th century and used for over three hundred years. Synthetic dyes, developed in the 1860s, provided a better colouring-matter, chemical drying-agents appeared, and by the early 20th century ink-making for various purposes had become a complicated industrial process. Ball-point pens use an oil-based ink that dries almost instantly.

Inkatha /ɪnˈkɑːtə/ (in full **Inkatha Freedom Party**; abbr. **IFP**) a mainly Zulu political party and organization in South Africa. It was founded in 1928 by the Zulu king Solomon as a cultural and social movement and revived by Chief Buthelezi in 1975. It has a professed aim of racial equality and universal franchise in South Africa, but progress towards political reform was obstructed by violent clashes between Inkatha factions and members of the rival ANC (an organization with a largely Xhosa membership) in the late 1980s and early 1990s. Inkatha participated in the general elections of April 1994, although it had previously threatened to boycott them amid its demands for Zulu autonomy. Winning 10.5 per cent of the vote, the party was allocated forty-three seats in the National Assembly, and three Cabinet posts. [Zulu *inKhata* crown of woven grass, a tribal emblem symbolizing the force unifying the Zulu nation]

inkling /ˈɪŋklɪŋ/ n. (often foll. by *of*) a slight knowledge or suspicion; a hint. [ME *inkle* utter in an undertone, of unkn. orig.]

inkstand /ˈɪŋkstænd/ n. a stand for one or more ink bottles, often incorporating a pen tray etc.

inky /ˈɪŋkɪ/ adj. (**inkier, inkiest**) of, as black as, or stained with ink. □ **inkiness** n.

INLA see IRISH NATIONAL LIBERATION ARMY.

inlaid past and past part. of INLAY.

inland adj., n., & adv. ● adj. /ˈɪnlənd, -lænd/ **1** situated in the interior of a country. **2** esp. *Brit.* carried on within the limits of a country; domestic (*inland trade*). ● n. /ˈɪnlənd, -lænd/ the parts of a country remote from the sea or frontiers; the interior. ● adv. /ɪnˈlænd/ in or towards the interior of a country. □ **inland duty** a tax payable on inland trade. □ **inlander** /ˈɪnləndə(r), -ˌlændə(r)/ n. **inlandish** adj.

Inland Revenue (in the UK) the government department responsible for assessing and collecting direct taxes on income and capital.

Inland Sea an almost landlocked arm of the Pacific Ocean, surrounded by the Japanese islands of Honshu, Shikoku, and Kyushu. Its chief port is Hiroshima.

in-law /ˈɪnlɔː/ n. (often in *pl.*) a relative by marriage.

inlay v. & n. ● v.tr. /ɪnˈleɪ/ (past and past part. **inlaid** /-ˈleɪd/) **1 a** (usu. foll. by *in*) embed (a thing in another) so that the surfaces are even. **b** (usu. foll. by *with*) ornament (a thing with inlaid work). **2** (as **inlaid** adj.) (of a piece of furniture etc.) ornamented by inlaying. **3** insert (a page, an illustration, etc.) in a space cut in a larger thicker page. ● n. /ˈɪnleɪ/ **1** inlaid work. **2** material inlaid. **3** a filling shaped to fit a tooth-cavity. □ **inlayer** /ɪnˌleɪə(r)/ n. [IN-² + LAY¹]

inlet /ˈɪnlet, -lɪt/ n. **1** a small arm of the sea, a lake, or a river. **2** a piece inserted, esp. in dressmaking etc. **3** a way of entry. [ME f. IN + LET¹ v.]

inlier /ˈɪnˌlaɪə(r)/ n. *Geol.* a structure or area of older rocks completely surrounded by newer rocks. [IN, after *outlier*]

in-line /ˈɪnlaɪn/ adj. **1** having parts arranged in a line. **2** constituting an integral part of a continuous sequence of operations or machines.

in loco parentis /ɪn ˌləʊkəʊ pəˈrentɪs/ adv. in the place or position of a parent (used of a teacher etc. responsible for children). [L]

inly /ˈɪnlɪ/ adv. poet. **1** inwardly; in the heart. **2** intimately; thoroughly. [OE *innlīce* (as IN, -LY²)]

inlying /ˈɪnˌlaɪɪŋ/ adj. situated within, or near a centre.

Inmarsat /ˈɪnmɑːˌsæt/ an international organization, founded in 1978 and with headquarters in London, that operates a system of satellites to provide telecommunication services, as well as distress and safety communication services, to the world's shipping, aviation, and offshore industries. [acronym, f. *International Maritime Satellite Organization*]

inmate /ˈɪnmeɪt/ n. (often foll. by *of*) **1** an occupant of a hospital, prison, institution, etc. **2** an occupant of a house etc., esp. one of several. [prob. orig. INN + MATE¹, assoc. with IN]

in medias res /ɪn ˌmiːdɪˌæs ˈreɪz/ adv. **1** into the midst of things. **2** into the middle of a story, without preamble. [L]

in memoriam /ˌɪn mɪˈmɔːrɪəm/ prep. & n. ● prep. in memory of (a dead person). ● n. a written article or notice etc. in memory of a dead person; an obituary. [L]

inmost /ˈɪnməʊst, -məst/ adj. **1** most inward. **2** most intimate; deepest. [OE *innemest* (as IN, -MOST)]

inn /ɪn/ n. **1** a public house providing alcoholic liquor for consumption on the premises, and sometimes accommodation etc. **2** *hist.* a house providing accommodation, esp. for travellers. [OE *inn* (as IN)]

innards /ˈɪnədz/ n.pl. colloq. **1** entrails. **2** internal workings (of an engine etc.). [dial. etc. pronunc. of *inwards* pl. of *inward* n.: see INWARD]

innate /ɪˈneɪt, ˈɪneɪt/ adj. **1** inborn; natural. **2** *Philos.* originating in the mind. □ **innately** adv. **innateness** n. [ME f. L *innatus* (as IN-², *natus* past part. of *nasci* be born)]

inner /ˈɪnə(r)/ adj. & n. ● adj. (usu. attrib.) **1** further in; inside; interior (*the inner compartment*). **2** (of thoughts, feelings, etc.) deeper; more secret. ● n. *Archery* **1** a division of the target next to the bull's-eye. **2** a shot that strikes this. □ **inner bar** *Brit. Law* Queen's or King's Counsel collectively. **inner city** the area near the centre of a city, esp. when densely populated (also (with hyphen) *attrib.: inner-city housing*). **inner-directed** *Psychol.* governed by standards formed in childhood. **inner ear** *Anat.* the semicircular canals and cochlea, which form the organs of balance and are embedded in the temporal bone. **inner man** (or **woman** or **person**) **1** the soul or mind. **2** *joc.* the stomach. **inner planet** *Astron.* a planet with an orbit inside the asteroid belt, i.e. Mercury, Venus, Earth, and Mars (cf. TERRESTRIAL adj. 3). **inner space 1** the region between the earth and outer space, or below the surface of the sea. **2** the part of the mind not normally accessible to consciousness. **inner-spring** *N. Amer.* = *interior-sprung*. **inner tube** a separate inflatable tube inside the cover of a pneumatic tyre. □ **innerly** adv. **innermost** adj. **innerness** n. [OE *innera* (adj.), compar. of IN]

Inner Hebrides see HEBRIDES, THE.

Inner Mongolia an autonomous region of northern China, on the border with Mongolia; capital, Hohhot.

Inner Temple one of the two Inns of Court on the site of the Temple in London (cf. MIDDLE TEMPLE; see also TEMPLE, THE 2).

innervate /ˈɪnəˌveɪt, ɪˈnɜːv-/ v.tr. supply (an organ etc.) with nerves. □ **innervation** /ˌɪnəˈveɪʃ(ə)n/ n. [IN-² + L *nervus* nerve + -ATE³]

inning /ˈɪnɪŋ/ n. *N. Amer.* each division of a game of baseball during which both sides have a turn at batting. [*in* (v.) go in (f. IN)]

innings /ˈɪnɪŋz/ n. (pl. same or colloq. **inningses**) **1** esp. *Cricket* **a** the part of a game during which a side is in or batting. **b** the play of or score achieved by a player during a turn at batting. **2** a period during which a government, party, cause, etc. is in office or effective. **3 a** a period during which a person can achieve something. **b** colloq. a person's life span (*had a good innings and died at 94*).

innkeeper /ˈɪnˌkiːpə(r)/ n. a person who keeps an inn.

innocent /ˈɪnəs(ə)nt/ adj. & n. ● adj. **1** free from moral wrong; sinless. **2** (usu. foll. by *of*) not guilty (of a crime etc.). **3** not responsible for or involved in an event, yet suffering circumstantially from it (*innocent bystander*). **4 a** simple; guileless; naive. **b** pretending to be guileless. **5** harmless. **6** (foll. by *of*) colloq. without, lacking (*appeared, innocent of shoes*). ● n. **1** an innocent person, esp. a young child. **2** (in *pl.*) the young children killed by Herod the Great after the birth of Jesus (Matt. 2:16). **3** a person involved by chance in a situation, esp. a victim of crime or war. □ **Innocents'** (or **Holy Innocents'**) **Day** the day, 28 December, commemorating the massacre of the innocents. □ **innocence** n. **innocency** n. **innocently** adv. [ME f. OF *innocent* or L *innocens innocent-* (as IN-¹, *nocere* hurt)]

innocuous /ɪˈnɒkjʊəs/ adj. **1** not injurious; harmless. **2** inoffensive. □ **innocuously** adv. **innocuousness** n. [L *innocuus* (as IN-¹, *nocuus* formed as INNOCENT)]

Inn of Court n. *Law* **1** each of the four legal societies having the exclusive right of admitting people to the English bar. **2** any of the sets of buildings in London belonging to these societies. **3** a similar society in Ireland.

innominate /ɪˈnɒmɪnət/ adj. not having a name; unnamed. □ **innominate bone** *Anat.* the bone formed from the fusion of the ilium, ischium, and pubis; the hip-bone. [LL *innominatus* (as IN-¹, NOMINATE)]

innovate /ˈɪnəˌveɪt/ v.intr. **1** bring in new methods, ideas, etc. **2** (often foll. by *in*) make changes. □ **innovator** n. **innovation** /ˌɪnəˈveɪʃ(ə)n/ n. **innovational** adj. **innovative** /ˈɪnəvətɪv, -ˌveɪtɪv/ adj. **innovativeness** n. **innovatory** adj. [L *innovare* make new, alter (as IN-², *novus* new)]

Innsbruck /ˈɪnzbrʊk/ a city in western Austria, capital of Tyrol; pop. (1991) 115,000.

Inns of Chancery n. Brit. hist. the buildings in London formerly used as hostels for law students.

innuendo /ˌɪnjʊˈendəʊ/ n. & v. ● n. (pl. **-oes** or **-os**) **1** an allusive or oblique remark or hint, usu. disparaging. **2** a remark with a double meaning, usu. suggestive. ● v.intr. (**-oes**, **-oed**) make innuendoes. [L, = by nodding at, by pointing to: ablat. gerund of innuere nod at (as IN-², nuere nod)]

Innuit var. of INUIT.

innumerable /ɪˈnjuːmərəb(ə)l/ adj. too many to be counted. □ **innumerably** adv. **innumerability** /ɪˌnjuːmərəˈbɪlɪtɪ/ n. [ME f. L innumerabilis (as IN-¹, NUMERABLE)]

innumerate /ɪˈnjuːmərət/ adj. having no knowledge of or feeling for mathematical operations; not numerate. □ **innumeracy** n. [IN-¹, NUMERATE]

innutrition /ˌɪnjuːˈtrɪʃ(ə)n/ n. lack of nutrition. □ **innutritious** adj.

inobservance /ˌɪnəbˈzɜːv(ə)ns/ n. **1** inattention. **2** (usu. foll. by of) non-observance (of a law etc.). [F inobservance or L inobservantia (as IN-¹, OBSERVANCE)]

inoculate /ɪˈnɒkjʊˌleɪt/ v.tr. **1 a** treat (a person or animal) with vaccine containing a dead or modified disease agent, usu. by injection, to promote immunity against the disease. (See also VACCINATE.) **b** introduce (an infective agent) into an organism. **c** introduce (cells or organisms) into a culture medium. **2** instil (a person) with ideas or opinions. □ **inoculator** n. **inoculable** /-ləb(ə)l/ adj. **inoculative** adj. **inoculation** /ɪˌnɒkjʊˈleɪʃ(ə)n/ n. [orig. in sense 'insert (a bud) into a plant': L inoculare inoculat- engraft (as IN-², oculus eye, bud)]

inoculum /ɪˈnɒkjʊləm/ n. (pl. **inocula** /-lə/) any substance used for inoculation. [mod.L (as INOCULATE)]

inodorous /ɪnˈəʊdərəs/ adj. having no smell; odourless.

in-off /ˈɪnɒf/ n. Billiards the act of pocketing a ball by bouncing it off another ball.

inoffensive /ˌɪnəˈfensɪv/ adj. not objectionable; harmless. □ **inoffensively** adv. **inoffensiveness** n.

inoperable /ɪnˈɒpərəb(ə)l/ adj. **1** Surgery that cannot suitably be operated on (inoperable cancer). **2** that cannot be operated; inoperative. **3** impractical, unworkable. □ **inoperably** adv. **inoperability** /-ˌɒpərəˈbɪlɪtɪ/ n. [F inopérable (as IN-¹, OPERABLE)]

inoperative /ɪnˈɒpərətɪv/ adj. not working or taking effect.

inopportune /ɪnˈɒpəˌtjuːn/ adj. not appropriate, esp. as regards time; unseasonable. □ **inopportunely** adv. **inopportuneness** n. [L inopportunus (as IN-¹, OPPORTUNE)]

inordinate /ɪnˈɔːdɪnət/ adj. **1** immoderate; excessive. **2** intemperate. **3** disorderly. □ **inordinately** adv. [ME f. L inordinatus (as IN-¹, ordinatus past part. of ordinare ORDAIN)]

inorganic /ˌɪnɔːˈgænɪk/ adj. **1** Chem. (of a compound) not organic, usu. of mineral origin (opp. ORGANIC). **2** without organized physical structure. **3** not arising by natural growth; extraneous. **4** Philol. not explainable by normal etymology. □ **inorganic chemistry** the chemistry of inorganic compounds. □ **inorganically** adv.

inosculate /ɪnˈɒskjʊˌleɪt/ v.intr. & tr. **1** join by running together. **2** join closely. □ **inosculation** /-ˌɒskjʊˈleɪʃ(ə)n/ n. [IN-² + L osculare provide with a mouth f. osculum dimin. of os mouth]

in-patient /ˈɪnˌpeɪʃ(ə)nt/ n. a patient who lives in hospital while under treatment.

in propria persona /ɪn ˌprəʊprɪə pɜːˈsəʊnə/ adv. in his or her own person. [L]

input /ˈɪnpʊt/ n. & v. ● n. **1** what is put in or taken in, or operated on by any process or system. **2** Electronics **a** a place where, or a device through which, energy, information, etc., enters a system (a tape recorder with inputs for microphone and radio). **b** energy supplied to a device or system; an electrical signal. **3** the information fed into a computer. **4** the action or process of putting in or feeding in. **5** a contribution of information etc. ● v.tr. (**inputting**; past and past part. **input** or **inputted**) (often foll. by into) **1** put in. **2** Computing supply (data, programs, etc., to a computer, program, etc.). □ **input-output** (or **input/output**) Computing etc. of, relating to, or for input and output. □ **inputter** n.

inquest /ˈɪnkwest/ n. **1** Law **a** Brit. an inquiry by a coroner's court into the cause of a death. **b** a judicial inquiry to ascertain the facts relating to an incident etc. **c** Brit. a coroner's jury. **2** colloq. a discussion analysing the outcome of a game, an election, etc. [ME f. OF enqueste (as ENQUIRE)]

inquietude /ɪnˈkwaɪɪˌtjuːd/ n. uneasiness of mind or body. [ME f. OF inquietude or LL inquietudo f. L inquietus (as IN-¹, quietus quiet)]

inquiline /ˈɪnkwɪˌlaɪn/ n. an animal living in the home of another; a commensal. [L inquilinus sojourner (as IN-², colere dwell)]

inquire /ɪnˈkwaɪə(r)/ v. **1** intr. seek information formally; make a formal investigation. **2** intr. & tr. = ENQUIRE. □ **inquirer** n. [var. of ENQUIRE]

inquiry /ɪnˈkwaɪərɪ/ n. (pl. **-ies**) **1** an investigation, esp. an official one. **2** = ENQUIRY 1. □ **inquiry agent** Brit. a private detective.

inquisition /ˌɪnkwɪˈzɪʃ(ə)n/ n. **1** usu. derog. an intensive search or investigation. **2** a judicial or official inquiry. □ **inquisitional** adj. [ME f. OF f. L inquisitio -onis examination (as INQUIRE)]

Inquisition, the an ecclesiastical court established by Pope Gregory IX c.1232 for the detection of heretics, at a time when certain heretical groups were regarded by the Church as enemies of society. It was active chiefly in northern Italy and southern France, becoming notorious for the use of torture; condemned heretics who refused to recant were handed over to the civil authorities and could be burned at the stake. In 1542 the papal Inquisition was reinstituted to combat Protestantism, eventually becoming an organ of papal government. The Spanish Inquisition was a separate body, established in 1478 and directed originally against converts from Judaism and Islam but later also against Protestants. It operated with great severity, especially under its first inquisitor, Torquemada, and was not suppressed until the early 19th century.

inquisitive /ɪnˈkwɪzɪtɪv/ adj. **1** unduly curious; prying. **2** seeking knowledge; inquiring. □ **inquisitively** adv. **inquisitiveness** n. [ME f. OF inquisitif -ive f. LL inquisitivus (as INQUISITION)]

inquisitor /ɪnˈkwɪzɪtə(r)/ n. **1** an official investigator. **2** hist. an officer of the Inquisition. □ **Grand Inquisitor** the director of the court of Inquisition in some countries. [F inquisiteur f. L inquisitor -oris (as INQUIRE)]

Inquisitor-General /ɪnˌkwɪzɪtəˈdʒenrəl/ n. the head of the Spanish Inquisition.

inquisitorial /ɪnˌkwɪzɪˈtɔːrɪəl/ adj. **1** of or like an inquisitor. **2** offensively prying. **3** Law (of a trial etc.) in which the judge has a prosecuting role (opp. ACCUSATORIAL). □ **inquisitorially** adv. [med.L inquisitorius (as INQUISITOR)]

inquorate /ɪnˈkwɔːreɪt/ adj. not constituting a quorum.

in re /ɪn ˈriː, ˈreɪ/ prep. = RE¹. [L, = in the matter of]

INRI abbr. Jesus of Nazareth, King of the Jews. [L Iesus Nazarenus Rex Iudaeorum]

inroad /ˈɪnrəʊd/ n. **1** (often in pl.) **a** (usu. foll. by on, into) an encroachment; a using up of resources etc. (makes inroads on my time). **b** (often foll. by in, into) progress, an advance (making inroads into a difficult market). **2** a hostile attack; a raid. [IN + ROAD¹ in sense 'riding']

inrush /ˈɪnrʌʃ/ n. a rushing in; an influx. □ **inrushing** adj. & n.

ins. abbr. **1** inches. **2** insurance.

insalubrious /ˌɪnsəˈluːbrɪəs, -ˈljuːbrɪəs/ adj. (of a climate or place) unhealthy. □ **insalubrity** n. [L insalubris (as IN-¹, SALUBRIOUS)]

insane /ɪnˈseɪn/ adj. **1** not of sound mind; mad. **2** colloq. extremely foolish; irrational. □ **insanely** adv. **insanity** /-ˈsænɪtɪ/ n. (pl. **-ies**). [L insanus (as IN-¹, sanus healthy)]

insanitary /ɪnˈsænɪtərɪ/ adj. not sanitary; dirty or germ-carrying.

insatiable /ɪnˈseɪʃəb(ə)l/ adj. **1** unable to be satisfied. **2** extremely greedy. □ **insatiably** adv. **insatiability** /-ˌseɪʃəˈbɪlɪtɪ/ n. [ME f. OF insaciable or L insatiabilis (as IN-¹, SATIATE)]

insatiate /ɪnˈseɪʃɪət/ adj. never satisfied. [L insatiatus (as IN-¹, SATIATE)]

inscape /ˈɪnskeɪp/ n. literary the unique inner quality or essence of an object etc. as shown in a work of art, esp. a poem. [coined by Gerard Manley Hopkins, perh. f. IN-² + -SCAPE]

inscribe /ɪnˈskraɪb/ v.tr. **1 a** (usu. foll. by in, on) write or carve (words etc.) on stone, metal, paper, a book, etc. **b** (usu. foll. by with) mark (a sheet, tablet, etc.) with characters. **2** (usu. foll. by to) write an informal dedication (to a person) in or on (a book etc.). **3** enter the name of (a person) on a list or in a book. **4** Geom. draw (a figure) within another so that some or all points of it lie on the boundary of the other (cf. CIRCUMSCRIBE). **5** (esp. as **inscribed** adj.) Brit. issue (stock etc.) in the form of shares with registered holders. □ **inscribable** adj. **inscriber** n. [L inscribere inscript- (as IN-², scribere write)]

inscription /ɪnˈskrɪpʃ(ə)n/ n. **1** words inscribed, esp. on a monument, coin, stone, or in a book etc. **2 a** the act of inscribing, esp. the informal dedication of a book etc. **b** an instance of this. □ **inscriptional** adj.

inscriptive /-ˈskrɪptɪv/ adj. [ME f. L inscriptio (as INSCRIBE)]

inscrutable /ɪn'skruːtəb(ə)l/ adj. wholly mysterious, impenetrable. □ **inscrutably** adv. **inscrutableness** n. **inscrutability** /-ˌskruːtə'bɪlɪtɪ/ n. [ME f. eccl.L inscrutabilis (as IN-[1], scrutari search: see SCRUTINY)]

insect /'ɪnsekt/ n. **1 a** a six-legged arthropod of the class Insecta, frequently having one or two pairs of wings. (See note below.) **b** (loosely) any small invertebrate animal with several pairs of legs. **2** an insignificant or contemptible person or creature. □ **insectile** /ɪn'sektaɪl/ adj. [L insectum (animal) notched (animal) f. insecare insect- (as IN-[2], secare cut)]

▪ Insects are the most numerous animals, with more than a million species described. They display great diversity of form and occur in all habitats except the sea, but all of them breathe air as adults. The insect body is divided into three sections — head, thorax, and abdomen — with the legs and wings borne on the thorax. Primitive wingless forms such as springtails and bristletails are sometimes considered not to be true insects. The more advanced forms have a complete metamorphosis, passing through distinct larval and pupal stages. Many are of great economic importance, some as pollinators of crops, or producers of honey or silk, others as pests of crops, or carriers of disease.

insectarium /ˌɪnsek'teərɪəm/ n. (also **insectary** /ɪn'sektərɪ/) (pl. **insectariums** or **insectaries**) a place for keeping insects.

insecticide /ɪn'sektɪˌsaɪd/ n. a substance used for killing insects. □ **insecticidal** /-ˌsektɪ'saɪd(ə)l/ adj.

insectivore /ɪn'sektɪˌvɔː(r)/ n. **1** an animal that feeds on insects, worms, etc.; esp. a small mammal of the order Insectivora. (See note below.) **2** a plant that captures and absorbs insects. □ **insectivorous** /ˌɪnsek'tɪvərəs/ adj. [F f. mod.L insectivorus (as INSECT, -VORE: see -VOROUS)]

▪ The members of the order Insectivora (or Lipotyphla) are considered to be among the most primitive of placental mammals, and they are all small but voracious predators. They comprise shrews, moles, and hedgehogs, together with the less familiar desmans, tenrecs, and solenodons. The tree shrews and elephant shrews are now usually categorized as belonging in separate orders.

insecure /ˌɪnsɪ'kjʊə(r)/ adj. **1** (of a person or state of mind) uncertain; lacking confidence. **2 a** unsafe; not firm or fixed. **b** (of ice, ground, etc.) liable to give way. **c** lacking security, unprotected. □ **insecurely** adv. **insecurity** n.

inselberg /'ɪns(ə)lˌbɜːg, 'ɪnz(ə)l-/ n. an isolated hill or mountain rising abruptly from its surroundings. [G, f. Insel island + Berg mountain]

inseminate /ɪn'semɪˌneɪt/ v.tr. **1** introduce semen into (a female) by natural or artificial means. **2** sow (seed etc.). □ **inseminator** n. **insemination** /-ˌsemɪ'neɪʃ(ə)n/ n. [L inseminare (as IN-[2], SEMEN)]

insensate /ɪn'senseɪt/ adj. **1** without physical sensation; unconscious. **2** without sensibility; unfeeling. **3** stupid. □ **insensately** adv. [eccl.L insensatus (as IN-[1], sensatus f. sensus SENSE)]

insensibility /ɪnˌsensɪ'bɪlɪtɪ/ n. **1** unconsciousness. **2** a lack of mental feeling or emotion; hardness. **3** (often foll. by to) indifference. [F insensibilité or LL insensibilitas (as INSENSIBLE)]

insensible /ɪn'sensɪb(ə)l/ adj. **1 a** without one's mental faculties; unconscious. **b** (of the extremities etc.) numb; without feeling. **2** (usu. foll. by of, to) unaware; indifferent (insensible of her needs). **3** without emotion; callous. **4** too small or gradual to be perceived; inappreciable. □ **insensibly** adv. [ME f. OF insensible or L insensibilis (as IN-[1], SENSIBLE)]

insensitive /ɪn'sensɪtɪv/ adj. (often foll. by to) **1** lacking mental or moral sensitivity; unsympathetic, unfeeling. **2** not sensitive to physical stimuli. □ **insensitively** adv. **insensitiveness** n. **insensitivity** /-ˌsensɪ'tɪvɪtɪ/ n.

insentient /ɪn'senʃ(ə)nt/ adj. not sentient; inanimate. □ **insentience** n. **insentiency** n.

inseparable /ɪn'sepərəb(ə)l/ adj. & n. ● adj. **1** (esp. of friends) unable or unwilling to be separated. **2** Gram. (of a prefix, or a verb in respect of it) unable to be used as a separate word, e.g.: dis-, mis-, un-. ● n. (usu. in pl.) an inseparable person or thing, esp. a friend. □ **inseparably** adv. **inseparability** /-ˌsepərə'bɪlɪtɪ/ n. [ME f. L inseparabilis (as IN-[1], SEPARABLE)]

insert v. & n. ● v.tr. /ɪn'sɜːt/ **1** (usu. foll. by in, into, between, etc.) place, fit, or thrust (a thing) into another. **2** (usu. foll. by in, into) introduce (a letter, word, article, advertisement, etc.) into a newspaper etc. **3** (as **inserted** adj.) Anat. etc. (of a muscle etc.) attached (at a specific point). ● n. /'ɪnsɜːt/ something inserted, e.g. a loose page in a magazine, a piece of cloth in a garment, a shot in a cinema film. □ **insertable** /ɪn'sɜːtəb(ə)l/ adj. **inserter** n. [L inserere (as IN-[2], serere sert- join)]

insertion /ɪn'sɜːʃ(ə)n/ n. **1** the act or an instance of inserting. **2** an amendment etc. inserted in writing or printing. **3** each appearance of an advertisement in a newspaper etc. **4** an ornamental section of needlework inserted into plain material (lace insertions). **5** the manner or place of attachment of a muscle, an organ, etc. **6** the placing of a spacecraft in an orbit. [LL insertio (as INSERT)]

in-service /'ɪnˌsɜːvɪs/ attrib.adj. (of training) intended for those actively engaged in the profession or activity concerned.

INSET /'ɪnset/ n. (also **Inset**) (in the UK) in-service term-time training for teachers in state schools. [acronym f. in-service education and training]

inset n. & v. ● n. /'ɪnset/ **1 a** an extra page or pages inserted in a folded sheet or in a book; an insert. **b** a small map, photograph, etc., inserted within the border of a larger one. **2** a piece let into a dress etc. ● v.tr. /ɪn'set/ (**insetting**; past and past part. **inset** or **insetted**) **1** put in as an inset. **2** decorate with an inset. □ **insetter** /'ɪnˌsetə(r)/ n.

inshallah /ɪn'ʃælə/ int. if Allah wills it. [Arab. in šā' Allah]

inshore /ɪn'ʃɔː(r), 'ɪnʃɔː(r)/ adv. & adj. at sea but close to the shore. □ **inshore of** nearer to shore than.

inside n., adj., adv., & prep. ● n. /ɪn'saɪd, 'ɪnsaɪd/ **1 a** the inner side or surface of a thing. **b** the inner part; the interior. **2 a** (of a path) the side next to the wall or away from the road. **b** Brit. (of a double-decker bus) the lower section. **3** (usu. in pl.) colloq. **a** the stomach and bowels (something wrong with my insides). **b** the operative part of a machine etc. **4** colloq. a position affording inside information (knows someone on the inside). **5** Football & Hockey **a** a position towards the centre of the field. **b** a player in this position. ● adj. /'ɪnsaɪd/ **1 a** situated on or in, or derived from, the inside. **b** (of information etc.) available only to those on the inside. **2** Hockey, (now less often) Football nearer to the centre of the field (inside forward; inside left; inside right). ● adv. /ɪn'saɪd/ **1** on, in, or to the inside. **2** sl. in prison. ● prep. /ɪn'saɪd, 'ɪnsaɪd/ **1** on the inner side of; within (inside the house). **2** in less than (inside an hour). □ **inside country** Austral. settled areas near the coast. **inside information** information not accessible to outsiders. **inside job** colloq. a crime committed by a person living or working on the premises burgled etc. **inside of** colloq. **1** in less than (a week etc.). **2** Brit. the middle part of. **inside out** with the inner surface turned outwards. **inside story** = inside information. **inside track 1** the track which is shorter, because of the curve. **2** a position of advantage. **know a thing inside out** know a thing thoroughly. **turn inside out 1** turn the inner surface of outwards. **2** colloq. ransack; cause confusion in. [IN + SIDE]

insider /ɪn'saɪdə(r)/ n. **1** a person who is within a society, organization, etc. (cf. OUTSIDER). **2** a person privy to a secret, esp. when using it to gain advantage. □ **insider dealing** Stock Exch. the illegal practice of trading to one's own advantage through having access to confidential information.

insidious /ɪn'sɪdɪəs/ adj. **1** proceeding or progressing inconspicuously but harmfully (an insidious disease). **2** treacherous; crafty. □ **insidiously** adv. **insidiousness** n. [L insidiosus cunning f. insidiae ambush (as IN-[2], sedere sit)]

insight /'ɪnsaɪt/ n. (usu. foll. by into) **1** the capacity of understanding hidden truths etc., esp. of character or situations. **2** an instance of this. □ **insightful** adj. **insightfully** adv. [ME, = 'discernment', prob. of Scand. & LG orig. (as IN-[2], SIGHT)]

insignia /ɪn'sɪgnɪə/ n. (treated as sing. or pl.; usu. foll. by of) **1** badges (wore his insignia of office). **2** distinguishing marks. [L, pl. of insigne neut. of insignis distinguished (as IN-[2], signis f. signum SIGN)]

insignificant /ˌɪnsɪg'nɪfɪkənt/ adj. **1** unimportant; trifling. **2** (of a person) undistinguished. **3** meaningless. □ **insignificance** n. **insignificancy** n. **insignificantly** adv.

insincere /ˌɪnsɪn'sɪə(r)/ adj. not sincere; not candid. □ **insincerely** adv. **insincerity** /-'serɪtɪ/ n. (pl. **-ies**) [L insincerus (as IN-[1], SINCERE)]

insinuate /ɪn'sɪnjʊˌeɪt/ v.tr. **1** (often foll. by that + clause) convey indirectly or obliquely; hint (insinuated that she was lying). **2** (often refl.; usu. foll. by into) **a** introduce (oneself, a person, etc.) into favour, office, etc., by subtle manipulation. **b** introduce (a thing, an idea, oneself, etc.) subtly or deviously into a place (insinuated himself into the Royal Box). □ **insinuatingly** adv. **insinuator** n. **insinuative** /-jʊətɪv/ adj. **insinuation** /-ˌsɪnjʊ'eɪʃ(ə)n/ n. [L insinuare insinuat- (as IN-[2], sinuare to curve)]

insipid /ɪn'sɪpɪd/ adj. **1** lacking vigour or interest; dull. **2** lacking flavour; tasteless. □ **insipidly** adv. **insipidness** n. **insipidity** /ˌɪnsɪ'pɪdɪtɪ/ n. [F insipide or LL insipidus (as IN-[1], sapidus SAPID)]

insist /ɪn'sɪst/ v.tr. (usu. foll. by that + clause; also absol.) maintain or demand positively and assertively (insisted that he was innocent; give me *the bag! I insist!*). □ **insist on** demand or maintain (*I insist on being present; insists on his suitability*). □ **insistingly** *adv.* [L *insistere* stand on, persist (as IN-², *sistere* stand)]

insistent /ɪnˈsɪstənt/ *adj.* **1** (often foll. by *on*) insisting; demanding positively or continually (*is insistent on taking me with him*). **2** obtruding itself on the attention (*the insistent rattle of the window frame*). □ **insistence** *n.* **insistency** *n.* **insistently** *adv.*

in situ /ɪn ˈsɪtjuː/ *adv.* **1** in its place. **2** in its original place. [L]

insobriety /ˌɪnsəˈbraɪɪti/ *n.* intemperance, esp. in drinking.

insofar /ˌɪnsəʊˈfɑː(r)/ *adv.* in so far (see FAR).

insolation /ˌɪnsəʊˈleɪʃ(ə)n/ *n.* exposure to the sun's rays, esp. for bleaching. [L *insolatio* f. *insolare* (as IN-², *solare* f. *sol* sun)]

insole /ˈɪnsəʊl/ *n.* **1** a removable sole worn in a boot or shoe for warmth etc. **2** the fixed inner sole of a boot or shoe.

insolent /ˈɪnsələnt/ *adj.* offensively contemptuous or arrogant; insulting. □ **insolence** *n.* **insolently** *adv.* [ME, = 'arrogant', f. L *insolens* (as IN-¹, *solens* pres. part. of *solere* be accustomed)]

insoluble /ɪnˈsɒljʊb(ə)l/ *adj.* **1** incapable of being solved. **2** incapable of being dissolved. □ **insolubilize** *v.tr.* (also **-ise**). **insolubly** *adv.* **insolubility** /-ˌsɒljʊˈbɪlɪti/ *n.* [ME f. OF *insoluble* or L *insolubilis* (as IN-¹, SOLUBLE)]

insolvable /ɪnˈsɒlvəb(ə)l/ *adj.* = INSOLUBLE.

insolvent /ɪnˈsɒlv(ə)nt/ *adj. & n.* ● *adj.* **1** unable to pay one's debts. **2** relating to insolvency (*insolvent laws*). ● *n.* an insolvent person. □ **insolvency** *n.*

insomnia /ɪnˈsɒmnɪə/ *n.* habitual sleeplessness; inability to sleep. □ **insomniac** /-nɪˌæk/ *n. & adj.* [L f. *insomnis* sleepless (as IN-¹, *somnus* sleep)]

insomuch /ˌɪnsəʊˈmʌtʃ/ *adv.* **1** (foll. by *that* + clause) to such an extent. **2** (foll. by *as*) inasmuch. [ME, orig. *in so much*]

insouciant /ɪnˈsuːsɪənt/ *adj.* carefree; unconcerned. □ **insouciance** *n.* **insouciantly** *adv.* [F (as IN-¹, *souciant* pres. part. of *soucier* care)]

inspan /ɪnˈspæn/ *v.* (**inspanned**, **inspanning**) *S. Afr.* **1** *tr.* (also *absol.*) **a** yoke (oxen etc.) in a team to a vehicle. **b** harness an animal or animals to (a wagon). **2** *tr.* harness (people or resources) into service. [Du. *inspannen* stretch (as IN-², SPAN²)]

inspect /ɪnˈspekt/ *v.tr.* **1** look closely at or into. **2** examine (a document etc.) officially. □ **inspection** /-ˈspekʃ(ə)n/ *n.* [L *inspicere inspect-* (as IN-², *specere* look at), or its frequent. *inspectare*]

inspector /ɪnˈspektə(r)/ *n.* **1** a person who inspects. **2** an official employed to supervise a service, a machine, etc., and make reports. **3** *Brit.* a police officer below a superintendent and above a sergeant in rank. □ **inspector general** a chief inspector. **inspector of taxes** (in the UK) an official of the Inland Revenue responsible for collecting taxes. □ **inspectorship** *n.* **inspectorate** /-tərət/ *n.* **inspectorial** /ˌɪnspekˈtɔːrɪəl/ *adj.* [L (as INSPECT)]

inspiration /ˌɪnspɪˈreɪʃ(ə)n/ *n.* **1 a** a supposed force or influence on poets, artists, musicians, etc., stimulating creative activity, exalted thoughts, etc. **b** a person, principle, faith, etc. that is a source of stimulation to creative activity etc. **c** a similar divine influence supposed to have led to the writing of Scripture etc. **2** a sudden brilliant, creative, or timely idea. **3** a drawing in of breath; inhalation. □ **inspirational** *adj.* [ME f. OF f. LL *inspiratio -onis* (as INSPIRE)]

inspirator /ˈɪnspɪˌreɪtə(r)/ *n.* an apparatus for drawing in air or vapour. [LL (as INSPIRE)]

inspire /ɪnˈspaɪə(r)/ *v.tr.* **1** (often foll. by *to* + infin.) stimulate or arouse (a person) to esp. creative activity (*inspired by God; inspired her to write*). **2 a** (usu. foll. by *with*) animate (a person) with a feeling. **b** (usu. foll. by *into*) instil (a feeling) into a person etc. **c** (usu. foll. by *in*) create (a feeling) in a person. **3** prompt; give rise to (*the poem was inspired by the autumn*). **4** (as **inspired** *adj.*) **a** (of a work of art etc.) as if prompted by or emanating from a supernatural source; characterized by inspiration (*an inspired speech*). **b** (of a guess) intuitive but accurate. **5** (also *absol.*) breathe in (air etc.); inhale. □ **inspirer** *n.* **inspiring** *adj.* **inspiringly** *adv.* **inspiratory** /-ˈspaɪərətərɪ/ *adj.* **inspiredly** /-ˈspaɪərɪdlɪ/ *adv.* [ME f. OF *inspirer* f. L *inspirare* breathe in (as IN-², *spirare* breathe)]

inspirit /ɪnˈspɪrɪt/ *v.tr.* (**inspirited**, **inspiriting**) **1** put life into; animate. **2** (usu. foll. by *to*, or *to* + infin.) encourage (a person). □ **inspiriting** *adj.* **inspiritingly** *adv.*

inspissate /ɪnˈspɪseɪt/ *v.tr. literary* thicken; condense. □ **inspissation** /ˌɪnspɪˈseɪʃ(ə)n/ *n.* [LL *inspissare inspissat-* (as IN-², L *spissus* thick)]

inspissator /ˈɪnspɪˌseɪtə(r)/ *n.* an apparatus for thickening serum etc. by heat.

inst. /ɪnst/ *abbr.* **1** = INSTANT *adj.* 4 (*the 6th inst.*). **2** institute. **3** institution.

instability /ˌɪnstəˈbɪlɪti/ *n.* (*pl.* **-ies**) **1** a lack of stability. **2** *Psychol.* unpredictability in behaviour etc. **3** an instance of instability. [ME f. F *instabilité* f. L *instabilitas -tatis* f. *instabilis* (as IN-¹, STABLE¹)]

install /ɪnˈstɔːl/ *v.tr.* (also **instal**) (**installed**, **installing**) **1** place (equipment, machinery, etc.) in position ready for use. **2** place (a person) in an office or rank with ceremony (*installed in the office of Chancellor*). **3** establish (oneself, a person, etc.) in a place, condition, etc. (*installed herself at the head of the table*). □ **installant** *adj. & n.* **installer** *n.* [med.L *installare* (as IN-², *stallare* f. *stallum* STALL¹)]

installation /ˌɪnstəˈleɪʃ(ə)n/ *n.* **1 a** the act or an instance of installing. **b** the process or an instance of being installed. **2 a** a large piece of equipment etc. installed for use. **b** a subsidiary military or industrial establishment. **3** an art exhibit constructed within a gallery etc. [med.L *installatio* (as INSTALL)]

instalment /ɪnˈstɔːlmənt/ *n.* (*US* **installment**) **1** a sum of money due as one of several usu. equal payments for something, spread over an agreed period of time. **2** any of several parts, esp. of a television or radio serial or a magazine story, published or shown in sequence at intervals. □ **instalment plan** payment by instalments, esp. hire purchase. [alt. f. obs. *estallment* f. AF *estalement* f. *estaler* fix: prob. assoc. with INSTALLATION]

instance /ˈɪnstəns/ *n. & v.* ● *n.* **1** an example or illustration of (*just another instance of his lack of determination*). **2** a particular case (*that's not true in this instance*). **3** *Law* a legal suit. ● *v.tr.* cite (a fact, case, etc.) as an instance. □ **at the instance of** at the request or suggestion of. **court of first instance** *Law* a court of primary jurisdiction. **for instance** as an example. **in the first** (or **second** etc.) **instance** in the first (or second etc.) place; at the first (or second etc.) stage of a proceeding. [ME f. OF f. L *instantia* (as INSTANT)]

instancy /ˈɪnstənsɪ/ *n.* **1** urgency. **2** pressing nature. [L *instantia*: see INSTANCE]

instant /ˈɪnstənt/ *adj. & n.* ● *adj.* **1** occurring immediately (*gives an instant result*). **2 a** (of food etc.) processed to allow quick preparation. **b** prepared hastily and with little effort (*I have no instant solution*). **3** urgent; pressing. **4** *Commerce* of the current month (*the 6th instant*). **5** *archaic* of the present moment. ● *n.* **1** a precise moment of time, esp. the present (*come here this instant; told you the instant I heard*). **2** a short space of time (*was there in an instant*). □ **instant replay** the immediate repetition of part of a filmed sports event, often in slow motion. [ME f. F f. L *instare instant-* be present, press upon (as IN-², *stare* stand)]

instantaneous /ˌɪnstənˈteɪnɪəs/ *adj.* **1** occurring or done in an instant or instantly. **2** *Physics* existing at a particular instant. □ **instantaneously** *adv.* **instantaneousness** *n.* **instantaneity** /-tə'niːɪtɪ/ *n.* [med.L *instantaneus* f. L *instans* (as INSTANT) after eccl.L *momentaneus*]

instanter /ɪnˈstæntə(r)/ *adv. archaic or joc.* immediately; at once. [L f. *instans* (as INSTANT)]

instantiate /ɪnˈstænʃɪˌeɪt/ *v.tr.* represent by an instance. □ **instantiation** /-ˌstænʃɪˈeɪʃ(ə)n/ *n.* [L *instantia*: see INSTANCE]

instantly /ˈɪnstəntlɪ/ *adv.* **1** immediately; at once. **2** *archaic* urgently; pressingly.

instar /ˈɪnstɑː(r)/ *n. Zool.* a stage in the development of an insect larva etc. between two periods of moulting. [L, = form]

instate /ɪnˈsteɪt/ *v.tr.* (often foll. by *in*) install; establish. [IN-² + STATE]

in statu pupillari /ɪn ˌstætjuː ˌpjuːpɪˈlɑːrɪ/ *adj.* **1** under guardianship, esp. as a pupil. **2** in a junior position at university; not having a master's degree. [L]

instauration /ˌɪnstɔːˈreɪʃ(ə)n/ *n. formal* **1** restoration; renewal. **2** an act of instauration. □ **instaurator** /ˈɪnstɔːˌreɪtə(r)/ *n.* [L *instauratio* f. *instaurare* (as IN-²: cf. RESTORE)]

instead /ɪnˈsted/ *adv.* **1** (foll. by *of*) as a substitute or alternative to; in place of (*instead of this one; stayed instead of going*). **2** as an alternative (*took me instead*) (cf. STEAD). [ME, f. IN + STEAD]

instep /ˈɪnstep/ *n.* **1** the inner arch of the foot between the toes and the ankle. **2** the part of a shoe etc. fitting over or under this. **3** a thing shaped like an instep. [16th c.: ult. formed as IN-² + STEP, but immediate orig. uncert.]

instigate /ˈɪnstɪˌgeɪt/ *v.tr.* **1** bring about by incitement or persuasion; provoke (*who instigated the inquiry?*). **2** (usu. foll. by *to*) urge on, incite (a person etc.) to esp. an evil act. □ **instigator** *n.* **instigation** /ˌɪnstɪˈgeɪʃ(ə)n/ *n.* [L *instigare instigat-*]

instil /ɪnˈstɪl/ *v.tr.* (*US* **instill**) (**instilled**, **instilling**) (often foll. by *in* or *into*) **1** introduce (a feeling, idea, etc.) into a person's mind etc. gradually. **2** put (a liquid) into something in drops. □ **instilment** n. **instillation** /ˌɪnstɪˈleɪʃ(ə)n/ n. [L *instillare* (as IN-², *stillare* drop): cf. DISTIL]

instinct n. & adj. ● n. /ˈɪnstɪŋkt/ **1 a** an innate, usu. fixed, pattern of behaviour in most animals in response to certain stimuli. **b** a similar propensity in human beings to act without conscious intention; innate impulsion. **2** (usu. foll. by *for*) unconscious skill; intuition. ● predic.adj. /ɪnˈstɪŋkt/ (foll. by *with*) imbued, filled (with life, beauty, force, etc.). □ **instinctual** /ɪnˈstɪŋktʃʊəl, -tjʊəl/ adj. **instinctually** adv. [ME, = 'impulse', f. L *instinctus* f. *instinguere* incite (as IN-², *stinguere* stinct- prick)]

instinctive /ɪnˈstɪŋktɪv/ adj. **1** relating to or prompted by instinct. **2** apparently unconscious or automatic (*an instinctive reaction*). □ **instinctively** adv.

institute /ˈɪnstɪˌtjuːt/ n. & v. ● n. **1 a** a society or organization for the promotion of science, education, etc. **b** a building used by an institute. **2** (usu. in pl.) Law a digest of the elements of a legal subject (*Institutes of Justinian*). **3** a principle of instruction. ● v.tr. **1** establish; found. **2 a** initiate (an inquiry etc.). **b** begin (proceedings) in a court. **3** (usu. foll. by *to*, *into*) appoint (a person) as a cleric in a church etc. [ME f. L *institutum* design, precept, neut. past part. of *instituere* establish, arrange, teach (as IN-², *statuere* set up)]

institution /ˌɪnstɪˈtjuːʃ(ə)n/ n. **1** the act or an instance of instituting. **2 a** a society or organization founded esp. for charitable, religious, educational, or social purposes. **b** a building used by an institution. **3** an established law, practice, or custom. **4** colloq. (of a person, a custom, etc.) a familiar object. **5** the establishment of a cleric etc. in a church. [ME f. OF f. L *institutio -onis* (as INSTITUTE)]

institutional /ˌɪnstɪˈtjuːʃən(ə)l/ adj. **1** of or like an institution. **2** typical of institutions, esp. in being regimented or unimaginative (*the food was dreadfully institutional*). **3** (of religion) expressed or organized through institutions (churches etc.). **4** (of advertising) intended to create prestige rather than immediate sales. □ **institutionalism** n. **institutionally** adv.

institutionalize /ˌɪnstɪˈtjuːʃənəˌlaɪz/ v.tr. (also **-ise**) **1** (as **institutionalized** adj.) **a** (of a prisoner, a long-term patient, etc.) subjected to institutional life, esp. for a period of time resulting in unfitness for life outside an institution. **b** existing in or treated as if in or belonging to an institution (*institutionalized racism*). **2** place or keep (a person) in an institution. **3** convert into an institution; make institutional. □ **institutionalization** /-ˌtjuːʃənəlaɪˈzeɪʃ(ə)n/ n.

Inst.P. abbr. (in the UK) Institute of Physics.

INSTRAW /ˈɪnstrɔː/ abbr. a United Nations agency, the International Research and Training Institute for the Advancement of Women.

instruct /ɪnˈstrʌkt/ v.tr. **1** (often foll. by *in*) teach (a person) a subject etc. (*instructed her in French*). **2** (usu. foll. by *to* + infin.) direct; command (*instructed him to fill in the hole*). **3** (often foll. by *of*, or *that* etc. + clause) inform (a person) of a fact etc. **4** Brit. **a** (of a client or solicitor) give information to (a solicitor or counsel). **b** authorize (a solicitor or counsel) to act for one. [ME f. L *instruere* instruct- build, teach (as IN-², *struere* pile up)]

instruction /ɪnˈstrʌkʃ(ə)n/ n. **1** (often in pl.) a direction; an order (*read the instructions; gave him his instructions*). **2** teaching; education (*took a course of instruction*). **3** (in pl.) Law directions to a solicitor or counsel. **4** Computing a direction in a computer program defining and effecting an operation. □ **instructional** adj. [ME f. OF f. LL *instructio -onis* (as INSTRUCT)]

instructive /ɪnˈstrʌktɪv/ adj. tending to instruct; conveying a lesson; enlightening (*found the experience instructive*). □ **instructively** adv. **instructiveness** n.

instructor /ɪnˈstrʌktə(r)/ n. (fem. **instructress** /-trɪs/) **1** a person who instructs; a teacher, demonstrator, etc. **2** N. Amer. a university teacher ranking below assistant professor. □ **instructorship** n.

instrument n. & v. ● n. /ˈɪnstrəmənt/ **1** a tool or implement, esp. for delicate or scientific work. **2** (in full **musical instrument**) a device for producing musical sounds by vibration, wind, percussion, electronic means, etc. **3 a** a thing used in performing an action (*the meeting was an instrument in his success*). **b** a person made use of (*is merely their instrument*). **4** a measuring-device, esp. one in an aeroplane etc. serving to determine the position or speed. **5** a formal, esp. legal, document. ● v.tr. /ˈɪnstrəˌment/ **1** arrange (music) for instruments. **2** equip with instruments (for measuring, recording, controlling, etc.). □ **instrument panel** (or **board**) a surface, esp. in a car or aeroplane, containing the dials etc. of monitoring devices. [ME f. OF *instrument* or L *instrumentum* (as INSTRUCT)]

instrumental /ˌɪnstrəˈment(ə)l/ adj. & n. ● adj. **1** (usu. foll. by *to*, *in*, or *in* + verbal noun) serving as an instrument or means (*was instrumental in finding the money*). **2** (of music) performed on instruments, without singing (cf. VOCAL adj. 5). **3** of, or arising from, an instrument (*instrumental error*). **4** Gram. of or in the instrumental. ● n. **1** a piece of music performed by instruments, not by the voice. **2** Gram. the case of nouns and pronouns (and words in grammatical agreement with them) indicating a means or instrument. □ **instrumentalist** n. (in sense 2 of adj.). **instrumentality** /-menˈtælɪtɪ/ n. [ME f. F f. med.L *instrumentalis* (as INSTRUMENT)]

instrumentation /ˌɪnstrəmenˈteɪʃ(ə)n/ n. **1 a** the arrangement or composition of music for a particular group of musical instruments. **b** the instruments used in any one piece of music. **2 a** the design, provision, or use of instruments in industry, science, etc. **b** such instruments collectively. [F f. *instrumenter* (as INSTRUMENT)]

insubordinate /ˌɪnsəˈbɔːdɪnət/ adj. disobedient; rebellious. □ **insubordinately** adv. **insubordination** /-ˌbɔːdɪˈneɪʃ(ə)n/ n.

insubstantial /ˌɪnsəbˈstænʃ(ə)l/ adj. **1** lacking solidity or substance. **2** not real. □ **insubstantially** adv. **insubstantiality** /-ˌstænʃɪˈælɪtɪ/ n. [LL *insubstantialis* (as IN-¹, SUBSTANTIAL)]

insufferable /ɪnˈsʌfərəb(ə)l/ adj. **1** intolerable. **2** unbearably arrogant or conceited etc. □ **insufferableness** n. **insufferably** adv.

insufficiency /ˌɪnsəˈfɪʃənsɪ/ n. **1** the condition of being insufficient. **2** Med. the inability of an organ to perform its normal function (*renal insufficiency*). [ME f. LL *insufficientia* (as INSUFFICIENT)]

insufficient /ˌɪnsəˈfɪʃ(ə)nt/ adj. not sufficient; inadequate. □ **insufficiently** adv. [ME f. OF f. LL *insufficiens* (as IN-¹, SUFFICIENT)]

insufflate /ˈɪnsəˌfleɪt/ v.tr. **1** Med. **a** blow or breathe (air, gas, powder, etc.) into a cavity of the body etc. **b** treat (the nose etc.) in this way. **2** (in Christian theology) blow or breathe on (a person) to symbolize spiritual influence. □ **insufflation** /ˌɪnsəˈfleɪʃ(ə)n/ n. [LL *insufflare* insufflat- (as IN-², *sufflare* blow upon)]

insufflator /ˈɪnsəˌfleɪtə(r)/ n. **1** a device for blowing powder on to a surface in order to make fingerprints visible. **2** an instrument for insufflating.

insular /ˈɪnsjʊlə(r)/ adj. **1 a** of or like an island. **b** separated or remote, like an island. **2** ignorant of or indifferent to cultures, peoples, etc., outside one's own experience; narrow-minded. **3** of a British variant of Latin handwriting current in the Middle Ages. **4** (of climate) equable. □ **insularly** adv. **insularism** n. **insularity** /ˌɪnsjʊˈlærɪtɪ/ n. [LL *insularis* (as INSULATE)]

insulate /ˈɪnsjʊˌleɪt/ v.tr. **1** prevent the passage of electricity, heat, or sound from (a thing, room, etc.) by interposing non-conductors. **2** detach (a person or thing) from its surroundings; isolate. **3** archaic make (land) into an island. □ **insulating tape** an adhesive tape used to cover exposed electrical wires etc. □ **insulation** /ˌɪnsjʊˈleɪʃ(ə)n/ n. [L *insula* island + -ATE³]

insulator /ˈɪnsjʊˌleɪtə(r)/ n. **1** a thing or substance used for insulation against electricity, heat, or sound. **2** an insulating device to support telegraph wires etc. **3** a device preventing contact between electrical conductors.

insulin /ˈɪnsjʊlɪn/ n. Biochem. a polypeptide hormone involved in carbohydrate metabolism in humans and some other vertebrates. Insulin, discovered by the Canadian scientists F. G. Banting and C. H. Best in 1921–2, is produced in the pancreas by the islets of Langerhans. Its effects include the removal of glucose from the blood and the promotion of protein synthesis and fat storage. Deficient production of insulin causes diabetes mellitus. [L *insula* island + -IN]

insult v. & n. ● v.tr. /ɪnˈsʌlt/ **1** speak to or treat with scornful abuse or indignity. **2** offend the self-respect or modesty of. ● n. /ˈɪnsʌlt/ **1** an insulting remark or gesture. **2** (often foll. by *to*) colloq. an action which is considered offensive by virtue of its implications (*an insult to his intelligence; the payment was an insult*). **3** Med. **a** an agent causing damage to the body. **b** such damage. □ **insulter** /ɪnˈsʌltə(r)/ n. **insultingly** adv. [F *insulte* or L *insultare* (as IN-², *saltare* frequent. of *salire* salt- leap)]

insuperable /ɪnˈsuːpərəb(ə)l, ɪnˈsjuː-/ adj. **1** (of a barrier) impossible to surmount. **2** (of a difficulty etc.) impossible to overcome. □ **insuperably** adv. **insuperability** /-ˌsuːpərəˈbɪlɪtɪ, -ˌsjuː-/ n. [ME f. OF *insuperable* or L *insuperabilis* (as IN-¹, SUPERABLE)]

insupportable /ˌɪnsəˈpɔːtəb(ə)l/ adj. **1** unable to be endured. **2** unjustifiable. □ **insupportableness** n. **insupportably** adv. [F (as IN-¹, SUPPORT)]

insurance /ɪnˈʃʊərəns/ n. **1 a** the act or an instance of insuring. **b** the business of providing this. (*See note below.*) **2 a** a sum paid for this; a premium. **b** a sum paid out as compensation for theft, damage, loss, etc. **3** = *insurance policy.* **4** a measure taken to provide for a possible contingency (*take an umbrella as insurance*). □ **insurance agent** *Brit.* a person employed to collect premiums door to door. **insurance company** *Brit.* a company engaged in the business of insurance. **insurance policy** *Brit.* **1** a contract of insurance. **2** a document detailing such a policy and constituting a contract. **insurance stamp** *Brit.* a stamp certifying the payment of a sum, usu. paid weekly, for National Insurance. [earlier *ensurance* f. OF *enseürance* (as ENSURE)]

▪ Insurance was known in ancient Greece and among the maritime peoples with whom the Greeks traded. It developed first as a means of spreading the huge risks attendant on early maritime enterprises and dates as a distinct contract from the 14th century, when it evolved in the commercial cities of Italy. It is found in the Admiralty Court in England in the 16th century; life and fire insurance developed later. Lloyd's of London began in the 17th century, and the first US company was organized by Benjamin Franklin in 1752. Since the mid-19th century insurance against other kinds of risk has developed greatly. (See also ASSURANCE.)

insure /ɪnˈʃʊə(r)/ *v.tr.* **1** (often foll. by *against, for*; also *absol.*) secure the payment of a sum of money in the event of loss or damage to (property, life, a person, etc.) by regular payments or premiums (*insured the house for £100,000; we are insured against flood damage; the car's not insured*) (cf. ASSURANCE). **2** (of the owner of a property, an insurance company, etc.) secure the payment of (a sum of money) in this way. **3** (usu. foll. by *against*) provide for (a possible contingency) (*insured themselves against the rain by taking umbrellas*). **4** *US* = ENSURE. □ **insurable** *adj.* **insurability** /-ˌʃʊərəˈbɪlɪtɪ/ *n.* [ME, var. of ENSURE]

insured /ɪnˈʃʊəd/ *adj. & n.* ● *adj.* covered by insurance. ● *n.* (usu. prec. by *the*) a person etc. covered by insurance.

insurer /ɪnˈʃʊərə(r)/ *n.* **1** a person or company offering insurance policies for premiums; an underwriter. **2** a person who takes out insurance.

insurgent /ɪnˈsɜːdʒənt/ *adj. & n.* ● *adj.* **1** rising in active revolt. **2** (of the sea etc.) rushing in. ● *n.* a rebel; a revolutionary. □ **insurgence** *n.* **insurgency** *n.* (pl. **-ies**). [F f. L *insurgere insurrect-* (as IN-², *surgere* rise)]

insurmountable /ˌɪnsəˈmaʊntəb(ə)l/ *adj.* unable to be surmounted or overcome. □ **insurmountably** *adv.*

insurrection /ˌɪnsəˈrekʃ(ə)n/ *n.* a rising in open resistance to established authority; a rebellion. □ **insurrectionary** *adj.* **insurrectionist** *n.* [ME f. OF f. LL *insurrectio -onis* (as INSURGENT)]

insusceptible /ˌɪnsəˈseptɪb(ə)l/ *adj.* (usu. foll. by *of, to*) not susceptible (of treatment, to an influence). □ **insusceptibility** /-ˌseptɪˈbɪlɪtɪ/ *n.*

in-swinger /ˈɪnˌswɪŋə(r)/ *n.* **1** *Cricket* a ball bowled with a swing towards the batsman. **2** *Football* a pass or kick that sends the ball curving towards the goal.

int. *abbr.* **1** interior. **2** internal. **3** international.

intact /ɪnˈtækt/ *adj.* **1** entire; unimpaired. **2** untouched. □ **intactness** *n.* [ME f. L *intactus* (as IN-¹, *tactus* past part. of *tangere* touch)]

intagliated /ɪnˈtælɪˌeɪtɪd/ *adj.* decorated with surface carving. [It. *intagliato* past part. of *intagliare* cut into]

intaglio /ɪnˈtælɪəʊ, -ˈtɑːlɪəʊ/ *n. & v.* ● *n.* (pl. **-os**) **1** a gem with an incised design (cf. CAMEO). **2** an engraved design. **3** a carving, esp. incised, in hard material. **4** a process of printing from an engraved design. ● *v.tr.* (**-oes, -oed**) **1** engrave (material) with a sunk pattern or design. **2** engrave (such a design). [It. (as INTAGLIATED)]

intake /ˈɪnteɪk/ *n.* **1 a** the action of taking in. **b** an instance of this. **2 a** a number (of people etc.) or the amount taken in or received (*this year's intake of students*). **b** such people etc. **3** a place where water is taken into a channel or pipe from a river, or fuel or air enters an engine etc. **4** an airway into a mine. **5** esp. *N. Engl.* land reclaimed from a moor etc.

intangible /ɪnˈtændʒɪb(ə)l/ *adj. & n.* ● *adj.* **1** unable to be touched; not solid. **2** unable to be grasped mentally. ● *n.* something that cannot be precisely measured or assessed. □ **intangibly** *adv.* **intangibility** /-ˌtændʒɪˈbɪlɪtɪ/ *n.* [F *intangible* or med.L *intangibilis* (as IN-¹, TANGIBLE)]

intarsia /ɪnˈtɑːsɪə/ *n.* **1** the craft of using wood inlays, esp. as practised in 15th-century Italy. **2** work of this kind. [It. *intarsio*]

integer /ˈɪntɪdʒə(r)/ *n.* **1** a whole number. **2** a thing complete in itself. [L (adj.) = untouched, whole: see ENTIRE]

integral *adj. & n.* ● *adj.* /ˈɪntɪɡrəl, disp. ɪnˈteɡrəl/ **1 a** of a whole or necessary to the completeness of a whole. **b** forming a whole (*integral design*). **c** whole, complete. **2** *Math.* **a** of or denoted by an integer. **b** involving only integers, esp. as coefficients of a function. ● *n.*

/ˈɪntɪɡrəl/ *Math.* a quantity of which a given function is the derivative, and which may express the area under the curve of a graph of the function; a function satisfying a given differential equation. □ **definite integral** *Math.* an integral expressed as the difference between the values of the integral at specified upper and lower limits of the independent variable. **indefinite integral** *Math.* an integral expressed without limits and hence containing an indeterminate additive constant. **integral calculus** *Math.* the part of calculus concerned with integrals and integration (cf. DIFFERENTIAL). □ **integrally** *adv.* **integrality** /ˌɪntɪˈɡrælɪtɪ/ *n.* [LL *integralis* (as INTEGER)]

integrand /ˈɪntɪˌɡrænd/ *n. Math.* a function that is to be integrated. [L *integrandus* gerundive of *integrare*: see INTEGRATE]

integrant /ˈɪntɪɡrənt/ *adj.* (of parts) making up a whole; component. [F *intégrant* f. *intégrer* (as INTEGRATE)]

integrate *v. & adj.* ● *v.* /ˈɪntɪˌɡreɪt/ **1** *tr.* **a** combine (parts) into a whole. **b** complete (an imperfect thing) by the addition of parts. **2** *tr. & intr.* bring or come into equal participation in or membership of society, a school, etc. **3** *tr.* desegregate, esp. racially (a school etc.). **4** *tr. Math.* **a** find the integral of. **b** (as **integrated** *adj.*) indicating the mean value or total sum of (temperature, an area, etc.). ● *adj.* /ˈɪntɪɡrət/ **1** made up of parts. **2** whole; complete. □ **integrated circuit** *Electronics* a small chip etc. of material replacing several separate components in a conventional electrical circuit. **integrated services digital network** a telecommunications network through which sound, images, and data can be transmitted as digitized signals. □ **integrable** /ˈɪntɪɡrəb(ə)l/ *adj.* **integrative** *adj.* **integrability** /ˌɪntɪɡrəˈbɪlɪtɪ/ *n.* [L *integrare integrat-* make whole (as INTEGER)]

integration /ˌɪntɪˈɡreɪʃ(ə)n/ *n.* **1** the act or an instance of integrating. **2** the intermixing of persons previously segregated. **3** *Psychol.* the combination of the diverse elements of perception etc. in a personality. □ **integrationist** *n.* [L *integratio* (as INTEGRATE)]

integrator /ˈɪntɪˌɡreɪtə(r)/ *n.* **1** an instrument for indicating or registering the total amount or mean value of some physical quality, as area, temperature, etc. **2** a person or thing that integrates.

integrity /ɪnˈteɡrɪtɪ/ *n.* **1** moral uprightness; honesty. **2** wholeness; soundness. [ME f. F *intégrité* or L *integritas* (as INTEGER)]

integument /ɪnˈteɡjʊmənt/ *n.* a natural outer covering, as a skin, husk, rind, etc. □ **integumental** /-ˌteɡjʊˈment(ə)l/ *adj.* **integumentary** *adj.* [L *integumentum* f. *integere* (as IN-², *tegere* cover)]

intellect /ˈɪntɪˌlekt/ *n.* **1 a** the faculty of reasoning, knowing, and thinking, as distinct from feeling. **b** the understanding or mental powers (of a particular person etc.) (*his intellect is not great*). **2 a** a clever or knowledgeable person. **b** the intelligentsia regarded collectively (*the combined intellect of four universities*). [ME f. OF *intellect* or L *intellectus* understanding (as INTELLIGENT)]

intellection /ˌɪntɪˈlekʃ(ə)n/ *n.* the action or process of understanding. □ **intellective** /-ˈlektɪv/ *adj.* [ME f. med.L *intellectio* (as INTELLIGENT)]

intellectual /ˌɪntɪˈlektʊəl, -tjʊəl/ *adj. & n.* ● *adj.* **1** of or relating to the intellect. **2** possessing a high level of understanding or intelligence. **3** requiring, appealing to, or engaging the intellect. ● *n.* a person possessing a highly developed intellect. □ **intellectual property** *Law* property that is the result of creativity and does not exist in tangible form, such as patents, copyright, trademarks, etc. □ **intellectualize** *v.tr. & intr.* (also **-ise**). **intellectually** *adv.* **intellectuality** /-ˌlektʃʊˈælɪtɪ, -ˌlektjʊ-/ *n.* [ME f. L *intellectualis* (as INTELLECT)]

intellectualism /ˌɪntɪˈlektʃʊəˌlɪz(ə)m, ˌɪntɪˈlektjʊ-/ *n.* **1** the exercise, esp. when excessive, of the intellect at the expense of the emotions. **2** *Philos.* the theory that knowledge is wholly or mainly derived from pure reason. □ **intellectualist** *n.*

intelligence /ɪnˈtelɪdʒəns/ *n.* **1 a** the intellect; the understanding. **b** (of a person or an animal) quickness of understanding; wisdom. **2 a** the collection of information, esp. of military or political value. **b** people employed in this. **c** information so collected. **3** an intelligent or rational being. □ **intelligence department** a usu. government department engaged in collecting esp. secret information. **intelligence quotient** see IQ. **intelligence test** a test designed to measure intelligence rather than acquired knowledge. □ **intelligential** /-ˌtelɪˈdʒenʃ(ə)l/ *adj.* [ME f. OF f. L *intelligentia* (as INTELLIGENT)]

intelligent /ɪnˈtelɪdʒənt/ *adj.* **1** having or showing intelligence, esp. of a high level. **2** quick of mind; clever. **3 a** (of a device or machine) able to vary its behaviour in response to varying situations and requirements and past experience. **b** (esp. of a computer terminal) having its own data-processing capability; incorporating a microprocessor (opp. DUMB 8). □ **intelligently** adv. [L intelligere intellect-understand (as INTER-, legere gather, pick out, read)]

intelligentsia /ɪnˌtelɪˈdʒentsɪə/ n. **1** the class of intellectuals regarded as possessing culture and political initiative. **2** people doing intellectual work; intellectuals. [Russ. orig. applied to disaffected intellectuals in pre-revolutionary Russia, f. Pol. inteligencja f. L intelligentia (as INTELLIGENT)]

intelligible /ɪnˈtelɪdʒɪb(ə)l/ adj. **1** (often foll. by to) able to be understood; comprehensible. **2** Philos. able to be understood only by the intellect, not by the senses. □ **intelligibility** /-ˌtelɪdʒɪˈbɪlɪtɪ/ n. **intelligibly** adv. [L intelligibilis (as INTELLIGENT)]

Intelpost /ˈintelˌpəʊst/ n. the international electronic transmission of messages and graphics by fax, telex, etc. [acronym, f. International Electronic Post]

Intelsat /ˈintelˌsæt/ an international organization of more than 100 countries, formed in 1964, which owns and operates the worldwide commercial communications satellite system. Its headquarters are in Washington, DC. [acronym, f. International Telecommunications Satellite Consortium]

intemperate /ɪnˈtempərət/ adj. **1** (of a person, person's conduct, or speech) immoderate; unbridled; violent (used intemperate language). **2 a** given to excessive indulgence in alcohol. **b** excessively indulgent in one's appetites. □ **intemperance** n. **intemperately** adv. **intemperateness** n. [ME f. L intemperatus (as IN-1, TEMPERATE)]

intend /ɪnˈtend/ v.tr. **1** have as one's purpose (an action etc.); propose (we intend to go; we intend going; we intend that it shall be done). **2** (foll. by for, as, or to + infin.) design or destine (a person or a thing) for a purpose (I intend him to go; I intend it as a warning). **3** mean (what does he intend by that?). **4** (in passive; foll. by for) **a** be meant for a person to have or use etc. (they are intended for the children). **b** be meant to represent (the picture is intended for you). **5** (as **intending** adj.) who intends to be (an intending visitor). [ME entende, intende f. OF entendre, intendre f. L intendere intent- or intens- strain, direct, purpose (as IN-2, tendere stretch, tend)]

intendant /ɪnˈtendənt/ n. **1** (esp. as a title of foreign officials) a superintendent or manager of a department of public business etc. **2** the administrator of an opera house or theatre. □ **intendancy** n. [F f. L intendere (as INTEND)]

intended /ɪnˈtendɪd/ adj. & n. ● adj. **1** done on purpose; intentional. **2** designed, meant. ● n. colloq. the person one intends to marry; one's fiancé or fiancée (is this your intended?). □ **intendedly** adv.

intense /ɪnˈtens/ adj. (**intenser, intensest**) **1** (of a quality, feeling, etc.) existing in a high degree; extreme, forceful (intense joy; intense cold). **2** (of a person) feeling, or apt to feel, strong emotion (very intense about her music). **b** expressing strong emotion (a deeply intense poem). **3** (of a colour) very strong or deep. **4** (of an action etc.) highly concentrated (intense preparation). □ **intensely** adv. **intenseness** n. [ME f. OF intens or L intensus (as INTEND)]

intensifier /ɪnˈtensɪˌfaɪə(r)/ n. **1** a person or thing that intensifies. **2** Gram. a word or prefix used to give force or emphasis.

intensify /ɪnˈtensɪˌfaɪ/ v. (**-ies, -ied**) **1** tr. & intr. make or become intense or more intense. **2** tr. Photog. increase the opacity of (a negative). □ **intensification** /-ˌtensɪfɪˈkeɪʃ(ə)n/ n.

intension /ɪnˈtenʃ(ə)n/ n. **1** Logic the internal content of a concept. **2** formal the intensity, or high degree, of a quality. **3** formal the strenuous exertion of the mind or will. □ **intensional** adj. **intensionally** adv. [L intensio (as INTEND)]

intensity /ɪnˈtensɪtɪ/ n. (pl. **-ies**) **1** the quality or an instance of being intense. **2** esp. Physics the measurable amount of some quality, e.g. force, brightness, a magnetic field, etc.

intensive /ɪnˈtensɪv/ adj. & n. ● adj. **1** thorough, vigorous; directed to a single point, area, or subject (intensive study; intensive bombardment). **2** of or relating to intensity as opposed to extent; producing intensity. **3** serving to increase production in relation to costs (intensive farming). **4** (usu. in comb.) Econ. making much use of (a labour-intensive industry). **5** Gram. (of an adjective, adverb, etc.) expressing intensity; giving force or emphasis, as really in my feet are really cold. ● n. Gram. an intensive adjective, adverb, etc.; an intensifier. □ **intensive care 1** medical treatment with constant monitoring etc. of a dangerously ill patient (also (with hyphen) attrib.: intensive-care unit). **2** a part of a hospital devoted to this. □ **intensively** adv. **intensiveness** n. [F intensif -ive or med.L intensivus (as INTEND)]

intent /ɪnˈtent/ n. & adj. ● n. (usu. without article) intention; a purpose (with intent to defraud; my intent to reach the top; with evil intent). ● adj. **1** (usu. foll. by on) **a** resolved; bent; determined (was intent on succeeding).

b attentively occupied (intent on his books). **2** (esp. of a look) earnest; eager; meaningful. □ **to all intents and purposes** practically; virtually. □ **intently** adv. **intentness** n. [ME entent f. OF f. L intentus (as INTEND)]

intention /ɪnˈtenʃ(ə)n/ n. **1** (often foll. by to + infin., or of + verbal noun) a thing intended; an aim or purpose (it was not his intention to interfere; have no intention of staying). **2** the act of intending (done without intention). **3** colloq. (usu. in pl.) a person's, esp. a man's, designs in respect to marriage (are his intentions strictly honourable?). **4** Logic a conception. □ **first intention** Med. the healing of a wound by natural contact of the parts. **first intentions** Logic one's primary conceptions of things, e.g. the conceptions of a tree, an oak. **intention tremor** Med. a trembling of a part of a body when commencing a movement. **second intention** Med. the healing of a wound by granulation. **second intentions** Logic one's secondary conceptions (e.g. difference, identity, species). **special** (or **particular**) **intention** RC Ch. a special aim or purpose for which a mass is celebrated, prayers are said, etc. □ **intentioned** adj. (usu. in comb.). [ME entencion f. OF f. L intentio stretching, purpose (as INTEND)]

intentional /ɪnˈtenʃən(ə)l/ adj. done with an aim or purpose; deliberate. □ **intentionality** /-ˌtenʃəˈnælɪtɪ/ n. **intentionally** adv. [F intentionnel or med.L intentionalis (as INTENTION)]

inter /ɪnˈtɜː(r)/ v.tr. (**interred, interring**) deposit (a corpse etc.) in the earth, a tomb, etc.; bury. [ME f. OF enterrer f. Rmc (as IN-2, L terra earth)]

inter. abbr. intermediate.

inter- /ˈintə(r)/ prefix **1** between, among (intercontinental). **2** mutually, reciprocally (interbreed). [OF entre- or L inter between, among]

interact /ˌɪntərˈækt/ v.intr. act reciprocally; act on each other. □ **interactant** adj. & n.

interaction /ˌɪntərˈækʃ(ə)n/ n. **1** reciprocal action or influence. **2** Physics the action of atomic and subatomic particles on each other. □ **interactional** adj.

interactive /ˌɪntərˈæktɪv/ adj. **1** reciprocally active; acting upon or influencing each other. **2** (of a computer or other electronic device) allowing a two-way flow of information between it and a user, responding to the user's input. □ **interactively** adv. [INTERACT, after active]

inter alia /ˌɪntər ˈeɪlɪə, ˈælɪə/ adv. among other things. [L]

inter-allied /ˌɪntərˈælaɪd/ adj. relating to two or more allies (in war etc.).

interarticular /ˌɪntərɑːˈtɪkjʊlə(r)/ adj. between the contiguous surfaces of a joint.

interatomic /ˌɪntərəˈtɒmɪk/ adj. between atoms.

interbank /ˈintəˌbæŋk/ adj. agreed, arranged, or operating between banks (interbank loan).

interbed /ˌɪntəˈbed/ v.tr. (**-bedded, -bedding**) embed (one thing) among others.

interblend /ˌɪntəˈblend/ v. **1** tr. (usu. foll. by with) mingle (things) together. **2** intr. blend with each other.

interbreed /ˌɪntəˈbriːd/ v.intr. & tr. (past and past part. **-bred** /-ˈbred/) breed or cause to breed with members of a different race, stock, or species to produce a hybrid.

intercalary /ɪnˈtɜːkələrɪ, ˌɪntəˈkælərɪ/ adj. **1 a** (of a day or a month) inserted in the calendar to harmonize it with the solar year, e.g. 29 Feb. in leap years. **b** (of a year) having such an addition. **2** interpolated; intervening. [L intercalari(u)s (as INTERCALATE)]

intercalate /ɪnˈtɜːkəˌleɪt/ v.tr. **1** (also absol.) insert (an intercalary day etc.). **2** interpose (anything out of the ordinary course). **3** (as **intercalated** adj.) (of strata etc.) interposed. □ **intercalation** /-ˌtɜːkəˈleɪʃ(ə)n/ n. [L intercalare intercalat- (as INTER-, calare proclaim)]

intercede /ˌɪntəˈsiːd/ v.intr. (usu. foll. by with) interpose or intervene on behalf of another; plead (they interceded with the king for his life). □ **interceder** n. [F intercéder or L intercedere intercess- intervene (as INTER-, cedere go)]

intercellular /ˌɪntəˈseljʊlə(r)/ adj. Biol. located or occurring between cells.

intercensal /ˌɪntəˈsens(ə)l/ adj. between two censuses.

intercept v. & n. ● v.tr. /ˌɪntəˈsept/ **1** seize, catch, or stop (a person, message, vehicle, ball, etc.) going from one place to another. **2** (usu. foll. by from) cut off (light etc.). **3** check or stop (motion etc.). **4** Math. mark off (a space) between two points etc. ● n. /ˈintəˌsept/ Math. the part of a line between two points of intersection with usu. the coordinate axes or other lines. □ **interception** /ˌɪntəˈsepʃ(ə)n/ n.

interceptive /-'septɪv/ adj. [L *intercipere intercept-* (as INTER-, *capere* take)]

interceptor /ˌɪntə'septə(r)/ n. **1** an aircraft used to intercept enemy raiders. **2** a person or thing that intercepts.

intercession /ˌɪntə'seʃ(ə)n/ n. **1** the act of interceding, esp. by prayer. **2** an instance of this, esp. a prayer on behalf of another. □ **intercessional** adj. **intercessor** /-'sesə(r)/ n. **intercessory** adj. **intercessorial** /-se'sɔːrɪəl/ adj. [F *intercession* or L *intercessio* (as INTERCEDE)]

interchange v. & n. ● v.tr. /ˌɪntə'tʃeɪndʒ/ **1** (of two people) exchange (things) with each other. **2** put each of (two things) in the other's place; alternate. ● n. /'ɪntə,tʃeɪndʒ/ **1** (often foll. by *of*) a reciprocal exchange between two people etc. **2** alternation (*the interchange of woods and fields*). **3** a road junction designed so that traffic streams do not intersect. □ **interchangeable** /ˌɪntə'tʃeɪndʒəb(ə)l/ adj. **interchangeably** adv. **interchangeableness** n. **interchangeability** /ˌɪntə,tʃeɪndʒə'bɪlɪtɪ/ n. [ME f. OF *entrechangier* (as INTER-, CHANGE)]

intercity /ˌɪntə'sɪtɪ/ adj. existing or travelling between cities.

inter-class /ˌɪntə'klɑːs/ adj. existing or conducted between different social classes.

intercollegiate /ˌɪntəkə'liːdʒət/ adj. existing or conducted between colleges or universities.

intercolonial /ˌɪntəkə'ləʊnɪəl/ adj. existing or conducted between colonies.

intercom /'ɪntə,kɒm/ n. colloq. **1** a system of intercommunication by radio or telephone between or within offices, aircraft, etc. **2** an instrument used in this. [abbr.]

intercommunicate /ˌɪntəkə'mjuːnɪ,keɪt/ v.intr. **1** communicate reciprocally. **2** (of rooms etc.) have free passage into each other; have a connecting door. □ **intercommunicative** /-kətɪv/ adj. **intercommunication** /-,mjuːnɪ'keɪʃ(ə)n/ n.

intercommunion /ˌɪntəkə'mjuːnɪən/ n. **1** mutual communion. **2** a mutual action or relationship, esp. between Christian denominations.

intercommunity /ˌɪntəkə'mjuːnɪtɪ/ n. **1** the quality of being common to various groups etc. **2** having things in common.

interconnect /ˌɪntəkə'nekt/ v.tr. & intr. connect with each other. □ **interconnection** /-'nekʃ(ə)n/ n.

intercontinental /ˌɪntə,kɒntɪ'nent(ə)l/ adj. connecting or travelling between continents. □ **intercontinentally** adv.

interconvert /ˌɪntəkən'vɜːt/ v.tr. & intr. convert into each other. □ **interconvertible** adj. **interconversion** /-'vɜːʃ(ə)n/ n.

intercooler /'ɪntə,kuːlə(r)/ n. an apparatus for cooling gas between successive compressions, esp. in a car or truck engine. □ **intercool** /ˌɪntə'kuːl/ v.tr.

intercorrelate /ˌɪntə'kɒrə,leɪt/ v.tr. & intr. correlate with one another. □ **intercorrelation** /-,kɒrə'leɪʃ(ə)n/ n.

intercostal /ˌɪntə'kɒst(ə)l/ adj. between the ribs (of the body or a ship). □ **intercostally** adv.

intercounty /ˌɪntə'kaʊntɪ/ adj. existing or conducted between counties.

intercourse /'ɪntə,kɔːs/ n. **1** communication or dealings between individuals, nations, etc. **2** = *sexual intercourse*. **3** communion between human beings and God. [ME f. OF *entrecours* exchange, commerce, f. L *intercursus* (as INTER-, *currere curs-* run)]

intercrop /ˌɪntə'krɒp/ v.tr. (also absol.) (**-cropped, -cropping**) raise (a crop) among plants of a different kind, usu. in the space between rows. □ **intercropping** n.

intercross /ˌɪntə'krɒs/ v. **1** tr. & intr. lay or lie across each other. **2 a** intr. (of animals) breed with each other. **b** tr. cause to do this.

intercrural /ˌɪntə'krʊərəl/ adj. between the legs.

intercurrent /ˌɪntə'kʌrənt/ adj. **1** (of a time or event) intervening. **2** Med. **a** (of a disease) occurring during the progress of another. **b** recurring at intervals. [L *intercurrere intercurrent-* (as INTERCOURSE)]

intercut /ˌɪntə'kʌt/ v.tr. (**-cutting**; past and past part. **-cut**) Cinematog. alternate (shots) with contrasting shots by cutting.

interdenominational /ˌɪntədɪ,nɒmɪ'neɪʃən(ə)l/ adj. concerning more than one (religious) denomination. □ **interdenominationally** adv.

interdepartmental /ˌɪntə,diːpɑːt'ment(ə)l/ adj. concerning more than one department. □ **interdepartmentally** adv.

interdepend /ˌɪntədɪ'pend/ v.intr. depend on each other. □ **interdependence** n. **interdependency** n. **interdependent** adj.

interdict n. & v. ● n. /'ɪntə,dɪkt, -,daɪt/ **1** an authoritative prohibition. **2** RC Ch. a sentence debarring a person, or esp. a place, from ecclesiastical functions and privileges. **3** Sc. Law an injunction. ● v.tr. /ˌɪntə'dɪkt, -'daɪt/ **1** prohibit (an action). **2** forbid the use of. **3** (usu. foll. by *from* + verbal noun) restrain (a person). **4** (usu. foll. by *to*) forbid (a thing) to a person. **5** US **a** Mil. impede (an enemy force), esp. by bombing lines of communication or supply. **b** intercept (a prohibited commodity); prevent (its movement). □ **interdiction** /ˌɪntə'dɪkʃ(ə)n/ n. **interdictory** /-'dɪktərɪ/ adj. [ME f. OF *entredit* f. L *interdictum* past part. of *interdicere* interpose, forbid by decree (as INTER-, *dicere* say)]

interdigital /ˌɪntə'dɪdʒɪt(ə)l/ adj. between the fingers or toes.

interdigitate /ˌɪntə'dɪdʒɪ,teɪt/ v.intr. interlock like clasped fingers. [INTER- + L *digitus* finger + -ATE³]

interdisciplinary /ˌɪntə,dɪsɪ'plɪnərɪ/ adj. of or between more than one branch of learning.

interest /'ɪntrəst, -trɪst/ n. & v. ● n. **1 a** a feeling of curiosity or concern (*have no interest in fishing*). **b** a quality exciting curiosity or holding the attention (*this magazine lacks interest*). **c** the power of an issue, action, etc. to hold the attention; noteworthiness, importance (*findings of no particular interest*). **2** a subject, hobby, etc., in which one is concerned (*his interests are gardening and sport*). **3** advantage or profit, esp. when financial (*it is in your interest to go; look after your own interests*). **4** money paid for the use of money lent, or for not requiring the repayment of a debt. **5** (usu. foll. by *in*) **a** a financial stake (in an undertaking etc.). **b** a legal concern, title, or right (in property). **6 a** a party or group having a common concern (*the brewing interest*). **b** a principle in which a party or group is concerned (*the selfish pursuit of one's own welfare, self-interest*). ● v.tr. **1** excite the curiosity or attention of (*your story interests me greatly*). **2** (usu. foll. by *in*) cause (a person) to take a personal interest or share (*can I interest you in a holiday abroad?*). **3** (as **interested** adj.) having a private interest; not impartial or disinterested (*an interested party*). □ **at interest** (of money borrowed) on the condition that interest is payable. **declare an** (or **one's**) **interest** make known one's financial etc. interests in an undertaking before it is discussed. **in the interest** (or **interests**) **of** as something that is advantageous to. **lose interest** become bored or boring. **with interest 1** with interest charged or paid. **2** with increased force etc. (*returned the blow with interest*). □ **interestedly** adv. **interestedness** n. [ME, earlier *interesse* f. AF f. med.L, alt. app. after OF *interest*, both f. L *interest*, 3rd sing. pres. of *interesse* matter, make a difference (as INTER-, *esse* be)]

interesting /'ɪntrəstɪŋ, -trɪstɪŋ/ adj. causing curiosity; holding the attention. □ **in an interesting condition** archaic pregnan\ □ **interestingly** adv. **interestingness** n.

interface /'ɪntə,feɪs/ n. & v. ● n. **1** esp. Physics a surface forming a common boundary between two regions. **2** a point where interaction occurs between two systems, processes, subjects, etc. (*the interface between psychology and education*). **3** esp. Computing an apparatus for connecting two pieces of equipment so that they can be operated jointly. ● v. (often foll. by *with*) **1** tr. & intr. connect (with another piece of equipment etc.) by an interface. **2** intr. interact (with another person etc.).

interfacial /ˌɪntə'feɪʃ(ə)l/ adj. **1** included between two faces of a crystal or other solid. **2** of or forming an interface.

interfacing /'ɪntə,feɪsɪŋ/ n. a stiffish material, esp. buckram, between two layers of fabric in collars etc.

interfemoral /ˌɪntə'femərəl/ adj. between the thighs.

interfere /ˌɪntə'fɪə(r)/ v.intr. **1** (usu. foll. by *with*) **a** (of a person) meddle; obstruct a process etc. **b** (of a thing) be a hindrance; get in the way. **2** (usu. foll. by *in*) take part or intervene, esp. without invitation or necessity. **3** (foll. by *with*) euphem. molest or assault sexually. **4** Physics (of light or other waves) combine so as to cause interference. **5** (of a horse) knock one leg against another. □ **interferer** n. **interfering** adj. **interferingly** adv. [OF *s'entreferir* strike each other (as INTER-, *ferir* f. L *ferire* strike)]

interference /ˌɪntə'fɪərəns/ n. **1** (usu. foll. by *with*) **a** the act of interfering. **b** an instance of this. **2** the fading or disturbance of received radio signals by the interference of waves from different sources, or esp. by atmospherics or unwanted signals. **3** Physics the combination of two or more wave motions to form a resultant wave in which the displacement is reinforced or cancelled. □ **interferential** /-fə'renʃ(ə)l/ adj.

interferometer /ˌɪntəfə'rɒmɪt(ə)r/ n. an instrument for measuring wavelengths etc. by means of interference phenomena. □ **interferometry** n. **interferometric** /-,ferə'metrɪk/ adj. **interferometrically** adv.

interferon /ˌɪntəˈfɪərɒn/ n. Biochem. any of a group of proteins released by animal cells, usually in response to the entry of a virus, which has the property of inhibiting virus replication. The first such protein was discovered by the virologists Alick Isaacs (1921–67) and Jean Lindenmann in the UK in 1957. [INTERFERE + -ON]

interfibrillar /ˌɪntəˈfɪbrɪlə(r)/ adj. between fibrils.

interfile /ˌɪntəˈfaɪl/ v.tr. **1** file (two sequences) together. **2** file (one or more items) into an existing sequence.

interflow v. & n. ● v.intr. /ˌɪntəˈfləʊ/ flow into each other. ● n. /ˈɪntəˌfləʊ/ the process or result of this.

interfluent /ˌɪntəˈfluːənt, ɪnˈtɜːfluənt/ adj. flowing into each other. [L interfluere interfluent- (as INTER-, fluere flow)]

interfuse /ˌɪntəˈfjuːz/ v. **1** tr. **a** (usu. foll. by with) mix (a thing) with; intersperse. **b** blend (things) together. **2** intr. (of two things) blend with each other. □ **interfusion** /-ˈfjuːʒ(ə)n/ n. [L interfundere interfus- (as INTER-, fundere pour)]

intergalactic /ˌɪntəɡəˈlæktɪk/ adj. of or situated between two or more galaxies. □ **intergalactically** adv.

interglacial /ˌɪntəˈɡleɪʃ(ə)l, -ˈɡleɪsɪəl/ adj. & n. ● adj. of or relating to a period of milder climate between glacial periods. ● n. such a period.

intergovernmental /ˌɪntəˌɡʌvənˈment(ə)l/ adj. concerning or conducted between two or more governments (intergovernmental talks). □ **intergovernmentally** adv.

intergradation /ˌɪntəɡrəˈdeɪʃ(ə)n/ n. the process of merging together by gradual change of the constituents.

intergrade v. & n. ● v.intr. /ˌɪntəˈɡreɪd/ pass into another form by intervening grades. ● n. /ˈɪntəˌɡreɪd/ such a grade.

intergrowth /ˈɪntəˌɡrəʊθ/ n. the growing of things into each other.

interim /ˈɪntərɪm/ n., adj., & adv. ● n. the intervening time (in the interim he had died). ● adj. intervening; provisional, temporary. ● adv. archaic meanwhile. □ **interim dividend** a dividend declared on the basis of less than a full year's results. [L, as INTER- + adv. suffix -im]

interior /ɪnˈtɪərɪə(r)/ adj. & n. ● adj. **1** inner (opp. EXTERIOR adj. 1a). **2** remote from the coast or frontier; inland. **3** internal; domestic (cf. HOME adj. 2b; opp. FOREIGN 2). **4** (usu. foll. by to) situated further in or within. **5** existing in the mind or soul; inward. **6** drawn, photographed, etc. within a building. **7** coming from inside. ● n. **1** the interior part; the inside. **2** the interior part of a country or region. **3 a** the home affairs of a country. **b** a department dealing with these (Minister of the Interior). **4** a representation of the inside of a building or a room (Dutch interior). **5** the inner nature; the soul. □ **interior angle** the angle between adjacent sides of a rectilinear figure. **interior decoration** (or **design**) the decoration or design of the interior of a building, a room, etc. **interior monologue** a form of writing expressing a character's inner thoughts. **interior-sprung** (of a mattress etc.) with internal springs. □ **interiorize** v.tr. (also **-ise**). **interiorly** adv. [L, compar. f. inter among]

interject /ˌɪntəˈdʒekt/ v.tr. **1** utter (words) abruptly or parenthetically. **2** interrupt with. □ **interjectory** adj. [L interjicere (as INTER-, jacere throw)]

interjection /ˌɪntəˈdʒekʃ(ə)n/ n. an exclamation, esp. as a part of speech (e.g. ah!, dear me!). □ **interjectional** adj. [ME f. OF f. L interjectio -onis (as INTERJECT)]

interknit /ˌɪntəˈnɪt/ v.tr. & intr. (**-knitting**; past and past part. **-knitted** or **-knit**) knit together; intertwine.

interlace /ˌɪntəˈleɪs/ v. **1** tr. bind intricately together; interweave. **2** tr. mingle, intersperse. **3** intr. cross each other intricately. □ **interlacement** n. [ME f. OF entrelacier (as INTER-, LACE v.)]

Interlaken /ˈɪntəˌlɑːkən/ the chief town of the Bernese Alps in central Switzerland, situated on the River Aare between Lake Brienz and Lake Thun; pop. (1980) 4,852.

interlanguage /ˈɪntəˌlæŋɡwɪdʒ/ n. a language or use of language having features of two others, often a pidgin or dialect form.

interlap /ˌɪntəˈlæp/ v.intr. (**-lapped, -lapping**) overlap.

interlard /ˌɪntəˈlɑːd/ v.tr. (usu. foll. by with) diversify or provide with things scattered or placed at intervals, esp. embellish (writing or speech) with foreign or technical vocabulary. [F entrelarder (as INTER-, LARD v.)]

interleaf /ˈɪntəˌliːf/ n. (pl. **-leaves**) an extra (usu. blank) leaf between the leaves of a book.

interleave /ˌɪntəˈliːv/ v.tr. insert (usu. blank) leaves between the leaves of (a book etc.).

interleukin /ˌɪntəˈluːkɪn/ n. Biochem. any of a class of glycoproteins produced by leucocytes for regulating immune responses. [INTER- + LEUCOCYTE]

interlibrary /ˌɪntəˌlaɪbrərɪ/ adj. between libraries (esp. interlibrary loan).

interline[1] /ˌɪntəˈlaɪn/ v.tr. **1** insert words between the lines of (a document etc.). **2** insert (words) in this way. □ **interlineation** /-ˌlɪnɪˈeɪʃ(ə)n/ n. [ME f. med.L interlineare (as INTER-, LINE[1])]

interline[2] /ˌɪntəˈlaɪn/ v.tr. put an extra lining between the ordinary lining and the fabric of (a garment).

interlinear /ˌɪntəˈlɪnɪə(r)/ adj. written or printed between the lines of a text. [ME f. med.L interlinearis (as INTER-, LINEAR)]

interlining /ˌɪntəˌlaɪnɪŋ/ n. material used to interline a garment.

interlink /ˌɪntəˈlɪŋk/ v.tr. & intr. link or be linked together.

interlobular /ˌɪntəˈlɒbjʊlə(r)/ adj. situated between lobes.

interlock v., adj., & n. ● v. /ˌɪntəˈlɒk/ **1** intr. engage with each other by overlapping or by the fitting together of projections and recesses. **2** tr. (usu. in passive) lock or clasp within each other. ● adj. /ˈɪntəˌlɒk/ (of a fabric) knitted with closely interlocking stitches. ● n. /ˈɪntəˌlɒk/ a device or mechanism for connecting or coordinating the function of different components. □ **interlocker** /ˈɪntəˌlɒkə(r)/ n.

interlocutor /ˌɪntəˈlɒkjʊtə(r)/ n. (fem. **interlocutrix** /-trɪks/) a person who takes part in a dialogue or conversation. □ **interlocution** /-ləˈkjuːʃ(ə)n/ n. [mod.L f. L interloqui interlocut- interrupt in speaking (as INTER-, loqui speak)]

interlocutory /ˌɪntəˈlɒkjʊtərɪ/ adj. **1** of dialogue or conversation. **2** Law (of a decree etc.) given provisionally in a legal action. [med.L interlocutorius (as INTERLOCUTOR)]

interloper /ˈɪntəˌləʊpə(r)/ n. **1** an intruder. **2** a person who interferes in others' affairs, esp. for profit. □ **interlope** v.intr. [INTER- + loper as in landloper vagabond f. MDu. landlooper]

interlude /ˈɪntəˌluːd, -ˌljuːd/ n. **1 a** a pause between the acts of a play. **b** something performed or done during this pause. **2 a** an intervening time, space, or event that contrasts with what goes before or after. **b** a temporary amusement or entertaining episode. **3** a piece of music played between other pieces, the verses of a hymn, etc. [ME, = a light dramatic item between the acts of a morality play, f. med.L interludium (as INTER-, ludus play)]

intermarriage /ˌɪntəˈmærɪdʒ/ n. **1** marriage between people of different races, castes, families, etc. **2** (loosely) marriage between near relations.

intermarry /ˌɪntəˈmærɪ/ v.intr. (**-ies, -ied**) (foll. by with) (of races, castes, families, etc.) become connected by marriage.

intermediary /ˌɪntəˈmiːdɪərɪ/ n. & adj. ● n. (pl. **-ies**) an intermediate person or thing, esp. a mediator. ● adj. acting as mediator; intermediate. [F intermédiaire f. It. intermediario f. L intermedius (as INTERMEDIATE)]

intermediate adj., n., & v. ● adj. /ˌɪntəˈmiːdɪət/ coming between two things in time, place, order, character, etc. ● n. /ˌɪntəˈmiːdɪət/ **1** an intermediate thing. **2** a chemical compound formed by one reaction and then used in another, esp. during synthesis. ● v.intr. /ˌɪntəˈmiːdɪˌeɪt/ (foll. by between) act as intermediary; mediate. □ **intermediate frequency** the frequency to which a radio signal is converted during heterodyne reception. **intermediate technology** technology suitable for use in developing countries, esp. by making use of locally available resources. □ **intermediacy** /-ˈmiːdɪəsɪ/ n. **intermediately** adv. **intermediateness** n. **intermediation** /-ˌmiːdɪˈeɪʃ(ə)n/ n. **intermediator** /-ˈmiːdɪˌeɪtə(r)/ n. [med.L intermediatus (as INTER-, medius middle)]

interment /ɪnˈtɜːmənt/ n. the burial of a corpse, esp. with ceremony.

intermesh /ˌɪntəˈmeʃ/ v.tr. & intr. make or become meshed together.

intermezzo /ˌɪntəˈmetsəʊ/ n. (pl. **intermezzi** /-sɪ/ or **-os**) **1 a** a short connecting instrumental movement in an opera or other musical work. **b** a similar piece performed independently. **c** a short piece for a solo instrument. **2** a short light dramatic or other performance inserted between the acts of a play. [It. f. L intermedium interval (as INTERMEDIATE)]

interminable /ɪnˈtɜːmɪnəb(ə)l/ adj. **1** endless. **2** tediously long or habitual. **3** with no prospect of an end. □ **interminableness** n. **interminably** adv. [ME f. OF interminable or LL interminabilis (as IN-[1], TERMINATE)]

intermingle /ˌɪntəˈmɪŋɡ(ə)l/ v.tr. & intr. (often foll. by with) mix together; mingle.

intermission /ˌɪntəˈmɪʃ(ə)n/ n. **1** a pause or cessation. **2** an interval between parts of a play, film, concert, etc. **3** a period of inactivity. [F *intermission* or L *intermissio* (as INTERMIT)]

intermit /ˌɪntəˈmɪt/ v. (**intermitted**, **intermitting**) **1** intr. esp. Med. stop or cease activity briefly (e.g. of a fever, or a pulse). **2** tr. suspend; discontinue for a time. [L *intermittere intermiss-* (as INTER-, *mittere* let go)]

intermittent /ˌɪntəˈmɪt(ə)nt/ adj. occurring at intervals; not continuous or steady. □ **intermittence** n. **intermittency** n. **intermittently** adv. [L *intermittere intermittent-* (as INTERMIT)]

intermix /ˌɪntəˈmɪks/ v.tr. & intr. mix together. □ **intermixable** adj. **intermixture** n. [back-form. f. *intermixed, intermixt* f. L *intermixtus* past part. of *intermiscere* mix together (as INTER-, *miscere* mix)]

intermolecular /ˌɪntəməˈlekjʊlə(r)/ adj. between molecules.

intern n. & v. ● n. /ˈɪntɜːn/ (also **interne**) esp. N. Amer. a recent graduate or advanced student receiving supervised training in a hospital and acting as an assistant physician or surgeon. ● v. **1** tr. /ɪnˈtɜːn/ confine; oblige (a prisoner, alien, etc.) to reside within prescribed limits. **2** intr. /ˈɪntɜːn/ esp. N. Amer. serve as an intern. □ **internment** /ɪnˈtɜːnmənt/ n. **internship** /ˈɪntɜːnˌʃɪp/ n. esp. N. Amer. [F *interne* f. L *internus* internal]

internal /ɪnˈtɜːn(ə)l/ adj. & n. ● adj. **1** of or situated in the inside or invisible part (opp. EXTERNAL). **2** relating or applied to the inside of the body (*internal injuries*). **3** of a nation's domestic affairs. **4** (of a student) attending a university etc. as well as taking its examinations. **5** used or applying within an organization. **6 a** of the inner nature of a thing; intrinsic. **b** of the mind or soul. ● n. (in pl.) intrinsic qualities. □ **internal energy** the energy in a system arising from the relative positions and interactions of its parts. **internal evidence** evidence derived from the contents of the thing discussed. **internal exile** see EXILE n. 1. **internal market 1** = SINGLE MARKET. **2** a system of decentralized funding in the National Health Service whereby hospital departments purchase each other's services contractually. **internal rhyme** a rhyme involving a word in the middle of a line and another at the end of the line or in the middle of the next. □ **internally** adv. **internality** /ˌɪntɜːˈnælɪtɪ/ n. [mod.L *internalis* (as INTERN)]

internal-combustion engine /ɪnˌtɜːn(ə)lkəmˈbʌstʃən/ n. an engine which generates motive power by the burning of fuel with air inside the engine, the hot gases produced being used as the working fluid. Although the gas turbine, for example, is an internal-combustion engine, the term usually denotes diesel and petrol engines in which the exhaust gases are made to drive a piston as they expand. Internal-combustion engines were pioneered in the early 19th century, but it was not until the development of the four-stroke cycle by N. A. Otto and others in the late 19th century that they became sufficiently powerful and reliable to challenge the dominance of steam. The chief advantage of such engines over steam engines is their compactness; all processes take place in a single mechanism, giving a high output of power for a low weight. The disadvantage is the need for very specific types of liquid or gaseous fuels.

internalize /ɪnˈtɜːnəˌlaɪz/ v.tr. (also **-ise**) **1** make internal. **2** make (attitudes, behaviour, etc.) part of one's nature by learning or unconscious assimilation. □ **internalization** /-ˌtɜːnəlaɪˈzeɪʃ(ə)n/ n.

internat. abbr. international.

International /ˌɪntəˈnæʃ(ə)n(ə)l/ n. any of various international socialist organizations. The First International was formed by Karl Marx in London in 1864 as an international working men's association and was dissolved twelve years later after internal wrangling between Marxists and anarchists. The Second International was formed in Paris in 1889 to celebrate the 100th anniversary of the French Revolution and, although gravely weakened by the First World War, still survives as a loose association of social democrats. The Third International, also known as the Comintern, was formed by the Bolsheviks in 1919 to further the cause of world revolution. Active if seldom effective between the wars, it was abolished in 1943 as a gesture towards the Soviet Union's war allies. The Fourth International, a body of Trotskyist organizations, was formed in 1938 in opposition to the policies of the Stalin-dominated Third International.

international /ˌɪntəˈnæʃ(ə)n(ə)l/ adj. & n. ● adj. **1** existing, involving, or carried on between two or more nations. **2** agreed on or used by all or many nations (*international driving licence*). ● n. **1** Brit. a contest, esp. in sport, between teams representing different countries. **2** a member of such a team. □ **international candle** see CANDLE n. 2. **international law** a body of rules established by custom or treaty and agreed as binding by nations in their relations with one another. **international unit** a standard quantity of a vitamin etc. □ **internationally** adv. **internationality** /-ˌnæʃəˈnælɪtɪ/ n.

International Atomic Energy Agency an international organization set up in 1957 to promote research into and development of atomic energy for peaceful purposes. Its headquarters are in Vienna.

International Bank for Reconstruction and Development (abbr. **IBRD**) (also called *World Bank*) an agency of the United Nations, established in 1945 and affiliated with the International Development Association and the International Finance Corporation. It exists to promote the economic development of member nations by facilitating the investment of capital for productive purposes, encouraging private foreign investment, and if necessary lending money from its own funds. Its headquarters are in Washington, DC.

International Brigade a group of volunteers (mainly from Europe and the US) raised internationally by foreign Communist parties, which fought on the Republican side in the Spanish Civil War.

International Civil Aviation Organization an agency of the United Nations, founded in 1947 to study problems of international civil aviation and establish standards and regulations etc. Its headquarters are in Montreal.

International Confederation of Free Trade Unions (abbr. **ICFTU**) an association formed in 1949 to promote free trade unionism worldwide. Its headquarters are in Brussels. (See also WORLD FEDERATION OF TRADE UNIONS.)

International Court of Justice a judicial court of the United Nations which replaced the Cour Permanente de Justice in 1945 and meets at The Hague.

International Criminal Police Commission see INTERPOL.

International Date Line see DATE LINE.

International Development Association (abbr. **IDA**) an affiliate of the International Bank for Reconstruction and Development (World Bank), established in 1960 to provide assistance primarily in the poorer developing countries. Its headquarters are in Washington, DC.

Internationale, the /ˌɪntəˌnæsjɒˈnɑːl/ **1** a revolutionary hymn composed (1888) by Pierre Degeyter of Lille to words written earlier (1871) by Eugène Pottier, a Parisian transport worker. It was adopted by French socialists and subsequently by others, and was the official anthem of the USSR until 1944. **2** = INTERNATIONAL. [F, fem. of *international* (adj.) f. INTERNATIONAL]

International Energy Agency (abbr. **IEA**) an agency founded in 1974, within the framework of the OECD, to coordinate energy supply and demand worldwide. Its headquarters are in Paris.

International Finance Corporation an organization established in 1956 as an affiliate of the International Bank for Reconstruction and Development (World Bank) to assist developing member countries by promoting the growth of the private sector of their economies.

International Fund for Agricultural Development (abbr. **IFAD**) an agency of the United Nations that began operations in 1977, with headquarters in Rome. Its purpose is to mobilize additional funds for agricultural and rural development in developing countries through programmes that directly benefit the poorest rural populations.

internationalism /ˌɪntəˈnæʃ(ə)nəˌlɪz(ə)m/ n. **1** the advocacy of a community of interests among nations. **2** (**Internationalism**) the principles of any of the Internationals. □ **internationalist** n.

internationalize /ˌɪntəˈnæʃ(ə)nəˌlaɪz/ v.tr. (also **-ise**) **1** make international. **2** bring under the protection or control of two or more nations. □ **internationalization** /-ˌnæʃ(ə)nəlaɪˈzeɪʃ(ə)n/ n.

International Labour Organization (abbr. **ILO**) an organization, established with the League of Nations in 1919, that became in 1946 the first specialized agency associated with the United Nations. Its headquarters are in Geneva. Its aim is to promote lasting peace through social justice, and to this end it works for better economic and social conditions everywhere. The organization was awarded the Nobel Peace Prize in 1969.

International Maritime Association an agency of the United Nations established in 1958 for cooperation and exchange of information among governments on matters relating to international shipping. Its headquarters are in London.

International Maritime Satellite Organization see INMARSAT.

International Monetary Fund (abbr. **IMF**) an international organization affiliated to the UN, with headquarters in Washington, DC. Established in 1945, it exists to promote international trade and monetary cooperation and the stabilization of exchange rates. Member countries contribute in gold and in their own currencies to provide a reserve on which they may draw to meet foreign obligations during periods of deficit in their international balance of payments. In recent years most of the recipients of payments from this reserve have been developing countries. Payments are usually made on the basis of the country's acceptance of stipulated measures for economic correction, which often entail cuts in public expenditure and an increased cost of living, and have frequently caused controversy.

International Organization for Standardization an organization founded in 1946 to standardize measurements etc. for international industrial, commercial, and scientific purposes. The British Standards Institution is a member.

International Phonetic Alphabet (abbr. **IPA**) a set of phonetic symbols for international use, developed in the late 19th century with the intention of enabling strict one-for-one correspondence between sound and symbol. It is based on the Roman and Greek alphabets, with the addition of some special symbols and diacritical marks.

International Society for Krishna Consciousness see HARE KRISHNA.

international style n. an architectural style of the 20th century, associated especially with Walter Gropius, Frank Lloyd Wright, and Le Corbusier, so called because it crossed national and cultural barriers. It is characterized by the use of new building materials (especially steel and reinforced concrete), wide windows, uninterrupted interior spaces, simple lines, and strict geometric forms.

International System of Units a system of physical units known as *SI units* (together with a set of prefixes indicating multiplication or division by a power of ten), based on the metre, kilogram, second, ampere, kelvin, candela, and mole as independent basic units, with each of the derived units defined in terms of these without any multiplying factor. It was instituted in 1957. [transl. F *Système International d'Unités*]

International Telecommunications Satellite Consortium see INTELSAT.

International Telecommunication Union (abbr. **ITU**) an organization whose purpose is to promote international cooperation in the use and improvement of telecommunications of all kinds. Founded at Paris in 1865 as the International Telegraph Union, it became an agency of the United Nations in 1947; its headquarters are in Geneva.

interne var. of INTERN n.

internecine /ˌɪntəˈniːsaɪn/ adj. mutually destructive. [orig. = deadly, f. L *internecinus* f. *internecio* massacre f. *internecare* slaughter (as INTER-, *necare* kill)]

internee /ˌɪntɜːˈniː/ n. a person interned.

Internet /ˈɪntəˌnet/ propr. an international computer network linking computers from educational institutions, government agencies, industry, etc. [INTER- + NETWORK]

internist /ɪnˈtɜːnɪst/ n. esp. N. Amer. Med. a specialist in internal diseases.

internode /ˈɪntəˌnəʊd/ n. **1** Bot. a part of a stem between two of the nodes or knobs from which leaves arise. **2** Anat. a slender part between two nodes or joints, e.g. the bone of a finger or toe.

internuclear /ˌɪntəˈnjuːklɪə(r)/ adj. between nuclei.

internuncial /ˌɪntəˈnʌnʃ(ə)l/ adj. Anat. (of nerves) communicating between different parts of the system. [*internuncio* ambassador f. It. *internunzio*]

interoceanic /ˌɪntərˌəʊʃɪˈænɪk, ˌɪntərˌəʊsɪ-/ adj. between or connecting two oceans.

interoceptive /ˌɪntərəʊˈseptɪv/ adj. Biol. relating to stimuli produced within an organism, esp. in the viscera. [irreg. f. L *internus* interior + RECEPTIVE]

interoperable /ˌɪntərˈɒpərəb(ə)l/ adj. able to operate in conjunction. □ **interoperability** /-ˌɒpərəˈbɪlɪtɪ/ n.

interosculate /ˌɪntərˈɒskjʊˌleɪt/ v.intr. = INOSCULATE.

interosseous /ˌɪntərˈɒsɪəs/ adj. Anat. between bones.

interparietal /ˌɪntəpəˈraɪət(ə)l/ adj. Anat. between the right and left parietal bones of the skull. □ **interparietally** adv.

interpellate /ɪnˈtɜːpeˌleɪt/ v.tr. (in a parliament) interrupt the order of the day by demanding an explanation from (the Minister concerned). □ **interpellator** n. **interpellation** /-ˌtɜːpeˈleɪʃ(ə)n/ n. [L *interpellare* interpellat- (as INTER-, *pellere* drive)]

interpenetrate /ˌɪntəˈpenɪˌtreɪt/ v. **1** intr. (of two things) penetrate each other. **2** tr. pervade; penetrate thoroughly. □ **interpenetrative** /-trətɪv/ adj. **interpenetration** /-ˌpenɪˈtreɪʃ(ə)n/ n.

interpersonal /ˌɪntəˈpɜːsən(ə)l/ adj. (of relations) occurring between persons, esp. reciprocally. □ **interpersonally** adv.

interphase /ˈɪntəˌfeɪz/ n. Biol. the resting phase between successive mitotic divisions of a cell, or between the first and second divisions of meiosis.

interplait /ˌɪntəˈplæt/ v.tr. & intr. plait together.

interplanetary /ˌɪntəˈplænɪtərɪ/ adj. **1** between planets. **2** relating to travel between planets.

interplay /ˈɪntəˌpleɪ/ n. **1** reciprocal action. **2** the operation of two things on each other.

interplead /ˌɪntəˈpliːd/ v. **1** intr. litigate with each other to settle a point concerning a third party. **2** tr. cause to do this. □ **interpleader** /ˈɪntəˌpliːdə(r)/ n. [ME f. AF *enterpleder* (as INTER-, PLEAD)]

Interpol /ˈɪntəˌpɒl/ the International Criminal Police Commission, an organization that coordinates investigations made by the police forces of member countries into crimes with an international dimension. It was founded in 1923, with headquarters in Vienna, and reconstituted in Paris after the Second World War. [contr. of International *police*]

interpolate /ɪnˈtɜːpəˌleɪt/ v.tr. **1 a** insert (words) in a book etc., esp. to give false impressions as to its date etc. **b** make such insertions in (a book etc.). **2** interject (a remark) in a conversation. **3** estimate (intermediate values) from surrounding known values. □ **interpolator** n. **interpolative** /-lətɪv/ adj. **interpolation** /-ˌtɜːpəˈleɪʃ(ə)n/ n. [L *interpolare* furbish up (as INTER-, *polire* POLISH)]

interpose /ˌɪntəˈpəʊz/ v. **1** tr. (often foll. by between) place or insert (a thing) between others. **2** tr. say (words) as an interruption. **3** tr. exercise or advance (a veto or objection) so as to interfere. **4** intr. (often foll. by between) intervene (between parties). [F *interposer* f. L *interponere* put (as INTER-, POSE[1])]

interposition /ˌɪntəpəˈzɪʃ(ə)n/ n. **1** the act of interposing. **2** a thing interposed. **3** an interference. [ME f. OF *interposition* or L *interpositio* (as INTER-, POSITION)]

interpret /ɪnˈtɜːprɪt/ v. (**interpreted, interpreting**) **1** tr. explain the meaning of (something mysterious or abstruse, foreign words, a dream, etc.). **2** tr. make out or bring out the meaning of (creative work, esp. musical composition). **3** intr. act as an interpreter, esp. of foreign languages. **4** tr. explain or understand (behaviour etc.) in a specified manner (*interpreted his gesture as mocking*). □ **interpretable** adj. **interpretability** /-ˌtɜːprɪtəˈbɪlɪtɪ/ n. **interpretation** /-ˌtɜːprɪˈteɪʃ(ə)n/ n. **interpretational** adj. **interpretative** /-ˈtɜːprɪtətɪv/ adj. **interpretive** adj. **interpretively** adv. [ME f. OF *interpreter* or L *interpretari* explain, translate f. *interpres -pretis* explainer]

interpreter /ɪnˈtɜːprɪtə(r)/ n. a person, esp. an official, who translates speech orally. [ME f. AF *interpretour*, OF *interpreteur* f. LL *interpretator -oris* (as INTERPRET)]

interprovincial /ˌɪntəprəˈvɪnʃ(ə)l/ adj. situated or carried on between provinces.

interracial /ˌɪntəˈreɪʃ(ə)l/ adj. existing between or affecting different races. □ **interracially** adv.

interregnum /ˌɪntəˈregnəm/ n. (pl. **interregnums** or **interregna** /-nə/) **1** an interval when the normal government is suspended, esp. between successive reigns or regimes. **2** an interval or pause. [L (as INTER-, *regnum* reign)]

interrelate /ˌɪntərɪˈleɪt/ v. **1** tr. relate (two or more things) to each other. **2** intr. (of two or more things) relate to each other. □ **interrelation** /-ˈleɪʃ(ə)n/ n. **interrelationship** n.

interrogate /ɪnˈterəˌgeɪt/ v.tr. ask questions of (a person) esp. closely, thoroughly, or formally. □ **interrogator** n. [ME f. L *interrogare interrogat-* ask (as INTER-, *rogare* ask)]

interrogation /ɪnˌterəˈgeɪʃ(ə)n/ n. **1** the act or an instance of interrogating; the process of being interrogated. **2** a question or enquiry. □ **interrogation point** (or **mark** etc.) = question mark. □ **interrogational** adj. [ME f. F *interrogation* or L *interrogatio* (as INTERROGATE)]

interrogative /ˌɪntəˈrɒgətɪv/ adj. & n. ● adj. **1 a** of or like a question; used in questions. **b** Gram. (of an adjective or pronoun) asking a question (e.g. *who?*, *which?*). **2** having the form or force of a question. **3** suggesting enquiry (*an interrogative tone*). ● n. an interrogative word (e.g. *what?*, *why?*). □ **interrogatively** adv. [LL *interrogativus* (as INTERROGATE)]

interrogatory /ˌɪntəˈrɒgətərɪ/ adj. & n. ● adj. questioning; of or suggesting enquiry (*an interrogatory eyebrow*). ● n. (pl. **-ies**) a formal set of questions, esp. Law one formally put to an accused person etc. [LL *interrogatorius* (as INTERROGATE)]

interrupt /ˌɪntəˈrʌpt/ *v.tr.* **1** act so as to break the continuous progress of (an action, speech, a person speaking, etc.). **2** obstruct (a person's view etc.). **3** break or suspend the continuity of. □ **interruptible** *adj.* **interruptive** *adj.* **interruptory** *adj.* **interruption** /-ˈrʌpʃ(ə)n/ *n.* [ME f. L *interrumpere interrupt-* (as INTER-, *rumpere* break)]

interrupter /ˌɪntəˈrʌptə(r)/ *n.* (also **interruptor**) **1** a person or thing that interrupts. **2** a device for interrupting, esp. an electric circuit.

intersect /ˌɪntəˈsekt/ *v.* **1** *tr.* divide (a thing) by passing or lying across it. **2** *intr.* (of lines, roads, etc.) cross or cut each other. [L *intersecare intersect-* (as INTER-, *secare* cut)]

intersection /ˌɪntəˈsekʃ(ə)n/ *n.* **1** the act of intersecting. **2** a place where two or more roads intersect. **3** a point or line common to lines or planes that intersect. □ **intersectional** *adj.* [L *intersectio* (as INTERSECT)]

interseptal /ˌɪntəˈseptəl/ *adj.* between septa or partitions.

intersex /ˈɪntəˌseks/ *n.* **1** the abnormal condition of being intermediate between male and female. **2** an individual in this condition.

intersexual /ˌɪntəˈseksjʊəl, -ˈsekʃʊəl/ *adj.* **1** existing between the sexes. **2** of or relating to intersex. □ **intersexuality** /-ˌseksjʊˈælɪtɪ, -ˌsekʃʊ-/ *n.*

interspace *n. & v.* ● *n.* /ˈɪntəˌspeɪs/ an interval of space or time. ● *v.tr.* /ˌɪntəˈspeɪs/ put interspaces between.

interspecific /ˌɪntəspɪˈsɪfɪk/ *adj.* Biol. **1** formed from different species. **2** occurring among individuals of different species.

intersperse /ˌɪntəˈspɜːs/ *v.tr.* **1** (often foll. by *between, among*) scatter; place here and there. **2** (foll. by *with*) diversify (a thing or things with others so scattered). □ **interspersion** /-ˈspɜːʃ(ə)n/ *n.* [L *interspergere interspers-* (as INTER-, *spargere* scatter)]

interspinal /ˌɪntəˈspaɪn(ə)l/ *adj.* (also **interspinous** /-nəs/) between spines or spinous processes.

interstadial /ˌɪntəˈsteɪdɪəl/ *adj. & n.* ● *adj.* of or relating to a minor period of ice retreat during a glacial period. ● *n.* such a period. [INTER- + L *stadium* stage]

interstate /ˈɪntəsteɪt/ *adj. & n.* ● *adj.* existing or carried on between states, esp. of the US. ● *n.* US each motorway of a system of motorways between states.

interstellar /ˌɪntəˈstelə(r)/ *adj.* occurring or situated between stars.

interstice /ɪnˈtɜːstɪs/ *n.* **1** an intervening space. **2** a chink or crevice. [L *interstitium* (as INTER-, *sistere stit-* stand)]

interstitial /ˌɪntəˈstɪʃ(ə)l/ *adj.* of, forming, or occupying interstices. □ **interstitially** *adv.*

intertextuality /ˌɪntəˌtekstʃʊˈælɪtɪ, ˌɪntəˌtekstjʊ-/ *n.* the relationship between esp. literary texts.

intertidal /ˌɪntəˈtaɪd(ə)l/ *adj.* of or relating to the area which is covered at high tide and uncovered at low tide.

intertribal /ˌɪntəˈtraɪb(ə)l/ *adj.* existing or occurring between different tribes.

intertrigo /ˌɪntəˈtraɪgəʊ/ *n.* (pl. **-os**) Med. inflammation from the rubbing of one area of skin on another. [L f. *intertrerere intertrit-* (as INTER-, *terere* rub)]

intertwine /ˌɪntəˈtwaɪn/ *v.* **1** *tr.* (often foll. by *with*) entwine (together). **2** *intr.* become entwined. □ **intertwinement** *n.*

intertwist /ˌɪntəˈtwɪst/ *v.tr.* twist together.

interval /ˈɪntəv(ə)l/ *n.* **1** an intervening time or space. **2** Brit. a pause or break, esp. between the parts of a theatrical or musical performance. **3** the difference in pitch between two sounds. **4** the distance between persons or things in respect of qualities. □ **at intervals** here and there; now and then. □ **intervallic** /ˌɪntəˈvælɪk/ *adj.* [ME ult. f. L *intervallum* space between ramparts, interval (as INTER-, *vallum* rampart)]

intervene /ˌɪntəˈviːn/ *v.intr.* (often foll. by *between, in*) **1** occur in time between events. **2** interfere; come between so as to prevent or modify the result or course of events. **3** be situated between things. **4** come in as an extraneous factor or thing. **5** *Law* interpose in a lawsuit as a third party. □ **intervener** *n.* **intervenient** *adj.* **intervenor** *n.* [L *intervenire* (as INTER-, *venire* come)]

intervention /ˌɪntəˈvenʃ(ə)n/ *n.* **1** the act or an instance of intervening. **2** interference, esp. by a state in another's affairs. **3** mediation. [ME f. F *intervention* or L *interventio* (as INTERVENE)]

interventionist /ˌɪntəˈvenʃənɪst/ *n.* a person who favours intervention. □ **interventionism** *n.*

intervertebral /ˌɪntəˈvɜːtɪbrəl/ *adj.* between vertebrae.

interview /ˈɪntəˌvjuː/ *n. & v.* ● *n.* **1** an oral examination of an applicant for employment, a college place, etc. **2** a conversation between a reporter etc. and a person of public interest, used as a basis of a broadcast or publication. **3** a meeting of persons face to face, esp. for consultation. **4** a session of formal questioning by the police. ● *v.* **1** *tr.* hold an interview with. **2** *tr.* question to discover the opinions or experience of (a person). **3** *intr.* participate in an interview; perform (well etc.) at interview. □ **interviewer** *n.* **interviewee** /ˌɪntəvjuːˈiː/ *n.* [F *entrevue* f. *s'entrevoir* see each other (as INTER-, *voir* f. L *videre* see: see VIEW)]

interwar /ˌɪntəˈwɔː(r)/ *attrib.adj.* existing in the period between two wars, esp. the two world wars.

interweave /ˌɪntəˈwiːv/ *v.tr.* (*past* **-wove** /-ˈwəʊv/; *past part.* **-woven** /-ˈwəʊv(ə)n/) **1** (often foll. by *with*) weave together. **2** blend intimately.

interwind /ˌɪntəˈwaɪnd/ *v.tr. & intr.* (*past* and *past part.* **-wound** /-ˈwaʊnd/) wind together.

interwork /ˌɪntəˈwɜːk/ *v.* **1** *intr.* work together or interactively. **2** *tr.* interweave.

intestate /ɪnˈtesteɪt, -tət/ *adj. & n.* ● *adj.* (of a person) not having made a will before death. ● *n.* a person who has died intestate. □ **intestacy** /-təsɪ/ *n.* [ME f. L *intestatus* (as IN-[1], *testari testat-* make a will f. *testis* witness)]

intestinal /ˌɪntesˈtaɪn(ə)l, ɪnˈtestɪn(ə)l/ *adj.* of, relating to, or found in the intestines. □ **intestinal flora** the symbiotic bacteria naturally inhabiting the gut.

intestine /ɪnˈtestɪn/ *n.* (in *sing.* or *pl.*) **1** (in vertebrates) the lower part of the alimentary canal from the end of the stomach to the anus. **2** (esp. in invertebrates) the whole alimentary canal from the mouth downward. ● **large intestine** the caecum, colon, and rectum collectively. **small intestine** the duodenum, jejunum, and ileum collectively. [L *intestinum* f. *intestinus* internal]

inthrall US var. of ENTHRAL.

inti /ˈɪntɪ/ *n.* (pl. same or **intis**) a former monetary unit of Peru, equal to 100 centimos. [Sp. f. Quechua *ɣnti* sun, the Inca sun-god]

intifada /ˌɪntɪˈfɑːdə/ *n.* a movement of Palestinian uprising in the Israeli-occupied West Bank and Gaza Strip, beginning in 1987. [Arab., = uprising]

intimacy /ˈɪntɪməsɪ/ *n.* (pl. **-ies**) **1** intimate friendship; close familiarity. **2** an instance of this, esp. a sexual act. **3** an intimate remark; an endearment.

intimate[1] /ˈɪntɪmət/ *adj. & n.* ● *adj.* **1** closely acquainted; familiar, close (*an intimate friend; an intimate relationship*). **2** private and personal (*intimate thoughts*). **3** (usu. foll. by *with*) having sexual relations. **4** (of knowledge) detailed, thorough. **5** (of a relationship between things) close. **6** (of mixing etc.) thorough. **7** essential, intrinsic. **8** (of a place etc.) friendly; promoting close personal relationships. ● *n.* a very close friend. □ **intimately** *adv.* [L *intimus* inmost]

intimate[2] /ˈɪntɪˌmeɪt/ *v.tr.* **1** (often foll. by *that* + clause) state or make known. **2** imply, hint. □ **intimation** /ˌɪntɪˈmeɪʃ(ə)n/ *n.* [LL *intimare* announce f. L *intimus* inmost]

intimidate /ɪnˈtɪmɪˌdeɪt/ *v.tr.* frighten or overawe, esp. to subdue or influence. □ **intimidating** *adj.* **intimidator** *n.* **intimidatory** *adj.* **intimidation** /-ˌtɪmɪˈdeɪʃ(ə)n/ *n.* [med.L *intimidare* (as IN-[2], *timidare* f. *timidus* TIMID)]

intinction /ɪnˈtɪŋkʃ(ə)n/ *n.* Eccl. the dipping of the Eucharistic bread in the wine so that the communicant receives both together. [LL *intinctio* f. L *intingere intinct-* (as IN-[2], TINGE)]

intitule /ɪnˈtɪtjuːl/ *v.tr.* Brit. entitle (an Act of Parliament). [OF *intituler* f. LL *intitulare* (as IN-[2], *titulare* f. *titulus* title]

into /ˈɪntuː, -tə/ *prep.* **1** expressing motion or direction to a point on or within (*walked into a tree; ran into the house*). **2** expressing direction of attention or concern (*will look into it*). **3** expressing a change of state (*turned into a dragon; separated into groups; forced into cooperation*). **4** after the beginning of (*five minutes into the game*). **5** *colloq.* interested in; knowledgeable about (*is really into art*). [OE *intō* (as IN, TO)]

intolerable /ɪnˈtɒlərəb(ə)l/ *adj.* that cannot be endured. □ **intolerableness** *n.* **intolerably** *adv.* [ME f. OF *intolerable* or L *intolerabilis* (as IN-[1], TOLERABLE)]

intolerant /ɪnˈtɒlərənt/ *adj.* not tolerant, esp. of views, beliefs, or behaviour differing from one's own. □ **intolerance** *n.* **intolerantly** *adv.* [L *intolerans* (as IN-[1], TOLERANT)]

intonate /ˈɪntəˌneɪt/ *v.tr.* intone. [med.L *intonare*: see INTONE]

intonation /ˌɪntəˈneɪʃ(ə)n/ n. **1** modulation of the voice; accent. **2** the act of intoning. **3** accuracy of pitch in playing or singing (has good intonation). **4** the opening phrase of a plainsong melody. □ **intonational** adj. [med.L intonatio (as INTONE)]

intone /ɪnˈtəʊn/ v.tr. **1** recite (prayers etc.) with prolonged sounds, esp. in a monotone. **2** utter with a particular tone. □ **intoner** n. [med.L intonare (as IN-², L tonus TONE)]

in toto /ɪn ˈtəʊtəʊ/ adv. completely. [L]

intoxicant /ɪnˈtɒksɪkənt/ adj. & n. ● adj. intoxicating. ● n. an intoxicating substance.

intoxicate /ɪnˈtɒksɪˌkeɪt/ v.tr. **1** make drunk. **2** excite or elate beyond self-control. □ **intoxication** /-ˌtɒksɪˈkeɪʃ(ə)n/ n. [med.L intoxicare (as IN-², toxicare poison f. L toxicum): see TOXIC]

intoxicating /ɪnˈtɒksɪˌkeɪtɪŋ/ adj. **1** liable to cause intoxication; alcoholic. **2** exhilarating, exciting. □ **intoxicatingly** adv.

intra- /ˈɪntrə/ prefix forming adjectives usu. from adjectives, meaning 'on the inside, within' (intracellular). [L intra inside]

intracellular /ˌɪntrəˈseljʊlə(r)/ adj. Biol. located or occurring within a cell or cells.

intracranial /ˌɪntrəˈkreɪnɪəl/ adj. within the skull. □ **intracranially** adv.

intractable /ɪnˈtræktəb(ə)l/ adj. **1** hard to control or deal with. **2** difficult, stubborn. □ **intractably** adv. **intractableness** n. **intractability** /-ˌtræktəˈbɪlɪtɪ/ n. [L intractabilis (as IN-¹, TRACTABLE)]

intrados /ɪnˈtreɪdɒs/ n. the lower or inner curve of an arch (opp. EXTRADOS). [F (as INTRA-, dos back f. L dorsum)]

intramolecular /ˌɪntrəməˈlekjʊlə(r)/ adj. within a molecule.

intramural /ˌɪntrəˈmjʊərəl/ adj. **1** situated or done within walls. **2 a** forming part of normal university or college studies. **b** taking place within a single (esp. educational) institution. □ **intramurally** adv.

intramuscular /ˌɪntrəˈmʌskjʊlə(r)/ adj. in or into a muscle or muscles.

intransigent /ɪnˈtrænsɪdʒənt, ɪnˈtrænzɪ-/ adj. & n. ● adj. uncompromising, stubborn. ● n. an intransigent person. □ **intransigence** n. **intransigency** n. **intransigently** adv. [F intransigeant f. Sp. los intransigentes extreme republicans in Cortes, ult. formed as IN-¹ + L transigere transigent- come to an understanding (as TRANS-, agere act)]

intransitive /ɪnˈtrænzɪtɪv, ɪnˈtrɑːnz-/ adj. (of a verb or sense of a verb) that does not take or require a direct object (whether expressed or implied), e.g. look in look at the sky (opp. TRANSITIVE). □ **intransitively** adv. **intransitivity** /-ˌtrænzɪˈtɪvɪtɪ, -ˌtrɑːnz-/ n. [LL intransitivus (as IN-¹, TRANSITIVE)]

intrauterine /ˌɪntrəˈjuːtəˌraɪn, -rɪn/ adj. within the womb. □ **intrauterine death** (abbr. **IUD**) death of the foetus before birth. **intrauterine device** (abbr. **IUD**) a contraceptive device placed in the womb.

intravenous /ˌɪntrəˈviːnəs/ adj. in or into a vein or veins. □ **intravenously** adv. [INTRA- + L vena vein]

in-tray /ˈɪntreɪ/ n. a tray for incoming documents, letters, etc.

intrepid /ɪnˈtrepɪd/ adj. fearless; very brave. □ **intrepidly** adv. **intrepidity** /ˌɪntrɪˈpɪdɪtɪ/ n. [F intrépide or L intrepidus (as IN-¹, trepidus alarmed)]

intricate /ˈɪntrɪkət/ adj. very complicated; perplexingly detailed or obscure. □ **intricacy** n. (pl. **-ies**). **intricately** adv. [ME f. L intricare intricat- (as IN-², tricare f. tricae tricks)]

intrigant /ˈɪntrɪgənt/ n. (fem. **intrigante** /ˌɪntrɪˈgɒnt/) an intriguer. [F intriguant f. intriguer: see INTRIGUE]

intrigue v. & n. ● v. /ɪnˈtriːg/ (**intrigues**, **intrigued**, **intriguing**) **1** intr. (foll. by with) **a** carry on an underhand plot. **b** use secret influence. **2** tr. arouse the curiosity of; fascinate. ● n. /ɪnˈtriːg, ˈɪntriːg/ **1** an underhand plot or plotting. **2** archaic a secret love affair. □ **intriguer** /ɪnˈtriːgə(r)/ n. **intriguing** adj. (esp. in sense 2 of v.). **intriguingly** adv. [F intrigue (n.), intriguer (v.) f. It. intrigo, intrigare f. L (as INTRICATE)]

intrinsic /ɪnˈtrɪnzɪk/ adj. inherent, essential; belonging naturally (intrinsic value). Opp. EXTRINSIC 1. □ **intrinsically** adv. [ME, = interior, f. F intrinsèque f. LL intrinsecus f. L intrinsecus (adv.) inwardly]

intro /ˈɪntrəʊ/ n. (pl. **-os**) colloq. an introduction. [abbr.]

intro- /ˈɪntrəʊ/ comb. form into (introgression). [L intro to the inside]

introduce /ˌɪntrəˈdjuːs/ v.tr. **1** (foll. by to) make (a person or oneself) known by name to another, esp. formally. **2** announce or present to an audience. **3 a** bring (a custom, idea, etc.) into use. **b** put on sale for

the first time. **4** bring (a piece of legislation) before a legislative assembly. **5** (foll. by to) draw the attention or extend the understanding of (a person) to a subject. **6** (often foll. by into) insert; add or incorporate (introduced a note of optimism). **7** bring in; usher in; bring forward. **8** begin; occur just before the start of. □ **introducer** n. **introducible** adj. [ME f. L introducere introduct- (as INTRO-, ducere lead)]

introduction /ˌɪntrəˈdʌkʃ(ə)n/ n. **1** the act or an instance of introducing; the process of being introduced. **2** a formal presentation of one person to another. **3** an explanatory section at the beginning of a book etc. **4** a preliminary section in a piece of music, often thematically different from the main section. **5** an introductory treatise on a subject. **6** a thing introduced. [ME f. OF introduction or L introductio (as INTRODUCE)]

introductory /ˌɪntrəˈdʌktərɪ/ adj. serving as an introduction; preliminary. [LL introductorius (as INTRODUCTION)]

introit /ˈɪntrɔɪt/ n. a psalm or antiphon sung or said while the priest approaches the altar for the Eucharist. [ME f. OF f. L introitus f. introire introit- enter (as INTRO-, ire go)]

introjection /ˌɪntrəʊˈdʒekʃ(ə)n/ n. the unconscious incorporation of external ideas into one's mind. [INTRO- after projection]

intromit /ˌɪntrəˈmɪt/ v.tr. (**intromitted**, **intromitting**) **1** archaic (foll. by into) let in, admit. **2** insert. □ **intromittent** adj. **intromission** /-ˈmɪʃ(ə)n/ n. [L intromittere intromiss- introduce (as INTRO-, mittere send)]

introspection /ˌɪntrəˈspekʃ(ə)n/ n. the examination or observation of one's own mental and emotional processes etc. □ **introspective** /-ˈspektɪv/ adj. **introspectively** adv. **introspectiveness** n. [L introspicere introspect- look inwards (as INTRO-, specere look)]

introvert n., adj., & v. ● n. /ˈɪntrəˌvɜːt/ **1** Psychol. a person predominantly concerned with his or her own thoughts and feelings rather than with external things. **2** a shy inwardly thoughtful person. ● adj. /ˈɪntrəˌvɜːt/ (also **introverted**) typical or characteristic of an introvert. ● v.tr. /ˌɪntrəˈvɜːt/ **1** Psychol. direct (one's thoughts or mind) inwards. **2** Zool. withdraw (an organ etc.) within its own tube or base, like the finger of a glove. □ **introversion** /ˌɪntrəˈvɜːʃ(ə)n/ n. **introversive** /-ˈvɜːsɪv/ adj. **introvertive** /ˈɪntrəˌvɜːtɪv/ adj. **introverted** /-ˌvɜːtɪd/ adj. [INTRO- + vert as in INVERT]

intrude /ɪnˈtruːd/ v. (foll. by on, upon, into) **1** intr. come uninvited or unwanted; force oneself abruptly on others. **2** tr. thrust or force (something unwelcome) on a person. [L intrudere intrus- (as IN-², trudere thrust)]

intruder /ɪnˈtruːdə(r)/ n. a person who intrudes, esp. into a building with criminal intent.

intrusion /ɪnˈtruːʒ(ə)n/ n. **1** the act or an instance of intruding. **2** an unwanted interruption etc. **3** Geol. an influx of molten rock between or through strata etc. but not reaching the surface. **4** the occupation of a vacant estate etc. to which one has no claim. [ME f. OF intrusion or med.L intrusio (as INTRUDE)]

intrusive /ɪnˈtruːsɪv/ adj. **1** that intrudes or tends to intrude. **2** characterized by intrusion. **3** Phonet. pronounced between words or syllables to facilitate pronunciation, e.g. intrusive r in saw a film (/ˌsɔːrəˈfɪlm/). □ **intrusively** adv. **intrusiveness** n.

intrust var. of ENTRUST.

intubate /ˈɪntjʊˌbeɪt/ v.tr. Med. insert a tube into the trachea for ventilation, usu. during anaesthesia. □ **intubation** /ˌɪntjʊˈbeɪʃ(ə)n/ n. [IN-² + L tuba tube]

intuit /ɪnˈtjuːɪt/ v. **1** tr. know by intuition. **2** intr. receive knowledge by direct perception. □ **intuitable** adj. [L intueri intuit- consider (as IN-², tueri look)]

intuition /ˌɪntjuːˈɪʃ(ə)n/ n. **1** immediate apprehension by the mind without reasoning. **2** immediate apprehension by a sense. **3** immediate insight. □ **intuitional** adj. [LL intuitio (as INTUIT)]

intuitionism /ˌɪntjuːˈɪʃəˌnɪz(ə)m/ n. (also **intuitionalism** /-ˈɪʃənəˌlɪz(ə)m/) Philos. the theory that the perception of truth is by intuition; the theory that in perception objects are known immediately by intuition; the theory that ethical principles are matters of intuition. □ **intuitionist** n.

intuitive /ɪnˈtjuːɪtɪv/ adj. **1** of, characterized by, or possessing intuition. **2** perceived by intuition. **3** Philos. proceeding by intuition (opp. DISCURSIVE). □ **intuitively** adv. **intuitiveness** n. [med.L intuitivus (as INTUIT)]

intuitivism /ɪnˈtjuːɪtɪˌvɪz(ə)m/ n. the doctrine that ethical principles can be established by intuition. □ **intuitivist** n.

intumesce /ˌɪntjuːˈmes/ v.intr. swell up. □ **intumescence** n.

intumescent adj. [L intumescere (as IN-², tumescere inceptive of tumere swell)]

intussusception /ˌɪntəsəˈsepʃ(ə)n/ n. **1** Med. the inversion of one portion of the intestine within another. **2** Bot. the deposition of new cellulose particles in a cell wall, to increase the surface area of the cell. [F intussusception or mod.L intussusceptio f. L intus within + susceptio f. suscipere take up]

intwine var. of ENTWINE.

Inuit /ˈɪnjuːɪt, ˈɪnʊɪt/ n. (also **Innuit**) (pl. same) **1 a** an Inupiaq-speaking Eskimo, esp. in Canada. **b** (in general use) an Eskimo. **2 a** the Inupiaq languages, esp. the language of Canadian Eskimos. **b** (in general use) the languages of the Eskimos. ¶ See usage note at ESKIMO. [Eskimo (Inuit) inuit people, pl. of inuk person; cf. INUK]

Inuk /ˈɪnʊk/ n. (pl. same) **1** an Eskimo of Canada or Greenland. **2** the language of these peoples. [Eskimo (Inuit), lit. 'person'; cf. INUIT]

Inuktitut /ɪˈnʊktɪˌtʊt/ n. an Inupiaq language, spoken by the Inuit of northern Canada. [Eskimo (Inuit), lit. 'the Eskimo way']

inundate /ˈɪnʌnˌdeɪt/ v.tr. (often foll. by with) **1** flood. **2** overwhelm (inundated with enquiries). □ **inundation** /ˌɪnʌnˈdeɪʃ(ə)n/ n. [L inundare flow (as IN-², unda wave)]

Inupiaq /ɪˈnuːpɪˌæk/ n. & adj. (also **Inupik** /-pɪk/, **Inupiat** /-pɪˌæt/) ● n. **1** the language of the Inuit, a major division of the Eskimo-Aleut family, comprising numerous dialects spoken in northern Alaska, the Canadian Arctic, and Greenland. **2** an Inupiaq-speaking Eskimo, esp. in north-western Alaska. ● adj. of or relating to the Inupiaqs or their language. [Eskimo (Inuit), f. inuk person (cf. INUK) + piaq genuine]

inure /ɪˈnjʊə(r)/ v. **1** tr. (often in passive; foll. by to) accustom (a person) to something esp. unpleasant. **2** intr. Law come into operation; take effect. □ **inurement** n. [ME f. AF eneurer f. phr. en eure (both unrecorded) in use or practice, f. en in + OF e(u)vre work f. L opera]

in utero /ɪn ˈjuːtəˌrəʊ/ adv. in the womb; before birth. [L]

in vacuo /ɪn ˈvækjʊˌəʊ/ adv. in a vacuum. [L]

invade /ɪnˈveɪd/ v.tr. (often absol.) **1** enter (a country etc.), esp. in a hostile manner and with armed force with intent to control or subdue it; attack. **2** swarm into or on to (fans invaded the football pitch). **3** (of a disease) attack (a body etc.). **4** encroach upon (a person's rights, esp. privacy). □ **invader** n. [L invadere invas- (as IN-², vadere go)]

invaginate /ɪnˈvædʒɪˌneɪt/ v.tr. **1** put in a sheath. **2** turn (a tube or tubular organ) inside out. □ **invagination** /-ˌvædʒɪˈneɪʃ(ə)n/ n. [IN-² + L vagina sheath]

invalid¹ n. & v. ● n. /ˈɪnvəˌliːd, -lɪd/ **1** a person enfeebled or disabled by illness or injury. **2** (attrib.) **a** of or for invalids (invalid car; invalid diet). **b** being an invalid (caring for her invalid mother). ● v. /ˈɪnvəˌliːd/ (**invalided, invaliding**) **1** tr. (often foll. by out etc.) remove from active service (a person who has become an invalid). **2** tr. (usu. in passive) disable (a person) by illness. **3** intr. become an invalid. □ **invalidism** n. [L invalidus weak, infirm (as IN-¹, VALID)]

invalid² /ɪnˈvælɪd/ adj. not valid, esp. having no legal force. □ **invalidly** adv. [L invalidus (as INVALID¹)]

invalidate /ɪnˈvælɪˌdeɪt/ v.tr. **1** make (esp. an argument etc.) invalid. **2** remove the validity or force of (a treaty, contract, etc.). □ **invalidation** /-ˌvælɪˈdeɪʃ(ə)n/ n. [med.L invalidare invalidat- (as IN-¹, validus VALID)]

invalidity /ˌɪnvəˈlɪdɪtɪ/ n. **1** lack of validity. **2** bodily infirmity. [F invalidité or med.L invaliditas (as INVALID¹)]

invaluable /ɪnˈvæljʊəb(ə)l/ adj. above valuation; inestimable. □ **invaluableness** n. **invaluably** adv.

Invar /ɪnˈvɑː(r)/ n. propr. an iron-nickel alloy with a negligible coefficient of expansion, used in the manufacture of clocks and scientific instruments. [abbr. of INVARIABLE]

invariable /ɪnˈveərɪəb(ə)l/ adj. **1** unchangeable. **2** always the same. **3** Math. constant, fixed. □ **invariably** adv. **invariableness** n. **invariability** /-ˌveərɪəˈbɪlɪtɪ/ n. [F invariable or LL invariabilis (as IN-¹, VARIABLE)]

invariant /ɪnˈveərɪənt/ adj. & n. ● adj. invariable. ● n. Math. a function which remains unchanged when a specified transformation is applied. □ **invariance** n.

invasion /ɪnˈveɪʒ(ə)n/ n. **1** the act of invading or process of being invaded. **2** an entry of a hostile army into a country. [F invasion or LL invasio (as INVADE)]

invasive /ɪnˈveɪsɪv/ adj. **1** (of weeds, cancer cells, etc.) tending to spread. **2** (of medical procedures etc.) involving the introduction of instruments into the body.

invective /ɪnˈvektɪv/ n. **1 a** strongly attacking words. **b** the use of these. **2** abusive rhetoric. [ME f. OF f. LL invectivus attacking (as INVEIGH)]

inveigh /ɪnˈveɪ/ v.intr. (foll. by against) speak or write with strong hostility. [L invehi go into, assail (as IN-², vehi passive of vehere vect- carry)]

inveigle /ɪnˈveɪg(ə)l, -ˈviːg(ə)l/ v.tr. (foll. by into, or to + infin.) entice; persuade by guile. □ **inveiglement** n. [earlier enve(u)gle f. AF envegler, OF aveugler to blind f. aveugle blind prob. f. Rmc ab oculis (unrecorded) without eyes]

invent /ɪnˈvent/ v.tr. **1** create by thought, devise; originate (a new method, an instrument, etc.). **2** concoct (a false story etc.). [ME, = discover, f. L invenire invent- find, contrive (as IN-², venire vent- come)]

invention /ɪnˈvenʃ(ə)n/ n. **1** the process of inventing. **2** a thing invented; a contrivance, esp. one for which a patent is granted. **3** a fictitious story. **4** inventiveness. **5** Mus. a short piece for keyboard, developing a simple idea. [ME f. L inventio (as INVENT)]

Invention of the Cross 1 in Christianity, the reputed finding of the Cross by Helena, mother of the emperor Constantine, in AD 326. **2** a festival commemorating this, held on 3 May.

inventive /ɪnˈventɪv/ adj. **1** able or inclined to invent; original in devising. **2** showing ingenuity of devising. □ **inventively** adv. **inventiveness** n. [ME f. F inventif -ive or med.L inventivus (as INVENT)]

inventor /ɪnˈventə(r)/ n. (fem. **inventress** /-trɪs/) a person who invents, esp. as an occupation.

inventory /ˈɪnvəntərɪ/ n. & v. ● n. (pl. **-ies**) **1** a complete list of goods in stock, house contents, etc. **2** the goods listed in this. **3** US the total of a firm's commercial assets. ● v.tr. (**-ies, -ied**) **1** make an inventory of. **2** enter (goods) in an inventory. [ME f. med.L inventorium f. LL inventarium (as INVENT)]

Invercargill /ˌɪnvəˈkɑːgɪl/ a city in New Zealand, capital of Southland region, South Island; pop. (1991) 51,980.

Inverness /ˌɪnvəˈnes/ a city in Scotland, administrative centre of Highland Region, situated at the mouth of the River Ness; pop. (1981) 40,000.

inverse /ˈɪnvɜːs, ɪnˈvɜːs/ adj. & n. ● adj. inverted in position, order, or relation. ● n. **1** the state of being inverted. **2** (often foll. by of) a thing that is the opposite or reverse of another. **3** Math. an element which, when combined with a given element in an operation, produces the identity element for that operation. □ **inverse proportion** (or **ratio**) a relation between two quantities such that one increases in proportion as the other decreases. **inverse square law** a law by which the intensity of an effect, such as gravitational force, illumination, etc., changes in inverse proportion to the square of the distance from the source. □ **inversely** adv. [L inversus past part. of invertere: see INVERT]

inversion /ɪnˈvɜːʃ(ə)n/ n. **1** the act of turning upside down or inside out. **2** the reversal of a normal order, position, or relation. **3** the reversal of the order of words, for rhetorical effect. **4** (in full **temperature inversion**) the reversal of the normal variation of air temperature with altitude. **5** the process or result of inverting. **6** the reversal of direction of rotation of a plane of polarized light. **7** homosexuality. □ **inversion layer** Meteorol. a layer of air in which temperature increases with height, often resulting in fog at ground level. **inversion temperature** Physics the temperature at which the Joule–Thomson effect for a given gas changes sign, so that the gas is neither heated nor cooled when allowed to expand without expending energy. □ **inversive** /-ˈvɜːsɪv/ adj. [L inversio (as INVERT)]

invert v. & n. ● v.tr. /ɪnˈvɜːt/ **1** turn upside down. **2** reverse the position, order, or relation of. **3** Mus. change the relative position of the notes of (a chord or interval) by placing the lowest note higher, usu. by an octave. **4** subject to inversion. ● n. /ˈɪnvɜːt/ **1** a homosexual. **2** an inverted arch, as at the bottom of a sewer. □ **inverted comma** = quotation mark. **inverted snob** a person who likes or takes pride in what a snob might be expected to disapprove of. **invert sugar** a mixture of glucose and fructose, obtained by the hydrolysis of sucrose. □ **inverter** /ɪnˈvɜːtə(r)/ n. **invertible** /ɪnˈvɜːtɪb(ə)l/ adj. **invertibility** /ɪnˌvɜːtɪˈbɪlɪtɪ/ n. [L invertere invers- (as IN-², vertere turn)]

invertase /ˈɪnvɜːˌteɪz, ɪnˈvɜːteɪz/ n. Biochem. an enzyme from yeast which catalyses the inversion of sucrose. [INVERT + -ASE]

invertebrate /ɪnˈvɜːtɪbrət, -ˌbreɪt/ adj. & n. ● adj. **1** (of an animal) not having a backbone. **2** lacking firmness of character. ● n. an invertebrate animal. (See note below.) [mod.L invertebrata (pl.) (as IN-¹, VERTEBRA)]

▪ The invertebrates constitute an artificial division of the animal kingdom, comprising as they do 95 per cent of all animals. There are about thirty phyla, including arthropods (the most numerous group), sponges, coelenterates, molluscs, echinoderms, and several groups of wormlike organisms, but many are minor phyla found only in the sea. The unicellular protozoans are now usually placed with the unicellar algae and fungi in the kingdom Protista.

invest /ɪn'vest/ v. **1** tr. (often foll. by *in*) apply or use (money), esp. for profit. **2** intr. (often foll. by *in*) devote (time, effort, etc.) to an enterprise. **3** intr. (foll. by *in*) **a** put money for profit (into stocks etc.). **b** colloq. buy (something useful) (*invested in a new car*). **4** tr. **a** (foll. by *with*) provide or credit (a person or thing with qualities, insignia, or rank). **b** (foll. by *in*) attribute or entrust (qualities or feelings to a person or thing). **5** tr. cover as a garment. **6** tr. lay siege to. □ **investable** adj. **investible** adj. **investor** n. [ME f. F *investir* or L *investire investit-* (as IN-², *vestire* clothe f. *vestis* clothing): sense 1 f. It. *investire*]

investigate /ɪn'vestɪˌgeɪt/ v. **1** tr. **a** inquire into; examine; study carefully. **b** make an official inquiry into. **2** intr. make a systematic inquiry or search. □ **investigator** n. **investigatory** /-gətəri/ adj. [L *investigare investigat-* (as IN-², *vestigare* track)]

investigation /ɪnˌvestɪˈgeɪʃ(ə)n/ n. **1** the process or an instance of investigating. **2** a formal examination or study. □ **investigational** adj.

investigative /ɪn'vestɪgətɪv/ adj. seeking or serving to investigate, esp. (of journalism) inquiring intensively into controversial issues.

investiture /ɪn'vestɪˌtjʊə(r)/ n. **1** the formal investing of a person with honours or rank, esp. a ceremony at which a sovereign confers honours. **2** (often foll. by *with*) the act of investing (with attributes). [ME f. med.L *investitura* (as INVEST)]

investment /ɪn'vestmənt/ n. **1** the act or process of investing. **2** money invested. **3** property etc. in which money is invested. **4** the act of besieging; a blockade. □ **investment bank** (in the US) a bank that fulfils many of the functions of a UK merchant bank. **investment trust** a trust that buys and sells shares in selected companies to make a profit for its members.

inveterate /ɪn'vetərət/ adj. **1** (of a person) confirmed in an (esp. undesirable) habit etc. (*an inveterate gambler*). **2 a** (of a habit etc.) long-established. **b** (of an activity, esp. an undesirable one) habitual. □ **inveteracy** n. **inveterately** adv. [ME f. L *inveterare inveterat-* make old (as IN-², *vetus veteris* old)]

invidious /ɪn'vɪdɪəs/ adj. (of an action, conduct, attitude, etc.) likely to excite resentment or indignation against the person responsible, esp. by real or seeming injustice (*an invidious position; an invidious task*). □ **invidiously** adv. **invidiousness** n. [L *invidiosus* f. *invidia* ENVY]

invigilate /ɪn'vɪdʒɪˌleɪt/ v.intr. Brit. supervise candidates at an examination. □ **invigilator** n. **invigilation** /-ˌvɪdʒɪˈleɪʃ(ə)n/ n. [orig. = keep watch, f. L *invigilare invigilat-* (as IN-², *vigilare* watch f. *vigil* watchful)]

invigorate /ɪn'vɪgəˌreɪt/ v.tr. give vigour or strength to. □ **invigorating** adj. **invigoratingly** adv. **invigorator** n. **invigorative** /-rətɪv/ adj. **invigoration** /-ˌvɪgəˈreɪʃ(ə)n/ n. [IN-² + med.L *vigorare vigorat-* make strong]

invincible /ɪn'vɪnsɪb(ə)l/ adj. unconquerable; that cannot be defeated. □ **invincibly** adv. **invincibleness** n. **invincibility** /-ˌvɪnsɪˈbɪlɪtɪ/ n. [ME f. OF f. L *invincibilis* (as IN-¹, VINCIBLE)]

inviolable /ɪn'vaɪələb(ə)l/ adj. not to be violated or profaned. □ **inviolably** adv. **inviolability** /-ˌvaɪələ'bɪlɪtɪ/ n. [F *inviolable* or L *inviolabilis* (as IN-¹, VIOLATE)]

inviolate /ɪn'vaɪələt/ adj. **1** not violated or profaned. **2** safe from violation or harm. □ **inviolacy** n. **inviolately** adv. **inviolateness** n. [ME f. L *inviolatus* (as IN-¹, *violare*, *violat-* treat violently)]

invisible /ɪn'vɪzɪb(ə)l/ adj. & n. ● adj. **1** not visible to the eye, either characteristically or because hidden. **2** too small to be seen or noticed. **3** artfully concealed (*invisible mending*). ● n. an invisible person or thing, esp. (in pl.) invisible exports and imports. □ **invisible exports** (or **imports**) items, esp. services, involving payment between countries but not constituting tangible commodities (cf. VISIBLE 3). □ **invisibly** adv. **invisibleness** n. **invisibility** /-ˌvɪzɪ'bɪlɪtɪ/ n. [ME f. OF *invisible* f. L *invisibilis* (as IN-¹, VISIBLE)]

invitation /ˌɪnvɪ'teɪʃ(ə)n/ n. **1 a** the process of inviting or fact of being invited, esp. to a social occasion. **b** the spoken or written form in which a person is invited. **2** the action or an act of enticing; attraction, allurement.

invite v. & n. ● v. /ɪn'vaɪt/ **1** tr. (often foll. by *to*, or *to* + infin.) ask (a person) courteously to come, or to do something (*were invited to lunch; invited them to reply*). **2** tr. make a formal courteous request for (*invited comments*). **3** tr. tend to call forth unintentionally (something unwanted) (*invited*

hostility). **4 a** tr. attract. **b** intr. be attractive. ● n. /'ɪnvaɪt/ colloq. an invitation. □ **invitee** /ˌɪnvaɪ'ti:/ n. **inviter** /ɪn'vaɪtə(r)/ n. [F *inviter* or L *invitare*]

inviting /ɪn'vaɪtɪŋ/ adj. **1** attractive. **2** enticing, tempting. □ **invitingly** adv.

in vitro /ɪn 'vi:trəʊ/ adv. Biol. (of processes or reactions) taking place in a test-tube, culture dish, or elsewhere outside a living organism (opp. IN VIVO). [L, = in glass]

in vivo /ɪn 'vi:vəʊ/ adv. Biol. (of processes) taking place in a living organism. [L, = in a living thing]

invocation /ˌɪnvə'keɪʃ(ə)n/ n. **1** the act or an instance of invoking, esp. in prayer. **2** an appeal to a supernatural being or beings, e.g. the Muses, for psychological or spiritual inspiration. **3** Eccl. the words 'In the name of the Father' etc. used as the preface to a sermon etc. □ **invocatory** /ɪn'vɒkətəri/ adj. [ME f. OF f. L *invocatio -onis* (as INVOKE)]

invoice /'ɪnvɔɪs/ n. & v. ● n. a list of goods shipped or sent, or services rendered, with prices and charges; a bill. ● v.tr. **1** make an invoice of (goods and services). **2** send an invoice to (a person). [earlier *invoyes* pl. of *invoy* = ENVOY²]

invoke /ɪn'vəʊk/ v.tr. **1** call on (a deity etc.) in prayer or as a witness. **2** appeal to (the law, a person's authority, etc.). **3** summon (a spirit) by charms. **4** ask earnestly for (vengeance, help, etc.). □ **invocable** adj. **invoker** n. [F *invoquer* f. L *invocare* (as IN-², *vocare* call)]

involucre /'ɪnvə,lu:kə(r), -,lju:kə(r)/ n. **1** a covering or envelope. **2** Anat. a membranous envelope. **3** Bot. a whorl of bracts surrounding an inflorescence. □ **involucral** /ˌɪnvə'lu:krəl, -'lju:krəl/ adj. [F *involucre* or L *involucrum* (as INVOLVE)]

involuntary /ɪn'vɒləntəri/ adj. **1** done without conscious control; unintentional. **2** (of a limb, muscle, or movement) not under the control of the will. □ **involuntarily** adv. **involuntariness** n. [LL *involuntarius* (as IN-¹, VOLUNTARY)]

involute /'ɪnvə,lu:t, -,lju:t/ adj. & n. ● adj. **1** involved, intricate. **2** curled spirally. **3** Bot. rolled inwards at the edges. ● n. Geom. the locus of a point fixed on a straight line that rolls without sliding on a curve and is in the plane of that curve (cf. EVOLUTE). [L *involutus* past part. of *involvere*: see INVOLVE]

involuted /'ɪnvə,lu:tɪd, -,lju:tɪd/ adj. **1** complicated, abstruse. **2** = INVOLUTE adj. 2.

involution /ˌɪnvə'lu:ʃ(ə)n, -'lju:ʃ(ə)n/ n. **1** the process of involving. **2** an entanglement. **3** intricacy. **4** curling inwards. **5** a part that curls inwards. **6** Math. the raising of a quantity to any power. **7** Physiol. the reduction in size of an organ in old age, or when its purpose has been fulfilled (esp. the uterus after childbirth). □ **involutional** adj. [L *involutio* (as INVOLVE)]

involve /ɪn'vɒlv/ v.tr. **1** (often foll. by *in*) cause (a person or thing) to participate, or share the experience or effect (of a situation, activity, etc.). **2** imply, entail, make necessary. **3** (foll. by *in*) implicate (a person in a charge, crime, etc.). **4** include or affect in its operations. **5** (as **involved** adj.) **a** (often foll. by *in*) concerned or interested. **b** complicated in thought or form. **c** amorously associated. [ME f. L *involvere involut-* (as IN-², *volvere* roll)]

involvement /ɪn'vɒlvmənt/ n. **1** (often foll. by *in*, *with*) the act or an instance of involving; the process of being involved. **2** financial embarrassment. **3** a complicated affair or concern.

invulnerable /ɪn'vʌlnərəb(ə)l/ adj. that cannot be wounded or hurt, physically or mentally. □ **invulnerably** adv. **invulnerability** /-ˌvʌlnərə'bɪlɪtɪ/ n. [L *invulnerabilis* (as IN-¹, VULNERABLE)]

inward /'ɪnwəd/ adj. & adv. ● adj. **1** directed toward the inside; going in. **2** situated within. **3** mental, spiritual. ● adv. (also **inwards** /-wədz/) **1** (of motion or position) towards the inside. **2** in the mind or soul. [OE *innanweard* (as IN, -WARD)]

inwardly /'ɪnwədlɪ/ adv. **1** on the inside. **2** in the mind or soul. **3** (of speaking) not aloud; inaudibly. [OE *inweardlīce* (as INWARD)]

inwardness /'ɪnwədnɪs/ n. **1** inner nature; essence. **2** the condition of being inward. **3** spirituality.

inwards var. of INWARD adv.

inweave /ɪn'wi:v/ v.tr. (also **enweave**) (past **-wove** /-'wəʊv/; past part. **-woven** /-'wəʊv(ə)n/) **1** weave (two or more things) together. **2** intermingle.

inwrap var. of ENWRAP.

inwreathe var. of ENWREATHE.

inwrought /ɪn'rɔːt/ adj. **1 a** (often foll. by *with*) (of a fabric) decorated (with a pattern). **b** (often foll. by *in*, *on*) (of a pattern) wrought (in or on a fabric). **2** closely blended.

Io /ˈaɪəʊ/ **1** *Gk Mythol.* a priestess of Hera who was loved by Zeus. Trying to protect her from the jealousy of Hera, he turned her into a heifer. Hera sent a gadfly to torture the heifer which then, driven mad by the stinging insect, fled across the world. She finally reached Egypt, where Zeus turned her back into human form. The Bosporus (= cow's passage) and the Ionian Sea are reputed to have been crossed by Io, and derive their names from her story. **2** *Astron.* satellite I of Jupiter, the fifth closest to the planet, and one of the Galilean moons (diameter 3,630 km). Its vivid red and yellow coloration is due to sulphur compounds on the surface; it is actively volcanic.

IOC *abbr.* International Olympic Committee.

iodic /aɪˈɒdɪk/ *adj. Chem.* containing iodine in chemical combination (*iodic acid*). □ **iodate** /ˈaɪəˌdeɪt/ *n.*

iodide /ˈaɪəˌdaɪd/ *n. Chem.* any compound of iodine with another element or group.

iodinate /aɪˈɒdɪˌneɪt, ˈaɪədɪ-/ *v.tr.* treat or combine with iodine. □ **iodination** /aɪˌɒdɪˈneɪʃ(ə)n, ˌaɪədɪ-/ *n.*

iodine /ˈaɪəˌdiːn, -ˌdaɪn, -dɪn/ *n.* a non-metallic chemical element (atomic number 53; symbol I). One of the halogens, iodine was discovered by the French chemist Bernard Courtois (1777–1838) in 1811. It forms black crystals which readily sublime to form a violet vapour. Iodine occurs chiefly as salts in sea water and brines; as a constituent of thyroid hormones it is required in small amounts in the body, and deficiency can lead to goitre. A solution of iodine in alcohol is a commonly used antiseptic; potassium iodide is used in photography. [F *iode* f. Gk *iōdēs* violet-like f. *ion* violet + -INE⁴]

iodism /ˈaɪəˌdɪz(ə)m/ *n. Med.* a condition caused by an overdose of iodides.

iodize /ˈaɪəˌdaɪz/ *v.tr.* (also **-ise**) treat or impregnate with iodine. □ **iodization** /ˌaɪədaɪˈzeɪʃ(ə)n/ *n.*

iodo- /aɪˈəʊdəʊ, -ˈɒdəʊ/ *comb. form* (usu. **iod-** before a vowel) *Chem.* iodine.

iodoform /aɪˈəʊdəˌfɔːm, -ˈɒdəˌfɔːm/ *n.* a pale yellow volatile sweet-smelling solid iodine compound (chem. formula: CHI_3) with antiseptic properties. [IODINE after *chloroform*]

IOM *abbr.* Isle of Man.

ion /ˈaɪən/ *n.* an atom, molecule, or group that has lost one or more electrons (= CATION), or gained one or more electrons (= ANION). □ **ion exchange** the exchange of ions of the same charge between a usu. aqueous solution and a solid, used in water-softening, chromatography, etc. **ion exchanger** a substance or equipment for this process. [Gk, neut. pres. part. of *eimi* go]

-ion /ən/ *suffix* (usu. as **-sion, -tion, -xion**; see -ATION, -ITION, -UTION) forming nouns denoting: **1** verbal action (*excision*). **2** an instance of this (*a suggestion*). **3** a resulting state or product (*vexation*; *concoction*). [from or after F *-ion* or L *-io -ionis*]

Iona /aɪˈəʊnə/ a small island in the Inner Hebrides, off the west coast of Mull. It is the site of a monastery founded by St Columba in about 563, which became the centre for Celtic Christian missions in Scotland.

Ionesco /ˌiːəˈneskəʊ/, Eugène (1912–94), Romanian-born French dramatist. A leading exponent of the Theatre of the Absurd, he achieved fame with his first play *The Bald Prima Donna* (1950), which blended a dialogue of platitudes with absurd logic and surrealist effects. In *Rhinoceros* (1960), he depicted a totalitarian society whose members eventually conform by turning into rhinoceroses.

Ionia /aɪˈəʊnɪə/ in classical times, the central part of the west coast of Asia Minor. In the 11th century BC peoples speaking the Ionic dialect of Greek (the Ionians) settled in the Aegean Islands and in the western part of Asia Minor, part of which was named Ionia after them. This area was also colonized by Greeks from the mainland from about the 8th century BC.

Ionian /aɪˈəʊnɪən/ *n. & adj.* ● *n.* a member of an ancient Hellenic people inhabiting Attica, parts of Asia Minor, the Aegean Islands, etc., after whom Ionia was named. (*See note below.*) ● *adj.* of or relating to Ionia or the Ionians. □ **Ionian mode** *Mus.* the mode represented by the natural diatonic scale C–C. [L *Ionius* f. Gk *Iōnios*]

 ▪ The Ionians were among the earliest Greek-speaking inhabitants of Greece. In the 11th or 12th century BC they are thought to have been dislodged by the Dorians. The Ionians retained their settlements in Attica, in particular its chief city, Athens, where they were responsible for some of the greatest of classical Greek art, architecture, and literature. They also established settlements in

western Asia Minor (see IONIA), the Aegean Islands, and those off the west coast of Greece.

Ionian Islands a chain of about forty Greek islands off the western coast of mainland Greece, in the Ionian Sea, including Corfu, Cephalonia, and Ithaca.

Ionian Sea the part of the Mediterranean Sea between western Greece and southern Italy, at the mouth of the Adriatic. According to legend it is named after Io.

Ionic /aɪˈɒnɪk/ *adj. & n.* ● *adj.* **1** of the order of Greek architecture characterized by a column with scroll-shapes on either side of the capital. **2** of the ancient Greek dialect used in Ionia. **3** of or relating to Ionia or the Ionians. ● *n.* the Ionic dialect. [L *Ionicus* f. Gk *Iōnikos*]

ionic /aɪˈɒnɪk/ *adj.* of, relating to, or using ions. □ **ionically** *adv.*

ionization /ˌaɪənaɪˈzeɪʃ(ə)n/ *n.* (also **-isation**) the process of producing ions as a result of solvation, heat, radiation, etc. □ **ionization chamber** an instrument for detecting ionizing radiation.

ionize /ˈaɪəˌnaɪz/ *v.tr. & intr.* (also **-ise**) convert or be converted into an ion or ions. □ **ionizing radiation** a radiation of sufficient energy to cause ionization in the medium through which it passes. □ **ionizable** *adj.*

ionizer /ˈaɪəˌnaɪzə(r)/ *n.* any thing which produces ionization, esp. a device used to improve the quality of the air in a room etc.

ionosphere /aɪˈɒnəˌsfɪə(r)/ *n.* an ionized region of the atmosphere above the stratosphere, extending to about 1,000 km above the earth's surface and able to reflect radio waves for long-distance transmission round the earth (cf. TROPOSPHERE). □ **ionospheric** /-ˌɒnəˈsferɪk/ *adj.*

-ior¹ /ɪə(r)/ *suffix* forming adjectives of comparison (*senior*; *ulterior*). [L]

-ior² var. of -IOUR.

iota /aɪˈəʊtə/ *n.* **1** the ninth letter of the Greek alphabet (*I, ι*). **2** (usu. with *neg.*) the smallest possible amount. [Gk *iōta*]

IOU /ˌaɪəʊˈjuː/ *n.* a signed document acknowledging a debt. [abbr., = I owe you]

-iour /ɪə(r)/ *suffix* (also **-ior**) forming nouns (*saviour*; *warrior*). [-I- (as a stem element) + -OUR², -OR¹]

-ious /ɪəs, əs/ *suffix* forming adjectives meaning 'characterized by, full of', often corresponding to nouns in *-ion* (*cautious*; *curious*; *spacious*). [from or after F *-ieux* f. L *-iosus*]

IOW *abbr.* Isle of Wight.

Iowa /ˈaɪəwə/ a state in the Middle West of the US, acquired as part of the Louisiana Purchase in 1803; pop. (1990) 2,776,770; capital, Des Moines. It became the 29th state of the US in 1846. □ **Iowan** *adj. & n.*

Iowa City a city in eastern Iowa; pop. (1990) 59,740. Founded in 1838, it was the state capital until replaced by Des Moines in 1858.

IPA see INTERNATIONAL PHONETIC ALPHABET.

Ipatieff /ɪˈpætɪˌef/, Vladimir Nikolaievich (1867–1952), Russian-born American chemist. He worked mainly on the catalysis of hydrocarbons, particularly the use of high-pressure catalysis and of metallic oxides as catalysts. These techniques became vitally important to the petrochemical industry, which he helped to establish in both pre- and post-revolutionary Russia. Ipatieff continued his research on catalysis after moving to the US in 1930.

IPCS *abbr. hist.* (in the UK) Institute of Professional Civil Servants.

ipecac /ˈɪpɪˌkæk/ *n. colloq.* ipecacuanha. [abbr.]

ipecacuanha /ˌɪpɪˌkækjʊˈɑːnə/ *n.* the root of a South American shrub, *Cephaelis ipecacuanha*, used as an emetic and expectorant. [Port. f. Tupi-Guarani *ipekaaguéne* emetic creeper]

Iphigenia /ˌɪfɪdʒɪˈnaɪə/ *Gk Mythol.* the daughter of Agamemnon, who was obliged to offer her as a sacrifice to Artemis when the Greek fleet was becalmed on its way to the Trojan War. The goddess snatched her away to Tauris in the Crimea, where she became a priestess until rescued by her brother Orestes.

Ipiros see EPIRUS.

IPMS *abbr.* (in the UK) Institution of Professionals, Managers, and Specialists.

Ipoh /ˈiːpəʊ/ the capital of the state of Perak in western Malaysia; pop. (1980) 300,700. It replaced Taiping as state capital in 1937.

ipomoea /ˌɪpəˈmiːə/ *n.* a twining plant of the genus *Ipomoea*, having trumpet-shaped flowers, e.g. the sweet potato and morning glory. [mod.L f. Gk *ips ipos* worm + *homoios* like]

ips *abbr.* (also **i.p.s.**) inches per second.

ipse dixit /ˌɪpsɪ ˈdɪksɪt/ *n.* a dogmatic statement resting merely on the speaker's authority. [L, he himself said it (orig. of Pythagoras)]

ipsilateral /ˌɪpsɪˈlætərəl/ *adj.* belonging to or occurring on the same side of the body. [irreg. f. L *ipse* self + LATERAL]

ipsissima verba /ɪpˌsɪsɪmə ˈvɜːbə/ *n.pl.* the precise words. [L]

ipso facto /ˌɪpsəʊ ˈfæktəʊ/ *adv.* **1** by that very fact or act. **2** thereby. [L]

Ipswich /ˈɪpswɪtʃ/ the county town of Suffolk, a port and industrial town on the estuary of the River Orwell; pop. (1991) 115,500.

IQ *abbr.* intelligence quotient, a number used to express a person's performance in intelligence tests. It represents the ratio of the subject's score to the statistical norm for a population group, an IQ of 100 being average. IQ has been the most widely used measure of intelligence, although how that should be defined is controversial and testing often involves cultural bias. The quotient was introduced by the German-born American psychologist William Louis Stern (1871–1938) in 1912.

Iqbal /ˈɪkbæl/, Sir Muhammad (1875–1938), Indian poet and philosopher, generally regarded as the father of Pakistan. Writing in both Persian and Urdu, he became a champion of an international Islamic community, eventually concluding that it could only find expression in the free association of Muslim states. As president of the Muslim League in 1930, he advocated the creation of a separate Muslim state in NW India; the demands of the Muslim League led ultimately to the establishment of Pakistan in 1947.

-ique /ɪk, iːk/ *archaic* var. of -IC.

Iquitos /ɪˈkiːtɒs/ a city in NE Peru, a river port on the west bank of the Amazon; pop. (est. 1988) 247,000. Situated in tropical rainforest, it is a centre for oil exploration.

IR *abbr.* infrared.

Ir *symb. Chem.* the element iridium.

ir- /ɪr/ *prefix* assim. form of IN-¹, IN-² before *r*.

IRA see IRISH REPUBLICAN ARMY.

irade /ɪˈrɑːdɪ/ *n. hist.* a written decree of the sultan of Turkey. [Turk. f. Arab. *'irāda* will]

Iráklion see HERAKLION.

Iran /ɪˈrɑːn, ɪˈræn/ a country in the Middle East, between the Caspian Sea and the Persian Gulf; pop. (est. 1990) 54,600,000; languages, Farsi (Persian) (official), Azerbaijani, Kurdish, Arabic, and others; capital, Tehran. Previously known as Persia, the country adopted the name Iran in 1935. Iran was a monarchy until 1979, when the shah was overthrown in a popular uprising, headed by Ayatollah Khomeini, which led soon after to the establishment of an Islamic republic. From 1980 to 1988 Iran was involved in war with its neighbour Iraq (see IRAN–IRAQ WAR). (See also PERSIA.)

Irangate /ɪˈrɑːngeɪt, ɪˈræn-/ (also called *Iran–Contra affair* or *Iran–Contra scandal*) a US political scandal of 1987, during the presidency of Ronald Reagan. It involved the covert supplying by the US of arms to Iran, at a time when official relations between the countries were suspended (and while Iran was at war with Iraq), and the subsequent release of American hostages held in the Middle East. The proceeds of the arms sales were used by officials to supply arms to the anti-Communist Contras in Nicaragua, despite Congressional prohibition. [IRAN + -GATE]

Iranian /ɪˈreɪnɪən/ *adj. & n.* ● *adj.* **1** of or relating to Iran. **2** of or relating to the Indo-European group of languages including Persian, Pashto, Avestan, and Kurdish. ● *n.* **1** a native or national of Iran. **2** a person of Iranian descent.

Iran–Iraq War /ɪˌrɑːnɪˈrɑːk, ɪˌrænɪˈræk/ (also called *Gulf War*) the war of 1980–8 between Iran and Iraq in the general area of the Persian Gulf. It ended inconclusively after great hardship and loss of life on both sides.

Iraq /ɪˈrɑːk, ɪˈræk/ a country in the Middle East, on the Persian Gulf; pop. (est. 1988) 17,583,450; official language, Arabic; capital, Baghdad. Iraq is traversed by the Tigris and Euphrates rivers, whose valley was the site of the ancient civilizations of Mesopotamia. It was conquered by Arabia in the 7th century (by which time it had become known as Iraq) and from 1534 formed part of the Ottoman Empire. After the First World War a kingdom was established, although the country was under British administration until 1932. In 1958 the monarchy was overthrown in a coup and the country became a republic. Saddam Hussein came to power as President in 1979. The country was at war with its eastern neighbour Iran in 1980–8. In Aug. 1990, Iraq invaded Kuwait in an attempt to obtain that country's wealth and oilfields and to secure its own access to the Persian Gulf; it was expelled by an international coalition of forces in the Gulf War of 1991. A large

Kurdish minority in the north has been the subject of military attacks by Iraqi forces, forcing many to flee into Turkey.

Iraqi /ɪˈrɑːkɪ, ɪˈrækɪ/ *adj. & n.* ● *adj.* of or relating to Iraq. ● *n.* (pl. **Iraqis**) **1 a** a native or national of Iraq. **b** a person of Iraqi descent. **2** the form of Arabic spoken in Iraq.

IRAS /ˈaɪəræs/ Infrared Astronomical Satellite, which was launched and operated successfully in 1983 to map the distribution of infrared radiation in the sky. [acronym]

irascible /ɪˈræsɪb(ə)l/ *adj.* irritable; hot-tempered. □ **irascibly** *adv.* **irascibility** /ɪˌræsɪˈbɪlɪtɪ/ *n.* [ME f. F f. LL *irascibilis* f. L *irasci* grow angry f. *ira* anger]

irate /aɪˈreɪt/ *adj.* angry, enraged. □ **irately** *adv.* **irateness** *n.* [L *iratus* f. *ira* anger]

IRBM *abbr.* intermediate-range ballistic missile.

ire /ˈaɪə(r)/ *n. literary* anger. □ **ireful** *adj.* [ME f. OF f. L *ira*]

Ireland /ˈaɪələnd/ an island of the British Isles, lying west of Great Britain. Approximately four-fifths of the area of Ireland forms the Republic of Ireland, with the remaining one-fifth forming Northern Ireland. Ireland was inhabited by Celts as early as the 6th century BC. Christianity reached Ireland probably by the 4th century, and in the Dark Ages the country was a leading cultural centre, with the monasteries fostering learning and missionary work. Irish Celtic society was based on clans, with a high king at Tara claiming nominal overlordship. English invasions began in the 12th century under Henry II, but English authority was largely confined to the Pale, around Dublin, until the Tudors succeeded in conquering the whole of the island. Revolts against English rule, and against the imposition of Protestantism, resulted in the Plantation of Ireland by English and Scottish families on confiscated land, in an attempt to anglicize the country and secure its allegiance. In parts of Ulster the descendants of Protestant settlers form a majority. After an unsuccessful rebellion in 1798, union of Britain and Ireland followed in 1801. A share of Britain's 19th-century industrial prosperity reached Protestant Ulster, but not the rest of the island. After the failure of the potato crop in the 1840s thousands died in a famine, and thousands more emigrated. In the late 19th century there was increasing agitation for Home Rule; a bill, passed in 1914, was not implemented because of the First World War. An armed uprising (see EASTER RISING) in Dublin at Easter, 1916, was suppressed. The Government of Ireland Act (1920) provided for two Irish Parliaments, one in the north and one in the south; the following year, in 1921, Ireland was partitioned by the Anglo-Irish Treaty, which gave dominion status to Ireland with the exception of six of the counties of Ulster (Northern Ireland), whose Protestant majority wished to preserve the union and which remained part of the United Kingdom. (For later history, see IRELAND, REPUBLIC OF and NORTHERN IRELAND.)

Ireland, Republic of (also **Irish Republic**) a country forming approximately four-fifths of Ireland; pop. (1991) 3,523,400; languages, Irish (official), English; capital, Dublin. The Republic of Ireland includes the three provinces of Munster, Leinster, and Connacht, and three of the counties of Ulster, although the constitution lays claim to the whole island. The Anglo-Irish Treaty of 1921 gave southern Ireland dominion status as the Irish Free State. The treaty was followed by civil war between the Free State government and republicans, led by Eamon de Valera, who rejected partition. The war ended in victory for the government in 1923. After de Valera became Prime Minister in 1932, the country began to sever its links with Britain. A new constitution as a sovereign state (Eire) was adopted in 1937. Eire remained neutral during the Second World War; in 1949 the country left the Commonwealth and became fully independent as the Republic of Ireland. The Republic of Ireland joined the EEC in 1973. (See also EIRE.)

Irenaeus, St /ˌaɪəriˈniːəs/ (AD c.130–c.200), Greek theologian. He became bishop of Lyons in Gaul in 177, and was the author of *Against Heresies* (c.180), a detailed attack on Gnosticism. Feast day, (in the Eastern Church) 23 Aug.; (in the Western Church) 28 June.

irenic /aɪˈriːnɪk/ *adj.* (also **irenical**, **eirenic**) *literary* aiming or aimed at peace. [Gk *eirēnikos*: see EIRENICON]

irenicon var. of EIRENICON.

Irgun /ɪəˈɡʊn/ a right-wing Zionist organization founded in 1931. During the period when it was active (1937–48) it carried out violent attacks on Arabs (sometimes also Britons) in its campaign to establish a Jewish state; it was disbanded after the creation of Israel in 1948. [mod.Heb. *'irgūn ṣĕbā'ī lĕ'ummī* = national military organization]

Irian Jaya /ˌɪrɪən ˈdʒaɪə/ (also **West Irian**) a province of eastern Indonesia comprising the western half of the island of New Guinea together with the adjacent small islands; capital, Jayapura. Until its incorporation into Indonesia in 1963 it was known as Dutch New Guinea.

iridaceous /ˌɪrɪˈdeɪʃəs/ *adj. Bot.* of or relating to the family Iridaceae of flowering plants growing from bulbs, corms, or rhizomes, e.g. iris, crocus, and gladiolus. [mod.L *iridaceus* (as IRIS)]

iridescent /ˌɪrɪˈdes(ə)nt/ *adj.* **1** showing rainbow-like luminous or gleaming colours. **2** changing colour with position. □ **iridescence** *n.* **iridescently** *adv.* [L IRIS + -ESCENT]

iridium /ɪˈrɪdɪəm/ *n.* a hard silvery-white metallic chemical element (atomic number 77; symbol **Ir**). A transition element and one of the platinum metals, iridium was discovered by the British chemist Smithson Tennant (1761–1815) in 1804. Iridium–platinum alloys are used in jewellery, electrical contacts, and other situations where hardness and resistance to corrosion are important; an alloy with osmium is used in fountain-pen nibs. Iridium is very rare in the earth's crust; relatively high levels of iridium in geological strata from the end of the Cretaceous period are interpreted as evidence of a major meteorite impact 65 million years ago. [L *iris* rainbow, from its forming compounds of many colours]

iridology /ˌɪrɪˈdɒlədʒɪ/ *n.* (in alternative medicine) diagnosis by examination of the iris of the eye.

Iris /ˈaɪərɪs/ *Gk Mythol.* the goddess of the rainbow, who acted as a messenger of the gods.

iris /ˈaɪərɪs/ *n.* **1** the flat circular coloured membrane behind the cornea of the eye, with a circular opening (pupil) in the centre. **2** a herbaceous plant of the genus *Iris* (family Iridaceae), usu. with tuberous roots, sword-shaped leaves, and showy flowers. **3** (in full **iris diaphragm**) an adjustable diaphragm of thin overlapping plates for regulating the size of a central hole, esp. for the admission of light to a lens. [ME f. L *iris iridis* f. Gk *iris iridos* rainbow, iris]

Irish /ˈaɪərɪʃ/ *adj. & n.* ● *adj.* **1** of or relating to Ireland. **2 a** of or like its people or culture. **b** (of a statement or expression) paradoxical, (apparently) illogical. ● *n.* **1** the Celtic language of Ireland, formerly also called *Erse*. (*See note below.*) **2** (prec. by *the*; treated as *pl.*) the people of Ireland. □ **Irish bull** = BULL³ 1. **Irish coffee** coffee mixed with a dash of Irish whiskey and served with cream on top. **Irish moss** dried carrageen. **Irish stew** a stew of mutton, potato, and onion. **Irish terrier** a rough-haired light reddish-brown breed of terrier. **Irish wolfhound** a large breed of hound with a rough coat, often greyish. □ **Irishness** *n.* [ME f. OE *Iras* the Irish]

▪ Irish belongs to the Celtic family of languages and is a distinct variety of Gaelic. It was brought to Ireland by Celtic invaders *c.*500 BC, and until the end of the 18th century was spoken by the great majority of the people, especially in areas other than the cities. Its earliest attestation is in inscriptions from the 4th century AD, written in the ogham script, and there has been a tradition of literature since the 6th century, with a large amount of material from the 9th to 19th centuries. In the 19th century English gained ground rapidly and Irish is now spoken regularly only in a few isolated areas in the west of Ireland. Since 1922 the Irish government has organized its revival and it is now taught in all state schools, but despite this active support and the establishment of Irish as an official language there are probably only about 120,000 people who use it on an everyday basis. It is the first official language of the Republic of Ireland (the second is English). A few English words are derived from Irish, e.g. *banshee, blarney, galore, leprechaun,* and *Tory.* (See also GAELIC.)

Irish Free State the name for southern Ireland from 1921, when it gained dominion status on the partition of Ireland, until 1937, when it became the sovereign state of Eire. (See IRELAND, REPUBLIC OF.)

Irishman /ˈaɪərɪʃmən/ *n.* (*pl.* **-men**) a man who is Irish by birth or descent.

Irish National Liberation Army (abbr. **INLA**) a small paramilitary organization seeking to achieve union between Northern Ireland and the Republic of Ireland. It was formed in the early 1970s, probably as an offshoot of the Provisional IRA.

Irish Republic see IRELAND, REPUBLIC OF.

Irish Republican Army (abbr. **IRA**) the military arm of Sinn Fein, formed during the struggle for independence from Britain in 1916–21, and aiming for union between the Republic of Ireland and Northern Ireland. After a bombing campaign in England in 1939 many IRA members were interned without trial during the Second World War. In 1969 the IRA split into Official and Provisional wings. The Official IRA became virtually inactive, while the Provisional IRA stepped up the level of violence against military and civilian targets in Northern

Ireland, Britain, and Europe. The IRA declared a ceasefire in August 1994.

Irish Republican Brotherhood see FENIAN.

Irish Sea the sea separating Ireland from England and Wales.

Irish Sweepstake (also **Irish Sweep**) a sweepstake on the results of certain major horse-races (especially the Derby and the Grand National), authorized since 1930 by the government of the Republic of Ireland in order to benefit Irish hospitals. It is the largest international lottery. Most of its revenue is derived from the US, though the buying and selling of sweepstake tickets is illegal there and they have to be smuggled into the country.

Irishwoman /ˈaɪərɪʃˌwʊmən/ *n.* (*pl.* **-women**) a woman who is Irish by birth or descent.

iritis /ˌaɪəˈraɪtɪs/ *n.* inflammation of the iris.

irk /ɜːk/ *v.tr.* (usu. prec. by *it* as subject) irritate, bore, annoy. [ME: orig. unkn.]

irksome /ˈɜːksəm/ *adj.* tedious, annoying, tiresome. □ **irksomely** *adv.* **irksomeness** *n.* [ME, = tired etc., f. IRK + -SOME¹]

Irkutsk /ɪəˈkʊtsk/ the chief city of Siberia, situated on the western shore of Lake Baikal in eastern Russia; pop. (1990) 635,000.

IRO *abbr.* **1** (in the UK) Inland Revenue Office. **2** International Refugee Organization.

iroko /ɪˈrəʊkəʊ/ *n.* (*pl.* **-os**) **1** an African tree of the genus *Chlorophora*, esp. *C. excelsa* or *C. regia.* **2** the light-coloured hardwood from these trees. [Ibo]

iron /ˈaɪən/ *n., adj., & v.* ● *n.* **1** a strong, hard, silvery-grey metallic chemical element (atomic number 26; symbol **Fe**). (*See note below.*) **2** this as a type of unyieldingness or a symbol of firmness (*man of iron*; *will of iron*). **3** a tool or implement made of iron (*branding iron*; *curling iron*). **4** a household, now usu. electrical, implement with a flat base which is heated to smooth clothes etc. **5** a golf club with an iron or steel sloping head which is angled in order to loft the ball (often in *comb.* with a number indicating the degree of angle: *seven-iron*). **6** (usu. in *pl.*) a fetter (*clapped in irons*). **7** (usu. in *pl.*) a stirrup. **8** (often in *pl.*) *Brit.* an iron support for a malformed leg. **9** a preparation of iron as a tonic or dietary supplement (*iron tablets*). ● *adj.* **1** made of iron. **2** very robust. **3** unyielding, merciless (*iron determination*). ● *v.tr.* **1** smooth (clothes etc.) with an iron. **2** furnish or cover with iron. **3** shackle with irons. □ **in irons** handcuffed, chained, etc. **iron-bound 1** bound with iron. **2** rigorous; hard and fast. **3** (of a coast) rock-bound. **with an iron hand** firmly and ruthlessly. **an iron hand in a velvet glove** firmness or strength cloaked in outward gentleness. **iron in the fire** an undertaking, opportunity, or commitment (usu. in *pl.*: *too many irons in the fire*). **ironing-board** a flat surface usu. on legs and of adjustable height on which clothes etc. are ironed. **iron lung** a rigid airtight case fitted over a patient's body, used for giving prolonged artificial respiration by means of mechanical pumps. **iron maiden** *hist.* an instrument of torture consisting of a coffin-shaped box lined with iron spikes. **iron man 1** a brave or robust man, esp. a powerful athlete. **2** a multi-event sporting contest demanding stamina, esp. a triathlon involving consecutively a 3.9 km (2.4 mile) swim, a 180 km (112 mile) cycle ride, and a 42.2 km (26.2 mile) run. **iron-mould** (*US* **-mold**) a spot caused by iron-rust or an ink-stain, esp. on fabric. **iron-on** able to be fixed to the surface of a fabric etc. by ironing. **iron out** remove or smooth over (difficulties etc.). **iron pyrites** see PYRITES. **iron ration** a small emergency supply of food. □ **ironer** *n.* **ironing** *n.* (in sense 1 of *v.*). **ironless** *adj.* **iron-like** *adj.* [OE *īren, īsern* f. Gmc, prob. f. Celt.]

▪ Iron is one of the most abundant elements, widely distributed throughout the earth's crust; the chief ores are haematite, magnetite, and siderite. It occurs in the free state in meteorites, and the earth's core is believed to consist largely of metallic iron and nickel. Iron is the most extensively used of all metals because of its properties of strength, ductility, malleability, and magnetism. Iron was known in Egypt *c.*3000 BC and became widely used in Europe between 1200 and 1000 BC (see IRON AGE). In the modern world iron is mostly used in the form of steel, an alloy of purified iron with carbon and other materials. Other important forms of the metal are cast iron and wrought iron. Chemically a transition metal, iron is essential to many types of animal life as a constituent of haemoglobin. The symbol Fe is from the Latin *ferrum*.

Iron Age a period of prehistory when weapons and tools were made of iron, following the Bronze Age. The Hittites in Anatolia were working iron by *c.*1400 BC, although on a small scale. By the 12th–13th centuries BC its use had spread more widely; the Hebrews and Greeks knew it by 1000 BC, but it was not in use on a large scale until two centuries later. The application of the term ends, in Europe, at the Roman period. (See PREHISTORY.)

ironbark /ˈaɪənˌbɑːk/ n. Austral. any of several eucalyptus trees with thick solid bark and hard dense timber, esp. *Eucalyptus paniculata* and *E. sideroxylon*.

Iron Chancellor the nickname of Bismarck after he used the phrase 'blood and iron' in a speech in 1862, referring to war as an instrument of foreign policy.

ironclad adj. & n. ● adj. /ˌaɪənˈklæd/ **1** clad or protected with iron. **2** impregnable; rigorous. ● n. /ˈaɪənˌklæd/ hist. a wooden battleship protected by armour plating.

Iron Cross a German military decoration for bravery, originally awarded in Prussia (instituted 1813), revived by Hitler on the invasion of Poland by German forces in Sept. 1939.

Iron Curtain the notional barrier to the passage of people and information which existed between the West and the countries of the former Soviet bloc until the decline of Communism following the political events in eastern Europe in 1989. The first reference to such a barrier in connection with Russia dates from 1920, but its use by Winston Churchill in 1946 fixed it in the English language.

Iron Duke the nickname of Wellington, said to have been first used by the magazine *Punch* in 1845.

Iron Gate (called in Romanian *Porţile de Fier*, Serbo-Croat *Gvozdena Vrata*) a gorge through which a section of the River Danube flows, forming part of the boundary between Romania and Serbia. Navigation was improved by means of a ship canal constructed through it in 1896, and a joint Yugoslav–Romanian project completed a dam and hydroelectric power plant in 1972.

ironic /aɪˈrɒnɪk/ adj. (also **ironical**) **1** using or displaying irony. **2** in the nature of irony. □ **ironically** adv. [F *ironique* or LL *ironicus* f. Gk *eirōnikós* dissembling (as IRONY[1])]

ironist /ˈaɪərənɪst/ n. a person who uses irony. □ **ironize** v.intr. (also **-ise**). [Gk *eirōn* dissembler + -IST]

Iron Lady the nickname first given by Soviet journalists to Margaret Thatcher while she was British Prime Minister.

ironmaster /ˈaɪənˌmɑːstə(r)/ n. a manufacturer of iron.

ironmonger /ˈaɪənˌmʌŋɡə(r)/ n. Brit. a dealer in hardware etc. □ **ironmongery** n. (pl. **-ies**).

Ironside[1] /ˈaɪənˌsaɪd/, Edmund, the nickname of Edmund II of England (see EDMUND).

Ironside[2] /ˈaɪənˌsaɪd/ a nickname given to Oliver Cromwell.

Ironsides /ˈaɪənˌsaɪdz/ n.pl. Cromwell's cavalry troopers in the English Civil War, so called by their Royalist opponents in allusion to their hardiness in battle.

ironstone /ˈaɪənˌstəʊn/ n. **1** any rock containing a substantial proportion of an iron compound. **2** a kind of hard white opaque stoneware.

ironware /ˈaɪənˌweə(r)/ n. articles made of iron, esp. domestic implements.

ironwork /ˈaɪənwɜːk/ n. **1** things made of iron. **2** work in iron.

ironworks /ˈaɪənwɜːks/ n. (as sing. or pl.) a place where iron is smelted or iron goods are made.

irony[1] /ˈaɪərənɪ/ n. (pl. **-ies**) **1** an expression of meaning, often humorous or sarcastic, by the use of language of a different or opposite tendency. **2** an ill-timed or perverse arrival of an event or circumstance that is in itself desirable. **3** (also **dramatic irony**) a literary technique in which the audience can perceive hidden meanings unknown to the characters. [L *ironia* f. Gk *eirōneia* simulated ignorance f. *eirōn* dissembler]

irony[2] /ˈaɪənɪ/ adj. of or like iron.

Iroquoian /ˌɪrəˈkwɔɪən/ n. & adj. ● n. **1** a language family of eastern North America, including Cherokee and Mohawk. **2** a member of the Iroquois. ● adj. of or relating to the Iroquois or the Iroquoian language family.

Iroquois /ˈɪrəˌkwɔɪ/ n. & adj. ● n. (pl. same) **1 a** a confederacy of American Indian peoples living in southern Ontario and Quebec and northern New York State. (See note below.) **b** a member of any of these peoples. **2** any of the Iroquoian languages of these peoples, including Mohawk, Oneida, Seneca, etc. ● adj. of or relating to the Iroquois or their languages. [F f. Algonquian]

▪ Iroquois is the collective designation of the League of Five (later Six) Nations of American Indian peoples (Mohawk, Oneida, Seneca, Onondaga, Cayuga, and later Tuscarora), which joined in a confederacy c.1570. A powerful force in early colonial history, the confederacy was split by conflict over support for the different sides in the War of American Independence, with half the League (the Cayugas, Mohawks, and Senecas) migrating to Canada after the war to live on land given to them as allies of the defeated loyalists. Traditional Iroquois society revolved around matrilineal residential and social organization.

irradiant /ɪˈreɪdɪənt/ adj. literary shining brightly. □ **irradiance** n.

irradiate /ɪˈreɪdɪˌeɪt/ v.tr. **1** subject to (any form of) radiation. **2** shine upon; light up. **3** throw light on (a subject). □ **irradiative** /-dɪətɪv/ adj. [L *irradiare irradiat-* (as IN-[2], *radiare* f. *radius* RAY[1])]

irradiation /ɪˌreɪdɪˈeɪʃ(ə)n/ n. **1** the process of irradiating; esp. the process of exposing food to gamma rays to kill micro-organisms. **2** shining, illumination. **3** the apparent extension of the edges of an illuminated object seen against a dark background. [F *irradiation* or LL *irradiatio* (as IRRADIATE)]

irrational /ɪˈræʃ(ə)n(ə)l/ adj. **1** illogical; unreasonable. **2** not endowed with reason. **3** Math. (of a root etc.) not rational; not commensurate with the natural numbers (e.g. a non-terminating decimal). □ **irrationalize** v.tr. (also **-ise**). **irrationally** adv. **irrationality** /ɪˌræʃəˈnælɪtɪ/ n. [L *irrationalis* (as IN-[1], RATIONAL)]

Irrawaddy /ˌɪrəˈwɒdɪ/ the principal river of Burma (Myanmar), 2,090 km (1,300 miles) long. It flows in a large delta into the eastern part of the Bay of Bengal.

irreclaimable /ˌɪrɪˈkleɪməb(ə)l/ adj. that cannot be reclaimed or reformed. □ **irreclaimably** adv.

irreconcilable /ɪˈrekənˌsaɪləb(ə)l/ adj. & n. ● adj. **1** implacably hostile. **2** (of ideas etc.) incompatible. ● n. **1** an uncompromising opponent of a political measure etc. **2** (usu. in pl.) any of two or more items, ideas, etc., that cannot be made to agree. □ **irreconcilably** adv. **irreconcilableness** n. **irreconcilability** /ɪˌrekənˌsaɪləˈbɪlɪtɪ/ n.

irrecoverable /ˌɪrɪˈkʌvərəb(ə)l/ adj. that cannot be recovered or remedied. □ **irrecoverably** adv.

irrecusable /ˌɪrɪˈkjuːzəb(ə)l/ adj. that must be accepted. [F *irrécusable* or LL *irrecusabilis* (as IN-[1], *recusare* refuse)]

irredeemable /ˌɪrɪˈdiːməb(ə)l/ adj. **1** that cannot be redeemed. **2** hopeless, absolute. **3 a** (of a government annuity) not terminable by repayment. **b** (of paper currency) for which the issuing authority does not undertake ever to pay coin. □ **irredeemably** adv. **irredeemability** /-ˌdiːməˈbɪlɪtɪ/ n.

irredentist /ˌɪrɪˈdentɪst/ n. **1** (**Irredentist**) an Italian nationalist of the late 19th century advocating the return to Italy of the Italian-speaking districts of the Austro-Hungarian empire. **2** a person holding similar views of other areas. □ **irredentism** n. [It. *irredentista* f. (*Italia*) *irredenta* unredeemed (Italy)]

irreducible /ˌɪrɪˈdjuːsɪb(ə)l/ adj. **1** that cannot be reduced or simplified. **2** (often foll. by to) that cannot be brought to a desired condition. □ **irreducibly** adv. **irreducibility** /-ˌdjuːsɪˈbɪlɪtɪ/ n.

irrefragable /ɪˈrefrəɡəb(ə)l/ adj. **1** (of a statement, argument, or person) unanswerable, indisputable. **2** (of rules etc.) inviolable. □ **irrefragably** adv. [LL *irrefragabilis* (as IN-[1], *refragari* oppose)]

irrefrangible /ˌɪrɪˈfrændʒɪb(ə)l/ adj. **1** inviolable. **2** Optics incapable of being refracted.

irrefutable /ɪˈrefjʊtəb(ə)l, ˌɪrɪˈfjuːt-/ adj. that cannot be refuted. □ **irrefutably** adv. **irrefutability** /ɪˌrefjʊtəˈbɪlɪtɪ, ˌɪrɪˌfjuːt-/ n. [LL *irrefutabilis* (as IN-[1], REFUTE)]

irregardless /ˌɪrɪˈɡɑːdlɪs/ adj. & adv. US dial. or joc. = REGARDLESS. [prob. blend of IRRESPECTIVE and REGARDLESS]

irregular /ɪˈreɡjʊlə(r)/ adj. & n. ● adj. **1** not regular in shape, unsymmetrical; varying in form. **2** (of a surface) uneven. **3** contrary to a rule, moral principle, or custom; abnormal. **4** uneven in duration, order, etc.; not occurring at regular intervals. **5** (of troops) not belonging to the regular army. **6** Gram. (of a verb, noun, etc.) not inflected according to the usual rules. **7** disorderly. **8** (of a flower) having unequal petals etc. ● n. (in pl.) irregular troops. □ **irregularly** adv. **irregularity** /ɪˌreɡjʊˈlærɪtɪ/ n. (pl. **-ies**). [ME f. OF *irreguler* f. LL *irregularis* (as IN-[1], REGULAR)]

irrelative /ɪˈrelətɪv/ adj. **1** (often foll. by to) unconnected, unrelated. **2** having no relations; absolute. **3** irrelevant. □ **irrelatively** adv.

irrelevant /ɪˈrelɪv(ə)nt/ adj. (often foll. by to) not relevant; not applicable (to a matter in hand). □ **irrelevance** n. **irrelevancy** n. (pl. **-ies**). **irrelevantly** adv.

irreligion /ˌɪrɪˈlɪdʒən/ n. disregard of or hostility to religion. □ **irreligionist** n. [F *irréligion* or L *irreligio* (as IN-¹, RELIGION)]

irreligious /ˌɪrɪˈlɪdʒəs/ adj. **1** indifferent or hostile to religion. **2** lacking a religion. □ **irreligiously** adv. **irreligiousness** n.

irremediable /ˌɪrɪˈmiːdɪəb(ə)l/ adj. that cannot be remedied. □ **irremediably** adv. [L *irremediabilis* (as IN-¹, REMEDY)]

irremissible /ˌɪrɪˈmɪsɪb(ə)l/ adj. **1** unpardonable. **2** unalterably obligatory. [ME f. OF *irremissible* or eccl.L *irremissibilis* (as IN-¹, REMISSIBLE)]

irremovable /ˌɪrɪˈmuːvəb(ə)l/ adj. that cannot be removed, esp. from office. □ **irremovably** adv. **irremovability** /-ˌmuːvəˈbɪlɪtɪ/ n.

irreparable /ɪˈrepərəb(ə)l, disp. ˌɪrɪˈpeərə-/ adj. (of an injury, loss, etc.) that cannot be rectified or made good. □ **irreparably** adv. **irreparableness** n. **irreparability** /ɪˌrepərəˈbɪlɪtɪ, disp. ˌɪrɪˌpeərə-/ n. [ME f. OF f. L *irreparabilis* (as IN-¹, REPARABLE)]

irreplaceable /ˌɪrɪˈpleɪsəb(ə)l/ adj. **1** that cannot be replaced. **2** of which the loss cannot be made good. □ **irreplaceably** adv.

irrepressible /ˌɪrɪˈpresɪb(ə)l/ adj. that cannot be repressed or restrained. □ **irrepressibly** adv. **irrepressibleness** n. **irrepressibility** /-ˌpresɪˈbɪlɪtɪ/ n.

irreproachable /ˌɪrɪˈprəʊtʃəb(ə)l/ adj. faultless, blameless. □ **irreproachably** adv. **irreproachableness** n. **irreproachability** /-ˌprəʊtʃəˈbɪlɪtɪ/ n. [F *irréprochable* (as IN-¹, REPROACH)]

irresistible /ˌɪrɪˈzɪstɪb(ə)l/ adj. **1** too strong or convincing to be resisted. **2** delightful; alluring. □ **irresistibly** adv. **irresistibleness** n. **irresistibility** /-ˌzɪstɪˈbɪlɪtɪ/ n. [med.L *irresistibilis* (as IN-¹, RESIST)]

irresolute /ɪˈrezəˌluːt, -ˌljuːt/ adj. **1** hesitant, undecided. **2** lacking in resoluteness. □ **irresolutely** adv. **irresoluteness** n. **irresolution** /ɪˌrezəˈluːʃ(ə)n, -ˈljuːʃ(ə)n/ n.

irresolvable /ˌɪrɪˈzɒlvəb(ə)l/ adj. **1** that cannot be resolved into its components. **2** (of a problem) that cannot be solved.

irrespective /ˌɪrɪˈspektɪv/ adj. (foll. by *of*) not taking into account; regardless of. □ **irrespectively** adv.

irresponsible /ˌɪrɪˈspɒnsɪb(ə)l/ adj. **1** acting or done without due sense of responsibility. **2** not responsible for one's conduct. □ **irresponsibly** adv. **irresponsibility** /-ˌspɒnsɪˈbɪlɪtɪ/ n.

irresponsive /ˌɪrɪˈspɒnsɪv/ adj. (often foll. by *to*) not responsive. □ **irresponsiveness** n.

irretrievable /ˌɪrɪˈtriːvəb(ə)l/ adj. that cannot be retrieved or restored. □ **irretrievably** adv. **irretrievability** /-ˌtriːvəˈbɪlɪtɪ/ n.

irreverent /ɪˈrevərənt/ adj. lacking reverence; disrespectful. □ **irreverence** n. **irreverently** adv. **irreverential** /ɪˌrevəˈrenʃ(ə)l/ adj. [L *irreverens* (as IN-¹, REVERENT)]

irreversible /ˌɪrɪˈvɜːsɪb(ə)l/ adj. not reversible or alterable. □ **irreversibly** adv. **irreversibility** /-ˌvɜːsɪˈbɪlɪtɪ/ n.

irrevocable /ɪˈrevəkəb(ə)l, disp. ˌɪrɪˈvəʊkə-/ adj. **1** unalterable. **2** gone beyond recall. □ **irrevocably** adv. **irrevocability** /ɪˌrevəkəˈbɪlɪtɪ, disp. ˌɪrɪˌvəʊkə-/ n. [ME f. L *irrevocabilis* (as IN-¹, REVOKE)]

irrigate /ˈɪrɪˌɡeɪt/ v.tr. **1 a** supply water to (land) by means of channels. **b** (of a stream etc.) supply (land) with water. **2** Med. apply a cleansing or cooling flow of water or medication to (a wound etc.). **3** refresh as with moisture. □ **irrigative** adj. **irrigator** n. **irrigable** /-ɡəb(ə)l/ adj. **irrigation** /ˌɪrɪˈɡeɪʃ(ə)n/ n. [L *irrigare* (as IN-², *rigare* moisten)]

irritable /ˈɪrɪtəb(ə)l/ adj. **1** easily annoyed or angered. **2** (of an organ etc.) very sensitive to contact. **3** Biol. responding actively to physical stimulus. □ **irritable bowel syndrome** Med. a stress-related condition involving abdominal pain and diarrhoea or constipation. □ **irritably** adv. **irritability** /ˌɪrɪtəˈbɪlɪtɪ/ n. [L *irritabilis* (as IRRITATE)]

irritant /ˈɪrɪt(ə)nt/ adj. & n. ● adj. causing irritation. ● n. an irritant substance. □ **irritancy** n.

irritate /ˈɪrɪˌteɪt/ v.tr. **1** excite to anger; annoy. **2** stimulate discomfort or pain in (a part of the body). **3** Biol. stimulate (an organ etc.) to an active response. □ **irritatedly** adv. **irritating** adj. **irritatingly** adv. **irritative** adj. **irritator** n. **irritation** /ˌɪrɪˈteɪʃ(ə)n/ n. [L *irritare irritat-*]

irrupt /ɪˈrʌpt/ v.intr. (foll. by *into*) enter forcibly or suddenly. [L *irrumpere irrupt-* (as IN-², *rumpere* break)]

irruption /ɪˈrʌpʃ(ə)n/ n. **1** the action of irrupting; a sudden incursion. **2** a sudden temporary increase in the local population of a migrant bird or animal species. □ **irruptive** /ɪˈrʌptɪv/ adj.

Irtysh /ɪəˈtɪʃ/ a river of central Asia, which rises in the Altai Mountains in northern China and flows westwards into NE Kazakhstan, where it

turns north-west into Russia, joining the River Ob near its mouth at the head of the Gulf of Ob. Its length is 4,248 km (2,655 miles).

Irving¹ /ˈɜːvɪŋ/, Sir Henry (born John Henry Brodribb) (1838–1905), English actor-manager. In 1874, he first played Hamlet at the Lyceum Theatre, and proceeded to manage the theatre from 1878 to 1902. During this period, he entered into a celebrated acting partnership with Ellen Terry; they were particularly noted for their performances in Irving's productions of Shakespeare.

Irving² /ˈɜːvɪŋ/, Washington (1783–1859), American writer. His first publication was a series of satirical essays (1807–8) entitled *Salmagundi*. He travelled extensively in Europe and also served as US ambassador to Spain 1842–6; he is best known for *The Sketch Book of Geoffrey Crayon, Gent* (1819–20), which contains such tales as 'Rip Van Winkle' and 'The Legend of Sleepy Hollow'. He also wrote the burlesque *History of New York* (1809), under the pretended name of Diedrich Knickerbocker; the word *knickerbocker* derives from this.

Is. abbr. **1 a** Island(s). **b** Isle(s). **2** (also **Isa.**) (in the Bible) Isaiah.

is 3rd sing. present of BE.

Isaac /ˈaɪzək/ a Hebrew patriarch, son of Abraham and Sarah and father of Jacob and Esau. He was nearly sacrificed by Abraham at God's command (Gen. 21:3 etc.).

Isabella I /ˌɪzəˈbelə/ (known as Isabella of Castile or Isabella the Catholic) (1451–1504), queen of Castile 1474–1504 and of Aragon 1479–1504. Her marriage to Ferdinand of Aragon in 1469 helped to join together the Christian kingdoms of Castile and Aragon, marking the beginning of the unification of Spain. As joint monarchs of Castile, they instituted the Spanish Inquisition (1478). Later, as rulers of Aragon as well as Castile, they supported Columbus's expedition in which he discovered the New World (1492). (See also FERDINAND.)

Isabella of France /ˌɪzəˈbelə/ (1292–1358), daughter of Philip IV of France. She was queen consort of Edward II of England from 1308, but returned to France in 1325. She and her lover Roger de Mortimer organized an invasion of England in 1326, forcing Edward to abdicate in favour of his son, who was crowned Edward III after his father's murder in 1327. Isabella and Mortimer acted as regents for Edward III until 1330, after which Edward took control of the kingdom and Isabella was banished.

isagogic /ˌaɪsəˈɡɒdʒɪk/ adj. introductory. [L *isagogicus* f. Gk *eisagōgikos* f. *eisagōgē* introduction f. *eis* into + *agōgē* leading f. *agō* lead]

isagogics /ˌaɪsəˈɡɒdʒɪks/ n. an introductory study, esp. of the literary and external history of the Bible.

Isaiah /aɪˈzaɪə/ **1** a Hebrew major prophet of Judah in the 8th century BC, who taught the supremacy of the God of Israel and emphasized the moral demands on worshippers. **2** a book of the Bible containing his prophecies.

isatin /ˈaɪsətɪn/ n. Chem. a red crystalline derivative of indole used in the manufacture of dyes. [L *isatis* woad f. Gk]

ISBN abbr. international standard book number, a ten-digit number assigned to every book before publication, recording such details as language, provenance, and publisher. The system was adopted in 1969.

ischaemia /ɪˈskiːmɪə/ n. (US **ischemia**) Med. a reduction of the blood supply to part of the body. □ **ischaemic** adj. [mod.L f. Gk *iskhaimos* f. *iskhō* keep back]

Ischia /ˈɪskɪə/ an island in the Tyrrhenian Sea off the west coast of Italy, about 26 km (16 miles) west of Naples.

ischium /ˈɪskɪəm/ n. (pl. **ischia** /-kɪə/) the curved bone forming the base of each half of the pelvis. □ **ischial** adj. [L f. Gk *iskhion* hip-joint: cf. SCIATIC]

ISDN abbr. integrated services digital network.

Ise /ˈiːseɪ/ a city in central Honshu island, Japan, on Ise Bay; pop. (1990) 104,000. It has several noted Shinto shrines, including one dedicated to the sun-goddess, from whom the Japanese royal family were once claimed to be descended. The city was known as Ujiyamada until 1956.

-ise¹ suffix var. of -IZE. ¶ See the note at *-ize*.

-ise² /aɪz, iːz/ suffix forming nouns of quality, state, or function (*exercise; expertise; franchise; merchandise*). [from or after F or OF *-ise* f. L *-itia* etc.]

-ise³ suffix var. of -ISH².

isentropic /ˌaɪsenˈtrɒpɪk/ adj. Physics having equal entropy. [ISO- + ENTROPY]

Iseult /ɪˈzuːlt, ɪˈsuːlt/ (also **Isolde** /ɪˈzɒld, ɪˈzəʊldə/) (in medieval legend) **1** in one account, the sister or daughter of the king of Ireland, and wife of King Mark of Cornwall, loved by Tristram. **2** in another account, the daughter of the king of Brittany and wife of Tristram.

Isfahan /ˌɪsfəˈhɑːn/ (also **Esfahan** /ˌesfə-/, **Ispahan** /ˌɪspə-/) an industrial city in central Iran, the country's third largest city; pop. (1991) 1,127,000. It was made the capital in 1598 and became one of the largest and most beautiful cities of this period until captured and destroyed by the Afghans in 1722.

-ish[1] /ɪʃ/ suffix forming adjectives: **1** from nouns, meaning: **a** having the qualities or characteristics of (boyish). **b** of the nationality of (Danish). **2** from adjectives, meaning 'somewhat' (thickish). **3** colloq. denoting an approximate age or time of day (fortyish; six-thirtyish). [OE -isc]

-ish[2] /ɪʃ/ suffix (also **-ise** /aɪz/) forming verbs (vanish; advertise). [from or after F -iss- (in extended stems of verbs in -ir) f. L -isc- inceptive suffix]

Isherwood /ˈɪʃəˌwʊd/, Christopher (William Bradshaw) (1904–86), British-born American novelist. His novels Mr Norris Changes Trains (1935) and Goodbye to Berlin (1939; filmed as Cabaret, 1972) vividly portray Germany on the eve of Hitler's rise to power and reflect Isherwood's experiences in Berlin from 1929 to 1933. He collaborated with W. H. Auden on three verse plays and emigrated with him to the US in 1939.

Ishiguro /ˌɪʃɪˈɡʊərəʊ/, Kazuo (b.1954), Japanese-born British novelist. In 1960 he moved with his family to Great Britain. He gained recognition with his novel An Artist of the Floating World (1986); his next novel, The Remains of the Day (1989), won the Booker Prize.

Ishmael /ˈɪʃmeɪəl/ (in the Bible) a son of Abraham and his wife Sarah's maid Hagar, driven away with his mother after the birth of Isaac (Gen. 16:12). He is the traditional ancestor of Muhammad and of the Arab peoples. □ **Ishmaelite** /ˈɪʃməˌlaɪt/ n.

Ishtar /ˈɪʃtɑː(r)/ Mythol. a Babylonian and Assyrian goddess whose name and functions correspond to those of Astarte.

Isidore of Seville, St /ˈɪzɪˌdɔː(r)/ (also called Isidorus Hispalensis) (c.560–636), Spanish archbishop and Doctor of the Church. He is noted for his Etymologies, an encyclopedic work used by many medieval authors. Feast day, 4 Apr.

isinglass /ˈaɪzɪŋˌɡlɑːs/ n. **1** a kind of gelatin obtained from fish, esp. sturgeon, and used in making jellies, glue, etc., and for fining beer. **2** mica. [corrupt. of obs. Du. huisenblas sturgeon's bladder, assim. to GLASS]

Isis /ˈaɪsɪs/ Egyptian Mythol. a nature-goddess, wife of Osiris and mother of Horus. Her worship spread to western Asia, Greece, and Rome, where she was identified with various local goddesses, and became the focus of one of the major mystery religions (see MYSTERY[1] 7a), involving enactment of the myth of the death and resurrection of Osiris.

Iskenderun /ɪsˈkendəˌruːn/ a port and naval base in southern Turkey, on the Mediterranean coast; pop. (1990) 158,930. Formerly named Alexandretta, it lies on or near the site of Alexandria ad Issum, founded by Alexander the Great in 333 BC. The port was an important outlet for goods from Persia, India, and eastern Asia before the development of sea routes round the Cape of Good Hope and later through the Suez Canal.

Islam /ˈɪzlɑːm, -læm, ɪzˈlɑːm/ n. the religion of the Muslims, a monotheistic faith regarded as revealed through Muhammad as the prophet of Allah. Founded in the Arabian peninsula in the 7th century AD, Islam is now the professed faith of nearly 1,000 million people worldwide, particularly in North Africa, the Middle East, and parts of Asia. To be a Muslim means both to affirm an individual surrender to God and to live as a member of a social community. The Muslim performs prescribed acts of worship and strives to fulfil good works within the group; expected duties (the Five Pillars of Islam) include profession of the faith in a prescribed form, observance of ritual prayer (five obligatory prayer sequences each day as well as non-obligatory prayers), giving alms to the poor, fasting during the month of Ramadan, and performing the pilgrimage (hajj) to Mecca. Prayer takes place in a mosque and is led by an imam; there is no organized priesthood, although respect is given to holy men and to the descendants of Muhammad. The ritual observances of Islam, as well as a code governing social behaviour, were said to have been given to Muhammad as a series of revelations, which were codified in the Koran. They are supplemented by the deeds and discourse of the prophet, as collected in the Hadith. Islam is regarded by its adherents as the last of the revealed religions (following Judaism and Christianity), and Muhammad is seen as the last of the prophets, building on and perfecting the examples and teachings of Abraham, Moses, and Jesus. There are two major branches in Islam, Sunni and Shia, and other schools including the mystical movement of Sufism. (See also KAABA.) □ **Islamic** /ɪzˈlæmɪk/ adj. **Islamism** /ˈɪzləˌmɪz(ə)m/ n. **Islamist** n.

Islamize v.tr. (also **-ise**). **Islamization** /ˌɪzləmaɪˈzeɪʃ(ə)n/ n. [Arab. islām submission (to God) f. aslama resign oneself]

Islamabad /ɪzˈlɑːməˌbæd/ the capital of Pakistan, a modern planned city in the north of the country, which replaced Rawalpindi as capital in 1967; pop. (1981) 201,000.

Islamic fundamentalism n. the belief that the revitalization of Islamic societies can come about only through a return to the fundamental principles and practices of early Islam. Although often regarded as a phenomenon characteristic of the late 20th century, Islamic fundamentalism appeared in the 18th and 19th centuries in the form of a number of differing revivalist movements, which emerged in the face of the disintegration of Islamic political and economic power and the ascendancy of the West. In the 20th century fundamentalist groups have become active in Egypt, Afghanistan, Pakistan, Sudan, and elsewhere, many gaining impetus from the 1979 revolution in Iran, which brought Ayatollah Khomeini to power. Such groups are characterized by the view that religion is central to both state and society and by advocacy of a return to a life patterned on the 7th-century community established by Muhammad at Medina, governed by the sharia (Islamic law) and supported if need be by jihad or holy war. Literal interpretation of the Koran and close observance of the penal code contained in the sharia are stressed, at least partially because of their importance in symbolizing cultural identity and resistance to westernization.

Islamic Jihad /dʒɪˈhæd, -ˈhɑːd/ (also **Jehad**) a Muslim fundamentalist terrorist group within the Shiite Hezbollah association. The group was involved in the holding of some Western hostages in Lebanon in the 1980s. [see JIHAD]

Island see ICELAND.

island /ˈaɪlənd/ n. **1** a piece of land surrounded by water. **2** anything compared to an island, esp. in being surrounded in some way. **3** = traffic island. **4 a** a detached or isolated thing. **b** Anat. a detached portion of tissue or group of cells (cf. ISLET 2). **5** Naut. a ship's superstructure, bridge, etc. □ **island area** each of three administrative areas in Scotland (Orkney, Shetland, Western Isles), consisting of groups of islands. [OE īgland f. īg island + LAND: first syllable infl. by ISLE]

islander /ˈaɪləndə(r)/ n. a native or inhabitant of an island.

Islands of the Blessed a mythical abode, often located near where the sun sets in the west, to which people in classical times believed the souls of the good were conveyed to a life of bliss.

Islay /ˈaɪleɪ/ an island in western Scotland, to the south of Jura. It is the southernmost of the Inner Hebrides.

isle /aɪl/ n. poet. (and in place-names) an island or peninsula, esp. a small one. [ME ile f. OF ile f. L insula: later ME & OF isle after L]

Isle of Man an island in the Irish Sea which is a British Crown possession having home rule, with its own legislature (the Tynwald) and judicial system; pop. (1991) 69,790; capital, Douglas. The island was part of the Norse kingdom of the Hebrides in the Middle Ages, passing into Scottish hands in 1266 for a time, until the English gained control in the early 15th century. Its ancient language, Manx, is still occasionally used for ceremonial purposes.

Isle of Wight /waɪt/ an island off the south coast of England, a county since 1974; pop. (1991) 126,600; administrative centre, Newport. It lies at the entrance to Southampton Water and is separated from the mainland by the Solent and Spithead.

islet /ˈaɪlɪt/ n. **1** a small island. **2** Anat. a portion of tissue structurally distinct from surrounding tissues. **3** an isolated place. [OF, dimin. of isle ISLE]

islets of Langerhans /ˈlæŋəˌhænz/ n.pl. Anat. groups of pancreatic cells secreting insulin and glucagon. [Paul Langerhans, Ger. anatomist (1847–88)]

ism /ˈɪz(ə)m/ n. colloq. usu. derog. any distinctive but unspecified doctrine or practice of a kind with a name in -ism.

-ism /ɪz(ə)m/ suffix forming nouns, esp. denoting: **1** an action or its result (baptism; organism). **2** a system, principle, or ideological movement (Conservatism; jingoism; feminism). **3** a state or quality (heroism; barbarism). **4** a basis of prejudice or discrimination (racism; sexism). **5** a peculiarity in language (Americanism). **6** a pathological condition (alcoholism; Parkinsonism). [from or after F -isme f. L -ismus f. Gk -ismos or -isma f. -izō -IZE]

Ismaili /ɪzˈmaɪlɪ, ˌɪzmɑːˈiːlɪ/ n. a member of a Shiite Muslim branch that seceded from the main group in the 8th century over the question of succession to the position of imam, upholding the claims of Ismail as the seventh imam, from whom the sect takes its name. The Ismaili movement developed an elaborate esoteric doctrine which diverged considerably from the rest of Islam and was organized as a hierarchical secret society. It eventually split into many sects, of which the best known is the Nizari sect, headed by the Aga Khan. Today Ismailis are found especially in India, Pakistan, and East Africa, with smaller groups in Syria, Iran, and some other countries.

isn't /'ɪz(ə)nt/ *contr.* is not.

ISO *abbr.* **1** (in the UK) Imperial Service Order. **2** International Organization for Standardization.

iso- /'aɪsəʊ/ *comb. form* **1** equal (*isometric*). **2** *Chem.* isomeric, esp. of a hydrocarbon with a branched chain of carbon atoms (*isobutane*). [Gk *isos* equal]

isobar /'aɪsəʊ̩bɑː(r)/ *n.* **1** a line on a map connecting positions having the same atmospheric pressure at a given time or on average over a given period. **2** a curve for a physical system at constant pressure. **3** each of two or more isotopes of different elements, with the same atomic weight. □ **isobaric** /̩aɪsəʊ'bærɪk/ *adj.* [Gk *isobarēs* of equal weight (as ISO-, *baros* weight)]

isocheim /'aɪsəʊ̩kaɪm/ *n.* a line on a map connecting places having the same average temperature in winter. [ISO- + Gk *kheima* winter weather]

isochromatic /̩aɪsəʊkrəʊ'mætɪk/ *adj.* of the same colour.

isochronous /aɪ'sɒkrənəs/ *adj.* **1** occurring at the same time. **2** occupying equal time. □ **isochronously** *adv.* [ISO- + Gk *khronos* time]

isoclinal /̩aɪsəʊ'klaɪn(ə)l/ *adj.* (also **isoclinic** /-'klɪnɪk/) **1** *Geol.* (of a fold) in which the two limbs are parallel. **2** corresponding to equal values of magnetic dip. [ISO- + CLINE]

isoclinic var. of ISOCLINAL.

Isocrates /aɪ'sɒkrə̩tiːz/ (436–338 BC), Athenian orator. His written speeches are amongst the earliest political pamplets; they advocate the union of Greeks under Philip II of Macedon and a pan-Hellenic crusade against Persia.

isocyanic acid /̩aɪsəsaɪ'ænɪk/ *n. Chem.* a volatile pungent liquid (chem. formula: HNCO), isomeric with cyanic acid. (See also FULMINIC ACID.) □ **isocyanate** /-'saɪə̩neɪt/ *n.* [ISO- + CYANIC ACID]

isodynamic /̩aɪsəʊdaɪ'næmɪk/ *adj.* corresponding to equal values of (magnetic) force.

isoelectric /̩aɪsəʊɪ'lektrɪk/ *adj.* having or involving no net electric charge or difference in electrical potential. □ **isoelectric focusing** *Biochem.* a form of high-resolution electrophoresis. [ISO- + ELECTRIC]

isoenzyme /'aɪsəʊ̩enzaɪm/ *n. Biochem.* each of two or more enzymes with identical function but different structure.

isogamy /aɪ'sɒgəmɪ/ *n. Biol.* sexual reproduction by fusion of similar gametes. [ISO- + Gk *-gamia* f. *gamos* marriage]

isogeotherm /̩aɪsəʊ'dʒiː̩əʊ̩θɜːm/ *n.* a line or surface connecting points in the interior of the earth having the same temperature. □ **isogeothermal** /-̩dʒiː̩əʊ'θɜːm(ə)l/ *adj.*

isogloss /'aɪsəʊ̩glɒs/ *n.* a line on a map marking an area having a distinct linguistic feature.

isogonic /̩aɪsəʊ'gɒnɪk/ *adj.* corresponding to equal values of magnetic declination.

isohel /'aɪsəʊ̩hel/ *n.* a line on a map connecting places having the same duration of sunshine. [ISO- + Gk *hēlios* sun]

isohyet /̩aɪsəʊ'haɪɪt/ *n.* a line on a map connecting places having the same amount of rainfall in a given period. [ISO- + Gk *huetos* rain]

isokinetic /̩aɪsəʊkɪ'netɪk, -kaɪ'netɪk/ *adj.* **1** characterized by or producing a constant speed. **2** *Physiol.* of or relating to muscular action with a constant rate of movement.

isolate /'aɪsə̩leɪt/ *v.tr.* **1 a** place apart or alone, cut off from society. **b** place (a patient thought to be contagious or infectious) in quarantine. **2 a** identify and separate for attention (*isolated the problem*). **b** *Chem.* prepare (a substance) in a pure form. **3** insulate (electrical apparatus). □ **isolator** *n.* **isolatable** *adj.* **isolable** /-ləb(ə)l/ *adj.* [orig. in past part., f. F *isolé* f. It. *isolato* f. LL *insulatus* f. L *insula* island]

isolated /'aɪsə̩leɪtɪd/ *adj.* **1** lonely; cut off from society or contact; remote (*feeling isolated; an isolated farmhouse*). **2** untypical, exceptional (*an isolated example*).

isolating /'aɪsə̩leɪtɪŋ/ *adj.* (of a language) having each element as an independent word without inflections.

isolation /̩aɪsə'leɪʃ(ə)n/ *n.* the act or an instance of isolating; the state of being isolated or separated. □ **in isolation** considered singly and

not relatively. **isolation hospital** (or **ward** etc.) a hospital, ward, etc., for patients with contagious or infectious diseases.

isolationism /̩aɪsə'leɪʃə̩nɪz(ə)m/ *n.* the policy of withdrawal from or non-participation in the affairs of other countries or groups, esp. in politics. □ **isolationist** *n. & adj.*

Isolde see ISEULT.

isoleucine /̩aɪsəʊ'luːsiːn/ *n. Biochem.* a hydrophobic amino acid that is a constituent of proteins and an essential nutrient. [G *Isoleucin* (see ISO-, LEUCINE)]

isomer /'aɪsəmə(r)/ *n.* **1** *Chem.* each of two or more compounds with the same molecular formula but a different arrangement of atoms and different properties. **2** *Physics* each of two or more atomic nuclei that have the same atomic number and the same mass number but different energy states. □ **isomeric** /̩aɪsə'merɪk/ *adj.* **isomerism** /aɪ'sɒmə̩rɪz(ə)m/ *n.* **isomerize** *v.* (also **-ise**). [G f. Gk *isomerēs* sharing equally (as ISO-, *meros* share)]

isomerous /aɪ'sɒmərəs/ *adj. Bot.* (of a flower) having the same number of petals in each whorl. [Gk *isomerēs*: see ISOMER]

isometric /̩aɪsəʊ'metrɪk/ *adj.* **1** of equal measure. **2** *Physiol.* (of muscle action) developing tension while the muscle is prevented from contracting. **3** (of a drawing etc.) with the plane of projection at equal angles to the three principal axes of the object shown. **4** *Math.* (of a transformation) without change of shape or size. □ **isometrically** *adv.* **isometry** /aɪ'sɒmɪtrɪ/ *n.* (in sense 4). [Gk *isometria* equality of measure (as ISO-, -METRY)]

isometrics /̩aɪsəʊ'metrɪks/ *n.pl.* a system of physical exercises in which muscles are caused to act against each other or against a fixed object.

isomorph /'aɪsəʊ̩mɔːf/ *n.* an isomorphic substance or organism. [ISO- + Gk *morphē* form]

isomorphic /̩aɪsəʊ'mɔːfɪk/ *adj.* (also **isomorphous** /-fəs/) **1** exactly corresponding in form and relations. **2** *Crystallog.* having the same form. □ **isomorphism** *n.*

-ison /ɪs(ə)n/ *suffix* forming nouns, = -ATION (*comparison; garrison; jettison; venison*). [OF *-aison* etc. f. L *-atio* etc.: see -ATION]

isophote /'aɪsəʊ̩fəʊt/ *n.* a line (imaginary or in a diagram) of equal brightness or illumination. [ISO- + Gk *phōs phōtos* light]

isopleth /'aɪsəʊ̩pleθ/ *n.* a line on a map connecting places having equal incidence of a meteorological feature. [ISO- + Gk *plēthos* fullness]

isopod /'aɪsəʊ̩pɒd/ *n. Zool.* a crustacean of the order Isopoda, having a flattened body with seven pairs of legs. The order includes woodlice and slaters, and many marine and freshwater species (some of which are parasitic). [F *isopode* f. mod.L *Isopoda* (as ISO-, Gk *pous podos* foot)]

isosceles /aɪ'sɒsɪ̩liːz/ *adj.* (of a triangle) having two sides equal. [LL f. Gk *isoskelēs* (as ISO-, *skelos* leg)]

isoseismal /̩aɪsəʊ'saɪzm(ə)l/ *adj. & n.* (also **isoseismic** /-mɪk/) ● *adj.* having equal strength of earthquake shock. ● *n.* a line on a map connecting places having an equal strength of earthquake shock.

isostasy /aɪ'sɒstəsɪ/ *n. Geol.* the general state of equilibrium of the earth's crust, which behaves as if it consists of blocks floating on the underlying mantle. A consequence of this is that the crust rises if material (e.g. an ice-cap) is removed and sinks if material is deposited. □ **isostatic** /̩aɪsəʊ'stætɪk/ *adj.* [ISO- + Gk *stasis* station]

isothere /'aɪsəʊ̩θɪə(r)/ *n.* a line on a map connecting places having the same average temperature in the summer. [ISO- + Gk *theros* summer]

isotherm /'aɪsəʊ̩θɜːm/ *n.* **1** a line on a map connecting places having the same temperature at a given time or on average over a given period. **2** a curve for changes in a physical system at a constant temperature. □ **isothermal** /̩aɪsəʊ'θɜːm(ə)l/ *adj.* **isothermally** *adv.* [F *isotherme* (as ISO-, Gk *thermē* heat)]

isotonic /̩aɪsəʊ'tɒnɪk/ *adj.* **1** having the same osmotic pressure. **2** *Physiol.* (of muscle action) taking place with normal contraction. □ **isotonically** *adv.* **isotonicity** /-tə'nɪsɪtɪ/ *n.* [Gk *isotonos* (as ISO-, TONE)]

isotope /'aɪsə̩təʊp/ *n. Chem.* each of two or more forms of the same element that contain equal numbers of protons but different numbers of neutrons in their nuclei, and hence differ in relative atomic mass but not in chemical properties. Most elements consist of one stable (non-radioactive) isotope; there are also numerous radioactive isotopes, the majority of them artificially created. □ **isotopic** /̩aɪsə'tɒpɪk/ *adj.* **isotopically** *adv.* **isotopy** /aɪ'sɒtəpɪ/ *n.* [ISO- + Gk *topos* place (i.e. in the periodic table of elements)]

isotropic /̩aɪsəʊ'trɒpɪk/ *adj.* having the same physical properties in all directions (opp. ANISOTROPIC). □ **isotropically** *adv.* **isotropy** /aɪˈsɒtrəpɪ/ *n.* [ISO- + -TROPIC]

Ispahan see ISFAHAN.

I-spy /aɪˈspaɪ/ *n.* a game in which players try to identify something observed by one of them and identified by its initial letter.

Israel[1] /ˈɪzreɪəl/ **1** (also **children of Israel**) the Hebrew nation or people. According to tradition they are descended from the patriarch Jacob (his alternative name was Israel), whose twelve sons became founders of the twelve tribes. **2** the northern kingdom of the Hebrews (c.930–721 BC), formed after the reign of Solomon, whose inhabitants were carried away to captivity in Babylon. (See also JUDAH 3.) [Heb. *yisrā'ēl* he that strives with God; see Gen. 32:28]

Israel[2] /ˈɪzreɪəl/ a country in the Middle East, on the Mediterranean Sea; pop. (est. 1990) 4,600,000; languages, Hebrew (official), English, Arabic; capital (not recognized as such by the UN), Jerusalem. The modern state of Israel was established as a Jewish homeland in 1948, on land that was at that time part of the British mandated territory of Palestine. Jewish settlement in the area had increased greatly during the Nazi persecution of the 1930s and 1940s. Britain had supported the creation of such a state in the Balfour Declaration of 1917, and the United Nations had voted for it in 1947. Israel was immediately attacked by the surrounding Arab states, which it defeated. The continuing conflict with the neighbouring Arabs, mainly over the rights of the Palestinian Arabs displaced from their homes or living under Israeli rule, has caused continual tension and intermittent terrorist and military activity. Further wars occurred in 1956, 1967, and 1973, which resulted in Israeli occupation of eastern Jerusalem, the West Bank, the Gaza Strip, and the Golan Heights. In 1993 Israel and the Palestine Liberation Organization signed an agreement for limited Palestinian autonomy in the West Bank and the Gaza Strip. (See also PALESTINE.) □ **Israeli** /ɪzˈreɪlɪ/ *adj. & n.*

Israelite /ˈɪzrəˌlaɪt/ *n. & adj.* ● *n.* a member of the ancient Hebrew nation or people, esp. an inhabitant of the northern kingdom of the Hebrews (c.930–721 BC). ● *adj.* of or relating to the Israelites. [ISRAEL[1]]

Israfel /ˈɪzrəˌfel/ (in Muslim tradition) the angel of music, who will sound the trumpet on the Day of Judgement.

Issa /ˈiːsɑː/ *n. & adj.* ● *n.* **1** a member of a Somali people living in the Republic of Djibouti. **2** the Cushitic language of this people. ● *adj.* of or relating to the Issas or their language. [Issa]

Issachar /ˈɪsəkə(r)/ **1** a Hebrew patriarch, son of Jacob and Leah (Gen. 30:18). **2** the tribe of Israel traditionally descended from him.

Issigonis /ˌɪsɪˈɡəʊnɪs/, Sir Alec (Arnold Constantine) (1906–88), Turkish-born British car designer. His most famous designs were the Morris Minor (1948), produced until 1971, and the Mini (1959).

ISSN *abbr.* international standard serial number.

issuant /ˈɪʃʊənt, ˈɪsjʊ-/ *adj.* Heraldry (esp. of a beast with only the upper part shown) rising from the bottom or top of a bearing.

issue /ˈɪʃuː, ˈɪsjuː/ *n. & v.* ● *n.* **1 a** a giving out or circulation of shares, notes, stamps, etc. **b** a quantity of coins, supplies, copies of a newspaper or book etc., circulated or put on sale at one time. **c** an item or amount given out or distributed. **d** each of a regular series of a magazine etc. (*the May issue*). **2 a** an outgoing, an outflow. **b** a way out, an outlet, esp. the place of the emergence of a stream etc. **3** a point in question; an important subject of debate or litigation. **4** a result; an outcome; a decision. **5** *Law* children, progeny (*without male issue*). **6** *archaic* a discharge of blood etc. ● *v.* (**issues, issued, issuing**) **1** *intr.* (often foll. by *out, forth*) *literary* go or come out. **2** *tr.* **a** send forth; publish; put into circulation. **b** supply, esp. officially or authoritatively (foll. by *to, with*: *issued passports to them; issued them with passports; issued orders to the staff*). **3** *intr.* **a** (often foll. by *from*) be derived or result. **b** (foll. by *in*) end, result. **4** *intr.* (foll. by *from*) emerge from a condition. □ **at issue 1** under discussion; in dispute. **2** at variance. **issue of fact** (or **law**) a dispute at law when the significance of a fact or facts is denied or when the application of the law is contested. **join** (or **take**) **issue** identify an issue for argument (foll. by *with, on*). **make an issue of** make a fuss about; turn into a subject of contention. □ **issuable** *adj.* **issuance** *n.* **issueless** *adj.* **issuer** *n.* [ME f. OF ult. f. L *exitus* past part. of *exire* EXIT]

-ist /ɪst/ *suffix* forming personal nouns (and in some senses related adjectives) denoting: **1** an adherent of a system etc. in -ism: see -ISM 2 (*Marxist; fatalist*). **2 a** a member of a profession (*pathologist*). **b** a person concerned with something (*tobacconist*). **3** a person who uses a thing (*violinist; balloonist; motorist*). **4** a person who does something expressed by a verb in -ize (*plagiarist*). **5** a person who subscribes to a prejudice or practises discrimination (*racist; sexist*). [OF -iste, L -ista f. Gk -istēs]

Istanbul /ˌɪstænˈbʊl/ a port in Turkey on the Bosporus, lying partly in Europe, partly in Asia; pop. (1990) 7,309,190. It was the capital of Turkey from 1453 until 1923, when it was replaced by Ankara. Formerly the Roman city of Constantinople (330–1453), it was built on the site of the ancient Greek city of Byzantium, founded in the 7th century BC. It was captured by the Ottoman Turks in 1453 and most of its characteristic buildings date from the Ottoman era. Its Turkish name is derived from Greek *eis tēn polin* meaning 'into the city'; it was formerly also known as Stamboul.

Isthmian /ˈɪsmɪən, ˈɪsθmɪ-/ *adj.* of or relating to the Isthmus of Corinth in southern Greece. □ **Isthmian games** games held by the ancient Greeks every other year near the Isthmus of Corinth.

isthmian /ˈɪsmɪən, ˈɪsθmɪ-/ *adj.* of or relating to an isthmus.

isthmus /ˈɪsməs, ˈɪsθməs/ *n.* **1** (*pl.* **isthmuses**) a narrow piece of land connecting two larger bodies of land. **2** (*pl.* **isthmi** /-maɪ/) *Anat.* a narrow part connecting two larger parts. [L f. Gk *isthmos*]

istle /ˈɪstlɪ/ *n.* a fibre used for cord, nets, etc., obtained from agave. [Mex. *ixtli*]

IT *abbr.* information technology.

It. *abbr.* Italian.

it[1] /ɪt/ *pron.* (*poss.* **its**; *pl.* **they**) **1** the thing (or sometimes the animal or child) previously named or in question (*took a stone and threw it*). **2** the person in question (*Who is it? It is I; is it a boy or a girl?*). **3** as the subject of an impersonal verb (*it is raining; it is winter; it is Tuesday; it is two miles to Bath*). **4** as a substitute for a deferred subject or object (*it is intolerable, this delay; it is silly to talk like that; I take it that you agree*). **5** as a substitute for a vague object (*brazen it out; run for it!*). **6** as the antecedent to a relative word (*it was an owl I heard*). **7** exactly what is needed (*absolutely it*). **8** the extreme limit of achievement. **9** *colloq.* sexual intercourse; sex appeal. **10** (in children's games) a player who has to perform a required feat, esp. to catch the others. □ **that's it** *colloq.* that is: **1** what is required. **2** the difficulty. **3** the end, enough. **this is it** *colloq.* **1** the expected event is at hand. **2** this is the difficulty. [OE *hit* neut. of HE]

it[2] /ɪt/ *n. colloq.* Italian vermouth (*gin and it*). [abbr.]

ITA *abbr.* **1** (also **i.t.a.**) initial teaching alphabet. **2** see INDEPENDENT TELEVISION AUTHORITY.

Itaipu /iːˈtaɪpuː/ a dam on the Paraná river in SW Brazil, one of the world's largest hydroelectric installations, formally opened in 1982.

ital. *abbr.* italic (type).

Italia see ITALY.

Italian /ɪˈtæljən/ *n. & adj.* ● *n.* **1 a** a native or national of Italy. **b** a person of Italian descent. **2** the official language of Italy. (*See note below.*) ● *adj.* of or relating to Italy or its people or language. □ **Italian vermouth** a sweet kind of vermouth. [ME f. It. *Italiano* f. *Italia* Italy]

▪ Italian is a Romance language which in many ways has remained closer to Latin than have the others of this group. It is spoken by some 60 million people in Italy and Switzerland and by large numbers in the US, Australia, and South America.

Italianate /ɪˈtæljəˌneɪt/ *adj.* of Italian style or appearance. [It. *Italianato*]

Italic /ɪˈtælɪk/ *adj. & n.* ● *adj.* **1** of ancient Italy. **2** of or relating to the branch of Indo-European languages including Latin, Oscan, Umbrian, and the Romance languages. ● *n.* the Italic group of languages. [L *italicus* f. Gk *italikos* Italian]

italic /ɪˈtælɪk/ *adj. & n.* ● *adj.* **1** of the sloping kind of letters now used esp. for emphasis or distinction and in foreign words. ● *n.* **1** a letter in italic type. **2** this type. [as ITALIC]

italicize /ɪˈtælɪˌsaɪz/ *v.tr.* (also **-ise**) print in italics. □ **italicization** /ɪˌtælɪsaɪˈzeɪʃ(ə)n/ *n.*

Italiot /ɪˈtælɪət/ *n. & adj.* ● *n.* an inhabitant of the Greek colonies in ancient Italy. ● *adj.* of or relating to the Italiots. [Gk *Italiōtēs* f. *Italia* Italy]

Italo- /ɪˈtæləʊ/ *comb. form* Italian; Italian and.

Italy /ˈɪtəlɪ/ (Italian **Italia** /iˈtaːlja/) a country in southern Europe; pop. (est. 1990) 57,746,160; official language, Italian; capital, Rome. Mainland Italy forms a peninsula extending south from the Alps into the Mediterranean Sea. There are also several offshore islands, of which the largest are Sicily and Sardinia. Italy was united under Rome from the 2nd century BC to the collapse of the empire in AD 476. In the Middle Ages, it was dominated by several city-states and the papacy, and was the centre of the Renaissance. Modern Italy was created by the nationalist movement of the mid-19th century, led by Garibaldi and the kingdom of Sardinia; the Sardinian monarch, Victor

Emmanuel II, became king of Italy in 1861. Italy entered the First World War on the Allied side in 1915. In 1922, the country was taken over by the Fascist dictator Mussolini; participation in support of Germany during the Second World War resulted in defeat and Mussolini's downfall. A republic was established by popular vote in 1946. The postwar period brought sustained economic growth, particularly in the north of the country, but also political instability and frequent changes of government. In 1993 the electoral system of proportional representation was replaced by a 'first past the post' method for three-quarters of parliamentary seats, and the political establishment was discredited following revelations of widespread corruption. Italy was a founder member of the EEC.

Itanagar /ˌiːtəˈnʌgə(r)/ a city in the far north-east of India, north of the Brahmaputra river, capital of the state of Arunachal Pradesh; pop. (1991) 17,300.

ITAR-Tass /ˌɪtɑːˈtæs/ see TASS.

ITC see INDEPENDENT TELEVISION COMMISSION.

itch /ɪtʃ/ n. & v. ● n. **1** an irritation in the skin. **2** an impatient desire; a hankering. **3** (prec. by *the*) (in general use) scabies. ● v.intr. **1** feel an irritation in the skin, causing a desire to scratch it. **2** (usu. foll. by *to* + infin.) (of a person) feel a desire to do something (*I'm itching to tell you the news*). □ **itching palm** avarice. **itch-mite** a parasitic mite, *Sarcoptes scabiei*, which burrows under the skin causing scabies. [OE *gycce*, *gyccan* f. WG]

itchy /ˈɪtʃɪ/ adj. (**itchier**, **itchiest**) having or causing an itch. □ **have itchy feet** colloq. **1** be restless. **2** have a strong urge to travel. □ **itchiness** n.

it'd /ˈɪtəd/ contr. colloq. **1** it had. **2** it would.

-ite[1] /aɪt/ suffix forming nouns meaning 'a person or thing connected with': **1** in names of persons: **a** as natives of a country (*Israelite*). **b** often derog. as followers of a movement etc. (*pre-Raphaelite*; *Trotskyite*). **2** in names of things: **a** fossil organisms (*ammonite*). **b** minerals (*graphite*). **c** constituent parts of a body or organ (*somite*). **d** explosives (*dynamite*). **e** commercial products (*ebonite*; *vulcanite*). **f** salts of acids having names in *-ous* (*nitrite*; *sulphite*). [from or after F *-ite* f. L *-ita* f. Gk *-itēs*]

-ite[2] /aɪt, ɪt/ suffix **1** forming adjectives (*erudite*; *favourite*). **2** forming nouns (*appetite*). **3** forming verbs (*expedite*; *unite*). [from or after L *-itus* past part. of verbs in *-ere*, and *-ire*]

item /ˈaɪtəm/ n. & adv. ● n. **1 a** any of a number of enumerated or listed things. **b** an entry in an account. **2** an article, esp. one for sale (*household items*). **3** a separate or distinct piece of news, information, etc. **4** colloq. a set of two people who live together; a couple. ● adv. archaic (introducing the mention of each item) likewise, also. [orig. as adv.: L, = in like manner, also]

itemize /ˈaɪtəˌmaɪz/ v.tr. (also **-ise**) state or list item by item. □ **itemizer** n. **itemization** /ˌaɪtəmaɪˈzeɪʃ(ə)n/ n.

iterate /ˈɪtəˌreɪt/ v.tr. repeat; state repeatedly. □ **iteration** /ˌɪtəˈreɪʃ(ə)n/ n. [L *iterare iterat-* f. *iterum* again]

iterative /ˈɪtərətɪv/ adj. Gram. = FREQUENTATIVE. □ **iteratively** adv.

Ithaca /ˈɪθəkə/ an island off the western coast of Greece in the Ionian Sea, the legendary home of Odysseus.

ithyphallic /ˌɪθɪˈfælɪk/ adj. Gk Hist. **1 a** of the phallus carried in Bacchic festivals. **b** (of a statue etc.) having an erect penis. **2** lewd, licentious. **3** (of a poem or metre) used for Bacchic hymns. [LL *ithyphallicus* f. Gk *ithuphallikos* f. *ithus* straight, *phallos* PHALLUS]

-itic /ˈɪtɪk/ suffix forming adjectives and nouns corresponding to nouns in *-ite*, *-itis*, etc. (*Semitic*; *arthritic*; *syphilitic*). [from or after F *-itique* f. L *-iticus* f. Gk *-itikos*: see -IC]

itinerant /aɪˈtɪnərənt, ɪˈtɪn-/ adj. & n. ● adj. travelling from place to place. ● n. an itinerant person; a tramp. □ **itinerant judge** (or **minister** etc.) a judge, minister, etc. travelling within a circuit. □ **itineracy** n. **itinerancy** n. [LL *itinerari* travel f. L *iter itiner-* journey]

itinerary /aɪˈtɪnərərɪ, ɪˈtɪn-/ n. & adj. ● n. (pl. **-ies**) **1** a detailed route. **2** a record of travel. **3** a guidebook. ● adj. of roads or travelling. [LL *itinerarius* (adj.), *-um* (n.) f. L *iter*: see ITINERANT]

itinerate /aɪˈtɪnəˌreɪt, ɪˈtɪn-/ v.intr. travel from place to place or (of a minister etc.) within a circuit. □ **itineration** /aɪˌtɪnəˈreɪʃ(ə)n, ɪˌtɪn-/ n. [LL *itinerari*: see ITINERANT]

-ition /ˈɪʃ(ə)n/ suffix forming nouns, = -ATION (*admonition*; *perdition*; *position*). [from or after F *-ition* or L *-itio -itionis*]

-itious[1] /ˈɪʃəs/ suffix forming adjectives corresponding to nouns in *-ition* (*ambitious*; *suppositious*). [L *-itio* etc. + -OUS]

-itious[2] /ˈɪʃəs/ suffix forming adjectives meaning 'related to, having the

nature of' (*adventitious*; *suppositious*). [L *-icius* + -OUS, commonly written with *t* in med.L manuscripts]

-itis /ˈaɪtɪs/ suffix forming nouns, esp.: **1** names of inflammatory diseases (*appendicitis*; *bronchitis*). **2** colloq. in extended uses with ref. to conditions compared to diseases (*electionitis*). [Gk *-itis*, forming fem. of adjectives in *-itēs* (with *nosos* 'disease' implied)]

-itive /ɪtɪv/ suffix forming adjectives, = -ATIVE (*positive*; *transitive*). [from or after F *-itif -itive* or L *-itivus* f. participial stems in *-it-*: see -IVE]

it'll /ˈɪt(ə)l/ contr. colloq. it will; it shall.

ITN abbr. (in the UK) Independent Television News.

ITO abbr. International Trade Organization.

Ito /ˈiːtəʊ/, Prince Hirobumi (1841–1909), Japanese statesman, Premier four times between 1884 and 1901. In 1889 he was prominent in drafting the Japanese constitution, and the following year helped to establish a bicameral national diet. He was assassinated by a member of the Korean independence movement.

-itor /ɪtə(r)/ suffix forming agent nouns, usu. from Latin words (sometimes via French) (*creditor*). (See also -OR[1].)

-itory /ɪtərɪ, ɪtrɪ/ suffix forming adjectives meaning 'relating to or involving (a verbal action)' (*inhibitory*). (See also -ORY[2].) [L *-itorius*]

-itous /ɪtəs/ suffix forming adjectives corresponding to nouns in *-ity* (*calamitous*; *felicitous*). [from or after F *-iteux* f. L *-itosus*]

its /ɪts/ poss.pron. of it; of itself (*can see its advantages*).

it's /ɪts/ contr. **1** it is. **2** it has.

itself /ɪtˈself/ pron. emphatic and refl. form of IT[1]. □ **by itself** apart from its surroundings, automatically, spontaneously. **in itself** viewed in its essential qualities (*not in itself a bad thing*). [OE f. IT[1] + SELF, but often treated as ITS + SELF (cf. *its own self*)]

itsy-bitsy /ˌɪtsɪˈbɪtsɪ/ adj. (also **itty-bitty** /ˌɪtɪˈbɪtɪ/) colloq. usu. derog. tiny, insubstantial, slight. [redupl. of LITTLE, infl. by BIT[1]]

ITU see INTERNATIONAL TELECOMMUNICATION UNION.

ITV abbr. (in the UK) Independent Television.

-ity /ɪtɪ/ suffix forming nouns denoting: **1** quality or condition (*authority*; *humility*; *purity*). **2** an instance or degree of this (*a monstrosity*; *capacity*). [from or after F *-ité* f. L *-itas -itatis*]

IU abbr. international unit.

IUD abbr. **1** an intrauterine device (to prevent pregnancy). **2** intrauterine death (of the foetus before birth).

-ium /ɪəm/ suffix forming nouns denoting esp.: **1** (also **-um**) names of metallic elements (*uranium*; *tantalum*). **2** a region of the body (*pericardium*; *hypogastrium*). **3** a biological structure (*mycelium*; *prothallium*). [from or after L *-ium* f. Gk *-ion*]

IUPAC /ˈjuːpæk/ abbr. International Union of Pure and Applied Chemistry.

IV abbr. intravenous.

Ivan /ˈaɪv(ə)n/ the name of six rulers of Russia, notably:

Ivan I (*c.*1304–41), grand duke of Muscovy 1328–40. He strengthened and enlarged the duchy, making Moscow the ecclesiastical capital in 1326.

Ivan III (known as Ivan the Great) (1440–1505), grand duke of Muscovy 1462–1505. He consolidated and enlarged his territory, defending it against a Tartar invasion in 1480, and adopting the title of 'Ruler of all Russia' in 1472.

Ivan IV (known as Ivan the Terrible) (1530–84), grand duke of Muscovy 1533–47 and first tsar of Russia 1547–84. His expansionist foreign policy resulted in the capture of Kazan (1552), Astrakhan (1556), and Siberia (1581). However, the Tartar siege of Moscow (1572) and Ivan's defeat by the Poles in the Livonian War (1558–82) left Russia weak and divided. He grew increasingly unpredictable and tyrannical; in 1581, he killed his eldest son Ivan in a fit of rage, the succession passing to his retarded second son Fyodor.

Ivan the Great, Ivan III of Russia (see IVAN).

Ivan the Terrible, Ivan IV of Russia (see IVAN).

I've /aɪv/ contr. I have.

-ive /ɪv/ suffix forming adjectives meaning 'tending to, having the nature of', and corresponding nouns (*suggestive*; *corrosive*; *palliative*; *coercive*; *talkative*). □ **-ively** suffix forming adverbs. **-iveness** suffix forming nouns. [from or after F *-if -ive* f. L *-ivus*]

Ives /aɪvz/, Charles (Edward) (1874–1954), American composer. Influenced by popular music and the sounds of everyday life, he developed the use of polyrhythms, polytonality, quarter-tones, note-clusters, and aleatoric techniques. He is noted for his second piano sonata *Concord* (1915), and his chamber work *The Unanswered Question* (1906), scored for two unsynchronized orchestras.

IVF *abbr. in vitro* fertilization.

ivied /ˈaɪvɪd/ *adj.* overgrown with ivy.

Ivory /ˈaɪvərɪ/, James (b.1928), American film director. He has made a number of films in partnership with the producer Ismail Merchant, including *Heat and Dust* (1983), *A Room with a View* (1986), *Maurice* (1987), *Howard's End* (1992), and *The Remains of the Day* (1993).

ivory /ˈaɪvərɪ/ *n.* (*pl.* **-ies**) **1** a hard creamy-white substance composing the main part of the tusks of the elephant, hippopotamus, walrus, and narwhal. **2** the colour of this. **3** (usu. in *pl.*) **a** an article made of ivory. **b** *sl.* anything made of or resembling ivory, esp. a piano key or a tooth. □ **fossil ivory** ivory from the tusks of a mammoth. **ivory black** black pigment from calcined ivory or bone. **ivory-nut** the seed of a corozo palm, *Phytelephas macrocarpa*, used as a source of vegetable ivory for carving (also called *corozo nut*). **ivory tower** a state of seclusion or separation from the ordinary world and the harsh realities of life. **vegetable ivory** a hard white material obtained from the endosperm of the ivory-nut. □ **ivoried** *adj.* [ME f. OF *yvoire* ult. f. L *ebur eboris*]

Ivory Coast (called in French *Côte d'Ivoire*) a country in West Africa, on the Gulf of Guinea; pop. (est. 1990) 12,000,000; languages, French (official), West African languages; capital, Yamoussoukro. The area was explored by the Portuguese in the late 15th century. Subsequently it was disputed by traders from various European countries, who mainly sought ivory and slaves. It was made a French protectorate in 1842, and became an autonomous republic within the French Community in 1958 and a fully independent republic outside it in 1960.

ivy /ˈaɪvɪ/ *n.* (*pl.* **-ies**) **1** a climbing evergreen shrub of the genus *Hedera*; esp. *H. helix*, with usu. dark green shiny five-angled leaves. **2** any climbing or trailing plant (*poison ivy*; *ground ivy*). [OE īfig]

Ivy League a group of long-established eastern US universities of high academic and social prestige, including Harvard, Yale, Princeton, and Columbia.

IWC *abbr.* International Whaling Commission.

Iwo Jima /ˌiːwəʊ ˈdʒiːmə/ a small volcanic island, the largest of the Volcano Islands in the western Pacific, 1,222 km (760 miles) south of Tokyo. During the Second World War it was the heavily fortified site of a Japanese air base, and its attack and capture in 1944–5 was one of the severest US campaigns. It was returned to Japan in 1968.

IWW see INDUSTRIAL WORKERS OF THE WORLD.

ixia /ˈɪksɪə/ *n.* a plant of the genus *Ixia* (iris family), native to southern Africa, with large showy flowers. [L f. Gk, a kind of thistle]

Ixion /ɪkˈsaɪən/ *Gk Mythol.* a king who was the first to murder one of his kin. He was purified by Zeus, but tried to seduce Hera, for which he was punished by being pinned to a fiery wheel that revolved unceasingly through the underworld.

Iyyar /ˈiːjɑː/ *n.* (in the Jewish calendar) the eighth month of the civil and second of the religious year, usually coinciding with parts of April and May. [Heb. *'iyyār*]

izard /ˈɪzɑːd/ *n.* the Pyrenean variety of the chamois. [F *isard*, of unkn. orig.]

-ize /aɪz/ *suffix* (also **-ise**) forming verbs, meaning: **1** make or become such (*Americanize*; *pulverize*; *realize*). **2** treat in such a way (*monopolize*; *pasteurize*). **3 a** follow a special practice (*economize*). **b** have a specified feeling (*sympathize*). **4** affect with, provide with, or subject to (*oxidize*; *hospitalize*). ¶ The form *-ize* has been in use in English since the 16th century; it is widely used in American English, but is not an Americanism. The alternative spelling *-ise* (reflecting a French influence) is in common use, esp. in British English, and is obligatory in certain cases: (*a*) where it forms part of a larger word-element, such as *-mise* (= sending) in *compromise*, and *-prise* (= taking) in *surprise*; and (*b*) in verbs corresponding to nouns with *-is-* in the stem, such as *advertise* and *televise*. □ **-ization** /aɪˈzeɪʃ(ə)n/ *suffix* forming nouns. **-izer** /ˌaɪzə(r)/ *suffix* forming agent nouns. [from or after F *-iser* f. LL *-izare* f. Gk *-izō*]

Izhevsk /ɪˈʒefsk/ an industrial city in central Russia, capital of the republic of Udmurtia; pop. (1990) 642,000. It was known as Ustinov from 1984 to 1987.

Izmir /ˈɪzmɪə(r)/ a seaport and naval base in western Turkey, on an inlet of the Aegean Sea; pop. (1990) 1,757,410. Formerly known as Smyrna, it is the third largest city in Turkey.

Izmit /ˈɪzmɪt/ a city in NW Turkey, situated on the Gulf of Izmit, an inlet of the Sea of Marmara; pop. (1990) 256,880.

Iznik /ˈɪznɪk/ a town in NW Turkey, situated to the south-east of the Sea of Marmara; pop. (1990) 17,230. Built on the site of ancient Nicaea, it has been a noted centre for the production of coloured tiles since the 16th century.

Izvestia /ɪzˈvestɪə/ (also ***Izvestiya***) a Russian daily newspaper founded in 1917 as the official organ of the Soviet government. It has continued to be published independently since the collapse of Communist rule and the breakup of the Soviet Union. [Russ. *Izvestiya* news]

Jj

J[1] /dʒeɪ/ n. (also **j**) (pl. **Js** or **J's**) **1** the tenth letter of the alphabet. **2** (as a Roman numeral) = i in a final position (*ij*; *vj*).

J[2] abbr. (also **J.**) **1** Judge. **2** Justice. **3** (in cards) jack.

J[3] symb. joule(s).

jab /dʒæb/ v. & n. ● v.tr. (**jabbed, jabbing**) **1 a** poke roughly. **b** stab. **2** (foll. by *into*) thrust (a thing) hard or abruptly. ● n. **1** an abrupt blow with one's fist or a pointed implement. **2** colloq. a hypodermic injection, esp. a vaccination. [orig. Sc. var. of JOB[2]]

Jabalpur /ˌdʒʌb(ə)l'pʊə(r)/ an industrial city and military post in Madhya Pradesh, central India; pop. (1991) 760,000.

jabber /'dʒæbə(r)/ v. & n. ● v. **1** intr. chatter volubly and incoherently. **2** tr. utter (words) fast and indistinctly. ● n. meaningless jabbering; a gabble. [imit.]

jabberwocky /'dʒæbəˌwɒkɪ/ n. (pl. **-ies**) a piece of nonsensical writing or speech, esp. for comic effect. [title of a poem in Lewis Carroll's *Through the Looking Glass* (1871)]

jabiru /'dʒæbɪˌruː/ n. a large black-necked stork of the genus *Ephippiorhynchus*; esp. *E. mycteria* of Central and South America, with mainly white plumage. [Tupi-Guarani *jabirú*]

jaborandi /ˌdʒæbə'rændɪ/ n. (pl. **jaborandis**) **1** a shrub of the genus *Pilocarpus*, of South America. **2** the dried leaflets of this, having diuretic and diaphoretic properties. [Tupi-Guarani *jaburandi*]

jabot /'ʒæbəʊ/ n. an ornamental frill or ruffle of lace etc. on the front of a shirt or blouse. [F, orig. = crop of a bird]

jacamar /'dʒækəˌmɑː(r)/ n. a small insect-eating bird with partly iridescent plumage, of the tropical South American family Galbulidae. [F, app. f. Tupi]

jacana /ˌdʒækənə, ˌdʒækəsə'nɑː/ n. (also **jaçana**) a small tropical wading bird of the family Jacanidae, with elongated toes and hind-claws which enable it to walk on floating leaves etc. (See also *lily-trotter*.) [Port. *jaçanã* f. Tupi-Guarani *jasaná*]

jacaranda /ˌdʒækə'rændə/ n. a tropical American tree with fragrant wood; esp. one of the genus *Jacaranda*, with showy trumpet-shaped blue flowers. [Tupi-Guarani *jacarandá*]

jacinth /'dʒæsɪnθ, 'dʒeɪs-/ n. a reddish-orange variety of zircon used as a gem. [ME *iacynt* etc. f. OF *iacinte* or med.L *jacint(h)us* f. L *hyacinthus* HYACINTH]

jack[1] /dʒæk/ n. & v. ● n. **1** a device for lifting heavy objects, esp. the axle of a vehicle off the ground while changing a wheel etc. **2** a court-card with a picture of a man, esp. a soldier, page, or knave, etc. **3** a ship's flag, esp. one flown from the bow and showing nationality. **4** a device using a single plug to connect an electrical circuit. **5** a small white ball in bowls, at which the players aim. **6 a** (in pl.) = JACKSTONE 1. **b** JACKSTONE 2. **7** (**Jack**) the familiar form of *John* esp. typifying the common man (*I'm all right, Jack*). **8** the figure of a man striking the bell on a clock. **9** sl. a detective; a policeman. **10** US sl. money. **11** N. Amer. colloq. = LUMBERJACK. **12** = STEEPLEJACK. **13** a device for turning a spit. **14** a marine perchlike fish of the family Carangidae (*amberjack*). **15** a device for plucking the string of a harpsichord etc., one being operated by each key. **16 a** the male of various animals (*jackass*). **b** a species etc. of animal smaller than other similar kinds (*jackdaw*; *jack snipe*). ● v.tr. **1** (usu. foll. by *up*) raise with or as with a jack (in sense 1). **2** (usu. foll. by *up*) colloq. raise e.g. prices. **3** (foll. by *off*) **a** go away, depart. **b** coarse sl. masturbate. □ **before you can say Jack Robinson** colloq. very quickly or suddenly. **every man jack** each and every person. **Jack**

Frost frost personified. **jack in** (or **up**) sl. abandon (an attempt etc.). **jack-in-the-box** a toy figure that springs out of a box when it is opened. **jack-in-office** a self-important minor official. **jack of all trades** a person who can do many different kinds of work. **jack-o'-lantern 1** a will-o'-the wisp. **2** a lantern made esp. from a pumpkin with holes for facial features. **jack plane** a medium-sized plane for use in rough joinery. **jack plug** a plug for use with a jack (see sense 4 of n.). **jack snipe** a small dark snipe, *Lymnocryptes minimus*. **Jack tar** a sailor. **Jack-the-lad** colloq. a brash, self-assured young man. **on one's jack** (or **Jack Jones**) sl. alone; on one's own. [ME *Iakke*, a pet-name for *John*, erron. assoc. with F *Jacques* James]

jack[2] /dʒæk/ n. **1** = BLACKJACK[2]. **2** hist. a sleeveless padded tunic worn by foot-soldiers. [ME f. OF *jaque*, of uncert. orig.]

jackal /'dʒæk(ə)l/ n. **1** a bushy-tailed wild dog of the genus *Canis*, native to Africa and southern Eurasia, usu. hunting or scavenging for food in packs; esp. *C. aureus*, the golden or common jackal. **2** colloq. **a** a person who does preliminary drudgery for another. **b** a person who assists another's immoral behaviour. [Turk. *çakal* f. Pers. *šagāl*]

jackanapes /'dʒækəˌneɪps/ n. archaic **1** a pert or insolent fellow. **2** a mischievous child. **3** a tame monkey. [earliest as *Jack Napes* (1450): supposed to refer to the Duke of Suffolk, whose badge was an ape's clog and chain]

jackaroo /ˌdʒækə'ruː/ n. (also **jackeroo**) Austral. colloq. a novice on a sheep-station or cattle-station. [JACK[1] + KANGAROO]

jackass /'dʒækæs/ n. **1** a male ass. **2** a stupid person.

jackboot /'dʒækbuːt/ n. **1** a large boot reaching above the knee, worn esp. by cavalry soldiers of the 17th–18th centuries and by German soldiers under the Nazi regime. **2** this as a symbol of fascism or military oppression. □ **jackbooted** adj.

Jack-by-the-hedge /ˌdʒækbaɪðə'hedʒ/ n. a white-flowered cruciferous plant, *Alliaria petiolata*, of shady places, with leaves smelling of garlic.

jackdaw /'dʒækdɔː/ n. a small grey-headed crow, *Corvus monedula*, often frequenting rooftops and nesting in tall buildings, and noted for its inquisitiveness (cf. DAW).

jackeroo var. of JACKAROO.

jacket /'dʒækɪt/ n. & v. ● n. **1 a** a sleeved short outer garment. **b** a thing worn esp. round the torso for protection or support (*life-jacket*). **2** a casing or covering, e.g. as insulation round a boiler. **3** = *dust-jacket*. **4** the skin of a potato, esp. when baked whole. **5** an animal's coat. ● v.tr. (**jacketed, jacketing**) cover with a jacket. □ **jacket potato** Brit. a baked potato served with the skin on. [ME f. OF *ja(c)quet* dimin. of *jaque* JACK[2]]

jackfish /'dʒækfɪʃ/ n. (pl. same) = PIKE[1].

jackfruit /'dʒækfruːt/ n. **1** a tropical Asian tree, *Artocarpus heterophyllus*, bearing fruit resembling breadfruit. **2** this fruit. [Port. *jaca* f. Malayalam *chakka* + FRUIT]

jackhammer /'dʒækˌhæmə(r)/ n. esp. N. Amer. a portable pneumatic hammer or drill.

Jack-in-the-pulpit /ˌdʒækɪnðə'pʊlpɪt/ n. a small woodland plant of the arum family, esp. cuckoo-pint.

jackknife /'dʒæknaɪf/ n. & v. ● n. (pl. **-knives**) **1** a large clasp-knife. **2** a dive in which the body is first bent at the waist and then straightened. ● v. (**-knifed, -knifing**) **1** intr. (of an articulated vehicle) fold against itself in an accidental skidding movement. **2** intr. & tr. fold like a jackknife. **3** *intr.* perform a jackknife dive. [f. *jack* (perh. f. JACK[1]) + KNIFE]

jackpot /ˈdʒækpɒt/ *n.* a large prize or amount of winnings, esp. accumulated in a game or lottery etc. □ **hit the jackpot** *colloq.* **1** win a large prize. **2** have remarkable luck or success. [JACK[1] *n.* 2 + POT[1]: orig. in a form of poker with two jacks as minimum to open the pool]

jackrabbit /ˈdʒækˌræbɪt/ *n.* N. *Amer.* a large North American hare of the genus *Lepus*, with very long ears and hind legs. [abbr. of *jackass-rabbit* (so called from its long ears)]

Jack Russell /dʒæk ˈrʌs(ə)l/ *n.* (also **Jack Russell terrier**) a small working terrier with short legs. [f. Revd John (*Jack*) *Russell* (1795–1883), Engl. clergyman and dog-breeder]

Jackson[1] /ˈdʒæks(ə)n/ the state capital of Mississippi; pop. (1990) 196,640. Founded as a trading post by the French-Canadian trader Louis Le Fleur, it was later named after President Andrew Jackson.

Jackson[2] /ˈdʒæks(ə)n/, Andrew (1767–1845), American general and Democratic statesman, 7th President of the US 1829–37. After waging several campaigns against American Indians, he defeated a British army at New Orleans (1815) and successfully invaded Florida (1818). As President, he replaced an estimated twenty per cent of those in public office with Democrat supporters, a practice that became known as the spoils system. His reputation for toughness gave rise to the nickname 'Old Hickory' (see HICKORY).

Jackson[3] /ˈdʒæks(ə)n/, Glenda (b.1936), English actress and politician. Throughout her career, she has appeared on stage and television, but is best known for her film work. She won Oscars for her performances in *Women in Love* (1969) and *A Touch of Class* (1973); other films include *Sunday Bloody Sunday* (1971) and *Turtle Diary* (1985). In 1992 she became Labour MP for the London constituency of Hampstead and Highgate.

Jackson[4] /ˈdʒæks(ə)n/, Jesse (Louis) (b.1941), American politician and clergyman. After working with Martin Luther King in the civil-rights struggle, he competed for but failed to win the Democratic Party's 1984 and 1988 presidential nominations.

Jackson[5] /ˈdʒæks(ə)n/, Michael (Joe) (b.1958), American singer and songwriter. In the 1970s he performed with his four older brothers in the pop group the Jackson Five. His full-time solo career began in 1979, when *Off the Wall* became the best-selling album to date by a black artist. Subsequent albums, including *Bad* (1987), confirmed him as the most commercially successful American star of the 1980s. His career suffered a setback in 1993 after allegations of child molestation concerning young boys. In 1994 he married Lisa Marie Presley (b.1968), the daughter of Elvis Presley.

Jackson[6] /ˈdʒæks(ə)n/, Thomas Jonathan (known as 'Stonewall Jackson') (1824–63), American general. During the American Civil War he made his mark as a commander at the first battle of Bull Run in 1861; a successful defensive stand there earned him his nickname. As the deputy of Robert E. Lee, he played an important part in the Confederate victories in Virginia in the first two years of the war.

Jacksonville /ˈdʒæks(ə)nˌvɪl/ an industrial city and port in NE Florida; pop. (1990) 672,970. It was named in honour of President Andrew Jackson.

jackstaff /ˈdʒækstɑːf/ *n.* Naut. **1** a staff at the bow of a ship for a jack. **2** a staff carrying the flag that is to show above the masthead.

jackstone /ˈdʒækstəʊn/ *n.* **1** (in *pl.*) a game played by tossing and catching small round pebbles or star-shaped pieces of metal. **2** a pebble or piece of metal used for this.

jackstraw /ˈdʒækstrɔː/ *n.* **1** a spillikin. **2** (in *pl.*, treated as *sing.*) a game of spillikins.

Jack the Ripper, unidentified (19th-century) English murderer. From August to November 1888, at least six prostitutes were brutally killed in the East End of London, the bodies being mutilated in a way that indicated a knowledge of anatomy. The authorities received taunting notes from a person calling himself Jack the Ripper and claiming to be the murderer, but the cases remain unsolved.

Jacky /ˈdʒækɪ/ *n.* (*pl.* **-ies**) (in full **Jacky Jacky**) *Austral. sl. offens.* an Aboriginal. [JACK[1]]

Jacob /ˈdʒeɪkəb/ a Hebrew patriarch, the younger of the twin sons of Isaac and Rebecca, who persuaded his brother Esau to sell him his birthright and tricked him out of his father's blessing (Gen. 25, 27). His twelve sons became the founders of the twelve tribes of ancient Israel. (See also ISRAEL[1].) [Heb. *ya 'aqōb* following after, supplanter]

Jacobean /ˌdʒækəˈbɪən/ *adj. & n.* ● *adj.* **1** of or relating to the reign of James I of England. **2** (of furniture) in the style prevalent during this period, esp. of the colour of dark oak. ● *n.* a person of the time of James I. [mod.L *Jacobaeus* f. eccl.L *Jacobus* James f. Gk *Iakōbos* Jacob]

Jacobi /dʒæˈkəʊbɪ/, Karl Gustav Jacob (1804–51), German mathematician. He worked on the theory of elliptic functions, in competition with Niels Abel. Jacobi also investigated number theory, mathematical analysis, geometry, and differential equations, and his work on determinants is important in dynamics and quantum mechanics.

Jacobin /ˈdʒækəbɪn/ *n.* **1** *hist.* a member of a political society existing during the French Revolution. (*See note below.*) **2** a radical or revolutionary. **3** *archaic* a Dominican friar. **4** (**jacobin**) a pigeon with reversed feathers on the back of its neck like a cowl. □ **Jacobinism** *n.* **Jacobinic** /ˌdʒækəˈbɪnɪk/ *adj.* **Jacobinical** *adj.* [orig. in sense 3 by assoc. with the Rue St Jacques in Paris: ME f. F f. med.L *Jacobinus* f. eccl.L *Jacobus*]

▪ Taking their name from the old Jacobin convent where they held their first meetings in 1789, the Jacobins were the most radical and ruthless of the political groups formed in the wake of the Revolution of 1789. In association with Robespierre they purged their more moderate rivals the Girondists, and instituted the Terror of 1793–4.

Jacobite[1] /ˈdʒækəˌbaɪt/ *n.* a follower of Jacobus Bardaeus, a 6th-century Syrian Monophysite monk. The Jacobites became the national Church of Syria, but declined in the 13th and 14th centuries with the Mongol invasions; a small Jacobite Church still exists.

Jacobite[2] /ˈdʒækəˌbaɪt/ *n.* a supporter of the deposed James II and his descendants in their claim to the British throne after the Revolution of 1688. The Jacobites drew most of their support from Catholic clans of the Scottish Highlands, backed by France only when it was politically convenient. A number of attempts were launched to regain the throne, including those in 1689–90, 1715, 1719, and 1745–6, but support finally collapsed when the clans were suppressed after the Battle of Culloden. □ **Jacobitism** *n.* **Jacobitical** /ˌdʒækəˈbɪtɪk(ə)l/ *adj.* [L *Jacobus* James: see JACOBEAN]

Jacobs /ˈdʒeɪkəbz/, William Wymark (1863–1943), English short-story writer. He is noted for his tales of the macabre, such as 'The Monkey's Paw' (1902).

Jacob's ladder *n.* **1** a plant, *Polemonium caeruleum*, with corymbs of blue or white flowers, and leaves suggesting a ladder. **2** a rope-ladder with wooden rungs. [f. Jacob's dream of a ladder reaching to heaven, as described in Gen. 28:12]

Jacob's staff *n.* **1** a surveyor's iron-shod rod used instead of a tripod. **2** an instrument for measuring distances and heights. [f. the staffs used by Jacob, as described in Gen. 30:37–43]

jaconet /ˈdʒækənɪt/ *n.* a cotton cloth like cambric, esp. a dyed waterproof kind for poulticing etc. [Urdu *jagannāthi* f. *Jagannath* (now Puri) in India, its place of origin: see JUGGERNAUT]

Jacopo della Quercia /ˈjækəˌpəʊ/ see DELLA QUERCIA.

jacquard /ˈdʒækɑːd/ *n.* (also **Jacquard**) **1** an apparatus with perforated cards, fitted to a loom to facilitate the weaving of figured fabrics. **2** (in full **Jacquard loom**) a loom fitted with this. **3** a fabric or article made with this, with an intricate variegated pattern. [J. M. *Jacquard*, Fr. inventor (1752–1834)]

jactitation /ˌdʒæktɪˈteɪʃ(ə)n/ *n.* **1** *Med.* **a** the restless tossing of the body in illness. **b** the twitching of a limb or muscle. **2** *archaic* the offence of falsely claiming to be a person's spouse. [med.L *jactitatio* false declaration f. L *jactitare* boast, frequent. of *jactare* throw: sense 1 f. earlier *jactation*]

Jacuzzi /dʒəˈkuːzɪ/ *n.* (*pl.* **Jacuzzis**) *propr.* a large round bath with underwater jets of water to massage the body. [name of the inventor and manufacturers]

jade[1] /dʒeɪd/ *n.* **1** a hard usu. green stone composed of silicates of calcium and magnesium, or of sodium and aluminium, used for ornaments and implements. **2** the green colour of jade. [F: *le jade* for *l'ejade* f. Sp. *piedra de ijada* stone of the flank, i.e. stone for colic (which it was believed to cure)]

jade[2] /dʒeɪd/ *n.* **1** an inferior or worn-out horse. **2** *derog.* a disreputable woman. [ME: orig. unkn.]

jaded /ˈdʒeɪdɪd/ *adj.* tired or worn out; surfeited. □ **jadedly** *adv.* **jadedness** *n.* [JADE[2]]

jadeite /ˈdʒeɪdaɪt/ *n.* a green, blue, or white sodium aluminium silicate form of jade.

j'adoube /ʒæˈduːb/ *int.* Chess a declaration by a player intending to adjust the placing of a piece without making a move with it. [F, = I adjust]

jaeger /ˈjeɪɡə(r)/ *n.* N. *Amer.* a seabird of the family Stercorariidae, a skua; esp. one of the smaller kinds, of the genus *Stercorarius*. (See also SKUA.) [G *Jäger* hunter f. *jagen* to hunt]

Jaffa[1] /ˈdʒæfə/ (Hebrew **Yafo** /ˈjɑːfɔː/; biblical name **Joppa** /ˈdʒɒpə/) a city and port on the Mediterranean coast of Israel, forming a southern suburb of the Tel Aviv conurbation, and since 1949 united with Tel Aviv. Inhabited since prehistoric times, Jaffa was a Byzantine bishopric until captured by the Arabs in 636; later, it was a stronghold of the Crusaders.

Jaffa[2] /ˈdʒæfə/ n. a large oval thick-skinned variety of orange. [f. JAFFA[1], near where it was first grown]

Jaffna /ˈdʒæfnə/ a city and port on the Jaffna peninsula at the northern tip of Sri Lanka; pop. (1990) 129,000. It has a predominantly Tamil population and until the 17th century was the capital of a Tamil monarchy.

jag[1] /dʒæg/ n. & v. ● n. **1** a sharp projection of rock etc. **2** Sc. a prickle, a thorn. ● v.tr. (**jagged, jagging**) **1** cut or tear unevenly. **2** make indentations in. □ **jagger** n. [ME, prob. imit.]

jag[2] /dʒæg/ n. sl. **1** a drinking-bout; a spree. **2** a period of indulgence in an activity, emotion, etc. [orig. 16th c., = load for one horse: orig. unkn.]

Jagannatha /ˌdʒæɡəˈnɑːθə/ see JUGGERNAUT.

jagged /ˈdʒæɡɪd/ adj. **1** with an unevenly cut or torn edge. **2** deeply indented; with sharp points. □ **jaggedly** adv. **jaggedness** n. [JAG[1]]

jaggy /ˈdʒæɡɪ/ adj. (**jaggier, jaggiest**) **1** = JAGGED. **2** (also **jaggie**) Sc. prickly.

jaguar /ˈdʒæɡjʊə(r)/ n. a large leopard-like spotted feline, *Panthera onca*, of Central and South America. [Tupi-Guarani *jaguara*]

jaguarundi /ˌdʒæɡwəˈrʌndɪ/ n. (pl. **jaguarundis**) an unspotted feline, *Felis yaguarondi*, of Central and South America, with a long slender body and tail, and short legs. [Tupi-Guarani]

jai alai /ˈhaɪ əˌlaɪ/ n. a game like pelota played with large curved wicker baskets. [Sp. f. Basque *jai* festival + *alai* merry]

jail /dʒeɪl/ n. & v. (also **gaol**) ● n. **1** a place to which persons are committed by a court for detention. **2** confinement in a jail. ● v.tr. put in jail. □ **jail-bait** sl. a girl under the age of consent. [ME *gayole* f. OF *jaiole, jeole* & ONF *gaole* f. Rmc dimin. of L *cavea* CAGE]

jailbird /ˈdʒeɪlbɜːd/ n. (also **gaolbird**) a prisoner or habitual criminal.

jailbreak /ˈdʒeɪlbreɪk/ n. (also **gaolbreak**) an escape from jail.

jailer /ˈdʒeɪlə(r)/ n. (also **gaoler**) a person in charge of a jail or of the prisoners in it.

Jain /dʒaɪn/ n. & adj. ● n. an adherent of Jainism. ● adj. of or relating to Jainism. [Hindi f. Skr. *jaina* saint, victor f. *jina* victorious; see JINA]

Jainism /ˈdʒaɪnɪz(ə)m/ n. a non-theistic religion founded in India in the 6th century BC by the Jina Vardhamana Mahavira as a reaction against the teachings of orthodox Brahmanism. One of its central doctrines is non-injury to living creatures. Salvation is attained by perfection of the soul through successive lives (see KARMA). Jainism survives in India today but, unlike Buddhism, it has never spread beyond it. There are two major sects: the white-robed Svetambaras and the Digambaras, who are generally naked. □ **Jainist** n. & adj.

Jaipur /dʒaɪˈpʊə(r)/ a city in western India, the capital of Rajasthan; pop. (1991) 1,455,000.

Jakarta see DJAKARTA.

jake /dʒeɪk/ adj. Austral. & NZ sl. all right; satisfactory. [20th c.: orig. uncert.]

Jakobson /ˈjækəbs(ə)n/, Roman (Osipovich) (1896–1982), Russian-born American linguist. In 1941 he emigrated to the US and from 1949 to 1967 was professor of Slavic languages and literature and general linguistics at Harvard University. In *Child Language, Aphasia, and Phonological Universals* (1941) he developed the hypothesis that there may be a universal sequence according to which speech sounds are learned. His *Fundamentals of Language* (1956) postulates a phonological system of twelve binary oppositions to cover all the permutations of sounds in the world's languages.

Jalalabad /dʒəˈlæləˌbæd/ a city in eastern Afghanistan, situated east of Kabul, near the border with Pakistan; pop. (est. 1984) 61,000.

Jalal ad-Din ar-Rumi /dʒəˌlæl ædˌdɪn ɑːˈruːmɪ/ (also called Mawlana) (1207–73), Persian poet and Sufi mystic. He was born in Balkh (in modern Afghanistan), but lived for most of his life at Konya in Anatolia, where he founded the order of whirling dervishes (see DERVISH). He wrote much lyrical poetry and an influential epic on Sufi mystical doctrine.

Jalandhar see JULLUNDUR.

jalap /ˈdʒæləp/ n. a purgative drug obtained esp. from the tuberous roots of a Mexican climbing plant, *Exogonium purga*. [F f. Sp. *jalapa* f. JALAPA]

Jalapa /həˈlɑːpə/ (in full **Jalapa Enríquez** /enˈriːkez/) a city in east central Mexico, capital of the state of Veracruz; pop. (1990) 288,330.

jalapeño /ˌhæləˈpeɪnjəʊ, -ˈpiːnəʊ/ n. (pl. **-os**) (in full **jalapeño pepper**) a very hot green chilli pepper. [Mex. Sp.]

Jalisco /həˈliːskəʊ/ a state of west central Mexico, on the Pacific coast; capital, Guadalajara.

jalopy /dʒəˈlɒpɪ/ n. (pl. **-ies**) colloq. a dilapidated old motor vehicle. [20th c.: orig. unkn.]

jalousie /ˈʒæluˌziː/ n. a blind or shutter made of a row of angled slats to keep out rain etc. and control the influx of light. [F (as JEALOUSY)]

Jam. abbr. **1** Jamaica. **2** James (New Testament).

jam[1] /dʒæm/ v. & n. ● v.tr. & intr. (**jammed, jamming**) **1 a** tr. (usu. foll. by *into*) squeeze or wedge into a space. **b** intr. become wedged. **2 a** tr. cause (machinery or a component) to become wedged or immovable so that it cannot work. **b** intr. become jammed in this way. **3** tr. push or cram together in a compact mass. **4** intr. (foll. by *in, on to*) push or crowd (they jammed on to the bus). **5** tr. **a** block (a passage, road, etc.) by crowding or obstructing. **b** (foll. by *in*) obstruct the exit of (we were jammed in). **6** tr. (usu. foll. by *on*) apply (brakes etc.) forcefully or abruptly. **7** tr. make (a radio transmission) unintelligible by causing interference. **8** intr. colloq. (in jazz etc.) extemporize with other musicians. ● n. **1** a squeeze or crush. **2** a crowded mass (traffic jam). **3** colloq. an awkward situation or predicament. **4** a stoppage (of a machine etc.) due to jamming. **5** (in full **jam session**) colloq. improvised playing by a group of jazz musicians. □ **jam-packed** colloq. full to capacity. □ **jammer** n. [imit.]

jam[2] /dʒæm/ n. & v. ● n. **1** a conserve of fruit and sugar boiled to a thick consistency. **2** Brit. colloq. something easy or pleasant (money for jam). ● v.tr. (**jammed, jamming**) **1** spread jam on. **2** make (fruit etc.) into jam. □ **jam tomorrow** a pleasant thing often promised but usu. never forthcoming. [perh. = JAM[1]]

Jamaica /dʒəˈmeɪkə/ an island country in the Caribbean Sea, south-east of Cuba; pop. (est. 1990) 2,500,000; official language, English; capital, Kingston. Visited by Columbus in 1494, Jamaica was colonized by the Spanish, who enslaved or killed the native people. Both the Spanish and the British, who took the island by force in 1655, imported slaves, mainly to work on sugar plantations. In the 17th century, Jamaica was a base for buccaneers preying on Spanish interests. British colonial rule was threatened by popular violence in the mid-19th century which led to the suspension of representative government for two decades. Self-government was achieved in 1944, and in 1962 Jamaica became an independent Commonwealth state. □ **Jamaican** adj. & n.

jamb /dʒæm/ n. Archit. a side post or surface of a doorway, window, or fireplace. [ME f. OF *jambe* ult. f. LL *gamba* hoof]

jambalaya /ˌdʒæmbəˈlaɪə/ n. a Cajun dish of rice with shrimps, chicken, etc. [Louisiana F f. mod. Prov. *jambalaia*]

jamberoo /ˌdʒæmbəˈruː/ n. Austral. a spree. [alt. of JAMBOREE]

jamboree /ˌdʒæmbəˈriː/ n. **1** a celebration or merrymaking. **2** a large rally of Scouts. [19th c.: orig. unkn.]

James[1] /dʒeɪmz/ the name of seven Stuart kings of Scotland:

James I (1394–1437), son of Robert III, reigned 1406–37. Captured by the English while a child, James remained a captive until 1424. He returned to a country divided by baronial feuds, but managed to restore some measure of royal authority. He was murdered in Perth by rebel nobles.

James II (1430–60), son of James I, reigned 1437–60. After ascending the throne as a minor, he eventually overthrew his regents and considerably strengthened the position of the Crown by crushing the powerful Douglas family (1452–5). He was killed during the siege of Roxburgh Castle.

James III (1451–88), son of James II, reigned 1460–88. He proved increasingly unable to control his nobles, who eventually raised an army against him in 1488, using his son, the future James IV, as a figurehead. The king was defeated and killed in battle at Sauchieburn near Stirling.

James IV (1473–1513), son of James III, reigned 1488–1513. He re-established royal power throughout the realm, notably in the Highlands. He took an active part in European alliance politics, forging a dynastic link with England through his marriage to Margaret Tudor, the daughter of Henry VII, and revitalizing the traditional pact with France. When England and France went to war in 1513, he supported the latter and invaded England at the head of a large army. He died along with many of his nobles when his army was defeated at Flodden.

James V (1512–42), son of James IV, reigned 1513–42. Both during his long minority and after his marriage to the French noblewoman, Mary of Guise, Scotland was dominated by French interests. Relations with England deteriorated in the later years of his reign, culminating in an invasion by Henry VIII's army and the defeat of James's troops near the border at Solway Moss in 1542.

James VI (1566–1625), James I of England (1603–25): see JAMES[2].

James VII the Scottish title of James II of England.

James[2] /dʒeɪmz/ the name of two kings of England, Ireland, and Scotland:

James I (1566–1625), son of Mary, Queen of Scots, king of Scotland (as James VI) 1567–1625, and of England and Ireland 1603–25. After his minority ended in 1583, he was largely successful in restoring royal authority in Scotland. He inherited the throne of England on the death of Elizabeth I, as great-grandson of Margaret Tudor, daughter of Henry VII. His declaration of the divine right of kings, his favouritism towards the Duke of Buckingham, and his intended alliance with Spain made him unpopular with Parliament. He was succeeded by his second son, Charles I.

James II (1633–1701), son of Charles I, king of England, Ireland, and (as James VII) Scotland 1685–8. His Catholic beliefs led to the rebellion of the Duke of Monmouth in 1685 and to his deposition in favour of William of Orange and Mary II three years later. Attempts to regain the throne resulted in James's defeat at the Battle of the Boyne in 1690. He died in exile in France, leaving the Jacobite claim to the throne in the hands of his son, James Stuart.

James[3] /dʒeɪmz/, C(yril) L(ionel) R(obert) (1901–89), Trinidadian historian, journalist, political theorist, and novelist. After working as a cricket columnist, he established a reputation as a historian with his study of the Haitian revolution, *Black Jacobins* (1938). A Trotskyist from 1934, he wrote a number of political works, including *World Revolution: 1917–1936* (published in 1937). He is also noted for *Beyond a Boundary* (1963), an analysis of cricket and anti-colonialism.

James[4] /dʒeɪmz/, Henry (1843–1916), American-born British novelist and critic. He settled in England in 1876, and in his early novels wrote about the relationship between European civilization and American life, notably in *The Portrait of a Lady* (1881). In *The Bostonians* (1886) he portrayed American society in its own right, before producing many novels of English life. He is also remembered for his ghost story *The Turn of the Screw* (1898), the subject of Benjamin Britten's opera by the same name (1954). He was the brother of William James.

James[5] /dʒeɪmz/, Jesse (Woodson) (1847–82), American outlaw. He joined with his brother Frank (1843–1915) and others to form a notorious band of outlaws which specialized in bank and train robberies and inspired many westerns.

James[6] /dʒeɪmz/, Dame P(hyllis) D(orothy) (b.1920), English writer of detective fiction. She is noted for her novels featuring the poet-detective Adam Dalgleish, including *Death of an Expert Witness* (1977) and *A Taste for Death* (1986).

James[7] /dʒeɪmz/, William (1842–1910), American philosopher and psychologist. Influenced by C. S. Peirce, James was a leading exponent of pragmatism, who sought a functional definition of truth rather than a depiction of a structural relation between ideas and reality. Major works include *The Will to Believe* (1907) and *The Meaning of Truth* (1909). In psychology, he is credited with introducing the concept of the stream of consciousness. He was the brother of Henry James.

James, St[1] (known as St James the Great), an Apostle, son of Zebedee and brother of John. He was put to death by Herod Agrippa I; afterwards, according to a Spanish tradition, his body was taken to Santiago de Compostela. Feast day, 25 July.

James, St[2] (known as St James the Less), an Apostle. Feast day (in the Eastern Church) 9 Oct.; (in the Western Church) 1 May.

James, St[3] **1** (known as St James the Just or 'the Lord's brother'), leader of the early Christian Church at Jerusalem. He was put to death by the Sanhedrin. Feast day, 1 May. **2** the epistle of the New Testament traditionally ascribed to St James.

James Bay a shallow southern arm of Hudson Bay, Canada. It was discovered in 1610 by Henry Hudson but was later named after Captain Thomas James (*c*.1593–*c*.1635), who explored the region in 1631.

Jameson Raid /ˈdʒeɪms(ə)n/ an abortive raid into Boer territory made in 1895–6 by pro-British extremists led by Dr L. S. Jameson in an attempt to incite an uprising among recent, non-Boer immigrants. The raid seriously heightened tension in South Africa and contributed to the eventual outbreak of the Second Boer War.

Jamestown /ˈdʒeɪmstaʊn/ **1** a British settlement established in Virginia in 1607, during the reign of King James I. Built on a marshy and unhealthy site, it was abandoned when the state capital of Virginia was moved to Williamsburg at the end of the 17th century. **2** the capital and chief port of the island of St Helena; pop. (1981) 1,500.

jamjar /ˈdʒæmdʒɑː(r)/ n. **1** a glass jar used for jam. **2** *rhyming sl.* a car.

Jammu /ˈdʒʌmuː/ a town in NW India; pop. (1991) 206,000. It is the winter capital of the state of Jammu and Kashmir.

Jammu and Kashmir a mountainous state of NW India, at the western end of the Himalayas; capitals, Srinagar (in summer) and Jammu (in winter). See also KASHMIR.

jammy /ˈdʒæmɪ/ adj. (**jammier**, **jammiest**) **1** covered with jam. **2** *Brit. colloq.* **a** lucky. **b** profitable.

Jamnagar /dʒʌmˈnʌgə(r)/ a port and walled city in the state of Gujarat, western India; pop. (1991) 325,000. It was famous in the past for its pearl fishing and for its tie-dyed fabrics.

Jamshedpur /ˌdʒʌmʃedˈpʊə(r)/ an industrial city in the state of Bihar, NE India; pop. (1991) 461,000.

Jamshid /dʒæmˈʃiːd/ a legendary early king of Persia, reputed inventor of the arts of medicine, navigation, and iron-working. According to legend he was king of the peris (or fairies) who was condemned to assume human form for boasting of his immortality, and ruled Persia for 700 years.

Jan. abbr. January.

Janáček /ˈjænəˌtʃek/, Leoš (1854–1928), Czech composer. Influenced at first by Dvořák, from 1885 he began to collect the Moravian folk-songs, pitch inflections, and rhythmic speech patterns that pervade his music. His works include nine operas, notably *Jenufa* (1904) and *The Cunning Little Vixen* (1924), the *Sinfonietta* (1926), and the *Glagolitic Mass* (1927).

Jane /dʒeɪn/ n. sl. a woman. [the name *Jane*]

JANET /ˈdʒænɪt/ n. (also **Janet**) (in the UK) Joint Academic Network. [acronym]

jangle /ˈdʒæŋg(ə)l/ v. & n. ● v. **1** intr. & tr. make, or cause (a bell etc.) to make, a harsh metallic sound. **2** tr. irritate (the nerves etc.) by discordant sound or speech etc. ● n. a harsh metallic sound. [ME f. OF *jangler*, of uncert. orig.]

Janglish /ˈdʒæŋglɪʃ/ n. = JAPLISH. [*Japanese* + *English*]

janitor /ˈdʒænɪtə(r)/ n. **1** a doorkeeper. **2** a caretaker of a building. □ **janitorial** /ˌdʒænɪˈtɔːrɪəl/ adj. [L f. *janua* door]

janizary /ˈdʒænɪzərɪ/ n. (also **janissary** /-ɪsərɪ/) (pl. **-ies**) **1** hist. a member of the Turkish infantry forming the sultan's guard and the main fighting force of the Turkish army from the late 14th to early 19th centuries. **2** a devoted follower or supporter. [ult. f. Turk. *yeniçeri* f. *yeni* new + *çeri* troops]

jankers /ˈdʒæŋkəz/ n. Mil. sl. punishment for defaulters. [20th c.: orig. unkn.]

Jan Mayen /jæn ˈmaɪən/ a barren and virtually uninhabited island in the Arctic Ocean between Greenland and Norway, annexed by Norway in 1929. It is named after a Dutch sea captain, Jan May, who claimed the island for his company and his country in 1614.

Jansen /ˈdʒæns(ə)n/, Cornelius Otto (1585–1638), Flemish Roman Catholic theologian and founder of Jansenism. A strong opponent of the Jesuits, he proposed a reform of Christianity through a return to St Augustine. To this end he produced his major work, *Augustinus* (1640), which was published by his followers after his death. The four-volume study followed St Augustine's teachings on grace, predestination, and free will, and formed the basis of Jansenism.

Jansenism /ˈdʒænsəˌnɪz(ə)m/ n. a Christian movement of the 17th and 18th centuries, based on the writings of Jansen and characterized by general harshness and moral rigour. Its most famous exponent was Pascal. The movement received papal condemnation and its adherents were persecuted in France (though tolerated in the Netherlands) during most of the 18th century. □ **Jansenist** n.

Jansens, Cornelius, see JOHNSON[3].

January /ˈdʒænjʊərɪ/ n. (pl. **-ies**) the first month of the year. [ME f. AF *Jenever* f. L *Januarius* (*mensis*) (month) of Janus]

Janus /ˈdʒeɪnəs/ Rom. Mythol. an ancient Italian deity, guardian of doorways, gates, and beginnings, and protector of the state in time of war. He is usually represented with two faces, so that he looks both forwards and backwards.

Jap /dʒæp/ n. & adj. colloq. often offens. = JAPANESE. [abbr.]

Japan /dʒə'pæn/ a country in eastern Asia, occupying a festoon of islands in the Pacific roughly parallel with the east coast of the Asiatic mainland; pop. (est. 1988) 122,626,000; official language, Japanese; capital, Tokyo. According to Japanese tradition the islands were brought together in 660 BC by the first emperor, Jimmu. After a long period of courtly rule centred on Kyoto, from the 12th century onwards the country was dominated by succeeding clans of warriors. With the restoration of imperial authority in 1867, Japan began a modernizing process which eventually made it into a major world power. This process was accelerated by wars against China (1894–5) and Russia (1904–5), and Japan fought on the Allied side in the First World War. After the war Japan adopted more militarist and expansionist policies, occupying Manchuria in 1931 and launching a full-scale invasion of China in 1937. In 1936 an alliance was formed with Germany and later with Italy. Japan entered the Second World War in 1941 with a surprise attack on Pearl Harbor and offensives in SE Asia. Their advance was eventually halted by the Allies, and Japan surrendered in 1945 after the dropping of atom bombs by the US on Hiroshima and Nagasaki. Under American occupation, Japan became a constitutional monarchy in 1946, the emperor Hirohito renouncing divine status. Japan is now the most highly industrialized country and the leading economic power in the region, with a range of manufacturing industries that includes electrical goods, motor vehicles, chemicals, and shipping. The Japanese name of the country is *Nippon*, literally 'rising sun'; *Japan* is a rendering of the Chinese form *Riben*.

japan /dʒə'pæn/ n. & v. ● n. **1** a hard usu. black varnish, esp. of a kind brought orig. from Japan. **2** work in a Japanese style. ● v.tr. (**japanned**, **japanning**) **1** varnish with japan. **2** make black and glossy as with japan. [JAPAN]

Japan, Sea of the sea between Japan and the mainland of Asia.

Japanese /ˌdʒæpə'niːz/ n. & adj. ● n. (pl. same) **1 a** a native or national of Japan. **b** a person of Japanese descent. **2** the official language of Japan, spoken by virtually the whole population of that country. (*See note below.*) ● adj. of or relating to Japan or its people or language. □ **Japanese beetle** a chafer, *Popillia japonica*, which is a plant pest in eastern North America. **Japanese cedar** = CRYPTOMERIA. **Japanese print** a colour print from woodblocks. **Japanese quince** = JAPONICA.

▪ Japanese contains many Chinese loan-words and has no genders, no article, and no number in nouns or verbs. It is written vertically, in a system that is partly ideographic and partly syllabic. The ideographs (known as *kanji*) were adopted from the Chinese in the early centuries of the Christian era and designate the chief meaningful elements of the language. They are supplemented by two groups of syllabic characters (*kana*), known as *hiragana* and *katakana*, for the agglutinative and inflectional endings. The exact relationship between Japanese and other languages is not clear, although it is almost certainly related to Korean, and possibly to the Altaic languages.

jape /dʒeɪp/ n. & v. ● n. a practical joke. ● v.intr. play a joke. □ **japery** /'dʒeɪpərɪ/ n. [ME: orig. uncert.]

Japheth /'dʒeɪfeθ/ (in the Bible) a son of Noah (Gen. 10:1), traditional ancestor of the peoples living round the Mediterranean. His name is probably to be connected with that of Iapetus, a Titan in Greek mythology.

Japlish /'dʒæplɪʃ/ n. a blend of Japanese and English, used in Japan. [*Japanese* + *English*]

japonica /dʒə'pɒnɪkə/ n. a plant native to Japan; esp. *Chaenomeles speciosa*, a shrub with bright red flowers and round white, green, or yellow edible fruits (also called *Japanese quince*). [mod.L, fem. of *japonicus* Japanese]

Jaques-Dalcroze /ˌʒækdæl'krəʊz/, Émile (1865–1950), Austrian-born Swiss music teacher and composer. While professor of harmony at the Geneva Conservatory he evolved the eurhythmics method of teaching music and dance. He first used the method with elementary pupils in 1905, before establishing a school for eurhythmics instruction in 1910.

jar¹ /dʒɑː(r)/ n. **1 a** a container of glass, earthenware, plastic, etc., usu. cylindrical. **b** the contents of this. **2** Brit. colloq. a glass of beer. □ **jarful** n. (pl. **-fuls**). [F *jarre* f. Arab. *jarra*]

jar² /dʒɑː(r)/ v. & n. ● v. (**jarred**, **jarring**) **1** intr. (often foll. by on) (of sound, words, manner, etc.) sound discordant or grating (on the nerves

etc.). **2 a** tr. (foll. by against, on) strike or cause to strike with vibration or a grating sound. **b** intr. (of a body affected) vibrate gratingly. **3** tr. send a shock through (a part of the body) (*the fall jarred his neck*). **4** intr. (often foll. by with) (of an opinion, fact, etc.) be at variance; be in conflict or in dispute. ● n. **1** a jarring sound or sensation. **2** a physical shock or jolt. **3** lack of harmony; disagreement. [16th c.: prob. imit.]

jar³ /dʒɑː(r)/ n. □ **on the jar** ajar. [late form of obs. *char* turn: see AJAR¹, CHAR²]

jardinière /ˌʒɑːdɪ'njeə(r)/ n. **1** an ornamental pot or stand for the display of growing plants. **2** a dish of mixed vegetables. [F]

jargon¹ /'dʒɑːgən/ n. **1** words or expressions used by a particular group or profession (*medical jargon*). **2** barbarous or debased language. **3** gibberish. □ **jargonize** v.tr. & intr. (also **-ise**). **jargonic** /dʒɑː'gɒnɪk/ adj. **jargonistic** /ˌdʒɑːgə'nɪstɪk/ adj. [ME f. OF: orig. unkn.]

jargon² /'dʒɑːgən/ n. (also **jargoon** /dʒɑː'guːn/) a translucent, colourless, or smoky variety of zircon. [F f. It. *giargone*, prob. ult. formed as ZIRCON]

jargonelle /ˌdʒɑːgə'nel/ n. an early-ripening variety of pear. [F, dimin. of JARGON²]

jarl /jɑːl/ n. hist. a Norse or Danish chief. [ON, orig. = man of noble birth, rel. to EARL]

jarrah /'dʒærə/ n. **1** the mahogany gum-tree, *Eucalyptus marginata*, of western Australia. **2** the durable timber of this. [Aboriginal *djarryl*]

Jarrow /'dʒærəʊ/ a town in NE England, on the Tyne estuary; pop. (1981) 31,310. From the 7th century until the Viking invasions its monastery was a centre of Northumbrian Christian culture; the Venerable Bede lived and worked there. Its name is associated with a series of hunger marches to London by the unemployed during the Depression of the 1930s.

Jarry /'dʒærɪ, 'ʒærɪ/, Alfred (1873–1907), French dramatist. His satirical farce *Ubu Roi* (1896) is widely claimed to have anticipated surrealism and the Theatre of the Absurd.

Jaruzelski /ˌjærʊ'zelskɪ/, Wojciech (b.1923), Polish general and Communist statesman, Prime Minister 1981–5, head of state 1985–9, and President 1989–90. After becoming Premier in 1981 he responded to Poland's economic crisis and the rise of Solidarity by imposing martial law and banning trade-union operation. Following the victory of Solidarity in the 1989 free elections, Jaruzelski supervised Poland's transition to a novel 'socialist pluralist' democracy.

Jas. abbr. James.

jasmine /'dʒæzmɪn/ n. (also **jasmin**, **jessamin**, **jessamine** /'dʒesəmɪn/) an ornamental shrub of the genus *Jasminum*, usu. with white or yellow fragrant flowers. □ **jasmine tea** a tea perfumed with dried jasmine blossom. [F *jasmin*, *jessemin* f. Arab. *yās(a)mīn* f. Pers. *yāsamīn*]

Jason /'dʒeɪs(ə)n/ Gk Mythol. the son of the king of Iolcos in Thessaly, and leader of the Argonauts in the quest for the Golden Fleece.

jaspé /'dʒæspeɪ/ adj. like jasper; randomly coloured (esp. of cotton fabric). [F, past part. of *jasper* marble f. *jaspe* JASPER]

jasper /'dʒæspə(r)/ n. **1** an opaque variety of quartz, usu. red, yellow, or brown in colour. **2** (in full **jasper-ware**) a kind of fine white stoneware invented by Josiah Wedgwood (1775), usu. stained powder blue and with white cameo decoration. Cf. WEDGWOOD¹. [ME f. OF *jasp(r)e* f. L *iaspis* f. Gk, of oriental orig.]

Jassy see IAŞI.

Jat /dʒɑːt/ n. a member of an Indic people widely distributed in NW India. [Hindi *jāṭ*]

Jataka /'dʒʌtəkə/ n. any of the various stories of the former lives of the Buddha found in Buddhist literature. [Skr. f. *jata* born]

jato /'dʒeɪtəʊ/ n. (pl. **-os**) Aeron. **1** jet-assisted take-off. **2** an auxiliary power unit providing extra thrust at take-off. [acronym]

jaundice /'dʒɔːndɪs/ n. & v. ● n. **1** Med. a condition with yellowing of the skin or whites of the eyes, often caused by obstruction of the bile duct or by liver disease. **2** disordered (esp. mental) vision. **3** envy, jealousy, resentment. ● v.tr. **1** affect with jaundice. **2** (esp. as **jaundiced** adj.) affect (a person) with envy, resentment, or jealousy. [ME *iaunes* f. OF *jaunice* yellowness f. *jaune* yellow]

jaunt /dʒɔːnt/ n. & v. ● n. a short excursion for enjoyment. ● v.intr. take a jaunt. □ **jaunting car** hist. a light two-wheeled horse-drawn vehicle formerly used in Ireland. [16th c.: orig. unkn.]

jaunty /'dʒɔːntɪ/ adj. (**jauntier**, **jauntiest**) **1** cheerful and self-confident; carefree. **2** lively, rakish, dashing. □ **jauntily** adv. **jauntiness** n. [earlier *jentee* f. F *gentil* GENTLE]

Java /'dʒɑːvə/ a large island in the Malay Archipelago, forming part of Indonesia; pop. (est. 1989) 107,513,800 (with Madura). The island was chiefly under Dutch rule from the 17th century until it was occupied by Japanese troops in 1942.

Java man n. an early form of the fossil hominid *Homo erectus*, originally named as *Pithecanthropus* from remains found in a river-bank in central Java in 1891. These probably occurred in deposits dating from c.1,000,000–800,000 years BP (the Middle Pleistocene period). (See also PEKING MAN.)

Javan /'dʒɑːv(ə)n/ n. & adj. = JAVANESE.

Javanese /,dʒɑːvə'niːz/ n. & adj. ● n. (pl. same) **1 a** a native or inhabitant of Java in Indonesia. **b** a person of Javanese descent. **2** the language of Java, which belongs to the Malayo-Polynesian group of languages and is spoken by about 70 million people. ● adj. of or relating to Java, its people, or its language.

Java Sea a sea in the Malay Archipelago of SE Asia, surrounded by the islands of Borneo, Java, and Sumatra.

Java sparrow n. a waxbill, *Padda oryzivora*, native to Java and Bali, often kept as a cage-bird.

javelin /'dʒævlɪn/ n. **1** a light spear thrown in a competitive sport or as a weapon. **2** the athletic event or sport of throwing the javelin. [F *javeline, javelot* f. Gallo-Roman *gabalottus*]

jaw /dʒɔː/ n. & v. ● n. **1 a** each of the upper and lower bony structures in vertebrates forming the framework of the mouth and containing the teeth. **b** the parts of certain invertebrates used for the ingestion of food. **2 a** (in pl.) the mouth with its bones and teeth. **b** the narrow mouth of a valley, channel, etc. **c** the gripping parts of a tool or machine. **d** gripping-power (*jaws of death*). **3** colloq. **a** talkativeness; tedious talk (*hold your jaw*). **b** a sermonizing talk; a lecture. ● v. colloq. **1** intr. speak esp. at tedious length. **2** tr. **a** persuade by talking. **b** admonish or lecture. □ **jaw-breaker** colloq. a word that is very long or hard to pronounce. □ **jawless** adj. [ME f. OF *joe* cheek, jaw, of uncert. orig.]

jawbone /'dʒɔːbəʊn/ n. a bone of the jaw, esp. that of the lower jaw (the mandible) or either half of this.

jay /dʒeɪ/ n. **1** a bird of the crow family, frequently with colourful plumage; esp. *Garrulus glandarius* of Europe, with a screeching call and pinkish-brown, blue, black, and white plumage. **2** a person who chatters impertinently. [ME f. OF f. LL *gaius, gaia*, perh. f. L praenomen *Gaius*: cf. *jackdaw, robin*]

jaywalk /'dʒeɪwɔːk/ v.intr. cross or walk in the street or road without regard to traffic. □ **jaywalker** n.

jazz /dʒæz/ n. & v. ● n. **1** a type of music of US black origin, characterized by its use of improvisation, syncopated phrasing, and a regular or forceful rhythm. (*See note below.*) **2** colloq. pretentious talk or behaviour, nonsensical stuff (*all that jazz*). ● v.intr. play or dance to jazz. □ **jazz up** brighten or enliven. □ **jazzer** n. [20th c.: orig. uncert.]

▪ Jazz was developed in the southern US early in the 20th century, blending West African rhythms with elements from ragtime, brass bands, spirituals, blues, and work-songs. Brass and woodwind instruments and piano are particularly associated with jazz, although guitar and occasionally violin are also used. In its earliest form – Dixieland – jazz was played by small groups, such as those led by Jelly Roll Morton and Louis Armstrong. During the 1920s bands became larger, leading to the swing of the 1930s, which emphasized pre-arranged orchestrations, as exemplified by the music of Benny Goodman. In the early 1940s the bebop of players like Dizzy Gillespie, Charlie Parker, and Thelonious Monk marked a return to smaller groups, playing complex melodies with extreme syncopation. More recent developments have included modal and atonal experimentation, notably by John Coltrane and Miles Davis, and attempts to fuse jazz with classical or rock music.

Jazz Age the 1920s in the US characterized as a period of carefree hedonism, wealth, freedom, and youthful exuberance, reflected in the novels of writers such as F. Scott Fitzgerald.

jazzman /'dʒæzmæn/ n. (pl. **-men**) a jazz musician.

jazzy /'dʒæzɪ/ adj. (**jazzier, jazziest**) **1** of or like jazz. **2** vivid, unrestrained, showy. □ **jazzily** adv. **jazziness** n.

JCB /,dʒeɪsiː'biː/ n. propr. a type of mechanical excavator with a shovel at the front and a digging arm at the rear. [J. C. Bamford, the makers]

JCL abbr. Computing job-control language.

JCR abbr. Brit. Junior Common (or Combination) Room.

jealous /'dʒeləs/ adj. **1** (often foll. by of) fiercely protective (of rights etc.). **2** afraid, suspicious, or resentful of rivalry in love or affection.

3 (often foll. by of) envious or resentful (of a person or a person's advantages etc.). **4** (of God) intolerant of disloyalty. **5** (of inquiry, supervision, etc.) vigilant. □ **jealously** adv. [ME f. OF *gelos* f. med.L *zelosus* ZEALOUS]

jealousy /'dʒeləsɪ/ n. (pl. **-ies**) **1** a jealous state or feeling. **2** an instance of this. [ME f. OF *gelosie* (as JEALOUS)]

jean /dʒiːn/ n. heavy twilled cotton cloth, esp. denim. (See also JEANS.) [ME, attrib. use of *Jene* f. OF *Janne* f. med.L *Janua* Genoa]

Jean Paul /ʒɒn 'pɔːl/ (pseudonym of Johann Paul Friedrich Richter) (1763–1825), German novelist. He is noted for his romantic novels, including *Hesperus* (1795), and for comic works such as *Titan* (1800–3).

Jeans /dʒiːnz/, Sir James Hopwood (1877–1946), English physicist and astronomer. He began as a mathematician, but his major contributions were in molecular physics and astrophysics. Jeans proposed a theory for the formation of the solar system, according to which the planets formed from natural material pulled out of the sun by the gravity of a passing star. He was the first to propose that matter is continuously created throughout the universe, one of the tenets of the steady-state theory. He also became a popularizer of science, especially as a radio lecturer.

jeans /dʒiːnz/ n.pl. hard-wearing trousers made of denim or another cotton fabric, for informal wear. [JEAN]

Jedburgh /'dʒedbərə/ a town in southern Scotland near the English border. Disputes arising between border peoples in this area gave rise to the expression *Jedburgh justice*, a summary procedure whereby a person is executed first and tried later.

Jeddah see JIDDAH.

Jeep /dʒiːp/ n. propr. a small sturdy esp. military motor vehicle with four-wheel drive. [orig. US, f. GP = general purposes, infl. by 'Eugene the Jeep', an animal in a comic strip]

jeepers /'dʒiːpəz/ int. (also **jeepers creepers**) N. Amer. sl. expressing surprise etc. [corrupt. of *Jesus*]

jeer /dʒɪə(r)/ v. & n. ● v. **1** intr. (usu. foll. by at) scoff derisively. **2** tr. scoff at; deride. ● n. a scoff or taunt. □ **jeeringly** adv. [16th c.: orig. unkn.]

Jeeves /dʒiːvz/ the resourceful and influential valet of Bertie Wooster in the novels of P. G. Wodehouse.

Jeez /dʒiːz/ int. sl. a mild expression of surprise, discovery, etc. (cf. GEE¹). [abbr. of *Jesus*]

Jefferies /'dʒefrɪz/, (John) Richard (1848–87), English writer and naturalist. He is renowned for his observation of English rural life. Important works include *Bevis* (1882), an evocation of his country childhood in Wiltshire, and his autobiography *The Story of my Heart* (1883).

Jefferson /'dʒefəs(ə)n/, Thomas (1743–1826), American Democratic Republican statesman, 3rd President of the US 1801–9. Jefferson was the principal drafter of the Declaration of Independence (1776) and played a key role in the American leadership during the War of Independence. He advocated decentralization and the restrained use of presidential power, in defiance of Alexander Hamilton. While President, Jefferson secured the Louisiana Purchase (1803).

Jefferson City the state capital of Missouri; pop. (1990) 35,480. It is named after President Thomas Jefferson.

Jeffreys /'dʒefrɪz/, George, 1st Baron (c.1645–89), Welsh judge. In 1683 he was made Chief Justice of the King's Bench and took part in the Popish Plot prosecutions, sitting at the trial of the plot's fabricator, Titus Oates. He later became infamous for his brutal sentencing at the Bloody Assizes.

jehad var. of JIHAD.

Jehovah /dʒɪ'həʊvə/ a form of the Hebrew name for God used in the Bible. The word was formed in medieval Latin from YHVH (the Hebrew name of God which was too sacred to pronounce) and the vowels of *adonai*, Hebrew for 'my lord', which was substituted for it in reading the Hebrew Bible. (See also YAHWEH.)

Jehovah's Witness n. a member of a millenarian fundamentalist Christian sect, the Watch Tower Bible and Tract Society, founded in the US c.1879 by Charles Taze Russell (1852–1916). Jehovah's Witnesses deny many of the traditional Christian doctrines and do not acknowledge the claims of the state when these conflict with the sect's principles; for example, they refuse to undertake military service or receive blood transfusions.

Jehovist /dʒɪ'həʊvɪst/ n. = YAHWIST.

Jehu /'dʒiːhjuː/ (842–815 BC), king of Israel. He was famous for driving his chariot furiously (2 Kings 9).

jejune /dʒɪˈdʒuːn/ adj. **1** intellectually unsatisfying; shallow. **2** puerile, childish; naive. **3** (of ideas, writings, etc.) meagre, scanty; dry and uninteresting. **4** (of land) barren, poor. □ **jejunely** adv. **jejuneness** n. [orig. = fasting, f. L jejunus]

jejunum /dʒɪˈdʒuːnəm/ n. (pl. **-na** /-nə/) Anat. the part of the small intestine between the duodenum and ileum. [L, neut. of jejunus fasting]

Jekyll /ˈdʒek(ə)l/ the hero of Robert Louis Stevenson's story The Strange Case of Dr Jekyll and Mr Hyde (1886). Dr Jekyll, a physician, discovers a drug which creates a separate personality (appearing in the character of Mr Hyde) into which Jekyll's evil impulses are channelled.

jell /dʒel/ v.intr. colloq. **1 a** set as a jelly. **b** (of ideas etc.) take a definite form. **2** (of people) readily cooperate or reach an understanding. [back-form. f. JELLY]

jellaba var. of DJELLABA.

Jellicoe /ˈdʒelɪˌkəʊ/, John Rushworth, 1st Earl (1859–1935), British admiral. He was commander of the Grand Fleet at the Battle of Jutland. After the war he was appointed Governor-General of New Zealand (1920–4).

jellify /ˈdʒelɪˌfaɪ/ v.tr. & intr. (**-ies, -ied**) turn into jelly; make or become like jelly. □ **jellification** /ˌdʒelɪfɪˈkeɪʃ(ə)n/ n.

jelly /ˈdʒelɪ/ n. & v. ● n. (pl. **-ies**) **1 a** a soft somewhat elastic semi-transparent preparation of boiled sugar and fruit juice or milk etc., often cooled in a mould and eaten as a dessert. **b** a similar preparation of fruit juice etc. for use as a jam or a condiment (redcurrant jelly). **c** a similar preparation derived from meat, bones, etc., and gelatin (marrowbone jelly). **2** any substance of a similar consistency. **3** Brit. sl. gelignite (cf. GELLY). ● v. (**-ies, -ied**) **1** intr. & tr. set or cause to set as a jelly, congeal. **2** tr. set (food) in a jelly (jellied eels). □ **jelly baby** Brit. a soft gelatinous sweet in the stylized shape of a baby. **jelly bag** a bag for straining juice for jelly. **jelly bean** a bean-shaped sweet with a gelatinous centre and a hard sugar coating. □ **jelly-like** adj. [ME f. OF gelee frost, jelly, f. Rmc gelata f. L gelare freeze f. gelu frost]

jellyfish /ˈdʒelɪˌfɪʃ/ n. (pl. usu. same) **1** a marine coelenterate of the class Scyphozoa, having an umbrella-shaped jelly-like body and stinging tentacles. **2** colloq. a feeble person.

jemmy /ˈdʒemɪ/ n. & v. (N. Amer. **jimmy** /ˈdʒɪmɪ/) ● n. (pl. **-ies**) a burglar's short crowbar, usu. made in sections. ● v.tr. (**-ies, -ied**) force open with a jemmy. [pet-form of the name James]

Jena /ˈjeɪnə/ a university town in central Germany, in Thuringia; pop. (1991) 100,970. It was the scene of a battle (1806) in which Napoleon defeated the Prussians. It is also noted as a manufacturing centre for optical and precision instruments.

je ne sais quoi /ˌʒə nə seɪ ˈkwʌ/ n. an indefinable something. [F, = I do not know what]

Jenkins's Ear, War of /ˈdʒeŋkɪnzɪz/ a war between England and Spain (1739), which derived its name from Robert Jenkins, a British sea captain who appeared before Parliament in 1738 to produce what he claimed was his ear, cut off by the Spanish while they were carrying out a search of his ship in the West Indies. His story was probably at least partially fabricated, but it caused great popular indignation and precipitated the naval war with Spain in the following year.

Jenner /ˈdʒenə(r)/, Edward (1749–1823), English physician, the pioneer of vaccination. A local belief that dairymaids who had had cowpox did not catch smallpox led Jenner to the idea of deliberately infecting people with cowpox in order to protect them from the more serious disease. The practice was eventually accepted throughout the world, leading to the widespread use of vaccination for other diseases and eventually to the eradication of smallpox in the late 20th century. In intervals between medical practice he indulged his keen interest in natural history, and wrote a paper on the habits of the cuckoo.

jennet /ˈdʒenɪt/ n. a small Spanish horse. [F genet f. Sp. jinete light horseman f. Arab. zenāta Berber tribe famous as horsemen]

jenny /ˈdʒenɪ/ n. (pl. **-ies**) **1** hist. = SPINNING-JENNY. **2** a female donkey or ass. **3** a locomotive crane. □ **jenny-wren** a popular name for a female wren. [pet-form of the name Janet]

jeon /dʒʌn/ n. a monetary unit of South Korea, equal to one-hundredth of a won. [Korean]

jeopardize /ˈdʒepəˌdaɪz/ v.tr. (also **-ise**) endanger; put into jeopardy.

jeopardy /ˈdʒepədɪ/ n. **1** danger, esp. of severe harm or loss. **2** Law danger resulting from being on trial for a criminal offence. [ME iuparti f. OF ieu parti divided (i.e. even) game, f. L jocus game + partitus past part. of partire divide f. pars partis part]

Jephthah /ˈdʒefθə/ (in the Bible) a judge of Israel who sacrificed his daughter in consequence of a vow that if victorious in battle he would sacrifice the first living thing that met him on his return (Judges 11, 12). A similar rash promise was made by Idomeneus in Greek mythology.

Jer. abbr. (in the Bible) Jeremiah.

Jerba see DJERBA.

jerbil var. of GERBIL.

jerboa /dʒɜːˈbəʊə/ n. a small desert rodent of the family Dipodidae, with long hind legs and the ability to make large jumps. [mod.L f. Arab. yarbū' flesh of loins, jerboa]

jeremiad /ˌdʒerɪˈmaɪæd/ n. a doleful complaint or lamentation; a list of woes. [F jérémiade f. Jérémie JEREMIAH[1]]

Jeremiah[1] /ˌdʒerɪˈmaɪə/ **1** a Hebrew major prophet (c.650 –c.585 BC), who foresaw the fall of Assyria, the conquest of his country by Egypt and then by Babylon, and the destruction of Jerusalem. The biblical Lamentations are traditionally ascribed to him. **2** a book of the Bible containing his prophecies.

Jeremiah[2] /ˌdʒerɪˈmaɪə/ n. a doleful or pessimistic person; a dismal prophet. [JEREMIAH[1]]

Jerez /heˈrez/ (in full **Jerez de la Frontera** /deɪ læ frɒnˈteərə/) a town in Andalusia, Spain; pop. (1991) 184,020. It is the centre of the sherry-making industry.

Jericho /ˈdʒerɪˌkəʊ/ a town in Palestine, in the West Bank north of the Dead Sea. It has been occupied from at least 9000 BC. According to the Bible, Jericho was a Canaanite city destroyed by the Israelites after they crossed the Jordan into the Promised Land; its walls were flattened by the shout of the army and the blast of the trumpets. Occupied by the Israelis since the Six Day War of 1967, in 1994 Jericho was the first area given partial autonomy under the PLO–Israeli peace accord.

jerk[1] /dʒɜːk/ n. & v. ● n. **1** a sharp sudden pull, twist, twitch, start, etc. **2** a spasmodic muscular twitch. **3** (in pl.) Brit. colloq. exercises (physical jerks). **4** sl. a fool; a stupid or contemptible person. **5** (in weightlifting) the rapid raising of a weight from shoulder-level to above the head by straightening the arms (and legs). ● v. **1** intr. move with a jerk. **2** tr. pull, thrust, twist, etc., with a jerk. **3** tr. throw with a suddenly arrested motion. **4** tr. (in weightlifting) raise (a weight) from shoulder-level to above the head. □ **jerk off** coarse sl. masturbate. □ **jerker** n. [16th c.: perh. imit.]

jerk[2] /dʒɜːk/ v.tr. cure (beef) by cutting it in long slices and drying it in the sun. [Amer. Sp. charquear f. charqui f. Quechua echarqui dried flesh]

jerkin /ˈdʒɜːkɪn/ n. **1** a sleeveless jacket. **2** hist. a man's close-fitting jacket, often of leather. [16th c.: orig. unkn.]

jerky /ˈdʒɜːkɪ/ adj. (**jerkier, jerkiest**) **1** having sudden abrupt movements. **2** spasmodic. □ **jerkily** adv. **jerkiness** n.

jeroboam /ˌdʒerəˈbəʊəm/ n. a wine bottle of four times the ordinary size. [Jeroboam king of Israel (1 Kings 11:28, 14:16)]

Jerome /dʒəˈrəʊm/, Jerome K(lapka) (1859–1927), English novelist and dramatist. He is chiefly remembered for his humorous novel Three Men in a Boat (1889).

Jerome, St (c.342–420), Doctor of the Church. Born in Dalmatia, he acted as secretary to Pope Damasus in Rome (382–5) before settling in Bethlehem, where he ruled a newly founded monastery and devoted his life to study. He is chiefly known for his compilation of the Vulgate. Feast day, 30 Sept.

Jerry /ˈdʒerɪ/ n. (pl. **-ies**) Brit. sl. **1** a German (esp. in military contexts). **2** the Germans collectively. [prob. alt. of German]

jerry[1] /ˈdʒerɪ/ n. (pl. **-ies**) Brit. sl. a chamber-pot.

jerry[2] /ˈdʒerɪ/ v.intr. (**-ies, -ied**) Austral. sl. understand, realize. [20th c.: orig. unkn.]

jerry-builder /ˈdʒerɪˌbɪldə(r)/ n. a builder of unsubstantial houses with poor-quality materials. □ **jerry-building** n. **jerry-built** /-ˌbɪlt/ adj. [19th c.: origin unkn.]

jerrycan /ˈdʒerɪˌkæn/ n. (also **jerrican**) a kind of (orig. German) petrol- or water-can. [JERRY + CAN[2]]

jerrymander var. of GERRYMANDER.

Jersey /ˈdʒɜːzɪ/ the largest of the Channel Islands; pop. (1990) 82,810; capital, St Helier.

jersey /ˈdʒɜːzɪ/ n. (pl. **-eys**) **1 a** a knitted usu. woollen pullover or similar garment. **b** a plain-knitted (orig. woollen) fabric. **2** (**Jersey**) a light brown breed of dairy cattle from Jersey, producing milk with a high fat content. [JERSEY]

Jersey City an industrial city in NE New Jersey, on the Hudson River opposite New York City; pop. (1990) 228,540.

Jerusalem /dʒəˈruːsələm/ the holy city of the Jews, sacred also to Christians and Muslims, lying in the Judaean hills about 30 km (20 miles) from the River Jordan; pop. (1987) 482,700. In early biblical times it was a Canaanite stronghold, captured by King David (c.1000 BC), who made it the capital of the Jewish state. After the building of the Temple by Solomon (957 BC) it became a religious as well as a political capital. Since then it has shared the troubled history of the area — destroyed by the Babylonians in 586 BC and by the Romans in AD 70, refounded by Hadrian as a Gentile city (AD 135) under the name of *Aelia Capitolina*, destroyed again by the Persians in 614, and fought over by Saracens and Crusaders in the Middle Ages. From 1099 it was the capital of a Crusader kingdom, which persisted as a political entity until 1291, even though Jerusalem itself was captured by Saladin in 1187. Suleiman the Magnificent rebuilt its walls (1542). From 1947 the city was divided between the states of Israel and Jordan until the Israelis occupied the whole city in June 1967. Its Christian history begins with the short ministry of Christ, culminating in the Crucifixion. For Muslims Jerusalem is the holiest city after Mecca and Medina, containing the Dome of the Rock, one of Islam's most sacred sites. After the Second World War it was envisaged by the United Nations as an international city, but was seized by Israel during the Six Day War of 1967 and proclaimed the capital of the state of Israel. (See also NEW JERUSALEM.)

Jerusalem artichoke n. **1** a kind of sunflower, *Helianthus tuberosus*, with edible underground tubers. **2** this tuber used as a vegetable. [corrupt. of It. *girasole* sunflower]

Jervis /ˈdʒɑːvɪs/, John, Earl St Vincent (1735-1823), British admiral. In 1795 he was put in command of the British fleet, and in 1797, with Nelson as his commodore, led his forces to victory over a Spanish fleet off Cape St Vincent; Jervis was created Earl St Vincent in recognition of this achievement.

Jespersen /ˈjespəs(ə)n/ (Jens) Otto (Harry) (1860–1943), Danish philologist, grammarian, and educationist. He promoted the use of the 'direct method' in language teaching with the publication of his theoretical work *How to Teach a Foreign Language* (1904). Other books include his seven-volume *Modern English Grammar* (1909–49).

jess /dʒes/ n. & v. ● n. a short strap of leather, silk, etc., put round the leg of a hawk in falconry. ● v.tr. put jesses on (a hawk etc.). [ME *ges* f. OF *ges*, *get* ult. f. L *jactus* a throw f. *jacere jact-* to throw]

jessamin (also **jessamine**) var. of JASMINE.

Jesse /ˈdʒesɪ/ (in the Bible) the father of David (1 Sam. 16), hence represented as the first in the genealogy of Jesus Christ. □ **Jesse window** a church window showing Jesus' descent from Jesse, usually in the form of a tree of Jesse. **tree of Jesse** an illustration of a tree showing the descent of Jesus from Jesse, with the intermediate descendants placed on scrolls of foliage branching out of each other.

jest /dʒest/ n. & v. ● n. **1 a** a joke. **b** fun. **2 a** raillery, banter. **b** an object of derision (*a standing jest*). ● v.intr. **1** joke; make jests. **2** fool about; play or act triflingly. □ **in jest** in fun. [orig. = exploit, f. OF *geste* f. L *gesta* neut. pl. past part. of *gerere* do]

jester /ˈdʒestə(r)/ n. hist. a professional joker or 'fool' at a medieval court etc., traditionally wearing a cap and bells.

Jesu /ˈdʒiːzjuː/ Jesus (as an archaic form of address).

Jesuit /ˈdʒezjʊɪt/ n. a member of the Society of Jesus, a Roman Catholic order of priests founded in 1534 in Paris by St Ignatius Loyola, Francis Xavier, and others to do missionary work throughout the world. The order took the lead in opposing the Reformation in Europe, and became known for the uncompromising zeal with which it spread Catholic beliefs. Despite periodic persecution the Jesuits have become a large and influential order noted for their educational work and establishments. They take a special oath of obedience to the pope, and wear no distinctive habit. [F *jésuite* or mod.L *Jesuita* f. *Jesus*: see JESUS]

Jesuitical /ˌdʒezjʊˈɪtɪk(ə)l/ adj. **1** of or concerning the Jesuits. **2** often offens. dissembling or equivocating, in the manner once associated with Jesuits. □ **Jesuitically** adv.

Jesus /ˈdʒiːzəs/ (also **Jesus Christ** or **Jesus of Nazareth**) the central figure of the Christian religion. A Jew, the son of Mary, he lived in Palestine at the beginning of the 1st century AD. In about AD 28–30 he conducted a mission of preaching and healing (with reported miracles) which is described in the New Testament, as are his arrest and death by crucifixion. His followers considered him to be the Christ or Messiah and the Son of the living God, and belief in his Resurrection

from the dead, as recorded in the Gospels, became a central tenet of Christianity.

JET /dʒet/ abbr. Joint European Torus, a machine for conducting experiments in nuclear fusion, established by Euratom at Culham in Oxfordshire as the centre for all such research in western Europe.

jet¹ /dʒet/ n. & v. ● n. **1** a stream of liquid, gas, or (more rarely) solid particles shot out, esp. from a small opening. **2** a spout or nozzle for emitting water etc. in this way. **3 a** a jet engine. **b** an aircraft powered by one or more jet engines. ● v. (**jetted, jetting**) **1** intr. spurt out in jets. **2** tr. & intr. colloq. send or travel by jet plane. □ **jet engine** an engine using jet propulsion for forward thrust, esp. of an aircraft. **jetfoil** a vessel that travels above the surface of the water on struts attached to underwater foils. **jet lag** extreme tiredness and other bodily effects felt after a long flight involving marked differences of local time. **jet-lagged** suffering from jet lag. **jet-propelled 1** having jet propulsion. **2** very fast. **jet set** colloq. wealthy people frequently travelling by air, esp. for pleasure. **jet-setter** colloq. a member of the jet set. **jet-setting** colloq. of or belonging to the jet set. [earlier as verb (in sense 1): F *jeter* throw ult. f. L *jactare* frequent. of *jacere jact-* throw]

jet² /dʒet/ n. **1 a** a hard black variety of lignite capable of being carved and highly polished. **b** (attrib.) made of this. **2** (in full **jet-black**) a deep glossy black colour. [ME f. AF *geet*, OF *jaiet* f. L *gagates* f. Gk *gagatēs* f. *Gagai* in Asia Minor]

jeté /ʒeˈteɪ/ n. Ballet a spring or leap with one leg forward and the other stretched backwards. [F, past part. of *jeter* throw: see JET¹]

jet propulsion n. ejection of a high-speed jet of gas (or liquid) as a source of propulsive power, especially for aircraft. Simple examples are the movement of a deflating balloon, and of squids and other cephalopods, which propel themselves by expelling water through a funnel from the mantle cavity. The principle of jet propulsion for aircraft was discussed in the 1860s, but its successful application awaited the development of the turbojet — a form of gas turbine — concurrently by Frank Whittle in England and Hans von Ohain (b.1911) in Germany. The first jet aeroplane flew in Germany in 1939. Developments of the turbojet include the turboprop (in which the turbine drives a propeller), and the turbofan (in which thrust is augmented by a fan which drives air backwards around the engine proper). Some use has been made of the ramjet, a very simple engine with no moving parts, efficient only at high speeds.

jetsam /ˈdʒetsəm/ n. discarded material washed ashore, esp. that thrown overboard to lighten a ship etc. (cf. FLOTSAM). [contr. of JETTISON]

jet stream n. **1** a narrow band of very strong winds encircling the globe several miles above the earth, blowing in an approximately horizontal direction predominantly from west to east. (See note below.) **2** the flow of vapour from a jet engine.

■ The existence of jet streams in the atmosphere was discovered by aviators during the Second World War. There are typically two or three in each hemisphere, associated with the boundaries between air masses of differing temperatures. Jet streams have speeds up to 500 k.p.h. (310 m.p.h.) and follow a variable, meandering course. Besides having practical effects on air travel, they strongly influence the movement of the lower atmosphere and the development of cyclones.

jettison /ˈdʒetɪs(ə)n, -ɪz(ə)n/ v. & n. ● v.tr. **1** a throw (esp. heavy material) overboard to lighten a ship, hot-air balloon, etc. **b** drop (goods) from an aircraft. **2** abandon; get rid of (something no longer wanted). ● n. the act of jettisoning. [ME f. AF *getteson*, OF *getaison* f. L *jactatio -onis* f. *jactare* throw: see JET¹]

jetton /ˈdʒet(ə)n/ n. a counter with a stamped or engraved design, esp. for insertion like a coin to operate a machine etc. [F *jeton* f. *jeter* throw, add up accounts: see JET¹]

jetty /ˈdʒetɪ/ n. (pl. **-ies**) **1** a pier or breakwater constructed to protect or defend a harbour, coast, etc. **2** a landing-pier. [ME f. OF *jetee*, fem. past part. of *jeter* throw: see JET¹]

jeu d'esprit /ˌʒɜː deˈspriː/ n. (pl. **jeux d'esprit** pronunc. same) a witty or humorous (usu. literary) trifle. [F, = game of the spirit]

jeunesse dorée /ʒɜːˌnes dɔːˈreɪ/ n. = gilded youth (see GILD¹). [F]

Jew /dʒuː/ n. **1** a person of Hebrew descent or whose religion is Judaism. (See note below.) **2** sl. offens. (as a stereotype) a person considered to be parsimonious or to drive a hard bargain in trading. [ME f. OF *giu* f. L *judaeus* f. Gk *ioudaios* ult. f. Heb. *yĕhûdî* f. *yĕhûdāh* Judah]

■ Jewish identity is a complex blend of racial and religious affiliation which has persisted through the many centuries of the Jewish

Diaspora. The term *Jew* may be applied to those of Hebrew descent (traditionally through the mother) even if they have abandoned the practice of Judaism, but also to other groups, such as the Falashas, who profess Judaism. The two main cultural divisions of European Jews are the Ashkenazim and the Sephardim. Jewish identity finds political expression in Zionism, and in the state of Israel, established as a Jewish homeland in 1948. (See also ANTI-SEMITISM, JUDAISM.)

jewel /ˈdʒuːəl/ *n. & v.* ● *n.* **1 a** a precious stone. **b** this as used for its hardness as a bearing in watchmaking. **2** a personal ornament containing a jewel or jewels. **3** a precious person or thing. ● *v.tr.* (**jewelled, jewelling**; *US* **jeweled, jeweling**) **1** (esp. as **jewelled** *adj.*) adorn or set with jewels. **2** (in watchmaking) set with jewels. □ **jewel-fish** a scarlet and green cichlid fish, *Hemichromis bimaculatus*, kept in aquaria. □ **jewelly** *adj.* [ME f. AF *juel, jeuel*, OF *joel*, of uncert. orig.]

jeweller /ˈdʒuːələ(r)/ *n.* (*US* **jeweler**) a maker of or dealer in jewels or jewellery. □ **jeweller's rouge** finely ground rouge for polishing metal. [ME f. AF *jueler*, OF *juelier* (as JEWEL)]

jewellery /ˈdʒuːəlrɪ/ *n.* (also **jewelry**) jewels or other ornamental objects, esp. for personal adornment, regarded collectively. [ME f. OF *juelerie* and f. JEWEL, JEWELLER]

Jewess /ˈdʒuːɪs/ *n.* often *offens.* a female Jew.

jewfish /ˈdʒuːfɪʃ/ *n.* (*pl.* same) **1** a grouper, *Epinephelus itajara*, of the Atlantic and Pacific coasts of North America. **2** a large Australian fish used as food, esp. the mulloway.

Jewish /ˈdʒuːɪʃ/ *adj.* **1** of or relating to Jews. **2** of Judaism.

Jewish calendar *n.* a complex ancient calendar in use among Jewish people. It is a lunar calendar adapted to the solar year, normally consisting of twelve months but having thirteen months in leap years, which occur seven times in every cycle of nineteen years. The years are reckoned from the Creation (which is placed at 3761 BC); the months are Nisan, Iyyar, Sivan, Thammuz, Ab, Elul, Tishri, Hesvan, Kislev, Tebet, Sebat, and Adar, with an intercalary month (First Adar) being added in leap years. The religious year begins with Nisan and ends with Adar, while the civil year begins with Tishri and ends with Elul.

Jewish New Year see ROSH HASHANA.

Jewry /ˈdʒʊərɪ/ *n.* (*pl.* **-ies**) **1** Jews collectively. **2** *hist.* a Jews' quarter in a town etc. [ME f. AF *juerie*, OF *juierie* (as JEW)]

Jew's ear *n.* a rubbery cup-shaped fungus, *Auricularia auricula-judae*, growing on trees. [mistransl. med.L *auricula Judae* 'Judas's ear', f. its shape and its occurrence on the elder, said to be the tree from which Judas hanged himself]

Jew's harp *n.* a musical instrument consisting of a small U-shaped metal frame held in the teeth while a springy metal clip joining its ends is twanged with a finger. The strip can produce only one note, but harmonics of this note can be produced by altering the shape of the mouth-cavity. The name of the instrument is an ancient one, but no connection with Jews has been established with certainty.

Jezebel /ˈdʒezəˌbel/ a Phoenician princess of the 9th century BC, traditionally the great-aunt of the legendary Dido and in the Bible the wife of Ahab king of Israel. She was denounced by Elijah for introducing the worship of Baal into Israel (1 Kings 16:31, 21:5–15, 2 Kings 9:30–7). Her use of make-up shocked Puritan England and led to the use of her name to mean a shameless or immoral woman.

Jhansi /ˈdʒɑːnsɪ/ a city in the state of Uttar Pradesh, northern India; pop. (1991) 301,000.

Jhelum /ˈdʒiːləm/ a river which rises in the Himalayas and flows for about 720 km (450 miles) through the Vale of Kashmir into the province of Punjab in Pakistan, where it meets the Chenab river. It is one of the five rivers that gave Punjab its name. In ancient times it was called the Hydaspes.

Jiang Jie Shi see CHIANG KAI-SHEK.

Jiangsu /dʒjæŋˈsuː/ (also **Kiangsu** /kjæŋ-/) a province of eastern China; capital, Nanjing. It includes much of the Yangtze delta.

Jiangxi /dʒjæŋˈʃiː/ (also **Kiangsi** /kjæŋˈsiː/) a province of SE China; capital, Nanchang.

jib[1] /dʒɪb/ *n. & v.* ● *n.* **1** a triangular staysail extending from the outer end of the jib-boom to the top of the foremast or from the bowsprit to the masthead. **2** the projecting arm of a crane. ● *v.tr. & intr.* (**jibbed, jibbing**) (of a sail etc.) pull or swing round from one side of the ship to the other; gybe. □ **jib-boom** a spar run out from the end of the bowsprit. [17th c.: orig. unkn.]

jib[2] /dʒɪb/ *v.intr.* (**jibbed, jibbing**) **1 a** (of an animal, esp. a horse) stop and refuse to go on; move backwards or sideways instead of going on. **b** (of a person) refuse to continue. **2** (foll. by *at*) show aversion to (a person or course of action). □ **jibber** *n.* [19th c.: orig. unkn.]

jibba /ˈdʒɪbə/ *n.* (also **jibbah**) a long coat worn by Muslim men. [Egypt. var. of Arab. *jubba*]

jibe[1] var. of GIBE.

jibe[2] *US* var. of GYBE.

jibe[3] /dʒaɪb/ *v.intr.* (usu. foll. by *with*) *US colloq.* agree; be in accord. [19th c.: orig. unkn.]

Jibuti see DJIBOUTI.

Jiddah /ˈdʒɪdə/ (also **Jeddah** /ˈdʒedə/) a seaport on the Red Sea coast of Saudi Arabia, near Mecca; pop. (est. 1986) 1,400,000.

jiff /dʒɪf/ *n.* (also **jiffy** /ˈdʒɪfɪ/, *pl.* **-ies**) *colloq.* a short time; a moment (*in a jiffy; half a jiff*). [18th c.: orig. unkn.]

Jiffy bag /ˈdʒɪfɪ/ *n. propr.* a type of padded envelope for postal use.

jig /dʒɪg/ *n. & v.* ● *n.* **1 a** a lively dance with leaping movements. **b** the music for this, usu. in triple time; = GIGUE 2. **2** a device that holds a piece of work and guides the tools operating on it. **3** *Fishing* a device for catching fish that is jerked up and down through the water. ● *v.* (**jigged, jigging**) **1** *intr.* dance a jig. **2** *tr. & intr.* move quickly and jerkily up and down. **3** *tr.* work on or equip with a jig or jigs. **4** *tr. & intr. Fishing* fish (for) or catch with a jig. □ **jig about** fidget. [16th c.: orig. unkn.]

jigger[1] /ˈdʒɪgə(r)/ *n. & v.* ● *n.* **1** *Naut.* **a** a small tackle consisting of a double and single block with a rope. **b** a small sail at the stern. **c** a small smack having this. **2** *sl.* a gadget. **3** *Golf* an iron club with a narrow face. **4** *Billiards* a cue-rest. **5 a** a measure of spirits etc. **b** a small glass holding this. **6** a person or thing that jigs. ● *v.tr.* **1** (usu. in phr. *I'll be jiggered*) *sl.* confound, damn. **2** exhaust; damage, break. [prob. euphem. for BUGGER *v.*]

jigger[2] /ˈdʒɪgə(r)/ *n.* (also **chigger** /ˈtʃɪg-/) **1** a tropical flea, *Tunga penetrans*, the females of which burrow beneath the skin causing painful sores. Also called *chigoe, sand-flea*. **2** a harvest mite of the genus *Leptotrombidium*, with parasitic larvae. [alt. of CHIGOE]

jiggery-pokery /ˌdʒɪgərɪˈpəʊkərɪ/ *n. Brit. colloq.* deceitful or dishonest dealing, trickery. [cf. Sc. *joukery-pawkery* f. *jouk* dodge, skulk]

jiggle /ˈdʒɪg(ə)l/ *v. & n.* ● *v.* (often foll. by *about* etc.) **1** *tr. & intr.* shake lightly; rock jerkily. **2** *intr.* fidget. ● *n.* a light shake. □ **jiggly** *adj.* [JIG or JOGGLE[1]]

jigsaw /ˈdʒɪgsɔː/ *n.* **1 a** (in full **jigsaw puzzle**) a puzzle consisting of a picture on board or wood etc. cut into irregular interlocking pieces to be reassembled for amusement. The first jigsaw puzzles (18th century) were maps, mounted on wood and cut into oddly shaped pieces. **b** a mental puzzle resolvable by assembling various pieces of information. **2** a machine saw with a fine blade enabling it to cut curved lines in a sheet of wood, metal, etc.

jihad /dʒɪˈhæd, -ˈhɑːd/ *n.* (also **jehad**) (in Islam) a holy war. A jihad may be undertaken to defend Islam against external threats or to spread the religion among non-believers; the term is also used, however, to mean effort or exertion in the cause of God and can encompass struggle against resistance to the rule of divine law within oneself. Prosecution of a jihad when summoned is prescribed as a religious duty by the Koran and by tradition. [Arab. *jihād* fight, struggle]

Jilin /dʒiːˈlɪn/ (also **Kirin** /kiːˈrɪn/) **1** a province of NE China; capital, Changchun. **2** an industrial city in Jilin province; pop. (1986) 3,974,000.

jill var. of GILL[4].

jilt /dʒɪlt/ *v. & n.* ● *v.tr.* abruptly reject or abandon (a lover etc.). ● *n.* a person (esp. a woman) who jilts a lover. [17th c.: orig. unkn.]

Jim Crow /dʒɪm ˈkrəʊ/ *n. US* **1** the policy of segregating and discriminating against blacks, esp. by laws passed in the southern states of the US in the late 19th century. The policy was not seriously challenged until after the Second World War. **2** *offens.* a black person. **3** an implement for straightening iron bars or bending rails by screw pressure. □ **Jim Crowism** *n.* (in sense 1). [eponymous black character in early 19th-c. plantation song 'Jim Crow']

Jiménez de Cisneros /hɪˌmenez deɪ sɪsˈneərɒs/ (also **Ximenes de Cisneros**), Francisco (1436–1517), Spanish statesman, regent of Spain 1516–17. He was made Cardinal in 1507 and served as Grand Inquisitor for Castile and Léon from 1507 to 1517, during which time he undertook a massive campaign against heresy, having some 2,500 alleged heretics put to death.

jim-jams /ˈdʒɪmdʒæmz/ *n.pl.* **1** *sl.* = delirium tremens. **2** *colloq.* a fit of depression or nervousness. **3** *colloq.* pyjamas. [fanciful redupl.]

Jimmu /ˈdʒɪmuː/ the legendary first emperor of Japan (660 BC), descendant of the sun-goddess Amaterasu and founder of the imperial dynasty.

jimmy *N. Amer. var. of* JEMMY.

jimmygrant /ˈdʒɪmɪgrənt/ *n. Austral. rhyming sl.* an immigrant.

Jimmy Woodser /ˌdʒɪmɪ ˈwʊdzə(r)/ *n. Austral. sl.* **1** a person who drinks alone. **2** a drink taken on one's own. [Jimmy *Wood*, name of a character in the poem of that name by Barcroft Boake]

jimson /ˈdʒɪms(ə)n/ *n.* (in full **jimson weed**) *US* a highly poisonous tall weed, *Datura stramonium*, with large trumpet-shaped flowers. [JAMESTOWN 1]

Jin /dʒɪn/ (also **Chin** /tʃɪn/) **1** a dynasty that ruled China AD 265–420, commonly divided into *Western Jin* (265–317) and *Eastern Jin* (317–420). **2** a dynasty that ruled Manchuria and northern China AD 1115–1234.

Jina /ˈdʒɪnə/ *n. Jainism* **1** a great teacher who has attained liberation from karma. **2** a representation in sculpture of such a teacher. [Skr. *jina* (see JAIN)]

Jinan /dʒiːˈnæn/ (also **Tsinan** /tsiː-/) a city in eastern China, the capital of Shandong province; pop. (1990) 2,290,000.

jingle /ˈdʒɪŋg(ə)l/ *n. & v.* ● *n.* **1** a mixed noise as of bells or light metal objects being shaken together. **2 a** a repetition of the same sound in words, esp. as an aid to memory or to attract attention. **b** a short verse of this kind used in advertising etc. ● *v.* **1** *intr. & tr.* make or cause to make a jingling sound. **2** *intr.* (of writing) be full of alliterations, rhymes, etc. □ **jingly** *adj.* [ME: imit.]

jingo /ˈdʒɪŋgəʊ/ *n.* (*pl.* **-oes**) a supporter of policy favouring war; a blustering patriot. □ **by jingo!** a mild oath. [a conjuror's word of 17th-c. origin: see JINGOISM]

jingoism /ˈdʒɪŋgəʊˌɪz(ə)m/ *n.* advocacy of a warlike policy; loud and blustering patriotism. The term derives from the use of 'jingo' as a nickname for a supporter of Disraeli's action in sending the British fleet to resist the advance of Russia in 1878. The word had been popularized by its use in a music-hall song of the time, 'We don't want to fight, yet by Jingo! if we do, We've got the ships, we've got the men, and got the money too'. (See JINGO.) □ **jingoist** *n.* **jingoistic** /ˌdʒɪŋgəʊˈɪstɪk/ *adj.*

jink /dʒɪŋk/ *v. & n.* ● *v.* **1** *intr.* move elusively; dodge. **2** *tr.* elude by dodging. ● *n.* an act of dodging or eluding. [orig. Sc.: prob. imit. of nimble motion]

jinker /ˈdʒɪŋkə(r)/ *n. & v. Austral.* ● *n.* **1** a wheeled conveyance for moving heavy logs. **2** a light two-wheeled cart. ● *v.tr.* convey by such a cart. [Sc. *janker* long pole on wheels used for carrying logs]

Jinnah /ˈdʒɪnə/, Muhammad Ali (1876–1948), Indian statesman and founder of Pakistan. He headed the Muslim League in its struggle with the Hindu-oriented Indian National Congress, and from 1928 onwards championed the rights of the Muslim minority at conferences on Indian independence. After 1937, when self-governing Hindu provinces began to be formed, his fear that Muslims would be excluded from office led him to campaign for a separate Muslim state. With the establishment of Pakistan in 1947 he became its first Governor-General.

jinnee /dʒɪˈniː, ˈdʒɪnɪ/ *n.* (also **jinn, djinn** /dʒɪn/) (*pl.* **jinn** or **djinn**) *Islamic Mythol.* a supernatural being similar to an angel but of a lower order, able to appear in human and animal form and to help or hinder human beings. [Arab. *jinnī*, pl. *jinn*: cf. GENIE]

jinx /dʒɪŋks/ *n. & v.* ● *n.* a person or thing that seems to cause bad luck. ● *v.tr.* (often in *passive*) subject (a person or thing) to bad luck; cast an evil spell on. [perh. var. of *jynx* wryneck, charm]

JIT *abbr.* = just-in-time.

jitter /ˈdʒɪtə(r)/ *n. & v. colloq.* ● *n.* (**the jitters**) extreme nervousness. ● *v.intr.* be nervous; act nervously. □ **jittery** *adj.* **jitteriness** *n.* [20th c.: orig. unkn.]

jitterbug /ˈdʒɪtəˌbʌɡ/ *n. & v.* ● *n.* **1** a nervous person. **2 a** a fast dance popular in the 1940s, performed esp. to swing music. **b** a person fond of dancing this. ● *v.intr.* (**-bugged, -bugging**) dance the jitterbug.

jiu-jitsu *var. of* JU-JITSU.

jive /dʒaɪv/ *n. & v.* ● *n.* **1** a jerky lively style of dance popular esp. in the 1950s, performed to jazz or rock and roll music. **2** music for this. **3** *US sl.* talk, conversation, esp. when misleading or pretentious. ● *v.intr.* **1** dance the jive. **2** play jive music. □ **jiver** *n.* [20th c.: orig. uncert.]

jizz /dʒɪz/ *n.* the characteristic impression given by an animal or plant. [20th c.: orig. unkn.]

Jnr. *abbr.* Junior.

jo /dʒəʊ/ *n.* (*pl.* **joes**) *Sc.* a sweetheart or beloved. [var. of JOY]

Joachim, St /ˈdʒəʊəˌkɪm/ (in Christian tradition) the husband of

St Anne and father of the Virgin Mary. He is first mentioned in an apocryphal work of the 2nd century, and then rarely referred to until much later times.

Joan of Arc, St /dʒəʊn, ɑːk/ (known as 'the Maid of Orleans') (c.1412–31), French national heroine. Inspired by 'voices' of St Catherine and St Margaret, she led the French armies against the English in the Hundred Years War, relieving besieged Orleans (1429) and ensuring that Charles VII could be crowned in previously occupied Reims. Captured by the Burgundians in 1430, she was handed over to the English, convicted of heresy, and burnt at the stake in Rouen. She was canonized in 1920. Feast day, 30 May.

João Pessoa /ˌʒwaʊ peˈsəʊə/ a city in NE Brazil, on the Atlantic coast, capital of the state of Paraíba; pop. (1990) 484,290.

Job /dʒəʊb/ **1** (in the Bible) a prosperous man whose patience and exemplary piety are tried by dire and undeserved misfortunes, and who, in spite of his bitter lamentations, remains finally confident in the goodness and justice of God. **2** a book of the Bible telling of Job.

job[1] /dʒɒb/ *n. & v.* ● *n.* **1** a piece of work, esp. one done for hire or profit. **2** a paid position of employment. **3 a** anything one has to do, a task. **b** *colloq.* a specified operation or other matter, esp. an operation involving plastic surgery (*a nose job*; *a respray job*). **4** *colloq.* a difficult task (*had a job to find them*). **5** *sl.* a product of work, esp. if well done; an example of its type. **6** *Computing* an item of work regarded separately. **7** *sl.* a crime, esp. a robbery. **8** a transaction in which private advantage prevails over duty or public interest. **9** esp. *Brit. colloq.* a state of affairs or set of circumstances (*is a bad job*). ● *v.* (**jobbed, jobbing**) **1 a** *intr.* do jobs; do piecework. **b** *tr.* (usu. foll. by *out*) let or deal with by contract; subcontract. **2 a** *intr.* deal in stocks. **b** *tr.* buy and sell (stocks or goods) as a middleman. **3 a** *intr.* turn a position of trust to private advantage. **b** *tr.* deal corruptly with (a matter). **4** *tr. US sl.* swindle. □ **job-control language** *Computing* a language enabling the user to determine the tasks to be undertaken by the operating system. **job-hunt** *colloq.* seek employment. **job lot** a miscellaneous group of articles, esp. as bought together for sale. **jobs for the boys** *colloq.* profitable situations etc. to reward one's supporters. **job-share** *v.intr.* be employed or work under a job-sharing arrangement. ● *n.* job-sharing; an instance of this. **job-sharing** an arrangement by which a full-time job is done jointly by two or more part-time employees who share the remuneration etc. **just the job** *colloq.* exactly what is wanted. **make a job** (or **good job**) **of** do thoroughly or successfully. **on the job** *colloq.* **1** at work; in the course of doing a piece of work. **2** *euphem.* engaged in sexual intercourse. **out of a job** unemployed. [16th c.: orig. unkn.]

job[2] /dʒɒb/ *v. & n.* ● *v.* (**jobbed, jobbing**) **1** *tr.* prod; stab slightly. **2** *intr.* (foll. by *at*) thrust. ● *n.* a prod or thrust; a jerk at a horse's bit. [ME, app. imit.: cf. JAB]

jobber /ˈdʒɒbə(r)/ *n.* **1** (in the UK) a principal or wholesaler dealing on the Stock Exchange. ¶ Up to Oct. 1986 permitted to deal only with brokers, not directly with the public. From Oct. 1986 the name has ceased to be in official use (see BROKER 2). **2** *US* **a** a wholesaler. **b** *derog.* a broker (see BROKER 2). **3** a person who jobs. [JOB[1]]

jobbery /ˈdʒɒbərɪ/ *n.* corrupt dealing.

jobbing /ˈdʒɒbɪŋ/ *adj.* working on separate or occasional jobs (esp. of a computer, gardener, or printer).

jobcentre /ˈdʒɒbˌsentə(r)/ *n.* (in the UK) a government office displaying information about available jobs.

jobless /ˈdʒɒblɪs/ *adj.* without a job; unemployed. □ **joblessness** *n.*

Job's comforter *n.* a person who under the guise of comforting aggravates distress (Job 16:2). [JOB]

jobsheet /ˈdʒɒbʃiːt/ *n.* a sheet for recording details of jobs done.

Job's tears *n.* a grass, *Coix lacryma-jobi*, of eastern Asia, with tearlike seeds used as beads. [JOB]

jobsworth /ˈdʒɒbzwɜːθ/ *n. Brit. colloq.* an official who upholds petty rules. [contr. of 'It's more than my *job's worth*' (as typical justification)]

Joburg /ˈdʒəʊbɜːɡ/ an informal name for JOHANNESBURG.

jobwork /ˈdʒɒbwɜːk/ *n.* work done and paid for by the job.

Jocasta /dʒəˈkæstə/ *Gk Mythol.* a Theban woman, the wife of Laius and mother and later wife of Oedipus.

Jock /dʒɒk/ *n. colloq. often offens.* a Scotsman. [Sc. form of the name *Jack* (see JACK[1])]

jock[1] /dʒɒk/ *n. colloq.* **1** a jockey. **2** a disc jockey. [abbr.]

jock[2] /dʒɒk/ *n. US sl.* = JOCKSTRAP 2. [abbr.]

jockey /ˈdʒɒkɪ/ *n. & v.* ● *n.* (*pl.* **-eys**) a rider in horse-races, esp. a professional one. ● *v.* (**-eys, -eyed**) **1** *tr.* **a** trick or cheat (a person).

b outwit. **2** *tr.* (foll. by *away*, *out*, *in*, etc.) draw (a person) by trickery. **3** *intr.* cheat. □ **jockey cap** a strengthened cap with a long peak, as worn by jockeys. **jockey for position** try to gain an advantageous position esp. by skilful manoeuvring or unfair action. □ **jockeydom** *n.* **jockeyship** *n.* [dimin. of JOCK]

Jockey Club an organization whose stewards are the central authority for the administration and control of horse-racing in Britain. It was founded in 1750.

jockstrap /ˈdʒɒkstræp/ *n.* **1** a support or protection for the male genitals, worn esp. by sportsmen. **2** *US sl.* an athletic (rather than intellectual or aesthetic) young man, esp. at a university. [sl. *jock* genitals + STRAP]

jocose /dʒəˈkəʊs/ *adj.* **1** playful in style. **2** fond of joking, jocular. □ **jocosely** *adv.* **jocoseness** *n.* **jocosity** /-ˈkɒsɪtɪ/ *n.* (*pl.* **-ies**). [L *jocosus* f. *jocus* jest]

jocular /ˈdʒɒkjʊlə(r)/ *adj.* **1** merry; fond of joking. **2** of the nature of a joke; humorous. □ **jocularly** *adv.* **jocularity** /ˌdʒɒkjʊˈlærɪtɪ/ *n.* (*pl.* **-ies**). [L *jocularis* f. *joculus* dimin. of *jocus* jest]

jocund /ˈdʒɒkənd/ *adj. literary* merry, cheerful, sprightly. □ **jocundly** *adv.* **jocundity** /dʒəˈkʌndɪtɪ/ *n.* (*pl.* **-ies**). [ME f. OF f. L *jocundus*, *jucundus* f. *juvare* delight]

Jodhpur /ˈdʒɒdpʊə(r)/ **1** a city in western India, in Rajasthan; pop. (1991) 649,000. **2** a former princely state of India, now part of Rajasthan.

jodhpurs /ˈdʒɒdpəz/ *n.pl.* long breeches for riding etc., close-fitting from the knee to the ankle. [JODHPUR]

Jodrell Bank /ˈdʒɒdrəl/ the site in Cheshire of the Nuffield Radio Astronomy Laboratory of Manchester University. It has one of the world's largest radio telescopes, with a fully steerable dish 76 m (250 ft) in diameter.

Joe Bloggs /dʒəʊ ˈblɒgz/ *n. Brit. colloq.* a hypothetical average man.

Joe Blow /dʒəʊ ˈbləʊ/ *n. N. Amer. colloq.* = JOHN DOE.

Joel /ˈdʒəʊəl/ **1** a Hebrew minor prophet of the 5th or possibly 9th century BC. **2** a book of the Bible containing his prophecies.

joey /ˈdʒəʊɪ/ *n.* (*pl.* **-eys**) *Austral.* **1** a young kangaroo. **2** a young animal. [19th c.: origin unkn.]

Joffre /ˈʒɒfrə/, Joseph Jacques Césaire (1852–1931), French Marshal. During the First World War he was Commander-in-Chief of the French army on the Western Front (1914–16). Joffre was chiefly responsible for the Allied victory in the first battle of the Marne, but resigned after the costly Battle of Verdun.

jog /dʒɒg/ *v. & n.* ● *v.* (**jogged**, **jogging**) **1** *intr.* run at a slow pace, esp. as physical exercise. **2** *intr.* (of a horse) move at a jogtrot. **3** *intr.* (often foll. by *on*, *along*) proceed laboriously; trudge. **4** *intr.* go on one's way. **5** *intr.* proceed; get through the time (*we must jog on somehow*). **6** *intr.* move up and down with an unsteady motion. **7** *tr.* nudge (a person), esp. to arouse attention. **8** *tr.* shake with a push or jerk. **9** *tr.* stimulate (a person's or one's own memory). ● *n.* **1** a shake, push, or nudge. **2** a slow walk or trot. [ME: app. imit.]

jogger /ˈdʒɒgə(r)/ *n.* a person who jogs, esp. one who runs for physical exercise.

joggle[1] /ˈdʒɒg(ə)l/ *v. & n.* ● *v.tr. & intr.* shake or move by or as if by repeated jerks. ● *n.* **1** a slight shake. **2** the act or action of joggling. [frequent. of JOG]

joggle[2] /ˈdʒɒg(ə)l/ *n. & v.* ● *n.* **1** a joint of two pieces of stone or timber, contrived to prevent their sliding on one another. **2** a notch in one of the two pieces, a projection in the other, or a small piece let in between the two, for this purpose. ● *v.tr.* join with a joggle. [perh. f. *jog* = JAG[1]]

Jogjakarta see YOGYAKARTA.

jogtrot /ˈdʒɒgtrɒt/ *n.* **1** a slow regular trot. **2** a monotonous progression.

Johannesburg /dʒəʊˈhænɪsˌbɜːg/ a city in South Africa, the capital of the province of Pretoria-Witwatersrand-Vereeniging; pop. (1985) 1,609,400. The largest city in South Africa and the centre of its gold-mining industry, it was founded in 1886 and was probably named after Johannes Meyer, its first mining commissioner.

John[1] /dʒɒn/ (known as John Lackland) (1165–1216), son of Henry II, king of England 1199–1216. He lost Normandy and most of his French possessions to Phillip II of France by 1205. His refusal to accept Stephen Langton as Archbishop of Canterbury caused an interdict to be placed on England in 1208, and led to his own excommunication the following year. In 1215 John was forced to sign Magna Carta by his

barons. When he ignored its provisions, civil war broke out and he died on campaign.

John[2] /dʒɒn/ the name of six kings of Portugal, notably:

John I (known as John the Great) (1357–1433), reigned 1385–1433. Reinforced by an English army, he won independence for Portugal with his victory over the Castilians at Aljubarrota (1385). He established an Anglo-Portuguese alliance (1386), married a daughter of John of Gaunt (1387), and presided over a long period of peace and prosperity which was notable for his encouragement of voyages of discovery.

John IV (known as John the Fortunate) (1604–56), reigned 1640–56. The founder of the Braganza dynasty, he expelled a Spanish usurper and proclaimed himself king. He defeated the Spanish at Montijo (1644) and drove the Dutch out of Brazil (1654).

John[3] /dʒɒn/, Augustus (Edwin) (1878–1961), Welsh painter. He is perhaps best known for *The Smiling Woman* (1908), a portrait of his second wife Dorelia, which portrayed a robust gypsy type of beauty; the gypsies of Wales were frequent subjects of his work. He was subsequently noted for his portraits of the wealthy and famous, particularly prominent writers such as Thomas Hardy, George Bernard Shaw, W. B. Yeats, James Joyce, and Dylan Thomas. He was the brother of Gwen John.

John[4] /dʒɒn/, Barry (b.1945), Welsh Rugby Union player. His international career, during which he played at half-back and scored a record ninety points for his country, lasted from 1966 until his retirement in 1972. He played a prominent part in the British Lions' victorious tour of New Zealand in 1971.

John[5] /dʒɒn/, Elton (Hercules) (born Reginald Kenneth Dwight) (b.1947), English pop and rock singer, pianist, and songwriter. He has written many hit songs, the majority of them with lyricist Bernie Taupin (b.1950); they include 'Your Song' (1970) and 'Nikita' (1985). He is noted for his good-humoured flamboyance, performing handstands at the piano and wearing outrageous costumes. In 1979 he became the first Western rock star to tour the Soviet Union.

John[6] /dʒɒn/, Gwen (1876–1939), Welsh painter. After studying with Whistler in Paris (1898), she settled in France and worked as Rodin's model (1904), becoming his devoted friend and mistress. She converted to Catholicism in 1913; her paintings, mainly watercolours, often depict nuns or girls in interior settings and are noted for their grey tonality. She was the sister of Augustus John.

john /dʒɒn/ *n.* **1** *N. Amer. colloq.* a lavatory. **2** *sl.* a prostitute's client. [the name *John*]

John III (known as John Sobieski) (1624–96), king of Poland 1674–96. He was elected king of Poland after a distinguished early career as a soldier. In 1683 he relieved Vienna when it was besieged by the Turks, thereby becoming the hero of the Christian world.

John, St 1 (known as St John the Evangelist or St John the Divine) an Apostle, son of Zebedee and brother of James. He has traditionally been credited with the authorship of the fourth Gospel, Revelation, and three epistles of the New Testament. Feast day, 27 Dec. **2** the fourth Gospel (see GOSPEL 2). **3** any of the three epistles of the New Testament attributed to St John.

John Bull a personification of England or the typical Englishman, represented as a stout red-faced farmer in a top hat and high boots. It was originally the name of a character representing the English nation in John Arbuthnot's satire *History of John Bull* (1712).

John Chrysostom, St see CHRYSOSTOM, ST JOHN.

John Doe /dʒɒn ˈdəʊ/ *n. N. Amer. colloq.* an ordinary or typical citizen; Joe Bloggs. [orig. in Engl. law, with ref. to an anonymous party in a legal action]

John Dory *n.* (*pl.* **-ies**) a European dory (fish), *Zeus faber*, with a laterally flattened body and a black spot on each side.

John Hop *n. Austral. sl.* a police officer. [rhyming sl. for *cop*]

John Lackland, King John of England (see JOHN[1]).

johnny /ˈdʒɒnɪ/ *n.* (*pl.* **-ies**) *Brit.* **1** *sl.* a condom. **2** *colloq.* a fellow; a man. □ **johnny-come-lately** *colloq.* a recently arrived person. [familiar form of the name *John*]

John of Damascus, St (*c.*675–*c.*749), Syrian theologian and Doctor of the Church. After championing image worship against the iconoclasts, he wrote his encyclopedic work on Christian theology, *The Fount of Wisdom*. Its last section summarized the teachings of the Greek Fathers of the Church on the principal mysteries of the Christian faith and was influential for centuries in both Eastern and Western Churches. Feast day, 4 Dec.

John of Gaunt /gɔːnt/ (1340–99), son of Edward III. Born in Ghent, he was created Duke of Lancaster in 1362. John of Gaunt headed the government during the final years of his father's reign and the minority of Richard II, and was effective ruler of England in this period. His son Henry Bolingbroke later became King Henry IV.

John of the Cross, St (born Juan de Yepis y Alvarez) (1542–91), Spanish mystic and poet. A Carmelite monk and priest, he joined with St Teresa of Ávila in trying to reassert the original Carmelite observance of austerity, and in 1568 founded the 'discalced' Carmelite order for monks. He also wrote mystical poems including 'The Dark Night of the Soul', describing the soul's purgation. Feast day, 14 Dec.

John o'Groats /ə'ɡrəʊts/ a village at the extreme NE point of the Scottish mainland. It is said to be named from a house built there in the 16th century by a Dutchman, Jan Groot. (Cf. LAND'S END.)

John Paul II (born Karol Jozef Wojtyla) (b.1920), Polish cleric, pope since 1978. The first non-Italian pope since 1522, he has travelled abroad extensively during his papacy, especially in Central and South America. He has upheld the Church's traditional opposition to artificial means of contraception and abortion, as well as condemning homosexuality, the ordination of women, and the relaxation of the rule of celibacy for priests. He has also discouraged priests from taking part in political activity.

Johns /dʒɒnz/, Jasper (b.1930), American painter, sculptor, and printmaker. A key figure in the development of pop art, he rebelled against abstract expressionism and depicted commonplace and universally recognized images such as the US flag. He is best known for his *Flags, Targets,* and *Numbers* series produced in the mid-1950s; in these, he was noted for his use of encaustic (wax-based) paint. In the late 1950s he produced sculptures of such objects as beer cans and light bulbs cast in bronze.

John Sobieski /sɒ'bjeskɪ/ see JOHN III.

Johnson[1] /'dʒɒns(ə)n/, Amy (1903–41), English aviator. In 1930 she became the first woman to fly solo to Australia, although her time of 19½ days was three days short of the record. She later set a record with her solo flight to Tokyo (1931) and broke the solo-flight record to Cape Town (1932). She joined the Auxiliary Air Force in 1939, but was lost when her plane disappeared in a flight over the Thames estuary.

Johnson[2] /'dʒɒns(ə)n/, Andrew (1808–75), American Democratic statesman, 17th President of the US 1865–9. His lenient policy towards the southern states after the American Civil War brought him into bitter conflict with the Republican majority in Congress, who impeached him (1868); he was acquitted by a single vote.

Johnson[3] /'dʒɒns(ə)n/, Cornelius (also **Jansens** /'dʒæns(ə)nz/ or **Janssen van Ceulen** /,dʒæns(ə)n væn 'kɜ:lən/) (1593–c.1661), English-born Dutch portrait painter. He painted for the court of Charles I, where he was influenced by Van Dyck and became noted for his individual portrait heads. After the outbreak of the English Civil War he emigrated to Holland (1643).

Johnson[4] /'dʒɒns(ə)n/, Jack (1878–1946), American boxer. In 1908 he took the world heavyweight title, becoming the first black holder of the title; he retained it until 1915.

Johnson[5] /'dʒɒns(ə)n/, Lyndon Baines (known as 'LBJ') (1908–73), American Democratic statesman, 36th President of the US 1963–9. His administration continued the programme of social and economic reform initiated by John F. Kennedy, passing the 1964 and 1965 Civil Rights Acts and legislating to reduce taxation. However, the increasing involvement of the US in the Vietnam War undermined his popularity and he refused to seek re-election.

Johnson[6] /'dʒɒns(ə)n/, Samuel (known as Dr Johnson) (1709–84), English lexicographer, writer, critic, and conversationalist. His principal works include his *Dictionary of the English Language* (1755), one of the first to use illustrative quotations, his edition of Shakespeare (1765), and *The Lives of the English Poets* (1777). A leading figure in the literary London of his day, he formed the Literary Club (1764), which numbered Edmund Burke, Oliver Goldsmith, Sir Joshua Reynolds, David Garrick, and Johnson's biographer Boswell among its members.

Johnsonian /dʒɒn'səʊnɪən/ adj. of or relating to Samuel Johnson, or typical of his style of writing.

John the Baptist, St, Jewish preacher and prophet, a contemporary of Jesus. In AD c.27 he preached on the banks of the River Jordan, demanding repentance and baptism from his hearers in view of the approach of God's judgement. Among those whom he baptized was Christ. He was imprisoned and beheaded by Herod Antipas after denouncing the latter's marriage to Herodias, the wife of Herod's brother Philip (Matt. 14:1–12). (See also SALOME.) Feast day, 24 June.

John the Evangelist, St see JOHN, ST 1.

John the Fortunate, John IV of Portugal (see JOHN[2]).

John the Great, John I of Portugal (see JOHN[2]).

Johor /dʒəʊ'hɔː(r)/ (also **Johore**) a state of Malaysia, at the southernmost point of mainland Asia, joined to Singapore by a causeway; capital, Johor Baharu.

Johor Baharu the capital of the state of Johor in Malaysia, situated at the southern tip of the Malay Peninsula, opposite the island of Singapore; pop. (1980) 246,400.

joie de vivre /,ʒwʌ də 'viːvrə/ n. a feeling of healthy and exuberant enjoyment of life. [F, = joy of living]

join /dʒɔɪn/ v. & n. ● v. **1** tr. (often foll. by *to, together*) put together; fasten, unite (one thing or person to another or several together). **2** tr. connect (points) by a line etc. **3** tr. become a member of (an association, society, organization, etc.). **4** tr. take one's place with or in (a company, group, procession, etc.). **5** tr. **a** come into the company of (a person). **b** (foll. by *in*) take part with (others) in an activity etc. (*joined me in condemnation of the outrage*). **c** (foll. by *for*) share the company of (for a specified occasion (*may I join you for lunch?*). **6** intr. (often foll. by *with, to*) come together; be united. **7** intr. (often foll. by *in*) take part with others in an activity etc. **8** tr. be or become connected or continuous with (*the Ruhr joins the Rhine near Duisburg*). ● n. a point, line, or surface at which two or more things are joined. □ **join battle** begin fighting. **join forces** combine efforts. **join hands 1 a** clasp each other's hands. **b** clasp one's hands together. **2** combine in an action or enterprise. **join in** (also *absol.*) take part in (an activity). **join up 1** enlist for military service. **2** (often foll. by *with*) unite, connect. □ **joinable** adj. [ME f. OF *joindre* (stem *joign-*) f. L *jungere junct-* join]

joinder /'dʒɔɪndə(r)/ n. Law the act of bringing together. [AF f. OF *joindre* to join]

joiner /'dʒɔɪnə(r)/ n. **1** a person who makes furniture and light woodwork. **2** colloq. a person who readily joins societies etc. □ **joinery** n. (in sense 1). [ME f. AF *joignour*, OF *joigneor* (as JOIN)]

joint /dʒɔɪnt/ n., adj., & v. ● n. **1 a** a place at which two things are joined together. **b** a point at which, or a contrivance by which, two parts of an artificial structure are joined. **2** a structure in an animal body by which two bones are fitted together. **3 a** any of the parts into which an animal carcass is divided for food. **b** any of the parts of which a body is made up. **4** sl. a place of meeting for drinking etc. **5** sl. a marijuana cigarette. **6** the part of a stem from which a leaf or branch grows. **7** a piece of flexible material forming the hinge of a book-cover. **8** Geol. a fissure in a mass of rock. ● adj. **1** held or done by, or belonging to, two or more persons etc. in conjunction (*a joint mortgage; joint action*). **2** sharing with another in some action, state, etc. (*joint author; joint favourite*). ● v.tr. **1** connect by joints. **2** divide (a body or member) at a joint or into joints. **3** fill up the joints of (masonry etc.) with mortar etc.; trim the surface of (a mortar joint). **4** prepare (a board etc.) for being joined to another by planing its edge. □ **joint account** a bank account held by more than one person, each of whom has the right to deposit and withdraw funds. **joint and several** (of an obligation etc.) undertaken and signed by two or more people, of whom each is liable for the whole obligation etc. **joint stock** capital held jointly; a common fund. **joint-stock company** one formed on the basis of a joint stock. **out of joint 1** (of a bone) dislocated. **2** out of order. □ **jointless** adj. **jointly** adv. [ME f. OF, past part. of *joindre* JOIN]

jointer /'dʒɔɪntə(r)/ n. **1 a** a plane for jointing. **b** a tool for jointing or pointing masonry. **2** a worker employed in jointing wires, pipes, etc.

jointress /'dʒɔɪntrɪs/ n. a widow who holds a jointure. [obs. *jointer* joint possessor]

jointure /'dʒɔɪntʃə(r)/ n. & v. ● n. an estate settled on a wife for the period during which she survives her husband. ● v.tr. provide (a wife) with a jointure. [ME f. OF f. L *junctura* (as JOIN)]

joist /dʒɔɪst/ n. each of a series of parallel supporting beams of timber, steel, etc., used in floors, ceilings, etc. □ **joisted** adj. [ME f. OF *giste* ult. f. L *jacere* lie]

jojoba /həʊ'həʊbə/ n. a desert shrub, *Simmondsia chinensis*, of Mexico and the southern US, with seeds yielding an oily extract used in cosmetics etc. [Mex. Sp.]

joke /dʒəʊk/ n. & v. ● n. **1 a** a thing said or done to excite laughter. **b** a witticism or jest. **2** a ridiculous thing, person, or circumstance. ● v. **1** intr. make jokes. **2** tr. poke fun at; banter. □ **no joke** colloq. a serious matter. □ **jokingly** adv. **joky** adj. (also **jokey**). **jokily** adv. **jokiness** n. [17th c. (*joque*), orig. sl.: perh. f. L *jocus* jest]

joker /ˈdʒəʊkə(r)/ n. **1** a person who jokes. **2** sl. a fellow; a man. **3** a playing card usu. with a figure of a jester, used in some games esp. as a wild card. **4** US a clause unobtrusively inserted in a bill or document and affecting its operation in a way not immediately apparent. **5** an unexpected factor or resource. □ **the joker in the pack** an unpredictable factor or participant.

jokesmith /ˈdʒəʊksmɪθ/ n. a skilled user or inventor of jokes.

jolie laide /ˌʒɒlɪˈled/ n. (pl. **jolies laides** pronunc. same) = BELLE LAIDE. [F f. jolie pretty + laide ugly]

Joliot /ˈʒɒlɪˌəʊ/, Jean-Frédéric (1900–58), French nuclear physicist. He gave up engineering to study radioactivity and became Madame Curie's assistant at the Radium Institute. There he worked with her daughter Irène (1897–1956), whom he married, taking the name Joliot-Curie; their joint discovery of artificial radioactivity earned them the 1935 Nobel Prize for chemistry. Shortly before the war Joliot demonstrated that a nuclear chain reaction was possible, and later he and his wife became involved with the establishment of the French atomic energy commission, only to be removed because of their Communism.

jollify /ˈdʒɒlɪˌfaɪ/ v.tr. & intr. (**-ies**, **-ied**) make or be merry, esp. in drinking. □ **jollification** /ˌdʒɒlɪfɪˈkeɪʃ(ə)n/ n.

jollity /ˈdʒɒlɪtɪ/ n. (pl. **-ies**) **1** merrymaking; festiveness. **2** (in pl.) festivities. [ME f. OF joliveté (as JOLLY[1])]

jollo /ˈdʒɒləʊ/ n. (pl. **-os**) Austral. colloq. a spree; a party.

jolly[1] /ˈdʒɒlɪ/ adj., adv., v., & n. ● adj. (**jollier**, **jolliest**) **1** cheerful and good-humoured; merry. **2** festive, jovial. **3** slightly drunk. **4** colloq. (of a person or thing) very pleasant, delightful (often iron.: a jolly shame). ● adv. colloq. very (they were jolly unlucky). ● v.tr. (**-ies**, **-ied**) **1** (usu. foll. by along) colloq. coax or humour (a person) in a friendly way. **2** chaff, banter. ● n. (pl. **-ies**) colloq. a party or celebration; an outing. □ **jollily** adv. **jolliness** n. [ME f. OF jolif gay, pretty, perh. f. ON jól YULE]

jolly[2] /ˈdʒɒlɪ/ n. (pl. **-ies**) (in full **jolly boat**) a clinker-built ship's boat smaller than a cutter. [18th c.: orig. unkn.: perh. rel. to YAWL]

Jolly Roger n. a pirates' black flag, usu. with the skull and crossbones.

Jolson /ˈdʒəʊls(ə)n/, Al (born Asa Yoelson) (1886–1950), Russian-born American singer, film actor, and comedian. He made the Gershwin song 'Swanee' his trademark, and appeared in the first full-length talking film The Jazz Singer (1927).

jolt /dʒəʊlt, dʒɒlt/ v. & n. ● v. **1** tr. disturb or shake from the normal position (esp. in a moving vehicle) with a jerk. **2** tr. give a mental shock to; perturb. **3** intr. (of a vehicle) move along with jerks, as on a rough road. ● n. **1** such a jerk. **2** a surprise or shock. □ **jolty** adj. [16th c.: orig. unkn.]

Jomon /ˈdʒəʊmən/ adj. **1** designating an early hand-made pottery of Japan, decorated with a characteristic cord-pattern. **2** the early neolithic or pre-neolithic culture (c.3000 BC) characterized by this pottery. [Jap. jōmon cord-pattern]

Jon. abbr. **1** (in the Bible) Jonah. **2** Jonathan.

Jonah /ˈdʒəʊnə/ **1** a Hebrew minor prophet. He was called by God to preach in Nineveh, but disobeyed and attempted to escape by sea; in a storm he was thrown overboard as a bringer of bad luck and swallowed by a great fish, only to be saved and finally succeed in his mission. **2** a book of the Bible telling of Jonah.

Jonathan /ˈdʒɒnəθən/ (in the Bible) the son of Saul and friend of David, killed at the battle of Mount Gilboa (1 Sam. 13 ff.).

Jones[1] /dʒəʊnz/, Daniel (1881–1967), British linguist and phonetician. From 1907 he developed the recently invented International Phonetic Alphabet at the first British department of phonetics, at University College, London. He went on to invent a system of cardinal vowels, used as reference points for transcribing all vowel sounds. In his English Pronouncing Dictionary (1917), he described the influential system of received pronunciation.

Jones[2] /dʒəʊnz/, Inigo (1573–1652), English architect and stage designer. He introduced the Palladian style to England and is best known as the architect of the Queen's House at Greenwich (1616) and the Banqueting Hall at Whitehall (1619). He also pioneered the use of the proscenium arch and movable stage scenery in England, and was for many years involved with costume design for court masques.

Jones[3] /dʒəʊnz/, John Paul (born John Paul) (1747–92), Scottish-born American admiral. He became famous for his raids off the northern coasts of Britain during the War of American Independence.

Jones[4] /dʒəʊnz/, Robert Tyre ('Bobby') (1902–71), American golfer. In a short competitive career (1923–30), and as an amateur, he won thirteen

major competitions out of twenty-seven, including four American and three British open championships.

Jong /jɒŋ/, Erica (Mann) (b.1942), American poet and novelist. She made her name with the award-winning poetry collection Fruits and Vegetables (1971). Her international reputation is based on the picaresque novels Fear of Flying (1973), recounting the sexual exploits of its heroine Isadora Wing, and Fanny (1980), written in a pseudo-18th-century style.

jongleur /ʒɒŋˈglɜː(r)/ n. hist. an itinerant minstrel. [F, var. of jougleur JUGGLER]

Jönköping /ˈjɜːnˌtʃɜːpɪŋ/ an industrial city in southern Sweden, at the south end of Lake Vättern; pop. (1990) 111,500.

jonquil /ˈdʒɒŋkwɪl/ n. a kind of narcissus, Narcissus jonquilla, with clusters of small fragrant yellow flowers. [mod.L jonquilla or F jonquille f. Sp. junquillo dimin. of junco: see JUNCO]

Jonson /ˈdʒɒns(ə)n/, Benjamin ('Ben') (1572–1637), English dramatist and poet. With his play Every Man in his Humour (1598) he established his 'comedy of humours', whereby each character is dominated by a particular obsession. His vigorous and often savage wit is evident in his comedies Volpone (1606), The Alchemist (1610), and Bartholomew Fair (1614). During the reign of James I his prestige and influence were unrivalled, and he became the first Poet Laureate in the modern sense.

Joplin /ˈdʒɒplɪn/, Scott (1868–1917), American pianist and composer. One of the creators of ragtime, he was the first to write down his compositions. Two of Joplin's best-known rags, 'Original Rags' and 'Maple Leaf Rag', were written in 1899. The latter was so successful that a million copies of the sheet music were sold. Joplin's music, including the rag 'The Entertainer', was featured in the film The Sting (1973).

Joppa see JAFFA[1].

Jordaens /jɔːˈdɑːns/, Jacob (1593–1678), Flemish painter. Influenced by Rubens, he painted in warm colours and is noted for his boisterous peasant scenes. His major works include The King Drinks (1638).

Jordan /ˈdʒɔːd(ə)n/ **1** a river flowing southward for 320 km (200 miles) from the Anti-Lebanon mountains through the Sea of Galilee into the Dead Sea. John the Baptist baptized Christ in the River Jordan. It is regarded as sacred not only by Christians but also by Jews and Muslims. **2** (in full **Hashemite Kingdom of Jordan**), a country in the Middle East east of the River Jordan; pop. (est. 1990) 4,000,000; official language, Arabic; capital, Amman. Romans, Arabs, and Crusaders dominated the area successively until it fell under Turkish rule in the 16th century. In 1916 the land east of the River Jordan was made a British protectorate, the Amirate of Transjordan; this became independent in 1946 and changed its name to the Hashemite Kingdom of Jordan in 1949. The country is almost completely landlocked, its only outlet to the sea being the port of Aqaba at the north-eastern end of the Red Sea. During the war of 1948–9 following the establishment of the state of Israel the Jordanians took over a large area on the west bank of the river, but were driven out by Israel in the Six Day War of 1967 (see WEST BANK); many Palestinian refugees then entered the country, and the Palestine Liberation Organization maintained its headquarters there until it was expelled following the short civil war of 1970. Jordan supported Iraq in the Gulf War of 1991, but has since repudiated this policy; a peace treaty with Israel was signed in 1994, ending an official state of war between the two countries. □ **Jordanian** /dʒɔːˈdeɪnɪən/ adj. & n.

jorum /ˈdʒɔːrəm/ n. **1** a large drinking-bowl. **2** its contents, esp. punch. [perh. f. Joram (2 Sam. 8:10)]

Jorvik /ˈjɔːvɪk/ (also **Yorvik**) the Viking name for YORK.

Jos. abbr. Joseph.

Joseph /ˈdʒəʊzɪf/ a Hebrew patriarch, son of Jacob. He was given a coat of many colours by his father, but was then sold by his jealous brothers into captivity in Egypt, where he attained high office (Gen. 30–50).

Joseph, St, carpenter of Nazareth, husband of the Virgin Mary. At the time of the Annunciation, he was betrothed to Mary. Feast day, 19 Mar.

Josephine /ˈdʒəʊzɪˌfiːn/, (born Marie Joséphine Rose Tascher de la Pagerie) (1763–1814), Empress of France 1796–1809. Born in the West Indies, she was married to the Viscount de Beauharnais before marrying Napoleon in 1796. Their marriage proved childless and Josephine was divorced by Napoleon in 1809.

Joseph of Arimathea /ˌærɪməˈθiːə/ a member of the council at Jerusalem who, after the Crucifixion, asked Pilate for Christ's body, which he buried. He is also known from the medieval story that he came to England with the Holy Grail and built the first church at Glastonbury.

Josephus /dʒəʊˈsiːfəs/, Flavius (born Joseph ben Matthias) (c.37–c.100), Jewish historian, general, and Pharisee. A leader of the Jewish revolt against the Romans from 66, he was captured in 67; his life was spared when he prophesied that Vespasian would become emperor. He subsequently received Roman citizenship and a pension, and is remembered as the author of the *Jewish War*, an eyewitness account of the events leading up to the revolt, and of *Antiquities of the Jews*, a history running from the Creation to 66.

Josh. *abbr.* (in the Bible) Joshua.

josh /dʒɒʃ/ n. & v. *colloq.* ● n. a good-natured or teasing joke. ● v. **1** tr. tease or banter. **2** *intr.* indulge in ridicule. □ **josher** n. [19th c.: orig. unkn.]

Joshua /ˈdʒɒʃʊə/ **1** the Israelite leader (probably 13th century BC) who succeeded Moses and led his people into the Promised Land. **2** the sixth book of the Bible, telling of the conquest of Canaan and its division among the twelve tribes of Israel.

Joshua tree n. a yucca, *Yucca brevifolia*, of arid parts of western North America. [app. f. JOSHUA, the plant being likened to a man brandishing a spear (*Joshua* 8:18)]

Josquin des Prez /ˈʒɒskæn/ see DES PREZ.

joss[1] /dʒɒs/ n. a Chinese idol. □ **joss-stick** a stick of fragrant tinder mixed with clay, burnt as incense. [perh. ult. f. Port. *deos* f. L *deus* god]

joss[2] /dʒɒs/ n. *Austral. sl.* a person of influence and importance. [Brit. dial.]

josser /ˈdʒɒsə(r)/ n. *sl.* **1** *Brit.* **a** a fool. **b** a fellow. **2** *Austral.* a clergyman.

jostle /ˈdʒɒs(ə)l/ v. & n. ● v. **1** tr. push against; elbow. **2** tr. (often foll. by *away*, *from*, etc.) push (a person) abruptly or roughly. **3** *intr.* (foll. by *against*) knock or push, esp. in a crowd. **4** *intr.* (foll. by *with*) struggle; have a rough exchange. ● n. **1** the act or an instance of jostling. **2** a collision. [ME: earlier *justle* f. JOUST + -LE⁴]

jot /dʒɒt/ v. & n. ● v.tr. (**jotted**, **jotting**) (usu. foll. by *down*) write briefly or hastily. ● n. (usu. with *neg.* expressed or implied) a very small amount (*not one jot*). [earlier as noun: L f. Gk *iōta*: see IOTA]

jotter /ˈdʒɒtə(r)/ n. a small pad or notebook for making notes etc.

jotting /ˈdʒɒtɪŋ/ n. (usu. in *pl.*) a note; something jotted down.

Jotun /ˈjəʊtʊn/ n. *Scand. Mythol.* a member of a race of giants, enemies of the gods.

Jotunheim /ˈjəʊtʊnˌhaɪm/ **1** *Scand. Mythol.* part of Asgard, inhabited by giants. **2** a mountain range in south central Norway. Its highest peak is Glittertind (2,472 m, 8,110 ft). [Norw., = giants' home]

Joule /dʒuːl/, James Prescott (1818–89), English physicist. Experimenting in his private laboratory and at the family's brewery, he established that all forms of energy were basically the same and interchangeable – the basic principle of what is now called the first law of thermodynamics. Among other things, he measured the thermal effects of an electric current due to the resistance of the wire, establishing the law governing this. In 1852 he and William Thomson, later Lord Kelvin, discovered the fall in temperature when gases expand (the Joule–Thomson effect), which led to the development of the refrigerator and to the science of cryogenics.

joule /dʒuːl/ n. *Physics* the SI unit of work or energy (symbol: **J**), equal to 10⁷ erg. One joule is the work done by a force of one newton when the point of application of the force moves one metre along its direction of action. [JOULE]

Joule–Thomson effect /dʒuːlˈtɒms(ə)n/ n. *Physics* the change of temperature of a gas when it is allowed to expand without doing any external work. The gas becomes cooler if it was initially below a certain temperature (the *inversion temperature*), or hotter if initially above it. [JOULE + William *Thomson* (see KELVIN)]

jounce /dʒaʊns/ v.tr. & intr. bump, bounce, jolt. [ME: orig. unkn.]

journal /ˈdʒɜːn(ə)l/ n. **1** a newspaper or periodical. **2** a daily record of events. **3** *Naut.* a logbook. **4** a book in which business transactions are entered, with a statement of the accounts to which each is to be debited and credited. **5** the part of a shaft or axle that rests on bearings. **6** (**the Journals**) *Parl.* a record of daily proceedings. [ME f. OF *jurnal* f. LL *diurnalis* DIURNAL]

journalese /ˌdʒɜːnəˈliːz/ n. a hackneyed style of language characteristic of some newspaper writing.

journalism /ˈdʒɜːnəˌlɪz(ə)m/ n. the business or practice of writing and producing newspapers or periodicals.

journalist /ˈdʒɜːnəlɪst/ n. a person employed to write for, edit, or report for a newspaper, periodical, or newscast. □ **journalistic** /ˌdʒɜːnəˈlɪstɪk/ adj. **journalistically** adv.

journalize /ˈdʒɜːnəˌlaɪz/ v.tr. (also **-ise**) record in a private journal.

journey /ˈdʒɜːnɪ/ n. & v. ● n. (pl. **-eys**) **1** an act of going from one place to another, esp. at a long distance. **2** the distance travelled in a specified time (*a day's journey*). **3** the travelling of a vehicle along a route at a stated time. ● v.intr. (**-eys**, **-eyed**) make a journey. □ **journeyer** n. [ME f. OF *jornee* day, day's work or travel, ult. f. L *diurnus* daily]

journeyman /ˈdʒɜːnɪmən/ n. (pl. **-men**) **1** a qualified mechanic or artisan who works for another. **2** *derog.* **a** a reliable but not outstanding worker. **b** a mere hireling. [JOURNEY in obs. sense 'day's work' + MAN]

journo /ˈdʒɜːnəʊ/ n. (pl. **-os**) *colloq.* a journalist. [shortened form of JOURNALIST + -O]

joust /dʒaʊst/ n. & v. *hist.* ● n. a combat between two knights on horseback with lances, esp. for exercise or sport. ● v.intr. engage in a joust. □ **jouster** n. [ME f. OF *juster* bring together ult. f. L *juxta* near]

Jove /dʒəʊv/ *Rom. Mythol.* Jupiter. □ **by Jove!** an exclamation of surprise or approval. [ME f. L *Jovis* genitive of OL *Jovis* used as genitive of JUPITER]

jovial /ˈdʒəʊvɪəl/ adj. **1** merry. **2** convivial. **3** hearty and good-humoured. □ **jovially** adv. **joviality** /ˌdʒəʊvɪˈælɪtɪ/ n. [F f. LL *jovialis* of Jupiter (as JOVE), with ref. to the supposed influence of the planet Jupiter on those born under it]

Jovian /ˈdʒəʊvɪən/ adj. **1** *Rom. Mythol.* of or like Jupiter. **2** *Astron.* of the planet Jupiter.

jowar /dʒaʊˈwɑː(r)/ n. = DURRA. [Hindi *jawār*]

jowl[1] /dʒaʊl/ n. **1** the jaw or jawbone. **2** the cheek (*cheek by jowl*). □ **-jowled** adj. (in comb.). [ME *chavel* jaw f. OE *ceafl*]

jowl[2] /dʒaʊl/ n. **1** the external loose skin on the throat or neck when prominent. **2** the dewlap of an ox, wattle of a bird, etc. □ **jowly** adj. [ME *cholle* neck f. OE *ceole*]

joy /dʒɔɪ/ n. & v. ● n. **1** (often foll. by *at*, *in*) a vivid emotion of pleasure; extreme gladness. **2** a thing that causes joy. **3** *Brit. colloq.* satisfaction, success (*got no joy*). ● v. esp. *poet.* **1** *intr.* rejoice. **2** tr. gladden. □ **wish a person joy of** *Brit. iron.* be gladly rid of (what that person has to deal with). □ **joyless** adj. **joylessly** adv. [ME f. OF *joie* ult. f. L *gaudium* f. *gaudere* rejoice]

Joyce /dʒɔɪs/, James (Augustine Aloysius) (1882–1941), Irish writer. He left Ireland in 1904 and thereafter lived in Trieste, Zurich, and Paris, becoming one of the most important writers of the modernist movement. His first major publication was *Dubliners* (1914), a collection of short stories depicting his native Dublin, which was followed by the semi-autobiographical novel *A Portrait of the Artist as a Young Man* (1914–15). His novel *Ulysses* (published in Paris in 1922 but banned in the UK and the US until 1936) revolutionized the form and structure of the modern novel and influenced the development of the stream of consciousness technique. *Finnegans Wake* (1939) pushed linguistic experimentation to the extreme.

Joycean /ˈdʒɔɪsɪən/ adj. & n. ● adj. of or characteristic of James Joyce. ● n. a specialist in or admirer of Joyce's works.

joyful /ˈdʒɔɪfʊl/ adj. full of, showing, or causing joy. □ **joyfully** adv. **joyfulness** n.

joyous /ˈdʒɔɪəs/ adj. (of an occasion, circumstance, etc.) characterized by pleasure or joy; joyful. □ **joyously** adv. **joyousness** n.

joyride /ˈdʒɔɪraɪd/ n. & v. *colloq.* ● n. a ride for pleasure in a motor car, esp. the crime of doing this (often recklessly) in a stolen car. ● v.intr. (past **-rode** /-rəʊd/; past part. **-ridden** /-ˌrɪd(ə)n/) go for a joyride. □ **joyrider** n.

joystick /ˈdʒɔɪstɪk/ n. **1** *colloq.* the control column of an aircraft. **2** a lever that can be moved in several directions to control the movement of an image on a VDU screen.

JP *abbr.* (in the UK) Justice of the Peace.

Jr. *abbr.* Junior.

jt. *abbr.* joint.

Juan Carlos /hwɑːn ˈkɑːlɒs/ (b.1938) (full name Juan Carlos Victor María de Borbón y Borbón), grandson of Alfonso XIII, king of Spain since 1975. He was nominated by Franco as his successor and became king when Franco died. His reign has seen Spain's increasing liberalization and its entry into NATO and the European Community.

Juan Fernandez Islands /ˌhwɑːn fəˈnændez/ a group of three almost uninhabited islands in the Pacific Ocean 640 km (400 miles) west of Chile. (See also SELKIRK.)

Juárez /ˈhwɑːrez/, Benito Pablo (1806–72), Mexican statesman, President 1861–4 and 1867–72. His refusal to repay Mexico's foreign debts led to the occupation of Mexico by Napoleon III and the establishment of Maximilian as emperor of Mexico in 1864. The withdrawal of the occupying French forces in 1867 prompted the execution of Maximilian and the rehabilitation of Juárez in the same year.

Juba /ˈdʒuːbə/ the capital of the southern region of Sudan, on the White Nile; pop. (est. 1990) 100,000. Since 1983 it has been virtually isolated by the civil war in Sudan.

Jubba /ˈdʒʊbə, ˈdʒuːbə/ (also **Juba**) a river in East Africa, rising in the highlands of central Ethiopia and flowing southwards for about 1,600 km (1,000 miles) through Somalia to the Indian Ocean.

jube /dʒuːb/ n. Austral. & NZ = JUJUBE 2. [abbr.]

jubilant /ˈdʒuːbɪlənt/ adj. exultant, rejoicing, joyful. □ **jubilance** n. **jubilantly** adv. [L jubilare jubilant- shout for joy]

jubilate /ˈdʒuːbɪˌleɪt/ v.intr. exult; be joyful. □ **jubilation** /ˌdʒuːbɪˈleɪʃ(ə)n/ n. [L jubilare (as JUBILANT)]

jubilee /ˈdʒuːbɪˌliː/ n. **1** a time or season of rejoicing. **2** an anniversary of an event, esp. the twenty-fifth or fiftieth. **3** Jewish Hist. a year of emancipation and restoration, kept every fifty years. **4** RC Ch. a period of remission from the penal consequences of sin, granted under certain conditions for a year usu. at intervals of twenty-five years. **5** exultant joy. [ME f. OF jubilé f. LL jubilaeus (annus) (year) of jubilee ult. f. Heb. yōbēl, orig. = ram, ram's-horn trumpet]

Jud. abbr. Judith (Apocrypha).

Judaea /dʒuːˈdiːə/ the southern part of ancient Palestine, corresponding to the former kingdom of Judah. The Jews returned to the region in 537 BC after the Babylonian Captivity, and in 165 the Maccabees again established it as an independent kingdom. It became a province of the Roman Empire in 63 BC, and was subsequently amalgamated with Palestine. (See also JUDAH 3.) □ **Judaean** adj.

Judaeo- /dʒuːˈdiːəʊ, -ˈdeɪəʊ/ comb. form (US **Judeo-**) Jewish; Jewish and. [L judaeus Jewish]

Judah /ˈdʒuːdə/ (in the Bible) **1** a Hebrew patriarch, the fourth son of Jacob. **2** the tribe of Israel traditionally descended from him, the most powerful of the twelve tribes of Israel. **3** the southern part of ancient Palestine, occupied by the tribe of Judah. After the reign of Solomon (c.930 BC) it formed a separate kingdom, which outlasted that of the northern tribes (see ISRAEL¹ 2). Judah was overrun by the Babylonians in 587 BC. The Jews returned to the region, which became known as Judaea, in 537 after the Babylonian Captivity. (See also JUDAEA.)

Judaic /dʒuːˈdeɪɪk/ adj. of or characteristic of the Jews or Judaism. [L Judaicus f. Gk Ioudaïkos f. Ioudaios JEW]

Judaism /ˈdʒuːdeɪ.ɪz(ə)m/ n. the religion of the Jews, a monotheistic faith having a basis in Mosaic and rabbinical teachings. Judaism was founded, according to the Bible, on the covenant made between God and Abraham, which ordained the Jewish people's special relationship with God. The first five books of the Bible, the Torah (or Pentateuch), reveal God's laws; next in importance to these is the Talmud, comprising the Mishnah and the Gemara. Judaism is a religion of much ritual, centred on the synagogue, although the home is also of great importance. The central time for worship is the sabbath (sunset on Friday to sunset on Saturday), while among annual festivals the most important are Yom Kippur, Hanukkah, Rosh Hashana, and Passover. □ **Judaist** n. [ME f. LL Judaismus f. Gk Ioudaïsmos (as JUDAIC)]

Judaize /ˈdʒuːdeɪ.aɪz/ v. (also **-ise**) **1** intr. follow Jewish customs or rites. **2** tr. a make Jewish. **b** convert to Judaism. □ **Judaization** /ˌdʒuːdeɪaɪˈzeɪʃ(ə)n/ n. [LL judaizare f. Gk ioudaïzō (as JUDAIC)]

Judas¹ /ˈdʒuːdəs/ n. **1** a person who betrays a friend. **2** (**judas**) a peephole in a door. [JUDAS ISCARIOT]

Judas² see JUDE, ST.

Judas Iscariot /ɪˈskærɪət/ an Apostle. He betrayed Christ to the Jewish authorities in return for thirty pieces of silver; the Gospels leave his motives uncertain. Overcome with remorse, he later committed suicide.

Judas Maccabaeus /ˌmækəˈbiːəs/ (died c.161 BC), Jewish leader. He led a Jewish revolt in Judaea against the Seleucid king Antiochus IV Epiphanes from around 167, and succeeded in recovering Jerusalem, dedicating the Temple anew, and protecting Judaism from Hellenization. He also features in the Apocrypha as the hero of the Maccabees.

Judas-tree /ˈdʒuːdəsˌtriː/ n. a leguminous Mediterranean tree, Cercis siliquastrum, with purple flowers appearing before the leaves.

judder /ˈdʒʌdə(r)/ v. & n. esp. Brit. ● v.intr. **1** (esp. of a mechanism) vibrate noisily or violently. **2** (of a singer's voice) oscillate in intensity. ● n. an instance of juddering. [imit.: cf. SHUDDER]

Jude, St /dʒuːd/ **1** (known as Judas) an Apostle, supposed brother of James. Thaddaeus (mentioned in St Matthew's Gospel) is traditionally identified with him. According to tradition, he was martyred in Persia with St Simon. Feast day (with St Simon), 28 Oct. **2** the last epistle of the New Testament, ascribed to St Jude.

Judeo- US var. of JUDAEO-.

Judg. abbr. (in the Bible) Judges.

judge /dʒʌdʒ/ n. & v. ● n. **1** a public officer appointed to hear and try causes in a court of justice. **2** a person appointed to decide in a competition, contest, or dispute. **3 a** a person who decides a question. **b** a person regarded in terms of capacity to decide on the merits of a thing or question (am no judge of that; a good judge of art). **4** hist. a leader having temporary authority in ancient Israel in the period between Joshua and the Kings (c.13th–11th centuries BC). ● v. **1** tr. a try (a cause) in a court of justice. **b** pronounce sentence on (a person). **2** tr. form an opinion about; estimate, appraise. **3** tr. act as a judge of (a dispute or contest). **4** tr. (often foll. by to + infin. or that + clause) conclude, consider, or suppose. **5** intr. a form a judgement. **b** act as judge. □ **judgeship** n. [ME f. OF juge (n.), juger (v.) f. L judex judicis f. jus law + -dicus speaking]

Judge Advocate General n. an officer in supreme control of the courts martial in the armed forces.

judgement /ˈdʒʌdʒmənt/ n. (also **judgment**) **1** the critical faculty; discernment (an error of judgement). **2** good sense. **3** an opinion or estimate (in my judgement). **4** the sentence of a court of justice; a decision by a judge. **5** often joc. a misfortune viewed as a deserved recompense (it is a judgement on you for getting up late). **6** criticism. □ **against one's better judgement** contrary to what one really feels to be advisable. **judgement by default** see DEFAULT. **judgement-seat** a judge's seat; a tribunal. [ME f. OF jugement (as JUDGE)]

judgemental /dʒʌdʒˈment(ə)l/ adj. (also **judgmental**) **1** of or concerning or by way of judgement. **2** condemning, critical. □ **judgementally** adv.

Judgement Day (in some beliefs) the day on which the Last Judgement is expected to take place.

Judgement of Solomon (in the Bible) the arbitration of King Solomon of Israel over a baby claimed by two women (I Kings 3:16–28). He proposed cutting the baby in half, and then gave it to the woman who showed concern for its life.

Judges /ˈdʒʌdʒɪz/ the seventh book of the Bible, describing the gradual conquest of Canaan under various leaders (known as judges) in an account that is parallel to that of the Book of Joshua and is probably more accurate historically.

Judges' Rules (in the UK) a set of rules concerning the mode of questioning of suspects by police. Although the rules have no force in law, if they are not observed a court may decline to admit an accused person's statements in evidence.

judicature /ˈdʒuːdɪkətʃə(r), dʒuːˈdɪkə-/ n. **1** the administration of justice. **2** a judge's office or term of office. **3** judges collectively. **4** a court of justice. [med.L judicatura f. L judicare to judge]

judicial /dʒuːˈdɪʃ(ə)l/ adj. **1** of, done by, or proper to a court of law. **2** having the function of judgement (a judicial assembly). **3** of or proper to a judge. **4** expressing a judgement; critical. **5** impartial. **6** regarded as a divine judgement. □ **judicial factor** Sc. an official receiver. **judicial separation** the separation of man and wife by decision of a court. □ **judicially** adv. [ME f. L judicialis f. judicium judgement f. judex JUDGE]

judiciary /dʒuːˈdɪʃɪərɪ/ n. (pl. **-ies**) the judges of a state collectively. [L judiciarius (as JUDICIAL)]

judicious /dʒuːˈdɪʃəs/ adj. **1** sensible, prudent. **2** sound in discernment and judgement. □ **judiciously** adv. **judiciousness** n. [F judicieux f. L judicium (as JUDICIAL)]

Judith /ˈdʒuːdɪθ/ **1** (in the Apocrypha) a rich Israelite widow who saved the town of Bethulia from Nebuchadnezzar's army by captivating the besieging general Holofernes and cutting off his head while he slept. **2** a book of the Apocrypha recounting the story of Judith.

judo /ˈdʒuːdəʊ/ n. a sport of unarmed combat that developed from jujitsu primarily in Japan, founded by Dr Jigoro Kano (1860–1938). Its aim is to train the body and cultivate the mind through practice of the methods of attack and defence. Degrees of skill in judo are indicated by the colour of belt worn, from white for a novice to black for an expert. □ **judoist** n. [Jap. f. jū gentle + dō way]

Judy /ˈdʒuːdɪ/ n. (pl. **-ies**) **1** the wife of Punch (see PUNCH AND JUDY). **2** (also **judy**) sl. a woman. [pet-form of the name *Judith*]

jug /dʒʌg/ n. & v. • n. **1** Brit. **a** a deep vessel for holding liquids, with a handle and often with a spout or lip shaped for pouring. **b** the contents of this; a jugful. **2** US a large vessel, esp. for liquids, with a narrow mouth. **3** sl. prison. **4** (in pl.) N. Amer. coarse sl. a woman's breasts. • v.tr. (**jugged**, **jugging**) **1** (usu. as **jugged** adj.) stew or boil (a hare or rabbit) in a covered vessel. **2** sl. imprison. □ **jugful** n. (pl. **-fuls**). [perh. f. *Jug*, pet-form of the name *Joan* etc.]

Jugendstil /ˈjuːgəntˌʃtiːl/ n. the German name for ART NOUVEAU. [G f. *Jugend* youth + *Stil* style]

Juggernaut /ˈdʒʌgəˌnɔːt/ (also called *Jagannatha*) Hinduism the form of Krishna worshipped in Puri, Orissa (India), where in the annual festival his image is dragged through the streets on a heavy chariot; devotees are said formerly to have thrown themselves under its wheels. [Hindi *Jagannath* f. Skr. *Jagannātha* lord of the world]

juggernaut /ˈdʒʌgəˌnɔːt/ n. **1** esp. Brit. a large heavy motor vehicle, esp. an articulated lorry. **2** a huge or overwhelming force or object; an institution or notion to which persons blindly sacrifice themselves or others. [JUGGERNAUT]

juggins /ˈdʒʌgɪnz/ n. Brit. sl. a simpleton. [perh. f. proper name *Juggins* (as JUG): cf. MUGGINS]

juggle /ˈdʒʌg(ə)l/ v. & n. • v. **1 a** intr. (often foll. by *with*) perform feats of dexterity, esp. by tossing objects in the air and catching them, keeping several in the air at the same time. **b** tr. perform such feats with. **2** tr. continue to deal with (several activities) at once, esp. with ingenuity. **3** intr. (foll. by *with*) & tr. **a** deceive or cheat. **b** misrepresent (facts). **c** rearrange adroitly. • n. **1** a piece of juggling. **2** a fraud. [ME, back-form. f. JUGGLER or f. OF *jogler*, *jugler* f. L *joculari* jest f. *joculus* dimin. of *jocus* jest]

juggler /ˈdʒʌglə(r)/ n. **1 a** a person who juggles. **b** a conjuror. **2** a trickster or impostor. □ **jugglery** n. [ME f. OF *jouglere* *-eor* f. L *joculator* *-oris* (as JUGGLE)]

Jugoslav var. of YUGOSLAV.

jugular /ˈdʒʌgjʊlə(r)/ adj. & n. • adj. **1** of the neck or throat. **2** (of fish) having ventral fins in front of the pectoral fins. • n. = *jugular vein*. □ **jugular vein** any of several large veins in the neck which carry blood from the head. [LL *jugularis* f. L *jugulum* collar-bone, throat, dimin. of *jugum* YOKE]

jugulate /ˈdʒʌgjʊˌleɪt/ v.tr. **1** kill by cutting the throat. **2** arrest the course of (a disease etc.) by a powerful remedy. [L *jugulare* f. *jugulum* (as JUGULAR)]

Jugurtha /dʒəˈgɜːθə/ (d.104 BC), joint king of Numidia *c.*118–104. His attacks on his royal partners prompted intervention by Rome and led to the outbreak of the Jugurthine War (112–105). He was eventually captured by the Roman general Marius and executed in Rome. □ **Jugurthine** /-θaɪn/ adj.

juice /dʒuːs/ n. & v. • n. **1** the liquid part of vegetables or fruits. **2** the fluid part of an animal body or substance, esp. a secretion (*gastric juice*). **3** the essence or spirit of anything. **4** colloq. petrol or electricity as a source of power. **5** US sl. alcoholic liquor. • v.tr. extract the juice from (a fruit etc.). □ **juiceless** adj. [ME f. OF *jus* f. L *jus* broth, juice]

juicer /ˈdʒuːsə(r)/ n. US sl. an alcoholic.

juicy /ˈdʒuːsɪ/ adj. (**juicier**, **juiciest**) **1** full of juice; succulent. **2** colloq. substantial or interesting; racy, scandalous. **3** colloq. profitable. □ **juicily** adv. **juiciness** n.

ju-jitsu /dʒuːˈdʒɪtsuː/ n. (also **jiu-jitsu**, **ju-jutsu**) a Japanese method of self-defence using throws, punches, kicks, arm-locks, etc., and seeking to utilize the opponent's strength and weight to his or her disadvantage. It began to take on a systematized form in the latter half of the 16th century and many different schools developed, each distinguished by its individual features. It fell into disrepute for various reasons (including its ruthlessness), and the expansion of judo further reduced its popularity. [Jap. *jūjutsu* f. *jū* gentle + *jutsu* skill]

ju-ju /ˈdʒuːdʒuː/ n. **1** a charm or fetish of some West African peoples. **2** a supernatural power attributed to this. [perh. f. F *joujou* toy]

jujube /ˈdʒuːdʒuːb/ n. **1 a** a tree of the genus *Zizyphus*, bearing edible acidic berry-like fruits. **b** this fruit. **2** a lozenge of gelatin etc. flavoured with or imitating this. [F *jujube* or med.L *jujuba* ult. f. Gk *zizuphon*]

jukebox /ˈdʒuːkbɒks/ n. a machine that automatically plays a selected musical recording when a coin is inserted. [Gullah *juke* disorderly + BOX[1]]

Jul. abbr. July.

julep /ˈdʒuːlɪp/ n. **1 a** a sweet drink, esp. as a vehicle for medicine. **b** a medicated drink as a mild stimulant etc. **2** US iced and flavoured spirits and water (*mint julep*). [ME f. OF f. Arab. *julāb* f. Pers. *gulāb* f. *gul* rose + *āb* water]

Julian[1] /ˈdʒuːlɪən/ adj. of or associated with Julius Caesar. [L *Julianus* f. *Julius*]

Julian[2] /ˈdʒuːlɪən/ (known as the Apostate) (full name Flavius Claudius Julianus) (AD *c.*331–63), Roman emperor 360–3, nephew of Constantine. He restored paganism as the state cult in place of Christianity, but this move was reversed after his death on campaign against the Persians.

Julian Alps an Alpine range in western Slovenia and NE Italy, rising to a height of 2,863 m (9,395 ft) at Triglav.

Julian calendar n. a calendar introduced under the authority of Julius Caesar in 46 BC and used in most of Christian Europe until its replacement by the Gregorian calendar from 1582, although retained in Russia until 1918. The Julian calendar instituted the modern system in which the ordinary year has 365 days and every fourth year is a leap year of 366 days. After some initial problems, an adjusted version of the calendar was instituted in 8 AD, in which the months had their modern lengths. Dates reckoned by the Julian calendar are sometimes designated 'Old Style'. (See also GREGORIAN CALENDAR.)

Julian of Norwich (*c.*1342–*c.*1413), English mystic. Her name probably derives from St Julian's Church, Norwich, outside which she is said to have lived as a religious recluse. She is chiefly associated with the *Revelations of Divine Love* (*c.*1393), which describe a series of visions she had at the end of a serious illness in 1373. In her account, she affirms the love of God and depicts the Holy Trinity as Father, Mother, and Lord.

julienne /ˌdʒuːlɪˈen/ n. & adj. • n. foodstuff, esp. vegetables, cut into short thin strips. • adj. cut into thin strips. [F f. the name *Jules* or *Julien*]

Juliet cap /ˈdʒuːlɪət/ n. a small network ornamental cap worn by brides etc. [the heroine of Shakespeare's *Romeo & Juliet*]

Julius Caesar /ˌdʒuːlɪəs ˈsiːzə(r)/, Gaius (100–44 BC), Roman general and statesman. He established the First Triumvirate with Pompey and Crassus (60), and became consul in 59, obtaining command of the provinces of Illyricum, Cisalpine Gaul, and Transalpine Gaul. Between 58 and 51 he fought the Gallic Wars, subjugating Transalpine Gaul, invading Britain (55–54), and acquiring immense power. Resentment at this on the part of Pompey and other powerful Romans led to civil war, which resulted in Pompey's defeat at Pharsalus (48). Caesar was made dictator of the Roman Empire and initiated a series of reforms, including the introduction of the Julian calendar. Hostility to Caesar's autocracy culminated in his murder on the Ides (15th) of March in a conspiracy led by Brutus and Cassius.

Jullundur /ˈdʒʌləndə(r)/ (also **Jalandhar**) a city in Punjab, NW India; pop. (1991) 520,000. It has long been an important communications centre and was capital of Punjab 1947–54.

July /dʒuːˈlaɪ/ n. (pl. **Julys**) the seventh month of the year. [ME f. AF *julie* f. L *Julius* (*mensis* month), named after Julius Caesar]

jumble /ˈdʒʌmb(ə)l/ v. & n. • v. **1** tr. (often foll. by *up*) confuse; mix up. **2** intr. move about in disorder. • n. **1** a confused state or heap; a muddle. **2** Brit. articles collected for a jumble sale. □ **jumble sale** Brit. a sale of miscellaneous usu. second-hand articles, esp. for charity. □ **jumbly** adj. [prob. imit.]

jumbo /ˈdʒʌmbəʊ/ n. & adj. colloq. • n. (pl. **-os**) **1** a large animal (esp. an elephant), person, or thing. **2** (in full **jumbo jet**) a large jet airliner with capacity for several hundred passengers; specifically, the US Boeing 747, introduced in 1970. • adj. **1** very large of its kind. **2** extra large (*jumbo packet*). [19th c. (orig. of a person): orig. unkn.: popularized as the name of a zoo elephant sold in 1882]

jumbuck /ˈdʒʌmbʌk/ n. Austral. colloq. a sheep. [19th-c. Austral. pidgin: origin unkn.]

Jumna /ˈdʒʌmnə/ (Hindi **Yamuna** /ˈjʌmʊnə/) a river of northern India, which rises in the Himalayas and flows over 1,370 km (850 miles) in a large arc southwards and south-eastwards, through Delhi, joining the Ganges below Allahabad. Its source (Yamunotri) and its confluence with the Ganges are both Hindu holy places.

jump /dʒʌmp/ v. & n. • v. **1** intr. move off the ground or other surface (usu. upward, at least initially) by sudden muscular effort in the legs. **2** intr. (often foll. by *up*, *from*, *in*, *out*, etc.) move suddenly or hastily in a specified way (*we jumped into the car*). **3** intr. give a sudden bodily movement from shock or excitement etc. **4** intr. undergo a rapid change, esp. an advance in status. **5** intr. (often foll. by *about*) change or move rapidly from one idea or subject to another. **6 a** intr. rise or increase suddenly (*prices jumped*). **b** *tr.* cause to do this. **7** *tr.* **a** pass over (an obstacle, barrier, etc.) by jumping. **b** move or pass over (an intervening thing) to a point beyond. **8** *tr.* skip or pass over (a passage in a book etc.). **9** *tr.* cause (a thing, or an animal, esp. a horse) to jump. **10** *intr.* (foll. by *to*, *at*) reach a conclusion hastily. **11** *tr.* (of a train) leave (the rails) owing to a fault or error. **12** *tr.* ignore and pass (a red traffic light etc.). **13** *tr.* get on or off (a train etc.) quickly, esp. illegally or dangerously. **14** *tr.* pounce on or attack (a person) unexpectedly. **15** *tr.* take summary possession of (a piece of land) after alleged abandonment or forfeiture by the former occupant. ● *n.* **1** the act or an instance of jumping. **2 a** a sudden bodily movement caused by shock or excitement. **b** (**the jumps**) *colloq.* extreme nervousness or anxiety. **3** an abrupt rise in amount, price, value, status, etc. **4** an obstacle to be jumped, esp. by a horse. **5 a** a sudden transition. **b** a gap in a series, logical sequence, etc. □ **get** (or **have**) **the jump on** *colloq.* get (or have) an advantage over (a person) by prompt action. **jump at** accept eagerly. **jump bail** see BAIL¹. **jump down a person's throat** *colloq.* reprimand or contradict a person fiercely. **jumped-up** *colloq.* upstart; presumptuously arrogant. **jump the gun** see GUN. **jumping-off place** (or **point**) the place or point of starting. **jump-jet** a jet aircraft that can take off and land vertically. **jump-lead** *Brit.* each of a pair of cables for conveying current from the battery of a motor vehicle to boost (or recharge) another. **jump-off** a deciding round in a showjumping competition. **jump on** *colloq.* attack or criticize severely and without warning. **jump out of one's skin** *colloq.* be extremely startled. **jump the queue 1** push forward out of one's turn. **2** take unfair precedence over others. **jump-rope** *N. Amer.* a skipping-rope. **jump seat** *US* a folding extra seat in a motor vehicle. **jump ship** (of a seaman) desert. **jump-start** *v.tr.* start (a motor vehicle) by pushing it or with jump-leads. ● *n.* the action of jump-starting. **jump suit** a one-piece garment for the whole body, of a kind orig. worn by paratroopers. **jump to it** *colloq.* act promptly and energetically. **one jump ahead** one stage further on than a rival etc. **on the jump** *colloq.* on the move; in a hurry. □ **jumpable** *adj.* [16th c.: prob. imit.]

jumper¹ /ˈdʒʌmpə(r)/ *n.* **1** a knitted pullover. **2** a loose outer jacket of canvas etc. worn by sailors. **3** *N. Amer.* a pinafore dress. [prob. f. (17th-c., now dial.) *jump* short coat perh. f. F *jupe* f. Arab. *jubba*]

jumper² /ˈdʒʌmpə(r)/ *n.* **1** a person or animal that jumps. **2** *Electr.* a short wire used to shorten a circuit or close it temporarily. **3** *Naut.* a rope made fast to keep a yard, mast, etc., from jumping. **4** a heavy chisel-ended iron bar for drilling blast-holes.

jumping bean *n.* the seed of a Mexican plant that jumps with the movement of the moth larva inside.

jumping jack *n.* **1** a small firework producing repeated explosions. **2** a toy figure of a man, with movable limbs.

jumpy /ˈdʒʌmpɪ/ *adj.* **1** nervous; easily startled. **2** making sudden movements, esp. of nervous excitement. □ **jumpily** *adv.* **jumpiness** *n.*

Jun. *abbr.* **1** June. **2** Junior.

jun /dʒʌn/ *n.* a monetary unit of North Korea, equal to one-hundredth of a won. [Korean]

junco /ˈdʒʌŋkəʊ/ *n.* (*pl.* **-os** or **-oes**) a small American bunting of the genus *Junco*. [Sp. f. L *juncus* rush plant]

junction /ˈdʒʌŋkʃ(ə)n/ *n.* **1** a point at which two or more things are joined. **2** a place where two or more railway lines or roads meet, unite, or cross. **3** the act or an instance of joining. **4** *Electronics* a region of transition in a semiconductor between regions where conduction is mainly by electrons and regions where it is mainly by holes. □ **junction box** a box containing a junction of electric cables etc. [L *junctio* (as JOIN)]

juncture /ˈdʒʌŋktʃə(r)/ *n.* **1** a critical convergence of events; a critical point of time (*at this juncture*). **2** a place where things join. **3** an act of joining. [ME f. L *junctura* (as JOIN)]

June /dʒuːn/ *n.* the sixth month of the year. □ **June bug** a beetle, esp. a North American chafer, appearing in June. [ME f. OF *juin* f. L *Junius* var. of *Junonius* sacred to Juno]

Juneau /ˈdʒuːnəʊ/ the state capital of Alaska, a seaport on an inlet of the Pacific Ocean in the south of the state; pop. (1990) 26,750. It developed following the discovery of gold there by Joseph Juneau in 1880.

June War see SIX DAY WAR.

Jung /jʊŋ/, Carl (Gustav) (1875–1961), Swiss psychologist. He collaborated with Sigmund Freud in the development of the psychoanalytic theory of personality, though he later divorced himself from Freud's viewpoint because of its preoccupation with sexuality as the determinant of personality. Jung originated the concept of introvert and extrovert personality, and of the four psychological functions of sensation, intuition, thinking, and feeling. In his major work, *The Psychology of the Unconscious* (1912), he proposed the existence of a collective unconscious, which he combined with a theory of archetypes for studying the history and psychology of religion.

Jungfrau /ˈjʊŋfraʊ/ a mountain in the Swiss Alps, 4,158 m (13,642 ft) high.

Jungian /ˈjʊŋɪən/ *adj. & n.* ● *adj.* of Carl Jung or his system of analytical psychology. ● *n.* a supporter of Jung or of his system.

jungle /ˈdʒʌŋg(ə)l/ *n.* **1 a** land overgrown with underwood or tangled vegetation, esp. in the tropics. **b** an area of such land. **2** a wild tangled mass. **3** a place of bewildering complexity or confusion, or of a struggle for survival (*blackboard jungle*). □ **jungle fever** a severe form of malaria. **law of the jungle** a state of ruthless competition. □ **jungled** *adj.* **jungly** *adj.* [Hindi *jangal* f. Skr. *jangala* desert, forest]

junior /ˈdʒuːnɪə(r)/ *adj. & n.* ● *adj.* **1** less advanced in age. **2** (foll. by *to*) inferior in age, standing, or position. **3** the younger (esp. appended to a name for distinction from an older person of the same name). **4** of less or least standing; of the lower or lowest position (*junior partner*). **5** *Brit.* (of a school) having pupils in a younger age-range, usu. 7–11. **6** *N. Amer.* of the year before the final year at university, high school, etc. ● *n.* **1** a junior person. **2** one's inferior in length of service etc. **3** a junior student. **4** a barrister who is not a QC. **5** *US colloq.* a young male child, esp. in relation to his family. □ **junior college** *US* a college offering a two-year course esp. in preparation for completion at senior college. **junior common** (or **combination**) **room** *Brit.* **1** a room for social use by the junior members of a college. **2** the junior members collectively. **junior lightweight** see LIGHTWEIGHT. **junior middleweight** see MIDDLEWEIGHT. □ **juniority** /ˌdʒuːnɪˈɒrɪtɪ/ *n.* [L, compar. of *juvenis* young]

juniper /ˈdʒuːnɪpə(r)/ *n.* an evergreen coniferous shrub or tree of the genus *Juniperus*; esp. *J. communis* of Europe, with prickly leaves and dark purple berries. □ **oil of juniper** oil from juniper berries, used in medicine and in flavouring gin etc. [ME f. L *juniperus*]

junk¹ /dʒʌŋk/ *n. & v.* ● *n.* **1** discarded articles; rubbish. **2** anything regarded as of little value. **3** *sl.* a narcotic drug, esp. heroin. **4** old cables or ropes cut up for oakum etc. **5** *Brit.* a lump or chunk. **6** *Naut.* hard salt meat. **7** a lump of fibrous tissue in a sperm whale's head, containing spermaceti. ● *v.tr.* discard as junk. □ **junk food** food with low nutritional value. **junk mail** unsolicited advertising matter sent by post. **junk shop** a shop selling cheap second-hand goods or antiques. [ME: orig. unkn.]

junk² /dʒʌŋk/ *n.* a flat-bottomed sailing vessel used in the China Sea, with a prominent stem and lugsails. [obs. F *juncque*, Port. *junco*, or Du. *jonk*, f. Javanese *djong*]

junk bond *n.* *US Commerce* a bond bearing high interest but judged to be a very risky investment, issued by a company seeking to raise a large amount of capital quickly, e.g. in order to finance a takeover. The term is applied when there is doubt about the company's ability to pay the interest from income generated by the assets purchased. The practice of issuing such bonds dates from the mid-1970s.

junker /ˈjʊŋkə(r)/ *n. hist.* **1** a young German nobleman. **2** a member of an exclusive Prussian aristocratic party. □ **junkerdom** *n.* [G, earlier *Junkher* f. OHG (as YOUNG, HERR)]

junket /ˈdʒʌŋkɪt/ *n. & v.* ● *n.* **1** a dish of sweetened and flavoured curds, often served with fruit or cream. **2** a feast. **3** a pleasure outing. **4** *N. Amer.* an official's tour at public expense. ● *v.intr.* (**junketed**, **junketing**) feast, picnic. □ **junketing** *n.* [ME *jonket* f. OF *jonquette* rush-basket (used to carry junket) f. *jonc* rush f. L *juncus*]

junkie /ˈdʒʌŋkɪ/ *n. sl.* a drug addict.

Juno /ˈdʒuːnəʊ/ **1** *Rom. Mythol.* the most important goddess of the Roman state, wife of Jupiter, identified with Hera. She was originally an ancient Italian goddess. **2** *Astron.* asteroid 3, discovered in 1804 (diameter 244 km).

Junoesque /ˌdʒuːnəʊˈesk/ *adj.* resembling the goddess Juno in stately beauty.

Junr. *abbr.* Junior.

junta /ˈdʒʌntə, ˈhʊntə/ *n.* **1 a** a political or military clique or faction taking power after a revolution or *coup d'état*. **b** a secretive group; a cabal. **2** a deliberative or administrative council in Spain or Portugal. [Sp. & Port. f. L *juncta*, fem. past part. (as JOIN)]

Jupiter /ˈdʒuːpɪtə(r)/ **1** *Rom. Mythol.* the chief god of the Roman state, giver of victory, identified with Zeus. He was originally a sky-god, associated with lightning and the thunderbolt; his wife was Juno. Also called *Jove*. **2** *Astron.* the fifth planet from the sun in the solar system, orbiting between Mars and Saturn at an average distance of 778 million km from the sun. It is the largest planet in the solar system, with an equatorial diameter of 142,800 km, and is the closest of the gas giants. It has a massive atmosphere, consisting mainly of hydrogen with swirling clouds of ammonia and methane, with a circulation system that results in a number of distinct latitudinal bands. The planet's famous Great Red Spot is a cyclonic weather system in the southern hemisphere extending over 10,000 km, and has persisted at least since the beginning of telescopic observations. There at least sixteen satellites, four of which (the Galilean moons) are visible through binoculars, and a faint ring system. A familiar planet to the ancients, Jupiter appears as one of the brightest objects in the night sky. [OL *Jovis pater* father of the bright heaven]

Jura[1] /ˈʒʊərə/ a system of mountain ranges on the border of France and Switzerland. It has given its name to the Jurassic period, when most of its rocks were laid down.

Jura[2] /ˈdʒʊərə/ an island of the Inner Hebrides, separated from the west coast of Scotland by the Sound of Jura.

jural /ˈdʒʊərəl/ *adj.* **1** of law. **2** of rights and obligations. [L *jus juris* law, right]

Jurassic /dʒʊəˈræsɪk/ *adj. & n. Geol.* ● *adj.* of or relating to the second period of the Mesozoic era. (*See note below.*) ● *n.* this era or the corresponding geological system. [F *jurassique* f. JURA[1]: cf. *Triassic*]

- The Jurassic lasted from about 213 to 144 million years ago, between the Triassic and Cretaceous periods. Dinosaurs and other reptiles attained their maximum size and were found on land, in the sea, and in the air. The first birds appeared towards the end of the period.

jurat[1] /ˈdʒʊəræt/ *n. Brit.* **1** a municipal officer (esp. of the Cinque Ports) holding a position similar to that of an alderman. **2** an honorary judge or magistrate in the Channel Islands. [ME f. med.L *juratus* past part. of L *jurare* swear]

jurat[2] /ˈdʒʊəræt/ *n.* a statement of the circumstances in which an affidavit was made. [L *juratum* neut. past part. (as JURAT[1])]

juridical /dʒʊəˈrɪdɪk(ə)l/ *adj.* **1** of judicial proceedings. **2** relating to the law. □ **juridically** *adv.* [L *juridicus* f. *jus juris* law + *-dicus* saying f. *dicere* say]

jurisconsult /ˈdʒʊərɪskən‚sʌlt, ‚dʒʊərɪsˈkɒnsʌlt/ *n.* a person learned in law; a jurist. [L *jurisconsultus* f. *jus juris* law + *consultus* skilled: see CONSULT]

jurisdiction /‚dʒʊərɪsˈdɪkʃ(ə)n/ *n.* **1** (often foll. by *over, of*) the administration of justice. **2 a** legal or other authority. **b** the extent of this; the territory it extends over. □ **jurisdictional** *adj.* [ME *jurisdiccioun* f. OF *jurediction, juridiction*, L *jurisdictio* f. *jus juris* law + *dictio* DICTION]

jurisprudence /‚dʒʊərɪsˈpruːd(ə)ns/ *n.* **1** the science or philosophy of law. **2** skill in law. □ **jurisprudent** *adj. & n.* **jurisprudential** /-pruːˈdenʃ(ə)l/ *adj.* [LL *jurisprudentia* f. L *jus juris* law + *prudentia* knowledge: see PRUDENT]

jurist /ˈdʒʊərɪst/ *n.* **1** an expert in law. **2** a legal writer. **3** *US* a lawyer. □ **juristic** /dʒʊəˈrɪstɪk/ *adj.* **juristical** *adj.* [F *juriste* or med.L *jurista* f. *jus juris* law]

juror /ˈdʒʊərə(r)/ *n.* **1** a member of a jury. **2** a person who takes an oath (cf. NONJUROR). [ME f. AF *jurour*, OF *jureor* f. L *jurator -oris* f. *jurare* jurat- swear]

jury /ˈdʒʊərɪ/ *n.* (*pl.* **-ies**) **1** a body of usu. twelve persons sworn to render a verdict on the basis of evidence submitted to them in a court of justice. **2** a body of persons selected to award prizes in a competition. □ **jury-box** the enclosure for the jury in a lawcourt. **the jury is out** a decision has not yet been reached. [ME f. AF & OF *juree* oath, inquiry, f. *jurata* fem. past part. of L *jurare* swear]

juryman /ˈdʒʊərɪmən/ *n.* (*pl.* **-men**) a member of a jury.

jury-rigged /ˈdʒʊərɪ‚rɪgd/ *adj. Naut.* having temporary makeshift rigging. [perh. ult. f. OF *ajurie* aid]

jurywoman /ˈdʒʊərɪ‚wʊmən/ *n.* (*pl.* **-women**) a female member of a jury.

Jussieu /ʒuːˈsjɜː/, Antoine Laurent de (1748–1836), French botanist. Jussieu came from a family of botanists whose home was a centre for plant collection and research. From extensive observation he grouped plants into families on the basis of common essential properties and, in *Genera Plantarum* (1789), developed the system on which modern plant classification is based.

jussive /ˈdʒʌsɪv/ *adj. Gram.* expressing a command. [L *jubere juss-* command]

just *adj. & adv.* ● *adj.* /dʒʌst/ **1** acting or done in accordance with what is morally right or fair. **2** (of treatment etc.) deserved (*a just reward*). **3** (of feelings, opinions, etc.) well grounded (*just resentment*). **4** right in amount etc.; proper. ● *adv.* /dʒʌst, dʒəst/ **1** exactly (*just what I need*). **2** exactly or at this or that moment; a little time ago (*I have just seen them*). **3** *colloq.* simply, merely (*we were just good friends; it just doesn't make sense*). **4** barely; no more than (*I just managed it; just a minute*). **5** *colloq.* positively (*it is just splendid*). **6** quite (*not just yet; it is just as well that I checked*). **7** *colloq.* really, indeed (*won't I just tell him!*). **8** in questions, seeking precise information (*just how did you manage?*). □ **just about** *colloq.* almost exactly; almost completely. **just in case 1** lest. **2** as a precaution. **just-in-time** (often *attrib.*) a factory system whereby materials are delivered immediately before they are required, so minimizing storage costs. **just now 1** at this moment. **2** a little time ago. **just so 1** exactly arranged (*they like everything just so*). **2** it is exactly as you say. □ **justly** /ˈdʒʌstlɪ/ *adv.* **justness** *n.* [ME f. OF *juste* f. L *justus* f. *jus* right]

justice /ˈdʒʌstɪs/ *n.* **1** just conduct. **2** fairness. **3** the exercise of authority in the maintenance of right. **4** judicial proceedings (*was duly brought to justice; the Court of Justice*). **5 a** a magistrate. **b** a judge, esp. (in England) of the Supreme Court of Judicature. □ **do justice to** treat fairly or appropriately; show due appreciation of. **do oneself justice** perform in a manner worthy of one's abilities. **in justice to** out of fairness to. **Mr** (or **Mrs**) **Justice** *Brit.* a form of address or reference to a Supreme Court Judge. **with justice** reasonably. □ **justiceship** *n.* (in sense 5). [ME f. OF f. L *justitia* (as JUST)]

Justice of the Peace *n.* an unpaid lay magistrate appointed to preserve the peace in a county, town, etc., hear minor cases, grant licences, etc.

justiciable /dʒʌˈstɪʃɪəb(ə)l/ *adj.* liable to legal consideration. [OF f. *justicier* bring to trial f. med.L *justitiare* (as JUSTICE)]

justiciary /dʒʌˈstɪʃɪərɪ/ *n. & adj.* ● *n.* (*pl.* **-ies**) an administrator of justice. ● *adj.* of the administration of justice. [med.L *justitiarius* f. L *justitia*: see JUSTICE]

justifiable /ˈdʒʌstɪ‚faɪəb(ə)l/ *adj.* that can be justified or defended. □ **justifiable homicide** killing regarded as lawful and without criminal guilt, esp. the execution of a death sentence. □ **justifiably** *adv.* **justifiableness** *n.* **justifiability** /‚dʒʌstɪ‚faɪəˈbɪlɪtɪ/ *n.* [F f. *justifier*: see JUSTIFY]

justification /‚dʒʌstɪfɪˈkeɪʃ(ə)n/ *n.* **1 a** the action of justifying. **b** that which justifies; a defence; a good reason or cause. **2** (in Christian theology) the action whereby humankind is justified or made righteous by God; the fact or condition of being so justified (*justification by faith*).

justify /ˈdʒʌstɪ‚faɪ/ *v.tr.* (**-ies, -ied**) **1** show the justice or rightness of (a person, act, etc.). **2** demonstrate the correctness of (an assertion etc.). **3** adduce adequate grounds for (conduct, a claim, etc.). **4 a** (esp. in *passive*) (of circumstances) be such as to justify. **b** vindicate. **5** (as **justified** *adj.*) just, right (*am justified in assuming*). **6** (in Christian theology) declare (a person) righteous. **7** *Printing* adjust (a line of type) to fill a space evenly. □ **justifier** *n.* **justificatory** /-fɪ‚keɪtərɪ/ *adj.* [ME f. F *justifier* f. LL *justificare* do justice to f. L *justus* JUST]

Justin, St /ˈdʒʌstɪn/ (known as St Justin the Martyr) (*c.*100–165), Christian philosopher. Born in Samaria, he became a Christian convert at Ephesus (*c.*130). He is remembered for his *Apologia* (*c.*150) defending Christianity. Tradition holds that he was martyred in Rome together with some of his followers. Feast day, 1 June.

Justinian /dʒʌˈstɪnɪən/ (Latin name Flavius Petrus Sabbatius Justinianus) (483–565), Byzantine emperor 527–65. He set out to recover the lost provinces of the Western Empire, and through his general Belisarius (*c.*505–65) succeeded in reclaiming North Africa from the Vandals, Italy from the Ostrogoths, and Spain from the Visigoths. Justinian's codification of Roman law in 529 had a significant impact on the course of law in European countries. He carried out an active building programme throughout the Empire and commissioned the construction of St Sophia at Constantinople in 532.

jut /dʒʌt/ *v. & n.* ● *v.intr.* (**jutted, jutting**) (often foll. by *out, into, through*, etc.) protrude, project. ● *n.* a projection; a protruding point. [var. of JET[1]]

Jute /dʒuːt/ *n.* a member of a Low German tribe that invaded southern England (according to legend under Horsa and Hengist) in the 5th century and set up a kingdom in Kent. □ **Jutish** *adj.* [repr. med.L *Jutae, Juti*, in OE *Eotas, Iotas* = Icel. *Iótar* people of Jutland in Denmark]

jute /dʒuːt/ *n.* **1** a rough fibre made from the bark of a jute plant, used for making twine and rope and woven into sacking, mats, etc. **2** an Asian plant of the genus *Corchorus* yielding this fibre. [Bengali *jhōṭo* f. Skr. *jūṭa* = *jaṭā* braid of hair]

Jutland /'dʒʌtlənd/ (Danish **Jylland** /'jylan/) a peninsula of NW Europe, forming the mainland of Denmark together with the north German state of Schleswig-Holstein.

Jutland, Battle of a major naval battle in the First World War, fought between the British Grand Fleet under Admiral Jellicoe and the German High Seas Fleet under Admiral Scheer in the North Sea west of Jutland on 31 May 1916. Although the battle itself was indecisive, the German fleet never again sought a full-scale engagement, and Allied control of the North Sea remained unshaken.

Juvenal /'dʒu:vɪn(ə)l/ (Latin name Decimus Junius Juvenalis) (*c.*60–*c.*140), Roman satirist. His sixteen verse satires present a savage attack on the vice and folly of Roman society, chiefly in the reign of the emperor Domitian. They deal variously with the hardship of poverty, the profligacy of the rich, and the futility of ambition.

juvenescence /ˌdʒu:vɪ'nes(ə)ns/ *n.* **1** youth. **2** the transition from infancy to youth. □ **juvenescent** *adj.* [L *juvenescere* reach the age of youth f. *juvenis* young]

juvenile /'dʒu:vɪˌnaɪl/ *adj. & n.* ● *adj.* **1 a** young, youthful. **b** of or for young persons. **2** suited to or characteristic of youth. **3** often *derog.* immature (*behaving in a very juvenile way*). ● *n.* **1** a young person. **2** *Commerce* a book intended for young people. **3** an actor playing the part of a youthful person. □ **juvenile court** a court for the trial of children under 17. **juvenile delinquency** habitual committing of offences or perpetration of antisocial behaviour by young persons. **juvenile delinquent** a young person behaving in such a way. □ **juvenilely** *adv.* **juvenility** /ˌdʒu:vɪ'nɪlɪtɪ/ *n.* [L *juvenilis* f. *juvenis* young]

juvenilia /ˌdʒu:vɪ'nɪlɪə/ *n.pl.* works produced by an author or artist in youth. [L, neut. pl. of *juvenilis* (as JUVENILE)]

juxtapose /ˌdʒʌkstə'pəʊz/ *v.tr.* **1** place (things) side by side. **2** (foll. by *to, with*) place (a thing) beside another. □ **juxtaposition** /-pə'zɪʃ(ə)n/ *n.* **juxtapositional** *adj.* [F *juxtaposer* f. L *juxta* next: see POSE¹]

Jylland see JUTLAND.

Jyväskylä /'ju:væsˌkʊlə/ a city in central Finland; pop. (1990) 66,530.

Kk

K[1] /keɪ/ *n.* (also **k**) (*pl.* **Ks** or **K's**) the eleventh letter of the alphabet.

K[2] *abbr.* (also **K.**) **1** King, King's. **2** Köchel (catalogue of Mozart's works). **3** (also **k**) (prec. by a numeral) **a** *Computing* a unit of 1,024 (i.e. 2^{10}) bytes or bits, or loosely 1,000. **b** thousand. [sense 3 as abbr. of KILO-]

K[3] *symb.* **1** *Chem.* the element potassium. **2** kelvin(s). [sense 1 f. L *kalium*]

k[1] *abbr.* knot(s).

k[2] *symb.* **1** kilo-. **2** *Math.* a constant.

K2 /keɪˈtuː/ (also called *Dapsang*) the highest mountain in the Karakoram range, on the border between Pakistan and China. It is the second highest peak in the world, rising to 8,611 m (28,250 ft). It was discovered in 1856 and named K2 because it was the second peak to be surveyed in the Karakoram range. It is also known as Mount Godwin-Austen after Col. H. H. Godwin-Austen, who first surveyed it.

ka /kɑː/ *n.* in ancient Egypt, the supposed spiritual part of an individual human being or god, which survived (with the soul) after death and could reside in a statue of the dead person.

Kaaba /ˈkɑːəbə/ (also **Caaba**) a building in the centre of the Great Mosque at Mecca, the Muslim Holy of Holies, containing a sacred black stone, in the direction of which Muslims must face when praying. The shrine is square-shaped and made of pre-Islamic granite and marble, and is said to have been constructed by Abraham on divine orders; it is regarded by Muslims as the 'navel' of the earth. The sacred stone is made of basalt, and is lodged in the eastern corner of the shrine. It is said that on Judgement Day the stone will speak as witness to the sins of humanity. [Arab. *Ka'ba*]

kabaddi /kəˈbɑːdɪ/ *n.* a game popular in northern India and Pakistan, played between two teams of nine players. It is a traditional team pursuit game, requiring members of each team to run and hold their breath in turn while repeating the word *kabaddi*. [20th c.: orig. uncert.]

Kabalega Falls /ˌkæbəˈleɪgə/ a waterfall on the lower Victoria Nile near Lake Albert, in NW Uganda. It is a central feature of the Kabalega National Park (established 1952) and is formed at a point where the river narrows to 6 m (20 ft) and drops 120 m (400 ft). It was formerly known as Murchison Falls.

Kabardino-Balkaria /ˌkæbəˈdiːnəʊ bælˈkɑːrɪə/ (also called **Kabarda-Balkar Republic** /ˌkæbədə bælˈkɑː(r)/) an autonomous republic of SW Russia, on the border with Georgia; pop. (1990) 768,000; capital, Nalchik.

kabbala var. of CABBALA.

kabuki /kəˈbuːkɪ/ *n.* a form of traditional Japanese drama with highly stylized song, mime, and dance. It originated in narrative dances performed by women in the early 17th century, but by the end of that century it was a seasonal commercial entertainment performed only by men, who played both male and female roles. Kabuki actors use their whole bodies to express complex emotions through stylized and exaggerated techniques. The plays are divided mainly into historical plays, domestic dramas, and dance pieces, a programme consisting of scenes or acts from several different plays. The kabuki theatre still flourishes. [Jap. f. *ka* song + *bu* dance + *ki* art]

Kabul /ˈkɑːbʊl/ the capital of Afghanistan; pop. (est. 1984) 1,179,300. It is situated in the north-east of the country, with a strategic position commanding the mountain passes through the Hindu Kush, especially the Khyber Pass. It has existed for more than 3,000 years and has been destroyed and rebuilt several times in its history. It was capital of the Mogul empire 1504-1738 and in 1773 replaced Kandahar as capital of an independent Afghanistan. It suffered severe damage in the conflict that followed the Soviet invasion of Afghanistan in 1979.

Kabwe /ˈkæbweɪ/ a town in central Zambia, situated to the north of Lusaka; pop. (1987) 190,750. One of the oldest mining towns in Zambia, it is the site of a cave which has yielded human fossils associated with the Upper Pleistocene period. It was known as Broken Hill from 1904 to 1965.

kachina /kəˈtʃiːnə/ *n.* **1** an American Indian ancestral spirit. **2** (in full **kachina dancer**) a person who represents a kachina in ceremonial dances. □ **kachina doll** a wooden doll representing a kachina. [Hopi, = supernatural]

Kádár /ˈkɑːdɑː(r)/, János (1912-89), Hungarian statesman, First Secretary of the Hungarian Socialist Workers' Party 1956-88 and Prime Minister 1956-8 and 1961-5. He replaced Imre Nagy as Premier after crushing the Hungarian uprising of 1956. Kádár consistently supported the Soviet Union, involving Hungarian troops in the 1968 invasion of Czechoslovakia, while retaining a degree of decentralization for the economy. His policy of 'consumer socialism' made Hungary the most affluent state in eastern Europe. He was removed as First Secretary following his resistance to the political reforms of the 1980s.

Kaddish /ˈkædɪʃ/ *n. Judaism* **1** a Jewish mourner's prayer. **2** a doxology in the synagogue service. [Aram. *ḳaddîš* holy]

kadi var. of CADI.

Kadiköy see CHALCEDON.

Kaffir /ˈkæfə(r)/ *n.* **1 a** *hist.* a member of the Xhosa-speaking peoples of South Africa. **b** the language of these peoples. **2** *S. Afr. offens.* any black African (now an actionable insult). [Arab. *kāfir* infidel f. *kafara* not believe]

kaffiyeh var. of KEFFIYEH.

Kafir /ˈkæfə(r)/ *n.* a native of the Hindu Kush mountains of NE Afghanistan. [formed as KAFFIR]

Kafka /ˈkæfkə/, Franz (1883-1924), Czech novelist, who wrote in German. A sense of guilt haunts Kafka's stories, and his work is characterized by its lack of scenic description and its portrayal of an enigmatic and nightmarish reality where the individual is perceived as lonely, perplexed, and threatened. *The Metamorphosis* (1917) was one of the few works published in Kafka's lifetime: his novels *The Trial* (1925) and *The Castle* (1926) were published posthumously by his friend the writer Max Brod (1884-1968), against the directions of his will.

Kafkaesque /ˌkæfkəˈesk/ *adj.* (of a situation, atmosphere, etc.) impenetrably oppressive, nightmarish, in a manner characteristic of the fictional world of Franz Kafka.

kaftan var. of CAFTAN.

Kagoshima /ˌkægɒˈʃiːmə/ a city and port in Japan; pop. (1990) 537,000. Situated on the southern coast of Kyushu island, on the Satsuma Peninsula, it is noted for its porcelain, Satsuma ware.

kai /kaɪ/ *n. NZ colloq.* food. [Maori]

Kaifeng /kaɪˈfeŋ/ a city in Henan province, eastern China, on the Yellow River; pop. (est. 1984) 619,200. Established in the 4th century BC, it is one of the oldest cities in China; from AD 907 to AD 960, as Pien Liang, it was the country's capital.

kail var. of KALE.

kailyard var. of KALEYARD.

Kairouan /ˌkaɪəruːˈɑːn/ a city in NE Tunisia; pop. (1984) 72,250. It is a Muslim holy city and a place of pilgrimage.

Kaiser /ˈkaɪzə(r)/, Georg (1878–1945), German dramatist. Author of some sixty plays, he is best known for his expressionist plays *The Burghers of Calais* (1914), and *Gas I* (1918) and *Gas II* (1920); the last two provide a gruesome interpretation of futuristic science ending with the extinction of all life by poisonous gas.

kaiser /ˈkaɪzə(r)/ *n. hist.* an emperor, esp. the German emperor, the emperor of Austria, or the head of the Holy Roman Empire. □ **kaisership** *n.* [in mod.Engl. f. G *Kaiser* and Du. *keizer*; in ME f. OE *cāsere* f. Gmc adoption (through Gk *kaisar*) of L *Caesar*: see CAESAR[1]]

Kaiserslautern /ˌkaɪzəsˈlaʊt(ə)n/ a city in western Germany, in Rhineland-Palatinate; pop. (1991) 100,540.

Kaiser Wilhelm, Wilhelm II of Germany (see WILHELM II).

kaizen /kaɪˈzen/ *n.* continuous improvement of working practices, personal efficiency, etc., as a business philosophy. [Jap., = improvement]

kaka /ˈkɑːkɑː/ *n.* (*pl.* **kakas**) a large New Zealand parrot, *Nestor meridionalis*, with olive-brown plumage. [Maori]

kakapo /ˈkɑːkəˌpaʊ/ *n.* (*pl.* **-os**) a flightless owl-like New Zealand parrot, *Strigops habroptilus*. [Maori, = night kaka]

kakemono /ˌkækɪˈməʊnəʊ/ *n.* (*pl.* **-os**) a vertical Japanese unframed wall-picture, usu. painted or inscribed on paper or silk and mounted on rollers. [Jap. f. *kake*- hang + *mono* thing]

Kalaallit Nunaat /kəˌlɑːlɪt nəˈnɑːt/ the Inuit name for GREENLAND.

kala-azar /ˌkɑːləəˈzɑː(r)/ *n.* a tropical disease caused by the parasitic protozoan *Leishmania donovani*, which is transmitted to humans by sandflies. [Assamese f. *kālā* black + *āzār* disease]

Kalahari /ˌkæləˈhɑːrɪ/ (in full **Kalahari Desert**) a high, vast, arid plateau in southern Africa north of the Orange River. It comprises most of Botswana with parts in Namibia and South Africa. It is the site of the Central Kalahari Game Reserve (established 1931) and the Gemsbok National Park (established 1971) in Botswana, and the Kalahari Gemsbok National Park (established 1931) in South Africa.

kalanchoe /ˌkælənˈkəʊɪ/ *n.* a succulent plant of the mainly African genus *Kalanchoe*, which includes several house plants, some producing miniature plants from the edges of the leaves. [mod.L f. F, ult. f. Chin. *gāláncài*]

kale /keɪl/ *n.* (also **kail**) **1** a variety of cabbage with leaves that do not form a compact head. **2** *N. Amer. sl.* money. □ **curly kale** a form of kale with leaves curled like parsley. [ME, northern form of COLE]

kaleidoscope /kəˈlaɪdəˌskəʊp/ *n.* **1** a tube containing mirrors and pieces of coloured glass or paper, whose reflections produce changing patterns when the tube is rotated. **2** a constantly changing group of bright or interesting objects. □ **kaleidoscopic** /-ˌlaɪdəˈskɒpɪk/ *adj.* **kaleidoscopically** *adv.* [Gk *kalos* beautiful + *eidos* form + -SCOPE]

kalends var. of CALENDS.

Kalevala /ˈkɑːlɪˌvɑːlə/ a group of popular Finnish legends transmitted orally until published in the 19th century and now regarded as the Finnish national epic. The stories are concerned with the myths of Finland and the conflicts of the Finns with the Lapps. [Finnish f. *kaleva* of a hero + -*la* home]

kaleyard /ˈkeɪljɑːd/ *n.* (also **kailyard**) *Sc.* a kitchen garden. [KALE + YARD[2]]

Kaleyard School a group of 19th-century Scottish writers of fiction including J. M. Barrie, who described local town life in a romantic vein and with much use of the vernacular.

Kalgan /kɑːlˈgɑːn/ the Mongolian name for ZHANGJIAKOU.

Kalgoorlie /kælˈɡʊəlɪ/ a gold-mining town in Western Australia; pop. (est. 1987) 11,100. Gold was discovered there in 1887, leading to a gold rush in the 1890s.

Kali /ˈkɑːlɪ/ *Hinduism* the most terrifying goddess, wife of Siva, often identified with Durga (see also PARVATI). She is usually depicted as black, naked, old, and hideous, with a necklace of skulls, a belt of severed hands, and a protruding blood-stained tongue. The infamous Thugs were her devotees. [Skr., = black]

kali /ˈkælɪ, ˈkeɪlɪ/ *n.* a glasswort, *Salsola kali*, with fleshy jointed stems, having a high soda content. [Arab. *ḳalī* ALKALI]

Kalidasa /ˌkælɪˈdɑːsə/, Indian poet and dramatist. He is best known for his drama *Sakuntala*, the love story of King Dushyanta and the maiden Sakuntala. Kalidasa probably lived in the 5th century AD, although there is some diversity of opinion on this point.

Kalimantan /ˌkælɪˈmæntæn/ a region of Indonesia, comprising the southern part of the island of Borneo.

Kalinin[1] /kəˈliːnɪn/, Mikhail Ivanovich (1875–1946), Soviet statesman, head of state of the USSR 1919–46. Born in Russia, he was a founder of the newspaper *Pravda* in 1912.

Kalinin[2] /kəˈliːnɪn/ the former name (1931–91) for TVER.

Kaliningrad /kəˈliːnɪnˌɡræd/ **1** a port on the Baltic coast of eastern Europe, capital of the Russian region of Kaliningrad; pop. (1990) 406,000. Founded in the 13th century, it was known by its German name of Königsberg until 1946, when it was ceded to the Soviet Union under the Potsdam Agreement and renamed in honour of Kalinin. Its port is ice-free all the year round and is a significant naval base for the Russian fleet. **2** a region of Russia, an enclave situated on the Baltic coast of eastern Europe; capital, Kaliningrad. It shares its borders with Lithuania and Poland and is separated from Russia by the intervening countries of Lithuania, Latvia, and Belarus.

Kalisz /ˈkɑːlɪʃ/ a city in central Poland; pop. (1990) 106,150.

Kalmar /ˈkælmɑː(r)/ a port in SE Sweden, on the Kalmar Sound opposite Öland; pop. (1990) 56,200.

Kalmar Sound a narrow strait between the mainland of SE Sweden and the island of Öland, in the Baltic Sea.

Kalmar, Union of the treaty which joined together the crowns of Denmark, Sweden, and Norway in 1397, dissolved in 1523.

kalmia /ˈkælmɪə/ *n.* an evergreen ericaceous shrub of the genus *Kalmia*, native to North America; esp. *K. latifolia*, with showy pink flowers. [mod.L f. Pehr *Kalm*, Swedish botanist (1716–79)]

Kalmuck /ˈkælmʌk/ *n. & adj.* (also **Kalmyk** /ˈkælmɪk/) ● *n.* **1** a member of a people living on the north-western shores of the Caspian Sea. They invaded Russia in the 17th and 18th centuries and settled along the lower Volga, many migrating to Chinese Turkestan in the 18th century. **2** the language of this people, a western form of Mongolian. ● *adj.* of or relating to the Kalmucks or their language. [Russ. *kalmyk*]

Kalmykia /kælˈmɪkɪə/ (official name **Republic of Kalmykia-Khalmg Tangch**) an autonomous republic in SW Russia, on the Caspian Sea; pop. (1990) 325,000; capital, Elista.

kalong /ˈkɑːlɒŋ/ *n.* an Asian fruit-bat of the family Pteropodidae; esp. *Pteropus edulis*, the common flying fox. [Malay]

kalpa /ˈkælpə/ *n. Hinduism & Buddhism* the period between the beginning and the end of the world considered as the day of Brahma (4,320 million human years). [Skr.]

Kaluga /kəˈluːɡə/ an industrial city and river port in European Russia, on the River Oka south-west of Moscow; pop. (1990) 314,000.

Kalyan /kʌlˈjɑːn/ a city on the west coast of India, in the state of Maharashtra, north-east of Bombay; pop. (1991) 1,014,000.

Kama /ˈkɑːmə/ *Hinduism* the god of sexual love, usually represented as a beautiful youth with a bow of sugar-cane, a bowstring of bees, and arrows of flowers. [Skr., = love]

Kama Sutra /ˈsuːtrə/ *n.* an ancient Sanskrit treatise on the art of love and sexual technique. [Skr., = love-treatise]

Kamchatka /kæmˈtʃætkə/ a vast mountainous peninsula of the NE coast of Siberian Russia, separating the Sea of Okhotsk from the Bering Sea; chief port, Petropavlovsk. It is a volcanically active zone containing twenty-two active volcanoes and many hot springs.

kame /keɪm/ *n.* a short ridge of sand and gravel deposited from the water of a melted glacier. [Sc. form of COMB]

Kamenskoe /kæˈmjenskɔɪjə/ a former name (until 1936) for DNIPRODZERZHINSK.

Kamensk-Uralsky /ˌkɑːmɪnskʊˈrælskɪ/ an industrial city in central Russia, in the eastern foothills of the Urals; pop. (1990) 208,000.

Kamerlingh Onnes /ˌkæməlɪŋ ˈɒnɪs/, Heike (1853–1926), Dutch physicist, who studied cryogenic phenomena. Using the Joule–Thomson effect, he succeeded in liquefying helium in 1908, and achieved a temperature of less than one degree above absolute zero. Onnes discovered the phenomenon of superconductivity in 1911, and was awarded the Nobel Prize for his work on low-temperature physics two years later.

kamikaze /ˌkæmɪˈkɑːzɪ/ *n. & adj.* ● *n.* a Japanese aircraft loaded with explosives and deliberately crashed on its target; the pilot of such an aircraft. (*See note below.*) ● *adj.* **1** of or relating to a kamikaze. **2** reckless, dangerous, potentially self-destructive. [Jap. f. *kami* divinity + *kaze* wind]

▪ During the Second World War kamikaze attacks were launched from 1944 by the Japanese naval command against advancing Allied ships, achieving a certain amount of success, particularly in the fighting at Okinawa. The name refers to a supposedly divine wind which arose and destroyed an invading Mongol fleet in 1281.

Kamilaroi /kə'mɪlə,rɔɪ/ n. & adj. ● n. (pl. same) **1** a member of a group of Australian Aboriginal peoples living between the Gwydir and Lachlan rivers in New South Wales. **2** the language of these peoples, now extinct. ● adj. of or relating to these peoples or their language. [Aboriginal]

Kampala /kæm'pɑːlə/ the capital of Uganda; pop. (1980) 458,400. It is situated on the northern shores of Lake Victoria and replaced Entebbe as capital when the country became independent in 1963.

kampong /'kæmpɒŋ/ n. a Malayan enclosure or village. [Malay: cf. COMPOUND[2]]

Kampuchea /,kæmpʊ'tʃiːə/ the former name (1976–89) for CAMBODIA.

Kampuchean /,kæmpʊ'tʃiːən/ n. & adj. = CAMBODIAN.

Kan. abbr. Kansas.

kana /'kɑːnə/ n. any of various Japanese syllabaries. [Jap.]

kanaka /kə'nækə, 'kænəkə/ n. a South Sea Islander, esp. hist. one employed in forced labour in Australia. [Hawaiian, = man]

Kanarese /,kænə'riːz/ n. & adj. (also **Canarese**) ● n. (pl. same) **1** a member of a people of Kanara, a district in SW India. **2** the language spoken by this people; = KANNADA. ● adj. of or relating to the Kanarese or their language.

kanban /'kænbæn/ n. **1** a card used for ordering parts etc. in a Japanese just-in-time manufacturing system. **2** (in full **kanban system**) a Japanese just-in-time manufacturing system in which parts etc. are ordered on cards. [Jap., = billboard, sign]

Kanchenjunga /,kæntʃen'dʒʌŋgə/ (also **Kangchenjunga**, **Kinchinjunga** /,kɪntʃɪn-/) a mountain in the Himalayas, on the border between Nepal and Sikkim. Rising to a height of 8,598 m (28,209 ft), it is the world's third-highest mountain. Its summit is split into five separate peaks, whence its name, which is Tibetan and means 'the five treasures of the snows'.

Kandahar /,kændə'hɑː(r)/ a city in southern Afghanistan; pop. (est. 1984) 203,000. It was the first capital of Afghanistan after the country became independent, from 1748 until replaced by Kabul in 1773.

Kandinsky /kæn'dɪnskɪ/, Wassily (1866–1944), Russian painter and theorist. He was a pioneer of abstract art, producing non-representational works as early as 1910. His treatise On the Spiritual in Art (1912) urged the expression of inner and essential feelings in art rather than the representation of surface appearances of the natural world. In 1911 he co-founded the Munich-based Blaue Reiter group of artists; he later taught at the Bauhaus (1922–33). His paintings from this time became almost wholly abstract, with their energy and movement being conveyed purely by colour, line, and shape.

Kandy /'kændɪ/ a city in the highlands of central Sri Lanka; pop. (1990) 104,000. It was the capital (1480–1815) of the former independent kingdom of Kandy and contains one of the most sacred Buddhist shrines, the Dalada Maligava (Temple of the Tooth).

kanga /'kæŋgə/ n. Austral. **1** colloq. = KANGAROO. **2** rhyming sl. a prison warder (from screw). [abbr.]

Kangar /'kæŋgɑː(r)/ the capital of the state of Perlis in northern Malaysia, near the west coast of the Malay Peninsula; pop. (1980) 12,950.

kangaroo /,kæŋgə'ruː/ n. a large plant-eating marsupial of the genus Macropus, native to Australia and New Guinea, with a long tail and strongly developed hind quarters enabling it to travel by jumping. □ **kangaroo closure** Brit. Parl. a closure involving the chairperson of a committee selecting some amendments for discussion and excluding others. **kangaroo court** an improperly constituted or illegal court held by strikers etc. **kangaroo mouse** a small rodent of the genus Microdipodops, native to North America, with long hind legs for hopping. **kangaroo paw** a plant of the genus Angiozanthos, with green and red woolly flowers. **kangaroo-rat** a burrowing rodent of the genus Dipodomys, of the southern US and Mexico, having elongated hind feet. **kangaroo vine** an evergreen climbing plant, Cissus antarctica, of Australia, with heart-shaped leaves. [ganurru name of a specific kind of kangaroo in an extinct Aboriginal language of N. Queensland]

Kangchenjunga see KANCHENJUNGA.

KaNgwane /,kɑːəŋ'gwɑːneɪ/ a former homeland established in South Africa for the Swazi people, now part of the province of Eastern Transvaal. (See also HOMELAND.)

kanji /'kændʒɪ/ n. (any of) the set of Chinese characters used in the Japanese writing system. [Jap. f. kan Chinese + ji character]

Kannada /'kænədə, 'kɑːn-/ n. the Dravidian language spoken by the Kanarese, closely allied to Telugu, with about 22 million speakers. Its alphabet is similar to that of Telugu, and developed from the Grantha script. [Kannada kannaḍa]

Kano /'kɑːnəʊ/ a city in northern Nigeria; pop. (est. 1986) 553,000.

kanoon /kə'nuːn/ n. an instrument like a zither, with fifty to sixty strings. [Pers. or Arab. ḳānūn]

Kanpur /kɑːn'pʊə(r)/ (also **Cawnpore** /kɔːn'pɔː(r)/) a city in Uttar Pradesh, northern India, on the River Ganges; pop. (1991) 2,100,000. It was the site of a massacre of British soldiers and European families in July 1857, during the Indian Mutiny.

Kans. abbr. Kansas.

Kansas /'kænzəs/ a state in the central US; pop. (est. 1990) 2,477,570; capital, Topeka. Acquired as part of the Louisiana Purchase in 1803, it became the 34th state of the US in 1861. □ **Kansan** adj. & n.

Kansas City each of two adjacent cities in the US, situated at the junction of the Missouri and Kansas rivers, one in NE Kansas (1990 pop. 149,770) and the other in NW Missouri (1990 pop. 435,150).

Kansu see GANSU.

Kant /kænt/, Immanuel (1724–1804), German philosopher. In the Critique of Pure Reason (1781) he countered Hume's empiricism by arguing that the human mind can neither confirm, deny, nor scientifically demonstrate the ultimate nature of reality. He claimed, however, that it can know the objects of experience, which it interprets with notions of space and time and orders according to twelve categories of thought, grouped under the classes of quantity, quality, reason, and modality. Kant's Critique of Practical Reason (1788) deals with ethics and affirms the existence of an absolute moral law — the categorical imperative — whose motivation is reason. His idealism left an important legacy for the philosophy of Hegel and Fichte. □ **Kantian** adj. & n. **Kantianism** n.

Kanto /kæn'təʊ/ a region of Japan, on the island of Honshu; capital, Tokyo.

KANU /'kɑːnuː/ see KENYA AFRICAN NATIONAL UNION.

Kaohsiung /kaʊ'ʃjʊŋ/ the chief port of Taiwan, on the SW coast; pop. (1990) 1,390,000.

kaolin /'keɪəlɪn/ n. a fine soft white clay produced by the decomposition of other clays or feldspar, used esp. for making porcelain and in medicines. Also called china clay. □ **kaolinize** v.tr. (also **-ise**). **kaolinic** /,keɪə'lɪnɪk/ adj. [F f. Chin. gaoling the name of a mountain f. gao high + ling hill]

kaon /'keɪɒn/ n. Physics a meson having a mass several times that of a pion. [ka repr. the letter K (as symbol for the particle) + -ON]

Kapachira Falls /,kæpə'tʃɪərə/ a waterfall on the Shire river in southern Malawi. It was formerly known as Murchison Falls.

kapellmeister /kə'pel,maɪstə(r)/ n. (pl. same) the conductor of an orchestra, opera, choir, etc., esp. in German contexts. [G f. Kapelle court orchestra f. It. cappella CHAPEL + Meister master]

Kap Farvel /,kab far'vɛl/ the Danish name for Cape Farewell (see FAREWELL, CAPE 1).

Kapil Dev /,kæpɪl 'dev/ (full name Kapil Dev Nikhanj) (b.1959), Indian cricketer. He made his début for India in 1978 as a medium-pace bowler, soon developing into an all-rounder. He had two spells as captain, in one of which he led India to victory in the 1983 World Cup. In 1994 he passed Richard Hadlee's world record of 431 test match wickets.

kapok /'keɪpɒk/ n. **1** a fine fibrous cotton-like substance found surrounding the seeds of the kapok tree, used for stuffing cushions, soft toys, etc. **2** the large tropical tree, Ceiba pentandra, yielding this. [ult. f. Malay kāpoq]

Kapoor /kæ'pʊə(r)/, (Prithvi) Raj (1924–88), Indian actor and film-maker. In 1944 he founded the Prithvi Theatres in Bombay, a company notable for the realism it brought to Hindi drama. Kapoor went on to direct a large number of films for the Indian market. Productions, in which Kapoor often played the lead, include Pathan (1946).

Kaposi's sarcoma /kə'pəʊsɪz/ n. Med. a form of cancer involving multiple tumours of the lymph nodes and skin, usu. occurring in people with depressed immune systems, e.g. as a consequence of Aids. [M. K. Kaposi, Hung. dermatologist (1837–1902)]

kappa /ˈkæpə/ n. the tenth letter of the Greek alphabet (K, κ). [Gk]

kaput /kæˈpʊt/ predic.adj. colloq. broken, ruined; done for. [G kaputt]

karabiner /ˌkærəˈbiːnə(r)/ n. a coupling link with safety closure, used by mountaineers. [G, lit. 'carbine']

Karachai-Cherkessia /ˌkærəˌtʃaɪtʃeəˈkesɪə/ (official name **Karachai-Cherkess Republic**) an autonomous republic in the northern Caucasus, SW Russia; pop. (1989) 418,000; capital, Cherkessk.

Karachi /kəˈrɑːtʃɪ/ a major city and port in Pakistan, capital of Sind province; pop. (est. 1991) 6,700,000. Situated on the Arabian Sea, it was the capital of Pakistan 1947–59 before being replaced by Rawalpindi.

Karadžić /ˈkærəˌdʒɪtʃ/, Vuk Stefanović (1787–1864), Serbian writer, grammarian, lexicographer, and folklorist. He modified the Cyrillic alphabet for Serbian written usage and compiled a Serbian dictionary in 1818. Widely claimed to be the father of modern Serbian literature, he undertook the task of collecting and publishing national folk stories and poems (1821–33).

Karafuto /ˌkærəˈfuːtəʊ/ the Japanese name for the southern part of the island of Sakhalin.

Karaganda see QARAGHANDY.

Karaite /ˈkeərəˌaɪt/ n. a member of a Jewish sect founded in the 8th century and located chiefly in the Crimea and neighbouring areas, and in Israel, which rejects rabbinical tradition in favour of a literal interpretation of the Scriptures. [Heb. qerāīm scripturalists f. qārā read]

Karaj /kæˈrɑːdʒ/ a city in northern Iran, to the west of Tehran; pop. (1991) 442,000.

Karajan /ˈkærəˌjæn/, Herbert von (1908–89), Austrian conductor. He is chiefly remembered as the principal conductor of the Berlin Philharmonic Orchestra (1955–89), although he was also associated with the Vienna State Opera (1957–64). Karajan was artistic director of the Salzburg Festival (1956–60; 1964) and founded the Salzburg Easter Festival of operas in 1967.

Karakoram /ˌkærəˈkɔːrəm/ a great mountain system of central Asia, extending over 480 km (300 miles) south-eastwards from NE Afghanistan to Kashmir and forming part of the borders of India and Pakistan with China. One of the highest mountain systems in the world, it consists of a group of parallel ranges, forming a westwards continuation of the Himalayas, with many peaks over 7,900 m (26,000 ft), the highest being K2. Virtually inaccessible, it also contains the highest passes in the world, at elevations over 4,900 m (16,000 ft), including Karakoram Pass and Khardungla Pass.

Karakorum /ˌkærəˈkɔːrəm/ an ancient city in central Mongolia, now ruined, which was the capital of the Mongol empire, established by Genghis Khan in 1220. The capital was later moved by Kublai Khan to Khanbaliq (modern Beijing) in 1267, and Karakorum was destroyed by Chinese forces in 1388.

karakul /ˈkærəˌkʊl/ n. (also **caracul**) **1** a variety of Asian sheep with a dark curled fleece when young. **2** fur made from or resembling this. Also called Persian lamb. [Russ.]

Kara Kum /ˌkærə ˈkuːm/ (Russian **Karakumy** /kərəˈkumij/) a desert in central Asia, to the east of the Caspian Sea, covering much of Turkmenistan.

karaoke /ˌkærɪˈəʊkɪ/ n. a form of entertainment in which people sing popular songs as soloists against a pre-recorded backing. □ **karaoke bar** (or **club**) a bar or club with this form of entertainment. [Jap., = empty orchestra]

Kara Sea /ˈkɑːrə/ an arm of the Arctic Ocean off the northern coast of Russia, bounded to the east by the islands of Severnaya Zemlya and to the west by Novaya Zemlya.

karat US var. of CARAT 2.

karate /kəˈrɑːtɪ/ n. a Japanese form of unarmed combat using the hands and feet as weapons. It involves a training of the mind as well as the body and seeks to concentrate the body's power at the point of impact. Modern karate is a product of the 20th century, but its roots are ancient. The ranking system involves the wearing of belts similar to those used in judo. [Jap. f. kara empty + te hand]

Karbala /ˈkɑːbələ/ a city in southern Iraq; pop. (est. 1985) 184,600. A holy city for Shiite Muslims, it is the site of the tomb of Husayn, grandson of Muhammad, who was killed there in AD 680.

Karelia /kəˈriːlɪə, -ˈreɪlɪə/ a region of NE Europe on the border between Russia and Finland. It was an independent state of Finnish-speaking people in medieval times; folk-tales from there were the source of the Finnish epic, the Kalevala. In the 16th century Karelia came under Swedish rule and in 1721 it was annexed by Russia. Following Finland's declaration of independence in 1917, part of Karelia became a region of Finland and part an autonomous republic of the Soviet Union. After the Russo-Finnish war of 1939–40 the greater part of Finnish Karelia was ceded to the Soviet Union and the area now constitutes the Republic of Karelia; pop. (1989) 792,000; capital, Petrozavodsk. The remaining part of Karelia constitutes a province of eastern Finland. □ **Karelian** adj. & n.

Karen /kəˈren/ n. & adj. ● n. **1** a member of a non-Burmese Mongoloid people, most of whom live in eastern Burma (Myanmar). **2** the language spoken by this people, which is probably of the Sino-Tibetan family. ● adj. of or relating to the Karens or their language. [Burmese ka-reng wild unclean man]

Karen State (also called Kawthoolay, Kawthulei) a state in SE Burma (Myanmar), on the border with Thailand; capital, Pa-an. Inaugurated in 1954 as an autonomous state of Burma, the state was given the traditional Karen name of Kawthoolay in 1964, but reverted to Karen after the 1974 constitution limited its autonomy. The people are engaged in armed conflict with the Burmese government in an attempt to gain independence.

Kariba, Lake /kəˈriːbə/ a large, man-made lake on the Zambia-Zimbabwe border in central Africa. It was created by the damming of the Zambezi river by the Kariba Dam. On its northern shore is the town of Kariba, which was originally built to house the 10,000 workers on the dam-building project and is now a resort town. Its name means 'where the waters have been trapped'.

Kariba Dam /kəˈriːbə/ a concrete arch dam on the Zambezi river, 385 km (240 miles) downstream from the Victoria Falls. It was built in 1955–9, creating Lake Kariba and providing a bridge over the Zambezi between Zambia and Zimbabwe. Its construction was a major engineering feat and it is the chief source of hydroelectric power for both Zimbabwe and Zambia.

Karl XII see CHARLES XII.

Karl-Marx-Stadt /kɑːlˈmɑːksʃtæt/ the former name (1953–1990) for CHEMNITZ.

Karloff /ˈkɑːlɒf/, Boris (born William Henry Pratt) (1887–1969), British-born American actor. His name is chiefly linked with horror films, such as Frankenstein (1931) and The Body Snatcher (1945).

Karlovy Vary /ˌkɑːləvɪ ˈvɑːrɪ/ (called in German Karlsbad) a spa town in the western Czech Republic; pop. (1991) 56,290. Founded in the 14th century by the Holy Roman emperor Charles IV, it is famous for its alkaline thermal springs.

Karlsbad /ˈkɑːlsbɑːt/ the German name for KARLOVY VARY.

Karlsruhe /ˈkɑːlzˌruːə/ an industrial town and port on the Rhine in western Germany; pop. (1991) 278,580.

karma /ˈkɑːmə/ n. **1** Hinduism & Buddhism the doctrine that the sum of a person's actions in previous states of existence controls his or her fate in future existences. The doctrine reflects the Hindu belief that life as a human is just one of a chain of successive existences by transmigration, each life's condition being a consequence of actions in a previous life. **2** Jainism subtle physical matter which binds the soul as a result of bad actions. □ **karmic** adj. [Skr., = action, fate]

Karnak /ˈkɑːnæk/ a village in Egypt, on the Nile near Luxor. It is the site of the northern complex of monuments of ancient Thebes, including the great temple of Amun.

Karnataka /kəˈnɑːtəkə/ a state in SW India; capital, Bangalore. It was known as Mysore until 1973.

Kärnten see CARINTHIA.

Karoo /kəˈruː/ n. (also **Karroo**) an elevated semi-desert plateau in South Africa. [Nama]

Karpov /ˈkɑːpɒf/, Anatoli (b.1951), Russian chess player. He was world champion from 1975 until defeated by Gary Kasparov in 1985.

karri /ˈkærɪ/ n. (pl. **karris**) **1** a tall western Australian tree, Eucalyptus diversicolor, with a hard red wood. **2** the timber from this. [Aboriginal]

Kars /kɑːs/ a city and province in NE Turkey; pop. (1990) 78,455. In the 9th and 10th centuries it was the capital of an independent Armenian principality.

karst /kɑːst/ n. a limestone region with underground drainage and many cavities and passages caused by the dissolution of the rock. [German Karst (Slovenian Kras), name of a limestone region of Slovenia, at the north-western end of the Dinaric Alps]

kart /kɑːt/ n. a small four-wheeled motor-racing vehicle usually consisting of a tubular frame with a rear-mounted engine, used in the sport of karting. [commercial alt. of cart]

karting /ˈkɑːtɪŋ/ n. a motor-racing sport using karts, begun in the US in 1956.

karyo- /ˈkærɪəʊ/ comb. form Biol. denoting the nucleus of a cell. [Gk karuon kernel]

karyokinesis /ˌkærɪəʊkɪˈniːsɪs, -kaɪˈniːsɪs/ n. Biol. the division of a cell nucleus during mitosis. [KARYO- + Gk kinēsis movement f. kineō move]

karyotype /ˈkærɪəˌtaɪp/ n. Biol. the number and visual appearance of the chromosomes in the nucleus of a cell.

kasbah /ˈkæzbɑː/ n. (also **casbah**) **1** the citadel of a North African city. **2** an Arab quarter near this. [F casbah f. Arab. kas(a)ba citadel]

Kashmir /kæʃˈmɪə(r)/ a region on the northern border of India and NE Pakistan. Formerly a state of India, it has been disputed between India and Pakistan since partition in 1947, with sporadic outbreaks of fighting. The north-western part is controlled by Pakistan, most of it forming the state of Azad Kashmir, while the remainder is incorporated into the Indian state of Jammu and Kashmir.

Kashmiri /kæʃˈmɪərɪ/ n. & adj. ● n. **1** a native or inhabitant of the territory of Kashmir. **2** the Indic language of the Kashmiris. ● adj. of or relating to the Kashmiris or their language.

Kasparov /ˈkæspəˌrɒf/, Gary (born Gary Weinstein) (b.1963), Azerbaijani chess player of Armenian-Jewish descent. At the age of 22 he became the youngest-ever world champion, defeating Anatoli Karpov in 1985. He has retained the title ever since, defending it against challenges from Karpov in 1986, 1987, and 1990.

Kassel /ˈkæs(ə)l/ a city in central Germany, in Hesse; pop. (1991) 196,830. Founded c.913, it was the capital of the kingdom of Westphalia (1807–13) and of the Prussian province of Hesse-Nassau (1866–1944).

Kassite /ˈkæsaɪt/ n. & adj. ● n. **1** a member of an ancient Elamite people from the Zagros mountains in western Iran, who ruled Babylonia from the 18th to the 12th century BC until overthrown by the Assyrians. **2** the language of this people. ● adj. of or relating to the Kassites or their language. [Assyr. Kaššu + -ITE¹]

Kasur /kəˈsʊə(r)/ a city in Punjab province, NE Pakistan; pop. (1981) 155,000.

katabatic /ˌkætəˈbætɪk/ adj. Meteorol. (of wind) caused by local downward motion of cold air (cf. ANABATIC). [Gk katabatikos f. katabainō go down]

katabolism var. of CATABOLISM.

katakana /ˌkætəˈkɑːnə/ n. an angular form of Japanese kana. [Jap., = side kana]

Katanga /kəˈtæŋgə/ the former name (until 1972) for SHABA.

katharevousa /ˌkæθəˈrevuːsə/ n. the literary form of modern Greek, as opposed to the form based on the spoken language (cf. DEMOTIC). [mod.Gk, ult. f. Gk katharos pure]

Kathiawar /ˌkætɪəˈwɑː(r)/ a peninsula on the western coast of India, in the state of Gujarat, separating the Gulf of Kutch from the Gulf of Cambay.

Kathmandu /ˌkætmænˈduː/ the capital of Nepal; pop. (1981) 235,000. It is situated in the Himalayas at an altitude of 1,370 m (4,450 ft).

kathode var. of CATHODE.

Katowice /ˌkætəˈviːtsə/ a city in SW Poland; pop. (1990) 349,360. It is the industrial centre of the Silesian coal-mining region.

Kattegat /ˈkætɪˌgæt/ a strait, 225 km (140 miles) in length, between Sweden and Denmark. It is linked to the North Sea by the Skagerrak and to the Baltic Sea by the Øresund.

katydid /ˈkeɪtɪˌdɪd/ n. a large green grasshopper of the family Tettigoniidae, native to North America. [imit. of the sound it makes]

Kauai /kaʊˈwɑːɪ/ an island in the state of Hawaii, separated from Oahu by the Kauai Channel; chief town, Lihue.

Kauffmann /ˈkaʊfmən/, Angelica (also **Kauffman**) (1740–1807), Swiss painter. She made her name in Rome with her portrait of Johann Winckelmann (1764). In London from 1766, she became well known for her neoclassical and allegorical paintings (e.g. Self-Portrait Hesitating Between the Arts of Music and Painting, 1791), and for her decorative wall-paintings in houses designed by Robert Adam and his brothers. She was a founder member of the Royal Academy of Arts (1768).

Kaunas /ˈkaʊnəs/ an industrial city and river port in southern Lithuania, at the confluence of the Vilnya and Neman rivers; pop. (1991) 430,000.

Kaunda /kɑːˈʊndə/, Kenneth (David) (b.1924), Zambian statesman, President 1964–91. He led the United National Independence Party to electoral victory in 1964, becoming Prime Minister and the first President of independent Zambia. As chairman of the Organization of African Unity (1970–1; 1987–8), he played a key role in the negotiations leading to Namibian independence in 1990.

kauri /ˈkaʊərɪ/ n. (pl. **kauris**) a coniferous New Zealand tree, Agathis australis, which produces valuable timber and a resin. □ **kauri-gum** this resin. [Maori]

kava /ˈkɑːvə/ n. **1** a Polynesian shrub, Piper methysticum, related to the peppers. **2** an intoxicating drink made from the crushed roots of this. [Polynesian]

Kaválla /kəˈvælə/ a port on the Aegean coast of NE Greece; pop. (1981) 56,375. Originally a Byzantine city and fortress controlling Macedonia, it was Turkish until 1912, when it was ceded to Greece. It occupies the site of Neapolis, the port of ancient Philippi.

Kaveri see CAUVERY.

Kawabata /ˌkɑːwəˈbɑːtə/, Yasunari (1899–1972), Japanese novelist. Known as an experimental writer in the 1920s, he reverted to traditional Japanese novel forms in the mid-1930s. His novels include The Izu Dancer (1925), Snow Country (1935–47), and The Sound of the Mountain (1949–54). He won the Nobel Prize for literature in 1968, the first Japanese writer to do so.

kawakawa /ˌkɑːwəˈkɑːwə/ n. a New Zealand shrub, Macropiper excelsum, with aromatic leaves. [Maori]

Kawasaki /ˌkɑːwəˈsɑːkɪ/ an industrial city on the SE coast of the island of Honshu, Japan; pop. (1990) 1,174,000.

Kawthoolay /ˌkɔːθuːˈleɪ/ (also **Kawthulei**) a former name (1964–74) for KAREN STATE.

kayak /ˈkaɪæk/ n. & v. ● n.**1** a light Eskimo one-person canoe consisting of a wooden framework covered with sealskins. **2** a small covered canoe resembling this. ● v.intr. (**kayaked, kayaking**) travel by kayak; paddle a kayak. [Eskimo (Inuit)]

Kaye /keɪ/, Danny (born David Daniel Kominski) (1913–87), American actor and comedian. After a successful Broadway career he made his first feature film in 1944, and went on to take a number of roles in which he became known for his mimicry, comic songs, and slapstick humour. His films include The Secret Life of Walter Mitty (1947), Hans Christian Andersen (1952), and The Court Jester (1956).

kayo /keɪˈəʊ/ v. & n. colloq. ● v.tr. (**-oes, -oed**) knock out; stun by a blow. ● n. (pl. **-os**) a knockout. [repr. pronunc. of KO]

Kayseri /ˈkaɪsərɪ/ a city in central Turkey, capital of a province of the same name; pop. (1990) 421,360. Known as Kayseri since the 11th century, it was formerly called Caesarea Mazaca and was the capital of Cappadocia.

Kazakh /kəˈzɑːk/ n. & adj. ● n. (pl. **Kazakhs, Kazakhi** /-ˈzɑːkɪ/) **1** a member of a largely pastoral people inhabiting Kazakhstan and parts of China, Mongolia, and Afghanistan. Traditionally nomadic, Kazakhs are predominantly Sunni Muslims. **2** the Turkic language of the Kazakhs. ● adj. of or relating to the Kazakhs or their language.

Kazakhstan /ˌkæzɑːkˈstɑːn/ a republic in central Asia; population (est. 1991) 16,899,000; languages, Kazakh (official), Russian; capital, Almaty. It is situated on the southern border of Russia, extending from the Caspian Sea eastwards to the Altai Mountains and China. The Turkic tribes of Kazakhstan were overrun by the Mongols in the 13th century, and the region was eventually absorbed into the Russian empire. Kazakhstan formed a constituent republic of the Soviet Union, becoming an independent republic within the Commonwealth of Independent States in 1991.

Kazan¹ /kəˈzæn, -ˈzɑːn/ a port situated on the River Volga to the east of Nizhni Novgorod in Russia, capital of the autonomous republic of Tatarstan; pop. (1990) 1,103,000.

Kazan² /kəˈzæn/, Elia (born Elia Kazanjoglous) (b.1909), Turkish-born American film and theatre director. In 1947 he co-founded the Actors' Studio, one of the leading centres of method acting. Kazan's stage productions include A Streetcar Named Desire (1947), which he made into a film four years later; both starred Marlon Brando. Other films include On the Waterfront (1954), again with Marlon Brando, and East of Eden (1955), starring James Dean.

kazoo /kəˈzuː/ n. a toy musical instrument into which the player sings or hums. [19th c., app. with ref. to the sound produced]

KB abbr. **1** (also **Kb**) Computing kilobyte(s). **2** (in the UK) King's Bench.

kb abbr. Biochem. kilobase(s).

KBE abbr. (in the UK) Knight Commander of the Order of the British Empire.

kbyte abbr. Computing kilobyte(s).

KC *abbr.* **1** King's College. **2** King's Counsel.

kc *abbr.* kilocycle(s).

KCB *abbr.* (in the UK) Knight Commander of the Order of the Bath.

KCMG *abbr.* (in the UK) Knight Commander of the Order of St Michael and St George.

kc/s *abbr.* kilocycles per second.

KCVO *abbr.* (in the UK) Knight Commander of the Royal Victorian Order.

KE *abbr.* kinetic energy.

kea /ˈkiːə, ˈkeɪə/ *n.* a large parrot, *Nestor notabilis*, found in the uplands of New Zealand, with brownish-green and red plumage. [Maori, imit.]

Kean /kiːn/, Edmund (1787–1833), English actor. He achieved fame with his performance as Shylock at London's Drury Lane theatre in 1814, and became particularly renowned for his interpretations of Shakespearian tragic roles, notably those of Macbeth and Iago.

Keating /ˈkiːtɪŋ/, Paul (John) (b.1944), Australian Labor statesman, Prime Minister since 1991. He entered politics in 1969 when he became a member of the House of Representatives. He served as federal treasurer (1983–91) and deputy Prime Minister (1990–1) under Bob Hawke, whom he replaced as Premier in 1991. His term of office has been notable for a vociferous republican campaign as well as for measures to combat high unemployment.

Keaton /ˈkiːt(ə)n/, Buster (born Joseph Francis Keaton) (1895–1966), American actor and director. His deadpan face and acrobatic skills made him one of the biggest comedy stars of the silent-film era. Major films include *Our Hospitality* (1923), *The Navigator* (1924), and *The General* (1926).

Keats /kiːts/, John (1795–1821), English poet. In 1818 he wrote his most famous poems, including 'Hyperion', 'The Eve of St Agnes', 'La Belle Dame sans Merci', 'Ode to a Nightingale', 'Ode on a Grecian Urn', and 'Ode to Autumn', all published in 1820. A principal figure of the romantic movement, Keats was noted for his spiritual and intellectual contemplation of beauty. He died in Rome of tuberculosis.

kebab /kɪˈbæb/ *n.* a dish consisting of pieces of meat, vegetables, etc., packed closely and cooked on a skewer or spit. [Urdu f. Arab. *kabāb*]

Keble /ˈkiːb(ə)l/, John (1792–1866), English churchman. His sermon on national apostasy (1833) is generally held to mark the beginning of the Oxford Movement. Politically, it failed to win support for its idea that the law of the land need not coincide with the Church's teaching; theologically, however, the work of Keble's followers did much to revive traditional Catholic teaching, as well as to define and mould the Church of England.

Kebnekaise /ˌkebnəˈkaɪsə/ the highest peak in Sweden, in the north of the country, rising to a height of 2,117 m (6,962 ft).

ked /ked/ *n.* a bloodsucking fly of the family Hippoboscidae; esp. *Melophagus ovinus*, a wingless flat-bodied fly infesting sheep. [16th c.: orig. unkn.]

Kedah /ˈkedə/ a state of NW Malaysia, on the west coast of the Malay Peninsula; capital, Alor Setar.

kedge /kedʒ/ *v. & n.* ● *v.* **1** *tr.* move (a ship) by means of a hawser attached to a small anchor. **2** *intr.* (of a ship) move in this way. ● *n.* (in full **kedge-anchor**) a small anchor for this purpose. [perh. a specific use of obs. *cagge*, dial. *cadge* bind, tie]

kedgeree /ˈkedʒəri, ˌkedʒəˈriː/ *n.* **1** an Indian dish of rice, split pulse, onions, eggs, etc. **2** a European dish of fish, rice, hard-boiled eggs, etc. [Hindi *khichṛī*, Skr. *k'rsara* dish of rice and sesame]

keek /kiːk/ *v. & n.* Sc. ● *v.intr.* peep. ● *n.* a peep. [ME *kike*: cf. MDu., MLG *kīken*]

keel[1] /kiːl/ *n. & v.* ● *n.* **1** the lengthwise timber or steel structure along the base of a ship, airship, or some aircraft, on which the framework of the whole is built up. **2** *poet.* a ship. **3** a ridge along the breastbone of many birds; a carina. **4** *Bot.* a prow-shaped pair of petals in a corolla etc. ● *v.* **1** (often foll. by *over*) **a** *intr.* turn over or fall down. **b** *tr.* cause to do this. **2** *Naut.* **a** *intr.* (of a boat) turn keel upwards. **b** *tr.* turn up the keel of (a boat). □ **keelless** /ˈkiːllɪs/ *adj.* [ME *kele* f. ON *kjölr* f. Gmc]

keel[2] /kiːl/ *n.* Brit. hist. **1** a flat-bottomed vessel, esp. of the kind formerly used on the rivers Tyne and Wear for loading coal-ships. **2** an amount carried by such a vessel. [ME *kele* f. MLG *kēl*, MDu. *kiel* ship, boat, f. Gmc]

keelhaul /ˈkiːlhɔːl/ *v.tr.* **1** *hist.* drag (a person) through the water under the keel of a ship as a punishment. **2** scold or rebuke severely.

Keeling Islands /ˈkiːlɪŋ/ an alternative name for the COCOS ISLANDS.

keelson /ˈkiːls(ə)n/ *n.* (also **kelson** /ˈkels-/) a line of timber fastening a ship's floor-timbers to its keel. [ME *kelswayn*, perh. f. LG *kielswīn* f. *kiel* KEEL[1] + (prob.) *swīn* SWINE used as the name of a timber]

keen[1] /kiːn/ *adj.* **1** (of a person, desire, or interest) eager, ardent (*a keen sportsman; keen to be involved*). **2** (foll. by *on*) much attracted by; fond of or enthusiastic about. **3 a** (of the senses) sharp; highly sensitive. **b** (of memory etc.) clear, vivid. **4** (of a person) intellectually acute; (of a remark etc.) quick, sharp, biting. **5 a** having a sharp edge or point. **b** (of an edge etc.) sharp. **6** (of a sound, light, etc.) penetrating, vivid, strong. **7** (of a wind, frost, etc.) piercingly cold. **8** (of a pain etc.) acute, bitter. **9** Brit. (of a price) competitive. **10** *colloq.* excellent. □ **keenly** *adv.* **keenness** /ˈkiːnnɪs/ *n.* [OE *cēne* f. Gmc]

keen[2] /kiːn/ *n. & v.* ● *n.* an Irish funeral song accompanied with wailing. ● *v.* **1** *intr.* utter the keen. **2** *tr.* bewail (a person) in this way. **3** *tr.* utter in a wailing tone. □ **keener** *n.* [Ir. *caoine* f. *caoinim* wail]

Keene /kiːn/, Charles Samuel (1823–91), English illustrator and caricaturist. He is remembered for his work in the weekly journal *Punch* from 1851.

keep /kiːp/ *v. & n.* ● *v.* (*past* and *past part.* **kept** /kept/) **1** *tr.* have continuous charge of; retain possession of; save or hold on to. **2** *tr.* (foll. by *for*) retain or reserve for a future occasion or time (*will keep it for tomorrow*). **3** *tr. & intr.* retain or remain in a specified condition, position, course, etc. (*keep cool; keep off the grass; keep them happy*). **4** *tr.* put or store in a regular place (*knives are kept in this drawer*). **5** *tr.* (foll. by *from*) cause to avoid or abstain from something (*will keep you from going too fast*). **6** *tr.* detain; cause to be late (*what kept you?*). **7** *tr.* **a** observe or pay due regard to (a law, custom, etc.) (*keep one's word*). **b** honour or fulfil (a commitment, undertaking, etc.). **c** respect the commitment implied by (a secret etc.). **d** act fittingly on the occasion of (*keep the sabbath*). **8** *tr.* own and look after (animals) for pleasure or profit (*keeps bees*). **9 a** provide for the sustenance of (a person, family, etc.). **b** (foll. by *in*) maintain (a person) with a supply of. **10** *tr.* manage (a shop, business, etc.). **11 a** *tr.* maintain (accounts, a diary, etc.) by making the requisite entries. **b** *tr.* maintain (a house) in proper order. **12** *tr.* have (a commodity) regularly on sale (*do you keep buttons?*). **13** *tr.* **a** confine or detain (a person, animal, etc.). **b** guard or protect (a person or place, a goal in football, etc.). **14** *tr.* preserve in being; continue to have (*keep order*). **15** *intr.* (foll. by verbal noun) continue or do repeatedly or habitually (*why do you keep saying that?*). **16** *tr.* continue to follow (a way or course). **17** *intr.* **a** (esp. of perishable commodities) remain in good condition. **b** (of news or information etc.) admit of being withheld for a time. **18** *tr.* remain in (one's bed, room, house, etc.). **19** *tr.* retain one's place in (a seat or saddle, one's ground, etc.) against opposition or difficulty. **20** *tr.* maintain (a person) in return for sexual favours. ● *n.* **1** maintenance or the essentials for this (esp. food) (*hardly earn your keep*). **2** charge or control (*is in your keep*). **3** *hist.* a tower or stronghold. □ **for keeps** *colloq.* (esp. of something received or won) permanently, indefinitely. **how are you keeping?** how are you? **keep at** persist or cause to persist with. **keep away** (often foll. by *from*) **1** avoid being near. **2** prevent from being near. **keep back 1** remain or keep at a distance. **2** retard the progress of. **3** conceal; decline to disclose. **4** retain, withhold (*kept back £50*). **keep one's balance 1** remain stable; avoid falling. **2** retain one's composure. **keep down 1** hold in subjection. **2** keep low in amount. **3** lie low; stay hidden. **4** manage not to vomit (food eaten). **keep one's feet** manage not to fall. **keep-fit** esp. Brit. regular exercises to promote personal fitness and health. **keep one's hair on** see HAIR. **keep one's hand in** see HAND. **keep in 1** confine or restrain (one's feelings etc.). **2** remain or confine indoors. **3** keep (a fire) burning. **keep in mind** take into account having remembered. **keep in with** remain on good terms with. **keep off 1** stay or cause to stay away from. **2** ward off; avert. **3** abstain from. **4** avoid (a subject) (*let's keep off religion*). **keep on 1** continue to do something; do continually (*kept on laughing*). **2** continue to use or employ. **3** (foll. by *at*) pester or harass. **keep open house** see OPEN. **keep out 1** keep or remain outside. **2** exclude. **keep state 1** maintain one's dignity. **2** be difficult of access. **keep to 1** adhere to (a course, schedule, etc.). **2** observe (a promise). **3** confine oneself to. **keep to oneself 1** avoid contact with others. **2** refuse to disclose or share. **keep together** remain or keep in harmony. **keep track of** see TRACK[1]. **keep under** hold in subjection. **keep up 1** maintain (progress etc.). **2** prevent (prices, one's spirits, etc.) from sinking. **3** keep in repair, in an efficient or proper state, etc. **4** carry on (a correspondence etc.). **5** prevent (a person) from going to bed, esp. when late. **6** (often foll. by *with*) manage not to fall behind. **keep up with the Joneses** strive to compete socially with one's neighbours. **keep one's word** see WORD. **kept woman** a woman maintained or supported in return for sexual favours. □ **keepable** *adj.* [OE *cēpan*, of unkn. orig.]

keeper /'ki:pə(r)/ n. **1** a person who keeps or looks after something or someone. **2** Brit. a custodian of a museum, art gallery, forest, etc. **3 a** = GAMEKEEPER. **b** a person in charge of animals in a zoo. **4 a** = wicket-keeper. **b** = GOALKEEPER. **5** a fruit etc. that remains in good condition. **6** a bar of soft iron placed across the poles of a horseshoe magnet to maintain its strength. **7 a** a plain ring to preserve a hole in a pierced ear lobe; a sleeper. **b** a ring worn to guard against the loss of a more valuable one.

keeping /'ki:pɪŋ/ n. **1** custody, charge (in safe keeping). **2** agreement, harmony (esp. in or out of keeping).

keepsake /'ki:pseɪk/ n. a thing kept for the sake of or in remembrance of the giver.

keeshond /'keɪshɒnd/ n. a Dutch breed of dog with long thick hair like a large Pomeranian. [Du.]

kef /kef/ n. (also **kif** /kɪf/) **1** a drowsy state induced by marijuana etc. **2** the enjoyment of idleness. **3** a substance smoked to produce kef, esp. marijuana. [Arab. kayf enjoyment, well-being]

Kefallinía see CEPHALONIA.

keffiyeh /kə'fi:jə/ n. (also **kaffiyeh**) a Bedouin Arab's kerchief worn as a head-dress. [Arab. keffiya, kūfiyya, perh. f. LL cofea COIF]

Keflavik /'keflə,vɪk/ a fishing port in SW Iceland; pop. (1990) 7,525. Iceland's international airport is located nearby.

keg /keg/ n. a small barrel, usu. of less than 10 gallons or (in the US) 30 gallons. □ **keg beer** Brit. beer to which carbon dioxide has been added, supplied from a sealed metal container. [ME cag f. ON kaggi, of unkn. orig.]

keister /'ki:stə/, 'kaɪs-/ n. US sl. **1** the buttocks. **2** a suitcase, satchel, handbag, etc. [19th c.: orig. unkn.]

Kekulé /'kekjʊ,leɪ/, Friedrich August, (full name Friedrich August Kekulé von Stradonitz) (1829–96), German chemist. He was one of the founders of structural organic chemistry, in which he was perhaps helped by his early training as an architect. He suggested in 1858 that carbon was tetravalent, and that carbon atoms could combine with others to form complex chains. Kekulé is best known for discovering the ring structure of benzene, the key to understanding many organic compounds.

Kelantan /kə'læntən/ a state of northern Malaysia, on the east coast of the Malay Peninsula; capital, Kota Baharu.

Keller /'kelə(r)/, Helen (Adams) (1880–1968), American writer, social reformer, and academic. Blind and deaf from the age of nineteen months, she learned how to read, type, and speak and went on to champion the cause of blind, deaf, and dumb people throughout the world. She is particularly remembered for her campaigning in aid of the American Foundation for the Blind.

Kellogg /'kelɒg/, Will Keith (1860–1951), American food manufacturer. He collaborated with his brother, a doctor, to develop a process of manufacturing a breakfast cereal for sanatorium patients consisting of crisp flakes of rolled and toasted wheat and corn. The product's success led to the establishment of the W. K. Kellogg company in 1906 and a subsequent revolution in Western eating habits.

Kellogg Pact (also called Kellogg–Briand Pact) a treaty renouncing war as an instrument of national policy, signed in Paris in 1928 by representatives of fifteen nations. It grew out of a proposal made by the French Premier Aristide Briand (1862–1932) to Frank B. Kellogg (1856–1937), US Secretary of State.

Kells, Book of /kelz/ an illuminated manuscript of the Gospels, perhaps made by Irish monks in Iona in the 8th or early 9th century, now kept at Trinity College, Dublin. [Kells, town in County Meath, Ireland, where formerly kept]

Kelly[1] /'kelɪ/, Edward ('Ned') (1855–80), Australian outlaw. He was the leader of a band of horse and cattle thieves and bank raiders operating in Victoria. A bushranger from 1878, Kelly was eventually hanged in Melbourne.

Kelly[2] /'kelɪ/, Gene (born Eugene Curran Kelly) (1912–96), American dancer and choreographer. He began his career on Broadway in 1938 and made a successful transition to film with For Me and My Girl (1942). He went on to perform in and choreograph many film musicals, including Anchors Aweigh (1945), An American in Paris (1951), and Singin' in the Rain (1952).

Kelly[3] /'kelɪ/, Grace (Patricia) (also called (from 1956) Princess Grace of Monaco) (1928–82), American film actress. Her first starring role came in the classic western High Noon (1952). She won an Oscar for her performance in The Country Girl (1954) and also made three Hitchcock films, including Rear Window (1954). Kelly retired from films in 1956

on her marriage to Prince Rainier III of Monaco (b.1923). She died in a road accident.

Kelly[4] /'kelɪ/, Petra (Karin) (1947–92), German political leader. Formerly a member of the German Social Democratic Party, she became disillusioned with their policies and in 1979 co-founded the Green Party, a broad alliance of environmentalists, feminists, and anti-nuclear activists. She became the Party's leading spokesperson and in 1983 was one of seventeen Green Party members elected to the West German Parliament. The cause of her death remains a subject of controversy.

keloid /'ki:lɔɪd/ n. fibrous tissue formed at the site of a scar or injury. [Gk khēlē claw + -OID]

kelp /kelp/ n. **1** a large broad-fronded brown seaweed; esp. of the genus Laminaria, suitable for use as manure, or Macrocystis, of the Pacific coast of America. **2** the calcined ashes of seaweed, formerly used in glass-making and soap manufacture because of their high content of sodium, potassium, and magnesium salts. [ME cülp(e), of unkn. orig.]

kelpie /'kelpɪ/ n. Sc. **1** a water-spirit, usu. in the form of a horse, reputed to delight in the drowning of travellers etc. **2** an Australian sheepdog orig. bred from a Scottish collie. [18th c.: orig. unkn.]

kelson var. of KEELSON.

kelt /kelt/ n. a salmon or sea trout after spawning. [ME: orig. unkn.]

kelter var. of KILTER.

Kelvin /'kelvɪn/, William Thomson, 1st Baron (1824–1907), British physicist, professor of natural philosophy at Glasgow from 1846 to 1895. He restated the second law of thermodynamics in 1850, and introduced the absolute scale of temperature. His concept of an electromagnetic field influenced Maxwell's electromagnetic theory of light, which Kelvin never accepted. He was involved in the laying of the first Atlantic cable, for which he invented several instruments, and he devised many scientific instruments for other purposes. Kelvin's calculation of an age for the earth, although a gross underestimate, showed that physics could be useful in helping to establish a geological time-scale.

kelvin /'kelvɪn/ n. the SI base unit of thermodynamic temperature (symbol: **K**). One kelvin is the fraction $1/_{273.16}$ of the thermodynamic temperature of the triple point of water, and is equal in magnitude to the degree Celsius. [KELVIN]

Kelvin scale a scale of temperature with absolute zero as zero and the triple point of water exactly 273.16 degrees. (See also KELVIN.)

Kemal Pasha /ke'mɑːl ,pɑːʃə/ see ATATÜRK.

Kemble[1] /'kemb(ə)l/, Frances Anne ('Fanny') (1809–93), English actress. The daughter of Charles Kemble and the niece of Sarah Siddons, she was a success in both Shakespearian comedy and tragedy, playing such parts as Portia, Beatrice, Juliet, and Lady Macbeth.

Kemble[2] /'kemb(ə)l/, John Philip (1757–1823), English actor-manager, brother of Sarah Siddons. He was noted for his performances in Shakespearian tragedy, notably as Hamlet and Coriolanus, and for his interpretations of historical roles such as Brutus in Julius Caesar. He was manager of Drury Lane (1788–1803) and Covent Garden (1803–17) theatres. His younger brother Charles (1775–1854) was also a successful actor-manager.

Kemerovo /'kemɪrəvə/ an industrial city in south central Russia, to the east of Novosibirsk; pop. (est. 1989) 520,000.

kemp /kemp/ n. coarse hair in wool. □ **kempy** adj. [ME f. ON kampr beard, whisker]

Kempis, Thomas à, see THOMAS À KEMPIS.

kempt /kempt/ adj. combed; neatly kept. [past part. of (now dial.) kemb COMB v. f. OE cemban f. Gmc]

ken /ken/ n. & v. ● n. range of sight or knowledge (it's beyond my ken). ● v.tr. (**kenning**; past and past part. **kenned** or **kent** /kent/) Sc. & N. Engl. **1** recognize at sight. **2** know. [OE cennan f. Gmc]

Kendal /'kend(ə)l/ a town in Cumbria, NW England; pop. (1981) 24,200.

Kendall /'kend(ə)l/, Edward Calvin (1886–1972), American biochemist. He was the first to isolate crystalline thyroxine from the thyroid gland. From the adrenal cortex he obtained a number of steroid hormones, one of which was later named cortisone, and several of which are now of great value in the treatment of rheumatic, allergic, and inflammatory diseases. He shared a Nobel Prize in 1950.

kendo /'kendəʊ/ n. a Japanese form of fencing with two-handed bamboo swords. Its origins go back more than 1,500 years; it developed from the need for non-lethal practice in the art of swordsmanship, which was an essential skill for the samurai of medieval Japan. [Jap., = sword-way]

Keneally /kəˈnælɪ, -ˈniːlɪ/, Thomas (Michael) (b.1935), Australian novelist. He first gained notice for *The Chant of Jimmy Blacksmith* (1972). Later works include war novels such as *Confederates* (1979), and the Booker Prize-winning *Schindler's Ark* (1982), the true story of the German industrialist Oskar Schindler, who helped more than 1,200 Jews to escape death in Nazi concentration camps; the book was filmed by Steven Spielberg in 1993 as *Schindler's List*.

Kennedy[1] /ˈkenədɪ/, Edward Moore ('Teddy') (b.1932), American Democratic politician. The brother of John F. Kennedy and Robert F. Kennedy, he was elected to the Senate in 1962. His subsequent political career was overshadowed by his involvement in a mysterious fatal car accident at Chappaquiddick Island (1969), although he remains a prominent Democratic spokesman.

Kennedy[2] /ˈkenədɪ/, John F(itzgerald) (known as 'JFK') (1917-63), American Democratic statesman, 35th President of the US 1961-3. A national war hero during the Second World War, Kennedy became, at 43, the youngest man ever to be elected President, as well as the first Catholic. He gained a popular reputation as an advocate of civil rights, although reforms were delayed by Congress until 1964. In foreign affairs he recovered from the Bay of Pigs fiasco to demand successfully the withdrawal of Soviet missiles from Cuba (see CUBAN MISSILE CRISIS), and negotiated the Test-Ban Treaty of 1963 with the USSR and the UK. Kennedy was assassinated while riding in a motorcade through Dallas, Texas, in Nov. 1963; Lee Harvey Oswald was charged with his murder (see OSWALD).

Kennedy[3] /ˈkenədɪ/, Robert F(rancis) (1925-68), American Democratic statesman. The brother of John F. Kennedy and Edward Kennedy, he closely assisted his brother John in domestic policy, serving as Attorney-General (1961-4), and was a champion of the civil-rights movement. Robert Kennedy stood as a prospective presidential candidate in 1968, but was assassinated during his campaign.

kennel /ˈken(ə)l/ *n. & v.* ● *n.* **1** a small shelter for a dog. **2** (in *pl.*) a breeding or boarding establishment for dogs. **3** a mean dwelling. ● *v.* (**kennelled, kennelling**; *US* **keneled, kenneling**) **1** *tr.* put into or keep in a kennel. **2** *intr.* live in or go to a kennel. [ME f. OF *chenil* f. med.L *canile* (unrecorded) f. L *canis* dog]

Kennelly /ˈkenəlɪ/, Arthur Edwin (1861-1939), American electrical engineer. His principal work was on the theory of alternating currents, and he also worked on the practical problems of electrical transmission. Kennelly independently discovered the layer in the atmosphere responsible for reflecting radio waves back to earth (see HEAVISIDE). He helped to develop electrical units and standards, and promoted the adoption of the metric system.

Kenneth I /ˈkenɪθ/ (known as Kenneth MacAlpin) (d.858), king of Scotland *c*.844-58. He is traditionally viewed as the founder of the kingdom of Scotland, which was established following Kenneth's defeat of the Picts in about 844.

kenning /ˈkenɪŋ/ *n.* a compound expression in Old English and Old Norse poetry, e.g. *oar-steed* = ship. [ME, = 'teaching' etc. f. KEN]

kenosis /kɪˈnəʊsɪs/ *n.* (in Christian theology) the renunciation of the divine nature, at least in part, by Christ in the Incarnation. □ **kenotic** /-ˈnɒtɪk/ *adj.* [Gk. *kenōsis* f. *kenoō* to empty f. *kenos* empty]

Kensington /ˈkenzɪŋtən/ a fashionable residential district in central London. Part of the borough of Kensington and Chelsea, it contains Kensington Palace (the birthplace of Queen Victoria), Kensington Gardens, and the Victoria and Albert Museum, Natural History Museum, and Science Museum.

kenspeckle /ˈkenˌspek(ə)l/ *adj.* Sc. conspicuous. [*kenspeck* of Scand. orig.: rel. to KEN]

Kent[1] /kent/ a county on the SE coast of England; county town, Maidstone.

Kent[2] /kent/, William (*c*.1685-1748), English architect and landscape gardener. He promoted the Palladian style of architecture in England and is renowned for such works as the Treasury (1733-7) and Whitehall (1734-6). Holkham Hall, begun in 1734, was one of the first English buildings to feature interiors and furniture designed by the architect. Kent is chiefly remembered, however, for his landscape gardens at Stowe House in Buckinghamshire (*c*.1730). His design principles overturned the formal taste of the time and anticipated the innovations of Capability Brown.

kent *past* and *past part.* of KEN.

Kentish /ˈkentɪʃ/ *adj.* of Kent in England. [OE *Centisc* f. *Cent* f. L *Cantium*]

Kentish fire *n. Brit.* a prolonged volley of rhythmic applause or a demonstration of dissent.

kentledge /ˈkentlɪdʒ/ *n. Naut.* pig-iron etc. used as permanent ballast. [F *quintelage* ballast, with assim. to *kentle* obs. var. of QUINTAL]

Kentucky /kenˈtʌkɪ/ a state in the south-eastern US; pop. (1990) 3,685,300; capital, Frankfort. Ceded by the French to the British in 1763, Kentucky entered the Union as the 15th state in 1792. It is also known as the Bluegrass state.

Kentucky Derby an annual horse-race for three-year-olds at Louisville, Kentucky. First held in 1875, it is the oldest horse-race in the US.

Kenya /ˈkenjə/ an equatorial country in East Africa, on the Indian Ocean; pop. (est. 1991) 25,016,000; languages, Swahili (official), English (official), Kikuyu; capital, Nairobi. Largely populated by Bantu-speaking peoples, Kenya was not exposed to European influence until the arrival of the British in the late 19th century. After the opening up of the interior it became a Crown Colony in 1920. The demands made on land by European settlers caused discontent, resulting in the Mau Mau rebellion of the 1950s. Kenya became an independent state within the Commonwealth in 1963, and a republic was established the following year. □ **Kenyan** *adj. & n.*

Kenya, Mount a mountain in central Kenya, just south of the equator, rising to a height of 5,200 m (17,058 ft). The second highest mountain in Africa, it gave its name to the country Kenya.

Kenya African National Union (abbr. **KANU**) a Kenyan political party formed in 1960 and led first by Jomo Kenyatta. KANU won the first Kenyan elections and took the country into independence in 1963; it has since dominated Kenyan politics, ruling as the sole legal party 1982-1991.

Kenyatta /kenˈjætə/, Jomo (*c*.1891-1978), Kenyan statesman, Prime Minister of Kenya 1963 and President 1964-78. He was imprisoned from 1952 to 1961 for alleged complicity in the Mau Mau uprising. On his release he was elected president of the Kenya African National Union and led his country to independence in 1963, subsequently serving as independent Kenya's first President.

kepi /ˈkepɪ, ˈkeɪpɪ/ *n.* (*pl.* **kepis**) a French military cap with a horizontal peak. [F *képi* f. Swiss G *Käppi* dimin. of *Kappe* cap]

Kepler /ˈkeplə(r)/, Johannes (1571-1630), German astronomer. He settled in Prague in 1599, becoming Tycho Brahe's assistant and later court mathematician to the emperor. His analysis of Brahe's planetary observations enabled him to discover the laws governing planetary motion, and foreshadowed the general application of scientific method to astronomy. The first two laws recognized the elliptical orbits of the planets; in *Harmonices Mundi* (1620) he expounded the third law of planetary dynamics, relating the distances of the planets from the sun to their orbital periods. Despite these advances, Kepler's approach remained medieval in spirit; he believed in the music of the spheres and sought an inner relationship between the planets that would express it.

kept *past* and *past part.* of KEEP.

Kerala /ˈkerələ/ a state on the coast of SW India; capital, Trivandrum. It was created in 1956 from the former state of Travancore-Cochin and part of Madras. □ **Keralite** *adj. & n.*

keratin /ˈkerətɪn/ *n.* any of a group of fibrous proteins which occur in hair, feathers, hooves, claws, horns, etc. [Gk *keras keratos* horn + -IN]

keratinize /ˈkerətɪˌnaɪz/ *v.tr. & intr.* (also **-ise**) cover or become covered with a deposit of keratin. □ **keratinization** /ˌkerətɪnaɪˈzeɪʃ(ə)n/ *n.*

keratitis /ˌkerəˈtaɪtɪs/ *n. Med.* inflammation of the cornea of the eye. [*kerat-* used to denote the cornea (f. Gk *kerat- keras* horn) + -ITIS]

keratotomy /ˌkerəˈtɒtəmɪ/ *n. Med.* a surgical operation involving cutting into the cornea of the eye, esp. (in full **radial keratotomy**) to correct myopia. [*kerato-* cornea (f. Gk *keras -atos* horn) + -TOMY]

kerb /kɜːb/ *n. Brit.* a stone edging to a pavement or raised path. □ **kerb-crawler** a (usu. male) person who indulges in kerb-crawling. **kerb-crawling** the practice of driving slowly along the edge of a road in an attempt to engage a prostitute or harass esp. female passers-by. **kerb drill** *Brit.* precautions, esp. looking to right and left, before crossing a road. [var. of CURB]

kerbstone /ˈkɜːbstəʊn/ *n. Brit.* each of a series of stones forming a kerb.

Kerch /kɜːtʃ/ a city in southern Ukraine, the chief port and industrial centre of the Crimea, at the eastern end of the Kerch peninsula; pop. (1990) 176,000. Founded in the 6th century BC, it was originally called Panticapaeum. Renamed Korchev by the Genoese in the 14th century, it was captured by the Crimean Tartars in 1475, who lost it to the Russian empire in 1771.

kerchief /ˈkɜːtʃiːf, -tʃɪf/ n. **1** a cloth used to cover the head. **2** poet. a handkerchief. □ **kerchiefed** adj. [ME curchef f. AF courchef, OF couvrechief f. couvrir COVER + CHIEF head]

kerf /kɜːf/ n. **1** a slit made by cutting, esp. with a saw. **2** the cut end of a felled tree. [OE cyrf f. Gmc (as CARVE)]

kerfuffle /kəˈfʌf(ə)l/ n. esp. Brit. colloq. a fuss or commotion. [Sc. curfuffle f. fuffle to disorder: imit.]

Kerguelen Islands /ˈkɜːgɪlɪn, kɜːˈgeɪlən/ a group of islands in the southern Indian Ocean, comprising the island of Kerguelen and some 300 small islets, forming part of French Southern and Antarctic Territories. The only settlement is a scientific base. They are named after the Breton navigator Yves-Joseph de Kerguélen-Trémarec, who discovered them in 1772.

Kérkira /ˈkɛrkira/ the modern Greek name for CORFU.

Kerkrade /ˈkɜːkˌrɑːdə/ a mining town in the southern Netherlands, on the German border; pop. (1991) 53,280. An international music competition is held there every four years.

Kermadec Islands /kɜːˈmædək/ a group of uninhabited islands in the western South Pacific, north of New Zealand, administered by New Zealand since 1887.

kermes /ˈkɜːmɪz/ n. **1** (in full **kermes oak**) a small evergreen oak, Quercus coccifera, of the Mediterranean region. **2 a** the female of a scale insect, Kermes ilicis, which forms berry-like galls on the kermes oak. **b** a red dye made from the dried bodies of these. **3** (in full **kermes mineral**) a bright red hydrous trisulphide of antimony. [F kermès f. Arab. & Pers. ḳirmiz: rel. to CRIMSON]

kermis /ˈkɜːmɪs/ n. **1** a periodical country fair, esp. in the Netherlands. **2** US a charity bazaar. [Du., orig. = mass on the anniversary of the dedication of a church, when yearly fair was held: f. kerk formed as CHURCH + mis, misse MASS²]

Kern /kɜːn/, Jerome (David) (1885–1945), American composer. He wrote several musical comedies, including Showboat (1927), which proved a major influence in the development of the musical. It also featured the song 'Ol' Man River', first sung by Paul Robeson.

kern¹ /kɜːn/ n. Printing the part of a metal type projecting beyond its body or shank. □ **kerned** adj. [perh. f. F carne corner f. OF charne f. L cardo cardinis hinge]

kern² /kɜːn/ n. (also **kerne**) **1** hist. a light-armed Irish foot-soldier. **2** archaic a rustic, a peasant; a boor. [ME f. Ir. ceithern]

kernel /ˈkɜːn(ə)l/ n. **1** a central, softer, usu. edible part within a hard shell of a nut, fruit stone, seed, etc. **2** the whole seed of a cereal. **3** the nucleus or essential part of anything. [OE cyrnel, dimin. of CORN¹]

kero /ˈkerəʊ/ n. Austral. = KEROSENE. [abbr.]

kerosene /ˈkerəˌsiːn/ n. (also **kerosine**) esp. US a fuel oil suitable for use in jet engines and domestic heating boilers; paraffin oil. [Gk kēros wax + -ENE]

Kerouac /ˈkeruˌæk/, Jack (born Jean-Louis Lebris de Kérouac) (1922–69), American novelist and poet, of French-Canadian descent. A leading figure in the beat generation, he is best known for his semi-autobiographical novel On the Road (1957). Other works include Big Sur (1962).

Kerry¹ /ˈkerɪ/ a county of the Republic of Ireland, on the SW coast in the province of Munster; county town, Tralee.

Kerry² /ˈkerɪ/ n. (pl. **-ies**) a breed of small black dairy cattle originating in County Kerry. □ **Kerry blue** a breed of terrier with a silky blue-grey coat.

kersey /ˈkɜːzɪ/ n. (pl. **-eys**) **1** a kind of coarse narrow cloth woven from long wool, usu. ribbed. **2** a variety of this. [ME, prob. f. Kersey in Suffolk]

kerseymere /ˈkɜːzɪˌmɪə(r)/ n. a twilled fine woollen cloth. [alt. of cassimere, var. of CASHMERE, assim. to KERSEY]

Kesey /ˈkiːzɪ/, Ken (Elton) (b.1935), American novelist. His best-known novel, One Flew over the Cuckoo's Nest (1962), is based on his experiences as a ward attendant in a mental hospital. Kesey's adventures with the Merry Pranksters, a group who pioneered the use of psychedelic drugs, are described in The Electric Kool-Aid Acid Test (1967) by Tom Wolfe (b.1931).

keskidee var. of KISKADEE.

Kesteven /keˈstiːv(ə)n/ one of three administrative divisions of Lincolnshire before the local government reorganization of 1974 (the others were Holland and Lindsey), comprising the inland, south-

western parts of the county. Since 1974 it has been administered as the districts of North and South Kesteven.

kestrel /ˈkestrəl/ n. a small falcon which hovers while searching for its prey; esp. Falco tinnunculus of the Old World. [ME castrell, perh. f. F dial. casserelle, F créc(er)elle, perh. imit. of its cry]

Keswick /ˈkezɪk/ a market town and tourist centre in Cumbria, NW England; pop. (1981) 5,645. It is situated on the northern shores of Derwent Water.

ketch /ketʃ/ n. a two-masted fore-and-aft rigged sailing-boat with a mizen-mast stepped forward of the rudder and smaller than its foremast. [ME catche, prob. f. CATCH]

ketchup /ˈketʃʌp/ n. (also **catchup** /ˈkætʃ-/) a spicy sauce made from tomatoes, mushrooms, vinegar, etc., used as a condiment. [Chin. dial. kōechiap pickled-fish brine]

ketone /ˈkiːtəʊn/ n. any of a class of organic compounds in which two hydrocarbon groups are linked by a carbonyl group, e.g. propanone (acetone). □ **ketone body** Biochem. each of three related compounds including acetone, produced during the metabolism of fats. □ **ketonic** /kɪˈtɒnɪk/ adj. [G Keton alt. of Aketon ACETONE]

ketonuria /ˌkiːtəʊˈnjʊərɪə/ n. Med. the excretion of abnormally large amounts of ketone bodies in the urine, characteristic of diabetes mellitus, starvation, etc.

ketosis /kɪˈtəʊsɪs/ n. Med. a condition characterized by raised levels of ketone bodies in the body, associated with fat metabolism and diabetes mellitus. □ **ketotic** /-ˈtɒtɪk/ adj.

Kettering /ˈketərɪŋ/, Charles Franklin (1876–1958), American automobile engineer. His first significant development was the electric starter (1912). He was leader of research at General Motors until 1947, discovering with his team tetraethyl lead as an antiknock agent and defining the octane rating of fuels. He did important work on two-stroke diesel engines, which came into widespread use for railway locomotives and road coaches, and was also responsible for the development of synchromesh gearboxes, automatic transmissions, and power steering.

kettle /ˈket(ə)l/ n. a vessel, usu. of metal with a lid, spout, and handle, for boiling water in. □ **kettle hole** a depression in the ground resulting from the melting of an ice block trapped in glacial deposits. **a different kettle of fish** a different matter altogether. **a pretty** (or **fine**) **kettle of fish** an awkward state of affairs. □ **kettleful** n. (pl. **-fuls**). [ME f. ON ketill ult. f. L catillus dimin. of catinus deep food-vessel]

kettledrum /ˈket(ə)lˌdrʌm/ n. a large drum shaped like a bowl with a membrane adjustable for tension (and so pitch) stretched across. □ **kettledrummer** n.

keV abbr. kilo-electronvolt.

Kevlar /ˈkevlə(r)/ n. propr. a synthetic fibre of high tensile strength used esp. as a reinforcing agent in the manufacture of rubber products, e.g. tyres and bulletproof vests.

Kew Gardens /kjuː/ the Royal Botanic Gardens at Kew, in Richmond, London. Developed by the mother of George III with the aid of Sir Joseph Banks, the gardens were presented to the nation in 1841 and are now an important botanical institution.

kewpie /ˈkjuːpɪ/ n. a small chubby doll with wings and a curl or topknot. [CUPID + -IE]

key¹ /kiː/ n., adj., & v. ● n. (pl. **keys**) **1** an instrument, usu. of metal, for moving the bolt of a lock forwards or backwards to lock or unlock. **2** a similar implement for operating a switch in the form of a lock. **3** an instrument for grasping screws, pegs, nuts, etc., esp. one for winding a clock etc. **4** a lever depressed by the finger in playing the organ, piano, flute, concertina, etc. **5** (often in pl.) each of several buttons for operating a typewriter, word processor, computer terminal, etc. **6** a thing that gives or precludes the opportunity for or access to something. **7** a place that by its position gives control of a sea, territory, etc. **8 a** a solution or explanation. **b** a word or system for solving a cipher or code. **c** an explanatory list of symbols used in a map, table, etc. **d** a book of solutions to mathematical problems etc. **e** a literal translation of a book written in a foreign language. **f** the first move in a chess-problem solution. **9** Mus. a group of notes based on a particular note and comprising a scale, regarded as a unit forming the tonal basis of a piece of music (a study in the key of C major). **10** a tone or style of thought or expression. **11** a piece of wood or metal inserted between others to secure them. **12** the part of a first coat of wall plaster that passes between the laths and so secures the rest. **13** the roughness of a surface, helping the adhesion of plaster etc. **14** the samara of a sycamore etc. **15** a mechanical device for making or breaking an electric circuit, e.g. in telegraphy. **16** *Basketball* the area around the basket outside which all players except the thrower must stand at a free throw. ● *adj.* essential; of vital importance (*the key element in the problem*). ● *v.tr.* (**keys, keyed**) **1** (foll. by *in, on*, etc.) fasten with a pin, wedge, bolt, etc. **2** (often foll. by *in*) enter (data) by means of a keyboard. **3** roughen (a surface) to help the adhesion of plaster etc. **4** (foll. by *to*) align or link (one thing to another). **5** regulate the pitch of the strings of (a violin etc.). **6** word (an advertisement in a particular periodical) so that answers to it can be identified (usu. by varying the form of address given). □ **key industry** an industry essential to the carrying on of others, e.g. coal mining or dyeing. **key map** a map in bare outline, to simplify the use of a full map. **key money** *Brit.* a payment demanded from an incoming tenant for the provision of a key to the premises. **key-ring** a ring for keeping keys on. **key signature** *Mus.* any of several combinations of sharps or flats after the clef at the beginning of each staff, indicating the key of a composition. **key up** (often foll. by *to*, or *to* + infin.) make (a person) nervous or tense; excite. □ **keyer** *n.* **keyless** *adj.* [OE *cǣg*, of unkn. orig.]

key[2] /kiː/ *n.* a low-lying island or reef, esp. in the West Indies (cf. CAY). [Sp. *cayo* shoal, reef, infl. by QUAY]

keyboard /ˈkiːbɔːd/ *n. & v.* ● *n.* **1** a set of keys on a typewriter, computer, piano, etc. **2** an electronic musical instrument with keys arranged as on a piano. ● *v.* **1** *tr.* enter (data) by means of a keyboard. **2** *intr.* work at a keyboard. □ **keyboarder** *n.* (in sense 1 of *n.*). **keyboardist** *n.* (in sense 2 of *n.*).

keyhole /ˈkiːhəʊl/ *n.* a hole by which a key is put into a lock. □ **keyhole surgery** minimally invasive surgery carried out through a very small incision.

Key Largo /ˈlɑːgəʊ/ a resort island off the south coast of Florida, the northernmost and the longest of the Florida Keys.

Keynes /keɪnz/, John Maynard, 1st Baron (1883–1946), English economist. Keynes served as an adviser to the Treasury during both world wars and was its representative at the Versailles peace conference (1919), subsequently becoming one of the most influential critics of the Treaty of Versailles. He laid the foundations of modern macroeconomics with *The General Theory of Employment, Interest and Money* (1936). In this he argued that full employment is not a natural condition but is determined by effective demand, requiring government spending on public works to stimulate this. His theories influenced Roosevelt's decision to introduce the American New Deal. □ **Keynesian** *adj. & n.* **Keynesianism** *n.*

keynote /ˈkiːnəʊt/ *n.* **1** a prevailing tone or idea (*the keynote of the whole occasion*). **2** (*attrib.*) intended to set the prevailing tone at a meeting or conference (*keynote address*). **3** *Mus.* the note on which a key is based.

keypad /ˈkiːpæd/ *n.* a miniature keyboard or set of buttons for operating a portable electronic device, telephone, etc.

keypunch /ˈkiːpʌntʃ/ *n. & v.* ● *n.* a device for transferring data by means of punched holes or notches on a series of cards or paper tape. ● *v.tr.* transfer (data) by means of a keypunch. □ **keypuncher** *n.*

Keys, House of see HOUSE OF KEYS.

Keystone /ˈkiːstəʊn/ a US film company formed in 1912. The company is remembered for its silent slapstick comedy films, many featuring the bumbling Keystone Kops police characters.

keystone /ˈkiːstəʊn/ *n.* **1** the central principle of a system, policy, etc., on which all the rest depends. **2** a central stone at the summit of an arch locking the whole together.

keystroke /ˈkiːstrəʊk/ *n.* a single depression of a key on a keyboard, esp. as a measure of work.

keyway /ˈkiːweɪ/ *n.* a slot for receiving a machined key.

Key West a city in southern Florida, at the southern tip of the Florida Keys; pop. (1990) 24,800. It is the southernmost city in continental USA.

keyword /ˈkiːwɜːd/ *n.* **1** the key to a cipher etc. **2 a** a word of great significance. **b** a word used in an information-retrieval system to indicate the content of a document etc.

KG *abbr.* (in the UK) Knight of the Order of the Garter.

kg *abbr.* kilogram(s).

KGB the Soviet secret police organization created on Stalin's death (1953) to take over state security, with responsibility for external espionage, internal counter-intelligence, and internal 'crimes against the state'. Its most famous chairman was Yuri Andropov. The KGB was dissolved in Oct. 1991. [abbr. of Russ. *Komitet Gosudarstvennoĭ Bezopasnosti* Committee of State Security]

Kgs. *abbr.* (in the Bible) Kings.

Khabarovsk /kəˈbɑːrɒfsk/ **1** a krai (administrative territory) on the east coast of Siberian Russia; pop. (1990) 1,840,000. **2** its capital, a city on the Amur river, on the Chinese border; pop. (1990) 608,000.

khaddar /ˈkædə(r)/ *n.* Indian homespun cloth. [Hindi]

Khakassia /kɑːˈkæsɪə/ an autonomous republic in south central Russia; pop. (est. 1990) 569,000; capital, Abakan.

khaki /ˈkɑːkɪ/ *adj. & n.* ● *adj.* dull brownish-yellow in colour. ● *n.* (*pl.* **khakis**) **1** khaki fabric of twilled cotton or wool, used esp. in military dress. **2** the dull brownish-yellow colour of this. [Urdu *kākī* dust-coloured f. *kāk* dust]

khalasi /kəˈlæsɪ/ *n.* (*pl.* **khalasis**) a servant or labourer in the Indian subcontinent, esp. one employed as a seaman. [Hind.]

Khalkís see CHALCIS.

Khama /ˈkɑːmə/, Sir Seretse (1921–80), Botswanan statesman, Prime Minister of Bechuanaland 1965 and President of Botswana 1966–80. An heir to the chieftainship of the ruling tribe in Bechuanaland, he was banished because of opposition to his marriage to an Englishwoman in 1948. He returned with his wife in 1956 and formed the Democratic Party in 1962, leading the party to a landslide victory in the elections of 1965; he became Botswana's first President the following year. A strong believer in multiracial democracy, he achieved nationwide free education.

Khambat, Gulf of see CAMBAY, GULF OF.

khamsin /ˈkæmsɪn/ *n.* (also **hamsin** /ˈhæm-/) an oppressive hot south or south-east wind occurring in Egypt for about fifty days in March, April, and May. [Arab. *ḵamsīn* f. *ḵamsūn* fifty]

Khan[1], Ayub, see AYUB KHAN.

Khan[2] /kɑːn/, Imran (full name Imran Ahmad Khan Niazi) (b.1952), Pakistani cricketer. He made his test début in 1970 and served as captain of his country in four periods between 1982 and 1992. A batsman and fast bowler, he also played county cricket for Worcestershire (1971–6) and Sussex (1977–88).

Khan[3] /kɑːn/, Jahangir (b.1963), Pakistani squash player. In 1979 he became world amateur champion at the age of 15; after turning professional he was world squash champion five consecutive times (1981–5), and again in 1988.

khan[1] /kɑːn, kæn/ *n.* **1** a title given to rulers and officials in central Asia, Afghanistan, etc. **2** *hist.* **a** the supreme ruler of the Turkish, Tartar, and Mongol tribes. **b** the emperor of China in the Middle Ages. □ **khanate** /-neɪt/ *n.* [Turkic *ḵān* lord]

khan[2] /kɑːn, kæn/ *n.* a caravanserai. [Arab. *ḵān* inn]

Khaniá see CHANIA.

Kharg Island /kɑːg/ a small island at the head of the Persian Gulf, site of Iran's principal deep-water oil terminal.

Kharkiv /ˈhɑːkɪv, Ukrainian ˈxarkiw/ (Russian **Kharkov** /ˈxarjkəf/) an industrial city in NE Ukraine, in the Donets basin; pop. (1990) 1,618,000. It was the first capital of the Ukrainian Soviet Socialist Republic, from 1919 until replaced by Kiev in 1934.

Khartoum /kɑːˈtuːm/ the capital of Sudan, situated at the junction of the Blue Nile and the White Nile; pop. (1983) 476,220. Originally established in 1821 as an Egyptian army camp, it developed into a garrison town. In 1885 a British and Egyptian force under the command of General Gordon was besieged there for ten months by the Mahdists, who eventually stormed the garrison, killing most of the defenders, including Gordon. It remained under the control of the Mahdists until they were defeated by the British in 1898 and the city was recaptured by General Kitchener. It was the capital of the Anglo-Egyptian government of Sudan until 1956, when it became capital of the independent Republic of Sudan.

khat /kɑːt/ *n.* **1** a shrub, *Catha edulis*, grown in Arabia. **2** the leaves of this shrub, chewed or infused as a stimulant. [Arab. *ḵāt*]

Khaylitsa /kaɪˈlɪtsə/ a township 40 km (25 miles) south-east of Cape Town, South Africa. Designed to accommodate 250,000 people, it was built in 1983 for black Africans from the squatter camps of Crossroads, Langa, and KTC.

Khedive /krˈdiːv/ *n.* *hist.* the title of the viceroy of Egypt under Turkish rule 1867–1914. □ **Khedival** *adj.* **Khedivial** *adj.* [F *khédive*, ult. f. Pers. *ḵadīv* prince]

Kherson /kɪəˈsɒn/ a port on the south coast of Ukraine, on the Dnieper estuary; pop. (1990) 361,000.

Khios see CHIOS.

Khitai see CATHAY.

Khmer /kmeə(r)/ n. **1** a native of the ancient kingdom of Khmer in SE Asia, which reached the peak of its power in the 11th century, when it ruled over the entire Mekong valley from the capital at Angkor, and was destroyed by Siamese conquests in the 12th and 14th centuries. **2** a native of the Khmer Republic or Cambodia. **3** the official language of Cambodia, spoken by about 7 million people. It is largely monosyllabic, and is the most important member of the Mon-Khmer group of the Austro-Asiatic family. Also called *Cambodian*. [Khmer]

Khmer Republic the former official name (1970–5) for CAMBODIA.

Khmer Rouge /ruːʒ/ a Communist guerrilla organization which opposed the Cambodian government in the 1960s and waged a civil war from 1970, taking power in 1975. Under Pol Pot the Khmer Rouge undertook a forced reconstruction of Cambodian society, involving mass deportations from the towns to the countryside and the execution of many thousands of Cambodians as 'bourgeois elements'. More than two million died before the regime was overthrown by the Vietnamese in 1979. Khmer Rouge forces have continued a programme of guerrilla warfare from bases in Thailand, and have resisted UN attempts to negotiate a lasting peace treaty.

Khoikhoi /ˈkɔɪkɔɪ/ n. see NAMA. [Nama, lit. 'men of men']

Khoisan /ˈkɔɪsaːn/ n. & adj. ● n. **1** a collective term for the Nama (Khoikhoi) and the San (Bushmen) of southern Africa. **2** a southern African language family, the smallest in Africa, spoken mainly by the Nama and the San. Khoisan languages are distinguished by clicks (made by suction with the tongue), which function as consonants. ● adj. of or relating to these languages or their speakers. [f. KHOIKHOI + SAN]

Khomeini /xɒˈmeɪnɪ/, Ruhollah (known as Ayatollah Khomeini) (1900–89), Iranian Shiite Muslim leader. After sixteen years in exile he returned to Iran in 1979 to lead an Islamic revolution which overthrew the shah. He established a fundamentalist Islamic republic, supported the seizure of the US embassy (1979) by Iranian students, and relentlessly pursued the Iran–Iraq War 1980–8. In 1989 he issued a fatwa condemning Salman Rushdie, author of *The Satanic Verses*, provoking criticism from the West.

Khonsu /ˈkɒnsuː/ *Egyptian Mythol.* a moon-god worshipped especially at Thebes, a member of a triad as the divine son of Amun and Mut. His name means 'he who crosses'.

Khorramshahr /ˌxɔːrəmˈʃɑː(r)/ an oil port on the Shatt al-Arab waterway in western Iran, known as Mohammerah until 1924. It was almost totally destroyed during the Iran–Iraq War of 1980–8.

khoum /kuːm/ n. a monetary unit of Mauritania, equal to one-fifth of an ouguiya. [Arab. *ḳums* one-fifth]

Khrushchev /ˈkruʃtʃɒf/, Nikita (Sergeevich) (1894–1971), Soviet statesman, Premier of the USSR 1958–64. Born in Ukraine, Khrushchev became First Secretary of the Communist Party of the USSR (1953–64) after the death of Stalin. He played a prominent part in the 'de-Stalinization' programme that began in 1956, denouncing the former leader in a historic speech, and went on to succeed Bulganin as Premier (Chairman of the Council of Ministers) in 1958. He came close to war with the US over the Cuban Missile Crisis in 1962 and clashed with China over economic aid and borders. He was ousted two years later by Brezhnev and Kosygin, largely because of his antagonism to China.

Khufu /ˈkuːfuː/ see CHEOPS.

Khulna /ˈkʊlnɑː/ an industrial city in southern Bangladesh, on the Ganges river; pop. (1991) 601,050.

Khunjerab Pass /ˈkʌnjəˌrɑːb/ a high-altitude pass through the Himalayas, on the Karakoram highway at a height of 4,900 m (16,088 ft), linking China and Pakistan.

Khyber Pass /ˈkaɪbə(r)/ a mountain pass in the Hindu Kush, on the border between Pakistan and Afghanistan at a height of 1,067 m (3,520 ft). The pass was for long of great commercial and strategic importance, the route by which successive invaders entered India, and was garrisoned by the British intermittently between 1839 and 1947.

kHz abbr. kilohertz.

kiang /kɪˈæŋ/ n. a wild ass, *Equus hemionus*, of a race native to Tibet, with a thick furry coat (cf. KULAN, ONAGER). [Tibetan *kyang*]

Kiangsi see JIANGXI.

Kiangsu see JIANGSU.

kibble¹ /ˈkɪb(ə)l/ v. & n. ● v.tr. grind or chop (dried corn, beans, etc.) coarsely. ● n. N. Amer. ground meal shaped into pellets esp. for pet food. [18th c.: orig. unkn.]

kibble² /ˈkɪb(ə)l/ n. Brit. an iron hoisting-bucket used in mines. [G *Kübel* (cf. OE *cyfel*) f. med.L *cupellus*, corn-measure, dimin. of *cuppa* cup]

kibbutz /kɪˈbʊts/ n. (pl. **kibbutzim** /ˌkɪbʊtˈsiːm/) a communal settlement in Israel, especially one engaged in farming, with collective holding of property and earnings, group involvement in decision-making, and communal rearing of children. Kibbutzim became established in Israel in the 1950s. [mod.Heb. *ḳibbūṣ* gathering]

kibbutznik /kɪˈbʊtsnɪk/ n. a member of a kibbutz. [Yiddish (as KIBBUTZ)]

kibe /kaɪb/ n. an ulcerated chilblain, esp. on the heel. [ME, prob. f. Welsh *cibi*]

kibitka /kɪˈbɪtkə/ n. **1** a type of Russian hooded sledge. **2 a** a Tartar's circular tent, covered with felt. **b** a Tartar household. [Russ. f. Tartar *kibitz*]

kibitz /ˈkɪbɪts/ v.intr. esp. N. Amer. colloq. act as a kibitzer. [Yiddish f. G *kiebitzen* (as KIBITZER)]

kibitzer /ˈkɪbɪtsə(r), kɪˈbɪt-/ n. esp. N. Amer. colloq. **1** an onlooker at cards etc., esp. one who offers unwanted advice. **2** a busybody, a meddler. [Yiddish *kibitser* f. G *Kiebitz* lapwing, busybody]

kiblah /ˈkɪblə/ n. (also **qibla**) **1** the direction of the Kaaba at Mecca, to which Muslims turn at prayer. **2** = MIHRAB. [Arab. *ḳibla* that which is opposite]

kibosh /ˈkaɪbɒʃ/ n. (also **kybosh**) sl. nonsense. □ **put the kibosh on** put an end to; finally dispose of. [19th c.: orig. unkn.]

kick¹ /kɪk/ v. & n. ● v. **1** tr. strike or propel forcibly with the foot or hoof etc. **2** intr. (usu. foll. by at, against) **a** strike out with the foot. **b** express annoyance at or dislike of (treatment, a proposal, etc.); rebel against. **3** tr. colloq. give up (a habit). **4** tr. (often foll. by out etc.) expel or dismiss forcibly. **5** refl. be annoyed with oneself (I'll kick myself if I'm wrong). **6** tr. Football score (a goal) by a kick. **7** intr. Cricket (of a ball) rise sharply from the pitch. **8** intr. (as **kicking** adj.) sl. lively, exciting; excellent. ● n. **1 a a** blow with the foot or hoof etc. **b** the delivery of such a blow. **2** colloq. **a** a sharp stimulant effect, esp. of alcohol (has some kick in it; a cocktail with a kick in it). **b** (often in pl.) a pleasurable thrill (did it just for kicks; got a kick out of flying). **3** strength, resilience (have no kick left). **4** colloq. a specified temporary interest or enthusiasm (on a jogging kick). **5** the recoil of a gun when discharged. **6** Football a player of specified kicking ability (is a good kick). □ **kick about** (or **around**) colloq. **1** a drift idly from place to place. **b** be unused or unwanted. **2 a** treat roughly or scornfully. **b** discuss (an idea) unsystematically. **kick against the pricks** see PRICK. **kick a person's ass** esp. N. Amer. sl. dominate, defeat, or punish a person. **kick ass** esp. N. Amer. sl. act forcefully or in a domineering manner. **kick-ass** adj. N. Amer. sl. forceful, aggressive, domineering. **kick the bucket** sl. die. **kick-down** a device for changing gear in a motor vehicle by full depression of the accelerator. **kick one's heels** see cool one's heels (see HEEL¹). **kick in 1** knock down (a door etc.) by kicking. **2** esp. US sl. contribute (esp. money); pay one's share. **3** become activated, start. **kick in the pants** (or **teeth**) colloq. a humiliating punishment or setback. **kick off 1 a** Football begin or resume a match. **b** begin. **2** remove (shoes etc.) by kicking. **kick-off 1** Football the start or resumption of a match. **2** (in **for a kick-off**) colloq. for a start (that's wrong for a kick-off). **kick over the traces** see TRACE². **kick-pleat** a pleat in a narrow skirt to allow freedom of movement. **kick-turn** a standing turn in skiing. **kick up** (or **kick up a fuss, dust**, etc.) colloq. create a disturbance; object or register strong disapproval. **kick up one's heels** frolic. **kick a person upstairs** colloq. remove a person from the scene of action by giving him or her ostensible promotion or a title. □ **kickable** adj. **kicker** n. [ME *kike*, of unkn. orig.]

kick² /kɪk/ n. an indentation in the bottom of a glass bottle. [19th c.: orig. unkn.]

kickback /ˈkɪkbæk/ n. **1** the force of a recoil. **2** colloq. usu. illegal payment for collaboration, esp. collaboration for profit.

kickshaw /ˈkɪkʃɔː/ n. **1** archaic, usu. derog. a fancy dish in cookery. **2** something elegant but insubstantial; a toy or trinket. [F *quelque chose* something]

kicksorter /ˈkɪkˌsɔːtə(r)/ n. colloq. a device for analysing electrical pulses according to amplitude.

kickstand /ˈkɪkstænd/ n. a rod attached to a bicycle or motorcycle and kicked into a vertical position to support the vehicle when stationary.

kick-start /ˈkɪkstɑːt/ n. & v. ● n. (also **kick-starter**) **1** a device to start the engine of a motorcycle etc. by the downward thrust of a pedal. **2** an act of starting a motorcycle etc. in this way. **3** an impetus given to get a thing started or restarted. ● v.tr. **1** start (a motorcycle etc.) in this way. **2** start or restart (a process etc.) by providing some initial impetus.

kid[1] /kɪd/ *n. & v.* ● *n.* **1** a young goat. **2** the leather made from its skin. **3** *colloq.* a child or young person. ● *v.intr.* (**kidded, kidding**) (of a goat) give birth. □ **handle with kid gloves** handle in a gentle, delicate, or gingerly manner. **kid brother** (or **sister**) *colloq.* a younger brother or sister. **kid-glove** (*attrib.*) dainty or delicate. **kids' stuff** *colloq.* something very simple. [ME *kide* f. ON *kith* f. Gmc]

kid[2] /kɪd/ *v.* (**kidded, kidding**) *colloq.* **1** *tr. & refl.* deceive, trick (*don't kid yourself*; *kidded his mother that he was ill*). **2** *tr. & intr.* tease (*only kidding*). □ **no kidding** (or **kid**) that is the truth. □ **kidder** *n.* **kiddingly** *adv.* [perh. f. KID[1]]

kid[3] /kɪd/ *n. hist.* a small wooden tub, esp. a sailor's mess tub for grog or rations. [perh. var. of KIT[1]]

Kidd /kɪd/, William (known as Captain Kidd) (1645–1701), Scottish pirate. Sent to the Indian Ocean in 1695 in command of an anti-pirate expedition, Kidd became a pirate himself. In 1699 he went to Boston in the hope of obtaining a pardon, but was arrested in the same year and hanged in London.

Kidderminster /'kɪdə‚mɪnstə(r)/ a town in west central England, in Hereford and Worcester, on the River Stour; pop. (1981) 50,750.

Kidderminster carpet *n.* a carpet made of two cloths of different colours woven together so that the carpet is reversible. [KIDDERMINSTER]

kiddie /'kɪdɪ/ *n.* (also **kiddy**) (*pl.* **-ies**) *colloq.* = KID[1] *n.* 3.

kiddle /'kɪd(ə)l/ *n.* **1** a barrier in a river having an opening fitted with nets etc. to catch fish. **2** an arrangement of fishing-nets hung on stakes along the seashore. [ME f. AF *kidel*, OF *quidel, guidel*]

kiddo /'kɪdəʊ/ *n.* (*pl.* **-os**) *colloq.* = KID[1] *n.* 3.

kidnap /'kɪdnæp/ *v.tr.* (**kidnapped, kidnapping**; *US* **kidnaped, kidnaping**) carry off (a person etc.) by illegal force or deception, esp. to obtain a ransom; abduct. □ **kidnapper** *n.* [back-form. f. *kidnapper* f. KID[1] + *nap* = NAB]

kidney /'kɪdnɪ/ *n.* (*pl.* **-eys**) **1** either of a pair of organs in the abdominal cavity of mammals, birds, and reptiles, which remove nitrogenous wastes from the blood and excrete urine. **2** the kidney of a sheep, ox, or pig as food. **3** temperament, nature, kind (*a man of that kidney*; *of the right kidney*). □ **kidney bean 1** a dwarf French bean. **2** a scarlet runner bean. **kidney dialysis** see DIALYSIS 2. **kidney dish** a kidney-shaped dish, esp. one used in surgery. **kidney machine** see ARTIFICIAL KIDNEY. **kidney-shaped** shaped like a kidney, with one side concave and the other convex. **kidney vetch** a yellow-flowered leguminous plant, *Anthyllis vulneraria*, found in grassland (also called *lady's finger*). [ME *kidnei*, pl. *kidneiren*, app. partly f. *ei* EGG[1]]

kidskin /'kɪdskɪn/ *n.* = KID[1] *n.* 2.

kiekie /'kiːkiː/ *n.* a New Zealand climbing plant with edible bracts, and leaves which are used for basket-making etc. [Maori]

Kiel /kiːl/ a naval port in northern Germany, capital of Schleswig-Holstein, on the Baltic Sea coast at the eastern end of the Kiel Canal; pop. (1991) 247,100.

Kiel Canal a man-made waterway, 98 km (61 miles) in length, in NW Germany, running westwards from Kiel to Brunsbüttel at the mouth of the Elbe. It connects the North Sea with the Baltic and was constructed in 1895 to provide the German navy with a shorter route between these two seas.

Kielce /'kjeltsə/ an industrial city in southern Poland; pop. (1990) 214,200.

Kierkegaard /'kɪəkə‚gɑːd/, Søren (Aabye) (1813–55), Danish philosopher. Viewed as one of the founders of existentialism, he opposed the prevailing Hegelian philosophy of the time by affirming the importance of individual experience and choice. Accordingly, he refused to subscribe to the possibility of an objective system of Christian doctrinal truths, and held that one could know God only through a 'leap of faith'. His philosophical works include *Either-Or* (1843); he is also noted for his religious writings, including *The Concept of Dread* (1844) and *The Sickness unto Death* (1849).

kieselguhr /'kiːz(ə)l‚gʊə(r)/ *n.* a soft friable porous form of diatomite used as a filter, filler, insulator, etc., in various manufacturing processes. Also called *diatomaceous earth*. [G f. *Kiesel* gravel + dial. *Guhr* earthy deposit]

Kiev /'kiːef/ the capital of Ukraine, an industrial city and port on the River Dnieper; pop. (est. 1989) 2,587,000. Founded in the 8th century, it was capital of the first Russian state and became capital of the Ukrainian Soviet Socialist Republic, following Kharkiv, in 1934. In 1991 it became capital of independent Ukraine.

kif var. of KEF.

Kigali /kɪ'gɑːlɪ/ the capital of Rwanda; pop. (1981) 156,650.

kike /kaɪk/ *n.* esp. *US sl. offens.* a Jew. [20th c.: orig. uncert.]

Kikládhes see CYCLADES.

Kikongo /kɪ'kɒŋgəʊ/ *n.* the Bantu language of the Kongo people, used in Congo, Zaire, and adjacent areas.

Kikuyu /kɪ'kuːjuː/ *n. & adj.* ● *n.* (*pl.* same or **Kikuyus**) **1** a member of a Bantu-speaking people forming the largest ethnic group in Kenya. The Kikuyu are traditionally agricultural, but have come to the fore in the professions and in Kenyan political life. (See also MAU MAU.) **2** the Bantu language of the Kikuyu. ● *adj.* of or relating to the Kikuyu or their language. [Bantu]

Kilauea /‚kiːlaʊ'eɪə/ a volcano with a crater roughly 8 km (5 miles) long by 5 km (3 miles) broad on the island of Hawaii, situated on the eastern flanks of Mauna Loa at an altitude of 1,247 m (4,090 ft). Kilauea is one of the most active volcanoes in the world.

Kildare /kɪl'deə(r)/ a county of the Republic of Ireland, in the east, in the province of Leinster; county town, Naas.

kilderkin /'kɪldəkɪn/ *n.* **1** a cask for liquids etc., usu. holding 18 imperial gallons (about 82 litres). **2** this measure. [ME, alt. of *kinderkin* f. MDu. *kinde(r)kin, kinneken*, dimin. of *kintal* QUINTAL]

kilim /'kɪlɪm/ *n.* (also **kelim** /'kiːlɪ-/) a pileless woven carpet, rug, etc., made in Turkey, Kurdistan, and neighbouring areas. [Turk. *Kılım*, f. Pers. *gelīm*]

Kilimanjaro, Mount /‚kɪlɪmən'dʒɑːrəʊ/ an extinct volcano in northern Tanzania. It has twin peaks, the higher of which, Kibo (5,895 m, 19,340 ft), is the highest mountain in Africa.

Kilkenny /kɪl'kenɪ/ **1** a county of the Republic of Ireland, in the southeast, in the province of Leinster. **2** its county town; pop. (1991) 8,510. It was the capital of the ancient kingdom of Ossory.

Kilkenny cats two cats which, according to legend, fought until only their tails remained.

kill[1] /kɪl/ *v. & n.* ● *v.tr.* **1 a** deprive of life or vitality; put to death; cause the death of. **b** (*absol.*) cause or bring about death (*must kill to survive*). **2** destroy; put an end to (feelings etc.) (*overwork killed my enthusiasm*). **3** *refl.* (often foll. by pres. part.) *colloq.* **a** overexert oneself (*don't kill yourself lifting them all at once*). **b** laugh heartily. **4** *colloq.* overwhelm (a person) with amusement, delight, etc. (*the things he says really kill me*). **5** switch off (a spotlight, engine, etc.). **6** *colloq.* delete (a line, paragraph, etc.) from a computer file. **7** *colloq.* cause pain or discomfort to (*my feet are killing me*). **8** pass (time, or a specified amount of it), usu. while waiting for a specific event (*had an hour to kill before the interview*). **9** defeat (a bill in Parliament). **10** *colloq.* consume the entire contents of (a bottle of wine etc.). **11 a** *Tennis* etc. hit (the ball) so skilfully that it cannot be returned. **b** stop (the ball) dead. **12** neutralize or render ineffective (taste, sound, colour, etc.) (*thick carpet killed the sound of footsteps*). ● *n.* **1** an act of killing (esp. an animal). **2** an animal or animals killed, esp. by a sportsman. **3** *colloq.* the destruction or disablement of an enemy aircraft, submarine, etc. □ **dressed to kill** dressed showily, alluringly, or impressively. **in at the kill** present at or benefiting from the successful conclusion of an enterprise. **kill off 1** get rid of or destroy completely (esp. a number of persons or things). **2** (of an author) bring about the death of (a fictional character). **kill or cure** (usu. *attrib.*) (of a remedy etc.) drastic, extreme. **kill two birds with one stone** achieve two aims at once. **kill with kindness** spoil (a person) with overindulgence. [ME *cülle, kille*, perh. ult. rel. to QUELL]

kill[2] /kɪl/ *n. US dial.* a stream, creek, or tributary river.

Killarney /kɪ'lɑːnɪ/ a town in the south-west of the Republic of Ireland, in County Kerry, famous for the beauty of the nearby lakes and mountains; pop. (1991) 7,250.

killdeer /'kɪldɪə(r)/ *n.* a large North American plover, *Charadrius vociferus*, with a plaintive call. [imit. of its cry]

killer /'kɪlə(r)/ *n.* **1 a** a person, animal, or thing that kills. **b** a murderer. **2** *colloq.* **a** an impressive, formidable, or excellent person or thing (*this one is quite difficult, but the next one is a real killer*). **b** a hilarious joke. **c** a decisive blow (*his brilliant header proved to be the killer*). □ **killer bee** an Africanized honey bee. **killer cell** *Physiol.* a white blood cell which destroys infected or cancerous cells. **killer instinct 1** an innate tendency to kill. **2** a ruthless streak. **killer whale** a predatory black and white whale, *Orcinus orca*, with a prominent dorsal fin (also called *grampus, orca*).

killick /ˈkɪlɪk/ *n.* **1** a heavy stone used by small craft as an anchor. **2** a small anchor. **3** *Brit. Naut. sl.* a leading seaman. [17th c.: orig. unkn.]

killifish /ˈkɪlɪˌfɪʃ/ *n.* a small freshwater or brackish-water fish of the family Cyprinodontidae or Poeciliidae, often brightly coloured and kept in aquariums; esp. one of the genus *Fundulus*, of eastern North America, or *Pterolebias*, of the upper Amazon. [perh. f. KILL² + FISH¹]

killing /ˈkɪlɪŋ/ *n.* & *adj.* ● *n.* **1 a** the causing of death. **b** an instance of this. **2** a great (esp. financial) success (*make a killing*). ● *adj. colloq.* **1** overwhelmingly funny. **2** exhausting; very strenuous. □ **killing-bottle** a bottle containing poisonous vapour to kill insects collected as specimens. □ **killingly** *adv.*

killjoy /ˈkɪldʒɔɪ/ *n.* a person who throws gloom over or prevents other people's enjoyment.

Kilmarnock /kɪlˈmɑːnək/ a town in Strathclyde region, west central Scotland; pop. (1981) 52,080.

kiln /kɪln/ *n.* a furnace or oven for burning, baking, or drying, esp. for calcining lime or firing pottery etc. [OE *cylene* f. L *culina* kitchen]

kiln-dry /ˈkɪlndraɪ/ *v.tr.* (**-ies, -ied**) dry in a kiln.

kilo /ˈkiːləʊ/ *n.* (*pl.* **-os**) **1** a kilogram. **2** a kilometre. [F: abbr.]

kilo- /ˈkɪləʊ/ *comb. form* (often **k**, or **K** in Computing) denoting a factor of 1,000 (esp. in metric units). [F f. Gk *khilioi* thousand]

kilobase /ˈkɪləˌbeɪs/ *n.* (abbr. **kb**) *Biochem.* (in expressing the lengths of nucleic acid molecules) one thousand bases. (See BASE¹ *n.* 6b.)

kilobyte /ˈkɪləˌbaɪt/ *n.* (abbr. **KB**) *Computing* 1,024 (i.e. 2¹⁰) bytes as a measure of data capacity.

kilocalorie /ˈkɪləˌkælərɪ/ *n.* = CALORIE 2.

kilocycle /ˈkɪləˌsaɪk(ə)l/ *n.* a former measure of frequency, equivalent to 1 kilohertz (symbol: **kc**).

kilogram /ˈkɪləˌgræm/ *n.* (also **-gramme**) a metric unit of mass (loosely, of weight) equal to 1,000 g, about 2.205 lb. The kilogram (symbol: **kg**) is the SI base unit of mass and is equivalent to the mass of a standard cylinder made of platinum–iridium, kept at Sèvres near Paris. [F *kilogramme* (as KILO, GRAM¹)]

kilohertz /ˈkɪləˌhɜːts/ *n.* a measure of frequency equivalent to 1,000 cycles per second (symbol: **kHz**).

kilojoule /ˈkɪləˌdʒuːl/ *n.* 1,000 joules, esp. as a measure of the energy value of foods (symbol: **kJ**).

kilolitre /ˈkɪləˌliːtə(r)/ *n.* (*US* **-liter**) 1,000 litres (equivalent to 220 imperial gallons) (symbol: **kl**).

kilometre /ˈkɪləˌmiːtə(r), kɪˈlɒmɪtə(r)/ *n.* (*US* **kilometer**) a metric unit of measurement equal to 1,000 metres (approx. 0.62 miles) (symbol: **km**). □ **kilometric** /ˌkɪləˈmetrɪk/ *adj.* [F *kilomètre* (as KILO-, METRE¹)]

kiloton /ˈkɪləˌtʌn/ *n.* (also **kilotonne**) a unit of explosive power equivalent to 1,000 tons of TNT.

kilovolt /ˈkɪləˌvəʊlt/ *n.* 1,000 volts (symbol: **kV**).

kilowatt /ˈkɪləˌwɒt/ *n.* 1,000 watts (symbol: **kW**).

kilowatt-hour /ˌkɪləwɒtˈaʊə(r)/ *n.* a unit of electrical energy equivalent to a power consumption of 1,000 watts for one hour (symbol: **kWh**).

Kilroy /ˈkɪlrɔɪ/ a mythical person, popularized by American servicemen in the Second World War, who left such inscriptions as 'Kilroy was here' on walls all over the world. There are many unverifiable accounts of the origin of the name.

kilt /kɪlt/ *n.* & *v.* ● *n.* **1** a skirtlike garment, usu. of pleated tartan cloth and reaching to the knees, as traditionally worn by Highland men. **2** a similar garment worn by women and children. ● *v.tr.* **1** tuck up (skirts) round the body. **2** (esp. as **kilted** *adj.*) gather in vertical pleats. □ **kilted** *adj.* [orig. as verb: ME, of Scand. orig.]

kilter /ˈkɪltə(r)/ *n.* (also **kelter** /ˈkel-/) good working order (esp. *out of kilter*). [17th c.: orig. unkn.]

kiltie /ˈkɪltɪ/ *n.* a wearer of a kilt, esp. a kilted Highland soldier.

Kimberley /ˈkɪmbəlɪ/ **1** a city in South Africa, in the province of Northern Cape; pop. (1985) 149,700. Named after the 1st Earl of Kimberley, a British Colonial Secretary, it has been a diamond-mining centre since the early 1870s and gave its name to kimberlite. **2** (also **the Kimberleys**) a plateau region in the far north of Western Australia. A mining and cattle-rearing region, it was the scene of a gold rush in 1885.

kimberlite /ˈkɪmbəˌlaɪt/ *n. Mineral.* a rare igneous blue-tinged rock sometimes containing diamonds, found in South Africa and Siberia. Also called *blue ground* (see BLUE¹). [KIMBERLEY 1]

Kim Il Sung /ˌkɪm ɪl ˈsʊŋ/ (born Kim Song Ju) (1912–94), Korean Communist statesman, first Premier of North Korea 1948–72 and President 1972–94. In the 1930s and 1940s he led the armed resistance to the Japanese domination of Korea; following the country's partition at the end of the Second World War he became Premier of the Democratic People's Republic of Korea (1948). He ordered his forces to invade South Korea in 1950, precipitating the Korean War (1950–3), and remained committed to the reunification of the country. He maintained a one-party state and created a personality cult around himself and his family; on his death he was quickly replaced in power by his son Kim Jong Il (b.1942).

kimono /kɪˈməʊnəʊ/ *n.* (*pl.* **-os**) **1** a long loose Japanese robe worn with a sash. **2** a European dressing-gown modelled on this. □ **kimonoed** *adj.* [Jap.]

kin /kɪn/ *n.* & *adj.* ● *n.* one's relatives or family. ● *predic.adj.* (of a person) related (*we are kin; he is kin to me*) (see also AKIN 1). □ **kith and kin** see KITH. **near of kin** closely related by blood, or in character. **next of kin** see NEXT. □ **kinless** *adj.* [OE *cynn* f. Gmc]

-kin /kɪn/ *suffix* forming diminutive nouns (*catkin; manikin*). [from or after MDu. -*kijn*, -*ken*, OHG -*chin*]

kina /ˈkiːnə/ *n.* the chief monetary unit of Papua New Guinea. [Papuan]

Kinabalu, Mount /ˌkɪnəbəˈluː/ a mountain in the state of Sabah in eastern Malaysia, on the north coast of Borneo. Rising to 4,094 m (13,431 ft), it is the highest peak of Borneo and of SE Asia.

kinaesthesia /ˌkɪnəsˈθiːzɪə/ *n.* (*US* **kinesthesia**) the brain's awareness of the position and movement of the body or limbs etc., by means of sensory nerves within the muscles and joints. □ **kinaesthetic** /-ˈθetɪk/ *adj.* [Gk *kineō* move + *aisthēsis* sensation]

Kincardineshire /kɪnˈkɑːdɪnˌʃɪə(r)/ a former county of eastern Scotland. In 1975 it became part of Grampian region.

Kinchinjunga see KANCHENJUNGA.

kincob /ˈkɪŋkɒb/ *n.* a rich Indian fabric embroidered with gold or silver. [Urdu f. Pers. *kamk̲āb* f. *kamk̲ā* damask]

kind¹ /kaɪnd/ *n.* **1 a** a race or species (*human kind*). **b** a natural group of animals, plants, etc. (*the wolf kind*). **2** class, type, sort, variety (*what kind of job are you looking for?*). ¶ In this sense *these* (or *those*) *kind* is often encountered when followed by a plural, as in *I don't like these kind of things*, but *this kind* and *these kinds* are usually preferred. **3** the manner or fashion natural to a person etc. (*act after their kind; true to kind*). □ **in kind 1** in the same form, likewise (*was insulted and replied in kind*). **2** (of payment) in goods or labour as opposed to money (*received their wages in kind*). **3** in character or quality (*differ in degree but not in kind*). **kind of** *colloq.* to some extent (*felt kind of sorry; I kind of expected it*). **a kind of** used to imply looseness, vagueness, exaggeration, etc., in the term used (*a kind of Jane Austen of our times; I suppose he's a kind of doctor*). **law of kind** *archaic* nature in general; the natural order. **nothing of the kind 1** not at all like the thing in question. **2** (expressing denial) not at all. **of its kind** within the limitations of its own class (*good of its kind*). **of a kind 1** *derog.* scarcely deserving the name (*a choir of a kind*). **2** similar in some important respect (*they're two of a kind*). **one's own kind** those with whom one has much in common. **something of the kind** something like the thing in question. [OE *cynd(e), gecynd(e)* f. Gmc: cf. KIND²]

kind² /kaɪnd/ *adj.* **1** of a friendly, generous, benevolent, or gentle nature. **2** (usu. foll. by *to*) showing friendliness, affection, or consideration. **3 a** affectionate. **b** *archaic* loving. [OE *gecynde* (as KIND¹): orig. = 'natural, native']

kinda /ˈkaɪndə/ *colloq.* = *kind of*. [corrupt.]

kinder /ˈkɪndə(r)/ *n. Austral.* = KINDERGARTEN. [abbr.]

kindergarten /ˈkɪndəˌgɑːt(ə)n/ *n.* an establishment for preschool learning, a nursery school. [G, = children's garden]

kind-hearted /kaɪndˈhɑːtɪd/ *adj.* of a kind disposition. □ **kind-heartedly** *adv.* **kind-heartedness** *n.*

kindle /ˈkɪnd(ə)l/ *v.* **1** *tr.* light or set on fire (a flame, fire, substance, etc.). **2** *intr.* catch fire, burst into flame. **3** *tr.* arouse or inspire (*kindle enthusiasm for the project; kindle jealousy in a rival*). **4** *intr.* (usu. foll. by *to*) respond, react (to a person, an action, etc.) (*kindle to his courage*). **5** *intr.* become animated, glow with passion etc. (*her imagination kindled*). **6** *tr.* & *intr.* make or become bright (*kindle the embers to a glow*). □ **kindler** *n.* [ME f. ON *kynda*, kindle: cf. ON *kindill* candle, torch]

kindling /ˈkɪndlɪŋ/ *n.* small sticks etc. for lighting fires.

kindly¹ /ˈkaɪndlɪ/ *adv.* **1** in a kind manner (*spoke to the child kindly*). **2** often *iron.* used in a polite request or demand (*kindly acknowledge this letter; kindly leave me alone*). □ **look kindly upon** regard sympathetically. **take a thing kindly** like or be pleased by it. **take**

kindly to be pleased by or endeared to (a person or thing). **thank kindly** thank very much. [OE *gecyndelīce* (as KIND²)]

kindly² /'kaɪndlɪ/ *adj.* (**kindlier, kindliest**) **1** kind, kind-hearted. **2** (of climate etc.) pleasant, genial. **3** *archaic* native-born (*a kindly Scot*). □ **kindlily** *adv.* **kindliness** *n.* [OE *gecyndelic* (as KIND¹)]

kindness /'kaɪndnɪs/ *n.* **1** the state or quality of being kind. **2** a kind act.

kindred /'kɪndrɪd/ *n. & adj.* ● *n.* **1** one's relations, referred to collectively. **2** a relationship by blood. **3** a resemblance or affinity in character. ● *adj.* **1** related by blood or marriage. **2** allied or similar in character (*other kindred symptoms*). □ **kindred spirit** a person whose character and outlook have much in common with one's own. [ME f. KIN + -red f. OE *ræden* condition]

kine /kaɪn/ *archaic pl.* of COW¹.

kinematics /ˌkɪnɪ'mætɪks, ˌkaɪn-/ *n.pl.* (usu. treated as *sing.*) the branch of mechanics concerned with the motion of objects without reference to the forces which cause the motion. □ **kinematic** *adj.* **kinematically** *adv.* [Gk *kinēma -matos* motion f. *kineō* move + -ICS]

kinematograph var. of CINEMATOGRAPH.

kinesics /kɪ'niːsɪks, kaɪ-/ *n.pl.* (usu. treated as *sing.*) **1** the study of body movements and gestures which contribute to communication. **2** these movements; body language. [Gk *kinēsis* motion (as KINETIC)]

kinesiology /kɪˌniːsɪ'ɒlədʒɪ, kaɪ-/ *n.* the study of the mechanics of body movements.

kinesis /kɪ'niːsɪs, kaɪ-/ *n.* **1** movement, motion. **2** *Biol.* undirected movement of an organism in response to a stimulus (cf. TAXIS 2). **3** *Zool.* mobility of the bones of the skull, as in some birds and reptiles. [Gk *kinēsis* movement]

kinesthesia US var. of KINAESTHESIA.

kinetic /kɪ'netɪk, kaɪ-/ *adj.* of or due to motion. □ **kinetic energy** energy which a body possesses by virtue of being in motion. □ **kinetically** *adv.* [Gk *kinētikos* f. *kineō* move]

kinetic art *n.* a form of art, especially sculpture, incorporating real or apparent movement. The term was first used by Naum Gabo and his brother Antoine Pevsner in 1920, but did not become established until the 1950s. It is associated with, for example, the mobiles of Alexander Calder.

kinetics /kɪ'netɪks, kaɪ-/ *n.pl.* (usu. treated as *sing.*) **1** = DYNAMICS 1a. **2** the branch of physical chemistry or biochemistry concerned with measuring and studying the rates of chemical or enzymic reactions.

kinetic theory *n.* a theory that explains the physical properties of matter in terms of the motions of its constituent particles. According to the kinetic theory the temperature of a solid, liquid, or gas is a measure of the mean kinetic energy distributed among its atoms or molecules. In solids the particles vibrate about a fixed point; in liquids they slide past each other but remain in close contact; in gases they move freely and independently. Evaporation from a liquid, for example, occurs when the more energetic, and hence faster-moving, molecules escape. Simple particle theories have existed from antiquity, but modern kinetic theory, based on statistics, was developed in the 19th century by Clausius, Maxwell, Boltzmann, and others. Kinetic theory has had most success in describing thermal properties and the behaviour of gases.

kinetin /'kaɪnɪtɪn/ *n. Biochem.* a synthetic kinin used to stimulate cell division in plants. [as KINETIC + -IN]

kinfolk N. Amer. var. of KINSFOLK.

King¹ /kɪŋ/, B. B. (born Riley B King) (b.1925), American blues singer and guitarist. He became an established blues performer in the 1950s and early 1960s, but only came to the notice of a wider audience in the late 1960s, when his style of guitar playing was imitated by rock musicians.

King² /kɪŋ/, Billie Jean (b.1943), American tennis player. She won a record twenty Wimbledon titles, including six singles titles (1966–8; 1972–3; 1975), ten doubles titles, and four mixed doubles titles. King retired in 1983.

King³ /kɪŋ/, Martin Luther (1929–68), American Baptist minister and civil-rights leader. King opposed discrimination against blacks by organizing non-violent resistance and peaceful mass demonstrations, notably the year-long black boycott of the local bus company in Montgomery, Alabama, in 1955 and the march on Washington involving 200,000 demonstrators in 1963. At the latter, King delivered his celebrated speech beginning 'I have a dream … '. He was awarded the Nobel Peace Prize in 1964. King was assassinated in Memphis in 1968.

King⁴ /kɪŋ/, William Lyon Mackenzie (1874–1950), Canadian Liberal statesman, Prime Minister 1921–6, 1926–30, and 1935–48. The grandson of William Lyon Mackenzie, he represented Canada at the imperial conferences in London (1923; 1926; 1927), where he played an important role in establishing the status of the self-governing nations of the Commonwealth. He went on to strengthen ties with the UK and the US and introduced a number of social reforms, including unemployment insurance (1940).

king /kɪŋ/ *n. & v.* ● *n.* **1** (as a title usu. **King**) a male sovereign, esp. the hereditary ruler of an independent state. **2** a person or thing pre-eminent in a specified field or class (*railway king*). **3** (*attrib.*) a large (or the largest) kind of plant, animal, etc. (*king penguin*). **4** *Chess* the piece on each side which the opposing side has to checkmate to win. **5** a piece in draughts with extra capacity for moving, made by crowning an ordinary piece that has reached the opponent's baseline. **6** a court-card bearing a representation of a king and usu. ranking next below an ace. **7** (**the King**) (in the UK) the national anthem when there is a male sovereign. ● *v.tr.* make (a person) king. □ **king cobra** a large and venomous hooded Indian snake, *Ophiophagus hannah*. **king-crab 1** = *horseshoe crab*. **2** US a large edible spider crab. **king it 1** play or act the king. **2** (usu. foll. by *over*) govern, control. **king of beasts** the lion. **king of birds** the eagle. **king of the castle** a children's game consisting of trying to displace a rival from a mound. **king-post** an upright post from the tie-beam of a roof to the apex of a truss. **king prawn** a large edible prawn of the genus *Penaeus*, common esp. in Australasia. **king's bishop, knight**, etc. *Chess* (of pieces which exist in pairs) the piece starting on the king's side of the board. **king's bounty** see BOUNTY. **King's Champion** see CHAMPION OF ENGLAND. **King's colour** see COLOUR. **King's evidence** see EVIDENCE. **king's evil** *hist.* scrofula, formerly held to be curable by the royal touch. **King's highway** see HIGHWAY. **king-size** (or **-sized**) larger than normal; very large. **king's pawn** *Chess* the pawn in front of the king at the beginning of a game. **king's ransom** a fortune. **King's speech** see SPEECH. □ **kinghood** *n.* **kingless** *adj.* **kinglike** *adj.* **kingly** *adj.* **kingliness** *n.* **kingship** *n.* [OE *cyning, cyng* f. Gmc]

kingbird /'kɪŋbɜːd/ *n.* a tyrant flycatcher of the genus *Tyrannus*, of North America, with olive-grey plumage and long pointed wings.

kingbolt /'kɪŋbəʊlt/ *n.* = KINGPIN.

King Charles spaniel *n.* a spaniel of a small black and tan breed.

kingcraft /'kɪŋkrɑːft/ *n. archaic* the skilful exercise of kingship.

kingcup /'kɪŋkʌp/ *n. Brit.* a marsh marigold.

kingdom /'kɪŋdəm/ *n.* **1** an organized community headed by a king. **2** the territory subject to a king. **3 a** the spiritual reign attributed to God (*Thy kingdom come*). **b** the sphere of this (*kingdom of heaven*). **4** a domain belonging to a person, animal, etc. **5** a province of nature (*the vegetable kingdom*). **6** a specified mental or emotional province (*kingdom of the heart*; *kingdom of fantasy*). **7** *Biol.* the highest category in taxonomic classification. □ **come into** (or **to**) **one's kingdom** achieve recognition or supremacy. **kingdom come** eternity; the next world. **till kingdom come** for ever. [OE *cyningdōm* (as KING)]

kingfish /'kɪŋfɪʃ/ *n.* a fish notable for its size, appearance, or value, e.g. the opah.

kingfisher /'kɪŋˌfɪʃə(r)/ *n.* a bird of the family Alcedinidae, with a long sharp beak, which dives for fish in rivers etc.; esp. *Alcedo atthis*, a small European bird with brightly coloured blue and orange plumage.

King James Bible (also **King James Version**) = AUTHORIZED VERSION.

King Kong /kɒŋ/ a huge apelike monster featured in the film *King Kong* (1933).

kinglet /'kɪŋlɪt/ *n.* **1** a petty king. **2** a tiny warbler of the genus *Regulus*, e.g. the goldcrest.

King Log *n.* a ruler noted for extreme laxity. In the fable of Jupiter and the frogs, the frogs wanted a king, and Jupiter gave them a log of wood; when they complained, he sent them a stork, which promptly gobbled them up. (See also KING STORK.)

kingmaker /'kɪŋˌmeɪkə(r)/ *n.* a person who makes kings, leaders, etc., through the exercise of political influence, orig. with ref. to the Earl of Warwick in the reigns of Henry VI and Edward IV of England.

King of Arms *n. Heraldry* (in the UK) a chief herald (at the College of Arms: Garter, Clarenceux, and Norroy and Ulster; in Scotland: Lyon).

King of Kings *n.* **1** God. **2** the title assumed by many ancient Eastern kings.

kingpin /ˈkɪŋpɪn/ n. **1 a** a main or large bolt in a central position. **b** a vertical bolt used as a pivot. **2** an essential person or thing, esp. in a complex system; the most important person in an organization.

Kings /kɪŋz/ either of two books of the Bible recording the history of Israel and Judah from the accession of Solomon to the destruction of the Temple in 586 BC.

King's Bench = QUEEN'S BENCH.

Kings Canyon National Park a national park in the Sierra Nevada, California, to the north of Sequoia National Park. Established in 1940, it preserves groves of ancient sequoia trees, including some of the largest in the world.

King's Counsel n. = QUEEN'S COUNSEL.

King's English see ENGLISH.

Kingsley /ˈkɪŋzlɪ/, Charles (1819–75), English novelist and clergyman. He is remembered for his historical novel *Westward Ho!* (1855) and for his classic children's story *The Water-Babies* (1863).

King's Messenger n. = QUEEN'S MESSENGER.

King's Proctor n. = QUEEN'S PROCTOR.

King's Scout see SCOUT[1].

Kingston /ˈkɪŋstən/ **1** the capital and chief port of Jamaica; pop. (1991) 588,000. Founded in 1693, it became capital in 1870. **2** a port in SE Canada, on Lake Ontario, at the head of the St Lawrence River; pop. (1991) 56,600; metropolitan area pop. 94,710.

Kingston-upon-Hull /ˌkɪŋstənəpɒnˈhʌl/ the official name for HULL.

King Stork n. a tyrannical ruler noted for extremes of oppression. (See KING LOG.)

Kingstown /ˈkɪŋztaʊn/ the capital and chief port of St Vincent in the West Indies; pop. (1989) 29,370.

kinin /ˈkaɪnɪn/ n. **1** any of a group of polypeptides formed in body tissue in response to injury, causing vasodilation and muscle contraction. **2** any of a group of compounds which promote cell division and inhibit ageing in plants. [Gk *kineō* move + -IN]

kink /kɪŋk/ n. & v. ● n. **1 a** a short backward twist in wire or tubing etc. such as may cause an obstruction. **b** a tight wave in human or animal hair. **2** a mental twist or quirk. ● v.intr. & tr. form or cause to form a kink. [MLG *kinke* (v.) prob. f. Du. *kinken*]

kinkajou /ˈkɪŋkəˌdʒuː/ n. a nocturnal fruit-eating mammal, *Potos flavus*, of Central and South America, related to the raccoon. It lives in trees and has a prehensile tail. [F *quincajou* f. Amer. Indian: cf. Algonquian *kwingwaage* wolverine]

Kinki /ˈkiːŋkiː/ a region of Japan, on the island of Honshu; capital, Osaka.

kinky /ˈkɪŋkɪ/ adj. (**kinkier, kinkiest**) **1** colloq. **a** given to or involving bizarre or unusual sexual behaviour. **b** (of clothing etc.) bizarre in a sexually provocative way. **2** strange, eccentric. **3** having kinks or twists. □ **kinkily** adv. **kinkiness** n. [KINK + -Y[1]]

kino /ˈkiːnəʊ/ n. (pl. **-os**) a gum produced by various tropical trees and used in medicine and tanning as an astringent. [W. Afr.]

Kinross-shire /kɪnˈrɒsʃɪə(r)/ a former county of east central Scotland. In 1975 it became part of Tayside region.

-kins /kɪnz/ suffix = -KIN, often with suggestions of endearment (*babykins*).

Kinsey /ˈkɪnzɪ/, Alfred Charles (1894–1956), American zoologist and sex researcher. He co-founded and directed the Institute for Sex Research at Indiana University, carrying out pioneering studies by interviewing large numbers of people. His best-known work, *Sexual Behaviour in the Human Male* (1948) (often referred to as the *Kinsey Report*), was controversial but highly influential. It was followed five years later by a companion volume, *Sexual Behaviour in the Human Female*.

kinsfolk /ˈkɪnzfəʊk/ n.pl. (also **kinfolk** /ˈkɪnfəʊk/) one's relations by blood.

Kinshasa /kɪnˈʃɑːsə, -ˈʃɑːzə/ the capital of Zaire, a port on the River Congo, in the south-west; pop. (est. 1984) 2,653,550. Founded in 1881 by the explorer Sir Henry Morton Stanley, it was known until 1966 as Léopoldville. It became capital of the Republic of Zaire in 1960.

kinship /ˈkɪnʃɪp/ n. **1** blood relationship. **2** the sharing of characteristics or origins.

kinsman /ˈkɪnzmən/ n. (pl. **-men**; fem. **kinswoman**, pl. **-women**) **1** a blood relation or (loosely) a relation by marriage. **2** a member of one's own tribe or people.

Kintyre /kɪnˈtaɪə(r)/ a peninsula on the west coast of Scotland, to the west of Arran, extending southwards for 64 km (40 miles) into the

North Channel and separating the Firth of Clyde from the Atlantic Ocean. Its southern tip is the Mull of Kintyre.

kiosk /ˈkiːɒsk/ n. **1** a light open-fronted booth or cubicle from which food, newspapers, tickets, etc., are sold. **2** a telephone box. **3** Austral. a building in which refreshments are served in a park, zoo, etc. **4** a light open pavilion in Turkey and Iran. [F *kiosque* f. Turk. *kiũshk* pavilion f. Pers. *guš*]

kip[1] /kɪp/ n. & v. Brit. sl. ● n. **1** a sleep or nap. **2** a bed or cheap lodging-house. **3** a brothel. ● v.intr. (**kipped, kipping**) **1** sleep, take a nap. **2** (foll. by *down*) lie or settle down to sleep. [cf. Danish *kippe* mean hut]

kip[2] /kɪp/ n. the hide of a young or small animal as used for leather. [ME: orig. unkn.]

kip[3] /kɪp/ n. (pl. same or **kips**) the chief monetary unit of Laos. [Thai]

kip[4] /kɪp/ n. Austral. sl. a small piece of wood from which coins are spun in the game of two-up. [perh. f. Engl. dial.: cf. *keper* a flat piece of wood preventing a horse from eating the corn, or Ir. dial. *kippeen* f. Ir. *cípín* a little stick]

Kipling /ˈkɪplɪŋ/, (Joseph) Rudyard (1865–1936), English novelist, short-story writer, and poet. He was born in India, where he worked as a journalist 1882–9, and set many of his writings in the India of the Raj. Hist best-known poems, such as 'The White Man's Burden', 'If', and 'Gunga Din', came to be regarded as epitomizing the British colonial spirit. Of his vast and varied output, Kipling is perhaps now primarily known for his tales for children, notably *The Jungle Book* (1894) and the *Just So Stories* (1902). In 1907 he became the first English writer to be awarded the Nobel Prize for literature.

kipper /ˈkɪpə(r)/ n. & v. ● n. **1** a kippered fish, esp. herring. **2** a male salmon in the spawning season. ● v.tr. cure (a herring etc.) by splitting open, salting, and drying in the open air or smoke. [ME: orig. uncert.]

kipsie /ˈkɪpsɪ/ n. (also **kipsy**) (pl. **-ies**) Austral. sl. a house, home, lean-to, or shelter. [perh. f. KIP[1]]

kir /kɪə(r), kɜː(r)/ n. propr. a drink made from dry white wine and crème de cassis. [Canon Felix *Kir* (1876–1968), said to have invented the recipe]

kirby-grip /ˈkɜːbɪˌɡrɪp/ n. (also **Kirbigrip** propr.) Brit. a type of sprung hairgrip. [*Kirby*, part of orig. manufacturer's name]

Kirchhoff /ˈkɪəxhɒf/, Gustav Robert (1824–87), German physicist, a pioneer in spectroscopy. Working with Bunsen, he developed a spectroscope which used the flame of Bunsen's gas burner in conjunction with a prism which he had designed. Using this, he discovered that solar absorption lines are specific to certain elements, developed the concept of black-body radiation, and discovered the elements caesium and rubidium. He also worked on electrical circuits and the flow of currents.

Kirchner /ˈkɪəxnə(r)/, Ernst Ludwig (1880–1938), German expressionist painter. In 1905 he was a founder of the first group of German expressionists, who sought inspiration from medieval German and primitive art and were influenced by Van Gogh and Gauguin. His paintings, such as *Five Women in the Street* (1913), are characterized by the use of bright contrasting colours and angular outlines, and often depict claustrophobic street scenes. He committed suicide in 1938 after condemnation of his work by the Nazis.

Kirghiz var. of KYRGYZ.

Kirghizia /kɪəˈɡɪzɪə/ an alternative name for KYRGYZSTAN.

Kiribati /ˈkɪrɪbæts/ a country in the SW Pacific including the Gilbert Islands, the Line islands, the Phoenix Islands, and Banaba (Ocean Island); pop. (est. 1991) 71,000; official languages, English and I-Kiribati (local Malayo-Polynesian language); capital, Bairiki (on Tarawa). Inhabited by Micronesian people, the islands were sighted by the Spaniards in the mid-16th century. The islands became a centre for whaling and, later, a source of phosphates (now exhausted). Britain declared a protectorate over the Gilbert and Ellice Islands in 1892, and they became a colony in 1915. Links with the Ellice Islands (now Tuvalu) having ended in 1975, in 1979 Kiribati became an independent republic within the Commonwealth.

Kirin see JILIN.

Kiritimati /kɪˈrɪsɪməs/ an island in the Pacific Ocean, one of the Line Islands of Kiribati; pop. (1990) 2,530. It is the largest atoll in the world. Known as Christmas Island until 1981, it was discovered by Captain James Cook on Christmas Eve 1777 and was British until it became part of an independent Kiribati in 1979. In the 1950s it was used as an operational base for US and British nuclear weapons testing.

kirk /kɜːk/ n. Sc. & N. Engl. **1** a church. **2** (**the Kirk** or **the Kirk of Scotland**) the Church of Scotland as distinct from the Church of

England or from the Episcopal Church in Scotland. [ME f. ON *kirkja* f. OE *cir(i)ce* CHURCH]

Kirkcaldy /kɜːˈkɔːdɪ/ an industrial town and port in Fife, SE Scotland, on the north shore of the Firth of Forth; pop. (1981) 46,500.

Kirkcudbright /kɜːˈkuːbrɪ/ a town in Dumfries and Galloway, SW Scotland, on the River Dee; pop. (1981) 3,400.

Kirkcudbrightshire /kɜːˈkuːbrɪˌʃɪə(r)/ a former county of SW Scotland. It became part of the region of Dumfries and Galloway in 1975.

kirkman /ˈkɜːkmən/ *n.* (*pl.* **-men**) *Sc. & N. Engl.* a member of the Church of Scotland.

Kirk-session /ˈkɜːkˌseʃ(ə)n/ *n.* **1** the lowest court in the Church of Scotland. **2** *hist.* the lowest court in other Presbyterian Churches, composed of ministers and elders.

Kirkuk /kɜːˈkʊk/ an industrial city in northern Iraq, centre of the oil industry in that region; pop. (1985) 208,000.

Kirkwall /ˈkɜːkwɔːl/ a port in the Orkney Islands; pop. (1981) 6,000. Situated on Mainland, it is the chief town of the islands.

Kirov /ˈkɪərɒf/ the former name (1934–92) for VYATKA.

Kirovabad /ˌkɪərəvəˈbæd/ a former name (1935–89) for GÄNCÄ.

kirsch /kɪəʃ/ *n.* (also **kirschwasser** /ˈkɪəˌvæsə(r)/) a brandy distilled from the fermented juice of cherries. [G *Kirsche* cherry, *Wasser* water]

kirtle /ˈkɜːt(ə)l/ *n. archaic* **1** a woman's gown or outer petticoat. **2** a man's tunic or coat. [OE *cyrtel* f. Gmc, ult. perh. f. L *curtus* short]

Kiruna /ˈkɪərʊnə/ the northernmost town of Sweden, situated in the Lapland iron-mining region; pop. (1990) 26,150.

Kirundi /kɪˈrʊndɪ/ *n.* a Bantu language, one of the official languages of Burundi.

Kisangani /ˌkɪsæŋˈɡɑːnɪ/ a city in northern Zaire, on the River Congo; pop. (1984) 282,650. Founded in 1882 by the explorer Sir Henry Morton Stanley, it was known as Stanleyville until 1966.

Kishinev see CHIŞINĂU.

kiskadee /ˌkɪskəˈdiː/ *n.* (also **keskidee** /ˌkeskɪˈdiː/) a tyrant flycatcher of Central and South America; esp. *Pitangus sulphuratus*, with brown and yellow plumage. [imit. of its cry]

Kislev /ˈkɪslef/ *n.* (also **Kislew**) (in the Jewish calendar) the third month of the civil and ninth of the religious year, usually coinciding with parts of November and December. [Heb. *kislēw*]

kismet /ˈkɪzmet/ *n.* destiny, fate. [Turk. f. Arab. *ḳisma(t)* f. *ḳasama* divide]

kiss /kɪs/ *v. & n.* ● *v.* **1** *tr.* touch with the lips, esp. as a sign of love, affection, greeting, or reverence. **2** *tr.* express (greeting or farewell) in this way. **3** *absol.* (of two persons) touch each others' lips in this way. **4** *tr.* (also *absol.*) (of a snooker ball etc. in motion) lightly touch (another ball). ● *n.* **1** a touch with the lips in kissing. **2** the slight impact when one snooker ball etc. lightly touches another. **3** a small sweetmeat or piece of confectionery. □ **kiss and tell** recount one's sexual exploits. **kiss a person's arse** *coarse sl.* act obsequiously towards a person. **kiss away** remove (tears etc.) by kissing. **kiss-curl** a small curl of hair on the forehead, at the nape, or in front of the ear. **kiss the dirt** (or **dust**) submit abjectly; be overthrown. **kiss goodbye to** accept the loss of. **kiss the ground** prostrate oneself as a token of homage. **kissing cousin** (or **kin** or **kind**) a distant relative (given a formal kiss on occasional meetings). **kissing-gate** *Brit.* a gate hung in a V- or U-shaped enclosure, letting one person through at a time. **kiss of death** an apparently friendly act which causes ruin. **kiss off** esp. *N. Amer. sl.* **1** dismiss, get rid of. **2** go away, die. **kiss of life** mouth-to-mouth resuscitation. **kiss of peace** *Eccl.* a ceremonial kiss, esp. during the Eucharist, as a sign of unity. **kiss the rod** accept chastisement submissively. □ **kissable** *adj.* [OE *cyssan* f. Gmc]

kisser /ˈkɪsə(r)/ *n.* **1** a person who kisses. **2** (orig. *Boxing*) *sl.* the mouth; the face.

Kissinger /ˈkɪsɪndʒə(r)/, Henry (Alfred) (b.1923), German-born American statesman and diplomat, Secretary of State 1973–7. As presidential assistant to Richard Nixon for national security affairs (1968–73), he helped to improve relations with both China and the Soviet Union. His role in negotiating the withdrawal of US troops from South Vietnam in 1973 was recognized when he was jointly awarded the Nobel Peace Prize that year. Later in 1973 he restored US diplomatic relations with Egypt in the wake of the Yom Kippur War and became known for his 'shuttle diplomacy' while subsequently mediating between between Israel and Syria.

kissogram /ˈkɪsəˌɡræm/ *n.* (also **Kissagram** *propr.*) a novelty telegram or greetings message sent through a commercial agency, usu. delivered with a kiss by a young woman.

kissy /ˈkɪsɪ/ *adj. colloq.* given to kissing (*not the kissy type*).

kist var. of CIST[1].

Kiswahili /ˌkɪswɑːˈhiːlɪ/ *n.* see SWAHILI. [Swahili *ki-* prefix for an abstract or inanimate object]

kit[1] /kɪt/ *n. & v.* ● *n.* **1** a set of articles, equipment, or clothing needed for a specific purpose (*first-aid kit*; *bicycle-repair kit*). **2** the clothing etc. needed for any activity, esp. sport (*football kit*). **3** a set of all the parts needed to assemble an item, e.g. a piece of furniture, a model, etc. **4** *Brit.* a wooden tub. ● *v.tr.* (**kitted**, **kitting**) (often foll. by *out*, *up*) equip with the appropriate clothing or tools. □ **the whole kit and caboodle** see CABOODLE. [ME f. MDu. *kitte* wooden vessel, of unkn. orig.]

kit[2] /kɪt/ *n.* **1** a kitten. **2** a young fox, badger, etc. [abbr.]

kit[3] /kɪt/ *n. hist.* a small fiddle esp. as used by a dancing-master. [perh. f. L *cithara*; see CITTERN]

Kitakyushu /ˌkiːtəˈkjuːʃuː/ a port in southern Japan, on the north coast of Kyushu island; pop. (1990) 1,026,470.

kitbag /ˈkɪtbæɡ/ *n.* a large, usu. cylindrical bag used for carrying equipment by a soldier, traveller, or sports player.

kit-cat /ˈkɪtkæt/ *n.* (in full **kit-cat portrait**) a portrait of less than half length, but including one hand; usu. 36 x 28 in. [named after a series of portraits of the members of the KIT-CAT CLUB]

Kit-Cat Club an association of prominent Whigs and literary figures founded in the early part of the 18th century. According to Alexander Pope its members included Steele, Addison, Congreve, and Vanbrugh.

kitchen /ˈkɪtʃɪn/ *n.* **1 a** the room or area where food is prepared and cooked. **b** kitchen fitments or units as sold together. **2** (*attrib.*) of or belonging to the kitchen (*kitchen knife*; *kitchen table*). **3** *sl.* the percussion section of an orchestra. □ **everything but the kitchen sink** everything imaginable. **kitchen cabinet** a group of unofficial advisers thought to be unduly influential. The name was first used in the US during the administration of President Andrew Jackson, *c.*1830. **kitchen garden** a garden where vegetables and sometimes fruit or herbs are grown. **kitchen midden** a prehistoric refuse-heap which marks an ancient settlement, chiefly containing bones, seashells, etc. **kitchen tea** *Austral. & NZ* a party held before a wedding to which female guests bring items of kitchen equipment as presents. [OE *cycene* f. L *coquere* cook]

Kitchener[1] /ˈkɪtʃɪnə(r)/ a city in Ontario, southern Canada; pop. (1991) 168,300; metropolitan area pop. 356,420. Settled by German Mennonites in 1806, as Dutch Sand Hills, it was renamed Berlin in 1830 and Kitchener in 1916, in honour of Field Marshal Kitchener.

Kitchener[2] /ˈkɪtʃɪnə(r)/, (Horatio) Herbert, 1st Earl Kitchener of Khartoum (1850–1916), British soldier and statesman. After defeating the Mahdist forces at Omdurman and reconquering Sudan in 1898, he served as Chief of Staff (1900–2) in the Second Boer War and Commander-in-Chief (1902–9) in India. At the outbreak of the First World War he was made Secretary of State for War, in which capacity he was responsible for organizing the large volunteer army which eventually fought the war on the Western Front. His commanding image appeared on recruiting posters urging 'Your country needs you!' He died when the ship taking him to Russia was sunk by a mine.

Kitchener bun *n. Austral.* a cream-filled bun coated with cinnamon and sugar. [KITCHENER[2]]

kitchenette /ˌkɪtʃɪˈnet/ *n.* a small kitchen or part of a room fitted as a kitchen.

kitchen-sink drama /ˌkɪtʃɪnˈsɪŋk/ *n.* a style of postwar British drama using working-class, often drab, domestic settings rather than the drawing-rooms of conventional middle-class drama; examples include John Osborne's *Look Back in Anger* (1956) and Arnold Wesker's *Roots* (1959).

kitchen-sink school /ˌkɪtʃɪnˈsɪŋk/ *n.* a group of postwar British realist painters whose work characteristically deals with drab and sordid themes.

kitchenware /ˈkɪtʃɪnˌweə(r)/ *n.* the utensils used in the kitchen.

kite /kaɪt/ *n. & v.* ● *n.* **1** a toy consisting of a light framework with thin material stretched over it, esp. a triangle with a semicircular base or a quadrilateral symmetrical about one diagonal, flown in the wind at the end of a long string. **2** a soaring bird of prey, esp. of the genus *Milvus*, with long wings and a forked tail. **3** *Brit. sl.* an aeroplane. **4** *sl.* a fraudulent cheque, bill, or receipt. **5** *Geom.* a quadrilateral figure symmetrical about one diagonal. **6** *sl.* a letter or note, esp. one that is illicit or surreptitious. **7** (in *pl.*) the highest sail of a ship, set only in a light wind. **8** *archaic* a dishonest person, a sharper. ● *v.* **1** *intr.* soar like a kite. **2** *tr.* (also *absol.*) *sl.* originate or pass (fraudulent cheques, bills, or receipts). **3** *tr.* (also *absol.*) *sl.* raise (money by dishonest means) (*kite a loan*). □ **kite balloon** a sausage-shaped captive balloon for military observations. **kite-flying** fraudulent practice. [OE *cȳta*, of unkn. orig.]

Kitemark /ˈkaɪtmɑːk/ *n.* (in the UK) an official mark in the shape of a kite, on goods approved by the British Standards Institution.

kit-fox /ˈkɪtfɒks/ *n.* *Zool.* a small fox, *Vulpes velox*, of North American prairies; esp. a form inhabiting the south-west US and Mexico, sometimes regarded as a distinct species (*V. macrotis*). [prob. f. KIT²]

kith /kɪθ/ *n.* □ **kith and kin** friends and relations. [OE *cȳthth* f. Gmc]

kitsch /kɪtʃ/ *n.* (often *attrib.*) garish, pretentious, or sentimental art, esp. as appreciated in a perverse or self-conscious way (*kitsch plastic models of the royal family*). □ **kitschy** *adj.* **kitschiness** *n.* [G]

kitten /ˈkɪt(ə)n/ *n. & v.* ● *n.* **1** a young cat. **2** a young ferret etc. ● *v.intr. & tr.* (of a cat etc.) give birth or give birth to. □ **have kittens** *Brit. colloq.* be extremely upset, anxious, or nervous. [ME *kito(u)n, ketoun* f. OF *chitoun, chetoun* dimin. of *chat* CAT¹]

kittenish /ˈkɪt(ə)nɪʃ/ *adj.* **1** like a young cat; playful and lively. **2** flirtatious, coy. □ **kittenishly** *adv.* **kittenishness** *n.* [KITTEN]

kittiwake /ˈkɪtɪˌweɪk/ *n.* a small gull of the genus *Rissa*, nesting on sea cliffs; esp. *R. tridactyla*, of the North Atlantic and Arctic Oceans. [imit. of its cry]

kittle /ˈkɪt(ə)l/ *adj.* (also **kittle-cattle** /-ˌkæt(ə)l/) **1** (of a person) capricious, rash, or erratic in behaviour. **2** difficult to deal with. [ME (now Sc. & dial.) *kittle* tickle, prob. f. ON *kitla*]

kitty¹ /ˈkɪtɪ/ *n.* (*pl.* **-ies**) **1** a fund of money for communal use. **2** the pool in some card-games. **3** the jack in bowls. [19th c.: orig. unkn.]

kitty² /ˈkɪtɪ/ *n.* (*pl.* **-ies**) a pet-name or a child's name for a kitten or cat.

Kitty Hawk /ˈkɪtɪ ˌhɔːk/ a town on a narrow sand peninsula on the Atlantic coast of North Carolina. It was there that, in 1903, the Wright brothers made the first powered airplane flight.

Kitwe /ˈkɪtweɪ/ a city in the Copperbelt mining region of northern Zambia; pop. (est. 1987) 449,400.

Kitzbühel /ˈkɪtsˌbjʊəl/ a town in the Tyrol, western Austria; pop. (1981) 7,840. Founded in the 12th century, it was a town formerly prosperous through copper and silver mining. It is now a popular resort centre for winter sports.

Kitzinger /ˈkɪtsɪŋə(r)/, Sheila (Helena Elizabeth) (b.1929), English childbirth educator. As a member of the advisory board of the National Childbirth Trust, and as one of the Trust's teachers, she has been a pioneer of natural childbirth and a leading advocate of breastfeeding. Her many books include *The Experience of Childbirth* (1962).

Kivu, Lake /ˈkiːvuː/ a lake in central Africa, on the Zaire–Rwanda frontier.

kiwi /ˈkiːwiː/ *n.* (*pl.* **kiwis**) **1** a flightless nocturnal New Zealand bird of the genus *Apteryx*, with hairlike feathers and a long bill. **2** (**Kiwi**) *colloq.* a New Zealander, esp. a soldier or member of a national sports team. □ **kiwi fruit** the fruit of a cultivated climbing plant, *Actinidia chinensis*, having a thin hairy skin, green flesh, and black seeds (also called *Chinese gooseberry*). [Maori]

kJ *abbr.* kilojoule(s).

KKK see KU KLUX KLAN.

KL *abbr. colloq.* Kuala Lumpur.

kl *abbr.* kilolitre(s).

Klagenfurt /ˈklɑːɡənˌfʊət/ a city in southern Austria, capital of Carinthia; pop. (1991) 89,500.

Klaipeda /ˈklaɪpədə/ (called in German *Memel*) a city and port in Lithuania, on the Baltic Sea; pop. (1991) 206,000. Founded in 1252 by the Teutonic Knights, it was known as Memel when under German control (1918–23 and 1941–4).

Klaproth /ˈklæprəʊt/, Martin Heinrich (1743–1817), German chemist, one of the founders of analytical chemistry. He discovered three new elements (zirconium, uranium, and titanium) in certain minerals, and contributed to the identification of others. A follower of Lavoisier, he helped to introduce the latter's new system of chemistry into Germany.

Klausenburg /ˈklaʊz(ə)nˌbɛrk, -ˌbʊrk/ the German name for CLUJ-NAPOCA.

Klaxon /ˈklæks(ə)n/ *n. propr.* a horn or warning hooter, orig. on a motor vehicle. [name of the manufacturing company]

Klee /kleɪ/, Paul (1879–1940), Swiss painter, resident in Germany from 1906. He began as a graphic artist and exhibited with Kandinsky, whose *Blaue Reiter* group he joined in 1912. He later concentrated on painting and developed an art of free fantasy, describing his drawing method as 'taking a line for a walk'. His paintings often have a childlike quality, as in *A Tiny Tale of a Tiny Dwarf* (1925). He taught at the Bauhaus (1920–33), after which he returned to Switzerland. Seventeen of his works appeared in the 'degenerate art' exhibition mounted by the Nazi regime in Munich in 1937.

Kleenex /ˈkliːneks/ *n.* (*pl.* same or **Kleenexes**) *propr.* an absorbent disposable paper tissue, used esp. as a handkerchief.

Klein¹ /klaɪn/, Calvin (Richard) (b.1942), American fashion designer. In 1968 he formed his own company, since when he has gained a reputation for his understated fashions for both men and women, including designer jeans. He is also known for his ranges of cosmetics and household linen.

Klein² /klaɪn/, Melanie (1882–1960), Austrian-born psychoanalyst. Klein was the first psychologist to specialize in the psychoanalysis of small children: she discovered surprising levels of aggression and sadism in young infants, and made an important contribution to the understanding of the more severe mental disorders found in children. She moved to London in 1926, becoming an influential member of the British Psychoanalytical Society.

Klein bottle *n.* *Math.* a closed surface with only one side, formed by passing the neck of a tube through the side of the tube to join the hole in the base. [Felix *Klein*, Ger. mathematician (1849–1925)]

Klemperer /ˈklempərə(r)/, Otto (1885–1973), German-born conductor and composer. While conductor at the Kroll Theatre in Berlin (1927–31), he was noted as a champion of new work and opera; his premières included Janáček's opera *From the House of the Dead*. He left Germany in 1933 and became an American citizen in 1937. Klemperer subsequently established a reputation as a conductor of symphonies by Beethoven, Brahms, and Mahler, and received particular acclaim for his Beethoven recordings in London during the 1950s. He adopted Israeli citizenship in 1970.

klepht /kleft/ *n.* **1** a member of the original body of Greeks who refused to submit to the Turks in the 15th century. **2** any of their descendants. **3** a brigand or bandit. [mod.Gk *klephtēs* f. Gk *kleptēs* thief]

kleptomania /ˌkleptəʊˈmeɪnɪə/ *n.* a recurrent urge to steal, usu. without regard for need or profit. □ **kleptomaniac** /-nɪˌæk/ *n. & adj.* [Gk *kleptēs* thief + -MANIA]

Klerk, F. W. de, see DE KLERK.

Klerksdorp /ˈklɜːksdɔːp/ a city in South Africa, in North-West Province, south-west of Johannesburg; pop. (1980) 238,865.

klieg /kliːɡ/ *n.* (also **klieg light**) a powerful lamp in a film studio etc. [A. T. & J. H. *Kliegl*, Amer. inventors (1872–1927, 1869–1959)]

Klimt /klɪmt/, Gustav (1862–1918), Austrian painter and designer. In 1897 he co-founded and became the first president of the Vienna Secession. His work combines stylized human forms with decorative and ornate clothing or backgrounds in elaborate mosaic patterns, often using gold leaf. He depicted mythological and allegorical subjects, and painted a number of portraits, chiefly of women; his works include *Judith I* (1901) and *The Kiss* (1908).

klipspringer /ˈklɪpˌsprɪŋə(r)/ *n.* a southern African dwarf antelope, *Oreotragus oreotragus*, which can bound up and down rocky slopes. [Afrik. f. *klip* rock + *springer* jumper]

Klondike /ˈklɒndaɪk/ a tributary of the Yukon river, in Yukon Territory, NW Canada, which rises in the Ogilvie mountains and flows 160 km (100 miles) westwards to join the Yukon at Dawson. It gave its name to the surrounding region, which became famous when gold was found in nearby Bonanza Creek in 1896. In the ensuing gold rush of 1897–8 thousands settled in the area to mine gold and the town of Dawson was established. Within ten years the area was exhausted and the population dramatically decreased.

kloof /kluːf/ *n.* a steep-sided ravine or valley in South Africa. [Du., = cleft]

Klosters /ˈkləʊstəz/ an Alpine winter-sports resort in eastern Switzerland, near the Austrian border.

kludge /klʌdʒ/ *n.* orig. *US sl.* **1** an ill-assorted collection of poorly matching parts. **2** *Computing* a machine, system, or program that has been badly put together. [20th c.: orig. uncert.]

klutz /klʌts/ *n.* *US sl.* **1** a clumsy awkward person. **2** a fool. □ **klutzy** *adj.* [Yiddish f. G *Klotz* wooden block]

klystron /ˈklaɪstrɒn/ *n.* *Electr.* an electron tube that generates or amplifies microwaves by velocity modulation. [Gk *kluzō klus-* wash over]

km *abbr.* kilometre(s).

K-meson /ˈkeɪˈmezɒn, -ˈmiːzɒn/ *n.* = KAON. [K (see KAON) + MESON]

kn. *abbr. Naut.* knot(s).

knack /næk/ *n.* **1** an acquired or intuitive faculty of doing a thing adroitly. **2** a trick or habit of action or speech etc. (*has a knack of offending people*). **3** *archaic* an ingenious device (see KNICK-KNACK). [ME, prob. identical with *knack* sharp blow or sound f. LG, ult. imit.]

knacker /ˈnækə(r)/ *n. & v. Brit.* ● *n.* **1** a buyer of useless horses, cattle, etc., for slaughter. **2** a buyer of old houses, ships, etc., for the materials. ● *v.tr. sl.* **1** kill. **2** (esp. as **knackered** *adj.*) exhaust, wear out. [19th c.: orig. unkn.]

knackery /ˈnækəri/ *n.* (*pl.* **-ies**) a knacker's yard or business.

knag /næg/ *n.* **1** a knot in wood; the base of a branch. **2** a short dead branch. **3** a peg for hanging things on. [ME, perh. f. LG *Knagge*]

knap[1] /næp/ *n. esp. dial.* the crest of a hill or of rising ground. [OE *cnæp(p)*, perh. rel. to ON *knappr* knob]

knap[2] /næp/ *v.tr.* (**knapped**, **knapping**) **1** break (stones for roads or building, flints, or *Austral.* ore) with a hammer. **2** *archaic* knock, rap, snap asunder. ● **knapper** *n.* [ME, imit.]

knapsack /ˈnæpsæk/ *n.* a soldier's or hiker's bag with shoulder-straps, carried on the back, and usu. made of canvas or weatherproof material. [MLG, prob. f. *knappen* bite + SACK[1]]

knapweed /ˈnæpwiːd/ *n.* a composite plant of the genus *Centaurea*, having thistle-like purple flowers. [ME, orig. *knopweed* f. KNOP + WEED]

knar /nɑː(r)/ *n.* a knot or protuberance in a tree-trunk, root, etc. [ME *knarre*, rel. to MLG, M.Du., MHG *knorre* knobbed protuberance]

knave /neɪv/ *n.* **1** a rogue, a scoundrel. **2** = JACK[1] *n.* 2. ● **knavery** /ˈneɪvəri/ *n.* (*pl.* **-ies**). **knavish** *adj.* **knavishly** *adv.* **knavishness** *n.* [OE *cnafa* boy, servant, f. WG]

knawel /ˈnɔːəl/ *n.* a low-growing plant of the genus *Scleranthus*, of the pink family. [G *Knauel*]

knead /niːd/ *v.tr.* **1 a** work (a yeast mixture, clay, etc.) into dough, paste, etc., by pummelling. **b** make (bread, pottery, etc.) in this way. **2** blend or weld together (*kneaded them into a unified group*). **3** massage (muscles etc.) as if kneading. ● **kneadable** *adj.* **kneader** *n.* [OE *cnedan* f. Gmc]

knee /niː/ *n. & v.* ● *n.* **1 a** (often *attrib.*) the joint between the thigh and the lower leg in humans. **b** the corresponding joint in other animals. **c** the area around this. **d** the upper surface of the thigh of a sitting person; the lap (*held her on his knee*). **2** the part of a garment covering the knee. **3** anything resembling a knee in shape or position, esp. a piece of wood or iron bent at an angle, a sharp turn in a graph, etc. ● *v.tr.* (**knees**, **kneed**, **kneeing**) **1** touch or strike with the knee (*kneed the ball past him; kneed him in the groin*). **2** *Brit. colloq.* cause (trousers) to bulge at the knee. □ **bend** (or **bow**) **the knee 1** kneel in submission, worship, or supplication. **2** submit. **bring to its** (or **his** or **her**) **knees** reduce (a thing or person) to a state of weakness or submission. **knee-bend** the action of bending the knee, esp. as a physical exercise in which the body is raised and lowered without the use of the hands. **knee-breeches** close-fitting trousers reaching to or just below the knee. **knee-deep 1** (usu. foll. by *in*) **a** immersed up to the knees. **b** deeply involved. **2** so deep as to reach the knees. **knee-high** so high as to reach the knees. **knee-high to a grasshopper** very small or very young. **knee-hole** (often *attrib.*) a space for the knees, esp. under a desk (*knee-hole desk*). **knee-jerk 1** a sudden involuntary kick caused by a blow on the tendon just below the knee. **2** (*attrib.*) predictable, automatic, stereotyped (*a knee-jerk reaction*). **knee-joint 1** = senses 1a, b of *n.* **2** a joint made of two pieces hinged together. **knee-length** reaching the knees. **knee-pan** the kneecap. **knees-up** *Brit. colloq.* a lively party or gathering. **knee-trembler** *sl.* an act of sexual intercourse between people in a standing position. **on** (or **on one's**) **bended knee** (or **knees**) kneeling, esp. in supplication, submission, or worship. [OE *cnēo(w)*]

kneecap /ˈniːkæp/ *n. & v.* ● *n.* **1** the convex bone in front of the knee-joint. **2** a protective covering for the knee. ● *v.tr.* (**-capped**, **-capping**) *colloq.* shoot (a person) in the knee or leg as a punishment, esp. for betraying a terrorist group. □ **kneecapping** *n.*

kneel /niːl/ *v.intr.* (*past and past part.* **knelt** /nelt/ or esp. *US* **kneeled**) fall or rest on the knees or a knee, esp. in reverence or submission. [OE *cnēowlian* (as KNEE)]

kneeler /ˈniːlə(r)/ *n.* **1** a hassock or cushion used for kneeling, esp. in church. **2** a person who kneels.

knell /nel/ *n. & v.* ● *n.* **1** the sound of a bell, esp. when rung solemnly for a death or funeral. **2** an announcement, event, etc., regarded as a solemn warning of disaster. ● *v.* **1** *intr.* **a** (of a bell) ring solemnly, esp. for a death or funeral. **b** make a doleful or ominous sound. **2** *tr.* proclaim by or as by a knell (*knelled the death of all their hopes*). □ **ring the knell of** announce or herald the end of. [OE *cnyll, cnyllan*: perh. infl. by *bell*]

knelt *past and past part.* of KNEEL.

Knesset /ˈknesɪt/ the Parliament of modern Israel, established in 1949. It consists of 120 members elected every four years. [Heb., lit. 'gathering']

knew *past* of KNOW.

knickerbocker /ˈnɪkəˌbɒkə(r)/ *n.* **1** (in *pl.*) loose-fitting breeches gathered at the knee or calf. **2** (**Knickerbocker**) **a** a New Yorker. **b** a descendant of the original Dutch settlers in New York. [Diedrich *Knickerbocker*, pretended author of Washington Irving's *History of New York* (1809)]

Knickerbocker Glory *n. Brit.* ice-cream served with other ingredients in a tall glass.

knickers /ˈnɪkəz/ *n.pl.* **1** *Brit.* a woman's or girl's undergarment covering the body from the waist or hips to the top of the thighs and having leg-holes or separate legs. **2** esp. *N. Amer.* **a** knickerbockers. **b** a boy's short trousers. **3** (as *int.*) *Brit. sl.* an expression of contempt. [abbr. of KNICKERBOCKER]

knick-knack /ˈnɪknæk/ *n.* **1** a useless and usu. worthless ornament; a trinket. **2** a small, dainty article of furniture, dress, etc. □ **knick-knackery** *n.* **knick-knackish** *adj.* [redupl. of *knack* in obs. sense 'trinket']

knife /naɪf/ *n. & v.* ● *n.* (*pl.* **knives** /naɪvz/) **1 a** a metal blade used as a cutting tool with usu. one long sharp edge fixed rigidly in a handle or hinged (cf. PENKNIFE). **b** a similar tool used as a weapon. **2** a cutting-blade forming part of a machine. **3** (as **the knife**) *colloq.* a surgical operation or operations. ● *v.* **1** *tr.* cut or stab with a knife. **2** *tr. sl.* bring about the defeat of (a person) by underhand means. **3** *intr.* (usu. foll. by *through*) cut or cut its way like a knife. □ **at knife-point** threatened with a knife or an ultimatum etc. **before you can say knife** *colloq.* very quickly or suddenly. **get one's knife into** treat maliciously or vindictively, persecute. **go under the knife** *colloq.* have surgery. **knife-board** a board on which knives are cleaned. **knife-edge 1** the edge of a knife. **2** a position of extreme danger or uncertainty. **3** a steel wedge on which a pendulum etc. oscillates. **4** = ARÊTE. **knife-grinder 1** a travelling sharpener of knives etc. **2** a person who grinds knives etc. during their manufacture. **knife-machine** a machine for cleaning knives. **knife-pleat** a narrow flat pleat on a skirt etc., usu. overlapping another. **knife-rest** a metal or glass support for a carving-knife or -fork at table. **knife-throwing** a circus etc. act in which knives are thrown at targets. **that one could cut with a knife** *colloq.* (of an accent, atmosphere, etc.) very obvious, oppressive, etc. □ **knifelike** *adj.* **knifer** *n.* [OE *cnif* f. ON *knífr* f. Gmc]

knight /naɪt/ *n. & v.* ● *n.* **1** a man awarded a non-hereditary title (*Sir*) by a sovereign in recognition of merit or service. **2** *hist.* **a** a man, usu. noble, raised esp. by a sovereign to honourable military rank after service as a page and squire. (*See note below.*) **b** a military follower or attendant, esp. of a lady as her champion in a war or tournament. **3** a man devoted to the service of a woman, cause, etc. **4** *Chess* a piece usu. shaped like a horse's head. **5 a** *Rom. Hist.* a member of the class of *equites*, orig. the cavalry of the Roman army. **b** *Gk Hist.* a citizen of the second class in Athens. **6** (in full **knight of the shire**) *hist.* a gentleman representing a shire or county in Parliament. ● *v.tr.* confer a knighthood on. □ **knight bachelor** (*pl.* **knights bachelor**) a knight not belonging to a special order. **knight commander** see COMMANDER 3. **knight errant** a medieval knight wandering in search of chivalrous adventures. **2** a man of a chivalrous or quixotic nature. **knight-errantry** the practice or conduct of a knight errant. **knight in shining armour** a chivalrous rescuer or helper, esp. of a woman. **knight marshal** *hist.* an officer of the royal household with judicial functions. **knight of the road** *colloq.* **1** a highwayman. **2** a commercial traveller. **3** a tramp. **4** a lorry driver or taxi driver. **knight-service** *hist.* the tenure of land by military service. □ **knighthood** *n.* **knightlike** *adj.* **knightly** *adj. & adv. poet.* **knightliness** *n.* [OE *cniht* boy, youth, hero f. WG]

▪ In early medieval England knights formed an élite cavalry force giving military service (usually forty days a year) in return for the lands they held under the feudal system. During the 12th century they gained in economic and social importance, becoming involved in local administration, and derived further impetus from the formation of the great religious and secular orders in the wake of the Crusades. The foremost religious organizations included the Knights Hospitallers and the Knights Templars; among British orders the oldest, and the only surviving medieval order, is the Order of the Garter. Men knighted by the sovereign today usually become knights bachelor rather than members of an order of knighthood.

knightage /ˈnaɪtɪdʒ/ n. **1** knights collectively. **2** a list and account of knights.

Knightsbridge /ˈnaɪtsbrɪdʒ/ a district in the West End of London, to the south of Hyde Park, noted for its fashionable and expensive shops.

Knights Hospitallers a military and religious order founded as the Knights of the Hospital of St John at Jerusalem in the 11th century. Originally protectors of pilgrims, they also undertook the care of the sick. During the Middle Ages they became a powerful and wealthy military force, with foundations in various European countries; their military power ended when Malta was surrendered to Napoleon (1798). In England, the order was revived in 1831 and was responsible for the foundation of the St John Ambulance Brigade in 1888 (see ST JOHN AMBULANCE).

Knights Templars a military and religious order founded in 1118 (as the Poor Knights of Christ and the Temple of Solomon) to protect pilgrims from bandits in the Holy Land, where the order was given quarters on the site of Solomon's temple in Jerusalem. The order became powerful and wealthy, particularly in the 13th century, but its members' arrogance towards rulers, together with their wealth and their rivalry with the Knights Hospitallers, led to their downfall; the order was suppressed in 1312, many of its possessions being given to the Hospitallers. The Inner and Middle Temple in London are on the site of the Templars' English headquarters.

kniphofia /nɪˈfəʊfɪə, naɪˈfəʊ-, nɪpˈhəʊ-/ n. a tall ornamental liliaceous plant of the genus *Kniphofia*, native to South and East Africa, with long spikes or dense racemes of red, yellow or orange flowers, esp. the red-hot poker. [mod.L f. Johann Hieronymus *Kniphof*, Ger. botanist (1704–63)]

knish /knɪʃ/ n. a dumpling of flaky dough filled with cheese etc. and baked or fried. [Yiddish f. Russ.]

knit /nɪt/ v. & n. ● v. (**knitting**; past and past part. **knitted** or (esp. in senses 2–4) **knit**) **1** tr. (also absol.) **a** make (a garment, blanket, etc.) by interlocking loops of esp. wool with knitting-needles. **b** make (a garment etc.) with a knitting machine. **c** make (a plain stitch) in knitting (*knit one, purl one*). **2 a** tr. contract (the forehead) in vertical wrinkles. **b** intr. (of the forehead) contract; frown. **3** tr. & intr. (often foll. by *together*) make or become close or compact, esp. through common interests etc. (*a close-knit group*). **4** intr. (often foll. by *together*) (of parts of a broken bone) become joined; heal. ● n. knitted material or a knitted garment. □ **knit up** make or repair by knitting. **2** conclude, finish, or end. □ **knitter** n. [OE *cnyttan* f. WG: cf. KNOT[1]]

knitting /ˈnɪtɪŋ/ n. **1** a garment etc. in the process of being knitted. **2 a** the act of knitting. **b** an instance of this. □ **knitting-machine** a machine used for knitting garments etc. **knitting-needle** a thin pointed rod of steel, wood, plastic, etc., used esp. in pairs for knitting.

knitwear /ˈnɪtweə(r)/ n. knitted garments.

knives pl. of KNIFE.

knob /nɒb/ n. & v. ● n. **1 a** a rounded protuberance, esp. at the end or on the surface of a thing. **b** a handle of a door, drawer, etc., shaped like a knob. **c** a knob-shaped attachment for pulling, turning, etc. (*press the knob under the desk*). **2** a small, usu. round, piece (of butter, coal, sugar, etc.). **3** esp. N. Amer. a prominent round hill. **4** coarse sl. the penis. ● v. (**knobbed, knobbing**) **1** tr. provide with knobs. **2** intr. (usu. foll. by *out*) bulge. □ **with knobs on** Brit. sl. that and more (used as a retort to an insult, in emphatic agreement, etc.) (*and the same to you with knobs on*). □ **knobby** adj. **knoblike** adj. [ME f. MLG *knobbe* knot, knob, bud: cf. KNOP, NOB[2], NUB]

knobble /ˈnɒb(ə)l/ n. a small knob. □ **knobbly** adj. [ME, dimin. of KNOB: cf. Du. & LG *knobbel*]

knobkerrie /ˈnɒbˌkerɪ/ n. a short stick with a knobbed head used as a weapon esp. by indigenous peoples of South Africa. [after Afrik. *knopkierie*]

knobstick /ˈnɒbstɪk/ n. **1** = KNOBKERRIE. **2** archaic = BLACKLEG.

knock /nɒk/ v. & n. ● v. **1 a** tr. strike (a hard surface) with an audible sharp blow (*knocked the table three times*). **b** intr. strike, esp. a door to gain admittance (*can you hear someone knocking?; knocked at the door*). **2** tr. make (a hole, a dent, etc.) by knocking (*knock a hole in the fence*). **3** tr. (usu. foll. by *in, out, off*, etc.) drive (a thing, a person, etc.) by striking (*knocked the ball into the hole; knocked those ideas out of his head; knocked her hand away*). **4** tr. colloq. criticize, disparage. **5** intr. **a** (of a motor or other engine) make a thumping or rattling noise, esp. as the result of a loose bearing. **b** = PINK[3]. **6** tr. Brit. sl. make a strong impression on, astonish. **7** tr. Brit. coarse sl. offens. = knock off 7. ● n. **1** an act of knocking. **2** a sharp rap, esp. at a door. **3** an audible sharp blow. **4** the sound of knocking in esp. a motor engine. **5** Cricket colloq. an innings. □ **knock about** (or **around**) **1** strike repeatedly; treat roughly (*knocked her about*). **2** lead a wandering adventurous life; wander aimlessly. **3** be present without design or volition (*there's a cup knocking about somewhere*). **4** (usu. foll. by *with*) be associated socially (*knocks about with his brother*). **knock against 1** collide with. **2** come across casually. **knock back 1** colloq. eat or drink, esp. quickly. **2** Brit. colloq. disconcert. **3** reverse the progress of. **4** colloq. refuse, rebuff. **knock-back** n. colloq. a refusal, a rebuff. **knock the bottom out of** see BOTTOM. **knock down 1** strike (esp. a person) to the ground with a blow. **2** demolish. **3** (usu. foll. by *to*) (at an auction) dispose of (an article) to a bidder by a knock with a hammer (*knocked the Picasso down to him for a million*). **4** colloq. lower the price of (an article). **5** take (machinery, furniture, etc.) to pieces for transportation. **6** US sl. steal. **7** Austral. & NZ sl. spend a pay cheque etc.) freely. **knock-down** attrib.adj. **1** (of a blow, misfortune, argument, etc.) overwhelming. **2** colloq. (of a price) very low. **3** (of a price at auction) reserve. **4** (of furniture etc.) easily dismantled and reassembled. **5** (of an insecticide's effect) immobilizing. ● n. Austral. & NZ sl. an introduction (to a person). **knock for knock agreement** an agreement between insurance companies by which each pays its own policyholder regardless of liability. **knock one's head against** come into collision with (unfavourable facts or conditions). **knocking-shop** Brit. sl. a brothel. **knock into a cocked hat** see COCK[1]. **knock into the middle of next week** colloq. send (a person) flying, esp. with a blow. **knock into shape** see SHAPE. **knock-kneed** having knock knees. **knock knees** an abnormal condition with the legs curved inwards at the knee. **knock off 1** strike off with a blow. **2** colloq. **a** finish work (*knocked off at 5.30*). **b** finish (work) (*knocked off work early*). **3** colloq. dispatch (business). **4** colloq. rapidly produce (a work of art, verses, etc.). **5** (often foll. by *from*) deduct (a sum) from a price, bill, etc. **6** sl. steal. **7** Brit. coarse sl. offens. have sexual intercourse with (a woman). **8** sl. kill. **knock on** Rugby drive (a ball) with the hand or arm towards the opponents' goal-line. **knock-on** n. an act of knocking on. **knock-on effect** a secondary, indirect, or cumulative effect. **knock on the head 1** stun or kill (a person) by a blow on the head. **2** colloq. put an end to (a scheme etc.). **knock on** (or **knock**) **wood** N. Amer. = touch wood. **knock out 1** make (a person) unconscious by a blow on the head. **2** knock down (a boxer) for a count of 10, thereby winning the contest. **3** defeat, esp. in a knockout competition. **4** colloq. please greatly, astonish. **5** (refl.) colloq. exhaust (*knocked themselves out swimming*). **6** colloq. make or write (a plan etc.) hastily. **7** empty (a tobacco-pipe) by tapping. **8** Austral., NZ, & US sl. earn. **knock sideways** colloq. disconcert; astonish. **knock spots off** defeat easily. **knock together** put together or assemble hastily or roughly. **knock under** submit. **knock up 1** make or arrange hastily. **2** drive upwards with a blow. **3 a** become exhausted or ill. **b** exhaust or make ill. **4** Brit. arouse (a person) by a knock at the door. **5** Cricket score (runs) rapidly. **6** esp. US sl. make pregnant. **7** practise a ball game before formal play begins. **knock-up** n. Brit. a warm-up at tennis etc. **take a** (or **the**) **knock** be hard hit financially or emotionally. [ME f. OE *cnocian*: prob. imit.]

knockabout /ˈnɒkəˌbaʊt/ adj. & n. ● attrib.adj. **1** (of comedy) boisterous; slapstick. **2** (of clothes) suitable for rough use. **3** Austral. of a farm or station handyman. ● n. **1** Austral. a farm or station handyman. **2** a knockabout performer or performance.

knocker /ˈnɒkə(r)/ n. **1** a metal or wooden instrument hinged to a door for knocking to call attention. **2** a person or thing that knocks. **3** (in pl.) coarse sl. a woman's breasts. **4** a person who buys or sells door to door. □ **knocker-up** Brit. hist. a person employed to rouse early workers by knocking at their doors or windows. **on the knocker 1 a** (buying or selling) from door to door. **b** (obtained) on credit. **2** Austral. & NZ colloq. promptly. **up to the knocker** Brit. sl. in good condition; to perfection.

knockout /ˈnɒkaʊt/ n. **1** the act of making unconscious by a blow. **2** Boxing etc. a blow that knocks an opponent out. **3** a competition in which the loser in each round is eliminated (also attrib.: *a knockout round*). **4** colloq. an outstanding or irresistible person or thing. □ **knockout drops** a drug added to a drink to cause unconsciousness.

knoll[1] /nəʊl/ n. a small hill or mound. [OE *cnoll* hilltop, rel. to MDu., MHG *knolle* clod, ON *knollr* hilltop]

knoll² /nəʊl/ v. & n. archaic ● v. **1** tr. & intr. = KNELL. **2** tr. summon by the sound of a bell. ● n. = KNELL. [ME, var. of KNELL: perh. imit.]

knop /nɒp/ n. **1** a knob, esp. ornamental. **2** an ornamental loop or tuft in yarn. **3** archaic a flower-bud. [ME f. MLG, MDu. *knoppe*]

knopkierie /ˈknɒpˌkɪərɪ/ n. S. Afr. = KNOBKERRIE. [Afrik.]

Knossos /ˈknɒsəs, ˈnɒs-/ the principal city of Minoan Crete, the remains of which are situated on the north coast of Crete. Excavations by Sir Arthur Evans from 1900 onwards revealed the remains of a luxurious and spectacularly decorated complex of buildings, which he named the Palace of Minos, with frescos of landscapes, animal life, and the sport of bull-leaping. The city site was occupied from neolithic times until c.1200 BC; in c.1450 BC Crete was overrun by the Mycenaeans, but the palace was not finally destroyed until the 14th or early 13th century BC.

knot¹ /nɒt/ n. & v. ● n. **1 a** an intertwining of a rope, string, tress of hair, etc., with another, itself, or something else to join or fasten together. **b** a set method of tying a knot (a reef knot). **c** a ribbon etc. tied as an ornament and worn on a dress etc. **d** a tangle in hair, knitting, etc. **2 a** a unit of a ship's or aircraft's speed equivalent to one nautical mile per hour (see nautical mile). **b** a division marked by knots on a log-line, as a measure of speed. **c** colloq. a nautical mile. **3** (usu. foll. by of) a group or cluster (a small knot of journalists at the gate). **4** something forming or maintaining a union; a bond or tie, esp. of wedlock. **5** a hard lump of tissue in an animal or human body. **6 a** a knob or protuberance in a stem, branch, or root. **b** a hard mass formed in a tree-trunk at the intersection with a branch. **c** a round cross-grained piece in timber where a branch has been cut through. **d** a node on the stem of a plant. **7** a difficulty; a problem. **8** a central point in a problem or the plot of a story etc. **9** (in full **porter's knot**) Brit. hist. a double shoulder-pad and forehead-loop used for carrying loads. ● v. (**knotted, knotting**) **1** tr. tie (a string etc.) in a knot. **2** tr. entangle. **3** tr. knit (the brows). **4** tr. unite closely or intricately (knotted together in intrigue). **5 a** intr. make knots for fringing. **b** tr. make (a fringe) with knots. □ **at a rate of knots** colloq. very fast. **get knotted!** colloq. an expression of disbelief, annoyance, etc. **knot-garden** an intricately designed formal garden. **knot-hole** a hole in a piece of timber where a knot has fallen out (sense 6c). **tie in knots** colloq. baffle or confuse completely. □ **knotless** adj. **knotter** n. **knotting** n. (esp. in sense 5 of v.). [OE cnotta f. WG]

knot² /nɒt/ n. a small short-billed sandpiper, Calidris canutus. [ME: orig. unkn.]

knotgrass /ˈnɒtɡrɑːs/ n. **1** a common weed, Polygonum aviculare, with jointed creeping stems and small pink flowers. **2** any other plant, esp. a grass, with jointed stems.

knotty /ˈnɒtɪ/ adj. (**knottier, knottiest**) **1** full of knots. **2** hard to explain; puzzling (a knotty problem). □ **knottily** adv. **knottiness** n.

knotweed /ˈnɒtwiːd/ n. a polygonum; esp. Fallopia japonica, a tall fast-growing Japanese plant that is widely naturalized.

knotwork /ˈnɒtwɜːk/ n. ornamental work representing or consisting of intertwined cords.

knout /naʊt, nuːt/ n. & v. hist. ● n. a scourge used in imperial Russia, often causing death. ● v.tr. flog with a knout. [F f. Russ. knut f. Icel. knútr, rel. to KNOT¹]

know /nəʊ/ v. & n. ● v. (past **knew** /njuː/; past part. **known** /nəʊn/) **1** tr. (often foll. by that, how, what, etc.) **a** have in the mind; have learned; be able to recall (knows a lot about cars; knows what to do). **b** (also absol.) be aware of (a fact) (he knows I am waiting; I think he knows). **c** have a good command of (a subject or language) (knew German; knows his subject). **2** tr. be acquainted or friendly with (a person or thing). **3** tr. **a** recognize; identify (I knew him at once; knew him for an American). **b** (foll. by to + infin.) be aware of (a person or thing) as being or doing what is specified (knew them to be rogues). **c** (foll. by from) be able to distinguish (one from another) (did not know him from Adam). **4** tr. be subject to (her joy knew no bounds). **5** tr. have personal experience of (fear etc.). **6** tr. (as **known** adj.) **a** publicly acknowledged (a known thief; a known fact). **b** Math. (of a quantity etc.) having a value that can be stated. **7** intr. have understanding or knowledge. **8** tr. archaic have sexual intercourse with. ● n. (in phr. **in the know**) colloq. well-informed; having special knowledge. □ **all one knows** (or **knows how**) **1** all one can (did all he knew to stop it). **2** adv. to the utmost of one's power (tried all she knew). **before one knows where one is** with baffling speed. **be not to know 1** have no way of learning (wasn't to know they'd arrive late). **2** be not to be told (she's not to know about the party). **don't I know it!** colloq. an expression of rueful assent. **don't you know** colloq. or joc. an expression used for emphasis (such a bore, don't you know). **for all** (or

aught) I **know** so far as my knowledge extends. **have been known to** have done on occasion (they have been known to not turn up). I **knew it!** I was sure that this would happen. I **know what** I have a new idea, suggestion, etc. **know about** have information about. **know-all** colloq. a person who seems to know everything. **know best** be or claim to be better informed etc. than others. **know better than** (foll. by that, or to + infin.) be wise, well-informed, or well-mannered enough to avoid (specified behaviour etc.). **know by name 1** have heard the name of. **2** be able to give the name of. **know by sight** recognize the appearance (only) of. **know how** know the way to do something. **know-how** n. **1** practical knowledge; technique, expertise. **2** natural skill or invention. **know-it-all** = know-all. **know-nothing 1** an ignorant person. **2** an agnostic. **know of** be aware of; have heard of (not that I know of). **know one's own mind** be decisive, not vacillate. **know the ropes** (or **one's stuff**) be fully knowledgeable or experienced. **know a thing or two** be experienced or shrewd. **know what's what** have adequate knowledge of the world, life, etc. **know who's who** be aware of who or what each person is. **not if I know it** only against my will. **not know that** … be fairly sure that … not (I don't know that I want to go). **not know what hit one** be suddenly injured, killed, disconcerted, etc. **not want to know** refuse to take any notice of. **what do you know** (or **know about that**)? an expression of surprise. **you know** colloq. **1** an expression implying something generally known or known to the hearer (you know, the pub on the corner). **2** an expression used as a gap-filler in conversation. **you know something** (or **what**)? I am going to tell you something. **you-know-what** (or **-who**) a thing or person unspecified but understood. **you never know** nothing in the future is certain. □ **knowable** adj. **knower** n. [OE (ge)cnāwan, rel. to CAN¹, KEN]

knowing /ˈnəʊɪŋ/ n. & adj. ● n. the state of being aware or informed of any thing. ● adj. **1** cunning; sly. **2** showing knowledge or awareness; shrewd. □ **there is no knowing** no one can tell. □ **knowingness** n.

knowingly /ˈnəʊɪŋlɪ/ adv. **1** consciously; intentionally (had never knowingly injured him). **2** in a knowing manner (smiled knowingly).

knowledge /ˈnɒlɪdʒ/ n. **1 a** (usu. foll. by of) awareness or familiarity gained by experience (of a person, fact, or thing) (have no knowledge of their character). **b** a person's range of information (is not within his knowledge). **c** specific information; facts or intelligence about something (received knowledge of their imminent departure). **2 a** (usu. foll. by of) a theoretical or practical understanding of a subject, language, etc. (has a good knowledge of Greek). **b** the sum of what is known (every branch of knowledge). **c** learning, scholarship. **3** Philos. true, justified belief; certain understanding, as opposed to opinion. **4** = carnal knowledge. □ **come to one's knowledge** become known to one. **to my knowledge 1** so far as I know. **2** as I know for certain. [ME knaulege, with earlier knawlechen (v.) formed as KNOW + OE -lǣcan f. lāc as in WEDLOCK]

knowledgeable /ˈnɒlɪdʒəb(ə)l/ adj. (also **knowledgable**) well-informed; intelligent. □ **knowledgeably** adv. **knowledgeableness** n. **knowledgeability** /ˌnɒlɪdʒəˈbɪlɪtɪ/ n.

known past part. of KNOW.

Knox¹ /nɒks/, John (c.1505–72), Scottish Protestant reformer. After early involvement in the Scottish Reformation he spent more than a decade preaching in Europe, during which time he stayed in Geneva and was influenced by Calvin. In 1559 he returned to Scotland and played a central part in the establishment of the Church of Scotland within a Scottish Protestant state. A fiery orator, he became the spokesman of the religious interests opposed to the Catholic Mary, Queen of Scots when she returned to rule in her own right in 1561.

Knox² /nɒks/, Ronald Arbuthnott (1888–1957), English theologian and writer. In 1917 he converted to Roman Catholicism and later served as Catholic chaplain at Oxford University (1926–39). His translation of the Bible from the Vulgate (1945–9) was accepted for use in the Roman Catholic Church. His literary output was varied and included detective fiction and humour.

Knoxville /ˈnɒksvɪl/ a port on the Tennessee river, in eastern Tennessee; pop. (1990) 165,120. It was twice the state capital, in 1796–1812 and 1817–19, and is now the headquarters of the Tennessee Valley Authority.

Knt. abbr. Knight.

knuckle /ˈnʌk(ə)l/ n. & v. ● n. **1** the bone at a finger-joint, esp. that adjoining the hand. **2 a** a projection of the carpal or tarsal joint of a quadruped. **b** a joint of meat consisting of this with the adjoining parts, esp. of bacon or pork. ● v.tr. strike, press, or rub with the knuckles. □ **go the knuckle** Austral. sl. fight, punch. **knuckle-bone 1** bone forming a knuckle. **2** the bone of a sheep or other animal corresponding to or resembling a knuckle. **3** a knuckle of meat. **knuckle-bones 1** animal knuckle-bones used in the game of jacks. **2** the game of jacks. **knuckle down** (often foll. by *to*) **1** apply oneself seriously (to a task etc.). **2** give in; submit. **knuckle sandwich** *sl.* a punch in the mouth. **knuckle under** give in; submit. **rap on** (or **over**) **the knuckles** see RAP[1]. □ **knuckly** *adj.* [ME *knokel* f. MLG, MDu. *knökel*, dimin. of *knoke* bone]

knuckleduster /'nʌk(ə)l,dʌstə(r)/ *n.* a metal guard worn over the knuckles in fighting, esp. to increase the effect of blows.

knucklehead /'nʌk(ə)l,hed/ *n. colloq.* a slow-witted or stupid person.

knur /nɜ:(r)/ *n.* (also **knurr**) **1** a hard excrescence on the trunk of a tree. **2** a hard concretion. [ME *knorre*, var. of KNAR]

knurl /nɜ:l/ *n.* a small projecting knob, ridge, etc. □ **knurled** /nɜ:ld/ *adj.* [KNUR]

Knut see CANUTE.

KO *abbr.* **1** knockout. **2** kick-off.

koa /'kəʊə/ *n.* **1** a Hawaiian tree, *Acacia koa*, which produces dark red wood. **2** this wood. [Hawaiian]

koala /kəʊ'ɑ:lə/ *n.* an Australian bearlike marsupial, *Phascolarctos cinereus*, having thick grey fur and feeding on eucalyptus leaves. ¶ The fuller form *koala bear* is now considered incorrect. [Dharuk *gula, gulawan*[y]]

koan /'kəʊæn/ *n.* a riddle used in Zen Buddhism to demonstrate the inadequacy of logical reasoning. [Jap., = public matter (for thought)]

Kobe /'kəʊbɪ/ a port in central Japan, on the island of Honshu; pop. (1990) 1,477,420. The city was severely damaged by an earthquake in Jan. 1995.

København see COPENHAGEN.

kobold /'kəʊbɒld, 'kɒbəʊld/ *n. Germanic Mythol.* **1** a familiar spirit; a brownie. **2** an underground spirit in mines etc. [G]

Koch /kɒx/, Robert (1843–1910), German bacteriologist. He successfully identified and cultured the bacillus causing anthrax in cattle, devised better methods for obtaining pure cultures, and identified the organisms causing tuberculosis and cholera. He also studied typhoid fever, malaria, and other tropical diseases, and formulated the conditions to be satisfied before a disease can be ascribed to a specific micro-organism. The techniques that Koch devised are the basis of modern bacteriological methods. He was awarded a Nobel Prize in 1905.

Köchel number /'kɜ:x(ə)l/ *n. Mus.* a number given to each of Mozart's compositions in the complete catalogue of his works compiled by the Austrian scientist Ludwig von Köchel (1800–77) and his successors.

KO'd /keɪ'əʊd/ *adj.* knocked out. [abbr.]

Kodály /'kəʊdaɪ/, Zoltán (1882–1967), Hungarian composer. He was influenced by Debussy's music while studying in Paris (1910), but his main source of inspiration was his native land; he was much involved in the collection and publication of Hungarian folk-songs. His best-known compositions include the choral work *Psalmus Hungaricus* (1923), the opera *Háry János* (1925–7), and the *Marosszék Dances* (1930).

Kodiak /'kəʊdɪ,æk/ *n.* (in full **Kodiak bear**) a very large Alaskan race of the brown bear. [*Kodiak* Island, Alaska]

koel /'kəʊəl/ *n.* a dark-coloured cuckoo of the genus *Eudynamys*, esp. *E. scolopacea* of Asia. [Hindi *kóïl* f. Skr. *kokila*]

Koestler /'kɜ:stlə(r)/, Arthur (1905–83), Hungarian-born British novelist and essayist. In 1940 he settled in Britain and published his best-known novel *Darkness at Noon*, which exposed the Stalinist purges of the 1930s. In later works such as *The Sleepwalkers* (1959), a study of the Copernican revolution in astronomy, he questioned some of the common assumptions of science. He became increasingly interested in parapsychology and left money in his will for a university chair in the subject, subsequently founded at Edinburgh. He and his wife committed suicide together.

Kohima /'kəʊ'hi:mə/ a city in the far north-east of India, capital of the state of Nagaland; pop. (1991) 53,000.

Koh-i-noor /'kəʊɪ,nʊə(r)/ a famous Indian diamond, one of the treasures belonging to Aurangzeb, which has a history going back to the 14th century. It passed into British possession on the annexation of Punjab in 1849, and was set in the queen's state crown for the coronation of George VI (1937). [Pers. *kōh-i nūr* mountain of light]

Kohl /kəʊl/, Helmut (b.1930), German statesman, Chancellor of the Federal Republic of Germany 1982–90, and of Germany since 1990. He became chairman of the Christian Democratic Party in 1973, and

was leader of the opposition until 1982, becoming Chancellor of the Federal Republic of Germany when the ruling coalition collapsed. As Chancellor he showed a strong commitment to NATO and to closer European union within the European Community. In 1990 he presided over the reunification of East and West Germany and was elected Chancellor of the united country later the same year.

kohl /kəʊl/ *n.* a black powder, usu. antimony sulphide or lead sulphide, used as eye make-up esp. in Eastern countries. [Arab. *kuḥl*]

kohlrabi /kəʊl'rɑ:bɪ/ *n.* (*pl.* **kohlrabies**) a variety of cabbage with an edible turnip-like swollen stem. [G f. It. *cavoli rape* (pl.] f. med.L *caulorapa* (as COLE, RAPE[2])]

Koil /kɔɪl/ see ALIGARH.

koine /'kɔɪni:/ *n.* **1** the common language of the Greeks from the close of the classical period to the Byzantine era. **2** a common language shared by various peoples; a lingua franca. [Gk *koinē* (*dialektos*) common (language)]

kola var. of COLA.

Kola Peninsula /'kəʊlə/ a peninsula on the NW coast of Russia, separating the White Sea from the Barents Sea. The port of Murmansk lies on its northern coast.

Kolhapur /ˌkəʊlhɑ:'pʊə(r)/ an industrial city in the state of Maharashtra, western India; pop. (1991) 405,000.

kolinsky /kə'lɪnskɪ/ *n.* (*pl.* **-ies**) **1** a Siberian weasel, *Mustela sibirica*, with a rich brown coat. **2** the fur of this. [f. *Kola* (cf. KOLA PENINSULA) + pseudo-Russ. ending -*insky*]

Kolkhis see COLCHIS.

kolkhoz /kɒl'kɒz, kɒl'hɒz/ *n.* a collective farm in the former USSR. [Russ. f. *kollektivnoe khozyaĭstvo* collective farm]

Köln see COLOGNE.

Kolozsvár /'kɒlɒʒ,vɑ:r/ the Hungarian name for CLUJ-NAPOCA.

Kolyma /ˌkɒlɪ'mɑ:/ a river of far eastern Siberia, which flows approximately 2,415 km (1,500 miles) northwards to the Arctic Ocean.

Komi /'kəʊmɪ/ an autonomous republic of NW Russia; pop. (1990) 1,265,000; capital, Syktyvkar.

komitadji (also **komitaji**) var. of COMITADJI.

Komodo /kə'məʊdəʊ/ a small island in Indonesia, in the Lesser Sunda Islands, situated between the islands of Sumbawa and Flores.

Komodo dragon *n.* (also **Komodo monitor**) a very large monitor lizard, *Varanus komodoensis*, found only on Komodo and a few neighbouring islands. It is the largest living lizard.

Komsomol /'kɒmsə,mɒl/ *n. hist.* **1** a Communist youth organization in the former Soviet Union. **2** a member of this. [Russ. f. *Kommunisticheskiĭ soyuz molodezhi* Communist League of Youth)]

Komsomolsk /ˌkɒmsə'mɒlsk/ (also **Komsomolsk-on-Amur** /-,ɒnə'mʊə(r)/) an industrial city in the far east of Russia, on the Amur river; pop. (1990) 318,000. It was built in 1932 by members of the Komsomol on the site of the village of Permskoya.

Kongo /'kɒŋgəʊ/ *n. & adj.* ● *n.* (*pl.* same or **Kongos**) **1** a member of a Bantu-speaking people inhabiting the region of the River Congo in west central Africa. **2** the language of this people, Kikongo. ● *adj.* of or relating to this people or their language.

Königgrätz /ˌkɜ:nɪç'grɛts/ the German name for HRADEC KRÁLOVÉ.

Königsberg /'kɜ:nɪçs,bɛrk/ the German name for KALININGRAD.

Kon-Tiki /kɒn'tɪkɪ/ the raft made of balsa logs in which the Norwegian anthropologist Thor Heyerdahl sailed from the western coast of Peru to the islands of Polynesia in 1947. It was named after an Inca god.

Konya /'kɒnjə/ a city in SW central Turkey; pop. (1990) 513,350. Originally settled by Phrygians in the 8th century BC, it was known in Roman times as Iconium. Towards the end of the 11th century AD it became the capital of the Seljuk sultans and was renamed Konya.

koodoo var. of KUDU.

kook /ku:k/ *n. & adj. US sl.* ● *n.* a crazy or eccentric person. ● *adj.* crazy; eccentric. [20th c.: prob. f. CUCKOO]

kookaburra /'kʊkə,bʌrə/ *n.* a large tree-living Australian kingfisher of the genus *Dacelo*; esp. *D. novaeguineae*, which has a strange laughing cry (also called *laughing jackass*). [Wiradhuri *guguburra*]

kooky /'ku:kɪ/ *adj.* (**kookier, kookiest**) *sl.* crazy, eccentric, strange. □ **kookily** *adv.* **kookiness** *n.*

Kooning, Willem de, see DE KOONING.

kop /kɒp/ *n.* **1** *S. Afr.* a prominent hill or peak. **2** (**Kop**) *Football* a high bank of terracing for standing spectators, esp. those supporting the home side. [Afrik. f. Du., = head (cf. COP²): sense 2 f. *Spion Kop*, site of a battle in the Second Boer War]

kopek (also **kopeck**) var. of COPECK.

kopi /'kəʊpɪ/ *n. Austral.* powdered gypsum. [Aboriginal]

koppie /'kɒpɪ/ *n.* (also **kopje**) *S. Afr.* a small hill. [Afrik. *koppie*, Du. *kopje*, dimin. of KOP]

koradji /kə'rædʒɪ/ *n.* (*pl.* **koradjis**) *Austral.* an Aboriginal medicine man. [Dharuk *garraaji* doctor]

Koran /kɔː'rɑːn/ *n.* (also **Qur'an**) the holy book of Islam, composed of the revelations (said to have come from the archangel Gabriel) to the prophet Muhammad. The Koran was written in Arabic during the Prophet's lifetime, from *c*.610 to his death in 632. The revelations are grouped into 114 units of varying lengths, known as *suras*; the first sura is said as part of the ritual prayer. The revelations touch upon all aspects of human existence, from the doctrinally focused revelations of Muhammad's early career in Mecca to those concerning social organization and legislation. Considered to be the direct and inimitable word of God, the Koran is held by Muslims to be untranslatable, although versions or interpretations in many other languages are available. □ **Koranic** /-'rænɪk, -'rɑːnɪk/ *adj.* [Arab. *ḳur'ān* recitation f. *ḳara'a* read]

Korbut /'kɔːbət/, Olga (b.1955), Soviet gymnast, born in Belarus. Her performances (especially at the 1972 Olympic Games, where she won two individual gold medals) greatly increased the popularity of the sport.

Korda /'kɔːdə/, Sir Alexander (born Sándor Kellner) (1893–1956), Hungarian-born British film producer and director. He settled in Britain in 1930 and founded London Film Productions two years later. His productions included *The Private Life of Henry VIII* (1933), which he also directed, *Sanders of the River* (1935), *Things to Come* (1936), and *The Third Man* (1949).

Kordofan /ˌkɔːdə'fɑːn/ a region of central Sudan.

Korea /kə'rɪə/ a region of eastern Asia forming a peninsula between the Sea of Japan and the Yellow Sea, now divided into the countries of North Korea and South Korea. Possessed of a distinct national and cultural identity and ruled from the 14th century by the Korean Yi dynasty, Korea has suffered as a result of its position between Chinese and Japanese spheres of influence. A period of Chinese domination was ended by the Sino-Japanese War (1894–5), and after the Russo-Japanese War a decade later the country was annexed by Japan (1910). After the Japanese surrender at the end of the Second World War, Korea was partitioned along the 38th parallel, the north being occupied by the Soviets and the south by the US. (See also KOREAN WAR, NORTH KOREA, SOUTH KOREA.)

Korea, Democratic People's Republic of see NORTH KOREA.

Korea, Republic of see SOUTH KOREA.

Korean /kə'rɪən/ *n.* & *adj.* ● *n.* **1** a native or national of Korea. **2** the language of Korea. (*See* note below.) ● *adj.* of or relating to Korea or its people or language.

▪ Korean is spoken by more than 60 million people in North and South Korea, and by large numbers in China, Russia, Japan, and North America. Its linguistic affiliations are uncertain, although it is almost certainly related to Japanese. Its vocabulary and orthography have been heavily influenced by Chinese; in South Korea Chinese characters are used in addition to the Korean alphabet, although in the North only the Korean alphabet is used.

Korean War the war of 1950–3 between North and South Korea, which began with the invasion of the South by North Korean forces. The United Nations voted to oppose the invasion, and UN troops, dominated by US forces, invaded and advanced to the border of North Korea with China. China intervened on the side of the North, and for a time there appeared to be a danger of global conflict; however, peace negotiations were begun in 1951, and the war ended two years later with the restoration of previous boundaries.

korfball /'kɔːfbɔːl/ *n.* a game like basketball played by two teams consisting of six men and six women each. [Du. *korfbal* f. *korf* basket + *bal* ball]

Kórinthos see CORINTH.

korma /'kɔːmə/ *n.* a mildly spiced Indian curry dish of meat or fish marinated in yoghurt or curds. [Urdu *ḳormā*, f. Turk *kavurma*]

Kortrijk /'kɔːtreɪk/ (French **Courtrai** /kurtrɛ/) a city in western Belgium, in West Flanders; pop. (1991) 76,140.

koruna /'kɒrʊnə/ *n.* the basic monetary unit of the Czech Republic and Slovakia, equal to 100 haleru. [Czech = crown]

Korup National Park /'kɒrəp/ a national park in western Cameroon, on the border with Nigeria. It was established in 1961 to protect a large area of tropical rainforest.

Kos /kɒs/ (also **Cos**) a Greek island in the SE Aegean, one of the Dodecanese group.

Kosciusko /ˌkɒsɪ'ʌskəʊ/, Thaddeus (or Tadeusz) (1746–1817), Polish soldier and patriot. A trained soldier, he fought for the American colonists during the War of American Independence, returning to Poland in 1784. Ten years later he led a nationalist uprising, defeating a large Russian force at Racławice. Captured and imprisoned by the Russians (1794–6), he eventually moved to France, where he devoted the rest of his life to the cause of Polish independence.

Kosciusko, Mount a mountain in SE Australia, in the Great Dividing Range in SE New South Wales. Rising to a height of 2,228 m (7,234 ft), it is the highest mountain in Australia. It was named by the explorer Sir Paul Edmund de Strzelecki (1797–1873) in honour of Thaddeus Kosciusko.

kosher /'kəʊʃə(r)/ *adj.* & *n.* ● *adj.* **1** (of food or premises in which food is sold, cooked, or eaten) fulfilling the requirements of Jewish law. (*See* note below.) **2** *colloq.* correct; genuine; legitimate. ● *n.* **1** kosher food. **2** a kosher shop. [Heb. *kāšēr* proper]

▪ Restrictions on the foods suitable for Jews are derived from the Mosaic prescriptions found in the books of Leviticus (11) and Deuteronomy (14). Animals must be slaughtered and prepared in the prescribed way, in which the blood is drained from the body, while certain creatures, such as pigs and shellfish, are forbidden altogether; meat and milk must not be cooked or consumed together, and separate utensils must be kept for each. Strict observance of these rules, which may well have originated for reasons of hygiene, is today confined mainly to Orthodox Jews.

Košice /'kɒʃɪtsə/ an industrial city in southern Slovakia; pop. (1991) 234,840.

Kosovo /'kɒsəvə/ an autonomous province of Serbia; pop. (1987) 1,850,000; capital, Priština. It borders on Albania and the majority of the people are of Albanian descent.

Kossuth /'kɒsuːθ, 'kɒʃuːt/, Lajos (1802–94), Hungarian statesman and patriot. Long an opponent of Habsburg domination of Hungary, he led the 1848 insurrection and was appointed governor of the country during the brief period of independence which followed. In 1849 he began a lifelong period of exile when the uprising was crushed, although he continued to strive for Hungarian independence.

Kostroma /ˌkɒstrə'mɑː/ an industrial city in European Russia, situated on the River Volga to the north-west of Nizhni Novgorod; pop. (1990) 280,000.

Kosygin /kɒ'siːgɪn/, Aleksei Nikolaevich (1904–80), Soviet statesman, Premier of the USSR 1964–80. Born in Russia, he became a Central Committee member in 1939 and held a series of ministerial posts, mostly concerned with finance and industry. He succeeded Khrushchev as Premier (Chairman of the Council of Ministers) in 1964, but devoted most of his attention to internal economic affairs, being gradually eased out of the leadership by Brezhnev. He resigned owing to ill health in 1980.

Kota /'kəʊtə/ an industrial city in Rajasthan state, in NW India, on the Chambal river; pop. (1991) 536,000.

Kota Baharu /ˌkəʊtə bə'hɑːruː/ a city in Malaysia, on the east coast of the Malay Peninsula, the capital of the state of Kelantan; pop. (1980) 170,600.

Kota Kinabalu /ˌkəʊtə ˌkɪnəbə'luː/ a port in Malaysia, on the north coast of Borneo, capital of the state of Sabah; pop. (1980) 56,000.

Kotka /'kɒtkə/ a port on the south coast of Finland; pop. (1990) 56,630.

koto /'kəʊtəʊ/ *n.* (*pl.* **-os**) a Japanese musical instrument with thirteen long esp. silk strings. [Jap.]

kotow var. of KOWTOW.

Kotzebue /'kɒtsəˌbjuː, -ˌbuː/, August von (1761–1819), German dramatist. His many plays were popular in both Germany and England; the tragedy *Menschenhass und Reue* (1789) was produced by Richard Sheridan as *The Stranger*, and *Das Kind der Liebe* (1790) was adapted in England as *Lovers' Vows*. He was a political informant to Tsar Alexander I and was assassinated by the Germans. His son, Otto von Kotzebue (1787–1846), was a navigator and explorer; he discovered an inlet of NW Alaska (Kotzebue Sound) now named after him.

koumiss /'kuːmɪs, 'kʊm-/ *n.* (also **kumiss**, **kumis**) a fermented liquor prepared from esp. mare's milk, used by Asian nomads and medicinally. [Tartar *kumiz*]

kouprey /ˈkuːpreɪ/ n. a rare grey wild ox, *Bos sauveli*, native to forests in Indo-China. [Cambodian]

kourbash /ˈkʊəbæʃ/ n. (also **kurbash**) a whip, esp. of hippopotamus hide, used as an instrument of punishment in Turkey and Egypt. [Arab. *kurbāj* f. Turk. *kırbāç* whip]

Kourou /kʊˈruː/ a town on the north coast of French Guiana; pop. (1990) 11,200. Nearby is a satellite-launching station of the European Space Agency, established in 1967.

kowhai /ˈkəʊaɪ/ n. a leguminous tree or shrub of the genus *Sophora*; esp. *S. tetraptera*, native to New Zealand and Chile, with pendent clusters of yellow flowers. [Maori]

Kowloon /kaʊˈluːn/ a densely populated peninsula on the SE coast of China, forming part of Hong Kong. It is separated from Hong Kong Island by Victoria Harbour.

kowtow /kaʊˈtaʊ/ n. & v. (also **kotow** /kəʊ-/) ● n. hist. the Chinese custom of kneeling and touching the ground with the forehead in worship or submission. ● v.intr. **1** (usu. foll. by *to*) act obsequiously. **2** hist. perform the kowtow. [Chin. *ketou* f. *ke* knock + *tou* head]

KP n. US Mil. colloq. **1** enlisted men detailed to help the cooks. **2** kitchen duty. [abbr. of *kitchen police*]

k.p.h. abbr. kilometres per hour.

Kr symb. Chem. the element krypton.

Kra, Isthmus of /krɑː/ the narrowest part of the Malay Peninsula, forming part of southern Thailand.

kraal /krɑːl/ n. S. Afr. **1** a village of huts enclosed by a fence. **2** an enclosure for cattle or sheep. [Afrik. f. Port. *curral* f. Nama]

Krafft-Ebing /ˈkræftˈeɪbɪŋ/, Richard von (1840–1902), German physician and psychologist. He is best known for establishing the relationship between syphilis and general paralysis, and for his *Psychopathia Sexualis* (1886), which pioneered the systematic study of aberrant sexual behaviour.

kraft /krɑːft/ n. (in full **kraft paper**) a kind of strong smooth brown wrapping paper. [G f. Sw., = strength]

Kragujevac /ˈkræɡʊjəˌvæts/ a city in central Serbia; pop. (1981) 164,820. It was the capital of Serbia 1818–39.

krai /kraɪ/ n. (also **kray**) an administrative territory of Russia. In pre-revolutionary times krais were each made up of a number of provinces, becoming in 1924 large administrative units in the Soviet territorial system. By the time of the breakup of the USSR in 1991 there were six krais. [Russ. = edge, border]

krait /kraɪt/ n. a brightly coloured venomous snake of the genus *Bungarus*, of India and SE Asia. [Hindi *karait*]

Krakatoa /ˌkrækəˈtəʊə/ a small volcanic island in Indonesia, lying between Java and Sumatra, scene of a great eruption in 1883 which destroyed most of the island.

kraken /ˈkrɑːkən/ n. a large mythical sea monster said to appear off the coast of Norway. [Norw.]

Kraków see CRACOW.

krans /krɑːns/ n. S. Afr. a precipitous or overhanging wall of rocks. [Afrik. f. Du. *krans* coronet]

Krasnodar /ˌkræsnəˈdɑː(r)/ **1** a krai (administrative territory) in the northern Caucasus, on the Black Sea in southern Russia; pop. (1990) 5,135,000. **2** its capital, a port on the lower Kuban river; pop. (1990) 627,000. It was known until 1922 as Yekaterinodar (Ekaterinodar).

Krasnoyarsk /ˌkræsnəˈjɑːsk/ **1** a krai (administrative territory) in central Siberian Russia; pop. (1990) 3,612,000. **2** its capital, a port on the Yenisei river; pop. (1990) 922,000.

Kraut /kraʊt/ n. sl. offens. a German. [shortening of SAUERKRAUT]

Krebs /krebz/, Sir Hans Adolf (1900–81), German-born British biochemist. While still in Germany he discovered the cycle of reactions by which urea is synthesized by the liver as a nitrogenous waste product. After moving to Britain he discovered the biochemical cycle that is now named after him, for which he shared a Nobel Prize in 1953.

Krebs cycle n. Biochem. a cyclic sequence of reactions by which most living cells generate energy during the process of aerobic respiration. It takes place in the mitochondria, using up oxygen and producing carbon dioxide and water as waste products, and ADP is converted to energy-rich ATP.

Krefeld /ˈkreɪfelt/ an industrial town and port on the Rhine in western Germany, in North Rhine-Westphalia; pop. (1991) 245,770.

Kreisler /ˈkraɪslə(r)/, Fritz (1875–1962), Austrian-born American violinist and composer. He made his first public appearances in the US in 1889 and became an American citizen in 1943. A noted interpreter of the standard classics, in 1910 he gave the first performance of Elgar's violin concerto, which was dedicated to him.

Kremenchuk /ˌkremənˈtʃuːk/ (Russian **Kremenchug** /ˌkrɪmɪnˈtʃuk/) an industrial city in east central Ukraine, on the River Dnieper; pop. (1990) 238,000.

kremlin /ˈkremlɪn/ n. a citadel within a Russian town. [F, f. Russ. *Kreml'*, of Tartar orig.]

Kremlin, the the citadel in Moscow, centre of administration of the Russian (and formerly the Soviet) government. A fort was first built on the site in 1156, and in the 15th century walls were added which formed the limits of Moscow. The Kremlin became the residence of the grand dukes of Moscow and the place where Russian emperors and empresses were crowned; in the 20th century the term has been used to signify the government of the USSR and, latterly, of Russia.

kriegspiel /ˈkriːɡspiːl, -ʃpiːl/ n. **1** a war-game in which blocks representing armies etc. are moved about on maps. **2** a form of chess with an umpire, in which each player has only limited information about the opponent's moves. [G f. *Krieg* war + *Spiel* game]

Kriemhild /ˈkriːmhɪlt/ (in the Nibelungenlied) a Burgundian princess, wife of Siegfried and later of Etzel (Attila the Hun), whom she marries in order to be revenged on her brothers for Siegfried's murder.

krill /krɪl/ n. tiny planktonic crustaceans found in the seas around the Antarctic and eaten by fish and some seals and whales. [Norw. *kril* tiny fish]

krimmer /ˈkrɪmə(r)/ n. a grey or black furry fleece obtained from the wool of young Crimean lambs. [G f. *Krim* Crimea]

kris /kriːs/ n. (also **crease, creese** /kriːs/) a Malay or Indonesian dagger with a wavy blade. [ult. f. Malay *k(i)rĭs*]

Krishna /ˈkrɪʃnə/ *Hinduism* one of the most popular gods, the eighth and most important avatar or incarnation of Vishnu. He is worshipped in several forms: as the child god whose miracles and pranks are extolled in the Puranas; as the divine cowherd whose erotic exploits, especially with his favourite, Radha, have produced both romantic and religious literature; and as the divine charioteer who preaches to Arjuna on the battlefield in the Bhagavadgita. [Skr., = black]

Krishnaism /ˈkrɪʃnəˌɪz(ə)m/ n. *Hinduism* the worship of Krishna as an incarnation of Vishnu.

Krishnamurti /ˌkrɪʃnəˈmʊətɪ/, Jiddu (1895–1986), Indian spiritual leader. He was originally associated with the Theosophical Society and was declared a World Teacher by Annie Besant. In 1929 he broke away from the society, advocating his own spiritual philosophy based on a rejection of organized religion and the attainment of self-realization by introspection. His teachings enjoyed a revival of interest in the 1960s and he settled in California in 1969.

Krishna River a river which rises in the Western Ghats of southern India and flows generally eastwards for 1,288 km (805 miles) to the Bay of Bengal.

Kristallnacht /ˈkrɪst(ə)lˌnɑːxt/ the occasion of concerted violence by Nazis throughout Germany and Austria against Jews and their property on the night of 9–10 Nov. 1938. Its name refers to the broken glass produced by the smashing of shop windows. The event marked an escalation of the persecution of Jews in the Third Reich. [G, = night of crystal or of (broken) glass]

Kristiania see CHRISTIANIA.

Kristiansand /ˈkrɪstʃənˌsænd/ a ferry port on the south coast of Norway, in the Skagerrak; pop. (1991) 65,690.

Kriti see CRETE.

Krivoy Rog see KRYVY RIH.

kromesky /krəˈmeskɪ/ n. (pl. **-ies**) a croquette of minced meat or fish, rolled in bacon and fried. [app. f. Pol. *kromeczka* small slice]

krona /ˈkrəʊnə/ n. **1** (pl. **kronor** /-nə(r)/) the chief monetary unit of Sweden. **2** (pl. **kronur** /-nə(r)/) the chief monetary unit of Iceland. [Sw. & Icel., = CROWN]

krone /ˈkrəʊnə/ n. (pl. **kroner** /-nə(r)/) the chief monetary unit of Denmark and Norway. [Danish & Norw., = CROWN]

Kronos var. of CRONUS.

Kronstadt /ˈkrɒnʃtat/ the German name for BRAŞOV.

Kroo var. of KRU.

kroon /kruːn/ n. the basic monetary unit of Estonia, equal to 100 sents. [Estonian = crown: cf. KRONA, KRONE.]

Kropotkin /krə'pɒtkɪn/, Prince Peter (1842–1921), Russian anarchist. He was a geographer who carried out explorations of Siberia, Finland, and Manchuria before devoting his life to political activities. He became an influential exponent of anarchism and was imprisoned in 1874. He escaped abroad two years later and only returned to Russia after the Russian Revolution in 1917. His works include *Modern Science and Anarchism* (1903).

Kru /kruː/ *n. & adj.* (also **Kroo**) ● *n.* (*pl.* same) **1** a member of a seafaring people of the coast of Liberia. **2** the language of this people. ● *adj.* of or relating to the Kru or their language. [W. Afr.]

Kru Coast a section of the coast of Liberia to the north-west of Cape Palmas, inhabited by the Kru people.

Kruger /'kruːɡə(r)/, Stephanus Johannes Paulus (known as 'Oom (= uncle) Paul') (1825–1904), South African soldier and statesman. He led the Afrikaners to victory in the First Boer War in 1881 and afterwards served as President of Transvaal from 1883 to 1899. His refusal to allow equal rights to non-Boer immigrants was one of the causes of the Second Boer War, during which Kruger was forced to flee the country. He died in exile in Switzerland.

Kruger National Park a national park in South Africa, in Eastern Transvaal on the Mozambique border. Designated a national park in 1926, it was originally a game reserve established in 1898 by President Kruger.

krugerrand /'kruːɡəˌrænd, -ˌrɑːnt/ *n.* a South African gold coin depicting President Kruger. [f. KRUGER + RAND[1]]

krummhorn var. of CRUMHORN.

Krupp /krʊp/, Alfred (1812–87), German arms manufacturer. In the 1840s he began to manufacture ordnance at the ironworks founded in Essen by his father, and built up the company to become the largest such manufacturer in Europe. Under the management of successive members of the family the Krupp Works played a pre-eminent part in German arms production through to the end of the Second World War.

krypton /'krɪptɒn/ *n.* an unreactive gaseous chemical element (atomic number 36; symbol **Kr**). One of the noble gases, krypton was discovered by William Ramsay and M. W. Travers in 1898. It is obtained by distillation of liquid air, and is used in some kinds of electric lamps and bulbs. Although chemically very inert, it can be made to combine with fluorine. [Gk *krupton* hidden, neut. adj. f. *kruptō* hide]

Kryvy Rih /krɪˌviː 'rɪx/ (Russian **Krivoy Rog** /krɪˌvɔj 'rɔk/) an industrial city in southern Ukraine, at the centre of an iron-ore mining region; pop. (est. 1989) 713,000.

KS *abbr.* **1** US Kansas (in official postal use). **2** (in the UK) King's Scholar.

Kshatriya /'kʃætrɪə, 'kʃɑːt-/ *n.* a member of the second of the four great Hindu classes or varnas, the warrior or baronial class. The traditional function of the Kshatriyas is to protect society by fighting in wartime and governing in peacetime. [Skr. f. *kshatra* rule]

K. St. J. *abbr.* Knight of the Order of St John.

KT *abbr.* **1** Knight Templar. **2** (in the UK) Knight of the Order of the Thistle.

Kt. *abbr.* Knight.

kt. *abbr.* knot.

K/T boundary /keɪ'tiː/ *n. Geol.* **1** the time at the end of the Cretaceous period and the start of the Tertiary, marked by the extinction of many groups of organisms, including dinosaurs. **2** a sedimentary deposit laid down at this time. (See also CRETACEOUS.) [symbols for *Cretaceous* and *Tertiary*]

Ku *symb. Chem.* the element kurchatovium.

Kuala Lumpur /ˌkwɑːlə 'lʊmpʊə(r)/ the capital of Malaysia, in the south-west of the Malay Peninsula; pop. (1990) 1,237,900. It is a major commercial centre in the middle of a rubber-growing and tin-mining region.

Kuala Trengganu /ˌkwɑːlə trəŋ'ɡænuː/ (also **Kuala Terengganu**) the capital of the state of Trengganu in Malaysia, on the east coast of the Malay Peninsula at the mouth of the Trengganu river; pop. (1980) 180,300.

Kuantan /kwɑːn'tɑːn/ the capital of the state of Pahang in Malaysia, on the east coast of the Malay Peninsula; pop. (1981) 131,550.

Kuan Yin /kwɑːn 'jɪn/ (in Chinese Buddhism) the goddess of compassion.

Kublai Khan /ˌkuːblaɪ 'kɑːn/ (1216–94), Mongol emperor of China, grandson of Genghis Khan. Between 1252 and 1259 he conquered southern China with his brother Mangu (then Mongol Khan). On Mangu's death in 1259 he was elected Khan himself, completing the conquest of China and founding the Yuan dynasty; he established his capital on the site of the modern Beijing. He successfully invaded Korea and Burma, but failed in attacks on Java and Japan.

Kubrick /'kjuːbrɪk/, Stanley (b.1928), American film director, producer, and writer. He first gained acclaim as a director with the thriller *The Killing* (1956). The coldly enigmatic science-fiction epic *2001: A Space Odyssey* (1968) set new standards for special effects. Other notable films include *Lolita* (1962), *Dr Strangelove* (1964), *A Clockwork Orange* (1971), and *The Shining* (1980).

Kuching /'kuːtʃɪŋ/ a port in Malaysia, on the Sarawak river near the NW coast of Borneo, capital of the state of Sarawak; pop. (1989) 157,000.

kudos /'kjuːdɒs/ *n.* glory; renown. [Gk]

kudu /'kuːduː/ *n.* (also **koodoo**) (*pl.* same or **kudus**) either of two large African antelopes of the genus *Tragelaphus*, with white stripes and spiral horns. [Afrik. *koedoe* f. Xhosa *i-qudu*]

kudzu /'kʊdzuː/ *n.* (in full **kudzu vine**) a quick-growing climbing leguminous plant, *Pueraria lobata*, with reddish-purple flowers. [Jap. *kuzu*]

Kufic /'kjuːfɪk/ *n. & adj.* (also **Cufic**) ● *n.* an early angular form of the Arabic alphabet found chiefly in decorative inscriptions. ● *adj.* of or in this type of script. [*Cufa*, a city south of Baghdad]

Kuibyshev /'kuːɪbɪˌʃef/ the former name (1935–91) for SAMARA.

Ku Klux Klan /ˌkuː klʌks 'klæn, ˌkjuː/ (abbr. **KKK**) an extremist right-wing secret society in the southern states of the US, founded originally to oppose the new rights being given to blacks during the Reconstruction after the Civil War. Members disguised themselves in white robes and hoods in order to carry out acts of terrorism and intimidation against blacks and their supporters. The original Klan was outlawed by Congress in 1870–1, but a similar, much more powerful, organization emerged in 1915, and by the 1920s had over 4 million members, this time being responsible for lynchings and murders of Jews, immigrants, Catholics, and trade unionists as well as blacks. A burning cross became the symbol of this new organization. Although temporarily disbanded in 1944, it emerged again in the 1950s and 1960s and continues at a local level. [perh. f. Gk *kuklos* circle + CLAN]

kukri /'kʊkrɪ/ *n.* (*pl.* **kukris**) a curved knife broadening towards the point, used by Gurkhas. [Hindi *kukrī*]

kulak /'kuːlæk/ *n. hist.* a peasant in Russia wealthy enough to own a farm, hire labour, and engage in money-lending. Kulaks emerged after the emancipation of the serfs in the 19th century, and remained important after the Revolution; when they resisted Stalin's forced collectivization of agriculture from 1929, however, they were destroyed as a class, with millions being arrested, sent to Siberia, or killed. [Russ., = fist, tight-fisted person]

kulan /'kuːlən/ *n.* a wild ass, *Equus hemionus*, of a race native to the central Asian steppes (cf. KIANG, ONAGER). [Tartar]

Kultur /kʊl'tʊə(r)/ *n. esp. derog.* German civilization and culture, seen as racist, authoritarian, and militaristic. [G f. L *cultura* CULTURE]

Kulturkampf /kʊl'tʊəkæmpf/ the conflict between the German government (headed by Bismarck) and the papacy for the control of schools and Church appointments between 1872 and 1887. Bismarck passed legislation in an attempt to break the authority and influence of the Catholic Church in the new German Empire, but he was later forced to repeal most of it. [G (as KULTUR, *Kampf* struggle)]

Kum see QOM.

Kumamoto /ˌkuːmə'məʊtəʊ/ a city in southern Japan, on the west coast of Kyushu island; pop. (1990) 579,300.

kumara /'kʊmərə/ *n. NZ* a sweet potato. [Maori]

Kumasi /kuː'mæsɪ/ a city in southern Ghana; pop. (1984) 376,250. It is the capital of the Ashanti region.

Kumayri see GYUMRI.

kumis (also **kumiss**) var. of KOUMISS.

kümmel /'kʊm(ə)l/ *n.* a sweet liqueur flavoured with caraway and cumin seeds. [G (as CUMIN)]

kumquat /'kʌmkwɒt/ *n.* (also **cumquat**) **1** a small orange-like citrus fruit with a sweet rind and acid pulp, used in preserves. **2** a shrub or small tree of the genus *Fortunella* producing this. [Cantonese var. of Chin. *kin kü* golden orange]

Kundera /'kʊndərə/, Milan (b.1929), Czech novelist. His books were proscribed in Czechoslovakia following the Soviet military invasion of 1968. He emigrated to France (1975), was stripped of his Czech citizenship in 1979, and became a French citizen two years later. Major novels include *Life is Elsewhere* (1973), *The Book of Laughter and Forgetting* (1979), and *The Unbearable Lightness of Being* (1984).

Kung /kʊŋ/ *n. & adj.* ● *n.* (*pl.* same) **1** a member of a San (Bushman) people of the Kalahari Desert in southern Africa, maintaining to some extent a nomadic way of life dependent on hunting and gathering. **2** the Khoisan language of the Kung. ● *adj.* of or relating to the Kung or their language.

kung fu /kʊŋ ˈfuː, kʌŋ/ *n.* a primarily unarmed Chinese martial art resembling karate, dating back to the Zhou dynasty (11th century BC–256 BC) or earlier. Kung fu became popular in the West during the 1970s, largely through the films of Bruce Lee. [Chin. *gongfu* f. *gong* merit + *fu* master]

Kung Fu-tzu /ˌkʊŋ fuːˈtsuː/ see CONFUCIUS.

Kunlun Shan /ˌkʊnlʊn ˈʃɑːn/ a range of mountains in western China, on the northern edge of the Tibetan plateau, extending eastwards for over 1,600 km (1,000 miles) from the Pamir Mountains. Its highest peak is Muztag, which rises to 7,723 m (25,338 ft).

Kunming /kʊnˈmɪŋ/ a city in SW China, capital of Yunnan province; pop. (est. 1989) 1,500,000.

Kuomintang /ˌkwəʊmɪnˈtæŋ/ (also **Guomindang** /ˌɡwəʊmɪnˈdæŋ/) a nationalist party founded in China under Sun Yat-sen in 1912, which held power from 1928 until the Communist Party took power in Oct. 1949. The Kuomintang subsequently formed the central administration of Taiwan. After Sun Yat-sen's death in 1925, it was led by Chiang Kai-shek. [Chin., = national people's party]

Kuopio /kuˈəʊpɪˌəʊ/ a city in southern Finland, capital of a province of the same name; pop. (1990) 80,610.

kurbash var. of KOURBASH.

kurchatovium /ˌkɜːtʃəˈtəʊvɪəm/ *n.* a name proposed in the former Soviet Union for the chemical element unnilquadium (cf. RUTHERFORDIUM). The corresponding symbol is **Ku**. [I. V. *Kurchatov*, Russian physicist (1903–60)]

Kurd /kɜːd/ *n.* a member of a mainly pastoral Muslim people living chiefly in eastern Turkey, northern Iraq, western Iran, and eastern Syria. (See KURDISTAN.) [Kurdish]

kurdaitcha /kəˈdaɪtʃə/ *n. Austral.* **1** the tribal use of a bone in spells intended to cause sickness or death. **2** a man empowered to point the bone at a victim. [Aboriginal]

Kurdish /ˈkɜːdɪʃ/ *n. & adj.* ● *n.* the language spoken by over 10 million people in the region of Kurdistan. It belongs to the Indo-Iranian language group and is generally written in an Arabic script. ● *adj.* of or relating to the Kurds or their language.

Kurdistan /ˌkɜːdɪˈstɑːn/ an extensive plateau and mountainous region in the Middle East, south of the Caucasus, including large parts of eastern Turkey, northern Iraq, western Iran, eastern Syria, Armenia, and Azerbaijan. For centuries it has been the traditional home of the Kurdish people. The creation of a separate state of Kurdistan was proposed by the Allies after the First World War, but this was abandoned in 1923 when Turkey reasserted its territorial authority in the region. In 1970 the Kurds in Iraq were allocated an autonomous region with limited powers of self-determination, but in 1988 came under attack by Saddam Hussein's troops, and many fled to Turkey and Iran. Following further persecution in the aftermath of the Gulf War of 1991, 'safe havens' were established for the Kurds in northern Iraq. Although not officially recognized as a state, this region is called Kurdistan by its inhabitants, elections have taken place, and a coalition government has been elected comprising the Kurdistan Party and the Patriotic Union of Kurdistan. There is still armed conflict and political instability in the area; Iran, Syria, and Turkey, as well as Iraq, oppose the creation of a Kurdish state.

Kure /kuˈreɪ/ a city in southern Japan, on the south coast of the island of Honshu, near Hiroshima; pop. (1990) 216,720.

Kurgan /kʊəˈɡɑːn/ a city in central Russia, commercial centre for an agricultural region; pop. (1990) 360,000.

Kurile Islands /kʊəˈriːl/ (also **Kuril Islands** or **Kurils**) a chain of 56 islands between the Sea of Okhotsk and the North Pacific, stretching from the southern tip of the Kamchatka peninsula to the north-eastern corner of the Japanese island of Hokkaido. The islands were given to Japan in exchange for the northern part of Sakhalin island in 1875, but were returned to the Soviet Union in 1945. They are a continuing source of dispute between the two countries.

Kurosawa /ˌkʊərəˈsɑːwə/, Akira (b.1910), Japanese film director. He first gained international acclaim with *Rashomon* (1950), and later

became known for his samurai films, such as *The Seven Samurai* (1954) and *Ran* (1985), a Japanese version of Shakespeare's *King Lear*. He also treats modern themes, mainly of social injustice, in films such as *Living* (1952), and has made adaptations of Dostoevsky (*The Idiot*, 1951) and Maxim Gorky (*The Lower Depths*, 1957).

kurrajong /ˈkʌrəˌdʒɒŋ/ *n.* (also **currajong**) any of various Australian trees, esp. *Brachychiton populneum*, which produce a tough bast fibre. [Dharuk *garrajung* fibre fishing-line]

kursaal /ˈkʊəzɑːl, ˈkɜːz(ə)l/ *n.* **1** a building for the use of visitors at a health resort, esp. at a German spa. **2** a casino. [G f. *Kur* CURE + *Saal* room]

Kursk /kʊəsk/ an industrial city in SW Russia; pop. (1990) 430,000. It was the scene of an important Soviet victory in the Second World War.

kurta /ˈkʊətə/ *n.* (also **kurtha**) a loose shirt or tunic worn by esp. Hindu men and women. [Hind.]

kurtosis /kɜːˈtəʊsɪs/ *n. Statistics* the sharpness of the peak of a frequency-distribution curve. [mod.L f. Gk *kurtōsis* bulging f. *kurtos* convex]

kuru /ˈkʊruː/ *n. Med.* a fatal infection of the brain occurring in some peoples in New Guinea, believed to be caused by either prions or virinos, and hence related to CJD, scrapie, and BSE. [name in New Guinea]

Kuşadasi /ˈkʊʃədəsɪ/ a resort town on the Aegean coast of western Turkey; pop. (1990) 31,910.

Kutaisi /ˌkʊtəˈiːsɪ/ an industrial city in central Georgia; pop. (1990) 236,100. One of the oldest cities in Transcaucasia, it has been the capital of various kingdoms, including Colchis and Abkhazia.

Kutch, Gulf of /kʌtʃ, kʊtʃ/ an inlet of the Arabian Sea on the west coast of India.

Kutch, Rann of /kʌtʃ, kʊtʃ, ræn/ a vast salt marsh on the eastern shores of the Arabian Sea, on the frontier between India and Pakistan. Most of it lies in the state of Gujarat in NW India, the remainder in the province of Sind in SE Pakistan.

Kuwait /kuˈweɪt/ a country on the NW coast of the Persian Gulf; pop. (est. 1991) 1,200,000; official language, Arabic; capital, Kuwait City. Kuwait has been an autonomous Arab sheikhdom, under the rule of an amir, from the 18th century, although the British established a protectorate from 1897 until 1961. One of the world's leading oil-producing countries, Kuwait was invaded by Iraq in Aug. 1990, the occupying forces being expelled in the Gulf War of Jan.–Feb. 1991. □ **Kuwaiti** *adj. & n.*

Kuwait City a port on the Persian Gulf, the capital city of Kuwait; pop. (1985) 44,335.

Kuzbass /kʊzˈbæs/ an alternative name for the KUZNETS BASIN.

Kuznets Basin /kʊzˈnjets/ (also **Kuznetsk** /-ˈnjetsk/; also called *Kuzbass*) an industrial region of southern Russia, situated in the valley of the Tom river, between Tomsk and Novokuznetsk. The region is rich in iron and coal deposits.

kV *abbr.* kilovolt(s).

kvass /kvɑːs/ *n.* a fermented beverage, low in alcohol, made from rye-flour or bread with malt in Russia etc. [Russ. *kvas*]

kvetch /kvetʃ/ *n. & v. US sl.* ● *n.* an objectionable person, esp. one who complains a great deal. ● *v.* complain, whine. □ **kvetcher** *n.* [Yiddish, f. G *quetschen* crush, press]

kW *abbr.* kilowatt(s).

Kwa /kwɑː/ *n. & adj.* (also **Qua**) ● *n.* (*pl.* same) **1** the group of related languages, spoken from Ivory Coast to Nigeria, which includes Ibo and Yoruba. **2** a member of a Kwa-speaking people. ● *adj.* of or relating to this group of languages. [Kwa]

KWAC /kwæk/ *n. Computing* etc. keyword and context. [abbr.]

kwacha /ˈkwɑːtʃə/ *n.* the chief monetary unit of Zambia and Malawi. [Bantu, = dawn]

KwaNdebele /ˌkwɑːəndəˈbiːlɪ/ a former homeland established in South Africa for the Ndebele people, now part of the province of Eastern Transvaal. (See also HOMELAND.)

Kwangchow see GUANGZHOU.

Kwangju /kwæŋˈdʒuː/ a city in SW South Korea; pop. (1990) 1,144,700.

Kwangsi Chuang see GUANGXI ZHUANG.

Kwangtung see GUANGDONG.

kwanza /ˈkwænzə/ *n.* (*pl.* same or **kwanzas**) the basic monetary unit of Angola, equal to 100 lweis. [perh. f. Swahili = first]

kwashiorkor /ˌkwɒʃɪˈɔːkɔː(r)/ *n.* a form of malnutrition caused by a protein deficiency of diet, esp. in young children in the tropics. [local name in Ghana]

KwaZulu /kwɑːˈzuːluː/ a former homeland established in South Africa for the Zulu people, now part of the province of KwaZulu/Natal. The general area was formerly known as Zululand. (See also HOMELAND.)

KwaZulu/Natal a province of eastern South Africa, on the Indian Ocean; capital, Pietermaritzburg. Formerly called Natal, it became one of the new provinces of South Africa following the democratic elections of 1994. (See also NATAL.)

Kweichow see GUIZHOU.

Kweilin see GUILIN.

Kweiyang see GUIYANG.

Kwesui /kweɪˈsweɪ/ a former name for HOHHOT.

kWh *abbr.* kilowatt-hour(s).

KWIC /kwɪk/ *n. Computing* etc. keyword in context. [abbr.]

KWOC /kwɒk/ *n. Computing* etc. keyword out of context. [abbr.]

KY *abbr. US* Kentucky (in official postal use).

Ky. *abbr.* Kentucky.

kyanite /ˈkaɪəˌnaɪt/ *n.* a blue crystalline mineral of aluminium silicate. □ **kyanitic** /ˌkaɪəˈnɪtɪk/ *adj.* [Gk *kuanos* dark blue]

kyanize /ˈkaɪəˌnaɪz/ *v.tr.* (also **-ise**) treat (wood) with a solution of corrosive sublimate to prevent decay. [J. H. *Kyan*, Irish inventor (1774–1850)]

kyat /kɪˈɑːt/ *n.* (*pl.* same or **kyats**) the basic monetary unit of Burma (Myanmar) since 1952, equal to 100 pyas. [Burmese]

kybosh var. of KIBOSH.

Kyd /kɪd/, Thomas (1558–94), English dramatist. His anonymously published *The Spanish Tragedy* (1592), an early example of revenge tragedy, was very popular on the Elizabethan stage. The only work published under his name was a translation of Robert Garnier's *Cornelia* (1594; reissued as *Pompey the Great*, 1595). Other works attributed to Kyd are *The Tragedy of Solyman and Perseda* (1592) and a lost pre-Shakespearian play on Hamlet.

kyle /kaɪl/ *n.* (in Scotland) a narrow channel between islands or between an island and the mainland. [Gael. *caol* strait]

kylie /ˈkaɪlɪ/ *n. W. Austral.* a boomerang. [Aboriginal]

kylin /kiːˈlɪn/ *n.* a mythical composite animal figured on Chinese and Japanese ceramics. [Chin. *qilin* f. *qi* male + *lin* female]

kyloe /ˈkaɪləʊ/ *n. Brit.* a breed of small usu. black long-horned Highland cattle. [Gael. *gaidhealach* Gaelic, Highland]

kymograph /ˈkaɪməˌɡrɑːf/ *n.* an instrument for recording variations in pressure, e.g. in sound waves or in blood within blood vessels. □ **kymographic** /ˌkaɪməˈɡræfɪk/ *adj.* [Gk *kuma* wave + -GRAPH]

Kyoto /kɪˈəʊtəʊ/ an industrial city in central Japan, on the island of Honshu; pop. (1990) 1,461,140. Founded in the 8th century, it was the imperial capital from 794 until 1868.

kyphosis /kaɪˈfəʊsɪs/ *n. Med.* excessive outward curvature of the spine, causing hunching of the back (opp. LORDOSIS). □ **kyphotic** /-ˈfɒtɪk/ *adj.* [mod.L f. Gk *kuphōsis* f. *kuphos* bent]

Kyrgyz /kɪəˈɡɪz, ˈkɜːɡɪz/ *n. & adj.* (also **Kirghiz**) ● *n.* (*pl.* same) **1** a member of a Mongol people living in central Asia. (*See note below.*) **2** the Turkic language of the Kyrgyz. ● *adj.* of or relating to the Kyrgyz or their language. [Kyrgyz]

▪ The Kyrgyz live between the Volga and the Irtysh rivers, chiefly in Kyrgyzstan but also in Tajikistan, Uzbekistan, and parts of China and Afghanistan. Traditionally they live a nomadic and pastoral existence, during the winter inhabiting individual family tents and coming together for the summer in larger settlements. They are Sunni Muslims.

Kyrgyz Republic an alternative name for KYRGYZSTAN.

Kyrgyzstan /ˌkɪrɡɪˈstɑːn/ (also called *Kirghizia*; *Kyrgyz Republic*) a mountainous country in central Asia, on the north-western border of China; population (est. 1991) 4,448,000; official language, Kyrgyz; capital, Bishkek. The region was annexed by Russia in 1864, and became a constituent republic of the Soviet Union. On the breakup of the USSR in 1991, Kyrgyzstan became an independent republic within the Commonwealth of Independent States.

Kyrie /ˈkɪərɪˌeɪ/ (in full **Kyrie eleison** /ɪˈleɪɪˌzɒn, -ˌsɒn/) *n.* a short repeated invocation (in Greek or translated) used in liturgies, esp. at the beginning of the Eucharist or as a response in a litany. [ME f. med.L f. Gk *Kurie eleēson* Lord, have mercy]

Kyushu /kɪˈuːʃuː/ the most southerly of the four main islands of Japan, constituting an administrative region; pop. (1990) 13,296,000; capital, Fukuoka.

Kyzyl /kəˈzɪl/ a city in south central Russia, on the Yenisei river, capital of the republic of Tuva; pop. (1989) 80,000.

Kyzyl Kum /kuːm/ an arid desert region in central Asia, extending eastwards from the Aral Sea to the Pamir Mountains and covering part of Uzbekistan and southern Kazakhstan.

Ll

L¹ /el/ n. (also **l**) (pl. **Ls** or **L's**) **1** the twelfth letter of the alphabet. **2** (as a Roman numeral) 50. **3** a thing shaped like an L, esp. a joint connecting two pipes at right angles.

L² abbr. (also **L.**) **1** Lake. **2** Brit. learner driver (cf. L-PLATE). **3** Liberal. **4** Biol. Linnaeus. **5** lire. **6** left.

l¹ abbr. (also **l.**) **1** left. **2** line. **3** liquid. **4** length. **5** archaic pound(s) (money).

l² symb. litre(s).

£ abbr. (preceding a numeral) pound or pounds (of money). [L libra]

LA abbr. **1** Library Association. **2** Los Angeles. **3** US Louisiana (in official postal use).

La symb. Chem. the element lanthanum.

La. abbr. Louisiana.

la var. of LAH.

laager /ˈlɑːgə(r)/ n. & v. ● n. **1** esp. S. Afr. a camp or encampment, esp. formed by a circle of wagons. **2** Mil. a park for armoured vehicles. ● v. **1** tr. a form (vehicles) into a laager. **b** encamp (people) in a laager. **2** intr. encamp. [Afrik. f. Du. leger: see LEAGUER²]

Laayoune see LA'YOUN.

Lab. abbr. **1** Labour. **2** Labrador.

lab /læb/ n. colloq. a laboratory. [abbr.]

Laban /ˈlɑːb(ə)n/, Rudolf von (1879–1958), Hungarian choreographer and dancer. He was a pioneer of the central European school of modern dance and is especially significant for his contribution to the theory of dance. In 1920 he published the first of several volumes outlining his system of dance notation (known as Labanotation). In 1938 he moved to England, where he concentrated on modern educational dance.

la Barca, Pedro Calderón de, see CALDERÓN DE LA BARCA.

labarum /ˈlæbərəm/ n. **1** a symbolic banner. **2** hist. Constantine the Great's imperial standard, an adaptation of the traditional Roman military standard with added Christian symbols. [LL: orig. unkn.]

labdanum var. of LADANUM.

labefaction /ˌlæbɪˈfækʃ(ə)n/ n. literary a shaking, weakening, or downfall. [L labefacere weaken f. labi fall + facere make]

label /ˈleɪb(ə)l/ n. & v. ● n. **1** a usu. small piece of paper, card, linen, metal, etc., for attaching to an object and giving its name, information about it, instructions for use, etc. **2** esp. derog. a short classifying phrase or name applied to a person, a work of art, etc. **3 a** a small fabric label sewn into a garment bearing the brand name, size, etc. **b** the logo, title, or trademark of esp. a fashion or recording company (brought it out under his own label). **c** the piece of paper in the centre of a gramophone record describing its contents etc. **d** a record-producing company or part of one. **4** an adhesive stamp on a parcel etc. **5** Biol. & Chem. a radioactive isotope, fluorescent dye, etc. used to label another substance. **6** a word placed before, after, or in the course of a dictionary definition etc. to specify its subject, register, nationality, etc. **7** Archit. a dripstone. **8** Heraldry the mark of an eldest son, consisting of a superimposed horizontal bar with usu. three downward projections. ● v.tr. (**labelled**, **labelling**) **1** attach a label to. **2** (usu. foll. by as) assign to a category (labelled them as irresponsible). **3** Biol. & Chem. make (a substance, molecule, or atom) recognizable by replacing an atom with one of a distinctive radioactive isotope, or by attaching a fluorescent dye to the molecule. □ **labeller** n. [ME f. OF, = ribbon, prob. f. Gmc (as LAP¹)]

labellum /ləˈbeləm/ n. (pl. **labella** /-lə/) **1** Zool. each of a pair of lobes at the tip of the proboscis in some insects. **2** Bot. a central petal at the base of an orchid flower, usu. large and unlike the others. [L, dimin. of labrum lip]

labia pl. of LABIUM.

labial /ˈleɪbɪəl/ adj. & n. ● adj. **1 a** of the lips. **b** Zool. of, like, or serving as a lip, a liplike part, or a labium. **2** Dentistry designating the surface of a tooth adjacent to the lips. **3** Phonet. requiring partial or complete closure of the lips (e.g. p, b, f, v, m, w; and vowels in which lips are rounded, e.g. oo in moon). ● n. Phonet. a labial sound. □ **labial pipe** Mus. an organ-pipe having lips; a flue-pipe. □ **labialize** v.tr. (also **-ise**). **labially** adv. [med.L labialis f. L labia lips]

labiate /ˈleɪbɪət/ n. & adj. ● n. Bot. a plant of the large family Labiatae. (See note below.) ● adj. **1** Bot. of or relating to the Labiatae. **2** Bot. & Zool. like a lip or labium. [mod.L labiatus (as LABIUM)]

▪ The labiates are typified by the dead-nettles, having square stems and a characteristic corolla and calyx divided into two parts that suggest a pair of lips. Many kinds, such as mint, sage, rosemary, and thyme, have aromatic foliage and are popular as culinary herbs.

labile /ˈleɪbaɪl, -bɪl/ adj. Chem. (of a compound) unstable; liable to displacement or change esp. if an atom or group is easily replaced by other atoms or groups. □ **lability** /ləˈbɪlɪtɪ/ n. [ME f. LL labilis f. labi to fall]

labio- /ˈleɪbɪəʊ/ comb. form of the lips. [as LABIUM]

labiodental /ˌleɪbɪəʊˈdent(ə)l/ adj. Phonet. (of a speech sound) made with the lips and teeth, e.g. f and v.

labiovelar /ˌleɪbɪəʊˈviːlə(r)/ adj. Phonet. (of a speech sound) made with the lips and soft palate, e.g. w.

labium /ˈleɪbɪəm/ n. (pl. **labia** /-bɪə/) **1** (usu. in pl.) Anat. each fold of skin of the two pairs that enclose the vulva. **2** Zool. the lower lip in the mouthparts of an insect or crustacean. **3** Bot. a lip, esp. the lower one of a labiate plant's corolla. □ **labia majora** /məˈdʒɔːrə/ the larger outer pair of labia (in sense 1). **labia minora** /mɪˈnɔːrə/ the smaller inner pair of labia (in sense 1). [L, = lip]

labor etc. US & Austral. var. of LABOUR etc.

laboratory /ləˈbɒrətərɪ/ n. (pl. **-ies**) a room or building fitted out for scientific experiments, research, teaching, or the manufacture of drugs and chemicals. [med.L laboratorium f. L laborare LABOUR]

laborious /ləˈbɔːrɪəs/ adj. **1** needing hard work or toil (a laborious task). **2** (esp. of literary style) showing signs of toil; pedestrian; not fluent. □ **laboriously** adv. **laboriousness** n. [ME f. OF laborieus f. L laboriosus (as LABOUR)]

Labor Party, Australian see AUSTRALIAN LABOR PARTY.

labour /ˈleɪbə(r)/ n. & v. (US, Austral. **labor**) ● n. **1 a** physical or mental work; exertion; toil. **b** such work considered as supplying the needs of a community. **2 a** workers, esp. manual, considered as a social class or political force (a dispute between capital and labour). **b** (**Labour**) the Labour Party. **3** the process of childbirth, esp. the period from the start of uterine contractions to delivery (has been in labour for three hours). **4** a particular task, esp. of a difficult nature. ● v. **1** intr. work hard; exert oneself. **2** intr. (usu. foll. by for, or to + infin.) strive for a purpose (laboured to fulfil his promise). **3** tr. a treat at excessive length; elaborate needlessly (I will not labour the point). **b** (as **laboured** adj.) done with great effort; not spontaneous or fluent. **4** intr. (often foll. by under) suffer under (a disadvantage or delusion). **5** intr. proceed with trouble or difficulty (laboured slowly up the hill). **6** intr. (of a ship) roll or pitch heavily. **7** tr. archaic or *poet.* till (the ground). □ **labour camp** a prison camp enforcing a regime of hard labour. **labour exchange** *Brit. colloq.* or *hist.* an employment exchange; a jobcentre. **labour force** the body of workers employed, esp. at a single plant. **labouring man** a labourer. **labour-intensive** (of a form of work) needing a large work force. **labour in vain** make a fruitless effort. **labour-market** the supply of labour with reference to the demand on it. **labour of Hercules** a task needing enormous strength or effort. **labour of love** a task done for pleasure, not reward. **labour-saving** (of an appliance etc.) designed to reduce or eliminate work. **labour theory of value** *Econ.* the theory that the value of a commodity should be determined by the amount of labour used in its production. **labour union** *US* a trade union. **lost labour** fruitless effort. [ME f. OF labo(u)r, labourer f. L labor, -oris, laborare]

Labour Day May 1 (or in the US and Canada the first Monday in September), celebrated in honour of working people.

labourer /ˈleɪbərə(r)/ *n.* (*US* **laborer**) **1** a person doing unskilled, usu. manual, work for wages. **2** a person who labours. [ME f. OF laboureur (as LABOUR)]

Labourite /ˈleɪbəˌraɪt/ *n.* (*US* **Laborite**) a member or follower of the Labour Party.

Labour Party a left-of-centre political party formed to represent the interests of ordinary working people. The Labour Party in Britain had its roots in the trade union movement, which combined with the Independent Labour Party, the Fabian Society, and other socialist groups in 1900 to form the Labour Representation Committee, changing its name to the Labour Party after electoral successes in 1906; Keir Hardie was its co-founder and first leader. (See also INDEPENDENT LABOUR PARTY.) After the First World War, Labour was reorganized as a true national party and replaced the Liberals as the country's second major political party. Having formed minority governments in 1924 and 1929–31 it entered Churchill's wartime coalition before forming its first majority government under Clement Attlee (1945–51). This government established the welfare state and carried out a substantial nationalization programme. Labour subsequently held power in 1964–70 and 1974–9, since when it has suffered disunity, competition from splinter groups and the Liberal Democrats, and the loss of its traditional industrial working class constituency. (See also AUSTRALIAN LABOR PARTY.)

labra *pl.* of LABRUM.

Labrador /ˈlæbrəˌdɔː(r)/ a coastal region of eastern Canada, which forms the mainland part of the province of Newfoundland and Labrador. Formerly disputed between Newfoundland and Quebec, the area was awarded to Newfoundland in 1927.

labrador /ˈlæbrəˌdɔː(r)/ *n.* (also **Labrador**) (in full **labrador dog** or **retriever**) a breed of retriever with a black or golden coat, often used as a gun dog or as a guide for a blind person.

Labrador Current a cold ocean current which flows southwards from the Arctic Ocean along the NE coast of North America. It meets the warm Gulf Stream in an area off the coast of Newfoundland which is noted for its dense fogs.

labradorite /ˌlæbrəˈdɔːraɪt, ˈlæbrədəˌraɪt/ *n. Mineral.* a kind of plagioclase feldspar, often showing iridescence from internal reflective planes. [LABRADOR + -ITE¹]

Labrador Peninsula (also **Labrador-Ungava** /ʊŋɡeɪvə, -ˈɡɑːvə/) a broad peninsula of eastern Canada, between Hudson Bay, the Atlantic, and the Gulf of St Lawrence. Consisting of the Ungava Peninsula and Labrador, it contains most of Quebec and the mainland part of the province of Newfoundland and Labrador. It is very sparsely inhabited.

labret /ˈlæbrɪt, ˈleɪbret/ *n.* a piece of shell, bone, etc., inserted in the lip as an ornament. [LABRUM]

labrum /ˈleɪbrəm/ *n.* (*pl.* **labra** /-brə/) the upper lip in the mouthparts of an insect. [L, = lip: rel. to LABIUM]

La Bruyère /ˌlæ bruːˈjeə(r)/, Jean de (1645–96), French writer and moralist. He is known for his *Caractères* (1688), a book consisting of two parts, one being a translation of the *Characters* of Theophrastus and the other a collection of portrait sketches modelled on this work. These often portray contemporary French figures with disguised names and expose the vanity and corruption of human behaviour by satirizing Parisian society. La Bruyère was a leading figure in the 'Ancient and Modern' dispute, which preoccupied the Académie française in his day; he spoke on behalf of the Ancients.

Labuan /ləˈbuːən/ a small Malaysian island off the north coast of Borneo; pop. (1980) 26,410; capital Victoria.

laburnum /ləˈbɜːnəm/ *n.* a small leguminous tree of the genus *Laburnum*, with hanging clusters of yellow flowers yielding poisonous seeds. Also called *golden chain*. [L]

labyrinth /ˈlæbəˌrɪnθ/ *n.* **1** a complicated irregular network of passages or paths etc.; a maze. The term was first used of the building constructed by Daedalus for King Minos of Crete, in which the Minotaur lived. **2** an intricate or tangled arrangement. **3** *Anat.* the complex arrangement of bony and membranous canals and chambers of the inner ear which constitute the organs of hearing and balance. □ **labyrinth fish** = GOURAMI. □ **labyrinthian** /ˌlæbəˈrɪnθɪən/ *adj.* **labyrinthine** /-ˈrɪnθaɪn/ *adj.* [F labyrinthe or L labyrinthus f. Gk laburinthos]

LAC *abbr.* Leading Aircraftman.

lac¹ /læk/ *n.* a resinous substance secreted as a protective covering by the lac insect, and used to make varnish and shellac. □ **lac insect** an Asian scale insect, *Laccifer lacca*, living in trees. [ult. f. Hind. lākh f. Prakrit lakkha f. Skr. lākṣā]

lac² var. of LAKH.

Lacan /læˈkɒn/, Jacques (1901–81), French psychoanalyst and writer. He founded the Freudian School in Paris (1964) and carried out influential work in reinterpreting Freudian psychoanalysis in the light of developments in structural linguistics and anthropology. His most significant contributions to the field concern his theory of the unconscious, which he saw as being structured like, and developing simultaneously with, language. A number of Lacan's articles and lectures are collected in *Écrits* (1966). (See also POST-STRUCTURALISM.) □ **Lacanian** /læˈkeɪnɪən/ *adj.*

Laccadive Islands /ˈlækədɪv/ one of the groups of islands forming the Indian territory of Lakshadweep in the Indian Ocean.

laccolith /ˈlækəlɪθ/ *n. Geol.* a lens-shaped intrusion of igneous rock which thrusts the overlying strata into a dome. [Gk lakkos reservoir + -LITH]

lace /leɪs/ *n. & v.* ● *n.* **1** a fine open fabric, esp. of cotton or silk, made by weaving thread in patterns and used esp. for trimming garments. **2** a cord or leather strip passed through eyelets or hooks on opposite sides of a shoe, the opening of a garment, etc., pulled tight and fastened. **3** braid used for trimming esp. dress uniform (*gold lace*). ● *v.* **1** *tr.* (usu. foll. by *up*) **a** fasten or tighten (a shoe, garment, etc.) with a lace or laces. **b** compress the waist of (a person) with a laced corset. **2** *tr.* (usu. foll. by *with*) **a** flavour or fortify (coffee, beer, etc.) with a dash of spirits. **b** enhance or adulterate (a foodstuff) by adding an extra ingredient (*laced the casserole with arsenic*). **3** *tr.* (usu. foll. by *with*) **a** streak (a sky etc.) with colour (*cheek laced with blood*). **b** interlace or embroider (fabric) with thread etc. **4** *tr.* & (foll. by *into*) *intr. colloq.* lash, beat, defeat. **5** *tr.* (often foll. by *through*) pass (a shoelace etc.) through. **6** *tr.* trim with lace. □ **lace-glass** Venetian glass with lacelike designs. **lace-pillow** a cushion placed on the lap and providing support in lacemaking. **lace-up** *n.* a shoe fastened with a lace. ● *attrib.adj.* (of a shoe etc.) fastened by a lace or laces. [ME f. OF laz, las, lacier ult. f. L laqueus noose]

La Ceiba /læ ˈseɪbə/ a seaport on the Caribbean coast of Honduras; pop. (1988) 68,200.

lacemaker /ˈleɪsˌmeɪkə(r)/ *n.* a person who makes lace, esp. professionally. □ **lacemaking** *n.*

lacerate /ˈlæsəˌreɪt/ *v.tr.* **1** mangle or tear (esp. flesh or tissue). **2** distress or cause pain to (the feelings, the heart, etc.). □ **lacerable** /-rəb(ə)l/ *adj.* **laceration** /ˌlæsəˈreɪʃ(ə)n/ *n.* [L lacerare f. lacer torn]

lacertilian /ˌlæsəˈtɪlɪən/ *n. & adj.* (also **lacertian** /ləˈsɜːtɪən, -ˈsɜːʃ(ə)n/, **lacertine** /ˈlæsəˌtaɪn/) *Zool.* ● *n.* a reptile of the suborder Lacertilia, which comprises the lizards. ● *adj.* of or relating to the Lacertilia; lizard-like, saurian. [L lacerta lizard]

lacewing /ˈleɪswɪŋ/ *n.* a neuropterous insect with delicate lacelike wings.

lacewood /ˈleɪswʊd/ *n.* the timber of the plane tree.

laches /ˈlætʃɪz, ˈleɪtʃ-/ *n. Law* delay in performing a legal duty, asserting a right, claiming a privilege, etc. [ME f. AF laches(se), OF laschesse f. lasche ult. f. L laxus loose]

Lachesis /ˈlækɪsɪs/ *Gk Mythol.* one of the three Fates. [Gk, = getting by lot]

Lachlan /ˈlæklən/ a river of New South Wales, Australia, which rises in the Great Dividing Range and flows some 1,472 km (920 miles) north-west then south-west to join the Murrumbidgee river near the border with Victoria. It is named after Lachlan Macquarie, the governor of New South Wales from 1810 to 1821.

lachryma Christi /ˌlækrɪmə ˈkrɪstɪ/ n. a white, red, or pink Italian wine originally from grapes grown near Mount Vesuvius, now also produced elsewhere in Italy. [L, = Christ's tear]

lachrymal /ˈlækrɪm(ə)l/ adj. & n. (also **lacrimal**, **lacrymal**) ● adj. **1** formal of or for tears. **2** (usu. as **lacrimal**) Anat. concerned in the secretion of tears (lacrimal canal; lacrimal duct). ● n. **1** = lachrymal vase. **2** (in pl.) (usu. as **lacrimals**) Anat. the lacrimal organs. □ **lachrymal vase** hist. a phial holding the tears of mourners at a funeral. [ME f. med.L lachrymalis f. L lacrima tear]

lachrymation /ˌlækrɪˈmeɪʃ(ə)n/ n. (also **lacrimation**, **lacrymation**) formal the flow of tears. [L lacrimatio f. lacrimare weep (as LACHRYMAL)]

lachrymator /ˈlækrɪˌmeɪtə(r)/ n. an agent irritating the eyes, causing tears.

lachrymatory /ˈlækrɪmətərɪ/ adj. & n. ● adj. formal of or causing tears. ● n. (pl. **-ies**) a phial of a kind found in ancient Roman tombs and thought to be a lachrymal vase.

lachrymose /ˈlækrɪˌməʊs, -ˌməʊz/ adj. formal **1** given to weeping; tearful. **2** intended to produce tears; sad, mournful. □ **lachrymosely** adv. [L lacrimosus f. lacrima tear]

lacing /ˈleɪsɪŋ/ n. **1** lace trimming, esp. on a uniform. **2** a laced fastening on a shoe or garment. **3** colloq. a beating. **4** a dash of spirits in a beverage. □ **lacing course** Building a strengthening course built into an arch or wall.

laciniate /ləˈsɪnɪət/ adj. (also **laciniated** /-nɪˌeɪtɪd/) Bot. & Zool. divided into deep narrow irregular segments; fringed. □ **laciniation** /-ˌsɪnɪˈeɪʃ(ə)n/ n. [L lacinia flap of a garment]

lack /læk/ n. & v. ● n. (usu. foll. by of) an absence, want, or deficiency (a lack of talent; felt the lack of warmth). ● v.tr. be without or deficient in (lacks courage). □ **for lack of** owing to the absence of (went hungry for lack of money). **lack for** lack. [ME lac, lacen, corresp. to MDu., MLG lak deficiency, MDu. laken to lack]

lackadaisical /ˌlækəˈdeɪzɪk(ə)l/ adj. **1** unenthusiastic; listless; idle. **2** feebly sentimental and affected. □ **lackadaisically** adv. **lackadaisicalness** n. [archaic lackaday, -daisy (int.): see ALACK]

lacker var. of LACQUER.

lackey /ˈlækɪ/ n. & v. (also **lacquey**) ● n. (pl. **-eys**) **1** derog. **a** a servile political follower. **b** an obsequious parasitical person. **2 a** a (usu. liveried) footman or manservant. **b** a servant. ● v.tr. (**-eys**, **-eyed**) archaic behave servilely to; dance attendance on. □ **lackey moth** a brown moth, Malacosoma neustria, having a brightly striped caterpillar. [F laquais, obs. alaquais f. Catalan alacay = Sp. ALCALDE]

lacking /ˈlækɪŋ/ adj. **1** absent or deficient (money was lacking; is lacking in determination). **2** colloq. deficient in intellect; mentally subnormal.

lackland /ˈlæklənd/ n. & adj. ● n. **1** a person having no land. **2** (**Lackland**) a nickname for King John of England. ● adj. having no land.

lacklustre /ˈlækˌlʌstə(r)/ adj. (US **lackluster**) **1** lacking in vitality, force, or conviction. **2** (of the eye) dull.

Lac Léman /lak lemɑ̃/ the French name for Lake Geneva (see GENEVA, LAKE).

Laclos /læˈkləʊ/, Pierre (-Ambroise-François) Choderlos de (1741–1803), French novelist. He is remembered for his epistolary novel Les Liaisons dangereuses (1782), which caused a scandal with its depiction of the corrupt, erotic schemes of an aristocratic couple.

Laconia /ləˈkəʊnɪə/ (also **Lakonia**) a modern department and ancient region of Greece, in the SE Peloponnese. Throughout the classical period the region was dominated by its capital, Sparta, which remains the administrative centre of the modern department. □ **Laconian** adj. & n.

laconic /ləˈkɒnɪk/ adj. **1** (of a style of speech or writing) brief; concise; terse. **2** (of a person) laconic in speech etc. □ **laconically** adv. **laconicism** /-nɪˌsɪz(ə)m/ n. [L f. Gk Lakōnikos f. Lakōn Spartan, the ancient Spartans being known for their terse speech]

laconism /ˈlækəˌnɪz(ə)m/ n. **1** brevity of speech. **2** a short pithy saying. [Gk lakōnismos f. lakōnizō behave like a Spartan: see LACONIC]

La Coruña see CORUNNA.

lacquer /ˈlækə(r)/ n. & v. (also **lacker**) ● n. **1** a sometimes coloured liquid made of shellac dissolved in alcohol, or of synthetic substances, that dries to form a hard protective coating for wood, brass, etc. **2** Brit. a chemical substance sprayed on hair to keep it in place. **3** the sap of the lacquer-tree used to varnish wood etc. **4** decorative ware made of

wood coated with lacquer. ● v.tr. coat with lacquer. □ **lacquer-tree** an eastern Asian tree, Rhus verniciflua, the sap of which is used as a hard-wearing varnish for wood. □ **lacquerer** n. [obs. F lacre sealing-wax, f. unexpl. var. of Port. laca LAC[1]]

lacquey var. of LACKEY.

lacrimal var. of LACHRYMAL.

lacrimation var. of LACHRYMATION.

lacrosse /ləˈkrɒs/ n. a field game played with a crosse or netted stick with which a small ball is driven or thrown, caught, and carried. Scoring is by goals. In the men's game there are ten players on each side, whereas women's lacrosse is a twelve-a-side game. Lacrosse was played by American Indians in southern Canada and parts of the US in the 17th century or earlier, and was taken up more widely around the mid-19th century, later spreading from America and Canada to Britain and Australia. [F f. la the + CROSSE]

lacrymal var. of LACHRYMAL.

lacrymation var. of LACHRYMATION.

lactase /ˈlækteɪz, -teɪs/ n. Biochem. any of a group of enzymes which catalyse the hydrolysis of lactose to glucose and galactose. [F f. lactose LACTOSE]

lactate[1] /lækˈteɪt/ v.intr. (of mammals) secrete milk. [as LACTATION]

lactate[2] /ˈlækteɪt/ n. Chem. a salt or ester of lactic acid.

lactation /lækˈteɪʃ(ə)n/ n. **1** the secretion of milk by the mammary glands. **2** the suckling of young. [L lactare suckle f. lac lactis milk]

lacteal /ˈlæktɪəl/ adj. & n. ● adj. **1** of milk. **2** Anat. (of a vessel) conveying chyle or other milky fluid. ● n. (in pl.) Anat. the lymphatic vessels of the small intestine which absorb digested fats. [L lacteus f. lac lactis milk]

lactescence /lækˈtes(ə)ns/ n. **1** a milky form or appearance. **2** a milky juice. [L lactescere f. lactere be milky (as LACTIC)]

lactescent /lækˈtes(ə)nt/ adj. **1** milky. **2** Bot. yielding a milky juice.

lactic /ˈlæktɪk/ adj. Chem. of, relating to, or obtained from milk. □ **lactic acid** a clear odourless syrupy carboxylic acid formed in sour milk, and produced in the muscle tissues during strenuous exercise. [L lac lactis milk]

lactiferous /lækˈtɪfərəs/ adj. yielding milk or milky fluid. [LL lactifer (as LACTIC)]

lacto- /ˈlæktəʊ/ comb. form milk. [L lac lactis milk]

lactobacillus /ˌlæktəʊbəˈsɪləs/ n. (pl. **-bacilli** /-laɪ/) Biol. a rod-shaped bacterium of the family Lactobacillaceae, producing lactic acid from the anaerobic fermentation of carbohydrates. [mod.L Lactobacillus genus name, f. as LACTO- + BACILLUS]

lactometer /lækˈtɒmɪtə(r)/ n. an instrument for testing the density of milk.

lactone /ˈlæktəʊn/ n. Chem. any of a class of cyclic esters formed by the elimination of water from a hydroxy-carboxylic acid. [G Lacton]

lactoprotein /ˌlæktəʊˈprəʊtiːn/ n. Chem. the albuminous constituent of milk.

lactose /ˈlæktəʊz, -təʊs/ n. Chem. a disaccharide sugar that occurs in milk, consisting of a glucose molecule linked to a galactose molecule. [as LACTO-]

lacuna /ləˈkjuːnə/ n. (pl. **lacunae** /-niː/ or **lacunas**) **1** a hiatus, blank, or gap. **2** a missing portion or empty page, esp. in an ancient manuscript, book, etc. **3** Anat. a cavity or depression, esp. in bone. □ **lacunal** adj. **lacunar** adj. **lacunary** adj. **lacunose** /-nəʊs/ adj. [L, = pool, f. lacus LAKE[1]]

lacustrine /ləˈkʌstraɪn/ adj. formal or Biol. **1** of or relating to lakes. **2** living or growing in or beside a lake. [L lacus LAKE[1], after palustris marshy]

LACW abbr. Leading Aircraftwoman.

lacy /ˈleɪsɪ/ adj. (**lacier**, **laciest**) of or resembling lace fabric. □ **lacily** adv. **laciness** n.

lad /læd/ n. **1 a** a boy or youth. **b** a young son. **2** (esp. in pl.) colloq. a man; a fellow, esp. a workmate, drinking companion, etc. (he's one of the lads). **3** colloq. a high-spirited fellow; a rogue (he's a bit of a lad). **4** Brit. a stable-worker (regardless of age). □ **lad's love** = SOUTHERNWOOD. [ME ladde, of unkn. orig.]

Ladakh /ləˈdɑːk/ a high-altitude region of NW India, Pakistan, and China, containing the Ladakh and Karakoram mountain ranges and the upper Indus valley; chief town, Leh (in India). It is one of the highest regions of the world.

ladanum /ˈlædənəm/ n. (also **labdanum** /ˈlæbdə-/) a gum resin from plants of the genus *Cistus*, used in perfumery etc. [L f. Gk *ladanon* f. *lēdon* mastic]

ladder /ˈlædə(r)/ *n. & v.* ● *n.* **1** a set of horizontal bars of wood or metal fixed between two uprights and used for climbing up or down. **2** *Brit.* a vertical strip of unravelled fabric in a stocking etc. resembling a ladder. **3 a** a hierarchical structure. **b** such a structure as a means of advancement, promotion, etc. ● *v. Brit.* **1** *intr.* (of a stocking etc.) develop a ladder. **2** *tr.* cause a ladder in (a stocking etc.). □ **ladder-back** an upright chair with a back resembling a ladder. **ladder-stitch** transverse bars in embroidery. **ladder tournament** a sporting contest with each participant listed and entitled to a higher place by defeating the one above. [OE *hlæd(d)er*, ult. f. Gmc: cf. LEAN[1]]

laddie /ˈlædɪ/ *n. colloq.* a young boy or lad.

lade /leɪd/ *v.* (*past part.* **laden** /ˈleɪd(ə)n/) **1** *tr.* **a** put cargo on board (a ship). **b** ship (goods) as cargo. **2** *intr.* (of a ship) take on cargo. **3** *tr.* (as **laden** *adj.*) (usu. foll. by *with*) **a** (of a vehicle, animal, person, tree, table, etc.) heavily loaded. **b** (of the conscience, spirit, etc.) painfully burdened with sin, sorrow, etc. [OE *hladan*]

la-di-da /ˌlɑːdɪˈdɑː/ *adj. & n. colloq.* ● *adj.* pretentious or snobbish, esp. in manner or speech. ● *n.* **1** a la-di-da person. **2** la-di-da speech or manners. [imit. of an affected manner of speech]

ladies *pl.* of LADY.

Ladies' Gallery a public gallery in the House of Commons, reserved for women.

ladify var. of LADYFY.

Ladin /ləˈdiːn/ *n.* the Rhaeto-Romanic dialect of the Engadine in Switzerland. [Romansh, f. L *latinus* LATIN]

lading /ˈleɪdɪŋ/ *n.* **1** a cargo. **2** the act or process of lading.

Ladino /ləˈdiːnəʊ/ *n.* (*pl.* **-os**) **1** a language based on old Spanish and written in modified Hebrew characters, used by some Sephardic Jews, esp. in Mediterranean countries. **2** a mestizo or Spanish-speaking white person in Central America. [Sp., orig. = Latin, f. L (as LADIN)]

ladino /ləˈdiːnəʊ/ *n.* a large variety of white clover, *Trifolium repens*, native to Italy and cultivated esp. in the US for fodder. [It.]

Ladislaus I /ˈlædɪsˌlɔːs/ (canonized as St Ladislaus) (*c*.1040–95), king of Hungary 1077–95. He conquered Croatia and Bosnia and extended Hungarian power into Transylvania, as well as establishing order in his kingdom and advancing the spread of Christianity. He was canonized in 1192. Feast day, 27 June.

Ladislaus II /ˈlædɪsˌlaʊs/ (Polish name Władysław /vwadˈiswaf/) (*c*.1351–1434), king of Poland 1386–1434. He was grand duke of Lithuania from 1377 to 1401, during which time he was known as Jogaila, and acceded to the Polish throne on his marriage with the Polish monarch, Queen Jadwiga (1374–99), thus uniting Lithuania and Poland. He converted Lithuania to Christianity and was the founder of the Jagiellon dynasty that ruled the two states until 1572.

ladle /ˈleɪd(ə)l/ *n. & v.* ● *n.* **1** a large long-handled spoon with a cup-shaped bowl used for serving esp. soups and gravy. **2** a vessel for transporting molten metal in a foundry. ● *v.tr.* (often foll. by *out*) transfer (liquid) from one receptacle to another. □ **ladle out** distribute, esp. lavishly. □ **ladleful** *n.* (*pl.* **-fuls**). **ladler** *n.* [OE *hlædel* f. *hladan* LADE]

Ladoga, Lake /ˈlɑːdəgə/ a large lake in NW Russia, north-east of St Petersburg, near the border with Finland. It is the largest lake in Europe, with an area of 17,700 sq. km (6,837 sq. miles).

lady /ˈleɪdɪ/ *n.* (*pl.* **-ies**) **1 a** a woman regarded as being of superior social status or as having the refined manners associated with this (cf. GENTLEMAN). **b** (**Lady**) a title used by peeresses, female relatives of peers, the wives and widows of knights, etc. **2** (often *attrib.*) a woman; a female person or animal (*ask that lady over there*; *lady butcher*; *lady dog*). **3** *colloq.* **a** a wife. **b** a man's girlfriend. **4** a ruling woman (*lady of the house*; *lady of the manor*). **5** (in *pl.* as a form of address) a female audience or the female part of an audience. **6** *hist.* a woman to whom a man, esp. a knight, is chivalrously devoted; a mistress. □ **find the lady** = *three-card trick*. **the ladies** (or **ladies'**) *Brit.* a women's public lavatory. **ladies' chain** a figure in a quadrille etc. **ladies' fingers** = OKRA (cf. *lady's finger*). **ladies'** (or **lady's**) **man** a man fond of female company; a seducer. **ladies' night** a function at a men's club etc. to which women are invited. **ladies' room** a women's lavatory in a hotel, office, etc. **Lady altar** the altar in a Lady chapel. **Lady chapel** a chapel in a large church or cathedral, usu. to the E. of the high altar, dedicated to the Virgin Mary. **lady-fern** a slender fern, *Athyrium filix-femina*. **lady-in-waiting** (*pl.* **ladies-in-waiting**) a lady attending a queen or princess. **lady-killer** a practised and habitual seducer. **lady-love** a man's sweetheart. **lady of the bedchamber** = *lady-in-waiting*. **lady of**

the night a prostitute. **lady of easy virtue** a sexually promiscuous woman; a prostitute. **lady's bedstraw** a yellow-flowered herbaceous plant, *Galium verum*. **lady's companion** a small case or bag containing implements for needlework. **lady's finger 1** = *kidney vetch*. **2** = LADYFINGER (cf. *ladies' fingers*). **lady's maid** a lady's personal maidservant. **lady's mantle** a rosaceous plant of the genus *Alchemilla*, with yellowish-green clustered flowers. **lady's slipper** an orchid of the genus *Cypripedium*, having a flower with a usu. yellow slipper-shaped lip. **lady's smock** = *cuckoo flower* 1. **lady's tresses** an orchid of the genus *Spiranthes*, with small white flowers. **my lady** a form of address used chiefly by servants etc. to holders of the title 'Lady'. **my lady wife** *joc.* my wife. **old lady** *colloq.* **1** a mother. **2** a wife or mistress. □ **ladyhood** *n.* [OE *hlæfdige* f. *hlāf* LOAF[1] + (unrecorded) *dig-* knead, rel. to DOUGH): in *Lady Day*. etc. f. OE genitive *hlæfdigan* (Our) Lady's]

ladybird /ˈleɪdɪˌbɜːd/ *n.* a small rounded beetle of the family Coccinellidae, with wing-cases often of a reddish colour with black spots.

Lady Bountiful *n.* a patronizingly generous lady of the manor etc. [a character in George Farquhar's *The Beaux' Stratagem*]

ladybug /ˈleɪdɪˌbʌg/ *n. dial. & N.Amer.* = LADYBIRD.

Lady Day *esp. Brit.* the feast of the Annunciation, 25 March.

ladyfinger /ˈleɪdɪˌfɪŋgə(r)/ *n. N.Amer.* a finger-shaped sponge cake.

ladyfy /ˈleɪdɪˌfaɪ/ *v.tr.* (also **ladify**) (**-ies, -ied**) *esp. Brit.* **1** make a lady of. **2** call (a person) 'lady'. **3** (as **ladyfied** *adj.*) having the manner of a fine lady.

ladylike /ˈleɪdɪˌlaɪk/ *adj.* **1 a** with the modesty, manners, etc., of a lady. **b** befitting a lady. **2** (of a man) effeminate.

Lady Mayoress *n.* the wife of a Lord Mayor.

Lady Muck *n. sl. derog.* a socially pretentious woman.

ladyship /ˈleɪdɪˌʃɪp/ *n. archaic* being a lady. □ **her** (or **your** or **their**) **ladyship** (or **ladyships**) **1** a respectful form of reference or address to a Lady or Ladies. **2** *iron.* a form of reference or address to a woman thought to be giving herself airs.

Ladysmith /ˈleɪdɪˌsmɪθ/ a town in eastern South Africa, in KwaZulu/Natal. It was founded in the early 19th century and named after the wife of the governor of Natal, Sir Harry Smith (1787–1860). It was subjected to a four-month siege by Boer forces during the Second Boer War and was finally relieved on 28 Feb. 1900 by Lord Frederick Roberts, who replaced General Sir Redvers Buller (1839–1908) as commander of the British forces.

Lady Superior *n.* the head of a convent or nunnery in certain orders.

Lae /ˈlɑːeɪ/ an industrial seaport on the east coast of Papua New Guinea, the country's second largest city; pop. (1990) 80,655.

laevo- /ˈliːvəʊ/ *comb. form* (also **levo-**) on or to the left. [L *laevus* left]

laevorotatory /ˌliːvəʊˈrəʊtətərɪ/ *adj.* (*US* **levorotatory**) *Chem.* having the property of rotating the plane of a polarized light ray to the left (anticlockwise facing the oncoming radiation). Cf. DEXTROROTATORY.

laevulose /ˈliːvjʊˌləʊz, -ˌləʊs/ *n.* (*US* **levulose**) *Chem.* = FRUCTOSE. [LAEVO- + -ULE + -OSE[2]]

Lafayette /ˌlæfaɪˈet/ (also **La Fayette**), Marie Joseph Paul Yves Roch Gilbert du Motier, Marquis de (1757–1834), French soldier and statesman. In 1777 he went to America and became one of the leaders of the French Expeditionary Force, which fought alongside the American colonists in the War of Independence. On his return he played a crucial part in the early phase of the French Revolution, commanding the National Guard (1789–91) and advocating moderate policies. He became an opposition leader in the Chamber of Deputies (1825–30) and participated in the Revolution of 1830.

La Fontaine /ˌlæ fɒnˈteɪn/, Jean de (1621–95), French poet. He is chiefly remembered for his *Fables* (1668–94), drawn from oriental, classical, and contemporary sources; they include such tales as 'The Cicada and the Ant' and 'The Crow and the Fox'. He also wrote *Contes et nouvelles* (1664–74), a collection of bawdy verse tales drawn from Ariosto, Boccaccio, and others.

lag[1] /læg/ *v. & n.* ● *v.intr.* (**lagged, lagging**) **1** (often foll. by *behind*) fall behind; not keep pace. **2** *US Billiards* make the preliminary strokes that decide which player shall begin. ● *n.* **1** a delay. **2** *Physics* **a** a retardation in a current or movement. **b** the amount of this. □ **lag of tide** the interval by which a tide falls behind mean time at the 1st and 3rd quarters of the moon (cf. PRIMING[2]). □ **lagger** *n.* [orig. = hindmost person, hang back: perh. f. a fanciful distortion of LAST[1] in a children's game (*fog, seg, lag*, = 1st, 2nd, last, in dial.)]

lag[2] /læg/ *v. & n.* ● *v.tr.* (**lagged, lagging**) enclose or cover in lagging.

● *n.* **1** the non-heat-conducting cover of a boiler etc.; lagging. **2** a piece of this. [prob. f. Scand.: cf. ON *lögg* barrel-rim, rel. to LAY¹]

lag³ /læg/ *n. & v. Brit. sl.* ● *n.* (esp. as **old lag**) a habitual convict. ● *v.tr.* (**lagged, lagging**) **1** send to prison. **2** apprehend; arrest. [19th c.: orig. unkn.]

lagan /ˈlægən/ *n.* goods or wreckage lying on the bed of the sea, sometimes with a marker buoy etc. for later retrieval. [OF, perh. of Scand. orig., f. root of LIE¹, LAY¹]

lager /ˈlɑːgə(r)/ *n.* a kind of beer, effervescent and light in colour and body. □ **lager lout** *Brit. colloq.* a youth who behaves badly as a result of excessive drinking. [G *Lagerbier* beer brewed for keeping f. *Lager* store]

Lagerlöf /ˈlɑːgəˌlɜːf/, Selma (Ottiliana Lovisa) (1858–1940), Swedish novelist. She made her name with *Gösta Berlings Saga* (1891), a book inspired by local legends and traditions, as were many of her later novels. She was awarded the Nobel Prize for literature in 1909, the first woman to win a Nobel Prize in any field.

lagerphone /ˈlɑːgəˌfəʊn/ *n. Austral.* an improvised musical instrument employing beer bottle tops.

laggard /ˈlægəd/ *n. & adj.* ● *n.* a dawdler; a person who lags behind. ● *adj.* dawdling; slow. □ **laggardly** *adj. & adv.* **laggardness** *n.* [LAG¹]

lagging /ˈlægɪŋ/ *n.* material providing heat insulation for a boiler, pipes, etc. [LAG²]

La Gioconda /ˌlæ dʒɪəˈkɒndə/ an alternative name for the MONA LISA.

lagomorph /ˈlægəˌmɔːf/ *n. Zool.* a mammal of the order Lagomorpha, which comprises the hares, rabbits, and pikas. [Gk *lagōs* hare + *morphē* form]

lagoon /ləˈguːn/ *n.* **1** a stretch of salt water separated from the sea by a low sandbank, coral reef, etc. **2** the enclosed water of an atoll. **3** *US, Austral., & NZ* a small freshwater lake near a larger lake or river. **4** an artificial pool for the treatment of effluent or to accommodate an overspill from surface drains during heavy rain. □ **lagoonal** *adj.* [F *lagune* or It. & Sp. *laguna* f. L *lacuna*: see LACUNA]

Lagos /ˈleɪgɒs/ the chief city of Nigeria, a port on the Gulf of Guinea; pop. (1986) 1,739,000. A centre of the slave trade between the 16th and 19th centuries, it was held by Britain from 1851 until 1960, when it became capital of the newly independent Nigeria. It was replaced as capital by Abuja in 1982.

Lagrange /læˈgrɒnʒ/, Joseph Louis, Comte de (1736–1813), Italian-born French mathematician. He is remembered for his proof that every positive integer can be expressed as a sum of at most four squares, and for his study of the solution of algebraic equations, which later provided the inspiration for the founding of the theory of groups and Galois theory. His most influential work, however, was the *Traité de mécanique analytique* (1788), which was the culmination of his extensive work on mechanics and its application to the description of planetary and lunar motion.

Lagrangian point /læˈgrɒnʒɪən/ *n. Astron.* each of five points in the plane of orbit of one body around another (e.g. the moon around the earth), at which a small third body can remain stationary with respect to both. [LAGRANGE]

lah /lɑː/ *n.* (also **la**) *Mus.* **1** (in tonic sol-fa) the sixth note of a major scale. **2** the note A in the fixed-doh system. [ME f. L *labii*: see GAMUT]

lahar /ˈlɑːhɑː(r)/ *n.* a mud-flow composed mainly of volcanic debris. [Javanese]

Lahore /ləˈhɔː(r)/ the capital of Punjab province and second largest city of Pakistan, situated near the border with India; pop. (est. 1991) 3,200,000. It was the capital of the former province of West Pakistan between 1955 and 1970.

Laibach /ˈlaɪbax/ the German name for LJUBLJANA.

laic /ˈleɪɪk/ *adj. & n.* ● *adj.* non-clerical; lay; secular; temporal. ● *n. formal* a lay person; a non-cleric. □ **laical** *adj.* **laically** *adv.* [LL f. Gk *laïkos* f. *laos* people]

laicity /leɪˈɪsɪtɪ/ *n.* the status or influence of the laity.

laicize /ˈleɪɪˌsaɪz/ *v.tr.* (also **-ise**) **1** make (an office etc.) tenable by lay people. **2** subject (a school or institution) to the control of lay people. **3** secularize. □ **laicization** /ˌleɪɪsaɪˈzeɪʃ(ə)n/ *n.*

laid *past* and *past part.* of LAY¹.

lain *past part.* of LIE¹.

Laing /læŋ/, R(onald) D(avid) (1927–89), Scottish psychiatrist. He became famous for his controversial views on madness and in particular on schizophrenia, in which he proposed that what society calls insanity is in fact a defensive façade in response to the tensions of the close-knit nuclear family. Major works include *The Divided Self* (1960) and *Sanity, Madness, and the Family* (1965).

lair¹ /leə(r)/ *n. & v.* ● *n.* **1 a** a wild animal's resting-place. **b** a person's hiding-place; a den (*tracked him to his lair*). **2** a place where domestic animals lie down. **3** *Brit.* a shed or cattle on the way to market. ● *v.* **1** *intr.* go to or rest in a lair. **2** *tr.* place (an animal) in a lair. □ **lairage** *n.* [OE *leger* f. Gmc: cf. LIE¹]

lair² /leə(r)/ *n. & v. Austral. sl.* ● *n.* a youth or man who dresses flashily and shows off. ● *v.intr.* (often foll. by *up*) behave or dress like a lair. □ **lairy** *adj.* [*lair* back-form. f. *lairy; lairy* alt. f. LEERY]

laird /leəd/ *n. Sc.* a landed proprietor. □ **lairdship** *n.* [Sc. form of LORD]

laissez-aller /ˌleseɪˈæleɪ/ *n.* (also **laisser-aller**) unconstrained freedom; an absence of constraint. [F, = let go]

laissez-faire /ˌleseɪˈfeə(r)/ *n.* (also **laisser-faire**) the theory or practice of governmental abstention from interference in the workings of the market etc. Laissez-faire ideas were influentially expounded by Adam Smith (although the phrase originated with the 18th-century French physiocrats). They were particularly popular in the mid-19th century, and have since become central to the theories of free-market economists. [F, = let act]

laissez-passer /ˌleseɪˈpæseɪ/ *n.* (also **laisser-passer**) a document allowing the holder to pass; a permit. [F, = let pass]

laity /ˈleɪɪtɪ/ *n.* (usu. prec. by *the*; usu. treated as *pl.*) **1** lay people, as distinct from the clergy. **2** non-professionals. [ME f. LAY² + -ITY]

Laius /ˈlaɪəs/ *Gk Mythol.* a king of Thebes, the father of Oedipus and husband of Jocasta.

lake¹ /leɪk/ *n.* a large body of water surrounded by land. □ **lake trout 1** a large pale partly migrating race of the trout, *Salmo trutta*, of North European lakes. **2** *US* a related fish, *Salvelinus namaycush*, of North American lakes. □ **lakeless** *adj.* **lakelet** *n.* [ME f. OF *lac* f. L *lacus* basin, pool, lake]

lake² /leɪk/ *n.* **1** a reddish colouring orig. made from lac (*crimson lake*). **2** a complex formed by the action of dye and mordants applied to fabric to fix colour. **3** any insoluble product of a soluble dye and mordant. [var. of LAC¹]

Lake Albert, Lake Baikal, etc. see ALBERT, LAKE; BAIKAL, LAKE, etc.

Lake District (or **the Lakes**) the region of the English lakes in Cumbria.

lake-dwelling /ˈleɪkˌdwelɪŋ/ *n.* a prehistoric hut built on piles driven into the bed or shore of a lake. Such dwellings occur in Switzerland and northern Italy from the neolithic to the Iron Age period (4th–1st millennium BC) and in other parts of temperate Europe in the Iron Age (7th–1st centuries BC). □ **lake-dweller** *n.*

Lakeland /ˈleɪklənd/ *n.* = LAKE DISTRICT.

Lakeland terrier *n.* a small stocky breed of terrier originating in the Lake District.

Lake of the Woods a lake on the border between Canada and the US, to the west of the Great Lakes. It is bounded to the west by Manitoba, to the north and east by Ontario, and to the south by Minnesota.

Lake Poets (also **Lake School**) the poets Samuel Taylor Coleridge, Robert Southey, and William Wordsworth, who lived in and were inspired by the Lake District.

lakeside /ˈleɪksaɪd/ *attrib.adj.* beside a lake.

lakh /læk, lɑːk/ *n.* (also **lac**) *Ind.* (usu. foll. by *of*) a hundred thousand (rupees etc.). [Hind. *lākh* f. Skr. *lakṣa*]

Lakonia see LACONIA.

Lakshadweep /lækˈʃædwiːp, ˌlʌkʃəˈdwiːp/ a group of islands off the Malabar Coast of SW India, constituting a Union Territory in India; pop. (1991) 51,680; capital, Kavaratti. The group consists of the Laccadive, Minicoy, and Amindivi Islands.

Lakshmi /ˈlʌkʃmɪ/ *Hinduism* the goddess of prosperity, consort of Vishnu. She assumes different forms (e.g. Radha, Sita) in order to accompany her husband in his various incarnations. [Skr., = prosperity]

Lallan /ˈlælən/ *n. & adj. Sc.* ● *n.* (now usu. **Lallans** /-lənz/) a Lowland Scots dialect, esp. as a literary language. ● *adj.* of or concerning the Lowlands of Scotland. [var. of LOWLAND]

lallation /læˈleɪʃ(ə)n/ *n.* **1** the pronunciation of *r* as *l*. **2** imperfect speech, esp. the repetition of meaningless sounds by young children. [L *lallare lallat-* sing a lullaby]

lallygag /ˈlælɪˌgæg/ *v.intr.* (**lallygagged, lallygagging**) *N.Amer. sl.* **1** loiter. **2** cuddle amorously. [20th c.: orig. unkn.]

La Louvière /læ ˈluːviˌeə(r)/ an industrial city in SW Belgium, in the province of Hainaut west of Charleroi; pop. (1991) 76,430.

Lam. abbr. (in the Bible) Lamentations.

lam[1] /læm/ v. (**lammed, lamming**) colloq. **1** tr. thrash; hit. **2** intr. (foll. by into) hit (a person etc.) hard with a stick etc. [perh. f. Scand.: cf. ON lemja beat so as to LAME]

lam[2] /læm/ n. □ **on the lam** N.Amer. sl. in flight, esp. from the police. [20th c.: orig. unkn.]

lama /ˈlɑːmə/ n. Buddhism **1** an honorific applied to a spiritual leader in Tibetan Buddhism, whether a reincarnate lama (such as the Dalai Lama) or one who has earned the title in life. **2** any Tibetan Buddhist monk. [Tibetan blama (with silent b) superior one]

Lamaism /ˈlɑːməˌɪz(ə)m/ n. the system of doctrine and observances inculcated and maintained by lamas; Tibetan Buddhism. (See DALAI LAMA, TIBETAN BUDDHISM.) □ **Lamaist** n.

Lamarck /læˈmɑːk/, Jean Baptiste de (1744–1829), French naturalist, an early proponent of organic evolution. He suggested that species could have evolved from each other by small changes in their structure, and that the mechanism of such change was that characteristics acquired in order to survive could be passed on to offspring. His theory found little favour in his lifetime (it was criticized notably by Georges Cuvier), but the concept of inheritance of acquired characteristics was revived by those who did not accept Darwin's later theory of natural selection. It is not usually accepted today.

Lamarckism /læˈmɑːkɪz(ə)m/ n. Biol. the theory of evolution devised by Lamarck, based on the inheritance of acquired characteristics. □ **Lamarckian** n. & adj.

Lamartine /ˌlæməˈtiːn/, Alphonse Marie Louis de (1790–1869), French poet, statesman, and historian. His first volume of poems, Méditations poétiques (1820), brought a fresh lyricism to French poetry and established him as a leading figure of French romanticism. During the 1830s he devoted more time to politics and spoke out on behalf of the working classes. He served as Minister of Foreign Affairs in the provisional government following the Revolution of February 1848, but was deposed in June. His writings include Histoire des Girondins (1847).

lamasery /ˈlɑːməsərɪ, ləˈmɑːsərɪ/ n. (pl. **-ies**) a monastery of lamas. [F lamaserie irreg. f. lama LAMA]

Lamb /læm/, Charles (1775–1834), English essayist and critic. He devoted much of his life to caring for his sister Mary, who suffered from a recurrent mental illness; together they wrote Tales from Shakespeare (1807). He also compiled Specimens of English Dramatic Poets (1808), an anthology of scenes and speeches from Elizabethan and Jacobean dramatists, with accompanying critical comments. His essays were published in leading periodicals; the best known are the semi-autobiographical Essays of Elia (published in a collected edition in 1823).

lamb /læm/ n. & v. ● n. **1** a young sheep. **2** the flesh of a lamb as food. **3** a mild or gentle person, esp. a young child. ● v. **1 a** tr. (in passive) (of a lamb) be born. **b** intr. give birth to lambs. **2** tr. tend (lambing ewes). □ **lamb's ears** a labiate garden plant, Stachys byzantina, with whitish woolly leaves. **lamb's fry** lamb's testicles or other offal as food. **lamb's lettuce** a plant, Valerianella locusta, used in salad (also called corn-salad). **lamb's-tails** Brit. catkins of the hazel. **like a lamb** meekly, obediently. □ **lamber** n. **lambhood** n. **lambkin** n. **lamblike** adj. [OE lamb f. Gmc]

lambada /læmˈbɑːdə/ n. a fast erotic Brazilian dance in which couples dance with their hips touching each other. [Port. = a beating]

lambaste /læmˈbeɪst/ v.tr. (also **lambast** /-ˈbæst/) colloq. **1** thrash; beat. **2** criticize severely. [LAM[1] + BASTE[3]]

lambda /ˈlæmdə/ n. **1** the eleventh letter of the Greek alphabet (Λ, λ). **2** (as λ) the symbol for wavelength. [ME f. Gk la(m)bda]

lambent /ˈlæmb(ə)nt/ adj. **1** (of a flame or a light) playing on a surface with a soft radiance but without burning. **2** (of the eyes, sky, etc.) softly radiant. **3** (of wit etc.) lightly brilliant. □ **lambency** n. **lambently** adv. [L lambere lambent- lick]

Lambert /ˈlæmbət/, (Leonard) Constant (1905–51), English composer, conductor, and critic. While still a student he was commissioned by Diaghilev to write the music for the ballet Romeo and Juliet (1926). Thereafter he took a leading part in the establishment of British ballet as musical director of Sadler's Wells (1930–47). His other works include The Rio Grande (1929), a work in a jazz idiom for orchestra, piano, and voices.

lambert /ˈlæmbət/ n. a former unit of luminance, equal to the emission or reflection of one lumen per square centimetre. [Johann Heinrich Lambert, Ger. physicist (1728–77)]

Lambeth /ˈlæmbəθ/ a borough of inner London, on the south bank of the Thames; pop. (1991) 220,100. It is the site of Lambeth Palace and the South Bank arts centre.

Lambeth Conference an assembly of bishops from the Anglican Communion, usually held every ten years (since 1867) at Lambeth Palace and presided over by the Archbishop of Canterbury.

Lambeth Palace a palace in the London borough of Lambeth, the residence of the Archbishop of Canterbury since 1197.

Lamb of God a name for Jesus (see John 1:29) (cf. AGNUS DEI).

lambrequin /ˈlæmbrɪˌkɪn, ˈlæmbəˌkɪn/ n. **1** US a short piece of drapery hung over the top of a door or a window or draped on a mantelpiece. **2** Heraldry = MANTLING. [F f. Du. (unrecorded) lamperkin, dimin. of lamper veil]

lambskin /ˈlæmskɪn/ n. a prepared skin from a lamb with the wool on or as leather.

lambswool /ˈlæmzwʊl/ n. & adj. (also **lamb's-wool**) ● n. soft fine wool from a young sheep used in knitted garments etc. ● adj. made of lambswool.

lame /leɪm/ adj. & v. ● adj. **1** disabled, esp. in the foot or leg; limping; unable to walk normally (lame in his right leg). **2 a** (of an argument, story, excuse, etc.) unconvincing; unsatisfactory; weak. **b** (of verse etc.) halting. ● v.tr. **1** make lame; disable. **2** harm permanently. □ **lame-brain** N.Amer. colloq. a stupid person. **lame duck 1** a disabled or weak person. **2** Brit. a defaulter on the Stock Exchange. **3** Brit. a firm etc. in financial difficulties. **4** US an official (esp. the President) in the final period of office, after the election of a successor. □ **lamely** adv. **lameness** n. **lamish** adj. [OE lama f. Gmc]

lamé /ˈlɑːmeɪ/ n. & adj. ● n. a fabric with gold or silver threads interwoven. ● adj. (of fabric, a dress, etc.) having such threads. [F]

lamella /ləˈmelə/ n. (pl. **lamellae** /-liː/) **1** a thin layer, membrane, scale, or platelike tissue or part, esp. in bone tissue. **2** Bot. a membranous fold in a chloroplast. □ **lamellar** adj. **lamelliform** adj. **lamellate** /ˈlæməˌleɪt, ləˈmeleɪt, -lət/ adj. **lamellose** adj. [L, dimin. of lamina: see LAMINA]

lamellibranch /ləˈmelɪˌbræŋk/ n. & adj. Zool. = BIVALVE. [mod.L f. LAMELLA + Gk bragkhia gills]

lamellicorn /ləˈmelɪˌkɔːn/ n. & adj. Zool. ● n. a beetle of the superfamily Scarabaeoidea (formerly Lamellicornia), having lamelliform antennae, including stag beetles, chafers, dung-beetles, etc. ● adj. of or relating to this superfamily. [mod.L lamellicornis f. L lamella (see LAMELLA) + cornu horn]

lament /ləˈment/ n. & v. ● n. **1** a passionate expression of grief. **2** a song or poem of mourning or sorrow. ● v.tr. (also absol.) **1** express or feel grief for or about; regret (lamented the loss of his ticket). **2** (as **lamented** adj.) a conventional expression referring to a recently dead person (your late lamented father). □ **lament for** (or **over**) mourn or regret. □ **lamenter** n. **lamentingly** adv. [L lamentum]

lamentable /ˈlæməntəb(ə)l, disp. ləˈment-/ adj. **1** (of an event, fate, condition, character, etc.) deplorable; regrettable. **2** archaic mournful. □ **lamentably** adv. [ME f. OF lamentable or L lamentabilis (as LAMENT)]

lamentation /ˌlæmənˈteɪʃ(ə)n/ n. **1** the act or an instance of lamenting. **2** a lament. [ME f. OF lamentation or L lamentatio (as LAMENT)]

Lamentations /ˌlæmənˈteɪʃ(ə)nz/ (in full **The Lamentations of Jeremiah**) a book of the Bible traditionally ascribed to Jeremiah but probably of a later period, telling of the desolation of Judah after the destruction of Jerusalem in 586 BC.

lamina /ˈlæmɪnə/ n. (pl. **laminae** /-ˌniː/) a thin plate or scale, e.g. of bone, stratified rock, or animal or vegetable tissue. □ **laminose** /-ˌnəʊz, -ˌnəʊs/ adj. [L]

laminar /ˈlæmɪnə(r)/ adj. **1** consisting of laminae. **2** Physics (of a flow) taking place along constant streamlines, not turbulent.

laminate v., n., & adj. ● v. /ˈlæmɪˌneɪt/ **1** tr. beat or roll (metal) into thin plates. **2** tr. overlay with metal plates, a plastic layer, etc. **3** tr. manufacture by placing layer on layer. **4** tr. & intr. split or be split into layers or leaves. ● n. /ˈlæmɪnət/ a laminated structure or material, esp. of layers fixed together to form rigid or flexible material. ● adj. /ˈlæmɪnət/ in the form of lamina or laminae. □ **laminator** /-ˌneɪtə(r)/ n. **lamination** /ˌlæmɪˈneɪʃ(ə)n/ n. [LAMINA + -ATE[2], -ATE[3]]

lamington /ˈlæmɪŋtən/ n. Austral. & NZ a square of sponge cake coated in chocolate icing and desiccated coconut. [Charles Wallace Baillie, Baron Lamington, Governor of Queensland (1860–1940)]

laminitis /ˌlæmɪˈnaɪtɪs/ n. inflammation of the sensitive laminae in the hoof of a horse, ox, etc.

Lammas /ˈlæməs/ (in full **Lammas Day**) the first day of August, formerly observed as an English harvest festival at which loaves made from the first ripe corn were consecrated; in Scotland, one of the quarter days. [OE hlāfmæsse (as LOAF¹, MASS²)]

lammergeier /ˈlæməˌɡaɪə(r)/ n. a large vulture, Gypaetus barbatus, with a very large wingspan (often of 3 m), and dark beardlike feathers on either side of its beak. It inhabits mountainous country in southern Europe, Africa, and Asia. [G Lämmergeier f. Lämmer lambs + Geier vulture]

lamp /læmp/ n. & v. ● n. **1** a device for producing a steady light, esp.: **a** an electric bulb, and usu. its holder and shade or cover (bedside lamp; bicycle lamp). **b** an oil lamp. **c** a usu. glass holder for a candle. **d** a gas-jet and mantle. **2** a source of spiritual or intellectual inspiration. **3** poet. the sun, the moon, or a star. **4** a device producing esp. ultraviolet or infrared radiation as a treatment for various complaints. ● v. **1** intr. poet. shine. **2** tr. supply with lamps; illuminate. **3** tr. US sl. look at. □ **lamp-chimney** a glass cylinder enclosing and making a draught for the flame of an oil-lamp. **lamp-holder** a device for supporting a lamp, esp. an electric one. **lamp shell** = BRACHIOPOD. **lamp standard** = LAMPPOST. □ **lampless** adj. [ME f. OF lampe f. LL lampada f. accus. of L lampas torch f. Gk]

lampblack /ˈlæmpblæk/ n. a pigment made from soot.

lampern /ˈlæmpən/ n. a lamprey, Lampetra fluviatilis, of rivers and coasts in NW Europe. [OF lampreion, dimin. of lampreie LAMPREY]

lamplight /ˈlæmplaɪt/ n. light given by a lamp or lamps. □ **lamplit** /-lɪt/ adj.

lamplighter /ˈlæmpˌlaɪtə(r)/ n. hist. **1** a person who lights street lamps. **2** US a spill for lighting lamps.

lampoon /læmˈpuːn/ n. & v. ● n. a satirical attack on a person etc. ● v.tr. satirize. □ **lampooner** n. **lampoonery** n. **lampoonist** n. [F lampon, conjectured to be f. lampons let us drink f. lamper gulp down f. laper LAP³]

lamppost /ˈlæmppəʊst/ n. a tall post supporting a street-light.

lamprey /ˈlæmprɪ/ n. (pl. **-eys**) an eel-like fish of the family Petromyzonidae, without scales, paired fins, or jaws, but having a sucker mouth with horny teeth and a rough tongue. [ME f. OF lampreie f. med.L lampreda: cf. LL lampetra perh. f. L lambere lick + petra stone]

lampshade /ˈlæmpʃeɪd/ n. a translucent cover for a lamp used to soften or direct its light.

LAN /læn/ n. Computing local area network. [acronym]

Lanarkshire /ˈlænəkˌʃɪə(r)/ a former county of south-west central Scotland. It became part of Strathclyde region in 1975.

Lancashire /ˈlæŋkəˌʃɪə(r)/ a county of NW England, on the Irish Sea; administrative centre, Preston. The county was noted for the production of textiles, especially cotton goods, between the 16th and 19th centuries.

Lancaster¹ /ˈlæŋˌkæstə(r)/ a city in Lancashire, on the estuary of the River Lune; pop. (1981) 44,450. The town developed around a castle and a Benedictine priory built in the 11th century on the site of a former Roman fortification. It was the county town and administrative centre of Lancashire until 1974.

Lancaster² /ˈlæŋˌkæstə(r)/ n. a type of four-engined British heavy bomber of the Second World War.

Lancaster, House of the English royal house descended from John of Gaunt, Duke of Lancaster, which ruled England from 1399 (Henry IV) until 1461 (the deposition of Henry VI) and again on Henry's brief restoration in 1470–1. The House of Lancaster (with the red rose as its emblem) fought the Wars of the Roses with the House of York, both houses being branches of the Plantagenet line. Their descendants, the Tudors, eventually prevailed through Henry VII's accession to the throne in 1485.

Lancaster House Agreement an agreement which brought about the establishment of the independent state of Zimbabwe, reached in September 1979 at Lancaster House in London between representatives of the British government, the Rhodesian government, and black nationalist groups.

Lancastrian /læŋˈkæstrɪən/ adj. & n. ● adj. **1** of Lancashire or Lancaster. **2** hist. of the family descended from John of Gaunt, Duke of Lancaster, or of its supporters during the Wars of the Roses (cf. YORKIST). ● n. **1** a native of Lancashire or Lancaster. **2** hist. a member or adherent of the house of Lancaster.

lance /lɑːns/ n. & v. ● n. **1 a** hist. a long weapon with a wooden shaft and a pointed steel head, used by a horseman in charging. **b** a similar weapon used for spearing a fish, killing a harpooned whale, etc. **2** a metal pipe supplying oxygen to burn metal. **3** = LANCER 1. ● v.tr. **1** Surgery prick or cut open with a lancet. **2** pierce with a lance. **3** poet. fling; launch. □ **lance-bombardier** a rank in the Royal Artillery corresponding to lance-corporal in the infantry. **lance-corporal** the lowest rank of NCO in the Army. **lance-jack** Brit. sl. a lance-corporal or lance-bombardier. **lance-sergeant** a corporal acting as sergeant. **lance-snake** = FER-DE-LANCE. [ME f. OF lancier f. L lancea: lance-corporal on analogy of obs. lancepesade lowest grade of NCO ult. f. It. lancia spezzata broken lance]

lancelet /ˈlɑːnslɪt/ n. a small non-vertebrate fishlike chordate of the family Branchiostomidae, that burrows in sand; esp. amphioxus. [LANCE n. + -LET, with ref. to its thin form]

Lancelot /ˈlɑːnsəlɒt/ (also **Launcelot**) (in Arthurian legend) the most famous of Arthur's knights, lover of Queen Guinevere and father of Galahad.

lanceolate /ˈlɑːnsɪələt/ adj. shaped like a lance-head, tapering to each end. [LL lanceolatus f. lanceola dimin. of lancea lance]

lancer /ˈlɑːnsə(r)/ n. **1** hist. a soldier of a cavalry regiment armed with lances. **2** (in pl.) **a** a quadrille for 8 or 16 pairs. **b** the music for this. [F lancier (as LANCE)]

lancet /ˈlɑːnsɪt/ n. a small broad two-edged surgical knife with a sharp point. □ **lancet arch** (or **light** or **window**) Archit. a narrow arch or window with a pointed head. □ **lanceted** adj. [ME f. OF lancette (as LANCE)]

lancewood /ˈlɑːnswʊd/ n. a tough elastic wood from a West Indian tree, Oxandra lanceolata, used for carriage-shafts, fishing-rods, etc.

Lanchow see LANZHOU.

Lancs. abbr. Lancashire.

Land /lænd, lɑːnt/ n. (pl. **Länder** /ˈlendə(r)/) a province of Germany or Austria. [G (as LAND)]

land /lænd/ n. & v. ● n. **1** the solid part of the earth's surface, as distinguished from the sea or water, or from the air. **2 a** an expanse of country; ground; soil. **b** such land in relation to its use, quality, etc., or (often prec. by the) as a basis for agriculture (building land; this is good land; works on the land). **3** a country, nation, or state (land of hope and glory). **4 a** landed property. **b** (in pl.) estates. **5** the space between the rifling-grooves in a gun. **6** Sc. a building containing several dwellings. **7** S. Afr. ground fenced off for tillage. **8** a strip of plough or pasture land parted from others by drain-furrows. ● v. **1 a** tr. & intr. set or go ashore. **b** intr. (often foll. by at) disembark (landed at the harbour). **2** tr. bring (an aircraft, its passengers, etc.) to the ground or the surface of water. **3** intr. (of an aircraft, bird, parachutist, etc.) alight on the ground or water. **4** tr. bring (a fish) to land, esp. with a hook or net. **5** tr. & intr. (also refl.; often foll. by up) bring to, reach, or find oneself in a certain situation, place, or state (landed himself in jail; landed up in France; landed her in trouble; landed up penniless). **6** tr. colloq. **a** deal (a person etc. a blow etc.) (landed him one in the eye). **b** (foll. by with) Brit. present (a person) with (a problem, job, etc.). **7** tr. set down (a person, cargo, etc.) from a vehicle, ship, etc. **8** tr. colloq. win or obtain (a prize, job, etc.) esp. against strong competition. □ **how the land lies** what the state of affairs is. **in the land of the living** joc. still alive. **land-agency 1** the stewardship of an estate. **2** an agency for the sale etc. of estates. **land-agent 1** the steward of an estate. **2** an agent for the sale of estates. **land-bank** a bank issuing banknotes on the securities of landed property. **land breeze** a breeze blowing towards the sea from the land, esp. at night. **land-bridge** a neck of land joining two large land masses. **land-crab** a crab of the tropical family Gecarcinidae, spending most of its time on land; esp. Cardisoma guanhumi, living in burrows and migrating to the sea to breed. **land force** (or **forces**) armies, not naval or air forces. **land-form** a natural feature of the earth's surface. **land-girl** Brit. a woman doing farm work, esp. in wartime. **land-grabber** an illegal seizer of land, esp. (hist) a person who took the land of an evicted Irish tenant. **land-law** (usu. in pl.) the law of landed property. **land-line** a means of telecommunication over land. **land mass** a large area of land. **land-mine 1** an explosive mine laid in or on the ground. **2** a parachute mine. **land of cakes** Scotland. **land office** US an office recording dealings in public land. **land-office business** US colloq. enormous trade. **land of the midnight sun** any of the most northerly European countries in which it never becomes fully dark during the summer months. **land of Nod** sleep (with pun on the phr. in Gen. 4:16). **land on one's feet** attain a good position, job, etc., by luck. **land-tax** hist. a tax assessed on landed property. **land-tie** a rod, beam, or piece of masonry securing or supporting a wall etc. by connecting it with the ground. **land-wind** a wind blowing seaward from the land. **land yacht** a vehicle with wheels and sails for recreational use on a beach etc. □ **landless** adj. **landward** adj. & adv. **landwards** adv. [OE f. Gmc]

Land Acts a series of British parliamentary acts concerning land tenure in Ireland, passed in 1870, 1881, 1903, and 1909, intended to give tenants greater security and further rights.

Landau /'lændɔː, Russian lanˈdaʊ/, Lev (Davidovich) (1908–68), Soviet theoretical physicist, born in Russia. He created an influential school of theoretical physics at Moscow State University, studying a wide range of problems. He contributed to thermodynamics, particle physics, quantum mechanics and electrodynamics, astrophysics, condensed matter physics, and several other areas. Landau was awarded the Nobel Prize for physics in 1962 for his work on the superfluidity and thermal conductivity of liquid helium.

landau /'lændɔː/ n. a four-wheeled enclosed carriage with a removable front cover and a back cover that can be raised and lowered. [*Landau* near Karlsruhe in Germany, where it was first made]

landaulet /ˌlændɔːˈlet/ n. **1** a small landau. **2** hist. a car with a folding hood over the rear seats.

landed /'lændɪd/ adj. **1** owning land (*landed gentry*). **2** consisting of, including, or relating to land (*landed property*).

Länder pl. of LAND.

lander /'lændə(r)/ n. **1** a person who lands or goes ashore. **2** a spacecraft designed to land on the surface of a planet or moon (*lunar lander, soft lander*) (cf. ORBITER).

landfall /'lændfɔːl/ n. the approach to land, esp. for the first time on a sea or air journey.

landfill /'lændfɪl/ n. **1** waste material etc. used to landscape or reclaim areas of ground. **2** the process of disposing of rubbish in this way. □ **landfill site** a place where rubbish is disposed of by burying it in the ground.

landgrave /'lændɡreɪv/ n. (fem. **landgravine** /-ɡrəˌviːn/) hist. **1** a count having jurisdiction over a territory. **2** the title of certain German princes. □ **landgraviate** /lændˈɡreɪvɪət/ n. [MLG *landgrave*, MHG *lantgrāve* (as LAND, G *Graf* COUNT²)]

landholder /'lændˌhəʊldə(r)/ n. the tenant or proprietor of land.

landing /'lændɪŋ/ n. **1 a** the act or process of coming to land. **b** an instance of this. **c** (also **landing-place**) a place where ships etc. land. **2 a** a platform between two flights of stairs, or at the top or bottom of a flight. **b** a passage leading to upstairs rooms. □ **landing-craft** any of several types of craft esp. designed for putting troops and equipment ashore. **landing-gear** the undercarriage of an aircraft. **landing-net** a net for landing a large fish which has been hooked. **landing pad** a small space on an airfield or heliport, where helicopters land and take off. **landing-stage** a platform, often floating, on which goods and passengers are disembarked. **landing-strip** an airstrip.

landlady /'lændˌleɪdɪ/ n. (pl. **-ies**) **1** a woman who lets land, a building, part of a building, etc., to a tenant. **2** a woman who keeps a public house, boarding-house, or lodgings.

Land League an Irish organization formed in 1879 to campaign for tenants' rights, in particular fair rents and security of tenure. Among its techniques was the use of the boycott against anyone taking on a farm from which the tenant had been evicted. The Land Act passed by Gladstone's government in 1881 met many of the League's demands. (See also BOYCOTT.)

ländler /'lendlə(r)/ n. **1** an Austrian dance in triple time, a precursor of the waltz. **2** the music for a ländler. [G f. *Landl* Upper Austria]

landlocked /'lændlɒkt/ adj. almost or entirely enclosed by land.

landloper /'lændˌləʊpə(r)/ n. esp. Sc. a vagabond. [MDu. *landlooper* (as LAND, *loopen* run, formed as LEAP)]

landlord /'lændlɔːd/ n. **1** a man who lets land, a building, part of a building, etc., to a tenant. **2** a man who keeps a public house, boarding-house, or lodgings.

landlubber /'lændˌlʌbə(r)/ n. a person unfamiliar with the sea or sailing.

landmark /'lændmɑːk/ n. **1 a** a conspicuous object in a district etc. **b** an object marking the boundary of an estate, country, etc. **2** an event, change, etc. marking a stage or turning-point in history etc. **3** attrib. serving as a landmark; signifying an important change, development, etc.

Landor /'lændɔː(r)/, Walter Savage (1775–1864), English poet and essayist. Among his poems is the exotic oriental epic *Gebir* (1798), which won him the admiration and friendship of the poet Robert Southey. During his long residence in Italy (1815–35) he wrote his best-known prose work, *Imaginary Conversations of Literary Men and Statesmen* (1824–8). In verse and prose his style shows a clear debt to classical forms and themes.

landowner /'lændˌəʊnə(r)/ n. an owner of land. □ **landownership** n. **landowning** adj. & n.

landrail /'lændreɪl/ n. = CORNCRAKE.

Landsat /'lændsæt/ n. a series of artificial satellites designed to monitor the earth's resources, the first of which was launched by the US in 1972. Images of the earth's surface (the whole of which is covered every few weeks) in several wavebands of the spectrum are used to provide information about agriculture, geology, ecological changes, etc.

landscape /'lændskeɪp/ n. & v. ● n. **1** natural or imaginary scenery, as seen in a broad view. **2** (often attrib.) a picture representing this; the genre of landscape painting. **3** (in graphic design etc.) a format in which the width of an illustration etc. is greater than the height (cf. PORTRAIT 4). ● v.tr. (also absol.) improve (a piece of land) by landscape gardening. □ **landscape gardener** (or **architect**) a person who plans the layout of landscapes, esp. extensive grounds. **landscape gardening** (or **architecture**) the art or practice of laying out ornamental grounds or grounds imitating natural scenery. **landscape-marble** marble with treelike markings. **landscape painter** an artist who paints landscapes. □ **landscapist** n. [MDu. *landscap* (as LAND, -SHIP)]

Landseer /'lændsɪə(r)/, Sir Edwin Henry (1802–73), English painter and sculptor. He was Queen Victoria's favourite painter and his works enjoyed great popular appeal. He is best known for his animal subjects, including scenes set in the Scottish Highlands (such as *The Monarch of the Glen*, 1851) and sentimental pictures of domestic pets. As a sculptor he is chiefly remembered for the bronze lions which he modelled in 1867 for the base of Nelson's Column in Trafalgar Square, London.

Land's End a rocky promontory in SW Cornwall, which forms the westernmost point of England. The approximate distance by road from Land's End to John o'Groats, at the north-east tip of Scotland, is 1,400 km (876 miles).

landslide /'lændslaɪd/ n. **1** the sliding down of a mass of land from a mountain, cliff, etc. **2** an overwhelming majority for one side in an election.

landslip /'lændslɪp/ n. = LANDSLIDE 1.

Landsmål /'læntsmɔːl/ n. a form of the Norwegian language (see NORWEGIAN). [Norw. f. *land* country + *mål* language]

landsman /'lændzmən/ n. (pl. **-men**) a non-sailor.

Landsteiner /'lændˌstaɪnə(r)/, Karl (1868–1943), Austrian-born American physician. His main interest was immunology, and he devised the system of classifying blood into four main immunological groups (A, B, AB, and O), which made it possible for blood transfusions to be carried out successfully. Landsteiner was awarded a Nobel Prize for this in 1930, and in 1940 he was the first to describe the rhesus factor in blood.

lane /leɪn/ n. **1** a narrow, often rural, road, street, or path. **2** a division of a road for a stream of traffic (*three-lane highway*). **3** a strip of track or water for a runner, rower, or swimmer in a race. **4** a path or course prescribed for or regularly followed by a ship, aircraft, etc. (*ocean lane*). **5** a gangway between crowds of people, objects, etc. □ **it's a long lane that has no turning** change is inevitable. [OE: orig. unkn.]

Lang /læŋ/, Fritz (1890–1976), Austrian-born film director. A pioneer of German cinema, during the 1920s he directed such notable silent films as the dystopian *Metropolis* (1927), making the transition to sound in 1931 with the thriller *M*. When *The Testament of Dr Mabuse* (1933) was banned by the Nazis, Lang left Germany. He eventually settled in the US and made a range of films, including westerns (such as *Rancho Notorious*, 1952) and *films noirs* (such as *The Big Heat*, 1953).

Langland /'læŋlənd/, William (c.1330–c.1400), English poet. A minor friar, he devoted much of his life to writing and rewriting *Piers Plowman* (c.1367–70), a long allegorical poem in alliterative verse; it takes the form of a spiritual pilgrimage, through which the narrator is guided by the Plowman and experiences a series of visions, with vivid vignettes of contemporary life, on his journey in search of Truth.

langlauf /'læŋlaʊf/ n. cross-country skiing; a cross-country skiing race. [G, = long run]

Langley /'læŋlɪ/, Samuel Pierpoint (1834–1906), American astronomer and aviation pioneer. He invented the bolometer (1879–81) and used it to study the radiant energy of the sun. His work on aerodynamics contributed to the design of early aeroplanes.

Langmuir /'læŋmjʊə(r)/, Irving (1881–1957), American chemist and physicist. His principal work was in surface chemistry, especially the phenomenon of adsorption and the application of this to catalysis. He worked on high-temperature electrical discharges in gases, introducing the use of inert gas in the tungsten lamp, developing an atomic-hydrogen welding torch capable of temperatures up to 3,000°C, and first using the term *plasma*. While studying atomic structure he introduced the terms *covalence* and *electrovalence*.

Lango /'læŋgəʊ/ *n. & adj.* ● *n.* (*pl.* same) **1** a member of a village-dwelling people inhabiting the Nile region of Uganda. **2** the Nilotic language of this people. ● *adj.* of or relating to the Lango or their language.

langouste /'lɒŋguːst/ *n.* a crawfish or spiny lobster. [F]

langoustine /'lɒŋguˌstiːn/ *n.* = NORWAY LOBSTER. [F]

lang syne /læŋ 'saɪn/ *adv. & n.* Sc. ● *adv.* in the distant past. ● *n.* the old days (cf. AULD LANG SYNE). [= long since]

Langton /'læŋtən/, Stephen (*c.*1150–1228), English prelate. His reputation rests mainly on his promotion of the interests of the English Church in the face of conflicting pressures from the papacy and the English throne. As Archbishop of Canterbury he defended the Church's interests against King John, was intermediary during the negotiations leading to the signing of Magna Carta, and protected the young Henry III against baronial domination.

Langtry /'læŋtrɪ/, Lillie (born Emilie Charlotte le Breton) (1853–1929), British actress. Born in Jersey and the daughter of the dean of the island, she was noted for her beauty and became known as 'the Jersey Lily' from the title of a portrait of her painted by Sir John Millais. She made her stage début in 1881 and was one of the first actresses from an aristocratic background. She became the mistress of the Prince of Wales, later Edward VII.

language /'læŋgwɪdʒ/ *n.* **1** the method of human communication, either spoken or written, consisting of the use of words in an agreed way. **2** the language of a particular community or country etc. (*speaks several languages*). **3 a** the faculty of speech. **b** a style or faculty of expression; the use of words, etc. (*his language was poetic; hasn't the language to express it*). **c** (also **bad language**) coarse, crude, or abusive speech (*didn't like his language*). **4** a system of symbols and rules for writing computer programs or algorithms. **5** any method of expression or communication (*the language of mime; sign language*). **6** a professional or specialized vocabulary. **7** literary style. □ **language laboratory** a room equipped with tape recorders etc. for learning a foreign language. **language of flowers** a set of symbolic meanings attached to different flowers. **speak the same language** have a similar outlook, manner of expression, etc. [ME f. OF *langage* ult. f. L *lingua* tongue]

langue de chat /ˌlɒŋg də 'ʃʌ/ *n.* (*pl.* **langues de chat** *pronunc.* same) a very thin finger-shaped crisp biscuit or piece of chocolate. [F, = cat's tongue]

Languedoc /lɒŋ'dɒk/ a former province of southern France, which extended from the Rhône valley to the northern foothills of the eastern Pyrenees. It united with the former province of Roussillon to form the administrative region of Languedoc-Roussillon. (See also LANGUE D'OC.)

langue d'oc /lɒŋ 'dɒk/ *n.* the vernacular language spoken in medieval France south of the Loire. The term comes from Old French and means 'language of *oc*', in reference to the use of the word *oc* (from Latin *hoc*) for 'yes', in contrast to *langue d'oïl*, the language spoken north of this region, where *oïl* was used. It includes a number of dialects, including Provençal; the latter is sometimes used synonymously with *langue d'oc*. (See also LANGUE D'OÏL.)

Languedoc-Roussillon /lɒŋgˌdɒkruːsiː'jɒn/ a region of southern France, on the Mediterranean coast, extending from the Rhône delta to the border with Spain.

langue d'oïl /lɒŋg 'dɔɪl/ *n.* the form of medieval French spoken north of the Loire, the basis of modern standard French. The term means 'language of oïl', and refers to the use of the word *oïl* (from Latin *hoc ille*) for 'yes'; *oïl* has developed into the modern French *oui*. (See also LANGUE D'OC.)

languid /'læŋgwɪd/ *adj.* **1** lacking vigour; idle; inert; apathetic. **2** (of ideas etc.) lacking force; uninteresting. **3** (of trade etc.) slow-moving; sluggish. **4** faint; weak. □ **languidly** *adv.* **languidness** *n.* [F *languide* or L *languidus* (as LANGUISH)]

languish /'læŋgwɪʃ/ *v.intr.* **1** be or grow feeble; lose or lack vitality. **2** put on a sentimentally tender or languid look. □ **languish for** droop or pine for. **languish under** suffer under (esp. depression, confinement, etc.). □ **languisher** *n.* **languishingly** *adv.* **languishment** *n.* [ME f. OF *languir*, ult. f. L *languere*, rel. to LAX]

languor /'læŋgə(r)/ *n.* **1** lack of energy or alertness; inertia; idleness; dullness. **2** faintness; fatigue. **3** a soft or tender mood or effect. **4** an oppressive stillness (of the air etc.). □ **languorous** *adj.* **languorously** *adv.* [ME f. OF f. L *languor -oris* (as LANGUISH)]

langur /lʌŋ'gʊə(r), 'læŋgə(r)/ *n.* a long-tailed monkey, esp. of the genus *Presbytis*, native to Asia (cf. *leaf-monkey*). [Hindi]

laniary /'læɪnɪərɪ/ *adj. & n.* ● *adj.* (of a tooth) adapted for tearing; canine. ● *n.* (*pl.* **-ies**) a laniary tooth. [L *laniarius* f. *lanius* butcher f. *laniare* to tear]

laniferous /lə'nɪfərəs/ *adj.* (also **lanigerous** /-'nɪdʒərəs/) wool-bearing. [L *lanifer*, *-ger* f. *lana* wool]

lank /læŋk/ *adj.* **1** (of hair, grass, etc.) long, limp, and straight. **2** thin and tall. **3** shrunken; spare. □ **lankly** *adv.* **lankness** *n.* [OE *hlanc* f. Gmc: cf. FLANK, LINK¹]

lanky /'læŋkɪ/ *adj.* (**lankier**, **lankiest**) (of limbs, a person, etc.) ungracefully thin and long or tall. □ **lankily** *adv.* **lankiness** *n.*

lanner /'lænə(r)/ *n.* a falcon, *Falco biarmicus*, native to Africa and the Mediterranean countries, esp. the female. [ME f. OF *lanier* perh. f. OF *lanier* cowardly, orig. = weaver f. L *lanarius* wool-merchant f. *lana* wool]

lanneret /'lænərɪt/ *n.* a male lanner, smaller than the female. [ME f. OF *laneret* (as LANNER)]

lanolin /'lænəlɪn/ *n.* a fat found naturally on sheep's wool and used purified as a basis for ointments. [G f. L *lana* wool + *oleum* oil]

Lansing /'lænsɪŋ/ the state capital of Michigan; pop. (1990) 127,320. First settled in 1847, the city expanded rapidly after the establishment there in 1887 of the motor industry.

lansquenet /'lænskənət/ *n.* **1** a card-game of German origin. **2** a German mercenary soldier in the 16th–17th centuries. [F f. G *Landsknecht* (as LAND, *Knecht* soldier f. OHG *kneht*: see KNIGHT)]

lantana /læn'teɪnə/ *n.* an evergreen subtropical shrub of the genus *Lantana*, with red, yellow, white, or varicoloured flowers. [mod.L]

Lantau /læn'taʊ/ (called in Chinese *Tai Yue Shan*) an island of Hong Kong, situated to the west of Hong Kong Island and separated from it by the Lamma Channel. It forms part of the New Territories, which were ceded to Britain by China in 1898.

lantern /'læntən/ *n.* **1 a** a lamp with a transparent usu. glass case protecting a candle flame etc. **b** a similar electric etc. lamp. **c** its case. **2 a** a raised structure on a dome, room, etc., glazed to admit light. **b** a similar structure for ventilation etc. **3** the light-chamber of a lighthouse. **4** = *magic lantern*. □ **lantern fish** a marine fish of the family Myctophidae, having small light organs on the head and body. **lantern-fly** (*pl.* **-flies**) a tropical homopterous insect of the family Fulgoridae, the head of which bears a large hollow projection formerly thought to be luminous. **lantern-jawed** having lantern jaws. **lantern jaws** long thin jaws and chin, giving a hollow look to the face. **lantern-slide** a slide for projection by a magic lantern etc. **lantern-wheel** a lantern-shaped gearwheel; a trundle. [ME f. OF *lanterne* f. L *lanterna* f. Gk *lamptēr* torch, lamp]

lanthanide /'lænθəˌnaɪd/ *n.* (also **lanthanoid** /-ˌnɔɪd/) Chem. any of the series of fifteen metallic elements from lanthanum to lutetium in the periodic table. They tend to be soft, quite reactive, and usually trivalent. (See also RARE EARTH.) [G *Lanthanid* (as LANTHANUM)]

lanthanum /'lænθənəm/ *n.* a silvery-white metallic chemical element (atomic number 57; symbol **La**). One of the rare-earth metals, lanthanum was first discovered (as the oxide) by the Swedish chemist C. G. Mosander in 1839. It is a component of misch metal and other alloys, and is also used in catalysts for oil refining; the oxide is used to make specialized types of glass. [Gk *lanthanō* escape notice, from having remained undetected in cerium oxide]

lanugo /lə'njuːgəʊ/ *n.* fine soft hair, esp. that which covers the body and limbs of a human foetus. [L, = down f. *lana* wool]

lanyard /'lænjəd, -jɑːd/ *n.* **1** a cord hanging round the neck or looped round the shoulder, esp. of a Scout or sailor etc., to which a knife, a whistle, etc., may be attached. **2** Naut. a short rope or line used for securing, tightening, etc. **3** a cord attached to a breech mechanism for firing a gun. [ME f. OF *laniere*, *lasniere*: assim. to YARD¹]

Lanzarote /ˌlænzə'rɒtɪ/ one of the Canary Islands, the most easterly island of the group; chief town, Arrecife. The island's landscape was dramatically altered after a series of volcanic eruptions in about 1730. It is noted for the black sand of its beaches and for the 'Mountains of Fire' in the south-west, an area of many volcanic cones.

Lanzhou /læn'dʒəʊ/ (also **Lanchow** /-'tʃaʊ/) a city in northern China, on the upper Yellow River, capital of Gansu province; pop. (1990) 1,480,000.

Laocoon /leɪˈɒkəʊˌɒn/ *Gk Mythol.* a Trojan priest who, with his two sons, was crushed to death by two great sea-serpents as a penalty for warning the Trojans against drawing the wooden horse of the Greeks into Troy. The incident is the subject of a passage in Virgil's *Aeneid* and of one of the most famous examples of ancient Hellenistic sculpture, a Roman reproduction of which was rediscovered in the Renaissance and is now on show in the Vatican Museum.

Laodicean /ˌleɪəʊdɪˈsɪən/ *adj. & n.* ● *adj.* lukewarm or half-hearted, esp. in religion or politics. ● *n.* such a person. [L *Laodicea* in Asia Minor (with ref. to the early Christians there: see Rev. 3:16)]

Laois /liːʃ/ (also **Laoighis**, **Leix**) a county of the Republic of Ireland, in the province of Leinster; county town, Portlaoise. It was formerly called Queen's County.

Laos /laʊs, ˈlɑːɒs/ a landlocked country in SE Asia; pop. (est. 1991) 4,279,000; official language, Laotian; capital, Vientiane. The kingdom of Laos was established in 1353 by Thai Buddhists who had migrated southwards from China. In decline from the 16th century, Laos came under French colonial influence and was incorporated in French Indo-China. Laos became independent in 1949, but for most of the next 25 years was torn by strife and civil war between the Communist Pathet Lao movement and government supporters. In 1975 the Pathet Lao achieved total control, the king abdicated, and a Communist republic was established.

Laotian /ˈlaʊʃɪən, lɑːˈəʊʃɪən/ *n. & adj.* ● *n.* **1** a member of a people inhabiting Laos and NE Thailand. **2** the Sino-Tibetan language of this people. ● *adj.* of or relating to the Laotian people or their language.

Lao-tzu /laʊˈtsuː/ (also **Laoze** /-ˈtseɪ/) **1** the legendary founder of Taoism (*fl.* 6th century BC) and traditional author of the Tao-te-Ching, its most sacred scripture. **2** the Tao-te-Ching. [Chin., = Lao the Master]

lap[1] /læp/ *n.* **1 a** the front of the body from the waist to the knees of a sitting person (*sat on her lap*; *caught it in his lap*). **b** the clothing, esp. a skirt, covering the lap. **c** the front of a skirt held up to catch or contain something. **2** a hollow among hills. **3** a hanging flap on a garment, a saddle, etc. □ **in** (or **on**) **a person's lap** as a person's responsibility. **in the lap of the gods** (of an event etc.) open to chance; beyond human control. **in the lap of luxury** in extremely luxurious surroundings. **lap-dog** a small pet dog. **lap robe** *N.Amer.* a travelling-rug. □ **lapful** *n.* (*pl.* **-fuls**). [OE *læppa* fold, flap]

lap[2] /læp/ *n. & v.* ● *n.* **1 a** one circuit of a racetrack etc. **b** a section of a journey etc. (*finally we were on the last lap*). **2 a** an amount of overlapping. **b** an overlapping or projecting part. **3 a** a layer or sheet (of cotton etc. being made) wound on a roller. **b** a single turn of rope, silk, thread, etc., round a drum or reel. **4** a rotating disc for polishing a gem or metal. ● *v.* (**lapped, lapping**) **1** *tr.* lead or overtake (a competitor in a race) by one or more laps. **2** *tr.* (often foll. by *about, round*) coil, fold, or wrap (a garment etc.) round esp. a person. **3** *tr.* (usu. foll. by *in*) enfold or swathe (a person) in wraps etc. **4** *tr.* (as **lapped** *adj.*) (usu. foll. by *in*) protectively encircled; enfolded caressingly. **5** *tr.* surround (a person) with an influence etc. **6** *intr.* (usu. foll. by *over*) project; overlap. **7** *tr.* cause to overlap. **8** *tr.* polish (a gem etc.) with a lap. □ **half-lap** = *lap joint*. **lap joint** the joining of rails, shafts, etc., by halving the thickness of each at the joint and fitting them together. **lap of honour** a ceremonial circuit of a football pitch, a track, etc., by a winner or winners. **lap-strake** *Naut. n.* a clinker-built boat. ● *adj.* clinker-built. **lap-weld** *v.tr.* weld with overlapping edges. ● *n.* such a weld. [ME, prob. f. LAP[1]]

lap[3] /læp/ *v. & n.* ● *v.* (**lapped, lapping**) **1** *tr.* (also *absol.*) (usu. of an animal) drink (liquid) with the tongue. **b** (usu. foll. by *up, down*) consume (liquid) greedily. **c** (usu. foll. by *up*) consume (gossip, praise, etc.) greedily. **2 a** *tr.* (of water) move or beat upon (a shore) with a rippling sound as of lapping. **b** *intr.* (of waves etc.) move in ripples; make a lapping sound. ● *n.* **1 a** the process or an act of lapping. **b** the amount of liquid taken up. **2** the sound of wavelets on a beach. **3** liquid food for dogs. **4** *sl.* **a** a weak beverage. **b** any liquor. [OE *lapian* f. Gmc]

La Palma /læ ˈpælmə, ˈpɑːmə/ one of the Canary Islands, the most north-westerly in the group. It is the site of an astronomical observatory that has several major telescopes, including the principal instruments of the Royal Greenwich Observatory.

laparoscope /ˈlæpərəˌskəʊp/ *n. Surgery* a fibre optic instrument inserted through the abdominal wall to give a view of the organs in

the abdomen. □ **laparoscopy** /ˌlæpəˈrɒskəpɪ/ *n.* (*pl.* **-ies**). [Gk *lapara* flank + -SCOPE]

laparotomy /ˌlæpəˈrɒtəmɪ/ *n.* (*pl.* **-ies**) a surgical incision into the abdominal cavity for exploration or diagnosis. [Gk *lapara* flank + -TOMY]

La Paz /læ ˈpæz/ **1** the capital of Bolivia, in the north-west of the country near the border with Peru; pop. (1990) 1,126,000. (The judicial capital is Sucre.) Situated in the Andes at an altitude of 3,660 m (12,000 ft), La Paz is the highest capital city in the world. It was founded by the Spanish in 1548 on the site of an Inca settlement. **2** a city in Mexico, near the southern tip of the Baja California peninsula, capital of the state of Baja California Sur; pop. (est. 1991) 1,050,000.

lapel /ləˈpel/ *n.* the part of a coat, jacket, etc., folded back against the front round the neck opening. □ **lapelled** *adj.* [LAP[1] + -EL]

lapicide /ˈlæpɪˌsaɪd/ *n.* a person who cuts or engraves on stone. [L *lapicida* irreg. f. *lapis -idis* stone: see -CIDE]

lapidary /ˈlæpɪdərɪ/ *adj. & n.* ● *adj.* **1** concerned with stone or stones. **2** engraved upon stone. **3** (of writing style) dignified and concise, suitable for inscriptions. ● *n.* (*pl.* **-ies**) a cutter, polisher, or engraver of gems. [ME f. L *lapidarius* f. *lapis -idis* stone]

lapilli /ləˈpɪlaɪ/ *n.pl.* stone fragments ejected from volcanoes. [It. f. L, pl. dimin. of *lapis* stone]

lapis lazuli /ˈlæpɪs/ *n.* **1** a blue mineral containing sodium aluminium silicate and sulphur, used as a gemstone. **2** a bright blue pigment formerly made from this. **3** its colour. [ME f. L *lapis* stone + med.L *lazuli* genitive of *lazulum* f. Pers. (as AZURE)]

Lapita /læˈpiːtə/ an ancient Oceanic culture centred on Melanesia around *c.*1500 BC. It is characterized by pottery distinctively stamped with a toothed instrument.

Lapith /ˈlæpɪθ/ *n. Gk Mythol.* a member of a Thessalian people who fought and defeated the centaurs.

Laplace /læˈplɑːs/, Pierre Simon, Marquis de (1749–1827), French applied mathematician and theoretical physicist. He devoted his greatest treatise, *Mécanique céleste* (1799–1825), to an extensive mathematical analysis of geophysical matters and of planetary and lunar motion. He demonstrated the long-term stability of planetary orbits, and added considerably to the earlier work of Newton. Laplace is also known for his innovative work on partial differential equations, for his contributions to probability theory, and for other mathematical discoveries.

Lapland /ˈlæplænd/ a region of northern Europe which extends from the Norwegian Sea to the White Sea and lies mainly within the Arctic Circle. It consists of the northern parts of Norway, Sweden, and Finland, and the Kola Peninsula of Russia. □ **Laplander** *n.*

La Plata /læ ˈplɑːtə/ a port in Argentina, on the River Plate (Río de la Plata) south-east of Buenos Aires; pop. (1991) 640,000.

Lapp /læp/ *n. & adj.* ● *n.* **1** a member of the people inhabiting the extreme north of Scandinavia. (Cf. SAMI.) (*See note below.*) **2** the Finno-Ugric language of this people. ● *adj.* of or relating to the Lapps or their language. [Sw. *Lapp*, perh. orig. a term of contempt: cf. MHG *lappe* simpleton]

▪ Originating in the region of Lake Onega in Russia, the Lapps moved westward to northern Scandinavia about 10,000 years ago. Although nominally under Swedish and Norwegian government since the Middle Ages, they were not Christianized until the 18th century. The language of the Lapps, of which there are several mutually unintelligible dialects, is related to Finnish. Today the majority of Lapps live in Norway and Sweden, with small communities in Finland and Russia. Traditionally associated with the domestication and herding of reindeer, few Lapps continue the nomadic herding of the animals, most now living in permanent settlements with year-round pasture or deriving their livelihood from a combination of fishing, hunting, trapping, and farming. Scandinavian industrialization — particularly hydroelectric schemes, mining, and new roads — has severely disrupted the Lapps' traditional lifestyle.

lappet /ˈlæpɪt/ *n.* **1** a small flap or fold of a garment etc. **2** a hanging or loose piece of flesh, such as a lobe or wattle. **3** (in full **lappet moth**) a usu. brown velvety moth which has a caterpillar with flaps along its sides; esp. *Gastropacha quercifolia*, of Eurasia. □ **lappeted** *adj.* [LAP[1] + -ET[1]]

Lappish /ˈlæpɪʃ/ *adj. & n.* ● *adj.* = LAPP *adj.* ● *n.* the Lapp language.

lapse /læps/ *n. & v.* ● *n.* **1** a slight error; a slip of memory etc. **2** a weak or careless decline into an inferior state. **3** (foll. by *of*) an interval or passage of time (*after a lapse of three years*). **4** *Law* the termination of a right or privilege through disuse or failure to follow appropriate procedures. ● *v.intr.* **1** fail to maintain a position or standard. **2** (foll. by *into*) fall back into an inferior or previous state. **3** (of a right or privilege etc.) become invalid because it is not used or claimed or renewed. **4** (as **lapsed** *adj.*) (of a person or thing) that has lapsed. □ **lapse rate** *Meteorol.* the rate at which the temperature falls with increasing altitude. [L *lapsus* f. *labi laps-* glide, slip, fall]

lapstone /ˈlæpstəʊn/ *n.* a shoemaker's stone held in the lap and used to beat leather on.

lapsus calami /ˌlæpsəs ˈkæləˌmaɪ/ *n.* (*pl.* same) a slip of the pen. [L: see LAPSE]

lapsus linguae /ˌlæpsəs ˈlɪŋgwaɪ/ *n.* (*pl.* same) a slip of the tongue. [L: see LAPSE]

Laptev Sea /ˈlæptef/ a part of the Arctic Ocean, which lies to the north of Russia between the Taimyr Peninsula and the New Siberian Islands.

laptop /ˈlæptɒp/ *n.* (often *attrib.*) a microcomputer that is portable and suitable for use while travelling.

lapwing /ˈlæpwɪŋ/ *n.* a plover, *Vanellus vanellus*, with dark green and white plumage, crested head, and a shrill cry. Also called *green plover*, *peewit*. [OE *hlēapewince* f. *hlēapan* LEAP + WINK: assim. to LAP[1], WING]

Lara /ˈlɑːrə/, Brian (Charles) (b.1969), West Indian cricketer. Born in Trinidad, he first played for the West Indies in 1990. Two years later Lara made his mark on international cricket with an innings of 277 against Australia in Sydney. In 1994 he scored 375 against England in Antigua, breaking the record test score previously set by Gary Sobers in 1957. A few weeks later, playing for Warwickshire against Durham, he scored 501 not out, a world record in first-class cricket.

Laramie /ˈlærəmɪ/ a city in SE Wyoming; pop. (1990) 26,690. It was first settled in 1868, during the construction of the Union Pacific Railroad. Its early years are associated with the lawlessness of the Wild West.

larboard /ˈlɑːbəd/ *n. & adj. Naut. archaic* = PORT[3]. [ME *lade-, ladde-, lathe-* (perh. = LADE + BOARD): later assim. to *starboard*]

larceny /ˈlɑːsənɪ/ *n.* (*pl.* **-ies**) the theft of personal property. ¶ In 1968 replaced as a statutory crime in English law by *theft*. □ **larcener** *n.* **larcenist** *n.* **larcenous** *adj.* [OF *larcin* f. L *latrocinium* f. *latro* robber, mercenary f. Gk *latreus*]

larch /lɑːtʃ/ *n.* **1** a deciduous coniferous tree of the genus *Larix*, with bright foliage and producing tough timber. **2** (in full **larchwood**) its wood. [MHG *larche* ult. f. L *larix -icis*]

lard /lɑːd/ *n. & v.* ● *n.* the internal fat of the abdomen of pigs, esp. when rendered and clarified for use in cooking and pharmacy. ● *v.tr.* **1** insert strips of fat or bacon in (meat etc.) before cooking. **2** (foll. by *with*) embellish (talk or writing) with foreign or technical terms. [ME f. OF *lard* bacon f. L *lardum, laridum*, rel. to Gk *larinos* fat]

larder /ˈlɑːdə(r)/ *n.* **1** a room or large cupboard for storing food. **2** a wild animal's store of food, esp. for winter. [ME f. OF *lardier* f. med.L *lardarium* (as LARD)]

lardon /ˈlɑːd(ə)n/ *n.* (also **lardoon** /lɑːˈduːn/) a strip of fat bacon used to lard meat. [ME f. F *lardon* (as LARD)]

lardy /ˈlɑːdɪ/ *adj.* like or with lard. □ **lardy-cake** *Brit.* a cake made with lard, currants, etc.

lares /ˈlɑːriːz/ *n.pl. Rom. Hist.* gods worshipped, together with the penates, by households in ancient Rome. They are probably originally deities of the farmland. □ **lares and penates** the home. [L]

large /lɑːdʒ/ *adj. & n.* ● *adj.* **1** of considerable or relatively great size or extent. **2** of the larger kind (*the large intestine*). **3** of wide range; comprehensive. **4** pursuing an activity on a large scale (*large farmer*). ● *n.* (**at large**) **1** at liberty. **2** as a body or whole (*popular with the people at large*). **3** (of a narration etc.) at full length and with all details. **4** without a specific target (*scatters insults at large*). **5** *US* representing a whole area and not merely a part of it (*congressman at large*). □ **in large** on a large scale. **large as life** see LIFE. **large-minded** liberal; not narrow-minded. **larger than life** see LIFE. **large-scale** made or occurring on a large scale or in large amounts. □ **largeness** *n.* **largish** *adj.* [ME f. OF f. fem. of L *largus* copious]

largely /ˈlɑːdʒlɪ/ *adv.* to a great extent; principally (*is largely due to laziness*).

largesse /lɑːˈʒes/ *n.* (also **largess**) **1** money or gifts freely given, esp. on a special occasion by a person in high position. **2** generosity, beneficence. [ME f. OF *largesse* ult. f. L *largus* copious]

larghetto /lɑːˈgetəʊ/ *adv., adj., & n. Mus.* ● *adv. & adj.* in a fairly slow tempo. ● *n.* (*pl.* **-os**) a larghetto passage or movement. [It., dimin. of LARGO]

largo /ˈlɑːgəʊ/ *adv., adj., & n. Mus.* ● *adv. & adj.* in a slow tempo and dignified in style. ● *n.* (*pl.* **-os**) a largo passage or movement. [It., = broad]

lari /ˈlɑːrɪ/ *n.* (*pl.* **laris** or same) a monetary unit of the Maldives, equal to one-hundredth of a rufiyaa. [Pers.]

lariat /ˈlærɪət/ *n.* **1** a lasso. **2** a tethering-rope, esp. used by cowboys. [Sp. *la reata* f. *reatar* tie again (as RE-, L *aptare* adjust f. *aptus* APT, fit)]

La Rioja /ˌlæ rɪˈɒhɑ/ an autonomous region of northern Spain, in the wine-producing valley of the River Ebro; capital, Logroño.

Larissa /ləˈrɪsə/ (Greek **Lárisa** /ˈlarisa/) a city in Greece, the chief town of Thessaly; pop. (1991) 113,000.

lark[1] /lɑːk/ *n.* **1** a small bird of the family Alaudidae, with brown plumage, elongated hind claws, and a tuneful song, esp. the skylark. **2** any similar bird, such as the meadowlark. [OE *lǣferce, lǣwerce*, of unkn. orig.]

lark[2] /lɑːk/ *n. & v. colloq.* ● *n.* **1** a frolic or spree; an amusing incident; a joke. **2** *Brit.* a type of activity, affair, etc. (*fed up with this digging lark*). ● *v.intr.* (foll. by *about*) play tricks; frolic. □ **larky** *adj.* **larkiness** *n.* [19th c.: orig. uncert.]

Larkin /ˈlɑːkɪn/, Philip (Arthur) (1922–85), English poet. His distinctive poetic voice first became apparent in *The Less Deceived* (1955), and was further developed in *The Whitsun Weddings* (1964) and *High Windows* (1974). His style is notable for its adaptation of contemporary speech rhythms and colloquial vocabulary to poetic metre; many poems are set in urban and suburban landscapes and are pervaded by an air of melancholy, bitterness, and stoic wit.

larkspur /ˈlɑːkspɜː(r)/ *n.* **1** a ranunculaceous plant of the genus *Consolida*, with spurred flowers. **2** a delphinium.

larn /lɑːn/ *v. colloq. or joc.* **1** *intr.* = LEARN. **2** *tr.* teach (*that'll larn you*). [dial. form of LEARN]

La Rochefoucauld /ˌlæ rɒʃfuːˈkəʊ/, François de Marsillac, Duc de (1613–80), French writer and moralist. He was a supporter of Marie de Médicis in plotting against Richelieu and later joined the uprising of the Fronde against Mazarin (1648–53). He returned to court on Mazarin's death in 1661. His chief work, *Réflexions, ou sentences et maximes morales* (1665), consists of 504 epigrammatic reflections analysing human conduct, and finding self-interest to be its driving force.

La Rochelle /ˌlæ rɒˈʃel/ a port on the Atlantic coast of western France; pop. (1990) 73,740.

Larousse /læˈruːs/, Pierre (1817–75), French lexicographer and encyclopedist. He edited the fifteen-volume *Grand dictionnaire universel du XIXᵉ siècle* (1866–76), which aimed to treat every area of human knowledge. In 1852 he co-founded the publishing house of Larousse, which continues to issue the dictionaries and reference works that bear its name.

larrikin /ˈlærɪkɪn/ *n. Austral.* a hooligan. [also Engl. dial.: perh. f. the name *Larry* (pet-form of *Lawrence*) + -KIN]

larrup /ˈlærəp/ *v.tr.* (**larruped, larruping**) *colloq.* thrash. [dial.: perh. f. LATHER]

Larry /ˈlærɪ/ *n.* □ **as happy as Larry** *colloq.* extremely happy. [20th c.: orig. uncert.: cf. LARRIKIN]

larva /ˈlɑːvə/ *n.* (*pl.* **larvae** /-viː/) **1** the active immature form of an insect, esp. one that differs greatly from the adult and is the stage between egg and pupa, e.g. a caterpillar or grub (cf. NYMPH 3). **2** an immature form of other animals that undergo some metamorphosis, e.g. a tadpole. □ **larval** *adj.* **larvicide** /-vɪˌsaɪd/ *n.* [L, = ghost, mask]

laryngeal /ləˈrɪndʒɪəl/ *adj.* **1** of or relating to the larynx. **2** *Phonet.* denoting a speech sound made in the larynx with the vocal cords partly closed.

laryngitis /ˌlærɪnˈdʒaɪtɪs/ *n.* inflammation of the larynx. □ **laryngitic** /-ˈdʒɪtɪk/ *adj.*

laryngoscope /ləˈrɪŋgəˌskəʊp/ *n.* an instrument for examining the larynx, or for inserting a tube through it.

laryngotomy /ˌlærɪŋˈgɒtəmɪ/ *n.* (*pl.* **-ies**) a surgical incision of the larynx, esp. to provide an air passage when breathing is obstructed.

larynx /ˈlærɪŋks/ *n.* (*pl.* **larynges** /ləˈrɪndʒiːz/) *Anat.* the hollow muscular organ forming an air passage to the lungs and holding the vocal cords in humans and other mammals. [mod.L f. Gk *larugx -ggos*]

lasagne /ləˈsænjə, -ˈsɑːnjə/ *n.* pasta in the form of sheets or wide ribbons, esp. as cooked and served with minced meat, cheese sauce, etc. [It., pl. of *lasagna* f. L *lasanum* cooking-pot]

La Salle /ˌlæ ˈsæl/, René-Robert Cavelier, Sieur de (1643–87), French explorer. A settler in French Canada, he sailed down the Ohio and

Mississippi rivers to the sea in 1682, naming the Mississippi basin Louisiana in honour of Louis XIV. In 1684 he led an expedition to establish a French colony on the Gulf of Mexico; over two years were wasted in fruitless searches for the Mississippi delta. La Salle eventually landed in Texas by mistake and was murdered when his followers mutinied.

La Scala /læ ˈskɑːlə/ an opera house in Milan, Italy, one of the most famous in the world, built 1776–8 on the site of the Church of Santa Maria della Scala.

Lascar /ˈlæskə(r)/ n. a sailor from India or SE Asia. [ult. f. Urdu & Pers. laškar army]

Lascaux /læˈskəʊ/ Archaeol. the site of a cave in the Dordogne, SW France, which is richly decorated with palaeolithic wall-paintings of animals, particularly horses, bison, cattle, and deer. Like those at Altamira, the paintings at Lascaux are dated to the Magdalenian period. Discovered in 1940, the cave was closed in 1963 to protect the paintings, a replica later being opened nearby.

lascivious /ləˈsɪvɪəs/ adj. **1** lustful. **2** inciting to or evoking lust. □ **lasciviously** adv. **lasciviousness** n. [ME f. LL lasciviosus f. L lascivia lustfulness f. lascivus sportive, wanton]

lase /leɪz/ v.intr. **1** function as or in a laser. **2** (of a substance) undergo the physical processes employed in a laser. [back-form. f. LASER]

laser /ˈleɪzə(r)/ n. a device that generates an intense beam of coherent monochromatic light (or other electromagnetic radiation) by stimulated emission from excited atoms or molecules. (See note below.) □ **laser printer** a printer in which a laser is used to form a pattern of dots on a photosensitive drum corresponding to the pattern of print required on a page. [light amplification by stimulated emission of radiation]

- A laser is basically a tube (with reflective ends) containing a suitable substance. The atoms or molecules of the lasing material are forced into an excited state by the application of energy in some form, e.g. light, microwaves, etc., from which they can return to normal by emitting light of a particular wavelength. In a laser the emission of one photon of light stimulates other excited atoms to emit similar photons. The light is reflected back and forth along the tube, building up in intensity; the laser beam is emitted through one (partially reflective) end of the tube. The first laser was built in 1960 by the US physicist Theodore Harold Maiman (b. 1927); it employed a ruby rod and emitted pulses of visible light. Many kinds of laser have been developed, emitting radiation in various regions of the spectrum, either as continuous or pulsed beams. The power and pinpoint accuracy of laser beams finds many applications, e.g. in drilling and cutting, alignment and guidance, and in surgery; the optical properties are exploited in holography, reading bar-codes, and in recording and playing compact discs. (See also MASER.)

laserdisc /ˈleɪzəˌdɪsk/ n. = DISC 4b.

LaserVision /ˈleɪzəˌvɪʒ(ə)n/ n. propr. a system for the reproduction of video signals recorded on a disc with a laser. [LASER + VISION, after TELEVISION]

lash /læʃ/ v. & n. ● v. **1** tr. beat with a whip, rope, etc. **2 a** intr. make a sudden whiplike movement with a limb or flexible instrument. **b** tr. move (a limb etc.) in this way (the stallion lashed his tail). **3** intr. pour or rush with great force. **4** intr. (foll. by at, against) strike violently. **5** tr. castigate in words. **6** tr. urge on as with a lash. **7** tr. (foll. by down, together, etc.) fasten with a cord, rope, etc. **8** tr. (of rain, wind, etc.) beat forcefully upon. ● n. **1 a** a sharp blow made by a whip, rope, etc. **b** (prec. by the) punishment by beating with a whip etc. **2** the flexible end of a whip. **3** (usu. in pl.) an eyelash. □ **lash out 1** (often foll. by at) speak or hit out angrily. **2** Brit. spend money extravagantly, be lavish. **lash-up** Brit. a makeshift or improvised structure or arrangement. □ **lasher** n. **lashless** adj. [ME: prob. imit.]

lashing /ˈlæʃɪŋ/ n. **1** a beating. **2** cord used for lashing.

lashings /ˈlæʃɪŋz/ n.pl. Brit. colloq. (foll. by of) plenty; an abundance.

Las Palmas /læs ˈpælmæs/ (in full **Las Palmas de Gran Canaria** /də ˌgræn kəˈnɑːrɪə/) a port and resort on the north coast of the island of Gran Canaria, capital of the Canary Islands; pop. (1991) 372,270.

La Spezia /læ ˈspetsɪə/ an industrial port in NW Italy; pop. (1990) 103,000. Since 1861 it has been Italy's chief naval station.

lass /læs/ n. esp. Sc. & N. Engl. or poet. a girl or young woman. [ME lasce ult. f. ON laskwa unmarried (fem.)]

Lassa fever /ˈlæsə/ n. Med. an acute viral disease, with fever, occurring in tropical Africa. It was first reported at the village of Lassa in Nigeria in 1969, and results in a high mortality rate.

lassie /ˈlæsɪ/ n. colloq. = LASS.

lassitude /ˈlæsɪˌtjuːd/ n. **1** languor, weariness. **2** disinclination to exert or interest oneself. [F lassitude or L lassitudo f. lassus tired]

lasso /læˈsuː, ˈlæsəʊ/ n. & v. ● n. (pl. **-os** or **-oes**) a rope with a noose at one end, used esp. in North America for catching cattle etc. ● v.tr. (**-oes**, **-oed**) catch with a lasso. □ **lassoer** n. [Sp. lazo LACE]

Lassus /ˈlæsəs/, Orlande de (Italian name Orlando di Lasso) (c.1532–94), Flemish composer. He was a notable composer of polyphonic music and wrote over 2,000 secular and sacred works, including masses, motets, madrigals, and settings of psalms.

last[1] /lɑːst/ adj., adv., & n. ● adj. **1** after all others; coming at or belonging to the end. **2** most recent; next before a specified time (last Christmas; last week). **b** preceding; previous in a sequence (got on at the last station). **3** only remaining (the last biscuit; our last chance). **4** (prec. by the) least likely or suitable (the last person I'd want; the last thing I'd have expected). **5** the lowest in rank (the last place). ● adv. **1** after all others; esp. in comb.: last-mentioned. **2** on the last occasion before the present (when did you last see him?). **3** (esp. in enumerating) lastly. ● n. **1** a person or thing that is last, last-mentioned, most recent, etc. **2** (prec. by the) the last mention or sight etc. (shall never hear the last of it). **3** the last performance of certain acts (breathed his last). **4** (prec. by the) **a** the end or last moment. **b** death. □ **at last** (or **long last**) in the end; after much delay. **last agony** the pangs of death. **last ditch** (often with hyphen) attrib.) a place of final desperate defence. **last man** Cricket the batsman who goes in to bat last. **last minute** (or **moment**) (often with hyphen) attrib.) the time just before an important event. **last name** surname. **last post** see POST[3]. **last rites** sacred rites for a person about to die. **the last straw** a slight addition to a burden or difficulty that makes it finally unbearable. **last thing** adv. very late, esp. as a final act before going to bed. **the last word 1** a final or definitive statement (always has the last word; is the last word on this subject). **2** (often foll. by in) the latest fashion. **on one's last legs** see LEG. **pay one's last respects** see PAY[1]. **to** (or **till**) **the last** till the end; esp. till death. [OE latost superl.: see LATE]

last[2] /lɑːst/ v.intr. **1** remain unexhausted or adequate or alive for a specified or considerable time; suffice (enough food to last us a week; the battery lasts and lasts). **2** continue for a specified time (the journey lasts an hour). □ **last out** remain adequate or in existence for the whole of a period previously stated or implied. [OE læstan f. Gmc]

last[3] /lɑːst/ n. a shoemaker's model for shaping or repairing a shoe or boot. □ **stick to one's last** not meddle with what one does not understand. [OE læste last, læst boot, lāst footprint f. Gmc]

lasting /ˈlɑːstɪŋ/ adj. **1** continuing, permanent. **2** durable. □ **lastingly** adv. **lastingness** n.

Last Judgement n. (in some beliefs) the judgement of humankind expected to take place at the end of the world.

lastly /ˈlɑːstlɪ/ adv. finally; in the last place.

Last Supper the final meal of Jesus with his Apostles on the night before the Crucifixion. Traditionally it has been regarded as the Passover meal, since the Synoptic Gospels appear to put it on the evening when the Passover celebrations began. The Gospel of St John, however, places the Crucifixion itself a few hours before the Passover meal. The Last Supper is commemorated by Christians in the sacrament of the Eucharist.

Las Vegas /læs ˈveɪgəs/ a city in southern Nevada; pop. (1990) 258,295. It is noted for its casinos and nightclubs.

lat. abbr. latitude.

Latakia /ˌlætəˈkiːə/ a seaport on the coast of western Syria, opposite the north-eastern tip of Cyprus; pop. (1981) 196,800. It is famous for its tobacco.

latch /lætʃ/ n. & v. ● n. **1** a bar with a catch and lever used as a fastening for a gate etc. **2** a spring-lock preventing a door from being opened from the outside without a key after being shut. ● v.tr. & intr. fasten or be fastened with a latch. □ **latch on** (often foll. by to) colloq. **1** attach oneself (to). **2** understand. **on the latch** fastened by the latch only, not locked. [prob. f. (now dial.) latch (v.) seize f. OE læccan f. Gmc]

latchkey /ˈlætʃkiː/ n. (pl. **-eys**) a key of an outer door. □ **latchkey child** a child who is alone at home after school until a parent returns from work.

late /leɪt/ adj. & adv. ● adj. **1** after the due or usual time; occurring or done after the proper time (late for dinner; a late milk delivery). **2 a** far on in the day or night or in a specified time or period. **b** far on in development. **3** flowering or ripening towards the end of the season (late strawberries). **4** (prec. by the or my, his, etc.) no longer alive or having the specified status (*my late husband*; *the late president*). **5** (esp. in *superl.*) of recent date (*the late storms*; *the latest songs*). **6** (as **latest**, prec. by *the*) fashionable, up to date. ● *adv.* **1** after the due or usual time (*arrived late*). **2** far on in time (*this happened later on*). **3** at or till a late hour. **4** at a late stage of development. **5** formerly but not now (*late of New York*). □ **at the latest** as the latest time envisaged (*will have done it by six at the latest*). **late in the day** at a late stage in the proceedings, esp. too late to be useful. **late Latin** see LATIN. **the latest 1** the most recent news etc. (*have you heard the latest?*). **2** the current fashion. □ **lateness** *n.* [OE læt (adj.), late (adv.) f. Gmc]

latecomer /ˈleɪtˌkʌmə(r)/ *n.* a person who arrives late.

lateen /ləˈtiːn/ *adj.* (of a ship) rigged with a lateen sail. □ **lateen sail** a triangular sail on a long yard at an angle of 45° to the mast. [F (*voile*) *latine* Latin (sail), because common in the Mediterranean]

lately /ˈleɪtlɪ/ *adv.* not long ago; recently; in recent times. [OE lætlīce (as LATE, -LY²)]

La Tène /læ ˈten/ *adj. Archaeol.* of, relating to, or denoting the second phase of the European Iron Age, dating from the mid-5th century BC until the coming of the Romans. It follows the Hallstatt period. The culture of this period, representing the height of early Celtic achievement, is characterized by hill-forts, developments in agriculture, rich and elaborate burials, and artefacts of excellent craftsmanship ornamented with the idiosyncratic Celtic style of swirling lines. [a district at the east end of Lake Neuchâtel, Switzerland, location of the type-site]

latent /ˈleɪt(ə)nt/ *adj.* **1** concealed, dormant. **2** existing but not developed or manifest. □ **latent heat** *Physics* the heat required to convert a solid into a liquid or vapour, or a liquid into a vapour, without change of temperature. **latent image** *Photog.* an image not yet made visible by developing. □ **latency** *n.* **latently** *adv.* [L *latere* latent- be hidden]

-later /lətə(r)/ *comb. form* denoting a person who worships a particular thing or person (*idolater*). [Gk: see LATRIA]

lateral /ˈlætərəl/ *adj. & n.* ● *adj.* **1** of, at, towards, or from the side or sides. **2** descended from a brother or sister of a person in direct line. ● *n.* a side part etc., esp. a lateral shoot or branch. □ **lateral line** *Zool.* a visible line along the side of a fish consisting of a series of sense-organs acting as vibration receptors. **lateral thinking** a method of solving problems indirectly or by apparently illogical methods. □ **laterally** *adv.* [L *lateralis* f. *latus lateris* side]

Lateran /ˈlætərən/ *the* site in Rome containing the basilica dedicated to St John the Baptist and St John the Evangelist, which is the cathedral church of Rome, and the Lateran Palace where the popes resided until the 14th century.

Lateran Council any of five general councils of the Western Church held in the Lateran Palace, Rome, in 1123, 1139, 1179, 1215, and 1512–17. The fourth council (1215) was the most important and involved the condemnation of the Albigenses as heretical as well as a clarification of Church doctrine on transubstantiation, the Trinity, and the Incarnation.

Lateran Treaty a treaty signed in 1929 in the Lateran Palace in Rome, a concordat between the kingdom of Italy (represented by Mussolini) and the Holy See (represented by Pope Pius XI), which recognized as fully sovereign and independent the papal state under the name Vatican City.

laterite /ˈlætəˌraɪt/ *n.* a red or yellow ferruginous clay, friable and hardening in air, used for making roads in the tropics. □ **lateritic** /ˌlætəˈrɪtɪk/ *adj.* [L *later* brick + -ITE¹]

latex /ˈleɪteks/ *n.* (pl. **latexes** or **latices** /ˈlætɪˌsiːz/) **1** a milky fluid of mixed composition found in various plants and trees, esp. the rubber tree, and used for commercial purposes. **2** a synthetic product resembling this. [L, = liquid]

lath /lɑːθ/ *n. & v.* ● *n.* (pl. **laths** /lɑːðz, lɑːθs/) **1** a thin flat strip of wood, esp. each of a series forming a framework or support for plaster etc. **2** (esp. in phr. **lath and plaster**) laths collectively as a building material, esp. as a foundation for supporting plaster. ● *v.tr.* attach laths to (a wall or ceiling). [OE lætt]

lathe /leɪð/ *n.* a machine for shaping wood, metal, etc., by means of a rotating drive which turns the piece being worked on against changeable cutting tools. [prob. rel. to Old Danish *lad* structure, frame, f. ON *hlath*, rel. to *hlatha* LADE]

lather /ˈlɑːðə(r), ˈlæð-/ *n. & v.* ● *n.* **1** a froth produced by agitating soap etc. and water. **2** frothy sweat, esp. of a horse. **3** a state of agitation. ● *v.* **1** *intr.* (of soap etc.) form a lather. **2** *tr.* cover with lather. **3** *intr.* (of a

horse etc.) develop or become covered with lather. **4** *tr. colloq.* thrash. □ **lathery** *adj.* [OE *lēathor* (n.), *lēthran* (v.)]

lathi /ˈlɑːtɪ/ *n.* (pl. **lathis**) (in India) a long heavy iron-bound bamboo stick used as a weapon, esp. by police. [Hindi *lāṭhī*]

latices *pl.* of LATEX.

Latimer /ˈlætɪmə(r)/, Hugh (*c.*1485–1555), English Protestant prelate and martyr. He became one of Henry VIII's chief advisers when the king formally broke with the papacy in 1534, and was made bishop of Worcester in 1535. Latimer's opposition to Henry's moves to restrict the spread of Reformation doctrines and practices led to his resignation in 1539. Under Mary I he was imprisoned for heresy and burnt at the stake with Nicholas Ridley at Oxford.

Latin /ˈlætɪn/ *n. & adj.* ● *n.* **1** the language of ancient Rome and its empire. (*See note below.*) **2** *Rom. Hist.* an inhabitant of ancient Latium in Italy. ● *adj.* **1** of or in Latin. **2** of the countries or peoples (e.g. France and Spain) using languages developed from Latin. **3** *Rom. Hist.* of or relating to ancient Latium or its inhabitants. **4** of the Roman Catholic Church. □ **Latin Church** the Western Church. **Latin cross** a plain cross with the lowest arm longer than the other three. □ **Latinism** *n.* **Latinist** *n.* [ME f. OF *Latin* or L *Latinus* f. *Latium*]

▪ Latin is an Indo-European language, inflected and with complex syntax, the ancestor of all the Romance languages. It was originally the dialect of the Latini or people of Latium. With the rise of Rome it spread as the official and literary language of the Roman Empire. In the Middle Ages it remained the international medium of communication in western Europe, the language of law, the sciences, and in particular of liturgy; it was the universal language of the Roman Catholic mass until the mid-20th century. Latin of the post-classical period is distinguished chronologically as *late Latin* (AD *c.*200–600), *medieval Latin* (*c.*600–1500), and *modern Latin* (since 1500); the latter is used for the scientific names of animals and plants and many Latin terms persist in legal and medical language. Other forms of Latin include *silver Latin*, which is the literary language and style of the century following the death of Augustus in AD 14, and *vulgar Latin*, a term applied to popular and provincial forms of Latin, especially those from which the Romance languages developed.

Latin America the parts of the Americas where Spanish or Portuguese is the main language. □ **Latin American** *adj. & n.*

Latin American Integration Association an economic grouping of South American countries instituted in 1981 in place of the Latin American Free Trade Area, having as its ultimate aim the creation of a common market. Its headquarters are in Montevideo.

Latinate /ˈlætɪˌneɪt/ *adj.* having the character of Latin.

Latinize /ˈlætɪˌnaɪz/ *v.* (also **-ise**) **1** *tr.* give a Latin or Latinate form to. **2** *tr.* translate into Latin. **3** *tr.* make conformable to the ideas, customs, etc., of the ancient Romans, Latin peoples, or Latin Church. **4** *intr.* use Latin forms, idioms, etc. □ **Latinizer** *n.* **Latinization** /ˌlætɪnaɪˈzeɪʃ(ə)n/ *n.* [LL *latinizare* (as LATIN)]

latish /ˈleɪtɪʃ/ *adj. & adv.* fairly late.

latitude /ˈlætɪˌtjuːd/ *n.* **1** *Geog.* **a** the angular distance on a meridian north or south of the equator, expressed in degrees and minutes. **b** (usu. in *pl.*) regions or climes, esp. with reference to temperature (*warm latitudes*). **2** freedom from narrowness; liberality of interpretation. **3** tolerated variety of action or opinion (*was allowed much latitude*). **4** *Astron.* the angular distance of a celestial object or point from the ecliptic. □ **high latitudes** regions near the poles. **low latitudes** regions near the equator. □ **latitudinal** /ˌlætɪˈtjuːdɪn(ə)l/ *adj.* **latitudinally** *adv.* [ME, = breadth, f. L *latitudo -dinis* f. *latus* broad]

latitudinarian /ˌlætɪˌtjuːdɪˈneərɪən/ *adj. & n.* ● *adj.* allowing latitude esp. in religion; showing no preference among varying creeds and forms of worship. (*See note below.*) ● *n.* a person with a latitudinarian attitude. □ **latitudinarianism** *n.* [L *latitudo -dinis* breadth + -ARIAN]

▪ The term was applied derogatorily in the 17th century to Anglican divines who, while remaining in the Church of England, attached relatively little importance to matters of dogma, ecclesiastical organization, and liturgical practice, and deprecated quarrels over these, regarding personal piety and morality as of more consequence.

Latium /ˈleɪʃɪəm/ an ancient region of west central Italy, west of the Apennines and south of the River Tiber. It was settled during the early part of the first millennium BC by a branch of the Indo-European people known as the Latini. By the end of the 4th century BC the region had become dominated by Rome. It is now part of the modern region of Lazio.

Latona /lə'təʊnə/ *Rom. Mythol.* the Roman name for Leto.

latria /lə'traɪə/ *n.* (in Christian theology) supreme worship allowed to God alone. [LL f. Gk *latreia* worship f. *latreuō* serve]

latrine /lə'triːn/ *n.* a communal lavatory, esp. in a camp, barracks, etc. [F f. L *latrina*, shortening of *lavatrina* f. *lavare* wash]

-latry /lətrɪ/ *comb. form* denoting worship (*idolatry*). [Gk *latreia*: see LATRIA]

latten /'læt(ə)n/ *n.* an alloy of copper and zinc, often rolled into sheets, and formerly used for monumental brasses and church articles. [ME *latoun* f. OF *laton, leiton*]

latter /'lætə(r)/ *adj.* **1 a** denoting the second-mentioned of two, or *disp.* the last-mentioned of three or more. **b** (prec. by *the*; usu. *absol.*) the second or last-mentioned person or thing. (Opp. FORMER.) **2** nearer to the end (*the latter part of the year*). **3** recent. **4** belonging to the end of a period, of the world, etc. □ **latter-day** modern, contemporary. [OE *lætra*, compar. of *læt* LATE]

Latter-day Saints /'lætə,deɪ/ *n.pl.* see MORMON.

latterly /'lætəlɪ/ *adv.* **1** in the latter part of life or of a period. **2** recently.

lattice /'lætɪs/ *n.* **1 a** a structure of crossed laths or bars with spaces between, used as a screen, fence, etc. **b** (in full **lattice-work**) laths arranged in lattice formation. **2** *Crystallog.* a regular periodic arrangement of atoms, ions, or molecules in a crystalline solid. □ **lattice frame** (or **girder**) a girder or truss made of top and bottom members connected by struts usu. crossing diagonally. **lattice window** a window with small panes set in diagonally crossing strips of lead. □ **latticed** *adj.* **latticing** *n.* [ME f. OF *lattis* f. *latte* lath f. WG]

Latvia /'lætvɪə/ a country on the eastern shore of the Baltic Sea, between Estonia and Lithuania; pop. (est. 1991) 2,693,000; languages, Latvian (official), Russian; capital, Riga. Latvia became a province of the Russian empire in the early 18th century, after having been under Polish and then Swedish rule. It was proclaimed an independent republic in 1918, but in 1940 was annexed by the Soviet Union as a constituent republic. In 1991, on the breakup of the USSR, Latvia became an independent republic once again.

Latvian /'lætvɪən/ *n. & adj.* ● *n.* **1 a** a native of Latvia. **b** a person of Latvian descent. **2** the language of Latvia, spoken by some 1,500,000 people. It is an Indo-European language, most closely related to Lithuanian, with which it constitutes the Baltic branch of languages. Formerly also called *Lettish*. ● *adj.* of or relating to Latvia or its people or language.

Laud /lɔːd/, William (1573–1645), English prelate. In 1633 he was appointed Archbishop of Canterbury and set about the suppression of the prevailing Calvinism in England and Presbyterianism in Scotland. His moves to impose liturgical uniformity by restoring pre-Reformation practices aroused great hostility; they led to war in Scotland and were a contributory cause of the English Civil War. In 1640 Laud was impeached and imprisoned; he was later executed for treason.

laud /lɔːd/ *v. & n.* ● *v.tr.* praise or extol, esp. in hymns. ● *n.* **1** *literary* praise; a hymn of praise. **2** (in *pl.*) the first religious service of the day in the Western Church. In the Book of Common Prayer parts of lauds and matins were combined to form the service of morning prayer. [ME: (n.) f. OF *laude*, (v.) f. L *laudare*, f. L *laus laudis* praise]

Lauda /'laʊdə/, Nikolaus Andreas ('Niki') (b.1949), Austrian motor-racing driver. He was world champion in 1975 and went on to win two more championships (1977 and 1984), despite suffering severe injuries in a crash in the 1976 German Grand Prix. He retired in 1985.

laudable /'lɔːdəb(ə)l/ *adj.* commendable, praiseworthy. □ **laudably** *adv.* **laudability** /,lɔːdə'bɪlɪtɪ/ *n.* [ME f. L *laudabilis* (as LAUD)]

laudanum /'lɔːdnəm, 'lɒd–/ *n.* a solution containing morphine and prepared from opium, formerly used as a narcotic painkiller. [mod.L, the name given by Paracelsus to a costly medicament, later applied to preparations containing opium: perh. var. of LADANUM]

laudation /lɔː'deɪʃ(ə)n/ *n. formal* praise. [L *laudatio -onis* (as LAUD)]

laudatory /'lɔːdətərɪ/ *adj.* (also **laudative** /-tɪv/) expressing praise.

laugh /lɑːf/ *v. & n.* ● *v.* **1** *intr.* make the spontaneous sounds and movements usual in expressing lively amusement, scorn, derision, etc. **2** *tr.* express by laughing. **3** *tr.* bring (a person) into a certain state by laughing (*laughed them into agreeing*). **4** *intr.* (foll. by *at*) ridicule, make fun of (*laughed at us for going*). **5** *intr.* (in phr. **be laughing**) *colloq.* be in a fortunate or successful position. **6** *intr.* esp. *poet.* make sounds reminiscent of laughing. ● *n.* **1** the sound or act or manner of laughing. **2** *colloq.* a comical or ridiculous person or thing. □ **have the last laugh** be ultimately the winner. **laugh in a person's face** show open scorn for a person. **laugh off** get rid of (embarrassment or humiliation) by joking. **laugh on the other side of one's face** change from enjoyment or amusement to displeasure, shame, apprehension, etc. **laugh out of court** deprive of a hearing by ridicule. **laugh up one's sleeve** be secretly or inwardly amused. □ **laugher** *n.* [OE *hlæhhan, hliehhan* f. Gmc]

laughable /'lɑːfəb(ə)l/ *adj.* ludicrous; highly amusing. □ **laughably** *adv.*

laughing /'lɑːfɪŋ/ *n. & adj.* ● *n.* laughter. ● *adj.* in senses of LAUGH *v.* □ **laughing-gas** nitrous oxide as an anaesthetic, formerly used without oxygen and causing an exhilarating effect when inhaled. **laughing hyena** see HYENA. **laughing jackass** = KOOKABURRA. **laughing-stock** a person or thing open to general ridicule. **no laughing matter** something serious. □ **laughingly** *adv.*

laughter /'lɑːftə(r)/ *n.* the act or sound of laughing. [OE *hleahtor* f. Gmc]

Laughton /'lɔːt(ə)n/, Charles (1899–1962), British-born American actor. He began his acting career on the English stage, turning to film in 1932. His appearance suited him for character roles such as Henry VIII (*The Private Life of Henry VIII*, 1933) and Captain Bligh (*Mutiny on the Bounty*, 1935); he also played Quasimodo in *The Hunchback of Notre Dame* (1939). He became an American citizen in 1950.

launce /lɑːns, læns/ *n.* a sand eel. [perh. f. LANCE: cf. *garfish*]

Launcelot var. of LANCELOT.

Launceston /'lɔːnsəstən/ a city in northern Tasmania, on the Tamar estuary, the second largest city of the island; pop. (1991) 66,750.

launch¹ /lɔːntʃ/ *v. & n.* ● *v.* **1** *tr.* set (a vessel) afloat. **2** *tr.* hurl or send forth (a weapon, rocket, etc.). **3** *tr.* start or set in motion (an enterprise, a person on a course of action, etc.). **4** *tr.* formally introduce (a new product) with publicity etc. **5** *intr.* (often foll. by *out, into*, etc.) **a** make a start, esp. on an ambitious enterprise. **b** plunge into a long recital, a tirade, etc. ● *n.* the act or an instance of launching. □ **launch** (or **launching**) **pad** a platform with a supporting structure, from which rockets are launched. **launch vehicle** a rocket-powered vehicle used to send artificial satellites or spacecraft into space. [ME f. AF *launcher*, ONF *lancher*, OF *lancier* LANCE]

launch² /lɔːntʃ/ *n.* **1** a large motor boat, used esp. for pleasure. **2** *hist.* a man-of-war's largest boat. [Sp. *lancha* pinnace perh. f. Malay *lancharan* f. *lanchār* swift]

launcher /'lɔːntʃə(r)/ *n.* a structure or device to hold a rocket during launching.

launder /'lɔːndə(r)/ *v. & n.* ● *v.tr.* **1** wash and iron (clothes, linen, etc.). **2** *colloq.* transfer (funds) to conceal a dubious or illegal origin. ● *n.* a channel for conveying liquids, esp. molten metal. □ **launderer** *n.* [ME *launder* (n.), washer of linen, contr. of *lavander* f. OF *lavandier* ult. f. L *lavanda* things to be washed, neut. pl. gerundive of *lavare* wash]

launderette /lɔːn'dret/ *n.* (also **laundrette**) an establishment with coin-operated washing-machines and driers for public use.

laundress /'lɔːndrɪs/ *n.* a woman who launders clothes, linen, etc., esp. professionally.

Laundromat /'lɔːndrə,mæt/ *n.* (also **laundromat**) esp. *US propr.* a launderette. [alt. f. as US *laundermat*]

laundry /'lɔːndrɪ/ *n.* (pl. **-ies**) **1 a** a room or building for washing clothes etc. **b** a firm washing clothes etc. commercially. **2** clothes or linen for laundering or newly laundered. **3** the action of laundering clothes etc. [contr. f. *lavendry* (f. OF *lavanderie*) after LAUNDER]

Laurasia /lɔː'reɪʒə, -'reɪʃə/ a vast continental area believed to have existed in the northern hemisphere and to have resulted from the breakup of Pangaea in Mesozoic times. It comprised the present North America, Greenland, Europe, and most of Asia north of the Himalayas. (See also CONTINENTAL DRIFT.) [*Laurentia*, name given to the ancient forerunner of North America + Eur*asia*]

laureate /'lɒrɪət, 'lɔːr–/ *adj. & n.* ● *adj.* **1** wreathed with laurel as a mark of honour. **2** consisting of laurel; laurel-like. ● *n.* **1** a person who is honoured for outstanding creative or intellectual achievement (*Nobel laureate*). **2** = POET LAUREATE. □ **laureateship** *n.* [L *laureatus* f. *laurea* laurel-wreath f. *laurus* laurel (from the Greek and Roman custom of honouring poets and military victors with laurel crowns)]

laurel /'lɒr(ə)l/ *n. & v.* ● *n.* **1** = BAY². **2 a** (in *sing.* or *pl.*) the foliage of the bay tree used as an emblem of victory or distinction in poetry usu. formed into a wreath or crown. **b** (in *pl.*) honour or distinction. **3** any plant with dark green glossy leaves like a bay tree, e.g. cherry-laurel, mountain laurel, spurge laurel. ● *v.tr.* (**laurelled, laurelling**; *US* **laureled, laureling**) wreathe with laurel. □ **look to one's laurels**

beware of losing one's pre-eminence. **rest on one's laurels** be satisfied with what one has done and not seek further success. [ME *lorer* f. OF *lorier* f. Prov. *laurier* f. *laur* f. L *laurus*]

Laurel and Hardy /'lɒrəl, 'hɑːdɪ/, Stan Laurel (born Arthur Stanley Jefferson) (1890–1965) and Oliver Hardy (1892–1957), American comedy duo. British-born Stan Laurel played the scatterbrained and often tearful innocent, Oliver Hardy ('Ollie') his pompous, overbearing, and frequently exasperated friend. They brought their distinctive slapstick comedy to many films from 1927 onwards.

Laurentian Plateau /lɒ'renʃ(ə)n/ an alternative name for the CANADIAN SHIELD.

Laurier /'lɔːrɪ,eɪ/, Sir Wilfrid (1841–1919), Canadian Liberal statesman, Prime Minister 1896–1911. He became the leader of the Liberal Party in 1891 and five years later was elected Canada's first French-Canadian and Roman Catholic Prime Minister. While in office he worked to achieve national unity in the face of cultural conflict; he also oversaw the building of a second transcontinental railway and the creation of the provinces of Alberta and Saskatchewan.

laurustinus /ˌlɒrə'staɪnəs/ n. an evergreen winter-flowering shrub, *Viburnum tinus*, with dense glossy green leaves and white or pink flowers. [mod.L f. L *laurus* laurel + *tinus* wild laurel]

Lausanne /ləʊ'zæn/ a town in SW Switzerland, on the north shore of Lake Geneva; pop. (1990) 122,600.

Lausitzer Neisse /ˌlaʊzɪtsə 'naɪsə/ an alternative German name for the River Neisse (see NEISSE 1).

lav /læv/ n. Brit. colloq. lavatory. [abbr.]

lava /'lɑːvə/ n. **1** the molten matter which flows from a volcano. **2** the solid substance which it forms on cooling. [It. f. *lavare* wash f. L]

lavabo /lə'vɑːbəʊ, -'veɪbəʊ/ n. (pl. **-os**) **1** RC Ch. **a** the ritual washing of the celebrant's hands at the offertory of the mass. **b** a towel or basin used for this. **2** a monastery washing-trough. **3** a wash-basin. [L, = I will wash, first word of Psalm 26:6]

lavage /'lævɪdʒ/ n. Med. the washing-out of a body cavity, such as the colon or stomach, with water or a medicated solution. [F f. *laver* wash: see LAVE]

lavation /lə'veɪʃ(ə)n/ n. formal washing. [L *lavatio* f. *lavare* wash]

lavatorial /ˌlævə'tɔːrɪəl/ adj. **1** of or relating to lavatories, esp. resembling the architecture or decoration of public lavatories. **2** (of humour etc.) scatalogical.

lavatory /'lævətərɪ/ n. (pl. **-ies**) **1** a receptacle into which a person can urinate or defecate, usu. with a flush mechanism as a means of disposing of the products. (*See note below.*) **2** a room or compartment containing one or more of these. □ **lavatory paper** Brit. = toilet paper. [ME, = washing vessel, f. LL *lavatorium* f. L *lavare lavat-* wash]

▪ Some ancient civilizations had elaborate systems of drainage and lavatories which in construction surpassed anything of later periods up to the 19th century. The first English flushing lavatory or water-closet was produced in Elizabethan times, but the innovation did not become established and was never patented; the first patent for a water-closet was taken out in 1755 by a Scottish watchmaker and engineer, Alexander Cumming (1733–1814). By mid-Victorian times most middle-class homes had indoor lavatories, but until the end of the 19th century these drained into cesspools, which were later connected by pipes to town sewers. Improvements on the basic design included ceramic pans to replace the earlier metal ones and the provision of an S-bend in the outlet pipe (in which trapped water prevents drain odours from rising).

lave /leɪv/ v.tr. literary **1** wash, bathe. **2** (of water) wash against; flow along. [ME f. OF *laver* f. L *lavare* wash, perh. coalescing with OE *lafian*]

lavender /'lævɪndə(r)/ n., adj., & v. ● n. **1 a** a small evergreen labiate shrub of the genus *Lavandula*, with narrow leaves and blue, purple, or pink aromatic flowers. **b** its flowers and stalks dried and used to scent linen, clothes, etc. **2** a pale blue colour with a trace of red. **3** colloq. effeminacy, homosexuality. ● adj. **1** of the colour or fragrance of lavender flowers. **2 a** refined, genteel, sentimental. **b** colloq. of or relating to homosexuality. **c** colloq. (of a man) effeminate, homosexual. ● v.tr. put lavender among (linen etc.). □ **lavender-water** a perfume made from distilled lavender, alcohol, and ambergris. [ME f. AF *lavendre*, ult. f. med.L *lavandula*]

Laver /'leɪvə(r)/, Rodney George ('Rod') (b.1938), Australian tennis player. In 1962 he became the second man (after Don Budge in 1938) to win the four major singles championships (British, American, French, and Australian) in one year; in 1969 he was the first to repeat this.

laver¹ /'leɪvə(r), 'lɑːv-/ n. an edible seaweed; esp. *Porphyra umbilicaulis*, having sheetlike fronds. □ **laver bread** a Welsh dish of laver which is boiled, dipped in oatmeal, and fried. [L]

laver² /'leɪvə(r)/ n. **1** Bibl. a large brass vessel for Jewish priests' ritual ablutions. **2** archaic a washing or fountain basin; a font. [ME *lavo(u)r* f. OF *laveo(i)r* f. LL (as LAVATORY)]

lavish /'lævɪʃ/ adj. & v. ● adj. **1** giving or producing in large quantities; profuse. **2** generous, unstinting. **3** excessive, over-abundant. ● v.tr. (often foll. by *on*) bestow or spend (money, effort, praise, etc.) abundantly. □ **lavishly** adv. **lavishness** n. [ME f. obs. *lavish*, *lavas* (n.) profusion f. OF *lavasse* deluge of rain f. *laver* wash]

Lavoisier /læ'vwʌzɪ,eɪ/, Antoine Laurent (1743–94), French scientist, regarded as the father of modern chemistry. He caused a revolution in chemistry by his description of the true nature of combustion, his rigorous methods of analysis, and his development of a new rational chemical nomenclature. He realized that it was Joseph Priestley's 'dephlogisticated air' that combined with substances during burning, and (believing it to be a constituent of acids) he renamed the gas *oxygen*. He held a number of important public offices, and his involvement in the collection of indirect taxes led to his death by guillotine in the French Revolution.

Law /lɔː/, (Andrew) Bonar (1858–1923), Canadian-born British Conservative statesman, Prime Minister 1922–3. Brought up in Scotland, he was a successful businessman before entering Parliament in 1900. He was leader of the Conservative Party (1911–21; 1922–3) and held several ministerial posts from 1915 until retiring in 1921. In October 1922 he returned from retirement to become Prime Minister following Lloyd George's resignation, but himself resigned six months later because of ill health.

law /lɔː/ n. **1 a** a rule enacted or customary in a community and recognized as enjoining or prohibiting certain actions and enforced by the imposition of penalties. **b** a body of such rules (*the law of the land*; *forbidden under Scots law*). **2** the controlling influence of laws; a state of respect for laws (*law and order*). **3** laws collectively as a social system or subject of study (*was reading law*). **4** (with defining word) any of the specific branches or applications of law (*commercial law*; *law of contract*). **5** binding force or effect (*their word is law*). **6** (prec. by *the*) **a** the legal profession. **b** colloq. the police. **7** the statute and common law (cf. EQUITY 2). **8** (in pl.) jurisprudence. **9 a** the judicial remedy; litigation. **b** the lawcourts as providing this (*go to law*). **10** a rule of action or procedure, e.g. in a game, social context, form of art, etc. **11** a regularity in natural occurrences, esp. as formulated or propounded in particular instances (*the laws of nature*; *the law of gravity*; *Parkinson's law*). **12 a** the body of divine commandments as expressed in the Bible or other sources. **b** (**Law of Moses**) the precepts of the Pentateuch. **c** (**the Law**) the Pentateuch as distinguished from the other parts of the Bible (the Prophets and the Writings). □ **at** (or **in**) **law** according to the laws. **be a law unto oneself** do what one feels is right; disregard custom. **go to law** take legal action; make use of the lawcourts. **law-abiding** obedient to the laws. **law-abidingness** obedience to the laws. **law agent** (in Scotland) a solicitor. **law centre** Brit. an independent publicly funded advisory service on legal matters. **Law Lord** a member of the House of Lords qualified to perform its legal work. **law of diminishing returns** see DIMINISH. **law of nature** = natural law. **laws of war** the limitations on belligerents' action recognized by civilized nations. **law term** a period appointed for the sitting of lawcourts. **lay down the law** be dogmatic or authoritarian. **take the law into one's own hands** redress a grievance by one's own means, esp. by force. [OE *lagu* f. ON *lag* something 'laid down' or fixed, rel. to LAY¹]

lawbreaker /'lɔː,breɪkə(r)/ n. a person who breaks the law. □ **lawbreaking** n. & adj.

Law Commission either of two bodies established in 1965 (one for England and Wales, another for Scotland) to consider changes to the law and removal of obsolete legislation. Made up of judges and lawyers, the bodies have influence but no power to make actual changes to the law.

lawcourt /'lɔːkɔːt/ n. a court of law.

lawful /'lɔːfʊl/ adj. conforming with, permitted by, or recognized by law; not illegal or illegitimate. □ **lawfully** adv. **lawfulness** n.

lawgiver /'lɔː,gɪvə(r)/ n. a person who lays down laws.

lawless /'lɔːlɪs/ adj. **1** having no laws or enforcement of them. **2** disregarding laws. **3** unbridled, uncontrolled. □ **lawlessly** adv. **lawlessness** n.

lawmaker /'lɔːˌmeɪkə(r)/ n. a legislator. □ **law-making** adj. & n.

lawman /'lɔːmæn/ n. (pl. **-men**) US a law-enforcement officer, esp. a sheriff or policeman.

lawn[1] /lɔːn/ n. & v. ● n. a piece of grass kept mown and smooth in a garden, park, etc. ● v.tr. turn into lawn, lay with lawn. [ME laund glade f. OF launde f. OCelt., rel. to LAND]

lawn[2] /lɔːn/ n. a fine linen or cotton fabric used for clothes. □ **lawny** adj. [ME, prob. f. Laon in France]

lawnmower /'lɔːnˌməʊə(r)/ n. a machine for cutting the grass on a lawn.

lawn tennis n. see TENNIS.

Lawrence[1] /'lɒrəns/, D(avid) H(erbert) (1885–1930), English novelist, poet, and essayist. Although he gained acclaim for Sons and Lovers (1913), his subsequent work met with hostile reactions; he left England permanently in 1919 and thereafter lived chiefly in Italy and Mexico. A moralist who believed that industrial society was causing man to become divorced from his basic instincts, he is chiefly remembered for his frank exploration of sexual relationships, as in The Rainbow (1915) and Women in Love (1921); Lady Chatterley's Lover (published in Italy, 1928) was only published in Britain in an unexpurgated form after an unsuccessful prosecution for obscenity in 1960.

Lawrence[2] /'lɒrəns/, Ernest Orlando (1901–58), American physicist. He developed the first circular particle accelerator, later called a cyclotron, capable of achieving very high electron voltages. This opened the way for the new science of high-energy physics, with the production of many new isotopes and elements, and Lawrence and his team investigated some of the subatomic particles generated. He also worked on providing fissionable material for the atom bomb. He received the Nobel Prize for physics in 1939.

Lawrence[3] /'lɒrəns/, Sir Thomas (1769–1830), English painter. He first achieved success with his full-length portrait (1789) of Queen Charlotte, the wife of King George III. Many portrait commissions followed and by 1810 he was recognized as the leading portrait painter of his time. In 1818 he was sent by the Prince Regent (later George IV) to paint the portraits of heads of state and military leaders after the allied victory over Napoleon.

Lawrence[4] /'lɒrəns/, T(homas) E(dward) (known as Lawrence of Arabia) (1888–1935), British soldier and writer. From 1916 onwards he helped to organize and lead the Arab revolt against the Turks in the Middle East. His campaign of guerrilla raids contributed to General Allenby's eventual victory in Palestine in 1918; Lawrence described this period in The Seven Pillars of Wisdom (1926). In 1922 he enlisted in the RAF under an assumed name to avoid attention and remained in the ranks of that service for most of the rest of his life. He was killed in a motorcycle accident.

Lawrence, St (Latin name Laurentius) (d.258), Roman martyr and deacon of Rome. According to tradition, Lawrence was ordered by the prefect of Rome to deliver up the treasure of the Church; when in response to this order he presented the poor people of Rome to the prefect, he was roasted to death on a gridiron. Feast day, 10 Aug.

lawrencium /lɒ'rensɪəm/ n. an artificial radioactive metallic chemical element (atomic number 103; symbol **Lr**, formerly **Lw**). A member of the actinide series, lawrencium was first produced by the US physicist Albert Ghiorso and his colleagues in 1961 by bombarding californium with boron nuclei. [LAWRENCE[2]]

Law Society the professional body responsible for regulating solicitors in England and Wales, established in 1825.

lawsuit /'lɔːsuːt, -sjuːt/ n. the process or an instance of making a claim in a lawcourt.

lawyer /'lɔɪə(r), 'lɔːjə(r)/ n. a member of the legal profession, esp. a solicitor. □ **lawyerly** adj. [ME law(i)er f. LAW]

lax /læks/ adj. **1** lacking care, concern, or firmness. **2** loose, relaxed; not compact. **3** Phonet. pronounced with the vocal muscles relaxed. □ **laxity** n. **laxly** adv. **laxness** n. [ME, = loose, f. L laxus: rel. to SLACK[1]]

laxative /'læksətɪv/ adj. & n. ● adj. tending to stimulate or facilitate evacuation of the bowels. ● n. a laxative medicine. [ME f. OF laxatif -ive or LL laxativus f. L laxare loosen (as LAX)]

lay[1] /leɪ/ v. & n. ● v. (past and past part. **laid** /leɪd/) **1** tr. place on a surface, esp. horizontally in a position of rest (laid the book on the table). **2** tr. put or bring into a certain or the required position or state (lay a carpet, a cable). **3** intr. dial. or erron. lie. ¶ This use, incorrect in standard English, is probably partly encouraged by confusion with lay as the past of lie, as in the dog lay on the floor which is correct; the dog is laying on the floor is not correct. **4** tr. make by laying (lay the foundations). **5** tr. (often absol.)

(of a hen bird) produce (an egg). **6** tr. **a** cause to subside or lie flat. **b** deal with to remove (a ghost, fear, etc.). **7** tr. place or present for consideration (a case, proposal, etc.). **8** tr. set down as a basis or starting-point. **9** tr. (usu. foll. by on) attribute or impute (blame etc.). **10** tr. locate (a scene etc.) in a certain place. **11** tr. prepare or make ready (a plan or a trap). **12** tr. prepare (a table) for a meal. **13** tr. place or arrange the material for (a fire). **14** tr. put down as a wager; stake. **15** tr. (foll. by with) coat or strew (a surface). **16** tr. sl. offens. have sexual intercourse with (esp. a woman). ● n. **1** the way, position, or direction in which something lies. **2** sl. offens. **a** a partner (esp. female) in sexual intercourse. **b** an act of sexual intercourse. **3** the direction or amount of twist in rope-strands. □ **in lay** (of a hen) laying eggs regularly. **laid-back** colloq. relaxed, unbothered, easygoing. **laid paper** paper with the surface marked in fine ribs. **laid up** confined to bed or the house. **lay about one 1** hit out on all sides. **2** criticize indiscriminately. **lay aside 1** put to one side. **2** cease to practise or consider. **3** save (money etc.) for future needs. ● n. the door of see DOOR. **lay back** cause to slope back from the vertical. **lay bare** expose, reveal. **lay a charge** make an accusation. **lay claim to** claim as one's own. **lay down 1** put on the ground or other surface. **2** relinquish; give up (an office). **3** formulate or insist on (a rule or principle). **4** pay or wager (money). **5** begin to construct (a ship or railway). **6** store (wine) in a cellar. **7** set down on paper. **8** sacrifice (one's life). **9** convert (land) into pasture. **10** record (esp. popular music). **lay down the law** see LAW. **lay hands on 1** seize or attack. **2** place one's hands on or over, esp. in confirmation, ordination, or spiritual healing. **lay one's hands on** obtain, acquire, locate. **lay hold of** seize or grasp. **lay in** provide oneself with a stock of. **lay into** colloq. attack violently with words or blows. **lay it on thick** (or **with a trowel**) colloq. flatter or exaggerate grossly. **lay low** overthrow, kill, or humble. **lay off 1** discharge (workers) temporarily or permanently because of a shortage of work; make redundant. **2** colloq. desist. **lay-off** n. **1** a temporary or permanent discharge of workers; a redundancy. **2** a period when this is in force. **lay on 1** esp. Brit. provide (a facility, amenity, etc.). **2** impose (a penalty, obligation, etc.). **3** inflict (blows). **4** spread on (paint etc.). **lay on the table** see TABLE. **lay open 1** break the skin of. **2** (foll. by to) expose (to criticism etc.). **lay out 1** spread out. **2** expose to view. **3** prepare (a corpse) for burial. **4** colloq. knock unconscious. **5** dispose (grounds etc.) according to a plan. **6** expend (money). **7** refl. (foll. by to + infin.) take pains (to do something) (laid themselves out to help). **lay store by** see STORE. **lay to rest** bury in a grave. **lay up 1** store, save. **2** put (a ship etc.) out of service. **lay waste** see WASTE. [OE lecgan f. Gmc]

lay[2] /leɪ/ adj. **1 a** non-clerical. **b** not ordained into the clergy. **2 a** not professionally qualified, esp. in law or medicine. **b** of or done by such persons. □ **lay brother** (or **sister**) a person who has taken the vows of a religious order but is not ordained and is employed in ancillary or manual work. **lay reader** a lay person licensed to conduct some religious services. [ME f. OF lai f. eccl.L laicus f. Gk laikos LAIC]

lay[3] /leɪ/ n. **1** a short lyric or narrative poem meant to be sung. **2** a song. [ME f. OF lai, Prov. lais, of unkn. orig.]

lay[4] past of LIE[1].

layabout /'leɪəˌbaʊt/ n. a habitual loafer or idler.

Layamon /'laɪəmən/ (late 12th century), English poet and priest. He wrote the verse chronicle known as the Brut, a history of England from the arrival of the legendary Brutus to the 7th-century king Cadwalader. One of the earliest major works in Middle English, the poem introduces for the first time in English the story of King Arthur and other figures prominent in later English literature.

lay-by /'leɪbaɪ/ n. (pl. **lay-bys**) **1** Brit. an area at the side of an open road where vehicles may stop. **2** a similar arrangement on a canal or railway. **3** Austral. & NZ a system of paying a deposit to secure an article for later purchase.

layer /'leɪə(r)/ n. & v. ● n. **1** a thickness of matter, esp. one of several, covering a surface. **2** a person or thing that lays. **3** (usu. with a qualifying word) a hen that lays eggs. **4** a shoot fastened down to take root while attached to the parent plant. ● v.tr. **1 a** arrange in layers. **b** cut (hair) in layers. **2** propagate (a plant) as a layer. □ **layer-out** a person who prepares a corpse for burial. □ **layered** adj. [ME f. LAY[1] + -ER[1]]

layette /leɪ'et/ n. a set of clothing, toilet articles, and bedclothes for a newborn child. [F, dimin. of OF laie drawer f. MDu. laege]

lay figure n. **1** a dummy or jointed figure of a human body used by artists for arranging drapery on etc. **2** an unrealistic character in a novel etc. **3** a person lacking in individuality. [lay f. obs. layman f. Du. leeman f. obs. led joint]

layman /'leɪmən/ n. (pl. **-men**; fem. **laywoman**, pl. **-women**) **1** any non-ordained member of a Church. **2** a person without professional or specialized knowledge in a particular subject.

La'youn /lɑːˈjuːn/ (also **Laayoune**, Arabic **El Aaiún** /ˌel aɪˈuːn/) the capital of Western Sahara; pop. (1982) 96,800.

layout /'leɪaʊt/ n. **1** the disposing or arrangement of a site, ground, etc. **2** the way in which plans, printed matter, etc., are arranged or set out. **3** something arranged or set out in a particular way. **4** the make-up of a book, newspaper, etc.

layover /'leɪˌəʊvə(r)/ n. a period of rest or waiting before a further stage in a journey etc.; a stopover.

layshaft /'leɪʃɑːft/ n. a second or intermediate transmission shaft in a machine.

lazar /'læzə(r)/ n. archaic a poor and diseased person, esp. a leper. [ME f. med.L lazarus f. the name in Luke 16:20]

lazaret /ˌlæzəˈret/ n. (also **lazaretto** /-ˈretəʊ/) (pl. **lazarets** or **lazarettos**) **1** a hospital for diseased people, esp. lepers. **2** a building or ship for quarantine. **3** the after part of a ship's hold, used for stores. [(F lazaret) f. It. lazzaretto f. lazzaro LAZAR]

Lazarist /'læzərɪst/ n. a member of a religious body, the Congregation of the Mission, established at the priory of St Lazare in Paris in 1625 by St Vincent de Paul. Set up to preach to the rural poor and train candidates for the priesthood, the Lazarists now have foundations worldwide. Also called *Vincentian*. [F, f. Lazarus (Luke 16:20); cf. LAZAR]

laze /leɪz/ v. & n. ● v. **1** intr. spend time lazily or idly. **2** tr. (often foll. by away) pass (time) in this way. ● n. a spell of lazing. [back-form. f. LAZY]

Lazio /'lætsɪˌəʊ/ an administrative region of west central Italy, on the Tyrrhenian Sea, including the ancient region of Latium; capital, Rome.

lazuli /'læzjʊlɪ, -ˌlaɪ/ n. = LAPIS LAZULI. [abbr.]

lazy /'leɪzɪ/ adj. (**lazier**, **laziest**) **1** disinclined to work, doing little work. **2** of or inducing idleness. **3** (of a river etc.) slow-moving. □ **lazily** adv. **laziness** n. [earlier laysie, lasie, laesy, perh. f. LG: cf. LG lasich idle]

lazybones /'leɪzɪˌbəʊnz/ n. (pl. same) colloq. a lazy person.

lb abbr. a pound or pounds (weight). [L libra]

l.b. abbr. Cricket leg-bye(s), leg-byed.

LBC abbr. London Broadcasting Company.

L/Bdr abbr. Lance-Bombardier.

l.b.w. abbr. Cricket leg before wicket.

l.c. abbr. **1** in the passage etc. cited. **2** lower case. **3** letter of credit. [sense 1 f. L loco citato]

LCC abbr. hist. London County Council.

LCD abbr. **1** liquid crystal display. **2** lowest (or least) common denominator.

LCM abbr. lowest (or least) common multiple.

L/Cpl abbr. Lance-Corporal.

LD abbr. Biol. lethal dose, usu. with a following numeral indicating the percentage of a group of animals, or cultured cells or micro-organisms, killed by such a dose (LD$_{50}$).

Ld. abbr. Lord.

Ldg. abbr. Leading (Seaman etc.).

LDL abbr. Biochem. = low-density lipoprotein.

L-dopa /el'dəʊpə/ n. Biochem. the laevorotatory form of dopa, used to treat Parkinson's disease. Also called *levodopa*. (See DOPA.)

LDS abbr. **1** Licentiate in Dental Surgery. **2** Latter Day Saints (see MORMON.)

-le¹ /(ə)l/ suffix forming nouns, esp.: **1** names of appliances or instruments (handle; thimble). **2** names of animals and plants (beetle; thistle). ¶ The suffix has ceased to be syllabic in fowl, snail, stile. [ult. from or repr. OE -el etc. f. Gmc, with many Indo-European cognates]

-le² /(ə)l/ suffix (also **-el**) forming nouns with (or orig. with) diminutive sense (angle; castle; mantle; syllable; novel; tunnel). [ME -el, -elle f. OF ult. f. L forms -ellus, -ella, -alis (cf. -AL¹) etc.]

-le³ /(ə)l/ suffix forming adjectives, often with (or orig. with) the sense 'apt or liable to' (brittle; fickle; little; nimble). [ME f. OE -el etc. f. Gmc, corresp. to L -ulus]

-le⁴ /(ə)l/ suffix forming verbs, esp. expressing repeated action or movement or having diminutive sense (bubble; crumple; wriggle). ¶ Examples from OE are handle, nestle, startle, twinkle. [OE -lian f. Gmc]

LEA abbr. (in the UK) Local Education Authority.

lea /liː/ n. poet. a piece of meadow or pasture or arable land. [OE lēa(h) f. Gmc]

Leach /liːtʃ/, Bernard (Howell) (1887–1979), British potter, born in Hong Kong. After studying in Japan he settled in Britain in 1920 and, with the Japanese potter Shoji Hamada, founded his pottery at St Ives in Cornwall. He practised and taught for more than fifty years, becoming a key figure in British 20th-century ceramics. His work amalgamated the ideas, methods, and traditions of Japanese and English pottery and his products were designed to combine beauty with functionality.

leach /liːtʃ/ v. **1** tr. make (a liquid) percolate through some material. **2** tr. subject (bark, ore, ash, or soil) to the action of percolating fluid. **3** tr. & intr. (foll. by away, out) remove (soluble matter) or be removed in this way. [prob. repr. OE leccan to water, f. WG]

Leacock /'liːkɒk/, Stephen (Butler) (1869–1944), Canadian humorist and economist. He was the head of the department of economics and political science at McGill University (1908–36) and published a number of books on these subjects. However, he is chiefly remembered for his many humorous short stories, parodies, and essays, among which are *Sunshine Sketches of a Little Town* (1912), an affectionate account of Canadian small-town life, and *Arcadian Adventures with the Idle Rich* (1914), a sharper and more satirical portrayal of an American city.

lead¹ /liːd/ v., n., & adj. ● v. (past and past part. **led** /led/) **1** tr. cause to go with one, esp. by guiding or showing the way or by going in front and taking a person's hand or an animal's halter etc. **2** tr. **a** direct the actions or opinions of. **b** (often foll. by to, or to + infin.) guide by persuasion or example or argument (what led you to that conclusion?; was led to think you may be right). **3** tr. (also absol.) provide access to; bring to a certain position or destination (this door leads you into a small room; the road leads to Lincoln; the path leads uphill). **4** tr. pass or go through (a life etc. of a specified kind) (led a miserable existence). **5** tr. **a** have the first place in (lead the dance; leads the world in sugar production). **b** (absol.) go first; be ahead in a race or game. **c** (absol.) be pre-eminent in some field. **6** tr. be in charge of (leads a team of researchers). **7** tr. **a** direct by example. **b** set (a fashion). **c** be the principal player of (a group of musicians). **8** tr. (also absol.) begin a round of play at cards by playing (a card) or a card of (a particular suit). **9** intr. (foll. by to) have as an end or outcome; result in (what does all this lead to?). **10** intr. (foll. by with) Boxing make an attack (with a particular blow). **11 a** intr. (foll. by with) (of a newspaper) use a particular item as the main story (led with the stock market crash). **b** tr. (of a story) be the main feature of (a newspaper or part of it) (the royal wedding will lead the front page). **12** tr. (foll. by through) make (a liquid, strip of material, etc.) pass through a pulley, channel, etc. ● n. **1** guidance given by going in front; example. **2 a** a leading place; the leadership (is in the lead; take the lead). **b** the amount by which a competitor is ahead of the others (a lead of ten yards). **3** a clue, esp. an early indication of the resolution of a problem (is the first real lead in the case). **4** a strap or cord for leading a dog etc. **5 a** Brit. a conductor (usu. a wire) conveying electric current from a source to an appliance. **b** US a conductor used in internal wiring. **6 a** the chief part in a play etc. **b** the person playing this. **c** (attrib.) the chief performer or instrument of a specified type (lead guitar). **7** (in full **lead story**) the item of news given the greatest prominence in a newspaper or magazine. **8 a** the act or right of playing first in a game or round of cards. **b** the card led. **9** the distance advanced by a screw in one turn. **10 a** an artificial watercourse, esp. one leading to a mill. **b** a channel of water in an ice-field. ● attrib.adj. leading, principal, first. □ **lead by the nose** cajole (a person) into compliance. **lead a person a dance** see DANCE. **lead-in 1** an introduction, opening, etc. **2** a wire leading in from outside, esp. from an aerial to a receiver or transmitter. **lead off 1** begin; make a start. **2** Brit. colloq. lose one's temper. **lead-off** n. an action beginning a process. **lead on 1** entice into going further than was intended. **2** mislead or deceive. **lead time** the time between the initiation and completion of a production process. **lead up the garden path** colloq. mislead. **lead up to 1** form an introduction to, precede; prepare for. **2** direct one's talk gradually or cautiously to a particular topic etc. **lead the way** see WAY. □ **leadable** adj. [OE lǣdan f. Gmc]

lead² /led/ n. & v. ● n. **1** a heavy grey soft metallic chemical element (atomic number 82; symbol **Pb**). (See note below.) **2 a** graphite. **b** a thin length of this for use in a pencil. **3** a lump of lead used in sounding water. **4** (in pl.) Brit. **a** strips of lead covering a roof. **b** a piece of lead-covered roof. **5** (in pl.) lead frames holding the glass of a lattice or stained-glass window. **6** Printing a blank space between lines of print (orig. with ref. to the metal strip used to give this space). **7** (attrib.) made of lead. ● v.tr. **1** cover, weight, or frame (a roof or window panes) with lead. **2** Printing separate lines of (printed matter) with leads. **3** (often as

leaded adj.) add a lead compound to (petrol etc.). □ **lead balloon** a failure, an unsuccessful venture. **lead-free** (of petrol) without added tetraethyl lead. **lead pencil** a pencil of graphite enclosed in wood. **lead poisoning** acute or chronic poisoning by absorption of lead into the body (also called *plumbism*). **lead shot** = SHOT[1] 3b. **lead tetraethyl** = TETRAETHYL LEAD. **lead wool** a fibrous form of lead, used for jointing water-pipes. □ **leaded** adj. **leadless** adj. [OE lēad f. WG]

■ Lead was known to the ancient Egyptians and Babylonians, and was used by the Romans for making water-pipes. It is durable, resistant to corrosion, and a poor conductor of electricity; it has been used in roofing, ammunition, damping sound and vibration, storage batteries, cable sheathing, and shielding radioactive material. Lead compounds have been used in crystal glass, as an antiknock agent in petrol, and were formerly used extensively in paints. Lead and its compounds can accumulate in the body as poisons. The symbol Pb is from the Latin *plumbum*.

leaden /'led(ə)n/ adj. **1** of or like lead. **2** heavy, slow, burdensome (*leaden limbs*). **3** inert, depressing (*leaden rule*). **4** lead-coloured (*leaden skies*). □ **leaden seal** a stamped piece of lead holding the ends of a wire used as a fastening. □ **leadenly** adv. **leadenness** /'led(ə)nnɪs/ n. [OE lēaden (as LEAD[2])]

leader /'liːdə(r)/ n. **1 a** a person or thing that leads. **b** a person followed by others. **2 a** the principal player in a music group; the first violin in an orchestra. **b** US a conductor of an orchestra. **3** Brit. = leading article (see LEADING[1]). **4** a short strip of non-functioning material at each end of a reel of film or recording tape for connection to the spool. **5** (in full **Leader of the House**) Brit. a member of the government officially responsible for initiating business in Parliament. **6** a shoot of a plant at the apex of a stem or of the main branch. **7** (in pl.) Printing a series of dots or dashes across the page to guide the eye, esp. in tabulated material. **8** the horse placed at the front in a team or pair. □ **leaderless** adj. **leadership** n. [OE lǣdere (as LEAD[1])]

leading[1] /'liːdɪŋ/ adj. & n. ● adj. chief; most important. ● n. guidance, leadership. □ **leading aircraftman** the rank above aircraftman in the RAF. **leading article** Brit. a newspaper article giving the editorial opinion. **leading counsel** the senior barrister of two or more in a case. **leading edge 1** the foremost edge of an aerofoil, esp. a wing or propeller blade. **2** the forefront of development, esp. in technology. **3** Electronics the part of a pulse in which the amplitude increases (opp. trailing edge (see TRAIL)). **leading lady** the actress playing the principal part in a play etc. **leading light** a prominent and influential person. **leading man** the actor playing the principal part in a play etc. **leading note** Mus. = SUBTONIC. **leading question** a question that prompts the answer wanted. **leading seaman** the rank next below NCO in the Royal Navy. **leading tone** US Mus. = leading note.

leading[2] /'ledɪŋ/ n. Printing = LEAD[2] n. 6.

leadwort /'ledwɜːt/ n. = PLUMBAGO 2.

leaf /liːf/ n. & v. ● n. (pl. **leaves** /liːvz/) **1 a** each of several flattened usu. green structures of a plant, usu. on the side of a stem or branch and the main organ of photosynthesis. **b** other similar plant structures, e.g. bracts, sepals, and petals (*floral leaf*). **2 a** foliage regarded collectively. **b** the state of having leaves out (*a tree in leaf*). **3** the leaves of tobacco or tea. **4** a single thickness of paper, esp. in a book with each side forming a page. **5** a very thin sheet of metal, esp. gold or silver. **6 a** the hinged part or flap of a door, shutter, table, etc. **b** an extra section inserted to extend a table. ● v. **1** intr. put forth leaves. **2** tr. (foll. by through) turn over the pages of (a book etc.). □ **leaf-beetle** a small beetle of the family Chrysomelidae, the larvae of which feed on leaves. **leaf green** n. the colour of green leaves. ● adj. (hyphenated when attrib.) of this colour. **leaf insect** a large tropical insect of the family Phylliidae, having the body and legs flattened and leaflike in appearance. **leaf-miner** an insect larva, esp. of some small moths and flies, that burrows between the cuticles of leaves. **leaf-monkey** a langur of the genus *Presbytis*. **leaf-mould** soil consisting chiefly of decayed leaves. **leaf spring** a spring made of strips of metal. **leaf-stalk** a petiole. □ **leafage** n. **leafed** adj. (also in comb.). **leafless** adj. **leaflessness** n. **leaflike** adj. [OE lēaf f. Gmc]

leafcutter /'liːfˌkʌtə(r)/ n. **1** (in full **leafcutter ant**) an ant of the mainly tropical American genus *Atta*, which cuts pieces from leaves in order to cultivate fungus for food. **2** (in full **leafcutter bee**) a solitary bee of the family Megachilidae, which lines its nest with leaf fragments.

leafhopper /'liːfˌhɒpə(r)/ n. a leaping homopterous insect of the

superfamily Cicadelloidea, which sucks the sap of plants and often causes damage and spreads disease.

leaflet /'liːflɪt/ n. & v. ● n. **1** a sheet of (usu. printed) paper (sometimes folded but not stitched) giving information, esp. for free distribution. **2** a young leaf. **3** Bot. a division of a compound leaf. ● v.tr. (**leafleted**, **leafleting**) distribute leaflets to.

leafy /'liːfɪ/ adj. (**leafier**, **leafiest**) **1 a** having many leaves. **b** (of a place) rich in foliage; verdant. **2** resembling a leaf. □ **leafiness** n.

league[1] /liːg/ n. & v. ● n. **1** a collection of people, countries, groups, etc., combining for a particular purpose, esp. mutual protection or cooperation. **2** an agreement to combine in this way. **3** a group of sports clubs which compete over a period for a championship. **4** a class of contestants etc. of comparable activity. ● v.intr. (**leagues**, **leagued**, **leaguing**) (often foll. by together) join in a league. □ **in league** allied, conspiring. **league football** Austral. Rugby League or Australian Rules football played in leagues. **league table 1** a listing of competitors as a league, showing their ranking according to performance. **2** any list of ranking order. [F ligue or It. liga, var. of lega f. legare bind f. L ligare]

league[2] /liːg/ n. archaic a varying measure of travelling-distance by land, usu. about 3 miles (5 km). [ME, ult. f. LL leuga, leuca, of Gaulish orig.]

League of Arab States (also called Arab League) an organization of Arab states, founded in 1945 with headquarters in Cairo, whose purpose is to ensure cooperation among its member states and protect their independence and sovereignty.

League of Nations an association of countries established in 1919 by the Treaty of Versailles to promote international cooperation and achieve international peace and security. Although the League accomplished much of value in postwar economic reconstruction, it failed in its prime purpose through the refusal of member nations to put international interests before national ones, and was powerless to stop Italian, German, and Japanese expansionism leading to the Second World War. After the war it was replaced by the United Nations.

leaguer[1] /'liːgə(r)/ n. esp. US a member of a league.

leaguer[2] /'liːgə(r)/ n. & v. = LAAGER. [Du. leger camp, rel. to LAIR[1]]

leak /liːk/ n. & v. ● n. **1 a** a hole in a vessel, pipe, or container etc. caused by wear or damage, through which matter, esp. liquid or gas, passes accidentally in or out. **b** the matter passing in or out through this. **c** the act or an instance of leaking. **2 a** a similar escape of electrical charge. **b** the charge that escapes. **3** the intentional disclosure of secret information. ● v. **1 a** intr. (of liquid, gas, etc.) pass in or out through a leak. **b** tr. lose or admit (liquid, gas, etc.) through a leak. **2** tr. intentionally disclose (secret information). **3** intr. (often foll. by out) (of a secret, secret information) become known. □ **have** (or **take**) **a leak** sl. urinate. □ **leaker** n. [ME prob. f. LG]

leakage /'liːkɪdʒ/ n. **1** the action or result of leaking. **2** what leaks in or out. **3** an intentional disclosure of secret information.

Leakey /'liːkɪ/, Louis (Seymour Bazett) (1903–72), British-born Kenyan archaeologist and anthropologist. Leakey is noted for his work on human origins in East Africa, where after the Second World War his excavations brought to light the remains of early hominids and their implements at Olduvai Gorge. His wife Mary (Douglas) Leakey (b.1913) is an anthropologist in her own right, discovering *Australopithecus* (or *Zinjanthropus*) *boisei* at Olduvai in 1959 and initiating work at the nearby Laetoli site in the mid-1970s. Their son Richard (Erskine) Leakey (b.1944) has continued his parents' work on early hominids; he was appointed director of the new Kenya Wildlife Service in 1989, but resigned in 1994 following a controversial political campaign to remove him.

leaky /'liːkɪ/ adj. (**leakier**, **leakiest**) **1** having a leak or leaks. **2** given to disclosing secrets. □ **leakiness** n.

leal /liːl/ adj. Sc. loyal, honest. [ME f. AF leal, OF leel, loial (as LOYAL)]

Leamington Spa /'lemɪŋtən/ (official name **Royal Leamington Spa**) a town in central England, in Warwickshire, south-east of Birmingham; pop. (1981) 57,350. Noted for its saline springs, it was granted the status of royal spa after a visit by Queen Victoria in 1838.

Lean /liːn/, Sir David (1908–91), English film director. He made his début as a director in 1942 and went on to make many notable films, including *Brief Encounter* (1945), *Great Expectations* (1946), *The Bridge on the River Kwai* (1957), *Lawrence of Arabia* (1962), and *A Passage to India* (1984).

lean[1] /liːn/ v. & n. ● v. (past and past part. **leaned** /liːnd, lent/ or **leant** /lent/) **1** intr. & tr. (often foll. by across, back, over, etc.) be or place in a sloping position; incline from the perpendicular. **2** intr. & tr. (foll. by *against, on, upon*) rest or cause to rest for support against etc. **3** *intr.* (foll. by *on, upon*) rely on; derive support from. **4** *intr.* (foll. by *to, towards*) be inclined or partial to; have a tendency towards. ● *n.* a deviation from the perpendicular; an inclination (*has a decided lean to the right*). □ **lean on** *colloq.* put pressure on (a person) to act in a certain way. **lean over backwards** see BACKWARDS. **lean-to** (*pl.* **-tos**) a building with its roof leaning against a larger building or a wall. [OE *hleonian, hlinian* f. Gmc]

lean² /liːn/ *adj.* & *n.* ● *adj.* **1** (of a person or animal) thin; having no superfluous fat. **2** (of meat) containing little fat. **3** a meagre; of poor quality (*lean crop*). **b** not nourishing (*lean diet*). **4** unremunerative. **5** (of various materials, as soil, mortar, fuel, etc.) low in essential or valuable elements or qualities. ● *n.* the lean part of meat. □ **lean years** years of scarcity. **lean-burn** of or relating to an internal-combustion engine designed to run on a lean mixture to reduce pollution. **lean mixture** vaporized fuel containing a high proportion of air. □ **leanly** *adv.* **leanness** /ˈliːnnɪs/ *n.* [OE *hlǣne* f. Gmc]

Leander /lɪˈændə(r)/ **1** *Gk Mythol.* a young man, the lover of the priestess Hero. He was drowned swimming across the Hellespont to visit her. **2** (in full **Leander Club**) the oldest amateur rowing club in the world, founded early in the 19th century. Originally based in London, by the end of the 19th century it had built its present boathouse at Henley-on-Thames. Membership is a mark of distinction in the rowing world, and a large proportion of its members are former Oxford and Cambridge oarsmen.

leaning /ˈliːnɪŋ/ *n.* a tendency or partiality.

leap /liːp/ *v.* & *n.* ● *v.* (*past* and *past part.* **leaped** /liːpt, lept/ or **leapt** /lept/) **1** *intr.* jump or spring forcefully. **2** *tr.* jump across. **3** *intr.* (of prices etc.) increase dramatically. **4** *intr.* hurry, rush; proceed without pausing for thought (*leapt to the wrong conclusion; leapt to their defence*). ● *n.* a forceful jump. □ **by leaps and bounds** with startlingly rapid progress. **leap at 1** rush towards, pounce upon. **2** accept eagerly. **leap in the dark** a daring step or enterprise whose consequences are unpredictable. **leap to the eye** be immediately apparent. **leap year** a year, occurring once in four, with 366 days, including 29th Feb. as an intercalary day (perh. so called because feast-days after Feb. in a leap year fall two days later, instead of the normal one day, than in the previous year). □ **leaper** *n.* [OE *hlȳp, hlēapan* f. Gmc]

leap-frog /ˈliːpfrɒg/ *n.* & *v.* ● *n.* a game in which players in turn vault with parted legs over others who are bending down. ● *v.* (**-frogged**, **-frogging**) **1** *intr.* (foll. by *over*) perform such a vault. **2** *tr.* vault over in this way. **3** *tr.* & *intr.* (of two or more people, vehicles, etc.) overtake alternately.

Lear¹ /lɪə(r)/ a legendary early king of Britain, the central figure in Shakespeare's tragedy *King Lear*. He is mentioned by the chronicler Geoffrey of Monmouth.

Lear² /lɪə(r)/, Edward (1812–88), English humorist and illustrator. He worked as a zoological draughtsman, and was especially noted as a bird artist; his *Illustrations of the Family of the Psittacidae* was published in 1832. He later came under the patronage of the 13th Earl of Derby, for whose grandchildren he wrote *A Book of Nonsense* (1845) with his own limericks and illustrations; subsequent collections of nonsense verses include *Laughable Lyrics* (1877). He also ,published illustrated accounts of his travels in Italy, Greece, and the Holy Land.

learn /lɜːn/ *v.* (*past* and *past part.* **learned** /lɜːnt, lɜːnd/ or **learnt** /lɜːnt/) **1** *tr.* gain knowledge of or skill in by study, experience, or being taught. **2** *tr.* (foll. by *to* + infin.) acquire or develop a particular ability (*learn to swim*). **3** *tr.* commit to memory (*will try to learn your names*). **4** *intr.* (foll. by *of*) be informed about. **5** *tr.* (foll. by *that, how*, etc. + clause) become aware of by information or from observation. **6** *intr.* receive instruction; acquire knowledge or skill. **7** *tr.* *archaic* or *sl.* teach. □ **learn one's lesson** see LESSON. □ **learnable** *adj.* **learnability** /ˌlɜːnəˈbɪlɪtɪ/ *n.* [OE *leornian* f. Gmc: cf. LORE¹]

learned /ˈlɜːnɪd/ *adj.* **1** having much knowledge acquired by study. **2** showing or requiring learning (*a learned work*). **3** studied or pursued by learned persons. **4** concerned with the interests of learned persons; scholarly (*a learned journal*). **5** *Brit.* as a courteous description of a lawyer in certain formal contexts (*my learned friend*). □ **learnedly** *adv.* **learnedness** *n.* [ME f. LEARN in the sense 'teach']

learner /ˈlɜːnə(r)/ *n.* **1** a person who is learning a subject or skill. **2** (in full **learner driver**) a person who is learning to drive a motor vehicle and has not yet passed a driving test.

learning /ˈlɜːnɪŋ/ *n.* knowledge acquired by study. [OE *leornung* (as LEARN)]

lease /liːs/ *n.* & *v.* ● *n.* an agreement by which the owner of a building or land allows another to use it for a specified time, usu. in return for payment. ● *v.tr.* grant or take on lease. □ **a new lease of** (*N. Amer.* **on**) **life** a substantially improved prospect of living, or of use after repair. □ **leasable** *adj.* **leaser** *n.* [ME f. AF *les*, OF *lais, leis* f. *lesser, laissier* leave f. L *laxare* make loose (*laxus*)]

leaseback /ˈliːsbæk/ *n.* the leasing of a property back to the vendor.

leasehold /ˈliːshəʊld/ *n.* & *adj.* ● *n.* **1** the holding of property by lease. **2** property held by lease. ● *adj.* held by lease. □ **leaseholder** *n.*

Lease-Lend /ˈliːslend/ see LEND-LEASE.

leash /liːʃ/ *n.* & *v.* ● *n.* **1** a dog's lead. **2** a restraint. ● *v.tr.* **1** put a leash on. **2** restrain. □ **straining at the leash** eager to begin. [ME f. OF *lesse, laisse* f. specific use of *laisser* let run on a slack lead: see LEASE]

least /liːst/ *adj., n.*, & *adv.* ● *adj.* **1** smallest, slightest, most insignificant. **2** (prec. by *the*; esp. with *neg.*) any at all (*it does not make the least difference*). **3** (of a species or variety) very small (*least tern*). ● *n.* the least amount. ● *adv.* in the least degree. □ **at least 1** at all events; anyway; even if there is doubt about a more extended statement. **2** (also **at the least**) not less than. **in the least** (or **the least**) (usu. with *neg.*) in the smallest degree; at all (*not in the least offended*). **least common denominator, multiple** see DENOMINATOR, MULTIPLE. **least squares** (in full **method** or **principle of least squares**) *Statistics* a method of estimating a quantity or fitting a graph to data, so as to minimize the sum of the squares of the differences between the observed values and the estimated values. **to say the least** (or **the least of it**) used to imply the moderation of a statement (*that is doubtful to say the least*). [OE *lǣst, lǣsest* f. Gmc]

leastways /ˈliːstweɪz/ *adv.* (also **leastwise** /-waɪz/) *dial.* or *colloq.* at least, or rather.

leat /liːt/ *n. Brit.* an open watercourse conducting water to a mill etc. [OE *-gelǣt* (as Y- + root of LET¹)]

leather /ˈleðə(r)/ *n.* & *v.* ● *n.* **1** a material made from the skin of an animal by tanning or a similar process. **b** (*attrib.*) made of leather. **2** a piece of leather for polishing with. **3** the leather part or parts of something. **4** *sl.* a cricket-ball or football. **5** (in *pl.*) leather clothes, esp. leggings, breeches, or clothes for wearing on a motorcycle. **6** a thong (*stirrup-leather*). ● *v.tr.* **1** cover with leather. **2** polish or wipe with a leather. **3** beat, thrash (orig. with a leather thong). □ **leather-neck** *Naut. sl.* a soldier or (esp. *US*) a marine (with reference to the leather stock formerly worn by them). [OE *lether* f. Gmc]

leatherback /ˈleðəbæk/ *n.* a large marine turtle, *Dermochelys coriacea*, having a thick leathery carapace.

leathercloth /ˈleðəklɒθ/ *n.* strong fabric coated to resemble leather.

leatherette /ˌleðəˈret/ *n.* imitation leather.

leatherjacket /ˈleðəˌdʒækɪt/ *n.* **1** *Brit.* a burrowing crane-fly larva with a tough skin. **2** a thick-skinned marine fish of the family Carangidae or Balistidae.

leathern /ˈleðən/ *n. archaic* made of leather.

leathery /ˈleðərɪ/ *adj.* **1** like leather. **2** (esp. of meat etc.) tough. □ **leatheriness** *n.*

leave¹ /liːv/ *v.* & *n.* ● *v.* (*past* and *past part.* **left** /left/) **1 a** *tr.* go away from; cease to remain in or on (*left him quite well an hour ago; leave the track; leave here*). **b** *intr.* (often foll. by *for*) depart (*we leave tomorrow; has just left for London*). **2** *tr.* cause to or let remain; depart without taking (*has left his gloves; left a slimy trail; left a bad impression; six from seven leaves one*). **3** *tr.* (also *absol.*) cease to reside at or attend or belong to or work for (*has left the school; I am leaving for another firm*). **4** *tr.* abandon, forsake, desert. **5** *tr.* have remaining after one's death (*leaves a wife and two children*). **6** *tr.* bequeath. **7** *tr.* (foll. by *to* + infin.) allow (a person or thing) to do something without interference or assistance (*leave the future to take care of itself*). **8** *tr.* (foll. by *to*) commit or refer to another person or agent (*leave that to me; nothing was left to chance*). **9** *tr.* **a** abstain from consuming or dealing with. **b** (in *passive*; often foll. by *over*) remain over. **10** *tr.* **a** deposit or entrust (a thing) to be attended to, collected, delivered, etc., in one's absence (*left a message with his secretary*). **b** depute (a person) to perform a function in one's absence. **11** *tr.* allow to remain or cause to be in a specified state or position (*left the door open; the performance left them unmoved; left nothing that was necessary undone*). **12** *tr.* pass (an object) so that it is in a specified relative direction (*leave the church on the left*). ● *n.* the position in which a player leaves the balls in billiards, croquet, etc. □ **be left with 1** retain (a feeling etc.). **2** be burdened with (a responsibility etc.). **be well left** be well provided for by a legacy etc. **get left** *colloq.* be deserted or worsted. **have left** have remaining (*has no friends left*). **leave alone 1** refrain from disturbing, not interfere with. **2** not have dealings with. **leave be** *colloq.* refrain from disturbing, not interfere with. **leave behind 1** go away without.

2 leave as a consequence or a visible sign of passage. **3** pass. **leave a person cold** (or **cool**) not impress or excite a person. **leave go** colloq. relax one's hold. **leave hold of** cease holding. **leave it at that** abstain from comment or further action. **leave much** (or **a lot** etc.) **to be desired** be highly unsatisfactory. **leave off 1** come to or make an end. **2** discontinue (leave off work; leave off talking). **3** not wear. **leave out** omit, not include. **leave over** Brit. leave to be considered, settled, or used later. **leave a person to himself** or **herself 1** not attempt to control a person. **2** leave a person solitary. **left at the post** beaten from the start of a race. **left for dead** abandoned as being beyond rescue. **left luggage** Brit. **1** luggage deposited for later retrieval. **2** (in full **left-luggage office**) a place, esp. at a railway station, where such luggage may be left for a small charge. □ **leaver** n. [OE læfan f. Gmc]

leave² /liːv/ n. **1** (often foll. by to + infin.) permission. **2 a** (in full **leave of absence**) permission to be absent from duty. **b** the period for which this lasts. □ **by** (or **with**) **your leave** often iron. an expression of apology for taking a liberty or making an unwelcome statement. **on leave** legitimately absent from duty. **take one's leave** bid farewell. **take one's leave of** bid farewell to. **take leave of one's senses** see SENSE. **take leave to** venture or presume to. [OE lēaf f. WG: cf. LIEF, LOVE]

leaved /liːvd/ adj. **1** having leaves. **2** (in comb.) having a leaf or leaves of a specified kind or number (four-leaved clover).

leaven /ˈlev(ə)n/ n. & v. ● n. **1** a substance added to dough to make it ferment and rise, esp. yeast, or fermenting dough reserved for the purpose. **2 a** a pervasive transforming influence (cf. Matt. 13:33). **b** (foll. by of) a tinge or admixture of a specified quality. ● v.tr. **1** ferment (dough) with leaven. **2 a** permeate and transform. **b** (foll. by with) modify with a tempering element. □ **the old leaven** traces of the unregenerate state (cf. 1 Cor. 5:6–8). [ME f. OF levain f. Gallo-Roman use of L levamen relief f. levare lift]

leaves pl. of LEAF.

leavings /ˈliːvɪŋz/ n.pl. things left over, esp. as worthless.

Leavis /ˈliːvɪs/, F(rank) R(aymond) (1895–1978), English literary critic. As a teacher of English at Cambridge from the 1920s and founder and editor of the quarterly Scrutiny (1932–53), he exerted a considerable influence on literary criticism. He regarded the rigorous critical study of English literature as central to preserving cultural continuity, which he considered to be under threat from technology and the mass media. He was a champion of D. H. Lawrence and led the way for a more serious appreciation of Charles Dickens. His books include The Great Tradition (1948) and The Common Pursuit (1952).

Lebanon /ˈlebənən/ a country in the Middle East with a coastline on the Mediterranean Sea; pop. (est. 1990) 2,700,000; official language, Arabic; capital, Beirut. The Lebanon region was the centre of the Phoenician culture (c.2700–450 BC). The mountains of Lebanon were a refuge for Christians from the 7th century, and were partly occupied by the Muslim Druze from the 11th century. The area was part of the Ottoman Empire from the early 16th century, becoming a French mandate after the First World War. Lebanon achieved independence in 1943 after the defeat of the Vichy French garrison by the Allies during the Second World War. Until the mid-1970s the country prospered, with Beirut the commercial capital of the Middle East. Friction between the previously dominant Christian community and the Muslims, the influx of Palestinian refugees, and repeated Middle Eastern wars, have chronically destabilized the country and led to the growth of armed militias and intermittent civil war. Syrian forces have been present in parts of the country since 1976, and an invasion by Israel in 1982 led to the evacuation of the Palestinians. Israeli forces later withdrew to a buffer zone on the southern border. The first general elections for twenty years were held in 1992. □ **Lebanese** /ˌlebəˈniːz/ adj. & n.

Lebanon Mountains a range of mountains in Lebanon. Running parallel to the Mediterranean coast, it rises to a height of 3,087 m (10,022 ft) at Qornet es Saouda. It is separated from the Anti-Lebanon range, on the border with Syria, by the Bekaa valley.

Lebensraum /ˈleɪbənzˌraʊm/ n. the territory which a state or nation believes is needed for its natural development. The term was used in the 1930s by the German Nazis in attempted justification of their claim to large areas of eastern Europe, which they said were needed to relieve pressure on a supposedly overpopulated Germany. [G, = living-space]

Leblanc /ləˈblɒŋk/, Nicolas (1742–1806), French surgeon and chemist. Leblanc became interested in the large-scale manufacture of soda because of the offer of a prize. He developed a process for making soda ash (sodium carbonate) from common salt, making possible the large-scale manufacture of glass, soap, paper, and other chemicals. The factory he set up with others was confiscated during the French Revolution, and he later committed suicide.

Lebowa /ləˈbəʊə/ a former homeland established in South Africa for the North Sotho people, now part of the province of Northern Transvaal. (See also HOMELAND.)

Lebrun /ləˈbrɜːn/, Charles (1619–90), French painter, designer, and decorator. He was prominent in the development and institutionalization of French art. In 1648 he helped to found the Royal Academy of Painting and Sculpture, becoming its director in 1663; from this position he sought to impose orthodoxy in artistic matters, laying the basis of academicism. His work for Louis XIV at Versailles (1661–83), which included painting, furniture, and tapestry design, established him as a leading exponent of 17th-century French classicism.

Le Carré /lə ˈkæreɪ/, John (pseudonym of David John Moore Cornwell) (b.1931), English novelist. His spy novels are characterized by their unromanticized view of espionage and frequently explore the moral dilemmas inherent in such work; they often feature the British agent George Smiley and include The Spy Who Came in from the Cold (1963) and Tinker, Tailor, Soldier, Spy (1974).

lech /letʃ/ v. & n. colloq. ● v.intr. feel lecherous; behave lustfully. ● n. **1** a strong desire, esp. sexual. **2** a lecher. [back-form. f. LECHER: (n.) perh. f. letch longing]

lecher /ˈletʃə(r)/ n. a lecherous man; a debauchee. [ME f. OF lecheor etc. f. lechier live in debauchery or gluttony f. Frank., rel. to LICK]

lecherous /ˈletʃərəs/ adj. lustful, having strong or excessive sexual desire. □ **lecherously** adv. **lecherousness** n. [ME f. OF lecheros etc. f. lecheur LECHER]

lechery /ˈletʃərɪ/ n. unrestrained indulgence of sexual desire. [ME f. OF lecherie f. lecheur LECHER]

lecithin /ˈlesɪθɪn/ n. Biochem. **1** any of a group of phospholipids found naturally in animals, egg-yolk, and some higher plants. **2** a preparation of this used to emulsify foods etc. [Gk lekithos egg-yolk + -IN]

Leclanché cell /ləˈklɒnʃeɪ/ n. a primary electrochemical cell having a zinc cathode in contact with zinc chloride, ammonium chloride (as a solution or a paste) as the electrolyte, and a carbon anode in contact with a mixture of manganese dioxide and carbon powder. [Georges Leclanché, Fr. chemist (1839–82)]

Leconte de Lisle /ləˌkɒnt də ˈliːl/ Charles Marie René (1818–94), French poet and leader of the Parnassians. He published a number of collections of poetry, including Poèmes antiques (1852) and Poèmes barbares (1862); his work often draws inspiration from mythology, biblical history, and exotic Eastern landscape.

Le Corbusier /ˌlə kɔːˈbjuːzɪˌeɪ/ (born Charles Édouard Jeanneret) (1887–1965), French architect and town planner, born in Switzerland. He was a pioneer of the international style and was influential both as an architect and as a theorist. In 1918 he set out his aesthetic manifesto with the launch of Purism; he later developed his theories on functionalism, the use of new materials and industrial techniques in architecture, and the Modulor, a modular system of standard-sized units, in books such as Towards a New Architecture (1923) and Le Modulor I (1948). His buildings include the block of flats in Marseilles known as the unité d'habitation ('living unit', 1945–50); he also planned the city of Chandigarh in India.

lectern /ˈlektɜːn/ n. **1** a stand for holding a book in a church or chapel, esp. for a bible from which lessons are to be read. **2** a similar stand for a lecturer etc. [ME lettorne f. OF let(t)run, med.L lectrum f. legere lect- read]

lectin /ˈlektɪn/ n. Biochem. any of a class of proteins, usu. of plant origin, that cause the agglutination of particular cell types. [L lect- legere choose, select + -IN]

lection /ˈlekʃ(ə)n/ n. a reading of a text found in a particular copy or edition. [L lectio reading (as LECTION)]

lectionary /ˈlekʃənərɪ/ n. (pl. **-ies**) **1** a list of portions of Scripture appointed to be read at divine service. **2** a book containing such portions. [ME f. med.L lectionarium (as LECTION)]

lector /ˈlektɔː(r)/ n. **1** a reader, esp. of lessons in a church service. **2** (fem. **lectrice** /lekˈtriːs/) a lecturer or reader, esp. one employed in a foreign university to give instruction in his or her native language. [L f. legere lect- read]

lecture /ˈlektʃə(r)/ n. & v. ● n. **1** a discourse giving information about a subject to a class or other audience. **2** a long serious speech esp. as a scolding or reprimand. ● v. **1** intr. (often foll. by on) deliver a lecture or lectures. **2** *tr.* talk seriously or reprovingly to (a person). **3** *tr.* instruct or entertain (a class or other audience) by a lecture. [ME f. OF *lecture* or med.L *lectura* f. L (as LECTOR)]

lecturer /ˈlektʃərə(r)/ *n.* a person who lectures, esp. as a teacher in higher education.

lectureship /ˈlektʃəˌʃɪp/ *n.* the office of lecturer. ¶ The form *lecturership*, which is strictly more regular, is in official use at Oxford University and elsewhere, but is not widely current.

lecythus /ˈlesɪθəs/ *n.* (*pl.* **lecythi** /-ˌθaɪ/) *Gk Antiq.* a thin narrow-necked vase or flask. [Gk *lēkuthos*]

LED *abbr.* light-emitting diode, a semiconductor diode which glows when a voltage is applied (often *attrib.*: *LED display*).

led *past* and *past part.* of LEAD¹.

Leda /ˈliːdə/ *Gk Mythol.* the wife of Tyndareus king of Sparta. She was loved by Zeus, who visited her in the form of a swan; among her children were the Dioscuri, Helen, and Clytemnestra.

lederhosen /ˈleɪdəˌhəʊz(ə)n/ *n.pl.* leather shorts as worn by men in Bavaria etc. [G, = leather trousers]

ledge /ledʒ/ *n.* **1** a narrow horizontal surface projecting from a wall etc. **2** a shelflike projection on the side of a rock or mountain. **3** a ridge of rocks, esp. below water. **4** *Mining* a stratum of metal-bearing rock. □ **ledged** *adj.* **ledgy** *adj.* [perh. f. ME *legge* LAY¹]

ledger /ˈledʒə(r)/ *n.* **1** a tall narrow book in which a firm's accounts are kept, esp. one which is the principal book of a set and contains debtor-and-creditor accounts. **2** a flat gravestone. **3** a horizontal timber in scaffolding, parallel to the face of the building. □ **ledger line** *Mus.* = LEGER LINE. **ledger-tackle** a kind of fishing tackle in which a lead weight keeps the bait on the bottom. [ME f. senses of Du. *ligger* and *legger* (f. *liggen* LIE¹, *leggen* LAY¹) & pronunc. of ME *ligge*, *legge*]

Led Zeppelin /led ˈzepəlɪn/ a British rock group formed in 1969, featuring vocalist Robert Plant (b. 1948) and guitarist Jimmy Page (b. 1944). Remembered as being one of the first groups to play heavy rock, Led Zeppelin were at first strongly influenced by blues but later broadened their style to take in elements of folk, funk, and Eastern music. Perhaps their best-known song is 'Stairway to Heaven' (1971).

Lee¹ /liː/, Bruce (born Lee Yuen Kam) (1941–73), American actor. He was an expert in kung fu and starred in a number of martial arts films featuring elaborately staged fight scenes, such as *Fists of Fury* (1972) and *Enter the Dragon* (1973).

Lee² /liː/, Christopher (Frank Carandini) (b.1922), English actor. Since the late 1940s he has made a variety of films, including thrillers and adventure films, but his reputation is chiefly based on the horror films that he made for the British film company Hammer. These include *Dracula* (1958) and seven sequels (the last in 1973), and *The Mummy* (1959).

Lee³ /liː/, Gypsy Rose (born Rose Louise Hovick) (1914–70), American striptease artist. In the 1930s she became famous on Broadway for her sophisticated striptease act, imbuing what was previously considered vulgar entertainment with a new artistic content and style.

Lee⁴ /liː/, Robert E(dward) (1807–70), American general. He was the commander of the Confederate army of Northern Virginia, leading it for most of the American Civil War. Although he did much to prolong Confederate resistance against the Union's greater manpower and resources, his invasion of the North was repulsed by General Meade at the Battle of Gettysburg (1863) and he eventually surrendered to General Grant in 1865.

lee /liː/ *n.* **1** shelter given by a neighbouring object (*under the lee of*). **2** (in full **lee side**) the sheltered side, the side away from the wind (opp. *weather side*). □ **lee-board** *Naut.* a plank frame fixed to the side of a flat-bottomed vessel and let down into the water to diminish leeway. **lee shore** the shore to leeward of a ship. [OE *hlēo* f. Gmc]

leech¹ /liːtʃ/ *n.* **1** an aquatic or terrestrial annelid worm of the class Hirudinea, having suckers at both ends; esp. *Hirudo medicinalis*, a bloodsucking parasite of vertebrates formerly much used medically. **2** a person who extorts profit from or sponges on others. □ **like a leech** persistently or clingingly present. [OE *lǣce*, assim. to LEECH²]

leech² /liːtʃ/ *n.* archaic or joc. a physician; a healer. [OE *lǣce* f. Gmc]

leech³ /liːtʃ/ *n.* *Naut.* **1** a perpendicular or sloping side of a square sail. **2** the side of a fore-and-aft sail away from the mast or stay. [ME, perh. rel. to ON *lik*, a nautical term of uncert. meaning]

leechcraft /ˈliːtʃkrɑːft/ *n.* archaic the art of healing. [OE *lǣcecræft* (as LEECH², CRAFT)]

Leeds /liːdz/ an industrial city in West Yorkshire; pop. (1991) 674,400.

It developed as a wool town in the Middle Ages, becoming a centre of the clothing trade in the Industrial Revolution.

leek /liːk/ *n.* **1** an allium, *Allium porrum*, with flat overlapping leaves forming an elongated cylindrical bulb, used as food. **2** this as a Welsh national emblem. [OE *lēac* f. Gmc]

leer¹ /lɪə(r)/ *v. & n.* ● *v.intr.* look slyly or lasciviously or maliciously. ● *n.* a leering look. □ **leeringly** *adv.* [perh. f. obs. *leer* cheek f. OE *hlēor*, as though 'to glance over one's cheek']

leer² var. of LEHR.

leery /ˈlɪərɪ/ *adj.* (**leerier**, **leeriest**) *sl.* **1** knowing, sly. **2** (foll. by *of*) wary. □ **leeriness** *n.* [perh. f. obs. *leer* looking askance f. LEER¹ + -Y¹]

lees /liːz/ *n.pl.* **1** the sediment of wine etc. (*drink to the lees*). **2** dregs, refuse. [pl. of ME *lie* f. OF *lie* f. med.L *lia* f. Gaulish]

leet¹ /liːt/ *n.* *Brit. hist.* **1** (in full **Court leet**) a yearly or half-yearly court of record that lords of certain manors might hold. **2** its jurisdiction or district. [ME f. AF *lete* (= AL *leta*), of unkn. orig.]

leet² /liːt/ *n.* *Sc.* a selected list of candidates for some office. □ **short leet** = *short list*. [ME *lite* etc., prob. f. AF & OF *lit(t)e*, var. of *liste* LIST¹]

Leeuwenhoek /ˈleɪv(ə)nˌhuːk/, Antoni van (1632–1723), Dutch naturalist. Apprenticed to a Delft cloth-merchant, he developed a lens for scientific purposes from those used to inspect cloth, and was the first to observe bacteria, protozoa, and yeast. He accurately described red blood cells, capillaries, striated muscle fibres, spermatozoa, and the crystalline lens of the eye. Being without Latin he was out of touch with the scientific community, and his original work on micro-organisms only became known through the Royal Society's translation and publication of his letters (1673–1723).

leeward /ˈliːwəd, *Naut.* ˈluːəd/ *adj., adv., & n.* ● *adj. & adv.* on or towards the side sheltered from the wind (opp. WINDWARD). ● *n.* the leeward region, side, or direction (*to leeward; on the leeward of*).

Leeward Islands /ˈliːwəd/ a group of islands in the West Indies, constituting the northern part of the Lesser Antilles. The group includes Guadeloupe, Antigua, St Kitts, and Montserrat. The name refers to their position further downwind, in terms of the prevailing south-easterly winds, than the Windward Islands.

leewardly /ˈliːwədlɪ, *Naut.* ˈluːədlɪ/ *adj.* (of a ship) apt to drift to leeward.

leeway /ˈliːweɪ/ *n.* **1** the sideways drift of a ship to leeward of the desired course. **2 a** allowable deviation or freedom of action. **b** margin of safety. □ **make up leeway** struggle out of a bad position, recover lost time, etc.

Le Fanu /ˈlefəˌnjuː, lə ˈfɑːnuː/, Joseph Sheridan (1814–73), Irish novelist. He is best known for his stories of mystery, suspense, and the supernatural; notable works include *The House by the Churchyard* (1861), *Uncle Silas* (1864), and the collection of ghost stories *In a Glass Darkly* (1872).

left¹ /left/ *adj., adv., & n.* (opp. RIGHT). ● *adj.* **1** on or towards the side of the human body which is to the west when facing north, and in a majority of persons has the less-used hand. **2** on or towards the part of an object which is analogous to a person's left side or (with opposite sense) which is nearer to an observer's left hand. **3** (also **Left**) *Polit.* of the Left. ● *adv.* on to the left side. ● *n.* **1** the left-hand part or region or direction. **2** *Boxing* **a** the left hand. **b** a blow with this. **3 a** (often **Left**) *Polit.* a group or section favouring radical socialism; such radicals collectively. (See also LEFT WING.) **b** the more advanced or innovative section of any group. **4** the side of a stage which is to the left of a person facing the audience. **5** (esp. in marching) the left foot. **6** the left wing of an army. □ **have two left feet** be clumsy. **left and right** = *right and left*. **left bank** the bank of a river on the left facing downstream. **left bower** see BOWER³. **left field** *Baseball* the part of the outfield to the left of the batter as he or she faces the pitcher. **left hand 1** the hand of the left side. **2** (usu. prec. by *at, on, to*) the region or direction on the left side of a person. **left-hand** *adj.* **1** on or towards the left side of a person or thing (*left-hand drive*). **2** done with the left hand (*left-hand blow*). **3 a** (of rope) twisted counter-clockwise. **b** (of a screw) = LEFT-HANDED 4c. **left turn** a turn that brings one's front to face as one's left side did before. **marry with the left hand** marry morganatically (see LEFT-HANDED 7). □ **leftish** *adj.* [ME *lüft, lift, left*, f. OE, orig. sense 'weak, worthless']

left² *past* and *past part.* of LEAVE¹.

Left Bank a district of the city of Paris, situated on the left bank of the River Seine, to the south of the river. The site of several educational and arts institutions, including the Sorbonne and the Académie des Beaux-Arts, it is an area noted for its intellectual and artistic life.

left-handed /left ˈhændɪd/ *adj.* **1** using the left hand by preference as more serviceable than the right. **2** (of a tool etc.) made to be used with the left hand. **3** (of a blow etc.) done or made with the left hand. **4 a** turning to the left; towards the left. **b** (of a racecourse) turning anticlockwise. **c** (of a screw) advanced by turning to the left (anticlockwise). **5** awkward, clumsy. **6 a** (of a compliment) ambiguous. **b** of doubtful sincerity or validity. **7** (of a marriage) morganatic (from a German custom by which the bridegroom gave the bride his left hand in such marriages). □ **left-handedly** adv. **left-handedness** n.

left-hander /left'hændə(r)/ n. **1** a left-handed person. **2** a left-handed blow.

leftie var. of LEFTY.

leftism /'leftɪz(ə)m/ n. Polit. the principles or policy of the left. □ **leftist** n. & adj.

leftmost /'leftməust/ adj. furthest to the left.

leftover /'left,əuvə(r)/ adj. & n. ● adj. remaining over; not used up or disposed of. ● n. (usu. in pl.) an item (esp. of food) remaining after the rest has been used.

leftward /'leftwəd/ adv. & adj. ● adv. (also **leftwards** /-wədz/) towards the left. ● adj. going towards or facing the left.

left wing n. & adj. ● n. **1** the radical, progressive, or socialist section of society, a political party, etc. The term originated with the National Assembly in France (1789–91), where the nobles sat on the president's right hand and the commons on the left. **2** the left side of a football etc. team on the field. **3** the left side of an army. ● adj. (also **left-wing**) **1** of or relating to the left wing. **2** Polit. socialist, radical, politically progressive. □ **left-winger** n.

lefty /'leftɪ/ n. (also **leftie**) (pl. **-ies**) colloq. **1** Polit. a left-winger. **2** a left-handed person.

leg /leg/ n. & v. ● n. **1 a** each of the limbs on which a person or animal walks and stands. **b** the part of this from the hip to the ankle. **2** a leg of an animal or bird as food. **3** an artificial leg (wooden leg). **4** a part of a garment covering a leg or part of a leg. **5 a** a support of a chair, table, bed, etc. **b** a long thin support or prop, esp. a pole. **6** Cricket the half of the field (as divided lengthways through the pitch) in which the striker's feet are placed (opp. OFF n. 1). **7 a** a section of a journey. **b** a section of a relay race. **c** a stage in a competition. **d** one of two or more games constituting a round. **e** a single game in darts. **8** one branch of a forked object. **9** Naut. a run made on a single tack. **10** archaic an obeisance made by drawing back one leg and bending it while keeping the front leg straight. ● v.tr. (**legged**, **legging**) propel (a boat) through a canal tunnel by pushing with one's legs against the tunnel sides. □ **feel** (or **find**) **one's legs** become able to stand or walk. **get one's leg over** coarse sl. (of a man) engage in sexual intercourse. **give a person a leg up** help a person to mount a horse etc. or get over an obstacle or difficulty. **have the legs of** be able to go further than. **have no legs** colloq. (of a golf ball etc.) have not enough momentum to reach the desired point. **keep one's legs** not fall. **leg before wicket** (abbr. **l.b.w.**) Cricket adj. & adv. (of a batsman) out because of illegally obstructing the ball with a part of the body. ● n. such a dismissal. **leg break** Cricket **1** a ball which deviates from the leg side after bouncing. **2** such deviation. **leg-bye** see BYE[1]. **leg-cutter** Cricket a fast leg break. **leg-iron** a shackle or fetter for the leg. **leg it** colloq. walk or run fast. **leg-of-mutton sail** a triangular mainsail. **leg-of-mutton sleeve** a sleeve which is full and loose on the upper arm but close-fitting on the forearm. **leg-pull** colloq. a hoax. **leg-rest** a support for a seated invalid's leg. **leg-room** space for the legs of a seated person. **leg-show** colloq. a theatrical performance by scantily dressed women. **leg slip** Cricket a fielding position close behind the batsman on the leg side. **leg spin** Cricket a type of spin which causes the ball to deviate from the leg side after bouncing. **leg stump** Cricket the stump on the leg side. **leg theory** Cricket bowling to leg with fielders massed on that side. **leg trap** Cricket a group of fielders near the wicket on the leg side. **leg warmer** either of a pair of tubular knitted garments covering the leg from ankle to thigh. **not have a leg to stand on** be unable to support one's argument by facts or sound reasons. **on one's last legs** near death or the end of one's usefulness etc. **on one's legs** Brit. **1** (also **on one's hind legs**) standing esp. to make a speech. **2** well enough to walk about. **take to one's legs** run away. □ **legged** /legd, 'legɪd/ adj. (also in comb.). [ME f. ON leggr f. Gmc]

legacy /'legəsɪ/ n. (pl. **-ies**) **1** a gift left in a will. **2** something handed down by a predecessor (legacy of corruption). □ **legacy-hunter** a person who is attentive to esp. old and rich people in the hope of obtaining a legacy. [ME f. OF legacie legateship f. med.L legatia f. L legare bequeath]

legal /'li:g(ə)l/ adj. **1** of or based on law; concerned with law; falling within the province of law. **2** appointed or required by law. **3** permitted

by law, lawful. **4** recognized by law, as distinct from equity. **5** (in Christian theology) **a** of the Mosaic Law. **b** of salvation by works rather than by faith. □ **legal aid** payment from public funds allowed, in cases of need, to help pay for legal advice or proceedings. **legal fiction** an assertion accepted as true (though probably fictitious) to achieve a useful purpose, esp. in legal matters. **legal holiday** US a public holiday established by law. **legal proceedings** see PROCEEDING 2. **legal separation** see SEPARATION 2. **legal tender** currency that cannot legally be refused in payment of a debt (usu. up to a limited amount for coins not made of gold). □ **legally** adv. [F légal or L legalis f. lex legis law: cf. LEAL, LOYAL]

legalese /,li:gə'li:z/ n. colloq. the technical language of legal documents.

legalism /'li:gə,lɪz(ə)m/ n. **1** excessive adherence to law or formula. **2** (in Christian theology) **a** adherence to the Mosaic Law rather than to the Gospel. **b** the doctrine of justification by works. □ **legalist** n. **legalistic** /,li:gə'lɪstɪk/ adj. **legalistically** adv.

legality /lɪ'gælɪtɪ/ n. (pl. **-ies**) **1** lawfulness. **2** legalism. **3** (in pl.) obligations imposed by law. [F légalité or med.L legalitas (as LEGAL)]

legalize /'li:gə,laɪz/ v.tr. (also **-ise**) **1** make lawful. **2** bring into harmony with the law. □ **legalization** /,li:gəlaɪ'zeɪʃ(ə)n/ n.

legate /'legət/ n. **1** a member of the clergy representing the pope. **2** Rom. Hist. **a** a deputy of a general. **b** a governor or deputy governor of a province. **3** archaic an ambassador or delegate. □ **legate a latere** /ɑː 'lætə,reɪ/ a papal legate of the highest class, with full powers. □ **legateship** n. **legatine** /-,taɪn/ adj. [OE f. OF legat f. L legatus past part. of legare depute, delegate]

legatee /,legə'ti:/ n. the recipient of a legacy. [as LEGATOR + -EE]

legation /lɪ'geɪʃ(ə)n/ n. **1** a body of deputies. **2 a** the office and staff of a diplomatic minister (esp. when not having ambassadorial rank). **b** the official residence of a diplomatic minister. **3** a legateship. **4** the sending of a legate or deputy. [ME f. OF legation or L legatio (as LEGATE)]

legato /lɪ'gɑːtəu/ adv., adj., & n. Mus. ● adv. & adj. in a smooth flowing manner, without breaks between notes (cf. STACCATO, TENUTO). ● n. (pl. **-os**) **1** a legato passage. **2** legato playing. [It., = bound, past part. of legare f. L ligare bind]

legator /,legə'tɔː(r)/ n. the giver of a legacy. [archaic legate bequeath f. L legare (as LEGACY)]

legend /'ledʒənd/ n. **1 a** a traditional story sometimes popularly regarded as historical but unauthenticated; a myth. **b** such stories collectively. **c** a popular but unfounded belief. **d** colloq. a person about whom unauthenticated tales are told; a famous or notorious person (became a legend in his own lifetime). **2 a** an inscription, esp. on a coin or medal. **b** Printing a caption. **c** wording on a map etc. explaining the symbols used. **3** hist. a the story of a saint's life. **b** a collection of lives of saints or similar stories. □ **legendry** n. [ME (in sense 3) f. OF legende f. med.L legenda what is to be read, neut. pl. gerundive of L legere read]

legendary /'ledʒəndərɪ/ adj. **1** of or connected with legends. **2** described in or based on a legend. **3** colloq. remarkable enough to be a subject of legend. □ **legendarily** adv. [med.L legendarius (as LEGEND)]

legerdemain /,ledʒədə'meɪn/ n. **1** sleight of hand; conjuring or juggling. **2** trickery, sophistry. [ME f. F léger de main light of hand, dexterous]

leger line /'ledʒə(r)/ n. Mus. a short line added for notes above or below the range of a staff. [var. of LEDGER]

legging /'legɪŋ/ n. **1** (in pl.) a close-fitting stretch garment, worn esp. by women and children, covering the legs and the lower part of the torso. **2** (usu. in pl.) a stout protective outer covering for the leg from the knee to the ankle.

leggy /'legɪ/ adj. (**leggier**, **leggiest**) **1 a** long-legged. **b** (of a woman) having attractively long legs. **2** long-stemmed. □ **legginess** n.

Leghari /leg'hɑːrɪ/, Farooq Ahmed (b.1940), Pakistani statesman, President since 1993.

Leghorn see LIVORNO.

leghorn n. **1** /'leghɔːn, 'legɔːn/ **a** fine plaited straw. **b** a hat of this. **2** (**Leghorn**) /lɪ'gɔːn/ a small hardy breed of domestic fowl. [Leghorn LIVORNO, from where the straw and fowls were imported]

legible /'ledʒɪb(ə)l/ adj. (of handwriting, print, etc.) clear enough to read; readable. □ **legibly** adv. **legibility** /,ledʒɪ'bɪlɪtɪ/ n. [ME f. LL legibilis f. legere read]

legion /'li:dʒən/ n. & adj. ● n. **1** a division of 3,000–6,000 men, including a complement of cavalry, in the ancient Roman army. **2** a large organized body. **3** a vast host, multitude, or number. ● predic.adj. great in number (*his good works have been legion*). [ME f. OF f. L *legio -onis* f. *legere* choose]

legionary /'li:dʒənərɪ/ adj. & n. ● adj. of a legion or legions. ● n. (pl. **-ies**) a member of a legion. [L *legionarius* (as LEGION)]

legioned /'li:dʒənd/ adj. poet. arrayed in legions.

legionella /ˌli:dʒə'nelə/ n. the bacterium *Legionella pneumophila*, which causes legionnaires' disease.

legionnaire /ˌli:dʒə'neə(r)/ n. **1** a member of a foreign legion. **2** a member of the American Legion or the Royal British Legion. □ **legionnaires' disease** *Med.* a form of bacterial pneumonia disseminated in water droplets from air-conditioning systems etc. and first identified after an outbreak at an American Legion meeting in 1976 (cf. LEGIONELLA). [F *légionnaire* (as LEGION)]

Legion of Honour n. a French order of distinction founded by Napoleon in 1802, conferred as a reward for civil or military services. [French *Légion d'honneur*]

legislate /'ledʒɪsˌleɪt/ v.intr. **1** make laws. **2** (foll. by *for*) make provision by law. [back-form. f. LEGISLATION]

legislation /ˌledʒɪs'leɪʃ(ə)n/ n. **1** the process of making laws. **2** laws collectively. [LL *legis latio* f. *lex legis* law + *latio* proposing f. *lat-* past part. stem of *ferre* bring]

legislative /'ledʒɪslətɪv/ adj. of or empowered to make legislation. □ **legislatively** adv.

legislator /'ledʒɪsˌleɪtə(r)/ n. **1** a member of a legislative body. **2** a lawgiver. [L (as LEGISLATION)]

legislature /'ledʒɪsˌleɪtʃə(r), -lətʃə(r)/ n. the legislative body of a state.

legit /lɪ'dʒɪt/ adj. & n. colloq. ● adj. legitimate. ● n. **1** legitimate drama. **2** an actor in legitimate drama. [abbr.]

legitimate adj. & v. ● adj. /lɪ'dʒɪtɪmət/ **1 a** (of a child) born of parents lawfully married to each other. **b** (of a parent, birth, descent, etc.) with, of, through, etc., a legitimate child. **2** lawful, proper, regular, conforming to the standard type. **3** logically admissible. **4 a** (of a sovereign's title) based on strict hereditary right. **b** (of a sovereign) having a legitimate title. **5** constituting or relating to serious drama as distinct from musical comedy, revue, etc. (*See note below.*) ● v.tr. /lɪ'dʒɪtɪˌmeɪt/ **1** make legitimate by decree, enactment, or proof. **2** justify, serve as a justification for. □ **legitimacy** /-məsɪ/ n. **legitimately** /-mətlɪ/ adv. **legitimation** /-ˌdʒɪtɪ'meɪʃ(ə)n/ n. [med.L *legitimare* f. L *legitimus* lawful f. *lex legis* law]

▪ The term was first applied to theatre in the 18th century during the struggle of the patent theatres, Covent Garden and Drury Lane, against the new theatres which were springing up all over London. It covered plays dependent entirely on acting, with little or no singing, dancing, or spectacle, which the new theatres included in order to circumvent legislation.

legitimatize /lɪ'dʒɪtɪməˌtaɪz/ v.tr. (also **-ise**) legitimize. □ **legitimatization** /-ˌdʒɪtɪmətaɪ'zeɪʃ(ə)n/ n.

legitimism /lɪ'dʒɪtɪˌmɪz(ə)m/ n. adherence to a sovereign or pretender whose claim is based on direct descent (esp. in French and Spanish history). □ **legitimist** n. & adj. [F *légitimisme* f. *légitime* LEGITIMATE]

legitimize /lɪ'dʒɪtɪˌmaɪz/ v.tr. (also **-ise**) **1** make legitimate. **2** serve as a justification for. □ **legitimization** /-ˌdʒɪtɪmaɪ'zeɪʃ(ə)n/ n.

legless /'leglɪs/ adj. **1** having no legs. **2** sl. drunk, esp. too drunk to stand.

legman /'legmæn/ n. (pl. **-men**) a person employed to go about gathering news or running errands etc.

Lego /'legəʊ/ n. propr. a construction toy consisting of interlocking plastic building blocks. [Danish *leg godt* play well f. *lege* to play]

legume /'legju:m/ n. **1** the seed pod of a leguminous plant. **2** any seed, pod, or other edible part of a leguminous plant used as food. [F *légume* f. L *legumen -minis* f. *legere* pick, because pickable by hand]

leguminous /lɪ'gju:mɪnəs/ adj. Bot. of, relating to, or like the family Leguminosae, which includes peas and beans. Plants of this family have seeds in pods, and often root nodules containing symbiotic bacteria able to fix nitrogen. [mod.L *leguminosus* (as LEGUME)]

legwork /'legwɜ:k/ n. work which involves a lot of walking, travelling, or physical activity.

Leh /leɪ/ a town in Jammu and Kashmir, northern India, to the east of Srinagar near the Indus river; pop. (est. 1991) 9,000. It is the chief town of the Himalayan region of Ladakh, and the administrative centre of Ladakh district in Jammu and Kashmir.

Lehár /'leɪhɑ:(r)/, Franz (Ferencz) (1870–1948), Hungarian composer.

He is chiefly known for his operettas, of which the most famous is *The Merry Widow* (1905).

Le Havre /lə 'hɑ:vrə/ a port in northern France, on the English Channel at the mouth of the Seine; pop. (1990) 197,220.

lehr /lɪə(r)/ n. (also **leer**) a furnace used for the annealing of glass. [17th c.: orig. unkn.]

lei[1] /'leɪi:, leɪ/ n. a Polynesian garland of flowers. [Hawaiian]

lei[2] pl. of LEU.

Leibniz /'laɪbnɪts/, Gottfried Wilhelm (1646–1716), German rationalist philosopher, mathematician, and logician. He spent his life in the diplomatic and political service and in 1700 was appointed first president of the Academy of Sciences in Berlin. Leibniz is chiefly known as an exponent of optimism; he believed that the world is fundamentally harmonious and good, being composed of single units (monads), each of which is self-contained but acts in harmony with every other; these form an ascending hierarchy culminating in God. Their harmony is ordained by God, who never acts except for a reason that requires it, and so this world is the best of all possible worlds (a view satirized in Voltaire's *Candide*). Leibniz made the important distinction between necessary (logical) truths and contingent (factual) truths, and proposed a universal logical language which would eliminate ambiguity. He also devised a method of calculus independently of Newton.

Leicester[1] /'lestə(r)/ a city in central England, on the River Soar, the county town of Leicestershire; pop. (1991) 270,600. It was founded as a Roman settlement where Fosse Way crosses the Soar (AD 50–100).

Leicester[2] /'lestə(r)/, Earl of, see DUDLEY.

Leicester[3] /'lestə(r)/ n. (often **Red Leicester**) a kind of mild firm cheese, usu. orange-coloured and orig. made in Leicestershire.

Leicestershire /'lestəˌʃɪə(r)/ a county of central England; county town, Leicester.

Leichhardt /'laɪkhɑ:t/, (Friedrich Wilhelm) Ludwig (1813–48), Australian explorer, born in Prussia. After emigrating to Australia in 1841, he began a series of geological surveys, crossing from Moreton Bay near Brisbane to Port Essington on the coast of Arnhem Land (1843–5); he disappeared without trace during another attempt at a transcontinental crossing in 1848.

Leics. abbr. Leicestershire.

Leiden /'laɪd(ə)n/ (also **Leyden**) a city in the west Netherlands, 15 km (9 miles) north-east of The Hague; pop. (1991) 111,950. It is the site of the country's oldest university, founded in 1575.

Leif Ericsson see ERICSSON[2].

Leigh /li:/, Vivien (born Vivian Mary Hartley) (1913–67), British actress, born in India. She made her screen début in 1934; major film roles include her Oscar-winning performances as Scarlett O'Hara in *Gone with the Wind* (1939) and Blanche du Bois in *A Streetcar Named Desire* (1951). From 1935 Leigh also pursued a successful career on stage, often playing opposite Laurence Olivier, to whom she was married from 1940 to 1961.

Leighton /'leɪt(ə)n/, Frederic, 1st Baron Leighton of Stretton (1830–96), English painter and sculptor. He first gained renown with the painting *Cimabue's Madonna Carried in Procession through the Streets of Florence* (1855). He became a leading exponent of Victorian neoclassicism and chiefly painted large-scale mythological and genre scenes. His sculptures include *Athlete Struggling with a Python* (1874–7).

Leinster /'lenstə(r)/ a province of the Republic of Ireland, in the south-east of the country, centred on Dublin.

Leipzig /'laɪpsɪg/ an industrial city in east central Germany; pop. (1991) 503,190. It is a centre of publishing and music. An annual trade fair has been held there since the 12th century.

leishmaniasis /ˌli:ʃmə'naɪəsɪs/ n. Med. a disease caused by parasitic protozoans of the genus *Leishmania*, transmitted by the bite of sandflies, occurring in several forms that affect the skin or various internal organs. [Sir William Boog *Leishman*, Brit. physician (1865–1926)]

leister /'li:stə(r)/ n. & v. ● n. a pronged salmon-spear. ● v.tr. pierce with a leister. [ON *ljóstr* f. *ljósta* to strike]

leisure /'leʒə(r)/ n. **1** free time; time at one's own disposal. **2** enjoyment of free time. **3** (usu. foll. by *for*, or *to* + infin.) opportunity afforded by free time. □ **at leisure 1** not occupied. **2** in an unhurried manner. **at one's leisure** when one has time. **leisure centre** a large public building with sports facilities, bars, etc. □ **leisureless** adj. [ME f. AF *leisour*, OF *leisir* ult. f. L *licere* be allowed]

leisured /ˈleʒəd/ adj. having ample leisure.

leisurely /ˈleʒəlɪ/ adj. & adv. ● adj. having leisure; acting or done at leisure; unhurried, relaxed. ● adv. without hurry. □ **leisureliness** n.

leisurewear /ˈleʒəˌweə(r)/ n. informal clothes, especially tracksuits and other sportswear.

leitmotif /ˈlaɪtməʊˌtiːf/ n. (also **leitmotiv**) a recurrent theme associated throughout a musical, literary, etc. composition with a particular person, idea, or situation. [G *Leitmotiv* (as LEAD¹, MOTIVE)]

Leitrim /ˈliːtrɪm/ a county of the Republic of Ireland, in the province of Connacht; county town, Carrick-on-Shannon.

Leix see LAOIS.

lek¹ /lek/ n. the chief monetary unit of Albania. [Albanian]

lek² /lek/ n. a patch of ground used by groups of certain birds, esp. black grouse, during the breeding season as a setting for the males' display and their meeting with the females. [perh. f. Sw. *leka* to play]

Lely /ˈliːlɪ/, Sir Peter (Dutch name Pieter van der Faes) (1618–80), Dutch portrait painter, resident in England from 1641. He became principal court painter to Charles II and consolidated the tradition of society portrait painting. By 1650 he had a large studio and produced hundreds of portraits of court figures in a baroque style. Notable works include his series of *Windsor Beauties*, painted during the 1660s.

LEM abbr. lunar excursion module.

leman /ˈlemən/ n. (pl. **lemans**) archaic **1** a lover or sweetheart. **2** an illicit lover, esp. a mistress. [ME *leofman* (as LIEF, MAN)]

Le Mans /lə ˈmɒn/ an industrial town in NW France; pop. (1990) 148,465. It is the site of a motor-racing circuit, on which a 24-hour endurance race (established in 1923) is held each summer.

Lemberg /ˈlɛmbɛrk/ the German name for LVIV.

lemma /ˈlemə/ n. **1** an assumed or demonstrated proposition used in an argument or proof. **2 a** a heading indicating the subject or argument of a literary composition, a dictionary entry, etc. **b** (pl. **lemmata** /-mətə/) a heading indicating the subject or argument of an annotation. **3** a motto appended to a picture etc. [L f. Gk *lēmma -matos* thing assumed, f. the root of *lambanō* take]

lemme /ˈlemɪ/ colloq. let me. [corrupt.]

lemming /ˈlemɪŋ/ n. a small vole-like rodent, esp. of the Arctic genus *Lemmus*. The Norwegian *L. lemmus* periodically undergoes mass migrations, during which it frequently attempts to cross large bodies of water. [Norw.]

Lemmon /ˈlemən/, Jack (born John Uhler) (b.1925), American actor. He made his name in comedy films such as *Some Like It Hot* (1959); he later played serious dramatic parts in such films as *Save the Tiger* (1973), for which he won an Oscar, *The China Syndrome* (1979), and *Missing* (1981).

Lemnos /ˈlemnɒs/ (Greek **Limnos** /ˈlimnɔs/) a Greek island in the northern Aegean Sea; chief town, Kástron.

lemon /ˈlemən/ n. **1 a** a pale-yellow thick-skinned oval citrus fruit with acidic juice. **b** a citrus tree, *Citrus limon*, which produces this fruit. **2 a** pale-yellow colour. **3** colloq. a person or thing regarded as feeble or unsatisfactory or disappointing. □ **lemon balm** a bushy labiate plant, *Melissa officinalis*, with leaves smelling and tasting of lemon. **lemon curd** (or **cheese**) a conserve made from lemons, butter, eggs, and sugar, with the consistency of cream cheese. **lemon drop** a boiled sweet flavoured with lemon. **lemon geranium** a lemon-scented pelargonium, *Pelargonium crispum*. **lemon grass** a fragrant tropical grass of the genus *Cymbopogon*, yielding an oil smelling of lemon. **lemon squash** Brit. a soft drink made from lemons and other ingredients, often sold in concentrated form. **lemon-squeezer** a device for extracting the juice from a lemon. **lemon thyme** a labiate herb, *Thymus citriodorus*, with lemon-scented leaves. **lemon verbena** (or **plant**) a shrub, *Aloysia triphylla*, with lemon-scented leaves. [ME f. OF *limon* f. Arab. *līma*: cf. LIME²]

lemonade /ˌleməˈneɪd/ n. **1** an effervescent or still drink made from lemon juice. **2** a synthetic substitute for this.

lemon sole n. a common flatfish, *Microstomus kitt*, of the flounder family. [F *limande*]

lemony /ˈlemənɪ/ adj. **1** tasting or smelling of lemons. **2** Austral. & NZ sl. irritable.

lempira /lemˈpɪərə/ n. the basic monetary unit of Honduras, equal to 100 centavos. [the name of a 16th-c. chieftain who opposed the Spanish conquest of Honduras]

lemur /ˈliːmə(r)/ n. a primitive tree-living primate of the family Lemuridae or a related family, found only in Madagascar, having a

pointed snout and long tail. [mod.L f. L *lemures* (pl.) spirits of the dead, from its spectre-like face]

Lena /ˈleɪnə/, Russian /ˈljɛnə/ a river in Siberia, which rises in the mountains on the western shore of Lake Baikal and flows generally north-east and north for 4,400 km (2,750 miles) into the Laptev Sea, a part of the Arctic Ocean. It is famous for the gold-fields in its basin.

Lenclos /lɒŋˈkləʊ/, Ninon de (born Anne de Lenclos) (1620–1705), French courtesan. She was a famous wit and beauty and numbered many prominent writers and nobles among her lovers. She advocated a form of Epicureanism and defended her philosophy and lifestyle in her book *La Coquette vengée* (1659). In later life she presided over one of the most distinguished literary salons of the age.

lend /lend/ v.tr. (past and past part. **lent** /lent/) **1** (usu. foll. by *to*) grant (to a person) the use of (a thing) on the understanding that it or its equivalent shall be returned. **2** allow the use of (money) at interest. **3** bestow or contribute (something temporary) (*lend assistance*; *lends a certain charm*). □ **lend an ear** (or **one's ears**) listen. **lend a hand** = give a hand (see HAND). **lending library** a library from which books may be temporarily taken away with or Brit. without direct payment. **lend itself to** (of a thing) be suitable for. **lend oneself to** accommodate oneself to (a policy or purpose). □ **lendable** adj. **lender** n. **lending** n. [ME, earlier *lēne(n)* f. OE *lǣnan* f. *lǣn* LOAN¹]

Lendl /ˈlend(ə)l/, Ivan (b.1960), Czech-born tennis player. He won many singles titles in the 1980s and early 1990s, including the US, Australian, and the French Open championships. He became an American citizen in 1992.

Lend-Lease /ˈlendliːs/ (also called *Lease-Lend*) an arrangement made in 1941 whereby the US supplied military equipment and armaments to the UK and her allies in the Second World War, originally as a loan in return for the use of British-owned military bases.

length /leŋθ, leŋkθ/ n. **1** measurement or extent from end to end; the greater of two or the greatest of three dimensions of a body. **2** extent in, of, or with regard to, time (*a stay of some length*; *the length of a speech*). **3** the distance a thing extends (*at arm's length*; *ships a cable's length apart*). **4** the length of a swimming-pool as a measure of the distance swum. **5** the length of a horse, boat, etc., as a measure of the lead in a race. **6** a long stretch or extent (*a length of hair*). **7** a degree of thoroughness in action (*went to great lengths*; *prepared to go to any length*). **8** a piece of material of a certain length (*a length of cloth*). **9** Prosody the quantity of a vowel or syllable. **10** Cricket **a** the distance from the batsman at which the ball pitches (*the bowler keeps a good length*). **b** the proper amount of this. **11** the extent of a garment in a vertical direction when worn. **12** the full extent of one's body. □ **at length 1** (also **at full** or **great** etc. **length**) in detail, without curtailment. **2** after a long time, at last. [OE *lengthu* f. Gmc (as LONG¹)]

lengthen /ˈleŋθən, ˈleŋkθ-/ v. **1** tr. & intr. make or become longer. **2** tr. make (a vowel) long. □ **lengthener** n.

lengthman /ˈleŋθmən, ˈleŋkθ-/ n. (pl. **-men**) Brit. a person employed to maintain a section of railway or road.

lengthways /ˈleŋθweɪz, ˈleŋkθ-/ adv. in a direction parallel with a thing's length.

lengthwise /ˈleŋθwaɪz, ˈleŋkθ-/ adv. & adj. ● adv. lengthways. ● adj. lying or moving lengthways.

lengthy /ˈleŋθɪ, ˈleŋkθɪ/ adj. (**lengthier, lengthiest**) **1** of unusual length. **2** (of speech, writing, style, a speaker, etc.) tedious, prolix. □ **lengthily** adv. **lengthiness** n.

lenient /ˈliːnɪənt/ adj. **1** merciful, tolerant, not disposed to severity. **2** (of punishment etc.) mild. **3** archaic emollient. □ **lenience** n. **leniency** n. **leniently** adv. [L *lenire lenit-* soothe f. *lenis* gentle]

Lenin /ˈlenɪn/, Vladimir Ilich (born Vladimir Ilich Ulyanov) (1870–1924), the principal figure in the Russian Revolution and first Premier (Chairman of the Council of People's Commissars) of the Soviet Union 1918–24. Lenin was the first political leader to attempt to put Marxist principles into practice, though, like Marx, he saw the need for a transitional period to full communism, during which there would be a 'dictatorship of the proletariat'. The policies that he pursued led ultimately to the establishment of Marxism-Leninism in the Soviet Union and, later, in China. Born in Russia, he lived in Switzerland from 1900, but was instrumental in the split between the Bolsheviks and Mensheviks in 1903, when he became leader of the more radical Bolsheviks. He returned to Russia in 1917, established Bolshevik control after the overthrow of the tsar, and in 1918 became head of state; he founded the Third International or Comintern (see INTERNATIONAL) the following year. With Trotsky's help, he defeated counter-revolutionary forces in the Russian Civil War, but was forced to moderate his socio-economic policies to allow the country to recover from the effects of war and revolution. During the last years of his life he denounced, but was unable to prevent, the concentration of power in the hands of Stalin.

Leninakan /ˌlenɪnəˈkɑːn/ a former name (1924–91) for GYUMRI.

Leningrad /ˈlenɪnˌɡræd/ a former name (1924–91) for ST PETERSBURG.

Leninism /ˈlenɪˌnɪz(ə)m/ n. Marxism as interpreted and applied by Lenin. □ **Leninist** n. & adj. **Leninite** n. & adj.

lenition /liːˈnɪʃ(ə)n/ n. (in Celtic languages) the process or result of a consonant's becoming weak or lost; weakening of articulation. [L lenis soft, after G Lenierung]

lenitive /ˈlenɪtɪv/ adj. & n. ● adj. Med. soothing. ● n. **1** Med. a soothing drug or appliance. **2** a palliative. [ME f. med.L lenitivus (as LENIENT)]

lenity /ˈlenɪtɪ/ n. (pl. **-ies**) literary **1** mercifulness, gentleness. **2** an act of mercy. [F lénité or L lenitas f. lenis gentle]

Lennon /ˈlenən/, John (1940–80), English pop and rock singer, guitarist, and songwriter. He was a founder member of the Beatles and wrote most of their songs in collaboration with Paul McCartney; he announced his intention to leave the group in 1969. His subsequent recording career, in which he often collaborated with his second wife Yoko Ono (b.1933), included the album *Imagine* (1971). He took up residency in the US and was assassinated outside his home in New York.

leno /ˈliːnəʊ/ n. (pl. **-os**) an open-work fabric with the warp threads twisted in pairs before weaving. [F linon f. lin flax f. L linum]

Le Nôtre /lə ˈnəʊtrə/, André (1613–1700), French landscape gardener. He designed many formal gardens, including the parks of Vaux-le-Vicomte and Versailles, begun in 1655 and 1662 respectively. These incorporated his ideas on architecturally conceived garden schemes: geometric formality and perfect equilibrium of all the individual elements — sculpture, fountains, parterres, and open spaces. His influence spread throughout the Continent and to England, where his style was imitated.

lens /lenz/ n. **1** a piece of a transparent substance with one or (usu.) both sides curved for concentrating or dispersing light-rays, esp. in optical instruments. **2** a combination of lenses used in photography etc. **3** Anat. = crystalline lens. **4** Physics a device for focusing or otherwise modifying the direction of movement of light, sound, electrons, etc. **5** = contact lens. **6** Astron. = gravitational lens. □ **lensed** adj. **lensless** adj. [L lens lentis lentil (from the similarity of shape)]

Lent /lent/ n. **1** Eccl. the period from Ash Wednesday to Holy Saturday, of which the 40 weekdays are devoted to fasting and penitence in commemoration of Jesus' fasting in the wilderness. **2** (in pl.) the boat races held at Cambridge in the Lent term. □ **Lent lily** Brit. a daffodil, esp. a wild one. **Lent term** Brit. a university etc. term in which Lent falls. [ME f. LENTEN]

lent past and past part. of LEND.

-lent /lənt/ suffix forming adjectives (pestilent; violent) (cf. -ULENT). [L -lentus -ful]

Lenten /ˈlentən/ adj. of, in, or appropriate to, Lent. □ **Lenten fare** food without meat. [orig. as noun, = spring, f. OE lencten f. Gmc, rel. to LONG¹, perh. with ref. to lengthening of the day in spring: now regarded as adj. f. LENT + -EN²]

lenticel /ˈlentɪˌsel/ n. Bot. any of the raised pores in the stems of woody plants that allow gas exchange between the atmosphere and the internal tissues. [mod.L lenticella dimin. of L lens: see LENS]

lenticular /lenˈtɪkjʊlə(r)/ adj. **1** shaped like a lentil or a biconvex lens. **2** of the lens of the eye. [L lenticularis (as LENTIL)]

lentigo /lenˈtaɪɡəʊ/ n. (pl. **lentigines** /-ˈtɪdʒɪˌniːz/) Med. **1** a condition marked by small brown patches on the skin, esp. in elderly people. **2** such a patch. [L, f. lens: see LENS]

lentil /ˈlentɪl/ n. **1** a leguminous plant, Lens culinaris, yielding edible biconvex seeds. **2** this seed, esp. used as food with the husk removed. [ME f. OF lentille f. L lenticula (as LENS)]

lento /ˈlentəʊ/ adj. & adv. Mus. ● adj. slow. ● adv. slowly. [It.]

lentoid /ˈlentɔɪd/ adj. = LENTICULAR 1. [L lens (see LENS) + -OID]

Leo¹ /ˈliːəʊ/ n. **1** Astron. a large constellation (the Lion), said to represent the lion slain by Hercules. It contains the bright stars Regulus and Denebola, and numerous galaxies. The stars outlining the 'head' form a group called the Sickle. **2** Astrol. **a** the fifth sign of the zodiac, which the sun enters about 21 July. **b** a person born when the sun is in this sign. [OE f. L, = LION]

Leo² /ˈliːəʊ/ the name of thirteen popes, notably:

Leo I (known as Leo the Great; canonized as St Leo I) (d.461), pope from 440 and Doctor of the Church. His statement of the doctrine of the Incarnation was accepted at the Council of Chalcedon (451). He extended and consolidated the power of the Roman see, claiming jurisdiction in Africa, Spain, and Gaul. He persuaded the Huns to retire beyond the Danube and secured concessions from the Vandals when they captured Rome. Feast day (in the Eastern Church) 18 Feb.; (in the Western Church) 11 Apr.

Leo X (born Giovanni de' Medici) (1475–1521), pope from 1513. He excommunicated Martin Luther and bestowed on Henry VIII of England the title of Defender of the Faith. He was a noted patron of learning and the arts.

Leo I, St Pope Leo I (see LEO²).

Leo III /ˈliːəʊ/ (c.680–741), Byzantine emperor 717–41. He repulsed several Muslim invasions and carried out an extensive series of financial, legal, administrative, and military reforms. In 726 he forbade the use of images in public worship; this policy, enforced by teams of iconoclasts, met with much popular opposition and a split with Rome.

León /leɪˈɒn/ **1** a city in northern Spain; pop. (1991) 146,270. It is the capital of the province and former kingdom of León, now part of Castilla-León region. **2** an industrial city in central Mexico; pop. (1990) 872,450. **3** a city in western Nicaragua, the second largest city in the country; pop. (1991) 158,577.

Leonardo da Vinci /ˌliːəˈnɑːdəʊ də ˈvɪntʃɪ/ (1452–1519), Italian painter, scientist, and engineer. He spent his early life in Florence; thereafter he worked in Milan, Florence, Rome, and France. Although Leonardo's paintings are relatively few in number, they are notable for their use of the technique of *sfumato* and reflect his studies of nature; they include *The Virgin of the Rocks* (1483–5), *The Last Supper* (1498), a tempera painting on the wall of the refectory at Santa Maria delle Grazie in Milan, and *Mona Lisa* (1504–5), his most famous easel painting, showing a woman with an enigmatic smile. In addition to painting, he devoted his mental energy to a wide range of subjects, from anatomy and biology to mechanics and hydraulics. His nineteen notebooks contain meticulously observed drawings of plants, clouds, skeletons, etc., studies of the human circulatory system, and plans for a type of aircraft (see HELICOPTER), an armoured tank, and a submarine.

leone /lɪˈəʊn, -ˈəʊnɪ/ n. the basic monetary unit of Sierra Leone, equal to 100 cents. [f. SIERRA LEONE]

Leonids /ˈliːənɪdz/ n.pl. Astron. an annual meteor shower with a radiant in the constellation of Leo, reaching a peak about 17 November. [L leo (see LEO¹) leonis + -ID³]

Leonine /ˈliːəˌnaɪn/ adj. & n. ● adj. of or relating to a pope or emperor named Leo. ● n. (in pl.) Leonine verse. □ **Leonine verse 1** medieval Latin verse in hexameter or elegiac metre with internal rhyme. **2** English verse with internal rhyme. [the name Leo (as LEONINE)]

leonine /ˈliːəˌnaɪn/ adj. **1** like a lion. **2** of or relating to lions. [ME f. OF leonin -ine or L leoninus f. leo leonis lion]

Leonine City the part of Rome in which the Vatican stands, walled and fortified by Pope Leo IV (d.855).

leopard /ˈlepəd/ n. (fem. **leopardess** /-dɪs/) **1** a large African and Asian feline, Panthera pardus, with either a black-spotted yellowish-fawn or all black coat. (See note below.) **2** Heraldry a lion passant guardant as in the arms of England. **3** (attrib.) spotted like a leopard (leopard moth). □ **leopard's bane** a composite plant of the genus Doronicum, with large yellow daisy-like flowers. [ME f. OF f. LL f. late Gk leopardos (as LION, PARD)]

▪ The name 'leopard' was originally given to the animal now called a cheetah (the 'hunting-leopard'), which was thought to be a cross between the lion and a 'pard' or panther (formerly supposed to be a more powerful leopard). A black form of the leopard is widely known as the 'black panther'. Unlike the cheetah and lion, the leopard is a solitary animal, hunting mainly at night, and is better able to climb trees.

Leopold I /ˈliːəˌpəʊld/ (1790–1865), first king of Belgium 1831–65. The fourth son of the Duke of Saxe-Coburg-Saalfield, Leopold was an uncle of Queen Victoria, whom he advised during the early part of her reign. In 1830 he refused the throne of Greece, but a year later accepted that of the newly independent Belgium, reigning peacefully thereafter.

Léopoldville /ˈliːəˌpəʊldvɪl/ the former name (until 1966) for KINSHASA.

leotard /ˈliːəˌtɑːd/ n. a close-fitting one-piece garment, usu. covering the torso and arms, worn by ballet-dancers, acrobats, etc. [Jules Léotard, Fr. trapeze artist (1830–70)]

Leo the Great Pope Leo I (see LEO[2]).

Lepanto, Battle of /lɪˈpæntəʊ/ a naval battle fought in 1571 close to the port of Lepanto (Greek *Návpaktos*) in Greece, at the entrance to the Gulf of Corinth. The Christian forces of Rome, Venice, and Spain, under the command of Don John of Austria, defeated a large Turkish fleet, ending for the time being Turkish naval domination in the eastern Mediterranean.

Lepanto, Gulf of /lɪˈpæntəʊ/ an alternative name for the Gulf of Corinth (see CORINTH, GULF OF).

leper /ˈlepə(r)/ n. **1** a person suffering from leprosy. **2** a person who is shunned for moral or social reasons. [ME, prob. attrib. use of *leper* leprosy f. OF *lepre* f. L *lepra* f. Gk, fem. of *lepros* scaly f. *lepos* scale]

Lepidoptera /ˌlepɪˈdɒptərə/ n.pl. Zool. an order of insects comprising the butterflies and moths, which have four large scale-covered wings that are often brightly coloured, and larvae that are caterpillars. (*See note below.*) □ **lepidopteran** n. & adj. **lepidopterist** n. **lepidopterous** adj. [mod.L f. Gk *lepis -idos* scale + *pteron* wing]

▪ There are many superfamilies within the Lepidoptera, the great majority of which are regarded as moths. The order is divided informally into the macrolepidoptera (the usually larger species favoured by collectors) and the microlepidoptera (the numerous smaller species studied only by specialists). (See also BUTTERFLY, MOTH.)

Lepidus /ˈlepɪdəs/, Marcus Aemilius (died *c.*13 BC), Roman statesman and triumvir. After supporting Julius Caesar in the civil war against Pompey, Lepidus was elected consul in 46. He was appointed one of the Second Triumvirate with Octavian and Antony in 43 as well as consul again in 42. Lepidus was given control over Africa after losing the provinces of Gaul and Spain to his two more powerful partners. He retired from public life following a failed revolt in Sicily against Octavian in 36.

leporine /ˈlepəˌraɪn/ adj. of or like hares. [L *leporinus* f. *lepus -oris* hare]

leprechaun /ˈleprəˌkɔːn/ n. a small mischievous sprite in Irish folklore. [OIr. *luchorpán* f. *lu* small + *corp* body]

leprosarium /ˌleprəˈseərɪəm/ n. a hospital for people with leprosy.

leprosy /ˈleprəsɪ/ n. **1** Med. a contagious bacterial disease that affects the skin, mucous membranes, and nerves, causing discoloration and lumps on the skin and, in severe cases, disfigurement and deformities. (*See note below.*) **2** moral corruption or contagion. [LEPROUS + -Y[3]]

▪ Leprosy is also called *Hansen's disease*, after the Norwegian Gerhard Hansen (1841–1912), who identified the bacillus, *Mycobacterium leprae*, which causes it. The disease can be controlled, but rarely cured, by drugs. Formerly more widespread, leprosy is now almost entirely confined to tropical Africa and Asia. In the past, and particularly in biblical translation, the word was used to refer to disfiguring skin diseases in general.

leprous /ˈleprəs/ adj. **1** suffering from leprosy. **2** like or relating to leprosy. [ME f. OF f. LL *leprosus* f. *lepra*: see LEPER]

lepta pl. of LEPTON[1].

Leptis Magna /ˌleptɪs ˈmægnə/ an ancient seaport and trading centre on the Mediterranean coast of North Africa, near present-day Al Khums in Libya. Founded by the Phoenicians (perhaps before 600 BC) it was settled from Carthage, becoming one of the three chief cities of Tripolitania. It became a Roman colony under Trajan. Most of its impressive remains date from the reign of Septimius Severus (AD 193–211), a native of the city.

lepto- /ˈleptəʊ/ comb. form small, narrow. [Gk *leptos* fine, small, thin, delicate]

leptocephalic /ˌleptəʊsɪˈfælɪk, -kɪˈfælɪk/ adj. (also **leptocephalous** /-ˈsefələs, -ˈkefələs/) narrow-skulled.

lepton[1] /ˈleptɒn/ n. (pl. **lepta**) a Greek monetary unit worth one-hundredth of a drachma. [Gk *lepton* (*nomisma* coin) neut. of *leptos* small]

lepton[2] /ˈleptɒn/ n. Physics a subatomic particle, such as an electron, muon, or neutrino, which does not take part in the strong interaction (cf. HADRON). □ **leptonic** /lepˈtɒnɪk/ adj. [LEPTO-]

leptospirosis /ˌleptəʊspaɪəˈrəʊsɪs/ n. Med. an infectious disease caused by spirochaete bacteria of the genus *Leptospira*, that occurs in rodents, dogs, and other mammals, and can be transmitted to humans (cf. WEIL'S DISEASE). [LEPTO- + SPIRO-[1] + -OSIS]

leptotene /ˈleptəˌtiːn/ n. Biol. the first stage of the prophase of meiosis in which each chromosome is apparent as two fine chromatids. [LEPTO- + Gk *tainia* band]

Lepus /ˈliːpəs, ˈlep-/ Astron. a small constellation (the Hare) at the foot

of Orion, said to represent the hare pursued by him. It has few bright stars. [L]

Lerwick /ˈlɜːwɪk/ the capital of the Shetland Islands, on the island of Mainland; pop. (1991) 7,220. The most northerly town in the British Isles, it is a fishing centre and a service port for the oil industry.

Lesage /ləˈsɑːʒ/, Alain-René (1668–1747), French novelist and dramatist. He is best known for the picaresque novel *Gil Blas* (1715–35).

lesbian /ˈlezbɪən/ n. & adj. ● n. a homosexual woman. ● adj. **1** of homosexuality in women; being a female homosexual. **2** (**Lesbian**) of Lesbos. □ **lesbianism** n. [L *Lesbius* f. Gk *Lesbios* f. LESBOS, home of SAPPHO]

Lesbos /ˈlezbɒs/ (Greek **Lésvos** /ˈlɛzvɒs/) a Greek island in the eastern Aegean, off the coast of NW Turkey; chief town, Mytilene. Its artistic golden age of the late 7th and early 6th centuries BC produced the poets Alcaeus and Sappho.

lese-majesty /liːz ˈmædʒɪstɪ/ n. (also **lèse-majesté** /leɪz ˈmæʒeˌsteɪ/) **1** treason. **2** an insult to a sovereign or ruler. **3** presumptuous conduct. [F *lèse-majesté* f. L *laesa majestas* injured sovereignty f. *laedere laes-* injure + *majestas* MAJESTY]

lesion /ˈliːʒ(ə)n/ n. **1** damage. **2** injury. **3** Med. a morbid change in the functioning or texture of an organ etc. [ME f. OF f. L *laesio -onis* f. *laedere laes-* injure]

Lesotho /ləˈsuːtuː/ a landlocked mountainous country forming an enclave in South Africa; pop. (est. 1991) 1,816,000; official languages, Sesotho and English; capital, Maseru. The region was settled by the Sotho people in the 16th century, coming under British rule (as Basutoland) in 1868. The country became an independent kingdom within the Commonwealth in 1966, changing its name to Lesotho.

less /les/ adj., adv., n., & prep. ● adj. **1** smaller in extent, degree, duration, number, etc. (*of less importance; in a less degree*). **2** of smaller quantity, not so much (opp. MORE) (*find less difficulty; eat less meat*). **3** disp. fewer (*eat less biscuits*). **4** of lower rank etc. (*no less a person than; James the Less*). ● adv. to a smaller extent, in a lower degree. ● n. a smaller amount or quantity or number (*cannot take less; for less than £10; is little less than disgraceful*). ● prep. minus (*made £1,000 less tax*). □ **in less than no time** joc. very quickly or soon. **less developed country** see DEVELOPING COUNTRY. **much** (or **still**) **less** with even greater force of denial (*do not suspect him of negligence, much less of dishonesty*). [OE *læssa* (adj.), *læs* (adv.), f. Gmc]

-less /lɪs/ suffix forming adjectives and adverbs: **1** from nouns, meaning 'not having, without, free from' (*doubtless; powerless*). **2** from verbs, meaning 'not affected by or doing the action of the verb' (*fathomless; tireless*). □ **-lessly** suffix forming adverbs. **-lessness** suffix forming nouns. [OE *-lēas* f. *lēas* devoid of]

lessee /leˈsiː/ n. (often foll. by *of*) a person who holds a property by lease. □ **lesseeship** n. [ME f. AF past part., OF *lessé* (as LEASE)]

lessen /ˈles(ə)n/ v.tr. & intr. make or become less, diminish.

Lesseps /ˈlesəps/, Ferdinand Marie, Vicomte de (1805–94), French diplomat. While in the consular service in Egypt he became aware of plans to link the Mediterranean and the Red Sea by means of a canal, and from 1854 onwards devoted himself to the project. Work began in 1859 and the Suez Canal was opened ten years later. In 1881 he embarked on the building of the Panama Canal, but had not anticipated the difficulties of this very different enterprise; the project was abandoned in 1889.

lesser /ˈlesə(r)/ adj. (usu. attrib.) not so great as the other or the rest (*the lesser evil; the lesser celandine*). [double compar., f. LESS + -ER[2]]

Lesser Antilles see ANTILLES.

Lesser Sunda Islands see SUNDA ISLANDS.

Lessing[1] /ˈlesɪŋ/, Doris (May) (b.1919), British novelist and short-story writer, brought up in Rhodesia. An active Communist in her youth, she frequently deals with social and political conflicts in her fiction, especially as they affect women; *The Golden Notebook* (1962) was hailed as a landmark by the women's movement. Other works include *The Grass is Singing* (1950) about interracial relationships in Africa, and a quintet of science-fiction novels collectively entitled *Canopus in Argus: Archives* (1979–83).

Lessing[2] /ˈlesɪŋ/, Gotthold Ephraim (1729–81), German dramatist and critic. He wrote tragedies such as *Emilia Galotti* (1772) and *Miss Sara Sampson* (1755), the latter considered to be the first significant domestic tragedy in German; a comedy, *Minna von Barnhelm* (1767); and the dramatic poem *Nathan der Weise* (1779), a plea for religious toleration. In his critical works, such as *Laokoon* (1766), he criticized the reliance of German literature on the conventions of the French classical school and suggested that German writers should look to Shakespeare and English literature instead.

Les Six /leɪ 'siːs/ (also **the Six**) a group of six Parisian composers (Louis Durey, Arthur Honegger, Darius Milhaud, Germaine Tailleferre, Georges Auric, and Francis Poulenc) formed after the First World War, whose music represents a reaction against romanticism and impressionism. [F, = the six]

lesson /'les(ə)n/ n. & v. ● n. **1 a** an amount of teaching given at one time. **b** the time assigned to this. **2** (in pl.; foll. by in) systematic instruction (gives lessons in dancing; took lessons in French). **3** a thing learned or to be learned by a pupil; an assignment. **4 a** an occurrence, example, rebuke, or punishment, that serves or should serve to warn or encourage (let that be a lesson to you). **b** a thing inculcated by experience or study. **5** a passage from the Bible read aloud during a church service, esp. either of two readings at morning and evening prayer in the Church of England. ● v.tr. archaic **1** instruct. **2** admonish, rebuke. □ **learn one's lesson** profit from or bear in mind a particular (usu. unpleasant) experience. **teach a person a lesson** punish a person, esp. as a deterrent. [ME f. OF leçon f. L lectio -onis: see LECTION]

lessor /le'sɔː(r)/ n. a person who lets a property by lease. [AF f. lesser: see LEASE]

lest /lest/ conj. **1** in order that not, for fear that (lest we forget). **2** that (afraid lest we should be late). [OE thÿ læs whereby less that, later the læste, ME lest(e)]

Lésvos see LESBOS.

let[1] /let/ v. & n. ● v. (**letting**; past and past part. **let**) **1** tr. **a** allow to, not prevent or forbid (we let them go). **b** cause to (let me know; let it be known). **2** tr. (foll. by into) **a** allow to enter. **b** make acquainted with (a secret etc.). **c** inlay in. **3** tr. Brit. grant the use of (rooms, land, etc.) for rent or hire (was let to the new tenant for a year). **4** tr. allow or cause (liquid or air) to escape (let blood). **5** tr. award (a contract for work). **6** aux. supplying the first and third persons of the imperative in exhortations (let us pray), commands (let it be done at once; let there be light), assumptions (let AB be equal to CD), and permission or challenge (let him do his worst). ● n. Brit. the act or an instance of letting a house, room, etc. (a long let). □ **let alone 1** not to mention, far less or more (hasn't got a television, let alone a video). **2** = let be. **let be** not interfere with, attend to, or do. **let down 1** lower. **2** fail to support or satisfy, disappoint. **3** lengthen (a garment). **4** deflate (a tyre). **let-down** n. **1** a disappointment. **2** the release of milk from a mammary gland, esp. a cow's udder. **let down gently** avoid humiliating abruptly. **let drop** (or **fall**) **1** drop (esp. a word or hint) intentionally or by accident. **2** (foll. by on, upon, to) Geom. draw (a perpendicular) from an outside point to a line. **let fly 1** (often foll. by at) attack physically or verbally. **2** discharge (a missile). **let go 1** release, set at liberty. **2 a** (often foll. by of) lose or relinquish one's hold. **b** lose hold of. **3** cease to think or talk about. **let oneself go 1** give way to enthusiasm, impulse, etc. **2** cease to take trouble, neglect one's appearance or habits. **let in 1** allow to enter (let the dog in; let in a flood of light; this would let in all sorts of evils). **2** (usu. foll. by for) involve (a person, often oneself) in loss or difficulty. **3** (foll. by on) allow (a person) to share privileges, information, etc. **4** inlay (a thing) in another. **let oneself in** enter a building by means of a latchkey. **let loose** release or unchain esp. something fierce or uncontrollable. **let me see** see SEE[1]. **let off 1 a** fire (a gun). **b** explode (a bomb or firework). **2** allow or cause (steam, liquid, etc.) to escape. **3** allow to alight from a vehicle etc. **4 a** not punish or compel. **b** (foll. by with) punish lightly. **5** Brit. let (part of a house etc.). **6** Brit. colloq. break wind. **let-off** n. being allowed to escape something. **let off steam** see STEAM. **let on** colloq. **1** reveal a secret. **2** pretend (let on that he had succeeded). **let out 1** allow to go out, esp. through a doorway. **2** release from restraint. **3** (often foll. by that + clause) reveal (a secret etc.). **4** make (a garment) looser esp. by adjustment at a seam. **5** put out to rent esp. to several tenants, or to contract. **6** exculpate. **7** give vent or expression to; emit (a sound etc.). **let-out** n. Brit. colloq. an opportunity to escape from an awkward situation etc. **let rip** see RIP[1]. **let slip** see SLIP[1]. **let through** allow to pass. **let up** colloq. **1** become less intense or severe. **2** relax one's efforts. **let-up** n. **1** a reduction in intensity. **2** a relaxation of effort. **to let** available for rent. [OE lǣtan f. Gmc, rel. to LATE]

let[2] /let/ n. & v. ● n. **1** Tennis etc. an obstruction of a ball or a player in certain ways, requiring the ball to be served again. **2** (archaic except in **without let or hindrance**) obstruction, hindrance. ● v.tr. (**letting**; past and past part. **letted** or **let**) archaic hinder, obstruct. [OE lettan f. Gmc, rel. to LATE]

-let /lɪt, lət/ suffix forming nouns, usu. diminutives (flatlet; leaflet) or denoting articles of ornament or dress (anklet). [orig. corresp. (in bracelet, crosslet, etc.) to F -ette added to nouns in -el]

lethal /'liːθəl/ adj. causing or sufficient to cause death. □ **lethal chamber** a chamber in which animals may be killed painlessly with gas. **lethal dose** the amount of a toxic compound or drug that causes death in humans, animals, micro-organisms, or cultured cells. □ **lethally** adv. **lethality** /lɪ'θælɪtɪ/ n. [L let(h)alis f. letum death]

lethargy /'leθədʒɪ/ n. **1** lack of energy or vitality; a torpid, inert, or apathetic state. **2** Med. morbid drowsiness or prolonged and unnatural sleep. □ **lethargic** /lɪ'θɑːdʒɪk/ adj. **lethargically** adv. [ME f. OF litargie f. LL lethargia f. Gk lēthargia f. lēthargos forgetful f. lēth-, lanthanomai forget]

Lethe /'liːθɪ/ Gk Mythol. one of the rivers of the underworld, whose water when drunk made the souls of the dead forget their life on earth. □ **Lethean** /lɪ'θiːən/ adj. [L, use of Gk lēthē forgetfulness (as LETHARGY)]

Leticia /lə'tiːsɪə/ a town and river port at the southern tip of Colombia, on the upper reaches of the Amazon on the border with Brazil and Peru; pop. (1985) 24,090.

Leto /'liːtəʊ/ Gk Mythol. the daughter of a Titan, mother (by Zeus) of Artemis and Apollo. The Roman Latona was identified with her.

let's /lets/ contr. let us (let's go now).

Lett /let/ n. archaic = LATVIAN n. [G Lette f. Latvian Latvi]

letter /'letə(r)/ n. & v. ● n. **1 a** a character representing one or more of the simple or compound sounds used in speech; any of the alphabetic symbols. **b** (in pl.) colloq. the initials of a degree etc. after the holder's name. **c** US a school or college initial as a mark of proficiency in games etc. **2 a** a written, typed, or printed communication, usu. sent by post or messenger. **b** (in pl.) an addressed legal or formal document for any of various purposes. **3** the precise terms of a statement, the strict verbal interpretation (opp. SPIRIT n. 6) (according to the letter of the law). **4** (in pl.) **a** literature. **b** acquaintance with books, erudition. **c** authorship (the profession of letters). **5** Printing **a** types collectively. **b** a font of type. ● v.tr. **1 a** inscribe letters on. **b** impress a title etc. on (a book-cover). **2** classify with letters. □ **letter-bomb** a terrorist explosive device in the form of a postal packet. **letter-box** esp. Brit. a box or slot into which letters are posted or delivered. **letter-card** Brit. a folded card with a gummed edge for posting as a letter. **letter-carrier** N. Amer. a postman or postwoman. **letter-heading** = LETTERHEAD. **letter of comfort** an assurance about a debt, short of a legal guarantee, given to a bank by a third party. **letter of credence** see CREDENCE. **letter of credit** see CREDIT. **letter-perfect** Theatr. knowing one's part perfectly. **letter-quality** of the quality of printing suitable for a business letter; producing print of this quality. **letters missive** see MISSIVE. **letters of administration** authority to administer the estate of an intestate. **letters of marque** see MARQUE[2]. **letters patent** see PATENT. **letter-writer 1** a person who writes letters. **2** a book giving guidance on writing letters. **man of letters** a scholar or author. **to the letter** with adherence to every detail. □ **letterless** adj. [ME f. OF lettre f. L litera, littera letter of alphabet, (in pl.) epistle, literature]

lettered /'letəd/ adj. well-read or educated.

letterhead /'letəˌhed/ n. **1** a printed heading on stationery. **2** stationery with this.

lettering /'letərɪŋ/ n. **1** the process of inscribing letters. **2** letters inscribed.

letterpress /'letəˌpres/ n. **1 a** the contents of an illustrated book other than the illustrations. **b** printed matter relating to illustrations. **2** printing from raised type, not from lithography or other planographic processes.

Lettic /'letɪk/ adj. & n. archaic ● adj. **1** = LATVIAN adj. **2** of or relating to the Baltic branch of languages. ● n. = LATVIAN n. 2.

Lettish /'letɪʃ/ adj. & n. archaic = LATVIAN.

lettuce /'letɪs/ n. **1** a composite plant, Lactuca sativa, with crisp edible leaves used in salads. **2** any plant resembling this. [ME letus(e), rel. to OF laituë f. L lactuca f. lac lactis milk, with ref. to its milky juice]

Letzeburgesch /'letsəˌbɜːɡɪʃ/ see LUXEMBURGISH.

leu /'leɪuː/ n. (pl. **lei** /leɪ/) the chief monetary unit of Romania and Moldova. [Romanian, = lion]

leucine /'luːsiːn/ n. Biochem. a hydrophobic amino acid present in protein and essential in the diet of vertebrates. [F f. Gk leukos white + -INE[4]]

leuco- /'luːkəʊ/ comb. form white. [Gk leukos white]

leucocyte /'luːkəˌsaɪt/ n. (also **leukocyte**) a colourless amoeboid cell in the blood, lymph, etc., containing a nucleus and important in fighting disease. Leucocytes may be classed according to the dyes that they take up (*acidophil, basophil, neutrophil,* and *eosinophil*), and they may take various forms (e.g. *lymphocyte, granulocyte, monocyte,* and *macrophage*). Also called *white* (*blood*) *cell* or *corpuscle*. (See also BLOOD.) □ **leucocytic** /ˌluːkəˈsɪtɪk/ *adj.*

leucoma /luːˈkəʊmə/ *n. Med.* a white opacity in the cornea of the eye.

leucorrhoea /ˌluːkəˈrɪə/ *n. Med.* a whitish or yellowish discharge of mucus from the vagina.

leucotomy /luːˈkɒtəmɪ/ *n.* (*pl.* **-ies**) *Med.* surgical cutting of tracts of white nerve fibres in the brain, orig. esp. prefrontal lobotomy, formerly used in psychosurgery.

leukaemia /luːˈkiːmɪə/ *n.* (*US* **leukemia**) *Med.* any of a group of malignant diseases in which the bone marrow and other blood-forming organs produce increased numbers of leucocytes. □ **leukaemic** *adj.* [mod.L f. G *Leukämie* f. Gk *leukos* white + *haima* blood]

leukocyte var. of LEUCOCYTE.

Leuven /ˈlɜːvən/ (French **Louvain** /luvɛ̃/) a town in Belgium, east of Brussels; pop. (1991) 85,020. From the 11th to the 15th centuries it was the capital of the former duchy (now province) of Brabant.

Lev. *abbr.* (in the Bible) Leviticus.

lev /lev, lef/ *n.* (also **leva** /ˈlevə/) (*pl.* **leva**, **levas**, or **levs**) the basic monetary unit of Bulgaria, equal to 100 stotinki. [Bulg., var. of *lăv* lion]

Levallois /ləˈvælwʌ/ *adj. Archaeol.* of, relating to, or denoting a flint-working technique first employed in the late Acheulean period in western Europe and associated with numerous Mousterian cultures throughout the world, in which a flint is trimmed so that a flake of predetermined size and shape can be struck from it. □ **Levalloisean** /ˌlevəˈlɔɪzɪən/ *adj.* [the type-site in northern Paris]

Levant /lɪˈvænt/ *n.* (prec. by *the*) *archaic* or *hist.* the eastern part of the Mediterranean with its islands and neighbouring countries. □ **Levant morocco** high-grade large-grained morocco leather. [F, pres. part. of *lever* rise, used as noun = point of sunrise, east]

levanter /lɪˈvæntə(r)/ *n.* **1** a strong easterly Mediterranean wind. **2** (**Levanter**) *archaic* or *hist.* a native or inhabitant of the Levant.

Levantine /lɪˈvæntaɪn, ˈlevən-/ *adj. & n. archaic* or *hist.* ● *adj.* of or trading to the Levant. ● *n.* a native or inhabitant of the Levant.

levator /lɪˈveɪtə(r)/ *n. Anat.* a muscle that lifts the structure into which it is inserted. [L, = one who lifts f. *levare* raise]

levee¹ /ˈlevɪ/ *n.* **1** *archaic* or *N. Amer.* an assembly of visitors or guests, esp. at a formal reception. **2** *hist.* (in the UK) an assembly held by the sovereign or sovereign's representative at which men only were received. **3** *hist.* a reception of visitors on rising from bed. [F *levé* var. of *lever* rising f. *lever* to rise: see LEVY]

levee² /ˈlevɪ, lɪˈviː/ *n. US* **1** an embankment against river floods. **2** a natural embankment built up by a river. **3** a landing-place, a quay. [F *levée* fem. past part. of *lever* raise: see LEVY]

level /ˈlev(ə)l/ *n., adj., & v.* ● *n.* **1** a horizontal line or plane. **2** a height or value reached, a position on a real or imaginary scale (*eye level; sugar level in the blood; danger level*). **3** a social, moral, or intellectual standard. **4** a plane of rank or authority (*discussions at Cabinet level*). **5 a** an instrument giving a line parallel to the plane of the horizon for testing whether things are horizontal. **b** *Surveying* an instrument for giving a horizontal line of sight. **6** a more or less level surface. **7** a flat tract of land. **8** a floor or storey in a building, ship, etc. ● *adj.* **1** having a flat and even surface; not bumpy. **2** horizontal; perpendicular to the plumb-line. **3** (often foll. by *with*) **a** on the same horizontal plane as something else. **b** having equality with something else. **c** (of a spoonful etc.) with the contents flat with the brim. **4** even, uniform, equable, or well-balanced in quality, style, temper, judgement, etc. **5** (of a race) having the leading competitors close together. ● *v.* (**levelled**, **levelling**; *US* **leveled**, **leveling**) **1** *tr.* make level, even, or uniform. **2** *tr.* (often foll. by *to* (or *with*) *the ground*) raze or demolish. **3** *tr.* (also *absol.*) aim (a missile or gun). **4** *tr.* (also *absol.*; foll. by *at, against*) direct (an accusation, criticism, or satire). **5** *tr.* abolish (distinctions). **6** *intr.* (usu. foll. by *with*) *sl.* be frank or honest. **7** *tr.* place on the same level. **8** *tr.* (also *absol.*) *Surveying* ascertain differences in the height of (land). □ **do one's level best** do one's utmost; make all possible efforts. **find one's level 1** reach the right social, intellectual, etc. place in relation to others. **2** (of a liquid) reach the same height in receptacles or regions which communicate with each other. **level crossing** *Brit.* a crossing of a railway and a road, or two railways, at the same level. **level down** bring down to a standard. **levelling-screw** a screw for adjusting parts of a machine etc. to an exact level. **level off** make or become level or smooth. **level out** make or become level, remove differences from. **level pegging** *Brit.* equality of scores or achievements. **level up** bring up to a standard. **on the level** *colloq. adv.* honestly, without deception. ● *adj.* honest, truthful. **on a level with 1** in the same horizontal plane as. **2** equal with. □ **levelly** *adv.* **levelness** *n.* [ME f. OF *livel* ult. f. L *libella* dimin. of *libra* scales, balance]

level-headed /ˌlev(ə)lˈhedɪd/ *adj.* mentally well-balanced, cool, sensible. □ **level-headedly** *adv.* **level-headedness** *n.*

leveller /ˈlevələ(r)/ *n.* (*US* **leveler**) **1** a person who advocates the abolition of social distinctions. **2** (**Leveller**) *hist.* an extreme radical dissenter in 17th-century England. (*See note below.*) **3** a person or thing that levels.
■ The Levellers arose in the London area in 1645–6, during the English Civil War, with a programme that included the abolition of the monarchy and the House of Lords, substantial extension of the franchise, social and agrarian reforms, and religious freedom. They found support within the New Model Army, which elected Leveller representatives from each regiment to join in debate with Oliver Cromwell and senior army officers. Lack of progress led to mutinies by Leveller elements in the army in 1647 and 1649; the suppression of these mutinies by Cromwell brought the movement to an end.

lever /ˈliːvə(r)/ *n. & v.* ● *n.* **1** a bar resting on a pivot, used to help lift a heavy or firmly fixed object. **2** *Mech.* a simple machine consisting of a rigid bar pivoted about a fulcrum (fixed point) which can be acted upon by a force (effort) in order to move a load. **3** a projecting handle moved to operate a mechanism. **4** a means of exerting moral pressure. ● *v.* **1** *intr.* use a lever. **2** *tr.* (often foll. by *away, out, up,* etc.) lift, move, or act on with a lever. □ **lever escapement** a mechanism in a watch connecting the escape wheel and the balance wheel using two levers. **lever watch** a watch with a lever escapement. [ME f. OF *levier, leveor* f. *lever* raise: see LEVY]

leverage /ˈliːvərɪdʒ/ *n.* **1** the action of a lever; a way of applying a lever. **2** the power of a lever; the mechanical advantage gained by use of a lever. **3** a means of accomplishing a purpose; power, influence. **4** a set or system of levers. **5** *US Commerce* gearing. □ **leveraged buyout** the buyout of a company by its management using outside capital.

leveret /ˈlevərɪt/ *n.* a young hare, esp. one in its first year. [ME f. AF, dimin. of *levre*, OF *lievre* f. L *lepus leporis* hare]

Leverhulme /ˈliːvəˌhjuːm/, 1st Viscount (born William Hesketh Lever) (1851–1925), English industrialist and philanthropist. He and his brother started the manufacture of soap from vegetable oil (instead of tallow) under the trade name Sunlight. In the 1880s they founded and built the industrial new town and factory complex of Port Sunlight in Cheshire (now in Merseyside); workers were accommodated in model housing and were entitled to medical care, pensions, and a form of profit-sharing. In the 20th century Leverhulme's company, Lever Bros., came to form the basis of the international corporation Unilever.

Leverkusen /ˈleɪvəˌkuːz(ə)n/ an industrial city in western Germany, in North Rhine-Westphalia, on the River Rhine north of Cologne; pop. (1991) 161,150.

Le Verrier /lə ˈverɪˌeɪ/, Urbain (1811–77), French mathematician. His analysis of the motions of the planets suggested that an unknown body was disrupting the orbit of Uranus, the same conclusion being reached almost simultaneously by John Couch Adams. Under the prompting of Le Verrier, the German astronomer Johann Galle (1812–1910) searched the region of the sky in which the mysterious object was predicted to lie, and discovered the planet Neptune on 23 Sept. 1846.

Levi¹ /ˈliːvaɪ/ **1** a Hebrew patriarch, son of Jacob and Leah (Gen. 29:34). **2** the tribe of Israel traditionally descended from him.

Levi² /ˈleɪvɪ/, Primo (1919–87), Italian novelist and poet, of Jewish descent. His experiences as a survivor of Auschwitz are recounted in his first book *If This is a Man* (1947); other books include *The Periodic Table* (1985), a collection of memoirs.

leviathan /lɪˈvaɪəθən/ *n.* **1** *Bibl.* a very large aquatic creature, a sea monster, identified in different passages with the whale and the crocodile (e.g. Job 41, Ps. 74:14). The name is also used for the Devil (after Isaiah 27:1). **2** anything very large or powerful, esp. a ship. **3** an autocratic monarch or state (in allusion to a book by Thomas Hobbes, 1651). [ME f. LL f. Heb. *liwyāṯān*]

levigate /ˈlevɪˌgeɪt/ *v.tr.* **1** reduce to a fine smooth powder. **2** make a smooth paste of. □ **levigation** /ˌlevɪˈgeɪʃ(ə)n/ *n.* [L *levigare levigat-* f. *levis* smooth]

levin /ˈlevɪn/ *n. archaic* **1** lightning. **2** a flash of lightning. [ME *leven(e)*, prob. f. ON]

levirate /ˈleviɪrət/ n. a custom of the ancient Hebrews and some other peoples by which a man is obliged to marry his brother's widow. □ **leviratic** /ˌlevɪˈrætɪk/ adj. **leviratical** adj. [L levir brother-in-law + -ATE[1]]

Levis /ˈliːvaɪz/ n.pl. (also **Levi's** propr.) a type of (orig. blue) denim jeans or overalls reinforced with rivets. [Levi Strauss (1829–1902), orig. Amer. manufacturer in 1860s]

Lévi-Strauss /ˌlevɪˈstraʊs/, Claude (b.1908), French social anthropologist. He was an influential pioneer in the use of a structuralist analysis to study cultural systems; he regarded language as an essential common denominator underlying cultural phenomena, a view which forms the basis for his theories concerning the relationships of such societal elements as religion, myth, and kinship. His books include the two-volume *Structural Anthropology* (1958; 1973).

levitate /ˈlevɪˌteɪt/ v. **1** intr. rise and float in the air (esp. with reference to spiritualism). **2** tr. cause to do this. □ **levitator** n. **levitation** /ˌlevɪˈteɪʃ(ə)n/ n. [L levis light, after GRAVITATE]

Levite /ˈliːvaɪt/ n. a member of the tribe of Levi, esp. of that part of it which provided assistants to the priests in the worship in the Jewish temple. [ME f. LL levita f. Gk leuitēs f. Leui f. Heb. lēwî LEVI[1]]

Levitical /lɪˈvɪtɪk(ə)l/ adj. **1** of the Levites or the tribe of Levi. **2** of the Levites' ritual. **3** of Leviticus. [LL leviticus f. Gk leuitikos (as LEVITE)]

Leviticus /lɪˈvɪtɪkəs/ the third book of the Bible, containing details of laws and ritual. [L, = (book) of the Levites]

levity /ˈlevɪtɪ/ n. **1** lack of serious thought, frivolity, unbecoming jocularity. **2** inconstancy. **3** undignified behaviour. **4** archaic lightness of weight. [L levitas f. levis light]

levo- US var. of LAEVO-.

levodopa /ˌliːvəˈdəʊpə/ n. Biochem. = L-DOPA.

levulose US var. of LAEVULOSE.

levy /ˈlevɪ/ v. & n. ● v.tr. (-ies, -ied) **1 a** impose (a rate or toll). **b** raise (contributions or taxes). **c** (also absol.) raise (a sum of money) by legal execution or process (*the debt was levied on the debtor's goods*). **d** seize (goods) in this way. **e** extort (*levy blackmail*). **2** enlist or enrol (troops etc.). **3** (usu. foll. by upon, against) wage, proceed to make (war). ● n. (pl. -ies) **1 a** the collecting of a contribution, tax, etc., or of property to satisfy a legal judgement. **b** a contribution, tax, etc., levied. **2 a** the act or an instance of enrolling troops etc. **b** (in pl.) troops enrolled. **c** a body of troops enrolled. **d** the number of troops enrolled. □ **leviable** adj. [ME f. OF levee fem. past part. of lever f. L levare raise f. levis light]

lewd /ljuːd/ adj. **1** lascivious. **2** indecent, obscene. □ **lewdly** adv. **lewdness** n. [OE lǣwede LAY[2], of unkn. orig.]

Lewes /ˈluːɪs/ a town in southern England, north-east of Brighton, the administrative centre of East Sussex; pop. (1981) 14,970. It was the site in 1264 of a battle in which the younger Simon de Montfort defeated Henry III.

Lewis[1] /ˈluːɪs/ the northern part of the island of Lewis with Harris, in the Outer Hebrides of Scotland.

Lewis[2] /ˈluːɪs/, Cecil Day, see DAY-LEWIS.

Lewis[3] /ˈluːɪs/, C(live) S(taples) (1898–1963), British novelist, religious writer, and literary scholar. A convert to Christianity from atheism, Lewis wrote and broadcast widely on religious and moral issues, producing books such as *The Screwtape Letters* (1942). He also wrote a trilogy of allegorical science-fiction novels, and a series of fantasies for children set in the imagined land of 'Narnia', beginning with *The Lion, the Witch, and the Wardrobe* (1950). His works on medieval and Renaissance literature include *The Allegory of Love* (1936). In 1957 he met and married an American writer, Joy Davidman (1915–60), who was dying of cancer; their relationship was the subject of the popular film *Shadowlands* (1993).

Lewis[4] /ˈluːɪs/, Frederick Carleton ('Carl') (b.1961), American athlete. In 1984 he won four Olympic gold medals (in the 100 and 200 metres, long jump, and 4 × 100 metre relay) and he repeated his victories in the 100 metres and long jump in the Olympics of 1988. He has broken the world record for the 100 metres on several occasions.

Lewis[5] /ˈluːɪs/, (Harry) Sinclair (1885–1951), American novelist. He gained recognition with *Main Street* (1920), a social satire on small-town life in the Midwest. His later novels, such as *Babbitt* (1922), *Elmer Gantry* (1927), and *Dodsworth* (1929), continued in a similar vein, using satire and caricature to attack targets such as the urban middle class and the Church. He was awarded the Nobel Prize for literature in 1930, the first American writer to achieve this.

Lewis[6] /ˈluːɪs/, Meriwether (1774–1809), American explorer. He was the joint leader, with William Clark, of an expedition to explore the newly acquired Louisiana Purchase. The Lewis and Clark expedition started out from St Louis in 1804 and crossed America, reaching the mouth of the Columbia river on the Pacific coast in 1805; they returned to St Louis in 1806.

Lewis[7] /ˈluːɪs/, (Percy) Wyndham (1882–1957), British novelist, critic, and painter, born in Canada. He was a leader of the vorticist movement, and with Ezra Pound edited the magazine *Blast* (1914–15). His satirical novels and polemical works include *The Apes of God* (1930) and the trilogy *The Human Age* (1928–55); he expounds his philosophical ideas in *Time and Western Man* (1927). He later aroused hostility for his Fascist sympathies and his satirical attacks on his contemporaries (especially the Bloomsbury Group).

lewis /ˈluːɪs/ n. an iron contrivance for gripping heavy blocks of stone or concrete for lifting. [18th c.: orig. unkn.]

Lewis gun n. a light air-cooled machine-gun with a magazine, operated by gas from its own firing. [Isaac Newton Lewis, Amer. soldier (1858–1931), its inventor]

lewisite /ˈluːɪˌsaɪt/ n. an irritant gas that produces blisters, developed for use in chemical warfare. [Winford Lee Lewis, Amer. chemist (1878–1943) + -ITE[1]]

Lewis with Harris (also **Lewis and Harris**) the largest and northernmost island of the Outer Hebrides in Scotland; chief town, Stornoway. The island, which is separated from the mainland by the Minch, consists of a northern part, Lewis, and a smaller but more mountainous southern part, Harris.

lex domicilii /ˌleks dɒmɪˈsɪlɪˌaɪ/ n. Law the law of the country in which a person is domiciled. [L]

lexeme /ˈleksiːm/ n. Linguistics a basic lexical unit of a language comprising one or several words, the elements of which do not separately convey the meaning of the whole. [LEXICON + -EME]

lex fori /ˌleks ˈfɔːraɪ/ n. Law the law of the country in which an action is brought. [L]

lexical /ˈleksɪk(ə)l/ adj. **1** of the words of a language. **2** of or as of a lexicon. □ **lexically** adv. [Gk lexikos, lexikon: see LEXICON]

lexicography /ˌleksɪˈkɒɡrəfɪ/ n. the compiling of dictionaries. □ **lexicographer** n. **lexicographic** /-kəˈɡræfɪk/ adj. **lexicographical** adj. **lexicographically** adv.

lexicology /ˌleksɪˈkɒlədʒɪ/ n. the study of the form, history, and meaning of words. □ **lexicologist** n. **lexicological** /-kəˈlɒdʒɪk(ə)l/ adj. **lexicologically** adv.

lexicon /ˈleksɪkən/ n. **1** a dictionary, esp. of Greek, Hebrew, Syriac, or Arabic. **2** the vocabulary of a person, language, branch of knowledge, etc. [mod.L f. Gk lexikon (biblion book), neut. of lexikos f. lexis word f. legō speak]

lexigraphy /lekˈsɪɡrəfɪ/ n. a system of writing in which each character represents a word, as in Chinese. [Gk lexis (see LEXICON) + -GRAPHY]

Lexington /ˈleksɪŋtən/ **1** a city in north central Kentucky; pop. (1990) 225,270 (with Fayette). It is a noted horse-breeding centre. **2** a residential town north-west of Boston, Massachusetts; pop. (1990) 28,970. It was the scene in 1775 of the first battle in the War of American Independence.

lexis /ˈleksɪs/ n. **1** words, vocabulary. **2** the total stock of words in a language. [Gk: see LEXICON]

lex loci /ˌleks ˈləʊsaɪ/ n. Law the law of the country in which a transaction is performed, a tort is committed, or a property is situated. [L]

lex talionis /ˌleks tælɪˈəʊnɪs/ n. Law the law of retaliation, whereby a punishment resembles the offence committed, in kind and degree. [L]

ley[1] /leɪ/ n. a field temporarily under grass. □ **ley farming** esp. Brit. alternate growing of crops and grass. [ME (orig. adj.), perh. f. OE, rel. to LAY[1], LIE[1]]

ley[2] /liː, leɪ/ n. (usu. **ley-line**) a supposed straight line connecting prehistoric sites etc., often regarded as the line of an ancient track and credited by some with paranormal properties. [var. of LEA]

Leyden see LEIDEN.

Leyden jar /ˈlaɪd(ə)n, ˈleɪd/ n. an early form of capacitor consisting of a glass jar with layers of metal foil on the outside and inside. It was invented at Leiden in 1745.

Leyte /ˈleɪtɪ/ an island in the central Philippines; pop. (1990) 1,362,050; chief town, Tacloban.

LF abbr. low frequency.

LH *abbr. Biochem.* luteinizing hormone.

l.h. *abbr.* left hand.

Lhasa /ˈlɑːsə/ the capital of Tibet; pop. (1986) 108,000. It is situated in the northern Himalayas at an altitude of 3,600 m (*c.*11,800 ft), on a tributary of the Brahmaputra. Its inaccessibility and the hostility of the Tibetan Buddhist priests to foreign visitors — to whom Lhasa was closed until the 20th century — earned it the title of the Forbidden City. The spiritual centre of Tibetan Buddhism, Lhasa was the seat of the Dalai Lama until 1959, when direct Chinese administration was imposed on the city.

LI *abbr.* **1** Light Infantry. **2** *US* Long Island.

Li *symb. Chem.* the element lithium.

liability /ˌlaɪəˈbɪlɪtɪ/ *n.* (*pl.* **-ies**) **1** the state of being liable. **2** a person or thing that is troublesome as an unwelcome responsibility; a handicap. **3** what a person is liable for, esp. (in *pl.*) debts or pecuniary obligations.

liable /ˈlaɪəb(ə)l/ *predic.adj.* **1** legally bound. **2** (foll. by *to*) subject to (a tax or penalty). **3** (foll. by *to* + infin.) under an obligation. **4** (foll. by *to*) exposed or open to (something undesirable). **5** (foll. by *to* + infin.) apt, likely (*it is liable to rain*). **6** (foll. by *for*) answerable. [ME perh. f. AF f. OF *lier* f. L *ligare* bind]

liaise /lɪˈeɪz/ *v.intr.* (foll. by *with*, *between*) establish cooperation, act as a link. [back-form. f. LIAISON]

liaison /lɪˈeɪzɒn/ *n.* **1** communication or cooperation, esp. between military forces or units. **2** an illicit sexual relationship. **3** the binding or thickening agent of a sauce. **4** *Phonet.* the sounding of an ordinarily silent final consonant before a word beginning with a vowel (or a mute *h* in French). □ **liaison officer** an officer acting as a link between allied forces or units of the same force. [F f. *lier* bind f. L *ligare*]

liana /lɪˈɑːnə/ *n.* (also **liane** /-ˈɑːn/) a climbing and twining plant occurring in tropical forests. [F *liane*, *lierne* clematis, of uncert. orig.]

Liao¹ /ljaʊ/ a river of NE China, which rises in Inner Mongolia and flows about 1,450 km (900 miles) east and south to the Gulf of Liaodong at the head of the gulf of Bo Hai.

Liao² /ljaʊ/ a dynasty which ruled much of Manchuria and part of NE China AD 947–1125.

Liaodong Peninsula /ljaʊˈdʊŋ/ a peninsula in NE China, which extends southwards into the Yellow Sea between Bo Hai and Korea Bay. It contains the major port and industrial centre of Luda, and forms part of Liaoning province.

Liaoning /ljaʊˈnɪŋ/ a province of NE China, bordered on the east by North Korea; capital, Shenyang.

liar /ˈlaɪə(r)/ *n.* a person who tells a lie or lies, esp. habitually. □ **liar dice** a game with poker dice in which the result of a throw may be announced falsely. [OE *lēogere* (as LIE², -AR⁴)]

lias /ˈlaɪəs/ *n.* **1** (**Lias**) *Geol.* the lower strata of the Jurassic system of rocks, consisting of shales and limestones rich in fossils. **2** a blue limestone rock found in SW England. □ **liassic** /laɪˈæsɪk/ *adj.* (in sense 1). [ME f. OF *liois* hard limestone, prob. f. Gmc]

Lib. *abbr.* Liberal.

lib /lɪb/ *n. colloq.* (in names of political movements etc.) liberation (*women's lib*). [abbr.]

libation /laɪˈbeɪʃ(ə)n, lɪ-/ *n.* **1 a** the pouring out of a drink-offering to a god. **b** such a drink-offering. **2** *joc.* a drink. [ME f. L *libatio* f. *libare* pour as offering]

libber /ˈlɪbə(r)/ *n. colloq.* an advocate of women's liberation.

libel /ˈlaɪb(ə)l/ *n. & v.* ● *n.* **1** *Law* **a** a published false statement damaging to a person's reputation (cf. SLANDER). **b** the act of publishing this. **2 a** a false and defamatory written statement. **b** (foll. by *on*) a thing that brings discredit by misrepresentation etc. (*the portrait is a libel on him; the book is a libel on human nature*). **3 a** (in civil and ecclesiastical law) the plaintiff's written declaration. **b** *Sc. Law* a statement of the grounds of a charge. ● *v.tr.* (**libelled**, **libelling**; *US* **libeled**, **libeling**) **1** defame by libellous statements. **2** accuse falsely and maliciously. **3** *Law* publish a libel against. **4** (in ecclesiastical law) bring a suit against. □ **criminal libel** *Law* a deliberate defamatory statement in a permanent form. **public libel** *Law* a published libel. □ **libeller** *n.* [ME f. OF f. L *libellus* dimin. of *liber* book]

libellous /ˈlaɪbələs/ *adj.* containing or constituting a libel. □ **libellously** *adv.*

Liberace /ˌlɪbəˈrɑːtʃɪ/ (full name Wladziu Valentino Liberace) (1919–87), American pianist and entertainer. He was known for his romantic arrangements of popular piano classics and for his flamboyant costumes. His television show ran for five years (1952–7) and he gave many live performances; recordings of his performances sold millions of copies.

liberal /ˈlɪbərəl, ˈlɪbrəl/ *adj. & n.* ● *adj.* **1** given freely; ample, abundant. **2** (often foll. by *of*) giving freely, generous, not sparing. **3** open-minded, not prejudiced. **4** not strict or rigorous; (of interpretation) not literal. **5** for general broadening of the mind, not professional or technical (*liberal studies*). **6** *Polit.* **a** favouring individual liberty, free trade, and moderate political and social reform. **b** (**Liberal**) of or characteristic of Liberals or a Liberal Party. **7** *Theol.* regarding many traditional beliefs as dispensable, invalidated by modern thought, or liable to change (*liberal Protestant; liberal Judaism*). ● *n.* **1** a person of liberal views. **2** (**Liberal**) *Polit.* a supporter or member of a Liberal Party. □ **liberal arts 1** esp. *N. Amer.* the arts as distinct from science and technology. **2** *hist.* the medieval trivium and quadrivium. □ **liberalism** *n.* **liberalist** *n.* **liberally** *adv.* **liberalness** *n.* **liberalistic** /ˌlɪbərəˈlɪstɪk, ˌlɪbrə-/ *adj.* [ME, orig. = befitting a free man, f. OF f. L *liberalis* f. *liber* free (man)]

Liberal Democrats (in the UK) a party formed from the Liberal Party and members of the Social Democratic Party. The party was founded in 1988 as the Social and Liberal Democrats, and changed its name in 1989.

liberality /ˌlɪbəˈrælɪtɪ/ *n.* **1** free giving, munificence. **2** freedom from prejudice, breadth of mind. [ME f. OF *liberalite* or L *liberalitas* (as LIBERAL)]

liberalize /ˈlɪbərəˌlaɪz, ˈlɪbrə-/ *v.tr. & intr.* (also **-ise**) make or become more liberal or less strict. □ **liberalizer** *n.* **liberalization** /ˌlɪbərəlaɪˈzeɪʃ(ə)n, ˌlɪbrə-/ *n.*

Liberal Party a political party advocating liberal policies. The British Liberal Party emerged in the 1860s from the old Whig Party, encompassing not only former Whigs but also Radicals and former Peelite Conservatives. Dominated by Gladstone until the 1890s, the Liberal Party advocated free trade, political reform, and a restrained foreign policy. With Gladstone's conversion to Home Rule for Ireland, the party was weakened by the defection of the Liberal Unionists and only returned to a position of strength under Henry Campbell-Bannerman and Herbert Asquith in the decade before the First World War. Lloyd George's revolt against Asquith's wartime administration fatally weakened the Liberals and after the war they lost their position as one of the two major parties to Labour. The Liberals have not regained their former eminence and have since gained few parliamentary seats. The name was discontinued in official use in 1988, when the party regrouped with elements of the Social Democratic Party to form the Social and Liberal Democrats, now known as the Liberal Democrats.

Liberal Party of Australia an Australian political party established in its modern form by Robert Menzies in 1944, in opposition to the Australian Labor Party. It first gained power in 1949, being in office until 1972 and then in 1975–83.

Liberal Party of Canada a Canadian political party generally taking a moderate, left-of-centre position. The party emerged in the mid-19th century, and during the twentieth century held power for most of the period 1963–84, for over half of that time under Pierre Trudeau; it was defeated by the Progressive Conservative Party in 1984.

Liberal Unionist *n.* a member of a group of British Liberal MPs who left the party in 1886 because of Gladstone's support for Irish Home Rule. Led by Joseph Chamberlain from 1891, the Liberal Unionists formed an alliance with the Conservative Party in Parliament, and merged officially with them in 1909 as the Conservative and Unionist Party.

liberate /ˈlɪbəˌreɪt/ *v.tr.* **1** (often foll. by *from*) set at liberty, set free. **2** free (a country etc.) from an oppressor or an enemy occupation. **3** (often as **liberated** *adj.*) free (a person) from rigid social conventions, esp. in sexual behaviour. **4** *sl.* steal. **5** *Chem.* release (esp. a gas) from a state of combination. □ **liberator** *n.* [L *liberare liberat-* f. *liber* free]

liberation /ˌlɪbəˈreɪʃ(ə)n/ *n.* the act or an instance of liberating; the state of being liberated. □ **liberationist** *n.* [ME f. L *liberatio* f. *liberare*: see LIBERATE]

liberation theology *n.* a movement in Christian theology developed in the 1960s, principally by Latin American Roman Catholics such as Gustavo Gutiérrez. Drawing to an extent on Marxism, liberation theology attempts to address the problems of poverty and social injustice as well as spiritual matters; it interprets liberation from social, political, and economic oppression as an anticipation of ultimate salvation. Liberation theology has attracted criticism from Pope John Paul II and other Catholic authorities.

Liberia /laɪˈbɪərɪə/ a country on the Atlantic coast of West Africa; pop. (est. 1991) 2,639,000; languages, English (official), English-based pidgin; capital, Monrovia. Liberia was founded in 1822 as a settlement for freed slaves from the US, and was proclaimed independent in 1847. Indigenous peoples, however, form the majority of the population. In 1980, following a military coup that overthrew the predominant Liberian–American élite, Samuel (Kanyon) Doe (1950–90) assumed power, but was himself overthrown and murdered during a civil war that began in 1990. After the war ended in July 1993 an interim government was installed, which scheduled multi-party elections for late 1994. □ **Liberian** adj. & n. [L liber free]

libertarian /ˌlɪbəˈteərɪən/ n. & adj. ● n. **1** an advocate of liberty. **2** Philos. a believer in free will (opp. NECESSITARIAN). **3** an adherent of libertarianism. ● adj. believing in free will or libertarianism.

libertarianism /ˌlɪbəˈteərɪəˌnɪz(ə)m/ n. a libertarian position; specifically, an extreme laissez-faire political philosophy, popular today especially in the US, advocating only minimal state intervention in the lives of citizens. Its adherents believe that private morality is not the state's affair, and that therefore activities such as drug use and prostitution that arguably harm no one but the participants should not be illegal. Libertarianism has some common ground with anarchism, although it is generally associated more with the political right; it differs from traditional liberalism in being less concerned with social justice.

libertine /ˈlɪbəˌtiːn, -tɪn, -ˌtaɪn/ n. & adj. ● n. **1** a dissolute or licentious person. **2** a free thinker on religion. **3** a person who follows his or her own inclinations. ● adj. **1** licentious, dissolute. **2** freethinking. **3** following one's own inclinations. □ **libertinage** n. **libertinism** n. [L libertinus freedman f. libertus made free f. liber free]

liberty /ˈlɪbətɪ/ n. (pl. **-ies**) **1 a** freedom from captivity, imprisonment, slavery, or despotic control. **b** a personification of this. **2 a** the right or power to do as one pleases. **b** (foll. by to + infin.) right, power, opportunity, permission. **c** Philos. freedom from control by fate or necessity. **3 a** (usu. in pl.) a right, privilege, or immunity, enjoyed by prescription or grant. **b** (in sing. or pl.) hist. an area having such privileges etc., esp. a district controlled by a city though outside its boundary or an area outside a prison where some prisoners might reside. **4** setting aside of rules or convention. □ **at liberty 1** free, not imprisoned. **2** (foll. by to + infin.) entitled, permitted. **3** available, disengaged. **liberty boat** Brit. Naut. a boat carrying liberty men. **liberty bodice** a close-fitting under-bodice. **liberty hall** a place where one may do as one likes. **liberty horse** a horse performing in a circus without a rider. **liberty man** Brit. Naut. a sailor with leave to go ashore. **liberty of the subject** the rights of a subject under constitutional rule. **take liberties 1** (often foll. by with) behave in an unduly familiar manner. **2** (foll. by with) deal freely or superficially with rules or facts. **take the liberty** (foll. by to + infin., or of + verbal noun) presume, venture. [ME f. OF liberté f. L libertas -tatis f. liber free]

Liberty, Statue of a statue on an island at the entrance to New York harbour, a symbol of welcome to immigrants, representing a draped female figure carrying a book of laws in her left hand and holding aloft a torch in her right. Dedicated in 1886, it was designed by the French sculptor Frédéric-Auguste Bartholdi (1834–1904) (with inner structure by Alexandre Eiffel) and was the gift of the French to the American people, commemorating the alliance of France and the US during the War of American Independence and marking its centenary.

Liberty Bell a large bell that is the traditional symbol of US freedom, bearing the legend 'Proclaim liberty throughout all the land unto all the inhabitants thereof' (Leviticus 25:10). First rung on 8 July 1776 in Philadelphia to celebrate the first public reading of the Declaration of Independence, it cracked irreparably when rung for George Washington's birthday in 1846. It is now housed near Independence Hall, Philadelphia.

Liberty ship n. hist. a prefabricated US-built freighter of the war of 1939–45.

libidinous /lɪˈbɪdɪnəs/ adj. lustful. □ **libidinously** adv. **libidinousness** n. [ME f. L libidinosus f. libido -dinis lust]

libido /lɪˈbiːdəʊ/ n. (pl. **-os**) Psychol. psychic drive or energy, esp. that associated with sexual desire. □ **libidinal** /-ˈbɪdɪn(ə)l/ adj. **libidinally** adv. [L: see LIBIDINOUS]

Lib-Lab /lɪbˈlæb/ adj. Brit. hist. Liberal and Labour. [abbr.]

Li Bo see LI PO.

Libra /ˈliːbrə, ˈlɪb-, ˈlaɪb-/ n. **1** Astron. a small constellation (the Scales or Balance), said to represent the pair of scales which is the symbol of justice. It contains no bright stars. **2** Astrol. **a** the seventh sign of the zodiac, which the sun enters at the northern autumnal equinox (about 22 Sept.). **b** a person born when the sun is in this sign. □ **Libran** n. & adj. [ME f. L, orig. = pound weight]

librarian /laɪˈbreərɪən/ n. a person in charge of, or an assistant in, a library. □ **librarianship** n. [L librarius: see LIBRARY]

library /ˈlaɪbrərɪ/ n. (pl. **-ies**) **1 a** a collection of books etc. for use by the public or by members of a group. (See note below.) **b** a person's collection of books. **2** a room or building containing a collection of books (for reading or reference rather than for sale). **3 a** a similar collection of films, records, computer routines, etc. **b** the place where these are kept. **4** a series of books issued by a publisher in similar bindings etc., usu. as a set. **5** a public institution charged with the care of a collection of books, films, etc. □ **library edition** a strongly bound edition. **library school** a college or a department in a university or polytechnic teaching librarianship. **library science** the study of librarianship. [ME f. OF librairie f. L libraria (taberna shop), fem. of librarius bookseller's, of books, f. liber libri book]

▪ Libraries have existed since ancient times, famously at Nineveh, Alexandria, and Athens, where the first public library opened in the 4th century BC. All ancient libraries were reference libraries, where books could be consulted but not borrowed; lending or circulating libraries did not appear until the 18th century, becoming accessible to the general public in the 19th century with the growth of literacy and the development of free public book-lending services supported by philanthropists such as Andrew Carnegie. Libraries have today become regarded as a public service and a vital source of knowledge and education; cataloguing systems such as Dewey decimal classification have been supplemented in recent years by the application of information technology, and in addition to books many libraries now collect videos and sound recordings. Most countries operate a system of copyright or legal deposit libraries which have a right to receive a copy of every work published in that country; there are five of these in Britain, including the Bodleian Library in Oxford, the British Library in London, and the National Library of Scotland in Edinburgh (see COPYRIGHT LIBRARY).

Library of Congress the US national library, in Washington, DC. It was established in 1800, originally for the benefit of members of the US Congress, and was at first housed in the Capitol, moving to its present site in 1897.

libration /laɪˈbreɪʃ(ə)n/ n. Astron. an apparent oscillation of a celestial body, esp. the moon, by which the parts near the edge of the disc are alternately in view and out of view. [L libratio f. librare f. libra balance]

libretto /lɪˈbretəʊ/ n. (pl. **libretti** /-tɪ/ or **-os**) the text of an opera or other long musical vocal work. □ **librettist** n. [It., dimin. of libro book f. L liber libri]

Libreville /ˈliːbrəˌvɪl/ the capital of Gabon, a port on the Atlantic coast at the mouth of the Gabon river; pop. (1983) 350,000.

Librium /ˈlɪbrɪəm/ n. propr. a white crystalline benzodiazepine drug used as a tranquillizer.

Libya /ˈlɪbɪə/ a country in North Africa; pop. (est. 1991) 4,714,000; official language, Arabic; capital, Tripoli. Much of Libya forms part of the Sahara Desert, with a narrow coastal plain bordering the Mediterranean; the country has major oil deposits. Having formed part of the Roman Empire, Libya was conquered by the Arabs, brought under Turkish domination in the 16th century, and annexed by Italy in 1912 and partially colonized. During the Second World War it was the scene of heavy fighting, and, after a period of French and British administration, became an independent kingdom in 1951. After prolonged unrest, the monarchy was overthrown in 1969 and the country emerged with a radical revolutionary leadership. Alleged support for international terrorism has brought Libya into conflict with Western countries, and the US launched a punitive air strike against Libyan targets in 1986.

Libyan /ˈlɪbɪən/ adj. & n. ● adj. **1** of or relating to modern Libya. **2** of ancient North Africa west of Egypt. **3** of or relating to the Berber group of languages. ● n. **1 a** a native or national of modern Libya. **b** a person of Libyan descent. **2** an ancient language of the Berber group.

lice pl. of LOUSE.

licence /ˈlaɪs(ə)ns/ n. (US **license**) **1** a permit from an authority to own or use something (esp. a dog, gun, television set, or vehicle), do something (esp. marry, print something, preach, or drive on a public road), or carry on a trade (esp. in alcoholic liquor). **2** leave, permission (have I your licence to remove the fence?). **3 a** liberty of action, esp. when excessive; disregard of law or propriety, abuse of freedom. **b** licentiousness. **4** a writer's or artist's irregularity in grammar, metre, perspective, etc., or deviation from fact, esp. for effect (*poetic licence*). **5** a university certificate of competence in a faculty. □ **license plate** *US* the number plate of a licensed vehicle. [ME f. OF f. L *licentia* f. *licere* be lawful: *-se* by confusion with LICENSE]

license /'laɪs(ə)ns/ *v.tr.* (also **licence**) **1** grant a licence to (a person). **2** authorize the use of (premises) for a certain purpose, esp. the sale and consumption of alcoholic liquor. **3** authorize the publication of (a book etc.) or the performance of (a play). **4** *archaic* allow. □ **licensed victualler** see VICTUALLER 1b. □ **licensable** *adj.* **licenser** *n.* **licensor** *n.* [ME f. LICENCE: *-se* on analogy of the verbs PRACTISE, PROPHESY, perh. after ADVISE, where the sound differs from the corresp. noun]

licensee /ˌlaɪs(ə)n'siː/ *n.* the holder of a licence, esp. to sell alcoholic liquor.

licentiate /laɪ'senʃɪət, -ʃət/ *n.* **1** a holder of a certificate of competence to practise a certain profession, or of a university licence. **2** a licensed preacher not yet having an appointment, esp. in a Presbyterian church. [ME f. med.L *licentiatus* past part. of *licentiare* f. L *licentia*: see LICENCE]

licentious /laɪ'senʃəs/ *adj.* **1** immoral, esp. in sexual relations. **2** *archaic* disregarding accepted rules or conventions. □ **licentiously** *adv.* **licentiousness** *n.* [L *licentiosus* f. *licentia*: see LICENCE]

lichee var. of LYCHEE.

lichen /'laɪkən, 'lɪtʃən/ *n.* **1** a composite plant organism, composed of a fungus and an alga in symbiotic association, usu. of green, grey, or yellow tint and growing on and colouring rocks, trees, roofs, walls, etc. (*See note below.*) **2** *Med.* any skin disease in which small round hard lesions occur close together. □ **lichened** *adj.* (in sense 1). **lichenous** *adj.* (in sense 2). **lichenology** /ˌlaɪkə'nɒlədʒɪ, ˌlɪtʃə-/ *n.* (in sense 1). [L f. Gk *leikhēn*]

▪ There are encrusting, leaflike, and branching forms of lichen, and they are now classified mainly by their fungal component. The alga provides nutrients for the fungus, and the fungus presumably supplies a suitable protective environment. Lichens are small, slow-growing, and very hardy, often growing on bare rock surfaces and in polar regions. Some form an important source of food for browsing animals, especially reindeer or caribou, in the tundra. Their economic uses include the production of dyes, and they are also sensitive pollution indicators.

Lichfield /'lɪtʃfiːld/ a town in central England, in Staffordshire north of Birmingham; pop. (1981) 25,740. It was the birthplace of Samuel Johnson.

lich-gate /'lɪtʃgeɪt/ *n.* (also **lych-gate**) a roofed gateway to a churchyard, formerly used for sheltering a coffin until the clergyman's arrival for a burial. [ME f. OE *līc* corpse f. Gmc + GATE[1]]

Lichtenstein /'lɪktənˌstaɪn/, Roy (b.1923), American painter and sculptor. A leading exponent of pop art, he chiefly based his work on images from commercial art or made parodies of the work of other painters. He became known in the 1960s for paintings inspired by comic strips; these paintings are made up of thick black outlines enclosing areas of dots in imitation of the half-tones used to print blocks of colour in comics. One of the best-known examples of this style is *Whaam!* (1963).

licit /'lɪsɪt/ *adj.* not forbidden; lawful. □ **licitly** *adv.* [L *licitus* past part. of *licere* be lawful]

lick /lɪk/ *v. & n.* ● *v.* **1** *tr.* pass the tongue over, esp. to taste, moisten, or (of animals) clean. **2** *tr.* bring into a specified condition or position by licking (*licked it all up; licked it clean*). **3** a *tr.* (of a flame, waves, etc.) touch; play lightly over. **b** *intr.* move gently or caressingly. **4** *tr. colloq.* **a** defeat, overcome. **b** excel, surpass. **c** surpass the comprehension of (*has got me licked*). **5** *tr. colloq.* thrash. ● *n.* **1** an act of licking with the tongue. **2** = *salt-lick.* **3** *colloq.* a fast pace (*at a lick; at full lick*). **4** *colloq.* a small amount, quick treatment with (foll. by *of: a lick of paint*). **5** a smart blow with a stick etc. □ **a lick and a promise** *colloq.* a hasty performance of a task, esp. of washing oneself. **lick a person's boots** (or **shoes**) toady; be servile. **lick into shape** see SHAPE. **lick one's lips** (or **chops**) **1** look forward with relish. **2** show one's satisfaction. **lick one's wounds** be in retirement after defeat. □ **licker** *n.* (also in comb.). [OE *liccian* f. WG]

lickerish /'lɪkərɪʃ/ *adj.* (also **liquorish**) **1** lecherous. **2 a** fond of fine food. **b** greedy, longing. [ME *lickerous* f. OF *lecheros*: see LECHER]

lickety-split /ˌlɪkɪtɪ'splɪt/ *adv. colloq.* at full speed; headlong. [prob. f. LICK *n.* 3 + SPLIT]

licking /'lɪkɪŋ/ *n. colloq.* **1** a thrashing. **2** a defeat.

lickspittle /'lɪkˌspɪt(ə)l/ *n.* a toady.

licorice var. of LIQUORICE.

lictor /'lɪktɔː(r)/ *n.* (usu. in *pl.*) *Rom. Hist.* an officer attending the consul or other magistrate, bearing the fasces, and executing sentence on offenders. [ME f. L, perh. rel. to *ligare* bind]

lid /lɪd/ *n.* **1** a hinged or removable cover, esp. for the top of a container. **2** = EYELID. **3** the operculum of a shell or a plant. **4** *sl.* a hat. □ **put the lid** (or **tin lid**) **on** *Brit. colloq.* **1** be the culmination of. **2** put a stop to. **take the lid off** *colloq.* expose (a scandal etc.). □ **lidded** *adj.* (also in comb.). **lidless** *adj.* [OE *hlid* f. Gmc]

Liddell /'lɪd(ə)l/, Eric (1902–45), British athlete and missionary, born in China. In the 1924 Olympic Games the heats of his own event (the 100 metres) were held on a Sunday and he withdrew on religious grounds; he ran in the 400 metres instead, winning the race in a world record time. His exploits were celebrated in the film *Chariots of Fire* (1981). Liddell went on to serve as a missionary in China and died there in a Japanese prisoner-of-war camp.

Liddell Hart /ˌlɪd(ə)l 'hɑːt/, Sir Basil Henry (1895–1970), British military historian and theorist. Appalled at the slaughter produced by trench warfare in the First World War, he formulated a strategy using an indirect approach, in which attacks would be made with tanks and aircraft to destroy enemy command centres, communications, and supply lines. His theories were particularly influential in Germany, where the idea of strategic penetration by tank divisions was successfully adopted in the Second World War.

Lidingö /'liːdɪŋɜː/ an island suburb of Stockholm in Sweden, in the north-east of the city; pop. (1990) 38,400.

Lido /'liːdəʊ/ **1** (in full **Lido di Malamocco** /dɪ ˌmæləˈmɒkəʊ/) an island reef off the coast of NE Italy, in the northern Adriatic. It separates the Lagoon of Venice from the Gulf of Venice. **2** (also **the Lido**) a town and beach resort in NE Italy, on the Lido reef opposite Venice; pop. (1980) 20,950.

lido /'liːdəʊ, 'laɪdəʊ/ *n.* (pl. **-os**) a public open-air swimming-pool or bathing-beach. [It. f. LIDO 2, f. L *litus* shore]

lidocaine /'lɪdəˌkeɪn/ *n. Pharm.* = LIGNOCAINE.

Lie /liː/, Trygve Halvdan (1896–1968), Norwegian Labour politician, first Secretary-General of the United Nations 1946–53.

lie[1] /laɪ/ *v. & n.* ● *v.intr.* (**lying**; *past* **lay** /leɪ/; *past part.* **lain** /leɪn/) **1** be in or assume a horizontal position on a supporting surface; be at rest on something. **2** (of a thing) rest flat on a surface (*snow lay on the ground*). **3** (of abstract things) remain undisturbed or undiscussed etc. (*let matters lie*). **4 a** be kept or remain or be in a specified, esp. concealed, state or place (*lie hidden; lie in wait; malice lay behind those words; they lay dying; the books lay unread; the money is lying in the bank*). **b** (of abstract things) exist, reside; be in a certain position or relation (foll. by *in, with,* etc.: *the answer lies in education; my sympathies lie with the family*). **5 a** be situated or stationed (*the village lay to the east; the ships are lying off the coast*). **b** (of a road, route, etc.) lead (*the road lies over mountains*). **c** be spread out to view (*the desert lay before us*). **6** (of the dead) be buried in a grave. **7** (foll. by *with*) *archaic* have sexual intercourse. **8** *Law* be admissible or sustainable (*the objection will not lie*). **9** (of a game bird) not rise. ● *n.* **1 a** the way or direction or position in which a thing lies. **b** *Golf* the position of a golf ball when about to be struck. **2** the place of cover of an animal or a bird. □ **as far as in me lies** to the best of my power. **let lie** not raise (a controversial matter etc.) for discussion etc. **lie about** (or **around**) be left carelessly out of place. **lie ahead** be going to happen; be in store. **lie back** recline so as to rest. **lie down** assume a lying position; have a short rest. **lie-down** *n.* a short rest. **lie down under** *Brit.* accept (an insult etc.) without protest. **lie heavy** cause discomfort or anxiety. **lie in 1** remain in bed in the morning. **2** *archaic* be brought to bed in childbirth. **lie-in** *n.* a prolonged stay in bed in the morning. **lie in state** (of a deceased person of high rank) be laid in a public place of honour before burial. **lie low 1** keep quiet or unseen. **2** be discreet about one's intentions. **lie off** *Naut.* stand some distance from shore or from another ship. **the lie of the land** the current state of affairs. **lie over** be deferred. **lie to** *Naut.* come almost to a stop facing the wind. **lie up** (of a ship) go into dock or be out of commission. **lie with** (often foll. by *to* + infin.) be the responsibility of (a person) (*it lies with you to answer*). **take lying down** (usu. with *neg.*) accept (defeat, rebuke, etc.) without resistance or protest etc. [OE *licgan* f. Gmc]

lie[2] /laɪ/ *n. & v.* ● *n.* **1** an intentionally false statement (*tell a lie; pack of lies*). **2** imposture; false belief (*live a lie*). ● *v.* (**lies, lied, lying**) **1** *intr.* **a** tell a lie or lies (*they lied to me*). **b** (of a thing) be deceptive (*the camera cannot lie*). **2** *tr.* (usu. *refl.*; foll. by *into, out of*) get (oneself) into or out of a situation by lying (*lied themselves into trouble; lied my way out of danger*).

☐ **give the lie to** serve to show the falsity of (a supposition etc.). [OE *lyge lēogan* f. Gmc]

Liebfraumilch /'liːbfraʊˌmɪlk, -ˌmɪlx/ n. a light white wine from the Rhine region. [G f. *Liebfrau* the Virgin Mary, the patroness of the convent where it was first made + *Milch* milk]

Liebig /'liːbɪx/, Justus von, Baron (1803–73), German chemist and teacher. With Friedrich Wöhler he discovered the benzoyl radical, and demonstrated that such radicals were groups of atoms that remained unchanged in many chemical reactions. He applied chemistry to physiology and to agriculture, stressed the importance of artificial fertilizers, and developed techniques for quantitative organic analysis.

Liechtenstein /'lɪktənˌstaɪn/ a small independent principality in the Alps, between Switzerland and Austria; pop. (1990) 28,880; official language, German; capital, Vaduz. The principality was created in 1719 within the Holy Roman Empire, becoming independent of the German confederation in 1866. Liechtenstein is economically integrated with Switzerland. ☐ **Liechtensteiner** n.

lied /liːd, liːt/ n. (pl. **lieder** /'liːdə(r)/) a type of German song, esp. of the romantic period, usu. for solo voice with piano accompaniment. [G]

lie-detector /'laɪdɪˌtektə(r)/ n. an instrument intended to determine whether a person is telling the truth by testing for physiological changes considered to be symptomatic of lying. First devised by John A. Larson in California in 1921, lie-detectors are generally not accepted for judicial purposes.

lief /liːf/ adv. archaic gladly, willingly (usu. **had lief, would lief**). [orig. as adj. f. OE *lēof* dear, pleasant, f. Gmc, rel. to LEAVE², LOVE]

Liège /lɪ'eɪʒ/ (called in Flemish *Luik*) **1** a province of eastern Belgium. Formerly ruled by independent prince-bishops, it became a part of the Netherlands in 1815 and of Belgium in 1830. **2** its capital city; pop. (1991) 194,600. Situated at the junction of the Meuse and Ourthe rivers, it is a major river port and industrial centre.

liege /liːdʒ/ adj. & n. usu. hist. ● adj. (of a superior) entitled to receive or (of a vassal) bound to give feudal service or allegiance. ● n. **1** (in full **liege lord**) a feudal superior or sovereign. **2** (usu. in pl.) a vassal or subject. [ME f. OF *lige, liege* f. med.L *laeticus*, prob. f. Gmc]

liegeman /'liːdʒmæn/ n. (pl. **-men**) hist. a sworn vassal; a faithful follower.

lien /'liːən/ n. Law a right over another's property to protect a debt charged on that property. [F f. OF *loien* f. L *ligamen* bond f. *ligare* bind]

lierne /lɪ'ɜːn/ n. Archit. (in vaulting) a short rib connecting the bosses and intersections of the principal ribs. [ME f. F: see LIANA]

lieu /ljuː/ n. ☐ **in lieu 1** instead. **2** (foll. by of) in the place of. [ME f. F f. L *locus* place]

Lieut. abbr. Lieutenant.

lieutenant /lef'tenənt/ n. **1** a deputy or substitute acting for a superior. **2 a** an army officer next in rank below captain. **b** a naval officer next in rank below lieutenant commander. **3** US a police officer next in rank below captain. ☐ **lieutenant colonel** an army officer ranking next below colonel. **lieutenant commander** a naval officer ranking below a commander and above a lieutenant. **lieutenant general** an army officer ranking below a general and above a major-general. **lieutenant-governor** the acting or deputy governor of a state, province, etc., under a governor or Governor-General. ☐ **lieutenancy** n. (pl. **-ies**). [ME f. OF (as LIEU, TENANT)]

Lieutenant of the Tower the acting commandant of the Tower of London.

life /laɪf/ n. (pl. **lives** /laɪvz/) **1** the condition which distinguishes active animals and plants from inorganic matter, including the capacity for growth, functional activity, and continual change preceding death. **2 a** living things and their activity (*insect life; is there life on Mars?*). **b** human presence or activity (*no sign of life*). **3 a** the period during which life lasts, or the period from birth to the present time or from the present time to death (*have done it all my life; will regret it all my life; life membership*). **b** the duration of a thing's existence or of its ability to function; validity, efficacy, etc. (*the battery has a life of two years*). **4 a** a person's state of existence as a living individual (*sacrificed their lives; took many lives*). **b** a living person (*many lives were lost*). **5 a** an individual's occupation, actions, or fortunes; the manner of one's existence (*that would make life easy; start a new life*). **b** a particular aspect of this (*love-life; private life*). **6** the active part of existence; the business and pleasures of the world (*travel is the best way to see life*). **7** humankind's earthly or supposed future existence (*this life and the next*). **8 a** energy, liveliness, animation (*full of life; put some life into it!*). **b** an animating influence

(*was the life of the party*). **c** (of an inanimate object) power, force; ability to perform its intended function. **9** the living, esp. nude, form or model (*taken from the life*). **10** a written account of a person's life; a biography. **11** colloq. a sentence of imprisonment for life (*they were all serving life*). **12** a chance; a fresh start (*cats have nine lives; gave the player three lives*). ☐ **come to life 1** emerge from unconsciousness or inactivity; begin operating. **2** (of an inanimate object) assume an imaginary animation. **for dear** (or **one's**) **life** as if or in order to escape death; as a matter of extreme urgency (*hanging on for dear life; run for your life*). **for life** for the rest of one's life. **for the life of** (foll. by pers. pron.) even if (one's) life depended on it (*cannot for the life of me remember*). **give one's life 1** (foll. by for) die; sacrifice oneself. **2** (foll. by to) dedicate oneself. **large as life** in person, esp. prominently (*stood there large as life*). **larger than life 1** exaggerated. **2** (of a person) having an exuberant or striking personality. **life-and-death** vitally important; desperate (*a life-and-death struggle*). **life cycle** the series of changes in the life of an organism including reproduction. **life expectancy** the average period that a person at a specified age may expect to live. **life-force** inspiration or a driving force or influence. **life-form** an organism. **life-giving** sustaining life; uplifting and revitalizing. **life history** the story of a person's life, esp. told at tedious length. **life insurance** insurance for a sum to be paid on the death of the insured person. **life-jacket** a buoyant or inflatable jacket for keeping a person afloat in water. **life peer** Brit. a peer whose title lapses on death. **life peerage** Brit. a title held by a life peer. **life-preserver 1** a short stick with a heavily loaded end. **2** a life-jacket etc. **life-raft** an inflatable or timber etc. raft for use in an emergency instead of a boat. **life-saver** colloq. **1** a thing that saves one from serious difficulty. **2** Austral. & NZ = LIFEGUARD. **life sciences** biology and related subjects. **life sentence 1** a sentence of imprisonment for life. **2** an illness or commitment etc. perceived as a continuing threat to one's freedom. **life-size** (or **-sized**) of the same size as the person or thing represented. **life-support** attrib.adj. (of equipment) allowing vital functions to continue in an adverse environment or during severe disablement. **life-support machine** Med. a ventilator or respirator. **life's-work** a task etc. pursued throughout one's lifetime. **lose one's life** be killed. **a matter of life and death** a matter of vital importance. **not on your life** colloq. most certainly not. **save a person's life 1** prevent a person's death. **2** save a person from serious difficulty. **take one's life in one's hands** take a crucial personal risk. **to the life** true to the original. [OE *līf* f. Gmc]

lifebelt /'laɪfbelt/ n. a belt of buoyant or inflatable material for keeping a person afloat in water.

lifeblood /'laɪfblʌd/ n. **1** the blood, as being necessary to life. **2** the vital factor or influence.

lifeboat /'laɪfbəʊt/ n. **1** a specially constructed boat for rescuing people in distress at sea, launched from the land. (*See note below.*) **2** a ship's small boat for use in emergency.

▪ The first lifeboat is often considered to be the converted fishing-boat operated by Lionel Lukin (1742–1834) off the coast of NE England in 1785, although Chinese vessels may have predated this. Lifeboats in Britain are now specially built self-righting vessels, operated by the Royal National Lifeboat Institution.

lifebuoy /'laɪfbɔɪ/ n. a buoyant support (usu. a ring) for keeping a person afloat in water.

lifeguard /'laɪfɡɑːd/ n. an expert swimmer employed to rescue bathers from drowning.

Life Guards n. (in the UK) a regiment of the royal Household Cavalry.

lifeless /'laɪflɪs/ adj. **1** lacking life; no longer living. **2** unconscious. **3** lacking movement or vitality. ☐ **lifelessly** adv. **lifelessness** n. [OE *līflēas* (as LIFE, -LESS)]

lifelike /'laɪflaɪk/ adj. closely resembling the person or thing represented. ☐ **lifelikeness** n.

lifeline /'laɪflaɪn/ n. **1 a** a rope etc. used for life-saving, e.g. that attached to a lifebuoy. **b** a diver's signalling line. **2** a sole means of communication or transport. **3** a fold in the palm of the hand, regarded as significant in palmistry. **4** an emergency telephone counselling service.

lifelong /'laɪflɒŋ/ adj. lasting a lifetime.

lifer /'laɪfə(r)/ n. sl. a person serving a life sentence.

lifestyle /'laɪfstaɪl/ n. the particular way of life of a person or group.

lifetime /'laɪftaɪm/ n. **1** the duration of a person's life. **2** the duration of a thing or its usefulness. **3** colloq. an exceptionally long time. ☐ **of a**

lifetime such as does not occur more than once in a person's life (*the chance of a lifetime; the journey of a lifetime*).

Liffe /ˈlɪfɪ, laɪf/ (also **LIFFE**) London International Financial Futures Exchange. [acronym]

Liffey /ˈlɪfɪ/ a river of eastern Ireland, which flows for 80 km (50 miles) from the Wicklow Mountains to Dublin Bay. The city of Dublin is situated at its mouth.

Lifford /ˈlɪfəd/ the county town of Donegal, in the Republic of Ireland; pop. (1986) 1,460.

lift /lɪft/ *v. & n.* ● *v.* **1** *tr.* (often foll. by *up, off, out*, etc.) raise or remove to a higher position. **2** *intr.* go up; be raised; yield to an upward force (*the window will not lift*). **3** *tr.* give an upward direction to (the eyes or face). **4** *tr.* **a** elevate to a higher plane of thought or feeling (*the news lifted their spirits*). **b** make less heavy or dull; add interest to (something esp. artistic). **c** enhance, improve (*lifted their game after half-time*). **5** *intr.* (of a cloud, fog, etc.) rise, disperse. **6** *tr.* remove (a barrier or restriction). **7** *tr.* transport (supplies, troops, etc.) by air. **8** *tr. colloq.* **a** steal. **b** plagiarize (a passage of writing etc.). **9** *Phonet.* **a** *tr.* make louder; raise the pitch of. **b** *intr.* (of the voice) rise. **10** *tr.* dig up (esp. potatoes etc. at harvest). **11** *intr.* (of a floor) swell upwards, bulge. **12** *tr.* hold or have on high (*the church lifts its spire*). **13** *tr.* hit (a cricket-ball) into the air. **14** *tr.* (usu. in *passive*) perform cosmetic surgery on (esp. the face or breasts) to reduce sagging. ● *n.* **1** the act of lifting or process of being lifted. **2** a free ride in another person's vehicle (*gave them a lift*). **3 a** *Brit.* = ELEVATOR 3a. **b** a similar apparatus for carrying persons up or down a mountain etc. (see *ski-lift*). **4 a** transport by air (see AIRLIFT *n.*). **b** a quantity of goods transported by air. **5** the upward pressure which air exerts on an aerofoil to counteract the force of gravity. **6** a supporting or elevating influence; a feeling of elation. **7** a layer of leather in the heel of a boot or shoe, esp. to correct shortening of a leg or increase height. **8 a** a rise in the level of the ground. **b** the extent to which water rises in a canal lock. □ **lift down** pick up and bring to a lower position. **lift a finger** (or **hand** etc.) (in *neg.*) make the slightest effort (*didn't lift a finger to help*). **lift off** (of a spacecraft or rocket) rise from the launching pad. **lift-off** *n.* the vertical take-off of a spacecraft or rocket. **lift up one's head** hold one's head high with pride. **lift up one's voice** sing out. □ **liftable** *adj.* **lifter** *n.* [ME f. ON *lypta* f. Gmc]

lig /lɪg/ *v.intr.* (**ligged, ligging**) *sl.* **1** idle or lie about. **2** sponge, freeload; attend or gatecrash parties. □ **ligger** *n.* [f. dial. var. of LIE¹]

ligament /ˈlɪgəmənt/ *n.* **1** *Anat.* **a** a short band of tough flexible fibrous connective tissue linking bones together. **b** any membranous fold keeping an organ in position. **2** *archaic* a bond of union. □ **ligamental** /ˌlɪgəˈment(ə)l/ *adj.* **ligamentary** *adj.* **ligamentous** *adj.* [ME f. L *ligamentum* bond f. *ligare* bind]

ligand /ˈlɪgənd/ *n.* **1** *Chem.* an ion or molecule attached to a metal atom by covalent bonding in which both electrons are supplied by one atom. **2** *Biochem.* a molecule that binds to another (usu. larger) molecule. [L *ligandus* gerundive of *ligare* bind]

ligate /lɪˈgeɪt/ *v.tr. Surgery* tie up (a bleeding artery etc.). □ **ligation** /-ˈgeɪʃ(ə)n/ *n.* [L *ligare* ligat-]

ligature /ˈlɪgətʃə(r)/ *n. & v.* ● *n.* **1** a tie or bandage, esp. in surgery for a bleeding artery etc. **2** *Mus.* a slur; a tie. **3** *Printing* two or more letters joined, e.g. æ. **4** a bond; a thing that unites. **5** the act of tying or binding. ● *v.tr.* bind or connect with a ligature. [ME f. LL *ligatura* f. L *ligare* ligat- tie, bind]

liger /ˈlaɪgə(r)/ *n.* the offspring of a lion and a tigress (cf. TIGON). [portmanteau word f. LION + TIGER]

Ligeti /ˈlɪgətɪ/, György Sándor (b.1923), Hungarian composer. His early works employ electronic instruments, but he made his name when he returned to the traditional orchestra with *Apparitions* (1958–9) and *Atmosphères* (1961). These compositions are remarkable for their use of slowly evolving complexes of polyphonic sound, dispensing with the formal elements of melody, harmony, and rhythm. His many other compositions include choral works, such as the *Requiem* (1963–5), string quartets, and pieces satirizing the work of other composers.

light¹ /laɪt/ *n., v., & adj.* ● *n.* **1** the natural agent that stimulates sight and makes things visible, and consists of electromagnetic radiation of wavelength between about 390 and 740 nm. (*See note below.*) **2** the medium or condition of the space in which this is present. **3** an appearance of brightness (*saw a distant light*). **4 a** a source of light, e.g. the sun, or a lamp, fire, etc. **b** (in *pl.*) illuminations. **5** (often in *pl.*) a traffic light (*went through a red light; stop at the lights*). **6 a** the amount or quality of illumination in a place (*bad light stopped play*). **b** one's fair or usual share of this (*you are standing in my light*). **7 a** a flame or spark serving to ignite (*struck a light*). **b** a device producing this (*have you got*

a light?). **8** the aspect in which a thing is regarded or considered (*appeared in a new light*). **9 a** mental illumination; elucidation, enlightenment. **b** hope, happiness; a happy outcome. **c** spiritual illumination by divine truth. **10** vivacity, enthusiasm, or inspiration visible in a person's face, esp. in the eyes. **11** (in *pl.*) a person's mental powers or ability (*according to one's lights*). **12** an eminent person (*a leading light*). **13 a** the bright part of a thing; a highlight. **b** the bright parts of a picture etc. esp. suggesting illumination (*light and shade*). **14 a** a window or opening in a wall to let light in. **b** *Archit.* the perpendicular division of a mullioned window. **c** a pane of glass esp. in the side or roof of a greenhouse. **15** *Brit.* (in a crossword etc.) each of the items filling a space and to be deduced from the clues. **16** *Law* the light falling on windows, the obstruction of which by a neighbour is illegal. ● *v.* (*past* **lit** /lɪt/; *past part.* **lit** or (*attrib.*) **lighted**) **1** *tr. & intr.* set burning or begin to burn; ignite. **2** *tr.* provide with light or lighting. **3** *tr.* show (a person) the way or surroundings with a light. **4** *intr.* (usu. foll. by *up*) (of the face or eyes) brighten with animation, pleasure, etc. ● *adj.* **1** well provided with light; not dark. **2** (of a colour) pale (*light blue; a light-blue ribbon*). □ **bring** (or **come**) **to light** reveal or be revealed. **festival of lights 1** = HANUKKAH. **2** = DIWALI. **in a good** (or **bad**) **light** giving a favourable (or unfavourable) impression. **in (the) light of** having regard to; drawing information from. **light bulb** a glass bulb containing an inert gas and a metal filament, providing light when an electric current is passed through. **light-emitting diode** see LED. **lighting-up time** the time during or after which vehicles on the road must show the prescribed lights. **light meter** an instrument for measuring the intensity of the light, esp. to show the correct photographic exposure. **light of day 1** daylight, sunlight. **2** general notice; public attention. **light of one's life** usu. *joc.* a much-loved person. **light-pen** (or **-gun**) **1** a penlike or gunlike photosensitive device held to the screen of a computer terminal for passing information on to it. **2** a light-emitting device used for reading barcodes. **light pollution** excessive illumination of the night sky by street lights etc. **light show** a display of changing coloured lights for entertainment. **light up 1** *colloq.* begin to smoke a cigarette etc. **2** switch on lights or lighting; illuminate a scene. **light-year 1** *Astron.* the distance light travels in one year, nearly 6 million million miles. **2** (in *pl.*) *colloq.* a long distance or great amount. **lit up** *colloq.* drunk. **out like a light** deeply asleep or unconscious. **throw** (or **shed**) **light on** help to explain. □ **lightish** *adj.* **lightless** *adj.* **lightness** *n.* [OE *lēoht*, *līht*, *līhtan* f. Gmc]

▪ Visible light is electromagnetic radiation whose wavelength falls within the range to which the human retina responds, i.e. between about 390 nanometres (violet light) and 740 nanometres (red). White light consists of a roughly equal mixture of all visible wavelengths, which can be separated to yield the colours of the spectrum, as was first demonstrated conclusively by Newton. The nature of light has been a subject of dispute since ancient times. The speed of light was shown to be finite in the 17th century, and at about the same time two rival theories, a corpuscular (particle) theory and a wave theory, were proposed. The former, advocated by Newton, predominated until about 1800, when experiments showed that light could be made to produce the interference patterns characteristic of waves. Following the work of James Maxwell later in the century it became clear that light was a form of electromagnetic radiation. Although the wave theory seemed well established, in the 20th century it has become apparent that light consists of energy quanta called photons which behave partly like waves and partly like particles. The velocity of light in a vacuum is 299,792 km per second. (See also SPECTRUM.)

light² /laɪt/ *adj., adv., & v.* ● *adj.* **1** of little weight; not heavy; easy to lift. **2 a** relatively low in weight, amount, density, intensity, etc. (*light arms; light traffic; light metal; light rain; a light breeze*). **b** deficient in weight (*light coin*). **c** *Physics* (of an isotope etc.) having not more than the usual mass. **3 a** carrying or suitable for small loads (*light aircraft; light railway*). **b** (of a ship) unladen. **c** carrying only light arms, armaments, etc. (*light brigade; light infantry*). **d** (of a locomotive) with no train attached. **4 a** (of food, a meal, etc.) small in amount; easy to digest (*had a light lunch*). **b** (of a foodstuff) low in fat, cholesterol, or sugar, etc. **c** (of drink) not heavy on the stomach or strongly alcoholic. **5 a** (of entertainment, music, etc.) intended for amusement, rather than edification; not profound. **b** frivolous, thoughtless, trivial (*a light remark*). **6** (of sleep or a sleeper) easily disturbed. **7** easily borne or done (*light duties*). **8** nimble; quick-moving (*a light step; light of foot; a light rhythm*). **9** (of a building etc.) graceful, elegant, delicate. **10** (of type) not heavy or bold. **11 a** free from sorrow; cheerful (*a light heart*). **b** giddy (*light in the head*). **12** (of soil) not dense; porous. **13** (of pastry, a sponge cake, etc.) fluffy and well-aerated during cooking and with the fat fully absorbed.

14 *archaic* (of a woman) unchaste or wanton; fickle. ● *adv.* **1** in a light manner (*tread light; sleep light*). **2** with a minimum load or minimum luggage (*travel light*). ● *v.intr.* (*past* and *past part.* **lit** /lɪt/ or **lighted**) **1** (foll. by *on, upon*) come upon or find by chance. **2** *archaic* **a** alight, descend. **b** (foll. by *on*) land on (shore etc.). □ **light air** *Meteorol.* a very light wind, force 1 on the Beaufort scale (1–3 m.p.h. or 0.3–1.5 metres per second). **lighter-than-air** (of an aircraft) weighing less than the air it displaces. **light-fingered** given to stealing. **light flyweight** see FLYWEIGHT. **light-footed** nimble. **light-footedly** nimbly. **light-headed** giddy, frivolous, delirious. **light-headedly** in a light-headed manner. **light-headedness** being light-headed. **light-hearted 1** cheerful. **2** (unduly) casual, thoughtless. **light-heartedly** in a light-hearted manner. **light-heartedness** being light-hearted. **light heavyweight** see HEAVYWEIGHT. **light industry** the manufacture of small or light articles. **light into** *colloq.* attack. **light middleweight** see MIDDLEWEIGHT. **light opera** a light-hearted or less serious opera; an operetta. **light out** *colloq.* depart. **light touch** delicate or tactful treatment. **light welterweight** see WELTERWEIGHT. **make light of** treat as unimportant. **make light work of** do a thing quickly and easily. □ **lightish** *adj.* **lightness** *n.* [OE *lēoht, līht, līhtan* f. Gmc, the verbal sense from the idea of relieving a horse etc. of weight]

Light Brigade, Charge of the see CHARGE OF THE LIGHT BRIGADE.

lighten[1] /ˈlaɪt(ə)n/ *v.* **1 a** *tr.* & *intr.* make or become lighter in weight. **b** *tr.* reduce the weight or load of. **2** *tr.* bring relief to (the heart, mind, etc.). **3** *tr.* mitigate (a penalty).

lighten[2] /ˈlaɪt(ə)n/ *v.* **1 a** *tr.* shed light on. **b** *tr.* & *intr.* make or grow lighter or brighter. **2** *intr.* **a** shine brightly; flash. **b** emit lightning (*it is lightening*).

lightening /ˈlaɪt(ə)nɪŋ/ *n.* *Med.* a sensation of reduced pressure on the upper part of the womb felt in the late stages of pregnancy as the head of the foetus engages in the pelvis.

lighter[1] /ˈlaɪtə(r)/ *n.* a device for lighting cigarettes etc.

lighter[2] /ˈlaɪtə(r)/ *n.* a boat, usu. flat-bottomed, for transferring goods from a ship to a wharf or another ship. [ME f. MDu. *lichter* (as LIGHT[2] in the sense 'unload')]

lighterage /ˈlaɪtərɪdʒ/ *n.* **1** the transference of cargo by means of a lighter. **2** a charge made for this.

lighterman /ˈlaɪtəmən/ *n.* (*pl.* **-men**) a person who works on a lighter.

lighthouse /ˈlaɪthaʊs/ *n.* a tower or other structure containing a beacon light to warn or guide ships at sea. One of the earliest was the Pharos off Alexandria, one of the Seven Wonders of the World, which used a fire of burning wood, as did later lighthouses until first oil lamps and then electric lamps came into use, giving greatly increased power. The intensity of the light from modern lighthouses is increased by the use of mirrors and lenses to concentrate the light into a narrow beam, which usually sweeps round in a horizontal circle at a fixed speed so that, to ships at sea, it appears to emit a regular pattern of flashes. Nowadays lighthouses are equipped also with radio beacons, and each has a characteristic pattern of light or radio signals which enables it to be identified.

lighting /ˈlaɪtɪŋ/ *n.* **1** equipment in a room or street etc. for producing light. (*See note below.*) **2** the arrangement or effect of lights.

▪ The first technical advance on a naked flame for lighting was the oil lamp with a wick and glass chimney, e.g. that patented by the Swiss Aimé Argand (1755–1803) in the late 18th century. Gas lighting, introduced in the early 19th century, was improved by the introduction of the incandescent mantle by Auer von Welsbach in 1885 (although not enough to compete with electricity). The first successful electric lamps were developed almost simultaneously by Joseph Swan in England and Thomas Edison in America in 1879. Gas discharge tubes containing neon were pioneered in the late 19th century; most street lighting is now of this type except that mercury or sodium vapour is usually used. Other forms of electric light include arc lamps, developed in the early 19th century and used where a high-intensity beam is needed, e.g. in searchlights and lighthouses, and fluorescent tubes, developed in the 1930s.

lightly /ˈlaɪtlɪ/ *adv.* in a light (esp. frivolous or unserious) manner. □ **get off lightly** escape with little or no punishment. **take lightly** not be serious about (a thing).

lightning /ˈlaɪtnɪŋ/ *n.* & *adj.* ● *n.* the sudden bright light produced by an electric discharge between clouds or between clouds and the ground. (*See note below.*) ● *attrib.adj.* very quick (*with lightning speed*). □ **lightning bug** *N. Amer.* a firefly. **lightning conductor** (or **rod**) a metal rod or wire fixed to an exposed part of a building or to a mast

to divert lightning into the earth or sea. **lightning strike** a strike by workers at short notice, esp. without official union backing. [ME, differentiated from *lightening*, verbal noun f. LIGHTEN[2]]

▪ Lightning is caused by a build-up of static electricity in clouds. Charges are formed by friction between the air and water droplets and ice particles in the violent air currents associated with thunderclouds. Complex charge separation processes tend to concentrate positive charge in the upper part of the cloud and negative charge lower down, with a smaller positive charge near the ground. A lightning flash occurs when the electrical potential is great enough (typically several hundred million volts) to overcome the resistance of the air and create a conducting path. The violent heating and expansion of the air by the passage of a current of 10,000–20,000 amperes produces a shock wave which is heard as thunder.

lightproof /ˈlaɪtpruːf/ *adj.* able to resist the harmful effects of (esp. excessive) light.

lights /laɪts/ *n.pl.* the lungs of sheep, pigs, bullocks, etc., used as a food esp. for pets. [ME, noun use of LIGHT[2]: cf. LUNG]

lightship /ˈlaɪtʃɪp/ *n.* a moored or anchored ship with a beacon light.

lightsome /ˈlaɪtsəm/ *adj.* **1** gracefully light; nimble. **2** merry. □ **lightsomely** *adv.* **lightsomeness** *n.*

lightweight /ˈlaɪtweɪt/ *adj.* & *n.* ● *adj.* **1** (of a person, animal, garment, etc.) of below average weight. **2** of little importance or influence. ● *n.* **1 a** lightweight person, animal, or thing. **2 a** a weight in certain sports intermediate between featherweight and welterweight, in the amateur boxing scale 57–60 kg but differing for professional boxers, wrestlers, and weightlifters. **b** a boxer etc. of this weight. □ **junior lightweight 1** a weight in professional boxing of 57.1–59 kg. **2** a professional boxer of this weight.

lightwood /ˈlaɪtwʊd/ *n.* **1** a tree with a light wood. **2** *US* wood or a tree with wood that burns with a bright flame.

ligneous /ˈlɪɡnɪəs/ *adj.* **1** (of a plant) woody (opp. HERBACEOUS). **2** of the nature of wood. [L *ligneus* (as LIGNI-)]

ligni- /ˈlɪɡnɪ/ *comb. form* wood. [L *lignum* wood]

lignify /ˈlɪɡnɪfaɪ/ *v.tr.* & *intr.* (**-ies, -ied**) *Bot.* make or become woody by the deposition of lignin.

lignin /ˈlɪɡnɪn/ *n.* *Bot.* a complex organic polymer deposited in the cell-walls of many plants making them rigid and woody. [as LIGNI- + -IN]

lignite /ˈlɪɡnaɪt/ *n.* a soft brown coal showing traces of plant structure, intermediate between bituminous coal and peat. □ **lignitic** /lɪɡˈnɪtɪk/ *adj.* [F (as LIGNI-, -ITE[1])]

lignocaine /ˈlɪɡnəkeɪn/ *n.* *Pharm.* a local anaesthetic for the gums, mucous membranes, or skin, usu. given by injection. Also called *lidocaine*. [*ligno-* (as LIGNI-) for XYLO- + COCA + -INE[4]]

lignum vitae /ˌlɪɡnəm ˈvaɪtɪ, ˈviːtaɪ/ *n.* = GUAIACUM 2a. [L, = wood of life]

ligroin /ˈlɪɡrəʊɪn/ *n.* *Chem.* a volatile hydrocarbon mixture obtained from petroleum and used as a solvent. [20th c.: orig. unkn.]

ligulate /ˈlɪɡjʊlət/ *adj.* *Bot.* having strap-shaped florets. [formed as LIGULE + -ATE[2]]

ligule /ˈlɪɡjuːl/ *n.* *Bot.* a narrow projection from the top of a leaf-sheath of a grass. [L *ligula* strap, spoon f. *lingere* lick]

Liguria /lɪˈɡjʊərɪə/ a coastal region of NW Italy, which extends along the Mediterranean coast from Tuscany to the border with France; capital, Genoa. In ancient times, prior to its occupation by the Celts and then the Romans, Liguria was much more extensive (see LIGURIAN *adj.* 2). A coastal strip around Genoa, designated the Ligurian Republic in 1797 after Napoleon's Italian campaign, was annexed to France between 1805 and 1815. [L, f. *Ligur* Ligurian f. Gk *Ligus*]

Ligurian /lɪˈɡjʊərɪən/ *adj.* & *n.* ● *adj.* **1** of or relating to ancient or modern Liguria, its people, or their language or dialect. **2** of or relating to an ancient people inhabiting NW Italy, Switzerland, and SE Gaul and speaking a pre-Italic Indo-European language. ● *n.* **1** a native or inhabitant of ancient or modern Liguria. **2** the language of the ancient Ligurians. **3** the dialect of modern Liguria.

Ligurian Sea a part of the northern Mediterranean, between Corsica and the NW coast of Italy.

ligustrum /lɪˈɡʌstrəm/ *n.* = PRIVET. [L]

likable var. of LIKEABLE.

like[1] /laɪk/ *adj., prep., adv., conj.,* & *n.* ● *adj.* (often governing a noun as if a transitive participle such as *resembling*) (**more like, most like**) **1 a** having some or all of the qualities of another or each other or an original; alike (*in like manner; as like as two peas; is very like her brother*). **b** resembling in some way, such as; in the same class as (*good writers like Dickens*). **c** (usu. in pairs correlatively) as one is so will the other be (*like mother, like daughter*). **2** characteristic of (*it is not like them to be late*). **3** in a suitable state or mood for (doing or having something) (*felt like working; felt like a cup of tea*). ● *prep.* in the manner of; to the same degree as (*drink like a fish; acted like hot cakes; acted like an idiot*). ● *adv.* **1** *archaic* likely (*they will come, like enough*). **2** *archaic* in the same manner (foll. by *as: sang like as a nightingale*). **3** *sl.* so to speak (*did a quick getaway, like; as I said, like, I'm no Shakespeare*). ● *conj. colloq. disp.* **1** as (*cannot do it like you do*). **2** as if (*ate like they were starving*). ● *n.* **1** a counterpart; an equal; a similar person or thing (*shall not see its like again; compare like with like*). **2** (prec. by *the*) a thing or things of the same kind (*will never do the like again*). □ **and the like** and similar things; et cetera (*music, painting, and the like*). **be nothing like** (usu. with complement) be in no way similar or comparable or adequate. **like anything** see ANYTHING. **like** (or **as like**) **as not** *colloq.* probably. **like-minded** having the same tastes, opinions, etc. **like-mindedly** in accordance with the same tastes etc. **like-mindedness** being like-minded. **like so** *colloq.* like this; in this manner. **the likes of** *colloq.* a person such as. **more like it** *colloq.* nearer what is required. **of like** (or **of a like**) **mind** = *like-minded*. **what is he** (or **she** or **it** etc.) **like?** what sort of a person is he (or she, or thing is it etc.)? [ME *līc, līk*, shortened form of OE *gelīc* ALIKE]

like² /laɪk/ *v. & n.* ● *v.tr.* **1 a** find agreeable or enjoyable or satisfactory (*like reading; like the sea; like to dance*). **b** be fond of (a person). **2 a** choose to have; prefer (*like my coffee black; do not like such things discussed*). **b** wish for or be inclined to (*would like a cup of tea; should like to come*). **3** (usu. in *interrog.*; prec. by *how*) feel about; regard (*how would you like it if it happened to you?*). ● *n.* (in *pl.*) the things one likes or prefers. □ **I like that!** *iron.* as an exclamation expressing affront. **like it or not** *colloq.* whether it is acceptable or not. [OE *līcian* f. Gmc]

-like /laɪk/ *comb. form* forming adjectives from nouns, meaning 'similar to, characteristic of' (*doglike; shell-like; tortoise-like*). ¶ In formations intended as nonce-words, or not generally current, the hyphen should be used. It may be omitted when the first element is of one syllable, but nouns in -*l* always require it.

likeable /ˈlaɪkəb(ə)l/ *adj.* (also **likable**) pleasant; easy to like. □ **likeably** *adv.* **likeableness** *n.* **likeability** /ˌlaɪkəˈbɪlɪtɪ/ *n.*

likelihood /ˈlaɪklɪˌhʊd/ *n.* probability; the quality or fact of being likely. □ **in all likelihood** very probably.

likely /ˈlaɪklɪ/ *adj. & adv.* ● *adj.* **1** probable; such as well might happen or be true (*it is not likely that they will come; the most likely place is London; a likely story*). **2** (foll. by *to* + infin.) to be reasonably expected (*he is not likely to come now*). **3** promising; apparently suitable (*three likely lads; this is a likely spot*). ● *adv.* probably (*is very likely true*). □ **as likely as not** probably. **not likely!** *colloq.* certainly not, I refuse. □ **likeliness** *n.* [ME f. ON *līkligr* (as LIKE¹, -LY¹)]

liken /ˈlaɪkən/ *v.tr.* (foll. by *to*) point out the resemblance of (a person or thing to another). [ME f. LIKE¹ + -EN¹]

likeness /ˈlaɪknɪs/ *n.* **1** (usu. foll. by *between, to*) resemblance. **2** (foll. by *of*) a semblance or guise (*in the likeness of a ghost*). **3** a portrait or representation (*is a good likeness*). [OE *gelīcnes* (as LIKE¹, -NESS)]

likewise /ˈlaɪkwaɪz/ *adv.* **1** also, moreover, too. **2** similarly (*do likewise*). [for *in like wise*]

liking /ˈlaɪkɪŋ/ *n.* **1** what one likes; one's taste (*is it to your liking?*). **2** (foll. by *for*) regard or fondness; taste or fancy (*had a liking for toffee*). [OE *līcung* (as LIKE², -ING¹)]

Likud /lɪˈkuːd/ *n.* a coalition of right-wing Israeli political parties, formed in 1973, that won power in the Israeli elections of 1977 and governed under Menachem Begin until 1984. Likud has since entered into coalitions with the Labour Party. [Heb., = consolidation, unity]

likuta /lɪˈkuːtə/ *n.* (*pl.* **makuta** /məˈkuːtə/) a monetary unit of Zaire, equal to one-hundredth of a zaire. [Kikongo]

lilac /ˈlaɪlək/ *n. & adj.* ● *n.* **1** a shrub or small tree of the genus *Syringa*; esp. *S. vulgaris*, with fragrant pale pinkish-violet or white blossoms. **2** a pale pinkish-violet colour. ● *adj.* of this colour. [obs. F f. Sp. f. Arab. *līlāk* f. Pers. *līlak*, var. of *nīlak* bluish f. *nīl* blue]

lilangeni /ˌliːlæŋˈɡeɪnɪ/ *n.* (*pl.* **emalangeni** /ˌɪˌmɑːlæŋ-/) the basic monetary unit of Swaziland, equal to 100 cents. [Bantu, f. *li-* sing. pref. (*ema-* pl. pref.) + -*langeni* member of royal family]

liliaceous /ˌlɪlɪˈeɪʃəs/ *adj.* **1** *Bot.* of or relating to the family Liliaceae, which comprises plants with elongated leaves growing from a corm, bulb, or rhizome, e.g. tulip, lily, or onion. **2** lily-like. [LL *liliaceus* f. L *lilium* lily]

Lilienthal /ˈliːlɪənˌtɑːl/, Otto (1848–96), German pioneer in the design and flying of gliders. Trained as an engineer, he invented a light steam motor and worked on marine signals. In his flying experiments he constructed wings connected to a tail, made of osier wands and covered with shirt fabric, fitted them to his shoulders, and took off by running downhill into the wind. In 1896 he experimented with a small motor to flap the wings. Working with his brother he made over 2,000 flights in various gliders before being killed in a crash. He also studied the science of bird flight, demonstrating the superiority of a curved over a flat wing.

Lilith /ˈlɪlɪθ/ a female demon of Jewish folklore, who tries to kill newborn children. In the Talmud she is the first wife of Adam, dispossessed by Eve. [Heb., = night-monster]

Lille /liːl/ an industrial city in northern France, near the border with Belgium; pop. (1990) 178,300.

Lillee /ˈlɪlɪ/, Dennis (Keith) (b.1949), Australian cricketer. He was a notable fast bowler who took 355 wickets in seventy matches during his career in test cricket (1971–84).

Lilliburlero /ˌlɪlɪbəˈleərəʊ/ the title and part of the refrain of a song ridiculing the Irish, popular at the end of the 17th century especially among soldiers and supporters of William III during the Revolution of 1688. With different words the song has remained associated with the Orange Party, as 'Protestant Boys'.

lilliputian /ˌlɪlɪˈpjuːʃ(ə)n/ *n. & adj.* ● *n.* a diminutive person or thing. ● *adj.* diminutive. [*Lilliput* in Swift's *Gulliver's Travels*]

Lilo /ˈlaɪləʊ/ *n.* (also **Li-lo** (*pl.* **-os**) *propr.*) a type of inflatable mattress. [f. *lie low*]

Lilongwe /lɪˈlɒŋweɪ/ the capital of Malawi; pop. (1987) 233,970.

lilt /lɪlt/ *n. & v.* ● *n.* **1 a** a light springing rhythm or gait. **b** a song or tune marked by this. **2** (of the voice) a characteristic cadence or inflection; a pleasant accent. ● *v.intr.* (esp. as **lilting** *adj.*) move or speak etc. with a lilt (*a lilting step; a lilting melody*). [ME *lilte, lülte*, of unkn. orig.]

lily /ˈlɪlɪ/ *n.* (*pl.* **-ies**) **1 a** a bulbous plant of the genus *Lilium* (family Liliaceae), with large trumpet-shaped often spotted flowers on a tall slender stem (*madonna lily, tiger lily*). **b** any plant of similar appearance (*Arum lily*). **c** the water lily. **2** a person or thing of special whiteness or purity. **3** a heraldic fleur-de-lis. **4** (*attrib.*) **a** delicately white (*a lily hand*). **b** pallid. □ **lily-livered** cowardly. **lily of the valley** a liliaceous plant of the genus *Convallaria*, with oval leaves in pairs and clusters of white bell-shaped fragrant flowers. **lily-pad** a floating leaf of a water lily. **lily-trotter** a jacana, esp. an African one. **lily-white 1** as white as a lily. **2** faultless. □ **lilied** *adj.* [OE *lilie* f. L *lilium* prob. f. Gk *leirion*]

Lima /ˈliːmə/ the capital of Peru; pop. (1990) 426,300. Founded in 1535 by Francisco Pizarro, it was the capital of the Spanish colonies in South America until the 19th century.

lima bean /ˈliːmə/ *n.* **1** a tropical American bean plant, *Phaseolus limensis*, having large flat white edible seeds. **2** the seed of this plant. [LIMA]

Limassol /ˈlɪməˌsɒl/ a port on the south coast of Cyprus, on Akrotiri Bay; pop. (1990) 135,400.

limb¹ /lɪm/ *n.* **1** a projecting part of a person's or animal's body such as an arm, leg, or wing. **2** a large branch of a tree. **3** a branch of a cross. **4** a spur of a mountain. **5** *Brit.* a clause of a sentence. □ **out on a limb 1** isolated, stranded. **2** at a disadvantage. **tear limb from limb** violently dismember. **with life and limb** (esp. escape) without grave injury. □ **limbed** *adj.* (also in *comb.*). **limbless** *adj.* [OE *lim* f. Gmc]

limb² /lɪm/ *n.* **1 a** *Astron.* a specified edge of the sun, moon, etc. (*eastern limb; lower limb*). **b** the graduated edge of a quadrant etc. **2** *Bot.* the broad part of a petal, sepal, or leaf. [F *limbe* or L *limbus* hem, border]

limber¹ /ˈlɪmbə(r)/ *adj. & v.* ● *adj.* **1** lithe, agile, nimble. **2** flexible. ● *v.* (usu. foll. by *up*) **1** *tr.* make (oneself or a part of the body etc.) supple. **2** *intr.* warm up in preparation for athletic etc. activity. □ **limberness** *n.* [16th c.: orig. uncert.]

limber² /ˈlɪmbə(r)/ *n. & v.* ● *n.* the detachable front part of a gun-carriage, consisting of two wheels, axle, pole, and ammunition-box. ● *v.* **1** *tr.* attach a limber to (a gun etc.). **2** *intr.* fasten together the two parts of a gun-carriage. [ME *limo(u)r*, app. rel. to med.L *limonarius* f. *limo -onis* shaft]

limbic /ˈlɪmbɪk/ *adj. Anat.* of or relating to a part of the brain concerned with basic emotions and instinctive actions. [F *limbique*, f. L *limbus* edge]

limbo[1] /'lɪmbəʊ/ n. (pl. **-os**) **1** (in medieval Christian theology) a region on the border of hell, the supposed abode of pre-Christian righteous persons and of unbaptized infants. **2** an intermediate state or condition of awaiting a decision etc. **3** prison, confinement. **4** a state of neglect or oblivion. [ME f. med.L phr. *in limbo*, f. *limbus*: see LIMB[2]]

limbo[2] /'lɪmbəʊ/ n. (pl. **-os**) a West Indian dance in which the dancer bends backwards to pass under a horizontal bar which is progressively lowered to a position just above the ground. [a West Indian word, perh. = LIMBER[1]]

Limburg /'lɪmbɜːɡ/ (French **Limbourg** /lɛ̃bur/) a former duchy of Lorraine, divided in 1839 between Belgium and the Netherlands. It now forms a province of NE Belgium (capital, Hasselt) and a province of the SE Netherlands (capital, Maastricht).

Limburger /'lɪmˌbɜːɡə(r)/ n. a soft white cheese with a characteristic strong smell, orig. made in Limburg. [Du. f. LIMBURG]

lime[1] /laɪm/ n. & v. ● n. **1** (in full **quicklime**) a white caustic alkaline substance (calcium oxide) obtained by heating limestone and used for making mortar or as a fertilizer or bleach etc. Also called *calx*. **2** (in full **slaked lime**) a white substance (calcium hydroxide) made by adding water to quicklime, used esp. in cement. **3** calcium or calcium salts, esp. calcium carbonate in soil etc. **4** *archaic* = BIRDLIME. ● v.tr. **1** treat (wood, skins, land, etc.) with lime. **2** *archaic* catch (a bird etc.) with birdlime. □ **lime water** an aqueous solution of calcium hydroxide used esp. to detect the presence of carbon dioxide. □ **limeless** adj. **limy** adj. (**limier**, **limiest**). [OE *līm* f. Gmc, rel. to LOAM]

lime[2] /laɪm/ n. **1 a** a rounded citrus fruit like a lemon but greener, smaller, and more acid. **b** the citrus tree, *Citrus aurantifolia*, bearing this. **2** (in full **lime juice**) the juice of limes as a drink and formerly given to prevent scurvy on long sea voyages. **3** (in full **lime-green**) a pale green colour like a lime. [F f. mod.Prov. *limo*, Sp. *lima* f. Arab. *līma*: cf. LEMON]

lime[3] /laɪm/ n. **1** (in full **lime tree**) a tree of the genus *Tilia*, with heart-shaped leaves; esp. *T. europaea*, with fragrant yellow blossom. Also called *linden*. **2** the wood of this tree. [alt. of *line* = OE *lind* = LINDEN]

limekiln /'laɪmkɪln/ n. a kiln for heating limestone to produce quicklime.

limelight /'laɪmlaɪt/ n. **1** an intense white light obtained by heating a cylinder of lime in an oxyhydrogen flame, used formerly in theatres. **2** (prec. by *the*) the full glare of publicity; the focus of attention.

limepit /'laɪmpɪt/ n. a pit containing lime for steeping hides to remove hair.

Limerick /'lɪmərɪk/ **1** a county of the Republic of Ireland, in the west in the province of Munster. **2** its county town, on the River Shannon; pop. (est. 1991) 52,040.

limerick /'lɪmərɪk/ n. a humorous or comic form of five-line poem with a rhyme-scheme *aabba*. [said to be from the chorus 'will you come up to Limerick?' sung between improvised verses at a gathering: f. LIMERICK]

limestone /'laɪmstəʊn/ n. *Geol.* a sedimentary rock composed mainly of calcium carbonate, used as building material and in the making of cement.

limewash /'laɪmwɒʃ/ n. a mixture of lime and water for coating walls.

Limey /'laɪmɪ/ n. (pl. **-eys**) N. Amer. sl. offens. a British person (orig. a sailor) or ship. [LIME[2], because of the former enforced consumption of lime juice in the British navy to prevent scurvy]

limit /'lɪmɪt/ n. & v. ● n. **1** a point, line, or level beyond which something does not or may not extend or pass. **2** (often in *pl.*) the boundary of an area. **3** the greatest or smallest amount permissible or possible (*upper limit*; *lower limit*). **4** *Math.* a quantity which a function or sum of a series can be made to approach as closely as desired. ● v.tr. (**limited**, **limiting**) **1** set or serve as a limit to. **2** (foll. by *to*) restrict. □ **be the limit** *colloq.* be intolerable or extremely irritating. **off limits** *US* out of bounds. **within limits** moderately; with some degree of freedom. **without limit** with no restriction. □ **limitable** adj. **limitative** adj. **limiter** n. [ME f. L *limes limitis* boundary, frontier]

limitary /'lɪmɪtərɪ/ adj. **1** subject to restriction. **2** of, on, or serving as a limit.

limitation /ˌlɪmɪˈteɪʃ(ə)n/ n. **1** the act or an instance of limiting; the process of being limited. **2** (often in *pl.*) a condition of limited ability (*know one's limitations*). **3** (often in *pl.*) a limiting rule or circumstance (*has its limitations*). **4** *Law* a legally specified period beyond which an action cannot be brought, or a property right is not to continue. [ME f. L *limitatio* (as LIMIT)]

limited /'lɪmɪtɪd/ adj. **1** confined within limits. **2** not great in scope or talents (*has limited experience*). **3 a** few, scanty, restricted (*limited accommodation*). **b** restricted to a few examples (*limited edition*). □ **limited** (or **limited liability**) **company** a company whose owners are legally responsible only to a limited amount for its debts. **limited liability** *Brit.* the status of being legally responsible only to a limited amount for debts of a trading company. □ **limitedly** adv. **limitedness** n.

limitless /'lɪmɪtlɪs/ adj. **1** extending or going on indefinitely (*a limitless expanse*). **2** unlimited (*limitless generosity*). □ **limitlessly** adv. **limitlessness** n.

limn /lɪm/ v.tr. **1** *archaic* paint (esp. a miniature portrait). **2** *hist.* illuminate (manuscripts). □ **limner** /'lɪmnə(r)/ n. [obs. *lumine* illuminate f. OF *luminer* f. L *luminare*: see LUMEN]

limnology /lɪmˈnɒlədʒɪ/ n. the study of the physical phenomena of lakes and other fresh waters. □ **limnologist** n. **limnological** /ˌlɪmnəˈlɒdʒɪk(ə)l/ adj. [Gk *limnē* lake + -LOGY]

Limnos see LEMNOS.

limo /'lɪməʊ/ n. (pl. **-os**) *colloq.* a limousine. [abbr.]

Limoges /lɪˈməʊʒ/ a city in west central France, the principal city of Limousin; pop. (1990) 136,400. Famous in the late Middle Ages for enamel work, it has been noted since the 18th century for the production of porcelain.

Limón /lɪˈmɒn/ (also **Puerto Limón** /'pwɜːtəʊ/) a port on the Caribbean coast of Costa Rica; pop. (1984) 64,400.

Limousin[1] /'lɪmuːˌzæn/ a region and former province of central France, centred on Limoges.

Limousin[2] /'lɪmuːˌzæn/ n. a breed of white beef cattle originating in Limousin.

limousine /'lɪməˌziːn/ n. a large luxurious motor car, often with a partition behind the driver. [F, orig. a caped cloak worn in LIMOUSIN[1]]

limp[1] /lɪmp/ v. & n. ● v.intr. **1** walk lamely. **2** (of a damaged ship, aircraft, etc.) proceed with difficulty. **3** (of verse) be defective. ● n. a lame walk. □ **limpingly** adv. [rel. to obs. *limphalt* lame, OE *lemp-healt*]

limp[2] /lɪmp/ adj. **1** not stiff or firm; easily bent. **2** without energy or will. **3** (of a book) having a soft cover. □ **limp-wristed** *sl.* homosexual or effeminate; weak, feeble. □ **limply** adv. **limpness** n. [18th c.: orig. unkn.: perh. rel. to LIMP[1] in the sense 'hanging loose']

limpet /'lɪmpɪt/ n. **1** a marine gastropod mollusc with a shallow conical shell and a broad muscular foot that sticks tightly to rocks; esp. the common limpet, *Patella vulgata*. **2** a clinging person. □ **limpet mine** a mine designed to be attached to a ship's hull and set to explode after a certain time. [OE *lempedu* f. med.L *lampreda* limpet, LAMPREY]

limpid /'lɪmpɪd/ adj. **1** (of water, eyes, etc.) clear, transparent. **2** (of writing) clear and easily comprehended. □ **limpidly** adv. **limpidness** n. **limpidity** /lɪmˈpɪdɪtɪ/ n. [F *limpide* or L *limpidus*, perh. rel. to LYMPH]

limpkin /'lɪmpkɪn/ n. a wading marsh bird, *Aramus guarauna*, of tropical America, resembling a large rail. [f. LIMP[1], with ref. to the bird's limping gait]

Limpopo /lɪmˈpəʊpəʊ/ a river of SE Africa. Rising as the Crocodile river near Johannesburg, it flows 1,770 km (1,100 miles) in a sweeping curve to the north and east to meet the Indian Ocean in Mozambique, north of Maputo. For much of its course it forms the boundary between NE South Africa and the neighbouring countries of Botswana and Zimbabwe.

Linacre /'lɪnəkə(r)/, Thomas (c.1460–1524), English physician and classical scholar. In 1518 he founded the College of Physicians in London, and became its first president. He wrote textbooks on Latin grammar, and his students of Greek included Thomas More and probably Erasmus. Linacre's translations of Galen's Greek works on medicine and philosophy into Latin brought about a revival of studies in anatomy, botany, and clinical medicine in Britain.

linage /'laɪnɪdʒ/ n. **1** the number of lines in printed or written matter. **2** payment by the line.

Lin Biao /lɪn ˈbjaʊ/ (also **Lin Piao** /'pjaʊ/) (1908–71), Chinese Communist statesman and general. After joining the Communists (1927), he became a commander of Mao Zedong's Red Army in the fight against the Kuomintang. He was appointed Minister of Defence (1959) and then Vice-Chairman under Mao (1966), later being nominated to become Mao's successor (1969). Having staged an unsuccessful coup in 1971, Lin Biao was reported to have been killed in an aeroplane crash while fleeing to the Soviet Union.

linchpin /'lɪntʃpɪn/ n. (also **lynchpin**) **1** a pin passed through an axle-end to keep a wheel in position. **2** a person or thing vital to an enterprise, organization, etc. [ME *linch* f. OE *lynis* + PIN]

Lincoln[1] /ˈlɪŋkən/ **1** the state capital of Nebraska; pop. (1990) 191,970. Founded as Lancaster in 1856, it was made state capital in 1867 and renamed in honour of Abraham Lincoln. **2** a city in eastern England, the county town of Lincolnshire; pop. (1991) 81,900. It was founded by the Romans as Lindum Colonia. Its 11th-century cathedral houses one of the four original copies of Magna Carta.

Lincoln[2] /ˈlɪŋkən/, Abraham (1809–65), American Republican statesman, 16th President of the US 1861–5. His election as President on an anti-slavery platform antipathetic to the interests of the southern states helped precipitate the American Civil War. He eventually managed to unite the Union side behind the anti-slavery cause and emancipation was formally proclaimed on New Year's Day, 1864. Lincoln won re-election in 1864, but was assassinated shortly after the surrender of the main Confederate army had ended the war. During his lifetime Lincoln was noted for his succinct, eloquent speeches, including the Gettysburg address of 1863.

Lincoln green *n.* a bright green cloth of a kind orig. made at Lincoln.

Lincoln Memorial a monument in Washington, DC to Abraham Lincoln, designed by Henry Bacon (1866–1924) and dedicated in 1922. Built in the form of a Greek temple, the monument houses a statue of Lincoln that is 6 m (19 ft) high.

Lincolnshire /ˈlɪŋkənˌʃɪə(r)/ a county on the east coast of England. The county town is Lincoln.

Lincoln's Inn one of the Inns of Court in London. Thomas de Lincoln, king's sergeant in the 14th century, may have been an early landlord.

Lincs. *abbr.* Lincolnshire.

linctus /ˈlɪŋktəs/ *n.* a syrupy medicine, esp. a soothing cough mixture. [L f. *lingere* lick]

Lind[1] /lɪnd/, James (1716–94), Scottish physician. After a period as a naval surgeon, he performed some famous experiments which demonstrated that sailors could be cured of scurvy by supplementing their diet with citrus fruit. It was not until just after his death that the Royal Navy officially adopted the practice of giving lime juice to sailors. His work was a major step towards the discovery of vitamins.

Lind[2] /lɪnd/, Jenny (born Johanna Maria Lind Goldschmidt) (1820–87), Swedish soprano. Known as 'the Swedish nightingale' for the purity and agility of her voice, she achieved international success with her performances in opera, oratorio, and concerts. She funded musical scholarships and other charitable causes in England and Sweden.

lindane /ˈlɪndeɪn/ *n.* a toxic colourless isomer of benzene hexachloride, used as an insecticide. Its use is now discouraged, as it is persistent in the environment and is highly toxic to fish etc. [Teunis van der *Linden*, Dutch chemist (b.1884)]

Lindbergh /ˈlɪndbɜːɡ/, Charles (Augustus) (1902–74), American aviator. In 1927 he made the first solo transatlantic flight, taking 33½ hours from New York to Paris, in a single-engined monoplane, *Spirit of St Louis*. Lindbergh moved to Europe with his wife to escape the publicity surrounding the kidnap and murder of his two-year-old son in 1932.

Lindemann /ˈlɪndəmən/, Frederick Alexander, see CHERWELL.

linden /ˈlɪndən/ *n.* a lime tree. [(orig. adj.) f. OE *lind* lime tree: cf. LIME[3]]

Lindisfarne /ˈlɪndɪsˌfɑːn/ (also called *Holy Island*) a small island off the coast of Northumberland, north of the Farne Islands. It is linked to the mainland by a causeway exposed only at low tide. It is the site of a church and monastery founded by St Aidan in 635, which was a missionary centre of the Celtic Church.

Lindsay /ˈlɪndzɪ/ a family of Australian artists. The most prominent members include Sir Lionel Lindsay (1874–1961), art critic, watercolour painter, and graphic artist, who did much to generate Australian interest in the collection of original prints. His brother, Norman Lindsay (1879–1969), was a graphic artist, painter, critic, and novelist.

Lindsey /ˈlɪndzɪ/ one of three administrative divisions of Lincolnshire before the local government reorganization of 1974 (the others were Holland and Kesteven), comprising the northern part of the county. An area of early settlement by the Angles, Lindsey was a separate Anglo-Saxon kingdom until the 8th century. In 1974, Lindsey was divided between Humberside and Lincolnshire, the portion within Lincolnshire being administered as the districts of East and West Lindsey.

Lindum Colonia /ˌlɪndəm kəˈləʊnɪə/ the Roman name for Lincoln (see LINCOLN[1] 2).

line[1] /laɪn/ *n. & v.* ● *n.* **1** a continuous mark or band made on a surface (*drew a line*). **2** use of lines in art, esp. draughtsmanship or engraving (*boldness of line*). **3** a thing resembling such a mark, esp. a furrow or wrinkle. **4** *Mus.* **a** each of (usu. five) horizontal marks forming a stave in musical notation. **b** a sequence of notes or tones forming an instrumental or vocal melody. **5 a** a straight or curved continuous extent of length without breadth. **b** the track of a moving point. **6 a** a contour or outline, esp. as a feature of design (*admired the yacht's clean lines; the pure line of a tailored jacket*). **b** a facial feature (*the cruel line of his mouth*). **7 a** (on a map or graph) a curve connecting all points having a specified common property. **b** (**the Line**) the equator. **8 a** a limit or boundary. **b** a mark limiting the area of play, the starting or finishing point in a race, etc. **c** the boundary between a credit and a debit in an account. **9 a** a row of persons or things. **b** a direction as indicated by them (*line of march*). **c** *N. Amer.* a queue. **10 a** a row of printed or written words. **b** a portion of verse written in one line. **11** (in *pl.*) **a** a piece of poetry. **b** the words of an actor's part. **c** a specified amount of text etc. to be written out as a school punishment. **12** a short letter or note (*drop me a line*). **13** (in *pl.*) = *marriage lines*. **14** a length of cord, rope, wire, etc., usu. serving a specified purpose, esp. a fishing-line or clothes-line. **15** a wire or cable for a telephone or telegraph. **b** a connection by means of this (*am trying to get a line*). **16 a** a single track of a railway. **b** one branch or route of a railway system, or the whole system under one management. **17 a** a regular succession of buses, ships, aircraft, etc., plying between certain places. **b** a company conducting this (*shipping line*). **18** a connected series of persons following one another in time (esp. several generations of a family); stock, succession (*a long line of craftsmen; next in line to the throne*). **19 a** a course or manner of procedure, conduct, thought, etc. (*did it along these lines; don't take that line with me*). **b** policy (*the party line*). **c** conformity (*bring them into line*). **20** a direction, course, or channel (*lines of communication*). **21** a department of activity; a province; a branch of business (*not in my line*). **22** a range of commercial goods (*a new line in hats*). **23** *colloq.* a false or exaggerated account or story; a dishonest approach (*gave me a line about missing the bus*). **24 a** a connected series of military fieldworks, defences, etc. (*behind enemy lines*). **b** an arrangement of soldiers or ships in a line; a line of battle (*ship of the line*). **c** (prec. by *the*) regular army regiments (not auxiliary forces or Guards). **25** each of the very narrow horizontal sections forming a television picture. **26** a narrow range of the spectrum that is noticeably brighter or darker than the adjacent parts. **27** the level of the base of most letters in printing and writing. **28** (as a measure) one twelfth of an inch. ● *v.* **1** *tr.* mark with lines. **2** *tr.* cover with lines (*a face lined with pain*). **3** *tr. & intr.* position or stand at intervals along (*crowds lined the route*). □ **all along the line** at every point. **bring into line** make conform. **come into line** conform. **end of the line** the point at which further effort is unproductive or one can go no further. **get a line on** *colloq.* learn something about. **in line for** likely to receive. **in line** in the course of (esp. duty). **in** (or **out of**) **line with** in (or not in) alignment or accordance with. **lay** (or **put**) **it on the line** speak frankly. **line-drawing** a drawing in which images are produced from variations of lines. **line of fire** the expected path of gunfire, a missile, etc. **line of force** *Physics* an imaginary line which represents the strength and direction of a magnetic, gravitational, or electric field at any point. **line of march** the route taken in marching. **line of vision** the straight line along which an observer looks. **line-out** (in rugby) parallel lines of opposing teams at right angles to the touchline for the throwing in of the ball. **line printer** a machine that prints output from a computer a line at a time rather than character by character. **line up 1** arrange or be arranged in a line or lines. **2** have ready; organize (*had a job lined up*). **line-up** *n.* **1** a line of people for inspection. **2** an arrangement of persons in a team or nations etc. in an alliance. **on the line 1** at risk (*put my reputation on the line*). **2** speaking on the telephone. **3** (of a picture in an exhibition) hung with its centre about level with the spectator's eye. **out of line 1** not in alignment; discordant. **2** failing to conform to a rule or convention, behaving inappropriately. **step out of line** behave inappropriately. [ME *line*, *ligne* f. OF *ligne* ult. f. L *linea* f. *linum* flax, & f. OE *līne* rope, series]

line[2] /laɪn/ *v.tr.* **1 a** cover the inside surface of (a garment, box, etc.) with a layer of usu. different material. **b** serve as a lining for. **2** cover as if with a lining (*shelves lined with books*). **3** *colloq.* fill, esp. plentifully. □ **line one's pocket** (or **purse**) make money, usu. by corrupt means. [ME f. obs. *line* flax, with ref. to the use of linen for linings]

lineage /ˈlɪnɪɪdʒ/ *n.* lineal descent; ancestry, pedigree. [ME f. OF *linage*, *lignage* f. Rmc f. L *linea* LINE[1]]

lineal /ˈlɪnɪəl/ *adj.* **1** in the direct line of descent or ancestry. **2** linear; of or in lines. □ **lineally** *adv.* [ME f. OF f. LL *linealis* (as LINE[1])]

lineament /ˈlɪnɪəmənt/ n. (usu. in pl.) a distinctive feature or characteristic, esp. of the face. [ME f. L lineamentum f. lineare make straight f. linea LINE[1]]

linear /ˈlɪnɪə(r)/ adj. **1 a** of or in lines; in lines rather than masses (linear development). **b** of length (linear extent). **2** long and narrow and of uniform breadth. **3** involving one dimension only. □ **linear accelerator** Physics a particle accelerator in which particles travel in straight lines, not in closed orbits. **linear equation** Algebra an equation between two variables that gives a straight line when plotted on a graph. **linear motor** a motor producing straight-line (not rotary) motion by means of a magnetic field. □ **linearize** v.tr. (also **-ise**). **linearly** adv. **linearity** /ˌlɪnɪˈærɪtɪ/ n. [L linearis f. linea LINE[1]]

Linear A the earlier of two related forms of writing (see also LINEAR B) discovered at Knossos in Crete by Sir Arthur Evans between 1894 and 1901, found on fewer than 400 tablets and stone vases dating from c.1700 to 1450 BC and still largely unintelligible.

Linear B the later of two related forms of writing discovered at Knossos (see also LINEAR A). Linear B is found on thousands of tablets, largely recording details of palace administration, dating from c.1400 to 1200 BC, and occurs also on the mainland of Greece. It was deciphered in 1952 by the archaeologist Michael Ventris (1922–56) and shown to be a syllabic script composed of linear signs (derived from Linear A and older Minoan hieroglyphic or pictographic script) and adapted to the writing of Mycenaean Greek.

lineation /ˌlɪnɪˈeɪʃ(ə)n/ n. **1** a marking with or drawing of lines. **2** a division into lines. [ME f. L lineatio f. lineare make straight]

linebacker /ˈlaɪnˌbækə(r)/ n. Amer. Football a defensive player stationed close to the line of scrimmage.

linefeed /ˈlaɪnfiːd/ n. **1** the action of advancing paper in a printing machine by the space of one line. **2** the analogous movement of text on a VDU screen.

Line Islands a group of eleven islands in the central Pacific, straddling the equator south of Hawaii. Eight of the islands, including Kiritimati (Christmas Island), form part of Kiribati; the remaining three are uninhabited dependencies of the US.

lineman /ˈlaɪnmən/ n. (pl. **-men**) **1 a** a person who repairs and maintains telephone or electrical etc. lines. **b** a person who tests the safety of railway lines. **2** Amer. Football a player in the line formed before a scrimmage.

linen /ˈlɪnɪn/ n. & adj. ● n. **1 a** cloth woven from flax. **b** a particular kind of this. **2** (collect.) articles made, or orig. made, of linen, e.g. sheets, cloths, shirts, undergarments, etc. ● adj. made of linen or flax (linen cloth). □ **linen basket** esp. Brit. a basket for soiled clothes. **wash one's dirty linen in public** be indiscreet about one's domestic quarrels etc. [OE līnen f. WG, rel. to obs. line flax]

linenfold /ˈlɪnɪnˌfəʊld/ n. (often attrib.) a carved or moulded ornament representing a fold or scroll of linen (linenfold panelling).

liner[1] /ˈlaɪnə(r)/ n. a ship or aircraft etc. carrying passengers on a regular line. □ **liner train** Brit. a fast goods train with detachable containers on permanently coupled wagons.

liner[2] /ˈlaɪnə(r)/ n. a removable lining.

-liner /ˈlaɪnə(r)/ comb. form (prec. by a numeral, usu. one or two) colloq. a spoken passage of a specified number of lines in a play etc. (a one-liner).

linesman /ˈlaɪnzmən/ n. (pl. **-men**) **1** (in games played on a pitch or court) an umpire's or referee's assistant who decides whether a ball falls within the playing area or not. **2** Brit. = LINEMAN 1.

ling[1] /lɪŋ/ n. a long slender marine fish, Molva molva, of the East Atlantic, related to the cod and used as food. [ME leng(e), prob. f. MDu, rel. to LONG[1]]

ling[2] /lɪŋ/ n. a plant of the heath family; esp. Calluna vulgaris, the common heather. □ **lingy** adj. [ME f. ON lyng]

-ling[1] /lɪŋ/ suffix **1** denoting a person or thing: **a** connected with (hireling; sapling). **b** having the property of being (weakling; underling) or undergoing (starveling). **2** denoting a diminutive (duckling), often derogatory (lordling). [OE (as -LE[1] + -ING[3]): sense 2 f. ON]

-ling[2] /lɪŋ/ suffix forming adverbs and adjectives (darkling; grovelling) (cf. -LONG). [OE f. Gmc]

linga /ˈlɪŋɡə/ n. (also **lingam** /-ɡəm/) a phallus, esp. as the Hindu symbol of Siva. [Skr. lingam, lit. 'mark']

Lingala /lɪŋˈɡɑːlə/ n. a Bantu language of Zaire, widely used as a lingua franca in the north and west of that country.

linger /ˈlɪŋɡə(r)/ v.intr. **1 a** be slow or reluctant to depart. **b** stay about a place. **c** (foll. by over, on, etc.) dally (lingered over dinner). **2** (usu. foll. by on) (of an action or condition) be protracted; drag on (his cold lingered on; the memory lingered on). **3** (foll. by on) (of a dying person or custom) be slow in dying; drag on feebly. □ **lingerer** n. **lingering** adj. **lingeringly** adv. [ME lenger, frequent. of leng f. OE lengan f. Gmc, rel. to LENGTHEN]

lingerie /ˈlɒnʒərɪ/ n. women's underwear and nightclothes. [F f. linge linen]

lingo /ˈlɪŋɡəʊ/ n. (pl. **-os** or **-oes**) colloq. **1** a foreign language. **2** the vocabulary of a special subject or group of people. [prob. f. Port. lingoa f. L lingua tongue]

lingua franca /ˌlɪŋɡwə ˈfræŋkə/ n. (pl. **lingua francas**) **1** a language adopted as a common language between speakers whose native languages are different. **2** a system for mutual understanding. **3** hist. a mixture of Italian with French, Greek, Arabic, and Spanish, used in the Levant. [It., = Frankish tongue]

lingual /ˈlɪŋɡwəl/ adj. **1** of or formed by the tongue. **2** of speech or languages. □ **lingualize** v.tr. (also **-ise**). **lingually** adv. [med.L lingualis f. L lingua tongue, language]

linguiform /ˈlɪŋɡwɪˌfɔːm/ adj. Bot., Zool., & Anat. tongue-shaped. [L lingua tongue + -FORM]

linguine /lɪŋˈɡwiːnɪ/ n.pl. pasta in the form of narrow ribbons. [It., pl. of linguina dim. of lingua tongue]

linguist /ˈlɪŋɡwɪst/ n. a person skilled in languages or linguistics. [L lingua language]

linguistic /lɪŋˈɡwɪstɪk/ adj. of or relating to language or the study of languages. □ **linguistically** adv.

linguistics /lɪŋˈɡwɪstɪks/ n. the scientific study of language and its structure. In India in the 4th century BC there was a flourishing tradition of grammatical analysis (exemplified by the work of Panini), although it did not make any impact on the West until the 19th century. In classical times the Greeks created a linguistic terminology and defined a number of categories (e.g. the parts of speech) used to analyse Greek. The application of these to Latin marked the beginning of the Western grammatical tradition, which continued in the Middle Ages, when the grammar of Priscian was a widely used standard for many centuries. The problem of the connection between linguistic and logical categories, between language and thought, was discussed at length in the 17th and 18th centuries in France and elsewhere. In the 19th century the focus shifted to historical and comparative linguistics, aiming at identifying language families, tracing the history of languages, and reconstructing lost proto-languages such as Indo-European. At the beginning of the 20th century the Swiss structuralist Saussure drew a number of important distinctions, e.g. between the diachronic (historical) and synchronic study of language, between signifier and signified, and between syntagmatic structures (words in context) and paradigmatic classes (words in sets like the parts of speech). American structural linguists such as Leonard Bloomfield insisted on the objective analysis of observed data. A turning point was the publication in 1957 of Noam Chomsky's Syntactic Structures, which led to the development of generative and transformational grammar, the objective of which is the statement of a set of rules capable of generating all and only the grammatical sentences of a language. [F linguistique or G Linguistik (as LINGUIST)]

linguodental /ˌlɪŋɡwəʊˈdent(ə)l/ adj. Phonet. = DENTAL 2. [L lingua tongue + DENTAL]

liniment /ˈlɪnɪmənt/ n. an embrocation, usu. made with oil. [LL linimentum f. L linire smear]

lining /ˈlaɪnɪŋ/ n. **1** a layer of material used to line a surface etc. **2** an inside layer or surface etc. (stomach lining).

link[1] /lɪŋk/ n. & v. ● n. **1** one loop or ring of a chain etc. **2 a** a connecting part, esp. a thing or person that unites or provides continuity; one in a series. **b** a state or means of connection. **3** a means of contact by radio or telephone between two points. **4** a means of travel or transport between two places. **5** = cuff-link (see CUFF[1]). **6** a measure equal to one-hundredth of a surveying chain (7.92 inches, approx. 20.1 cm). ● v. **1** tr. (foll. by together, to, with) connect or join (two things or one to another). **2** tr. clasp or intertwine (hands or arms). **3** intr. (foll. by on, to, in to) be joined; attach oneself to (a system, company, etc.). □ **link up** (foll. by with) connect or combine. **link-up** n. an act or result of linking up. [ME f. ON f. Gmc]

link[2] /lɪŋk/ n. hist. a torch of pitch and tow for lighting the way in dark streets. [16th c.: perh. f. med.L li(n)chinus wick f. Gk lukhnos light]

linkage /ˈlɪŋkɪdʒ/ n. **1** a connection. **2** a system of links; a linking or link. **3** the linking together of different political issues as a negotiating tactic.

linkman /ˈlɪŋkmæn/ n. (pl. **-men**) **1** a person providing continuity in a broadcast programme. **2** a player between the forwards and half-backs or strikers and backs in football etc.

Linköping /ˈlɪnˌtʃɜːpɪŋ/ an industrial town in SE Sweden; pop. (1990) 122,270. It was a noted cultural and ecclesiastical centre in the Middle Ages.

links /lɪŋks/ n.pl. **1** (treated as *sing.* or *pl.*) a golf course, esp. one having undulating ground, coarse grass, etc. **2** *Sc. dial.* level or undulating sandy ground near a seashore, with turf and coarse grass. [pl. of *link* 'rising ground' f. OE *hlinc*]

linn /lɪn/ n. *Sc.* **1 a** a waterfall. **b** a pool below this. **2** a precipice; a ravine. [Gael. *linne*]

Linnaean /lɪˈniːən, -ˈneɪən/ adj. & n. *Biol.* ● *adj.* of or relating to Linnaeus or his system of binary nomenclature in the classification of plants and animals. ● *n.* a follower of Linnaeus. ¶ Spelt *Linnean* in *Linnean Society.*

Linnaeus /lɪˈniːəs, -ˈneɪəs/, Carolus (Latinized name of Carl von Linné) (1707–78), Swedish botanist, founder of modern systematic botany and zoology. He devised a classification system for flowering plants based on stamen type and number of pistils, and became the authority to whom collectors all over the world sent specimens. He described over 7,000 plants, introducing binomial Latin names, although his classification was later superseded by that of Antoine Jussieu. His classification of animals was less satisfactory, as he paid little attention to internal anatomy. He set out his system in *Systema Naturae* (1735) and other works. The tenth edition of this (1758) and *Species Plantarum* (1753) are internationally recognized as the starting-points for zoological and botanical nomenclature respectively. Linnaeus also carried out early experiments in biological pest control.

linnet /ˈlɪnɪt/ n. a small finch, *Acanthis cannabina*, with mainly brown and grey plumage and a pinkish breast. [OF *linette* f. *lin* flax (the bird feeding on flax-seeds)]

lino /ˈlaɪnəʊ/ n. (pl. **-os**) linoleum. [abbr.]

linocut /ˈlaɪnəʊˌkʌt/ n. **1** a design or form carved in relief on a block of linoleum. **2** a print made from this. □ **linocutting** n.

linoleic acid /ˌlɪnəʊˈliːɪk, -ˈleɪɪk/ n. *Chem.* a polyunsaturated fatty acid occurring as a glyceride in linseed and other oils, and essential in the human diet. [L *linum* flax + OLEIC ACID]

linolenic acid /ˌlɪnəʊˈlenɪk, -ˈliːnɪk/ n. *Chem.* a polyunsaturated fatty acid with one more double bond than linoleic acid, occurring as a glyceride in linseed and other oils and essential in the human diet. [after LINOLEIC ACID with *-ene-* inserted]

linoleum /lɪˈnəʊlɪəm/ n. a material consisting of a canvas backing thickly coated with a preparation of linseed oil and powdered cork etc., used esp. as a floor-covering. □ **linoleumed** adj. [L *linum* flax + *oleum* oil]

Linotype /ˈlaɪnəʊˌtaɪp/ n. *Printing propr.* a composing-machine producing lines of words as single strips of metal, used esp. for newspapers. [= *line o' type*]

Lin Piao see LIN BIAO.

linsang /ˈlɪnsæŋ/ n. a civet-like mammal; esp. one of the genus *Prionodon* of SE Asia, or *Poiana* of Africa. [Javanese]

linseed /ˈlɪnsiːd/ n. the seed of flax. □ **linseed cake** pressed linseed used as cattle-food. **linseed meal** ground linseed. **linseed oil** oil extracted from linseed and used in paint and varnish. [OE *līnsǣd* f. *līn* flax + *sǣd* seed]

linsey-woolsey /ˌlɪnzɪˈwʊlzɪ/ n. a fabric of coarse wool woven on a cotton warp. [ME f. *linsey* coarse linen, prob. f. *Lindsey* in Suffolk + WOOL, with jingling ending]

linstock /ˈlɪnstɒk/ n. *hist.* a match-holder used to fire cannon. [earlier *lintstock* f. Du. *lontstok* f. *lont* match + *stok* stick, with assim. to LINT]

lint /lɪnt/ n. **1** a fabric, orig. of linen, with a raised nap on one side, used for dressing wounds. **2** fluff. **3** *Sc.* flax. □ **linty** adj. [ME *lyn(n)et*, perh. f. OF *linette* linseed f. *lin* flax]

lintel /ˈlɪnt(ə)l/ n. *Archit.* a horizontal supporting piece of timber, stone, etc., across the top of a door or window. □ **lintelled** adj. (*US* **linteled**). [ME f. OF *lintel* threshold f. Rmc *limitale* (unrecorded), infl. by LL *liminare* f. L *limen* threshold]

linter /ˈlɪntə(r)/ n. *US* **1** a machine for removing the short fibres from cotton seeds after ginning. **2** (in *pl.*) these fibres. [LINT + -ER¹]

liny /ˈlaɪnɪ/ adj. (**linier**, **liniest**) marked with lines; wrinkled.

Linz /lɪnts/ an industrial city in northern Austria, on the River Danube, capital of the state of Upper Austria; pop. (1991) 202,855.

lion /ˈlaɪən/ n. **1** (*fem.* **lioness** /-nɪs/) a large powerful feline, *Panthera leo*, of Africa and India, with a tawny coat and, in the male, a flowing shaggy mane. **2** (**the Lion**) the zodiacal sign or constellation Leo. **3** a brave or celebrated person. **4** the lion as a national emblem of Great Britain or as a representation in heraldry. □ **lion-heart** a courageous person (esp. as a sobriquet of Richard I of England). **lion-hearted** brave and generous. **the lion's share** the largest or best part. □ **lion-like** adj. [ME f. AF *liun* f. L *leo -onis* f. Gk *leōn leontos*]

lionize /ˈlaɪəˌnaɪz/ v.tr. (also **-ise**) treat as a celebrity. □ **lionizer** n. **lionization** /ˌlaɪənaɪˈzeɪʃ(ə)n/ n.

Lions (also **British Lions**) the Rugby Union team representing the British Isles, so called from the symbol on the official tie.

lip /lɪp/ n. & v. ● n. **1 a** either of the two fleshy parts forming the edges of the mouth-opening. **b** a thing resembling these. **c** = LABIUM. **2** the edge of a cup, container, etc., esp. the part shaped for pouring from. **3** *colloq.* impudent talk (*that's enough of your lip!*). ● v.tr. (**lipped**, **lipping**) **1 a** touch with the lips; apply the lips to. **b** touch lightly. **2** *Golf* **a** hit a ball just to the edge of (a hole). **b** (of a ball) reach the edge of (a hole) but fail to drop in. □ **bite one's lip** repress an emotion; stifle laughter, a retort, etc. **curl one's lip** express scorn. **hang on a person's lips** listen attentively to a person. **lick one's lips** see LICK. **lip-read** (*past* and *past part.* **-read** /-red/) (esp. of a deaf person) understand (speech) entirely from observing a speaker's lip-movements. **lip-reader** a person who lip-reads. **lip-service** an insincere expression of support etc. **lip-sync** (or **synch**) n. (in film acting etc.) movement of a performer's lips in synchronization with a pre-recorded soundtrack. ● v. perform, esp. a song, on film using this technique. **pass a person's lips** be eaten, drunk, spoken, etc. **smack one's lips** part the lips noisily in relish or anticipation, esp. of food. □ **lipless** adj. **liplike** adj. **lipped** adj. (also in *comb.*). [OE *lippa* f. Gmc]

Lipari Islands /ˈlɪpərɪ/ a group of seven volcanic islands in the Tyrrhenian Sea, off the NE coast of Sicily, and in Italian possession. Believed by the ancient Greeks to be the home of Aeolus, the god of the winds, the islands were formerly known as the Aeolian Islands.

lipase /ˈlaɪpeɪz, -peɪs/ n. *Biochem.* an enzyme that catalyses the breakdown of fats. [Gk *lipos* fat + -ASE]

Lipetsk /ˈlɪpɪtsk/ an industrial city in SW Russia, on the Voronezh river; pop. (1990) 455,000. It is a major centre of the iron and steel industries.

lipid /ˈlɪpɪd/ n. *Chem.* any of a group of organic compounds that are insoluble in water but soluble in organic solvents, including fatty acids, oils, waxes, and steroids. [F *lipide* (as LIPASE)]

lipidosis /ˌlɪpɪˈdəʊsɪs/ n. (also **lipoidosis** /ˌlɪpɔɪ-/) (pl. **-doses** /-siːz/) *Med.* any disorder of lipid metabolism in the body tissues.

Lipizzaner var. of LIPIZZANER.

Li Po /liː ˈpəʊ/ (also **Li Bo** /ˈbəʊ/, **Li T'ai Po** /ˌliː taɪ ˈpəʊ/) (AD 701–62), Chinese poet. He had a bohemian lifestyle at the emperor's court, alternating with long periods of wandering. Typical themes in his poetry are wine, women, and the beauties of nature.

lipography /lɪˈpɒɡrəfɪ/ n. the omission of letters or words in writing. [Gk *lip-* stem of *leipō* omit + -GRAPHY]

lipoid /ˈlɪpɔɪd/ adj. resembling fat.

lipoprotein /ˌlaɪpəʊˈprəʊtiːn/ n. *Biochem.* any of a group of soluble proteins that combine with and transport fats or other lipids in the blood plasma. [Gk *lipos* fat + PROTEIN]

liposome /ˈlaɪpəʊˌsəʊm/ n. *Biochem.* a minute artificial spherical sac usu. of a phospholipid membrane enclosing an aqueous core, often used to transport drugs to a target tissue. [G. *Liposom*: see LIPID]

liposuction /ˈlaɪpəʊˌsʌkʃ(ə)n/ n. a technique in cosmetic surgery in which excess fat is removed from under the skin using a vacuum pump. [Gk *lipos* fat + SUCTION]

Lippi¹ /ˈlɪpɪ/, Filippino (c.1457–1504), Italian painter, son of Fra Filippo Lippi. Having trained with his father and Botticelli he completed the fresco cycle on the life of St Peter in the Brancacci Chapel, Florence (c.1481–3), a project begun by Masaccio. His other works include the series of frescos in the Carafa Chapel in Rome (1488–93) and the painting *The Vision of St Bernard* (c.1486).

Lippi² /ˈlɪpɪ/, Fra Filippo (c.1406–69), Italian painter. He joined a Carmelite order, but later renounced his vows in order to marry; his son was the painter Filippino Lippi. In the early 1420s he became a pupil of Masaccio, whose influence can be seen in the fresco *The Relaxation of the Carmelite Rule* (c.1432). His characteristic later style is more decorative and less monumental than his early work; typical works depict the Madonna as the central feature, stressing the human aspect of the theme. His paintings influenced the Pre-Raphaelites.

Lippizaner /ˌlɪpɪˈtsɑːnə(r)/ n. (also **Lipizzaner**) a fine white breed of horse, used esp. in displays of dressage. [G f. *Lippiza* in Slovenia]

Lippmann /ˈlɪpmən/, Gabriel Jonas (1845–1921), French physicist. He is best known today for his production of the first fully orthochromatic colour photograph in 1893. His earlier research was on the effect of electricity on capillary tubing, using it to develop an electrometer that was sensitive to potential changes of a thousandth of a volt. Lippmann also designed a number of other instruments, notably a coelostat, and devised methods for the precise measurement of various units.

lippy /ˈlɪpɪ/ adj. (**lippier, lippiest**) colloq. **1** insolent, impertinent. **2** talkative.

lipsalve /ˈlɪpsælv/ n. **1** a preparation, usu. in stick form, to prevent or relieve sore lips. **2** flattery.

lipstick /ˈlɪpstɪk/ n. a small stick of cosmetic for colouring the lips.

liquate /lɪˈkweɪt/ v.tr. separate or purify (metals) by liquefying. □ **liquation** /-ˈkweɪʃ(ə)n/ n. [L *liquare* melt, rel. to LIQUOR]

liquefy /ˈlɪkwɪˌfaɪ/ v.tr. & intr. (also **liquify**) (**-ies, -ied**) Chem. make or become liquid. □ **liquefiable** adj. **liquefier** n. **liquefacient** /ˌlɪkwɪˈfeɪʃ(ə)nt/ adj. & n. **liquefaction** /-ˈfækʃ(ə)n/ n. **liquefactive** /-ˈfæktɪv/ adj. [F *liquéfier* f. L *liquefacere* f. *liquere* be liquid]

liquescent /lɪˈkwes(ə)nt/ adj. becoming or apt to become liquid. [L *liquescere* (as LIQUEFY)]

liqueur /lɪˈkjʊə(r)/ n. any of several strong sweet alcoholic spirits, variously flavoured, usu. drunk after a meal. [F, = LIQUOR]

liquid /ˈlɪkwɪd/ adj. & n. ● adj. **1** having a consistency like that of water or oil, flowing freely but of constant volume. **2** having the qualities of water in appearance; translucent (*liquid blue; a liquid lustre*). **3** (of a gas, e.g. air, hydrogen) reduced to a liquid state by intense cold. **4** (of sounds) clear and pure; harmonious, fluent. **5 a** (of assets) easily converted into cash. **b** having ready cash or liquid assets. **6** not fixed; fluid (*liquid opinions*). ● n. **1** a liquid substance. **2** Phonet. the sound of l or r. □ **liquid crystal** a turbid liquid with some order in its molecular arrangement. **liquid crystal display** a form of visual display in electronic devices, in which the reflectivity of a matrix of liquid crystals changes as a signal is applied. **liquid measure** a unit for measuring the volume of liquids. **liquid paper** = correction fluid. **liquid paraffin** Pharm. a colourless odourless oily liquid obtained from petroleum and used as a laxative. □ **liquidly** adv. **liquidness** n. [ME f. L *liquidus* f. *liquere* be liquid]

liquidambar /ˌlɪkwɪdˈæmbə(r)/ n. **1** a tree of the genus *Liquidambar*, yielding a resinous gum. **2** this gum. [mod.L app. f. L *liquidus* (see LIQUID) + med.L *ambar* amber]

liquidate /ˈlɪkwɪˌdeɪt/ v. **1 a** tr. wind up the affairs of (a company or firm) by ascertaining liabilities and apportioning assets. **b** intr. (of a company) be liquidated. **2** tr. clear or pay off (a debt). **3** tr. eliminate by killing, wipe out. [med.L *liquidare* make clear (as LIQUID)]

liquidation /ˌlɪkwɪˈdeɪʃ(ə)n/ n. **1** the process of liquidating a company etc. **2** elimination by killing. □ **go into liquidation** (of a company etc.) be wound up and have its assets apportioned.

liquidator /ˈlɪkwɪˌdeɪtə(r)/ n. a person called in to wind up the affairs of a company etc.

liquidity /lɪˈkwɪdɪtɪ/ n. (pl. **-ies**) **1** the state of being liquid. **2 a** availability of liquid assets. **b** (in pl.) liquid assets. [F *liquidité* or med.L *liquiditas* (as LIQUID)]

liquidize /ˈlɪkwɪˌdaɪz/ v.tr. (also **-ise**) reduce to a liquid or puréed state.

liquidizer /ˈlɪkwɪˌdaɪzə(r)/ n. (also **-iser**) a machine for liquidizing food.

liquify var. of LIQUEFY.

liquor /ˈlɪkə(r)/ n. & v. ● n. **1** an alcoholic (esp. distilled) drink. **2** water used in brewing. **3** other liquid, esp. that produced in cooking. **4** Pharm. a solution of a specified drug in water. ● v.tr. **1** dress (leather) with grease or oil. **2** steep (malt etc.) in water. [ME f. OF *lic(o)ur* f. L *liquor -oris* (as LIQUID)]

liquorice /ˈlɪkərɪs, -rɪʃ/ n. (also **licorice**) **1** a black root extract used as a sweet and in medicine. **2** the leguminous plant, *Glycyrrhiza glabra*, from which it is obtained. [ME f. AF *lycorys*, OF *licoresse* f. LL *liquiritia* f. Gk *glukurrhiza* f. *glukus* sweet + *rhiza* root]

liquorish /ˈlɪkərɪʃ/ adj. **1** = LICKERISH. **2** fond of or indicating a

fondness for liquor. □ **liquorishly** adv. **liquorishness** n. [var. of LICKERISH, misapplied]

lira /ˈlɪərə/ n. (pl. **lire** /-rɪ/) **1** the chief monetary unit of Italy and Malta. **2** the chief monetary unit of Turkey. [It. f. Prov. *liura* f. L *libra* pound (weight etc.)]

Lisbon /ˈlɪzbən/ (Portuguese **Lisboa** /liʒˈbɔə/) the capital and chief port of Portugal, on the Atlantic coast at the mouth of the River Tagus; pop. (1991) 677,790. Occupied by the Romans in AD 205 and the Moors in 714, it was captured by the newly independent kingdom of Portugal in 1147 and made the capital in the 13th century. It flourished during the period of Portuguese colonial expansion, but was devastated by an earthquake in 1755, after which much of the city had to be rebuilt.

Lisburn /ˈlɪzbɜːn/ a town in Northern Ireland, to the south-west of Belfast, on the border between Antrim and Down; pop. (1981) 40,390. It developed as a centre of linen production after the arrival in 1698 of French Huguenots.

Lisdoonvarna /ˌlɪsduːnˈvɑːnə/ a spa town in the Republic of Ireland, in County Clare; pop. (1981) 607. It is noted for its sulphurous water, its summer folk festival, and its October fair.

lisente pl. of SENTE.

lisle /laɪl/ n. (in full **lisle thread**) a fine smooth cotton thread for stockings etc. [*Lisle*, former spelling of LILLE, where orig. made]

lisp /lɪsp/ n. & v. ● n. **1** a speech defect in which s is pronounced like th in *thick* and z is pronounced like th in *this*. **2** a rippling of waters; a rustling of leaves. ● v.intr. & tr. speak or utter with a lisp. □ **lisper** n. **lispingly** adv. [OE *wlispian* (recorded in *āwlispian*) f. *wlisp* (adj.) lisping, of uncert. orig.]

lissom /ˈlɪsəm/ adj. (also **lissome**) lithe, supple, agile. □ **lissomly** adv. **lissomness** n. [ult. f. LITHE + -SOME¹]

list¹ /lɪst/ n. & v. ● n. **1** a number of connected items, names, etc., written or printed together usu. consecutively to form a record or aid to memory (*shopping list*). **2** (in pl.) a hist. palisades enclosing an area for a tournament. **b** the scene of a contest. **3** esp. Brit. **a** a selvedge or edge of cloth, usu. of different material from the main body. **b** such edges used as a material. ● v. **1** tr. **a** make a list of. **b** enumerate; name one by one as if in a list. **2** tr. enter in a list. **3** tr. (as **listed** adj.) **a** (of securities) approved for dealings on the Stock Exchange. **b** (of a building in the UK) officially designated as being of historical importance and having protection from demolition or major alterations. **4** tr. & intr. archaic enlist. □ **enter the lists** issue or accept a challenge. **list price** the price of something as shown in a published list. □ **listable** adj. [OE *liste* border, strip f. Gmc]

list² /lɪst/ v. & n. ● v.intr. (of a ship etc.) lean over to one side, esp. owing to a leak or shifting cargo (cf. HEEL²). ● n. the process or an instance of listing. [17th c.: orig. unkn.]

listen /ˈlɪs(ə)n/ v.intr. **1 a** make an effort to hear something. **b** attentively hear a person speaking. **2** (often foll. by to) give attention with the ear (*listened to my story*). **b** take notice of; respond to advice or a request or to the person expressing it. **3** (also **listen out**) (often foll. by for) seek to hear or be aware of by waiting alertly. □ **listen in 1** eavesdrop; tap a private conversation, esp. one by telephone. **2** use a radio receiving set. **listening-post 1 a** a point near an enemy's lines for detecting movements by sound. **b** a station for intercepting electronic communications. **2** a place for the gathering of information from reports etc. [OE *hlysnan* f. WG]

listenable /ˈlɪs(ə)nəb(ə)l/ adj. easy or pleasant to listen to. □ **listenability** /ˌlɪs(ə)nəˈbɪlɪtɪ/ n.

listener /ˈlɪs(ə)nə(r)/ n. **1** a person who listens. **2** a person receiving broadcast radio programmes.

Lister /ˈlɪstə(r)/, Joseph, 1st Baron (1827–1912), English surgeon, inventor of antiseptic techniques in surgery. In 1865 he became acquainted with Louis Pasteur's theory that putrefaction is due to micro-organisms, and realized its significance in connection with sepsis in wounds, a major cause of deaths in patients who had undergone surgery. In the same year Lister first used carbolic acid dressings, and later he used a carbolic spray in the operating theatre. After about 1883 aseptic rather than antiseptic techniques became popular, though Lister believed in the use of both.

lister /ˈlɪstə(r)/ n. US a plough with a double mould-board. [*list* prepare land for a crop + -ER¹]

listeria /lɪˈstɪərɪə/ n. a motile rodlike bacterium of the genus *Listeria*; esp. *L. monocytogenes*, infecting humans and animals eating contaminated food. [mod.L f. Joseph *Lister*, Engl. surgeon (1827–1912)]

listeriosis /lɪ‚stɪərɪ'əʊsɪs/ n. Med. infection with listeria; a disease resulting from this.

listing /'lɪstɪŋ/ n. **1** a list or catalogue (see LIST[1] 1). **2 a** the drawing up of a list. **b** an entry in a list or register. **3** Brit. selvedge (see LIST[1] n. 3).

listless /'lɪstlɪs/ adj. lacking energy or enthusiasm; disinclined for exertion. □ **listlessly** adv. **listlessness** n. [ME f. obs. list inclination + -LESS]

Liszt /lɪst/, Franz (1811–86), Hungarian composer and pianist. He was a key figure in the romantic movement and a virtuoso pianist; many of his piano compositions combine lyricism with great technical complexity. His orchestral works include the Faust and Dante Symphonies (1854–7; 1855–6); his twelve symphonic poems (1848–58) created a new musical form. He also composed masses, and oratorios such as Christus (1862–7). Apart from his own influence as a composer and teacher, Liszt was also significant as a champion of Wagner's work.

lit past and past part. of LIGHT[1], LIGHT[2].

Li T'ai Po see LI PO.

litany /'lɪtənɪ/ n. (pl. **-ies**) **1 a** a series of petitions for use in church services or processions, usu. recited by the clergy and responded to in a recurring formula by the people. **b** (**the Litany**) that contained in the Book of Common Prayer. **2** a tedious recital (a litany of woes). [ME f. OF letanie f. eccl.L litania f. Gk litaneia prayer f. litē supplication]

litchi var. of LYCHEE.

lite /laɪt/ n. & adj. colloq. ● n. **1** a light, esp. a courtesy light in a motor vehicle. **2** (also propr. **Lite**) a light beer with relatively few calories. ● adj.**1** (also propr. **Lite**) applied to lighter versions of various manufactured food or drink products, esp. to low-calorie beer. **2** US lacking in substance, facile, over-simplified. [var. of LIGHT[1] and LIGHT[2], now usu. a deliberate respelling]

-lite /laɪt/ suffix Geol. forming names of minerals and fossils (rhyolite; zeolite). [F f. Gk lithos stone]

liter US var. of LITRE.

literacy /'lɪtərəsɪ/ n. the ability to read and write. [LITERATE + -ACY after illiteracy]

literae humaniores /‚lɪtə‚raɪ hu:‚mænɪ'ɔ:rez/ n. Brit. the school of classics and philosophy at Oxford University; Greats (GREAT n. 2). [L, = the more humane studies]

literal /'lɪtərəl/ adj. & n. ● adj. **1** taking words in their usual or primary sense without metaphor or allegory (literal interpretation). **2** following the letter, text, or exact or original words (literal translation; a literal transcript). **3** (in full **literal-minded**) (of a person) prosaic; matter-of-fact. **4 a** not exaggerated (the literal truth). **b** so called without exaggeration (a literal extermination). **5** colloq. disp. so called with some exaggeration or using metaphor (a literal avalanche of mail). **6** of, in, or expressed by a letter or the letters of the alphabet. **7** Algebra not numerical. ● n. Printing a misprint of a letter. □ **literalize** v.tr. (also **-ise**). **literally** adv. **literalness** n. **literality** /‚lɪtə'rælɪtɪ/ n. [ME f. OF literal or LL litteralis f. L littera (as LETTER)]

literalism /'lɪtərə‚lɪz(ə)m/ n. insistence on a literal interpretation; adherence to the letter. □ **literalist** n. **literalistic** /‚lɪtərə'lɪstɪk/ adj.

literary /'lɪtərərɪ/ adj. **1** of, constituting, or occupied with books or literature or written composition, esp. of the kind valued for quality of form. **2** well-informed about literature. **3** (of a word or idiom) used chiefly in literary works or other formal writing. □ **literary executor** see EXECUTOR. **literary history** the history of the treatment of a subject in literature. □ **literarily** adv. **literariness** n. [L litterarius (as LETTER)]

literary criticism n. the art or practice of judging and commenting on the qualities and character of literary works. Literary criticism in the West was until the end of the 17th century much concerned with applying abstract rules derived from Aristotle, such as the three unities (see UNITY 5); modern criticism in English traces its origin to the more flexible approach of Samuel Johnson and Augustans such as John Dryden and Alexander Pope. In the 19th century Matthew Arnold argued that literature had an important moral and philosophical purpose, a view contradicted by the 'art for art's sake' ideal of the Aesthetic Movement. Criticism developed into an academic discipline in the 20th century, with a profusion of different schools of thought: these may be broadly divided into those approaching literary works as self-contained entities to be analysed in isolation from external factors (see NEW CRITICISM) and those discussing literature in the light of related spheres (such as biography, history, Marxism, feminism, etc.). Since the 1950s structuralism, post-structuralism, post-modernism,

and deconstruction have explored and questioned the fundamental concepts of authorship and meaning.

literate /'lɪtərət/ adj. & n. ● adj. able to read and write; educated. ● n. a literate person. □ **literately** adv. [ME f. L litteratus (as LETTER)]

literati /‚lɪtə'rɑ:tɪ/ n.pl. literary or scholarly people. [L, pl. of literatus (as LETTER)]

literatim /‚lɪtə'rɑ:tɪm/ adv. letter for letter; textually, literally. [med.L]

literation /‚lɪtə'reɪʃ(ə)n/ n. the representation of sounds etc. by a letter or group of letters. [L litera LETTER]

literature /'lɪtərətʃə(r), 'lɪtrə-/ n. **1** written works, esp. those whose value lies in beauty of language or in emotional effect. **2** the realm of letters. **3** the writings of a country or period. **4** literary production. **5** colloq. printed matter, leaflets, etc. **6** the material in print on a particular subject (there is a considerable literature on geraniums). [ME, = literary culture, f. L litteratura (as LITERATE)]

-lith /lɪθ/ suffix denoting types of stone (laccolith; monolith). [Gk lithos stone]

litharge /'lɪθɑ:dʒ/ n. lead monoxide (chem. formula: PbO), a red or yellow solid used as a pigment and in glass and ceramics. [ME f. OF litarge f. L lithargyrus f. Gk litharguros f. lithos stone + arguros silver]

lithe /laɪð/ adj. flexible, supple. □ **lithely** adv. **litheness** n. **lithesome** adj. [OE līthe f. Gmc]

lithia /'lɪθɪə/ n. lithium oxide (chem. formula: Li_2O), a white alkaline solid. □ **lithia water** mineral water containing lithium salts. [f. Gk neut. of litheios f. lithos stone, after soda etc.]

lithic /'lɪθɪk/ adj. **1** of, like, or made of stone. **2** Med. of a calculus. [Gk lithikos f. lithos stone]

lithium /'lɪθɪəm/ n. a soft silver-white metallic chemical element (atomic number 3; symbol **Li**). Lithium was discovered by the Swedish chemist J. A. Arfvedson (1792–1841) in 1817. It occurs widely as a minor component of silicate rocks, and in some rare minerals. It is the lightest and least reactive of the alkali metals. The element and its compounds now have numerous commercial uses in alloys, lubricating greases, chemical reagents, etc. [LITHIA]

litho /'laɪθəʊ/ n. & v. colloq. ● n. = LITHOGRAPHY. ● v.tr. (**-oes, -oed**) produce by lithography. [abbr.]

litho- /'lɪθəʊ, 'laɪθ-/ comb. form stone. [Gk lithos stone]

lithograph /'lɪθə‚grɑːf, 'laɪθ-/ n. & v. ● n. a lithographic print. ● v.tr. **1** print by lithography. **2** write or engrave on stone. [back-form. f. LITHOGRAPHY]

lithography /lɪ'θɒgrəfɪ/ n. a process of obtaining prints from a stone or metal surface treated so that the parts to be printed can be inked but the remaining area rejects ink. (See note below.) □ **lithographer** n. **lithographic** /‚lɪθə'græfɪk/ adj. **lithographically** adv. [G Lithographie (as LITHO-, -GRAPHY)]

▪ Lithography was invented in 1798 by a Bavarian dramatist, Aloys Senefelder (1771–1834), while experimenting with methods of duplicating his plays. He first drew images in greasy ink or crayon on a flat piece of limestone, which he etched with acid, treated with gum arabic, and then moistened with water. Ink then applied over the whole surface was retained only by the greasy images, from which a replica could be obtained by pressing a piece of paper on to the stone. In the modern printing press, thin aluminium plates are wrapped round a cylinder and damped and inked alternately by rollers. In offset lithography, the inked image is transferred (or set off) from the printing plate on to an intermediate rubber-covered transfer cylinder before being printed on to the chosen substrate. Photo-offset involves the photographic capture and deposition of the image on to a specially sensitized plate, which is then appropriately processed before use on the press.

lithology /lɪ'θɒlədʒɪ/ n. the science of the general physical characteristics of rocks (cf. PETROLOGY). □ **lithological** /‚lɪθə'lɒdʒɪk(ə)l/ adj.

lithophyte /'lɪθə‚faɪt/ n. Bot. a plant that grows on stone.

lithopone /'lɪθə‚pəʊn/ n. a white pigment of zinc sulphide, barium sulphate, and zinc oxide. [LITHO- + Gk ponos work]

lithosphere /'lɪθə‚sfɪə(r)/ n. Geol. the rigid outer part of the earth consisting of the crust and upper mantle. □ **lithospheric** /‚lɪθə'sferɪk/ adj.

lithotomy /lɪ'θɒtəmɪ/ n. (pl. **-ies**) Med. the surgical removal of a calculus from the bladder or urinary tract. □ **lithotomist** n. **lithotomize** v.tr. (also **-ise**). [LL f. Gk lithotomia (as LITHO-, -TOMY)]

lithotripsy /'lɪθə‚trɪpsɪ/ n. (pl. **-ies**) Med. a treatment using ultrasound to shatter a calculus in the bladder into small particles that can be passed through the urethra. □ **lithotripter** n. **lithotriptic** /ˌlɪθəˈtrɪptɪk/ adj. [LITHO- + Gk tripsis rubbing f. tribo rub]

Lithuania /ˌlɪθjuˈeɪnɪə, ˌlɪθuː-/ a country on the SE shore of the Baltic sea; pop. (est. 1991) 3,765,000; languages, Lithuanian (official), Russian; capital, Vilnius. Medieval Lithuania was an independent grand duchy which at its zenith, in the 14th century, extended to the shores of the Black Sea. It was united with Poland in 1386, and was absorbed into the Russian empire in 1795. Lithuania was declared an independent republic in 1918, but in 1940 was annexed by the Soviet Union as a constituent republic. In 1991, on the breakup of the USSR, Lithuania became an independent republic once again.

Lithuanian /ˌlɪθjuːˈeɪnɪən, ˌlɪθuː-/ n. & adj. ● n. **1 a** a native or inhabitant of Lithuania. **b** a person of Lithuanian descent. **2** the language of Lithuania, most closely related to Latvian, with which it constitutes the Baltic branch of Indo-European languages. ● adj. of or relating to Lithuania or its people or language.

litigant /ˈlɪtɪgənt/ n. & adj. ● n. a party to a lawsuit. ● adj. engaged in a lawsuit. [F (as LITIGATE)]

litigate /ˈlɪtɪˌgeɪt/ v. **1** intr. go to law; be a party to a lawsuit. **2** tr. contest (a point) in a lawsuit. □ **litigator** n. **litigable** /-gəb(ə)l/ adj. **litigation** /ˌlɪtɪˈgeɪʃ(ə)n/ n. [L litigare litigat- f. lis litis lawsuit]

litigious /lɪˈtɪdʒəs/ adj. **1** given to litigation; unreasonably fond of going to law. **2** disputable in a lawcourt; offering matter for a lawsuit. **3** of lawsuits. □ **litigiously** adv. **litigiousness** n. [ME f. OF litigieux or L litigiosus f. litigium litigation: see LITIGATE]

litmus /ˈlɪtməs/ n. a dye obtained from lichens that is red under acid conditions and blue under alkaline conditions. □ **litmus paper** a paper stained with litmus for use as a test for acids or alkalis. **litmus test 1** a test for acids and alkalis using litmus paper. **2** a simple test to establish true character. [ME f. ON litmosi f. litr dye + mosi moss]

litotes /laɪˈtəʊtiːz/ n. ironical understatement, esp. the expressing of an affirmative by the negative of its contrary (e.g. I shan't be sorry for I shall be glad). [LL f. Gk litotēs f. litos plain, meagre]

litre /ˈliːtə(r)/ n. (US **liter**) a metric unit of capacity equal to 1,000 cubic centimetres (about 1.75 pints) (symbol: l). The litre was formerly defined as the volume of one kilogram of water under standard conditions. □ **litreage** n. [F f. litron, an obs. measure of capacity, f. med.L f. Gk litra a Sicilian monetary unit]

Litt.D. abbr. Doctor of Letters. [L Litterarum Doctor]

litter /ˈlɪtə(r)/ n. & v. ● n. **1 a** refuse, esp. paper, discarded in an open or public place. **b** odds and ends lying about. **c** (attrib.) for disposing of litter (litter-bin). **2** a state of untidiness, disorderly accumulation of papers etc. **3** the young animals brought forth at a birth. **4** a vehicle containing a couch shut in by curtains and carried on men's shoulders or by beasts of burden. **5** a framework with a couch for transporting the sick and wounded. **6 a** straw, rushes, etc., as bedding, esp. for animals. **b** straw and dung in a farmyard. **7** granular absorbent material for lining a box for a cat to urinate and defecate in indoors. ● v.tr. **1** make (a place) untidy with discarded refuse. **2** scatter untidily and leave lying about. **3** give birth to (whelps etc.). **4** (often foll. by down) **a** provide (a horse etc.) with litter as bedding. **b** spread litter or straw on (a floor) or in (a stable). □ **litter-lout** = LITTERBUG. □ **littery** adj. (in senses 1, 2 of n.). [ME f. AF litere, OF litiere f. med.L lectaria f. L lectus bed]

littérateur /ˌliːtəraˈtɜː(r)/ n. a literary person. [F]

litterbug /ˈlɪtəˌbʌg/ n. colloq. a person who carelessly drops litter in a public place.

little /ˈlɪt(ə)l/ adj., n., & adv. ● adj. (**littler**, **littlest**; **less** or **lesser**; **least**) **1** small in size, amount, degree, etc.; not great or big: sometimes used to convey affectionate or emotional overtones, or condescension, not implied by small (a friendly little chap; a silly little fool; a nice little car). **2 a** short in stature (a little man). **b** of short distance or duration (will go a little way with you; wait a little while). **3** (prec. by a) a certain though small amount of (give me a little butter). **4** trivial; relatively unimportant (exaggerates every little difficulty). **5** not much; inconsiderable (gained little advantage from it). **6** operating on a small scale (the little shopkeeper). **7** as a distinctive epithet: **a** of a smaller or the smallest size etc. (little finger). **b** that is the smaller or smallest of the name (little auk; little grebe). **8** young or younger (a little boy; my little sister). **9** as of a child, evoking tenderness, condescension, amusement, etc. (we know their little ways). **10** mean, paltry, contemptible (you little sneak). ● n. **1** not much; only a small amount (got very little out of it; did what little I could). **2** (usu. prec. by a) **a** a certain but no great amount (knows a little of everything; every little helps). **b** a short time or distance (after a little).

● adv. (**less**, **least**) **1** to a small extent only (little-known authors; is little more than speculation). **2** not at all; hardly (they little thought). **3** (prec. by a) somewhat (is a little deaf). □ **in little** on a small scale. **little by little** by degrees; gradually. **little end** the smaller end of a connecting-rod, attached to the piston. **little finger** the smallest finger, at the outer end of the hand. **little grebe** a small Old World grebe, Tachybaptus ruficollis (also called dabchick). **little man 1** esp. joc. (as a form of address) a boy. **2** the average 'man in the street'. **little ones** young children or animals. **little or nothing** hardly anything. **little owl** a small owl, Athene noctua, of Africa and Eurasia, with speckled plumage. **the little people** fairies. **little slam** Bridge the winning of 12 tricks. **little theatre** a small playhouse, esp. one used for experimental productions. **the little woman** colloq. often derog. one's wife. **no little** considerable, a good deal of (took no little trouble over it). **not a little** n. much; a great deal. ● adv. extremely (not a little concerned). □ **littleness** n. [OE lȳtel f. Gmc]

Little Ararat see ARARAT, MOUNT.

Little Bighorn, Battle of /ˈbɪghɔːn/ a battle in which General George Custer and his forces were defeated by Sioux warriors on 25 June 1876, popularly known as Custer's Last Stand. It took place in the valley of the Little Bighorn river in Montana.

Little Corporal a nickname for Napoleon.

Little Englander n. esp. hist. a person opposed to an imperial role or policy for Britain.

Little Lord Fauntleroy see FAUNTLEROY.

Little Masters a group of 16th-century Nuremberg engravers, followers of Dürer, who worked small-dimension plates with biblical, mythological, and genre scenes. [mistransl. G Kleinmeister, 'Masters in Little']

Little Minch see MINCH, THE.

Little Ouse see OUSE 4.

Little Rock the state capital of Arkansas; pop. (1990) 175,800.

Little Russian n. & adj. hist. ● n. a Ukrainian. ● adj. Ukrainian.

Little St Bernard Pass see ST BERNARD PASS.

Little Tibet an alternative name for BALTISTAN.

Littlewood /ˈlɪt(ə)lˌwʊd/, (Maud) Joan (1914–91), English theatre director. She is best known for co-founding the Theatre Workshop (1945), which set out to present established plays in radical productions and to stage contemporary working-class plays. Memorable productions included Brendan Behan's The Quare Fellow (1956). Littlewood's name is most closely associated with the direction of the musical Oh, What a Lovely War (1963).

littoral /ˈlɪtərəl/ adj. & n. ● adj. of or on the shore of the sea, a lake, etc. ● n. a region lying along a shore. [L littoralis f. litus litoris shore]

Littré /ˈlɪtreɪ/, Émile (1801–81), French lexicographer and philosopher. He was the author of the major Dictionnaire de la langue française (1863–77) and also wrote a history of the French language (1862). He was a follower of Auguste Comte, and became the leading exponent of positivism after Comte's death.

liturgical /lɪˈtɜːdʒɪk(ə)l/ adj. of or related to liturgies or public worship. □ **liturgically** adv. [med.L f. Gk leitourgikos (as LITURGY)]

liturgy /ˈlɪtədʒɪ/ n. (pl. **-ies**) **1 a** a form of public worship, esp. in the Christian Church. **b** a set of formularies for this. **c** public worship in accordance with a prescribed form. **2** the Communion office of the Orthodox Church. **3** (**the Liturgy**) the Book of Common Prayer. **4** Gk Hist. a public office or duty performed voluntarily by a rich Athenian. □ **liturgist** n. [F liturgie or LL liturgia f. Gk leitourgia public service, worship of the gods f. leitourgos minister f. leit- public + ergon work]

Liuzhou /ljuːˈdʒəʊ/ (also **Liuchow** /-ˈtʃaʊ/) an industrial city in southern China, in Guangxi Zhuang province north-east of Nanning; pop. (1990) 740,000. It is the site of a major iron and steel complex.

livable var. of LIVEABLE.

live¹ /lɪv/ v. **1** intr. be or remain alive; have (esp. animal) life. **2** intr. (foll. by on) subsist or feed (lives on fruit). **3** intr. (foll. by on, off) depend for subsistence (lives off the family; lives on income from investments). **4** intr. (foll. by on, by) sustain one's position or repute (live on their reputation; lives by his wits). **5** tr. **a** (with complement) spend, pass, experience (lived a happy life). **b** express in one's life (was living a lie). **6** intr. spend or lead one's life; conduct oneself in a specified way (live quietly). **7** intr. arrange one's habits, expenditure, feeding, etc. (live modestly). **8** intr. make or have one's abode. **9** intr. (foll. by in) spend the daytime (the room does not seem to be lived in). **10** intr. (of a person or thing) survive. **11** intr. (of a ship) escape destruction. **12** intr. enjoy life intensely or to the full (you haven't *lived till you've drunk champagne).* □ **live and let live** condone others' failings so as to be similarly tolerated. **live down** (usu. with *neg.*) cause (past guilt, embarrassment, etc.) to be forgotten by different conduct over a period of time (*you'll never live that down!*). **live in** *Brit.* (of a domestic employee, student, etc.) reside on the premises of one's work, college, etc. **live-in** *attrib.adj.* **1** (of a sexual partner) cohabiting. **2** resident (*live-in nanny*). **live it up** *colloq.* live gaily and extravagantly. **live out 1** survive (a danger, difficulty, etc.). **2** (of a domestic employee, student, etc.) reside away from one's place of work, college, etc. **live through** survive; remain alive at the end of. **live to** survive and reach (*lived to a great age*). **live to oneself** live in isolation. **live together** (esp. of a man and woman not married to each other) share a home and have a sexual relationship. **live up to** honour or fulfil; put into practice (principles etc.). **live with 1** share a home with. **2** tolerate; find congenial. **long live** … **!** an exclamation of loyalty (to a person etc. specified). [OE *libban, lifian,* f. Gmc]

live² /laɪv/ *adj. & adv.* ● *adj.* **1** (*attrib.*) that is alive; living. **2** (of a performance) heard or seen at the time of its occurrence, not from a recording. **3** full of power, energy, or importance; not obsolete or exhausted (*disarmament is still a live issue*). **4** expending or still able to expend energy in various forms, esp.: **a** (of coals) glowing, burning. **b** (of a shell) unexploded. **c** (of a match) unkindled. **d** (of a wire, terminal, device, etc.) connected to a source of electrical power. **5** (of rock) not detached, seeming to form part of the earth's frame. **6 a** (of a wheel or axle etc. in machinery) moving or imparting motion. **b** in operation, fully operational (*the system has been live for two weeks now*). ● *adv.* in order to make a live broadcast; as a live performance (*going live now to the House of Commons; the show went out live*). □ **live bait** small fish used to entice prey. **live load** the weight of persons or goods in a building or vehicle. **live oak** an American evergreen tree, *Quercus virginiana*. **live wire** *colloq.* an energetic and forceful person. [aphetic form of ALIVE]

liveable /ˈlɪvəb(ə)l/ *adj.* (also **livable**) **1** (of a house, room, climate, etc.) fit to live in. **2** (of a life) worth living. **3** (usu. **liveable with**) *colloq.* (of a person) companionable; easy to live with. □ **liveableness** *n.* **liveability** /ˌlɪvəˈbɪlɪtɪ/ *n.*

livelihood /ˈlaɪvlɪˌhʊd/ *n.* a means of living; sustenance. [OE *līflād* f. *līf* LIFE + *lād* course (see LOAD): assim. to obs. *livelihood* liveliness]

livelong¹ /ˈlɪvlɒŋ/ *adj. poet. or rhet.* in its entire length or apparently so (*the livelong day*). [ME *lefe longe* (as LIEF, LONG¹): assim. to LIVE¹]

livelong² /ˈlɪvlɒŋ/ *n.* an orpine. [LIVE¹ + LONG¹]

lively /ˈlaɪvlɪ/ *adj.* (**livelier, liveliest**) **1** full of life; vigorous, energetic. **2** brisk (*a lively pace*). **3** vigorous, stimulating (*a lively discussion*). **4** vivacious, jolly, sociable. **5** esp. *Brit. joc.* exciting, dangerous, difficult (*the press is making things lively for them*). **6** (of a colour) bright and vivid. **7** lifelike, realistic (*a lively description*). **8** (of a boat etc.) rising lightly to the waves. □ **livelily** *adv.* **liveliness** *n.* [OE *līflic* (as LIFE, -LY¹)]

liven /ˈlaɪv(ə)n/ *v.tr. & intr.* (often foll. by *up*) *colloq.* make or become more lively.

liver¹ /ˈlɪvə(r)/ *n.* **1 a** a large lobed glandular organ in the abdomen of vertebrates, secreting bile. (*See note below.*) **b** a similar organ in other animals. **2** the flesh of an animal's liver as food. **3** (in full **liver-colour**) a dark reddish-brown. □ **liver chestnut** see CHESTNUT. **liver fluke** a parasitic fluke, esp. *Fasciola hepatica*, the adults of which live within the liver tissues of vertebrates, and the larvae within snails. **liver of sulphur** a liver-coloured mixture containing potassium sulphide etc., used as a lotion in skin disease. **liver salts** *Brit.* salts to cure dyspepsia or biliousness. **liver sausage** a sausage containing cooked liver etc. □ **liverless** *adj.* [OE *lifer* f. Gmc]

▪ The liver's main function is the chemical processing of digestive products into substances which will be useful to the body. It makes many of the proteins of blood plasma, converts glucose into glycogen for storage, neutralizes harmful substances absorbed from the intestine, and stores some minerals and vitamins. Its other major function is the secretion of bile, which is essential for the proper digestion and absorption of fats. The liver was anciently supposed to be the seat of love and of violent passion generally, whence expressions such as *lily-livered* (= cowardly).

liver² /ˈlɪvə(r)/ *n.* a person who lives in a specified way (*a clean liver*).

liverish /ˈlɪvərɪʃ/ *adj.* **1** suffering from a disorder of the liver. **2** peevish, glum. □ **liverishly** *adv.* **liverishness** *n.*

Liverpool¹ /ˈlɪvəˌpuːl/ a city and seaport in NW England, administrative centre of Merseyside; pop. (1991) 448,300. Situated at the east side of the mouth of the River Mersey, Liverpool developed as a port in the 17th century with the import of cotton from America

and the export of textiles produced in Lancashire and Yorkshire. In the 18th century it became an important centre of shipbuilding and engineering but these industries have declined in recent years.

Liverpool² /ˈlɪvəˌpuːl/, 2nd Earl of (title of Robert Banks Jenkinson) (1770–1828), British Tory statesman, Prime Minister 1812–27. His government opposed both parliamentary reform and Catholic Emancipation, and took repressive measures to deal with popular discontent at the time of the Peterloo massacre. Lord Liverpool was later influenced by more liberal figures such as Sir Robert Peel to introduce some important reforms.

Liverpudlian /ˌlɪvəˈpʌdliən/ *n. & adj.* ● *n.* a native of Liverpool in NW England. ● *adj.* of or relating to Liverpool. [joc. f. *Liverpool* + PUDDLE]

liverwort /ˈlɪvəˌwɜːt/ *n.* a small leafy or thalloid bryophyte of the class Hepaticae. [f. the lobed liver-shaped thallus of one kind]

livery¹ /ˈlɪvərɪ/ *n.* (*pl.* **-ies**) **1 a** a distinctive clothing worn by a member of a City Company, a servant, etc. **b** membership of a City livery company (see also LIVERY COMPANY). **2** a distinctive colour scheme in which vehicles, aircraft, etc., of a particular company or line are painted. **3** *US* a place where horses can be hired. **4** *hist.* a provision of food or clothing for retainers etc. **5** *Law* **a** the legal delivery of property. **b** a writ allowing this. □ **at livery** (of a horse) kept for the owner and fed and groomed for a fixed charge. **livery stable** a stable where horses are kept at livery or let out for hire. □ **liveried** *adj.* (esp. in senses 1, 2). [ME f. AF *liveré*, OF *livrée*, fem. past part. of *livrer* DELIVER]

livery² /ˈlɪvərɪ/ *adj.* **1** of the consistency or colour of liver. **2** *Brit.* (of soil) cohesive, heavy. **3** *colloq.* liverish.

livery company *n.* (in the UK) any of the London City companies, which formerly had distinctive costumes. Descended from medieval craft guilds, the companies are now largely social and charitable organizations; none is now a trading company, though some still have some involvement with the operation of their original trade; several support public schools (e.g. Merchant Taylors, Haberdashers), and collectively they are involved in various forms of technical education.

liveryman /ˈlɪvərɪmən/ *n.* (*pl.* **-men**) **1** *Brit.* a member of a livery company. **2** a keeper of or attendant in a livery stable.

lives *pl.* of LIFE.

livestock /ˈlaɪvstɒk/ *n.* (usu. treated as *pl.*) animals, esp. on a farm, regarded as an asset.

livid /ˈlɪvɪd/ *adj.* **1** *colloq.* furiously angry. **2 a** of a bluish leaden colour. **b** discoloured as by a bruise. □ **lividly** *adv.* **lividness** *n.* **lividity** /lɪˈvɪdɪtɪ/ *n.* [F *livide* or L *lividus* f. *livere* be bluish]

living /ˈlɪvɪŋ/ *n. & adj.* ● *n.* **1** a livelihood or means of maintenance (*made my living as a journalist; what does she do for a living?*). **2** *Brit. Eccl.* a position as a vicar or rector with an income or property. **3** (prec. by *the*; treated as *pl.*) those who are alive. ● *adj.* **1** contemporary; now existent (*the greatest living poet*). **2** (of a likeness or image of a person) exact. **3** (of a language) still in vernacular use. **4** (of water) perennially flowing. **5** (of rock etc.) = LIVE² 5. □ **living death** a state of hopeless misery. **living-room** a room for general day use. **living will** a written statement of a person's desire not to be kept alive by artificial means in the event of terminal illness or accident. **within living memory** within the memory of people still living.

Livingstone¹ /ˈlɪvɪŋstən/ the former name for MARAMBA.

Livingstone² /ˈlɪvɪŋstən/, David (1813–73), Scottish missionary and explorer. He first went to Bechuanaland as a missionary in 1841; on his extensive travels in the interior he discovered Lake Ngami (1849), the Zambezi river (1851), and the Victoria Falls (1855). In 1866 he led an expedition into central Africa in search of the source of the Nile; after many hardships he was eventually found in poor health by the explorer Sir Henry Morton Stanley on the eastern shore of Lake Tanganyika in 1871.

Livonia /lɪˈvəʊnɪə/ (German **Livland** /ˈliːflant/) a region on the east coast of the Baltic Sea, north of Lithuania, comprising most of present-day Latvia and Estonia. Formerly ruled by the Teutonic Knights, the region was the scene of the Livonian War (1558–82), in which Russia unsuccessfully fought an alliance of Poland, Sweden, and Lithuania for control of the territory. By the end of the 18th century, however, Livonia had been absorbed by Russia; it was divided between Estonia and Latvia in 1918. □ **Livonian** *adj. & n.*

Livorno /lɪˈvɔːnəʊ/ (also **Leghorn** /ˈleghɔːn/) a port in NW Italy, in Tuscany, on the Ligurian Sea; pop. (1990) 171,265. It is the site of the Italian Naval Academy.

Livy /ˈlɪvɪ/ (Latin name Titus Livius) (59 BC–AD 17), Roman historian. His history of Rome from its foundation to his own time contained

142 books, of which thirty-five survive (including the earliest history of the war with Hannibal). Livy is notable for his power of vivid historical reconstruction as he sought to give Rome a history that in conception and style should be worthy of her imperial rise and greatness.

lixiviate /lɪkˈsɪvɪˌeɪt/ *v.tr.* separate (a substance) into soluble and insoluble constituents by the percolation of liquid; leach. □ **lixiviation** /-ˌsɪvɪˈeɪʃ(ə)n/ *n.* [L *lixivius* made into lye f. *lix* lye]

lizard /ˈlɪzəd/ *n.* a reptile of the order Lacertilia (or Sauria), usu. having a long body and tail, four legs, and a scaly or granulated skin. Lizards differ from snakes in having movable eyelids, shedding their skin in pieces, and (except for one kind) lacking venom. Limbless forms adapted for burrowing have evolved in several families, especially the skinks and slow-worms. The largest living lizards are the monitors, from which the extinct marine mosasaurs evolved. [ME f. OF *lesard(e)* f. L *lacertus*]

Lizard, the a promontory in SW England, in Cornwall. Its southern tip, Lizard Point, is the southernmost point of the British mainland.

LJ *abbr.* (*pl.* **L JJ**) (in the UK) Lord Justice.

Ljubljana /luːˈbljɑːnə/ (called in German *Laibach*) the capital of Slovenia; pop. (1991) 267,000. The city was founded (as Emona) by the Romans in 34 BC. It was the capital of the Illyrian Provinces (later, Illyria) 1809–49.

LL *abbr.* Lord Lieutenant.

ll. *abbr.* lines.

'll /l, (ə)l/ *abbr.* (usu. after pronouns) shall, will (*I'll*; *that'll*).

llama /ˈlɑːmə/ *n.* **1** a domesticated South American ruminant, *Lama glama*, of the camel family, kept as a beast of burden and for its soft woolly fleece. **2** the wool from this animal, or cloth made from it. [Sp., prob. f. Quechua]

Llandudno /læɳˈdɪdnəʊ, hlæn-/ a resort town in northern Wales, on the Irish Sea; pop. (1981) 14,370.

llanero /lɑːˈneərəʊ, ljɑː-/ *n.* (*pl.* **-os**) an inhabitant of a llano. [Sp.]

llano /ˈlɑːnəʊ, ˈljɑː-/ *n.* (*pl.* **-os**) a treeless grassy plain or steppe, esp. in South America. [Sp. f. L *planum* plain]

LL B *abbr.* Bachelor of Laws. [L *legum baccalaureus*]

LL D *abbr.* Doctor of Laws. [L *legum doctor*]

Llewelyn /luːˈelɪn, hluː-/ (also **Llywelyn ap Gruffydd** /æp ˈgrɪfɪð/) (d.1282), prince of Gwynedd in North Wales. In 1258 he proclaimed himself prince of all Wales and four years later formed an alliance with Simon de Montfort, leader of the baronial opposition to Henry III. He later signed a treaty with Henry, which made him chief of the other Welsh princes but recognized Henry as his overlord. His refusal to pay homage to Edward I led the latter to invade and subjugate Wales (1277–84); Llewelyn died in battle after raising a rebellion against Edward's rule.

LL M *abbr.* Master of Laws. [L *legum magister*]

Llosa, Mario Vargas, see VARGAS LLOSA.

Lloyd /lɔɪd/, Marie (born Matilda Alice Victoria Wood) (1870–1922), English music-hall entertainer. She made her first stage appearance in 1885 and soon achieved fame for her risqué songs and extravagant costumes; she later took her act to the US, South Africa, and Australia.

Lloyd George /lɔɪd ˈdʒɔːdʒ/, David, 1st Earl Lloyd George of Dwyfor (1863–1945), British Liberal statesman, Prime Minister 1916–22. As Chancellor of the Exchequer (1908–15), he introduced old-age pensions (1908) and national insurance (1911). His 'People's Budget' (1909), intended to finance reform by raised death duties etc., was rejected by the Lords and led to a constitutional crisis which was eventually resolved by the Parliament Act of 1911. Supported by the Conservatives, he took over from Asquith as Prime Minister at the end of 1916 and led the coalition government for the remainder of the First World War. In the postwar period his administration was threatened by increasing economic problems and trouble in Ireland; he resigned in 1922 after the Conservatives withdrew their support.

Lloyd's /lɔɪdz/ an association of insurance underwriters in London incorporated by statute in 1871. Its members are private syndicates, elected after scrutiny of their finances and required to deposit a substantial sum as security against their underwriting activities. It is named after the coffee-house of Edward Lloyd (*fl.* 1688–1726), in which underwriters and merchants congregated and where a newspaper, *Lloyd's List*, was started in 1734, giving daily news of the movements of shipping. Originally Lloyd's dealt only in marine insurance, but it now undertakes most other kinds, and its business in marine and aircraft insurance is international. Lloyd's Register of Shipping is a separate

and independent society which surveys and classifies merchant ships over a certain tonnage, publishing details of these each year on the basis of reports by its surveyors. (See also LUTINE BELL.)

Lloyd Webber /lɔɪd ˈwebə(r)/, Sir Andrew (b.1948), English composer. He has written many successful musicals, several of them in collaboration with the lyricist Sir Tim Rice (b.1944); they include *Jesus Christ Superstar* (1970), *Evita* (1976), *Cats* (1981), *The Phantom of the Opera* (1986), and *Aspects of Love* (1989).

Llywelyn ap Gruffydd see LLEWELYN.

LM *abbr.* **1** long metre. **2** lunar module.

lm *abbr.* lumen(s).

LMS *abbr.* **1** local management of schools. **2** *hist.* London Midland and Scottish (Railway).

ln *abbr.* natural logarithm. [mod.L *logarithmus naturalis*]

LNER *abbr. hist.* London and North-Eastern Railway.

lo /ləʊ/ *int. archaic* calling attention to an amazing sight. □ **lo and behold** *joc.* a formula introducing a surprising or unexpected fact. [OE *lā* int. of surprise etc., & ME *lō* = *lōke* LOOK]

loa /ˈləʊə/ *n.* (*pl.* same or **loas**) a god in the voodoo cult of Haiti (see VOODOO). [Haitian *lwa* f. Yoruba *oluwa* lord, owner]

loach /ləʊtʃ/ *n.* a small slender freshwater fish of the family Cobitidae. [ME f. OF *loche*, of unkn. orig.]

load /ləʊd/ *n. & v.* ● *n.* **1 a** what is carried or is to be carried; a burden. **b** an amount usu. or actually carried (often in *comb.*: *a busload of tourists*; *a lorry-load of bricks*). **2** a unit of measure or weight of certain substances. **3** a burden or commitment of work, responsibility, care, grief, etc. **4** *colloq.* **a** (in *pl.*; often foll. by *of*) plenty; a lot. **b** (**a load of**) a quantity of (*a load of nonsense*). **5 a** the amount of electric power supplied by a generating system at any given time. **b** *Electronics* an impedance or circuit that receives or develops the output of a transistor or other device. **6** the weight or force borne by the supporting part of a structure. **7** a material object or force acting as a weight or clog. **8** the resistance of machinery to motive power. ● *v.* **1** *tr.* put a load on or aboard (a person, vehicle, ship, etc.). **b** place (a load or cargo) aboard a ship, on a vehicle, etc. **2** *intr.* (often foll. by *up*) (of a ship, vehicle, or person) take a load aboard, pick up a load. **3** *tr.* (often foll. by *with*) be a weight or burden upon; overburden, oppress; strain the bearing-capacity of. **4** *tr.* (also **load up**) (foll. by *with*) **a** a supply overwhelmingly (*loaded us with work*). **b** assail overwhelmingly (*loaded us with abuse*). **5** *tr.* charge (a firearm) with ammunition. **6** *tr.* insert (the required operating medium) in a device, e.g. film in a camera, magnetic tape in a tape recorder, a program into a computer, etc. **7** *tr.* add an extra charge to (an insurance premium) in the case of a poorer risk. **8** *tr.* **a** weight with lead. **b** give a bias to (dice, a roulette wheel, etc.) with weights. □ **get a load of** *sl.* listen attentively to; notice. **load-displacement** (or **-draught**) the displacement of a ship when laden. **load line** a Plimsoll line. [OE *lād* way, journey, conveyance, f. Gmc: rel. to LEAD[1], LODE]

loaded /ˈləʊdɪd/ *adj.* **1** bearing or carrying a load. **2** *sl.* **a** wealthy. **b** drunk. **c** *US* drugged. **3** (of dice etc.) weighted or given a bias. **4** (of a question or statement) charged with some hidden or improper implication.

loader /ˈləʊdə(r)/ *n.* **1** a loading-machine. **2** (in *comb.*) a gun, machine, lorry, etc., loaded in a specified way (*breech-loader*). **3** an attendant who loads guns at a shoot. □ **-loading** *adj.* (in *comb.*) (in sense 2).

loading /ˈləʊdɪŋ/ *n.* **1** *Electr.* the maximum current or power taken by an appliance. **2** an increase in an insurance premium due to a factor increasing the risk involved (see LOAD *v.* 7). **3** *Austral.* an increment added to a basic wage for special skills etc.

loadstar var. of LODESTAR.

loadstone var. of LODESTONE.

loaf[1] /ləʊf/ *n.* (*pl.* **loaves** /ləʊvz/) **1** a portion of baked bread, usu. of a standard size or shape. **2** a quantity of other food formed into a particular shape (*sugar loaf*; *meat loaf*). **3** *sl.* the head, esp. as a source of common sense (*use your loaf*). □ **loaf sugar** a sugar loaf as a whole or cut into lumps. [OE *hlāf* f. Gmc]

loaf[2] /ləʊf/ *v. & n.* ● *v.* **1** *intr.* (often foll. by *about*, *around*) spend time idly; hang about. **2** *tr.* (foll. by *away*) waste (time) idly (*loafed away the morning*). **3** *intr.* saunter. ● *n.* an act or spell of loafing. [prob. a back-form. f. LOAFER]

loafer /ˈləʊfə(r)/ *n.* **1** an idle person. **2** (**Loafer**) *propr.* a leather shoe shaped like a moccasin with a flat heel. [perh. f. G *Landläufer* vagabond]

loam /ləʊm/ *n.* **1** a fertile soil of clay and sand containing humus. **2** a paste of clay and water with sand, chopped straw, etc., used in making bricks, plastering, etc. □ **loamy** adj. **loaminess** n. [OE lām f. WG, rel. to LIME[1]]

loan[1] /ləʊn/ n. & v. ● n. **1** something lent, esp. a sum of money to be returned normally with interest. **2** the act of lending or state of being lent. **3** funds acquired by the state, esp. from individuals, and regarded as a debt. **4** a word, custom, etc., adopted by one people from another. ● v.tr. lend (esp. money). □ **loan shark** colloq. a person who lends money at exorbitant rates of interest. **loan-translation** an expression adopted by one language from another in a more or less literally translated form. **on loan** acquired or given as a loan. □ **loanable** adj. **loaner** n. **loanee** /ləʊˈniː/ n. [ME lan f. ON lán f. Gmc: cf. LEND]

loan[2] /ləʊn/ n. (also **loaning** /-nɪŋ/) Sc. **1** a lane. **2** an open place where cows are milked. [ME var. of LANE]

loanholder /ˈləʊnˌhəʊldə(r)/ n. **1** a person holding securities for a loan. **2** a mortgagee.

loanword /ˈləʊnwɜːd/ n. a word adopted, usu. with little modification, from a foreign language.

loath /ləʊθ/ predic.adj. (also **loth**) (usu. foll. by to + infin.) disinclined, reluctant, unwilling (was loath to admit it). □ **nothing loath** adj. quite willing. [OE lāth f. Gmc]

loathe /ləʊð/ v.tr. regard with disgust; abominate, detest. □ **loather** n. **loathing** n. [OE lāthian f. Gmc, rel. to LOATH]

loathsome /ˈləʊðsəm/ adj. arousing hatred or disgust; offensive, repulsive. □ **loathsomely** adv. **loathsomeness** n. [ME f. loath disgust f. LOATHE]

loaves pl. of LOAF[1].

lob /lɒb/ v. & n. ● v.tr. (**lobbed, lobbing**) **1** hit or throw (a ball or missile etc.) slowly or in a high arc. **2** send (an opponent) a lobbed ball. ● n. **1 a** a ball struck in a high arc. **b** a stroke producing this result. **2** Cricket a slow underarm ball. [earlier as noun, prob. f. LG or Du.]

Lobachevski /ˌlɒbəˈtʃefski/ Nikolai Ivanovich (1792–1856), Russian mathematician. At about the same time as Gauss in Germany and János Bolyai (1802–60) in Hungary, he independently discovered non-Euclidean geometry. His work was not widely recognized until the non-Euclidean nature of space–time was revealed by the general theory of relativity.

lobar /ˈləʊbə(r)/ adj. **1** of the lungs (lobar pneumonia). **2** of, relating to, or affecting a lobe.

lobate /ˈləʊbeɪt/ adj. Biol. having a lobe or lobes. □ **lobation** /ləʊˈbeɪʃ(ə)n/ n.

lobby /ˈlɒbɪ/ n. & v. ● n. (pl. **-ies**) **1** a porch, ante-room, entrance-hall, or corridor. **2 a** (in the House of Commons) a large hall used esp. for interviews between MPs and members of the public. **b** (also **division lobby**) either of two corridors to which MPs retire to vote. **3 a** a body of persons seeking to influence legislators on behalf of a particular interest (the anti-abortion lobby). **b** an organized attempt by members of the public to influence legislators (a lobby of MPs). **4** (prec. by the) (in the UK) a group of journalists who receive unattributable briefings from the government (lobby correspondent). ● v. (**-ies, -ied**) **1** tr. solicit the support of (an influential person). **2** tr. (of members of the public) seek to influence (the members of a legislature). **3** intr. frequent a parliamentary lobby. **4** tr. (foll. by through) get (a bill etc.) through a legislature, by interviews etc. in the lobby. □ **lobbyer** n. **lobbyist** n. [med.L lobia, lobium LODGE]

lobe /ləʊb/ n. **1** a roundish and flattish projecting or pendulous part, often each of two or more such parts divided by a fissure (lobes of the brain). **2** = ear lobe (see EAR[1]). □ **lobed** adj. **lobeless** adj. [LL f. Gk lobos lobe, pod]

lobectomy /ləˈbektəmɪ/ n. (pl. **-ies**) the surgical removal of a lobe of an organ such as the thyroid gland, lung, etc.

lobelia /ləˈbiːlɪə/ n. a plant of the genus Lobelia, with a deeply cleft corolla; esp. a cultivated species with blue, scarlet, white, or purple flowers. [Matthias de Lobel, Flemish botanist in England (1538–1616)]

Lobito /lʊˈbiːtəʊ/ a seaport and natural harbour on the Atlantic coast of Angola; pop. (est. 1983) 150,000. It is linked by rail to Zaire, Zambia, and the Pacific coast at Beira in Mozambique.

lobotomy /ləˈbɒtəmɪ/ n. (pl. **-ies**) Med. surgical incision into a lobe, esp. the prefrontal lobe of the brain (cf. LEUCOTOMY). □ **lobotomize** v.tr. (also **-ise**). [LOBE + -TOMY]

lobscouse /ˈlɒbskaʊs/ n. a sailor's dish of meat stewed with vegetables and ship's biscuit. [18th c.: orig. unkn.: cf. Du. lapskous, Danish, Norw., G Lapskaus]

lobster /ˈlɒbstə(r)/ n. & v. ● n. **1** a large marine crustacean of the family Nephropidae, with stalked eyes and the first pair of its ten limbs modified as pincer-like claws. **2** its flesh as food. ● v.intr. catch lobsters. □ **lobster-pot** a basket in which lobsters are trapped. **lobster thermidor** /ˈθɜːmɪˌdɔː(r)/ a mixture of lobster meat, mushrooms, cream, egg yolks, and sherry, cooked in a lobster shell. [OE lopustre, corrupt. of L locusta crustacean, locust: thermidor f. the name of the 11th month of the Fr. revolutionary calendar]

lobule /ˈlɒbjuːl/ n. esp. Anat. a small lobe. □ **lobular** adj. **lobulate** adj. [LOBE]

lobworm /ˈlɒbwɜːm/ n. **1** a large earthworm used as fishing-bait. **2** = LUGWORM. [LOB in obs. sense 'pendulous object']

local /ˈləʊk(ə)l/ adj. & n. ● adj. **1** belonging to or existing in a particular place or places. **2** peculiar to or only encountered in a particular place or places. **3** of or belonging to the neighbourhood (the local doctor). **4** of or affecting a part and not the whole, esp. of the body (local pain; a local anaesthetic). **5** (of a telephone call) to a nearby place and charged at a lower rate. **6** in regard to place. ● n. a local person or thing, esp.: **1** an inhabitant of a particular place regarded with reference to that place. **2** a local train, bus, etc. **3** (often prec. by the) Brit. colloq. a local public house. **4** a local anaesthetic. **5** N. Amer. a local branch of a trade union. □ **local area network** (abbr. **LAN**) a computer network in which computers in close proximity are able to communicate with each other and share resources. **local authority** Brit. an administrative body in local government. **local bus 1** a bus service operating over short distances. **2** Computing a bus that connects a microprocessor directly to an adjacent peripheral device such as a video system, enabling the transmission of data at a faster rate. **local Derby** see DERBY[3] 2. **local government** a system of administration of a county, district, parish, etc., by the elected representatives of those who live there. **local option** (or **veto**) a choice at local or district level as to whether to accept national legislation. **local preacher** a Methodist lay person authorized to conduct services in a particular circuit. **local time 1** time measured from the sun's transit over the meridian of a place. **2** time as reckoned in a particular region or time zone, esp. with reference to an event reported from it (we arrived at one o'clock local time). **local train** a train stopping at all the stations on its route. □ **locally** adv. **localness** n. [ME f. OF f. LL localis f. L locus place]

locale /ləʊˈkɑːl/ n. a scene or locality, esp. with reference to an event or occurrence taking place there. [F local (n.) (as LOCAL), respelt to indicate stress: cf. MORALE]

Local Group n. Astron. the cluster of galaxies of which our Galaxy is a member.

localism /ˈləʊkəˌlɪz(ə)m/ n. **1** preference for what is local. **2** a local idiom, custom, etc. **3 a** attachment to a place. **b** a limitation of ideas etc. resulting from this.

locality /ləʊˈkælɪtɪ/ n. (pl. **-ies**) **1** a district or neighbourhood. **2** the site or scene of something, esp. in relation to its surroundings. **3** the position of a thing; the place where it is. [F localité or LL localitas (as LOCAL)]

localize /ˈləʊkəˌlaɪz/ v.tr. (also **-ise**) **1** restrict or assign to a particular place. **2** invest with the characteristics of a particular place. **3** attach to districts; decentralize. □ **localizable** adj. **localization** /ˌləʊkəlaɪˈzeɪʃ(ə)n/ n.

Locarno /ləˈkɑːnəʊ/ a resort in southern Switzerland, at the northern end of Lake Maggiore; pop. (1990) 14,150.

Locarno Pact a series of agreements made in Locarno in 1925 between the UK, Germany, France, Belgium, Poland, and Czechoslovakia in an attempt to ensure the future peace of Europe. The Pact guaranteed the common borders of France, Germany, and Belgium and the demilitarization of the Rhineland, as specified by the Treaty of Versailles, but did not fully resolve the question of Germany's eastern boundaries. Its provisions were systematically breached by the Nazis after they came to power in Germany.

locate /ləʊˈkeɪt/ v. **1** tr. discover the exact place or position of (locate the enemy's camp). **2** tr. establish or install in a place or in its proper place. **3** tr. state the locality of. **4** tr. (in passive) be situated. **5** intr. (often foll. by in) N. Amer. take up residence or business (in a place). □ **locatable** adj. **locator** n. [L locare locat- f. locus place]

location /ləʊˈkeɪʃ(ə)n/ n. **1** a particular place; the place or position in which a person or thing is. **2** the act of locating or process of being located. **3** an actual place or natural setting featured in a film or broadcast, as distinct from a simulation in a studio (filmed entirely on location). **4** S. Afr. esp. hist. an area where blacks were obliged to live, usu. on the outskirts of a town or city. □ **locational** adj. [L locatio (as LOCATE)]

locative /'lɒkətɪv/ n. & adj. Gram. ● n. the case of nouns, pronouns, and adjectives, expressing location. ● adj. of or in the locative. [formed as LOCATE + -IVE, after vocative]

loc. cit. abbr. in the passage already cited. [L loco citato]

loch /lɒk, lɒx/ n. Sc. **1** a lake. **2** an arm of the sea, esp. when narrow or partially landlocked. [ME f. Gael.]

lochia /'lɒkɪə, 'ləʊk-/ n. Med. a (normal) discharge of material from the uterus after childbirth. □ **lochial** adj. [mod.L f. Gk lokhia neut. pl. of lokhios of childbirth]

Loch Ness /nes/ a deep lake in NW Scotland, in the Great Glen. Forming part of the Caledonian Canal, it is 38 km (24 miles) long, with a maximum depth of 230 m (755 ft).

Loch Ness monster a large creature alleged to live in the deep waters of Loch Ness. Reports of its existence date from the time of St Columba (6th century); the number of sightings increased after the construction of a major road alongside the loch in 1933, but, despite scientific expeditions, there is still no proof of its existence.

loci pl. of LOCUS.

loci classici pl. of LOCUS CLASSICUS.

lock[1] /lɒk/ n. & v. ● n. **1** a mechanism for fastening a door, lid, etc., with a bolt that requires a key of a particular shape, or a combination of movements (see combination lock), to work it. **2** a confined section of a canal or river where the level can be changed for raising and lowering boats between adjacent sections by the use of gates and sluices. **3 a** the turning of the front wheels of a vehicle to change its direction of motion. **b** (in full **full lock**) the maximum extent of this. **4** an interlocked or jammed state. **5** Wrestling a hold that keeps an opponent's limb fixed. **6** (in full **lock forward**) Rugby a player in the second row of a scrum. **7** an appliance to keep a wheel from revolving or slewing. **8** a mechanism for exploding the charge of a gun. **9** = AIRLOCK 2. ● v. **1 a** tr. fasten with a lock. **b** tr. (foll. by up) shut and secure (esp. a building) by locking. **c** intr. (of a door, window, box, etc.) have the means of being locked. **2 a** tr. (foll. by up, in, into) enclose (a person or thing) by locking or as if by locking. **b** (foll. by up) colloq. imprison (a person). **3** tr. (often foll. by up, away) store or allocate inaccessibly (capital locked up in land). **4** tr. hold fast (in sleep or enchantment etc.). **5** tr. (usu. in passive) (of land, hills, etc.) enclose. **6** tr. & intr. make or become rigidly fixed or immovable. **7** intr. & tr. (often foll. by into) become or cause to become jammed or caught (locked into a cycle of borrowing). **8** tr. (usu. in passive; foll. by in) entangle in an embrace or struggle (locked in a legal battle). **9** tr. provide (a canal etc.) with locks. **10** tr. (foll. by up, down) convey (a boat) through a lock. **11** intr. go through a lock on a canal etc. □ **lock-keeper** a keeper of a lock on a river or canal. **lock-knit** knitted with an interlocking stitch. **lock-nut** Mech. a nut screwed down on another to keep it tight. **lock on** to locate or cause to locate by radar etc. and then track. **lock out 1** keep (a person) out by locking the door. **2** (of an employer) submit (employees) to a lockout. **lock step** marching with each person as close as possible to the one in front. **lock stitch** a secure sewing machine stitch made by locking together two threads or stitches. **lock, stock, and barrel** n. the whole of a thing. ● adv. completely. **under lock and key** securely locked up. □ **lockable** adj. **lockless** adj. [OE loc f. Gmc]

lock[2] /lɒk/ n. **1 a** a portion of hair that coils or hangs together. **b** (in pl.) the hair of the head. **2** a tuft of wool or cotton. □ **-locked** adj. (in comb.). [OE locc f. Gmc]

lockage /'lɒkɪdʒ/ n. **1** the amount of rise and fall effected by canal locks. **2** a toll for the use of a lock. **3** the construction or use of locks. **4** locks collectively; the aggregate of locks constructed.

Locke[1] /lɒk/, John (1632–1704), English philosopher, a founder of empiricism and political liberalism. Both his major works were published in 1690. In Two Treatises of Government he justified the Revolution of 1688 by arguing that, contrary to the theory of the divine right of kings, the authority of rulers has a human origin and is limited. In An Essay concerning Human Understanding he denied that any ideas are innate, and argued instead for a central empiricist tenet that all knowledge is derived from sense-experience. He concluded that it is not possible to know everything of the world and that our limited knowledge must be reinforced by faith.

Locke[2] /lɒk/, Joseph (1805–60), English civil engineer. A major railway pioneer, he enjoyed a lifelong association with Thomas Brassey, during which time he built important lines in England, Scotland, and France.

locker /'lɒkə(r)/ n. **1** a small lockable cupboard or compartment, esp. each of several for public use. **2** Naut. a chest or compartment for clothes, stores, ammunition, etc. **3** a person or thing that locks.

□ **locker-room** a room containing lockers (in sense 1), esp. in a pavilion or sports centre.

Lockerbie /'lɒkəbɪ/ a town in SW Scotland, in Dumfries and Galloway; pop. (1981) 3,560. In 1988 the wreckage of an American airliner, destroyed by a terrorist bomb, crashed on the town. The disaster claimed the lives of all 259 people on the aircraft and eleven people on the ground.

locket /'lɒkɪt/ n. **1** a small ornamental case holding a portrait, lock of hair, etc., and usu. hung from the neck. **2** a metal plate or band on a scabbard. [OF locquet dimin. of loc latch, lock, f. WG (as LOCK[1])]

lockfast /'lɒkfɑːst/ adj. Sc. secured with a lock.

lockjaw /'lɒkdʒɔː/ n. = TRISMUS. ¶ Not in technical use.

lockout /'lɒkaʊt/ n. the exclusion of employees by their employer from their place of work until certain terms are agreed to.

locksman /'lɒksmən/ n. (pl. **-men**) a lock-keeper.

locksmith /'lɒksmɪθ/ n. a maker and mender of locks.

lock-up /'lɒkʌp/ n. & adj. ● n. **1** a house or room for the temporary detention of prisoners. **2** Brit. non-residential premises etc. that can be locked up, esp. a small shop or storehouse. **3 a** the locking up of premises for the night. **b** the time of doing this. **4 a** the unrealizable state of invested capital. **b** an amount of capital locked up. ● attrib.adj. Brit. that can be locked up (lock-up shop).

Lockyer /'lɒkjə(r)/, Sir (Joseph) Norman (1836–1920), English astronomer. Following the first observation of solar prominences during an eclipse, Lockyer demonstrated that they could be seen at other times with suitable equipment. His spectroscopic analysis of the sun led to his discovery of a new element, which he named helium. Lockyer also studied possible astronomical alignments in ancient monuments such as Stonehenge. He is perhaps best known today for founding both the Science Museum in London and the scientific journal Nature, which he edited for fifty years.

loco[1] /'ləʊkəʊ/ n. (pl. **-os**) colloq. a locomotive. [abbr.]

loco[2] /'ləʊkəʊ/ adj. & n. ● adj. sl. crazy. ● n. (pl. **-oes** or **-os**) (in full **loco-weed**) a poisonous leguminous plant of the US, esp. of the genus Astragalus, affecting the brains of cattle eating it. [Sp., = insane]

locomotion /ˌləʊkə'məʊʃ(ə)n/ n. **1** motion or the power of motion from one place to another. **2** travel; a means of travelling, esp. an artificial one. [L loco ablat. of locus place + motio MOTION]

locomotive /ˌləʊkə'məʊtɪv/ n. & adj. ● n. (in full **locomotive engine**) a vehicle powered by steam, diesel fuel, or electricity, used for pulling trains; a railway engine. ● adj. **1** of or relating to or effecting locomotion (locomotive power). **2** having the power of or given to locomotion; not stationary.

locomotor /ˌləʊkə'məʊtə(r)/ adj. of or relating to locomotion. [LOCOMOTION + MOTOR]

loculus /'lɒkjʊləs/ n. (pl. **loculi** /-ˌlaɪ/) Zool., Anat., & Bot. each of a number of small separate cavities. □ **locular** adj. [L, dimin. of locus: see LOCUS]

locum /'ləʊkəm/ n. colloq. = LOCUM TENENS. [abbr.]

locum tenens /ˌləʊkəm 'tiːnenz, 'ten-/ n. (pl. **locum tenentes** /tɪ'nentiːz/) a deputy acting esp. for a cleric or doctor. □ **locum tenency** /'tenənsɪ/ n. [med.L, one holding a place: see LOCUS, TENANT]

locus /'ləʊkəs, 'lɒk-/ n. (pl. **loci** /'ləʊsaɪ, 'lɒkiː/) **1** a position or point, esp. in a text, treatise, etc. **2** Math. a curve etc. formed by all the points satisfying a particular equation of the relation between coordinates, or by a point, line, or surface moving according to mathematically defined conditions. **3** Biol. the position of a gene, mutation, etc. on a chromosome. [L, = place]

locus classicus /ˌləʊkəs 'klæsɪkəs, ˌlɒk-/ n. (pl. **loci classici** /ˌləʊsaɪ 'klæsɪˌsaɪ, ˌlɒkiː 'klæsɪˌkiː/) the best-known or most authoritative passage on a subject. [L]

locus standi /ˌləʊkəs 'stændaɪ, ˌlɒk-/ n. a recognized or identifiable (esp. legal) status.

locust /'ləʊkəst/ n. **1** a large African and Asian grasshopper of the family Acrididae, migrating in swarms and destroying vegetation. **2** US a cicada. **3** (in full **locust bean**) a carob. **4** (in full **locust tree**) **a** a carob tree. **b** = ACACIA 2. **c** = KOWHAI. □ **locust-bird** (or **-eater**) a bird that feeds on locusts. [ME f. OF locuste f. L locusta lobster, locust]

locution /lə'kjuːʃ(ə)n/ n. **1** a word or phrase, esp. considered in regard to style or idiom. **2** style of speech. [ME f. OF locution or L locutio f. loqui locut- speak]

lode /ləʊd/ n. a vein of metal ore. [var. of LOAD]

loden /ˈləʊd(ə)n/ n. **1** a thick waterproof woollen cloth. **2** the dark green colour in which this is often made. [G]

lodestar /ˈləʊdstɑː(r)/ n. (also **loadstar**) **1** a star that a ship etc. is steered by, esp. the pole star. **2 a** a guiding principle. **b** an object of pursuit. [LODE in obs. sense 'way, journey' + STAR]

lodestone /ˈləʊdstəʊn/ n. (also **loadstone**) **1** magnetic oxide of iron, magnetite. **2 a** a piece of this used as a magnet. **b** a thing that attracts.

Lodge /lɒdʒ/, Sir Oliver (Joseph) (1851–1940), English physicist. He made important contributions to the study of electromagnetic radiation, and was a pioneer of radio-telegraphy. He also devised an ingenious experiment which demonstrated that the hypothetical ether did not exist, and he carried out intensive studies of psychic phenomena.

lodge /lɒdʒ/ n. & v. ● n. **1** a small house at the gates of a park or in the grounds of a large house, occupied by a gatekeeper, gardener, etc. **2** a large house or hotel, esp. in a resort. **3** a house occupied in the hunting or shooting season. **4 a** a porter's room or quarters at the gate of a college or other large building. **b** the residence of a head of a college, esp. at Cambridge. **5** the members or the meeting-place of a branch of a society such as the Freemasons. **6** a local branch of a trade union. **7** a beaver's or otter's lair. **8** a North American Indian tent or wigwam. ● v. **1** tr. deposit in court or with an official a formal statement of (complaint or information). **2** tr. deposit (money etc.) for security. **3** tr. bring forward (an objection etc.). **4** tr. (foll. by in, with) place (power etc.) in a person or group. **5** tr. & intr. make or become fixed or caught without further movement (the bullet lodged in his brain; the tide lodges mud in the cavities). **6** tr. provide (a person) with sleeping quarters; receive as a guest or inmate; establish as a resident in a house or room or rooms. **7** intr. reside or live, esp. as a guest paying for accommodation. **8** tr. serve as a habitation for; contain. **9** tr. (in passive; foll. by in) be contained in. **10 a** tr. (of wind or rain) flatten (crops). **b** intr. (of crops) be flattened in this way. [ME loge f. OF loge arbour, hut, f. med.L laubia, lobia (see LOBBY) f. Gmc]

lodgement /ˈlɒdʒmənt/ n. **1** the act of lodging or process of being lodged. **2** the depositing or a deposit of money. **3** an accumulation of matter intercepted in fall or transit. [F logement (as LODGE)]

lodger /ˈlɒdʒə(r)/ n. a person receiving accommodation in another's house for payment.

lodging /ˈlɒdʒɪŋ/ n. **1** temporary accommodation (a lodging for the night). **2** (in pl.) a room or rooms (other than in a hotel) rented for lodging in. **3** a dwelling-place. **4** (in pl.) the residence of a head of a college at Oxford. □ **lodging-house** a house in which lodgings are let.

lodicule /ˈlɒdɪˌkjuːl/ n. Bot. a small green or white scale below the ovary of a grass flower. [L lodicula dimin. of lodix coverlet]

Łódź /wʊtʃ/ an industrial city in central Poland, south-west of Warsaw, the second largest city in the country; pop. (1990) 848,260.

loess /ˈləʊɪs, lɜːs/ n. a deposit of fine light-coloured wind-blown dust found esp. in the basins of large rivers and very fertile when irrigated. □ **loessial** /ləʊˈesɪəl, ˈlɜːsɪəl/ adj. [G Löss f. Swiss G lösch loose f. lösen loosen]

Loewi /ˈləʊiː/, Otto (1873–1961), American pharmacologist and physiologist, born in Germany. He is chiefly remembered today for his contributions in the field of chemical transmission of nerve impulses. By means of experiments using a pair of isolated frog hearts, he was the first to show that a chemical neurotransmitter is produced at the junction between a parasympathetic nerve and a muscle, and later identified it as the substance acetylcholine. Loewi shared a Nobel Prize with Sir Henry Dale in 1936.

Lofoten Islands /ləˈfəʊt(ə)n/ an island group off the NW coast of Norway. They are situated within the Arctic Circle, south-west of the Vesterålen group.

loft /lɒft/ n. & v. ● n. **1** the space under the roof of a house, above the ceiling of the top floor; an attic. **2** a room over a stable, esp. for hay and straw. **3** a gallery in a church or hall (organ-loft). **4** US an upstairs room. **5** a pigeon-house. **6** Golf **a** a backward slope in a club-head. **b** a lofting stroke. ● v.tr. **1 a** send (a ball etc.) high up. **b** clear (an obstacle) in this way. **2** (esp. as **lofted** adj.) give a loft to (a golf club). [OE f. ON lopt air, sky, upper room, f. Gmc (as LIFT)]

lofter /ˈlɒftə(r)/ n. a golf club for lofting the ball.

lofty /ˈlɒftɪ/ adj. (**loftier, loftiest**) **1** literary (of things) of imposing height, towering, soaring (lofty heights). **2** consciously haughty, aloof, or dignified (lofty contempt). **3** exalted or noble; sublime (lofty ideals). □ **loftily** adv. **loftiness** n. [ME f. LOFT as in aloft]

log¹ /lɒg/ n. & v. ● n. **1** an unhewn log of a felled tree, or a similar rough mass of wood, esp. cut for firewood. **2 a** a float attached to a line wound on a reel for gauging the speed of a ship. **b** any other apparatus for the same purpose. **3** a record of events occurring during and affecting the voyage of a ship or aircraft (including the rate of a ship's progress shown by a log: see sense 2). **4** any systematic record of things done, experienced, etc. **5** = LOGBOOK. ● v.tr. (**logged, logging**) **1 a** enter (the distance made or other details) in a ship's logbook. **b** enter details about (a person or event) in a logbook. **c** (of a ship) achieve (a certain distance). **2 a** enter (information) in a regular record. **b** attain (a cumulative total of time etc. recorded in this way) (logged 50 hours on the computer). **3** cut into logs. □ **like a log 1** in a helpless or stunned state (fell like a log). **2** without stirring (slept like a log). **log cabin** a hut built of logs. **log in** = log on below. **log-jam 1** a crowded mass of logs in a river. **2** a deadlock. **log-line** a line to which a ship's log (see sense 2 a. of n.) is attached. **log on** (or **off**) go through the procedures to begin (or conclude) use of a computer system. [ME: orig. unkn.]

log² /lɒg/ n. a logarithm (esp. prefixed to a number or algebraic symbol whose logarithm is to be indicated). □ **log table** (usu. in pl.) a table of logarithms. [abbr.]

-log US var. of -LOGUE.

logₑ /ˈlɒgiː/ abbr. natural logarithm.

logan /ˈləʊgən/ n. (in full **logan-stone**) = rocking-stone (see ROCK²). [= logging f. dial. log to rock + STONE]

Logan, Mount /ˈləʊgən/ a mountain in SW Yukon Territory, Canada, near the border with Alaska. Rising to 6,054 m (19,850 ft), it is the highest peak in Canada and the second-highest peak in North America.

loganberry /ˈləʊgənbərɪ/ n. (pl. **-ies**) **1** a dull red soft fruit, apparently a hybrid of a raspberry and a dewberry. **2** the plant bearing this, Rubus loganbaccus. [James Harvey Logan, Amer. horticulturalist (1841–1928) + BERRY]

logarithm /ˈlɒgəˌrɪðəm/ n. (abbr. **log**) Math. the power to which a fixed number or base must be raised to produce a given number (the logarithm of 1,000 to base 10 is 3). (See note below.) □ **common logarithm** a logarithm to the base 10. **natural** (or **Napierian**) **logarithm** (abbr. **ln** or **log**ₑ) a logarithm to the base e (2.71828.....). □ **logarithmic** /ˌlɒgəˈrɪðmɪk/ adj. **logarithmically** adv. [mod.L logarithmus f. Gk logos reckoning, ratio + arithmos number]

▪ Logarithms were independently introduced by John Napier in Scotland and Joost Bürgi (1552–1632) in Germany in the early 17th century. Tables of logarithms were extensively published as aids to calculation, since the processes of multiplication and division could be replaced by the addition and subtraction of logarithms. Log tables lost their importance as aids to calculation with the invention of the electronic calculator.

logbook /ˈlɒgbʊk/ n. **1** a book containing a detailed record or log. **2** Brit. a document recording the registration details of a motor vehicle. ¶ Now officially called vehicle registration document.

loge /ləʊʒ/ n. a private box or enclosure in a theatre. [F]

-loger /lədʒə(r)/ comb. form forming nouns, = -LOGIST. [after astrologer]

logger /ˈlɒgə(r)/ n. N. Amer. a lumberjack.

loggerhead /ˈlɒgəˌhed/ n. **1** an iron instrument with a ball at the end heated for melting pitch etc. **2** (in full **loggerhead turtle**) a large-headed turtle, Caretta caretta, of warm seas worldwide. **3** archaic a blockhead or fool. □ **at loggerheads** (often foll. by with) disagreeing or disputing. **loggerhead shrike** a grey, black, and white shrike, Lanus ludovicianus, of the southern US and Mexico. [prob. f. dial. logger block of wood for hobbling a horse + HEAD]

loggia /ˈləʊdʒə, ˈlɒdʒə/ n. **1** an open-sided gallery or arcade. **2** an open-sided extension of a house. [It., = LODGE]

logging /ˈlɒgɪŋ/ n. the work of cutting and preparing forest timber.

logia pl. of LOGION.

logic /ˈlɒdʒɪk/ n. **1 a** the science of reasoning, proof, thinking, or inference. (See note below.) **b** a particular scheme of or treatise on this. **2 a** a chain of reasoning; reasoned argument (I don't follow your logic; is not governed by logic). **b** the correct or incorrect use of reasoning (your logic is flawed). **c** ability in reasoning (argues with great learning and logic). **3 a** the inexorable force or compulsion of a thing (the logic of events). **b** the necessary consequence of (an argument, decision, etc.). **4 a** a system or set of principles underlying the arrangements of elements in a computer or electronic device so as to perform a specified task. **b** logical operations collectively. □ **logic bomb** Computing instructions secretly incorporated into a program in such a way that if a particular logical condition is satisfied they will be carried out, usu. with deleterious results. □ **logician** /ləˈdʒɪʃ(ə)n/ *n.* [ME f. OF *logique* f. LL *logica* f. Gk *logikē (tekhnē)* (art) of reason (*logos* word, reason)]

■ Logic involves the systematic study of the patterns of argument, and in particular of those patterns of argument that are valid, i.e. such that if the premisses are true then of necessity the conclusion is true. The chief instrument of advance has been the development of symbols which make it possible to abstract from particular premisses and conclusions and consider only the patterns. The process was begun by Aristotle, who gave rules for which syllogisms are valid. Since the 19th century the formulation of such rules has drawn on the concepts and techniques of mathematics, using symbols to replace ordinary language; great advances were made by Gottlob Frege, who devised a simple method of handling 'all' and 'some', and by Bertrand Russell and A. N. Whitehead (see also MATHEMATICAL LOGIC).

-logic /ˈlɒdʒɪk/ *comb. form* (also **-logical**) forming adjectives corresponding esp. to nouns in *-logy* (*pathological*; *theological*). [from or after Gk *-logikos*: see -IC, -ICAL]

logical /ˈlɒdʒɪk(ə)l/ *adj.* **1** of logic or formal argument. **2** not contravening the laws of thought, correctly reasoned. **3** deducible or defensible on the ground of consistency; reasonably to be believed or done. **4** capable of correct reasoning. □ **logical necessity** the compulsion to believe that of which the opposite is inconceivable. □ **logically** *adv.* **logicality** /ˌlɒdʒɪˈkælɪtɪ/ *n.* [med.L *logicalis* f. LL *logica* (as LOGIC)]

logical atomism *n. Philos.* the theory, advanced by Bertrand Russell and developed by Wittgenstein, that all statements or propositions can be analysed into simple independent or atomic statements which correspond directly to facts about our experience of the world.

logical positivism *n.* the theories of the Vienna Circle, influenced by Wittgenstein and expounded by A. J. Ayer. The Circle aimed at evolving formal methods, similar to those of the mathematical sciences, for the verification of empirical questions and therefore eliminating metaphysical and other more speculative questions. Logical positivists held that a statement has meaning only if its truth or falsity can be tested empirically; logical and mathematical statements are tautologous because they are valid only within their own system; moral and value judgements are subjective and therefore 'meaningless'; metaphysical and religious speculation is logically ill-founded. □ **logical positivist** *n.*

logion /ˈlɒɡɪˌɒn/ *n.* (*pl.* **logia** /-ɡɪə/) a saying attributed to Christ, esp. one not recorded in the canonical Gospels. [Gk, = oracle f. *logos* word]

-logist /lədʒɪst/ *comb. form* forming nouns denoting a person skilled or involved in a branch of study etc. with a name in *-logy* (*archaeologist*; *etymologist*).

logistics /ləˈdʒɪstɪks/ *n.pl.* **1** the organization of moving, lodging, and supplying troops and equipment. **2** the detailed organization and implementation of a plan or operation. □ **logistic** *adj.* **logistical** *adj.* **logistically** *adv.* [F *logistique* f. *loger* lodge]

logo /ˈləʊɡəʊ, ˈlɒɡ-/ *n.* (*pl.* **-os**) an emblem or device used as the badge of an organization in display material. [abbr. of LOGOTYPE]

logogram /ˈlɒɡəˌɡræm/ *n.* a sign or character representing a word, as in shorthand or some ancient writing systems. [Gk *logos* word + -GRAM]

logographic /ˌlɒɡəˈɡræfɪk/ *adj.* denoting a writing system that consists of characters or signs, each of which represents a complete word. □ **logographically** *adv.*

logomachy /ləˈɡɒməkɪ/ *n.* (*pl.* **-ies**) *literary* a dispute about words; controversy turning on merely verbal points. [Gk *logomakhia* f. *logos* word + *makhia* fighting]

logorrhoea /ˌlɒɡəˈrɪə/ *n.* (*US* **logorrhea**) an excessive flow of words esp. in mental illness. □ **logorrhoeic** *adj.* [Gk *logos* word + *rhoia* flow]

Logos /ˈlɒɡɒs/ (in Christian theology) the Word of God, or second person of the Trinity, incarnate in Jesus Christ according to the fourth Gospel. [Gk, = word, reason (*legō* speak)]

logotype /ˈlɒɡəˌtaɪp/ *n.* **1** *Printing* a single piece of type that prints a word or group of separate letters or a logo. **2** = LOGO. [Gk *logos* word + TYPE]

logrolling /ˈlɒɡˌrəʊlɪŋ/ *n.* *US* **1** *colloq.* the practice of exchanging favours, esp. (in politics) of exchanging votes to mutual benefit. **2** a sport in which two contestants stand on a floating log and try to knock each other off. □ **logroller** *n.* [polit. sense f. phr. *you roll my log and I'll roll yours*]

Logroño /lɒˈɡrɒnjəʊ/ a market town in northern Spain, on the River Ebro, capital of La Rioja region; pop. (1991) 126,760.

-logue /lɒɡ/ *comb. form* (*US* **-log**) **1** forming nouns denoting talk (*dialogue*) or compilation (*catalogue*). **2** = -LOGIST (*ideologue*). [from or after F *-logue* f. Gk *-logos*, *-logon*]

logwood /ˈlɒɡwʊd/ *n.* **1** a leguminous West Indian tree, *Haematoxylon campechianum*. **2** the wood of this, producing a substance used in dyeing.

-logy /lədʒɪ/ *comb. form* forming nouns denoting: **1** (usu. as **-ology**) a subject of study or interest (*archaeology*; *zoology*). **2** a characteristic of speech or language (*tautology*). **3** discourse (*trilogy*). [F *-logie* or med.L *-logia* f. Gk (as LOGOS)]

Lohengrin /ˈləʊɪnˌɡrɪn/ (in medieval French and German romances) the son of Perceval (Parsifal). He was summoned from the temple of the Holy Grail and taken in a boat drawn by swans to Antwerp, where he consented to marry Elsa of Brabant on condition that she did not ask who he was. Elsa broke this condition and he was carried away again in the boat. Wagner's opera *Lohengrin* is based on this story.

loin /lɔɪn/ *n.* **1** (in *pl.*) **a** the part of the body on both sides of the spine between the false ribs and the hip-bones. **b** *literary* or *archaic* this as the source of reproductive power. **2** a joint of meat that includes the loin vertebrae. □ **gird one's loins** see GIRD[1]. [ME f. OF *loigne* ult. f. L *lumbus*]

loincloth /ˈlɔɪnklɒθ/ *n.* a cloth worn round the hips, esp. as a sole garment.

Loire /lwɑː(r)/ a river of west central France, which rises in the Massif Central and flows 1,015 km (630 miles) north and west to the Atlantic at St-Nazaire. Principal cities on its route are Orleans, Tours, and Nantes. The longest river in France, it is noted for the châteaux and vineyards that lie along its course.

loiter /ˈlɔɪtə(r)/ *v.* **1** *intr.* hang about; linger idly. **2** *intr.* travel indolently and with long pauses. **3** *tr.* (foll. by *away*) pass (time etc.) in loitering. □ **loiter with intent** *Brit.* hang about in order to commit a felony. □ **loiterer** *n.* [ME f. MDu. *loteren* wag about]

Loki /ˈləʊkɪ/ *Scand. Mythol.* a mischievous and sometimes evil god who contrived the death of Balder and was punished by being bound to a rock.

loll /lɒl/ *v.intr.* **1** (often foll. by *about*, *around*) stand, sit, or recline in a lazy attitude. **2** hang loosely (*a dog with its tongue lolling out*). □ **loller** *n.* [ME: prob. imit.]

Lolland /ˈlɒlɑːn/ a Danish island in the Baltic Sea, to the south of Zealand and west of Falster. Its name means 'low land'.

Lollard /ˈlɒləd/ *n.* any of a group of radical Christians in the Middle Ages who followed or held opinions similar to those of the 14th-century religious reformer John Wyclif. The Lollards believed that the Church should aid people to live a life of evangelical poverty and imitate Christ. The name itself was a derogatory term, derived from a Dutch word meaning 'mumbler'. Official attitudes to the Lollards varied considerably, but they were generally held to be heretics and often severely persecuted. Their ideas influenced the thought of John Huss, who in turn influenced Martin Luther. □ **Lollardism** *n.* **Lollardy** *n.* [MDu. *lollaerd* f. *lollen* mumble]

lollipop /ˈlɒlɪˌpɒp/ *n.* **1** *Brit.* a large usu. flat rounded boiled sweet on a small stick. **2** = *ice lolly.* □ **lollipop man** (or **lady** or **woman**) *Brit. colloq.* an official using a circular sign on a stick to stop traffic for children to cross the road, esp. near a school. [perh. f. dial. *lolly* tongue + POP[1]]

lollop /ˈlɒləp/ *v.intr.* (**lolloped**, **lolloping**) *colloq.* **1** flop about. **2** *Brit.* move or proceed in a lounging or ungainly way. [prob. f. LOLL, assoc. with TROLLOP]

lolly /ˈlɒlɪ/ *n.* (*pl.* **-ies**) **1** *colloq.* **a** a lollipop. **b** *Austral.* a sweet. **c** *Brit.* = *ice lolly.* **2** *Brit. sl.* money. [abbr. of LOLLIPOP]

Lombard /ˈlɒmbɑːd/ *n. & adj.* ● *n.* **1** a member of a Germanic people from the lower Elbe who invaded Italy in 568 and founded a kingdom (overthrown by Charlemagne in 774) in the valley of the Po. **2** a native of modern Lombardy. **3** the Italian dialect of modern Lombardy. ● *adj.* of or relating to Lombardy, the Lombards, or their dialect. □ **Lombardic** /lɒmˈbɑːdɪk/ *adj.* [ME f. OF *lombard* or MDu. *lombaerd*, f. It. *lombardo* f. med.L *Longobardus* f. L *Langobardus* f. Gmc]

Lombard Street a street in the City of London, formerly occupied by bankers from Lombardy and still containing many of the principal London banks.

Lombardy /ˈlɒmbədɪ/ (Italian **Lombardia** /ˌlombarˈdiːa/) a region of central northern Italy, between the Alps and the River Po; capital, Milan. Founded in the 6th century by the Germanic Lombards (also known as *Langobards* 'long beards'), it was taken by Spain in the 16th century, was ceded to Austria in 1713, and finally became a part of the kingdom of Italy in 1859.

Lombardy poplar *n.* a widely planted variety of the black poplar with a very tall slender form.

Lombok /ˈlɒmbɒk/ a volcanic island of the Lesser Sunda group in Indonesia, between Bali and Sumbawa; pop. (1991) 2,500,000; chief town, Mataram.

Lomé /ˈləʊmeɪ/ the capital and chief port of Togo, on the Gulf of Guinea; pop. (1983) 366,500.

Lomé Convention a trade agreement of 1975, reached in Lomé, between the EC and forty-six African, Caribbean, and Pacific Ocean states, aiming for technical cooperation and the provision of development aid. A second agreement was signed in 1979 by a larger group.

lomentum /ləʊˈmentəm/ *n.* (*pl.* **lomenta** /-tə/) (also **loment** /ˈləʊment/) *Bot.* the pod of some leguminous plants, breaking up when mature into one-seeded joints. □ **lomentaceous** /ˌləʊmənˈteɪʃəs/ *adj.* [L = bean-meal (orig. cosmetic) f. *lavare* wash]

Lomond, Loch /ˈləʊmənd/ a lake in west central Scotland, to the north-west of Glasgow. It is the largest freshwater lake in Scotland.

London[1] /ˈlʌndən/ **1** the capital of the United Kingdom, situated in SE England on the River Thames; pop. (1991) 6,378,600. It was settled as a river port and trading centre, called Londinium, shortly after the Roman invasion of AD 43. Since the Middle Ages London has been a flourishing commercial, business, and cultural centre. Devastated by the Great Plague (1665–6) and the Fire of London (1666), the city was rebuilt in the late 17th century, much of it, including St Paul's Cathedral, under the direction of the architect Sir Christopher Wren. Damage during the Second World War again led to substantial reconstruction. The more recent decline and redevelopment of the docklands in the East End have led to further changes in the city's landscape. London is divided administratively into the City of London, known as the Square Mile, which is the country's financial centre, and thirty-two boroughs. It is the seat, at Westminster, of the British government. **2** an industrial city in SE Ontario, Canada, situated to the north of Lake Erie; pop. (1991) 310,585.

London[2] /ˈlʌndən/, John Griffith ('Jack') (1876–1916), American novelist. He grew up in poverty, scratching a living in various ways and taking part in the Klondike gold rush of 1897–8 — experiences which provided the material for his works and made him a socialist. His most famous novel is *The Call of the Wild* (1903).

London clay *n.* a geological formation in the lower division of the Eocene in SE England. [LONDON[1]]

Londonderry /ˈlʌndənˌderɪ/ **1** one of the Six Counties of Northern Ireland, formerly an administrative area. **2** its chief town, a city and port on the River Foyle near its outlet on the north coast; pop. (1981) 62,700. Built on the site of an abbey founded by St Columba in AD 546, it was formerly called Derry, a name still used by many. In 1613 it was granted to the City of London for colonization and became known as Londonderry. In 1689 it resisted a siege by James II for 105 days before being relieved.

Londoner /ˈlʌndənə(r)/ *n.* a native or inhabitant of London.

London plane *n.* a hybrid plane tree, *Platanus* × *hispanica*, resistant to smoke and therefore often planted in streets. [LONDON[1]]

London pride *n.* a pink-flowered saxifrage, *Saxifraga urbium*. [LONDON[1]]

lone /ləʊn/ *attrib.adj.* **1** (of a person) solitary; without a companion or supporter. **2** (of a place) unfrequented, uninhabited, lonely. **3** unmarried, single (*lone parent*). **4** *literary* feeling or causing to feel lonely. □ **lone hand 1** a hand played, or a player playing, against the rest at quadrille and euchre. **2** a person or action without allies. **lone wolf** a person who prefers to act alone. [ME, f. ALONE]

lonely /ˈləʊnlɪ/ *adj.* (**lonelier, loneliest**) **1** (of a person) solitary, companionless, isolated. **2** (of a place) unfrequented; (of a thing) standing apart, isolated. **3 a** sad because without friends or company. **b** imparting a sense of loneliness; dreary. □ **lonely heart** a lonely person (in sense 3a). □ **loneliness** *n.*

loner /ˈləʊnə(r)/ *n.* a person or animal that prefers not to associate with others.

lonesome /ˈləʊnsəm/ *adj.* **1** solitary, lonely. **2** feeling lonely or forlorn. **3** causing such a feeling. □ **by** (or **on**) **one's lonesome** *colloq.* all alone. □ **lonesomely** *adv.* **lonesomeness** *n.*

long[1] /lɒŋ/ *adj.*, *n.*, & *adv.* ● *adj.* (**longer** /ˈlɒŋgə(r)/; **longest** /ˈlɒŋgɪst/) **1** measuring much from end to end in space or time; not soon traversed

or finished (*a long line; a long journey; a long time ago*). **2** (following a measurement) in length or duration (*2 metres long; the vacation is two months long*). **3** relatively great in extent or duration (*a long meeting*). **4 a** consisting of a large number of items (*a long list*). **b** seemingly more than the stated amount; tedious, lengthy (*ten long miles; tired after a long day*). **5** of elongated shape. **6 a** lasting or reaching far back or forward in time (*a long friendship*). **b** (of a person's memory) retaining things for a long time. **7** far-reaching; acting at a distance; involving a great interval or difference. **8** *Phonet.* & *Prosody* of a vowel or syllable: **a** having the greater of the two recognized durations. **b** (of a vowel in English) having the pronunciation shown in the name of the letter (as in *pile* and *cute* which have a long *i* and *u*, as distinct from *pill* and *cut*) (cf. SHORT *adj.* 6). **9** (of odds or a chance) reflecting or representing a low level of probability. **10** *Stock Exch.* **a** (of stocks) bought in large quantities in advance, with the expectation of a rise in price. **b** (of a broker etc.) buying etc. on this basis. **11** (of a bill of exchange) maturing at a distant date. **12** (of a cold drink) large and refreshing. **13** *colloq.* (of a person) tall. **14** (foll. by *on*) *colloq.* well supplied with. ● *n.* **1** a long interval or period (*shall not be away for long; it will not take long*). **2** *Phonet.* **a** a long syllable or vowel. **b** a mark indicating that a vowel is long. **3 a** long-dated stock. **b** a person who buys this. ● *adv.* (**longer** /ˈlɒŋgə(r)/; **longest** /ˈlɒŋgɪst/) **1** by or for a long time (*long before; long ago; long live the king!*). **2** (following nouns of duration) throughout a specified time (*all day long*). **3** (in *compar.*; with *neg.*) after an implied point of time (*shall not wait any longer*). □ **as** (or **so**) **long as 1** during the whole time that. **2** provided that; only if. **at long last** see *at last* (see LAST[1]). **before long** fairly soon (*shall see you before long*). **be long** (often foll. by *pres. part.* or *in* + verbal noun) take a long time; be slow (*was long finding it out; the chance was long in coming; I shan't be long*). **by a long chalk** see CHALK. **in the long run** over a long period. **2** eventually; finally. **long ago** in the distant past. **long-ago** *attrib.adj.* that is in the distant past. **the long and the short of it 1** all that can or need be said. **2** the eventual outcome. **long-case clock** a grandfather clock. **long-chain** *Chem.* (of a molecule) containing a chain of many carbon atoms. **long-dated** (of securities) not due for early payment or redemption. **long-day** (of a plant) needing a long daily period of light to cause flowering. **long-distance** *adj.* **1** (of a telephone call, public transport, etc.) between distant places. **2** *Brit.* (of a weather forecast) long-range. ● *adv.* (also **long distance**) between distant places (*phone long-distance*). ● *n.* (also **long distance**) (often *attrib.*) *Athletics* a race distance of 6 miles or 10,000 metres, or longer. **long division** division of numbers with details of the calculations written down. **long dozen** thirteen. **long-drawn** (or **-drawn-out**) prolonged, esp. unduly. **long face** a dismal or disappointed expression. **long-faced** with a long face. **long field** *Cricket* = *long off* (or *on*) below. **3** the part of the field behind the bowler. **long figure** (or **price**) a heavy cost. **long haul 1** the transport of goods or passengers over a long distance (hyphenated when *attrib.*: *long-haul flights*). **2** a prolonged effort or task. **long-headed** shrewd, far-seeing, sagacious. **long hop** a short-pitched easily hit ball in cricket. **long hundredweight** see HUNDREDWEIGHT 1. **long in the tooth** rather old (orig. of horses, from the recession of the gums with age). **long johns** *colloq.* underpants with full-length legs. **long jump** an athletic contest of jumping as far as possible along the ground in one leap. **long leg** *Cricket* a fielding position far behind the batsman on the leg side. **long-legged** speedy. **long-life** (of consumable goods) treated to preserve freshness. **long-lived** having a long life; durable. **long measure** a measure of length (metres, miles, etc.). **long metre** *Mus.* **1** a hymn stanza of four lines with eight syllables each. **2** a quatrain of iambic tetrameters with alternate lines rhyming. **long off** (or **on**) *Cricket* a fielding position far behind the bowler and towards the off (or on) side. **long-player** a long-playing record. **long-playing** (of a gramophone record) playing for about 20–30 minutes on each side. **long-range 1** (of a missile etc.) having a long range. **2** of or relating to a period of time far into the future. **long-running** continuing for a long time. **long shot 1** a wild guess or venture. **2** a bet at long odds. **3** *Cinematog.* a shot including objects at a distance. **long sight** the ability to see clearly only what is comparatively distant. **long-sleeved** with sleeves reaching to the wrist. **long-standing** that has long existed; not recent. **long-suffering** bearing provocation patiently. **long-sufferingly** in a long-suffering manner. **long suit 1** many cards of one suit in a hand (esp. more than 3 or 4 in a hand of 13). **2** a thing at which one excels. **long-tailed duck** an Arctic marine duck, *Clangula hyemalis*, the male of which has elongated tail feathers (also called (*N. Amer.*) *old squaw*). **long-term** occurring in or relating to a long period of time (*long-term plans*). **long-time** that has been such for a long time. **long ton** see TON[1] 1. **long tongue** loquacity. **long vacation** *Brit.* the summer vacation of lawcourts and universities. **long waist** a low or deep waist of a dress or body. **long-waisted** having a long waist. **long wave** a radio wave of frequency less than 300 kHz. **not by a long shot** by no means. □ **longish** adj. [OE long, lang]

long² /lɒŋ/ v.intr. (foll. by for or to + infin.) have a strong wish or desire for. [OE langian seem long to]

long. abbr. longitude.

-long /lɒŋ/ comb. form forming adjectives and adverbs: **1** for the duration of (lifelong). **2** = -LING² (headlong).

Long Beach a port and resort in California, situated on the south side of the Los Angeles conurbation; pop. (1990) 429,430.

longboard /'lɒŋbɔ:d/ n. US a type of surfboard.

longboat /'lɒŋbəʊt/ n. a sailing-ship's largest boat.

longbow /'lɒŋbəʊ/ n. a bow drawn by hand and shooting a long feathered arrow. Developed at the beginning of the 14th century, the longbow was the primary weapon of the English in the Hundred Years War, and proved influential at the battles of Crécy, Agincourt, and Poitiers. It was generally made of yew, with a shaft up to 1.8 m (6 ft) long and a range of over 180 m (200 yds).

longe var. of LUNGE².

longeron /'lɒndʒərən/ n. a longitudinal member of a plane's fuselage. [F, = girder]

longevity /lɒn'dʒevɪtɪ/ n. long life. [LL longaevitas f. L longus long + aevum age]

Longfellow /'lɒŋˌfeləʊ/, Henry Wadsworth (1807–82), American poet. His Ballads and other Poems (1841) contains such well-known pieces as 'The Wreck of the Hesperus' and 'The Village Blacksmith'. Longfellow's popularity increased with subsequent volumes, especially his narrative poems. His best-known work is The Song of Hiawatha (1855), which is based on American Indian stories concerning the life of the chieftain Hiawatha; its metre and novel subject-matter attracted many parodies and imitations.

Longford /'lɒŋfəd/ **1** a county of the Republic of Ireland, in the province of Leinster. **2** its county town; pop. (1991) 6,390.

longhair /'lɒŋheə(r)/ n. **1** a person characterized by the associations of long hair, esp. a hippie or intellectual. **2** a breed of long-haired cat.

longhand /'lɒŋhænd/ n. ordinary handwriting (as opposed to shorthand or typing or printing).

longhorn /'lɒŋhɔ:n/ n. **1** a breed of cattle with long horns. **2** an elongated beetle of the family Cerambycidae, with very long antennae.

longhouse /'lɒŋhaʊs/ n. (also **long house**) **1 a** hist. the traditional communal dwelling of the Iroquois and some other peoples of North America. **b** a building on an Iroquois reservation, used as a church and meeting hall. **2** a large communal village house in parts of Malaysia and Indonesia. **3** hist. a type of dwelling formerly built in Britain to house people and animals under the same roof.

longicorn /'lɒndʒɪˌkɔ:n/ n. a longhorn beetle. [mod.L longicornis f. L longus long + cornu horn]

longing /'lɒŋɪŋ/ n. & adj. ● n. a feeling of intense desire. ● adj. having or showing this feeling. □ **longingly** adv.

Longinus /lɒn'dʒaɪnəs/ (fl. 1st century AD), Greek scholar. He is the supposed author of a Greek literary treatise On the Sublime, a critical analysis of literary greatness showing concern with the moral function of literature and impatience with pedantry. After its translation into French in 1674 it became a very influential work with Augustan writers such as Dryden and Pope.

Long Island an island on the coast of New York State. Its western tip, comprising the New York districts of Brooklyn and Queens, is separated from Manhattan and the Bronx by the East River and is linked to Manhattan by the Brooklyn Bridge. The island extends 210 km (118 miles) eastwards roughly parallel to the coast of Connecticut, from which it is separated by Long Island Sound.

longitude /'lɒŋgɪˌtju:d, 'lɒndʒɪ-/ n. **1** Geog. the angular distance east or west from a standard meridian such as that of Greenwich to the meridian of any place (symbol: λ). **2** Astron. the angular distance of a celestial object north or south of the ecliptic measured along a great circle through the object and the poles of the ecliptic. [ME f. L longitudo -dinis f. longus long]

longitudinal /ˌlɒŋgɪˈtju:dɪn(ə)l, ˌlɒndʒɪ-/ adj. **1** of or in length. **2** running lengthwise. **3** of longitude. □ **longitudinal wave** a wave vibrating in the direction of propagation. □ **longitudinally** adv.

long knives, night of the n. see NIGHT OF THE LONG KNIVES.

Long March the epic withdrawal of the Chinese Communists from SE to NW China in 1934–5, over a distance of 9,600 km (6,000 miles).

The march was begun by 100,000 people evacuating the Communist rural base (the Jiangxi Soviet), which had been almost destroyed by the nationalist Kuomintang. Led by Mao Zedong during 1935, the Communists travelled through mountainous terrain cut by several major rivers, and in October some 6,000 survivors reached Yan'an in Shaanxi province. Other groups arrived later, but in all only about 20,000 survived the journey.

Long Parliament the English Parliament which sat from November 1640 to March 1653, was restored for a short time in 1659, and finally voted its own dissolution in 1660. It was summoned by Charles I and sat through the English Civil War and on into the interregnum which followed.

Longshan /lɒŋ'ʃæn/ an ancient civilization of the Yellow River valley in China c.2500–1700 BC, between the Yangshao and Shang periods, characterized by pottery kiln-fired to a uniform black colour, and by the establishment of towns.

longship /'lɒŋʃɪp/ n. a long narrow ship, powered by both oar and sail, as used by the Vikings. Its shallow draught enabled raiders to penetrate far inland up rivers.

longshore /'lɒŋʃɔ:(r)/ adj. **1** existing on or frequenting the shore. **2** directed along the shore. □ **longshore drift** the movement of material along a coast by waves which approach at an angle to the shore but recede directly away from it. [along shore]

longshoreman /'lɒŋˌʃɔ:mən/ n. (pl. **-men**) US a docker.

long-sighted /lɒŋ'saɪtɪd/ adj. **1** having long sight. **2** having foresight or imagination. □ **long-sightedly** adv. **long-sightedness** n.

longspur /'lɒŋspɜ:(r)/ n. a North American bunting of the genus Calcarius.

longstop /'lɒŋstɒp/ n. **1** Cricket a fielding position (now rarely used) directly behind the wicket-keeper. **2** an emergency precaution; a last resort.

longueur /lɒŋ'gɜ:(r)/ n. **1** a tedious passage in a book etc. **2** a tedious stretch of time. [F, = length]

longways /'lɒŋweɪz/ adv. (also **longwise** /-waɪz/) = LENGTHWAYS.

long-winded /lɒŋ'wɪndɪd/ adj. **1** (of speech or writing) tediously lengthy. **2** able to run a long distance without rest. □ **long-windedly** adv. **long-windedness** n.

lonicera /lə'nɪsərə/ n. **1** a dense evergreen shrub, Lonicera nitidum, much used as hedging. **2** = HONEYSUCKLE. [Adam Lonicer, Ger. botanist (1528–86)]

Lonsdale belt /'lɒnzdeɪl/ n. an ornate belt awarded to a professional boxer winning a British title fight. A fighter winning three title matches in one weight division is given a belt to keep. The 5th Earl of Lonsdale (Hugh Cecil Lowther, 1857–1944), presented the first belt in 1909.

loo¹ /lu:/ n. Brit. colloq. a lavatory. [20th c.: orig. uncert.]

loo² /lu:/ n. **1** a round card-game with penalties paid to the pool. **2** this penalty. □ **loo table** a circular table for playing loo on or resembling such a table. [abbr. of obs. lanterloo f. F lanturlu, refrain of a song]

loof var. of LUFF.

loofah /'lu:fə/ n. (also **luffa** /'lʌfə/) **1** a climbing gourdlike plant, Luffa cylindrica, native to Asia, producing edible marrow-like fruits. **2** the dried fibrous vascular system of this fruit used as a sponge. Also called vegetable sponge. [Egypt. Arab. lūfa, the plant]

look /lʊk/ v., n., & int. ● v. **1 a** intr. (often foll. by at) use one's sight; turn one's eyes in some direction. **b** tr. (foll. by adv.) turn one's eyes on; contemplate or examine (looked me in the eyes). **2** intr. **a** make a visual or mental search (I'll look in the morning). **b** (foll. by at) consider, examine (we must look at the facts). **3** intr. (foll. by for) **a** search for. **b** hope or be on the watch for. **c** expect. **4** intr. inquire (when one looks deeper). **5** intr. have a specified appearance; seem (look a fool; look foolish). **6** intr. (foll. by to) **a** consider; take care of; be careful about (look to the future). **b** rely on (a person or thing) (you can look to me for support). **c** expect. **7** intr. (foll. by into) investigate or examine. **8** tr. (foll. by what, where, etc. + clause) ascertain or observe by sight (look where we are). **9** intr. (of a thing) face or be turned, or have or afford an outlook, in a specified direction. **10** tr. express, threaten, or show (an emotion etc.) by one's looks. **11** intr. (foll. by that + clause) take care; make sure. **12** intr. (foll. by to + infin.) expect (am looking to finish this today). ● n. **1** an act of looking; the directing of the eyes to look at a thing or person; a glance (a scornful look). **2** (in sing. or pl.) the appearance of a face; a person's expression or personal aspect. **3** the (esp. characteristic) appearance of a thing (the place has a European look). **4** style, fashion (this year's look; the wet look). ● int. (also **look here!**)

calling attention, expressing a protest, etc. □ **look after 1** attend to; take care of. **2** follow with the eye. **3** seek for. **look one's age** appear to be as old as one really is. **look-alike** a person or thing closely resembling another (*a Prince Charles look-alike*). **look alive** (or **lively**) *colloq.* be brisk and alert. **look as if** suggest by appearance the belief that (*it looks as if he's gone*). **look back 1** (foll. by *on, upon, to*) turn one's thoughts to (something past). **2** (usu. with *neg.*) cease to progress (*since then we have never looked back*). **3** *Brit.* make a further visit later. **look before you leap** avoid precipitate action. **look daggers at** see DAGGER. **look down on** (or **upon** or **look down one's nose at**) regard with contempt or a feeling of superiority. **look for trouble** see TROUBLE. **look forward to** await (an expected event) eagerly or with specified feelings. **look in** make a short visit or call. **look-in** *n. colloq.* **1** an informal call or visit. **2** a chance of participation or success (*never gets a look-in*). **look a person in the eye** (or **eyes** or **face**) look directly and unashamedly at him or her. **look like 1** have the appearance of. **2** *Brit.* seem to be (*they look like winning*). **3** threaten or promise (*it looks like rain*). **4** indicate the presence of (*it looks like woodworm*). **look on 1** (often foll. by *as*) regard (*looks on you as a friend; looked on them with disfavour*). **2** be a spectator; avoid participation. **look oneself** appear in good health (esp. after illness etc.). **look out 1** direct one's sight or put one's head out of a window etc. **2** (often foll. by *for*) be vigilant or prepared. **3** (foll. by *on, over,* etc.) have or afford a specified outlook. **4** search for and produce (*shall look one out for you*). **look over 1** inspect or survey (*looked over the house*). **2** examine (a document etc.) esp. cursorily (*shall look it over*). **look round 1** look in every or another direction. **2** examine the objects of interest in a place (*you must come and look round sometime*). **3** examine the possibilities etc. with a view to deciding on a course of action. **look-see** *colloq.* a survey or inspection. **look sharp** act promptly; make haste (orig. = keep strict watch). **look small** see SMALL. **look through 1** examine the contents of, esp. cursorily. **2** penetrate (a pretence or pretender) with insight. **3** ignore by pretending not to see (*I waved, but you just looked through me*). **look up 1** search for (esp. information in a book). **2** *colloq.* go to visit (a person) (*had intended to look them up*). **3** raise one's eyes (*looked up when I went in*). **4** improve, esp. in price, prosperity, or well-being (*things are looking up all round*). **look a person up and down** scrutinize a person keenly or contemptuously. **look up to** respect or venerate. **not like the look of** find alarming or suspicious. □ **-looking** *adj.* (in comb.). [OE *lōcian* f. WG]

looker /'lʊkə(r)/ *n.* **1** a person having a specified appearance (*a good-looker*). **2** *colloq.* a good-looking person, esp. a woman. □ **looker-on** a person who is a mere spectator.

looking-glass /'lʊkɪŋˌglɑːs/ *n.* a mirror for looking at oneself.

lookout /'lʊkaʊt/ *n.* **1** a watch or looking out (*on the lookout for bargains*). **2 a** a post of observation. **b** a person or party or boat stationed to keep watch. **3** a view over a landscape. **4** a prospect of luck (*it's a bad lookout for them*). **5** *colloq.* a person's own concern.

loom[1] /luːm/ *n.* an apparatus for weaving yarn or thread into fabric. [ME *lōme* f. OE *gelōma* tool]

loom[2] /luːm/ *v. & n.* ● *v.intr.* (often foll. by *up*) **1** come into sight dimly, esp. as a vague and often magnified or threatening shape. **2** (of an event or prospect) be ominously close. ● *n.* a vague often exaggerated first appearance of land at sea etc. [prob. f. LG or Du.: cf. E Frisian *lōmen* move slowly, MHG *lüemen* be weary]

loon /luːn/ *n.* **1** *N. Amer.* = DIVER 3. **2** *colloq.* a crazy person (cf. LOONY). [alt. f. loom f. ON *lómr*]

loony /'luːnɪ/ *n. & adj. sl.* ● *n.* (*pl.* **-ies**) a mad or silly person; a lunatic. ● *adj.* (**loonier, looniest**) crazy, silly. □ **loony-bin** *sl.* a mental home or hospital. □ **looniness** *n.* [abbr. of LUNATIC]

loop /luːp/ *n. & v.* ● *n.* **1 a** a figure produced by a curve, or a doubled thread etc., that crosses itself. **b** anything forming this figure. **2** a similarly shaped attachment or ornament formed of cord or thread etc. and fastened at the crossing. **3** a ring or curved piece of material as a handle etc. **4** a contraceptive coil. **5** (in full **loop-line**) a railway or telegraph line that diverges from a main line and joins it again. **6** a manoeuvre in which an aeroplane describes a vertical loop. **7** *Skating* a manoeuvre describing a curve that crosses itself, made on a single edge. **8** *Electr.* a complete circuit for a current. **9** an endless strip of tape or film allowing continuous repetition. **10** *Computing* a programmed sequence of instructions that is repeated until or while a particular condition is satisfied. ● *v.* **1** *tr.* form (thread etc.) into a loop or loops. **2** *tr.* enclose with or as with a loop. **3** *tr.* (often foll. by *up, back, together*) fasten or join with a loop or loops. **4** *intr.* **a** form a loop. **b** move in looplike patterns. **5** *intr.* (also **loop the loop**) *Aeron.* perform the feat of circling in a vertical loop. [ME: orig. unkn.]

looper /'luːpə(r)/ *n.* **1** the twiglike caterpillar of a geometrid moth, which moves by looping and straightening the body as if measuring the ground. Also called *inchworm, measuring worm, (US) span-worm*. **2** a device for making loops.

loophole /'luːphəʊl/ *n. & v.* ● *n.* **1** a means of evading a rule etc. without infringing the letter of it. **2** a narrow vertical slit in a wall for shooting or looking through or to admit light or air. ● *v.tr.* make loopholes in (a wall etc.). [ME *loop* in the same sense + HOLE]

loopy /'luːpɪ/ *adj.* (**loopier, loopiest**) **1** *sl.* crazy. **2** having many loops.

loose /luːs/ *adj., n., & v.* ● *adj.* **1 a** not or no longer held by bonds or restraint. **b** (of an animal) not confined or tethered etc. **2** detached or detachable from its place (*has come loose*). **3** not held together or contained or fixed. **4** not specially fastened or packaged (*loose papers; had her hair loose*). **5** hanging partly free (*a loose end*). **6** slack, relaxed; not tense or tight. **7** not compact or dense (*loose soil*). **8** (of language, concepts, etc.) inexact; conveying only the general sense. **9** (preceding an agent noun) doing the expressed action in a loose or careless manner (*a loose thinker*). **10** morally lax; dissolute (*loose living*). **11** (of the tongue) likely to speak indiscreetly. **12** (of the bowels) tending to diarrhoea. **13** *Sport* **a** (of a ball) in play but not in any player's possession. **b** (of play etc.) with the players not close together. **14** *Cricket* **a** (of bowling) inaccurately pitched. **b** (of fielding) careless or bungling. **15** (in comb.) loosely (*loose-flowing; loose-fitting*). ● *n.* **1** a state of freedom or unrestrainedness. **2** loose play in football (*in the loose*). **3** free expression. ● *v.tr.* **1** release; set free; free from constraint. **2** untie or undo (something that constrains). **3** detach from moorings. **4** relax (*loosed my hold on it*). **5** discharge (a bullet or arrow etc.). □ **at a loose end** (*N. Amer.* **at loose ends**) (of a person) unoccupied, esp. temporarily. **loose box** *Brit.* a compartment for a horse, in a stable or vehicle, in which it can move about. **loose cannon** a person or thing causing damage unpredictably and indiscriminately. **loose change** money as coins in the pocket etc. for casual use. **loose cover** *Brit.* a removable cover for a chair or sofa etc. **loose forward** *Rugby* a forward at the back of the scrum. **loose-leaf** *adj.* (of a notebook, manual, etc.) with each leaf separate and removable. ● *n.* a loose-leaf notebook etc. **loose-limbed** having supple limbs. **loose order** an arrangement of soldiers etc. with wide intervals. **on the loose 1** escaped from captivity. **2** having a free enjoyable time. **play fast and loose** ignore one's obligations, be unreliable, trifle. □ **loosely** *adv.* **looseness** *n.* **loosish** *adj.* [ME *lōs* f. ON *lauss* f. Gmc]

loosen /'luːs(ə)n/ *v.* **1** *tr. & intr.* make or become less tight or compact or firm. **2** *tr.* make (a regime etc.) less severe. **3** *tr.* release (the bowels) from constipation. **4** *tr.* relieve (a cough) from dryness. □ **loosen a person's tongue** make a person talk freely. **loosen up** limber up (see LIMBER[1] *v.*). □ **loosener** *n.*

loosestrife /'luːsstraɪf/ *n.* **1** (in full **yellow loosestrife**) a tall waterside plant, *Lysimachia vulgaris*, of the primrose family, with spikes of yellow flowers. **2** (in full **purple loosestrife**) a tall waterside plant, *Lythrum salicaria*, with spikes of star-shaped purple flowers. [LOOSE + STRIFE, taking the Gk name *lusimakhion* (f. the personal name *Lusimakhos*) as if directly f. *luō* undo + *makhē* battle]

loot /luːt/ *n. & v.* ● *n.* **1** goods taken from an enemy; spoil. **2** booty; illicit gains made by an official. **3** *sl.* money. ● *v.tr.* **1** rob (premises) or steal (goods) left unprotected, esp. after riots or other violent events. **2** plunder or sack (a city, building, etc.). **3** carry off as booty. □ **looter** *n.* [Hindi *lūṭ*]

lop[1] /lɒp/ *v. & n.* ● *v.* (**lopped, lopping**) **1** *tr.* **a** (often foll. by *off, away*) cut or remove (a part or parts) from a whole, esp. branches from a tree. **b** remove branches from (a tree). **2** *tr.* (often foll. by *off, away*) remove (items) as superfluous. **3** *intr.* (foll. by *at*) make lopping strokes on (a tree etc.). ● *n.* parts lopped off, esp. branches and twigs of trees. □ **lop and top** (or **crop**) the trimmings of a tree. □ **lopper** *n.* [ME f. OE *loppian* (unrecorded): cf. obs. *lip* to prune]

lop[2] /lɒp/ *v.* (**lopped, lopping**) **1** *intr.* hang limply. **2** *intr.* (foll. by *about*) slouch, dawdle; hang about. **3** *intr.* move with short bounds. **4** *tr.* (of an animal) let (the ears) hang. □ **lop-ears** drooping ears. **lop-eared** (of an animal) having drooping ears. □ **loppy** *adj.* [prob. imit.]

lope /ləʊp/ *v. & n.* ● *v.intr.* (esp. of animals) run with a long bounding stride. ● *n.* a long bounding stride. [ME, var. of Sc. *loup* f. ON *hlaupa* LEAP]

lopho- /'ləʊfəʊ, 'lɒf-/ *comb. form Zool.* crested. [Gk *lophos* crest]

lophodont /'ləʊfəˌdɒnt, 'lɒf-/ *n. & adj. Zool.* ● *adj.* having transverse ridges on the grinding surface of molar teeth. ● *n.* an animal with these teeth. [LOPHO- + Gk *odous odont-* tooth]

lophophore /'ləʊfəˌfɔː(r), 'lɒf-/ *n. Zool.* a horseshoe-shaped structure of ciliated tentacles around the mouth of bryozoans and brachiopods.

Lop Nor /lɒp ˈnɔː(r)/ (also **Lop Nur** /ˈnʊə(r)/) a marshy depression in the arid basin of the Tarim river in NW China. Once a large salt lake, which dried up when the river changed its course, the remote area has been used since 1964 for nuclear testing.

lopolith /ˈlɒpəlɪθ/ n. Geol. a large saucer-shaped intrusion of igneous rock. [Gk lopas basin + -LITH]

lopsided /lɒpˈsaɪdɪd/ adj. with one side lower or smaller than the other; unevenly balanced. □ **lopsidedly** adv. **lopsidedness** n. [LOP² + SIDE]

loquacious /lɒˈkweɪʃəs/ adj. **1** talkative. **2** (of birds or water) chattering, babbling. □ **loquaciously** adv. **loquaciousness** n. **loquacity** /-ˈkwæsɪtɪ/ n. [L loquax -acis f. loqui talk]

loquat /ˈləʊkwɒt/ n. **1** a rosaceous tree, Eriobotrya japonica, bearing small yellow egg-shaped fruits. **2** this fruit. [Chin. dial. luh kwat rush orange]

loquitur /ˈlɒkwɪtə(r)/ v.intr. (he or she) speaks (with the speaker's name following, as a stage direction or to inform the reader). [L]

lor /lɔː(r)/ int. Brit. sl. an exclamation of surprise or dismay. [abbr. of LORD]

loran /ˈlɔːræn, ˈlɒr-/ n. a system of long-distance navigation in which position is determined from the intervals between signal pulses received from widely spaced radio transmitters. [long-range navigation]

Lorca /ˈlɔːkə/, Federico García (1898-1936), Spanish poet and dramatist. His volumes of verse include Gypsy Ballads (1928), strongly influenced by the folk poetry of his native Andalusia. However, he is particularly known for intense, poetic tragedies evoking the passionate emotions of Spanish life; they include Blood Wedding (1933), Yerma (1934), and The House of Bernada Alba (published posthumously in 1945). He was murdered by Nationalist partisans after the outbreak of the Spanish Civil War.

lord /lɔːd/ n., int., & v. ● n. **1** a master or ruler. **2** hist. a feudal superior, esp. of a manor. **3** a peer of the realm or a person entitled to the title Lord, esp. a marquess, earl, viscount, or baron. **4** (**Lord**) (often prec. by the) a name for God or Christ. **5** (**Lord**) **a** prefixed as the designation of a marquess, earl, viscount, or baron. **b** prefixed to the Christian name of the younger son of a duke or marquess. **c** (**the Lords**) = HOUSE OF LORDS. **6** Astrol. the ruling planet (of a sign, house, or chart). ● int. (**Lord**) expressing surprise, dismay, etc. ● v.tr. confer the title of Lord upon. □ **good Lord!** an exclamation of surprise, anger, etc. **live like a lord** live sumptuously. **lord it over** domineer over. **lord over** (usu. in passive) dominate, rule over. **lords and ladies** = cuckoo-pint. **Our Lord** a name for Christ. □ **lordless** adj. **lordlike** adj. [OE hlāford f. hlāfweard = bread-keeper (as LOAF¹, WARD)]

Lord Advocate n. the principal law-officer of the Crown in Scotland.

Lord Bishop n. the formal title of a bishop.

Lord Chamberlain n. (in full **Lord Chamberlain of the Household**) (in the UK) the official in charge of the royal household, formerly responsible for licensing plays.

Lord Chancellor n. (also **Lord High Chancellor**) **1** an officer presiding in the House of Lords, the Chancery Division, or the Court of Appeal. **2** hist. an officer of state acting as head of the judiciary and administrator of the royal household.

Lord Chief Justice n. (in the UK) the president of the Queen's Bench Division.

Lord Commissioner n. (or **Lord High Commissioner**) the representative of the Crown at the General Assembly of the Church of Scotland.

Lord Fauntleroy see FAUNTLEROY.

Lord High Commissioner see LORD COMMISSIONER.

Lord High Steward of England n. (in the UK) a high officer of state presiding at coronations.

Lord Howe Island /haʊ/ a volcanic island in the SW Pacific off the east coast of Australia, administered as part of New South Wales; pop. (1989) 320. It is named after Admiral Lord Howe (1726-99), who was First Lord of the Admiralty when it was first visited.

Lord Lieutenant n. **1** (in the UK) the chief executive authority and head of magistrates in each county. **2** hist. the viceroy of Ireland.

lordling /ˈlɔːdlɪŋ/ n. usu. derog. a minor lord.

lordly /ˈlɔːdlɪ/ adj. (**lordlier**, **lordliest**) **1** haughty, imperious. **2** suitable for a lord. □ **lordliness** n. [OE hlāfordlic (as LORD)]

Lord Mayor n. the title of the mayor in London and some other large cities.

Lord Ordinary n. see ORDINARY n. 5.

lordosis /lɔːˈdəʊsɪs/ n. Med. inward curvature of the spine (opp. KYPHOSIS). □ **lordotic** /-ˈdɒtɪk/ adj. [mod.L f. Gk lordōsis f. lordos bent backwards]

Lord President of the Council n. (in the UK) the Cabinet minister presiding at the Privy Council.

Lord Privy Seal n. (in the UK) a senior Cabinet minister without specified official duties.

Lord Provost n. the head of a municipal corporation or borough in certain Scottish cities.

Lord's a cricket ground in St John's Wood, north London, headquarters since 1814 of the MCC. It is named after the cricketer Thomas Lord (1755-1832).

Lords Commissioners n.pl. the members of a board performing the duties of a high state office put in commission.

Lord's Day Sunday.

lordship /ˈlɔːdʃɪp/ n. **1** (usu. **Lordship**) a title used in addressing or referring to a man with the rank of Lord or a judge or a bishop (Your Lordship; His Lordship). **2** (foll. by of, over) dominion, rule, or ownership. **3** the condition of being a lord. [OE hlāfordscipe (as LORD, -SHIP)]

Lords of Session n.pl. the judges of the Scottish Court of Session.

Lord's Prayer the prayer taught by Christ to his disciples (Matt. 6:9-13), beginning 'Our Father'.

Lords Spiritual n.pl. the bishops in the House of Lords.

Lord's Supper the Eucharist.

Lords Temporal n.pl. the members of the House of Lords other than the bishops.

Lordy /ˈlɔːdɪ/ int. = LORD int.

lore¹ /lɔː(r)/ n. a body of traditions and knowledge on a subject or held by a particular group (herbal lore; gypsy lore). [OE lār f. Gmc, rel. to LEARN]

lore² /lɔː(r)/ n. Zool. a straplike surface between the eye and upper mandible in birds, or between the eye and nostril in snakes. [L lorum strap]

Lorelei /ˈlɒrəˌlaɪ/ **1** a rock on the right bank of the Rhine, in the Rhine gorge near Sankt Goarshausen. It is held by legend to be the home of a siren whose song lures boatmen to destruction. **2** the siren herself.

Loren /ləˈren/, Sophia (born Sophia Scicolone) (b.1934), Italian actress. She has starred in many Italian and American films, ranging from the romantic melodrama The Black Orchid (1959) and the slapstick comedy The Millionairess (1960) to the wartime drama La Ciociara (1961), for which she won an Oscar.

Lorentz /ˈlɒrənts/, Hendrik Antoon (1853-1928), Dutch theoretical physicist. He worked on the forces affecting electrons, making substantial advances on the work of Maxwell and realizing that electrons and cathode rays were the same thing. Lorentz's name is applied to various concepts and phenomena which he described. For their work on electromagnetic theory, he and his pupil Pieter Zeeman (1865-1943) shared the 1902 Nobel Prize for physics.

Lorentz transformation n. Physics the set of equations which in Einstein's special theory of relativity relate the space and time coordinates of one frame of reference to those of another.

Lorenz /ˈlɒrənts/, Konrad (Zacharias) (1903-89), Austrian zoologist. He pioneered the science of ethology, emphasizing innate rather than learned behaviour or conditioned reflexes. His major studies were in ornithology, especially with geese (in which he discovered the phenomenon of imprinting), and with jackdaws. Lorenz extrapolated these studies to human behaviour patterns, and compared the ill effects of the domestication of animals to human civilizing processes. His popular books include King Solomon's Ring (1952) and On Aggression (1966); he shared a Nobel Prize in 1973 with Karl von Frisch and Nikolaas Tinbergen.

Lorenzo de' Medici /ləˈrenzəʊ/ (known as Lorenzo the Magnificent) (1449-92), Italian statesman and scholar. The grandson of Cosimo de' Medici, he came to power in Florence in 1469 following his father's death. He was a patron of the arts, promoted humanist learning and Neoplatonic philosophy, and was a noted poet and scholar in his own right; Botticelli, Leonardo da Vinci, and Michelangelo were among the artists who enjoyed his patronage.

Loreto /ləˈretəʊ/ a town in eastern Italy, near the Adriatic coast to the south of Ancona; pop. (1990) 10,640. It is the site of the 'Holy House', said to be the home of the Virgin Mary and to have been brought from Nazareth by angels in 1295.

lorgnette /lɔː'njet/ n. (in sing. or pl.) a pair of eyeglasses or opera-glasses held by a long handle. [F f. lorgner to squint]

loricate /'lɒrɪˌkeɪt/ adj. Zool. having a defensive armour of bone, plates, scales, etc. [L loricatus f. lorica breastplate f. lorum strap]

Lorient /'lɒrɪˌɒn/ a port in NW France, on the south coast of Brittany; pop. (1990) 61,630.

lorikeet /'lɒrɪˌkiːt/ n. a small lory (parrot) found in Australasia. [dimin. of LORY, after parakeet]

loris /'lɔːrɪs/ n. (pl. same) a small slow-moving nocturnal primate, with small ears and a very short tail; esp. the slender loris (Loris tardigradus) of southern India, and the slow loris (Nycticebus coucang) of SE Asia and Indonesia. [F perh. f. obs. Du. loeris clown]

lorn /lɔːn/ adj. literary desolate, forlorn, abandoned. [past part. of obs. leese f. OE -lēosan lose]

Lorrain, Claude see CLAUDE LORRAIN.

Lorraine /lə'reɪn/ a region of NE France, between Champagne and the Vosges mountains. The modern region corresponds to the southern part of the medieval kingdom of Lorraine, which extended from the North Sea to Italy. In 1871 the western part was annexed by Prussia, along with Alsace. It was restored to France after the First World War. [L Lotharingia f. Lothair name of king (825–69)]

Lorraine, Cross of n. a cross with two parallel crosspieces, originally a heraldic device. It was the symbol of Joan of Arc, and in the Second World War was adopted by the Free French forces of General de Gaulle.

lorry /'lɒrɪ/ n. Brit. (pl. -ies) a heavy road vehicle for transporting goods etc.; = TRUCK[1] 2. □ **fall off** (or **off the back of**) **a lorry** (of goods etc.) be acquired in dubious circumstances from an unspecified source. [19th c. orig. = long flat wagon: orig. uncert.]

lory /'lɔːrɪ/ n. (pl. -ies) a brightly coloured parrot of the subfamily Loriinae, found in SE Asia and Australasia. [Malay lūrī]

Los Alamos /lɒs 'æləˌmɒs/ a town in northern New Mexico; pop. (1990) 11,450. It has been a centre for nuclear research since the 1940s, when it was the site of the development of the first atomic and hydrogen bombs.

Los Angeles /lɒs 'ændʒɪ,liːz, -lɪs/ a city on the Pacific coast of southern California, the second largest city in the US; pop. (1990) 3,485,400. Founded by the Spanish in 1781, it developed after the arrival of the Southern Pacific Railroad and the discovery of oil in 1894. It has become a major centre of industry, film-making, and television in the 20th century, its metropolitan area having expanded to include towns such as Beverly Hills, Hollywood, Santa Monica, and Pasadena.

lose /luːz/ v. (past and past part. **lost** /lɒst/) **1** tr. be deprived of or cease to have, esp. by negligence or misadventure. **2** tr. **a** be deprived of (a person, esp. a close relative) by death. **b** suffer the loss of (a baby) in childbirth; miscarry (a baby). **3** tr. become unable to find; fail to keep in sight or follow or mentally grasp (lose one's way). **4** tr. let or have pass from one's control or reach (lose one's chance; lose one's bearings). **5** tr. be defeated in (a game, race, lawsuit, battle, etc.). **6** tr. evade; get rid of (lost our pursuers). **7** tr. fail to obtain, catch, or perceive (lose a train; lose a word). **8** tr. forfeit (a stake, deposit, right to a thing, etc.). **9** tr. spend (time, efforts, etc.) to no purpose (lost no time in raising the alarm). **10** intr. **a** suffer loss or detriment; incur a disadvantage. **b** be worse off, esp. financially. **11** tr. cause (a person) the loss of (will lose you your job). **12** intr. & tr. (of a timepiece) become slow; become slow by (a specified amount of time). **13** tr. (in passive) **a** disappear, perish; be dead (was lost in the war). **b** fall, sin; be damned (souls lost to drunkenness and greed). **14** (as **lost** adj.) **a** gone, stray, mislaid; forgotten (lost valuables; a lost art). **b** dead, destroyed (lost comrades). **c** damned, fallen (lost souls in hell). □ **be lost** (or **lose oneself**) **in** be engrossed in. **be lost on** colloq. be wasted on, or not noticed or appreciated by. **be lost to** be no longer affected by or accessible to (is lost to pity; is lost to the world). **be lost without** have great difficulty if deprived of (am lost without my diary). **get lost** sl. (usu. in imper.) go away. **lose one's balance 1** fail to remain stable; fall. **2** fail to retain one's composure. **lose one's cool** colloq. lose one's composure. **lose face** be humiliated; lose one's credibility. **lose ground** see give ground (see GROUND[1]). **lose one's head** see HEAD. **lose heart** be discouraged. **lose one's heart** see give one's heart (see HEART). **lose one's nerve** become timid or irresolute. **lose out** colloq. (often foll. by on) **1** be unsuccessful; not get a fair chance or advantage (in). **2** (often foll. by to) be beaten in competition or replaced by. **lose sleep over a thing** lie awake

worrying about a thing. **lose one's temper** become angry. **lose time** allow time to pass with something unachieved etc. **lose touch** see TOUCH. **lose track of** see TRACK[1]. **lose the** (or **one's**) **way** become lost; fail to reach one's destination. **losing battle** a contest or effort in which failure seems certain. **lost cause 1** an enterprise etc. with no chance of success. **2** a person one can no longer hope to influence. □ **losable** adj. [OE losian perish, destroy f. los loss]

loser /'luːzə(r)/ n. **1** a person or thing that loses or has lost (esp. a contest or game) (is a poor loser; the loser pays). **2** colloq. a person who regularly fails.

loss /lɒs/ n. **1 a** the act or an instance of losing; the state of being lost. **b** the fact of being deprived of a person by death, estrangement, etc. **2** a person, thing, or amount lost. **3** the detriment or disadvantage resulting from losing (that is no great loss). □ **at a loss** (sold etc.) for less than was paid for it. **be at a loss** be puzzled or uncertain. **be at a loss for words** not know what to say. **loss adjuster** an insurance agent who assesses the amount of compensation arising from a loss. **loss-leader** an item sold at a loss to attract customers. [ME los, loss prob. back-form. f. lost, past part. of LOSE]

lost past and past part. of LOSE.

lost generation n. a generation with many of its men killed in war, esp. the one maturing in 1915–25, which lost a great number in the First World War. The phrase was applied by Gertrude Stein to the disillusioned young US writers, such as Ernest Hemingway, Scott Fitzgerald, and Ezra Pound, who went to live in Paris in the 1920s.

Lost Tribes (also **Ten Lost Tribes of Israel**) the ten tribes of Israel taken away c.720 BC by Sargon II to captivity in Assyria (2 Kings 17:6), from which they are believed never to have returned, while the tribes of Benjamin and Judah remained. (See also TRIBES OF ISRAEL.)

Lot[1] /lɒt, ləʊ/ a river of southern France, which rises in the Auvergne and flows 480 km (300 miles) west to meet the Garonne south-east of Bordeaux.

Lot[2] /lɒt/ (in the Bible) the nephew of Abraham, who was allowed to escape from the destruction of Sodom (Gen. 19). His wife, who disobeyed orders and looked back, was turned into a pillar of salt.

lot /lɒt/ n. & v. ● n. **1** colloq. (prec. by a or in pl.) **a** a large number or amount (a lot of people; lots of chocolate). **b** colloq. much (a lot warmer; smiles a lot; is lots better). **2 a** each of a set of objects used in making a chance selection. **b** this method of deciding (chosen by lot). **3** a share, or the responsibility resulting from it. **4** a person's destiny, fortune, or condition. **5** esp. N. Amer. a plot; an allotment of land (parking lot). **6** an article or set of articles for sale at an auction etc. **7** a number or quantity of associated persons or things. ● v.tr. (**lotted, lotting**) divide into lots. □ **bad lot** a person of bad character. **cast** (or **draw**) **lots** decide by means of lots. **throw in one's lot with** decide to share the fortunes of. **the** (or **the whole**) **lot** esp. Brit. the whole number or quantity. **a whole lot** colloq. very much (is a whole lot better). [OE hlot portion, choice f. Gmc]

loth var. of LOATH.

Lothario /lə'θɑːrɪˌəʊ, -'θeərɪˌəʊ/ n. (pl. **-os**) a rake or libertine. [a character in Nicholas Rowe's play Fair Penitent (1703)]

Lothian /'ləʊðɪən/ a local government region in SE central Scotland, on the Firth of Forth; administrative centre, Edinburgh.

Loti /lɒ'tiː/, Pierre (pseudonym of Louis Marie Julien Viaud) (1850–1923), French novelist. His novels were written while he served as a naval officer and his voyages provided the background for his work. His fame chiefly rests on three novels: Mon frère Yves (1883), Pêcheur d'Islande (1886), and Matelot (1893). These tell of the struggles of sailors who leave Brittany to fish in the waters around Iceland and the heartbreak of those left behind.

loti /'ləʊtɪ, 'luːtɪ/ n. (pl. **maloti** /mə'ləʊtɪ, -'luːtɪ/) the basic monetary unit of Lesotho, equal to 100 lisente. [Sesotho]

lotion /'ləʊʃ(ə)n/ n. a medicinal or cosmetic liquid preparation applied externally. [ME f. OF lotion or L lotio f. lavare lot- wash]

lottery /'lɒtərɪ/ n. (pl. **-ies**) **1** a means of raising money by selling numbered tickets and giving prizes to the holders of numbers drawn at random. **2** an enterprise, process, etc., whose success is governed by chance (life is a lottery). [prob. f. Du. loterij (as LOT)]

Lotto /'lɒtəʊ/, Lorenzo (c.1480–1556), Italian painter. His early art reflects the influence of his training in the studio of Giovanni Bellini in Venice, while his more mature work incorporates a wide range of influences, from the art of northern Europe to Raphael and Titian. Although chiefly a painter of religious subjects, he also produced a number of notable portraits, such as A Lady as Lucretia (c.1533).

lotto /'lɒtəʊ/ n. **1** a game of chance like bingo, but with numbers drawn by the players instead of being called. **2** esp. US a lottery. [It.]

lotus /'ləʊtəs/ n. **1** Gk Mythol. a legendary plant inducing luxurious languor when eaten. **2 a** a water lily of the genus Nelumbo; esp. N. nucifera of India, with large pink flowers. **b** this flower used symbolically in Hinduism and Buddhism. **3** an Egyptian water lily, Nymphaea lotus, with white flowers. **4** a leguminous plant of the genus Lotus, e.g. bird's-foot trefoil. □ **lotus-eater** a person given to indolent enjoyment. **lotus-land** a place of indolent enjoyment. **lotus position** a cross-legged position of meditation with the feet resting on the thighs. [L f. Gk lōtos, of Semitic orig.]

Lotus Sutra Buddhism one of the most important texts in Mahayana Buddhism, significant particularly in China and Japan and given special veneration by the Nichiren sect.

Louangphrabang see LUANG PRABANG.

louche /luːʃ/ adj. disreputable, shifty. [F, = squinting]

loud /laʊd/ adj. & adv. ● adj. **1 a** strongly audible, esp. noisily or oppressively so. **b** able or liable to produce loud sounds (a loud engine). **c** clamorous, insistent (loud complaints). **2** (of colours, design, etc.) gaudy, obtrusive. **3** (of behaviour) aggressive and noisy. ● adv. in a loud manner. □ **loud hailer** an electronic device for amplifying the sound of the voice so that it can be heard at a distance. **loud-mouth** colloq. a loud-mouthed person. **loud-mouthed** colloq. noisily self-assertive; vociferous. **out loud 1** aloud. **2** loudly (laughed out loud). □ **louden** v.tr. & intr. **loudish** adj. **loudly** adv. **loudness** n. [OE hlūd f. WG]

loudspeaker /laʊd'spiːkə(r)/ n. an apparatus that converts electrical impulses into sound, esp. music and voice.

Lou Gehrig's disease see GEHRIG.

lough /lɒk, lɒx/ n. Ir. = LOCH. [Ir. loch LOCH, assim. to the related obs. ME form lough]

Loughborough /'lʌfbərə/ a town in Leicestershire, on the River Soar north of Leicester; pop. (1981) 46,120.

Louis[1] /'luːɪ/ the name of eighteen kings of France, notably:

Louis IX (canonized as St Louis) (1214–70), son of Louis VIII, reigned 1226–70. His reign was dominated by his two crusades to the Holy Land, neither of which proved successful: the first (1248–54) ended in disaster with his capture by the Egyptians, the second (1270–1) in his own death from plague in Tunis. Feast day, 25 Aug.

Louis XI (1423–83), son of Charles VII, reigned 1461–83. He continued his father's work in laying the foundations of a united France ruled by an absolute monarchy. His reign was dominated by his struggle with Charles the Rash, Duke of Burgundy. This ended with Charles's death in battle in 1477 and France's absorption of much of Burgundy's former territory along her border.

Louis XIII (1601–43), son of Henry IV of France, reigned 1610–43. During his minority, the country was ruled by his mother Marie de Médicis. Louis asserted his right to rule in 1617, but from 1624 he was heavily influenced in policy-making by his chief minister Cardinal Richelieu.

Louis XIV (1638–1715), son of Louis XIII, reigned 1643–1715. He is known as the 'Sun King' from the magnificence of his reign, which represented the high point of the Bourbon dynasty and of French power in Europe, and during which French art and literature flourished. However, his almost constant wars of expansion united Europe against him, and, despite the reforms of Colbert, gravely weakened France's financial position. The Peace of Utrecht (1713–14), which ended the War of the Spanish Succession, represented the ultimate failure of Louis's attempt at European hegemony, preventing as it did the union of the French and Spanish crowns.

Louis XVI (1754–93), grandson and successor of Louis XV, reigned 1774–93. He inherited a situation of growing political discontent and severe problems of debt in the state finances. When the French Revolution broke out, he took refuge in a series of half measures, such as constitutional reforms and concessions to the republicans, which proved disastrous to his cause. After Louis's unsuccessful attempt to flee the country (1791), the Revolution became progressively more extreme and, with foreign invaders massing on the borders, the monarchy was abolished and Louis and his wife Marie Antoinette were executed.

Louis XVIII (1755–1824), brother of Louis XVI, reigned 1814–24. Following the outbreak of the French Revolution, he went into exile in 1791; two years later he pronounced himself regent for his nephew Louis XVII (1785–95), who had become king of France in name on the execution of his parents Louis XVI and Marie Antoinette. After his nephew's death, Louis XVIII became titular king until the fall of

Napoleon in 1814, when he returned to Paris on the summons of Talleyrand and was officially restored to the throne. Louis introduced a constitutional monarchy in the same year, but was forced to flee the capital when Napoleon regained power briefly in 1815. After the latter's defeat at Waterloo, Louis returned to Paris and inaugurated parliamentary government.

Louis[2] /'luːɪs/, Joe (born Joseph Louis Barrow) (1914–81), American boxer. Known as the 'Brown Bomber', he was heavyweight champion of the world 1937–49, defending his title twenty-five times during that period.

louis /'luːɪ/ n. (pl. same /'luːɪz/) hist. (in full **louis d'or** /'dɔː(r)/) a former French gold coin worth about 20 francs. [Louis, the name of kings of France]

Louis I /'luːɪ/ (known as Louis the Great) (1326–82), king of Hungary 1342–82 and of Poland 1370–82. He fought two successful wars against Venice (1357–8; 1378–81), and the rulers of Serbia, Wallachia, Moldavia, and Bulgaria became his vassals. Under his rule Hungary became a powerful state, though Poland was troubled by revolts.

Louis, St, Louis IX of France (see LOUIS[1]).

Louisiana /luːˌiːzɪ'ænə/ a state in the southern US, on the Gulf of Mexico; pop. (1990) 4,219,970; capital, Baton Rouge. Louisiana originally denoted the large region of the Mississippi basin claimed for France by the explorer La Salle in 1682, named in honour of Louis XIV. It was sold by the French to the US in the Louisiana Purchase of 1803. The smaller area now known as Louisiana became the 18th state in 1812. □ **Louisianan**, **Louisianian** adj. & n.

Louisiana Purchase the territory sold by France to the US in 1803, comprising the western part of the Mississippi valley. The area had been explored by France, ceded to Spain in 1762, and returned to France in 1800.

Louis-Napoleon /ˌluːmɑː'pəʊliən/, Napoleon III of France (see NAPOLEON).

Louis Philippe /ˌluːɪ fiˈliːp/ (1773–1850), king of France 1830–48. As the Duc d'Orléans, Louis Philippe participated in the early, liberal phase of the French Revolution, but later went into exile abroad. Returning to France after the restoration of the Bourbons (see RESTORATION, THE 2), he became the focus for liberal discontent, and after the overthrow of Charles X in 1830 was made king. His bourgeois-style regime was popular at first but it was gradually undermined by radical discontent and overthrown in a brief uprising in 1848, with Louis once more going into exile.

Louis the Great, Louis I of Hungary (see LOUIS I).

Louisville /'luːɪˌvɪl/ an industrial city and river port in northern Kentucky, on the Ohio river just south of the border with Indiana; pop. (1990) 269,060. It is the site of the annual Kentucky Derby, which takes place on the nearby Churchill Downs racetrack.

lounge /laʊndʒ/ v. & n. ● v.intr. **1** recline comfortably and casually; loll. **2** stand or move about idly. ● n. **1** a place for lounging, esp.: **a** a public room (e.g. in a hotel). **b** a place in an airport etc. with seats for waiting passengers. **c** a sitting-room in a house. **2** a spell of lounging. □ **lounge bar** Brit. a more comfortable room for drinking in a public house. **lounge lizard** colloq. an idler in fashionable society. **lounge suit** Brit. a man's formal suit for ordinary day wear. [perh. f. obs. lungis lout]

lounger /'laʊndʒə(r)/ n. **1** a person who lounges. **2** a piece of furniture for relaxing on. **3** a casual garment for wearing when relaxing.

loupe /luːp/ n. a small magnifying glass used by jewellers etc. [F]

louping-ill /'laʊpɪŋˌɪl/ n. a viral disease of animals, esp. sheep, transmitted by ticks and causing staggering and jumping. [Sc. & N. dial. loup leap]

lour /'laʊə(r)/ v. & n. (also **lower**) ● v.intr. **1** frown; look sullen. **2** (of the sky etc.) look dark and threatening. ● n. **1** a scowl. **2** a gloomy look (of the sky etc.). □ **louringly** adv. **loury** adj. [ME loure, of unkn. orig.]

Lourdes /lʊəd/ a town in SW France, at the foot of the Pyrenees; pop. (1982) 17,619. It has been a major place of Roman Catholic pilgrimage since in 1858 a young peasant girl, Marie Bernarde Soubirous (St Bernadette), claimed to have had a series of visions of the Virgin Mary.

Lourenço Marques /ləˌrensəʊ 'mɑːks/ the former name (until 1976) for MAPUTO.

louse /laʊs/ n. & v. ● n. **1** (pl. lice /laɪs/) **a** (in full **head louse**, **body louse**) a parasitic insect, Pediculus humanus, infesting the human hair and skin and transmitting various diseases (cf. CRAB[1] 3). **b** any other insect of the order Anoplura or Mallophaga, parasitic on mammals and birds. (See note below.) **c** any other invertebrate that attaches itself parasitically to an animal, esp. an aquatic one, or that infests plants.

2 *sl.* (*pl.* **louses**) a contemptible or unpleasant person. ● *v.tr.* remove lice from. □ **louse up** *sl.* make a mess of. [OE *lūs*, pl. *lȳs*]

▪ The true lice are insects belonging to two main groups, sometimes placed together in the order Phthiraptera. The Anoplura (or Siphunculata) comprise the sucking lice, all of which are parasites of mammals. The Mallophaga comprise three suborders of biting or chewing lice, most of which are parasites of birds.

lousewort /ˈlaʊswɜːt/ *n.* a plant of the genus *Pedicularis*, with purple-pink flowers, found in marshes and wet places.

lousy /ˈlaʊzɪ/ *adj.* (**lousier, lousiest**) **1** infested with lice. **2** *colloq.* very bad; disgusting (also as a term of general disparagement). **3** *colloq.* (often foll. by *with*) well supplied, teeming (with). □ **lousily** *adv.* **lousiness** *n.*

lout /laʊt/ *n.* a rough, crude, or ill-mannered person (usu. a man). □ **loutish** *adj.* **loutishly** *adv.* **loutishness** *n.* [perh. f. archaic *lout* to bow]

Louth /laʊθ/ a county of the Republic of Ireland, on the east coast in the province of Leinster; county town, Dundalk.

Louvain see LEUVEN.

Louvre /ˈluːvrə/ the principal museum and art gallery of France, in Paris, housed in the former royal palace built by Francis I and later extended. When the court moved to Versailles in 1678 the conversion of the Louvre into a museum was begun. Francis I had set the pattern for royal collecting and patronage which continued until the Revolution, and the royal collections, greatly increased by Louis XIV, formed the nucleus of the national collection. Among the Louvre's paintings (which represent work up to the time of the Impressionists) is the Mona Lisa, while its other holdings include the Venus de Milo.

louvre /ˈluːvə(r)/ *n.* (also **louver**) **1** each of a set of overlapping slats designed to admit air and some light and exclude rain. **2** a domed structure on a roof with side openings for ventilation etc. □ **louvre-boards** the slats or boards making up a louvre. □ **louvred** *adj.* [ME f. OF *lover, lovier* skylight, prob. f. Gmc]

lovable /ˈlʌvəb(ə)l/ *adj.* (also **loveable**) inspiring or deserving love or affection. □ **lovably** *adv.* **lovableness** *n.* **lovability** /ˌlʌvəˈbɪlɪt/ *n.*

lovage /ˈlʌvɪdʒ/ *n.* **1** a southern European herb, *Levisticum officinale*, used for flavouring etc. **2** a white-flowered umbelliferous plant, *Ligusticum scoticum*. [ME *loveache* alt. f. OF *levesche* f. LL *levisticum* f. L *ligusticum* neut. of *ligusticus* Ligurian]

lovat /ˈlʌvət/ *n.* (also *attrib.*) a muted green colour found esp. in tweed and woollen garments. [*Lovat* in Scotland]

love /lʌv/ *n. & v.* ● *n.* **1** an intense feeling of deep affection or fondness for a person or thing; great liking. **2** sexual passion. **3** sexual relations. **4 a** a beloved one; a sweetheart (often as a form of address). **b** *Brit. colloq.* a familiar form of address regardless of affection. **5** *colloq.* a person of whom one is fond. **6** affectionate greetings (*give him my love*). **7** a formula for ending an affectionate letter etc. **8** (often **Love**) a representation of Cupid. **9** (in some games) no score; nil. ● *v.tr.* **1** (also *absol.*) feel love or deep fondness for. **2** delight in; admire; greatly cherish. **3** *colloq.* like very much (*loves books*). **4** (foll. by verbal noun, or *to* + infin.) be inclined, esp. as a habit; greatly enjoy; find pleasure in (*children love dressing up; loves to find fault*). □ **fall in love** (often foll. by *with*) develop a great (esp. sexual) love (for). **for love** for pleasure not profit. **for the love of** for the sake of. **in love** (often foll. by *with*) deeply enamoured (of). **2** an intense enthusiasm for liking for something. **love affair 1** a romantic or sexual relationship between two people in love. **2** an intense enthusiasm or liking for something. **love-apple** *archaic* a tomato. **love-bird 1** a small African or Madagascan parrot, esp. *Agapornis personata*. **2** (in *pl.*) *colloq.* an affectionate couple; lovers. **love-child** an illegitimate child. **love-feast 1** a meal affirming brotherly love among early Christians. **2** a religious service of Methodists, etc., imitating this. **love game** a game in which the loser makes no score. **love handle** (usu. in *pl.*) esp. *US sl.* excess fat at the waist. **love-hate relationship** an intensely emotional relationship in which one or each party has ambivalent feelings of love and hate for the other. **love-in-a-mist** a blue-flowered garden plant, *Nigella damascena*, with many delicate green bracts. **love-letter** a letter expressing feelings of (esp. romantic) love. **love-lies-bleeding** a garden plant, *Amaranthus caudatus*, with drooping spikes of purple-red blooms. **love-match** a marriage made for love's sake. **love-nest** a secluded retreat for (esp. illicit) lovers. **love-seat** an armchair or small sofa for two. **make love** (often foll. by *to*) **1** have sexual intercourse (with). **2** *archaic* pay amorous attention (to). **no love lost between** mutual dislike between (two people etc.). **not for love or money** *colloq.* not in any circumstances. **out of love** no longer in love. □ **loveworthy** *adj.* [OE *lufu* f. Gmc]

loveable var. of LOVABLE.

lovebite /ˈlʌvbaɪt/ *n.* a red mark on the skin, caused by biting or sucking during sexual play.

Lovelace[1] /ˈlʌvleɪs/, Countess of (title of Augusta Ada King) (1815–52), English mathematician. The daughter of Lord Byron, she was brought up by her mother, who encouraged her studies in mathematics and astronomy. In 1833 Lovelace met the mathematician and computer pioneer Charles Babbage, subsequently becoming his assistant. In 1843 she made a translation of an Italian paper on Babbage's computer or 'difference engine', to which she added significant and detailed notations as to how the machine could be programmed. The high-level computer programming language *Ada* is named after her.

Lovelace[2] /ˈlʌvleɪs/, Richard (1618–57), English poet. A Royalist, he was committed to prison in 1642, where he probably wrote the poem 'To Althea from Prison'. He rejoined Charles I in 1645 and was again imprisoned in 1648; during this time he prepared his collection of poetry *Lucasta*, which includes the lyric 'On going to the wars'.

loveless /ˈlʌvlɪs/ *adj.* without love; unloving or unloved or both. □ **lovelessly** *adv.* **lovelessness** *n.*

Lovell /ˈlʌv(ə)l/, Sir (Alfred Charles) Bernard (b.1913), English astronomer and physicist, and pioneer of radio astronomy. He became professor of radio astronomy at Manchester University in 1951, and founded the university's radio observatory at Jodrell Bank. He directed the construction of the large radio telescope there, now named after him.

lovelock /ˈlʌvlɒk/ *n.* a curl or lock of hair worn on the temple or forehead.

lovelorn /ˈlʌvlɔːn/ *adj.* pining from unrequited love.

lovely /ˈlʌvlɪ/ *adj. & n.* ● *adj.* (**lovelier, loveliest**) **1** exquisitely beautiful. **2** *colloq.* pleasing, delightful. ● *n.* (*pl.* **-ies**) *colloq.* a pretty woman. □ **lovely and** *colloq.* delightfully (*lovely and warm*). □ **lovelily** *adv.* **loveliness** *n.* [OE *luflic*]

lovemaking /ˈlʌvˌmeɪkɪŋ/ *n.* **1** amorous sexual activity, esp. sexual intercourse. **2** *archaic* courtship.

lover /ˈlʌvə(r)/ *n.* **1** a person in love with another. **2** a person with whom another is having sexual relations. **3** (in *pl.*) a couple in love or having sexual relations. **4** a person who likes or enjoys something specified (*a music lover; a lover of words*). □ **loverless** *adj.*

lovesick /ˈlʌvsɪk/ *adj.* languishing with romantic love. □ **lovesickness** *n.*

lovesome /ˈlʌvsəm/ *adj. literary* lovely, lovable.

lovey /ˈlʌvɪ/ *n.* (*pl.* **-eys**) *Brit. colloq.* love, sweetheart (esp. as a form of address).

lovey-dovey /ˌlʌvɪˈdʌvɪ/ *adj. colloq.* fondly affectionate, esp. unduly sentimental.

loving /ˈlʌvɪŋ/ *adj. & n.* ● *adj.* feeling or showing love; affectionate. ● *n.* affection; active love. □ **loving-cup** a two-handled drinking-cup passed round at banquets. **loving-kindness** tenderness and consideration. □ **lovingly** *adv.* **lovingness** *n.* [OE *lufiende* (as LOVE)]

low[1] /ləʊ/ *adj., n., & adv.* ● *adj.* **1** of less than average height; not high or tall or reaching far up (*a low wall*). **2 a** situated close to ground or sea level etc.; not elevated in position (*low altitude*). **b** (of the sun) near the horizon. **c** (of latitude) near the equator. **3** of or in humble rank or position (*of low birth*). **4** of small or less than normal amount or extent or intensity (*low price; low temperature; low in calories*). **5** small or reduced in quantity (*stocks are low*). **6** coming below the normal level (*a dress with a low neck*). **7 a** dejected; lacking vigour (*feeling low; in low spirits*). **b** poorly nourished; indicative of poor nutrition. **8** (of a sound) not shrill or loud or high-pitched. **9** not exalted or sublime; commonplace. **10** unfavourable (*a low opinion*). **11** abject, mean, vulgar (*low cunning; low slang*). ● *n.* **1** a low or the lowest level or number (*the dollar has reached a new low*). **2** an area of low barometric pressure; a depression. ● *adv.* **1** in or to a low position or state. **2** in a low tone (*speak low*). **3** (of a sound) at or to a low pitch. □ **low-born** of humble birth. **low-class** of low quality or social class. **low comedy** that in which the subject and the treatment border on farce. **low-cut** (of a dress etc.) made with a low neckline. **low-density lipoprotein** (abbr. **LDL**) the form of lipoprotein in which cholesterol is transported in the blood. **low-down** *adj.* abject, mean, dishonourable. ● *n. colloq.* (usu. foll. by *on*) the relevant information (about). **lowest common denominator, multiple** see DENOMINATOR, MULTIPLE. **low frequency** (in radio) 30–300 kilohertz. **low gear** see GEAR. **low-grade** of low quality or strength. **low-key** lacking intensity or prominence; restrained. **low-level** *Computing* (of a programming language) close in form to machine language. **low-loader** a lorry with a low floor and no sides, for heavy loads. **low-lying** at low altitude (above sea level etc.). **low mass** see MASS[2]. **low-pitched 1** (of a sound) low. **2** (of a roof) having only a slight slope. **low pressure 1** little demand for activity or exertion. **2** an atmospheric condition with pressure below average. **low profile** avoidance of attention or publicity. **low-profile** *adj.* **1** having a low profile. **2** (of a motor-vehicle tyre) having a greater width than usual in relation to height. **low relief** see RELIEF 6a. **low-rise** (of a building) having few storeys. **low season** the period of fewest visitors at a resort etc. **low-spirited** dejected, dispirited. **low-spiritedness** dejection, depression. **low spirits** dejection, depression. **low tide** (or **water**) the time or level of the tide at its lowest level. **low-water mark 1** the level reached at low water. **2** a minimum recorded level or value etc. □ **lowish** *adj.* **lowness** *n.* [ME *lāh* f. ON *lágr* f. Gmc]

low[2] /ləʊ/ *n. & v.* ● *n.* a sound made by cattle; a moo. ● *v.intr.* utter this sound. [OE *hlōwan* f. Gmc]

lowboy /ˈləʊbɔɪ/ *n. N. Amer.* a low chest or table with drawers and short legs.

lowbrow /ˈləʊbraʊ/ *adj. & n.* ● *adj.* not highly intellectual or cultured. ● *n.* a lowbrow person. □ **lowbrowed** *adj.*

Low Church the section of the Church of England which gives a relatively unimportant place to the episcopate, priesthood, and sacraments, and approximates to Protestant Nonconformism in its beliefs. Originally used of the latitudinarians, the term has been applied, since the time of the Oxford Movement, to evangelicals.

Low Countries a region of NW Europe, comprising the Netherlands, Belgium, and Luxembourg. (See also NETHERLANDS, THE.)

Lowell[1] /ˈləʊəl/, Amy (Lawrence) (1874–1925), American poet. After producing her first volume of relatively conventional poetry, she became influenced by imagism and while visiting England in 1913 and 1914 met Ezra Pound and other imagists. Her subsequent volumes, including *Men, Women and Ghosts* (1916), show her increasing allegiance to the imagist movement and her experiments in 'polyphonic prose'. Her love of New England is expressed in 'Lilacs' and 'Purple Grackles' (in *What's O'Clock*, 1925). She was the sister of the astronomer Percival Lowell.

Lowell[2] /ˈləʊəl/, James Russell (1819–91), American poet and critic. His works include volumes of verse, the satirical *Biglow Papers* (1848 and 1867; prose and verse), memorial odes after the Civil War, and various volumes of essays, including *Among my Books* (1870) and *My Study Window* (1871).

Lowell[3] /ˈləʊəl/, Percival (1855–1916), American astronomer. Lowell founded an observatory in Flagstaff, Arizona, which now bears his name. He inferred the existence of a ninth planet beyond Neptune, and when it was eventually discovered in 1930 it was given the name Pluto, with a symbol that also included Lowell's initials. He claimed to have seen the supposed canals on Mars, and was a devout believer in the existence of intelligent life on the planet. He was the brother of poet Amy Lowell.

Lowell[4] /ˈləʊəl/, Robert (Traill Spence) (1917–77), American poet. In 1940 he married and was converted to Roman Catholicism. His first volume, *Land of Unlikeness* (1944), reflects his conflicts with Catholicism and his Boston ancestry. His personal life was marked by recurring bouts of manic illness, alcoholism, and marital discord; his poetry is notable for its intense confessional nature and for its ambiguous complex imagery, as in the volumes *Life Studies* (1959), *For the Union Dead* (1964), and *The Dolphin* (1973).

lower[1] /ˈləʊə(r)/ *adj. & adv.* ● *adj.* (compar. of LOW[1]). **1** less high in position or status. **2** situated below another part (*lower lip*; *lower atmosphere*). **3 a** situated on less high ground or nearer to the coast (*Lower Egypt*). **b** situated to the south (*Lower California*). **4** *Geol. & Archaeol.* (of a stratigraphic division, deposit, or period) earlier or older, and usu. deeper (*Lower Jurassic*, *lower palaeolithic*). **5** (of an animal or plant) showing more primitive characteristics, e.g. a platypus or a fungus. ● *adv.* in or to a lower position, status, etc. □ **lower case** see CASE[2]. **lower class** the members of the working class. **lower-class** *adj.* of the lower class. **lower deck 1** the deck of a ship situated immediately over the hold. **2** the petty officers and men of a ship collectively. **lower regions** (or **world**) hell; the realm of the dead. □ **lowermost** *adj.*

lower[2] /ˈləʊə(r)/ *v.* **1** *tr.* let or haul down. **2** *tr. & intr.* make or become lower. **3** *tr.* reduce the height or pitch or elevation of (*lower your voice*; *lower one's eyes*). **4** *tr.* degrade. **5** *tr. & intr.* diminish. □ **lower the tone** diminish the cultural content, prestige, or moral character (of a conversation, place, etc.).

lower[3] var. of LOUR.

Lower Austria (called in German *Niederösterreich*) a state of NE Austria; capital, St Pölten.

Lower California an alternative name for Baja CALIFORNIA.

Lower Canada the mainly French-speaking region of Canada around the lower St Lawrence River, in what is now southern Quebec. It was a British colony from 1791 to 1841, when it was united with Upper Canada.

Lower House *n.* the larger and usu. elected body in a legislature, esp. in Britain, the House of Commons.

Lower Hutt /hʌt/ a city in New Zealand, near Wellington; pop. (1986) 63,860. It is the site of the Prime Minister's official residence, Vogel House.

Lower Saxony (called in German *Niedersachsen*) a state of NW Germany; capital, Hanover. It corresponds to the north-western part of the former kingdom of Saxony.

Lowestoft /ˈləʊɪstɒft/ a fishing port and resort town on the North Sea coast of eastern England, in NE Suffolk; pop. (1981) 59,875. It is the most easterly English town.

Low German *n. & adj.* (also called *Plattdeutsch*) ● *n.* the group of dialects of Germany which are not High German. Spoken in the lowland areas of northern Germany, Low German is most closely related to Dutch and Frisian. ● *adj.* of or relating to Low German.

lowland /ˈləʊlənd/ *n. & adj.* ● *n.* **1** (usu. in *pl.*) low-lying country. **2** (**Lowland**) (usu. in *pl.*) the region of Scotland lying south and east of the Highlands. ● *adj.* of or in lowland or (**Lowland**) the Scottish Lowlands. □ **lowlander** *n.* (also **Lowlander**).

Low Latin *n.* medieval and later forms of Latin.

lowlight /ˈləʊlaɪt/ *n.* **1** *colloq.* a monotonous or dull period; a feature of little prominence (*one of the lowlights of the evening*). **2** (usu. in *pl.*) a dark tint in the hair produced by dyeing. [after HIGHLIGHT]

lowly /ˈləʊlɪ/ *adj.* (**lowlier, lowliest**) **1** humble in feeling, behaviour, or status. **2** modest, unpretentious. **3** (of an organism) evolved to only a slight degree. □ **lowlily** *adv.* **lowliness** *n.*

low-minded /ləʊˈmaɪndɪd/ *adj.* vulgar or ignoble in mind or character. □ **low-mindedness** *n.*

Lowry[1] /ˈlaʊrɪ/, (Clarence) Malcolm (1909–57), English novelist. He lived in Mexico in the 1930s and his experiences provided the background for his symbolic semi-autobiographical novel *Under the Volcano* (1947); set in a Mexican town, it uses a complex narrative structure with many shifts of time sequence to trace the decline of an alcoholic British ex-consul.

Lowry[2] /ˈlaʊrɪ/, L(aurence) S(tephen) (1887–1976), English painter. He spent most of his life in Salford, near Manchester, which provided the characteristic industrial setting of his pictures. Deliberately adopting a childlike manner of visualization, he painted small matchstick figures set against the iron and brick expanse of urban and industrial landscapes, providing a wry perspective on life in the industrial North.

Low Sunday *n.* the Sunday after Easter.

Low Week *n.* the week beginning with Low Sunday.

lox[1] /lɒks/ *n.* liquid oxygen. [abbr.]

lox[2] /lɒks/ *n. N. Amer.* smoked salmon. [Yiddish *laks*]

loyal /ˈlɔɪəl/ *adj.* **1** (often foll. by *to*) true or faithful (to duty, love, or obligation). **2** steadfast in allegiance; devoted to the legitimate sovereign or government of one's country. **3** showing loyalty. □ **loyal toast** a toast to the sovereign. □ **loyally** *adv.* [F f. OF *loial* etc. f. L *legalis* LEGAL]

loyalist /ˈlɔɪəlɪst/ *n.* (often *attrib.*) **1** a person who remains loyal to the legitimate sovereign, a government, a cause, etc., esp. in the face of rebellion or usurpation. **2** (also **Loyalist**) a supporter or advocate of union between Great Britain and Northern Ireland or (before partition) the whole of Ireland; a Unionist. **3** (also **Loyalist**) *hist.* any of the colonists of the American revolutionary period who supported the British cause, of whom many went to Canada or Britain after the war. □ **loyalism** *n.*

loyalty /ˈlɔɪəltɪ/ *n.* (pl. **-ies**) **1** the state of being loyal. **2** (often in *pl.*) a feeling or application of loyalty.

Loyalty Islands a group of islands in the SW Pacific, forming part of the French overseas territory of New Caledonia; pop. (1989) 17,910. The group includes the three main islands of Maré, Lifou, and Uvéa in addition to a large number of small islets.

lozenge /ˈlɒzɪndʒ/ *n.* **1** a rhombus or diamond figure. **2** a small sweet or medicinal tablet, orig. lozenge-shaped, for dissolving in the mouth. **3** a lozenge-shaped pane in a window. **4** *Heraldry* a lozenge-shaped device. **5** the lozenge-shaped facet of a cut gem. □ **lozenged** adj. (in sense 4). **lozengy** adj. [ME f. OF losenge, ult. of Gaulish or Iberian orig.]

LP abbr. **1** long-playing (gramophone record). **2** low pressure.

LPG abbr. liquefied petroleum gas.

L-plate /'elpleɪt/ n. Brit. a sign bearing the letter L, attached to the front and rear of a motor vehicle to indicate that it is being driven by a learner.

LPO abbr. London Philharmonic Orchestra.

Lr symb. Chem. the element lawrencium.

LSD abbr. lysergic acid diethylamide, a potent hallucinogenic and psychedelic drug. LSD is synthesized from alkaloids found in ergot, a fungus which attacks cereal crops. It was first prepared in 1943 by the Swiss chemist Albert Hofmann (b.1906), who discovered its hallucinogenic properties by accident. It had some clinical use in psychotherapy in the 1960s, but this was discontinued.

l.s.d. /ˌeles'di:/ n. (also **£.s.d.**) Brit. **1** pounds, shillings, and pence (in former British currency). **2** money, riches. [L librae, solidi, denarii]

LSE abbr. London School of Economics.

LSO abbr. London Symphony Orchestra.

Lt. abbr. **1** Lieutenant. **2** light.

LTA abbr. Lawn Tennis Association.

Ltd. abbr. Limited.

Lu symb. Chem. the element lutetium.

Lualaba /ˌluːə'lɑːbə/ a river of central Africa, which rises near the southern border of Zaire and flows northwards for about 640 km (400 miles), joining the Lomami to form the River Congo.

Luanda /lu:'ændə/ the capital of Angola, a port on the Atlantic coast; pop. (est. 1988) 1,800,000. Founded by the Portuguese in 1575, it was a centre for the shipment of slaves to Brazil in the 17th and 18th centuries.

Luang Prabang /lu:ˌæŋ prə'bæŋ/ (also **Louangphrabang**) a city in NW Laos, on the Mekong river; pop. (est. 1984) 44,240. It was the capital of a kingdom of the same name from 1707 until the reorganization of 1946–7, when Vientiane became the administrative capital. Luang Prabang remained the royal residence and Buddhist religious centre of Laos until the end of the monarchy in 1975.

lubber /'lʌbə(r)/ n. a big clumsy fellow; a lout. □ **lubber line** Naut. a line marked on a compass, showing the ship's forward direction. □ **lubberlike** adj. **lubberly** adj. & adv. [ME, perh. f. OF lobeor swindler, parasite f. lober deceive]

Lubbock /'lʌbək/ a city in NW Texas; pop. (1990) 186,200. It is an agricultural trading centre.

Lübeck /'lu:bek/ a port in northern Germany, on the Baltic coast in Schleswig-Holstein, north-east of Hamburg; pop. (1991) 211,000. Between the 14th and 19th centuries it was an important city within the Hanseatic League.

Lublin /'lʊblɪn/ a manufacturing city in eastern Poland; pop. (1990) 351,350.

lubra /'lu:brə/ n. Austral. sometimes derog. an Aboriginal woman. [F loubra f. Tasmanian]

lubricant /'lu:brɪkənt/ n. & adj. ● n. a substance used to reduce friction. ● adj. lubricating.

lubricate /'lu:brɪˌkeɪt/ v.tr. **1** reduce friction in (machinery etc.) by applying oil or grease etc. **2** make slippery or smooth with oil or grease. □ **lubricator** n. **lubricative** /-kətɪv/ adj. **lubrication** /ˌlu:brɪ'keɪʃ(ə)n/ n. [L lubricare lubricat- f. lubricus slippery]

lubricious /lu:'brɪʃəs/ adj. (also **lubricous** /'lu:brɪkəs/) **1** slippery, smooth, oily. **2** lewd, prurient. **3** evasive. □ **lubricity** /-'brɪsɪtɪ/ n. [L lubricus slippery]

Lubumbashi /ˌlu:bʊm'bæʃɪ/ a city in SE Zaire, near the border with Zambia, capital of the region of Shaba; pop. (1984) 543,270. Founded by Belgian colonists in 1910, it is a copper-mining centre. Until 1966 it was called Elisabethville.

Lubyanka /lu:'bjæŋkə/ (also **Lubianka**) a building in Moscow used as a prison and as the headquarters of the KGB and other Russian secret-police organizations since the Russian Revolution.

Lucan[1] /'lu:kən/ (Latin name Marcus Annaeus Lucanus) (AD 39–65), Roman poet, born in Spain. At first held in esteem by Nero, he was forced to commit suicide after joining a conspiracy against the emperor. His major work, a hexametric epic in ten books known as the Pharsalia, deals with the civil war between Julius Caesar and Pompey;

Lucan's republican and Stoic ideals find expression in his description of Cato the Younger.

Lucan[2] /'lu:kən/ adj. of or relating to St Luke. [eccl.L Lucas f. Gk Loukas Luke]

Lucas /'lu:kəs/, George (b.1944), American film director, producer, and screenwriter. He is chiefly known as the director and writer of the science-fiction adventure film Star Wars (1977). He produced and wrote the screenplays for two further episodes in the saga, namely The Empire Strikes Back (1980) and Return of the Jedi (1983), as well as for Steven Spielberg's Raiders of the Lost Ark (1981).

Lucas van Leyden /ˌlu:kəs væn 'laɪd(ə)n/ (c.1494–1533), Dutch painter and engraver. He produced his most significant work as an engraver and was active in this field from an early age; his Muhammad and the Murdered Monk dates from 1508 and Ecce Homo from 1510. His later work was influenced by that of Dürer, whom he met in 1521. His paintings include portraits, genre scenes, and religious subjects, such as the triptych The Last Judgement (1526–7).

Lucca /'lu:kə/ a city in northern Italy, in Tuscany to the west of Florence; pop. (1990) 86,440. An ancient Ligurian settlement, it was occupied by the Romans in about 180 BC. An important Lombard and then Frankish city, it became a free commune in the 12th century, remaining independent until it fell to the French in 1799. It was ceded to Tuscany in 1847.

luce /lu:s/ n. a pike (fish), esp. when full-grown. [ME f. OF lus, luis f. LL lucius]

lucent /'lu:s(ə)nt/ adj. literary **1** shining, luminous. **2** translucent. □ **lucency** n. [L lucere shine (as LUX)]

Lucerne /lu:'sɜ:n/ (German **Luzern** /lu'tsɛrn/) a resort on the western shore of Lake Lucerne, in central Switzerland; pop. (1990) 59,370.

lucerne /lu:'sɜ:n/ n. (also **lucern**) Brit. = ALFALFA. [F luzerne f. mod. Prov. luzerno glow-worm, with ref. to its shiny seeds]

Lucerne, Lake (also **Lake of Lucerne**; also called the Lake of the Four Cantons, in German Vierwaldstättersee) a lake in central Switzerland, surrounded by the four cantons of Lucerne, Nidwalden, Uri, and Schwyz.

lucid /'lu:sɪd/ adj. **1** expressing or expressed clearly; easy to understand. **2** of or denoting intervals of sanity between periods of insanity or dementia. **3** Bot. with a smooth shining surface. **4** poet. bright. □ **lucidly** adv. **lucidness** n. **lucidity** /lu:'sɪdɪtɪ/ n. [L lucidus (perh. through F lucide or It. lucido) f. lucere shine (as LUX)]

Lucifer /'lu:sɪfə(r)/ n. **1** Satan. (See also DEVIL.) **2** poet. the morning star (the planet Venus). **3** (**lucifer**) archaic a friction match. [OE f. L, = light-bringing, morning-star (as LUX, -fer f. ferre bring)]

luck /lʌk/ n. & v. ● n. **1** chance regarded as the bringer of good or bad fortune. **2** circumstances of life (beneficial or not) brought by this. **3** good fortune; success due to chance (in luck; out of luck). ● v.intr. colloq. **1** (foll. by upon, up on) chance to find or meet with. **2** (foll. by into) esp. N. Amer. acquire by good fortune. **3** (foll. by out) esp. N. Amer. achieve success or advantage by good luck. □ **for luck** to bring good fortune. **good luck 1** good fortune. **2** an omen of this. **hard luck** see HARD. **no such luck** colloq. unfortunately not. **try one's luck** make a venture. **with luck** if all goes well. **worse luck** colloq. unfortunately. [ME f. LG luk f. MLG geluke]

luckily /'lʌkɪlɪ/ adv. **1** (qualifying a whole sentence or clause) fortunately (luckily there was enough food). **2** in a lucky or fortunate manner.

luckless /'lʌklɪs/ adj. having no luck; unfortunate. □ **lucklessly** adv. **lucklessness** n.

Lucknow /'lʌknaʊ/ a city in northern India, capital of the state of Uttar Pradesh; pop. (1991) 1,592,000. In 1775 it became the capital of the province of Oudh. In 1857, during the Indian Mutiny, its British residency was twice besieged by Indian insurgents.

lucky /'lʌkɪ/ adj. (**luckier**, **luckiest**) **1** having or resulting from good luck, esp. as distinct from skill or design or merit. **2** bringing good luck (a lucky mascot). **3** fortunate, appropriate (a lucky guess). □ **lucky dip** Brit. a tub containing different articles concealed in wrapping or bran etc., and chosen at random by participants. □ **luckiness** n.

lucrative /'lu:krətɪv/ adj. profitable, yielding financial gain. □ **lucratively** adv. **lucrativeness** n. [ME f. L lucrativus f. lucrari to gain]

lucre /'lu:kə(r)/ n. derog. financial profit or gain. □ **filthy lucre** see FILTHY. [ME f. F lucre or L lucrum]

Lucretia /lu:'kri:ʃə/ (in Roman legend) a woman who was raped by a son of Tarquinius Superbus and took her own life; this led to the expulsion of the Tarquins from Rome by a rebellion under Brutus.

Lucretius /luːˈkriːʃəs/ (full name Titus Lucretius Carus) (*c.*94–*c.*55 BC), Roman poet and philosopher. His didactic hexametric poem *On the Nature of Things* is an exposition of the atomist physics of Epicurus; it is based on a firmly materialistic view of the universe which is directed to the goal of giving humans peace of mind by showing that fear of the gods and of death is without foundation.

lucubrate /ˈluːkjʊˌbreɪt/ *v.intr. literary* **1** write or study, esp. by night. **2** express one's meditations in writing. □ **lucubrator** *n.* [L *lucubrare lucubrat-* work by lamplight (as LUX)]

lucubration /ˌluːkjʊˈbreɪʃ(ə)n/ *n. literary* **1** nocturnal study or meditation. **2** (usu. in *pl.*) literary writings, esp. of a pedantic or elaborate character. [L *lucubratio* (as LUCUBRATE)]

Lucullan /luːˈkʌlən/ *adj.* profusely luxurious. [Licinius *Lucullus*, Roman general of the 1st c. BC famous for his lavish banquets]

Lucy /ˈluːsɪ/ a nickname for a partial female skeleton of the fossil hominid *Australopithecus afarensis*, found at Hadar in the Afar region of Ethiopia in 1974. It is about 3.2 million years old, and would have been only about 1.2 m (4 ft) in height and 27 kg (60 lb) in weight. It is the most complete skeleton so far obtained of this species, which is regarded by many as the ancestor of all other *Australopithecus* and *Homo* species. Other specimens found at this site and at Laetoli near Olduvai Gorge show considerable variation, and new finds at Hadar in the early 1990s confirmed that they all belong to a single sexually dimorphic species that lived *c.*3.9–3.0 million years ago.

lud /lʌd/ *n. Brit.* □ **m'lud** (or **my lud**) a form of address to a judge in a court of law. [corrupt. of LORD]

Luda /luːˈdɑː/ an industrial conurbation and port in NE China, in the province of Liaoning at the south-eastern tip of the Liaodong Peninsula; pop. (est. 1986) 1,630,000. It comprises the cities of Lushun and Dalian.

Luddite /ˈlʌdaɪt/ *n. & adj.* ● *n.* **1** *hist.* a member of any of the 19th-century bands of English workers who destroyed machinery which they believed was threatening their jobs. (*See note below.*) **2** a person opposed to increased industrialization or new technology. ● *adj.* of the Luddites or their beliefs. □ **Luddism** *n.* **Ludditism** *n.*

▪ Luddites first appeared in 1811 in Nottingham, when knitters began wrecking machines used to make poor-quality stockings at prices which undercut skilled craftsmen. Their name came from a certain Ned Ludd, nicknamed 'King Ludd', said to have destroyed two stocking-frames *c.*1779, although whether or not he existed is uncertain. The movement spread rapidly, large groups storming cotton and woollen mills in Yorkshire and Lancashire, but was quickly put down by the government's harsh reprisals, which included making machine-breaking a capital offence. Although the Luddites were never well organized, they were considered a serious threat by the government of the day, which was haunted by the spectre of a popular uprising.

Ludendorff /ˈluːd(ə)nˌdɔːf/, Erich (1865–1937), German general. Shortly after the outbreak of the First World War he was appointed Chief of Staff to General von Hindenburg and they jointly directed the war effort until the final offensive failed (Sept. 1918). Ludendorff later joined the Nazi Party and served as an MP (1924–8).

Ludhiana /ˌlʊdrˈɑːnə/ a city in NW India, in Punjab south-east of Amritsar; pop. (1991) 1,012,000. Founded in 1480 by the rulers of the Muslim Lodi dynasty, it is now an industrial and agricultural centre and a major railway junction.

ludicrous /ˈluːdɪkrəs/ *adj.* absurd or ridiculous; laughable. □ **ludicrously** *adv.* **ludicrousness** *n.* [L *ludicrus* prob. f. *ludicrum* stage play]

ludo /ˈluːdəʊ/ *n. Brit.* a simple board game in which counters are moved round according to the throw of dice. [L, = I play]

Ludwig /ˈlʊdvɪɡ/ the name of three kings of Bavaria, notably:

Ludwig I (1786–1868), reigned 1825–48. His reactionary policies and lavish expenditure were the cause of radical protests in 1830; his domination by the dancer Lola Montez led to further unrest and he was forced to abdicate in favour of his son.

Ludwig II (1845–86), reigned 1864–86. He came increasingly under Prussian influence and his country eventually joined the new German Empire in 1871. A patron of the arts, in particular of Wagner, he later became a recluse and concentrated on building a series of elaborate castles. He was declared insane and deposed in 1886.

Ludwigshafen /ˈlʊdvɪɡzˌhɑːf(ə)n/ an industrial river port in west

central Germany, south-west of Mannheim, on the River Rhine in the state of Rhineland-Palatinate; pop. (1991) 165,370.

lues /ˈluːiːz/ *n.* (in full **lues venerea** /vɪˈnɪərɪə/) syphilis. □ **luetic** /luːˈetɪk/ *adj.* [L]

luff /lʌf/ *n. & v.* (also **loof** /luːf/) *Naut.* ● *n.* **1** the edge of the fore-and-aft sail next to the mast or stay. **2** *Brit.* the broadest part of the ship's bow where the sides begin to curve in. ● *v.tr.* (also *absol.*) **1** steer (a ship) nearer the wind. **2** turn (the helm) so as to achieve this. **3** obstruct (an opponent in yacht-racing) by sailing closer to the wind. **4** raise or lower (the jib of a crane or derrick). [ME lo(o)f f. OF *lof*, prob. f. LG]

luffa var. of LOOFAH.

Luftwaffe /ˈlʊftˌvæfə/ *n. hist.* the German Air Force before and during the Second World War. [G f. *Luft* air + *Waffe* weapon]

lug¹ /lʌɡ/ *v. & n.* ● *v.* (**lugged, lugging**) **1** *tr.* **a** drag or tug (a heavy object) with effort or violence. **b** (usu. foll. by *round, about*) carry (something heavy) around with one. **2** *tr.* (usu. foll. by *in, into*) introduce (a subject etc.) irrelevantly. **3** *tr.* (usu. foll. by *along, to*) force (a person) to join in an activity. **4** *intr.* (usu. foll. by *at*) pull hard. ● *n.* **1** a hard or rough pull. **2** (in *pl.*) *US sl.* affectation (*put on* lugs). [ME, prob. f. Scand.: cf. Sw. *lugga* pull a person's hair f. *lugg* forelock]

lug² /lʌɡ/ *n.* **1** *Sc.* or *colloq.* an ear. **2** a projection on an object by which it may be carried, fixed in place, etc. **3** esp. *N. Amer. sl.* a lout; a sponger; a stupid person. [prob. of Scand. orig.: cf. LUG¹]

lug³ /lʌɡ/ *n.* = LUGWORM. [17th c.: orig. unkn.]

lug⁴ /lʌɡ/ *n.* = LUGSAIL. [abbr.]

Lugano /luːˈɡɑːnəʊ/ a town in southern Switzerland, on the northern shore of Lake Lugano; pop. (1990) 26,010. It is a centre of international finance and a health and holiday resort.

Lugansk see LUHANSK.

Lugdunum /lʊɡˈduːnəm/ the Roman name for LYONS.

luge /luːʒ/ *n. & v.* ● *n.* a light toboggan for one or two people, ridden in a sitting or supine position. ● *v.intr.* ride on a luge. [Swiss F]

Luger /ˈluːɡə(r)/ *n.* a type of German automatic pistol. [Georg *Luger*, German firearms expert (1849–1923)]

luggage /ˈlʌɡɪdʒ/ *n.* suitcases, bags, etc. to hold a traveller's belongings. □ **luggage-van** *Brit.* a railway carriage for travellers' luggage. [LUG¹ + -AGE]

lugger /ˈlʌɡə(r)/ *n.* a small ship carrying two or three masts with a lugsail on each. [LUGSAIL + -ER¹]

lughole /ˈlʌɡhəʊl, ˈlʌɡəʊl/ *n. sl.* the ear orifice. [LUG² + HOLE]

Lugosi /ləˈɡəʊsɪ/, Bela (born Béla Ferenc Blasko) (1884–1956), Hungarian-born American actor. From 1904 he pursued a successful career as a classical actor in the Hungarian theatre, before emigrating to the US in 1921. Lugosi became famous with his performance in the title role of *Dracula* (1927) on Broadway; in 1931 he recreated the role for Hollywood in the first Dracula film. He subsequently appeared in a succession of horror films, including *Mark of the Vampire* (1935) and *The Wolf Man* (1940).

lugsail /ˈlʌɡseɪl, *Naut.* -s(ə)l/ *n. Naut.* a quadrilateral sail which is bent on and hoisted from a yard. [prob. f. LUG²]

lugubrious /luːˈɡuːbrɪəs/ *adj.* doleful, mournful, dismal. □ **lugubriously** *adv.* **lugubriousness** *n.* [L *lugubris* f. *lugere* mourn]

lugworm /ˈlʌɡwɜːm/ *n.* a polychaete worm of the genus *Arenicola*, living in muddy sand and leaving characteristic worm-casts on lower shores, and often used as bait by fishermen. [LUG³]

Luhansk /luːˈhænsk/ (Russian **Lugansk** /luˈɡansk/) an industrial city in eastern Ukraine, in the Donets Basin; pop. (1990) 501,000. From 1935 until 1991 it was known as Voroshilovgrad, in honour of the Soviet military and political leader Marshal Kliment Voroshilov (1881–1969).

Luik /lœjk/ the Flemish name for LIÈGE.

Lukács /ˈluːkætʃ/, György (1885–1971), Hungarian philosopher, literary critic, and politician. A major figure in Western Marxism, he is best known for his philosophical work *History and Class Consciousness* (1923), in which he stresses the central role of alienation in Marxist thought. His literary criticism is noted for its realist standpoint, notably in *The Theory of the Novel* (1916) and *The Historical Novel* (1955).

Luke, St /luːk/ **1** an evangelist, closely associated with St Paul and traditionally the author of the third Gospel and the Acts of the Apostles. A physician, he was possibly the son of a Greek freedman of Rome. Feast day, 18 Oct. **2** the third Gospel (see GOSPEL 2).

lukewarm /luːkˈwɔːm/ *adj.* **1** moderately warm; tepid.

2 unenthusiastic, indifferent. □ **lukewarmly** *adv.* **lukewarmness** *n.* [ME f. (now dial.) *luke, lew* f. OE]

lull /lʌl/ *v. & n.* ● *v.* **1** *tr.* soothe or send to sleep gently. **2** *tr.* (usu. foll. by *into*) deceive (a person) into confidence (*lulled into a false sense of security*). **3** *tr.* allay (suspicions etc.) usu. by deception. **4** *intr.* (of noise, a storm, etc.) abate or fall quiet. ● *n.* a temporary quiet period in a storm or in any activity. [ME, imit. of sounds used to quieten a child]

lullaby /ˈlʌləˌbaɪ/ *n. & v.* ● *n.* (*pl.* **-ies**) **1** a soothing song to send a child to sleep. **2** the music for this. ● *v.tr.* (**-ies, -ied**) sing to sleep. [as LULL + *-by* as in BYE-BYE²]

Lully /ˈluːlɪ/, Jean-Baptiste (Italian name Giovanni Battista Lulli) (1632–87), French composer, born in Italy. He lived in France from the age of 14 and entered the service of Louis XIV in 1653. From 1664 he collaborated with Molière, writing incidental music for a series of comedies, including *Le Bourgeois gentilhomme* (1670). In 1673 he turned to composing operas; his works, which include *Alceste* (1674) and *Armide* (1686), mark the beginning of the French operatic tradition.

lulu /ˈluːluː/ *n. sl.* a remarkable or excellent person or thing. [19th c., perh. f. *Lulu*, pet-form of *Louise*]

lumbago /lʌmˈbeɪɡəʊ/ *n.* rheumatic pain in the muscles of the lower back. [L f. *lumbus* loin]

lumbar /ˈlʌmbə(r)/ *adj. Anat. & Med.* relating to the loin, esp. the lower back area. □ **lumbar puncture** the withdrawal of spinal fluid from the lower back with a hollow needle, usu. for diagnosis. [med.L *lumbaris* f. L *lumbus* loin]

lumber¹ /ˈlʌmbə(r)/ *v.intr.* (usu. foll. by *along, past, by*, etc.) move in a slow clumsy noisy way. □ **lumbering** *adj.* [ME *lomere*, perh. imit.]

lumber² /ˈlʌmbə(r)/ *n. & v.* ● *n.* **1** *Brit.* disused articles of furniture etc. inconveniently taking up space. **2** useless or cumbersome objects. **3** *N. Amer.* partly or fully prepared timber. ● *v.* **1** *tr. Brit.* **a** (usu. foll. by *with*) leave (a person etc.) with something unwanted or unpleasant (*always lumbering me with the cleaning*). **b** (as **lumbered** *adj.*) in an unwanted or inconvenient situation (*afraid of being lumbered*). **2** *tr.* (usu. foll. by *together*) *Brit.* heap or group together carelessly. **3** *tr.* (usu. foll. by *up*) *Brit.* obstruct. **4** *intr.* cut and prepare forest timber for transport. □ **lumber-jacket** a jacket, usu. of warm checked material, of the kind worn by lumberjacks. **lumber-room** *Brit.* a room where disused or cumbrous things are kept. □ **lumberer** *n.* (in sense 4 of *v.*). **lumbering** *n.* (in sense 4 of *v.*). [perh. f. LUMBER¹: later assoc. with obs. *lumber* pawnbroker's shop]

lumberjack /ˈlʌmbəˌdʒæk/ *n.* (also **lumberman** /-mən/ *pl.* **-men**) esp. *N. Amer.* a person who fells, prepares, or conveys forest timber.

lumbersome /ˈlʌmbəsəm/ *adj.* unwieldy, awkward.

lumbrical muscle /ˈlʌmbrɪk(ə)l/ *n.* any of the muscles flexing the fingers or toes. [mod.L *lumbricalis* f. L *lumbricus* earthworm, with ref. to their shape]

lumen /ˈluːmɪn/ *n.* **1** *Physics* the SI unit of luminous flux (symbol: **lm**), equal to the amount of light emitted per second in a solid angle of one steradian from a uniform source of one candela. **2** (*pl.* **lumina** /-nə/) *Anat.* a cavity within a tube, cell, etc. □ **luminal** *adj.* [L *lumen luminis* a light, an opening]

Lumière /ˈluːmɪˌeə(r)/, Auguste Marie Louis Nicholas (1862–1954) and Louis Jean (1864–1948), French inventors and pioneers of cinema. In 1895 the brothers patented their 'Cinématographe', a cine-camera and projector in one; it had its first public demonstration later the same year. They also invented an improved process of colour photography.

Luminal /ˈluːmɪn(ə)l/ *n. propr.* phenobarbitone. [as LUMEN + -AL²]

luminance /ˈluːmɪnəns/ *n. Physics* the intensity of light emitted from a surface per unit area in a given direction. [L *luminare* illuminate (as LUMEN)]

luminary /ˈluːmɪnərɪ/ *n.* (*pl.* **-ies**) **1** *literary* a natural light-giving body, esp. the sun or moon. **2** a person as a source of intellectual light or moral inspiration. **3** a prominent member of a group or gathering (*a host of show-business luminaries*). [ME f. OF *luminarie* or LL *luminarium* f. L LUMEN]

luminescence /ˌluːmɪˈnes(ə)ns/ *n.* the emission of light by a substance other than as a result of incandescence. □ **luminescent** *adj.* [as LUMEN + -ESCENCE (see -ESCENT)]

luminiferous /ˌluːmɪˈnɪfərəs/ *adj.* producing or transmitting light.

luminous /ˈluːmɪnəs/ *adj.* **1** full of or shedding light; radiant, bright, shining. **2** phosphorescent, visible in darkness (*luminous paint*). **3** (esp. of a writer or a writer's work) throwing light on a subject. **4** of visible radiation (*luminous intensity*). □ **luminously** *adj.* **luminousness** *n.* **luminosity** /ˌluːmɪˈnɒsɪtɪ/ *n.* [ME f. OF *lumineux* or L *luminosus*]

lumme /ˈlʌmɪ/ *int. Brit. sl.* an expression of surprise or interest. [= (*Lord*) *love me*]

lummox /ˈlʌməks/ *n. N. Amer. colloq.* a clumsy or stupid person. [19th c. in US & dial.: orig. unkn.]

lump¹ /lʌmp/ *n. & v.* ● *n.* **1** a compact shapeless or unshapely mass. **2** *sl.* a quantity or heap. **3** a tumour, swelling, or bruise. **4** a heavy, dull, or ungainly person. **5** (prec. by *the*) *Brit.* casual workers in the building and other trades. ● *v.* **1** *tr.* (usu. foll. by *together, with, in with, under*, etc.) mass together or group indiscriminately. **2** *tr.* carry or throw carelessly (*lumping crates round the yard*). **3** *intr.* become lumpy. **4** *intr.* (usu. foll. by *along*) proceed heavily or awkwardly. **5** *intr.* (usu. foll. by *down*) sit down heavily. □ **in the lump** taking things as a whole; in a general manner. **lump in the throat** a feeling of pressure there, caused by emotion. **lump sugar** sugar shaped into lumps or cubes. **lump sum 1** a sum covering a number of items. **2** money paid down at once (opp. INSTALMENT). □ **lumper** *n.* (in sense 2 of *v.*). [ME, perh. of Scand. orig.]

lump² /lʌmp/ *v.tr. colloq.* endure or suffer (a situation) ungraciously. □ **like it or lump it** put up with something whether one likes it or not. [imit.: cf. *dump, grump*, etc.]

lumpectomy /lʌmˈpektəmɪ/ *n.* (*pl.* **-ies**) the surgical removal of a usu. cancerous lump from the breast.

lumpenproletariat /ˈlʌmpənˌprəʊlɪˌteərɪət/ *n.* (esp. in Marxist terminology) the unorganized and unpolitical lower orders of society, not interested in revolutionary advancement. □ **lumpen** *adj.* [G f. *Lumpen* rag, rogue: see PROLETARIAT]

lumpfish /ˈlʌmpfɪʃ/ *n.* a spiny-finned fish, *Cyclopterus lumpus*, of the North Atlantic, with modified pelvic fins for clinging to objects (cf. LUMPSUCKER). [MLG *lumpen*, MDu. *lumpe* (perh. = LUMP¹) + FISH¹]

lumpish /ˈlʌmpɪʃ/ *adj.* **1** heavy and clumsy. **2** stupid, lethargic. □ **lumpishly** *adv.* **lumpishness** *n.*

lumpsucker /ˈlʌmpˌsʌkə(r)/ *n.* a marine fish of the family Cyclopteridae, with a ventral sucker, esp. the lumpfish.

lumpy /ˈlʌmpɪ/ *adj.* (**lumpier, lumpiest**) **1** full of or covered with lumps. **2** (of water) cut up by the wind into small waves. □ **lumpily** *adv.* **lumpiness** *n.*

Luna /ˈluːnə/ a series of Soviet moon probes launched between 1959 and 1976. Luna 2 was the first probe to hit the moon (1959), Luna 3 took the first photographs of the far side (1959), and Luna 9 achieved the first soft landing (1966). [L, = moon]

lunacy /ˈluːnəsɪ/ *n.* (*pl.* **-ies**) **1** insanity (orig. of an intermittent kind supposedly due to changes of the moon); the state of being a lunatic. **2** *Law* such mental unsoundness as interferes with civil rights or transactions. **3** great folly or eccentricity; a foolish act.

luna moth /ˈluːnə/ *n.* a large North American moth, *Actias luna*, with crescent-shaped spots and long tails on its pale green wings. [L *luna*, = moon (from its markings)]

lunar /ˈluːnə(r)/ *adj.* **1** of, relating to, or determined by the moon. **2** concerned with travel to the moon and related research. **3** (of light, glory, etc.) pale, feeble. **4** crescent-shaped, lunate. **5** of or containing silver (from alchemists' use of *luna* (= moon) for 'silver'). □ **lunar caustic** silver nitrate, esp. in stick form. **lunar cycle** = METONIC CYCLE. **lunar distance** the angular distance of the moon from the sun, a planet, or a star, used in finding longitude at sea. **lunar eclipse** an eclipse of the moon (see ECLIPSE). **lunar module** (also **lunar excursion module**) a small craft used for travelling between the moon's surface and a spacecraft in orbit around the moon. **lunar month 1** the period of the moon's revolution, esp. the interval between new moons of about 29¹/₂ days. **2** (in general use) a period of four weeks. **lunar nodes** the points at which the moon's orbit cuts the ecliptic. **lunar observation** the finding of longitude by lunar distance. **lunar orbit 1** the orbit of the moon round the earth. **2** an orbit round the moon. **lunar year** a period of 12 lunar months. [L *lunaris* f. *luna* moon]

lunate /ˈluːneɪt/ *adj. & n.* ● *adj.* crescent-shaped. ● *n.* a crescent-shaped prehistoric implement etc. □ **lunate bone** a crescent-shaped bone in the wrist. [L *lunatus* f. *luna* moon]

lunatic /ˈluːnətɪk/ *n. & adj.* ● *n.* **1** an insane person. **2** someone foolish or eccentric. ● *adj.* mad, foolish. □ **lunatic asylum** *esp. hist.* a mental home or hospital. **lunatic fringe** an extreme or eccentric minority group. [ME f. OF *lunatique* f. LL *lunaticus* f. L *luna* moon]

lunation /luːˈneɪʃ(ə)n/ *n.* the interval between new moons, about 29¹/₂ days. [ME f. med.L *lunatio* (as LUNATIC)]

lunch /lʌntʃ/ n. & v. ● n. **1** the meal eaten in the middle of the day. **2** a light meal eaten at any time. ● v. **1** intr. eat one's lunch. **2** tr. provide lunch for. □ **lunch-box** a container for a packed meal. **lunch-hour** (or **-time**) a break from work, when lunch is eaten. **out to lunch** esp. N. Amer. sl. unaware; incompetent; crazy. □ **luncher** n. [LUNCHEON]

luncheon /'lʌntʃən/ n. formal lunch. □ **luncheon meat** a usu. tinned block of ground meat ready to cut and eat. **luncheon voucher** Brit. a voucher or ticket issued to employees and exchangeable for food at many restaurants and shops. [17th c.: orig. unkn.]

luncheonette /ˌlʌntʃə'net/ n. orig. US a small restaurant or snack bar serving light lunches.

Lund /lʊnd/ a city in SW Sweden, just north-east of Malmö; pop. (1991) 87,680. Its university was founded in 1666.

Lundy /'lʌndɪ/ **1** a granite island in the Bristol Channel, off the coast of north Devon. Acquired by the National Trust in 1969, it is the site of an important seabird colony. **2** a shipping forecast area covering the Bristol Channel and the eastern Celtic Sea.

lune /luːn/ n. Geom. a crescent-shaped figure formed on a sphere or plane by two arcs intersecting at two points. [F f. L luna moon]

lunette /luːˈnet/ n. **1** an arched aperture in a domed ceiling to admit light. **2** a crescent-shaped or semicircular space or alcove which contains a painting, statue, etc. **3** a watch-glass of flattened shape. **4** a ring through which a hook is placed to attach a vehicle to the vehicle towing it. **5** a temporary fortification with two faces forming a salient angle, and two flanks. **6** RC Ch. a holder for the consecrated host in a monstrance. [F, dimin. of lune (see LUNE)]

lung /lʌŋ/ n. each of the pair of respiratory organs, situated within the ribcage, which bring air into contact with the blood in humans and many other vertebrates. □ **lung-power** the power of one's voice. □ **lunged** adj. **lungful** n. (pl. **-fuls**). **lungless** adj. [OE lungen f. Gmc, rel. to LIGHT²]

lunge¹ /lʌndʒ/ n. & v. ● n. **1** a sudden movement forward. **2** a thrust with a sword etc., esp. the basic attacking move in fencing. **3** a movement forward by bending the front leg at the knee while keeping the back leg straight. ● v. **1** intr. make a lunge. **2** intr. (usu. foll. by at, out) deliver a blow from the shoulder in boxing. **3** tr. drive (a weapon etc.) violently in some direction. [earlier allonge f. F allonger lengthen f. à to + long LONG¹]

lunge² /lʌndʒ, ljuːndʒ/ n. & v. (also **longe** /lɒndʒ/) ● n. **1** a long rope on which a horse is held and made to move in a circle round its trainer. **2** a circular exercise-ground for training horses. ● v. tr. (**lungeing**) exercise (a horse) with or in a lunge. [F longe, allonge (as LUNGE¹)]

lungfish /'lʌŋfɪʃ/ n. a freshwater fish of the order Dipnoi, having gills and a modified swim bladder used as lungs, and able to aestivate to survive drought.

lungi /'lʊŋɡiː/ n. (pl. **lungis**) a length of cotton cloth, usu. worn as a loincloth in India, or as a skirt in Burma where it is the national dress for both sexes. [Urdu]

lungworm /'lʌŋwɜːm/ n. a nematode of the family Metastrongylidae, parasitic in the lungs of mammals, esp. farm and domestic animals.

lungwort /'lʌŋwɜːt/ n. **1** a herbaceous plant of the genus Pulmonaria, of the borage family; esp. P. officinalis, with white-spotted leaves likened to a diseased lung. **2** a tree lichen, Lobaria pulmonaria, formerly believed to be a remedy for lung disease.

lunisolar /ˌluːnɪˈsəʊlə(r)/ adj. of or concerning the sun and moon. □ **lunisolar period** a period of 532 years between the repetitions of both solar and lunar cycles. **lunisolar year** a year with divisions regulated by changes of the moon and an average length made to agree with the solar year. [L luna moon + sol sun]

lunula /'luːnjʊlə/ n. (pl. **lunulae** /-ˌliː/) **1** a crescent-shaped mark, esp. the white area at the base of the fingernail. **2** a crescent-shaped Bronze-Age ornament. [L, dimin. of luna moon]

Luo /'luːəʊ/ n. & adj. ● n. (pl. same or **Luos**) **1** a member of an East African people of Kenya and the upper Nile valley. After the Kikuyu they form the largest ethnic group in Kenya. **2** the Nilotic language of the Luo. ● adj. of or relating to the Luo or their language. [Luo]

Luoyang /laʊˈjæŋ/ (formerly called Honan) an industrial city in east central China, in Henan province on the Luo river; pop. (1990) 1,160,000. It was founded in the 12th century BC as the imperial capital of the Zhou dynasty and was the capital of several subsequent dynasties. Between the 4th and 6th centuries AD the construction of cave temples to the south of the city made it an important Buddhist centre.

Lupercalia /ˌluːpəˈkeɪlɪə/ an ancient Roman festival of purification and fertility, held on 15 February at a cave called the Lupercal.

lupin /'luːpɪn/ n. (also **lupine**) **1** a leguminous plant of the genus Lupinus, with long tapering spikes of blue, purple, pink, white, or yellow flowers. **2** (in pl.) seeds of the lupin. [ME f. L lupinus]

lupine /'luːpaɪn/ adj. of or like a wolf or wolves. [L lupinus f. lupus wolf]

lupus /'luːpəs/ n. an ulcerous skin disease, esp. tuberculosis of the skin. □ **lupus vulgaris** /vʌlˈɡɑːrɪs/ tuberculosis with dark red patches on the skin, usu. due to direct inoculation of the tuberculosis bacillus into the skin. □ **lupoid** adj. **lupous** adj. [L, = wolf]

lur /ljʊə(r), lʊə(r)/ n. (also **lure**) a bronze S-shaped trumpet dating from prehistoric times, still used in Scandinavia to call cattle. [Danish & Norw.]

lurch¹ /lɜːtʃ/ n. & v. ● n. a sudden unsteady movement or leaning, a stagger. ● v.intr. stagger, move suddenly and unsteadily. [orig. Naut., lee-lurch alt. of lee-latch drifting to leeward]

lurch² /lɜːtʃ/ n. □ **leave in the lurch** desert (a friend etc.) in difficulties. [orig. = a severe defeat in a game, f. F lourche (also the game itself, like backgammon)]

lurcher /'lɜːtʃə(r)/ n. **1** Brit. a cross-bred dog, usu. a retriever, collie, or sheepdog crossed with a greyhound, originally used esp. for hunting and by poachers. **2** archaic a petty thief, swindler, or spy. [f. obs. lurch (v.) var. of LURK]

lure¹ /ljʊə(r), lʊə(r)/ v. & n. ● v.tr. **1** (usu. foll. by away, into) entice (a person, an animal, etc.) usu. with some form of bait. **2** attract back again or recall (a person, animal, etc.) with the promise of a reward. ● n. **1** a thing used to entice. **2** (usu. foll. by of) the attractive or compelling qualities (of a pursuit etc.). **3** a falconer's apparatus for recalling a hawk, consisting of a bunch of feathers attached to a thong, within which the hawk finds food while being trained. □ **luring** adj. **luringly** adv. [ME f. OE luere f. Gmc]

lure² var. of LUR.

Lurex /'ljʊəreks, 'lʊə-/ n. propr. **1** a type of yarn which incorporates a glittering metallic thread. **2** fabric made from this yarn.

lurid /'ljʊərɪd, 'lʊə-/ adj. **1** vivid or glowing in colour; unnaturally glaring (lurid orange; lurid nocturnal brilliance). **2** sensational, horrifying, or terrible (lurid details). **3** showy, gaudy (paperbacks with lurid covers). **4** ghastly, wan. □ **cast a lurid light on** explain or reveal (facts or character) in a horrific, sensational, or shocking way. □ **luridly** adv. **luridness** n. [L luridus f. luror wan or yellow colour]

lurk /lɜːk/ v. & n. ● v.intr. **1** linger furtively or unobtrusively. **2 a** lie in ambush. **b** (usu. foll. by in, under, about, etc.) hide, esp. for sinister purposes. **3** (as **lurking** adj.) latent, semi-conscious (a lurking suspicion). ● n. Austral. colloq. a dodge, racket, or scheme; a method of profitable business. □ **lurker** n. [ME perh. f. LOUR with frequent. -k as in TALK]

Lusaka /luːˈsɑːkə/ the capital of Zambia; pop. (1989) 1,151,250. Founded in 1905, it was developed as a railway town, becoming capital of Northern Rhodesia in 1935.

luscious /'lʌʃəs/ adj. **1 a** richly sweet in taste or smell. **b** colloq. delicious. **2** (of literary style, music, etc.) over-rich in sound, imagery, or voluptuous suggestion. **3** voluptuously attractive. □ **lusciously** adv. **lusciousness** n. [ME, perh. alt. of obs. licious f. DELICIOUS]

lush¹ /lʌʃ/ adj. **1** (of vegetation, esp. grass) luxuriant and succulent. **2** luxurious. **3** (of colour, sound, etc.) rich, voluptuous **4** sl. good-looking, attractive. □ **lushly** adv. **lushness** n. [ME, perh. var. of obs. lash soft, f. OF lasche lax (see LACHES): assoc. with LUSCIOUS]

lush² /lʌʃ/ n. & v. esp. N. Amer. sl. ● n. **1** alcohol, liquor. **2** an alcoholic, a drunkard. ● v. **1** tr. & intr. drink (alcohol). **2** tr. ply with alcohol. [18th c.: perh. joc. use of LUSH¹]

Lushun /luːˈʃʊn/ a port on the Liaodong Peninsula in NE China, now part of the urban complex of Luda. It was leased by Russia for use as a Pacific naval port from 1898 until 1905, when it was known as Port Arthur. Between 1945 and 1955 it was jointly held by China and the Soviet Union.

Lusitania¹ /ˌluːsɪˈteɪnɪə/ an ancient Roman province in the Iberian peninsula, corresponding to modern Portugal. □ **Lusitanian** adj. & n.

Lusitania² /ˌluːsɪˈteɪnɪə/ a Cunard liner which was sunk by a German submarine in the Atlantic in May 1915 with the loss of over 1,000 lives. The anti-German feeling that this event generated in the US was a factor in bringing that country into the First World War.

lust /lʌst/ n. & v. ● n. **1** strong sexual desire. **2 a** (usu. foll. by for, of) a passionate desire for (a lust for power). **b** (usu. foll. by of) a passionate enjoyment of (the lust of battle). **3** (usu. in pl.) a sensuous appetite regarded as sinful (*the lusts of the flesh*). ● *v.intr.* (usu. foll. by *after, for*) have a strong or excessive (esp. sexual) desire. ☐ **lustful** *adj.* **lustfully** *adv.* **lustfulness** *n.* [OE f. Gmc]

luster *US* var. of LUSTRE[1].

lustra *pl.* of LUSTRUM.

lustral /ˈlʌstrəl/ *adj.* relating to or used in ceremonial purification. [L *lustralis* (as LUSTRUM)]

lustrate /ˈlʌstreɪt/ *v.tr.* purify by expiatory sacrifice, ceremonial washing, or other such rite. ☐ **lustration** /lʌˈstreɪʃ(ə)n/ *n.* [L *lustrare* (as LUSTRUM)]

lustre[1] /ˈlʌstə(r)/ *n. & v.* (*US* **luster**) ● *n.* **1** gloss, brilliance, or sheen. **2** a shining or reflective surface. **3 a** a thin metallic coating giving an iridescent glaze to ceramics. **b** = LUSTREWARE. **4** a radiance or attractiveness; splendour, glory, distinction (of achievements etc.) (*add lustre to*; *shed lustre on*). **5 a** a prismatic glass pendant on a chandelier etc. **b** a cut-glass chandelier or candelabra. **6 a** *Brit.* a thin dress-material with a cotton warp, woollen weft, and a glossy surface. **b** any fabric with a sheen or gloss. ● *v.tr.* put lustre on (pottery, a cloth, etc.). ☐ **lustreless** *adj.* (*US* **lusterless**). **lustrous** *adj.* **lustrously** *adv.* **lustrousness** *n.* [F f. It. *lustro* f. *lustrare* f. L *lustrare* illuminate]

lustre[2] /ˈlʌstə(r)/ *n.* (*US* **luster**) = LUSTRUM. [ME, anglicized f. LUSTRUM]

lustreware /ˈlʌstəˌweə(r)/ *n.* (*US* **lusterware**) ceramics with an iridescent glaze. [LUSTRE[1]]

lustrum /ˈlʌstrəm/ *n.* (*pl.* **lustra** /-rə/ or **lustrums**) a period of five years. [L, an orig. purificatory sacrifice after a quinquennial census]

lusty /ˈlʌsti/ *adj.* (**lustier, lustiest**) **1** healthy and strong. **2** vigorous or lively. ☐ **lustily** *adv.* **lustiness** *n.* [ME f. LUST + -Y[1]]

lusus /ˈluːsəs/ *n.* (*pl.* same or ***lususes***) (in full ***lusus naturae*** /nəˈtjʊəriː, -ˈtʊəraɪ/) a freak of nature. [L]

lutanist var. of LUTENIST.

lute[1] /luːt, ljuːt/ *n.* a plucked stringed musical instrument, fretted and with a round body resembling a halved pear. In Europe it was one of the most important solo instruments from the 16th to the 18th century, and with the 20th-century revival of interest in music of this period it has regained something of its former popularity. The lute was brought to Spain by the Moors at least as early as the 11th century, at that time having no frets and being plucked by a plectrum rather than the fingers. The word is also used for the many similar plucked instruments of Asia and eastern Europe, some dating back to the 2nd millennium BC. [ME f. F *lut, leüt,* prob. f. Prov. *laüt* f. Arab. *al-'ūd*]

lute[2] /luːt, ljuːt/ *n. & v.* ● *n.* **1** clay or cement used to stop a hole, make a joint airtight, coat a crucible, protect a graft, etc. **2** a rubber seal for a jar etc. ● *v.tr.* apply lute to. [ME f. OF *lut* f. L *lutum* mud, clay]

lutein /ˈluːtɪɪn/ *n. Chem.* a pigment of a deep yellow colour found in leaves, egg-yolk, and the corpus luteum. [L *luteum* yolk of egg, neut. of *luteus* yellow]

luteinizing hormone /ˈluːtəˌnaɪzɪŋ/ *n.* (abbr. **LH**) *Biochem.* a hormone secreted by the anterior pituitary gland that in females stimulates ovulation and in males stimulates the synthesis of androgen. [LUTEIN]

lutenist /ˈluːtənɪst, ˈljuː-/ *n.* (also **lutanist**) a lute-player. [med.L *lutanista* f. *lutana* LUTE[1]]

luteo- /ˈluːtɪəʊ/ *comb. form* orange-coloured. [as LUTEOUS + -O-]

luteofulvous /ˌluːtɪəʊˈfʌlvəs/ *adj.* orange-tawny.

luteous /ˈluːtɪəs/ *adj.* of a deep orange yellow or greenish-yellow. [L *luteus* f. *lutum* WELD[2]]

lutestring /ˈluːtstrɪŋ, ˈljuːt-/ *n. archaic* a glossy silk fabric. [app. f. *lustring* f. F *lustrine* or It. *lustrino* f. *lustro* LUSTRE[1]]

Lutetia /luːˈtiːʃə/ the Roman name for PARIS[1].

lutetium /luːˈtiːʃɪəm/ *n.* a silvery-white metallic chemical element (atomic number 71; symbol **Lu**). A member of the lanthanide series, lutetium was discovered by the French chemist Georges Urbain (1872–1938) in 1907. It has few commercial uses. [LUTETIA]

Luther /ˈluːθə(r)/, Martin (1483–1546), German Protestant theologian, the principal figure of the German Reformation. From 1508 he taught at the University of Wittenberg, latterly as professor of scripture (1512–46). He began to preach the doctrine of justification by faith rather than by works; his attack on the sale of indulgences with his ninety-five theses (1517) was followed by further attacks on papal authority; in 1521 Luther was condemned and excommunicated at the Diet of Worms. At a meeting with Swiss theologians at Marburg (1529), he opposed Zwingli and gave a defence of the doctrine of consubstantiation; the next year he gave his approval to Melanchthon's Augsburg Confession, which laid down the Lutheran

position. His translation of the Bible into High German (1522–34) contributed significantly to the spread of this form of the language and to the development of German literature in the vernacular.

Lutheran /ˈluːθərən/ *n. & adj.* ● *n.* **1** a follower of Martin Luther. **2** a member of the Lutheran Church. ● *adj.* of or characterized by the theology of Martin Luther. ☐ **Lutheranism** *n.* **Lutheranize** *v.tr. & intr.* (also **-ise**).

Lutheran Church the Protestant Church accepting the Augsburg Confession of 1530, with justification by faith alone as a cardinal doctrine. The Lutheran Church is the largest Protestant body, with substantial membership in Germany, Scandinavia, and the US.

Luthuli /luːˈtuːlɪ/, Albert John (also **Lutuli**) (*c.* 1898–1967), South African political leader. He inherited a Zulu chieftaincy in 1935, and was president of the African National Congress from 1952 to 1960. Luthuli's presidency was marked by the Defiance Campaign, his programme of civil disobedience. He was awarded the Nobel Peace Prize in 1960 for his commitment to non-violence as a means of opposing apartheid.

Lutine Bell /ˈluːtiːn/ the bell of HMS *Lutine*, which sank in 1799. The ship was carrying a large amount of coin and bullion, and the loss fell on the underwriters, who were members of Lloyd's of London. When the bell was recovered during salvage operations it was taken to Lloyd's, where it now hangs. It is rung (and business is halted) whenever there is an important announcement to be made to the underwriters. It was formerly rung once if a ship had sunk and twice for good news. [F, fem. of *lutin* spirit, imp]

luting /ˈluːtɪŋ, ˈljuː-/ *n.* = LUTE[2] *n.*

Luton /ˈluːt(ə)n/ an industrial town in Bedfordshire, north-west of London; pop. (1991) 167,300.

Lutosławski /ˌluːtəˈswɑːfskɪ/, Witold (1913–94), Polish composer. He is noted for his orchestral music, including three symphonies (1947; 1967; 1983), *Mi-parti* (1976), and *Chain 3* (1986). From the early 1960s, his works have been characterized by a blend of notational composition and aleatoric sections.

Lutuli see LUTHULI.

Lutyens[1] /ˈlʌtjənz/, (Agnes) Elizabeth (1906–83), English composer. She was one of the first English composers to use the twelve-note system; the Chamber Concerto No. 1 (1939) is an example of her interpretation of this technique. Her works include operas, orchestral and choral works, and chamber music; she also wrote many scores for films and radio, as well as incidental music for plays. She was the daughter of Sir Edwin Lutyens.

Lutyens[2] /ˈlʌtjənz/, Sir Edwin (Landseer) (1869–1944), English architect. He established his reputation designing country houses, moving from a romantic red-brick style to Palladian-influenced formal designs. Lutyens is best known for his plans for New Delhi (1912), where he introduced an open garden-city layout; his Viceroy's House (1915–30) combined classical features with decoration in the Indian idiom. He is also remembered for the Cenotaph in London (1919–21) and his unfulfilled design for the Roman Catholic cathedral in Liverpool (1929).

lutz /lʊts/ *n.* a jump in skating from the backward outside edge of one skate to the backward outside edge of the other, with a full turn in the air. [prob. f. Gustave *Lussi*, Swiss figure-skater (1898–1993), who invented it]

luvvy /ˈlʌvi/ *n.* (also **luvvie**) (*pl.* **-ies**) *Brit. colloq.* **1** (as a form of address) = LOVEY. **2** an actor or actress, esp. one who is particularly effusive or affected.

lux /lʌks/ *n.* (*pl.* same) *Physics* the SI unit of illuminance (symbol: **lx**), equivalent to one lumen per square metre. [L *lux lucis* light]

luxe /lʊks, lʌks/ *n.* luxury (cf. DE LUXE). [F f. L *luxus*]

Luxembourg /ˈlʌksəmˌbɜːɡ/ **1** a country in western Europe, situated between Belgium, Germany, and France; pop. (est. 1991) 378,000; official languages, Luxemburgish, French, and German; capital, Luxembourg. Luxembourg became a Habsburg possession in the 15th century, and was successively Spanish and then Austrian until annexed by France in 1795. As a result of the Treaty of Vienna in 1815 it became an independent grand duchy. It was occupied by Germany during both world wars. Luxembourg formed a customs union with Belgium in 1922, extended in 1948 into the Benelux Customs Union with the Netherlands. It was a founder member of the EEC in 1957. **2** the capital of the Grand Duchy of Luxembourg; pop. (1991) 75,620. It is the seat of the European Court of Justice and the secretariat of the Parliament of the EEC. **3** a province of SE Belgium, until 1839 a province of the Grand Duchy of Luxembourg; capital, Arlon. ☐ **Luxembourger** *n.*

Luxemburg /ˈlʌksəmˌbɜːg/, Rosa (1871–1919), Polish-born German revolutionary leader. She co-founded what became the Polish Communist Party (1893), before obtaining German citizenship in 1898. Imprisoned in 1915 for opposing the First World War, she founded the revolutionary and pacifist group known as the Spartacus League in 1916 with the German socialist Karl Liebknecht (1871–1919). After her release from prison in 1918, she co-founded the German Communist Party; the following year she and Liebknecht were assassinated after organizing an abortive Communist uprising in Berlin. (See also SPARTACIST.)

Luxemburgish /ˈlʌksəmˌbɜːgɪʃ/ n. & adj. (also called *Letzeburgesch*) ● n. a form of German spoken in Luxembourg. ● adj. of or relating to Luxemburgish.

Luxor /ˈlʌksɔː(r)/ (Arabic **El Uqsur** /el ˈʊksʊə(r)/, **Al Uqsur**) a city in eastern Egypt, on the east bank of the Nile; pop. (1991) 142,000. It is the site of the southern part of ancient Thebes and contains the ruins of the temple built by Amenhotep III between 1411 and 1375 BC and of monuments erected by Ramses II in the 13th century BC. [Arab. *al-uqsur* the castles]

luxuriant /lʌgˈʒʊərɪənt, -ˈzjʊərɪənt/ adj. **1** (of vegetation etc.) lush, profuse in growth. **2** prolific, exuberant, rank (*luxuriant imagination*). **3** (of literary or artistic style) florid, richly ornate. □ **luxuriance** n. **luxuriantly** adv. [L *luxuriare* grow rank f. *luxuria* LUXURY]

luxuriate /lʌgˈʒʊərɪˌeɪt, -ˈzjʊərɪˌeɪt/ v.intr. **1** (foll. by *in*) take self-indulgent delight in, enjoy in a luxurious manner. **2** take one's ease, relax in comfort.

luxurious /lʌgˈʒʊərɪəs, -ˈzjʊərɪəs/ adj. **1** characterized by luxury, sumptuous. **2** extremely comfortable. **3** fond of luxury, self-indulgent, voluptuous. □ **luxuriously** adv. **luxuriousness** n. [ME f. OF *luxurios* f. L *luxuriosus* (as LUXURY)]

luxury /ˈlʌkʃərɪ/ n. (pl. **-ies**) **1** choice or costly surroundings, possessions, food, etc.; luxuriousness (*a life of luxury*). **2** something desirable for comfort or enjoyment, but not indispensable. **3** (*attrib.*) providing great comfort, expensive (*a luxury flat; a luxury holiday*). [ME f. OF *luxurie*, *luxure* f. L *luxuria* f. *luxus* abundance]

Luzern see LUCERNE.

Luzon /luːˈzɒn/ the most northerly and the largest island in the Philippines. Its chief towns are Quezon City and Manila, the country's capital.

LV abbr. (in the UK) luncheon voucher.

Lviv /lvɪv, Ukrainian ljviw/ (Russian **Lvov** /ljvɒf/, Polish **Lwów** /lvuf/; called in German *Lemberg*) an industrial city in western Ukraine, near the border with Poland; pop. (1990) 798,000. Founded in the 13th century, it was an important trading town on a route linking the Black Sea with the Baltic. It belonged to Poland until 1772 and to Austria–Hungary until 1918.

Lw symb. Chem. a former symbol for the element lawrencium (cf. LR).

lwei /ləˈweɪ/ n. a monetary unit of Angola, equal to one-hundredth of a kwanza. [Angolan name]

LWM abbr. low-water mark.

Lwów see LVIV.

lx abbr. lux.

LXX symb. Septuagint. [Roman numeral for 70]

-ly[1] /lɪ/ suffix forming adjectives esp. from nouns, meaning: **1** having the qualities of (*princely; manly*). **2** recurring at intervals of (*daily; hourly*). [from or after OE *-lic* f. Gmc, rel. to LIKE[1]]

-ly[2] /lɪ/ suffix forming adverbs from adjectives, denoting esp. manner or degree (*boldly; happily; miserably; deservedly; amusingly*). [from or after OE *-līce* f. Gmc (as -LY[1])]

Lyallpur /ˌlaɪəlˈpʊə(r)/ the former name (until 1979) for FAISALABAD.

lycanthrope /ˈlaɪkənˌθrəʊp/ n. **1** a werewolf. **2** a person with lycanthropy. [mod.L *lycanthropus* f. Gk (as LYCANTHROPY)]

lycanthropy /laɪˈkænθrəpɪ/ n. **1** the mythical transformation of a person into a wolf (see also WEREWOLF). **2** a form of madness involving the delusion of being a wolf or other animal, with changed appetites, voice, etc. [mod.L *lycanthropia* f. Gk *lukanthrōpia* f. *lukos* wolf + *anthrōpos* man]

lycée /ˈliːseɪ/ n. (pl. **lycées** pronunc. same) a state secondary school in France. [F f. L (as LYCEUM)]

Lyceum /laɪˈsiːəm/ n. **1 a** the garden in Athens in which Aristotle taught philosophy. **b** Aristotelian philosophy and its followers. **2** (**lyceum**) US hist. a literary institution, lecture-hall, or teaching-place. **3** a theatre in London, built in 1834 and managed by Sir Henry

Irving 1878–1902. [L f. Gk *Lukeion* neut. of *Lukeios* epithet of Apollo (from whose neighbouring temple the Lyceum was named)]

lychee /ˈlaɪtʃɪ, laɪˈtʃiː/ n. (also **litchi, lichee**) **1** a sweet fleshy fruit with a thin spiny skin. **2** the tree, *Nephelium litchi*, orig. from China, bearing this. [Chin. *lizhi*]

lych-gate var. of LICH-GATE.

lychnis /ˈlɪknɪs/ n. a herbaceous plant of the genus *Lychnis*, of the pink family, e.g. ragged robin. [L f. Gk *lukhnis* a red flower f. *lukhnos* lamp]

Lycia /ˈlɪsɪə/ an ancient region on the coast of SW Asia Minor, between Caria and Pamphylia. Previously under Persian and Syrian rule, it was annexed by Rome as part of Pamphylia in AD 43, becoming a separate province in the 4th century. □ **Lycian** adj. & n.

lycopod /ˈlaɪkəˌpɒd/ n. a clubmoss, esp. of the genus *Lycopodium*. [anglicized form of LYCOPODIUM]

lycopodium /ˌlaɪkəˈpəʊdɪəm/ n. **1** = LYCOPOD. **2** a fine powder of spores from this, used as an absorbent in surgery, and in making fireworks etc. [mod.L f. Gk *lukos* wolf + *pous podos* foot]

Lycra /ˈlaɪkrə/ n. propr. an elastic polyurethane fibre or fabric used esp. for close-fitting sports clothing. [orig. unkn.]

Lycurgus /laɪˈkɜːgəs/ (9th century BC), Spartan lawgiver. He is traditionally held to have been the founder of the constitution and military regime of ancient Sparta.

Lydgate /ˈlɪdgeɪt/, John (c.1370–c.1450), English poet and monk. He is noted for his copious output of verse, often in Chaucerian style, and for translations. Of the latter, the best known are the *Troy Book* (1412–20), written at the request of Prince Henry (later Henry V), and *The Fall of Princes* (1431–8), based on a French version of a book on tragedy by Boccaccio.

Lydia /ˈlɪdɪə/ an ancient region of western Asia Minor, south of Mysia and north of Caria. It became a powerful kingdom in the 7th century BC but in 546 its final king, Croesus, was defeated by Cyrus and it was absorbed into the Persian empire. Lydia was probably the first realm to use coined money.

Lydian /ˈlɪdɪən/ adj. & n. ● n. **1** a native or inhabitant of ancient Lydia. **2** the language of the Lydians. ● adj. of or relating to the people of Lydia or their language. [L *Lydius* f. Gk *Ludios* of Lydia]

Lydian mode n. Mus. the mode represented by the natural diatonic scale F–F.

lye /laɪ/ n. **1** water that has been made alkaline by lixiviation of vegetable ashes. **2** any strong alkaline solution, esp. of potassium hydroxide used for washing or cleansing. [OE *lēag* f. Gmc: cf. LATHER]

Lyell /ˈlaɪəl/, Sir Charles (1797–1875), Scottish geologist. His textbook *Principles of Geology* (1830–3) influenced a generation of geologists. He held that the earth's features were shaped over a long period of time by natural processes, and not during short periodic upheavals as proposed by the catastrophist school of thought. In this he revived the theories of James Hutton, but his influence on geological opinion was much greater. Lyell's views cleared the way for Darwin's theory of evolution, which he accepted after some hesitation.

lying[1] /ˈlaɪɪŋ/ pres. part. of LIE[1]. ● n. a place to lie (*a dry lying*).

lying[2] /ˈlaɪɪŋ/ pres. part. of LIE[2]. ● adj. deceitful, false. □ **lyingly** adv.

lyke-wake /ˈlaɪkweɪk/ n. Brit. a night-watch over a dead body. [perh. f. ON: cf. LICH-GATE, WAKE[1]]

Lyly /ˈlɪlɪ/, John (c.1554–1606), English prose writer and dramatist. He is remembered for his prose romance in two parts: *Euphues, The Anatomy of Wit* (1578) and *Euphues and his England* (1580). Both were written in an elaborate style that became known as *euphuism*.

Lyme disease /laɪm/ n. Med. a form of arthritis caused by spirochaete bacteria transmitted by ticks. [*Lyme*, a town in Connecticut, US, where an outbreak occurred]

lymph /lɪmf/ n. **1** Physiol. a colourless fluid containing white blood cells, drained from the tissues and conveyed through the body in the lymphatic system. **2** this fluid used as a vaccine. **3** exudation from a sore etc. **4** poet. pure water. □ **lymph gland** (or **node**) a small mass of tissue in the lymphatic system where lymph is purified and lymphocytes are formed. □ **lymphoid** adj. **lymphous** adj. [F *lymphe* or L *lympha*, *limpa* water]

lymphatic /lɪmˈfætɪk/ adj. & n. ● adj. **1** of or secreting or conveying lymph. **2** (of a person) pale, flabby, or sluggish. ● n. a veinlike vessel conveying lymph. □ **lymphatic system** a network of vessels conveying lymph. [orig. = frenzied, f. L *lymphaticus* mad f. Gk *numpholēptos* seized by nymphs: now assoc. with LYMPH (on the analogy of *spermatic* etc.)]

lymphocyte /ˈlɪmfəˌsaɪt/ n. a form of leucocyte occurring in the blood, in lymph, etc. □ **lymphocytic** /ˌlɪmfəˈsɪtɪk/ adj.

lymphoma /lɪmˈfəʊmə/ n. (pl. **lymphomas** or **lymphomata** /-mətə/) Med. any malignant tumour of the lymph nodes, excluding leukaemia.

lyncean /lɪnˈsiːən/ adj. lynx-eyed, keen-sighted. [L lynceus f. Gk lugkeios f. lugx LYNX]

lynch /lɪntʃ/ v.tr. (of a group of people) put (a person) to death, esp. by hanging, for an alleged offence without a legal trial. The word is derived from the name of Captain William Lynch of Virginia, US, who led an irregular court set up to punish Loyalists during the War of American Independence (c.1780). □ **lynch law** the procedure of a self-constituted illegal court that punishes or executes. **lynch mob** a band of people intent on lynching someone. □ **lyncher** n. **lynching** n.

lynchet /ˈlɪntʃɪt/ n. (in the UK) a ridge or ledge formed by ancient ploughing on a slope. [linch f. OE hlinc: cf. LINKS]

lynchpin var. of LINCHPIN.

Lynn /lɪn/, Dame Vera (born Vera Margaret Lewis) (b.1917), English singer. During the Second World War she sang to the troops and became known as the 'Forces' Sweetheart'. She is mainly remembered for her rendering of such songs as 'We'll Meet Again' and 'White Cliffs of Dover'. She had a number of postwar successes, including 'Auf Wiederseh'n, Sweetheart' (1952).

lynx /lɪŋks/ n. **1** a medium-sized feline, Felis lynx, with short tail, spotted fur, and tufted ear-tips. **2** its fur. □ **lynx-eyed** keen-sighted. [ME f. L f. Gk lugx]

Lyon /ˈlaɪən/ n. (in full **Lord Lyon** or **Lyon King of Arms**) the chief herald of Scotland. □ **Lyon Court** the court over which the chief herald of Scotland presides. [archaic form of LION: named f. the lion on the royal shield]

Lyons /ˈliːɒn/ (French **Lyon** /ljɔ̃/) an industrial city and river port in SE France, situated at the confluence of the Rhône and Saône rivers; pop. (1990) 422,440. Founded by the Romans in AD 43 as Lugdunum, it was an important city of Roman Gaul. Today it is the principal town of the Rhône-Alpes region.

lyophilic /ˌlaɪəˈfɪlɪk/ adj. Chem. (of a colloid) readily dispersed by a solvent. [Gk luō loosen, dissolve + Gk philos loving]

lyophilize /laɪˈɒfɪˌlaɪz/ v.tr. (also **-ise**) freeze-dry.

lyophobic /ˌlaɪəˈfəʊbɪk/ adj. Chem. (of a colloid) not lyophilic. [Gk luō loosen, dissolve + -PHOBIC (see -PHOBIA)]

Lyra /ˈlaɪrə/ Astron. a small northern constellation (the Lyre), said to represent the lyre invented by Hermes. It contains the bright star Vega and the Ring Nebula, a bright planetary nebula. [L]

lyrate /ˈlaɪrət/ adj. Biol. lyre-shaped.

lyre /ˈlaɪə(r)/ n. a plucked stringed musical instrument in which strings are fixed to a crossbar supported by two arms. It was one of the most important instruments of ancient Greece, but is widespread now only in eastern Africa. □ **lyre-bird** an Australian bird of the family Menuridae, the male of which has a lyre-shaped tail. **lyre-flower** a bleeding heart. [ME f. OF lire f. L lyra f. Gk lura]

lyric /ˈlɪrɪk/ adj. & n. ● adj. **1** (of poetry) expressing the writer's emotions, usu. briefly and in stanzas or recognized forms. **2** (of a poet) writing in this manner. **3** of or for the lyre. **4** meant to be sung, fit to be expressed in song, songlike (lyric drama; lyric opera). ● n. **1** a lyric poem or verse. **2** (usu. in pl.) the words of a song. **3** (in pl.) lyric verses. [F lyrique or L lyricus f. Gk lurikos (as LYRE)]

lyrical /ˈlɪrɪk(ə)l/ adj. **1** = LYRIC adj. 1, 2. **2** resembling, couched in, or using language appropriate to lyric poetry. **3** colloq. highly enthusiastic (wax lyrical about). □ **lyrically** adv.

lyricism /ˈlɪrɪˌsɪz(ə)m/ n. **1** the character or quality of being lyric or lyrical. **2** a lyrical expression. **3** high-flown sentiments.

lyricist /ˈlɪrɪsɪst/ n. a person who writes the words to a song.

lyrist n. **1** /ˈlaɪərɪst/ a person who plays the lyre. **2** /ˈlɪrɪst/ a lyric poet. [L lyrista f. Gk luristēs f. lura lyre]

Lysander /laɪˈsændə(r)/ (d.395 BC) Spartan general. He commanded the Spartan fleet that defeated the Athenian navy in 405. Lysander captured Athens in 404, so bringing the Peloponnesian War to an end.

lyse /laɪz/ v.tr. & intr. Biol. bring about or undergo lysis. [back-form. f. LYSIS]

Lysenko /lɪˈsɛŋkəʊ/, Trofim Denisovich (1898–1976), Soviet biologist and geneticist. He was an adherent of Lamarck's theory of evolution by the inheritance of acquired characteristics. Since his ideas harmonized with Marxist ideology he was favoured by Stalin and dominated Soviet genetics for many years. Among other false claims, he stated that the process of vernalization — growing a plant, especially a food crop, in a cold climate — will adapt the plant genetically to resist low temperatures.

lysergic acid /laɪˈsɜːdʒɪk/ n. Chem. a crystalline polycyclic acid extracted from ergot or prepared synthetically. □ **lysergic acid diethylamide** see LSD. [hydrolysis + ergot + -IC]

lysin /ˈlaɪsɪn/ n. Biol. a substance, esp. an antibody, able to cause lysis of cells or bacteria. [G Lysine]

lysine /ˈlaɪsiːn/ n. Biochem. an amino acid present in protein and essential in the diet of vertebrates. [G Lysin, ult. f. LYSIS]

Lysippus /laɪˈsɪpəs/ (4th century BC), Greek sculptor. His name is associated with a series of bronze athletes, notably the Apoxyomenos (c.320–315), which represents a young male athlete scraping and cleaning his oil-covered skin. With such works Lysippus is said to have introduced a naturalistic scheme of proportions for the human body into Greek sculpture.

lysis /ˈlaɪsɪs/ n. (pl. **lyses** /-siːz/) Biol. the disintegration of a cell. [L f. Gk lusis loosening f. luō loosen]

-lysis /lɪsɪs/ comb. form forming nouns denoting disintegration or decomposition (electrolysis; haemolysis).

Lysol /ˈlaɪsɒl/ n. propr. a mixture of cresols and soft soap, used as a disinfectant. [LYSIS + -OL]

lysosome /ˈlaɪsəˌsəʊm/ n. Biol. a cytoplasmic organelle in eukaryotic cells containing degradative enzymes enclosed in a membrane. [LYSIS + -SOME³]

lysozyme /ˈlaɪsəˌzaɪm/ n. Biochem. an enzyme found in tears and egg-white which catalyses the destruction of cell walls of certain Gram-positive bacteria. [LYSIS + ENZYME]

lytic /ˈlɪtɪk/ adj. of, relating to, or causing lysis.

-lytic /ˈlɪtɪk/ comb. form forming adjectives corresponding to nouns in -lysis. [Gk lutikos (as LYSIS)]

Lytton /ˈlɪt(ə)n/, 1st Baron (born Edward George Earle Bulwer-Lytton) (1803–73), British novelist, dramatist, and statesman. His prolific literary output includes Pelham (1828), a novel of fashionable society with which he had his first success, many historical romances (such as The Last Days of Pompeii, 1834), and plays. He entered Parliament as an MP in 1831 and later served as Colonial Secretary in Lord Derby's government (1858–59).

Mm

M[1] /em/ n. (also **m**) (pl. **Ms** or **M's**) **1** the thirteenth letter of the alphabet. **2** (as a Roman numeral) 1,000. **3** *Printing* em.

M[2] abbr. (also **M.**) **1** *Monsieur.* **2** (in the UK in road designations) motorway. **3** *Chem.* molar.

M[3] symb. mega-.

m[1] abbr. (also **m.**) **1 a** masculine. **b** male. **2** married. **3** *Cricket* maiden (over). **4** mile(s). **5** million(s). **6** minute(s). **7** *Currency* mark(s). **8** mare.

m[2] symb. **1** metre(s). **2** milli-. **3** *Physics* mass.

m[3] /əm/ adj. = MY (*m'lud*).

'm /əm/ abbr. **1** (after *1st pers. pron.*) am (*I'm*). **2** colloq. madam (in *yes'm* etc.).

MA abbr. **1** Master of Arts. **2** *US* Massachusetts (in official postal use).

ma /mɑː/ n. colloq. mother. [abbr. of MAMMA[1]]

ma'am /mæm, mɑːm, məm/ n. madam (used esp. in addressing royalty). [contr.]

Maas see MEUSE.

Maastricht /ˈmɑːstrɪxt/ an industrial city in the Netherlands, capital of the province of Limburg, situated on the River Maas near the Belgian and German borders; pop. (1991) 117,420.

Maastricht Treaty the Treaty on European Union, which was agreed by the heads of government of the twelve member states of the EC at a summit meeting in Maastricht in Dec. 1991. The treaty was designed to pave the way to full economic and monetary union within the EC, but progress on this has been disrupted by difficulties with the Exchange Rate Mechanism and hesitation within some individual member states over ratifying the treaty; ratification was eventually completed in Oct. 1993, and the treaty came into effect on 1 Nov. (See also EUROPEAN MONETARY UNION.)

Maat /mɑːt/ *Egyptian Mythol.* the goddess of truth, justice, and cosmic order, daughter of Ra. She is depicted as a young and beautiful woman, standing or seated, with a feather on her head.

Mabinogion /ˌmæbɪˈnəʊɡɪən, -ˈnɒɡɪən/ a collection of Welsh prose tales of the 11th–13th centuries, dealing with Celtic legends and mythology. [Welsh *Mabinogi* instruction for young bards]

Mabuse /məˈbjuːz/ Jan (Flemish name Jan Gossaert) (*c.*1478–*c.*1533), Flemish painter. In 1508 he visited Italy, where the art of the High Renaissance made a lasting impression on him; he subsequently became one of the first artists to disseminate the Italian style in the Netherlands. His works are largely nudes, studies of the Virgin and Child, and commissioned portraits.

Mac /mæk/ n. colloq. **1** a Scotsman. **2** esp. *N. Amer.* a form of address to a male stranger. [*Mac-* as a patronymic prefix in many Scottish and Irish surnames]

mac /mæk/ n. (also **mack**) *Brit.* colloq. mackintosh. [abbr.]

macabre /məˈkɑːbrə/ adj. grim, gruesome. [ME f. OF *macabré* perh. f. *Macabé* a Maccabee, with ref. to a mystery play showing the slaughter of the Maccabees]

macadam /məˈkædəm/ n. **1** material for road-making with successive layers of compacted broken stone. **2** = TARMACADAM. □ **macadamize** v.tr. (also **-ise**). [John Loudon *McAdam*, Brit. surveyor (1756–1836), who advocated using this material]

macadamia /ˌmækəˈdeɪmɪə/ n. an evergreen Australian tree of the genus *Macadamia*; esp. *M. integrifolia* and *M. tetraphylla*, which bear edible nutlike seeds. [John *Macadam*, Austral. chemist (1827–65)]

MacAlpin /məˈkælpɪn/, Kenneth, see KENNETH I.

Macao /məˈkaʊ/ (Portuguese **Macau** /məˈkau/) a Portuguese dependency on the SE coast of China, on the west side of the Pearl River estuary opposite Hong Kong; pop. (est. 1991) 467,000; official languages, Portuguese and Cantonese; capital, Macao City. The colony comprises the Macao peninsula and two nearby islands. Visited by Vasco da Gama in 1497, Macao was developed by the Portuguese as a trading post, becoming in the 18th century the chief centre of trade between Europe and China. It was declared a free port by Portugal in 1849, but the development of Hong Kong in the 19th century led to a decline in its trade. Portugal's right of occupation was recognized by the Manchu government in 1887, but in 1987 it was agreed that sovereignty should pass to China in 1999. □ **Macanese** /mækəˈniːz/ adj. & n.

Macapá /ˌmækəˈpɑː/ a town in northern Brazil, on the Amazon delta, capital of the state of Amapá; pop. (1990) 166,750.

macaque /məˈkæk/ n. a medium-sized monkey of the Old World genus *Macaca*, including the rhesus monkey and Barbary ape and typically having a rather long face with cheek pouches. [F f. Port. *macaco* f. Fiot *makaku* some monkeys f. *kaku* monkey]

Macaronesia /ˌmækərəʊˈniːzɪə, -ˈniːʒə/ a biogeographical region comprising the Azores, Madeira, Canary Islands, and Cape Verde Islands in the eastern North Atlantic. □ **Macaronesian** adj. [Gk *makarōn nēsoi* Islands of the Blessed, identified with the Canaries]

macaroni /ˌmækəˈrəʊnɪ/ n. **1** a tubular variety of pasta. **2** (pl. **macaronies**) hist. an 18th-century British dandy affecting Continental fashions. [It. *maccaroni* f. late Gk *makaria* food made from barley]

macaronic /ˌmækəˈrɒnɪk/ n. & adj. ● n. (in pl.) burlesque verses containing Latin (or other foreign) words and vernacular words with Latin etc. terminations. ● adj. (of verse) of this kind. [mod.L *macaronicus* f. obs. It. *macaronico*, joc. formed as MACARONI]

macaroon /ˌmækəˈruːn/ n. a small light cake or biscuit made with white of egg, sugar, and ground almonds or coconut. [F *macaron* f. It. (as MACARONI)]

MacArthur /məˈkɑːθə(r)/, Douglas (1880–1964), American general. He was in command of US (later Allied) forces in the SW Pacific during the Second World War. He formally accepted Japan's surrender in 1945, and administered that country during the Allied occupation that followed. In 1950 he was put in charge of UN forces in Korea, but was relieved of his command the following year.

Macassar see MAKASSAR.

Macassarese see MAKASARESE.

Macassar oil /məˈkæsə(r)/ n. a kind of oil formerly used as a dressing for the hair, whose ingredients were said to come from Macassar.

Macau see MACAO.

Macaulay[1] /məˈkɔːlɪ/, Dame (Emilie) Rose (1881–1958), English novelist and essayist. After works such as *Dangerous Ages* (1921) and *They Were Defeated* (1932) she wrote no fiction for over a decade. Her return to the Anglican Church, from which she had long been estranged, was followed by her best-known novels *The World My Wilderness* (1950) and *The Towers of Trebizond* (1956).

Macaulay[2] /məˈkɔːlɪ/, Thomas Babington, 1st Baron (1800–59), English historian, essayist, and philanthropist. As a civil servant in India (1834–8) he established an English system of education and devised a new criminal code, before returning to Britain and devoting himself to literature and politics. Among his best-known works are *The Lays of Ancient Rome* (1842) and his *History of England* (1849–61), which covers the period from the accession of James II to the death of William III from a Whig standpoint.

macaw /məˈkɔː/ *n.* a large long-tailed brightly coloured parrot, esp. of the genus *Ara*, native to Central and South America. [Port. *macao*, of unkn. orig.]

Macbeth /məkˈbeθ/ (*c.*1005–57), king of Scotland 1040–57. He came to the throne after killing his cousin Duncan I in battle, and was himself defeated and killed by Malcolm III. He is chiefly remembered as the subject of Shakespeare's tragedy *Macbeth*, in which the historical events are considerably embroidered.

Macc. *abbr.* Maccabees (Apocrypha).

Maccabaeus, Judas, see JUDAS MACCABAEUS.

Maccabee /ˈmækəˌbiː/ *n.* **1** a member or supporter of a Jewish family which, headed by Judas Maccabaeus, led a religious revolt in Judaea against the Seleucid king, Antiochus IV Epiphanes, from around 167 BC, thus stemming the threatened destruction of Judaism by the advance of Hellenism. **2** (**the Maccabees**) four books of Jewish history and theology, of which the first two (whose hero is Judas Maccabaeus) are included in the Apocrypha. □ **Maccabean** /ˌmækəˈbiːən/ *adj.* [L *Maccabaeus* f. Gk, epithet of Judas, perh. f. Heb. *maqāb* hammer]

McCarthy[1] /məˈkɑːθɪ/, Joseph R(aymond) (1909–57), American Republican politician. Between 1950 and 1954, latterly as chairman of a government committee, he was the instigator of widespread investigations into alleged Communist infiltration in US public life. (See McCARTHYISM.)

McCarthy[2] /məˈkɑːθɪ/, Mary (Therese) (1912–89), American novelist and critic. Her novels are satirical social commentaries that draw on her experience of intellectual circles and academic life; they include *The Groves of Academe* (1952), which describes political persecution under McCarthyism, and *The Group* (1963), tracing the lives and careers of eight college girls.

McCarthyism /məˈkɑːθɪˌɪz(ə)m/ *n.* a vociferous campaign against alleged Communists in the US government and other institutions carried out by Senator Joseph R. McCarthy in the period 1950–4. Although most of those accused as a result of the 'witch-hunt' were not in fact members of the Communist Party, many of them were blacklisted, lost their jobs, or were otherwise discriminated against in a mood of hysteria which abated only after the public censure of McCarthy in Dec. 1954.

McCartney /məˈkɑːtnɪ/, (James) Paul (b.1942), English pop and rock singer, songwriter, and bass guitarist. He was a founder member of the Beatles and wrote most of their songs in collaboration with John Lennon. After the group broke up in 1970 he formed the band Wings, with whom he recorded hit singles such as 'Mull of Kintyre' (1977). Thereafter his musical career has included solo albums, film scores, and a classical composition, the *Liverpool Oratorio* (1991), written with the American composer Carl Davis (b.1936).

McCoy /məˈkɔɪ/ *n. colloq.* □ **the** (or **the real**) **McCoy** the real thing; the genuine article. [19th c.: orig. uncert.]

MacDiarmid /məkˈdɜːmɪd/, Hugh (pseudonym of Christopher Murray Grieve) (1892–1978), Scottish poet and nationalist. As a poet, he is chiefly remembered for his lyrics in a synthetic Scots that drew on the language of various regions and historical periods, such as the poems in the volume *A Drunk Man Looks at the Thistle* (1926). In the 1930s he wrote political poetry, including *First Hymn to Lenin and Other Poems* (1931). He was a founder member (1928) of the National Party of Scotland (later the Scottish National Party).

MacDonald[1] /məkˈdɒn(ə)ld/, Flora (1722–90), Scottish Jacobite heroine. She aided Charles Edward Stuart's escape from Culloden in 1746, by smuggling him over to the Isle of Skye in a small boat under the eyes of government forces.

MacDonald[2] /məkˈdɒn(ə)ld/, (James) Ramsay (1866–1937), British Labour statesman, Prime Minister 1924, 1929–31, and 1931–5. In 1922 he became leader of the Labour Party, and served as Britain's first Labour Prime Minister in the short-lived Labour government of 1924; he was elected Prime Minister again in 1929, but without an overall majority. Faced with economic crisis, and weakened by splits in his own party, he formed a national government with some Conservatives and Liberals; this led to his being expelled from the Labour Party.

Macdonald[3] /məkˈdɒn(ə)ld/, Sir John Alexander (1815–91), Scottish-born Canadian statesman, Prime Minister 1867–73 and 1878–91.

Entering politics in the 1840s, he became leader of the Conservatives and joint Premier (with George-Étienne Cartier) in 1856. Thereafter he played a leading role in the confederation of the Canadian provinces, and was appointed first Prime Minister of the Dominion of Canada in 1867.

MacDonnell Ranges /məkˈdɒn(ə)l/ a series of mountain ranges extending westwards from Alice Springs in Northern Territory, Australia. The highest peak is Mount Liebig, which rises to a height of 1,524 m (4,948 ft). Explored in 1860 by John McDouall Stuart, they were named after Sir Richard MacDonnell, governor of South Australia at that time.

Mace /meɪs/ *n. propr.* an irritant chemical preparation used in aerosol form as a disabling weapon. [prob. a use of MACE[1]]

mace[1] /meɪs/ *n.* **1** a staff of office, esp. the symbol of the Speaker's authority in the House of Commons. **2** *hist.* a heavy club usu. having a metal head and spikes. **3** a stick used in the game of bagatelle. **4** = *mace-bearer*. □ **mace-bearer** an official who carries a mace on ceremonial occasions. [ME f. OF *mace*, *masse* f. Rmc *mattea* (unrecorded) club]

mace[2] /meɪs/ *n.* the dried outer covering of the nutmeg, used as a spice. [ME *macis* (taken as pl.) f. OF *macis* f. L *macir* a red spicy bark]

macédoine /ˈmæsɪˌdwʌn/ *n.* mixed vegetables or fruit, esp. cut up small or in jelly. [F, = Macedonia, with ref. to the mixture of peoples there]

Macedonia /ˌmæsɪˈdəʊnɪə/ **1** (also **Macedon** /ˈmæsɪd(ə)n/) an ancient country in SE Europe, at the northern end of the Greek peninsula, including the coastal plain around Thessaloníki and the mountain ranges behind. In classical times it was a kingdom which under Philip II and Alexander the Great became a world power. The region is now divided between Greece, Bulgaria, and the republic of Macedonia (see sense 3). **2** a region in the north-east of modern Greece; capital, Thessaloníki. **3** a landlocked republic in the Balkans; pop. (est. 1991) 2,038,000; official language, Macedonian; capital, Skopje. Formerly a constituent republic of Yugoslavia, Macedonia became independent after a referendum in 1991.

Macedonian /ˌmæsɪˈdəʊnɪən/ *n. & adj.* ● *n.* **1** a native or inhabitant of ancient or present-day Macedonia. **2 a** the language of ancient Macedonia, usually regarded as a variety of Greek. **b** the Slavonic language of the republic of Macedonia and adjacent areas of Bulgaria and Greece. ● *adj.* of or relating to Macedonia or Macedonian.

Macedonian Wars a series of wars between Rome and Macedonia in the 3rd and 2nd centuries BC. The first (214–205 BC) occurred when Philip V of Macedonia formed an alliance against Rome. The second (200–196), in which Philip was defeated, resulted in the extension of Roman influence in Greece; Philip's son Perseus was defeated in the third war (171–168). Macedonia was then divided into four republics, and after a fourth war (149–148) became a Roman province.

Maceió /ˌmæseɪˈəʊ/ a port in eastern Brazil, on the Atlantic coast; pop. (1990) 699,760. It is the capital of the state of Alagoas.

McEnroe /ˈmækɪnˌrəʊ/, John (Patrick) (b.1959), American tennis player. A temperamental player, he dominated the game in the early 1980s; among his many titles are seven Wimbledon titles, three for the singles (1981, 1983–4) and four for the doubles (1979–84), six victories in the US Masters doubles (1978–83), and four US Open singles championships (1979–84).

macer /ˈmeɪsə(r)/ *n.* a mace-bearer, esp. *Sc.* an official keeping order in a lawcourt. [ME f. OF *massier* f. *masse*: see MACE[1]]

macerate /ˈmæsəˌreɪt/ *v.* **1** *tr. & intr.* make or become soft by soaking. **2** *intr.* waste away by fasting. □ **macerator** *n.* **maceration** /ˌmæsəˈreɪʃ(ə)n/ *n.* [L *macerare macerat-*]

Macgillicuddy's Reeks /məˌɡɪlɪˌkʌdɪz ˈriːks/ a range of hills in County Kerry in SW Ireland.

Mach[1] /mɑːk, mæk/, Ernst (1838–1916), Austrian physicist and philosopher of science. His belief that all knowledge of the physical world comes from sensations, and that science should be solely concerned with observables, inspired the logical positivist philosophers of the Vienna Circle in the 1920s. Mach also influenced scientists such as Einstein in the formulation of his theory of relativity, and Niels Bohr in quantum mechanics. In commemoration of his work on aerodynamics, his name has been preserved in the Mach number.

Mach[2] /mɑːk, mæk/ *n.* (in full **Mach number**) the ratio of the speed of a body to the speed of sound in the surrounding medium. □ **Mach one** (or **two** etc.) the speed (or twice the speed) of sound. [MACH[1]]

machete /məˈtʃetɪ, -ˈʃetɪ/ *n.* (also **matchet** /ˈmætʃɪt/) a broad heavy knife used esp. in Central America and the West Indies as an implement and weapon. [Sp. f. *macho* hammer f. LL *marcus*]

Machiavelli /ˌmækɪəˈvelɪ/, Niccolò di Bernardo dei (1469–1527), Italian statesman and political philosopher. After holding high office in Florence he was exiled by the Medicis on suspicion of conspiracy, but was subsequently restored to some degree of favour. His best-known work is *The Prince* (1532), a treatise on statecraft which advises rulers that the acquisition and effective use of power may necessitate unethical methods that are not in themselves desirable. He is thus often regarded as the originator of a political pragmatism in which 'the end justifies the means'.

machiavellian /ˌmækɪəˈvelɪən/ adj. elaborately cunning; scheming, unscrupulous. □ **machiavellianism** n. [MACHIAVELLI]

machicolate /məˈtʃɪkəˌleɪt/ v.tr. (usu. as **machicolated** adj.) provide (a parapet etc.) with openings between supporting corbels for dropping stones etc. on attackers. □ **machicolation** /-ˌtʃɪkəˈleɪʃ(ə)n/ n. [OF *machicoler*, ult. f. Prov. *machacol* f. *macar* crush + *col* neck]

machinable /məˈʃiːnəb(ə)l/ adj. capable of being cut by machine tools. □ **machinability** /-ˌʃiːnəˈbɪlɪtɪ/ n.

machinate /ˈmækɪˌneɪt, ˈmæʃɪ-/ v.intr. devise plots; intrigue. □ **machinator** n. **machination** /ˌmækɪˈneɪʃ(ə)n, ˌmæʃɪ-/ n. [L *machinari* contrive (as MACHINE)]

machine /məˈʃiːn/ n. & v. **1** an apparatus using or applying mechanical power, having several parts each with a definite function and together performing certain kinds of work. **2** a particular kind of machine, esp. a vehicle, a piece of electrical or electronic apparatus, etc. **3** an instrument that transmits a force or directs its application. **4** the controlling system of an organization etc. (*the party machine*). **5** a person who acts mechanically and with apparent lack of emotion. **6** (esp. in *comb.*) a coin-operated dispenser (*cigarette machine*) ● v.tr. make or operate on with a machine (esp. in sewing or manufacturing). □ **machine code** (or **language**) a computer language that a particular computer can respond to directly. **machine-readable** (of data, text, etc.) in a form that a computer can process. **machine tool** a mechanically operated tool for working on metal, wood, or plastics. **machine-tooled 1** shaped by a machine tool. **2** (of artistic presentation etc.) precise, slick, esp. excessively so. **machine translation** translation by a computer. [F f. L *machina* f. Gk *makhana* Doric form of *mēkhanē* f. *mēkhos* contrivance]

machine-gun /məˈʃiːngʌn/ n. & v. ● n. an automatic gun giving continuous fire. (See note below.) ● v.tr. (**-gunned**, **-gunning**) shoot at with a machine-gun. □ **machine-gunner** n.

▪ The first mechanically operated (hand-cranked) 'revolving battery gun' was the invention of a London lawyer, James Puckle, and was patented in 1718, but its flintlock mechanism proved unreliable. Other notable inventions include those of R. J. Gatling, B. B. Hotchkiss, I. N. Lewis, and J. M. Browning. The first satisfactory fully automatic weapon (using a water-cooling system) was that of Hiram S. Maxim. The machine-gun dominated the battlefields in the First World War and continued in use in the Second; technological advancements have made present-day machine-guns lighter, more rapid-firing, and more reliable.

machinery /məˈʃiːnərɪ/ n. (pl. **-ies**) **1** machines collectively. **2** the components of a machine; a mechanism. **3** (foll. by *of*) an organized system. **4** (foll. by *for*) the means devised or available (*the machinery for decision-making*).

machinist /məˈʃiːnɪst/ n. **1** a person who operates a machine, esp. a sewing machine or a machine tool. **2** a person who makes machinery.

machismo /məˈtʃɪzməʊ, məˈkɪz-/ n. exaggeratedly assertive manliness; a show of masculinity. [Sp. f. *macho* MALE f. L *masculus*]

Machmeter /ˈmɑːkˌmiːtə(r), ˈmæk-/ n. an instrument indicating air speed in the form of a Mach number.

macho /ˈmætʃəʊ/ adj. & n. ● adj. showily manly or virile. ● n. (pl. **-os**) **1** a macho man. **2** = MACHISMO. [MACHISMO]

Mach's principle n. *Physics* the hypothesis that a body's inertial mass results from its interaction with the rest of the matter in the universe. [MACH[1]]

Machtpolitik /ˈmɑːxtpɒlɪˌtiːk/ n. power politics. [G]

Machu Picchu /ˌmɑːtʃuː ˈpiːktʃuː/ a fortified Inca town in the Andes in Peru, which the invading Spaniards never found. Although it was not an important fortress it is famous for its dramatic position, perched high on a steep-sided ridge. It contains a palace, a temple to the sun, and extensive cultivation terraces. Discovered in 1911, it was named after the mountain that rises above it.

Macias Nguema /məˌsiːəs əŋˈgweɪmə/ a former name (1973–9) for BIOKO.

macintosh var. of MACKINTOSH.

mack var. of MAC.

Mackay /məˈkaɪ/ a port in NE Australia, on the coast of Queensland; pop. (est. 1991) 40,250. It was founded in 1862 and is named after Captain John MacKay, who explored the region in 1860.

Mackenzie[1] /məˈkenzɪ/, Sir Alexander (1764–1820), Scottish explorer of Canada. He entered the service of the North-West Company in 1779, undertaking explorations throughout NW Canada. He discovered the Mackenzie River in 1789 and in 1793 became the first European to reach the Pacific Ocean by land along a northern route.

Mackenzie[2] /məˈkenzɪ/, Sir (Edward Montague) Compton (1883–1972), English novelist, essayist, and poet. He produced essays, memoirs, poems, and biographies, but is best known as a novelist. His works include the semi-autobiographical *Sinister Street* (1913–14) and the comic novel *Whisky Galore* (1947).

Mackenzie[3] /məˈkenzɪ/, William Lyon (1795–1861), Scottish-born Canadian politician and journalist. Having emigrated to Canada in 1820, he became involved with the movement for political reform, at first as a radical journalist and later as a member of the provincial Parliament and mayor of Toronto (1834–6). In 1837 he led a short-lived rebellion, unsuccessfully attempting to set up a new government in Toronto. He fled to New York, returning to Canada in 1849.

Mackenzie River the longest river in Canada, flowing 1,700 km (1,060 miles) north-westwards from the Great Slave Lake to the Beaufort Sea, a part of the Arctic Ocean. It is named after Sir Alexander Mackenzie.

mackerel /ˈmækrəl/ n. (pl. same or **mackerels**) a North Atlantic marine fish, *Scomber scombrus*, with a greenish-blue body, used for food. □ **mackerel shark** a porbeagle. **mackerel sky** a sky dappled with rows of small white fleecy clouds, like the pattern on a mackerel's back. [ME f. AF *makerel*, OF *maquerel*]

McKinlay /məˈkɪnlɪ/, John (1819–72), Scottish-born explorer. Having emigrated to New South Wales in 1836, he was appointed in 1861 to lead an expedition to search for the missing explorers Burke and Wills. Although he found only traces of part of the Burke and Wills party, he carried out valuable exploratory work in the Australian interior and brought his entire party back safely despite tremendous hardships.

McKinley /məˈkɪnlɪ/, William (1843–1901), American Republican statesman, 25th President of the US 1897–1901. He supported US expansion into the Pacific, fighting the Spanish–American War of 1898 which resulted in the acquisition of Puerto Rico, Cuba, and the Philippines as well as the annexation of Hawaii. He was assassinated by an anarchist.

McKinley, Mount a mountain in south central Alaska. Rising to 6,194 m (20,110 ft), it is the highest mountain in North America. It is named after President William McKinley.

Mackintosh /ˈmækɪnˌtɒʃ/, Charles Rennie (1868–1928), Scottish architect and designer. He was a leading exponent of art nouveau and a precursor of several trends in 20th-century architecture. In particular, he pioneered the new concept of functionalism in architecture and interior design. His influence was very great abroad, especially in Austria and Germany, but less so in Britain. His fame chiefly rests on his Glasgow School of Art (1898–1909) and four Glasgow tearooms (1897–1912), designed with all their furniture and equipment.

mackintosh /ˈmækɪnˌtɒʃ/ n. (also **macintosh**) **1** *Brit.* a waterproof coat or cloak. **2** cloth waterproofed with rubber. [Charles *Macintosh*, Sc. inventor (1766–1843), who orig. patented the cloth]

mackle /ˈmæk(ə)l/ n. a blurred impression in printing. [F *macule* f. L *macula* blemish: see MACULA]

macle /ˈmæk(ə)l/ n. **1** a twin crystal. **2** a dark spot in a mineral. [F f. L (as MACKLE)]

Maclean[1] /məˈkleɪn/, Alistair (1922–87), Scottish novelist. His numerous thrillers and adventure stories, many of which were made into films, include *The Guns of Navarone* (1957), *Where Eagles Dare* (1967), and *Bear Island* (1971).

Maclean[2] /məˈkleɪn/, Donald Duart (1913–83), British Foreign Office official and Soviet spy. After acting as a Soviet agent from the late 1930s he fled to the USSR with Guy Burgess in 1951, following a warning from Kim Philby of impending proceedings against him.

Macleod /məˈklaʊd/, John James Rickard (1876–1935), Scottish physiologist. He specialized in carbohydrate metabolism and held various chairs in physiology, notably at the University of Toronto. He provided facilities there for the research on pancreatic extracts by F. G. Banting and C. H. Best, much of which was directed by Macleod personally, and which led to the discovery and isolation of insulin. Macleod shared a Nobel Prize with Banting in 1923.

McLuhan /məˈkluːən/, (Herbert) Marshall (1911–80), Canadian writer and thinker. He became famous in the 1960s for his theories on the role of the media and technology in society. He is particularly known for claiming that the world had become 'a global village' in its electronic interdependence, and that 'the medium is the message', because it is the characteristics of a particular medium rather than the information it disseminates which influence and control society. His books include *Understanding Media: The Extensions of Man* (1964).

Macmillan /məkˈmɪlən/, (Maurice) Harold, 1st Earl of Stockton (1894–1986), British Conservative statesman, Prime Minister 1957–63. His term of office saw the signing of the Test-Ban Treaty (1963) with the US and the USSR. He advocated the granting of independence to British colonies but his attempt to take Britain into the European Economic Community was blocked by the French President de Gaulle (1963). Macmillan resigned on grounds of ill health shortly after the scandal surrounding the Secretary of State for War, John Profumo.

McNaughten rules /məkˈnɔːt(ə)n/ *n.pl.* (also **M'Naghten rules**) *Brit.* rules governing the decision as to the criminal responsibility of a mentally ill person. [*McNaughten* or *McNaughtan*, name of a 19th-c. accused person]

MacNeice /məkˈniːs/, (Frederick) Louis (1907–63), Northern Irish poet. He was part of W. H. Auden's circle at Oxford, where he published his first volume of poetry in 1929; later volumes include *Collected Poems* (1966). His work is characterized by the use of assonance, internal rhythms, and ballad-like repetitions absorbed from the Irish background of his youth. He also wrote documentaries and plays for radio, notably the fantasy *The Dark Tower* (1947).

Macquarie /məˈkwɒri/, Lachlan (1762–1824), Scottish-born Australian colonial administrator. He served as governor of New South Wales 1809–21; the colony was chiefly populated by convicts, but during his term of office he improved its prosperity, expanded opportunities for former convicts, and promoted public works, further settlement, and exploration.

Macquarie River a river in New South Wales, Australia, rising on the western slopes of the Great Dividing Range and flowing 960 km (600 miles) north-west to join the River Darling, of which it is a headwater.

macramé /məˈkrɑːmɪ/ *n.* **1** the art of knotting cord or string in patterns to make decorative articles. **2** articles made in this way. [Turk. *makrama* bedspread f. Arab. *miḳrama*]

macro /ˈmækrəʊ/ *n.* (also **macro-instruction**) *Computing* a single instruction that expands automatically into a set of instructions to perform a particular task.

macro- /ˈmækrəʊ/ *comb. form* **1** long. **2** large, large-scale. [Gk *makro-* f. *makros* long, large]

macrobiotic /ˌmækrəʊbaɪˈɒtɪk/ *adj. & n.* ● *adj.* relating to or following a Zen Buddhist dietary system intended to prolong life by balancing the influences of *yin* and *yang*, comprising pure vegetable foods, brown rice, and little meat. ● *n.* (in *pl.*; treated as *sing.*) the use or theory of such a dietary system.

macrocarpa /ˌmækrəʊˈkɑːpə/ *n.* an evergreen coniferous tree, *Cupressus macrocarpa*, often cultivated for hedges or wind-breaks. [mod.L f. Gk MACRO- + *karpos* fruit]

macrocephalic /ˌmækrəʊsɪˈfælɪk, -kɪˈfælɪk/ *adj.* (also **macrocephalous** /-ˈsefələs, -ˈkefələs/) having a long or large head. □ **macrocephaly** *n.*

macrocosm /ˈmækrəʊˌkɒz(ə)m/ *n.* **1** the universe. **2** the whole of a complex structure. □ **macrocosmic** /ˌmækrəʊˈkɒzmɪk/ *adj.* **macrocosmically** *adv.*

macroeconomics /ˌmækrəʊˌiːkəˈnɒmɪks, -ˌekəˈnɒmɪks/ *n.* the branch of economics that deals with large-scale or general economic factors, such as government income and expenditure, the balance of payments, employment levels, inflation, national productivity, etc. Cf. MICROECONOMICS. □ **macroeconomic** *adj.*

macroevolution /ˌmækrəʊˌiːvəˈluːʃ(ə)n, -ˈljuːʃ(ə)n/ *n. Biol.* major evolutionary change, esp. over a long period. □ **macroevolutionary** *adj.*

macrolepidoptera /ˌmækrəʊˌlepɪˈdɒptərə/ *n.pl.* butterflies and moths of interest beyond the sphere of specialist entomology, generally comprising the larger kinds (cf. MICROLEPIDOPTERA). ¶ Not used in formal classification.

macromolecule /ˌmækrəʊˈmɒlɪˌkjuːl/ *n. Biochem.* a molecule containing a very large number of atoms, e.g. a protein or a nucleic acid. □ **macromolecular** /-məˈlekjʊlə(r)/ *adj.*

macron /ˈmækrɒn/ *n.* a written or printed mark (ˉ) over a long or stressed vowel. [Gk *makron* neut. of *makros* large]

macronutrient /ˌmækrəʊˈnjuːtrɪənt/ *n.* a chemical element required in relatively large amounts for the growth and development of living organisms.

macrophage /ˈmækrəʊˌfeɪdʒ/ *n. Physiol.* a large phagocytic white blood cell usu. occurring at points of infection.

macrophotography /ˌmækrəʊfəˈtɒgrəfɪ/ *n.* photography producing photographs larger than life.

macropod /ˈmækrəʊˌpɒd/ *n. Zool.* a plant-eating marsupial mammal of the family Macropodidae, native to Australia and New Guinea. The family comprises the kangaroos and wallabies. [f. MACRO- + Gk *pous podos* foot]

macroscopic /ˌmækrəʊˈskɒpɪk/ *adj.* **1** visible to the naked eye. **2** regarded in terms of large units. □ **macroscopically** *adv.*

macula /ˈmækjʊlə/ *n.* (*pl.* **maculae** /-ˌliː/) **1** a dark spot, esp. a permanent one, in the skin. **2** (in full **macula lutea** /ˈluːtɪə/) the region of greatest visual acuity in the retina. □ **macular** *adj.* **maculation** /ˌmækjʊˈleɪʃ(ə)n/ *n.* [L, = spot, mesh]

mad /mæd/ *adj. & v.* ● *adj.* (**madder, maddest**) **1** mentally ill or deranged; having a disordered mind. **2** (of a person, conduct, or an idea) wildly foolish. **3** (often foll. by *about, on*) wildly excited or infatuated (*mad about football; is chess-mad*). **4** (often foll. by *at, with*) *colloq.* angry. **5** (of an animal) rabid. **6** wildly light-hearted. ● *v.* (**madded, madding**) **1** *tr. US* make angry. **2** *intr. archaic* be mad; act madly (*the madding crowd*). □ **as mad as a hatter** see HATTER. **like mad** *colloq.* with great energy, intensity, or enthusiasm. **mad cow disease** *colloq.* = BSE. **mad keen** *colloq.* extremely eager. □ **madness** *n.* [OE *gemǣded* part. form f. *gemǣd* mad]

Madagascar /ˌmædəˈgæskə(r)/ an island country in the Indian Ocean, off the east coast of Africa; pop. (est. 1991) 12,016,000; official languages, Malagasy and French; capital, Antananarivo. Settled by peoples of mixed Indo-Melanesian and African descent, and later influenced by Arab traders, Madagascar was discovered by the Portuguese in 1500. Small rival kingdoms emerged, which were unified in the 19th century. The island resisted colonization until the French established control in 1896. It regained its independence as the Malagasy Republic in 1960, changing its name back to Madagascar in 1975. Madagascar is the fourth largest island in the world, and many of its plants and animals are not found elsewhere. □ **Madagascan** *adj. & n.*

madam /ˈmædəm/ *n.* **1** a polite or respectful form of address or mode of reference to a woman. **2** *Brit. colloq.* a conceited or precocious girl or young woman. **3** a woman brothel-keeper. [ME f. OF *ma dame* my lady]

Madame /məˈdɑːm, -ˈdæm, ˈmædəm/ *n.* **1** (*pl.* **Mesdames** /meɪˈdɑːm, -ˈdæm/) a title or form of address used of or to a French-speaking woman, corresponding to Mrs or madam. **2** (**madame**) = MADAM 1. [F (as MADAM)]

madcap /ˈmædkæp/ *adj. & n.* ● *adj.* **1** wildly impulsive. **2** undertaken without forethought. ● *n.* a wildly impulsive person.

madden /ˈmæd(ə)n/ *v.* **1** *tr. & intr.* make or become mad. **2** *tr.* irritate intensely. □ **maddening** *adj.* **maddeningly** *adv.*

madder /ˈmædə(r)/ *n.* **1** a herbaceous climbing plant, *Rubia tinctorum*, with whorled leaves and yellowish flowers. **2** a red dye obtained from the root of the madder, or its synthetic substitute. [OE *mædere*]

made /meɪd/ **1** past and past part. of MAKE. **2** *adj.* (usu. in *comb.*) **a** (of a person or thing) built or formed (*well-made; strongly made*). **b** successful (*a self-made man*). □ **have** (or **have got**) **it made** *colloq.* be sure of success. **made for** ideally suited to. **made of** consisting of. **made of money** *colloq.* very rich. **made-up 1** invented, fabricated. **2** wearing cosmetics or make-up. **3** prepared, assembled; ready-made.

Madeira¹ /məˈdɪərə/ **1** an island in the Atlantic Ocean off NW Africa, the largest of the Madeiras, a group of islands which constitutes an autonomous region of Portugal; pop. (est. 1986) 269,500; capital, Funchal. Encountered by the Portuguese in 1419, the islands were occupied by the Spanish 1580–1640 and the British 1807–14. **2** a river in NW Brazil, which rises on the Bolivian border and flows about 1,450 km (900 miles) to meet the Amazon east of Manaus. It is navigable to large ocean-going vessels as far as Pôrto Velho. [Port., = timber (f. L *materia* matter), from its dense woods]

Madeira[2] /məˈdɪərə/ n. **1** a fortified white wine from the island of Madeira. **2** (in full **Madeira cake**) a kind of rich sponge cake.

madeleine /ˈmædəˌleɪn/ n. a small fancy sponge cake. [F]

Mademoiselle /ˌmædmwəˈzel/ n. (pl. **Mesdemoiselles**) /ˌmeɪd-/ **1** a title or form of address used of or to an unmarried French-speaking woman, corresponding to Miss or madam. **2** (**mademoiselle**) **a** a young Frenchwoman. **b** a French governess. [F f. *ma* my + *demoiselle* DAMSEL]

madhouse /ˈmædhaʊs/ n. **1** archaic or colloq. a mental home or hospital. **2** colloq. a scene of extreme confusion or uproar.

Madhya Pradesh /ˌmʌdjə prəˈdeʃ/ a large state in central India, formed in 1956; capital, Bhopal.

Madison[1] /ˈmædɪs(ə)n/ the state capital of Wisconsin; pop. (1990) 191,260. It is named after President James Madison.

Madison[2] /ˈmædɪs(ə)n/, James (1751–1836), American Democratic Republican statesman, 4th President of the US 1809–17. Before taking office, he played a leading part in drawing up the US Constitution (1787) and proposed the Bill of Rights (1791). His presidency saw the US emerge successfully from the War of 1812 against Britain.

madly /ˈmædlɪ/ adv. **1** in a mad manner. **2** colloq. **a** passionately. **b** extremely.

madman /ˈmædmən/ n. (pl. **-men**) a man who is mad.

Madonna[1] /məˈdɒnə/ n. Eccl. **1** (prec. by the) a name for the Virgin Mary. **2** (usu. **madonna**) a picture or statue of the Madonna. □ **madonna lily** the white lily, *Lilium candidum*, as shown in many pictures of the Madonna. [It. f. *ma* = *mia* my + *donna* lady f. L *domina*]

Madonna[2] /məˈdɒnə/ (born Madonna Louise Ciccone) (b.1958), American pop singer and actress. She rose to international stardom in the mid-1980s through her records and accompanying videos, cultivating her image as a sex symbol and frequently courting controversy. Among her singles are 'Holiday' (1983), while her albums include *Like a Virgin* (1984) and *Erotica* (1992). Probably her best-known film is *Desperately Seeking Susan* (1985).

Madras[1] /məˈdrɑːs, -ˈdræs/ **1** a seaport on the east coast of India, capital of Tamil Nadu; pop. (1991) 3,795,000. **2** the former name (until 1968) for the state of Tamil Nadu.

Madras[2] /məˈdræs/ n. **1** (in full **Madras curry**) a hot curry dish usu. containing chicken or beef. **2** a strong cotton fabric with coloured or white stripes, checks, etc. [MADRAS[1]]

madrepore /ˈmædrɪˌpɔː(r)/ n. Zool. **1** a stony coral, esp. of the genus *Madrepora*. **2** the polyp producing this. □ **madreporic** /ˌmædrɪˈpɒrɪk/ adj. [F *madrépore* or mod.L *madrepora* f. It. *madrepora* f. *madre* mother + *poro* PORE[1]]

Madrid /məˈdrɪd/ the capital of Spain; pop. (1991) 2,984,600. Situated on a high plateau in the centre of the country, it replaced Valladolid as capital in 1561.

madrigal /ˈmædrɪɡ(ə)l/ n. **1** a part-song for several voices, usu. arranged in elaborate counterpoint and without instrumental accompaniment. (*See note below.*) **2** a short love poem. □ **madrigalist** n. **madrigalian** /ˌmædrɪˈɡeɪlɪən/ adj. [It. *madrigale* f. med.L *matricalis* mother (church), formed as MATRIX]

 • The term was used in the 14th century for a genre of Italian song for two or three voices, often comprising two or three stanzas, each of three lines. Today, however, it usually refers to late 16th and early 17th-century song for several voices but with no fixed form, in which the musical style is heavily dependent on the poem on which it is based. Favourite poets of madrigal composers include Petrarch and Boccaccio.

madroño /məˈdrəʊnjəʊ/ n. (also **madroña** /-njə/) (pl. **-os**) an evergreen ericaceous tree, *Arbutus menziesii*, of western North America, with white flowers, red berries, and glossy leaves. [Sp.]

Madura /məˈdʊərə/ an island of Indonesia, off the NE coast of Java. Its chief town is Pamekasan.

Madurai /ˈmædjʊˌraɪ/ a city in Tamil Nadu in southern India; pop. (1991) 952,000.

madwoman /ˈmædˌwʊmən/ n. (pl. **-women**) a woman who is mad.

Maeander /miːˈændə(r)/ the ancient name for the MENDERES.

Maecenas /maɪˈsiːnəs/, Gaius (c.70–8 BC), Roman statesman. He was a trusted adviser of Augustus but shunned official position. Himself a writer, he was a notable patron of poets such as Virgil and Horace (a role for which his name became a byword).

maelstrom /ˈmeɪlstrəm/ n. **1** a great whirlpool. **2** a state of confusion. [early mod.Du. f. *malen* grind, whirl + *stroom* STREAM]

maenad /ˈmiːnæd/ n. **1** a bacchante. **2** a frenzied woman. □ **maenadic** /miːˈnædɪk/ adj. [L *Maenas Maenad-* f. Gk *Mainas -ados* f. *mainomai* rave]

maestoso /maɪˈstəʊzəʊ/ adj., adv., & n. Mus. ● adj. & adv. to be performed majestically. ● n. (pl. **-os**) a piece of music to be performed in this way. [It.]

maestro /ˈmaɪstrəʊ/ n. (pl. **maestri** /-strɪ/ or **-os**) (often as a respectful form of address) **1** a distinguished musician, esp. a conductor or performer. **2** a great performer in any sphere, esp. artistic. [It., = master]

Maeterlinck /ˈmeɪtəˌlɪŋk/, Count Maurice (1862–1949), Belgian poet, dramatist, and essayist. He published a collection of symbolist poems in 1889, and became established as a leading figure in the symbolist movement with his prose dramas *La Princesse Maleine* (1889) and *Pelléas et Mélisande* (1892), the source of Debussy's opera of that name (1902). He also achieved great popularity in his day with the play *L'Oiseau bleu* (1908; *The Blue Bird*). His work draws on traditions of fairy tale and romance, and is characterized by an air of mystery and melancholy. He was awarded the Nobel Prize for literature in 1911.

Mae West /meɪ ˈwest/ n. sl. an inflatable life-jacket, orig. one issued to Royal Air Force men in the Second World War. [the name of the actress Mae West (see WEST[2]), with allusion to her large bust]

Mafeking /ˈmæfɪˌkɪŋ/ (also **Mafikeng**) a town in South Africa, in North-West Province. In 1899–1900, during the Second Boer War, a small British force under the command of Baden-Powell was besieged there by the Boers for 215 days. Although the town was of little strategic significance, its successful defence, at a time when the war was going very badly for the British, excited great interest, while its relief was hailed almost with a national sense of jubilation.

MAFF /mæf/ abbr. (in the UK) Ministry of Agriculture, Fisheries, and Food.

Mafia /ˈmæfɪə/ a secret society of criminals, highly organized and with a strong hierarchy, originating in Sicily in the 13th century but now operating internationally, especially in the US. In Sicily during the 18th–19th centuries the organization developed a complex and ruthless behavioural code (for example, *omertà*, the code of silence); in the US the Mafia developed amongst Italian immigrants, calling itself 'Cosa Nostra'. The latter became an integral part of organized crime that developed especially during the Prohibition era, and went on to be involved in gambling, moneylending, and drugs. [It. dial. (Sicilian), = bragging]

Mafioso /ˌmæfɪˈəʊsəʊ/ n. (pl. **Mafiosi** /-sɪ/) a member of the Mafia. [It. (as MAFIA)]

mag[1] /mæg/ n. colloq. a magazine or periodical. [abbr.]

mag[2] /mæg/ v. & n. esp. Austral. ● v.intr. chatter or talk incessantly. ● n. an excessively talkative person. [f. MAGPIE]

mag. abbr. **1** magnesium. **2** magneto. **3** magnetic.

Magadha /ˈmʌɡədə/ an ancient kingdom situated in the valley of the River Ganges in NE India (modern Bihar) which was the centre of several empires, notably those of the Mauryan and Gupta dynasties, between the 6th century BC and the 8th century AD.

Magadi, Lake /məˈɡɑːdɪ/ a salt lake in the Great Rift Valley, in southern Kenya, with extensive deposits of sodium carbonate and other minerals.

Magahi /ˈmʌɡəhɪ/ n. one of the Bihari group of languages, spoken by some 10 million people in central Bihar and West Bengal.

magazine /ˌmæɡəˈziːn/ n. **1** a periodical publication containing articles, stories, etc., usu. with photographs, illustrations, etc. **2** a chamber for holding a supply of cartridges to be fed automatically to the breech of a gun. **3** a similar device feeding a camera, slide projector, etc. **4** a store for arms, ammunition, and provisions for use in war. **5** a store for explosives. [F *magasin* f. It. *magazzino* f. Arab. *maḵāzin* pl. of *maḵzan* storehouse f. *ḵazana* store up]

magdalen /ˈmæɡdəlɪn/ n. **1** a reformed prostitute. **2** a home for reformed prostitutes. [f. eccl.L *Magdalena* f. Gk *Magdalēnē*; cf. MARY MAGDALENE, ST]

Magdalena /ˌmæɡdəˈleɪnə/ the principal river of Colombia, rising in the Andes and flowing northwards for about 1,600 km (1,000 miles) to enter the Caribbean at Barranquilla.

Magdalenian /ˌmæɡdəˈliːnɪən/ adj. & n. Archaeol. of, relating to, or denoting a culture of the late palaeolithic period in Europe, dated to *c.*17,000–11,000 BP. It is characterized by a range of bone and horn tools, including elaborate bone harpoons; cave art reached a zenith during this period (see ALTAMIRA 1). [F *Magdalénien* f. *La Madeleine*, the type-site in Dordogne, France]

Magdeburg /'mægdə,bɜːg/ an industrial city in Germany, the capital of Saxony-Anhalt, situated on the River Elbe and linked to the Rhine and Ruhr by the Mittelland Canal; pop. (est. 1990) 290,000.

Magdeburg hemispheres *n.pl.* a pair of copper hemispheres joined to form a hollow globe from which the air can be extracted, after which they cannot easily be pulled apart. They were devised by the 17th-century German physicist Otto von Guericke to demonstrate the effect of air pressure.

mage /meɪdʒ/ *n. archaic* **1** a magician. **2** a wise and learned person. [ME, anglicized f. MAGUS]

Magellan[1] /mə'gelən/, Ferdinand (Portuguese name Fernão Magalhães) (*c.*1480–1521), Portuguese explorer. In 1519, while in the service of Spain, he commanded five vessels on a voyage from Spain to the East Indies by the western route. He reached South America later that year, rounding the continent through the strait which now bears his name and emerging to become the first European to navigate the Pacific. He reached the Philippines in 1521, but soon after was killed in a skirmish on Cebu. The survivors, in the one remaining ship, sailed back to Spain round Africa, thereby completing the first circumnavigation of the globe (1522).

Magellan[2] /mə'gelən/ an American space probe launched in 1989 to map the surface of Venus. After orbiting the planet for three years, the probe was deliberately burned up in Venus's atmosphere in 1994.

Magellan, Strait of a passage separating Tierra del Fuego and other islands from mainland South America, connecting the Atlantic and Pacific Oceans. It is named after Ferdinand Magellan, who discovered it in Oct. 1520.

Magellanic cloud /,mædʒɪ'lænɪk/ *n. Astron.* either of two diffuse luminous regions of the southern sky, now known to be galaxies of irregular shape containing millions of stars. They are the closest galaxies to our own. [MAGELLAN[1]]

magenta /mə'dʒentə/ *n. & adj.* ● *n.* **1** a brilliant mauvish-crimson shade. **2** an aniline dye of this colour; fuchsine. ● *adj.* of or coloured with magenta. [*Magenta* in N. Italy, town near which the Austrians were defeated by the French under Napoleon III in 1859, shortly before the dye was discovered]

Maggiore, Lake /,mædʒɪ'ɔːrɪ/ the second largest of the lakes of northern Italy, extending into southern Switzerland.

maggot /'mægət/ *n.* **1** a soft-bodied legless larva, esp. that of a blowfly etc. found in decaying matter. **2** *archaic* a whimsical fancy. □ **maggoty** *adj.* [ME, perh. alt. f. *maddock*, earlier *mathek* f. ON *mathkr*: cf. MAWKISH]

Maghrib /'mægrɪb, 'mʌgrəb/ (also **Maghreb**) a region of North and NW Africa between the Atlantic Ocean and Egypt, comprising the coastal plain and Atlas Mountains of Morocco, together with Algeria, Tunisia, and sometimes also Tripolitania, forming a well-defined zone bounded by sea or desert. It formerly included Moorish Spain. (See also BARBARY.) [Arab., = west]

magi *pl.* of MAGUS.

Magi, the /'meɪdʒaɪ/ (in the Bible) the 'wise men' from the East who brought gifts to the infant Christ (Matt. 2:1). The Gospel does not mention their number (the later tradition that there were three probably arose from the three gifts of gold, frankincense, and myrrh). The tradition that they were kings appears first in Tertullian (2nd century); their names, Caspar, Melchior, and Balthasar, are first mentioned in the 6th century. In the Middle Ages they were venerated as saints, and what are claimed to be their relics lie in Cologne cathedral. [pl. of MAGUS]

magian /'meɪdʒɪən/ *adj. & n.* ● *adj.* of the magi or Magi. ● *n.* **1** a magus or Magus. **2** a magician. □ **magianism** *n.* [L *magus*: see MAGUS]

magic /'mædʒɪk/ *n., adj., & v.* ● *n.* **1 a** the supposed art of influencing the course of events by the occult control of nature or of the spirits. **b** witchcraft. **2** conjuring tricks. **3** an inexplicable or remarkable influence producing surprising results. **4** an enchanting quality or phenomenon. ● *adj.* **1** of or resulting from magic. **2** producing surprising results. **3** *colloq.* wonderful, exciting. ● *v.tr.* (**magicked**, **magicking**) change or create by magic, or apparently so. □ **like magic** very effectively or rapidly. **magic away** cause to disappear as if by magic. **magic bullet** *colloq.* any (usu. undiscovered) highly specific therapeutic agent. **magic carpet** a mythical carpet able to transport a person on it to any desired place. **magic eye 1** a photoelectric

device used in equipment for detection, measurement, etc. **2** a small cathode-ray tube used to indicate the correct tuning of a radio receiver. **magic lantern** a simple form of image-projector using photographic slides. **magic mushroom** a toadstool of the genus *Psilocybe*, containing psilocybin. **magic square** a square divided into smaller squares each containing a number such that the sum of the figures in each vertical, horizontal, or diagonal row is equal. [ME f. OF *magique* f. L *magicus* adj., LL *magica* n., f. Gk *magikos* (as MAGUS)]

magical /'mædʒɪk(ə)l/ *adj.* **1** of or relating to magic. **2** resembling magic; produced as if by magic. **3** wonderful, enchanting. □ **magically** *adv.*

magician /mə'dʒɪʃ(ə)n/ *n.* **1** a person skilled in or practising magic. **2** a conjuror. **3** a person with exceptional skill. [ME f. OF *magicien* f. LL *magica* (as MAGIC)]

magic realism *n.* a style in art and esp. literature in which realistic techniques such as narrative and naturalistic detail are combined with the unrealistic, imaginative elements of dream and fantasy. The term was originally used to describe the work of a group of German artists in the 1920s, but is now strongly associated with the writings of such Latin American authors as Jorge Luis Borges and Gabriel García Márquez as well as European writers such as Italo Calvino, Günter Grass, and, more recently, Angela Carter, Salman Rushdie, and John Fowles.

magilp var. of MEGILP.

Maginot Line /'mæʒɪ,nəʊ/ a line of defensive fortifications built along France's north-eastern frontier from Switzerland to Luxembourg, begun in 1920 as a defence against German invasion and completed in 1936. It was not extended along the Franco-Belgian border, however, and consequently, although the defences proved impregnable to frontal assault, the line could be outflanked, as happened when the Germans invaded France in the spring of 1940. It was named after the French Minister of War André Maginot (1877–1932).

magisterial /,mædʒɪ'stɪərɪəl/ *adj.* **1** imperious. **2** invested with authority. **3** of or conducted by a magistrate. **4** (of a work, opinion, etc.) highly authoritative. □ **magisterially** *adv.* [med.L *magisterialis* f. LL *magisterius* f. L *magister* MASTER]

magisterium /,mædʒɪ'stɪərɪəm/ *n. RC Ch.* the official teaching of a bishop or pope. [L, = the office of a master (as MAGISTERIAL)]

magistracy /'mædʒɪstrəsɪ/ *n.* (*pl.* **-ies**) **1** the office or authority of a magistrate. **2** magistrates collectively.

magistral /'mædʒɪstrəl, mə'dʒɪs-/ *adj.* **1** of a master or masters. **2** *Pharm.* (of a remedy etc.) devised and made up for a particular case (cf. OFFICINAL 1b). [F *magistral* or L *magistralis* f. *magister* MASTER]

magistrate /'mædʒɪstrət, -,streɪt/ *n.* **1** a civil officer administering the law. **2** an official conducting a court for minor cases and preliminary hearings (*magistrates' court*). □ **magistrateship** *n.* **magistrature** /-strə,tjʊə(r)/ *n.* [ME f. L *magistratus* (as MAGISTRAL)]

Maglemosian /,mæglə'məʊzɪən/ *adj. & n. Archaeol.* of, relating to, or denoting a mesolithic culture of northern Europe, dated to *c.*8300–6500 BC. [f. *Maglemose*, the type-site near Mullerup in Denmark]

maglev /'mæglev/ *n.* (usu. *attrib.*) a system in which trains glide above the track in a magnetic field. [abbr. of *magnetic levitation*]

magma /'mægmə/ *n.* **1** fluid or semifluid material under the earth's surface from which lava and other igneous rock is formed. **2** a crude pasty mixture of mineral or organic matter. □ **magmatic** /mæg'mætɪk/ *adj.* [ME, = a solid residue f. L f. Gk *magma -atos* f. the root of *massō* knead]

Magna Carta /,mægnə 'kɑːtə/ (also **Magna Charta** /'tʃɑː-/) the English political charter which King John was forced to sign by his rebellious barons at Runnymede in 1215. The barons, discontented with John's high-handedness, were led by Archbishop Langton to frame a charter which effectively redefined the limits of royal power. Although the charter was often violated by medieval kings, it eventually came to be seen as the seminal document of English constitutional practice. Among its provisions were that no freeman should be imprisoned or banished except by the law of the land. [med.L, = great charter]

Magna Graecia /,mægnə 'griːsɪə, 'griːʃə/ a group of ancient Greek cities in southern Italy, founded from *c.*750 BC by colonists from Euboea and from Sparta and other Greek cities. The cities thrived until after the 5th century BC; the Pythagorean and Eleatic systems of philosophy arose there. [L, = Great Greece]

magnanimous /mæg'nænɪməs/ *adj.* nobly generous; not petty in feelings or conduct. □ **magnanimously** *adv.* **magnanimity** /,mægnə'nɪmɪtɪ/ *n.* [L *magnanimus* f. *magnus* great + *animus* soul]

magnate /'mægneɪt, -nɪt/ n. a wealthy and influential person, esp. in business (*shipping magnate; financial magnate*). [ME f. LL *magnas -atis* f. L *magnus* great]

magnesia /mæg'niːʒə, -'niːʃə, -'niːzɪə/ n. **1** *Chem.* magnesium oxide. **2** (in general use) hydrated magnesium carbonate, a white powder used as an antacid and laxative. □ **magnesian** *adj.* [ME f. med.L f. Gk *Magnēsia (lithos)* (stone) of Magnesia in Asia Minor, orig. referring to loadstone]

magnesite /'mægnɪˌsaɪt/ n. a white or grey mineral form of magnesium carbonate.

magnesium /mæg'niːzɪəm/ n. a silvery metallic chemical element (atomic number 12; symbol **Mg**). Magnesium was identified as an element by Davy in 1807; it is an alkaline earth metal. Magnesium compounds are common constituents of rock-forming minerals such as magnesite and dolomite, and the element is important to life as the central atom of the chlorophyll molecule. Magnesium is used to make strong lightweight alloys, especially for the aerospace industry. It is also used in flash bulbs and pyrotechnics, as it burns with a brilliant white flame. Magnesium compounds have various uses, including as important reagents in organic chemistry; magnesium carbonate is the basis of many antacid and laxative preparations. □ **magnesium flare** (or **light**) a blinding white light produced by burning magnesium wire. [MAGNESIA]

magnet /'mægnɪt/ n. **1** a piece of iron, steel, alloy, ore, etc., usu. in the form of a bar or horseshoe, having properties of attracting or repelling iron. **2** a lodestone. **3** a person or thing that attracts. [ME f. L *magnes magnetis* f. Gk *magnēs = Magnēs -ētos (lithos)* (stone) of Magnesia: cf. MAGNESIA]

magnetic /mæg'netɪk/ adj. **1 a** having the properties of a magnet. **b** producing, produced by, or acting by magnetism. **2** capable of being attracted by or acquiring the properties of a magnet. **3** very attractive or alluring (*a magnetic personality*). □ **magnetic compass** = COMPASS 1. **magnetic disk** see DISC 4a. **magnetic equator** an imaginary line, passing round the earth near the equator, on which a magnetic needle shows no dip (see DIP n. 8). **magnetic field** a region of variable force around magnets, magnetic materials, or current-carrying conductors. **magnetic inclination** = DIP n. 8. **magnetic mine** a submarine mine detonated by the proximity of a magnetized body such as that of a ship. **magnetic moment** the property of a magnet that interacts with an applied field to give a mechanical moment. **magnetic needle** a piece of magnetized steel used as an indicator on the dial of a compass and in magnetic and electrical apparatus, esp. in telegraphy. **magnetic north** the point indicated by the north end of a compass needle. **magnetic pole 1** each of the points near the extremities of the axis of rotation of the earth or another body where a magnetic needle dips vertically (see also POLE²). **2** each of the regions of an artificial or natural magnet, from which the magnetic forces appear to originate. **magnetic resonance imaging** (abbr. **MRI**) *Med.* imaging of the internal structure of tissues etc. by means of nuclear magnetic resonance. **magnetic storm** a disturbance of the earth's magnetic field caused by charged particles from the sun etc. **magnetic tape** a tape coated with magnetic material for recording sound or pictures or for the storage of information. □ **magnetically** *adv.* [LL *magneticus* (as MAGNET)]

magnetism /'mægnɪˌtɪz(ə)m/ n. **1** magnetic phenomena and their study; the property of producing these phenomena. (*See note below.*) **2** attraction; personal charm. [mod.L *magnetismus* (as MAGNET)]

■ Magnetism has been known, and used in the form of the magnetic compass, since ancient times. William Gilbert, in *De Magnete* (1600), showed that the earth is a great magnet, and began the scientific study of magnetism. It was Ampère who first suggested that the powers of a magnet are due to the combined effect of tiny electric currents circulating in its atoms. In 1820 Oersted showed that an electric current gave rise to a magnetic field, and in 1831 Faraday demonstrated the converse of this, the induction of a current by a moving magnet. It is now generally accepted that all magnetism is due to circulating electric currents. In magnetic materials the magnetism is produced by electrons orbiting within the atoms; in most substances the magnetic effects of different electrons cancel each other out, but in some, such as iron, a net magnetic field can be induced. The earliest uses of magnets were as compasses; nowadays magnets, and especially electromagnets, are fundamental to the operation of electric generators and motors, loudspeakers, lifting magnets, transformers, etc. (See also FIELD.)

magnetite /'mægnɪˌtaɪt/ n. magnetic iron oxide. [G *Magnetit* (as MAGNET)]

magnetize /'mægnɪˌtaɪz/ v.tr. (also **-ise**) **1** give magnetic properties to. **2** make into a magnet. **3** attract as or like a magnet. □ **magnetizable** *adj.* **magnetizer** n. **magnetization** /ˌmægnɪtaɪ'zeɪʃ(ə)n/ n.

magneto /mæg'niːtəʊ/ n. (pl. **-os**) an electric generator using permanent magnets and producing high voltage, esp. for the ignition of an internal-combustion engine. [abbr. of MAGNETO-ELECTRIC]

magneto- /mæg'niːtəʊ/ comb. form indicating a magnet or magnetism. [Gk *magnēs*: see MAGNET]

magneto-electric /mægˌniːtəʊɪ'lektrɪk/ adj. (of an electric generator) using permanent magnets. □ **magneto-electricity** /-ˌɪlek'trɪsɪtɪ, -ˌel-/ n.

magnetograph /mæg'niːtəˌgrɑːf/ n. an instrument for recording measurements of magnetic quantities.

magnetohydrodynamics /mægˌniːtəʊˌhaɪdrəʊdaɪ'næmɪks/ n.pl. (usu. treated as *sing.*) the branch of physics that deals with the interaction of electrically conducting fluids (such as plasma or molten metal) and magnetic fields. Also called *hydromagnetics*. □ **magnetohydrodynamic** adj.

magnetometer /ˌmægnɪ'tɒmɪtə(r)/ n. an instrument measuring magnetic forces, esp. the earth's magnetism. □ **magnetometry** n.

magnetomotive /mægˌniːtəʊ'məʊtɪv/ adj. (of a force) being the sum of the magnetizing forces along a circuit.

magneton /'mægnɪˌtɒn/ n. a unit of magnetic moment in atomic and nuclear physics. [F *magnéton* (as MAGNETIC)]

magnetosphere /mæg'niːtəˌsfɪə(r)/ n. the region surrounding a planet, star, etc., in which its magnetic field is effective.

magnetron /'mægnɪˌtrɒn/ n. an electron tube for amplifying or generating microwaves, with the flow of electrons controlled by an external magnetic field. [MAGNET + -TRON]

Magnificat /mæg'nɪfɪˌkæt/ n. a canticle forming part of the Christian liturgy at evensong and vespers, and comprising the hymn of the Virgin Mary in Luke 1:46–55. It is named after the first word of the hymn in the Vulgate. [L, = magnifies (i.e. extols)]

magnification /ˌmægnɪfɪ'keɪʃ(ə)n/ n. **1** the act or an instance of magnifying; the process of being magnified. **2** the amount or degree of magnification. **3** the apparent enlargement of an object by a lens.

magnificent /mæg'nɪfɪs(ə)nt/ adj. **1** splendid, stately. **2** sumptuously or lavishly constructed or adorned. **3** *colloq.* fine, excellent. □ **magnificence** n. **magnificently** adv. [F *magnificent* or L *magnificus* f. *magnus* great]

magnifico /mæg'nɪfɪˌkəʊ/ n. (pl. **-oes**) a magnate or grandee. [It., = MAGNIFICENT: orig. with ref. to Venice]

magnify /'mægnɪˌfaɪ/ v.tr. (**-ies, -ied**) **1** make (a thing) appear larger than it is, as with a lens. **2** exaggerate. **3** intensify. **4** *archaic* extol, glorify. □ **magnifying glass** a lens used to produce an enlarged image. □ **magnifiable** adj. **magnifier** n. [ME f. OF *magnifier* or L *magnificare* (as MAGNIFICENT)]

magniloquent /mæg'nɪləkwənt/ adj. **1** grand or grandiose in speech. **2** boastful. □ **magniloquence** n. **magniloquently** adv. [L *magniloquus* f. *magnus* great + *-loquus* -speaking]

Magnitogorsk /ˌmægnɪtə'gɔːsk/ an industrial city in southern Russia, on the Ural river close to the border with Kazakhstan; pop. (1990) 443,000. Founded in 1921 near deposits of iron and magnetite, it has developed into a leading centre of metallurgy.

magnitude /'mægnɪˌtjuːd/ n. **1** largeness. **2 a** size. **b** a mathematical quantity. **3** importance. **4** *Astron.* a measure of the relative brightness of stars and other celestial objects. (*See note below.*) □ **absolute magnitude** *Astron.* a star's magnitude as it would be seen at a standard distance of 10 parsecs (32.6 light-years). **apparent magnitude** *Astron.* a star's magnitude as it is seen from the earth. **of the first magnitude** very important. [ME f. L *magnitudo* f. *magnus* great]

■ The astronomical scale of magnitude was based originally on one devised in the 2nd century BC by Hipparchus, who classed the brightest stars as first magnitude, down to the dimmest ones visible, which were classed as sixth magnitude. It is now quantified by a mathematical formula involving the logarithm of the measured energy, with the scale extended far beyond the range of unaided vision. A difference of five magnitudes represents a hundredfold difference in brightness. The brightest objects have negative magnitudes, with the planet Venus reaching −4.4, and the brightest star (Sirius) at −1.46.

magnolia /mæg'nəʊlɪə/ n. **1** a tree or shrub of the genus *Magnolia*, cultivated for its dark green foliage and large waxlike flowers in spring.

2 a pale creamy-white colour. [mod.L f. Pierre *Magnol*, Fr. botanist (1638–1715)]

magnox /ˈmægnɒks/ *n.* a magnesium-based alloy used to enclose uranium fuel elements in a nuclear reactor. [*magnesium no oxidation*]

magnum /ˈmægnəm/ *n.* (*pl.* **magnums**) **1** a wine bottle of about twice the standard size. **2 a** a cartridge or shell that is especially powerful or large. **b** (often *attrib.*) a cartridge or gun adapted so as to be more powerful than its calibre suggests. [L, neut. of *magnus* great]

magnum opus /ˌmægnəm ˈəʊpəs/ *n.* **1** a great and usu. large work of art, literature, etc. **2** the most important work of an artist, writer, etc. [L, = great work: see OPUS]

Magog see GOG AND MAGOG.

magpie /ˈmægpaɪ/ *n.* **1** a long-tailed crow; esp. *Pica pica*, of Europe and North America, with black and white plumage and a habit of collecting bright objects. **2** any bird with black and white plumage; esp. *Gymnorhina tibicen*, of Australia. **3** an idle chatterer. **4** a person who collects things indiscriminately. **5 a** the division of a circular target next to the outer one. **b** a rifle shot which strikes this. □ **magpie-lark** an Australian passerine bird of the family Grallinidae; esp. the common *Grallina cyanoleuca*, with black and white plumage and long legs (also called *peewee*). **magpie moth** a white geometrid moth, *Abraxas grossulariata*, with black and yellow spots, having caterpillars that feed on the leaves of fruit bushes. [*Mag*, abbr. of *Margaret* + PIE[2]]

Magritte /məˈgriːt/, René (François Ghislain) (1898–1967), Belgian painter. His paintings are typical examples of surrealism; they display startling or amusing juxtapositions of the ordinary, the strange, and the erotic, all depicted in a realist manner. He had a repertory of images which appear in incongruous settings, for example in *Threatening Weather* (1928), in which a chair, a table, and a torso hover like clouds over the sea.

magsman /ˈmægzmən/ *n.* (*pl.* **-men**) *Austral. sl.* **1** a confidence man. **2** a storyteller, a raconteur.

maguey /ˈmægweɪ/ *n.* an agave plant, esp. one yielding pulque. [Sp. f. Haitian]

magus /ˈmeɪgəs/ *n.* (*pl.* **magi** /ˈmeɪdʒaɪ/) **1** a member of a priestly caste of ancient Persia. **2** a magician, a sorcerer. (See also MAGI, THE.) [ME f. L f. Gk *magos* f. OPers. *magush*]

Magyar /ˈmægjɑː(r)/ *n. & adj.* ● *n.* **1** a member of a Uralic people now predominant in Hungary. **2** the language of this people; Hungarian. ● *adj.* of or relating to this people or language. [Hungarian]

Magyarország /ˈmɒdjɒrɒrˌsɑːg/ the Hungarian name for HUNGARY.

Mahabad /ˌmɑːhɑːˈbæd/ a city in NW Iran, near the Iraqi border, with a chiefly Kurdish population; pop. (1986) 63,000. Occupied by Soviet troops in 1941, Mahabad became the centre of a short-lived Soviet-supported Kurdish republic, which was overthrown by the Iranians in 1946.

Mahabharata /ˌmɑːhɑːˈbɑːrətə/ *n.* one of the two great Sanskrit epics of the Hindus (the other is the Ramayana), that evolved over centuries to reach its present form AD *c*.400. Containing almost 100,000 stanzas, it is probably the longest single poem in the world. The main story describes the civil war waged between the five Pandava brothers and their one hundred stepbrothers at Kuruksetra near modern Delhi, and there is much additional legendary, philosophical, and religious material; the numerous interpolated episodes include the Bhagavadgita. [Skr., = the great epic of the Bharata dynasty]

maharaja /ˌmɑːhɑːˈrɑːdʒə/ *n.* (also **maharajah**) *hist.* a title of some Indian princes. [Hindi *mahārāja* f. *mahā* great + RAJA]

maharanee /ˌmɑːhɑːˈrɑːniː/ *n.* (also **maharani**) *hist.* a maharaja's wife or widow. [Hindi *mahārānī* f. *mahā* great + RANEE]

Maharashtra /ˌmɑːhɑːˈræʃtrə, -ˈrɑːʃtrə/ a large state in western India bordering on the Arabian Sea, formed in 1960 from the south-eastern part of the former Bombay State; capital, Bombay. □ **Maharashtrian** *adj. & n.*

maharishi /ˌmɑːhɑːˈrɪʃɪ/ *n.* a great Hindu sage or spiritual leader. [Hindi f. *mahā* great + RISHI]

mahatma /məˈhætmə/ *n.* **1 a** (in India etc.) a person regarded with reverence. **b** a sage. **2** each of a class of persons in India and Tibet supposed by some to have preternatural powers. [Skr. *mahātman* f. *mahā* great + *ātman* soul]

Mahaweli /ˌmɑːhɑːˈweɪlɪ/ the largest river in Sri Lanka. Rising in the central highlands, it flows 330 km (206 miles) to the Bay of Bengal near Trincomalee.

Mahayana /ˌmɑːhɑːˈjɑːnə/ *n.* (also **Mahayana Buddhism**) one of the two major Buddhist traditions, now practised especially in China, Tibet, Japan, and Korea. It emerged around the 1st century AD and distinguishes itself from the more orthodox conservative tradition (see THERAVADA, HINAYANA). In particular, Mahayana Buddhism lays stress on the ideal of the bodhisattva (who postpones personal salvation for the love of others) rather than on personal enlightenment, and hence involves the belief that compassion is as important a virtue as wisdom. [Skr. f. *mahā* great + *yāna* vehicle]

Mahdi /ˈmɑːdɪ/ *n.* (*pl.* **Mahdis**) (in Muslim belief) a spiritual and temporal leader who will rule before the end of the world and restore religion and justice. Not part of orthodox doctrine, this concept was introduced into popular Islam through Sufi channels influenced by Christian doctrine. For Shiites, the title refers to the twelfth imam. The title has also been claimed by various leaders; the most widely known of these was Muhammad Ahmad of Dongola in Sudan (1843–85), who proclaimed himself Mahdi in 1881 and launched a political and revolutionary movement which captured Khartoum and overthrew the Egyptian regime. [Arab. *mahdīy* he who is rightly guided, past part. of *hadā* guide aright]

Mahdism /ˈmɑːdɪz(ə)m/ *n.* belief in or adherence to the Mahdi or a Mahdi, esp. Muhammad Ahmad of Dongola (see MAHDI). □ **Mahdist** *n. & adj.*

Mahfouz /mɑːˈfuːz/, Naguib (b.1911), Egyptian novelist and short-story writer. His novels include the Cairo Trilogy (1956–7), which monitors the stages of Egyptian nationalism up to the revolution of 1952, and *Miramar* (1967), an attack on Nasser's subsequent policies; his short stories of the late 1960s examine the aftermath of the 1967 war with Israel. In 1988 he became the first writer in Arabic to be awarded the Nobel Prize for literature.

Mahilyow /ˌmɑːhɪˈljəʊ/ (Russian **Mogilev** /məgɪˈljɒf/) an industrial city and railway centre in eastern Belarus, on the River Dnieper; pop. (1990) 363,000.

mah-jong /mɑːˈdʒɒŋ/ *n.* (also **mah-jongg**) an old Chinese game resembling certain card-games, introduced into Europe and America in the early 1920s, played usually by four persons using 136 or 144 pieces called tiles. [Chin. dial. *ma-tsiang*, lit. 'sparrows']

Mahler /ˈmɑːlə(r)/, Gustav (1860–1911), Austrian composer, conductor, and pianist. He was director (1897–1907) of the Vienna State Opera. His large-scale works include nine complete symphonies (1888–1910) and the symphonic song-cycle *Das Lied von der Erde* (1908). His music forms a link between the romantic tradition of the 19th century and the experimentalism of 20th-century composers such as Schoenberg; there has been a significant revival of interest in his work in the second half of the 20th century.

mahlstick var. of MAULSTICK.

mahogany /məˈhɒgənɪ/ *n.* (*pl.* **-ies**) **1 a** a hard reddish-brown wood used esp. for furniture. **b** the colour of this. **2** a tropical tree of the genus *Swietenia*, esp. *S. mahagoni*, yielding this wood. [17th c.: orig. unkn.]

Mahón /məˈhɒn/ (also **Mahon**, **Port Mahon**) the capital of the island of Minorca, a port on the SE coast; pop. (1991) 21,800. Thought to have been founded by the Carthaginians, the town was held by the Moors between the 8th and 13th centuries.

mahonia /məˈhəʊnɪə/ *n.* an evergreen shrub of the genus *Mahonia*, with prickly leaves and yellow bell-shaped or globular flowers. [F *mahonne*, Sp. *mahona*, It. *maona*, Turk. *māwuna*]

Mahore /məˈhɔː(r)/ an alternative name for MAYOTTE.

mahout /məˈhaʊt/ *n.* (in the Indian subcontinent) an elephant-driver or -keeper. [Hindi *mahāut* f. Skr. *mahāmātra* high official, lit. 'great in measure']

Mahratta var. of MARATHA.

Mahratti var. of MARATHI.

mahseer /ˈmɑːsɪə(r)/ *n.* a large Indian freshwater fish used for food; esp. *Barbus tor*, which resembles the barbel. [Hindi *mahāsir*]

Maia[1] /ˈmaɪə/ *Gk Mythol.* the daughter of Atlas and mother of Hermes. [Gk, = mother, nurse.]

Maia[2] /ˈmaɪə/ *Rom. Mythol.* a goddess associated (for unknown reasons) with Vulcan and also (by confusion with MAIA[1]) with Mercury (Hermes). She was worshipped on 1 May and 15 May; that month is named after her. [perh. f. L root *mag-* growth, increase]

maid /meɪd/ *n.* **1** a female domestic servant. **2** *archaic* or *poet.* a girl or young woman. □ **maid of honour 1** an unmarried lady attending a queen or princess. **2** a kind of small custard tart. **3** *N. Amer.* a principal bridesmaid. □ **maidish** *adj.* [ME, abbr. of MAIDEN]

maidan /maɪˈdɑːn/ *n. Anglo-Ind.* **1** an open space in or near a town. **2** a parade-ground. [Urdu f. Arab. *maydān*]

maiden /ˈmeɪd(ə)n/ *n.* **1 a** *archaic* or *poet.* a girl; a young unmarried woman. **b** (*attrib.*) unmarried (*maiden aunt*). **2** *Cricket* = maiden over. **3** (*attrib.*) (of a female animal) unmated. **4** (often *attrib.*) **a** a horse that has never won a race. **b** a race open only to such horses. **5** (*attrib.*) being or involving the first attempt or occurrence (*maiden speech; maiden voyage*). □ **maiden name** a wife's surname before marriage. **maiden over** *Cricket* an over in which no runs are scored off the bat. □ **maidenhood** *n.* **maidenish** *adj.* **maidenlike** *adj.* **maidenly** *adj.* [OE *mægden*, dimin. f. *mægeth* f. Gmc]

maidenhair /ˈmeɪd(ə)nˌheə(r)/ *n.* (in full **maidenhair fern**) a fern of the genus *Adiantum*, with fine hairlike stalks and delicate fronds; esp. *A. capillus-veneris*, often grown as a house plant. □ **maidenhair tree** = GINKGO.

Maidenhead /ˈmeɪd(ə)nhed/ a town in southern England, in Berkshire; pop. (1981) 60,460. It is situated to the west of London on the River Thames.

maidenhead /ˈmeɪd(ə)nhed/ *n.* **1** virginity. **2** the hymen.

maidservant /ˈmeɪdˌsɜːv(ə)nt/ *n.* a female domestic servant.

Maidstone /ˈmeɪdstəʊn/ a town in SE England, on the River Medway, the county town of Kent; pop. (1991) 133,200.

maieutic /meɪˈuːtɪk/ *adj. Philos.* (of the Socratic mode of enquiry) serving to bring a person's latent ideas into clear consciousness. [Gk *maieutikos* f. *maieuomai* act as a midwife f. *maia* midwife]

maigre /ˈmeɪgə(r)/ *adj. RC Ch.* **1** (of a day) on which abstinence from meat is ordered. **2** (of food) suitable for eating on maigre days. [F, lit. 'lean': cf. MEAGRE]

Maikop /maɪˈkɒp/ a city in SW Russia, capital of the republic of Adygea; pop. (est. 1986) 142,000.

mail[1] /meɪl/ *n. & v.* ● *n.* **1 a** letters and parcels etc. conveyed by post. **b** the postal system. **c** one complete delivery or collection of mail. **d** one delivery of letters to one place, esp. to a business on one occasion. **2** = *electronic mail*. **3** a vehicle carrying mail. **4** *hist.* a bag of letters for conveyance by post. ● *v.tr.* send (a letter etc.) by post. □ **mail-boat** a boat carrying mail. **mail carrier** *US* a postman or postwoman. **mail cart** *Brit. hist.* **1** a cart for carrying mail by road. **2** a light vehicle for carrying children. **mail coach** a railway coach or *hist.* stagecoach used for carrying mail. **mail drop** *N. Amer.* a receptacle for mail. **mailing list** a list of people to whom advertising matter, information, etc., is to be posted. **mail order** an order for goods sent by post. **mail-order firm** a firm doing business by post. **mail train** a train carrying mail. [ME f. OF *male* wallet f. WG]

mail[2] /meɪl/ *n. & v.* ● *n. hist.* **1** armour made of rings, chains, or plates, joined together flexibly. **2** the protective shell, scales, etc., of an animal. ● *v.tr.* clothe with or as if with mail. □ **coat of mail** *hist.* a jacket covered with mail or composed of mail. **mailed fist** physical force. □ **mailed** *adj.* [ME f. OF *maille* f. L *macula* spot, mesh]

mailable /ˈmeɪləb(ə)l/ *adj.* acceptable for conveyance by post.

mailbag /ˈmeɪlbæg/ *n.* a large sack or bag for carrying mail.

mailbox /ˈmeɪlbɒks/ *n. N. Amer.* a letter-box.

Mailer /ˈmeɪlə(r)/, Norman (b.1923), American novelist and essayist. He gained recognition with his first novel, *The Naked and the Dead* (1948), which drew on his experiences in the Second World War. The effect of war and violence on human relationships is a recurrent theme in his work, which frequently also includes an element of social criticism. His later novels, such as *The Presidential Papers* (1963) and *The Armies of the Night* (1968), combine journalism, autobiography, political commentary, and fictional passages in a wide range of styles.

maillot /ˈmaɪəʊ/ *n.* **1** tights for dancing, gymnastics, etc. **2** a woman's one-piece bathing-suit. **3** a jersey. [F]

mailman /ˈmeɪlmæn/ *n.* (pl. **-men**) *N. Amer.* a postman.

mailshot /ˈmeɪlʃɒt/ *n.* a dispatch of mail, esp. advertising and promotional material, to a large number of addresses.

maim /meɪm/ *v.tr.* **1** cripple, disable, mutilate. **2** harm, impair (*emotionally maimed by neglect*). [ME *maime* etc. f. OF *mahaignier* etc., of unkn. orig.]

Maimonides /maɪˈmɒnɪˌdiːz/ (born Moses ben Maimon) (1135–1204), Jewish philosopher and Rabbinic scholar, born in Spain. He eventually settled in Cairo, where he became head of the Jewish community. His writings include the *Guide for the Perplexed* (1190), which endeavoured to reconcile Talmudic scripture with the philosophy of Aristotle. His work had a great influence on medieval Christian thought.

Main /maɪn/ a river of SW Germany which rises in northern Bavaria and flows 500 km (310 miles) westwards, through Frankfurt, to meet the Rhine at Mainz.

main[1] /meɪn/ *adj. & n.* ● *adj.* **1** chief in size, importance, extent, etc.; principal (*the main part; the main point*). **2** exerted to the full (*by main force*). ● *n.* **1** a principal channel, duct, etc., for water, sewage, etc. (*water main*). **2** (usu. in *pl.*; prec. by *the*) **a** the central distribution network for electricity, gas, water, etc. **b** a domestic electricity supply as distinct from batteries. **3** *archaic* or *poet.* **a** the ocean or oceans (*the Spanish Main*). **b** the mainland. □ **in the main** for the most part. **main brace** *Naut.* the brace attached to the main yard. **the main chance** one's own interests. **main course 1** the chief course of a meal. **2** *Naut.* the mainsail. **main deck** *Naut.* **1** the deck below the spar-deck in a man-of-war. **2** the upper deck between the poop and the forecastle in a merchantman. **main line 1** a chief railway line. **2** *sl.* a principal vein, esp. as a site for a drug injection (cf. MAINLINE). **3** *US* a chief road or street. **main stem** *US colloq.* = *main street.* **main street 1** the principal street of a town. **2** esp. *US* mediocrity, parochialism, or materialism regarded as typical of small-town life (after Sinclair Lewis's novel, 1920). **main yard** *Naut.* the yard on which the mainsail is extended. **with might and main** with all one's force. [ME, partly f. ON *megenn*, *megn* (adj.), partly f. OE *mægen-* f. Gmc: (n.) orig. = physical force]

main[2] /meɪn/ *n.* **1** (in the game of hazard) a number (5, 6, 7, 8, or 9) called by a player before dice are thrown. **2** a match between fighting-cocks. [16th c.: prob. orig. *the main chance* (see MAIN[1])]

Maine /meɪn/ a north-eastern state of the US, on the Atlantic coast; pop. (1990) 1,227,930; capital, Augusta. Visited by John Cabot in 1498 and colonized from England in the 17th and 18th centuries, it became the 23rd state of the US in 1820.

mainframe /ˈmeɪnfreɪm/ *n.* **1** the central processing unit and primary memory of a computer. **2** (often *attrib.*) a large computer system.

Mainland /ˈmeɪnlənd/ **1** the largest island in Orkney. **2** the largest island in Shetland.

mainland /ˈmeɪnlənd/ *n.* a large continuous extent of land, excluding neighbouring islands etc. □ **mainlander** *n.*

mainline /ˈmeɪnlaɪn/ *v. sl.* **1** *intr.* take drugs intravenously. **2** *tr.* inject (drugs) intravenously. □ **mainliner** *n.*

mainly /ˈmeɪnlɪ/ *adv.* for the most part; chiefly.

mainmast /ˈmeɪnmɑːst, -məst/ *n. Naut.* the principal mast of a ship.

mainplane /ˈmeɪnpleɪn/ *n.* the principal supporting surface of an aircraft (cf. TAILPLANE).

mainsail /ˈmeɪnseɪl, *Naut.* -s(ə)l/ *n. Naut.* **1** (in a square-rigged vessel) the lowest sail on the mainmast. **2** (in a fore-and-aft rigged vessel) a sail set on the after part of the mainmast.

mainspring /ˈmeɪnsprɪŋ/ *n.* **1** the principal spring of a mechanical watch, clock, etc. **2** a chief motive power; an incentive.

mainstay /ˈmeɪnsteɪ/ *n.* **1** a chief support (*has been his mainstay since his trouble*). **2** *Naut.* a stay from the maintop to the foot of the foremast.

mainstream /ˈmeɪnstriːm/ *n.* **1** (often *attrib.*) the prevailing trend in opinion, fashion, etc. **2** a type of jazz based on the 1930s swing style and consisting esp. of solo improvisation on chord sequences. **3** the principal current of a river.

maintain /meɪnˈteɪn/ *v.tr.* **1** cause to continue; keep up, preserve (a state of affairs, an activity, etc.) (*maintained friendly relations*). **2** (often foll. by *in*; often *refl.*) support (life, a condition, etc.) by work, nourishment, expenditure, etc. (*maintained him in comfort; maintained themselves by fishing*). **3** (often foll. by *that* + clause) assert (an opinion, statement, etc.) as true (*maintained that she was the best; his story was true, he maintained*). **4** preserve or provide for the preservation of (a building, machine, road, etc.) in good repair. **5** give aid to (a cause, party, etc.). **6** provide means for (a garrison etc. to be equipped). □ **maintained school** *Brit.* a school supported from public funds. □ **maintainable** *adj.* **maintainability** /-ˌteɪnəˈbɪlɪtɪ/ *n.* [ME f. OF *maintenir* ult. f. L *manu tenere* hold in the hand]

maintainer /meɪnˈteɪnə(r)/ *n.* **1** a person or thing that maintains. **2** (also **maintainor**) *Law hist.* a person guilty of maintenance (see MAINTENANCE 3).

maintenance /ˈmeɪntənəns/ *n.* **1** the process of maintaining or being maintained. **2 a** the provision of the means to support life, esp. by work etc. **b** (also **separate maintenance**) a husband's or wife's provision for a spouse after separation or divorce; alimony. **3** *Law hist.* the offence of aiding a party in litigation without lawful cause. [ME f. OF f. *maintenir*: see MAINTAIN]

Maintenon /ˈmæntəˌnɒn/ Marquise de (title of Françoise d'Aubigné)

(1635–1719), mistress and later second wife of the French king Louis XIV. Already a widow, in 1669 she became the governess of Louis's children by his previous mistress, Madame de Montespan. In 1674, with the king's assistance, she bought the marquisate of Maintenon. She was married to Louis after his first wife's death in 1683.

maintop /ˈmeɪntɒp/ n. Naut. a platform above the head of the lower mainmast.

maintopmast /meɪnˈtɒpmɑːst, Naut. -məst/ n. Naut. a mast above the head of the lower mainmast.

Mainz /maɪnts/ a city in western Germany, capital of Rhineland-Palatinate, situated at the confluence of the Rhine and Main rivers; pop. (1991) 182,870.

maiolica var. of MAJOLICA.

maisonette /ˌmeɪzəˈnet/ n. (also **maisonnette**) **1** a part of a house, block of flats, etc., forming separate living accommodation, usu. on two floors and having a separate entrance. **2** a small house. [F maisonnette dimin. of maison house]

Maithili /ˈmaɪtɪlɪ/ n. one of the Bihari group of languages, spoken by some 24 million people in northern Bihar, elsewhere in India, and in Nepal. [Skr.]

maître d'hôtel /ˌmetrə dəʊˈtel/ n. (pl. **maîtres d'hôtel** pronunc. same) **1** the manager, head steward, etc., of a hotel. **2** a head waiter. [F, = master of (the) house]

maize /meɪz/ n. **1** a cereal plant, Zea mays, native to America, yielding large grains set in rows on a cob. Also called Indian corn. (See note below.) **2** the cobs or grains of this. [F maïs or Sp. maiz, of Carib orig.]

   ■ Maize had already been domesticated by about 5000 BC, probably in the area of Mexico. It is unclear, however, whether it was a cultivated form of a wild maize, now extinct, or developed from other native grasses. It is referred to in North America and Australia as 'corn', and is familiar as cornflakes, popcorn, and corn on the cob.

Maj. abbr. Major.

majestic /məˈdʒestɪk/ adj. showing majesty; stately and dignified; grand, imposing. □ **majestically** adv.

majesty /ˈmædʒɪstɪ/ n. (pl. **-ies**) **1** impressive stateliness, dignity, or authority, esp. of bearing, language, the law, etc. **2 a** royal power. **b** (**Majesty**) part of several titles given to a sovereign or a sovereign's wife or widow or used in addressing them (Your Majesty; Her Majesty the Queen Mother). **3** a picture of God or Christ enthroned within an aureole. □ **Her** (or **His**) **Majesty's** part of the title of several state institutions (Her Majesty's Stationery Office). [ME f. OF majesté f. L majestas -tatis (as MAJOR)]

Majlis /ˈmædʒlɪs/ n. the Parliament of various North African or Middle Eastern countries, esp. Iran. [Pers., = assembly]

majolica /məˈjɒlɪkə, məˈdʒɒl-/ n. (also **maiolica** /məˈjɒl-/) a fine kind of earthenware with coloured decoration on an opaque white glaze, originating in Italy during the Renaissance. [It., f. the former name of the island of MAJORCA, ships of which brought Spanish wares to Italy]

Major /ˈmeɪdʒ(ə)r/, John (b.1943), British Conservative statesman, Prime Minister since 1990. He became Prime Minister following the resignation of Margaret Thatcher, and in 1992 he was returned for a further term of office. His premiership has seen the abolition of the community charge, the negotiations leading to the signing of the Maastricht Treaty, and the joint 'Downing Street Declaration' of the UK and Irish governments, intended as the basis of a peace initiative in Northern Ireland.

major /ˈmeɪdʒə(r)/ adj., n., & v. ● adj. **1** important, large, serious, significant (a major road; a major war; the major consideration must be their health). **2** (of an operation) serious or life-threatening. **3** Mus. **a** (of a scale) having intervals of a semitone between the third and fourth, and seventh and eighth degrees. **b** (of an interval) greater by a semitone than a minor interval (major third). **c** (of a key) based on a major scale, tending to produce a bright or joyful effect (D major). **4** of full age. **5** Brit. (appended to a surname, esp. in public schools) indicating the elder of two brothers (Smith major). **6** Logic **a** (of a term) occurring in the predicate or conclusion of a syllogism. **b** (of a premiss) containing a major term. ● n. **1** Mil. **a** an army officer next below lieutenant colonel and above captain. **b** an officer in charge of a section of band instruments (drum major; pipe major). **2** a person of full age. **3** Mus. a major key etc. **4** US **a** a student's special subject or course. **b** a student specializing in a specified subject (a philosophy major). **5** Logic a major term or premiss. ● v.intr. (foll. by in) US study or qualify in as a special subject (majored in theology). □ **major axis** the axis of a conic, passing through its foci. **major-general** an officer next below a lieutenant

general. **major league** N. Amer. a league of the highest rank in baseball etc. **major part** (often foll. by of) the majority. **major piece** Chess a rook or queen. **major planet** Jupiter, Saturn, Uranus, or Neptune. **major prophet** see PROPHET. **major suit** Bridge spades or hearts. □ **majorship** n. [ME f. L, compar. of magnus great]

Majorca /məˈjɔːkə/ (Spanish **Mallorca** /maˈʎorka/) the largest of the Balearic Islands; pop. (1981) 561,200; capital, Palma.

major-domo /ˌmeɪdʒəˈdəʊməʊ/ n. (pl. **-os**) **1** the chief official of an Italian or Spanish princely household. **2** a house-steward; a butler. [orig. mayordome f. Sp. mayordomo, It. maggiordomo f. med.L major domus highest official of the household (as MAJOR, DOME)]

majorette /ˌmeɪdʒəˈret/ n. = drum majorette. [abbr.]

majority /məˈdʒɒrɪtɪ/ n. (pl. **-ies**) **1** (usu. foll. by of) the greater number or part. ¶ Strictly used only with countable nouns, e.g. a majority of people, and not with mass nouns, e.g. a majority of the work. **2** Polit. **a** the number by which the votes cast for one party, candidate, etc. exceed those of the next in rank (won by a majority of 151). **b** a party etc. receiving the greater number of votes. **3** full legal age (attained his majority). **4** the rank of major. □ **the great majority 1** much the greater number. **2** euphem. the dead (has joined the great majority). **in the majority** esp. Polit. belonging to or constituting a majority party etc. **majority rule** the principle that the greater number should exercise greater power. **majority verdict** a verdict given by more than half of the jury, but not unanimous. [F majorité f. med.L majoritas -tatis (as MAJOR)]

majuscule /ˈmædʒəˌskjuːl/ n. & adj. ● n. **1** a large letter, whether capital or uncial. **2** large lettering. ● adj. of, written in, or concerning majuscules. □ **majuscular** /məˈdʒʌskjʊlə(r)/ adj. [F f. L majuscula (littera letter), dimin. of MAJOR]

Makarios III /maˈkɑːrɪˌɒs/ (born Mikhail Christodolou Mouskos) (1913–77), Greek Cypriot archbishop and statesman, President of the republic of Cyprus 1960–77. He was primate and archbishop of the Greek Orthodox Church in Cyprus from 1950, and combined this position with a vigorous political role. He reorganized the movement for enosis (union of Cyprus with Greece) and was exiled (1956–9) by the British for allegedly supporting the EOKA terrorist campaign. He was elected first President of an independent Cyprus and, although forced briefly into exile by a Greek military coup (1974), continued in office until his death.

Makasarese /maˌkæsəˈriːz/ n. & adj. (also archaic **Macassarese**) ● n. (pl. same) **1** a member of a people of Makassar (now Ujung Pandang) in Indonesia. **2** the language of this people. ● adj. of or relating to this people or their language. [f. MAKASSAR + -ESE]

Makassar /maˈkæsə(r)/ (also **Macassar** or **Makasar**) the former name (until 1973) for UJUNG PANDANG.

Makassar Strait a stretch of water separating the islands of Borneo and Sulawesi (Celebes) and linking the Celebes Sea in the north with the Java Sea in the south.

make /meɪk/ v. & n. ● v. (past and past part. **made**) **1** tr. construct; create; form from parts or other substances (made a table; made it out of cardboard; made him a sweater). **2** tr. (usu. foll. by infin. with or without to) cause or compel (a person etc.) to do something (make him repeat it; was made to confess). **3** tr. **a** cause to exist; create; bring about (made a noise; made an enemy). **b** cause to become or seem (made an exhibition of myself; made him angry). **c** appoint; designate (made him a Cardinal). **4** tr. compose; prepare; draw up (made her will; made a film about Japan). **5** tr. constitute; amount to (makes a difference; 2 and 2 make 4; this makes the tenth time). **6** tr. **a** undertake or agree to (an aim or purpose) (made a promise; make an effort). **b** execute or perform (a bodily movement, a speech, etc.) (made a face; made a bow). **7** tr. gain, acquire, procure (money, a profit, etc.) (made £20,000 on the deal). **8** tr. prepare (tea, coffee, a dish, etc.) for consumption (made egg and chips). **9** tr. **a** arrange bedclothes tidily on (a bed) ready for use. **b** arrange and light materials for (a fire). **10** intr. **a** proceed (made towards the river). **b** (foll. by to + infin.) begin an action (he made to go). **11** tr. colloq. **a** arrive at (a place) or in time for (a train etc.) (made the border before dark; made the six o'clock train). **b** manage to attend; manage to attend on (a certain day) or at (a certain time) (couldn't make the meeting last week; can make any day except Friday). **c** achieve a place in (made the first eleven; made the six o'clock news). **d** esp. N. Amer. achieve the rank of (made colonel in three years). **12** tr. establish or enact (a distinction, rule, law, etc.). **13** tr. consider to be; estimate as (what do you make the time?; do you make that a 1 or a 7?). **14** tr. secure the success or advancement of (his mother made him; it made my day). **15** tr. accomplish (a distance, speed, score, etc.) (made 60 m.p.h. on the motorway). **16** tr. **a** become by development or training (made a great leader). **b** serve as (a log makes a useful seat). **17** tr. (usu. foll. by out) represent as; cause to appear as (makes him out a liar). **18** tr. form in the mind (I make no judgement). **19** tr. (foll. by it + complement) **a** determine, establish, or choose (let's make it Tuesday; made it my business to know). **b** bring to (a chosen value etc.) (decided to make it a dozen). **20** tr. sl. have sexual relations with. **21** tr. Cards **a** win (a trick). **b** play (a card) to advantage. **c** win the number of tricks that fulfils (a contract). **d** shuffle (a pack of cards) for dealing. **22** tr. Cricket score (runs). **23** tr. Electr. complete or close (a circuit) (opp. BREAK¹ v. 10). **24** intr. (of the tide) begin to flow or ebb. ● n. **1** (esp. of a product) a type, origin, brand, etc. of manufacture (different make of car; our own make). **2** a kind of mental, moral, or physical structure or composition. **3** an act of shuffling cards. **4** Electr. a making of contact. **b** the position in which this is made. □ **be made for** be ideally suited to. **be made of** consist of (cake made of marzipan). **have it made** colloq. be sure of success. **made dish** a dish prepared from several separate foods. **made man** a man who has attained success. **made of money** colloq. very rich. **made road** a properly surfaced road of tarmac, concrete, etc. **made to measure** (of a suit etc.) made to a specific customer's measurements. **made to order** see ORDER. **make after** archaic pursue. **make against** be unfavourable to. **make as if** (or **though**) (foll. by to + infin. or conditional) act as if the specified circumstances applied (made as if to leave; made as if he would hit me; made as if I had not noticed). **make away** (or **off**) depart hastily. **make away with 1** get rid of; kill. **2** squander. **3** see make off with. **make-believe** (or **-belief**) **1** pretence. **2** pretended. **make believe** pretend. **make conversation** talk politely. **make a day** (or **night** etc.) **of it** devote a whole day (or night etc.) to an activity. **make do 1** manage with the limited or inadequate means available. **2** (foll. by with) manage with (something) as an inferior substitute. **make an example of** punish as a warning to others. **make a fool of** see FOOL¹. **make for 1** tend to result in (happiness etc.). **2** proceed towards (a place). **3** assault; attack. **4** confirm (an opinion). **make friends** (often foll. by with) become friendly. **make fun of** see FUN. **make good** see GOOD. **make a habit of** see HABIT. **make a hash of** see HASH¹. **make hay** see HAY¹. **make head or tail of** see HEAD. **make headway** advance, progress. **make a House** Polit. secure the presence of enough members for a quorum or support in the House of Commons. **make it** colloq. succeed in reaching, esp. in time. **2** colloq. be successful. **3** (usu. foll. by with) sl. have sexual intercourse (with). **make it up 1** be reconciled, esp. after a quarrel. **2** fill in a deficit. **make it up to** remedy negligence, an injury, etc. to (a person). **make light of** see LIGHT². **make love** see LOVE. **make a meal of** see MEAL¹. **make merry** see MERRY. **make money** acquire wealth or an income. **make the most of** see MOST. **make much** (or **little** or **the best**) **of 1** derive much (or little etc.) advantage from. **2** give much (or little etc.) attention, importance, etc., to. **make a name for oneself** see NAME. **make no bones about** see BONE. **make nothing of 1** do without hesitation. **2** treat as a trifle. **3** be unable to understand, use, or deal with. **make of 1** construct from. **2** conclude to be the meaning or character of (can you make anything of it?). **make off** see make away. **make off** (or **away**) **with** carry away; steal. **make oneself scarce** see SCARCE. **make or break** (or **mar**) cause the success or ruin of. **make out 1 a** distinguish by sight or hearing. **b** decipher (handwriting etc.). **2** understand (can't make him out). **3** assert; pretend (made out he liked it). **4** colloq. make progress; fare (how did you make out?). **5** (usu. foll. by to, in favour of) draw up; write out (made out a cheque to her). **6** prove or try to prove (how do you make that out?). **7** (often foll. by with) esp. US colloq. **a** engage in sexual play or petting. **b** form a sexual relationship. **make over 1** transfer the possession of (a thing) to a person. **2** refashion (a garment etc.). **make a point of** see POINT. **make sail** Naut. **1** spread a sail or sails. **2** start a voyage. **make shift** see SHIFT. **make so bold as to** see BOLD. **make time 1** (usu. foll. by for or to + infin.) find an occasion when time is available. **2** (usu. foll. by with) N. Amer. sl. make sexual advances (to a person). **make-up 1** cosmetics for the face etc., either generally or to create an actor's appearance or disguise. **2** the appearance of the face etc. when cosmetics have been applied (his make-up was not convincing). **3** Printing the making up of a type. **4** Printing the type made up. **5** a person's character, temperament, etc. **6** the composition or constitution of a thing. **make up 1** serve or act to overcome (a deficiency). **2** complete (an amount, a party, etc.). **3** compensate. **4** be reconciled. **5** put together; compound; prepare (made up the medicine). **6** sew (parts of a garment etc.) together. **7** get (a sum of money, a company, etc.) together. **8** concoct (a story). **9** (of parts) compose (a whole). **10 a** apply cosmetics. **b** apply cosmetics to. **11** settle (a dispute). **12** prepare (a bed) for use with fresh sheets etc. **13** Printing arrange (type) in pages. **14** compile (a list, an account, a document, etc.). **15** arrange (a marriage etc.). **make up one's mind** decide, resolve. **make up to** curry favour with; court.

make water 1 urinate. **2** (of a ship) take in water. **make way 1** (often foll. by for) allow room for others to proceed. **2** achieve progress. **make one's way** proceed. **make with** US colloq. supply; perform; proceed with (made with the feet and left in a hurry). **on the make** colloq. **1** intent on gain. **2** looking for sexual partners. **self-made man** etc. a man etc. who has succeeded by his own efforts. □ **makable** adj. [OE macian f. WG: rel. to MATCH¹]

makeover /ˈmeɪkˌəʊvə(r)/ n. a complete transformation or remodelling.

maker /ˈmeɪkə(r)/ n. **1** (often in comb.) a person or thing that makes. **2** (**our, the**, etc. **Maker**) God. **3** archaic a poet.

makeshift /ˈmeɪkʃɪft/ adj. & n. ● adj. temporary; serving for the time being (a makeshift arrangement). ● n. a temporary substitute or device.

makeweight /ˈmeɪkweɪt/ n. **1** a small quantity or thing added to make up the full weight. **2** an unimportant extra person. **3** an unimportant point added to make an argument seem stronger.

Makgadikgadi Pans /ˌmæɡəˈdiːɡədɪ/ an extensive area of salt-pans in central Botswana. In prehistoric times it formed a large lake.

Makhachkala /ˌmæxətʃkəˈlɑː/ a port in SW Russia, on the Caspian Sea, capital of the autonomous republic of Dagestan; pop. (1990) 327,000. It was known as Port Petrovsk until 1922, when it was renamed after the Dagestani revolutionary Makhach (1882–1918).

making /ˈmeɪkɪŋ/ n. **1** in senses of MAKE v. **2** (in pl.) **a** earnings; profit. **b** (foll. by of) essential qualities or ingredients (has the makings of a general; we have the makings of a meal). **c** N. Amer. & Austral. colloq. paper and tobacco for rolling a cigarette. □ **be the making of** ensure the success or favourable development of. **in the making** in the course of being made or formed. [OE macung (as MAKE)]

Makkah see MECCA¹.

mako¹ /ˈmækəʊ/ n. (pl. **-os**) a large bluish shark, Isurus oxyrinchus, found worldwide. [Maori]

mako² /ˈmækəʊ/ n. (pl. **-os**) a small New Zealand tree, Aristotelia racemosa, with clusters of small pink flowers and dark red berries. Also called wineberry. [Maori]

Maksutov telescope /mækˈsuːtɒf, ˈmæksʊˌtɒf/ n. a type of catadioptric telescope having a deeply curved meniscus lens. A secondary mirror on the back of the lens brings the light to a focus just behind a hole in the primary mirror. [D. D. Maksutov, Soviet astronomer (1896–1964), who invented it]

makuta pl. of LIKUTA.

Mal. abbr. (in the Bible) Malachi.

mal- /mæl/ comb. form **1 a** bad, badly (malpractice; maltreat). **b** faulty, faultily (malfunction). **2** not (maladroit). [F mal badly f. L male]

Malabar Christians /ˈmæləˌbɑː(r)/ n.pl. a group of Christians of SW India tracing their origin to St Thomas, who according to their tradition landed on the Malabar Coast.

Malabar Coast /ˈmæləˌbɑː(r)/ the southern part of the west coast of India, including the coastal region of Karnataka and most of the state of Kerala. It is named from the Malabars, an ancient Dravidian people.

Malabo /məˈlɑːbəʊ/ the capital of Equatorial Guinea, on the island of Bioko; pop. (est. 1986) 10,000.

malabsorption /ˌmæləbˈsɔːpʃ(ə)n, -ˈzɔːpʃ(ə)n/ n. imperfect absorption of food material by the small intestine.

Malacca see MELAKA.

malacca /məˈlækə/ n. (in full **malacca cane**) a rich brown cane made from the stem of the palm tree Calamus scipionum, used for walking-sticks etc. [MALACCA]

Malacca, Strait of the channel between the Malay Peninsula and the Indonesian island of Sumatra, an important sea passage linking the Indian Ocean to the South China Sea. The ports of Melaka and Singapore lie on this strait.

Malachi /ˈmæləˌkaɪ/ a book of the Bible belonging to a period before Ezra and Nehemiah. Malachi is probably not a personal name. [Heb. mālāki my messenger]

malachite /ˈmæləˌkaɪt/ n. a bright-green mineral of hydrous copper carbonate, taking a high polish and used for ornament. [OF melochite f. L molochites f. Gk molokhitis f. molokhē = malakhē mallow]

malaco- /ˈmæləkəʊ/ comb. form soft. [Gk malakos soft]

malacology /ˌmæləˈkɒlədʒɪ/ n. the study of molluscs.

malacostracan /ˌmæləˈkɒstrəkən/ n. & adj. Zool. ● n. a crustacean of the large class Malacostraca, which includes crabs, shrimps, lobsters, isopods, and amphipods. ● *adj.* of or relating to this class. [mod.L f. MALACO- + Gk *ostrakon* shell]

maladaptive /ˌmæləˈdæptɪv/ *adj.* (of an individual, species, etc.) failing to adjust adequately to the environment, and undergoing emotional, behavioural, physical, or mental repercussions. □ **maladaptation** /ˌmælædæpˈteɪʃ(ə)n/ *n.*

maladjusted /ˌmæləˈdʒʌstɪd/ *adj.* **1** not correctly adjusted. **2** (of a person) unable to adapt to or cope with the demands of a social environment. □ **maladjustment** *n.*

maladminister /ˌmælədˈmɪnɪstə(r)/ *v.tr.* manage or administer inefficiently, badly, or dishonestly. □ **maladministration** /-ˌmɪnɪˈstreɪʃ(ə)n/ *n.*

maladroit /ˌmæləˈdrɔɪt/ *adj.* clumsy; bungling. □ **maladroitly** *adv.* **maladroitness** *n.* [F (as MAL-, ADROIT)]

malady /ˈmælədɪ/ *n.* (*pl.* **-ies**) **1** an ailment; a disease. **2** a morbid or depraved condition; something requiring a remedy. [ME f. OF *maladie* f. *malade* sick ult. f. L *male* ill + *habitus* past part. of *habere* have]

mala fide /ˌmeɪlə ˈfaɪdɪ/ *adj. & adv.* ● *adj.* acting or done in bad faith. ● *adv.* in bad faith. [L]

Malaga[1] /ˈmæləɡə/ (Spanish **Málaga** /ˈmalaɣa/) a seaport on the Andalusian coast of southern Spain; pop. (1991) 524,750.

Malaga[2] /ˈmæləɡə/ *n.* a sweet fortified wine from Malaga in Spain.

Malagasy /ˌmæləˈɡæsɪ/ *adj. & n.* ● *adj.* of or relating to Madagascar or its people or language. ● *n.* **1** a native or national of Madagascar. **2** the Malayo-Polynesian language of Madagascar, spoken by about 10 million people. [orig. *Malegass, Madegass* f. *Madagascar*]

Malagasy Republic the former name (1960–75) for MADAGASCAR.

malagueña /ˌmæləˈɡeɪnjə/ *n.* **1** a Spanish dance resembling the fandango. **2** a piece of music for or in the style of a fandango. [Sp. f. MALAGA[1]]

malaise /məˈleɪz/ *n.* **1** a nonspecific bodily discomfort not associated with the development of a disease. **2** a feeling of uneasiness. [F f. OF *mal* bad + *aise* EASE]

Malamud /ˈmæləməd/, Bernard (1914–86), American novelist and short-story writer. The son of Russian Jewish immigrants, he is perhaps best known for his novel *The Fixer* (1967), the story of a Jewish handyman or 'fixer' in tsarist Russia just before the First World War who is falsely accused of murder and turned into a scapegoat for anti-Semitic feeling.

malamute /ˈmæləˌmjuːt/ *n.* (also **malemute**) an Eskimo dog. [name of an Alaskan Inuit people]

malanders var. of MALLENDERS.

malapert /ˈmæləpɜːt/ *adj.* *n. archaic* ● *adj.* impudent; saucy. ● *n.* an impudent or saucy person. [ME f. OF (as MAL-, *apert* = *espert* EXPERT)]

malapropism /ˈmæləprɒˌpɪz(ə)m/ *n.* (also **malaprop** /-ˌprɒp/) the use of a word in mistake for one sounding similar, to comic effect, e.g. dance a *flamingo* (for *flamenco*). [Mrs *Malaprop* (f. MALAPROPOS) in Sheridan's *The Rivals* (1775)]

malapropos /ˌmæləprəˈpəʊ/ *adv., adj., & n.* ● *adv.* inopportunely; inappropriately. ● *adj.* inopportune; inappropriate. ● *n.* something inappropriately said, done, etc. [F *mal à propos* f. *mal* ill: see APROPOS]

malar /ˈmeɪlə(r)/ *adj. & n.* ● *adj.* of the cheek. ● *n.* a bone of the cheek. [mod.L *malaris* f. L *mala* jaw]

Mälaren /ˈmeləˌren/ a lake in SE Sweden, extending inland from the Baltic Sea. The city of Stockholm is situated at its outlet.

malaria /məˈleərɪə/ *n. Med.* an intermittent and remittent fever caused by a protozoan parasite of the genus *Plasmodium*, transmitted from infected persons by the bite of a female *Anopheles* mosquito after developing in the body of this insect. The frequency of episodes of fever and sweating varies in different forms of the disease; *malignant malaria* causes a nearly continuous fever and is associated with dangerous complications and death. Effective preventive and curative drugs have been developed (the earliest known was quinine), but the parasite is becoming resistant in some areas, and successful eradication of malaria depends on dealing with the mosquitoes and preventing their breeding. The disease remains a problem in many tropical and subtropical regions. □ **malarial** *adj.* **malarian** *adj.* **malarious** *adj.* [It. *mal'aria* bad air, the disease being formerly attributed to vapour given off by marshes]

malarkey /məˈlɑːkɪ/ *n. colloq.* humbug; nonsense. [20th c.: orig. unkn.]

malathion /ˌmæləˈθaɪən/ *n.* an insecticide containing phosphorus, with low toxicity to other animals. [diethyl *maleate* + *thio-* acid + -ON]

Malawi /məˈlɑːwɪ/ a country of south central Africa, in the Great Rift Valley; pop. (est. 1991) 8,796,000; official languages, English and Nyanja; capital, Lilongwe. Malawi is a landlocked country, heavily dependent on Mozambique for access to the sea, and much of its eastern border is formed by Lake Malawi. Malawi is the former Nyasaland, a British protectorate from 1891 (following Livingstone's exploration), and from 1953 to 1963 a part of the Federation of Rhodesia and Nyasaland; it became an independent Commonwealth state under President Hastings Banda in 1964 and a republic in 1966. □ **Malawian** *adj. & n.*

Malawi, Lake an alternative name for Lake Nyasa (see NYASA, LAKE).

Malay /məˈleɪ/ *n. & adj.* ● *n.* **1 a** a member of a people inhabiting Malaysia, Brunei, parts of Indonesia, and other areas. **b** a person of Malay descent. **2** the Malayo-Polynesian language of the Malays. (*See note below.*) ● *adj.* of or relating to this people or language. [Malay *malāyu*]

▪ Malay, which is almost identical to Indonesian, is spoken by about 20 million people, particularly in Malaysia, where it is the first language of about half the population. From the 14th century Malay was written in Arabic script but in the 19th century the British colonists devised a Roman-based alphabet which is in general use today.

Malaya /məˈleɪə/ a former country in SE Asia, consisting of the southern part of the Malay Peninsula and some adjacent islands (originally including Singapore), now forming the western part of the federation of Malaysia and known as West Malaysia. Malaya was dominated by the Buddhist kingdom of Srivijaya from the 9th to 14th centuries and by the Hindu kingdom of Majaphit in the 14th century. Islam was introduced with the rise of the princely states, especially Malacca, in the 15th century. The area was opened up by the Dutch and Portuguese, and eventually Britain became dominant, investing heavily in rubber plantations using much immigrant labour from China and India. The several Malay states federated under British control in 1896. Malaya was occupied by the Japanese from 1941 to 1945. After the war, Britain fought a successful campaign (the Malayan Emergency) from 1948 to 1960 against Communist guerrillas, mainly Chinese. The country became independent in 1957, the federation expanding into Malaysia in 1963. (See also MALAYSIA.)

Malayalam /ˌmæləˈjɑːləm/ *n.* the Dravidian language of the state of Kerala in southern India. [Malayalam *Malayālam* f. *mala* (Tamil *malai*) mountain + *āl* man]

Malayan /məˈleɪən/ *n. & adj.* ● *n.* **1** a Malay. **2** the Malay language. ● *adj.* of or relating to Malaya, Malaysia, the Malays, or the Malay language.

Malay Archipelago a very large group of islands, including Sumatra, Java, Borneo, the Philippines, and New Guinea, lying between SE Asia and Australia. They constitute the bulk of the area formerly known as the East Indies.

Malayo- /məˈleɪəʊ/ *comb. form* Malayan and (*Malayo-Chinese*). [MALAY]

Malayo-Polynesian /məˌleɪəʊˌpɒlɪˈniːzɪən, -ˈniːʒ(ə)n/ *n. & adj.* ● *n.* a family of languages (also called *Austronesian*) extending from Madagascar in the west to the Pacific islands in the east. (*See note below.*) ● *adj.* of or relating to the Malays and the Polynesians, or to Malayo-Polynesian.

▪ Malayo-Polynesian languages are spoken by about 140 million people, of whom all but 1 million speak a language of the Indonesian group, which includes Indonesian, Javanese, Tagalog, and Malagasy. The other groups are Micronesian, Melanesian, and Polynesian.

Malay Peninsula a peninsula in SE Asia separating the Indian Ocean from the South China Sea. It extends approximately 1,100 km (700 miles) southwards from the Isthmus of Kra and comprises the southern part of Thailand and the whole of Malaya (West Malaysia).

Malaysia /məˈleɪzɪə, -ˈleɪʒə/ a country in SE Asia; pop. (est. 1991) 18,294,000; official language, Malay; capital, Kuala Lumpur. Malaysia is a federation consisting of *East Malaysia* (the northern part of Borneo, including Sabah and Sarawak) and *West Malaysia* (the southern part of the Malay Peninsula). The two parts of Malaysia are separated from each other by 650 km (400 miles) of the South China Sea. West Malaysia is the world's leading producer of rubber and tin, while East Malaysia is an important exporter of oil. Malaysia federated as an independent Commonwealth state in 1963, although Singapore withdrew in 1965. The federation is a parliamentary monarchy ruled on a five-year rotation by each of the nine hereditary Malay rulers. (See also MALAYA.) □ **Malaysian** *adj. & n.*

Malcolm /ˈmælkəm/ the name of four kings of Scotland, notably:

Malcolm III (known as Malcolm Canmore, f. Gael. *Ceann-mor* great head) (*c.*1031–93), son of Duncan I, reigned 1058–93. He came to the throne after killing Macbeth in battle (1057). One of the monarchs most responsible for welding Scotland into an organized kingdom, Malcolm spent a large part of his reign involved in intermittent border warfare with the new Norman regime in England, eventually being killed in battle near Alnwick.

Malcolm IV (known as Malcolm the Maiden) (1141–65), grandson of David I, reigned 1153–65. His reign witnessed a progressive loss of power to Henry II of England; he died young and without an heir.

Malcolm X /ˈmælkəm 'eks/ (born Malcolm Little) (1925–65), American political activist. He joined the Black Muslims (Nation of Islam) in 1946 and during the 1950s and early 1960s became a vigorous campaigner against the exploitation of blacks. He advocated the use of violence for self-protection and was opposed to the cooperative approach that characterized the rest of the civil-rights movement. In 1964, after converting to orthodox Islam, he broke away from the Black Muslims and moderated his views on black separatism; he was assassinated the following year.

malcontent /ˈmælkənˌtent/ n. & adj. ● n. a discontented person; a rebel. ● adj. discontented or rebellious. [F (as MAL-, CONTENT¹)]

mal de mer /ˌmæl də 'meə(r)/ n. seasickness. [F, = sickness of (the) sea]

Maldives, the /ˈmɔːldaɪvz, -diːv/ a country consisting of a chain of coral islands in the Indian Ocean south-west of Sri Lanka; pop. (est. 1991) 221,000; official language, Maldivian; capital, Male. The islands were probably first settled from southern India and Sri Lanka in the 5th century BC, but later came under Arab influence. A British protectorate from 1887, the Maldives became independent within the Commonwealth under the rule of a sultan in 1965 and then a republic in 1968.

Maldivian /mɔːlˈdɪvɪən/ n. & adj. ● n. **1** a native or inhabitant of the Maldives. **2** the Sinhalese language spoken in the Maldives. ● adj. of or relating to the Maldives or Maldivian.

Male /ˈmɑːleɪ/ the capital of the Maldives; pop. (est. 1991) 55,130.

male /meɪl/ adj. & n. ● adj. **1** of the sex that can beget offspring by fertilization or insemination (*male child; male dog*). **2** of men or male animals, plants, etc.; masculine (*the male sex; a male-voice choir*). **3 a** (of plants or their parts) containing only fertilizing organs. **b** (of plants) thought of as male because of colour, shape, etc. **4** (of parts of machinery etc.) designed to enter or fill the corresponding female part (*a male screw*). ● n. a male person or animal. □ **male chauvinist** (**pig**) a man who is prejudiced against women or regards women as inferior.

male fern a common lowland fern, *Dryopteris filixmas*. **male menopause** a crisis of potency, confidence, etc., supposed to afflict men in middle life. □ **maleness** n. [ME f. OF ma(s)le, f. L *masculus* f. *mas* a male]

malediction /ˌmælɪˈdɪkʃ(ə)n/ n. **1** a curse. **2** the utterance of a curse. □ **maledictive** /-ˈdɪktɪv/ adj. **maledictory** adj. [ME f. L *maledictio* f. *maledicere* speak evil of f. *male* ill + *dicere dict-* speak]

malefactor /ˈmælɪˌfæktə(r)/ n. a criminal; an evildoer. □ **malefaction** /ˌmælɪˈfækʃ(ə)n/ n. [ME f. L f. *malefacere malefact-* f. *male* ill + *facere* do]

malefic /məˈlefɪk/ adj. *literary* (of magical arts etc.) harmful; baleful. [L *maleficus* f. *male* ill]

maleficent /məˈlefɪs(ə)nt/ adj. *literary* **1** (often foll. by *to*) hurtful. **2** criminal. □ **maleficence** n. [*maleficence* formed as MALEFIC after *malevolence*]

Malegaon /ˈmɑːləˌɡaʊn/ a city in western India, in Maharashtra north-east of Bombay; pop. (1991) 342,000.

maleic acid /məˈleɪɪk/ n. *Chem.* a colourless crystalline organic acid used in making synthetic resins. [F *maléique* (as MALIC ACID)]

malemute var. of MALAMUTE.

malevolent /məˈlevələnt/ adj. wishing evil to others. □ **malevolence** n. **malevolently** adv. [OF *malivolent* or f. L *malivolens* f. *male* ill + *volens* willing, part. of *velle*]

malfeasance /mælˈfiːz(ə)ns/ n. *Law* evildoing. □ **malfeasant** n. & adj. [AF *malfaisance* f. OF *malfaisant* (as MAL-, *faisant* part. of *faire* do f. L *facere*): cf. MISFEASANCE]

malformation /ˌmælfɔːˈmeɪʃ(ə)n/ n. faulty formation. □ **malformed** /-ˈfɔːmd/ adj.

malfunction /mælˈfʌŋkʃ(ə)n/ n. & v. ● n. a failure to function in a normal or satisfactory manner. ● v.intr. fail to function normally or satisfactorily.

Malherbe /mæˈleəb/ François de (1555–1628), French poet. An architect of classicism in poetic form and grammar, he sternly criticized excess of emotion and ornamentation and the use of Latin and dialectal forms.

Mali /ˈmɑːlɪ/ a landlocked country in West Africa, south of Algeria; pop. (est. 1991) 8,706,000; languages, French (official), other languages mainly of the Mande group; capital, Bamako. Apart from the north of the country, which is desert, Mali lies mostly within the Sahel. It takes its name from a mercantile empire which dominated the region from the 13th to the 16th centuries. Conquered by the French in the late 19th century, Mali became part of French West Africa, under the name French Sudan. It became a partner with Senegal in the Federation of Mali in 1959 and achieved full independence a year later, on the withdrawal of Senegal. □ **Malian** adj. & n.

mali /ˈmɑːlɪ/ n. (pl. **malis**) (in the Indian subcontinent) a member of a caste whose traditional occupation is gardening; a gardener. [Hindi]

Malibu /ˈmælɪˌbuː/ a resort on the Pacific coast of southern California, immediately to the west of Los Angeles. The home of a number of film stars, it is noted for its beaches and for the J. Paul Getty art museum, which is situated there.

malic acid /ˈmælɪk/ n. *Chem.* an organic acid found in unripe apples and other fruits. [F *malique* f. L *malum* apple]

malice /ˈmælɪs/ n. **1 a** the intention to do evil. **b** a desire to tease, esp. cruelly. **2** *Law* wrongful intention, esp. as increasing the guilt of certain offences. □ **malice aforethought** (or **prepense**) *Law* the intention to commit a crime, esp. murder. [ME f. OF f. L *malitia* f. *malus* bad]

malicious /məˈlɪʃəs/ adj. characterized by malice; intending or intended to do harm. □ **maliciously** adv. **maliciousness** n. [OF *malicius* f. L *malitiosus* (as MALICE)]

malign /məˈlaɪn/ adj. & v. ● adj. **1** (of a thing) injurious. **2** (of a disease) malignant. **3** malevolent. ● v.tr. speak ill of; slander. □ **maligner** n. **malignly** adv. **malignity** /-ˈlɪɡnɪtɪ/ n. (pl. **-ies**). [ME f. OF *malin maligne*, *malignier* f. LL *malignare* contrive maliciously f. L *malignus* f. *malus* bad: cf. BENIGN]

malignant /məˈlɪɡnənt/ adj. **1 a** (of a disease) very virulent or infectious (*malignant cholera*). **b** (of a tumour) tending to invade normal tissue and recur after removal; cancerous. **2** harmful; feeling or showing intense ill will. □ **malignant pustule** a form of anthrax. □ **malignancy** n. (pl. **-ies**). **malignantly** adv. [LL *malignare* (as MALIGN)]

Malines /malin/ the French name for MECHELEN.

malinger /məˈlɪŋɡə(r)/ v.intr. exaggerate or feign illness in order to escape duty, work, etc. □ **malingerer** n. [back-form. f. *malingerer* app. f. F *malingre*, perh. formed as MAL- + *haingre* weak]

Malin Head /ˈmælɪn/ a point on the coast of County Donegal, the northernmost point of Ireland. The shipping forecast area *Malin* covers the Atlantic north of Ireland and west of the southern half of Scotland.

Malinowski /ˌmælɪˈnɒfskɪ/, Bronisław Kaspar (1884–1942), Polish anthropologist. An influential teacher, from 1916 onwards he was chiefly based in England and the US. He initiated the technique of 'participant observation', which he first applied in his study of the people of the Trobriand Islands, conducted from 1915 to 1918. He also developed the functionalist approach to anthropology, especially in his studies of the Pueblo Indians in Mexico and Bantu-speaking peoples in Africa.

mall /mæl, mɔːl/ n. **1** a sheltered walk or promenade. **2** an enclosed shopping precinct. **3** *hist.* **a** = PALL-MALL. **b** an alley used for this. [var. of MAUL: applied to *The Mall* in London (orig. a pall-mall alley)]

mallard /ˈmælɑːd/ n. (pl. same or **mallards**) **1** a wild duck, *Anas platyrhynchos*, of the northern hemisphere, the ancestor of most domestic ducks. Also called *wild duck*. **2** the flesh of the mallard. [ME f. OF prob. f. *maslart* (unrecorded, as MALE)]

Mallarmé /ˌmælɑːˈmeɪ/, Stéphane (1842–98), French poet. His best-known poems include 'Hérodiade' (*c.*1871) and 'L'Après-midi d'un faune' (1876). He was a symbolist, who made use of elaborate symbols and metaphors in his work and experimented with rhythm and syntax by transposing words and omitting grammatical elements. These tendencies culminated in the poem 'Un Coup de dés jamais n'abolira le hasard' (1897), which makes revolutionary use of typographical possibilities to suggest a musical score.

malleable /ˈmælɪəb(ə)l/ adj. **1** (of metal etc.) able to be hammered or pressed permanently out of shape without breaking or cracking. **2** adaptable; pliable; flexible. □ **malleably** adv. **malleability** /ˌmælɪəˈbɪlɪtɪ/ n. [ME f. OF f. med.L *malleabilis* f. L *malleare* to hammer f. *malleus* hammer]

mallee /ˈmælɪ/ n. *Austral.* **1** a low-growing scrubby eucalyptus, flourishing in arid areas of Australia; esp. *Eucalyptus dumosa*. **2** scrub formed by this plant. □ **mallee-bird** (or **-fowl** or **-hen**) a megapode, *Leipoa ocellata*, resembling a turkey. [Aboriginal]

mallei *pl.* of MALLEUS.

mallemuck var. of MOLLYMAWK.

mallenders /ˈmæləndəz/ *n.pl.* (also **malanders**) a dry scabby eruption behind a horse's knee. [ME f. OF *malandre* (sing.) f. L *malandria* (pl.) neck-pustules]

malleolus /məˈliːələs/ *n.* (*pl.* **malleoli** /-ˌlaɪ/) *Anat.* a bone with the shape of a hammer-head, esp. each of those forming a projection on either side of the ankle. [L, dimin. of *malleus* hammer]

mallet /ˈmælɪt/ *n.* **1** a hammer, usu. of wood. **2** a long-handled wooden hammer for striking a croquet or polo ball. [ME f. OF *maillet* f. *mailler* to hammer f. *mail* hammer f. L *malleus*]

malleus /ˈmælɪəs/ *n.* (*pl.* **mallei** /-lɪˌaɪ/) *Anat.* a small bone in the middle ear transmitting the vibrations of the tympanum to the incus. Also called *hammer*. [L, = hammer]

Mallorca see MAJORCA.

mallow /ˈmæləʊ/ *n.* **1** a plant of the genus *Malva* (family Malvaceae), with hairy stems and leaves, and pink or purple flowers; esp. *M. sylvestris*. **2** any other plant of the family Malvaceae (*marsh mallow*; *tree mallow*). [OE *meal(u)we* f. L *malva*]

malm /mɑːm/ *n.* **1** a soft chalky rock. **2** a loamy soil produced by the disintegration of this rock. **3** a fine-quality brick made originally from malm, marl, or a similar chalky clay. [OE *mealm-* (in compounds) f. Gmc]

Malmö /ˈmɑːlmɜː/ a port and fortified city in SW Sweden, situated on the Øresund opposite Copenhagen; pop. (1990) 233,900.

malmsey /ˈmɑːmzi/ *n.* a strong sweet wine orig. from Greece, now chiefly from Madeira. [ME f. MDu., MLG *malmesie*, *-eye*, f. *Monemvasia* in S. Greece: cf. MALVOISIE]

malnourished /mælˈnʌrɪʃt/ *adj.* suffering from malnutrition.

malnourishment /mælˈnʌrɪʃmənt/ *n.* = MALNUTRITION.

malnutrition /ˌmælnjuːˈtrɪʃ(ə)n/ *n.* a dietary condition resulting from the absence of some foods or essential elements necessary for health; insufficient nutrition.

malocclusion /ˌmæləˈkluːʒ(ə)n/ *n.* imperfect positioning of the teeth when the jaws are closed.

malodorous /mælˈəʊdərəs/ *adj.* evil-smelling.

Malory /ˈmæləri/, Sir Thomas (d.1471), English writer. Although his exact identity is uncertain, it is probable that he was Sir Thomas Malory of Newbold Revel, Warwickshire. His major work, *Le Morte d'Arthur* (printed 1483), is a prose translation of a collection of the legends of King Arthur, selected from French and other sources. It was one of the earliest works to be printed by Caxton, and is the standard source for later versions in English of the Arthurian romances.

maloti *pl.* of LOTI.

Malpighi /mælˈpiːɡɪ/, Marcello (*c.*1628–94), Italian microscopist. Seeking a mechanical interpretation of animal bodies, he looked for and found visible structures underlying physiological functions. He discovered the alveoli and capillaries in the lungs and the fibres and red cells of clotted blood, and demonstrated the pathway of blood from arteries to veins. Malpighi began the study of embryology, and also investigated the structures of the kidney and the skin, the anatomy of the silkworm, the structure of plant cells, and the breathing system of animals.

Malpighian /mælˈpɪɡɪən/ *adj.* of, discovered by, or associated with Malpighi. □ **Malpighian layer** *Anat.* a layer of proliferating cells at the base of the epidermis. **Malpighian tubule** *Zool.* a tubular excretory organ, numbers of which open into the gut in insects and some other arthropods.

Malplaquet, Battle of /ˌmælplæˈkeɪ/ a battle in 1709 during the War of the Spanish Succession, near the village of Malplaquet in northern France, on the border with Belgium. A force of allied British and Austrian troops under the Duke of Marlborough won a victory over the French.

malpractice /mælˈpræktɪs/ *n.* **1** improper or negligent professional treatment, esp. by a medical practitioner. **2 a** criminal wrongdoing; misconduct. **b** an instance of this.

malt /mɔːlt, mɒlt/ *n.* & *v.* ● *n.* **1** barley or other grain that is steeped, germinated, and dried, esp. for brewing or distilling and vinegar-making. **2** *colloq.* malt whisky; malt liquor. ● *v.* **1** *tr.* convert (grain) into malt. **2** *intr.* (of seeds) become malt when germination is checked by drought. □ **malted milk 1** a hot drink made from dried milk and a

malt preparation. **2** the powdered mixture from which this is made.

malt-house a building used for preparing and storing malt. **malt liquor** alcoholic liquor made from malt by fermentation without distillation, e.g. beer. **malt whisky** whisky made from malted barley. [OE *m(e)alt* f. Gmc, rel. to MELT]

Malta /ˈmɔːltə, ˈmɒl-/ an island country in the central Mediterranean, about 100 km (60 miles) south of Sicily; pop. (est. 1991) 356,000; official languages, Maltese and English; capital, Valletta. Besides Malta itself, the country includes two other inhabited islands, Gozo and Comino. Malta's position is of great strategic importance, and the island was held in turn by Phoenicians, Greeks, Carthaginians, and Arabs, and in 1090 was conquered by Roger of Normandy. Given to the Knights Hospitallers by the Emperor Charles V in 1530, Malta successfully withstood a long siege by Turkish invaders and remained the Knights' headquarters until captured by the French in 1798. It was annexed by Britain in 1814 and was an important naval base until independence. During the Second World War the island was awarded the George Cross for its endurance under Axis air attack between 1940 and 1942. Malta became independent within the Commonwealth in 1964.

Maltese /mɔːlˈtiːz, mɒl-/ *n.* & *adj.* ● *n.* **1** (*pl.* same) **a** a native or national of Malta. **b** a person of Maltese descent. **2** the Semitic language of Malta, much influenced by Italian. ● *adj.* of or relating to Malta or its people or language. □ **Maltese cross** a cross with arms of equal length broadening from the centre, often indented at the ends. **Maltese dog** (or **terrier**) a small breed of spaniel or terrier.

maltha /ˈmælθə/ *n.* a cement made of pitch and wax or other ingredients. [L f. Gk]

Malthus /ˈmælθəs/, Thomas Robert (1766–1834), English economist and clergyman. He was a pioneer of the science of political economy and is known for his theory, as expressed in *Essay on Population* (1798), that the rate of increase of the population tends to be out of proportion to the increase of its means of subsistence; controls on population (by sexual abstinence or birth control) are therefore necessary to prevent catastrophe. □ **Malthusian** /mælˈθjuːzɪən/ *adj.*

malting /ˈmɔːltɪŋ, ˈmɒl-/ *n.* **1** the process or an instance of brewing or distilling with malt. **2** = *malt-house*.

maltose /ˈmɔːltəʊz, ˈmɒl-, -təʊs/ *n. Chem.* a disaccharide sugar consisting of two linked glucose molecules, produced by the hydrolysis of starch under the action of the enzymes in malt, saliva, etc. [MALT + -OSE²]

maltreat /mælˈtriːt/ *v.tr.* ill-treat. □ **maltreater** *n.* **maltreatment** *n.* [F *maltraiter* (as MAL-, TREAT)]

maltster /ˈmɔːltstə(r), ˈmɒl-/ *n.* a person who makes malt.

malty /ˈmɔːlti, ˈmɒl-/ *adj.* (**maltier, maltiest**) of, containing, or resembling malt. □ **maltiness** *n.*

Maluku see MOLUCCA ISLANDS.

malvaceous /mælˈveɪʃəs/ *adj. Bot.* of or relating to the family Malvaceae, which includes the mallows. [L *malvaceus* f. *malva* MALLOW]

Malvern Hills /ˈmɔːlv(ə)n/ (also **the Malverns**) a range of limestone hills in western England, in Hereford and Worcester. The highest point is Worcestershire Beacon (425 m; 1,394 ft).

malversation /ˌmælvəˈseɪʃ(ə)n/ *n. formal* **1** corrupt behaviour in a position of trust. **2** (often foll. by *of*) corrupt administration (of public money etc.). [F f. *malverser* f. L *male* badly + *versari* behave]

Malvinas, Islas /mælˈviːnəs, ˈiːzlæs/ the name by which the Falkland Islands are known in Argentina.

malvoisie /ˈmælvɔɪzɪ/ *n.* = MALMSEY. [ME f. OF *malvesie* f. F form of *Monemvasia*: see MALMSEY]

mam /mæm/ *n. colloq.* mother. [formed as MAMMA¹]

mama /ˈmæmə, məˈmɑː/ *n. colloq.* (esp. as a child's term) = MAMMA¹.

mamba /ˈmæmbə/ *n.* a large venomous African snake of the genus *Dendroaspis*; esp. the *green mamba* and *black mamba*, which are varieties of *D. angusticeps*. [Zulu *imamba*]

mambo /ˈmæmbəʊ/ *n.* & *v.* ● *n.* (*pl.* **-os**) **1** a Latin American dance like the rumba. **2** the music for this. ● *v.intr.* (**-oes, -oed**) perform the mambo. [Amer. Sp. prob. f. Haitian]

mamelon /ˈmæmələn/ *n.* a small rounded hillock. [F, = nipple f. *mamelle* breast f. L MAMILLA]

Mameluke /ˈmæməˌluːk/ *n.* a member of a group of Turkoman warriors who were originally brought to Egypt as slaves to act as a bodyguard for the caliphs and sultans. They became powerful, ruling as sultans in Egypt from 1250 and Syria from 1260 until conquered by the Ottoman Turks in 1517; in Egypt the Mamelukes continued as a ruling military caste under Ottoman sovereignty until they were finally defeated and massacred by the viceroy Muhammad Ali in 1811. [F *mameluk*, ult. f. Arab. *mamlūk* slave f. *malaka* possess]

Mamet /ˈmæmɪt/, David (b.1947), American dramatist, director, and screenwriter. In 1974 he co-founded the Chicago-based St Nicholas Theater Company, for whom he wrote and directed a succession of plays noted for their approach to social issues, including *Glengarry Glen Ross* (1984), which won the Pulitzer Prize. Other works include the film *House of Games* (1986), which he wrote and directed, and the play *Oleanna* (1992), whose portrayal of a case involving false accusations of sexual harassment caused much controversy.

mamilla /məˈmɪlə/ n. (US **mammilla**) (pl. **-llae** /-liː/) **1** the nipple of a woman's breast. **2** a nipple-shaped organ etc. □ **mamillary** /ˈmæmɪləri/ adj. **mamillate** /-ˌleɪt/ adj. [L, dimin. of MAMMA²]

mamma¹ /ˈmæmə/ n. (also **momma** /ˈmɒmə/) colloq. (esp. as a child's term) mother. [imit. of child's *ma, ma*]

mamma² /ˈmæmə/ n. (pl. **mammae** /-miː/) **1** a milk-secreting organ of female mammals. **2** a corresponding non-secretory structure in male mammals. □ **mammiform** /ˈmæmɪˌfɔːm/ adj. [OE f. L]

mammal /ˈmæm(ə)l/ n. a vertebrate of the class Mammalia, characterized by secretion of milk by the female to feed the young. Mammals evolved from reptiles about 200 million years ago, but remained relatively small and inconspicuous until the dinosaurs died out at the end of the Cretaceous period, when they evolved rapidly and became the dominant land vertebrates. They are a very diverse group with about 4,000 living species, of which nearly half are rodents and almost a quarter are bats. Among the distinctive features which help to account for their success are warm-bloodedness, hair and a layer of fat to conserve body heat, teeth evolved to suit a variety of diets, and a large brain permitting complex and adaptable behaviour. The three major subgroups are those of the monotremes, marsupials, and placental mammals: the latter includes most of the familiar types, including primates, carnivores, hoofed mammals, rodents, bats, whales, etc. □ **mammalian** /məˈmeɪlɪən/ adj. & n. **mammalogy** /-ˈmælədʒɪ/ n. [mod.L *mammalia* neut. pl. of L *mammalis* (as MAMMA²)]

mammaliferous /ˌmæməˈlɪfərəs/ adj. Geol. containing mammalian remains.

mammary /ˈmæməri/ adj. of the human female breasts or milk-secreting organs of other mammals. □ **mammary gland** the milk-producing gland of female mammals. [MAMMA² + -ARY¹]

mammee /mæˈmiː/ n. (in full **mammee apple**) a tropical American tree, *Mammea americana*, which has a large edible fruit with yellow aromatic flesh. [Sp. *mamei* f. Haitian]

mammilla US var. of MAMILLA.

mammography /mæˈmɒɡrəfɪ/ n. Med. an X-ray technique of diagnosing and locating abnormalities (esp. tumours) of the breasts. [MAMMA² + -GRAPHY]

Mammon /ˈmæmən/ n. **1** the god or devil of covetousness. (*See note below*.) **2** wealth regarded as an idol or as an evil influence. □ **Mammonism** n. **Mammonist** n. [ME f. LL *Mam(m)ona* f. Gk *mamōnas* f. Aram.]

■ 'Mammon' was the Aramaic word for 'riches' (*māmōn*) used in the Greek text of the New Testament in Matt. 6:24 and Luke 16:9–13, and retained in the Vulgate. It was taken by medieval writers as the name of the devil of covetousness, and this use was revived by Milton in *Paradise Lost*.

mammoth /ˈmæməθ/ n. & adj. ● n. a large extinct elephant of the genus *Mammuthus*, with a hairy coat and long curved tusks, found mainly during the Pleistocene ice ages, esp. the woolly mammoth. ● adj. huge. [Russ. *mamo(n)t*]

Mammoth Cave National Park a national park in west central Kentucky, site of the largest known cave system in the world. It consists of over 480 km (300 miles) of charted passageways and contains some spectacular rock formations.

mammy /ˈmæmɪ/ n. (pl. **-ies**) **1** colloq. (esp. as a child's term) mother. **2** US hist. a black nursemaid or nanny in charge of white children. [formed as MAMMA¹]

Mamoutzu /mæˈmuːtsuː/ the capital (since 1977) of Mayotte; pop. (1985) 12,120.

Man. abbr. Manitoba.

man /mæn/ n. & v. ● n. (pl. **men** /men/) **1** an adult human male, esp. as distinct from a woman or boy. **2 a** a creature of the genus *Homo*, distinguished from other animals by superior mental development, power of articulate speech, and upright stance (see HOMO); a human

being; a person (*no man is perfect*). **b** human beings in general; the human race, humankind (*man is mortal*). **3** a person showing characteristics traditionally associated with males (*she's more of a man than he is*). **4 a** a male worker or employee (*the manager spoke to the men*). **b** a manservant or valet. **c** hist. a vassal. **5 a** (usu. in pl.) soldiers, sailors, etc., esp. non-officers (*was in command of 200 men*). **b** an individual, usu. male, person (*fought to the last man*). **c** (usu. prec. by *the*, or poss. pron.) a person regarded as suitable or appropriate in some way; a person fulfilling requirements (*I'm your man; not the man for the job*). **6 a** a husband (*man and wife*). **b** colloq. a boyfriend or lover. **7 a** a human being of a specified historical period or character (*Renaissance man*). **b** a type of prehistoric human named after the place where the remains were found (*Peking man*). **8** any one of a set of pieces used in playing chess, draughts, etc. **9** (as second element in comb.) a man of a specified nationality, profession, skill, etc. (*Dutchman; clergyman; horseman; gentleman*). **10 a** an expression of impatience etc. used in addressing a male (*nonsense, man!*). **b** colloq. a general mode of address associated with hippies etc. (*blew my mind, man!*). **11** (prec. by *a*) a person; one (*what can a man do?*). **12** a person pursued; an opponent etc. (*the police have so far not caught their man*). **13** (**the Man**) US sl. **a** the police. **b** black sl. white people. **14** (in comb.) a ship of a specified type (*merchantman; Indiaman*). ● v.tr. (**manned, manning**) **1** supply (a ship, fort, factory, etc.) with a person or people for work or defence etc. **2** work or service or defend (a specified piece of equipment, a fortification, etc.) (*man the pumps*). **3** Naut. place men at (a part of a ship). **4** fill (a post or office). **5** (usu. refl.) fortify the spirits or courage of (*manned herself for the task*). □ **as one man** in unison; in agreement. **be a man** be courageous; not show fear. **be one's own man 1** be free to act; be independent. **2** be in full possession of one's faculties etc. **man about town** a fashionable man of leisure. **man and boy** from childhood. **man-at-arms** (pl. **men-at-arms**) archaic a soldier, esp. when heavily armed and mounted. **man enough** sufficiently manly **man Friday** see FRIDAY. **man-hour** (or **-day** etc.) an hour (or day etc.) regarded in terms of the amount of work that could be done by one person within this period. **man in the moon** the semblance of a face seen on the surface of a full moon. **man in** (US **on**) **the street** an ordinary average person, as distinct from an expert. **man-made** (esp. of a textile fibre) made by man, artificial, synthetic. **man of the cloth** a clergyman. **man of God 1** a clergyman. **2** a male saint. **man of honour** a man whose word can be trusted. **man of the house** the male head of a household. **man of letters** a scholar; an author. **man of the moment** a man of importance at a particular time. **man of straw 1** an insubstantial person; an imaginary person set up as an opponent. **2** a stuffed effigy. **3** a person undertaking a financial commitment without adequate means. **4** a sham argument set up to be defeated. **man-of-war 1** hist. an armed ship, esp. of a specified country. **2** a frigate bird. **3** see *Portuguese man-of-war*. **man of the world** see WORLD. **man-size** (or **-sized**) **1** of the size of a man; very large. **2** big enough for a man. **man to man** with candour; honestly. **men's** (or **men's room**) a usu. public lavatory for men. **my** (or **my good**) **man** a patronizing mode of address to a man. **separate** (or **sort out**) **the men from the boys** colloq. find those who are truly mature, competent, manly, etc. **to a man** all without exception. □ **manless** adj. [OE *man(n)*, pl. *menn*, *mannian*, f. Gmc]

Man, Isle of see ISLE OF MAN.

mana /ˈmɑːnə/ n. **1** esp. NZ power; authority; prestige. **2** Anthropol. supernatural or magical power. [Maori]

manacle /ˈmænək(ə)l/ n. & v. ● n. (usu. in pl.) **1** a fetter or shackle for the hand; a handcuff. **2** a restraint. ● v.tr. fetter with manacles. [ME f. OF *manicle* handcuff f. L *manicula* dimin. of *manus* hand]

manage /ˈmænɪdʒ/ v. & n. ● v. **1** tr. organize; regulate; be in charge of (a business, household, team, a person's career, etc.). **2** tr. (often foll. by *to* + infin.) succeed in achieving; contrive (*managed to arrive on time; managed a smile; managed to ruin the day*). **3** intr. **a** (often foll. by *with*) succeed in one's aim, esp. against heavy odds (*managed with one assistant*). **b** meet one's needs with limited resources etc. (*just about manages on a pension*). **4** tr. gain influence with or maintain control over (a person etc.) (*cannot manage their teenage son*). **5** tr. (also absol.; often prec. by *can, be able to*) **a** cope with; make use of (*couldn't manage another bite; can you manage by yourself?*). **b** be free to attend on (a certain day) or at (a certain time) (*can you manage Thursday?*). **6** tr. handle or wield (a tool, weapon, etc.). **7** tr. take or have charge or control of (an animal or animals, esp. cattle). ● n. archaic **1 a** the training of a horse. **b** the trained movements of a horse. **2** a riding-school (cf. MANÈGE). [It. *maneggiare*, *maneggio* ult. f. L *manus* hand]

manageable /ˈmænɪdʒəb(ə)l/ adj. able to be managed, controlled, or accomplished without great or excessive difficulty. □ **manageably** adv. **manageableness** n. **manageability** /ˌmænɪdʒəˈbɪlɪtɪ/ n.

management /ˈmænɪdʒmənt/ n. **1** the process or an instance of managing or being managed. **2 a** the professional administration of business concerns, public undertakings, etc. **b** the people engaged in this. **c** a governing body; a board of directors; the people in charge of running a business, regarded collectively. **3** (usu. foll. by of) Med. the technique of treating a disease etc. **4** trickery; deceit.

manager /ˈmænɪdʒə(r)/ n. **1** a person controlling or administering a business or part of a business. **2** a person controlling the affairs, training, etc., of a person or team in sports, entertainment, etc. **3** Brit. Parl. a member of either House of Parliament appointed with others for some duty in which both Houses are concerned. **4** a person regarded in terms of skill in household or financial or other management (a good manager). □ **managership** n. **managerial** /ˌmænɪˈdʒɪərɪəl/ adj. **managerially** adv.

manageress /ˌmænɪdʒəˈres/ n. a woman manager, esp. of a shop, hotel, theatre, etc.

managing /ˈmænɪdʒɪŋ/ adj. **1** (in comb.) having executive control or authority (managing director). **2** (attrib.) fond of controlling affairs etc. **3** archaic economical.

Managua /məˈnɑːgwə/ the capital of Nicaragua; pop. (1985) 682,100. The city was almost completely destroyed by an earthquake in 1972.

manakin /ˈmænəkɪn/ n. a small fruit-eating bird of the family Pipridae, of Central and South America, the male of which is often brightly coloured (cf. MANIKIN 4). [var. of MANIKIN]

Manama /məˈnɑːmə/ the capital of Bahrain; pop. (est. 1988) 151,500.

mañana /mæˈnjɑːnə/ adv. & n. ● adv. in the indefinite future (esp. to indicate procrastination). ● n. an indefinite future time. [Sp., = tomorrow]

Mana Pools National Park /ˈmɑːnə/ a national park in northern Zimbabwe, in the Zambezi valley north-east of Lake Kariba. It was established in 1963.

Manasseh /məˈnæsɪ, -ˈnæsə/ **1** a Hebrew patriarch, son of Joseph (Gen. 48:19). **2** the tribe of Israel traditionally descended from him.

Manasses see PRAYER OF MANASSES.

manatee /ˌmænəˈtiː/ n. an aquatic sirenian mammal of the genus Trichechus, of warm Atlantic coastal waters and adjacent rivers, with paddle-like forelimbs and a rounded tail fluke. Also called sea cow. [Sp. manati f. Carib manattoui]

Manaus /məˈnaʊs/ a city in NW Brazil, capital of the state of Amazonas; pop. (1990) 996,720. It is the principal commercial centre of the upper Amazon region.

Manawatu /ˌmænəˈwɑːtuː/ a river of North Island, New Zealand, flowing into Cook Strait.

Manchester /ˈmæntʃɪstə(r)/ an industrial city in NW England, administrative centre of the metropolitan county of Greater Manchester; pop. (1991) 397,400. Founded in Roman times, it developed in the 18th and 19th centuries as a centre of the English cotton industry.

Manchester Ship Canal a waterway in NW England, which links Manchester with the estuary of the River Mersey and the Irish Sea. Opened in 1894, it is 57 km (36 miles) long.

manchineel /ˌmæntʃɪˈniːl/ n. a West Indian tree, Hippomane mancinella, with a poisonous and caustic milky sap and acrid apple-like fruit. [F mancenille f. Sp. manzanilla dimin. of manzana apple]

Manchu /mænˈtʃuː/ n. & adj. ● n. **1** a member of a Tartar people of Manchuria who conquered China and founded the Qing dynasty (1644–1912). **2** the language of this people, which belongs to the Tungus group in the Altaic family of languages. At one time it was an official language of China, but it is now spoken only in parts of northern Manchuria. ● adj. of or relating to the Manchus or their language. [Manchu, = pure]

Manchuria /mænˈtʃʊərɪə/ a mountainous region forming the north-eastern portion of China, now comprising the provinces of Jilin, Liaoning, and Heilongjiang. In 1932 it was declared an independent state by Japan and renamed Manchukuo; it was restored to China in 1945.

manciple /ˈmænsɪp(ə)l/ n. an officer who buys provisions for a college, an Inn of Court, etc. [ME f. AF & OF f. L mancipium purchase f. manceps buyer f. manus hand + capere take]

Mancunian /mænˈkjuːnɪən/ n. & adj. ● n. a native or inhabitant of Manchester. ● adj. of or relating to Manchester. [L Mancunium Roman settlement on the site of Manchester]

-mancy /mænsɪ/ comb. form forming nouns meaning 'divination by' (geomancy; necromancy). □ **-mantic** comb. form forming adjectives. [OF -mancie f. LL -mantia f. Gk manteia divination]

Mandaean /mænˈdiːən/ n. & adj. ● n. **1** a member of a Gnostic sect still surviving in Iraq, who regard John the Baptist as the Messiah. **2** the Aramaic dialect in which the sacred books of this sect are written. ● adj. of or relating to the Mandaeans or Mandaean. [Aram. mandaia f. manda knowledge]

mandala /ˈmændələ/ n. **1** a symbolic circular figure representing the universe in various religions. **2** Psychol. such a symbol in a dream, representing the dreamer's search for completeness and self-unity. [Skr. máṇḍala disc]

Mandalay /ˌmændəˈleɪ/ a port on the Irrawaddy river in central Burma (Myanmar); pop. (1983) 533,000. Founded in 1857, it was the capital until 1885 of the Burmese kingdom. It is an important Buddhist religious centre.

mandamus /mænˈdeɪməs/ n. Law a judicial writ issued as a command to an inferior court, or ordering a person to perform a public or statutory duty. [L, = we command]

mandarin[1] /ˈmændərɪn/ n. **1** (**Mandarin**) the most widely spoken group of Chinese dialects, esp. the form that is used as the official language of China. **2** hist. a Chinese official in any of nine grades of the pre-Communist civil service. **3** a person of importance, esp. a government official or a reactionary or secretive bureaucrat. **4 a** a nodding Chinese figure, usu. of porcelain. **b** porcelain etc. decorated with Chinese figures in mandarin dress. □ **mandarin collar** a small close-fitting upright collar. **mandarin duck** a small duck, Aix galericulata, noted for its colourful plumage, native to eastern Asia and naturalized elsewhere. **mandarin sleeve** a wide loose sleeve. □ **mandarinate** n. [Port. mandarim f. Malay f. Hindi mantrī f. Skr. mantrin counsellor]

mandarin[2] /ˈmændərɪn/ n. (also **mandarine** /-ˌriːn/) (in full **mandarin orange**) **1** a small flattish deep-coloured orange with a loose skin. **2** the tree, Citrus reticulata, yielding this. (See also SATSUMA, TANGERINE.) [F mandarine (perh. as MANDARIN[1], with ref. to the official's yellow robes)]

mandatary /ˈmændətərɪ/ n. (pl. **-ies**) esp. hist. a person or state receiving a mandate. [LL mandatarius (as MANDATE)]

mandate /ˈmændeɪt/ n. & v. ● n. **1** an official command or instruction by an authority. **2** support for a policy or course of action, regarded by a victorious party, candidate, etc., as derived from the wishes of the people in an election. **3** a commission to act for another. **4** Law a commission by which a party is entrusted to perform a service, often gratuitously and with indemnity against loss by that party. **5** hist. a commission from the League of Nations to a member state to administer a territory. **6** a papal decree or decision. ● v.tr. **1** instruct (a delegate) to act or vote in a certain way. **2** (usu. foll. by to) hist. commit (a territory etc.) to a mandatary. [L mandatum, neut. past part. of mandare command f. manus hand + dare give: sense 2 of n. after F mandat]

mandatory /ˈmændətərɪ/ adj. & n. ● adj. **1** of or conveying a command. **2** compulsory. ● n. (pl. **-ies**) = MANDATARY. □ **mandatorily** adv. [LL mandatorius f. L (as MANDATE)]

Mande /ˈmɑːndeɪ/ n. & adj. ● n. **1** a member of a large group of peoples of West Africa. **2** the group of Niger-Congo languages spoken by these peoples. ● adj. of or relating to these peoples or their languages.

Mandela /mænˈdelə/, Nelson (Rolihlahla) (b.1918), South African statesman, President since 1994. From his twenties he was an activist for the African National Congress (ANC); he was first jailed in 1962 and was sentenced to life imprisonment in 1964. His authority as a moderate leader of black South Africans did not diminish while he was in detention, and he became a symbol of the struggle against apartheid. On his release in 1990 Mandela resumed his leadership of the ANC, and engaged in talks with President F. W. de Klerk on the introduction of majority rule. He shared the Nobel Peace Prize with President de Klerk in 1993, and in the country's first democratic elections was elected President the following year.

Mandelbrot set /ˈmænd(ə)lˌbrɒt/ n. Math. a particular set of complex numbers which has a highly convoluted fractal boundary when plotted. [B. B. Mandelbrot, Polish-born Amer. mathematician (b.1924)]

Mandelstam /ˈmænd(ə)lˌʃtæm/, Osip (Emilevich) (also **Mandelshtam**) (1891–1938), Russian poet. As one of the Acmeist group of poets (with Anna Akhmatova), Mandelstam favoured concrete detail, clarity, and precision of language as a reaction against the mysticism of contemporary Russian symbolist poetry. During the 1920s his poetry met with increasing official criticism; Mandelstam was sent into internal exile (1934-7) and eventually died in a prison camp. Major works include *Stone* (1913) and *Tristia* (1922).

Mandeville /ˈmændəˌvɪl/, Sir John (14th century), English nobleman. He is remembered as the reputed author of a book of travels and travellers' tales which takes the reader to Turkey, Tartary, Persia, Egypt, and India. Written in French and much translated, it was actually compiled by an unknown hand from the works of several writers.

mandible /ˈmændɪb(ə)l/ *n.* **1** the jaw, esp. the lower jaw in mammals and fishes. **2** the upper or lower part of a bird's beak. **3** the mouthpart of an arthropod used in biting and crushing. □ **mandibular** /-ˈdɪbjʊlə(r)/ *adj.* **mandibulate** /-lət/ *adj.* [ME f. OF *mandible* or LL *mandibula* f. *mandere* chew]

mandolin /ˌmændəˈlɪn/ *n.* a plucked stringed instrument of the lute family with metal strings tuned in pairs and a characteristic tremolo when sustaining long notes. It is played chiefly in folk music. □ **mandolinist** *n.* [F *mandoline* f. It. *mandolino* dimin. of *mandola*]

mandorla /mænˈdɔːlə/ *n.* = VESICA 2. [It., = almond]

mandragora /mænˈdrægərə/ *n. hist.* the mandrake, esp. as a type of narcotic (Shakespeare's *Othello* III. iii. 334). [OE f. med.L f. L f. Gk *mandragoras*]

mandrake /ˈmændreɪk/ *n.* a poisonous plant, *Mandragora officinarum*, with a short stem and whitish or purple flowers. It has emetic and narcotic properties, and a forked fleshy root once thought to resemble the human form and to shriek when plucked. [ME *mandrag(g)e*, prob. f. MDu. *mandrag(r)e* f. med.L (as MANDRAGORA): assoc. with MAN + *drake* dragon]

mandrel /ˈmændrəl/ *n.* **1 a** a shaft in a lathe to which work is fixed while being turned. **b** a cylindrical rod round which metal or other material is forged or shaped. **2** *Brit.* a miner's pick. [16th c.: orig. unkn.]

mandrill /ˈmændrɪl/ *n.* a large West African baboon, *Mandrillus sphinx*, the adult of which has a brilliantly coloured face and blue buttocks. [prob. f. MAN + DRILL³]

manducate /ˈmændjʊˌkeɪt/ *v.tr. literary* chew; eat. □ **manducation** /ˌmændjʊˈkeɪʃ(ə)n/ *n.* **manducatory** /ˈmændjʊkətərɪ, ˌmændjʊˈkeɪtərɪ/ *adj.* [L *manducare manducat-* chew f. *manduco* guzzler f. *mandere* chew]

mane /meɪn/ *n.* **1** a growth of long hair on the neck of a horse, lion, etc. **2** *colloq.* a person's long hair. □ **maned** *adj.* (also in *comb.*). **maneless** *adj.* [OE *manu* f. Gmc]

manège /mæˈneɪʒ/ *n.* (also **manege**) **1** a riding-school. **2** the movements of a trained horse. **3** horsemanship. [F *manège* f. It. (as MANAGE)]

manes /ˈmɑːneɪz, ˈmeɪniːz/ *n.pl.* **1** the deified souls of dead ancestors. **2** (as *sing.*) the revered ghost of a dead person. [ME f. L]

Manet /ˈmæneɪ/, Édouard (1832-83), French painter. He adopted a realist approach which greatly influenced the impressionists, and abandoned half-tones and shadings in favour of pure colour to give a direct unsentimental effect. Several of his paintings aroused outrage because of the frank and unidealized treatment of their subject-matter; *Olympia* (1865) depicted a nude woman with clear indications that she was a prostitute. Among other notable works are *Déjeuner sur l'herbe* (1863) and *A Bar at the Folies-Bergère* (1882).

Manetho /ˈmæˈneθəʊ/ (3rd century BC), Egyptian priest. He wrote a history of Egypt (*c.*280) from mythical times to 323. He arbitrarily divided the succession of rulers known to him into thirty dynasties, an arrangement which is still followed.

maneuver *US* var. of MANOEUVRE.

manful /ˈmænfʊl/ *adj.* brave; resolute. □ **manfully** *adv.* **manfulness** *n.*

mangabey /ˈmæŋgəˌbeɪ/ *n.* a small long-tailed monkey of the genus *Cercocebus*, of central and West Africa. [*Mangabey*, a region of Madagascar]

manganate /ˈmæŋgəˌneɪt/ *n. Chem.* a salt in which the anion contains both manganese and oxygen, especially one of the anion MnO_4^{2-}. (Cf. PERMANGANATE.)

manganese /ˌmæŋgəˈniːz/ *n.* a hard grey metallic chemical element (atomic number 25; symbol **Mn**). A transition metal, manganese was identified as an element by Scheele in 1774, although some of its compounds, notably the black dioxide (pyrolusite), had been known since antiquity. It is quite abundant and widely distributed in the earth's crust; there are various ores, and manganese-rich nodules

occur on the ocean bed. The metal is an important component of special steels and magnetic alloys. □ **manganic** /mænˈgænɪk/ *adj.* **manganous** /ˈmæŋgənəs/ *adj.* [F *manganèse* f. It. *manganese*, alt. f. MAGNESIA]

mange /meɪndʒ/ *n.* a skin disease in hairy and woolly animals, caused by a parasitic mite and occasionally communicated to humans. [ME *mangie, maniewe* f. OF *manjue, mangeue* itch f. *mangier manju-* eat f. L *manducare* chew]

mangel /ˈmæŋg(ə)l/ *n.* (also **mangold** /-gəʊld/) (in full **mangel-wurzel, mangold-wurzel** /-ˌwɜːz(ə)l/) a large kind of beet, *Beta vulgaris*, used as cattle food. [G *Mangoldwurzel* f. *Mangold* beet + *Wurzel* root]

manger /ˈmeɪndʒə(r)/ *n.* a long open box or trough in a stable etc., for horses or cattle to eat from. [ME f. OF *mangeoire, mangeure* ult. f. L (as MANDUCATE)]

mangetout /mɒnʒˈtuː/ *n.* (*pl.* same or **mangetouts** *pronunc.* same) the sugar pea. [F, = eat-all]

mangle¹ /ˈmæŋg(ə)l/ *n. & v. esp. Brit. hist.* ● *n.* a machine having two or more cylinders usu. turned by a handle, between which wet clothes etc. are squeezed and pressed. ● *v.tr.* press (clothes etc.) in a mangle. [Du. *mangel(stok)* f. *mangelen* to mangle, ult. f. Gk *magganon* + *stok* staff, STOCK]

mangle² /ˈmæŋg(ə)l/ *v.tr.* **1** hack, cut about, or mutilate by blows etc. **2** spoil (a quotation, text, etc.) by misquoting, mispronouncing, etc. **3** cut roughly so as to disfigure. □ **mangler** *n.* [AF *ma(ha)ngler*, app. frequent. of *mahaignier* MAIM]

mango /ˈmæŋgəʊ/ *n.* (*pl.* **-oes** or **-os**) **1** a fleshy yellowish-red fruit, eaten ripe or used green for pickles etc. **2** the Indian evergreen tree, *Mangifera indica*, bearing this. [Port. *manga* f. Malay *mangā* f. Tamil *mānkāy* f. *mān* mango-tree + *kāy* fruit]

mangold (also **mangold-wurzel**) var. of MANGEL.

mangonel /ˈmæŋgən(ə)l/ *n. Mil. hist.* a military engine for throwing stones etc. [ME f. OF *mangonel(le)*, f. med.L *manganellus* dimin. of LL *manganum* f. Gk *magganon*]

mangosteen /ˈmæŋgəˌstiːn/ *n.* **1** a white juicy-pulped fruit with a thick reddish-brown rind. **2** the Malaysian tree, *Garcinia mangostana*, bearing this. [Malay *manggustan*]

mangrove /ˈmæŋgrəʊv/ *n.* a tropical tree or shrub, esp. of the genus *Rhizophora* or *Bruguiera*, with interlacing aerial roots, forming dense thickets in muddy tidal swamps. [17th c.: orig. uncert.: assim. to GROVE]

mangy /ˈmeɪndʒɪ/ *adj.* (**mangier, mangiest**) **1** (esp. of a domestic animal) having mange. **2** squalid; shabby; inferior. □ **mangily** *adv.* **manginess** *n.*

manhandle /ˈmænˌhænd(ə)l/ *v.tr.* **1** move (heavy objects) by human effort. **2** *colloq.* handle (a person) roughly.

Manhattan /mænˈhæt(ə)n/ an island near the mouth of the Hudson River forming part of the city of New York. The site of the original Dutch settlement of New Amsterdam, it is now a borough containing the commercial and cultural centre of New York City and is famous for its skyscrapers. It takes its name from the Algonquin tribe from whom the Dutch settlers claimed to have bought the island in 1626.

manhattan /mænˈhæt(ə)n/ *n.* a cocktail made with vermouth and a spirit, as whisky or brandy, sometimes with a dash of bitters. [MANHATTAN]

Manhattan Project the code name for the American project set up in 1942 to develop an atom bomb. Based in New Mexico, the project culminated in 1945 with the detonation of the first nuclear weapon, at White Sands.

manhole /ˈmænhəʊl/ *n.* a covered opening in a floor, pavement, sewer, etc. for workmen to gain access.

manhood /ˈmænhʊd/ *n.* **1** the state of being a man rather than a child or woman. **2 a** manliness; courage. **b** a man's sexual potency. **c** *euphem.* the penis. **3** the men of a country etc. **4** the state of being human.

manhunt /ˈmænhʌnt/ *n.* an organized search for a person, esp. a criminal.

mania /ˈmeɪnɪə/ *n.* **1** *Psychol.* mental illness marked by periods of great excitement and violence. **2** (often foll. by *for*) excessive enthusiasm; an obsession (*has a mania for jogging*). [ME f. LL f. Gk, = madness f. *mainomai* be mad, rel. to MIND]

-mania /ˈmeɪnɪə/ *comb. form* **1** *Psychol.* denoting a special type of mental abnormality or obsession (*megalomania; nymphomania*). **2** denoting extreme enthusiasm or admiration (*bibliomania; Anglomania*).

maniac /ˈmeɪnɪˌæk/ *n. & adj.* ● *n.* **1** *colloq.* a person exhibiting extreme symptoms of wild behaviour etc.; a madman. **2** *colloq.* an obsessive enthusiast. **3** *Psychol. archaic* a person suffering from mania. ● *adj.* of or behaving like a maniac. □ **maniacal** /mə'naɪək(ə)l/ *adj.* **maniacally** *adv.* [LL *maniacus* f. late Gk *maniakos* (as MANIA)]

-maniac /'meɪnɪˌæk/ *comb. form* forming adjectives and nouns meaning 'affected with -mania' or 'a person affected with -mania' (*nymphomaniac*).

manic /'mænɪk/ *adj.* of or affected by mania. □ **manic depression** a mental disorder with alternating periods of elation and depression. **manic-depressive** *adj.* affected by or relating to manic depression. ● *n.* a manic-depressive person. □ **manically** *adv.*

Manicaland /mə'niːkəˌlænd/ a gold-mining province of eastern Zimbabwe; capital, Mutare.

Manichaean /ˌmænɪ'kiːən/ *adj. & n.* (also **Manichean**) ● *adj.* **1** of or relating to Manichaeism or its adherents. **2** *Philos.* of or relating to dualism. ● *n.* = MANICHEE.

Manichaeism /'mænɪkiːˌɪz(ə)m/ *n.* (also **Manicheism**) a dualistic religious system with Christian, Gnostic, and pagan elements, founded in Persia in the 3rd century by Manes (*c*.216–*c*.276). The system of Manichaean dualism was based on a supposed primeval conflict between light and darkness: matter was evil, but within each person's brain was imprisoned a particle of the divine 'light' which could be released by the practice of religion; Christ, Buddha, the Prophets, and Manes had been sent to help in this task. A strict ascetic regime was practised within the sect, which, although persecuted, spread widely in the Roman Empire and in Asia, surviving in Chinese Turkestan until the 13th century. [as MANICHEE]

Manichee /'mænɪˌkiː/ *n.* **1** a follower or adherent of Manichaeism. **2** *Philos.* a dualist (see DUALISM). [LL *Manichaeus* f. late Gk *Manikhaios*, f. *Manes* or *Manichaeus*]

manicure /'mænɪˌkjʊə(r)/ *n. & v.* ● *n.* **1** a usu. professional cosmetic treatment of the hands and fingernails. **2** = MANICURIST. ● *v.tr.* apply a manicure to (the hands or a person). [F f. L *manus* hand + *cura* care]

manicurist /'mænɪˌkjʊərɪst/ *n.* a person who manicures hands and fingernails.

manifest[1] /'mænɪˌfest/ *adj. & v.* ● *adj.* clear or obvious to the eye or mind (*his distress was manifest*). ● *v.* **1** *tr.* display or show (a quality, feeling, etc.) by one's acts etc. **2** *tr.* show plainly to the eye or mind. **3** *tr.* be evidence of; prove. **4** *intr. & refl.* (of a thing) reveal itself. **5** *intr.* (of a ghost) appear. □ **manifestly** *adv.* **manifestation** /ˌmænɪfe'steɪʃ(ə)n/ *n.* [ME f. OF *manifeste* (adj.), *manifester* (v.) or L *manifestus*, *manifestare* f. *manus* hand + *festus* (unrecorded) struck]

manifest[2] /'mænɪˌfest/ *n. & v.* ● *n.* **1** a cargo-list for the use of customs officers. **2** a list of passengers in an aircraft or of trucks etc. in a goods train. ● *v.tr.* record (names, cargo, etc.) in a manifest. [It. *manifesto*: see MANIFESTO]

manifesto /ˌmænɪ'festəʊ/ *n.* (*pl.* **-os**) a public declaration of policy and aims, esp. as issued before an election by a political party, candidate, government, etc. [It. f. *manifestare* f. L (as MANIFEST[1])]

manifold /'mænɪˌfəʊld/ *adj. & n.* ● *adj. literary* **1** many and various (*manifold vexations*). **2** having various forms, parts, applications, etc. **3** performing several functions at once. ● *n.* **1** a thing with many different forms, parts, applications, etc. **2** *Mech.* a pipe or chamber branching into several openings. □ **manifoldly** *adv.* **manifoldness** *n.* [OE *manigfeald* (as MANY, -FOLD)]

manikin /'mænɪkɪn/ *n.* (also **mannikin**) **1** a little man; a dwarf. **2** an artist's lay figure. **3** an anatomical model of the body. **4** (usu. **mannikin**) a small finchlike bird of the genus *Lonchura*, native to Africa and Australasia (cf. MANAKIN). [Du. *manneken*, dimin. of *man* MAN]

Manila[1] /mə'nɪlə/ the capital and chief port of the Philippines, on the island of Luzon; pop. (1990) 1,599,000. Founded in 1571, it was an important trade centre of the Spanish until taken by the US in 1898.

Manila[2] /mə'nɪlə/ *n.* (also **Manilla**) **1** a cigar or cheroot made in Manila. **2** (in full **Manila hemp**) the strong fibre of a Philippine tree, *Musa textilis*, used for rope etc. **3** (also **manila**) a strong brown paper made from Manila hemp or other material and used for wrapping paper, envelopes, etc.

manilla /mə'nɪlə/ *n.* a metal bracelet used by some African peoples as a medium of exchange. [Sp., prob. dimin. of *mano* hand f. L *manus*]

manille /mə'nɪl/ *n. Cards* the second-best trump or honour in ombre or quadrille. [F f. Sp. *malilla* dimin. of *mala* bad f. L *malus*]

Man in the Iron Mask a mysterious prisoner held in the Bastille and other prisons in 17th-century France. According to the novel of the same name by Alexandre Dumas the man was the twin brother of Louis XIV, and was considered a threat to Louis's position on the throne; his face was concealed by a mask so that he could not be recognized. Various other theories as to the identity of the prisoner have been advanced, but it is now considered most likely that he was an Italian agent, Count Matthioli, who had angered the king by his betrayal of secret negotiations for France to acquire the stronghold of Casale from the Duke of Mantua.

manioc /'mænɪˌɒk/ *n.* **1** cassava. **2** the flour made from it. [Tupi *mandioca*]

maniple /'mænɪp(ə)l/ *n.* **1** *Rom. Hist.* a subdivision of a legion, containing 120 or 60 men. **2** *Eccl.* a Eucharistic vestment consisting of a strip hanging from the left arm. [OF *maniple* or L *manipulus* handful, troop f. *manus* hand]

manipulate /mə'nɪpjʊˌleɪt/ *v.tr.* **1** handle, treat, or use, esp. skilfully (a tool, question, material, etc.). **2** manage (a person, situation, etc.) to one's own advantage, esp. unfairly or unscrupulously. **3** manually examine and treat (a part of the body). **4** *Computing* alter, edit, or move (text, data, etc.). **5** stimulate (the genitals). □ **manipulatable** *adj.* **manipulator** *n.* **manipulatory** /-lətərɪ/ *adj.* **manipulable** *adj.* **manipulability** /-ˌnɪpjʊlə'bɪlɪtɪ/ *n.* **manipulation** /-ˌnɪpjʊ'leɪʃ(ə)n/ *n.* [back-form. f. *manipulation* f. F *manipulation* f. mod.L *manipulatio* (as MANIPLE), after F *manipuler*]

manipulative /mə'nɪpjʊlətɪv/ *adj.* **1** characterized by unscrupulous exploitation of a situation, person, etc., for one's own ends. **2** of, relating to, or involving manipulation. □ **manipulatively** *adv.* **manipulativeness** *n.*

Manipur /ˌmʌnɪ'pʊə(r)/ a small state in the far east of India, east of Assam, on the border with Burma (Myanmar); capital, Imphal. □ **Manipuri** *adj. & n.*

Manit. *abbr.* Manitoba.

Manitoba /ˌmænɪ'təʊbə/ a province of central Canada, with a coastline on Hudson Bay; pop. (1991) 1,093,200; capital, Winnipeg. The area was part of Rupert's Land from 1670 until it was transferred to Canada by the Hudson's Bay Company and became a province in 1870. □ **Manitoban** *adj. & n.*

manitou /'mænɪˌtuː/ *n.* (among some American Indians) **1** a good or evil spirit as an object of reverence. **2** something regarded as having supernatural power. [Algonquian *manito*, *-tu* he has surpassed]

mankind *n.* **1** /mæn'kaɪnd/ the human species, humankind. **2** /'mænkaɪnd/ male people, as distinct from female.

manky /'mæŋkɪ/ *adj.* (**mankier**, **mankiest**) *colloq.* **1** bad, inferior, defective. **2** dirty. [obs. *mank* mutilated, defective]

Manley /'mænlɪ/, Michael (Norman) (b.1923), Jamaican statesman, Prime Minister 1972–80 and since 1989. He became the island's first Vice-President in 1955 and leader of the People's National Party in 1969. Elected Prime Minister on a socialist platform, he introduced policies to strengthen Jamaica's economy through the expansion of public works and the encouragement of local industry; he also introduced a system of free education.

manlike /'mænlaɪk/ *adj.* **1** having the qualities of a man. **2** (of a woman) mannish. **3** (of an animal, shape, etc.) resembling a human being.

manly /'mænlɪ/ *adj.* (**manlier**, **manliest**) **1** having good qualities traditionally associated with men, such as courage, frankness, etc. **2** (of a woman) mannish. **3** (of things, qualities, etc.) befitting a man. □ **manliness** *n.*

Mann /mæn/, Thomas (1875–1955), German novelist and essayist. He achieved recognition with his first novel *Buddenbrooks* (1901), which describes the decline of a merchant family and has strongly autobiographical features. The role and character of the artist in relation to society is a constant theme in his works, and is linked with the rise of Nazism in *Dr Faustus* (1947). Other notable works include the novella *Death in Venice* (1912). When Hitler came to power Mann was forced into exile; he became a US citizen in 1944 but later settled in Switzerland. He was awarded the Nobel Prize for literature in 1929.

manna /'mænə/ *n.* **1** (in the Bible) the substance miraculously supplied as food to the Israelites in the wilderness (Exod. 16). **2** an unexpected benefit (esp. *manna from heaven*). **3** spiritual nourishment, esp. the Eucharist. **4** the sweet dried juice from the manna-ash and other plants, used as a mild laxative. □ **manna-ash** a white-flowered ash tree, *Fraxinus ornus*, native to southern Europe. [OE f. LL f. Gk f. Aram. *mannā* f. Heb. *mān*, explained as = *mān hū*? what is it?, but prob. = Arab. *mann* exudation of common tamarisk (*Tamarix gallica*)]

Mannar /mæ'nɑː(r)/ **1** an island off the NW coast of Sri Lanka, linked to India by the chain of coral islands and shoals known as Adam's Bridge. **2** a town on this island; pop. (1981) 14,000.

Mannar, Gulf of an inlet of the Indian Ocean lying between NW Sri Lanka and the southern tip of India. It lies to the south of Adam's Bridge, which separates it from the Palk Strait.

manned /mænd/ adj. (of an aircraft, spacecraft, etc.) having a human crew. [past part. of MAN]

mannequin /'mænɪkɪn/ n. **1** a model employed by a dressmaker etc. to show clothes to customers. **2** a window dummy. [F, = MANIKIN]

manner /'mænə(r)/ n. **1** a way a thing is done or happens (always dresses in that manner). **2** (in pl.) **a** social behaviour (it is bad manners to stare). **b** polite or well-bred behaviour (he has no manners). **c** modes of life; conditions of society. **3** a person's outward bearing, way of speaking, etc. (has an imperious manner). **4 a** a style in literature, art, etc. (in the manner of Rembrandt). **b** = MANNERISM 2a. **5** archaic a kind or sort (what manner of man is he?). □ **all manner of** many different kinds of. **comedy of manners** satirical portrayal of social behaviour, esp. of the upper classes. **in a manner of speaking** in some sense; to some extent; so to speak. **manner of means** see MEANS. **to the manner born 1** colloq. naturally at ease in a specified job, situation, etc. **2** destined by birth to follow a custom or way of life (Shakespeare's Hamlet I. iv. 17). □ **mannerless** adj. (in sense 2b of n.). [ME f. AF manere, OF maniere ult. f. L manuarius of the hand (manus)]

mannered /'mænəd/ adj. **1** (in comb.) behaving in a specified way (ill-mannered; well-mannered). **2** (of a style, artist, writer, etc.) showing idiosyncratic mannerisms. **3** (of a person) eccentrically affected in behaviour.

mannerism /'mænə,rɪz(ə)m/ n. **1** a habitual gesture or way of speaking etc.; an idiosyncrasy. **2 a** excessive use of a distinctive style in art or literature. **b** a stylistic trick. **3** a style of 16th-century Italian art (c.1530–90) preceding the baroque. (See note below.) □ **mannerist** n. **manneristic** /,mænə'rɪstɪk/ adj. **manneristically** adv. [MANNER]

• The term *mannerism* has a vexed history in the literature of the arts. Popularized and used as a term of praise in the 16th century by Vasari, in the 17th century the term was applied negatively, as defining a highly artificial and overwrought style, symbolic of decadence, and regarded as a decline from artists of the High Renaissance such as Leonardo da Vinci and Raphael; it was not until the 20th century that mannerist art was looked at with a more sympathetic eye. Mannerist art is characterized by a self-conscious virtuosity and use of *trompe l'oeil*, unusual and often bizarre effects of scale, lighting, and perspective, and the use of bright, often lurid colours. Pontormo, Vasari, and the later Michelangelo are among the typical exponents of mannerism in painting; in architecture, it is associated with the buildings of Vignola and of Giulio Romano (1492/9–1546), and with Michelangelo's Laurentian Library in Florence (1524–34).

mannerly /'mænəlɪ/ adj. & adv. ● adj. well-mannered; polite. ● adv. politely. □ **mannerliness** n.

Mannheim /'mænhaɪm/ an industrial port at the confluence of the Rhine and the Neckar in Baden-Württemberg, SW Germany; pop. (1991) 314,685.

mannikin var. of MANIKIN.

mannish /'mænɪʃ/ adj. **1** usu. derog. (of a woman) masculine in appearance or manner. **2** characteristic of a man. □ **mannishly** adv. **mannishness** n. [OE mennisc f. (and assim. to) MAN]

Mano /'mɑːnəʊ/ a river of West Africa. It rises in NW Liberia and flows to the Atlantic, forming for part of its length the boundary between Liberia and Sierra Leone.

manoeuvre /mə'nuːvə(r)/ n. & v. (US **maneuver**) ● n. **1** a planned and controlled movement or series of moves. **2** (in pl.) a large-scale exercise of troops, warships, etc. **3 a** an often deceptive planned or controlled action designed to gain an objective. **b** a skilful plan. ● v. (-**ring**) **1** intr. & tr. perform or cause to perform a manoeuvre (manoeuvred the car into the space). **2** intr. & tr. perform or cause (troops etc.) to perform military manoeuvres. **3 a** tr. (usu. foll. by into, out, away) force, drive, or manipulate (a person, thing, etc.) by scheming or adroitness. **b** intr. use artifice; scheme. □ **manoeuvrer** /-'nuːvrə(r)/ n. **manoeuvrable** adj. **manoeuvrability** /-,nuːvrə'bɪlɪtɪ/ n. [F manœuvre, manœuvrer f. med.L manuoperare f. L manus hand + operari to work]

manometer /mə'nɒmɪtə(r)/ n. a pressure gauge for gases and liquids. □ **manometric** /,mænə'metrɪk/ adj. [F manomètre f. Gk manos thin]

ma non troppo see TROPPO[1].

manor /'mænə(r)/ n. **1** (also **manor-house**) **a** a large country house

with lands. **b** the house of the lord of the manor. **2** Brit. **a** a unit of land consisting of a lord's demesne and lands rented to tenants etc. **b** hist. a feudal lordship over lands. **3** Brit. colloq. the district covered by a police station. □ **manorial** /mə'nɔːrɪəl/ adj. [ME f. AF maner, OF maneir, f. L manere remain]

manpower /'mæn,paʊə(r)/ n. **1** the power generated by a man working. **2** the number of people available for work, service, etc.

manqué /'mɒŋkeɪ/ adj. (placed after noun) that might have been but is not; unfulfilled (a comic actor manqué). [F, past part. of manquer lack]

Man Ray see RAY[2].

Mans, Le see LE MANS.

mansard /'mænsɑːd/ n. a roof which has four sloping sides, each of which becomes steeper halfway down. The style was popularized by François Mansart but had been in existence earlier, for example in part of the roof of the Louvre dating from the mid-16th century. [F mansarde f. MANSART]

Mansart /'mɒnsɑː(r)/, François (1598–1666), French architect. His first major work was the rebuilding of part of the château of Blois, which incorporated the type of roof now named after him (see MANSARD). Other buildings include a number of town houses in Paris, the château of Maisons (1642–6), and the church of Val-de-Grâce (1645).

manse /mæns/ n. the house of a minister, esp. a Scottish Presbyterian. □ **son** (or **daughter**) **of the manse** the child of a Presbyterian etc. minister. [ME f. med.L mansus, -sa, -sum, house f. manere mans- remain]

Mansell /'mæns(ə)l/, Nigel (b.1954), English motor-racing driver. He won the Formula One world championship in 1992 and the Indy car championship in 1993, the only driver to win both titles.

manservant /'mæn,sɜːv(ə)nt/ n. (pl. **menservants**) a male servant.

Mansfield /'mænsfiːld/, Katherine (pseudonym of Kathleen Mansfield Beauchamp) (1888–1923), New Zealand short-story writer. Her stories show the influence of Chekhov and range from extended impressionistic evocations of family life to short sketches. Collections include In a German Pension (1911) and Bliss (1920). She married the English writer and critic John Middleton Murry (1889–1957) in 1918 and spent much of the remainder of her short life travelling in Europe in search of a cure for the tuberculosis from which she eventually died.

-manship /mənʃɪp/ suffix forming nouns denoting skill in a subject or activity (craftsmanship; gamesmanship).

mansion /'mænʃ(ə)n/ n. **1** a large house. **2** (usu. in pl.) Brit. a large building divided into flats. □ **mansion-house** Brit. the house of a lord mayor or a landed proprietor. [ME f. OF f. L mansio -onis a staying (as MANSE)]

Mansion House the official residence of the Lord Mayor of London.

manslaughter /'mæn,slɔːtə(r)/ n. **1** the killing of a human being. **2** Law the unlawful killing of a human being without malice aforethought.

Manson /'mæns(ə)n/, Sir Patrick (1844–1922), Scottish physician, pioneer of tropical medicine. Working for many years in China, he discovered the organism responsible for elephantiasis and established that it was spread by the bite of a mosquito. After returning to London he suggested a similar role for the mosquito in spreading malaria, and studied a number of tropical parasites and infections. He was the chief founder of the London School of Tropical Medicine.

mansuetude /'mænswɪ,tjuːd/ n. archaic meekness, docility, gentleness. [ME f. OF mansuetude or L mansuetudo f. mansuetus gentle, tame f. manus hand + suetus accustomed]

manta /'mæntə/ n. a very large plankton-eating ray of the family Mobulidae, esp. Manta birostris, having winglike pectoral fins, hornlike projections on the head, and a whiplike tail. Also called devilfish, devil ray. [Amer. Sp., = large blanket]

Mantegna /mæn'tenjə/, Andrea (1431–1506), Italian painter and engraver. He is noted especially for his frescos, which include those painted for the bridal chamber of the ducal family in Mantua, with both the ceiling and walls painted in an illusionistic style which extends the interior space and gives the impression that the room is open to the sky. His work reveals his knowledge of the artefacts and architecture of classical antiquity, as can be seen in his nine paintings depicting the Triumph of Caesar (begun 1486).

mantel /'mænt(ə)l/ n. **1** = MANTELPIECE 1. **2** = MANTELSHELF. [var. of MANTLE]

mantelet /'mæntəlɪt/ n. (also **mantlet** /'mæntlɪt/) **1** hist. a woman's short loose sleeveless mantle. **2** a bulletproof screen for gunners. [ME f. OF, dimin. of mantel MANTLE]

mantelpiece /'mænt(ə)l,piːs/ *n.* **1** a structure of wood, marble, etc., above and around a fireplace. **2** = MANTELSHELF.

mantelshelf /'mænt(ə)l,ʃelf/ *n.* a shelf above a fireplace.

mantic /'mæntɪk/ *adj. formal* of or concerning divination or prophecy. [Gk *mantikos* f. *mantis* prophet]

mantid /'mæntɪd/ *n.* = MANTIS.

mantilla /mæn'tɪlə/ *n.* a lace scarf worn by Spanish women over the hair and shoulders. [Sp., dimin. of *manta* MANTLE]

mantis /'mæntɪs/ *n.* (*pl.* same or **mantises**) a slender predatory insect of the family Mantidae, holding its forelegs raised and folded like hands in prayer while waiting for prey; esp. *Mantis religiosa*, of southern Europe. Also called *praying mantis*. [Gk, = prophet]

mantissa /mæn'tɪsə/ *n.* the part of a logarithm after the decimal point. [L, = makeweight]

mantle /'mænt(ə)l/ *n. & v.* ● *n.* **1** a loose sleeveless cloak, esp. of a woman. **2** a covering (*a mantle of snow*). **3** responsibility or authority, esp. as passing from one person to another (see 2 Kings 2:13). **4** a fragile lacelike tube fixed round a gas-jet to give an incandescent light. **5** *Zool.* an outer fold of skin enclosing a mollusc's viscera and secreting the shell. **6** a bird's back, scapulars, and wing-coverts, esp. if of a distinctive colour. **7** the region between the crust and the core of the earth. ● *v.* **1** *tr.* clothe in or as if in a mantle; cover, conceal, envelop. **2** *intr.* **a** (of the blood) suffuse the cheeks. **b** (of the face) glow with a blush. **3** *intr.* (of a liquid) become covered with a coating or scum. [ME f. OF f. L *mantellum* cloak]

mantlet var. of MANTELET.

mantling /'mæntlɪŋ/ *n. Heraldry* **1** ornamental drapery etc. behind and around a shield. **2** a representation of this. [MANTLE + -ING[1]]

mantra /'mæntrə/ *n. Buddhism & Hinduism* a syllable, word, or phrase used in meditation and believed to possess spiritual powers. It may be spoken aloud and as a chant, or can be repeated internally. In Hinduism, an important part of initiation involves the giving of a personal mantra by the guru. [Skr., = instrument of thought f. *man* think]

mantrap /'mæntræp/ *n.* a trap for catching people, esp. trespassers.

mantua /'mæntjʊə/ *n. hist.* a woman's loose gown of the 17th–18th centuries. [corrupt. of *manteau* (F, as MANTLE) after *Mantua* in Italy]

Manu /'mɑnuː/ the archetypal first man of Hindu mythology, survivor of the great flood and father of the human race. He is also the legendary author of one of the most famous codes of Hindu religious law, the *Manusmriti* (*Laws of Manu*), composed in Sanskrit and dating in its present form from the 1st century BC. [Skr., = man]

manual /'mænjʊəl/ *adj. & n.* ● *adj.* **1** of or done with the hands (*manual labour*). **2** (of a machine etc.) worked by hand, not automatically. ● *n.* **1 a** a book of instructions, esp. for operating a machine or learning a subject; a handbook (*a computer manual*). **b** any small book. **2** an organ keyboard played with the hands not the feet. **3** *colloq.* a vehicle with manual transmission. **4** *Mil.* an exercise in handling a rifle etc. **5** *hist.* a book of the forms to be used by priests in the administration of the sacraments. □ **manual alphabet** sign language. □ **manually** *adv.* [ME f. OF *manuel*, f. (and later assim. to) L *manualis* f. *manus* hand]

manufactory /,mænjʊ'fæktərɪ/ *n.* (*pl.* **-ies**) *archaic* = FACTORY. [MANUFACTURE, after *factory*]

manufacture /,mænjʊ'fæktʃə(r)/ *n. & v.* ● *n.* **1 a** the making of articles, esp. in a factory etc. **b** a branch of industry (*woollen manufacture*). **2** esp. *derog.* the merely mechanical production of literature, art, etc. ● *v.tr.* **1** make (articles), esp. on an industrial scale. **2** invent or fabricate (evidence, a story, etc.). **3** esp. *derog.* make or produce (literature, art, etc.) in a mechanical way. □ **manufacturer** *n.* **manufacturable** *adj.* **manufacturability** /-,fæktʃərə'bɪlɪtɪ/ *n.* [F f. It. *manifattura* & L *manufactum* made by hand]

manuka /'mæ'nuːkə, 'mɑːnəkə/ *n. Austral. & NZ* a small tree, *Leptospermum scoparium*, with aromatic leaves and hard timber. [Maori]

manumit /,mænjʊ'mɪt/ *v.tr.* (**manumitted, manumitting**) *hist.* set (a slave) free. □ **manumission** /-'mɪʃ(ə)n/ *n.* [ME f. L *manumittere manumiss-* f. *manus* hand + *emittere* send forth]

manure /mə'njʊə(r)/ *n. & v.* ● *n.* **1** animal dung, esp. *Brit.* of horses, used for fertilizing land. **2** any compost or artificial fertilizer. ● *v.tr.* (also *absol.*) apply manure to (land etc.). [ME f. AF *mainoverer* = OF *manouvrer* MANOEUVRE]

manuscript /'mænjʊ,skrɪpt/ *n. & adj.* ● *n.* **1** a book, document, etc., written by hand. **2** an author's handwritten or typed text, submitted for publication. **3** handwritten form (*produced in manuscript*). ● *adj.*

written by hand. [med.L *manuscriptus* f. *manu* by hand + *scriptus* past part. of *scribere* write]

Manutius, Aldus, see ALDUS MANUTIUS.

Manx /mæŋks/ *adj. & n.* ● *adj.* of or relating to the Isle of Man or its people or language. ● *n.* **1** the Celtic language of the Isle of Man, not now learned as a first language but still in use for ceremonial purposes. **2** (prec. by *the*; treated as *pl.*) the Manx people. □ **Manx cat** a tailless variety of domestic cat, originating in the Isle of Man. **Manx shearwater** a brownish-black and white shearwater, *Puffinus puffinus*, of Atlantic and Mediterranean waters. [ON f. OIr. *Manu* Isle of Man]

Manxman /'mæŋksmən/ *n.* (*pl.* **-men**; *fem.* **Manxwoman**, *pl.* **-women**) a native of the Isle of Man.

many /'menɪ/ *adj. & n.* ● *adj.* (**more; most**) great in number; numerous (*many times; many people; many a person; his reasons were many*). ● *n.* (as *pl.*) **1** a large number (*many like skiing; many went*). **2** (prec. by *the*) the majority of people, the multitude. □ **as many** the same number of (*six mistakes in as many lines*). **as many again** the same number additionally (*sixty here and as many again there*). **be too** (or **one too**) **many for** outwit, baffle. **a good** (or **great**) **many** a large number. **many-sided** having many sides, aspects, interests, capabilities, etc. **many-sidedness** *n.* the fact or state of being many-sided. **many's the time** often (*many's the time we saw it*). **many a time** many times. [OE *manig*, ult. f. Gmc]

manzanilla /,mænzə'nɪlə/ *n.* a pale very dry Spanish sherry. [Sp., lit. 'camomile']

manzanita /,mænzə'niːtə/ *n.* a North American bearberry; esp. *Arctostaphylos manzanita*, native to California. [Sp., dimin. of *manzana* apple]

Manzoni /mæn'zəʊnɪ/, Alessandro (1785–1873), Italian novelist, dramatist, and poet. He is remembered chiefly as the author of the historical novel *I promessi sposi* (1825–42). The novel is a powerfully characterized historical reconstruction of 17th-century Lombardy during the period of Spanish administration; it had great patriotic appeal at a time when Italy was seeking independence from Austrian rule.

Maoism /'maʊɪz(ə)m/ *n.* the Communist doctrines of Mao Zedong as formerly practised in China, having as a central idea permanent revolution and stressing the importance of the peasantry, of small-scale industry, and of agricultural collectivization. □ **Maoist** *n. & adj.*

Maori /'maʊrɪ/ *n. & adj.* ● *n.* (*pl.* same or **Maoris**) **1** a member of the Polynesian aboriginal people of New Zealand. (*See note below.*) **2** the Polynesian language of the Maoris, spoken by about 100,000 people. ● *adj.* of or relating to the Maoris or their language. [Maori]

▪ The Maoris arrived in New Zealand as part of a series of waves of migration from Tahiti, probably in the 9th century. Traditionally the way of life is hunting and gathering, and the tribal structure is based on common ancestry. In the 19th century colonization by the British, and, in particular, the threat to traditional Maori land rights, led to opposition and to a series of wars (see MAORI WARS), which finally ended in 1872. The Maoris lost large amounts of land as a result and the pace of colonization increased. The 20th century has seen a number of Maori cultural revivals, including the introduction of the Maori language into schools; although Maoris maintain their cultural identity, increasing urbanization and intermarriage has meant that in many practical ways they are now largely integrated with modern New Zealand society. The Maoris now number about 280,000.

Maori Wars a series of wars fought intermittently in 1845–8 and 1860–72 between Maoris and the colonial government of New Zealand over the enforced sale of Maori lands to Europeans, which was forbidden by the Treaty of Waitangi.

Mao Zedong /,maʊ dziˈdʊŋ/ (also **Mao Tse-tung** /tseɪˈtʊŋ/) (1893–1976), Chinese statesman, chairman of the Communist Party of the Chinese People's Republic 1949–76 and head of state 1949–59. After studying Marxism as a student he was among the founders of the Chinese Communist Party in 1921, becoming its effective leader following the Long March (1934–5). He eventually defeated both the occupying Japanese and rival Kuomintang nationalist forces to form the People's Republic of China, becoming its first head of state (1949). Although he initially adopted the Soviet Communist model, following Khrushchev's denunciation of Stalin (1956), Mao began to introduce his own measures (see MAOISM). A brief period of freedom of expression (see HUNDRED FLOWERS) ended with the introduction of the economically disastrous Great Leap Forward (1958–60). Mao resigned as head of state in 1959 but retained his position as chairman of the Communist Party, and as such remained China's most powerful politician. He was the instigator of the Cultural Revolution (1966–8), during which he became the focus of a powerful personality cult which lasted until his death.

map /mæp/ n. & v. ● n. **1 a** a usu. flat representation of the earth's surface, or part of it, showing physical features, cities, etc. (cf. GLOBE n. 2). **b** a diagrammatic representation of a route etc. (drew a map of the journey). **2** a two-dimensional representation of the stars, the heavens, etc., or of the surface of a planet, the moon, etc. **3** a diagram showing the arrangement or components of a thing, esp. Biol. of the gene sequence on a chromosome or the DNA base sequence of a gene. **4** sl. the face. ● v.tr. (**mapped, mapping**) **1** represent (a country etc.) on a map. **2** Math. associate each element of (a set) with an element of another set. □ **map out** arrange in detail; plan (a course of conduct etc.). **off the map** colloq. **1** of no account; obsolete. **2** very distant. **on the map** colloq. prominent, important. **wipe off the map** colloq. obliterate. □ **mapless** adj. **mappable** adj. **mapper** n. [L mappa napkin: in med.L mappa (mundi) map (of the world)]

maple /ˈmeɪp(ə)l/ n. **1** a tree or shrub of the genus Acer, with lobed leaves and winged seeds (see also sugar maple). **2** the wood of the maple. □ **maple-leaf** the leaf of the maple, used as an emblem of Canada. **maple sugar** a sugar produced by evaporating the sap of the sugar maple etc. **maple syrup** a syrup produced from the sap of the sugar maple etc. [ME mapul etc. f. OE mapeltrēow, mapulder]

Mappa Mundi /ˌmæpə ˈmʊndɪ/ a famous 13th-century map of the world, now in Hereford cathedral, England. The map is round and typical of similar maps of the time in that it depicts Jerusalem at its centre.

Maputo /məˈpuːtəʊ, -tuː/ the capital and chief port of Mozambique, on the Indian Ocean in the south of the country; pop. (1991) 1,098,000. Founded as a Portuguese fortress in the late 18th century, it became the capital of Mozambique in 1907. It was known as Lourenço Marques until 1976.

maquette /məˈket/ n. **1** a sculptor's small preliminary model in wax, clay, etc. **2** a preliminary sketch. [F f. It. machietta dimin. of macchia spot]

maquillage /ˌmækiːˈɑːʒ/ n. **1** make-up; cosmetics. **2** the application of make-up. [F f. maquiller make up f. OF masquiller stain]

Maquis /mæˈkiː/ n. (pl. same) **1** the French resistance movement during the German occupation in the Second World War (see RESISTANCE, THE). **2** a member of this. [F, = brushwood, f. Corsican It. macchia thicket]

Mar. abbr. March.

mar /mɑː(r)/ v.tr. (**marred, marring**) **1** impair the perfection of; spoil; disfigure. **2** archaic ruin. [OE merran hinder]

marabou /ˈmærəˌbuː/ n. (also **marabout**) **1** a large West African stork, Leptoptilos crumeniferus, feeding on carrion. **2** a tuft of down from the wing or tail of the marabou used as a trimming for hats etc. [F f. Arab. murābiṭ holy man (see MARABOUT), the stork being regarded as holy]

marabout /ˈmærəˌbuː, -ˌbuːt/ n. **1** a Muslim hermit or monk, esp. in North Africa. **2** a shrine marking a marabout's burial place. **3** var. of MARABOU. [F f. Port. marabuto f. Arab. murābiṭ holy man f. ribāṭ frontier station, where he acquired merit by combat against the infidel]

maraca /məˈrækə/ n. a hollow clublike gourd or gourd-shaped container filled with beans etc. and usu. shaken in pairs as a percussion instrument in Latin American music. [Port. maracá, prob. f. Tupi]

Maracaibo /ˌmærəˈkaɪbəʊ/ a city and port in NW Venezuela, situated on the channel linking the Gulf of Venezuela with Lake Maracaibo; pop. (1991) 1,400,640. Founded in 1571, the city expanded rapidly after the discovery of oil in the region in 1917. It is now the second largest city in the country.

Maracaibo, Lake a large lake in NW Venezuela, linked by a narrow channel to the Gulf of Venezuela and the Caribbean Sea.

Maradona /ˌmærəˈdɒnə/, Diego (Armando) (b.1960), Argentinian footballer. He was captain of the victorious Argentinian team in the 1986 World Cup, but aroused controversy with his apparent handball when scoring a goal in Argentina's quarter-final match against England. In 1984 he joined the Italian club Napoli, and subsequently contributed to that team's victories in the Italian championship (1987) and the UEFA Cup (1989). However, clashes with authority culminated in Maradona being suspended from football for fifteen months in 1991 for cocaine use, and then sent home from the 1994 World Cup after failing a drugs test.

Maramba /məˈræmbə/ a city in southern Zambia, about 5 km (3 miles)

from the Zambezi river and the Victoria Falls; pop. (1987) 94,640. Formerly called Livingstone in honour of the explorer David Livingstone, it was the capital of Northern Rhodesia from 1911 until Lusaka became capital in 1935.

Maranhão /ˌmærəˈnjaʊ/ a state of NE Brazil, on the Atlantic coast; capital, São Luís.

Marañón /ˌmærəˈnjɒn/ a river of northern Peru, which rises in the Andes and forms one of the principal headwaters of the Amazon.

maraschino /ˌmærəˈskiːnəʊ/ n. (pl. **-os**) a strong sweet liqueur made from small black Dalmatian cherries. □ **maraschino cherry** a cherry preserved in maraschino and used to decorate cocktails etc. [It. f. marasca small black cherry, for amarasca f. amaro bitter f. L amarus]

marasmus /məˈræzməs/ n. a wasting away of the body. □ **marasmic** adj. [mod.L f. Gk marasmos f. marainō wither]

Marat /ˈmærɑː/, Jean Paul (1743–93), French revolutionary and journalist. The founder of a radical newspaper, he became prominent during the French Revolution as a virulent critic of the moderate Girondists and was instrumental (with Danton and Robespierre) in their fall from power in 1793. Suffering from a skin disease, he spent much of his time in later life in his bath, where he was murdered by the Girondist Charlotte Corday. This was used as a pretext by Robespierre and the Jacobins to purge their Girondist rivals.

Maratha /məˈrɑːtə, -ˈrætə/ n. (also **Mahratta**) a member of a Hindu people native to the Indian state of Maharashtra. In the 17th century the Marathas rose to prominence under the leadership of Shivaji, rising in rebellion against the Muslim Moguls and establishing their own kingdom, of which Shivaji became raja in 1674. In the early 18th century they came to dominate much of southern and central India, but proved unable to stand together against expanding British power. [Hindi Marhaṭṭā f. Skr. Mahārāshṭra great kingdom]

Marathi /məˈrɑːtɪ, -ˈrætɪ/ n. (also **Mahratti**) the Indic language of the Marathas. [MARATHA]

marathon /ˈmærəθən/ n. **1** a long-distance running-race, now usu. of 26 miles 385 yards (42.195 km). (See note below.) **2** a long-lasting or difficult task, operation, etc. (often attrib.: a marathon shopping expedition). □ **marathoner** n.

• The marathon has been a principal event of the modern Olympic Games since 1896. It is named after Marathon in Greece, scene of an Athenian victory over an invading Persian army in 490 BC. Herodotus records that the herald Pheidippides ran the 150 miles (250 km) from Athens to Sparta to secure aid before the battle, but the race instituted in 1896 was based on a later tradition that a messenger ran all the way from Marathon to Athens (22 miles, 35 km) with news of victory, but fell dead on arrival. Similar races, open to applicants, also take place through the streets in certain cities, e.g. in Boston (annually from 1987) and London (from 1981). The present standardized distance dates from 1924.

maraud /məˈrɔːd/ v. **1** intr. **a** make a plundering raid. **b** pilfer systematically; plunder. **2** tr. plunder (a place). □ **marauder** n. [F marauder f. maraud rogue]

Marbella /mɑːˈbeɪjə/ a resort town on the Costa del Sol of southern Spain, in Andalusia; pop. (1991) 80,645.

marble /ˈmɑːb(ə)l/ n. & v. ● n. **1** limestone in a metamorphic crystalline (or granular) state, and capable of taking a polish, used in sculpture and architecture. **2** (often attrib.) **a** anything made of marble (a marble clock). **b** anything resembling marble in hardness, coldness, durability, etc. (her features were marble). **3 a** a small ball of marble, glass, clay, etc., used as a toy. **b** (in pl.) treated as sing.) a game using these. **4** (in pl.) sl. one's mental faculties (he's lost his marbles). **5** (in pl.) a collection of sculptures (Elgin Marbles). ● v.tr. **1** (esp. as **marbled** adj.) stain or colour (paper, the edges of a book, soap, etc.) to look like variegated marble. **2** (as **marbled** adj.) (of meat) streaked with alternating layers of lean and fat. □ **marble cake** a cake with a mottled appearance, made of light and dark sponge. **marbled white** a whitish satyrid butterfly with black markings, Melanargia galathea. □ **marbly** adj. [ME f. OF marbre, marble, f. L marmor f. Gk marmaros shining stone]

Marble Arch an arch with three gateways at the north-eastern corner of Hyde Park in London. Designed by John Nash, it was erected in 1827 in front of Buckingham Palace and moved in 1851 to its present site.

marbling /ˈmɑːblɪŋ/ n. **1** colouring or marking like marble. **2** streaks of fat in lean meat.

Marburg /ˈmɑːbʊrk/ **1** a city in the state of Hesse in west central Germany; pop. (1989) 71,358. It was the scene in 1529 of a debate between German and Swiss theologians, notably Martin Luther and

Ulrich Zwingli, on the doctrine of consubstantiation. **2** the German name for MARIBOR.

marc /maːk/ n. **1** the refuse of pressed grapes etc. **2** a brandy made from this. [F f. *marcher* tread, MARCH[1]]

Marcan /'maːkən/ adj. of or relating to St Mark. [L *Marcus* Mark]

marcasite /'maːkəˌsaɪt/ n. **1** a yellowish crystalline iron sulphide mineral. **2** bronze-yellow crystals of marcasite used in jewellery. [ME f. med.L *marcasita*, f. Arab. *marḳašīṭā* f. Pers.]

marcato /maːˈkɑːtəʊ/ adv. & adj. Mus. played with emphasis. [It., = marked]

Marceau /maːˈsəʊ/, Marcel (b.1923), French mime artist. He is known for his mimes in which he appears as the white-faced Bip, a character he developed from the French Pierrot character. He directed his own company between 1948 and 1964 and made many stage and television appearances. He also created a number of mime-dramas, including *Don Juan* (1964).

marcel /maːˈsel/ n. & v. hist. ● n. (in full **marcel wave**) a deep artificial wave in the hair. ● v.tr. (**marcelled**, **marcelling**) wave (hair) with a deep wave. [Marcel Grateau, Paris hairdresser (1852–1936), who invented the method]

marcescent /maːˈses(ə)nt/ adj. (of part of a plant) withering but not falling. □ **marcescence** n. [L *marcescere* inceptive of *marcere* wither]

March /maːtʃ/ n. the third month of the year. □ **March hare** a hare in the breeding season, characterized by excessive leaping, strange behaviour, etc. (*mad as a March hare*). [ME f. OF *march(e)*, dial. var. of *marz*, *mars*, f. L *Martius (mensis)* (month) of Mars, several of whose festivals were held in this month; it was orig. the first month of the year]

march[1] /maːtʃ/ v. & n. ● v. **1** intr. (usu. foll. by *away*, *off*, *out*, etc.) walk in a military manner with a regular measured tread. **2** tr. (often foll. by *away*, *on*, *off*, etc.) cause to march or walk (*marched the army to Moscow*; *marched him out of the room*). **3** intr. **a** walk or proceed steadily, esp. across country. **b** (of events etc.) continue unrelentingly (*time marches on*). **4** intr. take part in a protest march. ● n. **1 a** the act or an instance of marching. **b** the uniform step of troops etc. (*a slow march*). **2** a long difficult walk. **3** a procession as a protest or demonstration. **4** (usu. foll. by *of*) progress or continuity (*the march of events*). **5 a** a piece of music composed to accompany a march. **b** a composition of similar character and form. □ **marching order** Mil. equipment or a formation for marching. **marching orders 1** Mil. the direction for troops to depart for war etc. **2** a dismissal (*gave him his marching orders*). **march on 1** advance towards (a military objective). **2** proceed. **march past** n. the marching of troops past a saluting-point at a review. ● v.intr. (of troops) carry out a march past. **on the march 1** marching. **2** in steady progress. □ **marcher** n. [F *marche* (n.), *marcher* (v.), f. LL *marcus* hammer]

march[2] /maːtʃ/ n. & v. ● n. hist. **1** (usu. in pl.) a boundary, a frontier (see also MARCHES, THE). **2** a tract of often disputed land between two countries. ● v.intr. (foll. by *upon*, *with*) (of a country, an estate, etc.) have a common frontier with, border on. [ME f. OF *marche*, *marchir* ult. f. Gmc: cf. MARK[1]]

marcher /'maːtʃə(r)/ n. an inhabitant of a march or border district.

Marches, the /'maːtʃɪz/ **1** the parts of England along the borders with Wales and (formerly) Scotland. **2** (Italian **Marche** /'markeɪ/) a region of east central Italy, between the Apennines and the Adriatic Sea; capital, Ancona. [MARCH[2]]

marchioness /ˌmaːʃəˈnes, 'maːʃənɪs/ n. **1** the wife or widow of a marquess. **2** a woman holding the rank of marquess in her own right (cf. MARQUISE). [med.L *marchionissa* f. *marchio -onis* captain of the marches (as MARCH[2])]

marchpane archaic var. of MARZIPAN.

Marciano /ˌmaːsɪˈɑːnəʊ/, Rocky (born Rocco Francis Marchegiano) (1923–69), American boxer. In 1952 he became world heavyweight champion and successfully defended his title six times until he retired, undefeated, in 1956.

Marconi /maːˈkəʊnɪ/, Guglielmo (1874–1937), Italian electrical engineer, the father of radio. He appreciated the potential of earlier work on electric waves and their detection, and made a radio transmission over a distance of a mile at Bologna in 1895. He later transmitted a signal across the Atlantic from Cornwall to Newfoundland, and in 1912 produced a continuously oscillating wave, essential for the transmission of sound. Marconi went on to develop short-wave transmission over long distances, especially valuable to shipping, and set up successful companies to exploit his work. In 1909 he was awarded the Nobel Prize for physics.

Marco Polo /ˌmaːkəʊ ˈpəʊləʊ/ (c.1254–c.1324), Italian traveller. Between 1271 and 1275 he accompanied his father and uncle on a journey east from Acre into central Asia, eventually reaching China and the court of Kublai Khan. After service with the emperor and travelling widely in the empire for a decade and a half, Polo returned home (1292–5) via Sumatra, India, and Persia. His book recounting his travels gave considerable impetus to the European quest for the riches of the East.

Marcus Aurelius see AURELIUS.

Marcuse /maːˈkuːzə/, Herbert (1898–1979), German-born American philosopher. He was an associate at the Frankfurt Institute of Social Research (see FRANKFURT SCHOOL) until 1933, when he left Germany and eventually settled in the US. His works include *Eros and Civilization* (1955), *Soviet Marxism* (1958), a rejection of bureaucratic Communism which argues that revolutionary change can come only from alienated élites such as students, and *One-Dimensional Man* (1964).

Mar del Plata /ˌmaː del ˈplaːtə/ a fishing port and resort in Argentina, on the Atlantic coast south of Buenos Aires; pop. (1991) 520,000.

Mardi Gras /ˌmaːdɪ ˈɡraː/ n. **1 a** Shrove Tuesday in some Catholic countries. **b** merrymaking on this day. **2** the last day of a carnival etc. **3** Austral. a carnival or fair at any time. [F, = fat Tuesday]

Marduk /'maːdʊk/ Babylonian Mythol. the chief god of Babylon, who became lord of the gods of heaven and earth after conquering Tiamat, the monster of primeval chaos.

mardy /'maːdɪ/ adj. dial. sulky, whining, spoilt. [dial. *mard* spoilt, alt. of *marred* f. MAR]

Mare, Walter de la, see DE LA MARE.

mare[1] /meə(r)/ n. **1** the female of any equine animal, esp. the horse. **2** sl. derog. a woman. □ **mare's nest** an illusory discovery. **mare's tail 1** a tall slender marsh plant, *Hippuris vulgaris*. **2** (in pl.) long straight streaks of cirrus cloud. [ME f. OE *mearh* horse f. Gmc: cf. MARSHAL]

mare[2] /'maːreɪ/ n. (pl. **maria** /-rɪə/ or **mares**) **1** (in full **mare clausum** /'klaʊsʊm/) Law the sea under the jurisdiction of a particular country. **2** (in full **mare liberum** /'liːbərʊm/) Law the sea open to all nations. **3 a** any of a number of large dark flat areas on the surface of the moon, once thought to be seas. **b** a similar area on Mars. [L, = sea]

maremma /məˈremə/ n. (pl. **maremme** /-mɪ/) low marshy unhealthy land near a seashore. [It. f. L *maritima* (as MARITIME)]

Marengo, Battle of /məˈreŋɡəʊ/ a decisive French victory of Napoleon's campaign in Italy in 1800, close to the village of Marengo near Turin. After military reverses had all but destroyed French power in Italy, Napoleon crossed the Alps to defeat and capture an Austrian army, a victory which led to Italy coming under French control again.

Margaret, Princess /'maːɡrɪt/, Margaret Rose (b.1930), only sister of Elizabeth II. In 1960 she married Antony Armstrong-Jones (b.1930), who was later created Earl of Snowdon; the marriage was dissolved in 1978. Their two children are David, Viscount Linley (b.1961) and Lady Sarah Armstrong-Jones (b.1964).

Margaret, St /'maːɡrɪt/ (c.1046–93), Scottish queen, wife of Malcolm III. She exerted a strong influence over royal policy during her husband's reign, and was instrumental in the reform of the Scottish Church. Feast day, 16 Nov.

margarine /ˌmaːdʒəˈriːn, ˌmaːɡə-/ n. a substance used as a substitute for butter, made from vegetable oils or animal fats with milk etc. [F, misapplication of a chem. term, f. *margarique* f. Gk *margaron* pearl]

Margarita /ˌmaːɡəˈriːtə/ an island in the Caribbean Sea, off the coast of Venezuela. Visited by Columbus in 1498, it was used as a base by Simón Bolívar in 1816 in the struggle for independence from Spanish rule. The island has been a centre of pearl fishing for several hundred years.

margay /'maːɡeɪ/ n. a small Central and South American wild cat, *Felis wiedii*, with a marbled coat. [F f. Tupi *mbaracaïa*]

marge[1] /maːdʒ/ n. Brit. colloq. margarine. [abbr.]

marge[2] /maːdʒ/ n. poet. a margin or edge. [F f. L *margo* (as MARGIN)]

margin /'maːdʒɪn/ n. & v. ● n. **1** the edge or border of a surface. **2 a** the blank border on each side of the print on a page etc. **b** a line ruled esp. on exercise paper, marking off a margin. **3** an amount (of time, money, etc.) by which a thing exceeds, falls short, etc. (*won by a narrow margin*; *a margin of profit*). **4** the lower limit of possibility, success, etc. (*his effort fell below the margin*). **5** Austral. an increment to a basic wage, paid for skill. **6** a sum deposited with a stockbroker to cover the risk of loss on a transaction on account. ● v.tr. (**margined**, **margining**) provide with a margin or marginal notes. □ **margin of error** a usu. small difference allowed for miscalculation, change of circumstances, etc. **margin release** a device on a typewriter allowing a word to be typed beyond the margin normally set. [ME f. L *margo -ginis*]

marginal /ˈmɑːdʒɪn(ə)l/ *adj. & n.* ● *adj.* **1 a** of or written in a margin. **b** having marginal notes. **2 a** of or at the edge; not central. **b** not significant or decisive (*the work is of merely marginal interest*). **3** *Brit.* (of a parliamentary seat or constituency) having a small majority at risk in an election. **4** close to the limit, esp. of profitability. **5** (of the sea) adjacent to the shore of a state. **6** (of land) difficult to cultivate; unprofitable. **7** barely adequate; unprovided for. ● *n. Brit.* a marginal constituency or seat. □ **marginal cost** the cost added by making one extra copy etc. □ **marginally** *adv.* **marginality** /ˌmɑːdʒɪˈnælɪtɪ/ *n.* [med.L *marginalis* (as MARGIN)]

marginalia /ˌmɑːdʒɪˈneɪlɪə/ *n.pl.* marginal notes. [med.L, neut. pl. of *marginalis*]

marginalize /ˈmɑːdʒɪnəˌlaɪz/ *v.tr.* (also **-ise**) make or treat as insignificant. □ **marginalization** /ˌmɑːdʒɪnəlaɪˈzeɪʃ(ə)n/ *n.*

marginate *v. & adj.* ● *v.tr.* /ˈmɑːdʒɪˌneɪt/ **1** = MARGINALIZE. **2** provide with a margin or border. ● *adj.* /ˈmɑːdʒɪnət/ *Biol.* having a distinct margin or border. □ **margination** /ˌmɑːdʒɪˈneɪʃ(ə)n/ *n.*

margrave /ˈmɑːɡreɪv/ *n. hist.* the hereditary title of some princes of the Holy Roman Empire (orig. of a military governor of a border province). □ **margravate** /-ɡrəvət/ *n.* [MDu. *markgrave* border count (as MARK[1], *grave* count f. OLG *grēve*)]

margravine /ˈmɑːɡrəˌviːn/ *n. hist.* the wife of a margrave. [Du. *markgravin* (as MARGRAVE)]

marguerite /ˌmɑːɡəˈriːt/ *n.* an ox-eye daisy. [F f. L *margarita* f. Gk *margarītēs* f. *margaron* pearl]

Mari /ˈmɑːrɪ/ an ancient city on the west bank of the Euphrates, in Syria. Its strategic position commanding major trade routes ensured its rapid growth, and by about 2500 BC it was a thriving city, influenced by Sumerian culture. Its period of greatest importance was from the late 19th to the mid-18th centuries BC, when it was a kingdom with hegemony over the middle Euphrates valley. The vast palace of the last king, Zimrilim, has yielded an archive of 25,000 cuneiform tablets, which are the principal source for the history of northern Syria and Mesopotamia at that time. The city was sacked by Hammurabi of Babylon in 1759 BC.

maria *pl.* of MARE[2].

Maria de' Medici /məˈriːə/ see MARIE DE MÉDICIS.

mariage de convenance /marjaʒ də kɔ̃vənɑ̃s/ *n.* = marriage of convenience. [F]

Marian /ˈmɛərɪən/ *adj. RC Ch.* of or relating to the Virgin Mary (*Marian vespers*). [L *Maria* Mary]

Mariana Islands /ˌmærɪˈɑːnə/ (also **Marianas**) a group of islands in the western Pacific, comprising Guam and the Northern Marianas. Visited by Ferdinand Magellan in 1521, the islands were colonized by Spain in 1668 and named Las Marianas in honour of Maria Anna, widow of Philip IV. Guam was ceded to the US in 1898 and the remaining islands were sold to Germany the following year. Administered by Japan after the First World War, they became part of the Pacific Islands Trust Territory, administered by the US, in 1947. In 1975 the Northern Marianas voted to establish a commonwealth in union with the US, and became self-governing three years later. (See also NORTHERN MARIANAS.)

Mariana Trench an ocean trench to the south-east of the Mariana Islands in the western Pacific, with the greatest known ocean depth (11,034 m, 36,201 ft at the Challenger Deep).

Maria Theresa /məˌriːə təˈreɪzə/ (1717–80), Archduchess of Austria, queen of Hungary and Bohemia 1740–80. The daughter of the Emperor Charles VI, Maria Theresa married the future Emperor Francis I in 1736 and succeeded to the Habsburg dominions in 1740 by virtue of the Pragmatic Sanction. Her accession triggered the War of the Austrian Succession (1740–8), during which Silesia was lost to Prussia. She attempted but failed to regain Silesia from the Prussians in the Seven Years War (1756–63). After the death of Francis I in 1765 she ruled in conjunction with her son, the Emperor Joseph II.

Maribor /ˈmærɪˌbɔː(r)/ (called in German *Marburg*) an industrial city in NE Slovenia, on the River Drava near the border with Austria; pop. (1991) 103,900.

Marie Antoinette /ˌmærɪ ˌæntwəˈnet, mari ɑ̃twanɛt/ (1755–93), French queen, wife of Louis XVI. A daughter of Maria Theresa and the Emperor Francis I, she married the future Louis XVI of France in 1770, becoming queen four years later. She became a focus for opposition to

reform and won widespread unpopularity through her extravagant lifestyle. Like her husband she was imprisoned during the French Revolution and eventually executed.

Marie Byrd Land /ˌmɑːrɪ ˈbɜːd/ a region of Antarctica bordering the Pacific, between Ellsworth Land and the Ross Sea. It was explored in 1929 by the US naval commander Richard E. Byrd and named after his wife.

Marie de Médicis /ˌmɑːˌriː də ˌmeɪdiˈsiːs/ (Italian name Maria de' Medici) (1573–1642), queen of France. The second wife of Henry IV of France, she ruled as regent during the minority of her son Louis XIII (1610–17). She continued to exert a significant influence after her son came to power, and plotted against Richelieu, her former protégé, but was eventually exiled in 1631.

Mari El /ˌmɑːrɪ ˈel/ (also called **Mari Autonomous Republic**) an autonomous republic in European Russia, north of the Volga; pop. (1990) 754,000; capital, Yoshkar-Ola.

marigold /ˈmærɪˌɡəʊld/ *n.* a composite plant of the genus *Calendula* or *Tagetes*, with golden or bright yellow flowers. [ME f. *Mary* (prob. the Virgin) + dial. *gold*, OE *golde*, prob. rel. to GOLD]

marijuana /ˌmærɪˈwɑːnə/ *n.* (also **marihuana**) **1** the dried leaves, flowering tops, and stems of hemp, used as an intoxicating or hallucinogenic drug and usu. smoked in cigarettes; cannabis. **2** the plant yielding these (cf. HEMP). [Amer. Sp.]

marimba /məˈrɪmbə/ *n.* **1** a type of xylophone played in Africa and Central America. **2** a modern orchestral instrument derived from this. [Congo]

marina /məˈriːnə/ *n.* a specially designed harbour with moorings for pleasure-yachts etc. [It. & Sp. fem. adj. f. *marino* f. L (as MARINE)]

marinade *n. & v.* ● *n.* /ˌmærɪˈneɪd/ **1** a mixture of wine, vinegar, oil, spices, etc., in which meat, fish, etc., is soaked before cooking. **2** meat, fish, etc., soaked in such a mixture. ● *v.tr.* /ˈmærɪˌneɪd/ = MARINATE. [F f. Sp. *marinada* f. *marinar* pickle in brine f. *marino* (as MARINE)]

marinate /ˈmærɪˌneɪt/ *v.tr.* soak (meat, fish, etc.) in a marinade. □ **marination** /ˌmærɪˈneɪʃ(ə)n/ *n.* [It. *marinare* or F *mariner* (as MARINE)]

marine /məˈriːn/ *adj. & n.* ● *adj.* **1** of, found in, or produced by the sea. **2 a** of or relating to shipping or naval matters (*marine insurance*). **b** for use at sea. ● *n.* **1** a country's shipping, fleet, or navy (*mercantile marine; merchant marine*). **2** a member of a body of troops trained to serve on land or sea. (See ROYAL MARINES, US MARINES.) **3** a picture of a scene at sea. □ **marine stores** new or old ships' material etc. sold as merchandise. **marine trumpet** a large single-stringed viol with a trumpet-like tone. **tell that to the marines** (or **horse marines**) *colloq.* an expression of disbelief. [ME f. OF *marin marine* f. L *marinus* f. *mare* sea]

Mariner /ˈmærɪnə(r)/ a series of American space probes launched between 1962 and 1977 to investigate the inner planets. Mariners 2 and 5 made close-up observations of Venus, Mariners 4, 6, 7, and 9 were successful probes to Mars, and Mariner 10 visited Venus and Mercury.

mariner /ˈmærɪnə(r)/ *n.* a seaman, a sailor. □ **mariner's compass** a compass showing magnetic or true north and the bearings from it. [ME f. AF *mariner*, OF *marinier* f. med.L *marinarius* f. L (as MARINE)]

Marines, the see ROYAL MARINES, US MARINES.

Marinetti /ˌmærɪˈnetɪ/, Filippo Tommaso (1876–1944), Italian poet and dramatist. He launched the futurist movement with a manifesto published in the magazine *Le Figaro* (1909), which exalted technology, glorified war, and demanded revolution and innovation in the arts. In his poems he abandoned syntax and grammar; in the theatre he renounced verisimilitude and traditional methods of plot development and characterization and introduced the simultaneous staging of unrelated actions.

Mariolatry /ˌmɛərɪˈɒlətrɪ/ *n. derog.* idolatrous worship of the Virgin Mary. [L *Maria* Mary + -LATRY, after *idolatry*]

marionette /ˌmærɪəˈnet/ *n.* a puppet worked by strings. [F *marionnette* f. *Marion* dimin. of *Marie* Mary]

Marist /ˈmɛərɪst/ *n.* a member of a Roman Catholic missionary and teaching order, esp. the Society of Mary (founded in Lyons in the early 19th century). [F *Mariste* f. *Marie* Mary]

marital /ˈmærɪt(ə)l/ *adj.* **1** of marriage or the relations between husband and wife. **2** of or relating to a husband. □ **maritally** *adv.* [L *maritalis* f. *maritus* husband]

maritime /ˈmærɪˌtaɪm/ *adj.* **1** connected with the sea or seafaring (*maritime insurance*). **2** living or found near the sea. [L *maritimus* f. *mare* sea]

Maritime Provinces (also **Maritimes**) the Canadian provinces of New Brunswick, Nova Scotia, and Prince Edward Island, with coastlines on the Gulf of St Lawrence and the Atlantic. These provinces, with Newfoundland and Labrador, are also known as the Atlantic Provinces.

Maritsa /məˈrɪtsə/ (Turkish **Meriç** /məˈriːtʃ/; called in Greek *Évros*) a river of southern Europe, which rises in the Rila Mountains of SW Bulgaria and flows 480 km (300 miles) south to the Aegean Sea. It forms, for a small part of its length, the border between Bulgaria and Greece and then, for about 185 km (115 miles), that between Greece and Turkey. Its ancient name is the Hebros or Hebrus.

Mariupol /ˌmærɪˈuːpɒl/ an industrial port on the south coast of Ukraine, on the Sea of Azov; pop. (1989) 517,000. Between 1948 and 1989 it was named Zhdanov after the Soviet Politburo official Andrei Zhdanov, the defender of Leningrad during the siege of 1941–4.

Marius /ˈmæriəs/, Gaius (*c*.157–86 BC), Roman general and politician. Consul for the first time in 107, he established his dominance by victories over Jugurtha and invading Germanic tribes. He was subsequently involved in a struggle for power with Sulla and was expelled from Italy, only to return and take Rome by force in 87. He was again elected consul in 86 but died soon afterwards.

marjoram /ˈmɑːdʒərəm/ *n.* an aromatic labiate herb of the genus *Origanum*, the fresh or dried leaves of which are used as a flavouring in cookery. There are two species, *wild marjoram* (*O. vulgare*), used as the herb oregano, and *sweet marjoram* (*O. hortensis*). [ME & OF *majorane* f. med.L *majorana*, of unkn. orig.]

mark[1] /mɑːk/ *n. & v.* ● *n.* **1** a trace, sign, stain, scar, etc., on a surface, face, page, etc. **2** (esp. in *comb.*) **a** a written or printed symbol (*exclamation mark*; *question mark*). **b** a numerical or alphabetical award denoting excellence, conduct, proficiency, etc. (*got a good mark for effort*; *gave him a black mark*; *gained 46 marks out of 50*). **3** (usu. foll. by *of*) a sign or indication of quality, character, feeling, etc. (*took off his hat as a mark of respect*). **4 a** a sign, seal, etc., used for distinction or identification. **b** a cross etc. made in place of a signature by an illiterate person. **5 a a** target, object, goal, etc. (*missed the mark with his first play*). **b** a standard for attainment (*his work falls below the mark*). **6** a line etc. indicating a position; a marker. **7** (usu. **Mark**) (followed by a numeral) a particular design, model, etc., of a car, aircraft, etc. (*this is the Mark 2 model*). **8** a runner's starting-point in a race. **9** *Naut.* a piece of material etc. used to indicate a position on a sounding-line. **10 a** *Rugby* a heel-mark on the ground made by a player who has caught the ball direct from a kick, knock-on, or throw-forward by an opponent. **b** *Austral. Rules* the catching before it reaches the ground of a ball kicked at least ten metres; the spot from which the subsequent kick is taken. **11** *sl.* the intended victim of a swindler etc. **12** *Boxing* the pit of the stomach. **13** *hist.* a tract of land held in common by a Teutonic or medieval German village community. ● *v.tr.* **1 a** make a mark on (a thing or person), esp. by writing, cutting, scraping, etc. **b** put a distinguishing or identifying mark, initials, name, etc., on (clothes etc.) (*marked the tree with their initials*). **2 a** allot marks to; correct (a student's work etc.). **b** record (the points gained in games etc.). **3** attach a price to (goods etc.) (*marked the doll at 50p*). **4** (often foll. by *by*) show or manifest (displeasure etc.) (*marked his anger by leaving early*). **5** notice or observe (*she marked his agitation*). **6 a** characterize or be a feature of (*the day was marked by storms*). **b** acknowledge, recognize, celebrate (*marked the occasion with a toast*). **7** name or indicate (a place on a map, the length of a syllable, etc.) by a sign or mark. **8** characterize (a person or a thing) as (*marked them as weak*). **9 a** *Brit.* keep close to so as to prevent the free movement of (an opponent in sport). **b** *Austral. Rules* catch (the ball). **10** (as **marked** *adj.*) having natural marks (*is marked with silver spots*). **11** (of a graduated instrument) show, register (so many degrees etc.). **12** *Austral. & NZ* castrate (a lamb). □ **one's mark** *colloq.* **1** what one prefers. **2** an opponent, object, etc., of one's own size, calibre, etc. (*the little one's more my mark*). **beside** (or **off** or **wide of**) **the mark 1** not to the point; irrelevant. **2** not accurate. **make one's mark** attain distinction. **mark down 1** mark (goods etc.) at a lower price. **2** make a written note of. **3** choose (a person) as one's victim. **4** reduce the examination marks of. **mark-down** *n.* a reduction in price. **mark off** (often foll. by *from*) separate (one thing from another) by a boundary etc. (*marked off the subjects for discussion*). **mark of mouth** a depression in a horse's incisor indicating age. **mark out 1** plan (a course of action etc.). **2** destine (*marked out for success*). **3** trace out boundaries, a course, etc. **mark time 1** *Mil.* march on the spot, without moving forward. **2** act routinely; go through the motions. **3** await an opportunity to advance. **mark up 1** mark (goods etc.) at a higher price. **2** mark or

correct (text etc.) for typesetting or alteration. **mark-up** *n.* **1** the amount added to the cost price of goods to cover overhead charges, profit, etc. **2** the corrections made in marking up text. **mark you** please note (*without obligation, mark you*). **off the mark 1** having made a start. **2** see *beside the mark* above. **of mark** noteworthy. **on the mark** ready to start. **on your mark** (or **marks**) (as an instruction) get ready to start (esp. a race). **up to the mark** reaching the usual or normal standard, esp. of health. [OE *me(a)rc* (n.), *mearcian* (v.), f. Gmc]

mark[2] /mɑːk/ *n.* **1 a** = DEUTSCHMARK. **b** *hist.* = OSTMARK. **2** *hist.* **a** a denomination of weight for gold and silver. **b** English money of account. [OE *marc*, prob. rel. to med.L *marca, marcus*]

Mark, St /mɑːk/ **1** an Apostle, companion of St Peter and St Paul, traditional author of the second Gospel. Feast day, 25 Apr. **2** the second Gospel, the earliest in date (see GOSPEL 2).

Mark Antony see ANTONY.

marked /mɑːkt/ *adj.* **1** having a visible mark. **2** clearly noticeable; evident (*a marked difference*). **3** (of playing cards) having distinctive marks on their backs to assist cheating. □ **marked man 1** a person whose conduct is watched with suspicion or hostility. **2** a person destined to succeed. □ **markedly** /-kɪdlɪ/ *adv.* **markedness** *n.* [OE (past part. of MARK[1])]

marker /ˈmɑːkə(r)/ *n.* **1** a stone, post, etc., used to mark a position, place reached, etc. **2** a person or thing that marks. **3** a felt-tipped pen with a broad tip. **4** a person who records a score, esp. in billiards. **5** a flare etc. used to direct a pilot to a target. **6** a bookmark. **7** *US sl.* a promissory note; an IOU.

market /ˈmɑːkɪt/ *n. & v.* ● *n.* **1** the gathering of people for the purchase and sale of provisions, livestock, etc., esp. with a number of different vendors. **2** an open space or covered building used for this. **3** (often foll. by *for*) a demand for a commodity or service (*goods find a ready market*). **4** a place or group providing such a demand. **5** conditions as regards, or opportunity for, buying or selling. **6** the rate of purchase and sale, market value (*the market fell*). **7** (prec. by *the*) the trade in a specified commodity (*the corn market*). **8** the commercial world; the free market. ● *v.* (**marketed, marketing**) **1** *tr.* sell. **2** *tr.* offer for sale. **3** *intr.* buy or sell goods in a market. □ **be in the market for** wish to buy. **be on** (or **come into**) **the market** be offered for sale. **make a market** *Stock Exch.* induce active dealing in a stock or shares. **market cross** a structure erected in a market-place, orig. a stone cross, later an arcaded building. **market-day** a day on which a market is regularly held, usu. weekly. **market economy** an economy in which prices, incomes, industrial production, etc., are determined by the workings of the free market rather than by government planning. **market garden** esp. *Brit.* a place where vegetables and fruit are grown for the market etc. **market gardener** a person who owns or is employed in a market garden. **market maker** *Brit.* a member of the Stock Exchange granted certain privileges and trading to prescribed regulations. **market-place 1** an open space where a market is held in a town. **2** the commercial world; the trade in a particular commodity. **3** a place where ideas etc. are exchanged. **market price** the price in current dealings. **market research** the study of consumers' needs and preferences. **market town** *Brit.* a town where a market is held. **market value** value as a saleable thing (opp. *book value*). **put on the market** offer for sale. □ **marketer** *n.* [ME ult. f. L *mercatus* f. *mercari* buy: see MERCHANT]

marketable /ˈmɑːkɪtəb(ə)l/ *adj.* able or fit to be sold. □ **marketability** /ˌmɑːkɪtəˈbɪlɪtɪ/ *n.*

marketeer /ˌmɑːkɪˈtɪə(r)/ *n.* **1** a supporter of the EC and British membership of it. **2** a marketer.

marketing /ˈmɑːkɪtɪŋ/ *n.* **1** the action or business of promoting and selling products, including market research and advertising. **2** in senses of MARKET *v.*

markhor /ˈmɑːkɔː(r)/ *n.* a large spiral-horned wild goat, *Capra falconeri*, of central Asia. [Pers. *mār-ₖwār* f. *mār* serpent + *ₖwār* -eating]

marking /ˈmɑːkɪŋ/ *n.* (often in *pl.*) **1** an identification mark, esp. a symbol on an aircraft. **2** a mark or pattern of marks on an animal's fur, feathers, skin, etc. □ **marking-ink** indelible ink for marking linen etc.

markka /ˈmɑːkɑː/ *n.* the basic monetary unit of Finland, equal to 100 penniä. [Finnish]

Markova /mɑːˈkəʊvə, ˈmɑːkəvə/, Dame Alicia (born Lilian Alicia Marks) (b.1910), English ballet-dancer. In 1931 she joined the Vic-Wells Ballet, where she was the first English dancer to take the lead in *Giselle* and *Swan Lake*; she also created roles in new ballets such as Ashton's *Façade* (1931). She founded the Markova–Dolin Ballet with Anton Dolin in 1935; they later both joined the emergent London Festival Ballet (1950), with which Markova was prima ballerina until 1952.

Marks /mɑːks/, Simon, 1st Baron Marks of Broughton (1888–1964), English businessman. In 1907 he inherited the Marks and Spencer Penny Bazaars established by his father and Thomas Spencer (1851–1905). These became the nucleus of the successful company created in 1926 as Marks & Spencer, a chain of retail stores selling clothes, food, and household goods under the brand name 'St Michael'.

marksman /'mɑːksmən/ n. (pl. **-men**; fem. **markswoman**, pl. **-women**) a person skilled in shooting, esp. with a pistol or rifle. □ **marksmanship** n.

marl[1] /mɑːl/ n. & v. ● n. soil consisting of clay and lime, with fertilizing properties. ● v.tr. apply marl to (the ground). □ **marly** adj. [ME f. OF marle f. med.L margila f. L marga]

marl[2] /mɑːl/ n. **1** a mottled yarn of differently coloured threads. **2** the fabric made from this. [shortening of marbled: see MARBLE]

Marlborough /'mɔːlbərə/, 1st Duke of (title of John Churchill) (1650–1722), British general. He was appointed commander of British and Dutch troops in the War of the Spanish Succession and won a series of victories over the French armies of Louis XIV, most notably Blenheim (1704), Ramillies (1706), Oudenarde (1708), and Malplaquet (1709), which effectively ended Louis's attempts to dominate Europe. The building of Blenheim Palace, Marlborough's seat at Woodstock in Oxfordshire, was funded by Queen Anne as a token of the nation's gratitude for his victory.

Marley /'mɑːlɪ/, Robert Nesta ('Bob') (1945–81), Jamaican reggae singer, guitarist, and songwriter. In 1965 he formed the trio The Wailers, with whom he went on to become the first internationally acclaimed reggae musician. Bob Marley was a devout Rastafarian and supporter of black power whose lyrics frequently reflect his religious and political beliefs; by the 1970s he had gained the status of a cult hero. Among his albums are Burnin' (1973) and Exodus (1977).

marlin /'mɑːlɪn/ n. a large marine fish of the genus Makaira or Tetrapterus, with an elongated pointed upper jaw. [MARLINSPIKE, with ref. to its pointed snout]

marline /'mɑːlɪn/ n. Naut. a thin line of two strands. □ **marline-spike** = MÁRLINSPIKE. [ME f. Du. marlijn f. marren bind + lijn LINE[1]]

marlinspike /'mɑːlɪnˌspaɪk/ n. Naut. a pointed iron tool used to separate strands of rope or wire. [orig. app. marling-spike f. marl fasten with marline (f. Du. marlen frequent. of MDu. marren bind) + -ING[1] + SPIKE[1]]

marlite /'mɑːlaɪt/ n. a kind of marl that is not reduced to powder by the action of the air.

Marlowe /'mɑːləʊ/, Christopher (1564–93), English dramatist and poet. As a dramatist he brought a new strength and vitality to blank verse; major works include Tamburlaine the Great (1587–8), Doctor Faustus (c.1590), Edward II (1592), and The Jew of Malta (1592). His poems include 'Come live with me and be my love' (published in The Passionate Pilgrim, 1599) and the unfinished Hero and Leander (1598; completed by George Chapman). His work had a significant influence on Shakespeare's early historical plays. Marlowe was killed during a brawl in a tavern.

marmalade /'mɑːməˌleɪd/ n. a preserve of citrus fruit, usu. bitter oranges, made like jam. □ **marmalade cat** a cat with orange fur. [F marmelade f. Port. marmelada quince jam f. marmelo quince f. L melimelum f. Gk melimēlon f. meli honey + mēlon apple]

Marmara, Sea of /'mɑːmərə/ a small sea in NW Turkey. Connected by the Bosporus to the Black Sea and by the Dardanelles to the Aegean, it separates European Turkey from Asian Turkey. In ancient times it was known as the Propontis.

Marmite /'mɑːmaɪt/ n. **1** Brit. propr. a preparation made from yeast extract and vegetable extract, used in sandwiches and for flavouring. **2** (**marmite**) (also /mɑːˈmiːt/) an earthenware cooking vessel. [F, = cooking-pot]

marmoreal /mɑːˈmɔːrɪəl/ adj. poet. of or like marble. □ **marmoreally** adv. [L marmoreus (as MARBLE)]

marmoset /'mɑːməˌzet/ n. a small tropical American monkey of the family Callithricidae, having a long silky coat and a bushy tail. [OF marmouset grotesque image, of unkn. orig.]

marmot /'mɑːmət/ n. a burrowing colonial rodent of the genus Marmota, with a heavy-set body and short bushy tail. [F marmotte prob. f. Romansh murmont f. L murem (nominative mus) montis mountain mouse]

Marne /mɑːn/ a river of east central France, which rises in the Langres plateau north of Dijon and flows 525 km (328 miles) north and west to join the Seine near Paris. Its valley was the scene of two important battles in the First World War. The first battle (Sept. 1914) halted and repelled the German advance on Paris; the second (July 1918) ended the final German offensive.

marocain /'mærəˌkeɪn/ n. a dress-fabric of ribbed crêpe. [F, = Moroccan f. Maroc Morocco]

Maronite /'mærəˌnaɪt/ n. & adj. ● n. a member of a Christian sect of Syrian origin, living chiefly in Lebanon. (See note below.) ● adj. of or relating to the Maronites. [med.L Maronita f. Maro]

▪ Now numbering about 1 million, the Maronites claim to have been founded by St Maro, a friend of St John Chrysostom, but it seems certain that their origin is no earlier than the 7th century. Since 1181 they have been in communion with the Roman Catholic Church; their liturgy is in Syriac, and their head (under the pope) is the patriarch of Antioch. Maronites have been prominent in politics since the establishment of Lebanon as an independent state, traditionally holding the office of President.

maroon[1] /məˈruːn/ adj. & n. ● adj. brownish-crimson. ● n. **1** this colour. **2** an explosive device giving a loud report. [F marron chestnut f. It. marrone f. med.Gk maraon]

maroon[2] /məˈruːn/ v. & n. ● v.tr. **1** leave (a person) isolated in a desolate place (esp. an island). **2** (of a person or a natural phenomenon) cause (a person) to be unable to leave a place. ● n. **1** a person descended from a group of fugitive slaves in the remoter parts of Suriname and the West Indies. **2** a marooned person. [F marron f. Sp. cimarrón wild f. cima peak]

marque[1] /mɑːk/ n. a make of motor car, as distinct from a specific model (the Jaguar marque). [F, = MARK[1]]

marque[2] /mɑːk/ n. hist. reprisals. □ **letters of marque** (or **marque and reprisal**) **1** a licence to fit out an armed vessel and employ it in the capture of an enemy's merchant shipping. **2** (in sing.) a ship carrying such a licence. [ME f. F f. Prov. marca f. marcar seize as a pledge]

marquee /mɑːˈkiː/ n. **1** esp. Brit. a large tent used for social or commercial functions. **2** N.Amer. a rooflike projection over the entrance to a theatre, hotel, etc. [MARQUISE, taken as pl. & assim. to -EE]

Marquesas Islands /mɑːˈkeɪsəs/ a group of volcanic islands in the South Pacific, forming part of French Polynesia; pop. (1988) 7,540. The islands were annexed by France in 1842. They are described by the American writer Herman Melville in his novel Typee, written after he visited them in 1842. The largest island is Hiva Oa, on which the French painter Paul Gauguin spent the last two years of his life (1901–3).

marquess /'mɑːkwɪs/ n. a British hereditary nobleman ranking between a duke and an earl (cf. MARQUIS). The title was introduced into the peerage of England at the end of the 14th century and into that of Scotland in the late 15th century; it had been in use earlier in various European countries. □ **marquessate** /-kwɪzɪt/ n. [var. of MARQUIS]

marquetry /'mɑːkɪtrɪ/ n. (also **marqueterie**) inlaid work in wood, ivory, etc. [F marqueterie f. marqueter variegate f. MARQUE[1]]

Marquette /mɑːˈket/, Jacques (1637–75), French Jesuit missionary and explorer. He travelled to North America in 1666 and played a prominent part in the attempt to Christianize the American Indians there, especially during his mission among the Ottawa tribe. In 1673 he was a member of an expedition which explored the Wisconsin and Mississippi rivers as far as the mouth of the Arkansas.

Márquez, Gabriel García, see GARCÍA MÁRQUEZ.

marquis /'mɑːkwɪs, mɑːˈkiː/ n. a foreign nobleman ranking between a duke and a count (cf. MARQUESS). □ **marquisate** /'mɑːkwɪzɪt/ n. [ME f. OF marchis f. Rmc (as MARCH[2], -ESE)]

Marquis de Sade see SADE.

marquise /mɑːˈkiːz/ n. **1 a** the wife or widow of a marquis. **b** a woman holding the rank of marquis in her own right (cf. MARCHIONESS). **2** a finger-ring set with an oval pointed cluster of gems. **3** archaic = MARQUEE 1. [F, fem. of MARQUIS]

marquisette /ˌmɑːkɪˈzet/ n. a fine light cotton, rayon, or silk fabric for net curtains etc. [F, dimin. of MARQUISE]

Marrakesh /ˌmærəˈkeʃ/ (also **Marrakech**) a city in western Morocco, in the foothills of the High Atlas Mountains; pop. (1982) 439,700. Founded in 1062 as the capital of the Almoravids, it is now a centre of tourism and winter sports.

marram /'mærəm/ n. a shore grass, Ammophila arenaria, that binds sand with its tough rhizomes. [ON marálmr f. marr sea + hálmr HAULM]

marriage /'mærɪdʒ/ n. **1** the legal union of a man and a woman in order to live together and often to have children. **2** an act or ceremony establishing this union. **3** one particular union of this kind (*by a previous marriage*). **4** an intimate union (*the marriage of true minds*). **5** *Cards* the union of a king and queen of the same suit. □ **by marriage** as a result of a marriage (*related by marriage*). **in marriage** as husband or wife (*give in marriage; take in marriage*). **marriage bureau** an establishment arranging introductions between persons wishing to marry. **marriage certificate** a certificate certifying the completion of a marriage ceremony. **marriage guidance** counselling of couples who have problems in married life. **marriage licence** a licence to marry. **marriage lines** *Brit.* a marriage certificate. **marriage of convenience** a marriage concluded to achieve some practical purpose, esp. financial or political. **marriage settlement** an arrangement securing property between spouses. [ME f. OF *mariage* f. *marier* MARRY¹]

marriageable /ˈmærɪdʒəb(ə)l/ *adj.* **1** fit for marriage, esp. old or rich enough to marry. **2** (of age) fit for marriage. □ **marriageability** /ˌmærɪdʒəˈbɪlɪtɪ/ *n.*

Marriage of the Adriatic a ceremony formerly held on Ascension Day in Venice symbolizing the city's sea-power, during which the doge dropped a ring into the water from his official barge.

married /ˈmærɪd/ *adj. & n.* ● *adj.* **1** united in marriage. **2** of or relating to marriage (*married name; married life*). ● *n.* (usu. in *pl.*) a married person (*young marrieds*).

marron glacé /ˌmærɒn ˈɡlɑːseɪ/ *n.* (*pl.* **marrons glacés** *pronunc.* same) a chestnut preserved in and coated with sugar. [F, = iced chestnut: cf. GLACÉ]

marrow /ˈmærəʊ/ *n.* **1** (in full **vegetable marrow**) **a** a large usu. white-fleshed edible gourd used as food. **b** the plant, *Cucurbita pepo*, yielding this. **2** (in full **bone marrow**) **a** a soft fatty substance in the cavities of bones, important as the tissue in which red and white blood cells are produced. **b** this as typifying vitality and strength. □ **to the marrow** right through. □ **marrowless** *adj.* **marrowy** *adj.* [OE *mearg, mærg* f. Gmc]

marrowbone /ˈmærəʊbəʊn/ *n.* a bone containing edible marrow.

marrowfat /ˈmærəʊfæt/ *n.* a kind of large pea.

marry¹ /ˈmærɪ/ *v.* (**-ies, -ied**) **1** *tr.* **a** take as one's wife or husband in marriage. **b** (often foll. by *to*) (of a priest etc.) join (persons) in marriage. **c** (of a parent or guardian) give (a son, daughter, etc.) in marriage. **2** *intr.* **a** enter into marriage. **b** (foll. by *into*) become a member of (a family) by marriage. **3** *tr.* **a** unite intimately. **b** correlate (things) as a pair. **c** *Naut.* splice (rope-ends) together without increasing their girth. □ **marry off** find a wife or husband for. **marry up** (often foll. by *with*) link or join up. [ME f. OF *marier* f. L *maritare* f. *maritus* husband]

marry² /ˈmærɪ/ *int. archaic* expressing surprise, asseveration, indignation, etc. [ME, = (the Virgin) *Mary*]

Marryat /ˈmærɪət/, Frederick (known as Captain Marryat) (1792–1848), English novelist. In 1830 he resigned his commission in the navy to concentrate on writing. He produced a number of novels dealing with life at sea, such as *Peter Simple* (1833) and *Mr Midshipman Easy* (1836), while his books for children include the historical story *The Children of the New Forest* (1847).

marrying /ˈmærɪŋ/ *adj.* likely or inclined to marry (*not a marrying man*).

Mars /mɑːz/ **1** *Rom. Mythol.* the god of war and the most important Roman god after Jupiter. He is identified with Ares and was probably originally an agricultural god. The month of March is named after him. **2** *Astron.* the fourth planet from the sun in the solar system, orbiting between earth and Jupiter at an average distance of 228 million km from the sun, with an equatorial diameter of 6,740 km. Its characteristic red colour, clearly visible to the naked eye, arises from the iron-rich minerals covering its surface. A tenuous atmosphere of carbon dioxide whips up periodic dust storms which can cover the entire planet, causing erosion of surface features over long periods. Sinuous channels on the surface resemble river beds on earth and are perhaps associated with flows of liquid water (of which none remains), and extinct volcanoes hint at vigorous past geological activity. The polar caps are of frozen carbon dioxide, but may contain some ice. There is no evidence of the 'canals' reported by Percival Lowell in the early 20th century, and there is so far no unambiguous evidence for life on the planet. There are two small satellites, Phobos and Deimos.

Marsala /mɑːˈsɑːlə/ *n.* a dark sweet fortified dessert wine. [*Marsala* in Sicily, where orig. made]

Marseillaise /ˌmɑːseɪˈeɪz, ˌmɑːsəˈleɪz/ the national anthem of France, composed by a young engineer officer, Rouget de Lisle, in 1792, on the declaration of war against Austria, and first sung in Paris by Marseilles patriots. [F, fem. adj. f. *Marseille* MARSEILLES]

Marseilles /mɑːˈseɪ, -ˈseɪlz/ (French **Marseille** /marsɛj/) a city and port on the Mediterranean coast of southern France; pop. (1990) 807,725. It was settled as a Greek colony, called Massilia, in about 600 BC and became an ally of the Romans in their campaigns in Gaul in the first century BC. It was an important embarkation port during the Crusades of the 11th to the 14th centuries AD and in the 19th century served as a major port for French Algeria.

Marsh /mɑːʃ/, Dame Ngaio (Edith) (1899–1982), New Zealand writer of detective fiction. Her works include *Vintage Murder* (1937), *Surfeit of Lampreys* (1941), and *Final Curtain* (1947); many of the novels feature Chief Detective Inspector Roderick Alleyn.

marsh /mɑːʃ/ *n.* **1** low land flooded in wet weather and usu. watery at all times. **2** (*attrib.*) of or inhabiting marshland. □ **marsh fever** malaria. **marsh gas** methane. **marsh harrier** a large Old World harrier, *Circus aeruginosus*, frequenting reed-beds etc. **marsh hawk** *N. Amer.* = hen harrier. **marsh mallow** a shrubby herbaceous mallow, *Althaea officinalis*, the roots of which were formerly used to make marshmallow. **marsh marigold** a golden-flowered ranunculaceous plant, *Caltha palustris*, growing in moist meadows etc. (also called *kingcup*). **marsh tit** a grey tit, *Parus palustris*, inhabiting woods and hedges. **marsh trefoil** the buckbean. □ **marshy** *adj.* **marshiness** *n.* [OE *mer(i)sc* f. WG]

marshal /ˈmɑːʃ(ə)l/ *n. & v.* ● *n.* **1** (**Marshal**) **a** a high-ranking officer in the armed forces (*Air Marshal; Field Marshal; Marshal of France*). **b** a high-ranking officer of state (*Earl Marshal*). **2** an officer arranging ceremonies, controlling procedure at races, etc. **3** *US* the head of a police or fire department. **4** (in full **judge's marshal**) (in the UK) an official accompanying a judge on circuit, with secretarial and social duties. ● *v.* (**marshalled, marshalling**; *US* **marshaled, marshaling**) **1** *tr.* arrange (soldiers, facts, one's thoughts, etc.) in due order. **2** *tr.* (often foll. by *into, to*) conduct (a person) ceremoniously. **3** *tr.* Heraldry combine (coats of arms). **4** *intr.* take up positions in due arrangement. □ **marshalling yard** a railway yard in which goods trains etc. are assembled. **Marshal of the Royal Air Force** an officer of the highest rank in the Royal Air Force. □ **marshaller** *n.* **marshalship** *n.* [ME f. OF *mareschal* f. LL *mariscalcus* f. Gmc, lit. 'horse-servant']

Marshall /ˈmɑːʃ(ə)l/, George C(atlett) (1880–1959), American general and statesman. As US Secretary of State (1947–9) he initiated the programme of economic aid to European countries known as the Marshall Plan. He was awarded the Nobel Peace Prize in 1953.

Marshall Islands (also **Marshalls**) a country consisting of two chains of islands in the NW Pacific; pop. (1990) 43,420; languages, English (official), local Malayo-Polynesian languages; capital, Majuro. Visited by the Spanish in 1529 and by the English adventurer John Marshall in 1788, the Marshalls were not colonized until 1885, when a German protectorate was declared. After being under Japanese mandate following the First World War, they were administered by the US as part of the Pacific Islands Trust Territory from 1947 until 1986, when they became a republic in free association with the US.

Marshall Plan (formally called *European Recovery Program*) a programme of financial aid and other initiatives, sponsored by the US, designed to boost the economies of western European countries after the Second World War. The scheme was originally advocated by Secretary of State George C. Marshall and was passed by Congress in 1948; it was successful in helping renewal in major manufacturing industries such as engineering and iron and steel.

Marsh Arab *n.* a member of a semi-nomadic Arab people inhabiting marshland in southern Iraq, near the confluence of the Tigris and Euphrates rivers. Largely Shiite Muslims, they have suffered persecution under Saddam Hussein's Sunni regime.

marshland /ˈmɑːʃlənd/ *n.* land consisting of marshes.

marshmallow /mɑːʃˈmæləʊ/ *n.* a soft sweet made of sugar, albumen, gelatin, etc.

Marston Moor, Battle of /ˈmɑːstən/ a decisive battle of the English Civil War, fought in 1644 on Marston Moor near York. The combined Royalist armies of Prince Rupert and the Duke of Newcastle were defeated by the English and Scottish Parliamentary armies, a defeat which destroyed Royalist power in the north of England and fatally weakened Charles I's cause.

marsupial /mɑːˈsuːpɪəl/ *n. & adj.* ● *n.* a mammal of the order Marsupialia, usu. carrying the young in a pouch and including the kangaroo and opossum. (*See note below.*) ● *adj.* **1** of or belonging to this order. **2** of, like, or having a pouch (*marsupial muscle*). [mod.L *marsupialis* f. L *marsupium* f. Gk *marsupion* pouch, dimin. of *marsipos* purse]

▪ The young of marsupials are born in a very undeveloped state, and are nourished and complete their development attached to the teats of the milk-secreting glands on the mother's abdomen. The external pouch (marsupium) is found in many species, but it is not a universal feature. There are under 200 species of marsupials today, and, apart from the American opossums, they are confined to Australasia. There they have evolved to fill many of the niches occupied by placental mammals elsewhere, with marsupial counterparts of mice, moles, squirrels, rabbits, deer, cats, dogs, and others.

Marsyas /ˈmɑːsɪəs/ *Gk Mythol.* a satyr who took to flute-playing. He challenged Apollo to a musical contest and was flayed alive when he lost.

mart /mɑːt/ *n.* **1** a trade centre. **2** an auction-room. **3 a** a market. **b** a market-place. [ME f. obs. Du. *mart*, var. of *markt* MARKET]

Martaban, Gulf of /ˌmɑːtəˈbɑːn/ an inlet of the Andaman Sea, a part of the Indian Ocean, on the coast of SE Burma (Myanmar) east of Rangoon.

martagon /ˈmɑːtəgən/ *n.* a lily, *Lilium martagon*, with small purple turban-like flowers. [F f. Turk. *martagān* a form of turban]

Martel, Charles, see CHARLES MARTEL.

Martello /mɑːˈteləʊ/ *n.* (also **Martello tower**) (*pl.* **-os**) any of the small circular forts erected for defence purposes along the coasts of Britain during the Napoleonic Wars. They had massive walls and flat roofs on which guns were mounted. [alt. f. Cape *Mortella* in Corsica, where such a tower proved difficult for the British to capture in 1794]

marten /ˈmɑːtɪn/ *n.* a tree-living carnivore of the genus *Martes*, of the weasel family, having a bushy tail and valuable fur. [ME f. MDu. *martren* f. OF (*peau*) *martrine* marten (fur) f. *martre* f. WG]

martensite /ˈmɑːtɪnˌzaɪt/ *n.* the chief constituent of hardened steel. [A. *Martens*, German metallurgist (1850–1914) + -ITE[1]]

Martha /ˈmɑːθə/ (in the New Testament) the sister of Lazarus and Mary and friend of Jesus Christ (Luke 10:40). Her name is used allusively for an active or busy woman, much concerned with domestic affairs.

Martha's Vineyard a resort island off the coast of Massachusetts, to the south of Cape Cod. Settled by the English in 1642, it became an important centre of fishing and whaling in the 18th and 19th centuries.

Martial /ˈmɑːʃ(ə)l/ (Latin name Marcus Valerius Martialis) (AD *c.*40–*c.*104), Roman epigrammatist, born in Spain. His fifteen books of epigrams, in a variety of metres, reflect all facets of Roman life; they are witty, mostly satirical, and often coarse.

martial /ˈmɑːʃ(ə)l/ *adj.* **1** of or appropriate to warfare. **2** warlike, brave; fond of fighting. □ **martial arts** fighting sports such as judo and karate. **martial law** military government, involving the suspension of ordinary law. □ **martially** *adv.* [ME f. OF *martial* or L *martialis* of the Roman god Mars (see MARS 1)]

Martian /ˈmɑːʃ(ə)n/ *adj.* & *n.* ● *adj.* of the planet Mars. ● *n.* a hypothetical inhabitant of Mars. [ME f. OF *martien* or L *Martianus* f. *Mars*: see MARS]

martin /ˈmɑːtɪn/ *n.* a short-tailed bird of the swallow family (*house-martin*; *sand-martin*). [late ME, from the male forename *Martin*]

Martin, St /ˈmɑːtɪn/, (d.397), French bishop, a patron saint of France. While serving in the Roman army he gave half his cloak to a beggar and received a vision of Christ, after which he was baptized. He joined St Hilary at Poitiers and founded the first monastery in Gaul. On becoming bishop of Tours (371) he pioneered the evangelization of the rural areas. Feast day, 11 Nov. (See also MARTINMAS.)

martinet /ˌmɑːtɪˈnet/ *n.* a strict (esp. military or naval) disciplinarian. □ **martinettish** *adj.* (also **martinetish**). [J. *Martinet*, 17th-c. Fr. drill-master]

martingale /ˈmɑːtɪŋˌgeɪl/ *n.* **1** a strap, or set of straps, fastened at one end to the noseband or reins of a horse and at the other end to the girth, to prevent rearing etc. **2** *Naut.* a rope for holding down the jib-boom. **3** a gambling system of continually doubling the stakes in the hope of an eventual win that must yield a net profit. [F, of uncert. orig.]

Martini[1] /mɑːˈtiːnɪ/, Simone (*c.*1284–1344), Italian painter. His work is characterized by strong outlines and the use of rich colour. He worked in Siena and Assisi in Italy, and for the papal court at Avignon in France (*c.*1339–44). Notable works include *The Annunciation* (1333).

Martini[2] /mɑːˈtiːnɪ/ *n.* **1** *propr.* a type of vermouth. **2** a cocktail made of gin and French vermouth, and sometimes orange bitters etc. [*Martini* & *Rossi*, Italian firm selling vermouth]

Martinique /ˌmɑːtɪˈniːk/ a French island in the West Indies, in the Lesser Antilles group; pop. (1990) 359,570; capital, Fort-de-France. Its former capital, St Pierre, was completely destroyed by an eruption of Mount Pelée in 1902.

Martinmas /ˈmɑːtɪnməs/ St Martin's day, 11 Nov. It was formerly the usual time in England for hiring servants and for slaughtering cattle to be salted for the winter. [St *Martin* (see MARTIN, ST) + MASS[2]]

martlet /ˈmɑːtlɪt/ *n.* **1** *Heraldry* an imaginary footless bird borne as a charge. **2** *archaic* **a** a swift. **b** a house-martin. [F *martelet* alt. f. *martinet* dimin. f. MARTIN]

martyr /ˈmɑːtə(r)/ *n.* & *v.* ● *n.* **1 a** a person who is put to death for refusing to renounce a faith or belief. **b** a person who suffers for adhering to a principle, cause, etc. **c** a person who suffers or pretends to suffer in order to obtain sympathy or pity. **2** (foll. by *to*) a constant sufferer from (an ailment). ● *v.tr.* **1** put to death as a martyr. **2** torment. □ **make a martyr of oneself** accept or pretend to accept unnecessary discomfort etc. [OE *martir* f. eccl.L *martyr* f. Gk *martur*, *martus* *-uros* witness]

martyrdom /ˈmɑːtədəm/ *n.* **1** the sufferings and death of a martyr. **2** torment. [OE *martyrdōm* (as MARTYR, -DOM)]

martyrize /ˈmɑːtəˌraɪz/ *v.tr.* & *refl.* (also **-ise**) make a martyr of. □ **martyrization** /ˌmɑːtəraɪˈzeɪʃ(ə)n/ *n.*

martyrology /ˌmɑːtəˈrɒlədʒɪ/ *n.* (*pl.* **-ies**) **1** a list or register of martyrs. **2** the history of martyrs. □ **martyrologist** *n.* **martyrological** /-rəˈlɒdʒɪk(ə)l/ *adj.* [med.L *martyrologium* f. eccl.Gk *marturologion* (as MARTYR, *logos* account)]

martyry /ˈmɑːtərɪ/ *n.* (*pl.* **-ies**) a shrine or church erected in honour of a martyr. [ME f. med.L *martyrium* f. Gk *marturion* martyrdom (as MARTYR)]

Maruts /ˈmʌrʊts/ *Hinduism* (also called the *Rudras*) the sons of Rudra. In the Rig-veda they are the storm gods, Indra's helpers.

marvel /ˈmɑːv(ə)l/ *n.* & *v.* ● *n.* **1** a wonderful or astonishing person or thing. **2** (often foll. by *of*) a wonderful example (*a marvel of engineering*; *she's a marvel of patience*). ● *v.intr.* (**marvelled**, **marvelling**; US **marveled, marveling**) *literary* **1** (foll. by *at*, or *that* + clause) feel surprise or wonder. **2** (foll. by *how, why*, etc. + clause) wonder. □ **marvel of Peru** a showy garden plant, *Mirabilis jalapa*, with flowers opening at dusk. □ **marveller** *n.* [ME f. OF *merveille*, *merveiller* f. LL *mirabilia* neut. pl. of L *mirabilis* f. *mirari* wonder at: see MIRACLE]

Marvell /ˈmɑːv(ə)l, mɑːˈvel/, Andrew (1621–78), English poet. He served as an MP (1659–78) and was best known during his lifetime for his verse satires and pamphlets attacking Charles II and his ministers, particularly for corruption at court and in Parliament. Most of his poetry was published posthumously (1681) and did not achieve recognition until the early 20th century, when it was reappraised along with the work of other metaphysical poets such as John Donne. His best-known poems include 'To his Coy Mistress', 'Bermudas', and 'An Horatian Ode upon Cromwell's Return from Ireland'.

marvellous /ˈmɑːvələs/ *adj.* (US **marvelous**) **1** astonishing. **2** excellent. **3** extremely improbable. □ **marvellously** *adv.* **marvellousness** *n.* [ME f. OF *merveillos* f. *merveille*: see MARVEL]

Marx /mɑːks/, Karl Heinrich (1818–83), German political philosopher and economist, resident in England from 1849. The founder of modern communism with Engels, he collaborated with him in the writing of the *Communist Manifesto* (1848). Thereafter, Marx spent much of his time enlarging the theory of this pamphlet into a series of books, the most important being the three-volume *Das Kapital*. The first volume of this appeared in 1867 and the remainder was completed by Engels and published after Marx's death (1885; 1894). Marx was also a leading figure in the founding of the First International (1864). (See also MARXISM.)

Marx Brothers /mɑːks/ a family of American comedians, consisting of the brothers 'Chico' (Leonard, 1886–1961), 'Harpo' (Adolph Arthur, 1888–1964), 'Groucho' (Julius Henry, 1890–1977), and 'Zeppo' (Herbert, 1901–79). Their films, which are characterized by their anarchic humour, include *Horse Feathers* (1932), *Duck Soup* (1933), and *A Night at the Opera* (1935).

Marxism /ˈmɑːksɪz(ə)m/ *n.* the political and economic theories based on the writings of Karl Marx and Freidrich Engels, and later developed by their followers together with dialectical materialism to form the basis for the theory and practice of communism. At the centre of Marxist theory is an explanation of past and present societies and social change in terms of economic factors, in which labour and the means of production provide the economic *base* which influences or determines the political and ideological *superstructure*. The history of society shows progressive stages (ancient, feudal, capitalist) in the control and ownership of the means of production; in a projected future stage the imbalances inherent in these systems lead to change, and eventually to a class struggle in which the working classes overturn the capitalist class and capitalism. How an alternative society is then established was not detailed by Marx; it was assumed that a necessary interim period called the 'dictatorship of the proletariat' would be followed by the establishment of communism and the eventual 'withering away' of the state. (See also MARX.) □ **Marxist** *n.* & *adj.*

Marxism-Leninism /ˌmɑːksɪz(ə)mˈlenɪˌnɪz(ə)m/ *n.* the doctrines of Marx as interpreted and put into effect by Lenin; Communism as implemented in the Soviet Union and subsequently in China.

Mary[1] /ˈmeərɪ/ (known as the (Blessed) Virgin Mary, or St Mary, or Our Lady), mother of Jesus. According to the Gospels, she was a virgin betrothed to Joseph at the time of the Annunciation who conceived Jesus by the power of the Holy Spirit. She has been venerated by Catholic and Orthodox Churches from the earliest Christian times. The doctrines of her Immaculate Conception and Assumption are taught by some Churches. Feast days, 1 Jan. (RC Ch.), 25 Mar. (Annunciation), 15 Aug. (Assumption), 8 Sept. (Immaculate Conception).

Mary[2] /ˈmeərɪ/ the name of two queens of England:

Mary I (known as Mary Tudor) (1516–58), daughter of Henry VIII, reigned 1553–8. Having regained the throne after the brief attempt to install Lady Jane Grey in her place, Mary attempted to reverse the country's turn towards Protestantism, which had begun to gain momentum during the reign of her brother, Edward VI. She married Philip II of Spain, and after putting down several revolts began the series of religious persecutions which earned her the name of 'Bloody Mary'. She died childless and the throne passed to her Protestant sister, Elizabeth I.

Mary II (1662–94), daughter of James II, reigned 1689–94. Although her father was converted to Catholicism, Mary remained a Protestant, and was invited to replace him on the throne after his deposition in 1689. She insisted that her husband, William of Orange (William III; see WILLIAM), be crowned along with her and afterwards left most of the business of the kingdom to him, although she frequently had to act as sole head of state because of her husband's absence on campaigns abroad.

Mary, St see MARY[1].

Mary Celeste /sɪˈlest/ an American brig that set sail from New York for Genoa and was found in the North Atlantic in Dec. 1872 in perfect condition but abandoned. The fate of the crew was never discovered, and the abandonment of the ship remains one of the great mysteries of the sea.

Maryland /ˈmeərɪˌlænd/ a state of the eastern US, on the Atlantic coast, surrounding Chesapeake Bay; pop. (1990) 4,781,470; capital, Annapolis. Colonized from England in the 17th century and named after Queen Henrietta Maria, wife of Charles I, it was one of the original thirteen states of the Union (1788).

Mary Magdalene, St /ˈmægdəˌliːn/ (in the New Testament) a woman of Magdala in Galilee. She was a follower of Jesus, who cured her of evil spirits (Luke 8:2); she is also traditionally identified with the 'sinner' of Luke 7:37. Feast day, 22 July.

Mary, Queen of Scots (known as Mary Stuart) (1542–87), daughter of James V, queen of Scotland 1542–67. Mary was sent to France as an infant and was married briefly to Francis II, but after his death returned to Scotland in 1561 to resume personal rule. A devout Catholic, she was unable to control her Protestant lords; her position was made more difficult by her marriages to Lord Darnley and the Earl of Bothwell, and after the defeat of her supporters she fled to England in 1567. There she was imprisoned and became the focus of several Catholic plots against Elizabeth I; she was eventually beheaded.

Mary Rose a heavily armed ship, built for Henry VIII and named in honour of his sister. For some years from 1512 onwards the ship took part in Henry's wars with the French; in July 1545, when going out to engage the French fleet off Portsmouth, she was swamped and quickly sank with the loss of nearly all her company. The hull was discovered by skin-divers in 1968, and raised in 1982.

Mary Stuart see MARY, QUEEN OF SCOTS.

Mary Tudor, Mary I of England (see MARY[2]).

marzipan /ˈmɑːzɪˌpæn/ *n.* & *v.* (also *archaic* **marchpane** /ˈmɑːtʃpeɪn/) ● *n.* **1** a paste of ground almonds, sugar, etc., made up into small cakes etc. or used to coat large cakes. **2** a piece of marzipan. ● *v.tr.* (**marzipanned**, **marzipanning**) cover with or as with marzipan. [G f. It. *marzapane*]

Masaccio /mæˈsætʃɪˌəʊ/, (born Tommaso Giovanni di Simone Guidi) (1401–28), Italian painter. He was based chiefly in Florence and is remembered particularly for his frescos in the Brancacci Chapel (1424–7). He was the first artist to develop the laws of perspective (which he learned from Brunelleschi) and apply them to painting. His work is also notable for the use of light to define the construction of the body and its draperies and to unify a whole composition.

Masada /məˈsɑːdə/ the site, on a steep rocky hill on the SW shore of the Dead Sea, of the ruins of a palace and fortification built by Herod the Great in the 1st century BC. It was a Jewish stronghold in the Zealots' revolt against the Romans (AD 66–73) and was the scene in AD 73 of mass suicide by the Jewish defenders when the Romans breached the citadel after a siege of nearly two years.

Masai /ˈmɑːsaɪ/ *n.* & *adj.* ● *n.* (*pl.* same or **Masais**) **1** a member of a pastoral people inhabiting parts of Kenya and Tanzania. The Masai are traditionally credited with fierce raiding of neighbouring tribes, undertaken to protect their herds and prove prowess. **2** the Nilotic language of the Masai. ● *adj.* of or relating to the Masai or their language. [Masai]

masala /məˈsɑːlə/ *n.* **1** any of various spice mixtures ground into a paste or powder for use in Indian cookery. **2** a dish flavoured with this. [Urdu *maṣālaḥ*, ult. f. Arab. *maṣāliḥ* ingredients, materials]

Masaryk /ˈmæsərɪk/, Tomáš (Garrigue) (1850–1937), Czechoslovak statesman, President 1918–35. A founder of modern Czechoslovakia, during the First World War he was in London, where he founded the Czechoslovakian National Council with Edvard Beneš and promoted the cause of his country's independence. When this was achieved in 1918 he became Czechoslovakia's first President. In 1935 he retired in favour of Beneš.

Masbate /mæsˈbɑːtɪ/ **1** an island in the central Philippines; pop. (1990) 599,355. **2** its chief town.

Mascagni /mæˈskænjɪ/, Pietro (1863–1945), Italian composer and conductor. His compositions include operas and choral works; he is especially remembered for the opera *Cavalleria Rusticana* (1890).

mascara /mæˈskɑːrə/ *n.* a cosmetic for darkening the eyelashes. [It. *mascara, maschera* MASK]

Mascarene Islands /ˌmæskəˈriːn/ (also **Mascarenes**) a group of three islands in the western Indian Ocean, east of Madagascar, comprising Réunion, Mauritius, and Rodrigues. The group was named after the 16th-century Portuguese navigator Pedro de Mascarenhas.

mascarpone /ˌmæskəˈpəʊnɪ/ *n.* a soft mild Italian cream cheese. [It.]

mascle /ˈmæsk(ə)l/ *n.* Heraldry a lozenge voided, with a central lozenge-shaped aperture. [ME f. AF f. AL *ma(s)cula* f. L MACULA]

mascon /ˈmæskɒn/ *n.* Astron. a concentration of dense matter below the moon's surface, producing a gravitational pull. [*mass concentration*]

mascot /ˈmæskɒt/ *n.* a person, animal, or thing that is supposed to bring good luck. [F *mascotte* f. mod. Prov. *mascotto* fem. dimin. of *masco* witch]

masculine /ˈmæskjʊlɪn/ *adj.* & *n.* ● *adj.* **1** of or characteristic of men. **2** manly, vigorous. **3** (of a woman) having qualities considered appropriate to a man. **4** Gram. of or denoting the gender proper to words or grammatical forms classified as male. ● *n.* Gram. the masculine gender; a masculine word. □ **masculinely** *adv.* **masculinize** *v.tr.* (also **-ise**). **masculinity** /ˌmæskjʊˈlɪnɪtɪ/ *n.* [ME f. OF *masculin -ine* f. L *masculinus* (as MALE)]

Masefield /ˈmeɪsfiːld/, John (Edward) (1878–1967), English poet and novelist. His fascination with the sea is reflected in his first published book of poetry, *Salt-Water Ballads* (which contained 'I must go down to the sea again', 1902). His other works include several narrative poems and novels, and the children's story *The Midnight Folk* (1927). He was appointed Poet Laureate in 1930.

maser /ˈmeɪzə(r)/ *n.* a device using the stimulated emission of radiation by excited atoms to amplify or generate coherent monochromatic electromagnetic radiation in the microwave range (cf. LASER). [*microwave amplification by the stimulated emission of radiation*]

Maseru /məˈseəruː/ the capital of Lesotho; pop. (1986) 109,380. It is situated on the Caledon river near the border with the province of

Orange Free State in South Africa. It was established as the capital in 1869.

mash /mæʃ/ *n. & v.* ● *n.* **1** a soft mixture. **2** a mixture of boiled grain, bran, etc., given warm to horses etc. **3** *Brit. colloq.* mashed potatoes (*sausage and mash*). **4** a mixture of malt and hot water used to form wort for brewing. **5** a soft pulp made by crushing, mixing with water, etc. ● *v.* **1** *tr.* reduce (potatoes etc.) to a uniform mass by crushing. **2** *tr.* crush or pound to a pulp. **3** *tr.* mix (malt) with hot water to form wort. **4** *dial. a tr.* infuse, brew (tea). **b** *intr.* (of tea) infuse, draw, brew. □ **masher** *n.* [OE *māsc* f. WG, perh. rel. to MIX]

Mashhad /mæʃ'hæd/ (also **Meshed** /mə'ʃed/) a city in NE Iran, close to the border with Turkmenistan; pop. (1986) 1,463,500. The burial place in AD 809 of the Abbasid caliph Harun ar-Rashid and in 818 of the Shiite leader Ali ar-Rida, it is a holy city of the Shiite Muslims. It is the second largest city in Iran.

mashie /'mæʃɪ/ *n. Golf* an iron formerly used for lofting or for medium distances. [perh. f. F *massue* club]

Mashona /mə'ʃəʊnə/ *n. & adj.* ● *n.* (*pl.* same or **Mashonas**) a member of the Shona people (see SHONA). ● *adj.* of or relating to the Shona.

Mashonaland /mə'ʃəʊnə,lænd/ an area of northern Zimbabwe, occupied by the Shona people. A former province of Southern Rhodesia, it is now divided into the three provinces of Mashonaland East, West, and Central.

mask /mɑːsk/ *n. & v.* ● *n.* **1** a covering for all or part of the face: **a** worn as a disguise, or to appear grotesque and amuse or terrify. **b** made of wire, gauze, etc., and worn for protection (e.g. by a fencer) or by a surgeon to prevent infection of a patient. **c** worn to conceal the face at a ball etc. and usu. made of velvet or silk. **2** a respirator used to filter inhaled air or to supply gas for inhalation. **3** a likeness of a person's face, esp. one made by taking a mould from the face (*death-mask*). **4** a disguise or pretence (*throw off the mask*). **5** a hollow model of a human head worn by ancient Greek and Roman actors. **6** *Photog.* a screen used to exclude part of an image. **7** the face or head of an animal, esp. a fox. **8** = face-pack. **9** *archaic* a masked person. ● *v.tr.* **1** cover (the face etc.) with a mask. **2** disguise or conceal (a taste, one's feelings, etc.). **3** protect from a process. **4** *Mil.* **a** conceal (a battery etc.) from the enemy's view. **b** hinder (an army etc.) from action by observing with adequate force. **c** hinder (a friendly force) by standing in its line of fire. □ **masking tape** adhesive tape used in painting to cover areas on which paint is not wanted. □ **masker** *n.* [F *masque* f. It. *maschera* f. Arab. *maskara* buffoon f. *sakira* to ridicule]

masked /mɑːskt/ *adj.* wearing or disguised with a mask. □ **masked ball** a ball at which masks are worn.

maskinonge /'mæskɪ,nɒndʒ, ,mæskɪ'nɒndʒɪ/ *n.* = MUSKELLUNGE. [ult. f. Ojibwa, = great fish]

masochism /'mæsə,kɪz(ə)m/ *n.* **1** the condition or state of deriving esp. sexual gratification from one's own pain or humiliation (cf. SADISM). **2** *colloq.* the enjoyment of what appears to be painful or tiresome. □ **masochist** *n.* **masochistic** /,mæsə'kɪstɪk/ *adj.* **masochistically** *adv.* [Leopold von Sacher-Masoch, Austrian novelist (1835–95), who described cases of it]

Mason /'meɪs(ə)n/, A(lfred) E(dward) W(oodley) (1865–1948), English novelist. His work includes adventure stories (*The Four Feathers*, 1902), historical novels (*Musk and Amber*, 1942), detective novels featuring Inspector Hanaud, and several plays.

mason /'meɪs(ə)n/ *n. & v.* ● *n.* **1** a person who builds with stone. **2** (**Mason**) a Freemason. ● *v.tr.* build or strengthen with masonry. □ **mason's mark** a device carved on stone by the mason who dressed it. [ME f. OF *masson, maçonner*, ONF *machun*, prob. ult. f. Gmc]

Mason–Dixon Line /,meɪs(ə)n'dɪks(ə)n/ (also **Mason and Dixon Line**) the boundary line between Pennsylvania and Maryland, which was laid out in 1763–7 by the surveyors Charles Mason and Jeremiah Dixon. The name was later applied to the entire southern boundary of Pennsylvania, and in the years before the American Civil War it represented the division between the northern states and the slave-owning states of the South.

Masonic /mə'sɒnɪk/ *adj.* of or relating to Freemasons.

masonry /'meɪsənrɪ/ *n.* **1 a** the work of a mason. **b** stonework. **2** (**Masonry**) Freemasonry. [ME f. OF *maçonerie* (as MASON)]

Masorah /mə'sɔːrə/ *n.* (also **Massorah**) a body of traditional information and comment on the text of the Hebrew Bible, compiled by the Masoretes. [Heb. var. of *māsōret* bond]

Masorete /'mæsə,riːt/ *n.* (also **Massorete**) any of a group of Jewish scholars (6th–10th centuries AD) who were responsible for establishing a recognized text of the Hebrew Bible (called the *Masoretic text*), and who compiled a detailed commentary relating to it (Old Testament; see MASORAH). They also introduced vowel points and accents to indicate how the words of the text should be pronounced, at a time when Hebrew was ceasing to be a spoken language. □ **Masoretic** /,mæsə'retɪk/ *adj.* [F *Massoret* & mod.L *Massoreta*, orig. a misuse of Heb. (see MASORAH), assim. to -ETE]

masque /mɑːsk/ *n.* a form of amateur dramatic and musical entertainment, popular at court and amongst the nobility in 16th–17th-century England, consisting of dancing and acting performed by masked players, orig. in dumb show but later with metrical dialogue. The masque derived from a primitive folk ritual featuring the arrival of guests, usually in disguise, bearing gifts to a monarch or noble, who with his household then joined the visitors in a ceremonial dance. The Civil War put an end to the masque, which was never revived. □ **masquer** *n.* [var. of MASK]

masquerade /,mɑːskə'reɪd, ,mæs-/ *n. & v.* ● *n.* **1** a false show or pretence. **2** a masked ball. ● *v.intr.* (often foll. by *as*) appear in disguise, assume a false appearance. □ **masquerader** *n.* [F *mascarade* f. Sp. *mascarada* f. *máscara* mask]

Mass. *abbr.* Massachusetts.

mass¹ /mæs/ *n., v., & adj.* ● *n.* **1** a coherent body of matter of indefinite shape. **2** a dense aggregation of objects (*a mass of fibres*). **3** (in *sing.* or *pl.*; foll. by *of*) a large number or amount. **4** (usu. foll. by *of*) an unbroken expanse (of colour etc.). **5** (prec. by *a*; foll. by *of*) covered or abounding in (*was a mass of cuts and bruises*). **6** a main portion (of a painting etc.) as perceived by the eye. **7** (prec. by *the*) **a** the majority. **b** (in *pl.*) the ordinary people. **8** *Physics* the quantity of matter a body contains, measured in terms of resistance to acceleration by a force. (*See note below.*) **9** (*attrib.*) relating to, done by, or affecting large numbers of people or things; large-scale (*mass audience; mass action; mass murder*). ● *v.tr. & intr.* **1** assemble into a mass or as one body (*massed bands*). **2** *Mil.* (with ref. to troops) concentrate or be concentrated. □ **centre of mass** a point representing the mean position of matter in a body or system. **in the mass** in the aggregate. **law of mass action** the principle that the rate of a chemical reaction is proportional to the masses of the reacting substances. **mass defect** the difference between the mass of an isotope and its mass number. **mass energy** a body's ability to do work according to its mass. **mass media** = MEDIA¹ 2. **mass noun** *Gram.* a noun that is not countable and cannot be used with the indefinite article or in the plural (e.g. *bread*). **mass number** the total number of protons and neutrons in a nucleus. **mass observation** *Brit. esp. hist.* the study and recording of the social habits and opinions of ordinary people. **mass-produce** produce by mass production. **mass production** the production of large quantities of a standardized article by a standardized mechanical process. **mass spectrograph** a mass spectrometer in which the particles are detected photographically. **mass spectrum** the distribution of ions shown by the use of a mass spectrometer. □ **massless** *adj.* [ME f. OF *masse, masser* f. L *massa* f. Gk *maza* barley-cake: perh. rel. to *massō* knead]

▪ The scientific concept of mass as the quantity of matter began to be developed in the Middle Ages from the usual meaning of mass as an aggregation of matter. Newton introduced a more precise definition of mass based on inertia, the property possessed by all bodies of resisting any change in motion. For Newton all bodies with the same inertia had the same mass; he also demonstrated that all bodies with the same weight had the same inertia and, therefore, the same mass. Einstein, in the special theory of relativity, showed that mass increases with velocity; further, he suggested that the total energy content of every body was measured by the equation $E = mc^2$ (where E is energy, m is mass, and c is the speed of light), and that mass is convertible into energy — a great deal of energy, since c^2 is a very large number. The equivalence of mass and energy described by this equation has been demonstrated by, for example, the atom bomb and the nuclear reactor.

mass² /mæs, mɑːs/ *n.* (often **Mass**) **1** the Eucharist, esp. in the Roman Catholic Church. **2** a celebration of this. **3** the liturgy used in the mass. **4** a musical setting of parts of this. □ **high mass** mass with incense, music, and usu. the assistance of a deacon and subdeacon. **low mass** mass with no music and a minimum of ceremony. [OE *mæsse* f. eccl.L *missa* f. L *mittere miss-* dismiss, perh. f. the concluding dismissal *Ite, missa est* Go, it is the dismissal]

Massachusetts /,mæsə'tʃuːsɪts/ a state in the north-eastern US, on the Atlantic coast; pop. (1990) 6,016,425; capital, Boston. Settled by the Pilgrim Fathers in 1620, it was a centre of resistance to the British before and during the War of American Independence. It became one of the original thirteen states of the Union (1788).

Massachusetts Institute of Technology (abbr. **MIT**) a US institute of higher education, famous for scientific and technical research, founded in 1861 in Cambridge, Massachusetts.

massacre /ˈmæsəkə(r)/ n. & v. ● n. **1** a general slaughter (of persons, occasionally of animals). **2** colloq. an utter defeat or destruction. ● v.tr. **1** murder (esp. a large number of people) cruelly or violently. **2** colloq. defeat heavily, destroy. [OF, of unkn. orig.]

Massacre of St Bartholomew the massacre of Huguenots throughout France ordered by Charles IX at the instigation of his mother, Catherine de' Medici, and begun without warning on 24 Aug. (the feast of St Bartholomew) 1572.

massage /ˈmæsɑːʒ, -sɑːdʒ/ n. & v. ● n. **1** the rubbing, kneading, etc., of muscles and joints of the body with the hands, to stimulate their action, cure strains, etc. **2** an instance of this. **3** manipulation of figures or statistics (data massage). ● v.tr. **1** apply massage to. **2** manipulate (statistics) to give an acceptable result. **3** flatter (a person's ego etc.). □ **massage parlour 1** an establishment providing massage. **2** euphem. a brothel (operating under the guise of providing massage). □ **massager** n. [F f. masser treat with massage, perh. f. Port. amassar knead, f. massa dough: see MASS¹]

massasauga /ˌmæsəˈsɔːgə/ n. a small North American rattlesnake, Sistrurus catenatus. [irreg. f. Mississagi river, Ontario]

Massawa /məˈsɑːwə/ (also **Mitsiwa** /mɪˈtsiːwə/) the chief port of Eritrea, on the Red Sea; pop. (1984) 27,500.

massé /ˈmæsɪ/ n. (usu. attrib.) Billiards a stroke made with the cue held more or less vertical, imparting swerve to the ball. [F, past part. of masser make such a stroke (as MACE¹)]

masseter /mæˈsiːtə(r)/ n. (in full **masseter muscle**) Anat. either of two chewing-muscles which run from the temporal bone to the lower jaw. [Gk masētēr f. masaomai chew]

masseur /mæˈsɜː(r)/ n. (fem. **masseuse** /-ˈsɜːz/) a person who provides massage professionally. [F f. masser: see MASSAGE]

massicot /ˈmæsɪkət/ n. yellow lead monoxide, used as a pigment. [F, perh. rel. to It. marzacotto unguent prob. f. Arab. mashḥakūnyā]

massif /ˈmæsiːf, mæˈsiːf/ n. a compact group of mountain heights. [F massif used as noun: see MASSIVE]

Massif Central /ˌmæsiːf sɒnˈtrɑːl/ a mountainous plateau in south central France. Covering almost one-sixth of the country, it rises to a height of 1,887 m (6,188 ft) at Puy de Sancy in the Auvergne. It is bounded to the south-east by the Cévennes.

Massine /mæˈsiːn/, Léonide Fëdorovich (born Leonid Fëdorovich Myassin) (1895–1979), Russian-born choreographer and ballet-dancer. In 1914 he joined Diaghilev's Ballets Russes as a dancer; he took up choreography the following year and went on to create ballets such as Le Tricorne (1919). He was the originator of the symphonic ballet, including Les Présages (1933), which uses Tchaikovsky's Fifth Symphony. He also danced in and choreographed the film The Red Shoes (1948). He settled in Europe and became a French citizen in 1944.

Massinger /ˈmæsɪndʒə(r)/, Philip (1583–1640), English dramatist. He wrote many of his works in collaboration with other dramatists, most notably John Fletcher. His plays include tragedies (e.g. The Duke of Milan, 1621–2) and the social comedies A New Way to Pay Old Debts (1625–6) and The City Madam (1632).

massive /ˈmæsɪv/ adj. **1** large and heavy or solid. **2** (of the features, head, etc.) relatively large; of solid build. **3** exceptionally large (took a massive overdose). **4** substantial, impressive (a massive reputation). **5** Mineral. not visibly crystalline. **6** Geol. without structural divisions. □ **massively** adv. **massiveness** n. [ME f. F massif -ive f. OF massiz ult. f. L massa MASS¹]

Masson /ˈmæsɒn/, André (1896–1987), French painter and graphic artist. He joined the surrealists in the mid-1920s; his early works pioneered the use of 'automatic' drawing, a form of fluid, spontaneous composition intended to express images emerging from the unconscious. His later works are characterized by themes of psychic pain, violence, and eroticism, with near-abstract biomorphic forms; they include the painting Iroquois Landscape (1942).

Massorah var. of MASORAH.

Massorete var. of MASORETE.

mass spectrometer n. an apparatus for separating atoms, molecules, and molecular fragments according to mass by passing them in ionic form through electric and magnetic fields. The sample material is first vaporized and ionized; the ions are accelerated by an electric field and then deflected by a magnetic field into a curved trajectory which depends on their mass and charge. The ions are detected photographically or electrically as a mass spectrum. The first such apparatus was invented by F. W. Aston in 1920; it was actually a mass spectrograph, using photographic detection. Mass spectrometers have become important tools in nuclear physics, analytical chemistry, and organic chemistry.

mast¹ /mɑːst/ n. & v. ● n. **1** a long upright post of timber, iron, etc., set up on a ship's keel, esp. to support sails. **2** a post or lattice-work upright for supporting a radio or television aerial. **3** a flag-pole (half-mast). **4** (in full **mooring-mast**) a strong steel tower to the top of which an airship can be moored. ● v.tr. furnish (a ship) with masts. □ **before the mast** serving as an ordinary seaman (quartered in the forecastle). □ **masted** adj. (also in comb.). **master** n. (also in comb.). [OE mæst f. WG]

mast² /mɑːst/ n. the fruit of the beech, oak, chestnut, and other forest-trees, esp. as food for pigs, birds, etc. [OE mæst f. WG, prob. rel. to MEAT]

mastaba /ˈmæstəbə/ n. **1** an ancient Egyptian tomb, rectangular in shape with sloping sides and a flat roof, standing to a height of 5–6 m (17–20 ft). The interior is composed of two parts: an underground burial chamber, and rooms above it (at ground level) to store offerings etc. **2** in Islamic countries: a bench, usu. of stone, attached to a house. [Arab. maṣṭabah bench]

mast cell n. Physiol. a cell in connective tissue which releases histamine etc. during inflammatory and allergic reactions. [G Mast = fattening, feeding (cf. MAST²)]

mastectomy /mæˈstektəmɪ/ n. (pl. **-ies**) the surgical removal of a breast. [Gk mastos breast + -ECTOMY]

master /ˈmɑːstə(r)/ n., adj., & v. ● n. **1 a** a person having control of other persons or things. **b** an employer. **c** a male head of a household (master of the house). **d** the owner of a dog, horse, etc. **e** the owner of a slave. **f** Naut. the captain of a merchant ship. **g** Hunting the person in control of a pack of hounds etc. **2 a** male teacher or tutor, esp. a schoolmaster. **3 a** the head of a college, school, etc. **b** the presiding officer of a livery company, Masonic lodge, etc. **4** a person who has or gets the upper hand (we shall see which of us is master). **5 a** a person skilled in a particular trade and able to teach others (often attrib.: master carpenter). **b** a skilled practitioner (master of innuendo). **6** a holder of a university degree orig. giving authority to teach in the university (Master of Arts; Master of Science). **7 a** a revered teacher in philosophy etc. **b** (**the Master**) Christ. **8** a great artist. **9** Chess etc. a player of proved ability at international level. **10** an original version (e.g. of a film or gramophone record) from which a series of copies can be made. **11** (**Master**) **a** a title prefixed to the name of a boy not old enough to be called Mr (Master T. Jones; Master Tom). **b** archaic a title for a man of high rank, learning, etc. **12** (in England and Wales) an official of the Supreme Court. **13** a machine or device directly controlling another (cf. SLAVE n. 4). **14** (**Master**) a courtesy title of the heir apparent of a Scottish viscount or baron (the Master of Falkland). ● adj. **1** commanding, superior (a master spirit). **2** main, principal (master bedroom). **3** controlling others (master plan). ● v.tr. **1** overcome, defeat. **2** reduce to subjection. **3** acquire complete knowledge of (a subject) or facility in using (an instrument etc.). **4** rule as a master. □ **be master of 1** have at one's disposal. **2** know how to control. **be one's own master** be independent or free to do as one wishes. **make oneself master of** acquire a thorough knowledge of or facility in using. **master-at-arms** (pl. **masters-at-arms**) the chief police officer on a man-of-war or a merchant ship. **master-class** a class given by a person of distinguished skill, esp. in music. **master-hand 1** a person having commanding power or great skill. **2** the action of such a person. **master-key** a key that opens several locks, each of which also has its own key. **master mariner 1** the captain of a merchant ship. **2** a seaman certified competent to be captain. **master mason 1** a skilled mason, or one in business on his or her own account. **2** a fully qualified Freemason, who has passed the third degree. **master of ceremonies 1** (abbr. **MC**) a person introducing speakers at a banquet, or entertainers in a variety show. **2** a person in charge of ceremonies at a state or public occasion. **master-stroke** an outstandingly skilful act of policy etc. **master-switch** a switch controlling the supply of electricity etc. to an entire system. **master touch** a masterly manner of dealing with something. **master-work** a masterpiece. □ **masterdom** n. **masterhood** n. **masterless** adj. [OE mægester (later also f. OF maistre) f. L magister, prob. rel. to magis more]

Master Aircrew n. an RAF rank equivalent to warrant-officer.

masterful /ˈmɑːstəfʊl/ adj. **1** imperious, domineering. **2** masterly. ¶ Normally used of a person, whereas masterly is used of achievements, abilities, etc. □ **masterfully** adv. **masterfulness** n.

masterly /'mɑːstəlɪ/ adj. worthy of a master; very skilful (a masterly piece of work). □ **masterliness** n.

mastermind /'mɑːstə,maɪnd/ n. & v. ● n. **1 a** a person with an outstanding intellect. **b** such an intellect. **2** the person directing an intricate operation. ● v.tr. plan and direct (a scheme or enterprise).

Master of the Rolls n. (in England and Wales) a judge who presides over the Court of Appeal and was formerly in charge of the Public Record Office.

masterpiece /'mɑːstə,piːs/ n. **1** an outstanding piece of artistry or workmanship. **2** a person's best work.

mastership /'mɑːstə,ʃɪp/ n. **1** the position or function of a master, esp. a schoolmaster. **2** dominion, control.

mastersinger /'mɑːstə,sɪŋə(r)/ n. = MEISTERSINGER.

Masters Tournament a prestigious golf competition in which golfers (chiefly professionals) compete only by invitation on the basis of their past achievements. It has been held annually in the US since 1934 at Augusta, Georgia, and was instituted by the US golfer Bobby Jones.

mastery /'mɑːstərɪ/ n. **1** dominion, sway. **2** masterly skill. **3** (often foll. by of) comprehensive knowledge or use of a subject or instrument. **4** (prec. by the) the upper hand. [ME f. OF maistrie (as MASTER)]

masthead /'mɑːsthed/ n. & v. ● n. **1** the highest part of a ship's mast, esp. that of a lower mast as a place of observation or punishment. **2** the title of a newspaper etc. at the head of the front or editorial page. ● v.tr. **1** send (a sailor) to the masthead. **2** raise (a sail) to its position on the mast.

mastic /'mæstɪk/ n. **1** a gum or resin exuded from the bark of the mastic tree, used in making varnish. **2** (in full **mastic tree**) the evergreen Mediterranean tree, Pistacia lentiscus, yielding this. **3** a waterproof filler and sealant used in building. **4** a liquor flavoured with mastic gum. [ME f. OF f. LL mastichum f. L mastiche f. Gk mastikhē, perh. f. mastikhaō (see MASTICATE) with ref. to its use as chewing-gum]

masticate /'mæstɪ,keɪt/ v.tr. grind or chew (food) with one's teeth. □ **masticator** n. **masticatory** /-kətərɪ/ adj. **mastication** /,mæstɪ'keɪʃ(ə)n/ n. [LL masticare masticat- f. Gk mastikhaō gnash the teeth]

mastiff /'mæstɪf, 'mɑːs-/ n. a large strong breed of dog with drooping ears and pendulous lips. [ME ult. f. OF mastin ult. f. L mansuetus tame: see MANSUETUDE]

mastitis /mæ'staɪtɪs/ n. an inflammation of the mammary gland (the breast or udder). [Gk mastos breast + -ITIS]

mastodon /'mæstə,dɒn/ n. a large extinct elephant-like mammal of the genus Mammut. Its name refers to the nipple-shaped tubercles on the crowns of its molar teeth. □ **mastodontic** /,mæstə'dɒntɪk/ adj. [mod.L f. Gk mastos breast + odous odontos tooth]

mastoid /'mæstɔɪd/ adj. & n. ● adj. esp. Anat. shaped like a woman's breast. ● n. **1** Anat. = mastoid process. **2** colloq. mastoiditis. □ **mastoid process** a conical prominence on the temporal bone behind the ear, to which muscles are attached. [F mastoïde or mod.L mastoides f. Gk mastoeidēs f. mastos breast]

mastoiditis /,mæstɔɪ'daɪtɪs/ n. Med. inflammation of the mastoid process.

masturbate /'mæstə,beɪt/ v.intr. & tr. arouse oneself sexually or cause (another person) to be aroused by manual stimulation of the genitals. □ **masturbator** n. **masturbatory** adj. **masturbation** /,mæstə'beɪʃ(ə)n/ n. [L masturbari masturbat-]

Masuria /mə'sjʊərɪə/ (also **Masurian Lakes**) a low-lying and forested lakeland region of NE Poland. Formerly part of East Prussia, it was assigned to Poland after the Second World War. Extending from the Vistula to Poland's eastern borders, it contains some 2,700 lakes.

mat[1] /mæt/ n. & v. ● n. **1** a piece of coarse material for wiping shoes on, esp. a doormat. **2** a piece of cork, rubber, plastic, etc., to protect a surface from the heat or moisture of an object placed on it. **3** a piece of resilient material for landing on in gymnastics, wrestling, etc. **4** a piece of coarse fabric of plaited rushes, straw, etc., for lying on, packing furniture, etc. **5** a small rug. ● v. (**matted**, **matting**) **1 a** tr. (esp. as **matted** adj.) entangle in a thick mass (matted hair). **b** intr. become matted. **2** tr. cover or furnish with mats. □ **on the mat** sl. being reprimanded (orig. in the army, on the orderly-room mat before the commanding officer). [OE m(e)att(e) f. WG f. LL matta]

mat[2] var. of MATT.

mat[3] /mæt/ n. = MATRIX 1. [abbr.]

Matabele /,mætə'biːlɪ/ n. & adj. (pl. same or **Matabeles**) see NDEBELE.

Matabeleland /,mætə'biːlɪ,lænd/ a former province of Southern Rhodesia, lying between the Limpopo and Zambezi rivers and occupied by the Matabele people. The area is now divided into the two provinces of Matabeleland North and South, in southern Zimbabwe.

matador /'mætə,dɔː(r)/ n. **1** a bullfighter whose task is to kill the bull. **2** a principal card in ombre, quadrille, etc. **3** a domino game in which the piece played must make a total of seven. [Sp. f. matar kill f. Pers. māt dead]

Mata Hari /,mɑːtə 'hɑːrɪ/ (born Margaretha Geertruida Zelle) (1876–1917), Dutch dancer and secret agent. She became a professional dancer in Paris in 1905 and probably worked for both French and German intelligence services before being executed by the French in 1917. Her name derives from Malay mata eye and hari day.

match[1] /mætʃ/ n. & v. ● n. **1** a contest or game of skill etc. in which persons or teams compete against each other. **2 a** a person able to contend with another as an equal (meet one's match; be more than a match for). **b** a person equal to another in some quality (we shall never see his match). **c** a person or thing exactly like or corresponding to another. **3** a marriage. **4** a person viewed in regard to his or her eligibility for marriage, esp. as to rank or fortune (an excellent match). ● v. **1 a** tr. be equal to or harmonious with; correspond to in some essential respect (the curtains match the wallpaper). **b** intr. (often foll. by with) correspond; harmonize (his socks do not match; does the ribbon match with your hat?). **c** (as **matching** adj.) having correspondence in some essential respect (matching curtains). **2** tr. (foll. by against, with) place (a person etc.) in conflict, contest, or competition with (another). **3** tr. find material etc. that matches (another) (can you match this silk?). **4** tr. find (a person or thing) suitable for another (matching unemployed workers with vacant posts). **5** tr. prove to be a match for. **6** tr. Electronics produce an adjustment of (circuits) such that maximum power is transmitted between them. **7** tr. (usu. foll. by with) archaic join (a person) with another in marriage. □ **make a match** bring about a marriage. **match play** Golf play in which the score is reckoned by counting the holes won by each side (cf. stroke play). **match point 1** Tennis etc. **a** the state of a game when one side needs only one more point to win the match. **b** this point. **2** Bridge a unit of scoring in matches and tournaments. **match up** (often foll. by with) fit to form a whole; tally. **match up to** be as good as or equal to. **to match** corresponding in some essential respect with what has been mentioned (yellow dress with gloves to match). **well-matched** fit to contend with each other, live together, etc., on equal terms. □ **matchable** adj. [OE gemæcca mate, companion, f. Gmc]

match[2] /mætʃ/ n. **1** a short thin piece of wood, wax, etc., tipped with a composition that can be ignited by friction. **2** a piece of wick, cord, etc., designed to burn at a uniform rate, for firing a cannon etc. [ME f. OF mesche, meiche, perh. f. L myxa lamp-nozzle]

matchboard /'mætʃbɔːd/ n. a board with a tongue cut along one edge and a groove along another, so as to fit with similar boards.

matchbox /'mætʃbɒks/ n. a box for holding matches.

matchet var. of MACHETE.

matchless /'mætʃlɪs/ adj. without an equal, incomparable. □ **matchlessly** adv.

matchlock /'mætʃlɒk/ n. hist. **1** an old type of gun with a lock in which a match was placed for igniting the powder. **2** such a lock.

matchmaker /'mætʃ,meɪkə(r)/ n. **1** a person who arranges marriages. **2** a person who schemes to bring couples together. □ **matchmaking** n.

matchstick /'mætʃstɪk/ n. the stem of a match.

matchwood /'mætʃwʊd/ n. **1** wood suitable for matches. **2** minute splinters. □ **make matchwood of** smash utterly.

mate[1] /meɪt/ n. & v. ● n. **1** a friend or fellow worker. **2** colloq. a general form of address, esp. to another man. **3 a** each of a pair, esp. of birds. **b** colloq. a partner in marriage. **c** (in comb.) a fellow member or joint occupant of (team-mate; room-mate). **4** Naut. an officer on a merchant ship subordinate to the master. **5** an assistant to a skilled worker (plumber's mate). ● v. (often foll. by with) **1 a** tr. bring (animals or birds) together for breeding. **b** intr. (of animals or birds) come together for breeding. **2 a** tr. join (persons) in marriage. **b** intr. (of persons) be joined in marriage. **3** intr. Mech. fit well. □ **mateless** adj. [ME f. MLG mate f. gemate messmate f. WG, rel. to MEAT]

mate[2] /meɪt/ n. & v.tr. Chess = CHECKMATE. □ **fool's mate** a series of moves in which the first player is mated at the second player's second move. **scholar's mate** a series of moves in which the second player is mated at the first player's fourth move. [ME f. F mat(er): see CHECKMATE]

maté /'mæteɪ/ n. **1** an infusion of the leaves of a South American shrub,

Ilex paraguariensis. **2** this shrub, or its leaves. **3** a vessel in which these leaves are infused. [Sp. *mate* f. Quechua *mati*]

matelot /ˈmætləʊ/ *n.* (also **matlow, matlo**) *Brit. sl.* a sailor. [F]

matelote /ˈmætəˌləʊt/ *n.* a dish of fish etc. with a sauce of wine and onions. [F (as MATELOT)]

mater /ˈmeɪtə(r)/ *n. Brit. sl.* mother. ¶ Now only in jocular or affected use. [L]

materfamilias /ˌmeɪtəfəˈmɪlɪˌæs/ *n.* the woman head of a family or household (cf. PATERFAMILIAS). [L f. *mater* mother + *familia* FAMILY]

material /məˈtɪərɪəl/ *n. & adj.* ● *n.* **1** the matter from which a thing is made. **2** cloth, fabric. **3** (in *pl.*) things needed for an activity (*building materials; cleaning materials; writing materials*). **4** a person or thing of a specified kind or suitable for a purpose (*officer material*). **5** (in *sing.* or *pl.*) information etc. to be used in writing a book etc. (*experimental material; materials for a biography*). **6** (in *sing.* or *pl.*, often foll. by *of*) the elements or constituent parts of a substance. ● *adj.* **1** of matter; corporeal. **2** concerned with bodily comfort etc. (*material well-being*). **3** (of conduct, points of view, etc.) not spiritual. **4** (often foll. by *to*) important, essential, relevant (*at the material time; material witness*). **5** concerned with the matter, not the form, of reasoning. □ **materiality** /-ˌtɪərɪˈælɪtɪ/ *n.* [ME f. OF *materiel, -al,* f. LL *materialis* f. L (as MATTER)]

materialism /məˈtɪərɪəˌlɪz(ə)m/ *n.* **1** a tendency to consider material possessions and physical comfort to be more important than spiritual values. **2** *Philos.* the doctrine that nothing exists but matter and its movements and modifications (cf. SPIRITUALISM 2); the doctrine that consciousness and will are wholly due to material agency. (*See note below.*) **3** *Art* a tendency to lay stress on the material aspect of objects. □ **materialist** *n. & adj.* **materialistic** /-ˌtɪərɪəˈlɪstɪk/ *adj.* **materialistically** *adv.*

■ In saying that nothing exists but matter, materialism both denies the existence of minds and mental states and rejects the status of universals. Denial of the existence of minds has resulted in the assertion that the mechanistic workings of the brain are identical with the operations of the mind, as well as the notion that disembodied minds (e.g. God or the souls of the dead) cannot exist.

materialize /məˈtɪərɪəˌlaɪz/ *v.* (also **-ise**) **1** *intr.* become actual fact. **2** *tr.* cause (a spirit) to appear in bodily form. **b** *intr.* (of a spirit) appear in this way. **3** *intr. colloq.* appear or be present when expected. **4** *tr.* represent or express in material form. **5** *tr.* make materialistic. □ **materialization** /-ˌtɪərɪəlaɪˈzeɪʃ(ə)n/ *n.*

materially /məˈtɪərɪəlɪ/ *adv.* **1** substantially, considerably. **2** in respect of matter.

materia medica /məˌtɪərɪə ˈmedɪkə/ *n.* **1** the remedial substances used in the practice of medicine. **2** the study of the origin and properties of these substances. [mod.L, transl. Gk *hulē iatrikē* healing material]

matériel /məˌtɪərɪˈel/ *n.* available means, esp. materials and equipment in warfare (opp. PERSONNEL). [F (as MATERIAL)]

maternal /məˈtɜːn(ə)l/ *adj.* **1** of or like a mother. **2** motherly. **3** related through the mother (*maternal uncle*). **4** of the mother in pregnancy and childbirth. □ **maternally** *adv.* **maternalism** *n.* **maternalistic** /-ˌtɜːnəˈlɪstɪk/ *adj.* [ME f. OF *maternel* or L *maternus* f. *mater* mother]

maternity /məˈtɜːnɪtɪ/ *n.* **1** motherhood. **2** motherliness. **3** (*attrib.*) **a** for women during and just after childbirth (*maternity hospital; maternity leave*). **b** suitable for a pregnant woman (*maternity dress; maternity wear*). [F *maternité* f. med.L *maternitas -tatis* f. L *maternus* f. *mater* mother]

mateship /ˈmeɪtʃɪp/ *n. Austral.* companionship, fellowship.

matey /ˈmeɪtɪ/ *adj. & n.* (also **maty**) ● *adj.* (**matier, matiest**) (often foll. by *with*) sociable; familiar and friendly. ● *n. Brit.* (*pl.* **-eys**) *colloq.* (usu. as a form of address) mate, companion. □ **mateyness** *n.* (also **matiness**). **matily** *adv.*

math /mæθ/ *n. N. Amer.* mathematics (cf. MATHS). [abbr.]

mathematical /ˌmæθəˈmætɪk(ə)l/ *adj.* **1** of or relating to mathematics. **2** (of a proof etc.) rigorously precise. □ **mathematical induction** = INDUCTION 3b. **mathematical tables** tables of logarithms and trigonometric values etc. □ **mathematically** *adv.* [F *mathématique* or L *mathematicus* f. Gk *mathēmatikos* f. *mathēma -matos* science f. *manthanō* learn]

mathematical logic *n.* the part of mathematics concerned with the study of formal languages, formal reasoning, the nature of mathematical proof, provability of mathematical statements, computability, and other aspects of the foundations of mathematics. Its roots lie in Boole's algebraic description of logic. Its growth was enormously stimulated by *Principia Mathematica* (1910–13) in which

A. N. Whitehead and Bertrand Russell attempted to express all of mathematics in formal logical terms, and by Gödel's incompleteness theorem (1931) which, by showing that no such enterprise could ever succeed, has led to investigation of the boundary between what is possible in mathematics and what is not.

mathematics /ˌmæθəˈmætɪks/ *n.pl.* **1** (also treated as *sing.*) the abstract deductive science of number, quantity, arrangement, and space. (*See note below.*) **2** (as *pl.*) the use of mathematics in calculation etc. □ **mathematician** /-məˈtɪʃ(ə)n/ *n.* [prob. f. F *mathématiques* pl. f. L *mathematica* f. Gk *mathēmatika*: see MATHEMATICAL]

■ Mathematics is customarily divided into *pure mathematics,* those topics studied in their own right, and *applied mathematics,* the application of mathematical knowledge in science, technology, and other areas. The main parts of pure mathematics are algebra (including arithmetic), analysis, geometry (including topology), and logic. Applied mathematics includes the various forms of mechanics (hydrodynamics, elasticity, etc.), statistics, and many other newer areas of application to computing, economics, biological sciences, etc.

maths /mæθs/ *n. Brit.* mathematics (cf. MATH). [abbr.]

Matilda[1] /məˈtɪldə/ (known as 'the Empress Maud') (1102–67), English princess, daughter of Henry I. She was Henry's only legitimate child and was named his heir in 1127. In 1135 her father died and Matilda was forced to flee when his nephew Stephen seized the throne. Her claim was supported by King David I of Scotland, and she and her half-brother Robert, Earl of Gloucester (d.1147) invaded England in 1139. She waged an unsuccessful civil war against Stephen and eventually left England in 1148. Her son became Henry II.

Matilda[2] /məˈtɪldə/ *n. Austral. sl.* a bushman's bundle; a swag. □ **waltz** (or **walk**) **Matilda** carry a swag. [the name *Matilda*]

matinée /ˈmætɪˌneɪ/ *n.* (US **matinee**) an afternoon performance in the theatre, cinema, etc. □ **matinée coat** (or **jacket**) a baby's short coat. **matinée idol** a handsome actor admired chiefly by women. [F, = what occupies a morning f. *matin* morning (as MATINS)]

matins /ˈmætɪnz/ *n.* (also **mattins**) (as *sing* or *pl.*) **1 a** a service of morning prayer in the Church of England. **b** the office of one of the canonical hours of prayer, properly a night office, but also recited with lauds at daybreak or on the previous evening. **2** (also **matin**) *poet.* the morning song of birds. [ME f. OF *matines* f. eccl.L *matutinas,* accus. fem. pl. adj. f. L *matutinus* of the morning f. *Matuta* dawn-goddess]

Matisse /mæˈtiːs/, Henri (Emile Benoît) (1869–1954), French painter and sculptor. He was influenced by the impressionists, Gauguin, and oriental art. His use of non-naturalistic colour in works such as *Open Window Collioure* (1905) led him to be regarded as a leader of the fauvists. Large figure compositions, such as *The Dance* (1909), heralded a new style based on simple reductive line, giving a rhythmic decorative pattern on a flat ground of rich colour. Later works include abstracts made of cut-out coloured paper, such as *The Snail* (1953). His sculpture displays a similar trend towards formal simplification and abstraction.

matlo (also **matlow**) var. of MATELOT.

Matmata /mætˈmɑːtə/ a town in SE Tunisia, in the Matmata hills south of Gabès. In this region for many centuries the people have lived in underground dwellings hacked from the tufa.

Mato Grosso /ˌmætuː ˈɡrɒsəʊ/ **1** a high plateau region of SW Brazil, forming a watershed between the Amazon and Plate river systems. Its Portuguese name means 'dense forest'. The region is divided into two states: *Mato Grosso* and *Mato Grosso do Sul.* **2** a state of western Brazil, on the border with Bolivia; capital, Cuiabá.

Mato Grosso do Sul /duː ˈsʊl/ a state of SW Brazil, on the borders with Bolivia and Paraguay; capital, Campo Grande.

matrass /ˈmætrəs/ *n. hist.* a long-necked glass vessel with a round or oval body, used for distilling etc. [F *matras,* of uncert. orig.]

matriarch /ˈmeɪtrɪˌɑːk/ *n.* a woman who is the head of a family or tribe. □ **matriarchal** /ˌmeɪtrɪˈɑːk(ə)l/ *adj.* [L *mater* mother, on the false analogy of PATRIARCH]

matriarchy /ˈmeɪtrɪˌɑːkɪ/ *n.* (*pl.* **-ies**) a system of society, government, etc. ruled by a woman or women and with descent through the female line.

matric /məˈtrɪk/ *n. Brit. colloq. hist.* matriculation. [abbr.]

matrices *pl.* of MATRIX.

matricide /ˈmeɪtrɪˌsaɪd/ *n.* **1** the killing by a person of his or her own mother. **2** a person who does this. □ **matricidal** /ˌmeɪtrɪˈsaɪd(ə)l/ *adj.* [L *matricida, matricidium* f. *mater matris* mother]

matriculate /məˈtrɪkjʊˌleɪt/ *v.* **1** *intr.* be enrolled at a college or university. **2** tr. admit (a student) to membership of a college or university. □ **matriculatory** /-lətərɪ/ adj. [med.L matriculare matriculat-enrol f. LL matricula register, dimin. of L MATRIX]

matriculation /məˌtrɪkjʊˈleɪʃ(ə)n/ n. **1** the act or an instance of matriculating. **2** hist. an examination to qualify for this.

matrilineal /ˌmætrɪˈlɪnɪəl/ adj. of or based on kinship with the mother or the female line. □ **matrilineally** adv. [L mater matris mother + LINEAL]

matrilocal /ˌmætrɪˈləʊk(ə)l/ adj. of or denoting a custom in marriage where the husband goes to live with the wife's community. [L mater matris mother + LOCAL]

matrimony /ˈmætrɪmənɪ/ n. (pl. **-ies**) **1** the rite of marriage. **2** the state of being married. □ **matrimonial** /ˌmætrɪˈməʊnɪəl/ adj. **matrimonially** adv. [ME f. AF matrimonie, OF matremoi(g)ne f. L matrimonium f. mater matris mother]

matrix /ˈmeɪtrɪks/ n. (pl. **matrices** /-trɪˌsiːz/ or **matrixes**) **1** a mould in which a thing is cast or shaped, such as a gramophone record, printing type, etc. **2 a** an environment or substance in which a thing is developed. **b** a womb. **3** a mass of fine-grained rock in which gems, fossils, etc., are embedded. **4** Math. a rectangular array of elements in rows and columns that is treated as a single entity. **5** Biol. the substance between cells or in which structures are embedded. **6** Computing a gridlike array of interconnected circuit elements. □ **matrix printer** = dot matrix printer (see DOT[1]). [L, = breeding-female, womb, register f. mater matris mother]

matron /ˈmeɪtrən/ n. **1** a married woman, esp. a dignified and sober one. **2** a woman managing the domestic arrangements of a school etc. **3** Brit. a woman in charge of the nursing in a hospital. ¶ Now usu. called senior nursing officer. □ **matron of honour** a married woman attending the bride at a wedding. □ **matronhood** n. [ME f. OF matrone f. L matrona f. mater matris mother]

matronly /ˈmeɪtrənlɪ/ adj. like or characteristic of a matron, esp. in respect of staidness or portliness.

Matsuyama /ˌmætsuːˈjɑːmə/ a city in Japan, the capital and largest city of the island of Shikoku; pop. (1990) 443,320.

Matt. abbr. Matthew (esp. in the New Testament).

matt /mæt/ adj., n., & v. (also **mat**) ● adj. (of a colour, surface, etc.) dull, without lustre. ● n. **1** a border of dull gold round a framed picture. **2** (in full **matt paint**) paint formulated to give a dull flat finish (cf. GLOSS[1] n. 3). **3** the appearance of unburnished gold. ● v.tr. (**matted**, **matting**) **1** make (gilding etc.) dull. **2** frost (glass). [F mat, mater, identical with mat MATE[2]]

matte[1] /mæt/ n. an impure product of the smelting of sulphide ores, esp. those of copper or nickel. [F]

matte[2] /mæt/ n. Cinematog. a mask to obscure part of an image and allow another image to be superimposed, giving a combined effect. [F]

matter /ˈmætə(r)/ n. & v. ● n. **1 a** a physical substance in general, as distinct from mind and spirit. **b** Physics that which occupies space and possesses rest mass. (See note below.) **2** a particular substance (colouring matter). **3** (prec. by the; often foll. by with) the thing that is amiss (what is the matter?; there is something the matter with him). **4** material for thought or expression. **5 a** the substance of a book, speech, etc., as distinct from its manner or form. **b** Logic the particular content of a proposition, as distinct from its form. **6** a thing or things of a specified kind (printed matter; reading matter). **7** an affair or situation being considered, esp. in a specified way (a serious matter; a matter for concern; the matter of your overdraft). **8** Physiol. any substance in or discharged from the body (faecal matter; grey matter). **b** pus. **9** (foll. by of, for) what is or may be a good reason for (complaint, regret, etc.). **10** Printing the body of a printed work, as type or as printed sheets. ● v.intr. (often foll. by to) be of importance; have significance (it does not matter to me when it happened). □ **as a matter of fact** in reality (esp. to correct a falsehood or misunderstanding). **for that matter** (or **for the matter of that**) **1** as far as that is concerned. **2** and indeed also. **in the matter of** as regards. **a matter of 1** approximately (for a matter of 40 years). **2** a thing that relates to, depends on, or is determined by (a matter of habit; only a matter of time before they agree). **a matter of course** see COURSE. **a matter of fact 1** what belongs to the sphere of fact as distinct from opinion etc. **2** Law the part of a judicial inquiry concerned with the truth of alleged facts (see also MATTER-OF-FACT). **a matter of form** a mere routine. **a matter of law** Law the part of a judicial inquiry concerned with the interpretation of the law. **a matter of record** see RECORD. **no matter 1** (foll. by when, how, etc.) regardless of (will do it no matter what the consequences). **2** it is of no importance. **what is the matter with** surely there is no objection to. **what matter?** that need not worry us. [ME f. AF mater(i)e, OF matiere f. L materia timber, substance, subject of discourse]

▪ Until the 19th century, Western scientific thought regarded all bodies as forms of matter or as matter united to spirit. With the advent of Faraday's concept of electric and magnetic fields, and the discovery that light is a form of electromagnetic radiation, a new fundamental distinction gained prominence, that between matter and energy. In the 20th century, matter was shown to consist of protons, neutrons, electrons, and other particles. However, Einstein demonstrated the equivalence of matter and energy, and the transformation of matter into kinetic energy and electromagnetic radiation became the practical basis for nuclear weapons and power generation. (See also MASS[1].)

Matterhorn /ˈmætəˌhɔːn/ (called in French Mont Cervin, Italian Monte Cervino) a mountain in the Alps, on the border between Switzerland and Italy. Rising to 4,477 m (14,688 ft), it was first climbed in 1865 by the English mountaineer Edward Whymper.

matter-of-fact /ˌmætərəˈfækt/ adj. (see also MATTER). **1** unimaginative, prosaic. **2** unemotional. □ **matter-of-factly** adv. **matter-of-factness** n.

Matthew, St /ˈmæθjuː/ **1** an Apostle, a tax-gatherer from Capernaum in Galilee, traditional author of the first Gospel. Feast day, 21 Sept. **2** the first Gospel, written after AD 70 and based largely on that of St Mark (see GOSPEL 2).

Matthew Paris /ˌmæθjuː ˈpærɪs/ (c.1199–1259) English chronicler and Benedictine monk. His Chronica Majora, a history of the world from the Creation to the mid-13th century, is a valuable source for contemporary events.

Matthews /ˈmæθjuːz/, Sir Stanley (b.1915), English footballer. He played on the right wing for Stoke City and Blackpool, and was famous for his dribbling skill. His career in professional football lasted until he was 50, during which time he played for England fifty-four times.

Matthias, St /məˈθaɪəs/ an Apostle, chosen by lot after the Ascension to take the place left by Judas. Feast day (in the Western Church) 14 May; (in the Eastern Church) 9 Aug.

matting /ˈmætɪŋ/ n. **1** fabric of hemp, bast, grass, etc., for mats (coconut matting). **2** in senses of MAT[1] v.

mattins var. of MATINS.

mattock /ˈmætək/ n. an agricultural tool shaped like a pickaxe, with an adze and a chisel edge as the ends of the head. [OE mattuc, of unkn. orig.]

mattress /ˈmætrɪs/ n. a fabric case stuffed with soft, firm, or springy material, or a similar case filled with air or water, used on or as a bed. [ME f. OF materas f. It. materasso f. Arab. almatrah the place, the cushion f. taraha throw]

maturate /ˈmætjʊˌreɪt/ v.intr. Med. (of a boil etc.) come to maturation. [L maturatus (as MATURE v.)]

maturation /ˌmætjʊˈreɪʃ(ə)n/ n. **1 a** the act or an instance of maturing; the state of being matured. **b** the ripening of fruit. **2** Med. **a** the formation of purulent matter. **b** the causing of this. □ **maturational** adj. **maturative** /məˈtjʊərətɪv/ adj. [ME f. F maturation or med.L maturatio f. L (as MATURE v.)]

mature /məˈtjʊə(r)/ adj. & v. ● adj. (**maturer**, **maturest**) **1 a** with fully developed powers of body and mind; adult. **b** sensible, wise. **2** complete in natural development, ripe. **3** (of thought, intentions, etc.) duly careful and adequate. **4** (of a bill etc.) due for payment. ● v. **1 a** tr. & intr. develop fully. **b** tr. & intr. ripen. **c** intr. come to maturity. **2** tr. perfect (a plan etc.). **3** intr. (of a bill etc.) become due for payment. □ **mature student** esp. Brit. an adult student who is older than most students. □ **maturely** adv. **maturity** n. [ME f. L maturus timely, early]

matutinal /ˌmætjuːˈtaɪn(ə)l, məˈtjuːtɪn(ə)l/ adj. **1** of or occurring in the morning. **2** early. [LL matutinalis f. L matutinus: see MATINS]

maty var. of MATEY.

matzo /ˈmætsəʊ/ n. (pl. **-os** or **matzoth** /ˈmætsəʊt/) **1** a wafer of unleavened bread for the Passover. **2** such bread collectively. [Yiddish f. Heb. maṣṣāh]

maud /mɔːd/ n. **1** a Scots shepherd's grey striped plaid. **2** a travelling-rug like this. [18th c.: orig. unkn.]

maudlin /ˈmɔːdlɪn/ adj. & n. ● adj. weakly or tearfully sentimental, esp. in a tearful and effusive stage of drunkenness. ● n. weak or mawkish sentiment. [ME f. OF Madeleine f. eccl.L Magdalena Mary Magdalene, with ref. to pictures of the saint weeping]

Maugham /mɔːm/, (William) Somerset (1874–1965), British novelist, short-story writer, and dramatist. He was born in France, where he spent his childhood, returning to live on the Riviera in 1926. His life and wide travels often provide the background to his writing; his work for British intelligence during the First World War is reflected in the *Ashenden* short stories (1928), while a visit to Tahiti gave the setting for the novel *The Moon and Sixpence* (1919). Other works include the novels *Of Human Bondage* (1915) and *Cakes and Ale* (1930), and the play *East of Suez* (1922).

Maui /ˈmaʊɪ/ the second largest of the Hawaiian islands, lying to the north-west of the island of Hawaii.

maul /mɔːl/ v. & n. ● v.tr. **1** beat and bruise. **2** (of an animal) tear and mutilate (prey etc.). **3** handle roughly or carelessly. **4** damage by criticism. ● n. **1** *Rugby* a loose scrum with the ball off the ground. **2** a brawl. **3** a special heavy hammer, commonly of wood, esp. for driving piles. □ **mauler** n. [ME f. OF *mail* f. L *malleus* hammer]

maulstick /ˈmɔːlstɪk/ n. (also **mahlstick** /ˈmɑːl-/) a light stick with a padded leather ball at one end, held by a painter in one hand to support the other hand. [Du. *maalstok* f. *malen* to paint + *stok* stick]

Mau Mau /ˈmaʊ maʊ/ an African secret society originating among the Kikuyu and active in the 1950s, having as its aim the expulsion of European settlers and the ending of British rule in Kenya, to which end it was willing to use violence and terror. As the result of a well-organized counter-insurgency campaign the British were eventually able to subdue the organization, but went on to institute political and social reforms, eventually leading to Kenyan independence in 1963.

Mauna Kea /ˌmaʊnə ˈkeɪə/ an extinct volcano on the island of Hawaii, in the central Pacific. Rising to 4,205 m (13,796 ft), it is the highest peak in the Hawaiian islands. The summit area is the site of several large astronomical telescopes.

Mauna Loa /ˌmaʊnə ˈləʊə/ an active volcano on the island of Hawaii, to the south of Mauna Kea, rising to 4,169 m (13,678 ft). The volcano Kilauea is situated on its flanks.

maunder /ˈmɔːndə(r)/ v.intr. **1** talk in a dreamy or rambling manner. **2** move or act listlessly or idly. [perh. f. obs. *maunder* beggar, to beg]

Maundy /ˈmɔːndɪ/ originally, the ceremony of washing the feet of a number of poor people, performed by royal or other eminent persons, or by ecclesiastics, on the Thursday before Easter, and commonly followed by the distribution of clothing, food, or money. It was instituted in commemoration of Christ's washing of the Apostles' feet at the Last Supper (John 13). The distribution of Maundy money by the British monarch is now all that remains of this ceremony. From the time of Henry IV the number of recipients has corresponded to the number of years in the sovereign's age, and the face value in pence of the amount received by each now corresponds similarly. [ME f. OF *mandé* f. L *mandatum* MANDATE, commandment (see John 13:34)]

Maundy money n. specially minted silver coins distributed by the British sovereign to a selected number of poor people on Maundy Thursday. The first were issued in 1662.

Maundy Thursday n. the Thursday before Easter.

Maupassant /ˈməʊpæˌsɒn/, (Henri René Albert) Guy de (1850–93), French novelist and short-story writer. He embarked on a literary career after encouragement from Flaubert, joining Zola's circle of naturalist writers. He contributed the story 'Boule de Suif' to their collection *Les Soirées de Médan* (1880) and became an immediate celebrity. He wrote about 300 short stories and six novels, portraying a broad spectrum of society and embracing themes of war, mystery, hallucination, and horror; these are written in a simple direct narrative style. His best-known novels include *Une Vie* (1883) and *Bel-Ami* (1885).

Mauretania /ˌmɒrɪˈteɪnɪə/ an ancient region of North Africa, corresponding to the northern part of Morocco and western and central Algeria. Originally occupied by the Moors (Latin *Mauri*), it was annexed by Claudius in the mid-1st century AD and divided into two Roman provinces. It was conquered by the Arabs in the 7th century. □ **Mauretanian** adj. & n.

Mauriac /ˈmɒrɪˌæk/, François (1885–1970), French novelist, dramatist, and critic. His works include the novels *Thérèse Desqueyroux* (1927) and *Le Noeud de vipères* (1932), and the play *Asmodée* (1938). His stories, usually set in the country round Bordeaux, show the conflicts suffered by prosperous bourgeois people pulled in different directions by convention, religion, and human passions. He was awarded the Nobel Prize for literature in 1952.

Mauritania /ˌmɒrɪˈteɪnɪə/ a country in West Africa with a coastline on the Atlantic Ocean; pop. (est. 1991) 2,023,000; languages, Arabic (official), French; capital, Nouakchott. The north of the country is desert, while the south lies in the Sahel. Mauritania was a centre of Berber power in the 11th and 12th centuries, at which time Islam became established in the region. Later, nomadic Arab tribes became dominant, while on the coast European nations, especially France, established trading posts. A French protectorate from 1902 and a colony from 1920, Mauritania achieved full independence in 1961. Between 1978 and 1992 the country was a military dictatorship. □ **Mauritanian** adj. & n.

Mauritius /məˈrɪʃəs/ an island country in the Indian Ocean, about 850 km (550 miles) east of Madagascar; pop. (est. 1991) 1,083,000; languages, English (official), French Creole, Indian languages; capital, Port Louis. Previously uninhabited, Mauritius was discovered by the Portuguese in the early 16th century. It was held by the Dutch (who named it in honour of Prince Maurice) from 1598 to 1710 and then by the French until 1810, when it was ceded to Britain. The British abolished slavery, used in the sugar plantations, instead importing mainly Indian and Chinese labour. Mauritius became independent as a member of the Commonwealth in 1968. □ **Mauritian** adj. & n.

Maury /ˈmɔːrɪ/, Matthew Fontaine (1806–73), American oceanographer. He conducted the first systematic survey of oceanic winds and currents, publishing charts which were of great value to merchant shipping. Maury also produced the first bathymetric charts, including a transatlantic profile, and pilot charts which enabled voyages to be considerably shortened. His work on physical oceanography was flawed in the eyes of many scientists by the religious tone of his writing.

Maurya /ˈmaʊrɪə/ a dynasty which ruled northern India 321–*c*.184 BC. It was founded by Chandragupta Maurya, who introduced a centralized government and uniform script and developed a highway network which led to Mauryan control of most of the Indian subcontinent. The oldest extant Indian art dates from this era. □ **Mauryan** adj.

mausoleum /ˌmɔːsəˈliːəm/ n. (pl. **mausolea** /-ˈlɪə/ or **mausoleums**) a large and stately tomb or place of burial. The term was originally applied to the magnificent marble tomb (one of the Seven Wonders of the World) at Halicarnassus in Caria ordered for himself by Mausolus king of Caria (d.353 BC) and erected by his queen Artemisia. [L f. Gk *Mausōleion* f. *Mausōlos* Mausolus]

mauve /məʊv/ adj. & n. ● adj. pale purple. ● n. **1** this colour. **2** a bright but delicate pale purple dye made from aniline. (See also PERKIN.) □ **mauvish** adj. [F, lit. 'mallow', f. L *malva*]

maven /ˈmeɪv(ə)n/ n. N. Amer. colloq. an expert or connoisseur. [Heb. *mēḇīn*]

maverick /ˈmævərɪk/ n. **1** N. Amer. an unbranded calf or yearling. **2** an unorthodox or independent-minded person. [Samuel A. *Maverick*, Texas engineer and rancher (1803–70), who did not brand his cattle]

mavis /ˈmeɪvɪs/ n. poet. or dial. a song thrush. [ME f. OF *mauvis*, of uncert. orig.]

maw /mɔː/ n. **1 a** the stomach of an animal. **b** the jaws or throat of a voracious animal. **2** colloq. the stomach of a greedy person. [OE *maga* f. Gmc]

mawkish /ˈmɔːkɪʃ/ adj. **1** sentimental in a feeble or sickly way. **2** having a faint sickly flavour. □ **mawkishly** adv. **mawkishness** n. [obs. *mawk* maggot f. ON *mathkr* f. Gmc]

Mawlana /mɔːˈlɑːnə/ see JALAL AD-DIN AR-RUMI.

max. /mæks/ abbr. maximum. □ **to the max** US sl. to the utmost, to the fullest extent.

maxi /ˈmæksɪ/ n. (pl. **maxis**) colloq. a maxi-coat, -skirt, etc. [abbr.]

maxi- /ˈmæksɪ/ comb. form very large or long (*maxi-coat*). [abbr. of MAXIMUM: cf. MINI-]

maxilla /mækˈsɪlə/ n. (pl. **maxillae** /-liː/) **1** the jaw or jawbone, esp. the upper jaw in most vertebrates. **2** the mouthpart of an arthropod that is used in chewing. □ **maxillary** adj. [L, = jaw]

maxim /ˈmæksɪm/ n. a general truth or rule of conduct expressed in a sentence. [ME f. F *maxime* or med.L *maxima* (*propositio*), fem. adj. (as MAXIMUM)]

maxima pl. of MAXIMUM.

maximal /ˈmæksɪm(ə)l/ adj. being or relating to a maximum; the greatest possible in size, duration, etc. □ **maximally** adv.

maximalist /ˈmæksɪməlɪst/ n. a person who rejects compromise and expects a full response to (esp. political) demands. [MAXIMAL, after Russ. *maksimalist*]

Maxim gun /ˈmæksɪm/ *n.* the first fully automatic water-cooled machine-gun, designed in 1884 and firing ten shots a second from a 250-round cartridge belt. It was adopted as a standard weapon by the British army and those of other major powers, and was used with devastating effect in the First World War. [Hiram S. *Maxim*, American-born British engineer (1840–1916)]

Maximilian /ˌmæksɪˈmɪlɪən/ (full name Ferdinand Maximilian Joseph) (1832–67), emperor of Mexico 1864–7. Brother of the Austro-Hungarian emperor Franz Josef and Archduke of Austria, Maximilian was established as emperor of Mexico under French auspices in 1864. In 1867, however, Napoleon III was forced to withdraw his support as a result of American pressure, and Maximilian was confronted by a popular uprising led by Benito Juárez. His forces proved unable to resist the rebels and he was captured and executed.

maximize /ˈmæksɪˌmaɪz/ *v.tr.* (also **-ise**) increase or enhance to the utmost. □ **maximizer** *n.* **maximization** /ˌmæksɪmaɪˈzeɪʃ(ə)n/ *n.* [L *maximus*: see MAXIMUM]

maximum /ˈmæksɪməm/ *n. & adj.* ● *n.* (*pl.* **maxima** /-mə/ or **maximums**) the highest possible or attainable amount. ● *adj.* that is a maximum. [mod.L, neut. of L *maximus*, superl. of *magnus* great]

Maxwell[1] /ˈmækswel/, (Ian) Robert (born Jan Ludvik Hoch) (1923–91), Czech-born British publisher and media entrepreneur. In 1940 he came to Britain, founding Pergamon Press, the basis of his publishing empire, in 1951. His business interests expanded during the 1980s and he became the proprietor of Mirror Group Newspapers in 1984; he also moved into cable television and became chairman of two football clubs. He died in obscure circumstances while yachting off Tenerife; it subsequently emerged that he had misappropriated company pension funds.

Maxwell[2] /ˈmækswel/, James Clerk (1831–79), Scottish physicist. Maxwell contributed to thermodynamics, the kinetic theory of gases (showing the importance of the statistical approach), and colour vision (demonstrating one of the earliest colour photographs in 1861). His greatest achievement was to extend the ideas of Faraday and Kelvin into his field equations of electromagnetism, thus unifying electricity and magnetism, identifying the electromagnetic nature of light, and postulating the existence of other electromagnetic radiation.

maxwell /ˈmækswel/ *n.* a unit of magnetic flux in the c.g.s. system, equal to that induced through one square centimetre by a perpendicular magnetic field of one gauss. [MAXWELL[2]]

Maxwell's demon *n. Physics* a hypothetical being conceived as controlling a hole in a partition dividing a gas-filled vessel into two parts, and allowing only fast-moving molecules to pass in one direction, and slow-moving molecules in the other. This would result in one side of the vessel becoming warmer and the other colder, in violation of the second law of thermodynamics. James Clerk Maxwell first imagined the demon in his *Theory of Heat* (1871).

May /meɪ/ *n.* **1** the fifth month of the year. **2** (**may**) the hawthorn or its blossom. **3** *poet.* bloom, prime. □ **may-apple** an American herbaceous plant, *Podophyllum peltatum*, bearing a yellow egg-shaped fruit in May. **May-bug** = COCKCHAFER. **May queen** a girl chosen to preside over celebrations on May Day. **Queen of the May** = *May queen*. [ME f. OF *mai* f. L *Maius* (*mensis*) (month) of the goddess Maia (see MAIA[2]), who was worshipped in this month]

may /meɪ/ *v.aux.* (*3rd sing. present* **may**; *past* **might** /maɪt/) **1** (often foll. by *well* for emphasis) expressing possibility (*it may be true; I may have been wrong; you may well lose your way*). **2** expressing permission (*you may not go; may I come in?*). ¶ Both *can* and *may* are used to express permission; in more formal contexts *may* is usual since *can* also denotes capability (*can I move?* = am I physically able to move?; *may I move* = am I allowed to move?). **3** expressing a wish (*may he live to regret it*). **4** expressing uncertainty or irony in questions (*who may you be?; who are you, may I ask?*). **5** in purpose clauses and after *wish, fear,* etc. (*take such measures as may avert disaster; hope he may succeed*). □ **be that as it may** regardless of whether or not that is so. **may as well** = *might as well* (see MIGHT[1]). **that is as may be** that may or may not be so (implying that there are other factors). [OE *mæg* f. Gmc, rel. to MAIN[1], MIGHT[2]]

Maya /ˈmɑːjə/ *n. & adj.* ● *n.* **1** (*pl.* same or **Mayas**) a member of a native people of Yucatán and Central America. (*See note below.*) **2** the language of this people. ● *adj.* of or relating to the Maya or their language. □ **Mayan** *adj. & n.* [Sp. f. Maya]

▪ Mayan civilization developed over an extensive area of southern Mexico, Guatemala, and Belize from the second millennium BC, reaching its peak *c.*300–*c.*900 AD. The cultural achievements of this period can be seen in a highly ornamented and detailed artistic style,

in stone temples built on pyramids and ornamented with sculptures, in its system of pictorial writing (still largely undeciphered), and in an extremely accurate calendar system (still in use at the time of the Spanish conquest in the 16th century).

maya /ˈmɑːjə/ *n.* **1** *Hinduism* illusion, magic, the supernatural power wielded by gods and demons. **2** *Hinduism & Buddhism* the power by which the universe becomes manifest, the illusion or appearance of the phenomenal world. [Skr. *māyā*(*mā* create)]

Mayakovsky /ˌmaɪəˈkɒfskɪ/, Vladimir (Vladimirovich) (1893–1930), Soviet poet and dramatist, born in Georgia. From 1910 he aligned himself with the Russian futurists, signing the futurist manifesto 'A Slap in the Face for Public Taste' in 1912. His early poems are declamatory in tone and employ an aggressive avant-garde style. After 1917, Mayakovsky saw a clear political role for futurism in the Bolshevik revolution and the new society, altering his style to have a comic mass appeal. He fell from official favour by the end of the 1920s, and committed suicide in 1930; five years later his reputation was restored by Stalin.

maybe /ˈmeɪbiː/ *adv.* perhaps, possibly. [ME f. *it may be*]

May Day 1 May, a day of traditional springtime celebrations, probably associated with pre-Christian fertility rites. Traditional activities include processions carrying garlands, the naming of a May king or queen, and dancing round a maypole. May Day was designated an international labour day by the International Socialist congress of 1889.

mayday /ˈmeɪdeɪ/ *n.* an international radio distress-signal used esp. by ships and aircraft. [repr. pronunc. of F *m'aidez* help me]

Mayer /ˈmeɪə(r)/, Louis B(urt) (born Eliezer Mayer) (1885–1957), Russian-born American film executive. In 1907 he acquired a chain of cinemas, moving into production with Metro Films in 1915. He joined with Samuel Goldwyn to form Metro-Goldwyn-Mayer (MGM) in 1924; he was head of MGM until 1951 and the company was responsible for many successful films. He also helped establish the Academy of Motion Picture Arts and Sciences (1927), and received an honorary award from the Academy in 1950.

mayest /ˈmeɪɪst/ *archaic* = MAYST.

Mayflower /ˈmeɪˌflaʊə(r)/ the ship in which, in 1620, the Pilgrim Fathers sailed from Plymouth to establish the first permanent colony in New England. It arrived at Cape Cod (Massachusetts) on 21 Nov., after a voyage of sixty-six days.

mayflower /ˈmeɪˌflaʊə(r)/ *n.* a flower that blooms in May, esp. the trailing arbutus, *Epigaea repens*.

mayfly /ˈmeɪflaɪ/ *n.* (*pl.* **-flies**) **1** a fragile winged insect of the order Ephemeroptera, with an aquatic nymph, and a brief adult life in spring. **2** an imitation mayfly used by anglers.

mayhap /meɪˈhæp/ *adv. archaic* perhaps, possibly. [ME f. *it may hap*]

mayhem /ˈmeɪhem/ *n.* **1** violent or damaging action. **2** rowdy confusion, chaos. **3** *hist.* the crime of maiming a person so as to render him or her partly or wholly defenceless. [AF *mahem*, OF *mayhem* (as MAIM)]

maying /ˈmeɪɪŋ/ *n. & adj.* participation in May Day festivities. [ME f. MAY]

Maynooth /meɪˈnuːθ/ a village in County Kildare in the Republic of Ireland; pop. (1981) 1,300. It is the site of St Patrick's College, a Roman Catholic seminary founded in 1795.

mayn't /ˈmeɪənt/ *contr.* may not.

Mayo /ˈmeɪəʊ/ a county in the Republic of Ireland, in the north-west in the province of Connacht; county town, Castlebar.

mayonnaise /ˌmeɪəˈneɪz/ *n.* **1** a thick creamy dressing made of egg-yolks, oil, vinegar, etc. **2** a (usu. specified) dish dressed with this (*chicken mayonnaise*). [F, perh. f. *mahonnais -aise* of Port Mahon on Minorca]

mayor /meə(r)/ *n.* **1** the head of the municipal corporation of a city or borough. **2** (in England, Wales, and Northern Ireland) the head of a district council with the status of a borough. □ **mayoral** *adj.* **mayorship** *n.* [ME f. OF *maire* f. L (as MAJOR)]

mayoralty /ˈmeərəltɪ/ *n.* (*pl.* **-ies**) **1** the office of mayor. **2** a mayor's period of office. [ME f. OF *mairalté* (as MAYOR)]

mayoress /ˈmeərɪs/ *n.* **1** a woman holding the office of mayor. **2** the wife of a mayor. **3** a woman fulfilling the ceremonial duties of a mayor's wife.

Mayotte /mɑːˈjɒt/ (also called *Mahore*) an island to the east of the Comoros in the Indian Ocean; pop. (est. 1991) 95,000; languages, French (official), local Swahili dialect; capital, Mamoutzu. When the

Comoros became independent in 1974, Mayotte elected to remain an overseas territory of France. Since 1976 the island has had the special status of *collectivité territoriale*.

maypole /ˈmeɪpəʊl/ *n.* a pole painted and decked with flowers and ribbons, for dancing round on May Day.

mayst /meɪst/ *archaic 2nd sing. present of* MAY.

mayweed /ˈmeɪwiːd/ *n.* a wild camomile often found as a weed, esp. *stinking mayweed* (*Anthemis cotula*), and *scentless mayweed* (*Tripleurospermum inodorum*). [earlier *maidwede* f. obs. *maithe(n)* f. OE *magothe*, *mægtha* + WEED]

mazard /ˈmæzəd/ *n.* (also **mazzard**) **1** the wild sweet cherry, *Prunus avium*, of Europe. **2** *archaic* a head or face. [perh. alt. of MAZER in sense 'hardwood']

Mazar-e-Sharif /mæˌzɑːriːˈʃəˈriːf/ a city in northern Afghanistan; pop. (est. 1984) 118,000. The city, whose name means 'tomb of the saint', is the reputed burial place of Ali, son-in-law of Muhammad.

Mazarin /ˈmæzərɪn/, Jules (Italian name Giulio Mazzarino) (1602–61), Italian-born French statesman. In 1634 he was sent to Paris as the Italian papal legate; he became a naturalized Frenchman and entered the service of Louis XIII in 1639. He was made a cardinal in 1641 and the following year succeeded Richelieu as chief minister of France, which he governed during the minority of Louis XIV. His administration aroused such opposition as to provoke the civil wars of the Fronde (1648–53).

Mazarin Bible an alternative name for the Gutenberg Bible. It is so called because a copy was kept in the library of Cardinal Mazarin.

mazarine /ˌmæzəˈriːn/ *n. & adj.* a rich deep blue. [17th c., perh. f. the name of MAZARIN or the Duchesse de *Mazarin*, Fr. noblewoman (d.1699)]

Mazatlán /ˌmæzætˈlɑːn/ a seaport and resort in Mexico, on the Pacific coast in the state of Sinaloa; pop. (1990) 314,250. Founded in 1531, it developed as a centre of Spanish colonial trade with the Philippines. It is linked by ferry to the Baja California peninsula.

Mazdaism /ˈmæzdəˌɪz(ə)m/ *n.* Zoroastrianism. [Avestan *mazda*, the supreme god (AHURA MAZDA) of ancient Persian religion]

maze /meɪz/ *n. & v.* ● *n.* **1** a network of paths and hedges designed as a puzzle for those who try to penetrate it. **2** a complex network of paths or passages; a labyrinth. **3** confusion, a confused mass, etc. ● *v.tr.* (esp. as **mazed** *adj.*) bewilder, confuse. □ **mazy** *adj.* (**mazier**, **maziest**). [ME, orig. as *mased* (adj.): rel. to AMAZE]

mazer /ˈmeɪzə(r)/ *n. hist.* a hardwood drinking-bowl, usu. silver-mounted. [ME f. OF *masere* f. Gmc]

mazurka /məˈzɜːkə/ *n.* **1** a usu. lively Polish dance in triple time. **2** the music for this. [F *mazurka* or G *Masurka*, f. Pol. *mazurka* woman of the province *Mazovia*]

mazzard var. of MAZARD.

Mazzini /mætˈsiːnɪ/, Giuseppe (1805–72), Italian nationalist leader. While in exile in Marseilles he founded the patriotic movement Young Italy (1831) and thereafter became one of the Risorgimento's most committed leaders, planning attempted insurrections in a number of Italian cities during the 1850s. He continued to campaign for a republican Italy following the country's unification as a monarchy in 1861.

MB *abbr.* **1** Bachelor of Medicine. **2** (also **Mb**) *Computing* megabyte(s). [sense 1 f. L *Medicinae Baccalaureus*]

MBA *abbr.* Master of Business Administration.

Mbabane /ˌəmbɑːˈbɑːnɪ/ the capital of Swaziland; pop. (1986) 38,300.

MBE *abbr.* Member of the Order of the British Empire.

MC *abbr.* **1** see *master of ceremonies*. **2** (in the UK) Military Cross. **3** (in the US) Member of Congress. **4** music cassette (of pre-recorded audiotape).

Mc *abbr.* megacycle(s).

MCC *abbr.* Marylebone Cricket Club, founded in 1787, which has its headquarters at Lord's Cricket Ground in London. The MCC was the tacitly accepted governing body of cricket until 1969, when many of its powers were transferred to other bodies, and continues to have primary responsibility for the game's laws. The MCC was also formerly responsible for organizing overseas tours by the England team, which played under the name 'MCC' except in internationals.

M.Ch. *abbr.* (also **M.Chir.**) Master of Surgery. [L *Magister Chirurgiae*]

mCi *abbr.* millicurie(s).

M.Com. *abbr.* Master of Commerce.

MCP *abbr. colloq.* male chauvinist pig.

MCR *abbr. Brit.* Middle Common Room.

Mc/s *abbr.* megacycles per second.

MD *abbr.* **1** Doctor of Medicine. **2** Managing Director. **3** *US* Maryland (in official postal use). **4** mentally deficient. [sense 1 f. L *Medicinae Doctor*]

Md *symb. Chem.* the element mendelevium.

Md. *abbr.* Maryland.

MDMA *abbr.* methylenedioxymethamphetamine, an amphetamine-based drug (also called *Ecstasy*) that causes euphoric and hallucinatory effects, originally produced as an appetite suppressant.

MDT *abbr.* Mountain Daylight Time, one hour ahead of Mountain Standard Time.

ME *abbr.* **1** *US* Maine (in official postal use). **2** myalgic encephalomyelitis.

Me. *abbr.* **1** Maine. **2** *Maître* (title of a French advocate).

me[1] /miː, mɪ/ *pron.* **1** objective case of I² (*he saw me*). **2** = I² (*it's me all right; is taller than me*). ¶ Strictly, the pronoun following the verb *to be* should be in the subjective rather than the objective case, i.e. *is taller than I* rather than *is taller than me*. **3** *US colloq.* myself, to or for myself (*I got me a gun*). **4** *colloq.* used in exclamations (*ah me!; dear me!; silly me!*). □ **me and mine** me and my relatives. [OE *me*, *mē* accus. & dative of I² f. Gmc]

me[2] /miː/ *n.* (also **mi**) *Mus.* **1** (in tonic sol-fa) the third note of a major scale. **2** the note E in the fixed-doh system. [ME f. L *mira*: see GAMUT]

mea culpa /ˌmiːə ˈkʌlpə, ˌmeɪə ˈkʊlpə/ *n. & int.* ● *n.* an acknowledgement of one's fault or error. ● *int.* expressing such an acknowledgement. [L, = by my fault]

Mead /miːd/, Margaret (1901–78), American anthropologist and social psychologist. She worked in Samoa and the New Guinea area and wrote a number of specialized studies of primitive cultures, but was also concerned to relate her findings to current American life and its problems; her writings made anthropology accessible to a wide readership and demonstrated its relevance to Western society. Her books include *Male and Female* (1949), an examination of the extent to which male and female roles are shaped by social factors rather than heredity.

mead[1] /miːd/ *n.* an alcoholic drink of fermented honey and water. [OE *me(o)du* f. Gmc]

mead[2] /miːd/ *n. poet.* or *archaic* = MEADOW. [OE *mæd* f. Gmc, rel. to MOW¹]

meadow /ˈmedəʊ/ *n.* **1** a piece of grassland, esp. one used for hay. **2** a piece of low well-watered ground, esp. near a river. □ **meadow brown** a common brown butterfly, *Maniola jurtina*. **meadow-grass** a perennial creeping grass, *Poa pratensis*. **meadow pipit** a common Old World pipit, *Anthus pratensis*, inhabiting open country. **meadow rue** a ranunculaceous plant of the genus *Thalictrum*; esp. *T. flavum*, with small yellow flowers. **meadow saffron** a crocus-like meadow plant, *Colchicum autumnale*, that produces lilac flowers in the autumn while still leafless (also called *autumn crocus*). □ **meadowy** *adj.* [OE *mædwe*, oblique case of *mæd*: see MEAD²]

meadowlark /ˈmedəʊˌlɑːk/ *n. N. Amer.* a brown and yellow songbird of the genus *Sturnella*, native to North America.

meadowsweet /ˈmedəʊˌswiːt/ *n.* **1** a rosaceous plant, *Filipendula ulmaria*, common in meadows and damp places, with creamy-white fragrant flowers. **2** a rosaceous shrub of the genus *Spiraea*, native to North America.

meagre /ˈmiːgə(r)/ *adj.* (*US* **meager**) **1** lacking in amount or quality (*a meagre salary*). **2** (of literary composition, ideas, etc.) lacking fullness, unsatisfying. **3** (of a person or animal) lean, thin. □ **meagrely** *adv.* **meagreness** *n.* [ME f. AF *megre*, OF *maigre* f. L *macer*]

meal[1] /miːl/ *n.* **1** an occasion when food is eaten. **2** the food eaten on one occasion. □ **make a meal of 1** treat (a task etc.) too laboriously or fussily. **2** consume as a meal. **meals on wheels** *Brit.* a service by which meals are delivered to old people, invalids, etc. **meal-ticket 1** a ticket entitling one to a meal, esp. at a specified place with reduced cost. **2** *colloq.* a person or thing that is a source of food or income. [OE *mæl* mark, fixed time, meal f. Gmc]

meal[2] /miːl/ *n.* **1** the edible part of a grain or pulse (usu. other than wheat) ground to powder. **2** *Sc.* oatmeal. **3** *US* maize flour. **4** any powdery substance made by grinding. □ **meal-beetle** a beetle, *Tenebrio molitor*, infesting granaries etc. **meal-worm** the larva of the meal-beetle. [OE *melu* f. Gmc]

mealie /ˈmiːlɪ/ *n.* (also **mielie**) *S. Afr.* **1** (usu. in *pl.*) maize. **2** a corn-cob. [Afrik. *mielie* f. Port. *milho* maize, millet f. L *milium*]

mealtime /ˈmiːltaɪm/ *n.* a time at which a meal is usually eaten.

mealy /ˈmiːlɪ/ *adj.* (**mealier**, **mealiest**) **1 a** of or like meal; soft and powdery. **b** containing meal. **2** (of a complexion) pale. **3** (of a horse) spotty. **4** (in full **mealy-mouthed**) not outspoken; ingratiating; afraid to use plain expressions. □ **mealy bug** a scale insect of the family Pseudococcidae, infesting vines etc., and covered with a waxy powder. □ **mealiness** *n.*

mean¹ /miːn/ *v.tr.* (*past* and *past part.* **meant** /ment/) **1 a** (often foll. by *to* + infin.) have as one's purpose or intention; have in mind (*they really mean mischief; I didn't mean to break it*). **b** (foll. by *by*) have as a motive in explanation (*what do you mean by that?*). **2** (often in *passive*) design or destine for a purpose (*mean it to be used; mean it for a stopgap; is meant to be a gift*). **3** intend to convey or indicate or refer to (a particular thing or notion) (*I mean we cannot go; I mean Richmond in Surrey*). **4** entail, involve (*it means catching the early train*). **5** (often foll. by *that* + clause) portend, signify (*this means trouble; your refusal means that we must look elsewhere*). **6** (of a word) have as its explanation in the same language or its equivalent in another language. **7** (foll. by *to*) be of some specified importance to (a person), esp. as a source of benefit or object of affection etc. (*that means a lot to me*). □ **mean business** *colloq.* be in earnest. **mean it** not be joking or exaggerating. **mean to say** really admit (usu. in *interrog.: do you mean to say you have lost it?*). **mean well** (often foll. by *to, towards, by*) have good intentions. [OE *mǣnan* f. WG, rel. to MIND]

mean² /miːn/ *adj.* **1** niggardly; not generous or liberal. **2** ignoble; uncooperative, unkind, or unfair. **3** (of a person's capacity, understanding, etc.) inferior, poor. **4** (of housing) not imposing in appearance; shabby. **5 a** malicious, ill-tempered. **b** *US* vicious or aggressive in behaviour. **6** *colloq.* skilful, formidable (*is a mean fighter*). □ **no mean** a very good (*that is no mean achievement*). **mean white** = *poor white.* □ **meanly** *adv.* **meanness** /ˈmiːnnɪs/ *n.* [OE *mǣne, gemǣne* f. Gmc]

mean³ /miːn/ *n. & adj.* ● *n.* **1 a** a condition, quality, virtue, or course of action equally removed from two opposite (usu. unsatisfactory) extremes. **2** *Math.* **a** the term or one of the terms midway between the first and last terms of an arithmetical or geometrical etc. progression (*2 and 8 have the arithmetic mean 5 and the geometric mean 4*). **b** the quotient of the sum of several quantities and their number, the average. ● *adj.* **1** (of a quantity) equally far from two extremes. **2** calculated as a mean. □ **mean free path** the average distance travelled by a gas molecule etc. between collisions. **mean sea level** the sea level halfway between the mean levels of high and low water. **mean sun** an imaginary sun moving in the celestial sphere at the mean rate of the real sun, used in calculating solar time. **mean time** the time based on the movement of the mean sun. [ME f. AF *meen* f. OF *meien, moien* f. L *medianus* MEDIAN]

meander /mɪˈændə(r)/ *v. & n.* ● *v.intr.* **1** wander at random. **2** (of a stream) wind about. ● *n.* **1 a** a curve in a winding river etc. **b** a crooked or winding path or passage. **2** a circuitous journey. **3** an ornamental pattern of lines winding in and out; a fret. □ **meandering** *adj. & n.* [L *maeander*: see MENDERES]

meanie /ˈmiːnɪ/ *n.* (also **meany**) (*pl.* **-ies**) *colloq.* a mean, niggardly, or small-minded person.

meaning /ˈmiːnɪŋ/ *n. & adj.* ● *n.* **1** what is meant by a word, action, idea, etc. **2** significance. **3** importance. ● *adj.* expressive, significant (*a meaning glance*). □ **meaningly** *adv.*

meaningful /ˈmiːnɪŋfʊl/ *adj.* **1** full of meaning; significant. **2** *Logic* able to be interpreted. □ **meaningfully** *adv.* **meaningfulness** *n.*

meaningless /ˈmiːnɪŋlɪs/ *adj.* having no meaning or significance. □ **meaninglessly** *adv.* **meaninglessness** *n.*

means /miːnz/ *n.pl.* **1** (often treated as *sing.*) that by which a result is brought about (*a means of quick travel*). **2 a** money resources (*live beyond one's means*). **b** wealth (*a man of means*). □ **by all means** (or **all manner of means**) **1** certainly. **2** in every possible way. **3** at any cost. **by means of** by the agency or instrumentality of (a thing or action). **by no means** (or **no manner of means**) not at all; certainly not. **means test** an official inquiry to establish need before financial assistance from public funds is given. [pl. of MEAN³]

meant *past* and *past part.* of MEAN¹.

meantime /ˈmiːntaɪm/ *adv. & n.* ● *adv.* = MEANWHILE. ¶ Less common than *meanwhile.* ● *n.* the intervening period (esp. *in the meantime*). [MEAN³ + TIME]

meanwhile /ˈmiːnwaɪl/ *adv. & n.* ● *adv.* **1** in the intervening period of time. **2** at the same time. ● *n.* the intervening period (esp. *in the meanwhile*). [MEAN³ + WHILE]

meany var. of MEANIE.

measles /ˈmiːz(ə)lz/ *n.pl.* (also treated as *sing.*) **1 a** an acute infectious viral disease marked by red spots on the skin. **b** the spots of measles.

2 a disease of pigs etc. caused by infestation with encysted larvae of the human tapeworm. [ME *masele(s)* prob. f. MLG *masele*, MDu. *masel* pustule (cf. Du. *mazelen* measles), OHG *masala*: change of form prob. due to assim. to ME *meser* leper]

measly /ˈmiːzlɪ/ *adj.* (**measlier, measliest**) **1** *colloq.* inferior, contemptible, worthless. **2** *colloq. derog.* ridiculously small in size, amount, or value. **3** of or affected with measles. **4** (of pork etc.) infested with encysted larvae of the human tapeworm. [MEASLES + -Y¹]

measurable /ˈmeʒərəb(ə)l/ *adj.* that can be measured. □ **within a measurable distance of** getting near (something undesirable). □ **measurably** *adv.* **measurability** /ˌmeʒərəˈbɪlɪtɪ/ *n.* [ME f. OF *mesurable* f. LL *mensurabilis* f. L *mensurare* (as MEASURE)]

measure /ˈmeʒə(r)/ *n. & v.* ● *n.* **1** a size or quantity found by measuring. **2** a system of measuring (*liquid measure; linear measure*). **3** a rod or tape etc. for measuring. **4** a vessel of standard capacity for transferring or determining fixed quantities of liquids etc. (*a pint measure*). **5 a** the degree, extent, or amount of a thing. **b** (foll. by *of*) some degree of (*there was a measure of wit in her remark*). **6** a unit of capacity, e.g. a bushel (*20 measures of wheat*). **7** a factor by which a person or thing is reckoned or evaluated (*their success is a measure of their determination*). **8** (usu. in *pl.*) suitable action to achieve some end (*took measures to ensure a good profit*). **9** a legislative enactment. **10** a quantity contained in another an exact number of times. **11** a prescribed extent or quantity. **12** *Printing* the width of a page or column of type. **13 a** poetical rhythm; metre. **b** a metrical group of a dactyl or two iambuses, trochees, spondees, etc. **14** *US Mus.* a bar or the time-content of a bar. **15** *archaic* a dance. **16** a mineral stratum (*coal measures*). ● *v.* **1** *tr.* ascertain the extent or quantity of (a thing) by comparison with a fixed unit or with an object of known size. **2** *intr.* be of a specified size (*it measures six inches*). **3** *tr.* ascertain the size and proportion of (a person) for clothes. **4** *tr.* estimate (a quality, person's character, etc.) by some standard or rule. **5** *tr.* (often foll. by *off*) mark (a line etc. of a given length). **6** *tr.* (foll. by *out*) deal or distribute (a thing) in measured quantities. **7** *tr.* (foll. by *with, against*) bring (oneself or one's strength etc.) into competition with. **8** *tr. poet.* traverse (a distance). □ **beyond measure** excessively. **for good measure** as something beyond the minimum; as a finishing touch. **in a** (or **some**) **measure** partly. **made to measure** see MAKE. **measure up 1 a** determine the size etc. of by measurement. **b** take comprehensive measurements. **2** (often foll. by *to*) have the necessary qualifications (for). **measuring-jug** (or **-cup**) a jug (or cup) marked to measure its contents. **measuring tape** a tape marked to measure length. [ME f. OF *mesure* f. L *mensura* f. *metiri mens-* measure]

measured /ˈmeʒəd/ *adj.* **1** rhythmical; regular in movement (*a measured tread*). **2** (of language) carefully considered. □ **measuredly** *adv.*

measureless /ˈmeʒəlɪs/ *adj.* not measurable; infinite. □ **measurelessly** *adv.*

measurement /ˈmeʒəmənt/ *n.* **1** the act or an instance of measuring. **2** an amount determined by measuring. **3** (in *pl.*) detailed dimensions.

meat /miːt/ *n.* **1** the flesh of animals (esp. mammals) as food. **2** (foll. by *of*) the essence or chief part of. **3** the edible part of fruits, nuts, eggs, shellfish, etc. **4** *archaic* food of any kind. **b** a meal. □ **meat and drink to** a source of great pleasure to. **meat-axe** a butcher's cleaver. **meat-fly** (*pl.* **-flies**) a fly that breeds in meat. **meat loaf** minced or chopped meat moulded into the shape of a loaf and baked. **meat safe** a cupboard for storing meat, usu. of wire gauze etc. □ **meatless** *adj.* [OE *mete* food f. Gmc]

meatball /ˈmiːtbɔːl/ *n.* minced meat compressed into a small round ball.

Meath /miːθ/ a county in the eastern part of the Republic of Ireland, in the province of Leinster; county town, Navan.

meatus /mɪˈeɪtəs/ *n.* (*pl.* same or **meatuses**) *Anat.* a channel or passage in the body or its opening, esp. that leading into the ear. [L, = passage f. *meare* flow, run]

meaty /ˈmiːtɪ/ *adj.* (**meatier, meatiest**) **1** full of meat; fleshy. **2** of or like meat. **3** full of substance. □ **meatily** *adv.* **meatiness** *n.*

Mecca¹ /ˈmekə/ (Arabic **Makkah** /ˈmækɑː/) a city in western Saudi Arabia, an oasis town in the Red Sea region of Hejaz, east of Jiddah; pop. (est. 1986) 618,000. The birthplace in AD 570 of the prophet Muhammad, it was the scene of his early teachings and his expulsion to Medina in 622 (see HEGIRA). On Muhammad's return to Mecca in 630 it became the centre of the new Muslim faith. Considered by Muslims to be the holiest city of Islam, it is the site of the Great Mosque and the Kaaba, and is a centre of Islamic ritual, including the hajj pilgrimage which leads thousands of visitors to the city each year.

Mecca[2] /ˈmekə/ n. **1** a place which attracts people (*a Mecca for gardeners*). **2** the centre or birthplace of a faith, policy, pursuit, etc. [MECCA[1]]

mechanic /mɪˈkænɪk/ n. a skilled worker, esp. one who makes or uses or repairs machinery. [ME (orig. as adj.) f. OF *mecanique* or L *mechanicus* f. Gk *mēkhanikos* (as MACHINE)]

mechanical /mɪˈkænɪk(ə)l/ adj. & n. ● adj. **1** of or relating to machines or mechanisms. **2** working or produced by machinery. **3** (of a person or action) like a machine; automatic; lacking originality. **4 a** (of an agency, principle, etc.) belonging to mechanics. **b** (of a theory etc.) explaining phenomena by the assumption of mechanical action. **5** of or relating to mechanics as a science. ● n. (in pl.) = MECHANICS 3. □ **mechanical advantage** the ratio of exerted to applied force in a machine. **mechanical drawing** a scale drawing of machinery etc. done with precision instruments. **mechanical engineer** an expert in mechanical engineering. **mechanical engineering** the branch of engineering dealing with the design, construction, and repair of machines. **mechanical equivalent of heat** the conversion factor between heat energy and mechanical energy. □ **mechanicalism** n. (in sense 4). **mechanically** adv. **mechanicalness** n. [ME f. L *mechanicus* (as MECHANIC)]

mechanician /ˌmekəˈnɪʃ(ə)n/ n. a person skilled in constructing machinery.

mechanics /mɪˈkænɪks/ n.pl. (usu. treated as *sing.*) **1** the branch of applied mathematics dealing with motion and tendencies to motion. **2** the science of machinery. **3** the method of construction or routine operation of a thing (*the mechanics of a lawnmower; the mechanics of government*).

mechanism /ˈmekəˌnɪz(ə)m/ n. **1** the structure or adaptation of parts of a machine. **2** a system of mutually adapted parts working together in or as in a machine. **3** the mode of operation of a process. **4** a means (*defence mechanism; no mechanism for complaints*). **5** *Art* mechanical execution; technique. **6** *Philos.* the doctrine that all natural phenomena, including life, allow mechanical explanation by physics and chemistry. [mod.L *mechanismus* f. Gk (as MACHINE)]

mechanist /ˈmekənɪst/ n. **1** a mechanician. **2** an expert in mechanics. **3** *Philos.* a person who holds the doctrine of mechanism. □ **mechanistic** /ˌmekəˈnɪstɪk/ adj. **mechanistically** adv.

mechanize /ˈmekəˌnaɪz/ v.tr. (also **-ise**) **1** give a mechanical character to. **2** introduce machines in. **3** *Mil.* equip with tanks, armoured cars, etc. (orig. as a substitute for horse-drawn vehicles and cavalry). □ **mechanizer** n. **mechanization** /ˌmekənaɪˈzeɪʃ(ə)n/ n.

mechano- /ˈmekənəʊ/ comb. form mechanical. [Gk *mēkhano-* f. *mēkhanē* machine]

mechanoreceptor /ˌmekənəʊrɪˈseptə(r)/ n. *Biol.* a sensory receptor that responds to mechanical stimuli such as touch or sound.

mechatronics /ˌmekəˈtrɒnɪks/ n. technology combining electronics and mechanical engineering, esp. in developing new manufacturing techniques. [*mechanics* + *electronics*]

Mechelen /ˈmexələn/ (called in French *Malines*) a city in northern Belgium, north of Brussels; pop. (1991) 75,310. It is noted for its cathedral, which contains a painting of the Crucifixion by the Flemish painter Van Dyck (1599–1641). It was formerly known in English as Mechlin, lending its name to the Mechlin lace made there.

Mechlin /ˈmeklɪn/ (in full **Mechlin lace**) lace made at Mechlin (now Mechelen or Malines) in Belgium.

Mecklenburg /ˈmeklənˌbɜːg/ a former state of NE Germany, on the Baltic coast, now part of Mecklenburg-West Pomerania. Inhabited originally by Germanic tribes, it was occupied in about AD 600 by Slavonic peoples but was reclaimed by Henry the Lion, Duke of Saxony, in 1160. It was divided in the 16th and 17th centuries into two duchies, Mecklenburg-Schwerin and Mecklenburg-Strelitz, which were reunited as the state of Mecklenburg in 1934. The region was part of the German Democratic Republic between 1949 and 1990.

Mecklenburg-West Pomerania a state of NE Germany, on the coast of the Baltic Sea; capital, Schwerin. The modern state consists of the former state of Mecklenburg and the western part of Pomerania.

M.Econ. abbr. Master of Economics.

meconium /mɪˈkəʊnɪəm/ n. *Med.* a dark substance forming the first faeces of a newborn infant. [L, lit. 'poppy juice', f. Gk *mēkōnion* f. *mēkōn* poppy]

M.Ed. abbr. Master of Education.

Med /med/ n. *colloq.* the Mediterranean Sea. [abbr.]

med. abbr. medium.

medal /ˈmed(ə)l/ n. a piece of metal, usu. in the form of a disc, struck or cast with an inscription or device to commemorate an event etc., or awarded as a distinction to a soldier, scholar, athlete, etc., for services rendered, for proficiency, etc. □ **medal play** *Golf* = stroke play. □ **medalled** adj. **medallic** /mɪˈdælɪk/ adj. [F *médaille* f. It. *medaglia* ult. f. L *metallum* METAL]

medallion /mɪˈdæljən/ n. **1** a large medal. **2** a thing shaped like this, e.g. a decorative panel or tablet, portrait, etc. [F *médaillon* f. It. *medaglione* augment. of *medaglia* (as MEDAL)]

medallist /ˈmed(ə)lɪst/ n. (*US* **medalist**) **1** a recipient of a (specified) medal (*gold medallist*). **2** an engraver or designer of medals.

Medan /ˈmedɑːn/ a city in Indonesia, in NE Sumatra near the Strait of Malacca; pop. (1980) 1,378,950. It was established as a trading post by the Dutch in 1682, becoming the capital of the region and a leading commercial centre.

Medawar /ˈmedəwə(r)/, Sir Peter (Brian) (1915–87), English immunologist and author. He studied the biology of tissue transplantation, and his early work showed that the rejection of grafts was the result of an immune mechanism. His subsequent discovery of the acquired tolerance of grafts encouraged the early attempts at human organ transplantation. In later life he wrote a number of popular books on the philosophy of science, notably *The Limits of Science* (1985). Medawar shared a Nobel Prize in 1960.

meddle /ˈmed(ə)l/ v.intr. (often foll. by *with, in*) interfere in or busy oneself unduly with others' concerns. □ **meddler** n. [ME f. OF *medler*, var. of *mesler* ult. f. L *miscere* mix]

meddlesome /ˈmed(ə)lsəm/ adj. fond of meddling; interfering. □ **meddlesomely** adv. **meddlesomeness** n.

Mede /miːd/ n. a member of an ancient Indo-European people that inhabited Media. In the 7th–6th centuries BC the Medes ruled an empire that included most of modern Iran and extended to Cappadocia and Syria; it passed into Persian control after the defeat of King Astyages by Cyrus in 550 BC, although the Medes held a prominent position in Cyrus's Achaemenid empire. □ **Median** adj. [ME f. L *Medi* (pl.) f. Gk *Mēdoi*]

Medea /mɪˈdiːə/ *Gk Mythol.* a sorceress, daughter of Aeetes king of Colchis. She helped Jason to obtain the Golden Fleece, and married him, but was deserted in Corinth and avenged herself by killing their two children. [Gk, = cunning]

Medellín /ˌmedeɪˈjiːn/ a city in eastern Colombia, the second largest city in the country; pop. (1985) 1,468,000. A major centre of coffee production, it has in recent years gained a reputation as a centre for cocaine production and the hub of the Colombian drug trade.

Media /ˈmiːdɪə/ an ancient region of Asia to the south-west of the Caspian Sea, corresponding approximately to present-day Azerbaijan, NW Iran, and NE Iraq. The region is roughly the same as that inhabited today by the Kurds. Originally inhabited by the Medes, the region was conquered in 550 BC by Cyrus the Great of Persia.

media[1] /ˈmiːdɪə/ n.pl. **1** pl. of MEDIUM. **2** (usu. prec. by *the*) the main means of mass communication (esp. newspapers and broadcasting) regarded collectively. ¶ Use as a mass noun with a singular verb is common (e.g. *the media is on our side*), but is generally disfavoured (cf. AGENDA, DATA). □ **media event** an event primarily intended to attract publicity.

media[2] /ˈmiːdɪə/ n. (pl. **mediae** /-dɪˌiː/) **1** *Phonet.* a voiced stop, e.g. g, b, d. **2** *Anat.* a middle layer of the wall of an artery or other vessel. [L, fem. of *medius* middle]

mediaeval var. of MEDIEVAL.

medial /ˈmiːdɪəl/ adj. **1** situated in the middle. **2** of average size. □ **medially** adv. [LL *medialis* f. L *medius* middle]

median /ˈmiːdɪən/ adj. & n. ● adj. situated in the middle. ● n. **1** *Anat.* a median artery, vein, nerve, etc. **2** *Geom.* a straight line drawn from any vertex of a triangle to the middle of the opposite side. **3** *Math.* the middle value of a series of values arranged in order of size. □ **medianly** adv. [F *médiane* or L *medianus* (as MEDIAL)]

mediant /ˈmiːdɪənt/ n. *Mus.* the third note of a diatonic scale of any key. [F *médiante* f. It. *mediante* part. of obs. *mediare* come between, f. L (as MEDIATE)]

mediastinum /ˌmiːdɪəˈstaɪnəm/ n. (pl. **mediastina** /-nə/) *Anat.* a membranous middle septum, esp. between the lungs. □ **mediastinal**

adj. [mod.L f. med.L *mediastinus* medial, after L *mediastinus* drudge f. *medius* middle]

mediate *v. & adj.* ● *v.* /ˈmiːdɪˌeɪt/ **1** *intr.* (often foll. by *between*) intervene (between parties in a dispute) to produce agreement or reconciliation. **2** *tr.* be the medium for bringing about (a result) or for conveying (a gift etc.). **3** *tr.* form a connecting link between. ● *adj.* /ˈmiːdɪət/ **1** connected not directly but through some other person or thing. **2** involving an intermediate agency. □ **mediately** /-dɪətlɪ/ *adv.* **mediator** /-dɪˌeɪtə(r)/ *n.* **mediatory** /-dɪətərɪ/ *adj.* **mediation** /ˌmiːdɪˈeɪʃ(ə)n/ *n.* [LL *mediare mediat-* f. L *medius* middle]

medic[1] /ˈmedɪk/ *n. colloq.* a medical practitioner or student. [L *medicus* physician f. *mederi* heal]

medic[2] var. of MEDICK.

medicable /ˈmedɪkəb(ə)l/ *adj.* admitting of remedial treatment. [L *medicabilis* (as MEDICATE)]

Medicaid /ˈmedɪˌkeɪd/ *n.* (in the US) a Federal system of health insurance for those requiring financial assistance. (See also MEDICARE 1.) [MEDICAL + AID]

medical /ˈmedɪk(ə)l/ *adj. & n.* ● *adj.* **1** of or relating to the science of medicine in general. **2** of or relating to conditions requiring medical and not surgical treatment (*medical ward*). ● *n. colloq.* a medical examination. □ **medical certificate** a certificate of fitness or unfitness to work etc. **medical examination** an examination to determine a person's physical fitness. **medical jurisprudence** the law relating to medicine. **medical officer** *Brit.* a person in charge of the health services of a local authority or other organization. **medical practitioner** a physician or surgeon. □ **medically** *adv.* [F *médical* or med.L *medicalis* f. L *medicus*: see MEDIC[1]]

medicament /mɪˈdɪkəmənt, ˈmedɪ-/ *n.* a substance used for medical treatment. [F *médicament* or L *medicamentum* (as MEDICATE)]

Medicare /ˈmedɪˌkeə(r)/ **1** (in the US) a Federal system of health insurance for persons aged 65 and over, introduced in 1965 by President Johnson in a series of reforms which also included Medicaid. **2** (in Canada and Australia) a national health care scheme financed by taxation. [MEDICAL + CARE]

medicate /ˈmedɪˌkeɪt/ *v.tr.* **1** treat medically. **2** impregnate with a medicinal substance (*medicated shampoo*). □ **medicative** /-kətɪv/ *adj.* [L *medicari medicat-* administer remedies to f. *medicus*: see MEDIC[1]]

medication /ˌmedɪˈkeɪʃ(ə)n/ *n.* **1** a substance used for medical treatment. **2** treatment using drugs.

Medicean /ˌmedɪˈtʃiːən/ *adj.* of or relating to the Medici family. [mod.L *Mediceus* f. It. MEDICI]

Medici /ˈmedɪtʃɪ, məˈdiːtʃɪ/ (also **de' Medici** /də/) a powerful and influential Italian family of bankers and merchants whose members effectively ruled Florence for much of the 15th century and from 1569 were grand dukes of Tuscany. Cosimo and Lorenzo de' Medici were notable rulers and patrons of the arts in Florence; the family also provided four popes (including Leo X) and two queens of France (Catherine de' Medici and Marie de Médicis).

medicinal /mɪˈdɪsɪn(ə)l/ *adj. & n.* ● *adj.* (of a substance) having healing properties. ● *n.* a medicinal substance. □ **medicinally** *adv.* [ME f. OF f. L *medicinalis* (as MEDICINE)]

medicine /ˈmedsɪn, -dɪsɪn/ *n.* **1** the science or practice of the diagnosis, treatment, and prevention of disease, esp. as distinct from surgical methods. **2** a drug or preparation used for the treatment or prevention of disease, esp. one taken by mouth. **3** a spell, charm, or fetish which is thought to cure afflictions. □ **a dose** (or **taste**) **of one's own medicine** treatment such as one is accustomed to giving others. **medicine ball** a large heavy ball thrown and caught for exercise. **medicine chest** a box containing medicines etc. **medicine man** a person believed to have magical powers of healing, esp. among American Indians. **take one's medicine** submit to something disagreeable. [ME f. OF *medecine* f. L *medicina* f. *medicus*: see MEDIC[1]]

Médicis, Marie de, see MARIE DE MÉDICIS.

medick /ˈmedɪk/ *n.* (also **medic**) a leguminous plant of the genus *Medicago*, esp. alfalfa. [ME f. L *medica* f. Gk *Mēdikē poa* Median grass]

medico /ˈmedɪˌkəʊ/ *n.* (*pl.* **-os**) *colloq.* = MEDIC[1]. [It. f. L (as MEDIC[1])]

medico- /ˈmedɪˌkəʊ/ *comb. form* medical; medical and (*medico-legal*). [L *medicus* (as MEDIC[1])]

medieval /ˌmedɪˈiːv(ə)l/ *adj.* (also **mediaeval**) **1** of or in the style of the Middle Ages. **2** *colloq.* old-fashioned, archaic. □ **medieval history** the history of the 5th–15th centuries. **medieval Latin** Latin of about AD 600–1500. □ **medievalism** *n.* **medievalist** *n.* **medievalize** *v.tr.* &

intr. (also **-ise**). **medievally** *adv.* [mod.L *medium aevum* f. L *medius* middle + *aevum* age]

Medina /meˈdiːnə/ (Arabic **Al Madinah** /ˌæl mæˈdiːnɑː/) a city in western Saudi Arabia, around an oasis some 320 km (200 miles) north of Mecca; pop. (est. 1981) 500,000. Formerly known as Yathrib and controlled by Jewish settlers, in AD 622 it became the refuge of Muhammad's infant Muslim community after its expulsion from Mecca until its return there in 630. It was renamed Medina, meaning 'city', by Muhammad and made the capital of the new Islamic state until it was superseded by Damascus in 661. It was Muhammad's burial place and the site of the first Islamic mosque, constructed around his tomb. It is considered by Muslims to be the second most holy city after Mecca and a visit to the prophet's tomb at Medina forms a frequent sequel to the formal pilgrimage to Mecca.

medina /meˈdiːnə/ *n.* the old Arab or non-European quarter of a North African town. [Arab., = town]

mediocre /ˌmiːdɪˈəʊkə(r)/ *adj.* **1** of middling quality, neither good nor bad. **2** second-rate. [F *médiocre* or f. L *mediocris* of middle height or degree f. *medius* middle + *ocris* rugged mountain]

mediocrity /ˌmiːdɪˈɒkrɪtɪ/ *n.* (*pl.* **-ies**) **1** the state of being mediocre. **2** a mediocre person or thing.

meditate /ˈmedɪˌteɪt/ *v.* **1** *intr.* **a** exercise the mind in (esp. religious) contemplation. **b** (usu. foll. by *on*, *upon*) focus on a subject in this manner. **2** *tr.* plan mentally; design. □ **meditator** *n.* **meditation** /ˌmedɪˈteɪʃ(ə)n/ *n.* [L *meditari* contemplate]

meditative /ˈmedɪtətɪv/ *adj.* **1** inclined to meditate. **2** indicative of meditation. □ **meditatively** *adv.* **meditativeness** *n.*

Mediterranean /ˌmedɪtəˈreɪnɪən/ *adj. & n.* ● *adj.* of or characteristic of the Mediterranean Sea; of or characteristic of the countries bordering the Mediterranean Sea or their inhabitants. ● *n.* **1** (**the Mediterranean**) the Mediterranean Sea; the countries bordering on the Mediterranean Sea. **2** a native of a country bordering on the Mediterranean Sea. [L *mediterraneus* inland f. *medius* middle + *terra* land]

Mediterranean climate *n.* a climate characterized by hot dry summers and warm wet winters.

Mediterranean Sea an almost landlocked sea between southern Europe, the north coast of Africa, and SW Asia. It is connected with the Atlantic by the Strait of Gibraltar, with the Red Sea by the Suez Canal, and with the Black Sea by the Dardanelles, the Sea of Marmara, and the Bosporus.

medium /ˈmiːdɪəm/ *n. & adj.* ● *n.* (*pl.* **media** or **mediums**) **1** the middle quality, degree, etc. between extremes (*find a happy medium*). **2** the means by which something is communicated (*the medium of sound*; *the medium of television*). **3** the intervening substance through which impressions are conveyed to the senses etc. (*light passing from one medium into another*). **4** *Biol.* the physical environment or conditions of growth, storage, or transport of a living organism (*the shape of a fish is ideal for its fluid medium*; *growing mould on the surface of a medium*). **5** an agency or means of doing something (*the medium through which money is raised*). **6** the material or form used by an artist, composer, etc. (*language as an artistic medium*). **7** the liquid (e.g. oil or gel) with which pigments are mixed for use in painting. **8** (*pl.* **mediums**) a person claiming to be in contact with the spirits of the dead and to communicate between the dead and the living. ● *adj.* **1** between two qualities, degrees, etc. **2** average; moderate (*of medium height*). □ **medium bowler** *Cricket* a bowler who bowls at a medium pace. **medium dry** (of sherry, wine, etc.) having a flavour intermediate between dry and sweet. **medium frequency** a radio frequency between 300 kHz and 3 MHz. **medium of circulation** something that serves as an instrument of commercial transactions, e.g. coin. **medium-range** (of an aircraft, missile, etc.) able to travel a medium distance. **medium wave** a radio wave of medium frequency. □ **mediumism** *n.* (in sense 8 of *n.*). **mediumship** *n.* (in sense 8 of *n.*). **mediumistic** /ˌmiːdɪəˈmɪstɪk/ *adj.* (in sense 8 of *n.*). [L, = middle, neut. of *medius*]

medlar /ˈmedlə(r)/ *n.* **1** a rosaceous tree, *Mespilus germanica*, bearing small brown apple-like fruits. **2** the hard fruit of this tree, eaten when half-rotten. [ME f. OF *medler* f. L *mespila* f. Gk *mespilē*, *-on*]

medley /ˈmedlɪ/ *n., adj., & v.* ● *n.* (*pl.* **-eys**) **1** a varied mixture; a miscellany. **2** a collection of musical items from one work or various sources arranged as a continuous whole. ● *adj. archaic* mixed; motley. ● *v.tr.* (**-eys**, **-eyed**) *archaic* make a medley of; intermix. □ **medley relay** a relay race between teams in which each member runs a different distance, swims a different stroke, etc. [ME f. OF *medlee* var. of *meslee* f. Rmc (as MEDDLE)]

Medoc /'meɪ'dɒk, 'medɒk/ *n.* a fine red claret from the Médoc region of SW France.

medulla /mɪ'dʌlə/ *n.* **1** *Anat.* the inner region of certain organs or tissues usu. when it is distinguishable from the outer region or cortex, as in hair or a kidney. **2** *Bot.* the soft internal tissue of plants. □ **medulla oblongata** /ˌɒblɒŋ'gɑːtə/ the continuation of the spinal cord within the skull, forming the lowest part of the brain stem. □ **medullary** *adj.* [L, = pith, marrow, prob. rel. to *medius* middle]

Medusa /mɪ'djuːzə/ *Gk Mythol.* one of the gorgons, the only mortal one, slain by Perseus, who cut off her head. (See also GORGON.)

medusa /mɪ'djuːzə/ *n.* (*pl.* **medusae** /-ziː/ or **medusas**) **1** a jellyfish. **2** *Zool.* a free-swimming and sexual form of a coelenterate, having tentacles round the edge of a usu. umbrella-shaped jelly-like body, e.g. a jellyfish. □ **medusan** *adj.* [L f. Gk MEDUSA]

meed /miːd/ *n. literary* or *archaic* **1** reward. **2** merited portion (of praise etc.). [OE *mēd* f. WG, rel. to Goth. *mizdō*, Gk *misthos* reward]

meek /miːk/ *adj.* **1** humble and submissive; suffering injury etc. tamely. **2** piously gentle in nature. □ **meekly** *adv.* **meekness** *n.* [ME *me(o)c* f. ON *mjúkr* soft, gentle]

meerkat /'mɪəkæt/ *n.* a small African mongoose; esp. the *grey meerkat* (*Suricata suricatta*), with grey and black stripes, living gregariously in burrows (also called *suricate*). [Du., = sea-cat]

meerschaum /'mɪəʃəm/ *n.* **1** a soft white form of hydrated magnesium silicate, chiefly found in Turkey, which resembles clay. **2** a tobacco-pipe with the bowl made from this. [G, = sea-foam f. *Meer* sea + *Schaum* foam, transl. Pers. *kef-i-daryā*, with ref. to its frothiness]

Meerut /'mɪərət/ a city in northern India, in Uttar Pradesh north-east of Delhi; pop. (1991) 850,000. It was the scene in May 1857 of the first uprising against the British in the Indian Mutiny.

meet[1] /miːt/ *v. & n.* ● *v.* (*past* and *past part.* **met** /met/) **1 a** *tr.* encounter (a person or persons) by accident or design; come face to face with. **b** *intr.* (of two or more people) come into each other's company by accident or design (*decided to meet on the bridge*). **2** *tr.* go to a place to be present at the arrival of (a person, train, etc.). **3 a** *tr.* (of a moving object, line, feature of landscape, etc.) come together or into contact with (*where the road meets the flyover*). **b** *intr.* come together or into contact (*where the sea and the sky meet*). **4 a** *tr.* make the acquaintance of (*delighted to meet you*). **b** *intr.* (of two or more people) make each other's acquaintance. **5** *intr. & tr.* come together or come into contact with for the purposes of conference, business, worship, etc. (*the committee meets every week*; *the union met management yesterday*). **6** *tr.* **a** (of a person or a group) deal with or answer (a demand, objection, etc.) (*met the original proposal with hostility*). **b** satisfy or conform with (proposals, deadlines, a person, etc.) (*agreed to meet the new terms*; *did my best to meet them on that point*). **7** *tr.* pay (a bill etc.); provide the funds required by (a cheque etc.) (*meet the cost of the move*). **8** *tr.* & (foll. by *with*) *intr.* experience, encounter, or receive (success, disaster, a difficulty, etc.) (*met their death*; *met with many problems*). **9** *tr.* oppose in battle, contest, or confrontation. **10** *intr.* (of clothes, curtains, etc.) join or fasten correctly (*my jacket won't meet*). ● *n.* **1** the assembly of riders and hounds for a hunt. **2** the assembly of competitors for various sporting activities, esp. athletics. □ **make ends meet** see END. **meet the case** be adequate. **meet the eye** (or **the ear**) be visible (or audible). **meet a person's eye** check if another person is watching and look into his or her eyes in return. **meet a person half way** make a compromise, respond in a friendly way to the advances of another person. **meet up** (often foll. by *with*) *colloq.* meet or make contact, esp. by chance. **meet with 1** see sense 8 of *v.* **2** receive (a reaction) (*met with the committee's approval*). **3** esp. *US* = sense 1a of *v.* **more in it than meets the eye** hidden qualities or complications. [OE *mētan* f. Gmc: cf. MOOT]

meet[2] /miːt/ *adj. archaic* suitable, fit, proper. □ **meetly** *adv.* **meetness** *n.* [ME *(i)mete* repr. OE *gemǣte* f. Gmc, rel. to METE[1]]

meeting /'miːtɪŋ/ *n.* **1** in senses of MEET[1]. **2** an assembly of people, esp. the members of a society, committee, etc., for discussion or entertainment. **3** = *race meeting.* **4** an assembly (esp. of Quakers) for worship. **5** the persons assembled (*address the meeting*). □ **meeting-house** a place of worship, esp. of Quakers etc.

mega /'megə/ *adj. & adv. sl.* ● *adj.* **1** excellent. **2** enormous. ● *adv.* extremely. [Gk f. as MEGA-]

mega- /'megə/ *comb. form* (abbr. **M**) **1** large. **2** denoting a factor of 1 million (10^6) in the metric system of measurement. [Gk f. *megas* great]

megabuck /'megəˌbʌk/ *n. colloq.* **1** a million dollars. **2** (in *pl.*) a huge sum of money.

megabyte /'megəˌbaɪt/ *n.* (abbr. **Mb**, **MB**) *Computing* 1,048,576 (i.e. 2^{20}) bytes as a measure of data capacity, or loosely 1,000,000.

megadeath /'megəˌdeθ/ *n.* the death of one million people (esp. as a unit in estimating the casualties of war).

Megaera /mɪ'dʒɪərə/ *Gk Mythol.* one of the Furies. [Gk, perh. = she who bewitches]

megaflop /'megəˌflɒp/ *n.* **1** *Computing* a unit of computing speed equal to one million floating-point operations per second. **2** *sl.* a complete failure.

megahertz /'megəˌhɜːts/ *n.* (*pl.* same) one million hertz, esp. as a measure of frequency of radio transmissions (symbol: **MHz**).

megalith /'megəlɪθ/ *n. Archaeol.* a large stone, esp. one placed upright as a monument or part of one. [MEGA- + Gk *lithos* stone]

megalithic /ˌmegə'lɪθɪk/ *adj. Archaeol.* made of or marked by the use of large stones.

megalo- /'megələʊ/ *comb. form* great (*megalomania*). [Gk f. *megas* megal-great]

megalomania /ˌmegələ'meɪnɪə/ *n.* **1** a mental disorder producing delusions of grandeur. **2** a passion for grandiose schemes. □ **megalomaniac** /-'meɪnɪˌæk/ *adj. & n.* **megalomaniacal** /-mə'naɪək(ə)l/ *adj.* **megalomanic** /-'mænɪk/ *adj.*

megalopolis /ˌmegə'lɒpəlɪs/ *n.* **1** a great city or its way of life. **2** an urban complex consisting of a city and its environs. □ **megalopolitan** /-lə'pɒlɪt(ə)n/ *adj. & n.* [MEGA- + Gk *polis* city]

megalosaurus /ˌmegələ'sɔːrəs/ *n.* (also **megalosaur** /'megələˌsɔː(r)/) a large flesh-eating dinosaur of the genus *Megalosaurus*, with stout hind legs and small forelimbs. [mod.L f. MEGALO- + Gk *sauros* lizard]

megaphone /'megəˌfəʊn/ *n.* a large funnel-shaped device for amplifying the sound of the voice.

megapode /'megəˌpəʊd/ *n.* (also **megapod** /-ˌpɒd/) a bird of the family Megapodidae, native to Australasia, building a large mound of debris for the incubation of its eggs, e.g. the mallee fowl and brush turkey. [mod.L *Megapodius* (genus name) formed as MEGA- + Gk *pous podos* foot]

megaron /'megəˌrɒn/ *n. Archaeol.* the central hall of a large Mycenaean house. [Gk, = hall]

megaspore /'megəˌspɔː(r)/ *n. Bot.* the larger of the two kinds of spores produced by some ferns (cf. MICROSPORE).

megastar /'megəˌstɑː(r)/ *n.* a very famous person, esp. in the world of entertainment.

megaton /'megəˌtʌn/ *n.* (also **megatonne**) a unit of explosive power equal to 1 million tons of TNT.

megavolt /'megəˌvəʊlt/ *n.* one million volts, esp. as a unit of electromotive force (symbol: **MV**).

megawatt /'megəˌwɒt/ *n.* one million watts, esp. as a measure of electrical power as generated by power stations (symbol: **MW**).

Megger /'megə(r)/ *n. Electr. propr.* an instrument for measuring electrical insulation resistance. [cf. MEGOHM]

Meghalaya /meɪ'gɑːləjə/ a small state in the extreme north-east of India, on the northern border of Bangladesh; capital, Shillong. It was created in 1970 from part of Assam.

Megiddo /mə'gɪdəʊ/ an ancient city of NW Palestine, situated to the south-east of Haifa in present-day Israel. Founded in the 4th millennium BC, the city controlled an important route linking Syria and Mesopotamia with the Jordan valley, Jerusalem, and Egypt. Its commanding location made the city the scene of many early battles, and from its name the word *Armageddon* ('hill of Megiddo') is derived. It was the scene in 1918 of the defeat of Turkish forces by the British under General Allenby.

megilp /mə'gɪlp/ *n.* (also **magilp**) a mixture of mastic resin and linseed oil, added to oil paints, much used in the 19th century. [18th c.: orig. unkn.]

megohm /'megəʊm/ *n. Electr.* one million ohms. [MEGA- + OHM]

megrim[1] /'miːgrɪm/ *n.* **1** *archaic* migraine. **2** a whim, a fancy. **3** (in *pl.*) **a** depression; low spirits. **b** staggers, vertigo in horses etc. [ME *mygrane* f. OF MIGRAINE]

megrim[2] /'megrɪm/ *n.* either of two European deep-water flatfishes, *Lepidorhombus whiffiagonis* and *Arnoglossus laterna*. [19th c.: orig. unkn.]

Meiji Tenno /ˌmeɪdʒiː 'tenəʊ/ (born Mutsuhito) (1852–1912), emperor of Japan 1868–1912. He took the name Meiji Tenno when he became emperor. His accession saw the restoration of imperial power after centuries of control by the shoguns. During his reign he encouraged Japan's rapid process of modernization and political reform and laid the foundations for the country's emergence as a major world power; the feudal system was abolished, Cabinet government introduced, and a new army and navy were established.

meiosis /maɪˈəʊsɪs/ *n.* (*pl.* **meioses** /-siːz/) **1** *Biol.* a type of cell division that results in daughter cells with half the number of chromosomes of the parent cell (cf. MITOSIS). **2** = LITOTES. □ **meiotic** /-ˈɒtɪk/ *adj.* **meiotically** *adv.* [mod.L f. Gk *meiōsis* f. *meioō* lessen f. *meiōn* less]

Meir /meɪˈɪə(r)/, Golda (born Goldie Mabovich) (1898–1978), Israeli stateswoman, Prime Minister 1969–74. Born in Ukraine, she emigrated to the US in 1907 and in 1921 moved to Palestine, where she became active in the Labour movement. Following Israel's independence she served in ministerial posts (1949–66); having left government in 1966 to build up the Labour Party from disparate socialist factions, she was elected Prime Minister in 1969, retaining her position through coalition rule until her retirement in 1974.

Meissen[1] /ˈmaɪs(ə)n/ a city in eastern Germany, in Saxony, on the River Elbe north-west of Dresden; pop. (1981) 39,280. It is famous for its porcelain, which has been made there since 1710.

Meissen[2] /ˈmaɪs(ə)n/ *n.* a kind of hard-paste porcelain made at Meissen. In Britain it is often called Dresden china.

Meistersinger /ˈmaɪstəˌsɪŋə(r), -ˈzɪŋə(r)/ *n.* (*pl.* same) a member of one of the guilds of German lyric poets and musicians. They began to be organized from 1311 and continued to flourish until the 17th century. Their technique was elaborate and they were subject to rigid regulations, as depicted in Wagner's opera *Die Meistersinger von Nürnberg* (1868). [G f. *Meister* MASTER + *Singer* SINGER (see SING)]

Meitner /ˈmaɪtnə(r)/, Lise (1878–1968), Austrian-born Swedish physicist. She was the second woman to obtain a doctorate from Vienna University, and became interested in radiochemistry. She worked in Germany with Otto Hahn, discovering the element protactinium with him in 1917, but fled the Nazis in 1938 to continue her research in Sweden. Meitner formulated the concept of nuclear fission with her nephew Otto Frisch, but, unlike him, refused to work on nuclear weapons.

Mekele /mɪˈkeɪlɪ/ the capital of Tigray province in Ethiopia; pop. (est. 1984) 62,000.

Meknès /mekˈnes/ a city in northern Morocco, in the Middle Atlas mountains west of Fez; pop. (1982) 319,800. Founded in the 10th century, it developed around a citadel of the Berber Almoravids. In the 17th century it was the residence of the Moroccan sultan.

Mekong /miːˈkɒŋ/ a river of SE Asia, which rises in Tibet and flows south-east and south for 4,180 km (2,600 miles) through southern China, Laos, Cambodia, and Vietnam to its extensive delta on the South China Sea. For part of its course it forms the boundary between Laos and its western neighbours Burma and Thailand.

Melaka /məˈlækə/ (also **Malacca**) **1** a state of Malaysia, on the SW coast of the Malay Peninsula, on the Strait of Malacca. **2** its capital and chief port; pop. (1980) 88,070. Founded *c.*1400, it was conquered by the Portugese in 1511 and played an important role in the development of trade between Europe and the East, especially China.

melamine /ˈmeləmiːn/ *n.* **1** a white crystalline compound that can be copolymerized with formaldehyde to give thermosetting resins. **2** (in full **melamine resin**) a plastic made from melamine and used esp. for laminated coatings. [*melam* (arbitrary) + AMINE]

melancholia /ˌmelənˈkəʊlɪə/ *n.* **1** a mental illness marked by depression and ill-founded fears. **2** = MELANCHOLY 1, 2. [LL: see MELANCHOLY]

melancholy /ˈmelənkəlɪ/ *n. & adj.* ● *n.* (*pl.* **-ies**) **1** a pensive sadness; dejection, depression. **2** a habitual or constitutional tendency to this. **3** *hist.* one of the four humours; black bile (see HUMOUR *n.* 5). ● *adj.* (of a person) sad, gloomy; (of a thing) saddening, depressing; (of words, a tune, etc.) expressing sadness. □ **melancholic** /ˌmelənˈkɒlɪk/ *adj.* **melancholically** *adv.* [ME f. OF *melancolie* f. LL *melancholia* f. Gk *melagkholia* f. *melas melanos* black + *kholē* bile]

Melanchthon /məˈlæŋkθɒn/, Philipp (born Philipp Schwarzerd) (1497–1560), German Protestant reformer. In 1521 he succeeded Luther as leader of the Reformation movement in Germany. Professor of Greek at Wittenberg, he helped to systematize Luther's teachings in the *Loci Communes* (1521) and drew up the Augsburg Confession (1530).

Melanesia /ˌmeləˈniːzɪə, -ˈniːʒə/ a region of the western Pacific to the south of Micronesia and west of Polynesia. Lying south of the equator,

it contains the Bismarck Archipelago, the Solomon Islands, Vanuatu, New Caledonia, Fiji, and the intervening islands. [Gk *melas* black + *nēsos* island]

Melanesian /ˌmeləˈniːzɪən, -ˈniːʒ(ə)n/ *n. & adj.* ● *n.* **1** a member of the dominant Negroid people of Melanesia. **2** any of the Malayo-Polynesian languages of this people. ● *adj.* of or relating to this people or their languages.

mélange /ˈmeɪlɒnʒ/ *n.* a mixture, a medley. [F f. *mêler* mix (as MEDDLE)]

melanin /ˈmelənɪn/ *n.* a dark brown to black pigment occurring in the hair, skin, and iris of the eye, that is responsible for tanning of the skin when exposed to sunlight. [Gk *melas melanos* black + -IN]

melanism /ˈmeləˌnɪz(ə)m/ *n.* the unusual darkening of body tissues caused by excessive production of melanin, esp. as a form of colour variation in animals. □ **melanic** /mɪˈlænɪk/ *adj.*

melanoma /ˌmeləˈnəʊmə/ *n.* (*pl.* **melanomas** or **melanomata** /ˌmeləˈnəʊmətə/) *Med.* a usu. malignant tumour of melanin-forming cells, usu. in the skin. [MELANIN]

melanosis /ˌmeləˈnəʊsɪs/ *n.* **1** = MELANISM. **2** a disorder in the body's production of melanin. □ **melanotic** /-ˈnɒtɪk/ *adj.* [mod.L f. Gk (as MELANIN)]

Melba[1] /ˈmelbə/, Dame Nellie (born Helen Porter Mitchell) (1861–1931), Australian operatic soprano. She was born near Melbourne, from which city she took her professional name. Melba gained worldwide fame with her coloratura singing.

Melba[2] /ˈmelbə/ *n.* □ **do a Melba** *Austral. sl.* **1** return from retirement. **2** make several farewell appearances. **Melba sauce** a sauce made from puréed raspberries thickened with icing sugar. **Melba toast** very thin crisp toast. **peach Melba** a dish of ice-cream and peaches with liqueur or sauce. [MELBA[1]: the culinary items were all invented by Georges-Auguste Escoffier]

Melbourne[1] /ˈmelbən, -bɔːn/ the capital of Victoria, SE Australia, on the Bass Strait opposite Tasmania; pop. (1991) 2,762,000. Founded in 1835 and named after the British prime minister William Melbourne, it became state capital in 1851 and was capital of Australia from 1901 until 1927. It is a major port and the second largest city in Australia.

Melbourne[2] /ˈmelbən, -bɔːn/ William Lamb, 2nd Viscount (1779–1848), British Whig statesman, Prime Minister 1834 and 1835–41. He was appointed Home Secretary under Lord Grey in 1830, before becoming Premier in 1834. Out of office briefly that year, he subsequently became chief political adviser to Queen Victoria after her accession in 1837. His term was marked by Chartist and anti-Corn Laws agitation.

Melchior /ˈmelkɪˌɔː(r)/ the traditional name of one of the Magi, represented as a king of Nubia.

Melchite /ˈmelkaɪt/ *n.* originally, an Eastern Christian adhering to the Orthodox faith as defined by the councils of Ephesus (431) and Chalcedon (451), and as accepted by the Byzantine emperor (see also MONOPHYSITE). Now the term is applied to a member of a group of Orthodox or Uniat Christians principally in Syria and Egypt. [f. eccl.L *Melchitae* f. Byzantine Gk *Melkhitai* repr. Syriac *malkāyā* royalists, f. *malkā* king]

Melchizedek /melˈkɪzɪˌdek/ (in the Bible) a priest and king of Salem (which is usually identified with Jerusalem). He was revered by Abraham, who paid tithes to him (Gen. 14:18).

meld[1] /meld/ *v. & n.* (also *absol.*) (in rummy, canasta, etc.) lay down or declare (one's cards) in order to score points. ● *n.* a completed set or run of cards in any of these games. [G *melden* announce]

meld[2] /meld/ *v.tr. & intr.* orig. US merge, blend, combine. [perh. f. MELT + WELD[1]]

Meleager[1] /ˌmelɪˈeɪgə(r)/ *Gk Mythol.* a hero at whose birth the Fates declared that he would die when a brand then on the fire was consumed. His mother Althaea seized the brand and kept it, but threw it back into the fire when she quarrelled with and killed her brothers in a hunting expedition, whereupon he died.

Meleager[2] /ˌmelɪˈeɪgə(r)/ (fl. 1st century BC), Greek poet. He is best known as the compiler of *Stephanos*, one of the first large anthologies of epigrams. Meleager was also the author of many epigrams of his own and of short poems on love and death.

mêlée /ˈmeleɪ/ *n.* (US **melee**) **1** a confused fight, skirmish, or scuffle. **2** a muddle. [F (as MEDLEY)]

melic /ˈmelɪk/ *adj.* (of a poem, esp. a Greek lyric) meant to be sung. [L *melicus* f. Gk *melikos* f. *melos* song]

Melilla /meˈliːjə/ a Spanish enclave on the Mediterranean coast of

Morocco; pop. (1991) 56,600 (with Ceuta). It was occupied by Spain in 1497.

melilot /'melɪˌlɒt/ n. a leguminous plant of the genus *Melilotus*, with trifoliate leaves, small flowers, and a scent of hay when dried. [(O)F *mélilot* f. L *melilotus* f. Gk *melilōtos* 'honey lotus']

meliorate /'miːlɪəˌreɪt/ v.tr. & intr. literary improve (cf. AMELIORATE). □ **meliorative** /-rətɪv/ adj. **melioration** /ˌmiːlɪə'reɪʃ(ə)n/ n. [LL *meliorare* (as MELIORISM)]

meliorism /'miːlɪəˌrɪz(ə)m/ n. a doctrine that the world may be made better by human effort. □ **meliorist** n. [L *melior* better + -ISM]

melisma /mɪ'lɪzmə/ n. (pl. **melismata** /-mətə/ or **melismas**) Mus. a group of notes sung to one syllable of text. □ **melismatic** /ˌmelɪz'mætɪk/ adj. [Gk]

melliferous /mɪ'lɪfərəs/ adj. yielding or producing honey. [L *mellifer* f. *mel* honey]

mellifluous /mɪ'lɪflʊəs/ adj. (of a voice or words) pleasing, musical, flowing. □ **mellifluence** n. **mellifluent** adj. **mellifluously** adv. **mellifluousness** n. [ME f. OF *melliflue* or LL *mellifluus* f. *mel* honey + *fluere* flow]

Mellon /'melən/, Andrew W(illiam) (1855–1937), American financier and philanthropist. He donated his considerable art collection, together with funds, to establish the National Gallery of Art in Washington, DC in 1941.

mellow /'meləʊ/ adj. & v. ● adj. **1** (of sound, colour, light) soft and rich, free from harshness. **2** (of character) softened or matured by age or experience. **3** genial, jovial. **4** slightly drunk. **5** (of fruit) soft, sweet, and juicy. **6** (of wine) well-matured, smooth. **7** (of earth) rich, loamy. ● v.tr. & intr. make or become mellow. □ **mellow out** US sl. relax. □ **mellowly** adv. **mellowness** n. [ME, perh. f. attrib. use of OE *melu*, *melw-* MEAL²]

melodeon /mɪ'ləʊdɪən/ n. (also **melodion**) **1** a small organ popular in the 19th century, similar to the harmonium. **2** a small German accordion, played esp. by folk musicians. [in sense 1, alt. of *melodium*, f. MELODY + HARMONIUM; in sense 2, perh. f. MELODY + ACCORDION]

melodic /mɪ'lɒdɪk/ adj. **1** of or relating to melody. **2** having or producing melody. □ **melodic minor** a scale with the sixth and seventh degrees raised when ascending and lowered when descending. □ **melodically** adv. [F *mélodique* f. LL *melodicus* f. Gk *melōidikos* (as MELODY)]

melodious /mɪ'ləʊdɪəs/ adj. **1** of, producing, or having melody. **2** sweet-sounding. □ **melodiously** adv. **melodiousness** n. [ME f. OF *melodieus* (as MELODY)]

melodist /'melədɪst/ n. **1** a composer of melodies. **2** a singer.

melodize /'meləˌdaɪz/ v. (also **-ise**) **1** intr. make a melody or melodies; make sweet music. **2** tr. make melodious. □ **melodizer** n.

melodrama /'meləˌdrɑːmə/ n. **1** an extravagantly sensational dramatic piece with stock heroes and villains and a strong sentimental appeal. **2** the genre of drama of this type. (See note below.) **3** exaggerated language, behaviour, or an occurrence suggestive of this. **4** hist. a play with songs interspersed and with orchestral music accompanying the action. □ **melodramatic** /ˌmelədrə'mætɪk/ adj. **melodramatically** adv. **melodramatist** /-'dræmətɪst/ n. **melodramatize** v.tr. (also **-ise**). [earlier *melodrame* f. F *mélodrame* f. Gk *melos* music + F *drame* DRAMA]

▪ Melodrama developed in the late 18th century and was influenced by the Gothic novel and by German romantic writers such as Goethe and Schiller; the term derives from the use of incidental music in spoken dramas in Germany and from the French *mélodrame*, a dumb show accompanied by music. The music gradually became less important and the setting less Gothic, and during the 19th century there was more emphasis on creating elaborate sensational effects: the genre included domestic dramas, those based on real-life or legendary crime, including the anonymous *Maria Marten* in the 1830s, and dramatizations of popular novels (as Mrs Henry Wood's *East Lynne*, 1861).

melodramatics /ˌmelədrə'mætɪks/ n.pl. melodramatic behaviour, action, or writing.

melody /'melədɪ/ n. (pl. **-ies**) **1** an arrangement of single notes in a musically expressive succession. **2** the principal part in harmonized music. **3** a musical arrangement of words. **4** sweet music, tunefulness. [ME f. OF *melodie* f. LL *melodia* f. Gk *melōidia* f. *melos* song]

melon /'melən/ n. **1** the large round fruit of various plants of the gourd family, with sweet pulpy flesh and many seeds (*honeydew melon*; *water melon*). **2** the plant producing this. **3** Zool. a mass of waxy material in the head of some toothed whales, believed to focus acoustic signals.

□ **cut the melon 1** decide a question. **2** share abundant profits among a number of people. [ME f. OF f. LL *melo -onis* abbr. of L *melopepo* f. Gk *mēlopepōn* f. *mēlon* apple + *pepōn* gourd f. *pepōn* ripe]

Melos /'miːlɒs/ (Greek **Mílos** /miləs/) a Greek island in the Aegean Sea, in the south-west of the Cyclades group. It was the centre of a flourishing civilization in the Bronze Age and is the site of the discovery in 1820 of a Hellenistic marble statue of Aphrodite (see VENUS DE MILO).

Melpomene /mel'pɒmɪnɪ/ Gk & Rom. Mythol. the Muse of tragedy. [Gk, = singer]

melt /melt/ v. & n. ● v. **1** intr. become liquefied by heat. **2** tr. change to a liquid condition by heat. **3** tr. (as **molten** adj.) (usu. of materials that require a great deal of heat to melt them) liquefied by heat (*molten lava*; *molten lead*). **4** intr. & tr. soften or liquefy by the agency of moisture; dissolve. **5** intr. **a** (of a person, feelings, the heart, etc.) be softened as a result of pity, love, etc. **b** dissolve into tears. **6** tr. soften (a person, feelings, the heart, etc.) (*a look to melt a heart of stone*). **7** intr. (usu. foll. by *into*) change or merge imperceptibly into another form or state (*night melted into dawn*). **8** intr. (often foll. by *away*) (of a person) leave or disappear unobtrusively (*melted into the background*; *melted away into the crowd*). **9** intr. (usu. as **melting** adj.) (of sound) be soft and liquid (*melting chords*). **10** intr. colloq. (of a person) suffer extreme heat (*I'm melting in this thick jumper*). ● n. **1** liquid metal etc. **2** an amount melted at any one time. **3** the process or an instance of melting. □ **melt away** disappear or make disappear by liquefaction. **melt down 1** melt (esp. metal articles) in order to reuse the raw material. **2** become liquid and lose structure (cf. MELTDOWN). **melt in the mouth** (of food) be delicious and esp. very light. **melting-point** the temperature at which any given solid will melt. **melting-pot 1** a pot in which metals etc. are melted and mixed. **2** a place where races, theories, etc. are mixed, or an imaginary pool where ideas are mixed together. **melt water** water formed by the melting of snow and ice, esp. from a glacier. □ **meltable** adj. & n. **melter** n. **meltingly** adv. [OE *meltan*, *mieltan* f. Gmc, rel. to MALT]

meltdown /'meltdaʊn/ n. **1** the melting of (and consequent damage to) a structure, esp. the overheated core of a nuclear reactor. **2** a disastrous event, esp. a rapid fall in share prices.

melton /'melt(ə)n/ n. cloth with a close-cut nap, used for overcoats etc. [*Melton Mowbray* in central England]

Melville /'melvɪl/, Herman (1819–91), American novelist and short-story writer. After first going to sea in 1839, Melville made a voyage on a whaler to the South Seas in 1841; this experience formed the basis of several novels, including *Moby Dick* (1851). He is also remembered for his novella *Billy Budd* (first published in 1924), a symbolic tale of a mutiny, which inspired Benjamin Britten's opera of the same name (1951).

member /'membə(r)/ n. **1** a person belonging to a society, team, etc. **2** (**Member**) a person formally elected to take part in the proceedings of certain organizations (*Member of Parliament*; *Member of Congress*). **3** (also attrib.) a part or branch of a political body (*member state*; *a member of the Commonwealth*). **4** a constituent portion of a complex structure (*load-bearing member*). **5** a part of a sentence, equation, group of figures, mathematical set, etc. **6 a** part or organ of the body, esp. a limb. **b** (also **male member**) the penis (cf. MEMBRUM VIRILE). **7** used in the title awarded to a person admitted to (usu. the lowest grade of) certain honours (*Member of the British Empire*). □ **membered** adj. (also in comb.). **memberless** adj. [ME f. OF *membre* f. L *membrum* limb]

membership /'membəˌʃɪp/ n. **1** being a member. **2** the number of members. **3** the body of members.

membrane /'membreɪn/ n. **1** a pliable sheetlike structure acting as a boundary, lining, or partition in an organism (*mucous membrane*). **2** a thin pliable sheet or skin of various kinds. □ **membranaceous** /ˌmembrə'neɪʃəs/ adj. **membraneous** /mem'breɪnɪəs/ adj. **membranous** /'membrənəs/ adj. [L *membrana* skin of body, parchment (as MEMBER)]

membrum virile /ˌmembrəm vɪ'raɪlɪ/ n. archaic the penis. [L, = male member]

Memel /'meːm(ə)l/ **1** the German name for KLAIPEDA. **2** a former district of East Prussia, centred on the city of Memel (Klaipeda). After the First World War it was administered by France, becoming an autonomous region of Lithuania in 1924. In 1938 it was taken by Germany but was restored to Lithuania by the Soviet Union in 1945. **3** the River Neman in its lower course (see NEMAN).

memento /mɪ'mentəʊ/ n. (pl. **-oes** or **-os**) an object kept as a reminder or a souvenir of a person or an event. [L, imper. of *meminisse* remember]

memento mori /ˈmɔːrɪ, -raɪ/ n. a warning or reminder of death (e.g. a skull). [L, = remember you must die]

Memnon /ˈmemnɒn/ Gk Mythol. an Ethiopian king who went to Troy to help Priam, his uncle, and was killed.

memo /ˈmeməʊ/ n. (pl. **-os**) colloq. a memorandum. [abbr.]

memoir /ˈmemwɑː(r)/ n. **1** a historical account or biography written from personal knowledge or special sources. **2** (in pl.) an autobiography or a written account of one's memory of certain events or people. **3 a** an essay on a learned subject specially studied by the writer. **b** (in pl.) the proceedings or transactions of a learned society (Memoirs of the American Mathematical Society). □ **memoirist** n. [F mémoire (masc.), special use of mémoire (fem.) MEMORY]

memorabilia /ˌmemərəˈbɪlɪə/ n.pl. **1** souvenirs of memorable events, people, etc. (railway memorabilia). **2** archaic memorable or noteworthy things. [L, neut. pl. (as MEMORABLE)]

memorable /ˈmemərəb(ə)l/ adj. **1** worth remembering, not to be forgotten. **2** easily remembered. □ **memorably** adv. **memorableness** n. **memorability** /ˌmemərəˈbɪlɪtɪ/ n. [ME f. F mémorable or L memorabilis f. memorare bring to mind f. memor mindful]

memorandum /ˌmeməˈrændəm/ n. (pl. **memoranda** /-də/ or **memorandums**) **1** a note or record made for future use. **2** an informal written message, esp. in business, diplomacy, etc. **3** Law a document recording the terms of a contract or other legal details. [ME f. L neut. sing. gerundive of memorare: see MEMORABLE]

memorial /mɪˈmɔːrɪəl/ n. & adj. ● n. **1** an object, institution, or custom established in memory of a person or event (the Albert Memorial). **2** (often in pl.) hist. a statement of facts as the basis of a petition etc.; a record; an informal diplomatic paper. ● adj. intending to commemorate a person or thing (memorial service). □ **memorialist** n. [ME f. OF memorial or L memorialis (as MEMORY)]

Memorial Day (in the US) a day on which those who died on active service are remembered, usu. the last Monday in May.

memorialize /mɪˈmɔːrɪəˌlaɪz/ v.tr. (also **-ise**) **1** commemorate. **2** address a memorial to (a person or body).

memoria technica /mɪˌmɔːrɪə ˈteknɪkə/ n. a system or contrivance used to assist the memory. [mod.L, = artificial memory]

memorize /ˈmeməˌraɪz/ v.tr. (also **-ise**) commit to memory, learn by heart. □ **memorizable** adj. **memorizer** n. **memorization** /ˌmeməraɪˈzeɪʃ(ə)n/ n.

memory /ˈmemərɪ/ n. (pl. **-ies**) **1** the faculty by which things are recalled to or kept in the mind. **2 a** this faculty in an individual (my memory is beginning to fail). **b** one's store of things remembered (buried deep in my memory). **3** a recollection or remembrance (the memory of better times). **4 a** the part of a computer or other electronic device in which data or instructions can be stored and from which they can be retrieved. **b** capacity for storing information in this way. **5** the remembrance of a person or thing (his mother's memory haunted him). **6 a** the reputation of a dead person (his memory lives on). **b** in formulaic phrases used of a dead sovereign etc. (of blessed memory). **7** the length of time over which the memory or memories of any given person or group extends (within living memory; within the memory of anyone still working here). **8** the act of remembering (a deed worthy of memory). □ **commit to memory** learn (a thing) so as to be able to recall it. **from memory** without verification in books etc. **in memory of** to keep alive the remembrance of. **memory bank 1** the memory device of a computer etc. **2** colloq. a place where memories are stored, esp. in a person's brain. **memory board** a detachable storage device which can be connected to a computer. **memory lane** (usu. prec. by down, along) a succession of memories or past events deliberately and sentimentally recalled. **memory mapping** Computing a technique whereby a computer treats peripheral devices as if they were located in the main memory. [ME f. OF memorie, memoire f. L memoria f. memor mindful, remembering, rel. to MOURN]

Memphis /ˈmemfɪs/ **1** an ancient city of Egypt, whose ruins are situated on the Nile about 15 km (nearly 10 miles) south of Cairo. It is thought to have been founded as the capital of the Old Kingdom of Egypt c.3100 BC by Menes, the ruler of the first Egyptian dynasty, who united the kingdoms of Upper and Lower Egypt. Associated with the god Ptah, it remained one of Egypt's principal cities even after Thebes was made the capital of the New Kingdom c.1550 BC. It is the site of the pyramids of Saqqara and Giza and the Sphinx. **2** a river port on the Mississippi in the extreme south-west of Tennessee; pop. (1990) 610,340. Founded in 1819, it was named after the ancient city on the Nile because of its river location. It was the home in the late 19th century of blues music, the scene in 1968 of the assassination of Martin Luther King, and the childhood home and burial place of Elvis Presley.

memsahib /ˈmemsɑːɪb, -sɑːb/ n. Anglo-Ind. hist. a European married woman in India, as spoken of or to by Indians. [MA'AM + SAHIB]

men pl. of MAN.

menace /ˈmenɪs/ n. & v. ● n. **1** a threat. **2** a dangerous or obnoxious thing or person. **3** joc. a pest, a nuisance. ● v.tr. & intr. threaten, esp. in a malignant or hostile manner. □ **menacer** n. **menacing** adj. **menacingly** adv. [ME ult. f. L minax -acis threatening f. minari threaten]

ménage /meɪˈnɑːʒ/ n. the members of a household. [OF manaige ult. f. L (as MANSION)]

ménage à trois /meɪˌnɑːʒ æ ˈtrwʌ/ n. (pl. **ménages à trois** pronunc. same) an arrangement in which three people live together, usu. a married couple and the lover of one of them. [F, = household of three (as MÉNAGE)]

menagerie /mɪˈnædʒərɪ/ n. **1** a collection of wild animals in captivity for exhibition etc. **2** the place where these are housed. [F ménagerie (as MÉNAGE)]

Menai Strait /ˈmenaɪ/ a channel separating Anglesey from the mainland of NW Wales. It is spanned by two bridges, the first a suspension bridge built by Thomas Telford between 1819 and 1826. The second, built by Robert Stephenson between 1846 and 1850, originally carried railway tracks within gigantic box girders but was rebuilt to a more conventional design as a combined road and rail bridge after being damaged by fire in 1970.

Menander /məˈnændə(r)/ (c.342–292 BC), Greek dramatist. He is noted as the originator of New Comedy; his comic plays, set in contemporary Greece, deal with domestic situations and capture colloquial speech patterns. The Dyskolos is his sole complete extant play; others were adapted by Terence and Plautus. (See also COMEDY.)

menaquinone /ˌmenəˈkwɪnəʊn, -kwɪˈnəʊn/ n. one of the K vitamins, produced by bacteria found in the large intestine. It is essential for the blood-clotting process. Also called vitamin K_2. [chem. name methyl-naphthoquinone]

menarche /meˈnɑːkɪ/ n. the onset of first menstruation. [mod.L formed as MENO- + Gk arkhē beginning]

Mencius /ˈmenʃɪəs/ **1** (Latinized name of Meng-tzu or Mengzi = 'Meng the Master') (c.371–c.289 BC), Chinese philosopher. He is noted for developing Confucianism; two of his central doctrines were that rulers should provide for the welfare of the people and that human nature is intrinsically good. **2** one of the Four Books of Confucianism, containing the teachings of Mencius, which formed the basis of primary and secondary education in imperial China from the 14th century.

Mencken /ˈmeŋkən/, H(enry) L(ouis) (1880–1956), American journalist and literary critic. From 1908 he boldly attacked the political and literary Establishment, championing such diverse writers as G. B. Shaw, Nietzsche, and Mark Twain. He strongly opposed the dominance of European culture in America, and in his book The American Language (1919) defended the vigour and versatility of colloquial American usage.

mend /mend/ v. & n. ● v. **1** tr. restore to a sound condition; repair (a broken article, a damaged road, torn clothes, etc.). **2** intr. regain health. **3** tr. improve (mend matters). **4** tr. add fuel to (a fire). ● n. a darn or repair in material etc. (a mend in my shirt). □ **mend one's fences** make peace with a person. **mend one's manners** improve one's behaviour. **mend or end** improve or abolish. **mend one's pace** go faster; alter one's pace to another's. **mend one's ways** reform, improve one's habits. **on the mend** improving in health or condition. □ **mendable** adj. **mender** n. [ME f. AF mender f. amender AMEND]

mendacious /menˈdeɪʃəs/ adj. lying, untruthful. □ **mendaciously** adv. **mendaciousness** n. **mendacity** /-ˈdæsɪtɪ/ n. (pl. **-ies**). [L mendax -dacis perh. f. mendum fault]

Mendel /ˈmend(ə)l/, Gregor Johann (1822–84), Moravian monk, the father of genetics. From his experiments with peas, he demonstrated that parent plants showing different characters produced hybrids exhibiting the dominant parental character, and that the hybrids themselves produced offspring in which the parental characters re-emerged unchanged and in precise ratios. After the rediscovery of his work in 1900, Mendelism was often thought, wrongly, to be the antithesis of the Darwinian theory of natural selection; in fact, Mendel had demonstrated the primary source of variability in plants and animals, on which natural selection could then operate.

Mendeleev /ˌmendəˈleɪef/, Dmitri (Ivanovich) (1834–1907), Russian chemist. He developed the periodic table, in which the chemical elements are classified according to their atomic weights in groups with similar properties. This allowed him to systematize much chemical knowledge, pinpoint elements with incorrectly assigned atomic weights, and successfully predict the discovery of several new elements. His study of gases and liquids led him to the concept of critical temperature, independently of Thomas Andrews.

mendelevium /ˌmendəˈliːvɪəm/ n. an artificial radioactive metallic chemical element (atomic number 101; symbol **Md**). A member of the actinide series, mendelevium was first produced by Glenn Seaborg and his colleagues in 1955 by bombarding einsteinium with helium ions. [MENDELEEV]

Mendelian /menˈdiːlɪən/ adj. Biol. of or relating to Mendel's theory of heredity by genes.

Mendelism /ˈmendəˌlɪz(ə)m/ n. Biol. the theory that the inheritance of particular characters is controlled by the inheritance of discrete units. These units, now called genes, occur in pairs in somatic cells and separate independently of each other at meiosis. The main principles were first outlined by Mendel, although they have been modified by later discoveries.

Mendelssohn /ˈmend(ə)ls(ə)n/, Felix (full name Jakob Ludwig Felix Mendelssohn-Bartholdy) (1809–47), German composer and pianist. As a child prodigy, Mendelssohn first performed in public at the age of 9, proceeding to compose a String Octet at 16 and the overture to *A Midsummer Night's Dream* at 17. His romantic music is known for its elegance and lightness, as well as its melodic inventiveness. Other works include the overture *Fingal's Cave* (also called *The Hebrides*; 1830–2), his Violin Concerto (1844), and the oratorio *Elijah* (1846).

Menderes /ˌmendəˈres/ a river of SW Turkey. Rising in the Anatolian plateau, it flows for some 384 km (240 miles), entering the Aegean Sea south of the Greek island of Samos. Known in ancient times as the Maeander, and noted for its winding course, it gave its name to the verb *meander*.

mendicant /ˈmendɪkənt/ adj. & n. ● adj. **1** begging. **2** (of a friar) living solely on alms. ● n. **1** a beggar. **2** a mendicant friar. □ **mendicancy** n. **mendicity** /menˈdɪsɪtɪ/ n. [L *mendicare* beg f. *mendicus* beggar f. *mendum* fault]

mending /ˈmendɪŋ/ n. **1** the action of a person who mends. **2** things, esp. clothes, to be mended.

Mendip Hills /ˈmendɪp/ (also **Mendips**) a range of limestone hills in SW England, on the borders of Avon and Somerset.

Mendoza[1] /menˈdəʊzə/ a city in western Argentina, situated in the foothills of the Andes at the centre of a noted wine-producing region; pop. (1991) 121,700.

Mendoza[2] /menˈdəʊzə/, Antonio de (c.1490–1552), Spanish colonial administrator. He served as the first viceroy of New Spain from 1535 to 1550, and did much to improve relations between Spaniards and American Indians, fostering economic development (especially in mining) and educational opportunities for both groups. From 1551 he was viceroy of Peru.

Menelaus /ˌmenɪˈleɪəs/ Gk Mythol. king of Sparta, husband of Helen and brother of Agamemnon. Helen was stolen from him by Paris, an event which provoked the Trojan War. They were reunited after the fall of Troy.

Menes /ˈmiːniːz/, Egyptian pharaoh, reigned c.3100 BC. He founded the first dynasty that ruled ancient Egypt and is traditionally held to have united Upper and Lower Egypt with Memphis as its capital.

menfolk /ˈmenfəʊk/ n.pl. **1** men in general. **2** the men of one's family.

Meng-tzu /ˈmeŋˈtsuː/ see MENCIUS 1.

Mengzi /ˈmeŋˈziː/ see MENCIUS 1.

menhaden /menˈheɪd(ə)n/ n. a large herring-like fish of the genus *Brevoortia*, of the east coast of North America, yielding valuable oil and used to make guano. [Algonquian]

menhir /ˈmenhɪə(r)/ n. Archaeol. a tall upright usu. prehistoric monumental stone. [Breton *men* stone + *hir* long]

menial /ˈmiːnɪəl/ adj. & n. ● adj. **1** (esp. of unskilled domestic work) degrading, servile. **2** usu. derog. (of a servant) domestic. ● n. **1** a menial servant. **2** a servile person. □ **menially** adv. [ME f. OF *meinee* household]

meningitis /ˌmenɪnˈdʒaɪtɪs/ n. inflammation of the meninges due to infection by viruses or bacteria. □ **meningitic** /-ˈdʒɪtɪk/ adj.

meningococcus /mɪˌnɪŋɡəʊˈkɒkəs, mɪˌnɪndʒəʊ-/ n. (pl. **meningococci** /-ˈkɒksaɪ, -ˈkɒkaɪ/) a bacterium, *Neisseria meningitidis*, involved in some forms of meningitis and cerebrospinal infection. □ **meningococcal** adj. [f. *meninges* pl. of MENINX + COCCUS]

meninx /ˈmiːnɪŋks/ n. (pl. **meninges** /mɪˈnɪndʒiːz/) (usu. in pl.) Anat. each of three membranes (the dura mater, arachnoid, and pia mater) that line the skull and vertebral canal and enclose the brain and spinal cord . □ **meningeal** /mɪˈnɪndʒɪəl/ adj. [mod.L f. Gk *mēnigx -iggos* membrane]

meniscus /mɪˈnɪskəs/ n. (pl. **menisci** /-ˈnɪsaɪ/) **1** Physics the curved upper surface of a liquid in a tube, usually concave upwards when the walls are wetted and convex when they are dry, because of the effects of surface tension. **2** a lens that is convex on one side and concave on the other. **3** Anat. a thin fibrous cartilage between the surfaces of some joints, e.g. the knee. **4** Math. a crescent-shaped figure. □ **meniscoid** adj. [mod.L f. Gk *mēniskos* crescent, dimin. of *mēnē* moon]

Mennonite /ˈmenəˌnaɪt/ n. a member of a Protestant sect originating in Friesland in the 16th century, maintaining principles similar to those of the Anabaptists, being opposed to infant baptism, the taking of oaths, military service, and the holding of civic offices. In the following centuries many emigrated in search of political freedom, first to other European countries and to Russia, later to North and South America. (See also AMISH.) [*Menno* Simons, its founder, (1496–1561)]

meno- /ˈmenəʊ/ comb. form menstruation. [Gk *mēn mēnos* month]

menology /mɪˈnɒlədʒɪ/ n. (pl. **-ies**) a calendar, esp. that of the Greek Orthodox Church, with biographies of the saints. [mod.L *menologium* f. eccl.Gk *mēnologion* f. *mēn* month + *logos* account]

Menomini /mɪˈnɒmɪnɪ/ n. & adj. ● n. (pl. same) **1** a member of an American Indian people of Wisconsin. **2** the Algonquian language of this people. ● adj. of or relating to the Menomini or their language. [Algonquian]

menopause /ˈmenəˌpɔːz/ n. **1** the ceasing of menstruation. **2** the period in a woman's life (usu. between 45 and 50) when this occurs (see also *male menopause*). □ **menopausal** /ˌmenəˈpɔːz(ə)l/ adj. [mod.L *menopausis* (as MENO-, PAUSE)]

menorah /mɪˈnɔːrə/ n. a sacred candelabrum with seven branches, used in the Temple in Jerusalem, originally that made by the craftsman Bezalel and placed in the sanctuary of the Tabernacle (Exod. 37:17 ff). It has long been a symbol of Judaism; the menorah framed by two olive branches is the emblem of the state of Israel. A similar candelabrum (usually with eight branches) is used in Jewish worship, especially during Hanukkah. [Heb., = candlestick]

Menorca see MINORCA.

menorrhagia /ˌmenəˈreɪdʒɪə/ n. Med. abnormally heavy bleeding at menstruation. [MENO- + stem of Gk *rhēgnumi* burst]

menorrhoea /ˌmenəˈrɪə/ n. Med. ordinary flow of blood at menstruation. [MENO- + Gk *rhoia* f. *rheō* flow]

Mensa /ˈmensə/ an international organization whose members are admitted on the basis of high performance in IQ tests, founded in England in 1945. [L, = table, prob. with allusion to a round table at which all members have equal status]

menses /ˈmensiːz/ n.pl. **1** blood and other materials discharged from the uterus at menstruation. **2** the time of menstruation. [L, pl. of *mensis* month]

Menshevik /ˈmenʃəˌvɪk/ n. a member of a minority faction of the Russian Socialist Party who opposed the Bolshevik policies of non-cooperation with more moderate reformers as well as disagreeing with their advocacy of revolutionary action by a small political élite. The Mensheviks, along with the other revolutionary groups, were defeated by the Bolsheviks in the power struggle following the successful overthrow of the tsar in 1917. [Russ. *Men'shevik* a member of the minority (*men'she* less)]

mens rea /menz ˈriːə, ˈreɪə/ n. criminal intent; the knowledge of wrongdoing. [L, = guilty mind]

menstrual /ˈmenstrʊəl/ adj. of or relating to the menses or menstruation. □ **menstrual cycle** the process of ovulation and menstruation in female primates. [ME f. L *menstrualis* f. *mensis* month]

menstruate /ˈmenstrʊˌeɪt/ v.intr. undergo menstruation. [LL *menstruare menstruat-* (as MENSTRUAL)]

menstruation /ˌmenstrʊˈeɪʃ(ə)n/ n. the process of discharging blood and other material from the lining of the uterus in sexually mature non-pregnant women at intervals of about one lunar month until the menopause.

menstruous /ˈmenstrʊəs/ adj. **1** of or relating to the menses. **2** menstruating. [ME f. OF *menstrueus* or LL *menstruosus* (as MENSTRUAL)]

menstruum /ˈmenstrʊəm/ n. (pl. **menstrua** /-strʊə/) archaic a solvent.

[ME f. L, neut. of *menstruus* monthly f. *mensis* month f. the alchemical parallel between transmutation into gold and the supposed action of menses on the ovum]

mensurable /ˈmensjʊrəb(ə)l/ *adj.* **1** measurable, having fixed limits. **2** *Mus.* = MENSURAL 2. [F *mensurable* or LL *mensurabilis* f. *mensurare* to measure f. L *mensura* MEASURE]

mensural /ˈmensjʊrəl/ *adj.* **1** of or involving measure. **2** *Mus.* of or involving a fixed rhythm or notes of definite duration (cf. PLAINSONG). [L *mensuralis* f. *mensura* MEASURE]

mensuration /ˌmensjʊˈreɪʃ(ə)n/ *n.* **1** measuring. **2** *Math.* the measuring of geometric magnitudes such as the lengths of lines, areas of surfaces, and volumes of solids. [LL *mensuratio* (as MENSURABLE)]

menswear /ˈmenzweə(r)/ *n.* clothes for men.

-ment /mənt/ *suffix* **1** forming nouns expressing the means or result of the action of a verb (*abridgement*; *embankment*). **2** forming nouns from adjectives (*merriment*; *oddment*). [from or after F f. L *-mentum*]

mental /ˈment(ə)l/ *adj.* **1** of or in the mind. **2** done by the mind. **3** *colloq.* **a** insane. **b** mad, crazy, angry, fanatical (*is mental about pop music*). **4** (*attrib.*) of or relating to disorders or illnesses of the mind (*mental hospital*). □ **mental age** the degree of a person's mental development expressed as an age at which the same degree is attained by an average person. **mental arithmetic** arithmetic performed in the mind. **mental cruelty** the infliction of suffering on another's mind, esp. *Law* as grounds for divorce. **mental defective** *archaic* or *offens.* a person with a mental handicap. **mental deficiency** *archaic* or *offens.* the condition of having a mental handicap. **mental illness** a disorder of the mind. **mental nurse** a nurse dealing with mentally ill patients. **mental patient** a sufferer from mental illness. **mental reservation** a qualification tacitly added in making a statement etc. □ **mentally** *adv.* [ME f. OF *mental* or LL *mentalis* f. L *mens mentis* mind]

mentalism /ˈmentəˌlɪz(ə)m/ *n.* **1** *Philos.* the theory that physical and psychological phenomena are ultimately explicable only as aspects or functions of the mind. **2** *Psychol.* the primitive tendency to personify in spirit form the forces of nature, or endow inert objects with the quality of 'soul'. □ **mentalist** *n.* **mentalistic** /ˌmentəˈlɪstɪk/ *adj.*

mentality /menˈtælɪtɪ/ *n.* (*pl.* **-ies**) **1** mental character or disposition. **2** kind or degree of intelligence. **3** what is in or of the mind.

mentation /menˈteɪʃ(ə)n/ *n.* **1** mental action. **2** state of mind. [L *mens mentis* mind]

menthol /ˈmenθɒl/ *n.* a mint-tasting organic alcohol found in oil of peppermint etc., used as a flavouring and to relieve local pain. [G f. L *mentha* MINT[1]]

mentholated /ˈmenθəˌleɪtɪd/ *adj.* treated with or containing menthol.

mention /ˈmenʃ(ə)n/ *v. & n.* ● *v.tr.* **1** refer to briefly. **2** specify by name. **3** reveal or disclose (*do not mention this to anyone*). **4** (in dispatches) award (a person) a minor honour for meritorious, usu. gallant, military service. ● *n.* **1** a reference, esp. by name, to a person or thing. **2** (in dispatches) a military honour awarded for outstanding conduct. □ **don't mention it** said in polite dismissal of an apology or thanks. **make mention** (or **no mention**) of refer (or not refer) to. **not to mention** introducing a fact or thing of secondary or (as a rhetorical device) of primary importance. □ **mentionable** *adj.* [OF f. L *mentio -onis* f. the root of *mens* mind]

mentor /ˈmentɔ:(r)/ *n.* an experienced and trusted adviser. [F f. L f. Gk *Mentōr* adviser of the young Telemachus in Homer's *Odyssey* and Fénelon's *Télémaque*]

menu /ˈmenju:/ *n.* **1 a** a list of dishes available in a restaurant etc. **b** a list of items to be served at a meal. **2** *Computing* a list of options (usu. displayed onscreen) showing the commands or facilities available. □ **menu-driven** (of a program or computer) used by making selections from menus. [F, = detailed list, f. L *minutus* MINUTE[2]]

Menuhin /ˈmenjʊɪn/, Sir Yehudi (b.1916), American-born British violinist. His performing career began as a child prodigy in 1924, when he played the Mendelssohn Violin Concerto. Menuhin received critical acclaim for his interpretations of Bach, Beethoven, and Mozart, as well as for his 1932 performance of Elgar's violin concerto, conducted by the composer. He also became a noted performer of contemporary music, including Bartók's solo violin sonata (1942). Having settled in England, he founded a school of music, named after him, in Surrey in 1962.

Menzies /ˈmenzɪz/, Sir Robert Gordon (1894–1978), Australian Liberal statesman, Prime Minister 1939–41 and 1949–66. Australia's longest-serving Prime Minister, he implemented policies resulting in fast industrial growth in the 1950s and gave impetus to the development of Australian universities. Menzies was noted for his anti-Communism, making an abortive attempt to abolish the Australian Communist Party in 1951 and actively supporting the US in the Vietnam War.

meow var. of MIAOW.

MEP *abbr.* Member of the European Parliament.

mepacrine /ˈmepəkrɪn/ *n. Brit.* quinacrine. [*methyl* + *paludism* (malaria) + *acridine*]

Mephistopheles /ˌmefɪˈstɒfɪˌli:z/ an evil spirit to whom Faust, in the German legend, sold his soul. □ **Mephistophelean** /-stəˈfi:lɪən/ *adj.* **Mephistophelian** *adj.*

mephitis /mɪˈfaɪtɪs/ *n.* **1** a noxious emanation, esp. from the earth. **2** a foul-smelling or poisonous stench. □ **mephitic** /-ˈfɪtɪk/ *adj.* [L]

-mer /mə(r)/ *comb. form* denoting a compound or molecule of a specified class, esp. a polymer (*dimer*; *isomer*; *tautomer*). [Gk *meros* part, share]

meranti /məˈrӕntɪ/ *n.* a white, red, or yellow hardwood timber from a Malaysian or Indonesian tree of the genus *Shorea*. [Malay]

mercantile /ˈmɜ:kənˌtaɪl/ *adj.* **1** of trade, trading. **2** commercial. **3** mercenary, fond of bargaining. □ **mercantile marine** shipping employed in commerce not war. [F f. It. f. *mercante* MERCHANT]

mercantilism /ˈmɜ:kəntɪˌlɪz(ə)m/ *n.* the economic theory that trade generates wealth and is stimulated by the accumulation of bullion. The theory was prevalent between 1500 and 1800, mainly in England and France. It held that exports created wealth for a nation, imports diminished it, and that imports of manufactured goods should be restricted. An implication of this theory is that trade benefits one country at the expense of another; it became discredited when Adam Smith and others showed that trade can benefit both sides. □ **mercantilist** *n.*

mercaptan /mɜ:ˈkӕptən/ *n. Chem.* = THIOL. [mod.L *mercurium captans* capturing mercury]

Mercator /mɜ:ˈkeɪtə(r)/, Gerardus (Latinized name of Gerhard Kremer) (1512–94), Flemish geographer and cartographer, resident in Germany from 1552. He is best known for inventing the system of map projection that is named after him. His world map of 1569 showed a navigable North-west Passage between Asia and America, and a large southern continent. Mercator is also credited with introducing the term *atlas* to refer to a book of maps, following the publication of his *Atlas* of part of Europe (1585).

Mercator projection *n.* (also **Mercator's projection**) a projection of a map of the world on to a cylinder so that all the parallels of latitude have the same length as the equator, first published in 1569 and used esp. for marine charts and certain climatological maps.

mercenary /ˈmɜ:sɪnərɪ/ *adj. & n.* ● *adj.* primarily concerned with money or other reward (*mercenary motives*). ● *n.* (*pl.* **-ies**) **1** a hired soldier in foreign service. **2** *derog.* someone offering his or her services to anyone prepared to pay. □ **mercenariness** *n.* [ME f. L *mercenarius* f. *merces -edis* reward]

mercer /ˈmɜ:sə(r)/ *n. Brit.* a dealer in textile fabrics, esp. silk and other costly materials. □ **mercery** *n.* (*pl.* **-ies**). [ME f. AF *mercer*, OF *mercier* ult. f. L *merx mercis* goods]

mercerize /ˈmɜ:səˌraɪz/ *v.tr.* (also **-ise**) treat (cotton fabric or thread) under tension with caustic alkali to give greater strength and impart lustre. [John *Mercer*, English dyer and chemist (1791–1866)]

merchandise /ˈmɜ:tʃənˌdaɪz/ *n. & v.* ● *n.* goods for sale. ● *v.* **1** *intr.* trade, traffic. **2** *tr.* trade or traffic in. **3** *tr.* **a** put on the market, promote the sale of (goods etc.). **b** advertise, publicize (an idea or person). □ **merchandisable** *adj.* **merchandiser** *n.* [ME f. OF *marchandise* f. *marchand*: see MERCHANT]

Merchant /ˈmɜ:tʃənt/, Ismail (b.1936), Indian film producer. In 1961 he became a partner with James Ivory in Merchant Ivory Productions and is noted for a number of films made in collaboration with Ivory, including *Shakespeare Wallah* (1965), *The Europeans* (1979), *The Bostonians* (1984), and *Howard's End* (1992).

merchant /ˈmɜ:tʃənt/ *n.* **1** a wholesale trader, esp. with foreign countries. **2** esp. *US & Sc.* a retail trader. **3** *colloq.* usu. *derog.* a person showing a partiality for a specified activity or practice (*speed merchant*). □ **merchant bank** esp. *Brit.* a bank dealing in commercial loans and finance. **merchant banker** a member of a merchant bank. **merchant marine** a nation's commercial shipping, esp. that of the US. **merchant navy** a nation's commercial shipping, esp. that of the UK. **merchant prince** a wealthy merchant. **merchant ship** = MERCHANTMAN. [ME f. OF *marchand*, *marchant* ult. f. L *mercari* trade f. *merx mercis* merchandise]

merchantable /'mɜːtʃəntəb(ə)l/ adj. saleable, marketable. [ME f. merchant (v.) f. OF marchander f. marchand: see MERCHANT]

Merchant Adventurers an English trading guild involved in trade overseas, principally with the Netherlands (and later Germany) during the 15th–18th centuries. It was established by charter in 1407 and mainly engaged in the lucrative business of exporting woollen cloth; in the 16th century it reached its peak as one of the most wealthy and influential trading guilds. It was formally disbanded in 1806.

merchantman /'mɜːtʃəntmən/ n. (pl. **-men**) a ship conveying merchandise.

Mercia /'mɜːʃɪə, 'mɜːsɪə/ a former kingdom of central England. It was established by invading Angles in the 6th century AD in the border areas between the new Anglo-Saxon settlements in the east and the Celtic regions in the west. Becoming dominant in the 8th century under King Offa, it expanded to cover an area stretching from the Humber to the south coast. Its decline began after Offa's death in 796 and in 926, when Athelstan became king of all England, it finally lost its separate identity. The name Mercia is sometimes revived in names of organizations to refer to parts of the English midlands. □ **Mercian** adj. & n. [L f. OE, = people of the marches (i.e. borders, with ref. to the border with Wales)]

merciful /'mɜːsɪfʊl/ adj. having or showing or feeling mercy. □ **mercifulness** n.

mercifully /'mɜːsɪfʊlɪ/ adv. **1** in a merciful manner. **2** (qualifying a whole sentence) fortunately (mercifully, the sun came out).

merciless /'mɜːsɪlɪs/ adj. **1** pitiless. **2** showing no mercy. □ **mercilessly** adv. **mercilessness** n.

Merckx /mɜːks/, Eddy (b.1945), Belgian racing cyclist. During his professional career he won the Tour de France five times (1969–72 and 1974). He also gained five victories in the Tour of Italy between 1968 and 1974.

mercurial /mɜːˈkjʊərɪəl/ adj. & n. ● adj. **1** (of a person) sprightly, ready-witted, volatile. **2** of or containing mercury. **3** (**Mercurial**) of the planet Mercury. ● n. a drug containing mercury. □ **mercurially** adv. **mercuriality** /-ˌkjʊərɪˈælɪtɪ/ n. [ME f. OF mercuriel or L mercurialis (as MERCURY)]

Mercury /'mɜːkjʊrɪ/ **1** Rom. Mythol. the god of eloquence, skill, trading, and thieving, herald and messenger of the gods, who was identified with Hermes. **2** Astron. the innermost planet of the solar system, orbiting 57.9 million km from the sun. With a diameter of 4,878 km it is somewhat larger than the moon, which it resembles in having a heavily cratered surface. Early theories that its rotation period was the same as its orbital period, so that it always kept the same face turned towards the sun, are now known to be incorrect; its 'day' of 58.65 days is precisely two-thirds the length of its 'year' of 87.97 days. Daytime temperatures average 170°C. There is no atmosphere, nor has the planet any satellites. [L Mercurius f. merx mercis merchandise]

mercury /'mɜːkjʊrɪ/ n. **1** a heavy silvery-white metallic chemical element (atomic number 80; symbol **Hg**), liquid at ordinary temperatures. Also called quicksilver. (See note below.) **2** a green-flowered plant of the genus Mercurialis; esp. dog's mercury, M. perenne. □ **mercury vapour lamp** a lamp in which light is produced by an electric discharge through mercury vapour. □ **mercurous** adj. **mercuric** /mɜːˈkjʊərɪk/ adj. [ME f. L Mercurius MERCURY]

▪ The element mercury was known in ancient times and was regarded by the alchemists as one of the five elementary principles of which all substances were supposed to be compounded. A rare element, its main ore is cinnabar, although it can occur in the uncombined state. The main use of mercury is in batteries, switches, lamps, and other electrical equipment. It is also used in thermometers and barometers, and in alloys (amalgams) with other metals. Mercury compounds were much used in early medicine, but this declined as their toxicity became recognized. The symbol Hg is from the Latin hydrargyrum.

mercy /'mɜːsɪ/ n. & int. ● n. (pl. **-ies**) **1** compassion or forbearance shown to enemies or offenders in one's power. **2** the quality of compassion. **3** an act of mercy. **4** (attrib.) administered or performed out of mercy or pity for a suffering person (mercy killing). **5** something to be thankful for (small mercies). ● int. expressing surprise or fear. □ **at the mercy of 1** wholly in the power of. **2** liable to danger or harm from. **have mercy on** (or **upon**) show mercy to. **mercy flight** the transporting by air of an injured or sick person from a remote area to a hospital. [ME f. OF merci f. L merces -edis reward, in LL pity, thanks]

mere[1] /mɪə(r)/ attrib.adj. (**merest**) that is solely or no more or better than what is specified (a mere boy; no mere theory). □ **mere right** Law a

right in theory. □ **merely** adv. [ME f. AF meer, OF mier f. L merus unmixed]

mere[2] /mɪə(r)/ n. archaic or poet. a lake or pond. [OE f. Gmc]

mere[3] /'merɪ/ n. a Maori war-club, esp. one made of greenstone. [Maori]

Meredith /'merəˌdɪθ/, George (1828–1909), English novelist and poet. His semi-autobiographical verse collection Modern Love (1862) describes the disillusionment of married love. Meredith's reputation rests chiefly on his novels, particularly The Egoist (1879). He is noted for his control of narrative, sharp psychological characterization, and deliberately intricate style.

merengue /məˈreŋɡeɪ/ n. (also **meringue** /məˈræŋ/) **1** a dance of Dominican and Haitian origin, with alternating long and short stiff-legged steps. **2** a piece of music for this dance, usu. in duple and triple time. [Amer. Sp., f. Haitian Creole méringue lit. 'meringue' f. F]

meretricious /ˌmerɪˈtrɪʃəs/ adj. **1** (of decorations, literary style, etc.) showily but falsely attractive. **2** of or befitting a prostitute. □ **meretriciously** adv. **meretriciousness** n. [L meretricius f. meretrix -tricis prostitute f. mereri be hired]

merganser /mɜːˈɡænsə(r)/ n. a diving fish-eating northern duck of the genus Mergus, with a long narrow serrated hooked bill (cf. GOOSANDER). Also called sawbill. [mod.L f. L mergus diver f. mergere dive + anser goose]

merge /mɜːdʒ/ v. **1** tr. & intr. (often foll. by with) **a** combine or be combined. **b** join or blend gradually. **2** intr. & tr. (foll. by in) lose or cause to lose character and identity in (something else). **3** tr. (foll. by in) embody (a title or estate) in (a larger one). □ **mergence** n. [L mergere mers- dip, plunge, partly through legal AF merger]

merger /'mɜːdʒə(r)/ n. **1** the combining of two commercial companies etc. into one. **2** a merging, esp. of one estate in another. **3** Law the absorbing of a minor offence in a greater one. [AF (as MERGE)]

Meriç see MARITSA.

Mérida /'merɪdə/ **1** a city in western Spain, on the Guadiana river, capital of Extremadura region; pop. (1991) 49,830. Founded by the Romans in 25 BC, it became the capital of the province of Lusitania. **2** a city in SE Mexico, capital of the state of Yucatán; pop. (1990) 557,340. Founded in 1542 on the site of the ancient Mayan city of T'ho, it developed in the 19th century as a centre of the trade in hemp.

meridian /məˈrɪdɪən/ n. & adj. ● n. **1** a circle passing through the celestial poles and the zenith of a given place on the earth's surface. **2 a** a circle of constant longitude, passing through a given place and the terrestrial poles. **b** the corresponding line on a map. **3** archaic the point at which a sun or star attains its highest altitude. **4** prime; full splendour. ● adj. **1** of noon. **2** of the period of greatest splendour, vigour, etc. [ME f. OF meridien or L meridianus (adj.) f. meridies midday f. medius middle + dies day]

meridional /məˈrɪdɪən(ə)l/ adj. & n. ● adj. **1** of or in the south (esp. of Europe). **2** of or relating to a meridian. ● n. an inhabitant of the south (esp. of France). [ME f. OF f. LL meridionalis irreg. f. L meridies: see MERIDIAN]

meringue[1] /məˈræŋ/ n. **1** a confection of sugar, the white of eggs, etc., baked crisp. **2** a small cake or shell of this, usu. decorated or filled with whipped cream etc. [F, of unkn. orig.]

meringue[2] var. of MERENGUE.

merino /məˈriːnəʊ/ n. (pl. **-os**) **1** (in full **merino sheep**) a variety of sheep with long fine wool. **2** a soft woollen or wool-and-cotton material like cashmere, orig. of merino wool. **3** a fine woollen yarn. [Sp., of uncert. orig.]

Merionethshire /ˌmerɪˈɒnɪθˌʃɪə(r)/ a former county of NW Wales. It became a part of Gwynedd in 1974.

meristem /'merɪˌstem/ n. Bot. a plant tissue consisting of actively dividing cells forming new tissue. □ **meristematic** /ˌmerɪstɪˈmætɪk/ adj. [Gk meristos divisible f. merizō divide f. meros part, after xylem]

merit /'merɪt/ n. & v. ● n. **1** the quality of deserving well. **2** excellence, worth. **3** (usu. in pl.) **a** a thing that entitles one to reward or gratitude. **b** esp. Law intrinsic rights and wrongs (the merits of a case). **4** (in Christian theology) good deeds as entitling to a future reward. ● v.tr. (**merited, meriting**) deserve or be worthy of (reward, punishment, consideration, etc.). □ **make a merit of** regard or represent (one's own conduct) as praiseworthy. **on its merits** with regard only to its intrinsic worth. [ME f. OF merite f. L meritum price, value, = past part. of mereri earn, deserve]

meritocracy /ˌmerɪˈtɒkrəsɪ/ n. (pl. **-ies**) **1** government by persons selected competitively according to merit. **2** a group of persons selected in this way. **3** a society governed by meritocracy. □ **meritocratic** /-təˈkrætɪk/ adj.

meritorious /ˌmerɪˈtɔːrɪəs/ adj. (of a person or act) having merit; deserving reward, praise, or gratitude. □ **meritoriously** adv. **meritoriousness** n. [ME f. L meritorius f. mereri merit- earn]

merle /mɜːl/ n. Sc. or archaic a blackbird. [ME f. F f. L merula]

Merlin /ˈmɜːlɪn/ (in Arthurian legend) a magician who aided and supported King Arthur. He is associated with various figures in Celtic mythology: Geoffrey of Monmouth (12th century) identified him with a boy (Ambrosius) who, according to Nennius in his Historia Britonum (c.796), interpreted an omen for the British king Vortigern.

merlin /ˈmɜːlɪn/ n. a small European and North American falcon, Falco columbarius, that hunts small birds. [ME f. AF merilun f. OF esmerillon augment. f. esmeril f. Frank.]

merlon /ˈmɜːlən/ n. the solid part of an embattled parapet between two embrasures. [F f. It. merlone f. merlo battlement]

Merlot /ˈmɜːləʊ/ n. **1** a variety of black grape used in wine-making. **2** a red wine made from these grapes. [F]

mermaid /ˈmɜːmeɪd/ n. a legendary sea-creature with a woman's head and upper body and a fish's tail. (See note below.) □ **mermaid's purse** the horny egg-case of a skate, ray, or shark; a sea purse. [ME f. MERE² in obs. sense 'sea' + MAID]

▪ In early use the mermaid is often identified with the siren of classical mythology, a beautiful feminine being who beguiles unsuspecting sailors to their destruction; until the 18th century there were reports of sightings in many parts of the world. The mermaid is conventionally depicted (esp. in heraldry) with long flowing golden hair and a comb in the left hand and a mirror in the right.

merman /ˈmɜːmæn/ n. (pl. **-men**) the male equivalent of a mermaid.

mero- /ˈmerəʊ/ comb. form partly, partial. [Gk meros part]

Meroe /ˈmerəʊɪ/ an ancient city on the Nile, in present-day Sudan north-east of Khartoum. Founded in c.750 BC, it was the capital of the ancient kingdom of Cush from c.590 BC until it fell to the invading Aksumites in the early 4th century AD.

meronymy /mɪəˈrɒnəmɪ/ n. (in semantics) use of a term denoting part of something to refer to the whole of it (e.g. I see several familiar faces present, where faces means people). [f. MERO- after SYNONYMY]

-merous /mərəs/ comb. form esp. Bot. having so many parts (dimerous; 5-merous). [Gk (as MERO-)]

Merovingian /ˌmerəʊˈvɪndʒɪən/ adj. & n. ● adj. of or relating to the Frankish dynasty reigning in Gaul and Germany c.500–750, whose territory was substantially extended under Clovis. The Merovingian rulers were overthrown and replaced by the Carolingians. ● n. a member of this dynasty. [F mérovingien f. med.L Merovingi f. L Meroveus name of the reputed founder]

merriment /ˈmerɪmənt/ n. **1** exuberant enjoyment; being merry. **2** mirth, fun.

merry /ˈmerɪ/ adj. (**merrier, merriest**) **1 a** joyous. **b** full of laughter or gaiety. **2** Brit. colloq. slightly drunk. □ **make merry 1** be festive; enjoy oneself. **2** (foll. by over) make fun of. **merry andrew** hist. a mountebank's assistant; a clown or buffoon. **merry thought** esp. Brit. the wishbone of a fowl. **play merry hell with** see HELL. □ **merrily** adv. **merriness** n. [OE myrige f. Gmc]

merry-go-round /ˈmerɪɡəʊˌraʊnd/ n. **1** a revolving machine with wooden horses or cars for riding on at a fair etc. **2** a cycle of bustling activities.

merrymaking /ˈmerɪˌmeɪkɪŋ/ n. festivity, fun. □ **merrymaker** n.

Mersa Matruh /ˈmɜːsə məˈtruː/ a town on the Mediterranean coast of Egypt, 250 km (156 miles) west of Alexandria; pop. (1990) 112,770.

Mersey /ˈmɜːzɪ/ a river in NW England, which rises in the Peak District of Derbyshire and flows 112 km (70 miles) to the Irish Sea near Liverpool.

Merseyside /ˈmɜːzɪˌsaɪd/ a metropolitan county of NW England; administrative centre, Liverpool.

Mersin /mɜːˈsiːn/ an industrial port in southern Turkey, on the Mediterranean south-west of Adana; pop. (1990) 422,360.

Merthyr Tydfil /ˌmɜːθə ˈtɪdvɪl/ a town in Mid Glamorgan, in the South Wales coalfield; pop. (1990) 59,300.

mesa /ˈmeɪsə/ n. an isolated flat-topped hill with steep sides, found in landscapes with horizontal strata. [Sp., lit. 'table', f. L mensa]

mésalliance /meˈzælɪəns, ˌmeɪzəˈlaɪəns/ n. a marriage with a person of a lower social position. [F (as MIS-², ALLIANCE)]

mescal /ˈmeskæl/ n. **1 a** a maguey. **b** liquor obtained from this. **2** a peyote cactus. □ **mescal button** disc-shaped dried top from a peyote cactus, eaten or chewed as an intoxicant. [Sp. mezcal f. Nahuatl mexcalli]

mescaline /ˈmeskəˌliːn/ n. (also **mescalin** /-lɪn/) a hallucinogenic alkaloid present in mescal buttons.

Mesdames pl. of MADAME.

Mesdemoiselles pl. of MADEMOISELLE.

mesembryanthemum /mɪˌzembrɪˈænθɪməm/ n. a succulent southern African plant of the genus Mesembryanthemum, having brightly coloured daisy-like flowers that only open in sunshine. [mod.L f. Gk mesēmbria noon + anthemon flower]

mesencephalon /ˌmesenˈsefəˌlɒn, -ˈkefəˌlɒn/ n. Anat. = MIDBRAIN. [Gk mesos middle + ENCEPHALON]

mesentery /ˈmesəntərɪ/ n. (pl. **-ies**) Anat. a double layer of peritoneum attaching the stomach, small intestine, pancreas, spleen, and other abdominal organs to the posterior wall of the abdomen. □ **mesenteric** /ˌmesənˈterɪk/ adj. **mesenteritis** /ˌmesˌentəˈraɪtɪs/ n. [med.L mesenterium f. Gk mesenterion (as MESO-, enteron intestine)]

mesh /meʃ/ n. & v. ● n. **1** a network fabric or structure. **2** each of the open spaces or interstices between the strands of a net or sieve etc. **3** (in pl.) **a** a network. **b** a snare. **4** (in pl.) Anat. an interlaced structure. ● v. **1** intr. (often foll. by with) (of the teeth of a wheel) be engaged (with others). **2** intr. be harmonious. **3** tr. catch in or as in a net. □ **in mesh** (of the teeth of wheels) engaged. [earlier meish etc. f. MDu. maesche f. Gmc]

Meshed see MASHHAD.

mesial /ˈmiːzɪəl/ adj. Anat. of, in, or directed towards the middle line of a body. □ **mesially** adv. [irreg. f. Gk mesos middle]

Mesmer /ˈmezmə(r)/, Franz Anton (1734–1815), Austrian physician. He had a successful practice in Vienna, where he used a number of novel treatments. Mesmer is chiefly remembered for introducing hypnotism — formerly known as animal magnetism or mesmerism — as a therapeutic technique. However, it was much steeped in archaic ideas, mysticism, and sensationalism, and Mesmer effectively retired following the critical report of a royal commission in 1784.

mesmerism /ˈmezməˌrɪz(ə)m/ n. **1** esp. hist. **a** the practice of deliberately inducing hypnosis in a person; hypnotism. **b** belief in this as a therapeutic technique. **2** fascination. □ **mesmerist** n. **mesmeric** /mezˈmerɪk/ adj. **mesmerically** adv. [MESMER]

mesmerize /ˈmezməˌraɪz/ v.tr. (also **-ise**) **1** esp. hist. hypnotize; exercise mesmerism on. **2** fascinate, spellbind. □ **mesmerizer** n. **mesmerizingly** adv. **mesmerization** /ˌmezmərəˈzeɪʃ(ə)n/ n.

mesne /miːn/ adj. Law intermediate. □ **mesne lord** hist. a lord holding an estate from a superior feudal lord. **mesne process** proceedings in a suit intervening between a primary and final process. **mesne profits** profits received from an estate by a tenant between two dates. [ME f. law F, var. of AF meen, MEAN³: cf. DEMESNE]

meso- /ˈmesəʊ, ˈmezəʊ, ˈmiːsəʊ, ˈmiːzəʊ/ comb. form middle, intermediate. [Gk mesos middle]

Meso-America /ˌmesəʊəˈmerɪkə, ˌmez-, ˌmiːs-, ˌmiːz-/ the central region of America, from central Mexico to Nicaragua, especially as a region of ancient civilizations and native cultures before the arrival of the Spanish. □ **Meso-American** adj. & n.

mesoderm /ˈmesəʊˌdɜːm, ˈmez-, ˈmiːs-, ˈmiːz-/ n. Zool. the middle layer of an embryo in early development. □ **mesodermal** /ˌmesəʊˈdɜːm(ə)l/ adj. [MESO- + Gk derma skin]

mesolithic /ˌmesəʊˈlɪθɪk, ˌmez-, ˌmiːs-, ˌmiːz-/ adj. & n. Archaeol. of, relating to, or denoting the transitional period between the palaeolithic and neolithic periods, especially in Europe, where the mesolithic falls between the end of the last glacial period (mid-9th millennium BC) and the beginnings of agriculture (see NEOLITHIC). The period is characterized by the use of microliths and the first domestication of an animal (the dog). [MESO- + Gk lithos stone]

Mesolóngion see MISSOLONGHI.

mesomorph /ˈmesəʊˌmɔːf, ˈmez-, ˈmiːs-, ˈmiːz-/ n. a person with a compact and muscular build of body (cf. ECTOMORPH, ENDOMORPH). □ **mesomorphic** /ˌmesəʊˈmɔːfɪk, ˌmez-, ˌmiːs-, ˌmiːz-/ adj. [MESO- + Gk morphē form]

meson /ˈmiːzɒn, ˈmez-/ n. Physics a subatomic particle which is intermediate in mass between electron and proton and transmits the strong interaction binding nucleons together in the atomic nucleus. □ **mesic** adj. **mesonic** /mɪˈzɒnɪk/ adj. [MESO-]

mesopause /ˈmesəʊˌpɔːz, ˈmez-, ˈmiːs-, ˈmiːz-/ n. the boundary between the mesosphere and the thermosphere in the earth's atmosphere, where the temperature stops decreasing with increasing height and starts to increase.

mesophyll /ˈmesəʊˌfɪl, ˈmez-, ˈmiːs-, ˈmiːz-/ n. Bot. the inner tissue of a leaf, containing chloroplasts. [MESO- + Gk phullon leaf]

mesophyte /ˈmesəʊˌfaɪt, ˈmez-, ˈmiːs-, ˈmiːz-/ n. a plant needing only a moderate amount of water.

Mesopotamia /ˌmesəpəˈteɪmɪə/ an ancient region of SW Asia in present-day Iraq, lying between the rivers Tigris and Euphrates. Its alluvial plains were the site of the civilizations of Akkad, Sumer, Babylonia, and Assyria. □ **Mesopotamian** adj. & n. [Gk, f. mesos middle, potamos river]

mesosphere /ˈmesəʊˌsfɪə(r), ˈmez-, ˈmiːs-, ˈmiːz-/ n. the region of the atmosphere extending about 80 km from the top of the stratosphere.

Mesozoic /ˌmesəʊˈzəʊɪk, ˌmez-, ˌmiːs-, ˌmiːz-/ adj. & n. Geol. of, relating to, or denoting the geological era between the Palaeozoic and Cenozoic, comprising the Triassic, Jurassic, and Cretaceous periods. The Mesozoic lasted from about 248 to 65 million years ago, and was a time of abundant vegetation; reptiles were dominant, although by its close they were being rapidly replaced by the mammals. The Mesozoic corresponds to the Secondary period. [MESO- + Gk zōē life]

mesquite /ˈmeskiːt/ n. (also **mesquit**) a thorny leguminous shrub or tree of the genus Prosopis, found in arid parts of North America, esp. P. glandulosa. □ **mesquite bean** a pod from the mesquite, used as fodder. [Mex. Sp. mezquite]

mess /mes/ n. & v. ● n. **1** a dirty or untidy state of things (the room is a mess). **2** a state of confusion, embarrassment, or trouble. **3** something causing a mess, e.g. spilt liquid. **4** a domestic animal's excreta. **5 a** a company of persons who take meals together, esp. in the armed forces. **b** a place where such meals or recreation take place communally. **c** a meal taken there. **6** derog. a disagreeable concoction or medley. **7** a liquid or mixed food for hounds etc. **8** a portion of liquid or pulpy food. ● v. **1** tr. (often foll. by up) **a** make a mess of; dirty. **b** muddle; make into a state of confusion. **2** intr. (foll. by with) interfere with. **3** intr. take one's meals. **4** intr. colloq. (esp. of an animal or infant) defecate. □ **make a mess of** bungle (an undertaking). **mess about** (or **around**) **1** act desultorily. **2** colloq. make things awkward for; cause arbitrary inconvenience to (a person). **mess-hall** a military dining area. **mess-jacket** a short close-fitting coat worn at the mess. **mess kit** a soldier's cooking and eating utensils. **mess of pottage** a material comfort etc. for which something higher is sacrificed (Gen. 25:29–34). **mess tin** a small container as part of a mess kit. [ME f. OF mes portion of food f. LL missus course at dinner, past part. of mittere send]

message /ˈmesɪdʒ/ n. & v. ● n. **1** an oral or written communication sent by one person to another. **2 a** an inspired or significant communication from a prophet, writer, or preacher. **b** the central import or meaning of an artistic work etc. **3** a mission or errand. **4** (in pl.) Sc. & N. Engl. things bought; shopping. ● v.tr. **1** send as a message. **2** transmit (a plan etc.) by signalling etc. □ **get the message** colloq. understand what is meant. **message stick** Austral. a stick carved with significant marks, carried as identification by Aboriginal messengers. [ME f. OF ult. f. L mittere miss- send]

Messalina /ˌmesəˈliːnə/, Valeria (also **Messallina**) (AD c.22–48), Roman empress, third wife of Claudius. She married her second cousin Claudius in about 39, and became notorious in Rome for the murders she instigated in court and for her extramarital affairs. She was executed on Claudius' orders, after the disclosure of her secret marriage with one of his political opponents.

Messeigneurs pl. of MONSEIGNEUR.

messenger /ˈmesɪndʒə(r)/ n. **1** a person who carries a message. **2** a person employed to carry messages. □ **messenger RNA** (abbr. **mRNA**) Biochem. the form of RNA in which genetic information is transcribed from DNA as a sequence of bases and transferred to a ribosome. [ME & OF messager (as MESSAGE): -n- as in harbinger, passenger, etc.]

Messiaen /ˈmesjɒn/, Olivier (Eugène Prosper Charles) (1908–92), French composer. Messiaen was organist of the church of La Trinité in Paris for more than forty years. His music shows many influences, including Greek and Hindu rhythms, birdsong, the music of Stravinsky and Debussy, and the composer's Roman Catholic faith. His Quartet for the End of Time for violin, clarinet, cello, and piano (1941) was written and first performed in a prison camp in Silesia during the Second World War. Other major works include La Nativité du Seigneur for organ (1935), the Turangalîla Symphony for large orchestra (1946–8), Catalogues

d'oiseaux for piano (1956–8), and La Transfiguration de Notre Seigneur Jésus-Christ (1969) for chorus and orchestra.

Messiah /mɪˈsaɪə/ n. **1** the promised deliverer of the Jewish nation prophesied in the Hebrew Bible; Jesus regarded by Christians as the saviour of humankind. **2** an actual or expected liberator of an oppressed people or country; a leader or saviour of a particular group, cause, etc. □ **Messiahship** n. [ME f. OF Messie ult. f. Heb. māšīaḥ anointed]

Messianic /ˌmesɪˈænɪk/ adj. **1** of the Messiah. **2** inspired by hope or belief in a Messiah. □ **Messianism** /mɪˈsaɪəˌnɪz(ə)m/ n. [F messianique (as MESSIAH) after rabbinique rabbinical]

Messieurs pl. of MONSIEUR.

Messina /meˈsiːnə/ a city in NE Sicily; pop. (1990) 274,850. Founded in 730 BC by the Greeks, it is situated on the Strait of Messina.

Messina, Strait of a channel separating the island of Sicily from the 'toe' of Italy. It forms a link between the Tyrrhenian and Ionian seas. The strait, which is 32 km (20 miles) in length, is noted for the strength of its currents. It is traditionally identified as the location of the legendary sea monster Scylla and the whirlpool Charybdis.

messmate /ˈmesmeɪt/ n. a person with whom one regularly takes meals, esp. in the armed forces.

Messrs /ˈmesəz/ pl. of MR. [abbr. of MESSIEURS]

messuage /ˈmeswɪdʒ/ n. Law a dwelling-house with outbuildings and land assigned to its use. [ME f. AF: perh. an alternative form of mesnage dwelling]

messy /ˈmesɪ/ adj. (**messier**, **messiest**) **1** untidy or dirty. **2** causing or accompanied by a mess. **3** difficult to deal with; full of awkward complications (a messy divorce). □ **messily** adv. **messiness** n.

mestizo /meˈstiːzəʊ/ n. (pl. **-os**; fem. **mestiza** /-zə/, pl. **-as**) a Spaniard or Portuguese of mixed race, esp. the offspring of a Spaniard and an American Indian. [Sp. ult. f. L mixtus past part. of miscere mix]

met[1] past and past part. of MEET[1].

met[2] /met/ adj. colloq. **1** meteorological. **2** metropolitan. **3** (**the Met**) **a** (in full **Met Office**) (in the UK) the Meteorological Office. **b** the Metropolitan Police in London. **c** the Metropolitan Opera House in New York. [abbr.]

meta- /ˈmetə/ comb. form (usu. **met-** before a vowel or h) **1** denoting change of position or condition (metabolism). **2** denoting position: **a** behind. **b** after or beyond (metaphysics; metacarpus). **c** of a higher or second-order kind (metalanguage). **3** Chem. **a** relating to two carbon atoms separated by one other carbon atom in a benzene ring. **b** relating to a compound formed by dehydration (metaphosphate). [Gk meta-, met-, meth- f. meta with, after]

metabolism /mɪˈtæbəˌlɪz(ə)m/ n. the chemical processes that occur in a living organism in order to maintain life. Two kinds of metabolism can be distinguished: constructive metabolism or anabolism, the synthesis of the proteins, carbohydrates, and fats which form tissue and store energy, and destructive metabolism or catabolism, the breakdown of complex substances and the consequent production of energy and waste matter. □ **metabolic** /ˌmetəˈbɒlɪk/ adj. **metabolically** adv. [Gk metabolē change (as META-, bolē f. ballō throw)]

metabolite /mɪˈtæbəˌlaɪt/ n. Physiol. a substance formed in or necessary for metabolism.

metabolize /mɪˈtæbəˌlaɪz/ v.tr. & intr. (also **-ise**) process or be processed by metabolism. □ **metabolizable** adj.

metacarpus /ˌmetəˈkɑːpəs/ n. (pl. **metacarpi** /-paɪ/) **1** the set of five bones of the hand that connects the wrist to the fingers. **2** this part of the hand. □ **metacarpal** adj. [mod.L f. Gk metakarpon (as META-, CARPUS)]

metacentre /ˈmetəˌsentə(r)/ n. (US **metacenter**) the point of intersection between a line (vertical in equilibrium) through the centre of gravity of a floating body and a vertical line through the centre of pressure after a slight angular displacement, which must be above the centre of gravity to ensure stability. □ **metacentric** /ˌmetəˈsentrɪk/ adj. [F métacentre (as META-, CENTRE)]

metage /ˈmiːtɪdʒ/ n. **1** the official measuring of a load of coal etc. **2** the duty paid for this. [METE[1] + -AGE]

metagenesis /ˌmetəˈdʒenɪsɪs/ n. Biol. the alternation of generations between sexual and asexual reproduction. □ **metagenetic** /-dʒɪˈnetɪk/ adj. [mod.L (as META-, GENESIS)]

metal /ˈmet(ə)l/ n., adj., & v. ● n. **1 a** any of a class of substances which are in general lustrous, malleable, ductile, fusible solids, and good conductors of heat and electricity. (See note below.) **b** material of this kind. **2** material used for making glass, in a molten state. **3** *Heraldry* gold or silver as tincture. **4** (in *pl.*) the rails of a railway line. **5** = *road-metal* (see ROAD[1]). ● *adj.* made of metal. ● *v.tr.* (**metalled, metalling**; *US* **metaled, metaling**) **1** provide or fit with metal. **2** *Brit.* make or mend (a road) with road-metal. □ **metal detector** an electronic device giving a signal when it locates metal. **metal fatigue** fatigue (see FATIGUE *n.* 2) in metal. [ME f. OF *metal* or L *metallum* f. Gk *metallon* mine]

▪ About three-quarters of the hundred or so known chemical elements are metals, of which only seven were known to the ancients – gold, silver, copper, iron, lead, tin, and mercury. Their properties, and those of their many alloys, vary quite considerably and no single combination of characteristics distinguishes them from all other substances. Metallic character arises from a distinctive kind of chemical bonding, in which each atom contributes one or more of its electrons to a so-called 'sea' of electrons able to move freely throughout the mass.

metalanguage /ˈmetəˌlæŋɡwɪdʒ/ *n.* **1** a form of language used to discuss a language. **2** a system of propositions about propositions.

metallic /mɪˈtælɪk/ *adj.* **1** of, consisting of, or characteristic of metal or metals. **2** sounding sharp and ringing, like struck metal. **3** having the sheen or lustre of metals. □ **metallically** *adv.* [L *metallicus* f. Gk *metallikos* (as METAL)]

metalliferous /ˌmetəˈlɪfərəs/ *adj.* bearing or producing metal. [L *metallifer* (as METAL, -FEROUS)]

metallize /ˈmetəˌlaɪz/ *v.tr.* (also **-ise**; *US* **metalize**) **1** render metallic; give a metallic form or appearance to. **2** coat with a thin layer of metal. □ **metallization** /ˌmetəlaɪˈzeɪʃ(ə)n/ *n.*

metallography /ˌmetəˈlɒɡrəfɪ/ *n.* the descriptive science of the structure and properties of metals. □ **metallographic** /mɪˌtæləˈɡræfɪk/ *adj.* **metallographical** *adj.* **metallographically** *adv.*

metalloid /ˈmetəˌlɔɪd/ *adj. & n.* ● *adj.* having the form or appearance of a metal. ● *n.* an element intermediate in properties between metals and non-metals, e.g. boron, silicon, and germanium.

metallurgy /mɪˈtælədʒɪ, ˈmetəˌlɜːdʒɪ/ *n.* the science concerned with the production, purification, and properties of metals and their application. □ **metallurgist** *n.* **metallurgic** /ˌmetəˈlɜːdʒɪk/ *adj.* **metallurgical** *adj.* **metallurgically** *adv.* [Gk *metallon* metal + *-ourgia* working]

metalwork /ˈmet(ə)lˌwɜːk/ *n.* **1** the art of working in metal. **2** metal objects collectively. □ **metalworker** *n.* **metalworking** *n. & attrib.adj.*

metamere /ˈmetəˌmɪə(r)/ *n. Zool.* each of several similar body segments containing the same internal structures, e.g. in an earthworm. [META- + Gk *meros* part]

metameric /ˌmetəˈmerɪk/ *adj.* **1** *Chem.* having the same proportional composition and molecular weight, but different functional groups and chemical properties. **2** *Zool.* of or relating to metameres. □ **metamerically** *adv.* **metamer** /ˈmetəmə(r)/ *n.* **metamerism** /mɪˈtæməˌrɪz(ə)m/ *n.*

metamorphic /ˌmetəˈmɔːfɪk/ *adj.* **1** of or marked by metamorphosis. **2** *Geol.* (of rock) that has undergone transformation by natural agencies such as heat and pressure. (*See note below.*) □ **metamorphism** *n.* [META- + Gk *morphē* form]

▪ Metamorphic rocks are formed on a large scale in the folding and deformation of strata (as in the formation of gneiss and schist), or on a smaller scale in the vicinity of an igneous intrusion (as in the transformation of limestone into marble, or granite into china clay). The main effect of metamorphism is recrystallization of the constituents of the rock; crushing and shearing forces can also cause structural breakdown. The term does not include alterations caused by weathering or consolidation of sediments. (See also IGNEOUS, SEDIMENTARY.)

metamorphose /ˌmetəˈmɔːfəʊz/ *v.tr.* **1** change in form. **2** (foll. by *to, into*) **a** turn (into a new form). **b** change the nature of. [F *métamorphoser* f. *métamorphose* METAMORPHOSIS]

metamorphosis /ˌmetəˈmɔːfəsɪs, -mɔːˈfəʊsɪs/ *n.* (*pl.* **metamorphoses** /-ˌsiːz/) **1** a change of form (by natural or supernatural means). **2** a changed form. **3** a change of character, conditions, etc. **4** *Zool.* the transformation (or transformations) that some animals undergo on the course to maturity, e.g. from a larva to a pupa, from a pupa to an adult insect, or from a tadpole to a frog. [L f. Gk *metamorphōsis* f. *metamorphoō* transform (as META-, *morphoō* f. *morphē* form)]

metaphase /ˈmetəˌfeɪz/ *n. Biol.* the stage of meiotic or mitotic cell division when the chromosomes become attached to the spindle fibres.

metaphor /ˈmetəˌfɔː(r)/ *n.* **1 a** the application of a name or descriptive term or phrase to an object or action to which it is imaginatively but not literally applicable (e.g. *a glaring error*). **b** an instance of this. **2** (usu. foll. by *for* or *of*) something considered as representing or symbolizing another (usu. abstract) thing. □ **metaphoric** /ˌmetəˈfɒrɪk/ *adj.* **metaphorical** *adj.* **metaphorically** *adv.* [F *métaphore* or L *metaphora* f. Gk *metaphora* f. *metapherō* transfer]

metaphrase /ˈmetəˌfreɪz/ *n. & v.* ● *n.* literal translation. ● *v.tr.* put into other words. □ **metaphrastic** /ˌmetəˈfræstɪk/ *adj.* [mod.L *metaphrasis* f. Gk *metaphrasis* f. *metaphrazō* translate]

metaphysic /ˌmetəˈfɪzɪk/ *n.* a system of metaphysics.

metaphysical /ˌmetəˈfɪzɪk(ə)l/ *adj.* **1** of or relating to metaphysics. **2** based on abstract general reasoning. **3** excessively subtle or theoretical. **4** incorporeal; supernatural. **5** visionary. □ **metaphysically** *adv.*

metaphysical poets a group of 17th-century poets whose work is characterized by the use of complex and elaborate images or conceits, often using an intellectual form of argumentation to express emotional states. Members of the group include John Donne, George Herbert, Henry Vaughan, Andrew Marvell, and Thomas Traherne. Their reputation dwindled after the Restoration with a new taste for clarity and impatience with figurative language, and it was not until after the First World War that they were reappraised.

metaphysics /ˌmetəˈfɪzɪks/ *n.pl.* (usu. treated as *sing.*) **1** the branch of philosophy that deals with the first principles of things, including such concepts as being, knowing, substance, essence, cause, identity, time, and space. (*See note below.*) **2** the philosophy of mind. **3** *colloq.* abstract or subtle talk; mere theory. □ **metaphysician** /-fɪˈzɪʃ(ə)n/ *n.* **metaphysicize** /-ˈfɪzɪˌsaɪz/ *v.intr.* (also **-ise**). [ME *metaphysic* f. OF *metaphysique* f. med.L *metaphysica* ult. f. Gk *ta meta ta phusika* the things after physics]

▪ Metaphysics, as an investigation using rational argument to determine what really exists, has two main strands: that which holds that what exists lies beyond experience (as argued e.g. by Plato), and that which holds that objects of experience constitute the only reality (as argued by Kant, the logical positivists, and Hume). Metaphysics has also concerned itself with a discussion of whether what exists is made of one substance or many, and whether what exists is inevitable or driven by chance.

metaplasia /ˌmetəˈpleɪzɪə/ *n. Physiol.* an abnormal change in the nature of a tissue. □ **metaplastic** /-ˈplæstɪk/ *adj.* [mod.L f. G *Metaplase* f. Gk *metaplasis* (as META-, *plasis* f. *plassō* to mould)]

metapsychology /ˌmetəsaɪˈkɒlədʒɪ/ *n.* the study of the nature and functions of the mind beyond what can be studied experimentally. □ **metapsychological** /-kəˈlɒdʒɪk(ə)l/ *adj.*

metastable /ˌmetəˈsteɪb(ə)l/ *adj.* **1** (of a state of equilibrium) stable only under small disturbances. **2** (of a substance etc.) technically unstable but so long-lived as to be stable for all practical purposes. □ **metastability** /-stəˈbɪlɪtɪ/ *n.*

metastasis /mɪˈtæstəsɪs/ *n.* (*pl.* **metastases** /-ˌsiːz/) *Med.* **1** the transfer of a disease etc. from one part of the body to another, esp. the development of secondary tumours at a distance from the primary site of cancer. **2** a secondary tumour. □ **metastasize** *v.intr.* (also **-ise**). **metastatic** /ˌmetəˈstætɪk/ *adj.* [LL f. Gk f. *methistēmi* change]

metatarsus /ˌmetəˈtɑːsəs/ *n.* (*pl.* **metatarsi** /-saɪ/) **1** the part of the foot between the ankle and the toes. **2** the set of bones in this. □ **metatarsal** *adj.* [mod.L (as META-, TARSUS)]

metatherian /ˌmetəˈθɪərɪən/ *n. & adj. Zool.* ● *n.* a mammal of the infraclass Metatheria, comprising the marsupials. ● *adj.* of or relating to this infraclass. [mod.L *Metatheria* f. Gk META- + Gk *thēr* wild beast]

metathesis /mɪˈtæθɪsɪs/ *n.* (*pl.* **metatheses** /-ˌsiːz/) **1** *Gram.* the transposition of sounds or letters in a word. **2** *Chem.* = *double decomposition*. **3** an instance of either of these. □ **metathetic** /ˌmetəˈθetɪk/ *adj.* **metathetical** *adj.* [LL f. Gk *metatithēmi* transpose]

metazoan /ˌmetəˈzəʊən/ *n. & adj. Zool.* ● *n.* an animal of the subkingdom Metazoa, having a multicellular body and differentiated tissues and comprising all animals except protozoa and sponges. ● *adj.* of or relating to this subkingdom. [mod.L *Metazoa* f. Gk META- + *zōia* pl. of *zōion* animal]

mete[1] /miːt/ *v.tr.* **1** (usu. foll. by *out*) *literary* apportion or allot (a punishment or reward). **2** *poet.* or *Bibl.* measure. [OE *metan* f. Gmc., rel. to MEET[1]]

mete[2] /miːt/ *n.* a boundary or boundary stone. [ME f. OF f. L *meta* boundary, goal]

metempsychosis /ˌmetəmsaɪˈkəʊsɪs/ n. (pl. **-psychoses** /-siːz/) **1** the supposed transmigration of the soul of a human being or animal at death into a new body of the same or a different species. **2** an instance of this. □ **metempsychosist** n. [LL f. Gk metempsukhōsis (as META-, EN-[2], psukhē soul)]

meteor /ˈmiːtɪə(r)/ n. Astron. a small body of rock or metal from space that enters the earth's atmosphere, becoming incandescent as a result of friction and appearing as a streak of light. Also called *shooting star*. □ **meteor shower** a group of meteors appearing to come from a particular point in the sky (the radiant), usually occurring at a particular date each year. [ME f. mod.L meteorum f. Gk meteōron neut. of meteōros lofty, (as META-, aeirō raise)]

Meteora /ˌmetiˈɔːrə/ a group of monasteries in north central Greece, in the region of Thessaly. The monasteries, built between the 12th and the 16th centuries, are perched on the summits of curiously shaped rock formations.

meteoric /ˌmiːtɪˈɒrɪk/ adj. **1 a** of, relating to, or derived from the atmosphere (*meteoric water*). **b** (of a plant) dependent on atmospheric conditions. **2** of meteors or meteorites. **3** rapid like a meteor; dazzling, transient (*meteoric rise to fame*). □ **meteorically** adv.

meteorite /ˈmiːtɪəˌraɪt/ n. a fallen meteor, of sufficient size to reach the surface of the earth without completely burning up in the atmosphere. Meteors are rarely more than a few kilograms in weight, although a small number have been much larger and have formed a crater, of which perhaps the most famous is Meteor Crater near Flagstaff, Arizona. Meteorites are broadly classified according to their composition: over 90 per cent are of rock while the remainder consist partly or wholly of iron and nickel. □ **meteoritic** /ˌmiːtɪəˈrɪtɪk/ adj.

meteorograph /ˈmiːtɪərəˌɡrɑːf/ n. an apparatus that records several meteorological phenomena at the same time. [F météorographe (as METEOR, -GRAPH)]

meteoroid /ˈmiːtɪəˌrɔɪd/ n. Astron. a small body moving through space, being a potential meteor in the event of an encounter with the earth's atmosphere. □ **meteoroidal** /ˌmiːtɪəˈrɔɪd(ə)l/ adj.

Meteorological Office /ˌmiːtɪərəˈlɒdʒɪk(ə)l/ (in the UK) a government department providing weather forecasts etc. Its headquarters are in Bracknell, Berkshire.

meteorology /ˌmiːtɪəˈrɒlədʒɪ/ n. **1** the study of the processes and phenomena of the atmosphere, esp. as a means of forecasting the weather. **2** the atmospheric character of a region. □ **meteorologist** n. **meteorological** /-rəˈlɒdʒɪk(ə)l/ adj. **meteorologically** adv. [Gk meteōrologia (as METEOR)]

meter[1] /ˈmiːtə(r)/ n. & v. ● n. **1** a person or thing that measures, esp. an instrument for recording a quantity of gas, electricity, etc. supplied, present, or needed. **2** = *parking-meter* (see PARK). ● v.tr. measure by means of a meter. [ME f. METE[1] + -ER[1]]

meter[2] US var. of METRE[1].

meter[3] US var. of METRE[2].

-meter /mɪtə(r), miːtə(r)/ comb. form **1** forming nouns denoting measuring instruments (*barometer*). **2** Prosody forming nouns denoting lines of poetry with a specified number of measures (*pentameter*).

methadone /ˈmeθəˌdəʊn/ n. a potent narcotic analgesic drug used to relieve severe pain, as a linctus to suppress coughs, and as a substitute for morphine or heroin. [chem. name 6-dimethylamino-4,4-diphenyl-3-heptan*one*]

methamphetamine /ˌmeθæmˈfetəmɪn, -ˌmiːn/ n. an amphetamine derivative with quicker and longer action, used as a stimulant. [METHYL + AMPHETAMINE]

methanal /ˈmeθəˌnæl/ n. Chem. = FORMALDEHYDE. [METHANE + -AL[2]]

methane /ˈmeθeɪn, ˈmiːθ-/ n. Chem. a colourless odourless inflammable gaseous hydrocarbon (chem. formula: CH_4), the simplest in the alkane series, and the main constituent of natural gas. Also called *marsh gas*. □ **methanoate** /-ˈnəʊeɪt/ n. [METHYL + -ANE[2]]

methanoic acid /ˌmeθəˈnəʊɪk/ n. Chem. = FORMIC ACID. □ **methanoate** /-ˈnəʊeɪt/ n. [METHANE + -IC]

methanol /ˈmeθəˌnɒl/ n. Chem. a colourless volatile inflammable liquid alcohol (chem. formula: CH_3OH), used as a solvent. Also called *methyl alcohol*. [METHANE + -OL]

methinks /mɪˈθɪŋks/ v.intr. (past **methought** /-ˈθɔːt/) archaic it seems to me. [OE mē thyncth f. mē dative of ME[1] + thyncth 3rd sing. of thyncan seem, THINK]

methionine /meˈθaɪəˌniːn/ n. Biochem. an amino acid which contains sulphur and is an important constituent of proteins. [METHYL + Gk theion sulphur]

metho /ˈmeθəʊ/ n. (pl. **-os**) Austral. sl. **1** methylated spirit. **2** a person addicted to drinking methylated spirit. [abbr.]

method /ˈmeθəd/ n. **1** a special form of procedure esp. in any branch of mental activity. **2** orderliness; regular habits. **3** the orderly arrangement of ideas. **4** a scheme of classification. **5** Theatr. = METHOD ACTING. □ **method in one's madness** sense in what appears to be foolish or strange behaviour. [F méthode or L methodus f. Gk methodos pursuit of knowledge (as META-, hodos way)]

method acting n. an acting theory and technique in which an actor aspires to complete emotional identification with a part. Based on the system evolved by Stanislavsky, it came into prominence in the US in the 1930s, and was developed in institutions such as the Actors' Studio in New York City (founded 1947), notably by Elia Kazan and Lee Strasberg. It was very successful in the work of modern American dramatists such as Tennessee Williams, and is particularly associated with actors such as Marlon Brando and Dustin Hoffman.

methodical /mɪˈθɒdɪk(ə)l/ adj. (also **methodic**) characterized by method or order. □ **methodically** adv. [LL methodicus f. Gk methodikos (as METHOD)]

Methodist /ˈmeθədɪst/ n. & adj. ● n. a member of a Christian Protestant denomination originating in the 18th-century evangelistic movement founded by John and Charles Wesley at Oxford. (*See note below.*) ● adj. of or relating to Methodists or Methodism. □ **Methodism** n. **Methodistic** /ˌmeθəˈdɪstɪk/ adj. **Methodistical** adj. [mod.L methodista (as METHOD)]

■ Methodism has a strong tradition of missionary work and concern with social welfare, and emphasizes the believer's personal relationship with God without an insistence on theological details. The movement was strongly influenced by Arminian ideas, and is usually Presbyterian in government. Methodism is particularly strong in the US and now constitutes one of the largest Protestant denominations worldwide, with more than 30 million members. The movement grew out of a religious society established within the Church of England and soon spread beyond Britain to North America; it formally separated from the Church of England in 1791. Doctrinal conflicts gave rise to a number of divisions within both British and American Methodism in the 19th century, but a degree of reunification was achieved in the early 20th century. The origin of the name is uncertain: it is generally supposed to have originated as a derisory reference to 'methodical' Bible study and weekday meetings.

Methodius, St /mɪˈθəʊdɪəs/ the brother of St Cyril (see CYRIL, ST).

methodize /ˈmeθəˌdaɪz/ v.tr. (also **-ise**) **1** reduce to order. **2** arrange in an orderly manner. □ **methodizer** n.

methodology /ˌmeθəˈdɒlədʒɪ/ n. (pl. **-ies**) **1** the science of method. **2** a body of methods used in a particular branch of activity. □ **methodologist** n. **methodological** /-dəˈlɒdʒɪk(ə)l/ adj. **methodologically** adv. [mod.L methodologia or F méthodologie (as METHOD)]

methought past of METHINKS.

meths /meθs/ n. Brit. colloq. methylated spirit. [abbr.]

Methuselah /mɪˈθjuːzələ/ (in the Bible) a patriarch, grandfather of Noah, said to have lived 969 years (Gen. 5:27).

methuselah /mɪˈθjuːzələ/ n. **1** a very old person or thing. **2** a wine bottle of about eight times the standard size. [METHUSELAH]

methyl /ˈmiːθaɪl, ˈmeθɪl/ n. Chem. the univalent hydrocarbon radical $-CH_3$, present in many organic compounds. □ **methyl alcohol** = METHANOL. **methyl benzene** = TOLUENE. □ **methylic** /mɪˈθɪlɪk/ adj. [G Methyl or F méthyle, back-form. f. G Methylen, F méthylène: see METHYLENE]

methylate /ˈmeθɪˌleɪt/ v.tr. **1** mix or impregnate with methanol. **2** introduce a methyl group into (a molecule or compound). □ **methylated spirit** (or **spirits**) alcohol impregnated with methanol to make it unfit for drinking and exempt from duty. □ **methylation** /ˌmeθɪˈleɪʃ(ə)n/ n.

methylene /ˈmeθɪˌliːn/ n. Chem. the highly reactive divalent group of atoms CH_2. [F méthylène f. Gk methu wine + hulē wood + -ENE]

metic /ˈmetɪk/ n. Gk Antiq. an alien living in a Greek city with some privileges of citizenship. [irreg. f. Gk metoikos (as META-, oikos dwelling)]

metical /ˌmetɪˈkɑːl/ n. the basic monetary unit of Mozambique, equal to 100 centavos. [Port. matical, f. Arab. mitkāl Arabian unit of weight, f. ṯakala weigh]

meticulous /məˈtɪkjʊləs/ adj. **1** giving great or excessive attention to details. **2** very careful and precise. □ **meticulously** *adv.* **meticulousness** *n.* [L *meticulosus* f. *metus* fear]

métier /ˈmetɪˌeɪ/ *n.* **1** one's trade, profession, or department of activity. **2** one's forte. [F ult. f. L *ministerium* service]

Metis /ˈmeɪˈtiːs, -ˈtiː, ˈmeɪtɪ/ *n.* (also **Métis**) (*pl.* **Metis** *pronunc.* same) a person of mixed descent, esp. (in Canada) the offspring or a descendant of a white person and an American Indian. [F *métis*, OF *mestis* f. Rmc, rel. to MESTIZO]

metol /ˈmiːtɒl/ *n.* a white soluble phenol derivative used as a photographic developer. [G, arbitrary name]

Metonic cycle /mɪˈtɒnɪk/ *n.* a period of 19 years (235 lunar months), after which the new and full moons return to the same day of the year. It was the basis of the ancient Greek calendar, and is still used for calculating movable feasts such as Easter. [Gk *Metōn*, Athenian astronomer (5th c. BC)]

metonym /ˈmetənɪm/ *n.* a word used in metonymy. [back-form. f. METONYMY, after *synonym*]

metonymy /mɪˈtɒnɪmɪ/ *n.* the substitution of the name of an attribute or adjunct for that of the thing meant (e.g. *Crown* for *king*, *the turf* for *horse-racing*). □ **metonymic** /ˌmetəˈnɪmɪk/ *adj.* **metonymical** *adj.* **metonymically** *adv.* [LL *metonymia* f. Gk *metōnumia* (as META-, *onoma*, *onuma* name)]

metope /ˈmetəʊp, ˈmetəpɪ/ *n. Archit.* a square space between triglyphs in a Doric frieze. [L *metopa* f. Gk *metopē* (as META-, *opē* hole for a beam-end)]

metre[1] /ˈmiːtə(r)/ *n.* (*US* **meter**) a unit of length equal to about 39.4 inches (symbol: **m**). (*See note below.*) □ **metre-kilogram-second** (abbr. **mks**) denoting a system of measure using these as the basic units of length, mass, and time. □ **metreage** *n.* [F *mètre* f. Gk *metron* measure]

▪ The metre is the fundamental unit of length in the metric system, and the SI base unit of linear measure. When the metric system was standardized, the metre was defined as one ten millionth part of the length of a quadrant of the meridian. Since then, various other official definitions have been used; in 1983, the metre was defined as the length of the path travelled by light in a vacuum during a time interval of $^1/_{299\,792\,458}$ of a second.

metre[2] /ˈmiːtə(r)/ *n.* (*US* **meter**) **1 a** any form of poetic rhythm, determined by the number and length of feet in a line. **b** a metrical group or measure. **2** the basic pulse and rhythm of a piece of music. [OF *metre* f. L *metrum* f. Gk *metron* MEASURE]

metric /ˈmetrɪk/ *adj. & n.* ● *adj.* **1** of or based on the metre. **2** of or relating to measurement. ● *n.* **1** a system or standard of measurement. **2** a mathematical function based on distances or quantities treated as analogous to distances for the purpose of analysis. □ **metric ton** (or **tonne**) 1,000 kilograms (2205 lb). [F *métrique* (as METRE[1])]

-metric /ˈmetrɪk/ *comb. form* (also **-metrical** /-k(ə)l/) forming adjectives corresponding to nouns in *-meter* and *-metry* (*thermometric*; *geometric*). □ **-metrically** *comb. form* forming adverbs. [from or after F *-métrique* f. L (as METRICAL)]

metrical /ˈmetrɪk(ə)l/ *adj.* **1** of, relating to, or composed in metre (*metrical psalms*). **2** of or involving measurement (*metrical geometry*). □ **metrically** *adv.* [ME f. L *metricus* f. Gk *metrikos* (as METRE[2])]

metricate /ˈmetrɪˌkeɪt/ *v.intr. & tr.* change or adapt to a metric system of measurement. □ **metricize** /-trɪˌsaɪz/ *v.tr.* (also **-ise**). **metrication** /ˌmetrɪˈkeɪʃ(ə)n/ *n.*

metric system *n.* a decimal measuring system with the metre, litre, and gram as the units of length, capacity, and weight or mass. The system was first proposed by the French astronomer and mathematician Gabriel Mouton (1618–94) in 1670, and was standardized in France under the Republican government in the 1790s, its use being made compulsory there in 1801. Napoleon's conquests facilitated the spread of the system throughout Europe, and the metric system has gained ground elsewhere at the expense of imperial units. The UK and other countries began the process of metrication in the 1960s, but metric units have not yet been generally adopted in the US. An important development of the metric system is the International System of Units (SI), used by the international scientific community. (*See also* DECIMALIZATION.)

metritis /mɪˈtraɪtɪs/ *n. Med.* inflammation of the womb. [Gk *mētra* womb + -ITIS]

metro /ˈmetrəʊ/ *n.* (*pl.* **-os**) an underground railway system in a city, esp. Paris. [F *métro*, abbr. of *métropolitain* METROPOLITAN]

metrology /mɪˈtrɒlədʒɪ/ *n.* the scientific study of measurement.

□ **metrologic** /ˌmetrəˈlɒdʒɪk/ *adj.* **metrological** *adj.* [Gk *metron* measure + -LOGY]

metronome /ˈmetrəˌnəʊm/ *n. Mus.* a device marking time at a selected rate by giving a regular tick. □ **metronomic** /ˌmetrəˈnɒmɪk/ *adj.* [Gk *metron* measure + *nomos* law]

metronymic /ˌmetrəˈnɪmɪk/ *adj. & n.* ● *adj.* (of a name) derived from the name of a mother or female ancestor. ● *n.* a metronymic name. [Gk *mētēr mētros* mother, after *patronymic*]

metropolis /mɪˈtrɒpəlɪs/ *n.* **1** the chief city of a country; a capital city. **2** a metropolitan bishop's see. **3** a city or town which is a local centre of activity. [LL f. Gk *mētropolis* parent state f. *mētēr mētros* mother + *polis* city]

metropolitan /ˌmetrəˈpɒlɪt(ə)n/ *adj. & n.* ● *adj.* **1** of or relating to a metropolis, esp. as distinct from its environs (*metropolitan New York*). **2** belonging to, forming or forming part of, a mother country as distinct from its colonies etc. (*metropolitan France*). **3** of an ecclesiastical metropolis. ● *n.* **1** (in full **metropolitan bishop**) a bishop having authority over the bishops of a province, in the Western Church equivalent to archbishop, in the Orthodox Church ranking above archbishop and below patriarch. **2** an inhabitant of a metropolis. □ **metropolitan county** each of the six units of English local government centred on a large urban area (Greater London excepted) established in 1974, with functions analogous to those of counties. Their councils were abolished in 1986. **metropolitan magistrate** *Brit.* a paid professional magistrate in London (cf. *stipendiary magistrate*). □ **metropolitanate** /-nət/ *n.* (in sense 1 of *n.*). **metropolitanism** *n.* [ME f. LL *metropolitanus* f. Gk *mētropolitēs* (as METROPOLIS)]

Metropolitan Museum of Art an important museum of art and archaeology in New York, founded in 1870.

metrorrhagia /ˌmiːtrəʊˈreɪdʒɪə/ *n. Med.* abnormal bleeding from the womb. [mod.L f. Gk *mētra* womb + *-rrhage* as HAEMORRHAGE]

-metry /mɪtrɪ/ *comb. form* forming nouns denoting procedures and systems corresponding to instruments in *-meter* (*calorimetry*; *thermometry*). [after *geometry* etc. f. Gk *-metria* f. *-metrēs* measurer]

Metternich /ˈmetənɪx/, Klemens Wenzel Nepomuk Lothar, Prince of Metternich-Winneburg-Beilstein (1773–1859), Austrian statesman. As Foreign Minister (1809–48), he was one of the organizers of the Congress of Vienna (1814–15), which devised the settlement of Europe after the Napoleonic Wars. He pursued policies which reflected his reactionary conservatism at home and abroad until forced to resign during the revolutions of 1848.

mettle /ˈmet(ə)l/ *n.* **1** the quality of a person's disposition or temperament (*a chance to show your mettle*). **2** natural ardour. **3** spirit, courage. □ **on one's mettle** incited to do one's best. □ **mettled** *adj.* (also in comb.). **mettlesome** *adj.* [var. of METAL *n.*]

Metz /mets/ a city in Lorraine, NE France, on the Moselle river; pop. (1990) 123,920. Formerly the capital of the medieval Frankish kingdom of Austrasia, the city grew to prosperity in the 13th century when it was a free town within the Holy Roman Empire, ruled by a virtually independent bishop. It was annexed in 1552 by Henry II of France and formally ceded to France in 1648. Metz fell to the Prussians in 1870 and was annexed to the German Empire in 1871. It was restored to France in 1918, after the First World War.

meu /mjuː/ *n.* (also **meum** /ˈmiːəm/) = BALDMONEY. [irreg. f. L *meum* f. Gk *mēon*]

meunière /mɜːˈnjeə(r)/ *adj.* (esp. of fish) cooked or served in lightly browned butter with lemon juice and parsley (*sole meunière*). [F (*à la*) *meunière* (in the manner of) a miller's wife]

Meuse /mɜːz/ (Flemish and Dutch **Maas** /maːs/) a river of western Europe, which rises in NE France and flows 950 km (594 miles) through Belgium and the Netherlands to the North Sea south of Dordrecht.

MeV *abbr.* mega-electronvolt(s).

mew[1] /mjuː/ *v. & n.* ● *v.intr.* (of a cat, gull, etc.) utter its characteristic cry. ● *n.* this sound, esp. of a cat. [ME: imit.]

mew[2] /mjuː/ *n. & v.* ● *n.* a cage for hawks, esp. while moulting. ● *v.tr.* **1** put (a hawk) in a cage. **2** (often foll. by *up*) shut up; confine. [ME f. OF *mue* f. *muer* moult f. L *mutare* change]

mew gull *n.* esp. *N. Amer.* the common gull, *Larus canus*. [OE *mæw* gull, f. Gmc]

mewl /mjuːl/ *v.intr.* (also **mule**) **1** cry feebly; whimper. **2** mew like a cat. [imit.: cf. MIAUL]

mews /mjuːz/ *n. Brit.* **1** a set of stabling round an open yard or along a lane. **2** such a set of buildings converted into dwellings; a row of houses in the style of a mews. [pl. (now used as sing.) of MEW², orig. of the royal stables on the site of hawks' mews at Charing Cross]

Mexicali /ˌmeksɪˈkɑːlɪ/ the capital of the state of Baja California, in NW Mexico; pop. (1990) 602,400.

Mexican /ˈmeksɪk(ə)n/ n. & adj. ● n. **1 a** a native or national of Mexico. **b** a person of Mexican descent. **2** an indigenous language of Mexico, esp. Nahuatl. ● adj. of or relating to Mexico or its people or indigenous languages. □ **Mexican wave** an effect like a moving wave produced by successive sections of the crowd in a stadium standing up, raising their arms, lowering them, and sitting down again (so called from being popularized at the Association Football World Cup competition in Mexico in 1986).

Mexico /ˈmeksɪˌkəʊ/ **1** a country in North America, with extensive coastlines on the Gulf of Mexico and the Pacific Ocean, bordered by the US to the north; pop. (1990) 81,140,920; official language, Spanish; capital, Mexico City. The centre of both Mayan and Aztec civilizations, Mexico was conquered and colonized by the Spanish in the early 16th century. The Aztecs are said to have fulfilled an ancient prophecy when they saw an eagle perched on a cactus eating a snake; on this site they founded a city, and the symbol of the eagle, snake, and cactus is the national emblem of Mexico. Mexico remained under Spanish rule until independence was achieved in 1821; a republic was established three years later. Texas rebelled and broke away in 1836, while all the remaining territory north of the Rio Grande was lost to the US in the Mexican War of 1846–8. Half a century of political instability, including a brief French occupation and imperial rule by Maximilian (1864–7), ended with the establishment of Porfirio Díaz as President in the 1870s. Civil war broke out again in 1910–20, leading to partial political reform. Mexico is a federal republic headed by a President, who is elected for a six-year term. **2** a state of central Mexico, to the west of Mexico City; capital, Toluca de Lerdo. [*Mextili*, Aztec war-god]

Mexico, Gulf of a large extension of the western Atlantic Ocean. Bounded in a sweeping curve by the US to the north, by Mexico to the west and south, and by Cuba to the south-east, it is linked to the Atlantic by the Straits of Florida and to the Caribbean Sea by the Yucatán Channel.

Mexico City the capital of Mexico; pop. (1990) 13,636,130. Founded in about 1300 as the Aztec capital Tenochtitlán, it was taken in 1519 by the Spanish conquistador Cortés, who destroyed the old city in 1521 and rebuilt it as the capital of New Spain. It is one of the most populous cities in the world.

Meyerbeer /ˈmaɪəˌbɪə(r)/, Giacomo (born Jakob Liebmann Beer) (1791–1864), German composer. He made his mark as a pianist before achieving success as an opera composer during a stay in Italy (1816–24). He then settled in Paris, establishing himself as a leading exponent of French grand opera with a series of works including *Robert le diable* (1831), *Les Huguenots* (1836), and *L'Africaine* (1865).

Meyerhof /ˈmaɪəˌhɒf/, Otto Fritz (1884–1951), German-born American biochemist. He worked in Germany on the biochemical processes involved in muscle action, including the production of lactic acid and heat as by-products, and provided the basis for understanding the process by which glucose is broken down to provide energy. He shared a Nobel Prize in 1922, and fled the Nazis in 1938 to continue his work in America.

mezereon /mɪˈzɪərɪən/ n. a small European and Asian shrub, *Daphne mezereum*, with fragrant purplish-red flowers and red berries. [med.L f. Arab. *māzaryūn*]

mezuzah /meˈzʊzə/ n. (pl. **mezuzoth** /ˌmezuˈzəʊt/) a parchment inscribed with religious texts and attached in a case to the doorpost of a Jewish house as a sign of faith. [Heb. *mᵉzûzāh* doorpost]

mezzanine /ˈmetsəˌniːn, ˈmezə-/ n. **1** a low storey between two others (usu. between the ground and first floors) (often attrib.: *mezzanine floor*). **2** Brit. Theatr. **a** a floor or space beneath the stage. **b** US a dress circle. **3** (attrib.) Finance of or involving unsecured high-interest loans subordinate to secured loans but ranking above equity. [F f. It. *mezzanino* dimin. of *mezzano* middle f. L *medianus* MEDIAN]

mezza voce /ˌmetsə ˈvəʊtʃeɪ/ adv. Mus. with less than the full strength of the voice or sound. [It., = half voice]

mezzo /ˈmetsəʊ/ adv. & n. Mus. ● adv. half, moderately. ● n. (in full **mezzo-soprano**) (pl. **-os**) **1 a** a female singing voice between soprano and contralto. **b** a singer with this voice. **2** a part written for mezzo-soprano. □ **mezzo forte** fairly loud. **mezzo piano** fairly soft. [It., f. L *medius* middle]

Mezzogiorno /ˌmetsəʊˈdʒɔːnəʊ/ the southern part of Italy, including Sicily and Sardinia. [It., = midday (cf. MIDI)]

mezzo-relievo /ˌmetsəʊrɪˈliːvəʊ, -ˈljeɪvəʊ/ n. (also ***mezzo-rilievo***) (pl. **-os**) a raised surface in the form of half-relief, in which the figures project half their true proportions. [It. *mezzo-rilievo* = half-relief]

mezzotint /ˈmetsəʊˌtɪnt/ n. & v. ● n. **1** a method of engraving a copper or steel plate in which the surface is partially roughened and partially scraped smooth, giving shaded areas where it has been left rough and light areas where it has been scraped smooth. (*See note below.*) **2** a print produced by this process. ● v.tr. engrave in mezzotint. □ **mezzotinter** n. [It. *mezzotinto* f. *mezzo* half + *tinto* tint]

▪ The mezzotint technique was invented by Ludwig von Siegen of Utrecht c.1640 and introduced into England by Prince Rupert. It was much used in the 17th, 18th, and early 19th centuries for the reproduction of paintings (in particular, oil paintings), but was superseded by photography in the 19th century.

MF abbr. medium frequency.

mf abbr. mezzo forte.

MFH abbr. Brit. Master of Foxhounds.

MG abbr. **1** machine-gun. **2** Morris Garages (as a make of car).

Mg symb. Chem. the element magnesium.

mg abbr. milligram(s).

MGB the Ministry of State Security in the former USSR (see MVD). [Russ. abbr.]

MGM abbr. Metro-Goldwyn-Mayer, a film company formed in 1924 by Samuel Goldwyn and Louis B. Mayer.

Mgr. abbr. **1** Manager. **2** *Monseigneur*. **3** Monsignor.

mho /məʊ/ n. (pl. **-os**) Electr. a unit of conductance, now usually called the siemens. [OHM reversed]

MHR abbr. (in the US and Australia) Member of the House of Representatives.

MHz abbr. megahertz.

MI abbr. **1** US Michigan (in official postal use). **2** Brit. hist. Military Intelligence.

mi var. of ME².

mi. abbr. mile(s).

MI5 /ˌemaɪˈfaɪv/ (in the UK) the secret governmental agency responsible for dealing with internal security and counter-intelligence on British territory. It was formed in 1909. The agency was officially named the Security Service in 1964, but the name MI5 is still in popular use. (See also MI6.) [Military Intelligence section 5]

MI6 /ˌemaɪˈsɪks/ (in the UK) the secret governmental agency responsible for dealing with matters of internal security and counter-intelligence overseas. It was formed in 1912. The agency was officially named the Secret Intelligence Service in 1964, but the name MI6 remains in popular use. (See also MI5.) [Military Intelligence section 6]

Miami /maɪˈæmɪ/ a city and port on the coast of SE Florida; pop. (1990) 358,550. Its subtropical climate and miles of beaches make this and the resort island of Miami Beach, separated from the mainland by Biscayne Bay, a year-round holiday resort.

miaow /mɪˈaʊ/ n. & v. (also **meow**) ● n. the characteristic cry of a cat. ● v.intr. make this cry. [imit.]

miasma /mɪˈæzmə, maɪ-/ n. (pl. **miasmata** /-mətə/ or **miasmas**) archaic an infectious or noxious vapour. □ **miasmal** adj. **miasmic** adj. **miasmically** adv. **miasmatic** /ˌmiːæzˈmætɪk, ˌmaɪ-/ adj. [Gk, = defilement, f. *miainō* pollute]

miaul /mɪˈaʊl/ v.intr. cry like a cat; mew. [F *miauler*: imit.]

Mic. abbr. (in the Bible) Micah.

mica /ˈmaɪkə/ n. any of a group of silicate minerals with a layered structure, esp. muscovite. □ **mica-schist** (or **slate**) a fissile rock containing quartz and mica. □ **micaceous** /maɪˈkeɪʃəs/ adj. [L, = crumb]

Micah /ˈmaɪkə/ **1** a Hebrew minor prophet. **2** a book of the Bible bearing his name, foretelling the destruction of Samaria and of Jerusalem.

Micawber /mɪˈkɔːbə(r)/ the name of a character, Mr Wilkins Micawber, in Dickens's novel *David Copperfield* (1850), apparently based on Dickens's own father. He is characterized as an eternal optimist, dreaming up elaborate schemes for making money which never materialize, but still remaining undaunted. □ **Micawberish** adj. **Micawberism** n.

mice pl. of MOUSE.

micelle /mɪˈsel, maɪ-/ n. Chem. an aggregate of molecules in a colloidal solution, as formed by detergents and other surfactants. [mod.L *micella* dimin. of L *mica* crumb]

Mich. *abbr.* **1** Michaelmas. **2** Michigan.

Michael, St /ˈmaɪk(ə)l/ one of the archangels, usually represented slaying a dragon (see Rev. 12:7). Feast day, 29 Sept. (Michaelmas Day).

Michaelmas /ˈmɪk(ə)lməs/ *n.* the feast of St Michael, 29 Sept. □ **Michaelmas daisy** an autumn-flowering aster. **Michaelmas term** *Brit.* a term in some universities and the law beginning soon after Michaelmas. [OE *sancte Micheles mæsse* Saint Michael's mass: see MASS²]

Michelangelo /ˌmaɪk(ə)lˈændʒəˌləʊ/ (full name Michelangelo Buonarroti) (1475–1564), Italian sculptor, painter, architect, and poet. A leading figure during the High Renaissance, Michelangelo established his reputation in Rome with statues such as the *Pietà* (c.1497–1500) and then in Florence with his marble *David* (1501–4). Under papal patronage, Michangelo decorated the ceiling of the Sistine Chapel in Rome (1508–12) and painted the fresco, *The Last Judgement* (1536–41), both important works in the development of the mannerist style. His architectural achievements include the design of the Laurentian Library in Florence (1524–34), as well as the completion of St Peter's in Rome, including the great dome (1546–64).

Michelin /ˈmɪtʃəlɪn, French miʃlɛ̃/, André (1853–1931) and Édouard (1859–1940), French industrialists. In 1888 they founded the Michelin Tyre Company, and they pioneered the use of pneumatic tyres on motor vehicles in the 1890s. The company also introduced steel-belted radial tyres in 1948.

Michelozzo /ˌmiːkeˈlɒtsəʊ/ (full name Michelozzo di Bartolommeo) (1396–1472), Italian architect and sculptor. In partnership with Ghiberti and Donatello, he led a revival of interest in Roman architecture. His most famous building is the Palazzo Medici-Riccardi in Florence (1444–59), one of the most influential palace designs of the early Renaissance.

Michelson /ˈmaɪk(ə)ls(ə)n/, Albert Abraham (1852–1931), American physicist. He specialized in precision measurement in experimental physics, and became in 1907 the first American to be awarded a Nobel Prize. He performed a number of accurate determinations of the velocity of light. His crucial experiment demonstrating that the hypothetical ether did not exist was repeated in 1887 with E. W. Morley, using improved apparatus. The Michelson–Morley result contradicted Newtonian physics, and was eventually resolved by Einstein's special theory of relativity.

Michigan /ˈmɪʃɪgən/ a state in the northern US, bordered on the west, north, and east by Lakes Michigan, Superior, Huron, and Erie; pop. (1990) 9,295,300; capital, Lansing. Explored by the French in the 17th century, it was ceded to Britain in 1763 and acquired by the US in 1783, becoming the 26th state in 1837.

Michigan, Lake one of the five Great Lakes of North America. Bordered by Michigan, Wisconsin, Illinois, and Indiana, it is the only one of the Great Lakes to lie wholly within the US. The cities of Milwaukee and Chicago are on its shores.

Michoacán /ˌmiːtʃəʊəˈkɑːn/ a state of western Mexico, on the Pacific coast; capital, Morelia.

mick /mɪk/ *n. sl. offens.* **1** an Irishman. **2** a Roman Catholic. [pet-form of the name *Michael*]

mickery /ˈmɪkərɪ/ *n. Austral.* (also **mickerie**) a water-hole or excavated well, esp. in a dry river bed. [Aboriginal *migri*]

mickey /ˈmɪkɪ/ *n.* (also **micky**) □ **take the mickey** (often foll. by *out of*) *sl.* tease, mock, ridicule. [20th c.: orig. uncert.]

Mickey Finn /ˌmɪkɪ ˈfɪn/ *n. sl.* **1** a drink adulterated with a narcotic or laxative. **2** the adulterant itself. [20th c.: orig. uncert.]

Mickey Mouse /ˌmɪkɪ ˈmaʊs/ a Walt Disney cartoon character, who first appeared as Mortimer Mouse in 1927, becoming Mickey in 1928. During the 1930s he became established as the central Disney character, with Disney himself speaking the soundtrack for Mickey's voice.

mickle /ˈmɪk(ə)l/ *adj. & n.* (also **muckle** /ˈmʌk-/) *archaic or Sc.* ● *adj.* much, great. ● *n.* a large amount. □ **many a little makes a mickle** (orig. *erron.* **many a mickle makes a muckle**) many small amounts accumulate to make a large amount. [ME f. ON *mikell* f. Gmc]

Mick the Miller /mɪk/ a racing greyhound who won many races in the UK from 1928 to 1931 and later starred in the film *Wild Boy* (1935).

micky var. of MICKEY.

Micmac /ˈmɪkmæk/ *n. & adj.* (also **Mi'kmaq**) ● *n.* (*pl.* same or **Micmacs**) **1** a member of an American Indian people inhabiting the Maritime Provinces of Canada. **2** the language spoken by the Micmac. ● *adj.* of or relating to this people or their language. [F f. Micmac]

micro /ˈmaɪkrəʊ/ *n.* (*pl.* **-os**) *colloq.* **1** = MICROCOMPUTER. **2** = MICROPROCESSOR.

micro- /ˈmaɪkrəʊ/ *comb. form* **1** small (*microchip*). **2** denoting a factor of one millionth (10⁻⁶) (*microgram*) (symbol: μ). [Gk *mikro-* f. *mikros* small]

microanalysis /ˌmaɪkrəʊəˈnælɪsɪs/ *n.* the quantitative analysis of chemical compounds using a sample of a few milligrams.

microbe /ˈmaɪkrəʊb/ *n.* a minute living being; a micro-organism (esp. a bacterium causing disease and fermentation). □ **microbial** /maɪˈkrəʊbɪəl/ *adj.* **microbic** *adj.* [F f. Gk *mikros* small + *bios* life]

microbiology /ˌmaɪkrəʊbaɪˈɒlədʒɪ/ *n.* the scientific study of micro-organisms, e.g. bacteria, viruses, and fungi. □ **microbiologist** *n.* **microbiological** /-ˌbaɪəˈlɒdʒɪk(ə)l/ *adj.* **microbiologically** *adv.*

microburst /ˈmaɪkrəʊˌbɜːst/ *n.* a particularly strong wind shear, esp. during a thunderstorm.

microcephaly /ˌmaɪkrəʊˈsefəlɪ, -ˈkefəlɪ/ *n.* an abnormal smallness of the head in relation to the rest of the body. □ **microcephalous** *adj.* **microcephalic** /-sɪˈfælɪk, -kɪˈfælɪk/ *adj. & n.*

microchip /ˈmaɪkrəʊˌtʃɪp/ *n.* a tiny wafer of semiconducting material used to make an integrated circuit.

microcircuit /ˈmaɪkrəʊˌsɜːkɪt/ *n.* a minute electric circuit, esp. an integrated circuit. □ **microcircuitry** *n.*

microclimate /ˈmaɪkrəʊˌklaɪmɪt/ *n.* the climate of a small local area or enclosed space, esp. where this differs from that of the surroundings. □ **microclimatic** /ˌmaɪkrəʊklaɪˈmætɪk/ *adj.* **microclimatically** *adv.*

microcode /ˈmaɪkrəʊˌkəʊd/ *n.* **1** = MICROINSTRUCTION. **2** = MICROPROGRAM.

microcomputer /ˈmaɪkrəʊkəmˌpjuːtə(r)/ *n.* a small computer that contains a microprocessor as its central processor.

microcopy /ˈmaɪkrəʊˌkɒpɪ/ *n. & v.* ● *n.* (*pl.* **-ies**) a copy of printed matter that has been reduced by microphotography. ● *v.tr.* (**-ies**, **-ied**) make a microcopy of.

microcosm /ˈmaɪkrəˌkɒz(ə)m/ *n.* **1** (often foll. by *of*) a miniature representation. **2** humankind viewed as the epitome of the universe. **3** any community or complex unity viewed in this way. □ **microcosmic** /ˌmaɪkrəˈkɒzmɪk/ *adj.* **microcosmically** *adv.* [ME f. F *microcosme* or med.L *microcosmus* f. Gk *mikros kosmos* little world]

microdot /ˈmaɪkrəʊˌdɒt/ *n.* a microphotograph of a document etc. reduced to the size of a dot.

microeconomics /ˌmaɪkrəʊˌiːkəˈnɒmɪks, -ˌekəˈnɒmɪks/ *n.* the branch of economics that deals with small-scale economic factors such as individual commodities, producers, consumers, etc. Cf. MACROECONOMICS. □ **microeconomic** *adj.*

microelectronics /ˌmaɪkrəʊˌɪlekˈtrɒnɪks, -ˌelekˈtrɒnɪks/ *n.* the design, manufacture, and use of microchips and microcircuits. □ **microelectronic** *adj.*

microevolution /ˌmaɪkrəʊˌiːvəˈluːʃ(ə)n, -ˈljuːʃ(ə)n/ *n. Biol.* evolutionary change within a species or small group of organisms, esp. over a short period. □ **microevolutionary** *adj.*

microfiche /ˈmaɪkrəʊˌfiːʃ/ *n.* (*pl.* same or **microfiches**) a flat rectangular piece of film bearing microphotographs of the pages of a printed text or document.

microfilm /ˈmaɪkrəʊˌfɪlm/ *n. & v.* ● *n.* a length of film bearing microphotographs of documents etc. ● *v.tr.* photograph (a document etc.) on microfilm.

microfloppy /ˈmaɪkrəʊˌflɒpɪ/ *n.* (*pl.* **-ies**) (in full **microfloppy disk**) *Computing* a floppy disk with a diameter of less than 5¼ inches, 13.3 cm (usu. 3½ inches, 8.9 cm).

microform /ˈmaɪkrəʊˌfɔːm/ *n.* microphotographic reproduction on film or paper of a manuscript etc.

microgram /ˈmaɪkrəʊˌgræm/ *n.* one-millionth of a gram.

micrograph /ˈmaɪkrəʊˌgrɑːf/ *n.* a photograph taken by means of a microscope.

microgravity /ˈmaɪkrəʊˌgrævɪtɪ/ *n.* very weak gravity, as in an orbiting spacecraft.

microgroove /ˈmaɪkrəʊˌgruːv/ *n.* a very narrow groove on a long-playing gramophone record.

microinstruction /ˌmaɪkrəʊɪnˈstrʌkʃ(ə)n/ *n.* a machine-code instruction that effects a basic operation in a computer system.

microlepidoptera /ˌmaɪkrəʊˌlepɪˈdɒptərə/ *n.pl.* the numerous smaller moths that are of interest only to specialists (cf. MACROLEPIDOPTERA). ¶ Not used in formal classification.

microlight /ˈmaɪkrəʊˌlaɪt/ n. a very small, light, low-speed aircraft with an open frame.

microlith /ˈmaɪkrəʊˌlɪθ/ n. Archaeol. a minute worked flint usu. as part of a composite tool. □ **microlithic** /ˌmaɪkrəʊˈlɪθɪk/ adj.

micromesh /ˈmaɪkrəʊˌmeʃ/ n. (often attrib.) material, esp. nylon, consisting of a very fine mesh.

micrometer /maɪˈkrɒmɪtə(r)/ n. a gauge for accurately measuring small distances, thicknesses, etc. □ **micrometry** n.

micrometre /ˈmaɪkrəʊˌmiːtə(r)/ n. one-millionth of a metre.

microminiaturization /ˌmaɪkrəʊˌmɪnɪtʃəraɪˈzeɪʃ(ə)n/ n. (also **-isation**) the manufacture of very small electronic devices by using integrated circuits.

micron /ˈmaɪkrɒn/ n. one-millionth of a metre. [Gk mikron neut. of mikros small: cf. MICRO-]

Micronesia /ˌmaɪkrəʊˈniːzɪə, -ˈniːʒə/ **1** a region of the western Pacific to the north of Melanesia and north and west of Polynesia. It includes the Mariana, Caroline, and Marshall island groups and Kiribati. **2** (in full **Federated States of Micronesia**) a group of associated island states comprising the 600 islands of the Caroline Islands, in the western Pacific to the north of the equator; pop. (est. 1990) 107,900; languages, English (official), Malayo-Polynesian languages; capital, Kolonia (on Pohnpei). The group, which includes the islands of Yap, Kosrae, Truk, and Pohnpei, was administered by the US as part of the Pacific Islands Trust Territory from 1947 until 1986, when it entered into free association with the US as an independent state. [MICRO- + Gk nēsos island]

Micronesian /ˌmaɪkrəʊˈniːzɪən, -ˈniːʒ(ə)n/ adj. & n. ● adj. of or relating to Micronesia or its people, or their group of languages. ● n. **1** a native or inhabitant of Micronesia. **2** the group of Malayo-Polynesian languages spoken in Micronesia.

micronutrient /ˌmaɪkrəʊˈnjuːtrɪənt/ n. a chemical element or substance required in trace amounts for the growth and development of living organisms.

micro-organism /ˌmaɪkrəʊˈɔːɡəˌnɪz(ə)m/ n. a microscopic organism, esp. a bacterium or virus.

microphone /ˈmaɪkrəˌfəʊn/ n. an instrument for converting sound waves into electrical energy variations which may be reconverted into sound after transmission by wire or radio or after recording. (See note below.) □ **microphonic** /ˌmaɪkrəˈfɒnɪk/ adj.

 ▪ Different types of microphones are used in different applications, and most of them are either of the electrodynamic or the electrostatic type. The electrodynamic, dynamic, or moving-coil microphone has a diaphragm attached to a small coil that moves in a magnetic field, producing a varying electric signal. A modification of this is the ribbon microphone, which uses a thin corrugated metal ribbon suspended in a magnetic field and is highly directional. The electrostatic, capacitor, or condenser microphone uses a flimsy foil diaphragm close to a fixed plate, which together form a variable capacitor. A variety of different microphones are used in telephones, ranging from the carbon microphone patented by Edison in 1886 to modifications of the electrostatic type with full electronic control.

microphotograph /ˌmaɪkrəʊˈfəʊtəˌɡrɑːf/ n. a photograph reduced to a very small size.

microphyte /ˈmaɪkrəʊˌfaɪt/ n. a microscopic plant.

microprocessor /ˌmaɪkrəʊˈprəʊsesə(r)/ n. an integrated circuit that contains all the functions of a central processing unit of a computer.

microprogram /ˌmaɪkrəʊˈprəʊɡræm/ n. a microinstruction program that controls the functions of a central processing unit of a computer.

micropyle /ˈmaɪkrəʊˌpaɪl/ n. Bot. a small opening in the surface of an ovule, through which pollen passes. [MICRO- + Gk pulē gate]

microscope /ˈmaɪkrəˌskəʊp/ n. an instrument magnifying small objects by means of a lens or lenses so as to reveal details invisible to the naked eye. Simple microscopes of one lens (such as those used by Antoni van Leeuwenhoek, or the modern magnifying glass) were used as early as the 15th century. The compound microscope, which permits greater magnification, first appeared at the beginning of the 17th century. It has at least two lenses: an objective lens, placed near the specimen, which produces an enlarged image of it, and an eyepiece lens which magnifies this image. Early instruments produced distortions of colour or shape in the image which were not fully eliminated until the 19th century. Optical microscopes cannot discriminate between points which are closer together than the

wavelength of light (a few hundred nanometres). (See also ELECTRON MICROSCOPE.) [mod.L microscopium (as MICRO-, -SCOPE)]

microscopic /ˌmaɪkrəˈskɒpɪk/ adj. **1** so small as to be visible only with a microscope. **2** extremely small. **3** regarded in terms of small units. **4** of the microscope. □ **microscopical** adj. (in sense 4). **microscopically** adv.

microscopy /maɪˈkrɒskəpɪ/ n. the use of the microscope. □ **microscopist** n.

microsecond /ˈmaɪkrəʊˌsekənd/ n. one-millionth of a second.

microsome /ˈmaɪkrəʊˌsəʊm/ n. Biol. a small particle of organelle fragments obtained by centrifugation of homogenized cells. [MICRO- + -SOME³]

microspore /ˈmaɪkrəʊˌspɔː(r)/ n. Bot. the smaller of the two kinds of spore produced by some ferns (cf. MEGASPORE).

microstructure /ˈmaɪkrəʊˌstrʌktʃə(r)/ n. (in a metal or other material) the arrangement of crystals etc. which can be made visible and examined with a microscope.

microsurgery /ˈmaɪkrəʊˌsɜːdʒərɪ/ n. intricate surgery performed using microscopes, enabling the tissue to be operated on with miniaturized precision instruments. □ **microsurgical** /ˌmaɪkrəʊˈsɜːdʒɪk(ə)l/ adj.

microswitch /ˈmaɪkrəʊˌswɪtʃ/ n. a switch that can be operated rapidly by a small movement.

microtome /ˈmaɪkrəʊˌtəʊm/ n. an instrument for cutting extremely thin sections of material for examination under a microscope. [MICRO- + -TOME]

microtone /ˈmaɪkrəʊˌtəʊn/ n. Mus. an interval smaller than a semitone.

microtubule /ˌmaɪkrəʊˈtjuːbjuːl/ n. Biol. a minute protein filament occurring in cytoplasm and involved in forming the spindles during cell division etc.

microwave /ˈmaɪkrəʊˌweɪv/ n. & v. ● n. **1** an electromagnetic wave with a wavelength in the range 0.001–0.3 m. (See note below.) **2** (in full **microwave oven**) an oven that uses microwaves to heat food. ● v.tr. (**-ving**) cook in a microwave oven.

 ▪ Microwaves have wavelengths shorter than those of normal radio waves but longer than those of infrared radiation. They are used in radar, in communications, and for heating in microwave ovens and in various industrial processes. Microwave cooking depends largely on the moisture present in food. Water molecules absorb microwaves, the energy taken up causing the molecules to spin; collisions then convert this rotation into faster motion and hence a rise in temperature, and the food is rapidly cooked.

micrurgy /ˈmaɪkrɜːdʒɪ/ n. the manipulation of individual cells etc. under a microscope. [MICRO- + Gk -ourgia work]

micturition /ˌmɪktjʊəˈrɪʃ(ə)n/ n. formal or Med. urination. [L micturire micturit-, desiderative f. mingere mict- urinate]

mid /mɪd/ prep. poet. = AMID. [abbr. f. AMID]

mid- /mɪd/ comb. form **1** that is the middle of (in mid-air; from mid-June to mid-July). **2** that is in the middle; medium, half. **3** Phonet. (of a vowel) pronounced with the tongue neither high nor low. [OE midd (recorded only in oblique cases), rel. to L medius, Gk mesos]

Midas /ˈmaɪdəs/ Gk Mythol. a king of Phrygia, who, according to one story, was given by Dionysus the power of turning everything he touched into gold. Unable to eat or drink, he prayed to be relieved of the gift and was instructed to wash in the River Pactolus, which since then has had golden sands. According to another story, he declared Pan a better flute-player than Apollo, who thereupon gave him ass's ears. Midas tried to hide them but his barber whispered the secret to some reeds, which repeat it whenever they rustle in the wind.

Midas touch n. the ability to turn one's activities to financial advantage.

Mid-Atlantic Ridge /ˌmɪdətˈlæntɪk/ a submarine ridge system extending the length of the Atlantic Ocean from the Arctic to the Antarctic. It is seismically and (in places) volcanically active; the islands of Iceland, the Azores, Ascension, St Helena, and Tristan da Cunha are situated on it. (See also MID-OCEAN RIDGE.)

midbrain /ˈmɪdbreɪn/ n. Anat. a small central part of the brainstem, developing from the middle of the primitive or embryonic brain. Also called mesencephalon. (See also BRAIN.)

midday /mɪdˈdeɪ/ n. the middle of the day; noon. [OE middæg (as MID, DAY)]

midden /ˈmɪd(ə)n/ n. **1** a dunghill. **2** a refuse heap near a dwelling.

3 = kitchen midden. [ME *myddyng*, of Scand. orig.: cf. Danish *mødding* muck heap]

middle /'mɪd(ə)l/ *adj., n., & v.* ● *attrib.adj.* **1** at an equal distance from the extremities of a thing. **2** (of a member of a group) so placed as to have the same number of members on each side. **3** intermediate in rank, quality, etc. **4** average (*of middle height*). **5** (of a language) of the period between the old and modern forms. **6** *Gram.* designating the voice of (esp. Greek) verbs that expresses reciprocal or reflexive action. ● *n.* **1** (often foll. by *of*) the middle point or position or part. **2** a person's waist. **3** *Gram.* the middle form or voice of a verb. **4** = middle term. ● *v.tr.* **1** place in the middle. **2** *Football* return (the ball) from the wing to the midfield. **3** *Cricket* strike (the ball) with the middle of the bat. **4** *Naut.* fold in the middle. □ **in the middle of** (often foll. by *verbal noun*) in the process of; during. **middle age** the period between youth and old age, about 45 to 60. **middle-aged** in middle age. **middle-age** (or **-aged**) **spread** the increased bodily girth often associated with middle age. **middle C** *Mus.* the C near the middle of the piano keyboard, the note between the treble and bass staves, at about 260 Hz. **middle class** the class of society between the upper and the lower, including professional and business workers and their families. **middle-class** *adj.* of the middle class. **middle common room** *Brit.* a common room for the use of graduate members of a college who are not fellows. **middle course** a compromise between two extremes. **middle distance 1** (in a painted or actual landscape) the part between the foreground and the background. **2** *Athletics* a race distance intermediate between that of a sprint and a long-distance race, now usually 400, 800, or 1500 metres. **middle ear** the cavity of the central part of the ear behind the drum. **middle finger** the finger next to the forefinger. **middle game** the central phase of a chess game, when strategies are developed. **middle name 1** a person's name placed after the first name and before the surname. **2** a person's most characteristic quality (*sobriety is my middle name*). **middle-of-the-road 1** (of a person, course of action, etc.) moderate; avoiding extremes. **2** (of music) intended to appeal to a wide audience, deliberately unadventurous. **middle passage** the sea journey between West Africa and the West Indies (with ref. to the slave trade). **middle school** *Brit.* a school for children from about 9 to 13 years old. **middle-sized** of medium size. **middle term** *Logic* the term common to both premisses of a syllogism. **middle watch** *Naut.* the watch from midnight to 4 a.m. **middle way 1** = middle course. **2** the eightfold path of Buddhism between indulgence and asceticism. [OE *middel* f. Gmc]

Middle Ages the period of European history from the fall of the Roman Empire in the West (5th century) to the fall of Constantinople (1453). The earlier part of the period (*c*.500–*c*.1100), or *early Middle Ages*, is sometimes distinguished as the *Dark Ages*, while the later part (*c*.1100–1453), or *high* or *late Middle Ages*, is often thought of as the Middle Ages proper. In general terms the whole period is characterized by the emergence of separate kingdoms, the growth of trade and urban life, and the growth in power of monarchies; the later period was dominated by power struggles and war. The Church became immensely powerful during the late Middle Ages, with the pope in many ways operating like a secular ruler; the influence of the Church and religious life was also felt in more general terms, as is shown by the Crusades, the popularity of pilgrimage, the building of monasteries, and the flowering of scholarship associated with them. The growth of interest in classical models within art and scholarship in the 15th century is seen as marking the transition to the Renaissance period and the end of the Middle Ages.

Middle America 1 Mexico and Central America. **2** the middle class in the US, esp. as a conservative political force. □ **Middle American** *adj. & n.*

middlebrow /'mɪd(ə)l,braʊ/ *adj. & n. colloq.* ● *adj.* claiming to be or regarded as only moderately intellectual. ● *n.* a middlebrow person.

Middle Congo see CONGO 2.

Middle East a term loosely applied to an extensive area of southwest Asia and northern Africa, stretching from the Mediterranean to Pakistan, including the Arabian peninsula, having a predominantly Muslim population.

Middle England the middle classes in England outside London, esp. as representative of conservative political views. □ **Middle Englander** *n.*

Middle English *n.* the English language from *c*.1150 to 1500. (See ENGLISH.)

Middle Kingdom a period of ancient Egyptian history (*c*.2040–1640 BC). (See EGYPT.)

middleman /'mɪd(ə)l,mæn/ *n.* (*pl.* **-men**) **1** a trader who handles a commodity between its producer and its consumer. **2** an intermediary.

Middlesbrough /'mɪd(ə)lzbrə/ a port in NE England, on the estuary of the River Tees; pop. (1991) 141,100. It is the administrative centre of Cleveland.

Middlesex /'mɪd(ə)l,seks/ a former county of SE England, situated to the north of London. In 1965 it was divided between Hertfordshire, Surrey, and Greater London.

Middle Temple one of the two Inns of Court on the site of the Temple in London (cf. INNER TEMPLE).

Middleton /'mɪd(ə)lt(ə)n/, Thomas (*c*.1570–1627), English dramatist. After collaborating with Dekker on the first part of *The Honest Whore* (1604), Middleton wrote the two tragedies for which he is best known, namely *The Changeling* (1622), written with William Rowley (*c*.1585–1626), and *Women Beware Women* (1620–7).

middleweight /'mɪd(ə)l,weɪt/ *n.* **1** a weight in certain sports intermediate between welterweight and light heavyweight, in the amateur boxing scale 71–5 kg but differing for professionals, wrestlers, and weightlifters. **2** a sportsman of this weight. □ **junior middleweight 1** a weight in professional boxing of 66.7–69.8 kg. **2** a professional boxer of this weight. **light middleweight 1** a weight in amateur boxing of 67–71 kg. **2** an amateur boxer of this weight.

Middle West (also **Midwest**) a part of the US occupying the northern half of the Mississippi basin, containing the states of Ohio, Indiana, Illinois, Michigan, Wisconsin, Iowa, and Minnesota.

middling /'mɪdlɪŋ/ *adj., n., & adv.* ● *adj.* **1 a** a moderately good (esp. *fair to middling*). **b** second-rate. **2** (of goods) of the second of three grades. ● *n.* (in *pl.*) middling goods, esp. flour of medium fineness. ● *adv.* **1** fairly or moderately (*middling good*). **2** *colloq.* fairly well (esp. in health). □ **middlingly** *adv.* [ME, of Sc. orig.: prob. f. MID + -LING²]

Middx. *abbr.* Middlesex.

middy¹ /'mɪdɪ/ *n.* (*pl.* **-ies**) **1** *colloq.* a midshipman. **2** (in full **middy blouse**) a woman's or child's loose blouse with a collar like that worn by sailors.

middy² /'mɪdɪ/ *n.* (*pl.* **-ies**) *Austral. sl.* a measure of beer of varying size. [20th c.: orig. unkn.]

Mideast /mɪd'i:st/ *n. US* = MIDDLE EAST.

midfield /mɪd'fi:ld/ *n. Football* the central part of the pitch, away from the goals. □ **midfielder** *n.*

Midgard /'mɪdgɑ:d/ *Scand. Mythol.* the region, encircled by the sea, in which human beings live; the earth. □ **Midgard serpent** a monstrous serpent, the offspring of Loki, thrown by Odin into the sea, where, with its tail in its mouth, it encircled the earth.

midge /mɪdʒ/ *n.* **1 a** a small gnatlike dipterous fly of the family Chironomidae, frequently found in dancing swarms near water. **b** a similar dipterous fly, esp. a tiny biting one of the family Ceratopogonidae, with piercing mouthparts for sucking blood. **2** *colloq.* a small or insignificant person. [OE *mycg(e)* f. Gmc]

midget /'mɪdʒɪt/ *n.* **1** an extremely small person or thing. **2** (*attrib.*) very small. [MIDGE + -ET¹]

Mid Glamorgan a county of South Wales; administrative centre, Cardiff. It was formed in 1974 from parts of Breconshire, Glamorgan, and Monmouthshire.

midgut /'mɪdgʌt/ *n. Zool.* the middle part of the alimentary canal, including (in vertebrates) the small intestine.

MIDI /'mɪdɪ/ *n.* a system for using combinations of electronic equipment, esp. audio and computer equipment. [acronym of *musical instrument digital interface*]

Midi /'mi:dɪ/ the south of France. [F, = midday (cf. MEZZOGIORNO)]

midi /'mɪdɪ/ *n.* (*pl.* **midis**) a garment of medium length, usu. reaching to mid-calf. [MID after MINI]

midibus /'mɪdɪ,bʌs/ *n.* a bus seating up to about 25 passengers.

midinette /,mɪdɪ'net/ *n.* a Parisian shop-girl, esp. a milliner's assistant. [F f. *midi* midday + *dînette* light dinner]

Midi-Pyrénées /,mi:dɪ,pɪreɪ'neɪ/ a region of southern France, between the Pyrenees and the Massif Central, centred on Toulouse.

midiron /'mɪd,aɪən/ *n. Golf* an iron with a medium degree of loft.

midland /'mɪdlənd/ *n. & adj.* ● *n.* **1** (**the Midlands**) the inland counties of central England. **2** the middle part of a country. ● *adj.* **1** (also **Midland**) of or in the midland or Midlands. **2** Mediterranean. □ **midlander** *n.*

mid-life /'mɪdlaɪf/ *n.* middle age (often *attrib.*: *mid-life planning*).

☐ **mid-life crisis** an emotional crisis of self-confidence that can occur in early middle age.

midline /ˈmɪdlaɪn/ n. a median line, or plane of bilateral symmetry.

Midlothian /mɪdˈləʊðɪən/ a former county of central Scotland, centred on Edinburgh. It became a part of Lothian region in 1975.

midmost /ˈmɪdməʊst/ adj. & adv. in the very middle.

midnight /ˈmɪdnaɪt/ n. **1** the middle of the night; 12 o'clock at night. **2** intense darkness. ☐ **midnight blue** a very dark blue. **midnight sun** the sun visible at midnight during the summer in polar regions. [OE midniht (as MID, NIGHT)]

mid-ocean ridge /mɪdˈəʊʃ(ə)n/ n. Geol. a long, seismically active, submarine ridge system rising in the middle of an ocean basin and marking the site of the upwelling of magma associated with sea-floor spreading. An example is the Mid-Atlantic Ridge.

mid-off /mɪdˈɒf/ n. Cricket a fielding position near the bowler on the off side.

mid-on /mɪdˈɒn/ n. Cricket a fielding position near the bowler on the on side.

Midrash /ˈmɪdræʃ/ n. (pl. **Midrashim** /ˌmɪdræˈʃɪm/) an ancient commentary on part of the Hebrew Bible, attached to the biblical text. The earliest Midrashim come from the 2nd century AD, although much of their content is older. [Bibl. Heb. midrāš commentary]

midrib /ˈmɪdrɪb/ n. the central rib of a leaf.

midriff /ˈmɪdrɪf/ n. **1 a** the region of the front of the body between the thorax and abdomen. **b** the diaphragm. **2** a garment or part of a garment covering the abdomen. [OE midhrif (as MID, hrif belly)]

midship /ˈmɪdʃɪp/ n. the middle part of a ship or boat.

midshipman /ˈmɪdˌʃɪpmən/ n. (pl. **-men**) **1** Brit. a naval officer of rank between naval cadet and sub-lieutenant. **2** US a naval cadet.

midships /ˈmɪdʃɪps/ adv. = AMIDSHIPS.

midst /mɪdst/ prep. & n. ● prep. poet. amidst. ● n. middle (now only in phrases as below). ☐ **in the midst of** among; in the middle of. **in our** (or **your** or **their**) **midst** among us (or you or them). [ME middest, middes f. in middes, in middan (as IN, MID)]

midsummer /mɪdˈsʌmə(r)/ n. the period of or near the summer solstice, about 21 June in the northern hemisphere. ☐ **midsummer madness** extreme folly. [OE midsumor (as MID, SUMMER[1])] ·

Midsummer Day (also **Midsummer's Day**) 24 June.

midtown /ˈmɪdtaʊn/ n. US the central part of a city between the downtown and uptown areas.

midway /ˈmɪdweɪ/ adv. in or towards the middle of the distance between two points.

Midway Islands two small islands with a surrounding coral atoll, in the central Pacific in the western part of the Hawaiian chain. The islands, which lie outside the state of Hawaii, were annexed by the US in 1867 and remain a US territory and naval base. They were the scene in 1942 of the decisive Battle of Midway, in which the US navy repelled a Japanese invasion fleet, sinking four aircraft carriers. This defeat marked the end of Japanese expansion in the Pacific during the Second World War.

Midwest /mɪdˈwest/ n. = MIDDLE WEST.

midwicket /mɪdˈwɪkɪt/ n. Cricket a fielding position on the leg side opposite the middle of the pitch.

midwife /ˈmɪdwaɪf/ n. (pl. **-wives**) a person (usu. a woman) trained to assist women in childbirth. ☐ **midwife toad** a European toad, Alytes obstetricans, the male of which carries the developing eggs wrapped round his hind legs. ☐ **midwifery** /-ˌwɪfrɪ/ n. [ME, prob. f. obs. prep. mid with + WIFE woman, in the sense of 'a person who is with the mother']

midwinter /mɪdˈwɪntə(r)/ n. the period of or near the winter solstice, about 22 Dec. in the northern hemisphere. [OE (as MID, WINTER)]

mielie var. of MEALIE.

mien /miːn/ n. literary a person's look or bearing, as showing character or mood. [prob. f. obs. demean f. DEMEAN[2], assim. to F mine expression]

Mies van der Rohe /ˌmiːz væn də ˈrəʊə/, Ludwig (1886–1969), German-born architect and designer. In the 1920s he produced unexecuted designs for glass skyscrapers and designed the German pavilion at the 1929 International Exhibition at Barcelona; the latter is regarded as a classic example of pure geometrical architecture. He is also noted for his design of tubular steel furniture, notably the 'Barcelona Chair'. He succeeded Gropius as director of the Bauhaus

1930–3. He emigrated to the US in 1937; his most celebrated American design is probably the Seagram Building in New York (1954–8).

mifepristone /ˌmɪfeˈprɪstəʊn/ n. Pharm. a synthetic steroid that inhibits the action of progesterone, given orally in early pregnancy to induce abortion. [AMINO + fe- (representing PHENOL) + PROPANE + ist- (representing oest- as in OESTRADIOL) + -ONE]

miff /mɪf/ v. & n. colloq. ● v.tr. (usu. in passive) put out of humour; offend. ● n. **1** a petty quarrel. **2** a huff. [perh. imit.: cf. G muff, exclam. of disgust]

might[1] /maɪt/ past of MAY, used esp. **1** in reported speech, expressing possibility (said he might come) or permission (asked if I might leave) (cf. MAY 1, 2). **2** (foll. by perfect infin.) expressing a possibility based on a condition not fulfilled (if you'd looked you might have found it; but for the radio we might not have known). **3** (foll. by present infin. or perfect infin.) expressing complaint that an obligation or expectation is not or has not been fulfilled (you might help instead of just standing there; they might have asked; you might have known they wouldn't come). **4** expressing a request (you might call in at the butcher's). **5** colloq. **a** = MAY 1 (it might be true). **b** (in tentative questions) = MAY 2 (might I have the pleasure of this dance?). **c** = MAY 4 (who might you be?). ☐ **might as well** expressing that it is probably at least as desirable to do a thing as not to do it (finished the work and decided they might as well go to lunch; won't win but might as well try). **might-have-been** colloq. **1** a past possibility that no longer applies. **2** a person who could have been more eminent.

might[2] /maɪt/ n. **1** great bodily or mental strength. **2** power to enforce one's will (usu. in contrast with right). ☐ **with all one's might** to the utmost of one's power. **with might and main** see MAIN[1]. [OE miht, mieht f. Gmc, rel. to MAY]

mightn't /ˈmaɪt(ə)nt/ contr. might not.

mighty /ˈmaɪtɪ/ adj. & adv. ● adj. (**mightier, mightiest**) **1** powerful or strong, in body, mind, or influence. **2** massive, bulky. **3** colloq. great, considerable. ● adv. colloq. very (a mighty difficult task). ☐ **mightily** adv. **mightiness** n. [OE mihtig (as MIGHT[2])]

mignonette /ˌmɪnjəˈnet/ n. **1 a** a plant of the genus Reseda; esp. R. odorata, with spikes of fragrant grey-green flowers. **b** the colour of these. **2** a light fine narrow pillow-lace. [F mignonnette dimin. of mignon small]

migraine /ˈmiːɡreɪn, ˈmaɪɡ-/ n. a recurrent throbbing headache that usually affects one side of the head, often accompanied by nausea and disturbance of vision. ☐ **migrainous** adj. [F f. LL hemicrania f. Gk hēmikrania (as HEMI-, CRANIUM): orig. of a headache confined to one side of the head]

migrant /ˈmaɪɡrənt/ adj. & n. ● adj. that migrates. ● n. a migrant person or animal, esp. a bird.

migrate /maɪˈɡreɪt/ v.intr. **1** (of people) move from one place of abode to another, esp. in a different country. **2** (of an animal, esp. a bird or fish) change its area of habitation with the seasons. **3** move under natural forces. ☐ **migrator** n. **migration** /-ˈɡreɪʃ(ə)n/ n. **migrational** adj. **migratory** /ˈmaɪɡrətərɪ/ adj. [L migrare migrat-]

Mihailović /mɪˈhaɪləˌvɪtʃ/, Dragoljub ('Draža') (1893–1946), Yugoslav soldier. Born in Serbia, he was leader of the Chetniks during the Second World War, resisting the German occupation of Yugoslavia. In 1943 he became Minister of War for the Yugoslav government in exile, but as Chetnik relations with the Communist partisans grew more strained, Allied support went to Tito and Mihailović was forced to go into hiding. After the war, he was tried and shot for collaboration and war crimes.

mihrab /ˈmiːrɑːb/ n. a niche or slab in a mosque, used to show the direction of Mecca. [Arab. miḥrāb praying-place]

mikado /mɪˈkɑːdəʊ/ n. (pl. **-os**) hist. the emperor of Japan. [Jap. f. mi august + kado gate]

Mike /maɪk/ n. sl. ☐ **for the love of Mike** an exclamation of entreaty or dismay. [abbr. of the name Michael]

mike[1] /maɪk/ n. colloq. a microphone. [abbr.]

mike[2] /maɪk/ v. & n. Brit. sl. ● v.intr. shirk work; idle. ● n. an act of shirking. [19th c.: orig. unkn.]

Míkonos see MYKONOS.

mil /mɪl/ n. one-thousandth of an inch, as a unit of measure for the diameter of wire etc. [L millesimum thousandth f. mille thousand]

milady /mɪˈleɪdɪ/ n. (pl. **-ies**) **1** an English noblewoman or great lady. **2** a form used in speaking of or to such a person. [F f. Engl. my lady: cf. MILORD]

milage var. of MILEAGE.

Milan /mɪˈlæn/ (Italian **Milano** /miˈlaːno/) an industrial city in NW Italy, capital of Lombardy region; pop. (1990) 1,432,180. Settled by the Gauls in about 600 BC, it was taken by the Romans in 222 BC, becoming the second city, after Rome, of the Western Empire. Although devastated by Attila the Hun in AD 452, it regained its powerful status, particularly from the 13th to the 15th centuries as a duchy under the Visconti and Sforza families. From the 16th century it was contested by the Habsburgs and the French, finally becoming a part of Italy in 1860. The city is today a leading financial and commercial centre and is known for its large opera house, La Scala.

Milan, Edict of an edict made by the Roman emperor Constantine in 313 which recognized Christianity and gave freedom of worship in the Roman Empire.

Milanese /ˌmɪləˈniːz/ adj. & n. ● adj. of or relating to Milan. ● n. (pl. same) a native of Milan. □ **Milanese silk** a finely woven silk or rayon.

milch /mɪltʃ/ adj. (of a domestic mammal) giving or kept for milk. □ **milch cow** a source of easy profit, esp. a person. [ME m(i)elche repr. OE mielce (unrecorded) f. Gmc: see MILK]

mild /maɪld/ adj. & n. ● adj. **1** (esp. of a person) gentle and conciliatory. **2** (of a rule, punishment, illness, feeling, etc.) moderate; not severe. **3** (of the weather, esp. in winter) moderately warm. **4** (of food, tobacco, etc.) not sharp or strong in taste etc. **5** (of medicine) operating gently. **6** tame, feeble; lacking energy or vivacity. ● n. Brit. beer not strongly flavoured with hops (cf. BITTER). □ **mild steel** steel containing a small percentage of carbon, strong and tough but not readily tempered. □ **milden** v.tr. & intr. **mildish** adj. **mildness** n. [OE milde f. Gmc]

mildew /ˈmɪldjuː/ n. & v. ● n. **1** a destructive growth of minute fungi on plants. **2** a similar growth on paper, leather, etc. exposed to damp. ● v.tr. & intr. taint or be tainted with mildew. □ **mildewy** adj. [OE mildēaw f. Gmc]

mildly /ˈmaɪldlɪ/ adv. in a mild fashion. □ **to put it mildly** as an understatement (implying the reality is more extreme).

mile /maɪl/ n. **1** (also **statute mile**) a unit of linear measure equal to 1,760 yards (approx. 1.609 kilometres). (See note below.) **2** hist. a Roman measure of 1,000 paces (approx. 1,620 yards). **3** (in pl.) colloq. a great distance or amount (miles better; beat them by miles). **4** a race extending over a mile. [OE mīl ult. f. L mil(l)ia pl. of mille thousand (see sense 2)]

▪ The mile was originally the Roman measure of 1,000 paces. Its length as a unit has varied considerably at different periods and in different localities, chiefly owing to the influence of the agricultural system of measures with which it was brought into relation (see FURLONG). The nautical mile, used in navigation, is a unit of 2,025 yards (1.852 km).

mileage /ˈmaɪlɪdʒ/ n. (also **milage**) **1 a** a number of miles travelled, used, etc. **b** the number of miles travelled by a vehicle per unit of fuel. **2** travelling expenses (per mile). **3** colloq. benefit, profit, advantage.

milepost /ˈmaɪlpəʊst/ n. a post one mile from the finishing-post of a race etc.

miler /ˈmaɪlə(r)/ n. colloq. a person or horse qualified or trained specially to run a mile.

milestone /ˈmaɪlstəʊn/ n. **1** a stone set up beside a road to mark a distance in miles. **2** a significant event or stage in a life, history, project, etc.

Miletus /maɪˈliːtəs/ an ancient city of the Ionian Greeks in SW Asia Minor. In the 7th and 6th centuries BC it was a powerful port, from which more than 60 colonies were founded on the shores of the Black Sea and in Italy and Egypt. In the same period it was the home of the philosophers Thales, Anaximander, and Anaximenes. It was conquered by the Persians in 494 BC. By the 6th century AD its harbours had become silted up by the alluvial deposits of the Menderes river.

milfoil /ˈmɪlfɔɪl/ n. **1** the common yarrow, Achillea millefolium. **2** (in full **water milfoil**) an aquatic plant of the genus Myriophyllum, with whorls of finely divided leaves. [ME f. OF f. L millefolium f. mille thousand + folium leaf, after Gk muriophullon]

Milhaud /ˈmiːjəʊ/, Darius (1892–1974), French composer. After travelling in Brazil, he became a member of the group known as Les Six. Milhaud composed the music to Cocteau's ballet Le Boeuf sur le toit (1919), and his contact with Latin American music formed the inspiration for his two dance suites Saudades do Brasil (1920–1). Much of his music was polytonal, and, after 1920, influenced by jazz.

miliary /ˈmɪlɪərɪ/ adj. **1** like a millet-seed in size or form. **2** (of a disease) having as a symptom a rash with lesions resembling millet-seed. [L miliarius f. milium millet]

milieu /ˈmiːljɜː, ˈmiːljɜː/ n. (pl. **milieux** or **milieus** /-ˈljɜːz/) one's environment or social surroundings. [F f. mi MID + lieu place]

Militant /ˈmɪlɪt(ə)nt/ n. a Trotskyite political organization in Britain, often referred to as the Militant Tendency, which publishes the weekly newspaper Militant. During the early 1980s, its tactics included widely publicized attempts to infiltrate and express its views from within the Labour Party; after 1983 the Labour Party leadership attempted systematically to expel or exclude Militant sympathizers from the party.

militant /ˈmɪlɪt(ə)nt/ adj. & n. ● adj. **1** combative; aggressively active esp. in support of a (usu. political) cause. **2** engaged in warfare. ● n. **1** a militant person, esp. a political activist. **2** a person engaged in warfare. □ **militancy** n. **militantly** adv. [ME f. OF f. L (as MILITATE)]

militarism /ˈmɪlɪtəˌrɪz(ə)m/ n. **1** the spirit or tendencies of a professional soldier. **2** undue prevalence of the military spirit or ideals. □ **militaristic** /ˌmɪlɪtəˈrɪstɪk/ adj. **militaristically** adv. [F militarisme (as MILITARY)]

militarist /ˈmɪlɪtərɪst/ n. **1** a person dominated by militaristic ideas. **2** a student of military science.

militarize /ˈmɪlɪtəˌraɪz/ v.tr. (also **-ise**) **1** equip with military resources. **2** make military or warlike. **3** imbue with militarism. □ **militarization** /ˌmɪlɪtəraɪˈzeɪʃ(ə)n/ n.

military /ˈmɪlɪtərɪ/ adj. & n. ● adj. of, relating to, or characteristic of soldiers or armed forces. ● n. (as sing. or pl.; prec. by the) members of the armed forces, as distinct from civilians and the police. □ **military honours** marks of respect paid by troops at the burial of a soldier, to royalty, etc. **military police** a corps responsible for police and disciplinary duties in the army. **military policeman** a member of the military police. □ **militarily** adv. **militariness** n. [F militaire or L militaris f. miles militis soldier]

militate /ˈmɪlɪˌteɪt/ v.intr. (usu. foll. by against) (of facts or evidence) have force or effect (what you say militates against our opinion). ¶ Often confused with mitigate. [L militare militat- f. miles militis soldier]

militia /mɪˈlɪʃə/ n. a military force, esp. one raised from the civil population and supplementing a regular army in an emergency. [L, = military service f. miles militis soldier]

militiaman /mɪˈlɪʃəmən/ n. (pl. **-men**) a member of a militia.

milk /mɪlk/ n. & v. ● n. **1** an opaque white fluid secreted by female mammals for the nourishment of their young. **2** the milk of cows, goats, or sheep as food. **3** the milklike juice of plants, e.g. in the coconut. **4** a milklike preparation of herbs, drugs, etc. ● v.tr. **1** draw milk from (a cow, ewe, goat, etc.). **2 a** exploit (a person) esp. financially. **b** get all possible advantage from (a situation). **3** extract sap, venom, etc. from. **4** sl. tap (telegraph or telephone wires etc.). □ **cry over spilt milk** lament an irremediable loss or error. **milk and honey** abundance, prosperity. **milk and water** a feeble or insipid or mawkish discourse or sentiment. **milk bar** a snack bar selling milk drinks and other refreshments. **milk chocolate** chocolate for eating, made with milk. **milk fever 1** an acute illness in female cows, goats, etc. that have just produced young, caused by calcium deficiency. **2** Med. a fever in women caused by infection after childbirth (formerly supposed to be due to the swelling of the breasts with milk). **milk float** Brit. a small usu. electric vehicle used in delivering milk. **milk-leg** a painful swelling, esp. of the legs, after childbirth. **milk-loaf** a loaf of bread made with milk. **milk of human kindness** kindness regarded as natural to humanity. **milk of sulphur** the amorphous powder of sulphur formed by precipitation. **milk-powder** milk dehydrated by evaporation. **milk pudding** a pudding of rice, sago, tapioca, etc., baked with milk in a dish. **milk round 1** a fixed route on which milk is delivered regularly. **2** colloq. a regular trip or tour involving calls at several places. **milk run** colloq. a routine expedition or service journey. **milk shake** a drink of milk, flavouring, etc., mixed by shaking or whisking. **milk sugar** lactose. **milk tooth** a temporary tooth in young mammals. **milk-vetch** a yellow-flowered leguminous plant of the genus Astragalus. **milk-white** white like milk. □ **milker** n. [OE milc, milcian f. Gmc]

milkmaid /ˈmɪlkmeɪd/ n. a girl or woman who milks cows or works in a dairy.

milkman /ˈmɪlkmən/ n. (pl. **-men**) a person who sells or delivers milk.

Milk of Magnesia n. Brit. propr. a white suspension of hydrated magnesium carbonate usu. in water as an antacid or laxative.

milksop /ˈmɪlksɒp/ n. a spiritless man or youth.

milkweed /ˈmɪlkwiːd/ n. **1** a plant with milky sap, esp. a poisonous

North American plant of the genus *Asclepias*. **2** (in full **milkweed butterfly**) = MONARCH 4.

milkwort /ˈmɪlkwɜːt/ *n.* a small flowering plant of the genus *Polygala*, formerly supposed to increase women's milk.

milky /ˈmɪlkɪ/ *adj.* (**milkier**, **milkiest**) **1** of, like, or mixed with milk. **2** (of a gem or liquid) cloudy; not clear. **3** effeminate; weakly amiable. □ **milkiness** *n.*

Milky Way *Astron.* a faint band of light crossing the sky, clearly visible on dark moonless nights and discovered by Galileo to be made up of vast numbers of faint stars. It corresponds to the plane of our Galaxy, in which most of its stars are located. (See also GALAXY.)

Mill /mɪl/, J(ohn) S(tuart) (1806–73), English philosopher and economist. He won recognition as a philosopher with his defence of empiricism in *System of Logic* (1843). Mill is best known, however, for his political and moral works, especially *On Liberty* (1859), which argued for the importance of individuality, and *Utilitarianism* (1861), which developed Bentham's theory, considering explicitly the relation between utilitarianism and justice. In other works he advocated representative democracy, criticized the contemporary treatment of married women, and claimed that an end to economic growth was desirable as well as inevitable.

mill¹ /mɪl/ *n. & v.* ● *n.* **1 a** a building fitted with a mechanical apparatus for grinding corn. **b** such an apparatus. **2** an apparatus for grinding a solid substance to powder or pulp (*pepper-mill*). **3 a** a building fitted with machinery for manufacturing processes etc. (*cotton-mill*). **b** such machinery. **4** *sl.* a boxing-match; a fist-fight. ● *v.* **1** *tr.* grind (a substance) in a mill, esp. grind (corn), produce (flour), or hull (seeds) in a mill. **2** *tr.* produce regular ribbed markings on the edge of (a coin). **3** *tr.* cut or shape (metal) with a rotating tool. **4** *intr.* (often foll. by *about*, *around*) (of people or animals) move in an aimless manner, esp. in a confused mass. **5** *tr.* thicken (cloth etc.) by fulling. **6** *tr.* beat (chocolate etc.) to froth. **7** *tr. sl.* beat, strike, fight. □ **go** (or **put**) **through the mill** undergo (or cause to undergo) intensive work or training etc. **mill-dam** a dam put across a stream to make it usable by a mill. **mill-hand** a worker in a mill or factory. **mill-race** a current of water that drives a mill-wheel. **mill-wheel** a wheel used to drive a water-mill. □ **millable** *adj.* [OE *mylen* ult. f. LL *molinum* f. L *mola* grindstone, mill f. *molere* grind]

mill² /mɪl/ *n. N. Amer.* one-thousandth of a dollar as money of account. [L *millesimum* thousandth: cf. CENT]

Millais /ˈmɪleɪ/, Sir John Everett (1829–96), English painter. A founder member of the Pre-Raphaelite Brotherhood, he initially adhered to a pure vision of nature with unidealized figures in such paintings as *Christ in the House of his Parents* (1850). However, with the success of *The Blind Girl* (1856) and other paintings, he gradually departed from the moral and aesthetic rigour of the early Pre-Raphaelites, going on to produce lavishly painted portraits, landscapes, and sentimental genre pictures, notably *Bubbles* (1886).

millboard /ˈmɪlbɔːd/ *n.* stout pasteboard for bookbinding etc.

Mille, Cecil B. de, see DE MILLE.

millefeuille /ˈmiːlˌfɜːɪ/ *n.* a rich confection of puff pastry split and filled with jam, cream, etc. [F, = thousand-leaf]

millenarian /ˌmɪlɪˈneərɪən/ *adj. & n.* ● *adj.* **1** of or related to the millennium. **2** believing in the millennium or a millennium. ● *n.* a person who believes in the millennium or a millennium. (See also MILLENARIANISM.) [as MILLENARY]

millenarianism /ˌmɪlɪˈneərɪəˌnɪz(ə)m/ *n.* in some Christian groups, the doctrine of or belief in the coming of the millennium. Millenarianist belief includes both those who believe the second coming of Christ will coincide with the beginning of the 1,000-year period of the millennium, and those who believe it will come at the end. There was a revival of interest in millenarianism during the 19th century, especially in the US; present-day millenarianist groups include the Mormons, Seventh-Day Adventists, and Jehovah's Witnesses. The term has also been used more recently in the social sciences to refer to any religious and political group seeking solutions to present crises through rapid and radical transformation of politics and society. □ **millenarianist** *n. & adj.*

millenary /mɪˈlenərɪ/ *n. & adj.* ● *n.* (*pl.* **-ies**) **1** a period of 1,000 years. **2** the festival of the 1,000th anniversary of a person or thing. **3** a person who believes in the millennium or a millennium. ● *adj.* of or relating to a millenary. [LL *millenarius* consisting of a thousand f. *milleni* distributive of *mille* thousand]

millennium /mɪˈlenɪəm/ *n.* (*pl.* **millenniums** or **millennia** /-nɪə/) **1** a

period of 1,000 years. **2** the period of 1,000 years during which, according to one interpretation, Christ will reign in person on earth (Rev. 20:1–5). According to another interpretation, the second coming of Christ will take place at the end of the 1,000-year period. (See MILLENARIANISM.) **2** a period of good government, great happiness, and prosperity. □ **millennial** *adj.* **millennialist** *n. & adj.* [mod.L f. L *mille* thousand after BIENNIUM]

millepede var. of MILLIPEDE.

millepore /ˈmɪlɪˌpɔː(r)/ *n. Zool.* a reef-building coral of the order Milleporina, with polyps protruding through minute pores in the calcareous skeleton. [F *millépore* or mod.L *millepora* f. L *mille* thousand + *porus* PORE¹]

Miller¹ /ˈmɪlə(r)/, (Alton) Glenn (1904–44), American jazz trombonist and band-leader. From 1938 he led his celebrated swing big band, with whom he recorded his signature tune 'Moonlight Serenade'. He joined the US army in 1942 and died when his aircraft disappeared on a routine flight across the English Channel.

Miller² /ˈmɪlə(r)/, Arthur (b.1915), American dramatist. He established his reputation with *Death of a Salesman* (1949), before writing *The Crucible* (1953), which used the Salem witch trials of 1692 as an allegory for McCarthyism in America in the 1950s. Miller was married to Marilyn Monroe from 1955 to 1961.

Miller³ /ˈmɪlə(r)/, Henry (Valentine) (1891–1980), American novelist. From 1930 to 1940 he lived in France, where he published the autobiographical novels *Tropic of Cancer* (1934), about his life in Paris, and *Tropic of Capricorn* (1939), dealing with his youth in New York. Their frank depiction of sex and use of obscenities caused them to be banned in the US until the 1960s. Other works include *The Air-Conditioned Nightmare* (1945), reflections on a return to the US.

miller /ˈmɪlə(r)/ *n.* **1** the proprietor or tenant of a corn-mill. **2** a person who works or owns a mill. □ **miller's thumb** a small spiny freshwater fish, *Cottus gobio*, with a large head (cf. BULLHEAD). [ME *mylnere*, prob. f. MLG, MDu. *molner*, *mulner*, OS *mulineri* f. LL *molinarius* f. *molina* MILL¹, assim. to MILL¹]

millesimal /mɪˈlesɪm(ə)l/ *adj. & n.* ● *adj.* **1** thousandth. **2** of or belonging to a thousandth. **3** of or dealing with thousandths. ● *n.* a thousandth part. □ **millesimally** *adv.* [L *millesimus* f. *mille* thousand]

millet /ˈmɪlɪt/ *n.* **1** a cereal plant, esp. *Panicum miliaceum*, bearing a large crop of small nutritious seeds. **2** the seed of this. □ **millet-grass** a tall woodland grass, *Milium effusum*. [ME f. F, dimin. of *mil* f. L *milium*]

milli- /ˈmɪlɪ/ *comb. form* (abbr. **m**) a thousand, esp. denoting a factor of one-thousandth. [L *mille* thousand]

milliammeter /ˌmɪlɪˈæmɪtə(r)/ *n.* an instrument for measuring electrical current in milliamperes.

milliampere /ˌmɪlɪˈæmpeə(r)/ *n.* one-thousandth of an ampere, a measure for small electrical currents.

milliard /ˈmɪljəd, -jɑːd/ *n. Brit.* one thousand million. ¶ Now largely superseded by *billion*. [F f. *mille* thousand]

millibar /ˈmɪlɪˌbɑː(r)/ *n.* one-thousandth of a bar, the cgs unit of atmospheric pressure equivalent to 100 pascals.

milligram /ˈmɪlɪˌgræm/ *n.* (also **-gramme**) one-thousandth of a gram.

Millikan /ˈmɪlɪkən/, Robert Andrews (1868–1953), American physicist. He was the first to give an accurate figure for the electric charge on an electron. Progressing from this to study the photoelectric effect, he confirmed the validity of Einstein's equation and gave an accurate figure for Planck's constant. Millikan also worked on the spectrometry of the lighter elements, and investigated cosmic rays. He was awarded the 1923 Nobel Prize for physics, and did much to establish the scientific reputation of the California Institute of Technology.

millilitre /ˈmɪlɪˌliːtə(r)/ *n.* (US **milliliter**) one-thousandth of a litre (0.002 pint).

millimetre /ˈmɪlɪˌmiːtə(r)/ *n.* (US **millimeter**) one-thousandth of a metre (0.039 in.).

milliner /ˈmɪlɪnə(r)/ *n.* a person who makes or sells women's hats. □ **millinery** *n.* [orig. = vendor of goods from Milan]

million /ˈmɪljən/ *n. & adj.* ● *n.* (*pl.* same or (in sense 2) **millions**) (in *sing.* prec. by *a* or *one*) **1** a thousand thousand (1,000,000). **2** (in *pl.*) *colloq.* a very large number (*millions of years*). **3** (prec. by *the*) the bulk of the population. **4 a** *Brit.* a million pounds. **b** *N. Amer.* a million dollars. ● *adj.* that amount to a million. □ **gone a million** *Austral. sl.* completely defeated. □ **millionfold** *adj. & adv.* **millionth** *adj. & n.* [ME f. OF, prob. f. It. *millione* f. *mille* thousand + *-one* augment. suffix]

millionaire /ˌmɪljəˈneə(r)/ *n.* (*fem.* **millionairess** /-rɪs/) **1** a person whose assets are worth at least a million pounds, dollars, etc. **2** a person of great wealth. [F *millionnaire* (as MILLION)]

millipede /'mɪlɪˌpiːd/ *n*. (also **millepede**) a herbivorous arthropod of the class Diplopoda, having an elongated segmented body with two pairs of legs on each segment. [L *millepeda* wood-louse f. *mille* thousand + *pes pedis* foot]

millisecond /'mɪlɪˌsekənd/ *n*. one-thousandth of a second.

millpond /'mɪlpɒnd/ *n*. a pool of water retained by a mill-dam for the operation of a mill. □ **like a millpond** (of a stretch of water) very calm.

Mills /mɪlz/, Sir John (b.1908), English actor. He appeared in war films such as *This Happy Breed* (1944), and also in classics such as *Great Expectations* (1946), where he played the adult Pip, and adventure films such as *Scott of the Antarctic* (1948). Mills won an Oscar for his portrayal of a village idiot in *Ryan's Daughter* (1971). His daughters Juliet Mills (b.1941) and Hayley Mills (b.1946) have also had acting careers.

Mills bomb *n*. an oval hand-grenade. [Sir William Mills, Engl. engineer (1856–1932)]

millstone /'mɪlstəʊn/ *n*. **1** either of two circular stones used for grinding corn. **2** a heavy burden or responsibility (cf. Matt. 18:6).

millwright /'mɪlraɪt/ *n*. a person who designs or builds mills.

Milne /mɪln/, A(lan) A(lexander) (1882–1956), English writer of stories and poems for children. He is remembered for his series of nursery stories written for his son Christopher Robin (b.1920), namely *Winnie-the-Pooh* (1926) and *The House at Pooh Corner* (1928). He also wrote two collections of verse for children: *When We Were Very Young* (1924) and *Now We Are Six* (1927).

milometer /maɪˈlɒmɪtə(r)/ *n*. an instrument for measuring the number of miles travelled by a vehicle.

milord /mɪˈlɔːd/ *n. hist*. an Englishman travelling in Europe in aristocratic style. [F f. Engl. *my lord*: cf. MILADY]

Milos see MELOS.

milt /mɪlt/ *n*. **1** the spleen in mammals. **2** an analogous organ in other vertebrates. **3 a** a sperm-filled reproductive gland of a male fish. **b** the semen of a male fish. [OE *milt(e)* f. Gmc, perh. rel. to MELT]

milter /'mɪltə(r)/ *n*. a male fish in spawning-time.

Milton /'mɪlt(ə)n/, John (1608–74), English poet. Milton's prolific early writings include the masque *Comus* (1637) and the elegy 'Lycidas' (1638). He became politically active during the Civil War, publishing the *Areopagitica* (1644), which demanded a free press, and writing a defence of republicanism on the eve of the Restoration (1660). His three major poems, all completed after he had gone blind (1652), have biblical subjects and show his skilful use of blank verse: they are *Paradise Lost* (1667, revised 1674), an epic on the fall of man, *Paradise Regained* (1671), on Christ's temptations, and *Samson Agonistes* (1671), on Samson's final years.

Milton Keynes /ˌmɪlt(ə)n 'kiːnz/ a town in south central England, in Buckinghamshire; pop. (1991) 172,300. It was established as a new town in the late 1960s, and is the site of the headquarters of the Open University.

Milwaukee /mɪlˈwɔːkiː/ an industrial port and city in SE Wisconsin, on the west shore of Lake Michigan; pop. (1990) 628,090. The city developed from the early 1830s, attracting large numbers of settlers from Germany in the 1840s and from Poland and Italy later in the century. It is noted for its brewing industry and is an important port on the St Lawrence Seaway.

Mimas /'maɪmæs, -məs/ **1** *Gk Mythol*. one of the giants, killed by Mars. **2** *Astron*. satellite I of Saturn, the seventh closest to the planet, discovered by William Herschel in 1789 (diameter 390 km). It is heavily cratered, with one crater a third the diameter of the whole satellite, and is probably composed entirely of ice.

mimbar /'mɪmbɑː(r)/ *n*. (also **minbar** /'mɪnbɑː(r)/) a stepped platform for preaching in a mosque. [Arab. *minbar*]

mime /maɪm/ *n*. & *v*. ● *n*. **1** the theatrical technique of suggesting action, character, etc. by gesture and expression without using words. **2** a theatrical performance using this technique. **3** *Gk & Rom. Antiq*. a simple farcical drama including mimicry. **4** (also **mime artist**) a practitioner of mime. ● *v*. **1** *tr*. (also *absol*.) convey (an idea or emotion) by gesture without words. **2** *intr*. (often foll. by *to*) (of singers etc.) mouth the words of a song etc. along with a soundtrack (*mime to a record*). □ **mimer** *n*. [L *mimus* f. Gk *mimos*]

mimeograph /'mɪmɪəˌɡrɑːf/ *n*. & *v*. ● *n*. **1** (often *attrib*.) a duplicating machine which produces copies from a stencil. **2** a copy produced in

this way. ● *v.tr*. reproduce (text or diagrams) by this process. [irreg. f. Gk *mimeomai* imitate: see -GRAPH]

mimesis /mɪˈmiːsɪs, maɪ-/ *n. Biol*. a close external resemblance of an animal to another that is distasteful or harmful to predators of the first. [Gk *mimēsis* imitation]

mimetic /mɪˈmetɪk/ *adj*. **1** relating to or habitually practising imitation or mimicry. **2** *Biol*. of or exhibiting mimesis. □ **mimetically** *adv*. [Gk *mimētikos* imitation (as MIMESIS)]

mimic /'mɪmɪk/ *v*, *n*., & *adj*. ● *v.tr*. (**mimicked, mimicking**) **1** imitate (a person, gesture, etc.) esp. to entertain or ridicule. **2** copy minutely or servilely. **3** (of a thing) resemble closely. ● *n*. a person skilled in imitation. ● *adj*. having an aptitude for mimicry; imitating; imitative of a thing, esp. for amusement. □ **mimicker** *n*. [L *mimicus* f. Gk *mimikos* (as MIME)]

mimicry /'mɪmɪkrɪ/ *n*. (*pl*. **-ies**) **1** the act or art of mimicking. **2** a thing that mimics another. **3** *Biol*. a close external resemblance of an animal (or part of an animal) to another animal, or to a plant or inanimate object; a similar resemblance in a plant. (See also BATESIAN MIMICRY, MÜLLERIAN MIMICRY.)

miminy-piminy /ˌmɪmɪnɪˈpɪmɪnɪ/ *adj*. overrefined, finicky (cf. NIMINY-PIMINY, NAMBY-PAMBY). [imit.]

mimosa /mɪˈməʊzə/ *n*. **1** a leguminous shrub of the genus *Mimosa*; esp. the sensitive plant, *M. pudica*, having globular usu. yellow flowers and sensitive leaflets which droop when touched. **2** an acacia with showy yellow flowers. [mod.L, app. f. L (as MIME, because the leaves imitate animals in their sensitivity) + -osa fem. suffix]

mimulus /'mɪmjʊləs/ *n*. a flowering plant of the genus *Mimulus*, including musk and the monkey flower. [mod.L, app. dimin. of L (as MIME, perh. with ref. to its masklike flowers)]

Min /mɪn/ *n*. any of the Chinese languages or dialects spoken in the Fukien province in SE China. [Chin.]

Min. *abbr*. **1** Minister. **2** Ministry.

min. *abbr*. **1** minute(s). **2** minimum. **3** minim (fluid measure).

mina var. of MYNAH.

Minaean /mɪˈniːən/ *n*. & *adj*. ● *n*. **1** a native or inhabitant of Ma'in, an ancient kingdom of southern Arabia (*c*.400 BC), which was absorbed into the Sabaean kingdom by the late 1st century BC. **2** the Semitic language of this people. ● *adj*. of the Minaeans or their language. [L *Minaeus* f. Arab. *Ma'in*]

minaret /ˌmɪnəˈret/ *n*. a slender turret connected with a mosque and having a balcony from which the muezzin calls at hours of prayer. □ **minareted** *adj*. [F *minaret* or Sp. *minarete* f. Turk. *minare* f. Arab. *manār(a)* lighthouse, minaret f. *nār* fire, light]

Minas Gerais /ˌmiːnəs ʒeˈraɪs/ a state of SE Brazil; capital, Belo Horizonte. It has major deposits of iron ore, coal, gold, and diamonds.

minatory /'mɪnətərɪ/ *adj*. threatening, menacing. [LL *minatorius* f. *minari minat-* threaten]

minbar var. of MIMBAR.

mince /mɪns/ *v*. & *n*. ● *v*. **1** *tr*. cut up or grind (esp. meat) into very small pieces. **2** *tr*. (usu. with *neg*.) restrain (one's words etc.) within the bounds of politeness. **3** *intr*. (usu. as **mincing** *adj*.) speak or walk with an affected delicacy. ● *n*. esp. *Brit*. minced meat. □ **mince matters** (usu. with *neg*.) use polite expressions etc. **mince pie** a usu. small round pie containing mincemeat. □ **mincer** *n*. **mincingly** *adv*. (in sense 3 of *v*.). [ME f. OF *mincier* ult. f. L (as MINUTIA)]

mincemeat /'mɪnsmiːt/ *n*. a mixture of currants, raisins, sugar, apples, candied peel, spices, and often suet. □ **make mincemeat of** utterly defeat (a person, argument, etc.).

Minch, the /mɪntʃ/ (also **the Minches**) a channel of the Atlantic, between the mainland of Scotland and the Outer Hebrides. The northern stretch is called the *North Minch*, the southern stretch, north-west of Skye, is called the *Little Minch*.

mind /maɪnd/ *n*. & *v*. ● *n*. **1 a** the seat of awareness, thought, volition, and feeling. **b** attention, concentration (*my mind keeps wandering*). **2** the intellect; intellectual powers; aptitude. **3** remembrance, memory (*it went out of my mind; I can't call it to mind*). **4** one's opinion (*we're of the same mind*). **5** a way of thinking or feeling (*shocking to the Victorian mind*). **6** the focus of one's thoughts or desires (*put one's mind to it*). **7** the state of normal mental functioning (*lose one's mind; in one's right mind*). **8** a person as embodying mental faculties (*a great mind*). ● *v.tr*. **1** (usu. with *neg*. or *interrog*.) object to (*do you mind if I smoke?; I don't mind your being late*). **2 a** remember; take care to (*mind you come on time*). **b** (often foll. by *out*) take care; be careful. **3** have charge of temporarily (*mind the* *house while I'm away*). **4** apply oneself to, concern oneself with (business, affairs, etc.) (*I try to mind my own business*). **5** give heed to; notice (*mind the step; don't mind the expense; mind how you go*). **6** *N. Amer. & Ir.* be obedient to (*mind what your mother says*). □ **be in** (or esp. *N. Amer.* **of**) **two minds** be undecided. **be of a mind** (often foll. by *to* + infin.) be prepared or disposed. **cast one's mind back** think back; recall an earlier time. **come into a person's mind** be remembered. **come** (or **spring**) **to mind** (of a thought, idea, etc.) suggest itself. **don't mind me** *iron.* do as you please. **do you mind!** *iron.* an expression of annoyance. **give a person a piece of one's mind** scold or reproach a person. **have a good** (or **great** or **half a**) **mind to** (often as a threat, usu. unfulfilled) feel tempted to (*I've a good mind to report you*). **have** (**it**) **in mind** intend. **have a mind of one's own** be capable of independent opinion or action. **have on one's mind** be troubled by the thought of. **in one's mind's eye** in one's imagination or mental view. **mind-bending** *colloq.* (esp. of a psychedelic drug) influencing or altering one's state of mind. **mind-blowing** *sl.* **1** confusing, shattering. **2** (esp. of drugs etc.) inducing hallucinations. **mind-boggling** *colloq.* overwhelming, startling. **mind out for** *Brit.* guard against, avoid. **mind over matter** the power of the mind asserted over the physical universe. **mind one's Ps & Qs** be careful in one's behaviour. **mind-read** discern the thoughts of (another person). **mind-reader** a person capable of mind-reading. **mind-set** habits of mind formed by earlier events. **mind the shop** have charge of affairs temporarily. **mind you** an expression used to qualify a previous statement (*I found it quite quickly; mind you, it wasn't easy*). **mind your back** (or **backs**) *colloq.* an expression to indicate that a person wants to get past. **never mind 1** an expression used to comfort or console. **2** (also **never you mind**) an expression used to evade a question. **3** disregard (*never mind the cost*). **open** (or **close**) **one's mind to** be receptive (or unreceptive) to (changes, new ideas, etc.). **out of one's mind** crazy. **put a person in mind of** remind a person of. **put** (or **set**) **a person's mind at rest** reassure a person. **put a person or thing out of one's mind** deliberately forget. **read a person's mind** discern a person's thoughts. **spring to mind** = *come to mind*. **to my mind** in my opinion. [ME *mynd* f. OE *gemynd* f. Gmc]

Mindanao /ˌmɪndəˈnaʊ/ the second largest island in the Philippines, in the south-east of the group. Its chief town is Davao.

minded /ˈmaɪndɪd/ *adj.* **1** (in *comb.*) **a** inclined to think in some specified way (*mathematically minded; fair-minded*). **b** having a specified kind of mind (*high-minded*). **c** interested in or enthusiastic about a specified thing (*car-minded*). **2** (usu. foll. by *to* + infin.) disposed or inclined (to an action).

minder /ˈmaɪndə(r)/ *n.* **1** (esp. in *comb.*) a person whose job it is to attend to a person or thing (*child-minder; machine-minder*). **2** *sl.* **a** a bodyguard, esp. a person employed to protect a criminal. **b** a person employed to protect anyone, esp. from the media.

mindful /ˈmaɪndfʊl/ *adj.* (often foll. by *of*) taking heed or care; being conscious. □ **mindfully** *adv.* **mindfulness** *n.*

mindless /ˈmaɪndlɪs/ *adj.* **1** lacking intelligence; stupid. **2** not requiring thought or skill (*totally mindless work*). **3** (usu. foll. by *of*) heedless of (advice etc.). □ **mindlessly** *adv.* **mindlessness** *n.*

Mindoro /mɪnˈdɔːrəʊ/ an island in the Philippines. It is situated to the south-west of Luzon.

mine[1] /maɪn/ *poss.pron.* **1** the one or ones belonging to or associated with me (*it is mine; mine are over there*). **2** (*attrib.* before a vowel) *archaic* = MY (*mine eyes have seen; mine host*). □ **of mine** of or belonging to me (*a friend of mine*). [OE *mīn* f. Gmc]

mine[2] /maɪn/ *n. & v.* ● *n.* **1** an excavation in the earth for extracting metal, coal, salt, etc. **2** an abundant source (of information etc.). **3** a receptacle filled with explosive and placed in the ground or in the water for destroying enemy personnel, ships, etc. **4 a** a subterranean gallery in which explosive is placed to blow up fortifications. **b** *hist.* a subterranean passage under the wall of a besieged fortress. ● *v.tr.* **1** obtain (metal, coal, etc.) from a mine. **2** (also *absol.* often foll. by *for*) dig in (the earth etc.) for ore etc. **3 a** dig or burrow in (usu. the earth). **b** delve into (an abundant source) for information etc. **c** make (a hole, passage, etc.) underground. **4** lay explosive mines under or in. **5** = UNDERMINE. □ **mine-detector** an instrument for detecting the presence of military mines. □ **mining** *n.* [ME f. OF *mine, miner*, perh. f. Celt.]

minefield /ˈmaɪnfiːld/ *n.* **1** an area planted with explosive mines. **2** a subject or situation presenting unseen hazards.

minelayer /ˈmaɪnˌleɪə(r)/ *n.* a ship or aircraft for laying mines.

miner /ˈmaɪnə(r)/ *n.* **1** a person who works in a mine. **2** any burrowing insect or grub. □ **miner's right** *Austral.* a licence to dig for gold etc. on private or public land. [ME f. OF *minëor, minour* (as MINE[2])]

mineral /ˈmɪnərəl/ *n. & adj.* ● *n.* **1** any of the species into which inorganic substances (especially those occurring naturally) are classified, having a definite chemical composition and usually a characteristic crystalline structure. **2** a substance obtained by mining. **3** (often in *pl.*) *Brit.* an artificial mineral water or other effervescent drink. ● *adj.* **1** of or containing a mineral or minerals. **2** obtained by mining. □ **mineral oil** petroleum or one of its distillation products (see also OIL). **mineral water 1** water found in nature with some dissolved salts present. **2** an artificial imitation of this, esp. soda water. **3** any effervescent non-alcoholic drink. **mineral wax** a fossil resin, esp. ozocerite. **mineral wool** a wool-like substance made from inorganic material, used for packing etc. [ME f. OF *mineral* or med.L *mineralis* f. *minera* ore f. OF *miniere* mine]

mineralize /ˈmɪnərəˌlaɪz/ *v.* (also **-ise**) **1** *v.tr. & intr.* change wholly or partly into a mineral. **2** *v.tr.* impregnate (water etc.) with a mineral substance.

mineralogy /ˌmɪnəˈrælədʒɪ/ *n.* the scientific study of minerals. □ **mineralogist** *n.* **mineralogical** /-rəˈlɒdʒɪk(ə)l/ *adj.*

Minerva /mɪˈnɜːvə/ *Rom. Mythol.* the goddess of handicrafts, widely worshipped and regularly identified with Athene, which led to her being regarded also as the goddess of war.

minestrone /ˌmɪnɪˈstrəʊnɪ/ *n.* a soup containing vegetables, pasta, and beans. [It.]

minesweeper /ˈmaɪnˌswiːpə(r)/ *n.* a ship for clearing away floating and submarine mines.

minever var. of MINIVER.

mineworker /ˈmaɪnˌwɜːkə(r)/ *n.* a person who works in a mine, esp. a coal mine.

Ming /mɪŋ/ a Chinese dynasty founded in 1368 by Zhu Yuanzhang (1328–98) after the collapse of Mongol authority in China, and ruling until succeeded by the Manchus in 1644 (see QING). The Ming period was one of expansion and exploration, with lasting contact established in the 16th century between China and Europe, and one in which the arts flourished. The capital was established at Beijing in 1421. The term is particularly used to describe the porcelain made in China during this period, characterized by elaborate designs and vivid colours. [Chin., lit. 'bright, clear']

mingle /ˈmɪŋɡ(ə)l/ *v.* **1** *tr. & intr.* mix, blend. **2** *intr.* (often foll. by *with*) (of a person) move about, associate. □ **mingle their** etc. **tears** *literary* weep together. [ME *mengel* f. obs. *meng* f. OE *mengan*, rel. to AMONG]

Mingus /ˈmɪŋɡəs/, Charles (1922–79), American jazz bassist and composer. After studying the double bass in Los Angeles with Louis Armstrong, he became part of the 1940s jazz scene alongside Dizzy Gillespie, Thelonious Monk, and Charlie Parker. Well-known compositions of that time included 'Goodbye Porkpie Hat' and 'The Black Saint and the Sinner Lady'. Mingus's music reflects his experiments with atonality, as well as the influence of gospel and blues.

mingy /ˈmɪndʒɪ/ *adj.* (**mingier, mingiest**) *Brit. colloq.* mean, stingy. □ **mingily** *adv.* [perh. f. MEAN[2] and STINGY]

Minho see MIÑO.

mini /ˈmɪnɪ/ *n.* (*pl.* **minis**) **1** *colloq.* a miniskirt, minidress, etc. **2** (**Mini**) *propr.* a make of small car. [abbr.]

mini- /ˈmɪnɪ/ *comb. form* miniature; very small or minor of its kind (*minibus; mini-budget*). [abbr. of MINIATURE]

miniature /ˈmɪnɪtʃə(r)/ *adj., n., & v.* ● *adj.* **1** much smaller than normal. **2** represented on a small scale. ● *n.* **1** any object reduced in size. **2** a small-scale minutely finished portrait. **3** this branch of painting. **4** a picture or decorated letters in an illuminated manuscript. ● *v.tr.* represent on a smaller scale. □ **in miniature** on a small scale. **miniature camera** a camera producing small negatives. □ **miniaturist** *n.* (in senses 2 and 3 of *n.*). [It. *miniatura* f. med.L *miniatura* f. L *miniare* rubricate, illuminate f. L *minium* red lead, vermilion]

miniaturize /ˈmɪnɪtʃəˌraɪz/ *v.tr.* (also **-ise**) produce in a smaller version; make small. □ **miniaturization** /ˌmɪnɪtʃəraɪˈzeɪʃ(ə)n/ *n.*

minibus /ˈmɪnɪbʌs/ *n.* a small bus for about twelve passengers.

minicab /ˈmɪnɪˌkæb/ *n. Brit.* a car used as a taxi, but not licensed to ply for hire.

minicomputer /ˈmɪnɪkəmˌpjuːtə(r)/ *n.* a computer of medium power, more than a microcomputer but less than a mainframe.

Minicoy Islands /ˈmɪnɪˌkɔɪ/ one of the groups of islands forming the Indian territory of Lakshadweep in the Indian Ocean.

minikin /ˈmɪnɪkɪn/ adj. & n. ● adj. **1** diminutive. **2** affected, mincing. ● n. a diminutive person or thing. [obs. Du. minneken f. minne love + -ken, -kijn -KIN]

minim /ˈmɪnɪm/ n. **1** Mus. a note having the time value of two crotchets or half a semibreve and represented by a hollow ring with a stem. Also called half-note. **2** one-sixtieth of a fluid drachm, about a drop. **3** an object or portion of the smallest size or importance. **4** a single downstroke of the pen. [ME f. L minimus smallest]

minima pl. of MINIMUM.

minimal /ˈmɪnɪm(ə)l/ adj. **1** very minute or slight. **2** being or related to a minimum. **3** the least possible in size, duration, etc. **4 a** Art & Archit. characterized by the use of simple or elemental (often geometric) forms or structures. (See MINIMALISM 2.) **b** Mus. characterized by the repetition of short phrases. (See MINIMALISM 3.) □ **minimally** adv. (in senses 1–3). [L minimus smallest]

minimalism /ˈmɪnɪməˌlɪz(ə)m/ n. **1** the advocacy or practice of a minimalist approach. **2** a trend in painting, and especially sculpture, which arose during the 1950s, and used simple, often massive, geometric forms. It is particularly associated with the US and figures such as Carl Andre; its impersonality can be seen as a reaction against the emotiveness of abstract expressionism. **3** an avant-garde movement in music characterized by the repetition of very short phrases which change gradually as the music proceeds, producing a hypnotic effect. Minimalism originated in the 1960s in the work of such musicians as Philip Glass, Steve Reich, and Michael Nyman (b. 1944), influenced by Erik Satie and John Cage but also by Balinese, African, and Indian music.

minimalist /ˈmɪnɪməlɪst/ n. & adj. ● n. **1** a person advocating minor or moderate reform in politics (opp. MAXIMALIST). **2** hist. = MENSHEVIK. **3** a person who advocates or practises minimalism in art or music. ● adj. **1 a** advocating moderate policies. **b** hist. of or relating to the Mensheviks. **2** of or relating to minimal art or music.

minimax /ˈmɪnɪˌmæks/ n. **1** Math. the lowest of a set of maximum values. **2** (usu. attrib.) **a** a strategy that minimizes the greatest risk to a participant in a game etc. **b** the theory that in a game with two players, a player's smallest possible maximum loss is equal to the same player's greatest possible minimum gain. [MINIMUM + MAXIMUM]

minimize /ˈmɪnɪˌmaɪz/ v. (also **-ise**) **1** tr. reduce to, or estimate at, the smallest possible amount or degree. **2** tr. estimate or represent at less than the true value or importance. **3** intr. attain a minimum value. □ **minimizer** n. **minimization** /ˌmɪnɪmaɪˈzeɪʃ(ə)n/ n.

minimum /ˈmɪnɪməm/ n. & adj. ● n. (pl. **minima** /-mə/ or **minimums**) the least possible or attainable amount (reduced to a minimum). ● adj. that is a minimum. □ **minimum lending rate** the announced minimum percentage at which a central bank will discount bills (cf. base rate (see BASE¹)). ¶ Abolished in the UK in 1981. **minimum wage** the lowest wage permitted by law or special agreement. [L, neut. of minimus least]

minion /ˈmɪnjən/ n. derog. **1** a servile agent; a slave. **2** a favourite servant, animal, etc. **3** a favourite of a sovereign etc. [F mignon, OF mignot, of Gaulish orig.]

minipill /ˈmɪnɪˌpɪl/ n. a contraceptive pill containing progestogen only (not oestrogen).

miniseries /ˈmɪnɪˌsɪərɪz/ n. (pl. same) a short series of television programmes on a common theme.

miniskirt /ˈmɪnɪˌskɜːt/ n. a very short skirt.

minister /ˈmɪnɪstə(r)/ n. & v. ● n. **1** (often **Minister**) a head of a government department. **2** (in full **minister of religion**) a member of the clergy, esp. in the Presbyterian and Nonconformist Churches. **3** a diplomatic agent, usu. ranking below an ambassador. **4** (usu. foll. by of) a person employed in the execution of (a purpose, will, etc.) (a minister of justice). **5** (in full **minister general**) the superior of some religious orders. ● v. **1** intr. (usu. foll. by to) render aid or service (to a person, cause, etc.). **2** tr. archaic furnish, supply, etc. □ **ministering angel** a kind-hearted person, esp. a woman, who nurses or comforts others (with ref. to Mark 1:13). **Minister of the Crown** Brit. Parl. a member of the Cabinet. **Minister of State** a government minister, in the UK usu. regarded as holding a rank below that of Head of Department. **Minister without Portfolio** a government minister who has Cabinet status, but is not in charge of a specific Department of State. □ **ministership** n. **ministrable** /-strəb(ə)l/ adj. [ME f. OF ministre f. L minister servant f. minus less]

ministerial /ˌmɪnɪˈstɪərɪəl/ adj. **1** of a minister of religion or a minister's office. **2** instrumental or subsidiary in achieving a purpose (ministerial in bringing about a settlement). **3 a** of a government minister. **b** supporting the government against the opposition. □ **ministerialist** n. (in sense 3b). **ministerially** adv. [F ministériel or LL ministerialis f. L (as MINISTRY)]

ministration /ˌmɪnɪˈstreɪʃ(ə)n/ n. **1** (usu. in pl.) aid or service (the kind ministrations of his neighbours). **2** ministering, esp. in religious matters. **3** (usu. foll. by of) the supplying (of help, justice, etc.). □ **ministrant** /ˈmɪnɪstrənt/ adj. & n. **ministrative** /-strətɪv/ adj. [ME f. OF ministration or L ministratio (as MINISTER)]

ministry /ˈmɪnɪstrɪ/ n. (pl. **-ies**) **1** (often **Ministry**) **a** a government department headed by a minister. **b** the building which it occupies (the Ministry of Defence). **2 a** (prec. by the) the vocation or profession of a religious minister (called to the ministry). **b** the office of a religious minister, priest, etc. **c** the period of tenure of this. **3** (prec. by the) the body of ministers of a government or of a religion. **4** a period of government under one Prime Minister. **5** ministering, ministration. [ME f. L ministerium (as MINISTER)]

miniver /ˈmɪnɪvə(r)/ n. (also **minever**) plain white fur used in ceremonial costume. [ME f. AF menuver, OF menu vair (as MENU, VAIR)]

mink /mɪŋk/ n. **1** either of two semiaquatic stoatlike animals of the genus Mustela, M. vison of North America, and M. intreola of Europe. **2** the thick brown fur of the American mink. **3** a coat made of this. [cf. Sw. mänk, menk]

minke /ˈmɪŋkɪ, -kə/ n. a small baleen whale, Balaenoptera acutorostrata, with a pointed snout. [prob. f. Meincke, the name of a Norw. whaler]

Minkowski /mɪŋˈkɒfskɪ/, Hermann (1864–1909), Russian-born German mathematician. He studied the theory of quadratic forms, and contributed to the understanding of the geometrical properties of sets in multidimensional space. Minkowski was the first to suggest the concept of four-dimensional space-time, which was the basis for Einstein's later work on the general theory of relativity.

Minn. abbr. Minnesota.

Minneapolis /ˌmɪnɪˈæpəlɪs/ an industrial city and port on the Mississippi in SE Minnesota; pop. (1990) 368,380. A leading lumbering centre in the 19th century, it is now a major agricultural centre of the upper Midwest.

Minnesinger /ˈmɪnɪˌsɪŋə(r)/ n. (pl. same) any of the aristocratic German poet-musicians who performed songs of courtly love in the 12th–14th centuries. [G (mod. Sänger), = love-singer]

Minnesota /ˌmɪnɪˈsəʊtə/ a state in the north central US, on the Canadian border; pop. (1990) 4,375,100; capital, St Paul. Part of it was ceded to Britain by the French in 1763 and acquired by the US in 1783, the remainder forming part of the Louisiana Purchase in 1803. It became the 32nd state of the US in 1858.

minnow /ˈmɪnəʊ/ n. a small freshwater fish of the carp family, esp. Phoxinus phoxinus. [late ME menow, perh. repr. OE mynwe (unrecorded), myne: infl. by ME menuse, menise f. OF menuise, ult. rel. to MINUTIA]

Miño /ˈmiːnjəʊ/ (Portuguese **Minho** /ˈmiːɲu/) a river which rises in NW Spain and flows south to the Portuguese border, which it follows before entering the Atlantic north of Viana do Castelo.

Minoan /mɪˈnəʊən/ adj. & n. Archaeol. ● adj. of or relating to the Bronze Age civilization centred on Crete (c.3000–1100 BC) or its people or language. (See note below.) ● n. **1** an inhabitant of Minoan Crete or the Minoan world. **2** the language or scripts associated with the Minoans. [named after King Minos, to whom the palace excavated at Knossos was attributed]

▪ The Minoan civilization was first revealed by the excavations of Sir Arthur Evans, who gave it its name. It had reached its zenith by the beginning of the late Bronze Age; impressive remains, especially in Crete, reveal the existence of large urban centres dominated by palaces at Knossos, Mallia, Phaistos, and Zakro. The civilization is also noted for its Linear A script and distinctive art and architecture. The Minoan civilization greatly influenced the Mycenaeans, whose presence in Crete is attested from the 16th century BC and who succeeded the Minoans in control of the Aegean c.1400 BC. The precise reasons for the collapse of Minoan civilization remain controversial.

minor /ˈmaɪnə(r)/ adj., n., & v. ● adj. **1** lesser or comparatively small in size or importance (minor poet; minor operation). **2** Mus. **a** (of a scale) having intervals of a semitone between the second and third, fifth and sixth, and seventh and eighth degrees. **b** (of an interval) less by a semitone than a major interval. **c** (of a key) based on a minor scale, tending to produce a melancholy effect. **3** Brit. (appended to a surname, esp. in public schools) indicating the younger of two brothers (*Smith minor*). **4** *Logic* **a** (of a term) occurring as the subject of the conclusion of a categorical syllogism. **b** (of a premiss) containing the minor term in a categorical syllogism. ● *n.* **1** a person under the legal age limit or majority (*no unaccompanied minors*). **2** *Mus.* a minor key etc. **3** *US* a student's subsidiary subject or course. **4** *Logic* a minor term or premiss. ● *v.intr.* (foll. by *in*) *US* (of a student) undertake study in (a subject) as a subsidiary to a main subject. □ **in a minor key** (of novels, events, people's lives, etc.) understated, uneventful. **minor axis** *Geom.* (of a conic) the axis perpendicular to the major axis. **minor canon** a cleric who assists in daily cathedral services but is not a member of the chapter. **minor league** *N. Amer.* (in baseball, football, etc.) a league of professional clubs other than the major leagues. **minor orders** *see* ORDER. **minor piece** *Chess* a bishop or a knight. **minor planet** = ASTEROID 1. **minor prophet** *see* PROPHET. **minor suit** *Bridge* diamonds or clubs. [L, = smaller, less, rel. to *minuere* lessen]

Minorca /mɪˈnɔːkə/ (Spanish **Menorca** /meˈnorka/) the most easterly and second largest of the Balearic Islands; pop. (1981) 58,700; capital, Mahón.

Minorite /ˈmaɪnəˌraɪt/ *n.* a Franciscan friar or Friar Minor.

minority /maɪˈnɒrɪtɪ, mɪ-/ *n.* (*pl.* **-ies**) **1** (often foll. by *of*) a smaller number or part, esp. within a political party or structure. **2** the number of votes cast for this (*a minority of two*). **3** the state of having less than half the votes or of being supported by less than half of the body of opinion (*in the minority*). **4** a relatively small group of people differing from others in the society of which they are a part in race, religion, language, political persuasion, etc. **5** (*attrib.*) relating to or done by the minority (*minority interests*). **6 a** the state of being under full legal age. **b** the period of this. [F *minorité* or med.L *minoritas* f. L *minor*: see MINOR]

Minos /ˈmaɪnɒs/ *Gk. Mythol.* a legendary king of Crete, son of Zeus and Europa. The proliferation of traditions associated with him led some classical writers to assume that the name refers to several kings rather than one. According to one story, Poseidon revenged himself on Minos by causing his wife Pasiphaë to give birth to the Minotaur, a cruel tyrant, Minos later exacted tribute from Athens in the form of youths and maidens to be devoured by the monster.

Minos, Palace of a building (excavated and reconstructed by Sir Arthur Evans) identified with the largest Minoan palace of Knossos, which yielded local coins portraying the labyrinth as the city's symbol and a Linear B religious tablet which, when eventually deciphered, was found to refer to the 'lady of the labyrinth'.

Minotaur /ˈmaɪnəˌtɔː(r), ˈmɪn-/ *Gk Mythol.* the creature half-man, half-bull, offspring of Pasiphaë and a bull with which she fell in love, confined in Crete in a labyrinth made by Daedalus and fed on human flesh. It was eventually slain by Theseus. [ME f. OF f. L *Minotaurus* f. Gk *Minōtauros* f. MINOS + *tauros* bull]

Minsk /mɪnsk/ the capital of Belarus, an industrial city in the central region of the country; pop. (1990) 1,613,000. Formerly held by Lithuania and Poland, it passed to Russia in 1793, becoming capital of the newly independent Belarus in 1991.

minster /ˈmɪnstə(r)/ *n.* **1** a large or important church (*York Minster*). **2** the church of a monastery. [OE *mynster* f. eccl.L *monasterium* f. Gk *monastērion* MONASTERY]

minstrel /ˈmɪnstrəl/ *n.* **1** *hist.* a medieval singer or musician, esp. singing or reciting poetry. **2** *hist.* a person who entertained patrons with singing, buffoonery, etc. **3** (usu. in *pl.*) a member of a band of public entertainers with blackened faces etc., performing songs and music ostensibly of Negro origin. [ME f. OF *menestral* entertainer, servant, f. Prov. *menest(ai)ral* officer, employee, musician, f. LL *ministerialis* official, officer: see MINISTERIAL]

minstrelsy /ˈmɪnstrəlsɪ/ *n.* (*pl.* **-ies**) **1** the minstrel's art. **2** a body of minstrels. **3** minstrel poetry. [ME f. OF *menestralsie* (as MINSTREL)]

mint[1] /mɪnt/ *n.* **1** an aromatic plant of the genus *Mentha* (family Labiatae), esp. garden spearmint. **2** a peppermint sweet or lozenge. □ **mint julep** *US* a sweet iced alcoholic drink of bourbon flavoured with mint. **mint sauce** chopped spearmint in vinegar and sugar, usu. eaten with lamb. □ **minty** *adj.* (**mintier**, **mintiest**). [OE *minte* ult. f. L *ment(h)a* f. Gk *minthē*]

mint[2] /mɪnt/ *n.* & *v.* ● *n.* **1** a place where money is coined, usu. under state authority. (See ROYAL MINT.) **2** a vast sum of money (*making a mint*). **3** a source of invention etc. (*a mint of ideas*). ● *v.tr.* **1** make (coin) by stamping metal. **2** invent, coin (a word, phrase, etc.). □ **in mint condition** (or **state**) freshly minted; (of books etc.) as new. **mint-mark** a mark on a coin to indicate the mint at which it was struck. **mint-master** the superintendent of coinage at a mint. **mint par** (in

full **mint parity**) **1** the ratio between the gold equivalents of currency in two countries. **2** their rate of exchange based on this. □ **mintage** *n.* [OE *mynet* f. WG f. L *moneta* MONEY]

Minton /ˈmɪntən/ *n.* a kind of pottery made at Stoke-on-Trent, from 1793 onwards, by Thomas Minton (1766–1836) and his successors.

minuend /ˈmɪnjʊˌend/ *n. Math.* a quantity or number from which another is to be subtracted. [L *minuendus* gerundive of *minuere* diminish]

minuet /ˌmɪnjʊˈet/ *n.* & *v.* ● *n.* **1** a slow stately ballroom dance in triple time. **2** *Mus.* the music for this, or music in the same rhythm and style, often as a movement in a suite, sonata, or symphony. ● *v.intr.* (**minueted**, **minueting**) dance a minuet. [F *menuet*, orig. adj. = fine, delicate, dimin. of *menu*: see MENU]

minus /ˈmaɪnəs/ *prep.*, *adj.*, & *n.* ● *prep.* **1** with the subtraction of (*7 minus 4 equals 3*) (symbol: −). **2** (of temperature) below zero (*minus 2°*). **3** lacking; deprived of (*returned minus their dog*). ● *adj.* **1** *Math.* negative. **2** *Electronics* having a negative charge. ● *n.* **1** = minus sign. **2** *Math.* a negative quantity. **3** a disadvantage. □ **minus sign** the symbol −, indicating subtraction or a negative value. [L, neut. of *minor* less]

minuscule /ˈmɪnəˌskjuːl/ *n.* & *adj.* ● *n.* **1** a kind of cursive script developed in the 7th century. **2** a lower-case letter. ● *adj.* **1** lower-case. **2** *colloq.* extremely small or unimportant. □ **minuscular** /mɪˈnʌskjʊlə(r)/ *adj.* [F f. L *minuscula* (*littera* letter) dimin. of *minor*: see MINOR]

minute[1] /ˈmɪnɪt/ *n.* & *v.* ● *n.* **1** a sixtieth of an hour (symbol: ′). **2** a distance covered in one minute (*twenty minutes from the station*). **3 a** a moment; an instant; a point of time (*expecting her any minute; the train leaves in a minute*). **b** (prec. by *the*) *colloq.* the present time (*what are you doing at the minute?*). **c** (foll. by *clause*) as soon as (*call me the minute you get back*). **4** a sixtieth of a degree of angular distance (symbol: ′). **5** (in *pl.*) a brief summary of the proceedings at a meeting. **6** an official memorandum authorizing or recommending a course of action. ● *v.tr.* **1** record (proceedings) in the minutes. **2** send the minutes of a meeting to (a person). □ **just** (or **wait**) **a minute 1** a request to wait for a short time. **2** as a prelude to a query or objection. **minute-gun** a gun fired at intervals of a minute at funerals etc. **minute-hand** the hand on a watch or clock which indicates minutes. **minute steak** a thin slice of steak to be cooked quickly. **up to the minute** completely up to date. [ME f. OF f. LL *minuta* (n.), f. fem. of *minutus* MINUTE[2]: senses 1 & 4 of noun f. med.L *pars minuta prima* first minute part (cf. SECOND[2]): senses 5 & 6 perh. f. med.L *minuta scriptura* draft in small writing]

minute[2] /maɪˈnjuːt/ *adj.* (**minutest**) **1** very small. **2** trifling, petty. **3** (of an inquiry, inquirer, etc.) accurate, detailed, precise. □ **minutely** *adv.* **minuteness** *n.* [ME f. L *minutus* past part. of *minuere* lessen]

Minuteman /ˈmɪnɪtˌmæn/ *n.* (*pl.* **-men**) *US* **1** a political watchdog or activist. **2** a type of American three-stage intercontinental ballistic missile. **3** *hist.* an American militiaman of the revolutionary period (ready to march at a minute's notice).

minutia /maɪˈnjuːʃɪə, mɪ-/ *n.* (*pl.* **-iae** /-ʃɪˌiː/) (usu. in *pl.*) a precise, trivial, or minor detail. [L, = smallness, in pl. trifles f. *minutus*: see MINUTE[2]]

minx /mɪŋks/ *n.* a pert, sly, or playful girl. □ **minxish** *adj.* [16th c.: orig. unkn.]

Miocene /ˈmaɪəˌsiːn/ *adj.* & *n. Geol.* of, relating to, or denoting the fourth epoch of the Tertiary period, between the Oligocene and the Pliocene. The Miocene lasted from about 24.6 to 5.1 million years ago, and was a period of great earth movements during which the Alps and Himalayas were being formed. [irreg. f. Gk *meiōn* less + *kainos* new]

miosis /maɪˈəʊsɪs/ *n.* (also **myosis**) *Med.* excessive constriction of the pupil of the eye. □ **miotic** /-ˈɒtɪk/ *adj.* [Gk *muō* shut the eyes + -OSIS]

MIPS /mɪps/ *n. Computing* a unit of computing speed equivalent to a million instructions per second. [acronym]

Miquelon see ST PIERRE AND MIQUELON.

Mir /mɪə(r)/ a Soviet space station, launched in 1986 and designed to be permanently manned. [Russ., = peace]

Mira /ˈmaɪərə/ *Astron.* a star in the constellation of Cetus, regarded as the prototype of long-period variable stars. [L, = wonderful]

Mirabeau /ˈmɪrəˌbəʊ/, Honoré Gabriel Riqueti, Comte de (1749–91), French revolutionary politician. Mirabeau rose to prominence in the early days of the French Revolution, when he became deputy of the Third Estate in the States General. His moderate political stance led him to press for a form of constitutional monarchy; he was made President of the National Assembly in 1791, but died shortly afterwards.

mirabelle /ˈmɪrəˌbel/ *n.* **1 a** a European variety of plum tree, bearing small round yellow fruit. **b** a fruit from this tree. **2** a liqueur distilled from this fruit. [F]

miracidium /ˌmaɪərəˈsɪdɪəm/ n. (pl. **miracidia** /-dɪə/) Zool. a free-swimming ciliated larval stage in which a parasitic fluke passes from the egg to its first host (esp. a snail) (cf. CERCARIA). [Gk *meirakidion* dimin. of *meirakion* boy, stripling]

miracle /ˈmɪrək(ə)l/ n. **1** an extraordinary event attributed to some supernatural agency. **2 a** any remarkable occurrence. **b** a remarkable development in some specified area (*an economic miracle; the German miracle*). **3** (usu. foll. by *of*) a remarkable or outstanding specimen (*the plan was a miracle of ingenuity*). □ **miracle drug** a drug which represents a breakthrough in medical science. **miracle play** see MYSTERY PLAY. [ME f. OF f. L *miraculum* object of wonder f. *mirari* wonder f. *mirus* wonderful]

miraculous /mɪˈrækjʊləs/ adj. **1** of the nature of a miracle. **2** supernatural. **3** remarkable, surprising. □ **miraculously** adv. **miraculousness** n. [F *miraculeux* or med.L *miraculosus* f. L (as MIRACLE)]

mirador /ˌmɪrəˈdɔː(r)/ n. a turret or tower etc. attached to a building, and commanding an excellent view. [Sp. f. *mirar* to look]

mirage /ˈmɪrɑːʒ/ n. **1** an optical illusion caused by atmospheric conditions, esp. the appearance of a sheet of water in a desert or on a hot road from the reflection of light. **2** an illusory thing. [F f. *se mirer* be reflected, f. L *mirare* look at]

Miranda /mɪˈrændə/ **1** the daughter of Prospero in Shakespeare's *The Tempest*. **2** Astron. satellite V of Uranus, the eleventh closest to the planet, discovered in 1948 (diameter 480 km). It has a complex terrain of cratered areas and tracts of grooves and ridges, suggesting that it was once shattered by an impact.

MIRAS /ˈmaɪəræs/ abbr. (also **Miras**) mortgage interest relief at source.

mire /ˈmaɪə(r)/ n. & v. ● n. **1** a stretch of swampy or boggy ground. **2** mud, dirt. ● v. **1** tr. & intr. plunge or sink in a mire. **2** tr. involve in difficulties. □ **miry** adj. **in the mire** in difficulties. [ME f. ON *mýrr* f. Gmc, rel. to MOSS]

mirepoix /mɪəˈpwʌ/ n. a mixture of sautéed diced vegetables, used in sauces etc. [F, f. Duc de *Mirepoix*, Fr. general (1699–1757)]

mirid /ˈmɪrɪd, ˈmaɪər-/ n. & adj. Zool. ● n. a heteropteran bug of the family Miridae (formerly Capsidae), which includes numerous plant pests. Also called *capsid*. ● adj. of or relating to this family. [mod.L f. *Miris* genus of bugs, f. *mirus* wonderful]

mirk var. of MURK.

mirky var. of MURKY.

Miró /mɪˈrəʊ/, Joan (1893–1983), Spanish painter. From 1919 he spent much of his time in Paris, returning to live in Spain in 1940. One of the most prominent figures of surrealism, he painted a brightly coloured fantasy world of variously spiky and amoebic calligraphic forms against plain backgrounds. Major works include *Harlequinade* (1924–5).

mirror /ˈmɪrə(r)/ n. & v. ● n. **1** a polished surface, usu. of amalgam-coated glass or metal, which reflects an image; a looking-glass. **2** anything regarded as giving an accurate reflection or description of something else. ● v.tr. reflect as in a mirror. □ **mirror carp** a breed of carp with large shiny scales. **mirror finish** a reflective surface. **mirror image** an identical image, but with the structure reversed, as in a mirror. **mirror symmetry** symmetry as of an object and its reflection. **mirror writing** reversed writing, like ordinary writing reflected in a mirror. [ME f. OF *mirour* ult. f. L *mirare* look at]

mirth /mɜːθ/ n. merriment, laughter. □ **mirthful** adj. **mirthfully** adv. **mirthfulness** n. **mirthless** adj. **mirthlessly** adv. **mirthlessness** n. [OE *myrgth* (as MERRY)]

MIRV /mɜːv/ abbr. multiple independently targeted re-entry vehicle (a type of missile).

MIS abbr. Computing management information system.

mis-[1] /mɪs/ prefix added to verbs and verbal derivatives: meaning 'amiss', 'badly', 'wrongly', 'unfavourably' (*mislead; misshapen; mistrust*). [OE f. Gmc]

mis-[2] /mɪs/ prefix occurring in a few words adopted from French meaning 'badly', 'wrongly', 'amiss', 'ill-', or having a negative force (*misadventure; mischief*). [OF *mes-* ult. f. L *minus* (see MINUS): assim. to MIS-[1]]

misaddress /ˌmɪsəˈdres/ v.tr. **1** address (a letter etc.) wrongly. **2** address (a person) wrongly, esp. impertinently.

misadventure /ˌmɪsədˈventʃə(r)/ n. **1** Law an accident without concomitant crime or negligence (*death by misadventure*). **2** bad luck.

3 a misfortune. [ME f. OF *mesaventure* f. *mesavenir* turn out badly (as MIS-[2], ADVENT: cf. ADVENTURE)]

misalign /ˌmɪsəˈlaɪn/ v.tr. give the wrong alignment to. □ **misalignment** n.

misalliance /ˌmɪsəˈlaɪəns/ n. an unsuitable alliance, esp. an unsuitable marriage. □ **misally** /-ˈlaɪ/ v.tr. (**-ies**, **-ied**) [MIS-[1] + ALLIANCE, after MÉSALLIANCE]

misanthrope /ˈmɪz(ə)nˌθrəʊp, ˈmɪs(ə)n-/ n. (also **misanthropist** /mɪˈzænθrəpɪst/) **1** a person who hates humankind. **2** a person who avoids human society. □ **misanthropic** /ˌmɪz(ə)nˈθrɒpɪk, ˌmɪs(ə)n-/ adj. **misanthropical** adj. **misanthropically** adv. **misanthropize** /mɪˈzænθrəˌpaɪz/ v.intr. (also **-ise**). **misanthropy** n. [F f. Gk *misanthrōpos* f. *misos* hatred + *anthrōpos* man]

misapply /ˌmɪsəˈplaɪ/ v.tr. (**-ies**, **-ied**) apply (esp. funds) wrongly. □ **misapplication** /-ˌæplɪˈkeɪʃ(ə)n/ n.

misapprehend /ˌmɪsæprɪˈhend/ v.tr. misunderstand (words, a person). □ **misapprehension** /-ˈhenʃ(ə)n/ n. **misapprehensive** /-ˈhensɪv/ adj.

misappropriate /ˌmɪsəˈprəʊprɪˌeɪt/ v.tr. apply (usu. another's money) to one's own use, or to a wrong use. □ **misappropriation** /-ˌprəʊprɪˈeɪʃ(ə)n/ n.

misbegotten /ˌmɪsbɪˈgɒt(ə)n/ adj. **1** illegitimate, bastard. **2** contemptible, disreputable.

misbehave /ˌmɪsbɪˈheɪv/ v.intr. & refl. (of a person or machine) behave badly. □ **misbehaviour** n.

misbelief /ˌmɪsbɪˈliːf/ n. **1** wrong or unorthodox religious belief. **2** a false opinion or notion.

misc. abbr. miscellaneous.

miscalculate /mɪsˈkælkjʊˌleɪt/ v.tr. (also absol.) calculate (amounts, results, etc.) wrongly. □ **miscalculation** /ˌmɪskælkjʊˈleɪʃ(ə)n/ n.

miscall /mɪsˈkɔːl/ v.tr. **1** call by a wrong or inappropriate name. **2** archaic or dial. call (a person) names.

miscarriage /ˈmɪsˌkærɪdʒ/ n. **1** a spontaneous abortion, esp. before the 28th week of pregnancy. **2** the failure (of a plan etc.) to reach completion. **3** Brit. the failure (of a letter etc.) to reach its destination. □ **miscarriage of justice** any failure of the judicial system to attain the ends of justice. [MISCARRY, after CARRIAGE]

miscarry /mɪsˈkærɪ/ v.intr. (**-ies**, **-ied**) **1** (of a woman) have a miscarriage. **2** Brit. (of a letter etc.) fail to reach its destination. **3** (of a business, plan, etc.) fail, be unsuccessful.

miscast /mɪsˈkɑːst/ v.tr. (past and past part. **-cast**) allot an unsuitable part to (an actor).

miscegenation /ˌmɪsɪdʒɪˈneɪʃ(ə)n/ n. the interbreeding of races, esp. of whites and non-whites. [irreg. f. L *miscere* mix + *genus* race]

miscellanea /ˌmɪsəˈleɪnɪə/ n.pl. **1** a literary miscellany. **2** a collection of miscellaneous items. [L neut. pl. (as MISCELLANEOUS)]

miscellaneous /ˌmɪsəˈleɪnɪəs/ adj. **1** of mixed composition or character. **2** (foll. by pl. noun) of various kinds. **3** (of a person) many-sided. □ **miscellaneousness** n. [L *miscellaneus* f. *miscellus* mixed f. *miscere* mix]

miscellany /mɪˈselənɪ/ n. (pl. **-ies**) **1** a mixture, a medley. **2** a book containing a collection of stories etc., or various literary compositions. □ **miscellanist** n. [F *miscellanées* (fem. pl.) or L MISCELLANEA]

mischance /mɪsˈtʃɑːns/ n. **1** bad luck. **2** an instance of this. [ME f. OF *mesch(e)ance* f. *mescheoir* (as MIS-[2], CHANCE)]

mischief /ˈmɪstʃɪf/ n. **1** conduct which is troublesome, but not malicious, esp. in children. **2** pranks, scrapes (*get into mischief; keep out of mischief*). **3** playful malice, archness, satire (*eyes full of mischief*). **4** harm or injury caused by a person or thing. **5** a person or thing responsible for harm or annoyance (*that loose connection is the mischief*). **6** (prec. by *the*) the annoying part or aspect (*the mischief of it is that etc.*). □ **do a person a mischief** wound or kill a person. **get up to** (or **make**) **mischief** create discord. **mischief-maker** a person who encourages discord, esp. by gossip etc. [ME f. OF *meschief* f. *meschever* (as MIS-[2], *chever* come to an end f. *chef* head: see CHIEF]

mischievous /ˈmɪstʃɪvəs/ adj. **1** (of a person) disposed to mischief. **2** (of conduct) playfully malicious. **3** (of a thing) having harmful effects. □ **mischievously** adv. **mischievousness** n. [ME f. AF *meschevous* f. OF *meschever*: see MISCHIEF]

misch metal /mɪʃ/ n. an alloy of lanthanide metals, usu. added to iron to improve its malleability. [G *mischen* mix + *Metall* metal]

miscible /ˈmɪsɪb(ə)l/ adj. (often foll. by *with*) capable of being mixed. □ **miscibility** /ˌmɪsɪˈbɪlɪtɪ/ n. [med.L *miscibilis* f. L *miscere* mix]

misconceive /ˌmɪskən'siːv/ v. **1** intr. (often foll. by of) have a wrong idea or conception. **2** tr. (as **misconceived** adj.) badly planned, organized, etc. **3** tr. misunderstand (a word, person, etc.). □ **misconceiver** n. **misconception** /-'sepʃ(ə)n/ n.

misconduct n. & v. ● n. /mɪs'kɒndʌkt/ **1** improper or unprofessional behaviour. **2** bad management. ● v. /ˌmɪskən'dʌkt/ **1** refl. misbehave. **2** tr. mismanage.

misconstrue /ˌmɪskən'struː/ v.tr. (**-construes**, **-construed**, **-construing**) **1** interpret (a word, action, etc.) wrongly. **2** mistake the meaning of (a person). □ **misconstruction** /-'strʌkʃ(ə)n/ n.

miscopy /mɪs'kɒpɪ/ v.tr. (**-ies**, **-ied**) copy (text etc.) incorrectly.

miscount v. & n. ● v.tr. /mɪs'kaʊnt/ (also absol.) count wrongly. ● n. /'mɪskaʊnt/ a wrong count.

miscreant /'mɪskrɪənt/ n. & adj. ● n. **1** a wretch, a villain. **2** archaic a heretic. ● adj. **1** depraved, villainous. **2** archaic heretical. [ME f. OF mescreant (as MIS-², creant part. of croire f. L credere believe)]

miscue /mɪs'kjuː/ n. & v. ● n. (in snooker etc.) the failure to strike the ball properly with the cue. ● v.intr. (**-cues**, **-cued**, **-cueing** or **-cuing**) make a miscue.

misdate /mɪs'deɪt/ v.tr. date (an event, a letter, etc.) wrongly.

misdeal /mɪs'diːl/ v. & n. ● v.tr. (also absol.) (past and past part. **-dealt** /-'delt/) make a mistake in dealing (cards). ● n. **1** a mistake in dealing cards. **2** a misdealt hand.

misdeed /mɪs'diːd/ n. an evil deed, a wrongdoing; a crime. [OE misdǣd (as MIS-¹, DEED)]

misdemeanant /ˌmɪsdɪ'miːnənt/ n. a person convicted of a misdemeanour or guilty of misconduct. [archaic misdemean misbehave]

misdemeanour /ˌmɪsdɪ'miːnə(r)/ n. (US **misdemeanor**) **1** an offence, a misdeed. **2** Law an indictable offence, (in the UK formerly) less heinous than a felony. (See also FELONY.)

misdiagnose /mɪsˌdaɪəg'nəʊz/ v.tr. diagnose incorrectly. □ **misdiagnosis** /-'nəʊsɪs/ n.

misdial /mɪs'daɪəl/ v.tr. (also absol.) (**-dialled**, **-dialling**; US **-dialed**, **-dialing**) dial (a telephone number etc.) incorrectly.

misdirect /ˌmɪsdaɪ'rekt, -dɪ'rekt/ v.tr. **1** direct (a person, letter, blow, etc.) wrongly. **2** (of a judge) instruct (the jury) wrongly. □ **misdirection** /-'rekʃ(ə)n/ n.

misdoing /mɪs'duːɪŋ/ n. a misdeed.

misdoubt /mɪs'daʊt/ v.tr. **1** have doubts or misgivings about the truth or existence of. **2** be suspicious about; suspect that.

miseducation /ˌmɪsedjʊ'keɪʃ(ə)n/ n. wrong or faulty education. □ **miseducate** /-'edjʊˌkeɪt/ v.tr.

mise en scène /ˌmiːz ɒn 'sen/ n. **1** Theatr. the scenery and properties of a play. **2** the setting or surroundings of an event. [F]

misemploy /ˌmɪsɪm'plɔɪ/ v.tr. employ or use wrongly or improperly. □ **misemployment** n.

miser /'maɪzə(r)/ n. **1** a person who hoards wealth and lives miserably. **2** an avaricious person. [L, = wretched]

miserable /'mɪzərəb(ə)l/ adj. **1** wretchedly unhappy or uncomfortable (felt miserable). **2** contemptible, inadequate, mean (a miserable attempt). **3** causing wretchedness or discomfort (miserable in a suit). **4** Sc., Austral., & NZ (of a person) stingy, mean. **5** colloq. (of a person) gloomy, morose. □ **miserableness** n. **miserably** adv. [ME f. F misérable f. L miserabilis pitiable f. miserari to pity f. miser wretched]

misère /mɪ'zeə(r)/ n. Cards (in solo whist etc.) a declaration undertaking to win no tricks. [F, = poverty, MISERY]

miserere /ˌmɪzə'reərɪ, -'rɪərɪ/ n. **1** a cry for mercy. **2** = MISERICORD 1. [ME f. L, imper. of miserere have mercy (as MISER); first word of Ps. 51 in Latin]

misericord /mɪ'zerɪˌkɔːd/ n. **1** a shelving projection on the under side of a hinged seat in a choir stall serving (when the seat is turned up) to help support a person standing. **2** an apartment in a monastery in which some relaxations of discipline are permitted. **3** hist. a dagger for dealing the death stroke. [ME f. OF misericorde f. L misericordia f. misericors compassionate f. stem of misereri pity + cor cordis heart]

miserly /'maɪzəlɪ/ adj. like a miser, niggardly. □ **miserliness** n. [MISER]

misery /'mɪzərɪ/ n. (pl. **-ies**) **1** a wretched state of mind, or of outward circumstances. **2** a thing causing this. **3** colloq. a constantly depressed or discontented person. **4** = MISÈRE. □ **put out of its** etc. **misery 1** release (a person, animal, etc.) from suffering or suspense. **2** kill (an animal in pain). [ME f. OF misere or L miseria (as MISER)]

misfeasance /mɪs'fiːz(ə)ns/ n. Law a transgression, esp. the wrongful exercise of lawful authority. [ME f. OF mesfaisance f. mesfaire misdo (as MIS-², faire do f. L facere): cf. MALFEASANCE]

misfield v. & n. ● v.tr. /mɪs'fiːld/ (also absol.) (in cricket, baseball, etc.) field (the ball) badly. ● n. /'mɪsfiːld/ an instance of this.

misfire v. & n. ● v.intr. /mɪs'faɪə(r)/ **1** (of a gun, motor engine, etc.) fail to go off or start or function regularly. **2** (of an action etc.) fail to have the intended effect. ● n. /'mɪsˌfaɪə(r)/ a failure of function or intention.

misfit /'mɪsfɪt/ n. **1** a person unsuited to a particular kind of environment, occupation, etc. **2** a garment etc. that does not fit. □ **misfit stream** Geog. a stream not corresponding in size to its valley.

misfortune /mɪs'fɔːtjuːn, -'fɔːtʃuːn/ n. **1** bad luck. **2** an instance of this.

misgive /mɪs'gɪv/ v.tr. (past **-gave** /-'geɪv/; past part. **-given** /-'gɪv(ə)n/) (often foll. by about, that) (of a person's mind, heart, etc.) fill (a person) with suspicion or foreboding.

misgiving /mɪs'gɪvɪŋ/ n. (usu. in pl.) a feeling of mistrust or apprehension.

misgovern /mɪs'gʌv(ə)n/ v.tr. govern (a state etc.) badly. □ **misgovernment** n.

misguide /mɪs'gaɪd/ v.tr. **1** (as **misguided** adj.) mistaken in thought or action. **2** mislead, misdirect. □ **misguidance** n. **misguidedly** adv. **misguidedness** n.

mishandle /mɪs'hænd(ə)l/ v.tr. **1** deal with incorrectly or ineffectively. **2** handle (a person or thing) roughly or rudely; ill-treat.

mishap /'mɪshæp/ n. an unlucky accident.

mishear /mɪs'hɪə(r)/ v.tr. (past and past part. **-heard** /-'hɜːd/) hear incorrectly or imperfectly.

mishit v. & n. ● v.tr. /mɪs'hɪt/ (**-hitting**; past and past part. **-hit**) hit (a ball etc.) faultily. ● n. /'mɪshɪt/ a faulty or bad hit.

mishmash /'mɪʃmæʃ/ n. a confused mixture. [ME, reduplication of MASH]

Mishnah /'mɪʃnə/ n. an authoritative collection of exegetical material embodying the oral tradition of Jewish law. Written in Hebrew and traditionally attributed to Rabbi Judah ha-Nasi (AD c.200), it forms the first part of the Talmud, and has had an influence on Judaism second only to that of the Hebrew Bible. □ **Mishnaic** /mɪʃ'neɪɪk/ adj. [Heb. mišnāh (teaching by) repetition]

misidentify /ˌmɪsaɪ'dentɪˌfaɪ/ v.tr. (**-ies**, **-ied**) identify erroneously. □ **misidentification** /-ˌdentɪfɪ'keɪʃ(ə)n/ n.

misinform /ˌmɪsɪn'fɔːm/ v.tr. give wrong information to, mislead. □ **misinformation** /-fə'meɪʃ(ə)n/ n.

misinterpret /ˌmɪsɪn'tɜːprɪt/ v.tr. (**-interpreted**, **-interpreting**) **1** interpret wrongly. **2** draw a wrong inference from. □ **misinterpreter** n. **misinterpretation** /-ˌtɜːprɪ'teɪʃ(ə)n/ n.

misjudge /mɪs'dʒʌdʒ/ v.tr. (also absol.) **1** judge wrongly. **2** have a wrong opinion of. □ **misjudgement** n. (also **misjudgment**).

miskey /mɪs'kiː/ v.tr. (**-keys**, **-keyed**) key (data) wrongly.

miskick v. & n. ● v.tr. /mɪs'kɪk/ (also absol.) kick (a ball etc.) badly or wrongly. ● n. /'mɪskɪk/ an instance of this.

Miskito /mɪ'skiːtəʊ/ n. & adj. (also **Mosquito** /mɒ'skiː-/) ● n. (pl. same or **-os**) **1** a member of a native people living on the Atlantic coast of Nicaragua and Honduras. **2** the language of this people. ● adj. of or relating to the Miskito or their language. [Amer. Indian name: see also MOSQUITO COAST]

Miskolc /'miːʃkɒlts/ a city in NE Hungary; pop. (est. 1989) 208,000. It is one of Hungary's major industrial centres.

mislay /mɪs'leɪ/ v.tr. (past and past part. **-laid** /-'leɪd/) **1** unintentionally put (a thing) where it cannot readily be found. **2** euphem. lose.

mislead /mɪs'liːd/ v.tr. (past and past part. **-led** /-'led/) **1** cause (a person) to go wrong, in conduct, belief, etc. **2** lead astray or in the wrong direction. □ **misleader** n.

misleading /mɪs'liːdɪŋ/ adj. causing to err or go astray; imprecise, confusing. □ **misleadingly** adv. **misleadingness** n.

mislike /mɪs'laɪk/ v.tr. archaic dislike. [OE mislīcian (as MIS-¹, LIKE²)]

mismanage /mɪs'mænɪdʒ/ v.tr. manage badly or wrongly. □ **mismanagement** n.

mismarriage /mɪs'mærɪdʒ/ n. an unsuitable marriage or alliance. [MIS-¹ + MARRIAGE]

mismatch v. & n. ● v.tr. /mɪs'mætʃ/ (usu. as **mismatched** adj.) match unsuitably or incorrectly, esp. in marriage. ● n. /'mɪsmætʃ/ a bad match.

mismated /mɪsˈmeɪtɪd/ adj. **1** (of people) not suited to each other, esp. in marriage. **2** (of objects) not matching.

mismeasure /mɪsˈmeʒə(r)/ v.tr. measure or estimate incorrectly. □ **mismeasurement** n.

misname /mɪsˈneɪm/ v.tr. = MISCALL.

misnomer /mɪsˈnəʊmə(r)/ n. **1** a name or term used wrongly. **2** the wrong use of a name or term. [ME f. AF f. OF mesnom(m)er (as MIS-², nommer name f. L nominare formed as NOMINATE)]

miso /ˈmiːsəʊ/ n. a paste made from fermented soya beans and barley or rice malt, used in Japanese cookery. [Jap.]

misogamy /mɪˈsɒgəmɪ/ n. the hatred of marriage. □ **misogamist** n. [Gk misos hatred + gamos marriage]

misogyny /mɪˈsɒdʒɪnɪ/ n. the hatred of women. □ **misogynist** n. **misogynous** adj. **misogynistic** /-ˌsɒdʒɪˈnɪstɪk/ adj. [Gk misos hatred + gunē woman]

mispickel /ˈmɪsˌpɪk(ə)l/ n. Mineral. an arsenide and sulphide of iron which is a major source of arsenic compounds. [G]

misplace /mɪsˈpleɪs/ v.tr. **1** put in the wrong place. **2** bestow (affections, confidence, etc.) on an inappropriate object. **3** time (words, actions, etc.) badly. □ **misplacement** n.

misplay v. & n. ● v.tr. /mɪsˈpleɪ/ play (a ball, card, etc.) in a wrong or ineffective manner. ● n. /ˈmɪspleɪ/ an instance of this.

misprint n. & v. ● n. /ˈmɪsprɪnt/ a mistake in printing. ● v.tr. /mɪsˈprɪnt/ print wrongly.

misprision¹ /mɪsˈprɪʒ(ə)n/ n. Law **1** (in full **misprision of a felony** or **of treason**) the deliberate concealment of one's knowledge of a crime, treason, etc. **2** a wrong action or omission. [ME f. AF mesprisioun f. OF mesprison error f. mesprendre to mistake (as MIS-², prendre take)]

misprision² /mɪsˈprɪʒ(ə)n/ n. **1** a misreading, misunderstanding, etc. **2** (usu. foll. by of) a failure to appreciate the value of a thing. **3** archaic contempt. [MISPRIZE after MISPRISION¹]

misprize /mɪsˈpraɪz/ v.tr. literary despise, scorn; fail to appreciate. [ME f. OF mesprisier (as MIS-¹, PRIZE¹)]

mispronounce /ˌmɪsprəˈnaʊns/ v.tr. pronounce (a word etc.) wrongly. □ **mispronunciation** /-ˌnʌnsɪˈeɪʃ(ə)n/ n.

misquote /mɪsˈkwəʊt/ v.tr. quote wrongly. □ **misquotation** /ˌmɪskwəʊˈteɪʃ(ə)n/ n.

misread /mɪsˈriːd/ v.tr. (past and past part. **-read** /-ˈred/) read or interpret (text, a situation, etc.) wrongly.

misremember /ˌmɪsrɪˈmembə(r)/ v.tr. remember imperfectly or incorrectly.

misreport /ˌmɪsrɪˈpɔːt/ v. & n. ● v.tr. give a false or incorrect report of. ● n. a false or incorrect report.

misrepresent /ˌmɪsreprɪˈzent/ v.tr. represent wrongly; give a false or misleading account or idea of. □ **misrepresentative** adj. **misrepresentation** /-zenˈteɪʃ(ə)n/ n.

misrule /mɪsˈruːl/ n. & v. ● n. bad government; disorder. ● v.tr. govern badly.

Miss. abbr. Mississippi.

miss¹ /mɪs/ v. & n. ● v. **1** tr. (also absol.) fail to hit, reach, find, catch, etc. (an object or goal). **2** tr. fail to catch (a bus, train, etc.). **3** tr. fail to experience, see, or attend (an occurrence or event). **4** tr. fail to meet (a person); fail to keep (an appointment). **5** tr. fail to seize (an opportunity etc.) (I missed my chance). **6** tr. fail to hear or understand (I'm sorry, I missed what you said). **7** tr. a regret the loss or absence of (a person or thing) (did you miss me while I was away?). **b** notice the loss or absence of (an object or a person) (bound to miss the key if it isn't there). **8** tr. avoid (go early to miss the traffic). **9** tr. = miss out 1. **10** intr. (of an engine etc.) fail, misfire. ● n. **1** a failure to hit, reach, attain, connect, etc. **2** colloq. = MISCARRIAGE 1. □ **be missing** not have, lack (see also MISSING adj.). **give (a thing) a miss** Brit. avoid, leave alone (gave the party a miss). **miss the boat** (or **bus**) lose an opportunity. **miss fire** (of a gun) fail to go off or hit the mark (cf. MISFIRE). **a miss is as good as a mile** the fact of failure or escape is not affected by the narrowness of the margin. **miss out 1** Brit. omit, leave out (missed out my name from the list). **2** (usu. foll. by on) colloq. fail to get or experience (always misses out on the good times). **not miss much** be alert. **not miss a trick** colloq. never fail to seize an opportunity, advantage, etc. □ **missable** adj. [OE missan f. Gmc]

miss² /mɪs/ n. **1** a girl or unmarried woman. **2** (**Miss**) **a** the title of an unmarried woman or girl, or of a married woman retaining her maiden name for professional purposes. **b** the title of a beauty queen (Miss World). **3** usu. derog. or joc. a girl, esp. a schoolgirl, with implications

of silliness etc. **4** the title used to address a female schoolteacher, shop assistant, etc. □ **missish** adj. (in sense 3). [abbr. of MISTRESS]

missal /ˈmɪs(ə)l/ n. RC Ch. **1** a book containing the texts used in the service of the Mass throughout the year. As a liturgical book, the missal appeared from about the 10th century, combining in one book the devotions that had previously appeared in several. **2** a book of prayers, esp. an illuminated one. [ME f. med.L missale neut. of eccl.L missalis of the mass f. missa MASS²]

missel thrush var. of MISTLE THRUSH.

misshape /mɪsˈʃeɪp/ v.tr. give a bad shape or form to; distort.

misshapen /mɪsˈʃeɪp(ə)n/ adj. ill-shaped, deformed, distorted. □ **misshapenly** adv. **misshapenness** /-p(ə)nnɪs/ n.

missile /ˈmɪsaɪl/ n. **1** an object or weapon suitable for throwing at a target or for discharge from a machine. **2** a weapon, esp. a nuclear weapon, directed by remote control or automatically. □ **missilery** /-saɪlrɪ/ n. [L missilis f. mittere miss- send]

missing /ˈmɪsɪŋ/ adj. **1** not in its place; lost. **2** (of a person) not yet traced or confirmed as alive but not known to be dead. **3** not present.

missing link n. **1** a thing lacking to complete a series. **2** a hypothetical intermediate type, especially between humans and apes. The missing link was a Victorian concept, arising from a simplistic picture of human evolution, and represented either a common evolutionary ancestor for both humans and apes, or, in popular thought, some kind of ape-man through which humans had evolved from the other higher primates. It is now clear that human evolution has been much more complex; even in the 2 million years' history of the genus Homo, several distinct hominid forms have existed.

mission /ˈmɪʃ(ə)n/ n. **1 a** a particular task or goal assigned to a person or group. **b** a journey undertaken as part of this. **c** a person's vocation (mission in life). **2** a military or scientific operation or expedition for a particular purpose. **3** a body of persons sent, esp. to a foreign country, to conduct negotiations etc. **4 a** a body sent to propagate a religious faith. **b** a field of missionary activity. **c** a missionary post or organization. **d** a place of worship attached to a mission. **5** a particular course or period of preaching, services, etc., undertaken by a parish or community. [F mission or L missio f. mittere miss- send]

missionary /ˈmɪʃənərɪ/ adj. & n. ● adj. of, concerned with, or characteristic of, religious missions. ● n. (pl. **-ies**) a person doing missionary work. □ **missionary position** colloq. a position for sexual intercourse with the woman lying on her back and the man lying on top and facing her. [mod.L missionarius f. L (as MISSION)]

missioner /ˈmɪʃənə(r)/ n. **1** a missionary. **2** a person in charge of a religious mission.

missis /ˈmɪsɪz/ n. (also **missus**) sl. or joc. **1** a form of address to a woman. **2** a wife. □ **the missis** my or your wife. [corrupt. of MISTRESS: cf. MRS]

Mississauga /ˌmɪsɪˈsɔːgə/ a town in southern Ontario, on the western shores of Lake Ontario; pop. (1991) 463,400. It forms a southern suburb of Toronto.

Mississippi /ˌmɪsɪˈsɪpɪ/ **1** a major river of North America, which rises in Minnesota near the Canadian border and flows south to a delta on the Gulf of Mexico. With its chief tributary, the Missouri, it is 5,970 km (3,710 miles) long. In the second half of the 17th century it provided a route south through the centre of the continent for French explorers from Canada. From the 1830s onwards it was famous for the sternwheeler steamboats which plied between New Orleans, St Louis, and other northern cities. **2** a state of the southern US, on the Gulf of Mexico, bounded to the west by the lower Mississippi river; pop. (1990) 2,573,200; capital, Jackson. A French colony in the first half of the 18th century, it was ceded to Britain in 1763 and to the US in 1783, becoming the 20th state in 1817.

missive /ˈmɪsɪv/ n. **1** joc. a letter, esp. a long and serious one. **2** an official letter. □ **letter** (or **letters**) **missive** a letter from a sovereign to a dean and chapter nominating a person to be elected bishop. [ME f. med.L missivus f. L (as MISSION)]

Missolonghi /ˌmɪsəˈlɒŋgɪ/ (Greek **Mesolóngion** /ˌmɛsəˈlɒŋgiˌɔn/) a city in western Greece, on the north shore of the Gulf of Patras; pop. (1981) 10,150. It resisted the Turkish forces in the War of Greek Independence (1821–9) and is noted as the place where the poet Byron, who had joined the fight, died of malaria in 1824.

Missouri /mɪˈzʊərɪ/ **1** a major river of North America, one of the main tributaries of the Mississippi. It rises in the Rocky Mountains in Montana and flows 3,736 km (2,315 miles) to meet the Mississippi just north of St Louis. **2** a state of the US, bounded on the east by the

Mississippi river; pop. (1990) 5,117,070; capital, Jefferson City. It was acquired as part of the Louisiana Purchase in 1803, becoming the 24th state of the US in 1821.

misspell /mɪsˈspel/ *v.tr.* (*past* and *past part.* **-spelt** /-ˈspelt/ or **-spelled**) spell wrongly. □ **misspelling** *n.*

misspend /mɪsˈspend/ *v.tr.* (*past* and *past part.* **-spent** /-ˈspent/) (esp. as **misspent** *adj.*) spend amiss or wastefully.

misstate /mɪsˈsteɪt/ *v.tr.* state wrongly or inaccurately. □ **misstatement** *n.*

misstep /mɪsˈstep/ *n.* **1** a wrong step or action. **2** a faux pas.

missus var. of MISSIS.

missy /ˈmɪsɪ/ *n.* (*pl.* **-ies**) an affectionate or derogatory form of address to a young girl.

mist /mɪst/ *n.* & *v.* ● *n.* **1 a** water vapour near the ground in minute droplets limiting visibility. **b** condensed vapour settling on a surface and obscuring glass etc. **2** dimness or blurring of the sight caused by tears etc. **3** a cloud of particles resembling mist. ● *v.tr.* & *intr.* (usu. foll. by *up*, *over*) cover or become covered with mist or as with mist. [OE f. Gmc]

mistake /mɪˈsteɪk/ *n.* & *v.* ● *n.* **1** an incorrect idea or opinion; a thing incorrectly done or thought. **2** an error of judgement. ● *v.tr.* (*past* **mistook** /-ˈstʊk/; *past part.* **mistaken** /-ˈsteɪkən/) **1** misunderstand the meaning or intention of (a person, a statement, etc.). **2** (foll. by *for*) wrongly take or identify (*mistook me for you*). **3** choose wrongly (*mistake one's vocation*). □ **and** (or **make**) **no mistake** undoubtedly. **by mistake** accidentally; in error. **there is no mistaking** one is sure to recognize (a person or thing). □ **mistakable** *adj.* **mistakably** *adv.* [ME f. ON *mistaka* (as MIS-[1], TAKE)]

mistaken /mɪˈsteɪkən/ *adj.* **1** wrong in opinion or judgement. **2** based on or resulting from this (*mistaken loyalty; mistaken identity*). □ **mistakenly** *adv.* **mistakenness** /-kənnɪs/ *n.*

misteach /mɪsˈtiːtʃ/ *v.tr.* (*past* and *past part.* **-taught** /-ˈtɔːt/) teach wrongly or incorrectly.

mister /ˈmɪstə(r)/ *n.* **1** a man without a title of nobility etc. (*a mere mister*). **2** *sl.* or *joc.* a form of address to a man. [weakened form of MASTER in unstressed use before a name: cf. MR]

mistigris /ˈmɪstɪˌɡrɪs, -ˌɡriː/ *n. Cards* **1** a blank card used as a wild card in a form of draw poker. **2** this game. [F *mistigri* jack of clubs]

mistime /mɪsˈtaɪm/ *v.tr.* say or do at the wrong time. [OE *mistīmian* (as MIS-[1], TIME)]

mistitle /mɪsˈtaɪt(ə)l/ *v.tr.* give the wrong title or name to.

mistle thrush /ˈmɪs(ə)l/ *n.* (also **missel thrush**) a large thrush, *Turdus viscivorus*, with a spotted breast, often feeding on mistletoe berries. [OE *mistel* basil, mistletoe, of unkn. orig.]

mistletoe /ˈmɪs(ə)lˌtəʊ/ *n.* **1** a parasitic plant, *Viscum album*, growing on apple and other trees and bearing white glutinous berries in winter. **2** a related plant of the genus *Phoradendron*, native to North America. [OE *misteltān* (as MISTLE THRUSH, *tān* twig)]

mistook *past* of MISTAKE.

mistral /ˈmɪstrɑːl/ *n.* a cold northerly wind that blows down the Rhône valley and southern France into the Mediterranean. [F & Prov. f. L (as MAGISTRAL)]

mistranslate /ˌmɪstrænzˈleɪt, ˌmɪstrɑːnz-/ *v.tr.* translate incorrectly. □ **mistranslation** /-ˈleɪʃ(ə)n/ *n.*

mistreat /mɪsˈtriːt/ *v.tr.* treat badly. □ **mistreatment** *n.*

mistress /ˈmɪstrɪs/ *n.* **1** a female head of a household. **2 a** a woman in authority over others. **b** the female owner of a pet. **3** a woman with power to control etc. (often foll. by *of*: *mistress of the situation*). **4** *Brit.* **a** a female teacher (*music mistress*). **b** a female head of a college etc. **5 a** a woman (other than his wife) with whom a married man has a (usu. prolonged) sexual relationship. **b** *archaic* or *poet.* a woman loved and courted by a man. **6** *archaic* or *dial.* (as a title) = MRS. [ME f. OF *maistresse* f. *maistre* MASTER]

Mistress of the Robes *n.* (in the UK) a lady in charge of the queen's wardrobe.

mistrial /mɪsˈtraɪəl/ *n.* **1** a trial rendered invalid through some error in the proceedings. **2** *US* a trial in which the jury cannot agree on a verdict.

mistrust /mɪsˈtrʌst/ *v.* & *n.* ● *v.tr.* **1** be suspicious of. **2** feel no confidence in (a person, oneself, one's powers, etc.). ● *n.* **1** suspicion. **2** lack of confidence.

mistrustful /mɪsˈtrʌstfʊl/ *adj.* **1** (foll. by *of*) suspicious. **2** lacking confidence or trust. □ **mistrustfully** *adv.* **mistrustfulness** *n.*

misty /ˈmɪstɪ/ *adj.* (**mistier, mistiest**) **1** of or covered with mist. **2** indistinct or dim in outline. **3** obscure, vague (*a misty idea*). □ **mistily** *adv.* **mistiness** *n.* [OE *mistig* (as MIST)]

mistype /mɪsˈtaɪp/ *v.tr.* type wrongly. [MIS-[1] + TYPE]

misunderstand /ˌmɪsʌndəˈstænd/ *v.tr.* (*past* and *past part.* **-understood** /-ˈstʊd/) **1** fail to understand correctly. **2** (usu. as **misunderstood** *adj.*) misinterpret the words or actions of (a person).

misunderstanding /ˌmɪsʌndəˈstændɪŋ/ *n.* **1** a failure to understand correctly. **2** a slight disagreement or quarrel.

misusage /mɪsˈjuːsɪdʒ/ *n.* **1** wrong or improper usage. **2** ill-treatment.

misuse *v.* & *n.* ● *v.tr.* /mɪsˈjuːz/ **1** use wrongly; apply to the wrong purpose. **2** ill-treat. ● *n.* /mɪsˈjuːs/ wrong or improper use or application. □ **misuser** /-ˈjuːzə(r)/ *n.*

MIT see MASSACHUSETTS INSTITUTE OF TECHNOLOGY.

Mitanni /mɪˈtænɪ/ *n.* & *adj.* ● *n.* (*pl.* same) **1** a member of the predominant people of a largely Hurrian kingdom centred on the Khabur and Upper Euphrates rivers which flourished in the 15th and early 14th centuries BC. **2** the language of this people. ● *adj.* of or relating to this people or their language. □ **Mitannian** *adj.* & *n.* (Mitanni or Hurrian)

Mitchell[1] /ˈmɪtʃəl/, Joni (born Roberta Joan Anderson) (b.1943), Canadian singer and songwriter. After making her name as a singer and guitarist in Toronto clubs, she became known as a songwriter in the late 1960s. Mitchell's career as a recording artist began in 1968; her many albums, often highly personal in their lyrics, reflect her move from a folk style to a fusion of folk, jazz, and rock. They include *Blue* (1971), *Mingus* (1979), and *Dog Eat Dog* (1986).

Mitchell[2] /ˈmɪtʃəl/, Margaret (1900–49), American novelist. She is famous as the author of the best-selling novel *Gone with the Wind* (1936), set during the American Civil War. It was awarded the Pulitzer Prize, as well as being made into a successful film (1939).

Mitchell[3] /ˈmɪtʃəl/, R(eginald) J(oseph) (1895–1937), English aeronautical engineer. He is best known for designing the Spitfire fighter aircraft; 19,000 aeroplanes based on his 1936 prototype were used by the RAF during the Second World War.

mite[1] /maɪt/ *n.* a small arachnid of the order Acarina, having four pairs of legs when adult. Most mites are tiny or microscopic, and are abundant in the soil and on plants. Some are parasitic, though these are much smaller than the related ticks. [OE *mīte* f. Gmc]

mite[2] /maɪt/ *n.* & *adv.* ● *n.* **1** *hist.* a Flemish copper coin of small value. **2** any small monetary unit. **3** a small object or person, esp. a child. **4** a modest contribution; the best one can do (*offered my mite of comfort*). ● *adv.* (usu. prec. by *a*) *colloq.* somewhat (*is a mite shy*). [ME f. MLG, MDu. *mīte* f. Gmc: prob. the same as MITE[1]]

miter *US* var. of MITRE.

Mitford /ˈmɪtfəd/, Nancy (Freeman) (1904–73) and her sister Jessica (Lucy) (b.1917), English writers. They were born into an aristocratic family; their sisters included Unity (1914–48), who was an admirer of Hitler, and Diana (b.1910), who married Sir Oswald Mosley in 1936. Nancy achieved fame with her comic novels, including *The Pursuit of Love* (1945) and *Love in a Cold Climate* (1949). She was also the editor of *Noblesse Oblige* (1956), which popularized the terms *U* and *Non-U* (used by the British linguist Alan Ross) to categorize the speech and behaviour of the upper class as opposed to other social classes. Jessica became an American citizen in 1944 and was a member of the American Communist Party in the 1940s and 1950s. She is best known for her works on American culture, notably *The American Way of Death* (1963) and *The American Way of Birth* (1992).

Mithras /ˈmɪθræs/ *Rom. Mythol.* a god of light, truth, and the plighted word, probably of Persian origin. He was the central figure of a cult, centred on bull-sacrifice and also associated with merchants and the protection of warriors, which was popular among Roman soldiers of the later empire. The cult was the principal rival of Christianity in the first three centuries AD. □ **Mithraic** /mɪθˈreɪɪk/ *adj.* **Mithraism** /ˈmɪθreɪˌɪz(ə)m/ *n.* **Mithraist** *n.* [L *Mithras* f. Gk *Mithras* f. OPers. *Mithra* f. Skr. *Mitra*]

Mithridates VI /ˌmɪθrɪˈdeɪtiːz/ (also **Mithradates VI**; known as **Mithridates the Great**) (*c.*132–63 BC), king of Pontus 120–63. His expansionist policies led to a war with Rome (88–85), in which he occupied most of Asia Minor and much of Greece, until driven back by Sulla. Two further wars followed; he was defeated by the Roman general Lucullus (*c.*110–*c.*57) and finally by Pompey (in 66).

mithridatize /mɪˈθrɪdəˌtaɪz, ˌmɪθrɪˈdeɪtaɪz/ *v.tr.* (also **-ise**) render proof against a poison by administering gradually increasing doses of it. □ **mithridatic** /ˌmɪθrɪˈdætɪk/ adj. **mithridatism** /ˈmɪθrɪdeɪˌtɪz(ə)m/ n. [f. *mithridate* a supposed universal antidote attributed to MITHRIDATES VI]

mitigate /ˈmɪtɪˌgeɪt/ v.tr. make milder or less intense or severe; moderate (*your offer certainly mitigated their hostility*). ¶ Often confused with *militate*. □ **mitigating circumstances** Law circumstances permitting greater leniency. □ **mitigator** n. **mitigatory** adj. **mitigable** /-gəb(ə)l/ adj. **mitigation** /ˌmɪtɪˈgeɪʃ(ə)n/ n. [ME f. L *mitigare* *mitigat-* f. *mitis* mild]

Mitilíni see MYTILENE.

Mitla /ˈmiːtlə/ an ancient city in southern Mexico, to the east of the city of Oaxaca, now a noted archaeological site. Believed to have been established as a burial site by the Zapotecs some centuries before the arrival of the Spanish, it was eventually overrun by the Mixtecs in about AD 1000. Its Nahuatl name means 'place of the dead'.

mitochondrion /ˌmaɪtəˈkɒndrɪən/ n. (pl. **mitochondria** /-drɪə/) Biol. an organelle found in most eukaryotic cells, containing enzymes for respiration and energy production. [mod.L f. Gk *mitos* thread + *khondrion* dimin. of *khondros* granule]

mitosis /mɪˈtəʊsɪs, maɪ-/ n. Biol. a type of cell division that results in two daughter cells each having the same number and kind of chromosomes as the parent nucleus (cf. MEIOSIS). □ **mitotic** /-ˈtɒtɪk/ adj. [mod.L f. Gk *mitos* thread]

mitral /ˈmaɪtrəl/ adj. of or like a mitre. □ **mitral valve** Anat. a two-cusped valve between the left atrium and the left ventricle of the heart. [mod.L *mitralis* f. L *mitra* girdle]

mitre /ˈmaɪtə(r)/ n. & v. (US **miter**) ● n. **1** a tall deeply cleft head-dress worn by a bishop or abbot, esp. as a symbol of episcopal office, forming in outline the shape of a pointed arch, and often made of embroidered white satin or linen. **2** (in full **mitre-joint**) **a** the joint of two pieces of wood or other material at an angle of 90°, such that the line of junction bisects this angle. **b** any joint in which the angle made by the sides of the joined pieces is bisected by the line of junction. **3** a diagonal join of two pieces of fabric that meet at a corner, made by folding. ● v. **1** tr. bestow the mitre on (a bishop or abbot). **2** tr. & intr. join with a mitre. □ **mitre-block** (or **-board** or **-box**) a guide for a saw in cutting mitre-joints. **mitre-wheels** a pair of bevelled cog-wheels with teeth set at 45° and axes at right angles. □ **mitred** adj. [ME f. OF f. L *mitra* f. Gk *mitra* girdle, turban]

Mitsiwa see MASSAWA.

mitt /mɪt/ n. **1** = MITTEN 1. **2** a glove leaving the fingers and thumb-tip exposed. **3** sl. a hand or fist. **4** a baseball glove for catching the ball. [abbr. of MITTEN]

Mittelland Canal /ˈmɪt(ə)lˌlænd/ a canal in NW Germany, which was constructed between 1905 and 1930. It is part of an inland waterway network linking the rivers Rhine and Elbe.

mitten /ˈmɪt(ə)n/ n. **1** a glove with two sections, one for the thumb and the other for all four fingers. **2** sl. (in pl.) boxing gloves. □ **mittened** adj. [ME f. OF *mitaine* ult. f. L *medietas* half: see MOIETY]

Mitterrand /ˈmiːtəˌrɒn/, François (Maurice Marie) (1916–96), French statesman, President 1981–95. After working to strengthen the Left alliance, Mitterrand became First Secretary of the new Socialist Party in 1971. As President he initially moved to raise basic wages, increase social benefits, nationalize key industries, and decentralize government. After the Socialist Party lost its majority vote in the 1986 general election, Mitterrand asked the right-wing politician Jacques Chirac to serve as Prime Minister. He was re-elected as President in 1988.

mittimus /ˈmɪtɪməs/ n. a warrant committing a person to prison. [ME f. L, = we send]

Mitty, Walter /ˈmɪtɪ, ˈwɔːltə/ the hero of a story (by James Thurber) who indulged in extravagant day-dreams of his own triumphs.

mitzvah /ˈmɪtsvə/ n. (pl. **mitzvahs** or **mitzvoth** /ˌmɪtsˈvəʊt/) in Judaism: **1** a precept or commandment. **2** a good deed done from religious duty. [Heb. *miṣwāh* commandment]

mix /mɪks/ v. & n. ● v. **1** tr. combine or put together (two or more substances or things) so that the constituents of each are diffused among those of the other(s). **2** tr. prepare (a compound, cocktail, etc.) by combining the ingredients. **3** tr. combine (an activity etc.) with another simultaneously (*mix business and pleasure*). **4** intr. **a** join, be mixed, or combine, esp. readily (*oil and water will not mix*). **b** be compatible. **c** be sociable (*must learn to mix*). **5** intr. **a** (foll. by *with*) (of a person) be harmonious or sociable with; have regular dealings with.

b (foll. by *in*) participate in. **6** tr. drink different kinds of (alcoholic liquor) in close succession. **7** tr. combine (two or more sound signals) into one. ● n. **1 a** the act or an instance of mixing; a mixture. **b** the proportion of materials etc. in a mixture. **2** a group of persons of different types (*social mix*). **3** the ingredients prepared commercially for making a cake etc. or for a process such as making concrete. **4** the merging of film pictures or sound. □ **be mixed up in** (or **with**) be involved in or with (esp. something undesirable). **mix and match** select from a range of alternative combinations. **mix in** be harmonious or sociable. **mix it** (US **mix it up**) colloq. start fighting. **mix up 1** mix thoroughly. **2** confuse; mistake the identity of. **mix-up** n. a confusion, misunderstanding, or mistake. □ **mixable** adj. [back-form. f. MIXED (taken as past part.)]

mixed /mɪkst/ adj. **1** of diverse qualities or elements. **2** containing persons from various backgrounds etc. **3** for or involving persons of both sexes (*a mixed school; mixed bathing*). □ **mixed bag** (or **bunch**) a diverse assortment of things or persons. **mixed blessing** a thing having advantages and disadvantages. **mixed crystal** a crystal formed from more than one substance. **mixed doubles** Tennis a doubles game with a man and a woman as partners on each side. **mixed economy** an economic system combining private and state enterprise. **mixed farming** farming of both crops and livestock. **mixed feelings** a mixture of pleasure and dismay about something. **mixed grill** a dish of various grilled meats and vegetables etc. **mixed marriage** a marriage between persons of different races or religions. **mixed metaphor** a combination of inconsistent metaphors (e.g. *this tower of strength will forge ahead*). **mixed number** Math. an integer and a proper fraction. **mixed-up** mentally or emotionally confused; socially ill-adjusted. □ **mixedness** /ˈmɪksɪdnɪs/ n. [ME *mixt* f. OF *mixte* f. L *mixtus* past part. of *miscere* mix]

mixer /ˈmɪksə(r)/ n. **1** a device for mixing foods etc. or for processing other materials. **2** a person who manages socially in a specified way (*a good mixer*). **3** a (usu. soft) drink to be mixed with another. **4** Broadcasting & Cinematog. **a** a device for merging input signals to produce a combined output in the form of sound or pictures. **b** a person who operates this. □ **mixer tap** a tap through which mixed hot and cold water is drawn by means of separate controls.

Mixtec /ˈmɪkstek/ n. & adj. ● n. (pl. same or **Mixtecs**) **1** a member of a people of southern Mexico, noted for their skill in pottery and metallurgy. **2** the language of this people. ● adj. of or relating to this people or their language. [Sp., f. Nahuatl *mixtecah* person from a cloudy place]

mixture /ˈmɪkstʃə(r)/ n. **1** the process of mixing or being mixed. **2** the result of mixing; something mixed; a combination. **3** Chem. the product of the random distribution of one substance through another without any chemical reaction taking place between the components, as distinct from a chemical compound. **4** ingredients mixed together to produce a substance, esp. a medicine (*cough mixture*). **5** a person regarded as a combination of qualities and attributes. **6** gas or vaporized petrol or oil mixed with air, forming an explosive charge in an internal-combustion engine. □ **the mixture as before** the same treatment repeated. [ME f. F *mixture* or L *mixtura* (as MIXED)]

mizen /ˈmɪz(ə)n/ n. (also **mizzen**) (in full **mizen-sail**) Naut. the lowest fore-and-aft sail of a fully rigged ship's mizen-mast. □ **mizen-mast** the mast next aft of the mainmast. **mizen yard** the yard on which the mizen is extended. [ME f. F *misaine* f. It. *mezzana* mizen-sail, fem. of *mezzano* middle: see MEZZANINE]

Mizoram /mɪˈzəʊˌræm/ a state in the far north-east of India, lying between Bangladesh and Burma (Myanmar); capital, Aizawl. Separated from Assam in 1972, it was administered as a Union Territory in India until 1986, when it became a state.

mizzle¹ /ˈmɪz(ə)l/ n. & v.intr. esp. dial. drizzle. □ **mizzly** adj. [ME, prob. f. LG *miseln*: cf. MDu. *miezelen*]

mizzle² /ˈmɪz(ə)l/ v.intr. Brit. sl. run away; decamp. [18th c.: orig. unkn.]

Mk. abbr. **1** the German mark. **2** Mark (esp. in the New Testament).

mks abbr. metre-kilogram-second.

Mkt. abbr. Market.

ml abbr. **1** millilitre(s). **2** mile(s).

MLA abbr. **1** Member of the Legislative Assembly. **2** Modern Language Association (of America).

MLC abbr. Member of the Legislative Council.

MLD abbr. minimum lethal dose.

MLF abbr. multilateral nuclear force.

M.Litt. abbr. Master of Letters. [L *Magister Litterarum*]

Mlle *abbr.* (*pl.* **Mlles**) Mademoiselle.

MLR *abbr.* minimum lending rate.

MM *abbr.* **1** *Messieurs.* **2** (in the UK) Military Medal. **3** Maelzel's metronome (an indication of tempo in music, from the original metronome invented by Johann Nepomuk *Maelzel* (1772–1838)).

mm *abbr.* millimetre(s).

Mmabatho /məˈbɑːtəʊ, ˌəmmə-/ the capital of North-West Province in South Africa, near the border with Botswana; pop. (1985) 28,000.

Mme *abbr.* (*pl.* **Mmes**) Madame.

m.m.f. *abbr.* magnetomotive force.

M.Mus. *abbr.* Master of Music.

MN *abbr.* **1** *Brit.* Merchant Navy. **2** *US* Minnesota (in official postal use).

Mn *symb. Chem.* the element manganese.

M'Naghten rules var. of MCNAUGHTEN RULES (see at MACN-).

mnemonic /nɪˈmɒnɪk/ *adj. & n.* ● *adj.* of or designed to aid the memory. ● *n.* a mnemonic device. □ **mnemonically** *adv.* **mnemonist** /ˈniːmənɪst/ *n.* [med.L *mnemonicus* f. Gk *mnēmonikos* f. *mnēmōn* mindful]

mnemonics /nɪˈmɒnɪks/ *n.pl.* (usu. treated as *sing.*) **1** the art of improving memory. **2** a system for this.

Mnemosyne /niːˈmɒzɪnɪ/ *Gk Mythol.* the mother of the Muses. [Gk *mnēmosunē* memory]

MO *abbr.* **1** Medical Officer. **2** money order. **3** *US* Missouri (in official postal use).

Mo *symb. Chem.* the element molybdenum.

Mo. *abbr.* Missouri.

mo[1] /məʊ/ *n.* (*pl.* **mos**) *colloq.* a moment (*wait a mo*). [abbr.]

mo[2] /məʊ/ *n. Austral. & NZ colloq.* (*pl.* **-os**) = MOUSTACHE. [abbr.]

mo. *abbr. US* month.

m.o. *abbr. modus operandi.*

moa /ˈməʊə/ *n.* (*pl.* **moas**) a large extinct flightless bird of the family Dinornithidae, resembling the emu. Several kinds were found in New Zealand until they were exterminated by humans, and some possibly survived to the 19th century. [Maori]

Moabite /ˈməʊəˌbaɪt/ *adj. & n.* ● *adj.* of Moab, an ancient region by the Dead Sea, or its people. ● *n.* a member of a Semitic people traditionally descended from Lot, living in Moab.

Moabite Stone a monument erected by Mesha, king of Moab, *c.*850 BC which describes (in an early form of the Hebrew language) the campaign between Moab and ancient Israel (2 Kings 3), and furnishes an early example of an inscription in the Phoenician alphabet. It is now in the Louvre in Paris.

moan /məʊn/ *n. & v.* ● *n.* **1** a long murmur expressing physical or mental suffering or pleasure. **2** a low plaintive sound of wind etc. **3** a complaint; a grievance. ● *v.* **1** *intr.* make a moan or moans. **2** *intr.* complain or grumble. **3** *tr.* **a** utter with moans. **b** lament. □ **moaner** *n.* **moanful** *adj.* **moaningly** *adv.* [ME f. OE *mān* (unrecorded) f. Gmc]

moat /məʊt/ *n. & v.* ● *n.* a deep defensive ditch round a castle, town, etc., usu. filled with water. ● *v.tr.* surround with or as with a moat. [ME *mot(e)* f. OF *mote, motte* mound]

mob /mɒb/ *n. & v.* ● *n.* **1** a disorderly crowd; a rabble. **2** (prec. by *the*) usu. *derog.* the populace. **3** *colloq.* **a** *Brit.* a gang; an associated group of persons. **b** (**the Mob**) the Mafia or a similar criminal organization. **4** *Austral.* a flock or herd. ● *v.tr. & intr.* (**mobbed, mobbing**) **1** *tr.* **a** crowd round in order to attack or admire. **b** (of a mob) attack. **c** *US* crowd into (a building). **2** *intr.* assemble in a mob. □ **mob law** (or **rule**) law or rule imposed and enforced by a mob. □ **mobber** *n.* [abbr. of *mobile*, short for L *mobile vulgus* excitable crowd: see MOBILE]

mob-cap /ˈmɒbkæp/ *n. hist.* a woman's large indoor cap covering all the hair, worn in the 18th and early 19th centuries. [obs. (18th-c.) *mob,* orig. = slut + CAP]

Mobile /məʊˈbiːl/ an industrial city and port on the coast of southern Alabama; pop. (1990) 196,280. It is situated at the head of Mobile Bay, an inlet of the Gulf of Mexico.

mobile /ˈməʊbaɪl/ *adj. & n.* ● *adj.* **1** movable; not fixed; free or able to move or flow easily. **2** (of the face etc.) readily changing its expression. **3** (of a shop, library, etc.) accommodated in a vehicle so as to serve various places. **4** (of a person) able to change his or her social status. ● *n.* a decorative structure that may be hung so as to turn freely. □ **mobile home** a large caravan permanently parked and used as a residence. **mobile sculpture** a sculpture having moving parts. □ **mobility** /məˈbɪlɪtɪ/ *n.* [ME f. F f. L *mobilis* f. *movere* move]

mobilize /ˈməʊbɪˌlaɪz/ *v.* (also **-ise**) **1 a** *tr.* organize for service or action (esp. troops in time of war). **b** *intr.* be organized in this way. **2** *tr.* render movable; bring into circulation. □ **mobilizable** *adj.* **mobilizer** *n.* **mobilization** /ˌməʊbɪlaɪˈzeɪʃ(ə)n/ *n.* [F *mobiliser* (as MOBILE)]

Möbius strip /ˈmɜːbɪəs/ *n. Math.* a one-sided surface formed by joining the ends of a rectangle after twisting one end through 180°. [August Ferdinand *Möbius*, Ger. mathematician (1790–1868)]

mobocracy /mɒˈbɒkrəsɪ/ *n.* (*pl.* **-ies**) *colloq.* **1** rule by a mob. **2** a ruling mob.

mobster /ˈmɒbstə(r)/ *n. sl.* a gangster.

Mobutu /məˈbuːtuː/, Sese Seko (full name Mobutu Sese Seko Kuku Ngbendu Wa Za Banga) (b.1930), Zairean statesman, President since 1965. After seizing power in a military coup, he changed his original name (Joseph-Désiré Mobutu) and that of his country (then known as the Belgian Congo) as part of his policy of Africanizing names. He has remained in power despite opposition from tribal groups and small farmers, as well as invasions from Angolan rebels.

Mobutu Sese Seko, Lake /məˌbuːtuː ˌseɪseɪ ˈseɪkəʊ/ the name given to Lake Albert in 1973 by Zaire. (See ALBERT, LAKE.)

moccasin /ˈmɒkəsɪn/ *n.* **1** a type of soft leather slipper or shoe with combined sole and heel, as orig. worn by North American Indians. **2** a venomous American snake of the genus *Agkistrodon*; esp. (in full **water moccasin**) the semiaquatic *A. piscivorus*, native to the south-eastern US. [Algonquian etc. *mockasin, makisin*]

mocha /ˈmɒkə/ *n.* **1** a coffee of fine quality. **2** a beverage or flavouring made with this, often with chocolate added. **3** a soft kind of leather made from sheepskin. [*Mocha*, a port on the Red Sea, from where the coffee first came]

Mochica /məˈtʃiːkə/ *n. & adj.* ● *n.* (*pl.* same) **1** a member of a pre-Inca people living on the coast of Peru AD 100–800. **2** the language of this people. ● *adj.* of or relating to this people or their language. [Sp., f. an Amer. Indian word: cf. *Moche,* archaeological site and valley on NW coast of Peru]

mock /mɒk/ *v., adj., & n.* ● *v.* **1 a** *tr.* ridicule; scoff at. **b** *intr.* (foll. by *at*) act with scorn or contempt for. **2** *tr.* mimic contemptuously. **3** *tr.* jeer, defy, or delude contemptuously. ● *attrib.adj.* sham, imitation (esp. without intention to deceive); pretended (*a mock battle; mock cream*). ● *n.* **1 a** a thing deserving scorn. **2** (in *pl.*) *colloq.* mock examinations. □ **make mock** (or **a mock**) **of** ridicule. **mock-heroic** *adj.* (of a literary style) burlesquing a heroic style. ● *n.* such a style. **mock moon** = PARASELENE. **mock orange** a white-flowered heavy-scented shrub, *Philadelphus coronarius* (also called *syringa*). **mock sun** = PARHELION. **mock turtle soup** soup made from a calf's head etc. to resemble turtle soup. **mock-up** an experimental model or replica of a proposed structure etc. □ **mockable** *adj.* **mockingly** *adv.* [ME *mokke, mocque* f. OF *mo(c)quer* deride f. Rmc]

mocker[1] /ˈmɒkə(r)/ *n.* a person who mocks. □ **put the mockers on** *sl.* **1** bring bad luck to. **2** put a stop to. [20th c.: orig. unkn.]

mocker[2] /ˈmɒkə(r)/ *n. Austral. & NZ sl.* clothing, dress. [orig. unkn.]

mockery /ˈmɒkərɪ/ *n.* (*pl.* **-ies**) **1 a** derision, ridicule. **b** a subject or occasion of this. **2** (often foll. by *of*) a counterfeit or absurdly inadequate representation. **3** a ludicrously or insultingly futile action etc. [ME f. OF *moquerie* (as MOCK)]

mockingbird /ˈmɒkɪŋˌbɜːd/ *n.* a long-tailed songbird of the American family Mimidae, noted for mimicking other birds' calls etc., esp. *Mimus polyglottos*.

MOD *abbr.* (in the UK) Ministry of Defence.

mod[1] /mɒd/ *adj. & n. colloq.* ● *adj.* modern, esp. in style of dress. ● *n.* (in the UK) a young person (esp. in the early 1960s) belonging to a subculture characterized by the cultivation of a smart stylish appearance, the riding of motor scooters, and a liking for soul music. □ **mod cons** modern conveniences. [abbr.]

mod[2] /mɒd/ *prep. Math.* = MODULO. [abbr.]

mod[3] /mɒd/ *n.* a Highland Gaelic meeting for music and poetry. [Gael. *mòd*]

modal /ˈməʊd(ə)l/ *adj.* **1** of or relating to mode or form as opposed to substance. **2** *Gram.* **a** of or denoting the mood of a verb. **b** (of an auxiliary verb, e.g. *would*) used to express the mood of another verb. **c** (of a particle) denoting manner. **3** *Statistics* of or relating to a mode; occurring most frequently in a sample or population. **4** *Mus.* denoting a style of music using a particular mode. **5** *Logic* (of a proposition) in which the predicate is affirmed of the subject with some qualification, or which involves the affirmation of possibility, impossibility, necessity, or contingency. □ **modally** *adv.* [med.L *modalis* f. L (as MODE)]

modality /məˈdælɪtɪ/ n. (pl. **-ies**) **1** the state of being modal. **2** (in *sing.* or *pl.*) a prescribed method of procedure. [med.L *modalitas* (as MODAL)]

mode /məʊd/ n. **1** a way or manner in which a thing is done; a method of procedure. **2** a prevailing fashion or custom. **3** *Computing* a way of operating or using a system (*print mode*). **4** *Statistics* the value that occurs most frequently in a given set of data. **5** *Mus.* **a** each of the scale systems that result when the white notes of the piano are played consecutively over an octave (*Lydian mode*). **b** each of the two main modern scale systems, the major and minor (*minor mode*). **6** *Logic* **a** the character of a modal proposition. **b** = MOOD[2] 2. **7** *Physics* any of the distinct kinds or patterns of vibration of an oscillating system. **8** *N. Amer. Gram.* = MOOD[2] 1. [F *mode* and L *modus* measure]

model /ˈmɒd(ə)l/ n. & v. ● n. **1** a representation in three dimensions of an existing person or thing or of a proposed structure, esp. on a smaller scale (often *attrib.: a model train*). **2** a simplified (often mathematical) description of a system etc., to assist calculations and predictions. **3** a figure in clay, wax, etc., to be reproduced in another material. **4** a particular design or style of a structure or commodity, esp. of a car. **5 a** an exemplary person or thing (*a model of self-discipline*). **b** (*attrib.*) ideal, exemplary (*a model student*). **6 a** a person employed to pose for an artist, sculptor, photographer, etc. **b** a person employed to display clothes etc. by wearing them. **7** (often foll. by *for*) an actual person, place, etc., on which a fictional character, location, etc., is based. **8** a garment etc. by a well-known designer, or a copy of this. ● v. (**modelled, modelling**; *US* **modeled, modeling**) **1** *tr.* **a** fashion or shape (a figure) in clay, wax, etc. **b** (foll. by *after, on*, etc.) form (a thing in imitation of). **2 a** *intr.* act or pose as a model. **b** *tr.* (of a person acting as a model) display (a garment). **3** *tr.* devise a (usu. mathematical) model of (a phenomenon, system, etc.). **4** *tr.* (in drawing, painting, etc.) cause to appear three-dimensional. □ **modeller** n. [F *modelle* f. It. *modello* ult. f. L *modulus*: see MODULUS]

Model, the (also called *the New Model*) the plan for the reorganization of the Parliamentary army, passed by the House of Commons in 1644–5 (see NEW MODEL ARMY).

Model Parliament a parliament summoned by Edward I in November 1295 to obtain financial aid for his wars. It was termed this in the 19th century because it was more representative than any previous Parliament, although it did not provide an enduring pattern for future Parliaments.

modem /ˈməʊdem/ n. a combined device for modulation and demodulation, e.g. between a computer and a telephone line. [*modulator* + *demodulator*]

Modena /ˈmɒdɪnə/ a city in northern Italy, north-west of Bologna; pop. (1990) 177,500. A city-state from the 12th century, it was ruled from the late 13th century by the Este family, until it became a part of the kingdom of Italy in 1859.

moderate adj., n., & v. ● adj. /ˈmɒdərət/ **1** avoiding extremes; temperate in conduct or expression. **2** fairly or tolerably large or good. **3** (of the wind) of medium strength. **4** (of prices) fairly low. ● n. /ˈmɒdərət/ a person who holds moderate views, esp. in politics. ● v. /ˈmɒdəˌreɪt/ **1** *tr. & intr.* make or become less violent, intense, rigorous, etc. **2** *tr.* (also *absol.*) act as a moderator of or to. **3** *tr. Physics* retard (neutrons) with a moderator. □ **moderately** /-rətlɪ/ adv. **moderateness** n. **moderatism** n. [ME f. L *moderatus* past part. of *moderare* reduce, control: rel. to MODEST]

moderation /ˌmɒdəˈreɪʃ(ə)n/ n. **1** the process or an instance of moderating. **2** the quality of being moderate. **3** *Physics* the retardation of neutrons by a moderator (see MODERATOR 5). **4** (in pl.) (**Moderations**) the first public examination in some faculties for the Oxford BA degree. □ **in moderation** in a moderate manner or degree. [ME f. OF f. L *moderatio -onis* (as MODERATE)]

moderato /ˌmɒdəˈrɑːtəʊ/ adj., adv., & n. *Mus.* ● adj. & adv. performed at a moderate pace. ● n. (pl. **-os**) a piece of music to be performed in this way. [It. (as MODERATE)]

moderator /ˈmɒdəˌreɪtə(r)/ n. **1** an arbitrator or mediator. **2** a presiding officer. **3** *Eccl.* a Presbyterian minister presiding over an ecclesiastical body. **4** an examiner for Moderations. **5** *Physics* a substance used in a nuclear reactor to retard neutrons. □ **moderatorship** n. [ME f. L (as MODERATE)]

modern /ˈmɒd(ə)n/ adj. & n. ● adj. **1** of the present and recent times, as opposed to the remote past. **2** in current fashion; not antiquated. ● n. (usu. in pl.) a person living in modern times. □ **modern English** English from about 1500 onwards (see ENGLISH). **modern history** history from the end of the Middle Ages to the present day.

modernly adv. **modernness** n. **modernity** /məˈdɜːnɪtɪ/ n. [F *moderne* or LL *modernus* f. L *modo* just now]

modern dance n. a form of theatrical dance which arose in the early 20th century as a reaction to the formality and narrative purpose of classical ballet, emphasizing instead freedom of expression and experimental techniques. It was pioneered by figures such as the dancer Isadora Duncan and choreographer Rudolf von Laban, and developed principally by American dancers, such as Martha Graham and Doris Humphrey (1895–1958), who in turn influenced the increasingly abstract dance of Merce Cunningham. In recent years elements not usually associated with dance, such as speech and film, have been introduced, although there has also been some return to balletic technique.

modernism /ˈmɒdəˌnɪz(ə)m/ n. **1** modern ideas or methods. **2** a term or expression characteristic of modern times. **3** (also **Modernism**) a style or movement in the arts that aims to break with classical and traditional forms. (*See note below.*) **4** a movement towards modifying traditional beliefs in accordance with modern ideas, esp. in the Roman Catholic Church in the late 19th and early 20th centuries. □ **modernist** n. **modernistic** /ˌmɒdəˈnɪstɪk/ adj. **modernistically** adv.

▪ Modernism is a broad term used to refer to a whole range of individual artistic movements and artists, mostly in the first half of the twentieth century. The emphasis tends to be on form rather than content, and represents a deliberate programme to challenge traditional (often nineteenth-century) forms of expression associated with narrative and representation, at the same time often questioning basic tenets relating to Western civilization and human progress. In literature, it is associated with writers such as James Joyce, Virginia Woolf, Ezra Pound, and T. S. Eliot, and with movements such as imagism and vorticism; in the visual arts, with cubism, futurism, Dadaism, and surrealism; in music, with Schoenberg and Webern; and in architecture, with Frank Lloyd Wright.

modernize /ˈmɒdəˌnaɪz/ v. (also **-ise**) **1** *tr.* make modern; adapt to modern needs or habits. **2** *intr.* adopt modern ways or views. □ **modernizer** n. **modernization** /ˌmɒdənaɪˈzeɪʃ(ə)n/ n.

modest /ˈmɒdɪst/ adj. **1** having or expressing a humble or moderate estimate of one's own merits or achievements. **2** diffident, bashful, retiring. **3** decorous in manner and conduct. **4** moderate or restrained in amount, extent, severity, etc.; not excessive or exaggerated (*a modest sum*). **5** (of a thing) unpretentious in appearance etc. □ **modestly** adv. [F *modeste* f. L *modestus* keeping due measure]

modesty /ˈmɒdɪstɪ/ n. the quality of being modest.

modicum /ˈmɒdɪkəm/ n. (foll. by *of*) a small quantity. [L, = short distance or time, neut. of *modicus* moderate f. *modus* measure]

modification /ˌmɒdɪfɪˈkeɪʃ(ə)n/ n. **1** the act or an instance of modifying or being modified. **2** a change made. [F or f. L *modificatio* (as MODIFY)]

modifier /ˈmɒdɪˌfaɪə(r)/ n. **1** a person or thing that modifies. **2** *Gram.* a word, esp. an adjective or noun used attributively, that qualifies the sense of another word (e.g. *good* and *family* in *a good family house*).

modify /ˈmɒdɪˌfaɪ/ v.tr. (**-ies, -ied**) **1** make less severe or extreme; tone down (*modify one's demands*). **2** make partial changes in; make different. **3** *Gram.* qualify or expand the sense of (a word etc.). **4** *Phonet.* change (a vowel) by umlaut. **5** *Chem.* change or replace all the substituent radicals of a polymer, thereby changing its physical properties such as solubility etc. (*modified starch*). □ **modifiable** adj. **modificatory** /-fɪˌkeɪtərɪ/ adj. [ME f. OF *modifier* f. L *modificare* (as MODE)]

Modigliani /ˌmɒdɪˈljɑːnɪ/, Amedeo (1884–1920), Italian painter and sculptor, resident in France from 1906. Influenced by Botticelli and other 14th-century artists, his portraits and nudes are noted for their elongated forms, linear qualities, and earthy colours. Modigliani's works include the sculpture *Head of a Woman* (1910–13) and the portrait *Jeanne Hébuterne* (1919).

modillion /məˈdɪljən/ n. *Archit.* a projecting bracket under the corona of a cornice in the Corinthian and other orders. [F *modillon* f. It. *modiglione* ult. f. L *mutulus* mutule]

modish /ˈməʊdɪʃ/ adj. fashionable. □ **modishly** adv. **modishness** n.

modiste /mɒˈdiːst/ n. a milliner; a dressmaker. [F (as MODE)]

Mods /mɒdz/ n.pl. colloq. Moderations (see MODERATION 4). [abbr.]

modular /ˈmɒdjʊlə(r)/ adj. of or consisting of modules or moduli. □ **modularity** /ˌmɒdjʊˈlærɪtɪ/ n. [mod.L *modularis* f. L *modulus*: see MODULUS]

modulate /'mɒdjʊ,leɪt/ v. **1** tr. **a** regulate or adjust. **b** moderate. **2** tr. adjust or vary the tone or pitch of (the speaking voice). **3** tr. Electronics alter the amplitude or frequency of (a wave) by a wave of a lower frequency to convey a signal. **4** intr. & tr. (often foll. by from, to) Mus. change or cause to change from one key to another. □ **modulator** n. **modulation** /,mɒdjʊ'leɪʃ(ə)n/ n. [L modulari modulat- to measure f. modus measure]

module /'mɒdjuːl/ n. **1** a standardized part or independent unit used in construction, esp. of furniture, a building, or an electronic system. **2** an independent self-contained unit of a spacecraft (lunar module). **3** a unit or period of training or education. **4 a** a standard or unit of measurement. **b** Archit. a unit of length for expressing proportions, e.g. the semidiameter of a column at the base. [F module or L modulus: see MODULUS]

modulo /'mɒdjʊ,ləʊ/ prep. & adj. Math. using, or with respect to, a modulus (see MODULUS 2). [L, ablat. of MODULUS]

modulus /'mɒdjʊləs/ n. (pl. **moduli** /-,laɪ/) Math. **1 a** the magnitude of a real number without regard to its sign. **b** the positive square root of the sum of the squares of the real and imaginary parts of a complex number. **2** a constant factor or ratio. **3** (in number theory) a number used as a divisor for considering numbers in sets giving the same remainder when divided by it. **4** a constant indicating the relation between a physical effect and the force producing it. [L, = measure, dimin. of modus]

modus operandi /,məʊdəs ,ɒpə'rændɪ/ n. (pl. **modi operandi** /,məʊdɪ/) **1** the particular way in which a person performs a task or action. **2** the way a thing operates. [L, = way of operating: see MODE]

modus vivendi /,məʊdəs vɪ'vendɪ/ n. (pl. **modi vivendi** /,məʊdɪ/) **1** a way of living or coping. **2 a** an arrangement whereby those in dispute can carry on pending a settlement. **b** an arrangement between people who agree to differ. [L, = way of living: see MODE]

Moesia /'miːsɪə, 'miːʃə/ an ancient country of southern Europe, corresponding to parts of modern Bulgaria and Serbia. Lying south of the lower Danube, it was bounded to the west by the River Drina, to the east by the Black Sea, and to the south by the Balkan Mountains. It became a province of Rome in AD 15, remaining part of the Roman Empire until the 7th century.

mofette /mɒ'fet/ n. **1** a fumarole. **2** an exhalation of vapour from this. [F mofette or Neapolitan It. mofetta]

mog /mɒg/ n. (also **moggie** /'mɒgɪ/) Brit. sl. a cat. [20th c.: of dial. orig.]

Mogadishu /,mɒgə'dɪʃuː/ (also **Muqdisho** /mʊk'dɪʃəʊ/, Italian **Mogadiscio** /,mɒga'diʃʃo/) the capital of Somalia, a port on the Indian Ocean; pop. (1982) 377,000. Founded by the Arabs in the 10th century, it was leased to the Italians in 1892 and sold to them in 1905. It was the capital of Italian Somalia from 1892 to 1960.

Mogadon /'mɒgə,dɒn/ n. propr. a benzodiazepine drug used to treat insomnia.

Mogilev see MAHILYOW.

Mogul /'məʊg(ə)l/ n. (also **Moghul** or **Mughal** /'mʊg-/) **1** a member of a Mongolian Muslim dynasty in India whose empire was consolidated, after the conquests of Tamerlane, by his descendant Babur (reigned c.1525–30) and greatly extended by Akbar (reigned 1556–1605). Gradually broken up by wars and revolts, and faced by European commercial expansion, the Mogul empire did not finally disappear until after the Indian Mutiny (1857). **2** (often **the Great Mogul**) any of the emperors of Delhi in the 16th–19th centuries. **3** (**mogul**) an important or influential person. [Pers. muġul MONGOL]

MOH abbr. Medical Officer of Health.

Mohács /'məʊhaːtʃ/ a river port and industrial town on the Danube in southern Hungary, close to the borders with Croatia and Serbia; pop. (est. 1984) 21,000. It was the site of a battle in 1526 in which the Hungarians were defeated by a Turkish force under Suleiman I, as a result of which Hungary became part of the Ottoman Empire. A site nearby was the scene of a further decisive battle fought in 1687 during the campaign that swept the Turks out of Hungary.

mohair /'məʊheə(r)/ n. **1** the hair of the angora goat. **2** a yarn or fabric from this, either pure or mixed with wool or cotton. [ult. f. Arab. muḵayyar, lit. 'choice, select']

Mohammed see MUHAMMAD[1].

Mohammedan var. of MUHAMMADAN.

Mohammerah /mə'hæmərə/ the former name (until 1924) for KHORRAMSHAHR.

Mohave Desert see MOJAVE DESERT.

Mohawk /'məʊhɔːk/ n. & adj. ● n. (pl. same or **Mohawks**) **1 a** a member of an Iroquois people inhabiting parts of New York State, Ontario, and Quebec. **b** the language of this people. **2** Skating a step from either edge of the skate to the same edge on the other foot in the opposite direction. **3** esp. N. Amer. a Mohican haircut. ● adj. of or relating to this people or their language. [Narragansett mohowawog lit. 'man-eaters']

Mohegan /məʊ'hiːgən/ adj. & n. (also **Mohican**) ● adj. of or relating to a member of an Algonquian people formerly inhabiting the western parts of Connecticut and Massachusetts, or their language. ● n. **1** a member of the Mohegan people. **2** the language of this people. ¶ Mohegan rather than Mohican is now the preferred form. [Mohegan]

Mohenjo-Daro /mə,hendʒəʊ'daːrəʊ/ an ancient city of the civilization of the Indus valley (c.2600–1700 BC), now a major archaeological site in Pakistan, south-west of Sukkur.

Mohican /məʊ'hiːkən/ adj. & n. ● adj. **1** = MOHEGAN adj. **2** designating a hairstyle in which the head is shaved except for a strip of hair, often spiked, from the forehead to the back of the neck (similar to the depictions of a deer-hair topknot worn by American Indian men in James Fenimore Cooper's novel The Last of the Mohicans). ● n. **1** = MOHEGAN n. **2** a Mohican haircut. [Mohegan]

moho /'məʊhəʊ/ n. Geol. the boundary surface between the earth's crust and the mantle, known more fully as the Mohorovičić discontinuity. The moho is situated at a depth of about 10–12 km under the ocean beds and 40–50 km under the continents. It was identified in 1909 by the Croatian seismologist Andrija Mohorovičić (1857–1936), who detected a change in the velocity of earthquake shock waves at those depths.

Moholy-Nagy /,məʊhɔɪ'nɒdʒ/, László (1895–1946), Hungarian-born American painter, sculptor, and photographer. Based in Berlin from 1920, he became identified with the constructivist school, pioneering the experimental use of plastic materials, light, photography, and film. Moholy-Nagy taught with Walter Gropius at the Bauhaus (1923–9), later heading the new Bauhaus school in Chicago from 1937.

Mohorovičić discontinuity /,məʊhə'rəʊvɪ,tʃɪtʃ/ n. Geol. see MOHO.

Mohs' scale /məʊz/ n. Mineral. a scale of hardness used in classifying minerals, ranging in increasing hardness from one to ten. The reference minerals used to define the scale are talc 1, gypsum 2, calcite 3, fluorspar 4, apatite 5, orthoclase 6, quartz 7, topaz 8, corundum 9, and diamond 10. Position on the scale depends on a mineral's ability to scratch those rated lower. [Friedrich Mohs, Ger. mineralogist (1773–1839)]

moidore /'mɔɪdɔː(r)/ n. hist. a Portuguese gold coin, current in England in the 18th century. [Port. moeda d'ouro money of gold]

moiety /'mɔɪətɪ/ n. (pl. **-ies**) Law or literary **1** a half. **2** each of the two parts into which a thing is divided. [ME f. OF moité, moitié f. L medietas -tatis middle f. medius (adj.) middle]

moil /mɔɪl/ v. & n. archaic ● v.intr. drudge (esp. toil and moil). ● n. drudgery. [ME f. OF moillier moisten, paddle in mud, ult. f. L mollis soft]

Moirai /'mɔɪraɪ/ Gk Mythol. the Greek name for the Fates (see FATES, THE).

moire /mwaː(r)/ n. (in full **moire antique**) watered fabric, orig. mohair, now usu. silk. [F earlier mouaire) f. MOHAIR]

moiré /'mwaːreɪ/ adj. & n. ● adj. **1** (of silk) watered. **2** (of metal) having a patterned appearance like watered silk. ● n. **1** this patterned appearance. **2** = MOIRE. [F, past part. of moirer (as MOIRE)]

Moissan /'mwʌsɒn/, Ferdinand Frédéric Henri (1852–1907), French chemist. In 1886 he succeeded in isolating the very reactive element fluorine. In 1892 he invented the electric arc furnace that bears his name, in which he claimed to have synthesized diamonds. This is now in doubt, but the very high temperatures achieved in the furnace made it possible to reduce some uncommon metals from their ores. An influential teacher, he was appointed professor of inorganic chemistry at the Sorbonne in 1900. Moissan was awarded the Nobel Prize for chemistry in 1906.

moist /mɔɪst/ adj. **1 a** slightly wet; damp. **b** (of the season etc.) rainy. **2** (of a disease) marked by a discharge of matter etc. **3** (of the eyes) wet with tears. □ **moistly** adv. **moistness** n. [ME f. OF moiste, ult. from or rel. to L mucidus (see MUCUS) and musteus fresh (see MUST[2])]

moisten /'mɔɪs(ə)n/ v.tr. & intr. make or become moist.

moisture /'mɔɪstʃə(r)/ n. water or other liquid diffused in a small quantity as vapour, or within a solid, or condensed on a surface. □ **moistureless** adj. [ME f. OF moistour (as MOIST)]

moisturize /'mɔɪstʃə,raɪz/ v.tr. (also **-ise**) make less dry (esp. the skin by use of a cosmetic). □ **moisturizer** n.

Mojave Desert /məʊˈhɑːvɪ/ (also **Mohave**) a desert in southern California, to the south-east of the Sierra Nevada and north and east of Los Angeles.

moke /məʊk/ *n. sl.* **1** *Brit.* a donkey. **2** *Austral.* a very poor horse. [19th c.: orig. unkn.]

moksha /ˈmɒkʃə/ *n. Hinduism & Jainism* release from the chain of births impelled by the law of karma; the transcendent state attained by this liberation (cf. NIRVANA). [Skr. *mokṣa*]

mol /məʊl/ *abbr. Chem.* = MOLE⁴.

molal /ˈməʊlæl/ *adj. Chem.* (of a solution) containing one mole of solute per kilogram of solvent. □ **molality** /məˈlælɪtɪ/ *n.* [MOLE⁴ + -AL¹]

molar¹ /ˈməʊlə(r)/ *adj. & n.* ● *adj.* (usu. of a mammal's back teeth) serving to grind. ● *n.* a molar tooth. [L *molaris* f. *mola* millstone]

molar² /ˈməʊlə(r)/ *adj.* **1** of or relating to mass. **2** acting on or by means of large masses or units. [L *moles* mass]

molar³ /ˈməʊlə(r)/ *adj. Chem.* **1** of or relating to one mole of a substance. **2** (of a solution) containing one mole of solute per litre of solvent. □ **molarity** /məˈlærɪtɪ/ *n.* [MOLE⁴ + -AR¹]

molasses /məˈlæsɪz/ *n.pl.* (treated as *sing.*) **1** uncrystallized syrup extracted from raw sugar during refining. **2** *N. Amer.* treacle. [Port. *melaço* f. LL *mellaceum* MUST² f. *mel* honey]

Mold /məʊld/ the county town of Clwyd; pop. (1981) 8,589.

mold *US* var. of MOULD¹, MOULD², MOULD³.

Moldau /ˈmɔldaʊ/ the German name for the VLTAVA.

Moldavia /mɒlˈdeɪvɪə/ **1** a former principality of SE Europe. Formerly a part of the Roman province of Dacia, the region became a principality in the 14th century, coming under Turkish rule in the 16th century. The province of Bukovina in the north-west was ceded to Austria in 1777 and Bessarabia, in the south-east, was ceded to Russia in 1812. In 1861 Moldavia united with Wallachia to form Romania. **2** see MOLDOVA.

Moldavian /mɒlˈdeɪvɪən/ *n. & adj.* ● *n.* **1** a native or inhabitant of Moldavia or Moldova. **2** the Romanian language as spoken and written (in the Cyrillic alphabet) in Moldavia or Moldova. ● *adj.* of or relating to Moldavia, Moldova, or Moldavian.

molder *US* var. of MOULDER.

molding *US* var. of MOULDING.

Moldova /mɒlˈdəʊvə, mɒlˈdɒvə/ (also **Moldavia**) a landlocked country in SE Europe, between Romania and Ukraine; pop. (1991) 4,384,000; languages, Moldavian (official), Russian; capital, Chişinău. A former constituent republic of the USSR, Moldova was formed from territory ceded by Romania in 1940. It became independent as a member of the Commonwealth of Independent States in 1991. □ **Moldovan** *adj. & n.*

moldy *US* var. of MOULDY.

mole¹ /məʊl/ *n.* **1** a small burrowing insect-eating mammal of the family Talpidae; esp. *Talpa europaea*, with dark velvety fur and very small eyes. **2** *colloq.* **a** a spy established deep within an organization and usu. dormant for a long period while attaining a position of trust. **b** a betrayer of confidential information. □ **mole cricket** a large burrowing nocturnal cricket-like insect of the family Gryllotalpidae. **mole rat** a subterranean burrowing rodent, with reduced eyes; esp. one of the family Bathyergidae, native to Africa, often living communally. [ME *molle*, prob. f. MDu. *moll(e)*, *mol*, MLG *mol*, *mul*]

mole² /məʊl/ *n.* a small often slightly raised dark blemish on the skin caused by a high concentration of melanin. [OE *māl* f. Gmc]

mole³ /məʊl/ *n.* **1** a massive structure serving as a pier, breakwater, or causeway. **2** an artificial harbour. [F *môle* f. L *moles* mass]

mole⁴ /məʊl/ *n. Chem.* that amount of a given substance or species having a mass numerically equal to its molecular or atomic weight (symbol: **mol**). The SI base unit of amount of substance, the mole is equivalent to the quantity containing as many specified elementary units (atoms, molecules, ions, etc.) as there are atoms in 0.012 kg of carbon-12. [G *Mol* f. *Molekül* MOLECULE]

mole⁵ /məʊl/ *n. Med.* an abnormal mass of tissue in the uterus. [F *môle* f. L *mola* millstone]

molecular /məˈlekjʊlə(r)/ *adj.* of, relating to, or consisting of molecules. □ **molecular biology** the study of the structure and function of large molecules associated with living organisms. **molecular sieve** a crystalline substance with pores of molecular dimensions which permit the passage of molecules below a certain size. **molecular weight** = *relative molecular mass*. □ **molecularly** *adv.* **molecularity** /-ˌlekjʊˈlærɪtɪ/ *n.*

molecule /ˈmɒlɪˌkjuːl/ *n.* **1** *Chem.* the smallest fundamental unit (usu.

a group of atoms) of a chemical compound that can take part in a chemical reaction. **2** (in general use) a small particle. [F *molécule* f. mod.L *molecula* dimin. of L *moles* mass]

molehill /ˈməʊlhɪl/ *n.* a small mound thrown up by a mole in burrowing. □ **make a mountain out of a molehill** exaggerate the importance of a minor difficulty.

moleskin /ˈməʊlskɪn/ *n.* **1** the skin of a mole used as fur. **2 a** a kind of cotton fustian with its surface shaved before dyeing. **b** (in *pl.*) clothes, esp. trousers, made of this.

molest /məˈlest/ *v.tr.* **1** annoy or pester (a person) in a hostile or injurious way. **2** attack or interfere with (a person), esp. sexually. □ **molester** *n.* **molestation** /ˌmɒleˈsteɪʃ(ə)n, ˌməʊl-/ *n.* [OF *molester* or L *molestare* annoy f. *molestus* troublesome]

Molière /ˈmɒlɪˌeə(r)/ (pseudonym of Jean-Baptiste Poquelin) (1622–73), French dramatist. In 1658 he established himself in Paris as an actor, dramatist, and manager of his own troupe. His high reputation as a writer of French comedy is based on more than twenty plays, which took the vices and follies of contemporary France as their subject, simultaneously adopting and developing stock characters from Italian *commedia dell'arte*. Major works include *Tartuffe* (1664), *Don Juan* (1665), *Le Misanthrope* (1666), and *Le Malade imaginaire* (1673). Molière also collaborated with the composer Lully for *Le Bourgeois gentilhomme* (1670).

moline /məˈlaɪn/ *adj. Heraldry* (of a cross) having each extremity broadened and curved back. [prob. f. AF *moliné* f. *molin* MILL¹, because of the resemblance to the iron support of a millstone]

Molise /mɒˈliːzeɪ/ a region of eastern Italy, on the Adriatic coast; capital, Campobasso.

moll /mɒl/ *n. sl.* **1** a gangster's female companion. **2** a prostitute. [pet-form of the name *Mary*]

mollify /ˈmɒlɪˌfaɪ/ *v.tr.* (**-ies, -ied**) **1** appease, pacify. **2** reduce the severity of; soften. □ **mollifier** *n.* **mollification** /ˌmɒlɪfɪˈkeɪʃ(ə)n/ *n.* [ME f. F *mollifier* or L *mollificare* f. *mollis* soft]

mollusc /ˈmɒləsk/ *n.* (*US* **mollusk**) an invertebrate animal of the phylum Mollusca, having a soft body and usu. a hard shell, including snails, cuttlefish, oysters, etc. (*See note below.*) □ **molluscan** /məˈlʌskən/ *adj.* **molluscoid** *adj.* **molluscous** *adj.* [mod.L *mollusca* neut. pl. of L *molluscus* f. *mollis* soft]

▪ The molluscs form one of the largest and most advanced of invertebrate phyla, and have diverged along several evolutionary pathways. Some have retained a single coiled shell (the gastropods), and others have a flattened body between two valves of a hinged shell (the bivalves). The cephalopods usually have an internal shell, and a number of tentacles. The majority are marine and breathe by means of gills, though a small number (all gastropods) have solved the problems of life on land, and some of these have returned to live in fresh water.

molly /ˈmɒlɪ/ *n.* (also **mollie**) a small live-bearing freshwater fish of the genus *Poecilia* (formerly *Mollienisia*), native to America; esp. *P. sphenops*, bred in many colour varieties and kept in aquariums. [abbr. of mod.L *Mollienisia*, f. Count Nicolas-François *Mollien*, Fr. statesman (1758–1850)]

mollycoddle /ˈmɒlɪˌkɒd(ə)l/ *v. & n.* ● *v.tr.* coddle, pamper. ● *n.* an effeminate man or boy; a milksop. [formed as MOLL + CODDLE]

mollymawk /ˈmɒlɪˌmɔːk/ *n.* (also **mallemuck** /ˈmælɪˌmʌk/) a fulmar, petrel, or similar bird; esp. a smaller albatross of the genus *Diomedea*. [Du. *mallemok* f. *mal* foolish + *mok* gull]

Moloch /ˈməʊlɒk/ *n.* **1 a** a Canaanite idol to whom children were sacrificed. **b** a tyrannical object of sacrifices. **2** (**moloch**) a spiny slow-moving grotesque lizard, *Moloch horridus*, native to Australia. [LL f. Gk *Molokh* f. Heb. *mōleḵ*]

Molotov¹ /ˈmɒləˌtɒf/ a former name (1940–57) for PERM.

Molotov² /ˈmɒləˌtɒf/, Vyacheslav (Mikhailovich) (born Vyacheslav Mikhailovich Skryabin) (1890–1986), Soviet statesman. Born in Russia, he was an early member of the Bolsheviks and a staunch supporter of Stalin after Lenin's death. As Commissar (later Minister) for Foreign Affairs (1939–49; 1953–6), he negotiated the non-aggression pact with Nazi Germany (1939) and after 1945 represented the Soviet Union at meetings of the United Nations, where his frequent exercise of the veto helped to prolong the cold war. He was expelled from his party posts in 1956 after quarrelling with Khrushchev.

Molotov cocktail *n.* a homemade incendiary grenade, usu. consisting of a bottle or other breakable container holding inflammable liquid and with a means of ignition; the production of similar grenades was organized by Vyacheslav Molotov during the Second World War.

molt US var. of MOULT.

molten /ˈməʊlt(ə)n/ adj. melted, esp. made liquid by heat. [past part. of MELT]

molto /ˈmɒltəʊ/ adv. Mus. very (molto sostenuto; allegro molto). [It. f. L multus much]

Molucca Islands /məˈlʌkə/ (also **Moluccas**; Indonesian **Maluku** /məˈluːkuː/) an island group in Indonesia, between Sulawesi and New Guinea; capital, Amboina. Settled by the Portuguese in the early 16th century, the islands were taken a century later by the Dutch, who controlled the lucrative trade in the spices produced on the islands. The islands were formerly known as the Spice Islands.

moly /ˈməʊlɪ/ n. (pl. **-ies**) **1** an allium, Allium moly, with small yellow flowers. **2** a mythical herb with white flowers and black roots, endowed with magic properties. [L f. Gk mōlu]

molybdenite /məˈlɪbdɪˌnaɪt/ n. Mineral. molybdenum disulphide as an ore.

molybdenum /məˈlɪbdənəm/ n. a brittle silver-grey metallic chemical element (atomic number 42; symbol **Mo**). A transition metal, molybdenum was discovered by Carl Scheele in 1782. The chief ore is molybdenite. The metal is used in steels to give strength and corrosion resistance; molybdenum disulphide is used as a lubricant in engine oils. [earlier molybdena, orig. = molybdenite, lead ore: L molybdena f. Gk molubdaina plummet f. molubdos lead]

mom /mɒm/ n. N. Amer. colloq. mother. [abbr. of MOMMA]

Mombasa /mɒmˈbæsə/ a seaport and industrial city in SE Kenya, on the Indian Ocean; pop. (est. 1984) 425,630. Established by the Portuguese as a fortified trading post in 1593, it is the leading port and second largest city of Kenya.

moment /ˈməʊmənt/ n. **1** a very brief portion of time; an instant. **2** a short period of time (wait a moment) (see also MINUTE[1]). **3** an exact or particular point of time (at last the moment arrived; I came the moment you called). **4** importance (of no great moment). **5** Physics & Mech. etc. **a** the turning effect produced by a force acting at a distance on an object. **b** this effect expressed as the product of the force and the distance from its line of action to a point. □ **at the moment** at this time; now. **in a moment 1** very soon. **2** instantly. **man** (or **woman** etc.) **of the moment** the one of importance at the time in question. **moment of inertia** Physics the quantity by which the angular acceleration of a body must be multiplied to give corresponding torque. **moment of truth** a time of crisis or test (orig. the final sword-thrust in a bullfight). **not for a** (or **one**) **moment** never; not at all. **this moment** immediately; at once (come here this moment). [ME f. OF f. L momentum: see MOMENTUM]

momenta pl. of MOMENTUM.

momentarily /ˈməʊməntərɪlɪ, ˌməʊmənˈterɪlɪ/ adv. **1** for a moment. **2** N. Amer. **a** at any moment. **b** instantly.

momentary /ˈməʊməntərɪ/ adj. **1** lasting only a moment. **2** short-lived; transitory. □ **momentariness** n. [L momentarius (as MOMENT)]

momently /ˈməʊməntlɪ/ adv. literary **1** from moment to moment. **2** every moment. **3** for a moment.

momentous /məˈmentəs/ adj. having great importance. □ **momentously** adv. **momentousness** n.

momentum /məˈmentəm/ n. (pl. **momenta** /-tə/) **1** Physics the quantity of motion of a moving body, measured as a product of its mass and velocity. **2** the impetus gained by movement. **3** strength or continuity derived from an initial effort. [L f. movimentum f. movere move]

momma /ˈmɒmə/ n. var. of MAMMA[1].

Mommsen /ˈmɒmz(ə)n/, Theodor (1817–1903), German historian. He is noted for his three-volume History of Rome (1854–6; 1885) and his treatises on Roman constitutional law (1871–88). Mommsen was also editor of the Corpus Inscriptionum Latinarum (1863) for the Berlin Academy. He was awarded the Nobel Prize for literature in 1902.

mommy /ˈmɒmɪ/ n. (pl. **-ies**) esp. US colloq. = MUMMY[1].

Mon /məʊn/ n. & adj. ● n. (pl. same or **Mons**) **1** a member of a people of Indo-Chinese origin, now inhabiting eastern Burma (Myanmar) and western Thailand but having their ancient capital at Pegu in southern Burma. **2** the Austro-Asiatic language of this people (see MON-KHMER). ● adj. of or relating to this people or their language. [Mon]

Mon. abbr. Monday.

Monaco /ˈmɒnəˌkəʊ/ a principality forming an enclave within French territory, on the Mediterranean coast near the Italian frontier; pop.

(1990) 29,880; official language, French. It includes the resort of Monte Carlo. Ruled by the Genoese from medieval times and by the Grimaldi family from 1297, Monaco was under French occupation from 1793 to 1814 and then was a Sardinian protectorate until 1861. It became a constitutional monarchy in 1911. The smallest sovereign state in the world apart from the Vatican, Monaco is almost entirely dependent on the tourist trade and maintains a customs union with France.

monad /ˈmɒnæd, ˈməʊn-/ n. **1** the number one; a unit. **2** Philos. any ultimate unit of being (e.g. a soul, an atom, a person, God). **3** Biol. a simple organism, e.g. one assumed as the first in the genealogy of living beings. □ **monadism** n. **monadic** /məˈnædɪk/ adj. (in sense 2). [F monade or LL monas monad- f. Gk monas -ados unit f. monos alone]

monadelphous /ˌmɒnəˈdelfəs/ adj. Bot. **1** (of stamens) having filaments united into one bundle. **2** (of a plant) with such stamens. [Gk monos one + adelphos brother]

monadnock /məˈnædnɒk/ n. a steep-sided isolated hill resistant to erosion and rising above a plain. [Mount Monadnock in New Hampshire, US]

Monaghan /ˈmɒnəhən/ **1** a county of the Republic of Ireland, part of the old province of Ulster. **2** its county town; pop. (1991) 5,750.

Mona Lisa /ˌməʊnə ˈliːzə/ a painting (now in the Louvre in Paris) executed 1503–6 by Leonardo da Vinci. The painting is known also as La Gioconda, the sitter being the wife of Francesco di Bartolommeo del Giocondo di Zandi. Her enigmatic smile has become one of the most famous images in Western art.

monandry /məˈnændrɪ/ n. **1** the custom of having only one husband at a time. **2** Bot. the state of having a single stamen. □ **monandrous** adj. [MONO- after polyandry]

monarch /ˈmɒnək/ n. **1** a sovereign with the title of king, queen, emperor, empress, or the equivalent. **2** a supreme ruler. **3** a powerful or pre-eminent person. **4** (in full **monarch butterfly**) a large orange and black butterfly, Danaus plexippus, found mainly in the Americas. Also called milkweed. **5** (in full **monarch flycatcher**) a flycatcher of the Old World family Monarchidae, esp. of the genus Monarcha. □ **monarchal** /məˈnɑːk(ə)l/ adj. **monarchic** adj. **monarchical** adj. **monarchically** adv. [ME f. F monarque or LL monarcha f. Gk monarkhēs, -os, f. monos alone + arkhō to rule]

monarchism /ˈmɒnəˌkɪz(ə)m/ n. the advocacy of or the principles of monarchy. □ **monarchist** n. [F monarchisme (as MONARCHY)]

monarchy /ˈmɒnəkɪ/ n. (pl. **-ies**) **1** a form of government with a monarch at the head. **2** a state with this. □ **monarchial** /məˈnɑːkɪəl/ adj. [ME f. OF monarchie f. LL monarchia f. Gk monarkhia the rule of one (as MONARCH)]

Monash /ˈmɒnæʃ/, Sir John (1865–1931), Australian general. After commanding the 4th Australian Brigade at Gallipoli (1915), he served with distinction as commander of the 3rd Australian Division in France (1916).

monastery /ˈmɒnəstərɪ/ n. (pl. **-ies**) the residence of a religious community, esp. of monks living in seclusion. [ME f. eccl.L monasterium f. eccl.Gk monastērion f. monazō live alone f. monos alone]

monastic /məˈnæstɪk/ adj. & n. ● adj. **1** of or relating to monasteries or the religious communities living in them. **2** resembling these or their way of life; solitary and celibate. ● n. a monk or other follower of a monastic rule. □ **monastically** adv. **monasticism** /-tɪˌsɪz(ə)m/ n. **monasticize** v.tr. (also **-ise**). [F monastique or LL monasticus f. Gk monastikos (as MONASTERY)]

monatomic /ˌmɒnəˈtɒmɪk/ adj. Chem. **1** (esp. of a molecule) consisting of one atom. **2** having one replaceable atom or radical.

monaural /mɒˈnɔːrəl/ adj. **1** = MONOPHONIC. **2** of or involving one ear. □ **monaurally** adv. [MONO- + AURAL[1]]

monazite /ˈmɒnəˌzaɪt/ n. Mineral. a phosphate mineral containing rare-earth elements and thorium. [G Monazit f. Gk monazō live alone (because of its rarity)]

Mönchengladbach /ˌmʌnx(ə)nˈɡlædbæx/ a city in NW Germany; pop. (1991) 262,580. It is the site of the NATO headquarters for northern Europe.

Monck /mʌŋk/, George, 1st Duke of Albemarle (1608–70), English general. Although initially a Royalist in the Civil War, he became a supporter of Oliver Cromwell, and later completed the suppression of the Royalists in Scotland (1651). He subsequently campaigned against the Dutch at sea (1652–4), and from 1654 to 1660 he was commander-in-chief in Scotland. Concerned at the growing unrest following the death of Cromwell in 1658, Monck eventually marched his army south to London, persuaded his fellow generals that the only alternative to anarchy was the restoration of the monarchy, and negotiated the return of Charles II (1660).

mondaine /mɒnˈden/ adj. & n. ● adj. **1** of the fashionable world. **2** worldly. ● n. a worldly or fashionable woman. [F, fem. of *mondain*: see MUNDANE]

Monday /ˈmʌndeɪ, -dɪ/ n. & adv. ● n. the second day of the week, following Sunday. ● adv. colloq. **1** on Monday. **2** (**Mondays**) on Mondays; each Monday. [OE *mōnandæg* day of the moon, transl. LL *lunae dies*]

Mondrian /ˈmɒndrɪˌɑːn/, Piet (born Pieter Cornelis Mondriaan) (1872–1944), Dutch painter. He was a co-founder of the De Stijl movement and the originator of neo-plasticism, one of the earliest and strictest forms of geometrical abstract painting. His use of vertical and horizontal lines, rectangular shapes, and primary colours is typified by such paintings as *Composition with Red, Yellow, and Blue* (1921).

Monégasque /ˌmɒnəˈɡæsk/ n. & adj. ● n. a native or inhabitant of Monaco. ● adj. of or relating to Monaco or its inhabitants.

Monel /mɒˈnel/ n. (in full **Monel metal**) propr. a nickel-copper alloy with high tensile strength and resisting corrosion. [Ambrose *Monell*, Amer. businessman (1873–1921)]

Monet /ˈmɒneɪ/, Claude (1840–1926), French painter. He was a founder member of the impressionists; his early painting *Impression: soleil levant* (1874) gave the movement its name. Of all the group, Monet remained the most faithful to the impressionist principles of painting directly from the subject (often out of doors) and giving primacy to transient visual perception. His fascination with the play of light on objects led to a series of paintings of single subjects painted at different times of day and under different weather conditions, notably the *Haystacks* series (1890–1), *Rouen Cathedral* (1892–5), and the *Water-lilies* sequence (1899–1906; 1916 onwards).

monetarism /ˈmʌnɪtəˌrɪz(ə)m/ n. the theory or practice of controlling the supply of money as the chief method of stabilizing an economy. This economic theory is based on the belief that the supply of money is the crucial determinant of the level of demand (in opposition to the Keynsian view), and that strict government control of the money supply is necessary to prevent high inflation. An early exponent was the American economist Milton Friedman, who holds that governments lack the political will to control inflation by cutting government spending, since this move will increase unemployment.

monetarist /ˈmʌnɪtərɪst/ n. & adj. ● n. an advocate of monetarism. ● adj. in accordance with the principles of monetarism.

monetary /ˈmʌnɪtərɪ/ adj. **1** of the currency in use. **2** of or consisting of money. □ **monetarily** adv. [F *monétaire* or LL *monetarius* f. L (as MONEY)]

monetize /ˈmʌnɪˌtaɪz/ v.tr. (also **-ise**) **1** give a fixed value as currency. **2** put (a metal) into circulation as money. □ **monetization** /ˌmʌnɪtaɪˈzeɪʃ(ə)n/ n. [F *monétiser* f. L (as MONEY)]

money /ˈmʌnɪ/ n. **1 a** a current medium of exchange in the form of coins and banknotes. **b** a particular form of this (*silver money*). **2** (pl. **-eys** or **-ies**) (in pl.) sums of money. **3 a** wealth; property viewed as convertible into money. **b** wealth as giving power or influence (*money speaks*). **c** a rich person or family (*has married into money*). **4 a** money as a resource (*time is money*). **b** profit, remuneration (*in it for the money*). □ **for my money** in my opinion or judgement; for my preference (*is too aggressive for my money*). **have money to burn** see BURN[1]. **in the money** colloq. having or winning a lot of money. **money box** a box for saving money dropped through a slit. **money-changer** a person whose business it is to change money, esp. at an official rate. **money for jam** (or **old rope**) colloq. profit for little or no trouble. **money-grubber** a person greedily intent on amassing money. **money-grubbing** n. the greedy amassing of money. ● adj. given to this practice. **money market** Stock Exch. trade in short-term stocks, loans, etc. **money of account** see ACCOUNT. **money order** an order for payment of a specified sum, issued by a bank or Post Office. **money spider** a very small household spider supposed to bring financial luck, esp. one of the family Linyphiidae. **money-spinner** esp. Brit. a thing that brings in a profit. **money-spinning** very profitable. **money's-worth** (prec. by *your, my, one's*, etc.) good value for one's money. **put money into** invest in. □ **moneyless** adj. [ME f. OF *moneie* f. L *moneta* mint, money, orig. a title of Juno, in whose temple at Rome money was minted]

moneybags /ˈmʌnɪˌbæɡz/ n.pl. (treated as sing.) colloq. usu. derog. a wealthy person.

moneyed /ˈmʌnɪd/ adj. **1** having much money; wealthy. **2** consisting of money (*moneyed assistance*).

moneylender /ˈmʌnɪˌlendə(r)/ n. a person who lends money, esp. as a business, at interest. □ **moneylending** n. & adj.

moneymaker /ˈmʌnɪˌmeɪkə(r)/ n. **1** a person who earns much money. **2** a thing, idea, etc., that produces much money. □ **moneymaking** n. & adj.

moneywort /ˈmʌnɪˌwɜːt/ n. a trailing evergreen plant, *Lysimachia nummularia*, with round glossy leaves and yellow flowers. Also called creeping jenny.

monger /ˈmʌŋɡə(r)/ n. (usu. in comb.) **1** a dealer or trader (*fishmonger; ironmonger*). **2** usu. derog. a person who promotes or deals in something specified (*warmonger; scaremonger*). [OE *mangere* f. *mangian* to traffic f. Gmc, ult. f. L *mango* dealer]

mongo /ˈmɒŋɡəʊ/ n. (pl. **mongos** or same) a monetary unit of Mongolia, equal to one-hundredth of a tugrik. [Mongolian *möngö* silver]

Mongol /ˈmɒŋɡ(ə)l/ adj. & n. ● adj. **1** of or relating to the Asian people now inhabiting Mongolia. **2** resembling this people, esp. in appearance. **3** (**mongol**) often offens. suffering from Down's syndrome. ● n. **1** a Mongolian. **2** (**mongol**) often offens. a person suffering from Down's syndrome. [Mongolian: perh. f. *mong* brave]

Mongolia /mɒŋˈɡəʊlɪə/ a large and sparsely populated country of eastern Asia, bordered by Siberian Russia and China; pop. (est. 1991) 2,184,000; official language, Mongolian; capital, Ulan Bator. To the north of the Gobi Desert, which occupies much of the southern half of the country, lies a fertile tableland at an altitude of 900–1,500 m (3,000–5,000 ft). The centre of the medieval Mongol empire, Mongolia subsequently became a Chinese province, achieving de facto independence in 1911. In 1924 it became a Communist state after the Soviet model. Free elections were held in 1990 and a new democratic constitution was introduced in 1992. It was formerly known as *Outer Mongolia* to distinguish it from Inner Mongolia, which remains a province of China.

Mongolian /mɒŋˈɡəʊlɪən/ adj. & n. ● adj. **1** of Mongolia or its people or language. **2** = MONGOLOID adj. 1. ● n. **1** a native or national of Mongolia. **2** the language of Mongolia, usually considered a member of the Altaic family. It is spoken by about 2 million people in Mongolia, while another form is used by some 2.7 million in China. **3** = MONGOLOID n. 1.

mongolism /ˈmɒŋɡəˌlɪz(ə)m/ n. often offens. = DOWN'S SYNDROME. [MONGOL]

Mongoloid /ˈmɒŋɡəˌlɔɪd/ adj. & n. ● adj. **1** of or relating to the division of humankind including the indigenous peoples of eastern Asia, SE Asia, and the Arctic region of North America. Characteristic features are dark eyes, straight dark hair, pale ivory to dark skin, and little facial or bodily hair. **2** (also **mongoloid**) often offens. affected with Down's syndrome. ● n. **1** a person of Mongoloid physical type. **2** (also **mongoloid**) often offens. a person affected with Down's syndrome.

mongoose /ˈmɒŋɡuːs/ n. (pl. **mongooses**) a small flesh-eating civet-like mammal of the family Viverridae (or Herpestidae), found in tropical Africa and Asia, and noted for its ability to kill venomous snakes. [Marathi *mangūs*]

mongrel /ˈmʌŋɡrəl, ˈmɒŋ-/ n. & adj. ● n. **1** a dog of no definable type or breed. **2** any other animal or plant resulting from the crossing of different breeds or types. **3** derog. a person of mixed race. ● adj. of mixed origin, nature, or character. □ **mongrelism** n. **mongrelize** v.tr. (also **-ise**). **mongrelization** /ˌmʌŋɡrəlaɪˈzeɪʃ(ə)n, ˌmɒŋ-/ n. [earlier *meng-, mang-* f. Gmc: prob. rel. to MINGLE]

'mongst /mʌŋst/ poet. amongst (see AMONG).

monial /ˈməʊnɪəl/ n. Archit. a mullion. [ME f. OF *moinel* middle f. *moien* MEAN[3]]

Monica, St /ˈmɒnɪkə/ (332–c.387), mother of St Augustine of Hippo. Born in North Africa, she is often seen as the model of Christian mothers for her patience with her son's spiritual crises, ending with his conversion in 386. She became the object of a cult in the late Middle Ages and is frequently chosen as the patron of associations of Christian mothers. Feast day, 27 Aug. (formerly 4 May).

monicker var. of MONIKER.

monies see MONEY 2.

moniker /ˈmɒnɪkə(r)/ n. (also **monicker, monniker**) sl. a name. [19th c.: orig. unkn.]

moniliform /məˈnɪlɪˌfɔːm/ adj. esp. Anat. & Zool. with a form suggesting a string of beads. [F *moniliforme* or mod.L *moniliformis* f. L *monile* necklace]

monism /ˈmɒnɪz(ə)m, ˈməʊn-/ n. **1** Philos. a theory denying the duality of matter and mind (opp. DUALISM 2). (See note below.) **2** Philos. or Theol. the doctrine that there is only one supreme being (opp. PLURALISM 3). □ **monist** n. **monistic** /məˈnɪstɪk/ adj. [mod.L monismus f. Gk monos single]

▪ The basic premiss of monism is that both physical and mental phenomena can be explained or analysed in terms of a single common substance; the arrangement differs but the basic underlying reality remains the same. The principal proponents of this view (called more specifically neutral monism) were William James and Bertrand Russell. (See also IDEALISM, MATERIALISM.)

monition /məˈnɪʃ(ə)n/ n. **1** (foll. by of) literary a warning (of danger). **2** Eccl. a formal notice from a bishop or ecclesiastical court admonishing a person not to commit an offence. [ME f. OF f. L monitio -onis (as MONITOR)]

monitor /ˈmɒnɪtə(r)/ n. & v. ● n. **1** any of various persons or devices for checking or warning about a situation, operation, etc. **2** a school pupil with disciplinary or other special duties. **3 a** a television receiver used in a studio to select or verify the picture being broadcast. **b** = visual display unit. **4** a person who listens to and reports on foreign broadcasts etc. **5** a detector of radioactive contamination. **6** (in full **monitor lizard**) a large tropical Old World lizard of the genus Varanus, formerly reputed to give warning of crocodiles. **7** hist. a heavily armed shallow-draught warship. ● v.tr. **1** act as a monitor of. **2** maintain regular surveillance over. **3** regulate the strength of (a radio transmission, television signal, etc.). □ **monitorship** n. **monitorial** /ˌmɒnɪˈtɔːrɪəl/ adj. [L f. monere monit- warn]

monitory /ˈmɒnɪtəri/ adj. & n. ● adj. literary giving or serving as a warning. ● n. (pl. **-ies**) Eccl. a letter of admonition from the pope or a bishop. [L monitorius (as MONITION)]

Monk /mʌŋk/, Thelonious (Sphere) (1917–82), American jazz pianist and composer. In the early 1940s he played alongside Dizzy Gillespie and Charlie Parker in Harlem, becoming one of the founders of the bebop style. He achieved popularity in the late 1950s, as the new style of 'cool' jazz reached a wider audience. Memorable compositions include 'Round Midnight', 'Straight, No Chaser', and 'Well, You Needn't'.

monk /mʌŋk/ n. a member of a religious community of men living under certain vows esp. of poverty, chastity, and obedience. □ **monkish** adj. [OE munuc ult. f. Gk monakhos solitary f. monos alone]

monkey /ˈmʌŋkɪ/ n. & v. ● n. (pl. **-eys**) **1** a primate, esp. a long-tailed one, and usually excluding the prosimians and the apes. (See note below.) **2** a mischievous person, esp. a child (young monkey). **3** Brit. sl. £500. **4** (in full **monkey engine**) a machine hammer for pile-driving etc. ● v. (**-eys, -eyed**) **1** tr. mimic or mock. **2** intr. (often foll. by with) tamper or play mischievous tricks. **3** intr. (foll. by around, about) fool around. □ **have a monkey on one's back** sl. be a drug addict. **make a monkey of** humiliate by making appear ridiculous. **monkey bread** the baobab tree or its fruit. **monkey business** colloq. mischief. **monkey flower** a mimulus; esp. Mimulus guttatus, with bright yellow flowers. **monkey-jacket** a short close-fitting jacket worn by sailors etc. or at a mess. **monkey-nut** a peanut. **monkey-puzzle** a coniferous tree, Araucaria araucana, native to Chile, with downward-pointing branches and small close-set spiky leaves. **monkey-suit** colloq. evening dress. **monkey tricks** colloq. mischief. **monkey wrench** a wrench with an adjustable jaw. **monkey-wrench** v.tr. sabotage esp. as a means of environmentalist protest. **monkey-wrenching** sabotage esp. of industrial sites etc. by ecological activists. □ **monkeyish** adj. [16th c.: orig. unkn. (perh. LG)]

▪ The majority of monkeys are agile tree-dwellers. Two familes are confined to the New World: the Callithricidae, comprising the small marmosets and tamarins, and the Cebidae, characterized by their widely spaced side-facing nostrils and usually prehensile tails. The latter include the capuchins, sakis, and howler, spider, and woolly monkeys. The family Cercopithecidae comprises the Old World monkeys, which have non-prehensile tails. These include the macaques, mangabeys, baboons, guenons, colobuses, and langurs, and represent the immediate ancestors of apes and humans. (See also PRIMATE.)

monkeyshine /ˈmʌŋkɪˌʃaɪn/ n. (usu. in pl.) US colloq. = monkey tricks.

monkfish /ˈmʌŋkfɪʃ/ n. **1** an angler fish; esp. Lophius piscatorius, often used as food. **2** a bottom-dwelling shark, Squatina squatina, with a flattened body and large pectoral fins. Also called angel-shark.

Mon-Khmer /məʊnˈkmeə(r)/ n. a group of Austro-Asiatic languages spoken in SE Asia, of which the most important are Mon and Khmer.

monkshood /ˈmʌŋkshʊd/ n. a poisonous ranunculaceous garden plant Aconitum napellus, with hood-shaped blue or purple flowers. (See also ACONITE.)

Monmouth /ˈmɒnməθ/, Duke of (title of James Scott) (1649–85), English claimant to the throne of England. The illegitimate son of Charles II, he became the focus for Whig supporters of a Protestant succession. In 1685 he led a rebellion against the Catholic James II; he proclaimed himself king at Taunton in Somerset, but his force was defeated at the Battle of Sedgemoor and he was executed.

Monmouthshire /ˈmɒnməθˌʃɪə(r)/ a former county of SE Wales, on the border with England. The major part of it was incorporated into Gwent in 1974.

monniker var. of MONIKER.

mono /ˈmɒnəʊ/ adj. & n. ● adj. monophonic. ● n. (pl. **-os**) a monophonic record, reproduction, etc. [abbr.]

mono- /ˈmɒnəʊ/ comb. form (usu. **mon-** before a vowel) **1** one, alone, single. **2** Chem. (forming names of compounds) containing one atom or group of a specified kind. [Gk f. monos alone]

monoacid /ˌmɒnəʊˈæsɪd/ adj. Chem. (of a base) having one replaceable hydroxide ion.

monobasic /ˌmɒnəʊˈbeɪsɪk/ adj. Chem. (of an acid) having one replaceable hydrogen atom.

monocarpic /ˌmɒnəʊˈkɑːpɪk/ adj. (also **monocarpous** /-ˈkɑːpəs/) Bot. bearing fruit only once. [MONO- + Gk karpos fruit]

monocausal /ˌmɒnəʊˈkɔːz(ə)l/ adj. in terms of a sole cause.

monocephalous /ˌmɒnəʊˈsefələs, -ˈkefələs/ adj. Bot. having only one head.

monochord /ˈmɒnəˌkɔːd/ n. Mus. an instrument with a single string and a movable bridge, used esp. to determine intervals. [ME f. OF monocorde f. LL monochordon f. Gk monokhordon (as MONO-, CHORD[1])]

monochromatic /ˌmɒnəkrəˈmætɪk/ adj. **1** Physics (of light or other radiation) of a single wavelength or frequency. **2** containing only one colour. □ **monochromatically** adv.

monochromatism /ˌmɒnəʊˈkrəʊməˌtɪz(ə)m/ n. complete colour-blindness in which all colours appear as shades of one colour.

monochrome /ˈmɒnəˌkrəʊm/ n. & adj. ● n. a photograph or picture done in one colour or different tones of this, or in black and white only. ● adj. having or using only one colour or in black and white only. □ **monochromic** /ˌmɒnəˈkrəʊmɪk/ adj. [ult. f. Gk monokhrōmatos (as MONO-, khrōmatos f. khrōma colour)]

monocle /ˈmɒnək(ə)l/ n. a single eyeglass. □ **monocled** adj. [F, orig. adj. f. LL monoculus one-eyed (as MONO-, oculus eye)]

monocline /ˈmɒnəˌklaɪn/ n. Geol. a bend in rock strata that are otherwise uniformly dipping or horizontal. □ **monoclinal** /ˌmɒnəˈklaɪn(ə)l/ adj. [MONO- + Gk klinō lean, slope]

monoclinic /ˌmɒnəʊˈklɪnɪk/ adj. (of a crystal) having one axial intersection oblique. [MONO- + Gk klinō lean, slope]

monoclonal /ˌmɒnəʊˈkləʊn(ə)l/ adj. forming a single clone; derived from a single individual or cell. □ **monoclonal antibodies** antibodies produced by a single clone of cells or a single cell line, consisting of identical antibody molecules.

monocoque /ˈmɒnəˌkɒk/ n. esp. Aeron. an aircraft or vehicle structure in which the chassis is integral with the body. [F (as MONO-, coque shell)]

monocot /ˈmɒnəˌkɒt/ n. Bot. = MONOCOTYLEDON. [abbr.]

monocotyledon /ˌmɒnəˌkɒtɪˈliːd(ə)n/ n. Bot. a flowering plant with an embryo that bears a single cotyledon. The monocotyledons belong to one of the two great divisions of flowering plants, Monocotyledoneae (or Monocotyledones), and typically have long narrow leaves with parallel veins (cf. DICOTYLEDON). □ **monocotyledonous** adj.

monocracy /məˈnɒkrəsɪ/ n. (pl. **-ies**) government by one person only. □ **monocratic** /ˌmɒnəˈkrætɪk/ adj.

monocular /məˈnɒkjʊlə(r)/ adj. with or for one eye. □ **monocularly** adv. [LL monoculus having one eye]

monoculture /ˈmɒnəˌkʌltʃə(r)/ n. the cultivation of a single crop.

monocycle /ˈmɒnəˌsaɪk(ə)l/ n. = UNICYCLE.

monocyte /ˈmɒnəˌsaɪt/ n. Biol. a large leucocyte with a single kidney-shaped nucleus, migrating to tissue where it becomes a macrophage.

monodactylous /ˌmɒnəʊˈdæktɪləs/ adj. having one finger, toe, or claw.

monodrama /ˈmɒnəˌdrɑːmə/ n. a dramatic piece for one performer.

monody /ˈmɒnədɪ/ n. (pl. **-ies**) **1** an ode sung by a single actor in a

Greek tragedy. **2** a poem lamenting a person's death. **3** *Mus.* a composition with only one melodic line. □ **monodist** *n.* **monodic** /məˈnɒdɪk/ *adj.* [LL *monodia* f. Gk *monōidia* f. *monōidos* singing alone (as MONO-, ODE)]

monoecious /məˈniːʃəs/ *adj.* **1** *Bot.* with unisexual male and female organs on the same plant. **2** *Zool.* hermaphrodite. [mod.L *Monoecia* the class of such plants (Linnaeus) f. Gk *monos* single + *oikos* house]

monofilament /ˈmɒnəˌfɪləmənt/ *n.* **1** a single strand of man-made fibre. **2** a type of fishing line using this.

monogamy /məˈnɒgəmɪ/ *n.* **1** the practice or state of being married to one person at a time. **2** *Zool.* the habit of having only one mate at a time. □ **monogamist** *n.* **monogamous** *adj.* **monogamously** *adv.* [F *monogamie* f. eccl.L f. Gk *monogamia* (as MONO-, *gamos* marriage)]

monogenesis /ˌmɒnəʊˈdʒɛnɪsɪs/ *n.* (also **monogeny** /məˈnɒdʒɪnɪ/) **1** the theory of the development of all beings from a single cell. **2** the theory that humankind descended from one pair of ancestors. □ **monogenetic** /-dʒɪˈnɛtɪk/ *adj.*

monoglot /ˈmɒnəˌglɒt/ *adj.* & *n.* ● *adj.* using only one language. ● *n.* a monoglot person.

monogram /ˈmɒnəˌgræm/ *n.* two or more letters, esp. a person's initials, interwoven as a device. □ **monogrammed** *adj.* **monogrammatic** /ˌmɒnəgrəˈmætɪk/ *adj.* [F *monogramme* f. LL *monogramma* f. Gk (as MONO-, -GRAM)]

monograph /ˈmɒnəˌgrɑːf/ *n.* & *v.* ● *n.* a separate treatise on a single subject or an aspect of it. ● *v.tr.* write a monograph on. □ **monographer** /məˈnɒgrəfə(r)/ *n.* **monographist** *n.* **monographic** /ˌmɒnəˈgræfɪk/ *adj.* [earlier *monography* f. mod.L *monographia* f. *monographus* writer on a single genus or species (as MONO-, -GRAPH, -GRAPHY)]

monogynous /məˈnɒdʒɪnəs/ *adj. Bot.* having only one pistil.

monogyny /məˈnɒdʒɪnɪ/ *n.* the custom of having only one wife at a time.

monohull /ˈmɒnəˌhʌl/ *n.* a boat with a single hull.

monohybrid /ˌmɒnəʊˈhaɪbrɪd/ *n.* & *adj. Biol.* ● *n.* a hybrid that is heterozygous for alleles of one gene. ● *adj.* of or relating to inheritance of alleles of one gene.

monohydric /ˌmɒnəʊˈhaɪdrɪk/ *adj. Chem.* containing one hydroxyl group.

monokini /ˌmɒnəʊˈkiːnɪ/ *n.* a woman's one-piece beach-garment equivalent to the lower half of a bikini. [MONO- + BIKINI, by false assoc. with BI-]

monolayer /ˈmɒnəˌleɪə(r)/ *n.* **1** *Chem.* a layer with one molecule in thickness. **2** *Biol.* & *Med.* a cell culture consisting of a layer one cell thick.

monolingual /ˌmɒnəʊˈlɪŋgwəl/ *adj.* speaking or using only one language.

monolith /ˈmɒnəlɪθ/ *n.* **1** a single block of stone, esp. shaped into a pillar or monument. **2** a person or thing like a monolith in being massive, immovable, or solidly uniform. **3** a large block of concrete. □ **monolithic** /ˌmɒnəˈlɪθɪk/ *adj.* [F *monolithe* f. Gk *monolithos* (as MONO-, *lithos* stone)]

monologue /ˈmɒnəˌlɒg/ *n.* **1 a** a scene in a drama in which a person speaks alone. **b** a dramatic composition for one performer. **2** a long speech by one person in a conversation etc. □ **monologuist** *n.* (also **monologist**). **monologic** /ˌmɒnəˈlɒdʒɪk/ *adj.* **monological** *adj.* **monologically** *adv.*

monomania /ˌmɒnəʊˈmeɪnɪə/ *n.* obsession of the mind by one idea or interest. □ **monomaniac** /-ˈmeɪnɪˌæk/ *n.* & *adj.* **monomaniacal** /-məˈnaɪək(ə)l/ *adj.* [F *monomanie* (as MONO-, -MANIA)]

monomer /ˈmɒnəmə(r)/ *n. Chem.* **1** a unit in a dimer, trimer, or polymer. **2** a molecule or compound that can be polymerized. □ **monomeric** /ˌmɒnəˈmɛrɪk/ *adj.*

monomial /məˈnəʊmɪəl/ *adj.* & *n. Math.* ● *adj.* (of an algebraic expression) consisting of one term. ● *n.* a monomial expression. [MONO- after *binomial*]

monomolecular /ˌmɒnəʊməˈlɛkjʊlə(r)/ *adj. Chem.* (of a layer) only one molecule in thickness.

monomorphic /ˌmɒnəʊˈmɔːfɪk/ *adj.* (also **monomorphous** /-ˈmɔːfəs/) *Biochem.* not changing form during development. □ **monomorphism** *n.*

mononucleosis /ˌmɒnəʊˌnjuːklɪˈəʊsɪs/ *n.* an abnormally high proportion of monocytes in the blood, esp. = *glandular fever.* [MONO- + NUCLEO- + -OSIS]

monopetalous /ˌmɒnəʊˈpɛt(ə)ləs/ *adj. Bot.* having the corolla in one piece, or the petals united into a tube.

monophonic /ˌmɒnəʊˈfɒnɪk/ *adj.* **1** (of sound-reproduction) using only one channel of transmission (cf. STEREOPHONIC). **2** *Mus.* homophonic. □ **monophonically** *adv.* [MONO- + Gk *phōnē* sound]

monophthong /ˈmɒnəfˌθɒŋ/ *n. Phonet.* a single vowel sound. □ **monophthongal** /ˌmɒnəfˈθɒŋg(ə)l/ *adj.* [Gk *monophthoggos* (as MONO-, *phthoggos* sound)]

monophyletic /ˌmɒnəfaɪˈlɛtɪk/ *adj. Biol.* (of a group of organisms) descended from a common evolutionary ancestor or ancestral group, esp. one not shared with any other group.

Monophysite /məˈnɒfɪˌsaɪt, ˈmɒnəʊfɪ-/ *n.* (in Christian theology) an adherent of the doctrine that there is only one inseparable nature in the person of Jesus, contrary to a declaration of the Council of Chalcedon (451). Despite attempts at reconciliation, the declaration occasioned a split in the 6th century, and some Eastern Churches, such as the Coptic, Ethiopian, Armenian, and Jacobite Churches, continue to reject the Council of Chalcedon's ruling. (See also MELCHITE.) [eccl.L *monophysita* f. eccl.Gk *monophusitēs* (as MONO-, *phusis* nature)]

monoplane /ˈmɒnəˌpleɪn/ *n.* an aeroplane with one set of wings (cf. BIPLANE).

monopole[1] /ˈmɒnəˌpəʊl/ *n.* **1** *Physics* a single electric charge or magnetic pole, esp. a hypothetical isolated magnetic pole. **2** a radio aerial, pylon, etc. consisting of a single pole or rod. [MONO- + (in sense 1) POLE², (in sense 2) POLE¹]

monopole[2] /ˈmɒnəˌpəʊl/ *n.* (also **Monopole**) a champagne exclusive to one shipper. [F, = MONOPOLY]

Monopolies and Mergers Commission a UK government body (originally set up in 1948) designed to investigate and report on activities relating to the setting up of trading monopolies, company mergers, and takeovers, in which the public has an interest.

monopolist /məˈnɒpəlɪst/ *n.* a person who has or advocates a monopoly. □ **monopolistic** /-ˌnɒpəˈlɪstɪk/ *adj.*

monopolize /məˈnɒpəˌlaɪz/ *v.tr.* (also **-ise**) **1** obtain exclusive possession or control of (a trade or commodity etc.). **2** dominate or prevent others from sharing in (a conversation, person's attention, etc.). □ **monopolizer** *n.* **monopolization** /-ˌnɒpəlaɪˈzeɪʃ(ə)n/ *n.*

Monopoly /məˈnɒpəlɪ/ proprietary name for a board game (invented in the US by Charles Darrow *c.*1935) in which players use imitation money to engage in simulated property and financial dealings, the board representing streets and other locations in a large city which can be acquired and developed. □ **Monopoly money** money that has no real existence or value.

monopoly /məˈnɒpəlɪ/ *n.* (pl. **-ies**) **1 a** the exclusive possession or control of the trade in a commodity or service. **b** this conferred as a privilege by the state. **2 a** a commodity or service that is subject to a monopoly. **b** a company etc. that possesses a monopoly. **3** (foll. by *of, on*) exclusive possession, control, or exercise. [L *monopolium* f. Gk *monopōlion* (as MONO-, *pōleō* sell)]

monorail /ˈmɒnəˌreɪl/ *n.* a railway in which the track consists of a single rail, usu. elevated with the train units suspended from it.

monosaccharide /ˌmɒnəʊˈsækəˌraɪd/ *n. Chem.* a sugar that cannot be hydrolysed to give a simpler sugar, e.g. glucose.

monosodium glutamate /ˌmɒnəʊˌsəʊdɪəm ˈgluːtəˌmeɪt/ *n.* (abbr. **MSG**) a sodium salt of glutamic acid, used as a flavour enhancer in food. Monosodium glutamate, which occurs naturally as a product of the breakdown of proteins, was originally obtained in the Far East from seaweed, but is now mainly made from bean and cereal protein. Although itself tasteless, it has the property of enhancing the flavour of some foods, particularly meat and vegetables. It is a traditional ingredient in oriental cooking, and has been widely used as a food additive in the West. Excessive ingestion of MSG has been claimed to cause flushing, headaches, and other symptoms.

monospermous /ˌmɒnəʊˈspɜːməs/ *adj. Bot.* having one seed. [MONO- + Gk *sperma* seed]

monostichous /məˈnɒstɪkəs/ *adj. Bot.* & *Zool.* arranged in or consisting of one layer or row. [MONO- + Gk *stikhos* row]

monosyllabic /ˌmɒnəsɪˈlæbɪk/ *adj.* **1** (of a word) having one syllable. **2** (of a person or statement) using or expressed in monosyllables. □ **monosyllabically** *adv.*

monosyllable /ˈmɒnəˌsɪləb(ə)l/ *n.* a word of one syllable. □ **in monosyllables** in simple direct words.

monotheism /'mɒnəθɪ,ɪz(ə)m/ n. the doctrine that there is only one God. □ **monotheist** n. **monotheistic** /,mɒnəθɪ'ɪstɪk/ adj. **monotheistically** adv. [MONO- + Gk theos god]

Monothelite /məˈnɒθə,laɪt/ n. (in Christian theology) an adherent of the doctrine that Jesus had only one (divine) will. The theory (condemned as heresy in 681) was put forward as an attempt to reconcile the orthodox and Monophysite doctrines, but it led only to fresh controversy. [eccl. L monothelita f. eccl. Gk monothelētēs (as MONO-, thelēma will)]

monotint /'mɒnə,tɪnt/ n. = MONOCHROME.

monotone /'mɒnə,təʊn/ n. & adj. ● n. **1** a sound or utterance continuing or repeated on one note without change of pitch. **2** sameness of style in writing. ● adj. without change of pitch. [mod.L monotonus f. late Gk monotonos (as MONO-, TONE)]

monotonic /,mɒnə'tɒnɪk/ adj. **1** uttered in a monotone. **2** Math. (of a function or quantity) varying in such a way that it either never decreases or never increases. □ **monotonically** adv.

monotonous /məˈnɒtənəs/ adj. **1** lacking in variety; tedious through sameness. **2** (of a sound or utterance) without variation in tone or pitch. □ **monotonize** v.tr. (also **-ise**). **monotonously** adv. **monotonousness** n.

monotony /məˈnɒtənɪ/ n. **1** the state of being monotonous. **2** dull or tedious routine.

monotreme /'mɒnə,triːm/ n. Zool. a primitive toothless mammal of the order Monotremata, including only the platypus and the echidnas, native to Australia and New Guinea. They lay large yolky eggs, and have a common opening or cloaca for the passage of urine, faeces, and eggs or sperm. [MONO- + Gk trēma -matos hole]

monotype /'mɒnə,taɪp/ n. **1** (**Monotype**) Printing propr. a typesetting machine that casts and sets up types in individual characters. **2** an impression on paper made from an inked design painted on glass or metal.

monotypic /,mɒnəʊ'tɪpɪk/ adj. having only one type or representative.

monounsaturated /,mɒnəʊʌn'sætʃə,reɪtɪd, -'sætjʊ,reɪtɪd/ adj. Chem. (of a compound, esp. a fat or oil molecule) containing one double bond.

monovalent /,mɒnəʊ'veɪlənt/ adj. Chem. having a valency of one; univalent.

monoxide /məˈnɒksaɪd/ n. Chem. an oxide containing one oxygen atom (carbon monoxide). [MONO- + OXIDE]

Monroe[1] /mənˈrəʊ/, James (1758–1831), American Democratic Republican statesman, 5th President of the US 1817–25. In 1803, while minister to France under President Jefferson, he negotiated and ratified the Louisiana Purchase; he is chiefly remembered, however, as the originator of the Monroe doctrine.

Monroe[2] /mənˈrəʊ/, Marilyn (born Norma Jean Mortenson, later Baker) (1926–62), American actress. After a career as a photographer's model, she starred in a series of comedy films, including Gentlemen Prefer Blondes (1953) and Some Like it Hot (1959), emerging as the definitive Hollywood sex symbol. Her last film was The Misfits (1961), written for her by her third husband, the dramatist Arthur Miller. She is thought to have died of an overdose of sleeping-pills, although there is continuing controversy over the cause of her death.

Monroe doctrine n. a principle of US foreign policy that any intervention by external powers in the politics of the Americas is a potentially hostile act against the US. It was first expressed in President Monroe's address to Congress in 1823 against a background of continued involvement and threat of expansion from European colonial powers in South America. [MONROE[1]]

Monrovia /mɒnˈrəʊvɪə/ the capital and chief port of Liberia; pop. (est. 1985) 500,000.

Mons /mɒnz/ (called in Flemish Bergen) a town in southern Belgium, capital of the province of Hainaut; pop. (1991) 91,730. In August 1914 it was the scene of the first major battle between British and German forces during the First World War.

Monseigneur /,mɒnseɪˈnjɜː(r)/ n. (pl. **Messeigneurs** /,meɪseɪ-/) a title given to an eminent French person, esp. a prince, cardinal, archbishop, or bishop. [F f. mon my + seigneur lord]

Monsieur /məˈsjɜː(r)/ n. (pl. **Messieurs** /meɪˈsjɜː(r)/) **1** the title or form of address used of or to a French-speaking man, corresponding to Mr or sir. **2** a Frenchman. [F f. mon my + sieur lord]

Monsignor /mɒnˈsiːnjə(r), ,mɒnsɪˈnjɔː(r)/ n. (pl. **Monsignori** /,mɒnsɪˈnjɔːrɪ/) the title of various Roman Catholic prelates, officers of the papal court, etc. [It., after MONSEIGNEUR: see SIGNOR]

monsoon /mɒnˈsuːn/ n. **1** a seasonal wind in southern Asia, esp. in the Indian Ocean, blowing from the south west in summer (wet monsoon) and the north east in winter (dry monsoon). (See note below.) **2** a rainy season accompanying a wet monsoon. **3** any other wind with periodic alternations. □ **monsoonal** adj. [obs. Du. monssoen f. Port. monção f. Arab. mawsim fixed season f. wasama to mark]

▪ The wet monsoon blows in the period May to September, bringing generally rainy weather to the Indian subcontinent and SE Asia. It results from the northward movement of the earth's wind belts in the northern hemisphere summer, partly due to prevailing low pressure over the warm Eurasian land mass. In winter, there is a corresponding high over central Asia, and cold air spreads out over southern and eastern Asia, bringing prevailing dry weather from October to April. Monsoon winds are generally steady rather than strong.

mons pubis /mɒnz 'pjuːbɪs/ n. a rounded mass of fatty tissue lying over the joints of the pubic bones (cf. MONS VENERIS). [L, = mount of the pubes]

monster /'mɒnstə(r)/ n. **1** an imaginary creature, usu. large and frightening, compounded of incongruous elements. **2** an inhumanly cruel or wicked person. **3** a misshapen animal or plant. **4** a large usu. ugly animal or thing (e.g. a building). **5** (attrib.) huge; extremely large of its kind. [ME f. OF monstre f. L monstrum portent, monster f. monere warn]

monstera /mɒnˈstɪərə/ n. a climbing plant of the genus Monstera, esp. the Swiss cheese plant. [mod.L, perh. f. L monstrum monster (from the odd appearance of its leaves)]

monstrance /'mɒnstrəns/ n. RC Ch. an open or transparent receptacle in which the consecrated Host is exposed for veneration. [ME, = demonstration, f. med.L monstrantia f. L monstrare show]

monstrosity /mɒnˈstrɒsɪtɪ/ n. (pl. **-ies**) **1** a huge or outrageous thing. **2** monstrousness. **3** = MONSTER 3. [LL monstrositas (as MONSTROUS)]

monstrous /'mɒnstrəs/ adj. **1** like a monster; abnormally formed. **2** huge. **3 a** outrageously wrong or absurd. **b** atrocious. □ **monstrously** adv. **monstrousness** n. [ME f. OF monstreux or L monstrosus (as MONSTER)]

mons Veneris /mɒnz 'venərɪs/ n. the rounded mass of fatty tissue on a woman's abdomen above the vulva. [L, = mount of Venus]

Mont. abbr. Montana.

montage /mɒnˈtɑːʒ/ n. **1** (in cinematography) **a** a combination of images in quick succession to compress background information or provide atmosphere. **b** a system of editing in which the narrative is modified or interrupted to include images that are not necessarily related to the dramatic development. The technique was used notably by Eisenstein. **2 a** the technique of producing a new composite whole from fragments of pictures, words, music, etc. **b** a composition produced in this way. [F f. monter MOUNT[1]]

Montagna /mɒnˈtɑːnjə/, Bartolommeo Cincani (c.1450–1523), Italian painter. He settled in Vicenza and helped establish it as a centre of art. He is noted for his altarpiece Sacra Conversazione (1499).

Montagu's harrier /'mɒntə,gjuːz/ n. a slender migratory Eurasian bird of prey, Circus pygargus. [George Montagu, Br. naturalist (1751–1815)]

Montaigne /mɒnˈteɪn/, Michel (Eyquem) de (1533–92), French essayist. Often regarded as the originator of the modern essay, he wrote about prominent personalities and ideas of his age in his sceptical Essays (1580; 1588). Translated by John Florio in 1603, they were an influence on Shakespeare, Bacon, and others.

Montana[1] /mɒnˈtænə/ a state in the western US, on the Canadian border to the east of the Rocky Mountains; pop. (1990) 799,065; capital, Helena. Acquired from France as part of the Louisiana Purchase in 1803, it became the 41st state of the US in 1889. □ **Montanan** adj. & n.

Montana[2] /mɒnˈteɪnə/, Joe ('Cool Joe') (b.1956), American football player. He joined the San Francisco 49ers as quarterback in 1980 and played in four winning Super Bowls (1982; 1985; 1989; 1990). He retired in 1995 after two seasons with the Kansas City Chiefs.

montane /'mɒnteɪn/ adj. of or inhabiting mountainous country. [L montanus (as MOUNT[2], -ANE[1])]

Montanism /'mɒntə,nɪz(ə)m/ n. the beliefs of a heretical millenarian Christian movement, founded in Phrygia in Asia Minor by Montanus in the 2nd century, which reacted against the growing secularism of the Church and desired stricter adherence to the principles of primitive Christianity. □ **Montanist** n. & adj.

Mont Blanc /mɒn 'blɒŋk/ a peak in the Alps on the border between

France and Italy, rising to 4,807 m (15,771 ft). It is the highest peak in the Alps and the highest in western Europe.

montbretia /mɒnˈbriːʃə/ *n.* a cultivated plant of the genus *Crocosmia*, of the iris family, with bright orange-yellow trumpet-shaped flowers. [mod.L f. A. F. E. Coquebert de *Montbret*, Fr. botanist (1780–1801)]

Montcalm /mɒnˈkɑːm/, Louis Joseph de Montcalm-Gozon, Marquis de (1712–59), French general. He defended Quebec against British troops under General Wolfe, but was defeated and mortally wounded in the battle on the Plains of Abraham.

Mont Cervin /mɔ̃ sɛrvɛ̃/ the French name for the MATTERHORN.

monte /ˈmɒntɪ/ *n. Cards* **1** a Spanish game of chance, played with 45 cards. **2** (in full **three-card monte**) a game of Mexican origin played with three cards, similar to three-card trick. [Sp., = mountain, heap of cards]

Monte Albán /ˌmɒntɪ ælˈbɑːn/ an ancient city, now in ruins, in Oaxaca, southern Mexico. Occupied from the 8th century BC, it became a centre of the Zapotec culture from about the 1st century BC to the 8th century AD, after which it was occupied by the Mixtecs until the Spanish conquest in the 16th century.

Monte Carlo[1] /ˌmɒntɪ ˈkɑːləʊ/ a resort in Monaco, forming one of the four communes of the principality; pop. (1985) 12,000. It is famous as a gambling resort and as the terminus of the annual Monte Carlo rally.

Monte Carlo[2] /ˌmɒntɪ ˈkɑːləʊ/ *adj.* designating statistical methods which use the random sampling of numbers in order to estimate the solutions to numerical problems.

Monte Cassino /ˌmɒntɪ kəˈsiːnəʊ/ a hill in central Italy near the town of Cassino, midway between Rome and Naples. It is the site of the principal monastery of the Benedictines, founded by St Benedict *c.*529. The monastery, demolished and rebuilt several times in its history, was almost totally destroyed during the Second World War, but has since been restored. In 1944 Allied forces advancing towards Rome were halted by German defensive positions in which Monte Cassino played a major part. The Allies succeeded in capturing the site only after four months of bitter fighting and the destruction of the town and the monastery.

Monte Cervino /ˌmonte tʃerˈviːno/ the Italian name for the MATTERHORN.

Montego Bay /mɒnˈtiːgəʊ/ a free port and tourist resort on the north coast of Jamaica; pop. (1982) 70,265.

Montenegro /ˌmɒntɪˈniːgrəʊ/ a mountainous landlocked republic in the Balkans; pop. (1988) 632,000; official language, Serbo-Croat; capital, Podgorica. Joined with Serbia before the Turkish conquest of 1355, Montenegro became independent in 1851. In 1918 it became part of the federation of Yugoslavia, of which it remains, with Serbia, a nominal constituent. □ **Montenegrin** *adj. & n.*

Monterey /ˌmɒntəˈreɪ/ a city and fishing port on the coast of California, founded by the Spanish in the 18th century; pop. (1990) 31,950. The Monterey Jazz Festival has been held there annually since 1958. The harbour is the site of the Monterey Bay Aquarium, a centre of marine biological conservation and research.

Monterrey /ˌmɒntəˈreɪ/ an industrial city in NE Mexico, capital of the state of Nuevo León; pop. (1990) 2,521,700.

Montespan /ˈmɒntɪˌspɒn/, Marquise de (title of Françoise-Athénaïs de Rochechouart) (1641–1707), French noblewoman. She was mistress of Louis XIV from 1667 to 1679, and had seven illegitimate children by him. She subsequently fell from favour when the king became attracted to their governess, Madame de Maintenon.

Montesquieu /ˈmɒntəˌskjɜː, -ˌsjkuː/, Baron de La Brède et de (title of Charles Louis de Secondat) (1689–1755), French political philosopher. A former advocate, he became known with the publication of his *Lettres Persanes* (1721), a satire of French society from the perspective of two Persian travellers visiting Paris. Montesquieu's reputation rests chiefly on *L'Esprit des lois* (1748), a comparative study of political systems in which he championed the separation of judicial, legislative, and executive powers as being most conducive to individual liberty, holding up the English state as a model. His theories were highly influential in Europe in the late 18th century, as they were in the drafting of the American Constitution.

Montessori /ˌmɒntɪˈsɔːrɪ/, Maria (1870–1952), Italian educationist. Her success with mentally retarded children led her, in 1907, to apply similar methods to younger children of normal intelligence. Montessori's system, set out in her book *The Montessori Method* (1909), advocates a child-centred approach, in which the pace is largely set by

the child and play is free but guided, using a variety of sensory materials. Her ideas have since become an integral part of modern nursery and infant-school education.

Monteverdi /ˌmɒntɪˈveədɪ/, Claudio (1567–1643), Italian composer. From the 1580s he published many books of madrigals, noted for their use of harmonic dissonance. Monteverdi is chiefly remembered for his opera *Orfeo* (1607), which introduced a sustained dramatic focus and more fully defined characters, interweaving the instrumental accompaniment and the singing with the drama. His other baroque operas include *The Return of Ulysses* (1641) and *The Coronation of Poppea* (1642). As a composer of sacred music, he is particularly associated with the *Vespers* (1610).

Montevideo /ˌmɒntɪvɪˈdeɪəʊ/ the capital and chief port of Uruguay, on the River Plate; pop. (1985) 1,247,900.

Montez /ˈmɒntez/, Lola (born Marie Dolores Eliza Rosanna Gilbert) (1818–61), Irish dancer. While performing in Munich in 1846 she came to the notice of Ludwig I of Bavaria; she became his mistress and exercised great influence on his ruling of the country until banished the following year.

Montezuma II /ˌmɒntɪˈzuːmə/ (1466–1520), Aztec emperor 1502–20. The last ruler of the Aztec empire in Mexico, he was defeated and imprisoned by the Spanish conquistadors under Cortés in 1519. Montezuma was killed while trying to pacify some of his former subjects during the Aztec uprising against his captors.

Montezuma's revenge *n. sl.* diarrhoea suffered by travellers, esp. visitors to Mexico. [f. MONTEZUMA II]

Montfort[1] /ˈmɒntfət/, Simon de (*c.*1165–1218), French soldier. From 1209 he was leader of the Albigensian Crusade against the Cathars in southern France. He died while besieging the city of Toulouse. He was the father of Simon de Montfort, Earl of Leicester (see MONTFORT[2]).

Montfort[2] /ˈmɒntfət/, Simon de, Earl of Leicester (*c.*1208–65), English soldier, born in Normandy. He was the son of the French soldier Simon de Montfort (see MONTFORT[1]). As leader of the baronial opposition to Henry III, he campaigned against royal encroachment on the privileges gained through Magna Carta, and defeated the king at Lewes, Sussex, in 1264. The following year he summoned a Parliament which included not only barons, knights, and clergymen, but also two citizens from every borough in England. He was defeated and killed by reorganized royal forces under Henry's son (later Edward I) at Evesham.

Montgolfier /mɒnˈgɒlfɪˌeɪ, -fɪə(r)/, Joseph Michel (1740–1810) and Jacques Étienne (1745–99), French inventors. Sons of a paper manufacturer, they pioneered experiments in hot-air ballooning. In 1782 they built a large balloon from linen and paper, lit a fire on the ground, and with the rising hot air successfully lifted a number of animals; the first human ascents followed in 1783.

Montgomery[1] /mɒntˈgʌmərɪ, -ˈgɒmərɪ/ the state capital of Alabama; pop. (1990) 187,100. From February 1861 until its capture in July it was the capital of the Confederate states of America. The capital was subsequently moved to Richmond, Virginia.

Montgomery[2] /mɒntˈgʌmərɪ, -ˈgɒmərɪ/, Bernard Law, 1st Viscount Montgomery of Alamein ('Monty') (1887–1976), British Field Marshal. In 1942 he commanded the 8th Army in the Western Desert, where his victory at El Alamein proved the first significant Allied success in the Second World War. He was later given command of the Allied ground forces in the invasion of Normandy in 1944 and accepted the German surrender on 7 May, 1945.

Montgomery[3] /mɒntˈgʌmərɪ, -ˈgɒmərɪ/, Lucy Maud (1874–1942), Canadian novelist. She is chiefly remembered for her first novel *Anne of Green Gables* (1908), the story of a spirited orphan girl brought up by an elderly couple. It became an instant best seller and was followed by seven sequels.

Montgomeryshire /mɒntˈgʌmərɪˌʃɪə(r), -ˈgɒmərɪˌʃɪə(r)/ a former county of central Wales. It became a part of Powys in 1974.

month /mʌnθ/ *n.* **1** (in full **calendar month**) **a** each of usu. twelve periods into which a year is divided. **b** a period of time between the same dates in successive calendar months. (*See note below.*) **2** a period of twenty-eight days or of four weeks. **3** = *lunar month.* □ **month of Sundays** a very long period. [OE *mōnath* f. Gmc, rel. to MOON]

• The primitive calendar month began on the day of a new moon or the day after, and thus coincided (except for fractions of a day) with the lunar month. Among many peoples, however, it was from a very early period found desirable that the calendar year should be divisible into whole numbers of the smaller periods used in ordinary reckoning, and the lunar months were superseded by a series of twelve periods each having a fixed number of days. The names of the months in general use in the West date back to Roman times, and some are named after ancient Roman gods and goddesses. (See also JULIAN CALENDAR, GREGORIAN CALENDAR, JEWISH CALENDAR.)

monthly /ˈmʌnθlɪ/ *adj., adv., & n.* ● *adj.* done, produced, or occurring once a month. ● *adv.* once a month; from month to month. ● *n.* (*pl.* **-ies**) **1** a monthly periodical. **2** (in *pl.*) *colloq.* a menstrual period.

monticule /ˈmɒntɪˌkjuːl/ *n.* **1** a small hill. **2** a small mound caused by a volcanic eruption. [F f. LL *monticulus* dimin. of *mons* MOUNT²]

Montmartre /mɒnˈmɑːtrə/ a district in northern Paris, on a hill above the Seine, much frequented by artists in the late 19th and early 20th centuries when it was a village separated from Paris. Many of its buildings have artistic associations, e.g. the Moulin de la Galette, which was painted by Renoir.

montmorillonite /ˌmɒntməˈrɪləˌnaɪt/ *n. Mineral.* any of a group of clay minerals which undergo reversible expansion on absorbing water, and the main constituent of fuller's earth and bentonite. [*Montmorillon*, a town in France]

Montparnasse /ˌmɒnpɑːˈnæs/ a district of Paris, on the left bank of the River Seine. Noted for its cafés, it was frequented in the late 19th century by writers and artists and is traditionally associated with Parisian cultural life.

Montpelier /mɒntˈpiːljə(r)/ the state capital of Vermont; pop. (1990) 8,250.

Montpellier /mɒnˈpelɪˌeɪ/ a city in southern France, near the Mediterranean coast, capital of Languedoc-Roussillon; pop. (1990) 210,870. It developed in the 10th century as a trading centre on the spice route from the Near East and was a stopover on the pilgrimage to Santiago de Compostela. A distinguished medical school and university, world famous in medieval times, was founded there in 1221.

Montreal /ˌmɒntrɪˈɔːl/ (French **Montréal** /mɔ̃real/) a port on the St Lawrence in Quebec, SE Canada; pop. (1991) 1,017,700; metropolitan area pop. (1991) 3,127,240. Founded in 1642, it was under French rule until 1763; almost two-thirds of its present-day population are French-speaking.

Montreux /mɒnˈtrɜː/ a resort town in SW Switzerland, at the east end of Lake Geneva; pop. (1990) 19,850. It is the site of an international jazz festival, which has been held there annually every July since 1967, and an annual television festival, which has been held there every spring since 1961.

Montrose /mɒnˈtrəʊz/, James Graham, 1st Marquis of (1612–50), Scottish general. Montrose supported Charles I when Scotland entered the English Civil War and, commanding a small army of Irish and Scottish irregulars, inflicted a dramatic series of defeats on the stronger Covenanter forces in the north (1644–5) before being defeated at Philiphaugh. After several years in exile, he returned to Scotland in 1650 in a bid to restore the new king Charles II, but was betrayed to the Covenanters and hanged.

Montserrat /ˌmɒntsəˈræt/ an island in the West Indies, one of the Leeward Islands; pop. (est. 1988) 12,000; capital, Plymouth. It was visited by Columbus in 1493 and named after a Benedictine monastery on the mountain of Montserrat in Catalonia, NE Spain. It was colonized by Irish settlers in 1632 and is now a British dependency.

Mont St Michel /ˌmɒn sæn miːˈʃel/ a rocky islet off the coast of Normandy, NW France. An island only at high tide, it is surrounded by sandbanks and linked to the mainland, since 1875, by a causeway. It is crowned by a magnificent medieval Benedictine abbey-fortress.

Monty Python's Flying Circus /ˈmɒntɪ/ an influential British television comedy series (1969–74), noted especially for its absurdist or surrealist style of humour and starring, amongst others, John Cleese.

monument /ˈmɒnjʊmənt/ *n.* **1** anything enduring that serves to commemorate or make celebrated, esp. a structure or building. **2** a stone or other structure placed over a grave or in a church etc. in memory of the dead. **3** an ancient building or site etc. that has survived or been preserved. **4** (foll. by *of, to*) a typical or outstanding example (*a monument of indiscretion*). **5** a lasting reminder. **6** a written record. [ME f. F f. L *monumentum* f. *monere* remind]

Monument, the a Doric column in the City of London, designed by Robert Hooke and Sir Christopher Wren (1671–7) to commemorate the Great Fire of 1666, which broke out in Pudding Lane nearby.

monumental /ˌmɒnjʊˈment(ə)l/ *adj.* **1a** extremely great; stupendous (*a monumental achievement*). **b** (of a literary work) massive and permanent. **2** of or serving as a monument. **3** *colloq.* (as an intensifier)

very great; calamitous (*a monumental blunder*). □ **monumental mason** a maker of tombstones etc. □ **monumentally** *adv.* **monumentality** /-menˈtælɪtɪ/ *n.*

monumentalize /ˌmɒnjʊˈmentəˌlaɪz/ *v.tr.* (also **-ise**) record or commemorate by or as by a monument.

-mony /mənɪ/ *suffix* forming nouns esp. denoting an abstract state or quality (*acrimony; testimony*). [L *-monia, -monium*, rel. to -MENT]

moo /muː/ *v. & n.* ● *v.intr.* (**moos**, **mooed**) make the characteristic vocal sound of cattle; = LOW². ● *n.* (*pl.* **moos**) this sound. □ **moo-cow** a childish name for a cow. [imit.]

mooch /muːtʃ/ *v. colloq.* **1** *intr. Brit.* loiter or saunter desultorily. **2** *tr. esp. N. Amer.* **a** steal. **b** beg. □ **moocher** *n.* [ME, prob. f. OF *muchier* hide, skulk]

mood¹ /muːd/ *n.* **1** a state of mind or feeling. **2** a fit of melancholy or bad temper. **3** (*attrib.*) inducing a particular mood (*mood music*). □ **in the** (or **no**) **mood** (foll. by *for*, or *to* + infin.) inclined (or disinclined) (*was in no mood to agree*). [OE *mōd* mind, thought, f. Gmc]

mood² /muːd/ *n.* **1** *Gram.* **a** a form or set of forms of a verb serving to indicate whether it is to express fact, command, wish, etc. (*subjunctive mood*). **b** the distinction of meaning expressed by different moods. **2** *Logic* any of the classes into which each of the figures of a valid categorical syllogism is subdivided. [var. of MODE, assoc. with MOOD¹]

moody /ˈmuːdɪ/ *adj. & n.* ● *adj.* (**moodier**, **moodiest**) given to changes of mood; gloomy, sullen. ● *n.* (*pl.* **-ies**) *colloq.* a bad mood; a tantrum. □ **moodily** *adv.* **moodiness** *n.* [OE *mōdig* brave (as MOOD¹)]

Moog /muːg, məʊg/ *n.* (in full **Moog synthesizer**) a type of electronic synthesizer with a keyboard, used esp. in 1970s rock music. [Robert Arthur *Moog*, Amer. engineer (b.1934), who invented it]

moolah /ˈmuːlə/ *n. sl.* money. [20th c.: orig. unkn.]

mooli /ˈmuːlɪ/ *n.* (also **muli**) a long white radish, the root of *Raphanus sativus* (variety *longipinnatus*) used esp. in Eastern cooking. [Hindi *mūlī* f. Skr. *mūlika*, f. *mūla* root]

moolvi /ˈmuːlvɪ/ *n.* (also **moolvie**) **1** a Muslim doctor of the law. **2** a learned person or teacher (esp. as a term of respect among Muslims in India). [Urdu *mulvī* f. Arab. *mawlawīy* judicial: cf. MULLAH]

Moon /muːn/, Sun Myung (b.1920), Korean industrialist and religious leader. In 1954 he founded the Holy Spirit Association for the Unification of World Christianity, which became known as the Unification Church.

moon /muːn/ *n. & v.* ● *n.* **1a** the natural satellite of the earth, orbiting it monthly, illuminated by the sun and reflecting some light to the earth. (See note below.) **b** this regarded in terms of its waxing and waning in a particular month (*new moon*). **c** the moon when visible (*there is no moon tonight*). **2** a satellite of any planet. **3** (prec. by *the*) *colloq.* something desirable but unattainable (*promised them the moon*). **4** *poet.* a month. ● *v.* **1** *intr.* (often foll. by *about, around*, etc.) move or look listlessly. **2** *tr.* (foll. by *away*) spend (time) in a listless manner. **3** *intr.* (foll. by *over*) act aimlessly or inattentively from infatuation for (a person). **4** *intr. sl.* expose one's buttocks. □ **many moons ago** a long time ago. **moon boot** a thickly padded boot designed for low temperatures. **moon-faced** having a round face. **over the moon** extremely happy or delighted. □ **moonless** *adj.* [OE *mōna* f. Gmc, rel. to MONTH]

▪ The moon orbits earth at an average distance of some 384,000 km. It is 3,476 km in diameter, and thus is large in proportion to the earth. The bright and dark features which outline the face of 'the Man in the Moon' are seen through a telescope to be highland and lowland regions, the former heavily pockmarked by craters due to the impact of millions of meteorites in the distant past. The same side is always presented to the earth. The moon has been visited by the Apollo astronauts, and the samples of rock and dust which they collected are being analysed. The moon lacks an atmosphere.

moonbeam /ˈmuːnbiːm/ *n.* a ray of moonlight.

mooncalf /ˈmuːnkɑːf/ *n.* (*pl.* **-calves** /-kɑːvz/) a born fool.

moonfish /ˈmuːnfɪʃ/ *n.* = OPAH.

Moonie /ˈmuːnɪ/ *n. colloq. offens.* a member of the Unification Church. [MOON]

moonlight /ˈmuːnlaɪt/ *n. & v.* ● *n.* **1** the light of the moon. **2** (*attrib.*) lighted by the moon. ● *v.intr.* (**-lighted**) *colloq.* have two paid occupations, esp. one by day and one by night. □ **moonlight flit** a hurried departure by night, esp. to avoid paying a debt. □ **moonlighter** *n.*

moonlit /ˈmuːnlɪt/ *adj.* lighted by the moon.

moonquake /ˈmuːnkweɪk/ *n.* a tremor of the moon's surface.

moonrise /ˈmuːnraɪz/ *n.* **1** the rising of the moon. **2** the time of this.

moonscape /ˈmuːnskeɪp/ n. **1** the surface or landscape of the moon. **2** an area resembling this; a wasteland.

moonset /ˈmuːnset/ n. **1** the setting of the moon. **2** the time of this.

moonshee /ˈmuːnʃiː/ n. (also **munshi**) a secretary or language-teacher in India. [Urdu munshī f. Arab. munšī writer]

moonshine /ˈmuːnʃaɪn/ n. **1** foolish or unrealistic talk or ideas. **2** esp. N. Amer. illicitly distilled or smuggled alcoholic liquor. **3** moonlight.

moonshiner /ˈmuːnˌʃaɪnə(r)/ n. N. Amer. an illicit distiller or smuggler of alcoholic liquor.

moonshot /ˈmuːnʃɒt/ n. the launching of a spacecraft to the moon.

moonstone /ˈmuːnstəʊn/ n. feldspar of pearly appearance.

moonstruck /ˈmuːnstrʌk/ adj. mentally deranged.

moony /ˈmuːnɪ/ adj. (**moonier**, **mooniest**) **1** listless; stupidly dreamy. **2** of or like the moon.

Moor /mʊə(r), mɔː(r)/ n. a member of a Muslim people of mixed Berber and Arab descent, inhabiting NW Africa, who in the 8th century conquered the Iberian peninsula. The Moors were finally driven out of their last stronghold in Spain, Granada, at the end of the 15th century. They left an important architectural legacy, which includes the Alhambra palace in Granada (begun 785). [ME f. OF More f. L Maurus f. Gk Mauros inhabitant of Mauretania]

moor[1] /mʊə(r), mɔː(r)/ n. **1 a** an area of open uncultivated land on acid peaty soil, usu. high-lying and covered with heather, coarse grasses, etc. (cf. HEATH 1a). **b** a tract of such land preserved for shooting. **2** dial. & US a fen. □ **moorish** adj. **moory** adj. [OE mōr waste land, marsh, mountain, f. Gmc]

moor[2] /mʊə(r), mɔː(r)/ v. **1** tr. make fast (a boat, buoy, etc.) by attaching a cable etc. to a fixed object. **2** intr. (of a boat) be moored. □ **moorage** n. [ME more, prob. f. LG or MLG mōren]

moorcock /ˈmʊəkɒk, ˈmɔːkɒk/ n. a male moorfowl.

Moore[1] /mʊə(r), mɔː(r)/, Francis (1657–c.1715), English physician, astrologer, and schoolmaster. In 1699 he published an almanac containing weather predictions in order to promote the sale of his pills, and in 1700 one with astrological observations. There are now several almanacs called 'Old Moore', and predictions range far beyond the weather.

Moore[2] /mʊə(r), mɔː(r)/, George (Augustus) (1852–1933), Irish novelist. From about 1870 he lived in Paris, where he studied painting and gained a knowledge of the works of writers such as Balzac and Zola. On his return to London (1880) Moore embarked on a career as a novelist; influenced by Zola, he experimented with naturalistic techniques in works such as A Mummer's Wife (1885), set in the Potteries in North Staffordshire, and Esther Waters (1894). He became involved in the Irish literary revival and collaborated in the planning of the Irish National Theatre, established in 1899.

Moore[3] /mʊə(r), mɔː(r)/, G(eorge) E(dward) (1873–1958), English philosopher. He led the revolt against the Hegelianism prevalent at the turn of the century, objecting that it was inapplicable to the familiar world of 'tables and chairs'. In his best-known work Principia Ethica (1903), he argued that good was a simple, indefinable, unanalysable, and non-natural property, but that it was still possible to identify certain things as pre-eminently good. These he declared to be 'personal affection and aesthetic enjoyments', values seized upon by several of his associates in the Bloomsbury Group.

Moore[4] /mʊə(r), mɔː(r)/, Henry (Spencer) (1898–1986), English sculptor and draughtsman. In the 1920s he rejected modelling in favour of direct carving in stone and wood, and allowed natural qualities such as texture and grain to influence form. Moore subsequently received major commissions for architectural sculpture, including large reclining figures for the UNESCO building in Paris (1957–8). Prominent themes for the postwar period were upright figures, family groups, and two and three-piece semi-abstract reclining forms; Moore was keen that these sculptures should be viewed in the open air.

Moore[5] /mʊə(r), mɔː(r)/, Sir John (1761–1809), British general. From 1808 he commanded the British army during the Peninsular War, conducting a successful 250-mile retreat to Corunna in mid-winter before being mortally wounded by his French pursuers. His burial was the subject of a famous poem by the Irish poet Charles Wolfe (1791–1823).

Moore[6] /mʊə(r), mɔː(r)/, Robert Frederick ('Bobby') (1941–93), English footballer. He is chiefly remembered as the captain of the English team that won the World Cup in 1966.

Moore[7] /mʊə(r), mɔː(r)/, Thomas (1779–1852), Irish poet and musician.

He wrote patriotic and nostalgic songs, which he set to Irish tunes, and collected in Irish Melodies (1807–34). His most famous songs include 'The Harp that once through Tara's Halls' and 'The Minstrel Boy'. He is also noted for his oriental romance Lalla Rookh (1817).

moorfowl /ˈmʊəfaʊl, ˈmɔːfaʊl/ n. a red grouse.

moorhen /ˈmʊəhen, ˈmɔːhen/ n. **1** a small aquatic bird, Gallinula chloropus, of the rail family, with mainly blackish plumage and a short red and yellow bill. **2** a female moorfowl.

mooring /ˈmʊərɪŋ, ˈmɔːr-/ n. **1 a** a fixed object to which a boat, buoy, etc., is moored. **b** (often in pl.) a place where a boat etc. is moored. **2** (in pl.) a set of permanent anchors and chains laid down for ships to be moored to.

Moorish /ˈmʊərɪʃ, ˈmɔːr-/ adj. of or relating to the Moors. □ **Moorish idol** a tropical reef fish, Zanclus cornutus, of the Indian and Pacific oceans, with a deep black and white striped body.

moorland /ˈmʊələnd, ˈmɔːlənd/ n. an extensive area of moor.

moose /muːs/ n. (pl. same) esp. N. Amer. = ELK 1. [Narragansett moos]

moot /muːt/ adj., v., & n. ● adj. (orig. the noun used attrib.) **1** debatable, undecided (a moot point). **2** US Law having no practical significance. ● v.tr. raise (a question) for discussion. ● n. **1** hist. an assembly, esp. one in England before the Norman Conquest for legislative or judicial purposes. **2** Law a discussion of a hypothetical case as an academic exercise. [OE mōt, and mōtian converse, f. Gmc, rel. to MEET[1]]

mop[1] /mɒp/ n. & v. ● n. **1** a wad or bundle of cotton or synthetic material fastened to the end of a stick, for cleaning floors etc. **2** a similarly shaped large or small implement for various purposes. **3** anything resembling a mop, e.g. a thick mass of hair. **4** an act of mopping or being mopped (gave it a mop). ● v.tr. (**mopped**, **mopping**) **1** wipe or clean with or as with a mop. **2 a** wipe tears or sweat etc. from (one's face or brow etc.). **b** wipe away (tears etc.). □ **mop up 1** wipe up with or as with a mop. **2** colloq. absorb (profits etc.). **3** dispatch; make an end of. **4** Mil. **a** complete the occupation of (a district etc.) by capturing or killing enemy troops left there. **b** capture or kill (stragglers). □ **moppy** adj. [ME mappe, perh. ult. rel. to L mappa napkin]

mop[2] /mɒp/ n. Brit. hist. an autumn fair or gathering at which farm-hands and servants were formerly hired. [perh. = mop-fair, at which a mop was carried by a maidservant seeking employment]

mope /məʊp/ v. & n. ● v.intr. **1** be gloomily depressed or listless; behave sulkily. **2** wander about listlessly. ● n. **1** a person who mopes. **2** (**the mopes**) low spirits. □ **moper** n. **mopish** adj. **mopy** adj. (**mopier**, **mopiest**). **mopily** adv. **mopiness** n. [16th c.: prob. rel. to mope, mopp(e) fool]

moped /ˈməʊped/ n. a motorized bicycle with an engine capacity below 50 cc. [Sw., perh. contr. of motorvelociped]

mophead /ˈmɒphed/ n. a person with thick matted hair.

mopoke /ˈməʊpəʊk/ n. (also **morepork** /ˈmɔːpɔːk/) **1** a boobook owl. **2** an Australian frogmouth, Podargus strigoides. [imit. of the birds' cry]

moppet /ˈmɒpɪt/ n. colloq. (esp. as a term of endearment) a baby or small child. [obs. moppe baby, doll]

Mopti /ˈmɒptɪ/ a city in central Mali, at the junction of the Niger and Bani rivers; pop. (1976) 53,900.

moquette /mɒˈket/ n. a thick pile or looped material used for carpets and upholstery. [F, perh. f. obs. It. mocaiardo mohair]

MOR abbr. middle-of-the-road (popular music).

mor /mɔː(r)/ n. humus formed under acid conditions. [Danish]

Moradabad /ˌmɔːrɑːdəˈbæd/ a city and railway junction in northern India, in Uttar Pradesh; pop. (1991) 417,000.

moraine /məˈreɪn/ n. an area covered by rocks and debris carried down and deposited by a glacier. □ **morainal** adj. **morainic** adj. [F f. It. dial. morena f. F dial. mor(re) snout f. Rmc]

moral /ˈmɒrəl/ adj. & n. ● adj. **1 a** concerned with goodness or badness of human character or behaviour, or with the distinction between right and wrong. **b** concerned with accepted rules and standards of human behaviour. **2 a** conforming to accepted standards of general conduct. **b** capable of moral action (man is a moral agent). **3** (of rights or duties etc.) founded on moral law. **4 a** concerned with morals or ethics (moral philosophy). **b** (of a literary work etc.) dealing with moral conduct. **5** concerned with or leading to a psychological effect associated with confidence in a right action (moral courage; moral support; moral victory). ● n. **1 a** a moral lesson (esp. at the end) of a fable, story, event, etc. **b** a moral maxim or principle. **2** (in pl.) moral behaviour, e.g. in sexual conduct. □ **moral certainty** probability so great as to allow no reasonable doubt. **moral law** the conditions to be satisfied by any right course of action. **moral majority** the majority of people, regarded as favouring firm moral standards (orig. *Moral Majority*, a right-wing US movement). **moral philosophy** the branch of philosophy concerned with ethics. **moral pressure** persuasion by appealing to a person's moral sense. **moral science** systematic knowledge as applied to morals. **moral sense** the ability to distinguish right and wrong. □ **morally** adv. [ME f. L *moralis* f. *mos moris* custom, pl. *mores* morals]

morale /məˈrɑːl/ n. the mental attitude or bearing of a person or group, esp. as regards confidence, discipline, etc. [F *moral* respelt to preserve the pronunciation]

moralism /ˈmɒrəˌlɪz(ə)m/ n. **1** a natural system of morality. **2** religion regarded as moral practice.

moralist /ˈmɒrəlɪst/ n. **1** a person who practises or teaches morality. **2** a person who follows a natural system of ethics. □ **moralistic** /ˌmɒrəˈlɪstɪk/ adj. **moralistically** adv.

morality /məˈrælɪtɪ/ n. (pl. **-ies**) **1** the degree of conformity of an idea, practice, etc., to moral principles. **2** right moral conduct. **3** a lesson in morals. **4** the science of morals. **5** a particular system of morals (*commercial morality*). **6** (in pl.) moral principles; points of ethics. **7** see MORALITY PLAY. [ME f. OF *moralité* or LL *moralitas* f. L (as MORAL)]

morality play n. any of the medieval allegorical dramas, teaching a moral lesson, in which the main characters are personified human qualities. They were popular chiefly in England, Scotland, France, and the Netherlands in the 15th and early 16th centuries. Among the most notable examples in English are *Everyman* (c.1510); *Ane Satire of the Thrie Estaitis* (c.1515) by Sir David Lyndsay (c.1486–1555); *Magnificence* (1516) by John Skelton; *Mankind* (c.1405); and *The Castle of Perseverance* (c.1405).

moralize /ˈmɒrəˌlaɪz/ v. (also **-ise**) **1** intr. (often foll. by *on*) indulge in moral reflection or talk. **2** tr. interpret morally; point the moral of. **3** tr. make moral or more moral. □ **moralizer** n. **moralizingly** adv. **moralization** /ˌmɒrəlaɪˈzeɪʃ(ə)n/ n. [F *moraliser* or med.L *moralizare* f. L (as MORAL)]

Moral Re-Armament n. **1** = OXFORD GROUP. **2** the beliefs of this organization, esp. as applied to international relations.

Morar, Loch /ˈmɔːrə(r)/ a loch in western Scotland. At 310 m (1,017 ft), it is the deepest loch in the country.

morass /məˈræs/ n. **1** an entanglement; a disordered situation, esp. one impeding progress. **2** *literary* a bog or marsh. [Du. *moeras* (assim. to *moer* MOOR[1]) f. MDu. *marasch* f. OF *marais* marsh f. med.L *mariscus*]

moratorium /ˌmɒrəˈtɔːrɪəm/ n. (pl. **moratoriums** or **moratoria** /-rɪə/) **1** (often foll. by *on*) a temporary prohibition or suspension (of an activity). **2 a** a legal authorization to debtors to postpone payment. **b** the period of this postponement. [mod.L, neut. of LL *moratorius* delaying f. L *morari morat-* to delay f. *mora* delay]

Moravia /məˈreɪvɪə/ a region of the Czech Republic, situated between Bohemia in the west and the Carpathians in the east; chief town, Brno. Formerly the western part of the medieval Slavonic kingdom of Greater Moravia, it became a province of Bohemia in the 11th century. It was made an Austrian province in 1848, becoming a part of Czechoslovakia in 1918. Traversed by the Morava river, it is a fertile agricultural region and is rich in mineral resources.

Moravian /məˈreɪvɪən/ n. & adj. ● n. **1** a native of Moravia. **2** a member of a Protestant Church holding Hussite doctrines, founded in Saxony in 1722 by emigrants from Moravia. ● adj. **1** of, relating to, or characteristic of Moravia or its people. **2** of or relating to the Moravian Church or its members.

Moray /ˈmʌrɪ/ (also **Morayshire** /ˈmʌrɪˌʃɪə(r)/) a former county of northern Scotland, bordered on the north by the Moray Firth. It was made a district of Grampian region in 1975.

moray /ˈmɒreɪ/ n. a voracious eel-like fish of the family Muraenidae, found in warm seas; esp. *Muraena helena*, of the Mediterranean and eastern Atlantic. [Port. *moreia* f. L f. Gk *muraina*]

Moray Firth a deep inlet of the North Sea on the NE coast of Scotland. The city of Inverness is near its head.

morbid /ˈmɔːbɪd/ adj. **1 a** (of the mind, ideas, etc.) macabre; unwholesome, sickly. **b** given to morbid feelings. **2** *colloq.* melancholy. **3** *Med.* of the nature of or indicative of disease. □ **morbid anatomy** the anatomy of diseased organs, tissues, etc. **morbidly** adv. **morbidness** n. **morbidity** /mɔːˈbɪdɪtɪ/ n. [L *morbidus* f. *morbus* disease]

morbific /mɔːˈbɪfɪk/ adj. causing disease. [F *morbifique* or mod.L *morbificus* f. L *morbus* disease]

morbilli /mɔːˈbɪlaɪ/ n.pl. *Med.* **1** measles. **2** the spots characteristic of measles. [L, pl. of *morbillus* pustule f. *morbus* disease]

mordant /ˈmɔːd(ə)nt/ n. & adj. ● n. **1** a substance that enables a dye or stain to become fixed in a fabric etc. **2** an adhesive compound for fixing gold leaf. **3** a corrosive liquid used to etch the lines on a printing plate. ● adj. **1** (of sarcasm etc.) caustic, biting. **2** pungent, smarting. **3** (of a substance) having the property of a mordant. **4** (of a dye) needing a mordant. □ **mordancy** n. **mordantly** adv. [ME f. F, part. of *mordre* bite f. L *mordere*]

mordent /ˈmɔːd(ə)nt/ n. *Mus.* **1** an ornament consisting of one rapid alternation of a written note with the note immediately below it. **2** a pralltriller. [G f. It. *mordente* part. of *mordere* bite]

Mordred /ˈmɔːdrəd/ (in Arthurian legend) the nephew of King Arthur.

Mordvinia /mɔːˈdvɪnɪə/ (also called **Mordvinian Autonomous Republic**) an autonomous republic in European Russia, south-east of Nizhni Novgorod; pop. (1990) 964,000; capital, Saransk.

More /mɔː(r)/, Sir Thomas (canonized as St Thomas More) (1478–1535), English scholar and statesman, Lord Chancellor 1529–32. From the time of the accession of Henry VIII (1509), More held a series of public offices, but was forced to resign as Lord Chancellor when he opposed the king's divorce from Catherine of Aragon. He was imprisoned in 1534 after refusing to take the oath on the Act of Succession, sanctioning Henry's marriage to Anne Boleyn. After opposing the Act of Supremacy in the same year, More was beheaded. Regarded as one of the leading humanists of the Renaissance, he owed his reputation largely to his *Utopia* (1516), describing an ideal city-state. Feast day, 22 June.

more /mɔː(r)/ adj., n., & adv. ● adj. **1** existing in a greater or additional quantity, amount, or degree (*more problems than last time; bring some more water*). **2** greater in degree (*more's the pity; the more fool you*). ● n. a greater quantity, number, or amount (*more than three people; more to it than meets the eye*). ● adv. **1** in a greater degree (*do it more carefully*). **2** to a greater extent (*people like to walk more these days*). **3** forming the comparative of adjectives and adverbs, esp. those of more than one syllable (*more absurd; more easily*). **4** again (*once more; never more*). **5** moreover. □ **more and more** in an increasing degree. **more like it** see LIKE[1]. **more of** to a greater extent (*more of a poet than a musician*). **more or less 1** in a greater or smaller degree. **2** approximately; as an estimate. **more so** of the same kind to a greater degree. [OE *māra* f. Gmc]

Morecambe Bay /ˈmɔːkəm/ an inlet of the Irish Sea, on the NW coast of England between Cumbria and Lancashire. The name was derived in the 18th century from a reference in a work of the 2nd-century Greek geographer Ptolemy to *morī kambē*, from the old Celtic name for the Lune estuary, *mori cambo* 'great bay'. The resort town of Morecambe on its eastern shore was named after the bay in 1870.

moreen /mɒˈriːn/ n. a strong ribbed woollen or cotton material for curtains etc. [perh. fanciful f. MOIRE]

moreish /ˈmɔːrɪʃ/ adj. (also **morish**) *colloq.* pleasant to eat, causing a desire for more.

morel[1] /məˈrel/ n. an edible fungus, *Morchella esculenta*, with a brown honeycombed cap. [F *morille* f. Du. *morilje*]

morel[2] /məˈrel/ n. nightshade. [ME f. OF *morele* fem. of *morel* dark brown ult. f. L *Maurus* MOOR]

Morelia /məˈreɪlɪə/ a city in central Mexico, capital of the state of Michoacán; pop. (1990) 489,760. Founded in 1541, it was known as Valladolid until 1828, when it was renamed in honour of José María Morelos y Pavón (1765–1815), a key figure in Mexico's independence movement.

morello /məˈreləʊ/ n. (pl. **-os**) a sour kind of dark cherry. [It. *morello* blackish f. med.L *morellus* f. L (as MOREL[2])]

Morelos /məˈreɪlɒs/ a state of central Mexico, to the west of Mexico City; capital, Cuernavaca.

moreover /mɔːˈrəʊvə(r)/ adv. (introducing or accompanying a new statement) further, besides.

morepork var. of MOPOKE.

mores /ˈmɔːreɪz, -riːz/ n.pl. customs or conventions regarded as essential to or characteristic of a community. [L, pl. of *mos* custom]

Moresco var. of MORISCO.

Moresque /mɒˈresk/ adj. (of art or architecture) Moorish in style or design. [F f. It. *moresco* f. *Moro* MOOR]

Morgan[1] /ˈmɔːgən/, J(ohn) P(ierpont) (1837–1913), American financier, philanthropist, and art collector. From 1871 he was a partner in a New York banking firm (which became J. P. Morgan and Company in 1895); during the 1870s and 1880s he became a powerful financier, acquiring interests in a number of US railways and instigating their reorganization. Morgan created General Electric (1891) from two smaller concerns, and, in 1901, merged several companies to form the United States Steel Corporation. In 1893 and 1907 his personal influence was sufficient to stabilize critically unbalanced financial markets. He built up one of the leading art collections of his day, bequeathing it to the Museum of Modern Art in New York.

Morgan[2] /'mɔːgən/, Thomas Hunt (1866–1945), American zoologist. He is best known for showing the mechanism in the animal cell responsible for inheritance. His studies with the rapidly reproducing fruit fly *Drosophila* showed that the genetic information was carried by genes arranged along the length of the chromosomes. Though this work was not widely accepted initially, Morgan was awarded a Nobel Prize in 1933.

morganatic /ˌmɔːgə'nætɪk/ adj. **1** (of a marriage) between a person of high rank and another of lower rank, the spouse and children having no claim to the possessions or title of the person of higher rank. **2** (of a wife) married in this way. □ **morganatically** adv. [F *morganatique* or G *morganatisch* f. med.L *matrimonium ad morganaticam* 'marriage with a morning gift', the husband's gift to the wife after consummation being his only obligation in such a marriage]

Morgan le Fay /ˌmɔːgən lə 'feɪ/ (in Arthurian legend) 'Morgan the Fairy', a magician, sister of King Arthur. (See FATA MORGANA.)

morgue /mɔːg/ n. **1** a mortuary. **2** (in a newspaper office) a room or file of miscellaneous information, esp. for future obituaries. [F, orig. the name of a Paris mortuary]

moribund /'mɒrɪˌbʌnd/ adj. **1** at the point of death. **2** lacking vitality. **3** on the decline, stagnant. □ **moribundity** /ˌmɒrɪ'bʌndɪtɪ/ n. [L *moribundus* f. *mori* die]

Morisco /mə'rɪskəʊ/ n. & adj. (also **Moresco** /-'reskəʊ/) ● n. (pl. **-os** or **-oes**) **1** a Moor, esp. (in Spain) one who accepted Christian baptism. (See note below.) **2** a morris dance. ● adj. Moorish. [Sp. f. *Moro* MOOR]

▪ When the Christians reconquered Muslim Spain in the 11th–15th centuries the Muslim religion and customs of the Moors were at first tolerated (see MUDÉJAR), but after the fall of Granada in 1492 Islam was officially prohibited and Muslims were forced to observe allegiance to the Christian monarch (when they were known as Moriscos) or go into exile. Many who remained practised their own religion in private in spite of persecution; in 1609 they were expelled, mainly to Africa.

morish var. of MOREISH.

Morisot /ˌmɒrɪˌzəʊ/, Berthe (Marie Pauline) (1841–95), French painter. A pupil of Corot, she was the first woman to join the impressionists, exhibiting with them from 1874. Her works typically depicted women and children, as in *The Cradle* (1872), and waterside scenes, notably *A Summer's Day* (1879).

Morland /'mɔːlənd/, George (1763–1804), English painter. Although indebted to Dutch and Flemish genre painters such as David Teniers the Younger, he drew his inspiration for his pictures of taverns, cottages, and farmyards from local scenes, as with *Inside a Stable* (1791). His art achieved widespread popularity through the engravings of William Ward (1766–1826).

Morley /'mɔːlɪ/, Edward Williams (1838–1923), American chemist. He specialized in accurate quantitative measurements, such as those of the combining weights of hydrogen and oxygen. He is best known, however, for his collaboration with Albert Michelson in their 1887 experiment to determine the speed of light (see MICHELSON).

Mormon /'mɔːmən/ n. a member of the Church of Jesus Christ of Latter-Day Saints, a millenarian religious movement founded in New York State (1830) by Joseph Smith. The central scripture (apart from the Bible) is called the Book of Mormon, which tells the history of a group of Hebrews who migrated to America c.600 BC; Smith claimed to have found and translated the Book of Mormon through divine revelation. A further revelation led him to institute polygamy, a practice which brought the Mormons into conflict with the US government and was officially abandoned in 1890. Smith was succeeded as leader by Brigham Young, who moved the Mormon headquarters to Salt Lake City, Utah, in 1847. Mormons believe in the second coming of Christ; the movement has no professional clergy, self-help is emphasized, and tithing and missionary work are demanded from its members. □ **Mormonism** n.

morn /mɔːn/ n. poet. morning. [OE *morgen* f. Gmc]

mornay /'mɔːneɪ/ n. a cheese-flavoured white sauce. [20th c.: orig. uncert.]

morning /'mɔːnɪŋ/ n. & int. ● n. **1** the early part of the day, esp. from sunrise to noon (*this morning*; *during the morning*; *morning coffee*). **2** this time spent in a particular way (*had a busy morning*). **3** sunrise, daybreak. **4** a time compared with the morning, esp. the early part of one's life etc. ● int. good morning (see GOOD adj. 14). □ **in the morning 1** during or in the course of the morning. **2** colloq. tomorrow. **morning after** colloq. a hangover. **morning-after pill** a contraceptive pill effective when taken some hours after intercourse. **morning coat** a coat with tails, and with the front cut away below the waist. **morning dress** a man's morning coat and striped trousers. **morning glory** a twining plant of the genus *Ipomoea*, native to tropical America, with trumpet-shaped flowers. **morning-room** a sitting-room for the morning. **morning sickness** nausea felt in the morning in pregnancy. **morning star** a planet, esp. Venus, when visible in the east before sunrise. **morning watch** Naut. the 4–8 a.m. watch. [ME *mor(we)ning* f. *morwen* MORN + -ING[1] after *evening*]

Moro /'mɔːrəʊ/ n. (pl. **-os**) a Muslim living in the Philippines. [Sp., = MOOR]

Morocco /mə'rɒkəʊ/ a country in NW Africa, with coastlines on the Mediterranean Sea and Atlantic Ocean; pop. (est. 1991) 25,731,000; languages, Arabic (official), Berber; capital, Rabat. Conquered by the Arabs in the 7th century, Morocco was penetrated by the Portuguese in the 15th century and later fell under French and Spanish influence, each country establishing protectorates in the early 20th century. It became an independent monarchy after the withdrawal of the colonial powers in 1956, the sultan becoming king. □ **Moroccan** adj. & n.

morocco /mə'rɒkəʊ/ n. (pl. **-os**) **1** a fine flexible leather made (orig. in Morocco) from goatskins tanned with sumac, used esp. in bookbinding and shoemaking. **2** an imitation of this in grained calf etc.

moron /'mɔːrɒn/ n. **1** colloq. a very stupid or foolish person. **2** Psychol. an adult with a mental age of about 8–12. □ **moronic** /mə'rɒnɪk/ adj. **moronically** adv. [Gk *mōron*, neut. of *mōros* foolish]

Moroni /mə'rəʊni/ the capital of Comoros, on the island of Grande Comore; pop. (1980) 20,100.

morose /mə'rəʊs/ adj. sullen and ill-tempered. □ **morosely** adv. **moroseness** n. [L *morosus* peevish etc. f. *mos moris* manner]

Morpeth /'mɔːpəθ/ a town in NE England, the county town of Northumberland; pop. (1981) 15,000.

morph[1] /mɔːf/ n. = ALLOMORPH. [back-form.]

morph[2] /mɔːf/ n. Biol. a variant form of an animal or plant. [Gk *morphē* form]

morpheme /'mɔːfiːm/ n. Linguistics **1** a morphological element considered in respect of its functional relations in a linguistic system. **2** a meaningful morphological unit of a language that cannot be further divided (e.g. in, come, -ing, forming *incoming*). □ **morphemic** /mɔː'fiːmɪk/ adj. **morphemically** adv. [F *morphème* f. Gk *morphē* form, after PHONEME]

morphemics /mɔː'fiːmɪks/ n.pl. (usu. treated as sing.) Linguistics the study of word structure.

Morpheus /'mɔːfɪəs/ Rom. Mythol. the son of Somnus (god of sleep), god of dreams, and, in later writings, also god of sleep.

morphia /'mɔːfɪə/ n. (in general use) = MORPHINE.

morphine /'mɔːfiːn/ n. an analgesic and narcotic drug obtained from opium and used medicinally to relieve pain. □ **morphinism** /-fɪˌnɪz(ə)m/ n. [G *Morphin* & mod.L *morphia* f. MORPHEUS]

morphing /'mɔːfɪŋ/ n. a computer graphics technique used in film-making, whereby an image is apparently transformed into another by a smooth progression; the act or process of changing one image into another using this technique. [shortened f. METAMORPHOSIS + -ING[1]]

morphogenesis /ˌmɔːfə'dʒenɪsɪs/ n. Biol. the development of form in organisms. □ **morphogenetic** /-dʒɪ'netɪk/ adj. [mod.L f. Gk *morphē* form + GENESIS]

morphology /mɔː'fɒlədʒɪ/ n. the study of the forms of things, esp.: **1** Biol. the study of the forms of organisms. **2** Philol. **a** the study of the forms of words. **b** the system of forms in a language. □ **morphologist** n. **morphological** /ˌmɔːfə'lɒdʒɪk(ə)l/ adj. **morphologically** adv. [Gk *morphē* form + -LOGY]

Morris[1] /'mɒrɪs/, William (1834–96), English designer, craftsman, poet, and socialist writer. He was a leading figure in the Arts and Crafts Movement, and in 1861 he established Morris & Company, an association of craftsmen whose members included Edward Burne-Jones and Dante Gabriel Rossetti, to produce hand-crafted goods for the home. Morris's Kelmscott Press, founded in 1890, printed limited editions of fine books using his own type designs and ornamental borders, and was an important influence on English book design. He is also noted for his poetry and many prose romances, especially *News from Nowhere* (1891), which portrays a socialist Utopia.

Morris[2] /ˈmɒrɪs/, William Richard, see NUFFIELD.

Morris chair *n.* a type of plain easy chair with an adjustable back. [MORRIS[1]]

morris dance /ˈmɒrɪs/ *n.* a lively traditional English folk-dance performed by groups of people in distinctive costume. There are many local variations in performance and dress: some dancers use handkerchiefs and bells, others sticks. Such dances are referred to in the 14th century but are probably much earlier, dating back to pre-Christian pagan rituals. Morris dancing underwent a revival at the beginning of the 20th century, due largely to the work of Cecil Sharp, and they survive today. □ **morris dancer** *n.* **morris dancing** *n.* [*morys*, var. of MOORISH]

Morrison[1] /ˈmɒrɪs(ə)n/, Toni (full name Chloe Anthony Morrison) (b.1931), American novelist. She is noted for her novels depicting the black American experience and heritage, often focusing on rural life in the South, as in *The Bluest Eye* (1970). Other works include *Song of Solomon* (1976), *Tar Baby* (1979), and the Pulitzer Prize-winning *Beloved* (1987), a tale of a runaway slave who commits infanticide in mid-19th-century Kentucky. Morrison was awarded the Nobel Prize for literature in 1993, becoming the first black woman writer to receive the prize.

Morrison[2] /ˈmɒrɪs(ə)n/, Van (full name George Ivan Morrison) (b.1945), Northern Irish singer, instrumentalist, and songwriter. He has developed a distinctive personal style from a background of blues, soul, folk music, and rock. Among his albums are *Astral Weeks* (1968), *Moondance* (1970), and *Irish Heartbeat* (1989).

morrow /ˈmɒrəʊ/ *n.* (usu. prec. by *the*) *literary* **1** the following day. **2** the time following an event. [ME *morwe*, *moru* (as MORN)]

Morse[1] /mɔːs/, Samuel F(inley) B(reese) (1791–1872), American inventor. His early career was as a painter, but he became interested in electricity and from 1832 pioneered the development of the electric telegraph. In 1837 he extended the range and capabilities of his working model by means of electromagnetic relays, though similar work was being done concurrently in England by Charles Wheatstone and William Cooke. The US Congress gave financial support for an experimental line from Washington, DC, to Baltimore, over which Morse sent the famous message 'What hath God wrought' (in his new code) on 24 May 1844.

Morse[2] /mɔːs/ *n. & v.* ● *n.* = MORSE CODE. ● *v.tr. & intr.* signal by Morse code.

Morse code *n.* an alphabet or code in which letters are represented by combinations of long and short light or sound signals. Invented by Samuel F. B. Morse, it was adopted internationally in wireless or radio-telegraphy when this superseded wire telegraphy.

morsel /ˈmɔːs(ə)l/ *n.* a mouthful; a small piece (esp. of food). [ME f. OF, dimin. of *mors* a bite f. *mordere mors-* to bite]

mort /mɔːt/ *n. Hunting* a note sounded when the quarry is killed. [ME f. OF f. L *mors mortis* death]

mortadella /ˌmɔːtəˈdelə/ *n.* (*pl.* **mortadelle** /-ˈdelɪ/) a large spiced pork sausage. [It. dimin., irreg. f. L *murtatum* seasoned with myrtle berries]

mortal /ˈmɔːt(ə)l/ *adj. & n.* ● *adj.* **1 a** (of a living being, esp. a human) subject to death. **b** (of material or earthly existence) temporal, ephemeral. **2** (often foll. by *to*) causing death; fatal. **3** (of a battle) fought to the death. **4** associated with death (*mortal agony*). **5** (of an enemy) implacable. **6** (of pain, fear, an affront, etc.) intense, very serious. **7** *colloq.* **a** very great (*in a mortal hurry*). **b** long and tedious (*for two mortal hours*). **8** *colloq.* conceivable, imaginable (*every mortal thing*; *of no mortal use*). ● *n.* **1** a mortal being, esp. a human. **2** *joc.* a person described in some specified way (*a thirsty mortal*). □ **mortal sin** (in Catholic theology) a grave sin that is regarded as depriving the soul of divine grace. □ **mortally** *adv.* [ME f. OF *mortal*, *mortel* or L *mortalis* f. *mors mortis* death]

mortality /mɔːˈtælɪtɪ/ *n.* (*pl.* **-ies**) **1** the state of being subject to death. **2** loss of life on a large scale. **3 a** the number of deaths in a given period etc. **b** (in full **mortality rate**) a death rate. [ME f. OF *mortalité* f. L *mortalitas -tatis* (as MORTAL)]

mortar /ˈmɔːtə(r)/ *n. & v.* ● *n.* **1** a mixture of lime with cement, sand, and water, used in building to bond bricks or stones. **2** a short large-bore cannon for firing bombs at high angles. **3** a contrivance for firing a lifeline or firework. **4** a usu. cup-shaped receptacle made of hard material, in which ingredients are pounded with a pestle. ● *v.tr.*

1 plaster or join with mortar. **2** fire on with mortars. □ **mortarless** *adj.* (in sense 1). **mortary** *adj.* (in sense 1). [ME f. AF *morter*, OF *mortier* f. L *mortarium*: partly from LG]

mortarboard /ˈmɔːtəˌbɔːd/ *n.* **1** an academic cap with a stiff flat square top. **2** a flat board with a handle on the undersurface, for holding mortar in bricklaying etc.

mortgage /ˈmɔːɡɪdʒ/ *n. & v.* ● *n.* **1 a** a conveyance of property by a debtor to a creditor as security for a debt (esp. one incurred by the purchase of the property), on the condition that it shall be returned on payment of the debt within a certain period. **b** a deed effecting this. **2 a** a debt secured by a mortgage. **b** a loan resulting in such a debt. ● *v.tr.* **1** convey (a property) by mortgage. **2** (often foll. by *to*) pledge (oneself, one's powers, etc.). □ **mortgage rate** the rate of interest charged by a mortgagee. □ **mortgageable** *adj.* [ME f. OF, = dead pledge f. *mort* f. L *mortuus* dead + *gage* GAGE[1]]

mortgagee /ˌmɔːɡɪˈdʒiː/ *n.* the creditor in a mortgage, usu. a bank or building society.

mortgager /ˈmɔːɡɪdʒə(r)/ *n.* (also **mortgagor** /ˌmɔːɡɪˈdʒɔː(r)/) the debtor in a mortgage.

mortice var. of MORTISE.

mortician /mɔːˈtɪʃ(ə)n/ *n. N. Amer.* an undertaker; a manager of funerals. [L *mors mortis* death + -ICIAN]

mortify /ˈmɔːtɪˌfaɪ/ *v.* (**-ies, -ied**) **1** *tr.* **a** cause (a person) to feel shamed or humiliated. **b** wound (a person's feelings). **2** *tr.* bring (the body, the flesh, the passions, etc.) into subjection by self-denial or discipline. **3** *intr.* (of flesh) be affected by gangrene or necrosis. □ **mortifying** *adj.* **mortifyingly** *adv.* **mortification** /ˌmɔːtɪfɪˈkeɪʃ(ə)n/ *n.* [ME f. OF *mortifier* f. eccl.L *mortificare* kill, subdue f. *mors mortis* death]

Mortimer /ˈmɔːtɪmə(r)/, Roger de, 8th Baron of Wigmore and 1st Earl of March (*c.*1287–1330), English noble. In 1326 he invaded England with his lover Isabella of France, forcing her husband Edward II to abdicate in favour of her son, the future Edward III. Mortimer and Isabella acted as regents for the young Edward until 1330, when the monarch assumed royal power and had Mortimer executed.

mortise /ˈmɔːtɪs/ *n. & v.* (also **mortice**) ● *n.* a hole in a framework designed to receive the end of another part, esp. a tenon. ● *v.tr.* **1** join securely, esp. by mortise and tenon. **2** cut a mortise in. □ **mortise lock** a lock recessed into a mortise in the frame of a door or window etc. [ME f. OF *mortoise* f. Arab. *murtazz* fixed in]

mortmain /ˈmɔːtmeɪn/ *n. Law* **1** the status of lands or tenements held inalienably by an ecclesiastical or other corporation. **2** the land or tenements themselves. [ME f. AF, OF *mortemain* f. med.L *mortua manus* dead hand, prob. in allusion to impersonal ownership]

Morton[1] /ˈmɔːt(ə)n/, Jelly Roll (born Ferdinand Joseph La Menthe Morton) (1885–1941), American jazz pianist, composer, and bandleader. He was one of the principal links between ragtime and New Orleans jazz, and formed his own band, the Red Hot Peppers, in 1926. For the next four years he and his band made a series of notable jazz recordings, but Morton's popularity waned during the 1930s.

Morton[2] /ˈmɔːt(ə)n/, John (*c.*1420–1500), English prelate and statesman. He rose to become Henry VII's chief adviser, being appointed Archbishop of Canterbury in 1486 and Chancellor a year later. He is traditionally associated with the Crown's stringent taxation policies, which made the regime in general and Morton in particular widely unpopular.

Morton's Fork the argument used by John Morton in demanding gifts for the royal treasury: if a man lived well he was obviously rich, and if he lived frugally then he must have savings. [MORTON[2]]

mortuary /ˈmɔːtjʊərɪ/ *n. & adj.* ● *n.* (*pl.* **-ies**) a room or building in which dead bodies may be kept until burial or cremation. ● *adj.* of or concerning death or burial. [ME f. AF *mortuarie* f. med.L *mortuarium* f. L *mortuarius* f. *mortuus* dead]

morula /ˈmɒrʊlə/ *n.* (*pl.* **morulae** /-liː/) *Zool.* a solid ball of cells that results from division of an ovum, and from which a blastula is formed. [mod.L, dimin. of L *morum* mulberry]

morwong /ˈmɔːwɒŋ/ *n.* a marine fish of the family Cheilodactylidae, native to Australasia, used as food. [Aboriginal]

Mosaic /məʊˈzeɪɪk/ *adj.* of or associated with Moses (in the Hebrew Bible). □ **Mosaic Law** the laws attributed to Moses and listed in the Pentateuch. [F *mosaïque* or mod.L *Mosaicus* f. *Moses* f. Heb. *Mōšeh*]

mosaic /məʊˈzeɪɪk/ *n. & v.* ● *n.* **1 a** a picture or pattern produced by an arrangement of small variously coloured pieces of glass or stone etc. **b** work of this kind as an art form. (*See note below.*) **2** a diversified thing. **3** an arrangement of photosensitive elements in a television camera.

4 *Biol.* a chimera. **5** (in full **mosaic disease**) a viral disease causing leaf-mottling in plants, esp. tobacco, maize, and sugar cane. **6** (*attrib.*) **a** of or like a mosaic. **b** diversified. ● *v.tr.* (**mosaicked, mosaicking**) **1** adorn with mosaics. **2** combine into or as into a mosaic. □ **mosaic gold 1** tin disulphide. **2** an alloy of copper and zinc used in cheap jewellery etc. □ **mosaicist** /-ˈzeɪɪsɪst/ *n.* [ME f. F *mosaïque* f. It. *mosaico* f. med.L *mosaicus, musaicus* f. Gk *mous(e)ion* mosaic work f. *mousa* MUSE[1]]

▪ The first extensive use of mosaic was by the Romans; a crucial technical advance was their invention of a durable and waterproof cement in which the fragments were fixed, and as a result well-preserved examples survive from all parts of the Roman world. The classical Roman mosaic tradition was continued in the Christian period, and survived longest and most splendidly in the Byzantine world. In the West, the stylized mosaic technique was largely superseded from the 13th century by the new naturalistic possibilities of fresco painting.

Mosander /mɒˈsændə(r)/, Carl Gustaf (1797–1858), Swedish chemist. He succeeded Berzelius in Stockholm and continued his work on the rare-earth elements, isolating new elements successively from preparations that turned out to be mixtures. In 1839 Mosander discovered and named the element lanthanum, which was present as the oxide in a mineral that had also yielded cerium. Four years later he announced the discovery of the new elements erbium and terbium, and the supposed element of didymium.

mosasaur /ˈməʊzəˌsɔː(r)/ *n.* (also **mosasaurus** /ˌməʊzəˈsɔːrəs/) a large extinct marine reptile, esp. of the genus *Mosasaurus*, with a long slender body and paddle-like limbs. They were predators of the late Cretaceous seas, related to the monitor lizards. [mod.L f. *Mosa* River MEUSE (near which it was first discovered) + Gk *sauros* lizard]

moschatel /ˌmɒskəˈtel/ *n.* a small plant, *Adoxa moschatellina*, with pale green flowers and a musky smell. [F *moscatelle* f. It. *moscatella* f. *moscato* musk]

Moscow /ˈmɒskəʊ/ (Russian **Moskva** /mas'kva/) the capital of Russia, situated at the centre of the vast plain of European Russia, on the River Moskva; pop. (1990) 9,000,000. First mentioned in medieval chronicles in 1147, it soon became the chief city of the increasingly powerful Muscovite princes. In the 16th century, when Ivan the Terrible proclaimed himself the first tsar of Russia, Moscow became the capital of the new empire, its central position giving it supreme military and economic value. Though Peter the Great moved his capital to St Petersburg in 1712, Moscow remained the heart of Russia and the centre of the Russian Orthodox Church. In 1812 it was attacked and occupied by Napoleon and three-quarters of the city was destroyed by fire. By the mid-19th century Moscow had become a large and growing industrial city. After the Bolshevik Revolution of 1917 it was made the capital of the USSR and seat of the new Soviet government, with its centre in the Kremlin, the ancient citadel of the 15th-century city. It is a major industrial and cultural centre, with world-famous theatres and museums, and is home to the Bolshoi Ballet.

Moscow Art Theatre a theatre group established in Moscow in 1898 as an experimental, amateur company. Under its founding artistic director Konstantin Stanislavsky, the originator of method acting, it was the first group to produce the plays of Chekhov; it went on to become internationally influential after foreign tours in the 1920s, and remains one of Russia's foremost artistic institutions.

Mosel /ˈməʊz(ə)l/ (also **Moselle** /məʊˈzel/) a river of western Europe, which rises in the Vosges mountains of NE France and flows 550 km (346 miles) north-east through Luxembourg and Germany to meet the Rhine at Koblenz.

mosel var. of MOSELLE.

Moseley /ˈməʊzlɪ/, Henry Gwyn Jeffreys (1887–1915), English physicist. While working under Rutherford, he discovered that there is a relationship between the atomic numbers of elements and the wavelengths of the X-rays they emit. He demonstrated experimentally that nuclear charge and atomic number are connected, that the element's chemical properties are determined by this number, and that there are only 92 naturally occurring elements. He was killed in action in the 1914–18 war.

Moselle see MOSEL.

moselle /məʊˈzel/ *n.* (also **mosel** /ˈməʊz(ə)l/) a light medium-dry white wine produced in the valley of the Mosel river.

Moses[1] /ˈməʊzɪz/ (*fl.* c.14th–13th centuries BC), Hebrew prophet and lawgiver. According to the biblical account, he was born in Egypt and led the Israelites away from servitude there, across the desert towards the Promised Land. During the journey he was inspired by God on Mount Sinai to write down the Ten Commandments on tablets of stone (Exod. 20). He was the brother of Aaron.

Moses[2] /ˈməʊzɪz/, Anna Mary (known as Grandma Moses) (1860–1961), American painter. She lived as a farmer's wife until widowed in 1927, when she took up painting as a hobby; her work began to appear in exhibitions from the late 1930s. Grandma Moses produced more than a thousand naive paintings, principally colourful scenes of American rural life.

mosey /ˈməʊzɪ/ *v.intr.* (**-eys, -eyed**) (often foll. by *along*) *sl.* walk in a leisurely or aimless manner. [19th c.: orig. unkn.]

mosh /mɒʃ/ *v.intr. sl.* dance to rock music in a violent manner involving colliding with others and headbanging. □ **mosh-pit** an area where moshing occurs, esp. in front of the stage at a rock concert. [perh. alt. of MASH *v.*]

moshav /məʊˈʃɑːv/ *n.* (pl. **moshavim** /ˌməʊʃɑːˈvɪm/) a cooperative association of Israeli smallholders. [Heb. *mošāb̲*, lit. 'dwelling']

Moskva see MOSCOW.

Moslem var. of MUSLIM.

Mosley /ˈməʊzlɪ/, Sir Oswald (Ernald), 6th Baronet (1896–1980), English Fascist leader. Mosley sat successively as a Conservative, Independent, and Labour MP before founding and leading the British Union of Fascists (1932), also known as the Blackshirts. The party was effectively destroyed by the Public Order Act of 1936 and Mosley was interned from 1940 to 1943. In 1948 he founded the right-wing Union Movement.

mosque /mɒsk/ *n.* a Muslim place of worship. The first mosques were modelled on the place of worship of Muhammad; the dominant form of architecture today consists of a courtyard, forming a place for communal gathering, a covered area, frequently domed, used for worship, on the side facing Mecca, and a fountain for ablutions, obligatory before prayer. No music or singing is allowed, and since representations of the human form are forbidden, decoration is by geometric designs and Arabic calligraphy. [F *mosquée* f. It. *moschea* f. Arab. *masjid*]

Mosquito var. of MISKITO.

mosquito /mɒˈskiːtəʊ/ *n.* (pl. **-oes**) a slender dipterous biting fly of the family Culicidae, esp. of the genus *Culex, Anopheles,* or *Aedes*. The female uses a long proboscis to puncture the skin of victims to suck blood, transmitting diseases such as filariasis and malaria. □ **mosquito-boat** *US* a motor torpedo-boat. **mosquito-net** a net to keep off mosquitoes. [Sp. & Port., dimin. of *mosca* f. L *musca* fly]

Mosquito Coast a sparsely populated coastal strip of swamp, lagoon, and tropical forest comprising the Caribbean coast of Nicaragua and NE Honduras, occupied by the Miskito people after whom it is named. The British maintained a protectorate over the area intermittently from the 17th century until the mid-19th century, when it briefly became an autonomous state. In 1894 Nicaragua appropriated the territory and in 1960 the northern part was awarded to Honduras.

Moss /mɒs/, Stirling (b.1929), English motor-racing driver. He was especially successful in the 1950s, winning various Grands Prix and other competitions, though the world championship always eluded him.

moss /mɒs/ *n. & v.* ● *n.* **1** a small leafy bryophyte of the class Musci, growing in dense clusters on the surface of the ground, in bogs, on trees, stones, etc. **2** *Sc. & N. Engl.* a bog, esp. a peatbog. ● *v.tr.* cover with moss. □ **moss agate** agate with mosslike dendritic markings. **moss-grown** overgrown with moss. **moss-hag** *Sc.* broken ground from which peat has been taken. **moss-stitch** alternate plain and purl in knitting. □ **mosslike** *adj.* [OE *mos* bog, moss f. Gmc]

Mossad /mɒˈsæd/ **1** the Supreme Institution for Intelligence and Special Assignments, founded in 1951 and the principal secret intelligence service of the state of Israel. **2** the Institution for the Second Immigration, an earlier organization formed in 1938 for the purpose of bringing Jews from Europe to Palestine. [Heb. *mōsād̲* institution]

mossie /ˈmɒzɪ/ *n.* esp. *Austral. sl.* = MOSQUITO.

mosso /ˈmɒsəʊ/ *adv. Mus.* with animation or speed. [It., past part. of *muovere* move]

mosstrooper /ˈmɒsˌtruːpə(r)/ *n. hist.* a freebooter of the Scottish Border in the 17th century.

mossy /ˈmɒsɪ/ *adj.* (**mossier, mossiest**) **1** covered in or resembling moss. **2** *US sl.* antiquated, old-fashioned. □ **mossiness** *n.*

most /məʊst/ *adj., n., & adv.* ● *adj.* **1** existing in the greatest quantity or degree (*you have made most mistakes*; *see who can make the most noise*). **2** the majority of; nearly all of (*most people think so*). ● *n.* **1** the greatest quantity or number (*this is the most I can do*). **2** (**the most**) *sl.* the best of all. **3** the majority (*most of them are missing*). ● *adv.* **1** in the highest degree (*this is most interesting*; *what most annoys me*). **2** the superlative of adjectives and adverbs, esp. those of more than one syllable (*most certain*; *most easily*). **3** *US colloq.* almost. □ **at most** no more or better than (*this is at most a makeshift*). **at the most 1** as the greatest amount. **2** not more than. **for the most part 1** as regards the greater part. **2** usually. **make the most of 1** employ to the best advantage. **2** represent at its best or worst. [OE *mæst* f. Gmc]

-most /məʊst/ *suffix* forming superlative adjectives and adverbs from prepositions and other words indicating relative position (*foremost*; *uttermost*). [OE *-mest* f. Gmc]

Mostar /mɒsˈtɑː(r)/ a largely Muslim city in Bosnia–Herzegovina south-west of Sarajevo, the chief town of Herzegovina; pop. (1981) 110,380. Its chief landmark, an old Turkish bridge across the River Neretva, was destroyed during the siege of the city by Serb forces in 1993.

Most High, the a name for God.

Most Honourable *n.* a title given to marquesses and to members of the Privy Council and the Order of the Bath.

mostly /ˈməʊstlɪ/ *adv.* **1** as regards the greater part. **2** usually.

Most Reverend *n.* a title given to archbishops and to Roman Catholic bishops.

Mosul /ˈməʊsʊl/ a city in northern Iraq, on the River Tigris, opposite the ruins of Nineveh; pop. (est. 1985) 570,900. It gives its name to muslin, a cotton fabric first produced there.

MOT *abbr.* **1** *hist.* (in the UK) Ministry of Transport. **2** (in full **MOT test**) a compulsory annual test of motor vehicles of more than a specified age.

mot /məʊ/ *n.* (*pl.* **mots** *pronunc.* same) a witty saying. □ **mot juste** /ʒuːst/ (*pl.* **mots justes** *pronunc.* same) the most appropriate expression. [F, = word, ult. f. L *muttum* uttered sound f. *muttire* murmur]

mote /məʊt/ *n.* a speck of dust. [OE *mot*, corresp. to Du. *mot* dust, sawdust, of unkn. orig.]

motel /məʊˈtel/ *n.* a roadside hotel providing accommodation for motorists and parking for their vehicles. [portmanteau word f. MOTOR + HOTEL]

motet /məʊˈtet/ *n.* *Mus.* a short sacred choral composition. [ME f. OF, dimin. of *mot*: see MOT]

moth /mɒθ/ *n.* **1** a usu. nocturnal lepidopterous insect with often drab coloration. (*See note below.*) **2** (in full **clothes-moth**) a small lepidopterous insect of the family Tineidae, breeding in cloth etc., on which its larva feeds. □ **moth-eaten 1** damaged or destroyed by moths. **2** antiquated, time-worn. [OE *moththe*]

▪ The moths are generally distinguished from butterflies by their nocturnal behaviour, antennae without clubs, stout bodies, and the folded and flattened position of the wings when at rest. They are, however, not taxonomically distinct from butterflies, and they constitute the great majority of Lepidoptera. There are a few groups of day-flying moths, which are often brightly coloured. (See also LEPIDOPTERA.)

mothball /ˈmɒθbɔːl/ *n.* & *v.* ● *n.* a ball of naphthalene etc. placed in stored clothes to keep away moths. ● *v.tr.* **1** place in mothballs. **2** leave unused. □ **in mothballs** stored unused for a considerable time.

mother /ˈmʌðə(r)/ *n.* & *v.* ● *n.* **1 a** a woman in relation to a child or children to whom she has given birth. **b** (in full **adoptive mother**) a woman who has continuous care of a child, esp. by adoption. **2** any female animal in relation to its offspring. **3** a quality or condition etc. that gives rise to another (*necessity is the mother of invention*). **4** (in full **Mother Superior**) the head of a female religious community. **5** *archaic* (esp. as a form of address) an elderly woman. **6** (*attrib.*) **a** designating an institution etc. regarded as having maternal authority (*Mother Church*; *mother earth*). **b** designating the main ship, spacecraft, etc., in a convoy or mission (*the mother craft*). ● *v.tr.* **1** give birth to; be the mother of. **2** protect as a mother. **3** give rise to; be the source of. **4** acknowledge or profess oneself the mother of. □ **mother country** a country in relation to its colonies. **mother-figure** an older woman who is regarded as a source of nurture, support, etc. **mother-in-law** (*pl.* **mothers-in-law**) the mother of one's husband or wife. **mother-in-law's tongue** an African plant, *Sansevieria trifasciata*, of the agave family, with long erect pointed leaves. **mother-lode** *Mining* the main vein of a system. **mother-naked** stark naked. **mother-of-pearl** a

smooth iridescent substance forming the inner layer of the shell of some molluscs. **mother's ruin** *colloq.* gin. **mother's son** *colloq.* a man (*every mother's son of you*). **mother tongue 1** one's native language. **2** a language from which others have evolved. **mother wit** native wit; common sense. □ **motherless** *adj.* **motherlessness** *n.* **motherlike** *adj.* & *adv.* [OE *mōdor* f. Gmc]

motherboard /ˈmʌðəbɔːd/ *n.* *Computing* a printed circuit board containing the principal components of a microcomputer etc.

Mother Carey's chicken *n.* = storm petrel.

mothercraft /ˈmʌðəkrɑːft/ *n.* skill in or knowledge of looking after children as a mother.

mother goddess *n.* (also called *Great Mother*) a mother-figure deity, goddess of the entire complex of birth and growth, commonly a central figure of early nature-cults where maintenance of fertility was of prime religious importance. Such goddesses were widespread and common; in the eastern Mediterranean, for example, Isis, Astarte, Cybele, and Demeter were all traditional mother goddesses.

Mother Goose rhyme *n.* *N. Amer.* a nursery rhyme.

motherhood /ˈmʌðəhʊd/ *n.* **1** the condition or fact of being a mother. **2** (*attrib.*) *US & Austral.* (of an issue, report, etc.) protective, withholding the worst aspects.

Mothering Sunday /ˈmʌðərɪŋ/ *n.* *Brit.* the fourth Sunday in Lent, traditionally a day for honouring mothers with gifts.

motherland /ˈmʌðəlænd/ *n.* one's native country.

motherly /ˈmʌðəlɪ/ *adj.* **1** like or characteristic of a mother in affection, care, etc. **2** of or relating to a mother. □ **motherliness** *n.* [OE *mōdorlic* (as MOTHER)]

Mother's Day 1 *Brit.* = MOTHERING SUNDAY. **2** *N. Amer.* an equivalent day on the second Sunday in May.

Mother Teresa see TERESA, MOTHER.

mothproof /ˈmɒθpruːf/ *adj.* & *v.* ● *adj.* (of clothes) treated so as to repel moths. ● *v.tr.* treat (clothes) in this way.

mothy /ˈmɒθɪ/ *adj.* (**mothier**, **mothiest**) infested with moths.

motif /məʊˈtiːf/ *n.* **1** a distinctive feature or dominant idea in artistic or literary composition. **2** *Mus.* = FIGURE *n.* 10. **3** a decorative design or pattern. **4** an ornament of lace etc. sewn separately on a garment. **5** *Brit.* an ornament on a vehicle identifying the maker, model, etc. [F (as MOTIVE)]

motile /ˈməʊtaɪl/ *adj.* *Zool.* & *Bot.* capable of motion. □ **motility** /məʊˈtɪlɪtɪ/ *n.* [L *motus* motion (as MOVE)]

motion /ˈməʊʃ(ə)n/ *n.* & *v.* ● *n.* **1** the act or process of moving or of changing position. **2** a particular manner of moving the body in walking etc. **3** a change of posture. **4** a gesture. **5** a formal proposal put to a committee, legislature, etc. **6** *Law* an application for a rule or order of court. **7** *Brit.* **a** an evacuation of the bowels. **b** (in *sing.* or *pl.*) faeces. **8** a piece of moving mechanism. ● *v.* (often foll. by *to* + infin.) **1** *tr.* direct (a person) by a sign or gesture. **2** *intr.* (often foll. by *to* a person) make a gesture directing (*motioned to me to leave*). □ **go through the motions 1** make a pretence; do something perfunctorily or superficially. **2** simulate an action by gestures. **in motion** moving; not at rest. **motion picture** (often with hyphen *attrib.*) a cinema film. **put** (or **set**) **in motion** set going or working. **motion sickness** nausea induced by motion, esp. by travelling in a vehicle. □ **motional** *adj.* **motionless** *adj.* **motionlessly** *adv.* [ME f. OF f. L *motio -onis* (as MOVE)]

motivate /ˈməʊtɪveɪt/ *v.tr.* **1** supply a motive to; be the motive of. **2** cause (a person) to act in a particular way. **3** stimulate the interest of (a person in an activity). □ **motivator** *n.* **motivation** /ˌməʊtɪˈveɪʃ(ə)n/ *n.* **motivational** *adj.* **motivationally** *adv.*

motive /ˈməʊtɪv/ *n.*, *adj.*, & *v.* ● *n.* **1** a factor or circumstance that induces a person to act in a particular way. **2** a motif in art, literature, or music. ● *adj.* **1** tending to initiate movement. **2** concerned with movement. ● *v.tr.* = MOTIVATE. □ **motive power** a moving or impelling power, esp. a source of energy used to drive machinery. □ **motiveless** *adj.* **motivelessly** *adv.* **motivelessness** *n.* **motivity** /məʊˈtɪvɪtɪ/ *n.* [ME f. OF *motif* (adj. & n.) f. LL *motivus* (adj.) (as MOVE)]

motley /ˈmɒtlɪ/ *adj.* & *n.* ● *adj.* (**motlier**, **motliest**) **1** diversified in colour. **2** of varied character (*a motley crew*). ● *n.* **1** an incongruous mixture. **2** *hist.* the particoloured costume of a jester. □ **wear motley** play the fool. [ME *mottelay*, perh. ult. rel. to MOTE]

motmot /ˈmɒtmɒt/ *n.* a bird of the tropical American family Momotidae, some members of which have two long tail feathers like rackets. [Amer. Sp., imit.]

moto-cross /ˈməʊtəʊˌkrɒs/ n. cross-country racing on motorcycles. [MOTOR + CROSS-]

moto perpetuo /ˌməʊtəʊ pəˈpetjʊˌəʊ/ n. Mus. a usu. fast-moving instrumental composition consisting mainly of notes of equal value. [It., = perpetual motion]

motor /ˈməʊtə(r)/ n. & v. ● n. **1** a thing that imparts motion. **2** a machine (esp. one using electricity or internal combustion) supplying motive power for a vehicle etc. or for some other device with moving parts. **3** Brit. a motor car. **4** (attrib.) **a** giving, imparting, or producing motion. **b** driven by a motor (motor-mower). **c** of or for motor vehicles. **d** Anat. relating to muscular movement or the nerves activating it. ● v.intr. & tr. Brit. go or convey in a motor vehicle. □ **motor area** Anat. the part of the frontal lobe of the brain associated with the initiation of muscular action. **motor bicycle** a motorcycle or moped. **motor bike** colloq. = MOTORCYCLE. **motor boat** a motor-driven boat. **motor car** = CAR 1. **motor mouth** esp. N. Amer. sl. a person who talks incessantly and trivially. **motor nerve** Anat. a nerve carrying impulses from the brain or spinal cord to a muscle. **motor neurone disease** Med. a progressive disease involving degeneration of the motor neurones and wasting of the muscles. **motor scooter** see SCOOTER n. 2. **motor vehicle** a road vehicle powered by an internal-combustion engine. □ **motory** adj. (in sense 4a of n.). **motorial** /məʊˈtɔːrɪəl/ adj. (in sense 4a of n.). [L, = mover (as MOVE)]

motorable /ˈməʊtərəb(ə)l/ adj. (of a road) that can be used by motor vehicles.

motorcade /ˈməʊtəˌkeɪd/ n. a procession of motor vehicles. [MOTOR, after cavalcade]

motorcycle /ˈməʊtəˌsaɪk(ə)l/ n. a two-wheeled motor-driven road vehicle, now usually one without pedal propulsion. The first practical motorcycle (setting aside early experiments with steam) was built in 1885 by Daimler, who fitted an internal-combustion engine into a wooden cycle frame. This was also the origin of the motor car, but the motorcycle continued as a distinct vehicle, being lighter, cheaper, and more stable on rough ground. After the Second World War it lost sales to the scooter and the moped, but was given a new lease of life when Japan entered the market with a large range of models and quickly became the world's largest supplier. □ **motorcycling** n. **motorcyclist** n.

motorist /ˈməʊtərɪst/ n. the driver of a motor car.

motorize /ˈməʊtəˌraɪz/ v.tr. (also **-ise**) **1** equip (troops etc.) with motor transport. **2** provide with a motor for propulsion etc. □ **motorization** /ˌməʊtəraɪˈzeɪʃ(ə)n/ n.

motorman /ˈməʊtəˌmæn/ n. (pl. **-men**) the driver of an underground train, tram, etc.

motor-racing /ˈməʊtəˌreɪsɪŋ/ n. racing of motorized vehicles, especially cars, as a sport. The earliest motor-races were held on public roads; the first, held in 1894 between Paris and Rouen, was won at an average speed of ten miles per hour. The first race on a closed circuit took place in France in 1898, while organized Grand Prix racing was instituted in 1906 at the Le Mans circuit. A world championship for Grand Prix drivers began in 1950, points being awarded for performances over the season. Formula One Grand Prix cars can now average speeds of more than 150 m.p.h. (240 k.p.h.) on a circuit, and have little in common with any mass-produced model of car; modified sports cars take part in competitions such as the Le Mans 24-hour race, whereas saloon cars also compete on public roads and over rough terrain in rallies such as the Paris–Dakar and the Monte Carlo Rally. Indy racing, a form of competition in which cars similar to those used in Grand Prix races compete over long distances on banked tracks, has recently attracted particular attention due to the decision of former Formula One champion Nigel Mansell to take up the sport.

motorway /ˈməʊtəˌweɪ/ n. Brit. a main road with separate carriageways and limited access, specially constructed and controlled for fast motor traffic.

Motown /ˈməʊtaʊn/ **1** a nickname for DETROIT. **2** (also **Tamla Motown** /ˈtæmlə/) propr. the first black-owned record company in the US, founded in 1959 in Detroit by Berry Gordy (b.1929) and important in popularizing soul music. Throughout the 1960s the label produced music by artists such as the Supremes, the Four Tops, Stevie Wonder, and Marvin Gaye (1939–84). Its influence declined somewhat after Gordy moved the company's operations to Los Angeles in 1971. [abbr. of motor town]

motte /mɒt/ n. a mound forming the site of a castle, camp, etc. [ME f. OF mote (as MOAT)]

mottle /ˈmɒt(ə)l/ v. & n. ● v.tr. (esp. as **mottled** adj.) mark with spots or smears of colour. ● n. **1** an irregular arrangement of spots or patches of colour. **2** any of these spots or patches. [prob. back-form. f. MOTLEY]

motto /ˈmɒtəʊ/ n. (pl. **-oes**) **1** a maxim adopted as a rule of conduct. **2** a phrase or sentence accompanying a coat of arms or crest. **3** a sentence inscribed on some object and expressing an appropriate sentiment. **4** verses etc. in a paper cracker. **5** a quotation prefixed to a book or chapter. **6** Mus. a recurrent phrase having some symbolical significance. [It. (as MOT)]

moue /muː/ n. = POUT[1] n. [F]

mouflon /ˈmuːflɒn/ n. (also **moufflon**) a wild mountain sheep, Ovis orientalis, of southern Europe. [F mouflon f. It. muflone f. Rmc]

mouillé /ˈmuːjeɪ/ adj. Phonet. (of a consonant) palatalized. [F, = wetted]

moujik var. of MUZHIK.

mould[1] /məʊld/ n. & v. (US **mold**) ● n. **1** a hollow container into which molten metal etc. is poured or soft material is pressed to harden into a required shape. **2 a** a metal or earthenware vessel used to give shape to puddings etc. **b** a pudding etc. made in this way. **3** a form or shape, esp. of an animal body. **4** Archit. a moulding or group of mouldings. **5** a frame or template for producing mouldings. **6** character or disposition (in heroic mould). ● v.tr. **1** make (an object) in a required shape or from certain ingredients (was moulded out of clay). **2** give a shape to. **3** influence the formation or development of (consultation helps to mould policies). **4** (esp. of clothing) fit closely to (the gloves moulded his hands). □ **mouldable** adj. **moulder** n. [ME mold(e), app. f. OF modle f. L modulus: see MODULUS]

mould[2] /məʊld/ n. (US **mold**) a woolly or furry growth of minute fungi occurring esp. in moist warm conditions. [ME prob. f. obs. mould adj.; past part. of moul grow mouldy f. ON mygla]

mould[3] /məʊld/ n. (US **mold**) **1** loose earth. **2** the upper soil of cultivated land, esp. when rich in organic matter. □ **mould-board** the board in a plough that turns over the furrow-slice. [OE molde f. Gmc., rel. to MEAL[2]]

moulder /ˈməʊldə(r)/ v.intr. (US **molder**) **1** decay to dust. **2** (foll. by away) rot or crumble. **3** deteriorate. [perh. f. MOULD[3], but cf. Norw. dial. muldra crumble]

moulding /ˈməʊldɪŋ/ n. (US **molding**) **1 a** an ornamentally shaped outline as an architectural feature, esp. in a cornice. **b** a strip of material in wood or stone etc. for use as moulding. **2** similar material in wood or plastic etc. used for other decorative purposes, e.g. in picture-framing.

mouldy /ˈməʊldɪ/ adj. (US **moldy**) (**-ier**, **-iest**) **1** covered with mould. **2** stale; out of date. **3** colloq. (as a general term of disparagement) dull, miserable, boring. □ **mouldiness** n.

moulin /ˈmuːlæn/ n. a nearly vertical shaft in a glacier, formed by surface water percolating through a crack in the ice. [F, lit. 'mill']

Moulin Rouge /ˌmuːlæn ˈruːʒ/ a cabaret in Montmartre, Paris, a favourite resort of poets and artists around the turn of the century. Toulouse-Lautrec immortalized its dancers in his posters.

Moulmein /maʊlˈmeɪn/ a port in SE Burma (Myanmar); pop. (1983) 220,000.

moult /məʊlt/ v. & n. (US **molt**) ● v. **1** intr. shed feathers, hair, a shell, skin, etc., in the process of renewing plumage or acquiring a new growth. **2** tr. (of an animal) shed (feathers, hair, etc.). ● n. the act or an instance of moulting (is in moult once a year). [ME moute f. OE mutian (unrecorded) f. L mutare change: -l- after fault etc.]

mound[1] /maʊnd/ n. & v. ● n. **1** a raised mass of earth, stones, or other compacted material. **2** a heap or pile. **3** a hillock. **4** Baseball the slight elevation on which a pitcher stands. ● v.tr. **1** heap up in a mound or mounds. **2** enclose with mounds. [16th c. (orig. = hedge or fence): orig. unkn.]

mound[2] /maʊnd/ n. Heraldry a ball of gold etc. representing the earth, and usu. surmounting a crown. [ME f. OF monde f. L mundus world]

mount[1] /maʊnt/ v. & n. ● v. **1** tr. ascend or climb (a hill, stairs, etc.). **2** tr. **a** get up on (an animal, esp. a horse) to ride it. **b** set (a person) on horseback. **c** provide (a person) with a horse. **d** (as **mounted** adj.) serving on horseback (mounted police). **3** tr. go up or climb on to (a raised surface). **4** intr. a move upwards. **b** (often foll. by up) increase, accumulate. **c** (of a feeling) become stronger or more intense (excitement was mounting). **d** (of the blood) rise into the cheeks. **5** tr. (esp. of a male animal) get on to (a female) to copulate. **6** tr. (often foll. by on) place (an object) on an elevated support. **7** tr. **a** set in or attach to a backing, setting, or other support. **b** attach (a picture etc.) to a mount or frame. **c** fix (an object for viewing) on a microscope slide. **8** tr. **a** arrange (a play, exhibition, etc.) or present for public view or display.

b take action to initiate (a programme, campaign, etc.). **9** *tr.* prepare (specimens) for preservation. **10** *tr.* **a** bring into readiness for operation. **b** raise (guns) into position on a fixed mounting. **11** *intr.* rise to a higher level of rank, power, etc. ● *n.* **1 a** a backing, setting, or other support on which something is set for display. **b** the base on which an instrument etc. is supported. **c** the fitting to which an interchangeable camera lens is attached. **d** the margin surrounding a picture or photograph. **e** = *stamp-hinge* (see HINGE). **2 a** a horse available for riding. **b** an opportunity to ride a horse, esp. as a jockey. □ **mount guard** (often foll. by *over*) perform the duty of guarding; take up sentry duty. □ **mountable** *adj.* **mounter** *n.* [ME f. OF *munter*, *monter* ult. f. L (as MOUNT²)]

mount² /maʊnt/ *n. archaic* (except before a name): mountain, hill (*Mount Everest*; *Mount of Olives*). [ME f. OE *munt* & OF *mont* f. L *mons montis* mountain]

mountain /ˈmaʊntɪn/ *n.* **1** a large natural elevation of the earth's surface rising abruptly from the surrounding level; a large or high and steep hill. **2** a large heap or pile; a huge quantity (*a mountain of work*). **3** a very large surplus stock of a commodity (*butter mountain*). □ **make a mountain out of a molehill** see MOLEHILL. **mountain ash 1** a small rosaceous tree of the genus *Sorbus*; esp. *S. aucuparia*, with delicate pinnate leaves and scarlet berries (also called *rowan*). **2 a** tall Australian eucalyptus. **mountain bike** a bicycle with a light sturdy frame, broad deep-treaded tyres, and multiple gears, originally designed for riding on mountainous terrain. **mountain chain** a connected series of mountains. **mountain goat 1** a goat which lives on mountains, proverbial for agility. **2** = ROCKY MOUNTAIN GOAT. **mountain laurel** an evergreen ericaceous shrub, *Kalmia latifolia*, native to North America. **mountain lion** a puma. **mountain panther 1** = *snow leopard*. **2** = PUMA. **mountain range** a line of mountains connected by high ground. **mountain sickness** a sickness caused by the rarefaction of the air at great heights. **move mountains 1** achieve spectacular results. **2** make every possible effort. □ **mountainy** *adj.* [ME f. OF *montaigne* ult. f. L (as MOUNT²)]

mountaineer /ˌmaʊntɪˈnɪə(r)/ *n. & v.* ● *n.* **1** a person skilled in mountain-climbing. **2** a person living in an area of high mountains. ● *v.intr.* climb mountains as a sport. (See MOUNTAINEERING.)

mountaineering /ˌmaʊntɪˈnɪərɪŋ/ *n.* the sport of mountain-climbing. Mountaineering was first organized as a sport by British climbers in the decade 1855–65. Many peaks and high passes in the Alps were then climbed or crossed for the first time (notably the Matterhorn in 1865, by Edward Whymper), before climbers turned their attention to more distant ranges in South America, New Zealand, Africa, and the Himalayas. Aided by the introduction of equipment such as pitons, fixed ropes, and later oxygen, expeditions have climbed most of the world's highest mountains (the highest, Mount Everest, was eventually conquered in 1953), and now tend to concentrate on untried routes.

mountainous /ˈmaʊntɪnəs/ *adj.* **1** (of a region) having many mountains. **2** huge.

mountainside /ˈmaʊntɪnˌsaɪd/ *n.* the slope of a mountain below the summit.

Mountain Standard Time (also **Mountain Time**) (abbr. **MST**) the standard time in a zone including parts of Canada and the US in or near the Rocky Mountains, seven hours behind GMT.

Mount Ararat, Mount Carmel, etc. see ARARAT, MOUNT; CARMEL, MOUNT, etc.

Mountbatten /maʊntˈbæt(ə)n/, Louis (Francis Albert Victor Nicholas), 1st Earl Mountbatten of Burma (1900–79), British admiral and administrator. A great-grandson of Queen Victoria, Mountbatten served in the Royal Navy before rising to become supreme Allied commander in SE Asia (1943–5). As the last viceroy (1947) and first Governor-General of India (1947–8), he oversaw the independence of India and Pakistan. He was killed by an IRA bomb while on his yacht in Ireland.

mountebank /ˈmaʊntɪˌbæŋk/ *n.* **1** a swindler; a charlatan. **2** a clown. **3** *hist.* an itinerant quack appealing to an audience from a platform. □ **mountebankery** *n.* [It. *montambanco* = *monta in banco* climb on bench: see MOUNT¹, BENCH]

Mount Garmo /ˈgɑːməʊ/ a former name (until 1933) for COMMUNISM PEAK.

Mountie /ˈmaʊntɪ/ *n. colloq.* a member of the Royal Canadian Mounted Police.

mounting /ˈmaʊntɪŋ/ *n.* **1** = MOUNT¹ *n.* 1. **2** in senses of MOUNT¹ *v.* □ **mounting-block** a block of stone placed to help a rider mount a horse.

Mount Isa /ˈaɪzə/ a lead and silver-mining town in NE Australia, in western Queensland; pop. (est. 1987) 24,200.

Mount of Olives see OLIVES, MOUNT OF.

Mount Vernon /ˈvɜːnən/ a property in NE Virginia, about 24 km (15 miles) from Washington, DC, on a site overlooking the Potomac river. Built in 1743, it was the home of George Washington from 1747 until his death in 1799.

mourn /mɔːn/ *v.* **1** *tr.* & (foll. by *for*) *intr.* feel or show deep sorrow or regret for (a dead person, a lost thing, a past event, etc.). **2** *intr.* show conventional signs of grief for a period after a person's death. [OE *murnan*]

Mourne Mountains /mɔːn/ a range of hills in SE Northern Ireland, in County Down.

mourner /ˈmɔːnə(r)/ *n.* **1** a person who mourns, esp. at a funeral. **2** a person hired to attend a funeral.

mournful /ˈmɔːnfʊl/ *adj.* **1** doleful, sad, sorrowing. **2** expressing or suggestive of mourning. □ **mournfully** *adv.* **mournfulness** *n.*

mourning /ˈmɔːnɪŋ/ *n.* **1** the expression of deep sorrow, esp. for a dead person, by the wearing of solemn dress. **2** the clothes worn in mourning. □ **in mourning** assuming the signs of mourning, esp. in dress. **mourning-band** a band of black crape etc. round a person's sleeve or hat as a token of mourning. **mourning cloak** *N. Amer.* = CAMBERWELL BEAUTY. **mourning dove** a North American dove, *Zenaida macroura*, with a plaintive call. **mourning-paper** notepaper with a black edge. **mourning-ring** a ring worn as a memorial of a deceased person.

mousaka var. of MOUSSAKA.

Mousalla see MUSALA, MOUNT.

mouse /maʊs/ *n. & v.* ● *n.* (*pl.* **mice** /maɪs/) **1 a** a small rodent, esp. of the family Muridae, usu. having a pointed snout and relatively large ears and eyes. (*See note below.*) **b** any similar small mammal, such as a shrew or vole. **2** a timid or feeble person. **3** (*pl.* also **mouses**) *Computing* a small hand-held device which controls the cursor on a VDU screen. **4** *sl.* a black eye. ● *v.intr.* (also /maʊz/) **1** (esp. of a cat, owl, etc.) hunt for or catch mice. **2** (foll. by *about*) search industriously; prowl about as if searching. □ **mouse-coloured 1** dark grey with a yellow tinge. **2** nondescript light brown. **mouse deer** a chevrotain. **mouse hare** a pika. □ **mouselike** *adj. & adv.* **mouser** *n.* [OE *mūs*, pl. *mȳs* f. Gmc]

▪ There are several hundred species of mice in the large family Muridae, being generally distinguished from rats by their smaller size, and from voles by their pointed snouts and prominent ears. A number of small rodents of other families, such as dormice, are not regarded as true mice. The house mouse, *Mus musculus*, is the world's most cosmopolitan mammal, and has been used extensively as an experimental animal.

mousetrap /ˈmaʊstræp/ *n.* **1** a sprung trap with bait for catching and usu. killing mice. **2** (often *attrib.*) cheese of poor quality.

moussaka /mʊˈsɑːkə/ *n.* (also **mousaka**) a Greek dish of minced meat, aubergine, etc. with a cheese sauce. [mod.Gk or Turk.]

mousse /muːs/ *n.* **1 a** a dessert of whipped cream, eggs, etc., usu. flavoured with fruit or chocolate. **b** a meat or fish purée made with whipped cream etc. **2** a preparation applied to the hair enabling it to be styled more easily. **3** a mixture of oil and sea water which forms a froth on the surface of the water after an oil-spill. [F, = moss, froth]

mousseline /muːˈsliːn/ *n.* **1** a muslin-like fabric of silk etc. **2 a** a soft light mousse. **b** hollandaise sauce made frothy with whipped cream or egg white. [F: see MUSLIN]

Moussorgsky see MUSSORGSKY.

moustache /məˈstɑːʃ/ *n.* (US **mustache** /ˈmʌstæʃ/) **1** hair left to grow on a man's upper lip. **2** a similar growth round the mouth of some animals. □ **moustache cup** a cup with a partial cover to protect the moustache when drinking. □ **moustached** *adj.* [F f. It. *mostaccio* f. Gk *mustax -akos*]

Mousterian /muːˈstɪərɪən/ *adj. & n. Archaeol.* of, relating to, or denoting a middle palaeolithic culture following the Acheulean and preceding the Aurignacian, typified by flints worked on one side only, and dated mainly to c.80,000–35,000 BP. The period is associated with that of Neanderthal man. [F *moust(i)érien* f. *Le Moustier*, a cave and type-site in Dordogne, France]

mousy /ˈmaʊsɪ/ *adj.* (also **mousey**) (**mousier**, **mousiest**) **1** of or like a mouse. **2** (of a person) shy or timid; ineffectual. **3** = *mouse-coloured*. □ **mousily** *adv.* **mousiness** *n.*

mouth *n. & v.* ● *n.* /maʊθ/ (*pl.* **mouths** /maʊðz/) **1 a** an external opening in the head, through which most animals admit food and emit communicative sounds. **b** (in humans and some animals) the cavity behind it containing the means of biting and chewing and the vocal organs. **2 a** the opening of a container such as a bag or sack. **b** the opening of a cave, volcano, etc. **c** the open end of a woodwind or brass instrument. **d** the muzzle of a gun. **3** the place where a river enters the sea. **4** *colloq.* **a** talkativeness. **b** impudent talk; cheek. **5** an individual regarded as needing sustenance (*an extra mouth to feed*). **6 a** horse's readiness to feel and obey the pressure of the bit. **7** an expression of displeasure; a grimace ● *v.* /maʊð/ **1** *tr. & intr.* utter or speak solemnly or with affectations; rant, declaim (*mouthing platitudes*). **2** *tr.* utter very distinctly. **3** *intr.* **a** move the lips silently. **b** grimace. **4** *tr.* say (words) with movement of the mouth but no sound. **5** *tr.* take (food) in the mouth. **6** *tr.* touch with the mouth. **7** *tr.* train the mouth of (a horse). □ **give mouth** (of a dog) bark, bay. **keep one's mouth shut** *colloq.* not reveal a secret. **mouth-organ** = HARMONICA. **mouth-to-mouth** (of resuscitation) in which a person breathes into a subject's lungs through the mouth. **mouth-watering 1** (of food etc.) having a delicious smell or appearance. **2** tempting, alluring. **put words into a person's mouth** represent a person as having said something in a particular way. **take the words out of a person's mouth** say what another was about to say. □ **mouthed** /maʊðd/ *adj.* (also in *comb.*). **mouther** /ˈmaʊðə(r)/ *n.* **mouthless** /ˈmaʊθlɪs/ *adj.* [OE *mūth* f. Gmc]

mouthbrooder /ˈmaʊθˌbruːdə(r)/ *n.* a fish which protects its eggs (and sometimes its young) by carrying them in its mouth.

mouthful /ˈmaʊθfʊl/ *n.* (*pl.* **-fuls**) **1** a quantity, esp. of food, that fills the mouth. **2** a small quantity. **3** a long or complicated word or phrase. **4** *US colloq.* something important said.

mouthpart /ˈmaʊθpɑːt/ *n.* *Zool.* one of the (usu. paired) organs surrounding the mouth of an insect or other arthropod, adapted for feeding.

mouthpiece /ˈmaʊθpiːs/ *n.* **1 a** the part of a musical instrument placed between or against the lips. **b** the part of a telephone for speaking into. **c** the part of a tobacco-pipe placed between the lips. **2 a** a person who speaks for another or others. **b** *colloq.* a lawyer. **3** a part attached as an outlet.

mouthwash /ˈmaʊθwɒʃ/ *n.* **1** a liquid antiseptic etc. for rinsing the mouth or gargling. **2** *colloq.* nonsense.

mouthy /ˈmaʊðɪ/ *adj.* (**mouthier**, **mouthiest**) **1** ranting, railing. **2** bombastic.

movable /ˈmuːvəb(ə)l/ *adj. & n.* (also **moveable**) ● *adj.* **1** that can be moved. **2** *Law* (of property) of the nature of a chattel, as distinct from land or buildings. **3** (of a feast or festival) variable in date from year to year. ● *n.* **1** an article of furniture that may be removed from a house, as distinct from a fixture. **2** (in *pl.*) personal property. □ **movable-doh** *Mus.* applied to a system of sight-singing in which doh is the keynote of any major scale (cf. *fixed-doh* (see FIX)). □ **movably** *adv.* **movableness** *n.* **movability** /ˌmuːvəˈbɪlɪtɪ/ *n.* [ME f. OF (as MOVE)]

move /muːv/ *v. & n.* ● *v.* **1** *intr. & tr.* change one's position or posture, or cause to do this. **2** *tr. & intr.* put or keep in motion; rouse, stir. **3 a** *intr.* make a move in a board-game. **b** *tr.* change the position of (a piece) in a board-game. **4** *intr.* (often foll. by *about, away*, etc.) go or pass from place to place. **5** *intr.* take action, esp. promptly (*moved to reduce unemployment*). **6** *intr.* make progress (*the project is moving fast*). **7 a** *intr.* change one's place of residence. **b** *intr.* (of a business etc.) change to new premises. **c** *tr.* change (one's place of residence or to new premises). **8** *intr.* (foll. by *in*) live or be socially active in (a specified place or group etc.) (*moves in the best circles*). **9** *tr.* affect (a person) with (usu. tender or sympathetic) emotion. **10 a** *tr.* (foll. by *to*) stimulate (laughter, anger, etc., in a person). **b** (foll. by *to*) provoke (a person to laughter etc.). **11** *tr.* (foll. by *to*, or *to* + *infin.*) prompt or incline (a person to a feeling or action). **12** *tr. & intr.* (cause to) change in attitude or opinion (*nothing can move me on this issue*). **13 a** *tr.* cause (the bowels) to be evacuated. **b** *intr.* (of the bowels) be evacuated. **14** *tr.* (often foll. by *that* + clause) propose in a meeting, deliberative assembly, etc. **15** *intr.* (foll. by *for*) make a formal request or application. **16 a** *intr.* (of merchandise) be sold. **b** *tr.* sell. ● *n.* **1** the act or an instance of moving. **2** a change of house, business premises, etc. **3** a step taken to secure some action or effect; an initiative. **4 a** the changing of the position of a piece in a board-game. **b** a player's turn to do this. □ **get a move on** *colloq.* **1** hurry up. **2** make a start. **make a move** take action. **move along** (or **on**) change to a new position, esp. to avoid crowding, getting in the way, etc. **move heaven and earth** see HEAVEN. **move in 1** take possession of a new house. **2** get into a position of influence, interference, etc. **3** (often foll. by *on*) get into a position of readiness or proximity (for an offensive action etc.). **move in with** start to share accommodation with (an existing

resident). **move mountains** see MOUNTAIN. **move out 1** leave one's home; change one's place of residence. **2** leave a position, job, etc. **move over** (or **up**) adjust one's position to make room for another. **on the move 1** progressing. **2** moving about. [ME f. AF *mover*, OF *moveir* f. L *movere mot-*]

moveable var. of MOVABLE.

movement /ˈmuːvmənt/ *n.* **1** the act or an instance of moving or being moved. **2 a** the moving parts of a mechanism (esp. a clock or watch). **b** a particular group of these. **3 a** a body of persons with a common object (*the peace movement*). **b** a campaign undertaken by such a body. **4** (usu. in *pl.*) a person's activities and whereabouts, esp. at a particular time. **5** *Mus.* a principal division of a longer musical work, self-sufficient in terms of key, tempo, structure, etc. **6** the progressive development of a poem, story, etc. **7** motion of the bowels. **8 a** an activity in a market for some commodity. **b** a rise or fall in price. **9** a mental impulse. **10** a development of position by a military force or unit. **11** a prevailing tendency in the course of events or conditions; trend. [ME f. OF f. med.L *movimentum* (as MOVE)]

mover /ˈmuːvə(r)/ *n.* **1** a person or thing that moves. **2** a person who moves a proposition. **3** *N. Amer.* a remover of furniture. **4** the author of a fruitful idea.

movie /ˈmuːvɪ/ *n.* *colloq.* **1** esp. *N. Amer.* a motion-picture film. **2** (in full **movie-house**) *US* a cinema.

moving /ˈmuːvɪŋ/ *adj.* **1** that moves or causes to move. **2** affecting with emotion. □ **moving-coil** (of an electrical device) containing a coil of wire suspended in a magnetic field, so that either the coil moves when a current is passed through it or else a current is generated when the coil is caused to move. **moving pavement** *Brit.* a structure like a conveyor belt for pedestrians. **moving picture** a continuous picture of events obtained by projecting a sequence of photographs taken at very short intervals. **moving staircase** *Brit.* an escalator. □ **movingly** *adv.* (in sense 2).

mow[1] /maʊ/ *v.tr.* (*past part.* **mowed** or **mown** /maʊn/) **1** cut down (grass, hay, etc.) with a scythe or machine. **2** cut down the produce of (a field) or the grass etc. of (a lawn) by mowing. □ **mow down** kill or destroy randomly or in great numbers. □ **mower** *n.* [OE *māwan* f. Gmc, rel. to MEAD[2]]

mow[2] /maʊ/ *n.* *N. Amer.* or *dial.* **1** a stack of hay, corn, etc. **2** a place in a barn where hay etc. is heaped. [OE *mūga*]

moxa /ˈmɒksə/ *n.* a downy substance from the dried leaves of an Asian plant, *Crossostephium artemisioides*, burnt near the skin in oriental medicine as a counterirritant. [Jap. *mogusa* f. *moe kusa* burning herb]

moxie /ˈmɒksɪ/ *n.* *US sl.* energy, courage, daring. [trade name of a drink]

Mozambique /ˌməʊzæmˈbiːk/ a country on the east coast of southern Africa; pop. (est. 1991) 16,142,000; languages, Portuguese (official), Bantu languages; capital, Maputo. First visited by Vasco da Gama, Mozambique was subdued and colonized by the Portuguese in the early 16th century. It was a centre of the slave trade in the 17th and 18th centuries. Mozambique became an independent republic in 1975, after a ten-year armed struggle by the Frelimo liberation movement. Support for guerrilla campaigns being waged in South Africa and Rhodesia (Zimbabwe) led to repeated incursions by troops of those countries, and civil war between the Frelimo government and the Renamo opposition continued for many years; a cease-fire was arranged in 1992, overseen by the UN. □ **Mozambican** *adj. & n.*

Mozambique Channel an arm of the Indian Ocean separating the eastern coast of mainland Africa from the island of Madagascar.

Mozart /ˈməʊtsɑːt/, (Johann Chrysostom) Wolfgang Amadeus (1756–91), Austrian composer. A child prodigy as a harpsichordist, pianist, and composer, he was taken on tours of western Europe by his father Leopold (1719–87). While in Vienna, he collaborated with the librettist Da Ponte in the composition of his three comic operas, *The Marriage of Figaro* (1786), *Don Giovanni* (1787), and *Così fan tutte* (1790). His use of music to aid characterization in these works marked an important advance in the development of opera. Early influenced by Haydn, Mozart came to epitomize classical music in its purity of form and melody. A prolific composer, he wrote more than forty symphonies, nearly thirty piano concertos, over twenty string quartets, and a vast quantity of other instrumental and orchestral music.

mozz /mɒz/ *n.* *Austral. colloq.* a jinx, a malign influence (esp. *put the mozz on*). [abbr. of MOZZLE]

mozzarella /ˌmɒtsəˈrelə/ *n.* an Italian curd cheese orig. of buffalo milk. [It.]

mozzle /'mɒz(ə)l/ n. Austral. colloq. luck, fortune. [Heb. mazzā]

MP abbr. **1** member of parliament. **2 a** military police. **b** military policeman.

mp abbr. mezzo piano.

m.p. abbr. melting-point.

m.p.g. abbr. miles per gallon.

m.p.h. abbr. miles per hour.

M.Phil. abbr. Master of Philosophy.

MPLA abbr. the Popular Movement for the Liberation of Angola, a Marxist organization founded in the 1950s that emerged as the ruling party in Angola after independence from Portugal in 1975. Once in power the MPLA fought UNITA and other rival groups for many years, supported at first by the USSR and Cuba. [Port. abbr., f. *Movimento Popular de Libertação de Angola*]

MR abbr. Master of the Rolls.

Mr /'mɪstə(r)/ n. (pl. **Messrs**) **1** the title of a man without a higher title (*Mr Jones*). **2** a title prefixed to a designation of office etc. (*Mr President*; *Mr Speaker*). □ **Mr Big** US sl. the head of an organization; any important person. **Mr Right** joc. a woman's destined husband. [abbr. of MISTER]

MRA abbr. Moral Re-Armament.

MRBM abbr. medium-range ballistic missile.

MRC abbr. (in the UK) Medical Research Council.

MRCA abbr. multi-role combat aircraft.

MRI abbr. magnetic resonance imaging.

mRNA abbr. Biol. messenger RNA.

MRPharmS abbr. Member of the Royal Pharmaceutical Society.

Mrs /'mɪsɪz/ n. (pl. same or **Mesdames**) the title of a married woman without a higher title (*Mrs Jones*). [abbr. of MISTRESS: cf. MISSIS]

MS abbr. **1** manuscript. **2** Master of Science. **3** Master of Surgery. **4** US Mississippi (in official postal use). **5** US motor ship. **6** multiple sclerosis.

Ms /mɪz, məz/ n. the title of a woman without a higher title, used regardless of marital status. [combination of MRS, MISS[2]]

MSC abbr. (in the UK) Manpower Services Commission.

M.Sc. abbr. Master of Science.

MS-DOS /ˌemes'dɒs/ abbr. Computing propr. Microsoft disk operating system.

MSF abbr. (in the UK) Manufacturing, Science, and Finance (Union).

Msgr. abbr. US **1** Monseigneur. **2** Monsignor.

MSS /em'esɪz/ abbr. manuscripts.

MST abbr. Mountain Standard Time.

MT abbr. **1** mechanical transport. **2** US Montana (in official postal use).

Mt. abbr. Mount.

MTB abbr. motor torpedo-boat.

M.Tech. abbr. Master of Technology.

mu /mju:/ n. **1** the twelfth letter of the Greek alphabet (*M, μ*). **2** (*μ*, as a symbol) = MICRO- 2. □ **mu-meson** = MUON. [Gk]

Mubarak /mu:'bɑːræk/, (Muhammad) Hosni (Said) (b.1928), Egyptian statesman, President since 1981. Appointed head of the Egyptian air force in 1972, Mubarak became Vice-President in 1975 and succeeded President Sadat following the latter's assassination. Although he did much to establish closer links between Egypt and other Arab nations, including distancing himself from Israel when it invaded Lebanon in 1982, he risked division by aligning Egypt against Saddam Hussein in the Gulf War of 1991. After the resurgence of militant Islamic fundamentalism in Egypt in 1992, Mubarak's National Democratic Party government adopted harsh measures to suppress activists.

much /mʌtʃ/ adj., n., & adv. ● adj. **1** existing or occurring in a great quantity (*much trouble; not much rain; too much noise*). **2** (prec. by *as*, *how*, *that*, etc.) with relative rather than distinctive sense (*I don't know how much money you want*). ● n. **1** a great quantity (*much of that is true*). **2** (prec. by *as*, *how*, *that*, etc.) with relative rather than distinctive sense (*we do not need that much*). **3** (usu. in *neg.*) a noteworthy or outstanding example (*not much to look at; not much of a party*). ● adv. **1 a** in a great degree (*much to my surprise; is much the same*). **b** (qualifying a verb or past participle) greatly (*they much regret the mistake; I was much annoyed*). ¶ *Much* implies a strong verbal element in the participle, whereas *very* implies a strong adjectival element: compare the second example above with *I was very annoyed*. **c** qualifying a comparative or superlative adjective (*much better; much the most likely*). **2** for a large part of one's time (*is much away from home*). □ **as much** the extent or quantity just specified; the idea just mentioned (*I thought as much; as much as that?*).

a bit much colloq. somewhat excessive or immoderate. **make much of** see MAKE. **much as** even though (*cannot come, much as I would like to*). **much less** see LESS. **much obliged** see OBLIGE. **not much** colloq. iron. very much. **not much in it** (or **to it**) see *nothing in it*. **too much** colloq. an intolerable situation etc. (*that really is too much*). **too much for 1** more than a match for. **2** beyond what is endurable by. □ **muchly** adv. joc. [ME f. *muchel* MICKLE: for loss of *el* cf. BAD, WENCH]

Mucha /'muːkə/, Alphonse (born Alfons Maria) (1860–1939), Czech painter and designer. Based in Paris from 1889, he was a leading figure in the art nouveau movement. He is noted for his flowing poster designs, often featuring the actress Sarah Bernhardt, as in *Gismonda* (1894); with the success of this poster, Mucha was given a six-year commission to design further posters, sets, costumes, and jewellery for the actress.

Muchinga Mountains /muː'tʃɪŋɡə/ a range of mountains in eastern Zambia.

muchness /'mʌtʃnɪs/ n. greatness in quantity or degree. □ **much of a muchness** very nearly the same or alike.

mucilage /'mjuːsɪlɪdʒ/ n. **1** a viscous or gelatinous solution obtained from plant roots, seeds, etc., used in medicines and adhesives. **2** N. Amer. a solution of gum. **3** a viscous secretion or bodily fluid, e.g. mucus. □ **mucilaginous** /ˌmjuːsɪ'lædʒɪnəs/ adj. [ME f. F f. LL *mucilago -ginis* musty juice (MUCUS)]

muck /mʌk/ n. & v. ● n. **1** farmyard manure. **2** colloq. dirt or filth; anything disgusting. **3** colloq. a mess. ● v.tr. **1** (usu. foll. by *up*) colloq. bungle (a job), spoil, ruin. **2** (foll. by *out*) remove muck from. **3** make dirty or untidy. **4** manure with muck. □ **make a muck of** colloq. bungle. **muck about** (or **around**) Brit. colloq. **1** potter or fool about. **2** (foll. by *with*) fool or interfere with. **muck in** Brit. (often foll. by *with*) share tasks etc. equally. **muck sweat** Brit. colloq. a profuse sweat. [ME *muk* prob. f. Scand.: cf. ON *myki* dung, rel. to MEEK]

mucker /'mʌkə(r)/ n. sl. **1** Brit. a friend or companion. **2** US a rough or coarse person. **3** Brit. a heavy fall. [prob. f. *muck in*: see MUCK]

muckle var. of MICKLE.

muckrake /'mʌkreɪk/ v.intr. search out and reveal scandal, esp. among famous people. □ **muckraker** n. **muckraking** n.

mucky /'mʌkɪ/ adj. (**muckier**, **muckiest**) **1** covered with muck. **2** dirty. □ **muckiness** n.

muco- /'mjuːkəʊ/ comb. form Biochem. mucus, mucous.

mucopolysaccharide /ˌmjuːkəʊˌpɒlɪ'sækəˌraɪd/ n. Biochem. any of a group of polysaccharides whose molecules contain sugar residues and are often found as components of connective tissue.

mucosa /mjuː'kəʊsə/ n. (pl. **mucosae** /-siː/) Anat. a mucous membrane. [mod.L, fem. of *mucosus*: see MUCOUS]

mucous /'mjuːkəs/ adj. of or covered with mucus. □ **mucous membrane** Anat. a mucus-secreting epithelial tissue lining many body cavities and tubular organs. □ **mucosity** /mjuː'kɒsɪtɪ/ n. [L *mucosus* (as MUCUS)]

mucro /'mjuːkrəʊ/ n. (pl. **mucrones** /mjuː'krəʊniːz/) Bot. & Zool. a sharp-pointed part or organ. □ **mucronate** /-krənət/ adj. [L *mucro -onis* sharp point]

mucus /'mjuːkəs/ n. **1** a slimy substance, usu. not miscible with water, secreted by a mucous membrane or gland. **2** a gummy substance found in plants. [L]

mud /mʌd/ n. **1** wet soft earthy matter. **2** hard ground from the drying of an area of this. **3** what is worthless or polluting. □ **as clear as mud** colloq. not at all clear. **drag through the mud** publish unpleasant information or allegations about. **fling** (or **sling** or **throw**) **mud** speak disparagingly or slanderously. **here's mud in your eye!** colloq. a drinking-toast. **mud-bath 1** a bath in the mud of mineral springs, esp. to relieve rheumatism etc. **2** a muddy scene or occasion. **mud-brick** a brick made from baked mud. **mud-flat** a stretch of muddy land left uncovered at low tide. **mud pack** a cosmetic paste applied thickly to the face. **mud pie** mud made into a pie shape by a child. **mud puppy** US a large neotenous aquatic salamander, esp. *Necturus maculosus* of the eastern US. **mud-slinger** colloq. one given to making abusive or disparaging remarks. **mud-slinging** colloq. abuse, disparagement. **mud volcano** a volcano discharging mud. **one's name is mud** one is unpopular or in disgrace. [ME *mode*, *mudde*, prob. f. MLG *mudde*, MHG *mot* bog]

muddle /'mʌd(ə)l/ v. & n. ● v. **1** tr. (often foll. by *up*, *together*) bring into disorder. **2** tr. bewilder, confuse. **3** tr. mismanage (an affair). **4** tr. US crush and mix (the ingredients for a drink). **5** intr. (often foll. by *with*) busy oneself in a confused and ineffective way. ● n. **1** disorder. **2** a muddled condition. □ **make a muddle of 1** bring into disorder. **2** bungle. **muddle along** (or **on**) progress in a haphazard way. **muddle-headed** stupid, confused. **muddle-headedness** stupidity; a confused state. **muddle through** succeed by perseverance rather than skill or efficiency. **muddle up** confuse (two or more things). □ **muddler** n. **muddlingly** adv. [perh. f. MDu. *moddelen*, frequent. of *modden* dabble in mud (as MUD)]

muddy /ˈmʌdɪ/ adj. & v. ● adj. (**muddier, muddiest**) **1** like mud. **2** covered in or full of mud. **3** (of liquid) turbid. **4** mentally confused. **5** (of light) dull. **6** (of colour) impure. ● v.tr. (**-ies, -ied**) make muddy. □ **muddily** adv. **muddiness** n.

Mudéjar /muːˈdeɪhɑː(r)/ n. & adj. ● n. (pl. **Mudéjares** /-hɑːˌres/) a subject Muslim during the Christian reconquest of the Iberian peninsula from the Moors (11th–15th centuries) who was allowed to retain Islamic laws and religion in return for loyalty to a Christian monarch. After 1492 such people were treated with less toleration, dubbed Moriscos (see MORISCO), and forced to accept the Christian faith or leave the country. ● adj. of or relating to a style of architecture and decorative art of the 12th–15th centuries produced by Mudéjares. The style combines Islamic and Gothic elements; examples can be seen in the churches and palaces of the Spanish cities Toledo, Cordoba, Seville, and Valencia. [Sp. f. Arab. *mudajjan* permitted to remain]

mudfish /ˈmʌdfɪʃ/ n. any fish that burrows in mud, esp. the bowfin.

mudflap /ˈmʌdflæp/ n. a flap hanging behind the wheel of a vehicle, to catch mud and stones etc. thrown up from the road.

mudguard /ˈmʌdɡɑːd/ n. a curved strip or cover over a wheel of a bicycle or motorcycle to reduce the amount of mud etc. thrown up from the road.

mudlark /ˈmʌdlɑːk/ n. Brit. hist. **1** a person who scavenges in river mud for objects of value. **2** a street urchin.

mudskipper /ˈmʌdˌskɪpə(r)/ n. a small goby of the family Periophthalmidae, able to scramble over mud etc., found on coasts of the Old World tropics.

mudstone /ˈmʌdstəʊn/ n. a dark clay rock.

muesli /ˈmuːzlɪ, ˈmjuːz-/ n. (pl. **-is**) a breakfast food of crushed cereals, dried fruits, nuts, etc., eaten with milk. [Swiss G]

muezzin /muːˈezɪn/ n. a Muslim crier who proclaims the hours of prayer usu. from a minaret. [Arab. *muʾaddin* part. of *ʾaddana* proclaim]

muff[1] /mʌf/ n. a fur or other covering, usu. in the form of a tube with an opening at each end for the hands to be inserted for warmth. [Du. *mof*, MDu. *moffel, muffel* f. med.L *muff(u)la*, of unkn. orig.]

muff[2] /mʌf/ v. & n. ● v.tr. **1** bungle; deal clumsily with. **2** fail to catch or receive (a ball etc.). **3** blunder in (a theatrical part etc.). ● n. **1** esp. Brit. a person who is awkward or stupid, orig. in some athletic sport. **2** a failure, esp. to catch a ball at cricket etc. [19th c.: orig. unkn.]

muffin /ˈmʌfɪn/ n. **1** Brit. a circular flat cake made from yeast dough, eaten toasted and buttered. **2** N. Amer. a small spongy cake made with eggs and baking powder. □ **muffin-man** Brit. hist. a seller of muffins in the street. [18th c.: orig. unkn.]

muffle[1] /ˈmʌf(ə)l/ v. & n. ● v.tr. **1** (often foll. by *up*) wrap or cover for warmth. **2** cover or wrap up (a source of sound) to reduce its loudness. **3** (usu. as **muffled** adj.) stifle (an utterance, e.g. a curse). **4** prevent from speaking. ● n. **1** a receptacle in a furnace where substances may be heated without contact with combustion products. **2** a similar chamber in a kiln for baking painted pottery. [ME: (n.) f. OF *moufle* thick glove; (v.) perh. f. OF *enmoufler* f. *moufle*]

muffle[2] /ˈmʌf(ə)l/ n. the thick part of the upper lip and nose of ruminants and rodents. [F *mufle*, of unkn. orig.]

muffler /ˈmʌflə(r)/ n. **1** a wrap or scarf worn for warmth. **2** any of various devices used to deaden sound in musical instruments. **3** N. Amer. the silencer of a motor vehicle.

mufti[1] /ˈmʌftɪ/ n. a Muslim legal expert empowered to give rulings on religious matters. [Arab. *muftī*, part. of *ʾaftā* decide a point of law]

mufti[2] /ˈmʌftɪ/ n. plain clothes worn by a person who also wears (esp. military) uniform (*in mufti*). [19th c.: perh. f. MUFTI[1]]

mug[1] /mʌɡ/ n. & v. ● n. **1 a** a drinking-vessel, usu. cylindrical and with a handle and used without a saucer. **b** its contents. **2** sl. the face or mouth of a person. **3** Brit. sl. **a** a simpleton. **b** a gullible person. **4** US sl. a hoodlum or thug. ● v. (**mugged, mugging**) **1** tr. rob (a person) with violence, esp. in a public place. **2** intr. sl. make faces, esp. before an audience, a camera, etc. □ **a mug's game** Brit. colloq. a foolish or unprofitable activity. **mug shot** sl. a photograph of a face, esp. for official purposes. □ **mugger** n. (esp. in sense 1 of v.). **mugful** n. (pl.

-**fuls**). **mugging** n. (in sense 1 of v.). [prob. f. Scand.: sense 2 of n. prob. f. the representation of faces on mugs, and sense 3 prob. from this]

mug[2] /mʌɡ/ v.tr. (**mugged, mugging**) Brit. (usu. foll. by *up* or *up on*) sl. learn (a subject) by concentrated study. [19th c.: orig. unkn.]

Mugabe /muˈɡɑːbɪ/, Robert (Gabriel) (b.1924), Zimbabwean statesman, Prime Minister 1980–7 and President since 1987. In 1963 he co-founded the Zimbabwe African National Union (ZANU) and in 1975 became its leader; the following year he formed the Patriotic Front with the leader of the Zimbabwe African People's Union (ZAPU), Joshua Nkomo. Mugabe was declared Prime Minister in 1980 after ZANU won a landslide victory in the country's first post-independence elections. In 1982 he ousted Nkomo from his Cabinet; ZANU and ZAPU agreed to merge in 1987 and Mugabe became President.

mugger[1] see MUG[1].

mugger[2] /ˈmʌɡə(r)/ n. a broad-nosed Indian crocodile, *Crocodylus palustris*, venerated by many Hindus. [Hindi *magar*]

muggins /ˈmʌɡɪnz/ n. (pl. same or **mugginses**) **1** colloq. **a** a simpleton. **b** a person who is easily outwitted (often with allusion to oneself: *so muggins had to pay*). **2** a card-game like snap. [perh. the surname *Muggins*, with allusion to MUG[1]]

muggy /ˈmʌɡɪ/ adj. (**muggier, muggiest**) (of the weather, a day, etc.) oppressively damp and warm; humid. □ **mugginess** n. [dial. *mug* mist, drizzle f. ON *mugga*]

Mughal var. of MOGUL.

mugwort /ˈmʌɡwɜːt/ n. a composite plant of the genus *Artemisia*; esp. *A. vulgaris*, with silver-grey aromatic foliage. [OE *mucgwyrt* (as MIDGE, WORT)]

mugwump /ˈmʌɡwʌmp/ n. N. Amer. **1** a great man; a boss. **2** a person who holds aloof, esp. from party politics. [Algonquian *mugquomp* great chief]

Muhammad[1] /məˈhæmɪd/ (also **Mohammed**) (c.570–632), Arab prophet and founder of Islam. He was born in Mecca, where in c.610 he received the first of a series of revelations (see KORAN) which became the doctrinal and legislative basis of Islam; his sayings (the Hadith) and the accounts of his daily practice (the Sunna) constitute the other major sources of guidance for most Muslims. In the face of opposition to his preaching he and his small group of supporters were forced to flee to Medina in 622 (see HEGIRA). After consolidation of the community there, Muhammad led his followers into a series of battles which resulted in the capitulation of Mecca in 630. He died two years later, having successfully united tribal factions of the Hejaz region into a force which would expand the frontiers of Islam. He was buried in Medina. (See also CALIPH, ISLAM.)

Muhammad[2] /məˈhæmɪd/, Mahathir (b.1925), Malaysian statesman, Prime Minister since 1981.

Muhammad Ahmad /ˈɑːmæd/ see MAHDI.

Muhammad Ali[1] /məˈhæmɪd ˈɑːlɪ, ɑːˈliː/ (1769–1849), Ottoman viceroy and pasha of Egypt 1805–49, possibly of Albanian descent. As a commander in the Ottoman army, he had overthrown the Mamelukes by 1811. Although technically the viceroy of the Ottoman sultan, he was effectively an independent ruler and modernized Egypt's infrastructure, making it the leading power in the eastern Mediterranean. In 1841 he and his family were given the right to become hereditary rulers of Egypt, and the dynasty survived until 1952.

Muhammad Ali[2] /məˌhæmɪd ˈɑːlɪ, ɑːˈliː/ (born Cassius Marcellus Clay) (b.1942), American boxer. He first won the world heavyweight title in 1964 and regained it in 1974 and 1978, becoming the only boxer to be world champion three times. After his retirement in 1981 it was confirmed that he was suffering from Parkinson's disease.

Muhammadan /məˈhæmɪd(ə)n/ n. & adj. (also **Mohammedan**) = MUSLIM. ¶ A term not used or favoured by Muslims, and often regarded as offens. □ **Muhammadanism** n. [MUHAMMAD[1]]

Mühlhausen see MULHOUSE.

Muir[1] /mjʊə(r)/, Edwin (1887–1959), Scottish poet and translator. His collections of poems include *The Labyrinth* (1949). He is also remembered for his translations of Kafka, done in collaboration with his wife, the novelist Willa Anderson (1890–1970); these appeared in the 1930s and established Kafka's reputation in Britain.

Muir[2] /mjʊə(r)/, John (1838–1914), Scottish-born American naturalist, a pioneer of environmental conservation. Devoting himself to nature after being injured in an industrial accident, Muir campaigned vigorously for the protection of unspoilt wilderness areas and was largely responsible for the establishment of Yosemite and Sequoia

National Parks in California (1890). He wrote several books about the American wilderness, such as *The Mountains of California* (1894).

mujahidin /ˌmʊdʒəˈhiːdiːn/ *n.pl.* (also **mujahedin**, **mujahideen**) guerrilla fighters in Islamic countries, esp. supporting Islamic fundamentalism. [Pers. & Arab. *mujāhidīn* pl. of *mujāhid* one who fights a JIHAD]

Mujibur Rahman /mʊˌdʒiːbʊə rəˈmɑːn/ (known as Sheikh Mujib) (1920–75), Bangladeshi statesman, Prime Minister 1972–5 and President 1975. In 1949 he co-founded the Awami (People's) League, which advocated autonomy for East Pakistan. He led the party to victory in the 1970 elections, but was imprisoned in 1971 when civil war broke out. Released in 1972, he became the first Prime Minister of independent Bangladesh. After his failure to establish parliamentary democracy, he assumed dictatorial powers in 1975. He and his family were assassinated in a military coup.

Mukalla /mʊˈkælə/ a port on the south coast of Yemen, in the Gulf of Aden; pop. (1987) 154,360.

Mukden /ˈmʊkdən/ a former name for SHENYANG.

mulatto /mjuːˈlætəʊ/ *n. & adj.* ● *n.* (pl. **-os** or **-oes**) a person of mixed white and black parentage. ● *adj.* of the colour of mulattos; tawny. [Sp. *mulato* young mule, *mulatto*, irreg. f. *mulo* MULE[1]]

mulberry /ˈmʌlbəri/ *n.* (pl. **-ies**) **1** a deciduous ornamental Asian tree of the genus *Morus*; esp. the *white mulberry* (*M. alba*), whose leaves are used to feed silkworms, and the *black mulberry* (*M. nigra*), grown for its fruit. **2** its white or dark red berry. **3** a dark red or purple colour. [OE *mōrberie*, f. L *morum* + BERRY, with dissimilation of r-r in l-r in ME]

mulch /mʌltʃ, mʌlʃ/ *n. & v.* ● *n.* a mixture of wet straw, leaves, etc., spread around or over a plant to enrich or insulate the soil. ● *v.tr.* treat with mulch. [prob. use as noun of *mulsh* soft: cf. dial. *melsh* mild f. OE *melsc*]

mulct /mʌlkt/ *v. & n.* ● *v.tr.* **1** extract money from by fine or taxation. **2 a** (often foll. by *of*) deprive by fraudulent means; swindle. **b** obtain by swindling. ● *n.* a fine. [earlier *mult*(*e*) f. L *multa*, *mulcta*: (v.) through F *mulcter* & L *mulctare*]

Muldoon /mʌlˈduːn/, Sir Robert (David) (1921–92), New Zealand statesman, Prime Minister 1975–84. He became a National Party MP in 1960, serving as deputy Prime Minister for a brief period in 1972 and as leader of the opposition from 1973 to 1974 before becoming Premier the following year. He was chairman of the board of governors for the IMF and World Bank (1979–80) and chairman of the ministerial council for the OECD (1982). His term of office was marked by domestic measures to tackle low economic growth and high inflation.

mule[1] /mjuːl/ *n.* **1** the offspring (usu. sterile) of a male donkey and a female horse, or (in general use) of a female donkey and a male horse (cf. HINNY[1]), used as a beast of burden. **2** a stupid or obstinate person. **3** (often *attrib.*) a hybrid and usu. sterile plant or animal (*mule canary*). **4** = SPINNING MULE. □ **mule deer** a long-eared black-tailed deer, *Odocoileus hemionus*, of western North America. [ME f. OF *mul*(*e*) f. L *mulus mula*]

mule[2] /mjuːl/ *n.* a light shoe or slipper without a back. [F]

mule[3] var. of MEWL.

muleteer /ˌmjuːlɪˈtɪə(r)/ *n.* a mule-driver. [F *muletier* f. *mulet* dimin. of OF *mul* MULE[1]]

mulga /ˈmʌlɡə/ *n. Austral.* **1** a small spreading leguminous tree, *Acacia aneura*. **2** the wood of this tree. **3** scrub or bush. **4** *colloq.* the outback. [Aboriginal]

Mulhacén /ˌmuːləˈsen/ a mountain in southern Spain, south-east of Granada, in the Sierra Nevada range. Rising to 3,482 m (11,424 ft), it is the highest mountain in the country.

Mülheim /ˈmuːlhaɪm/ (in full **Mülheim an der Ruhr** /ˌæn deə ˈrʊə(r)/) an industrial city in western Germany, in North Rhine-Westphalia south-west of Essen; pop. (1991) 177,040.

Mulhouse /mʊˈluːz/ (German **Mühlhausen** /ˈmyːlˌhaʊz(ə)n/) an industrial city in NE France, in Alsace; pop. (1990) 109,900. Founded in 803, it was a free imperial city until it joined the French Republic in 1798. In 1871, after the Franco-Prussian War, the city became part of the German Empire until it was reunited with France in 1918.

muliebrity /ˌmjuːlɪˈebrɪtɪ/ *n. literary* **1** womanhood. **2** the normal characteristics of a woman. **3** softness, effeminacy. [LL *muliebritas* f. L *mulier* woman]

mulish /ˈmjuːlɪʃ/ *adj.* **1** like a mule. **2** stubborn. □ **mulishly** *adv.* **mulishness** *n.*

Mull /mʌl/ an island of the Inner Hebrides, off the west coast of Scotland; chief town, Tobermory. It is separated from the mainland by the Sound of Mull.

mull[1] /mʌl/ *v.tr. & intr.* (often foll. by *over*) ponder or consider. [perh. f. *mull* grind to powder, ME *mul* dust f. MDu.]

mull[2] /mʌl/ *v.tr.* (usu. as **mulled** *adj.*) warm (wine or beer) with added sugar, spices, etc. [17th c.: orig. unkn.]

mull[3] /mʌl/ *n. Sc.* a promontory. [ME: cf. Gael. *maol*, Icel. *múli*]

mull[4] /mʌl/ *n.* humus formed under non-acid conditions. [G f. Danish *muld*]

mull[5] /mʌl/ *n.* a thin soft plain muslin. [abbr. of *mulmull* f. Hindi *malmal*]

mullah /ˈmʌlə, ˈmʊlə/ *n.* a Muslim learned in Islamic theology and sacred law. [Pers., Turk., Urdu *mullā* f. Arab. *mawlā*]

mullein /ˈmʌlɪn/ *n.* a tall herbaceous plant of the genus *Verbascum*, with woolly leaves and long spikes of yellow flowers. [ME f. OF *moleine* f. Gaulish]

Muller /ˈmʌlə(r)/, Hermann Joseph (1890–1967), American geneticist. Realizing that natural mutations were both rare and detrimental, he discovered that he could use X-rays to induce mutations in the genetic material of the fruit fly *Drosophila*, enabling him to carry out many more genetic studies with it. He also recognized the danger of X-radiation to living things, and was concerned about the build-up of genetic mutations in the human population. Muller was awarded a Nobel Prize in 1946.

Müller[1] /ˈmʊlə(r)/, (Friedrich) Max (1823–1900), German-born British philologist. He is remembered for his edition of the Sanskrit *Rig-veda* (1849–75; see RIG-VEDA); he also promoted the comparative study of Indo-European languages, as well as exploring comparative mythology and religion.

Müller[2] /ˈmʊlə(r)/, Johannes Peter (1801–58), German anatomist and zoologist. He was a pioneer of comparative and microscopical methods in biology. His investigations in physiology included respiration in the foetus, the nervous and sensory systems, the glandular system, and locomotion. Müller also studied the classification of marine animals.

Müller[3] /ˈmʊlə(r)/, Paul Hermann (1899–1965), Swiss chemist. Searching for an effective chemical for use in pest control, he synthesized DDT in 1939 and soon patented it as an insecticide. It was immediately successful, especially in controlling lice and mosquitoes, but was withdrawn by most countries in the 1970s when its environmental persistence and toxicity in higher animals was realized. He was awarded a Nobel Prize in 1948.

muller /ˈmʌlə(r)/ *n.* a stone or other heavy weight used for grinding material on a slab. [ME, perh. f. AF *moldre* grind]

Müllerian mimicry /mʊˈlɪərɪən/ *n. Zool.* a form of mimicry in which two or more noxious animals develop similar patterns of coloration etc. as a shared protective device (cf. BATESIAN MIMICRY). [J. F. T. *Müller*, Ger. zoologist (1821–97)]

mullet /ˈmʌlɪt/ *n.* a fish of the family Mullidae (see *red mullet*) or (esp. *US*) Mugilidae (see *grey mullet*). [ME f. OF *mulet* dimin. of L *mullus* red mullet f. Gk *mollos*]

mulligatawny /ˌmʌlɪɡəˈtɔːnɪ/ *n.* a highly seasoned soup orig. from India. [Tamil *milagutannir* pepper-water]

Mullingar /ˌmʌlɪŋˈɡɑː(r)/ the county town of Westmeath, in the Republic of Ireland; pop. (1981) 7,470.

mullion /ˈmʌljən/ *n.* (also **munnion** /ˈmʌnjən/) *Archit.* a vertical bar dividing the lights in a window (cf. TRANSOM). □ **mullioned** *adj.* [prob. an altered form of MONIAL]

mullock /ˈmʌlək/ *n.* **1** *Austral.* or *dial.* refuse, rubbish. **2** *Austral.* **a** rock containing no gold. **b** refuse from which gold has been extracted. **3** *Austral.* ridicule. [ME dimin. of *mul* dust, rubbish, f. MDu.]

mulloway /ˈmʌləˌweɪ/ *n.* a large Australian marine fish, *Argyrosomos hololepidotus*, used as food. (See also JEWFISH 2.) [Aboriginal *malowe*]

Mulroney /mʌlˈruːnɪ/, (Martin) Brian (b.1939), Canadian Progressive Conservative statesman, Prime Minister 1984–93. After becoming leader of the Progressive Conservative Party in 1983, he won a landslide victory in the 1984 election. He was re-elected in 1988 on a ticket of free trade with the US, but stood down in 1993 after the Canadian recession caused his popularity to slump in the opinion polls.

Multan /mʊlˈtɑːn/ a commercial city in Punjab province, east central Pakistan; pop. (1991) 980,000.

multangular /mʌlˈtæŋɡjʊlə(r)/ *adj.* having many angles. [med.L *multangularis* (as MULTI-, ANGULAR)]

multi- /ˈmʌltɪ/ *comb. form* many; more than one. [L f. *multus* much, many]

multi-access /ˈmʌltɪˌæksɛs/ n. (often attrib.) the simultaneous connection to a computer of a number of terminals.

multiaxial /ˌmʌltɪˈæksɪəl/ adj. of or involving several axes.

multicellular /ˌmʌltɪˈsɛljʊlə(r)/ adj. Biol. having many cells.

multichannel /ˈmʌltɪˌtʃæn(ə)l/ attrib.adj. employing or possessing many communication or television channels.

multicolour /ˈmʌltɪˌkʌl(ə)r/ adj. (also **multicoloured**) (US **-color**, **-colored**) of many colours.

multicultural /ˌmʌltɪˈkʌltʃərəl/ adj. of or relating to or constituting several cultural or ethnic groups within a society. □ **multiculturalism** n. **multiculturalist** n. & adj. **multiculturally** adv.

multidimensional /ˌmʌltɪdaɪˈmɛnʃən(ə)l, -dɪˈmɛnʃən(ə)l/ adj. of or involving more than three dimensions. □ **multidimensionally** adv. **multidimensionality** /-ˌmɛnʃəˈnælɪtɪ/ n.

multidirectional /ˌmʌltɪdaɪˈrɛkʃən(ə)l, -dɪˈrɛkʃən(ə)l/ adj. of, involving, or operating in several directions.

multifaceted /ˌmʌltɪˈfæsɪtɪd/ adj. having several facets.

multifarious /ˌmʌltɪˈfɛərɪəs/ adj. **1** (foll. by pl. noun) many and various. **2** having great variety. □ **multifariously** adv. **multifariousness** n. [L multifarius]

multifid /ˈmʌltɪfɪd/ adj. Bot. & Zool. divided into many parts. [L multifidus (as MULTI-, fid- stem of findere cleave)]

multifoil /ˈmʌltɪˌfɔɪl/ n. Archit. an ornament consisting of more than five foils.

multiform /ˈmʌltɪˌfɔːm/ adj. **1** having many forms. **2** of many kinds. □ **multiformity** /ˌmʌltɪˈfɔːmɪtɪ/ n.

multifunctional /ˌmʌltɪˈfʌŋkʃən(ə)l/ adj. (also **multifunction** /ˈmʌltɪˌfʌŋkʃ(ə)n/) having or fulfilling several functions.

multigrade /ˈmʌltɪˌgreɪd/ n. (usu. attrib.) an engine oil etc. meeting the requirements of several standard grades.

multilateral /ˌmʌltɪˈlætərəl/ adj. **1 a** (of an agreement, treaty, conference, etc.) in which three or more parties participate. **b** performed by more than one party (multilateral disarmament). **2** having many sides. □ **multilateralism** n. **multilateralist** n. & adj. **multilaterally** adv.

multilingual /ˌmʌltɪˈlɪŋgwəl/ adj. in or using several languages. □ **multilingualism** n. **multilingually** adv.

multimedia /ˌmʌltɪˈmiːdɪə/ adj. & n. ● adj. using more than one medium of communication or expression. ● n. = HYPERMEDIA.

multimillion /ˈmʌltɪˌmɪljən/ attrib.adj. costing or involving several million (pounds, dollars, etc.) (multimillion dollar fraud).

multimillionaire /ˌmʌltɪˌmɪljəˈnɛə(r)/ n. a person with a fortune of several millions.

multinational /ˌmʌltɪˈnæʃ(ə)n(ə)l/ adj. & n. ● adj. **1** (of a business organization) operating in several countries. **2** relating to or including several nationalities or ethnic groups. ● n. a multinational company. □ **multinationally** adv.

multinomial /ˌmʌltɪˈnəʊmɪəl/ adj. & n. Math. = POLYNOMIAL. [MULTI-, after binomial]

multiparous /mʌlˈtɪpərəs/ adj. **1** bringing forth many young at a birth. **2** having borne more than one child. [MULTI- + -PAROUS]

multipartite /ˌmʌltɪˈpɑːtaɪt/ adj. divided into many parts.

multiphase /ˈmʌltɪˌfeɪz/ n. Electr. = POLYPHASE.

multiple /ˈmʌltɪp(ə)l/ adj. & n. ● adj. **1** having several or many parts, elements, or individual components. **2** (foll. by pl. noun) many and various. ● n. **1** a number that may be divided by another a certain number of times without a remainder (56 is a multiple of 7). **2** a multiple shop or store. □ **least** (or **lowest**) **common multiple** the least quantity that is a multiple of two or more given quantities. **multiple-choice** (of a question in an examination) accompanied by several possible answers from which the correct one has to be chosen. **multiple personality** Psychol. two or more distinct personalities apparently existing in one individual. **multiple sclerosis** see SCLEROSIS 2. **multiple shop** (or **store**) Brit. a shop or store with branches in several places. **multiple standard** see STANDARD. **multiple star** several stars so close as to seem one, esp. when forming a connected system. □ **multiply** /-plɪ/ adv. [F f. LL multiplus f. L (as MULTIPLEX)]

multiplex /ˈmʌltɪˌplɛks/ adj., n., & v. ● adj. **1** manifold; of many elements. **2** involving simultaneous transmission of several messages along a single channel of communication. **3** of or relating to a cinema complex incorporating two or more cinemas on a single site. ● n. **1** a multiplex system or signal. **2** a multiplex cinema. ● v.tr. incorporate into a multiplex signal or system. □ **multiplexer** n. (also **multiplexor**). **multiplexing** n. [L (as MULTI-, -plex -plicis -fold)]

multipliable /ˈmʌltɪˌplaɪəb(ə)l/ adj. that can be multiplied.

multiplicable /ˈmʌltɪˌplɪkəb(ə)l/ adj. = MULTIPLIABLE. [OF multiplicable or med.L multiplicabilis f. L (as MULTIPLY)]

multiplicand /ˌmʌltɪplɪˈkænd/ n. a quantity to be multiplied by a multiplier. [med.L multiplicandus gerundive of L multiplicare (as MULTIPLY)]

multiplication /ˌmʌltɪplɪˈkeɪʃ(ə)n/ n. **1** the arithmetical process of multiplying. **2** the act or an instance of multiplying. □ **multiplication sign** the sign (×) to indicate that one quantity is to be multiplied by another, as in $2 \times 3 = 6$. **multiplication table** a list of multiples of a particular number, usu. from 1 to 12. □ **multiplicative** /-ˈplɪkətɪv/ adj. [ME f. OF multiplication or L multiplicatio (as MULTIPLY)]

multiplicity /ˌmʌltɪˈplɪsɪtɪ/ n. (pl. **-ies**) **1** manifold variety. **2** (foll. by of) a great number. [LL multiplicitas (as MULTIPLEX)]

multiplier /ˈmʌltɪˌplaɪə(r)/ n. **1** a quantity by which a given number is multiplied. **2** Econ. a factor by which an increment of income exceeds the resulting increment of saving or investment. **3** Electr. an instrument for increasing by repetition the intensity of a current, force, etc.

multiply /ˈmʌltɪˌplaɪ/ v. (**-ies**, **-ied**) **1** tr. (also absol.) obtain from (a number) another that is a specified number of times its value (multiply 6 by 4 and you get 24). **2** intr. increase in number esp. by procreation. **3** tr. produce a large number of (instances etc.). **4** tr. **a** breed (animals). **b** propagate (plants). [ME f. OF multiplier f. L multiplicare (as MULTIPLEX)]

multipolar /ˌmʌltɪˈpəʊlə(r)/ adj. having many poles (see POLE²).

multiprocessing /ˌmʌltɪˈprəʊsɛsɪŋ/ n. Computing processing by a number of processors sharing a common memory and common peripherals.

multiprogramming /ˌmʌltɪˈprəʊgræmɪŋ/ n. Computing the execution of two or more independent programs concurrently.

multi-purpose /ˈmʌltɪˌpɜːpəs/ attrib.adj. having several purposes.

multiracial /ˌmʌltɪˈreɪʃ(ə)l/ adj. relating to or made up of many human races. □ **multiracially** adv.

multi-role /ˈmʌltɪˌrəʊl/ attrib.adj. having several roles or functions.

multi-stage /ˈmʌltɪˌsteɪdʒ/ attrib.adj. (of a rocket etc.) having several stages of operation.

multi-storey /ˈmʌltɪˌstɔːrɪ/ attrib.adj. (of a building) having several (esp. similarly designed) storeys.

multitasking /ˈmʌltɪˌtɑːskɪŋ/ n. Computing the execution of a number of tasks at the same time. □ **multitask** v.tr. & intr.

multitude /ˈmʌltɪˌtjuːd/ n. **1** (often foll. by of) a great number. **2** a large gathering of people; a crowd. **3** (**the multitude**) the common people. **4** the state of being numerous. [ME f. OF f. L multitudo -dinis f. multus many]

multitudinous /ˌmʌltɪˈtjuːdɪnəs/ adj. **1** very numerous. **2** consisting of many individuals or elements. **3** (of an ocean etc.) vast. □ **multitudinously** adv. **multitudinousness** n. [L (as MULTITUDE)]

multi-user /ˈmʌltɪˌjuːzə(r)/ attrib.adj. (of a computer system) having a number of simultaneous users (cf. MULTI-ACCESS).

multivalent /ˌmʌltɪˈveɪlənt/ adj. **1** Chem. **a** having a valency of more than two. **b** having a variable valency. **2** having many applications, meanings, or values. □ **multivalency** n.

multivalve /ˈmʌltɪˌvælv/ attrib.adj. having several valves.

multivariate /ˌmʌltɪˈvɛərɪət/ adj. Statistics involving two or more variable quantities.

multiversity /ˌmʌltɪˈvɜːsɪtɪ/ n. (pl. **-ies**) a large university with many different departments. [MULTI- + UNIVERSITY]

multivocal /mʌlˈtɪvək(ə)l/ adj. having many possible meanings. [after EQUIVOCAL]

mum¹ /mʌm/ n. Brit. colloq. mother. [abbr. of MUMMY¹]

mum² /mʌm/ adj. colloq. silent (keep mum). □ **mum's the word** say nothing. [ME: imit. of closed lips]

mum³ /mʌm/ v.intr. (**mummed**, **mumming**) act in a traditional masked mime or a mummers' play. [cf. MUM² and MLG mummen]

Mumbai /mʊmˈbaɪ/ the Hindi name (official from 1995) for BOMBAY.

mumble /ˈmʌmb(ə)l/ v. & n. ● v. **1** intr. & tr. speak or utter indistinctly. **2** tr. bite or chew with or as with toothless gums. ● n. an indistinct utterance. □ **mumbler** n. □ **mumblingly** adv. [ME momele, as MUM²: cf. LG mummelen]

mumbo-jumbo /ˌmʌmbəʊˈdʒʌmbəʊ/ n. (pl. **-jumbos**) **1** meaningless or ignorant ritual. **2** language or action intended to mystify or confuse. **3** an object of senseless veneration. [*Mumbo Jumbo*, a supposed African idol]

mummer /ˈmʌmə(r)/ n. **1** an actor in a traditional masked mime or a mummers' play. **2** *archaic* or *derog.* an actor in the theatre. [ME f. OF *momeur* f. *momer* MUM³]

mummers' play n. (also called *mumming play*) the best-known type of English folk-play, mainly associated with Christmas and popular especially from the late 18th until the mid-19th century. In most versions a hero is killed and then brought back to life. The plays derive from a version of the legend of St George dating from the late 16th century, but it is likely that they also have links with primitive ceremonies marking stages in the agricultural year.

mummery /ˈmʌməri/ n. (pl. **-ies**) **1** ridiculous (esp. religious) ceremonial. **2** a performance by mummers. [OF *momerie* (as MUMMER)]

mummify /ˈmʌmɪˌfaɪ/ v.tr. (**-ies, -ied**) **1** embalm and preserve (a body) in the form of a mummy (see MUMMY²). **2** (usu. as **mummified** adj.) shrivel or dry up (tissues etc.). □ **mummification** /ˌmʌmɪfɪˈkeɪʃ(ə)n/ n.

mummy¹ /ˈmʌmi/ n. (pl. **-ies**) *Brit. colloq.* mother. □ **mummy's boy** a boy or man who is excessively influenced by or attached to his mother. [imit. of a child's pronunc.: cf. MAMMA¹]

mummy² /ˈmʌmi/ n. (pl. **-ies**) **1** a body of a human being or animal embalmed for burial, esp. in ancient Egypt. (*See note below.*) **2** a dried-up body. **3** a pulpy mass (*beat it to a mummy*). **4** a rich brown pigment. [F *momie* f. med.L *mumia* f. Arab. *mūmiyā* f. Pers. *mūm* wax]

▪ The concept of mummification was based on the belief that the preservation of the body was necessary for the soul and the *ka* to meet and live with it again. The practice developed in the predynastic period with the change from simple desert burials, which had dehydrated and preserved the body naturally, to burial in coffins. The essential steps in mummification were removing the viscera, dehydrating the body with natron, treating it with resin, bandaging, and decorating it. The ceremonial aspects were considered crucial, and the procedure could take seventy days.

mumps /mʌmps/ n.pl. **1** (treated as *sing.*) a contagious and infectious viral disease with swelling of the parotid salivary glands in the face, and sometimes causing sterility in adult males. **2** a fit of sulks. □ **mumpish** adj. (in sense 2). [archaic *mump* be sullen]

Munch /mʊŋk/, Edvard (1863–1944), Norwegian painter and engraver. One of the chief sources of German expressionism, he infused his subjects with an intense emotionalism and explored the use of violent colour and linear distortion to express feelings about life and death. Major works include his *Frieze of Life* sequence, incorporating *The Scream* (1893).

munch /mʌntʃ/ v.tr. eat steadily with a marked action of the jaws. [ME, imit.: cf. CRUNCH]

Munchausen, Baron /ˈmʌntʃaʊz(ə)n, German ˈmʏnçˌhaʊz(ə)n/ the hero of a book of fantastic travellers' tales (1785) written in English by a German, Rudolph Erich Raspe. The original Baron Munchausen is said to have lived 1720–97, to have served in the Russian army against the Turks, and to have related extravagant tales of his prowess.

Munchausen's syndrome n. *Med.* a mental abnormality in which a person persistently pretends to have a dramatic or severe illness in order to obtain hospital treatment. [MUNCHAUSEN, BARON]

München see MUNICH.

Munda /ˈmʊndə/ n. & adj. ● n. (pl. same or **Mundas**) **1** a member of an ancient Indian people surviving in NE India. **2** the group of languages which includes the dialects of the Mundas. ● adj. of or relating to this people or their language. [Munda *Muṇḍā*]

mundane /mʌnˈdeɪn/ adj. **1** dull, routine. **2** of this world; worldly. □ **mundanely** adv. **mundaneness** n. **mundanity** /-ˈdænɪti/ n. (pl. **-ies**). [ME f. OF *mondain* f. LL *mundanus* f. L *mundus* world]

mung /mʌŋ/ n. (in full **mung bean**) a leguminous Asian plant of the genus *Vigna*, yielding a small bean used as food. See also GRAM². [Hindi *mūng*]

mungo /ˈmʌŋgəʊ/ n. the short fibres recovered from heavily felted material. [19th c.: orig. uncert.]

Munich /ˈmjuːnɪk/ (German **München** /ˈmʏnç(ə)n/) a city in SE Germany, capital of Bavaria; pop. (1991) 1,229,050. Munich is noted for the architecture of the medieval city centre and as an artistic and cultural centre. It is the home of the Bavarian State Opera Company and Munich Philharmonic.

Munich Agreement (also called *Munich Pact*) an agreement between

Britain, France, Germany, and Italy, signed at Munich on 29 Sept. 1938, under which part of Czechoslovakia (the Sudetenland) was ceded to Germany. It is remembered as an act of appeasement, illustrating the inadequacy of such action in the face of a fiercely expansionist and powerful state such as Germany.

municipal /mjuːˈnɪsɪp(ə)l/ adj. of or concerning a municipality or its self-government. □ **municipally** adv. **municipalize** v.tr. (also **-ise**).

municipalization /-ˌnɪsɪpəlaɪˈzeɪʃ(ə)n/ n. [L *municipalis* f. *municipium* free city f. *municeps -cipis* citizen with privileges f. *munia* civic offices + *capere* take]

municipality /mjuːˌnɪsɪˈpælɪti/ n. (pl. **-ies**) **1** a town or district having local government. **2** the governing body of this area. [F *municipalité* f. *municipal* (as MUNICIPAL)]

munificent /mjuːˈnɪfɪs(ə)nt/ adj. (of a giver or a gift) splendidly generous, bountiful. □ **munificence** n. **munificently** adv. [L *munificent-*, var. stem of *munificus* f. *munus* gift]

muniment /ˈmjuːnɪmənt/ n. (usu. in pl.) **1** a document kept as evidence of rights or privileges etc. **2** an archive. [ME f. OF f. L *munimentum* defence, in med.L title-deed f. *munire munit-* fortify]

munition /mjuːˈnɪʃ(ə)n/ n. & v. ● n. (usu. in pl.) military weapons, ammunition, equipment, and stores. ● v.tr. supply with munitions. [F f. L *munitio -onis* fortification (as MUNIMENT)]

munitioner /mjuːˈnɪʃənə(r)/ n. a person who makes or supplies munitions.

munnion var. of MULLION.

Munro /mənˈrəʊ/, H(ector) H(ugh), see SAKI.

munshi var. of MOONSHEE.

Munster /ˈmʌnstə(r)/ a province of the Republic of Ireland, in the south-west of the country.

Münster /ˈmʊnstə(r)/ a city in NW Germany; pop. (1991) 249,900. It was formerly the capital of Westphalia; the Treaty of Westphalia, ending the Thirty Years War, was signed simultaneously there and at Osnabrück in 1648.

munt /mʊnt/ n. *S. Afr. sl. offens.* a black African. [Bantu *umuntu* person]

muntjac /ˈmʌntdʒæk/ n. (also **muntjak**) a small deer of the genus *Muntiacus*, native to SE Asia, the male of which has tusks and small antlers. [Sundanese *minchek*]

Muntz metal /mʌnts/ n. an alloy (60% copper, 40% zinc) used for sheathing ships etc. [George Frederick *Muntz*, Engl. manufacturer (1794–1857)]

muon /ˈmjuːɒn/ n. *Physics* an unstable subatomic particle like an electron, but with a much greater mass. □ **muonic** /mjuːˈɒnɪk/ adj. [μ (MU), as the symbol for it]

Muqdisho see MOGADISHU.

murage /ˈmjʊərɪdʒ/ n. *hist.* a tax levied for building or repairing the walls of a town. [ME f. OF, in med.L *muragium* f. OF *mur* f. L *murus* wall]

mural /ˈmjʊərəl/ n. & adj. ● n. a painting executed directly on a wall. ● adj. **1** of or like a wall. **2** on a wall. □ **mural crown** *Rom. Antiq.* a crown or garland given to the soldier who was first to scale the wall of a besieged town. □ **muralist** n. [F f. L *muralis* f. *murus* wall]

Murat /mjʊəˈrɑː/, Joachim (c.1767–1815), French general, king of Naples 1808–15. One of Napoleon's marshals, Murat made his name as a cavalry commander in the Italian campaign (1800). After he was made king of Naples by Napoleon, he made a bid to become king of all Italy in 1815, but was captured in Calabria and executed.

Murchison Falls /ˈmɜːtʃɪs(ə)n/ **1** the former name for the KABALEGA FALLS. **2** the former name for the KAPACHIRA FALLS.

Murcia /ˈmʊəsɪə/ **1** an autonomous region in SE Spain. In the Middle Ages, along with Albacete, it formed an ancient Moorish kingdom. **2** its capital city; pop. (1991) 328,840.

murder /ˈmɜːdə(r)/ n. & v. ● n. **1** the unlawful premeditated killing of a human being by another (cf. MANSLAUGHTER). **2** *colloq.* an unpleasant, troublesome, or dangerous state of affairs (*it was murder here on Saturday*). ● v.tr. **1** kill (a human being) unlawfully, esp. wickedly or inhumanly. **2** *Law* kill (a human being) unlawfully with a premeditated motive. **3** *colloq.* **a** utterly defeat. **b** spoil by a bad performance, mispronunciation, etc. (*murdered the soliloquy in the second act*). □ **cry** (or **scream**) **blue** (*N. Amer.* **bloody**) **murder** *colloq.* make an extravagant outcry. **get away with murder** *colloq.* do whatever one wishes and escape punishment. **murder will out** murder cannot remain undetected. □ **murderer** n. **murderess** /-rɪs/ n. [OE *morthor* & OF *murdre* f. Gmc]

murderous /ˈmɜːdərəs/ adj. **1** (of a person, weapon, action, etc.)

capable of, intending, or involving murder or great harm. **2** *colloq.* extremely arduous or unpleasant. □ **murderously** *adv.* **murderousness** *n.*

Murdoch[1] /ˈmɜːdɒk/, Dame (Jean) Iris (b.1919), British novelist and philosopher, born in Ireland. The author of several philosophical works, Murdoch is primarily known for her novels; many of these portray complex sexual relationships, as in *The Sandcastle* (1957) and *The Red and the Green* (1965). Others simultaneously explore the quest for the spiritual life, particularly *The Sea, The Sea* (1978), which won the Booker Prize. More recent novels include *The Good Apprentice* (1985) and *The Message to the Planet* (1989).

Murdoch[2] /ˈmɜːdɒk/, (Keith) Rupert (b.1931), Australian-born American publisher and media entrepreneur. As the founder and head of the News International Communications empire, he owns major newspapers in Australia, Britain, and the US, together with film and television companies and the publishing firm HarperCollins.

mure /mjʊə(r)/ *v.tr. archaic* **1** immure. **2** (foll. by *up*) wall up or shut up in an enclosed space. [ME f. OF *murer* f. *mur*: see MURAGE]

murex /ˈmjʊəreks/ *n.* (*pl.* **murices** /-rɪˌsiːz/ or **murexes**) a spiny-shelled gastropod mollusc, esp. of the genus *Murex*, yielding a purple dye. [L]

muriatic acid /ˌmjʊərɪˈætɪk/ *n. Chem.* = HYDROCHLORIC ACID. □ **muriate** /ˈmjʊərɪət/ *n.* ¶ Regarded as archaic in technical use. [L *muriaticus* f. *muria* brine]

Murillo /mjʊəˈrɪləʊ/, Bartolomé Esteban (*c.*1618–82), Spanish painter. He is noted both for his genre scenes of urchins and peasants and for his devotional pictures, which are characterized by delicate colour and ethereal form. Major works include *Two Boys Eating a Pie* (*c.*1665–75) and the *Soult Immaculate Conception* (1678).

murine /ˈmjʊəraɪn/ *adj.* of or like a mouse or mice. [L *murinus* f. *mus muris* mouse]

murk /mɜːk/ *n. & adj.* (also **mirk**) ● *n.* **1** darkness, poor visibility. **2** air obscured by fog etc. ● *adj. archaic* (of night, day, place, etc.) = MURKY. [prob. f. Scand.: cf. ON *myrkr*]

murky /ˈmɜːkɪ/ *adj.* (also **mirky**) (**-ier, -iest**) **1** dark, gloomy. **2** (of darkness, liquid, etc.) thick, dirty. **3** suspiciously obscure (*murky past*). □ **murkily** *adv.* **murkiness** *n.*

Murmansk /mʊəˈmænsk/ a port in NW Russia, on the northern coast of the Kola Peninsula, in the Barents Sea; pop. (1990) 472,000. It is the largest city north of the Arctic Circle and its port is ice-free throughout the year.

murmur /ˈmɜːmə(r)/ *n. & v.* ● *n.* **1** a subdued continuous sound, as made by waves, a brook, etc. **2** a softly spoken or nearly inarticulate utterance. **3** *Med.* a recurring sound heard in the auscultation of the heart and usu. indicating abnormality. **4** a subdued expression of discontent. ● *v.* **1** *intr.* make a subdued continuous sound. **2** *tr.* utter (words) in a low voice. **3** *intr.* (usu. foll. by *at, against*) complain in low tones, grumble. □ **murmurer** *n.* **murmuringly** *adv.* **murmurous** *adj.* [ME f. OF *murmurer* f. L *murmurare*: cf. Gk *mormurō* (of water) roar, Skr. *marmaras* noisy]

murphy /ˈmɜːfɪ/ *n.* (*pl.* **-ies**) *sl.* a potato. [Irish surname]

Murphy's Law /ˈmɜːfɪz/ *n. joc.* the maxim that anything that can go wrong will go wrong. [Irish surname; orig. of phrase uncertain]

murrain /ˈmʌrɪn/ *n.* **1** an infectious disease of cattle, carried by parasites. **2** *archaic* a plague, esp. the potato blight during the Irish famine in the mid-19th century. [ME f. AF *moryn*, OF *morine* f. *morir* f. L *mori* die]

Murray[1] /ˈmʌrɪ/, (George) Gilbert (Aimé) (1866–1957), Australian-born British classical scholar. He is remembered for his rhymed verse translations of Greek dramatists, particularly Euripides. His translations of the latter's *Medea, Bacchae*, and *Electra* were staged in London from 1902, and helped to revive contemporary interest in Greek drama. Murray was also a founder of the League of Nations and later a joint president of the United Nations.

Murray[2] /ˈmʌrɪ/, Sir James (Augustus Henry) (1837–1915), Scottish lexicographer. He was chief editor of the largest of all dictionaries in English, the *Oxford English Dictionary*. Murray did not live to see the dictionary completed; he died after finishing a section of the letter T, two years short of his 80th birthday. (See OXFORD ENGLISH DICTIONARY.)

Murray River the principal river of Australia, which rises in the Great Dividing Range in New South Wales and flows 2,590 km (1,610 miles) generally north-westwards, forming part of the border between the states of Victoria and New South Wales, before turning south-wards in South Australia to empty into the Indian Ocean south-east of Adelaide.

murre /mɜː(r)/ *n. esp. N. Amer.* an auk or guillemot. [16th c.: orig. unkn.]

murrelet /ˈmɜːlɪt/ *n.* a small auk of the genus *Brachyramphus* or *Synthliboramphus*, of the North Pacific.

murrey /ˈmʌrɪ/ *n. & adj. archaic* ● *n.* the colour of a mulberry; a deep red or purple. ● *adj.* of this colour. [ME f. OF *moré* f. med.L *moratus* f. *morum* mulberry]

Murrumbidgee /ˌmʌrəmˈbɪdʒiː/ a river of SE Australia, in New South Wales. Rising in the Great Dividing Range, it flows 1,759 km (1,099 miles) westwards to join the Murray, of which it is a major tributary.

murther /ˈmɜːðə(r)/ *archaic* var. of MURDER.

Mururoa /ˌmʊərʊˈrəʊə/ a remote South Pacific atoll in the Tuamotu archipelago, in French Polynesia, used as a nuclear testing site since 1966.

Musala, Mount /muːˈsɑːlə/ (also **Mousalla**) the highest peak in Bulgaria, in the Rila Mountains, rising to 2,925 m (9,596 ft).

Mus.B. *abbr.* (also **Mus. Bac.**) Bachelor of Music. [L *Musicae Baccalaureus*]

muscadel var. of MUSCATEL.

Muscadet /ˈmʌskəˌdeɪ/ *n.* **1** a white wine from the Loire region of France. **2** a variety of grape from which the wine is made. [F f. *muscade* nutmeg f. *musc* MUSK + -ET[1]]

muscadine /ˈmʌskədɪn, -ˌdaɪn/ *n.* a variety of grape with a musk flavour, used chiefly in wine-making. [perh. Engl. form f. Prov. MUSCAT]

muscarine /ˈmʌskərɪn/ *n.* a poisonous alkaloid from the fly agaric fungus, *Amanita muscaria*. [L *muscarius* f. *musca* fly]

Muscat /ˈmʌskæt/ the capital of Oman, a port on the SE coast of the Arabian peninsula; pop. (1990) 380,000.

muscat /ˈmʌskæt/ *n.* **1** a sweet fortified white wine made from muscadines. **2** a muscadine. [F f. Prov. *muscat muscade* (adj.) f. *musc* MUSK]

Muscat and Oman the former name (until 1970) for OMAN.

muscatel /ˌmʌskəˈtel/ *n.* (also **muscadel** /-ˈdel/) **1** = MUSCAT. **2** a raisin from a muscadine grape. [ME f. OF f. Prov. dimin. of *muscat*: see MUSCAT]

muscle /ˈmʌs(ə)l/ *n. & v.* ● *n.* **1** a fibrous tissue with the ability to contract, producing movement in or maintaining the position of an animal body. (*See note below.*) **2** the part of an animal body that is composed of muscles. **3** physical power or strength. ● *v.intr.* (usu. foll. by *in*) *colloq.* force oneself on others; intrude by forceful means. □ **muscle-bound** with muscles stiff and inelastic through excessive exercise or training. **muscle-man** a man with highly developed muscles, esp. one employed as an intimidator. **not move a muscle** be completely motionless. □ **muscled** *adj.* (usu. in *comb.*). **muscleless** *adj.* **muscly** *adj.* [F f. L *musculus* dimin. of *mus* mouse, from the fancied mouselike form of some muscles]

  ▪ All animals except protozoans and most sponges possess muscle cells. These cells are very small, but when grouped may form conspicuous columns, bands, or sheets of muscle. In vertebrates there are three main types of muscle tissue. Skeletal or striated muscles, whose tissue appears striated under the microscope, are under voluntary control; they are attached to bones by tendons and move the joints. Cardiac muscle, found only in the heart, is not under direct voluntary control; it is notable for its ability to contract spontaneously and rhythmically. Smooth or visceral muscle, which is also controlled involuntarily, is found chiefly in the walls of hollow organs such as the stomach and intestines. Muscle cells convert chemical into mechanical energy; besides making movement and stance possible, their chemical reactions produce most of the heat which maintains body temperature in warm-blooded animals.

muscology /mʌˈskɒlədʒɪ/ *n.* the study of mosses. □ **muscologist** *n.* [mod.L *muscologia* f. L *muscus* moss]

muscovado /ˌmʌskəˈvɑːdəʊ/ *n.* (*pl.* **-os**) an unrefined sugar made from the juice of sugar cane by evaporation and draining off the molasses. [Sp. *mascabado* (sugar) of the lowest quality]

Muscovite /ˈmʌskəˌvaɪt/ *n. & adj.* ● *n.* **1** a native or inhabitant of Moscow. **2** *archaic* a Russian. ● *adj.* **1** of or relating to Moscow. **2** *archaic* of or relating to Russia. [mod.L *Muscovita* f. *Muscovia* = MUSCOVY]

muscovite /ˈmʌskəˌvaɪt/ *n.* a silver-grey form of mica with a sheetlike crystalline structure that is used in the manufacture of electrical equipment etc. [obs. MUSCOVY *glass* (in the same sense) + -ITE[1]]

Muscovy /ˈmʌskəvɪ/ a medieval principality in west central Russia, centred on Moscow, which formed the nucleus of modern Russia. It was founded in the late 13th century by Daniel, son of Alexander Nevsky, and gradually expanded despite repeated Tartar depredations, finally overcoming the Tartars in 1380. Princes of Muscovy became the rulers of Russia; in 1472 Ivan III, grand duke of Muscovy, completed the unification, overcoming rivalry from the principalities of Novgorod, Tver, and Vladimir and adopting the title 'Ruler of all Russia'. In 1547 Ivan IV (Ivan the Terrible) became the first tsar of the new empire, and made Moscow its capital. [obs. F *Muscovie* f. mod.L *Moscovia* f. Russ. *Moskva* Moscow]

Muscovy duck *n.* a large tropical American duck, *Cairina moschata*, with a slight smell of musk (and no connection with Russia). Also called *musk duck*.

muscular /'mʌskjʊlə(r)/ *adj.* **1** of or affecting the muscles. **2** having well-developed muscles. **3** robust. □ **muscular Christianity** a Christian life of cheerful physical activity as described in the writings of Charles Kingsley. **muscular dystrophy** see DYSTROPHY. **muscular rheumatism** = MYALGIA. **muscular stomach** the gizzard of a bird etc. □ **muscularly** *adv.* **muscularity** /ˌmʌskjʊˈlærɪtɪ/ *n.* [earlier *musculous* (as MUSCLE)]

musculature /'mʌskjʊlətʃə(r)/ *n.* the muscular system of a body or organ. [F f. L (as MUSCLE)]

Mus.D. *abbr.* (also **Mus. Doc.**) Doctor of Music. [L *Musicae Doctor*]

muse[1] /mjuːz/ *n.* **1** (**Muse**) *Gk & Rom. Mythol.* any of the goddesses who presided over the arts and sciences. They were the daughters of Zeus and Mnemosyne, traditionally nine in number (Calliope, Clio, Euterpe, Terpsichore, Erato, Melpomene, Thalia, Polyhymnia, and Urania), though their functions and names vary considerably between different sources. **2** (usu. prec. by *the*) **a** a poet's inspiring goddess or woman. **b** a poet's genius or characteristic style. [ME f. OF *muse* or L *musa* f. Gk *mousa*]

muse[2] /mjuːz/ *v. & n.* ● *v.* **1** *intr.* **a** (usu. foll. by *on, upon*) ponder, reflect. **b** (usu. foll. by *on*) gaze meditatively (on a scene etc.). **2** *tr.* say meditatively. ● *n. archaic* a fit of abstraction. □ **musingly** *adv.* [ME f. OF *muser* to waste time f. Rmc perh. f. med.L *musum* muzzle]

musette /mjuːˈzet/ *n.* **1 a** a kind of small bagpipe with bellows, common in the French court in the 17th and 18th centuries. **b** a tune imitating the sound of this. **2** a small oboe-like double-reed instrument of 19th-century France. **3** a dance popular in the courts of Louis XIV and XV of France, a rustic form of the gavotte. Its name derives from the music's bagpipe-like bass drone. **4** *US* a small knapsack. [ME f. OF, dimin. of *muse* bagpipe]

museum /mjuːˈzɪəm/ *n.* a building used for storing and exhibiting objects of historical, scientific, or cultural interest. □ **museum piece 1** a specimen of art etc. fit for a museum. **2** *derog.* an old-fashioned or quaint person or object. □ **museology** /ˌmjuːzɪˈɒlədʒɪ/ *n.* [L f. Gk *mouseion* seat of the Muses: see MUSE[1]]

mush[1] /mʌʃ/ **1** soft pulp. **2** feeble sentimentality. **3** *N. Amer.* maize porridge. **4** /mʊʃ/ *sl.* the mouth; the face. □ **mushy** *adj.* (**mushier, mushiest**) **mushily** *adv.* **mushiness** *n.* [app. var. of MASH]

mush[2] /mʌʃ/ *v. & n. N. Amer.* ● *v.intr.* **1** (in *imper.*) used as a command to dogs pulling a sledge to urge them forward. **2** go on a journey across snow with a dog-sledge. ● *n.* a journey across snow with a dog-sledge. [prob. corrupt. f. F *marchons* imper. of *marcher* advance]

mushroom /'mʌʃrʊm, -ruːm/ *n. & v.* ● *n.* **1** the usu. edible spore-producing body of various fungi; esp. *Agaricus campestris*, with a stem and domed cap, proverbial for its rapid growth (cf. TOADSTOOL). **2** the pinkish-brown colour of this. **3** any item resembling a mushroom in shape (*darning mushroom*). **4** (usu. *attrib.*) something that appears or develops suddenly or is ephemeral; an upstart. ● *v.intr.* **1** appear or develop rapidly. **2** expand and flatten like a mushroom cap. **3** gather mushrooms. □ **mushroom cloud** a cloud suggesting the shape of a mushroom, esp. from a nuclear explosion. **mushroom growth 1** a sudden development or expansion. **2** anything undergoing this. □ **mushroomy** *adj.* [ME f. OF *mousseron* f. LL *mussirio -onis*]

music /'mjuːzɪk/ *n.* **1** the art of combining vocal or instrumental sounds (or both) to produce beauty of form, harmony, and expression of emotion. **2** the sounds so produced. **3** musical compositions. **4** the written or printed score of a musical composition. **5** certain pleasant sounds, e.g. birdsong, the sound of a stream, etc. □ **music box** *N. Amer.* = *musical box*. **music centre** *Brit.* equipment combining radio, record-player, tape recorder, etc. **music drama** Wagnerian-type opera without formal arias etc. and governed by dramatic considerations (cf. *number opera*). **music of the spheres** see SPHERE. **music-paper** paper printed with staves for writing music. **music stand** a rest or frame

on which sheet music or a score is supported. **music stool** a stool for a pianist, usu. with adjustable height. **music theatre** in late 20th-century music, a combination of elements from music and drama in new forms distinct from traditional opera, esp. as designed for small groups of performers. **music to one's ears** something very pleasant to hear. [ME f. OF *musique* f. L *musica* f. Gk *mousikē* (*tekhnē* art) of the Muses (*mousa* Muse: see MUSE[1])]

musical /'mjuːzɪk(ə)l/ *adj. & n.* ● *adj.* **1** of or relating to music. **2** (of sounds, a voice, etc.) melodious, harmonious. **3** fond of or skilled in music (*the musical one of the family*). **4** set to or accompanied by music. ● *n.* a film or theatrical piece (not opera or operetta) in which music is an essential element. (*See note below.*) □ **musical box** *Brit.* a mechanical musical instrument in a box, typically incorporating a toothed cylinder which plucks a row of tuned metal strips as it revolves. **musical bumps** *Brit.* a game similar to musical chairs, with players sitting on the floor and the one left standing eliminated. **musical chairs 1** a party game in which the players compete in successive rounds for a decreasing number of chairs. **2** a series of changes or political manoeuvring etc. after the manner of the game. **musical comedy** a musical; a light dramatic entertainment of songs, dialogue, and dancing, usu. without a strong plot. (*See note below.*) **musical film** a film in which music is an important feature. **musical glasses** an instrument in which notes are produced by rubbing graduated glass bowls or tubes. **musical saw** a bent saw played with a violin bow. □ **musicalize** *v.tr.* (also **-ise**). **musically** *adv.* **musicalness** *n.* **musicality** /ˌmjuːzɪˈkælɪtɪ/ *n.* [ME f. OF f. med.L *musicalis* f. L *musica*: see MUSIC]

▪ The musical as a genre includes both stage and film productions, and evolved from light opera at the end of the 19th century, using a combination of spoken dialogue and songs. The musicals of the earlier period, which reached its peak in the US in the work of such composers as Jerome Kern, Irving Berlin, Cole Porter, George Gershwin, and Richard Rodgers, are sometimes distinguished by the name *musical comedy*: they have tuneful scores and often elaborate choreography, but tend not to have strong plots. During and after the 1940s the genre developed stronger story-lines and often more serious themes; notable examples include Rodgers and Hammerstein's *Oklahoma!* (1943) and *South Pacific* (1949), the film *Singin' in the Rain* (1953), Leonard Bernstein's *West Side Story* (1957), and, more recently, Andrew Lloyd Webber's *Jesus Christ Superstar* (1970) and *Evita* (1978).

musicale /ˌmjuːzɪˈkɑːl/ *n. US* a musical party. [F fem. adj. (as MUSICAL)]

music-hall /'mjuːzɪkˌhɔːl/ *n.* a type of popular entertainment which flourished in Britain during the second half of the 19th and in the early 20th century; also, a place for performing this. Music-hall developed from tavern performances and by the early 19th century had come to be performed in a separate building, often an elaborate two or three-tier auditorium. The 'turns' on a normal evening's bill — sometimes as many as twenty-five — ranged from acrobats and jugglers to comedians and singers, all presided over by a chairman. In its heyday music-hall was extremely popular, but it was killed by the advent of the cinema, radio, and above all television.

musician /mjuːˈzɪʃ(ə)n/ *n.* a person who plays a musical instrument, esp. professionally, or is otherwise musically gifted. □ **musicianly** *adj.* **musicianship** *n.* [ME f. OF *musicien* f. *musique* (as MUSIC, -ICIAN)]

musicology /ˌmjuːzɪˈkɒlədʒɪ/ *n.* the study of music other than that directed to proficiency in performance or composition. □ **musicologist** *n.* **musicological** /-kəˈlɒdʒɪk(ə)l/ *adj.* [F *musicologie* or MUSIC + -LOGY]

Musil /'muːzɪl/, Robert (1880–1942), Austrian novelist. He is best known for his unfinished novel *The Man Without Qualities* (1930–43), a complex experimental work without a conventional plot, depicting the disintegration of traditional Austrian society and culture just before the outbreak of the First World War.

musique concrète /muːˌziːk kɒnˈkret/ *n.* (also called *concrete music*) a form of experimental music constructed from recorded natural sounds, as of machinery, running water, rustling leaves, etc. One of the earliest examples of music produced by electronic means, it was pioneered in the late 1940s by Pierre Schaeffer (b.1910) and developed by Edgar Varèse. [F]

musk /mʌsk/ *n.* **1** a strong-smelling reddish-brown substance produced by a gland in the male musk deer, used as an ingredient in perfumes. **2** a yellow-flowered plant, *Mimulus moschatus*, with pale-green ovate leaves (orig. with a smell of musk which is no longer perceptible in modern varieties). □ **musk deer** a small Asian deer of the genus

Moschus, the male of which has long protruding canine teeth but no antlers. **musk duck 1** an Australian duck, *Biziura lobata*, having a musky smell. **2** the Muscovy duck. **musk melon** the common yellow or green melon, *Cucumis melo*, usu. with a raised network of markings on the skin. **musk ox** a large goat-antelope, *Ovibos moschatus*, found esp. on the tundra of Canada and Greenland, with a thick shaggy coat and small curved horns. **musk-rose** a rambling rose, *Rosa moschata*, with large white flowers smelling of musk. **musk thistle** a nodding thistle, *Carduus nutans*, whose flowers have a musky fragrance. **musk-tree** (or **-wood**) an Australian tree, *Olearia argyrophylla*, with a musky smell. □ **musky** *adj.* **muskiness** *n.* [ME f. LL *muscus* f. Pers. *mušk*, perh. f. Skr. *muṣka* scrotum (from the shape of the musk deer's gland)]

muskeg /ˈmʌskeg/ *n.* a level swamp or bog in Canada. [Cree]

muskellunge /ˈmʌskəˌlʌndʒ/ *n.* a large North American pike, *Esox masquinongy*, found in the Great Lakes. Also called *maskinonge*. [Algonquian]

musket /ˈmʌskɪt/ *n. hist.* an infantryman's (esp. smooth-bored) light gun, often supported on the shoulder. □ **musket-shot 1** a shot fired from a musket. **2** the range of this shot. [F *mousquet* f. It. *moschetto* crossbow bolt f. *mosca* fly]

musketeer /ˌmʌskɪˈtɪə(r)/ *n. hist.* a soldier armed with a musket.

musketry /ˈmʌskɪtrɪ/ *n.* **1** muskets, or soldiers armed with muskets, referred to collectively. **2** the knowledge of handling muskets.

Muskogean /ˌmʌskəˈgiːən, mʌˈskəʊgɪən/ *n. & adj.* ● *n.* a family of American Indian languages in south-eastern North America, including Creek and Choctaw. ● *adj.* of or relating to this language family. [*Muskogee* its speakers, perh. of Algonquian origin]

muskrat /ˈmʌskræt/ *n.* **1** a large aquatic rodent, *Ondatra zibethicus*, native to North America, resembling a small beaver and having a musky smell. Also called *musquash*. **2** the fur of this.

Muslim /ˈmʊzlɪm, ˈmʌz-/ *n. & adj.* (also **Moslem** /ˈmɒzləm/) ● *n.* a follower of the Islamic religion. ● *adj.* of or relating to the Muslims or their religion. [Arab. *muslim*, part. of *aslama*: see ISLAM]

Muslim Brotherhood an Islamic religious and political organization, founded in Egypt in 1928 and dedicated to the establishment of a nation based on Islamic principles. The group at first engaged mostly in educational initiatives, but after 1938 it became more radical, opposed to any sort of Western influence and involved in terrorist activity. It has been banned by the Egyptian government on several occasions but continues to operate under cover in Egypt and other Sunni countries.

Muslim League one of the main political parties in Pakistan. It was formed in 1906 in India to represent the rights of Indian Muslims; its demands in 1940 for an independent Muslim state led ultimately to the establishment of Pakistan.

muslin /ˈmʌzlɪn/ *n.* **1** a fine delicately woven cotton fabric. **2** *US* a cotton cloth in plain weave. □ **muslined** *adj.* [F *mousseline* f. It. *mussolina* f. *Mussolo* Mosul in Iraq, where it was made]

musmon /ˈmʌzmən/ *n. Zool.* = MOUFLON. [L *musimo* f. Gk *mousmōn*]

muso /ˈmjuːzəʊ/ *n.* (*pl.* **-os**) *Brit. sl.* a musician, esp. a professional. [abbr.]

musquash /ˈmʌskwɒʃ/ *n.* = MUSKRAT. [Algonquian]

muss /mʌs/ *v. & n. N. Amer. colloq.* ● *v.tr.* (often foll. by *up*) disarrange; throw into disorder. ● *n.* a state of confusion; untidiness, mess. □ **mussy** *adj.* [app. var. of MESS]

mussel /ˈmʌs(ə)l/ *n.* **1** a marine bivalve mollusc of the genus *Mytilus*; esp. *M. edulis*, often found clustered on rocks etc. and used for food. **2** a similar mollusc of the genus *Margaritifer* or *Anodonta*, found in fresh water and forming pearls. [ME f. OE *mus(c)le* & MLG *mussel*, ult. rel. to L *musculus* (as MUSCLE)]

Mussolini /ˌmʊsəˈliːnɪ/, Benito (Amilcaro Andrea) (known as 'Il Duce' = the leader) (1883–1945), Italian Fascist statesman, Prime Minister 1922–43. Originally a socialist, Mussolini founded the Italian Fascist Party in 1919. Three years later he orchestrated the march on Rome by the Blackshirts and was created Prime Minister, proceeding to organize his government along dictatorial lines. He annexed Abyssinia in 1936 and entered the Second World War on Germany's side in 1940. Mussolini was forced to resign after the Allied invasion of Sicily in 1943; he was rescued from imprisonment by German paratroopers, but was captured and executed by Italian Communist partisans in 1945, a few weeks before the end of the war.

Mussorgsky /məˈsɔːgskɪ/, Modest (Petrovich) (also **Moussorgsky**) (1839–81), Russian composer. Most of his best-known works are vocal, his interest in speech rhythms combining with the lyricism of his songs. They include the opera *Boris Godunov* (1874) and *Songs and Dances*

of Death (1875–7). He is also noted for the piano suite *Pictures at an Exhibition* (1874). After his death many of his works were completed and altered by Rimsky-Korsakov and others, but recently there has been a tendency to return to Mussorgsky's original scoring.

Mussulman /ˈmʌs(ə)lmən/ *n. & adj. archaic* ● *n.* (*pl.* **-mans** or **-men**) a Muslim. ● *adj.* of or concerning Muslims. [Pers. *musulmān* orig. adj. f. *muslim* (as MUSLIM)]

must[1] /mʌst/ *v. & n.* ● *v.aux.* (*3rd sing. present* **must**; *past* **had to** or in indirect speech **must**) (foll. by infin., or *absol.*) **1 a** be obliged to (*you must go to school; must we leave now?; said he must go; I must away*). ¶ The negative (i.e. lack of obligation) is expressed by *not have to* or *need not*; *must not* denotes positive forbidding, as in *you must not smoke*. **b** in ironic questions (*must you slam the door?*). **2** be certain to (*we must win in the end; you must be her sister; he must be mad; they must have left by now; seemed as if the roof must blow off*). **3** ought to (*we must see what can be done; it must be said that*). **4** expressing insistence (*I must ask you to leave*). **5** (foll. by *not* + infin.) **a** not be permitted to, be forbidden to (*you must not smoke*). **b** ought not; need not (*you mustn't think he's angry; you must not worry*). **c** expressing insistence that something should not be done (*they must not be told*). **6** (as past or historic present) expressing the perversity of destiny (*what must I do but break my leg*). ● *n. colloq.* a thing that cannot or should not be overlooked or missed (*if you go to London St Paul's is a must*). □ **I must say** often *iron.* I cannot refrain from saying (*I must say he made a good attempt; a fine way to behave, I must say*). **must needs** see NEEDS. [OE *mōste* past of *mōt* may]

must[2] /mʌst/ *n.* grape juice before fermentation is complete. [OE f. L *mustum* neut. of *mustus* new]

must[3] /mʌst/ *n.* mustiness, mould. [back-form. f. MUSTY]

must[4] /mʌst/ *adj. & n.* (also **musth**) ● *adj.* (of a male elephant or camel) in a state of frenzy associated with the breeding season. ● *n.* this state. [Urdu f. Pers. *mast* intoxicated]

mustache *US* var. of MOUSTACHE.

mustachio /məˈstaːʃɪˌəʊ/ *n.* (*pl.* **-os**) (often in *pl.*) *archaic* a moustache. □ **mustachioed** *adj.* [Sp. *mostacho* & It. *mostaccio* (as MOUSTACHE)]

mustang /ˈmʌstæŋ/ *n.* a small wild horse native to Mexico and California. □ **mustang grape** a grape from the wild vine *Vitis candicans*, of the southern US, used for making wine. [Sp. *mestengo* f. *mesta* company of graziers, & Sp. *mostrenco*]

mustard /ˈmʌstəd/ *n.* **1 a** a plant of the genus *Brassica*, with slender pods and yellow flowers, esp. *B. nigra*. **b** a plant of the genus *Sinapis*; esp. *S. alba*, eaten at the seedling stage, often with cress. **2** the seeds of these, which are crushed, made into a paste, and used as a spicy condiment. **3** the brownish-yellow colour of this condiment. **4** *sl.* a thing which adds piquancy or zest. □ **mustard gas** a colourless oily liquid, whose vapour is a powerful irritant and vesicant. **mustard plaster** a poultice made with mustard. **mustard seed 1** the seed of the mustard plant. **2** a small thing capable of great development (Matt. 13:31). [ME f. OF *mo(u)starde*: orig. the condiment as prepared with MUST[2]]

mustelid /ˈmʌstɪlɪd, məˈstelɪd/ *n. & adj. Zool.* ● *n.* a flesh-eating mammal of the family Mustelidae, which includes weasels, stoats, badgers, skunks, martens, etc. ● *adj.* of or relating to this family. [mod.L *Mustelidae* f. L *mustela* weasel]

muster /ˈmʌstə(r)/ *v. & n.* ● *v.* **1** *tr.* collect (orig. soldiers) for inspection, to check numbers, etc. **2** *tr. & intr.* collect, gather together. **3** *tr.* (often foll. by *up*) summon up (courage, strength, etc.). **4** *tr. Austral.* round up (livestock). ● *n.* **1** the assembly of persons for inspection. **2** an assembly, a collection. **3** *Austral.* a rounding up of livestock. **4** *Austral. sl.* the number of people attending (a meeting, etc.) (*had a good muster*). □ **muster-book** a book for registering military personnel. **muster in** *US* enrol (recruits). **muster out** *US* discharge (soldiers etc.). **muster-roll** an official list of officers and men in a regiment or ship's company. **pass muster** be accepted as adequate. □ **musterer** *n.* (in sense 3 of n. & 4 of v.). [ME f. OF *mo(u)stre* ult. f. L *monstrare* show]

musth var. of MUST[4].

Mustique /mʊˈstiːk/ a small resort island in the northern Grenadines, in the Caribbean to the south of St Vincent.

mustn't /ˈmʌs(ə)nt/ *contr.* must not.

musty /ˈmʌstɪ/ *adj.* **1** mouldy. **2** of a mouldy or stale smell or taste. **3** stale, antiquated (*musty old books*). □ **mustily** *adv.* **mustiness** *n.* [perh. alt. f. *moisty* (MOIST) by assoc. with MUST[2]]

Mut /mʊt/ *Egyptian Mythol.* a goddess who was the wife of Amun and mother of Khonsu. Her name means 'the mother'.

mutable /'mjuːtəb(ə)l/ *adj. literary* **1** liable to change. **2** fickle. □ **mutability** /ˌmjuːtə'bɪlɪtɪ/ *n.* [L *mutabilis* f. *mutare* change]

mutagen /'mjuːtədʒən/ *n.* an agent promoting mutation, e.g. radiation. □ **mutagenic** /ˌmjuːtə'dʒenɪk/ *adj.* **mutagenesis** *n.* [MUTATION + -GEN]

mutant /'mjuːt(ə)nt/ *adj. & n.* ● *adj.* resulting from mutation. ● *n.* a mutant form. [L *mutant-* part. f. *mutare* change]

Mutare /muːˈtɑːrɪ/ an industrial town in the eastern highlands of Zimbabwe; pop. (1982) 69,620. It was known as Umtali until 1982.

mutate /mjuːˈteɪt/ *v.intr. & tr.* undergo or cause to undergo mutation. [back-form. f. MUTATION]

mutation /mjuːˈteɪʃ(ə)n/ *n.* **1** the process or an instance of change or alteration. **2** a genetic change which, when transmitted to offspring, gives rise to heritable variations. **3** a mutant. **4 a** an umlaut. **b** (in a Celtic language) a change of a consonant etc. determined by a preceding word. □ **mutational** *adj.* **mutationally** *adv.* [ME f. L *mutatio* f. *mutare* change]

mutatis mutandis /muːˌtɑːtɪs muːˈtændɪs, mjuː-/ *adv.* (in comparing cases) making the necessary alterations. [L]

mutch /mʌtʃ/ *n. dial.* a woman's or child's linen cap. [ME f. MDu. *mutse* MHG *mütze* f. med.L *almucia* AMICE[2]]

mute /mjuːt/ *adj., n., & v.* ● *adj.* **1** silent, refraining from or temporarily bereft of speech. **2** not emitting articulate sound. **3** (of a person or animal) dumb. **4** not expressed in speech (*mute protest*). **5 a** (of a letter) not pronounced. **b** (of a consonant) plosive. **6** (of hounds) not giving tongue. ● *n.* **1** a dumb person (*a deaf mute*). **2** *Mus.* **a** a clamp for damping the resonance of the strings of a violin etc. **b** a pad or cone for damping the sound of a wind instrument. **3** an unsounded consonant. **4** an actor whose part is in a dumb show. **5** a dumb servant in oriental countries. **6** a hired mourner. ● *v.tr.* **1** deaden, muffle, or soften the sound of (a thing, esp. a musical instrument). **2 a** tone down, make less intense. **b** (as **muted** *adj.*) (of colours etc.) subdued (*a muted green*). □ **mute button** a device on a telephone etc. to temporarily prevent the caller from hearing what is being said at the receiver's end. **mute swan** the commonest Eurasian swan, *Cygnus olor*, having white plumage and an orange-red bill with a swollen black base. □ **mutely** *adv.* **muteness** *n.* [ME f. OF *muet*, dimin. of *mu* f. L *mutus*, assim. to L]

mutilate /'mjuːtɪˌleɪt/ *v.tr.* **1 a** deprive (a person or animal) of a limb or organ. **b** destroy the use of (a limb or organ). **2** render (a book etc.) imperfect by excision or some act of destruction. □ **mutilator** *n.* **mutilation** /ˌmjuːtɪ'leɪʃ(ə)n/ *n.* [L *mutilare* f. *mutilus* maimed]

mutineer /ˌmjuːtɪ'nɪə(r)/ *n.* a person who mutinies. [F *mutinier* f. *mutin* rebellious f. *muete* movement ult. f. L *movere* move]

mutinous /'mjuːtɪnəs/ *adj.* rebellious; tending to mutiny. □ **mutinously** *adv.* [obs. *mutine* rebellion f. F *mutin*: see MUTINEER]

mutiny /'mjuːtɪnɪ/ *n. & v.* ● *n.* (*pl.* **-ies**) an open revolt against constituted authority, esp. by soldiers or sailors against their officers. ● *v.intr.* (**-ies**, **-ied**) (often foll. by *against*) revolt; engage in mutiny. [obs. *mutine* (as MUTINOUS)]

mutism /'mjuːtɪz(ə)m/ *n.* muteness; silence; dumbness. [F *mutisme* f. L (as MUTE)]

muton /'mjuːtɒn/ *n. Biol.* the smallest element of genetic material capable of giving rise to a mutant individual.

Mutsuhito /ˌmʊtsuːˈhiːtəʊ/ see MEIJI TENNO.

mutt /mʌt/ *n.* **1** *sl.* an ignorant, stupid, or blundering person. **2** *derog.* or *joc.* a dog, esp. a mongrel. [abbr. of *mutton-head*]

mutter /'mʌtə(r)/ *v. & n.* ● *v.* **1** *intr.* speak low in a barely audible manner. **2** *intr.* (often foll. by *against, at*) murmur or grumble about. **3** *tr.* utter (words etc.) in a low tone. **4** *tr.* say in secret. ● *n.* **1** muttered words or sounds. **2** muttering. □ **mutterer** *n.* **mutteringly** *adv.* [ME, rel. to MUTE]

mutton /'mʌt(ə)n/ *n.* **1** the flesh of sheep used for food. **2** *joc.* a sheep. □ **mutton-bird** *Austral. & NZ* **1** a shearwater of the genus *Puffinus*, esp. *P. tenuirostris*. **2** an Antarctic petrel of the genus *Pterodroma*. **mutton chop 1** a piece of mutton, usu. the rib and half vertebra to which it is attached. **2** (in full **mutton chop whisker**) a side whisker shaped like this. **mutton dressed as lamb** *colloq.* a usu. middle-aged or elderly woman dressed or made up to appear younger. **mutton-head** *colloq.* a dull, stupid person. **mutton-headed** *colloq.* dull, stupid. □ **muttony** *adj.* [ME f. OF *moton* f. med.L *multo -onis* prob. f. Gaulish]

mutual /'mjuːtʃʊəl, 'mjuːtjʊ-/ *adj.* **1** (of feelings, actions, etc.) experienced or done by each of two or more parties with reference to the other or others (*mutual affection*). **2** common to two or more persons (*a mutual friend; a mutual interest*). **3** standing in (a specified) relation to each other (*mutual well-wishers; mutual beneficiaries*). □ **mutual fund** *N. Amer.* a unit trust. **mutual inductance** the property of an electric circuit that causes an electromotive force to be generated in it by change in the current flowing through a magnetically linked circuit. **mutual induction** the production of an electromotive force between adjacent circuits that are magnetically linked. **mutual insurance** insurance in which some or all of the profits are divided among the policyholders. □ **mutually** *adv.* **mutuality** /ˌmjuːtʃʊ'ælɪtɪ, ˌmjuːtjʊ-/ *n.* [ME f. OF *mutuel* f. L *mutuus* mutual, borrowed, rel. to *mutare* change]

mutualism /'mjuːtʃʊəˌlɪz(ə)m, 'mjuːtjʊ-/ *n.* **1** the doctrine that mutual dependence is necessary to social well-being. **2** mutually beneficial symbiosis. □ **mutualist** *n. & adj.* **mutualistic** /ˌmjuːtʃʊə'lɪstɪk, ˌmjuːtjʊ-/ *adj.* **mutualistically** *adv.*

mutuel /'mjuːtʃʊəl, 'mjuːtjʊ-/ *n. esp. US* a totalizator; a *pari-mutuel*. [abbr. of PARI-MUTUEL]

mutule /'mjuːtjuːl/ *n. Archit.* a block derived from the ends of wooden beams projecting under a Doric cornice. [F f. L *mutulus*]

muu-muu /'muːmuː/ *n.* a woman's loose brightly coloured dress. [Hawaiian]

Muzaffarabad /ˌmʊzəˌfærə'bæd/ a town in NE Pakistan, the administrative centre of Azad Kashmir.

Muzak /'mjuːzæk/ *n.* **1** *propr.* a system of music transmission for playing in public places. **2** (**muzak**) recorded light background music. [alt. f. MUSIC]

muzhik /'muːʒɪk/ *n.* (also **moujik**) *hist.* a Russian peasant. [Russ. *muzhik*]

Muztag /muːs'tɑːg/ a mountain in western China, on the north Tibetan border close to the Karamiran Shankou pass. Rising to 7,723 m (25,338 ft), it is the highest peak in the Kunlun Shan range.

muzzle /'mʌz(ə)l/ *n. & v.* ● *n.* **1** the projecting part of an animal's face, including the nose and mouth. **2** a guard, usu. made of straps or wire, fitted over an animal's nose and mouth to stop it biting or feeding. **3** the open end of a firearm. ● *v.tr.* **1** put a muzzle on (an animal etc.). **2** impose silence upon. **3** *Naut.* take in (a sail). □ **muzzle-loader** a gun that is loaded through the muzzle. **muzzle velocity** the velocity with which a projectile leaves the muzzle of a gun. [ME f. OF *musel* ult. f. med.L *musum*: cf. MUSE[2]]

muzzy /'mʌzɪ/ *adj.* (**muzzier**, **muzziest**) **1 a** mentally hazy; dull, spiritless. **b** dazed from drinking alcohol. **2** blurred, indistinct. □ **muzzily** *adv.* **muzziness** *n.* [18th c.: orig. unkn.]

MV *abbr.* **1** motor vessel. **2** muzzle velocity. **3** megavolt(s).

MVD the Ministry of Internal Affairs, the secret police of the former USSR which, together with the MGB (Ministry of State Security), replaced the NKVD in 1946. On Stalin's death (1953) the MVD and MGB were merged under Lavrenti Beria and remained in charge of internal security, but the organization's responsibilities were gradually downgraded in favour of the KGB. [Russ. abbr.]

MVO *abbr.* (in the UK) Member of the Royal Victorian Order.

MVP *abbr. US Sport* most valuable player.

MW *abbr.* **1** megawatt(s). **2** medium wave.

mW *abbr.* milliwatt(s).

Mx. *abbr.* **1** maxwell(s). **2** Middlesex.

MY *abbr.* motor yacht.

my /maɪ/ *poss.pron.* (*attrib.*) **1** of or belonging to me or myself (*my house; my own business*). **2** as a form of address in affectionate, sympathetic, jocular, or patronizing contexts (*my dear boy*). **3** in various expressions of surprise (*my God!; oh my!*). **4** *Brit. colloq.* indicating the speaker's husband, wife, child, etc. (*my Johnny's ill again*). □ **my Lady** (or **Lord**) the form of address to certain titled persons. [ME *mī*, reduced f. *mīn* MINE[1]]

my- *comb. form* var. of MYO-.

myalgia /maɪ'ældʒə/ *n.* pain in a muscle or group of muscles. [mod.L f. Gk *mus* muscle]

myalgic /maɪ'ældʒɪk/ *adj.* relating to or involving myalgia. □ **myalgic encephalomyelitis** (abbr. **ME**) a condition of unknown cause, with fever, aching, prolonged tiredness, and depression, occurring esp. after a viral infection.

myalism /'maɪəˌlɪz(ə)m/ *n.* a kind of sorcery akin to obeah, practised esp. in the West Indies. [*myal*, prob. of W.Afr. orig.]

myall /'maɪəl/ *n.* **1 a** a leguminous tree of the genus *Acacia*, esp. *A. pendula*, native to Australia. **b** the hard scented wood of this, used for fences and tobacco-pipes. **2** an Australian Aboriginal living in a traditional way. [Dharuk *mayal, miyal* stranger, person of another tribe]

Myanmar /ˌmaɪæn'mɑ:(r), 'mjæmɑ:(r)/ the official name (since 1989) for BURMA.

myasthenia /ˌmaɪəs'θi:nɪə/ n. Med. a condition causing abnormal weakness of certain muscles; esp. (in full **myasthenia gravis** /'ɡrævɪs/) a rare chronic autoimmune disease marked by muscular weakness without atrophy. [mod.L f. Gk mus muscle: cf. ASTHENIA]

mycelium /maɪ'si:lɪəm/ n. (pl. **mycelia** /-lɪə/) the vegetative part of a fungus, consisting of microscopic threadlike hyphae. □ **mycelial** adj. [f. MYCO- after EPITHELIUM]

Mycenae /maɪ'si:ni:/ an ancient city in Greece, situated near the coast in the NE Peloponnese, on a site dominating various land and sea routes. The capital of King Agamemnon, it was the centre of the late Bronze Age Mycenaean civilization. Its period of greatest prosperity was c.1400–1200 BC, which saw construction of the palace and the massive walls of Cyclopean masonry, including the 'Lion Gate', the entrance to the citadel (c.1250 BC). The city was destroyed in about 1100 BC by invading Dorians. Systematic excavation of the site began in 1840.

Mycenaean /ˌmaɪsɪ'ni:ən/ adj. & n. ● adj. of or relating to the late Bronze Age civilization in Greece depicted in the Homeric poems and represented by finds at Mycenae and other ancient cities of the Peloponnese. (See note below.) ● n. an inhabitant of Mycenae or the Mycenaean world. [L Mycenaeus]

▪ The Mycenaeans inherited control of the Aegean after the collapse of the Minoan civilization c.1400 BC. Their cities, such as those of Mycenae, Tiryns, and Pylos, were well populated and prosperous, with fortified citadels enclosing impressive palaces. The language they used was a form of Greek; in writing they used the Linear B script. Trade with other Mediterranean countries flourished, and Mycenaean products are found from southern Italy to Palestine and Syria. The end of Mycenaean power c.1100 BC coincided with a period of general upheaval and migrations at the end of the Bronze Age in the Mediterranean.

-mycin /'maɪsɪn/ comb. form used to form the names of antibiotic compounds derived from fungi. [MYCO- + -IN]

myco- /'maɪkəʊ/ comb. form fungus. [Gk mukēs]

mycology /maɪ'kɒlədʒɪ/ n. **1** the study of fungi. **2** the fungi of a particular region. □ **mycologist** n. **mycological** /ˌmaɪkə'lɒdʒɪk(ə)l/ adj. **mycologically** adv.

mycoplasma /ˌmaɪkəʊ'plæzmə/ n. (pl. **mycoplasmas**, **mycoplasmata** /-mətə/) any of a group of mainly parasitic micro-organisms smaller than bacteria and without a cell wall. [mod.L f. MYCO- + PLASMA]

mycoprotein /ˌmaɪkə,prəʊti:n/ n. protein derived from fungi, esp. as produced for human consumption.

mycorrhiza /ˌmaɪkə'raɪzə/ n. (pl. **mycorrhizae** /-zi:/) a symbiotic association of a fungus and the roots of a plant. □ **mycorrhizal** adj. [MYCO- + Gk rhiza root]

mycosis /maɪ'kəʊsɪs/ n. Med. a disease caused by a fungus, e.g. ringworm and athlete's foot. □ **mycotic** /-'kɒtɪk/ adj.

mycotoxin /ˌmaɪkə'tɒksɪn/ n. any toxic substance produced by a fungus.

mycotrophy /maɪ'kɒtrəfɪ/ n. the condition of a plant which has mycorrhizae and is perhaps helped to assimilate nutrients as a result. [MYCO- + Gk trophē nourishment]

mydriasis /mɪ'draɪəsɪs/ n. excessive dilation of the pupil of the eye. [L f. Gk mudriasis]

myelin /'maɪɪlɪn/ n. a white fatty substance which forms an insulating sheath around certain nerve-fibres. □ **myelination** /ˌmaɪɪlɪ'neɪʃ(ə)n/ n. [Gk muelos marrow + -IN]

myelitis /maɪ'laɪtɪs/ n. inflammation of the spinal cord. [mod.L f. Gk muelos marrow]

myeloid /'maɪɪ,lɔɪd/ adj. of or relating to bone marrow or the spinal cord. [Gk muelos marrow]

myeloma /ˌmaɪɪ'ləʊmə/ n. (pl. **myelomas** or **myelomata** /-mətə/) Med. a malignant tumour of the bone marrow. [as MYELITIS + -OMA]

Mykolayiv /ˌmɪkə'laɪf/ (Russian **Nikolaev** /ˌnɪkə'laɪf/) an industrial city in southern Ukraine, on the Southern Bug river near its confluence with the Dnieper and their joint estuaries on the northern shores of the Black Sea; pop. (1990) 507,900.

Mykonos /'mɪkə,nɒs/ (Greek **Míkonos** /'mikɒnɔs/) a Greek island in the Aegean, one of the Cyclades.

mylodon /'maɪləd(ə)n/ n. an extinct giant ground sloth of the genus Mylodon, with cylindrical teeth, living in South America during the Pleistocene ice ages. [mod.L f. Gk mulē mill, molar + odous odontos tooth]

Mymensingh /ˌmaɪmən'sɪŋ/ a port on the Brahmaputra river in central Bangladesh; pop. (1991) 186,000.

mynah /'maɪnə/ n. (also **myna**, **mina**) a SE Asian bird related to the starling; esp. Gracula religiosa, able to mimic the human voice. [Hindi mainā]

myo- /'maɪəʊ/ comb. form (also **my-** before a vowel) muscle. [Gk mus muos muscle]

myocardium /ˌmaɪəʊ'kɑ:dɪəm/ n. (pl. **myocardia** /-dɪə/) the muscular tissue of the heart. □ **myocardial** adj. [MYO- + Gk kardia heart]

myofibril /ˌmaɪəʊ'faɪbrɪl/ n. any of the elongated contractile threads found in striated muscle cells.

myogenic /ˌmaɪə'dʒenɪk/ adj. originating in muscle tissue.

myoglobin /ˌmaɪəʊ'ɡləʊbɪn/ n. an oxygen-carrying protein containing iron and found in muscle cells.

myology /maɪ'ɒlədʒɪ/ n. the study of the structure and function of muscles.

myope /'maɪəʊp/ n. a short-sighted person. [F f. LL myops f. Gk muōps f. muō shut + ōps eye]

myopia /maɪ'əʊpɪə/ n. **1** short-sightedness. **2** lack of imagination or intellectual insight. □ **myopic** /-'ɒpɪk/ adj. **myopically** adv. [mod.L (as MYOPE)]

myosin /'maɪəsɪn/ n. Biochem. a protein which with actin forms the contractile filaments of muscle. [MYO- + -OSE² + -IN]

myosis var. of MIOSIS.

myosotis /ˌmaɪə'səʊtɪs/ n. (also **myosote** /'maɪə,səʊt/) a plant of the genus Myosotis, with small blue, pink, or white flowers, esp. a forget-me-not. [L f. Gk muosōtis f. mus muos mouse + ous ōtos ear]

myotonia /ˌmaɪə'təʊnɪə/ n. the inability to relax voluntary muscle after vigorous effort. □ **myotonic** /-'tɒnɪk/ adj. [MYO- + Gk tonos tone]

myriad /'mɪrɪəd/ n. & adj. literary ● n. **1** an indefinitely great number. **2** ten thousand. ● adj. of an indefinitely great number. [LL mirias miriad- f. Gk murias -ados f. murioi 10,000]

myriapod /'mɪrɪə,pɒd/ n. & adj. Zool. ● n. an elongated land-living arthropod of the class Myriapoda, with numerous leg-bearing segments, including the centipedes and millipedes. ● adj. of or relating to this class. [mod.L Myriapoda (as MYRIAD, Gk pous podos foot)]

myrmecology /ˌmɜ:mɪ'kɒlədʒɪ/ n. the scientific study of ants. □ **myrmecologist** n. **myrmecological** /-kə'lɒdʒɪk(ə)l/ adj. [Gk murmēk- murmēx ant]

myrmidon /'mɜ:mɪd(ə)n/ n. **1** a hired ruffian. **2** a lowly servant. [L Myrmidones (pl.) f. Gk Murmidones, warlike Thessalian people who went with Achilles to Troy]

myrobalan /ˌmaɪə'rɒbələn/ n. **1** (in full **myrobalan plum**) = cherry plum. **2** (in full **myrobalan nut**) the fruit of an Asian tree, Terminalia chebula, used in medicines, for tanning leather, and to produce inks and dyes. [F myrobolan or L myrobolanum f. Gk murobalanos f. muron unguent + balanos acorn]

Myron /'maɪərən/ (fl. c.480–440 BC), Greek sculptor. Only two certain copies of his work survive, the best known being the Discobolus (c.450 BC), a figure of a man throwing the discus, which demonstrates a remarkable interest in symmetry and movement.

myrrh¹ /mɜ:(r)/ n. a gum resin from trees of the genus Commiphora, used, esp. in the Near East, in perfumery, medicine, incense, etc. □ **myrrhic** adj. **myrrhy** adj. [OE myrra, myrre f. L myrr(h)a f. Gk murra, of Semitic orig.]

myrrh² /mɜ:(r)/ n. = sweet cicely. [L myrris f. Gk murris]

myrtaceous /mɜ:'teɪʃ(ə)s/ adj. Bot. of or relating to the plant family Myrtaceae, which includes the myrtles and eucalypts.

myrtle /'mɜ:t(ə)l/ n. **1** an evergreen shrub of the genus Myrtus, with aromatic foliage and white flowers; esp. M. communis of southern Europe, bearing purple-black ovoid berries. **2** US = PERIWINKLE¹. [ME f. med.L myrtilla, -us dimin. of L myrta, myrtus f. Gk murtos]

myself /maɪ'self/ pron. **1** emphat. form of I² or ME¹ (I saw it myself; I like to do it myself). **2** refl. form of ME¹ (I was angry with myself; able to dress myself; as bad as myself). **3** in my normal state of body and mind (I'm not myself today). **4** poet. = I². □ **by myself** see by oneself. **I myself** I for my part (I myself am doubtful). [ME¹ + SELF: my- partly after herself with her regarded as poss. pron.]

Mysia /ˈmɪsɪə/ an ancient region of NW Asia Minor, on the Mediterranean coast south of the Sea of Marmara. □ **Mysian** adj. & n.

Mysore /maɪˈsɔː(r)/ **1** a city in the Indian state of Karnataka; pop. (1991) 480,000. It was the former capital of the princely state of Mysore and is noted for the production of silk, incense, and sandalwood oil. **2** the former name (until 1973) for KARNATAKA.

mysterious /mɪˈstɪərɪəs/ adj. **1** full of or wrapped in mystery. **2** (of a person) delighting in mystery. □ **mysteriously** adv. **mysteriousness** n. [F mystérieux f. mystère f. OF (as MYSTERY¹)]

mystery¹ /ˈmɪstərɪ/ n. (pl. **-ies**) **1** a secret, hidden, or inexplicable matter (the reason remains a mystery). **2** secrecy or obscurity (wrapped in mystery). **3** (attrib.) secret, undisclosed (mystery guest). **4** the practice of making a secret of (esp. unimportant) things (engaged in mystery and intrigue). **5** (in full **mystery story**) a fictional work dealing with a puzzling event, esp. a crime (a well-known mystery writer). **6 a** (in Christian theology) a religious truth divinely revealed, esp. one beyond human reason. **b** RC Ch. a decade of the rosary. **7** (in pl.) **a** the secret religious rites of the ancient Greeks, Romans, etc. (See note below.) **b** archaic the Eucharist. □ **make a mystery of** treat as an impressive secret. **mystery tour** Brit. a pleasure excursion to an unspecified destination. [ME f. OF mistere or L mysterium f. Gk mustērion, rel. to MYSTIC]

- The mysteries or mystery religions were secret forms of worship, in contradistinction to the very public nature of most Greek and Roman religious observances, and were available only to people who had been specially initiated. They involved ritual purification, and probably an enactment of the myth of Demeter and Persephone; initiates were promised a happy existence in another world after death. The Eleusinian mysteries, celebrated at Eleusis, were the most famous, and attracted visitors from the whole of Greece.

mystery² /ˈmɪstərɪ/ n. (pl. **-ies**) archaic a handicraft or trade, esp. as referred to in indentures etc. (art and mystery). [ME f. med.L misterium contr. of ministerium MINISTRY, assoc. with MYSTERY¹]

mystery play n. (also called miracle play) a medieval drama based on a religious story and performed in the vernacular. During the 13th century trade guilds in Europe started producing plays based on biblical stories; originally these were performed in churches, but they became increasingly secular and began to be staged elsewhere; later productions introduced apocryphal elements and were often satirical. In England they were performed on temporary stages or on wagons which were trundled along an established route, stopping at fixed points where the audience awaited them. Individual dramas merged into a cycle of plays: the best known are those of York, Chester, Coventry, and Wakefield.

mystic /ˈmɪstɪk/ n. & adj. ● n. a person who seeks by contemplation and self-surrender to obtain union with or absorption into God, or who believes in the spiritual apprehension of truths that are beyond the understanding. ● adj. **1** mysterious and awe-inspiring. **2** spiritually allegorical or symbolic. **3** occult, esoteric. **4** of hidden meaning. □ **mysticism** /-tɪˌsɪz(ə)m/ n. [ME f. OF mystique or L mysticus f. Gk mustikos f. mustēs initiated person f. muō close the eyes or lips, initiate]

mystical /ˈmɪstɪk(ə)l/ adj. of mystics or mysticism. □ **mystically** adv.

mystify /ˈmɪstɪˌfaɪ/ v.tr. (**-ies**, **-ied**) **1** bewilder, confuse. **2** hoax, take advantage of the credulity of. **3** wrap up in mystery. □ **mystification** /ˌmɪstɪfɪˈkeɪʃ(ə)n/ n. **mystifying** adj. **mystifyingly** adv. [F mystifier (irreg. formed as MYSTIC or MYSTERY¹)]

mystique /mɪˈstiːk/ n. **1** an atmosphere of mystery and veneration attending some activity or person. **2** any skill or technique impressive or mystifying to the layman. [F f. OF (as MYSTIC)]

myth /mɪθ/ n. **1** a traditional narrative usu. involving supernatural or imaginary persons and often embodying popular ideas on natural or social phenomena etc. **2** such narratives collectively. **3** a widely held but false notion. **4** a fictitious person, thing, or idea. **5** an allegory (the Platonic myth). □ **mythic** adj. **mythical** adj. **mythically** adv. [mod.L mythus f. LL mythos f. Gk muthos]

mythi pl. of MYTHUS.

mythicize /ˈmɪθɪˌsaɪz/ v.tr. (also **-ise**) treat (a story etc.) as a myth; interpret mythically. □ **mythicism** n. **mythicist** n.

mytho- /ˈmɪθəʊ/ comb. form myth.

mythogenesis /ˌmɪθəʊˈdʒenɪsɪs/ n. the production of myths.

mythographer /mɪˈθɒɡrəfə(r)/ n. a compiler of myths.

mythography /mɪˈθɒɡrəfɪ/ n. the representation of myths in plastic art.

mythology /mɪˈθɒlədʒɪ/ n. (pl. **-ies**) **1** a body of myths (Greek mythology). **2** the study of myths. □ **mythologist** n. **mythologize** v.tr. & intr. (also **-ise**). **mythologizer** n. **mythological** /ˌmɪθəˈlɒdʒɪk(ə)l/ adj. **mythologically** adv. [ME f. F mythologie or LL mythologia f. Gk muthologia (as MYTHO-, -LOGY)]

mythomania /ˌmɪθəʊˈmeɪnɪə/ n. an abnormal tendency to exaggerate or tell lies. □ **mythomaniac** /-nɪˌæk/ n. & adj.

mythopoeia /ˌmɪθəʊˈpiːə/ n. the making of myths. □ **mythopoeic** adj. (also **mythopoetic** /-pəʊˈetɪk/).

mythus /ˈmɪθəs/ n. (pl. **mythi** /-θaɪ/) literary a myth. [mod.L: see MYTH]

Mytilene /ˌmɪtɪˈliːnɪ/ (Greek **Mitilíni** /mitiˈlini/) the chief town of the Greek island of Lesbos; pop. (1980) 24,115.

myxo- /ˈmɪksəʊ/ comb. form (also **myx-** before a vowel) mucus. [Gk muxa mucus]

myxoedema /ˌmɪksəʊˈdiːmə/ n. (US **myxedema**) a syndrome caused by hypothyroidism, resulting in thickening of the skin, weight gain, mental dullness, loss of energy, and sensitivity to cold.

myxoma /mɪkˈsəʊmə/ n. (pl. **myxomas** or **myxomata** /-mətə/) Med. a benign tumour of mucous or gelatinous tissue. □ **myxomatous** /-ˈsɒmətəs/ adj.

myxomatosis /ˌmɪksəmɑˈtəʊsɪs/ n. an infectious usu. fatal viral disease in rabbits, causing swelling of the mucous membranes.

myxomycete /ˌmɪksəʊˈmaɪsiːt/ n. an acellular slime mould of the class Myxomycetes, usu. consisting of a multinucleate plasmodium. [mod.L Myxomycetes f. MYXO- + pl. of Gk mukēs -ētos fungus]

myxovirus /ˈmɪksəʊˌvaɪərəs/ n. any of a group of viruses including the influenza virus.

Nn

N¹ /en/ n. (also **n**) (pl. **Ns** or **N's**) **1** the fourteenth letter of the alphabet. **2** *Printing* en.

N² abbr. (also **N.**) **1** North; Northern. **2** *Chess* knight. **3** New. **4** nuclear.

N³ symb. **1** *Chem.* the element nitrogen. **2** newton(s).

n¹ abbr. (also **n.**) **1** name. **2** neuter. **3** noon. **4** note. **5** noun.

n² symb. **1** *Math.* an indefinite number. **2** nano-. □ **to the nth** (or **nth degree**) **1** *Math.* to any required power. **2** to any extent; to the utmost.

'n /ən/ conj. (also **'n'**) colloq. and. [abbr.]

-n¹ suffix see -EN².

-n² suffix see -EN³.

Na symb. *Chem.* the element sodium.

na /nə/ adv. Sc. (in comb.; usu. with an auxiliary verb) = NOT (*I canna do it; they didna go*).

n/a abbr. **1** not applicable. **2** not available.

NAACP see NATIONAL ASSOCIATION FOR THE ADVANCEMENT OF COLORED PEOPLE.

NAAFI /'næfı/ abbr. Brit. **1** Navy, Army, and Air Force Institutes. **2** a canteen for members of the armed forces run by the NAAFI.

naan var. of NAN².

Naas /neıs/ the county town of Kildare in the Republic of Ireland; pop. (1991) 11,140.

nab /næb/ v.tr. (**nabbed**, **nabbing**) colloq. **1** arrest; catch in wrongdoing. **2** seize, grab. [17th c., also *napp*, as in KIDNAP: orig. unkn.]

Nabataean /ˌnæbəˈtiːən/ n. & adj. ● n. **1** a member of an ancient Arabian people. (*See note below.*) **2** the language of this people, a form of Aramaic strongly influenced by Arabic. ● adj. of or relating to the Nabataeans or their language. [L *Nabataeus*, Gk *Nabataios* (*Nebatu* native name of country)]

- Probably originally a nomadic Arab tribe, the Nabataeans formed an independent state 312–63 BC and prospered from control of the trade which converged at their capital Petra (now in Jordan). From AD 63 they became allies and vassals of Rome, and in AD 106 their kingdom was transformed into the Roman province of Arabia. Their speech, religion, art, and architecture reflected Babylonian, Arab, Greek, and Roman influence.

Nabeul /næˈbɜːl/ a resort town in NE Tunisia, on the Cape Bon peninsula; pop. (1984) 39,500.

Nabi Group /'nɑːbı/ a group of late 19th-century French painters, largely symbolist in their approach and heavily indebted to Gauguin. Members of the group included Maurice Denis, Pierre Bonnard, and Edouard Vuillard. [Heb. *nābī* prophet]

Nablus /'nɑːbləs/ a town in the West Bank; pop. (est. 1984) 80,000. It is close to the site of the Canaanite city of Shechem, important in ancient times because of its position on an east–west route through the mountains of Samaria.

nabob /'neıbɒb/ n. hist. a Muslim official or governor under the Mogul empire. **2** a person of conspicuous wealth or high rank, esp. hist. one returned from India with a fortune. [Port. *nababo* or Sp. *nabab*, f. Urdu (as NAWAB)]

Nabokov /'næbəˌkɒf, nəˈbəʊkɒf/, Vladimir (Vladimorovich) (1899–1977), Russian-born American novelist and poet. After writing a number of novels in Russian, Nabokov turned to writing in English in 1941. He is best known for *Lolita* (1958), his novel about a middle-aged European man's obsession with a twelve-year-old American girl. Other works include *Pale Fire* (1962) and *Ada: A Family Chronicle* (1969). He was also a keen lepidopterist and wrote a number of papers on the subject.

Nacala /nəˈkɑːlə/ a deep-water port on the east coast of Mozambique; pop. (1990) 104,300. It is linked by rail with landlocked Malawi.

nacarat /'nækəˌræt/ n. a bright orange-red colour. [F, perh. f. Sp. & Port. *nacardo* (*nacar* NACRE)]

nacelle /nəˈsel/ n. **1** the outer casing of the engine of an aircraft. **2** the car of an airship. [F, f. LL *navicella* dimin. of L *navis* ship]

nacho /'nætʃəʊ/ n. (pl. **-os**) (usu. in pl.) a tortilla chip, usu. topped with melted cheese and spices etc. [20th c.: orig. uncert.]

NACODS /'neıkɒdz/ abbr. (in the UK) National Association of Colliery Overmen, Deputies, and Shotfirers.

nacre /'neıkə(r)/ n. mother-of-pearl from any shelled mollusc. □ **nacreous** /-krıəs/ adj. [F]

NAD abbr. Biochem. nicotinamide adenine dinucleotide, a coenzyme important in many biological oxidation reactions.

Nader /'neıdə(r)/, Ralph (b.1934), American lawyer and reformer. He initiated a campaign on behalf of public safety that gave impetus to the consumer rights movement of the 1960s onwards. His views on defective car design, set out in *Unsafe at Any Speed* (1965), led to Federal legislation on safety standards. Nader was also a moving force behind legislation concerning radiation hazards, food packaging, and the use of insecticides.

nadir /'neıdıə(r), 'næd-/ n. **1** Astron. the part of the celestial sphere directly below an observer (opp. ZENITH). **2** the lowest point in one's fortunes; a time of deep despair. [ME f. OF f. Arab. *naẓīr* (*as-samt*) opposite (to the zenith)]

naevus /'niːvəs/ n. (US **nevus**) (pl. **naevi** /-vaı/) **1** a birthmark in the form of a raised red patch on the skin. **2** = MOLE². [L]

naff¹ /næf/ v.intr. Brit. sl. **1** (in imper., foll. by off) go away. **2** (as **naffing** adj.) used as an intensive to express annoyance etc. [prob. euphem. for FUCK: cf. EFF]

naff² /næf/ adj. Brit. sl. **1** unfashionable, tasteless. **2** worthless, rubbishy. [20th c.: orig. unkn.]

Naffy /'næfı/ n. sl. = NAAFI. [phonet. sp.]

NAFTA /'næftə/ (also **Nafta**) see NORTH AMERICAN FREE TRADE AGREEMENT.

nag¹ /næg/ v. & n. ● v. (**nagged**, **nagging**) **1 a** tr. annoy or irritate (a person) with persistent fault-finding or continuous urging. **b** intr. (often foll. by at) find fault, complain, or urge, esp. persistently. **2** intr. (of a pain) ache dully but persistently. **3 a** tr. worry or preoccupy (a person, the mind, etc.) (*his mistake nagged him*). **b** intr. (often foll. by at) worry or gnaw. **c** (as **nagging** adj.) persistently worrying or painful. ● n. a persistently nagging person. □ **nagger** n. **naggingly** adv. [of dial., perh. Scand. or LG, orig.: cf. Norw. & Sw. *nagga* gnaw, irritate, LG (g)*naggen* provoke]

nag² /næg/ n. **1** colloq. a horse. **2** a small riding-horse or pony. [ME: orig. unkn.]

Naga /'nɑːgə/ n. & adj. ● n. **1** a member of a group of peoples living in or near the Naga Hills of Burma (Myanmar) and in the state of Nagaland in NE India. The Nagas live by practising shifting cultivation and some hunting and gathering; although many are now Christian, a majority retain traditional beliefs. **2** the language of this people, a member of the Tibetan language group. ● adj. of or relating to this people or their language. [as NAGA]

naga /ˈnɑːgə/ n. *Hinduism* a member of a race of semi-divine creatures, half-snake half-human. [Skr., = serpent]

Nagaland /ˈnɑːgəˌlænd/ a state in the far north-east of India, on the border with Burma (Myanmar); capital, Kohima. It was created in 1962 from parts of Assam and is inhabited mainly by the Naga people.

Nagasaki /ˌnægəˈsɑːkɪ/ a city and port in SW Japan, on the west coast of Kyushu island; pop. (1990) 444,620. It was the target of the second atom bomb, dropped by the United States on 9 Aug. 1945, which resulted in the deaths of about 75,000 people and devastated one-third of the city.

Nagorno-Karabakh /nəˌgɔːnəʊˌkærəˈbæx/ a region of Azerbaijan in the southern foothills of the Caucasus, which has long had a majority Armenian population; pop. (1990) 192,000; capital, Xankändi. Formerly a khanate, it was absorbed into the Russian empire in the 19th century, later becoming an autonomous region of the Soviet Union within Azerbaijan. Since 1985 it has been the scene of armed conflict between Azerbaijan and Armenia, with the majority Armenian population desiring to be separated from Muslim Azerbaijan and united with Armenia. After the region declared unilateral independence in 1991 the Azerbaijani government imposed direct rule and abolished the region's autonomous status.

Nagoya /nəˈgɔɪə/ a city in central Japan, on the south coast of the island of Honshu, capital of Chubu region; pop. (1990) 2,154,660.

Nagpur /næɡˈpʊə(r)/ a city in central India, in the state of Maharashtra; pop. (1991) 1,622,000.

Nagy /nɒdʒ/, Imre (1896–1958), Hungarian Communist statesman, Prime Minister 1953–5 and 1956. During his first term in office he introduced liberal policies, pushing for greater availability of consumer goods and less collectivization, but was forced to resign. Back in power in 1956, he announced Hungary's withdrawal from the Warsaw Pact and sought a neutral status for his country. When the Red Army moved in later that year to crush the uprising Nagy was removed from office and executed by the new regime under János Kádár.

Nah. abbr. (in the Bible) Nahum.

Naha /ˈnɑːhə/ a port in southern Japan, capital of Okinawa island; pop. (1990) 304,900.

Nahuatl /ˈnɑːwɑːt(ə)l/ n. & adj. ● n. **1** a member of a group of peoples native to southern Mexico and Central America, including the Aztecs. **2** the language of these peoples. ● adj. of or relating to the Nahuatl peoples or language. □ **Nahuatlan** adj. [Sp. f. Nahuatl]

Nahum /ˈneɪhəm/ **1** a Hebrew minor prophet. **2** a book of the Bible containing his prophecy of the fall of Nineveh (early 7th century BC).

naiad /ˈnaɪæd/ n. (pl. **naiads** or **-des** /ˈnaɪəˌdiːz/) **1** *Gk & Rom. Mythol.* a water-nymph. **2** the nymph of a dragonfly etc. **3** an aquatic plant of the genus *Najas*, with narrow leaves and small flowers. [L *Naïas Naïad-* f. Gk *Naias -ados* f. *naō* flow]

nail /neɪl/ n. & v. ● n. **1** a small usu. sharpened metal spike with a broadened flat head, driven in with a hammer to join things together or to serve as a peg, protection (cf. HOBNAIL), or decoration. **2 a** a horny covering on the upper surface of the tip of the finger and toe in humans and other primates. **b** a claw or talon. **c** a hard growth on the upper mandible of some soft-billed birds. **3** *hist.* a measure of cloth length (equal to $2\frac{1}{4}$ inches). ● v.tr. **1** fasten with a nail or nails (*nailed it to the beam; nailed the planks together*). **2** fix or keep (a person, attention, etc.) fixed. **3 a** secure, catch, or get hold of (a person or thing). **b** expose or discover (a lie or a liar). □ **hard as nails 1** callous; unfeeling. **2** in good physical condition. **nail-biting** causing severe anxiety or tension. **nail-brush** a small brush for cleaning the nails. **nail one's colours to the mast** persist; refuse to give in. **nail down 1** bind (a person) to a promise etc. **2** define precisely. **3** fasten (a thing) with nails. **nail enamel** *N. Amer.* = nail polish. **nail-file** a roughened metal or emery strip used for smoothing the nails. **nail-head** *Archit.* an ornament like the head of a nail. **nail in a person's coffin** something thought to increase the risk of death. **nail polish** a varnish applied to the nails to colour them or make them shiny. **nail-punch** (or **-set**) a tool for sinking the head of a nail below a surface. **nail-scissors** small curved scissors for trimming the nails. **nail up 1** close (a door etc.) with nails. **2** fix (a thing) at a height with nails. **nail varnish** *Brit.* = nail polish. **on the nail** (esp. of payment) without delay (*cash on the nail*). □ **nailed** adj. (also in comb.). **nailless** /ˈneɪllɪs/ adj. [OE *nægel, næglan* f. Gmc]

nailer /ˈneɪlə(r)/ n. a nail-maker. □ **nailery** n.

nainsook /ˈneɪnsʊk/ n. a fine soft cotton fabric, orig. Indian. [Hindi *nainsukh* f. *nain* eye + *sukh* pleasure]

Naipaul /ˈnaɪpɔːl/, V(idiadhar) S(urajprasad) (b.1932), Trinidadian novelist and travel writer of Indian descent, resident in Britain since 1950. He is best known for his satirical novels, mostly set in Trinidad, such as *A House for Mr Biswas* (1961). Naipaul's *In a Free State* (1971) won the Booker Prize for its sharp analysis of issues of nationality and identity. His travel books include *An Area of Darkness* (1964), about a visit to India.

naira /ˈnaɪərə/ n. the chief monetary unit of Nigeria. [contr. of NIGERIA]

Nairnshire /ˈneənʃɪə(r)/ a former county of NE Scotland, on the Moray Firth. It became a part of Highland Region, as the district of Nairn, in 1975.

Nairobi /naɪˈrəʊbɪ/ the capital of Kenya; pop. (est. 1987) 1,288,700. It is situated on the central Kenyan plateau at an altitude of 1,680 m (5,500 ft) and has been capital since 1905.

naive /nɑːˈiːv, naɪˈiːv/ adj. (also **naïve**) **1** artless; innocent; unaffected. **2** foolishly credulous; simple. **3** (in art) straightforward in style; avoiding subtlety or conventional technique. (*See note below*.) □ **naively** adv. **naiveness** n. [F, fem. of *naïf* f. L *nativus* NATIVE]

▪ Naive art involves the deliberate rejection of conventional or sophisticated artistic techniques such as perspective, and the use of bold, bright colours and strong lines; the general style and vision resemble the freshness and directness of a child's drawing. Adherents of the style include Henri Rousseau and Grandma Moses.

naivety /nɑːˈiːvtɪ, naɪˈiːvtɪ, -ˈiːvətɪ/ n., **naïveté** /ˌnɑːiːvˈteɪ, ˌnaɪːv-/ (pl. **-ies** or **naïvetés** pronunc. same) **1** the state or quality of being naive. **2** a naive action. [F *naïveté* (as NAIVE)]

Najaf /ˈnædʒæf/ (also **An Najaf** /æn/) a city in southern Iraq, on the Euphrates; pop. (est. 1985) 242,600. It contains the shrine of Ali, the prophet Muhammad's son-in-law, and is a holy city of the Shiite Muslims.

naked /ˈneɪkɪd/ adj. **1** without clothes; nude. **2** plain; undisguised; exposed (*the naked truth; his naked soul*). **3** (of a light, flame, etc.) unprotected from the wind etc.; unshaded. **4** defenceless. **5** without addition, comment, support, evidence, etc. (*his naked word; naked assertion*). **6 a** (of landscape) barren; treeless. **b** (of rock) exposed; without soil etc. **7** (of a sword etc.) unsheathed. **8** (usu. foll. by *of*) devoid; without. **9** without leaves, hairs, scales, shell, etc. **10** (of a room, wall, etc.) without decoration, furnishings, etc.; empty, plain. □ **the naked eye** unassisted vision, e.g. without a telescope, microscope, etc. **naked ladies** (or **boys**) the meadow saffron. □ **nakedly** adv. **nakedness** n. [OE *nacod* f. Gmc]

naker /ˈneɪkə(r)/ n. hist. a kettledrum. [ME f. OF *nacre nacaire* f. Arab. *nakkāra* drum]

Nakhichevan see NAXÇIVAN.

Nakuru /næˈkuːruː/ an industrial city in western Kenya; pop. (1986) 112,000. Nearby is Lake Nakuru, which is famous for its spectacular flocks of flamingos.

Nalchik /ˈnæltʃɪk/ a city in the Caucasus, SW Russia, capital of the republic of Kabardino-Balkaria; pop. (1990) 237,000.

NALGO /ˈnælɡəʊ/ abbr. hist. (in the UK) National and Local Government Officers' Association, merged with COHSE and NUPE to form UNISON in 1993.

Nam /næm/ *US colloq.* Vietnam. [abbr.]

Nama /ˈnɑːmə/ n. & adj. ● n. (pl. same or **Namas**) (also called *Khoikhoi* and formerly *Hottentot*) **1** a member of a people now living chiefly in western South Africa and SW Namibia (Namaqualand), related to the San or Bushmen. Traditionally nomadic hunter-gatherers, they formerly occupied the region near the Cape but were largely dispossessed by Dutch settlers. **2** the Khoisan language of this people. ● adj. of or relating to this people or their language. [Nama]

Namangan /ˌnæməˈŋɡɑːn/ a city in eastern Uzbekistan, near the border with Kyrgyzstan; pop. (1990) 312,000.

Namaqualand /nəˈmɑːkwəˌlænd/ a region of south-western Africa, the homeland of the Nama people of SW Namibia and South Africa. *Little Namaqualand* lies immediately to the south of the Orange River and is in the South African province of Northern Cape, while *Great Namaqualand* lies to the north of the river and is in Namibia.

namby-pamby /ˌnæmbɪˈpæmbɪ/ adj. & n. ● adj. **1** lacking vigour or drive; weak. **2** insipidly pretty or sentimental. ● n. (pl. **-ies**) **1** a namby-pamby person. **2** namby-pamby talk. [fanciful formulation on name of *Ambrose* Philips, Engl. pastoral writer (1674–1749)]

name /neɪm/ *n. & v.* ● *n.* **1 a** the word by which an individual person, animal, place, or thing is known, spoken of, etc. (*mentioned him by name*; *her name is Joanna*). **b** all who go under one name; a family, clan, or people in terms of its name (*all the clans hostile to the name of Campbell*). **2 a** a usu. abusive term used of a person etc. (*called him names*). **b** a word denoting an object or esp. a class of objects, ideas, etc. (*what is the name of that kind of vase?*; *that sort of behaviour has no name*). **3** a famous person (*many great names were there*). **4** a reputation, esp. a good one (*has a name for honesty*; *their name is guarantee enough*). **5** something existing only nominally (opp. FACT, REALITY). **6** (*attrib.*) widely known (*a name brand of shampoo*). **7** an underwriter of a Lloyd's syndicate. ● *v.tr.* **1** give a usu. specified name to (*named the dog Spot*). **2** call (a person or thing) by the right name (*named the man in the photograph*). **3** mention; specify; cite (*named his requirements*). **4** nominate, appoint, etc. (*was named the new chairman*). **5** specify as something desired (*named it as her dearest wish*). **6** *Brit. Parl.* (of the Speaker) mention (an MP) as disobedient to the chair. □ **by name** called (*Tom by name*). **have to one's name** possess. **in all but name** virtually. **in name** (or **name only**) as a mere formality; hardly at all (*is the leader in name only*). **in a person's name** = *in the name of.* **in the name of** calling to witness; invoking (*in the name of goodness*). **in one's own name** independently; without authority. **make a name for oneself** become famous. **name after** (*N.Amer.* also **for**) call (a person) by the name of (a specified person) (*named him after his uncle Roger*). **name-calling** abusive language. **name-child** (usu. foll. by *of*) one named after another person. **name-day 1** the feast-day of a saint after whom a person is named. **2** = *ticket-day*. **name the day** arrange a date (esp. of a woman fixing the date for her wedding). **name-drop** (**-dropped**, **-dropping**) indulge in name-dropping. **name-dropper** a person who name-drops. **name-dropping** the familiar mention of famous people as a form of boasting. **name names** mention specific names, esp. in accusation. **name of the game** *colloq.* the purpose or essence of an action etc. **name-part** the title role in a play etc. **name-plate** a plate or panel bearing the name of an occupant of a room etc. **name-tape** a tape fixed to a garment etc. and bearing the name of the owner. **of** (or **by**) **the name of** called. **put one's name down for 1** apply for. **2** promise to subscribe (a sum). **what's in a name?** names are arbitrary labels. **you name it** *colloq.* no matter what; whatever you like. □ **nameable** *adj.* [OE *nama*, *noma*, (ge)*namian* f. Gmc, rel. to L *nomen*, Gk *onoma*]

nameless /ˈneɪmlɪs/ *adj.* **1** having no name or name-inscription. **2** inexpressible; indefinable (*a nameless sensation*). **3** unnamed; anonymous, esp. deliberately (*our informant, who shall be nameless*). **4** too loathsome or horrific to be named (*nameless vices*). **5** obscure; inglorious. **6** illegitimate. □ **namelessly** *adv.* **namelessness** *n.*

namely /ˈneɪmlɪ/ *adv.* that is to say; in other words.

Namen see NAMUR.

namesake /ˈneɪmseɪk/ *n.* a person or thing having the same name as another (*was her aunt's namesake*). [prob. f. phr. *for the name's sake*]

Namib Desert /ˈnɑːmɪb/ a desert of SW Africa. It extends for 1,900 km (1,200 miles) along the Atlantic coast, from the Curoca river in SW Angola through Namibia to the border between Namibia and South Africa.

Namibia /nəˈmɪbɪə/ a country in southern Africa, with a coastline on the Atlantic Ocean; pop. (est. 1991) 1,834,000; languages, English (official), various Bantu languages, Khoisan languages, Afrikaans; capital, Windhoek. Namibia is an arid country, with large tracts of desert along the coast and in the east. German missionaries explored the area in the 19th century, and it became the protectorate of German South West Africa in 1884. In 1920 it was mandated to South Africa by the League of Nations, becoming known as South West Africa. In 1946 South Africa's request to the UN for permission to incorporate South West Africa into the Union of South Africa was denied; despite increasing international pressure and the ending of the UN mandate in 1964, the country continued to be administered by South Africa. Eventually, after several years of fighting against SWAPO guerrillas, South Africa agreed to withdraw. Free elections were held in 1989, and Namibia became independent the next year. □ **Namibian** *adj. & n.*

namma var. of GNAMMA.

Namur /nəˈmʊə(r)/ (Flemish **Namen** /ˈnaːmə/) **1** a province in central Belgium. It was the scene of the last German offensive in the Ardennes in 1945. **2** the capital of this province, at the junction of the Meuse and Sambre rivers; pop. (1991) 103,440.

nan[1] /næn/ *n.* (also **nana**, **nanna** /ˈnænə/) *Brit. colloq.* grandmother. [childish pronunc.]

nan[2] /nɑːn, næn/ *n.* (also **naan**) (in Indian cookery) a type of unleavened bread cooked esp. in a clay oven. [Pers. & Urdu *nān*]

nana /ˈnɑːnə/ *n. sl.* a silly person; a fool. [perh. f. BANANA]

Nanaimo /næˈnaɪməʊ/ a port on the east coast of Vancouver Island in British Columbia, Canada; pop. (1991) 63,730.

Nanak /ˈnɑːnək/ (known as Guru Nanak) (1469–1539), Indian religious leader and founder of Sikhism. He was born into a Hindu family in a village near Lahore. Many Sikhs believe that he was in a state of enlightenment at birth and that he was destined from then to be God's messenger. He learned about both Hinduism and Islam as a child, and at the age of 30 he underwent a religious experience which prompted him to become a wandering preacher. He eventually settled in Kartarpur, in what is now Punjab province, Pakistan; there he built the first Sikh temple. Nanak sought neither to unite the Hindu and Muslim faiths nor to create a new religion, preaching rather that spiritual liberation could be achieved through practising an inward and disciplined meditation on the name of God. His teachings are contained in a number of hymns which form part of the Adi Granth.

Nanchang /nænˈtʃæŋ/ the capital of Jiangxi province in SE China; pop. (1990) 1,330,000.

Nancy /ˈnænsɪ/ a city in NE France, chief town of Lorraine; pop. (1990) 102,410.

nancy /ˈnænsɪ/ *n. & adj.* (also **nance** /næns/) *sl.* ● *n.* (*pl.* **-ies**) (in full **nancy boy**) an effeminate man, esp. a homosexual. ● *adj.* effeminate. [pet-form of the name *Ann*]

Nandi /ˈnʌndɪ/ *Hinduism* a bull which serves as the mount of Siva and symbolizes fertility. [Skr., = the happy one]

Nanga Parbat /ˌnʌŋɡə ˈpɑːbʌt/ a mountain in northern Pakistan, in the western Himalayas. It rises to 8,126 m (26,660 ft).

Nanjing /nænˈdʒɪŋ/ (also **Nanking** /-ˈkɪŋ/) a city in eastern China, on the Yangtze river, capital of Jiangsu province; pop. (1990) 2,470,000. It was the capital of various ruling dynasties and capital of China from 1368 until replaced by Beijing in 1421.

nankeen /nænˈkiːn/ *n.* **1** a yellowish cotton cloth. **2** a yellowish buff colour. **3** (in *pl.*) trousers of nankeen. [*Nanking* (NANJING), where orig. made]

nanna var. of NAN[1].

Nanning /nænˈnɪŋ/ the capital of Guangxi Zhuang autonomous region in southern China; pop. (1990) 1,070,000.

nanny /ˈnænɪ/ *n. & v.* ● *n.* (*pl.* **-ies**) **1 a** a child's nurse. **b** an unduly protective person, institution, etc. (*the nanny state*). **2** = NAN[1]. **3** (in full **nanny-goat**) a female goat. ● *v.tr.* (**-ies**, **-ied**) be unduly protective towards. [formed as NANCY]

nano- /ˈnænəʊ, ˈnænə-/ *comb. form* denoting a factor of 10^{-9} (*nanosecond*). [L f. Gk *nanos* dwarf]

nanometre /ˈnænəʊˌmiːtə(r)/ *n.* one thousand-millionth of a metre (symbol: **nm**).

nanosecond /ˈnænəʊˌsekənd/ *n.* one thousand-millionth of a second (symbol: **ns**).

nanotechnology /ˌnænəʊtekˈnɒlədʒɪ/ *n.* the branch of technology that deals with dimensions and tolerances of less than 100 nanometres, esp. the manipulation of individual atoms and molecules. □ **nanotechnologist** *n.* **nanotechnological** /-ˌteknəˈlɒdʒɪk(ə)l/ *adj.*

Nansen /ˈnæns(ə)n/, Fridtjof (1861–1930), Norwegian Arctic explorer. In 1888 he led the first expedition to cross the Greenland ice-fields. Five years later he sailed north of Siberia on board the *Fram*, intending to reach the North Pole by allowing the ship to become frozen in the ice and letting the current carry it towards Greenland. By 1895, it had drifted as far north as 84° 4′; Nansen then made for the Pole on foot, reaching a latitude of 86° 14′, the furthest north anyone had been at that time. Nansen became increasingly involved in affairs of state, serving as Norwegian minister in London (1906–8). In 1922 he was awarded the Nobel Peace Prize for organizing relief work among victims of the Russian famine.

Nantes /nɒnt/ a city in western France, on the Loire, chief town of Pays de la Loire region; pop. (1990) 252,030. In 1598 the Edict of Nantes was signed there.

Nantes, Edict of an edict of 1598 signed by Henry IV of France and granting toleration to Catholics. It was revoked by Louis XIV in 1685.

Nantucket /nænˈtʌkɪt/ an island off the coast of Massachusetts, south of Cape Cod and east of Martha's Vineyard. First visited by the English in 1602, it was settled by the Quakers in 1659. It became an important whaling centre in the 18th and 19th centuries.

naos /ˈneɪɒs/ n. (pl. **naoi** /ˈneɪɔɪ/) Gk Antiq. the inner part of a temple. [Gk, = temple]

nap[1] /næp/ v. & n. ● v.intr. (**napped**, **napping**) sleep lightly or briefly. ● n. a short sleep or doze, esp. by day (took a nap). □ **catch a person napping 1** find a person asleep or off guard. **2** detect in negligence or error. [OE hnappian, rel. to OHG (h)naffezan to slumber]

nap[2] /næp/ n. & v. ● n. **1** the raised pile on textiles, esp. velvet. **2** a soft downy surface. **3** Austral. colloq. blankets, bedding, swag. ● v.tr. (**napped**, **napping**) raise a nap on (cloth). □ **napless** adj. [ME noppe f. MDu., MLG noppe nap, noppen trim nap from]

nap[3] /næp/ n. & v. ● n. **1 a** a form of whist in which players declare the number of tricks they expect to take, up to five. **b** a call of five in this game. **2 a** the betting of all one's money on one horse etc. **b** a tipster's choice for this. ● v.tr. (**napped**, **napping**) name (a horse etc.) as a probable winner. □ **go nap 1** attempt to take all five tricks in nap. **2** risk everything in one attempt. **3** win all the matches etc. in a series. **nap hand** a good winning position worth risking in a venture. **not go nap on** Austral. colloq. not be too keen on; not care much for. [abbr. of orig. name of game NAPOLEON]

napa var. of NAPPA.

napalm /ˈneɪpɑːm/ n. & v. ● n. **1** a thickening agent produced from naphthenic acid, other fatty acids, and aluminium. **2** a jellied petrol made from this, used in incendiary bombs. ● v.tr. attack with napalm bombs. [NAPHTHENIC + PALMITIC ACID]

nape /neɪp/ n. the back of the neck. [ME: orig. unkn.]

napery /ˈneɪpərɪ/ n. Sc. or archaic household linen, esp. table linen. [ME f. OF naperie f. nape (as NAPKIN)]

Naphtali /ˈnæftəˌlaɪ/ **1** a Hebrew patriarch, son of Jacob and Bilhah (Gen. 30:7–8). **2** the tribe of Israel traditionally descended from him.

naphtha /ˈnæfθə/ n. an inflammable oil obtained by the dry distillation of organic substances such as coal, shale, or petroleum. [L f. Gk, = inflammable volatile liquid issuing from the earth, of oriental origin]

naphthalene /ˈnæfθəˌliːn/ n. a white crystalline aromatic substance produced by the distillation of coal tar and used in mothballs and the manufacture of dyes etc. □ **naphthalic** /næfˈθælɪk/ adj. [NAPHTHA + -ENE]

naphthene /ˈnæfθiːn/ n. Chem. any of a group of cycloalkanes (inc. cyclohexane) present in or obtained from petroleum. [NAPHTHA + -ENE]

naphthenic /næfˈθiːnɪk/ adj. Chem. of a naphthene or its radical. □ **naphthenic acid** any carboxylic acid resulting from the refining of petroleum.

Napier[1] /ˈneɪpɪə(r)/ a seaport on Hawke Bay, North Island, New Zealand; pop. (1991) 52,470. Originally a whaling port, the town was first settled by missionaries in 1844 and named after the British general and colonial administrator Sir Charles Napier (1809–54).

Napier[2] /ˈneɪpɪə(r)/, John (1550–1617), Scottish mathematician. Napier was the inventor (independently of the German mathematician Joost Bürgi (1552–1632)) of logarithms. His tables, modified and republished by Henry Briggs, had an immediate and lasting influence on mathematics.

Napierian logarithm /neɪˈpɪərɪən/ n. see LOGARITHM. [NAPIER[2]]

Napier's bones n.pl. Math. slips of ivory or other material divided into sections marked with digits, formerly used to facilitate multiplication and division, according to a method devised by John Napier.

napkin /ˈnæpkɪn/ n. **1** (in full **table napkin**) a square piece of linen, paper, etc., used for wiping the lips, fingers, etc. at meals, or serving fish etc. on; a serviette. **2** Brit. a baby's nappy. **3** a small towel. **4** N. Amer. a sanitary towel. □ **napkin-ring** a ring used to hold (and distinguish) a person's table napkin when not in use. [ME f. OF nappe tablecloth f. L mappa MAP]

Naples /ˈneɪp(ə)lz/ (Italian **Napoli** /ˈnɑːpoli/) a city and port on the west coast of Italy, capital of Campania region; pop. (1990) 1,206,000. It was formerly the capital of the kingdom of Naples and Sicily (1816–60). [L Neapolis f. Gk f. neos new + polis city]

Napoleon /nəˈpəʊlɪən/ the name of three rulers of France, notably:

Napoleon I (known as Napoleon; full name Napoleon Bonaparte) (1769–1821), emperor 1804–14 and 1815. Born in Corsica, he was appointed general of the French Army in Italy in 1796, where he led successful campaigns against Sardinia and Austria to establish a French-controlled republic in northern Italy. Thwarted in his attempt (1798) to create a French empire overseas by Nelson, who defeated the French fleet at the Battle of Aboukir Bay, Napoleon returned to France

(1799) and joined a conspiracy which overthrew the Directory. As First Consul he became the supreme ruler of France and over the next four years began his reorganization of the French legal and education systems. He declared himself emperor in 1804, embarking on a series of campaigns known as the Napoleonic Wars, winning such battles as those at Austerlitz (1805) and Jena (1806), and establishing a French empire stretching from Spain to Poland. However, his plans to invade England resulted in the destruction of the French fleet at Trafalgar (1805); then, after the failure of his attack on Russia in 1812, his conquests were gradually lost to a coalition of all his major opponents. Forced into exile in 1814 (to the island of Elba), he returned briefly to power a year later, but, after his defeat at Waterloo (1815), he was once again exiled, this time to the island of St Helena.

Napoleon III (full name Charles Louis Napoleon Bonaparte; known as Louis-Napoleon) (1808–73), emperor 1852–70. A nephew of Napoleon I, Napoleon III came to power after the 1848 revolution, when he was elected President of the Second Republic. In late 1851 he staged a coup, dissolving the Legislative Assembly and establishing a new constitution which was approved by plebiscite, after which the empire was restored and he was confirmed emperor. As emperor, he was noted for his aggressive foreign policy, which included intervention in Mexico, participation in the Crimean War, and war against Austria in Italy. He abdicated in 1870 after the defeat at Sedan in the Franco-Prussian War.

napoleon /nəˈpəʊlɪən/ n. **1** hist. a gold twenty-franc piece minted in the reign of Napoleon I. **2** hist. a 19th-century high boot. **3** = NAP[3] n. 1a. **4** N. Amer. = MILLEFEUILLE. □ **double napoleon** hist. a forty-franc piece. [F napoléon f. NAPOLEON]

Napoleonic /nəˌpəʊlɪˈɒnɪk/ adj. of, relating to, or characteristic of Napoleon I or his time.

Napoleonic Wars a series of campaigns (1800–15) of French armies under Napoleon I against Austria, Russia, Great Britain, Portugal, Prussia, and other European powers. They ended with Napoleon's defeat at the Battle of Waterloo.

Napoli see NAPLES.

nappa /ˈnæpə/ n. (also **napa**) a soft leather made by a special process from the skin of sheep or goats. [Napa in California]

nappe /næp/ n. Geol. a sheet of rock that has moved sideways over neighbouring strata, usu. as a result of overthrust. [F nappe tablecloth]

napper /ˈnæpə(r)/ n. Brit. sl. the head. [18th c.: orig. uncert.]

nappy /ˈnæpɪ/ n. (pl. **-ies**) Brit. a piece of towelling or other absorbent material wrapped round a baby to absorb or retain urine and faeces. □ **nappy rash** inflammation of a baby's skin, caused by prolonged contact with a damp nappy. [abbr. of NAPKIN]

Nara /ˈnɑːrə/ a city in central Japan, on the island of Honshu; pop. (1990) 349,360. It was the first capital of Japan (710–84) and an important centre of Japanese Buddhism.

Narayan /nəˈrɑːjən/, R(asipuram) K(rishnaswamy) (b.1906), Indian novelist and short-story writer. His best-known novels are set in the imaginary small Indian town of Malgudi, and portray its inhabitants in an affectionate yet ironic manner; they include Swami and Friends (1935), The Man-Eater of Malgudi (1961), and The Painter of Signs (1977). His short stories include Under the Banyan Tree and Other Stories (1985).

Narayanganj /nəˈrɑːjənˌɡʌndʒ/ a river port in Bangladesh, on the Ganges delta south-east of Dhaka; pop. (1991) 406,000.

Narbonne /nɑːˈbɒn/ a city in southern France, in Languedoc-Roussillon, just inland from the Mediterranean; pop. (1990) 47,090. It was founded by the Romans in 118 BC, the first Roman colony in Transalpine Gaul. Known as Narbo Martius, it became the capital of the Roman province of Gallia Narbonensis. It was a prosperous port of medieval France until its harbour silted up in the 14th century.

narceine /ˈnɑːsiˌiːn, ˈnɑːsiːɪn/ n. Pharm. a narcotic alkaloid obtained from opium. [F narcéine f. Gk narkē numbness]

narcissism /ˈnɑːsɪˌsɪz(ə)m/ n. Psychol. excessive or erotic interest in oneself, one's physical features, etc. □ **narcissist** n. **narcissistic** /ˌnɑːsɪˈsɪstɪk/ adj. **narcissistically** adv. [NARCISSUS]

Narcissus /nɑːˈsɪsəs/ Gk Mythol. a beautiful youth who rejected the nymph Echo and fell in love with his own reflection in a pool, eventually pining away and being changed into the flower that bears his name.

narcissus /nɑːˈsɪsəs/ n. (pl. **narcissi** /-saɪ/ or **narcissuses**) an ornamental bulbous plant of the genus Narcissus; esp. N. poeticus, which bears a heavily scented single flower with a shallow undivided corona edged with crimson and yellow. [L f. Gk *narkissos*, perh. f. *narkē* numbness, with ref. to its narcotic effects: see NARCISSUS]

narcolepsy /'nɑːkə,lepsɪ/ *n. Med.* a disease with fits of sleepiness and drowsiness. □ **narcoleptic** /,nɑːkə'leptɪk/ *adj. & n.* [Gk *narkoō* make numb, after EPILEPSY]

narcosis /nɑː'kəʊsɪs/ *n.* **1** *Med.* the working or effects of soporific narcotics. **2** a state of insensibility. [Gk *narkōsis* f. *narkoō* make numb]

narcoterrorism /,nɑːkəʊ'terə,rɪz(ə)m/ *n.* violent crime associated with illicit drugs. □ **narcoterrorist** *adj. & n.* [NARCOTIC + TERRORISM]

narcotic /nɑː'kɒtɪk/ *adj. & n.* ● *adj.* **1** (of a substance) inducing drowsiness, sleep, stupor, or insensibility. **2** (of a drug etc.) affecting the mind. **3** of or involving narcosis. **4** soporific. ● *n.* a narcotic substance, drug, or influence. □ **narcotically** *adv.* **narcotism** /'nɑːkə,tɪz(ə)m/ *n.* **narcotize** *v.tr.* (also **-ise**). **narcotization** /,nɑːkətaɪ'zeɪʃ(ə)n/ *n.* [ME f. OF *narcotique* or med.L f. Gk *narkōtikos* (as NARCOSIS)]

nard /nɑːd/ *n.* **1** an aromatic plant of the valerian family; esp. the *Celtic nard* (*Valeriana celtica*), of European mountains. **2** = SPIKENARD 1. [ME f. L *nardus* f. Gk *nardos* ult. f. Skr.]

nardoo /nɑː'duː/ *n.* **1** an aquatic clover-like plant, *Marsilea quadrifolia*, native to Australia and related to the ferns. **2** a food made from the spores of this plant. [Aboriginal]

nares /'neərɪːz/ *n.pl. Anat.* the nostrils. □ **narial** *adj.* [pl. of L *naris*]

narghile /'nɑːgɪlɪ/ *n.* an oriental tobacco-pipe with the smoke drawn through water; a hookah. [Pers. *nārgīleh* (*nārgīl* coconut)]

nark /nɑːk/ *n. & v. sl.* ● *n.* **1** *Brit.* a police informer or decoy. **2** *Austral.* an annoying person or thing. ● *v.tr.* (usu. in *passive*) *Brit.* annoy; infuriate (*was narked by their attitude*). □ **nark it!** stop that! [Romany *nāk* nose]

narky /'nɑːkɪ/ *adj.* (**narkier, narkiest**) *Brit. sl.* bad-tempered, irritable. [NARK]

Narmada /'nɜːmədə/ a river which rises in Madhya Pradesh, central India, and flows generally westwards for 1,245 km (778 miles) to the Gulf of Cambay. It is regarded by Hindus as sacred.

Narragansett /,nærə'gænsɪt/ *n. & adj.* ● *n.* (*pl.* same) **1** a member of an American Indian people originally living in Rhode Island. Their power was destroyed by war with the British in the late 17th century, and today only a few Narragansett remain in New England. **2** the Algonquian language of this people. ● *adj.* of or relating to this people or their language. [Algonquian = people of the small point (of land)]

narrate /nə'reɪt/ *v.tr.* (also *absol.*) **1** give a continuous story or account of. **2** provide a spoken commentary or accompaniment for (a film etc.). □ **narratable** *adj.* **narration** /-'reɪʃ(ə)n/ *n.* [L *narrare narrat-*]

narrative /'nærətɪv/ *n. & adj.* ● *n.* **1** a spoken or written account of connected events in order of happening. **2** the practice or art of narration. ● *adj.* in the form of, or concerned with, narration (*narrative verse*). □ **narratively** *adv.* [F *narratif -ive* f. LL *narrativus* (as NARRATE)]

narrator /nə'reɪtə(r)/ *n.* **1** a person who narrates, esp. a character who recounts the events of a novel or narrative poem. **2** an actor, announcer, etc. who delivers a commentary in a film, broadcast, etc. [L (as NARRATE)]

narrow /'nærəʊ/ *adj., n., & v.* ● *adj.* (**narrower, narrowest**) **1 a** of small width in proportion to length; lacking breadth. **b** confined or confining; constricted (*within narrow bounds*). **2** of limited scope; restricted (*in the narrowest sense*). **3** with little margin (*a narrow escape*). **4** searching; precise; exact (*a narrow examination*). **5** = NARROW-MINDED. **6** *Phonet.* (of a vowel) tense. **7** of small size. ● *n.* (usu. in *pl.*) **1** the narrow part of a strait, river, sound, etc. **2** a narrow pass or street. ● *v.* **1** *intr.* become narrow; diminish; contract; lessen. **2** *tr.* make narrow; constrict; restrict. □ **narrow boat** *Brit.* a canal boat, esp. one less than 2.1 m (7 ft) wide. **narrow cloth** cloth less than 52 inches wide. **narrow gauge** a railway track that has a smaller gauge than the standard one. **narrow seas** the English Channel and the Irish Sea. **narrow squeak 1** a narrow escape. **2** a success barely attained. □ **narrowish** *adj.* **narrowly** *adv.* **narrowness** *n.* [OE *nearu nearw-* f. Gmc]

narrowcast /'nærəʊ,kɑːst/ *v. & n.* esp. *US* ● *v.intr. & tr.* transmit (a television programme etc.), esp. by cable, to an audience targeted by interests or location. ● *n.* a transmission or programme of this kind. □ **narrowcaster** *n.*

narrow-minded /,nærəʊ'maɪndɪd/ *adj.* rigid or restricted in one's views, intolerant, prejudiced, illiberal. □ **narrow-mindedly** *adv.* **narrow-mindedness** *n.*

narthex /'nɑːθeks/ *n. Archit.* **1** a railed-off antechamber or porch etc.

at the western entrance of some early Christian churches, used by catechumens, penitents, etc. **2** a similar antechamber in a modern church. [L f. Gk *narthēx* giant fennel, stick, casket, narthex]

Narvik /'nɑːvɪk/ an ice-free port on the NW coast of Norway, north of the Arctic Circle; pop. (1990) 18,640. It is linked by rail to the iron-ore mines of northern Sweden.

narwhal /'nɑːwəl/ *n.* a small Arctic whale, *Monodon monoceros*, the male of which has a long straight spirally twisted tusk developed from one of its teeth. [Du. *narwal* f. Danish *narhval* f. *hval* whale: cf. ON *náhvalr* (perh. f. *nár* corpse, with ref. to its skin-colour)]

nary /'neərɪ/ *adj. colloq.* or *dial.* not a; no (*nary a one*). [f. *ne'er a*]

NAS *abbr. Brit.* Noise Abatement Society.

NASA /'næsə/ (also **Nasa**) (in the US) National Aeronautics and Space Administration, a body (set up in 1958) responsible for carrying out the space research conducted by the US. [acronym]

nasal /'neɪz(ə)l/ *adj. & n.* ● *adj.* **1** of, for, or relating to the nose. **2** *Phonet.* (of a speech sound) pronounced with the breath passing through the nose, e.g. *m*, *n*, *ng*, or French *en*, *un*. **3** (of the voice or speech) having an intonation caused by breathing through the nose. ● *n.* **1** *Phonet.* a nasal sound. **2** *hist.* a nose-piece on a helmet. □ **nasally** *adv.* **nasalize** *v.intr. & tr.* (also **-ise**). **nasalization** /,neɪzəlaɪ'zeɪʃ(ə)n/ *n.* **nasality** /neɪ'zælɪtɪ/ *n.* [F *nasal* or med.L *nasalis* f. L *nasus* nose]

nascent /'næs(ə)nt, 'neɪs-/ *adj.* **1** in the act of being born. **2** just beginning to be; not yet mature. **3** *Chem.* just being formed and therefore unusually reactive (*nascent hydrogen*). □ **nascency** *n.* [L *nasci nascent-* be born]

naseberry /'neɪzbərɪ/ *n.* (*pl.* **-ies**) a sapodilla. [Sp. & Port. *néspera* medlar f. L (see MEDLAR): assim. to BERRY]

Naseby, Battle of /'neɪzbɪ/ a major battle of the English Civil War, which took place in 1645 near the village of Naseby in Northamptonshire. It was the last battle of the main phase of the war, in which the last Royalist army in England, commanded by Prince Rupert and King Charles I, was decisively defeated by the larger and better-organized New Model Army under General Thomas Fairfax and Oliver Cromwell. Following the destruction of his army Charles I's cause collapsed completely.

Nash[1] /næʃ/, (Frederic) Ogden (1902–71), American poet. He is noted for his sophisticated light verse, comprising puns, epigrams, highly asymmetrical lines, and other verbal eccentricities. His verse appeared in many collections from 1931 onwards.

Nash[2] /næʃ/, John (1752–1835), English town planner and architect. Under the patronage of the Prince Regent (later George IV), he planned the layout of Regent's Park (1811–25), Regent Street (1826–c.1835; subsequently rebuilt), Trafalgar Square (1826–c.1835), and many other parts of London. He also began the reconstruction of Buckingham Palace (c.1821–30), for which he designed the Marble Arch, again for George IV.

Nash[3] /næʃ/, Paul (1889–1946), English painter and designer. He won renown for his paintings as an official war artist in the First World War; these recorded scenes of devastation in a modernist style. From 1928 the influence of surrealism resulted in a number of enigmatic pictures based on dreams or suggestive landscape motifs, as in *Equivalents for the Megaliths* (1935). A war artist again in the Second World War, he depicted the Battle of Britain, notably in *Totes Meer* (1940–1).

Nash[4] /næʃ/, Richard (known as Beau Nash) (1674–1762), Welsh dandy. Master of Ceremonies in Bath from 1704, he established the city as the centre of fashionable society and was an arbiter of fashion and etiquette in the early Georgian age.

Nashe /næʃ/, Thomas (1567–1601), English pamphleteer, prose writer, and dramatist. After writing several anti-Puritan pamplets, he wrote his best-known work *The Unfortunate Traveller* (1594), a medley of picaresque narrative and pseudo-historical fantasy.

Nashville /'næʃvɪl/ the state capital of Tennessee; pop. (1990) 510,780. The city is noted for its music industry and is the site of Opryland USA (a family entertainment complex) and the Country Music Hall of Fame.

Nasik /'nɑːsɪk/ a city in western India, in Maharashtra, on the Godavari river north-east of Bombay; pop. (1991) 647,000.

Nasmyth /'neɪsmɪθ/, James (1808–90), British engineer. He designed and built a number of steam-driven machines, and went on to manufacture railway locomotives. He is best known, however, for his invention of the steam hammer (1839), a major innovation for the forging industry. Nasmyth also became interested in astronomy, particularly the moon, producing a map of it in 1851, and a book twenty-three years later.

naso- /'neɪzəʊ/ *comb. form* nose. [L *nasus* nose]

naso-frontal /ˌneɪzəʊˈfrʌnt(ə)l/ *adj.* of or relating to the nose and forehead.

nasogastric /ˌneɪzəʊˈgæstrɪk/ *adj. Med.* supplying the stomach via the nose (*nasogastric tube*).

Nassau 1 /'næsaʊ/ a former duchy of western Germany, centred on the small town of Nassau, from which a branch of the House of Orange arose. It was annexed to the Prussian province of Hesse-Nassau in 1866 and corresponds to parts of the present-day states of Hesse and Rhineland-Palatinate. **2** /'næsɔː/ a port on the island of New Providence, capital of the Bahamas; pop. (1990) 172,000.

Nasser /'nɑːsə(r), 'næs-/, Gamal Abdel (1918–70), Egyptian colonel and statesman, Prime Minister 1954–6 and President 1956–70. He was the leader of a successful military coup to depose King Farouk in 1952, after which a republic was declared with Muhammad Neguib (1901– 84) as its President. Nasser deposed Neguib in 1954, declaring himself Prime Minister; two years later he announced a new one-party constitution, becoming President shortly afterwards. His nationalization of the Suez Canal brought war with Britain, France, and Israel in 1956 (see SUEZ CRISIS); he also led Egypt in two unsuccessful wars against Israel (1956 and 1967). With considerable Soviet aid, he launched a programme of domestic modernization, including the building of the High Dam at Aswan.

Nasser, Lake a lake in SE Egypt created in the 1960s by the building of the two dams on the Nile at Aswan. It is named after the former President of Egypt Gamal Nasser.

nastic /'næstɪk/ *adj. Bot.* (of the movement of plant parts) caused by an external stimulus but unaffected in direction by it. [Gk *nastos* squeezed together f. *nassō* to press]

nasturtium /nəˈstɜːʃəm/ *n.* **1** a trailing plant, *Tropaeolum majus*, with rounded edible pungent-tasting leaves and bright orange, yellow, or red flowers. **2** *Bot.* a cruciferous plant of the genus *Nasturtium*, including watercress. [L]

nasty /'nɑːstɪ/ *adj. & n.* ● *adj.* (**nastier, nastiest**) **1 a** highly unpleasant (*a nasty experience*). **b** annoying; objectionable (*the car has a nasty habit of breaking down*). **2** difficult to negotiate; dangerous, serious (*a nasty fence; a nasty question; a nasty illness*). **3** (of a person or animal) ill-natured, ill-tempered, spiteful; violent, offensive (*nasty to her mother; turns nasty when he's drunk*). **4** (of the weather) foul, wet, stormy. **5 a** disgustingly dirty, filthy. **b** unpalatable; disagreeable (*nasty smell*). **c** (of a wound) septic. **6 a** obscene. **b** delighting in obscenity. ● *n.* (*pl.* **-ies**) *colloq.* **1** a nasty person, object, or event. **2** a video nasty. □ **a nasty bit** (or **piece**) **of work** *Brit. colloq.* an unpleasant or contemptible person. **a nasty one 1** a rebuff; a snub. **2** an awkward question. **3** a disabling blow etc. □ **nastily** *adv.* **nastiness** *n.* [ME: orig. unkn.]

NASUWT *abbr.* (in the UK) National Association of Schoolmasters and Union of Women Teachers.

Nat. *abbr.* **1** National. **2** Nationalist. **3** Natural.

Natal /nəˈtæl, -ˈtɑːl/ **1** a former province of South Africa, situated on the east coast. It was first settled by British traders in 1823, then by Boers, becoming a Boer republic in 1838. Annexed by the British to Cape Colony in 1845, it became a separate colony in 1856 and acquired internal self-government in 1893. It became a province of the Union of South Africa in 1910. It was named *Terra Natalis* – meaning 'land of the day of birth' in Latin – by Vasco da Gama in 1497, because he sighted the entrance to what is now Durban harbour on Christmas Day. The province was renamed KwaZulu/Natal after democratic elections in 1994. **2** a port on the Atlantic coast of NE Brazil, capital of the state of Rio Grande do Norte; pop. (1990) 606,280.

natal /'neɪt(ə)l/ *adj.* of or from one's birth. [ME f. L *natalis* (as NATION)]

natality /nəˈtælɪtɪ/ *n.* (*pl.* **-ies**) birth rate. [F *natalité* (as NATAL)]

natation /nəˈteɪʃ(ə)n/ *n. formal* or *literary* the act or art of swimming. [L *natatio* f. *natare* swim]

natatorial /ˌneɪtəˈtɔːrɪəl, ˌnæt-/ *adj.* (also **natatory** /'neɪtətərɪ, nəˈteɪtərɪ/) *formal* **1** swimming. **2** of or concerning swimming. [LL *natatorius* f. L *natator* swimmer (as NATATION)]

natatorium /ˌneɪtəˈtɔːrɪəm/ *n. N. Amer.* a swimming-pool, esp. indoors. [LL neut. of *natatorius* (see NATATORIAL)]

natch /nætʃ/ *adv. colloq.* = NATURALLY. [abbr.]

nates /'neɪtiːz/ *n.pl. Anat.* the buttocks. [L]

NATFHE *abbr.* (in the UK) National Association of Teachers in Further and Higher Education.

nathless /'neɪθlɪs/ *adv.* (also **natheless**) *archaic* nevertheless. [ME f. OE *nā* not (f. *ne* not + *ā* ever) + THE + *lǣs* LESS]

nation /'neɪʃ(ə)n/ *n.* **1** a community of people of mainly common descent, history, language, etc., forming a state or inhabiting a territory. **2** a tribe or confederation of tribes of North American Indians. □ **law of nations** *Law* international law. □ **nation-state** a sovereign state, most of the citizens or subjects of which are also united by factors such as language, common descent, etc., which define a nation. □ **nationhood** *n.* [ME f. OF f. L *natio -onis* f. *nasci* nat- be born]

national /'næʃ(ə)n(ə)l/ *adj. & n.* ● *adj.* **1** of or common to a nation or the nation. **2** peculiar to or characteristic of a particular nation. ● *n.* **1** a citizen of a specified country, usu. entitled to hold that country's passport (*French nationals*). **2** a fellow countryman. **3** (**the National**) = GRAND NATIONAL. □ **national anthem** a song adopted by a nation, expressive of its identity etc. and intended to inspire patriotism. **national bank** *US* a bank chartered under the Federal government. **national convention** *US* a convention of a major political party, nominating candidates for the presidency etc. **national football** *Austral.* Australian Rules football. **national grid** *Brit.* **1** the network of high-voltage electric power lines between major power stations. **2** the metric system of geographical coordinates used in maps of the British Isles. **national income** the total money earned within a nation. **national park** an area of natural beauty protected by the state for the use of the general public. **national service** *Brit. hist.* service in the army etc. under conscription. □ **nationally** *adv.* [F (as NATION)]

National Assembly *n.* **1** an elected house of legislature in various countries. **2** *hist.* the elected legislature in France 1789–91.

National Assistance *n. hist.* **1** (in Britain) the former official name for supplementary benefits under National Insurance. **2** such benefits.

National Association for the Advancement of Colored People (abbr. **NAACP**) a US civil-rights organization set up to oppose racial segregation and discrimination by non-violent means. Formed in 1909 through the merging of a group of white liberals with a black organization, it maintains a moderate stance and a commitment to integration.

National Curriculum a curriculum that state schools in England and Wales have been required to follow since 1990, involving the teaching of specified subjects and assessment at specified ages.

national debt *n.* the total amount of a state's borrowings. A government may raise money by means such as the selling of interest-bearing bonds to the public or borrowing from foreign creditors. Debt has traditionally been incurred to finance wars, but more recently has also been undertaken to support the national currency, pay for social programmes, or avoid the need to raise taxes. The permanent national debt in Britain was first established as a result of borrowing to finance the wars against Louis XIV of France at the end of the 17th century, and has been managed by the Bank of England since 1750.

National Front (abbr. **NF**) an extreme right-wing British political party, formed in 1967, holding reactionary views on immigration. The party gained a very small amount of support in the late 1970s, but was greatly weakened in the 1980s by the formation of various splinter groups, the most prominent of which is the British National Party.

National Gallery an art gallery in Trafalgar Square, London, holding one of the chief national collections of pictures. The collection began in 1824 when Parliament voted money for the purchase of 38 pictures from the J. J. Angerstein collection. The present main building, built in 1833–7 and at first shared with the Royal Academy, was opened in 1838; it has been extended several times, most recently by the addition of the Sainsbury wing (opened 1991).

National Guard *n.* **1** (a member of) an armed force existing in France at various times between 1789 and 1871, first commanded by the Marquis de Lafayette. **2** (in the US) a militia recruited on a state-by-state basis but serving as the primary reserve force of the US army, and available for Federal use in emergency.

National Health Service (abbr. **NHS**) (in the UK) a system of national medical care paid for mainly by taxation and started by the Labour government in 1948. Largely created by the Health Minister Aneurin Bevan, it sought to provide free medical, dental, and optical treatment, but successive governments have introduced charges for some items and services, although medical consultations and hospital care remain free. Since the early 1980s the Conservative government has introduced measures aimed at reducing costs, while increasing numbers of people now use private health insurance schemes.

National Insurance (in the UK) the system of compulsory payments by employed persons (supplemented by employers) to provide state assistance in sickness, unemployment, retirement, etc.

nationalism /'næʃnə,lɪz(ə)m/ n. **1 a** patriotic feeling, principles, etc. **b** an extreme form of this; chauvinism. **2** a policy of national independence.

nationalist /'næʃ(ə)nəlɪst/ n. & adj. ● n. an adherent or advocate of nationalism; an advocate of national independence; (also **Nationalist**) a member of a Nationalist party. ● adj. of or relating to nationalists or nationalism; (also **Nationalist**) designating or relating to a party or movement espousing nationalism or advocating independence for a particular country. □ **nationalistic** /,næʃ(ə)nə'lɪstɪk/ adj. **nationalistically** adv.

nationality /,næʃə'nælɪtɪ/ n. (pl. -**ies**) **1 a** the status of belonging to a particular nation (what is your nationality?; has British nationality). **b** a nation (people of all nationalities). **2** the condition of being national; distinctive national qualities. **3** an ethnic group forming a part of one or more political nations. **4** existence as a nation; nationhood. **5** patriotic sentiment.

nationalize /'næʃ(ə)nə,laɪz/ v.tr. (also -**ise**) **1** take over (railways, coal mines, the steel industry, land, etc.) from private ownership on behalf of the state. **2 a** make national. **b** make into a nation. **3** naturalize (a foreigner). □ **nationalizer** n. **nationalization** /,næʃnəlaɪ'zeɪʃ(ə)n/ n. [F nationaliser (as NATIONAL)]

National Party an Australian political party established in 1914 (as the Country Party) to represent agricultural and rural interests. In recent times it has been in coalition with the Liberal Party of Australia.

National Party of South Africa a political party that held power in South Africa from 1948 until the country's first democratic elections in 1994. Formed in 1914 as an Afrikaner party, it favoured racial segregation and was responsible for instituting apartheid; in 1982 the right wing of the party broke away to form the Conservative Party in protest at measures being taken towards liberalizing racial legislation. The party came second to the ANC in the 1994 elections, gaining just over 20 per cent of the vote and being allocated eighty-two seats in the National Assembly.

National Portrait Gallery an art gallery in London holding the national collection of portraits of eminent or well-known British men and women. Founded in 1856, it moved to its present site next to the National Gallery in 1896.

National Security Agency (abbr. **NSA**) (in the US) a secret body established after the Second World War to gather intelligence, deal with coded communications from around the world, and safeguard US transmissions.

National Security Council (abbr. **NSC**) (in the US) a body created by Congress after the Second World War to advise the President on issues relating to national security in domestic, foreign, and military policy. It is chaired by the President and includes senior figures from the government, military, and CIA.

National Socialism n. hist. the doctrines of nationalism, racial purity, etc., adopted by the Nazis; Nazism. (See NAZI.)

National Socialist n. hist. a member of the party implementing National Socialism in Germany, 1933–45; a Nazi.

National Trust (abbr. **NT**) a trust for the preservation of places of historic interest or natural beauty in England, Wales, and Northern Ireland, founded in 1895 and supported by endowment and private subscription. The National Trust for Scotland is a Scottish institution (founded in 1931) with similar aims.

Nation of Islam see BLACK MUSLIM.

nationwide adj. /'neɪʃ(ə)n,waɪd/ & adv. /,neɪʃ(ə)n'waɪd/ extending over the whole nation.

native /'neɪtɪv/ n. & adj. ● n. **1 a** (usu. foll. by of) a person born in a specified place, or whose parents are domiciled in that place at the time of the birth (a native of Bristol). **b** a local inhabitant. **2** often offens. a member of a non-white indigenous people, as regarded by the colonial settlers. **3** (usu. foll. by of) an indigenous animal or plant. **4** an oyster reared in British waters, esp. in artificial beds (a Whitstable native). **5** Austral. a white person born in Australia. ● adj. **1** (usu. foll. by to) belonging to a person or thing by nature; inherent; innate (spoke with the facility native to him). **2** of one's birth or birthplace (native dress; native country). **3** belonging to one by right of birth. **4** (usu. foll. by to) belonging to a specified place (the anteater is native to South America). **5 a** (esp. of a non-European) indigenous; born in a place. **b** of the natives of a place (native customs). **6** in a natural state; unadorned;

simple. **7** Geol. (of metal etc.) found in a pure or uncombined state. **8** Austral. & NZ resembling an animal or plant familiar elsewhere (native cat; native oak). □ **go native** (of a settler) adopt the local way of life, esp. in a non-European country. **native bear** Austral. = KOALA. **native rock** rock in its original place. □ **natively** adv. **nativeness** n. [ME (earlier as adj.) f. OF natif -ive or L nativus f. nasci nat- be born]

Native American n. & adj. (also called American Indian) ● n. a member of a group of indigenous peoples of North and South America and the Caribbean Islands. (See note below.) ● adj. of or relating to these peoples.

  ▪ The ancestors of the Native American peoples arrived from Siberia via the Bering Strait during the last glacial period (between 20,000 and 35,000 years ago). They were originally called Indians by the error of Columbus and other Europeans in the 15th–16th centuries, who thought they had reached part of India by a new route. Their cultures and civilizations are extremely diverse; they include the Aztecs, Maya, and Incas of Central and South America, and the Apache, Cherokee, Cheyenne, Iroquois, Mohawk, Navajo, Sioux, Cree, and Ojibwa in North America. Since the arrival of European settlers in the 16th century, the lives and culture of most Native Americans (except those in the most remote areas) have been badly affected: their numbers declined as a result both of fighting and the transmission of European diseases. Since the 1880s most North America Indians have lived on reservations, although in more recent years there has been increasing political activity to reclaim land and rights.

nativism /'neɪtɪ,vɪz(ə)m/ n. Philos. the doctrine that at least some of our concepts are innate rather than acquired through experience. Among famous nativists have been Plato, Descartes, Kant, and Noam Chomsky. □ **nativist** n.

nativity /nə'tɪvɪtɪ/ n. (pl. -**ies**) **1** (esp. **the Nativity**) **a** the birth of Jesus. **b** the festival of Jesus' birth; Christmas. **2** a picture of the Nativity. **3** birth. **4** the horoscope at a person's birth. **5 a** the birth of the Virgin Mary or St John the Baptist. **b** the festival of the nativity of the Virgin (8 Sept.) or St John (24 June). □ **nativity play** a play, usu. performed by children at Christmas, dealing with the birth of Jesus. [ME f. OF nativité f. LL nativitas -tatis f. L (as NATIVE)]

NATO, Nato /'neɪtəʊ/ see NORTH ATLANTIC TREATY ORGANIZATION.

natron /'neɪtrən/ n. a mineral form of hydrated sodium salts found in dried lake beds. [F f. Sp. natrón f. Arab. naṭrūn f. Gk nitron NITRE]

Natron, Lake /'neɪtrən/ a lake in northern Tanzania, on the border with Kenya, containing large deposits of salt and soda.

NATSOPA /næt'səʊpə/ abbr. hist. (in the UK) National Society of Operative Printers, Graphical and Media Personnel (orig. Printers and Assistants).

natter /'nætə(r)/ v. & n. Brit. colloq. ● v.intr. **1** chatter idly. **2** grumble; talk fretfully. ● n. **1** aimless chatter. **2** grumbling talk. □ **natterer** n. [orig. Sc., imit.]

natterjack /'nætə,dʒæk/ n. a small Eurasian toad, Bufo calamita, with a light yellow stripe down its back, and moving by running. [perh. f. NATTER, from its loud croak, + JACK[1]]

nattier blue /'nætɪə(r)/ n. & adj. ● n. a soft shade of blue. ● adj. of this colour. [much used by Jean Marc Nattier, Fr. painter (1685–1766)]

natty /'nætɪ/ adj. (**nattier**, **nattiest**) colloq. **1** smartly or neatly dressed, dapper. **b** spruce; trim; smart (a natty blouse). **2** deft. □ **nattily** adv. **nattiness** n. [orig. sl., perh. rel. to NEAT[1]]

Natufian /nɑː'tuːfɪən/ adj. & n. Archaeol. of, relating to, or denoting a late mesolithic culture of the Middle East, which provides evidence for the first settled villages and is characterized by the use of microliths and of bone for implements. [f. Wadi an-Natuf, the type-site, a cave near Jerusalem]

natural /'nætʃrəl/ adj. & n. ● adj. **1 a** existing in or caused by nature; not artificial (natural landscape). **b** uncultivated; wild (existing in its natural state). **2** in the course of nature; not exceptional or miraculous (died of natural causes; a natural occurrence). **3** (of human nature etc.) not surprising; to be expected (natural for her to be upset). **4 a** (of a person or a person's behaviour) unaffected, easy, spontaneous. **b** (foll. by to) spontaneous, easy (friendliness is natural to him). **5 a** (of qualities etc.) inherent; innate (a natural talent for music). **b** (of a person) having such qualities (a natural linguist). **6** not disguised or altered (as by make-up etc.). **7** lifelike; as if in nature (the portrait looked very natural). **8** likely by its or their nature to be such (natural enemies; the natural antithesis). **9** having a physical existence as opposed to what is spiritual, intellectual, etc. (the natural world). **10 a** having genetically the specified familial relationship; actually begotten, not adopted (her *natural son*). **b** illegitimate. **11** based on the innate moral sense; instinctive (*natural justice*). **12** *Mus.* **a** (of a note) not sharpened or flattened (*B natural*). **b** (of a scale) not containing any sharps or flats. **13** without spiritual enlightenment. ● *n.* **1** (usu. foll. by *for*) a person or thing naturally suitable, adept, expert, etc. (*a natural for the championship*). **2** *archaic* a person mentally handicapped from birth. **3** *Mus.* **a** a sign (♮) denoting a return to natural pitch after a sharp or a flat. **b** a natural note. **c** a white key on a piano. **4 a** *Cards* a hand making 21 in the first deal in pontoon. **b** a throw of 7 or 11 at craps. **5** a pale fawn colour. □ **natural-born** having a character or position by birth. **natural childbirth** *Med.* childbirth with minimal medical or technological intervention. **natural classification** a scientific classification according to natural features. **natural death** death by age or disease, not by accident, poison, violence, etc. **natural food** food without preservatives etc. **natural gas** an inflammable mainly methane gas found in the earth's crust, not manufactured. **natural historian** a writer or expert on natural history. **natural history 1** the study of animals or plants, esp. as set forth for popular use. **2** an aggregate of the facts concerning the flora and fauna etc. of a particular place or class (*a natural history of the Isle of Wight*). **natural language 1** language that has developed naturally, as opposed to artificial language or code. **2** a language of this kind. **natural law 1** *Philos.* unchanging moral principles common to all people by virtue of their nature as human beings. **2** a correct statement of an invariable sequence between specified conditions and a specified phenomenon. **3** the laws of nature; regularity in nature (*where they saw chance, we see natural law*). **natural life** the duration of one's life on earth. **natural logarithm** see LOGARITHM. **natural magic** magic involving nature spirits, healing, the use of herbs, etc. **natural note** *Mus.* a note that is neither sharp nor flat. **natural numbers** *Math.* the integers 1, 2, 3, etc. **natural philosopher** *hist.* a physicist. **natural philosophy** *hist.* physical science. **natural religion** a religion based on reason (opp. *revealed religion*); deism. **natural resources** materials or conditions occurring in nature and capable of economic exploitation. **natural science** the sciences used in the study of the physical world, e.g. physics, chemistry, geology, biology, botany. **natural theology** the knowledge of God as gained by the light of natural reason. **natural uranium** unenriched uranium. **natural virtues** *Philos.* justice, prudence, temperance, fortitude. **natural year** the time taken by one revolution of the earth round the sun, 365 days 5 hours 48 minutes. □ **naturalness** *n.* [ME f. OF *naturel* f. L *naturalis* (as NATURE)]

Natural History Museum a museum of zoological, botanical, palaeontological, and mineralogical items in South Kensington, London. The museum was originally formed from the natural history collections of the British Museum, where it was housed until 1881, and did not become fully independent until 1963. The Geological Museum merged with it in 1985.

naturalism /'nætʃrə,lɪz(ə)m/ *n.* **1** the theory or practice in art and literature of representing nature, character, etc. realistically and in great detail. (*See note below.*) **2 a** *Philos.* a theory of the world that excludes the supernatural or spiritual. **b** any moral or religious system based on this theory. **3** action based on natural instincts. **4** indifference to conventions. [NATURAL, in Philos. after F *naturalisme*]

▪ Naturalism is a very broad term, but in its simplest usage it means the representation of form as found in nature, with avoidance of stylized or conceptual forms. Within these general limits, there are many shades of meaning: the art of classical Greece or the Italian Renaissance is naturalistic in mirroring natural beauty; the 17th-century followers of Caravaggio, however, copy nature faithfully even when it seems ugly. Naturalism was later applied specifically to an artistic and literary movement of the 19th century whose adherents were influenced by contemporary ideas of science and society such as those of Darwin and Comte, and which was concerned to adopt a strictly scientific and objective approach to art. It is characterized by a refusal to idealize experience and by the conviction that human life is strictly subject to natural laws. The new naturalism can be detected in minutely detailed and closely observed painting (the Barbizon School for example) and in the rise of the naturalistic novel, Zola's *Le Roman expérimental* (1880) being regarded as the manifesto of the movement. Among writers influenced by naturalism are Gerhart Hauptmann, Strindberg, Ibsen, Chekhov, and Theodore Dreiser.

naturalist /'nætʃrəlɪst/ *n.* & *adj.* ● *n.* **1** an expert in natural history. **2** a person who believes in or practises naturalism. ● *adj.* = NATURALISTIC.

naturalistic /,nætʃrə'lɪstɪk/ *adj.* **1** imitating nature closely; lifelike.

2 of or according to naturalism. **3** of natural history. □ **naturalistically** *adv.*

naturalize /'nætʃrə,laɪz/ *v.* (also **-ise**) **1** *tr.* admit (a foreigner) to the citizenship of a country. **2** *tr.* introduce (an animal, plant, etc.) into another region so that it flourishes in the wild. **3** *tr.* adopt (a foreign word, custom, etc.). **4** *intr.* become naturalized. **5** *tr.* *Philos.* exclude from the miraculous; explain naturalistically. **6** *tr.* free from conventions; make natural. **7** *tr.* cause to appear natural. **8** *intr.* study natural history. □ **naturalization** /,nætʃrəlaɪ'zeɪʃ(ə)n/ *n.* [F *naturaliser* (as NATURAL)]

naturally /'nætʃrəlɪ/ *adv.* **1** in a natural manner. **2** as a natural result. **3** (qualifying a whole sentence) as might be expected; of course.

natural selection *n.* *Biol.* an evolutionary process proposed by Darwin, whereby those organisms that are better adapted to their environment are more likely to survive and have offspring. A simple example of this is the predominance of dark-coloured moths in industrially polluted areas, where they are less visible against dark tree-trunks than the light-coloured forms, of which a higher proportion are eaten by predators. (See also EVOLUTION.)

nature /'neɪtʃə(r)/ *n.* **1** a thing's or person's innate or essential qualities or character (*not in their nature to be cruel*; *is the nature of iron to rust*). **2** (often **Nature**) **a** the physical power causing all the phenomena of the material world (*Nature is the best physician*). **b** these phenomena, including plants, animals, landscape, etc. **3** a kind, sort, or class (*things of this nature*). **4** = *human nature*. **5 a** a specified element of human character (*the rational nature*; *our animal nature*). **b** a person of a specified character (*even strong natures quail*). **6 a** an uncultivated or wild area, condition, community, etc. **b** the countryside, esp. when picturesque. **7** inherent impulses determining character or action. **8** heredity as an influence on or determinant of personality (opp. NURTURE *n.* 3). **9** a living thing's vital functions or needs (*such a diet will not support nature*). □ **against** (or **contrary to**) **nature 1** unnatural; immoral. **2** miraculous; miraculously. **back to nature** returning to a pre-civilized or natural state. **by nature** innately. **from nature** *Art* using natural objects as models. **human nature** see HUMAN. **in nature 1** actually existing. **2** anywhere; at all. **in the nature of things** inevitable; inevitably. **in** (or **of**) **the nature of** characteristically resembling or belonging to the class of (*the answer was in the nature of an excuse*). **in a state of nature 1** in an uncivilized or uncultivated state. **2** totally naked. **3** in an unregenerate state. **law of nature** = *natural law* 2. **nature cure** = NATUROPATHY. **nature-printing** a method of producing a print of leaves etc. by pressing them on a prepared plate. **nature reserve** a tract of land managed so as to preserve its flora, fauna, physical features, etc. **nature study** the practical study of plant and animal life etc. as a school subject. **nature trail** a signposted path through the countryside designed to draw attention to natural phenomena. [ME f. OF f. L *natura* f. *nasci nat-* be born]

natured /'neɪtʃəd/ *adj.* (in *comb.*) having a specified disposition (*good-natured*; *ill-natured*).

naturism /'neɪtʃə,rɪz(ə)m/ *n.* **1** nudism. **2** naturalism in regard to religion. **3** the worship of natural objects. □ **naturist** *n.* & *adj.*

naturopathy /,neɪtʃə'rɒpəθɪ/ *n.* **1** the treatment of disease etc. without drugs, usu. involving diet, exercise, massage, etc. **2** this regimen used preventively. □ **naturopath** /'neɪtʃərə,pæθ/ *n.* **naturopathic** /,neɪtʃərə'pæθɪk/ *adj.*

naught /nɔːt/ *n.* & *adj.* ● *n.* **1** *archaic* or *literary* nothing, nought. **2** *US* = NOUGHT 1. ● *adj.* (usu. *predic.*) *archaic* or *literary* worthless; useless. □ **bring to naught** ruin; baffle. **come to naught** be ruined or baffled. **set at naught** disregard; despise. [OE *nāwiht*, *-wuht* f. *nā* (see NO²) + *wiht* WIGHT]

naughty /'nɔːtɪ/ *adj.* (**naughtier**, **naughtiest**) **1** (esp. of children) disobedient; badly behaved. **2** *colloq.* *joc.* indecent (*a naughty postcard*). **3** *archaic* wicked. □ **naughtily** *adv.* **naughtiness** *n.* [ME f. NAUGHT + -Y¹]

nauplius /'nɔːplɪəs/ *n.* (*pl.* **nauplii** /-plɪ,aɪ/) *Zool.* the first larval stage of some crustaceans. [L, = a kind of shellfish, or f. Gk *Nauplios* son of Poseidon]

Nauru /nɑː'uːruː/ an island country in the SW Pacific, near the equator; pop. (1989) 9,000; official languages, Nauruan and English; no official capital. Inhabited by Polynesians, Nauru was encountered by the British in 1798. It was annexed by Germany in 1888 and became a British mandate after the First World War. Since 1968 it has been an independent republic with a limited form of membership of the Commonwealth. Its economy is heavily dependent upon the mining of phosphates, of which it has the world's richest deposits.

Nauruan /nɑː'uːrʊən/ *n.* & *adj.* ● *n.* **1** a native or inhabitant of Nauru.

2 the Malayo-Polynesian language of Nauru. ● *adj.* of or relating to Nauru or Nauruan.

nausea /ˈnɔːzɪə, ˈnɔːsɪə/ *n.* **1** a feeling of sickness with an inclination to vomit. **2** loathing; revulsion. [L f. Gk *nausia* f. *naus* ship]

nauseate /ˈnɔːzɪˌeɪt, ˈnɔːsɪ-/ *v.* **1** *tr.* affect with nausea; disgust (*was nauseated by the smell*). **2** *intr.* (usu. foll. by *at*) loathe food, an occupation, etc.; feel sick. □ **nauseating** *adj.* **nauseatingly** *adv.* [L *nauseare* (as NAUSEA)]

nauseous /ˈnɔːzɪəs, ˈnɔːsɪ-/ *adj.* **1** affected with nausea; sick (*felt nauseous all day*). **2** causing nausea; offensive to the taste or smell. **3** disgusting; loathsome. □ **nauseously** *adv.* **nauseousness** *n.* [L *nauseosus* (as NAUSEA)]

nautch /nɔːtʃ/ *n.* a performance of professional Indian dancing-girls. □ **nautch-girl** a professional Indian dancing-girl. [Urdu (Hindi) *nāch* f. Prakrit *nachcha* f. Skr. *nṛtja* dancing]

nautical /ˈnɔːtɪk(ə)l/ *adj.* of or concerning sailors or navigation; naval; maritime. □ **nautical almanac** a yearbook containing astronomical and tidal information for navigators etc. **nautical mile** a unit of distance of approx. 2,025 yards (1,852 metres); see also *sea mile*. □ **nautically** *adv.* [F *nautique* or f. L *nauticus* f. Gk *nautikos* f. *nautēs* sailor f. *naus* ship]

Nautilus /ˈnɔːtɪləs/ the first nuclear-powered submarine, launched in 1954. This US navy vessel was capable of prolonged submersion, making a historic journey (1–5 Aug. 1958) from Alaska to the Greenland Sea, passing under the thick polar ice-cap of the North Pole. The name *Nautilus* had previously been given to two historic submarines, especially that of Robert Fulton, as well as to the fictitious one in Jules Verne's *Twenty Thousand Leagues under the Sea.*

nautilus /ˈnɔːtɪləs/ *n.* (*pl.* **nautiluses** or **nautili** /-ˌlaɪ/) **1** (in full **pearly** or **chambered nautilus**) a cephalopod of the genus *Nautilus,* of Indo-Pacific waters, having a light coiled chambered shell with nacreous septa, and numerous tentacles; esp. the common *N. pompilius.* **2** (in full **paper nautilus**) a small floating octopus of the genus *Argonauta,* the female of which secretes a thin chalky shell for use as an egg-case and has webbed sail-like arms. [L f. Gk *nautilos,* lit. 'sailor' (as NAUTICAL)]

Navajo /ˈnævəˌhəʊ/ *n.* & *adj.* (also **Navaho**) ● *n.* (*pl.* same or **Navajos**) **1** a member of an American Indian people of New Mexico and Arizona. **2** the language of this people. ● *adj.* of this people or their language. [Sp., = pueblo]

naval /ˈneɪv(ə)l/ *adj.* **1** of, in, for, etc. the navy or a navy. **2** of or concerning ships (*a naval battle*). □ **naval academy** a college for training naval officers. **naval architect** a designer of ships. **naval architecture** the designing of ships. **naval officer** an officer in a navy. **naval stores** all materials used in shipping. □ **navally** *adv.* [L *navalis* f. *navis* ship]

Navan /ˈnæv(ə)n/ the county town of Meath, in the Republic of Ireland; pop. (1991) 3,410.

Navanagar /ˌnʌvəˈnʌgə(r)/ a former princely state of NW India, centred on the city of Jamnagar. It was ruled between 1907 and 1933 by the cricketer Ranjitsinhji Vibhaji. It is now part of the modern state of Gujarat.

navarin /ˈnævərɪn/ *n.* a casserole of mutton or lamb with vegetables. [F]

Navarino, Battle of /ˌnævəˈriːnəʊ/ a decisive naval battle in the Greek struggle for independence from the Ottoman Empire, fought in 1827 in the Bay of Navarino off Pylos in the Peloponnese, in which Britain, Russia, and France sent a combined fleet which destroyed the Egyptian and Turkish fleet.

Navarre /nəˈvɑː(r)/ (Spanish **Navarra** /naˈβarra/) an autonomous region of northern Spain, on the border with France; capital, Pamplona. A former Franco-Spanish kingdom in the Pyrenees, Navarre achieved independence in the 10th century under Sancho III (*c.*992–1035), but during the Middle Ages it fell at various times under French or Spanish domination. The southern part was conquered by Ferdinand in 1512 and attached to Spain, while the northern part passed to France in 1589 through inheritance by Henry IV.

nave[1] /neɪv/ *n.* the central part of a church, usu. from the west door to the chancel and excluding the side aisles. [med.L *navis* f. L *navis* ship]

nave[2] /neɪv/ *n.* the hub of a wheel. [OE *nafu, nafa* f. Gmc, rel. to NAVEL]

navel /ˈneɪv(ə)l/ *n.* **1** a rounded knotty depression in the centre of the belly caused by the detachment of the umbilical cord. **2** a central point. □ **navel orange** a large seedless orange with a navel-like formation at the top. [OE *nafela* f. Gmc, rel. to NAVE[2]]

navelwort /ˈneɪv(ə)lˌwɜːt/ *n.* a plant, *Umbilicus rupestris,* with round fleshy leaves, growing on rocks and walls. Also called (*wall*) *pennywort.*

navicular /nəˈvɪkjʊlə(r)/ *adj.* & *n.* ● *adj.* boat-shaped. ● *n.* (in full **navicular bone**) *Anat.* a boat-shaped bone in the foot or hand. □ **navicular disease** an inflammatory disease of the navicular bone in horses, causing lameness. [F *naviculaire* or LL *navicularis* f. L *navicula* dimin. of *navis* ship]

navigable /ˈnævɪgəb(ə)l/ *adj.* **1** (of a river, the sea, etc.) affording a passage for ships. **2** (of a ship etc.) seaworthy (*in navigable condition*). **3** (of a balloon, airship, etc.) steerable. □ **navigability** /ˌnævɪgəˈbɪlɪtɪ/ *n.* [F *navigable* or L *navigabilis* (as NAVIGATE)]

navigate /ˈnævɪˌgeɪt/ *v.* **1 a** *tr.* manage or direct the course of (a ship, aircraft, etc.). **b** *intr.* find one's way; steer the correct course. **2** *tr.* **a** sail on or across (a sea, river, etc.). **b** travel or fly through (the air). **3** *intr.* (of a passenger in a vehicle) assist the driver by map-reading etc. **4** *intr.* sail a ship; sail in a ship. **5** *tr.* (often *refl.*) *colloq.* steer (oneself, a course, etc.) through a crowd etc. [L *navigare* f. *navis* ship + *agere* drive]

navigation /ˌnævɪˈgeɪʃ(ə)n/ *n.* **1** the act or process of navigating. **2** any of several methods of determining or planning a ship's or aircraft's position and course by geometry, astronomy, radio signals, etc. (*See note below.*) **3** a voyage. □ **inland navigation** communication by canals and rivers. **navigation light** a light on a ship or aircraft at night, indicating its position and direction. **navigation satellite** an artificial satellite whose orbit is accurately known and made available, so that signals from it may be used for navigational purposes. □ **navigational** *adj.* [F or f. L *navigatio* (as NAVIGATE)]

▪ In ancient times navigation depended on observation of landmarks and the positions of the stars. The magnetic compass was introduced in the 12th–13th centuries, and was followed by the astrolabe, quadrant, and sextant, successively used to ascertain latitude. The problem of determining longitude was not solved until the 18th century, with the development of accurate marine chronometers. In the 20th century it has become possible to monitor position constantly, owing to the development of the gyrocompass, radar, sonar, navigation satellites, radio time signals, etc.

navigator /ˈnævɪˌgeɪtə(r)/ *n.* **1** a person skilled or engaged in navigation. **2** an explorer by sea. [L (as NAVIGATE)]

Navratilova /ˌnævˌvrætɪˈləʊvə/, Martina (b.1956), Czech-born American tennis player. Her major successes include nine Wimbledon singles titles (1978–9; 1982–7; 1990), two world championships (1980; 1984), and eight successive grand slam doubles titles.

navvy /ˈnævɪ/ *n.* & *v. Brit.* ● *n.* (*pl.* **-ies**) a labourer employed in building or excavating roads, canals, etc. ● *v.intr.* (**-ies, -ied**) work as a navvy. [abbr. of NAVIGATOR]

navy /ˈneɪvɪ/ *n.* (*pl.* **-ies**) **1** (often **Navy**) **a** the whole body of a state's ships of war, including crews, maintenance systems, etc. (See also ROYAL NAVY.) **b** the officers and men of a navy. **2** (in full **navy blue**, or **navy-blue** when *attrib.*) a dark blue colour as used in naval uniform. **3** *poet.* a fleet of ships. □ **navy yard** *US* a government shipyard with civilian labour. [ME, = fleet f. OF *navie* ship, fleet f. Rmc & pop.L *navia* ship f. L *navis*]

Navy Department (in the US) the government department in charge of the navy.

Navy List (in the UK) an official list of commissioned officers.

nawab /nəˈwɑːb, -ˈwɔːb/ *n.* **1** the title of a distinguished Muslim in Pakistan. **2** *hist.* the title of a governor or nobleman in India. [Urdu *nawwāb* pl. f. Arab. *nā'ib* deputy: cf. NABOB]

Naxçivan /ˌnæxtʃɪˈvɑːn/ (Russian **Nakhichevan** /nəxitʃɪˈvan/) **1** a predominantly Muslim Azerbaijani autonomous republic, situated on the borders of Turkey and northern Iran and separated from the rest of Azerbaijan by a narrow strip of Armenia; pop. (est. 1987) 300,000. Persian from the 13th to the 19th century, it became part of the Russian empire in 1828 and an autonomous republic of the Soviet Union, administratively part of Azerbaijan, in 1924. In 1990 it was the first Soviet territory to declare unilateral independence since the 1917 revolution. It has a predominantly Azerbaijani population and, along with Nagorno-Karabakh, is the focal point of conflict between Armenia and Azerbaijan. **2** the capital city of this republic; pop. (est. 1987) 51,000. It is an ancient settlement, perhaps dating from *c.*1500 BC and believed in Armenian tradition to have been founded by Noah.

Naxos /ˈnæksɒs/ a Greek island in the southern Aegean, the largest of the Cyclades.

nay /neɪ/ *adv.* & *n.* ● *adv.* **1** or rather; and even; and more than that (*impressive, nay, magnificent*). **2** *archaic* = NO² *adv.* **1.** ● *n.* **1** the word nay. **2** a negative vote (*counted 16 nays*). [ME f. ON *nei* f. *ne* not + *ei* AYE²]

Nayarit /ˌnɑːjɑːˈriːt/ a state of western Mexico, on the Pacific coast; capital, Tepic.

naysay /ˈneɪseɪ/ *v.* (*3rd sing. present* **-says** /-sez/; *past* and *past part.* **-said** /-sed/) esp. *US* **1** *intr.* utter a denial or refusal. **2** *tr.* refuse or contradict. □ **naysayer** *n.*

Nazarene /ˌnæzəˈriːn/ *n. & adj.* ● *n.* **1 a** (prec. by *the*) Jesus. **b** (esp. in Jewish or Muslim use) a Christian. **2** a native or inhabitant of Nazareth. **3** a member of an early Jewish-Christian sect living in Syria, who continued to obey much of the Jewish Law although they were otherwise orthodox Christians. They used a version of the Gospel in Aramaic. **4** a member of a group of German painters called the Brotherhood of St Luke. (*See note below.*) ● *adj.* of or concerning Nazareth or Nazarenes. [ME f. LL *Nazarenus* f. Gk *Nazarēnos* f. *Nazaret* NAZARETH]

▪ The Nazarenes aspired to revive religious painting by a return to the art and working practices of medieval Germany and early Renaissance Italy. Founded in 1809, from 1810 they lived and worked in a disused monastery in Rome. Their style was marked by strong outlines and pure colour reminiscent of quattrocento art, and was a major influence on the British Pre-Raphaelite movement. Leading participants were Peter von Cornelius (1783–1867), Friedrich Overbeck (1789–1869), and Franz Pforr (1788–1812).

Nazareth /ˈnæzərəθ/ a historic town in lower Galilee in present-day northern Israel; pop. (1982) 39,000. It was mentioned in the Gospels as the home of Mary and Joseph, is closely associated with the childhood of Jesus, and is a centre of Christian pilgrimage.

Nazarite /ˈnæzəˌraɪt/ *n.* (also **Nazirite**) *hist.* any of the Israelites specially consecrated to the service of God who were under vows to abstain from wine, let their hair grow, and avoid the defilement of contact with a dead body (Num. 6). [LL *Nazaraeus* f. Heb. *nāzīr* f. *nāzar* to separate or consecrate oneself]

Nazca Lines /ˈnæzkə/ a group of huge abstract designs, outline drawings of birds and animals, and straight lines on the coastal plain north of Nazca in southern Peru, made by cleaning and aligning the surface stones to expose the underlying sand. They belong to a pre-Inca culture of that region, and their purpose is uncertain; some hold the designs to represent a vast calendar or astronomical information. Virtually indecipherable at ground level, the lines are clearly visible from the air; they have been preserved by the extreme dryness of the region.

Nazi /ˈnɑːtsɪ/ *n. & adj.* ● *n.* (*pl.* **Nazis**) **1** *hist.* a member of the National Socialist Party in Germany, led by Adolf Hitler. (*See note below.*) **2** *derog.* a person holding extreme racist or authoritarian views or behaving brutally. **3** a person belonging to any organization similar to the Nazis. ● *adj.* of or concerning the Nazis, Nazism, etc. □ **Nazidom** *n.* **Nazify** *v.tr.* (**-ies, -ied**). **Naziism** *n.* **Nazism** *n.* [repr. pronunc. of *Nati-* in G *Nationalsozialist*]

▪ The National Socialist German Workers' Party was formed in Munich soon after the First World War to espouse a right-wing brand of nationalist authoritarianism. It was dominated from its early days by Hitler, who exploited feelings of anti-Semitism and resentment of the Treaty of Versailles and added to them his own beliefs in the racial superiority of 'Aryan' Germans. The initial popularity of Nazism (as witnessed by an increase in party membership from 176,000 in 1929 to 2,000,000 in early 1933) can be partly ascribed to Hitler's charismatic appeal and to the dreadful conditions prevalent in Germany during the Depression. After his election as Chancellor in January 1933, Hitler built up a Nazi dictatorship in which the party effectively controlled the state at all levels; Nazi Germany rearmed extensively and precipitated the Second World War with its expansionist foreign policies. The Nazi Party collapsed at the end of the Second World War and was formally outlawed by the new West German constitution.

Nazirite var. of NAZARITE.

NB *abbr.* **1** New Brunswick. **2** no ball. **3** Scotland (North Britain). **4** *nota bene*.

Nb *symb. Chem.* the element niobium.

NBA see NET BOOK AGREEMENT.

NBC *abbr.* (in the US) National Broadcasting Company.

n.b.g. *abbr. colloq.* no bloody good.

N. by E. *abbr.* North by East.

N. by W. *abbr.* North by West.

NC *abbr.* North Carolina (also in official postal use).

NCB *abbr. hist.* (in the UK) National Coal Board. ¶ Since 1987 officially called *British Coal*.

NCO *abbr.* non-commissioned officer.

NCU *abbr.* (in the UK) National Communications Union.

ND *abbr. US* North Dakota (in official postal use).

Nd *symb. Chem.* the element neodymium.

n.d. *abbr.* no date.

-nd¹ *suffix* forming nouns (*fiend; friend*). [OE *-ond*, orig. part. ending]

-nd² *suffix see* -AND, -END.

N.Dak. *abbr.* North Dakota.

Ndebele /ˌəndəˈbiːlɪ/ *n. & adj.* ● *n.* (*pl.* same or **Indebele** /ˌɪndə-/, or **Ndebeles**) **1** a member of a Nguni people. (*See note below.*) **2** the Bantu language of this people, Sindebele. ● *adj.* of or relating to this people or their language. [Bantu, f. Ndebele *n-* sing. pref. + Sesotho (*lè*)*tèbèlè* Nguni, f. *le-* pref. + *tèbèla* drive away]

▪ Branches of the Ndebele are found in Zimbabwe (where they are better known as *Matabele*) and in north-eastern South Africa. Traditionally the Ndebele lived in kraals, rearing cattle and growing maize, but many lost their lands under colonial rule and became domestic or industrial workers.

N'Djamena /ˌəndʒæˈmeɪnə/ the capital of Chad; pop. (1992) 688,000. Founded by the French in 1900, it was known until 1973 as Fort Lamy.

Ndola /ənˈdəʊlə/ a city in the Copperbelt region of central Zambia; pop. (1987) 418,100.

NE *abbr.* **1** north-east. **2** north-eastern. **3** *US* Nebraska (in official postal use).

Ne *symb. Chem.* the element neon.

né /neɪ/ *adj.* born (indicating a man's previous name) (*Lord Beaconsfield, né Benjamin Disraeli*). [F, past part. of *naître* be born: cf. NÉE]

Neagh, Lough /neɪ/ a shallow lake in Northern Ireland, the largest freshwater lake in the British Isles.

Neanderthal man /nɪˈændəˌtɑːl/ *n.* a palaeolithic fossil hominid, usually regarded as a race of *Homo sapiens*, named from remains found in 1856 in Neanderthal (now Neandertal), a valley in western Germany. Neanderthal man was stocky, with a long low skull, prominent brow ridges, and a jutting face, but not as brutish in appearance as was formerly thought. The race probably evolved in Europe, though remains have also been found in the Near East and elsewhere in the Old World, and dates range from *c.*120,000–35,000 BP (the Middle and Upper Pleistocene). In western Europe Neanderthal remains are associated with the middle palaeolithic Mousterian stone-tool industries, and it appears that the Neanderthals may have lived for up to 10,000 years alongside the modern humans who eventually displaced them (see CRO-MAGNON MAN).

neap /niːp/ *n. & v.* ● *n.* (in full **neap tide**) a tide just after the first and third quarters of the moon when there is least difference between high and low water. ● *v.* **1** *intr.* (of a tide) tend towards or reach the highest point of a neap tide. **2** *tr.* (in *passive*) (of a ship) be kept aground, in harbour, etc., by a neap tide. [OE *nēpflōd* (cf. FLOOD), of unkn. orig.]

Neapolitan /nɪəˈpɒlɪt(ə)n/ *n. & adj.* ● *n.* a native or inhabitant of Naples. ● *adj.* of or relating to Naples. □ **Neapolitan ice-cream** ice-cream made in layers of different colours and flavours. **Neapolitan violet** a sweet-scented double viola. [ME f. L *Neapolitanus* f. L *Neapolis* NAPLES f. Gk f. *neos* new + *polis* city]

near /nɪə(r)/ *adv., prep., adj., & v.* ● *adv.* **1** (often foll. by *to*) to or at a short distance in space or time; close by (*the time drew near; dropped near to them*). **2** closely (*as near as one can guess*). **3** *archaic* almost, nearly (*very near died*). **4** *archaic* parsimoniously; meanly. ● *prep.* (compar. & superl. also used) **1** to or at a short distance (in space, time, condition, or resemblance) from (*stood near the back; occurs nearer the end; the sun is near setting*). **2** (in *comb.*) **a** that is almost (*near-hysterical; a near-Communist*). **b** intended as a substitute for; resembling (*near-beer*). ● *adj.* **1** (usu. *predic.*) close at hand; close to, in place or time (*the end is near; in the near future*). **2 a** closely related (*a near relation*). **b** intimate (*a near friend*). **3** (of a part of a vehicle, animal, or road) left (*the near fore leg; near side front wheel*) (orig. of the side from which one mounted) (opp. OFF *adj.* 2). **4** close; narrow (*a near escape; a near guess*). **5** (of a road or way) direct. **6** similar (to) (*is nearer the original*). **7** niggardly, mean. ● *v.* **1** *tr.* approach; draw near to (*neared the harbour*). **2** *intr.* draw near (*could distinguish them as they neared*). □ **come** (or **go**) **near** (foll. by verbal noun, or *to* + verbal noun) be on the point of, almost succeed in (*came near to falling*). **go near** (foll. by *to* + infin.) narrowly fail. **near at hand 1** within easy reach. **2** in the immediate future. **nearest and dearest** one's closest friends and relatives collectively. **near go** *colloq.* a narrow escape. **near the knuckle** *colloq.* verging on the indecent. **near miss 1** a bomb etc. falling close to the target. **2** a narrowly avoided collision. **3** an attempt that is almost but not quite successful. **near sight** esp. *US* = *short sight*. **near thing** a narrow escape. **near upon** *archaic* not far in time from. □ **nearish** *adj.* **nearness** *n.* [ME f. ON *nær*, orig. compar. of *ná* = OE *nēah* NIGH]

nearby /ˈnɪəbaɪ/ *adj. & adv.* ● *adj.* /ˈnɪəbaɪ/ situated in a near position (*a nearby hotel*). ● *adv.* /nɪəˈbaɪ/ (also **near by**) close; not far away.

Nearctic /nɪˈɑːktɪk/ *adj. & n.* (also **nearctic**) *Zool.* ● *adj.* of or relating to the Arctic and temperate parts of North America as a zoogeographical region. ● *n.* the Nearctic region. [NEO- + ARCTIC]

Near East a term originally applied to the Balkan states of SE Europe but now generally applied to the countries of SW Asia between the Mediterranean and India. It includes the area known as the Middle East.

nearly /ˈnɪəlɪ/ *adv.* **1** almost (*we are nearly there*). **2** closely (*they are nearly related*). □ **not nearly** nothing like; far from (*not nearly enough*).

nearside /ˈnɪəsaɪd/ *n.* (often *attrib.*) esp. *Brit.* the left side of a vehicle, animal, etc. (cf. OFFSIDE *n.*).

near-sighted /nɪəˈsaɪtɪd/ *adj.* esp. *US* = SHORT-SIGHTED. □ **near-sightedly** *adv.* **near-sightedness** *n.*

neat[1] /niːt/ *adj.* **1** tidy and methodical. **2** elegantly simple in form etc.; well-proportioned. **3** (of language, style, etc.) brief, clear, and pointed; epigrammatic. **4 a** cleverly executed (*a neat piece of work*). **b** deft; dexterous. **5** (of esp. alcoholic liquor) undiluted. **6** *N.Amer. sl.* (as a general term of approval) good, pleasing, excellent. □ **neatly** *adv.* **neatness** *n.* [F *net* f. L *nitidus* shining f. *nitere* shine]

neat[2] /niːt/ *n. archaic* **1** a bovine animal. **2** (as *pl.*) cattle. □ **neat's-foot oil** oil made from boiled cow-heel and used to dress leather. [OE *nēat* f. Gmc]

neaten /ˈniːt(ə)n/ *v.tr.* make neat.

Neath /niːθ/ an industrial town in South Wales, in West Glamorgan on the River Neath; pop. (1981) 49,130.

neath /niːθ/ *prep. poet.* beneath. [BENEATH]

NEB *abbr.* **1** (in the UK) National Enterprise Board. **2** see NEW ENGLISH BIBLE.

Neb. *abbr.* Nebraska.

neb /neb/ *n. Sc. & N. Engl.* **1** a beak or bill. **2** a nose; a snout. **3** a tip, spout, or point. [OE *nebb* ult. f. Gmc: cf. NIB]

nebbish /ˈnebɪʃ/ *n. & adj. colloq.* ● *n.* a submissive or timid person. ● *adj.* submissive; timid. [Yiddish *nebach* poor thing!]

Neblina, Pico da see PICO DA NEBLINA.

Nebr. *abbr.* Nebraska.

Nebraska /nɪˈbræskə/ a state in the central US to the west of the Missouri; pop. (1990) 1,578,385; capital, Lincoln. It was acquired as part of the Louisiana Purchase in 1803 and became the 37th state of the US in 1867. □ **Nebraskan** *adj. & n.*

Nebuchadnezzar /ˌnebjʊkədˈnezə(r)/ (*c.*630–562 BC), king of Babylon 605–562 BC. He rebuilt the city with massive fortification walls, a huge temple, and a ziggurat, and extended his rule over ancient Palestine and neighbouring countries. In 586 BC he captured and destroyed Jerusalem and deported many Israelites (see BABYLONIAN CAPTIVITY).

nebuchadnezzar /ˌnebjʊkədˈnezə(r)/ *n.* a wine bottle of about twenty times the standard size. [NEBUCHADNEZZAR]

nebula /ˈnebjʊlə/ *n.* (*pl.* **nebulae** /-ˌliː/ or **nebulas**) **1** *Astron.* a cloud of gas or dust in space. (*See note below.*) **2** *Med.* a clouded spot on the cornea causing defective vision. [L, = mist]

▪ Astronomers originally applied the term *nebula* to any indistinct cloudlike patches seen in the telescope, including distant galaxies, but it is now usually restricted to objects situated within a galaxy, usually our own. Many nebulae appear bright, either due to their own light or to illumination by stars; some appear as dark silhouettes against other glowing matter.

nebular /ˈnebjʊlə(r)/ *adj.* of or relating to a nebula or nebulae. □ **nebular theory** (or **hypothesis**) the theory that the solar and stellar systems were developed from a primeval nebula.

nebulous /ˈnebjʊləs/ *adj.* **1** cloudlike. **2 a** formless, clouded. **b** hazy, indistinct, vague (*put forward a few nebulous ideas*). **3** *Astron.* of or like a nebula or nebulae. □ **nebulous star** *Astron.* a small cluster of indistinct stars, or a star in a luminous haze. □ **nebulously** *adv.* **nebulousness** *n.* **nebulosity** /ˌnebjʊˈlɒsɪtɪ/ *n.* [ME f. F *nébuleux* or L *nebulosus* (as NEBULA)]

nebuly /ˈnebjʊlɪ/ *adj.* Heraldry wavy in form; cloudlike. [F *nébulé* f. med.L *nebulatus* f. L NEBULA]

NEC *abbr.* National Exhibition Centre (in Birmingham).

necessarian /ˌnesɪˈseərɪən/ *n. & adj. Philos.* = NECESSITARIAN. □ **necessarianism** *n.*

necessarily /ˈnesɪsərɪlɪ, ˌnesɪˈserɪlɪ/ *adv.* as a necessary result; inevitably.

necessary /ˈnesɪsərɪ/ *adj. & n.* ● *adj.* **1** requiring to be done, achieved, etc.; requisite, essential (*it is necessary to work; lacks the necessary documents*). **2** determined, existing, or happening by natural laws, predestination, etc., not by free will; inevitable (*a necessary evil*). **3** *Philos.* (of a concept or a mental process) inevitably resulting from or produced by the nature of things etc., so that the contrary is impossible. **4** *Philos.* (of an agent) having no independent volition. ● *n.* (*pl.* **-ies**) (usu. in *pl.*) any of the basic requirements of life, such as food, warmth, etc. □ **the necessary** *colloq.* **1** money. **2** an action, item, etc., needed for a purpose (*they will do the necessary*). [ME f. OF *necessaire* f. L *necessarius* f. *necesse* needful]

necessitarian /nɪˌsesɪˈteərɪən/ *n. & adj. Philos.* ● *n.* a person who holds that all action is predetermined and free will is impossible. ● *adj.* of or concerning such a person or theory (opp. LIBERTARIAN). □ **necessitarianism** *n.*

necessitate /nɪˈsesɪˌteɪt/ *v.tr.* **1** make necessary (esp. as a result) (*will necessitate some sacrifice*). **2** *US* (usu. foll. by *to* + infin.) force or compel (a person) to do something. [med.L *necessitare* compel (as NECESSITY)]

necessitous /nɪˈsesɪtəs/ *adj.* poor; needy. [F *nécessiteux* or f. NECESSITY + -OUS]

necessity /nɪˈsesɪtɪ/ *n.* (*pl.* **-ies**) **1 a** an indispensable thing; a necessary (*central heating is a necessity*). **b** (usu. foll. by *of*) indispensability (*the necessity of a warm overcoat*). **2** a state of things or circumstances enforcing a certain course (*there was a necessity to hurry*). **3** imperative need (*necessity is the mother of invention*). **4** want; poverty; hardship (*stole because of necessity*). **5** constraint or compulsion regarded as a natural law governing all human action. □ **of necessity** unavoidably. [ME f. OF *necessité* f. L *necessitas -tatis* f. *necesse* needful]

Nechtansmere, Battle of /ˈnektənzˌmɪə(r)/ a battle which took place in 685 at Nechtansmere, near Forfar, Scotland, in which the Northumbrians were decisively defeated by the Picts. Their expansion northward was permanently checked and they were forced to withdraw south of the Firth of Forth.

neck /nek/ *n. & v.* ● *n.* **1 a** the part of the body connecting the head to the shoulders. **b** the part of a shirt, dress, etc. round or close to the neck. **2 a** something resembling a neck, such as the narrow part of a cavity or vessel, a passage, channel, pass, isthmus, etc. **b** the narrow part of a bottle near the mouth. **3** the part of a violin etc. bearing the finger-board. **4** the length of a horse's head and neck as a measure of its lead in a race. **5** the flesh of an animal's neck (*neck of lamb*). **6** *Geol.* solidified lava or igneous rock in an old volcano crater or pipe. **7** *Archit.* the lower part of a capital. **8** *sl.* impudence (*you've got a neck, asking that*). ● *v.* **1** *intr. & tr. colloq.* kiss and caress amorously. **2 a** *tr.* form a narrowed part in. **b** *intr.* form a narrowed part. □ **get it in the neck** *colloq.* **1** receive a severe reprimand or punishment. **2** suffer a fatal or severe blow. **neck and neck** running level in a race etc. **neck of the woods** *colloq.* a community or locality. **neck or nothing** risking everything on success. **up to one's neck** (often foll. by *in*) *colloq.* very deeply involved; very busy. □ **necked** *adj.* (also in *comb.*). **necker** *n.* (in sense 1 of *v.*). **neckless** *adj.* [OE *hnecca* ult. f. Gmc]

Neckar /ˈnekɑː(r)/ a river of western Germany, which rises in the Black Forest and flows 367 km (228 miles) north and west through Stuttgart to meet the Rhine at Mannheim.

neckband /ˈnekbænd/ *n.* a strip of material round the neck of a garment.

neckcloth /ˈneklɒθ/ *n. hist.* a cravat.

Necker /ˈnekə(r)/, Jacques (1732–1804), Swiss-born banker. He began work as a bank clerk in Switzerland, moving to his firm's headquarters in Paris in 1750. He rose to hold the office of director-general of French finances on two occasions. During his first term in this post (1777–81), Necker's social and administrative reform programmes aroused the hostility of the court and led to his forced resignation. While in office for a second time (1788–9), he recommended summoning the States General, resulting in his dismissal on 11 July 1789. News of this angered the people and was one of the factors which resulted in the storming of the Bastille three days later.

neckerchief /'nekə‚tʃɪf, -‚tʃiːf/ n. a square of cloth worn round the neck.

necking /'nekɪŋ/ n. **1** in senses of NECK v. **2** Archit. = NECK n. 7.

necklace /'neklɪs/ n. & v. ● n. **1** a chain or string of beads, precious stones, links, etc., worn as an ornament round the neck. **2** S. Afr. a tyre soaked or filled with petrol, placed round a victim's neck, and set alight. ● v.tr. S. Afr. kill with a 'necklace'.

necklet /'neklɪt/ n. **1** = NECKLACE n. 1. **2** a strip of fur worn round the neck.

neckline /'neklaɪn/ n. the edge or shape of the opening of a garment at the neck (a square neckline).

necktie /'nektaɪ/ n. esp. US = TIE n. 2. □ **necktie party** sl. a lynching or hanging.

neckwear /'nekweə(r)/ n. collars, ties, etc.

necro- /'nekrəʊ/ comb. form corpse. [from or after Gk nekro- f. nekros corpse]

necrobiosis /‚nekrəʊbaɪ'əʊsɪs/ n. Med. decay in the tissues of the body, esp. swelling of the collagen bundles in the dermis. □ **necrobiotic** /-'ɒtɪk/ adj.

necrolatry /ne'krɒlətrɪ/ n. worship of, or excessive reverence towards, the dead.

necrology /ne'krɒlədʒɪ/ n. (pl. **-ies**) **1** a list of recently dead people. **2** an obituary notice. □ **necrological** /‚nekrə'lɒdʒɪk(ə)l/ adj.

necromancy /'nekrəʊ‚mænsɪ/ n. **1** the prediction of the future by the supposed communication with the dead. **2** witchcraft. □ **necromancer** n. **necromantic** /‚nekrəʊ'mæntɪk/ adj. [ME f. OF nigromancie f. med.L nigromantia changed (by assoc. with L niger nigri black) f. LL necromantia f. Gk nekromanteia (as NECRO-, -MANCY)]

necrophilia /‚nekrə'fɪlɪə/ n. (also **necrophily** /ne'krɒfɪlɪ/) a morbid and esp. erotic attraction to corpses. □ **necrophile** /'nekrə‚faɪl/ n. **necrophiliac** /‚nekrə'fɪlɪ‚æk/ n. **necrophilic** adj. **necrophilism** /ne'krɒfɪ‚lɪz(ə)m/ n. **necrophilist** n. [NECRO- + Gk -philia loving]

necrophobia /‚nekrə'fəʊbɪə/ n. an abnormal fear of death or dead bodies.

necropolis /ne'krɒpəlɪs/ n. **1** an ancient cemetery or burial place. **2** a cemetery, esp. a large one in or near a city.

necropsy /'nekrɒpsɪ/ n. (also **necroscopy** /ne'krɒskəpɪ/) (pl. **-ies**) = AUTOPSY 1. [NECRO- after AUTOPSY, or + -SCOPY]

necrosis /ne'krəʊsɪs/ n. Med. & Physiol. the death of tissue caused by disease or injury, esp. as one of the symptoms of gangrene or pulmonary tuberculosis. □ **necrotic** /-'krɒtɪk/ adj. **necrotize** /'nekrə‚taɪz/ v.intr. (also **-ise**). [mod.L f. Gk nekrōsis (as NECRO-, -OSIS)]

nectar /'nektə(r)/ n. **1** a sugary substance produced by plants to attract pollinating insects and made into honey by bees. **2** (in Greek and Roman mythology) the drink of the gods. **3** a drink compared to this. □ **nectarous** adj. **nectarean** /nek'teərɪən/ adj. **nectareous** adj. **nectariferous** /‚nektə'rɪfərəs/ adj. [L f. Gk nektar]

nectarine /'nektərɪn, -‚riːn/ n. **1** a variety of peach with a thin brightly coloured smooth skin and firm flesh. **2** the tree bearing this. [orig. as adj., = nectar-like, f. NECTAR + -INE⁴]

nectary /'nektərɪ/ n. (pl. **-ies**) Bot. the nectar-secreting organ of a flower or plant. [mod.L nectarium (as NECTAR)]

NEDC abbr. hist. (in the UK) National Economic Development Council.

neddy /'nedɪ/ n. (pl. **-ies**) Brit. colloq. **1** a donkey. **2** (**Neddy**) hist. = NEDC. [dimin. of Ned, pet-form of the name Edward]

Nederland see NETHERLANDS, THE.

Ned Kelly /ned 'kelɪ/ n. Austral. a person of reckless courage or unscrupulous business dealings. [the name of the most famous Australian bushranger; see KELLY¹]

née /neɪ/ adj. (US **nee**) (used in adding a married woman's maiden name after her surname) born (Mrs Ann Smith, née Jones). [F, fem. past part. of naître be born]

need /niːd/ v. & n. ● v. **1** tr. stand in want of; require (needs a new coat). **2** tr. (foll. by to + infin.; 3rd sing. present neg. or interrog. **need** without to) be under the necessity or obligation (it needs to be done carefully; he need not come; need you ask?). **3** intr. archaic be necessary. ● n. **1 a** a want or requirement (my needs are few; the need for greater freedom). **b** a thing wanted (my greatest need is a car). **2** circumstances requiring some course of action; necessity (there is no need to worry; if need arise). **3** destitution; poverty. **4** a crisis; an emergency (failed them in their need). □ **at need** in time of need. **had need** archaic ought to (had need remember). **have need of** require; want. **have need to** require to (has need to be warned).

in need requiring help. **in need of** requiring. **need not have** did not need to (but did). [OE nēodian, nēd f. Gmc]

needful /'niːdfʊl/ adj. **1** requisite; necessary; indispensable. **2** (prec. by the) **a** what is necessary. **b** colloq. money or action needed for a purpose. □ **needfully** adv. **needfulness** n.

needle /'niːd(ə)l/ n. & v. ● n. **1 a** a very thin small piece of smooth steel etc. pointed at one end and with a slit (eye) for thread at the other, used in sewing. **b** a larger plastic, wooden, etc. slender stick without an eye, used in knitting. **c** a slender hooked stick used in crochet. **2** a pointer on a dial (see magnetic needle). **3** any of several small thin pointed instruments, esp.: **a** a surgical instrument for stitching. **b** the end of a hypodermic syringe. **c** = STYLUS 1a. **d** an etching tool. **e** a steel pin exploding the cartridge of a breech-loading gun. **4 a** an obelisk (Cleopatra's Needle). **b** a pointed rock or peak. **5** the leaf of a fir or pine tree. **6** a beam used as a temporary support during underpinning. **7** (usu. prec. by the) Brit. sl. a fit of bad temper or nervousness (got the needle while waiting). ● v.tr. **1** colloq. incite or irritate; provoke (refused to be needled by him). **2** sew, pierce, or operate on with a needle. □ **needle game** (or **match** etc.) Brit. colloq. a contest that is very close or arouses personal grudges. **needle in a haystack** something almost impossible to find because it is concealed by so many other similar things. **needle-lace** lace made with needles not bobbins. **needle-point 1** a very sharp point. **2** = needle-lace. **3** = GROS POINT, PETIT POINT. **needle's eye** (or **eye of a needle**) the least possible aperture, esp. with ref. to Matt. 19:24. **needle time** an agreed maximum allowance of time for broadcasting recorded music. **needle valve** a valve closed by a thin tapering part. [OE nǽdl f. Gmc]

needlecord /'niːd(ə)l‚kɔːd/ n. Brit. a fine-ribbed corduroy fabric.

needlecraft /'niːd(ə)l‚krɑːft/ n. skill in needlework.

needlefish /'niːd(ə)l‚fɪʃ/ n. = GARFISH 1.

needleful /'niːd(ə)l‚fʊl/ n. (pl. **-fuls**) the length of thread etc. put into a needle at one time.

Needles, the a group of rocks in the sea off the west coast of the Isle of Wight, in southern England.

needless /'niːdlɪs/ adj. **1** unnecessary. **2** uncalled-for; gratuitous. □ **needless to say** of course; it goes without saying. □ **needlessly** adv. **needlessness** n.

needlewoman /'niːd(ə)l‚wʊmən/ n. (pl. **-women**) **1** a seamstress. **2** a woman or girl with specified sewing skill (a good needlewoman).

needlework /'niːd(ə)l‚wɜːk/ n. sewing or embroidery.

needs /niːdz/ adv. archaic (usu. preceded or foll. by must) of necessity (needs must decide). [OE nēdes (as NEED, -s³)]

needy /'niːdɪ/ adj. (**needier, neediest**) **1** (of a person) poor; destitute. **2** (of circumstances) characterized by poverty. □ **neediness** n.

neem /niːm/ n. a tree, Azadirachta indica (mahogany family), whose leaves and bark are used medicinally in the Indian subcontinent. [Hindi nīm]

neep /niːp/ n. Sc. & N. Engl. a turnip, or (esp.) a swede. [OE nǽp f. L napus]

ne'er /neə(r)/ adv. poet. = NEVER. □ **ne'er-do-well** n. a good-for-nothing person. ● adj. good-for-nothing. [ME contr. of NEVER]

nefarious /nɪ'feərɪəs/ adj. wicked; iniquitous. □ **nefariously** adv. **nefariousness** n. [L nefarius f. nefas wrong f. ne- not + fas divine law]

Nefertiti /‚nefə'tiːtɪ/ (also **Nofretete** /‚nɒfrə-/) (fl. 14th century BC), Egyptian queen, wife of Akhenaten. She initially supported her husband's religious reforms, although it is a matter of dispute whether she persisted in her support or withdrew it in favour of the new religion promoted by her half-brother Tutankhamen. Nefertiti is frequently represented beside Akhenaten, with their six daughters, on reliefs from Tell el-Amarna; however, she is best known from the painted limestone bust of her, now in Berlin (c.1350).

neg. abbr. negative.

negate /nɪ'geɪt/ v.tr. **1** nullify; invalidate. **2** imply, involve, or assert the non-existence of. **3** be the negation of. □ **negator** n. [L negare negat- deny]

negation /nɪ'geɪʃ(ə)n/ n. **1** the absence or opposite of something actual or positive. **2 a** the act of denying. **b** an instance of this. **3** (usu. foll. by of) a refusal, contradiction, or denial. **4** a negative statement or doctrine. **5** a negative or unreal thing; a nonentity. **6** Logic the assertion that a certain proposition is false. □ **negatory** /'negətərɪ/ adj. [F negation or L negatio (as NEGATE)]

negative /'negətɪv/ adj., n., & v. ● adj. **1** expressing or implying denial, prohibition, or refusal (a negative vote; a negative answer). **2** (of a person or attitude): **a** lacking positive attributes; apathetic; pessimistic.

b opposing or resisting; uncooperative. **3** marked by the absence of qualities (*a negative reaction*; *a negative result from the test*). **4** of the opposite nature to a thing regarded as positive (*debt is negative capital*). **5** less than zero, to be subtracted from others or from zero (opp. POSITIVE *adj.* 9). **6 a** of the kind of electric charge carried by electrons (opp. POSITIVE *adj.* 10). (See also ELECTRICITY.) **b** containing or producing such a charge. ● *n.* **1** a negative statement, reply, or word (*hard to prove a negative*). **2** *Photog.* **a** an image with black and white reversed or colours replaced by complementary ones, from which positive pictures are obtained. **b** a developed film or plate bearing such an image. **3** a negative quality; an absence of something. **4** (prec. by *the*) a position opposing the affirmative. **5** *Logic* = NEGATION 6. ● *v.tr.* **1** refuse to accept or countenance; veto; reject. **2** disprove (an inference or hypothesis). **3** contradict (a statement). **4** neutralize (an effect). □ **in the negative** with negative effect; so as to reject a proposal etc.; no (*the answer was in the negative*). **negative equity** the indebtedness arising when the market value of a property falls to below the outstanding amount of a mortgage secured on it. **negative evidence** (or **instance**) evidence of the non-occurrence of something. **negative feedback 1** a counter-productive or negative response to an experiment, questionnaire, etc. **2** *Electronics* the return of part of an output signal to the input, tending to decrease the amplification etc. **3** esp. *Biol.* feedback that tends to diminish or counteract the process giving rise to it. **negative geotropism** see GEOTROPISM. **negative income tax** a scheme whereby those on the lowest incomes receive benefit payments direct from the income tax system. **negative pole** the south-seeking pole of a magnet. **negative proposition** *Logic* = NEGATION 6. **negative quantity** *joc.* nothing. **negative sign** a symbol (-) indicating subtraction or a value less than zero. **negative virtue** abstention from vice. □ **negatively** *adv.* **negativeness** *n.* **negativity** /ˌnegəˈtɪvɪtɪ/ *n.* [ME f. OF *negatif* *-ive* or LL *negativus* (as NEGATE)]

negativism /ˈnegətɪˌvɪz(ə)m/ *n.* **1** a negative position or attitude; extreme scepticism, criticism, etc. **2** denial of accepted beliefs. □ **negativist** *n.* **negativistic** /ˌnegətɪˈvɪstɪk/ *adj.*

Negev, the /ˈnegev/ an arid region forming most of southern Israel, between Beersheba and the Gulf of Aqaba, on the Egyptian border. Large-scale irrigation projects have greatly increased the fertility of the region.

neglect /nɪˈglekt/ *v.* & *n.* ● *v.tr.* **1** fail to care for or to do; be remiss about (*neglected their duty*; *neglected his children*). **2** (foll. by verbal noun, or *to* + infin.) fail; overlook or forget the need to (*neglected to inform them*; *neglected telling them*). **3** not pay attention to; disregard (*neglected the obvious warning*). ● *n.* **1** lack of caring; negligence (*the house suffered from neglect*). **2 a** the act of neglecting. **b** the state of being neglected (*the house fell into neglect*). **3** (usu. foll. by *of*) disregard. □ **neglectful** *adj.* **neglectfully** *adv.* **neglectfulness** *n.* [L *neglegere* neglect- f. *neg-* not + *legere* choose, pick up]

negligée /ˈneglɪˌʒeɪ/ *n.* (also **negligee**, **négligé**) **1** (usu. **negligee**) a woman's dressing-gown of thin fabric. **2** unceremonious or informal attire. [F, past part. of *négliger* NEGLECT]

negligence /ˈneglɪdʒəns/ *n.* **1 a** a lack of proper care and attention; carelessness. **b** an act of carelessness. **2** *Law* = contributory negligence. **3** *Art* freedom from restraint or artificiality. □ **negligent** *adj.* **negligently** *adv.* [ME f. OF *negligence* or L *negligentia* f. *negligere* = *neglegere*: see NEGLECT]

negligible /ˈneglɪdʒɪb(ə)l/ *adj.* not worth considering; insignificant. □ **negligible quantity** a person etc. that need not be considered. □ **negligibly** *adv.* **negligibility** /ˌneglɪdʒɪˈbɪlɪtɪ/ *n.* [obs. F f. *négliger* NEGLECT]

Negombo /nɪˈgɒmbəʊ/ a port and resort on the west coast of Sri Lanka; pop. (1981) 60,700.

negotiable /nɪˈgəʊʃəb(ə)l, -ˈgəʊsɪəb(ə)l/ *adj.* **1** open to discussion or modification. **2** able to be negotiated. □ **negotiability** /-ˌgəʊʃəˈbɪlɪtɪ, -ˌgəʊsɪə-/ *n.*

negotiate /nɪˈgəʊʃɪˌeɪt, -ˈgəʊsɪˌeɪt/ *v.* **1** *intr.* (usu. foll. by *with*) confer with others in order to reach a compromise or agreement. **2** *tr.* arrange (a matter) or bring about (a result) by negotiating (*negotiated a settlement*). **3** *tr.* find a way over, through, etc. (an obstacle, difficulty, fence, etc.). **4** *tr.* **a** transfer (a cheque etc.) to another for a consideration. **b** convert (a cheque etc.) into cash or notes. **c** get or give value for (a cheque etc.) in money. □ **negotiator** *n.* **negotiant** /-ʃənt, -sɪənt/ *n.* **negotiation** /-ˌgəʊʃɪˈeɪʃ(ə)n, -ˌgəʊsɪ-/ *n.* [L *negotiari* f. *negotium* business f. *neg-* not + *otium* leisure]

Negress /ˈniːgrɪs/ *n.* a female Negro. ¶ Now often considered *offens.*; the term *black* is usually preferred.

Negrillo /nɪˈgrɪləʊ/ *n.* (pl. **-os**) a member of a small-statured Negroid people of central and southern Africa. [Sp., dimin. of NEGRO]

Negri Sembilan /ˌnegri semˈbiːlən/ a state of Malaysia, on the SW coast of the Malay Peninsula; capital, Seremban.

Negrito /nɪˈgriːtəʊ/ *n.* (pl. **-os**) a member of a small-statured Negroid people of the Malayo-Polynesian region. (See also PYGMY.) [Sp., as NEGRILLO]

Negritude /ˈniːgrɪˌtjuːd, ˈneg-/ *n.* **1** the quality or state of being black. **2** the affirmation or consciousness of the value of black culture. [F *négritude* NIGRITUDE]

Negro /ˈniːgrəʊ/ *n.* & *adj.* ● *n.* (pl. **-oes**) a member of the black or dark-skinned group of human populations that exist or originated in Africa south of the Sahara, now distributed around the world. ¶ Now often considered *offens.*; the term *black* is usually preferred. ● *adj.* **1** of or concerning Negroes or blacks. **2** (as **negro**) *Zool.* black or dark (*negro ant*). □ **Negro minstrel** = MINSTREL 3. **Negro spiritual** = SPIRITUAL *n.* [Sp. & Port., f. L *niger nigri* black]

Negroid /ˈniːgrɔɪd/ *adj.* & *n.* ● *adj.* **1** (of features etc.) characterizing a member of the black group of human populations, esp. in having dark skin, tightly curled hair, and a broad flattish nose. **2** of or concerning blacks. ● *n.* a black person. [NEGRO]

Negros /ˈneɪgrɒs/ an island, the fourth largest of the Philippines; pop. (1991) 3,182,180; chief city, Bacolod.

Negus /ˈniːgəs/ *n.* *hist.* the title of the ruler of Ethiopia. [Amh. *n'gus* king]

negus /ˈniːgəs/ *n.* a hot drink of port, sugar, lemon, and spice. [Col. Francis *Negus* (d.1732), its inventor]

Neh. *abbr.* (in the Bible) Nehemiah.

Nehemiah /ˌniːəˈmaɪə/ **1** a Hebrew leader (5th century BC) who supervised the rebuilding of the walls of Jerusalem (*c.*444) and introduced moral and religious reforms (*c.*432). His work was continued by Ezra. **2** a book of the Bible telling of this rebuilding and of the reforms.

Nehru /ˈneəruː/, Jawaharlal (known as Pandit Nehru) (1889–1964), Indian statesman, Prime Minister 1947–64. An early associate of Mahatma Gandhi, Nehru was elected leader of the Indian National Congress, succeeding his father Pandit Motilal Nehru (1861–1931), in 1929. Imprisoned nine times by the British for his nationalist campaigns during the 1930s and 1940s, he eventually played a major part in the negotiations preceding independence. Nehru subsequently became the first Prime Minister of independent India.

neigh /neɪ/ *n.* & *v.* ● *n.* **1** the high whinnying sound of a horse. **2** any similar sound, e.g. a laugh. ● *v.* **1** *intr.* make such a sound. **2** *tr.* say, cry, etc. with such a sound. [OE *hnægan*, of imit. orig.]

neighbour /ˈneɪbə(r)/ *n.* & *v.* (US **neighbor**) ● *n.* **1** a person living next door to or near or nearest another (*my next-door neighbour*; *his nearest neighbour is 12 miles away*; *they are neighbours*). **2 a** a person regarded as having the duties or claims of friendliness, consideration, etc., of a neighbour. **b** a fellow human being, esp. as having claims on friendship. **3** a person or thing near or next to another (*my neighbour at dinner*). **4** (*attrib.*) neighbouring. ● *v.* **1** *tr.* border on; adjoin. **2** *intr.* (often foll. by *on, upon*) border; adjoin. □ **neighbouring** *adj.* **neighbourless** *adj.* **neighbourship** *n.* [OE *nēahgebūr* (as NIGH: *gebūr*, cf. BOOR)]

neighbourhood /ˈneɪbəˌhʊd/ *n.* (US **neighborhood**) **1 a** a district, esp. one forming a community within a town or city. **b** the people of a district; one's neighbours. **2** neighbourly feeling or conduct. □ **in the neighbourhood of** roughly; about (*paid in the neighbourhood of £100*). **neighbourhood watch** systematic local vigilance by householders to discourage crime, esp. against property.

neighbourly /ˈneɪbəlɪ/ *adj.* (US **neighborly**) characteristic of a good neighbour; friendly; kind. □ **neighbourliness** *n.*

Neill /niːl/, A(lexander) S(utherland) (1883–1973), Scottish teacher and educationist. He is best known as the founder of the progressive school Summerhill, established in Dorset in 1924 and based in Suffolk from 1927. Since its inception, the school has attracted both admiration and hostility for its anti-authoritarian ethos.

Neisse /ˈnaɪsə/ (Polish **Nysa** /ˈnɪsa/) **1** a river in central Europe, which rises in the north of the Czech Republic and flows over 225 km (140 miles) generally northwards, forming the southern part of the border between Germany and Poland (the Oder–Neisse Line), joining the River Oder north-east of Cottbus. It is also known in German as the *Lausitzer Neisse*. **2** a river of southern Poland, which rises near the border with the Czech Republic and flows 195 km (120 miles) generally north-eastwards, through the town of Nysa, joining the River Oder south-east of Wrocław. It is also known in German as the *Glatzer Neisse*.

neither /'naɪðə(r), 'niː-ð-/ *adj., pron., adv.,* & *conj.* ● *adj.* & *pron.* (foll. by sing. verb) **1** not the one nor the other (of two things); not either (*neither of the accusations is true; neither of them knows; neither wish was granted; neither went to the fair*). **2** *disp.* none of any number of specified things. ● *adv.* **1** not either; not on the one hand (foll. by *nor*; introducing the first of two or more things in the negative: *neither knowing nor caring; would neither come in nor go out; neither the teachers nor the parents nor the children*). **2** not either; also not (*if you do not, neither shall I*). **3** (with *neg.*) *colloq. disp.* either (*I don't know that neither*). ● *conj. archaic* nor yet; nor (*I know not, neither can I guess*). [ME *naither, neither* f. OE *nowther* contr. of *nōhwæther* (as NO[2], WHETHER): assim. to EITHER]

Nejd /nedʒd/ an arid plateau region in central Saudi Arabia, north of the Rub' al Khali desert, at an altitude of about 1,500 m (5,000 ft).

nek /nek/ *n. S. Afr.* = COL 1. [Du., = NECK]

nekton /'nektən/ *n. Zool.* aquatic animals that are able to swim and move independently (cf. BENTHOS, PLANKTON). [G f. Gk *nēkton* neut. of *nēktos* swimming f. *nēkhō* swim]

Nellore /ne'lɔː(r)/ a city and river port in SE India, in Andhra Pradesh, on the River Penner; pop. (1991) 316,000. Situated close to the mouth of the river, it is one of the chief ports of the Coromandel Coast.

nelly /'nelɪ/ *n.* (*pl.* **-ies**) a silly or effeminate person. □ **not on your nelly** *Brit. sl.* certainly not. [perh. f. the name *Nelly*: idiom f. rhyming sl. *Nelly Duff* = puff = breath: cf. *not on your life*]

Nelson[1] /'nels(ə)n/ a port in New Zealand, on the north coast of South Island; pop. (1991) 47,390. It was founded in 1841 by the New Zealand Company and named after the British admiral Lord Nelson.

Nelson[2] /'nels(ə)n/, Horatio, Viscount Nelson, Duke of Bronte (1758–1805), British admiral. Nelson's victories at sea during the early years of the Napoleonic Wars made him a national hero. His unorthodox independent tactics (as a commodore under Admiral Jervis) led to the defeat of a Spanish fleet off Cape St Vincent in 1797. In 1798 Nelson virtually destroyed the French fleet in the Battle of Aboukir Bay; he began his notorious affair with Lady Hamilton shortly afterwards. He proceeded to rout the Danes at Copenhagen in 1801, but is best known for his decisive victory over a combined French and Spanish fleet at the Battle of Trafalgar in 1805; Nelson was mortally wounded in the conflict.

nelson /'nels(ə)n/ *n.* a wrestling-hold in which one arm is passed under the opponent's arm from behind and the hand is applied to the neck (**half nelson**), or both arms and hands are applied (**full nelson**). [app. f. the name *Nelson*]

Nelson's Column a memorial to Lord Nelson in Trafalgar Square, London, consisting of a column 58 m (170 ft) high surmounted by a statue of Nelson.

Nelspruit /'nelsprɔɪt/ a town in eastern South Africa, the capital of the province of Eastern Transvaal, situated on the Crocodile River to the west of Kruger National Park.

nelumbo /nɪ'lʌmbəʊ/ *n.* (*pl.* **-os**) a water lily of the genus *Nelumbo*, native to India and China, bearing small pink flowers. (See also LOTUS.) [mod.L f. Sinh. *nelum(bu)*]

Neman /'nemən/ (also **Nemunas** /'njæmʊnəs/) a river of eastern Europe, which rises south of Minsk in Belarus and flows 955 km (597 miles) west and north to the Baltic Sea. Its lower course, which forms the boundary between Lithuania and the Russian enclave of Kaliningrad, is called the Memel.

nematic /nɪ'mætɪk/ *adj.* & *n. Physics* ● *adj.* designating or involving a state of a liquid crystal in which the molecules are oriented in parallel but not arranged in well-defined planes (cf. SMECTIC). ● *n.* a nematic substance. [Gk *nēma nemat-* thread]

nematocyst /nɪ'mætə,sɪst, 'nemətə-/ *n. Zool.* a specialized cell in the tentacles of jellyfish and other coelenterates, containing a barbed or venomous coiled thread that can be ejected in defence or to capture prey. [as NEMATIC + CYST]

nematode /'nemə,təʊd/ *n. Zool.* a parasitic or free-living worm of the phylum Nematoda, with a slender unsegmented cylindrical shape. (See also ROUNDWORM.) [as NEMATIC + -ODE[1]]

Nembutal /'nembjʊ,tɑːl/ *n. propr.* a sodium salt of pentobarbitone, used as a sedative and anticonvulsant. [Na (= sodium) + 5-*ethyl*-5-(1-*methylbutyl*) barbiturate + -AL[2]]

nem. con. /nem 'kɒn/ *abbr.* with no one dissenting. [L *nemine contradicente*]

nemertean /nɪ'mɜːtɪən/ *n.* & *adj.* (also **nemertine** /'nemə,taɪn/) *Zool.* ● *n.* an aquatic worm of the phylum Nemertea, often very long, brightly coloured, and tangled into knots, found esp. in coastal waters of Europe and the Mediterranean. Also called *ribbon worm.* ● *adj.* of or relating to this phylum. [mod.L *Nemertes* f. Gk *Nēmertēs* name of a sea nymph]

nemesia /nɪ'miːʒə/ *n.* a southern African plant of the genus *Nemesia*, cultivated for its variously coloured and irregular flowers. [mod.L f. Gk *nemesion*, the name of a similar plant]

Nemesis /'nemɪsɪs/ *Gk Mythol.* a goddess usually portrayed as the agent of divine punishment for wrongdoing or presumption (hubris). She is often little more than the personification of retribution or righteous indignation, although she is occasionally seen as a deity pursued amorously by Zeus and taking various non-human forms to evade him. [Gk: see NEMESIS]

nemesis /'nemɪsɪs/ *n.* (*pl.* **nemeses** /-,siːz/) **1** retributive justice. **2 a** a downfall caused by this. **b** an agent of such a downfall. [Gk, = retribution, righteous indignation, f. *nemō* give what is due: see NEMESIS]

Nemunas see NEMAN.

Nennius /'nenɪəs/ (*fl. c.*800), Welsh chronicler. He is traditionally credited with the compilation or revision of the *Historia Britonum*, a collection of historical and geographical information about Britain, including one of the earliest known accounts of King Arthur.

neo- /'niːəʊ/ *comb. form* **1** new, modern. **2** a new or revived form of. [Gk f. *neos* new]

neoclassical /,niːəʊ'klæsɪk(ə)l/ *adj.* (also **neoclassic**) of or relating to a revival of a classical style or treatment in art, literature, music, etc.; characteristic of neoclassicism.

neoclassicism /,niːəʊ'klæsɪ,sɪz(ə)m/ *n.* (also **Neoclassicism**) an aesthetic movement and artistic style which originated in Rome in the mid-18th century, a form of classicism combining a reaction against the excesses of the late baroque and rococo with a new interest in the antique. It received stimulus from discoveries at Herculaneum and Pompeii. Noted neoclassicists are the sculptor Canova and the architect Sir John Soane. In music, the term refers to a trend which developed in the 1920s when several composers, especially Stravinsky and Hindemith, wrote works in 17th and 18th-century forms and styles as a reaction against the elaborate orchestration of the late 19th-century romantics. □ **neoclassicist** *n.*

neocolonialism /,niːəʊkə'ləʊnɪə,lɪz(ə)m/ *n.* the use of economic, political, or other pressures to control or influence other countries, esp. former dependencies. □ **neocolonialist** *n.* & *adj.*

Neo-Darwinian /,niːəʊdɑː'wɪnɪən/ *adj.* & *n.* ● *adj.* of or relating to the modern version of Darwin's theory of evolution by natural selection, incorporating the findings of genetics. ● *n.* an adherent of this theory. □ **Neo-Darwinism** /-'dɑː,wɪ,nɪz(ə)m/ *n.* **Neo-Darwinist** *n.* & *adj.*

neodymium /,niːəʊ'dɪmɪəm/ *n.* a silvery-white metallic chemical element (atomic number 60; symbol **Nd**). Neodymium was identified by the Austrian chemist Auer von Welsbach in 1885, when he showed that the supposed rare-earth element didymium was in fact a mixture (see also PRASEODYMIUM). Neodymium is a member of the lanthanide series. The metal is a component of misch metal and some other alloys, and its compounds are used in colouring glass and ceramics. [NEO- + DIDYMIUM]

neo-Fascist /,niːəʊ'fæʃɪst/ *n.* & *adj.* ● *n.* an adherent or supporter of a new or revived form of Fascism. (Cf. NEO-NAZI.) ● *adj.* of or relating to neo-Fascists or neo-Fascism. □ **neo-Fascism** *n.*

neo-impressionism /,niːəʊɪm'preʃə,nɪz(ə)m/ *n.* a late 19th-century movement in French painting which marked both an extension and critique of impressionism. The neo-impressionists saw themselves as refining and improving upon impressionist style, and are particularly associated with pointillism and with a precise, scientific approach to form and colour. The movement's leading adherents were Georges Seurat, Paul Signac, and Camille Pissarro. □ **neo-impressionist** *adj.* & *n.*

neolithic /,niːə'lɪθɪk/ *adj.* & *n. Archaeol.* of, relating to, or denoting the later part of the Stone Age, when ground or polished stone weapons and implements prevailed. This period, which saw the introduction of agriculture and the domestication of animals, turned humankind from being dependent on nature to controlling it at least partially and indirectly. The change led to the establishment of settled communities, accumulation of food and wealth, and heavier growth of population. In the Old World, agriculture had begun in the Near

East by the 8th millennium BC and had spread to northern Europe by the 4th millennium BC. [NEO- + Gk *lithos* stone]

neologism /nɪˈɒləˌdʒɪz(ə)m/ n. **1** a new word or expression. **2** the coining or use of new words. □ **neologist** n. **neologize** v.intr. (also **-ise**). [F *néologisme* (as NEO-, -LOGY, -ISM)]

neomycin /ˌniːəʊˈmaɪsɪn/ n. an antibiotic related to streptomycin.

neon /ˈniːɒn/ n. an inert gaseous chemical element (atomic number 10; symbol **Ne**). One of the noble gases, neon was discovered by the chemists William Ramsay and Morris William Travers (1872–1961) in 1898. It is obtained by the distillation of liquid air, and is mainly used in fluorescent lamps and advertising signs as it emits a reddish-orange glow when electricity is passed through it in a sealed low-pressure tube. It has no known chemical compounds. [Gk, neut. of *neos* new]

neonate /ˈniːəˌneɪt/ n. a newborn child or mammal. □ **neonatal** /ˌniːəˈneɪt(ə)l/ adj. [mod.L neonatus (as NEO-, L nasci nat- be born)]

neo-Nazi /ˌniːəʊˈnɑːtsɪ/ n. & adj. ● n. an adherent or supporter of a new or revived form of Nazism. (Cf. NEO-FASCIST.) ● adj. of or relating to neo-Nazis or neo-Nazism. □ **neo-Nazism** n.

neophyte /ˈniːəˌfaɪt/ n. **1** a new convert, esp. to a religious faith. **2** RC Ch. **a** a novice of a religious order. **b** a newly ordained priest. **3** a beginner; a novice. [eccl.L *neophytus* f. NT Gk *neophutos* newly planted (as NEO- *phuton* plant)]

neoplasm /ˈniːəʊˌplæz(ə)m/ n. Med. a new and abnormal growth of tissue in some part of the body, esp. a tumour. □ **neoplastic** /ˌniːəʊˈplæstɪk/ adj. [NEO- + Gk *plasma* formation: see PLASMA]

neo-plasticism /ˌniːəʊˈplæstɪˌsɪz(ə)m/ n. a term coined by Piet Mondrian in 1917 to describe the style of his painting, in which form was reduced to the geometric simplicity of horizontals and verticals, and colour was reduced to primary colours and black, white, and grey. It is closely related to the theories of the De Stijl group.

Neoplatonism /ˌniːəʊˈpleɪtəˌnɪz(ə)m/ n. a religious and philosophical system based on Platonic ideas and originating with Plotinus in the 3rd century AD. It was a synthesis of elements from the philosophies of Plato, Pythagoras, Aristotle, and the Stoics, with overtones of Eastern mysticism. It was the dominant philosophy of the pagan world from the mid-3rd century AD until the closing of the pagan schools by Justinian in 529, and also strongly influenced medieval and Renaissance thought. Plotinus, who gave Neoplatonism its abiding shape, postulated a hierarchy of being, at the summit of which is the transcendent One, immaterial and indescribable. The human soul aspires to knowledge of this One through ascetic virtue and sustained contemplation, and in doing so rises above the imperfection and multiplicity of the material world. □ **Neoplatonist** n. **Neoplatonic** /-pləˈtɒnɪk/ adj.

neoprene /ˈniːəʊˌpriːn/ n. a synthetic rubber-like polymer. [NEO- + *chloroprene* etc. (perh. f. PROPYL +-ENE)]

Neoptolemus /ˌniːɒpˈtɒlɪməs/ Gk Mythol. the son of Achilles and killer of Priam after the fall of Troy. [Gk, = young warrior]

neo-realism /ˌniːəʊˈrɪəlɪz(ə)m/ n. a naturalistic Italian movement in literature and cinema that emerged in the 1940s. Exponents include the writers Alberto Moravia (1907–91), Carlo Levi (1902–75), and Italo Calvino. Neo-realist cinema is marked by an almost documentary style, sometimes featuring non-professional actors; prominent neo-realist directors are Federico Fellini, Roberto Rossellini, and Vittorio De Sica. □ **neo-realist** adj. & n.

neoteny /nɪˈɒtɪnɪ/ n. Zool. **1** the retention of juvenile features in adult life. **2** sexual maturity of an animal still in a larval stage, e.g. an axolotl. □ **neotenous** adj. **neotenic** /ˌniːəʊˈtenɪk/ adj. [G *Neotenie* (as NEO- + Gk *teinō* extend)]

neoteric /ˌniːəˈterɪk/ adj. literary recent; newfangled; modern. [LL *neotericus* f. Gk *neōterikos* (*neōteros* compar. of *neos* new)]

neotropical /ˌniːəʊˈtrɒpɪk(ə)l/ adj. (also **Neotropical**) Zool. of or relating to Central and South America as a zoogeographical region.

Nepal /nɪˈpɔːl/ a mountainous landlocked country in southern Asia, in the Himalayas (and including Mount Everest); pop. (est. 1991) 19,406,000; official language, Nepali; capital, Kathmandu. The country was conquered by the Gurkhas in the 18th century and has maintained its independence despite border defeats by the British in the 19th century. Nepal was for long an absolute monarchy, but in 1990 democratic elections were held under a new constitution. Nepal was the birthplace of Gautama Buddha.

Nepalese /ˌnepəˈliːz/ adj. & n. (pl. same) = NEPALI.

Nepali /nɪˈpɔːlɪ/ n. & adj. ● n. (pl. same or **Nepalis**) **1 a** a native or national of Nepal. **b** a person of Nepali descent. **2** the official language

of Nepal, spoken also in parts of NE India. It belongs to the Indic branch of the Indo-European family of languages. ● adj. of or relating to Nepal or its people or language.

nepenthe /nɪˈpenθɪ/ n. literary = NEPENTHES 1. [var. of NEPENTHES, after It. *nepente*]

nepenthes /nɪˈpenθiːz/ n. **1** literary a drug causing forgetfulness of grief. **2** a pitcher-plant of the genus *Nepenthes*. [L f. Gk *nēpenthes* (*pharmakon* drug), neut. of *nēpenthēs* f. *nē-* not + *penthos* grief]

nephelometer /ˌnefəˈlɒmɪtə(r)/ n. an instrument for measuring the size and concentration of particles suspended in liquid or gas, esp. by means of the light scattered. [Gk *nephelē* cloud + -METER]

nephew /ˈnevjuː, ˈnefjuː/ n. a son of one's brother or sister, or of one's brother-in-law or sister-in-law. [ME f. OF *neveu* f. L *nepos nepotis* grandson, nephew]

nephology /nɪˈfɒlədʒɪ/ n. the study of clouds. [Gk *nephos* cloud + -LOGY]

nephrite /ˈnefraɪt/ n. a green, yellow, or white calcium magnesium silicate form of jade. [G *Nephrit* f. Gk *nephros* kidney, with ref. to its supposed efficacy in treating kidney disease]

nephritic /nɪˈfrɪtɪk/ adj. **1** of or in the kidneys; renal. **2** of or relating to nephritis. [LL *nephriticus* f. Gk *nephritikos* (as NEPHRITIS)]

nephritis /nɪˈfraɪtɪs/ n. Med. inflammation of the kidneys. Also called *Bright's disease*. [LL f. Gk *nephros* kidney]

nephro- /ˈnefrəʊ/ comb. form (usu. **nephr-** before a vowel) kidney. [Gk f. *nephros* kidney]

Nepia /ˈniːpɪə/, George (1905–86), New Zealand Rugby Union player. While playing for the All Blacks 1929–30, he created a New Zealand record by playing thirty-eight consecutive matches for the country.

ne plus ultra /ˌneɪ plʊs ˈʊltrɑː/ n. **1** the furthest attainable point. **2** the culmination, acme, or perfection. [L, = not further beyond, the supposed inscription on the Pillars of Hercules (the Strait of Gibraltar) prohibiting passage by ships]

nepotism /ˈnepəˌtɪz(ə)m/ n. favouritism shown to relatives or friends in conferring offices or privileges. □ **nepotist** n. **nepotistic** /ˌnepəˈtɪstɪk/ adj. [F *népotisme* f. It. *nepotismo* f. *nepote* NEPHEW: orig. with ref. to popes with illegitimate sons called nephews]

Neptune /ˈneptjuːn/ **1** Rom. Mythol. the god of water and of the sea, identified with Poseidon. **2** Astron. the eighth planet from the sun in the solar system, orbiting between Uranus and Pluto at an average distance of 4,497 million km from the sun (but temporarily outside the orbit of Pluto 1979–99). It is the fourth largest planet (but third in mass), with an equatorial diameter of 48,600 km, and the most remote of the gas giants. It was discovered in 1846 by the German astronomer Johann Galle (1812–1910) at the Berlin Observatory, close to the position predicted by the French mathematician Urbain Le Verrier. Voyager 2 (1989) showed the planet to be predominantly blue, with an upper atmosphere mainly of hydrogen and helium with some methane, and recorded some visible cloud features including a large dark spot. There are at least eight satellites, the largest of which is Triton, and a faint ring system. [ME f. F *Neptune* or L *Neptunus*]

Neptunian /nepˈtjuːnɪən/ adj. **1** Geol. produced by the action of water; maintaining that the action of water played a principal part in the formation of certain rocks (opp. *Plutonian*). (See also NEPTUNIST). **2** of the planet Neptune.

Neptunist /ˈneptjuːnɪst/ n. Geol. hist. an advocate of the Neptunian theory, which held that rocks such as granite were formed by crystallization from the waters of a primeval ocean. The controversy between the Neptunists, of whom the principal figure was the German geologist Abraham Gottlob Werner (1749–1817), and the Plutonists or Vulcanists, who correctly maintained the igneous origin of such rocks, dominated geology for several decades before it was finally settled in the early 19th century. □ **Neptunism** n.

neptunium /nepˈtjuːnɪəm/ n. a radioactive metallic chemical element (atomic number 93; symbol **Np**). Neptunium was discovered by the American physicists Edwin McMillan (1907–91) and Philip Abelson (b.1913) in 1940 as a product of the bombardment of uranium with neutrons. A member of the actinide series, it occurs only in trace amounts in nature. [NEPTUNE, after *uranium*, as the next planet beyond Uranus]

NERC abbr. (in the UK) Natural Environment Research Council.

nerd /nɜːd/ n. (also **nurd**) esp. US sl. a foolish, feeble, or uninteresting person. □ **nerdy** adj. [20th c.: orig. uncert.]

Nereid /ˈnɪərɪɪd/ Astron. satellite II of Neptune, the furthest from the planet, discovered in 1949. It has an irregular shape (diameter about

340 km), and a highly eccentric orbit which takes it 5.5 million km from Neptune. [NEREID]

nereid /ˈnɪərɪd/ n. **1** Gk Mythol. any of the sea-nymphs, daughters of Nereus; they include Thetis, mother of Achilles. **2** Zool. a carnivorous marine polychaete worm of the family Nereidae (cf. RAGWORM). [L Nereïs Nereïd- f. Gk Nēreïs -idos daughter of Nereus]

Nereus /ˈnɪərɪəs/ Gk Mythol. an old sea-god, the father of the nereids. Like Proteus he had the power of assuming various forms.

nerine /nɪˈraɪnɪ/ n. a southern African plant of the genus Nerine, bearing flowers with usu. six narrow strap-shaped petals, often crimped and twisted. [mod.L f. the L name of a water-nymph]

Nernst /neənst/, Hermann Walther (1864–1941), German physical chemist. He made a number of contributions to physical chemistry, chiefly in electrochemistry and thermodynamics. He is best known for his discovery of the third law of thermodynamics (also called Nernst's heat theorem: see THERMODYNAMICS) and devoted many years of low-temperature research to verify it. Nernst also investigated photochemistry, the diffusion of electrolytic ions, and the distribution of solutes in immiscible liquids, and devised an electric lamp. He was awarded the Nobel Prize for chemistry in 1920.

Nero /ˈnɪərəʊ/ (full name Nero Claudius Caesar Augustus Germanicus) (AD 37–68), Roman emperor 54–68. The adopted son and successor of Claudius, he became infamous for his cruelty following his ordering of the murder of his mother Agrippina in 59. His reign was marked by wanton executions of leading Romans and witnessed a fire which destroyed half of Rome in 64. A wave of uprisings in 68 led to his flight from Rome and his eventual suicide.

neroli /ˈnɪərəlɪ/ n. (in full **neroli oil**) an essential oil from the flowers of the Seville orange, used in perfumery. [F néroli f. It. neroli, perh. f. the name of an Italian princess]

Neruda /nəˈruːdə/, Pablo (born Ricardo Eliezer Neftalí Reyes) (1904–73), Chilean poet and diplomat. He adopted the name Neruda as his pseudonym in 1920 after the Czech poet Jan Neruda (1834–91), later changing his name by deed poll (1946). From 1927 to 1952 he spent much of his life abroad, either serving in diplomatic posts or as a result of his membership of the Chilean Communist party, which was outlawed in 1948. His major work, Canto General (completed 1950), was originally conceived as an epic on Chile and was later expanded to cover the history of all the Americas from their ancient civilizations to their modern wars of liberation. He was awarded the Nobel Prize for literature in 1971.

Nerva /ˈnɜːvə/, Marcus Cocceius (AD c.30–98), Roman emperor 96–8. Appointed emperor by the Senate after the murder of Domitian, he returned to a liberal and constitutional form of rule after the autocracy of his predecessor.

nervation /nɜːˈveɪʃ(ə)n/ n. Bot. the arrangement of nerves in a leaf. [NERVE + -ATION]

nerve /nɜːv/ n. & v. ● n. **1 a** a fibre or bundle of fibres that transmits impulses of sensation or motion between the brain or spinal cord and other parts of the body. **b** the material constituting these. **2 a** coolness in danger; bravery; assurance. **b** colloq. impudence, audacity (they've got a nerve). **3** (in pl.) **a** the bodily state in regard to physical sensitiveness and the interaction between the brain and other parts. **b** a state of heightened nervousness or sensitivity; a condition of mental or physical stress (need to calm my nerves). **4** Bot. a rib of a leaf, esp. the midrib. **5** poet. or archaic a sinew or tendon. ● v.tr. **1** (usu. refl.) brace (oneself) to face danger, suffering, etc. **2** give strength, vigour, or courage to. □ **get on a person's nerves** irritate or annoy a person. **have nerves of iron** (or **steel**) (of a person etc.) be not easily upset or frightened. **hit** (or **touch**) **a nerve** remark on or draw attention to a sensitive subject or point. **nerve cell** an elongated branched cell transmitting impulses in nerve tissue. **nerve-centre 1** a group of closely connected nerve cells associated in performing some function. **2** the centre of control of an organization etc. **nerve gas** a poisonous gas affecting the nervous system. **nerve-racking** stressful, frightening; straining the nerves. □ **nerved** adj. (also in comb.). [ME, = sinew, f. L nervus, rel. to Gk neuron]

nerveless /ˈnɜːvlɪs/ adj. **1** inert, lacking vigour or spirit. **2** confident; not nervous. **3** (of literary or artistic style) diffuse. **4** Bot. & Zool. without nervures. **5** Anat. & Zool. without nerves. □ **nervelessly** adv. **nervelessness** n.

Nervi /ˈneəvɪ/, Pier Luigi (1891–1979), Italian engineer and architect. He is noted as a pioneer of new technology and materials, especially reinforced concrete. Nervi co-designed the UNESCO building in Paris

(1953), as well as designing the Pirelli skyscraper in Milan (1958) and San Francisco cathedral (1970).

nervine /ˈnɜːvaɪn, -viːn/ adj. & n. Med. ● adj. relieving nerve disorders. ● n. a nervine drug. [F nervin (as NERVE)]

nervo- /ˈnɜːvəʊ/ comb. form (also **nerv-** before a vowel) a nerve or the nerves.

nervous /ˈnɜːvəs/ adj. **1** having delicate or disordered nerves. **2** timid or anxious. **3 a** excitable; highly strung; easily agitated. **b** resulting from this temperament (nervous tension; a nervous headache). **4** affecting or acting on the nerves. **5** (foll. by of + verbal noun) reluctant, afraid (am nervous of meeting them). □ **nervous breakdown** a period of mental illness, usu. resulting from severe depression or anxiety. **nervous wreck** a person suffering from mental stress, exhaustion, etc. □ **nervously** adv. **nervousness** n. [ME f. L nervosus (as NERVE)]

nervous system n. a network of specialized cells which transmits nerve impulses between parts of an animal's body; the body's nerves and nerve-centres as a whole. In humans and other vertebrates, several distinct parts of the system are distinguished: the autonomic nervous system, the part controlling or influencing involuntary functions (e.g. heartbeat, digestive processes); the central nervous system, the brain and spinal cord; the peripheral nervous system, the part outside the central nervous system, consisting of the autonomic nervous system, the cranial nerves (supplying the face and head), and the spinal nerves (distributed to the limb and trunk muscles and skin). Many invertebrates have a simple ventral nerve cord with segmentally arranged ganglia. (See also PARASYMPATHETIC, SYMPATHETIC.)

nervure /ˈnɜːvjʊə(r)/ n. **1** Zool. each of the hollow tubes that form the framework of an insect's wing; a venule. **2** Brit. the principal vein of a leaf. [F nerf nerve]

nervy /ˈnɜːvɪ/ adj. (**nervier, nerviest**) **1** esp. Brit. nervous; easily excited or disturbed. **2** N. Amer. colloq. bold, impudent. **3** poet. or archaic sinewy, strong. □ **nervily** adv. **nerviness** n.

Nesbit /ˈnezbɪt/, E(dith) (1858–1924), English novelist. She is best known for her children's books, including The Story of the Treasure Seekers (1899), Five Children and It (1902), and The Railway Children (1906). Nesbit was also a founder member of the Fabian Society.

nescient /ˈnesɪənt/ adj. literary (foll. by of) lacking knowledge; ignorant. □ **nescience** n. [LL nescientia f. L nescire not know f. ne- not + scire know]

ness /nes/ n. a headland or promontory. [OE næs, rel. to OE nasu NOSE]

-ness /nɪs/ suffix forming nouns from adjectives and sometimes other words, expressing: **1** state or condition, or an instance of this (bitterness; conceitedness; happiness; a kindness). **2** something in a certain state (wilderness). [OE -nes, -ness f. Gmc]

Ness, Loch see LOCH NESS.

nest /nest/ n. & v. ● n. **1** a structure or place where a bird lays eggs and shelters its young. **2** an animal's or insect's breeding-place or lair. **3** a snug or secluded retreat or shelter. **4** (often foll. by of) a place fostering something undesirable (a nest of vice). **5** a brood or swarm. **6** a group or set of similar objects, often of different sizes and fitting together for storage (a nest of tables). ● v. **1** intr. use or build a nest. **2** intr. take wild birds' nests or eggs. **3** intr. (of objects) fit together or one inside another. **4** tr. (usu. as **nested** adj.) establish in or as in a nest. □ **nest egg 1** a sum of money saved for the future. **2** a real or artificial egg left in a nest to induce hens to lay eggs there. □ **nestful** n. (pl. **-fuls**). **nesting** n. (in sense 2 of v.). **nestlike** adj. [OE nest]

nestle /ˈnes(ə)l/ v. **1** intr. (often foll. by down, in, etc.) settle oneself comfortably. **2** intr. press oneself against another in affection etc. **3** tr. (foll. by in, into, etc.) push (a head or shoulder etc.) affectionately or snugly. **4** intr. lie half hidden or embedded. [OE nestlian (as NEST)]

nestling /ˈneslɪŋ, ˈnestlɪŋ/ n. a bird that is too young to leave its nest.

Nestor /ˈnestə(r)/ Gk Mythol. a king of Pylos in the Peloponnese, who in old age led his subjects to the Trojan War, where his wisdom and eloquence were proverbial.

Nestorianism /neˈstɔːrɪəˌnɪz(ə)m/ n. an early Christian doctrine holding that there were two separate persons, one human and one divine, in the incarnate Christ, as opposed to the orthodox teaching that Christ was a single person, both God and man. The doctrine takes its name from Nestorius, patriarch of Constantinople from 428 until deposed and banished in 431. His supporters in Syria and Persia gradually constituted themselves into a separate Nestorian Church, which became active in missionary work in India, Arabia, and China before suffering drastic losses in the Mongol invasions of Persia in the

14th century. A small Nestorian Church still exists in Iraq. □ **Nestorian** adj. & n.

net[1] /net/ n. & v. ● n. **1** an open-meshed fabric of cord, rope, fibre, etc.; a structure resembling this. **2** a piece of net used esp. to restrain, contain, or delimit, or to catch fish or other animals. **3** a structure with net to enclose an area of ground, esp. in sport. **4 a** a structure with net used in various games, esp. forming the goal in football, netball, etc., and dividing the court in tennis etc. **b** (often in pl.) a practice-ground in cricket, surrounded by nets. **5** a system or procedure for catching or entrapping a person or persons. **6 a** esp. Computing a network. **b** (**the Net**) colloq.=INTERNET. ● v. (**netted**, **netting**) **1** tr. **a** cover, confine, or catch with a net. **b** obtain or acquire as with a net. **2** tr. hit (a ball) into the net, esp. of a goal. **3** intr. make netting. **4** tr. make (a purse, hammock, etc.) by knotting etc. threads together to form a net. **5** tr. fish with nets, or set nets, in (a river). **6** tr. (usu. as **netted** adj.) mark with a netlike pattern; reticulate. □ **netful** n. (pl. **-fuls**). [OE net, nett]

net[2] /net/ adj. & v. (also **nett**) ● adj. **1** (esp. of money) remaining after all necessary deductions, or free from deductions. **2** (of a price) to be paid in full; not reducible. **3** (of a weight) excluding that of the packaging or container etc. **4** (of an effect, result, etc.) ultimate, effective. ● v.tr. (**netted**, **netting**) gain or yield (a sum) as net profit. □ **net profit** the effective profit; the actual gain after working expenses have been paid. **net ton** see TON[1] 5b. [F net NEAT[1]]

netball /'netbɔːl/ n. a seven-a-side game in which the object is to score goals by throwing a ball so that it falls through an elevated horizontal ring from which a net hangs. It was introduced into England from the US in 1895 as a development of basketball, and is played almost entirely by girls and women, mainly in English-speaking countries. Players may not run with or bounce the ball, and on receiving it must stand still until they have passed the ball to another player.

net book agreement (abbr. **NBA**) (in the UK) an agreement set up in 1900 between booksellers and publishers, by which booksellers will not offer books to the public at a price below that marked on the book's cover. Exceptions were made for school textbooks, remaindered books, and national book sales. The agreement effectively collapsed in Sept. 1995 when several major UK publishers withdrew their support.

nether /'neðə(r)/ adj. archaic = LOWER[1]. □ **nether regions** (or **world**) hell; the underworld. □ **nethermost** adj. [OE nithera etc. f. Gmc]

Netherlands, the /'neðələndz/ (Dutch **Nederland** /'neːdər,lant/) **1** a country in western Europe, on the North Sea, often called Holland; pop. (1991) 15,010,445; official language, Dutch; capital, Amsterdam; seat of government, The Hague. (See note below.) **2** hist. the Low Countries. □ **Netherlander** /-,lændə(r)/ n. **Netherlandish** adj. [Du. (as NETHER, LAND)]

▪ The region of the Low Countries was occupied by Celts and Frisians who came under Roman rule from the 1st century BC until the 4th century AD. It was then overrun by German tribes, with the Franks establishing an ascendancy in the 5th to 8th centuries. During the Middle Ages it was divided among various principalities. Part of the Spanish Habsburg empire in the 16th century, the northern (Dutch) part revolted against Spanish attempts to crush the Protestant faith and won independence in a series of wars lasting into the 17th century. Independence was declared in 1581 and international recognition, as the Protestant Republic of the United Provinces of the Netherlands, was gained in 1648. In this period the Netherlands became a leading maritime power, building up a vast commercial empire in the East Indies, southern Africa, and South America. In 1814 north and south were united under a monarchy (William I became king of the United Netherlands the following year), but the south revolted in 1830 and became an independent kingdom, Belgium, in 1839. The Netherlands stayed neutral in the First World War but was occupied by the Germans in the Second. In 1948 it formed the Benelux Customs Union with Belgium and Luxembourg, becoming a founder member of the EEC in 1957. The name Holland strictly refers only to the western coastal provinces of the country.

Netherlands Antilles two widely separated groups of Dutch islands in the Caribbean, in the Lesser Antilles; capital, Willemstad, on Curaçao; pop. (1991) 200,000. The southernmost group, situated just off the north coast of Venezuela, comprises the islands of Bonaire and Curaçao. The other group, about 800 km (500 miles) to the north, situated at the northern end of the Lesser Antilles, comprises the islands of St Eustatius, St Martin, and Saba. The islands were originally visited in the 16th century by the Spanish, who wiped out the native Arawak and Carib population. They were settled by the Dutch between 1634 and 1648 and until 1986 also included Aruba. In 1954 the islands were granted self-government and became an autonomous region of the Netherlands.

Netherlands Reformed Church the largest Protestant Church in the Netherlands, established in 1816 as the successor to the Dutch Reformed Church.

netsuke /'netsʊkɪ/ n. (pl. same or **netsukes**) (in Japan) a carved button-like ornament, esp. of ivory or wood, formerly worn to suspend articles from a girdle. [Jap.]

nett var. of NET[2].

netting /'netɪŋ/ n. **1** netted fabric. **2** a piece of this.

nettle /'net(ə)l/ n. & v. ● n. **1** a plant of the genus Urtica, with jagged leaves covered with stinging hairs, esp. the common stinging nettle. **2** a plant resembling this (dead-nettle). ● v.tr. **1** irritate, provoke, annoy. **2** sting with nettles. □ **nettle-rash** Med. a rash of itching red spots on the skin caused by allergic reaction to food etc. Also called urticaria. [OE netle, netele]

nettlesome /'net(ə)lsəm/ adj. **1** awkward, difficult. **2** causing annoyance.

network /'netwɜːk/ n. & v. ● n. **1** an arrangement of intersecting horizontal and vertical lines, like the structure of a net. **2** a complex system of railways, roads, canals, etc. **3** a group of people who exchange information, contacts, and experience for professional or social purposes. **4** a chain of interconnected computers, machines, or operations. **5** a system of connected electrical conductors. **6** a group of broadcasting stations connected for a simultaneous broadcast of a programme. ● v. **1** tr. Brit. broadcast on a network. **2** intr. establish a network. **3** tr. link (machines, esp. computers) to operate interactively. **4** intr. communicate with other people as a member of a professional or social network to exchange information etc.

networker /'net,wɜːkə(r)/ n. **1** Computing a member of an organization or computer network who operates from home or from an external office. **2** a member of a professional or social network.

Neuchâtel, Lake /,nɜːʃæ'tel/ the largest lake lying wholly within Switzerland, situated at the foot of the Jura Mountains in western Switzerland.

Neumann /'nɔɪmən/, John von (1903–57), Hungarian-born American mathematician and computer pioneer. His contributions ranged from pure logic and set theory to the most practical areas of application. He analysed the mathematics of quantum mechanics, founding a new area of mathematical research (algebras of operators in Hilbert space), and also established the mathematical theory of games (see GAME THEORY). Neumann also helped to develop the US hydrogen bomb, but perhaps his most influential contribution was in the design and operation of electronic computers.

neume /njuːm/ n. (also **neum**) Mus. a sign in plainsong indicating a note or group of notes to be sung to a syllable. [ME f. OF neume f. med.L neu(p)ma f. Gk pneuma breath]

neural /'njʊərəl/ adj. of or relating to a nerve or the central nervous system. □ **neural network** (or **net**) Computing a computer system modelled on the human brain and nervous system. □ **neurally** adv. [Gk neuron nerve]

neuralgia /njʊə'rældʒə/ n. Med. an intense intermittent pain along the course of a nerve, esp. in the head or face. □ **neuralgic** adj. [as NEURAL + -ALGIA]

neurasthenia /,njʊərəs'θiːnɪə/ n. a general term for fatigue, anxiety, listlessness, etc. (not in medical use). □ **neurasthenic** /-'θenɪk/ adj. & n. [Gk neuron nerve + ASTHENIA]

neuritis /njʊə'raɪtɪs/ n. Med. inflammation of a nerve or nerves. □ **neuritic** /-'rɪtɪk/ adj. [formed as NEURO- + -ITIS]

neuro- /'njʊərəʊ/ comb. form a nerve or the nerves. [Gk neuron nerve]

neuroanatomy /,njʊərəʊə'nætəmɪ/ n. the anatomy of the nervous system. □ **neuroanatomist** n. **neuroanatomical** /-,ænə'tɒmɪk(ə)l/ adj.

neurogenesis /,njʊərəʊ'dʒenɪsɪs/ n. the growth and development of nervous tissue.

neurogenic /,njʊərəʊ'dʒenɪk/ adj. caused by or arising in nervous tissue.

neuroglia /njʊə'rɒglɪə/ n. Anat. the connective tissue supporting the central nervous system. [NEURO- + Gk glia glue]

neurohormone /,njʊərəʊ'hɔːməʊn/ n. a hormone produced by nerve cells and secreted into the circulation.

neurology /njʊə'rɒlədʒɪ/ n. the scientific study of nerve systems.

□ **neurologist** *n.* **neurological** /ˌnjʊərəˈlɒdʒɪk(ə)l/ *adj.* **neurologically** *adv.* [mod.L *neurologia* f. mod.Gk (as NEURO-, -LOGY)]

neuroma /njʊəˈrəʊmə/ *n.* (*pl.* **neuromas** or **neuromata** /-mətə/) *Med.* a tumour on a nerve or in nerve-tissue. [Gk *neuron* nerve + -OMA]

neuromuscular /ˌnjʊərəʊˈmʌskjʊlə(r)/ *adj.* of or relating to nerves and muscles.

neuron /ˈnjʊərɒn/ *n.* (also **neurone** /-rəʊn/) a specialized cell transmitting nerve impulses; a nerve cell. □ **neuronal** /njʊəˈrəʊn(ə)l/ *adj.* **neuronic** /-ˈrɒnɪk/ *adj.* [Gk *neuron* nerve]

neuropath /ˈnjʊərəʊˌpæθ/ *n. Med.* a person affected by nervous disease, or with an abnormally sensitive nervous system. □ **neuropathic** /ˌnjʊərəʊˈpæθɪk/ *adj.*

neuropathology /ˌnjʊərəʊpəˈθɒlədʒɪ/ *n.* the pathology of the nervous system. □ **neuropathologist** *n.*

neurophysiology /ˌnjʊərəʊˌfɪzɪˈɒlədʒɪ/ *n.* the physiology of the nervous system. □ **neurophysiologist** *n.* **neurophysiological** /-ˌfɪzɪəˈlɒdʒɪk(ə)l/ *adj.*

Neuroptera /njʊəˈrɒptərə/ *n.pl. Zool.* an order of predatory flying insects comprising the lacewings, alder-flies, snake flies, and ant-lions, which have four finely veined membranous wings. □ **neuropteran** *n.* & *adj.* **neuropterous** *adj.* [mod.L f. NEURO- + Gk *pteron* wing]

neuroscience /ˌnjʊərəʊˈsaɪəns/ *n.* any or all of the sciences dealing with the structure and function of the nervous system and brain. □ **neuroscientist** /-ˈsaɪəntɪst/ *n.*

neurosis /njʊəˈrəʊsɪs/ *n.* (*pl.* **neuroses** /-siːz/) a mild mental illness involving symptoms of stress (e.g. depression, anxiety, obsessive behaviour) without loss of contact with reality, and not caused by organic disease. [mod.L (as NEURO-, -OSIS)]

neurosurgery /ˌnjʊərəʊˈsɜːdʒərɪ/ *n.* surgery performed on the nervous system, esp. the brain and spinal cord. □ **neurosurgeon** *n.* **neurosurgical** *adj.*

neurotic /njʊəˈrɒtɪk/ *adj.* & *n.* ● *adj.* **1** caused by or relating to neurosis. **2** (of a person) suffering from neurosis. **3** *colloq.* abnormally sensitive or obsessive. ● *n.* a neurotic person. □ **neurotically** *adv.* **neuroticism** /-tɪˌsɪz(ə)m/ *n.*

neurotomy /njʊəˈrɒtəmɪ/ *n.* (*pl.* **-ies**) *Med.* the operation of cutting a nerve, esp. to produce sensory loss.

neurotoxin /ˌnjʊərəʊˈtɒksɪn/ *n.* any poison which acts on the nervous system.

neurotransmitter /ˌnjʊərəʊtrænzˈmɪtə(r), -trɑːnzˈmɪtə(r)/ *n. Biochem.* a chemical substance released from a nerve fibre that effects the transfer of an impulse to another nerve or muscle.

Neusiedler See /ˈnɔɪzɪːdlə ˌzeɪ/ (called in Hungarian *Fertö Tó*) a shallow steppe lake straddling the frontier between eastern Austria and NW Hungary.

neuter /ˈnjuːtə(r)/ *adj., n.,* & *v.* ● *adj.* **1** *Gram.* (of a noun etc.) neither masculine nor feminine. **2** (of a plant) having neither pistils nor stamen. **3** (of an insect, animal, etc.) sexually undeveloped; castrated or spayed. ● *n.* **1** *Gram.* a neuter word. **2 a** a non-fertile insect, esp. a worker bee or ant. **b** a castrated animal. ● *v.tr.* castrate or spay. [ME f. OF *neutre* or L *neuter* neither f. *ne-* not + *uter* either]

neutral /ˈnjuːtr(ə)l/ *adj.* & *n.* ● *adj.* **1** not helping or supporting either of two opposing sides, esp. states at war or in dispute; impartial. **2** belonging to a neutral party, state, etc. (*neutral ships*). **3** indistinct, vague, indeterminate. **4** (of a gear) in which the engine is disconnected from the driven parts. **5** (of colours) not strong or positive; grey or beige. **6** *Chem.* neither acid nor alkaline. **7** *Electr.* neither positive nor negative. **8** *Biol.* sexually undeveloped; asexual. ● *n.* **1 a** a neutral state or person. **b** a subject of a neutral state. **2** a neutral gear. □ **neutral monism** *Philos.* see MONISM 1. □ **neutrally** *adv.* **neutrality** /njuːˈtrælɪtɪ/ *n.* [ME f. obs. F *neutral* or L *neutralis* of neuter gender (as NEUTER)]

neutralism /ˈnjuːtrəˌlɪz(ə)m/ *n.* a policy of political neutrality. □ **neutralist** *n.*

neutralize /ˈnjuːtrəˌlaɪz/ *v.tr.* (also **-ise**) **1** make neutral. **2** counterbalance; render ineffective by an opposite force or effect. **3** exempt or exclude (a place) from the sphere of hostilities. **4** *euphem.* make (a person) harmless or ineffective; kill. □ **neutralizer** *n.* **neutralization** /ˌnjuːtrəlaɪˈzeɪʃ(ə)n/ *n.* [F *neutraliser* f. med.L *neutralizare* (as NEUTRAL)]

neutrino /njuːˈtriːnəʊ/ *n. Physics* a neutral subatomic particle with a mass close to zero and half-integral spin. The existence of neutrinos was postulated in 1931 by Wolfgang Pauli, who deduced that they were carrying away some of the energy released in beta decay; they were experimentally detected in 1956. Neutrinos travel at the speed of light and have an extremely low probability of interaction with normal matter. Three kinds of neutrinos are known, associated with the electron, muon, and tau particle. [It., dimin. of *neutro* neutral (as NEUTER)]

neutron /ˈnjuːtrɒn/ *n. Physics* a subatomic particle of similar mass to a proton but with no electric charge. Neutrons, discovered by James Chadwick in 1932, are present in all atomic nuclei except those of ordinary hydrogen. However, free neutrons are unstable, decaying with a half-life of about 1,000 seconds into a proton, an electron, and a neutrino. Because they are electrically neutral, a beam of neutrons can penetrate the nucleus of an atom, inducing radioactive fission in some nuclei. The fission reaction usually produces more neutrons, so that a chain reaction is possible. Neutron-induced fission is the basis of atomic power; in nuclear reactors a controlled flux of slow neutrons, produced in the fission reaction, also maintains that reaction. □ **neutron bomb** a nuclear bomb producing neutrons and little blast, causing damage to life but little destruction to property. [NEUTRAL + -ON]

neutron star *n. Astron.* an object of very small radius (typically 30 km) and very high density, composed predominantly of closely packed neutrons. Neutron stars are thought to form by the gravitational collapse of the remnant of a massive star after a supernova explosion, provided that the star is insufficiently massive to produce a black hole. Pulsars are believed to be rotating neutron stars.

neutrophil /ˈnjuːtrəˌfɪl/ *n. Physiol.* a white blood cell readily stained only by neutral dyes.

Nev. *abbr.* Nevada.

Neva /ˈniːvə, njɪˈvɑː/ a river in NW Russia which flows 74 km (46 miles) westwards from Lake Ladoga to the Gulf of Finland, passing through St Petersburg. Alexander Nevsky took his name from this river after defeating the Swedes there in 1240.

Nevada /nɪˈvɑːdə/ a state of the western US; pop. (1990) 1,201,830; capital, Carson City. Acquired from Mexico in 1848, it became the 36th state of the US in 1864. It lies on an arid plateau, almost totally in the Great Basin area. □ **Nevadan** *adj.* & *n.*

névé /ˈneveɪ/ *n.* an expanse of granular snow not yet compressed into ice at the head of a glacier. [Swiss F, = glacier, ult. f. L *nix nivis* snow]

never /ˈnevə(r)/ *adv.* **1 a** at no time; on no occasion; not ever (*have never been to Paris; never saw them again*). **b** *colloq.* as an emphatic negative (*I never heard you come in; never fear*). **2** not at all (*never fear*). **3** *colloq.* (expressing surprise) surely not (*you never left the key in the lock!*). □ **never-ending** eternal, undying; immeasurable. **never-never** (often prec. by *the*) *Brit. colloq.* hire purchase. **never-never land** an imaginary utopian place (from J. M. Barrie's *Peter Pan*; see also NEVER-NEVER). **never a one** none. **never say die** see DIE[1]. **well I never!** expressing great surprise. [OE *næfre* f. *ne* not + *æfre* EVER]

nevermore /ˌnevəˈmɔː(r)/ *adv.* at no future time.

Never-Never /ˌnevəˈnevə(r)/ **1** the unpopulated desert country of the interior of Australia; the remote outback. **2** (**Never-Never Land** (or **Country**)) a region of Northern Territory, Australia, south-east of Darwin; chief town, Katherine.

Nevers /nəˈveə(r)/ a city in central France, on the Loire; pop. (1990) 43,890. It was capital of the former province of Nivernais.

nevertheless /ˌnevəðəˈles/ *adv.* in spite of that; notwithstanding; all the same.

Neville /ˈnevɪl/, Richard, see WARWICK[2].

Nevis /ˈniːvɪs/ one of the Leeward Islands in the West Indies, part of St Kitts and Nevis; capital, Charlestown. (See also ST KITTS AND NEVIS.) □ **Nevisian** /niːˈvɪsɪən/ *n.* & *adj.*

Nevsky see ALEXANDER NEVSKY.

nevus *US* var. of NAEVUS.

new /njuː/ *adj.* & *adv.* ● *adj.* **1 a** of recent origin or arrival. **b** made, invented, discovered, acquired, or experienced recently or now for the first time (*a new star; has many new ideas*). **2** in original condition; not worn or used. **3 a** renewed or reformed (*a new life; the new order*). **b** reinvigorated (*felt like a new person*). **4** different from a recent previous one (*has a new job*). **5** in addition to others already existing (*have you been to the new supermarket?*). **6** (often foll. by *to*) unfamiliar or strange (*a new sensation; the idea was new to me*). **7** (often foll. by *at*) (of a person) inexperienced, unaccustomed (to doing something) (*am new at this business*). **8** (usu. prec. by *the*) often *derog.* **a** later, modern (*the new morality*). **b** newfangled. **c** recently affected by social change (*the new rich*). **9** (often prec. by *the*) advanced in method or theory (*the new* *formula*). **10** (in place-names) discovered or founded later than and named after (*New York; New Zealand*). ● *adv.* (usu. in *comb.*) **1** newly, recently (*new-found; new-baked*). **2** anew, afresh. □ **new birth** (in Christian theology) spiritual regeneration. **new broom** see BROOM. **new deal** new arrangements or conditions, esp. when better than the earlier ones. (See also NEW DEAL.) **new-laid** (of an egg) freshly laid. **new look 1** a new or revised appearance or presentation, esp. of something familiar. **2** (often **New Look**) a style of women's clothing introduced after the war of 1939–45, featuring long wide skirts and a generous use of material in contrast to wartime austerity. **new-look** *attrib.adj.* having a new image; restyled. **new mathematics** (or **maths**) a system of teaching mathematics to children, with emphasis on investigation by them and on set theory. **new moon 1** the moon when first seen as a crescent after conjunction with the sun. **2** the time of its appearance. **a new one** (often foll. by *on*) *colloq.* an account or idea not previously encountered (by a person). **new potatoes** the earliest potatoes of a new crop. **new star** a nova. **new-style** *attrib.adj.* having a new style. **new year 1** the calendar year just begun or about to begin. **2** the first few days of a year. □ **newish** *adj.* **newness** *n.* [OE *nīwe* f. Gmc]

New Age *n.* a broad movement characterized by alternative approaches to traditional Western culture, with interest in spiritual matters, mysticism, holistic ideas, environmentalism, etc. Dating from the early 1970s, the movement has its roots in the hippy subculture of the 1960s. The term is derived from the astrological Age of Aquarius. (See also AQUARIUS, AGE OF; ARIES, FIRST POINT OF.) □ **New Age music** a style of chiefly instrumental modern music characterized by light melodic harmonies, improvisation, and sounds reproduced from the natural world, intended to promote serenity. **New Age traveller** see TRAVELLER 4.

New Amsterdam the original name (until 1664) for NEW YORK.

Newark /'njuːək/ an industrial city in New Jersey; pop. (1990) 275,220.

newborn /njuː'bɔːn/ *adj.* **1** (of a child etc.) recently born. **2** spiritually reborn; regenerated.

New Britain a mountainous island in the South Pacific, administratively part of Papua New Guinea, lying off the NE coast of New Guinea; pop. (1990) 311,955; capital, Rabaul.

New Brunswick a maritime province on the SE coast of Canada; pop. (1991) 726,000; capital, Fredericton. It was first settled by the French and ceded to Britain in 1713. It became one of the original four provinces in the Dominion of Canada in 1867.

New Brutalism see BRUTALISM.

New Caledonia (called in French *Nouvelle-Calédonie*) an island in the South Pacific, east of Australia; pop. (1989) 164,170; capital, Nouméa. Since 1946 it has formed, with its dependencies, a French overseas territory. It was inhabited by Melanesians for at least 3,000 years before the arrival of Captain James Cook in 1774, who named the main island after the Roman name for Scotland. The French annexed the island in 1853, and after the discovery of nickel in 1863 it assumed some economic importance for France. There has been a growing movement for independence, with a referendum on the matter scheduled for 1998. □ **New Caledonian** *n. & adj.*

New Carthage see CARTAGENA 1.

Newcastle[1] /'njuːˌkɑːs(ə)l/ an industrial port on the SE coast of Australia, in New South Wales; pop. (1991) 262,160.

Newcastle[2] /'njuːˌkɑːs(ə)l/, 1st Duke of (title of Thomas Pelham-Holles) (1693–1768), British Whig statesman, Prime Minister 1754–6 and 1757–62. Newcastle succeeded his brother Henry Pelham as Prime Minister on the latter's death in 1754. During his second term in office, he headed a coalition with William Pitt the Elder, until Pitt's resignation in 1761.

Newcastle disease *n.* an acute infectious viral fever affecting birds, esp. poultry. [NEWCASTLE-UPON-TYNE]

Newcastle-under-Lyme /ˌnjuːkɑːs(ə)lˌʌndə'laɪm/ an industrial town in Staffordshire, just south-west of Stoke-on-Trent; pop. (1991) 117,400.

Newcastle-upon-Tyne /ˌnjuːkɑːs(ə)ləpɒn'taɪn/ an industrial city in NE England, a port on the River Tyne, administrative centre of Tyne and Wear; pop. (1991) 263,000.

New Comedy *n.* a style of ancient Greek comedy (see COMEDY).

Newcomen /'njuːˌkʌmən/, Thomas (1663–1729), English engineer, developer of the first practical steam engine. He designed a beam engine to operate a pump for the removal of water from mines; the first such steam engine was erected in Worcestershire in 1712. To avoid infringing a patent held by Thomas Savery for his pumping engine, Newcomen went into partnership with him. Newcomen's engine was later greatly improved by James Watt.

newcomer /'njuːˌkʌmə(r)/ *n.* **1** a person who has recently arrived. **2** a beginner in some activity.

New Commonwealth those countries which have achieved self-government within the Commonwealth since 1945.

New Criticism *n.* an influential movement in literary criticism particularly in the US, which stressed the importance of focusing on the text itself rather than being concerned with external biographical or social considerations. The term was first used in 1941 by John Crowe Ransom (1888–1974) in discussing such critics as T. S. Eliot and William Empson, and was later applied to I. A. Richards (1893–1979), Cleanth Brooks (1906–94), and others.

New Deal the economic measures introduced by Franklin D. Roosevelt as President of the US in 1933 to counteract the effects of the Great Depression. The New Deal depended largely on a massive public works programme, complemented by the large-scale granting of loans, and succeeded in reducing unemployment by between 7 and 10 million.

New Delhi see DELHI.

newel /'njuːəl/ *n.* **1** the supporting central post of winding stairs. **2** the top or bottom supporting post of a stair-rail. [ME f. OF *noel, nouel*, knob f. med.L *nodellus* dimin. of L *nodus* knot]

New England an area on the NE coast of the US, comprising the states of Maine, New Hampshire, Vermont, Massachusetts, Rhode Island, and Connecticut. The name was given to it by the English explorer John Smith (1580–1631) in 1614. □ **New Englander** *n.*

New English Bible a translation of the Bible into modern English, made under the auspices of the Oxford and Cambridge University Presses from 1948. The New Testament was published in 1961, and the complete Bible appeared in 1970; a revision, the Revised English Bible, was made available in 1989.

newfangled /njuː'fæŋg(ə)ld/ *adj. derog.* different from what one is used to; objectionably new. [ME *newfangle* (now dial.) liking what is new f. *newe* NEW *adv.* + *-fangel* f. OE *fangol* (unrecorded) inclined to take]

New Forest an area of heath and woodland in southern Hampshire. It has been reserved as Crown property since 1079, originally by William I as a royal hunting area. William II was killed by an arrow when hunting there in 1100.

Newfoundland[1] /'njuːfəndlənd, njuː'faʊndlənd/ a large island off the east coast of Canada, at the mouth of the St Lawrence River. It was discovered in 1497 by John Cabot and named 'the New Isle', 'Terra Nova', or more commonly Newfoundland. In 1949 it was united with Labrador (as Newfoundland and Labrador) to form a province of Canada. □ **Newfoundlander** /njuː'faʊndləndə(r)/ *n.*

Newfoundland[2] /njuː'faʊndlənd/ *n.* (in full **Newfoundland dog**) a very large breed of dog with a thick coarse coat. [NEWFOUNDLAND[1]]

Newfoundland and Labrador a province of Canada, comprising the island of Newfoundland and the sparsely inhabited Labrador coast of eastern Canada; pop. (1991) 573,500; capital, St John's. It joined the confederation of Canada in 1949.

Newgate /'njuːgeɪt/ a former London prison, originally the gatehouse of the main west gate to the city, first used as a prison in the early Middle Ages and rebuilt and enlarged with funds left to the city by Sir Richard Whittington. Its unsanitary conditions became notorious in the 18th century before the building was burnt down in the anti-Catholic riots of 1780. A new edifice was erected on the same spot soon after but was demolished in 1902 to make way for the Central Criminal Court. □ **Newgate Calendar** a publication issued from *c.*1774 until the mid-19th century that dealt with notorious crimes. **Newgate novel** a novel of a picaresque 19th-century genre involving criminal characters.

New Guinea an island in the western South Pacific, off the north coast of Australia, the second largest island in the world (following Greenland). It is divided into two parts; the western half comprises part of Irian Jaya, a province of Indonesia, the eastern half forms part of the country of Papua New Guinea. □ **New Guinean** *n. & adj.*

New Hampshire a state in the north-eastern US, on the Atlantic coast; pop. (1990) 1,109,250; capital, Concord. It was settled from England in the 17th century and was one of the original thirteen states of the Union (1788).

New Hebrides the former name (until 1980) for VANUATU.

Ne Win /neɪ 'wɪn/ (b.1911), Burmese general and socialist statesman, Prime Minister 1958-60, head of state 1962-74, and President 1974-81. An active nationalist in the 1930s, Ne Win was appointed Chief of Staff in Aung San's Burma National Army in 1943. He led a military coup in 1962, after which he established a military dictatorship and formed a one-party state, governed by the Burma Socialist Programme Party (BSSP). He stepped down from the presidency in 1981 and retired as leader of the BSSP after riots in Rangoon in 1988.

New Ireland an island in the South Pacific, administratively part of Papua New Guinea, lying to the north of New Britain; pop. (1990) 87,190; capital, Kavieng.

New Jersey a state in the north-eastern US, on the Atlantic coast; pop. (1990) 7,730,190; capital, Trenton. Colonized by Dutch settlers and ceded to Britain in 1664, it became one of the original thirteen states of the Union (1787). □ **New Jerseyan** n. & adj.

New Jerusalem n. **1** (in Christian theology) the abode of the blessed in heaven. **2** an ideal place or situation.

New Jerusalem Church a Christian sect instituted by followers of Emanuel Swedenborg. It was founded in London in 1787.

New Kingdom a period of ancient Egyptian history (c.1550-1070 BC). (See EGYPT.)

Newlands /'njuːləndz/, John Alexander Reina (1837-98), English industrial chemist. He proposed a periodic table shortly before Dmitri Mendeleev, based on his *law of octaves*. Newlands observed that if elements were arranged in order of atomic weight, similar chemical properties appeared in every eighth element, a pattern he likened to the musical scale. The significance of his idea was not understood until Mendeleev's periodic table had been accepted. Newlands claimed priority, but his rigid scheme had a number of inadequacies.

newly /'njuːlɪ/ adv. **1** recently (*a friend newly arrived*; *a newly discovered country*). **2** afresh, anew (*newly painted*). **3** in a new or different manner (*newly arranged*). □ **newly-wed** a recently married person.

Newman[1] /'njuːmən/, Barnett (1905-70), American painter. A seminal figure in colour-field painting from the late 1940s, he is noted for his vast canvases, which achieve dramatic effects from their juxtaposition of large blocks of uniform colour with narrow marginal strips of contrasting colours. His paintings include *Who's Afraid of Red, Yellow, and Blue III* (1966-7).

Newman[2] /'njuːmən/, John Henry (1801-91), English prelate and theologian. In 1833, while vicar of St Mary's Church in Oxford, he founded the Oxford Movement together with John Keble and Edward Pusey. Newman was a leading influence within the Movement and wrote twenty-four of the *Tracts for the Times*. However, his Tract 90 (1841) on the compatibility of the Thirty-nine Articles with Roman Catholic theology aroused much hostility and he withdrew from the Movement. In 1845 he was received into the Roman Catholic Church, becoming a cardinal in 1879. His works include *Apologia pro Vita Sua* (1864) and the poem *The Dream of Gerontius* (1865), describing the soul's journey to God.

Newman[3] /'njuːmən/, Paul (b.1925), American actor and film director. Among his many films are *Butch Cassidy and the Sundance Kid* (1969), *The Sting* (1973), and *The Color of Money* (1987), for which he won an Oscar. He has also directed several films, including *Rachel, Rachel* (1968) and *The Glass Menagerie* (1987).

Newmarket[1] /'njuː.mɑːkɪt/ a town in eastern England, in Suffolk; pop. (1981) 16,130. It is a noted horse-racing centre and headquarters of the Jockey Club.

Newmarket[2] /'njuː.mɑːkɪt/ n. a card-game in which players seek to play cards that match four cards ('horses') displayed on the table, betting on which cards will be played. [NEWMARKET[1]]

New Mexico a state in the south-western US, on the border with Mexico; pop. (1990) 1,515,070; capital, Santa Fe. It was obtained from Mexico in 1848 and 1854, and in 1912 it became the 47th state of the US. □ **New Mexican** adj. & n.

New Model see MODEL, THE.

New Model Army an army created in 1645 by Oliver Cromwell to fight for the Parliamentary cause in the English Civil War. Led by Thomas Fairfax, it was a disciplined and well-trained army that proved too strong for the Royalists, winning notable victories at Marston Moor and Naseby. It later came to possess considerable political influence.

New Orleans /ɔːˈliːnz, ɔːˈliːnz/ a city and port in SE Louisiana, on the Mississippi; pop. (1990) 496,940. It was founded by the French in 1718 and named after the Duc d'Orléans, regent of France. It is noted for its annual Mardi Gras celebrations and for its association with the development of blues and jazz.

New Plymouth a port in New Zealand, on the west coast of North Island; pop. (1991) 48,520.

Newport /'njuːpɔːt/ an industrial town and port in South Wales, in Gwent on the Bristol Channel; pop. (1991) 129,900.

Newport News a city in SE Virginia, at the mouth of the James river on the Hampton Roads estuary; pop. (1990) 170,045. Settled around 1620, it is now a major seaport and shipbuilding centre.

Newry /'njʊərɪ/ a port in the south-east of Northern Ireland, in County Down; pop. (1981) 19,400.

news /njuːz/ n.pl. (usu. treated as *sing.*) **1** information about important or interesting recent events, esp. when published or broadcast. **2** (prec. by *the*) a broadcast report of news. **3** newly received or noteworthy information. **4** (foll. by *to*) *colloq.* information not previously known (to a person) (*that's news to me*). □ **news agency** an organization that collects and distributes news items. **news bulletin** a collection of items of news, esp. for broadcasting. **news conference** a press conference. **news-gatherer** a person who researches news items, esp. for broadcast or publication. **news-gathering** this process. **news room** a room in a newspaper or broadcasting office where news is processed. **news-sheet** a simple form of newspaper; a newsletter. **news-stand** a stall for the sale of newspapers. **news-vendor** *Brit.* a newspaper-seller. □ **newsless** adj. [ME, pl. of NEW after OF *noveles* or med.L *nova* neut. pl. of *novus* new]

newsagent /'njuːz.eɪdʒənt/ n. *Brit.* a seller of or shop selling newspapers and usu. related items, e.g. stationery.

newsboy /'njuːzbɔɪ/ n. a boy who sells or delivers newspapers.

newsbrief /'njuːzbriːf/ n. a short item of news, esp. on television; a newsflash.

newscast /'njuːzkɑːst/ n. a radio or television broadcast of news reports.

newscaster /'njuːz.kɑːstə(r)/ n. = NEWSREADER.

newsflash /'njuːzflæʃ/ n. a single item of important news broadcast separately and often interrupting other programmes.

newsgirl /'njuːzɡɜːl/ n. a girl who sells or delivers newspapers.

newsletter /'njuːz.letə(r)/ n. an informal printed report issued periodically to the members of a society, business, organization, etc.

newsman /'njuːzmæn/ n. (pl. **-men**) a newspaper reporter; a journalist.

newsmonger /'njuːz.mʌŋɡə(r)/ n. a gossip.

New South Wales a state of SE Australia; pop. (1990) 5,827,400; capital, Sydney. First colonized from Britain in 1788, it was federated with the other states of Australia in 1901.

New Spain a former Spanish viceroyalty established in Central and North America in 1535, centred on present-day Mexico City. It comprised all the land under Spanish control north of the Isthmus of Panama, including parts of the southern US. It also came to include the Spanish possessions in the Caribbean and the Philippines. The viceroyalty was abolished in 1821, when Mexico achieved independence.

newspaper /'njuːs.peɪpə(r)/ n. a printed publication (usu. daily or weekly) containing news, advertisements, correspondence, etc. Daily publication of news dates back to ancient Rome, where public announcements about social and political matters were written up on a large whitened board in the city centre. For centuries news was circulated to important persons by letters, or by handwritten and then printed news-sheets. The first regular public newspapers appeared in Europe in the mid-17th century, and the first English daily paper, the *Daily Courant*, in 1702. Early newspapers were subject to government censorship and controls and to a great many prosecutions for libel; their editors were also frequently corrupt. During the 18th century, however, increasing revenue from advertisements encouraged editors to resist bribes, and reporting became more independent. During the 19th century improvements in printing technology, the setting up of international news agencies, and the spread of literacy all contributed to the general expansion of newspapers, and from the 1890s a more popular approach, offering amusing or sensational items of news, was introduced from America. The first generation of twentieth-century press barons, Beaverbrook, Northcliffe, and Hearst, used their papers to propagate their political opinions, a practice that continues to a certain extent with publishers like Rupert Murdoch. The 20th century has also seen many newspapers merging into large groups, and a number have ceased publication. The split between 'popular' and 'quality' newspapers has widened, while the introduction of computer technology has transformed established methods of production.

Newspeak /'nju:spi:k/ *n.* ambiguous euphemistic language used esp. in political propaganda. [an artificial official language in George Orwell's *Nineteen Eighty-four* (1949)]

newsprint /'nju:zprint/ *n.* a type of low-quality paper on which newspapers are printed.

newsreader /'nju:z,ri:də(r)/ *n.* a person who reads out broadcast news bulletins.

newsreel /'nju:zri:l/ *n.* a short cinema film of recent events. The regular issue of newsreels, typically consisting of several short items grouped together, began in France in 1908, introduced by the Pathé brothers. During the 1930s newsreels lasting about fifteen minutes appeared once or twice a week in Britain and America and formed a familiar part of cinema programmes. Newsreels played an important role in the Second World War, but after the war the greater topicality of television caused them to be abandoned.

New Style *n.* dating reckoned by the Gregorian calendar (superseding the Julian calendar in 1752 in England and Wales).

newsworthy /'nju:z,wɜ:ði/ *adj.* topical; noteworthy as news. □ **newsworthiness** *n.*

newsy /'nju:zi/ *adj.* (**newsier, newsiest**) *colloq.* full of news.

newt /nju:t/ *n.* a small chiefly aquatic amphibian of the family Salamandridae, with a well-developed tail; esp. one of the common Eurasian genus *Triturus*. [ME f. *ewt*, with *n* from *an* (cf. NICKNAME): var. of *evet* EFT]

New Territories part of the territory of Hong Kong on the south coast of mainland China, lying to the north of the Kowloon peninsula and including the islands of Lantau, Tsing Yi, and Lamma. It comprises 92 per cent of the land area of Hong Kong. Under the 1898 Convention of Peking the New Territories were leased to Britain by China for a period of 99 years. (See HONG KONG.)

New Testament the twenty-seven books of the Bible recording the life, ministry, Crucifixion, and Resurrection of Christ and the teaching of some of his earliest followers, written originally in Greek. At an early date the Church came to regard some of its own writings, especially those held to be of apostolic origin, as of equal authority and inspiration to those received from Judaism. The canon of the New Testament, based on the four Gospels and the Epistles of St Paul, came into existence largely without definition. It was probably formally fixed at Rome in 382, when the Christian Old Testament was also defined.

Newton /'nju:t(ə)n/, Sir Isaac (1642–1727), English mathematician and physicist, the greatest single influence on theoretical physics until Einstein. His most productive period was in 1665–7, when he retreated temporarily from Cambridge to his isolated home in Lincolnshire during the Great Plague. He discovered the binomial theorem, and made several other contributions to mathematics, notably differential calculus and its relationship with integration. A bitter quarrel with Leibniz ensued as to which of them had discovered calculus first. In his major treatise, *Principia Mathematica* (1687), Newton gave a mathematical description of the laws of mechanics and gravitation, and applied these to planetary and lunar motion. For most purposes Newtonian mechanics has survived even the introduction of relativity theory and quantum mechanics, to both of which it stands as a good approximation. Another influential work was *Opticks* (1704), which gave an account of Newton's optical experiments and theories, including the discovery that white light is made up of a mixture of colours. In 1699 Newton was appointed Master of the Mint; he entered Parliament as MP for Cambridge University in 1701, and in 1703 was elected president of the Royal Society.

newton /'nju:t(ə)n/ *n. Physics* the SI unit of force (symbol: **N**), equal to 100,000 dynes. One newton is that force which would give a mass of one kilogram an acceleration of one metre per second per second in the direction of the force. [NEWTON]

Newtonian /nju:'təυniən/ *adj.* of or devised by Sir Isaac Newton. □ **Newtonian mechanics** the system of mechanics which relies on Newton's laws of motion concerning the relations between forces acting and motions occurring. **Newtonian telescope** a reflecting telescope in which light reflected from a small flat 45° secondary mirror is brought to a focus just outside the side of the telescope.

Newton's laws of motion the three laws of motion on which Newtonian mechanics are based: (1) that a body continues in a state of rest or uniform motion in a straight line unless it is acted on by an external force; (2) that the rate of change of momentum of a moving body is proportional to the force acting to produce the change; (3) that if one body exerts a force on another, there is an equal and opposite force (or reaction) exerted by the second body on the first.

new town *n.* a planned urban centre created in an undeveloped or rural area, often with government sponsorship. In the UK such towns were built in the late 19th century by industrialists needing homes for their workers, examples being Bournville, near Birmingham (built by George Cadbury), and Viscount Leverhulme's Port Sunlight near Liverpool. After the Second World War a programme was instituted to rehouse expanding or war-displaced populations from the cities in moderately sized new towns, among which were Basildon, Cumbernauld, Harlow, and Hemel Hempstead. The towns were criticized by some as disrupting family and community ties, and a static population and cuts in government spending meant that Milton Keynes was the last British new town to be started, in 1967.

new wave *n.* **1** = NOUVELLE VAGUE. **2** a style of rock music popular in the late 1970s, deriving from punk but generally more sophisticated in sound and less aggressive in performance.

New World North and South America regarded collectively in relation to Europe. The term was first applied to the Americas (also to other areas, e.g. Australia) after the early voyages of European explorers. (Cf. OLD WORLD.)

New Year's Day 1 January.

New Year's Eve 31 December.

New York 1 a state in the north-eastern US; pop. (1990) 17,990,445; capital, Albany. It stretches from the Canadian border and Lake Ontario in the northwest to the Atlantic in the east, with the Adirondacks in the north and the Catskills in the south. Originally settled by the Dutch, it was surrendered to the British in 1664 and was one of the original thirteen states of the Union (1788). **2** a city in the south-east of New York state, a major port of the US, situated on the Atlantic coast at the mouth of the Hudson River; pop. (1990) 7,322,560. The Hudson River and Manhattan Island were discovered by European explorers in 1609, and in 1629 Dutch colonists purchased Manhattan Island from the American Indians for 24 dollars' worth of trinkets, establishing a settlement there which they called New Amsterdam. In 1664 it was captured by the British, who renamed it in honour of the Duke of York (later James II), who was at that time Lord High Admiral of England. It is situated mainly on islands, linked by bridges, in New York harbour and comprises five boroughs: Manhattan, Brooklyn, the Bronx, Queens, and Staten Island. Manhattan is the economic and cultural heart of the city, containing the country's financial centre, the Stock Exchange in Wall Street. Because of the island's small area and population density Manhattan has been built upwards, with skyscrapers that give its characteristic skyline. New York is the country's leading commercial and industrial city. It is the headquarters of the United Nations and has world-famous art galleries, museums, and theatres and is one of the most ethnically diverse and culturally complex cities in the world. □ **New Yorker** *n.*

New Zealand /'zi:lənd/ (called in Maori *Aotearoa*) an island country in the South Pacific about 1,900 km (1,200 miles) east of Australia; pop. (1991) 3,434,950; languages, English (official), Maori; capital, Wellington. New Zealand consists of two major islands (North and South Islands) separated by Cook Strait, and several smaller islands. The original discoverers and first colonists of the country were the Maoris. The first European to sight New Zealand was the Dutch navigator Abel Tasman in 1642, when it was named after the Netherlands province of Zeeland. The islands were circumnavigated by Captain James Cook in 1769–70 and came under British sovereignty in 1840; colonization led to the Maori Wars of the mid-19th century. Full dominion status was granted in 1907, and independence in 1931. New Zealand is a member of the Commonwealth and played a major role on the side of the Allies in each of the two world wars. (See also MAORI.) □ **New Zealander** *n.*

next /nekst/ *adj., adv., n., & prep.* ● *adj.* **1** (often foll. by *to*) being or positioned or living nearest (*in the next house; the chair next to the fire*). **2** the nearest in order of time; the first or soonest encountered or considered (*next Friday; ask the next person you see*). ● *adv.* **1** (often foll. by *to*) in the nearest place or degree (*put it next to mine; came next to last*). **2** on the first or soonest occasion (*when we next meet*). ● *n.* the next person or thing. ● *prep. colloq.* next to. □ **next-best** the next in order of preference. **next door** see DOOR. **next of kin** the closest living relative or relatives. **next to** almost (*next to nothing left*). **the next world** see WORLD. [OE *nēhsta* superl. (as NIGH)]

nexus /'neksəs/ n. (pl. **nexuses**) **1** a connected group or series. **2** a bond; a connection. [L f. *nectere nex-* bind]

Ney /neɪ/, Michel (1768–1815), French marshal. He was one of Napoleon's leading generals, and after the Battle of Borodino (1812) became known as 'the bravest of the brave'. He commanded the French cavalry at Waterloo (1815), but after Napoleon's defeat and final overthrow was executed by the Bourbons despite attempts by Wellington and other allied leaders to intervene on his behalf.

NF see NATIONAL FRONT.

Nfld abbr. (also **NF**) Newfoundland.

NFU abbr. (in the UK) National Farmers' Union.

n.g. abbr. no good.

NGA abbr. Brit. hist. National Graphical Association. (See GPMU.)

ngaio /'naɪəʊ/ n. (pl. **-os**) a small New Zealand tree, *Myoporum laetum*, with edible fruit and light white timber. [Maori]

Ngaliema, Mount /əŋˌɡɑːlɪ'eɪmə/ the Zairean name for Mount Stanley (see STANLEY, MOUNT).

Ngamiland /əŋ'ɡɑːmɪˌlænd/ a region in NW Botswana, north of the Kalahari Desert. It includes the Okavango marshes and Lake Ngami.

Ngata /'nɑːtə/, Sir Apirana Turupa (1874–1950), New Zealand Maori leader and politician. As Minister for Native Affairs he devoted much time to Maori resettlement. Believing firmly in the continuing individuality of the Maori people, he sought to preserve the characteristic elements of their life and culture, including tribal customs and folklore, and emphasized pride in Maori traditions and history.

Ngbandi /əŋ'bændɪ/ n. & adj. ● n. a Niger-Congo language of northern Zaire. Cf. SANGO. ● adj. of or relating to this language. [Ngbandi]

NGO abbr. non-governmental organization.

Ngorongoro /əŋˌɡɔːrəŋ'ɡɔːrəʊ/ a huge extinct volcanic crater in the Great Rift Valley in NE Tanzania, 326 sq. km (126 sq. miles) in area. It is the centre of a wildlife conservation area, established in 1959, which includes the Olduvai Gorge.

Nguni /əŋ'ɡuːnɪ/ n. & adj. ● n. (pl. same) **1** a member of a group of Bantu-speaking peoples living mainly in southern Africa. **2** the group of closely related Bantu languages, including Xhosa, Zulu, Swazi, and Sindebele, spoken by this group of peoples. ● adj. of or relating to this group of peoples or their languages.

NH abbr. US New Hampshire (also in official postal use).

NHI abbr. (in the UK) National Health Insurance.

NHS see NATIONAL HEALTH SERVICE.

Nhulunbuy /ˌnjuːlən'baɪ/ a bauxite-mining centre on the NE coast of Arnhem Land in Northern Territory, Australia; pop. (1986) 3,800.

NI abbr. **1** (in the UK) National Insurance. **2** Northern Ireland.

Ni symb. Chem. the element nickel.

niacin /'naɪəsɪn/ n. = NICOTINIC ACID. [nicotinic *acid* + -IN]

Niagara Falls /naɪ'ægərə/ **1** the waterfalls on the Niagara River, consisting of two principal parts separated by Goat Island: the Horseshoe Falls adjoining the west (Canadian) bank, which fall 47 m (158 ft), and the American Falls adjoining the east (American) bank, which fall 50 m (167 ft). They are a popular tourist venue and an attraction for various stunts. In 1859 Charles Blondin walked across on a tightrope and in 1901 Annie Edson Taylor was the first person to go over in a barrel. **2** a city in upper New York State situated on the right bank of the Niagara River beside the Falls; pop. (1990) 61,840. **3** a city in Canada, in southern Ontario, situated on the left bank of the Niagara River beside the Falls, opposite the city of Niagara Falls, US, to which it is linked by bridges; pop. (1991) 75,400.

Niagara River a river of North America. Flowing northwards for 56 km (35 miles) from Lake Erie to Lake Ontario, it forms part of the border between Canada and the US. The Niagara Falls are situated halfway along its course. It is a major source of hydroelectric power.

Niamey /nja:'meɪ/ the capital of Niger, a port on the River Niger; pop. (1988) 398,265.

nib /nɪb/ n. & v. ● n. **1** the point of a pen, which touches the writing surface. **2** (in pl.) shelled and crushed coffee or cocoa beans. **3** the point of a tool etc. ● v. (**nibbed, nibbing**) **1** tr. provide with a nib. **2** tr. mend the nib of. **3** tr. & intr. nibble. [prob. f. MDu. *nib* or MLG *nibbe*, var. of *nebbe* NEB]

nibble /'nɪb(ə)l/ v. & n. ● v. **1** tr. & (foll. by at) intr. **a** take small bites at. **b** eat in small amounts. **c** bite at gently or cautiously or playfully. **2** intr. (foll. by at) show cautious interest in. ● n. **1** an instance of nibbling.

2 (usu. in pl.) a morsel or titbit of food. **3** Computing half a byte, i.e. 4 bits. □ **nibbler** n. [prob. of LG or Du. orig.: cf. LG *nibbeln* gnaw]

Nibelung /'niːbəˌlʊŋ/ n. (pl. **Nibelungs, Nibelungen** /-ˌlʊŋən/) Germanic Mythol. **1** a member of a Scandinavian race of dwarfs, owners of a hoard of gold and magic treasures, who were ruled by Nibelung, king of Nibelheim (land of mist). **2** (in the Nibelungenlied) any of the supporters of Siegfried, the subsequent possessor of the hoard; any of the Burgundians who stole it from him.

Nibelungenlied /'niːbəˌlʊŋənˌliːd, -ˌliːt/ a 13th-century German poem, embodying a story found in the Edda, telling of the life and death of Siegfried, a prince of the Netherlands. Siegfried kills the dragon Fafner to seize the treasure of the Nibelungs; he then marries the Burgundian princess Kriemhild and uses trickery to help her brother Gunther win Brunhild, but is killed by Gunther's retainer Hagen. His wife Kriemhild agrees to marry Etzel (Attila the Hun) in order to be revenged, and beheads Hagen herself. There have been many adaptations of the story, including Wagner's epic music drama *Der Ring des Nibelungen*. [NIBELUNG + G *lied* song]

niblick /'nɪblɪk/ n. Golf an iron with a large round heavy head, used esp. for playing out of bunkers. [19th c.: orig. unkn.]

nibs /nɪbz/ n. □ **his nibs** joc. colloq. a mock title used with reference to an important or self-important person. [19th c.: orig. unkn. (cf. earlier *nabs*)]

nicad /'naɪkæd/ n. (often attrib.) a battery, often rechargeable, with a nickel anode and a cadmium cathode. [NICKEL + CADMIUM]

Nicaea /naɪ'siːə/ an ancient city in Asia Minor, on the site of modern Iznik, which was important in Roman and Byzantine times. It was the site of two ecumenical councils of the early Christian Church. The first, the Council of Nicaea in 325, condemned Arianism and produced the Nicene Creed. The second, in 787, condemned the iconoclasts.

Nicam /'naɪkæm/ n. (also **NICAM**) a digital system used in British television to provide video signals with high-quality stereo sound. [acronym f. *near instantaneously companded audio multiplex*]

Nicaragua /ˌnɪkə'rægjʊə, -'rægwə/ the largest country in Central America, with a coastline on both the Atlantic and the Pacific Ocean; pop. (est. 1991) 3,975,000; official language, Spanish; capital, Managua. Columbus sighted the eastern coast in 1502 and the country was soon colonized by the Spaniards, who faced little opposition from the indigenous peoples. Nicaragua broke away from Spain in 1821 and, after brief membership of the United Provinces of Central America, became an independent republic in 1838. In 1979 the dictator Anastasio Somoza, whose family had held power almost continuously since 1937, was overthrown by a popular revolution after several years of civil war. The new left-wing Sandinista regime then faced a counter-revolutionary guerrilla campaign by the US-backed Contras. In the 1990 election the Sandinistas lost power to an opposition coalition. □ **Nicaraguan** adj. & n.

Nicaragua, Lake a lake near the west coast of Nicaragua, the largest lake in Central America.

Nice /niːs/ a resort city on the French Riviera, near the border with Italy; pop. (1990) 345,670.

nice /naɪs/ adj. **1** pleasant, agreeable, satisfactory. **2** (of a person) kind, good-natured. **3** iron. bad or awkward (*a nice mess you've made*). **4 a** fine or subtle (*a nice distinction*). **b** requiring careful thought or attention (*a nice problem*). **5** fastidious; delicately sensitive. **6** punctilious, scrupulous (*were not too nice about their methods*). **7** (foll. by an adj., often with *and*) satisfactory or adequate in terms of the quality described (*a nice long time; nice and warm*). □ **nice one** int. colloq. expressing approval or commendation. **nice work** often iron. a task well done. □ **nicely** adv. **niceness** n. **nicish** adj. (also **niceish**). [ME, = stupid, wanton f. OF, = silly, simple f. L *nescius* ignorant (as *nescience*: see NESCIENT)]

Nicene Creed /naɪ'siːn, 'naɪsiːn/ a formal statement of Christian belief that appears in the Thirty-nine Articles, based on that adopted at the first Council of Nicaea in 325. [*Nicene* ME f. LL *Nicenus* of NICAEA]

nicety /'naɪsɪtɪ/ n. (pl. **-ies**) **1** a subtle distinction or detail. **2** precision, accuracy. **3** intricate or subtle quality (*a point of great nicety*). **4** (in pl.) **a** minutiae; fine details. **b** refinements, trimmings. □ **to a nicety** with exactness. [ME f. OF *niceté* (as NICE)]

niche /nɪtʃ, niːʃ/ n. & v. ● n. **1** a shallow recess, esp. in a wall to contain a statue etc. **2** a comfortable or suitable position in life or employment. **3** (often attrib.) a specialized but profitable segment of a commercial market. **4** Ecol. a position or role taken by a kind of organism within its community. ● v.tr. (often as **niched** adj.) **1** place in a niche.

2 ensconce (esp. oneself) in a recess or corner. [F f. *nicher* make a nest, ult. f. L *nidus* nest]

Nichiren /'nɪʃərən/ *n.* (also **Nichiren Buddhism**) a Japanese Buddhist sect founded by the religious teacher Nichiren (1222–82). Somewhat critical of Zen and other Buddhist schools, Nichiren has as its central scripture the Lotus Sutra. There are more than 30 million followers in more than forty subsects, the largest now being Nichiren-Shoshu, which is connected with the religious and political organization Soka Gakkai.

Nicholas /'nɪkələs/ the name of two tsars of Russia:

Nicholas I (1796–1855), brother of Alexander I, reigned 1825–55. He pursued rigidly conservative policies, maintaining serfdom and building up a large secret police force to suppress radical reformers. He was largely concerned with keeping the peace in Europe, but his expansionist policies in the Near East led to the Crimean War, during which he died.

Nicholas II (1868–1918), son of Alexander III, reigned 1894–1917. He proved incapable of coping with the dangerous political legacy left by his father and was criticized for allowing his wife Alexandra (and her favourites such as Rasputin) too much influence. Previously resistant to reform, after the disastrous war with Japan (1904–5) the tsar pursued a less reactionary line, but the programme of reforms which was introduced was not sufficient to prevent the disintegration of the tsarist regime under the strain of fresh military disasters during the First World War. Nicholas was forced to abdicate after the Russian Revolution in 1917 and was shot along with his family a year later.

Nicholas, St (4th century), Christian prelate. Little is known of his life, but he is said to have been bishop of Myra in Lycia; his supposed remains were taken to Bari in SE Italy in 1087. He became the subject of many legends and is patron saint of children, sailors, and the countries of Greece and Russia. The cult of Santa Claus (a corruption of his name) arose in North America in the 17th century from the Dutch custom of giving gifts to children on his feast day (6 Dec.), a practice now usually transferred to Christmas. (See also FATHER CHRISTMAS.)

Nicholson[1] /'nɪk(ə)ls(ə)n/, Ben (1894–1982), English painter. A pioneer of British abstract art, he met Piet Mondrian in 1933 and from that time produced painted reliefs with circular and rectangular motifs. These became his main output, together with purely geometrical paintings and still lifes.

Nicholson[2] /'nɪk(ə)ls(ə)n/, Jack (b.1937), American actor. He made his film début in 1958, but it was not until he appeared in *Easy Rider* (1969) that he gained wide recognition. He went on to act in such diverse films as *Five Easy Pieces* (1970), *The Shining* (1980), and *A Few Good Men* (1992), and won Oscars for *One Flew Over the Cuckoo's Nest* (1975) and *Terms of Endearment* (1983).

Nichrome /'naɪkrəʊm/ *n. propr.* a group of nickel-chromium alloys used for making wire in heating elements etc. [NICKEL + CHROME]

nick[1] /nɪk/ *n. & v.* ● *n.* **1** a small cut or notch. **2** *Brit. sl.* **a** a prison. **b** a police station. **3** (prec. by *in* with adj.) *Brit. colloq.* condition (in *reasonable nick*). **4** the junction between the floor and walls in a squash court. ● *v.tr.* **1** make a nick or nicks in. **2** *Brit. sl.* **a** a steal. **b** arrest, catch. □ **in the nick of time** only just in time; just at the right moment. [ME: orig. uncert.]

nick[2] /nɪk/ *v.intr. Austral. sl.* (foll. by *off*, *in*, etc.) move quickly or furtively. [19th c.: orig. uncert. (cf. NIP[1] 4)]

nickel /'nɪk(ə)l/ *n. & v.* ● *n.* **1** a silvery-white metallic chemical element (atomic number 28; symbol **Ni**. (See note below.) **2** *N. Amer. colloq.* a five-cent coin. ● *v.tr.* (**nickelled, nickelling**; *US* **nickeled, nickeling**) coat with nickel. □ **nickel brass** an alloy of copper, zinc, and a small amount of nickel. **nickel-plated** coated with nickel by plating. **nickel silver** = German silver. **nickel steel** a type of stainless steel with chromium and nickel. □ **nickelous** *adj.* **nickelic** /nɪ'kelɪk/ *adj.* [abbr. of G *Kupfernickel* copper-coloured ore, from which nickel was first obtained, f. *Kupfer* copper + *Nickel* demon, with ref. to the ore's failure to yield copper (cf. COBALT)]

▪ A transition metal, nickel was isolated by the Swedish mineralogist Axel Frederik Cronstedt (1712–65) in 1751, although nickel ores had long been known to German miners. The unalloyed metal is used in electroplating and as a catalyst in the hydrogenation of oils to form fats, notably in the manufacture of margarine. Nickel's chief use is in alloys, especially with iron, to which it imparts strength and resistance to corrosion, and with copper for coinage. The earth's core is believed to consist largely of metallic iron and nickel, and the metal is a component of many meteorites.

nickelodeon /ˌnɪkə'ləʊdɪən/ *n. US colloq.* a jukebox. [NICKEL + MELODEON]

nicker /'nɪkə(r)/ *n.* (*pl.* same) *Brit. sl.* a pound (in money). [20th c.: orig. unkn.]

Nicklaus /'nɪklaʊs, -ləs/, Jack William (b.1940), American golfer. Since the start of his professional career in 1962, he has won more than eighty tournaments: major titles include six wins in the PGA championship, four in the US Open, and three in the British Open.

nick-nack var. of KNICK-KNACK.

nickname /'nɪkneɪm/ *n. & v.* ● *n.* a familiar or humorous name given to a person or thing instead of or as well as the real name. ● *v.tr.* **1** give a nickname to. **2** call (a person or thing) by a nickname. [ME f. *ekename*, with *n* from *an* (cf. NEWT): *eke* = addition, f. OE *ēaca* (as EKE)]

Nicobar Islands see ANDAMAN AND NICOBAR ISLANDS.

nicol /'nɪk(ə)l/ *n.* (in full **nicol prism**) *Optics* a device for producing plane-polarized light, consisting of two pieces of cut calcite cemented together with Canada balsam. [William *Nicol*, Sc. physicist (1768–1851), its inventor]

Nicosia /ˌnɪkə'sɪə/ the capital of Cyprus; pop. (est. 1990) 171,000. Since 1974 it has been divided into Greek and Turkish sectors.

nicotiana /nɪˌkəʊʃɪ'ɑːnə, nɪˌkɒtɪ-/ *n.* = tobacco plant 2. [mod.L *nicotiana* (*herba*) tobacco plant, f. Jean *Nicot*, Fr. diplomat (1530–1600) who introduced tobacco to France]

nicotinamide /ˌnɪkə'tɪnəˌmaɪd/ *n. Biochem.* the amide of nicotinic acid, having a similar role in the diet, and important as a constituent of NAD. □ **nicotinamide adenine dinucleotide** see NAD.

nicotine /'nɪkəˌtiːn/ *n.* a colourless poisonous alkaloid present in tobacco. □ **nicotinism** *n.* **nicotinize** *v.tr.* (also **-ise**). [F f. NICOTIANA]

nicotinic acid /ˌnɪkə'tɪnɪk/ *n. Biochem.* a vitamin of the B complex, found in milk, liver, and yeast, a deficiency of which causes pellagra. Also called *niacin*.

nictitate /'nɪktɪˌteɪt/ *v.intr.* esp. *Zool.* close and open the eyes; blink or wink. □ **nictitating membrane** a clear membrane forming a third eyelid in amphibians, birds, and some other animals, that can be drawn across the eye to give protection without loss of vision. □ **nictitation** /ˌnɪktɪ'teɪʃ(ə)n/ *n.* [med.L *nictitare* frequent. of L *nictare* blink]

nide /naɪd/ *n.* a brood of pheasants. [F *nid* or L *nidus*: see NIDUS]

nidificate /'nɪdɪfɪˌkeɪt/ *v.intr.* = NIDIFY.

nidify /'nɪdɪˌfaɪ/ *v.intr.* (**-ies, -ied**) (of a bird) build a nest. □ **nidification** /ˌnɪdɪfɪ'keɪʃ(ə)n/ *n.* [L *nidificare* f. NIDUS]

nidus /'naɪdəs/ *n.* (*pl.* **nidi** /-daɪ/ or **niduses**) **1** a place in which an insect etc. deposits its eggs, or in which spores or seeds develop. **2** a place in which something is nurtured or developed. [L, rel. to NEST]

niece /niːs/ *n.* a daughter of one's brother or sister, or of one's brother-in-law or sister-in-law. [ME f. OF ult. f. L *neptis* granddaughter]

Niederösterreich /'niːdərˌøːstəraɪç/ the German name for LOWER AUSTRIA.

Niedersachsen /'niːdərˌzaks(ə)n/ the German name for LOWER SAXONY.

niello /nɪ'eləʊ/ *n.* (*pl.* **nielli** /-lɪ/ or **-os**) **1** a black composition of sulphur with silver, lead, or copper, for filling engraved lines in silver or other metal. **2 a** such ornamental work. **b** an object decorated with this. □ **nielloed** *adj.* [It. f. L *nigellus* dimin. of *niger* black]

nielsbohrium /niːlz'bɔːrɪəm/ *n.* a name proposed in the former Soviet Union for the chemical element unnilpentium (cf. HAHNIUM). The corresponding symbol is **Ns**. [Niels BOHR]

Nielsen /'niːls(ə)n/, Carl August (1865–1931), Danish composer. A major figure in the development of modern Scandinavian music, he gained his first success in 1888 with his *Little Suite* for string orchestra. His six symphonies (1890–1925) form the core of his achievement; other major works include the opera *Maskerade* (1906), concertos, and the organ work *Commotio* (1931).

Niemeyer /'niːˌmaɪə(r)/, Oscar (b.1907), Brazilian architect. He was an early exponent of modernist architecture in Latin America and was influenced by Le Corbusier, with whom he worked as part of the group which designed the Ministry of Education in Rio de Janeiro (1937–43). His most significant individual achievement was the design of the main public buildings of Brasilia (1950–60) within the master plan drawn up by Lúcio Costa.

Niemöller /'niːˌmɜːlə(r)/, Martin (1892–1984), German Lutheran pastor. He was ordained in 1924 after serving as a U-boat commander in the First World War. During the 1930s he was an outspoken opponent of Nazism and organized resistance to Hitler's attempts to control the German Church. Despite a prohibition he continued to preach and was eventually imprisoned in Sachsenhausen and Dachau concentration camps (1937–45). After the war he became known as an advocate of a united neutral Germany and nuclear disarmament.

Nietzsche /ˈniːtʃə/, Friedrich Wilhelm (1844–1900), German philosopher. His main period of creativity began in 1872 with the publication of his first book *The Birth of Tragedy* and lasted until 1889, when his mental instability developed into permanent insanity; other major works include *Thus Spake Zarathustra* (1883–5) and *Beyond Good and Evil* (1886). His work is open to different interpretations, but his influence can be seen in existentialism and in the work of such varied figures as Michel Foucault, Martin Heidegger, and George Bernard Shaw. The principal features of his writings are contempt for Christianity, with its compassion for the weak, and exaltation of the 'will to power' and of the *Übermensch* (superman), superior to ordinary morality, who will replace the Christian ideal. He divided humankind into a small, dominant 'master-class' and a large, dominated 'herd' – a thesis which was taken up in a debased form by the Nazis after Nietzsche's death.

niff /nɪf/ n. & v. Brit. colloq. ● n. a smell, esp. an unpleasant one. ● v.intr. smell, stink. □ **niffy** adj. (**niffier, niffiest**). [orig. dial.]

Niflheim /ˈnɪv(ə)lˌheɪm, -ˌhaɪm/ Scand. Mythol. the underworld, a place of eternal cold, darkness, and mist inhabited by those who died of old age or illness. [ON *Niflheimr,* = world of mist]

nifty /ˈnɪftɪ/ adj. (**niftier, niftiest**) colloq. **1** clever, adroit. **2** smart, stylish. □ **niftily** adv. **niftiness** n. [19th c.: orig. uncert.]

nigella /naɪˈdʒelə/ n. a ranunculaceous plant of the genus *Nigella*, with showy flowers and finely cut leaves; esp. love-in-a-mist, *N. damascena*. [mod.L f. L *nigellus*, dimin. of *niger* black]

Niger /ˈnaɪdʒə(r)/ **1** a river in NW Africa, which rises on the north-east border of Sierra Leone and flows in a great arc for 4,100 km (2,550 miles) north-east to Mali, then south-east through western Niger and Nigeria, before turning southwards to empty through a great delta into the Gulf of Guinea. **2** /French niʒer/ a landlocked country in West Africa, on the southern edge of the Sahara; pop. (est. 1991) 7,909,000: languages, French (official), Hausa, and other West African languages: capital, Niamey. The region was not explored by Europeans until the 18th and 19th centuries. A French colony (part of French West Africa) from 1922, it became an autonomous republic within the French Community in 1958 and fully independent in 1960.

Niger-Congo /ˌnaɪdʒəˈkɒŋɡəʊ/ n. the largest group of languages in Africa, named after the rivers Niger and Zaire (Congo). It includes the languages spoken by most of the indigenous peoples of western, central, and southern Africa: the important Bantu group, the Mande group (West Africa), the Voltaic group (Burkina), and the Kwa group (Nigeria), which includes Yoruba and Ibo.

Nigeria /naɪˈdʒɪərɪə/ a country on the coast of West Africa, bordered by the River Niger to the north; pop. (1991) 88,514,500; languages, English (official), Hausa, Ibo, Yoruba, and others; capital, Abuja. The country was the site of highly developed kingdoms in the Middle Ages, and the coast was explored by the Portuguese in the 15th century. The area of the Niger delta came gradually under British influence, particularly during the period between the annexation of Lagos in 1861 and the unification of Nigeria into a single colony in 1914. Independence came in 1960 and the state became a federal republic within the Commonwealth in 1963. A civil war with the breakaway eastern area of Biafra (1967–70) ended in victory for the federal forces. Since the discovery of oil in the 1960s and 70s Nigeria has become a major exporter of oil. In 1995, Nigeria was suspended from the Commonwealth following the military government's execution (for murder) of human rights campaigners protesting against the exploitation of land for oil extraction. □ **Nigerian** adj. & n.

niggard /ˈnɪɡəd/ n. & adj. ● n. a mean or stingy person. ● adj. archaic = NIGGARDLY. [ME, alt. f. earlier (obs.) *nigon*, prob. f. Scand. orig.: cf. NIGGLE]

niggardly /ˈnɪɡədlɪ/ adj. & adv. ● adj. **1** stingy, parsimonious. **2** meagre, scanty. ● adv. in a stingy or meagre manner. □ **niggardliness** n.

nigger /ˈnɪɡə(r)/ n. offens. **1** a black person. **2** a dark-skinned person. □ **a nigger in the woodpile** a hidden cause of trouble or inconvenience. [earlier *neger* f. F *nègre* f. Sp. *negro* NEGRO]

niggle /ˈnɪɡ(ə)l/ v. & n. ● v. **1** intr. be over-attentive to details. **2** intr. find fault in a petty way. **3** tr. colloq. irritate; nag pettily. ● n. a trifling complaint or criticism; a worry or annoyance. [app. of Scand. orig.: cf. Norw. *nigla*]

niggling /ˈnɪɡlɪŋ/ adj. **1** troublesome or irritating in a petty way. **2** trifling or petty. □ **nigglingly** adv.

nigh /naɪ/ adv., prep., & adj. archaic or dial. near. □ **nigh on** nearly, almost. [OE *nēh, nēah*]

night /naɪt/ n. **1** the period of darkness between one day and the next; the time from sunset to sunrise. **2** nightfall (*shall not reach home before night*). **3** the darkness of night (*as black as night*). **4** a night or evening appointed for some activity, or spent or regarded in a certain way (*last night of the Proms; a great night out*). □ **night-blindness** = NYCTALOPIA. **night fighter** an aeroplane used for interception at night. **night heron** a nocturnal heron of the genus *Nycticorax* or *Gorsachius*. **night-life** entertainment available at night in a town. **night-light** a dim light kept burning in a bedroom at night. **night-long** throughout the night. **night-monkey** = DOUROUCOULI. **night nurse** a nurse on duty during the night. **night-owl** colloq. a person habitually active at night. **night safe** a safe with access from the outer wall of a bank for the deposit of money etc. when the bank is closed. **night school** an institution providing evening classes for those working by day. **night shift** a shift of workers employed during the night. **night-soil** excrement removed at night from cesspools etc., esp. for use as manure. **night-time** the time of darkness. **night-watchman 1** a person whose job is to keep watch by night. **2** Cricket an inferior batsman sent in near the close of a day's play. □ **nightless** adj. [OE *neaht, niht* f. Gmc]

nightbird /ˈnaɪtbɜːd/ n. a person habitually active at night.

nightcap /ˈnaɪtkæp/ n. **1** hist. a cap worn in bed. **2** a hot or alcoholic drink taken at bedtime.

nightclothes /ˈnaɪtkləʊðz/ n. clothes worn in bed.

nightclub /ˈnaɪtklʌb/ n. a club that is open at night and provides refreshment and entertainment.

nightdress /ˈnaɪtdres/ n. a woman's or child's loose garment worn in bed.

nightfall /ˈnaɪtfɔːl/ n. the onset of night; the end of daylight.

nightgown /ˈnaɪtɡaʊn/ n. **1** = NIGHTDRESS. **2** hist. a dressing-gown.

nighthawk /ˈnaɪthɔːk/ n. **1** a nocturnal prowler, esp. a thief. **2** an American nightjar, esp. of the genus *Chordeiles*.

nightie /ˈnaɪtɪ/ n. colloq. a nightdress. [abbr.]

Nightingale /ˈnaɪtɪŋˌɡeɪl/, Florence (1820–1910), English nurse and medical reformer. She became famous during the Crimean War for her attempts to publicize the state of the army's medical arrangements and improve the standard of care. In 1854 she took a party of nurses to the army hospital at Scutari, where she improved sanitation and medical procedures, thereby achieving a dramatic reduction in the mortality rate; she became known as the 'Lady of the Lamp' for her nightly rounds. She returned to England in 1856 and devoted the rest of her life to attempts to improve public health and hospital care.

nightingale /ˈnaɪtɪŋˌɡeɪl/ n. a small brownish migratory thrush, *Luscinia megarhynchos*, the male of which has a melodious song that is often heard at night. [OE *nihtegala* (whence obs. *nightgale*) f. Gmc: for -n- cf. FARTHINGALE]

nightjar /ˈnaɪtdʒɑː(r)/ n. a nocturnal insectivorous bird of the family Caprimulgidae, with cryptic grey-brown plumage; esp. *Caprimulgus europaeus*, which has a distinctive churring call. Also called *goatsucker*.

Night Journey Islam the journey through the air made by Muhammad, guided by the archangel Gabriel. They flew first to Jerusalem, where Muhammad prayed with earlier prophets including Abraham, Moses, and Jesus, before ascending to heaven and entering the presence of Allah.

nightly /ˈnaɪtlɪ/ adj. & adv. ● adj. **1** happening, done, or existing in the night. **2** recurring every night. ● adv. every night. [OE *nihtlic* (as NIGHT)]

nightmare /ˈnaɪtmeə(r)/ n. **1** a frightening or unpleasant dream. **2** colloq. a terrifying or very unpleasant experience or situation. **3** a haunting or obsessive fear. □ **nightmarish** adj. **nightmarishly** adv. [an evil spirit (incubus) once thought to lie on and suffocate sleepers: OE *mære* incubus]

night of the long knives n. **1** a treacherous massacre, e.g. (according to legend) of the Britons by Hengist in 472. **2** the massacre of Ernst Röhm and his Brownshirt associates on Hitler's orders on 29–30 June 1934. **3** a similar ruthless or decisive action.

nightshade /ˈnaɪtʃeɪd/ n. a poisonous plant of the family Solanaceae; esp. one of the genus *Solanum*, which includes *black nightshade* (*S. nigrum*), with black berries, and *woody nightshade* (*S. dulcamara*), with red berries. □ **deadly nightshade** a highly poisonous solanaceous plant, *Atropa belladonna*, with drooping purple flowers and black cherry-like fruit (also called *belladonna*, *dwale*). [OE *nihtscada* app. formed as NIGHT + SHADE, prob. with ref. to its poisonous properties]

nightshirt /'naɪtʃɜːt/ n. a long shirt worn in bed.

nightspot /'naɪtspɒt/ n. a nightclub.

nightstick /'naɪtstɪk/ n. US a policeman's truncheon.

nigrescent /nɪ'gres(ə)nt/ adj. blackish; becoming black. □ **nigrescence** n. [L *nigrescere* grow black f. *niger nigri* black]

nigritude /'nɪgrɪˌtjuːd/ n. blackness. [L *nigritudo* (as NIGRESCENT)]

nihilism /'naɪˌlɪz(ə)m, 'naɪhɪ-/ n. **1** negative doctrines or the rejection of all religious and moral principles, often involving a general sense of despair coupled with the belief that life is devoid of meaning. **2** *Philos.* an extreme form of scepticism maintaining that nothing in the world has a real existence. **3** the doctrine of a Russian extreme revolutionary party in the 19th and early 20th centuries, finding nothing to approve of in the established order or social and political institutions. □ **nihilist** n. **nihilistic** /ˌnaɪɪ'lɪstɪk, ˌnaɪhɪ-/ adj. [L *nihil* nothing]

nihility /naɪ'hɪlɪtɪ/ n. (pl. **-ies**) **1** non-existence, nothingness. **2** a mere nothing; a trifle. [med.L *nihilitas* (as NIHILISM)]

nihil obstat /ˌnaɪhɪl 'ɒbstæt/ n. **1** *RC Ch.* a certificate that a book is not open to objection on doctrinal or moral grounds. **2** an authorization or official approval. [L, = nothing hinders]

Niigata /ˌniːɪ'gɑːtə/ an industrial port in central Japan, on the NW coast of the island of Honshu; pop. (1990) 486,090.

Nijinsky /nɪ'dʒɪnskɪ/, Vaslav (Fomich) (1890–1950), Russian ballet-dancer and choreographer. From 1909 he was the leading dancer with Diaghilev's Ballets Russes, giving celebrated performances in the classics and Fokine's ballets, including *Le Spectre de la rose* (1911). He was encouraged by Diaghilev to take up choreography, resulting in productions of Debussy's *L'Après-midi d'un faune* (1912) and *Jeux* (1913) and Stravinsky's *The Rite of Spring* (1913). These ballets foreshadowed many developments of later avant-garde choreography. Nijinsky's career declined after his separation from Diaghilev and was brought to a premature end by schizophrenia.

Nijmegen /'naɪˌmeɪgən/ an industrial town in the eastern Netherlands, south of Arnhem; pop. (1991) 145,780.

-nik /nɪk/ suffix forming nouns denoting a person associated with a specified thing or quality (*beatnik*; *refusenik*). [Russ. (as SPUTNIK) and Yiddish]

Nike /'naɪkɪ/ *Gk Mythol.* the goddess of victory. (See also WINGED VICTORY.) [Gk *nikē* victory]

Nikkei index /'nɪkeɪ/ a figure indicating the relative price of a representative set of shares on the Tokyo Stock Exchange, calculated since 1974 by Japan's principal financial daily newspaper. The index was originally calculated (from 1949) by the Tokyo Stock Exchange itself. [Jap. acronym, f. *Nihon Kezai Shimbun* Japanese Economic Journal]

Nikolaev see MYKOLAYIV.

nil /nɪl/ n. nothing; no number or amount (esp. *Brit.* as a score in games). [L, = *nihil* nothing]

nil desperandum /ˌnɪl despəˈrændəm/ int. do not despair; never despair. [L *nil desperandum* (*Teucro duce*) no need to despair (with Teucer as your leader), f. Horace *Odes*]

Nile[1] /naɪl/ a river in eastern Africa, the longest river in the world, which rises in east central Africa near Lake Victoria and flows 6,695 km (4,160 miles) generally northwards through Uganda, Sudan, and Egypt to empty through a large delta into the Mediterranean. It flows from Lake Victoria to Lake Albert, in Uganda, as the *Victoria Nile*. As the *Albert Nile* it flows onwards to the Ugandan–Sudanese border, where it is known as the *White Nile*. At Khartoum, after its confluence with the *Blue Nile* (which arises in NW Ethiopia), it continues northwards as the Nile. Seasonal flooding in Egypt has been controlled by the construction of a dam at Aswan, which also provides hydro-electric power.

Nile[2] /naɪl/ n. & adj. ● n. (in full **Nile-blue**, **Nile-green**) pale greenish-blue or green. ● adj. (hyphenated when *attrib.*) of this colour.

Nile, Battle of the see ABOUKIR BAY, BATTLE OF.

nilgai /'nɪlgaɪ/ n. a large Indian antelope, *Boselaphus tragocamelus*, the male of which has short horns. [Hindi *nīlgāī* f. *nīl* blue + *gāī* cow]

Nilgiri Hills /'nɪlgɪrɪ/ a range of hills in southern India, in western Tamil Nadu. They form a branch of the Western Ghats.

Nilotic /naɪ'lɒtɪk/ adj. **1** of or relating to the Nile or the Nile region of Africa or its inhabitants. **2** of or relating to a group of languages spoken in Egypt and Sudan, and further south in Kenya and Tanzania.

The western group includes Luo (Kenya), Dinka (Sudan), and Lango (Uganda); the eastern group includes Masai (Kenya and Tanzania), and Turkana (Kenya). [L *Niloticus* f. Gk *Neilōtikos* f. *Neilos* Nile]

Nilsson /'nɪls(ə)n/, (Märta) Birgit (b.1918), Swedish operatic soprano. She made her Swedish début in 1946, gaining international success in the 1950s. She was particularly noted for her interpretation of Wagnerian roles, and sang at the Bayreuth Festivals between 1953 and 1970. Her repertoire also included the operas of Richard Strauss and Verdi.

nim /nɪm/ n. a game in which two players must alternately take one or more objects from one of several heaps and seek either to avoid taking or to take the last remaining object. [20th c.: perh. f. archaic *nim* take (as NIMBLE), or G *nimm* imper. of *nehmen* take]

nimble /'nɪmb(ə)l/ adj. (**nimbler**, **nimblest**) **1** quick and light in movement or action; agile. **2** (of the mind) quick to comprehend; clever, versatile. □ **nimbleness** n. **nimbly** adv. [OE *nǣmel* quick to seize f. *niman* take f. Gmc, with *-b-* as in THIMBLE]

nimbostratus /ˌnɪmbəʊ'strɑːtəs, -'streɪtəs/ n. (often *attrib.*) *Meteorol.* a cloud type forming a dense grey layer at low altitude, from which precipitation commonly falls. [NIMBUS + STRATUS]

nimbus /'nɪmbəs/ n. (pl. **nimbi** /-baɪ/ or **nimbuses**) **1 a** a bright cloud surrounding a deity, person, or thing. **b** the halo of a saint etc. **2** *Meteorol.* a rain-cloud. □ **nimbused** adj. [L, = cloud, aureole]

Nîmes /niːm/ a city in southern France; pop. (1990) 133,600. It is noted for its many well-preserved Roman remains. It also gave its name to the fabric denim, which was originally produced there.

niminy-piminy /ˌnɪmɪnɪ'pɪmɪnɪ/ adj. feeble, affected; lacking in vigour. [cf. MIMINY-PIMINY, NAMBY-PAMBY]

Nimrod /'nɪmrɒd/ n. a great hunter or sportsman. [Heb. *Nimrōd* valiant: see Gen. 10:8–9]

Nimrud /'nɪmrʊd/ an ancient Mesopotamian city, situated on the eastern bank of the Tigris south of Nineveh, near the modern city of Mosul. Founded in 1250 BC, it was inaugurated as the Assyrian capital in 879 BC by Ashurnasirpal II (883–859 BC). It was later supplanted by Khorsabad with the accession of Sargon II (722–705 BC) and finally destroyed by the Medes in 612 BC. British excavations under Sir Austen Henry Layard (1817–94) and later Sir Max Mallowan (1904–78) uncovered palaces of the Assyrian kings with monumental sculptured reliefs, carved ivory furniture inlays, and metalwork. The city was known in biblical times as Calah (Gen. 10:11).

nincompoop /'nɪŋkəmˌpuːp/ n. a simpleton; a fool. [17th c.: orig. unkn.]

nine /naɪn/ n. & adj. ● n. **1** one more than eight, or one less than ten; the sum of five units and four units. **2** a symbol for this (9, ix, IX). **3** a size etc. denoted by nine. **4** a set or team of nine individuals. **5** nine o'clock. **6** a card with nine pips. **7** (**the Nine**) the nine muses. ● adj. that amount to nine. □ **dressed to** (or *Brit.* **up to**) **the nines** dressed very elaborately. **nine days' wonder** a person or thing that is briefly famous. **nine times out of ten** nearly always. **nine to five** a designation of typical office hours. [OE *nigon* f. Gmc]

ninefold /'naɪnfəʊld/ adj. & adv. **1** nine times as much or as many. **2** consisting of nine parts.

ninepin /'naɪnpɪn/ n. **1** (in *pl.*; usu. treated as *sing.*) the usual form of skittles, using nine pins. (See BOWLING.) **2** a pin used in this game.

nineteen /naɪn'tiːn/ n. & adj. ● n. **1** one more than eighteen, nine more than ten. **2** the symbol for this (19, xix, XIX). **3** a size etc. denoted by nineteen. ● adj. that amount to nineteen. □ **talk nineteen to the dozen** see DOZEN. □ **nineteenth** adj. & n. [OE *nigontÿne*]

ninety /'naɪntɪ/ n. & adj. ● n. (pl. **-ies**) **1** the product of nine and ten. **2** a symbol for this (90, xc, XC). **3** (in *pl.*) the numbers from 90 to 99, esp. the years of a century or of a person's life. ● adj. that amount to ninety. □ **ninety-first, -second**, etc. the ordinal numbers between ninetieth and hundredth. **ninety-one, -two**, etc. the cardinal numbers between ninety and a hundred. □ **ninetieth** adj. & n. **ninetyfold** adj. & adv. [OE *nigontig*]

Nineveh /'nɪnɪvə/ an ancient city located on the east bank of the Tigris, opposite the modern city of Mosul. It was the oldest city of the ancient Assyrian empire and its capital during the reign of Sennacherib until it was destroyed by a coalition of Babylonians and Medes in 612 BC. A famous archaeological site, it was first excavated by the French in 1820 and later by the British and is noted for its monumental Neo-Assyrian palace, library, and statuary as well as for its crucial sequence of prehistoric pottery.

Ningxia /nɪŋˈʃjɑ:/ (also **Ningsia**) an autonomous region of north central China; capital, Yinchuan.

Ninian, St /ˈnɪnɪən/ (*c.*360–*c.*432), Scottish bishop and missionary. According to Bede he founded a church at Whithorn in SW Scotland (*c.*400) and from there evangelized the southern Picts.

ninja /ˈnɪndʒə/ *n.* a person skilled in ninjutsu. [Jap.]

ninjutsu /nɪnˈdʒʊtsu:/ *n.* one of the Japanese martial arts, characterized by stealthy movement and camouflage. [Jap.]

ninny /ˈnɪnɪ/ *n.* (*pl.* **-ies**) a foolish or simple-minded person. [perh. f. *innocent*]

ninon /ˈni:nɒn/ *n.* a lightweight silk dress fabric. [F]

ninth /namθ/ *n.* & *adj.* ● *n.* **1** the position in a sequence corresponding to the number 9 in the sequence 1–9. **2** something occupying this position. **3** each of nine equal parts of a thing. **4** *Mus.* **a** an interval or chord spanning nine consecutive notes in the diatonic scale (e.g. C to D an octave higher). **b** a note separated from another by this interval. ● *adj.* that is the ninth. □ **ninthly** *adv.*

Niobe /ˈnaɪəbɪ/ *Gk Mythol.* the daughter of Tantalus, and mother of a large family. Apollo and Artemis, enraged because she boasted herself superior to their mother Leto, slew her children. Niobe herself was turned into a stone, and her tears into streams that trickled from it.

niobium /naɪˈəʊbɪəm/ *n.* a silver-grey metallic chemical element (atomic number 41; symbol **Nb**). Niobium was discovered in an American mineral in 1801, but not until 1844 was it clearly distinguished from tantalum by the German chemist Heinrich Rose (1795–1864). It was known by the alternative name *columbium* for many years, especially in the US. A transition metal, niobium is chiefly found in the minerals columbite and tantalite. The metal is used in superconducting alloys. [NIOBE: so called because found in TANTALITE]

Nip /nɪp/ *n. sl. offens.* a Japanese person. [abbr. of NIPPONESE]

nip[1] /nɪp/ *v.* & *n.* ● *v.* (**nipped**, **nipping**) **1** *tr.* pinch, squeeze, or bite sharply. **2** *tr.* (often foll. by *off*) remove by pinching etc. **3** *tr.* (of the cold, frost, etc.) cause pain or harm to. **4** *intr.* (foll. by *in, out*, etc.) *Brit. colloq.* go nimbly or quickly. **5** *tr. US sl.* steal, snatch. ● *n.* **1 a** a pinch, a sharp squeeze. **b** a bite. **2 a** a biting cold. **b** a check to vegetation caused by this. □ **nip and tuck** *US* neck and neck. **nip in the bud** suppress or destroy (esp. an idea) at an early stage. □ **nipping** *adj.* [ME, prob. of LG or Du. orig.]

nip[2] /nɪp/ *n.* & *v.* ● *n.* a small quantity of spirits. ● *v.intr.* (**nipped**, **nipping**) drink spirits. [prob. abbr. of *nipperkin* small measure: cf. LG, Du. *nippen* to sip]

nipa /ˈni:pə/ *n.* **1** a tropical Asian and Australian palm tree, *Nipa fruticans*, with a creeping trunk and large feathery leaves. **2** an alcoholic drink made from its sap. [Sp. & Port. f. Malay *nīpah*]

nipper /ˈnɪpə(r)/ *n.* **1** a person or thing that nips. **2** the claw of a crab, lobster, etc. **3** *Brit. colloq.* a young child. **4** (in *pl.*) any tool for gripping or cutting, e.g. forceps or pincers. **5** *Austral.* a burrowing prawn of the order Thalassinidea, used as bait. Also called *yabby*.

nipple /ˈnɪp(ə)l/ *n.* **1 a** a small projection in which the mammary ducts of a female mammal terminate and from which milk is secreted for the young. **b** the analogous structure in the male. **2** the teat of a feeding-bottle. **3** a device like a nipple in function, e.g. the tip of a grease-gun. **4** a nipple-like protuberance. **5** *US* a short section of pipe with a screw-thread at each end for coupling. [16th c., also *neble, nible*, perh. dimin. f. *neb*]

nipplewort /ˈnɪp(ə)l,wɜ:t/ *n.* a slender composite weed, *Lapsana communis*, with small yellow flowers.

Nipponese /ˌnɪpəˈni:z/ *n.* & *adj.* ● *n.* (*pl.* same) a Japanese person. ● *adj.* Japanese. [Jap. *Nippon* Japan, f. *nip* sun + *pon* origin, lit. 'place the sun comes from' (i.e. land of the rising sun)]

nippy /ˈnɪpɪ/ *adj.* (**nippier, nippiest**) *colloq.* **1** quick, nimble, active. **2** chilly, cold. □ **nippily** *adv.* [NIP[1] + -Y[1]]

NIREX /ˈnaɪəreks/ *abbr.* (in the UK) Nuclear Industry Radioactive Waste Executive.

Niro, Robert De, see DE NIRO.

nirvana /nɜ:ˈvɑ:nə, nɪəˈvɑ:nə/ *n.* the final goal of Buddhism, a transcendent state in which there is neither suffering, desire, nor sense of self (*atman*), with release from the effects of karma. [Skr. *nirvāṇa* extinction (i.e. the extinction of illusion, since suffering, desire, and self are all illusions) f. *nirvā* be extinguished f. *nis* out + *vā-* to blow]

Niš /ni:ʃ/ (also **Nish**) an industrial city in SE Serbia, on the Nišava river near its confluence with the Morava; pop. (1981) 230,710. The city,

commanding the principal route by river between Europe and the Aegean, was for centuries a strategic stronghold. Dominated by the Turks for 500 years, it fell to the Serbs in 1877. It was the birthplace of Constantine the Great (*c.*274 AD).

Nisan /ˈnɪs(ə)n, ˈnɪsɑ:n/ *n.* (in the Jewish calendar) the seventh month of the civil and first of the religious year, usually coinciding with parts of March and April. [Heb. *nīsān*]

nisei /ni:ˈseɪ/ *n. US* an American whose parents were immigrants from Japan. [Jap., lit. 'second generation']

Nish see NIŠ.

nisi /ˈnaɪsaɪ/ *adj. Law* that takes effect only on certain conditions (*decree nisi*). [L, = 'unless']

Nissen hut /ˈnɪs(ə)n/ *n.* a tunnel-shaped hut of corrugated iron with a cement floor. [Lt.-Col. Peter Norman *Nissen*, British military engineer (1871–1930), its inventor]

nit[1] /nɪt/ *n.* **1** the egg or young form of a louse or other parasitic insect, esp. of human head lice or body lice. **2** *Brit. sl.* a stupid person. □ **nit-pick** *colloq.* indulge in nit-picking. **nit-picker** *colloq.* a person who nit-picks. **nit-picking** *n.* & *adj. colloq.* fault-finding in a petty manner. [OE *hnitu* f. WG]

nit[2] /nɪt/ *int. Austral. sl.* used as a warning that someone is approaching. □ **keep nit** keep watch; act as guard. [19th c.: orig. unkn.: cf. NIX[3]]

niter *US* var. of NITRE.

Niterói /ˌni:təˈrɔɪ/ an industrial port on the coast of SE Brazil, on Guanabara Bay opposite the city of Rio de Janeiro; pop. (1990) 455,200.

nitinol /ˈnɪtɪ,nɒl/ *n.* an alloy of nickel and titanium. [*Ni* + *Ti* + *Naval Ordnance Laboratory*, Maryland, US]

nitrate *n.* & *v.* ● *n.* /ˈnaɪtreɪt/ **1** *Chem.* a salt or ester of nitric acid. **2** potassium or sodium nitrate when used as a fertilizer. ● *v.tr.* /naɪˈtreɪt/ *Chem.* treat, combine, or impregnate with nitric acid. □ **nitration** /naɪˈtreɪʃ(ə)n/ *n.* [F (as NITRE, -ATE[1])]

nitre /ˈnaɪtə(r)/ *n.* (*US* **niter**) saltpetre, potassium nitrate. [ME f. OF f. L *nitrum* f. Gk *nitron*, of Semitic orig.]

nitric /ˈnaɪtrɪk/ *adj. Chem.* of or containing nitrogen, esp. in the pentavalent state. □ **nitric acid** a corrosive poisonous acid (chem. formula: HNO_3), which is a colourless or pale yellow fuming liquid in the pure state. **nitric oxide** a colourless gas (chem. formula: NO). [F *nitrique* (as NITRE)]

nitride /ˈnaɪtraɪd/ *n. Chem.* a binary compound of nitrogen with a more electropositive element. [NITRE + -IDE]

nitrify /ˈnaɪtrɪ,faɪ/ *v.tr.* (**-ies, -ied**) **1** impregnate with nitrogen. **2** convert (nitrogen, usu. in the form of ammonia) into nitrites or nitrates. □ **nitrifiable** *adj.* **nitrification** /ˌnaɪtrɪfɪˈkeɪʃ(ə)n/ *n.* [F *nitrifier* (as NITRE)]

nitrile /ˈnaɪtrɪl, -traɪl/ *n. Chem.* an organic compound consisting of an alkyl radical bound to a cyanide radical.

nitrite /ˈnaɪtraɪt/ *n. Chem.* a salt or ester of nitrous acid.

nitro- /ˈnaɪtrəʊ/ *comb. form* of or containing nitric acid, nitre, or nitrogen, esp. as the monovalent group -NO_2. [Gk (as NITRE)]

nitrobenzene /ˌnaɪtrəʊˈbenzi:n/ *n. Chem.* a yellow oily liquid (chem. formula: $C_6H_5NO_2$) made by the nitration of benzene and used to make aniline etc.

nitrocellulose /ˌnaɪtrəʊˈseljʊˌləʊz, -ˌləʊs/ *n.* a highly flammable material made by treating cellulose with concentrated nitric acid, used in the manufacture of explosives and celluloid. Also called *cellulose nitrate*.

nitrogen /ˈnaɪtrədʒən/ *n.* a colourless odourless unreactive gaseous chemical element (atomic number 7; symbol **N**). (See note below.) □ **nitrogen dioxide** a reddish-brown poisonous gas (chem. formula: NO_2). **nitrogen fixation** a chemical process in which atmospheric nitrogen is assimilated into organic compounds, esp. naturally by certain organisms as part of the nitrogen cycle. □ **nitrogenous** /naɪˈtrɒdʒɪnəs/ *adj.* [F *nitrogène* (as NITRO-, -GEN)]

▪ Nitrogen forms about 78 per cent of the earth's atmosphere, and was discovered by the British chemist and botanist Daniel Rutherford (1749–1819) in 1772. Liquid nitrogen (made by distilling liquid air) is used as a coolant; the gas is used to provide an inert atmosphere. Nitrogen forms many important compounds including nitric acid, ammonia, and nitrogenous fertilizers. Nitrogen is essential to life, being a major constituent of proteins and other biological molecules.

nitrogen cycle *n. Ecol.* the cycle of processes whereby nitrogen is absorbed from and replaced in the atmosphere. Nitrogen in air is converted to ammonia and then into other compounds (fixation) mainly by the action of certain bacteria. These compounds are taken up from the soil by plants, which are eaten by animals. Nitrogen is returned to the atmosphere when animals die and their tissues decay.

nitroglycerine /ˌnaɪtrəʊˈglɪsəriːn/ n. (also **nitroglycerin** /-rɪn/) an explosive yellow liquid made by reacting glycerol with a mixture of concentrated sulphuric and nitric acids.

nitrosamine /naɪˈtrəʊsəˌmiːn/ n. Chem. any of a group of carcinogenic organic substances containing the chemical group :N–N:O.

nitrous /ˈnaɪtrəs/ adj. Chem. of, like, or impregnated with nitrogen, esp. in the trivalent state. □ **nitrous acid** a weak acid (chem. formula: HNO_2), existing only in solution and in the gas phase. **nitrous oxide** a colourless gas (chem. formula: N_2O) used as an anaesthetic (cf. *laughing-gas*) and as an aerosol propellant. [L *nitrosus* (as NITRE), partly through F *nitreux*]

nitty-gritty /ˌnɪtɪˈgrɪtɪ/ n. sl. the realities or practical details of a matter. [20th c.: orig. uncert.]

nitwit /ˈnɪtwɪt/ n. colloq. a stupid person. □ **nitwittery** /nɪtˈwɪtərɪ/ n. [perh. f. NIT[1] + WIT[1]]

nitwitted /ˈnɪtˌwɪtɪd/ adj. stupid. □ **nitwittedness** /nɪtˈwɪtɪdnɪs/ n.

Niue /nɪˈuːeɪ/ an island territory in the South Pacific to the east of Tonga; pop. (1986) 2,530; languages, English (official), local Malayo-Polynesian; capital, Alofi. Annexed by New Zealand in 1901, the island achieved self-government in free association with New Zealand in 1974. Niue is the largest coral island in the world.

Nivernais /ˌniːvəˈneɪ/ a former duchy and province of central France. Its capital was the city of Nevers.

nix[1] /nɪks/ n. & v. sl. ● n. **1** nothing. **2** a denial or refusal. ● v.tr. **1** cancel. **2** reject. [G, colloq. var. of *nichts* nothing]

nix[2] /nɪks/ n. (fem. **nixie** /ˈnɪksɪ/) a water-elf. [G fem. *Nixe*]

nix[3] /nɪks/ int. Brit. sl. giving warning to confederates etc. that a person in authority is approaching. [19th c.: perh. = NIX[1]]

Nixon /ˈnɪks(ə)n/, Richard (Milhous) (1913–94), American Republican statesman, 37th President of the US 1969–74. He served as Vice-President under Eisenhower (1953–61), narrowly losing to John F. Kennedy in the 1960 presidential election. In his first term of office he sought to resolve the Vietnam War; the negotiations were brought to a successful conclusion by his Secretary of State, Henry Kissinger, in 1973. Nixon also restored Sino-American diplomatic relations by his visit to China in 1972. He was elected for a second term in November of that year, but it soon became clear that he was implicated in the Watergate scandal, and in 1974 he became the first President to resign from office, taking this action shortly before impeachment proceedings began.

Nizari /nɪˈzɑːrɪ/ n. a member of a Muslim sect that split from the Ismaili branch in 1094 over disagreement about the succession to the caliphate. Based in Persia, the Nizaris were known to those whom they opposed as *Hashshishin* or 'Assassins' (see ASSASSIN). Their power was broken in the 13th century, and the community survived as a minor sect until reorganized under the Aga Khan in the 19th century. The majority then moved to the Indian subcontinent, although a much smaller remnant of the original Nizaris remains independent in Syria.

Nizhni Novgorod /ˌniːʒnɪ ˈnɒvɡəˌrɒd/ a river port in European Russia on the Volga; pop. (1990) 1,443,000. Between 1932 and 1991 it was named Gorky after the writer Maxim Gorky, who was born there.

Nizhni Tagil /ˌniːʒnɪ təˈɡiːl/ an industrial and metal-mining city in central Russia, in the Urals north of Ekaterinburg; pop. (1990) 440,000.

NJ abbr. US New Jersey (also in official postal use).

Nkomo /əŋˈkəʊməʊ/, Joshua (Mqabuko Nyongolo) (b.1917), Zimbabwean statesman. In 1961 he became the leader of ZAPU; in 1976 he formed the Patriotic Front with Robert Mugabe, leader of ZANU, and was appointed to a Cabinet post in Mugabe's government of 1980. Dismissed from his post in 1982, he returned to the Cabinet in 1988, when ZANU and ZAPU agreed to merge, and became Vice-President in 1990.

Nkrumah /əŋˈkruːmə/, Kwame (1909–72), Ghanaian statesman, Prime Minister 1957–60, President 1960–66. The leader of the non-violent struggle for the Gold Coast's independence, he became first Prime Minister of the country when it gained independence as Ghana in 1957. He declared Ghana a republic in 1960 and proclaimed himself President for life in 1964, banning all opposition parties; Nkrumah's dictatorial methods seriously damaged Ghana's economy and eventually led to his overthrow in a military coup.

NKVD the secret police agency in the former USSR responsible from 1934 for internal security and the labour prison camps, having absorbed the functions of the former OGPU. Mainly concerned with political offenders, it was notably used for Stalin's purges, and merged with the MVD in 1946. [Russ. abbr., = People's Commissariat of Internal Affairs]

NM abbr. US New Mexico (in official postal use).

nm abbr. **1** nautical mile. **2** nanometre.

N.Mex. abbr. New Mexico.

NMR abbr. (also **nmr**) nuclear magnetic resonance.

NNE abbr. north-north-east.

NNW abbr. north-north-west.

No[1] symb. Chem. the element nobelium.

No[2] var. of NOH.

No. abbr. **1** number. **2** US North. [sense 1 f. L *numero*, ablat. of *numerus* number]

no[1] /nəʊ/ adj. **1** not any (*there is no excuse; no circumstances could justify it; no two of them are alike*). **2** not a, quite other than (*is no fool; is no part of my plan; caused no slight inconvenience*). **3** hardly any (*is no distance; did it in no time*). **4** used elliptically as a slogan, notice, etc., to forbid, reject, or deplore the thing specified (*no parking; no surrender*). □ **by no means** see MEANS. **no-account** unimportant, worthless. **no-ball** Cricket n. an unlawfully delivered ball (counting one as an extra to the batting side if not otherwise scored from). ● v.tr. pronounce (a bowler) to have bowled a no-ball. **no-claim** (or **-claims**) **bonus** a reduction of the insurance premium charged when the insured has not made a claim under the insurance during an agreed preceding period. **no date** (of a book etc.) not bearing a date of publication etc. **no dice** see DICE. **no doubt** see DOUBT. **no end** see END. **no entry** (of a notice) prohibiting vehicles or persons from entering a road or place. **no-fault** esp. N. Amer. involving no fault or blame, esp. designating an insurance policy that is valid regardless of whether the policyholder was at fault. **no fear** see FEAR. **no-frills** lacking ornament or embellishment. **no go** impossible, hopeless. **no-go area** an area forbidden to unauthorized people. **no good** see GOOD. **no-good** see GOOD. **no-hitter** Baseball a match in which a pitcher yields no hits. **no-hoper** sl. a useless person. **no joke** see JOKE. **no joy** see JOY n. 3. **no little** see LITTLE. **no man** no person, nobody. **no man's land 1** Mil. the space between two opposing armies. **2** an area not assigned to any owner. **3** an area not clearly belonging to any one subject etc. **no-no** colloq. a thing not possible or acceptable. **no-nonsense** serious, without flippancy. **no place** US nowhere. **no-show** a person who has reserved a seat etc. but neither uses it nor cancels the reservation. **no side** Rugby **1** the end of a game. **2** the referee's announcement of this. **no small** see SMALL. **no sweat** colloq. no bother, no trouble. **no thoroughfare** an indication that passage along a street, path, etc., is blocked or prohibited. **no time** see TIME. **no trumps** (or **trump**) Bridge a declaration or bid involving playing without a trump suit. **no-trumper** Bridge a hand on which a no-trump bid can suitably be, or has been, made. **no way** colloq. **1** it is impossible. **2** I will not agree etc. **no whit** see WHIT. **no-win** of or designating a situation in which success is impossible. **no wonder** see WONDER. ... **or no** ... regardless of the ... (*rain or no rain, I shall go out*). **there is no** ... **ing** it is impossible to ... (*there is no accounting for tastes; there was no mistaking what he meant*). [ME f. *nān*, *nōn* NONE[1], orig. only before consonants]

no[2] /nəʊ/ adv. & n. ● adv. **1** equivalent to a negative sentence: the answer to your question is negative, your request or command will not be complied with, the statement made or course of action intended or conclusion arrived at is not correct or satisfactory, the negative statement made is correct. **2** (foll. by *compar.*) by no amount; not at all (*no better than before*). **3** Sc. not (*will ye no come back again?*). ● n. (pl. **noes**) **1** an utterance of the word no. **2** a denial or refusal. **3** a negative vote. □ **is no more** has died or ceased to exist. **no better than one should be** archaic of doubtful moral character, esp. sexually promiscuous. **no can do** colloq. I am unable to do it. **the noes have it** the negative voters are in the majority. **no less** (often foll. by *than*) **1** as much (*gave me £50, no less; gave me no less than £50; is no less than a scandal; a no less fatal victory*). **2** as important (*no less a person than the President*). **3** disp. no fewer (*no less than ten people have told me*). **no longer** not now or henceforth as formerly. **no more** n. nothing further (*have no more to say; want no more of it*). ● adj. not any more (*no more wine?*). ● adv. **1** no longer. **2** never again. **3** to no greater extent (*is no more a lord than I am; could no more do it than fly in the air*). **4** just as little, neither (*you did not come, and no more did he*). **no, no** an emphatic equivalent of a negative sentence (cf. sense 1 of adv.). **no-see-em** (or **-um**) N. Amer. a small bloodsucking insect, esp. a biting midge. **no sooner** ... **than** see

SOON. **not take no for an answer** persist in spite of refusals. **or no** or not (*pleasant or no, it is true*). **whether or no 1** in either case. **2** (as an indirect question) which of a case and its negative (*tell me whether or no*). [OE nō, nā f. ne not + ð, ā ever]

n.o. abbr. Cricket not out.

Noah /ˈnəʊə/ a Hebrew patriarch represented as tenth in descent from Adam. According to the story in Genesis he made the ark which saved his family and specimens of every animal from the flood sent by God to destroy the world, and his sons Ham, Shem, and Japheth were regarded as ancestors of all the races of humankind (Gen. 5–10). The tradition of a great flood in very early times is found also in other countries (see DEUCALION and GILGAMESH).

Noah's ark n. **1 a** the ship in which (according to the Bible) Noah, his family, and specimens of all the world's animals were saved. **b** an imitation of this as a child's toy. **2** a large, cumbrous, or old-fashioned trunk or vehicle. **3** a small bivalve mollusc, *Arca tetragona*, with a boat-shaped shell.

nob[1] /nɒb/ n. Brit. sl. a person of wealth or high social position. [orig. Sc. knabb, nab; 18th c., of unkn. orig.]

nob[2] /nɒb/ n. sl. the head. □ **his nob** Cribbage a score of one point for holding the jack of the same suit as a card turned up by the dealer. [perh. var. of KNOB]

nobble /ˈnɒb(ə)l/ v.tr. Brit. sl. **1** tamper with (a racehorse) to prevent its winning. **2** try to influence (e.g. a judge), esp. unfairly. **3** steal (money). **4** seize. [prob. = dial. knobble, knubble knock, beat, f. KNOB]

nobbler /ˈnɒblə(r)/ n. Austral. sl. a glass or drink of liquor. [19th c.: orig. unkn.]

Nobel /nəʊˈbel/, Alfred Bernhard (1833–96), Swedish chemist and engineer. He was interested in the use of high explosives, and after accidents with nitroglycerine he invented the much safer dynamite (1866), followed by other new explosives. He took out a large number of patents in a variety of disciplines, making a large fortune which enabled him to endow the prizes that bear his name.

Nobelist /nəʊˈbelɪst/ n. US a winner of a Nobel Prize.

nobelium /nəʊˈbiːlɪəm/ n. an artificial radioactive metallic chemical element (atomic number 102; symbol **No**). A member of the actinide series, nobelium was first produced by scientists in Sweden in 1958 by bombarding curium with carbon nuclei. [NOBEL]

Nobel Prize n. any of the prizes awarded annually to the person or persons adjudged by Swedish learned societies to have done the most significant recent work in physics, chemistry, physiology or medicine, and literature, and to the person who is adjudged by the Norwegian Parliament to have rendered the greatest service to the cause of peace. They were established by the will of Alfred Nobel, and are traditionally awarded on 10 December, the anniversary of his death. A Nobel Prize for economic sciences was added in 1969, financed by the Swedish National Bank.

nobiliary /nəˈbɪljərɪ/ adj. of the nobility. □ **nobiliary particle** a preposition forming part of a title of nobility (e.g. French de, German von). [F nobiliaire (as NOBLE)]

nobility /nəʊˈbɪlɪtɪ/ n. (pl. **-ies**) **1** nobleness of character, mind, birth, or rank. **2** (prec. by a, the) a class of nobles, an aristocracy. [ME f. OF nobilité or L nobilitas (as NOBLE)]

noble /ˈnəʊb(ə)l/ adj. & n. ● adj. (**nobler, noblest**) **1** belonging by rank, title, or birth to the aristocracy. **2** of excellent character; having lofty ideals; free from pettiness and meanness, magnanimous. **3** of imposing appearance, splendid, magnificent, stately. **4** excellent, admirable (*noble horse; noble cellar*). ● n. **1** a nobleman or noblewoman. **2** hist. a former English gold coin first issued in 1351. □ **noble metal** a metal (e.g. gold, silver, or platinum) that resists chemical action, does not corrode or tarnish in air or water, and is not easily attacked by acids. **the noble science** boxing. □ **nobleness** n. **nobly** adv. [ME f. OF f. L (g)nobilis, rel. to KNOW]

noble gas n. Chem. any of the gaseous elements helium, neon, argon, krypton, xenon, and radon, occupying Group 0 (18) of the periodic table. Apart from radon, they occur in small amounts in air. Also known as inert gases or rare gases, they were assumed to be totally unreactive, but in 1962 the British chemist Neil Bartlett succeeded in making the first xenon compounds; compounds of krypton and radon are now also known.

nobleman /ˈnəʊb(ə)lmən/ n. (pl. **-men**) a man of noble rank or birth, a peer.

noble savage n. primitive man idealized as being morally superior to modern humankind. The concept of the noble savage, symbolizing the innate goodness of humanity when free from the corrupting influence of civilization, is associated particularly with the romantic writers of the 18th and 19th centuries, particularly Jean-Jacques Rousseau, but the term was first used by Dryden in 1672; authors such as James Fenimore Cooper embodied the idea in the North American Indian.

noblesse /nəʊˈbles/ n. the class of nobles (esp. of a foreign country). □ **noblesse oblige** /əʊˈbliːʒ/ privilege entails responsibility. [ME = nobility, f. OF (as NOBLE)]

noblewoman /ˈnəʊb(ə)lˌwʊmən/ n. (pl. **-women**) a woman of noble rank or birth, a peeress.

nobody /ˈnəʊbədɪ/ pron. & n. ● pron. no person. ● n. (pl. **-ies**) a person of no importance, authority, or position. □ **like nobody's business** see BUSINESS. **nobody's fool** see no fool (see FOOL[1]). [ME f. NO[1] + BODY (= person)]

nock /nɒk/ n. & v. ● n. **1** a notch at either end of a bow for holding the string. **2 a** a notch at the butt-end of an arrow for receiving the bowstring. **b** a notched piece of horn serving this purpose. ● v.tr. set (an arrow) on the string. [ME, perh. = nock forward upper corner of some sails, f. MDu. nocke]

noctambulist /nɒkˈtæmbjʊlɪst/ n. a sleepwalker. □ **noctambulism** n. [L nox noctis night + ambulare walk]

noctuid /ˈnɒktjʊɪd/ n. & adj. Zool. ● n. a moth of the large family Noctuidae, comprising mainly large brownish moths, often with an eyespot on the forewings. ● adj. of or relating to this family. [mod.L Noctua genus name, f. L = night-owl]

noctule /ˈnɒktjuːl/ n. a large Eurasian bat, Nyctalus noctula. [F f. It. nottola bat]

nocturn /ˈnɒktɜːn/ n. RC Ch. a part of matins orig. said at night. [ME f. OF nocturne or eccl.L nocturnum neut. of L nocturnus: see NOCTURNAL]

nocturnal /nɒkˈtɜːn(ə)l/ adj. of or in the night; done or active by night. □ **nocturnal emission** involuntary emission of semen during sleep. □ **nocturnally** adv. [LL nocturnalis f. L nocturnus of the night f. nox noctis night]

nocturne /ˈnɒktɜːn/ n. **1** Mus. a short composition of a romantic nature, usu. for piano. **2** a picture of a night scene. [F (as NOCTURN)]

nocuous /ˈnɒkjʊəs/ adj. literary noxious, harmful. [L nocuus f. nocere hurt]

nod /nɒd/ v. & n. ● v. (**nodded, nodding**) **1** intr. incline one's head slightly and briefly in greeting, assent, or command. **2** intr. let one's head fall forward in drowsiness; be drowsy. **3** tr. incline (one's head). **4** tr. signify (assent etc.) by a nod. **5** intr. (of flowers, plumes, etc.) bend downwards and sway, or move up and down. **6** intr. make a mistake due to a momentary lack of alertness or attention. ● n. a nodding of the head, esp. as a sign to proceed etc. □ **get the nod** N. Amer. be chosen or approved. **nodding acquaintance** (usu. foll. by with) a very slight acquaintance with a person or subject. **nod off** colloq. fall asleep. **nod through** colloq. **1** approve on the nod. **2** Brit. Parl. formally count (a member of parliament) as if having voted when unable to do so. **on the nod** Brit. colloq. **1** with merely formal assent and no discussion. **2** on credit. [ME nodde, of unkn. orig.]

noddle[1] /ˈnɒd(ə)l/ n. colloq. the head. [ME nodle, of unkn. orig.]

noddle[2] /ˈnɒd(ə)l/ v.tr. nod or wag (one's head). [NOD + -LE[4]]

Noddy /ˈnɒdɪ/ a character in the writings of Enid Blyton, a toy figure of a boy whose head is fixed in such a way that he has to nod when he speaks.

noddy /ˈnɒdɪ/ n. (pl. **-ies**) **1** a simpleton. **2** a tropical tern of the genus Anous or Procelsterna, with usu. dark plumage. [prob. f. obs. noddy foolish, which is perh. f. NOD]

node /nəʊd/ n. **1** Bot. **a** the part of a plant stem from which one or more leaves emerge. **b** a knob on a root or branch. **2** Anat. **a** a small mass of differentiated tissue, esp. a lymph gland. **b** an interruption of the myelin sheath of a nerve. **3** Astron. either of two points at which a planet's orbit intersects the plane of the ecliptic or the celestial equator. **4** Physics a point of minimum disturbance in a standing wave system. **5** Electr. a point of zero current or voltage. **6** Math. **a** a point at which a curve intersects itself. **b** a vertex in a graph. **7** a component in a computer network. □ **nodal** adj. **nodical** /ˈnəʊdɪk(ə)l, ˈnɒd-/ adj. (in sense 3). [L nodus knot]

nodi pl. of NODUS.

nodose /ˈnəʊdəʊs, ˈnəʊdəs/ adj. knotty, knotted. □ **nodosity** /nəˈdɒsɪtɪ/ n. [L nodosus (as NODE)]

nodule /ˈnɒdjuːl/ n. **1** a small rounded lump of anything, e.g. flint in chalk, carbon in cast iron, or a mineral on the seabed. **2** a small swelling or aggregation of cells, e.g. a small tumour, node, or ganglion, or a swelling on a root of a legume containing bacteria. □ **nodular** adj. **nodulated** adj. **nodulose** adj. **nodulous** adj. **nodulation** /ˌnɒdjʊˈleɪʃ(ə)n/ n. [L nodulus dimin. of nodus: see NODUS]

nodus /ˈnəʊdəs/ n. (pl. **nodi** /-daɪ/) a knotty point, a difficulty, a complication in the plot of a story etc. [L, = knot]

Noel /nəʊˈel/ n. (also **Noël**) Christmas (esp. as a refrain in carols). [F f. L (as NATAL)]

Noether /ˈnɜːtə(r)/, Emmy (1882–1935), German mathematician. She simplified and extended the work of her predecessors, particularly Hilbert and Dedekind, on the properties of rings. She lacked status in what was then a man's world, and her position at Göttingen remained insecure and unsalaried until terminated by the anti-Semitic laws of 1933. Nevertheless, she exercised an enormous influence, and inaugurated the modern period in algebraic geometry and abstract algebra.

noetic /nəʊˈetɪk, -ˈiːtɪk/ adj. & n. ● adj. **1** of the intellect. **2** purely intellectual or abstract. **3** given to intellectual speculation. ● n. (in sing. or pl.) the science of the intellect. [Gk noētikos f. noētos intellectual f. noeō apprehend]

Nofretete see NEFERTITI.

nog[1] /nɒg/ n. & v. ● n. **1** a small block or peg of wood. **2** a snag or stump on a tree. **3** nogging. ● v.tr. (**nogged, nogging**) **1** secure with nogs. **2** build in the form of nogging. [17th c.: orig. unkn.]

nog[2] /nɒg/ n. Brit. **1** a strong beer brewed in East Anglia. **2** an egg-flip. [17th c.: orig. unkn.]

noggin /ˈnɒgɪn/ n. **1** a small mug. **2** a small measure, usu. ¹/₄ pint, of spirits. **3** sl. the head. [17th c.: orig. unkn.]

nogging /ˈnɒgɪŋ/ n. brickwork or timber braces in a timber frame. [NOG[1] + -ING[1]]

Noh /nəʊ/ n. (also **No**) a form of traditional Japanese drama, evolved from Shinto rites and dating from the 14th and 15th centuries. It flourished in the 17th–19th centuries and has since been revived. The plays were short, and five made up a complete programme. The general tone was noble, the language honorific and sonorous, and the subject-matter taken mainly from Japan's classical literature. The players were all male, the chorus playing a passive narrative role and even chanting the lines of the principal characters while the latter executed certain dances. About two hundred Noh plays are extant; they are comparable in various respects with early Greek drama. [Jap. nō]

nohow /ˈnəʊhaʊ/ adv. **1** US in no way; by no means. **2** dial. out of order; out of sorts.

noil /nɔɪl/ n. (in sing. or pl.) short wool-combings. [perh. f. OF noel f. med.L nodellus dimin. of L nodus knot]

noise /nɔɪz/ n. & v. ● n. **1** a sound, esp. a loud or unpleasant or undesired one. **2** a series of loud sounds, esp. shouts; a confused sound of voices and movements. **3** irregular fluctuations accompanying a transmitted signal but not relevant to it. **4** (in pl.) conventional remarks, or speechlike sounds without actual words (made sympathetic noises). ● v. **1** tr. (usu. in passive) make public; spread abroad (a person's fame or a fact). **2** intr. archaic make much noise. □ **make a noise 1** (usu. foll. by about) talk or complain much. **2** be much talked of; attain notoriety. **noise-maker** a device for making a loud noise at a festivity etc. **noise pollution** harmful or annoying noise. **noises off** sounds made offstage to be heard by the audience of a play. [ME f. OF, = outcry, disturbance, f. L nausea: see NAUSEA]

noiseless /ˈnɔɪzlɪs/ adj. **1** silent. **2** making no avoidable noise. □ **noiselessly** adv. **noiselessness** n.

noisette /nwʌˈzet/ n. **1** a small round piece of meat etc. **2** a chocolate made with hazelnuts. [F, dimin. of noix nut]

noisome /ˈnɔɪsəm/ adj. literary **1** harmful, noxious. **2** evil-smelling. **3** objectionable, offensive. □ **noisomeness** n. [ME f. obs. noy f. ANNOY]

noisy /ˈnɔɪzɪ/ adj. (**noisier, noisiest**) **1** full of or attended with noise. **2** making or given to making much noise. **3** clamorous, turbulent. **4** (of a colour, garment, etc.) loud, conspicuous. □ **noisily** adv. **noisiness** n.

Nok /nɒk/ n. an ancient civilization of northern Nigeria c.400 BC–AD 200, characterized by the production of distinctive terracotta figurines and significant for its development of iron-working.

Nolan /ˈnəʊlən/, Sir Sidney Robert (1917–93), Australian painter. He is chiefly known for his paintings of famous characters and events from Australian history, especially his 'Ned Kelly' series (begun in 1946). He

has also painted landscapes and themes from classical mythology, such as his 'Leda and the Swan' series of 1960.

nolens volens /ˌnəʊlenz ˈvəʊlenz/ adv. literary willy-nilly, perforce. [L participles, = unwilling, willing]

nolle prosequi /ˌnɒlɪ ˈprɒsɪˌkwaɪ/ n. Law **1** the relinquishment by a plaintiff or prosecutor of all or part of a suit. **2** the entry of this on record. [L, = refuse to pursue]

nom. abbr. nominal.

nomad /ˈnəʊmæd/ n. & adj. ● n. **1** a member of a pastoral people roaming from place to place for fresh pasture. **2** a wanderer. ● adj. **1** living as a nomad. **2** wandering. □ **nomadism** n. **nomadize** v.intr. (also **-ise**). **nomadic** /nəʊˈmædɪk/ adj. **nomadically** adv. [F nomade f. L nomas nomad- f. Gk nomas -ados f. nemō to pasture]

nombril /ˈnɒmbrɪl/ n. Heraldry the point halfway between fess point and the base of the shield. [F, = navel]

nom de guerre /ˌnɒm də ˈgeə(r)/ n. (pl. **noms de guerre** pronunc. same) an assumed name under which a person fights, plays, writes, etc. [F, = war-name]

nom de plume /ˌnɒm də ˈpluːm/ n. (pl. **noms de plume** pronunc. same) an assumed name under which a person writes. [formed in Engl. of F words, = pen-name, after NOM DE GUERRE]

Nome /nəʊm/ a city in western Alaska, on the south coast of the Seward Peninsula. Founded in 1896 as a gold-mining camp, it became a centre of the Alaskan gold rush at the turn of the century.

nomen /ˈnəʊmen/ n. an ancient Roman's second name, indicating the gens, as in Marcus Tullius Cicero. [L, = name]

nomenclature /nəʊˈmenklətʃə(r), ˈnəʊmənˌkleɪtʃə(r)/ n. **1** a person's or community's system of names for things. **2** the terminology of a science etc. **3** systematic naming. **4** a catalogue or register. □ **nomenclative** /nəʊˈmenklətɪv, ˈnəʊmənˌkleɪtɪv/ adj. **nomenclatural** /ˌnəʊmənˈklætʃərəl/ adj. [F f. L nomenclatura f. nomen + calare call]

nominal /ˈnɒmɪn(ə)l/ adj. **1** existing in name only; not real or actual (nominal and real prices; nominal ruler). **2** (of a sum of money, rent, etc.) virtually nothing; much below the actual value of a thing. **3** of or in names (nominal and essential distinctions). **4** consisting of or giving the names (nominal list of officers). **5** of or as or like a noun. □ **nominal definition** a statement of all that is connoted in the name of a concept. **nominal value** the face value (of a coin, shares, etc.). □ **nominally** adv. [ME f. F nominal or L nominalis f. nomen -inis name]

nominalism /ˈnɒmɪnəˌlɪz(ə)m/ n. Philos. the doctrine that universals or general ideas are mere names without any corresponding reality. Only particular, usually physical, objects exist, and properties, numbers, sets, etc., are not further things in the world but merely features of the way of considering the things that do exist. Important in medieval scholastic thought, nominalism is associated particularly with William of Occam. (See also REALISM.) □ **nominalist** n. **nominalistic** /ˌnɒmɪnəˈlɪstɪk/ adj. [F nominalisme (as NOMINAL)]

nominalize /ˈnɒmɪnəˌlaɪz/ v.tr. (also **-ise**) form a noun from (a verb, adjective, etc.), e.g. output, truth, from put out, true. □ **nominalization** /ˌnɒmɪnəlaɪˈzeɪʃ(ə)n/ n.

nominate /ˈnɒmɪˌneɪt/ v.tr. **1** propose (a candidate) for election. **2** appoint to an office (a board of six nominated and six elected members). **3** name or appoint (a date or place). **4** mention by name. **5** call by the name of, designate. □ **nominator** n. [L nominare nominat- (as NOMINAL)]

nomination /ˌnɒmɪˈneɪʃ(ə)n/ n. **1** the act or an instance of nominating; the state of being nominated. **2** the right of nominating for an appointment (have a nomination at your disposal). [ME f. OF nomination or L nominatio (as NOMINATE)]

nominative /ˈnɒmɪnətɪv/ n. & adj. ● n. Gram. **1** the case of nouns, pronouns, and adjectives expressing the subject of a verb. **2** a word in this case. ● adj. **1** Gram. of or in this case. **2** /-ˌneɪtɪv/ of, or appointed by, nomination (as distinct from election). □ **nominatival** /ˌnɒmɪnəˈtaɪv(ə)l/ adj. [ME f. OF nominatif -ive or L nominativus (as NOMINATE), transl. Gk onomastikē (ptōsis case)]

nominee /ˌnɒmɪˈniː/ n. **1** a person who is nominated for an office or as the recipient of a grant etc. **2** Commerce a person (not necessarily the owner) in whose name a stock etc. is registered. [NOMINATE]

nomogram /ˈnɒməˌgræm, ˈnəʊm-/ n. (also **nomograph** /-ˌgrɑːf/) a graphical presentation of relations between quantities whereby the value of one may be found by simple geometrical construction (e.g. drawing a straight line) from those of others. □ **nomographic** /ˌnɒməˈgræfɪk/ adj. **nomographically** adv. **nomography** /nəʊˈmɒgrəfɪ/ n. [Gk nomo- f. nomos law + -GRAM]

nomothetic /ˌnɒməˈθetɪk, ˌnəʊm-/ adj. **1** stating (esp. scientific) laws. **2** legislative. [obs. *nomothete* legislator f. Gk *nomothetēs*]

-nomy /nəmi/ comb. form denoting an area of knowledge or the laws governing it (*aeronomy*; *economy*).

non- /nɒn/ prefix giving the negative sense of words with which it is combined, esp.: **1** not doing or having or involved with (*non-attendance*; *non-payment*; *non-productive*). **2 a** not of the kind or class described (*non-alcoholic*; *non-member*; *non-event*). **b** forming terms used adjectivally (*non-union*; *non-party*). **3** a lack of (*non-access*). **4** (with adverbs) not in the way described (*non-aggressively*). **5** forming adjectives from verbs, meaning 'that does not' or 'that is not meant to (or to be)' (*non-skid*; *non-iron*). **6** used to form a neutral negative sense when a form in *in-* or *un-* has a special sense or (usu. unfavourable) connotation (*non-controversial*; *non-effective*; *non-human*). ¶ The number of words that can be formed with this prefix is unlimited; consequently only a selection, considered the most current or semantically noteworthy, can be given here. [from or after ME *no*(u)*n-* f. AF *noun-*, OF *non-*, *nom-* f. L *non* not]

nona- /ˈnɒnə/ comb. form nine. [L f. *nonus* ninth]

non-abstainer /ˌnɒnəbˈsteɪnə(r)/ n. a person who does not abstain (esp. from alcohol).

non-acceptance /ˌnɒnəkˈseptəns/ n. a lack of acceptance.

non-access /nɒnˈækses/ n. a lack of access.

non-addictive /ˌnɒnəˈdɪktɪv/ adj. (of a drug, habit, etc.) not causing addiction.

nonage /ˈnəʊnɪdʒ, ˈnɒn-/ n. **1** hist. the state of being under full legal age, minority. **2** a period of immaturity. [ME f. AF *nounage*, OF *nonage* (as NON-, AGE)]

nonagenarian /ˌnəʊnədʒɪˈneərɪən, ˌnɒn-/ n. & adj. ● n. a person from 90 to 99 years old. ● adj. of this age. [L *nonagenarius* f. *nonageni* distributive of *nonaginta* ninety]

non-aggression /ˌnɒnəˈɡreʃ(ə)n/ n. lack of or restraint from aggression (often attrib.: *non-aggression pact*).

nonagon /ˈnɒnəɡən/ n. a plane figure with nine sides and angles. [L *nonus* ninth, after HEXAGON]

non-alcoholic /ˌnɒnælkəˈhɒlɪk/ adj. (of a drink etc.) not containing alcohol.

non-aligned /ˌnɒnəˈlaɪnd/ adj. (of states etc.) not aligned with another (esp. major) power. □ **non-alignment** n.

Non-Aligned Movement a grouping of chiefly developing countries pursuing a policy of neutrality towards the superpowers (i.e. the US and formerly the USSR) in world politics. The term was first used by the Indian Prime Minister Nehru, an instigator of the movement in the early 1960s along with Tito of Yugoslavia and Nasser of Egypt. Membership in 1994 was 109 countries following the entry of South Africa.

non-allergic /ˌnɒnəˈlɜːdʒɪk/ adj. not causing allergy; not allergic.

non-ambiguous /ˌnɒnæmˈbɪɡjʊəs/ adj. not ambiguous. ¶ Neutral in sense: see NON- 6, UNAMBIGUOUS.

non-appearance /ˌnɒnəˈpɪərəns/ n. failure to appear or be present.

non-art /nɒnˈɑːt/ n. something that avoids the normal forms of art.

nonary /ˈnəʊnəri/ adj. & n. ● adj. Math. (of a scale of notation) having nine as its base. ● n. (pl. **-ies**) a group of nine. [L *nonus* ninth]

non-Aryan /nɒnˈeərɪən/ adj. & n. ● adj. (of a person or language) not Aryan or of Aryan descent. ● n. a non-Aryan person.

non-attached /ˌnɒnəˈtætʃt/ adj. that is not attached. ¶ Neutral in sense: see NON- 6, UNATTACHED.

non-attendance /ˌnɒnəˈtendəns/ n. failure to attend.

non-attributable /ˌnɒnəˈtrɪbjʊtəb(ə)l/ adj. that cannot or may not be attributed to a particular source etc. □ **non-attributably** adv.

non-availability /ˌnɒnəˌveɪləˈbɪlɪti/ n. a state of not being available.

non-believer /ˌnɒnbɪˈliːvə(r)/ n. a person who does not believe or has no (esp. religious) faith.

non-belligerency /ˌnɒnbəˈlɪdʒərənsi/ n. a lack of belligerency.

non-belligerent /ˌnɒnbəˈlɪdʒərənt/ adj. & n. ● adj. not engaged in hostilities. ● n. a non-belligerent nation, state, etc.

non-biological /ˌnɒnbaɪəˈlɒdʒɪk(ə)l/ adj. not concerned with biology or living organisms.

non-black /nɒnˈblæk/ adj. & n. ● adj. **1** (of a person) not black. **2** of or relating to non-black people. ● n. a non-black person.

non-breakable /nɒnˈbreɪkəb(ə)l/ adj. not breakable.

non-capital /nɒnˈkæpɪt(ə)l/ adj. (of an offence) not punishable by death.

non-Catholic /nɒnˈkæθəlɪk, -ˈkæθlɪk/ adj. & n. ● adj. not Roman Catholic. ● n. a non-Catholic person.

nonce /nɒns/ n. □ **for the nonce** for the time being; for the present occasion. **nonce-word** a word coined for one occasion. [ME *for than anes* (unrecorded) = for the one, altered by wrong division (cf. NEWT)]

nonchalant /ˈnɒnʃələnt/ adj. calm and casual, unmoved, unexcited, indifferent. □ **nonchalance** n. **nonchalantly** adv. [F, part. of *nonchaloir* f. *chaloir* be concerned]

non-Christian /nɒnˈkrɪstɪən, -ˈkrɪstʃən/ adj. & n. ● adj. not Christian. ● n. a non-Christian person.

non-citizen /nɒnˈsɪtɪz(ə)n/ n. a person who is not a citizen (of a particular state, town, etc.).

non-classified /nɒnˈklæsɪˌfaɪd/ adj. (esp. of information) that is not classified. ¶ Neutral in sense: see NON- 6, UNCLASSIFIED.

non-clerical /nɒnˈklerɪk(ə)l/ adj. not doing or involving clerical work.

non-collegiate /ˌnɒnkəˈliːdʒət/ adj. **1** not attached to a college. **2** not having colleges.

non-com /ˈnɒnkɒm/ n. colloq. a non-commissioned officer. [abbr.]

non-combatant /nɒnˈkɒmbət(ə)nt/ n. a person not fighting in a war, esp. a civilian, army chaplain, etc.

non-commissioned /ˌnɒnkəˈmɪʃ(ə)nd/ adj. Mil. (of an officer) not holding a commission.

noncommittal /ˌnɒnkəˈmɪt(ə)l/ adj. avoiding commitment to a definite opinion or course of action. □ **noncommittally** adv.

non-communicant /ˌnɒnkəˈmjuːnɪkənt/ n. a person who is not a communicant (esp. in the religious sense).

non-communicating /ˌnɒnkəˈmjuːnɪˌkeɪtɪŋ/ adj. that does not communicate.

non-communist /nɒnˈkɒmjʊnɪst/ adj. & n. (also **non-Communist** with ref. to a particular party) ● adj. not advocating or practising communism. ● n. a non-communist person.

non-compliance /ˌnɒnkəmˈplaɪəns/ n. failure to comply; a lack of compliance.

non compos mentis /ˌnɒn kɒmpɒs ˈmentɪs/ adj. (also **non compos**) not in one's right mind. [L, = not having control of one's mind]

non-conductor /ˌnɒnkənˈdʌktə(r)/ n. a substance that does not conduct heat or electricity. □ **non-conducting** adj.

non-confidential /ˌnɒnkɒnfɪˈdenʃ(ə)l/ adj. not confidential. □ **non-confidentially** adv.

nonconformist /ˌnɒnkənˈfɔːmɪst/ n. & adj. ● n. **1** a person who does not conform to the doctrine or discipline of an established Church; esp. (**Nonconformist**) a member of a (usu. Protestant) sect dissenting from the Anglican Church. **2** a person who does not conform to a prevailing principle. ● adj. of or relating to nonconformists or Nonconformists. □ **nonconformism** n.

nonconformity /ˌnɒnkənˈfɔːmɪti/ n. **1 a** nonconformists as a body; esp. (**Nonconformity**) Protestants dissenting from the Anglican Church. **b** the principles or practice of nonconformists; esp. (**Nonconformity**) Protestant dissent. **2** (usu. foll. by *to*) failure to conform to a rule etc. **3** lack of correspondence between things.

non-contagious /ˌnɒnkənˈteɪdʒəs/ adj. not contagious.

non-content /ˈnɒnkənˌtent/ n. Brit. a negative voter in the House of Lords.

non-contentious /ˌnɒnkənˈtenʃəs/ adj. not contentious.

non-contributory /ˌnɒnkənˈtrɪbjʊtri/ adj. not contributing or (esp. of a pension scheme) involving contributions.

non-controversial /ˌnɒnkɒntrəˈvɜːʃ(ə)l/ adj. not controversial. ¶ Neutral in sense: see NON- 6, UNCONTROVERSIAL.

non-cooperation /ˌnɒnkəʊˌɒpəˈreɪʃ(ə)n/ n. failure to cooperate; a lack of cooperation.

non-delivery /ˌnɒndɪˈlɪvəri/ n. failure to deliver.

non-denominational /ˌnɒndɪˌnɒmɪˈneɪʃən(ə)l/ adj. not restricted as regards religious denomination.

nondescript /ˈnɒndɪˌskrɪpt/ adj. & n. ● adj. lacking distinctive characteristics, not easily classified, neither one thing nor another. ● n. a nondescript person or thing. □ **nondescriptly** adv. **nondescriptness** n. [NON- + *descript* described f. L *descriptus* (as DESCRIBE)]

non-destructive /ˌnɒndɪˈstrʌktɪv/ adj. that does not involve destruction or damage.

non-drinker /nɒnˈdrɪŋkə(r)/ n. a person who does not drink alcoholic liquor.

non-driver /nɒnˈdraɪvə(r)/ n. a person who does not drive a motor vehicle.

none[1] /nʌn/ pron., adj., & adv. ● pron. **1** (foll. by of) **a** not any of (none of this concerns me; none of them have found it; none of your impudence!). **b** not any one of (none of them has come). ¶ The verb following none in this sense can be singular or plural according to the sense. **2 a** no persons (none but fools have ever believed it). **b** no person (none can tell). ● adj. (usu. with a preceding noun implied) **1** no; not any (you have money and I have none; would rather have a bad reputation than none at all). **2** not to be counted in a specified class (his understanding is none of the clearest; if a linguist is wanted, I am none). ● adv. (foll. by the + compar., or so, too) by no amount; not at all (am none the wiser; are none too fond of him). □ **none the less** (also **nonetheless**) nevertheless. **none other** (usu. foll. by than) no other person. **none-so-pretty** London Pride. [OE nān f. ne not + ān ONE]

none[2] /nəʊn/ n. (also in pl.) **1** the office of the fifth of the canonical hours of prayer, orig. said at the ninth hour (3 p.m.). **2** this hour. [F f. L nona fem. sing. of nonus ninth: cf. NOON]

non-earning /nɒnˈɜːnɪŋ/ adj. not earning (esp. a regular wage or salary).

non-effective /ˌnɒnɪˈfektɪv/ adj. that does not have an effect. ¶ Neutral in sense: see NON- 6, INEFFECTIVE.

non-ego /nɒnˈiːɡəʊ/ n. Philos. all that is not the conscious self.

nonentity /nɒˈnentɪtɪ/ n. (pl. -ies) **1** a person or thing of no importance. **2 a** non-existence. **b** a non-existent thing, a figment. [med.L nonentitas non-existence]

nones /nəʊnz/ n.pl. in the ancient Roman calendar, the ninth day before the ides by inclusive reckoning, i.e. the 7th day of March, May, July, October, the 5th of other months. [OF nones f. L nonae fem. pl. of nonus ninth]

non-essential /ˌnɒnɪˈsenʃ(ə)l/ adj. & n. ● adj. not essential. ● n. a non-essential thing. ¶ Neutral in sense: see NON- 6, INESSENTIAL.

nonesuch var. of NONSUCH.

nonet /nəʊˈnet, nɒˈnet/ n. **1** Mus. **a** a composition for nine voices or instruments. **b** the performers of such a piece. **2** a group of nine. [It. nonetto f. nono ninth f. L nonus]

nonetheless var. of none the less.

non-Euclidean /ˌnɒnjuːˈklɪdɪən/ adj. denying or going beyond Euclidean principles in geometry, esp. contravening the postulate that only one line through a given point can be parallel to a given line.

non-European /nɒnˌjʊərəˈpɪən/ adj. & n. ● adj. not European. ● n. a non-European person.

non-event /ˌnɒnɪˈvent/ n. an unimportant or anticlimactic occurrence.

non-existent /ˌnɒnɪɡˈzɪstənt/ adj. not existing. □ **non-existence** n.

non-explosive /ˌnɒnɪkˈspləʊsɪv/ adj. (of a substance) that does not explode.

non-fattening /nɒnˈfæt(ə)nɪŋ/ adj. (of food) that does not make the consumer fat.

nonfeasance /nɒnˈfiːz(ə)ns/ n. failure to perform an act required by law. [NON-: see MISFEASANCE]

non-ferrous /nɒnˈferəs/ adj. (of a metal) other than iron or steel.

non-fiction /nɒnˈfɪkʃ(ə)n/ n. literary work other than fiction, including biography and reference books. □ **non-fictional** adj.

non-flam /nɒnˈflæm/ adj. = NON-FLAMMABLE.

non-flammable /nɒnˈflæməb(ə)l/ adj. not inflammable.

non-fulfilment /ˌnɒnfʊlˈfɪlmənt/ n. failure to fulfil (an obligation).

non-functional /nɒnˈfʌŋkʃən(ə)l/ adj. not having a function.

nong /nɒŋ/ n. Austral. sl. a foolish or stupid person. [20th c.: orig. unkn.]

non-governmental /ˌnɒnɡʌvənˈment(ə)l/ adj. not belonging to or associated with a government.

non-human /nɒnˈhjuːmən/ adj. & n. ● adj. (of a being) not human. ● n. a non-human being. ¶ Neutral in sense: see NON- 6, INHUMAN, UNHUMAN.

non-infectious /ˌnɒnɪnˈfekʃəs/ adj. (of a disease) not infectious.

non-inflected /ˌnɒnɪnˈflektɪd/ adj. (of a language) not having inflections.

non-interference /ˌnɒnɪntəˈfɪərəns/ n. a lack of interference.

non-intervention /ˌnɒnɪntəˈvenʃ(ə)n/ n. the principle or practice of not becoming involved in others' affairs, esp. by one state in regard to another. □ **non-interventionist** adj. & n.

non-intoxicating /ˌnɒnɪnˈtɒksɪˌkeɪtɪŋ/ adj. (of drink) not causing intoxication.

non-iron /nɒnˈaɪən/ adj. (of a fabric) that needs no ironing.

nonjoinder /nɒnˈdʒɔɪndə(r)/ n. Law the failure of a partner etc. to become a party to a suit.

nonjuror /nɒnˈdʒʊərə(r)/ n. a person who refuses to take an oath, esp. hist. a member of the clergy refusing to take the oath of allegiance to William and Mary in 1689. □ **nonjuring** adj.

non-jury /nɒnˈdʒʊərɪ/ adj. (of a trial) without a jury.

non-linear /nɒnˈlɪnɪə(r)/ adj. not linear, esp. with regard to dimension.

non-literary /nɒnˈlɪtərərɪ/ adj. (of writing, a text, etc.) not literary in character.

non-logical /nɒnˈlɒdʒɪk(ə)l/ adj. not involving logic. ¶ Neutral in sense: see NON- 6, ILLOGICAL. □ **non-logically** adv.

non-magnetic /ˌnɒnmæɡˈnetɪk/ adj. (of a substance) not magnetic.

non-member /nɒnˈmembə(r)/ n. a person who is not a member (of a particular association, club, etc.). □ **non-membership** n.

non-metal /nɒnˈmet(ə)l/ adj. not made of metal. □ **non-metallic** /-mɪˈtælɪk/ adj.

non-militant /nɒnˈmɪlɪt(ə)nt/ adj. not militant.

non-military /nɒnˈmɪlɪtərɪ/ adj. not military; not involving armed forces, civilian.

non-ministerial /ˌnɒnmɪnɪˈstɪərɪəl/ adj. not ministerial (esp. in political senses).

non-moral /nɒnˈmɒrəl/ adj. not concerned with morality. ¶ Neutral in sense: see NON- 6, AMORAL, IMMORAL. □ **non-morally** adv.

non-natural /nɒnˈnætʃrəl/ adj. not involving natural means or processes. ¶ Neutral in sense: see NON- 6, UNNATURAL.

non-negotiable /ˌnɒnnɪˈɡəʊʃəb(ə)l, -ˈɡəʊsɪəb(ə)l/ adj. that cannot be negotiated (esp. in financial senses).

non-net /nɒnˈnet/ adj. (of a book) not subject to a minimum selling price.

non-nuclear /nɒnˈnjuːklɪə(r)/ adj. **1** not involving nuclei or nuclear energy. **2** (of a state etc.) not having nuclear weapons.

non-objective /ˌnɒnəbˈdʒektɪv/ adj. **1** not objective. **2** Art abstract.

non-observance /ˌnɒnəbˈzɜːv(ə)ns/ n. failure to observe (esp. an agreement, requirement, etc.).

non-operational /ˌnɒnɒpəˈreɪʃən(ə)l/ adj. **1** that does not operate. **2** out of order.

non-organic /ˌnɒnɔːˈɡænɪk/ adj. not organic. ¶ Neutral in sense: see NON- 6, INORGANIC.

nonpareil /ˈnɒnpərəl, ˌnɒnpəˈreɪl/ adj. & n. ● adj. unrivalled or unique. ● n. such a person or thing. [F f. pareil equal f. pop.L pariculus dimin. of L par]

non-participating /ˌnɒnpɑːˈtɪsɪˌpeɪtɪŋ/ adj. not taking part.

non-partisan /ˌnɒnpɑːtɪˈzæn/ adj. not partisan.

non-party /nɒnˈpɑːtɪ/ adj. independent of political parties.

non-payment /nɒnˈpeɪmənt/ n. failure to pay; a lack of payment.

non-person /ˈnɒnˌpɜːs(ə)n/ n. a person regarded as non-existent or insignificant (cf. UNPERSON).

non-personal /nɒnˈpɜːsən(ə)l/ adj. not personal. ¶ Neutral in sense: see NON- 6, IMPERSONAL.

non-physical /nɒnˈfɪzɪk(ə)l/ adj. not physical. □ **non-physically** adv.

non placet /nɒn ˈpleɪset/ n. a negative vote in a Church or university assembly. [L, = it does not please]

non-playing /nɒnˈpleɪɪŋ/ adj. that does not play or take part (in a game etc.).

nonplus /nɒnˈplʌs/ v. & n. ● v.tr. (**nonplussed**, **nonplussing**) completely perplex. ● n. a state of perplexity, a standstill (at a nonplus; reduce to a nonplus). [L non plus not more]

non-poisonous /nɒnˈpɔɪz(ə)nəs/ adj. (of a substance, plant, etc.) not poisonous.

non-political /ˌnɒnpəˈlɪtɪk(ə)l/ adj. not political; not involved in politics.

non-porous /nɒnˈpɔːrəs/ adj. (of a substance) not porous.

non possumus /nɒn ˈpɒsjʊməs/ *n.* a statement of inability to act in a matter. [L, = we cannot]

non-productive /ˌnɒnprəˈdʌktɪv/ *adj.* not productive. ¶ Neutral in sense: see NON- 6, UNPRODUCTIVE. □ **non-productively** *adv.*

non-professional /ˌnɒnprəˈfeʃən(ə)l/ *adj.* not professional (esp. in status). ¶ Neutral in sense: see NON- 6, UNPROFESSIONAL.

non-profit /nɒnˈprɒfɪt/ *adj.* not involving or making a profit.

non-profit-making /nɒnˈprɒfɪtˌmeɪkɪŋ/ *adj. Brit.* (of an enterprise) not conducted primarily to make a profit.

non-proliferation /ˌnɒnprəˌlɪfəˈreɪʃ(ə)n/ *n.* the prevention of an increase in something, esp. possession of nuclear weapons (usu. *attrib.*: *non-proliferation agreement*).

non-racial /nɒnˈreɪʃ(ə)l/ *adj.* not involving race or racial factors.

non-reader /nɒnˈriːdə(r)/ *n.* a person who cannot or does not read.

non-resident /nɒnˈrezɪd(ə)nt/ *adj. & n.* ● *adj.* **1** not residing in a particular place, esp. (of a member of the clergy) not residing where his or her duties require. **2** (of a post) not requiring the holder to reside at the place of work. ● *n.* a non-resident person, esp. a person using some of the facilities of a hotel. □ **non-residence** *n.* **non-residential** /-ˌrezɪˈdenʃ(ə)l/ *adj.*

non-resistance /ˌnɒnrɪˈzɪstəns/ *n.* failure to resist; a lack of resistance.

non-returnable /ˌnɒnrɪˈtɜːnəb(ə)l/ *adj.* that may or need not be returned.

non-rigid /nɒnˈrɪdʒɪd/ *adj.* (esp. of materials) not rigid.

non-scientific /ˌnɒnsaɪənˈtɪfɪk/ *adj.* not involving science or scientific methods. ¶ Neutral in sense: see NON- 6, UNSCIENTIFIC. □ **non-scientist** /-ˈsaɪəntɪst/ *n.*

non-sectarian /ˌnɒnsekˈteərɪən/ *adj.* not sectarian.

nonsense /ˈnɒns(ə)ns/ *n.* **1 a** (often as *int.*) absurd or meaningless words or ideas; foolish or extravagant conduct. **b** an instance of this. **2** a scheme, arrangement, etc., that one disapproves of. **3** (often *attrib.*) a form of literature meant to amuse by absurdity (*nonsense verse*). □ **nonsensical** /nɒnˈsensɪk(ə)l/ *adj.* **nonsensically** *adv.* **nonsensicality** /ˌnɒnsensɪˈkælɪti/ *n.* (*pl.* **-ies**).

non sequitur /nɒn ˈsekwɪtə(r)/ *n.* a conclusion that does not logically follow from the premisses. [L, = it does not follow]

non-sexual /nɒnˈseksjʊəl, -ˈsekʃʊəl/ *adj.* not based on or involving sex. □ **non-sexually** *adv.*

non-skid /nɒnˈskɪd/ *adj.* **1** that does not skid. **2** that inhibits skidding.

non-slip /nɒnˈslɪp/ *adj.* **1** that does not slip. **2** that inhibits slipping.

non-smoker /nɒnˈsməʊkə(r)/ *n.* **1** a person who does not smoke. **2** a train compartment etc. in which smoking is forbidden. □ **non-smoking** *adj. & n.*

non-soluble /nɒnˈsɒljʊb(ə)l/ *adj.* (esp. of a substance) not soluble. ¶ Neutral in sense: see NON- 6, INSOLUBLE.

non-specialist /nɒnˈspeʃəlɪst/ *n.* a person who is not a specialist (in a particular subject).

non-specific /ˌnɒnspɪˈsɪfɪk/ *adj.* not specific in cause, extent, effect, etc. □ **non-specific urethritis** (abbr. **NSU**) *Med.* inflammation of the urethra due to infection other than by gonococci, esp. chlamydiae.

non-standard /nɒnˈstændəd/ *adj.* not standard.

non-starter /nɒnˈstɑːtə(r)/ *n.* **1** a person or animal that does not start in a race. **2** *colloq.* a person or thing that is unlikely to succeed or be effective.

non-stick /nɒnˈstɪk/ *adj.* **1** that does not stick. **2** that does not allow things to stick to it.

non-stop /nɒnˈstɒp/ *adj., adv., & n.* ● *adj.* **1** (of a train etc.) not stopping at intermediate places. **2** (of a journey, performance, etc.) done without a stop or intermission. ● *adv.* without stopping or pausing. ● *n.* a non-stop train etc.

non-subscriber /ˌnɒnsəbˈskraɪbə(r)/ *n.* a person who is not a subscriber.

nonsuch /ˈnʌnsʌtʃ/ *n.* (also **nonesuch**) **1** a person or thing that is unrivalled, a paragon. **2** a leguminous plant, *Medicago lupulina*, with black pods. [NONE[1] + SUCH, usu. now assim. to NON-]

nonsuit /nɒnˈsuːt, -ˈsjuːt/ *n. & v. Law* the stoppage of a suit by the judge when the plaintiff fails to make out a legal case or to bring sufficient evidence. ● *v.tr.* subject (a plaintiff) to a nonsuit. [ME f. AF *no(u)nsuit*]

non-swimmer /nɒnˈswɪmə(r)/ *n.* a person who cannot swim.

non-technical /nɒnˈteknɪk(ə)l/ *adj.* **1** not technical. **2** without technical knowledge.

non-toxic /nɒnˈtɒksɪk/ *adj.* not toxic.

non-transferable /ˌnɒntrænsˈfɜːrəb(ə)l, ˌnɒntrɑːns-/ *adj.* that may not be transferred.

non-U /nɒnˈjuː/ *adj. colloq.* not characteristic of the upper class. [NON- + U[2]]

non-uniform /nɒnˈjuːnɪˌfɔːm/ *adj.* not uniform.

non-union /nɒnˈjuːnɪən/ *adj.* **1** not belonging to a trade union. **2** not done or produced by members of a trade union.

non-usage /nɒnˈjuːsɪdʒ/ *n.* failure to use.

non-use /nɒnˈjuːs/ *n.* failure to use. □ **non-user** /-ˈjuːzə(r)/ *n.*

non-verbal /nɒnˈvɜːb(ə)l/ *adj.* not involving words or speech. □ **non-verbally** *adv.*

non-vintage /nɒnˈvɪntɪdʒ/ *adj.* (of wine etc.) not vintage.

non-violence /nɒnˈvaɪələns/ *n.* the avoidance of violence, esp. as a principle. □ **non-violent** *adj.*

non-volatile /nɒnˈvɒləˌtaɪl/ *adj.* (esp. of a substance) not volatile.

non-voting /nɒnˈvəʊtɪŋ/ *adj.* not having or using a vote. □ **non-voter** *n.*

non-white /nɒnˈwaɪt/ *adj. & n.* ● *adj.* **1** (of a person) not white. **2** of or relating to non-white people. ● *n.* a non-white person.

non-word /ˈnɒnwɜːd/ *n.* an unrecorded or unused word.

noodle[1] /ˈnuːd(ə)l/ *n.* a strip or ring of pasta. [G *Nudel*]

noodle[2] /ˈnuːd(ə)l/ *n. sl.* **1** a simpleton. **2** the head. [18th c.: orig. unkn.]

nook /nʊk/ *n.* a corner or recess; a secluded place. □ **in every nook and cranny** everywhere. [ME *nok(e)* corner, of unkn. orig.]

nooky /ˈnʊki/ *n.* (also **nookie**) *sl.* sexual activity. [20th c.: perh. f. NOOK]

noon /nuːn/ *n.* **1** twelve o'clock in the day, midday. **2** the culminating point. [OE *nōn* f. L *nona (hora)* ninth hour: orig. = 3 p.m. (cf. NONE[2])]

noonday /ˈnuːndeɪ/ *n.* midday.

no one /ˈnəʊ wʌn/ *n.* no person; nobody.

noontide /ˈnuːntaɪd/ *n.* (also **noontime** /-taɪm/) midday.

noose /nuːs/ *n. & v.* ● *n.* **1** a loop with a running knot, tightening as the rope or wire is pulled, esp. in a snare, lasso, or hangman's halter. **2** a snare or bond. **3** *joc.* the marriage tie. ● *v.tr.* **1** catch with or enclose in a noose, ensnare. **2 a** make a noose on (a cord). **b** (often foll. by *round*) arrange (a cord) in a noose. □ **put one's head in a noose** bring about one's own downfall. [ME *nose*, perh. f. OF *no(u)s* f. L *nodus* knot]

Nootka /ˈnuːtkə, ˈnʊt-/ *n. & adj.* (also **Nuu-chah-nulth** /ˈnuːtʃɑːˈnʊlθ/) ● *n.* (*pl.* same or **Nootkas**) **1** a member of an American Indian people of Vancouver Island. **2** the language of the Nootka. ● *adj.* of or relating to this people or their language. □ **Nootkan** *adj. & n.*

nootropic /ˌnəʊəˈtrəʊpɪk, -ˈtrɒpɪk/ *adj. & n.* ● *adj.* (of a drug) enhancing cognitive function, esp. memory, used to treat some cases of dementia. ● *n.* a nootropic drug. [Gk *noos* mind + -TROPIC]

nopal /ˈnəʊp(ə)l/ *n.* an American cactus, *Nopalea cochinellifera*, related to the prickly pear, grown in plantations for breeding cochineal. [F & Sp. f. Nahuatl *nopalli* cactus]

nope /nəʊp/ *adv. colloq.* = NO[2] 1. [NO[2]]

nor /nɔː(r), nə(r)/ *conj.* **1** and not; and not either (*neither one thing nor the other; not a man nor a child was to be seen; I said I had not seen it, nor had I; all that is true, nor must we forget …; can neither read nor write*). **2** and no more; neither ('*I cannot go*' – '*Nor can I*'). □ **nor … nor** … *poet.* or *archaic* neither … nor … [ME, contr. f. obs. *nother* f. OE *nawther*, *nāhwæther* (as NO[2], WHETHER)]

nor' /nɔː(r)/ *n., adj., & adv.* (esp. in compounds) = NORTH (*nor'ward; nor'wester*). [abbr.]

noradrenalin /ˌnɔːrəˈdrenəlɪn/ *n.* (also **noradrenaline**) *Biochem.* a hormone released by the adrenal medulla and by sympathetic nerve endings as a neurotransmitter. Also called *norepinephrine*. [normal + ADRENALIN]

Nordic /ˈnɔːdɪk/ *adj. & n.* ● *adj.* **1** of or relating to a physical type of northern Germanic peoples characterized by tall stature and fair colouring. **2** of and relating to Scandinavia or Finland. **3** (of skiing) involving cross-country work and jumping. ● *n.* a Nordic person, esp. a native of Scandinavia or Finland. [F *nordique* f. *nord* north]

Nordkapp see NORTH CAPE.

Nordkyn /ˈnʊətʃʊn/ a promontory on the north coast of Norway, to the east of North Cape. At 71° 8′ N, it is the northernmost point of the European mainland, North Cape being on an island.

Nord-Pas-de-Calais /ˌnɔːpɑːdəˈkæleɪ/ a region of northern France, on the border with Belgium.

Nordrhein-Westfalen see NORTH RHINE-WESTPHALIA.

norepinephrine /ˌnɔːrepɪˈnefrɪn, -riːn/ n. Biochem. = NORADRENALIN. [*normal* + EPINEPHRINE]

Norfolk /ˈnɔːfək/ a county on the east coast of England, east of the Wash; county town, Norwich.

Norfolk Island an island in the Pacific Ocean, off the east coast of Australia, administered since 1913 as an external territory of Australia; pop. (1986) 1,977. Discovered by Captain James Cook in 1774, it was occupied from 1788 to 1814 as a penal colony. The island was settled in 1856 by some of the descendants of the mutineers from the *Bounty*, to ease overcrowding on Pitcairn Island.

Norfolk jacket n. a man's loose belted jacket, with box pleats. [NORFOLK]

Norge see NORWAY.

Noriega /ˌnɒriˈeɪgə/, Manuel (Antonio Morena) (b.1940), Panamanian statesman and general, head of state 1983–9. He was Panama's head of intelligence (1970), becoming Chief of Staff and de facto head of state in 1983. He was charged with drug trafficking by a US grand jury in 1988; US support for his regime was withdrawn, relations worsened, and a year later President George Bush sent US troops into the country to arrest Noriega. He eventually surrendered and was brought to trial and convicted in 1992.

nork /nɔːk/ n. (usu. in *pl.*) Austral. sl. a woman's breast. [20th c.: orig. uncert.]

norland /ˈnɔːlənd/ n. Brit. a northern region. [contr. of NORTHLAND]

norm /nɔːm/ n. **1** a standard or pattern or type. **2** a standard quantity to be produced or amount of work to be done. **3** customary behaviour etc. [L *norma* carpenter's square]

normal /ˈnɔːm(ə)l/ adj. & n. ● adj. **1** conforming to a standard; regular, usual, typical. **2** free from mental or emotional disorder. **3** Geom. (of a line) at right angles, perpendicular. **4** Chem. (of a solution) containing one gram-equivalent of solute per litre. ● n. **1 a** the normal value of a temperature etc., esp. blood-heat. **b** the usual state, level, etc. **2** Geom. a line at right angles. □ **normal distribution** Statistics a function that represents the distribution of many random variables as a symmetrical bell-shaped graph (also called *Gaussian distribution*). **normal school** (in the US, France, etc.) a school or college for training teachers. □ **normalcy** n. esp. N. Amer. **normality** /nɔːˈmælɪti/ n. [F *normal* or L *normalis* (as NORM)]

normalize /ˈnɔːməˌlaɪz/ v. (also **-ise**) **1** tr. make normal. **2** intr. become normal. **3** tr. cause to conform. □ **normalizer** n. **normalization** /ˌnɔːməlaɪˈzeɪʃ(ə)n/ n.

normally /ˈnɔːməli/ adv. **1** in a normal manner. **2** usually.

Norman[1] /ˈnɔːmən/ n. & adj. ● n. **1** a native or inhabitant of Normandy. **2** a descendant of the people inhabiting Normandy in early medieval times. (*See note below.*) **3** Norman French. **4** Archit. the style of Romanesque architecture developed by the Normans and employed in England after the Conquest, with rounded arches and heavy pillars. **5** any of the English kings from William I to Stephen. ● adj. **1** of or relating to the Normans. **2** of or relating to the Norman style of architecture. □ **Normanism** n. **Normanize** v.tr. & intr. (also **-ise**). **Normanesque** /ˌnɔːməˈnesk/ adj. [OF *Normans* pl. of *Normant* f. ON *Northmathr* (as NORTH, MAN)]

• The Normans were of mixed Frankish and Scandinavian origin, a group of Norsemen under their chief Rollo (c.860–c.932) having settled in Normandy in 912. They were the dominant military power in western Europe in the 11th century, conquering England in 1066 and, through the activities of adventurers such as Robert Guiscard (c.1015–85), setting up Norman kingdoms as far afield as Sicily, Greece, and the Holy Land.

Norman[2] /ˈnɔːmən/, Gregory John ('Greg') (b.1955), Australian golfer. He turned professional in 1976 and subsequently won the world match-play championship three times (1980; 1983; 1986) and the British Open twice (1986; 1993).

Norman[3] /ˈnɔːmən/, Jessye (b.1945), American operatic soprano. She made her début in Berlin (1969) and subsequently performed in the major European opera houses, appearing in New York for the first time in 1973. Her repertoire includes both opera and concert music and she has given notable interpretations of the works of Wagner, Schubert, and Mahler.

Norman Conquest (also **the Conquest**) the conquest of England by William of Normandy (William the Conqueror) after the Battle of Hastings in 1066. William was crowned William I, founding a Norman dynasty that ruled until 1154, and had crushed most resistance by 1071. Under the Normans England prospered commercially and expanded its population, but most of the Saxon nobles were dispossessed or killed and the population was heavily taxed (the Domesday Book was compiled in 1086). Norman institutions and customs (such as feudalism) were introduced, and Anglo-French and Latin adopted as the languages of literature, law, and government.

Normandy /ˈnɔːməndi/ a former province of NW France with its coastline on the English Channel, now divided into the two regions of Lower Normandy (Basse-Normandie) and Upper Normandy (Haute-Normandie); chief town, Rouen. In 912 it was ceded by Charles III of France to Rollo (c.860–c.932), first Duke of Normandy. It was contested between France and England throughout the Middle Ages until finally conquered by France in the mid-15th century.

Norman English n. English as spoken or influenced by the Normans.

Norman French n. French as spoken by the Normans or (after 1066) in English lawcourts.

normative /ˈnɔːmətɪv/ adj. of or establishing a norm. □ **normatively** adv. **normativeness** n. [F *normatif -ive* f. L *norma* (see NORM)]

Norn /nɔːn/ n. & adj. ● n. a form of Norwegian formerly spoken in Orkney and Shetland. ● adj. of or relating to this language. [ON *norroen* adj., *norr oenna* n., f. *norðr* north]

Norns /nɔːnz/ n.pl. Scand. Mythol. the three virgin goddesses of destiny (Urd or Urdar, Verdandi, and Skuld), who sit by the well of fate at the base of the ash tree Yggdrasil and spin the web of fate. [ON: orig. unkn.]

Norrköping /ˈnɔːˌtʃɜːpɪŋ/ an industrial city and seaport on an inlet of the Baltic Sea in SE Sweden; pop. (1990) 120,520.

Norroy /ˈnɒrɔɪ/ n. (in full **Norroy and Ulster**) Heraldry (in the UK) the title given to the third King of Arms, with jurisdiction north of the Trent and (since 1943) in Northern Ireland (cf. CLARENCEUX, KING OF ARMS). [ME f. AF *norroi* (unrecorded) f. OF *nord* north, *roi* king]

Norse /nɔːs/ n. & adj. ● n. **1 a** the Norwegian language. **b** the Scandinavian language-group. **2** (prec. by *the*; treated as *pl.*) **a** the Norwegians. **b** the Vikings. ● adj. **1** of or relating to Norway or Norwegian. **2** of or relating to ancient Scandinavia. □ **Norseman** n. (*pl.* **-men**). [Du. *noor(d)sch* f. *noord* north]

North /nɔːθ/, Frederick, Lord (1732–92), British Whig statesman, Prime Minister 1770–82. He sought to avoid the War of American Independence, but was regarded as responsible for the loss of the American colonies. This, together with allegations that his ministry was dominated by the influence of George III, led to his resignation in 1782.

north /nɔːθ/ n., adj., & adv. ● n. **1 a** the point of the horizon 90° anticlockwise from east. **b** the compass point corresponding to this. **c** the direction in which this lies. **2** (usu. **the North**) the part of the world or a country or a town lying to the north; esp. **a** = *north country*. **b** the states in the north of the US, esp. as forming the Union side in the American Civil War. **c** the Arctic. **d** the industrialized nations. **3** (**North**) Bridge a player occupying the position designated 'north'. ● adj. **1** towards, at, near, or facing north. **2** coming from the north (*north wind*). ● adv. **1** towards, at, or near the north. **2** (foll. by *of*) further north than. □ **north and south** lengthwise along a line from north to south. **north by east** (or **west**) between north and north-north-east (or north-north-west). **north country** the northern part of England (north of the Humber). **north-countryman** (*pl.* **-men**) a native of the north country. **north light** good natural light without direct sun, esp. as desired by painters and in factory design. **north pole** see POLE[2] 1. **to the north** (often foll. by *of*) in a northerly direction. [OE f. Gmc]

North Africa the northern part of the African continent, especially the countries bordering the Mediterranean and the Red Sea.

Northallerton /nɔːˈθælət(ə)n/ a town in the north of England, the administrative centre of North Yorkshire; pop. (1981) 13,860.

North America a continent comprising the northern half of the American land mass, connected to South America by the Isthmus of Panama. It contains Canada, the United States, Mexico, and the countries of Central America. (See also AMERICA.) □ **North American** adj. & n.

North American English see AMERICAN ENGLISH.

North American Free Trade Agreement (abbr. **NAFTA**) an agreement which came into effect in Jan. 1994 between the US,

Canada, and Mexico to remove barriers to trade between the three countries over a ten-year period.

Northampton /nɔː'θæmptən/ a town in SE central England, on the River Nene, the county town of Northamptonshire; pop. (1991) 178,200.

Northamptonshire /nɔː'θæmptən,ʃɪə(r)/ a county of central England; county town, Northampton.

Northants abbr. Northamptonshire.

North Atlantic Drift a continuation of the Gulf Stream across the Atlantic Ocean and along the coast of NW Europe, where it has a significant warming effect on the climate.

North Atlantic Ocean see ATLANTIC OCEAN.

North Atlantic Treaty Organization (abbr. **NATO, Nato**) an association of European and North American states, formed in 1949 for the defence of Europe and the North Atlantic against the perceived threat of Soviet aggression. Under strong influence from its most powerful member, the US, it now includes all the major non-neutral Western powers, although France withdrew from the military side of the alliance in 1966. With the collapse of NATO's main notional adversary, the Warsaw Pact, NATO forces have been reduced.

northbound /'nɔːθbaʊnd/ adj. travelling or leading northwards.

North Cape (Norwegian **Nordkapp** /'nuːrkap/) a promontory on Magerøya, an island off the north coast of Norway. Situated on the edge of the Barents Sea, North Cape is the northernmost point of the world accessible by road.

North Carolina /,kærə'laɪnə/ a state of the east central US, on the Atlantic coast; pop. (1990) 6,628,640; capital, Raleigh. Originally settled by the English, it was named from *Carolus*, the Latin name of Charles I, and was one of the original thirteen states of the Union (1788). (See also SOUTH CAROLINA.)

North Channel the stretch of sea separating SW Scotland from Northern Ireland and connecting the Irish Sea to the Atlantic Ocean.

Northcliffe /'nɔːθklɪf/, 1st Viscount (title of Alfred Charles William Harmsworth) (1865–1922), British newspaper proprietor. With his younger brother Harold (later Lord Rothermere, 1868–1940), Northcliffe built up a large newspaper empire in the years preceding the First World War, including *The Times*, the *Daily Mail*, and the *Daily Mirror*. During the war he used his press empire to exercise a strong influence over British war policy; despite his criticism of Lloyd George and Lord Kitchener, he worked for the British government in charge of propaganda in enemy countries in 1917–18.

North Dakota an agricultural state in the north central US, on the border with Canada; pop. (1990) 638,800; capital, Bismarck. Acquired partly by the Louisiana Purchase in 1803 and partly from Britain by treaty in 1818, it became the 39th state of the US in 1889.

north-east /nɔː'iːst, Naut. nɔːr'iːst/ n., adj., & adv. ● n. **1** the point of the horizon midway between north and east. **2** the compass point corresponding to this. **3** the direction in which this lies. **4** (usu. **the North-East**) the part of a country or town lying to the north-east. ● adj. of, towards, or coming from the north-east. ● adv. towards, at, or near the north-east. □ **north-easterly** adj. & adv. **north-eastern** adj.

northeaster /nɔː'θiːstə(r), Naut. nɔːr'iːst-/ n. a north-east wind.

North-east Passage a passage for ships along the northern coast of Europe and Asia, from the Atlantic to the Pacific via the Arctic Ocean, sought for many years as a possible trade route to the East. It was first navigated in 1878–9 by the Swedish Arctic explorer Baron Nordenskjöld (1832–1901).

norther /'nɔːðə(r)/ n. US a strong cold north wind blowing in autumn and winter over Texas, Florida, and the Gulf of Mexico.

northerly /'nɔːðəlɪ/ adj., adv., & n. ● adj. & adv. **1** in a northern position or direction. **2** (of wind) blowing from the north. ● n. (pl. **-ies**) (usu. in pl.) a wind blowing from the north.

northern /'nɔːðən/ adj. **1** of or in the north; inhabiting the north. **2** lying or directed towards the north. □ **northern harrier** N. Amer. = hen harrier. **northern lights** the aurora borealis. **northern hemisphere** the half of the earth north of the equator. □ **northernmost** adj. [OE *northerne* (as NORTH, -ERN)]

Northern Cape a province of western South Africa, formerly part of Cape Province; capital, Kimberley.

Northern Circars /'sɜːkɑːz/ a former name for the coastal region of eastern India between the Krishna River and Orissa, now in Andhra Pradesh.

northerner /'nɔːðənə(r)/ n. a native or inhabitant of the north.

Northern Ireland a province of the United Kingdom occupying the north-eastern part of Ireland; pop. (1991) 1,569,790; capital, Belfast. Northern Ireland, which comprises six of the counties of Ulster, was established as a self-governing province in 1920, having refused to be part of the Irish Free State which was established the following year. Northern Ireland has always been dominated by Unionist parties, which represent the Protestant majority. Many members of the Roman Catholic minority favour union with the Republic of Ireland. Discrimination against the latter group in local government, employment, and housing led to violent conflicts and (from 1969) the presence of British army units in an attempt to keep the peace. Terrorism and sectarian violence by the Provisional IRA and other paramilitary groups, both Republican and Loyalist, resulted in the imposition of direct rule from Westminster in 1972. Attempts to end terrorism and to organize an agreed and permanent system of government have included initiatives such as the power-sharing Northern Ireland Executive (1973–4), while in August 1994 the IRA declared a ceasefire. Closer cooperation has also developed between Britain and the Republic of Ireland over Northern Irish affairs, especially through the Anglo-Irish Agreement of 1985 and the 'Downing Street Declaration' of 1993.

Northern Marianas a self-governing territory in the western Pacific, comprising the Mariana Islands with the exception of the southernmost, Guam; pop. (1990) 43,345; languages, English (official), Malayo-Polynesian languages; capital, Chalan Kanoa (on Saipan). The Northern Marianas are constituted as a self-governing commonwealth in union with the United States. (See also MARIANA ISLANDS.)

Northern Paiute see PAIUTE.

Northern Rhodesia the former name (until 1964) for ZAMBIA.

Northern Territory a state of north central Australia; pop. (1990) 158,400; capital, Darwin. The territory was annexed by the state of South Australia in 1863, and administered by the Commonwealth of Australia from 1911. It became a self-governing territory in 1978 and a full state of Australia in 1995.

Northern Transvaal a province of northern South Africa, formerly part of Transvaal; capital, Pietersburg.

North Frisian Islands see FRISIAN ISLANDS.

northing /'nɔːθɪŋ, 'nɔːðɪŋ/ n. Naut. the distance travelled or measured northward.

North Island the northernmost of the two main islands of New Zealand, separated from South Island by Cook Strait.

North Korea (official name **Democratic People's Republic of Korea**) a country in the Far East, occupying the northern part of the peninsula of Korea; pop. (est. 1992) 22,227,000; official language, Korean; capital, Pyongyang. North Korea was formed in 1948 when Korea was partitioned along the 38th parallel. In 1950 North Korean forces invaded the south, but were forced back to more or less the previous border (see KOREAN WAR). A Communist state which was long dominated by the personality of Kim Il Sung, its leader from 1948 to 1994, North Korea has always sought Korean reunification. After decades of hostility, North and South Korea signed a series of preliminary accords in 1991; suspended in late 1992, talks between the countries were resumed in 1994. In 1993 North Korea refused to allow officials from the International Atomic Energy Agency (IAEA) to inspect nuclear facilities, amid international fears that it had developed nuclear weapons; in the following year it withdrew from the IAEA. □ **North Korean** adj. & n.

Northland /'nɔːðlənd/ n. poet. the northern lands; the northern part of a country. [OE (as NORTH, LAND)]

Northman /'nɔːθmən/ n. (pl. **-men**) a native of Scandinavia, esp. of Norway. [OE]

North Minch see MINCH, THE.

north-north-east /,nɔːθnɔː'iːst, Naut. ,nɔːnɔːr'iːst/ n., adj., & adv. midway between north and north-east.

north-north-west /,nɔːθnɔː'θwest, Naut. ,nɔːnɔː'west/ n., adj., & adv. midway between north and north-west.

North Ossetia an autonomous republic of Russia, in the Caucasus on the border with Georgia; pop. (1990) 638,000; capital, Vladikavkaz. (See also OSSETIA.)

North Rhine-Westphalia /,raɪnwest'feɪlɪə/ (German **Nordrhein-Westfalen** /,nɔːtraɪnvest'faːlən/) a state of western Germany; capital, Düsseldorf.

North Sea an arm of the Atlantic Ocean lying between the mainland of Europe and the east coast of Britain. Since 1959 it has become important for the exploitation of oil and gas deposits under the seabed.

North Star the pole star (see POLARIS).

North Uist see UIST.

Northumb. abbr. Northumberland.

Northumberland /nɔːˈθʌmbələnd/ a county in NE England, on the Scottish border; county town, Morpeth.

Northumbria /nɔːˈθʌmbrɪə/ **1** an ancient Anglo-Saxon kingdom in NE England extending from the Humber to the Forth. **2** (loosely) an area of NE England comprising Northumberland, Durham, and Tyne and Wear. □ **Northumbrian** adj. & n. [obs. Northumber, persons living beyond the Humber, f. OE Northhymbre]

North Utsire see UTSIRE.

northward /ˈnɔːθwəd/ Naut. ˈnɔːðəd/ adj., adv., & n. ● adj. & adv. (also **northwards**) towards the north. ● n. a northward direction or region.

north-west /nɔːθˈwest/ Naut. nɔːˈwest/ n., adj., & adv. ● n. **1** the point of the horizon midway between north and west. **2** the compass point corresponding to this. **3** the direction in which this lies. **4** (usu. **the North-West**) the part of a country or town lying to the north-west. ● adj. of, towards, or coming from the north-west. ● adv. towards, at, or near the north-west. □ **north-westerly** adj. & adv. **north-western** adj.

northwester /nɔːθˈwestə(r)/ Naut. nɔːˈwest-/ n. a north-west wind.

North-West Frontier Province a province of NW Pakistan, on the border with Afghanistan; capital, Peshawar.

North-west Passage a sea passage along the northern coast of the American continent, through the Canadian Arctic from the Atlantic to the Pacific. It was sought for centuries as a possible trading route by many explorers, including Sebastian Cabot, Sir Francis Drake, Martin Frobisher, and Henry Hudson. In 1850–4 it was first completed on foot by the Irish naval officer Robert McClure (1807–73), and was finally successfully navigated in 1903–6 by Roald Amundsen.

North-West Province a province of northern South Africa, formed in 1994 from the north-eastern part of Cape Province and SW Transvaal; capital, Mmabatho.

Northwest Territories a territory of northern Canada extending northwards from the 60th parallel and westwards from Hudson Bay to the Rocky Mountains; pop. (1991) 57,650; capital, Yellowknife. For the most part it consists of vast, inhospitable, and sparsely inhabited forests and tundra. From 1670 the southern part of this territory was part of Rupert's Land and administered by the Hudson's Bay Company. The remainder was under nominal British rule until 1870, when both parts were ceded to Canada.

Northwest Territory a region and former territory of the US lying between the Mississippi and Ohio rivers and the Great Lakes. It was acquired in 1783 after the War of American Independence and now forms the states of Indiana, Ohio, Michigan, Illinois, and Wisconsin.

North Yorkshire a county in NE England; administrative centre, Northallerton. It was formed in 1974 from parts of the former North, East, and West Ridings of Yorkshire.

Norway /ˈnɔːweɪ/ (Norwegian **Norge** /ˈnɔrgə/) a mountainous European country on the northern and western coastline of Scandinavia; pop. (1991) 4,249,830; official language, Norwegian; capital, Oslo. An independent if often divided kingdom in Viking and early medieval times, Norway was united with Denmark and Sweden by the Union of Kalmar in 1397, but after Sweden's withdrawal in 1523 became subject to Denmark. Ceded to Sweden in 1814, Norway emerged as an independent kingdom in 1905. It was occupied by German forces 1940–5. An invitation to join the EEC was rejected after a referendum in 1972; an application to join the European Union twenty years later was accepted by the European Parliament but failed to win approval in a referendum (1994).

Norway lobster n. a long slender European lobster, Nephrops norvegicus, eaten as scampi. Also called Dublin Bay prawn.

Norway rat n. the common brown rat, Rattus norvegicus.

Norwegian /nɔːˈwiːdʒən/ n. & adj. ● n. **1 a** a native or national of Norway. **b** a person of Norwegian descent. **2** the official language of Norway, which belongs to the Scandinavian language group. (See note below.) ● adj. of or relating to Norway or its people or language. [med.L Norvegia f. ON Norvegr (as NORTH, WAY), assim. to Norway]

▪ Norwegian today exists in two forms, a situation which dates from Norway's domination by Denmark in the Middle Ages, when Danish was the language of the upper classes and Norwegian was spoken by the peasants. The forms are Bokmål, the more widely used (also called Riksmål or Dano-Norwegian), a modified form of Danish, and Nynorsk (= 'new Norwegian', formerly called Landsmål), a literary form

devised in the 19th century from the country dialects most closely descended from Old Norse, and considered to be a purer form of the language than Bokmål.

Norwegian Sea a sea which lies between Iceland and Norway and links the Arctic Ocean with the NE Atlantic.

nor'-wester /nɔːˈwestə(r)/ n. **1** a northwester. **2** a glass of strong liquor. **3** an oilskin hat, a sou'wester. [contr.]

Norwich /ˈnɒrɪdʒ, -rɪtʃ/ a city in eastern England, the county town of Norfolk; pop. (1991) 121,000.

Norwich School an English regional school of landscape painting associated with the Norwich Society of Artists, an institution founded in 1803. Leading members of the School were John Sell Cotman and John Crome.

Nos. abbr. numbers. [cf. No.]

nose /nəʊz/ n. & v. ● n. **1** an organ above the mouth on the face or head of a human or animal, containing nostrils and used for smelling and breathing. **2 a** the sense of smell (dogs have a good nose). **b** the ability to detect a particular thing (a nose for scandal). **3** the odour or perfume of wine, tea, tobacco, hay, etc. **4** the open end or nozzle of a tube, pipe, pair of bellows, retort, etc. **5 a** the front end or projecting part of a thing, e.g. of a car or aircraft. **b** = NOSING. **6** sl. an informer of the police. ● v. **1** tr. (often foll. by out) **a** perceive the smell of, discover by smell. **b** detect. **2** tr. thrust or rub one's nose against or into, esp. in order to smell. **3** intr. (usu. foll. by about, around, etc.) pry or search. **4 a** intr. make one's way cautiously forward. **b** tr. make (one's or its way). □ **as plain as the nose on your face** easily seen. **by a nose** by a very narrow margin (won the race by a nose). **count noses** count those present, one's supporters, etc.; decide a question by mere numbers. **cut off one's nose to spite one's face** disadvantage oneself in the course of trying to disadvantage another. **get up a person's nose** sl. annoy a person. **keep one's nose clean** sl. stay out of trouble, behave properly. **keep one's nose to the grindstone** see GRINDSTONE. **nose-cone** the cone-shaped nose of a rocket etc. **nose-flute** a musical instrument blown with the nose in Fiji etc. **nose leaf** a fleshy part on the nostrils of some bats, used for echolocation. **nose-monkey** the proboscis monkey. **nose-piece 1** = NOSEBAND. **2** the part of a helmet etc. protecting the nose. **3** the part of a microscope to which the objective lenses are attached. **nose-rag** sl. a pocket handkerchief. **nose tackle** Amer. Football the defensive player who lines up in the centre of the formation. **nose-to-tail** (of vehicles) moving or stationary one close behind another, esp. in heavy traffic. **nose-wheel** a landing-wheel under the nose of an aircraft. **on the nose 1** N. Amer. sl. precisely. **2** Austral. sl. annoying. **3** Austral. sl. stinking. **poke** (or **stick**) **one's nose into** colloq. pry or intrude into (esp. a person's affairs). **put a person's nose out of joint** colloq. embarrass, disconcert, frustrate, or supplant a person. **rub a person's nose in it** see rub it in (see RUB[1]). **see no further than one's nose** be short-sighted, esp. in foreseeing the consequences of one's actions etc. **speak through one's nose** pronounce words with a nasal twang. **turn up one's nose** (usu. foll. by at) colloq. show disdain. **under a person's nose** colloq. right before a person (esp. of defiant or unnoticed actions). **with one's nose in the air** haughtily. □ **nosed** adj. (also in comb.). **noseless** adj. [OE nosu]

nosebag /ˈnəʊzbæg/ n. a bag containing fodder, hung on a horse's head.

noseband /ˈnəʊzbænd/ n. the lower band of a bridle, passing over the horse's nose.

nosebleed /ˈnəʊzbliːd/ n. an instance of bleeding from the nose.

nosedive /ˈnəʊzdaɪv/ n. & v. ● n. **1** a steep downward plunge by an aeroplane. **2** a sudden plunge or drop. ● v.intr. make a nosedive.

nosegay /ˈnəʊzgeɪ/ n. a bunch of flowers, esp. a sweet-scented posy. [NOSE + GAY in obs. use = ornament]

nosepipe /ˈnəʊzpaɪp/ n. a piece of piping used as a nozzle.

nosering /ˈnəʊzrɪŋ/ n. a ring fixed in the nose of an animal (esp. a bull) for leading it, or of a person for ornament.

nosey var. of NOSY.

nosh /nɒʃ/ v. & n. sl. ● v.tr. & intr. **1** eat or drink. **2** N. Amer. eat between meals. ● n. **1** food or drink. **2** N. Amer. a snack. □ **nosh-up** Brit. a large meal. [Yiddish]

noshery /ˈnɒʃərɪ/ n. (pl. **-ies**) sl. a restaurant or snack bar.

nosing /ˈnəʊzɪŋ/ n. a rounded edge of a step, moulding, etc., or a metal shield for it.

nosography /nəˈsɒɡrəfɪ/ n. Med. the systematic description of diseases. [Gk nosos disease + -GRAPHY]

nosology /nəˈsɒlədʒɪ/ n. the branch of medical science dealing with the classification of diseases. □ **nosological** /ˌnɒsəˈlɒdʒɪk(ə)l/ adj. [Gk nosos disease + -LOGY]

nostalgia /nɒˈstældʒə/ n. **1** (often foll. by for) sentimental yearning for a period of the past; regretful or wistful memory of an earlier time. **2** a thing or things which evoke a former era. **3** severe homesickness. □ **nostalgic** adj. **nostalgically** adv. [mod.L f. Gk nostos return home]

nostoc /ˈnɒstɒk/ n. a unicellular blue-green alga of the genus Nostoc, forming gelatinous masses and able to fix atmospheric nitrogen. [name invented by Paracelsus]

Nostradamus /ˌnɒstrəˈdɑːməs, -ˈdeɪməs/ (Latinized name of Michel de Nostredame) (1503–66), French astrologer and physician. His predictions, in the form of rhymed quatrains, appeared in two collections (1555; 1558). Cryptic and apocalyptic in tone, they were given extensive credence at the French court, where Nostradamus was for a time personal physician to Charles IX. Their interpretation has continued to be the subject of controversy into the 20th century.

nostril /ˈnɒstrɪl/ n. either of two external openings of the nasal cavity in vertebrates that admit air to the lungs and smells to the olfactory nerves. □ **nostrilled** adj. (also in comb.). [OE nosthyrl, nosterl f. nosu NOSE + thȳr(e)l hole: cf. THRILL]

nostrum /ˈnɒstrəm/ n. **1** a quack remedy, a patent medicine, esp. one prepared by the person recommending it. **2** a pet scheme, esp. for political or social reform. [L, neut. of noster our, used in sense 'of our own make']

nosy /ˈnəʊzɪ/ adj. & n. (also **nosey**) ● adj. (**nosier, nosiest**) **1** colloq. inquisitive, prying. **2** having a large nose. **3** having a distinctive (good or bad) smell. ● n. (pl. **-ies**) a person with a large nose. □ **nosily** adv. **nosiness** n.

Nosy Parker n. esp. Brit. colloq. a busybody.

not /nɒt/ adv. expressing negation, esp.: **1** (also **n't** joined to a preceding verb) following an auxiliary verb or be or (in a question) the subject of such a verb (I cannot say; she isn't there; didn't you tell me?; am I not right?; aren't we smart?). ¶ Use with other verbs is now archaic (I know not; fear not), except with participles and infinitives (not realizing her mistake; we asked them not to come). **2** used elliptically for a negative sentence or verb or phrase (Is she coming? – I hope not; Do you want it? – Certainly not!). **3** used to express the negative of other words (not a single one was left; Are they pleased? – Not they; he is not my cousin, but my nephew). **4** sl. following and emphatically negating an affirmative statement (great party ... not!). □ **not at all** (in polite reply to thanks) there is no need for thanks. **not but what** archaic **1** all the same; nevertheless (I cannot do it; not but what a stronger man might). **2** not such ... or so ... that ... not (not such a fool but what he can see it). **not half** see HALF. **not least** with considerable importance, notably. **not much** see MUCH. **not quite 1** almost (am not quite there). **2** noticeably not (not quite proper). **not that** (foll. by clause) it is not to be inferred that (if he said so – not that he ever did – he lied). **not a thing** nothing at all. **not very** see VERY. [ME contr. of NOUGHT]

nota bene /ˌnəʊtə ˈbeneɪ/ v.tr. (abbr. **NB**) (as imper.) observe what follows, take notice (usu. drawing attention to a following qualification of what has preceded). [L, = note well]

notability /ˌnəʊtəˈbɪlɪtɪ/ n. (pl. **-ies**) **1** the state of being notable (names of no historical notability). **2** a prominent person. [ME f. OF notabilité or LL notabilitas as NOTABLE]

notable /ˈnəʊtəb(ə)l/ adj. & n. ● adj. worthy of note; striking, remarkable, eminent. ● n. an eminent person. □ **notably** adv. [ME f. OF f. L notabilis (as NOTE)]

notarize /ˈnəʊtəˌraɪz/ v.tr. (also **-ise**) N. Amer. certify (a document) as a notary.

notary /ˈnəʊtərɪ/ n. (pl. **-ies**) (in full **notary public**) a person authorized to perform certain legal formalities, esp. to draw up or certify contracts, deeds, etc. □ **notarial** /nəʊˈteərɪəl/ adj. **notarially** adv. [ME f. L notarius secretary (as NOTE)]

notate /nəʊˈteɪt/ v.tr. write in notation. [back-form. f. NOTATION]

notation /nəʊˈteɪʃ(ə)n/ n. **1 a** the representation of numbers, quantities, pitch and duration of musical notes, etc. by symbols. **b** any set of such symbols. **2** a set of symbols used to represent chess moves, dance steps, etc. **3** US **a** a note or annotation. **b** a record. **4** = scale of notation (see SCALE[3] n. 5). □ **notational** adj. [F notation or L notatio (as NOTE)]

notch /nɒtʃ/ n. & v. ● n. **1** a V-shaped indentation on an edge or surface. **2** a nick made on a stick etc. in order to keep count. **3** colloq. a step or degree (move up a notch). **4** N. Amer. a deep narrow mountain pass. ● v.tr.

1 make notches in. **2** (foll. by up) record or score with or as with notches. **3** secure or insert by notches. □ **notched** adj. **notcher** n. **notchy** adj. (**notchier, notchiest**). [AF noche perh. f. a verbal form nocher (unrecorded), of uncert. orig.]

note /nəʊt/ n. & v. ● n. **1** a brief record of facts, topics, thoughts, etc., as an aid to memory, for use in writing, public speaking, etc. (often in pl.: make notes; spoke without notes). **2** an observation, usu. unwritten, of experiences etc. (compare notes). **3** a short or informal letter. **4** a formal diplomatic or parliamentary communication. **5** a short annotation or additional explanation in a book etc.; a footnote. **6 a** Brit. = BANKNOTE (a five-pound note). **b** a written promise or notice of payment of various kinds. **7 a** notice, attention (worthy of note). **b** distinction, eminence (a person of note). **8 a** a written sign representing the pitch and duration of a musical sound. **b** a single tone of definite pitch made by a musical instrument, the human voice, etc. **c** a key of a piano etc. **9 a** a bird's song or call. **b** a single tone in this. **10** a quality or tone of speaking, expressing mood or attitude etc.; a hint or suggestion (sound a note of warning; ended on a note of optimism). **11** a characteristic; a distinguishing feature. ● v.tr. **1** observe, notice; give or draw attention to. **2** (often foll. by down) record as a thing to be remembered or observed. **3** (in passive; often foll. by for) be famous or well known (for a quality, activity, etc.) (were noted for their generosity). □ **hit** (or **strike**) **the right note** speak or act in exactly the right manner. **note cluster** Mus. a group of neighbouring notes played simultaneously. **of note** important, distinguished (a person of note). **take note** (often foll. by of) observe; pay attention (to). □ **noted** adj. (in sense 3 of v.). **noteless** adj. [ME f. OF note (n.), noter (v.) f. L nota mark]

notebook /ˈnəʊtbʊk/ n. **1** a small book for making or taking notes. **2** Computing a portable computer smaller than a laptop.

notecase /ˈnəʊtkeɪs/ n. Brit. a wallet for holding banknotes.

notelet /ˈnəʊtlɪt/ n. a small folded sheet of paper, usu. with a decorative design, for an informal letter.

notepaper /ˈnəʊtˌpeɪpə(r)/ n. paper for writing letters.

noteworthy /ˈnəʊtˌwɜːðɪ/ adj. worthy of attention; remarkable. □ **noteworthiness** n.

nothing /ˈnʌθɪŋ/ n. & adv. ● n. **1** not anything (nothing has been done; have nothing to do). **2** no thing (often foll. by complement: I see nothing that I want; can find nothing useful). **3 a** a person or thing of no importance or concern; a trivial event or remark (was nothing to me; the little nothings of life). **b** (attrib.) colloq. of no value; indeterminate (a nothing sort of day). **4** non-existence; what does not exist. **5** (in calculations) no amount; nought (a third of nothing is nothing). ● adv. **1** not at all, in no way (helps us nothing; is nothing like enough). **2** US colloq. not at all (Is he ill? – Ill nothing, he's dead.). □ **be nothing to 1** not concern. **2** not compare with. **be** (or **have**) **nothing to do with 1** have no connection with. **2** not be involved or associated with. **for nothing 1** at no cost; without payment. **2** to no purpose. **have nothing on 1** be naked. **2** have no engagements. **no nothing** colloq. (concluding a list of negatives) nothing at all. **nothing doing** colloq. **1 a** there is no prospect of success or agreement. **b** I refuse. **2** nothing (is) happening. **nothing** (or **nothing else**) **for it** (often foll. by but to + infin.) no alternative (nothing for it but to pay up). **nothing** (or **not much**) **in it** (or **to it**) **1** untrue or unimportant. **2** simple to do. **3** no (or little) advantage to be seen in one possibility over another. **nothing less than** at least (nothing less than a disaster). **think nothing of it** do not apologize or feel bound to show gratitude. [OE nān thing (as NO[1], THING)]

nothingness /ˈnʌθɪŋnɪs/ n. **1** non-existence; the non-existent. **2** worthlessness, triviality, insignificance.

notice /ˈnəʊtɪs/ n. & v. ● n. **1** attention, observation (it escaped my notice). **2** a displayed sheet etc. bearing an announcement or other information. **3 a** an intimation or warning, esp. a formal one to allow preparations to be made (give notice; at a moment's notice). **b** (often foll. by to + infin.) a formal announcement or declaration of intention to end an agreement or leave employment at a specified time (hand in one's notice; notice to quit). **4** a short published review or comment about a new play, book, etc. ● v.tr. **1** (often foll. by that, how, etc. + clause) perceive, observe; take notice of. **2** remark upon; speak of. □ **at short** (or **a moment's**) **notice** with little warning. **notice-board** Brit. a board for displaying notices. **put a person on notice** US alert or warn a person. **take notice** (or **no notice**) show signs (or no signs) of interest. **take notice of 1** observe; pay attention to. **2** act upon. **under notice** served with a formal notice. [ME f. OF f. L notitia being known f. notus past part. of noscere know]

noticeable /ˈnəʊtɪsəb(ə)l/ adj. **1** easily seen or noticed; perceptible. **2** noteworthy. □ **noticeably** adv.

notifiable /ˈnəʊtɪˌfaɪəb(ə)l/ adj. (of a disease, crop pest, etc.) that must be notified to the appropriate authorities.

notify /ˈnəʊtɪˌfaɪ/ v.tr. (**-ies**, **-ied**) **1** (often foll. by of, or that + clause) inform or give notice to (a person). **2** make known; announce or report (a thing). □ **notification** /ˌnəʊtɪfɪˈkeɪʃ(ə)n/ n. [ME f. OF notifier f. L notificare f. notus known: see NOTICE]

notion /ˈnəʊʃ(ə)n/ n. **1 a** a concept or idea; a conception (it was an absurd notion). **b** an opinion (has the notion that people are honest). **c** a vague view or understanding (have no notion what you mean). **2** an inclination, impulse, or intention (has no notion of conforming). **3** (in pl.) small, useful articles, esp. haberdashery. [L notio idea f. notus past part. of noscere know]

notional /ˈnəʊʃən(ə)l/ adj. **1 a** hypothetical, imaginary. **b** (of knowledge etc.) speculative; not based on experiment etc. **2** Gram. (of a verb) conveying its own meaning, not auxiliary. □ **notionally** adv. [obs. F notional or med.L notionalis (as NOTION)]

notochord /ˈnəʊtəˌkɔːd/ n. Zool. a cartilaginous skeletal rod supporting the body in all embryo and some adult chordate animals. [Gk nōton back + CHORD[2]]

notorious /nəʊˈtɔːrɪəs/ adj. well known, esp. unfavourably (a notorious criminal; notorious for its climate). □ **notoriously** adv. **notoriety** /ˌnəʊtəˈraɪətɪ/ n. [med.L notorius f. L notus (as NOTION)]

notornis /nəˈtɔːnɪs/ n. the takahe. [mod.L Notornis, former genus name, f. Gk notos south + ornis bird]

Notre-Dame /ˌnɒtrəˈdɑːm/ the Gothic cathedral church of Paris, dedicated to the Virgin Mary, on the Île de la Cité (an island in the Seine), begun in 1163 and effectively finished by 1250. It is especially noted for its innovatory flying buttresses, sculptured façade, and great rose windows with 13th-century stained glass. [F, = our lady]

Notre Dame University /ˌnəʊtə ˈdeɪm/ a university in Notre Dame, near South Bend, northern Indiana. It was founded in 1842 by a French religious community affiliated with the Roman Catholic Church and was a men's college until 1972. It is noted for the success of its football team.

Nottingham /ˈnɒtɪŋəm/ a city in east central England, the county town of Nottinghamshire; pop. (1991) 261,500. It was the scene in 1642 of the outbreak of the English Civil War.

Nottinghamshire /ˈnɒtɪŋəmˌʃɪə(r)/ a county in central England; county town, Nottingham.

Notting Hill /ˌnɒtɪŋ ˈhɪl/ a district of NW central London, scene of an annual street carnival.

Notts. abbr. Nottinghamshire.

notwithstanding /ˌnɒtwɪθˈstændɪŋ, -wɪð'stændɪŋ/ prep., adv., & conj. ● prep. in spite of; without prevention by (notwithstanding your objections; this fact notwithstanding). ● adv. nevertheless; all the same. ● conj. (usu. foll. by that + clause) although. [ME, orig. absol. part. f. NOT + WITHSTAND + -ING[2]]

Nouadhibou /ˌnwædɪˈbuː/ the principal port of Mauritania, on the Atlantic coast at the border with Western Sahara; pop. (1976) 22,000. It was formerly known as Port Étienne.

Nouakchott /nwækˈʃɒt/ the capital of Mauritania, situated on the Atlantic coast; pop. (est. 1985) 500,000.

nougat /ˈnuːgɑː/ n. a sweet made from sugar or honey, nuts, and egg-white. [F f. Prov. nogat f. noga nut]

nought /nɔːt/ n. **1** the digit 0; a cipher. **2** poet. or archaic (in certain phrases) nothing (cf. NAUGHT n. 1). [OE nōwiht f. ne not + ōwiht var. of āwiht AUGHT[1]]

noughts and crosses n.pl. esp. Brit. a game in which players seek to complete a row of three noughts or three crosses drawn alternately in the spaces of a grid of nine squares. Also called (N. Amer.) tick-tack-toe.

Nouméa /nuːˈmeɪə/ the capital of the island of New Caledonia; pop. (1989) 65,110. It was formerly called Port de France.

noun /naʊn/ n. Gram. a word (other than a pronoun) or group of words used to name or identify any of a class of persons, places, or things (common noun), or a particular one of these (proper noun). □ **nounal** adj. [ME f. AF f. L nomen name]

nourish /ˈnʌrɪʃ/ v.tr. **1 a** sustain with food. **b** enrich; promote the development of (the soil etc.). **c** provide with intellectual or emotional sustenance or enrichment. **2** foster or cherish (a feeling etc.). □ **nourisher** n. [ME f. OF norir f. L nutrire]

nourishing /ˈnʌrɪʃɪŋ/ adj. (esp. of food) containing much nourishment; sustaining. □ **nourishingly** adv.

nourishment /ˈnʌrɪʃmənt/ n. sustenance, food.

nous /naʊs/ n. **1** colloq. common sense; gumption. **2** Philos. the mind or intellect. [Gk]

nouveau riche /ˌnuːvəʊ ˈriːʃ/ n. (pl. **nouveaux riches** pronunc. same) a person who has recently acquired (usu. ostentatious) wealth. [F, = new rich]

nouveau roman /ˌnuːvəʊ rəʊˈmɑːn/ n. a style of avant-garde French novel that came to prominence in the 1950s. It rejected the plot, characters, and omniscient narrator central to the traditional novel in an attempt to reflect more faithfully the sometimes random nature of experience. Writers of such novels include Alain Robbe-Grillet and Claude Simon (b.1913). [F, = new novel]

Nouvelle-Calédonie /nuvɛlkaledɔni/ the French name for NEW CALEDONIA.

nouvelle cuisine /ˌnuːvel kwɪˈziːn/ n. a modern style of cookery avoiding heaviness and emphasizing presentation. [F, = new cookery]

nouvelle vague /ˌnuːvel ˈvɑːg/ n. a grouping of French film directors in the late 1950s and 1960s who reacted against established French cinema and sought to make more individualistic and stylistically innovative films. Foremost among the directors were Claude Chabrol, Jean-Luc Godard, Alain Resnais, and François Truffaut. [F, fem. of nouveau new + vague wave]

Nov. abbr. November.

nova /ˈnəʊvə/ n. (pl. **novae** /-viː/ or **novas**) Astron. a star showing a sudden large increase in brightness and then slowly returning to its original state over a few months. Novae are believed to occur in binary star systems. Material falling from a companion star to the surface of a white dwarf collects over a long period, increasing in density and temperature, until there is a sudden runaway thermonuclear reaction which blasts much of the accumulated material into space. Novae are often visible in small telescopes, and sometimes to the naked eye (cf. SUPERNOVA). [L, fem. of novus new, because orig. thought to be a new star]

Nova Lisboa /ˌnəʊvə lɪzˈbəʊə/ the former name (until 1978) for HUAMBO.

Nova Scotia /ˌnəʊvə ˈskəʊʃə/ **1** a peninsula on the SE coast of Canada, projecting into the Atlantic Ocean and separating the Bay of Fundy from the Gulf of St Lawrence. **2** a province of eastern Canada, comprising the peninsula of Nova Scotia and the adjoining Cape Breton Island; pop. (1991) 900,600; capital, Halifax. It was settled by the French in the early 18th century, who named it Acadia. It changed hands several times between the French and English before being awarded to Britain in 1713, when it was renamed and the French settlers expelled. It became one of the original four provinces in the Dominion of Canada in 1867. □ **Nova Scotian** adj. & n.

Novaya Zemlya /ˌnəʊvəjə zɪmˈljɑː/ two large uninhabited islands in the Arctic Ocean off the north coast of Siberian Russia. The name means 'new land'.

novel[1] /ˈnɒv(ə)l/ n. **1** a fictitious prose story of book length. (See note below.) **2** (prec. by the) this type of literature. [It. novella (storia story) fem. of novello new f. L novellus f. novus]

■ Although there are precedents in the ancient world, such as the Greek Daphnis and Chloë of Longus (2nd–3rd centuries BC) and The Golden Ass by the Roman Apuleius (2nd century AD), the fully developed novel in the West of relatively recent origin: it is usually dated back to Don Quixote (1605–15) by Cervantes. The first significant novel in English was Aphra Behn's Oroonoko (1688), but Defoe's Robinson Crusoe (1719) is often regarded as the foundation of the modern tradition. The classic period of the novel was the 19th century, the age of Austen, Balzac, Dickens, Melville, Flaubert, George Eliot, Tolstoy, Dostoevsky, and Zola. The 20th century has seen radical experimentation at the hands of writers such as Joyce, Woolf, Faulkner, Vonnegut, and Martin Amis, and also increased diversification into historical novels, detective fiction, science fiction, etc. The novel is bound by no rules of structure, style, or subject-matter, a flexibility which has helped to make it the most important and popular literary genre in the modern age.

novel[2] /ˈnɒv(ə)l/ adj. of a new kind or nature; strange; previously unknown. □ **novelly** adv. [ME f. OF f. L novellus f. novus new]

novelese /ˌnɒvəˈliːz/ n. derog. a style characteristic of inferior novels.

novelette /ˌnɒvəˈlet/ n. **1 a** a short novel. **b** Brit. derog. a light romantic novel. **2** Mus. a piano piece in free form with several themes.

novelettish /ˌnɒvəˈletɪʃ/ adj. derog. in the style of a light romantic novel; sentimental.

novelist /ˈnɒvəlɪst/ n. a writer of novels. □ **novelistic** /ˌnɒvəˈlɪstɪk/ adj.

novelize /ˈnɒvəˌlaɪz/ v.tr. (also **-ise**) make into a novel. □ **novelization** /ˌnɒvəlaɪˈzeɪʃ(ə)n/ n.

novella /nəˈvelə/ n. (pl. **novellas**) a short novel or narrative story; a tale. [It.: see NOVEL¹]

Novello /nəˈveləʊ/, Ivor (born David Ivor Davies) (1893–1951), Welsh composer, actor, and dramatist. In 1914 he wrote 'Keep the Home Fires Burning', which became one of the most popular songs of the First World War. Later he composed and acted in a series of musicals, including *Glamorous Night* (1935), *The Dancing Years* (1939), and *King's Rhapsody* (1949).

novelty /ˈnɒv(ə)ltɪ/ n. & adj. ● n. (pl. **-ies**) **1 a** newness; new character. **b** originality. **2** a new or unusual thing or occurrence. **3** a small toy or decoration etc. of novel design. **4** (attrib.) having novelty (*novelty toys*). [ME f. OF *novelté* (as NOVEL²)]

November /nəʊˈvembə(r)/ n. the eleventh month of the year. [ME f. OF *novembre* f. L *November* f. *novem* nine (orig. the ninth month of the Roman year)]

novena /nəˈviːnə/ n. RC Ch. a devotion consisting of special prayers or services on nine successive days. [med.L f. L *novem* nine]

Noverre /nɒˈveə(r)/, Jean-Georges (1727–1810), French choreographer and dance theorist. A great reformer, he stressed the importance of dramatic motivation in ballet and was critical of the overemphasis hitherto placed on technical virtuosity. His work, especially as set out in *Lettres sur la danse et sur les ballets* (1760), had a significant influence on the development of ballet.

Novgorod /ˈnɒvgəˌrɒd/ a city in NW Russia, on the Volkhov river at the northern tip of Lake Ilmen; pop. (1990) 232,000. Russia's oldest city, first chronicled in 859, it was settled by the Varangian chief Rurik in 862, becoming the first Russian principality. It was a major commercial and cultural centre of medieval eastern Europe, developing important trading links with Constantinople, the Baltic, Asia, and the rest of Europe. It was ruled by Alexander Nevsky between 1238 and 1263. A rival with Moscow for supremacy during the 14th and 15th centuries, it was finally defeated by Ivan the Great in the 1470s. From the 17th century the city became less prominent.

novice /ˈnɒvɪs/ n. **1 a** a probationary member of a religious order, before the taking of vows. **b** a new convert. **2** a beginner; an inexperienced person. **3** a horse, dog, etc. that has not won a major prize in a competition. [ME f. OF f. L *novicius* f. *novus* new]

noviciate /nəˈvɪʃɪət/ n. (also **novitiate**) **1** the period of being a novice. **2** a religious novice. **3** novices' quarters. [F *noviciat* or med.L *noviciatus* (as NOVICE)]

Novi Sad /ˌnəʊvɪ ˈsæd/ an industrial city in Serbia, on the River Danube, capital of the autonomous province of Vojvodina; pop. (1991) 178,800.

Novocaine /ˈnəʊvəˌkeɪn/ n. (also **novocaine**) propr. a local anaesthetic derived from benzoic acid. [L *novus* new + COCAINE]

Novokuznetsk /ˌnəʊvəkʊzˈnjetsk/ an industrial city in the Kuznets Basin in south central Siberian Russia; pop. (1990) 601,000.

Novosibirsk /ˌnəʊvəsɪˈbɪəsk/ a city in central Siberian Russia, to the west of the Kuznets Basin, on the River Ob; pop. (1990) 1,443,000. The university, educational institutes, and several science research institutes are located in the satellite town of Akademgorodok.

Novotný /ˈnɒvɒtniː/, Antonín (1904–75), Czechoslovak Communist statesman, President 1957–68. In 1921 he became a founder member of the Czechoslovak Communist Party, rising to prominence when he played a major part in the Communist seizure of power in 1948. He was appointed First Secretary of the Czechoslovak Communist Party in 1953, and became President four years later. A committed Stalinist whose policies caused much resentment, he was ousted by the reform movement in 1968.

now /naʊ/ adv., conj., & n. ● adv. **1** at the present or mentioned time. **2** immediately (*I must go now*). **3** by this or that time (*it was now clear*). **4** under the present circumstances (*I cannot now agree*). **5** on this further occasion (*what do you want now?*). **6** in the immediate past (*just now*). **7** (esp. in a narrative or discourse) then, next (*the police now arrived*; *now to consider the next point*). **8** (without reference to time, giving various tones to a sentence) surely, I insist, I wonder, etc. (*now what do you mean by that?*; *oh come now!*). ● conj. (often foll. by *that* + clause) as a consequence of the fact (*now that I am older*; *now you mention it*). ● n. this time; the present (*should be there by now*; *has happened before now*). □ **as of now** from or at this time. **for now** until a later time (*goodbye for now*). **now**

and again (or **then**) from time to time; intermittently. **now or never** an expression of urgency. [OE *nū*]

nowadays /ˈnaʊəˌdeɪz/ adv. & n. ● adv. at the present time or age; in these times. ● n. the present time.

noway /ˈnəʊweɪ/ adv. = NOWISE (see *no way*).

Nowel /ˈnəʊel/ (also **Nowell**) archaic var. of NOEL.

nowhere /ˈnəʊweə(r)/ adv. & pron. ● adv. in or to no place. ● pron. no place. □ **be** (or **come in**) **nowhere** be unplaced in a race or competition. **come from nowhere** be suddenly evident or successful. **get nowhere** make or cause to make no progress. **in the middle of nowhere** colloq. remote from urban life. **nowhere near** not nearly. [OE *nāhwǣr* (as NO¹, WHERE)]

nowise /ˈnəʊwaɪz/ adv. in no manner; not at all.

nowt /naʊt/ n. colloq. or dial. nothing. [var. of NOUGHT]

noxious /ˈnɒkʃəs/ adj. harmful, unwholesome. □ **noxiously** adv. **noxiousness** n. [f. L *noxius* f. *noxa* harm]

noyau /ˈnwaɪəʊ/ n. (pl. **noyaux** /-əʊz/) a liqueur of brandy flavoured with fruit-kernels. [F, = kernel, ult. f. L *nux nucis* nut]

nozzle /ˈnɒz(ə)l/ n. a spout on a hose etc. from which a jet issues. [NOSE + -LE²]

NP abbr. Notary Public.

Np symb. Chem. the element neptunium.

n.p. abbr. **1** new paragraph. **2** no place of publication.

NPA abbr. (in the UK) Newspaper Publishers' Association.

NPL abbr. (in the UK) National Physical Laboratory.

nr. abbr. near.

NS abbr. **1** New Style. **2** new series. **3** Nova Scotia.

Ns symb. Chem. the element nielsbohrium.

ns abbr. nanosecond.

NSA see NATIONAL SECURITY AGENCY.

NSB abbr. (in the UK) National Savings Bank.

NSC see NATIONAL SECURITY COUNCIL.

NSF abbr. (in the US) National Science Foundation.

NSPCC abbr. (in the UK) National Society for the Prevention of Cruelty to Children.

NSU abbr. Med. non-specific urethritis.

NSW abbr. New South Wales.

NT abbr. **1** New Testament. **2** Northern Territory (of Australia). **3** National Trust. **4** no trumps.

n't /(ə)nt/ adv. (in comb.) = NOT (usu. with *is*, *are*, *have*, *must*, and the auxiliary verbs *can*, *do*, *should*, *would*: *isn't*; *mustn't*) (see also CAN'T, DON'T, WON'T). [contr.]

Nth. abbr. North.

nth /enθ/ see N².

NTP abbr. normal temperature and pressure.

nu /njuː/ n. the thirteenth letter of the Greek alphabet (*N*, *ν*). [Gk]

nuance /ˈnjuːɒns/ n. & v. ● n. a subtle difference in or shade of meaning, feeling, colour, etc. ● v.tr. give a nuance or nuances to. [F f. *nuer* to shade, ult. f. L *nubes* cloud]

nub /nʌb/ n. **1** the point or gist (of a matter or story). **2** a small lump, esp. of coal. **3** a stub; a small residue. □ **nubby** adj. [app. var. of *knub*, f. MLG *knubbe*, *knobbe* KNOB]

nubble /ˈnʌb(ə)l/ n. a small knob or lump. □ **nubbly** adj. [dimin. of NUB]

Nubia /ˈnjuːbɪə/ an ancient region of southern Egypt and northern Sudan, including the Nile valley between Aswan and Khartoum and the surrounding area. Nubia fell under ancient Egyptian rule from the time of the Middle Kingdom, soon after 2000 BC, and from about the 15th century BC was ruled by an Egyptian viceroy. The country was Egyptianized, and trade (especially in gold) flourished. By the 8th century BC, however, as Egypt's centralized administration disintegrated, an independent Nubian kingdom emerged, and for a brief period extended its power over Egypt. Much of Nubia is now drowned by the waters of Lake Nasser, formed by the building of the two dams at Aswan. Nubians constitute an ethnic minority group in Egypt. □ **Nubian** adj. & n.

nubile /ˈnjuːbaɪl/ adj. (of a woman) marriageable or sexually attractive. □ **nubility** /njuːˈbɪlɪtɪ/ [L *nubilis* f. *nubere* become the wife of]

nuchal /ˈnjuːk(ə)l/ adj. Anat. & Zool. of or relating to the nape of the neck.

[*nucha* nape f. med.L *nucha* medulla oblongata f. Arab. *nuk̲a'* spinal marrow]

nuci- /'njuːsɪ/ *comb. form* nut. [L *nux nucis* nut]

nuciferous /njuːˈsɪfərəs/ *adj. Bot.* bearing nuts.

nuclear /'njuːklɪə(r)/ *adj.* **1** of, relating to, or constituting a nucleus. **2** using nuclear energy (*nuclear reactor*). **3** having nuclear weapons. □ **nuclear bomb** a bomb using the release of energy by nuclear fission or fusion or both. **nuclear disarmament** the gradual or total reduction by a state of its nuclear weapons. **nuclear energy** energy obtained by nuclear fission or fusion. **nuclear family** a couple and their children, regarded as a basic social unit. **nuclear force** a strong attractive force between nucleons in the atomic nucleus that holds the nucleus together. **nuclear-free** free from nuclear weapons, power, etc. **nuclear fuel** a substance that will sustain a fission chain reaction so that it can be used as a source of nuclear energy. **nuclear physics** the physics of atomic nuclei and their interactions, esp. in the generation of nuclear energy. **nuclear power 1** electric or motive power generated by a nuclear reactor. **2** a country that has nuclear weapons. **nuclear-powered** powered by a nuclear reactor. **nuclear umbrella** supposed protection afforded by an alliance with a country possessing nuclear weapons. **nuclear warfare** warfare in which nuclear weapons are used. **nuclear waste** radioactive waste material e.g. from the use or reprocessing of nuclear fuel. [NUCLEUS + -AR¹]

nuclear fission *n.* the splitting of an atomic nucleus into smaller nuclei of roughly equal size, with consequent release of energy. The phenomenon was discovered by Otto Hahn and Fritz Strassmann (1902–80) in Germany in 1938, and can occur either spontaneously or after the impact of another particle. Fission in isotopes of uranium and plutonium is the basis of nuclear power and atomic weapons.

nuclear fusion *n.* the union of atomic nuclei to form heavier nuclei. With nuclei of lighter elements this process can result in an enormous release of energy, although it can occur only at very high temperatures. Heavy nuclei can undergo fusion in extreme conditions (e.g. in supernovae), but the process consumes energy. The sun's energy arises mainly from the conversion of hydrogen into helium by fusion, a process which is also the basis of the hydrogen bomb. Research continues into harnessing fusion so that it can be used in power stations, the major problem being how to contain the reacting substances, whose temperatures can rise to millions of degrees Celsius.

nuclear magnetic resonance *n.* (abbr. **NMR, nmr**) the absorption of electromagnetic radiation by a nucleus having a magnetic moment when in an external magnetic field, used mainly as an analytical technique and in body imaging for diagnosis.

nuclear reactor *n.* a device in which a nuclear fission chain reaction is sustained and controlled, usually in order to produce energy. The first nuclear reactor, then called a pile, was built by the Italian physicist Enrico Fermi at the University of Chicago in 1942. Since then several countries have built nuclear reactors to generate electric power, and they have been used to power submarines and other vessels. The main components are a suitable fissile material as fuel, moderators, which control the rate of reaction by absorbing neutrons, and a fluid coolant, to carry away the heat to a boiler to generate electrical power. Spent fuel and other materials, which will remain radioactive for hundreds of years, have to be securely and safely stored. Some reactors are involved in the production of radioactive material for medical and other purposes.

nuclear weapons *n.pl.* weapons whose destructive power comes from the rapid release of nuclear energy. Nuclear weapons utilizing fission were first exploded by the US in 1945, and fusion weapons followed in 1952 (see also ATOM BOMB, HYDROGEN BOMB). In the following decades the superpowers built up arsenals of tactical and strategic weapons, with a variety of delivery systems, sufficient to annihilate their enemies several times over. The UK, France, China, and India also acquired nuclear weapons, while some other countries (e.g. Israel, South Africa, and Pakistan) are believed or rumoured to possess nuclear capability. International treaties relating to nuclear weapons include the Test-Ban Treaty of 1963 (see TEST-BAN TREATY), and the Non-Proliferation Treaty of 1968, designed to prevent the spread of nuclear weapons to countries which do not already possess them. The US and the Soviet Union made significant cuts in the numbers of nuclear weapons they possessed with the ending of the cold war and the signing of the START treaty in 1991. After the breakup of the Soviet Union, the remaining Soviet warheads were divided between Russia, Ukraine, Belarus, and Kazakhstan; all but Russia agreed in 1993 to disarm.

nuclear winter *n.* a period of abnormal cold and darkness on the earth predicted to follow a nuclear war, due to an atmospheric layer of smoke and dust blocking the sun's rays. The idea of a nuclear winter, with extensive fires causing disastrous effects on the earth's ecosystem, was put forward by Carl Sagan and other American scientists in 1983. It remains controversial, although there is evidence for a similar (non-nuclear) catastrophe after an asteroid impact at the end of the Cretaceous period 65 million years ago (see CRETACEOUS).

nuclease /'njuːklɪˌeɪz/ *n. Biochem.* an enzyme that catalyses the breakdown of nucleic acids.

nucleate *adj. & v.* ● *adj.* /'njuːklɪət/ having a nucleus. ● *v.intr. & tr.* /'njuːklɪˌeɪt/ form or form into a nucleus. □ **nucleation** /ˌnjuːklɪˈeɪʃ(ə)n/ *n.* [LL *nucleare nucleat-* form a kernel (as NUCLEUS)]

nuclei *pl.* of NUCLEUS.

nucleic acid /njuːˈkliːɪk, -ˈkleɪɪk/ *n. Biochem.* either of two complex organic substances (DNA and RNA), with molecules consisting of many nucleotides linked in a long chain, and present in all living cells.

nucleo- /'njuːklɪəʊ/ *comb. form* nucleus; nucleic acid (*nucleo-protein*).

nucleolus /njuːˈkliːələs, ˌnjuːklɪˈəʊləs/ *n.* (*pl.* **nucleoli** /-ˌlaɪ/) *Biol.* a small dense spherical structure within the nucleus of a non-dividing cell during interphase. □ **nucleolar** *adj.* [LL, dimin. of L *nucleus*: see NUCLEUS]

nucleon /'njuːklɪˌɒn/ *n. Physics* a proton or neutron.

nucleonics /ˌnjuːklɪˈɒnɪks/ *n.pl.* (treated as *sing.*) the branch of science and technology concerned with atomic nuclei and nucleons, esp. the exploitation of nuclear power. □ **nucleonic** *adj.* [NUCLEAR, after *electronics*]

nucleoprotein /ˌnjuːklɪəʊˈprəʊtiːn/ *n. Biochem.* a complex of nucleic acid and protein.

nucleoside /'njuːklɪəˌsaɪd/ *n. Biochem.* an organic compound consisting of a purine or pyrimidine base linked to a sugar, e.g. adenosine.

nucleosynthesis /ˌnjuːklɪəʊˈsɪnθɪsɪs/ *n. Astron.* the cosmic formation of atoms more complex than the hydrogen atom, esp. as a result of nuclear reactions in the cores of stars and of supernova explosions. □ **nucleosynthetic** /-sɪnˈθetɪk/ *adj.*

nucleotide /'njuːklɪəˌtaɪd/ *n. Biochem.* an organic compound consisting of a nucleoside linked to a phosphate group.

nucleus /'njuːklɪəs/ *n.* (*pl.* **nuclei** /-lɪˌaɪ/) **1 a** the central part or thing round which others are collected. **b** the kernel of an aggregate or mass. **2** an initial part meant to receive additions. **3** *Astron.* the solid part of a comet's head. **4** *Physics* the positively charged central core of an atom that contains most of its mass. (*See note below.*) **5** *Biol.* a large dense organelle in a eukaryotic cell, containing the genetic material. **6** *Anat.* a discrete mass of grey matter in the central nervous system. [L, = kernel, inner part, dimin. of *nux nucis* nut]

▪ Ernest Rutherford, in experiments from 1911 onwards, established that the positive charge and most of the mass of an atom is concentrated in a nucleus of approximate diameter 10^{-13} cm, about $^1/_{10\,000}$ of that of the atom as a whole. The nucleus consists of protons and neutrons, bound together by very powerful forces which do not extend significantly beyond the nucleus itself. Nuclei vibrate, rotate, and alter their shape, properties which can be recognized through energy changes and other phenomena.

nuclide /'njuːklaɪd/ *n. Physics* a distinct kind of atom or nucleus, characterized by the number of protons and neutrons in the nucleus. □ **nuclidic** /njuːˈklɪdɪk/ *adj.* [NUCLEUS + Gk *eidos* form]

nuddy /'nʌdɪ/ *n. sl.* nude (esp. *in the nuddy*). [prob. f. NUDE + -Y²]

nude /njuːd/ *adj. & n.* ● *adj.* naked, bare, unclothed. ● *n.* **1** a painting, sculpture, photograph, etc. of a nude human figure; such a figure. **2** a nude person. **3** (prec. by *the*) **a** an unclothed state. **b** the representation of an undraped human figure as a genre in art. [L *nudus*]

nudge /nʌdʒ/ *v. & n.* ● *v.tr.* **1** prod gently with the elbow to attract attention. **2** push gently or gradually. **3** give a gentle reminder or encouragement to (a person). ● *n.* the act or an instance of nudging; a gentle push. □ **nudger** *n.* [17th c.: orig. unkn.: cf. Norw. dial. *nugga, nyggja* to push, rub]

nudibranch /'njuːdɪˌbræŋk/ *n. & adj. Zool.* ● *n.* a marine gastropod mollusc of the order Nudibranchia, with exposed gills and a vestigial shell; a sea slug. ● *adj.* of or relating to this order. □ **nudibranchiate** /ˌnjuːdɪˈbræŋkɪət/ *n. & adj.* [mod.L *Nudibranchia*, f. L *nudus* NUDE + BRANCHIA]

nudist /'njuːdɪst/ n. a person who advocates or practises going unclothed. □ **nudism** n.

nudity /'njuːdɪtɪ/ n. the state of being nude; nakedness.

nuée ardente /ˌnjuːeɪ ɑːˈdɒnt/ n. Geol. a hot cloud of gas, ash, and lava fragments ejected from a volcano, usu. accompanying a pyroclastic flow. [F, lit. 'burning cloud']

Nuer /nʊə(r)/ n. & adj. ● n. (pl. same) **1** a member of an African people living in SE Sudan and Ethiopia. Much disrupted now by civil war and famine, the traditional Nuer way of life is pastoral and dominated by the rearing and trading of cattle. **2** the Nilotic language of this people. ● adj. of or relating to this people or their language. [African name]

Nuevo León /ˌnweɪvəʊ leɪˈɒn/ a state of NE Mexico, on the border with the US; capital, Monterrey.

Nuffield /'nʌfiːld/, 1st Viscount (title of William Richard Morris) (1877–1963), British motor manufacturer and philanthropist. Working in Oxford, he started by building bicycles. In 1912 he opened the first Morris automobile factory there and launched his first car the following year. Morris Motors Limited was formed in 1926, the year of the first MG (Morris Garage) models. In later life he devoted his considerable fortune to philanthropic purposes; these include the endowment of Nuffield College, Oxford (1937) and the creation of the Nuffield Foundation (1943) for medical, social, and scientific research.

nugatory /'njuːɡətərɪ/ adj. **1** futile, trifling, worthless. **2** inoperative; not valid. [L nugatorius f. nugari to trifle f. nugae jests]

nugget /'nʌɡɪt/ n. **1 a** a lump of gold, platinum, etc., as found in the earth. **b** a lump of anything compared to this. **2** something valuable for its size (often abstract in sense: a little nugget of information). [app. f. dial. nug lump etc.]

nuisance /'njuːs(ə)ns/ n. **1** a person, thing, or circumstance causing trouble or annoyance. **2** anything harmful or offensive to the community or a member of it and for which a legal remedy exists. □ **nuisance value** an advantage resulting from the capacity to harass or frustrate. [ME f. OF, = hurt, f. nuire nuis- f. L nocere to hurt]

NUJ abbr. (in the UK) National Union of Journalists.

nuke /njuːk/ n. & v. colloq. ● n. a nuclear weapon. ● v.tr. bomb or destroy with nuclear weapons. [abbr.]

Nuku'alofa /ˌnuːkuːəˈlɔːfə/ the capital of Tonga, situated on the island of Tongatapu; pop. (1986) 28,900.

null /nʌl/ adj. & n. ● adj. **1** (esp. **null and void**) invalid; not binding. **2** non-existent; amounting to nothing. **3** having or associated with the value zero. **4** Computing **a** empty; having no elements (null list). **b** all the elements of which are zeros (null matrix). **5** without character or expression. ● n. a dummy letter in a cipher. □ **null character** Computing a character denoting nothing, usu. represented by a zero. **null hypothesis** Statistics a hypothesis suggesting that the difference between samples does not imply a difference between populations. **null instrument** an instrument used by adjustment to give a reading of zero. **null link** Computing a reference incorporated into the last item in a list to indicate there are no further items in the list. [F nul nulle or L nullus none f. ne not + ullus any]

nullah /'nʌlə/ n. Anglo-Ind. a dry river bed or ravine. [Hindi nālā]

nulla-nulla /ˌnʌləˈnʌlə/ n. (also **nulla**) Austral. a hardwood club used by Aboriginals. [Dharuk ngalla-ngalla]

Nullarbor Plain /'nʌləˌbɔː(r)/ a vast arid plain in SW Australia, stretching inland from the Great Australian Bight. It contains no surface water, with sparse vegetation, and is almost uninhabited. [L nullus arbor no tree]

nullifidian /ˌnʌlɪˈfɪdɪən/ n. & adj. (a person) having no religious faith or belief. [med.L nullifidius f. L nullus none + fides faith]

nullify /'nʌlɪˌfaɪ/ v.tr. (**-ies, -ied**) make null; neutralize, invalidate, cancel. □ **nullifier** n. **nullification** /ˌnʌlɪfɪˈkeɪʃ(ə)n/ n.

nullipara /nʌˈlɪpərə/ n. Med. a woman who has never borne a child. □ **nulliparous** adj. [mod.L f. L nullus none + -para fem. of -parus f. parere bear children]

nullipore /'nʌlɪˌpɔː(r)/ n. Bot. a red alga that secretes lime. [L nullus none + PORE¹]

nullity /'nʌlɪtɪ/ n. (pl. **-ies**) **1** Law **a** being null; invalidity, esp. of marriage. **b** an act, document, etc., that is null. **2 a** nothingness. **b** a mere nothing; a nonentity. [F nullité or med.L nullitas f. L nullus none]

NUM abbr. (in the UK) National Union of Mineworkers.

Num. abbr. (in the Bible) Numbers.

Numa Pompilius /ˌnjuːmə pɒmˈpɪlɪəs/ the legendary second king of Rome, successor to Romulus, revered by the ancient Romans as the founder of nearly all their religious institutions.

numb /nʌm/ adj. & v. ● adj. (often foll. by with) deprived of feeling or the power of motion (numb with cold). ● v.tr. **1** make numb. **2** stupefy, paralyse. □ **numb-fish** = electric ray. □ **numbly** adv. **numbness** n. [ME nome(n) past part. of nim take: for -b cf. THUMB]

numbat /'nʌmbæt/ n. a small termite-eating marsupial, Myrmecobius fasciatus, native to Australia, with a bushy tail and white bars on the face and rump. [Aboriginal]

number /'nʌmbə(r)/ n. & v. ● n. **1 a** an arithmetical value representing a particular quantity and used in counting and making calculations. (See note below.) **b** a word, symbol, or figure representing this; a numeral. **c** an arithmetical value showing position in a series, esp. for identification, reference, etc. (registration number). **2** (often foll. by of) the total count or aggregate (the number of accidents has decreased; twenty in number). **3 a** the study of the behaviour of numbers; numerical reckoning (the laws of number). **b** (in pl.) arithmetic (not good at numbers). **4 a** (in sing. or pl.) a quantity or amount; a total; a count (a large number of people; only in small numbers). **b** (**a number of**) several (of), some (of). **c** (in pl.) numerical preponderance (force of numbers; there is safety in numbers). **5 a** a person or thing having a place in a series, esp. a single issue of a magazine, an item in a programme, etc. **b** a song, dance, musical item, etc. **c** Mus. a separate section of an opera, oratorio, etc., such as an aria, duet, or piece of recitative, or a combination of these. **6** company, collection, group (among our number). **7** Gram. **a** the classification of words by their singular or plural forms. **b** a particular form so classified. **8** colloq. a person or thing regarded familiarly or affectionately (usu. qualified in some way: an attractive little number). ● v.tr. **1** include (I number you among my friends). **2** assign a number or numbers to. **3** have or amount to (a specified number). **4 a** count. **b** comprise (numbering forty thousand men). □ **any number of 1** any particular whole quantity of. **2** colloq. a large unspecified number of. **by numbers** following simple instructions (as if) identified by numbers. **one's days are numbered** one does not have long to live. **have a person's number** colloq. understand a person's real motives, character, etc. **have a person's number on it** (of a bomb, bullet, etc.) be destined to hit a specified person. **number cruncher** Computing & Math. sl. a machine capable of performing complex and laborious calculations. **number crunching** Computing & Math. sl. the act or process of making complex and laborious calculations. **one's number is up** colloq. one is finished or doomed to die. **number one** n. colloq. oneself (always takes care of number one). ● adj. most important (the number one priority). **number opera** in which the arias and other sections are clearly separable (cf. music drama). **number-plate** a plate on a vehicle displaying its registration number. **numbers game 1** usu. derog. action involving only arithmetical work. **2** N. Amer. a lottery based on the occurrence of unpredictable numbers in the results of races etc. **number two** a second in command. **without number** innumerable. [ME f. OF nombre (n.), nombrer (v.) f. L numerus, numerare]

▪ Mathematicians distinguish several kinds of number. The natural numbers are positive integers 1, 2, 3, … (sometimes including zero) used in counting. The integers are the natural numbers, zero, and their negatives −1, −2, −3, … . Rational numbers are the fractions p/q (where p and q are integers and q is not zero). Real numbers are the numbers used in measurement; they are limits of sequences of rational numbers and can be described by decimal expansions that may go on for ever. Complex numbers are numbers of the form a + bi in which a and b are real numbers and i is an imagined square root of −1. All natural numbers are integers, all integers are rational numbers, all rational numbers are real numbers, all real numbers are complex numbers. Algebraic numbers are complex numbers that are roots of polynomial equations with integer coefficients; transcendental numbers are the complex numbers that are not algebraic. Transfinite numbers are the cardinal and ordinal numbers assigned to infinite sets in Cantor's set theory.

numberless /'nʌmbəlɪs/ adj. innumerable.

Numbers /'nʌmbəz/ the fourth book of the Bible, relating the experiences of the Israelites under Moses during their wanderings in the desert. The English title is explained by its two records of a census; its Hebrew title means 'in the wilderness'.

Number Ten 10 Downing Street, the official London home of the British Prime Minister.

numbles /'nʌmb(ə)lz/ n.pl. Brit. archaic a deer's entrails. [ME f. OF numbles, nombles loin etc., f. L lumbulus dimin. of lumbus loin: cf. UMBLES]

numbskull var. of NUMSKULL.

numdah /ˈnʌmdɑː/ n. an embroidered felt rug from India etc. [Urdu *namdā* f. Pers. *namad* carpet]

numen /ˈnjuːmen/ n. (pl. **numina** /-mɪnə/) a presiding deity or spirit. [L *numen -minis*]

numerable /ˈnjuːmərəb(ə)l/ adj. that can be counted. □ **numerably** adv. [L *numerabilis* f. *numerare* NUMBER v.]

numeral /ˈnjuːmərəl/ n. & adj. ● n. a word, figure, or group of figures denoting a number. ● adj. of or denoting a number. [LL *numeralis* f. L (as NUMBER)]

numerate /ˈnjuːmərət/ adj. acquainted with the basic principles of mathematics. □ **numeracy** n. [L *numerus* number + -ATE[2] after *literate*]

numeration /ˌnjuːməˈreɪʃ(ə)n/ n. **1 a** a method or process of numbering or computing. **b** calculation. **2** the expression in words of a number written in figures. [ME f. L *numeratio* payment, in LL numbering (as NUMBER)]

numerator /ˈnjuːməˌreɪtə(r)/ n. **1** the number above the line in a vulgar fraction showing how many of the parts indicated by the denominator are taken (e.g. 2 in $^2/_3$). **2** a person or device that numbers. [F *numérateur* or LL *numerator* (as NUMBER)]

numerical /njuːˈmerɪk(ə)l/ adj. (also **numeric**) of or relating to a number or numbers (*numerical superiority*). □ **numerical analysis** the branch of mathematics that deals with the development and use of numerical methods for solving problems. □ **numerically** adv. [med.L *numericus* (as NUMBER)]

numerology /ˌnjuːməˈrɒlədʒɪ/ n. the study of the supposed occult significance of numbers. □ **numerologist** n. **numerological** /-rəˈlɒdʒɪk(ə)l/ adj. [L *numerus* number + -LOGY]

numerous /ˈnjuːmərəs/ adj. **1** (with pl.) great in number (*received numerous gifts*). **2** consisting of many (*a numerous family*). □ **numerously** adv. **numerousness** n. [L *numerosus* (as NUMBER)]

Numidia /njuːˈmɪdɪə/ an ancient kingdom, later a Roman province, situated in North Africa in an area north of the Sahara corresponding roughly to present-day Algeria. □ **Numidian** adj. & n.

numina pl. of NUMEN.

numinous /ˈnjuːmɪnəs/ adj. **1** indicating the presence of a divinity. **2** spiritual. **3** awe-inspiring. [L *numen*: see NUMEN]

numismatic /ˌnjuːmɪzˈmætɪk/ adj. of or relating to coins or medals. □ **numismatically** adv. [F *numismatique* f. L *numisma* f. Gk *nomisma -atos* current coin f. *nomizō* use currently]

numismatics /ˌnjuːmɪzˈmætɪks/ n.pl. (usu. treated as sing.) the study of coins or medals. □ **numismatist** /-ˈmɪzmətɪst/ n.

numismatology /njuːˌmɪzməˈtɒlədʒɪ/ n. = NUMISMATICS.

nummulite /ˈnʌmjʊˌlaɪt/ n. a disc-shaped fossil skeleton of a foraminiferous protozoan found in Tertiary strata. [L *nummulus* dimin. of *nummus* coin]

numnah /ˈnʌmnɑː/ n. a saddle-cloth or pad placed under a saddle. [Urdu *namdā*: see NUMDAH]

numskull /ˈnʌmskʌl/ n. (also **numbskull**) a stupid or foolish person. [NUMB + SKULL]

nun /nʌn/ n. a member of a community of women living apart under religious vows. □ **nunhood** n. **nunlike** adj. **nunnish** adj. [ME f. OE *nunne* and OF *nonne* f. eccl.L *nonna* fem. of *nonnus* monk, orig. a title given to an elderly person]

nunatak /ˈnʌnəˌtæk/ n. an isolated peak of rock projecting above a surface of inland ice or snow. [Eskimo]

nun-buoy /ˈnʌnbɔɪ/ n. a buoy circular in the middle and tapering to each end. [obs. *nun* child's top + BUOY]

Nunc Dimittis /ˌnʌŋk dɪˈmɪtɪs/ n. a canticle beginning *Nunc Dimittis* (Lord, now let (your servant) depart), the first words of the Song of Simeon (Luke 2:29). [L]

nunciature /ˈnʌnsɪətʃə(r)/ n. RC Ch. the office or tenure of a nuncio. [It. *nunziatura* (as NUNCIO)]

nuncio /ˈnʌnʃɪəʊ, ˈnʌnsɪ-/ n. (pl. **-os**) RC Ch. a papal ambassador. [It. f. L *nuntius* messenger]

nuncupate /ˈnʌŋkjʊˌpeɪt/ v.tr. declare (a will or testament) orally, not in writing. □ **nuncupation** /ˌnʌŋkjʊˈpeɪʃ(ə)n/ n. **nuncupative** /ˈnʌŋkjʊˌpeɪtɪv, nʌŋˈkjuːpətɪv/ adj. [L *nuncupare nuncupat-* name]

Nuneaton /nʌˈniːt(ə)n/ a town in north Warwickshire in central England, near Coventry; pop. (1981) 81,880.

nunnery /ˈnʌnərɪ/ n. (pl. **-ies**) a religious house of nuns; a convent.

NUPE /ˈnjuːpɪ/ abbr. hist. (in the UK) National Union of Public Employees, merged with COHSE and NALGO to form UNISON in 1993.

nuptial /ˈnʌpʃ(ə)l/ adj. & n. ● adj. of or relating to marriage or weddings. ● n. (usu. in pl.) a wedding. [F *nuptial* or L *nuptialis* f. *nuptiae* wedding f. *nubere nupt-* wed]

NUR abbr. (in the UK) National Union of Railwaymen. ¶ In 1990 merged with the NUS (sense 1) to form the RMT.

nurd var. of NERD.

Nuremberg /ˈnjʊərəmˌbɜːg/ (German **Nürnberg** /ˈnyrnbɛrk/) a city in southern Germany, in Bavaria; pop. (1991) 497,500. It was a leading cultural centre in the 15th and 16th centuries and was the home of Albrecht Dürer and Hans Sachs. In the 1930s the Nazi Party congresses and the annual Nazi Party rallies were held there and in 1945–6 it was the scene of the Nuremberg war trials, in which Nazi war criminals were tried by international military tribunal. After the war the city centre was carefully reconstructed, as its cobbled streets and timbered houses had been reduced to rubble by Allied bombing.

Nureyev /nəˈreɪef, ˈnjʊərɪˌef/, Rudolf (1939–93), Russian-born ballet-dancer and choreographer. He defected to the West in 1961, joining the Royal Ballet in London the following year; it was there he began his noted partnership with Margot Fonteyn. Thereafter he danced the leading roles of the classical and standard modern repertory and choreographed many others, including *La Bayadère* (1963). He became a naturalized Austrian citizen in 1982, and was artistic director of the Paris Opéra Ballet 1983–9.

Nürnberg see NUREMBERG.

nurse[1] /nɜːs/ n. & v. ● n. **1** a person trained to care for the sick or infirm. **2** (formerly) a person employed or trained to take charge of young children. **3** archaic = wet-nurse. **4** Forestry a tree planted as a shelter to others. **5** Zool. a worker bee, ant, etc., caring for a young brood. ● v. **1 a** intr. work as a nurse. **b** tr. attend to (a sick person). **c** tr. give medical attention to (an illness or injury). **2** tr. & intr. feed or be fed at the breast. **3** tr. (in passive; foll. by in) be brought up in (a specified condition) (*nursed in poverty*). **4** tr. hold or treat carefully or caressingly (*sat nursing my feet*). **5** tr. **a** foster; promote the development of (the arts, plants, etc.). **b** harbour or nurture (a grievance, hatred, etc.). **c** pay special attention to (*nursed the voters*). **6** tr. Billiards keep (the balls) together for a series of cannons. [reduced f. ME and OF *norice, nurice* f. LL *nutricia* fem. of L *nutricius* f. *nutrix -icis* f. *nutrire* NOURISH]

nurse[2] /nɜːs/ n. a dogfish or shark; esp. the *nurse hound* (*Scyliorhinus stellaris*), a large dogfish of the NE Atlantic, and the *nurse shark* (*Ginglymostoma cirratuma*), a slow-swimming brownish shark of warm Atlantic waters. [orig. *nusse*, perh. derived by wrong division of *an huss* (cf. NEWT)]

nurseling var. of NURSLING.

nursemaid /ˈnɜːsmeɪd/ n. **1** a woman in charge of a child or children. **2** a person who watches over or guides another carefully.

nursery /ˈnɜːsərɪ/ n. (pl. **-ies**) **1 a** a room or place equipped for young children. **b** = day nursery. **2** a place where plants, trees, etc., are reared for sale or transplantation. **3** any sphere or place in or by which qualities or types of people are fostered or bred. **4** Billiards **a** grouped balls (see NURSE[1] v. 6). **b** (in full **nursery cannon**) a cannon on three close balls. □ **nursery nurse** a person trained to take charge of babies and young children. **nursery rhyme** a simple traditional song or story in rhyme for children. **nursery school** a school for children between the ages of three and five. **nursery slopes** Skiing gentle slopes suitable for beginners. **nursery stakes** a race for two-year-old horses.

nurseryman /ˈnɜːsərɪmən/ n. (pl. **-men**) an owner of or worker in a plant nursery.

nursing /ˈnɜːsɪŋ/ n. **1** the practice or profession of caring for the sick as a nurse. **2** (attrib.) concerned with or suitable for nursing the home or elderly etc. (*nursing home; nursing sister*). □ **nursing officer** a senior nurse (see *senior nursing officer*).

nursling /ˈnɜːslɪŋ/ n. (also **nurseling**) an infant that is being suckled.

nurture /ˈnɜːtʃə(r)/ n. & v. ● n. **1** the process of bringing up or training (esp. children); fostering care. **2** nourishment. **3** sociological factors as an influence on or determinant of personality (opp. NATURE 8). ● v.tr. **1** bring up; rear. **2** nourish. □ **nurturer** n. [ME f. OF *nour(e)ture* (as NOURISH)]

NUS abbr. **1** (in the UK) National Union of Seamen. (See RMT.) **2** (in the UK) National Union of Students.

NUT abbr. (in the UK) National Union of Teachers.

Nut /nʊt/ Egyptian Mythol. the sky-goddess, thought to swallow the sun at night and give birth to it in the morning. She is usually depicted as a naked woman with her body arched above the earth, which she touches with her feet and hands.

nut /nʌt/ n. & v. ● n. **1 a** a fruit consisting of a hard or tough shell around an edible kernel. **b** this kernel. **2** a pod containing hard seeds. **3** a small usu. square or hexagonal flat piece of metal or other material with a threaded hole through it for screwing on the end of a bolt to secure it. **4** sl. a person's head. **5** sl. **a** a crazy or eccentric person. **b** an obsessive enthusiast or devotee (a health-food nut). **6** a small lump of coal, butter, etc. **7** (as second element in comb.) a small rounded biscuit or cake (ginger-nut; doughnut). **8 a** a device fitted to the bow of a violin for adjusting its tension. **b** the fixed ridge on the neck of a stringed instrument over which the strings pass. **9** (in pl.) coarse sl. the testicles. ● v. (**nutted, nutting**) **1** intr. seek or gather nuts (go nutting). **2** tr. sl. butt with the head. □ **do one's nut** Brit. sl. be extremely angry or agitated. **for nuts** Brit. colloq. even tolerably well (cannot sing for nuts). **nut cutlet** Brit. a cutlet-shaped portion of meat-substitute, made from nuts etc. **nut-house** sl. a mental home or hospital. **nut-oil** an oil obtained from hazelnuts and walnuts and used in paints and varnishes. **nuts and bolts** colloq. the practical details. **nut-tree** a tree bearing nuts, esp. a hazel. **off one's nut** sl. crazy. □ **nutlike** adj. [OE hnutu f. Gmc]

nutant /ˈnjuːt(ə)nt/ adj. Bot. nodding, drooping. [L nutare nod]

nutation /njuːˈteɪʃ(ə)n/ n. **1** the act or an instance of nodding. **2** Astron. a periodic oscillation of the earth's poles. **3** oscillation of a spinning top. **4** Bot. the spiral movement of a plant organ during growth. [L nutatio (as NUTANT)]

nutcase /ˈnʌtkeɪs/ n. sl. a crazy or foolish person.

nutcracker /ˈnʌtˌkrækə(r)/ n. **1** (usu. in pl.) a device for cracking nuts. **2** a small crow of the genus Nucifraga, feeding mainly on seeds of conifers.

Nutcracker man n. a nickname for the fossil hominid Australopithecus (or Zinjanthropus) boisei, especially the original specimen discovered by Mary Leakey at Laetoli near Olduvai Gorge in 1959. The jaws and cheek teeth are exceptionally massive, suggesting a tough vegetarian diet, and similar remains have since been found in South Africa. (See AUSTRALOPITHECUS.)

nutgall /ˈnʌtɡɔːl/ n. a gall found on dyer's oak, used in dyeing.

nuthatch /ˈnʌthætʃ/ n. a small songbird of the genus Sitta and family Sittidae, climbing up and down tree-trunks and feeding on nuts, insects, etc; esp. the Eurasian S. europaea. [NUT + hatch rel. to HATCH²]

nutlet /ˈnʌtlɪt/ n. a small nut or nutlike fruit.

nutmeg /ˈnʌtmeɡ/ n. **1** an evergreen tree, Myristica fragrans, native to the Moluccas, yielding a hard aromatic spheroidal seed. **2** the seed of this grated and used as a spice. □ **nutmeg-apple** the fruit of this tree, yielding both mace and nutmeg. [ME: partial transl. of OF nois mug(u)ede ult. f. L nux nut + LL muscus MUSK]

nutria /ˈnjuːtrɪə/ n. the skin or fur of a coypu. [Sp., = otter]

nutrient /ˈnjuːtrɪənt/ n. & adj. ● n. a substance that provides essential nourishment for the maintenance of life. ● adj. serving as or providing nourishment. [L nutrire nourish]

nutriment /ˈnjuːtrɪmənt/ n. **1** nourishing food. **2** an intellectual or artistic etc. nourishment or stimulus. □ **nutrimental** /ˌnjuːtrɪˈment(ə)l/ adj. [L nutrimentum (as NUTRIENT)]

nutrition /njuːˈtrɪʃ(ə)n/ n. **1 a** the process of providing or receiving nourishing substances. **b** food, nourishment. **2** the study of nutrients and nutrition. □ **nutritional** adj. **nutritionally** adv. [F nutrition or LL nutritio (as NUTRIENT)]

nutritionist /njuːˈtrɪʃənɪst/ n. a person who studies or is an expert on the processes of human nourishment.

nutritious /njuːˈtrɪʃəs/ adj. efficient as food; nourishing. □ **nutritiously** adv. **nutritiousness** n. [L nutritius (as NURSE¹)]

nutritive /ˈnjuːtrɪtɪv/ adj. & n. ● adj. **1** of or concerned in nutrition. **2** serving as nutritious food. ● n. a nutritious article of food. [ME f. F nutritif -ive f. med.L nutritivus (as NUTRIENT)]

nuts /nʌts/ adj. & int. ● predic.adj. sl. crazy, mad, eccentric. ● int. sl. an expression of contempt or derision (nuts to you). □ **be nuts about** (or Brit. **on**) colloq. be enthusiastic about or very fond of.

nutshell /ˈnʌtʃel/ n. the hard woody exterior covering of a nut. □ **in a nutshell** in a few words.

nutter /ˈnʌtə(r)/ n. Brit. sl. a crazy or eccentric person.

nutty /ˈnʌtɪ/ adj. (**nuttier, nuttiest**) **1 a** full of nuts. **b** tasting like nuts. **2** sl. = NUTS adj. □ **nuttiness** n.

Nuu-chah-nulth see NOOTKA.

Nuuk /nuːk/ the capital of Greenland, a port on the Davis Strait; pop. (1990) 12,220. It was known by the Danish name Godthåb until 1979.

nux vomica /nʌks ˈvɒmɪkə/ n. **1** a southern Asian tree, Strychnos nux-vomica, yielding a poisonous fruit. **2** the seeds of this tree, containing strychnine. [med.L f. L nux nut + vomicus f. vomere vomit]

nuzzle /ˈnʌz(ə)l/ v. **1** tr. prod or rub gently with the nose. **2** intr. (foll. by into, against, up to) press the nose gently. **3** tr. (also refl.) nestle; lie snug. [ME f. NOSE + -LE⁴]

NV abbr. US Nevada (in official postal use).

NW abbr. **1** north-west. **2** north-western.

NY abbr. US New York (also in official postal use).

nyala /ˈnjɑːlə/ n. a large antelope, Tragelaphus angasii, native to southern Africa, with curved spiral horns. [Zulu i-nyala]

Nyanja /ˈnjændʒə/ n. & adj. ● n. **1** (pl. same or **Nyanjas**) a member of a people of Malawi. **2** the Bantu language of this people. ● adj. of or relating to the Nyanja or their language.

Nyasa, Lake /naɪˈæsə/ (also called Lake Malawi) a lake in east central Africa, the third largest lake in Africa. About 580 km (360 miles) long, it forms most of the eastern border of Malawi with Mozambique and Tanzania. Its name means 'lake'.

Nyasaland /naɪˈæsəˌlænd/ the former name of Malawi until it gained independence in 1966.

NYC abbr. New York City.

nyctalopia /ˌnɪktəˈləʊpɪə/ n. Med. inability to see in dim light or at night. Also called night-blindness. [LL f. Gk nuktalōps f. nux nuktos night + alaos blind + ōps eye]

nyctitropic /ˌnɪktɪˈtrəʊpɪk, -ˈtrɒpɪk/ adj. Bot. (of plant movements) occurring at night and caused by changes in light and temperature. [Gk nukti- comb. form of nux nuktos night + -TROPIC]

nye /naɪ/ n. dial. a brood of pheasants. [OF ni f. L nidus nest; cf. NIDE]

Nyerere /njeˈreərɪ/, Julius Kambarage (b.1922), Tanzanian statesman, President of Tanganyika 1962-4 and of Tanzania 1964-85. He was an active campaigner for the nationalist movement in the 1950s, forming the Tanganyika African National Union (1954). He served as Prime Minister of Tanganyika following its independence in 1961 and became President a year later. In 1964 he successfully negotiated a union with Zanzibar and remained President of the new state of Tanzania until his retirement.

nylon /ˈnaɪlɒn/ n. **1** a synthetic polymer composed of linear polyamide molecules. (See note below.) **2** a nylon fabric. **3** (in pl.) stockings made of nylon. [invented word, after cotton and rayon]

▪ Nylon was developed in the US by Wallace Carothers of the du Pont chemical company in the 1930s. The name now covers a range of thermoplastic polyamides which are tough, lightweight, and resistant to heat and chemicals. Originally developed and most familiar as a textile fibre, nylon is increasingly used in other forms, e.g. moulded objects, sheeting, and strengthening filaments in composite materials.

nymph /nɪmf/ n. **1** a mythological semi-divine spirit regarded as a maiden and associated with aspects of nature, esp. rivers and woods. **2** poet. a beautiful young woman. **3** the immature form of an insect that does not undergo full metamorphosis, e.g. the aquatic larva of a dragonfly or mayfly, or the larva of a locust. □ **nymphal** adj. **nymphlike** adj. **nymphean** /nɪmˈfiːən, ˈnɪmfɪən/ adj. [ME f. OF nimphe f. L nympha f. Gk numphē]

nymphae /ˈnɪmfiː/ n.pl. Anat. the labia minora. [L, pl. of nympha: see NYMPH]

nymphalid /nɪmˈfælɪd/ n. & adj. Zool. ● n. a butterfly of the large family Nymphalidae, which includes many of the familiar temperate species such as the tortoiseshells, red admiral, and peacock. The nymphalids have short, bristly forelegs which are not used in walking. ● adj. of or relating to this family. [mod.L Nymphalis genus name, f. L nympha NYMPH]

nymphet /ˈnɪmfɪt, nɪmˈfet/ n. **1** a young nymph. **2** colloq. a sexually attractive (esp. precocious) young woman.

nympho /ˈnɪmfəʊ/ n. (pl. **-os**) colloq. a nymphomaniac. [abbr.]

nympholepsy /ˈnɪmfəˌlepsɪ/ n. ecstasy or frenzy caused by desire of the unattainable. [NYMPHOLEPT after epilepsy]

nympholept /ˈnɪmfəˌlept/ n. a person inspired by violent enthusiasm esp. for an ideal. □ **nympholeptic** /ˌnɪmfəˈleptɪk/ adj. [Gk numpholēptos caught by nymphs (as NYMPH, lambanō take)]

nymphomania /ˌnɪmfəˈmeɪnɪə/ n. excessive sexual desire in a woman. □ **nymphomaniac** /-nɪˌæk/ n. & adj. [mod.L (as NYMPH, -MANIA)]

Nynorsk /ˈnjuːnɔːsk/ *n.* a form of the Norwegian language (see Norwegian). [Norw., f. *ny* new + *Norsk* Norwegian]

Nysa see Neisse.

nystagmus /nɪˈstægməs/ *n. Med.* rapid involuntary movements of the eyes. □ **nystagmic** *adj.* [Gk *nustagmos* nodding f. *nustazō* nod]

nystatin /ˈnaɪstətɪn/ *n. Pharm.* a yellow antibiotic used to treat fungal infections. [*New York State*, where developed]

Nyx /nɪks/ *Gk Mythol.* the female personification of Night, daughter of Chaos, a primeval deity seldom worshipped.

NZ *abbr.* New Zealand.

Oo

O[1] /əʊ/ *n.* (also **o**) (*pl.* **Os** or **O's**) **1** the fifteenth letter of the alphabet. **2** (**0**) nought, zero (in a sequence of numerals esp. when spoken). **3** a human blood type of the ABO system.

O[2] *abbr.* (also **O.**) Old.

O[3] *symb. Chem.* the element oxygen.

O[4] /əʊ/ *int.* **1** var. of OH[1]. **2** prefixed to a name in the vocative (*O God*). [ME, natural excl.]

O' /əʊ, ə/ *prefix* of Irish patronymic names (*O'Connor*). [Ir. *ó*, *ua*, descendant]

o' /ə/ *prep.* of, on (esp. in phrases: *o'clock; will-o'-the-wisp*). [abbr.]

-o /əʊ/ *suffix* forming usu. *sl.* or *colloq.* variants or derivatives (*beano; wino*). [perh. OH[1] as joc. suffix]

-o- /əʊ/ *suffix* the terminal vowel of combining forms (*spectro-*; *chemico-*; *Franco-*). ¶ Often elided before a vowel, as in *neuralgia*. [orig. Gk]

oaf /əʊf/ *n.* (*pl.* **oafs**) **1** an awkward lout. **2** a stupid person. □ **oafish** *adj.* **oafishly** *adv.* **oafishness** *n.* [orig. = elf's child, var. of obs. *auf* f. ON *álfr* elf]

Oahu /əʊˈɑːhuː/ the third largest of the Hawaiian islands; pop. (1988) 838,500. Its principal town, Honolulu, is the state capital of Hawaii. It is the site of Pearl Harbor, a US naval base (see PEARL HARBOR).

oak /əʊk/ *n.* **1** a tree or shrub of the genus *Quercus*, usu. having lobed leaves and bearing acorns. (*See note below.*) **2** the durable wood of this tree, used esp. for furniture, and (formerly) in buildings and ships. **3** (*attrib.*) made of oak (*oak table*). **4** a heavy outer door of a set of university college rooms. □ **oak-apple** a spongy apple-like gall formed by the gall-wasp *Biorhiza pallida*, found on oak trees. **oak eggar** see EGGAR. □ **oaken** *adj.* [OE *āc* f. Gmc]

▪ Oak woodland is the dominant natural vegetation of many north temperate countries, though relatively small remnants survive today. The chief species in western Europe are deciduous (the pedunculate and durmast oaks), whereas those in the Mediterranean region are usually evergreen (e.g. the cork, holm, and kermes oaks). North American species such as the turkey oaks and red oak are often grown as ornamental trees for their red autumn foliage.

Oak-apple Day /ˈəʊkˌæp(ə)l/ 29 May, the day of Charles II's Restoration in 1660, on which oak-apples or oak-leaves were worn in memory of the royal-oak incident (see *royal oak*).

Oakland /ˈəʊklənd/ an industrial port on the east side of San Francisco Bay in California; pop. (1990) 372,240.

Oakley /ˈəʊklɪ/, Annie (full name Phoebe Anne Oakley Mozee) (1860–1926), American markswoman. In 1885 she joined Buffalo Bill's Wild West Show, of which she became a star attraction for the next seventeen years, often working with her husband, the marksman Frank E. Butler. The musical *Annie Get Your Gun* (1946), with music by Irving Berlin, was based on her life.

Oaks, the an annual flat horse-race for three-year-old fillies run on Epsom Downs in England, over the same course as the Derby and three days after it. It was first run in 1779, and is named after the 12th Earl of Derby's shooting-box, 'The Oaks'.

oakum /ˈəʊkəm/ *n.* a loose fibre obtained by picking old rope to pieces and used esp. in caulking. [OE *ǣcumbe*, *ācumbe*, lit. 'off-combings']

O. & M. *abbr.* organization and methods.

OAP *abbr. Brit.* old-age pensioner.

OAPEC see ORGANIZATION OF ARAB PETROLEUM EXPORTING COUNTRIES.

oar /ɔː(r)/ *n.* **1** a pole with a blade used for rowing or steering a boat by leverage against the water. **2** a rower. □ **put** (or **stick**) **one's oar in** interfere, meddle. **rest** (*US* **lay**) **on one's oars** relax one's efforts. □ **oared** *adj.* (also in *comb.*). **oarless** *adj.* [OE *ār* f. Gmc, perh. rel. to Gk *eretmos* oar]

oarfish /ˈɔːfɪʃ/ *n.* a large, very long, deep-water ribbonfish, *Regalecus glesne*.

oarlock /ˈɔːlɒk/ *n. N. Amer.* a rowlock.

oarsman /ˈɔːzmən/ *n.* (*pl.* **-men**; *fem.* **oarswoman**, *pl.* **-women**) a rower. □ **oarsmanship** *n.*

oarweed /ˈɔːwiːd/ *n.* (also **oreweed**) a large marine alga, esp. of the genus *Laminaria*, often growing along rocky shores.

OAS *abbr.* **1** see ORGANIZATION OF AMERICAN STATES. **2** on active service.

oasis /əʊˈeɪsɪs/ *n.* (*pl.* **oases** /-siːz/) **1** a fertile spot in a desert, where water is found. **2** an area or period of calm in the midst of turbulence. [LL f. Gk, app. of Egypt. orig.]

oast /əʊst/ *n.* a kiln for drying hops. □ **oast-house** a building containing an oast. [OE *āst* f. Gmc]

oat /əʊt/ *n.* **1 a** a cereal plant, *Avena sativa*, cultivated in cool climates. **b** (in *pl.*) the grain yielded by this, used as food. **2** a related grass, esp. the *wild oat* (*Avena fatua*). **3** *poet.* the oat-stem used as a musical pipe by shepherds etc., usu. in pastoral or bucolic poetry. **4** (in *pl.*) *sl.* sexual gratification. □ **feel one's oats** *colloq.* **1** be lively. **2** *US* feel self-important. **oat-grass** a grass, esp. of the genus *Arrhenatherum* or *Helictotrichon*, resembling the oat. **off one's oats** *colloq.* not hungry. **sow one's oats** (or **wild oats**) indulge in youthful excess or promiscuity. □ **oaten** *adj.* **oaty** *adj.* [OE *āte*, pl. *ātan*, of unkn. orig.]

oatcake /ˈəʊtkeɪk/ *n.* a thin unleavened biscuit-like food made of oatmeal, common in Scotland and northern England.

Oates /əʊts/, Titus (1649–1705), English clergyman and conspirator. He is remembered as the fabricator of the Popish Plot; convicted of perjury in 1685, Oates was imprisoned in the same year, but subsequently released and granted a pension.

oath /əʊθ/ *n.* (*pl.* **oaths** /əʊðz/) **1** a solemn declaration or undertaking (often naming God) as to the truth of something or as a commitment to future action. **2** a statement or promise contained in an oath (*oath of allegiance*). **3** a profane or blasphemous utterance; a curse. □ **on** (or **under**) **oath** having sworn a solemn oath. **take** (or **swear**) **an oath** make such a declaration or undertaking. [OE *āth* f. Gmc]

oatmeal /ˈəʊtmiːl/ *n.* **1** meal made from ground oats used esp. in porridge and oatcakes. **2** a greyish-fawn colour flecked with brown.

OAU see ORGANIZATION OF AFRICAN UNITY.

Oaxaca /wəˈhɑːkə/ **1** a state of southern Mexico. **2** (in full **Oaxaca de Juárez** /deɪ ˈhwɑːrez/) its capital city; pop. (1990) 212,940.

OB *abbr. Brit.* outside broadcast.

Ob /ɒb/ the principal river of the western Siberian lowlands and one of the largest rivers in Russia. Rising in the Altai Mountains, it flows generally north and west for 5,410 km (3,481 miles) before entering the Gulf of Ob (or Ob Bay), an inlet of the Kara Sea, a part of the Arctic Ocean.

ob. *abbr.* he or she died. [L *obiit*]

ob- /ɒb, əb/ *prefix* (also **oc-** before *c*, **of-** before *f*, **op-** before *p*) occurring mainly in words of Latin origin, meaning: **1** exposure, openness (*object; obverse*). **2** meeting or facing (*occasion; obvious*). **3** direction (*oblong; offer*).

4 opposition, hostility, or resistance (*obstreperous*; *opponent*; *obstinate*). **5** hindrance, blocking, or concealment (*obese*; *obstacle*; *occult*). **6** finality or completeness (*obsolete*; *occupy*). **7** (in modern technical words) inversely; in a direction or manner contrary to the usual (*obconical*; *obovate*). [L f. *ob* towards, against, in the way of]

Obad. *abbr.* (in the Bible) Obadiah.

Obadiah /ˌɒbəˈdaɪə/ **1** a Hebrew minor prophet. **2** the shortest book of the Bible, bearing his name.

Oban /ˈəʊb(ə)n/ a port and tourist resort on the west coast of Scotland, in Strathclyde region, opposite the island of Mull; pop. (1981) 8,110. It is a major ferry terminal for boats to the Hebrides.

obbligato /ˌɒblɪˈɡɑːtəʊ/ *n.* (*pl.* **-os**) *Mus.* an accompaniment, usu. special and unusual in effect, forming an integral part of a composition (*with violin obbligato*). [It., = obligatory, f. L *obligatus* past part. (as OBLIGE)]

obconical /ɒbˈkɒnɪk(ə)l/ *adj.* (also **obconic**) in the form of an inverted cone.

obcordate /ɒbˈkɔːdeɪt/ *adj. Biol.* in the shape of a heart, with the pointed end serving as the base or point of attachment.

obdurate /ˈɒbdjʊrət/ *adj.* **1** stubborn. **2** hardened against persuasion or influence. □ **obduracy** *n.* **obdurately** *adv.* **obdurateness** *n.* [ME f. L *obduratus* past part. of *obdurare* (as OB-, *durare* harden f. *durus* hard)]

OBE *abbr.* (in the UK) Officer of the Order of the British Empire.

obeah /ˈəʊbɪə/ *n.* (also **obi** /ˈəʊbɪ/) a kind of sorcery practised esp. in the West Indies. [Twi]

obeche /əʊˈbiːtʃɪ/ *n.* **1** a West African tree, *Triplochiton scleroxylon*. **2** the light-coloured timber from this. [Nigerian name]

obedience /əʊˈbiːdɪəns/ *n.* **1** the action or practice of obeying; the fact or quality of being obedient. **2** submission to another's rule or authority. **3** compliance with a law or command. **4** *Eccl.* **a** compliance with a monastic rule. **b** a sphere of authority (*the Roman obedience*). □ **in obedience to** actuated by or in accordance with. [ME f. OF f. L *obedientia* (as OBEY)]

obedient /əʊˈbiːdɪənt/ *adj.* **1** obeying or ready to obey. **2** (often foll. by *to*) submissive to another's will; dutiful (*obedient to the law*). □ **obediently** *adv.* [ME f. OF f. L *obediens -entis* (as OBEY)]

obeisance /əʊˈbeɪs(ə)ns/ *n.* **1** a bow, curtsy, or other respectful or submissive gesture (*make an obeisance*). **2** homage, submission, deference (*pay obeisance*). □ **obeisant** *adj.* [ME f. OF *obeissance* (as OBEY)]

obeli *pl.* of OBELUS.

obelisk /ˈɒbəlɪsk/ *n.* **1 a** a tapering usu. four-sided stone pillar set up as a monument or landmark etc. **b** a mountain, tree, etc., of similar shape. **2** = OBELUS. [L *obeliscus* f. Gk *obeliskos* dimin. of *obelos* SPIT²]

obelize /ˈɒbəˌlaɪz/ *v.tr.* (also **-ise**) mark with an obelus as spurious etc. [Gk *obelizō* f. *obelos*: see OBELISK]

obelus /ˈɒbələs/ *n.* (*pl.* **obeli** /-ˌlaɪ/) **1** a symbol (†) used as a reference mark in printed matter or to indicate that a person is deceased. **2** a mark (- or ÷) used in ancient manuscripts to mark a word or passage, esp. as spurious. [L f. Gk *obelos* SPIT²]

Oberammergau /ˌəʊbərˈæməˌɡaʊ/ a village in the Bavarian Alps of SW Germany; pop. (1983) 4,800. It is the site of the most famous of the few surviving passion-plays, which has been performed every tenth year (with few exceptions) from 1634 as a result of a vow made during an epidemic of plague. It is entirely amateur, the villagers dividing the parts among themselves and being responsible also for the production, music, costumes, and scenery.

Oberhausen /ˈəʊbəˌhaʊz(ə)n/ an industrial city in western Germany, in the Ruhr valley of North Rhine-Westphalia; pop. (1991) 224,560.

Oberon /ˈəʊbəˌrɒn/ **1** king of the fairies in Shakespeare's *A Midsummer Night's Dream*. **2** *Astron.* satellite IV of Uranus, the furthest from the planet, discovered by William Herschel in 1787 (diameter 1,550 km). It is heavily cratered, some of the craters having bright rays surrounding a dark centre.

Oberösterreich /ˈoːbərˌøːstəraɪç/ the German name for UPPER AUSTRIA.

obese /əʊˈbiːs/ *adj.* very fat; corpulent. □ **obeseness** *n.* **obesity** *n.* [L *obesus* (as OB-, *edere* eat)]

obey /əʊˈbeɪ/ *v.* **1** *tr.* **a** carry out the command of (*you will obey me*). **b** carry out (a command) (*obey orders*). **2** *intr.* do what one is told to do. **3** *tr.* be actuated by (a force or impulse). □ **obeyer** *n.* [ME f. OF *obeir* f. L *obedire* (as OB-, *audire* hear)]

obfuscate /ˈɒbfʌˌskeɪt/ *v.tr.* **1** obscure or confuse (a mind, topic, etc.).

2 stupefy, bewilder. □ **obfuscatory** *adj.* **obfuscation** /ˌɒbfʌˈskeɪʃ(ə)n/ *n.* [LL *obfuscare* (as OB-, *fuscus* dark)]

obi¹ var. of OBEAH.

obi² /ˈəʊbɪ/ *n.* (*pl.* **obis**) a broad sash worn with a Japanese kimono. [Jap. *obi* belt]

obit /ˈɒbɪt, ˈəʊb-/ *n. colloq.* an obituary. [abbr.]

obiter dictum /ˌɒbɪtə ˈdɪktəm/ *n.* (*pl.* **obiter dicta** /-tə/) **1** a judge's expression of opinion uttered in court or giving judgement, but not essential to the decision and therefore without binding authority. **2** an incidental remark. [L f. *obiter* by the way + *dictum* a thing said]

obituary /əˈbɪtjʊərɪ/ *n.* (*pl.* **-ies**) **1** a notice of a death or deaths esp. in a newspaper. **2** an account of the life of a deceased person. **3** (*attrib.*) of or serving as an obituary. □ **obituarist** *n.* **obituarial** /əˌbɪtjʊˈeərɪəl/ *adj.* [med.L *obituarius* f. L *obire obit-* die (as OB-, *ire* go)]

object *n. & v.* ● *n.* /ˈɒbdʒɪkt/ **1** a material thing that can be seen or touched. **2** (foll. by *of*) a person or thing to which action or feeling is directed (*the object of attention*; *the object of our study*). **3** a thing sought or aimed at; a purpose. **4** *Gram.* a noun or its equivalent governed by an active transitive verb or by a preposition. **5** *Philos.* a thing external to the thinking mind or subject. **6** *derog.* a person or thing of esp. a pathetic or ridiculous appearance. **7** *Computing* a package of information and a description of its manipulation. ● *v.* /əbˈdʒekt/ **1** *intr.* (often foll. by *to*, *against*) express or feel opposition, disapproval, or reluctance; protest (*I object to being treated like this*; *objecting against government policies*). **2** *tr.* (foll. by *that* + clause) state as an objection (*objected that they were kept waiting*). **3** *tr.* (foll. by *to*, *against*, or *that* + clause) adduce (a quality or fact) as contrary or damaging (to a case). □ **no object** not forming an important or restricting factor (*money no object*). **object-ball** *Billiards* the ball at which a player aims the cue-ball. **object-glass** = OBJECTIVE *n.* 3. **object language 1** a language described by means of another language (see METALANGUAGE). **2** *Computing* a language into which a program is translated by means of a compiler or assembler. **object-lesson** a striking practical example of some principle. **object of the exercise** the main point of an activity. □ **objectless** /ˈɒbdʒɪktlɪs/ *adj.* **objector** /əbˈdʒektə(r)/ *n.* [ME f. med.L *objectum* thing presented to the mind, past part. of L *objicere* (as OB-, *jacere ject-* throw)]

objectify /əbˈdʒektɪˌfaɪ/ *v.tr.* (**-ies, -ied**) **1** make objective; express in a concrete form. **2** present as an object of perception. □ **objectification** /-ˌdʒektɪfɪˈkeɪʃ(ə)n/ *n.*

objection /əbˈdʒekʃ(ə)n/ *n.* **1** an expression or feeling of opposition or disapproval. **2** the act of objecting. **3** an adverse reason or statement. [ME f. OF *objection* or LL *objectio* (as OBJECT)]

objectionable /əbˈdʒekʃənəb(ə)l/ *adj.* **1** open to objection. **2** unpleasant, offensive. □ **objectionableness** *n.* **objectionably** *adv.*

objective /əbˈdʒektɪv/ *adj. & n.* ● *adj.* **1** external to the mind; actually existing; real. **2** (of a person, writing, art, etc.) dealing with outward things or exhibiting facts uncoloured by feelings or opinions; not subjective. **3** *Gram.* (of a case or word) constructed as or appropriate to the object of a transitive verb or preposition (cf. ACCUSATIVE). **4** aimed at (*objective point*). **5** (of symptoms) observed by another and not only felt by the patient. ● *n.* **1** something sought or aimed at; an objective point. **2** *Gram.* the objective case. **3** (in full **objective lens**) the lens in a telescope, microscope, etc. nearest to the object observed. Also called *object-glass*. □ **objectively** *adv.* **objectiveness** *n.* **objectivize** *v.tr.* (also **-ise**). **objectivization** /-ˌdʒektɪvaɪˈzeɪʃ(ə)n/ *n.* **objectival** /ˌɒbdʒekˈtaɪv(ə)l/ *adj.* **objectivity** /-ˈtɪvɪtɪ/ *n.* [med.L *objectivus* (as OBJECT)]

objectivism /əbˈdʒektɪˌvɪz(ə)m/ *n.* **1** the tendency to lay stress on what is objective. **2** *Philos.* the doctrine that knowledge, morality, perception, etc. exist apart from human knowledge or perception of them (opp. SUBJECTIVISM). □ **objectivist** *n.* **objectivistic** /-ˌdʒektɪˈvɪstɪk/ *adj.*

objet d'art /ˌɒbʒeɪ ˈdɑː(r)/ *n.* (*pl.* **objets d'art** pronunc. same) a small decorative object. [F, lit. 'object of art']

objurgate /ˈɒbdʒəˌɡeɪt/ *v.tr. literary* chide or scold. □ **objurgation** /ˌɒbdʒəˈɡeɪʃ(ə)n/ *n.* **objurgatory** /ɒbˈdʒɜːɡətərɪ/ *adj.* [L *objurgare* *objurgat-* (as OB-, *jurgare* quarrel f. *jurgium* strife)]

oblanceolate /ɒbˈlɑːnsɪələt/ *adj. Bot.* (esp. of leaves) lanceolate with the more pointed end at the base.

oblast /ˈɒblæst/ *n.* an administrative division or region in Russia and the former Soviet Union, and in some constituent republics of the former Soviet Union. [Russ.]

oblate¹ /ˈɒbleɪt/ *n.* a person dedicated to a monastic or religious life or work. [F f. med.L *oblatus* f. L *offerre oblat-* offer (as OB-, *ferre* bring)]

oblate[2] /'ɒbleɪt/ adj. Geom. (of a spheroid) flattened at the poles (cf. PROLATE). [mod.L oblatus (as OBLATE[1])]

oblation /ɒ'bleɪʃ(ə)n/ n. Relig. **1** a thing offered to a divine being. **2** the presentation of bread and wine to God in the Eucharist. □ **oblational** adj. **oblatory** /'ɒblətərɪ/ adj. [ME f. OF oblation or LL oblatio (as OBLATE[1])]

obligate v. & adj. ● v.tr. /'ɒblɪ.geɪt/ **1** (usu. in passive; foll. by to + infin.) bind (a person) legally or morally. **2** US commit (assets) as security. ● adj. /'ɒblɪgət/ Biol. that has to be as described (obligate parasite). □ **obligator** /-.geɪtə(r)/ n. [L obligare obligat- (as OBLIGE)]

obligation /.ɒblɪ'geɪʃ(ə)n/ n. **1** the constraining power of a law, precept, duty, contract, etc. **2** a duty; a burdensome task. **3** a binding agreement, esp. one enforceable under legal penalty; a written contract or bond. **4 a** a service or benefit; a kindness done or received (repay an obligation). **b** indebtedness for this (be under an obligation). □ **day of obligation** Eccl. a day on which all are required to attend Mass or Communion. **of obligation** obligatory. □ **obligational** adj. [ME f. OF f. L obligatio -onis (as OBLIGE)]

obligatory /ə'blɪgətərɪ/ adj. **1** legally or morally binding. **2** compulsory and not merely permitted. **3** constituting an obligation. □ **obligatorily** adv. [ME f. LL obligatorius (as OBLIGE)]

oblige /ə'blaɪdʒ/ v. **1** tr. (foll. by to + infin.) constrain, compel. **2** tr. be binding on. **3** tr. a make indebted by conferring a favour. **b** (foll. by with, or by + verbal noun; also absol.) gratify, perform a service for (oblige me by leaving; obliged with a song; will you oblige?). **4** tr. (in passive; foll. by to) be indebted (am obliged to you for your help). **5** tr. (foll. by to, or to + infin.) archaic or Law bind by oath, promise, contract, etc. □ **much obliged** an expression of thanks. □ **obliger** n. [ME f. OF obliger f. L obligare (as OB-, ligare bind)]

obligee /.ɒblɪ'dʒiː/ n. Law a person to whom another is bound by contract or other legal procedure (cf. OBLIGOR).

obliging /ə'blaɪdʒɪŋ/ adj. courteous, accommodating; ready to do a service or kindness. □ **obligingly** adv. **obligingness** n.

obligor /.ɒblɪ'gɔː(r)/ n. Law a person who is bound to another by contract or other legal procedure (cf. OBLIGEE).

oblique /ə'bliːk/ adj., n., & v. ● adj. **1 a** slanting; declining from the vertical or horizontal. **b** diverging from a straight line or course. **2** not going straight to the point; roundabout, indirect. **3** Geom. **a** (of a line, plane figure, or surface) inclined at other than a right angle. **b** (of an angle) acute or obtuse. **c** (of a cone, cylinder, etc.) with an axis not perpendicular to the plane of its base. **4** Anat. neither parallel nor perpendicular to the long axis of a body or limb. **5** Bot. (of a leaf) with unequal sides. **6** Gram. denoting any case other than the nominative or vocative. ● n. **1** an oblique stroke (/). **2** an oblique muscle. ● v.intr. (**obliques**, **obliqued**, **obliquing**) esp. Mil. advance obliquely. □ **oblique oration** (or **speech**) = reported speech (see REPORT). **oblique sphere** see SPHERE. □ **obliquely** adv. **obliqueness** n. **obliquity** /ə'blɪkwɪtɪ/ n. [ME f. F f. L obliquus]

obliterate /ə'blɪtə.reɪt/ v.tr. **1 a** blot out; efface, erase, destroy. **b** leave no clear traces of. **2** deface (a postage stamp etc.) to prevent further use. □ **obliterator** n. **obliterative** /-rətɪv/ adj. **obliteration** /ə.blɪtə'reɪʃ(ə)n/ n. [L obliterare (as OB-, litera LETTER)]

oblivion /ə'blɪvɪən/ n. **1 a** the state of having forgotten. **b** disregard; forgetfulness. **2** the state or condition of being forgotten. **3** an amnesty or pardon. □ **fall into oblivion** be forgotten or disused. [ME f. OF f. L oblivio -onis f. oblivisci forget]

oblivious /ə'blɪvɪəs/ adj. **1** (often foll. by of) forgetful, unmindful. **2** (foll. by to, of) unaware or unconscious of. □ **obliviously** adv. **obliviousness** n. [ME f. L obliviosus (as OBLIVION)]

oblong /'ɒblɒŋ/ adj. & n. ● adj. **1** deviating from a square form by having one long axis, esp. rectangular with adjacent sides unequal. **2** greater in breadth than in height. ● n. an oblong figure or object. [ME f. L oblongus longish (as OB-, longus long)]

obloquy /'ɒbləkwɪ/ n. **1** the state of being generally ill spoken of. **2** abuse, detraction. [ME f. LL obloquium contradiction f. L obloqui deny (as OB-, loqui speak)]

obnoxious /əb'nɒkʃəs/ adj. offensive, objectionable, disliked. □ **obnoxiously** adv. **obnoxiousness** n. [orig. = vulnerable (to harm), f. L obnoxiosus or obnoxius (as OB-, noxa harm: assoc. with NOXIOUS)]

oboe /'əʊbəʊ/ n. **1 a** a woodwind instrument with a double-reed mouthpiece. (See note below.) **b** a player of this instrument. **2** an organ stop with a quality resembling an oboe. □ **oboe d'amore** /dæ'mɔː.reɪ/ an oboe with a pear-shaped bell and mellow tone, pitched a minor third below a normal oboe, commonly used in baroque music.

oboist /'əʊbəʊɪst/ n. [It. oboe or F hautbois f. haut high + bois wood: d'amore = of love]

▪ Generally resembling the clarinet in size and appearance, the oboe is an instrument of treble pitch and plaintive incisive tone, with a range of over two-and-a-half octaves. It was developed in France in the 17th century.

obol /'ɒb(ə)l/ n. an ancient Greek coin, equal to one-sixth of a drachma. [L obolus f. Gk obolos, var. of obelos OBELUS]

Obote /ə'bəʊtɪ/, (Apollo) Milton (b.1924), Ugandan statesman, Prime Minister 1962–6, President 1966–71 and 1980–5. After founding the Uganda People's Congress in 1960, he became the first Prime Minister of independent Uganda. Overthrown by Idi Amin in 1971, he returned from exile nine years later and was re-elected President. Obote established a multi-party democracy, but was removed in a second military coup in 1985.

obovate /ɒb'əʊveɪt/ adj. Bot. (of a leaf) ovate with the narrower end at the base.

O'Brien[1] /əʊ'braɪən/, Edna (b.1932), Irish novelist and short-story writer. Her novels include the trilogy The Country Girls (1960), The Lonely Girl (1962), and Girls in Their Married Bliss (1964), which follows the fortunes of two Irish girls from their rural, convent-educated early years to new lives in Dublin and later in London. Among her collections of short stories is Lantern Slides (1990).

O'Brien[2] /əʊ'braɪən/, Flann (pseudonym of Brian O'Nolan) (1911–66), Irish novelist and journalist. He gained recognition as a novelist with his first book At Swim-Two-Birds (1939), an exploration of Irish life combining naturalism and farce, employing an experimental narrative structure much influenced by James Joyce. Writing under the name of Myles na Gopaleen, O'Brien contributed a satirical column to the Irish Times for nearly twenty years.

obscene /əb'siːn/ adj. **1** offensively or repulsively indecent, esp. by offending accepted sexual morality. **2** colloq. highly offensive or repugnant (an obscene accumulation of wealth). **3** Brit. Law (of a publication) tending to deprave or corrupt. □ **obscenely** adv. [F obscène or L obsc(a)enus ill-omened, abominable]

obscenity /əb'senɪtɪ/ n. (pl. **-ies**) **1** the state or quality of being obscene. **2** an obscene word or gesture. [L obscaenitas (as OBSCENE)]

obscurantism /.ɒbskjʊə'ræntɪz(ə)m/ n. opposition to knowledge and enlightenment. □ **obscurantist** n. & adj. **obscurant** /əb'skjʊərənt/ n. [obscurant f. G f. L obscurans f. obscurare: see OBSCURE]

obscure /əb'skjʊə(r)/ adj. & v. ● adj. **1** not clearly expressed or easily understood; vague, uncertain. **2** dark, dim. **3** indistinct, faint. **4** hidden; remote from observation. **5 a** unnoticed. **b** (of a person or a person's origins etc.) undistinguished, hardly known; humble. **6** (of a colour) dingy, dull, indefinite. ● v.tr. **1** make obscure, dark, indistinct, or unintelligible. **2** dim the glory of; outshine. **3** conceal from sight. □ **obscure vowel** = indeterminate vowel. □ **obscurely** adv. **obscuration** /.ɒbskjʊə'reɪʃ(ə)n/ n. [ME f. OF obscur f. L obscurus dark]

obscurity /əb'skjʊərɪtɪ/ n. (pl. **-ies**) **1** the state of being obscure. **2** an obscure person or thing. [F obscurité f. L obscuritas (as OBSCURE)]

obsecration /.ɒbsɪ'kreɪʃ(ə)n/ n. earnest entreaty. [ME f. L obsecratio f. obsecrare entreat (as OB-, sacrare f. sacer sacri sacred)]

obsequies /'ɒbsɪkwɪz/ n.pl. **1** funeral rites. **2** a funeral. □ **obsequial** /əb'siːkwɪəl/ adj. [ME, pl. of obs. obsequy f. AF obsequie, OF obseque f. med.L obsequiae f. L exsequiae funeral rites (see EXEQUIES): assoc. with obsequium (see OBSEQUIOUS)]

obsequious /əb'siːkwɪəs/ adj. servilely obedient or attentive. □ **obsequiously** adv. **obsequiousness** n. [ME f. L obsequiosus f. obsequium compliance (as OB-, sequi follow)]

observance /əb'zɜːv(ə)ns/ n. **1** the act or process of keeping or performing a law, duty, custom, ritual, etc. **2** an act of a religious or ceremonial character; a customary rite. **3** the rule of a religious order. **4** archaic respect, deference. **5** archaic the act of observing or watching; observation. [ME f. OF f. L observantia (as OBSERVE)]

observant /əb'zɜːv(ə)nt/ adj. & n. ● adj. **1 a** acute or diligent in taking notice. **b** (often foll. by of) carefully particular; heedful. **2** attentive in esp. religious observances (an observant Jew). ● n. (**Observant**) a member of the branch of the Franciscan order that observes the strict rule. □ **observantly** adv. [F (as OBSERVE)]

observation /.ɒbzə'veɪʃ(ə)n/ n. **1** the act or an instance of noticing; the condition of being noticed. **2** perception; the faculty of taking notice. **3** a remark or statement, esp. one that is of the nature of a comment. **4 a** the accurate watching and noting of an object, organism, or phenomenon for the purpose of scientific investigation.

b a measurement or other result so obtained. **c** the noting of the symptoms of a patient, the behaviour of a suspect, etc. **5** the taking of the sun's or another celestial body's altitude to find a latitude or longitude. **6** *Mil.* the watching of a fortress or hostile position or movements. □ **observation car** esp. *N. Amer.* a carriage in a train built so as to afford good views. **observation post** *Mil.* a post for watching the effect of artillery fire etc. **under observation** (esp. of a suspect or hospital patient) being watched methodically. □ **observational** *adj.* **observationally** *adv.* [ME f. L *observatio* (as OBSERVE)]

observatory /əbˈzɜːvətərɪ/ *n.* (pl. **-ies**) a room or building equipped for the observation of natural, esp. astronomical or meteorological, phenomena. [mod.L *observatorium* f. L *observare* (as OBSERVE)]

observe /əbˈzɜːv/ *v.* **1** *tr.* (often foll. by *that, how* + clause) perceive; note; take notice of; become conscious of. **2** *tr.* watch carefully. **3** *tr.* **a** follow or adhere to (a law, command, method, principle, etc.). **b** keep or adhere to (an appointed time). **c** maintain (silence). **d** duly perform (a rite). **e** celebrate (an anniversary). **4** *tr.* examine and note (phenomena) without the aid of experiment. **5** *tr.* (often foll. by *that* + clause) say, esp. by way of comment. **6** *intr.* (foll. by *on*) make a remark or remarks about. □ **observable** *adj.* **observably** *adv.* [ME f. OF *observer* f. L *observare* watch (as OB-, *servare* keep)]

observer /əbˈzɜːvə(r)/ *n.* **1** a person who observes. **2** an interested spectator. **3** a person who attends a conference etc. to note the proceedings but does not participate. **4 a** a person trained to notice and identify aircraft. **b** a person carried in an aeroplane to note the enemy's position etc.

obsess /əbˈses/ *v.tr.* (often in *passive*) preoccupy, haunt; fill the mind of (a person) continually. □ **obsessive** *adj.* & *n.* **obsessively** *adv.* **obsessiveness** *n.* [L *obsidere obsess-* (as OB-, *sedere* sit)]

obsession /əbˈseʃ(ə)n/ *n.* **1** the act of obsessing or the state of being obsessed. **2** a persistent idea or thought dominating a person's mind. **3** a condition in which such ideas are present. □ **obsessional** *adj.* **obsessionalism** *n.* **obsessionally** *adv.* [L *obsessio* (as OBSESS)]

obsidian /əbˈsɪdɪən/ *n.* a dark glassy volcanic rock formed from hardened lava. [L *obsidianus*, error for *obsianus* f. *Obsius*, the name (in Pliny) of the discoverer of a similar stone]

obsolescent /ˌɒbsəˈles(ə)nt/ *adj.* becoming obsolete; going out of use or date. □ **obsolescence** *n.* [L *obsolescere obsolescent-* (as OB-, *solere* be accustomed)]

obsolete /ˈɒbsəliːt/ *adj.* **1** disused, discarded, antiquated. **2** *Biol.* less developed than formerly or than in a cognate species; rudimentary. □ **obsoletely** *adv.* **obsoleteness** *n.* **obsoletism** *n.* [L *obsoletus* past part. (as OBSOLESCENT)]

obstacle /ˈɒbstək(ə)l/ *n.* a person or thing that obstructs progress. □ **obstacle-race** a race in which various obstacles have to be negotiated. [ME f. OF f. L *obstaculum* f. *obstare* impede (as OB-, *stare* stand)]

obstetric /əbˈstetrɪk/ *adj.* (also **obstetrical** /-k(ə)l/) of or relating to childbirth and associated processes. □ **obstetrically** *adv.* **obstetrician** /ˌɒbstəˈtrɪʃ(ə)n/ *n.* [mod.L *obstetricus* for L *obstetricius* f. *obstetrix* midwife f. *obstare* be present (as OB-, *stare* stand)]

obstetrics /əbˈstetrɪks/ *n.pl.* (usu. treated as *sing.*) the branch of medicine and surgery concerned with childbirth and midwifery.

obstinate /ˈɒbstɪnət/ *adj.* **1** stubborn, intractable. **2** firmly adhering to one's chosen course of action or opinion despite dissuasion. **3** inflexible, self-willed. **4** unyielding; not readily responding to treatment etc. □ **obstinacy** *n.* **obstinately** *adv.* [ME f. L *obstinatus* past part. of *obstinare* persist (as OB-, *stare* stand)]

obstreperous /əbˈstrepərəs/ *adj.* **1** turbulent, unruly; noisily resisting control. **2** noisy, vociferous. □ **obstreperously** *adv.* **obstreperousness** *n.* [L *obstreperus* f. *obstrepere* (as OB-, *strepere* make a noise)]

obstruct /əbˈstrʌkt/ *v.tr.* **1** block up; make hard or impossible to pass along or through. **2** prevent or retard the progress of; impede. □ **obstructor** *n.* [L *obstruere obstruct-* (as OB-, *struere* build)]

obstruction /əbˈstrʌkʃ(ə)n/ *n.* **1** the act or an instance of blocking; the state of being blocked. **2** the act of making or the state of becoming more or less impassable. **3** an obstacle or blockage. **4** the retarding of progress by deliberate delays, esp. of Parliamentary business. **5** *Sport* the act of unlawfully obstructing another player. **6** *Med.* a blockage in a bodily passage, esp. in an intestine. □ **obstructionism** *n.* (in sense 4). **obstructionist** *n.* (in sense 4). [L *obstructio* (as OBSTRUCT)]

obstructive /əbˈstrʌktɪv/ *adj.* & *n.* ● *adj.* causing or intended to cause an obstruction. ● *n.* an obstructive person or thing. □ **obstructively** *adv.* **obstructiveness** *n.*

obtain /əbˈteɪn/ *v.* **1** *tr.* acquire, secure; have granted to one. **2** *intr.* be prevalent or established or in vogue. □ **obtainer** *n.* **obtainment** *n.* **obtainable** *adj.* **obtainability** /-ˌteɪnəˈbɪlɪtɪ/ *n.* **obtention** /-ˈtenʃ(ə)n/ *n.* [ME f. OF *obtenir* f. L *obtinere obtent-* keep (as OB-, *tenere* hold)]

obtrude /əbˈtruːd/ *v.* **1** *intr.* be or become obtrusive. **2** *tr.* (often foll. by *on, upon*) thrust forward (oneself, one's opinion, etc.) importunately. □ **obtruder** *n.* **obtrusion** /-ˈtruːʒ(ə)n/ *n.* [L *obtrudere obtrus-* (as OB-, *trudere* push)]

obtrusive /əbˈtruːsɪv/ *adj.* **1** unpleasantly or unduly noticeable. **2** obtruding oneself. □ **obtrusively** *adv.* **obtrusiveness** *n.* [as OBTRUDE]

obtund /əbˈtʌnd/ *v.tr.* blunt or deaden (a sense or faculty). [ME f. L *obtundere obtus-* (as OB-, *tundere* beat)]

obtuse /əbˈtjuːs/ *adj.* **1** dull-witted; slow to understand. **2** of blunt form; not sharp-pointed or sharp-edged. **3** *Geom.* (of an angle) more than 90° and less than 180°. **4** (of pain or the senses) dull; not acute. □ **obtusely** *adv.* **obtuseness** *n.* **obtusity** *n.* [L *obtusus* past part. (as OBTUND)]

obverse /ˈɒbvɜːs/ *n.* & *adj.* ● *n.* **1 a** the side of a coin or medal etc. bearing the head or principal design. **b** this design. Opp. REVERSE *n.* 6. **2** the front or proper or top side of a thing. **3** the counterpart of a fact or truth. ● *adj.* **1** *Biol.* narrower at the base or point of attachment than at the apex or top (see OB- 7). **2** answering as the counterpart to something else. □ **obversely** *adv.* [L *obversus* past part. (as OBVERT)]

obvert /əbˈvɜːt/ *v.tr. Logic* alter (a proposition) so as to infer another proposition with a contradictory predicate, e.g. *no men are immortal* to *all men are mortal.* □ **obversion** /-ˈvɜːʒ(ə)n/ *n.* [L *obvertere obvers-* (as OB-, *vertere* turn)]

obviate /ˈɒbvɪˌeɪt/ *v.tr.* get round or do away with (a need, inconvenience, etc.). □ **obviation** /ˌɒbvɪˈeɪʃ(ə)n/ *n.* [LL *obviare* oppose (as OB-, *via* way)]

obvious /ˈɒbvɪəs/ *adj.* easily seen or recognized or understood; palpable, indubitable. □ **obviously** *adv.* **obviousness** *n.* [L *obvius* f. *ob viam* in the way]

OC *abbr.* Officer Commanding.

oc- /ɒk, ək/ *prefix* assim. form of OB- before *c*.

ocarina /ˌɒkəˈriːnə/ *n.* a small egg-shaped ceramic (usu. terracotta) or metal wind instrument with holes for the fingers. Invented *c.*1860, it is used as a toy and as a folk instrument. [It. f. *oca* goose (from its shape)]

OCAS see ORGANIZATION OF CENTRAL AMERICAN STATES.

O'Casey /əʊˈkeɪsɪ/, Sean (1880–1964), Irish dramatist. Encouraged by W. B. Yeats, he wrote a number of plays, including *The Shadow of a Gunman* (1923) and *Juno and the Paycock* (1924), which were successfully staged at the Abbey Theatre, Dublin. They deal with the lives of the Irish poor before and during the civil war that followed the establishment of the Irish Free State.

Occam, William of see WILLIAM OF OCCAM.

Occam's razor /ˈɒkəmz/ *n.* (also **Ockham's**) the principle, attributed to the English philosopher William of Occam, that in explaining a thing no more assumptions should be made than are necessary.

occasion /əˈkeɪʒ(ə)n/ *n.* & *v.* ● *n.* **1 a** a special or noteworthy event or happening (*dressed for the occasion*). **b** the time or occurrence of this (*on the occasion of their marriage*). **2** (often foll. by *for*, or *to* + infin.) a reason, ground, or justification (*there is no occasion to be angry*). **3** a juncture suitable for doing something; an opportunity. **4** an immediate but subordinate or incidental cause (*the assassination was the occasion of the war*). ● *v.tr.* **1** be the occasion or cause of; bring about esp. incidentally. **2** (foll. by *to* + infin.) cause (a person or thing to do something). □ **on occasion** now and then; when the need arises. **rise to the occasion** produce the necessary will, energy, ability, etc., in unusually demanding circumstances. **take occasion** (foll. by *to* + infin.) make use of the opportunity. [ME f. OF *occasion* or L *occasio* juncture, reason, f. *occidere occas-* go down (as OB-, *cadere* fall)]

occasional /əˈkeɪʒən(ə)l/ *adj.* **1** happening irregularly and infrequently. **2 a** made or meant for, or associated with, a special occasion (*occasional wear*). **b** (of furniture etc.) made or adapted for infrequent and varied use. **3** acting on a special occasion. □ **occasional cause** a secondary cause; an occasion (see OCCASION *n.* 4). **occasional table** a small table for infrequent and varied use. □ **occasionally** *adv.* **occasionality** /əˌkeɪʒəˈnælɪtɪ/ *n.*

Occident /ˈɒksɪd(ə)nt/ n. poet. or rhet. (prec. by the) **1** the West. **2** western Europe. **3** Europe, America, or both, as distinct from the Orient. **4** European or Western in contrast to oriental civilization. [ME f. OF f. L occidens -entis setting, sunset, west (as OCCASION)]

occidental /ˌɒksɪˈdent(ə)l/ adj. & n. ● adj. **1** of or characteristic of the West or of Western nations. **2** western, westerly. ● n. (**Occidental**) a native or inhabitant of the West. □ **occidentalism** n. **occidentalist** n. **occidentalize** v.tr. (also **-ise**). **occidentally** adv. [ME f. OF occidental or L occidentalis (as OCCIDENT)]

occipito- /ɒkˈsɪpɪtəʊ/ comb. form the back of the head. [as OCCIPUT]

occiput /ˈɒksɪˌpʌt/ n. the back of the head. □ **occipital** /ɒkˈsɪpɪt(ə)l/ adj. [ME f. L occiput (as OB-, caput head)]

Occitan /ˈɒksɪt(ə)n/ n. **1** = PROVENÇAL. **2** = LANGUE D'OC. □ **Occitanian** /ˌɒksɪˈteɪnɪən/ n. & adj. [F: cf. LANGUE D'OC]

occlude /əˈkluːd/ v.tr. **1** stop up or close (pores or an orifice). **2** Chem. absorb and retain (gases or impurities). □ **occluded front** Meteorol. a front resulting from occlusion. [L occludere occlus- (as OB-, claudere shut)]

occlusion /əˈkluːʒ(ə)n/ n. **1** the act or process of occluding. **2** Meteorol. a phenomenon in which the cold front of a depression overtakes the warm front, causing upward displacement of warm air between them. **3** Dentistry the position of the teeth when the jaws are closed. **4** the blockage or closing of a hollow organ etc. (coronary occlusion). **5** Phonet. the momentary closure of the vocal passage. □ **occlusive** /əˈkluːsɪv/ adj.

occult adj. & v. ● adj. /ɒˈkʌlt, ˈɒkʌlt/ **1** involving the supernatural; mystical, magical. **2** kept secret; esoteric. **3** recondite, mysterious; beyond the range of ordinary knowledge. **4** Med. not obvious on inspection. ● v.tr. /ɒˈkʌlt/ Astron. (of a celestial body) conceal an apparently smaller body from view by passing or being in front of it. □ **the occult** occult phenomena generally. **occulting light** a lighthouse light that is cut off at regular intervals. □ **occultism** /ɒˈkʌltɪz(ə)m, ˈɒkʌlˌtɪz(ə)m/ n. **occultist** /ɒˈkʌltɪst, ˈɒkʌltɪst/ n. **occultly** /ɒˈkʌltlɪ/ adv. **occultness** /ɒˈkʌltnɪs/ n. **occultation** /ˌɒkʌlˈteɪʃ(ə)n/ n. [L occulere occult- (as OB-, celare hide)]

occupant /ˈɒkjʊp(ə)nt/ n. **1** a person who occupies, resides in, or is in a place etc. (both occupants of the car were unhurt). **2** a person holding property, esp. land, in actual possession. **3** a person who establishes a title by taking possession of something previously without an established owner. □ **occupancy** n. (pl. **-ies**). [F occupant or L occupans -antis (as OCCUPY)]

occupation /ˌɒkjʊˈpeɪʃ(ə)n/ n. **1** what occupies one; a means of passing one's time. **2** a person's temporary or regular employment; a business, calling, or pursuit. **3** the act of occupying or state of being occupied. **4 a** the act of taking or holding possession of (a country, district, etc.) by military force. **b** the state or time of this. **5** tenure, occupancy. **6** (attrib.) for the sole use of the occupiers of the land concerned (occupation road). [ME f. AF ocupacioun, OF occupation f. L occupatio -onis (as OCCUPY)]

occupational /ˌɒkjʊˈpeɪʃən(ə)l/ adj. **1** of or in the nature of an occupation or occupations. **2** (of a disease, hazard, etc.) rendered more likely by one's occupation. □ **occupational therapist** a person skilled in supervising occupational therapy. **occupational therapy** mental or physical activity designed to assist recovery from disease or injury; a programme of this.

occupier /ˈɒkjʊˌpaɪə(r)/ n. Brit. a person residing in a property as its owner or tenant.

occupy /ˈɒkjʊˌpaɪ/ v.tr. (**-ies, -ied**) **1** reside in; be the tenant of. **2** take up or fill (space or time or a place). **3** hold (a position or office). **4** take military possession of (a country, region, town, strategic position). **5** place oneself in (a building etc.) forcibly or without authority. **6** (usu. in passive or refl.; often foll. by in, with) keep busy or engaged. [ME f. OF occuper f. L occupare seize (as OB-, capere take)]

occur /əˈkɜː(r)/ v.intr. (**occurred, occurring**) **1** come into being as an event or process at or during some time; happen. **2** exist or be encountered in some place or conditions. **3** (foll. by to, prec. by it as subject; usu. foll. by that + clause) come into the mind of, esp. as an unexpected or casual thought (it never occurred to him to ask their opinion). [L occurrere go to meet, present itself (as OB-, currere run)]

occurrence /əˈkʌrəns/ n. **1** the act or an instance of occurring. **2** an incident or event. □ **of frequent occurrence** often occurring. [occurrent that occurs f. F f. L occurrens -entis (as OCCUR)]

ocean /ˈəʊʃ(ə)n/ n. **1 a** a large expanse of sea, esp. each of the main areas called the Atlantic, Pacific, Indian, Arctic, and Antarctic Oceans. **b** these regarded cumulatively as the body of water surrounding the land of the globe. **2** (usu. prec. by the) the sea. **3** (often in pl.) a very large expanse or quantity of anything (oceans of room). □ **ocean-going** (of a ship) designed to cross oceans. **ocean tramp** a merchant ship, esp. a steamer, running on no regular line or route. □ **oceanward** adv. (also **-wards**). [ME f. OF occean f. L oceanus f. Gk ōkeanos the great river encircling the earth, the Atlantic]

oceanarium /ˌəʊʃəˈneərɪəm/ n. (pl. **oceanariums** or **oceanaria** /-rɪə/) a large sea-water aquarium for keeping sea animals. [OCEAN + -ARIUM, after aquarium]

Oceania /ˌəʊʃɪˈɑːnɪə, ˌəʊsɪ-/ the islands of the Pacific and adjacent seas. □ **Oceanian** adj. & n. [mod.L f. F Océanie f. L (as OCEAN)]

oceanic /ˌəʊʃɪˈænɪk, ˌəʊsɪ-/ adj. **1** of, like, or near the ocean. **2** (of a climate) governed by the ocean. **3** of the part of the ocean distant from the continents. **4** (**Oceanic**) of Oceania.

Oceanid /əʊˈsiːənɪd/ n. (pl. **Oceanids** or **Oceanides** /ˌəʊsɪˈænɪˌdiːz/) Gk Mythol. an ocean nymph. [Gk ōkeanis -idos daughter of Oceanus]

Ocean Island an alternative name for BANABA.

oceanography /ˌəʊʃəˈnɒɡrəfɪ/ n. the study of the oceans. □ **oceanographer** n. **oceanographic** /-nəˈɡræfɪk/ adj. **oceanographical** adj.

Oceanus /əʊˈsiːənəs, ˌəʊsɪˈeɪn-, ˌəʊsɪˈɑːn-/ Gk Mythol. the son of Uranus (Heaven) and Gaia (Earth), and father of the ocean nymphs (Oceanids) and river gods. He is the personification of the great river encircling the whole world.

ocellus /ɒˈseləs/ n. (pl. **ocelli** /-laɪ/) **1** Zool. **a** = simple eye. **b** = eye-spot 1a. **2** a spot of colour surrounded by a ring of a different colour on the wing of a butterfly etc. □ **ocellar** adj. **ocellate** /ˈɒsɪlət, -ˌleɪt/ adj. **ocellated** /-ˌleɪtɪd/ adj. [L, dimin. of oculus eye]

ocelot /ˈɒsɪˌlɒt/ n. **1** a medium-sized feline, Felis pardalis, native to Central and South America, having a tan coat with light black-edged patches. **2** its fur. [F, abbr. by Fr. naturalist G.–L. Buffon f. Nahuatl tlalocelotl jaguar of the field]

och /ɒx/ int. Sc. & Ir. expressing surprise or regret. [Gael. & Ir.]

oche /ˈɒkɪ/ n. (also **hockey** /ˈɒkɪ, ˈhɒkɪ/) (in darts etc.) the line behind which the players stand when throwing. [20th c.: orig. uncert. (perh. connected with OF ocher cut a deep notch in)]

ocher US var. of OCHRE.

ochlocracy /ɒkˈlɒkrəsɪ/ n. (pl. **-ies**) mob rule. □ **ochlocrat** /ˈɒkləˌkræt/ n. **ochlocratic** /ˌɒkləˈkrætɪk/ adj. [F ochlocratie f. Gk okhlokratia f. okhlos mob]

ochone /ɒˈxəʊn/ int. (also **ohone**) Sc. & Ir. expressing regret or lament. [Gael. & Ir. ochóin]

ochre /ˈəʊkə(r)/ n. (US **ocher**) **1** a mineral of clay and ferric oxide, used as a pigment varying from light yellow to brown or red. **2** a pale brownish-yellow. □ **ochreish** /-kərɪʃ/ adj. **ochreous** /-krɪəs/ adj. **ochrous** /-krəs/ adj. **ochry** /-krɪ, -kərɪ/ adj. [ME f. OF ocre f. L ochra f. Gk ōkhra yellow ochre]

-ock /ək/ suffix forming nouns orig. with diminutive sense (hillock; bullock). [from or after OE -uc, -oc]

ocker /ˈɒkə(r)/ n. Austral. sl. a boorish or aggressive Australian (esp. as a stereotype). [20th c.: orig. uncert.]

Ockham's razor see OCCAM'S RAZOR.

Ockham, William of see WILLIAM OF OCCAM.

o'clock /əˈklɒk/ adv. of the clock (used to specify the hour) (6 o'clock).

O'Connell /əʊˈkɒn(ə)l/, Daniel (known as 'the Liberator') (1775–1847), Irish nationalist leader and social reformer. His election to Parliament in 1828 forced the British government to grant Catholic Emancipation in order to enable him to take his seat in the House of Commons, for which Roman Catholics were previously ineligible. In 1839 he established the Repeal Association to abolish the union with Britain; O'Connell was arrested and briefly imprisoned for sedition in 1844.

ocotillo /ˌəʊkəˈtiːjəʊ/ n. (pl. **-os**) a spiny scarlet-flowered desert shrub, Fouquiera splendens, of Mexico and the south-western US. [Amer. Sp.]

OCR abbr. optical character recognition.

Oct. abbr. October.

oct. abbr. octavo.

oct- /ɒkt/ comb. form assim. form of OCTA-, OCTO- before a vowel.

octa- /ˈɒktə/ comb. form (also **oct-** before a vowel) eight. [Gk okta- f. oktō eight]

octad /ˈɒktæd/ n. a group of eight. [LL octas octad- f. Gk oktas -ados f. oktō eight]

octagon /ˈɒktəɡən/ n. **1** a plane figure with eight sides and angles.

2 an object or building with this cross-section. □ **octagonal** /ɒkˈtægən(ə)l/ adj. **octagonally** adv. [L octagonos f. Gk octagōnos (as OCTA-, -GON)]

octahedron /ˌɒktəˈhiːdrən/ n. (pl. **octahedrons** or **octahedra** /-drə/) **1** a solid figure contained by eight (esp. triangular) plane faces. **2** a body, esp. a crystal, in the form of a regular octahedron. □ **regular octahedron** an octahedron contained by equal and equilateral triangles. □ **octahedral** adj. [Gk oktaedron (as OCTA-, -HEDRON)]

octal /ˈɒktəl/ adj. reckoning or proceeding by eights (octal scale).

octamerous /ɒkˈtæmərəs/ adj. **1** esp. Bot. having eight parts. **2** Zool. having organs arranged in eights.

octane /ˈɒkteɪn/ n. a colourless inflammable hydrocarbon of the alkane series (chem. formula: C_8H_{18}). □ **high-octane** (of fuel used in internal-combustion engines) having good antiknock properties, not detonating readily during the power stroke. **octane number** (or **rating**) a figure indicating the antiknock properties of a fuel. [OCT- + -ANE²]

octant /ˈɒktənt/ n. **1** an arc of a circle equal to one eighth of the circumference. **2** such an arc with two radii, forming an area equal to one eighth of the circle. **3** each of eight parts into which three planes intersecting (esp. at right angles) at a point divide the space or the solid body round it. **4** an instrument in the form of a graduated eighth of a circle, used in astronomy and navigation. [L octans octant- half-quadrant f. octo eight]

octaroon var. of OCTOROON.

octastyle /ˈɒktəˌstaɪl/ adj. & n. Archit. ● adj. having eight columns at the end or in front. ● n. an octastyle portico or building. [L octastylus f. Gk oktastulos (as OCTA- + stulos pillar)]

octavalent /ˌɒktəˈveɪlənt/ adj. Chem. having a valency of eight.

octave /ˈɒktɪv/ n. **1** Mus. **a** a series of eight notes occupying the interval between (and including) two notes, one having twice or half the frequency of vibration of the other. **b** this interval. **c** each of the two notes at the extremes of this interval. **d** these two notes sounding together. **2** a group or stanza of eight lines; an octet. **3 a** the seventh day after a festival. **b** a period of eight days including a festival and its octave. **4** a group of eight. **5** the last of eight parrying positions in fencing. **6** Brit. a wine-cask holding an eighth of a pipe. □ **law of octaves** Chem. hist. the principle according to which, when elements are arranged in order of atomic weight, similar properties appear in every eighth element (see NEWLANDS). [ME f. OF f. L octava dies eighth day (reckoned inclusively)]

Octavian /ɒkˈteɪvɪən/ see AUGUSTUS.

octavo /ɒkˈteɪvəʊ, -ˈtɑːvəʊ/ n. (pl. **-os**) (abbr. **8vo**) **1** a size of book or page given by folding a standard sheet three times to form a quire of eight leaves. **2** a book or sheet of this size. [L in octavo in an eighth f. octavus eighth]

octennial /ɒkˈtenɪəl/ adj. **1** lasting eight years. **2** occurring every eight years. [LL octennium period of eight years (as OCT-, annus year)]

octet /ɒkˈtet/ n. (also **octette**) **1** Mus. **a** a composition for eight voices or instruments. **b** the performers of such a piece. **2** a group of eight. **3** the first eight lines of a sonnet. **4** Chem. a stable group of eight electrons. [It. ottetto or G Oktett: assim. to OCT-, DUET, QUARTET]

octo- /ˈɒktəʊ/ comb. form (also **oct-** before a vowel) eight. [L octo or Gk oktō eight]

October /ɒkˈtəʊbə(r)/ n. the tenth month of the year. [OE f. L (as OCTO-): cf. DECEMBER, SEPTEMBER]

October Revolution see RUSSIAN REVOLUTION.

October War see YOM KIPPUR WAR.

Octobrist /ɒkˈtəʊbrɪst/ n. hist. a member of the moderate party in the Russian Duma, supporting the Imperial Constitutional Manifesto of 30 Oct. 1905. [OCTOBER, after Russ. oktyabríst]

octocentenary /ˌɒktəʊsenˈtiːnərɪ/ n. & adj. ● n. (pl. **-ies**) **1** an eight-hundredth anniversary. **2** a celebration of this. ● adj. of or relating to an octocentenary.

octodecimo /ˌɒktəʊˈdesɪˌməʊ/ n. (pl. **-os**) **1** a size of book or page given by folding a standard sheet into eighteen leaves. **2** a book or sheet of this size. [in octodecimo f. L octodecimus eighteenth]

octogenarian /ˌɒktəʊdʒɪˈneərɪən/ n. & adj. ● n. a person from 80 to 89 years old. ● adj. of this age. [L octogenarius f. octogeni distributive of octoginta eighty]

octopod /ˈɒktəˌpɒd/ n. Zool. an octopus or other cephalopod of the suborder Octopoda. [Gk oktṓpous -podos f. oktō eight + pous foot]

octopus /ˈɒktəpəs/ n. (pl. **octopuses**) **1** a cephalopod mollusc of the suborder Octopoda, and esp. of the genus Octopus, having a soft saclike body, beaklike jaws, and eight arms bearing suckers. **2** an organized and usu. harmful ramified power or influence. [Gk oktṓpous: see OCTOPOD]

octoroon /ˌɒktəˈruːn/ n. (also **octaroon**) the offspring of a quadroon and a white person, a person of one-eighth black blood. [OCTO-, after QUADROON]

octosyllabic /ˌɒktəʊsɪˈlæbɪk/ adj. & n. ● adj. having eight syllables. ● n. an octosyllabic verse. [LL octosyllabus (as OCTO-, SYLLABLE)]

octosyllable /ˌɒktəʊˈsɪləb(ə)l/ n. & adj. ● n. an octosyllabic verse or word. ● adj. = OCTOSYLLABIC.

octroi /ˈɒktrwʌ/ n. **1** a duty levied in some European countries on goods entering a town. **2 a** the place where this is levied. **b** the officials by whom it is levied. [F f. octroyer grant, f. med.L auctorizare: see AUTHORIZE]

octuple /ˈɒktjʊp(ə)l, ɒkˈtjuːp-/ adj., n., & v. ● adj. eightfold. ● n. an eightfold amount. ● v.tr. & intr. multiply by eight. [F octuple or L octuplus (adj.) f. octo eight: cf. DOUBLE]

ocular /ˈɒkjʊlə(r)/ adj. & n. ● adj. of or connected with the eyes or sight; visual. ● n. the eyepiece of an optical instrument. □ **ocular spectrum** see SPECTRUM 6. □ **ocularly** adv. [F oculaire f. LL ocularis f. L oculus eye]

ocularist /ˈɒkjʊlərɪst/ n. a maker of artificial eyes. [F oculariste (as OCULAR)]

oculate /ˈɒkjʊlət/ adj. = OCELLATE (see OCELLUS). [L oculatus f. oculus eye]

oculist /ˈɒkjʊlɪst/ n. a person who specializes in the medical treatment of eye disorders or defects. □ **oculistic** /ˌɒkjʊˈlɪstɪk/ adj. [F oculiste f. L oculus eye]

oculo- /ˈɒkjʊləʊ/ comb. form eye (oculo-nasal). [L oculus eye]

OD¹ abbr. ordnance datum.

OD² /əʊˈdiː/ n. & v. esp. N. Amer. sl. ● n. an overdose, esp. of a narcotic drug. ● v.intr. (**OD's**, **OD'd**, **OD'ing**) take an overdose. [abbr.]

od¹ /ɒd/ n. a hypothetical power once thought to pervade nature and account for various scientific phenomena. [arbitrary term coined in G by Baron von Reichenbach, Ger. scientist (1788–1869)]

od² /ɒd/ n. (as int. or in oaths) archaic God. [corruption]

o.d. abbr. outer diameter.

odal var. of UDAL.

odalisque /ˈəʊdəˌlɪsk/ n. hist. an Eastern female slave or concubine, esp. in the Turkish sultan's seraglio. [F f. Turk. odalik f. oda chamber + lik function]

odd /ɒd/ adj. & n. ● adj. **1** extraordinary, strange, queer, remarkable, eccentric. **2** casual, occasional, unconnected (odd jobs; odd moments). **3** not normally noticed or considered; unpredictable (in some odd corner; picks up odd bargains). **4** additional; beside the reckoning (a few odd pence). **5 a** (of numbers such as 3 and 5) not integrally divisible by two. **b** (of things or persons numbered consecutively) bearing such a number (no parking on odd dates). **6** left over when the rest have been distributed or divided into pairs (have got an odd sock). **7** detached from a set or series (a few odd volumes). **8** (appended to a number, sum, weight, etc.) somewhat more than (forty odd; forty-odd people). **9** by which a round number, given sum, etc., is exceeded (we have 102 — what shall we do with the odd 2?). ● n. Golf a handicap of one stroke at each hole. □ **odd job** a casual isolated piece of work. **odd job man** (or **odd jobber**) Brit. a person who does odd jobs. **odd man out 1** a person or thing differing from all the others in a group in some respect. **2** a method of selecting one of three or more persons e.g. by tossing a coin. □ **oddish** adj. **oddly** adv. **oddness** n. [ME f. ON odda- in odda-mathr third man, odd man, f. oddi angle]

oddball /ˈɒdbɔːl/ n. colloq. **1** an odd or eccentric person. **2** (attrib.) strange, bizarre.

Oddfellow /ˈɒdˌfeləʊ/ n. a member of a fraternity similar to the Freemasons, founded in the 18th century.

oddity /ˈɒdɪtɪ/ n. (pl. **-ies**) **1** a strange person, thing, or occurrence. **2** a peculiar trait. **3** the state of being odd.

oddment /ˈɒdmənt/ n. **1** an odd article; something left over. **2** (in pl.) miscellaneous articles. **3** (in pl.) Brit. Printing matter other than the main text.

odds /ɒdz/ n.pl. **1** the ratio between the amounts staked by the parties to a bet, based on the expected probability either way. **2** the chances or balance of probability in favour of or against some result (the odds are against it; the odds are that it will rain). **3** the balance of advantage (the odds are in your favour; won against all the odds). **4** an equalizing allowance to a weaker competitor. **5** a difference giving an advantage (*makes no odds*). □ **at odds** (often foll. by *with*) in conflict or at variance. **by all odds** certainly. **lay** (or **give**) **odds** offer a bet with odds favourable to the other better. **odds and ends** miscellaneous articles or remnants. **odds and sods** *colloq.* = odds and ends. **odds-on** a state when success is more likely than failure, esp. as indicated by the betting odds. **over the odds** above a generally agreed price etc. **take odds** offer a bet with odds unfavourable to the other better. **what's the odds?** *colloq.* what does it matter? [app. pl. of ODD *n.*: cf. NEWS]

ode¹ /əʊd/ *n.* a lyric poem often in the form of an address and now usually rhymed. In classical Greek times odes were poems meant to be sung, as exemplified by the grand odes to athletic victories written by Pindar. These were based on the odes sung by the chorus (choral odes) in Greek tragedy. Pindaric odes are generally dignified or exalted in subject and style, whereas those written in Latin by Horace provide a simpler, more intimate model. Keats is among later poets to have written odes drawing on Horace, while more elevated odes have been produced by Dryden, Thomas Gray, Wordsworth, and Coleridge. [F f. LL *oda* f. Gk *ōidē* Attic form of *aoidē* song f. *aeidō* sing]

-ode¹ /əʊd/ *suffix* forming nouns meaning 'thing of the nature of' (*geode*; *trematode*). [Gk *-ōdēs* adj. ending]

-ode² /əʊd/ *comb. form Electr.* forming names of electrodes, or devices having them (*cathode*; *diode*). [Gk *hodos* way]

Odense /ˈəʊd(ə)nsə/ a port in eastern Denmark, on the island of Fyn; pop. (1991) 177,640.

Oder /ˈəʊdə(r)/ (Czech and Polish **Odra** /ˈɔdra/) a river of central Europe which rises in the mountains in the west of the Czech Republic and flows 907 km (567 miles) northwards through western Poland to meet the River Neisse, then continues northwards forming the northern part of the border between Poland and Germany before flowing into the Baltic Sea. This frontier, known as the Oder–Neisse Line, was adopted at the Potsdam Conference in 1945.

Odessa /əʊˈdesə/ (Ukrainian **Odesa** /ɔˈdɛsa/) a city and port on the south coast of Ukraine, on the Black Sea; pop. (1990) 1,106,000.

Odets /əʊˈdets/, Clifford (1906–63), American dramatist. He was a founder member in 1931 of the avant-garde Group Theatre, which followed the naturalistic methods of the Moscow Art Theatre and staged his best-known play, *Waiting for Lefty* (1935). His plays of the 1930s (especially *The Golden Boy*, 1937) reflect the experiences of the Depression, often displaying a strong sense of social issues.

odeum /ˈəʊdɪəm/ *n.* (*pl.* **odeums** or **odea** /-dɪə/) a building for musical performances, esp. among the ancient Greeks and Romans. [F *odéum* or L *odeum* f. Gk *ōideion* (as ODE)]

Odin /ˈəʊdɪn/ (also called *Woden* or *Wotan*) *Scand. Mythol.* the supreme god and creator, god of victory and the dead, married to Frigga and usually represented as a one-eyed old man of great wisdom. Wednesday is named after him.

odious /ˈəʊdɪəs/ *adj.* hateful, repulsive. □ **odiously** *adv.* **odiousness** *n.* [ME f. OF *odieus* f. L *odiosus* (as ODIUM)]

odium /ˈəʊdɪəm/ *n.* a general or widespread dislike or reprobation incurred by a person or associated with an action. [L, = hatred f. *odi* to hate]

odometer /əʊˈdɒmɪtə(r)/ *n.* (also **hodometer** /hɒˈdɒm-/) esp. *US* an instrument for measuring the distance travelled by a wheeled vehicle. □ **odometry** *n.* [F *odomètre* f. Gk *hodos* way: see -METER]

Odonata /ˌəʊdəˈnɑːtə/ *n.pl. Zool.* an order of insects comprising the dragonflies and damselflies, which have a long slender body, two pairs of long transparent wings, and predatory aquatic larvae. □ **odonate** /ˈəʊdəˌneɪt/ *n. & adj.* [mod.L f. Gk *odous odont-* tooth]

odonto- /əʊˈdɒntəʊ/ *comb. form* tooth. [Gk *odous odont-* tooth]

odontoglossum /əʊˌdɒntəˈɡlɒsəm/ *n.* an epiphytic orchid of the genus *Odontoglossum*, native to Central and South America, bearing flowers with jagged edges like tooth-marks. [mod.L f. ODONTO- + Gk *glōssa* tongue]

odontoid /əʊˈdɒntɔɪd/ *adj.* toothlike. □ **odontoid process** a projection from the second cervical vertebra. [Gk *odontoeidēs* (as ODONTO- + Gk *eidos* form)]

odontology /ˌəʊdɒnˈtɒlədʒɪ/ *n.* the scientific study of the structure and diseases of teeth. □ **odontologist** *n.* **odontological** /əʊˌdɒntəˈlɒdʒɪk(ə)l/ *adj.*

odor *US* var. of ODOUR.

odoriferous /ˌəʊdəˈrɪfərəs/ *adj.* diffusing a scent, esp. an agreeable one; fragrant. □ **odoriferously** *adv.* [ME f. L *odorifer* (as ODOUR)]

odorous /ˈəʊdərəs/ *adj.* **1** having a scent. **2** = ODORIFEROUS. □ **odorously** *adv.* [L *odorus* fragrant (as ODOUR)]

odour /ˈəʊdə(r)/ *n.* (*US* **odor**) **1** the property of a substance that has an effect on the nasal sense of smell. **2** a lasting quality or trace attaching to something (*an odour of intolerance*). **3** regard, repute (*in bad odour*). □ **odourless** *adj.* (in sense 1). [ME f. AF *odour*, OF *odor* f. L *odor -oris* smell, scent]

Odra see ODER.

Odysseus /əˈdɪsɪəs/ *Gk Mythol.* the king of Ithaca and central figure of the *Odyssey*, called Ulysses by the Romans. Renowned for his cunning and resourcefulness, he survived the Trojan War but was kept from home by Poseidon for ten years while his wife Penelope waited. (See ODYSSEY.)

Odyssey /ˈɒdɪsɪ/ a Greek hexameter epic poem in twenty-four books, traditionally ascribed to Homer. It tells of the travels of Odysseus during his years of wandering after the sack of Troy, and of his eventual return home to Ithaca and his slaying of the rapacious suitors of his faithful wife Penelope. His adventures include amorous liaisons with Calypso and the witch Circe, hospitality at the court of the pleasure-loving Phaeacians, the evocation of the famous dead from the underworld, and encounters with a number of fabulous monsters, including the Cyclops Polyphemus and Scylla and Charybdis. □ **Odyssean** /ˌɒdɪˈsiːən/ *adj.* [L *Odyssea* f. Gk *Odusseia*]

odyssey /ˈɒdɪsɪ/ *n.* (*pl.* **-eys**) a series of wanderings; a long adventurous journey. [ODYSSEY]

Oe *symb.* oersted(s).

Oea /ˈiːə/ the ancient name for Tripoli in Libya (see TRIPOLI 1).

OECD see ORGANIZATION FOR ECONOMIC COOPERATION AND DEVELOPMENT.

OED see OXFORD ENGLISH DICTIONARY.

oedema /ɪˈdiːmə/ *n.* (*US* **edema**) *Med.* a condition characterized by an excess of watery fluid collecting in the cavities or tissues of the body. Also called *dropsy.* □ **oedematous** /ɪˈdemətəs/ *adj.* [LL f. Gk *oidēma -atos* f. *oideō* swell]

Oedipus /ˈiːdɪpəs/ *Gk Mythol.* the son of Jocasta and of Laius, king of Thebes. Left to die on a mountain by Laius, who had been told by an oracle that he would be killed by his own son, the infant Oedipus was saved by a shepherd. Returning eventually to Thebes, Oedipus solved the riddle of the sphinx, but unwittingly killed his father and married Jocasta. On discovering what he had done he put out his own eyes in a fit of madness and left Thebes as an outcast, while Jocasta hanged herself. [Gk = swollen foot, from the story that Laius ran a spike through the infant's feet before leaving it to die]

Oedipus complex *n.* (in Freudian psychoanalysis) the complex of emotions aroused in a young (esp. male) child by a subconscious sexual desire for the parent of the opposite sex and wish to exclude the parent of the same sex. □ **Oedipal** *adj.*

oenology /iːˈnɒlədʒɪ/ *n.* (*US* **enology**) the study of wines. □ **oenologist** *n.* **oenological** /ˌiːnəˈlɒdʒɪk(ə)l/ *adj.* [Gk *oinos* wine]

Oenone /iːˈnəʊnɪ/ *Gk Mythol.* a nymph of Mount Ida and lover of Paris, who deserted her for Helen.

oenophile /ˈiːnəˌfaɪl/ *n.* a connoisseur of wines. □ **oenophilist** /iːˈnɒfɪlɪst/ *n.* [as OENOLOGY]

o'er /ˈəʊə(r)/ *adv. & prep. poet.* = OVER. [contr.]

Oersted /ˈɜːsted/, Hans Christian (1777–1851), Danish physicist, discoverer of the magnetic effect of an electric current. He had postulated the existence of electromagnetism earlier, but it was not demonstrated until 1820, when he carried out an experiment during a lecture and noticed the deflection of a compass needle placed below a wire carrying a current. He also worked on the compressibility of gases and liquids, and on diamagnetism.

oersted /ˈɜːsted/ *n.* (*pl.* same or **oersteds**) *Physics* the cgs unit of magnetic field strength (symbol: **Oe**), equivalent to 79.58 amperes per metre. [OERSTED]

oesophagus /iːˈsɒfəɡəs/ *n.* (*US* **esophagus**) (*pl.* **oesophagi** /iːˈsɒfədʒaɪ, -ˌɡaɪ/ or **-guses**) the part of the alimentary canal from the mouth to the stomach; the gullet. □ **oesophageal** /iːˌsɒfəˈdʒiːəl, ˌiːsəˈfædʒɪəl/ *adj.* [ME f. Gk *oisophagos*]

oestradiol /ˌiːstrəˈdaɪɒl, ˌest-/ *n. Biochem.* a major oestrogen produced in the ovaries. [OESTRUS + DI-¹ + -OL]

oestrogen /ˈiːstrədʒən/ *n.* (*US* **estrogen**) **1** any of a class of steroid hormones developing and maintaining female characteristics of the body. **2** this hormone produced artificially for use in oral contraceptives etc. □ **oestrogenic** /ˌiːstrəˈdʒenɪk/ *adj.* **oestrogenically** *adv.* [OESTRUS + -GEN]

oestrus /ˈiːstrəs/ *n.* (also **oestrum** /-rəm/, *US* **estrus**, **estrum**) a recurring period of sexual receptivity in many female mammals; heat. □ **oestrous** *adj.* [Gk *oistros* gadfly, frenzy]

oeuvre /ˈɜːvrə/ *n.* the works of an author, painter, composer, etc., esp. regarded collectively. [F, = work, f. L *opera*]

of /ɒv, əv/ *prep.* connecting a noun (often a verbal noun) or pronoun with a preceding noun, adjective, adverb, or verb, expressing a wide range of relations broadly describable as follows: **1** origin, cause, or authorship (*paintings of Turner*; *people of Rome*; *died of malnutrition*). **2** the material or substance constituting or identifying a thing (*a house of cards*; *was built of bricks*). **3** belonging, connection, or possession (*a thing of the past*; *articles of clothing*; *the head of the business*; *the tip of the iceberg*). **4** identity or close relation (*the city of Rome*; *a pound of apples*; *a fool of a man*). **5** removal, separation, or privation (*north of the city*; *got rid of them*; *robbed us of £1,000*). **6** reference, direction, or respect (*beware of the dog*; *suspected of lying*; *very good of you*; *short of money*; *the selling of goods*). **7** objective relation (*love of music*; *in search of peace*). **8** partition, classification, or inclusion (*no more of that*; *part of the story*; *a friend of mine*; *this sort of book*; *some of us will stay*). **9** description, quality, or condition (*the hour of prayer*; *a person of tact*; *a girl of ten*; *on the point of leaving*). **10** *N. Amer.* time in relation to the following hour (*a quarter of three*). □ **be of** possess intrinsically; give rise to (*is of great interest*). **of all** designating the (nominally) least likely or expected example (*you of all people!*). **of all the nerve** (or **cheek** etc.) an exclamation of indignation at a person's impudence etc. **of an evening** (or **morning** etc.) *colloq.* **1** on most evenings (or mornings etc.). **2** at some time in the evenings (or mornings etc.). **of late** recently; long ago. [OE, unaccented form of *æf*, f. Gmc]

of- /ɒf, əf/ *prefix* assim. form of OB- before *f*.

ofay /ˈəʊfeɪ/ *n. US sl. offens.* a white person (esp. used by blacks). [20th c.: prob. of Afr. orig.]

Off. *abbr.* **1** Office. **2** Officer.

off /ɒf/ *adv., prep., adj., & n.* ● *adv.* **1 a** away; at or to a distance (*drove off*; *is three miles off*). **b** distant or remote in fact, nature, likelihood, etc. **2** out of position; not on or touching or attached; loose, separate, gone (*has come off*; *take your coat off*). **3** so as to be rid of (*sleep it off*). **4** so as to break continuity or continuance; discontinued, stopped (*turn off the radio*; *take a day off*; *the game is off*). **5** not available as a choice, e.g. on a menu (*chips are off*). **6** to the end; entirely; so as to be clear (*clear off*; *finish off*; *pay off*). **7** situated as regards money, supplies, etc. (*is badly off*; *is not very well off*). **8** off-stage (*noises off*). **9 a** *Brit. colloq.* in bad condition or form; unwell (*am feeling a bit off*). **b** (of food etc.) beginning to decay. **c** *Brit. colloq.* (of behaviour etc.) unacceptable, ill-mannered (*thought it was a bit off*). **10** (with preceding numeral) denoting a quantity produced or made at one time (esp. *one-off*). **11** away or free from a regular commitment (*How about tomorrow? I'm off then*). ● *prep.* **1 a** from; away or down or up from (*fell off the chair*; *took something off the price*; *jumped off the edge*). **b** not on (*was already off the pitch*). **2 a** (temporarily) relieved of or abstaining from (*off duty*; *am off my diet*). **b** not attracted by for the time being (*off their food*; *off smoking*). **c** not achieving or doing one's best in (*off form*; *off one's game*). **3** using as a source or means of support (*live off the land*). **4** leading from; not far from (*a street off the Strand*). **5** at a short distance to sea from (*sank off Cape Horn*). ● *adj.* **1** far, further (*the off side of the wall*). **2** (of a part of a vehicle, animal, or road) right (opp. NEAR *adj.* 3) (*the off front wheel*). **3** *Cricket* designating the half of the field (as divided lengthways through the pitch) to which the striker's feet are pointed. ● *n.* **1** *Cricket* the off side (opp. LEG *n.* 6). **2** *Brit. colloq.* the start of a race; the beginning, the departure (*ready for the off*). □ **off and on** intermittently; now and then. **off-break** *Cricket* a ball bowled so that, on pitching, it changes direction towards the leg side. **off-centre** not quite coinciding with a central position. **the off chance** see CHANCE. **off colour 1** not in good health. **2** *US* somewhat indecent. **off the cuff** see CUFF[1]. **off-day** a day when one is not at one's best. **off-drive** *Cricket v.tr.* drive (the ball) to the off side. ● *n.* a drive to the off side. **off one's feet** see FOOT. **off form** see FORM. **off guard** see GUARD. **off one's hands** see HAND. **off one's head** see HEAD. **off-key 1** out of tune. **2** not quite suitable or fitting. **off-licence** *Brit.* **1** a shop selling alcoholic drink for consumption elsewhere. **2** a licence for this. **off limits** see LIMIT. **off-line** *Computing adj.* not directly controlled by or connected to a central processor. ● *adv.* while not directly controlled by or connected to a central processor. **off of** *sl. disp.* = OFF *prep.* (*picked it off of the floor*). **off-peak** *adj.* used or for use at times other than those of greatest demand. ● *adv.* at times other than those of greatest demand. **off the peg** see PEG. **off-piste**

Skiing away from prepared ski runs. **off the point** *adj.* irrelevant. ● *adv.* irrelevantly. **off-putting** disconcerting; repellent. **off the record** see RECORD. **off-road** *attrib.adj.* **1** away from the road, on rough terrain. **2** (of a vehicle etc.) designed for rough terrain or for cross-country driving. **off-roading** driving on dirt tracks and other unmetalled surfaces as a sport or leisure activity. **off-season** a time when business etc. is slack (often *attrib.*: *off-season prices*). **off-stage** *adj. & adv.* not on the stage and so not visible or audible to the audience. **off-street** (esp. of parking vehicles) other than on a street. **off-time** a time when business etc. is slack. **off the wall** see WALL. **off-white** white with a grey or yellowish tinge. [orig. var. of OF, to distinguish the sense]

Offa /ˈɒfə/ (d.796), king of Mercia 757–96. After seizing power in Mercia in 757, he expanded his territory to become overlord of most of England south of the Humber. Offa is chiefly remembered for constructing the frontier earthworks called Offa's Dyke.

offal /ˈɒf(ə)l/ *n.* **1** the less valuable edible parts of a carcass, esp. the entrails and internal organs. **2** refuse or waste stuff. **3** carrion; putrid flesh. [ME f. MDu. *afval* f. *af* OFF + *vallen* FALL]

Offaly /ˈɒfəlɪ/ a county in the central part of the Republic of Ireland, in the province of Leinster; county town, Tullamore.

Offa's Dyke a series of earthworks marking the traditional boundary between England and Wales, running from near the mouth of the Wye to near the mouth of the Dee, originally built or repaired by Offa in the second half of the 8th century to mark the boundary established by his wars with the Welsh.

offbeat /ˈɒfbiːt/ *adj. & n.* ● *adj.* **1** *Mus.* not coinciding with the beat. **2** eccentric, unconventional. ● *n. Mus.* any of the normally unaccented beats in a bar.

offcut /ˈɒfkʌt/ *n.* a remnant of timber, paper, etc., after cutting.

Offenbach /ˈɒf(ə)nˌbɑːx/, Jacques (born Jacob Offenbach) (1819–80), German composer, resident in France from 1833. Offenbach is associated with the rise of the operetta, whose style was typified by his *Orpheus in the Underworld* (1858). He is also noted for his opera *The Tales of Hoffmann* (1881), based on the stories of E. T. A. Hoffmann and first produced after Offenbach's death.

offence /əˈfens/ *n.* (*US* **offense**) **1** an illegal act; a transgression or misdemeanour. **2** a wounding of the feelings (*no offence was meant*). **3** resentment or umbrage. **4** the act of attacking or taking the offensive; aggressive action. □ **give offence** cause hurt feelings. **take offence** suffer hurt feelings. □ **offenceless** *adj.* [orig. = stumbling, stumbling-block: ME & OF *offens* f. L *offensus* annoyance, and ME & F *offense* f. L *offensa* a striking against, hurt, displeasure, both f. *offendere* (as OB-, *fendere fens-* strike)]

offend /əˈfend/ *v.* **1** *tr.* cause offence to or resentment in; wound the feelings of. **2** *tr.* displease or anger. **3** *intr.* commit an illegal act. **4** *intr.* (often foll. by *against*) do wrong; transgress. □ **offendedly** *adv.* **offender** *n.* **offending** *adj.* [ME f. OF *offendre* f. L (as OFFENCE)]

offense *US* var. of OFFENCE.

offensive /əˈfensɪv/ *adj. & n.* ● *adj.* **1** giving or meant or likely to give offence; insulting (*offensive language*). **2** disgusting, foul-smelling, nauseous, repulsive. **3 a** aggressive, attacking. **b** (of a weapon) meant for use in attack. ● *n.* **1** an aggressive action or attitude (*take the offensive*). **2** an attack, an offensive campaign or stroke. **3** aggressive or forceful action in pursuit of a cause (*a peace offensive*). □ **offensively** *adv.* **offensiveness** *n.* [F *offensif -ive* or med.L *offensivus* (as OFFENCE)]

Offer /ˈɒfə(r)/ (also **OFFER**) Office of Electricity Regulation, a regulatory body supervising the operation of the British electricity industry. [acronym]

offer /ˈɒfə(r)/ *v. & n.* ● *v.* **1** *tr.* present for acceptance or refusal or consideration (*offered me a drink*; *was offered a lift*; *offer one's services*; *offer no apology*). **2** *intr.* (foll. by *to* + infin.) express readiness or show intention (*offered to take the children*). **3** *tr.* provide; give an opportunity for. **4** *tr.* make available for sale. **5** *tr.* (of a thing) present to one's attention or consideration (*each day offers new opportunities*). **6** *tr.* present (a sacrifice, prayer, etc.) to a deity. **7** *intr.* present itself; occur (*as opportunity offers*). **8** *tr.* give an opportunity for (battle) to an enemy. **9** *tr.* attempt, or try to show (violence, resistance, etc.). ● *n.* **1** an expression of readiness to do or give if desired, or to buy or sell (for a certain amount). **2** an amount offered. **3** a proposal (esp. of marriage). **4** a bid. □ **on offer** *Brit.* for sale at a certain (esp. reduced) price. □ **offerer** *n.* **offeror** *n.* [OE *offrian* in religious sense, f. L *offerre* (as OB-, *ferre* bring)]

offering /ˈɒfərɪŋ/ *n.* **1** a contribution, esp. of money, to a Church. **2** a thing offered as a religious sacrifice or token of devotion. **3** anything, esp. money, contributed or offered.

offertory /ˈɒfətərɪ/ *n.* (*pl.* **-ies**) **1** *Eccl.* **a** the offering of the bread and wine at the Eucharist. **b** an anthem accompanying this. **2 a** the collection of money at a religious service. **b** the money collected. [ME f. eccl.L *offertorium* offering f. LL *offert-* for L *oblat-* past part. stem of *offerre* OFFER]

offhand /ɒfˈhænd/ *adj.* & *adv.* ● *adj.* curt or casual in manner. ● *adv.* **1** in an offhand manner. **2** without preparation or premeditation. □ **offhanded** *adj.* **offhandedly** *adv.* **offhandedness** *n.*

office /ˈɒfɪs/ *n.* **1** a room or building used as a place of business, esp. for clerical or administrative work. **2** a room or department or building for a particular kind of business (*ticket office; post office*). **3** the local centre of a large business (*our London office*). **4** *N. Amer.* the consulting-room of a professional person. **5** a position with duties attached to it; a place of authority or trust or service, esp. of a public nature. **6** tenure of an official position, esp. that of a Minister of State or of the party forming the government (*hold office; out of office for thirteen years*). **7** (**Office**) *Brit.* the quarters or staff or collective authority of a government department etc. (*Foreign Office*). **8** a duty attaching to one's position; a task or function. **9** (usu. in *pl.*) a piece of kindness or attention; a service (esp. *through the good offices of*). **10** *Eccl.* **a** an authorized form of worship (*Office for the Dead*). **b** (in full **divine office**) the daily service of the Roman Catholic breviary (*say the office*). **11** a ceremonial duty. **12** (in *pl.*) *Brit.* the parts of a house devoted to household work, storage, etc. **13** *sl.* a hint or signal. □ **the last offices** rites due to the dead. **office-bearer** an official or officer. **office block** a large building designed to contain business offices. **office boy** (or **girl**) a young man (or woman) employed to do minor jobs in a business office. **office hours** the hours during which business is normally conducted. **office of arms** the College of Arms, or a similar body in another country. **office-worker** an employee in a business office. [ME f. OF f. L *officium* performance of a task (in med.L also office, divine service), f. *opus* work + *facere fic-* do]

officer /ˈɒfɪsə(r)/ *n.* & *v.* ● *n.* **1** a person holding a position of authority or trust, esp. one with a commission in the armed services, in the mercantile marine, or on a passenger ship. **2** a policeman or policewoman. **3** a holder of a post in a society (e.g. the president or secretary). **4** a holder of a public, civil, or ecclesiastical office; a sovereign's minister; an appointed or elected functionary (usu. with a qualifying word: *medical officer; probation officer; returning officer*). **5** a bailiff (*the sheriff's officer*). **6** a member of the grade below commander in the Order of the British Empire etc. ● *v.tr.* **1** provide with officers. **2** act as the commander of. □ **officer of arms** a herald or pursuivant. [ME f. AF *officer*, OF *officier* f. med.L *officiarius* f. L *officium*: see OFFICE]

official /əˈfɪʃ(ə)l/ *adj.* & *n.* ● *adj.* **1** of or relating to an office (see OFFICE *n.* 5, 6) or its tenure or duties. **2** esp. *derog.* characteristic of officials and bureaucracy. **3** emanating from or attributable to a person in office; properly authorized. **4** holding office; employed in a public capacity. **5** *Med.* according to the pharmacopoeia, officinal. ● *n.* **1** a person holding office or engaged in official duties. **2** (in full **official principal**) the presiding officer or judge of an archbishop's, bishop's, or esp. archdeacon's court. □ **official birthday** (in the UK) a day in June chosen for the observance of the sovereign's birthday. **official secrets** (in the UK) confidential information involving national security. □ **officialdom** *n.* **officialism** *n.* **officially** *adv.* [ME (as noun) f. OF f. L *officialis* (as OFFICE)]

officialese /əˌfɪʃəˈliːz/ *n.* *derog.* the formal precise language characteristic of official documents.

officiant /əˈfɪʃɪənt/ *n.* a person who officiates at a religious ceremony.

officiate /əˈfɪʃɪˌeɪt/ *v.intr.* **1** act in an official capacity, esp. on a particular occasion. **2** perform a divine service or ceremony. □ **officiator** *n.* **officiation** /əˌfɪʃɪˈeɪʃ(ə)n/ *n.* [med.L *officiare* perform a divine service (*officium*): see OFFICE]

officinal /ˌɒfɪˈsaɪn(ə)l, əˈfɪsɪn-/ *adj.* **1 a** (of a medicine) kept ready for immediate dispensing. **b** made from the pharmacopoeia recipe (cf. MAGISTRAL 2). **c** (of a name) adopted in the pharmacopoeia. **2** (of a herb or drug) used in medicine. □ **officinally** *adv.* [med.L *officinalis* f. L *officina* workshop]

officious /əˈfɪʃəs/ *adj.* **1** asserting one's authority aggressively; domineering. **2** intrusive or excessively enthusiastic in offering help etc.; meddlesome. **3** (in diplomacy) informal, unofficial. □ **officiously** *adv.* **officiousness** *n.* [L *officiosus* obliging f. *officium*: see OFFICE]

offing /ˈɒfɪŋ/ *n.* the more distant part of the sea in view. □ **in the offing** not far away; likely to appear or happen soon. [perh. f. OFF + -ING[1]]

offish /ˈɒfɪʃ/ *adj. colloq.* inclined to be aloof. □ **offishly** *adv.* **offishness** *n.* [OFF: cf. *uppish*]

offload /ˈɒfləʊd/ *v.tr.* **1** unload. **2** get rid of (esp. something unpleasant) by giving it to someone else.

off-price /ˈɒfpraɪs/ *adj. N. Amer.* involving merchandise sold at a lower price than that recommended by the manufacturer.

offprint /ˈɒfprɪnt/ *n.* a printed copy of an article etc. originally forming part of a larger publication.

offscreen *adj.* & *adv.* ● *adj.* /ˈɒfskriːn/ **1** not appearing on a cinema, television, or VDU screen. **2** (*attrib.*) in one's private life or in real life as opposed to a film or television role. ● *adv.* /ɒfˈskriːn/ **1** without use of a screen. **2** outside the view presented by a cinema-film scene. **3** in one's private life or in real life as opposed to a film or television role.

offset /ˈɒfset/ *n.* & *v.* ● *n.* **1** a side-shoot from a plant serving for propagation. **2** an offshoot or scion. **3** a compensation; a consideration or amount diminishing or neutralizing the effect of a contrary one. **4** *Archit.* a sloping ledge in a wall etc. where the thickness of the part above is diminished. **5** a mountain-spur. **6** a bend in a pipe etc. to carry it past an obstacle. **7** (often *attrib.*) a method of printing in which ink is transferred from a plate or stone to a uniform rubber surface and from there to paper etc. (*offset litho*). **8** *Surveying* a short distance measured perpendicularly from the main line of measurement. ● *v.tr.* (**-setting**; *past* and *past part.* **-set**) **1** counterbalance, compensate. **2** place out of line. **3** print by the offset process.

offshoot /ˈɒfʃuːt/ *n.* **1 a** a side-shoot or branch. **b** a collateral branch or descendant of a family. **2** a thing which originated as a branch of something else; a derivative.

offshore *adj.* & *adv.* ● *adj.* /ˈɒfʃɔː(r)/ **1** situated at sea some distance from the shore. **2** (of the wind) blowing seawards. **3** (of goods, funds, etc.) made or registered abroad. ● *adv.* /ɒfˈʃɔː(r)/ away from or at a distance from the land.

offside *adj.* & *n.* ● *adj.* /ɒfˈsaɪd/ *Sport* (of a player in a field game) in a position, usu. ahead of the ball, that is not allowed if it affects play. ● *n.* /ˈɒfsaɪd/ (often *attrib.*) esp. *Brit.* the right side of a vehicle, animal, etc. (cf. NEARSIDE).

offsider /ɒfˈsaɪdə(r)/ *n. Austral. colloq.* a partner, assistant, or deputy.

offspring /ˈɒfsprɪŋ/ *n.* (*pl.* same) **1** a person's child or children or descendant(s). **2** an animal's young or descendant(s). **3** a result. [OE *ofspring* f. OF from + *springan* SPRING *v.*]

Ofgas /ˈɒfgæs/ (also **OFGAS**) Office of Gas Supply, a regulatory body supervising the operation of the British gas industry. [acronym]

oft /ɒft/ *adv.* usu. *archaic* or *literary* often (usu. in *comb.*: *oft-recurring, oft-quoted*). □ **oft-times** often. [OE]

Oftel /ˈɒftel/ (also **OFTEL**) Office of Telecommunications, a regulatory body supervising the operation of the British telecommunications industry. [acronym]

often /ˈɒf(ə)n, ˈɒft(ə)n/ *adv.* (**oftener, oftenest**) **1 a** frequently; many times. **b** at short intervals. **2** in many instances. □ **as often as not** in roughly half the instances. [ME: extended f. OFT, prob. after *selden* = SELDOM]

Ofwat /ˈɒfwɒt/ (also **OFWAT**) Office of Water Services, a regulatory body supervising the operation of the British water industry. [acronym]

Ogaden, the /ˌɒgəˈden/ a desert region in SE Ethiopia, largely inhabited by Somali nomads. It has been claimed by successive governments of neighbouring Somalia.

ogam var. of OGHAM.

Ogbomosho /ˌɒgbəˈməʊʃəʊ/ a city and agricultural market in SW Nigeria, north of Ibadan; pop. (1983) 527,000.

ogdoad /ˈɒgdəʊˌæd/ *n.* a group of eight. [LL *ogdoas ogdoad-* f. Gk *ogdoas -ados* f. *ogdoos* eighth f. *oktō* eight]

ogee /ˈəʊdʒiː/ *adj.* & *n. Archit.* ● *adj.* showing in section a double continuous S-shaped curve. ● *n.* an S-shaped line or moulding. □ **ogee arch** an arch with two ogee curves meeting at the apex. □ **ogee'd** *adj.* [app. f. OGIVE, as being the usual moulding in groin-ribs]

ogham /ˈɒgəm/ *n.* (also **ogam**) **1** an ancient British and Irish alphabet of twenty characters formed by parallel strokes on either side of or across a continuous line. **2** an inscription in this alphabet. **3** each of its characters. [OIr. *ogam*, referred to *Ogma*, its mythical inventor]

ogive /ˈəʊdʒaɪv/ *n.* **1** a pointed or Gothic arch. **2** one of the diagonal groins or ribs of a vault. **3** an S-shaped line. **4** *Statistics* a cumulative frequency graph. □ **ogival** /əʊˈdʒaɪv(ə)l/ *adj.* [ME f. F, of unkn. orig.]

ogle /ˈəʊg(ə)l/ *v.* & *n.* ● *v.* **1** *tr.* eye amorously or lecherously. **2** *intr.* look amorously. ● *n.* an amorous or lecherous look. □ **ogler** *n.* [prob. LG or Du.: cf. LG *oegeln*, frequent. of *oegen* look at]

OGPU /ˈɒgpuː/ (also **Ogpu**) an organization for investigating and combating counter-revolutionary activities in the USSR, which existed from 1922 (1922–3 as the GPU) to 1934, replacing the Cheka. After 1928 it also operated to enforce the collectivization of farming. It was absorbed into the NKVD in 1934. [Russ. abbr., = United State Political Directorate]

O grade /əʊ/ *n.* = ordinary grade. [abbr.]

ogre /ˈəʊgə(r)/ *n.* (*fem.* **ogress** /-grɪs/) **1** a man-eating giant in folklore etc. **2** a terrifying person. □ **ogreish** /-grɪʃ, -gərɪʃ/ *adj.* (also **ogrish**). [F, first used by Charles Perrault in 1697, of unkn. orig.]

OH *abbr.* US Ohio (in official postal use).

oh[1] /əʊ/ *int.* (also **O**) expressing surprise, pain, entreaty, etc. (*oh, what a mess; oh for a holiday*). □ **oh boy** expressing surprise, excitement, etc. **oh well** expressing resignation. [var. of O[4]]

oh[2] /əʊ/ *n.* = O[1] 2.

o.h.c. *abbr.* overhead camshaft.

O'Higgins /əʊˈhɪgɪnz/, Bernardo (*c.*1778–1842), Chilean revolutionary leader and statesman, head of state 1817–23. The son of a Spanish officer of Irish origin, he was educated in England, where he first became involved in nationalist politics. On his return to Chile he led the independence movement and, with the help of José de San Martín, liberator of Argentina, led the army which triumphed over Spanish forces in 1817 and paved the way for Chilean independence the following year. For the next six years he was head of state (supreme director) of Chile, but then fell from power and lived in exile in Peru for the remainder of his life.

Ohio /əʊˈhaɪəʊ/ a state in the north-eastern US, bordering on Lake Erie; pop. (1990) 10,847,115; capital, Columbus. Acquired by Britain from France in 1763 and by the US in 1783, it became the 17th state of the US in 1803. □ **Ohioan** *adj.* & *n.*

Ohm /əʊm/, Georg Simon (1789–1854), German physicist. He published two major papers in 1826, which between them contained the law that is named after him. This states that the electric current flowing in a conductor is directly proportional to the potential difference (voltage), and inversely proportional to the resistance. Applying this to a wire of known diameter and conductivity, the current is inversely proportional to length. The units *ohm* and *mho* are also named after him.

ohm /əʊm/ *n. Electr.* the SI unit of resistance (symbol: Ω), equal to the resistance acting when a current of one ampere flows through a potential difference of one volt. □ **ohmic** *adj.* [OHM]

ohmmeter /ˈəʊmˌmiːtə(r)/ *n.* an instrument for measuring electrical resistance.

OHMS *abbr.* on Her (or His) Majesty's Service.

oho /əʊˈhəʊ/ *int.* expressing surprise or exultation. [ME f. O[4] + HO]

-oholic *comb. form* var. of -AHOLIC.

ohone var. of OCHONE.

OHP *abbr.* overhead projector.

Ohrid, Lake /ˈɒxrɪd/ a lake in SE Europe, on the border between Macedonia and Albania.

o.h.v. *abbr.* overhead valve.

oi /ɔɪ/ *int.* calling attention or expressing alarm etc. [var. of HOY[1]]

-oic /əʊɪk/ *suffix Chem.* forming names of carboxylic acids (*ethanoic acid*).

-oid /ɔɪd/ *suffix* forming adjectives and nouns, denoting form or resemblance (*asteroid; rhomboid; thyroid*). □ **-oidal** *suffix* forming adjectives. **-oidally** *suffix* forming adverbs. [mod.L *-oides* f. Gk *-oeidēs* f. *eidos* form]

oidium /əʊˈɪdɪəm/ *n.* (*pl.* **oidia** /-dɪə/) any of several kinds of fungal spore, formed by the breaking up of fungal hyphae into cells. [mod.L f. Gk *ōion* egg + *-idion* dimin. suffix]

oik /ɔɪk/ *n. colloq.* an uncouth or obnoxious person; an idiot. [20th c.: orig. unkn.]

oil /ɔɪl/ *n.* & *v.* ● *n.* **1** a thick, viscous, usu. inflammable liquid insoluble in water but soluble in organic solvents. (*See note below.*) **2** petroleum. **3** (in *comb.*) using oil as fuel (*oil-heater*). **4 a** (usu. in *pl.*) = *oil-paint.* **b** *colloq.* a picture painted in oil-paints. **5** (in *pl.*) = OILSKIN 2b. ● *v.* **1** *tr.* apply oil to; lubricate. **2** *tr.* impregnate or treat with oil (*oiled silk*). **3** *tr.* & *intr.* supply with or take on oil as fuel. **4** *tr.* & *intr.* make (butter, grease, etc.) into or (of butter etc.) become an oily liquid. □ **oil-bird** a nocturnal fruit-

eating bird, *Steatornis caripensis*, which resembles a nightjar and inhabits caves in Central and South America (also called *guacharo*). **oil drum** a metal drum used for transporting oil. **oiled silk** silk made waterproof with oil. **oil engine** an engine driven by the explosion of vaporized oil mixed with air. **oil-fired** using oil as fuel. **oil a person's hand** (or **palm**) bribe a person. **oil-immersion** (of a microscope lens or technique) using a thin film of oil between the objective lens and the specimen, allowing the use of higher magnification. **oil-lamp** a lamp using oil as fuel. **oil-meal** ground oilcake. **oil of vitriol** see VITRIOL. **oil-paint** (or **-colour**) a mix of ground colour pigment and oil. **oil-painting 1** the art of painting in oil-paints. **2** a picture painted in oil-paints. **3** (in phr. **be no oil-painting**) (of a person) be physically unattractive. **oil-palm** a West African palm tree, *Elaeis guineensis*, from which palm oil is extracted. **oil-pan** an engine sump. **oil-paper** paper made transparent or waterproof by soaking in oil. **oil platform** a structure designed to stand on the seabed to provide a stable base above water from which oil or gas wells can be drilled or regulated. **oil-press** an apparatus for pressing oil from seeds etc. **oil rig** a structure with equipment for drilling an oil well; an oil platform. **oil-sand** a stratum of porous rock yielding petroleum. **oil-seed** the seed from any cultivated crop yielding oil, e.g. rape, peanut, or cotton. **oil-shale** a fine-grained rock from which oil can be extracted. **oil-slick** a smooth patch of oil, esp. one on the sea. **oil-tanker** a ship designed to carry oil in bulk. **oil one's tongue** *archaic* say flattering or glib things. **oil well** a well from which mineral oil is drawn. **oil the wheels** help make things go smoothly. **well oiled** *colloq.* very drunk. □ **oilless** /ˈɔɪlɪs/ *adj.* [ME *oli, oile* f. AF, ONF *olie* = OF *oile* etc. f. L *oleum* (olive) oil f. *olea* olive]

▪ Oils are classified as: (i) non-volatile *fatty* or *fixed oils* of animal or vegetable origin, chemically identical with fats, and used as varnishes, lubricants, illuminants, and soap constituents; (ii) *essential* or *volatile oils*, chiefly of vegetable origin, responsible for the scent of plants, used in medicine and perfumery; (iii) *mineral oils*, consisting mainly of hydrocarbons, thought to be the remains of tiny living organisms deposited millions of years ago and trapped in suitable geological formations, and detectable by seismic surveys and by drilling. The mineral oil industry dates from 1859, when a well was bored at Titusville in Pennsylvania, although this was not the first strike: oil had been found in Ohio in 1841 during drilling for salt.

oilcake /ˈɔɪlkeɪk/ *n.* a mass of compressed linseed etc. left after oil has been extracted, used as fodder or manure.

oilcan /ˈɔɪlkæn/ *n.* a can containing oil, esp. one with a long nozzle for oiling machinery.

oilcloth /ˈɔɪlklɒθ/ *n.* **1** a fabric waterproofed with oil. **2** an oilskin. **3** a canvas coated with linseed or other oil and used to cover a table or floor.

oiler /ˈɔɪlə(r)/ *n.* **1** an oilcan for oiling machinery. **2** an oil-tanker. **3** US **a** an oil well. **b** (in *pl.*) oilskin.

oilfield /ˈɔɪlfiːld/ *n.* an area yielding mineral oil.

oilman /ˈɔɪlmən/ *n.* (*pl.* **-men**) a person who deals in oil.

oilskin /ˈɔɪlskɪn/ *n.* **1** cloth waterproofed with oil. **2 a** a garment made of this. **b** (in *pl.*) a suit made of this.

oilstone /ˈɔɪlstəʊn/ *n.* a fine-grained flat stone used with oil for sharpening flat tools, e.g. chisels, planes, etc. (cf. WHETSTONE).

oily /ˈɔɪlɪ/ *adj.* (**oilier, oiliest**) **1** of, like, or containing much oil. **2** covered or soaked with oil. **3** (of a manner etc.) fawning, insinuating, unctuous. □ **oilily** *adv.* **oiliness** *n.*

oink /ɔɪŋk/ *v.intr.* (of a pig) make its characteristic grunt. [imit.]

ointment /ˈɔɪntmənt/ *n.* a smooth greasy healing or cosmetic preparation for the skin. [ME *oignement*, ointment, f. OF *oignement* ult. f. L (as UNGUENT): *oint-* after obs. *oint* anoint f. OF, past part. of *oindre* ANOINT]

Oireachtas /ˈerəxtəs/ *n.* the legislature of the Republic of Ireland, composed of the President, Dáil, and Seanad. [Ir.]

Oirot-Tura /ˌɔɪrɒtˈtuːrə/ a former name (1932–48) for GORNO-ALTAISK.

Oisin /ˈəʊʃiːn/ see OSSIAN.

Ojibwa /əʊˈdʒɪbweɪ/ *n.* & *adj.* ● *n.* (also **Ojibway**) (*pl.* same) **1** a member of an Algonquian people inhabiting the lands round Lake Superior and certain adjacent areas. **2** the Algonquian language of this people. ● *adj.* of or relating to the Ojibwa or their language. [Ojibwa, f. root meaning 'puckered', with ref. to their moccasins]

OK[1] /əʊˈkeɪ/ *adj., adv., n., & v.* (also **okay**) *colloq.* ● *adj.* (often as *int.* expressing agreement or acquiescence) all right; satisfactory. (*See note below.*) ● *adv.* well, satisfactorily (*that worked out OK*). ● *n.* (*pl.* **OKs**) approval, sanction. ● *v.tr.* (**OK's, OK'd, OK'ing**) give an OK to; approve, sanction.

▪ OK originated in America, probably as an abbreviation for *orl* (or *oll*) *korrect*, a humorous form of 'all correct'; it became well known in the American presidential election of 1840, when used as a slogan for Martin Van Buren, who was nicknamed Old Kinderhook. There are many other suggestions about the expression's origin, none of them supported by good evidence.

OK² *abbr. US* Oklahoma (in official postal use).

okapi /əʊˈkɑːpɪ/ *n.* (*pl.* same or **okapis**) a large ruminant mammal, *Okapia johnstoni*, of the giraffe family. It lives in the dense forests of Zaire, and has a dark chestnut coat with transverse stripes on the hindquarters and upper legs. [Afr. name]

Okara /əʊˈkɑːrə/ a commercial city in NE Pakistan, in Punjab province; pop. (1981) 154,000.

Okavango /ˌəʊkəˈvæŋɡəʊ/ (also **Cubango** /kjuːˈbæŋɡəʊ/) a river of SW Africa which rises in central Angola and flows 1,600 km (1,000 miles) south-eastwards to Namibia, where it turns eastwards to form part of the border between Angola and Namibia before entering Botswana, where it drains into the extensive Okavango marshes of Ngamiland.

okay var. of OK¹.

Okayama /ˌəʊkəˈjɑːmə/ an industrial city and major railway junction in SW Japan, on the SW coast of the island of Honshu; pop. (1990) 593,740.

Okeechobee, Lake /ˌəʊkəˈtʃəʊbiː/ a lake in southern Florida. Fed by the Kissimmee river from the north, it drains into the Everglades in the south, this drainage being controlled by embankments and canals. It forms part of the Okeechobee Waterway, which crosses the Florida peninsula from west to east, linking the Gulf of Mexico with the Atlantic.

O'Keeffe /əʊˈkiːf/, Georgia (1887–1986), American painter. She was a pioneer of modernism in America and her early work is largely abstract. In 1916 her work was exhibited by the photographer Alfred Stieglitz, whose circle in New York she then joined; she married him in 1924. In the 1920s she adopted a more figurative style, producing her best-known paintings; they depict enlarged studies, particularly of flowers, and are often regarded as being sexually symbolic (e.g. *Black Iris*, 1926). She also painted notable landscapes of New Mexico, where she settled after her husband's death in 1946.

Okefenokee Swamp /ˌəʊkəfəˈnəʊkiː/ an area of swampland in SE Georgia and NE Florida. It extends over 1,555 sq. km (600 sq. miles).

okey-dokey /ˌəʊkɪˈdəʊkɪ/ *adj.* & *adv.* (also **okey-doke** /-ˈdəʊk/) *sl.* = OK¹. [redupl.]

Okhotsk, Sea of /əʊˈxɒtsk/ an inlet of the northern Pacific Ocean on the east coast of Russia, between the Kamchatka peninsula and the Kurile Islands.

Okinawa /ˌəʊkɪˈnɑːwə/ **1** a region in southern Japan, in the southern Ryukyu Islands; capital, Naha. **2** the largest of the Ryukyu Islands, in southern Japan. It was captured from the Japanese in the Second World War by a US assault in April–June 1945. With its bases commanding the approaches to Japan, it was a key objective, defended by the Japanese almost to the last man, with kamikaze air attacks inflicting substantial damage on US ships. After the war it was retained under US administration until 1972.

Okla. *abbr.* Oklahoma.

Oklahoma /ˌəʊkləˈhəʊmə/ a state in the south central US, north of Texas; pop. (1990) 3,145,585; capital, Oklahoma City. In 1803 it was acquired from the French as part of the Louisiana Purchase. In 1834–89 it was declared Indian territory, in which Europeans were forbidden to settle; after 1889 parts of it were opened to European settlement, and it became the 46th state of the US in 1907.

Oklahoma City the state capital of Oklahoma; pop. (1990) 444,720. It expanded rapidly after the discovery in 1928 of the oilfield on which it lies.

okra /ˈəʊkrə, ˈɒk-/ *n.* **1** an African plant, *Abelmoschus esculentus* (mallow family), yielding long ridged seed-pods. **2** the seed-pods eaten as a vegetable and used to thicken soups and stews. Also called *gumbo*, *ladies' fingers*. [app. W.Afr.: cf. Ibo *okuro* okra, Twi *krakra* broth]

okta /ˈɒktə/ *n.* (*pl.* **oktas** or same) *Meteorol.* a unit of cloud cover, equal to one eighth of the sky. [alt. of OCTA-]

-ol /ɒl/ *suffix Chem.* forming names of organic compounds, esp. alcohols,

phenols, and oils (*methanol*; *cresol*; *benzol*). [f. ALCOHOL; also as var. of -OLE]

Olaf /ˈəʊlæf/ the name of five kings of Norway, notably:

Olaf I Tryggvason (969–1000), reigned 995–1000. According to legend he was brought up in Russia, being converted to Christianity and carrying out extensive Viking raids before returning to Norway to be accepted as king. He jumped overboard and was lost after his fleet was defeated by the combined forces of Denmark and Sweden at the battle of Svöld, but his exploits as a warrior and his popularity as sovereign made him a national legend.

Olaf II Haraldsson (canonized as St Olaf) (*c*.995–1030), reigned 1016–30. Notable for his attempts to spread Christianity in his kingdom, Olaf was forced into exile by a rebellion in 1028 and killed in battle at Stiklestad while attempting to return. He is the patron saint of Norway. Feast day, 29 July.

Öland /ˈɜːlænd/ a narrow island in the Baltic Sea off the SE coast of Sweden, separated from the mainland by Kalmar Sound.

Olbers' paradox /ˈɒlbəz/ *n. Astron.* the apparent paradox that, if stars are distributed evenly in an infinite universe, the sky should be as bright by night as by day. More distant stars, although faint, would be so numerous that every line of sight from an observer would reach a star. The paradox is resolved by the observation that the universe is of finite age, and the light from the more distant stars is dimmed because they are receding from the observer as the universe expands. [H. W. M. *Olbers*, Ger. astronomer (1758–1840)]

old /əʊld/ *adj.* (**older, oldest**) (cf. ELDER¹, ELDEST). **1 a** advanced in age; far on in the natural period of existence. **b** not young or near its beginning. **2** made long ago. **3** long in use. **4** worn or dilapidated or shabby from the passage of time. **5** having the characteristics (experience, feebleness, etc.) of age (*the child has an old face*). **6** practised, inveterate (*an old offender*; *old in crime*). **7** belonging only or chiefly to the past; lingering on; former (*old times*; *haunted by old memories*). **8** dating from far back; long established or known; ancient, primeval (*old as the hills*; *old friends*; *an old family*). **9** (appended to a period of time) **a** (often in *comb.*) of age (*is four years old*; *a four-year-old boy*). **b** (in *comb.*, as noun) a person or animal of the age specified (*our four-year-old is ill*). **10** (of language) as used in former or earliest times. **11** *colloq.* as a term of affection or casual reference (*good old Charlie*; *old mate*). **12** the former or first of two or more similar things (*our old house*; *wants his old job back*). □ **old age** the later part of normal life. **old-age pension** = *retirement pension*. **old-age pensioner** a person receiving this. **old bird** *sl.* a wary person. **old boy 1** a former male pupil of a school. **2** *colloq.* **a** an elderly man. **b** an affectionate form of address to a boy or man. **old boy network** *Brit. colloq.* preferment in employment of those from a similar social background, esp. fellow ex-pupils of public schools. **the old country** the native country of colonists etc. **old-fashioned** in or according to a fashion or tastes no longer current; antiquated. **old fustic** see FUSTIC. **old girl 1** a former female pupil of a school. **2** *colloq.* **a** an elderly woman. **b** an affectionate term of address to a girl or woman. **old gold** a dull brownish-gold colour. **old guard** the original or past or conservative members of a group. **old hand** a person with much experience. **old hat** *colloq. adj.* tediously familiar or out of date. ● *n.* something tediously familiar or out of date. **old lady** *colloq.* one's mother or wife. **old lag** see LAG³. **old maid 1** *derog.* an elderly unmarried woman. **2** a prim and fussy person. **3** a card-game in which players try not to be left with an unpaired queen. **old-maidish** like an old maid. **old man** *colloq.* **1** one's husband or father. **2** one's employer or other person in authority over one. **3** an affectionate form of address to a boy or man. **old man's beard** traveller's joy, esp. in seed. **old master 1** a great artist of former times, esp. of the 13th–17th centuries in Europe. **2** a painting by such a painter. **old moon** the moon in its last quarter, before the new moon. **an old one** a familiar joke. **old retainer** see RETAINER 3b. **old school 1** traditional attitudes. **2** people having such attitudes. **old school tie** esp. *Brit.* **1** a necktie with a characteristic pattern worn by the pupils of a particular (usu. public) school. **2** the group loyalty and traditionalism associated with wearing such a tie. **old soldier** an experienced person, esp. in an arduous activity. **old squaw** *N. Amer.* = *long-tailed duck*. **old stager** an experienced person, an old hand. **old-time** belonging to former times. **old-timer** a person with long experience or standing. **old wives' tale** a foolish or unscientific tradition or belief. **old woman** *colloq.* **1** one's wife or mother. **2** a fussy or timid man. **old-womanish** fussy and timid. **old-world** belonging to or associated with old times. **old year** the year just ended or about to end. □ **oldish** *adj.* **oldness** *n.* [OE *ald* f. WG]

Old Bailey /ˈbeɪlɪ/ the Central Criminal Court in England, formerly standing in an ancient bailey of the London city wall. The present court, which chiefly tries those offences committed in the City and the Greater London area, was built in 1903–6 on the site of Newgate Prison.

Old Bill *n. Brit. sl.* the police.

Old Contemptibles the veterans of the British Expeditionary Force sent to France in the First World War (1914), so named because of a supposed German reference to the 'contemptible little army' facing them.

Old Delhi see DELHI.

olden /ˈəʊld(ə)n/ *adj. archaic* of old; of a former age (esp. *in olden times*).

Old English *n.* the English language up to *c.*1150 (see ENGLISH).

Old French *n.* the French language of the period before *c.*1400.

Old Glory *n. US* the US national flag.

Oldham /ˈəʊldəm/ an industrial town in NW England, in NE Greater Manchester; pop. (1991) 100,000.

Old Hickory a nickname given to Andrew Jackson (see JACKSON²).

Old High German *n.* High German up to *c.*1200. (See HIGH GERMAN.)

oldie /ˈəʊldɪ/ *n. colloq.* an old person or thing.

Old Kingdom a period of ancient Egyptian history (*c.*2575–2134 BC). (See EGYPT.)

Old Lady of Threadneedle Street the nickname of the Bank of England, which stands in this street. The term dates from the late 18th century. (See THREADNEEDLE STREET.)

Old Nick the Devil. [prob. f. a pet-form of the name *Nicholas*]

Old Norse *n.* the North Germanic language of Norway and its colonies, or of Scandinavia, until the 14th century. It is the ancestor of the modern Scandinavian languages and is most clearly preserved in the saga literature of Iceland.

Old Pals Act *n. Brit.* the principle that friends should always help one another.

Old Pretender, the James Stuart, son of James II of England and Ireland (James VII of Scotland) (see STUART³).

Old Sarum /ˈseərəm/ a hill in southern England 3 km (2 miles) north of Salisbury. It was the site of an ancient Iron Age settlement and hill fort which was later occupied by the Normans, who built a castle and a town. It became a bishopric but fell into decline after 1220, when the new cathedral and town of Salisbury were established to the south. The site of Old Sarum became deserted and the original cathedral was demolished in 1331.

oldster /ˈəʊldstə(r)/ *n.* an old person. [OLD + -STER, after *youngster*]

Old Style *n.* dating reckoned by the Julian calendar.

Old Testament the writings comprising the first thirty-nine books of the Christian Bible, the Hebrew Bible containing an account of the Creation, the origin of humankind, God's covenant with the Jews, and their early history. The Old Testament corresponds approximately to the Hebrew Bible, differing largely in its division of some books and in the acceptance (by the Roman Catholic Church) of the Apocrypha. Most of the books were written in Hebrew, some in Aramaic, *c.*1200–100 BC. They were classified by the Hebrews into three groups: the Law (i.e. the Pentateuch), the Prophets (Joshua, Judges, 1 & 2 Samuel, 1 & 2 Kings, Isaiah, Jeremiah, Ezekiel, Daniel, and the twelve minor prophets, Hosea to Malachi), and the Writings, comprising the remaining books. The canon of the Hebrew Bible is believed to have been settled by about AD 100.

Olduvai Gorge /ˈɒldʊˌvaɪ/ a gorge in northern Tanzania, 48 km (30 miles) long and up to 90 metres (300 ft) deep. The exposed strata contain numerous fossils spanning the full range of the Pleistocene period. Work by the Leakeys in the Gorge has provided evidence of the longest sequence of hominid presence and activity yet discovered anywhere in the world, with fossils, stone-tool industries, and other evidence of hominid activities that date from *c.*2.1–1.7 million years BP for the oldest deposits to *c.*22,000 BP for the most recent. The hominids found at individual sites within the Gorge include *Australopithecus* (or *Zinjanthropus*) *boisei, Homo habilis,* and *Homo erectus.*

Old Vic /vɪk/ the popular name of a London theatre, opened in 1818 as the Royal Coburg and renamed the Royal Victoria Theatre in honour of Princess (later Queen) Victoria in 1833. Under the management of Lilian Baylis from 1912 it gained an enduring reputation for its Shakespearian productions.

Old World Europe, Asia, and Africa, regarded collectively as the part of the world known before the discovery of the Americas. (Cf. NEW WORLD.)

-ole /əʊl/ *comb. form Chem.* forming names of esp. heterocyclic compounds (*indole*). [L *oleum* oil: cf. -OL]

oleaceous /ˌəʊlɪˈeɪʃəs/ *adj. Bot.* of the plant family Oleaceae, which includes olive, jasmine, lilac, privet, ash, etc. [mod.L *Oleaceae* f. L *olea* olive tree]

oleaginous /ˌəʊlɪˈædʒɪnəs/ *adj.* **1** having the properties of or producing oil. **2** oily, greasy. **3** obsequious, ingratiating. [F *oléagineux* f. L *oleaginus* f. *oleum* oil]

oleander /ˌəʊlɪˈændə(r)/ *n.* an evergreen poisonous shrub, *Nerium oleander,* native to the Mediterranean and bearing clusters of white, pink, or red flowers. [med.L]

oleaster /ˌəʊlɪˈæstə(r)/ *n.* a tree of the genus *Elaeagnus,* often thorny and with evergreen leathery foliage; esp. *E. angustifolia,* which bears olive-shaped yellowish fruits (also called *Russian olive*). [ME f. L f. *olea* olive tree: see -ASTER]

olecranon /əʊˈlekrəˌnɒn, ˌəʊlɪˈkreɪnən/ *n.* a bony prominence on the upper end of the ulna at the elbow. [Gk *ōle(no)kranon* f. *ōlenē* elbow + *kranion* head]

olefin /ˈəʊlɪfɪn/ *n.* (also **olefine** /-ˌfiːn/) *Chem.* = ALKENE. [F *oléfiant* oil-forming (with ref. to oily ethylene dichloride)]

oleic acid /əʊˈliːɪk/ *n. Chem.* an unsaturated fatty acid present in many fats and soaps. □ **oleate** /ˈəʊlɪət/ *n.* [L *oleum* oil]

oleiferous /ˌəʊlɪˈɪfərəs/ *adj.* yielding oil. [L *oleum* oil + -FEROUS]

oleo- /ˈəʊlɪəʊ/ *comb. form* oil. [L *oleum* oil]

oleograph /ˈəʊlɪəˌɡrɑːf/ *n.* a print made to resemble an oil-painting.

oleomargarine /ˌəʊlɪəʊˌmɑːdʒəˈriːn, -ˌmɑːɡəˈriːn/ *n.* **1** a fatty substance extracted from beef fat and often used in margarine. **2** *US* margarine.

oleometer /ˌəʊlɪˈɒmɪtə(r)/ *n.* an instrument for determining the density and purity of oils.

oleoresin /ˌəʊlɪəʊˈrezɪn/ *n.* a natural or artificial mixture of essential oils and a resin, e.g. balsam.

oleum /ˈəʊlɪəm/ *n.* concentrated sulphuric acid containing excess sulphur trioxide in solution forming a dense corrosive liquid. [L, = oil]

O level /əʊ/ *n. hist.* = *ordinary level.* [abbr.]

olfaction /ɒlˈfækʃ(ə)n/ *n.* the act or capacity of smelling; the sense of smell. □ **olfactive** /-ˈfæktɪv/ *adj.* [L *olfactus* a smell f. *olere* to smell + *facere* fact- make]

olfactory /ɒlˈfæktərɪ/ *adj.* of or relating to the sense of smell (*olfactory nerves*). [L *olfactare* frequent. of *olfacere* (as OLFACTION)]

olibanum /ɒˈlɪbənəm/ *n.* = FRANKINCENSE. [ME f. med.L f. LL *libanus* f. Gk *libanos* frankincense, of Semitic orig.]

oligarch /ˈɒlɪˌɡɑːk/ *n.* a member of an oligarchy. [Gk *oligarkhēs* f. *oligoi* few + *arkhō* to rule]

oligarchy /ˈɒlɪˌɡɑːkɪ/ *n.* (*pl.* **-ies**) **1** government by a small group of people. **2** a state governed in this way. **3** the members of such a government. □ **oligarchic** /ˌɒlɪˈɡɑːkɪk/ *adj.* **oligarchical** *adj.* **oligarchically** *adv.* [F *oligarchie* or med.L *oligarchia* f. Gk *oligarkhia* (as OLIGARCH)]

oligo- /ˈɒlɪɡəʊ/ *comb. form* **1** few, slight. **2** *Chem.* denoting the presence of a relatively small number of radicals etc. of a particular kind in a molecule, esp. a polymer (*oligosaccharide*). [Gk *oligos* small, *oligoi* few]

Oligocene /ˈɒlɪɡəˌsiːn/ *adj. & n. Geol.* of, relating to, or denoting the third epoch of the Tertiary period, between the Eocene and the Miocene. The Oligocene lasted from about 38 to 24.6 million years ago, and was a time of falling temperatures. [as OLIGO- + Gk *kainos* new]

oligochaete /ˈɒlɪɡəˌkiːt/ *n. & adj. Zool.* ● *n.* an annelid worm of the division Oligochaeta, which includes the earthworms, having a small number of bristles on each segment. ● *adj.* of or relating to this division. [mod.L, f. as OLIGO- + Gk *khaitē* long hair (taken as 'bristle')]

oligopoly /ˌɒlɪˈɡɒpəlɪ/ *n.* (*pl.* **-ies**) a state of limited competition between a small number of producers or sellers. □ **oligopolist** *n.* **oligopolistic** /-ˌɡɒpəˈlɪstɪk/ *adj.* [OLIGO-, after MONOPOLY]

oligosaccharide /ˌɒlɪɡəʊˈsækəˌraɪd/ *n.* any carbohydrate whose molecules are composed of a relatively small number of monosaccharide units.

oligotrophic /ˌɒlɪɡəʊˈtrəʊfɪk, -ˈtrɒfɪk/ *adj. Ecol.* (of a lake etc.) relatively poor in plant nutrients. □ **oligotrophy** /ˌɒlɪˈɡɒtrəfɪ/ *n.*

olio /ˈəʊlɪəʊ/ *n.* (*pl.* **-os**) **1** a mixed dish; a stew of various meats and vegetables. **2** a hotchpotch or miscellany. [Sp. *olla* stew f. L *olla* cooking-pot]

olivaceous /ˌɒlɪˈveɪʃəs/ adj. olive-green; of a dusky yellowish-green.

olivary /ˈɒlɪvərɪ/ adj. Anat. olive-shaped; oval. [L olivarius (as OLIVE)]

olive /ˈɒlɪv/ n. & adj. ● n. **1** (in full **olive tree**) an evergreen tree of the genus *Olea*, having dark green lance-shaped leathery leaves with silvery undersides; esp. *O. europaea* of the Mediterranean, and *O. africana* of South Africa. **2** the small oily oval fruit of this, having a hard stone and bitter flesh, green when unripe and bluish-black when ripe. **3** (in full **olive-green**) the greyish-green colour of an unripe olive. **4** the wood of the olive tree. **5** Anat. each of a pair of olive-shaped swellings in the medulla oblongata. **6 a** an olive-shaped gastropod mollusc of the genus *Oliva*. **b** the shell of this. **7** a slice of beef or veal made into a roll with stuffing inside and stewed. **8** a metal ring or fitting which is tightened under a threaded nut to form a seal, as in a compression joint. ● adj. **1** (in full **olive-green**) coloured like an unripe olive. **2** (of the complexion) yellowish-brown, sallow. □ **olive branch 1** the branch of an olive tree as a symbol of peace (with allusion to Gen. 8:11). **2** a gesture of reconciliation or friendship. **olive crown** a garland of olive leaves as a sign of victory. **olive drab** the dull olive colour of US army uniforms. **olive oil** an oil extracted from olives, used esp. in cookery. [ME f. OF f. L *oliva* f. Gk *elaia* f. *elaion* oil]

Oliver /ˈɒlɪvə(r)/ the companion of Roland in the *Chanson de Roland*. (See ROLAND.)

Olives, Mount of the highest point in the range of hills to the east of Jerusalem. It is a holy place for both Judaism and Christianity and is frequently mentioned in the Bible. The Garden of Gethsemane is located nearby. Its slopes have been a sacred Jewish burial ground for centuries.

Olivier /əˈlɪvɪˌeɪ/, Laurence (Kerr), Baron Olivier of Brighton (1907–89), English actor and director. He made his professional début in 1924 and subsequently performed all the major Shakespearian roles; he was also director of the National Theatre (1963–73). His films include *Wuthering Heights* (1939) and *Rebecca* (1940), as well as adaptations of Shakespeare; he produced, co-directed, and starred in *Henry V* (1944) and directed and starred in *Hamlet* (1948) and *Richard III* (1955). The Olivier Theatre, part of the National Theatre, is named in his honour. He was married to Vivien Leigh from 1940 to 1961 and to Joan Plowright (b.1929) from 1961 until his death.

olivine /ˈɒlɪˌviːn/ n. Mineral. a naturally occurring form of magnesium-iron silicate, usu. olive-green and found in igneous rocks.

olla podrida /ˌɒlə pɒˈdriːdə/ n. = OLIO. [Sp., lit. 'rotten pot' (as OLIO + L *putridus*: cf. PUTRID)]

olm /ɒlm/ n. a blind cave-dwelling salamander, *Proteus anguinus*, native to Austria, usu. transparent but turning brown in light and having external gills. [G]

Olmec /ˈɒlmek/ n. (pl. same) **1** a member of a prehistoric people inhabiting the coast of Veracruz and western Tabasco on the Gulf of Mexico *c.*1200–100 BC, who established what was probably the first developed civilization of Meso-America. They are noted for their sculptures, especially massive stone-hewn heads with realistic features and round helmets, and small jade carvings featuring a jaguar. **2** a native people living in the same general area during the 15th and 16th centuries. [Nahuatl, = people of the rubber(-tree) country]

Olmos /ˈɒlmɒs/ a small town on the eastern edge of the Sechura Desert in NW Peru, which gave its name to a major irrigation project initiated in 1926 for the purpose of increasing cotton and sugar production in the arid lowlands of this region.

-ology /ˈɒlədʒɪ/ comb. form see -LOGY.

Olomouc /ˈɒləˌməʊts/ an industrial city on the Morava river in northern Moravia in the Czech Republic; pop. (1991) 105,690.

oloroso /ˌɒləˈrəʊsəʊ/ n. (pl. **-os**) a heavy dark medium-sweet sherry. [Sp., lit. 'fragrant']

Olsztyn /ˈɒlʃtɪn/ (German **Allenstein** /ˈalənˌʃtaɪn/) a city in northern Poland, in the lakeland area of Masuria; pop. (1990) 163,935. It was founded in 1348 by the Teutonic Knights. Ceded to Poland in 1466, it became a part of Prussia in 1772, but was returned to Poland in 1945.

Olympia /əˈlɪmpɪə/ **1** a plain in Greece, in the western Peloponnese. In ancient Greece it was the site of the chief sanctuary of the god Zeus, the place where the original Olympic Games were held, after which the site is named. **2** the capital of the state of Washington, a port on Puget Sound; pop. (1990) 33,840.

Olympiad /əˈlɪmpɪˌæd/ n. **1 a** a period of four years between Olympic Games, used by the ancient Greeks in dating events. **b** a four-yearly celebration of the ancient Olympic Games. **2** a celebration of the modern Olympic Games. **3** a regular international contest in chess etc.

[ME f. F *Olympiade* f. L *Olympias Olympiad-* f. Gk *Olumpias Olumpiad-* f. *Olumpios*: see OLYMPIC]

Olympian /əˈlɪmpɪən/ adj. & n. ● adj. **1 a** of or associated with Mount Olympus in Greece, traditionally the home of the Greek gods. **b** celestial, godlike. **2** (of manners etc.) magnificent, condescending, superior. **3 a** of or relating to ancient Olympia in southern Greece. **b** = OLYMPIC. ● n. **1** any of the pantheon of twelve gods regarded as living on Olympus. **2** a person of great attainments or of superhuman calm and detachment. **3** a competitor in the Olympic Games. [L *Olympus* or *Olympia*: see OLYMPUS]

Olympic /əˈlɪmpɪk/ adj. & n. ● adj. of ancient Olympia or the Olympic Games. ● n.pl. (**the Olympics**) the Olympic Games. [L *Olympicus* f. Gk *Olumpikos* of Olympus or Olympia (the latter being named from the games in honour of Zeus of *Olympus*)]

Olympic Games 1 a festival of sport held at Olympia (traditionally from 776 BC, every fourth year) by the ancient Greeks, with athletic, musical, and literary competitions, until abolished by the Roman emperor Theodosius I in AD 393. **2** (also **the Olympics**) a modern international sports festival inspired by this. The games were revived by the Frenchman Baron de Coubertin (1863–1937) as an amateur championship of world sport. Since the first of the modern games was held in Athens in 1896, the Olympics have taken place at different venues every four years, interrupted only by the First and Second World Wars; athletes representing nearly 150 countries now compete for gold, silver, and bronze medals in more than twenty sports. The Olympics are a major televised worldwide event, and continue despite political problems which have included national boycotts and a massacre by Arab terrorists of Israeli participants at the 1972 Munich Olympics; there have also been scandals over the use of drugs to enhance athletes' performances. Since 1924 there have been separate Winter Olympics, which are now held at a two-year interval from the summer games.

Olympus /əˈlɪmpəs/ Gk Mythol. the home of the twelve greater gods and the court of Zeus, identified in later antiquity with Mount Olympus in Greece. [pre-Gk word, = mountain]

Olympus, Mount 1 a mountain in northern Greece, at the eastern end of the range dividing Thessaly from Macedonia; height 2,917 m (9,570 ft). (See also OLYMPUS.) **2** a mountain in Cyprus, in the Troodos range. Rising to 1,951 m (6,400 ft), it is the highest peak on the island.

OM abbr. (in the UK) Order of Merit.

om /əʊm/ n. Hinduism & Tibetan Buddhism a mystic syllable, considered the most sacred mantra. It appears at the beginning and end of most Sanskrit recitations, prayers, and texts. [Skr., a universal affirmation]

-oma /ˈəʊmə/ suffix forming nouns denoting tumours and other abnormal growths (*carcinoma*). [mod.L f. Gk *-ōma* suffix denoting the result of verbal action]

Omagh /əʊˈmɑː, ˈəʊmə/ a town in Northern Ireland, principal town of County Tyrone; pop. (1981) 14,600.

Omaha[1] /ˈəʊməˌhɑː/ a city in eastern Nebraska, on the Missouri river; pop. (1990) 335,795.

Omaha[2] /ˈəʊməˌhɑː/ n. (pl. same or **Omahas**) **1** a member of a Sioux people of NE Nebraska. **2** the language of this people. [Omaha *umonhon* upstream people]

Oman /əʊˈmɑːn/ a country at the eastern corner of the Arabian peninsula; pop. (est. 1991) 1,600,000; official language, Arabic; capital, Muscat. An independent sultanate, known as Muscat and Oman until 1970, Oman was the most influential power in the region in the 19th century, controlling Zanzibar and other territory. Since the late 19th century Oman has had strong links with Britain, which has on occasions provided military aid to support the authority of the sultan. The economy is dependent on oil, discovered in 1964. □ **Omani** adj. & n.

Oman, Gulf of an inlet of the Arabian Sea, connected by the Strait of Hormuz to the Persian Gulf.

Omar I /ˈəʊmɑː(r)/ (*c.*581–644), Muslim caliph 634–44. In early life an opponent of Muhammad, Omar was converted to Islam in 617. After becoming caliph he began an extensive series of conquests, adding Syria, Palestine, and Egypt to his empire. He was assassinated by a Persian slave.

Omar Khayyám /ˌəʊmɑː kaɪˈɑːm/ (d.1123), Persian poet, mathematician, and astronomer. He is remembered for his *rubáiyát* (quatrains), translated and adapted by Edward Fitzgerald in *The Rubáiyát of Omar Khayyám* (1859); the work contains meditations on the mysteries of existence, expressing scepticism regarding divine providence and a consequent celebration of the sensuous and fleeting pleasures of the earthly world.

omasum /əʊˈmeɪsəm/ n. (pl. **omasa** /-sə/) the third stomach of a ruminant. Also called *psalterium*. [L, = bullock's tripe]

Omayyad var. of UMAYYAD.

ombre /ˈɒmbə(r)/ n. a card-game for three, popular in Europe in the 17th–18th centuries. [Sp. *hombre* man, with ref. to one player seeking to win the pool]

ombré /ˈɒmbreɪ/ adj. (of a fabric etc.) having gradual shading of colour from light to dark. [F, past part. of *ombrer* to shadow (as UMBER)]

ombro- /ˈɒmbrəʊ/ comb. form atmospheric moisture (*ombrogenous*). [Gk *ombros* rain-shower]

ombrogenous /ɒmˈbrɒdʒɪnəs/ Ecol. (of a bog etc.) dependent on rain for its formation.

ombudsman /ˈɒmbʊdzmən/ n. (pl. **-men**) an official appointed to investigate complaints by individuals against maladministration, especially that of public authorities; (in the UK) the Parliamentary Commissioner for Administration. Such an office was first introduced in Sweden, in 1809; there an ombudsman is a deputy of a group, particularly a trade union or a business concern, appointed to handle the legal affairs of the group and protect its interests generally. In the UK the first ombudsman took office in 1967. [Sw., = legal representative]

Omdurman /ˌɒmdɜːˈmɑːn/ a city in central Sudan, on the Nile opposite Khartoum; pop. (1983) 526,290. In 1885, following the victory of the Mahdi (Muhammad Ahmad) and his forces over the British, it was made the capital of the Mahdist state of Sudan. In 1898 it was recaptured by the British after the Battle of Omdurman, Kitchener's decisive victory over the Mahdi's successor, which marked the end of the uprising.

-ome /əʊm/ suffix forming nouns denoting objects or parts of a specified nature (*rhizome*; *trichome*). [var. of -OMA]

omega /ˈəʊmɪɡə/ n. **1** the last (24th) letter of the Greek alphabet (Ω, ω). **2** the last of a series; the final development. [Gk, *ō mega* = great O]

omelette /ˈɒmlɪt/ n. (also **omelet**) a dish of beaten eggs cooked in a frying-pan and served plain or with a savoury or sweet filling. [F *omelette*, obs. *amelette* by metathesis f. *alumette* var. of *alumelle* f. *lemele* knife-blade f. L *lamella*: see LAMELLA]

omen /ˈəʊmən, -men/ n. & v. ● n. **1** an occurrence or object regarded as portending good or evil. **2** prophetic significance (*of good omen*). ● v.tr. (usu. in *passive*) portend; foreshow. □ **omened** adj. (also in comb.). [L *omen ominis*]

omentum /əʊˈmentəm/ n. (pl. **omenta** /-tə/) Anat. a fold of peritoneum connecting the stomach with other abdominal organs. □ **omental** adj. [L]

omertà /ˌəʊmeəˈtaː/ n. a code of silence, esp. as practised by the Mafia. [It., = conspiracy of silence]

omicron /əʊˈmaɪkrən/ n. the fifteenth letter of the Greek alphabet (O, o). [Gk, *o mikron* = small o]

ominous /ˈɒmɪnəs/ adj. **1** threatening; indicating disaster or difficulty. **2** of evil omen; inauspicious. **3** giving or being an omen. □ **ominously** adv. **ominousness** n. [L *ominosus* (as OMEN)]

omission /əˈmɪʃ(ə)n/ n. **1** the act or an instance of omitting or being omitted. **2** something that has been omitted or overlooked. □ **omissive** /əˈmɪsɪv/ adj. [ME f. OF *omission* or LL *omissio* (as OMIT)]

omit /əˈmɪt/ v.tr. (**omitted, omitting**) **1** leave out; not insert or include. **2** leave undone. **3** (foll. by verbal noun or *to* + infin.) fail or neglect (*omitted saying anything*; *omitted to say*). □ **omissible** /əˈmɪsɪb(ə)l/ adj. [ME f. L *omittere omiss-* (as OB-, *mittere* send)]

ommatidium /ˌɒməˈtɪdɪəm/ n. (pl. **ommatidia** /-dɪə/) Zool. a structural element in the compound eye of an insect. [mod.L f. Gk *ommatidion* dimin. of *omma ommat-* eye]

omni- /ˈɒmnɪ/ comb. form **1** all; of all things. **2** in all ways or places. [L f. *omnis* all]

omnibus /ˈɒmnɪbəs/ n. & adj. ● n. **1** formal = BUS. **2** a volume, television or radio programme, etc. containing several novels, programmes, etc. previously published or broadcast separately. ● adj. **1** serving several purposes at once. **2** comprising several items. [F f. L (dative pl. of *omnis*), = for all]

omnicompetent /ˌɒmnɪˈkɒmpɪt(ə)nt/ adj. **1** able to deal with all matters. **2** having jurisdiction in all cases. □ **omnicompetence** n.

omnidirectional /ˌɒmnɪdaɪˈrekʃ(ə)n(ə)l, -dɪˈrekʃ(ə)n(ə)l/ adj. (of an aerial etc.) receiving or transmitting in all directions.

omnifarious /ˌɒmnɪˈfeərɪəs/ adj. of all sorts or varieties. [LL *omnifarius* (as OMNI-): cf. MULTIFARIOUS]

omnipotent /ɒmˈnɪpət(ə)nt/ adj. **1** having great or absolute power. **2** having great influence. □ **omnipotence** n. **omnipotently** adv. [ME f. OF f. L *omnipotens* (as OMNI-, POTENT[1])]

omnipresent /ˌɒmnɪˈprez(ə)nt/ adj. **1** present everywhere at the same time. **2** widely or constantly encountered. □ **omnipresence** n. [med.L *omnipraesens* (as OMNI-, PRESENT[1])]

omniscient /ɒmˈnɪsɪənt, -ˈnɪʃɪənt/ adj. knowing everything or much. □ **omniscience** n. **omnisciently** adv. [med.L *omnisciens -entis* (as OMNI-, *scire* know)]

omnium gatherum /ˌɒmnɪəm ˈɡæðərəm/ n. colloq. a miscellany or strange mixture. [mock L f. L *omnium* of all + GATHER]

omnivorous /ɒmˈnɪvərəs/ adj. **1** feeding on many kinds of food, esp. on both plants and flesh. **2** making use of everything available. □ **omnivorously** adv. **omnivorousness** n. **omnivore** /ˈɒmnɪˌvɔː(r)/ n. [L *omnivorus* (as OMNI-, -VOROUS)]

omphalo- /ˈɒmfələʊ/ comb. form navel. [Gk (as OMPHALOS)]

omphalos /ˈɒmfəˌlɒs/ n. **1** Gk Antiq. a conical stone at Delphi in Greece, believed in ancient times to be the navel of the earth. **2** Gk Antiq. a boss on a shield. **3** a centre or hub. [Gk, = navel, boss, hub]

Omsk /ɒmsk/ a city in south central Russia, on the Irtysh river; pop. (1990) 1,159,000.

on /ɒn/ prep., adv., adj., & n. ● prep. **1** (so as to be) supported by or attached to or covering or enclosing (*sat on a chair*; *stuck on the wall*; *rings on her fingers*; *leaned on his elbow*). **2** carried with; about the person of (*have you a pen on you?*). **3** (of time) exactly at; during; contemporaneously with (*on 29 May*; *on schedule*; *working on Tuesday*). **4** immediately after or before (*I saw them on my return*). **5** as a result of (*on further examination I found this*). **6** having, or so as to have, membership etc. of, or residence at or in (*she is on the board of directors*; *lives on the Continent*). **7** supported financially by (*lives on £50 a week*; *lives on his wits*). **8** close to; just by (*a house on the sea*; *lives on the main road*). **9** in the direction of; against. **10** so as to threaten; touching or striking (*advanced on him*; *pulled a knife on me*; *a punch on the nose*). **11** having as an axis or pivot (*turned on his heels*). **12** having as a basis or motive (*works on a ratchet*; *arrested on suspicion*). **13** having as a standard, confirmation, or guarantee (*had it on good authority*; *did it on purpose*; *I promise on my word*). **14** concerning or about (*writes on frogs*). **15** using or engaged with (*is on the pill*; *here on business*). **16** so as to affect (*walked out on her*). **17** at the expense of (*the drinks are on me*; *the joke is on him*). **18** added to (*disaster on disaster*; *ten pence on a pint of beer*). **19** in a specified manner or style (often foll. by *the* + adj. or noun: *on the cheap*; *on the run*). ● adv. **1** (so as to be) covering or in contact with something, esp. of clothes (*put your boots on*). **2** in the appropriate direction; towards something (*look on*). **3** further forward; in an advanced position or state (*time is getting on*; *it happened later on*). **4** with continued movement or action (*went plodding on*; *keeps on complaining*). **5** in operation or activity (*the light is on*; *the chase was on*). **6** due to take place as planned (*is the party still on?*). **7** colloq. **a** (of a person) willing to participate or approve, or make a bet. **b** (of an idea, proposal, etc.) practicable or acceptable (*that's just not on*). **8** being shown or performed (*a good film on tonight*). **9** (of an actor) on stage. **10** (of an employee) on duty. **11** forward (*head on*). ● adj. Cricket designating the part of the field on the striker's side and in front of the wicket. ● n. Cricket the on side. □ **be on about** refer to or discuss esp. tediously or persistently (*what are they on about?*). **be on at** colloq. nag or grumble at. **be on to 1** realize the significance or intentions of. **2** get in touch with (esp. by telephone). **on and off** intermittently; now and then. **on and on** continually; at tedious length. **on-line** Computing adj. (of equipment or a process) directly controlled by or connected to a central processor. ● adv. while thus controlled or connected. **on-off 1** (of a switch) having two positions, 'on' and 'off'. **2** = *on and off*. **on-stage** adj. & adv. on the stage; visible to the audience. **on-street** attrib.adj. (with ref. to parking vehicles) at the side of a street. **on time** punctual, punctually, on the dot, in good time. **on to** to a position or state on or in contact with (cf. ONTO). [OE *on, an* f. Gmc]

-on /ɒn/ suffix Physics, Biochem., & Chem. forming nouns denoting: **1** subatomic particles (*meson*; *neutron*). **2** quanta (*photon*). **3** molecular units (*codon*). **4** substances (*interferon*; *parathion*). [ION, orig. in *electron*]

onager /ˈɒnədʒə(r)/ n. **1** a wild ass, esp. *Equus hemionus* of a race native to central Asia (cf. KIANG, KULAN). **2** hist. an ancient military engine for throwing rocks. [ME f. L f. Gk *onagros* f. *onos* ass + *agrios* wild]

onanism /ˈəʊnəˌnɪz(ə)m/ n. **1** masturbation. **2** coitus interruptus. □ **onanist** n. **onanistic** /ˌəʊnəˈnɪstɪk/ adj. [F *onanisme* or mod.L *onanismus* f. *Onan*, son of Judah, who practised coitus interruptus (Gen. 38:9)]

Onassis[1] /əʊ'næsɪs/, Aristotle (Socrates) (1906–75), Greek shipping magnate and international businessman. The owner of a substantial shipping empire, he was also the founder of the Greek national airline, Olympic Airways (1957). In 1968 he married Jacqueline Bouvier Kennedy, the widow of John F. Kennedy.

Onassis[2] /əʊ'næsɪs/, Jacqueline Lee Bouvier Kennedy (known as 'Jackie O') (1929–94), American First Lady. She worked as a photographer before marrying John F. Kennedy in 1953. Her term as First Lady, which began in 1961, was cut short by the President's assassination in 1963. She married Aristotle Onassis in 1968 and after being widowed for a second time in 1975 pursued a career in publishing.

ONC *abbr. hist.* (in the UK) Ordinary National Certificate.

once /wʌns/ *adv., conj., & n.* ● *adv.* **1** on one occasion or for one time only (*once is not enough; have read it once*). **2** at some point or period in the past (*could once play chess*). **3** ever or at all (*if you once forget it*). **4** multiplied by one; by one degree. ● *conj.* as soon as (*once they have gone we can relax*). ● *n.* one time or occasion (*just the once*). □ **all at once 1** without warning; suddenly. **2** all together. **at once 1** immediately. **2** simultaneously. **for once** on this (or that) occasion, even if at no other. **once again** (or **more**) another time. **once and for all** (or **once for all**) (done) in a final or conclusive manner, esp. so as to end hesitation or uncertainty. **once** (or **every once**) **in a while** from time to time; occasionally. **once or twice** a few times. **once-over** *colloq.* **1** a rapid preliminary inspection or piece of work. **2** an appraising glance. **once upon a time 1** at some vague time in the distant past. **2** formerly. [ME *ānes, ōnes*, genitive of ONE]

oncer /'wʌnsə(r)/ *n.* **1** *Brit. hist. sl.* a one-pound note. **2** *colloq.* a thing that occurs only once. **3** *Austral. colloq.* **a** an election of an MP likely to serve only one term. **b** such an MP.

onchocerciasis /ˌɒŋkəʊsɜː'kaɪəsɪs, -'saɪəsɪs/ *n. Med.* infestation with parasitic threadworms of the genus *Onchocerca*; esp. river blindness, caused by *O. volvulus* and transmitted by the bite of blackflies. [mod.L *Onchocerca* genus name (f. Gk *ogkos* barb + *kerkos* tail) + -IASIS]

onco- /'ɒŋkəʊ/ *comb. form Med.* tumour. [Gk *ogkos* mass]

oncogene /'ɒŋkəˌdʒiːn/ *n. Biol.* a gene which can transform a cell into a tumour cell.

oncogenic /ˌɒŋkə'dʒɛnɪk/ *adj. Med.* causing the development of a tumour or tumours. □ **oncogenicity** /-dʒə'nɪsɪti/ *n.*

oncology /ɒŋ'kɒlədʒi/ *n. Med.* the study and treatment of tumours. □ **oncologist** *n.*

oncoming /'ɒnˌkʌmɪŋ/ *adj. & n.* ● *adj.* approaching from the front. ● *n.* an approach or onset.

oncost /'ɒnkɒst/ *n. Brit.* an overhead expense.

OND *abbr. hist.* (in the UK) Ordinary National Diploma.

on dit /ɒn 'diː/ *n.* (*pl.* **on dits** *pronunc.* same) a piece of gossip or hearsay. [F, = they say]

one /wʌn/ *adj., n., & pron.* ● *adj.* **1** single and integral in number. **2** (with a noun implied) a single person or thing of the kind expressed or implied (*one of the best; a nasty one*). **3 a** a particular but undefined, esp. as contrasted with another (*that is one view; one thing after another*). **b** *colloq.* (as an emphatic) a noteworthy example of (*that is one difficult question*). **4** only such (*the one man who can do it*). **5** forming a unity (*one and undivided*). **6** identical; the same (*of one opinion*). ● *n.* **1 a** the lowest cardinal number. **b** a symbol for this (1, i, I). **2** unity; a unit (*one is half of two; came in ones and twos*). **3** a single thing or person or example (often referring to a noun previously expressed or implied: *the big dog and the small one*). **4** one o'clock. **5** *colloq.* an alcoholic drink (*have a quick one; have one on me*). **6** a story or joke (*the one about the frog*). ● *pron.* **1** a person of a specified kind (*loved ones; like one possessed*). **2** any person, as representing people in general (*one is bound to lose in the end*). **3** I, me (*one would like to help*). ¶ Often regarded as an affectation. □ **all one** (often foll. by *to*) a matter of indifference. **at one** in agreement. **for one** being one, even if the only one (*I for one do not believe it*). **for one thing** as a single consideration, ignoring others. **one and all** everyone. **one and only 1** unique. **2** superb, unequalled. **one and the same** the same; (the) identical. **one another** each the other or others (as a formula of reciprocity: *love one another*). **one-armed bandit** *colloq.* a fruit machine worked by a long handle at the side. **one by one** singly, successively. **one day 1** on an unspecified day. **2** at some unspecified future date. **one-horse 1** using a single horse. **2** *colloq.* small, poorly equipped. **one-liner** *colloq.* a single brief sentence,

often witty or apposite. **one-man** involving, done, or operated by only one man. **one-night stand 1** a single performance of a play etc. in a place. **2** *colloq.* a sexual liaison lasting only one night. **one-off** *colloq.* *adj.* made or done as the only one; not repeated. ● *n.* the only example of a manufactured product; something not repeated. **one or two** see OR[1]. **one-piece** (of a bathing-suit etc.) made as a single garment. **one-sided 1** favouring one side in a dispute; unfair, partial. **2** having or occurring on one side only. **3** larger or more developed on one side. **one-sidedly** in a one-sided manner. **one-sidedness** the act or state of being one-sided. **one-time** former. **one-to-one** with one member of one group corresponding to one of another. **one-track mind** a mind preoccupied with one subject. **one-two** *colloq.* **1** *Boxing* the delivery of two punches in quick succession. **2** *Football* etc. a series of reciprocal passes between two advancing players. **one-up** *colloq.* having a particular advantage. **one-upmanship** *colloq.* the art of gaining or maintaining a psychological advantage. **one-way** allowing movement or travel in one direction only. [OE *ān* f. Gmc]

-one /əʊn/ *suffix Chem.* forming nouns denoting various compounds, esp. ketones (*acetone*). [Gk *-ōnē* fem. patronymic]

onefold /'wʌnfəʊld/ *adj.* consisting of only one member or element; simple.

Onega, Lake /ə'njeɪgə/ a lake in NW Russia, near the border with Finland, the second largest European lake.

Oneida /əʊ'naɪdə/ *n. & adj.* ● *n.* (*pl.* same or **Oneidas**) **1** an American Indian people formerly inhabiting upper New York State, one of the five peoples comprising the original Iroquois confederacy. **2** the Iroquoian language of this people. ● *adj.* of or relating to the Oneida or their language. [Oneida *oneriyote* erected stone]

Oneida Community a religious community, founded in New York State in 1848. Originally embracing primitive Christian beliefs, communal economic practices, and polygamous marriage, the Oneida Community proved a considerable economic success. Although it gradually relinquished its radical social and economic ideas, it has continued to flourish since becoming a joint-stock company (1881), carrying on various industries, especially the manufacture of silver plate, as commercial ventures.

O'Neill /əʊ'niːl/, Eugene (Gladstone) (1888–1953), American dramatist. He achieved recognition with his first full-length play, *Beyond the Horizon* (1920), which won a Pulitzer Prize. Among his many other plays are: the trilogy *Mourning Becomes Electra* (1931), in which he adapted the theme of Aeschylus' *Oresteia* to portray the aftermath of the American Civil War; *The Iceman Cometh* (1946), a tragedy about a collection of bar-room derelicts; and *Long Day's Journey into Night* (performed and published posthumously in 1956), a semi-autobiographical tragedy portraying mutually destructive family relationships. He was awarded the Nobel Prize for literature in 1936.

oneiric /ə'naɪərɪk/ *adj.* of or relating to dreams or dreaming. [Gk *oneiros* dream]

oneiro- /ə'naɪərəʊ/ *comb. form* dream. [Gk *oneiros* dream]

oneiromancy /ə'naɪərəˌmænsɪ/ *n.* the interpretation of dreams.

oneness /'wʌnnɪs/ *n.* **1** the fact or state of being one; singleness. **2** uniqueness. **3** agreement; unity of opinion. **4** identity, sameness.

oner /'wʌnə(r)/ *n. Brit. sl.* **1** one pound (of money). **2** a remarkable person or thing.

onerous /'ɒnərəs, 'əʊn-/ *adj.* **1** burdensome; causing or requiring trouble. **2** *Law* involving heavy obligations. □ **onerously** *adv.* **onerousness** *n.* [ME f. OF *onereus* f. L *onerosus* f. *onus oneris* burden]

oneself /wʌn'sɛlf/ *pron.* the reflexive and (in apposition) emphatic form of *one* (*kill oneself; one has to do it oneself*).

onestep /'wʌnstɛp/ *n.* a vigorous kind of foxtrot in duple time.

onflow /'ɒnfləʊ/ *n.* an onward flow.

onglaze /'ɒnɡleɪz/ *adj.* (of painting etc.) done on a glazed surface.

ongoing /'ɒnˌɡəʊɪŋ/ *adj.* **1** continuing to exist or be operative etc. **2** that is or are in progress (*ongoing discussions*). □ **ongoingness** *n.*

onion /'ʌnjən/ *n.* **1** an alliaceous plant, *Allium cepa*, having a short stem and bearing greenish-white flowers. **2** the pungent swollen bulb of this, with many concentric skins, used in cooking, pickling, etc. □ **know one's onions** be fully knowledgeable or experienced. **onion dome** a dome which bulges in the middle and rises to a point, used esp. in Russian church architecture. **onion-skin 1** the brown outermost skin or any outer skin of an onion. **2** thin smooth translucent paper. □ **oniony** *adj.* [ME f. AF *union*, OF *oignon* ult. f. L *unio -onis*]

onkus /'ɒŋkəs/ *adj. Austral. colloq.* **1** unpleasant. **2** disorganized. [20th c.: orig. unkn.]

onlooker /'ɒn,lʊkə(r)/ *n.* a non-participating observer; a spectator. □ **onlooking** *adj.*

only /'əʊnlɪ/ *adv., adj., & conj.* ● *adv.* **1** solely, merely, exclusively; and no one or nothing more besides (*I only want to sit down; will only make matters worse; needed six only; is only a child*). **2** no longer ago than (*saw them only yesterday*). **3** not until (*arrives only on Tuesday*). **4** with no better result than (*hurried home only to find her gone*). ¶ In informal English *only* is usually placed between the subject and verb regardless of what it refers to (e.g. *I only want to talk to you*); in more formal English it is often placed more exactly, esp. to avoid ambiguity (e.g. *I want to talk only to you*). In speech, intonation usually serves to clarify the sense. ● *attrib.adj.* **1** existing alone or their kind (*their only son*). **2** best or alone worth knowing (*the only place to eat*). ● *conj. colloq.* **1** except that; but for the fact that (*I would go, only I feel ill*). **2** but then (as an extra consideration) (*he always makes promises, only he never keeps them*). □ **only-begotten** *literary* begotten as the only child. **only too** extremely (*is only too willing*). [OE *ānlic*, *ǣnlic*, ME *onliche* (as ONE, -LY²)]

o.n.o. *abbr. Brit.* or near offer.

onomastic /,ɒnə'mæstɪk/ *adj.* relating to names or nomenclature. [Gk *onomastikos* f. *onoma* name]

onomastics /,ɒnə'mæstɪks/ *n.pl.* (treated as *sing.*) the study of the origin and formation of (esp. personal) proper names.

onomatopoeia /,ɒnə,mætə'piːə/ *n.* **1** the formation of a word from a sound associated with what is named (e.g. *cuckoo*, *sizzle*). **2** the use of such words. □ **onomatopoeic** *adj.* **onomatopoeically** *adv.* [LL f. Gk *onomatopoiia* word-making f. *onoma -matos* name + *poieō* make]

Onondaga /,ɒnən'dɑːgə/ *n. & adj.* ● *n.* (pl. **Onondagas** or same) **1** a member of an Iroquois people, one of the five comprising the original Iroquois confederacy, formerly inhabiting an area near Syracuse, New York. **2** the language of this people. ● *adj.* of or relating to the Onondaga or their language. [Onondaga]

onrush /'ɒnrʌʃ/ *n.* an onward rush.

onscreen *adj. & adv.* ● *attrib.adj.* /'ɒnskriːn/ appearing on a cinema, television, or VDU screen. ● *adv.* /ɒn'skriːn/ **1** on or by means of a screen. **2** within the view presented by a cinema-film scene.

onset /'ɒnset/ *n.* **1** an attack. **2** a beginning, esp. an energetic or determined one.

onshore *adj. & adv.* ● *adj.* /'ɒnʃɔː(r)/ **1** on the shore. **2** (of the wind) blowing from the sea towards the land. ● *adv.* /ɒn'ʃɔː(r)/ on or towards the land.

onside /ɒn'saɪd/ *adj. & adv.* (of a player in a field game) in a lawful position; not offside.

onslaught /'ɒnslɔːt/ *n.* a fierce attack. [earlier *anslaight* f. MDu. *aenslag* f. *aen* on + *slag* blow, with assim. to obs. *slaught* slaughter]

Ont. *abbr.* Ontario.

-ont /ɒnt/ *comb. form Biol.* denoting an individual of a specified type (*symbiont*). [Gk *ōn ont-* being]

Ontario /ɒn'teərɪ,əʊ/ a province of eastern Canada, between Hudson Bay and the Great Lakes; pop. (1991) 9,914,200; capital, Toronto. It was settled by the French and English in the 17th century, ceded to Britain in 1763, and became one of the original four provinces in the Dominion of Canada in 1867. □ **Ontarian** *adj. & n.*

Ontario, Lake the smallest and most easterly of the Great Lakes, lying on the US–Canadian border between Ontario and New York State. It is linked to Lake Erie in the south by the Niagara River and to the Atlantic Ocean by the St Lawrence Seaway.

onto /'ɒntu:, -tə/ *prep. disp.* to a position or state on or in contact with (cf. *on to*). ¶ The form *onto* is still not fully accepted in the way that *into* is, although it is in wide use. It is, however, useful in distinguishing sense as between *we drove on to the beach* (i.e. in that direction) and *we drove onto the beach* (i.e. in contact with it).

ontogenesis /,ɒntə'dʒenɪsɪs/ *n.* the origin and development of an individual (cf. PHYLOGENESIS). □ **ontogenetic** /-dʒɪ'netɪk/ *adj.* **ontogenetically** *adv.* [formed as ONTOGENY + Gk *genesis* birth]

ontogeny /ɒn'tɒdʒənɪ/ *n.* = ONTOGENESIS. □ **ontogenic** /,ɒntə'dʒenɪk/ *adj.* **ontogenically** *adv.* [Gk *ōn ont-* being, pres. part. of *eimi* be + -GENY]

ontology /ɒn'tɒlədʒɪ/ *n.* the branch of metaphysics dealing with the nature of being. □ **ontologist** *n.* **ontological** /,ɒntə'lɒdʒɪk(ə)l/ *adj.* **ontologically** *adv.* [mod.L *ontologia* f. Gk *ōn ont-* being + -LOGY]

onus /'əʊnəs/ *n.* (pl. **onuses**) a burden, duty, or responsibility. [L]

onward /'ɒnwəd/ *adv. & adj.* ● *adv.* (also **onwards** /-wədz/) **1** further on. **2** towards the front. **3** with advancing motion. **4** into the future. ● *adj.* directed onwards.

onychophoran /,ɒnɪ'kɒfərən/ *n. & adj. Zool.* ● *n.* a soft-bodied arthropod of the class Onychophora, with a long body and stubby legs, sometimes regarded as intermediate between annelids and arthropods. ● *adj.* of or relating to this class. [mod.L f. Gk *onux onukh-* nail, claw + *-phoros* bearing]

onyx /'ɒnɪks/ *n.* a semiprecious variety of agate with different colours in layers. □ **onyx marble** banded calcite etc. used as a decorative material. [ME f. OF *oniche*, *onix* f. L f. Gk *onux* fingernail, onyx]

oo- /'əʊə/ *comb. form* (US **oö-**) *Biol.* egg, ovum. [Gk *ōion* egg]

oocyte /'əʊə,saɪt/ *n. Biol.* an immature ovum in an ovary.

oodles /'uːd(ə)lz/ *n.pl. colloq.* a very great amount. [19th-c. US: orig. unkn.]

oof /uːf/ *n. sl.* money, cash. [Yiddish *ooftisch*, G *auf dem Tische* on the table (of money in gambling)]

oofy /'uːfɪ/ *adj. sl.* rich, wealthy. □ **oofiness** *n.*

oogamous /əʊ'ɒgəməs/ *adj. Biol.* reproducing by the union of mobile male and immobile female cells. □ **oogamy** *n.*

oogenesis /,əʊə'dʒenɪsɪs/ *n. Biol.* the production or development of an ovum.

ooh /uː/ *int.* expressing surprise, delight, pain, etc. [natural exclam.]

oolite /'əʊə,laɪt/ *n.* **1** a sedimentary rock, usu. limestone, consisting of rounded grains made up of concentric layers. **2** = OOLITH. □ **oolitic** /,əʊə'lɪtɪk/ *adj.* [F *oölithe* (as OO-, -LITE)]

oolith /'əʊəlɪθ/ *n.* any of the rounded grains making up oolite.

oology /əʊ'ɒlədʒɪ/ *n.* the study or collecting of birds' eggs. □ **oologist** *n.* **oological** /,əʊə'lɒdʒɪk(ə)l/ *adj.*

oolong /'uːlɒŋ/ *n.* a dark kind of cured China tea. [Chin. *wulong* black dragon]

oomiak var. of UMIAK.

oompah /'ʊmpɑː/ *n. colloq.* the rhythmical sound of deep-toned brass instruments in a band. [imit.]

oomph /ʊmf/ *n. sl.* **1** energy, enthusiasm. **2** attractiveness, esp. sexual appeal. [20th c.: orig. uncert.]

-oon /uːn/ *suffix* forming nouns, orig. from French words in stressed -*on* (*balloon; buffoon*). ¶ Replaced by -*on* in recent borrowings and those with unstressed -*on* (*baron*). [L -*o -onis*, sometimes via It. -*one*]

oophorectomy /,əʊəfə'rektəmɪ/ *n.* (pl. **-ies**) *Med.* the surgical removal of one or both ovaries. [mod.L *oophoron* ovary (f. Gk *ōophoros* egg-bearing) + -ECTOMY]

oops /uːps, ʊps/ *int. colloq.* expressing surprise or apology, esp. on making an obvious mistake. [natural exclam.]

Oort /ɔːt/, Jan Hendrik (1900–92), Dutch astronomer. His early measurements of the proper motion of stars enabled him to prove that the Galaxy is rotating, and to determine the position and orbital period of the sun within it. Oort was director of the observatory at Leiden for twenty-five years, during which time he was involved in the discovery of the wavelength of radio emission from interstellar hydrogen, and noted the strong polarization of light from the Crab Nebula. He also proposed the existence of a cloud of incipient comets beyond the orbit of Pluto, now named after him.

Oort cloud *n. Astron.* a spherical cloud of small rocky and icy bodies postulated to orbit the sun beyond the orbit of Pluto and up to 1.5 light years from the sun, and to act as a reservoir of comets. It is thought likely that billions of these objects exist, though only a very few have so far been visually confirmed. [OORT]

Oostende see OSTEND.

ooze¹ /uːz/ *v. & n.* ● *v.* **1** *intr.* (of fluid) pass slowly through the pores of a body. **2** *intr.* trickle or leak slowly out. **3** *intr.* (of a substance) exude moisture. **4** *tr.* exude or exhibit (a feeling) liberally (*oozed sympathy*). ● *n.* **1** a sluggish flow or exudation. **2** an infusion of oak-bark or other vegetable matter, used in tanning. □ **oozy** *adj.* [orig. as noun (sense 2), f. OE *wōs* juice, sap]

ooze² /uːz/ *n.* **1** a deposit of wet mud or slime, esp. at the bottom of a river, lake, or estuary. **2** a bog or marsh; soft muddy ground. □ **oozy** *adj.* [OE *wāse*]

OP *abbr.* **1** RC Ch. *Ordo Praedicatorum*, Order of Preachers (Dominican). **2** observation post. **3** opposite prompt.

op /ɒp/ *n. colloq.* operation (in surgical and military senses).

op. /ɒp/ *abbr.* **1** *Mus.* opus. **2** operator.

op- /ɒp, əp/ *prefix* assim. form of OB- before *p*.

o.p. *abbr.* **1** out of print. **2** overproof.

opacify /əʊˈpæsɪˌfaɪ/ *v.tr. & intr.* (**-ies, -ied**) make or become opaque. □ **opacifier** *n*.

opacity /əʊˈpæsɪtɪ/ *n*. **1** the state of being opaque. **2** obscurity of meaning. **3** obtuseness of understanding. [F *opacité* f. L *opacitas -tatis* (as OPAQUE)]

opah /ˈəʊpə/ *n*. a large rare deep-sea fish, *Lampris guttatus*, usu. having a silver-blue back with white spots and crimson fins. Also called *kingfish*, *moonfish*. [W. Afr. name]

opal /ˈəʊp(ə)l/ *n*. a quartzlike form of hydrated silica, usu. white or colourless and sometimes showing changing colours, often used as a gemstone. □ **opal glass** a semi-translucent white glass. [F *opale* or L *opalus* prob. ult. f. Skr. *upalas* precious stone]

opalescent /ˌəʊpəˈles(ə)nt/ *adj.* showing changing colours like an opal. □ **opalesce** *v.intr.* **opalescence** *n*.

opaline /ˈəʊpəˌlaɪn/ *adj. & n.* ● *adj.* opal-like, opalescent, iridescent. ● *n.* opal glass.

opaque /əʊˈpeɪk/ *adj. & n.* ● *adj.* (**opaquer, opaquest**) **1** not transmitting light. **2** impenetrable to sight. **3** obscure; not lucid. **4** obtuse, dull-witted. ● *n.* **1** an opaque thing or substance. **2** a substance for producing opaque areas on negatives. □ **opaquely** *adv.* **opaqueness** *n*. [ME *opak* f. L *opacus*: spelling now assim. to F]

op art /ɒp/ *n*. (in full **optical art**) a form of abstract art developed in the 1960s, in which optical effects are used to provide illusions of movement in the patterns produced, or designs in which conflicting patterns emerge and overlap. Bridget Riley and Victor Vasarely are its most famous exponents. [abbr. *optical art*, after POP ART]

op. cit. *abbr.* in the work already quoted. [L *opere citato*]

OPEC see ORGANIZATION OF THE PETROLEUM EXPORTING COUNTRIES.

Opel /ˈəʊp(ə)l/, Wilhelm von (1871–1948), German motor manufacturer. In 1898 he and his brothers converted their grandfather's bicycle and sewing-machine factory to car production, launching their first original model in 1902. After the First World War the company became the first in Germany to introduce assembly-line production and manufactured more than 1 million cars. Opel sold control of the company to the US manufacturer General Motors in 1929.

open /ˈəʊp(ə)n/ *adj., v., & n.* ● *adj.* **1** not closed or locked or blocked up; allowing entrance or passage or access. **2 a** (of a room, field, or other area) having its door or gate in a position allowing access, or part of its confining boundary removed. **b** (of a container) not fastened or sealed; in a position or with the lid etc. in a position allowing access to the inside part. **3** unenclosed, unconfined, unobstructed (*the open road*; *open views*). **4 a** uncovered, bare, exposed (*open drain*; *open wound*). **b** *Sport* (of a goal mouth or other object of attack) unprotected, vulnerable. **5** undisguised, public, manifest; not exclusive or limited (*open scandal*; *open hostilities*). **6** expanded, unfolded, or spread out (*had the map open on the table*). **7** (of a fabric) not close; with gaps or intervals. **8 a** (of a person) frank and communicative. **b** (of the mind) accessible to new ideas; unprejudiced or undecided. **c** generous. **9 a** (of an exhibition, shop, etc.) accessible to visitors or customers; ready for business. **b** (of a meeting) admitting all, not restricted to members etc. **10 a** (of a race, competition, scholarship, etc.) unrestricted as to who may compete. **b** (of a champion, scholar, etc.) having won such a contest. **11** (of government) conducted in an informative manner receptive to enquiry, criticism, etc., from the public. **12** (foll. by *to*) **a** willing to receive (*is open to offers*). **b** (of a choice, offer, or opportunity) still available (*there are three courses open to us*). **c** likely to suffer from or be affected by (*open to abuse*). **13 a** (of the mouth) with lips apart, esp. in surprise or incomprehension. **b** (of the ears or eyes) eagerly attentive. **14** *Mus.* **a** (of a string) allowed to vibrate along its whole length. **b** (of a pipe) unstopped at each end. **c** (of a note) sounded from an open string or pipe. **15** (of an electrical circuit) having a break in the conducting path. **16** (of the bowels) not constipated. **17** (of a return ticket) not restricted as to day of travel. **18** (of a cheque) not crossed. **19** (of a boat) without a deck. **20** (of a river or harbour) free of ice. **21** (of the weather or winter) free of frost. **22** *Phonet.* **a** (of a vowel) produced with a relatively wide opening of the mouth. **b** (of a syllable) ending in a vowel. **23** (of a town, city, etc.) not defended even if attacked. ● *v.* **1** *tr. & intr.* make or become open or more open. **2 a** *tr.* change from a closed or fastened position so as to allow access (*opened the door*; *opened the box*). **b** *intr.* (of a door, lid, etc.) have its position changed to allow access (*the door opened slowly*). **3** *tr.* remove the sealing or fastening element of (a container) to get access to the contents

(*opened the envelope*). **4** *intr.* (foll. by *into*, *on to*, etc.) (of a door, room, etc.) afford access as specified (*opened on to a large garden*). **5 a** *tr.* start or establish or set going (a business, activity, etc.). **b** *intr.* be initiated; make a start (*the session opens tomorrow*; *the story opens with a murder*). **c** *tr.* (of a counsel in a lawcourt) make a preliminary statement in (a case) before calling witnesses. **6** *tr.* spread out or unfold (a map, newspaper, etc.). **7** *intr.* (often foll. by *with*) (of a person) begin speaking, writing, etc. (*he opened with a warning*). **8** *intr.* (of a prospect) come into view; be revealed. **9** *tr.* **a** reveal or communicate (one's feelings, intentions, etc.). **b** make available, provide. **10** *tr.* make (one's mind, heart, etc.) more sympathetic or enlightened. **11** *tr.* ceremonially declare (a building etc.) to be completed and in use. **12** *tr.* break up (ground) with a plough etc. **13** *tr.* cause evacuation of (the bowels). **14** *Naut.* **a** *tr.* get a view of by change of position. **b** *intr.* come into full view. ● *n.* **1** (prec. by *the*) **a** open space or country or air. **b** public notice or view; general attention (esp. *into the open*). **2** an open championship, competition, or scholarship (*the French Open*). □ **be open with** speak frankly to. **keep open house** provide general hospitality. **open air** (usu. prec. by *the*) a free or unenclosed space outdoors. **open-air** (*attrib.*) out of doors. **open-and-shut** (of an argument, case, etc.) straightforward and conclusive. **open-armed** cordial; warmly receptive. **open college** a college offering training and vocational courses mainly by correspondence. **open book** a person who is easily understood. **open day** a day when the public may visit a place normally closed to them. **open door** free admission of foreign trade and immigrants. **open-door** *adj.* open, accessible, public. **open the door to** see DOOR. **open-ended** having no predetermined limit or boundary. **open a person's eyes** see EYE. **open-eyed 1** with the eyes open. **2** alert, watchful. **open-faced** having a frank or ingenuous expression. **open-handed** generous. **open-handedly** generously. **open-handedness** generosity. **open-hearted** frank and kindly. **open-heartedness** an open-hearted quality. **open-heart surgery** surgery with the heart exposed and the blood made to bypass it. **open house** welcome or hospitality for all visitors. **open ice** ice through which navigation is possible. **open letter** a letter, esp. of protest, addressed to an individual and published in a newspaper or journal. **open market** an unrestricted market with free competition of buyers and sellers. **open-minded** accessible to new ideas; unprejudiced. **open-mindedly** in an open-minded manner. **open-mindedness** the quality of being open-minded. **open-mouthed** *adv. & adj.* with the mouth open, esp. in surprise. **open out 1** unfold; spread out. **2** develop, expand. **3** become communicative. **4** accelerate. **open-plan** (usu. *attrib.*) (of a house, office, etc.) having large undivided rooms. **open prison** a prison with the minimum of physical restraints on prisoners. **open question** a matter on which differences of opinion are legitimate. **open-reel** (of a tape recorder) having reels of tape requiring individual threading, as distinct from a cassette. **open sandwich** a sandwich without a top slice of bread. **open sea** an expanse of sea away from land. **open season** the season when restrictions on the killing of game etc. are lifted. **2** a time when something is unrestricted. **open secret** a supposed secret that is known to many people. **open sesame** see SESAME. **open shop 1** a business etc. where employees do not have to be members of a trade union (opp. *closed shop*). **2** this system. **open society** a society with wide dissemination of information and freedom of belief. **open up 1** unlock (premises). **2** make accessible. **3** reveal; bring to notice. **4** become less reticent; talk or speak openly. **5** accelerate esp. a motor vehicle. **6** begin shooting or sounding. **open verdict** a verdict affirming that a crime has been committed but not specifying the criminal or (in case of violent death) the cause. **with open arms** see ARM[1]. □ **openable** *adj.* **openness** /ˈəʊp(ə)nnɪs/ *n*. [OE *open*]

Open Brethren *n.pl.* one of the two principal divisions of the Plymouth Brethren (the other is the Exclusive Brethren), formed in 1849 as a result of doctrinal and other differences; collectively, the members of this division. The Open Brethren are less rigorous and less exclusive in matters such as conditions for membership and contact with outsiders than the Exclusive Brethren.

opencast /ˈəʊp(ə)nˌkɑːst/ *adj. Brit.* (of a mine or mining) with removal of the surface layers and working from above, not from shafts.

opener /ˈəʊp(ə)nə(r)/ *n*. **1** a device for opening tins, bottles, etc. **2** *colloq.* the first item on a programme etc. **3** *Cricket* an opening batsman. **4** *Cards* the player who opens the betting or bidding. □ **for openers** *colloq.* to start with.

open-field system /ˌəʊp(ə)nˈfiːld/ *n*. the traditional medieval system of farming in England (see ENCLOSURE).

open-hearth process /ˌəʊp(ə)nˈhɑːθ/ *n*. a steel-making process in which scrap iron or steel, limestone, and molten or cold pig-iron are melted together in a shallow reverberatory furnace. The mixture is heated from above using gaseous fuel and air which oxidizes impurities in the iron. Developed by Sir Charles Siemens and others in the mid-19th century, the process produced most of the world's steel until the 1950s. Although it was not as rapid as the Bessemer process, it enabled closer control of the finished product. (See also STEEL.)

opening /'əʊp(ə)nɪŋ/ *n. & adj.* ● *n.* **1** an aperture or gap, esp. allowing access. **2** a favourable situation or opportunity. **3** a beginning; an initial part. **4** *Chess* a recognized sequence of moves at the beginning of a game. **5** a counsel's preliminary statement of a case in a lawcourt. ● *adj.* initial, first. □ **opening-time** *Brit.* the time at which public houses may legally open for custom.

openly /'əʊp(ə)nlɪ/ *adv.* **1** frankly, honestly. **2** publicly; without concealment. [OE *openlīce* (as OPEN, -LY²)]

Open University (in the UK) a university, opened in 1971, providing instruction by a combination of television, radio, and correspondence courses and by audiovisual centres and summer schools. There are no formal academic requirements for entry to courses leading to first degrees. The headquarters are at Milton Keynes.

openwork /'əʊp(ə)nˌwɜːk/ *n.* a pattern with intervening spaces in metal, leather, lace, etc.

opera¹ /'ɒpərə, 'ɒprə/ *n.* **1 a** a dramatic work in one or more acts, set to music for singers (usu. in costume) and instrumentalists. **b** this as a genre. (*See note below.*) **2** a building for the performance of opera. □ **opera-glasses** small binoculars for use at the opera or theatre. **opera-hat** a man's tall collapsible hat. **opera-house** a theatre for the performance of opera. [It. f. L, = labour, work]

▪ The concept of a continuously sung dramatic work originated in Florence in the late 16th century; Claudio Monteverdi's *Orfeo* (1607) is generally considered the first great extant opera. The first public opera house was opened in Venice in 1637, and opera quickly spread through Italy and into France and Germany. During the 17th century melodic arias increasingly supplemented simple declaimed recitative; in the 18th century the primacy of dramatic action was reasserted by Gluck and musical characterization was developed by Mozart, both areas later being explored by the 19th-century composers Verdi and Wagner. Wagner dispensed with aria and recitative in favour of a seamless web of music in a series of scenes; in the 20th century Stravinsky's *The Rake's Progress* returned to a structure based on aria and recitative. Opera has flourished in the 20th century, the melodies of Puccini being perhaps especially popular (with Britten providing a significant contribution to the form in English), while from the light opera or operetta of the late 19th century developed the new genres of musical comedy and the musical.

opera² *pl.* of OPUS.

operable /'ɒpərəb(ə)l/ *adj.* **1** that can be operated. **2** suitable for treatment by surgical operation. □ **operability** /ˌɒpərə'bɪlɪtɪ/ *n.* [LL *operabilis* f. L (as OPERATE)]

opera buffa /'buːfə/ *n.* (esp. Italian) comic opera, esp. with characters drawn from everyday life. [It.]

opéra comique /ˌɒpeɪˌrɑː kɒ'miːk/ *n.* (esp. French) opera on a light-hearted theme, with spoken dialogue. [F]

operand /'ɒpəˌrænd/ *n. Math.* the quantity etc. on which an operation is to be done. [L *operandum* neut. gerundive of *operari*: see OPERATE]

opera seria /'sɪərɪə/ *n.* (esp. 18th-century Italian) opera on a serious, usu. classical or mythological theme. [It.]

operate /'ɒpəˌreɪt/ *v.* **1** *tr.* manage, work, control; put or keep in a functional state. **2** *intr.* be in action; function. **3** *intr.* produce an effect; exercise influence (*the tax operates to our disadvantage*). **4** *intr.* (often foll. by *on*) **a** perform a surgical operation. **b** conduct a military or naval action. **c** be active in business etc., esp. dealing in stocks and shares. **5** *intr.* (foll. by *on*) influence or affect (feelings etc.). **6** *tr.* bring about; accomplish. □ **operating system** the basic software that enables the running of a computer system. **operating table** a table on which the patient is placed during an operation. **operating theatre** (or **room**) a room for surgical operations. [L *operari* to work f. *opus operis* work]

operatic /ˌɒpə'rætɪk/ *adj.* **1** of or relating to opera. **2** resembling or characteristic of opera. □ **operatically** *adv.* [irreg. f. OPERA¹, after *dramatic*]

operatics /ˌɒpə'rætɪks/ *n.pl.* the production and performance of operas.

operation /ˌɒpə'reɪʃ(ə)n/ *n.* **1 a** the action or process or method of working or operating. **b** the state of being active or functioning (*not yet in operation*). **c** the scope or range of effectiveness of a thing's activity. **2** an active process; a discharge of a function (*the operation of breathing*). **3** a piece of work, esp. one in a series (often in *pl.: begin operations*). **4** an act of surgery performed on a patient. **5 a** a strategic movement of troops, ships, etc. for military action. **b** preceding a code-name (*Operation Overlord*). **6** a financial transaction. **7** *Math.* the subjection of a number or quantity or function to a process affecting its value or form, e.g. multiplication, differentiation. □ **operations research** = *operational research*. [ME f. OF f. L *operatio -onis* (as OPERATE)]

operational /ˌɒpə'reɪʃən(ə)l/ *adj.* **1 a** of or used for operations. **b** engaged or involved in operations. **2** able or ready to function. □ **operational research** the application of scientific principles to business management, providing a quantitative basis for complex decisions. □ **operationally** *adv.*

operative /'ɒpərətɪv/ *adj. & n.* ● *adj.* **1** in operation; having effect. **2** having the principal relevance ('*may*' *is the operative word*). **3** of or by surgery. **4** *Law* expressing an intent to perform a transaction. ● *n.* **1** a worker, esp. a skilled one. **2** *US* a private detective or secret agent. □ **operatively** *adv.* **operativeness** *n.* [LL *operativus* f. L (as OPERATE)]

operator /'ɒpəˌreɪtə(r)/ *n.* **1** a person operating a machine etc., esp. making connections of lines in a telephone exchange. **2** a person operating or engaging in business. **3** *colloq.* a person acting in a specified way (*a smooth operator*). **4** *Math.* a symbol or function denoting an operation (e.g. ×, +). [LL f. L *operari* (as OPERATE)]

operculum /əʊ'pɜːkjʊləm/ *n.* (*pl.* **opercula** /-lə/) **1** *Zool.* **a** a flaplike structure covering the gills in a fish. **b** a platelike structure closing the aperture of a gastropod mollusc's shell when the organism is retracted. **c** any other part covering or closing an aperture, such as a flap over the nostrils in some birds. **2** *Bot.* a lidlike structure of the spore-containing capsule of mosses. □ **opercular** *adj.* **operculate** /-lət/ *adj.* **operculi-** /-lɪ/ *comb. form*. [L f. *operire* cover]

operetta /ˌɒpə'retə/ *n.* a small-scale opera, usually on a light or humorous theme and originally consisting of only one act. Operettas range in complexity from simple musical plays to works that differ from operas chiefly in their use of spoken dialogue rather than recitative (melodic speech); they are similar also to the musical comedy of the early 20th century and the modern musical, the difference largely consisting in period style and the precise amount and weight of music involved. Among the notable composers of operettas are Offenbach, Johan Strauss, Franz Lehár, and Gilbert and Sullivan. [It., dimin. of *opera*: see OPERA¹]

operon /'ɒpərɒn/ *n. Biol.* a unit of linked genes which is believed to regulate other genes responsible for protein synthesis. [F *opéron* f. *opérer* effect, work]

ophicleide /'ɒfɪˌklaɪd/ *n.* **1** an obsolete usu. bass brass wind instrument developed from the serpent. **2** a powerful organ reed-stop. [F *ophicléide* f. Gk *ophis* serpent + *kleis kleidos* key]

ophidian /əʊ'fɪdɪən/ *n. & adj. Zool.* ● *n.* a reptile of the suborder Serpentes (formerly Ophidia), comprising snakes. ● *adj.* **1** of or relating to this suborder. **2** snakelike. [mod.L *Ophidia* f. Gk *ophis* snake]

ophio- /'ɒfɪəʊ/ *comb. form* snake. [Gk *ophis* snake]

Ophir /'əʊfə(r)/ (in the Bible) an unidentified region, perhaps in SE Arabia, famous for its fine gold and precious stones.

Ophiuchus /ɒ'fjuːkəs/ *Astron.* a large constellation (the Serpent Bearer or Holder), said to represent a man (probably identified with Aesculapius, god of medicine) in the coils of a snake. The ecliptic passes through it, but it is not counted among the signs of the zodiac. [L f. Gk *Ophioukhos*, f. *ophis* serpent + *ekhein* hold]

ophthalmia /ɒf'θælmɪə/ *n. Med.* inflammation of the eye, esp. conjunctivitis. [LL f. Gk f. *ophthalmos* eye]

ophthalmic /ɒf'θælmɪk/ *adj.* of or relating to the eye and its diseases. □ **ophthalmic optician** see OPTICIAN 3. [L *ophthalmicus* f. Gk *ophthalmikos* (as OPHTHALMIA)]

ophthalmo- /ɒf'θælməʊ/ *comb. form Optics* denoting the eye. [Gk *ophthalmos* eye]

ophthalmology /ˌɒfθæl'mɒlədʒɪ/ *n.* the scientific study of the eye. □ **ophthalmologist** *n.* **ophthalmological** /-mə'lɒdʒɪk(ə)l/ *adj.*

ophthalmoscope /ɒf'θælməˌskəʊp/ *n.* an instrument for inspecting the retina and other parts of the eye. □ **ophthalmoscopic** /-ˌθælmə'skɒpɪk/ *adj.*

-opia /'əʊpɪə/ *comb. form* denoting a visual disorder (*myopia*). [Gk f. *ōps* eye]

opiate _adj., n.,_ & _v._ ● _adj._ /ˈəʊpɪət/ **1** containing, derived from, or resembling opium. **2** narcotic, soporific. ● _n._ /ˈəʊpɪət/ **1** a drug containing, derived from, or resembling opium, usu. to ease pain or induce sleep. **2** a thing which soothes or stupefies. ● _v.tr._ /ˈəʊpɪˌeɪt/ **1** mix with opium. **2** stupefy. [med.L _opiatus, -um, opiare_ f. L _opium:_ see OPIUM]

Opie /ˈəʊpɪ/, John (1761–1807), English painter. His work includes portraits of contemporary figures such as Mary Wollstonecraft and history paintings such as _The Murder of Rizzio_ (1787).

opine /əʊˈpaɪn/ _v.tr._ (often foll. by _that_ + clause) hold or express as an opinion. [L _opinari_ think, believe]

opinion /əˈpɪnjən/ _n._ **1** a belief or assessment based on grounds short of proof. **2** a view held as probable. **3** (often foll. by _on_) what one thinks about a particular topic or question (_my opinion on capital punishment_). **4 a** a formal statement of professional advice (_will get a second opinion_). **b** _Law_ a formal statement of reasons for a judgement given. **5** an estimation (_had a low opinion of it_). □ **be of the opinion that** believe or maintain that. **in one's opinion** according to one's view or belief. **a matter of opinion** a disputable point. **opinion poll** = GALLUP POLL. **public opinion** views generally prevalent, esp. on moral questions. [ME f. OF f. L _opinio -onis_ (as OPINE)]

opinionated /əˈpɪnjəˌneɪtɪd/ _adj._ conceitedly assertive or dogmatic in one's opinions. [obs. _opinionate_ in the same sense f. OPINION]

opioid /ˈəʊpɪˌɔɪd/ _n._ & _adj. Pharm._ & _Biochem._ ● _n._ any compound resembling cocaine and morphine in its addictive properties or physiological effects. ● _adj._ of or relating to such a compound. [OPIUM + -OID]

opium /ˈəʊpɪəm/ _n._ **1** a reddish-brown heavy-scented addictive drug prepared from the juice of the opium poppy, used in medicine as an analgesic and narcotic. **2** anything regarded as soothing or stupefying. □ **opium den** a haunt of opium-smokers. **opium poppy** a poppy, _Papaver somniferum,_ native to Europe and eastern Asia, with white, red, pink, or purple flowers. [ME f. L f. Gk _opion_ poppy juice f. _opos_ juice]

Opium War either of two wars, that between Britain and China (1839–42) and that involving Britain and France against China (1856–60), following China's attempt to prohibit the (illegal) importation of opium from British India into China, and Chinese restrictions on foreign trade. Defeat of the Chinese resulted in the ceding of Hong Kong to Britain and the opening of five 'treaty ports' to traders.

opopanax /əʊˈpɒpəˌnæks/ _n._ **1 a** an umbelliferous European plant, _Opopanax chironium,_ with yellow flowers. **b** a fetid gum resin obtained from the roots of this plant. **2 a** a gum resin obtained from the tree _Commiphora kataf._ **b** a perfume made from this. **3** = _sponge tree._ [ME f. L f. Gk f. _opos_ juice + _panax_ formed as PANACEA]

Oporto /əʊˈpɔːtuː/ (Portuguese **Porto** /ˈpɔrtu/) the principal city and port of northern Portugal, near the mouth of the River Douro, famous for port wine; pop. (1991) 310,640.

opossum /əˈpɒsəm/ _n._ **1** a mainly tree-living marsupial of the family Didelphidae, native to America, having a hairless prehensile tail and hind feet with an opposable thumb; esp. _Didelphis virginiana,_ of North America. **2** _Austral._ = POSSUM 2. [Virginia Algonquian _opassom,_ f. _op_ white + _assom_ dog]

opp. _abbr._ opposite.

Oppenheimer /ˈɒp(ə)nˌhaɪmə(r)/, Julius Robert (1904–67), American theoretical physicist. He showed in 1928 that the positron should exist. Fourteen years later he was appointed director of the laboratory at Los Alamos which designed and built the first atom bomb. After the Second World War he opposed development of the hydrogen bomb and — like many intellectuals of his day — was investigated for alleged un-American activities (see McCARTHYISM). His security clearance was withdrawn in 1953 and his advisory activities stopped; with the passing of the McCarthy era his public standing was restored.

oppo /ˈɒpəʊ/ _n._ (pl. **-os**) _Brit. colloq._ a colleague or friend. [abbr. of _opposite number_]

opponent /əˈpəʊnənt/ _n._ & _adj._ ● _n._ a person who opposes or belongs to an opposing side. ● _adj._ opposing, contrary, opposed. □ **opponent muscle** a muscle enabling the thumb to be placed front to front against a finger of the same hand. □ **opponency** _n._ [L _opponere opponent-_ (as OB-, _ponere_ place)]

opportune /ˈɒpəˌtjuːn/ _adj._ **1** (of a time) well-chosen or especially favourable or appropriate (_an opportune moment_). **2** (of an action or event) well-timed; done or occurring at a favourable or useful time. □ **opportunely** _adv._ **opportuneness** _n._ [ME f. OF _opportun -une_ f. L

opportunus (as OB-, _portus_ harbour), orig. of the wind driving towards the harbour]

opportunism /ˌɒpəˈtjuːnɪz(ə)m/ _n._ **1** the adaptation of policy or judgement to circumstances or opportunity, esp. regardless of principle. **2** the seizing of opportunities when they occur. □ **opportunist** _n._ & _adj._ [OPPORTUNE after It. _opportunismo_ and F _opportunisme_ in political senses]

opportunistic /ˌɒpətjʊˈnɪstɪk/ _adj._ **1** of or relating to an opportunist. **2** _Ecol._ (of a species) able to spread quickly in unexploited or newly formed habitats. **3** _Med._ **a** (of a micro-organism) causing disease only in unusual circumstances, esp. in patients with depressed immune systems. **b** (of an infection) caused by such a micro-organism. □ **opportunistically** _adv._

opportunity /ˌɒpəˈtjuːnɪtɪ/ _n._ (pl. **-ies**) **1** a good chance; a favourable occasion. **2** a chance or opening offered by circumstances. **3** good fortune. □ **opportunity knocks** an opportunity occurs. [ME f. OF _opportunité_ f. L _opportunitas -tatis_ (as OPPORTUNE)]

opposable /əˈpəʊzəb(ə)l/ _adj._ **1** able to be opposed. **2** _Zool._ (of the thumb in primates) capable of facing and touching the other digits on the same hand.

oppose /əˈpəʊz/ _v.tr._ (often _absol._) **1** set oneself against; resist, argue against. **2** be hostile to. **3** take part in a game, sport, etc., against (another competitor or team). **4** (foll. by _to_) place in opposition or contrast. □ **as opposed to** in contrast with. □ **opposer** _n._ [ME f. OF _opposer_ f. L _opponere:_ see OPPONENT]

opposite /ˈɒpəzɪt/ _adj., n., adv.,_ & _prep._ ● _adj._ **1** (often foll. by _to_) having a position on the other or further side, facing or back to back. **2** (often foll. by _to, from_) **a** of a contrary kind; diametrically different. **b** being the other of a contrasted pair. **3** _Geom._ (of angles) between opposite sides of the intersection of two lines. **4** _Bot._ (of leaves etc.) placed at the same height on the opposite sides of the stem, or placed straight in front of another organ. ● _n._ an opposite thing or person or term. ● _adv._ in an opposite position (_the tree stands opposite_). ● _prep._ **1** in a position opposite to (_opposite the house is a tree_). **2** (of a leading theatrical etc. part) in a complementary role to (another performer). □ **opposite number** a person holding an equivalent position in another group or organization. **opposite prompt** _Brit._ the side of a theatre stage usually to an actor's right. **the opposite sex** women in relation to men or vice versa. □ **oppositely** _adv._ **oppositeness** _n._ [ME f. OF f. L _oppositus_ past part. of _opponere:_ see OPPONENT]

opposition /ˌɒpəˈzɪʃ(ə)n/ _n._ **1** resistance, antagonism. **2** the state of being hostile or in conflict or disagreement. **3** contrast or antithesis. **4 a** a group or party of opponents or competitors. **b** (usu. **the Opposition**) _Brit._ the principal parliamentary party opposed to that in office. **5** the act of opposing or placing opposite. **6 a** diametrically opposite position. **b** _Astron._ & _Astrol._ the position of two celestial bodies when their longitude differs by 180°, as seen from the earth. □ **oppositional** _adj._ [ME f. OF f. L _oppositio_ (as OB-, POSITION)]

oppress /əˈpres/ _v.tr._ **1** keep in subservience by coercion. **2** govern or treat harshly or with cruel injustice. **3** weigh down (with cares or unhappiness). □ **oppressor** _n._ [ME f. OF _oppresser_ f. med.L _oppressare_ (as OB-, PRESS¹)]

oppression /əˈpreʃ(ə)n/ _n._ **1** the act or an instance of oppressing; the state of being oppressed. **2** prolonged harsh or cruel treatment or control. **3** mental distress. [OF f. L _oppressio_ (as OPPRESS)]

oppressive /əˈpresɪv/ _adj._ **1** oppressing; harsh or cruel. **2** difficult to endure. **3** (of weather) close and sultry. □ **oppressively** _adv._ **oppressiveness** _n._ [F _oppressif -ive_ f. med.L _oppressivus_ (as OPPRESS)]

opprobrious /əˈprəʊbrɪəs/ _adj._ (of language) severely scornful; abusive. □ **opprobriously** _adv._ [ME f. LL _opprobriosus_ (as OPPROBRIUM)]

opprobrium /əˈprəʊbrɪəm/ _n._ **1** disgrace or bad reputation attaching to some act or conduct. **2** a cause of this. [L f. _opprobrum_ (as OB-, _probrum_ disgraceful act)]

oppugn /əˈpjuːn/ _v.tr. literary_ call into question; controvert. □ **oppugner** _n._ [ME f. L _oppugnare_ attack, besiege (as OB-, L _pugnare_ fight)]

oppugnant /əˈpʌgnənt/ _adj. literary_ antagonistic; opposing. □ **oppugnance** _n._ **oppugnancy** _n._ **oppugnation** /ˌɒpʌgˈneɪʃ(ə)n/ _n._

opsimath /ˈɒpsɪˌmæθ/ _n. literary_ a person who learns only late in life. □ **opsimathy** /ɒpˈsɪməθɪ/ _n._ [Gk _opsimathēs_ f. _opse_ late + _math-_ learn]

opsonin /ˈɒpsənɪn/ _n. Biol._ & _Med._ a substance, esp. an antibody, which attaches itself to foreign organisms and makes them more susceptible to phagocytosis. □ **opsonic** /ɒpˈsɒnɪk/ _adj._ [Gk _opsōnion_ victuals + -IN]

opt /ɒpt/ _v.intr._ (usu. foll. by _for, between_) exercise an option; make a choice. □ **opt out** (often foll. by _of_) **1** choose not to participate (_opted out of the_ *race*). **2** (in the UK) (of a school or hospital) decide to withdraw from the control of a local authority. [F *opter* f. L *optare* choose, wish]

optant /'ɒptənt/ n. **1** a person who may choose one of two nationalities. **2** a person who chooses or has chosen.

optative /ɒp'teɪtɪv, 'ɒptətɪv/ adj. & n. Gram. ● adj. expressing a wish. ● n. the optative mood. □ **optative mood** a set of verb-forms expressing a wish etc., distinct esp. in Sanskrit and Greek. □ **optatively** adv. [F *optatif -ive* f. LL *optativus* (as OPT)]

optic /'ɒptɪk/ adj. & n. ● adj. of or relating to the eye or vision (*optic nerve*). ● n. **1** a lens etc. in an optical instrument. **2** archaic or joc. the eye. **3** (**Optic**) Brit. propr. a device fastened to the neck of an inverted bottle for measuring out spirits etc. □ **optic angle** the angle formed by notional lines from the extremities of an object to the eye, or by lines from the eyes to a given point. **optic axis 1** a line passing through the centre of curvature of a lens or spherical mirror and parallel to the axis of symmetry. **2** the direction in a doubly refracting crystal for which no double refraction occurs. **optic lobe** the dorsal lobe in the brain from which the optic nerve arises. **optic nerve** each of the second pair of cranial nerves, transmitting impulses to the brain from the retina at the back of the eye. [F *optique* or med.L *opticus* f. Gk *optikos* f. *optos* seen]

optical /'ɒptɪk(ə)l/ adj. **1** of sight; visual. **2 a** of or concerning sight or light in relation to each other. **b** belonging to optics. **3** (esp. of a lens) constructed to assist sight or on the principles of optics. □ **optical activity** Chem. the property of rotating the plane of polarization of plane-polarized light. **optical art** = OP ART. **optical brightener** any fluorescent substance used to produce a whitening effect on laundry. **optical character recognition** the identification of printed characters using photoelectric devices. **optical disk** see DISC 4b. **optical fibre** thin glass fibre through which light can be transmitted. **optical glass** a very pure kind of glass used for lenses etc. **optical illusion 1** a thing having an appearance so resembling something else as to deceive the eye. **2** an instance of mental misapprehension caused by this. **optical isomer** Chem. each of two or more forms of a chemical substance which have the same structure but a different spatial arrangement of atoms, and so usu. differ in optical activity. **optical microscope** a microscope using the direct perception of light (cf. ELECTRON MICROSCOPE). □ **optically** adv.

optician /ɒp'tɪʃ(ə)n/ n. **1** a maker or seller of optical instruments. **2** (in full **dispensing optician**) a person qualified to make and supply spectacles and contact lenses. **3** (in full **ophthalmic optician**) a person trained in the detection and correction of poor eyesight; an optometrist. [F *opticien* f. med.L *optica* (as OPTIC)]

optics /'ɒptɪks/ n.pl. (treated as sing.) the scientific study of sight and the behaviour of light, or of other radiation or particles (*electron optics*).

optima pl. of OPTIMUM.

optimal /'ɒptɪm(ə)l/ adj. best or most favourable, esp. under a particular set of circumstances. □ **optimally** adv. [L *optimus* best]

optimism /'ɒptɪˌmɪz(ə)m/ n. **1** an inclination to hopefulness and confidence. **2** Philos. **a** the doctrine, esp. as set forth by Leibniz, that this world is the best of all possible worlds. **b** the theory that good must ultimately prevail over evil in the universe. Opp. PESSIMISM. □ **optimist** n. **optimistic** /ˌɒptɪ'mɪstɪk/ adj. **optimistically** adv. [F *optimisme* f. L OPTIMUM]

optimize /'ɒptɪˌmaɪz/ v. (also **-ise**) **1** tr. make the best or most effective use of (a situation, an opportunity, etc.). **2** intr. be an optimist. □ **optimization** /ˌɒptɪmaɪ'zeɪʃ(ə)n/ n. [L *optimus* best]

optimum /'ɒptɪməm/ n. & adj. ● n. (pl. **optima** /-mə/ or **optimums**) **1 a** the most favourable conditions (for growth, reproduction, etc.). **b** the best or most favourable situation. **2** the best possible compromise between opposing tendencies. ● adj. = OPTIMAL. [L, neut. (as n.) of *optimus* best]

option /'ɒpʃ(ə)n/ n. **1 a** the act or an instance of choosing; a choice. **b** a thing that is or may be chosen (*those are the options*). **2** the liberty of choosing; freedom of choice. **3** Stock Exch. etc. the right, obtained by, or as, payment, to buy, sell, etc. specified stocks etc. at a specified price within a set time. □ **have no option but to** must. **keep** (or **leave**) **one's options open** not commit oneself. [F or f. L *optio*, stem of *optare* choose]

optional /'ɒpʃən(ə)l/ adj. being an option only; not obligatory. □ **optionally** adv. **optionality** /ˌɒpʃə'nælɪtɪ/ n.

optoelectronics /ˌɒptəʊˌɪlek'trɒnɪks/ n. the branch of technology concerned with the combined use of electronics and light. □ **optoelectronic** adj.

optometer /ɒp'tɒmɪtə(r)/ n. an instrument for testing the refractive power of the eye. [Gk *optos* seen + -METER]

optometrist /ɒp'tɒmɪtrɪst/ n. esp. US a person who practises optometry; an ophthalmic optician.

optometry /ɒp'tɒmɪtrɪ/ n. the measurement of the refractive power and other properties of the eyes and the prescription of corrective lenses; the occupation of an ophthalmic optician. □ **optometric** /ˌɒptə'metrɪk/ adj.

optophone /'ɒptəˌfəʊn/ n. an instrument converting light into sound, and so enabling the blind to read print etc. by ear. [Gk *optos* seen + -PHONE]

opulent /'ɒpjʊlənt/ adj. **1** ostentatiously rich; wealthy. **2** luxurious (*opulent surroundings*). **3** abundant; profuse. □ **opulence** n. **opulently** adv. [L *opulens, opulent-* f. *opes* wealth]

opuntia /əʊ'pʌnʃɪə/ n. a cactus of the genus *Opuntia*, with flattened elliptical segments and barbed spines. See also prickly pear. [L plant-name f. *Opus -untis*, a city in Locris in ancient Greece]

opus /'əʊpəs, 'ɒp-/ n. (pl. **opuses** or **opera** /'ɒpərə/) **1** Mus. **a** a separate musical composition or set of compositions of any kind. **b** (also **op.**) used before a number given to a composer's work, usu. indicating the order of publication (*Beethoven, op. 15*). **2** any artistic work (cf. MAGNUM OPUS). [L, = work]

opuscule /ə'pʌskjuːl/ n. (also **opusculum** /ə'pʌskjʊləm/) (pl. **opuscules** or **opuscula** /-lə/) a minor (esp. musical or literary) work. [F f. L *opusculum* dimin. of OPUS]

opus Dei /'deɪiː/ n. Eccl. **1** liturgical worship regarded as the primary duty to God. **2** (**Opus Dei**) a Roman Catholic organization of laymen and priests founded in Spain in 1928 with the aim of re-establishing Christian ideals in society. [med.L]

OR abbr. **1** operational research. **2** US Oregon (in official postal use). **3** Mil. other ranks (non-commissioned officers and ordinary soldiers, seamen, etc.).

or[1] /ɔː(r), ə(r)/ conj. **1 a** introducing the second of two alternatives (*white or black*). **b** introducing all but the first, or only the last, of any number of alternatives (*white or grey or black; white, grey, or black*). **2** (often prec. by *either*) introducing the only remaining possibility or choice given (*take it or leave it; either come in or go out*). **3** (prec. by *whether*) introducing the second part of an indirect question or conditional clause (*ask him whether he was there or not; I don't know whether I like or dislike it*). **4** introducing a synonym or explanation of a preceding word etc. (*suffered from vertigo or giddiness*). **5** introducing a significant afterthought (*he must know — or is he bluffing?*). **6** = or else 1. **7** poet. each of two; either (*or in the heart or in the head*). □ **not A or B** not A, and also not B. **one or two** (or **two or three** etc.) colloq. a few. **or else 1** otherwise (*do it now, or else you will have to do it tomorrow*). **2** colloq. expressing a warning or threat (*hand over the money or else*). **or rather** introducing a rephrasing or qualification of a preceding statement etc. (*he was there, or rather I heard that he was*). **or so** (after a quantity or a number) or thereabouts (*send me ten or so*). [reduced form of obs. *other* conj. (which superseded OE *oththe* or), of uncert. orig.]

or[2] /ɔː(r)/ n. & adj. Heraldry ● n. a gold or yellow colour. ● adj. (usu. following noun) gold or yellow (*a crescent or*). [F f. L *aurum* gold]

-or[1] /ə(r)/ suffix forming nouns denoting a person or thing performing the action of a verb, or an agent more generally (*actor; escalator; tailor*) (see also -ATOR, -ITOR). [L *-or, -ator*, etc., sometimes via AF *-eour*, OF *-ëor, -ëur*]

-or[2] /ə(r)/ suffix forming nouns denoting state or condition (*error; horror*). [L *-or -oris*, sometimes via (or after) OF *-or, -ur*]

-or[3] /ə(r)/ suffix forming adjectives with comparative sense (*major; senior*). [AF *-our* f. L *-or*]

-or[4] /ə(r)/ suffix US = -OUR[1].

orache /'ɒrɪtʃ/ n. (also **orach**) an edible plant, *Atriplex hortensis*, with red, yellow, or green leaves sometimes used as a substitute for spinach or sorrel. Also called *saltbush*. [ME *arage* f. AF *arasche* f. L *atriplex* f. Gk *atraphaxus*]

oracle /'ɒrək(ə)l/ n. **1 a** a place at which advice or prophecy was sought from the gods in classical antiquity. **b** the usu. ambiguous or obscure response given at an oracle. **c** a prophet or prophetess at an oracle. **2 a** a person or thing regarded as an infallible guide to future action etc. **b** a saying etc. regarded as infallible guidance. **3** divine inspiration or revelation. [ME f. OF f. L *oraculum* f. *orare* speak]

oracular /ə'rækjʊlə(r)/ adj. **1** of or concerning an oracle or oracles. **2** (esp. of advice etc.) mysterious or ambiguous. **3** prophetic. □ **oracularly** adv. **oracularity** /əˌrækjʊ'lærɪtɪ/ n. [L (as ORACLE)]

oracy /ˈɔːrəsɪ/ n. the ability to express oneself fluently in speech. [L *os oris* mouth, after *literacy*]

Oradea /ɒˈrɑːdɪə/ an industrial city in western Romania, near the border with Hungary; pop. (1989) 225,420.

oral /ˈɔːrəl/ adj. & n. ● adj. **1** by word of mouth; spoken or sung; not written (*the oral tradition*). **2** done or taken by the mouth (*oral contraceptive*). **3** of or relating to the mouth. **4** *Psychol.* of or concerning a supposed stage of infant emotional and sexual development, in which the mouth is of central interest. ● n. a spoken examination, test, etc. □ **oral sex** sexual activity in which the genitals of one partner are stimulated by the mouth of the other. **oral society** a society that has not reached the stage of literacy. □ **orally** adv. [LL *oralis* f. L *os oris* mouth]

Oran /ɔːˈrɑːn/ a port on the Mediterranean coast of Algeria; pop. (1989) 664,000.

Orange[1] /ˈɒrɪndʒ, French ɔrɑ̃ʒ/ a town in southern France, on the Rhône. It was a small principality in the 16th century in the possession of the House of Orange, whose descendants became rulers of the Netherlands. (See ORANGE, HOUSE OF.)

Orange[2] /ˈɒrɪndʒ/ adj. of, relating to, or denoting a Protestant society or order (the Orange Order) in Ireland, especially in Northern Ireland, formed in 1795 (as the Association of Orangemen) for the defence of Protestantism and maintenance of Protestant ascendancy in Ireland. It was probably named from the wearing of orange badges etc. as a symbol of adherence to William III (William of Orange), who defeated the Catholic James II at the Battle of the Boyne in 1690. The Orange Order, organized as Freemasons' Lodges, spread beyond Ireland to Britain and many parts of the former British Empire; in the early 20th century it was strengthened in the north of Ireland in its campaign to resist the Home Rule bill and has continued to form a core of Protestant Unionist opinion since. □ **Orangeism** n.

orange /ˈɒrɪndʒ/ n. & adj. ● n. **1 a** a large roundish juicy citrus fruit with a bright reddish-yellow tough rind. **b** a tree or shrub of the genus *Citrus*, esp. *C. sinensis* or *C. aurantium*, bearing fragrant white flowers and yielding this fruit. **2** a fruit or plant resembling this. **3 a** the reddish-yellow colour of an orange. **b** orange pigment. ● adj. orange-coloured; reddish-yellow. □ **orange blossom** the flowers of the orange tree, traditionally worn by the bride at a wedding. **orange flower water** a solution of neroli in water. **orange peel 1** the skin of an orange. **2** a rough surface resembling this. **orange pekoe** a type of black tea made from very small leaves. **orange squash** *Brit.* a soft drink made from oranges and other ingredients, sold in concentrated form. **orange-stick** a thin stick, pointed at one end and usu. of orange-wood, for manicuring the fingernails. **orange-wood** the wood of the orange tree. [ME f. OF *orenge*, ult. f. Arab. *nāranj* f. Pers. *nārang*]

Orange, House of the Dutch royal house. The House of Orange was originally a princely dynasty of the principality centred on the town of Orange in France. After a marital alliance with the duchy of Nassau, however, the family became a major force in the politics of the Netherlands, members holding the position of stadtholder or magistrate from the mid-16th until the late 18th century; the son of the last stadtholder became King William I of the United Netherlands in 1815. William of Orange became King William III of Great Britain and Ireland in 1689.

Orange, William of William III of Great Britain and Ireland (see WILLIAM).

orangeade /ˌɒrɪndʒˈeɪd/ n. a usu. fizzy non-alcoholic drink flavoured with orange.

Orange Free State a province in central South Africa, situated to the north of the Orange River; capital, Bloemfontein. First settled by Boers after the Great Trek from Cape Colony (1836–8), it was annexed by Britain in 1848 but restored in 1854 to the Boers, who established the Orange Free State Republic. It was re-annexed by Britain in 1900 as the Orange River Colony, was given internal self-government in 1907, and became a province of the Union of South Africa in 1910 as Orange Free State. In 1994 it became one of the new provinces of South Africa, following democratic elections in April 1994. [*House of Orange* (see ORANGE, HOUSE OF)]

Orange Lodge the Orange Order (see ORANGE[2]).

Orangeman /ˈɒrɪndʒmən/ n. (pl. **-men**) a member of the Orange Order (see ORANGE[2]).

Orange Order a Protestant political society in Ireland (see ORANGE[2]).

Orange River the longest river in South Africa, which rises in the Drakensberg Mountains in NE Lesotho and flows generally westward for 1,859 km (1,155 miles) to the Atlantic across almost the whole breadth of the continent, forming the border between Namibia and South Africa in its lower course.

orangery /ˈɒrɪndʒərɪ/ n. (pl. **-ies**) a place, esp. a special structure, where orange trees are cultivated.

orang-utan /ɔːˌræŋuːˈtæn/ n. (also **orangutan**, **orang-outang** /-ˈtæŋ/) a large red long-haired tree-living ape, *Pongo pygmaeus*, native to Borneo and Sumatra, with characteristic long arms and hooked hands and feet. [Malay *orang utan* forest person]

Oranjestad /ɒˈrænjəˌstɑːt/ the capital of the Dutch island of Aruba in the West Indies; pop. (1988) 20,000.

Oraşul Stalin /ɒˌræʃul ˈstɑːlɪn/ a former name for BRAŞOV.

orate /ɔːˈreɪt/ v.intr. esp. joc. or derog. make a speech or speak, esp. pompously or at length. [back-form. f. ORATION]

oration /ɔːˈreɪʃ(ə)n/ n. **1** a formal speech, discourse, etc., esp. when ceremonial. **2** *Gram.* a way of speaking; language. [ME f. L *oratio* discourse, prayer f. *orare* speak, pray]

orator /ˈɒrətə(r)/ n. **1 a** a person making a speech. **b** an eloquent public speaker. **2** (in full **public orator**) an official speaking for a university on ceremonial occasions. □ **oratorial** /ˌɒrəˈtɔːrɪəl/ adj. [ME f. AF *oratour*, OF *orateur* f. L *orator -oris* speaker, pleader (as ORATION)]

oratorio /ˌɒrəˈtɔːrɪˌəʊ/ n. (pl. **-os**) a semi-dramatic work for orchestra and voices esp. on a sacred theme, performed without costume, scenery, or action. The form arose at the beginning of the 17th century, and was taken to its highest level by Bach and Handel, notably in his *Messiah*. □ **oratorial** adj. [It. f. eccl.L *oratorium*, orig. of musical services at church of the Oratory of St Philip Neri in Rome]

oratory /ˈɒrətərɪ/ n. (pl. **-ies**) **1** the art or practice of formal speaking, esp. in public. **2** exaggerated, eloquent, or highly coloured language. **3** a small chapel, esp. for private worship. **4** (**Oratory**) *RC Ch.* **a** a religious society of priests without vows founded in Rome in 1564 and providing plain preaching and popular services. **b** a branch of this in England etc. □ **oratorian** /ˌɒrəˈtɔːrɪən/ adj. & n. **oratorical** /-ˈtɒrɪk(ə)l/ adj. [senses 1 and 2 f. L *ars oratoria* art of speaking; senses 3 and 4 ME f. AF *oratorie*, OF *oratoire* f. eccl.L *oratorium*: both f. L *oratorius* f. *orare* pray, speak]

orb /ɔːb/ n. & v. ● n. **1** a globe surmounted by a cross, esp. carried by a sovereign at a coronation. **2** a sphere; a globe. **3** *poet.* a celestial body. **4** *poet.* an eyeball; an eye. ● v. usu. *poet.* **1** tr. enclose in (an orb); encircle. **2** intr. form or gather into an orb. [L *orbis* ring]

orbicular /ɔːˈbɪkjʊlə(r)/ adj. *formal* **1** circular and flat; disc-shaped; ring-shaped. **2** spherical; globular; rounded. **3** forming a complete whole. □ **orbicularly** adv. **orbicularity** /ɔːˌbɪkjʊˈlærɪtɪ/ n. [ME f. LL *orbicularis* f. L *orbiculus* dimin. of *orbis* ring]

orbiculate /ɔːˈbɪkjʊlət/ adj. *Bot.* (of a leaf etc.) circular.

Orbison /ˈɔːbɪs(ə)n/, Roy (1936–88), American singer and composer. He began by writing country-music songs for other artists, establishing himself as a singer with the ballad 'Only the Lonely' (1960). Further hits followed, including 'Crying' (1961), 'Blue Bayou' (1963), and 'Oh, Pretty Woman' (1964), which was one of the best-selling singles of the 1960s.

orbit /ˈɔːbɪt/ n. & v. ● n. **1 a** the curved, usu. closed course of a planet, satellite, etc. **b** (prec. by *in*, *into*, *out of*, etc.) the state of motion in an orbit. **c** one complete passage around an orbited body. **2** the path of an electron round an atomic nucleus. **3** a range or sphere of action. **4 a** the eye socket. **b** the area around the eye of a bird or insect. ● v. (**orbited**, **orbiting**) **1** intr. **a** (of a satellite etc.) go round in orbit. **b** fly in a circle. **2** tr. move in orbit round. **3** tr. put into orbit. [L *orbita* course, track (in med.L eye-cavity): fem. of *orbitus* circular f. *orbis* ring]

orbital /ˈɔːbɪt(ə)l/ adj. & n. ● adj. **1** *Anat.*, *Astron.*, & *Physics* of an orbit or orbits. **2** (of a road) passing round the outside of a town. ● n. **1** *Physics* a state or function representing the possible motion of an electron round an atomic nucleus. **2** an orbital road. □ **orbital sander** a sander having a circular and not oscillating motion.

orbiter /ˈɔːbɪtə(r)/ n. a spacecraft designed to remain in orbit (*lunar orbiter*) (cf. LANDER 2).

orca /ˈɔːkə/ n. the killer whale. [F *orque* or L *orca* a kind of whale]

Orcadian /ɔːˈkeɪdɪən/ adj. & n. ● adj. of or relating to the Orkney Islands. ● n. a native of the Orkney Islands. [L *Orcades* Orkney Islands]

Orcagna /ɔːˈkɑːnjə/ (born Andrea di Cione) (c.1308–68), Italian painter, sculptor, and architect. His painting represents a return to a devotional, brightly coloured style in opposition to Giotto's naturalism; his work includes frescos and an altarpiece in the church of Santa Maria Novella, Florence (1357). His only known sculpture is the tabernacle in the church of Or San Michele, Florence, notable for its reliefs depicting scenes from the life and the Assumption of the Virgin (1359).

orch. *abbr.* **1** orchestrated by. **2** orchestra.

orchard /'ɔːtʃəd/ *n.* a piece of enclosed land with fruit-trees. □ **orchardist** *n.* [OE *ortgeard* f. L *hortus* garden + YARD²]

orcharding /'ɔːtʃədɪŋ/ *n.* the cultivation of fruit-trees.

orchardman /'ɔːtʃədmən/ *n.* (*pl.* **-men**) a fruit-grower.

orchestra /'ɔːkɪstrə/ *n.* **1 a** usu. large group of instrumentalists, esp. combining strings, woodwinds, brass, and percussion (*symphony orchestra*). (*See note below.*) **2 a** (in full **orchestra pit**) the part of a theatre, opera house, etc., where the orchestra plays, usu. in front of the stage and on a lower level. **b** *N. Amer.* the stalls in a theatre. **3** the semicircular space in front of an ancient Greek theatre-stage where the chorus danced and sang. □ **orchestra stalls** *Brit.* the front of the stalls. □ **orchestral** /ɔː'kestrəl/ *adj.* **orchestrally** *adv.* [L f. Gk *orkhēstra* f. *orkheomai* to dance (see sense 3)]

- In its modern form the orchestra dates from the 18th century, when the forces required for orchestral music became established as four main categories with standard members: woodwind (flutes, clarinets, oboes, bassoons), brass (horns, trumpets), percussion (drums, cymbals), and strings (first and second violins, violas, cellos, double basses). In the 19th century the number of instruments in the orchestra was increased, and some new instruments, such as the trombone, were added, but in the 20th century a return has frequently been made to smaller ensembles.

orchestrate /'ɔːkɪˌstreɪt/ *v.tr.* **1** arrange, score, or compose for orchestral performance. **2** carefully direct or coordinate the elements of (a plan, situation, etc.). □ **orchestrator** *n.* **orchestration** /ˌɔːkɪ'streɪʃ(ə)n/ *n.*

orchid /'ɔːkɪd/ *n.* **1** a plant of the family Orchidaceae, bearing complex flowers with one petal that is larger than the others and variously modified. (*See note below.*) **2** a flower of one of these plants. □ **orchidist** *n.* **orchidaceous** /ˌɔːkɪ'deɪʃəs/ *adj.* **orchidology** /-'dɒlədʒɪ/ *n.* [mod.L *Orchidaceae* irreg. f. L *orchis*: see ORCHIS]

- The orchids constitute a very large family, reaching their greatest diversity in the tropics, where they are often epiphytic. The flowers have one of the petals (the 'lip') larger than the others and often highly modified, with various spurs, lobes, pouches, and other contrivances. These are all adaptations for pollination, with some species (e.g. of the genus *Ophrys*) having lips that both look and smell like female insects in order to attract the male insect. Most temperate orchids are relatively inconspicuous, but the exotic colours and shapes of many tropical species make them popular with collectors. They have tiny seeds, and most kinds require a mycorrhizal association in order to germinate and thrive.

orchido- /'ɔːkɪdəʊ/ *comb. form* (also **orchid-** /'ɔːkɪd/ *before a vowel*) *Med.* of a testicle or the testicles (*orchidectomy*). [mod.L f. Gk *orkhis* testicle]

orchil /'ɔːtʃɪl/ *n.* (also **orchilla** /ɔː'tʃɪlə/, **archil** /'ɑːtʃɪl/) **1** a red or violet dye obtained from a lichen, often used in litmus. **2** a tropical lichen, esp. *Roccella tinctoria*, yielding this. [ME f. OF *orcheil* etc. perh. ult. f. L *herba urceolaris* a plant for polishing glass pitchers]

orchis /'ɔːkɪs/ *n.* **1** an orchid of (or formerly of) the genus *Orchis*, having a tuberous root and an erect fleshy stem with a spike of usu. purple or red flowers. **2** any wild orchid. [L f. Gk *orkhis*, orig. = testicle (with ref. to the shape of its tuber)]

orchitis /ɔː'kaɪtɪs/ *n. Med.* inflammation of the testicles. [mod.L f. Gk *orkhis* testicle]

orcin /'ɔːsɪn/ *n.* (also **orcinol** /'ɔːsɪˌnɒl/) a crystalline substance, becoming red in air, extracted from certain lichens and used to make dyes. [mod.L *orcina* f. It. *orcello* orchil]

Orczy /'ɔːtsɪ/, Baroness Emmusca (1865–1947), Hungarian-born British novelist. Her best-known novel is *The Scarlet Pimpernel* (1905), telling of the adventures of an English nobleman smuggling aristocrats out of France during the French Revolution.

ord. *abbr.* ordinary.

ordain /ɔː'deɪn/ *v.tr.* **1** confer holy orders on; appoint to the Christian ministry (*ordained him priest; was ordained in 1970*). **2 a** (often foll. by *that* + clause) decree (*ordained that he should go*). **b** (of God, fate, etc.) destine; appoint (*has ordained us to die*). □ **ordainer** *n.* **ordainment** *n.* [ME f. AF *ordeiner*, OF *ordein-* stressed stem of *ordener* f. L *ordinare* f. *ordo -inis* order]

ordeal /ɔː'diːl/ *n.* **1** a painful or horrific experience; a severe trial. **2** *hist.*

an ancient test of guilt or innocence esp. used by Germanic peoples, in which the accused was subjected to severe pain or torture, survival of which was taken as divine proof of innocence. [OE *ordāl, ordēl* f. Gmc: cf. DEAL¹]

order /'ɔːdə(r)/ *n. & v.* ● *n.* **1 a** the condition in which every part, unit, etc. is in its right place; tidiness (*restored some semblance of order*). **b** a usu. specified sequence, succession, etc. (*alphabetical order; the order of events*). **2** (in *sing.* or *pl.*) an authoritative command, direction, instruction, etc. (*only obeying orders; gave orders for it to be done; the judge made an order*). **3** a state of peaceful harmony under a constituted authority (*order was restored; law and order*). **4** (esp. in *pl.*) a social class, rank, etc., constituting a distinct group in society (*the lower orders; the order of baronets*). **5** a kind; a sort (*talents of a high order*). **6 a** a usu. written direction to a manufacturer, tradesman, waiter, etc. to supply something. **b** the quantity of goods etc. supplied. **7** the constitution or nature of the world, society, etc. (*the moral order; the order of things*). **8** *Biol.* a taxonomic rank below a class and above a family. **9** (also **Order**) a fraternity of monks and friars, or formerly of knights, bound by a common rule of life (*the Franciscan order; the order of Templars*). **10 a** any of the grades of the Christian ministry. **b** (in *pl.*) the status of a member of the clergy (*Anglican orders*). **11 a** any of the styles of ancient architecture distinguished by the type of column used. The principal classical orders are the Doric, Ionic, Corinthian, Tuscan, and Composite, the first three being Greek in origin and the others Roman. **b** any style or mode of architecture subject to uniform established proportions. **12** (esp. **Order**) **a** a company of distinguished people instituted esp. by a sovereign to which appointments are made as an honour or reward (*Order of the Garter; Order of Merit*). **b** the insignia worn by members of an order. **13** *Math.* **a** a degree of complexity of a differential equation (*equation of the first order*). **b** the order of the highest derivative in the equation. **14** *Math.* **a** the size of a matrix. **b** the number of elements of a finite group. **15** *Eccl.* the stated form of divine service (*the order of confirmation*). **16** the principles of procedure, decorum, etc., accepted by a meeting, legislative assembly, etc. or enforced by its president (*review order*). **17** *Mil.* **a** a style of dress and equipment (*review order*). **b** (prec. by *the*) the position of a company etc. with arms ordered (see *order arms*). **18** a Masonic or similar fraternity. **19** in Christian theology, any of the nine grades of angelic beings (seraphim, cherubim, thrones, dominations, principalities, powers, virtues, archangels, angels). **20** a pass admitting the bearer to a theatre, museum, private house, etc. free or cheap or as a privilege. ● *v.tr.* **1** (usu. foll. by *to* + infin., or *that* + clause) command; bid; prescribe (*ordered him to go; ordered that they should be sent*). **2** command or direct (a person) to a specified destination (*was ordered to Singapore; ordered them home*). **3** direct a manufacturer, waiter, tradesman, etc. to supply (*ordered a new suit; ordered dinner*). **4** put in order; regulate (*ordered her affairs*). **5** (of God, fate, etc.) ordain (*fate ordered it otherwise*). **6** *US* command (a thing) done or (a person) dealt with (*ordered it settled; ordered him expelled*). □ **by order** according to the proper authority. **holy orders** the status of a member of the clergy, esp. the grades of bishop, priest, and deacon. **in bad** (or **good** etc.) **order** not working (or working properly etc.). **in order 1** one after another according to some principle. **2** ready or fit for use. **3** according to the rules (of procedure at a meeting etc.). **in order that** with the intention; so that. **in order to** with the purpose of doing; with a view to. **keep order** enforce orderly behaviour. **made to order 1** made according to individual requirements, measurements, etc. (opp. READY-MADE *adj.*). **2** exactly what is wanted. **minor orders** *RC Ch. hist.* the grades of members of the clergy below that of deacon. **not in order** not working properly. **of** (or **in** or **on**) **the order of 1** approximately. **2** having the order of magnitude specified by (*of the order of one in a million*). **on order** (of goods etc.) ordered but not yet received. **order about 1** dominate; command officiously. **2** send hither and thither. **order arms** *Mil.* hold a rifle with its butt on the ground close to one's right side. **order book 1** a book in which a tradesman enters orders. **2** the level of incoming orders. **order-form** a printed form in which details are entered by a customer. **order of the day 1** the prevailing state of things. **2** a principal topic of action or a procedure decided upon. **3** business set down for treatment; a programme. **order of magnitude** a class in a system of classification determined by size, usu. by powers of 10. **Order! Order!** *Parl.* a call for silence or calm, esp. by the Speaker of the House of Commons. **order-paper** esp. *Parl.* a written or printed order of the day; an agenda. **order to view** a house-agent's request for a client to be allowed to inspect premises. **out of order 1** not working properly. **2** not according to the rules (of a meeting, organization, etc.). **3** *colloq.* (of a person) not behaving acceptably; (of behaviour) objectionable, unacceptable. **4** not in proper sequence. **take orders 1** accept commissions. **2** accept and carry out commands. **3** (also **take holy orders**) be ordained. [ME f. OF *ordre* f. L *ordo ordinis* row, array, degree, command, etc.]

Order in Council *n. Brit.* a sovereign's order on an administrative matter given by the advice of the Privy Council.

orderly /ˈɔːdəlɪ/ *adj. & n.* ● *adj.* **1** methodically arranged; regular. **2** obedient to discipline; well-behaved; not unruly. **3** *Mil.* **a** of or concerned with orders. **b** charged with the conveyance or execution of orders. ● *n.* (*pl.* **-ies**) **1** an esp. male cleaner in a hospital. **2** a soldier who carries orders for an officer etc. □ **orderly book** *Brit. Mil.* a regimental or company book for entering orders. **orderly officer** *Brit. Mil.* the officer of the day. **orderly room** *Brit. Mil.* a room in a barracks used for company business. □ **orderliness** *n.*

Order of Merit (in the UK) an order, founded in 1902, whose members are admitted for distinguished achievement.

Order of the Bath see BATH, ORDER OF THE.

Order of the British Empire (in the UK) an order of knighthood instituted in 1917 and divided into five classes, each with military and civilian divisions. The classes are: Knight or Dame Grand Cross of the Order of the British Empire (GBE), Knight or Dame Commander (KBE/DBE), Commander (CBE), Officer (OBE), and Member (MBE). The two highest classes entail the awarding of a knighthood.

Order of the Garter see GARTER 2.

Order of the Thistle see THISTLE, ORDER OF THE.

ordinal /ˈɔːdɪn(ə)l/ *n. & adj.* ● *n.* **1** (in full **ordinal number**) a number defining a thing's position in a series, e.g. 'first', 'second', 'third', etc. (cf. *cardinal numbers*). **2** *Eccl.* a service-book, esp. one with the forms of service used at ordinations. ● *adj.* **1 a** of or relating to an ordinal number. **b** defining a thing's position in a series etc. **2** *Biol.* of or concerning an order (see ORDER *n.* 8). [ME f. LL *ordinalis* & med.L *ordinale* neut. f. L (as ORDER)]

ordinance /ˈɔːdɪnəns/ *n.* **1** an authoritative order; a decree. **2** an enactment by a local authority. **3** a religious rite. **4** archaic = ORDONNANCE. [ME f. OF *ordenance* f. med.L *ordinantia* f. L *ordinare*: see ORDAIN]

ordinand /ˈɔːdɪnænd/ *n. Eccl.* a candidate for ordination. [L *ordinandus*, gerundive of *ordinare* ORDAIN]

ordinary /ˈɔːdɪnərɪ, ˈɔːd(ə)nrɪ/ *adj. & n.* ● *adj.* **1 a** regular, normal, customary, usual (*in the ordinary course of events*). **b** boring; commonplace (*an ordinary little man*). **2** *Brit. Law* (esp. of a judge) having immediate or *ex officio* jurisdiction, not deputed. ● *n.* (*pl.* **-ies**) **1** *Brit. Law* a person, esp. a judge, having immediate or *ex officio* jurisdiction. **2** (**the Ordinary**) **a** an archbishop in a province. **b** a bishop in a diocese. **3** (usu. **Ordinary**) *RC Ch.* **a** those parts of a service, esp. the mass, which do not vary from day to day. **b** a rule or book laying down the order of divine service. **4** *Heraldry* a charge of the earliest, simplest, and commonest kind (esp. chief, pale, bend, fess, bar, chevron, cross, saltire). **5** (**Ordinary**) (also **Lord Ordinary**) any of the judges of the Court of Session in Scotland, constituting the Outer House. **6** an ungeared bicycle of an early type, with a large front wheel and small rear wheel (see BICYCLE). **7** *Brit. hist.* **a** a public meal provided at a fixed time and price at an inn etc. **b** an establishment providing this. **8** *US* a tavern. **9** (*prec. by the*) *colloq.* the customary or usual condition, course, or degree. □ **in ordinary** *Brit.* by permanent appointment (esp. to the royal household) (*physician in ordinary*). **in the ordinary way** if the circumstances are or were not exceptional. **ordinary grade** (in Scotland) the lower of the two main levels of examination leading to the Scottish Certificate of Education. (See also HIGHER.) **ordinary level** *hist.* (in the UK except Scotland) the lower of the two main levels of the GCE examination. **ordinary scale** = *decimal scale.* **ordinary seaman** a sailor of the lowest rank, that below able-bodied seaman. **ordinary shares** *Brit.* shares entitling holders to a dividend from net profits (cf. *preference shares*). **out of the ordinary** unusual. □ **ordinarily** *adv.* **ordinariness** *n.* [ME f. L *ordinarius* orderly (as ORDER)]

ordinate /ˈɔːdɪnət/ *n. Math.* a straight line from any point drawn parallel to one coordinate axis and meeting the other, usually a coordinate measured parallel to the vertical (cf. ABSCISSA). [L *linea ordinata applicata* line applied parallel f. *ordinare*: see ORDAIN]

ordination /ˌɔːdɪˈneɪʃ(ə)n/ *n.* **1 a** the act of conferring holy orders, esp. on a priest or deacon. **b** the admission of a priest etc. to church ministry. **2** the arrangement of things etc. in ranks; classification. **3** the act of decreeing or ordaining. [ME f. OF *ordination* or L *ordinatio* (as ORDAIN)]

ordnance /ˈɔːdnəns/ *n.* **1** mounted guns; cannon. **2** a branch of government service dealing esp. with military stores and materials. [ME var. of ORDINANCE]

ordnance datum *n.* mean sea level as defined for Ordnance Survey.

Ordnance map *n.* a map produced by Ordnance Survey.

Ordnance Survey (in the UK) an official survey organization, orig. under the Master of the Ordnance, preparing large-scale detailed maps of the whole country.

ordonnance /ˈɔːdənəns/ *n.* the systematic arrangement esp. of literary or architectural work. [F f. OF *ordenance* (as ORDINANCE)]

Ordovician /ˌɔːdəˈvɪsɪən, -ˈvɪʃɪən/ *adj. & n. Geol.* ● *adj.* of or relating to the second period of the Palaeozoic era. ● *n.* this period or the corresponding geological system. (*See note below.*) [L *Ordovices* ancient British tribe in N. Wales]

▪ The Ordovician lasted from about 505 to 438 million years ago, between the Cambrian and Silurian periods. It saw the diversification of many invertebrate groups and the appearance of the first vertebrates (jawless fish).

ordure /ˈɔːdjʊə(r)/ *n.* **1** excrement; dung. **2** obscenity; filth; foul language. [ME f. OF f. *ord* foul f. L *horridus*: see HORRID]

Ordzhonikidze /ˌɔːdʒɒnˈkɪdzɪ/ the former name (1954–93) for VLADIKAVKAZ.

Ore. *abbr.* Oregon.

ore¹ /ɔː(r)/ *n.* a naturally occurring solid material from which metal or other valuable minerals may be extracted. [OE *ōra* unwrought metal, *ār* bronze, rel. to L *aes* crude metal, bronze]

ore² /ˈɜːrə/ *n.* (also **øre**, **öre**) a Scandinavian monetary unit equal to one-hundredth of a krona or krone. [Danish, Norw. *øre*, Sw. *öre*]

oread /ˈɔːrɪˌæd/ *n. Gk & Roman Mythol.* a mountain nymph. [ME f. L *oreas -ados* f. Gk *oreias* f. *oros* mountain]

Örebro /ˌɜːrəˈbruː/ an industrial city in south central Sweden; pop. (1990) 120,940.

orectic /əˈrektɪk/ *adj. Philos. & Med.* of or concerning desire or appetite. [Gk *orektikos* f. *oregō* stretch out]

Oreg. *abbr.* Oregon.

oregano /ˌɒrɪˈɡɑːnəʊ/ *n.* the dried leaves of wild marjoram used as a culinary herb (cf. MARJORAM). [Sp., = ORIGAN]

Oregon /ˈɒrɪɡən/ a state in the north-western US, on the Pacific coast; pop. (1990) 2,842,320; capital, Salem. British claims to Oregon were formally ceded to the US in 1846 and it became the 33rd state in 1859. □ **Oregonian** /ˌɒrɪˈɡəʊnɪən/ *adj. & n.*

Oregon Trail a route across the central US, from the Missouri to Oregon, some 3,000 km (2,000 miles) in length. It was used chiefly in the 1840s by settlers moving west.

Orel /ɒˈrel, Russian ɐˈrjɒl/ an industrial city in SW Russia; pop. (1990) 342,000. It was founded in the 16th century as a fortress town to protect the southern frontiers of Muscovy against the Crimean Tartars.

Ore Mountains an alternative name for the ERZGEBIRGE.

Orenburg /ˈɒrənˌbɜːɡ/ a city in southern Russia, on the Ural river; pop. (1990) 552,000. Founded as a frontier fortress on the site of present-day Orsk in 1735, it was moved some 240 km (150 miles) downstream to its present site in 1743. It was known as Chkalov from 1938 to 1957.

Oreo /ˈɔːrɪəʊ/ *n.* (*pl.* **-os**) **1** *US propr.* a chocolate biscuit with a white cream filling. **2** *US sl. derog.* an American black who is seen, esp. by other blacks, as part of the white Establishment.

oreography var. of OROGRAPHY.

Orestes /ɒˈrestiːz/ *Gk Mythol.* the son of Agamemnon and Clytemnestra. He killed his mother and her lover Aegisthus to avenge the murder of Agamemnon.

Øresund /ˌɜːrəˈsʊnd/ (also called *the Sound*) a narrow channel between Sweden and the Danish island of Zealand.

oreweed var. of OARWEED.

orfe /ɔːf/ *n.* a silvery freshwater fish, *Leuciscus idus*, of the carp family, fished commercially in northern Eurasia. Also called *ide.* □ **golden orfe** a yellow variety of the orfe, domesticated as an ornamental fish. [G & F: cf. L *orphus* f. Gk *orphos* sea-perch]

Orff /ɔːf/, Carl (1895–1982), German composer. His early career was devoted to evolving a system of musical education for children, beginners, and amateurs, based on Jaques-Dalcroze eurhythmics. He set out his theory in *Schulwerk* (1930–3), stressing the value of simplicity and controlled improvisation in music. Orff is best known for his secular cantata *Carmina Burana* (1937), based on a collection of medieval Latin poems; although originally conceived as a dramatic work, it is more often performed on the concert platform. Among his other works is an operatic trilogy, including *Antigone* (1949).

organ /ˈɔːgən/ *n.* **1 a** a usu. large musical instrument having pipes supplied with air from bellows, sounded by keys, and distributed into sets or stops which form partial organs, each with a separate keyboard (*choir organ*; *pedal organ*). (*See note below.*) **b** a smaller instrument without pipes, producing similar sounds electronically. **c** a smaller keyboard wind instrument with metal reeds; a harmonium. **d** = BARREL-ORGAN. **2 a** a usu. self-contained part of an organism having a special vital function (*vocal organs*; *digestive organs*). **b** esp. *joc.* the penis. **3** a medium of communication, esp. a newspaper or periodical which serves as the mouthpiece of a movement, political party, etc. **4** *archaic* a professionally trained singing voice. **5** *hist.* a region of the brain formerly held to be the seat of a particular faculty. □ **organ-blower** a person or mechanism working the bellows of an organ. **organ-grinder** the player of a barrel-organ. **organ-loft** a gallery in a church or concert-room for an organ. **organ of Corti** see CORTI. **organ-pipe** any of the pipes on an organ. **organ-screen** an ornamental screen usu. between the choir and the nave of a church, cathedral, etc., on which the organ is placed. **organ-stop 1** a set of pipes of a similar tone in an organ. **2** the handle of the mechanism that brings it into action. [ME f. OE *organa* & OF *organe*, f. L *organum* f. Gk *organon* tool]

▪ The organ is thought to have been invented in the 3rd century BC in Alexandria, and was known to the Romans and Byzantines. Organs began to be constructed in western Europe after one was brought from Constantinople to France in the 8th century and they have been associated with the Christian Church for nearly a millennium. By the 17th century the organ was very similar to the instrument of today, and was at its most popular during the era of baroque music; J. S. Bach, in particular, wrote many compositions for the organ. Technical innovations of the 19th century made the organ louder and less difficult to play, and also incorporated a rich variety of orchestral stops.

organdie /ˈɔːgəndɪ, ɔːˈgændɪ/ *n.* (*US* **organdy**) (*pl.* **-ies**) a fine translucent cotton muslin, usu. stiffened. [F *organdi*, of unkn. orig.]

organelle /ˌɔːgəˈnɛl/ *n. Biol.* a specialized structure within a cell or unicellular organism, e.g. a nucleus. [mod.L *organella* dimin.; see ORGAN, -LE²]

organic /ɔːˈgænɪk/ *adj.* **1 a** *Physiol.* of or relating to a bodily organ or organs. **b** *Med.* (of a disease) affecting the structure of an organ. **2** (of a plant or animal) having organs or an organized physical structure. **3** *Agriculture* produced or involving production without the use of chemical fertilizers, pesticides, etc. (*organic crop*; *organic farming*). **4** *Chem.* (of a compound etc.) containing carbon (opp. INORGANIC). **5 a** structural, inherent. **b** constitutional, fundamental. **6** organized, systematic, coordinated (*an organic whole*). **7** characterized by or designating continuous or natural development (*the company expanded through organic growth rather than acquisitions*). □ **organic chemistry** the chemistry of carbon compounds. **organic law** a law stating the formal constitution of a country. □ **organically** *adv.* [F *organique* f. L *organicus* f. Gk *organikos* (as ORGAN)]

organism /ˈɔːgəˌnɪz(ə)m/ *n.* **1** a living individual consisting of a single cell or of a group of interdependent parts sharing the life processes. **2 a** an individual live plant or animal. **b** the material structure of this. **3** a whole with interdependent parts compared to a living being. [F *organisme* (as ORGANIZE)]

organist /ˈɔːgənɪst/ *n.* the player of an organ.

organization /ˌɔːgənaɪˈzeɪʃ(ə)n/ *n.* (also **-isation**) **1** the act or an instance of organizing; the state of being organized. **2** an organized body, esp. a business, government department, charity, etc. **3** systematic arrangement; tidiness. □ **organization man** a man who subordinates his individuality and his personal life to the organization he serves. □ **organizational** *adj.* **organizationally** *adv.*

Organization for Economic Cooperation and Development (abbr. **OECD**) an organization formed in 1961 (replacing the Organization for European Economic Cooperation) to assist the economy of its member nations and to promote world trade. Its members include the industrialized countries of western Europe together with Australia, Japan, New Zealand, and the US. Its headquarters are in Paris.

Organization for European Economic Cooperation see ORGANIZATION FOR ECONOMIC COOPERATION AND DEVELOPMENT.

Organization of African Unity (abbr. **OAU**) an association of African states founded in 1963 for mutual cooperation and the elimination of colonialism in Africa. It is based in Addis Ababa.

Organization of American States (abbr. **OAS**) an association including most of the countries of North and South America. Originally founded in 1890 for largely commercial purposes, it adopted its present name and charter in 1948. Its aims are to work for peace and prosperity in the region and to uphold the sovereignty of member nations. Its headquarters are in Washington, DC.

Organization of Arab Petroleum Exporting Countries (abbr. **OAPEC**) an association of Arab countries, founded in 1968 to promote economic cooperation and safeguard its members' interests and to ensure the supply of oil to consumer markets.

Organization of Central American States (abbr. **OCAS**) an association of Central American countries founded in 1951 for economic and political cooperation. Its members (Guatemala, Honduras, El Salvador, and Costa Rica) succeeded in establishing the Central American Common Market in 1960, but have cooperated in little else.

Organization of the Petroleum Exporting Countries (abbr. **OPEC** /ˈəʊpɛk/) an association of the thirteen major oil-producing countries, founded in 1960 to coordinate policies. Its headquarters are in Vienna.

organize /ˈɔːgəˌnaɪz/ *v.tr.* (also **-ise**) **1 a** give an orderly structure to, systematize. **b** bring the affairs of (another person or oneself) into order; make arrangements for (a person). **2 a** arrange for or initiate (a scheme etc.). **b** provide; take responsibility for (*organized some sandwiches*). **3** (often *absol.*) **a** enrol (new members) in a trade union, political party, etc. **b** form (a trade union or other political group). **4 a** form (different elements) into an organic whole. **b** form (an organic whole). **5** (esp. as **organized** *adj.*) make organic; make into a living being or tissue. □ **organizable** *adj.* **organizer** *n.* [ME f. OF *organiser* f. med.L *organizare* f. L (as ORGAN)]

organo- /ˈɔːgənəʊ, ɔːˈgæn-/ *comb. form* **1** esp. *Biol.* organ. **2** *Chem.* organic; esp. in naming classes of organic compounds containing a particular element (*organochlorine*; *organophosphorus*). [Gk (as ORGAN)]

organoleptic /ˌɔːgənəʊˈlɛptɪk/ *adj. Biol.* affecting the organs of sense. [ORGANO- + Gk *lēptikos* disposed to take f. *lambanō* take]

organometallic /ˌɔːgænəʊmɪˈtælɪk/ *adj. Chem.* (of a compound) organic and containing a metal.

organon /ˈɔːgəˌnɒn/ *n.* (also **organum** /-nəm/) an instrument of thought, esp. a means of reasoning or a system of logic. [Gk *organon* & L *organum* (as ORGAN): *Organon* was the title of Aristotle's logical writings, and *Novum* (new) *Organum* that of Bacon's]

organotherapy /ˌɔːgənəʊˈθɛrəpɪ/ *n. Med.* the treatment of disease using extracts from animal organs, esp. glands.

organza /ɔːˈgænzə/ *n.* a thin stiff transparent silk or synthetic dress fabric. [prob. f. *Lorganza* (US trade name)]

organzine /ˈɔːgənˌziːn, ɔːˈgænziːn/ *n.* a silk thread in which the main twist is in a contrary direction to that of the strands. [F *organsin* f. It. *organzino*, of unkn. orig.]

orgasm /ˈɔːgæz(ə)m/ *n. & v.* ● *n.* **1 a** the climax of sexual excitement, esp. during sexual intercourse. **b** an instance of this. **2** violent excitement; rage. ● *v.intr.* experience a sexual orgasm. □ **orgasmic** /ɔːˈgæzmɪk/ *adj.* **orgasmically** *adv.* **orgastic** /ɔːˈgæstɪk/ *adj.* **orgastically** *adv.* [F *orgasme* or mod.L f. Gk *orgasmos* f. *orgaō* swell, be excited]

orgeat /ˈɔːdʒɪˌæt, ˈɔːʒɑː/ *n.* a cooling drink made from barley or almonds and orange-flower water. [F f. Prov. *orjat* f. *ordi* barley f. L *hordeum*]

orgiastic /ˌɔːdʒɪˈæstɪk/ *adj.* of or resembling an orgy. □ **orgiastically** *adv.* [Gk *orgiastikos* f. *orgiastēs* agent-noun f. *orgiazō* hold an orgy]

orgulous /ˈɔːgjʊləs/ *adj. archaic* haughty; splendid. [ME f. OF *orguillus* f. *orguill* pride f. Frank.]

orgy /ˈɔːdʒɪ/ *n.* (*pl.* **-ies**) **1** a wild drunken festivity, esp. one at which indiscriminate sexual activity takes place. **2** excessive indulgence in an activity. **3** (usu. in *pl.*) *Gk & Rom. Hist.* secret rites used in the worship of esp. Bacchus, celebrated with dancing, drunkenness, singing, etc. [orig. pl., f. F *orgies* f. L *orgia* f. Gk *orgia* secret rites]

oribi /ˈɒrɪbɪ/ *n.* (*pl.* same or **oribis**) a small southern African grazing antelope, *Ourebia ourebi*, having a reddish-fawn back and white underparts. [prob. Khoisan]

oriel /ˈɔːrɪəl/ *n.* **1** a large polygonal recess built out usu. from an upper storey and supported from the ground or on corbels. **2** (in full **oriel**

window) **a** any of the windows in an oriel. **b** the projecting window of an upper storey. [ME f. OF *oriol* gallery, of unkn. orig.]

orient *n., adj., & v.* ● *n.* /'ɔːrɪənt, 'ɒr-/ **1** (**the Orient**) **a** *poet.* the east. **b** the countries east of the Mediterranean, esp. eastern Asia. **2** an orient pearl. ● *adj.* /'ɔːrɪənt/ **1** *poet.* eastern. **2** (of precious stones and esp. the finest pearls coming orig. from the East) lustrous; sparkling; precious. **3** *archaic* **a** radiant. **b** (of the sun, daylight, etc.) rising. ● *v.* /'ɔːrɪˌent, 'ɒr-/ **1** *tr.* **a** place or exactly determine the position of with the aid of a compass; settle or find the bearings of. **b** (often foll. by *towards*) bring (oneself, different elements, etc.) into a clearly understood position or relationship; direct. **2** *tr.* **a** place or build (a church, building, etc.) facing towards the east. **b** bury (a person) with the feet towards the east. **3** *intr.* turn eastward or in a specified direction. □ **orient oneself** determine how one stands in relation to one's surroundings. [ME f. OF *orient, orienter* f. L *oriens -entis* rising, sunrise, east, f. *oriri* rise]

oriental /ˌɔːrɪ'ent(ə)l, ˌɒr-/ *adj. & n.* ● *adj.* **1** (often **Oriental**) **a** of or characteristic of Eastern civilization etc. **b** of or concerning the East, esp. eastern Asia. **2** (of a pearl etc.) orient. ● *n.* (esp. **Oriental**) **1** a native of the Orient, esp. the Far East. **2** a person of Oriental esp. Asian descent. □ **orientalism** *n.* **orientalist** *n.* **orientalize** *v.intr. & tr.* (also **-ise**) **orientally** *adv.* [ME f. OF *oriental* or L *orientalis* (as ORIENT)]

orientate /'ɔːrɪenˌteɪt, 'ɒr-/ *v.tr. & intr.* = ORIENT *v.* [prob. back-form. f. ORIENTATION]

orientation /ˌɔːrɪen'teɪʃ(ə)n, ˌɒr-/ *n.* **1** the act or an instance of orienting; the state of being oriented. **2 a** a relative position. **b** a person's attitude or adjustment in relation to circumstances, esp. politically or psychologically. **3** an introduction to a subject or situation; a briefing. **4** the faculty by which birds etc. find their way home from a distance. □ **orientation course** esp. *N. Amer.* a course giving information to newcomers to a university etc. □ **orientational** *adj.* [app. f. ORIENT]

orienteering /ˌɔːrɪen'tɪərɪŋ, ˌɒr-/ *n.* a competitive sport in which runners find their way across open country with a map and compass. Although 'chart and compass' races were held in the sports meetings of British army units in the early 20th century, orienteering is generally recognized as being of Scandinavian origin, first introduced as a sport for young people in Sweden in 1918. □ **orienteer** *n. & v.intr.* [Sw. *orientering*]

Orient Express a train which ran between Paris and Istanbul and other Balkan cities, via Vienna, from 1883 to 1961. Since 1961 the name has been used for various trains running over parts of the old route.

orifice /'ɒrɪfɪs/ *n.* an opening, esp. the mouth of a cavity, a bodily aperture, etc. [F f. LL *orificium* f. *os oris* mouth + *facere* make]

oriflamme /'ɒrɪˌflæm/ *n.* **1** *hist.* the sacred scarlet silk banner of St Denis given to early French kings by the abbot of St Denis on setting out for war. **2** a standard, a principle, or an ideal as a rallying-point in a struggle. **3** a bright conspicuous object, colour, etc. [ME f. OF f. L *aurum* gold + *flamma* flame]

origami /ˌɒrɪ'gɑːmɪ/ *n.* the Japanese art of folding paper into decorative shapes and figures. [Jap. f. *ori* fold + *kami* paper]

origan /'ɒrɪgən/ *n.* (also **origanum** /ˌɒrɪ'gɑːnəm, ə'rɪgən-/) any plant of the genus *Origanum*, esp. wild marjoram (see MARJORAM). [(ME f. OF *origan*) f. L *origanum* f. Gk *origanon*]

Origen /'ɒrɪdʒən/ (*c.*185–*c.*254), Christian scholar and theologian, probably born in Alexandria. Of his numerous works the most famous was the *Hexapla*, an edition of the Old Testament with six or more parallel versions. He recognized literal, moral, and allegorical interpretations of Scripture, preferring the last. His teachings are important for their introduction of Neoplatonist elements into Christianity but were later rejected by Church orthodoxy.

origin /'ɒrɪdʒɪn/ *n.* **1** a beginning or starting-point; a derivation; a source (*a word of Latin origin*). **2** (often in *pl.*) a person's ancestry (*what are his origins?*). **3** *Anat.* **a** a place at which a muscle is firmly attached. **b** a place where a nerve or blood vessel begins or branches from a main nerve or blood vessel. **4** *Math.* a fixed point from which coordinates are measured. [F *origine* or f. L *origo -ginis* f. *oriri* rise]

original /ə'rɪdʒɪn(ə)l/ *adj. & n.* ● *adj.* **1** existing from the beginning; innate. **2** novel; inventive; creative (*has an original mind*). **3** serving as a pattern; not derivative or imitative; firsthand (*in the original Greek; has an original Rembrandt*). ● *n.* **1** an original model, pattern, picture, etc. from which another is copied or translated (*kept the copy and destroyed the original*). **2** an eccentric or unusual person. **3 a** a garment specially designed for a fashion collection. **b** a copy of such a garment made to order. □ **original instrument** a musical instrument, or a copy of one,

dating from the time the music played on it was composed. **original print** a print made directly from an artist's own woodcut, etching, etc., and printed under the artist's supervision. □ **originally** *adv.* [ME f. OF *original* or L *originalis* (as ORIGIN)]

originality /əˌrɪdʒɪ'nælɪtɪ/ *n.* (*pl.* **-ies**) **1** the power of creating or thinking creatively. **2** newness or freshness (*this vase has originality*). **3** an original act, thing, trait, etc.

original sin *n.* (in Christian theology) the tendency to evil supposedly innate in all human beings, held to be inherited from Adam in consequence of the Fall. The concept of original sin is not explicitly set out in the Bible, but was firmly established by the writings of St Augustine. The view of some early theologians that the human will is capable of good without the help of divine grace was branded a heresy and the doctrine of original sin was widely accepted throughout the Middle Ages. Jesus, and, according to Roman Catholics, the Virgin Mary, are believed to be free of original sin.

originate /ə'rɪdʒɪˌneɪt/ *v.* **1** *tr.* cause to begin; initiate. **2** *intr.* (usu. foll. by *from, in, with*) have as an origin; begin. □ **originator** *n.* **originative** /-nətɪv/ *adj.* **origination** /əˌrɪdʒɪ'neɪʃ(ə)n/ *n.* [med. L *originare* (as ORIGIN)]

orimulsion /ˌɒrɪ'mʌlʃ(ə)n/ *n.* (also **Orimulsion**) *propr.* an emulsion of bitumen in water, used as a fuel. [f. *Orinoco* in Venezuela (the original source of the bitumen) + EMULSION]

orinasal /ˌɔːrɪ'neɪz(ə)l/ *adj.* (esp. of French nasalized vowels) sounded with both the mouth and the nose. [L *os oris* mouth + NASAL]

o-ring /'əʊrɪŋ/ *n.* a gasket in the form of a ring with a circular cross-section.

Orinoco /ˌɒrɪ'nəʊkəʊ/ a river in northern South America, which rises in SE Venezuela and flows 2,060 km (1,280 miles) in a great arc through Venezuela, entering the Atlantic Ocean through a vast delta. For part of its length it forms the border between Colombia and Venezuela.

oriole /'ɔːrɪˌəʊl/ *n.* **1** an Old World bird of the family Oriolidae, esp. the genus *Oriolus*, many of which have brightly coloured plumage; e.g. the golden oriole. **2** a New World bird of the genus *Icterus*, with similar coloration, related to the grackles. [med.L *oriolus* f. OF *oriol* f. L *aureolus* dimin. of *aureus* golden f. *aurum* gold]

Orion /ə'raɪən/ *n.* **1** *Gk Mythol.* a giant and hunter who was changed into a constellation at his death. The association of his name with the constellation occurs in Homer, making it an exceptionally early star-myth. **2** *Astron.* a conspicuous constellation (the Hunter), said to represent a hunter holding a club and shield. It contains many bright stars, including Rigel and Betelgeuse, and a prominent line of three forms 'Orion's Belt'. Others form the 'Sword of Orion', which contains the Great Nebula and the multiple star Theta Orionis.

orison /'ɒrɪz(ə)n/ *n.* (usu. in *pl.*) *archaic* a prayer. [ME f. AF *ureison*, OF *oreison* f. L (as ORATION)]

Orissa /ə'rɪsə/ a state in eastern India, on the Bay of Bengal; capital, Bhubaneswar.

-orium /'ɔːrɪəm/ *suffix* forming nouns denoting a place for a particular function (*auditorium; crematorium*). [L, neut. of adjectives in *-orius*: see -ORY¹]

Oriya /ɒ'riːə/ *adj. & n.* ● *adj.* **1** of or relating to Odra, an ancient region corresponding to the modern Indian state of Orissa. **2** of or relating to Orissa. ● *n.* (*pl.* same or **Oriyas**) **1** a native or inhabitant of Orissa or Odra. **2** the Indic or Indo-Aryan language of Odra (later Orissa), which is descended from Sanskrit and closely related to Bengali, spoken today by some 20 million people. [Hindi, ult. f. Skr. *Oḍra*]

Orkney Islands /'ɔːknɪ/ (also **Orkney, Orkneys**) a group of more than seventy islands off the north-east tip of Scotland, constituting an administrative region of Scotland; pop. (1991) 19,450; chief town, Kirkwall. Colonized by the Vikings in the 9th century, they were ruled by Norway and Denmark until 1472, when they came into Scottish possession (together with Shetland) as security against the unpaid dowry of Margaret of Denmark after her marriage to James III of Scotland.

Orlando /ɔː'lændəʊ/ a city in central Florida; pop. (1990) 164,690. A popular tourist resort, it is situated near the John F. Kennedy Space Center and Disney World.

orle /ɔːl/ *n. Heraldry* a narrow band or border of charges near the edge of a shield. [F *o(u)rle* f. *ourler* to hem, ult. f. L *ora* edge]

Orleanist /'ɔːlɪənɪst, ɔː'liːə-/ *n. hist.* a supporter of the claim to the French throne of the branch of the Bourbon dynasty descended from the Duc d'Orléans, younger brother of Louis XIV. Louis Philippe, king of France 1830–48, belonged to this branch.

Orleans /ɔːˈliːənz/ (French **Orléans** /ɔrleɑ̃/) a city in central France, on the Loire; pop. (1990) 107,965. It was conquered by Julius Caesar in 52 BC and by the 10th century was one of the most important cities in France. In 1429 it was the scene of Joan of Arc's first victory over the English during the Hundred Years War.

Orlon /ˈɔːlɒn/ n. propr. a man-made fibre and fabric for textiles and knitwear. [invented word, after NYLON]

orlop /ˈɔːlɒp/ n. the lowest deck of a ship with three or more decks. [ME f. MDu. *overloop* covering f. *overloopen* run over (as OVER-, LEAP)]

Ormazd /ˈɔːmæzd/ see AHURA MAZDA.

ormer /ˈɔːmə(r)/ n. an edible abalone (gastropod mollusc); esp. *Haliotis tuberculata*, used as food in the Channel Islands. Also called *sea-ear*. [Channel Islands F f. F *ormier* f. L *auris maris* ear of sea]

ormolu /ˈɔːmə‚luː/ n. **1** (often attrib.) gilded bronze; gold-coloured alloy of copper, zinc, and tin used to decorate furniture, make ornaments, etc. **2** articles made of or decorated with this. [F *or moulu* powdered gold (for use in gilding)]

Ormuz see HORMUZ.

ornament n. & v. ● n. /ˈɔːnəmənt/ **1 a** a thing used or serving to adorn, esp. a small trinket, vase, ˈfigure, etc. (*a mantelpiece crowded with ornaments; her only ornament was a brooch*). **b** a quality or person conferring adornment, grace, or honour (*an ornament to her profession*). **2** decoration added to embellish esp. a building (*a tower rich in ornament*). **3** (in pl.) Mus. embellishments and decorations made to a melody. **4** (usu. in pl.) Eccl. the accessories of worship, e.g. the altar, chalice, sacred vessels, etc. ● v.tr. /ˈɔːnə‚ment/ adorn; beautify. □ **ornamentation** /ˌɔːnəmenˈteɪʃ(ə)n/ n. [ME f. AF *urnement*, OF *o(u)rnement* f. L *ornamentum* equipment f. *ornare* adorn]

ornamental /ˌɔːnəˈment(ə)l/ adj. & n. ● adj. serving as an ornament; decorative. ● n. a thing considered to be ornamental, esp. a cultivated plant. □ **ornamentalism** n. **ornamentalist** n. **ornamentally** adv.

ornate /ɔːˈneɪt/ adj. **1** elaborately adorned; highly decorated. **2** (of literary style) convoluted; flowery. □ **ornately** adv. **ornateness** n. [ME f. L *ornatus* past part. of *ornare* adorn]

ornery /ˈɔːnərɪ/ adj. N. Amer. colloq. **1** cantankerous; unpleasant. **2** of poor quality. □ **orneriness** n. [var. of ORDINARY]

ornithic /ɔːˈnɪθɪk/ adj. of or relating to birds. [Gk *ornithikos* birdlike (as ORNITHO-)]

ornithischian /ˌɔːnɪˈθɪʃɪən, -ˈθɪskɪən/ adj. & n. ● adj. of or relating to the order Ornithischia, which includes dinosaurs having a pelvic structure like that of birds. (See DINOSAUR.) ● n. a dinosaur of this order. [mod.L f. Gk *ornis ornithos* bird + *iskhion* hip joint]

ornitho- /ˈɔːnɪθəʊ/ comb. form bird. [Gk f. *ornis ornithos* bird]

ornithology /ˌɔːnɪˈθɒlədʒɪ/ n. the scientific study of birds. □ **ornithologist** n. **ornithological** /-θəˈlɒdʒɪk(ə)l/ adj. **ornithologically** adv. [mod.L *ornithologia* f. Gk *ornithologos* treating of birds (as ORNITHO-, -LOGY)]

ornithopod /ˈɔːnɪθə‚pɒd/ n. & adj. ● n. a bipedal herbivorous ornithischian dinosaur of the suborder Ornithopoda. ● adj. of or relating to this suborder. [mod.L f. Gk *ornis ornith-* bird + *pous pod-* foot]

ornithorhynchus /ˌɔːnɪθəʊˈrɪŋkəs/ n. = PLATYPUS. [ORNITHO- + Gk *rhugkhos* bill]

oro- /ˈɔːrəʊ/ comb. form mountain. [Gk *oros* mountain]

orogeny /ɔːˈrɒdʒɪnɪ/ n. (also **orogenesis** /ˌɔːrəʊˈdʒenɪsɪs/) the process of the formation of mountains. □ **orogenetic** /ˌɔːrəʊdʒɪˈnetɪk/ adj. **orogenic** /-ˈdʒenɪk/ adj.

orography /ɔːˈrɒgrəfɪ/ n. (also **oreography** /ˌɔːrɪˈɒg-/) the branch of physical geography dealing with mountains. □ **orographic** /ˌɔːrəˈgræfɪk/ adj. **orographical** adj.

Orontes /əˈrɒntiːz/ a river in SW Asia which rises near Baalbek in northern Lebanon and flows 571 km (355 miles) through western and northern Syria before turning west through southern Turkey to enter the Mediterranean. It is an important source of water for irrigation, especially in Syria.

orotund /ˈɒrə‚tʌnd, ˈɔːr-/ adj. **1** (of the voice or phrasing) full, round; imposing. **2** (of writing, style, expression, etc.) pompous; pretentious. [L *ore rotundo* with rounded mouth]

orphan /ˈɔːf(ə)n/ n. & v. ● n. (often attrib.) **1** a child bereaved of a parent or usu. both parents. **2** a person bereft of previous protection, advantages, etc. **3** Printing the first line of a paragraph at the foot of a page or column. ● v.tr. bereave (a child) of its parents or a parent. □ **orphanhood** n. **orphanize** v.tr. (also **-ise**). [ME f. LL *orphanus* f. Gk *orphanos* bereaved]

orphanage /ˈɔːfənɪdʒ/ n. **1** a usu. residential institution for the care and education of orphans. **2** orphanhood.

Orphean /ɔːˈfiːən/ adj. of or like the music of Orpheus; melodious; entrancing. [L *Orpheus* (adj.) f. Gk *Orpheios*; cf. ORPHEUS]

Orpheus /ˈɔːfɪəs/ Gk Mythol. a poet who sang and played with his lyre so wonderfully that wild beasts were spellbound by his music. When his wife Eurydice died from a snakebite he visited the underworld to plead for her release from the dead. This was granted, but he lost her because he failed to obey the condition that he must not look back at her until they had reached the world of the living. He was also regarded as the founder of Orphism.

Orphic /ˈɔːfɪk/ adj. **1** of or concerning Orpheus or the mysteries, doctrines, etc. associated with him; oracular; mysterious. **2** = ORPHEAN. [L *Orphicus* f. Gk *Orphikos*; cf. ORPHEUS]

Orphism /ˈɔːfɪz(ə)m/ n. **1** a mystic religion of ancient Greece, originating in the 7th or 6th century BC and based on poems (now lost) attributed to Orpheus, emphasizing the mixture of good and evil in human nature and the necessity for individuals to rid themselves of the evil part by ritual and moral purification throughout a series of reincarnations. It had declined by the 5th century BC, but was an important influence on Pindar and Plato. **2** a short-lived art movement (c.1912) within cubism, pioneered by a group of French painters (including Robert Delaunay, Sonia Delaunay-Terk, and Fernand Léger (1881–1955)) who called themselves *La Section d'Or* (Golden Section) and emphasized the lyrical use of colour rather than the austere intellectual cubism of Picasso, Braque, and Gris.

orphrey /ˈɔːfrɪ/ n. (pl. **-eys**) an ornamental stripe or border or separate piece of ornamental needlework, esp. on ecclesiastical vestments. [ME *orfreis* (taken as pl.) (gold) embroidery f. OF f. med.L *aurifrisium* etc. f. L *aurum* gold + *Phrygius* Phrygian, also 'embroidered']

orpiment /ˈɔːpɪmənt/ n. **1** a mineral form of arsenic trisulphide, formerly used as a dye and artist's pigment. Also called *yellow arsenic*. **2** (in full **red orpiment**) = REALGAR. [ME f. OF f. L *auripigmentum* f. *aurum* gold + *pigmentum* pigment]

orpine /ˈɔːpaɪn/ n. (also **orpin** /-pɪn/) a succulent herbaceous purple-flowered plant, *Sedum telephium*. Also called *livelong*. [ME f. OF *orpine*, prob. alt. of ORPIMENT, orig. of a yellow-flowered species of the same genus]

orra /ˈɒrə/ adj. Sc. **1** not matched; odd. **2** occasional; extra. [18th c.: orig. unkn.]

orrery /ˈɒrərɪ/ n. (pl. **-ies**) a clockwork model of the solar system. [Charles Boyle, 4th Earl of *Orrery* (1678–1731), for whom one was made]

orris /ˈɒrɪs/ n. **1** a plant of the genus *Iris*; esp. *I. florentina* or *I. pallida*, grown for their fragrant rhizomes. **2** = ORRISROOT. □ **orris-powder** powdered orrisroot. [16th c.: app. an unexpl. alt. of IRIS]

orrisroot /ˈɒrɪs‚ruːt/ n. the fragrant rootstock of the orris, used in perfumery and formerly in medicine.

Orsk /ɔːsk/ a city in southern Russia, in the Urals on the Ural river near the border with Kazakhstan; pop. (1990) 271,000. It was founded in 1735 as the fortress of Orenburg, which was moved down river in 1743, and developed as a mining town. Orsk is now a major industrial centre and railway junction.

ortanique /ˌɔːtəˈniːk/ n. a citrus fruit produced by crossing an orange and a tangerine. [ORANGE + TANGERINE, after *unique*]

Ortega /ɔːˈteɪgə/, Daniel (full surname Ortega Saavedra) (b.1945), Nicaraguan statesman, President 1985–90. He joined the Sandinista National Liberation Front (FSLN) in 1963, becoming its leader in 1966. After a period of imprisonment he played a major role in the revolution which overthrew Anastasio Somoza in 1979. He headed a provisional socialist government from this date, later gaining the presidency after the Sandinista victory in the 1984 elections, but his regime was constantly under attack from the US-backed Contras. He lost power to an opposition coalition following elections in 1990.

ortho- /ˈɔːθəʊ/ comb. form **1 a** straight, rectangular, upright. **b** right, correct. **2** Chem. **a** relating to two adjacent carbon atoms in a benzene ring. **b** relating to acids and salts (e.g. *orthophosphates*) giving *meta-* compounds on removal of water. [Gk *orthos* straight]

orthocephalic /ˌɔːθəʊsɪˈfælɪk, -kɪˈfælɪk/ adj. having a head with a medium ratio of breadth to height.

orthochromatic /ˌɔːθəʊkrəʊˈmætɪk/ adj. giving fairly correct relative intensity to colours in photography by being sensitive to all except red.

orthoclase /ˈɔːθəʊ‚kleɪs, -‚kleɪz/ n. a common alkali feldspar usu. occurring as variously coloured crystals, used in ceramics and glass-making. [ORTHO- + Gk *klasis* breaking]

orthodontics /ˌɔːθəˈdɒntɪks/ *n.pl.* (treated as *sing.*) (also **orthodontia** /-ˈdɒntɪə/) the treatment of irregularities in the teeth and jaws. □ **orthodontic** *adj.* **orthodontist** *n.* [ORTHO- + Gk *odous odont-* tooth]

orthodox /ˈɔːθəˌdɒks/ *adj.* **1 a** holding correct or currently accepted opinions, esp. on religious doctrine, morals, etc. **b** not independent-minded; unoriginal; unheretical. **2** (of religious doctrine, standards of morality, etc.) generally accepted as right or true; authoritatively established; conventional. **3** (also **Orthodox**) of or relating to Orthodox Judaism. **4** (**Orthodox**) of or relating to the Orthodox Church. □ **orthodoxly** *adv.* [eccl. L *orthodoxus* f. Gk *orthodoxos* (as ORTHO-, *doxa*) opinion)]

Orthodox Church *n.* a Christian Church or federation of Churches acknowledging the authority of the patriarch of Constantinople, originating in the Greek-speaking Church of the Byzantine Empire and its missions in eastern Europe and parts of Asia. The Orthodox Churches include the national Churches of Greece, Russia, Bulgaria, Romania, etc. The central act of worship is the Liturgy (Eucharist), celebrated with much ritual in a way which is in many respects unchanged from the 6th century. Separation from the mainly Latin-speaking Western Church began in the 4th century as doctrine and ritual diverged and papal authority in the West increased, and took formal and lasting effect with the mutual excommunication of the patriarch of Constantinople and the papal legate in 1054. Many Orthodox Churches were subjected to centuries of Muslim rule. The Orthodox Churches are strongly traditional, but have participated in ecumenical dialogue during the 20th century, and though opposition to papal supremacy is undiminished, the excommunications of 1054 were annulled in 1965.

Orthodox Judaism *n.* a major branch within Judaism which teaches strict adherence to rabbinical interpretation of Jewish law and its traditional observances. There are more than 600 rules or *mitzvahs*, governing moral behaviour, dress, diet, work, observance of the sabbath, religious customs, and personal hygiene. Unlike adherents of modern movements such as Conservative and Reform Judaism, Orthodox Jews maintain the separation of the sexes in synagogue worship.

orthodoxy /ˈɔːθəˌdɒksɪ/ *n.* (*pl.* **-ies**) **1** the state of being orthodox. **2 a** the orthodox practice of a religion, esp. Judaism or the Christian Orthodox Church. **b** the body of Orthodox Jews or Orthodox Christians. **3** esp. *Relig.* an authorized or generally accepted theory, doctrine, etc. [LL *orthodoxia* f. late Gk *orthodoxia* sound doctrine (as ORTHODOX)]

orthoepy /ˈɔːθəʊˌiːpɪ, ɔːˈθəʊɪpɪ/ *n.* the study of the (correct) pronunciation of words. □ **orthoepist** *n.* **orthoepic** /ˌɔːθəʊˈepɪk/ *adj.* [Gk *orthoepeia* correct speech (as ORTHO-, *epos* word)]

orthogenesis /ˌɔːθəʊˈdʒenɪsɪs/ *n.* a theory of evolution which proposes that variations follow a defined direction and are not merely sporadic and fortuitous. □ **orthogenetic** /-dʒɪˈnetɪk/ *adj.* **orthogenetically** *adv.*

orthognathous /ɔːˈθɒɡnəθəs/ *adj.* (of mammals, including man) having a jaw which does not project forwards and a facial angle approaching a right angle. [ORTHO- + Gk *gnathos* jaw]

orthogonal /ɔːˈθɒɡən(ə)l/ *adj.* of or involving right angles; at right angles (*orthogonal to each other*). □ **orthogonal map** see PETERS PROJECTION. [F f. *orthogone* (as ORTHO-, -GON)]

orthography /ɔːˈθɒɡrəfɪ/ *n.* (*pl.* **-ies**) **1 a** correct or conventional spelling; spelling according to accepted usage. **b** the study or science of spelling. **2 a** perspective projection used in maps and elevations in which the projection lines are parallel. **b** a map etc. so projected. □ **orthographer** *n.* **orthographic** /ˌɔːθəˈɡræfɪk/ *adj.* **orthographical** *adj.* **orthographically** *adv.* [ME f. OF *ortografie* f. L *orthographia* f. Gk *orthographia* (as ORTHO-, -GRAPHY)]

orthopaedics /ˌɔːθəˈpiːdɪks/ *n.pl.* (treated as *sing.*) (*US* **-pedics**) the branch of medicine dealing with the correction of deformities of bones or muscles, orig. in children. □ **orthopaedic** *adj.* **orthopaedist** *n.* [F *orthopédie* (as ORTHO-, *pédie* f. Gk *paideia* rearing of children)]

Orthoptera /ɔːˈθɒptərə/ *n.pl. Zool.* an order of insects comprising the grasshoppers, locusts, and crickets. They have narrow forewings, hind legs modified for jumping, and chirping or rasping calls. The cockroaches, mantids, and stick insects were formerly also included in this order. □ **orthopteran** *n.* & *adj.* **orthopterous** *adj.* [mod.L f. ORTHO- + Gk *pteros* wing]

orthoptic /ɔːˈθɒptɪk/ *adj.* relating to the correct or normal use of the eyes. □ **orthoptist** *n.* [ORTHO- + Gk *optikos* of sight: see OPTIC]

orthoptics /ɔːˈθɒptɪks/ *n. Med.* the study or treatment of irregularities of the eyes, esp. with reference to the eye-muscles.

orthorhombic /ˌɔːθəʊˈrɒmbɪk/ *adj. Crystallog.* (of a crystal) characterized by three mutually perpendicular axes which are unequal in length, as in topaz and talc.

orthotone /ˈɔːθəˌtəʊn/ *adj.* & *n.* ● *adj.* (of a word) having an independent stress pattern, not enclitic nor proclitic. ● *n.* a word of this kind.

ortolan /ˈɔːtələn/ *n.* (in full **ortolan bunting**) a small European bird, *Emberiza hortulana*, the male of which has a greenish head. It was formerly eaten as a delicacy. [F f. Prov., lit. 'gardener', f. L *hortulanus* f. *hortulus* dimin. of *hortus* garden]

Orton[1] /ˈɔːt(ə)n/, Arthur (known as 'the Tichborne claimant') (1834–98), English butcher. In 1852 he emigrated to Australia, but returned to England in 1866 claiming to be the heir to the rich Tichborne estate; he asserted that he was the eldest son of the 10th baronet, who was presumed lost at sea, and convinced the lost heir's mother that he was her son. After a long trial he lost his claim and was tried and imprisoned for perjury.

Orton[2] /ˈɔːt(ə)n/, Joe (born John Kingsley Orton) (1933–67), English dramatist. He wrote a number of unconventional black comedies, notable for their examination of corruption, sexuality, and violence; they include *Entertaining Mr Sloane* (1964), *Loot* (1965), and the posthumously performed *What the Butler Saw* (1969). Orton was murdered by his homosexual lover, who then committed suicide.

Oruro /əˈruərəʊ/ a city in western Bolivia; pop. (1990) 208,700. It is the centre of an important mining region, with rich deposits of tin, zinc, silver, copper, and gold.

Orvieto /ˌɔːvɪˈeɪtəʊ/ a town in Umbria, central Italy; pop. (1990) 21,575. It has been inhabited since Etruscan times and lies at the centre of a wine-producing area.

Orwell /ˈɔːwel/, George (pseudonym of Eric Arthur Blair) (1903–50), British novelist and essayist. Born in Bengal, he returned to Europe in 1928. His work is characterized by his concern with social injustice; after living as a vagrant in London, he described his experiences in *Down and Out in Paris and London* (1933); he also wrote about the plight of the unemployed in *The Road to Wigan Pier* (1937), and fought for the Republicans in the Spanish Civil War. His most famous works are *Animal Farm* (1945), a satire on Communism as it developed under Stalin, and *Nineteen Eighty-four* (1949), a dystopian account of a future state in which every aspect of life is controlled by Big Brother.

Orwellian /ɔːˈwelɪən/ *adj.* of or characteristic of the writings of George Orwell, esp. with reference to the totalitarian state as depicted in *Nineteen Eighty-four* and *Animal Farm*.

-ory[1] /ərɪ/ *suffix* forming nouns denoting a place for a particular function (*dormitory; refectory*). □ **-orial** /ˈɔːrɪəl/ *suffix* forming adjectives. [L *-oria, -orium*, sometimes via ONF and AF *-orie*, OF *-oire*]

-ory[2] /ərɪ/ *suffix* forming adjectives (and occasionally nouns) relating to or involving a verbal action (*accessory; compulsory; directory*). [L *-orius*, sometimes via AF *-ori(e)*, OF *-oir(e)*]

oryx /ˈɒrɪks/ *n.* a large straight-horned antelope of the genus *Oryx*, native to arid parts of Africa and Arabia. [ME f. L f. Gk *orux* stonemason's pickaxe, f. its pointed horns]

OS *abbr.* **1** Old Style. **2** ordinary seaman. **3** (in the UK) Ordnance Survey. **4** outsize. **5** out of stock.

Os *symb. Chem.* the element osmium.

Osage orange /ˈəʊseɪdʒ/ *n.* **1** a hardy thorny tree, *Maclura pomifera*, of North America, bearing inedible wrinkled orange-like fruit. **2** the durable orange-coloured timber from this. [name of an Amer. Indian people]

Osaka /əʊˈsɑːkə/ a port and commercial city in central Japan, on the island of Honshu, capital of Kinki region; pop. (1990) 2,642,000.

Osborne /ˈɒzbɔːn/, John James (1929–94), English dramatist. His first play, *Look Back in Anger* (1956), ushered in a new era of kitchen-sink drama; its hero Jimmy Porter was seen as the archetype of contemporary disillusioned youth, the so-called 'angry young man'. Later plays include *The Entertainer* (1957), *A Patriot for Me* (1965), and *Déjà vu* (1991).

Oscan /ˈɒskən/ *n.* & *adj.* ● *n.* the ancient language of Campania in Italy, related to Latin and surviving only in inscriptions in an alphabet derived from Etruscan. ● *adj.* relating to or written in Oscan. [L *Oscus*]

Oscar /ˈɒskə(r)/ *n.* the nickname for an Academy award. The award is a gold statuette; the nickname is said to have arisen because the statue reminded Margaret Herrick, one-time librarian and later executive director of the Academy of Motion Picture Arts and Sciences, of her uncle Oscar.

oscillate /ˈɒsɪˌleɪt/ v. **1** intr. & tr. **a** swing to and fro like a pendulum. **b** move to and fro between points. **2** intr. vacillate; vary between extremes of opinion, action, etc. **3** intr. Physics move with periodic regularity. **4** intr. Electr. (of a current) undergo high-frequency alternations as across a spark-gap or in a valve-transmitter circuit. **5** intr. (of a radio receiver) radiate electromagnetic waves owing to faulty operation. □ **oscillator** n. **oscillatory** /ˈɒsɪlətərɪ, -ˌleɪtərɪ/ adj. **oscillation** /ˌɒsɪˈleɪʃ(ə)n/ n. [L oscillare oscillat- swing]

oscillo- /əˈsɪləʊ/ comb. form of electric current.

oscillogram /əˈsɪləˌɡræm/ n. a record obtained from an oscillograph.

oscillograph /əˈsɪləˌɡrɑːf/ n. a device for recording oscillations. □ **oscillographic** /əˌsɪləˈɡræfɪk/ adj. **oscillography** /ˌɒsɪˈlɒɡrəfɪ/ n.

oscilloscope /əˈsɪləˌskəʊp/ n. a device for viewing oscillations by a display on the screen of a cathode-ray tube. □ **oscilloscopic** /əˌsɪləˈskɒpɪk/ adj.

oscine /ˈɒsɪn, ˈɒsaɪn/ adj. (also **oscinine** /ˈɒsɪˌniːn, -ˌnaɪn/) Zool. of or relating to the suborder Oscines of passerine birds, comprising the songbirds. [L oscen -cinis songbird (as OB-, canere sing)]

oscitation /ˌɒsɪˈteɪʃ(ə)n/ n. formal **1** yawning; drowsiness. **2** inattention; negligence. [L oscitatio f. oscitare gape f. os mouth + citare move]

oscula pl. of OSCULUM.

oscular /ˈɒskjʊlə(r)/ adj. **1** of or relating to the mouth. **2** of or relating to kissing. [L osculum mouth, kiss, dimin. of os mouth]

osculate /ˈɒskjʊˌleɪt/ v. **1** tr. Math. (of a curve or surface) have contact of at least the second order with; have two branches with a common tangent, with each branch extending in both directions of the tangent. **2** v.intr. & tr. joc. kiss. **3** intr. Biol. (of a species etc.) be related through an intermediate species; have common characteristics with another or with each other. □ **osculant** adj. **osculatory** /-lətərɪ/ adj. **osculation** /ˌɒskjʊˈleɪʃ(ə)n/ n. [L osculari kiss (as OSCULAR)]

osculum /ˈɒskjʊləm/ n. (pl. **oscula** /-lə/) Zool. a pore or orifice, esp. a large opening in a sponge through which water is expelled. [L: see OSCULAR]

-ose[1] /əʊs, əʊz/ suffix forming adjectives denoting possession of a quality (grandiose; verbose). □ **-osely** suffix forming adverbs. **-oseness** suffix forming nouns (cf. -OSITY). [from or after L -osus]

-ose[2] /əʊs, əʊz/ suffix Chem. forming names of carbohydrates (cellulose; sucrose). [after GLUCOSE]

Osh /ɒʃ/ a city in western Kyrgyzstan, near the border with Uzbekistan; pop. (1990) 236,200. It was, until the 15th century, an important post on an ancient trade route to China and India.

Oshawa /ˈɒʃəwə/ a city in Ontario, on the northern shores of Lake Ontario east of Toronto; pop. (1991) 174,010.

osier /ˈəʊzɪə(r)/ n. **1** a willow, esp. Salix viminalis, with long flexible shoots used in basketwork. **2** a shoot of a willow. □ **osier-bed** a place where osiers are grown. [ME f. OF: cf. med.L auseria osier-bed]

Osijek /ˈɒsɪˌjek/ a city in eastern Croatia, on the River Drava; pop. (1991) 104,700.

Osiris /əˈsaɪrɪs/ Egyptian Mythol. a god originally connected with fertility, husband of Isis and father of Horus. He is known chiefly through the story of his death at the hands of his brother Seth and his subsequent restoration to a new life as ruler of the afterlife. Under Ptolemy I his cult was combined with that of Apis to produce the cult of Serapis.

-osis /ˈəʊsɪs/ suffix (pl. **-oses** /ˈəʊsiːz/) denoting a process or condition (apotheosis; metamorphosis), esp. a pathological state (acidosis; neurosis; thrombosis). [L f. Gk -ōsis suffix of verbal nouns]

-osity /ˈɒsɪtɪ/ suffix forming nouns from adjectives in -ose (see -OSE[1]) and -ous (verbosity; curiosity). [F -osité or L -ositas -ositatis: cf. -ITY]

Osler /ˈəʊzlə(r), ˈəʊslə(r)/, Sir William (1849–1919), Canadian-born physician and classical scholar. He was professor of medicine at four universities, and his Principles and Practice of Medicine (1892) became the chosen clinical textbook for medical students. At Johns Hopkins University, Baltimore, he instituted a model teaching unit in which clinical observation was combined with laboratory research.

Oslo /ˈɒzləʊ/ the capital and chief port of Norway, on the south coast at the head of Oslofjord; pop. (1990) 458,360. Founded in the 11th century, it was known as Christiania (or Kristiania) from 1624 until 1924 in honour of Christian IV of Norway and Denmark (1577–1648), who rebuilt the city after it had been destroyed by fire in 1624.

Osman I /ˈɒzmən/ (also **Othman** /ˈɒθmən/) (1259–1326), Turkish conqueror, founder of the Ottoman (Osmanli) dynasty and empire. After succeeding his father as leader of the Seljuk Turks in 1288,

Osman reigned as sultan, conquering NW Asia Minor. He assumed the title of emir in 1299.

Osmanli /ɒzˈmænlɪ, ɒsˈmæn-/ adj. & n. (pl. same or **Osmanlis**) = OTTOMAN. [Turk. f. Osman (see OSMAN I) f. Arab. 'uṯmān (see OTTOMAN) + -li adj. suffix]

osmic[1] /ˈɒzmɪk/ adj. of or relating to odours or the sense of smell. □ **osmically** adv. [Gk osmē smell, odour]

osmic[2] /ˈɒzmɪk/ adj. Chem. containing osmium.

osmium /ˈɒzmɪəm/ n. a hard silvery-white metallic chemical element (atomic number 76; symbol **Os**). A transition element and one of the platinum metals, osmium was discovered by the British chemist Smithson Tennant (1761–1815) in 1804. Osmium is thought to be the densest non-radioactive element. Its main use is in alloys with platinum and related metals, where it contributes hardness; an alloy with iridium is used in fountain-pen nibs. Osmium tetroxide is used for staining tissues in electron microscopy. [Gk osmē smell (from the pungent smell of its tetroxide)]

osmoregulation /ˌɒzməʊˌreɡjʊˈleɪʃ(ə)n/ n. Biol. the maintenance of constant osmotic pressure in the fluids of an organism by control of water and salt levels etc.

osmosis /ɒzˈməʊsɪs/ n. **1** Biol. & Chem. the passage of a solvent from a less concentrated solution into a more concentrated one through a semi-permeable partition. (See note below.) **2** any process by which something is acquired by absorption. [orig. osmose, after F f. Gk ōsmos push]

▪ Osmosis, first studied by the French physicist Jean Nollet (1700–70), is an important mechanism in plant and animal physiology and plays a vital role in the passage of water into and out of living cells. It also has applications in the desalination of water. It occurs partly because of interactions between the solute molecules and the solvent, and partly because the presence of the solute reduces the number of molecules of solvent per unit area on one side of the partition. Osmosis continues until the osmotic pressure difference between the solutions on either side of the partition is zero.

osmotic /ɒzˈmɒtɪk/ adj. of or relating to osmosis. □ **osmotic pressure** Biol. & Chem. the pressure that would have to be applied to prevent pure solvent from passing into a solution by osmosis, often used to express the concentration of the solution. □ **osmotically** adv.

osmund /ˈɒzmənd/ n. (also **osmunda** /ɒzˈmʌndə/) = royal fern. [ME f. AF, of uncert. orig.]

Osnabrück /ˈɒznəˌbrʊk/ a city in NW Germany, in Lower Saxony; pop. (1991) 165,140. In 1648 the Treaty of Westphalia, ending the Thirty Years War, was signed there and in Münster.

osprey /ˈɒspreɪ, -prɪ/ n. (pl. **-eys**) **1** a large bird of prey, Pandion haliaetus, with a brown back and white markings, feeding on fish. Also called fish-hawk. **2** an egret's plume on a woman's hat. [ME f. OF ospres app. ult. f. L ossifraga osprey f. os bone + frangere break]

Ossa, Mount /ˈɒsə/ **1** a mountain in Thessaly, NE Greece, south of Mount Olympus, rising to a height of 1,978 m (6,489 ft). In Greek mythology the giants were said to have piled Mount Olympus and Mount Ossa on to Mount Pelion in an attempt to reach heaven and destroy the gods. **2** the highest mountain on the island of Tasmania, rising to a height of 1,617 m (5,305 ft).

ossein /ˈɒsiɪn/ n. the collagen of bones. [L osseus (as OSSEOUS)]

osseous /ˈɒsɪəs/ adj. **1** consisting of bone. **2** having a bony skeleton. **3** ossified. [L osseus f. os ossis bone]

Ossete /ˈɒsiːt/ n. & adj. ● n. a native or inhabitant of Ossetia. ● adj. of or relating to Ossetia or the Ossetes. [Russ.]

Ossetia /ɒˈsiːʃə/ a region of the central Caucasus. It is divided by the boundary between Russia and Georgia into two parts, North Ossetia and South Ossetia. From 1989 the region was the scene of ethnic conflict and unrest over Ossetian demands for South Ossetia to secede from Georgian control and join North Ossetia as part of Russia; the fighting was halted in 1992 with the arrival of a peacekeeping force.

Ossetian /ɒˈsiːʃ(ə)n/ n. & adj. ● n. **1** the Iranian language of the Ossetes. **2** = OSSETE n. ● adj. of or relating to the Ossetes or their language.

Ossian /ˈɒsɪən, ˈɒsɪ-/ the anglicized form of Oisin, a legendary Irish warrior and bard, whose name became well known in 1760–3 when the Scottish poet James Macpherson (1736–96) published what was later discovered to be his own verse as an alleged translation of 3rd-century Gaelic tales.

ossicle /ˈɒsɪk(ə)l/ n. **1** Anat. any small bone, esp. of the middle ear. **2** a small piece of bonelike substance. [L ossiculum dimin. (as OSSEOUS)]

Ossie var. of AUSSIE.

ossify /ˈɒsɪˌfaɪ/ v.tr. & intr. (-**ies**, -**ied**) **1** turn into bone; harden. **2** make or become rigid, callous, or unprogressive. □ **ossific** /ɒˈsɪfɪk/ adj. **ossification** /ˌɒsɪfɪˈkeɪʃ(ə)n/ n. [F ossifier f. L os ossis bone]

osso bucco /ˌɒsəʊ ˈbʊkəʊ/ n. shin of veal containing marrowbone stewed in wine with vegetables. [It., = marrowbone]

ossuary /ˈɒsjʊərɪ/ n. (pl. -**ies**) **1** a receptacle for the bones of the dead; a charnel-house; a bone-urn. **2** a cave in which ancient bones are found. [LL ossuarium irreg. f. os ossis bone]

Ostade /ɒˈstɑːdə/, Adriaen van (1610–85), Dutch painter and engraver. He is thought to have been a pupil of Frans Hals, and his work chiefly depicts lively genre scenes of peasants carousing or brawling in crowded taverns or barns. His brother and pupil, Isack (1621–49), was also a painter, particularly of winter landscapes and genre scenes of peasants outside cottages or taverns.

osteitis /ˌɒstɪˈaɪtɪs/ n. Med. inflammation of the substance of a bone. [Gk osteon bone + -ITIS]

Ostend /ɒˈstend/ (Flemish **Oostende** /oːstˈɛndə/, French **Ostende** /ɔstɑ̃d/) a port on the North Sea coast of NW Belgium, in West Flanders; pop. (1991) 68,500. It is a major ferry port with links to Dover.

ostensible /ɒˈstensɪb(ə)l/ adj. apparent but not necessarily real; professed (his ostensible function was that of interpreter). □ **ostensibly** adv. [F f. med.L ostensibilis f. L ostendere ostens- stretch out to view (as OB-, tendere stretch)]

ostensive /ɒˈstensɪv/ adj. **1** directly demonstrative. **2** (of a definition) indicating by direct demonstration that which is signified by a term. □ **ostensively** adv. **ostensiveness** n. [LL ostensivus (as OSTENSIBLE)]

ostensory /ɒˈstensərɪ/ n. (pl. -**ies**) RC Ch. a receptacle for displaying the host to the congregation; a monstrance. [med.L ostensorium (as OSTENSIBLE)]

ostentation /ˌɒstenˈteɪʃ(ə)n/ n. **1** pretentious and vulgar display esp. of wealth and luxury. **2** the attempt or intention to attract notice; showing off. □ **ostentatious** adj. **ostentatiously** adv. [ME f. OF f. L ostentatio -onis f. ostentare frequent. of ostendere: see OSTENSIBLE]

osteo- /ˈɒstɪəʊ/ comb. form bone. [Gk osteon]

osteoarthritis /ˌɒstɪəʊɑːˈθraɪtɪs/ n. degeneration of joint cartilage, esp. from middle age onward, causing pain and stiffness. □ **osteoarthritic** /-ˈθrɪtɪk/ adj.

osteogenesis /ˌɒstɪəʊˈdʒenɪsɪs/ n. the formation of bone. □ **osteogenetic** /-dʒɪˈnetɪk/ adj.

osteology /ˌɒstɪˈɒlədʒɪ/ n. the study of the structure and function of the skeleton and bony structures. □ **osteologist** n. **osteological** /-tɪəˈlɒdʒɪk(ə)l/ adj. **osteologically** adv.

osteomalacia /ˌɒstɪəʊməˈleɪʃɪə/ n. softening of the bones, often through a deficiency of vitamin D and calcium. □ **osteomalacic** /-ˈlæsɪk/ adj. [mod.L (as OSTEO-, Gk malakos soft)]

osteomyelitis /ˌɒstɪəʊmaɪəˈlaɪtɪs/ n. Med. inflammation of the bone or of bone marrow, usu. due to infection.

osteopathy /ˌɒstɪˈɒpəθɪ/ n. a system of healing involving the treatment of disease through the manipulation of bones (especially the spine) and muscles. Osteopathy is based on the belief that the displacement of bones and muscles is a cause of many other disorders. It was founded by an American doctor, Andrew Taylor Still (1828–1917), who, after failing to persuade medical schools to accept his ideas, established his own school in Missouri in 1892. □ **osteopath** /ˈɒstɪəˌpæθ/ n. **osteopathic** /ˌɒstɪəˈpæθɪk/ adj.

osteoporosis /ˌɒstɪəʊpəˈrəʊsɪs/ n. a condition of brittle and fragile bones caused by loss of bony tissue, esp. as a result of hormonal changes, or deficiency of calcium or vitamin D. [OSTEO- + Gk poros passage, pore]

Österreich /ˈøːstəˌraɪç/ the German name for AUSTRIA.

Ostia /ˈɒstɪə/ an ancient city and harbour which was situated on the western coast of Italy at the mouth of the River Tiber. It was the first colony founded by ancient Rome and was a major port and commercial centre. Now located about 6 km (4 miles) inland, the original city was buried and its ruins were preserved by the gradual silting up of the River Tiber.

ostinato /ˌɒstɪˈnɑːtəʊ/ n. (pl. -**os**) (often attrib.) Mus. a persistent phrase or rhythm repeated through all or part of a piece. [It., = OBSTINATE]

ostium /ˈɒstɪəm/ n. (pl. **ostia** /-tɪə/) Anat. & Zool. an opening into a vessel or body cavity. [L, = door, opening]

ostler /ˈɒslə(r)/ n. Brit. hist. a stableman at an inn. [f. earlier HOSTLER, hosteler f. AF hostiler, OF (h)ostelier (as HOSTEL)]

Ostmark /ˈɒstmɑːk/ n. hist. the chief monetary unit of the German Democratic Republic. [G, = east mark: see MARK²]

Ostpolitik /ˈɒstpɒlɪˌtiːk/ n. hist. the foreign policy (esp. of détente) of many western European countries with reference to the former Communist bloc. The term was used in the Federal Republic of Germany (West Germany) to describe the opening of relations with the Eastern bloc in the 1960s — a reversal of West Germany's refusal to recognize the legitimacy of the German Democratic Republic (East Germany). The policy was pursued with particular vigour by the West German Chancellor Willy Brandt. [G f. Ost east + Politik politics]

ostracize /ˈɒstrəˌsaɪz/ v.tr. (also -**ise**) **1** exclude (a person) from a society, favour, common privileges, etc.; refuse to associate with. **2** (esp. in ancient Athens) banish (a powerful or unpopular citizen) for five or ten years by popular vote. □ **ostracism** n. [Gk ostrakizō f. ostrakon shell, potsherd (used to write a name on in voting)]

Ostrava /ˈɒstrəvə/ an industrial city in the Moravian lowlands of the NE Czech Republic; pop. (1991) 327,550. It is situated in the coal-mining region of Silesia; coal was first mined there in 1767.

ostrich /ˈɒstrɪtʃ/ n. **1** a very large swift-running flightless bird, Struthio camelus, native to Africa, with long legs and two toes on each foot. **2** a person who refuses to accept facts (from the belief that ostriches bury their heads in the sand when pursued). □ **ostrich-farm** a place that breeds ostriches for their feathers. **ostrich-plume** a feather or bunch of feathers of an ostrich. [ME f. OF ostric(h)e f. L avis bird + LL struthio f. Gk strouthiōn ostrich f. strouthos sparrow, ostrich]

Ostrogoth /ˈɒstrəˌgɒθ/ n. hist. a member of the eastern branch of the Goths, who conquered Italy in the 5th–6th centuries AD. □ **Ostrogothic** /ˌɒstrəˈgɒθɪk/ adj. [LL Ostrogothi (pl.) f. Gmc austro- (unrecorded) east + LL Gothi Goths: see GOTH]

Ostwald /ˈɒstvælt/, Friedrich Wilhelm (1853–1932), German physical chemist. He did much to establish physical chemistry as a separate discipline, and is particularly remembered for his pioneering work on catalysis. He also worked on chemical affinities, the hydrolysis of esters, and electrolytic conductivity and dissociation. After retiring, Ostwald studied colour science, and developed a new quantitative colour theory. He was awarded the Nobel Prize for chemistry in 1909.

Oswald /ˈɒzwəld/, Lee Harvey (1939–63), American alleged assassin of John F. Kennedy. In Nov. 1963 he was arrested in Dallas, Texas shortly after the assassination of President Kennedy and charged with his murder. He denied the charge, but was murdered by Jack Ruby (1911–67), a Dallas nightclub owner, before he could be brought to trial. Oswald was said to be the sole gunman by the Warren Commission (1964), but the House of Representatives Assassinations Committee (1979) concluded that more than one gunman had been involved; the affair remains the focus for a number of conspiracy theories.

Oswald of York, St /ˈɒzwəld/ (d.992), English prelate and Benedictine monk. He rose first to become bishop of Worcester and then Archbishop of York. He founded several Benedictine monasteries and, along with St Dunstan, was responsible for the revival of the Church and of learning in 10th-century England. Feast day, 28 Feb.

OT abbr. Old Testament.

-ot¹ /ət/ suffix forming nouns, orig. diminutives (ballot; chariot; parrot). [F]

-ot² /ət/ suffix forming nouns denoting persons (patriot), e.g. natives of a place (Cypriot). [F -ote, L -ota, Gk -ōtēs]

Otago /ɒˈtɑːgəʊ/ a region of New Zealand, on the SE coast of South Island. Formerly a province, it was centred on the settlement of Dunedin, which had been founded by Scottish settlers in 1848. It was named after a Maori village.

OTC abbr. (in the UK) Officers' Training Corps.

other /ˈʌðə(r)/ adj., n. or pron., & adv. ● adj. **1** not the same as one or some already mentioned or implied; separate in identity or distinct in kind (other people; use other means; I assure you, my reason is quite other). **2 a** further; additional (a few other examples). **b** alternative of two (open your other eye) (cf. every other). **3** (prec. by the) that remains after all except the one or ones in question have been considered, eliminated, etc. (must be in the other pocket; where are the other two?; the other three men left). **4** (foll. by than) apart from; excepting (any person other than you). ● n. or pron. (orig. an ellipt. use of the adj., now with pl. in -s) **1** an additional, different, or extra person, thing, example, etc. (one or other of us will be there; some others have come) (see also ANOTHER pron., each other). **2** (in pl.; prec. by the) the ones remaining (where are the others?). **3** (prec. by the) sl. sexual intercourse (a bit of the other). ● adv. (usu. foll. by than) disp. otherwise (cannot react other than angrily). ¶ In this sense otherwise is standard except in less formal use. □ **no other** *archaic* nothing else (*I can do no other*). **of all others** out of the many possible or likely (*on this night of all others*). **on the other hand** see HAND. **the other day** (or **night** or **week** etc.) a few days etc. ago (*heard from him the other day*). **other-directed** governed by external circumstances and trends. **other half** *colloq.* one's wife or husband. **the other place 1** hell (as opposed to heaven). **2** the House of Lords as regarded from the House of Commons and vice versa. **other ranks** soldiers other than commissioned officers. **the other thing** esp. *joc.* an unexpressed alternative (*if you don't like it, do the other thing*). **other things being equal** if conditions are or were alike in all but the point in question. **the other woman** (or **man**) the lover of a married or similarly attached man (or woman). **the other world** see WORLD. **someone** (or **something** or **somehow** etc.) **or other** some unspecified person, thing, manner, etc. [OE *ōther* f. Gmc]

otherness /ˈʌðənɪs/ *n.* **1** the state of being different; diversity. **2** a thing or existence other than the thing mentioned and the thinking subject.

otherwhere /ˈʌðəˌweə(r)/ *adj. archaic* or *poet.* elsewhere.

otherwise /ˈʌðəˌwaɪz/ *adv. & adj.* ● *adv.* **1** else; or else; in the circumstances other than those considered etc. (*bring your umbrella, otherwise you will get wet*). **2** in other respects (*he is untidy, but otherwise very suitable*). **3** (often foll. by *than*) in a different way (*could not have acted otherwise; cannot react otherwise than angrily*). **4** as an alternative (*otherwise known as Jack*). ● *adj.* **1** (*predic.*) in a different state (*the matter is quite otherwise*). **2** *archaic* that would otherwise exist (*their otherwise dullness*). □ **and** (or **or**) **otherwise** the negation or opposite (of a specified thing) (*the merits or otherwise of the Bill; experiences pleasant and otherwise*). [OE *on ōthre wisan* (as OTHER, WISE²)]

other-worldly /ˌʌðəˈwɜːldlɪ/ *adj.* **1** unworldly; impractical. **2** not of this world; of or relating to an imaginary other world. **3** concerned with life after death etc. □ **other-worldliness** *n.*

Othman see OSMAN I.

Otho /ˈəʊθəʊ/, Marcus Salvius (AD 32–69), Roman emperor Jan.–Apr. 69. He was proclaimed emperor after he had procured the death of Galba in a conspiracy of the praetorian guard. Otho was not recognized as emperor by the German legions; led by their candidate, Vitellius, they defeated his troops and Otho committed suicide.

otic /ˈəʊtɪk, ˈɒt–/ *adj.* of or relating to the ear. [Gk *ōtikos* f. *ous ōtos* ear]

-otic /ˈɒtɪk/ *suffix* forming adjectives and nouns corresponding to nouns in *-osis*, meaning 'affected with or producing or resembling a condition in *-osis*' or 'a person affected with this' (*narcotic; neurotic; osmotic*). □ **-otically** *suffix* forming adverbs. [from or after F *-otique* f. L f. Gk *-ōtikos* adj. suffix]

otiose /ˈəʊʃɪəʊs, ˈəʊtɪ–, –əʊz/ *adj.* **1** serving no practical purpose; not required; functionless. **2** *archaic* indolent; futile. □ **otiosely** *adv.* **otioseness** *n.* [L *otiosus* f. *otium* leisure]

Otis /ˈəʊtɪs/, Elisha Graves (1811–61), American inventor and manufacturer. He produced the first efficient elevator with a safety device in 1852; it consisted of a mechanical hoist for carrying machinery to the upper floors of a factory, with a device to prevent it from falling even if the lifting cable broke. In 1857 he installed the first public elevator for passengers in a New York department store. (See also ELEVATOR.)

otitis /əʊˈtaɪtɪs/ *n. Med.* inflammation of the ear, esp. (in full **otitis media**) of the middle ear. [mod.L as OTO-)]

oto- /ˈəʊtəʊ/ *comb. form* ear. [Gk *ōto-* f. *ous ōtos* ear]

otolaryngology /ˌəʊtəˌlærɪŋˈɡɒlədʒɪ/ *n.* the study of diseases of the ear and throat. □ **otolaryngologist** *n.* **otolaryngological** /–gəˈlɒdʒɪk(ə)l/ *adj.*

otolith /ˈəʊtəlɪθ/ *n. Anat. & Zool.* any of the several small calcareous particles in the inner ear of vertebrates, important as sensors of gravity and acceleration. □ **otolithic** /ˌəʊtəˈlɪθɪk/ *adj.*

otology /əʊˈtɒlədʒɪ/ *n.* the study of the anatomy and diseases of the ear. □ **otologist** *n.* **otological** /ˌəʊtəˈlɒdʒɪk(ə)l/ *adj.*

Otomi /ˌəʊtəˈmiː/ *n.* (*pl.* same) **1** a member of a native people inhabiting parts of central Mexico. **2** the language of this people. [Amer. Sp. f. Nahuatl *otomih* lit. 'unknown']

otorhinolaryngology /ˌəʊtəˌraɪnəʊˌlærɪŋˈɡɒlədʒɪ/ *n.* the study of diseases of the ear, nose, and throat.

otoscope /ˈəʊtəˌskəʊp/ *n.* an apparatus for examining the eardrum and the passage leading to it from the ear. □ **otoscopic** /ˌəʊtəˈskɒpɪk/ *adj.*

Otranto, Strait of /ɒˈtræntəʊ/ a channel linking the Adriatic Sea with the Ionian Sea and separating the 'heel' of Italy from Albania.

OTT *abbr. colloq.* over-the-top.

ottava rima /ɒˌtɑːvə ˈriːmə/ *n.* a stanza of eight lines of ten or eleven syllables, rhyming *abababcc*. [It. = eighth rhyme]

Ottawa /ˈɒtəwə, –ˌwɑː/ the federal capital of Canada, on the Ottawa river (a tributary of the St Lawrence); pop. (1991) 313,990; metropolitan area pop. 920,860. Founded in 1827, it was named Bytown after Colonel John By (1779–1836). It received its present name in 1854 and was chosen as capital (of the United Provinces of Canada) in 1858.

otter /ˈɒtə(r)/ *n.* **1 a** a semiaquatic fish-eating mammal of the genus *Lutra* or a related genus, of the weasel family, having webbed feet and dense fur; esp. *L. lutra*, of European rivers. **b** its fur or pelt. **2** = *sea otter*. **3** a piece of board used to carry fishing-bait in water. **4** a type of paravane, esp. as used on non-naval craft. □ **otter-board** a device for keeping the mouth of a trawl-net open. **otter-dog** (or **-hound**) a breed of dog originally used in otter-hunting. [OE *otr*, *ot(t)or* f. Gmc]

Otto /ˈɒtəʊ/, Nikolaus August (1832–91), German engineer, whose name is given to the four-stroke cycle on which most internal-combustion engines work. Otto's patent of 1876 was invalidated ten years later when it was found that Alphonse-Eugène Beau de Rochas (1815–93) had described the successful cycle earlier, so enabling other manufacturers to adopt it.

otto var. of ATTAR.

Otto I /ˈɒtəʊ/ (known as Otto the Great) (912–73), king of the Germans 936–73, Holy Roman emperor 962–73. As king of the Germans he carried out a policy of eastward expansion from his Saxon homeland and defeated the invading Hungarians in 955. He was crowned Holy Roman emperor in 962 and began to establish a strong imperial presence in Italy to rival that of the papacy.

Ottoman /ˈɒtəmən/ *adj. & n.* ● *adj. hist.* **1 a** of or concerning the dynasty of Osman I (Othman I). **b** of or relating to the branch of the Turks to which Osman belonged. **c** of or relating to the empire ruled by the descendants of Osman. **2** Turkish. ● *n.* (*pl.* **Ottomans**) an Ottoman person; a Turk. □ **Ottoman Porte** see PORTE. [F f. Arab. *'uṯmānī* adj. of Othman (*'uṯmān*)]

ottoman /ˈɒtəmən/ *n.* (*pl.* **ottomans**) **1 a** an upholstered seat, usu. square and without a back or arms, sometimes a box with a padded top. **b** a footstool of similar design. **2** a heavy silken fabric with a mixture of cotton or wool. [F *ottomane* fem. (as OTTOMAN)]

Ottoman Empire the Turkish empire, established in northern Anatolia by Osman I at the end of the 13th century and expanded by his successors to include all of Asia Minor and much of SE Europe. Ottoman power received a severe check with the invasion of the Mongol ruler Tamerlane in 1402, but expansion resumed several decades later, resulting in the capture of Constantinople in 1453. The empire reached its zenith under Suleiman in the mid-16th century, dominating the eastern Mediterranean and threatening central Europe, but thereafter it began to decline. Still powerful in the 17th century, it had greatly declined by the 19th century, eventually collapsing after the First World War.

Otto the Great, Otto I of the Germans and Holy Roman emperor (see OTTO I).

Otway /ˈɒtweɪ/, Thomas (1652–85), English dramatist. After failing as an actor he wrote for the stage and achieved success with his second play, *Don Carlos* (1676), a tragedy in rhymed verse. He is now chiefly remembered for his two blank verse tragedies, *The Orphan* (1680) and *Venice Preserved* (1682).

OU *abbr. Brit.* **1** Open University. **2** Oxford University.

ouabain /ˈwɑːbeɪn, wɑːˈbeɪn/ *n. Chem.* a form of strophanthin used as a very rapid heart stimulant. [F f. Somali *wabayo* tree yielding arrow-poison containing this]

Ouagadougou /ˌwɑːɡəˈduːɡuː/ the capital of Burkina; pop. (1985) 442,220.

oubliette /ˌuːblɪˈet/ *n.* a secret dungeon with access only through a trapdoor. [F f. *oublier* forget]

ouch /aʊtʃ/ *int.* expressing pain. [imit.: cf. G *autsch*]

Oudenarde, Battle of /ˈuːdənɑːd/ a battle which took place in 1708 during the War of the Spanish Succession, near the town of Oudenarde (Flemish name Oudenaarde, French name Audenarde) in eastern Flanders, Belgium. A force of allied British and Austrian troops under the Duke of Marlborough and the Austrian general Prince Eugene defeated the French.

Oudh /aʊd/ (also **Audh, Awadh** /ˈʌwəd/) a region of northern India. It was an independent kingdom until it fell under Muslim rule in the 11th century, but acquired a measure of independence again in the early 18th century. After annexation by Britain in 1856 it became the centre of the Indian Mutiny of 1857–8. In 1877 it joined with Agra and formed the United Provinces of Agra and Oudh in 1902. This was renamed Uttar Pradesh in 1950.

ought[1] /ɔːt/ *v.aux.* (usu. foll. *by* to + infin.; present and past indicated by the following infin.) **1** expressing duty or rightness (*we ought to love our neighbours*). **2** expressing shortcoming (*it ought to have been done long ago*). **3** expressing advisability or prudence (*you ought to go for your own good*). **4** expressing esp. strong probability (*he ought to be there by now*). □ **ought not** the negative form of *ought* (*he ought not to have stolen it*). [OE *āhte*, past of *āgan* OWE]

ought[2] /ɔːt/ *n.* (also **aught**) *colloq.* a figure denoting nothing; nought. [perh. f. *an ought* for a NOUGHT; cf. ADDER]

ought[3] var. of AUGHT[1].

oughtn't /ˈɔːt(ə)nt/ *contr.* ought not.

ouguiya /uːˈɡiːjə/ *n.* (also **ougiya**) the basic monetary unit of Mauritania, equal to five khoums. [F f. Mauritanian Arab. *ūgiyya*, ult. f. L *uncia* ounce]

Ouida /ˈwiːdə/ (pseudonym of Marie Louise de la Ramée) (1839–1908), English novelist. She lived mostly in Italy and wrote forty-five novels, often set in a fashionable world far removed from reality and showing a spirit of rebellion against the moral ideals that were prevalent in much of the fiction of the time. Her books include *Under Two Flags* (1867), *Folle-Farine* (1871), and *Two Little Wooden Shoes* (1874).

Ouija /ˈwiːdʒə/ *n.* (in full **Ouija board**) *propr.* a board having letters or signs at its rim to which a planchette, movable pointer, or upturned glass supposedly points in answer to questions from attenders at a seance etc. [F *oui* yes + G *ja* yes]

Oulu /ˈaʊluː/ (called in Swedish *Uleåborg*) a city in central Finland, on the west coast, capital of a province of the same name; pop. (1990) 101,380.

ounce[1] /aʊns/ *n.* **1 a** a unit of weight of one-sixteenth of a pound avoirdupois, approximately 28 grams (symbol: **oz**). **b** a unit of one-twelfth of a pound troy or apothecaries' measure, equal to 480 grains (approx. 31 grams). **2** a small quantity. □ **fluid ounce 1** *Brit.* a unit of capacity equal to one-twentieth of a pint (approx. 0.028 litre). **2** *US* a unit of capacity equal to one-sixteenth of a pint (approx. 0.034 litre). [ME & OF *unce* f. L *uncia* twelfth part of pound or foot: cf. INCH[1]]

ounce[2] /aʊns/ *n.* = snow leopard. [ME f. OF *once* (earlier *lonce*) = It. *lonza* ult. f. L *lynx*: see LYNX]

OUP *abbr.* Oxford University Press.

our /ˈaʊə(r)/ *poss.pron.* (*attrib.*) **1** of or belonging to us or ourselves (*our house; our own business*). **2** of or belonging to all people (*our children's future*). **3** (esp. as **Our**) of Us, the king or queen, emperor or empress, etc. (*given under Our seal*). **4** of us, the editorial staff of a newspaper etc. (*a foolish adventure in our view*). **5** *Brit. colloq.* indicating a relative, acquaintance, or colleague of the speaker (*our Barry works there*). [OE *ūre* orig. genitive pl. of 1st pers. pron. = of us, later treated as possessive adj.]

-our[1] /ə(r)/ *suffix* var. of -OR[2] surviving in some nouns (*ardour; colour; valour*).

-our[2] /ə(r)/ *suffix* var. of -OR[1] (*saviour*).

Our Father 1 the Lord's Prayer. **2** God.

Our Lady the Virgin Mary.

Our Lord 1 Jesus Christ. **2** God.

ours /ˈaʊəz/ *poss.pron.* the one or ones belonging to or associated with us (*it is ours; ours are over there*). □ **of ours** of or belonging to us (*a friend of ours*).

Our Saviour Jesus Christ.

ourself /ˌaʊəˈself/ *pron. archaic* = MYSELF (used in place of *ourselves*, esp. when *we* refers to an individual; cf. OUR 3, 4).

ourselves /ˌaʊəˈselvz/ *pron.* **1 a** *emphat.* form of WE or US (*we ourselves did it; made it ourselves; for our friends and ourselves*). **b** *refl.* form of US (*are pleased with ourselves*). **2** in our normal state of body or mind (*not quite ourselves today*). □ **be ourselves** act in our normal unconstrained manner. **by ourselves** see *by oneself*.

-ous /əs/ *suffix* **1** forming adjectives meaning 'abounding in, characterized by, of the nature of' (*envious; glorious; mountainous; poisonous*). **2** *Chem.* denoting a lower valency or oxidation state than the corresponding word in *-ic* (*ferrous; sulphurous*). □ **-ously** *suffix* forming adverbs. **-ousness** *suffix* forming nouns. [from or after AF *-ous*, OF *-eus*, f. L *-osus*]

Ouse /uːz/ **1** (also **Great Ouse**) a river of eastern England, which rises in Northamptonshire and flows 257 km (160 miles) eastwards then northwards through East Anglia to the Wash near King's Lynn. **2** a river of NE England, formed at the confluence of the Ure and Swale in North Yorkshire and flowing 92 km (57 miles) south-eastwards through York to the Humber estuary. **3** a river of SE England, which rises in the Weald of West Sussex and flows 48 km (30 miles) south-eastwards to the English Channel. **4** (also **Little Ouse**) a river of East Anglia, which forms a tributary of the Great Ouse. For much of its length it marks the border between Norfolk and Suffolk.

ousel var. of OUZEL.

oust /aʊst/ *v.tr.* **1** (usu. foll. by *from*) drive out or expel, esp. by forcing oneself into the place of. **2** (usu. foll. by *of*) *Law* put (a person) out of possession; deprive. [AF *ouster*, OF *oster* take away, f. L *obstare* oppose, hinder (as OB-, *stare* stand)]

ouster /ˈaʊstə(r)/ *n.* **1** ejection as a result of physical action, judicial process, or political upheaval. **2** esp. *N. Amer.* dismissal, expulsion.

out /aʊt/ *adv., prep., n., adj., int., & v.* ● *adv.* **1** away from or not in or at a place etc. (*keep him out; get out of here; my son is out in Canada*). **2** (forming part of phrasal verbs) **a** indicating dispersal away from a centre etc. (*hire out; share out; board out*). **b** indicating coming or bringing into the open for public attention etc. (*call out; send out; shine out; stand out*). **c** indicating a need for attentiveness (*watch out; look out; listen out*). **3 a** not in one's house, office, etc. (*went out for a walk*). **b** attending a social event (*out with friends*). **c** in prison. **4** to or at an end; completely (*tired out; die out; out of bananas; fight it out; typed it out*). **5** (of a fire, candle, light, etc.) not burning, not alight. **6** in error (*was 3% out in my calculations*). **7** *colloq.* unconscious (*she was out for five minutes*). **8 a** (of a tooth) extracted. **b** (of a joint, bone, etc.) dislocated (*put his shoulder out*). **9** (of a party, politician, etc.) not in office. **10** (of a jury) considering its verdict in secrecy. **11** (of workers) on strike. **12** (of a secret) revealed. **13** (of a flower) blooming, open. **14** (of a book) published. **15** (of a star) visible after dark. **16** unfashionable (*turn-ups are out*). **17** (of a batsman, batter, etc.) no longer taking part as such, having been caught, stumped, etc. **18 a** not worth considering; rejected (*that idea is out*). **b** not allowed. **19** *colloq.* (prec. by *superl.*) known to exist (*the best game out*). **20** (of a stain, mark, etc.) not visible, removed (*painted out the sign*). **21** (of time) not spent working (*took five minutes out*). **22** (of a rash, bruise, etc.) visible. **23** (of the tide) at the lowest point. **24** *Boxing* unable to rise from the floor (*out for the count*). **25** *archaic* (of a young upper-class woman) introduced into society. **26** (in a radio conversation etc.) transmission ends (*over and out*). ● *prep.* **1** out of (*looked out the window*). **2** *archaic* outside; beyond the limits of. ● *n.* **1** *colloq.* a way of escape; an excuse. **2** (**the outs**) the political party out of office. **3** *Baseball* the action or an act of putting a player out. ● *adj.* **1** *Brit.* (of a match) played away. **2** (of an island) away from the mainland. ● *int.* a peremptory dismissal, reproach, etc. (*out, you scoundrel!*). ● *v.* **1** *tr.* **a** put out. **b** *colloq.* eject forcibly. **2** *intr.* come or go out; emerge (*murder will out*). **3** *tr.* *Boxing* knock out. **4** *tr. colloq.* expose the homosexuality of (a prominent person). □ **at outs** at variance or enmity. **come out of the closet** see CLOSET. **not out** *Cricket* (of a side or a batsman) not having been caught, bowled, etc. **out and about** (of a person, esp. after an illness) engaging in normal activity. **out and away** by far. **out and out 1** thorough; surpassing. **2** thoroughly; surpassingly. **out at elbows** see ELBOW. **out for** having one's interest or effort directed to; intent on. **out of 1** from within (*came out of the house*). **2** not within (*I was never out of England*). **3** from among (*nine people out of ten; must choose out of these*). **4** beyond the range of (*is out of reach*). **5** without or so as to be without (*was swindled out of his money; out of breath; out of sugar*). **6** from (*get money out of him*). **7** owing to; because of (*asked out of curiosity*). **8** by the use of (material) (*what did you make it out of?*). **9** at a specified distance from (*a town, port, etc.*) (*seven miles out of Liverpool*). **10** beyond (*something out of the ordinary*). **11** *Horse-racing* (of an animal, esp. a horse) born of. **out of bounds** see BOUND[2]. **out of date** see DATE[1]. **out of doors** see DOOR. **out of drawing** see DRAWING. **out of hand** see HAND. **out of it 1** not included; forlorn. **2** *sl.* extremely drunk. **out of order** see ORDER. **out of pocket** see POCKET. **out of the question** see QUESTION. **out of sorts** see SORT. **out of temper** see TEMPER. **out of this world** see WORLD. **out of the way** see WAY. **out** to keenly striving to do. **out to lunch** see LUNCH. **out with** an exhortation to expel or dismiss (an unwanted person). **out with it** say what you are thinking. [OE *ūt*, OHG *ūz*, rel. to Skr. *ud*-]

out- /aʊt/ *prefix* added to verbs and nouns, meaning: **1** so as to surpass or exceed (*outdo*; *outnumber*). **2** external, separate (*outline*; *outhouse*; *outdoors*). **3** out of; away from; outward (*outspread*; *outgrowth*).

out-act /aʊt'ækt/ *v.tr.* surpass in acting or performing.

outage /'aʊtɪdʒ/ *n.* a period of time during which a power-supply etc. is not operating.

out-and-outer /ˌaʊtənd'aʊtə(r)/ *n. sl.* **1** a thorough or supreme person or thing. **2** an extremist.

outback /'aʊtbæk/ *n. esp. Austral.* the remote and usu. uninhabited inland districts. □ **outbacker** *n.*

outbalance /aʊt'bæləns/ *v.tr.* **1** count as more important than. **2** outweigh.

outbid /aʊt'bɪd/ *v.tr.* (**-bidding**; *past* and *past part.* **-bid**) **1** bid higher than (another person) at an auction. **2** surpass in exaggeration etc.

outblaze /aʊt'bleɪz/ *v.* **1** *intr.* blaze out or outwards. **2** *tr.* blaze more brightly than.

outboard /'aʊtbɔːd/ *adj., adv., & n.* ● *adj.* **1** (of a motor) portable and attachable to the outside of the stern of a boat. **2** (of a boat) having an outboard motor. ● *adj. & adv.* on, towards, or near the outside of esp. a ship, an aircraft, etc. ● *n.* **1** an outboard engine. **2** a boat with an outboard engine.

outbound /'aʊtbaʊnd/ *adj.* outward bound.

outbrave /aʊt'breɪv/ *v.tr.* **1** outdo in bravery. **2** face defiantly.

outbreak /'aʊtbreɪk/ *n.* **1** a usu. sudden eruption of war, disease, rebellion, etc. **2** an outcrop.

outbreeding /'aʊt,briːdɪŋ/ *n.* the theory or practice of breeding from animals not closely related. □ **outbreed** *v.intr. & tr.* (*past* and *past part.* **-bred** /-bred/).

outbuilding /'aʊt,bɪldɪŋ/ *n.* a detached shed, barn, garage, etc. within the grounds of a main building; an outhouse.

outburst /'aʊtbɜːst/ *n.* **1** an explosion of anger etc., expressed in words. **2** an act or instance of bursting out. **3** an outcrop.

outcast /'aʊtkɑːst/ *n. & adj.* ● *n.* **1** a person cast out from or rejected by his or her home, country, society, etc. **2** a tramp or vagabond. ● *adj.* rejected; homeless; friendless.

outcaste *n. & v.* ● *n.* /'aʊtkɑːst/ (also *attrib.*) **1** a person who has no caste, esp. in Hindu society. **2** a person who has lost his or her caste. ● *v.tr.* /aʊt'kɑːst/ cause (a person) to lose his or her caste.

outclass /aʊt'klɑːs/ *v.tr.* **1** belong to a higher class than. **2** defeat easily.

outcome /'aʊtkʌm/ *n.* a result; a visible effect.

outcrop /'aʊtkrɒp/ *n. & v.* ● *n.* **1 a** the emergence of a stratum, vein, or rock at the surface. **b** a stratum etc. emerging. **2** a noticeable manifestation or occurrence. ● *v.intr.* (**-cropped**, **-cropping**) appear as an outcrop; crop out.

outcry /'aʊtkraɪ/ *n.* (*pl.* **-ies**) **1** the act or an instance of crying out. **2** an uproar. **3** a noisy or prolonged public protest.

outdance /aʊt'dɑːns/ *v.tr.* surpass in dancing.

outdare /aʊt'deə(r)/ *v.tr.* **1** outdo in daring. **2** overcome by daring.

outdated /aʊt'deɪtɪd/ *adj.* out of date; obsolete.

outdistance /aʊt'dɪstəns/ *v.tr.* leave (a competitor) behind completely.

outdo /aʊt'duː/ *v.tr.* (*3rd sing. present* **-does** /-'dʌz/; *past* **-did** /-'dɪd/; *pres. part.* **-doing**; *past part.* **-done** /-'dʌn/) exceed or excel in doing or performance; surpass.

outdoor /'aʊtdɔː(r)/ *attrib.adj.* **1** done, existing, or used out of doors. **2** fond of the open air (*an outdoor type*). □ **outdoor pursuits** outdoor sporting or leisure activities, such as climbing, sailing, canoeing, orienteering, caving, fishing, etc. **outdoor relief** *hist.* (in Britain) financial assistance given by the state to poor people not living in a workhouse or almshouse. The Poor Law Amendment Act of 1834 attempted to end such assistance (see POOR LAW).

outdoors /aʊt'dɔːz/ *adv. & n.* ● *adv.* in or into the open air; out of doors. ● *n.* the world outside buildings; the open air.

outer /'aʊtə(r)/ *adj. & n.* ● *adj.* **1** outside; external (*pierced the outer layer*). **2** farther from the centre or inside; relatively far out. **3** objective or physical, not subjective or psychical. ● *n.* **1 a** the division of a target furthest from the bull's-eye. **b** a shot that strikes this. **2** an outer garment or part of one. **3** *Austral. sl.* the part of a racecourse outside the enclosure. **4** an outer container for transport or display. □ **the outer Bar** see BAR[1]. **outer garments** clothes worn over other clothes or outdoors. **the outer man** (or **woman**) personal appearance; dress. **outer planet** *Astron.* a planet with an orbit outside the asteroid belt, i.e. Jupiter, Saturn, Uranus, Neptune, and Pluto. **outer space** see

SPACE *n.* 3a. **the outer world** people outside one's own circle. [ME f. OUT, replacing UTTER[1]]

Outer Hebrides see HEBRIDES, THE.

Outer House *Sc. Law* the hall where judges of the Court of Session sit singly.

Outer Mongolia see MONGOLIA.

outermost /'aʊtə,məʊst/ *adj.* furthest from the inside; the most far out.

outerwear /'aʊtə,weə(r)/ *n.* = outer garments.

outface /aʊt'feɪs/ *v.tr.* disconcert or defeat by staring or by a display of confidence.

outfall /'aʊtfɔːl/ *n.* the mouth of a river, drain, etc., where it empties into the sea etc.

outfield /'aʊtfiːld/ *n.* **1** the outer part of a cricket or baseball field. **2** outlying land. □ **outfielder** *n.*

outfight /aʊt'faɪt/ *v.tr.* fight better than; beat in a fight.

outfit /'aʊtfɪt/ *n. & v.* ● *n.* **1** a set of clothes worn or esp. designed to be worn together. **2** a complete set of equipment etc. for a specific purpose. **3** *colloq.* a group of people regarded as a unit, organization, etc.; a team. ● *v.tr.* (also *refl.*) (**-fitted**, **-fitting**) provide with an outfit, esp. of clothes.

outfitter /'aʊt,fɪtə(r)/ *n.* a supplier of equipment, esp. of conventional styles of clothing.

outflank /aʊt'flæŋk/ *v.tr.* **1 a** extend one's flank beyond that of (an enemy). **b** outmanoeuvre (an enemy) in this way. **2** get the better of; confound (an opponent).

outflow /'aʊtfləʊ/ *n.* **1** an outward flow. **2** the amount that flows out.

outfly /aʊt'flaɪ/ *v.tr.* (**-flies**; *past* **-flew** /-'fluː/; *past part.* **-flown** /-'fləʊn/) **1** surpass in flying. **2** fly faster or further than.

outfox /aʊt'fɒks/ *v.tr. colloq.* outwit.

outgas /aʊt'gæs/ *v.* (**outgases, outgassed, outgassing**) **1** *intr.* release or give off a dissolved or adsorbed gas or vapour. **2** *tr.* **a** release or give off (a substance) as a gas or vapour. **b** drive off a gas or vapour from.

outgeneral /aʊt'dʒenrəl/ *v.tr.* (**-generalled, -generalling**; US **-generaled, -generaling**) **1** outdo in generalship. **2** get the better of by superior strategy or tactics.

outgo *v. & n.* ● *v.tr.* /aʊt'gəʊ/ (*3rd sing. present* **-goes**; *past* **-went** /-'went/; *past part.* **-gone** /-'gɒn/) *archaic* go faster than; surpass. ● *n.* /'aʊtgəʊ/ (*pl.* **-goes**) expenditure of money, effort, etc.

outgoing /'aʊt,gəʊɪŋ/ *adj. & n.* ● *adj.* **1** friendly; sociable; extrovert. **2** retiring from office. **3** going out or away. ● *n.* **1** (in *pl.*) expenditure. **2** the act or an instance of going out.

outgrow /aʊt'grəʊ/ *v.tr.* (*past* **-grew** /-'gruː/; *past part.* **-grown** /-'grəʊn/) **1** grow too big for (one's clothes). **2** leave behind (a childish habit, taste, ailment, etc.) as one matures. **3** grow faster or taller than (a person, plant, etc.). □ **outgrow one's strength** become lanky and weak through too rapid growth.

outgrowth /'aʊtgrəʊθ/ *n.* **1** something that grows out. **2** an offshoot; a natural product. **3** the process of growing out.

outguess /aʊt'ges/ *v.tr.* guess correctly what is intended by (another person).

outgun /aʊt'gʌn/ *v.tr.* (**-gunned, -gunning**) **1** surpass in military or other power or strength. **2** shoot better than.

outhouse /'aʊthaʊs/ *n. & v.* ● *n.* **1** a building, esp. a shed, lean-to, barn, etc. built next to or in the grounds of a house. **2** *N. Amer.* an outdoor lavatory. ● *v.tr.* store (books etc.) away from the main collection.

outing /'aʊtɪŋ/ *n.* **1** a short holiday away from home, esp. of one day or part of a day; a pleasure-trip, an excursion. **2** any brief journey from home. **3** an appearance in an outdoor match, race, etc. **4** *colloq.* the practice or policy of exposing the homosexuality of a prominent person. [OUT *v.* = put out, go out + -ING[1]]

outjockey /aʊt'dʒɒkɪ/ *v.tr.* (**-eys, -eyed**) outwit by adroitness or trickery.

outjump /aʊt'dʒʌmp/ *v.tr.* surpass in jumping.

outlander /'aʊt,lændə(r)/ *n.* a foreigner, alien, or stranger.

outlandish /aʊt'lændɪʃ/ *adj.* **1** looking or sounding foreign. **2** bizarre, strange, unfamiliar. □ **outlandishly** *adv.* **outlandishness** *n.* [OE ūtlendisc f. ūtland foreign country f. OUT + LAND]

outlast /aʊt'lɑːst/ *v.tr.* last longer than (a person, thing, or duration) (*outlasted its usefulness*).

outlaw /ˈaʊtlɔː/ n. & v. ● n. **1** a fugitive from the law. **2** hist. a person deprived of the protection of the law. ● v.tr. **1** declare (a person) an outlaw. **2** make illegal; proscribe (a practice etc.). □ **outlaw strike** an unofficial strike. □ **outlawry** n. [OE ūtlaga, ūtlagian f. ON útlagi f. útlagr outlawed, rel. to OUT, LAW]

outlay /ˈaʊtleɪ/ n. what is spent on something.

outlet /ˈaʊtlet, -lɪt/ n. **1** a means of exit or escape. **2** (usu. foll. by for) a means of expression (of a talent, emotion, etc.) (find an outlet for tension). **3** an agency, distributor, or market for goods (a new retail outlet in China). **4** US a power point. [ME f. OUT- + LET¹]

outlier /ˈaʊtˌlaɪə(r)/ n. **1** (also attrib.) an outlying part or member. **2** Geol. a younger rock formation isolated in older rocks. **3** Statistics a result differing greatly from others in the same sample.

outline /ˈaʊtlaɪn/ n. & v. ● n. **1** a rough draft of a diagram, plan, proposal, etc. **2 a** a précis of a proposed novel, article, etc. **b** a verbal description of essential parts only; a summary. **3** a sketch containing only contour lines. **4** (in sing. or pl.) **a** lines enclosing or indicating an object (the outline of a shape under the blankets). **b** a contour. **c** an external boundary. **5** (in pl.) the main features or general principles (the outlines of a plan). **6** the representation of a word in shorthand. ● v.tr. **1** draw or describe in outline. **2** mark the outline of. □ **in outline** sketched or represented as an outline.

outlive /aʊtˈlɪv/ v.tr. **1** live longer than (another person). **2** live beyond (a specified date or time). **3** live through (an experience).

outlook /ˈaʊtlʊk/ n. **1** the prospect for the future (the outlook is bleak). **2** one's mental attitude or point of view (narrow in their outlook). **3** what is seen on looking out.

outlying /ˈaʊtˌlaɪɪŋ/ attrib.adj. situated far from a centre; remote.

outmanoeuvre /ˌaʊtməˈnuːvə(r)/ v.tr. (US **-maneuver**) **1** use skill and cunning to secure an advantage over (a person). **2** outdo in manoeuvring.

outmatch /aʊtˈmætʃ/ v.tr. be more than a match for (an opponent etc.); surpass.

outmeasure /aʊtˈmeʒə(r)/ v.tr. exceed in quantity or extent.

outmoded /aʊtˈməʊdɪd/ adj. **1** no longer in fashion. **2** obsolete. □ **outmodedly** adv. **outmodedness** n.

outmost /ˈaʊtməʊst/ adj. **1** outermost, furthest. **2** uttermost. [ME, var. of utmest UTMOST]

outnumber /aʊtˈnʌmbə(r)/ v.tr. exceed in number.

outpace /aʊtˈpeɪs/ v.tr. **1** go faster than. **2** outdo in a contest.

out-patient /ˈaʊtˌpeɪʃ(ə)nt/ n. a hospital patient who is resident at home but attends regular appointments in hospital.

outperform /ˌaʊtpəˈfɔːm/ v.tr. **1** perform better than. **2** surpass in a specified field or activity. □ **outperformance** n.

outplacement /ˈaʊtˌpleɪsmənt/ n. the act or process of finding new employment for esp. executive workers who have been dismissed or made redundant.

outplay /aʊtˈpleɪ/ v.tr. surpass in playing; play better than.

outpoint /aʊtˈpɔɪnt/ v.tr. (in various sports, esp. boxing) score more points than.

outport /ˈaʊtpɔːt/ n. **1** a subsidiary port. **2** Can. a small remote fishing village.

outpost /ˈaʊtpəʊst/ n. **1** a detachment set at a distance from the main body of an army, esp. to prevent surprise. **2** a distant branch or settlement. **3** the furthest territory of an (esp. the British) empire.

outpouring /ˈaʊtˌpɔːrɪŋ/ n. **1** (usu. in pl.) a copious spoken or written expression of emotion. **2** what is poured out.

output /ˈaʊtpʊt/ n. & v. ● n. **1** the product of a process, esp. of manufacture, or of mental or artistic work. **2** the quantity or amount of this. **3** the printout, results, etc. supplied by a computer. **4** the power etc. delivered by an apparatus. **5** a place where energy, information, etc. leaves a system. ● v.tr. (**-putting**; past and past part. **-put** or **-putted**) **1** put or send out. **2** (of a computer) supply (results etc.).

outrage /ˈaʊtreɪdʒ/ n. & v. ● n. **1** an extreme or shocking violation of others' rights, sentiments, etc. **2** a gross offence or indignity. **3** fierce anger or resentment (a feeling of outrage). ● v.tr. **1** subject to outrage. **2** injure, insult, etc. flagrantly. **3** shock and anger. [ME f. OF outrage f. outrer exceed f. outre f. L ultra beyond]

outrageous /aʊtˈreɪdʒəs/ adj. **1** immoderate. **2** shocking. **3** grossly cruel. **4** immoral, offensive. □ **outrageously** adv. **outrageousness** n. [ME f. OF outrageus (as OUTRAGE)]

outran past of OUTRUN.

outrange /aʊtˈreɪndʒ/ v.tr. (of a gun or its user) have a longer range than.

outrank /aʊtˈræŋk/ v.tr. **1** be superior in rank to. **2** take priority over.

outré /ˈuːtreɪ/ adj. **1** outside the bounds of what is usual or proper. **2** eccentric or indecorous. [F, past part. of outrer: see OUTRAGE]

outreach v. & n. ● v.tr. /aʊtˈriːtʃ/ **1** reach further than. **2** surpass. **3** poet. stretch out (one's arms etc.). ● n. /ˈaʊtriːtʃ/ **1 a** any organization's involvement with or influence in the community, esp. in the context of social welfare. **b** the extent of this. **2** the extent or length of reaching out (an outreach of 38 metres).

out-relief /ˈaʊtrɪˌliːf/ n. Brit. hist. = outdoor relief.

outride /aʊtˈraɪd/ v.tr. (past **-rode** /-ˈrəʊd/; past part. **-ridden** /-ˈrɪd(ə)n/) **1** ride better, faster, or further than. **2** (of a ship) come safely through (a storm etc.).

outrider /ˈaʊtˌraɪdə(r)/ n. **1** a mounted attendant riding ahead of, or with, a carriage etc. **2** a motorcyclist acting as a guard in a similar manner. **3** US a herdsman keeping cattle within bounds.

outrigged /ˈaʊtrɪgd/ adj. (of a boat etc.) having outriggers.

outrigger /ˈaʊtˌrɪgə(r)/ n. **1** a beam, spar, or framework, rigged out and projecting from or over a ship's side for various purposes. **2** a similar projecting beam etc. in a building. **3** a log etc. fixed parallel to a canoe to stabilize it. **4 a** an extension of the splinter-bar of a carriage etc. to enable another horse to be harnessed outside the shafts. **b** a horse harnessed in this way. **5 a** an iron bracket bearing a rowlock attached horizontally to a boat's side to increase the leverage of the oar. **b** a boat fitted with these. **6** a chassis extension supporting the body of a car etc. [OUT- + RIG¹: perh. partly after obs. (Naut.) outligger]

outright adv. & adj. ● adv. /aʊtˈraɪt/ **1** altogether, entirely (proved outright). **2** not gradually, nor by degrees, nor by instalments (bought it outright). **3** without reservation, openly (denied the charge outright). ● attrib.adj. /ˈaʊtraɪt/ **1** downright, direct, complete (their resentment turned to outright anger). **2** undisputed, clear (the outright winner). □ **outrightness** /ˈaʊtˌraɪtnɪs/ n.

outrival /aʊtˈraɪv(ə)l/ v.tr. (**-rivalled**, **-rivalling**; US **-rivaled**, **-rivaling**) outdo as a rival.

outrode past of OUTRIDE.

outrun v. & n. ● v.tr. /aʊtˈrʌn/ (**-running**; past **-ran** /-ˈræn/; past part. **-run**) **1 a** run faster or further than. **b** escape from. **2** go beyond (a specified point or limit). ● n. /ˈaʊtrʌn/ Austral. a sheep-run distant from its homestead.

outrush /ˈaʊtrʌʃ/ n. **1** a rushing out. **2** a violent overflow.

outsail /aʊtˈseɪl/ v.tr. sail better or faster than.

outsat past and past part. of OUTSIT.

outsell /aʊtˈsel/ v.tr. (past and past part. **-sold** /-ˈsəʊld/) **1** sell more than. **2** be sold in greater quantities than.

outset /ˈaʊtset/ n. the start, beginning. □ **at** (or **from**) **the outset** at or from the beginning.

outshine /aʊtˈʃaɪn/ v.tr. (past and past part. **-shone** /-ˈʃɒn/) shine brighter than; surpass in ability, excellence, etc.

outshoot /aʊtˈʃuːt/ v.tr. (past and past part. **-shot** /-ˈʃɒt/) **1** shoot better or further than (another person). **2** esp. US score more goals, points, etc. than (another player or team).

outside n., adj., adv., & prep. ● n. /aʊtˈsaɪd, ˈaʊtsaɪd/ **1** the external side or surface; the outer parts (painted blue on the outside). **2** the external appearance; the outward aspect of a building etc. **3** (of a path) the side away from the wall or next to the road. **4** (also attrib.) all that is without; the world as distinct from the thinking subject (learn about the outside world; viewed from the outside the problem is simple). **5** a position on the outer side (the gate opens from the outside). **6** colloq. the highest computation (it is a mile at the outside). **7** an outside player in football etc. **8** (in pl.) the outer sheets of a ream of paper. ● adj. /ˈaʊtsaɪd/ **1** of or on or nearer the outside; outer. **2 a** not of or belonging to some circle or institution (outside help; outside work). **b** (of a broker) not a member of the Stock Exchange. **3** (of a chance etc.) remote; very unlikely. **4** (of an estimate etc.) the greatest or highest possible (the outside price). **5** (of a player in football etc.) positioned nearest to the edge of the field. ● adv. /aʊtˈsaɪd/ **1** on or to the outside. **2** in or to the open air. **3** not within or enclosed or included. **4** sl. not in prison. ● prep. /aʊtˈsaɪd, ˈaʊtsaɪd/ (also disp. foll. by of) **1** not in; to or at the exterior of (meet me outside the post office). **2** external to, not included in, beyond the limits of (outside the law). **3** colloq. other than; apart from. □ **at the outside** (of an estimate etc.) at the most. **get outside of**

sl. eat or drink. **outside and in** outside and inside. **outside broadcast** *Brit.* a broadcast made on location and not in a studio. **outside edge** (on an ice-skate) each of the edges facing outwards when both feet are together. **outside in** = *inside out.* **outside interest** a hobby; an interest not connected with one's work or normal way of life. **outside track** the outside lane of a sports track etc. which is longer because of the curve.

outsider /aʊtˈsaɪdə(r)/ *n.* **1 a** a non-member of some circle, party, profession, etc. (cf. INSIDER). **b** an uninitiated person; a layman. **2** a person without special knowledge, breeding, etc., or not fit to mix with good society. **3** a competitor, applicant, etc. thought to have little chance of success.

outsit /aʊtˈsɪt/ *v.tr.* (**-sitting**; *past* and *past part.* **-sat** /-ˈsæt/) sit longer than (another person or thing).

outsize /ˈaʊtsaɪz/ *adj. & n.* ● *adj.* **1** unusually large. **2** (of garments etc.) of an exceptionally large size. ● *n.* an exceptionally large person or thing, esp. a garment. □ **outsizeness** *n.*

outskirts /ˈaʊtskɜːts/ *n.pl.* the outer border or fringe of a town, district, subject, etc.

outsmart /aʊtˈsmɑːt/ *v.tr. colloq.* outwit, be cleverer than.

outsold *past* and *past part.* of OUTSELL.

outsource /aʊtˈsɔːs/ *v.tr. Commerce* **1** obtain (goods etc.) by contract from an outside source. **2** contract (work) out. □ **outsourcing** /ˈaʊtˌsɔːsɪŋ/ *n.*

outspan /ˈaʊtspæn/ *v. & n. S. Afr.* ● *v.* (**-spanned, -spanning**) **1** *tr.* (also *absol.*) unharness (animals) from a cart, plough, etc. **2** *intr.* break a wagon journey. ● *n.* a place for grazing or encampment. [S. Afr. Du. *uitspannen* unyoke]

outspend /aʊtˈspend/ *v.tr.* (*past* and *past part.* **-spent** /-ˈspent/) spend more than (one's resources or another person).

outspoken /aʊtˈspəʊkən/ *adj.* given to or involving plain speaking; frank in stating one's opinions. □ **outspokenly** *adv.* **outspokenness** /-kənnɪs/ *n.*

outspread /aʊtˈspred/ *adj. & v.* ● *adj.* spread out; fully extended or expanded. ● *v.tr. & intr.* (*past* and *past part.* **-spread**) spread out; expand.

outstanding /aʊtˈstændɪŋ/ *adj.* **1 a** conspicuous, eminent, esp. because of excellence. **b** (usu. foll. by *at, in*) remarkable in (a specified field). **2** (esp. of a debt) not yet settled (*£200 still outstanding*). □ **outstandingly** *adv.*

outstare /aʊtˈsteə(r)/ *v.tr.* **1** outdo in staring. **2** abash by staring.

outstation /ˈaʊtˌsteɪʃ(ə)n/ *n.* **1** a branch of an organization, enterprise, or business in a remote area or at a considerable distance from headquarters. **2** esp. *Austral. & NZ* part of a farming estate separate from the main estate.

outstay /aʊtˈsteɪ/ *v.tr.* **1** stay beyond the limit of (one's welcome, invitation, etc.). **2** stay or endure longer than (another person etc.).

outstep /aʊtˈstep/ *v.tr.* (**-stepped, -stepping**) step outside or beyond.

outstretch /aʊtˈstretʃ/ *v.tr.* **1** (usu. as **outstretched** *adj.*) reach out or stretch out (esp. one's hands or arms). **2** reach or stretch further than.

outstrip /aʊtˈstrɪp/ *v.tr.* (**-stripped, -stripping**) **1** pass in running etc. **2** surpass in competition or relative progress or ability.

out-swinger /ˈaʊtˌswɪŋə(r)/ *n. Cricket* a ball that swings away from the batsman.

out-take /ˈaʊtteɪk/ *n.* a length of film or tape rejected in editing.

out-talk /aʊtˈtɔːk/ *v.tr.* outdo or overcome in talking.

out-think /aʊtˈθɪŋk/ *v.tr.* (*past* and *past part.* **-thought** /-ˈθɔːt/) outwit; outdo in thinking.

out-thrust *adj., v., & n.* ● *adj.* /ˈaʊtθrʌst/ extended; projected (*ran forward with out-thrust arms*). ● *v.tr.* /aʊtˈθrʌst/ (*past* and *past part.* **-thrust**) thrust out. ● *n.* /ˈaʊtθrʌst/ **1** the act or an instance of thrusting forcibly outward. **2** the act or an instance of becoming prominent or noticeable.

out-top /aʊtˈtɒp/ *v.tr.* (**-topped, -topping**) surmount, surpass in height, extent, etc.

out-tray /ˈaʊttreɪ/ *n.* a tray for outgoing documents, letters, etc.

out-turn /ˈaʊttɜːn/ *n.* **1** the quantity produced. **2** the result of a process or sequence of events.

outvalue /aʊtˈvælju:/ *v.tr.* (**-values, -valued, -valuing**) be of greater value than.

outvote /aʊtˈvəʊt/ *v.tr.* defeat by a majority of votes.

outwalk /aʊtˈwɔːk/ *v.tr.* **1** outdo in walking. **2** walk beyond.

outward /ˈaʊtwəd/ *adj., adv., & n.* ● *adj.* **1** situated on or directed towards the outside. **2** going out (*on the outward voyage*). **3** bodily, external, apparent, superficial (*in all outward respects*). **4** *archaic* outer (*the outward man*). ● *adv.* (also **outwards** /-wədz/) in an outward direction; towards the outside. ● *n.* the outward appearance of something; the exterior. □ **outward bound 1** (of a ship, passenger, etc.) going away from home. **2** (**Outward Bound**) (in the UK) a movement to provide adventure training, naval training, and other outdoor activities for young people. **outward form** appearance. **outward things** the world around us. **to outward seeming** apparently. □ **outwardly** *adv.* [OE *ūtweard* (as OUT, -WARD)]

outwardness /ˈaʊtwədnɪs/ *n.* **1** external existence; objectivity. **2** an interest or belief in outward things, objective-mindedness.

outwards var. of OUTWARD *adv.*

outwash /ˈaʊtwɒʃ/ *n.* the material carried from a glacier by melt water and deposited beyond the moraine.

outwatch /aʊtˈwɒtʃ/ *v.tr.* **1** watch more than or longer than. **2** *archaic* keep awake beyond the end of (night etc.).

outwear *v. & n.* ● *v.tr.* /aʊtˈweə(r)/ (*past* **-wore** /-ˈwɔː(r)/; *past part.* **-worn** /-ˈwɔːn/) **1** exhaust; wear out; wear away. **2** live or last beyond the duration of. **3** (as **outworn** *adj.*) out of date, obsolete. ● *n.* /ˈaʊtweə(r)/ outer clothing.

outweigh /aʊtˈweɪ/ *v.tr.* exceed in weight, value, importance, or influence.

outwent *past* of OUTGO.

outwit /aʊtˈwɪt/ *v.tr.* (**-witted, -witting**) be too clever or crafty for; deceive by greater ingenuity.

outwith /aʊtˈwɪθ/ *prep. Sc.* outside, beyond.

outwore *past* of OUTWEAR.

outwork /ˈaʊtwɜːk/ *n.* **1** an advanced or detached part of a fortification. **2** work done outside the shop or factory which supplies it. □ **outworker** *n.* (in sense 2).

outworn *past part.* of OUTWEAR.

ouzel /ˈuːz(ə)l/ *n.* (also **ousel**) *archaic* a blackbird. □ **ring ouzel** see RING[1]. **water ouzel** = DIPPER 1. [OE *ōsle* blackbird, of unkn. orig.]

ouzo /ˈuːzəʊ/ *n.* (*pl.* **-os**) a Greek aniseed-flavoured spirit. [mod.Gk]

ova *pl.* of OVUM.

oval /ˈəʊv(ə)l/ *adj. & n.* ● *adj.* **1** egg-shaped, ellipsoidal. **2** having the outline of an egg, elliptical. ● *n.* **1** an egg-shaped or elliptical closed curve. **2** an object with an oval outline. **3** a sports ground with an oval field, esp. (*Austral.*) for Australian Rules football. □ **ovally** *adv.* **ovalness** *n.* **ovality** /əʊˈvælɪtɪ/ *n.* [med.L *ovalis* (as OVUM)]

Oval Office the office of the US President in the White House.

Ovambo /əʊˈvæmbəʊ/ *n. & adj.* ● *n.* (*pl.* same or **Ovambos**) **1** a member of a Bantu-speaking people inhabiting northern Namibia. **2** the language of the Ovambo. ● *adj.* of or relating to the Ovambo or their language.

Ovamboland /əʊˈvæmbəʊˌlænd/ a semi-arid region of northern Namibia, the homeland of the Ovambo people.

ovary /ˈəʊvərɪ/ *n.* (*pl.* **-ies**) **1** each of the female reproductive organs in which ova are produced. **2** the hollow base of the carpel of a flower, containing one or more ovules. □ **ovarian** /əʊˈveərɪən/ *adj.* **ovariectomy** /ˌəʊvərɪˈektəmɪ/ *n.* (*pl.* **-ies**) (in sense 1). **ovariotomy** /-rɪˈɒtəmɪ/ *n.* (*pl.* **-ies**) (in sense 1). **ovaritis** /-ˈraɪtɪs/ *n.* (in sense 1). [mod.L *ovarium* (as OVUM)]

ovate /ˈəʊveɪt/ *adj.* esp. *Biol.* egg-shaped as a solid or in outline; oval. [L *ovatus* (as OVUM)]

ovation /əʊˈveɪʃ(ə)n/ *n.* **1** an enthusiastic reception, esp. spontaneous and sustained applause. **2** *Rom. Antiq.* a lesser form of triumph. □ **standing ovation** a period of prolonged applause during which the crowd or audience rise to their feet. □ **ovational** *adj.* [L *ovatio* f. *ovare* exult]

oven /ˈʌv(ə)n/ *n.* **1** an enclosed compartment of brick, stone, or metal for cooking food. **2** a chamber for heating or drying. **3** a small furnace or kiln used in chemistry, metallurgy, etc. □ **oven-ready** (of food) prepared before sale so as to be ready for immediate cooking in the oven. [OE *ofen* f. Gmc]

ovenbird /ˈʌv(ə)nˌbɜːd/ *n.* a bird that makes a domed nest; esp. one of the family Furnariidae, native to Central and South America.

ovenproof /ˈʌv(ə)nˌpruːf/ *adj.* suitable for use in an oven; heat-resistant.

ovenware /ˈʌv(ə)nˌweə(r)/ n. dishes that can be used for cooking food in the oven.

over /ˈəʊvə(r)/ adv., prep., n., & adj. ● adv. expressing movement or position or state above or beyond something stated or implied: **1** outward and downward from a brink or from any erect position (knocked the man over). **2** so as to cover or touch a whole surface (paint it over). **3** so as to produce a fold, or reverse a position; with the effect of being upside down. **4 a** across a street or other space (decided to cross over; came over from America). **b** for a visit etc. (invited them over last night). **5** with transference or change from one hand or part to another (went over to the enemy; swapped them over). **6** with motion above something; so as to pass across something (climb over; fly over; boil over). **7 a** from beginning to end with repetition or detailed concentration (think it over; did it six times over). **b** esp. US again, once more. **8** in excess; more than is right or required (left over). **9** for or until a later time (hold it over). **10** at an end; settled (the crisis is over; all is over between us). **11** (in full **over to you**) (as int.) (in radio conversations etc.) said to indicate that it is the other person's turn to speak. **12** (as int.) Cricket an umpire's call to change ends. ● prep. **1** above, in, or to a position higher than; upon. **2** out and down from; down from the edge of (fell over the cliff). **3** so as to cover (a hat over his eyes). **4** above and across; so as to clear (flew over the North Pole; a bridge over the Thames). **5** concerning; engaged with; as a result of; while occupied with (laughed over a good joke; fell asleep over the newspaper). **6 a** in superiority of; superior to; in charge of (a victory over the enemy; reign over three kingdoms). **b** in preference to. **7** divided by (22 over 7). **8 a** throughout; covering the extent of (travelled over most of Africa; a blush spread over his face). **b** so as to deal with completely (went over the plans). **9 a** for the duration of (stay over Saturday night). **b** at any point during the course of (I'll do it over the weekend). **10** beyond; more than (bids of over £50; are you over 18?). **11** transmitted by (heard it over the radio). **12** in comparison with (gained 20% over last year). **13** having recovered from (am now over my cold; will get over it in time). ● n. Cricket a sequence of balls (now usu. six), bowled from one end of the pitch; the portion of the game containing this (a maiden over). ● adj. (see also over-). **1** upper, outer. **2** superior. **3** extra. □ **begin** (or **start** etc.) **over** N. Amer. begin again. **get it over with** do or undergo something unpleasant etc. so as to be rid of it. **give over** (usu. as int.) colloq. stop talking, desist. **not over** not very; not at all (not over friendly). **over again** once again, again from the beginning. **over against** in an opposite situation to; adjacent to, in contrast with. **over-age** over a certain age limit. **over all** taken as a whole. **over and above** in addition to; not to mention (£100 over and above the asking price). **over and over** so that the same thing or the same point comes up again and again (said it over and over; rolled it over and over). **over the fence** Austral. & NZ sl. unreasonable; unfair; indecent. **over one's head** see HEAD. **over the hill** see HILL. **over the moon** see MOON. **over-the-top** colloq. (esp. of behaviour, dress, etc.) outrageous, excessive. **over the way** (in a street etc.) facing or opposite. [OE ofer f. Gmc]

over- /ˈəʊvə(r)/ prefix added to verbs, nouns, adjectives, and adverbs, meaning: **1** excessively; to an unwanted degree (overheat; overdue). **2** upper, outer, extra (overcoat; overtime). **3** 'over' in various senses (overhang; overshadow). **4** completely, utterly (overawe; overjoyed).

over-abundant /ˌəʊvərəˈbʌndənt/ adj. in excessive quantity. □ **over-abundance** n. **over-abundantly** adv.

overachieve /ˌəʊvərəˈtʃiːv/ v. **1** intr. do more than might be expected (esp. scholastically). **2** tr. achieve more than (an expected goal or objective etc.). □ **overachievement** n. **overachiever** n.

overact /ˌəʊvərˈækt/ v.tr. & intr. act (a role) in an exaggerated manner.

over-active /ˌəʊvərˈæktɪv/ adj. excessively active. □ **over-activity** /-ækˈtɪvɪtɪ/ n.

overage /ˈəʊvərɪdʒ/ n. a surplus or excess, esp. an amount greater than estimated.

overall adj., adv., & n. ● adj. /ˈəʊvərˌɔːl/ **1** total, inclusive of all (overall cost). **2** taking everything into account, general (overall improvement). **3** from end to end (overall length). ● adv. /ˌəʊvərˈɔːl/ in all parts; taken as a whole; from end to end (overall, the performance was excellent; length overall). ● n. /ˈəʊvərˌɔːl/ **1** Brit. an outer garment worn to keep out dirt, wet, etc. **2** (in pl.) protective trousers, dungarees, or a combination suit, worn by workmen etc. **3** Brit. close-fitting trousers worn as part of army uniform. □ **overalled** /ˈəʊvərˌɔːld/ adj.

overambitious /ˌəʊvəræmˈbɪʃəs/ adj. excessively ambitious. □ **overambition** n. **overambitiously** adv.

over-anxious /ˌəʊvərˈæŋkʃəs/ adj. excessively anxious. □ **over-anxiously** adv. **over-anxiety** /-æŋˈzaɪɪtɪ/ n.

overarch /ˌəʊvərˈɑːtʃ/ v.tr. form an arch over. □ **overarching** adj.

overarm /ˈəʊvərˌɑːm/ adj. & adv. **1** Cricket & Tennis etc. with the hand above the shoulder (bowl it overarm; an overarm service). **2** Swimming with one or both arms lifted out of the water during a stroke.

overate past of OVEREAT.

overawe /ˌəʊvərˈɔː/ v.tr. **1** restrain by awe. **2** keep in awe.

overbalance v. & n. ● v. /ˌəʊvəˈbæləns/ **1** tr. cause (a person or thing) to lose its balance and fall. **2** intr. fall over, capsize. **3** tr. outweigh. ● n. /ˈəʊvəˌbæləns/ **1** an excess. **2** the amount of this.

overbear /ˌəʊvəˈbeə(r)/ v.tr. (past **-bore** /-ˈbɔː(r)/; past part. **-borne** /-ˈbɔːn/) **1** (as **overbearing** adj.) **a** domineering, masterful. **b** overpowering. **2** bear down; upset by weight, force, or emotional pressure. **3** put down or repress by power or authority. **4** surpass in importance; outweigh. □ **overbearingly** adv. **overbearingness** n.

overbid v. & n. ● v. /ˌəʊvəˈbɪd/ (**-bidding**; past and past part. **-bid**) **1** tr. **a** make a higher bid than. **b** offer more than the value of. **2** tr. (also absol.) Cards **a** bid more on (one's hand) than warranted. **b** overcall. ● n. /ˈəʊvəˌbɪd/ a bid that is higher than another, or higher than is justified. □ **overbidder** /ˈəʊvəˌbɪdə(r)/ n.

overbite /ˈəʊvəˌbaɪt/ n. Dentistry the overlapping of the lower teeth by the upper.

overblouse /ˈəʊvəˌblaʊz/ n. a garment like a blouse, but worn without tucking it into a skirt or trousers.

overblown /ˌəʊvəˈbləʊn/ adj. **1** excessively inflated or pretentious. **2** (of a flower or a woman's beauty etc.) past its prime.

overboard /ˈəʊvəˌbɔːd/ adv. from on a ship into the water (fall overboard). □ **go overboard 1** be highly enthusiastic. **2** behave immoderately; go too far. **throw overboard** abandon, discard.

overbold /ˌəʊvəˈbəʊld/ adj. excessively bold.

overbook /ˌəʊvəˈbʊk/ v.tr. (also absol.) make more bookings for (an aircraft, hotel, group of people, etc.) than there are places available.

overboot /ˈəʊvəˌbuːt/ n. a boot worn over another boot or shoe.

overbore past of OVERBEAR.

overborne past part. of OVERBEAR.

overbought past and past part. of OVERBUY.

overbrim /ˌəʊvəˈbrɪm/ v. (**-brimmed, -brimming**) **1** tr. flow over the brim of. **2** intr. (of a vessel or liquid) overflow at the brim.

overbuild /ˌəʊvəˈbɪld/ v.tr. (past and past part. **-built** /-ˈbɪlt/) **1** build over or upon. **2** place too many buildings on (land etc.).

overburden /ˌəʊvəˈbɜːd(ə)n/ v. & n. ● v.tr. burden (a person, thing, etc.) to excess. ● n. **1** rock etc. that must be removed prior to mining the mineral deposit beneath it. **2** an excessive burden. □ **overburdensome** adj.

Overbury /ˈəʊvəbərɪ/, Sir Thomas (1581–1613), English poet and courtier. He is remembered for his 'Characters', portrait sketches on the model of those of Theophrastus, published posthumously in 1614. On the pretext of his refusal of a diplomatic post he was sent to the Tower of London. There he was fatally poisoned by the agents of Frances Howard, Lady Essex, whose marriage to his patron Robert Carr (afterwards Earl of Somerset) (d.1645) he had opposed.

overbusy /ˌəʊvəˈbɪzɪ/ adj. excessively busy.

overbuy /ˌəʊvəˈbaɪ/ v.tr. & intr. (past and past part. **-bought** /-ˈbɔːt/) buy (a commodity etc.) in excess of immediate need.

overcall v. & n. ● v.tr. /ˌəʊvəˈkɔːl/ (also absol.) Cards **1** make a higher bid than (a previous bid or opponent). **2** Brit. = OVERBID v. 2a. ● n. /ˈəʊvəˌkɔːl/ an act or instance of overcalling.

overcame past of OVERCOME.

overcapacity /ˌəʊvəkəˈpæsɪtɪ/ n. a state of saturation or an excess of productive capacity.

overcapitalize /ˌəʊvəˈkæpɪtəˌlaɪz/ v.tr. (also **-ise**) fix or estimate the capital of (a company etc.) too high.

overcareful /ˌəʊvəˈkeəfʊl/ adj. excessively careful. □ **overcarefully** adv.

overcast adj., v., & n. ● adj. /ˈəʊvəˌkɑːst/ **1** (of the sky, weather, etc.) covered with cloud; dull and gloomy. **2** (in sewing) edged with stitching to prevent fraying. ● v.tr. /ˌəʊvəˈkɑːst/ (past and past part. **-cast**) **1** cover (the sky etc.) with clouds or darkness. **2** stitch over (a raw edge etc.) to prevent fraying. ● n. /ˈəʊvəˌkɑːst/ cloud covering part of the sky.

overcautious /ˌəʊvəˈkɔːʃəs/ adj. excessively cautious. □ **overcaution** n. **overcautiously** adv. **overcautiousness** n.

overcharge v. & n. ● v.tr. /ˌəʊvəˈtʃɑːdʒ/ **1 a** charge too high a price to (a person) for a thing. **b** charge (a specified sum) beyond the right price.

2 put too much charge into (a battery, gun, etc.). **3** put exaggerated or excessive detail into (a description, picture, etc.). ● *n.* /'əʊvəˌtʃɑːdʒ/ an excessive charge (of explosive, money, etc.).

overcheck /'əʊvəˌtʃek/ *n.* **1** a combination of two different-sized check patterns. **2** a cloth with this pattern.

overcloud /ˌəʊvə'klaʊd/ *v.tr.* **1** cover with cloud. **2** mar, spoil, or dim, esp. as the result of anxiety etc. (*overclouded by uncertainties*). **3** make obscure.

overcoat /'əʊvəˌkəʊt/ *n.* **1** a heavy coat, esp. one worn over indoor clothes for warmth outdoors in cold weather. **2** a protective coat of paint etc.

overcome /ˌəʊvə'kʌm/ *v.* (*past* **-came** /-'keɪm/; *past part.* **-come**) **1** *tr.* prevail over, master, conquer. **2** *tr.* (as **overcome** *adj.*) **a** exhausted, made helpless. **b** (usu. foll. by *with, by*) affected by (emotion etc.). **3** *intr.* be victorious. [OE *ofercuman* (as OVER-, COME)]

overcompensate /ˌəʊvə'kɒmpenˌseɪt/ *v.* **1** *tr.* (usu. foll. by *for*) compensate excessively for (something). **2** *intr. Psychol.* strive for power etc. in an exaggerated way, esp. to make allowance or amends for a real or fancied grievance, defect, handicap, etc. □ **overcompensation** /-ˌkɒmpen'seɪʃ(ə)n/ *n.* **overcompensatory** /-kɒm'pensətərɪ, -ˌkɒmpen'seɪtərɪ/ *adj.*

overconfident /ˌəʊvə'kɒnfɪd(ə)nt/ *adj.* excessively confident. □ **overconfidence** *n.* **overconfidently** *adv.*

overcook /ˌəʊvə'kʊk/ *v.tr.* cook too much or for too long. □ **overcooked** *adj.*

overcritical /ˌəʊvə'krɪtɪk(ə)l/ *adj.* excessively critical; quick to find fault.

overcrop /ˌəʊvə'krɒp/ *v.tr.* (**-cropped, -cropping**) exhaust (the land) by the continuous growing of crops.

overcrowd /ˌəʊvə'kraʊd/ *v.tr.* (often as **overcrowded** *adj.*) fill (a space, object, etc.) beyond what is usual or comfortable. □ **overcrowding** *n.*

over-curious /ˌəʊvə'kjʊərɪəs/ *adj.* excessively curious. □ **over-curiosity** /-ˌkjʊərɪ'ɒsɪtɪ/ *n.*

over-delicate /ˌəʊvə'delɪkət/ *adj.* excessively delicate. □ **over-delicacy** *n.*

overdevelop /ˌəʊvədɪ'veləp/ *v.tr.* (**-developed, -developing**) **1** develop too much. **2** *Photog.* treat with developer for too long.

overdo /ˌəʊvə'duː/ *v.tr.* (*3rd sing. present* **-does** /-'dʌz/; *past* **-did** /-'dɪd/; *past part.* **-done** /-'dʌn/) **1** carry to excess, take too far, exaggerate (*I think you overdid the sarcasm*). **2** (esp. as **overdone** *adj.*) overcook. □ **overdo it** (or **things**) exhaust oneself. [OE *oferdōn* (as OVER-, DO¹)]

overdose *n. & v.* ● *n.* /'əʊvəˌdəʊs/ an excessive dose (of a drug etc.). ● *v.* **1** *tr.* /ˌəʊvə'dəʊs/ give an overdose of a drug etc. to (a person). **2** *intr.* (often foll. by *on*) take an overdose of a drug. □ **overdosage** /ˌəʊvə'dəʊsɪdʒ/ *n.*

overdraft /'əʊvəˌdrɑːft/ *n.* **1** a deficit in a bank account caused by drawing more money than is credited to it. **2** the amount of this.

overdraw /ˌəʊvə'drɔː/ *v.* (*past* **-drew** /-'druː/; *past part.* **-drawn** /-'drɔːn/) **1** *tr.* **a** draw a sum of money in excess of the amount credited to (one's bank account). **b** (as **overdrawn** *adj.*) having overdrawn one's account. **2** *tr.* overdraw one's account. **3** *tr.* exaggerate in describing or depicting. □ **overdrawer** /'əʊvəˌdrɔːə(r)/ *n.* (in senses 1 & 2)

overdress *v. & n.* ● *v.* /ˌəʊvə'dres/ **1** *tr.* dress with too much display or formality. **2** *intr.* overdress oneself. ● *n.* /'əʊvəˌdres/ a dress worn over another dress or a blouse etc.

overdrink /ˌəʊvə'drɪŋk/ *v.intr. & refl.* (*past* **-drank** /-'dræŋk/; *past part.* **-drunk** /-'drʌŋk/) drink too much.

overdrive /'əʊvəˌdraɪv/ *n.* **1 a** a mechanism in a motor vehicle providing a gear ratio higher than that of the usual gear. **b** an additional speed-increasing gear. **2** (usu. prec. by *in, into*) a state of high or excessive activity.

overdub *v. & n.* ● *v.tr.* /ˌəʊvə'dʌb/ (**-dubbed, -dubbing**) (also *absol.*) impose (additional sounds) on an existing recording. ● *n.* /'əʊvəˌdʌb/ the act or an instance of overdubbing.

overdue /ˌəʊvə'djuː/ *adj.* **1** past the time when due or ready. **2** not yet paid, arrived, born, etc., though after the expected time. **3** (of a library book etc.) retained longer than the period allowed.

overeager /ˌəʊvər'iːgə(r)/ *adj.* excessively eager. □ **overeagerly** *adv.* **overeagerness** *n.*

overeat /ˌəʊvər'iːt/ *v.intr. & refl.* (*past* **-ate** /-'et, -'eɪt/; *past part.* **-eaten** /-'iːt(ə)n/) eat too much.

overelaborate *adj. & v.* ● *adj.* /ˌəʊvərɪ'læbərət/ excessively elaborate. ● *v.tr.* /ˌəʊvərɪ'læbəˌreɪt/ elaborate in excessive detail.

overelaborately *adv.* **overelaborateness** /-rətnɪs/ *n.* **overelaboration** /-ˌlæbə'reɪʃ(ə)n/ *n.*

over-emotional /ˌəʊvərɪ'məʊʃən(ə)l/ *adj.* excessively emotional. □ **over-emotionally** *adv.*

overemphasis /ˌəʊvər'emfəsɪs/ *n.* excessive emphasis. □ **overemphasize** *v.tr. & intr.* (also **-ise**).

overenthusiasm /ˌəʊvərɪn'θjuːzɪˌæz(ə)m, -'θuːzɪˌæz(ə)m/ *n.* excessive enthusiasm. □ **overenthusiastic** /-ˌθjuːzɪ'æstɪk, -ˌθuːzɪ-/ *adj.* **overenthusiastically** *adv.*

overestimate *v. & n.* ● *v.tr.* /ˌəʊvər'estɪˌmeɪt/ (also *absol.*) form too high an estimate of (a person, ability, cost, etc.). ● *n.* /ˌəʊvər'estɪmət/ too high an estimate. □ **overestimation** /-ˌestɪ'meɪʃ(ə)n/ *n.*

overexcite /ˌəʊvərɪk'saɪt/ *v.tr.* excite excessively. □ **overexcitement** *n.* **overexcitable** *adj.*

over-exercise /ˌəʊvər'eksəˌsaɪz/ *v. & n.* ● *v.* **1** *tr.* use or exert (a part of the body, one's authority, etc.) too much. **2** *intr.* take too much exercise; overexert oneself. ● *n.* excessive exercise.

overexert /ˌəʊvərɪg'zɜːt/ *v.tr. & refl.* exert too much. □ **overexertion** /-'zɜːʃ(ə)n/ *n.*

overexpose /ˌəʊvərɪk'spəʊz/ *v.tr.* (also *absol.*) **1** expose too much, esp. to the public eye. **2** *Photog.* expose (film) for too long a time. □ **overexposure** /-'spəʊʒə(r)/ *n.*

overextend /ˌəʊvərɪk'stend/ *v.tr.* **1** extend (a thing) too far. **2** (also *refl.*) take on (oneself) or impose on (another person) an excessive burden of work.

overfall /'əʊvəˌfɔːl/ *n.* **1** a turbulent stretch of sea etc. caused by a strong current or tide over a submarine ridge, or by a meeting of currents. **2** a place provided on a dam, weir, etc. for the overflow of surplus water.

overfamiliar /ˌəʊvəfə'mɪlɪə(r)/ *adj.* excessively familiar. □ **overfamiliarity** /-ˌmɪlɪ'ærɪtɪ/ *n.*

overfatigue /ˌəʊvəfə'tiːg/ *n.* excessive fatigue.

overfeed /ˌəʊvə'fiːd/ *v.tr.* (*past and past part.* **-fed** /-'fed/) feed excessively.

overfill /ˌəʊvə'fɪl/ *v.tr. & intr.* fill to excess or to overflowing.

overfine /ˌəʊvə'faɪn/ *adj.* excessively fine; too precise.

overfish /ˌəʊvə'fɪʃ/ *v.tr.* **1** deplete (a stream etc.) by too much fishing. **2** deplete the stock of (a fish) by excessive fishing.

overflow *v. & n.* ● *v.* /ˌəʊvə'fləʊ/ **1** *tr.* **a** flow over (the brim, limits, etc.). **b** flow over the brim or limits of. **2** *intr.* **a** (of a receptacle etc.) be so full that the contents overflow it (*until the cup was overflowing*). **b** (of contents) overflow a container. **3** *tr.* (of a crowd etc.) extend beyond the limits of (a room etc.). **4** *tr.* flood (a surface or area). **5** *intr.* (foll. by *with*) be full of. **6** *intr.* (of kindness, a harvest, etc.) be very abundant. ● *n.* /'əʊvəˌfləʊ/ (also *attrib.*) **1** what overflows or is superfluous (*mop up the overflow; put the overflow audience in another room*). **2** an instance of overflowing (*overflow occurs when both systems are run together*). **3** (esp. in a bath or sink) an outlet for excess water etc. **4** *Computing* the generation of a number having more digits than the assigned location. □ **overflow meeting** a meeting for those who cannot be accommodated at the main gathering. [OE *oferflōwan* (as OVER-, FLOW)]

overfly /ˌəʊvə'flaɪ/ *v.tr.* (**-flies**; *past* **-flew** /-'fluː/; *past part.* **-flown** /-'fləʊn/) fly over or beyond (a place or territory). □ **overflight** /'əʊvəˌflaɪt/ *n.*

overfold /'əʊvəˌfəʊld/ *n. Geol.* a series of strata folded so that the middle part is upside down.

overfond /ˌəʊvə'fɒnd/ *adj.* (often foll. by *of*) having too great an affection or liking (for a person or thing) (*overfond of chocolate; an overfond parent*). □ **overfondly** *adv.* **overfondness** *n.*

overfulfil /ˌəʊvəfʊl'fɪl/ *v.tr.* (*US* **-fulfill**) (**-fulfilled, -fulfilling**) fulfil (a plan, quota, etc.) beyond expectation or before the appointed time. □ **overfulfilment** *n.*

overfull /ˌəʊvə'fʊl/ *adj.* filled excessively or to overflowing.

overgeneralize /ˌəʊvə'dʒenrəˌlaɪz/ *v.* (also **-ise**) **1** *intr.* draw general conclusions from inadequate data etc. **2** *intr.* argue more widely than is justified by the available evidence, by circumstances, etc. **3** *tr.* draw an excessively general conclusion from (data, circumstances, etc.). □ **overgeneralization** /-ˌdʒenərəlaɪ'zeɪʃ(ə)n/ *n.*

overgenerous /ˌəʊvə'dʒenərəs/ *adj.* excessively generous. □ **overgenerously** *adv.*

overglaze /'əʊvəˌgleɪz/ *n. & adj.* ● *n.* **1** a second glaze applied to ceramic ware. **2** decoration on a glazed surface. ● *adj.* (of painting etc.) done on a glazed surface.

overground /ˈəʊvəˌɡraʊnd/ *adj.* **1** raised above the ground. **2** not underground.

overgrow /ˌəʊvəˈɡrəʊ/ *v.tr.* (*past* **-grew** /-ˈɡruː/; *past part.* **-grown** /-ˈɡrəʊn/) **1** (as **overgrown** *adj.*) **a** abnormally large, often through retaining immature characteristics beyond the usual age (*a great overgrown child*). **b** wild; grown over with vegetation (*an overgrown pond*). **2** grow over, overspread, esp. so as to choke (*nettles have overgrown the pathway*). **3** grow too big for (one's strength etc.). □ **overgrowth** /ˈəʊvəˌɡrəʊθ/ *n.*

overhand /ˈəʊvəˌhænd/ *adj. & adv.* **1** (in cricket, tennis, baseball, etc.) thrown or played with the hand above the shoulder. **2** *Swimming* = OVERARM 2. **3 a** with the palm of the hand downward or inward. **b** with the hand above the object held. □ **overhand knot** a simple knot made by forming a loop and passing the free end through it.

overhang *v. & n.* ● *v.* /ˌəʊvəˈhæŋ/ (*past* and *past part.* **-hung** /-ˈhʌŋ/) **1** *tr. & intr.* project or hang over. **2** *tr.* menace, preoccupy, threaten. ● *n.* /ˈəʊvəˌhæŋ/ **1** an instance of overhanging. **2** the overhanging part of a structure or rock-formation. **3** the amount by which a thing overhangs.

overhaste /ˌəʊvəˈheɪst/ *n.* excessive haste. □ **overhasty** *adj.* **overhastily** *adv.*

overhaul *v. & n.* ● *v.tr.* /ˌəʊvəˈhɔːl/ **1 a** take to pieces in order to examine. **b** examine the condition of (and repair if necessary). **2** overtake. ● *n.* /ˈəʊvəˌhɔːl/ a thorough examination, with repairs if necessary. [orig. Naut., = release (rope-tackle) by slackening]

overhead *adv., adj., & n.* ● *adv.* /ˌəʊvəˈhed/ **1** above one's head. **2** in the sky or in the storey above. ● *adj.* /ˈəʊvəˌhed/ **1** placed overhead. **2** (of a driving mechanism etc.) above the object driven. **3** (of expenses) arising from general running costs, as distinct from particular business transactions. ● *n.* /ˈəʊvəˌhed/ (in *pl.* or *N. Amer.* in *sing.*) overhead expenses. □ **overhead projector** a device that projects an enlarged image of a transparency on to a surface above and behind the user.

overhear /ˌəʊvəˈhɪə(r)/ *v.tr.* (*past* and *past part.* **-heard** /-ˈhɜːd/) (also *absol.*) hear as an eavesdropper or as an unperceived or unintentional listener.

overheat /ˌəʊvəˈhiːt/ *v.* **1** *tr. & intr.* make or become too hot; heat to excess. **2** cause marked inflation in (a country's economy) by placing excessive pressure on resources at a time of expanding demand. **3** *tr.* (as **overheated** *adj.*) too passionate about a matter.

Overijssel /ˌəʊvərˈaɪs(ə)l/ a province of the east central Netherlands, north of the IJssel river, on the border with Germany; capital, Zwolle.

overindulge /ˌəʊvərɪnˈdʌldʒ/ *v.tr. & intr.* indulge to excess. □ **overindulgence** *n.* **overindulgent** *adj.*

overinsure /ˌəʊvərɪnˈʃʊə(r)/ *v.tr.* insure (property etc.) for more than its real value; insure excessively. □ **overinsurance** *n.*

overissue *v. & n.* ● *v.tr.* /ˌəʊvərˈɪʃuː, -ˈɪsjuː/ (**-issues, -issued, -issuing**) issue (notes, shares, etc.) beyond the authorized amount, or the ability to pay. ● *n.* /ˌəʊvərˌɪʃuː, -ˌɪsjuː/ the notes, shares, etc., or the amount so issued.

overjoyed /ˌəʊvəˈdʒɔɪd/ *adj.* (often foll. by *at, to hear,* etc.) filled with great joy.

overkill *n. & v.* ● *n.* /ˈəʊvəˌkɪl/ **1** the amount by which destruction or the capacity for destruction exceeds what is necessary for victory or annihilation. **2** excess; excessive behaviour, unwarranted thoroughness of treatment. ● *v.tr. & intr.* /ˌəʊvəˈkɪl/ kill or destroy to a greater extent than necessary.

overladen /ˌəʊvəˈleɪd(ə)n/ *adj.* bearing or carrying too large a load.

overlaid *past* and *past part.* of OVERLAY[1].

overlain *past part.* of OVERLIE.

overland /ˈəʊvəˌlænd/ *adj., adv., & v.* ● *adj. & adv.* (also /ˌəʊvəˈlænd/) **1** by land. **2** not by sea. ● *v.* *Austral.* **1** *tr.* drive (livestock) overland. **2** *intr.* go a long distance overland.

overlander /ˈəʊvəˌlændə(r)/ *n. Austral. & NZ* **1** a person who drives livestock overland (orig. one who drove stock from New South Wales to the colony of South Australia). **2** *sl.* a tramp, a sundowner.

overlap *v. & n.* ● *v.* /ˌəʊvəˈlæp/ (**-lapped, -lapping**) **1** *tr.* (of part of an object) partly cover (another object). **2** *tr.* cover and extend beyond. **3** *intr.* (of two things) partly coincide; not be completely separate (*where psychology and philosophy overlap*). ● *n.* /ˈəʊvəˌlæp/ **1** an instance of overlapping. **2** the amount of this.

over-large /ˌəʊvəˈlɑːdʒ/ *adj.* too large.

overlay[1] *v. & n.* ● *v.tr.* /ˌəʊvəˈleɪ/ (*past* and *past part.* **-laid** /-ˈleɪd/) **1** lay over. **2** (foll. by *with*) cover the surface of (a thing) with (a coating etc.). **3** overlie. ● *n.* /ˈəʊvəˌleɪ/ **1** a thing laid over another. **2** (in printing,

map-reading, etc.) a transparent sheet to be superimposed on another sheet. **3** *Computing* **a** the process of transferring a block of data etc. to replace what is already stored. **b** a section so transferred. **4** a coverlet, small tablecloth, etc.

overlay[2] *past* of OVERLIE.

overleaf /ˌəʊvəˈliːf/ *adv.* on the other side of the leaf (of a book) (*the diagram overleaf*).

overleap /ˌəʊvəˈliːp/ *v.tr.* (*past* and *past part.* **-leaped** or **-leapt** /-ˈlept/) **1** leap over, surmount. **2** omit, ignore. [OE *oferhlēapan* (as OVER, LEAP)]

overlie /ˌəʊvəˈlaɪ/ *v.tr.* (**-lying** /-ˈlaɪɪŋ/; *past* **-lay** /-ˈleɪ/; *past part.* **-lain** /-ˈleɪn/) **1** lie on top of. **2** smother (a child etc.) by lying on top.

overload *v. & n.* ● *v.tr.* /ˌəʊvəˈləʊd/ load excessively; force (a person, thing, etc.) beyond normal or reasonable capacity. ● *n.* /ˈəʊvəˌləʊd/ an excessive quantity; a demand etc. which surpasses capability or capacity.

over-long /ˌəʊvəˈlɒŋ/ *adj. & adv.* too or excessively long.

overlook *v. & n.* ● *v.tr.* /ˌəʊvəˈlʊk/ **1** fail to notice; ignore, condone (an offence etc.). **2** have a view from above, be higher than. **3** supervise, oversee. **4** bewitch with the evil eye. ● *n.* /ˈəʊvəˌlʊk/ *US* a commanding position or view. □ **overlooker** /ˈəʊvəˌlʊkə(r)/ *n.*

overlord /ˈəʊvəˌlɔːd/ *n.* a supreme lord. □ **overlordship** *n.*

overly /ˈəʊvəlɪ/ *adv.* excessively; too.

overlying *pres. part.* of OVERLIE.

overman *v. & n.* ● *v.tr.* /ˌəʊvəˈmæn/ (**-manned, -manning**) provide with too large a crew, staff, etc. ● *n.* /ˈəʊvəˌmæn/ (*pl.* **-men**) **1** an overseer in a colliery. **2** = ÜBERMENSCH.

overmantel /ˈəʊvəˌmænt(ə)l/ *n.* ornamental shelves etc. over a mantelpiece.

over-many /ˌəʊvəˈmenɪ/ *adj.* too many; an excessive number.

overmaster /ˌəʊvəˈmɑːstə(r)/ *v.tr.* master completely, conquer. □ **overmastering** *adj.* **overmastery** *n.*

overmatch /ˌəʊvəˈmætʃ/ *v.tr.* be more than a match for; defeat by superior strength etc.

overmeasure /ˈəʊvəˌmeʒə(r)/ *n.* an amount beyond what is proper or sufficient.

over-much /ˌəʊvəˈmʌtʃ/ *adv. & adj.* ● *adv.* to too great an extent; excessively. ● *adj.* excessive; superabundant.

over-nice /ˌəʊvəˈnaɪs/ *adj.* excessively fussy, punctilious, particular, etc. □ **over-niceness** *n.* **over-nicety** /-ˈnaɪsɪtɪ/ *n.*

overnight *adv. & adj.* ● *adv.* /ˌəʊvəˈnaɪt/ **1** for the duration of a night (*stay overnight*). **2** during the course of a night. **3** suddenly, immediately (*the situation changed overnight*). ● *adj.* /ˈəʊvəˌnaɪt/ **1** for use overnight (*an overnight bag*). **2** done etc. overnight (*an overnight stop*).

overnighter /ˌəʊvəˈnaɪtə(r)/ *n.* **1** a person who stops at a place overnight. **2** an overnight bag.

overpaid *past* and *past part.* of OVERPAY.

overparted /ˌəʊvəˈpɑːtɪd/ *adj. Theatr.* having too demanding a part to play; cast beyond one's ability.

over-particular /ˌəʊvəpəˈtɪkjʊlə(r)/ *adj.* excessively particular or fussy.

overpass *n. & v.* ● *n.* /ˈəʊvəˌpɑːs/ a road or railway line that passes over another by means of a bridge. ● *v.tr.* /ˌəʊvəˈpɑːs/ **1** pass over or across or beyond. **2** get to the end of; surmount. **3** (as **overpassed** or **overpast** *adj.* /-ˈpɑːst/) that has gone by, past.

overpay /ˌəʊvəˈpeɪ/ *v.tr.* (*past* and *past part.* **-paid** /-ˈpeɪd/) recompense (a person, service, etc.) too highly. □ **overpayment** *n.*

overpitch /ˌəʊvəˈpɪtʃ/ *v.tr.* **1** (often *absol.*) *Cricket* bowl (a ball) so that it pitches or would pitch too near the stumps. **2** exaggerate.

overplay /ˌəʊvəˈpleɪ/ *v.tr.* play (a part) to excess; give undue importance to; overemphasize. □ **overplay one's hand 1** be unduly optimistic about one's capabilities. **2** spoil a good case by exaggerating its value.

overplus /ˈəʊvəˌplʌs/ *n.* a surplus, a superabundance. [ME, partial transl. of AF *surplus* or med.L *su(pe)rplus*]

overpopulated /ˌəʊvəˈpɒpjʊˌleɪtɪd/ *adj.* having too large a population. □ **overpopulation** /-ˌpɒpjʊˈleɪʃ(ə)n/ *n.*

overpower /ˌəʊvəˈpaʊə(r)/ *v.tr.* **1** reduce to submission, subdue. **2** make (a thing) ineffective or imperceptible by greater intensity. **3** (of heat, emotion, etc.) be too intense for, overwhelm. □ **overpowering** *adj.* **overpoweringly** *adv.*

overprice /ˌəʊvəˈpraɪs/ *v.tr.* price (a thing) too highly.

overprint *v. & n.* ● *v.tr.* /ˌəʊvəˈprɪnt/ **1** print further matter on (a surface already printed, esp. a postage stamp). **2** print (further matter) in this way. **3** *Photog.* print (a positive) darker than was intended. **4** (also *absol.*) print too many copies of (a work). ● *n.* /ˈəʊvəˌprɪnt/ **1** the words etc. overprinted. **2** an overprinted postage stamp.

overproduce /ˌəʊvəprəˈdjuːs/ *v.tr.* (usu. *absol.*) **1** produce more of (a commodity) than is wanted. **2** produce to an excessive degree. □ **overproduction** /-ˈdʌkʃ(ə)n/ *n.*

overproof /ˈəʊvəˌpruːf/ *adj.* containing more alcohol than proof spirit does.

overprotective /ˌəʊvəprəˈtektɪv/ *adj.* excessively protective, esp. of a person in one's charge.

overqualified /ˌəʊvəˈkwɒlɪˌfaɪd/ *adj.* too highly qualified (esp. for a particular job etc.).

overran *past* of OVERRUN.

overrate /ˌəʊvəˈreɪt/ *v.tr.* (often as **overrated** *adj.*) assess or value too highly.

overreach /ˌəʊvəˈriːtʃ/ *v.tr.* circumvent, outwit; get the better of by cunning or artifice. □ **overreach oneself 1** strain oneself by reaching too far. **2** defeat one's object by going too far.

overreact /ˌəʊvərɪˈækt/ *v.intr.* respond more forcibly etc. than is justified. □ **overreaction** /-ˈækʃ(ə)n/ *n.*

overrefine /ˌəʊvərɪˈfaɪn/ *v.tr.* (also *absol.*) **1** refine too much. **2** make too subtle distinctions in (an argument etc.). □ **over-refinement** *n.*

override *v. & n.* ● *v.tr.* /ˌəʊvəˈraɪd/ (*past* **-rode** /-ˈrəʊd/; *past part.* **-ridden** /-ˈrɪd(ə)n/) **1** (often as **overriding** *adj.*) have or claim precedence or superiority over (*an overriding consideration*). **2 a** intervene and make ineffective. **b** interrupt the action of (an automatic device) esp. to take manual control. **3 a** trample down or underfoot. **b** supersede arrogantly. **4** extend over, esp. (of a part of a fractured bone) overlap (another part). **5** ride over (enemy country). **6** exhaust (a horse etc.) by hard riding. ● *n.* /ˈəʊvəˌraɪd/ **1** the action or process of suspending an automatic function. **2** a device for this.

overrider /ˈəʊvəˌraɪdə(r)/ *n. Brit.* each of a pair of projecting pieces on the bumper of a car.

overripe /ˌəʊvəˈraɪp/ *adj.* (esp. of fruit etc.) past its best; excessively ripe; full-blown.

overrode *past* of OVERRIDE.

overruff *v. & n. Cards* ● *v.tr.* /ˌəʊvəˈrʌf/ (also *absol.*) overtrump. ● *n.* /ˈəʊvəˌrʌf/ an instance of this.

overrule /ˌəʊvəˈruːl/ *v.tr.* **1** set aside (a decision, argument, proposal, etc.) by exercising a superior authority. **2** annul a decision by or reject a proposal of (a person) in this way.

overrun *v. & n.* ● *v.tr.* /ˌəʊvəˈrʌn/ (**-running**; *past* **-ran** /-ˈræn/; *past part.* **-run**) **1** (of vermin, weeds, etc.) swarm or spread over. **2** conquer or ravage (territory) by force. **3** (of time, expenditure, production, etc.) exceed (a fixed limit). **4** *Printing* carry over (a word etc.) to the next line or page. **5** *Mech.* rotate faster than. **6** (of water) flood (land). ● *n.* /ˈəʊvəˌrʌn/ **1** an instance of overrunning. **2** the amount of this. **3** the movement of a vehicle at a speed greater than is imparted by the engine. [OE *oferyrnan* (as OVER-, RUN)]

oversailing /ˌəʊvəˈseɪlɪŋ/ *adj.* (of a part of a building) projecting beyond what is below. [OVER + F *saillir* SALLY¹]

oversaw *past* of OVERSEE.

overscrupulous /ˌəʊvəˈskruːpjʊləs/ *adj.* excessively scrupulous or particular.

overseas *adv. & adj.* ● *adv.* /ˌəʊvəˈsiːz/ (also **oversea**) abroad (*was sent overseas for training; came back from overseas*). ● *attrib.adj.* /ˈəʊvəˌsiːz/ (also **oversea**) **1** foreign; across or beyond the sea. **2** of or connected with movement or transport over the sea (*overseas postage rates*).

oversee /ˌəʊvəˈsiː/ *v.tr.* (**-sees**; *past* **-saw** /-ˈsɔː/; *past part.* **-seen** /-ˈsiːn/) officially supervise (workers, work, etc.). [OE *ofersēon* look at from above (as OVER-, SEE)]

overseer /ˈəʊvəˌsiːə(r)/ *n.* a person who supervises others, esp. workers. □ **overseer of the poor** *Brit. hist.* a parish official who administered funds to the poor. [OVERSEE]

oversell /ˌəʊvəˈsel/ *v.tr.* (*past* and *past part.* **-sold** /-ˈsəʊld/) (also *absol.*) **1** sell more of (a commodity etc.) than one can deliver. **2** exaggerate the merits of.

over-sensitive /ˌəʊvəˈsensɪtɪv/ *adj.* excessively sensitive; easily hurt by, or too quick to react to, outside influences. □ **over-sensitiveness** *n.* **over-sensitivity** /-ˌsensɪˈtɪvɪti/ *n.*

overset /ˌəʊvəˈset/ *v.tr.* (**-setting**; *past* and *past part.* **-set**) **1** overturn, upset. **2** *Printing* set up (type) in excess of the available space.

oversew /ˈəʊvəˌsəʊ/ *v.tr.* (*past part.* **-sewn** /-ˌsəʊn/ or **-sewed**) **1** sew (two edges) with every stitch passing over the join. **2** join the sections of (a book) by a stitch of this type.

oversexed /ˌəʊvəˈsekst/ *adj.* having unusually strong sexual desires.

overshadow /ˌəʊvəˈʃædəʊ/ *v.tr.* **1** appear much more prominent or important than. **2 a** cast into the shade; shelter from the sun. **b** cast gloom over; mar, spoil. [OE *ofersceadwian* (as OVER-, SHADOW)]

overshoe /ˈəʊvəˌʃuː/ *n.* a shoe of rubber, felt, etc., worn over another as protection from wet, cold, etc.

overshoot *v. & n.* ● *v.tr.* /ˌəʊvəˈʃuːt/ (*past* and *past part.* **-shot** /-ˈʃɒt/) **1** pass or send beyond (a target or limit). **2** (of an aircraft) fly beyond or taxi too far along (the runway) when landing or taking off. ● *n.* /ˈəʊvəˌʃuːt/ **1** the act of overshooting. **2** the amount of this. □ **overshoot the mark** go beyond what is intended or proper; go too far. **overshot wheel** a water-wheel operated by the weight of water falling into buckets attached to its periphery.

overside /ˌəʊvəˈsaɪd/ *adv.* over the side of a ship (into a smaller boat, or into the sea).

oversight /ˈəʊvəˌsaɪt/ *n.* **1** a failure to notice something. **2** an inadvertent mistake. **3** supervision.

oversimplify /ˌəʊvəˈsɪmplɪˌfaɪ/ *v.tr.* (**-ies, -ied**) (also *absol.*) distort (a problem etc.) by stating it in too simple terms. □ **oversimplification** /-ˌsɪmplɪfɪˈkeɪʃ(ə)n/ *n.*

oversize /ˈəʊvəˌsaɪz/ *adj.* (also **-sized** /-ˌsaɪzd/) of more than the usual size.

overskirt /ˈəʊvəˌskɜːt/ *n.* an outer or second skirt.

overslaugh /ˈəʊvəˌslɔː/ *n. & v.* ● *n. Brit. Mil.* the passing over of one's turn of duty. ● *v.tr.* **1** *Brit. Mil.* pass over (one's duty) in consideration of another duty that takes precedence. **2** *US* pass over in favour of another. **3** *US* omit to consider. [Du. *overslag* (n.) f. *overslaan* omit (as OVER, *slaan* strike)]

oversleep /ˌəʊvəˈsliːp/ *v.intr. & refl.* (*past* and *past part.* **-slept** /-ˈslept/) **1** continue sleeping beyond the intended time of waking. **2** sleep too long.

oversleeve /ˈəʊvəˌsliːv/ *n.* a protective sleeve covering an ordinary sleeve.

oversold *past* and *past part.* of OVERSELL.

oversolicitous /ˌəʊvəsəˈlɪsɪtəs/ *adj.* excessively worried, anxious, eager, etc. □ **oversolicitude** *n.*

oversoul /ˈəʊvəˌsəʊl/ *n.* God as a spirit animating the universe and including all human souls. The concept is associated esp. with the Transcendentalism of Ralph Waldo Emerson.

overspecialize /ˌəʊvəˈspeʃəˌlaɪz/ *v.intr.* (also **-ise**) concentrate too much on one aspect or area. □ **overspecialization** /-ˌspeʃəlaɪˈzeɪʃ(ə)n/ *n.*

overspend *v. & n.* ● *v.* /ˌəʊvəˈspend/ (*past* and *past part.* **-spent** /-ˈspent/) **1** *intr. & refl.* spend too much. **2** *tr.* spend more than (a specified amount). ● *n.* /ˈəʊvəˌspend/ **1** the act of overspending a limit. **2** an instance of this. **3** the amount by which a limit is overspent.

overspill /ˈəʊvəˌspɪl/ *n.* **1** what is spilt over or overflows. **2** (often *attrib.*) the surplus population leaving a country or city to live elsewhere.

overspread /ˌəʊvəˈspred/ *v.tr.* (*past* and *past part.* **-spread**) **1** become spread or diffused over. **2** cover or occupy the surface of. **3** (as **overspread** *adj.*) (usu. foll. by *with*) covered (*high mountains overspread with trees*). [OE *ofersprædan* (as OVER-, SPREAD)]

overstaff /ˌəʊvəˈstɑːf/ *v.tr.* provide with too large a staff.

overstate /ˌəʊvəˈsteɪt/ *v.tr.* **1** state (esp. a case or argument) too strongly. **2** exaggerate. □ **overstatement** *n.*

overstay /ˌəʊvəˈsteɪ/ *v.tr.* stay longer than (one's welcome, a time-limit, etc.).

oversteer *v. & n.* ● *v.intr.* /ˌəʊvəˈstɪə(r)/ (of a motor vehicle) have a tendency to turn more sharply than was intended. ● *n.* /ˈəʊvəˌstɪə(r)/ this tendency.

overstep /ˌəʊvəˈstep/ *v.tr.* (**-stepped, -stepping**) **1** pass beyond (a boundary or mark). **2** violate (certain standards of behaviour etc.). □ **overstep the mark** violate conventions of behaviour.

overstock /ˌəʊvəˈstɒk/ *v.tr.* stock excessively.

overstrain /ˌəʊvəˈstreɪn/ *v.tr. & intr.* strain too much or too hard.

overstress /ˌəʊvəˈstres/ *v. & n.* ● *v.tr.* stress too much. ● *n.* an excessive degree of stress.

overstretch /ˌəʊvəˈstretʃ/ v.tr. **1** stretch too much. **2** (esp. as **overstretched** adj.) make excessive demands on (resources, a person, etc.).

overstrung adj. **1** /ˌəʊvəˈstrʌŋ/ (of a person, disposition, etc.) intensely strained, highly strung. **2** /ˈəʊvəˌstrʌŋ/ (of a piano) with strings in sets crossing each other obliquely.

overstudy /ˌəʊvəˈstʌdɪ/ v.tr. (**-ies**, **-ied**) **1** study beyond what is necessary or desirable. **2** (as **overstudied** adj.) excessively deliberate; affected.

overstuff /ˌəʊvəˈstʌf/ v.tr. **1** stuff more into than is necessary. **2** (as **overstuffed** adj.) (of furniture) made soft and comfortable by thick upholstery.

oversubscribe /ˌəʊvəsəbˈskraɪb/ v.tr. (usu. as **oversubscribed** adj.) subscribe for more than the amount available of (a commodity offered for sale etc.) (*the offer was oversubscribed*).

oversubtle /ˌəʊvəˈsʌt(ə)l/ adj. excessively subtle; not plain or clear.

oversupply /ˌəʊvəsəˈplaɪ/ v. & n. ● v.tr. (**-ies**, **-ied**) supply with too much. ● n. an excessive supply.

oversusceptible /ˌəʊvəsəˈseptɪb(ə)l/ adj. too susceptible or vulnerable.

overt /əʊˈvɜːt, ˈəʊvɜːt/ adj. unconcealed; done openly. □ **overtly** adv. **overtness** n. [ME f. OF past part. of *ovrir* open f. L *aperire*]

overtake /ˌəʊvəˈteɪk/ v.tr. (past **-took** /-ˈtʊk/; past part. **-taken** /-ˈteɪkən/) **1** (also absol.) catch up with and pass while travelling in the same direction. **2** (of a storm, misfortune, etc.) come suddenly or unexpectedly upon. **3** become level with and exceed (a compared value etc.).

overtask /ˌəʊvəˈtɑːsk/ v.tr. **1** give too heavy a task to. **2** be too heavy a task for.

overtax /ˌəʊvəˈtæks/ v.tr. **1** make excessive demands on (a person's strength etc.). **2** tax too heavily.

overthrow v. & n. ● v.tr. /ˌəʊvəˈθrəʊ/ (past **-threw** /-ˈθruː/; past part. **-thrown** /-ˈθrəʊn/) **1** remove forcibly from power. **2** put an end to (an institution etc.). **3** conquer, overcome. **4** knock down, upset. ● n. /ˈəʊvəˌθrəʊ/ **1** a defeat or downfall. **2** Cricket a fielder's return of the ball, not stopped near the wicket and so allowing further runs. **b** such a run. **3** Archit. a panel of decorated wrought-iron work in an arch or gateway.

overthrust /ˈəʊvəˌθrʌst/ n. Geol. the thrust of esp. lower strata on one side of a fault over those on the other side.

overtime /ˈəʊvəˌtaɪm/ n. & adv. ● n. **1** the time during which a person works at a job in addition to the regular hours. **2** payment for this. **3** N. Amer. Sport = extra time. ● adv. in addition to regular hours.

overtire /ˌəʊvəˈtaɪə(r)/ v.tr. & refl. exhaust or wear out (esp. an invalid etc.).

overtone /ˈəʊvəˌtəʊn/ n. **1** Mus. any of the tones above the lowest in a harmonic series. **2** a subtle or elusive quality or implication (*sinister overtones*). [OVER- + TONE, after G *Oberton*]

overtop /ˌəʊvəˈtɒp/ v.tr. (**-topped**, **-topping**) **1** be or become higher than. **2** surpass.

overtrain /ˌəʊvəˈtreɪn/ v.tr. & intr. subject to or undergo too much (esp. athletic) training with a consequent loss of proficiency.

overtrick /ˈəʊvəˌtrɪk/ n. Bridge a trick taken in excess of one's contract.

overtrump /ˌəʊvəˈtrʌmp/ v.tr. (also absol.) Cards play a higher trump than (another player).

overture /ˈəʊvəˌtjʊə(r)/ n. **1** an orchestral piece opening an opera etc. **2** a one-movement composition in this style. **3** (usu. in pl.) an opening of negotiations; a preliminary proposal or offer (esp. *make overtures to*). **4** the beginning of a poem etc. [ME f. OF f. L *apertura* APERTURE]

overturn v. & n. ● v. /ˌəʊvəˈtɜːn/ **1** tr. cause to fall down or over; upset. **2** tr. reverse; subvert; abolish; invalidate. **3** intr. fall down; fall over. ● n. /ˈəʊvəˌtɜːn/ a subversion, an act of upsetting.

overuse v. & n. ● v.tr. /ˌəʊvəˈjuːz/ use too much. ● n. /ˌəʊvəˈjuːs/ excessive use.

overvalue /ˌəʊvəˈvæljuː/ v.tr. (**-values**, **-valued**, **-valuing**) value too highly; have too high an opinion of.

overview /ˈəʊvəˌvjuː/ n. a general survey.

overweening /ˌəʊvəˈwiːnɪŋ/ adj. arrogant, presumptuous, conceited, self-confident. □ **overweeningly** adv. **overweeningness** n. [WEEN]

overweight adj., n., & v. ● adj. /ˌəʊvəˈweɪt/ beyond an allowed or suitable weight; (esp. of a person) weighing more than is normal or desirable, corpulent. ● n. /ˈəʊvəˌweɪt/ excessive or extra weight; preponderance. ● v.tr. /ˌəʊvəˈweɪt/ (usu. foll. by *with*) load unduly.

overwhelm /ˌəʊvəˈwelm/ v.tr. **1** overpower with emotion. **2** (usu. foll. by *with*) overpower with an excess of business etc. **3** bring to sudden ruin or destruction; crush. **4** bury or drown beneath a huge mass; submerge utterly.

overwhelming /ˌəʊvəˈwelmɪŋ/ adj. irresistible by force of numbers, influence, amount, etc. □ **overwhelmingly** adv.

overwind v. & n. ● v.tr. /ˌəʊvəˈwaɪnd/ (past and past part. **-wound** /-ˈwaʊnd/) wind (a mechanism, esp. a watch) beyond the proper stopping point. ● n. /ˈəʊvəˌwaɪnd/ an instance of this.

overwinter /ˌəʊvəˈwɪntə(r)/ v. **1** intr. (usu. foll. by *at, in*) spend the winter. **2** intr. (of insects, fungi, etc.) live through the winter. **3** tr. keep (animals, plants, etc.) alive through the winter.

overwork /ˌəʊvəˈwɜːk/ v. & n. ● v. **1** intr. work too hard. **2** tr. cause (another person) to work too hard. **3** tr. weary or exhaust with too much work. **4** tr. (esp. as **overworked** adj.) make excessive use of. **5** tr. (as **overworked** adj.) = OVERWROUGHT 2. ● n. excessive work.

overwound past and past part. of OVERWIND.

overwrite /ˌəʊvəˈraɪt/ v. (past **-wrote** /-ˈrəʊt/; past part. **-written** /-ˈrɪt(ə)n/) **1** tr. write on top of (other writing). **2** tr. Computing destroy (data) in (a file etc.) by entering new data. **3** intr. (as **overwritten** adj.) write too elaborately or too ornately. **4** intr. & refl. write too much; exhaust oneself by writing. **5** tr. write too much about. **6** intr. (esp. as **overwriting** n.) in shipping insurance, accept more risk than the premium income limits allow.

overwrought /ˌəʊvəˈrɔːt/ adj. **1** overexcited, nervous, distraught. **2** overdone; too elaborate.

overzealous /ˌəʊvəˈzeləs/ adj. too zealous in one's attitude, behaviour, etc.; excessively enthusiastic.

ovi-[1] /ˈəʊvɪ/ comb. form egg, ovum. [L *ovum* egg]

ovi-[2] /ˈəʊvɪ/ comb. form sheep. [L *ovis* sheep]

ovibovine /ˌəʊvɪˈbəʊvaɪn/ adj. & n. Zool. ● adj. having characteristics intermediate between those of a sheep and an ox. ● n. such an animal, e.g. a musk-ox.

Ovid /ˈɒvɪd/ (full name Publius Ovidius Naso) (43 BC–AD c. 17), Roman poet. He was a major poet of the Augustan period, particularly known for his elegiac love-poems (such as the *Amores* and the *Ars Amatoria*) and for the *Metamorphoses*, a hexameter epic which retells Greek and Roman myths in roughly chronological order. His irreverent attitudes offended Augustus and in AD 8 he was exiled to Tomis (modern Constanţa) on the Black Sea, where he continued to write elegiac poems describing his plight; these are collected in the *Tristia*.

oviduct /ˈəʊvɪˌdʌkt/ n. Anat. the tube through which an ovum passes from the ovary. □ **oviducal** /ˌəʊvɪˈdjuːk(ə)l/ adj. **oviductal** /-ˈdʌktəl/ adj.

Oviedo /ˌɒvɪˈeɪdəʊ/ a city in NW Spain, capital of the Asturias; pop. (1991) 203,190.

oviform /ˈəʊvɪˌfɔːm/ adj. egg-shaped.

ovine /ˈəʊvaɪn/ adj. of or like sheep. [LL *ovinus* f. L *ovis* sheep]

oviparous /əʊˈvɪpərəs/ adj. Zool. producing young by means of eggs expelled from the body before they are hatched (cf. VIVIPAROUS 1, OVOVIVIPAROUS). □ **oviparously** adv. **oviparity** /ˌəʊvɪˈpærɪtɪ/ n.

oviposit /ˌəʊvɪˈpɒzɪt/ v.intr. (**oviposited**, **ovipositing**) Zool. lay an egg or eggs, esp. with an ovipositor. □ **oviposition** /-pəˈzɪʃ(ə)n/ n. [OVI-[1] + L *ponere posit-* to place]

ovipositor /ˌəʊvɪˈpɒzɪtə(r)/ n. Zool. a pointed tubular organ with which a female insect deposits her eggs. [mod.L f. OVI-[1] + L *positor* f. *ponere posit-* to place]

ovoid /ˈəʊvɔɪd/ adj. & n. ● adj. **1** (of a solid or of a surface) egg-shaped. **2** oval, with one end more pointed than the other. ● n. an ovoid body or surface. [F *ovoïde* f. mod.L *ovoides* (as OVUM)]

ovolo /ˈəʊvəˌləʊ/ n. (pl. **ovoli** /-ˌlaɪ/) Archit. a rounded convex moulding. [It. dimin. of *ovo* egg f. L OVUM]

ovotestis /ˌəʊvəˈtestɪs/ n. (pl. **-testes** /-tiːz/) Zool. an organ producing both ova and spermatozoa. [OVUM + TESTIS]

ovoviviparous /ˌəʊvəʊvɪˈvɪpərəs/ adj. Zool. producing young by means of eggs hatched within the body (cf. OVIPAROUS, VIVIPAROUS 1). □ **ovoviviparity** /-ˌvɪvɪˈpærɪtɪ/ n. [OVUM + VIVIPAROUS]

ovulate /ˈɒvjʊˌleɪt/ v.intr. produce ova or ovules, or discharge them from the ovary. □ **ovulatory** *adj.* **ovulation** /ˌɒvjʊˈleɪʃ(ə)n/ *n.* [mod.L *ovulum* (as OVULE)]

ovule /ˈɒvjuːl/ *n. Bot.* the part of the ovary of seed plants that contains the germ cell; an unfertilized seed. □ **ovular** *adj.* [F f. med.L *ovulum*, dimin. of OVUM]

ovum /ˈəʊvəm/ *n.* (*pl.* **ova** /-və/) **1** a mature reproductive cell of female animals, produced by the ovary. **2** *Bot.* the egg cell of plants. [L, = egg]

ow /aʊ/ *int.* expressing sudden pain. [natural exclam.]

owe /əʊ/ *v.tr.* **1 a** be under obligation (to a person etc.) to pay or repay (money etc.) (*we owe you five pounds; owe more than I can pay*). **b** (*absol.*, usu. foll. by *for*) be in debt (*still owe for my car*). **2** (often foll. by *to*) be under obligation to render (gratitude etc., a person honour, gratitude, etc.) (*owe grateful thanks to*). **3** (usu. foll. by *to*) be indebted to a person or thing for (*we owe to Newton the principle of gravitation*). □ **owe a person a grudge** cherish resentment against a person. **owe it to oneself** (often foll. by *to* + infin.) need (to do) something to protect one's own interests. [OE *āgan* (see OUGHT¹) f. Gmc]

Owen¹ /ˈəʊɪn/, David (Anthony Llewellyn), Baron Owen of the City of Plymouth (b.1938), British politician. After serving as Foreign Secretary (1977–9) in the Labour government, he became increasingly dissatisfied with the Labour Party's policies, and in 1981 broke away to become a founding member of the Social Democratic Party (SDP) (see also GANG OF FOUR 2). He led the SDP from 1983 to 1987, resigning to form a breakaway SDP when the main party decided to merge with the Liberals; he eventually disbanded this party in 1990. In 1992 he was appointed the EC's chief mediator in attempts to solve the crisis in the former Yugoslavia.

Owen² /ˈəʊɪn/, Sir Richard (1804–92), English anatomist and palaeontologist. A qualified surgeon, Owen was superintendent of natural history at the British Museum for 28 years and planned the new Natural History Museum in South Kensington. He made important contributions to the evolution and taxonomy of monotremes and marsupials, flightless birds, and fossil reptiles, and coined the word *dinosaur* in 1841. Owen is chiefly remembered for his opposition to Darwinism and to its defender T. H. Huxley, because he did not accept that natural selection was sufficient to explain evolution.

Owen³ /ˈəʊɪn/, Robert (1771–1858), Welsh social reformer and industrialist. A pioneer socialist thinker, he believed that character is a product of the social environment. He founded a model industrial community centred on his cotton mills at New Lanark in Scotland; this was organized on principles of mutual cooperation, with improved working conditions and housing together with educational institutions provided for workers and their families. He went on to found a series of other cooperative communities; although these did not always succeed, his ideas had an important long-term effect on the development of British socialist thought and on the practice of industrial relations.

Owen⁴ /ˈəʊɪn/, Wilfred (1893–1918), English poet. He fought in the First World War and his experiences inspired his best-known works, most of which he wrote after a meeting with Siegfried Sassoon in 1917. Only five of Owen's poems appeared in his lifetime and his reputation has grown following editions of his poems by Sassoon in 1920 and Edmund Blunden in 1931; among the most famous are 'Strange Meeting' and 'Anthem for Doomed Youth'. Owen's poetry is characterized by its bleak realism, its indignation at the horrors of war, and its pity for the victims – of whom he became one, killed in action in the last hours of the war.

Owens /ˈəʊɪnz/, Jesse (born James Cleveland Owens) (1913–80), American athlete. In 1935 he equalled or broke six world records in 45 minutes, and in 1936 won four gold medals (100 and 200 metres, long jump, and 4 × 100 metres relay) at the Olympic Games in Berlin. The success in Berlin of Owens, as a black man, outraged Hitler, who was conspicuously absent when Owens's medals were presented.

owing /ˈəʊɪŋ/ *predic.adj.* **1** owed; yet to be paid (*the balance owing*). **2** (foll. by *to*) **a** caused by; attributable to (*the cancellation was owing to ill health*). **b** (as *prep.*) because of (*trains are delayed owing to bad weather*).

owl /aʊl/ *n.* **1** a nocturnal bird of prey of the order Strigiformes, with large eyes and a hooked beak. (*See note below.*) **2** *colloq.* a person compared to an owl, esp. in looking solemn or wise. □ **owl-light** dusk, twilight. **owl-monkey** (*pl.* **-eys**) = DOUROUCOULI. □ **owlery** *n.* (*pl.* **-ies**). **owlish** *adj.* **owlishly** *adv.* **owlishness** *n.* (in sense 2). **owl-like** *adj.* [OE *ūle* f. Gmc]

■ The owls belong to two families, Tytonidae (barn owls and their relatives) and Strigidae (tawny owls, eagle owls, snowy owls, little owls, etc.). They have a number of adaptations for hunting at night,

with large forward-facing eyes, a prominent facial disc, highly acute hearing, and silent flight. Owls appear to have a large head, short neck, and small beak, but a thick layer of soft feathers hides the real shape — the neck is unusually flexible, and the gape is large enough to swallow prey whole. Their calls are often highly distinctive.

owlet /ˈaʊlɪt/ *n.* **1** a young owl. **2** a small owl, esp. of the genus *Glaucidium*, *Xenoglaux*, or *Athene*.

own /əʊn/ *adj. & v.* ● *adj.* (prec. by possessive) **1 a** belonging to oneself or itself; not another's (*saw it with my own eyes*). **b** individual, peculiar, particular (*has its own charm*). **2** used to emphasize identity rather than possession (*cooks his own meals*). **3** (*absol.*) **a** private property (*is it your own?*). **b** kindred (*among my own*). ● *v.* **1** *tr.* have as property; possess. **2 a** *tr.* confess; admit as valid, true, etc. (*own their faults; owns he did not know*). **b** *intr.* (foll. by *to*) confess to (*owned to a prejudice*). **3** *tr.* acknowledge paternity, authorship, or possession of. □ **come into one's own 1** receive one's due. **2** achieve recognition. **get one's own back** (often foll. by *on*) *colloq.* get revenge. **hold one's own** maintain one's position; not be defeated or lose strength. **of one's own** belonging to oneself alone. **on one's own 1** alone. **2** independently, without help. **own-brand** (of goods) manufactured specially for a retailer and bearing the retailer's name. **own goal 1** a goal scored by mistake against the scorer's own side. **2** *colloq.* an act or initiative that unintentionally harms one's own interests. **own up** (often foll. by *to*) confess frankly. □ **-owned** *adj.* (in *comb.*). [OE *āgen*, *āgnian*: see OWE]

owner /ˈəʊnə(r)/ *n.* **1** a person who owns something. **2** *sl.* the captain of a ship. □ **owner-occupied** (of a house etc.) lived in by its owner. **owner-occupier** a person who owns the house etc. he or she lives in. □ **ownerless** *adj.* **ownership** *n.*

owt /aʊt, ɔʊt/ *n. colloq.* or *dial.* anything. [var. of AUGHT¹]

ox /ɒks/ *n.* (*pl.* **oxen** /ˈɒks(ə)n/) **1** a bovine animal; esp. a large usu. horned domesticated species of cattle, *Bos taurus*, used esp. for supplying milk, and for eating as meat. (See also CATTLE.) **2** a castrated male of this, used esp. as a draught animal. □ **ox-eyed** with large protuberant eyes. **ox-fence** a strong fence for keeping in cattle, consisting of railings, a hedge, and often a ditch. **ox-pecker** an African bird of the genus *Buphagus*, feeding on skin parasites living on large animals. [OE *oxa* f. Gmc]

ox- var. of OXY-².

oxalic acid /ɒkˈsælɪk/ *n. Chem.* a poisonous, sour, crystalline acid (chem. formula: $(COOH)_2$), found in sorrel and rhubarb leaves. □ **oxalate** /ˈɒksəˌleɪt/ *n.* [F *oxalique* f. L *oxalis* f. Gk *oxalis* wood sorrel]

oxalis /ˈɒksəlɪs, ɒkˈsɑːl-/ *n.* a plant of the genus *Oxalis*, with trifoliate leaves and white, yellow, or pink flowers, e.g. wood sorrel. [L f. Gk f. *oxus* sour]

oxbow /ˈɒksbəʊ/ *n.* **1** a U-shaped collar of an ox-yoke. **2 a** a loop formed by a horseshoe bend in a river. **b** a lake formed when the river cuts across the narrow end of the loop.

Oxbridge /ˈɒksbrɪdʒ/ *n. Brit.* **1** (also *attrib.*) Oxford and Cambridge universities regarded together, esp. in contrast to newer institutions. **2** (often *attrib.*) the characteristics of these universities. [portmanteau word f. Ox(ford) + (Cam)bridge]

oxen *pl.* of OX.

oxer /ˈɒksə(r)/ *n.* an ox-fence.

ox-eye /ˈɒksaɪ/ *n.* a composite plant with large flowers with conspicuous rays, e.g. the ox-eye daisy. □ **ox-eye daisy** *n.* a daisy, *Leucanthemum vulgare*, having large white flowers with yellow centres (also called *marguerite*, *white ox-eye*).

Oxf. *abbr.* Oxford.

Oxfam /ˈɒksfæm/ *abbr.* Oxford Committee for Famine Relief, a British charity founded in Oxford in 1942, dedicated to helping victims of famine and natural disasters as well as raising living standards in developing countries.

Oxford /ˈɒksfəd/ a city in central England, on the River Thames, the county town of Oxfordshire; pop. (1991) 109,000. Oxford University is located there.

Oxford bags *n.* wide baggy trousers.

Oxford blue *n.. & adj.* ● *n.* **1** a dark blue, sometimes with a purple tinge. **2** a blue (BLUE¹ *n.* 3a) of Oxford University. ● *adj.* of this colour.

Oxford English Dictionary (abbr. **OED**) the largest dictionary of the English language, prepared in Oxford and originally issued in instalments between 1884 and 1928 under the title *A New English Dictionary on Historical Principles* (NED). It was published under the present title in twelve volumes with a supplement in 1933. Based on historical principles, it was edited until his death in 1915 by Sir James Murray. Preparation for the dictionary was begun by the Philological Society of London in 1857, and editorial work (originally for a four-volume dictionary, to be completed in ten years) started in earnest in 1879 under an agreement with Oxford University Press. A second edition, incorporating post-1933 supplements edited by Robert Burchfield (b.1923), was published in 1989, and a third edition is being prepared.

Oxford Group a Christian movement founded at Oxford in 1921, with discussion of personal problems by groups. Also called *Moral Re-Armament*.

Oxford Movement a Christian movement (c.1833–45) based at Oxford and led by John Keble, John Henry Newman, and Edward Pusey, which aimed at restoring traditional Catholic teaching within the Church of England; its principles were set out in a series of pamphlets (1833–41) called *Tracts for the Times* (hence an alternative name *Tractarianism*). The Movement emphasized ceremonial, set up the first Anglican religious communities, and contributed to social work and scholarship. At first it met with much hostility, but it eventually had a profound effect on the Anglican Church, forming the basis for the development of Anglo-Catholicism.

Oxfordshire /ˈɒksfəd‚ʃɪə(r)/ a county of south central England; county town, Oxford.

Oxford University the oldest English university, established at Oxford soon after 1167, perhaps as a result of a migration of students from Paris. The university comprises a federation of 36 colleges, the first of which, University College, was formally founded in 1249. The first women's college, Lady Margaret Hall, was founded in 1878.

oxherd /ˈɒkshɜːd/ n. a cowherd.

oxhide /ˈɒkshaɪd/ n. **1** the hide of an ox. **2** leather made from this.

oxidant /ˈɒksɪd(ə)nt/ n. an oxidizing agent. [F, part. of *oxider* (as OXIDE)]

oxidation /‚ɒksɪˈdeɪʃ(ə)n/ n. the process or result of oxidizing or being oxidized. □ **oxidation number** (or **state**) *Chem.* **1** a number assigned to an element representing the number of electrons lost (or, when negative, gained) by an atom of that element when chemically combined, on the assumption that bonding in the compound in question is purely ionic. **2** the state represented by a particular value of this. ¶ In chemical nomenclature, oxidation number is represented by roman numerals, e.g. in iron(III) oxide (= ferric oxide). □ **oxidational** adj. **oxidative** /ˈɒksɪ‚deɪtɪv/ adj. [F, f. *oxider* (see OXIDANT, -ATION)]

oxide /ˈɒksaɪd/ n. a binary compound of oxygen. [F f. *oxygène* OXYGEN + -*ide* after *acide* ACID]

oxidize /ˈɒksɪ‚daɪz/ v. (also -**ise**) **1** intr. & tr. combine or cause to combine with oxygen. **2** tr. & intr. cover (metal) or (of metal) become covered with a coating of oxide etc.; make or become rusty or tarnished. **3** intr. & tr. *Chem.* undergo or cause to undergo a loss of electrons (opp. REDUCE 12b). □ **oxidizing agent** *Chem.* a substance that brings about oxidation by being reduced and gaining electrons. □ **oxidizable** adj. **oxidized** adj. **oxidizer** n. **oxidization** /‚ɒksɪdaɪˈzeɪʃ(ə)n/ n.

oxlip /ˈɒkslɪp/ n. **1** a woodland primula, *Primula elatior*, with the flowers clustered on a stalk. **2** (in full **false oxlip**) a natural hybrid between a primrose and a cowslip.

Oxon. /ˈɒks(ə)n/ abbr. **1** Oxfordshire. **2** (esp. in degree titles) of Oxford University. [abbr. of med.L *Oxoniensis* f. *Oxonia*: see OXONIAN]

Oxonian /ɒkˈsəʊnɪən/ adj. & n. ● adj. of or relating to Oxford or Oxford University. ● n. **1** a member of Oxford University. **2** a native or inhabitant of Oxford. [*Oxonia* Latinized name of *Oxford*]

oxtail /ˈɒksteɪl/ n. the tail of an ox, esp. as an ingredient in soup.

oxter /ˈɒkstə(r)/ n. Sc. & N. Engl. the armpit. [OE *ōhsta*, *ōxta*]

oxtongue /ˈɒkstʌŋ/ n. **1** the tongue of an ox, esp. cooked as food. **2** a composite plant of the genus *Picris*, with bright yellow flowers.

Oxus /ˈɒksəs/ the ancient name for the AMU DARYA.

oxy-¹ /ˈɒksɪ/ comb. form denoting sharpness (*oxytone*). [Gk *oxu-* f. *oxus* sharp]

oxy-² /ˈɒksɪ/ comb. form (also **ox-** /ɒks/) *Chem.* oxygen (*oxyacetylene*). [abbr.]

oxyacetylene /‚ɒksɪəˈsetɪ‚liːn/ adj. of or using oxygen and acetylene, esp. in cutting or welding metals (*oxyacetylene burner*).

oxyacid /ˈɒksɪ‚æsɪd/ n. *Chem.* an acid containing oxygen.

oxygen /ˈɒksɪdʒən/ n. a colourless odourless gaseous chemical element (atomic number 8; symbol **O**). (*See note below.*) □ **oxygen mask** a mask placed over the nose and mouth to supply oxygen for breathing. **oxygen tent** a tentlike enclosure supplying a patient with air rich in

oxygen. □ **oxygenous** /ɒkˈsɪdʒɪnəs/ adj. [F *oxygène* acidifying principle (as OXY-¹): it was at first held to be the essential principle in the formation of acids]

▪ Oxygen forms about 20 per cent of the earth's atmosphere, and was discovered by Joseph Priestley in 1774 (see also PHLOGISTON). It is the most abundant element in the earth's crust, mainly in the form of oxides, silicates, and carbonates; water is a compound of oxygen and hydrogen. Oxygen is essential to plant and animal life and is a constituent of most organic compounds. The processes of respiration and combustion involve the ability of oxygen to combine readily with other elements. Industrially, pure oxygen is obtained by distillation of liquefied air; it is used in the steel-making and chemical industries, and in medicine. Although oxygen usually consists of diatomic molecules, a triatomic form, ozone, is an important component of the upper atmosphere.

oxygenate /ˈɒksɪdʒə‚neɪt, ɒkˈsɪ-/ v.tr. supply or treat with oxygen, esp. charge or enrich (blood, water, etc.) with oxygen. □ **oxygenation** /‚ɒksɪdʒəˈneɪʃ(ə)n/ n. [F *oxygéner* (as OXYGEN)]

oxygenator /ˈɒksɪdʒə‚neɪtə(r)/ n. **1** an apparatus for oxygenating the blood. **2** an aquatic plant which enriches the surrounding water with oxygen.

oxygenize /ˈɒksɪdʒə‚naɪz, ɒkˈsɪ-/ (also -**ise**) v.tr. = OXYGENATE.

oxyhaemoglobin /‚ɒksɪ‚hiːməˈgləʊbɪn/ n. *Biochem.* a bright red complex formed when haemoglobin combines with oxygen.

oxymoron /‚ɒksɪˈmɔːrɒn/ n. *rhet.* a figure of speech in which apparently contradictory terms appear in conjunction (e.g. *faith unfaithful kept him falsely true*). [Gk *oxumōron* neut. of *oxumōros* pointedly foolish f. *oxus* sharp + *mōros* foolish]

oxytetracycline /‚ɒksɪ‚tetrəˈsaɪkliːn/ n. an antibiotic related to tetracycin.

oxytocin /‚ɒksɪˈtəʊsɪn/ n. **1** *Biochem.* a hormone released by the pituitary gland that causes increased contraction of the womb during labour and stimulates the ejection of milk into the ducts of the breasts. **2** a synthetic form of this used to induce labour etc. [*oxytocic* accelerating parturition f. Gk *oxutokia* sudden delivery (as OXY-¹, *tokos* childbirth)]

oxytone /ˈɒksɪ‚təʊn/ adj. & n. ● adj. (esp. in ancient Greek) having an acute accent on the last syllable. ● n. a word of this kind. [Gk *oxutonos* (as OXY-¹, *tonos* tone)]

oyer and terminer /ˈɔɪə(r), ˈtɜːmɪnə(r)/ n. hist. a commission issued to judges on a circuit to hold courts. [ME f. AF *oyer et terminer* f. L *audire* hear + *et* and + *terminare* determine]

oyez /əʊˈjes, -ˈjez/ int. (also **oyes**) uttered, usu. three times, by a public crier or a court officer to command silence and attention. [ME f. AF, OF *oiez*, *oyez*, imper. pl. of *oïr* hear f. L *audire*]

oyster /ˈɔɪstə(r)/ n. **1** a bivalve mollusc of the family Ostreidae or Aviculidae; esp. an edible kind, *Ostrea edulis*, of European waters. **2** an oyster-shaped morsel of meat in a fowl's back. **3** something regarded as containing all that one desires (*the world is my oyster*). **4** (in full **oyster-white**) a white colour with a grey tinge. □ **oyster-bed** a part of the sea-bottom where oysters breed or are bred. **oyster-farm** an area of the seabed used for breeding oysters. **oyster-plant 1** = SALSIFY. **2** a blue-flowered plant, *Mertensia maritima*, of the borage family, growing on beaches. [ME & OF *oistre* f. L *ostrea*, *ostreum* f. Gk *ostreon*]

oystercatcher /ˈɔɪstə‚kætʃə(r)/ n. a wading shorebird of the genus *Haematopus*, with black and white or all-black plumage, a long red bill, and red legs; esp. the common *H. ostralegus* of Europe.

Oz /ɒz/ adj. & n. Austral. sl. ● adj. Australian. ● n. **1** Australia. **2** an Australian. [abbr. of the pronunc.]

oz abbr. ounce(s). [It. f. *onza* ounce]

Ozark Mountains /ˈəʊzɑːk/ (also **Ozarks**) a heavily forested highland plateau dissected by rivers, valleys, and streams, lying between the Missouri and Arkansas rivers and within the states of Missouri, Arkansas, Oklahoma, Kansas, and Illinois.

Ozawa /əˈzɑːwə/, Seiji (b.1935), Japanese conductor. In 1959 he won an international conducting competition and since then has been based chiefly in North America; he was the conductor of the Toronto Symphony Orchestra (1965–70) and in 1973 became music director and conductor of the Boston Symphony Orchestra.

ozocerite /əʊˈzəʊkə‚raɪt/ n. (also **ozokerite**) a waxlike fossil paraffin used for candles, insulation, etc. [G *Ozokerit* f. Gk *ozō* smell + *kēros* wax]

ozone /ˈəʊzəʊn/ n. **1** *Chem.* a colourless unstable toxic gas (chem. formula: O_3) with a pungent odour and powerful oxidizing properties, formed from normal oxygen by the action of electrical discharges or ultraviolet light. (See also OZONE LAYER.) **2** *colloq.* **a** invigorating air at the seaside etc. **b** exhilarating influence. □ **ozone depletion** a reduction of ozone concentration in the stratosphere, believed to be due to atmospheric pollution. **ozone-friendly** (of manufactured articles) containing chemicals that are not destructive to the ozone layer. **ozone hole** an area of the ozone layer in which depletion has occurred. □ **ozonic** /əʊˈzɒnɪk/ *adj.* [G *Ozon* f. Gk, neut. pres. part. of *ozō* smell]

ozone layer *n.* a layer in the earth's stratosphere at an altitude of about 10 km (6.2 miles), containing a high concentration of ozone. The ozone is formed by the action of the sun's ultraviolet radiation on oxygen. Ozone itself absorbs ultraviolet radiation and the layer therefore prevents much ultraviolet light from reaching the ground. In recent years depletion of ozone in this layer has been observed, particularly at high latitudes. This has been attributed to reaction of the ozone with atmospheric pollutants, notably CFCs. Increased ultraviolet light at ground level, besides being harmful in itself, also makes a contribution to global warming.

ozonize /ˈəʊzəˌnaɪz/ *v.tr.* (also **-ise**) **1** convert into ozone. **2** treat with ozone. □ **ozonizer** *n.* **ozonization** /ˌəʊzənaɪˈzeɪʃ(ə)n/ *n.*

Ozzie var. of AUSSIE.

Pp

P¹ /piː/ *n.* (also **p**) (*pl.* **Ps** or **P's**) the sixteenth letter of the alphabet.

P² *abbr.* (also **P.**) **1** (on road signs) parking. **2** *Chess* pawn. **3** (also Ⓟ) proprietary.

P³ *symb.* **1** *Chem.* the element phosphorus. **2** *Physics* poise (unit).

p *abbr.* (also **p.**) **1** *Brit.* penny, pence. **2** page. **3** pico-. **4** *Mus.* piano (softly).

PA *abbr.* **1** personal assistant. **2** public address (esp. *PA system*). **3** Press Association (Ltd.). **4** *US* Pennsylvania (in official postal use).

Pa *symb. Chem.* the element protactinium.

pa /pɑː/ *n. colloq.* father. [abbr. of PAPA]

p.a. *abbr.* per annum.

pa'anga /pɑːˈɑːŋgə/ *n.* the basic monetary unit of Tonga, equal to 100 seniti. [Tongan]

Paarl /pɑːl/ a town in SW South Africa, in the province of Western Cape, north-east of Cape Town; pop. (1980) 71,300. It is at the centre of a noted wine-producing region.

pabulum /ˈpæbjʊləm/ *n.* **1** food, esp. for the mind. **2** bland or insipid intellectual fare, entertainment, etc.; pap. [L f. *pascere* feed]

PABX *abbr. Brit.* private automatic branch exchange.

PAC see PAN-AFRICANIST CONGRESS.

paca /ˈpækə/ *n.* a tailless rodent of the genus *Agouti*, native to Central and South America; esp. *A. paca*, hunted for food. [Sp. & Port., f. Tupi]

pace¹ /peɪs/ *n. & v.* ● *n.* **1 a** a single step in walking or running. **b** the distance covered in this (about 75 cm or 30 in.). **c** the distance between two successive stationary positions of the same foot in walking. **2** speed in walking or running. **3** *Theatr. & Mus.* speed or tempo in theatrical or musical performance (*played with great pace*). **4 a** the rate at which something progresses (*the pace of technological change*). **b** the speed at which life is led (*the pace of city life*). **5** a manner of walking or running; a gait, esp. of a trained horse etc. (*rode at an ambling pace*). ● *v.* **1** *intr.* **a** walk (esp. repeatedly or methodically) with a slow or regular pace (*pacing up and down*). **b** (of a horse) = AMBLE. **2** *tr.* traverse by pacing. **3** *tr.* set the pace for (a rider, runner, etc.). **4** *tr.* (often foll. by *out*) measure (a distance) by pacing. □ **keep pace** (often foll. by *with*) advance at an equal rate (as). **pace bowler** *Cricket* a fast bowler. **pace-setter 1** a leader. **2** = PACEMAKER 1. **put a person through his or her paces** test a person's qualities in action etc. **set the pace** determine the speed, esp. by leading. **stand** (or **stay**) **the pace** be able to keep up with others. □ **-paced** *adj.* **pacer** *n.* [ME f. OF *pas* f. L *passus* f. *pandere pass-* stretch]

pace² /ˈpɑːteɪ, ˈpeɪsɪ/ *prep.* (in stating a contrary opinion) with due deference to (the person named). [L, ablat. of *pax* peace]

pacemaker /ˈpeɪsˌmeɪkə(r)/ *n.* **1** a competitor who sets the pace in a race. **2** a natural or artificial device for stimulating the heart muscle and regulating its contractions.

pacha var. of PASHA.

pachinko /pəˈtʃɪŋkəʊ/ *n.* a Japanese form of pinball. [Jap.]

pachisi /pəˈtʃiːzɪ/ *n.* a four-handed Indian board-game with six cowries used like dice. [Hindi, = of 25 (the highest throw)]

Pachuca de Soto /pəˌtʃuːkə deɪ ˈsəʊtəʊ/ (also **Pachuca**) a city in Mexico, capital of the state of Hidalgo; pop. (1990) 179,440.

pachyderm /ˈpækɪˌdɜːm/ *n.* a thick-skinned mammal, esp. an elephant or rhinoceros. □ **pachydermatous** /ˌpækɪˈdɜːmətəs/ *adj.* [F *pachyderme* f. Gk *pakhudermos* f. *pakhus* thick + *derma -matos* skin]

pachytene /ˈpækɪˌtiːn/ *n. Biol.* a stage during the prophase of meiosis when the chromosomes thicken and may exchange genes by pairing and crossing over. [Gk *pakhus* thick + *tainia* band]

pacific /pəˈsɪfɪk/ *adj. & n.* ● *adj.* **1** characterized by or tending to peace; tranquil. **2** (**Pacific**) of or adjoining the Pacific Ocean. ● *n.* **1** (**the Pacific**) the Pacific Ocean. **2** a steam locomotive of 4-6-2 wheel arrangement. □ **pacifically** *adv.* [F *pacifique* or L *pacificus* f. *pax pacis* peace]

pacification /ˌpæsɪfɪˈkeɪʃ(ə)n/ *n.* the act of pacifying or the process of being pacified. □ **pacificatory** /pəˈsɪfɪkətərɪ/ *adj.* [F f. L *pacificatio -onis* (as PACIFY)]

Pacific Ocean the world's largest ocean, covering one-third of the earth's surface (181 million sq. km, 70 million sq. miles). It separates Asia and Australia from North and South America and extends from Antarctica in the south to the Bering Strait (which links it to the Arctic Ocean) in the north. It was named by its first European navigator, Ferdinand Magellan, because he experienced calm weather there.

Pacific Standard Time (also **Pacific Time**) (abbr. **PST**) the standard time in a zone including the Pacific coastal region of Canada and the US, eight hours behind GMT.

pacifier /ˈpæsɪˌfaɪə(r)/ *n.* **1** a person or thing that pacifies. **2** *US* a baby's dummy.

pacifism /ˈpæsɪˌfɪz(ə)m/ *n.* the belief that war and violence are morally unjustified and that all disputes should be settled by peaceful means. □ **pacifist** *n. & adj.* [F *pacifisme* f. *pacifier* PACIFY]

pacify /ˈpæsɪˌfaɪ/ *v.tr.* (**-ies, -ied**) **1** appease (a person, anger, etc.). **2** bring (a country etc.) to a state of peace. [ME f. OF *pacifier* or L *pacificare* (as PACIFIC)]

pack¹ /pæk/ *n. & v.* ● *n.* **1 a** a collection of things wrapped up or tied together for carrying. **b** = BACKPACK *n.* **2** a set of items packaged for use or disposal together. **3** usu. *derog.* a lot or set (of similar things or persons) (*a pack of lies; a pack of thieves*). **4** *Brit.* a set of playing cards. **5 a** a group of hounds esp. for fox-hunting. **b** a group of wild animals, esp. wolves, hunting together. **6** an organized group of Cub Scouts or Brownies. **7 a** *Rugby* a team's forwards. **b** *Sport* the main body of competitors following the leader or leaders esp. in a race. **8 a** a medicinal or cosmetic substance applied to the skin; = face-pack. **b** a hot or cold pad of absorbent material for treating a wound etc. **9** (also **ice-pack**) an area of pack ice. **10** a quantity of fish, fruit, etc., packed in a season etc. **11** *Med.* **a** the wrapping of a body or part of a body in a wet sheet etc. **b** a sheet etc. used for this. ● *v.* **1** *tr.* (often foll. by *up*) **a** fill (a suitcase, bag, etc.) with clothes and other items. **b** put (things) together in a bag or suitcase, esp. for travelling. **2** *intr. & tr.* come or put closely together; crowd or cram (*packed a lot into a few hours; passengers packed like sardines*). **3** *tr.* (in passive; often foll. by *with*) be filled (with); contain extensively (*the restaurant was packed; the book is packed with information*). **4** *tr.* fill (a hall, theatre, etc.) with an audience etc. **5** *tr.* cover (a thing) with something pressed tightly round. **6** *intr.* be suitable for packing. **7** *tr. colloq.* **a** carry (a gun etc.). **b** be capable of delivering (a punch) with skill or force. **8** *intr.* (also foll. by *in, down*) *Rugby* (of forwards) form or take their places in the scrum. □ **pack-animal** an animal for carrying packs. **pack-drill** a military punishment of marching up and down carrying full equipment. **packed lunch** a lunch carried in a bag, box, etc., esp. to work or school. **packed out** *colloq.* full of people, crowded. **pack ice** large pieces of floating ice crowded together in a large expanse in the sea. **pack in** *colloq.* stop, give up (*packed in his job*). **pack it in** (or **up**) *colloq.* end or stop it. **pack off** *colloq.* send (a person) away, esp. abruptly or promptly. **pack rat 1** a North American wood rat; esp. *Neotoma cinerea*, with a long bushy tail. **2** a person who hoards things. **pack-saddle** a saddle adapted for supporting packs. **pack up** *colloq.* **1** (esp. of a machine) stop functioning; break down. **2** retire from an activity, contest, etc. **send packing** *colloq.* dismiss (a person) summarily. □ **packable** *adj.* [ME f. MDu., MLG *pak, pakken*, of unkn. orig.]

pack[2] /pæk/ *v.tr.* select (a jury etc.) or fill (a meeting) so as to secure a decision in one's favour. [prob. f. obs. verb *pact* f. PACT]

package /ˈpækɪdʒ/ *n. & v.* ● *n.* **1 a** a bundle of things packed. **b** a parcel, box, etc., in which things are packed. **2** (in full **package deal**) a set of proposals or items offered or agreed to as a whole. **3** *Computing* a piece of software suitable for various applications rather than one which is custom-built. **4** *colloq.* = package holiday. ● *v.tr.* make up into or enclose in a package. □ **package holiday** (or **tour** etc.) a holiday or tour etc. with all arrangements made at an inclusive price. □ **packager** *n.* [PACK[1] + -AGE]

packaging /ˈpækɪdʒɪŋ/ *n.* **1** a wrapping or container for goods. **2** the process of packing goods.

Packer /ˈpækə(r)/, Kerry (Francis Bullmore) (b.1937), Australian media entrepreneur. He launched a number of Australian sport initiatives, notably the 'World Series Cricket' tournaments (1977–9), for which he claimed exclusive television coverage rights. As part of these tournaments, Packer engaged many of the world's leading cricketers in defiance of the wishes of cricket's ruling bodies, precipitating a two-year schism in international cricket.

packer /ˈpækə(r)/ *n.* a person or thing that packs, esp. a dealer who prepares and packs food for transportation and sale.

packet /ˈpækɪt/ *n.* **1** a small package. **2** *colloq.* a large sum of money won, lost, or spent. **3** (in full **packet-boat**) *hist.* a mail-boat or passenger ship. □ **packet switching** a method of data transmission in which parts of a message are sent independently by the optimum route for each part and then reassembled. [PACK[1] + -ET[1]]

packhorse /ˈpækhɔːs/ *n.* a horse for carrying loads.

packing /ˈpækɪŋ/ *n.* **1** the act or process of packing. **2** material used to fill up a space round or in something, esp. to protect a fragile article in transit. **3** material used to seal a join or assist in lubricating an axle. □ **packing-case** a case (usu. wooden) or framework for packing goods in.

packthread /ˈpækθred/ *n.* stout thread for sewing or tying up packs.

pact /pækt/ *n.* an agreement or a treaty. [ME f. OF *pact(e)* f. L *pactum*, neut. past part. of *pacisci* agree]

pad[1] /pæd/ *n. & v.* ● *n.* **1** a piece of soft material used to reduce friction or jarring, fill out hollows, hold or absorb liquid, etc. **2** a number of sheets of blank paper fastened together at one edge, for writing or drawing on. **3** = ink-pad. **4** the fleshy underpart of an animal's foot or of a human finger. **5** a guard for the leg and ankle in sports. **6** a flat surface for helicopter take-off or rocket-launching. **7** *colloq.* a lodging, esp. a bedsitter or flat. **8** the floating leaf of a water lily. ● *v.tr.* (**padded, padding**) **1** provide with a pad or padding; stuff. **2** (foll. by *out*) lengthen or fill out (a book etc.) with unnecessary material. □ **padded cell** a room with padded walls in a mental hospital. [prob. of LG or Du. orig.]

pad[2] /pæd/ *v. & n.* ● *v.* (**padded, padding**) **1** *intr.* walk with a soft dull steady step. **2 a** *tr.* tramp along (a road etc.) on foot. **b** *intr.* travel on foot. ● *n.* the sound of soft steady steps. [LG *padden* tread, *pad* PATH]

Padang /pəˈdæŋ/ a seaport of Indonesia, the largest city on the west coast of Sumatra; pop. (1980) 480,920.

padding /ˈpædɪŋ/ *n.* soft material to pad or stuff with.

paddle[1] /ˈpæd(ə)l/ *n. & v.* ● *n.* **1** a short broad-bladed oar used without a rowlock. **2** a paddle-shaped instrument. **3** *Zool.* a fin or flipper. **4** each of the boards fitted round the circumference of a paddle-wheel or mill-wheel. **5** the action or a spell of paddling. ● *v.* **1** *intr. & tr.* move on water or propel a boat by means of paddles. **2** *intr. & tr.* row gently. **3** *tr.* esp. *US colloq.* spank. □ **paddle-boat** (or **-steamer** etc.) a boat, steamer, etc., propelled by a paddle-wheel. **paddle-wheel** a wheel for propelling a ship, with boards round the circumference so as to press backwards against the water. □ **paddler** *n.* [15th c.: orig. unkn.]

paddle[2] /ˈpæd(ə)l/ *v. & n.* ● *v.intr.* walk barefoot in shallow water or dabble the feet or hands in it. ● *n.* the action or a spell of paddling. □ **paddler** *n.* [prob. of LG or Du. orig.: cf. LG *paddeln* tramp about]

paddock /ˈpædək/ *n.* **1** a small field, esp. for keeping horses in. **2** a turf enclosure adjoining a racecourse where horses or cars are

gathered before a race. **3** *Austral. & NZ* a field; a plot of land. [app. var. of (now dial.) *parrock* (OE *pearruc*): see PARK]

Paddy /ˈpædɪ/ *n.* (pl. **-ies**) *colloq. often offens.* an Irishman. [pet-form of the Irish name *Padraig* (= Patrick)]

paddy[1] /ˈpædɪ/ *n.* (pl. **-ies**) **1** (in full **paddy-field**) a field where rice is grown. **2** rice before threshing or in the husk. [Malay *pādī*]

paddy[2] /ˈpædɪ/ *n.* (pl. **-ies**) *Brit. colloq.* a rage; a fit of temper. [PADDY]

pademelon /ˈpædɪˌmelən/ *n.* a small wallaby of the genus *Thylogale*, inhabiting the coastal scrub of Australia and New Guinea. [earlier *paddymelon*, prob. alteration of Dharuk *badimaliyan*]

Paderewski /ˌpædəˈrefskɪ/, Ignacy Jan (1860–1941), Polish pianist, composer, and statesman, Prime Minister 1919. He became one of the most famous international pianists of his time and also received acclaim for his compositions, which include the opera *Manru* (1901). He was the first Prime Minister of independent Poland, but resigned after only ten months in office and resumed his musical career. In 1939 he served briefly as head of the Polish government in Paris, before emigrating to the US in 1940, when France surrendered to Germany.

padlock /ˈpædlɒk/ *n. & v.* ● *n.* a detachable lock hanging by a pivoted hook on the object fastened. ● *v.tr.* secure with a padlock. [ME f. LOCK[1]: first element unexpl.]

Padma /ˈpædmə/ a river of southern Bangladesh, formed by the confluence of the Ganges and the Brahmaputra near Rajbari.

padouk /pəˈduːk/ *n.* **1** a leguminous timber tree of the genus *Pterocarpus*, native to Africa and Asia. **2** the wood of this tree, resembling rosewood. [Burmese]

Padova see PADUA.

padre /ˈpɑːdrɪ, -dreɪ/ *n.* a chaplain in any of the armed services. [It., Sp., & Port., = father, priest, f. L *pater patris* father]

padsaw /ˈpædsɔː/ *n.* a saw with a narrow blade, for cutting curves.

Padua /ˈpædjʊə/ (Italian **Padova** /ˈpaːdova/) a city in NE Italy; pop. (1990) 218,190. The city, first mentioned in 302 BC as Patavium, was the birthplace in 59 BC of the Roman historian Livy. A leading city from the 11th century AD, it was ruled by the Carrara family from 1318 until 1405, when it passed to Venice. Galileo taught at its university from 1592 to 1610.

paean /ˈpiːən/ *n.* (*US* **pean**) a song of praise or triumph. [L f. Doric Gk *paian* hymn of thanksgiving to Apollo (under the name of *Paian*)]

paederast var. of PEDERAST.

paederasty var. of PEDERASTY.

paediatrics /ˌpiːdɪˈætrɪks/ *n.pl.* (treated as *sing.*) (*US* **pediatrics**) the branch of medicine dealing with children and their diseases. □ **paediatric** *adj.* **paediatrician** /-dɪəˈtrɪʃ(ə)n/ *n.* [PAEDO- + Gk *iatros* physician]

paedo- /ˈpiːdəʊ/ *comb. form* (*US* **pedo-**) child. [Gk *pais paid-* child]

paedophile /ˈpiːdəˌfaɪl/ *n.* (*US* **pedophile**) a person who displays paedophilia.

paedophilia /ˌpiːdəˈfɪlɪə/ *n.* (*US* **pedophilia**) sexual desire directed towards children.

paella /paɪˈelə/ *n.* a Spanish dish of rice, saffron, chicken, seafood, etc., cooked and served in a large shallow pan. [Catalan f. OF *paele* f. L *patella* pan]

paeon /ˈpiːən/ *n.* a metrical foot of one long syllable and three short syllables in any order. □ **paeonic** /piːˈɒnɪk/ *adj.* [L f. Gk *paiōn*, the Attic form of *paian* PAEAN]

paeony var. of PEONY.

Pagalu /ˌpɑːɡəˈluː/ a former name (1973–9) for ANNOBÓN.

Pagan /pəˈɡɑːn/ a town in Burma, situated on the Irrawaddy south-east of Mandalay. It is the site of an ancient city, founded in about AD 849, which was the capital of a powerful Buddhist dynasty from the 11th to the end of the 13th centuries.

pagan /ˈpeɪɡən/ *n. & adj.* ● *n.* a person not subscribing to any of the main religions of the world, esp. formerly regarded by Christians as unenlightened or heathen. ● *adj.* **1 a** of or relating to or associated with pagans. **b** irreligious. **2** identifying divinity or spirituality in nature; pantheistic. □ **paganish** *adj.* **paganism** *n.* **paganize** *v.tr. & intr.* (also **-ise**). [ME f. L *paganus* villager, rustic f. *pagus* country district: in Christian L = civilian, heathen]

Paganini /ˌpæɡəˈniːnɪ/, Niccolò (1782–1840), Italian violinist and composer. His virtuoso violin recitals established him as an almost legendary figure of the romantic movement and radically changed the violin technique of the day. Paganini's technical innovations, such as widespread use of pizzicato and harmonics as well as new styles of fingering, were exhibited in his best-known composition, the twenty-four *Capricci* (1820). His technical innovations had a major influence on the work of Liszt.

Page /peɪdʒ/, Sir Frederick Handley (1885–1962), English aircraft designer. In 1909 he founded Handley Page Ltd., the first British aircraft manufacturing company. He is noted for designing the first twin-engined bomber (1915), as well as the Halifax heavy bombers of the Second World War.

page[1] /peɪdʒ/ n. & v. • n. **1 a** a leaf of a book, periodical, etc. **b** each side of this. **c** what is written or printed on this. **2** *Computing* a section of the stored data, esp. as much as can be displayed on a screen at one time. **3 a** an episode that might fill a page in written history etc.; a record. **b** a memorable event. • v. **1** tr. paginate. **2** intr. **a** (foll. by *through*) leaf through (a book etc.). **b** (foll. by *through*, *up*, *down*) *Computing* display (text etc.) one page at a time. [F f. L *pagina* f. *pangere* fasten]

page[2] /peɪdʒ/ n. & v. • n. **1** a boy or man, usu. in livery, employed to run errands, attend to a door, etc. **2** a boy employed as a personal attendant of a person of rank, a bride, etc. **3** *hist.* a boy in training for knighthood and attached to a knight's service. • v.tr. **1** (in hotels, airports, etc.) summon by making an announcement or by sending a messenger. **2** summon by means of a pager. □ **page-boy 1** = PAGE[2] n. 2. **2** a woman's hairstyle with the hair reaching to the shoulder and rolled under at the ends. [ME f. OF, perh. f. It. *paggio* f. Gk *paidion*, dimin. of *pais paidos* boy]

pageant /ˈpædʒənt/ n. **1 a** a brilliant spectacle, esp. an elaborate parade. **b** a spectacular procession, or play performed in the open, illustrating historical events. **c** a tableau etc. on a fixed stage or moving vehicle. **2** an empty or specious show. [ME *pagyn*, of unkn. orig.]

pageantry /ˈpædʒəntrɪ/ n. (pl. **-ies**) **1** elaborate or sumptuous show or display. **2** an instance of this.

pager /ˈpeɪdʒə(r)/ n. a radio device with a bleeper, activated from a central point to alert the person wearing it.

paginal /ˈpædʒɪn(ə)l/ adj. **1** of pages (of books etc.). **2** corresponding page for page. □ **paginary** adj. [LL *paginalis* (as PAGE[1])]

paginate /ˈpædʒɪˌneɪt/ v.tr. assign numbers to the pages of a book etc. □ **pagination** /ˌpædʒɪˈneɪʃ(ə)n/ n. [F *paginer* f. L *pagina* PAGE[1]]

pagoda /pəˈɡəʊdə/ n. **1** a Hindu or Buddhist temple or sacred building, esp. a many-tiered tower, in India and the Far East. **2** an ornamental imitation of this. □ **pagoda-tree** a leguminous Chinese tree, *Sophora japonica*, with hanging clusters of cream flowers. [Port. *pagode*, prob. ult. f. Pers. *butkada* idol temple]

pah /pɑː/ int. expressing disgust or contempt. [natural utterance]

Pahang /pəˈhæŋ/ a mountainous forested state of Malaysia, on the east coast of the Malay Peninsula; capital, Kuantan.

Pahlavi[1] /ˈpɑːləvɪ/ n. (also **Pehlevi** /ˈpeɪl-/) **1 a** the language of Persia spoken from the 3rd century BC until the 10th century AD, the official language of the Sassanian empire. It is closely related to Avestan, and is evidenced mainly in Zoroastrian texts and commentaries. **b** the alphabet in which Pahlavi was written, developed from Aramaic script. **2** a member of the Iranian dynasty founded by Reza Khan (see PAHLAVI[3]). [Pers. *pahlawī* f. *pahlav* f. *parthava* Parthia]

Pahlavi[2] /ˈpɑːləvɪ/, Muhammad Reza (also known as Reza Shah) (1919–80), shah of Iran 1941–79. His early years as shah were dominated by conflict with the nationalist Prime Minister Muhammad Mosaddeq (1880–1967). From 1953 he assumed direct control over all aspects of Iranian life and, with US support, embarked on a national development plan promoting public works, industrial development, and social and land reform. Opposition to his regime culminated in the Islamic revolution of 1979 under Ayatollah Khomeini; Reza Shah was forced into exile and died in the US.

Pahlavi[3] /ˈpɑːləvɪ/, Reza (born Reza Khan) (1878–1944), shah of Iran 1925–41. An army officer, he took control of the Persian government after a coup in 1921. In the absence of the reigning monarch, Reza Khan was elected Shah by the National Assembly in 1925. He abdicated in 1941, following the occupation of Iran by British and Soviet forces, passing the throne to his son Muhammad Reza Pahlavi.

pahoehoe /pəˈhəʊɪˌhəʊɪ/ n. *Geol.* lava forming smooth undulating or ropy masses (cf. AA). [Hawaiian]

paid *past* and *past part.* of PAY[1].

Paignton /ˈpeɪntən/ a resort town in SW England, on the south coast of Devon; pop. (1981) 40,820.

pail /peɪl/ n. **1** a bucket. **2** an amount contained in this. □ **pailful** n. (pl.

-fuls). [OE *pægel* gill (cf. MDu. *pegel* gauge), assoc. with OF *paelle*: see PAELLA]

Pailin /ˈpeɪlɪn/ a ruby-mining town in western Cambodia, close to the border with Thailand.

paillasse var. of PALLIASSE.

paillette /pæˈljet, paɪˈjet/ n. **1** a piece of bright metal used in enamel painting. **2** a spangle. [F, dimin. of *paille* f. L *palea* straw, chaff]

pain /peɪn/ n. & v. • n. **1 a** a strongly unpleasant bodily sensation such as is produced by illness, injury, or other harmful physical contact etc.; the condition of hurting. **b** a particular kind or instance of this (often in *pl.*: *suffering from stomach pains*). **2** mental suffering or distress. **3** (in *pl.*) careful effort; trouble taken (*take pains; got nothing for my pains*). **4** (also **pain in the neck** etc.) *colloq.* a troublesome person or thing; a nuisance. • v.tr. **1** cause pain to. **2** (as **pained** adj.) expressing pain (*a pained expression*). □ **be at** (or **take**) **pains** (usu. foll. by to + infin.) take great care in doing something. **in pain** suffering pain. **on** (or **under**) **pain of** with (death etc.) as the penalty. [ME f. OF *peine* f. L *poena* penalty]

Paine /peɪn/, Thomas (1737–1809), English political writer. After emigrating to the US in 1774, he wrote the pamphlet *Common Sense* (1776), which called for American independence and laid the ground for the Declaration of Independence. On returning to England in 1787, he published *The Rights of Man* (1791), defending the French Revolution in response to Burke's *Reflections on the Revolution in France* (1790). His radical views prompted the British government to indict him for treason and he fled to France. There he supported the Revolution but opposed the execution of Louis XVI. He was imprisoned for a year, during which time he wrote *The Age of Reason* (1794), an attack on orthodox Christianity.

Paine Towers /ˈpaɪnɪ/ a group of spectacular granite peaks in southern Chile, rising to a height of 2,668 m (8,755 ft).

painful /ˈpeɪnfʊl/ adj. **1** causing bodily or mental pain or distress. **2** (esp. of part of the body) suffering pain. **3** causing trouble or difficulty; laborious (*a painful climb*). □ **painfully** adv. **painfulness** n.

painkiller /ˈpeɪnˌkɪlə(r)/ n. a medicine or drug for alleviating pain. □ **painkilling** adj.

painless /ˈpeɪnlɪs/ adj. not causing or suffering pain. □ **painlessly** adv. **painlessness** n.

painstaking /ˈpeɪnzˌteɪkɪŋ/ adj. careful, industrious, thorough. □ **painstakingly** adv. **painstakingness** n.

paint /peɪnt/ n. & v. • n. **1 a** a colouring matter, esp. in liquid form for imparting colour to a surface. **b** this as a dried film or coating (*the paint peeled off*). **2** *joc.* or *archaic* cosmetic make-up, esp. rouge or nail varnish. • v.tr. **1 a** cover the surface of (a wall, object, etc.) with paint. **b** apply paint of a specified colour to (*paint the door green*). **2** depict (an object, scene, etc.) with paint; produce (a picture) by painting. **3** describe vividly as if by painting (*painted a gloomy picture of the future*). **4** apply a liquid or (*joc.* or *archaic*) cosmetics to (the face, skin, etc.). **5** apply (a liquid) to a surface with a brush etc. □ **painted lady** a migratory orange butterfly, *Cynthia cardui*, with black and white markings. **paint out** efface with paint. **paint shop** the part of a factory where goods are painted, esp. by spraying. **paint-stick** a stick of water-soluble paint used like a crayon. **paint the town red** *colloq.* enjoy oneself flamboyantly. □ **paintable** adj. [ME f. *peint* past part. of OF *peindre* f. L *pingere pict-* paint]

paintball /ˈpeɪntbɔːl/ n. **1** an outdoor game simulating military combat, in which players use guns that fire capsules of paint which break on impact, any player hit being eliminated from further play. **2** a capsule of paint for use in this game.

paintbox /ˈpeɪntbɒks/ n. a box holding dry paints for painting pictures.

paintbrush /ˈpeɪntbrʌʃ/ n. a brush for applying paint.

painter[1] /ˈpeɪntə(r)/ n. a person who paints, esp. an artist or decorator. [ME f. OF *peintour* ult. f. L *pictor* (as PAINT)]

painter[2] /ˈpeɪntə(r)/ n. a rope attached to the bow of a boat for tying it to a quay etc. [ME, prob. f. OF *penteur* rope from a masthead: cf. G *Pentertakel* f. *pentern* fish the anchor]

painterly /ˈpeɪntəlɪ/ adj. **1 a** using paint well; artistic. **b** characteristic of a painter or paintings. **2** (of a painting) lacking clearly defined outlines.

painting /ˈpeɪntɪŋ/ n. **1** the process or art of using paint. **2** a painted picture.

paintwork /ˈpeɪntwɜːk/ *n.* **1** a painted surface or area in a building etc. **2** the work of painting.

painty /ˈpeɪntɪ/ *adj.* (**paintier, paintiest**) **1** of or covered in paint. **2** (of a picture etc.) overcharged with paint.

pair /peə(r)/ *n. & v.* ● *n.* **1** a set of two persons or things used together or regarded as a unit (*a pair of gloves; a pair of eyes*). **2** an article (e.g. scissors, trousers, or tights) consisting of two joined or corresponding parts not used separately. **3 a** an engaged or married couple. **b** a mated couple of animals. **4** two horses harnessed side by side (*a coach and pair*). **5** the second member of a pair in relation to the first (*cannot find its pair*). **6** two playing cards of the same denomination. **7** either or both of two members of a legislative assembly on opposite sides absenting themselves from voting by mutual arrangement. ● *v.tr. & intr.* **1** (often foll. by *off*) arrange or be arranged in couples. **2 a** join or be joined in marriage. **b** (of animals) mate. **3** form a pair in a legislative assembly (see PAIR *n.* 7). □ **in pairs** in twos. **pair production** *Physics* the conversion of a radiation quantum into an electron and a positron. **pair royal** a set of three cards of the same denomination. [ME f. OF *paire* f. L *paria* neut. pl. of *par* equal]

paisa /ˈpaɪzɑ/ *n.* (*pl.* **paise** /-zeɪ, -zə/) a coin and monetary unit of India, Pakistan, and Nepal, equal to one-hundredth of a rupee. [Hindi]

Paisley[1] /ˈpeɪzlɪ/ a town in Strathclyde region, central Scotland, to the west of Glasgow; pop. (1981) 84,800. It developed around a Cluniac abbey founded in 1163. A centre of hand-weaving by the 18th century, it became famous for its distinctive shawls, woven in imitation of the highly prized shawls imported from Kashmir in the late 18th and early 19th centuries.

Paisley[2] /ˈpeɪzlɪ/, Ian (Richard Kyle) (b.1926), Northern Irish clergyman and politician. He was ordained as a minister of the Free Presbyterian Church in 1946, becoming its leader in 1951. Paisley first became politically active in the 1960s and was elected MP for North Antrim in 1970. A co-founder of the Ulster Democratic Unionist Party (1972), he has been a vociferous and outspoken defender of the Protestant Unionist position in Northern Ireland. Paisley became a Member of the European Parliament in 1979.

paisley /ˈpeɪzlɪ/ *n.* (often *attrib.*) **1** a distinctive intricate pattern of curved feather-shaped figures. **2** a soft woollen material having this pattern; a garment made from it. [PAISLEY[1]]

Paiute /ˈpaɪuːt/ *n. & adj.* ● *n.* (*pl.* same or **Paiutes**) **1** a member of either of two culturally similar but geographically separate and linguistically distinct American Indian peoples (the *Southern Paiute* and the *Northern Paiute*) of the south-western US. **2** either of the Shoshonean languages of these peoples. ● *adj.* of or relating to the Paiute or their languages. [Sp. *Payuchi, Payuta*, infl. by UTE]

pajamas *US var. of* PYJAMAS.

pakapoo /ˈpækəˌpuː, ˌpækəˈpuː/ *n.* (also **pakapu**) *Austral.* a Chinese form of lottery played with slips of paper marked with columns of characters. □ **pakapoo ticket** a piece of writing etc. that is illegible or difficult to decipher. [Chin. *bái gē piào* lit. 'white pigeon ticket', perh. referring to a Cantonese competition which involved releasing pigeons]

pakeha /ˈpɑːkɪˌhɑː/ *n. & adj.* ● *n. NZ* a white person as opposed to a Maori. ● *adj.* of or relating to white people. [Maori]

Paki /ˈpækɪ/ *n.* (*pl.* **Pakis**) *Brit. sl. offens.* a Pakistani, esp. an immigrant in Britain. [abbr.]

Pakistan /ˌpɑːkɪˈstɑːn, ˌpæk-/ a country in the Indian subcontinent; pop. (est. 1991) 115,588,000; languages, Urdu (official), Panjabi, Sindhi, Pashto; capital, Islamabad. Pakistan was created as a separate country in 1947, following the British withdrawal from India. It originally comprised two territories, respectively to the east and west of India, in which the population was predominantly Muslim. The country's history has seen long periods of military rule and an unresolved dispute with India over the territory of Kashmir. Civil war in East Pakistan over local claims for autonomy led to Indian intervention in Dec. 1971 and the establishment of the independent state of Bangladesh in 1972. Pakistan withdrew from the Commonwealth in 1972 as a protest against international recognition of Bangladesh, but rejoined in 1989. □ **Pakistani** *adj. & n.* [Punjab, Afghan Frontier, Kashmir, Baluchi*stan*, lands where Muslims predominated]

Pakistan People's Party (abbr. **PPP**) one of the main political parties in Pakistan. It was founded in 1967 by Zulfikar Ali Bhutto, and has been led since 1984 by his daughter Benazir Bhutto.

pakora /pəˈkɔːrə/ *n.* a piece of cauliflower, carrot, or other vegetable, coated in seasoned batter and deep-fried. [Hind.]

Pakse /ˈpækseɪ/ (also **Pakxe**) a town in southern Laos, on the Mekong river; pop. (est. 1990) 25,000. The 7th-century ruins of the ancient Khmer capital of Wat Phou lie to the south.

pal /pæl/ *n. & v.* ● *n. colloq.* a friend, mate, or comrade. ● *v.intr.* (**palled, palling**) (usu. foll. by *up*) associate; form a friendship. [Romany = brother, mate, ult. f. Skr. *bhrātr* BROTHER]

palace /ˈpælɪs/ *n.* **1** the official residence of a sovereign, president, archbishop, or bishop. **2** a splendid mansion; a spacious building. □ **palace revolution** (or **coup**) the (usu. non-violent) overthrow of a sovereign, government, etc. at the hands of senior officials. [ME f. OF *palais* f. L *Palatium* Palatine (hill) in Rome where the house of the emperor was situated]

Palace of Westminster see WESTMINSTER, PALACE OF.

paladin /ˈpælədɪn/ *n. hist.* **1** each of the twelve peers of Charlemagne's court, of whom the Count Palatine was the chief. **2** a knight errant; a champion. [F *paladin* f. It. *paladino* f. L *palatinus*: see PALATINE[1]]

Palaearctic /ˌpælɪˈɑːktɪk/ *adj. & n.* (also **Palearctic, palaearctic**) *Zool.* ● *adj.* of or relating to the Arctic and temperate parts of the Old World as a zoogeographical region. ● *n.* the Palaearctic region. [PALAEO- + ARCTIC]

palaeo- /ˈpælɪəʊ, ˈpeɪl-/ *comb. form* (*US* **paleo-**) ancient, old; of ancient (esp. prehistoric) times. [Gk *palaios* ancient]

palaeoanthropology /ˌpælɪəʊˌænθrəˈpɒlədʒɪ, ˌpeɪl-/ *n.* (*US* **paleoanthropology**) the branch of anthropology concerned with fossil hominids. □ **palaeoanthropologist** *n.* **palaeoanthropological** /-pəˈlɒdʒɪk(ə)l/ *adj.*

palaeobotany /ˌpælɪəʊˈbɒtənɪ, ˌpeɪl-/ *n.* (*US* **paleobotany**) the study of fossil plants. □ **palaeobotanist** *n.* **palaeobotanical** /-bəˈtænɪk(ə)l/ *adj.*

Palaeocene /ˈpælɪəˌsiːn, ˈpeɪl-/ *adj. & n.* (*US* **Paleocene**) *Geol.* of, relating to, or denoting the earliest epoch of the Tertiary period, between the Cretaceous period and the Eocene epoch. The Palaeocene lasted from about 65 to 55 million years ago, and saw a sudden diversification of mammals, possibly consequent on the mass extinctions (notably of the dinosaurs) which preceded it. (See also CRETACEOUS.) [PALAEO- + Gk *kainos* new]

palaeoclimatology /ˌpælɪəʊˌklaɪməˈtɒlədʒɪ, ˌpeɪl-/ *n.* (*US* **paleoclimatology**) the study of the climate in geologically past times. □ **palaeoclimatologist** *n.* **palaeoclimatological** /-təˈlɒdʒɪk(ə)l/ *adj.*

palaeoecology /ˌpælɪəʊɪˈkɒlədʒɪ, ˌpeɪl-/ *n.* (*US* **paleoecology**) the ecology of extinct and prehistoric organisms. □ **palaeoecologist** *n.* **palaeoecological** /-kəˈlɒdʒɪk(ə)l/ *adj.*

palaeogeography /ˌpælɪəʊdʒɪˈɒɡrəfɪ, ˌpeɪl-/ *n.* (*US* **paleogeography**) the study of the geographical features at periods in the geological past. □ **palaeogeographer** *n.* **palaeogeographical** /-ˌdʒiːəˈɡræfɪk(ə)l/ *adj.*

palaeography /ˌpælɪˈɒɡrəfɪ, ˌpeɪl-/ *n.* (*US* **paleography**) the study of writing and documents from the past. □ **palaeographer** *n.* **palaeographic** /-lɪəˈɡræfɪk/ *adj.* **palaeographical** /-lɪəˈɡræfɪk(ə)l/ *adj.* **palaeographically** *adv.* [F *paléographie* f. mod.L *palaeographia* (as PALAEO-, -GRAPHY)]

palaeolithic /ˌpælɪəʊˈlɪθɪk, ˌpeɪl-/ *adj. & n.* (*US* **paleolithic**) *Archaeol.* of, relating to, or denoting the early part of the Stone Age, when primitive stone implements were used. This period extends from the first appearance of artefacts, some 2.5 million years ago, to the end of the last glacial period or ice age *c*.13,500 BP. It has been divided into the *lower palaeolithic*, with the earliest forms of humankind and the presence of hand-axe industries, ending *c*.120,000 BP, *middle palaeolithic*, the era of Neanderthal man, ending *c*.35,000 BP, and *upper palaeolithic*, during which only modern *Homo sapiens* is known to have existed. [PALAEO- + Gk *lithos* stone]

palaeomagnetism /ˌpælɪəʊˈmæɡnɪˌtɪz(ə)m, ˌpeɪl-/ *n.* (*US* **paleomagnetism**) the study of the magnetism remaining in rocks. □ **palaeomagnetic** /-mæɡˈnetɪk/ *adj.*

palaeontology /ˌpælɪɒnˈtɒlədʒɪ, ˌpeɪl-/ *n.* (*US* **paleontology**) the branch of science that deals with extinct and fossil animals and plants. □ **palaeontologist** *n.* **palaeontological** /-ˌɒntəˈlɒdʒɪkəl/ *adj.* [PALAEO- + Gk *onta* neut. pl. of *ōn* being, part. of *eimi* be + -LOGY]

Palaeozoic /ˌpælɪəʊˈzəʊɪk, ˌpeɪl-/ *adj. & n.* (*US* **Paleozoic**) *Geol.* of, relating to, or denoting the geological era between the Precambrian and the Mesozoic. The Palaeozoic, comprising the Cambrian, Ordovician, Silurian, Devonian, Carboniferous, and Permian periods, lasted from about 590 to 248 million years ago. The earliest hard-shelled fossils are from this era, which ended with the rise to dominance of reptiles. The Palaeozoic corresponds to the Primary period. [PALAEO- + Gk *zōē* life]

palaestra /pə'li:strə, -'laɪstrə/ *n.* (also **palestra** /-'lestrə/) *Gk & Rom. Antiq.* a wrestling-school or gymnasium. [ME f. L *palaestra* f. Gk *palaistra* f. *palaiō* wrestle]

palais /'pæleɪ/ *n. colloq.* a public hall for dancing. [F *palais* (*de danse*) (dancing) hall]

Palais de l'Elysée /palɛ də lelize/ the French name for the ELYSÉE PALACE.

palanquin /ˌpælən'ki:n/ *n.* (also **palankeen**) (in India and the East) a covered litter for one passenger. [Port. *palanquim*: cf. Hindi *pālkī* f. Skr. *palyanka* bed, couch]

palatable /'pælətəb(ə)l/ *adj.* **1** pleasant to taste. **2** (of an idea, suggestion, etc.) acceptable, satisfactory. □ **palatably** *adv.* **palatableness** *n.* **palatability** /ˌpælətə'bɪlɪti/ *n.*

palatal /'pælət(ə)l/ *adj. & n.* ● *adj.* **1** of the palate. **2** (of a sound) made by placing the surface of the tongue against the hard palate (e.g. *y* in *yes*). ● *n.* a palatal sound. □ **palatally** *adv.* **palatalize** *v.tr.* (also **-ise**). **palatalization** /ˌpælətəlaɪ'zeɪʃ(ə)n/ *n.* [F (as PALATE)]

palate /'pælət/ *n.* **1** the roof of the mouth of a vertebrate, separating the oral and nasal cavities. **2** the sense of taste. **3** a mental taste or inclination; liking. [ME f. L *palatum*]

palatial /pə'leɪʃ(ə)l/ *adj.* (of a building) like a palace, esp. spacious and splendid. □ **palatially** *adv.* [L (as PALACE)]

palatinate /pə'lætɪnət/ *n.* a territory under the jurisdiction of a Count Palatine.

palatine[1] /'pælətaɪn/ *adj.* (also **Palatine**) *hist.* **1** (of an official or feudal lord) having local authority that elsewhere belongs only to a sovereign (see COUNT PALATINE). **2** (of a territory) subject to this authority (see COUNTY PALATINE). [ME f. F *palatin -ine* f. L *palatinus* of the PALACE]

palatine[2] /'pælətaɪn/ *adj. & n.* ● *adj.* of or connected with the palate. ● *n.* (in full **palatine bone**) either of two bones forming the hard palate. [F *palatin -ine* (as PALATE)]

Palau /pə'laʊ/ (also **Belau** /bə-/) a group of islands in the western Pacific Ocean, part of the US Trust Territory of the Pacific Islands from 1947 and internally self-governing since 1980; pop. (est. 1988) 14,100; capital, Koror.

palaver /pə'lɑːvə(r)/ *n. & v.* ● *n.* **1** fuss and bother, esp. prolonged. **2** profuse or idle talk. **3** cajolery. **4** *colloq.* a prolonged or tiresome business. **5** *hist.* a parley between African or other natives and traders. ● *v.* **1** *intr.* talk profusely. **2** *tr.* flatter, wheedle. [Port. *palavra* word f. L (as PARABLE)]

Palawan /pə'lɑːwən/ a long, narrow island in the western Philippines, separating the Sulu Sea from the South China Sea. Its chief town is Puerto Princesa.

pale[1] /peɪl/ *adj. & v.* ● *adj.* **1** (of a person or complexion) of a whitish or ashen appearance. **2 a** (of a colour) faint; not dark or deep. **b** faintly coloured. **3** of faint lustre; dim. **4** lacking intensity, vigour, or strength (*pale imitation*). ● *v.* **1** *intr.* & *tr.* grow or make pale. **2** *intr.* (often foll. by *before*, *beside*) become feeble in comparison (with). □ **palely** *adv.* **paleness** *n.* **palish** *adj.* [ME f. OF *pale*, *palir* f. L *pallidus* f. *pallere* be pale]

pale[2] /peɪl/ *n.* **1** a pointed piece of wood for fencing etc.; a stake. **2** a boundary. **3** an enclosed area, often surrounded by a palisade or ditch. **4** *Heraldry* a vertical stripe in the middle of a shield. □ **beyond the pale** outside the bounds of acceptable behaviour. **in pale** *Heraldry* arranged vertically. [ME f. OF *pal* f. L *palus* stake]

Pale, the see ENGLISH PALE 2.

palea /'peɪlɪə/ *n.* (*pl.* **paleae** /-lɪ,i:/) *Bot.* a chafflike bract, esp. in the flower of a grass. [L, = chaff]

Palearctic see PALAEARCTIC.

paled /peɪld/ *adj.* having palings.

paleface /'peɪlfeɪs/ *n.* a name supposedly used by North American Indians for a white person.

Palembang /ˌpɑːləm'bɑːŋ, pɑː'lembɑːŋ/ a city in Indonesia, in the south-eastern part of the island of Sumatra, a river port on the Musi river; pop. (1980) 787,190. Formerly the capital of the Buddhist kingdom of Srivijaya, which flourished in Malaya between the 7th and 13th centuries, it became a sultanate in the late 15th century. From 1616 it was developed as a trading post by the Dutch, who abolished the sultanate in 1825.

Palenque /pə'leŋkeɪ/ the site of a former Mayan city in SE Mexico,

south-east of present-day Villahermosa. The well-preserved ruins of the city, which existed from about AD 300 to 900, include notable examples of Mayan architecture and extensive hieroglyphic texts. The city's ancient name has been lost and it is now named after a neighbouring village.

paleo- *US var. of* PALAEO-.

Paleocene *US var. of* PALAEOCENE.

Paleozoic *US var. of* PALAEOZOIC.

Palermo /pə'leəməʊ/ the capital of the Italian island of Sicily, a port on the north coast; pop. (1990) 734,240. Founded as a trading post by the Phoenicians in the 8th century BC and later settled by the Carthaginians, it was taken by the Romans in 254 BC. It was conquered by the Arabs in AD 831, becoming in 1072 the capital of Sicily, which was then a Norman kingdom.

Palestine /'pælɪˌstaɪn/ a territory in the Middle East on the eastern coast of the Mediterranean Sea. It has had many changes of frontier and status in the course of history, and contains several places that are sacred to Christians, Jews, and Muslims. In biblical times Palestine comprised the kingdoms of Israel (see ISRAEL[1]) and Judah. The land was controlled at various times by the Egyptian, Assyrian, Persian, and Roman empires before being conquered by the Arabs in AD 634. It remained in Muslim hands, except for a period during the Crusades (1098–1197), until the First World War, being part of the Ottoman Empire from 1516 to 1918, when Turkish and German forces were defeated by the British at Megiddo. The name Palestine was used as the official political title for the land west of the Jordan mandated to Britain in 1920. Jewish immigration, encouraged by the Balfour Declaration of 1917 and intensified by Nazi persecution in Europe, increased dramatically during the period 1920–45, and in 1948 the state of Israel was established. The name Palestine continues to be used, however, to describe a geographical entity, particularly in the context of the struggle for territory and political rights of Palestinian Arabs displaced when Israel was established. In 1993 an agreement was signed between Israel and the Palestine Liberation Organization relating to the Gaza Strip and the West Bank, by which Israeli troops withdrew in 1994 and a Palestinian administration – the Palestine National Authority (PNA) – and police force were set up. (See also PALESTINE LIBERATION ORGANIZATION.) [Gk *Palaistinē* (used in early Christian writing), L (*Syria*) *Palaestina* (name of Roman province), f. *Philistia* land of the Philistines]

Palestine Liberation Organization (abbr. **PLO**) a political and military organization formed in 1964 to unite various Palestinian Arab groups and ultimately to bring about an independent state of Palestine. Since 1967 it has been dominated by the Al Fatah group led by Yasser Arafat. The PLO was soon active in guerrilla activities against Israel, originally from a base in Jordan but after the Jordanian civil war in 1970 from Beirut in Lebanon. In 1974 the organization was recognized by the Arab nations as the representative of all Palestinians and in 1976 it was invited to take part in a UN debate. The Israeli invasion of Lebanon in 1982 disorganized the PLO's military power and caused serious strains within its political superstructure, with the moderate Al Fatah group being opposed by more extreme factions. In 1988 Arafat renounced terrorism, and in the same year the PLO officially recognized Israel's right to exist; in 1993 an agreement was signed with Israel giving some autonomy to the West Bank and the Gaza Strip (see PALESTINE).

Palestinian /ˌpælɪ'stɪnɪən/ *adj. & n.* ● *adj.* of or relating to Palestine or the Palestinians. ● *n.* a native or inhabitant of Palestine, esp. an Arab born in (or descended from someone born in) the former mandated territory of Palestine.

palestra *var. of* PALAESTRA.

Palestrina /ˌpælə'stri:nə/, Giovanni Pierluigi da (*c.*1525–94), Italian composer. He composed several madrigals, but is chiefly known for his sacred music, notably 105 masses and over 250 motets. His music is characterized by its control of counterpoint; major works include the *Missa Papae Marcelli* (1567).

palette /'pælɪt/ *n.* **1** a thin board or slab or other surface, usu. with a hole for the thumb, on which an artist lays and mixes colours. **2** the range of colours used by an artist. □ **palette-knife 1** a thin steel blade with a handle for mixing colours or applying or removing paint. **2** a kitchen knife with a long blunt round-ended flexible blade. [F, dimin. of *pale* shovel f. L *pala* spade]

palfrey /'pɔ:lfrɪ/ *n.* (*pl.* **-eys**) *archaic* a horse for ordinary riding, esp. for women. [ME f. OF *palefrei* f. med.L *palefredus*, LL *paraveredus* f. Gk *para* beside, extra, + L *veredus* light horse, of Gaulish orig.]

Pali /ˈpɑːlɪ/ n. & adj. ● n. an Indic language, closely related to Sanskrit, in which the sacred texts of Theravada Buddhism are written. (See note below.) ● adj. of or in this language. [Pali pāli-bhāsā f. pāli text + bhāsā language]

▪ Pali developed in northern India in the 5th–2nd centuries BC. As the language of a large part of the Buddhist scriptures it was brought to Sri Lanka and Burma (Myanmar), and, though not spoken there, became the vehicle of a large literature of commentaries and chronicles.

palimony /ˈpælɪmənɪ/ n. esp. US colloq. an allowance made by one member of an unmarried couple to the other after separation. [PAL + ALIMONY]

palimpsest /ˈpælɪmpˌsest/ n. **1** a piece of writing material or manuscript on which the original writing has been effaced to make room for other writing. **2** a monumental brass turned and re-engraved on the reverse side. [L palimpsestus f. Gk palimpsēstos f. palin again + psēstos rubbed smooth]

palindrome /ˈpælɪnˌdrəʊm/ n. a word or phrase that reads the same backwards as forwards (e.g. rotator, nurses run). □ **palindromist** n. **palindromic** /ˌpælɪnˈdrɒmɪk/ adj. [Gk palindromos running back again f. palin again + drom- run]

paling /ˈpeɪlɪŋ/ n. **1** a fence of pales. **2** a pale.

palingenesis /ˌpælɪnˈdʒenɪsɪs/ n. Biol. the exact reproduction of ancestral characteristics in ontogenesis. □ **palingenetic** /-dʒɪˈnetɪk/ adj. [Gk palin again + genesis birth, GENESIS]

palinode /ˈpælɪˌnəʊd/ n. **1** a poem in which the writer retracts a view or sentiment expressed in a former poem. **2** a recantation. [F palinode or LL palinodia f. Gk palinōidia f. palin again + ōidē song]

palisade /ˌpælɪˈseɪd/ n. & v. ● n. **1 a** a fence of pales or iron railings. **b** a strong pointed wooden stake used in a close row for defence. **2** (in pl.) US a line of high cliffs. ● v.tr. enclose or provide with a palisade. □ **palisade layer** Bot. a layer of elongated cells below the epidermis of a leaf. [F palissade f. Prov. palissada f. palissa paling ult. f. L palus stake]

Palisades, the a ridge of high basalt cliffs on the west bank of the Hudson River, in NE New Jersey.

Palissy /ˈpælɪsɪ/, Bernard (c.1510–90), French potter. From the late 1550s he became famous for richly coloured earthenware decorated with reliefs of plants and animals. From about 1565 he enjoyed royal patronage and was employed by the court.

Palk Strait /pɔːlk/ an inlet of the Bay of Bengal separating northern Sri Lanka from the coast of Tamil Nadu in India. It lies to the north of Adam's Bridge, which separates it from the Gulf of Mannar.

pall¹ /pɔːl/ n. **1** a cloth spread over a coffin, hearse, or tomb. **2** a shoulder-band with pendants, worn as an ecclesiastical vestment and sign of authority. **3** a dark covering (a pall of darkness; a pall of smoke). **4** Heraldry a Y-shaped bearing charged with crosses representing the front of an ecclesiastical pall. [OE pæll, f. L pallium cloak]

pall² /pɔːl/ v. **1** intr. (often foll. by on) become uninteresting (to). **2** tr. satiate, cloy. [ME, f. APPAL]

palladia pl. of PALLADIUM².

Palladian /pəˈleɪdɪən/ adj. Archit. in the neoclassical style of Palladio. The term is applied to a phase of English architecture from c.1715, when there was a revival of interest in the ideas and designs of Palladio and his English follower, Inigo Jones, and a reaction against the baroque. □ **Palladianism** n.

Palladio /pəˈlɑːdɪˌəʊ/, Andrea (1508–80), Italian architect. He led a revival of classical architecture in 16th-century Italy, in particular promoting the Roman ideals of harmonic proportions and symmetrical planning. He designed many villas, palaces, and churches; major buildings include the church of San Giorgio Maggiore in Venice (1566 onwards). His theoretical work Four Books on Architecture (1570) was the main source of inspiration for the English Palladian movement.

palladium¹ /pəˈleɪdɪəm/ n. a silvery-white metallic chemical element (atomic number 46; symbol **Pd**). A rare element and one of the platinum metals, palladium was discovered by W. H. Wollaston in 1803. The metal and its alloys are used in electrical contacts, precision instruments, and jewellery, and it is also used as a catalyst in the chemical industry. [f. PALLAS 2 (discovered just previously)]

palladium² /pəˈleɪdɪəm/ n. (pl. **palladia** /-dɪə/) a safeguard or source of protection. [ME f. L f. Gk palladion image of Pallas (Athene), a protecting deity]

Pallas /ˈpæləs/ **1** Gk Mythol. one of the names (of unknown meaning) of

Athene. **2** Astron. asteroid 2, discovered in 1802. It is the second largest (diameter 523 km), and its surface appears to be rich in carbon.

pallbearer /ˈpɔːlˌbeərə(r)/ n. a person helping to carry or officially escorting a coffin at a funeral.

pallet¹ /ˈpælɪt/ n. **1** a straw mattress. **2** a crude or makeshift bed. [ME pailet, paillet f. AF paillete straw f. OF paille f. L palea]

pallet² /ˈpælɪt/ n. **1** a portable platform for transporting and storing loads. **2** a flat wooden blade with a handle, used in ceramics to shape clay. **3** = PALETTE. **4** a projection transmitting motion from an escapement to a pendulum etc. **5** a projection on a machine-part, serving to change the mode of motion of a wheel. □ **palletize** v.tr. (also **-ise**) (in sense 1). [F palette: see PALETTE]

pallia pl. of PALLIUM.

palliasse /ˈpælɪˌæs/ n. (also **paillasse**) a straw mattress. [F paillasse f. It. pagliaccio ult. f. L palea straw]

palliate /ˈpælɪˌeɪt/ v.tr. **1** alleviate (disease) without curing it. **2** excuse, extenuate. □ **palliator** n. **palliation** /ˌpælɪˈeɪʃ(ə)n/ n. [LL palliare to cloak f. pallium cloak]

palliative /ˈpælɪətɪv/ n. & adj. ● n. something used to alleviate pain, anxiety, etc. ● adj. serving to alleviate. □ **palliatively** adv. [F palliatif -ive or med.L palliativus (as PALLIATE)]

pallid /ˈpælɪd/ adj. pale, esp. from illness. □ **pallidly** adv. **pallidness** n. **pallidity** /pəˈlɪdɪtɪ/ n. [L pallidus PALE¹]

pallium /ˈpælɪəm/ n. (pl. **palliums** or **pallia** /-lɪə/) **1** an ecclesiastical pall, esp. that sent by the pope to an archbishop as a symbol of authority. **2** hist. a man's large rectangular cloak esp. as worn in antiquity. **3** Zool. the mantle of a mollusc or brachiopod. [L]

pall-mall /pælˈmæl, pelˈmel/ n. a 16th and 17th-century game in which a boxwood ball was driven with a mallet through an iron ring suspended at the end of a long alley. The street Pall Mall in London was on the site of a pall-mall alley. [obs. F pallemaille f. It. pallamaglio f. palla ball + maglio mallet]

pallor /ˈpælə(r)/ n. pallidness, paleness. [L f. pallere be pale]

pally /ˈpælɪ/ adj. (**pallier**, **palliest**) colloq. like a pal; friendly.

palm¹ /pɑːm/ n. **1** (in full **palm tree**) a tree of the mainly tropical family Palmae, with no branches and a mass of large pinnate or fan-shaped leaves at the top. **2** the leaf of this tree as a symbol of victory. **3 a** supreme excellence. **b** a prize for this. **4** a branch of various trees used instead of a palm in non-tropical countries, esp. in celebrating Palm Sunday. □ **palm oil** oil from the fruit of various palms, esp. Elaeis guineensis of West Africa. **palm wine** an alcoholic drink made from fermented palm sap. □ **palmaceous** /pælˈmeɪʃəs/ adj. [OE palm(a) f. Gmc f. L palma PALM², its leaf being likened to a spread hand]

palm² /pɑːm/ n. & v. ● n. **1** the inner surface of the hand between the wrist and fingers. **2** the part of a glove that covers this. **3** the palmate part of an antler. ● v.tr. conceal in the hand; steal by concealing in the hand. □ **in the palm of one's hand** under one's control or influence. **palm off 1** (often foll. by on) **a** impose or thrust fraudulently (on a person). **b** cause a person to accept reluctantly or unknowingly (palmed my old typewriter off on him). **2** (often foll. by with) cause (a person) to accept reluctantly or unknowingly (palmed him off with my old typewriter). □ **palmed** adj. **palmful** n. (pl. **-fuls**). **palmar** /ˈpælmə(r)/ adj. [ME paume f. OF paume f. L palma: later assim. to L]

Palma /ˈpælmə, ˈpɑːmə/ n. (in full **Palma de Mallorca**) the capital of the Balearic Islands, an industrial port and resort on the island of Majorca; pop. (1991) 308,620.

Palmas /ˈpælmæs/ a town in central Brazil, on the Tocantins river, capital of the state of Tocantins; pop. (1990) 5,750.

palmate /ˈpælmeɪt/ adj. **1** shaped like an open hand. **2** having lobes etc. like spread fingers. [L palmatus (as PALM²)]

Palm Beach a resort town in SE Florida, situated on an island just off the coast; pop. (1990) 9,810.

Palme /ˈpɑːlmə/, (Sven) Olof (Joachim) (1927–86), Swedish statesman, Prime Minister 1969–76 and 1982–6. He became leader of the Social Democratic Socialist Workers' Party in 1969. During his first term of prime ministerial office, Palme was a critic of US intervention in the Vietnam War and granted asylum to US army deserters. His electoral defeat in 1976 marked the end of forty-four continuous years in power for the Social Democratic Party. He was killed by an unknown assassin during his second term of office.

Palmer /ˈpɑːmə(r)/, Arnold (Daniel) (b.1929), American golfer. His many championship victories include the Masters (1958; 1960; 1962; 1964), the US Open (1960), and the British Open (1961–2).

palmer /ˈpɑːmə(r)/ n. **1** hist. **a** a pilgrim returning from the Holy Land with a palm branch or leaf. **b** an itinerant monk under a vow of poverty. **2** a hairy artificial fly used in angling. **3** (in full **palmerworm**) a caterpillar covered in stinging hairs; esp. one of the European moth *Euproctis chrysorrhoea*, a fruit pest. [ME f. AF *palmer*, OF *palmier* f. med.L *palmarius* pilgrim]

Palmerston /ˈpɑːməstən/, Henry John Temple, 3rd Viscount (1784–1865), British Whig statesman, Prime Minister 1855–8 and 1859–65. He left the Tory Party in 1830 to serve with the Whigs as Foreign Secretary (1830–4; 1835–41; 1846–51). In his foreign policy Palmerston was single-minded in his promotion of British interests, declaring the second Opium War against China in 1856, and overseeing the successful conclusion of the Crimean War in 1856 and the suppression of the Indian Mutiny in 1858. He maintained British neutrality during the American Civil War.

Palmerston North a city in the south-western part of North Island, New Zealand; pop. (1990) 69,300. Founded in 1866, it was named after the British Prime Minister Lord Palmerston. Located on the Manawatu river, it is the chief town of the agricultural region of Manawatu.

palmette /pælˈmet/ n. *Archaeol.* an ornament of radiating petals like a palm-leaf. [F, dimin. of *palme* PALM¹]

palmetto /pælˈmetəʊ/ n. (pl. **-os**) a small palm tree, esp. a fan palm of the genus *Sabal* or *Chamaerops*. [Sp. *palmito*, dimin. of *palma* PALM¹, assim. to It. words in *-etto*]

palmiped /ˈpælmɪped/ adj. & n. (also **palmipede** /-ˌpiːd/) • adj. web-footed. • n. a web-footed bird. [L *palmipes -pedis* (as PALM², *pes pedis* foot)]

palmistry /ˈpɑːmɪstrɪ/ n. a method of divination by examination of the lines and swellings of the hand. The parts of the hand are held to correspond to various parts of the body, traits of personality, and to the celestial bodies. Palmistry can be traced back to ancient China and flourished in classical Greece, also enjoying a considerable vogue in medieval and Renaissance Europe. After a long decline it enjoyed a certain revival in the 19th century, linked to the renewed interest in physiognomy. □ **palmist** n. [ME, orig. *palmestry* f. PALM²: second element unexpl.]

palmitic acid /pælˈmɪtɪk/ n. *Chem.* a saturated fatty acid, solid at room temperature, found in palm oil and other vegetable and animal fats. □ **palmitate** /ˈpælmɪˌteɪt/ n. [F *palmitique*, f. *palme* PALM¹]

Palm Springs a resort city in the desert area of southern California, east of Los Angeles, noted for its hot mineral springs; pop. (1990) 40,180.

Palm Sunday the Sunday before Easter, on which Christ's entry into Jerusalem is celebrated by processions in which branches of palms are carried.

palmtop /ˈpɑːmtɒp/ n. a computer small and light enough to be held in one hand.

palmy /ˈpɑːmɪ/ adj. (**palmier**, **palmiest**) **1** of or like or abounding in palms. **2** triumphant, flourishing (*palmy days*).

Palmyra /pælˈmaɪərə/ an ancient city of Syria, an oasis in the Syrian desert north-east of Damascus on the site of present-day Tadmur. First mentioned in the 19th century BC, Palmyra was an independent state in the 1st century BC, becoming a dependency of Rome between the 1st and 3rd centuries AD. A flourishing city on a trade route between Damascus and the Euphrates, it regained its independence briefly under Zenobia, who became queen of Palmyra in 267, until it was taken by the Emperor Aurelian in 272. The name Palmyra is the Greek form of the city's modern and ancient pre-Semitic name Tadmur or Tadmor, meaning 'city of palms'.

palmyra /pælˈmaɪərə/ n. a tall fan palm, *Borassus flabellifer*, native to tropical Asia, with leaves used for thatch, matting, etc. [Port. *palmeira* palm tree, assim. to PALMYRA]

Palo Alto /ˌpæləʊ ˈæltəʊ/ a city in western California, south of San Francisco; pop. (1990) 55,900. It is a noted centre for electronics and computer technology, and the site of Stanford University.

Palomar, Mount /ˈpæləˌmɑː(r)/ a mountain in southern California, north-east of San Diego, rising to a height of 1,867 m (6,126 ft). It is the site of an astronomical observatory, which contains a 5-metre (200-inch) reflecting telescope developed by the American astronomer George Ellery Hale. Its name means 'place of the pigeons' in Spanish.

palomino /ˌpæləˈmiːnəʊ/ n. (pl. **-os**) a golden or cream-coloured horse with a light-coloured mane and tail, orig. bred in the south-western US. [Amer. Sp. f. Sp. *palomino* young pigeon f. *paloma* dove f. L *palumba*]

paloverde /ˌpæləʊˈvɜːdɪ/ n. a thorny leguminous tree of the genus *Cercidium* or *Parkinsonia*, native to Arizona etc., with yellow flowers. [Amer. Sp., = green tree]

palp /pælp/ n. (also **palpus** /ˈpælpəs/) (pl. **palps** or **palpi** /-paɪ/) a segmented sense-organ at the mouth of an arthropod; a feeler. □ **palpal** adj. [L *palpus* f. *palpare* feel]

palpable /ˈpælpəb(ə)l/ adj. **1** that can be touched or felt. **2** readily perceived by the senses or mind. □ **palpably** adv. **palpability** /ˌpælpəˈbɪlɪtɪ/ n. [ME f. LL *palpabilis* (as PALPATE)]

palpate /ˈpælpeɪt/ v.tr. examine (esp. medically) by touch. □ **palpation** /pælˈpeɪʃ(ə)n/ n. [L *palpare palpat-* touch gently]

palpebral /ˈpælpɪbrəl/ adj. of or relating to the eyelids. [LL *palpebralis* f. L *palpebra* eyelid]

palpitate /ˈpælpɪˌteɪt/ v.intr. **1** (of the heart) beat strongly and rapidly; undergo palpitation. **2** throb, tremble. □ **palpitant** adj. [L *palpitare* frequent. of *palpare* touch gently]

palpitation /ˌpælpɪˈteɪʃ(ə)n/ n. **1** throbbing, trembling. **2** (often in pl.) a noticeably rapid, strong, or irregular heartbeat due to exertion, agitation, or disease. [L *palpitatio* (as PALPITATE)]

palpus var. of PALP.

palsgrave /ˈpɔːlzgreɪv/ n. a Count Palatine. [Du. *paltsgrave* f. *palts* palatinate + *grave* count]

palstave /ˈpɔːlsteɪv/ n. *Archaeol.* a type of chisel made of bronze etc. shaped to fit into a split handle. [Danish *paalstav* f. ON *pálstafr* f. *páll* hoe (cf. L *palus* stake) + *stafr* STAFF¹]

palsy /ˈpɔːlzɪ, ˈpɒl-/ n. & v. • n. (pl. **-ies**) **1** paralysis, esp. with involuntary tremors. **2 a** a condition of utter helplessness. **b** a cause of this. • v.tr. (**-ies, -ied**) **1** affect with palsy. **2** render helpless. [ME *pa(r)lesi* f. OF *paralisie* ult. f. L *paralysis*: see PARALYSIS]

palter /ˈpɔːltə(r), ˈpɒl-/ v.intr. **1** haggle or equivocate. **2** trifle. □ **palterer** n. [16th c.: orig. unkn.]

paltry /ˈpɔːltrɪ, ˈpɒl-/ adj. (**paltrier**, **paltriest**) worthless, contemptible, trifling. □ **paltriness** n. [16th c.: f. *paltry* trash app. f. *palt*, *pelt* rubbish + -RY (cf. *trumpery*): cf. LG *paltrig* ragged]

paludal /pəˈljuːd(ə)l, ˈpæljʊd-/ adj. **1** of a marsh. **2** malarial. □ **paludism** /ˈpæljʊˌdɪz(ə)m/ n. (in sense 2). [L *palus -udis* marsh + -AL¹]

paly /ˈpeɪlɪ/ adj. *Heraldry* divided into equal vertical stripes. [OF *palé* f. *pal* PALE²]

palynology /ˌpælɪˈnɒlədʒɪ/ n. the study of pollen grains and other spores, esp. as found in archaeological or geological deposits etc. Pollen grains are resistant to decay and can usually be easily recognized: pollen extracted from sediments may be used for radiocarbon dating and for studying past climates and environments by identifying plants then growing. Another application of palynology is in the investigation of the causative agents of allergies. □ **palynologist** n. **palynological** /-nəˈlɒdʒɪk(ə)l/ adj. [Gk *palunō* sprinkle + -LOGY]

Pamir Mountains /pəˈmɪə(r)/ (also **Pamirs**) a mountain system of central Asia, centred in Tajikistan and extending into Kyrgyzstan, Afghanistan, Pakistan, and western China. It is the centre of the system including the Karakoram, Hindu Kush, Tien Shan, and Kunlun Shan mountain ranges. The highest peak in the Pamirs, Communism Peak in Tajikistan, rises to 7,495 m (24,590 ft).

pampas /ˈpæmpəs/ n.pl. large treeless plains in South America. □ **pampas-grass** a tall South American grass, *Cortaderia selloana*, with silky flowering plumes. [Sp. f. Quechua *pampa* plain]

pamper /ˈpæmpə(r)/ v.tr. **1** overindulge (a person, taste, etc.), cosset. **2** spoil (a person) with luxury. □ **pamperer** n. [ME, prob. of LG or Du. orig.]

pampero /pæmˈpeərəʊ/ n. (pl. **-os**) a strong cold SW wind in South America, blowing from the Andes towards the Atlantic. [Sp. (as PAMPAS)]

pamphlet /ˈpæmflɪt/ n. & v. • n. a small usu. unbound booklet or leaflet containing information or a short treatise. • v.tr. (**pamphleted**, **pamphleting**) distribute pamphlets to. [ME f. *Pamphilet*, the familiar name of the 12th-c. Latin love poem *Pamphilus seu de Amore*]

pamphleteer /ˌpæmflɪˈtɪə(r)/ n. & v. • n. a writer of (esp. political) pamphlets. • v.intr. write pamphlets.

Pamphylia /pæmˈfɪlɪə/ an ancient coastal region of southern Asia Minor, between Lycia and Cilicia, to the east of the modern port of Antalya. It became a Roman province in the reign of Augustus, between 31 BC and AD 14. □ **Pamphylian** adj. & n.

Pamplona /pæmˈpləʊnə/ a city in northern Spain, capital of the former kingdom and modern region of Navarre; pop. (1991) 191,110. It is noted for the fiesta of San Fermín, held there in July, which is celebrated with the running of bulls through the streets of the city.

Pan /pæn/ *Gk Mythol.* a god of flocks and herds, native to Arcadia, usually represented with the horns, ears, and legs of a goat on a man's body. He was thought of as loving mountains, caves, and lonely places and as playing on the pan-pipes. His sudden appearance was supposed to cause terror similar to that of a frightened and stampeding herd, and the word *panic* is derived from his name. This probably meant 'the feeder' (i.e. herdsman), although it was regularly associated with Greek *pas* or *pan* (= all), giving rise to his identification as a god of nature or the universe.

pan[1] /pæn/ *n. & v.* ● *n.* **1 a** a vessel of metal, earthenware, or plastic, usu. broad and shallow, used for cooking and other domestic purposes. **b** the contents of this. **2** a panlike vessel in which substances are heated etc. **3** any similar shallow container such as the bowl of a pair of scales or that used for washing gravel etc. to separate gold. **4** *Brit.* the bowl of a lavatory. **5** part of the lock that held the priming in old guns. **6** a hollow in the ground (*salt-pan*). **7** a hard substratum of soil (*hardpan*). **8** *US sl.* the face. **9 a** a metal drum in a steel band. **b** steel-band music and the associated culture. ● *v.* (**panned, panning**) **1** *tr. colloq.* criticize severely. **2** *tr. sl.* hit or punch (a person). **3 a** *tr.* (often foll. by *off, out*) wash (gold-bearing gravel) in a pan. **b** *intr.* search for gold by panning gravel. **c** *intr.* (foll. by *out*) (of gravel) yield gold. □ **pan out** (of an action etc.) turn out well or in a specified way. □ **panful** *n.* (*pl.* **-fuls**). **panlike** *adj.* [OE *panne*, perh. ult. f. L *patina* dish]

pan[2] /pæn/ *v. & n.* ● *v.* (**panned, panning**) **1** *tr.* swing (a camera) horizontally to give a panoramic effect or to follow a moving object. **2** *intr.* (of a camera) be moved in this way. ● *n.* a panning movement. [abbr. of PANORAMA]

pan[3] /paːn/ *n.* **1** a leaf of the betel. **2** this enclosing lime and areca-nut parings, chewed in India etc. [Hindi f. Skr. *parna* feather, leaf]

pan- /pæn/ *comb. form* **1** all; the whole of. **2** relating to the whole or all the parts of a continent, racial group, religion, etc. (*pan-American*; *pan-African*; *pan-Hellenic*; *pan-Anglican*). [Gk f. *pan* neut. of *pas* all]

panacea /ˌpænəˈsɪə/ *n.* a universal remedy; something resorted to in every case of difficulty. □ **panacean** *adj.* [L f. Gk *panakeia* f. *panakēs* all-healing (as PAN-, *akos* remedy)]

panache /pəˈnæʃ/ *n.* **1** flamboyant confidence of style or manner. **2** *hist.* a tuft or plume of feathers, esp. as a head-dress or on a helmet. [F f. It. *pennacchio* f. LL *pinnaculum* dimin. of *pinna* feather]

panada /pəˈnɑːdə/ *n.* **1** a thick paste of flour etc. **2** bread boiled to a pulp and flavoured. [Sp. ult. f. L *panis* bread]

Pan-Africanist Congress /ˌpænˈæfrɪkənɪst/ (in full **Pan-Africanist Congress of Azania**) (abbr. **PAC**) a South African political movement formed in 1959 as a militant offshoot of the African National Congress. It was outlawed in 1960 after demonstrations which culminated in the Sharpeville massacre, but continued its armed opposition to the South African government from bases outside the country until it was legalized in 1990. The PAC took part in the elections of 1994, winning 1.2 per cent of the vote and seven seats.

Panaji /ˈpʌnədʒɪ/ (also **Panjim** /ˈpʌnʒɪm/) a city in western India, a port on the Arabian Sea; pop. (1991) 85,200. It is the capital of the state of Goa.

Panama /ˈpænəˌmɑː, ˌpænəˈmɑː/ a country in Central America; pop. (1990) 2,329,330; official language, Spanish; capital, Panama City. Panama occupies the isthmus connecting North and South America; much of the country is low-lying tropical rainforest. Colonized by Spain in the early 16th century, Panama was freed from imperial control in 1821, becoming a Colombian province. It gained full independence in 1903, although the construction of the Panama Canal and the leasing of the zone around it to the US (until 1977) split the country in two. In 1989 US troops invaded Panama in a successful attempt to arrest the country's President, Gen. Manuel Noriega, so that he could face trial for drug trafficking. □ **Panamanian** /ˌpænəˈmeɪnɪən/ *adj. & n.*

panama /ˈpænəˌmɑː/ *n.* a hat of strawlike material, orig. made from the leaves of a palm tree. [PANAMA]

Panama Canal a canal about 80 km (50 miles) long, across the Isthmus of Panama, connecting the Atlantic and Pacific Oceans. Its construction, begun by Ferdinand de Lesseps in 1881 but abandoned in 1889, was completed by the US between 1904 and 1914. The surrounding territory, the (Panama) Canal Zone, was administered by the US until 1979, when it was returned to the control of Panama. Control of the canal itself remains with the US until 1999, at which date it is due to be ceded to Panama.

Panama City the capital of Panama, situated on the Pacific coast close to the Panama Canal; pop. (1990) 584,800. The old city, founded

by the Spanish in 1519 on the site of a native fishing village, was destroyed by the Welsh buccaneer Sir Henry Morgan (*c*.1635–88) in 1671. The new city, built in 1674 on a site a little to the west, became capital of Panama in 1903, when the republic gained its independence from Colombia. The city developed rapidly after the opening of the Panama Canal in 1914.

panatella /ˌpænəˈtelə/ *n.* a long thin cigar. [Amer. Sp. *panatela*, = long thin biscuit f. It. *panatella* dimin. of *panata* (as PANADA)]

Panay /pæˈnaɪ/ an island in the central Philippines; chief town, Iloilo.

pancake /ˈpænkeɪk/ *n. & v.* ● *n.* **1** a thin flat cake of batter usu. fried and turned in a pan and rolled up with a filling. **2** a flat cake of make-up etc. ● *v.* **1** *intr.* make a pancake landing. **2** *tr.* cause (an aircraft) to pancake. □ **flat as a pancake** completely flat. **pancake landing** an emergency landing by an aircraft with its undercarriage still retracted. [ME f. PAN[1] + CAKE]

Pancake Day Shrove Tuesday (on which pancakes are traditionally eaten).

panchayat /pʌnˈtʃaɪət/ *n.* a village council in India. [Hindi f. Skr. *pancha* five]

Panchen lama /ˈpæntʃən/ *n.* a Tibetan lama ranking next after the Dalai Lama. [Tibetan *panchen* great learned one]

panchromatic /ˌpænkrəʊˈmætɪk/ *adj. Photog.* (of a film etc.) sensitive to all colours of the visible spectrum.

pancreas /ˈpæŋkrɪəs/ *n. Anat.* a gland near the stomach supplying the duodenum with digestive enzymes and secreting insulin into the blood. □ **pancreatic** /ˌpæŋkrɪˈætɪk/ *adj.* **pancreatitis** /-krɪəˈtaɪtɪs/ *n.* [mod.L f. Gk *pagkreas* (as PAN-, *kreas -atos* flesh)]

pancreatin /ˈpæŋkrɪətɪn/ *n.* a digestive extract containing pancreatic enzymes, prepared from animal pancreases.

panda /ˈpændə/ *n.* **1** (in full **giant panda**) a large herbivorous bearlike mammal, *Ailuropoda melanoleuca*, found in limited mountainous forested areas of China and Tibet, having characteristic black and white markings. (*See note below*.) **2** (in full **red** or **lesser panda**) a raccoon-like mammal, *Ailurus fulgens*, native to the Himalayas and eastern Asia, with reddish-brown fur and a long bushy tail. □ **panda car** *Brit.* a police patrol car (orig. white with black stripes on the doors). [Nepali name]

■ The giant panda was regarded as a bear until its relationship to the red panda was believed established, both animals then being placed in the raccoon family on the basis of their teeth. However, recent anatomical, biochemical, and chromosome studies of the giant panda have shown that it does indeed belong in the bear family. Its unusual characteristics appear to have resulted from its specialized diet, and its total dependence on bamboo has made it highly vulnerable. The red panda is related to both bears and raccoons, and is usually placed in its own family.

pandanus /pænˈdeɪnəs, -ˈdænəs/ *n.* **1** a tropical tree or shrub of the genus *Pandanus*, with a twisted stem, aerial roots, spiral tufts of long narrow leaves at the top, and conelike fruits. Also called *screw pine*. **2** fibre made from pandanus leaves, or material woven from this.

Pandarus /ˈpændərəs/ *Gk Mythol.* a Lycian fighting on the side of the Trojans, described in the *Iliad* as breaking the truce with the Greeks by wounding Menelaus with an arrow. The role as the lovers' go-between that he plays in Chaucer's (and later Shakespeare's) story of Troilus and Cressida originated with Boccaccio and is also the origin of the word *pander*.

pandect /ˈpændekt/ *n.* (usu. in *pl.*) **1** a complete body of laws. **2** *hist.* a compendium in 50 books of the Roman civil law made by order of Justinian in the 6th century. [F *pandecte* or L *pandecta pandectes* f. Gk *pandektēs* all-receiver (as PAN-, *dektēs* f. *dekhomai* receive)]

pandemic /pænˈdemɪk/ *adj. & n.* ● *adj.* (of a disease) prevalent over a whole country or the world. ● *n.* an outbreak of such a disease. [Gk *pandēmos* (as PAN-, *dēmos* people)]

pandemonium /ˌpændɪˈməʊnɪəm/ *n.* **1** uproar; utter confusion. **2** a scene of this. [mod.L (place of all demons in Milton's *Paradise Lost*) f. PAN- + Gk *daimōn* DEMON[1]]

pander /ˈpændə(r)/ *v. & n.* ● *v.intr.* (foll. by *to*) gratify or indulge a person, a desire or weakness, etc. ● *n.* **1** a go-between in illicit love affairs; a procurer. **2** a person who encourages coarse desires. [ME, f. PANDARUS]

Pandit /ˈpʌndɪt/, Vijaya (Lakshmi) (1900–90), Indian politician and diplomat. After joining the Indian National Congress, led by her brother Jawaharlal Nehru, she was imprisoned three times by the British (1932; 1941; 1942) for nationalist activities. Following independence, she led the Indian delegation to the United Nations

(1946–8; 1952–3) and was the first woman to serve as president of the United Nations General Assembly (1953–4).

pandit var. of PUNDIT 1.

P. & O. *abbr.* Peninsular and Oriental (Steamship Company).

Pandora /pæn'dɔːrə/ *Gk Mythol.* the first mortal woman. In one story she was created by Zeus and sent to earth with a jar or box of evils in revenge for Prometheus' having brought the gift of fire back to the world. Prometheus' simple brother Epimetheus married her despite his brother's warnings, and Pandora let out all the evils from the jar to infect the earth; hope alone remained to assuage the lot of humankind. In another account the jar contained all the blessings which would have been preserved for the world had they not been allowed to escape. [Gk *Pandōra* all-gifted (as PAN-, *dōron* gift)]

Pandora's box *n.* a source of many and unforeseen evils; something which once started will generate many unmanageable problems (*risked opening a Pandora's box*).

p. & p. *abbr. Brit.* postage and packing.

pane /peɪn/ *n.* **1** a single sheet of glass in a window or door. **2** a rectangular division of a chequered pattern etc. [ME f. OF *pan* f. L *pannus* piece of cloth]

panegyric /ˌpænɪ'dʒɪrɪk/ *n.* a laudatory discourse; a eulogy. □ **panegyrical** *adj.* [F *panégyrique* f. L *panegyricus* f. Gk *panēgurikos* of public assembly (as PAN-, *ēguris = agora* assembly)]

panegyrize /'pænɪdʒɪˌraɪz/ *v.tr.* (also **-ise**) speak or write in praise of; eulogize. □ **panegyrist** /ˌpænɪ'dʒɪrɪst/ *n.* [Gk *panēgurizō* (as PANEGYRIC)]

panel /'pæn(ə)l/ *n. & v.* ● *n.* **1 a** a distinct, usu. rectangular, section of a surface (e.g. of a wall, door, or vehicle). **b** a control panel (see CONTROL *n.* 5). **c** = *instrument panel.* **2** a strip of material as part of a garment. **3** a group of people brought together for a purpose, esp. to report, advise, or adjudicate officially on some matter. **4** *Brit.* a list of medical practitioners registered in a district as accepting patients under the National Health Service. **5 a** a list of available jurors; a jury. **b** *Sc.* a person or persons accused of a crime. ● *v.tr.* (**panelled, panelling**, *US* **paneled, paneling**) **1** fit or provide with panels. **2** cover or decorate with panels. □ **panel-beater** *Brit.* a person whose job is to beat out the metal panels of motor vehicles. **panel game** a broadcast quiz etc. played by a group of participants. **panel heating** the heating of rooms by panels in the wall etc. containing the sources of heat. **panel pin** a thin nail with a very small head. **panel saw** a saw with small teeth for cutting thin wood for panels. **panel truck** *US* a small enclosed delivery truck. [ME & OF, = piece of cloth, ult. f. L *pannus*: see PANE]

panelling /'pænəlɪŋ/ *n.* (*US* **paneling**) **1** panelled work. **2** wood for making panels.

panellist /'pænəlɪst/ *n.* (*US* **panelist**) a member of a panel (esp. in broadcasting).

panettone /ˌpænɪ'təʊnɪ/ *n.* (*pl.* **panettoni** *pronunc.* same) a rich Italian bread made with eggs, dried fruit, and butter. [It. f. *panetto* cake]

panforte /pæn'fɔːtɪ/ *n.* a hard spicy Sienese cake made with nuts, candied peel, and honey. [It. f. *pane* bread + *forte* strong]

pang /pæŋ/ *n.* (often in *pl.*) a sudden sharp pain or painful emotion (*hunger pangs; pangs of remorse*). [16th c.: var. of earlier *prange* pinching f. Gmc]

panga /'pæŋgə/ *n.* a bladed African tool like a machete. [East African name]

Pangaea /pæn'dʒiːə/ a vast continental area or supercontinent comprising all the continental crust of the earth, which is postulated to have existed in late Palaeozoic and Mesozoic times before breaking up into Gondwana and Laurasia. (See also CONTINENTAL DRIFT.) [PAN- + Gk *gaia* land, earth]

pangolin /pæŋ'gəʊlɪn/ *n.* an anteating mammal of the family Manidae, native to Asia (genus *Manis*) and Africa (genus *Phataginus*). Pangolins are covered with large horny scales, and have a small head with an elongated snout, and a tapering tail. Also called *scaly anteater*. [Malay *peng-gōling* roller (from its habit of rolling itself up)]

panhandle /'pænˌhænd(ə)l/ *n. & v. US* ● *n.* a narrow strip of territory projecting from the main part of a state etc. (*the Oklahoma panhandle*). ● *v.tr. & intr. colloq.* beg for money in the street. □ **panhandler** *n.*

panic[1] /'pænɪk/ *n. & v.* ● *n.* **1** sudden uncontrollable fear or alarm. **2** infectious apprehension or fright esp. in commercial dealings. **3** (*attrib.*) characterized or caused by panic (*panic buying*). ● *v.tr. & intr.* (**panicked, panicking**) (often foll. by *into*) affect or be affected with panic (*was panicked into buying*). □ **panic button** a button (often

imaginary) for summoning help in an emergency. **panic-monger** a person who fosters a panic. **panic stations** a state of emergency. **panic-stricken** (or **-struck**) affected with panic; very apprehensive. □ **panicky** *adj.* [F *panique* f. mod.L *panicus* f. Gk *panikos* f. PAN]

panic[2] /'pænɪk/ *n.* a grass of the genus *Panicum*, which includes millet and other cereals. [OE f. L *panicum* f. *panus* thread on bobbin, millet-ear f. Gk *pēnos* web]

panicle /'pænɪk(ə)l/ *n. Bot.* a loose branching cluster of flowers, as in oats. □ **panicled** *adj.* [L *paniculum* dimin. of *panus* thread]

Panini /'pɑːnɪnɪ/, Indian grammarian. Little is known about his life; sources vary as to when he lived, with dates ranging from the 4th to the 7th century BC. He is noted as the author of the *Eight Lectures*, a grammar of Sanskrit, outlining rules for the derivation of grammatical forms.

Panjabi var. of PUNJABI.

panjandrum /pæn'dʒændrəm/ *n.* **1** a mock title for an important person. **2** a pompous or pretentious official etc. [app. invented in a nonsense composition by the English dramatist Samuel Foote (1720–77)]

Panjim see PANAJI.

Pankhurst /'pæŋkhɜːst/, Mrs Emmeline (1858–1928), Christabel (1880–1958), and (Estelle) Sylvia (1882–1960), English suffragettes. In 1903 Emmeline and her daughters Christabel and Sylvia founded the Women's Social and Political Union, with the motto 'Votes for Women'. Following the imprisonment of Christabel in 1905, Emmeline initiated the militant suffragette campaign and was responsible for keeping the suffragette cause in the public eye until the outbreak of the First World War.

Panmunjom /ˌpænmʊn'dʒɒm/ a village in the demilitarized zone between North and South Korea. It was there that the armistice ending the Korean War was signed on 27 July 1953.

panne /pæn/ *n.* (in full **panne velvet**) a velvet-like fabric of silk or rayon with a flattened pile. [F]

pannier /'pænɪə(r)/ *n.* **1** a basket, esp. each of a pair carried by a beast of burden. **2** each of a pair of bags or boxes on either side of the rear wheel of a bicycle or motorcycle. **3** *hist.* **a** part of a skirt looped up round the hips. **b** a frame supporting this. [ME f. OF *panier* f. L *panarium* bread-basket f. *panis* bread]

pannikin /'pænɪkɪn/ *n.* **1** *Brit.* a small metal drinking-cup. **2** *Brit.* the contents of this. **3** *Austral. sl.* the head (esp. *off one's pannikin*). □ **pannikin boss** *Austral. sl.* a minor overseer or foreman. [PAN[1] + -KIN, after *cannikin*]

Pannonia /pə'nəʊnɪə/ an ancient country of southern Europe lying south and west of the Danube, in present-day Austria, Hungary, Slovenia, and Croatia. It was occupied by the Romans from 35 BC, becoming a separate province in AD 6. It lost its separate identity after the Romans withdrew at the end of the 4th century.

panoply /'pænəplɪ/ *n.* (*pl.* **-ies**) **1** a complete or splendid array (often *full panoply*). **2** a complete suit of armour. □ **panoplied** *adj.* [F *panoplie* or mod.L *panoplia* full armour f. Gk (as PAN-, *oplia* f. *hopla* arms)]

panoptic /pæn'ɒptɪk/ *adj.* showing or seeing the whole at one view. [Gk *panoptos* seen by all, *panoptēs* all-seeing]

panorama /ˌpænə'rɑːmə/ *n.* **1** an unbroken view of a surrounding region. **2** a complete survey or presentation of a subject, sequence of events, etc. **3** a picture or photograph containing a wide view. **4** a continuous passing scene. □ **panoramic** /-'ræmɪk/ *adj.* **panoramically** *adv.* [PAN- + Gk *horama* view f. *horaō* see]

pan-pipes /'pænpaɪps/ *n.pl.* a musical instrument originally associated with the Greek rural god Pan, made from three or more tubes of different lengths joined in a row (or in some areas from a block of wood with tubes drilled down into it) with mouthpieces in line, and sounded by blowing across the top. Pan-pipes are known to have been used in neolithic times, though probably not in the West until about the 6th century BC.

pansy /'pænzɪ/ *n.* (*pl.* **-ies**) **1** a garden plant of the genus *Viola*, with flowers of various rich colours. **2** *colloq. derog.* **a** an effeminate man. **b** a male homosexual. [F *pensée* thought, pansy f. *penser* think f. L *pensare* frequent. of *pendere pens-* weigh]

pant /pænt/ *v. & n.* ● *v.* **1** *intr.* breathe with short quick breaths. **2** *tr.* (often foll. by *out*) utter breathlessly. **3** *intr.* (often foll. by *for*) yearn or crave. **4** *intr.* (of the heart etc.) throb violently. ● *n.* **1** a panting breath. **2** a throb. □ **pantingly** *adv.* [ME f. OF *pantaisier* ult. f. Gk *phantasioō* cause to imagine (as FANTASY)]

pantalets /ˌpæntə'lets/ *n.pl.* (also **pantalettes**) *hist.* **1** long underpants worn by women and girls in the 19th century, with a frill at the bottom of each leg. **2** women's cycling trousers. [dimin. of PANTALOONS]

Pantaloon /ˌpæntə'lu:n/ a Venetian character in Italian *commedia dell'arte* represented as a foolish old man wearing pantaloons, spectacles, and slippers. [*Pantalone* It. perh. f. *San Pantalone*, favourite Venetian saint in former times]

pantaloons /ˌpæntə'lu:nz/ *n.pl.* **1** *hist.* men's close-fitting breeches fastened below the calf or at the foot. **2** *colloq.* trousers. **3** baggy trousers (esp. for women) gathered at the ankles. [F *pantalon* f. It. *pantalone* PANTALOON]

Pantanal /ˌpæntə'nɑ:l/ a vast region of tropical swampland in the upper reaches of the Paraguay river in SW Brazil.

pantechnicon /pæn'teknɪkən/ *n. Brit.* a large van for transporting furniture. [PAN- + TECHNIC orig. as the name of a bazaar and then a furniture warehouse]

Pantelleria /ˌpæntelə'rɪə/ a volcanic Italian island in the Mediterranean, situated between Sicily and the coast of Tunisia. It was used as a place of exile by the ancient Romans, who called it Cossyra. Heavily fortified by the Italians during the Second World War, it was attacked and taken by Allied forces in 1943.

Panthalassa /ˌpænθə'læsə/ a universal sea or single ocean, such as would have surrounded Pangaea. [PAN- + Gk *thalassa* sea]

pantheism /'pænθɪ,ɪz(ə)m/ *n.* **1** the belief that God is identifiable with the forces of nature and with natural substances. **2** worship that admits or tolerates all gods. □ **pantheist** *n.* **pantheistic** /ˌpænθɪ'ɪstɪk/ *adj.* **pantheistical** *adj.* **pantheistically** *adv.* [PAN- + Gk *theos* god]

pantheon /'pænθɪən/ *n.* **1** a temple dedicated to all the gods, especially (**Pantheon**) the circular one still standing in Rome, erected in the early 2nd century AD probably on the site of an earlier one built by Agrippa in 27 BC. **2** the deities of a people collectively. **3** (**Pantheon**) a building in which the illustrious dead are buried or have memorials, especially the former church of St Geneviève (the Panthéon) in Paris, which in some respects resembles the Pantheon in Rome. [ME f. L f. Gk *pantheion* (as PAN-, *theion* holy f. *theos* god)]

panther /'pænθə(r)/ *n.* **1** a leopard, esp. with black fur. **2** *US* a puma. [ME f. OF *pantere* f. L *panthera* f. Gk *panthēr*]

pantie-girdle /'pæntɪ,gɜ:d(ə)l/ *n.* a woman's girdle with a crotch shaped like pants.

panties /'pæntɪz/ *n.pl. colloq.* short-legged or legless underpants worn by women and girls. [dimin. of PANTS]

pantihose /'pæntɪ,həʊz/ *n.* (*US* **panty hose**) (usu. treated as *pl.*) women's tights. [PANTIES + HOSE]

pantile /'pæntaɪl/ *n.* a roof-tile curved to form an S-shaped section, fitted to overlap. [PAN¹ + TILE]

panto /'pæntəʊ/ *n.* (*pl.* **-os**) *Brit. colloq.* = PANTOMIME 1. [abbr.]

panto- /'pæntəʊ/ *comb. form* all, universal. [Gk *pas pantos* all]

pantograph /'pæntə,grɑ:f/ *n.* **1** an instrument for copying a plan or drawing etc. on a different scale by a system of jointed rods. **2** a jointed framework conveying a current to an electric vehicle from overhead wires. □ **pantographic** /ˌpæntə'græfɪk/ *adj.* [PANTO- + Gk *-graphos* writing]

pantomime /'pæntə,maɪm/ *n.* **1** *Brit.* a theatrical entertainment usually based on a fairy tale. (*See note below.*) **2** the use of gestures and facial expression to convey meaning, esp. in drama and dance. **3** *colloq.* an absurd or outrageous piece of behaviour. □ **pantomimic** /ˌpæntə'mɪmɪk/ *adj.* [F *pantomime* or L *pantomimus* f. Gk *pantomimos* (as PANTO-, MIME)]

▪ Pantomime derives its name from the *pantomimus* of ancient Rome, in which a player represented in mime the different characters in a short scene based on history or mythology. In its modern form, however, pantomime is a British form of theatrical entertainment derived from the harlequinade, primarily for children and associated with Christmas. It is based on the dramatization of a fairy tale or nursery story, and includes songs and topical jokes, buffoonery and slapstick, and standard characters such as a pantomime dame played by a man, a principal boy played by a woman, and a pantomime animal (e.g. a horse, cat, or goose) played by actors dressed in a comic costume, with some regional variations. Pantomimes remain a feature of the Christmas season in Britain, and now often feature performers well known from television. (See also HARLEQUINADE.)

pantothenic acid /ˌpæntə'θenɪk/ *n. Biochem.* a vitamin of the B complex, found in rice, bran, and many other foods, and essential for the oxidation of fats and carbohydrates. □ **pantothenate** /-neɪt/ *n.* [Gk *pantothen* from every side]

pantry /'pæntrɪ/ *n.* (*pl.* **-ies**) **1** a small room or cupboard in which crockery, cutlery, table linen, etc., are kept. **2** a larder. [ME f. AF *panetrie*, OF *paneterie* f. *panetier* baker ult. f. LL *panarius* bread-seller f. L *panis* bread]

pantryman /'pæntrɪmən/ *n.* (*pl.* **-men**) a butler or a butler's assistant.

pants /pænts/ *n.pl.* **1** *Brit.* underpants or knickers. **2** *US* trousers or slacks. □ **bore** (or **scare** etc.) **the pants off** *colloq.* bore, scare, etc., to an intolerable degree. **pants** (or **pant**) **suit** esp. *US* a trouser suit. **with one's pants down** *colloq.* in an embarrassingly unprepared state. [abbr. of PANTALOONS]

panty hose *US* var. of PANTIHOSE.

panzer /'pæntsə(r), 'pænzə(r)/ *n.* esp. *hist.* **1** (in *pl.*) armoured units or troops, esp. of the German army in the Second World War. **2** (*attrib.*) of panzers; heavily armoured (*panzer division*). [G, = coat of mail]

Paolozzi /paʊ'lɒtsɪ/, Eduardo (Luigi) (b.1924), Scottish artist and sculptor, of Italian descent. He was a key figure in the development of pop art in Britain in the 1950s. His work is typified by mechanistic sculptures in a figurative style, often surfaced with cog wheels and machine parts, as in *Japanese War God* (1958).

pap¹ /pæp/ *n.* **1 a** a soft or semi-liquid food for infants or invalids. **b** a mash or pulp. **2** light or trivial reading matter; nonsense. □ **pappy** *adj.* [ME prob. f. MLG, MDu. *pappe*, prob. ult. f. L *pappare* eat]

pap² /pæp/ *n. archaic* or *dial.* the nipple of a breast. [ME, of Scand. orig.: ult. imit. of sucking]

papa /pə'pɑ:/ *n. archaic* father (esp. as a child's word). [F f. LL f. Gk *papas*]

papabile /pə'pɑ:bɪ,leɪ/ *adj.* suitable for high office. [It., = suitable to be pope, f. L *papa* pope]

papacy /'peɪpəsɪ/ *n.* (*pl.* **-ies**) **1** a pope's office or tenure. **2** the papal system. [ME f. med.L *papatia* f. *papa* pope]

papain /pə'peɪɪn/ *n.* a protein-digesting enzyme obtained from unripe pawpaws, used to tenderize meat and as a food supplement to aid digestion. [PAPAYA + -IN]

papal /'peɪp(ə)l/ *adj.* of or relating to a pope or to the papacy. □ **papally** *adv.* [ME f. OF f. med.L *papalis* f. eccl.L *papa* POPE¹]

Papal States a part of central Italy held between 756 and 1870 by the Catholic Church, corresponding to the modern regions of Emilia-Romagna, Marche, Umbria, and Lazio. Taken from the Lombards by the Frankish king Pepin III, father of Charlemagne, the states were given to the papacy as a strategy to undermine Lombard expansionism. Greatly extended by Pope Innocent III in the 13th century and Pope Julius II in the 16th, they were incorporated into the newly unified Italy in 1860 and 1870. Their annexation to Italy deprived the papacy of its temporal powers until the Lateran Treaty of 1929 recognized the sovereignty of the Vatican City.

paparazzo /ˌpæpə'rætsəʊ/ *n.* (*pl.* **paparazzi** /-sɪ/) a freelance photographer who pursues celebrities to get photographs of them. [It., from the name of a film character]

papaveraceous /pə,peɪvə'reɪʃəs/ *adj. Bot.* of the plant family Papaveraceae, which includes the poppy. [f. mod.L *Papaveraceae* f. L *papaver* poppy: see -ACEOUS]

papaw var. of PAWPAW.

papaya var. of PAWPAW 1.

Papeete /ˌpɑ:pɪ'eɪtɪ, -'i:tɪ/ the capital of French Polynesia, situated on the NW coast of Tahiti; pop. (1988) 78,800.

paper /'peɪpə(r)/ *n. & v.* ● *n.* **1** a material manufactured in thin sheets from the pulp of wood or other fibrous substances, used for writing or drawing or printing on, or as wrapping material etc. (*See note below.*) **2** (*attrib.*) **a** made of or using paper. **b** flimsy like paper. **3** = NEWSPAPER. **4 a** a document printed on paper. **b** (in *pl.*) documents attesting identity or credentials. **c** (in *pl.*) documents belonging to a person or relating to a matter. **5** *Commerce* a negotiable document, e.g. bills of exchange. **b** (*attrib.*) recorded on paper though not existing (*paper profits*). **6 a** a set of questions to be answered at one session in an examination. **b** the written answers to these. **7** = WALLPAPER *n.* 1. **8** an essay or dissertation, esp. one read to a learned society or published in a learned journal. **9** a piece of paper, esp. as a wrapper etc. **10** *Theatr. sl.* free tickets or the people admitted by them (*the house is full of paper*). ● *v.tr.* **1** apply paper to, esp. decorate (a wall etc.) with wallpaper. **2** (foll. by *over*) **a** cover (a hole or blemish) with paper. **b** disguise or try to hide (a fault etc.). **3** *Theatr. sl.* fill (a theatre) by giving free passes. □ **on paper 1** in writing. **2** in theory; to judge from written or printed evidence. **paper-boy** (or **-girl**) a boy or girl who delivers or sells newspapers. **paper-chase** a cross-country run in which the runners follow a trail marked by torn-up paper. **paper-clip** a clip of bent wire or of plastic for holding several sheets of paper together. **paper-hanger** a person who decorates with wallpaper, esp. professionally. **paper-knife** a blunt knife for opening letters etc. **paper-mill** a mill in which paper is made. **paper money** money in the form of banknotes. **paper mulberry** a small Asiatic tree, *Broussonetia papyrifera*, of the mulberry family, whose bark is used for making paper and cloth. **paper nautilus** see NAUTILUS 2. **paper round 1** a job of regularly delivering newspapers. **2** a route taken doing this. **paper tape** tape made of paper, esp. having holes punched in it for conveying data or instructions to a computer etc. **paper tiger** an apparently threatening, but ineffectual, person or thing. □ **paperer** n. **paperless** adj. [ME f. AF papir, = OF papier f. L papyrus: see PAPYRUS]

▪ The essence of paper-making is the compact interlacing of natural fibres. It originated in China about 2,000 years ago, when the fibres used were bamboo, rags, and old fishing-nets. In the 8th century the Arabs learned the process from Chinese prisoners captured in war; they used mostly flax as fibre. Paper-making spread through Europe from Moorish Spain during the 12th to 15th centuries. Rags were the main source of fibre until the 19th century, when wood fibres came into use; these are now the main constituent, though rags, straw, esparto grass, and waste paper are important ingredients for mixing with wood fibres. The fibres are pulped, washed, bleached, and dried before passing to a paper mill, where the pulp is mixed with water and beaten to a uniform consistency, often with additives such as size, dye, or filler. A thin stream is made to flow on to a wire mesh belt which allows water to drain away, leaving a felt of sufficient strength to pass between heated rollers which compress and dry the paper. Special papers are coated with, for example, white clay to give an especially smooth surface for fine printing of illustrations. Paper is now generally made from fast-growing trees which are replaced as they are felled, and used paper is increasingly recycled.

paperback /ˈpeɪpəˌbæk/ adj. & n. ● adj. (of a book) bound in stiff paper not boards. ● n. a paperback book.

paperweight /ˈpeɪpəˌweɪt/ n. a small heavy object for keeping loose papers in place.

paperwork /ˈpeɪpəˌwɜːk/ n. **1** routine clerical or administrative work. **2** documents, esp. for a particular purpose.

papery /ˈpeɪpərɪ/ adj. like paper in thinness or texture.

Paphlagonia /ˌpæfləˈɡəʊnɪə/ an ancient region of northern Asia Minor, on the Black Sea coast between Bithynia and Pontus, to the north of Galatia. It was incorporated into Roman Bithynia and Galatia between 65 and 6 BC. □ **Paphlagonian** adj. & n.

papier mâché /ˌpæpjeɪ ˈmæʃeɪ/ n. paper pulp used for moulding into boxes, trays, etc. [F, = chewed paper]

papilionaceous /pəˌpɪljəˈneɪʃəs/ adj. Bot. (of a plant or flower) having a corolla resembling a butterfly, as in many leguminous plants. [mod.L papilionaceus f. L papilio -onis butterfly]

papilla /pəˈpɪlə/ n. (pl. **papillae** /-liː/) **1** a small nipple-like protuberance of a part or organ of the body. **2** Bot. a small fleshy projection on a plant. □ **papillary** adj. **papillate** /ˈpæpɪˌleɪt/ adj. **papillose** /-ˌləʊs, -ˌləʊz/ adj. [L, = nipple, dimin. of papula: see PAPULE]

papilloma /ˌpæpɪˈləʊmə/ n. (pl. **papillomas** or **papillomata** /-mətə/) Med. a wartlike usu. benign tumour.

papillon /ˈpæpɪˌlɒn/ n. a breed of toy dog with ears suggesting the shape of a butterfly's wings. [F, = butterfly, f. L papilio -onis]

Papineau /ˈpæpɪˌnəʊ/, Louis Joseph (1786–1871), French-Canadian politician. The leader of the French-Canadian party in Lower Canada (later Quebec province), he served as speaker of the House of Assembly for Lower Canada from 1815 to 1837. During this time, he campaigned against British proposals for the union of Lower and Upper Canada (later Ontario), and pressed for greater French-Canadian autonomy in government. He was forced to flee the country after leading an abortive French rebellion against British rule in Lower Canada in 1837.

papist /ˈpeɪpɪst/ n. & adj. often derog. ● n. **1** a Roman Catholic. **2** hist. an advocate of papal supremacy. ● adj. of or relating to Roman Catholics. □ **papistry** n. **papistic** /pəˈpɪstɪk/ adj. **papistical** adj. [F papiste or mod.L papista f. eccl.L papa POPE[1]]

papoose /pəˈpuːs/ n. a young North American Indian child. [Algonquian]

pappardelle /ˌpæpəˈdeli/ n.pl. pasta in the form of broad flat ribbons. [It. f. pappare eat greedily]

Pappus /ˈpæpəs/ (known as Pappus of Alexandria) (fl. c.300–350 AD), Greek mathematician. Little is known of his life, but his Collection of six

books (another two are missing) is the principal source of knowledge of the mathematics of his predecessors. They are particularly strong on geometry, to which Pappus himself made major contributions. Fragments of other works survive.

pappus /ˈpæpəs/ n. (pl. **pappi** /-paɪ/) a group of hairs on the fruit of thistles, dandelions, etc. □ **pappose** /-pəʊs, -pəʊz/ adj. [L f. Gk pappos]

paprika /ˈpæprɪkə, pəˈpriːkə/ n. **1** a red pepper. **2** a condiment made from it. [Hungarian]

Pap test /pæp/ n. (also **pap**) Med. a test carried out on a cervical smear to detect cancer of the cervix or womb. [abbr. of George N. Papanicolaou, Greek-born Amer. scientist (1883–1962)]

Papua /ˈpæpjʊə, ˈpæpwə/ the south-eastern part of the island of New Guinea, now part of the independent state of Papua New Guinea. Papua was visited in 1526–7 by a Portuguese navigator, who named it from a Malay word meaning 'woolly-haired'. (See also PAPUA NEW GUINEA.)

Papuan /ˈpæpʊən, ˈpæpwən/ n. & adj. ● n. **1** a native or inhabitant of Papua. **2** a language group consisting of around 750 languages spoken by some 3 million people in New Guinea and neighbouring islands. ● adj. of or relating to Papua or its people or to Papuan.

Papua New Guinea a country in the western Pacific comprising the eastern half of the island of New Guinea together with some neighbouring islands; pop. (1990) 3,529,540; languages, English (official), pidgin, and several hundred native Malayo-Polynesian and Papuan languages; capital, Port Moresby. Papua New Guinea was formed from the administrative union, in 1949, of Papua, an Australian Territory since 1906, and the Trust Territory of New Guinea (NE New Guinea), formerly under German control and an Australian trusteeship since 1921. It was the scene of fierce jungle fighting 1942–5, when parts of the territory were occupied by Japan. In 1975 Papua New Guinea became an independent state within the Commonwealth. □ **Papua New Guinean** adj. & n.

papule /ˈpæpjuːl/ n. (also **papula** /-jʊlə/; pl. **papulae** /-ˌliː/ or **papulas**) **1** a pimple. **2** a small fleshy projection on a plant. □ **papular** /-jʊlə(r)/ adj. **papulose** /-ləʊs, -ləʊz/ adj. **papulous** adj. [L papula]

papyrology /ˌpæpɪˈrɒlədʒɪ/ n. the study of ancient papyri. □ **papyrologist** n. **papyrological** /-rəˈlɒdʒɪk(ə)l/ adj.

papyrus /pəˈpaɪrəs/ n. (pl. **papyri** /-raɪ/) **1** a tall aquatic sedge, *Cyperus papyrus*, with dark green stems topped with fluffy inflorescences. **2 a** a writing material prepared in ancient Egypt from the pithy stem of this. (See note below.) **b** a document written on this. [ME f. L papyrus f. Gk papuros]

▪ In ancient Egypt production and marketing of papyrus as a writing material was a royal monopoly, and the secret of its preparation was jealously guarded. Sheets of papyrus were a major export of Egypt until the end of the Greco-Roman period. Papyrus has also been used in the region to make articles such as rope, sandals, and boats.

par[1] /pɑː(r)/ n. **1** the average or normal amount, degree, condition, etc. (be up to par). **2** equality; an equal status or footing (on a par with). **3** Golf the number of strokes a first-class player should normally require for a hole or course. **4** Stock Exch. the face value of stocks and shares etc. (at par). **5** (in full **par of exchange**) the recognized value of one country's currency in terms of another's. □ **above par 1** better than usual. **2** Stock Exch. at a premium. **at par** Stock Exch. at face value. **below par 1** less good than usual, esp. in health. **2** Stock Exch. at a discount. **par for the course** colloq. what is normal or expected in any given circumstances. [L (adj. & n.) = equal, equality]

par[2] /pɑː(r)/ n. Brit. esp. Journalism colloq. paragraph. [abbr.]

par. /pɑː(r)/ abbr. (also **para.** /ˈpærə/) paragraph.

par- /pə(r), pæ(r), pɑː(r)/ prefix var. of PARA-[1] before a vowel or h (paraldehyde; parody; parhelion).

Pará /pəˈrɑː/ a state in northern Brazil, on the Atlantic coast at the delta of the Amazon; capital, Belém. It is a region of dense rainforest.

para /ˈpærə/ n. colloq. **1** a paratrooper. **2** a paragraph. [abbr.]

para-[1] /ˈpærə/ prefix (also **par-**) **1** beside (paramilitary). **2** beyond (paranormal). **3** Chem. **a** modification of. **b** relating to diametrically opposite carbon atoms in a benzene ring (paradichlorobenzene). [from or after Gk para- f. para beside, past, beyond]

para-[2] /ˈpærə/ comb. form protect, ward off (parachute; parasol). [F f. It. f. L parare defend]

parabiosis /ˌpærəbaɪˈəʊsɪs/ n. Biol. the anatomical joining of two organisms, either naturally or surgically. □ **parabiotic** /-ˈɒtɪk/ adj. [mod.L, formed as PARA-[1] + Gk biōsis mode of life f. bios life]

parable /ˈpærəb(ə)l/ n. **1** a narrative of imagined events used to illustrate a moral or spiritual lesson. **2** an allegory. [ME f. OF *parabole* f. LL sense 'allegory, discourse' of L *parabola* comparison]

parabola /pəˈræbələ/ n. (pl. **parabolas** or **parabolae** /-ˌliː/) a symmetrical open plane curve formed by the intersection of a cone with a plane parallel to its side, resembling the path of a projectile under the action of gravity. [mod.L f. Gk *parabolē* placing side by side, comparison (as PARA-[1], *bolē* a throw f. *ballō*)]

parabolic /ˌpærəˈbɒlɪk/ adj. **1** of or expressed in a parable. **2** of or like a parabola. □ **parabolically** adv. [LL *parabolicus* f. Gk *parabolikos* (as PARABOLA)]

parabolical /ˌpærəˈbɒlɪk(ə)l/ adj. = PARABOLIC 1.

paraboloid /pəˈræbəˌlɔɪd/ n. **1** (in full **paraboloid of revolution**) a solid generated by the rotation of a parabola about its axis of symmetry. **2** a solid having two or more non-parallel parabolic cross-sections. □ **paraboloidal** /-ˌræbəˈlɔɪd(ə)l/ adj.

Paracel Islands /ˌpærəˈsel/ (also **Paracels**) a group of about 130 small barren coral islands and reefs in the South China Sea to the south-east of the Chinese island of Hainan. The islands, which lie close to deposits of oil, are claimed by both China and Vietnam.

Paracelsus /ˌpærəˈselsəs/ (born Theophrastus Phillipus Aureolus Bombastus von Hohenheim) (c.1493–1541), Swiss physician. He developed a new approach to medicine and philosophy condemning medical teaching that was not based on observation and experience. He introduced chemical remedies to replace traditional herbal ones, and gave alchemy a wider perspective. Paracelsus saw illness as having a specific external cause rather than being caused by an imbalance of the humours in the body, although this progressive view was offset by his overall occultist perspective. His study of a disease of miners was one of the first accounts of an occupational disease.

paracetamol /ˌpærəˈsetəˌmɒl, -ˈsiːtəˌmɒl/ n. a drug used to relieve pain and reduce fever. Also called (N. Amer.) *acetaminophen*. [*para-acetylaminophenol*]

parachronism /pəˈrækrəˌnɪz(ə)m/ n. an error in chronology, esp. by assigning too late a date. [PARA-[1] + Gk *khronos* time, perh. after *anachronism*]

parachute /ˈpærəˌʃuːt/ n. & v. ● n. **1** a rectangular or umbrella-shaped apparatus allowing a person or heavy object attached to it to descend slowly from a height, esp. from an aircraft, or to retard motion in other ways. (*See note below.*) **2** (attrib.) dropped or to be dropped by parachute (*parachute troops*; *parachute flare*). ● v.tr. & intr. convey or descend by parachute. [F (as PARA-[2], CHUTE[1])]

▪ The principle of the parachute had been noted by Leonardo da Vinci. The first man to demonstrate one in action was Frenchman Louis-Sebastien Lenormand (1757–1839), jumping from a high tower (1783). Another pioneer was balloonist Jean Pierre François Blanchard, and in the 19th century jumps from balloons were often made as stunts. The first jump from an aircraft was made in 1912; by the end of the First World War parachutes were becoming standard equipment in military aeroplanes. The Second World War saw parachutes used to drop troops as well as supplies and clandestine agents, while since the war parachuting (including skydiving) has become a popular sport.

parachutist /ˈpærəˌʃuːtɪst/ n. **1** a person who uses a parachute. **2** (in pl.) parachute troops.

Paraclete /ˈpærəˌkliːt/ n. the Holy Spirit as advocate or counsellor (John 14:16, 26, etc.). [ME f. OF *paraclet* f. LL *paracletus* f. Gk *paraklētos* called in aid (as PARA-[1], *klētos* f. *kaleō* call)]

parade /pəˈreɪd/ n. & v. ● n. **1 a** a formal or ceremonial muster of troops for inspection. **b** = parade-ground. **2** a public procession. **3** ostentatious display (*made a parade of their wealth*). **4** a public square, promenade, or row of shops. ● v. **1** intr. assemble for parade. **2 a** tr. march through (streets etc.) in procession. **b** intr. march ceremonially. **3** tr. display ostentatiously. □ **on parade 1** taking part in a parade. **2** on display. **parade-ground** a place for the muster of troops. □ **parader** n. [F, = show, f. Sp. *parada* and It. *parata* ult. f. L *parare* prepare, furnish]

paradiddle /ˈpærəˌdɪd(ə)l/ n. a pattern of strokes in drumming, consisting of four beats played in the sequence left right left left, or right left right right. [imit.]

paradigm /ˈpærəˌdaɪm/ n. **1** a representative example or pattern, esp. one underlying a theory or viewpoint. **2** a set of the inflections of a noun, verb, etc. □ **paradigm shift** a fundamental change, esp. in science or philosophy. □ **paradigmatic** /ˌpærədɪɡˈmætɪk/ adj.

paradigmatically adv. [LL *paradigma* f. Gk *paradeigma* f. *paradeiknumi* show side by side (as PARA-[1], *deiknumi* show)]

paradise /ˈpærəˌdaɪs/ n. **1** (in some religions) heaven as the ultimate abode of the just. **2** a place or state of complete happiness. **3** (in full **earthly paradise**) the abode of Adam and Eve in the biblical account of the Creation; the Garden of Eden. □ **paradisal** adj. **paradisaical** /ˌpærədɪˈseɪk(ə)l/ adj. **paradisiacal** /-dɪˈsaɪək(ə)l/ adj. **paradisical** /-ˈdɪsɪk(ə)l/ adj. [ME f. OF *paradis* f. LL *paradisus* f. Gk *paradeisos* f. Avestan *pairidaēza* park]

parador /ˈpærəˌdɔː(r)/ n. (pl. **paradors** or **paradores** /ˌpærəˈdɔːrez/) a hotel owned and administered by the Spanish government. [Sp., = inn]

parados /ˈpærəˌdɒs, -ˌdəʊ/ n. an elevation of earth behind a fortified place as a protection against attack from the rear, esp. a mound along the back of a trench. [F (as PARA-[2], *dos* back f. L *dorsum*)]

paradox /ˈpærəˌdɒks/ n. **1 a** a seemingly absurd or contradictory statement, even if actually well-founded. **b** a self-contradictory or essentially absurd statement. **2** a person or thing conflicting with a preconceived notion of what is reasonable or possible. **3** a paradoxical quality or character. [orig. = a statement contrary to accepted opinion, f. LL *paradoxum* f. Gk *paradoxon* neut. adj. (as PARA-[1], *doxa* opinion)]

paradoxical /ˌpærəˈdɒksɪk(ə)l/ adj. **1** of or like or involving paradox. **2** fond of paradox. □ **paradoxically** adv.

paraffin /ˈpærəfɪn/ n. **1** an inflammable waxy or oily substance obtained by distillation from petroleum or shale, used in liquid form (also **paraffin oil**) esp. as a fuel. **2** Chem. = ALKANE. □ **paraffin wax** paraffin in its solid form. [G f. L *parum* little + *affinis* related, from its low reactivity]

paraglider /ˈpærəˌɡlaɪdə(r)/ n. a wide parachute-like canopy attached to the body by a harness, allowing a person to be hauled to or jump from a height and glide to the ground. □ **paraglide** v.intr. **paragliding** n. [PARACHUTE + GLIDER]

paragoge /ˌpærəˈɡəʊdʒɪ/ n. the addition of a letter or syllable to a word in some contexts or as a language develops (e.g. *t* in *peasant*). □ **paragogic** /-ˈɡɒdʒɪk/ adj. [LL f. Gk *paragōgē* derivation (as PARA-[1], *agōgē* f. *agō* lead)]

paragon /ˈpærəɡən/ n. **1 a** a model of excellence. **b** a supremely excellent person or thing. **2** (foll. by *of*) a model (of virtue etc.). **3** a perfect diamond of 100 carats or more. [obs. F f. It. *paragone* touchstone, f. med.Gk *parakonē* whetstone]

paragraph /ˈpærəˌɡrɑːf/ n. & v. ● n. **1** a distinct section of a piece of writing, beginning on a new usu. indented line. **2** a symbol (usu. ¶) used to mark a new paragraph, and also as a reference mark. **3** a short item in a newspaper, usu. of only one paragraph. ● v.tr. arrange (a piece of writing) in paragraphs. □ **paragraphic** /ˌpærəˈɡræfɪk/ adj. [F *paragraphe* or med.L *paragraphus* f. Gk *paragraphos* short stroke marking a break in sense (as PARA-[1], *graphō* write)]

Paraguay /ˈpærəˌɡwaɪ/ a landlocked country in central South America; pop. (est. 1991) 4,441,000; languages, Spanish (official), Guarani; capital, Asunción. The territory, of which more than half is part of the lowland plains of the Gran Chaco, was occupied by semi-nomadic Guarani peoples before Spanish rule was established in the 16th century. Paraguay achieved independence in 1811. It was devastated, losing more than half of its population, in war against Brazil, Argentina, and Uruguay in 1865–70, but gained land to the west in the Chaco War with Bolivia in 1932–5. The country was ruled by the military dictator Alfredo Stroessner (b.1912) from 1954 to 1989, when he was overthrown. □ **Paraguayan** /ˌpærəˈɡwaɪən/ adj. & n.

Paraíba /ˌpærəˈiːbə/ a state of eastern Brazil, on the Atlantic coast; capital, João Pessoa.

parakeet /ˈpærəˌkiːt/ n. (US also **parrakeet**) a small, usu. long-tailed parrot. [OF *paroquet*, It. *parrocchetto*, Sp. *periquito*, perh. ult. f. dimin. of *Pierre* etc. Peter: cf. PARROT]

paralanguage /ˈpærəˌlæŋɡwɪdʒ/ n. elements or factors in communication that are ancillary to language proper, e.g. intonation and gesture.

paraldehyde /pəˈrældɪˌhaɪd/ n. Chem. & Pharm. a cyclic polymer of acetaldehyde, used as a narcotic and sedative. [PARA-[1] (PAR-) + ALDEHYDE]

paralegal /ˌpærəˈliːɡ(ə)l/ adj. & n. esp. US ● adj. of or relating to auxiliary aspects of the law. ● n. a person trained in subsidiary legal matters. [PARA-[1] + LEGAL]

paralipomena /ˌpærəlɪˈpɒmɪnə/ n.pl. (also **-leipomena** /-laɪˈpɒmɪnə/) **1** things omitted from a work and added as a supplement. **2** Bibl. the books of Chronicles in the Old Testament, containing particulars omitted from Kings. [ME f. eccl.L f. Gk *paraleipomena* f. *paraleipō* omit (as PARA-¹, *leipō* leave)]

paralipsis /ˌpærəˈlɪpsɪs/ *n.* (also **-leipsis** /-ˈlaɪpsɪs/) (*pl.* **-ses** /-siːz/) *Rhet.* **1** the device of giving emphasis by professing to say little or nothing of a subject, as in *not to mention their unpaid debts of several millions*. **2** an instance of this. [LL f. Gk *paraleipsis* passing over (as PARA-¹, *leipsis* f. *leipō* leave)]

parallax /ˈpærəˌlæks/ *n.* **1** the apparent difference in the position or direction of an object when viewed from different positions, e.g. the difference between the images in the viewfinder and the lens of a camera. **2** the angular amount of this. □ **parallactic** /ˌpærəˈlæktɪk/ *adj.* [F *parallaxe* f. mod.L *parallaxis* f. Gk *parallaxis* change f. *parallassō* to alternate (as PARA-¹, *allassō* exchange f. *allos* other)]

parallel /ˈpærəˌlel/ *adj., n., & v.* ● *adj.* **1 a** (of lines or planes) side by side and having the same distance continuously between them. **b** (foll. by *to, with*) (of a line or plane) having this relation (to another). **2** (of circumstances etc.) precisely similar, analogous, or corresponding. **3 a** (of processes etc.) occurring or performed simultaneously. **b** *Computing* involving the simultaneous performance of operations. ● *n.* **1** a person or thing precisely analogous or equal to another. **2** a comparison (*drew a parallel between the two situations*). **3** (in full **parallel of latitude**) *Geog.* **a** each of the imaginary parallel circles of constant latitude on the earth's surface. **b** a corresponding line on a map (*the 49th parallel*). **4** *Printing* two parallel lines (‖) as a reference mark. ● *v.tr.* (**paralleled, paralleling**) **1** be parallel to; correspond to. **2** represent as similar; compare. **3** adduce as a parallel instance. □ **in parallel** (of electric circuits) arranged so as to join at common points at each end. **parallel bars** a pair of parallel rails on posts for gymnastics. □ **parallelism** *n.* [F *parallèle* f. L *parallelus* f. Gk *parallēlos* (as PARA-¹, *allēlos* one another)]

parallelepiped /ˌpærəˈlepɪˌped, -ˌleləˈpaɪped/ *n.* *Geom.* a solid body of which each face is a parallelogram. [Gk *parallēlepipedon* (as PARALLEL, *epipedon* plane surface)]

parallelogram /ˌpærəˈleləˌgræm/ *n.* a four-sided plane rectilinear figure with opposite sides parallel. □ **parallelogram of forces 1** a parallelogram illustrating the theorem that if two forces acting at a point are represented in magnitude and direction by two sides of a parallelogram meeting at that point, their resultant is represented by the diagonal drawn from that point. **2** this theorem. [F *parallélogramme* f. LL *parallelogrammum* f. Gk *parallēlogrammon* (as PARALLEL, *grammē* line)]

paralogism /pəˈrælədʒɪz(ə)m/ *n.* *Logic* **1** a fallacy. **2** illogical reasoning (esp. of which the reasoner is unconscious). □ **paralogist** *n.* **paralogize** *v.intr.* (also **-ise**). [F *paralogisme* f. LL *paralogismus* f. Gk *paralogismos* f. *paralogizomai* reason falsely f. *paralogos* contrary to reason (as PARA-¹, *logos* reason)]

Paralympics /ˌpærəˈlɪmpɪks/ *n.pl.* (also **Paralympic Games**) an international sports competition for paraplegic and other disabled athletes, modelled on the Olympic Games. [*paraplegic* (PARAPLEGIA) + OLYMPIC]

paralyse /ˈpærəˌlaɪz/ *v.tr.* (*US* **paralyze**) **1** affect with paralysis. **2** render powerless; cripple. **3** bring to a standstill. □ **paralysingly** *adv.* **paralysation** /ˌpærəlaɪˈzeɪʃ(ə)n/ *n.* [F *paralyser* f. *paralysie*: cf. PALSY]

paralysis /pəˈrælɪsɪs/ *n.* (*pl.* **paralyses** /-ˌsiːz/) **1** a nervous condition with impairment or loss of esp. the motor function of the nerves. **2** a state of utter powerlessness. [L f. Gk *paralusis* f. *paraluō* disable (as PARA-¹, *luō* loosen)]

paralytic /ˌpærəˈlɪtɪk/ *adj. & n.* ● *adj.* **1** affected by paralysis. **2** *sl.* very drunk. ● *n.* a person affected by paralysis. □ **paralytically** *adv.* [ME f. OF *paralytique* f. L *paralyticus* f. Gk *paralutikos* (as PARALYSIS)]

paramagnetic /ˌpærəmægˈnetɪk/ *adj.* *Physics* (of a body or substance) tending to become weakly magnetized so as to lie parallel to a magnetic field force. □ **paramagnetism** /-ˈmægnɪˌtɪz(ə)m/ *n.*

Paramaribo /ˌpærəˈmærɪˌbəʊ/ the capital of Suriname, a port on the Atlantic coast; pop. (est. 1988) 192,110.

paramatta var. of PARRAMATTA.

paramecium /ˌpærəˈmiːsɪəm/ *n.* (also *Brit.* **paramoecium**) a freshwater protozoan of the genus *Paramecium*, of a characteristic slipper-like shape, covered with cilia. [mod.L f. Gk *paramēkēs* oval (as PARA-¹, *mēkos* length)]

paramedic /ˌpærəˈmedɪk/ *n.* a paramedical worker.

paramedical /ˌpærəˈmedɪk(ə)l/ *adj.* (of services etc.) supplementing and supporting medical work.

parameter /pəˈræmɪtə(r)/ *n.* **1** *Math.* a quantity constant in the case

considered but varying in different cases. **2 a** an (esp. measurable or quantifiable) characteristic or feature. **b** (loosely) a constant element or factor, esp. serving as a limit or boundary. □ **parametrize** *v.tr.* (also **-ise**). **parametric** /ˌpærəˈmetrɪk/ *adj.* [mod.L f. Gk *para* beside + *metron* measure]

paramilitary /ˌpærəˈmɪlɪtərɪ/ *adj. & n.* ● *adj.* (of forces) organized similarly to military forces. ● *n.* (*pl.* **-ies**) a member of an unofficial paramilitary organization, esp. in Northern Ireland.

paramnesia /ˌpærəmˈniːzɪə/ *n.* *Psychol.* = DÉJÀ VU 1. [PARA-¹ + AMNESIA]

paramo /ˈpærəˌməʊ/ *n.* (*pl.* **-os**) a high treeless plateau in tropical South America. [Sp. & Port. f. L *paramus*]

paramoecium *Brit.* var. of PARAMECIUM.

Paramount /ˈpærəˌmaʊnt/ a US film production and distribution company established in 1914. A major studio of the silent era, Paramount acted as an outlet for many of the films of Cecil B. De Mille and helped to create stars such as Mary Pickford and Rudolph Valentino; notable later successes included the *Road* films of Bob Hope and Billy Wilder's *Sunset Boulevard* (1950). The company's trademark has always been a snow-capped mountain encircled by stars.

paramount /ˈpærəˌmaʊnt/ *adj.* **1** supreme; requiring first consideration; pre-eminent (*of paramount importance*). **2** in supreme authority. □ **paramountcy** *n.* **paramountly** *adv.* [AF *paramont* f. OF *par* by + *amont* above: cf. AMOUNT]

paramour /ˈpærəˌmʊə(r)/ *n.* archaic or derog. an illicit lover of a married person. [ME f. OF *par amour* by love]

Paraná /ˌpærəˈnɑː/ **1** a river of South America, which rises in SE Brazil and flows some 3,300 km (2,060 miles) southwards to the River Plate estuary in Argentina. For part of its length it forms the SE border of Paraguay. **2** a river port in eastern Argentina, on the Paraná river; pop. (1991) 276,000. It was the capital of Argentina between 1853 and 1862. **3** a state of southern Brazil, on the Atlantic coast; capital, Curitiba.

parang /ˈpæræŋ/ *n.* a large heavy Malayan knife used for clearing vegetation etc. [Malay]

paranoia /ˌpærəˈnɔɪə/ *n.* **1** a mental disorder esp. characterized by delusions of persecution and self-importance. **2** an abnormal tendency to suspect and mistrust others. □ **paranoiac** /-ˈnɔɪk/ *adj. & n.* **paranoiacally** *adv.* **paranoic** /-ˈnəʊɪk, -ˈnɔɪk/ *adj.* **paranoically** *adv.* **paranoid** /ˈpærəˌnɔɪd/ *adj. & n.* [mod.L f. Gk f. *paranoos* distracted (as PARA-¹, *noos* mind)]

paranormal /ˌpærəˈnɔːm(ə)l/ *adj.* beyond the scope of normal objective investigation or explanation. □ **paranormally** *adv.*

Paranthropus /pəˈrænθrəpəs/ *n.* a genus name sometimes applied to robust forms of the fossil hominid *Australopithecus*, named from remains of *A. robustus* found in South Africa in 1938. (See AUSTRALOPITHECUS.) [mod.L f. Gk *para* near + *anthrōpos* man]

parapet /ˈpærəpɪt/ *n.* **1** a low wall at the edge of a roof, balcony, etc., or along the sides of a bridge. **2** a defence of earth or stone to conceal and protect troops. □ **parapeted** *adj.* [F *parapet* or It. *parapetto* breast-high wall (as PARA-², *petto* breast f. L *pectus*)]

paraph /ˈpærəf/ *n.* a flourish after a signature, orig. as a precaution against forgery. [ME f. F *paraphe* f. med.L *paraphus* for *paragraphus* PARAGRAPH]

paraphernalia /ˌpærəfəˈneɪlɪə/ *n.pl.* (also treated as *sing.*) miscellaneous belongings, items of equipment, accessories, etc. [orig. = property owned by a married woman, f. med.L *paraphernalia* f. LL *parapherna* f. Gk *parapherna* personal articles which a woman could keep after marriage, as opposed to her dowry which went to her husband (as PARA-¹, *pherna* f. *phernē* dower)]

paraphrase /ˈpærəˌfreɪz/ *n. & v.* ● *n.* a free rendering or rewording of a passage. ● *v.tr.* express the meaning of (a passage) in other words. □ **paraphrastic** /ˌpærəˈfræstɪk/ *adj.* [F *paraphrase* or L *paraphrasis* f. Gk *paraphrasis* f. *paraphrazō* (as PARA-¹ *phrazō* tell)]

paraplegia /ˌpærəˈpliːdʒə/ *n.* paralysis of the legs and part or the whole of the trunk. □ **paraplegic** /-dʒɪk/ *adj. & n.* [mod.L f. Gk *paraplēgia* f. *paraplēssō* (as PARA-¹, *plēssō* strike)]

parapsychology /ˌpærəsaɪˈkɒlədʒɪ/ *n.* the study of mental phenomena outside the sphere of ordinary psychology (hypnosis, telepathy, etc.). □ **parapsychologist** *n.* **parapsychological** /-ˌsaɪkəˈlɒdʒɪk(ə)l/ *adj.*

paraquat /ˈpærəˌkwɒt/ *n.* a highly toxic quick-acting contact herbicide, rapidly deactivated by the soil. [PARA-¹ + QUATERNARY (from the position of the bond between the two parts of the molecule relative to a quaternary nitrogen atom)]

parasailing /ˈpærəˌseɪlɪŋ/ n. a sport in which participants wearing open parachutes are towed behind a motor boat. □ **parasailer** n. **parasailor** n.

parascending /ˈpærəˌsendɪŋ/ n. a sport in which participants wearing open parachutes are towed behind a vehicle or motor boat to gain height for a conventional descent, usu. towards a target. □ **parascender** n.

paraselene /ˌpærəsɪˈliːnɪ/ n. (pl. **paraselenae** /-niː/) a bright spot, esp. an image of the moon, on a lunar halo. Also called *mock moon*. [mod.L (as PARA-¹, Gk *selēnē* moon)]

parasite /ˈpærəˌsaɪt/ n. 1 an organism living in or on another and benefiting at the expense of the other. (*See note below*.) 2 a person who lives off or exploits another or others. 3 *Philol*. an inorganic sound or letter developing from an adjacent one. □ **parasitism** n. **parasiticide** /ˌpærəˈsɪtɪˌsaɪd/ n. [L *parasitus* f. Gk *parasitos* one who eats at another's table (as PARA-¹, *sitos* food)]

▪ The parasites living on the body surface of larger animals (ectoparasites) are mostly arthropods, and include mites, ticks, lice, and fleas. Others live inside the host's body (endoparasites), and many of these are worms belonging to several unrelated groups. Insects that parasitize other insects are sometimes not regarded as true parasites, because the host is always killed eventually (cf. PARASITOID).

parasitic /ˌpærəˈsɪtɪk/ adj. 1 of, relating to, or characteristic of parasites; of the nature of a parasite. 2 (foll. by *in*, *on*) living in or on an organism as a parasite. □ **parasitic jaeger** N. Amer. = Arctic skua. □ **parasitical** adj. **parasitically** adv.

parasitize /ˈpærəsɪˌtaɪz/ v.tr. (also **-ise**) infest as a parasite. □ **parasitization** /ˌpærəˌsaɪtɪˈzeɪʃ(ə)n/ n.

parasitoid /ˈpærəsɪˌtɔɪd/ n. & adj. Zool. ● n. an insect whose larvae live as parasites which eventually kill their hosts, e.g. an ichneumon wasp. ● adj. of, relating to, or denoting such an insect. [PARASITE + -OID]

parasitology /ˌpærəsaɪˈtɒlədʒɪ/ n. the branch of biology and medicine that deals with parasites. □ **parasitologist** n.

parasol /ˈpærəˌsɒl/ n. 1 a light umbrella used to give shade from the sun. 2 (in full **parasol mushroom**) a tall fungus of the genus *Lepiota*, with a broad shaggy domed cap, esp. the edible *L. procera*. [F f. It. *parasole* (as PARA-², *sole* sun f. L *sol*)]

parasuicide /ˌpærəˈsuːɪˌsaɪd, -ˈsjuːɪˌsaɪd/ n. 1 apparent attempted suicide without the actual intention of killing oneself. 2 a person who has carried out or is considered likely to carry out such action.

parasympathetic /ˌpærəˌsɪmpəˈθetɪk/ adj. Anat. (of a nerve) belonging to the parasympathetic nervous system. □ **parasympathetic nervous system** one of the two divisions of the autonomic nervous system, consisting of nerves from the brain and the lower part of the spinal cord that supply the internal organs, blood vessels, and glands, balancing the action of the sympathetic nervous system. [PARA-¹ + SYMPATHETIC, because some of these nerves run alongside sympathetic nerves]

parasynthesis /ˌpærəˈsɪnθɪsɪs/ n. Philol. a derivation from a compound, e.g. *black-eyed* from *black eye*(s) + *-ed*. □ **parasynthetic** /-sɪnˈθetɪk/ adj. [Gk *parasunthesis* (as PARA-¹, SYNTHESIS)]

parataxis /ˌpærəˈtæksɪs/ n. Gram. the placing of clauses etc. one after another, without words to indicate coordination or subordination, e.g. *Tell me, how are you?* □ **paratactic** /-ˈtæktɪk/ adj. **paratactically** adv. [Gk *parataxis* (as PARA-¹, *taxis* arrangement f. *tassō* arrange)]

parathion /ˌpærəˈθaɪən/ n. a highly toxic organic compound of phosphorus and sulphur, used as an agricultural insecticide. [PARA-¹ + THIO- + -ON]

parathyroid /ˌpærəˈθaɪrɔɪd/ n. & adj. Anat. ● n. a gland next to the thyroid, secreting a hormone that regulates calcium levels in the blood. ● adj. of or associated with this gland.

paratroop /ˈpærəˌtruːp/ n. 1 (in pl.) troops equipped to be dropped by parachute from aircraft. 2 (attrib.) of or consisting of paratroops (*paratroop regiment*).[contr. of PARACHUTE + TROOP]

paratrooper /ˈpærəˌtruːpə(r)/ n. a member of a body of paratroops.

paratyphoid /ˌpærəˈtaɪfɔɪd/ n. & adj. Med. ● n. a fever resembling typhoid but caused by various different though related bacteria. ● adj. of, relating to, or caused by this fever.

paravane /ˈpærəˌveɪn/ n. a torpedo-shaped device towed at a depth regulated by vanes or planes, esp. for cutting the moorings of mines.

par avion /paːr ˈævɪˌɒn/ adv. by airmail. [F, = by aeroplane]

parboil /ˈpaːbɔɪl/ v.tr. partly cook by boiling. [ME f. OF *parbo(u)illir* f. LL *perbullire* boil thoroughly (as PER-, *bullire* boil: confused with PART)]

parbuckle /ˈpaːˌbʌk(ə)l/ n. & v. ● n. a rope arranged like a sling, for raising or lowering casks and cylindrical objects. ● v.tr. raise or lower with this. [earlier *parbunkle*, of unkn. orig.: assoc. with BUCKLE]

Parcae /ˈpaːkaɪ, ˈpaːsiː/ Rom. Mythol. the Roman name for the Fates (see FATES, THE).

parcel /ˈpaːs(ə)l/ n. & v. ● n. 1 a goods etc. wrapped up in a single package. b a bundle of things wrapped up, usu. in paper. 2 a piece of land, esp. as part of an estate. 3 a quantity dealt with in one commercial transaction. 4 archaic part. ● v.tr. (**parcelled, parcelling**; US **parceled, parceling**) 1 (foll. by *up*) wrap as a parcel. 2 (foll. by *out*) divide into portions. 3 cover (rope) with strips of canvas. □ **parcel post** the branch of the postal service dealing with parcels. **part and parcel** see PART. [ME f. OF *parcelle* ult. f. L *particula* (as PART)]

parch /paːtʃ/ v. 1 tr. & intr. make or become hot and dry. 2 tr. roast (peas, corn, etc.) slightly. [ME *perch*, *parche*, of unkn. orig.]

parched /paːtʃt/ adj. 1 hot and dry; dried out with heat. 2 colloq. thirsty.

parchment /ˈpaːtʃmənt/ n. 1 a an animal skin, esp. that of a sheep or goat, prepared as a writing or painting surface. b a manuscript written on this. 2 (in full **vegetable parchment**) high-grade paper made to resemble parchment. [ME f. OF *parchemin*, ult. a blend of LL *pergamina* writing material from Pergamum with *Parthica pellis* Parthian skin (leather)]

parclose /ˈpaːkləʊz/ n. a screen or railing in a church, separating a side chapel. [ME f. OF *parclos -ose* past part. of *parclore* enclose]

pard /paːd/ n. archaic or poet. a leopard. [ME f. OF f. L *pardus* f. Gk *pardos*]

pardalote /ˈpaːdəˌləʊt/ n. a small brightly coloured Australian bird of the genus *Pardalotus*, with spotted plumage. Also called *diamond-bird*. [mod.L *Pardalotus* f. Gk *pardalōtos* spotted like a leopard (as PARD)]

pardner /ˈpaːdnə(r)/ n. US colloq. a partner or comrade. [corrupt.]

pardon /ˈpaːd(ə)n/ n., v., & int. ● n. 1 the act of excusing or forgiving an offence, error, etc. 2 (in full **free pardon**) a remission of the legal consequences of a crime or conviction. 3 RC Ch. an indulgence. ● v.tr. 1 release from the consequences of an offence, error, etc. 2 forgive or excuse a person for (an offence etc.). 3 make (esp. courteous) allowances for; excuse. ● int. (also **pardon me** or **I beg your pardon**) 1 a formula of apology or disagreement. 2 a request to the speaker to repeat something said. □ **pardonable** adj. **pardonably** adv. [ME f. OF *pardun*, *pardoner* f. med.L *perdonare* concede, remit (as PER-, *donare* give)]

pardoner /ˈpaːdənə(r)/ n. hist. a person licensed to sell papal pardons or indulgences. [ME f. AF (as PARDON)]

pare /peə(r)/ v.tr. 1 a trim or shave (esp. fruit and vegetables) by cutting away the surface or edge. b (often foll. by *off*, *away*) cut off (the surface or edge). 2 (often foll. by *away*, *down*) diminish little by little. □ **parer** n. [ME f. OF *parer* adorn, peel (fruit), f. L *parare* prepare]

paregoric /ˌpærɪˈgɒrɪk/ n. (in full **paregoric elixir**) hist. a camphorated tincture of opium used to reduce pain. [LL *paregoricus* f. Gk *parēgorikos* soothing (as PARA-¹, *-agoros* speaking f. *agora* assembly)]

pareira /pəˈreərə/ n. a drug from the root of a Brazilian shrub, *Chondrodendron tomentosum*, used as a muscle relaxant in surgery etc. [Port. *parreira* vine trained against a wall]

parenchyma /pəˈreŋkɪmə/ n. 1 Anat. the functional part of an organ as distinguished from the connective and supporting tissue. 2 Bot. the cellular material, usu. soft and succulent, found esp. in the softer parts of leaves, pulp of fruits, bark and pith of stems, etc. □ **parenchymal** adj. **parenchymatous** /ˌpærəŋˈkɪmətəs/ adj. [Gk *paregkhuma* something poured in besides (as PARA-¹, *egkhuma* infusion f. *egkheō* pour in)]

parent /ˈpeərənt/ n. & v. ● n. 1 a person who has begotten or borne offspring; a father or mother. 2 a person who holds the position or exercises the functions of such a parent. 3 archaic a forefather. 4 an animal or plant from which others are derived. 5 a source or origin. 6 an initiating organization or enterprise. ● v.tr. (also absol.) be a parent of. □ **parent company** a company of which other companies are subsidiaries. **parent–teacher association** a local organization of parents and teachers for promoting closer relations and improving educational facilities at a school. □ **parenthood** n. **parental** /pəˈrent(ə)l/ adj. **parentally** adv. [ME f. OF f. L *parens parentis* f. *parere* bring forth]

parentage /ˈpeərəntɪdʒ/ n. lineage; descent from or through parents (*their parentage is unknown*). [ME f. OF (as PARENT)]

parenteral /pəˈrentərəl/ adj. Med. administered or occurring elsewhere in the body than in the alimentary canal. □ **parenterally** adv. [PARA-¹ + Gk enteron intestine]

parenthesis /pə'renθəsɪs/ n. (pl. **parentheses** /-ˌsiːz/) **1 a** a word, clause, or sentence inserted as an explanation or afterthought into a passage which is grammatically complete without it, and usu. marked off by brackets or dashes or commas. **b** (in pl.) a pair of round brackets () used for this. **2** an interlude or interval. □ **in parenthesis** as a parenthesis or afterthought. [LL f. Gk parenthesis f. parentithēmi put in beside]

parenthesize /pə'renθəˌsaɪz/ v.tr. (also **-ise**) **1** (also absol.) insert as a parenthesis. **2** put into brackets or similar punctuation.

parenthetic /ˌpærən'θetɪk/ adj. **1** of or by way of a parenthesis. **2** interposed. □ **parenthetical** adj. **parenthetically** adv. [PARENTHESIS after synthesis, synthetic, etc.]

parenting /'peərəntɪŋ/ n. the occupation or concerns of parents.

parergon /pə'rɜːgən/ n. (pl. **parerga** /-gə/) **1** work subsidiary to one's main employment. **2** an ornamental accessory. [L f. Gk parergon (as PARA-¹, ergon work)]

paresis /pə'riːsɪs, 'pærɪsɪs/ n. (pl. **pareses** /-siːz/) Med. partial paralysis. □ **paretic** /pə'retɪk/ adj. [mod.L f. Gk f. pariēmi let go (as PARA-¹, hiēmi let go)]

par excellence /paːr ˌeksə'lɒns/ adv. as having special excellence; being the supreme example of its kind (the short story par excellence). [F, = by excellence]

parfait /'paːfeɪ/ n. **1** a rich iced pudding of whipped cream, eggs, etc. **2** layers of ice-cream, meringue, etc., served in a tall glass. [F parfait PERFECT adj.]

pargana /pə'gʌnə/ n. (also **pergunnah, pergana**) (in India) a group of villages or a subdivision of a district. [Urdu pargana district]

parget /'paːdʒɪt/ v. & n. ● v.tr. (**pargeted, pargeting**) **1** plaster (a wall etc.) esp. with an ornamental pattern. **2** roughcast. ● n. **1** plaster applied in this way; ornamental plasterwork. **2** roughcast. [ME f. OF pargeter, parjeter f. par all over + jeter throw]

parhelion /paː'hiːlɪən/ n. (pl. **parhelia** /-lɪə/) a bright spot on the solar halo. Also called mock sun, sun-dog. □ **parheliacal** adj. **parheliacal** /ˌpaːhɪ'laɪək(ə)l/ adj. [L parelion f. Gk (as PARA-¹, hēlios sun)]

pariah /pə'raɪə, 'pærɪə/ n. **1** a social outcast. **2** hist. a member of a low caste or of no caste in southern India. □ **pariah-dog** = PYE-DOG. [Tamil paṛaiyar pl. of paṛaiyan hereditary drummer f. paṛai drum]

Parian /'peərɪən/ adj. & n. ● adj. **1** of or relating to the Greek island of Paros or the fine-textured white marble quarried there, much used by sculptors in antiquity. **2** denoting or made of a fine white unglazed porcelain resembling Parian marble. ● n. **1** a native or inhabitant of Paros. **2** Parian marble. **3** Parian porcelain.

parietal /pə'raɪət(ə)l/ adj. **1** Anat. of the wall of the body or any of its cavities. **2** Bot. of the wall of a hollow structure etc. **3** US relating to residence within a college. □ **parietal bone** Anat. either of a pair of bones forming the central part of the sides and top of the skull. **parietal lobe** Anat. either of the paired lobes of the brain at the top of the head, including areas concerned with the reception and correlation of sensory information. [F pariétal or LL parietalis f. L paries -etis wall]

pari-mutuel /ˌpærɪ'mjuːtjʊəl, -'mjuːtjʊəl/ n. **1** a form of betting in which those backing the first three places divide the losers' stakes (less the operator's commission). **2** a totalizator. [F, = mutual stake]

paring /'peərɪŋ/ n. a strip or piece cut off.

pari passu /ˌpaːrɪ 'pæsuː, ˌpærɪ/ adv. **1** with equal speed. **2** simultaneously and equally. [L]

Paris¹ /'pærɪs, French pari/ the capital of France, on the River Seine; pop. (1990) 2,175,200. An early settlement on the small island in the Seine, known now as the Île de la Cité, was inhabited by a Gallic people called the Parisii. It was taken by the Romans, who called it Lutetia, in 52 BC. In the 5th century AD it fell to the Frankish king Clovis, who made it his seat of power. It declined under the succeeding Merovingians, but was finally established as the capital in 987 under Hugh Capet. The city was extensively developed during the reign of Philippe-Auguste (1180–1223) and organized into three parts: the Île de la Cité, the Right Bank, and the Left Bank. During the reign of Francis I (1515–47) the city expanded again, its architecture showing the influence of the Italian Renaissance. The city's neoclassical architecture characterizes the modernization of the Napoleonic era. This continued under Napoleon III, when the bridges and boulevards of the modern city were built. Paris was occupied by the German army

in 1940, being liberated by the Allies when the Germans withdrew (Aug. 1944).

Paris² /'pærɪs/ Gk Mythol. a Trojan prince, the son of Priam and Hecuba. Appointed by the gods to decide who among the three goddesses Hera, Athene, and Aphrodite should win a prize for beauty, he awarded it to Aphrodite, who promised him the fairest woman in the world — Helen, wife of Menelaus king of Sparta. He abducted Helen, bringing about the Trojan War, in which he killed Achilles but was later himself killed. The Judgement of Paris has been a favourite theme in art from the mid-7th century BC.

Paris³, Matthew, see MATTHEW PARIS.

Paris Commune see COMMUNE, THE.

Paris green n. a vivid green poisonous compound containing copper and arsenic, used as a pigment and insecticide. [PARIS¹]

parish /'pærɪʃ/ n. **1** an area having its own church and clergy. **2** (in full **civil parish**) a district constituted for purposes of local government. (See note below.) **3** the inhabitants of a parish. **4** US a county in Louisiana. □ **parish clerk** an official performing various duties concerned with the church. **parish council** Brit. the administrative body in a civil parish. **parish pump** (often attrib.) a symbol of a parochial or restricted outlook. **parish register** a book recording christenings, marriages, and burials, at a parish church. [ME paroche, parosse f. OF paroche, paroisse f. eccl.L parochia, paroechia f. Gk paroikia sojourning f. paroikos (as PARA-¹, -oikos -dwelling f. oikeō dwell)]

   ▪ From the 17th century parishes and churchwardens in England were entrusted with local administration, chiefly relating to Poor Law and highways. Urban parishes were abolished as a unit of local government in 1933, but rural parishes have retained limited powers since the local government reorganization of 1974.

parishioner /pə'rɪʃənə(r)/ n. an inhabitant of a parish. [obs. parishen f. ME f. OF parossien, formed as PARISH]

Parisian /pə'rɪzɪən/ adj. & n. ● adj. of or relating to Paris in France. ● n. **1** a native or inhabitant of Paris. **2** the kind of French spoken in Paris. [F parisien]

parison /'pærɪs(ə)n/ n. a rounded mass of glass formed by rolling immediately after removal from the furnace. [F paraison f. parer prepare f. L parare]

parity¹ /'pærɪtɪ/ n. **1** equality or equivalence, esp. as regards status or pay. **2** parallelism or analogy (parity of reasoning). **3** equivalence of one currency with another; being at par. **4** (of a number) the fact of being even or odd. **5** Physics (of a quantity) the fact of changing its sign or remaining unaltered under a given transformation of coordinates etc. [F parité or LL paritas (as PAR¹)]

parity² /'pærɪtɪ/ n. Med. **1** the fact or condition of having borne children. **2** the number of children previously borne. [formed as -PAROUS + -ITY]

Park /paːk/, Mungo (1771–1806), Scottish explorer. He undertook a series of explorations in West Africa (1795–7), among them being the navigation of the Niger. His experiences were recorded in his Travels in the Interior of Africa (1799). He drowned on a second expedition to the Niger (1805–6).

park /paːk/ n. & v. ● n. **1** a large public garden in a town, for recreation. **2** a large enclosed piece of ground, usu. with woodland and pasture, attached to a country house etc. **3 a** a large area of land kept in its natural state for public recreational use. **b** a large enclosed area of land used to accommodate wild animals in captivity (wildlife park). **4** an area for motor vehicles etc. to be left in (car park). **5** the gear position or function in automatic transmission in which the gears are locked, preventing the vehicle's movement. **6** an area devoted to a specified purpose (industrial park). **7 a** US a sports ground. **b** colloq. (usu. prec. by the) a football pitch. ● v.tr. **1** (also absol.) leave (a vehicle) usu. temporarily, in a car park, by the side of the road, etc. **2** colloq. deposit and leave, usu. temporarily. □ **parking-light** a small light on a vehicle, for use when the vehicle is parked at night. **parking-lot** US an outdoor area for parking vehicles, a car park. **parking-meter** a coin-operated meter which receives fees for vehicles parked in the street and indicates the time available. **parking-ticket** a notice, usu. attached to a vehicle, of a penalty imposed for parking illegally. **park oneself** colloq. sit down. [ME f. OF parc f. med.L parricus f. Gmc orig., rel. to pearruc: see PADDOCK]

parka /'paːkə/ n. **1** a skin jacket with a hood, worn by Eskimos. **2** a warm weatherproof coat with a (often fur-trimmed) hood. [Aleutian]

Park Chung Hee /ˌpaːk tʃʊŋ 'hiː/ (1917–79), South Korean statesman, President 1963–79. In 1961 he staged a military coup that ousted the country's democratic government. Two years later he was elected President, assuming dictatorial powers in 1971. Under Park's presidency, South Korea emerged as a leading industrial nation, with one of the world's highest rates of economic growth. He was assassinated by Kim Jae Kyu (1926–80), the head of the Korean Central Intelligence Agency.

Parker[1] /ˈpɑːkə(r)/, Charles Christopher ('Charlie'; known as 'Bird' or 'Yardbird') (1920–55), American saxophonist. From 1944 he was based in New York, where he played with Thelonious Monk and Dizzy Gillespie, and became one of the key figures of the bebop movement. He is noted especially for his recordings with Miles Davis in 1945.

Parker[2] /ˈpɑːkə(r)/, Dorothy (Rothschild) (1893–1967), American humorist, literary critic, short-story writer, and poet. She was a leading member of the Algonquin Round Table, a circle of writers and humorists that met in the 1920s and included James Thurber. From 1927 Parker wrote book reviews and short stories for the *New Yorker* magazine, becoming one of its legendary wits. As a poet, she made her name with the best-selling verse collection *Enough Rope* (1927).

parkin /ˈpɑːkɪn/ n. Brit. a cake or biscuit made with oatmeal, ginger, and treacle or molasses. [perh. f. the name *Parkin*, dimin. of Peter]

Parkinsonism /ˈpɑːkɪnsəˌnɪz(ə)m/ n. = PARKINSON'S DISEASE.

Parkinson's disease /ˈpɑːkɪns(ə)nz/ n. a progressive disease of the nervous system with tremor, muscular rigidity, and slow, imprecise movement, chiefly affecting middle-aged and elderly people. It is associated with degeneration of the basal ganglia of the brain and a deficiency of the neurotransmitter dopamine. The condition can be induced by known factors such as drugs or wider nervous-system disease, but often the cause is unknown. Also called *Parkinsonism*. (See also DOPA.) [James *Parkinson*, Engl. surgeon (1755–1824)]

Parkinson's law /ˈpɑːkɪns(ə)nz/ n. the notion that work expands to fill the time available for its completion. [Cyril Northcote *Parkinson*, Engl. writer (1909–93)]

parkland /ˈpɑːklænd/ n. open grassland with clumps of trees etc.

parkway /ˈpɑːkweɪ/ n. **1** US an open landscaped highway. **2** Brit. a railway station with extensive parking facilities (esp. attrib. or in names: *Didcot Parkway*).

parky /ˈpɑːkɪ/ adj. (**parkier, parkiest**) Brit. colloq. chilly. [19th c.: orig. unkn.]

Parl. abbr. Brit. **1** Parliament. **2** Parliamentary.

parlance /ˈpɑːləns/ n. a particular way of speaking, esp. as regards choice of words, idiom, etc. [OF f. *parler* speak, ult. f. L *parabola* (see PARABLE): in LL = 'speech']

parlay /ˈpɑːleɪ/ v. & n. US ● v.tr. **1** use (money won on a bet) as a further stake. **2** increase in value by or as if by parlaying. ● n. **1** an act of parlaying. **2** a bet made by parlaying. [F *paroli* f. It. f. *paro* like f. L *par* equal]

parley /ˈpɑːlɪ/ n. & v. ● n. (pl. **-eys**) a conference for debating points in a dispute, esp. a discussion of terms for an armistice etc. ● v.intr. (**-leys, -leyed**) (often foll. by *with*) hold a parley. [perh. f. OF *parlee*, fem. past part. of *parler* speak: see PARLANCE]

parliament /ˈpɑːləmənt/ n. **1** (**Parliament**) (in the UK) **a** the highest legislature, consisting of the Sovereign, the House of Lords, and the House of Commons. (*See note below.*) **b** the members of this legislature for a particular period, esp. between one dissolution and the next. **2** a similar legislature in other nations and states. [ME f. OF *parlement* speaking (as PARLANCE)]

▪ The British Parliament emerged in medieval times as the assembly of the king and his Lords, meeting irregularly at the king's bidding to discuss judicial and other matters of general importance, especially finances. From the 13th century knights and burgesses representing the shires and boroughs were occasionally summoned to these assemblies, and under Edward III these 'Commons' began to meet separately from the 'Lords', although they did not gain their own meeting chamber until the 16th century. Free speech, regular meetings, and control over taxation were not established as rights until after the Civil War and the Bill of Rights following the 1688 Revolution. After the passing of the first Reform Act (1832) the traditional influence of the landed aristocracy began to decline, and the Parliament Act of 1911 reduced the power of the House of Lords, allowing them to delay but not veto legislation; this Act also introduced payment for MPs. A further Act in 1949 reduced the maximum period of delay and ended the right completely for financial legislation, leaving the House of Commons unequivocally the more powerful and important body. The building in which

Parliament meets was designed by Sir Charles Barry and erected after the destruction by fire of the original Palace of Westminster, its meeting-place, in 1834.

parliamentarian /ˌpɑːləmenˈteərɪən/ n. & adj. ● n. **1** a member of a parliament, esp. one well-versed in its procedures. **2** (**Parliamentarian**) an adherent of the Parliamentary party in the English Civil War (1642–9). Also called *Roundhead*. ● adj. = PARLIAMENTARY.

parliamentary /ˌpɑːləˈmentərɪ/ adj. **1** of or relating to a parliament. **2** enacted or established by a parliament. **3** (of language) admissible in a parliament; polite. □ **parliamentary private secretary** a member of parliament assisting a government minister.

Parliamentary Commissioner for Administration n. the official name of the ombudsman in the UK.

parlour /ˈpɑːlə(r)/ n. (US **parlor**) **1** a sitting-room in a private house. **2** a room in a hotel, convent, etc., for the private use of residents. **3** esp. US a shop providing specified goods or services (*beauty parlour*; *ice-cream parlour*). **4** a room or building equipped for milking cows. **5** (attrib.) derog. denoting support for political views by those who do not try to practise them (*parlour socialist*). □ **parlour game** an indoor game, esp. a word game. **parlour-maid** hist. a maid who waits at table. [ME f. AF *parlur*, OF *parleor*, *parleur*: see PARLANCE]

parlous /ˈpɑːləs/ adj. & adv. archaic or joc. ● adj. **1** dangerous or difficult. **2** hard to deal with. ● adv. extremely. □ **parlously** adv. **parlousness** n. [ME, = PERILOUS]

Parma /ˈpɑːmə/ **1** a province of northern Italy, south of the River Po in Emilia-Romagna. With neighbouring Piacenza, it was detached from the Papal States and made a duchy by Pope Paul III (Alessandro Farnese) in 1545. It was ruled by the Farnese family until 1731, when it passed to the Spanish. It became a part of united Italy in 1861. **2** its capital; pop. (1990) 193,990. Founded by the Romans in 183 BC, it became a bishopric in the 9th century AD and capital of the duchy of Parma and Piacenza in about 1547.

Parma ham n. a type of ham which is eaten uncooked.

Parma violet n. a variety of sweet violet with heavy scent and lavender-coloured flowers.

Parmenides /pɑːˈmenɪˌdiːz/ (fl. 5th century BC), Greek philosopher. Born in Elea in SW Italy, he founded the Eleatic school of philosophers and was noted for the philosophical work *On Nature*, written in hexameter verse. In this, he maintained that the apparent motion and changing forms of the universe are in fact manifestations of an unchanging and indivisible reality.

Parmesan /ˌpɑːmɪˈzæn/ n. a kind of hard dry cheese made orig. at Parma and used esp. in grated form. [F f. It. *parmegiano* of Parma]

Parmigianino /ˌpɑːmɪdʒaˈniːnəʊ/ (also **Parmigiano** /-ˈdʒaːnəʊ/) (born Girolano Francesco Maria Mazzola) (1503–40), Italian painter. A follower of Correggio, he made an important contribution to early mannerism with the graceful figure style of his frescos and portraits. His works include *Self-Portrait in a Convex Mirror* (1524) and *Madonna with the Long Neck* (1534).

Parnassian /pɑːˈnæsɪən/ adj. & n. ● adj. **1** of Parnassus. **2** poetic. **3** of or relating to the Parnassians. ● n. a member of a group of French poets in the late 19th century, notably Leconte de Lisle, who emphasized strictness of form. They took their name from the anthology *Le Parnasse contemporain* (1866).

Parnassus, Mount /pɑːˈnæsəs/ (Greek **Parnassós** /ˌparnaˈsɔs/) a mountain in central Greece, just north of Delphi, rising to a height of 2,457 m (8,064 ft). Held to be sacred by the ancient Greeks, as was the spring of Castalia on its southern slopes, it was associated with Apollo and the Muses and regarded as a symbol of poetry.

Parnell /pɑːˈnel/, Charles Stewart (1846–91), Irish nationalist leader. Elected to Parliament in 1875, Parnell became leader of the Irish Home Rule faction in 1880, and, through his obstructive parliamentary tactics, successfully raised the profile of Irish affairs. In 1886 he supported Gladstone's Home Rule bill, following the latter's conversion to the cause. He was forced to retire from public life in 1890 after the public exposure of his adultery with Mrs Katherine ('Kitty') O'Shea (1840–1905). □ **Parnellite** adj. & n.

parochial /pəˈrəʊkɪəl/ adj. **1** of or concerning a parish. **2** (of affairs, views, etc.) merely local, narrow or restricted in scope. □ **parochialism** n. **parochially** adv. **parochiality** /-ˌrəʊkɪˈælɪtɪ/ n. [ME f. AF *parochiel*, OF *parochial* f. eccl.L *parochialis* (as PARISH)]

parody /ˈpærədɪ/ n. & v. ● n. (pl. **-ies**) **1 a** humorous exaggerated imitation of an author, literary work, style, etc. **b** a work of this kind.

2 a feeble imitation; a travesty. ● *v.tr.* (**-ies, -ied**) **1** compose a parody of. **2** mimic humorously. □ **parodist** *n.* **parodic** /pə'rɒdɪk/ *adj.* [LL *parodia* or Gk *parōidia* burlesque poem (as PARA-¹, *ōidē* ode)]

parol /pə'rəʊl/ *adj. & n. Law* ● *adj.* **1** given orally. **2** (of a document) not given under seal. ● *n.* an oral declaration. [OF *parole* (as PAROLE)]

parole /pə'rəʊl/ *n. & v.* ● *n.* **1 a** the release of a prisoner temporarily for a special purpose or completely before the expiry of a sentence, on the promise of good behaviour. **b** such a promise. **2** a word of honour. ● *v.tr.* put (a prisoner) on parole. □ **on parole** released on the terms of parole. □ **parolee** /-rəʊ'li:/ *n.* [F, = word: see PARLANCE]

paronomasia /ˌpærənə'meɪzɪə/ *n.* a play on words; a pun. [L f. Gk *paronomasia* (as PARA-¹, *onomasia* naming f. *onomazō* to name f. *onoma* a name)]

paronym /'pærənɪm/ *n.* **1** a word cognate with another. **2** a word formed from a foreign word. □ **paronymous** /pə'rɒnɪməs/ *adj.* [Gk *parōnumon*, neut. of *parōnumos* (as PARA-¹, *onuma* name)]

Paros /'pærɒs, 'peərɒs/ a Greek island in the southern Aegean, in the Cyclades. It is noted for the translucent white Parian marble which has been quarried there since the 6th century BC.

parotid /pə'rɒtɪd/ *adj. & n. Anat.* ● *adj.* situated near the ear. ● *n.* (in full **parotid gland**) a salivary gland in front of the ear. □ **parotid duct** a duct opening from the parotid gland into the mouth. [F *parotide* or L *parotis parotid-* f. Gk *parōtis -idos* (as PARA-¹, *ous ōtos* ear)]

parotitis /ˌpærə'taɪtɪs/ *n. Med.* **1** inflammation of the parotid gland. **2** mumps. [PAROTID + -ITIS]

-parous /pərəs/ *comb. form* bearing offspring of a specified number or kind (*multiparous; viviparous*). [L *-parus* -bearing f. *parere* bring forth]

Parousia /pə'ru:zɪə/ *n.* (in Christian theology) the supposed second coming of Christ. [Gk, = presence, coming]

paroxysm /'pærək̩sɪz(ə)m/ *n.* **1** (often foll. by *of*) a sudden attack or outburst (of rage, laughter, etc.). **2** a fit of disease. □ **paroxysmal** /ˌpærək'sɪzm(ə)l/ *adj.* [F *paroxysme* f. med.L *paroxysmus* f. Gk *paroxusmos* f. *paroxunō* exasperate (as PARA-¹, *oxunō* sharpen f. *oxus* sharp)]

paroxytone /pə'rɒksɪˌtəʊn/ *adj. & n.* ● *adj.* (of a word) having an acute accent on the last syllable but one. ● *n.* a word of this kind. [mod.L f. Gk *paroxutonos* (as PARA-¹, OXYTONE)]

parpen /'pɑ:p(ə)n/ *n.* a stone passing through a wall from side to side, with two smooth vertical faces. [ME f. OF *parpain*, prob. ult. f. L *per* through + *pannus* piece of cloth, in Rmc 'piece of wall']

parquet /'pɑ:keɪ/ *n. & v.* ● *n.* **1** (often *attrib.*) a flooring of wooden blocks arranged in a pattern. **2** *US* the stalls of a theatre. ● *v.tr.* (**parqueted** /-keɪd/; **parqueting** /-keɪɪŋ/) furnish (a room) with a parquet floor. [F, = small compartment, floor, dimin. of *parc* PARK]

parquetry /'pɑ:kɪtrɪ/ *n.* the use of wooden blocks to make floors or inlay for furniture.

Parr /pɑ:(r)/, Catherine (1512–48), sixth and last wife of Henry VIII. Having married the king in 1543, she influenced his decision to restore the succession to his daughters Mary and Elizabeth (later Mary I and Elizabeth I respectively).

parr /pɑ:(r)/ *n.* a young salmon with blue-grey finger-like markings on its sides, younger than a smolt. [18th c.: orig. unkn.]

parrakeet *US* var. of PARAKEET.

parramatta /ˌpærə'mætə/ *n.* (also **paramatta**) a light dress fabric of wool and silk or cotton. [*Parramatta* in New South Wales, Australia]

parricide /'pærɪˌsaɪd/ *n.* **1** the killing of a near relative, esp. of a parent. **2** an act of parricide. **3** a person who commits parricide. □ **parricidal** /ˌpærɪ'saɪd(ə)l/ *adj.* [F *parricide* or L *parricida* (= sense 3), *parricidium* (= sense 1), of uncert. orig., assoc. in L with *pater* father and *parens* parent]

parrot /'pærət/ *n. & v.* ● *n.* **1** a bird of the mainly tropical order Psittaciformes, with a short hooked bill, often having vivid plumage and sometimes able to mimic the human voice. **2** a person who mechanically repeats the words or actions of another. ● *v.tr.* (**parroted, parroting**) repeat mechanically. □ **parrot-fashion** (learning or repeating) mechanically without understanding. **parrot-fish** a fish with a strong beaklike mouth; esp. one of the genus *Scarus*, which forms a protective mucous cocoon against predators. [prob. f. obs. or dial. F *perrot* parrot, dimin. of *Pierre* Peter: cf. PARAKEET]

Parry /'pærɪ/, Sir (Charles) Hubert (Hastings) (1848–1918), English composer. He is noted for his choral music, including the cantata *Blest Pair of Sirens* (1887). Parry's best-known work, however, is his setting of William Blake's poem 'Jerusalem' (1916), which has acquired the status of a national song.

parry /'pærɪ/ *v. & n.* ● *v.tr.* (**-ies, -ied**) **1** avert or ward off (a weapon or attack), esp. with a countermove. **2** deal skilfully with (an awkward question etc.). ● *n.* (*pl.* **-ies**) an act of parrying. [prob. repr. F *parez* imper. of *parer* f. It. *parare* ward off]

parse /pɑ:z/ *v.tr.* **1** describe (a word in context) grammatically, stating its inflection, relation to the sentence, etc. **2** resolve (a sentence) into its component parts and describe them grammatically. **3** *Computing* analyse (a string) into components, esp. to test conformability to a grammar. □ **parser** *n.* (esp. sense 3). [perh. f. ME *pars* parts of speech f. OF *pars*, pl. of *part* PART, infl. by L *pars* part]

parsec /'pɑ:sek/ *n. Astron.* a unit of distance, equal to about 3.25 light years (3.08×10^{16} metres), the distance at which the mean radius of the earth's orbit subtends an angle of one second of arc. [PARALLAX + SECOND²]

Parsee /pɑ:'si:, 'pɑ:si:/ *n.* **1** an adherent of Zoroastrianism, esp. a descendant of those Zoroastrians who fled to India from Muslim persecution in Persia during the 7th–8th centuries. They are found in isolated areas of Iran and in India, but numbers are declining on account of their refusal to accept converts. **2** = PAHLAVI¹ 1. □ **Parseeism** *n.* [Pers. *pārsī* Persian f. *pārs* Persia]

Parsifal var. of PERCEVAL¹.

parsimony /'pɑ:sɪmənɪ/ *n.* **1** carefulness in the use of money or other resources. **2** meanness, stinginess. □ **law of parsimony** the assertion that no more causes or forces should be assumed than are necessary to account for the facts. □ **parsimonious** /ˌpɑ:sɪ'məʊnɪəs/ *adj.* **parsimoniously** *adv.* **parsimoniousness** *n.* [ME f. L *parsimonia, parcimonia* f. *parcere pars-* spare]

parsley /'pɑ:slɪ/ *n.* a biennial umbelliferous herb, *Petroselinum crispum*, with white flowers and crinkly aromatic leaves, used for seasoning and garnishing food. □ **parsley fern** a fern, *Cryptogramma crispa*, with leaves like parsley. **parsley-piert** a dwarf annual rosaceous plant, *Aphanes arvensis*, with fan-shaped leaves. [ME *percil, per(e)sil* f. OF *peresil*, and OE *petersilie* ult. f. L *petroselinum* f. Gk *petroselinon*; *parsley-piert* prob. corrupt. of F *perce-pierre* pierce stone]

parsnip /'pɑ:snɪp/ *n.* **1** a biennial umbelliferous plant, *Pastinaca sativa*, with yellow flowers and a large pale yellow tapering root. **2** this root eaten as a vegetable. [ME *pas(se)nep* (with assim. to *nep* turnip) f. OF *pasnaie* f. L *pastinaca*]

parson /'pɑ:s(ə)n/ *n.* **1** a rector. **2** a vicar or any beneficed member of the clergy. **3** *colloq.* any (esp. Protestant) member of the clergy. □ **parson's nose** the piece of fatty flesh at the rump of a fowl. □ **parsonical** /pɑ:'sɒnɪk(ə)l/ *adj.* [ME *person(e), parson* f. OF *persone* f. L *persona* PERSON (in med.L rector)]

parsonage /'pɑ:sənɪdʒ/ *n.* a church house provided for a parson.

Parsons /'pɑ:s(ə)nz/, Sir Charles (Algernon) (1854–1931), British engineer, scientist, and manufacturer. He patented and built the first practical steam turbine in 1884, a 7.5-kW engine designed to drive electricity generators. Many such machines were installed in power stations, and their output was later increased by adding a condenser and using superheated steam. Parsons also developed steam turbines for marine propulsion, the experimental vessel *Turbinia* creating a sensation by its unscheduled appearance at a British naval review in 1897. He was also interested in optics, manufacturing searchlight reflectors, large reflecting telescopes, and optical glass.

part /pɑ:t/ *n., v, & adv.* ● *n.* **1** some but not all of a thing or number of things. **2** an essential member or constituent of anything (*part of the family; a large part of the job*). **3** a component of a machine etc. (*spare parts; needs a new part*). **4 a** a portion of a human or animal body. **b** (in *pl.*) *colloq.* = private parts. **5** a division of a book, broadcast serial, etc., esp. as much as is issued or broadcast at one time. **6 a** each of several equal portions of a whole (*the recipe has 3 parts sugar to 2 parts flour*). **b** (prec. by ordinal number) a specified fraction of a whole (*each received a fifth part*). **7 a** a portion allotted; a share. **b** a person's share in an action or enterprise (*will have no part in it*). **c** one's duty (*was not my part to interfere*). **8 a** a character assigned to an actor. **b** the words spoken by an actor. **c** a copy of these. **9** *Mus.* **a** a melody or other constituent of harmony assigned to a particular voice or instrument (often in comb.: *four-part harmony*). **b** a copy of the music for a particular musician. **10** each of the sides in an agreement or dispute. **11** (in *pl.*) a region or district (*am not from these parts*). **12** (in *pl.*) abilities (*a man of many parts*). **13** *US* = PARTING 2. ● *v.* **1** *tr. & intr.* divide or separate into parts (*the crowd parted to let them through*). **2** *intr.* **a** leave one another's company (*they parted the best of friends*). **b** (foll. by *from*) say goodbye to. **3** *tr.* cause to separate (*they fought hard and had to be parted*). **4** *intr.* (foll. by *with*) give up possession of; hand over. **5** *tr.* separate (the hair of the head on either side of the parting) with a comb. ● *adv.* to some extent; partly (*is part iron and part wood; a lie that is part truth*). □ **for the most part** see MOST. **for one's part** as far as one is concerned. **in part** (or **parts**) to some extent; partly. **look the part** appear suitable for a role. **on the part of** on the behalf or initiative of (*no objection on my part*). **part and parcel** (usu. foll. by *of*) an essential part. **part company** see COMPANY. **part-exchange** *n.* a transaction in which goods are given as part of the payment for other goods, with the balance in money. ● *v.tr.* give (goods) in such a transaction. **part of speech** each of the categories to which words are assigned in accordance with their grammatical and semantic functions (in English esp. noun, pronoun, adjective, adverb, verb, preposition, conjunction, and interjection). **part-song** a song with three or more voice-parts, often without accompaniment, and harmonic rather than contrapuntal in character. **part time** less than the full time required by an activity. **part-time** *adj.* occupying or using only part of the usual working week. **part-timer** a person employed in part-time work. **part-work** *Brit.* a publication appearing in several parts over a period of time. **play a part 1** be significant or contributory. **2** act deceitfully. **3** perform a dramatic role. **take in good part** see GOOD. **take part** (often foll. by *in*) assist or have a share (in). **take the part of 1** support; back up. **2** perform the role of. **three parts** three-quarters. [ME f. OF f. L *pars partis* (n.), *partire, partiri* (v.)]

partake /pɑːˈteɪk/ *v.intr.* (*past* **partook** /-ˈtʊk/; *past part.* **partaken** /-ˈteɪkən/) **1** (foll. by *of, in*) take a share or part. **2** (foll. by *of*) eat or drink some or *colloq.* all (of a thing). **3** (foll. by *of*) have some (of a quality etc.) (*their manner partook of insolence*). □ **partakable** *adj.* **partaker** *n.* [16th c.: back-form. f. *partaker, partaking* = part-taker etc.]

parterre /pɑːˈteə(r)/ *n.* **1** a level space in a garden occupied by flower-beds arranged formally. **2** *US* the ground floor of a theatre auditorium, esp. the pit overhung by balconies. [F, = *par terre* on the ground]

parthenogenesis /ˌpɑːθɪnəʊˈdʒenɪsɪs/ *n. Biol.* reproduction from an ovum without fertilization, esp. as a normal process in invertebrates and lower plants. □ **parthenogenetic** /-dʒɪˈnetɪk/ *adj.* **parthenogenetically** *adv.* [mod.L f. Gk *parthenos* virgin + *genesis* as GENESIS]

Parthenon /ˈpɑːθɪnən/ the temple of Athene Parthenos (= the virgin), built on the Acropolis at Athens in 447–432 BC by Pericles to honour the city's patron goddess and to commemorate the recent Greek victory over the Persians. Designed by the architects Ictinus and Callicrates with sculptures by Phidias, including a colossal gold and ivory statue of Athene (known from descriptive accounts) and the 'Elgin marbles' now in the British Museum, the Parthenon was partly financed by tribute from the league of Greek states led by Athens, and housed the treasuries of Athens and the league. It remains standing, despite being severely damaged by Venetian bombardment in 1687.

Parthian /ˈpɑːθɪən/ *n. & adj.* ● *n.* a native or inhabitant of the ancient Asian kingdom of Parthia, which lay SE of the Caspian Sea in present-day Iran. (*See note below*.) ● *adj.* of or relating to the Parthians or Parthia. □ **Parthian shot** a telling remark reserved for the moment of departure, so called from the trick used by Parthians of shooting arrows while in real or pretended flight.

▪ From *c.*250 BC to AD *c.*230 the Parthians ruled an empire stretching from the Euphrates to the Indus, with Ecbatana as its capital. Established by the Parthians' rebellion against the Seleucids, the empire reached the peak of its power around the 2nd century BC and was eventually eclipsed by the Sassanians. The Parthians were skilled horsemen, with a culture that contained a mixture of Greek and Persian elements.

partial /ˈpɑːʃ(ə)l/ *adj. & n.* ● *adj.* **1** not complete; forming only part (*a partial success*). **2** biased, unfair. **3** (foll. by *to*) having a liking for. ● *n. Mus.* any of the constituents of a musical sound. □ **partial eclipse** an eclipse in which only part of the luminary is covered or darkened. **partial pressure** *Physics* the pressure that would be exerted by one of the gases in a mixture if it occupied the same volume on its own. **partial verdict** a verdict finding a person guilty of part of a charge. □ **partially** *adv.* **partialness** *n.* [ME f. OF *parcial* f. LL *partialis* (as PART)]

partiality /ˌpɑːʃɪˈælɪtɪ/ *n.* **1** bias, favouritism. **2** (foll. by *for*) fondness. [ME f. OF *parcialité* f. med.L *partialitas* (as PARTIAL)]

participant /pɑːˈtɪsɪp(ə)nt/ *n.* a participator. □ **participant observation** (in the social sciences) a method of research involving the use of a researcher who, while appearing to be a member of the group under observation, is in fact gathering information on it.

participate /pɑːˈtɪsɪˌpeɪt/ *v.intr.* **1** (often foll. by *in*) take part or share (in). **2** (foll. by *of*) *literary* or *formal* have a certain quality (*the speech participated of wit*). □ **participator** *n.* **participatory** /-ˌtɪsɪˈpeɪtərɪ, -ˈtɪsɪpətərɪ/ *adj.* **participation** /-ˌtɪsɪˈpeɪʃ(ə)n/ *n.* [L *participare* f. *particeps -cipis* taking part, formed as PART + *-cip-* = *cap-* stem of *capere* take]

participle /ˈpɑːtɪˌsɪp(ə)l, pɑːˈtɪsɪ-/ *n. Gram.* a word formed from a verb (e.g. *going, gone, being, been*) and used in compound verb-forms (e.g. *is going, has been*) or as an adjective (e.g. *working woman, burnt toast*). □ **participial** /ˌpɑːtɪˈsɪpɪəl/ *adj.* **participially** *adv.* [ME f. OF, by-form of *participe* f. L *participium* (as PARTICIPLE)]

particle /ˈpɑːtɪk(ə)l/ *n.* **1** a minute portion of matter. (See also SUBATOMIC PARTICLE.) **2** the least possible amount (*not a particle of sense*). **3** *Gram.* **a** a minor part of speech, a short indeclinable one. **b** a common prefix or suffix such as *in-, -ness*. □ **particle physics** the branch of physics concerned with the properties and interactions of subatomic particles. [ME f. L *particula* (as PART)]

particle accelerator *n. Physics* an apparatus for accelerating subatomic particles to high velocities by means of electric or electromagnetic fields. The accelerated particles are generally made to collide with other particles, either as a research technique or for the generation of high-energy X-rays and gamma rays. The earliest successful experiments with accelerators were those in which the British physicists Cockcroft and Walton 'split the atom' (by bombarding atoms with accelerated electrons) in 1932. The search for higher energies led to the use of circular, rather than linear, accelerators such as the cyclotron, betatron, and synchrotron. Storage rings, introduced in 1956, in which high-energy particles of opposite charge are made to collide with each other, have produced the highest particle energies — tens of gigaelectronvolts — so far obtained.

particoloured /ˈpɑːtɪˌkʌləd/ *adj.* (*US* **particolored**) partly of one colour, partly of another or others. [PARTY² + COLOURED]

particular /pəˈtɪkjʊlə(r)/ *adj. & n.* ● *adj.* **1** relating to or considered as one thing or person as distinct from others; individual (*in this particular instance*). **2** more than is usual; special, noteworthy (*took particular trouble*). **3** scrupulously exact; fastidious. **4** detailed (*a full and particular account*). **5** *Logic* (of a proposition) in which something is asserted of some but not all of a class (opp. UNIVERSAL *adj.* 2). ● *n.* **1** a detail; an item. **2** (in *pl.*) points of information; a detailed account. **3** *Philos.* according to Platonic theory, a concrete thing in the material world which exemplifies an abstract concept or universal. □ **in particular** especially, specifically. [ME f. OF *particuler* f. L *particularis* (as PARTICLE)]

particularism /pəˈtɪkjʊləˌrɪz(ə)m/ *n.* **1** exclusive devotion to one party, sect, etc. **2** the principle of leaving political independence to each state in an empire or federation. **3** the theological doctrine of individual election or redemption. □ **particularist** *n.* [F *particularisme*, mod.L *particularismus*, and G *Partikularismus* (as PARTICULAR)]

particularity /pəˌtɪkjʊˈlærɪtɪ/ *n.* (*pl.* **-ies**) **1** the quality of being individual or particular. **2** fullness or minuteness of detail in a description. **3** (usu. in *pl.*) detail, particular.

particularize /pəˈtɪkjʊləˌraɪz/ *v.tr.* (also **-ise**) (also *absol.*) **1** name specially or one by one. **2** specify (items). □ **particularization** /-ˌtɪkjʊləraɪˈzeɪʃ(ə)n/ *n.* [F *particulariser* (as PARTICULAR)]

particularly /pəˈtɪkjʊləlɪ/ *adv.* **1** especially, very. **2** specifically (*they particularly asked for you*). **3** in a particular or fastidious manner.

particulate /pəˈtɪkjʊˌleɪt, -lət/ *adj. & n.* ● *adj.* in the form of separate particles. ● *n.* (often in *pl.*) matter in this form. [L *particula* PARTICLE]

parting /ˈpɑːtɪŋ/ *n.* **1** a leave-taking or departure (often *attrib.*: *parting words*). **2** *Brit.* the dividing line of combed hair. **3** a division; an act of separating. □ **parting shot** = *Parthian shot*.

parti pris /ˌpɑːtɪ ˈpriː/ *n. & adj.* ● *n.* (*pl.* **partis pris** *pronunc.* same) a preconceived view; a bias. ● *adj.* prejudiced, biased. [F, = side taken]

partisan /ˈpɑːtɪˌzæn/ *n. & adj.* (also **partizan**) ● *n.* **1** a strong, esp. unreasoning, supporter of a party, cause, etc. **2** *Mil.* a guerrilla in wartime. ● *adj.* **1** of or characteristic of partisans. **2** loyal to a particular cause; biased. □ **partisanship** *n.* [F f. It. dial. *partigiano* etc. f. *parte* PART]

partita /pɑːˈtiːtə/ *n.* (*pl.* **partite** /-tɪ/) *Mus.* **1** a suite. **2** an air with variations. [It., fem. past part. of *partire* divide, formed as PART]

partite /ˈpɑːtaɪt/ *adj.* **1** divided (esp. in *comb.*: *tripartite*). **2** *Bot. & Zool.* divided to or nearly to the base. [L *partitus* past part. of *partiri* PART v.]

partition /pɑːˈtɪʃ(ə)n/ *n. & v.* ● *n.* **1** division into parts, esp. of a country with separate areas of government. **2** a structure dividing a space into two parts, esp. a light interior wall. ● *v.tr.* **1** divide into parts. **2** (foll. by *off*) separate (part of a room etc.) with a partition. □ **partitioned** *adj.* **partitioner** *n.* **partitionist** *n.* [ME f. OF f. L *partitio -onis* (as PARTITE)]

partitive /ˈpɑːtɪtɪv/ *adj. & n. Gram.* ● *adj.* (of a word, form, etc.) denoting part of a collective group or quantity. ● *n.* a partitive word (e.g. *some, any*) or form. □ **partitive genitive** a genitive used to indicate a whole divided into or regarded in parts, expressed in English by *of* as in *most of us.* □ **partitively** *adv.* [F *partitif -ive* or med.L *partitivus* (as PARTITE)]

partizan var. of PARTISAN.

partly /ˈpɑːtlɪ/ *adv.* **1** with respect to a part or parts. **2** to some extent.

partner /ˈpɑːtnə(r)/ *n. & v.* ● *n.* **1** a person who shares or takes part with another or others, esp. in a business firm with shared risks and profits. **2** a companion in dancing. **3** a player (esp. one of two) on the same side in a game. **4** either member of a married couple, or of an established unmarried couple. ● *v.tr.* **1** be the partner of. **2** associate as partners. □ **partnerless** *adj.* [ME, alt. of *parcener* joint heir, after PART]

partnership /ˈpɑːtnəˌʃɪp/ *n.* **1** the state of being a partner or partners. **2** a joint business. **3** a pair or group of partners.

Parton /ˈpɑːt(ə)n/, Dolly (Rebecca) (b.1946), American singer and songwriter. She is best known as a country-music singer; in the mid-1960s she moved to Nashville and had her first hit in 1967 with 'Dumb Blonde'. Her other hits include 'Joshua' (1971) and 'Jolene' (1974). She has also had a number of film roles.

partook *past* of PARTAKE.

partridge /ˈpɑːtrɪdʒ/ *n.* (*pl.* same or **partridges**) **1** a game bird of the Eurasian genus *Perdix*, smaller than a pheasant, esp. the *grey partridge* (*P. perdix*). **2** a similar bird of the family Phasianidae (*snow partridge*). [ME *partrich* etc. f. OF *perdriz* etc. f. L *perdix -dicis*: for *-dge* cf. CABBAGE]

parturient /pɑːˈtjʊərɪənt/ *adj.* about to give birth. [L *parturire* be in labour, inceptive of *parere part-* bring forth]

parturition /ˌpɑːtjʊˈrɪʃ(ə)n/ *n.* Med. the act of bringing forth young; childbirth. [LL *parturitio* (as PARTURIENT)]

party[1] /ˈpɑːtɪ/ *n. & v.* ● *n.* (*pl.* **-ies**) **1** a social gathering, usu. of invited guests. **2** a body of persons engaged in an activity or travelling together (*fishing party; search party*). **3** a group of people united in a cause, opinion, etc., esp. a political group organized on a national basis. **4** a person or persons forming one side in an agreement or dispute. **5** (foll. by *to*) Law an accessory (to an action). **6** *colloq.* a person. ● *v.tr. & intr.* (**-ies, -ied**) entertain at or attend a party or parties. □ **party animal** *colloq.* a person who enjoys giving or attending parties. **party line 1** the policy adopted by a political party. **2** a telephone line shared by two or more subscribers. **party pooper** *sl.* a person whose manner or behaviour inhibits other people's enjoyment; a killjoy. **party-wall** a wall common to two adjoining buildings or rooms. [ME f. OF *partie* ult. f. L *partire*: see PART]

party[2] /ˈpɑːtɪ/ *adj.* Heraldry divided into parts of different colours. [ME f. OF *parti* f. L (as PARTY[1])]

Parvati /ˈpɑːvətɪ/ *Hinduism* a benevolent goddess, wife of Siva, mother of Ganesha and Skanda. She is often identified with Uma, Sati, Devi, and Sakti, and in her malevolent aspect with Durga and Kali. [Skr., = daughter of the mountain]

parvenu /ˈpɑːvəˌnuː, -ˌnjuː/ *n. & adj.* ● *n.* (*fem.* **parvenue**) **1** a person of obscure origin who has gained wealth or position. **2** an upstart. ● *adj.* **1** associated with or characteristic of such a person. **2** upstart. [F, past part. of *parvenir* arrive f. L *pervenire* (as PER-, *venire* come)]

parvis /ˈpɑːvɪs/ *n.* (also **parvise**) **1** an enclosed area in front of a cathedral, church, etc. **2** a room over a church porch. [ME f. OF *parvis* ult. f. LL *paradisus* PARADISE, a court in front of St Peter's, Rome]

parvovirus /ˈpɑːvəʊˌvaɪərəs/ *n.* any of a class of small viruses affecting vertebrate animals, esp. one which causes contagious disease in dogs. [L *parvus* small + VIRUS]

pas /pɑː/ *n.* (*pl.* same) a step in dancing, esp. in classical ballet. □ **pas de chat** /də ˈʃɑ/ a leap in which each foot in turn is raised to the opposite knee. **pas de deux** /də ˈdɜː/ a dance for two persons. **pas glissé** see GLISSÉ. **pas seul** /sɜːl/ a solo dance. [F, = step]

Pasadena /ˌpæsəˈdiːnə/ a city in California, in the San Gabriel Mountains on the north-eastern side of the Los Angeles conurbation; pop. (1990) 131,590. It is the site of the Rose Bowl stadium, venue for the American Football Super Bowl.

Pascal /pæˈskɑːl/, Blaise (1623-62), French mathematician, physicist, and religious philosopher. A child prodigy, before the age of 16 he had proved an important theorem in the projective geometry of conics, and at 19 constructed the first mechanical calculator to be offered for sale. He discovered that air has weight, confirmed that vacuum could exist, and derived the principle that the pressure of a fluid at rest is transmitted equally in all directions. He also founded the theory of probabilities, and developed a forerunner of integral calculus. He later entered a Jansenist convent, where he wrote two classics of French

devotional thought, the *Lettres Provinciales* (1656-7), directed against the casuistry of the Jesuits, and *Pensées* (1670), a defence of Christianity.

pascal /ˈpæsk(ə)l/ *n. Physics* **1** the SI unit of pressure (symbol **Pa**), equal to one newton per square metre, about 1.45×10^{-4} p.s.i. **2** (**Pascal**) a computer language used esp. in training. [PASCAL]

paschal /ˈpæsk(ə)l/ *adj.* **1** of or relating to the Jewish Passover. **2** of or relating to Easter. □ **paschal lamb 1** a lamb sacrificed at Passover. **2** Christ. [ME f. OF *pascal* f. eccl.L *paschalis* f. *pascha* f. Gk *paskha* f. Aram. *pasḥa*, rel. to Heb. *pesaḥ* PASSOVER]

pash /pæʃ/ *n. sl.* a brief infatuation. [abbr. of PASSION]

pasha /ˈpɑːʃə/ *n.* (also **pacha**) *hist.* the title (placed after the name) of a Turkish officer of high rank, e.g. a military commander, the governor of a province, etc. [Turk. *paşa*, prob. = *başa* f. *baş* head, chief]

pashm /ˈpæʃəm/ *n.* the under-fur of some Tibetan animals, esp. that of goats as used for Cashmere shawls.

Pashto /ˈpʌʃtəʊ/ *n. & adj.* ● *n.* the Indo-Iranian language of the Pathans, the official language of Afghanistan and spoken also in NW Pakistan. It is written in a form of the Arabic script. ● *adj.* of or in this language. [Pashto]

Pašić /ˈpæʃɪtʃ/, Nikola (1845-1926), Serbian statesman, Prime Minister of Serbia five times between 1891 and 1918, and of the Kingdom of Serbs, Croats, and Slovenes 1921-4 and 1924-6. As Prime Minister of the Serbian government in exile during the First World War, he signed the Corfu Declaration (1917), which set down a blueprint for a postwar unified Yugoslavia. Although Pašić was reluctant for the Serbs to share power with the South Slavs of Austria-Hungary, he was a party to the formation of the Kingdom of Serbs, Croats, and Slovenes (called Yugoslavia from 1929) in 1918.

Pasiphaë /pəˈsɪfɪˌiː/ *Gk Mythol.* the wife of Minos and mother of the Minotaur.

paso doble /ˌpæsəʊ ˈdəʊbleɪ/ *n.* **1** a ballroom dance based on a Latin American style of marching. **2** this style of marching. [Sp., = double step]

pasque-flower /ˈpæskˌflaʊə(r)/ *n.* a spring-flowering plant, *Pulsatilla vulgaris*, related to anemones, with bell-shaped purple flowers and fernlike foliage. [earlier *passe-flower* f. F *passe-fleur*: assim. to *pasque* = obs. *pasch* (as PASCHAL), Easter]

pasquinade /ˌpæskwɪˈneɪd/ *n.* a lampoon or satire, orig. one displayed in a public place. [It. *pasquinata* f. *Pasquino*, a statue in Rome on which abusive Latin verses were annually posted in the 16th c.]

pass[1] /pɑːs/ *v. & n.* ● *v.* (*past part.* **passed**) (see also PAST). **1** *intr.* (often foll. by *along, by, down, on,* etc.) move onward; proceed, esp. past some point of reference (*saw the procession passing*). **2** *tr.* **a** go past; leave (a thing etc.) on one side or behind in proceeding. **b** overtake, esp. in a vehicle. **c** go across (a frontier, mountain range, etc.). **3** *intr. & tr.* be transferred or cause to be transferred from one person or place to another (*pass the butter; the title passes to his son*). **4** *tr.* surpass; be too great for (*it passes my comprehension*). **5** *intr.* get through; effect a passage. **6** *intr.* **a** be accepted as adequate; go uncensured (*let the matter pass*). **b** (foll. by *as, for*) be accepted or currently known as. **c** *US* (of a person with some black ancestry) be accepted as white. **7** *tr.* move; cause to go (*passed her hand over her face; passed a rope round it*). **8 a** *intr.* (of a candidate in an examination) be successful. **b** *tr.* be successful in (an examination). **c** *tr.* (of an examiner) judge the performance of (a candidate) to be satisfactory. **9 a** *tr.* (of a bill) be examined and approved by (a parliamentary body or process). **b** *tr.* cause or allow (a bill) to proceed to further legislative processes. **c** *intr.* (of a bill or proposal) be approved. **10** *intr.* **a** occur, elapse (*the remark passed unnoticed; time passes slowly*). **b** happen; be done or said (*heard what passed between them*). **11 a** *intr.* circulate; be current. **b** *tr.* put into circulation (*was passing forged cheques*). **12** *tr.* spend or use up (a certain time or period) (*passed the afternoon reading*). **13** *tr.* (also *absol.*) (in field games) send (the ball) to another player of one's own side. **14** *intr.* forgo one's turn or chance in a game etc. **15** *intr.* (foll. by *to, into*) change from one form to another. **16** *intr.* come to an end. **17** *tr.* discharge from the body as or with excreta. **18** *tr.* (foll. by *on, upon*) **a** utter (criticism) about. **b** pronounce (a judicial sentence) on. **19** *intr.* (often foll. by *on, upon*) adjudicate. **20** *tr.* not declare or pay (a dividend). **21** *tr.* cause (troops etc.) to go by esp. ceremonially. ● *n.* **1** an act or instance of passing. **2 a** a success in an examination. **b** *Brit.* the status of a university degree without honours. **3** written permission to pass into or out of a place, or to be absent from quarters. **4 a** a ticket or permit giving free entry or access etc. **b** = *free pass*. **5** (in field games) a transference of the ball to another player on the same side. **6** a thrust in fencing. **7** a juggling trick. **8** an act of passing the hands over anything, as in conjuring or hypnotism. **9** a critical position (*has come to a fine pass*). □ **in passing 1** by the way. **2** in the course of speech, conversation, etc. **make a pass at** *colloq.* make amorous or sexual advances to. **pass away 1** *euphem.* die. **2** cease to exist; come to an end. **pass by 1** go past. **2** disregard, omit. **passed pawn** *Chess* a pawn that has advanced beyond the pawns on the other side. **pass one's eye over** see EYE. **pass muster** see MUSTER. **pass off 1** (of feelings etc.) disappear gradually. **2** (of proceedings) be carried through (in a specified way). **3** (foll. by *as*) misrepresent (a person or thing) as something else. **4** evade or lightly dismiss (an awkward remark etc.). **pass on 1** proceed on one's way. **2** *euphem.* die. **3** transmit to the next person in a series. **pass out 1** become unconscious. **2** *Brit. Mil.* complete one's training as a cadet. **3** distribute. **pass over 1** omit, ignore, or disregard. **2** ignore the claims of (a person) to promotion or advancement. **3** *euphem.* die. **pass round 1** distribute. **2** send or give to each of a number in turn. **pass the time of day** see TIME. **pass up** *colloq.* refuse or neglect (an opportunity etc.). **pass water** urinate. □ **passer** *n.* [ME f. OF *passer* ult. f. L *passus* PACE[1]]

pass[2] /pɑːs/ *n.* **1** a narrow passage through mountains. **2** a navigable channel, esp. at the mouth of a river. □ **sell the pass** betray a cause. [ME, var. of PACE[1], infl. by F *pas* and by PASS[1]]

passable /ˈpɑːsəb(ə)l/ *adj.* **1** barely satisfactory; just adequate. **2** (of a road, pass, etc.) that can be passed. □ **passableness** *n.* **passably** *adv.* [ME f. OF (as PASS[1])]

passacaglia /ˌpæsəˈkɑːlɪə/ *n. Mus.* an instrumental piece usu. with a ground bass. [It. f. Sp. *pasacalle* f. *pasar* pass + *calle* street: orig. often played in the streets]

passage[1] /ˈpæsɪdʒ/ *n.* **1** the process or means of passing; transit. **2** = PASSAGEWAY. **3** the liberty or right to pass through. **4 a** the right of conveyance as a passenger by sea or air. **b** a journey by sea or air. **5** a transition from one state to another. **6 a** a short extract from a book etc. **b** a section of a piece of music. **7** the passing of a bill etc. into law. **8** (in *pl.*) an interchange of words etc. **9** *Anat.* a duct etc. in the body. □ **passage of** (or **at**) **arms** a fight or dispute. **work one's passage** earn a right (orig. of passage) by working for it. [ME f. OF (as PASS[1])]

passage[2] /ˈpæsɪdʒ/ *v.* **1** *intr.* (of a horse or rider) move sideways, by the pressure of the rein on the horse's neck and of the rider's leg on the opposite side. **2** *tr.* make (a horse) do this. [F *passager*, earlier *passéger* f. It. *passeggiare* to walk, pace f. *passeggio* walk f. L *passus* PACE[1]]

passageway /ˈpæsɪdʒˌweɪ/ *n.* a narrow way for passing along, esp. with walls on either side; a corridor.

passant /ˈpæs(ə)nt/ *adj. Heraldry* (of an animal) walking and looking to the dexter side, with three paws on the ground and the right forepaw raised. [ME f. OF, part. of *passer* PASS[1]]

passband /ˈpɑːsbænd/ *n.* a frequency band within which signals are transmitted by a filter without attenuation.

passbook /ˈpɑːsbʊk/ *n.* a book issued by a bank or building society etc. to an account-holder recording sums deposited and withdrawn.

Passchendaele, Battle of /ˈpæʃ(ə)nˌdeɪl/ (also **Passendale** /ˈpæs(ə)n-/) a prolonged and indecisive episode of trench warfare during the First World War in 1917, near the village of Passchendaele in western Belgium. The village was the furthest point of an Allied offensive involving appalling loss of life in a sea of mud, for no eventual strategic gain. The battle is also known as the third Battle of Ypres (see also YPRES, BATTLE OF).

passé /ˈpæseɪ/ *adj.* no longer fashionable or topical, outdated. [F, past part. of *passer* PASS[1]]

passementerie /ˈpæsmɒntrɪ/ *n.* a trimming of gold or silver lace, braid, beads, etc. [F f. *passement* gold lace etc. f. *passer* PASS[1]]

passenger /ˈpæsɪndʒə(r)/ *n.* **1** a traveller in or on a public or private conveyance (other than the driver, pilot, crew, etc.). **2** *colloq.* a member of a team, crew, etc., who does no effective work. **3** (*attrib.*) for the use of passengers (*passenger seat*). □ **passenger-mile** one mile travelled by one passenger, as a unit of traffic. **passenger-pigeon** an extinct wild pigeon of North America, noted for migrating in huge flocks, hunted to extinction by 1914. [ME f. OF *passager* f. OF *passager* (adj.) passing (as PASSAGE[1]): -n- as in *messenger* etc.]

passe-partout /ˌpæspɑːˈtuː, ˌpɑːs-/ *n.* **1** a master-key. **2** a simple picture-frame (esp. for mounted photographs), esp. one consisting of a piece of glass stuck to a backing by adhesive tape along the edges. **3** adhesive tape or paper used for this. [F, = passes everywhere]

passer-by /ˌpɑːsəˈbaɪ/ *n.* (*pl.* **passers-by**) a person who goes past, esp. by chance.

passerine /ˈpæsəˌriːn, -ˌraɪn/ *n. & adj. Zool.* ● *n.* a bird of the large order Passeriformes, which comprises the perching birds, having feet with three toes pointing forward and one pointing backwards. (*See note below*.) ● *adj.* **1** of or relating to this order. **2** of the size of a sparrow. [L *passer* sparrow]

▪ The passerines constitute only one order of birds, but they are often regarded informally as being one of two major divisions (the remainder being the 'non-passerines'). The order is itself divided into two groupings, the Deutero-Oscines (ovenbirds, manakins, tyrant flycatchers, etc.) and the Oscines (larks, thrushes, warblers, finches, crows, etc.). The Oscines are the songbirds, possessing a syrinx, and since all the passerines in Europe belong to this group the terms *perching bird* and *songbird* are effectively synonymous there.

passible /ˈpæsɪb(ə)l/ *adj.* (in Christian theology) capable of feeling or suffering. □ **passibility** /ˌpæsɪˈbɪlɪtɪ/ *n.* [ME f. OF *passible* or LL *passibilis* f. L *pati pass-* suffer]

passim /ˈpæsɪm/ *adv.* (of allusions or references in a published work) to be found at various places throughout the text. [L f. *passus* scattered f. *pandere* spread]

passing /ˈpɑːsɪŋ/ *adj. & n.* ● *adj.* **1** in senses of PASS[1] *v.* **2** transient, fleeting (*a passing glance*). **3** cursory, incidental (*a passing reference*). ● *n.* **1** in senses of PASS[1] *v.* **2** *euphem.* the death of a person (*mourned his passing*). □ **passing note** *Mus.* a note not belonging to the harmony but interposed to secure a smooth transition. **passing shot** *Tennis* a shot aiming the ball beyond and out of reach of the other player. □ **passingly** *adv.*

passion /ˈpæʃ(ə)n/ *n.* **1** strong barely controllable emotion. **2** an outburst of anger (*flew into a passion*). **3 a** intense sexual love. **b** a person arousing this. **4 a** strong enthusiasm (*has a passion for football*). **b** a thing arousing this. **5** (**Passion**) **a** the suffering of Christ during his last days. **b** a narrative of this from the Gospels. **c** a musical setting of any of these narratives. □ **passion-flower** a climbing plant of the genus *Passiflora*, with a flower that was believed to suggest the crown of thorns and other things associated with the Passion of Christ. **passion-fruit** the edible fruit of some species of passion-flower, esp. *Passiflora edulis* (also called *granadilla*). □ **passionless** *adj.* [ME f. OF f. LL *passio -onis* f. L *pati pass-* suffer]

passional /ˈpæʃ(ə)n(ə)l/ *adj. & n.* ● *adj. literary* of or marked by passion. ● *n.* a book of the sufferings of saints and martyrs.

passionate /ˈpæʃənət/ *adj.* **1** dominated by or easily moved to strong feeling, esp. love or anger. **2** showing or caused by passion. □ **passionately** *adv.* **passionateness** *n.* [ME f. med.L *passionatus* (as PASSION)]

passion-play /ˈpæʃ(ə)nˌpleɪ/ *n.* a medieval religious drama dealing with the events of Christ's Passion from the Last Supper to the Crucifixion. The establishment of the feast of Corpus Christi in the 13th century gave a great impetus to the enactment of passion-plays throughout Europe, but the tradition of performance of them mostly died out during the 15th century. The best known of the few plays that survive is that performed at Oberammergau.

Passion Sunday the fifth Sunday in Lent.

Passiontide /ˈpæʃ(ə)nˌtaɪd/ the last two weeks of Lent.

Passion Week 1 the week between Passion Sunday and Palm Sunday. **2** = HOLY WEEK.

passivate /ˈpæsɪˌveɪt/ *v.tr.* make (esp. metal) passive. □ **passivation** /ˌpæsɪˈveɪʃ(ə)n/ *n.*

passive /ˈpæsɪv/ *adj.* **1** suffering action; acted upon. **2** offering no opposition; submissive. **3 a** not active; inert. **b** (of a metal) abnormally unreactive through having a superficial oxide coating. **4** *Gram.* designating the voice in which the subject undergoes the action of the verb (e.g. in *they were killed*). **5** (of a debt) incurring no interest payment. □ **passive obedience 1** surrender to another's will without cooperation. **2** compliance with commands irrespective of their nature. **passive resistance** non-violent refusal to comply with esp. legal requirements, as a form of political protest. **passive smoking** the involuntary inhaling, esp. by a non-smoker, of smoke from others' cigarettes etc. □ **passively** *adv.* **passiveness** *n.* **passivity** /pæˈsɪvɪtɪ/ *n.* [ME f. OF *passif -ive* or L *passivus* (as PASSION)]

passkey /ˈpɑːskiː/ *n.* **1** a private key to a gate etc. for special purposes. **2** a master-key.

passmark /ˈpɑːsmɑːk/ *n.* the minimum mark needed to pass an examination.

Passos, John Dos, see DOS PASSOS.

Passover /ˈpɑːsˌəʊvə(r)/ *n.* **1** the Jewish festival celebrated each spring, held from 14 to 21 Nisan and commemorating the liberation of the Israelites from slavery in Egypt (see EXODUS). **2** = *paschal lamb*. [*pass over*

= pass without touching, with ref. to the exemption of the Israelites from the death of the first-born (Exod. 12)]

passport /'pɑːspɔːt/ n. **1** an official document issued by a government certifying the holder's identity and citizenship, and entitling the holder to travel under its protection to and from foreign countries. **2** (foll. by *to*) a thing that ensures admission or attainment (*a passport to success*). [F *passeport* (as PASS[1], PORT[1])]

password /'pɑːswɜːd/ n. a selected word, phrase, or string of characters securing recognition, admission, access to a computing system, etc., when used by those to whom it is disclosed.

past /pɑːst/ adj., n., prep., & adv. ● adj. **1** gone by in time and no longer existing (*in past years*; *the time is past*). **2** recently completed or gone by (*the past month*; *for some time past*). **3** relating to a former time (*past president*). **4** *Gram.* expressing a past action or state. ● n. **1** (prec. by *the*) **a** past time. **b** what has happened in past time (*cannot undo the past*). **2** a person's past life or career, esp. if discreditable (*a man with a past*). **3** a past tense or form. ● prep. **1** beyond in time or place (*is past two o'clock*; *ran past the house*). **2** beyond the range, duration, or compass of (*past belief*; *past endurance*). ● adv. so as to pass by (*hurried past*). □ **not put it past a person** believe it possible of a person. **past it** colloq. old and useless. **past master 1** a person who is especially adept or expert in an activity, subject, etc. **2** a person who has been a master in a guild, Freemason's lodge, etc. **past perfect** = PLUPERFECT. [past part. of PASS[1] v.]

pasta /'pæstə/ n. **1** flour dough (often dried) used in various shapes in cooking (e.g. lasagne, spaghetti). **2** a dish made from this. [It., = PASTE]

paste /peɪst/ n. & v. ● n. **1** a moist fairly stiff mixture, esp. of powder and liquid. **2** a dough of flour with fat, water, etc., used in baking. **3** an adhesive of flour, water, etc., esp. for sticking paper and other light materials. **4** an easily spread preparation of ground meat, fish, etc. (*anchovy paste*). **5 a** a hard vitreous composition used in making imitation gems. **b** imitation jewellery made of this. **6** a mixture of clay, water, etc., used in making ceramic ware, esp. a mixture of low plasticity used in making porcelain. ● v.tr. **1** fasten or coat with paste. **2** sl. **a** beat or thrash. **b** bomb or bombard heavily. □ **paste-up** a document prepared for copying etc. by combining and pasting various sections on a backing. □ **pasting** n. (esp. in sense 2 of v.). [ME f. OF f. LL *pasta* small square medicinal lozenge f. Gk *pastē* f. *pastos* sprinkled]

pasteboard /'peɪstbɔːd/ n. **1** a sheet of stiff material made by pasting together sheets of paper. **2** (*attrib.*) **a** flimsy, unsubstantial. **b** fake.

pastel /'pæst(ə)l/ n. & adj. ● n. **1** a crayon consisting of powdered pigments bound with a gum solution. **2** a work of art in pastel. **3** a light and subdued shade of a colour. ● adj. of a light and subdued shade or colour. □ **pastelist** n. **pastellist** n. [F *pastel* or It. *pastello*, dimin. of *pasta* PASTE]

pastern /'pæstən/ n. **1** the part of a horse's foot between the fetlock and the hoof. **2** a corresponding part in other animals. [ME *pastron* f. OF *pasturon* f. *pasture* hobble ult. f. L *pastorius* a shepherd: see PASTOR]

Pasternak /'pæstəˌnæk/, Boris (Leonidovich) (1890–1960), Russian poet, novelist, and translator. On the eve of the Russian Revolution in 1917 he wrote the lyric poems *My Sister, Life* (1922), which established his reputation when published. In the 1930s he started work on the novel *Doctor Zhivago* (1957), a testament to the experience of the Russian intelligentsia before, during, and after the Revolution. It was banned in the Soviet Union and first published in Italian and then, with equal success, in other languages; in 1958 Pasternak was awarded the Nobel Prize for literature, but was forced to turn it down under pressure from the Soviet authorities.

Pasteur /pæ'stɜː(r)/, Louis (1822–95), French chemist and bacteriologist. His early work, in which he discovered the existence of dextrorotatory and laevorotatory forms of sugars, was of fundamental importance in chemistry, but he is popularly remembered for his 'germ theory' (1865) – that each fermentation process could be traced to a specific living micro-organism. Following the success of his introduction of pasteurization, he developed an interest in diseases. Pasteur isolated bacteria infecting silkworms, finding methods of preventing the disease from spreading. He then isolated the bacteria causing anthrax and chicken cholera, made vaccines against them, and pioneered vaccination against rabies using attenuated virus.

pasteurize /'pɑːstjəˌraɪz, 'pæs-, -tʃəˌraɪz/ v.tr. (also **-ise**) subject (milk etc.) to the process of partial sterilization by heating. □ **pasteurizer** n. **pasteurization** /ˌpɑːstjəraɪˈzeɪʃ(ə)n, ˌpæs-, -tʃəraɪ-/ n. [PASTEUR]

pasticcio /pæ'stɪtʃəʊ/ n. (pl. **-os**) = PASTICHE. [It.: see PASTICHE]

pastiche /pæ'stiːʃ/ n. **1** a medley, esp. a picture or a musical

composition, made up from or imitating various sources. **2** a literary or other work of art composed in the style of a well-known writer, artist, etc. [F f. It. *pasticcio* ult. f. LL *pasta* PASTE]

pastille /'pæstɪl/ n. **1** a small sweet or lozenge. **2** a small roll of aromatic paste burnt as a fumigator etc. □ **pastille-burner** an ornamental ceramic container in which an aromatic pastille may be burnt. [F f. L *pastillus* little loaf, lozenge f. *panis* loaf]

pastime /'pɑːstaɪm/ n. **1** a recreation or hobby. **2** a sport or game. [PASS[1] + TIME]

pastis /'pæstɪs, pæ'stiːs/ n. an aniseed-flavoured aperitif. [F]

pastor /'pɑːstə(r)/ n. **1** a minister in charge of a church or a congregation. **2** a person exercising spiritual guidance. **3** (in full **rose-coloured pastor**) a pink and black starling, *Sturnus roseus*, native to eastern Europe and Asia. □ **pastorship** n. [ME f. AF & OF *pastour* f. L *pastor -oris* shepherd f. *pascere past-* feed, graze]

pastoral /'pɑːstərəl/ adj. & n. ● adj. **1** of, relating to, or associated with shepherds or flocks and herds. **2** (of land) used for pasture. **3** (of a poem, picture, etc.) portraying country life, usu. in a romantic or idealized form. **4** of or appropriate to a pastor. **5** *Education* of or relating to a teacher's responsibility for the general well-being of pupils or students. ● n. **1** a pastoral poem, play, picture, etc. **2** a letter from a pastor (esp. a bishop) to the clergy or people. □ **pastoral staff** a bishop's crosier. **pastoral theology** that considering religious truth in relation to spiritual needs. □ **pastoralism** n. **pastorally** adv. **pastorality** /ˌpɑːstəˈrælɪtɪ/ n. [ME f. L *pastoralis* (as PASTOR)]

pastorale /ˌpæstəˈrɑːl, -ˈrɑːlɪ/ n. (pl. **pastorales** or **pastorali** /-lɪ/) **1** a slow instrumental composition in compound time, usu. with drone notes in the bass. **2** a simple musical play with a rural subject. [It. (as PASTORAL)]

pastoralist /'pɑːstərəlɪst/ n. *Austral.* a farmer of sheep or cattle.

pastorate /'pɑːstərət/ n. **1** the office or tenure of a pastor. **2** a body of pastors.

pastrami /pæ'strɑːmɪ/ n. seasoned smoked beef. [Yiddish]

pastry /'peɪstrɪ/ n. (pl. **-ies**) **1** a dough of flour, fat, and water baked and used as a base and covering for pies etc. **2 a** food, esp. cake, made wholly or partly of this. **b** a piece or item of this food. □ **pastry-cook** a cook who specializes in pastry, esp. for public sale. [PASTE after OF *pastaierie*]

pasturage /'pɑːstʃərɪdʒ/ n. **1** land for pasture. **2** the process of pasturing cattle etc. [OF (as PASTURE)]

pasture /'pɑːstʃə(r)/ n. & v. ● n. **1** land covered with grass etc. suitable for grazing animals, esp. cattle or sheep. **2** herbage for animals. ● v. **1** tr. put (animals) to graze in a pasture. **2** intr. & tr. (of animals) graze. [ME f. OF f. LL *pastura* (as PASTOR)]

pasty[1] /'pæstɪ/ n. (pl. **-ies**) a pastry case with a sweet or savoury filling, baked without a dish to shape it. [ME f. OF *pasté* ult. f. LL *pasta* PASTE]

pasty[2] /'peɪstɪ/ adj. (**pastier, pastiest**) **1** of or like or covered with paste. **2** unhealthily pale (esp. in complexion) (*pasty-faced*). □ **pastily** adv. **pastiness** n.

Pat /pæt/ n. a nickname for an Irishman. [abbr. of the name *Patrick*]

Pat. abbr. Patent.

pat[1] /pæt/ v. & n. ● v. (**patted, patting**) **1** tr. strike gently with the hand or a flat surface. **2** tr. flatten or mould by patting. **3** tr. strike gently with the inner surface of the hand, esp. as a sign of affection, sympathy, or congratulation. **4** intr. (foll. by *on, upon*) beat lightly. ● n. **1** a light stroke or tap, esp. with the hand in affection etc. **2** the sound made by this. **3** a small mass (esp. of butter) formed by patting. □ **pat-a-cake** a child's game with the patting of hands (the first words of a nursery rhyme). **pat on the back** a gesture of approval or congratulation. **pat a person on the back** congratulate a person. [ME, prob. imit.]

pat[2] /pæt/ adj. & adv. ● adj. **1** known thoroughly and ready for any occasion. **2** apposite or opportune, esp. unconvincingly so (*gave a pat answer*). ● adv. **1** in a pat manner. **2** appositely, opportunely. □ **have off** (or **down**) **pat** know or have memorized perfectly. **stand pat** esp. US **1** stick stubbornly to one's opinion or decision. **2** *Poker* retain one's hand as dealt; not draw other cards. □ **patly** adv. **patness** n. [16th c.: rel. to PAT[1]]

pat[3] /pæt/ n. □ **on one's pat** *Austral. sl.* on one's own. [*Pat Malone*, rhyming slang for *own*]

patagium /pəˈteɪdʒɪəm/ n. (pl. **patagia** /-dʒɪə/) *Zool.* **1** the wing-membrane of a bat or similar animal. **2** a scale covering the wing-joint in moths and butterflies. [med.L use of L *patagium* f. Gk *patageion* gold edging]

Patagonia /ˌpætəˈgəʊnɪə/ a region of South America, in southern Argentina and Chile. Consisting largely of a dry barren plateau, it extends from the Colorado river in central Argentina to the Strait of Magellan and from the Andes to the Atlantic coast. □ **Patagonian** adj. & n. [obs. Patagon member of a native people alleged by travellers of the 17th and 18th centuries to be the tallest known people]

Pataliputra /ˌpɑːtlɪˈpʊtrə/ the ancient name for PATNA.

Patavium /pəˈteɪvɪəm/ the Latin name for PADUA.

patball /ˈpætbɔːl/ n. **1** a simple game of ball played between two players. **2** derog. lawn tennis, esp. when slow.

patch /pætʃ/ n. & v. ● n. **1** a piece of material or metal etc. used to mend a hole or as reinforcement. **2** a pad or shield worn to protect an injured eye. **3** a dressing etc. put over a wound. **4** a large or irregular distinguishable area. **5** colloq. a period of time in terms of its characteristic quality (went through a bad patch). **6** a piece of ground. **7** colloq. an area assigned to or patrolled by an authorized person, esp. a police officer. **8** a number of plants growing in one place (brier patch). **9** a scrap or remnant. **10 a** a temporary electrical connection. **b** Computing a small piece of code inserted to correct or enhance a program. **11** hist. a small disc etc. of black silk attached to the face, worn esp. by women in the 17th–18th centuries for adornment. **12** Mil. a piece of cloth on a uniform as the badge of a unit. ● v.tr. **1** (often foll. by up) repair with a patch or patches; put a patch or patches on. **2** (of material) serve as a patch to. **3** (often foll. by up) put together, esp. hastily or in a makeshift way. **4** (foll. by up) settle (a quarrel etc.) esp. hastily or temporarily. **5** (also absol.; foll. by through, into, etc.) make a temporary electrical, radio, etc. connection enabling (a person) to communicate with others. **6** Computing insert a patch in (a program etc.). □ **not a patch on** colloq. greatly inferior to. **patch cord** an insulated lead with a plug at each end, for use with a patchboard. **patch panel** = PATCHBOARD. **patch pocket** one made of a piece of cloth sewn on a garment. **patch test** a test for allergy by applying to the skin patches containing allergenic substances. □ **patcher** n. [ME pacche, patche, perh. var. of peche f. OF pieche dial. var. of piece PIECE]

patchboard /ˈpætʃbɔːd/ n. a board with electrical sockets linked to enable changeable permutations of connection.

patchouli /pəˈtʃuːlɪ, ˈpætʃʊlɪ/ n. **1** an Indo-Malayan labiate shrub of the genus Pogostemon, with strongly scented leaves. **2** the perfume obtained from these leaves. [Tamil pacculi]

patchwork /ˈpætʃwɜːk/ n. & adj. ● n. **1** a piece of needlework consisting of small pieces of fabric, differing in colour and pattern, sewn together to form one article, esp. a quilt. **2** a thing composed of various small pieces or fragments. ● adj. formed out of patchwork; having a pattern similar to patchwork (patchwork fields).

patchy /ˈpætʃɪ/ adj. (**patchier, patchiest**) **1** uneven in quality. **2** having or existing in patches. □ **patchily** adv. **patchiness** n.

pate /peɪt/ n. archaic or colloq. the head, esp. representing the seat of intellect. [ME: orig. unkn.]

pâte /pɑːt/ n. the paste of which porcelain is made. [F, = PASTE]

pâté /ˈpæteɪ/ n. a rich paste or spread of mashed and spiced meat or fish etc. □ **pâté de foie gras** /də fwʌ ˈɡrɑː/ a paste of fatted goose liver. [F f. OF pasté (as PASTY¹)]

patella /pəˈtelə/ n. (pl. **patellae** /-liː/) the kneecap. □ **patellar** adj. **patellate** adj. [L, dimin. of patina: see PATEN]

paten /ˈpæt(ə)n/ n. **1** Eccl. a shallow dish used for the bread at the Eucharist. **2** a thin circular plate of metal. [ME ult. f. OF patene or L patena, patina shallow dish f. Gk patanē a plate]

patent /ˈpeɪt(ə)nt, ˈpæt-/ n., adj., & v. ● n. **1** a government authority to an individual or organization conferring a right or title, esp. the sole right to make or use or sell some invention. (See note below.) **2** a document granting this authority. **3** an invention or process protected by it. ● adj. **1** /ˈpeɪt(ə)nt/ obvious, plain. **2** conferred or protected by patent. **3 a** made and marketed under a patent; proprietary. **b** to which one has a proprietary claim. **4** such as might be patented; ingenious, well-contrived. **5** (of an opening etc.) allowing free passage. ● v.tr. obtain a patent for (an invention). □ **letters patent** an open document from a sovereign or government conferring a patent or other right. **patent leather** leather with a glossy varnished surface. **patent medicine** medicine made and marketed under a patent and available without prescription. **patent office** an office from which patents are issued. **patent roll** (in the UK) a list of patents issued in a year. **patent theatre** a theatre established by royal patent, (in London) the theatres of Covent Garden and Drury Lane, whose patents were granted by

Charles II in 1662. □ **patency** n. **patentable** adj. **patently** /ˈpeɪt(ə)ntlɪ/ adv. (in sense 1 of adj.). [ME f. OF patent and L patere lie open]

▪ Patents for inventions seem to have been introduced in Italy in the 15th century and their use spread to other European states. In England, Elizabeth I and James I granted monopolies to favourites by means of 'letters patent', open letters relating not only to new inventions etc. but also to known commodities. General dissatisfaction with this system led to the passing of the Statute of Monopolies (1623), which declared such grants void but allowed future patents to confer exclusive rights on an inventor for a period of 14 years. In most countries patents are now regulated by statute, and are generally granted for periods of 15–20 years. Recently there has been much controversy over whether patents should be given for new strains of animals and plants produced by genetic engineering.

patentee /ˌpeɪtənˈtiː, ˌpæt-/ n. **1** a person who takes out or holds a patent. **2** a person for the time being entitled to the benefit of a patent.

patentor /ˈpeɪtəntə(r), ˈpæt-, ˌpeɪtənˈtɔː(r), ˌpæt-/ n. a person or body that grants a patent.

Pater /ˈpeɪtə(r)/, Walter (Horatio) (1839–94), English essayist and critic. He came to fame with Studies in the History of the Renaissance (1873), which incorporated his essays on the then neglected Botticelli and on Leonardo da Vinci's Mona Lisa; it had a major impact on the development of the Aesthetic Movement. Pater's other works include Marius the Epicurean (1885), which develops his ideas on 'art for art's sake'.

pater /ˈpeɪtə(r)/ n. Brit. sl. father. ¶ Now only in jocular or affected use. [L]

paterfamilias /ˌpeɪtəfəˈmɪlɪˌæs/ n. the male head of a family or household. [L, = father of the family]

paternal /pəˈtɜːn(ə)l/ adj. **1** of or like or appropriate to a father. **2** fatherly. **3** related through the father. **4** (of a government etc.) that pursues a policy of paternalism. □ **paternally** adv. [LL paternalis f. L paternus f. pater father]

paternalism /pəˈtɜːnəˌlɪz(ə)m/ n. the claim or attempt by a government, company, etc., to take responsibility for the welfare of its people or to regulate their life and limit their responsibilities for their benefit. □ **paternalist** n. **paternalistic** /-ˌtɜːnəˈlɪstɪk/ adj. **paternalistically** adv.

paternity /pəˈtɜːnɪtɪ/ n. **1** fatherhood. **2** one's paternal origin. **3** the source or authorship of a thing. □ **paternity suit** a lawsuit held to determine whether a certain man is the father of a certain child. **paternity test** a blood test to determine whether a man may be or cannot be the father of a particular child. [ME f. OF paternité or LL paternitas]

paternoster /ˌpætəˈnɒstə(r)/ n. **1 a** the Lord's Prayer, esp. in Latin. **b** a rosary bead indicating that this is to be said. **2** a lift consisting of a series of linked doorless compartments moving continuously on a circular belt. [OE f. L pater noster our father]

path /pɑːθ/ n. (pl. **paths** /pɑːðz/) **1** a way or track laid down for walking or made by continual treading. **2** the line along which a person or thing moves (flight path). **3** a course of action or conduct. **4** a sequence of movements or operations taken by a system. □ **pathless** adj. [OE pæth f. WG]

-path /pæθ/ comb. form forming nouns denoting: **1** a practitioner of curative treatment (homeopath; osteopath). **2** a person who suffers from a disease (psychopath). [back-form. f. -PATHY, or f. Gk -pathēs -sufferer (as PATHOS)]

Pathan /pəˈtɑːn/ n. a member of a Pashto-speaking people inhabiting NW Pakistan and SE Afghanistan. [Hindi]

Pathé /ˈpæθeɪ/, Charles (1863–1957), French film pioneer. In 1896 he and his brothers founded a company which dominated the production and distribution of films in the early 20th century, and which initiated the system of leasing (rather than selling) copies of films. The firm also became internationally known for its newsreels, the first of which were introduced in France in 1909. After Charles Pathé's retirement in 1929 the company continued to produce Pathé newsreels until the mid-1950s.

pathetic /pəˈθetɪk/ adj. **1** arousing pity or sadness or contempt. **2** Brit. colloq. miserably inadequate. **3** archaic of the emotions. □ **pathetic fallacy** the attribution of human feelings and responses to inanimate things, esp. in art and literature. □ **pathetically** adv. [F pathétique f. LL patheticus f. Gk pathētikos (as PATHOS)]

pathfinder /ˈpɑːθˌfaɪndə(r)/ n. **1** a person who explores new territory, investigates a new subject, etc. **2** an aircraft or its pilot sent ahead to locate and mark the target area for bombing.

patho- /ˈpæθəʊ/ *comb. form* disease. [Gk *pathos* suffering: see PATHOS]

pathogen /ˈpæθədʒən/ *n.* an agent causing disease. □ **pathogenic** /ˌpæθəˈdʒɛnɪk/ *adj.* **pathogenous** /pəˈθɒdʒənəs/ *adj.* [PATHO- + -GEN]

pathogenesis /ˌpæθəˈdʒɛnɪsɪs/ *n.* (also **pathogeny** /pəˈθɒdʒənɪ/) the manner of development of a disease. □ **pathogenetic** /-dʒɪˈnɛtɪk/ *adj.*

pathological /ˌpæθəˈlɒdʒɪk(ə)l/ *adj.* **1** of pathology. **2** of or caused by a physical or mental disorder (*a pathological fear of spiders*). □ **pathologically** *adv.*

pathology /pəˈθɒlədʒɪ/ *n.* **1** the science of bodily diseases. **2** the symptoms of a disease. □ **pathologist** *n.* [F *pathologie* or mod.L *pathologia* (as PATHO-, -LOGY)]

pathos /ˈpeɪθɒs/ *n.* a quality in speech, writing, events, etc., that excites pity or sadness. [Gk *pathos* suffering, rel. to *paskhō* suffer, *penthos* grief]

pathway /ˈpɑːθweɪ/ *n.* **1** a path or its course. **2** *Biochem.* etc. a sequence of reactions undergone in a living organism.

-pathy /pəθɪ/ *comb. form* forming nouns denoting: **1** curative treatment (*allopathy; homeopathy*). **2** feeling (*telepathy*). [Gk *patheia* suffering]

patience /ˈpeɪʃ(ə)ns/ *n.* **1** calm endurance of hardship, provocation, pain, delay, etc. **2** tolerant perseverance or forbearance. **3** the capacity for calm self-possessed waiting. **4** esp. *Brit.* a game for one player in which cards taken in random order have to be arranged in certain groups or sequences. □ **have no patience with 1** be unable to tolerate. **2** be irritated by. [ME f. OF f. L *patientia* (as PATIENT)]

patient /ˈpeɪʃ(ə)nt/ *adj. & n.* ● *adj.* having or showing patience. ● *n.* a person receiving or registered to receive medical treatment. □ **patiently** *adv.* [ME f. OF f. L *patiens -entis* pres. part. of *pati* suffer]

patina /ˈpætɪnə/ *n.* (*pl.* **patinas**) **1** a film, usu. green, formed on the surface of old bronze. **2** a similar film on other surfaces. **3** a gloss produced by age on woodwork. □ **patinated** /-ˌneɪtɪd/ *adj.* **patination** /ˌpætɪˈneɪʃ(ə)n/ *n.* [It. f. L *patina* dish]

patio /ˈpætɪəʊ/ *n.* (*pl.* **-os**) **1** a paved usu. roofless area adjoining and belonging to a house. **2** an inner court open to the sky in a Spanish or Spanish-American house. [Sp.]

patisserie /pəˈtiːsərɪ/ *n.* **1** a shop where pastries are made and sold. **2** pastries collectively. [F *pâtisserie* f. med.L *pasticium* pastry f. *pasta* PASTE]

Patmore /ˈpætmɔː(r)/, Coventry (Kersey Dighton) (1823–96), English poet. His most important work is *The Angel in the House* (1854–63), a sequence of poems in praise of married love.

Patmos /ˈpætmɒs/ a Greek island in the Aegean Sea, one of the Dodecanese group. It is believed that St John was living there in exile (from AD 95) when he had the visions described in Revelation.

Patna /ˈpætnə/ a city in NE India, on the Ganges, capital of the state of Bihar; pop. (1991) 917,000. Originally known as Pataliputra, it was the capital between the 5th and 1st centuries BC of the Magadha kingdom and in the 4th century AD of the Gupta dynasty. After this it declined and had become deserted by the 7th century. It was refounded in 1541 by the Moguls, becoming a prosperous city and viceregal capital. It lies in the centre of a rice-growing region.

Patna rice *n.* a variety of long-grained rice, orig. that of Patna, now also grown elsewhere, esp. in the US.

patois /ˈpætwɑː/ *n.* (*pl.* same /-wɑːz/) the dialect of the common people in a region, differing fundamentally from the literary language. [F, = rough speech, perh. f. OF *patoier* treat roughly f. *patte* paw]

Paton /ˈpeɪt(ə)n/, Alan (Stewart) (1903–88), South African writer and politician. He is best known for his novel *Cry, the Beloved Country* (1948), a passionate indictment of the apartheid system. Paton helped found the South African Liberal Party in 1953, later becoming its president until it was banned in 1968.

Patras /pəˈtræs, ˈpætrəs/ (Greek **Pátrai** /ˈpatrɛ/) an industrial port in the NW Peloponnese, on the Gulf of Patras; pop. (1991) 155,000. Taken by the Turks in the 18th century, it was the site in 1821 of the outbreak of the Greek war of independence. It was finally freed in 1828.

patrial /ˈpeɪtrɪəl/ *adj. & n. Brit. hist.* ● *adj.* having the right to live in the UK through the British birth of a parent or a grandparent. ● *n.* a person with this right. □ **patriality** /ˌpeɪtrɪˈælɪtɪ/ *n.* [obs. F *patrial* or med.L *patrialis* f. L *patria* fatherland f. *pater* father]

patriarch /ˈpeɪtrɪˌɑːk/ *n.* **1** a man who is the head of a family or tribe. **2** (often in *pl.*) *Bibl.* any of those regarded as fathers of the human race, esp. the sons of Jacob, or Abraham, Isaac, and Jacob, and their forefathers. **3** *Eccl.* **a** the title of a chief bishop, esp. those presiding over the Churches of Antioch, Alexandria, Constantinople, and (formerly) Rome; now also the title of the heads of certain autocephalous Orthodox Churches. **b** (in the Roman Catholic Church) a bishop ranking next above primates and metropolitans, and immediately below the pope. **c** the head of a Uniat community. **4 a** the founder of an order, science, etc. **b** a venerable old man. **c** the oldest member of a group. □ **patriarchal** /ˌpeɪtrɪˈɑːk(ə)l/ *adj.* **patriarchally** *adv.* [ME f. OF *patriarche* f. eccl.L *patriarcha* f. Gk *patriarkhēs* f. *patria* family f. *patēr* father + -*arkhēs* -ruler]

patriarchate /ˈpeɪtrɪˌɑːkət/ *n.* **1** the office, see, or residence of an ecclesiastical patriarch. **2** the rank of a tribal patriarch. [med.L *patriarchatus* (as PATRIARCH)]

patriarchy /ˈpeɪtrɪˌɑːkɪ/ *n.* (*pl.* **-ies**) a form of social organization or government etc. in which a man or men rule and descent is reckoned through the male line. □ **patriarchism** *n.* [med.L *patriarchia* f. Gk *patriarkhia* (as PATRIARCH)]

patrician /pəˈtrɪʃ(ə)n/ *n. & adj.* ● *n.* **1** *hist.* a member of the ancient Roman nobility (cf. PLEBEIAN). **2** *hist.* a nobleman in some Italian republics. **3** an aristocrat. ● *adj.* **1** noble, aristocratic. **2** *hist.* of the ancient Roman nobility. [ME f. OF *patricien* f. L *patricius* having a noble father f. *pater patris* father]

patriciate /pəˈtrɪʃət/ *n.* **1** a patrician order; an aristocracy. **2** the rank of patrician. [L *patriciatus* (as PATRICIAN)]

patricide /ˈpætrɪˌsaɪd/ *n.* = PARRICIDE (esp. with reference to the killing of one's father). □ **patricidal** /ˌpætrɪˈsaɪd(ə)l/ *adj.* [LL *patricida, patricidium*, alt. of L *parricida, parricidium* (see PARRICIDE) after *pater* father]

Patrick, St /ˈpætrɪk/ (5th century), Apostle and patron saint of Ireland. His *Confession* is the chief source for the events of his life. Of Romano-British parentage, he was captured at the age of 16 by raiders and shipped to Ireland as a slave; there he experienced a religious conversion. Escaping after six years, probably to Gaul, he was ordained and returned to Ireland in about 432. Many of the details of his mission are uncertain but it is known that he founded the archiepiscopal see of Armagh in about 454. Feast day, 17 March.

patrilineal /ˌpætrɪˈlɪnɪəl/ *adj.* of or relating to, or based on kinship with, the father or descent through the male line. [L *pater patris* father + LINEAL]

patrimony /ˈpætrɪmənɪ/ *n.* (*pl.* **-ies**) **1** property inherited from one's father or ancestor. **2** a heritage. **3** the endowment of a church etc. □ **patrimonial** /ˌpætrɪˈməʊnɪəl/ *adj.* [ME *patrimoigne* f. OF *patrimoine* f. L *patrimonium* f. *pater patris* father]

patriot /ˈpætrɪət, ˈpeɪt-/ *n.* a person who is devoted to and ready to support or defend his or her country. □ **patriotism** *n.* **patriotic** /ˌpætrɪˈɒtɪk, ˌpeɪt-/ *adj.* **patriotically** *adv.* [F *patriote* f. LL *patriota* f. Gk *patriōtēs* f. *patrios* of one's fathers f. *patris* fatherland]

patristic /pəˈtrɪstɪk/ *adj.* of the early Christian writers (the Fathers of the Church) or their work. [G *patristisch* f. L *pater patris* father]

patristics /pəˈtrɪstɪks/ *n.pl.* (usu. treated as *sing.*) the branch of Christian theology that deals with the early Christian theologians or their writings. [as PATRISTIC]

Patroclus /pəˈtrɒkləs/ *Gk Mythol.* a Greek hero of the Trojan War, the close friend of Achilles. The *Iliad* describes how Patroclus' death at the hands of Hector led Achilles to return to battle.

patrol /pəˈtrəʊl/ *n. & v.* ● *n.* **1** the act of walking or travelling around an area, esp. at regular intervals, in order to protect or supervise it. **2** one or more persons or vehicles assigned or sent out on patrol, esp. a detachment of guards, police, etc. **3 a** a detachment of troops sent out to reconnoitre. **b** such reconnaissance. **4** a routine operational voyage of a ship or aircraft. **5** a routine monitoring of astronomical or other phenomena. **6** *Brit.* an official controlling traffic where children cross the road. **7** a unit of six to eight Scouts or Guides. ● *v.* (**patrolled**, **patrolling**) **1** *tr.* carry out a patrol of. **2** *intr.* act as a patrol. □ **patrol car** a police car used in patrolling roads and streets. **patrol wagon** esp. *US* a police van for transporting prisoners. □ **patroller** *n.* [F *patrouiller* paddle in mud f. *patte* paw: (n.) f. G *Patrolle* f. F *patrouille*]

patrolman /pəˈtrəʊlmən/ *n.* (*pl.* **-men**) *US* a policeman of the lowest rank.

patrology /pəˈtrɒlədʒɪ/ *n.* (*pl.* **-ies**) **1** patristics. **2** a collection of the writings of the early Christian theologians. □ **patrologist** *n.* **patrological** /ˌpætrəˈlɒdʒɪk(ə)l/ *adj.* [Gk *patēr patros* father]

patron /ˈpeɪtrən/ *n.* (*fem.* **patroness** /-nɪs/) **1** a person who gives financial or other support to a person, cause, work of art, etc., esp. one who buys works of art, or takes an honorary position in a charity etc. **2** a usu. regular customer of a shop etc. **3** *Rom. Antiq.* **a** the former owner of a freed slave. **b** the protector of a client. **4** *Brit.* a person or institution with the right to present a member of the clergy to a benefice. □ **patron saint** the protecting or guiding saint of a person, place, etc. [ME f. OF f. L *patronus* protector of clients, defender f. *pater patris* father]

patronage /ˈpætrənɪdʒ/ *n.* **1** the support, promotion, or encouragement given by a patron. **2** the control of appointments to office, privileges, etc. **3** a patronizing or condescending manner. **4** a customer's support for a shop etc. **5** *Rom. Antiq.* the rights and duties or position of a patron. [ME f. OF (as PATRON)]

patronal /pəˈtrəʊn(ə)l/ *adj.* of or relating to a patron saint (*the patronal festival*). [F *patronal* or LL *patronalis* (as PATRON)]

patronize /ˈpætrənaɪz/ *v.tr.* (also **-ise**) **1** treat condescendingly. **2** act as a patron towards (a person, cause, artist, etc.); support; encourage. **3** frequent (a shop etc.) as a customer. □ **patronizer** *n.* **patronizing** *adj.* **patronizingly** *adv.* **patronization** /ˌpætrənaɪˈzeɪʃ(ə)n/ *n.* [obs. F *patroniser* or med.L *patronizare* (as PATRON)]

patronymic /ˌpætrəˈnɪmɪk/ *n. & adj.* ● *n.* a name derived from the name of a father or ancestor, e.g. *Johnson, O'Brien, Ivanovich.* ● *adj.* (of a name) so derived. [LL *patronymicus* f. Gk *patrōnumikos* f. *patrōnumos* f. *patēr patros* father + *onuma, onoma* name]

patroon /pəˈtruːn/ *n.* *US hist.* a landowner with manorial privileges under the Dutch governments of New York and New Jersey. [Du., = PATRON]

patsy /ˈpætsɪ/ *n.* (*pl.* **-ies**) esp. *US sl.* a person who is deceived, ridiculed, tricked, etc. [20th c.: orig. unkn.]

Pattaya /pæˈtaɪə/ a resort on the coast of southern Thailand, southeast of Bangkok.

pattée /ˈpæteɪ, -tɪ/ *adj.* (of a cross) having almost triangular arms becoming very broad at the ends so as to form a square. [F f. *patte* paw]

patten /ˈpæt(ə)n/ *n.* *hist.* a shoe or clog with a raised sole or set on an iron ring, for walking in mud etc. [ME f. OF *patin* f. *patte* paw]

patter[1] /ˈpætə(r)/ *v. & n.* ● *v.* **1** *intr.* make a rapid succession of taps, as of rain on a window-pane. **2** *intr.* run with quick short steps. **3** *tr.* cause (water etc.) to patter. ● *n.* a rapid succession of taps, short light steps, etc. [PAT[1]]

patter[2] /ˈpætə(r)/ *n. & v.* ● *n.* **1 a** the rapid speech used by a comedian or introduced into a song. **b** the words of a comic song. **2** the words used by a person selling or promoting a product; a sales pitch. **3** the special language or jargon of a profession, class, etc. **4** *colloq.* mere talk; chatter. ● *v.* **1** *tr.* repeat (prayers etc.) in a rapid mechanical way. **2** *intr.* talk glibly or mechanically. [ME f. *pater* = PATERNOSTER]

pattern /ˈpæt(ə)n/ *n. & v.* ● *n.* **1** a repeated decorative design on wallpaper, cloth, a carpet, etc. **2** a regular or logical form, order, or arrangement of parts (*behaviour pattern; the pattern of one's daily life*). **3** a model or design, e.g. of a garment, from which copies can be made. **4** an example of excellence; an ideal; a model (*a pattern of elegance*). **5** a wooden or metal figure from which a mould is made for a casting. **6** a sample (of cloth, wallpaper, etc.). **7** the marks made by shots, bombs, etc. on a target or target area. **8** a random combination of shapes or colours. ● *v.tr.* **1** (usu. foll. by *after, on*) model (a thing) on a design etc. **2** decorate with a pattern. □ **pattern bombing** bombing over a large area, not on a single target. [ME *patron* (see PATRON): differentiated in sense and spelling since the 16th–17th c.]

patty /ˈpætɪ/ *n.* (*pl.* **-ies**) **1** a little pie or pastry. **2** a small flat cake of minced meat etc. **3** esp. *N. Amer.* a small round flat sweet. [F *pâté* PASTY[1]]

pattypan /ˈpætɪˌpæn/ *n.* a pan for baking a patty.

patulous /ˈpætjʊləs/ *adj.* **1** esp. *literary* (of branches etc.) spreading. **2** *formal* open; expanded. [L *patulus* f. *patere* be open]

paua /ˈpaʊə/ *n.* **1** a large edible abalone (gastropod mollusc), *Haliotis iris*, native to New Zealand. **2** its ornamental shell. **3** a fish-hook made from this. [Maori]

paucity /ˈpɔːsɪtɪ/ *n.* smallness of number or quantity. [ME f. OF *paucité* or f. L *paucitas* f. *paucus* few]

Paul /pɔːl/ (Les (born Lester Polfus) (b.1915), American jazz guitarist. In 1946 he invented the solid-body electric guitar for which he is best known (see GUITAR); it was first promoted in 1952 as the Gibson Les Paul guitar. Paul was also among the first to use such recording techniques as overdubbing. In the 1950s he wrote and recorded a number of hit songs with his wife, Mary Ford (1928–77), such as 'Mockin' Bird Hill' (1951).

Paul III /pɔːl/ (born Alessandro Farnese) (1468–1549), Italian pope 1534–49. He excommunicated Henry VIII of England in 1538, instituted the order of the Jesuits in 1540, and initiated the Council of Trent in 1545. Paul III was also a keen patron of the arts, commissioning Michelangelo to paint the fresco of the *Last Judgement* for the Sistine Chapel and to design the dome of St Peter's in Rome.

Paul, St (known as Paul the Apostle, or Saul of Tarsus, or 'the Apostle of the Gentiles') (died *c.*64), missionary of Jewish descent. He was brought up as a Pharisee and at first opposed the followers of Jesus, assisting at the martyrdom of St Stephen. On a mission to Damascus, he was converted to Christianity after a vision and became one of the first major Christian missionaries and theologians. His missionary journeys are described in the Acts of the Apostles, and his epistles form part of the New Testament. He was martyred in Rome. Feast day, 29 June.

Pauli /ˈpaʊlɪ/, Wolfgang (1900–58), Austrian-born American physicist who worked chiefly in Switzerland. He is best known for the *exclusion principle*, according to which only two electrons in an atom could occupy the same quantum level, provided they had opposite spins. This made it easier to understand the structure of the atom and the chemical properties of the elements, and was later extended to a whole class of subatomic particles, the fermions, which includes the electron. In 1931 he postulated the existence of the neutrino, later discovered by Enrico Fermi. He was awarded the 1945 Nobel Prize for physics.

Pauline /ˈpɔːlaɪn/ *adj.* of or relating to St Paul (*the Pauline epistles*). [ME f. med.L *Paulinus* f. L *Paulus* Paul]

Pauling /ˈpɔːlɪŋ/, Linus Carl (1901–94), American chemist. He is particularly renowned for his study of molecular structure and chemical bonding, especially of complex biological macromolecules, for which he received the 1954 Nobel Prize for chemistry. His suggestion of the helix as a possible structure for proteins formed the foundation for the later elucidation of the structure of DNA. Pauling also proved that sickle-cell anaemia is caused by a defect in haemoglobin at the molecular level. After the war he became increasingly involved with attempts to ban nuclear weapons, for which he was awarded the Nobel Peace Prize in 1962.

Paul Jones *n.* a ballroom dance during which the dancers change partners after circling in concentric rings of men and women. The dance is named after the Scottish-born American admiral John Paul Jones (1747–92).

paulownia /pɔːˈləʊnɪə/ *n.* a Chinese tree of the genus *Paulownia*, with fragrant blue or lilac flowers. [Russ. *pavlovniya*, f. Anna *Pavlovna*, Russian princess (1795–1865)]

Paul Pry *n.* an inquisitive person. [a character in a US song of 1820]

paunch /pɔːntʃ/ *n. & v.* ● *n.* **1** the belly or stomach, esp. when protruding. **2** *Naut.* a thick strong mat used to give protection from chafing, esp. on a mast or spar. ● *v.tr.* disembowel (an animal). □ **paunchy** *adj.* (**paunchier, paunchiest**). **paunchiness** *n.* [ME f. AF *pa(u)nche*, ONF *panche* ult. f. L *pantex panticis* bowels]

pauper /ˈpɔːpə(r)/ *n.* **1** a person without means; a beggar. **2** *hist.* a recipient of poor-law relief. **3** *Law* a person who may sue *in forma pauperis*. □ **pauperdom** *n.* **pauperism** *n.* **pauperize** *v.tr.* (also **-ise**). **pauperization** /ˌpɔːpəraɪˈzeɪʃ(ə)n/ *n.* [L, = poor]

Pausanias /pɔːˈseɪnɪəs/ (2nd century), Greek geographer and historian. His *Description of Greece* (also called the *Itinerary of Greece*) is a guide to the topography and remains of ancient Greece and is still considered an invaluable source of information.

pause /pɔːz/ *n. & v.* ● *n.* **1** an interval of inaction, esp. when due to hesitation; a temporary stop. **2** a break in speaking or reading; a silence. **3** *Mus.* a mark (⌢) over a note or rest that is to be lengthened by an unspecified amount. **4** a control allowing the interruption of the operation of a tape recorder etc. ● *v.* **1** *intr.* make a pause; wait. **2** *intr.* (usu. foll. by *upon*) linger over (a word etc.). **3** *tr.* cause to hesitate or pause. □ **give pause to** cause (a person) to hesitate. [ME f. OF *pause* or L *pausa* f. Gk *pausis* f. *pauō* stop]

pavage /ˈpeɪvɪdʒ/ *n.* **1** paving. **2** *hist.* a tax or toll towards the paving of streets. [ME f. OF f. *paver* PAVE]

pavane /pəˈvɑːn, ˈpæv(ə)n/ *n.* (also **pavan**) **1** *hist.* a stately dance in slow duple time, performed in elaborate clothing, and popular in the 16th century. **2** the music for this. [F *pavane* f. Sp. *pavana*, perh. f. *pavon* peacock]

Pavarotti /ˌpævəˈrɒtɪ/, Luciano (b.1935), Italian operatic tenor. He made his début as Rudolfo in Puccini's *La Bohème* in 1961, and achieved rapid success in this and a succession of other leading roles, including Edgardo in Donizetti's *Lucia di Lammermoor* and the Duke in Verdi's

Rigoletto (both 1965). Widely acclaimed for his bel canto singing, Pavarotti has appeared in concerts and on TV throughout the world, and has made many recordings.

pave /peɪv/ *v.tr.* **1** cover (a street, floor, etc.) with asphalt, stone, etc. **2** cover or strew (a floor etc.) with anything (*paved with flowers*). □ **pave the way for** prepare for; facilitate. **paving-stone** a large flat usu. rectangular piece of stone etc. for paving. □ **paver** *n.* **paving** *n.* **pavior** /ˈpeɪvjə(r)/ *n.* (also **paviour**). [ME f. OF *paver*, back-form. (as PAVEMENT)]

pavé /ˈpæveɪ/ *n.* **1** a paved street, road, or path. **2** a setting of jewels placed closely together. [F, past part. of *paver*: see PAVE]

pavement /ˈpeɪvmənt/ *n.* **1** Brit. a paved path for pedestrians at the side of a road a little higher than a road. **2** the covering of a street, floor, etc., made of tiles, wooden blocks, asphalt, and esp. of rectangular stones. **3** US a roadway. **4** *Zool.* a pavement-like formation of close-set teeth, scales, etc. □ **pavement artist 1** Brit. an artist who draws on paving-stones or paper laid on a pavement with coloured chalks, hoping to be given money by passers-by. **2** US an artist who displays paintings for sale on a pavement. [ME f. OF f. L *pavimentum* f. *pavire* beat, ram]

Pavese /pæˈveɪsɪ/, Cesare (1908–50), Italian novelist, poet, and translator. He is best known for his last novel *La Luna e i falò* (1950), in which he portrays isolation and the failure of communication as a general human predicament. Pavese also made many important translations of works written in English, including novels by Herman Melville, James Joyce, and William Faulkner. He committed suicide in 1950.

pavilion /pəˈvɪljən/ *n. & v.* ● *n.* **1** Brit. a building at a cricket or other sports ground used for changing, refreshments, etc. **2** a summerhouse or other decorative building in a garden. **3** a tent, esp. a large one with crenellated decorations at a show, fair, etc. **4** a building used for entertainments. **5** a temporary stand at an exhibition. **6** a detached building at a hospital. **7** *Archit.* a usu. highly decorated projecting subdivision of a building. **8** the part of a cut gemstone below the girdle. ● *v.tr.* enclose in or provide with a pavilion. [ME f. OF *pavillon* f. L *papilio -onis* butterfly, tent]

Pavlov /ˈpævlɒf/, Ivan (Petrovich) (1849–1936), Russian physiologist. He was awarded a Nobel Prize in 1904 for his work on digestion, but is best known for his later studies on the conditioned reflex. He showed by experiment with dogs how the secretion of saliva can be stimulated not only by food but also by the sound of a bell associated with the presentation of food, and that this sound comes to elicit salivation when presented alone. Pavlov applied his findings to show the importance of such reflexes in human and animal behaviour. His experiments form the basis for much current research in the field of conditioning. □ **Pavlovian** /pævˈləʊvɪən/ *adj.*

Pavlova /ˈpævləvə, pævˈləʊvə/ Anna (Pavlovna) (1881–1931), Russian dancer, resident in Britain from 1912. As the prima ballerina of the Russian Imperial Ballet, she toured Russia and northern Europe in 1907 and 1908. Her highly acclaimed solo dance *The Dying Swan* was created for her by Michel Fokine in 1905. After brief appearances with the Ballets Russes in 1909, Pavlova made her New York and London débuts the following year. On settling in Britain, she formed her own company and embarked on numerous tours which made her a pioneer of classical ballet all over the world.

pavlova /pævˈləʊvə/ *n.* a meringue cake with cream and fruit. [PAVLOVA]

pavonine /ˈpævəˌnaɪn/ *adj.* of or like a peacock. [L *pavoninus* f. *pavo -onis* peacock]

paw /pɔː/ *n. & v.* ● *n.* **1** a foot of an animal having claws or nails. **2** *colloq.* a person's hand. ● *v.* **1** *tr.* strike or scrape with a paw or foot. **2** *intr.* scrape the ground with a paw or hoof. **3** *tr. colloq.* fondle awkwardly or indecently. [ME *pawe, powe* f. OF *poue* etc. ult. f. Frank.]

pawky /ˈpɔːkɪ/ *adj.* (**pawkier, pawkiest**) Sc. & dial. **1** drily humorous. **2** shrewd. □ **pawkily** *adv.* **pawkiness** *n.* [Sc. & N. Engl. dial. *pawk* trick, of unkn. orig.]

pawl /pɔːl/ *n. & v.* ● *n.* **1** a lever with a catch for the teeth of a wheel or bar. **2** *Naut.* a short bar used to lock a capstan, windlass, etc., to prevent it from recoiling. ● *v.tr.* secure (a capstan etc.) with a pawl. [perh. f. LG & Du. *pal*, rel. to *pal* fixed]

pawn[1] /pɔːn/ *n.* **1** Chess a piece of the smallest size and value. **2** a person used by others for their own purposes. [ME f. AF *poun*, OF *peon* f. med.L *pedo -onis* foot-soldier f. L *pes pedis* foot: cf. PEON]

pawn[2] /pɔːn/ *v. & n.* ● *v.tr.* **1** deposit an object, esp. with a pawnbroker, as security for money lent. **2** pledge or wager (one's life, honour, word, etc.). ● *n.* **1** an object left as security for money etc. lent. **2** anything or any person left with another as security etc. □ **in** (or **at**) **pawn** (of an object etc.) held as security. [ME f. OF *pan, pand, pant*, pledge, security f. WG]

pawnbroking /ˈpɔːnˌbrəʊkɪŋ/ *n.* the lending of money at interest on the security of personal property deposited with the pawnbroker. Such a practice existed in China more than 2,000 years ago, but in Europe it dates from the Middle Ages. It was introduced into England by the Lombards in the 13th century, and remained important for poorer people though is now less common owing to the easier availability of credit and hire purchase. □ **pawnbroker** *n.*

pawnshop /ˈpɔːnʃɒp/ *n.* a shop where pawnbroking is conducted.

pawpaw /ˈpɔːpɔː/ *n.* (also **papaw**) **1** (also **papaya** /pəˈpaɪə/) **a** a tropical fruit shaped like an elongated melon, with edible orange flesh and small black seeds. **b** a tropical American tree, *Carica papaya*, bearing this fruit and producing a milky sap from which papain is obtained. **2** US a North American tree, *Asimina triloba*, with purple flowers and edible fruit. [earlier *papay(a)* f. Sp. & Port. *papaya*, of Carib orig.]

PAX *abbr.* private automatic (telephone) exchange.

pax /pæks/ *n.* **1** the kiss of peace. **2** (as *int.*) Brit. sl. a call for a truce (used esp. by schoolchildren). [ME f. L, = peace]

Paxton /ˈpækstən/ Sir Joseph (1801–65), English gardener and architect. He became head gardener to the Duke of Devonshire at Chatsworth House in Derbyshire in 1826, and designed a series of glass-and-iron greenhouses. He later reworked these, making the first known use of prefabricated materials, in his design for the Crystal Palace (1851).

pay[1] /peɪ/ *v. & n.* ● *v.tr.* (past and past part. **paid** /peɪd/) **1** (also *absol.*) give (a person etc.) what is due for services done, goods received, debts incurred, etc. (*paid him in full; I assure you I have paid*). **2 a** give (a usu. specified amount) for work done, a debt, a ransom, etc. (*they pay £6 an hour*). **b** (foll. by *to*) hand over the amount of (a debt, wages, recompense, etc.) to (*paid the money to the assistant*). **3 a** give, bestow, or express (attention, respect, a compliment, etc.) (*paid them no heed*). **b** make (a visit, a call, etc.) (*paid a visit to their uncle*). **4** (also *absol.*) (of a business, undertaking, attitude, etc.) be profitable or advantageous to (a person etc.). **5** reward or punish (*can never pay you for what you have done for us; I shall pay you for that*). **6** (usu. as **paid** *adj.*) recompense (work, time, etc.) (*paid holiday*). **7** (usu. foll. by *out, away*) let out (a rope) by slackening it. ● *n.* wages; payment. □ **in the pay of** employed by. **paid holidays** an agreed holiday period for which wages are paid as normal. **paid-up member** (esp. of a trade-union member) a person who has paid the subscriptions in full. **pay-as-you-earn** Brit. (abbr. **PAYE**) the deduction of income tax from wages at source. **pay back 1** return (money). **2** take revenge on (a person). **pay-bed** a hospital bed for private patients. **pay-claim** a demand for an increase in pay, esp. by a trade union. **pay day** a day on which payment, esp. of wages, is made or expected to be made. **pay dearly** (usu. foll. by *for*) **1** obtain at a high cost, great effort, etc. **2** suffer for a wrongdoing etc. **pay dirt** (or **gravel**) US **1** *Mineral.* ground worth working for ore. **2** a financially promising situation. **pay envelope** US = *pay-packet*. **pay for 1** hand over the price of. **2** bear the cost of. **3** suffer or be punished for (a fault etc.). **pay in** pay (money) into a bank account. **paying guest** a boarder. **pay its** (or **one's**) **way** cover costs; not be indebted. **pay one's last respects** show respect towards a dead person by attending the funeral. **pay off 1** dismiss (workers) with a final payment. **2** *colloq.* yield good results; succeed. **3** pay (a debt) in full. **4** (of a ship) turn to leeward through the movement of the helm. **pay-off** *n.* **1** an act of payment. **2** return on investment or on a bet. **3** a final outcome. **4** *colloq.* a bribe; bribery. **pay out** (or **back**) punish or be revenged on. **pay-packet** Brit. a packet or envelope containing an employee's wages. **pay party** a private party at which guests pay an entrance fee. **pay phone** a coin-box telephone. **pay the piper and call the tune** pay for, and therefore have control over, a proceeding. **pay one's respects** make a polite visit. **pay station** US = *pay phone*. **pay through the nose** *colloq.* pay much more than a fair price. **pay up** pay the full amount, or the full amount of. **put paid to** *colloq.* **1** deal effectively with (a person). **2** terminate (hopes etc.). □ **payer** *n.* [ME f. OF *paie, payer* f. L *pacare* appease f. *pax pacis* peace]

pay[2] /peɪ/ *v.tr.* (past and past part. **payed**) *Naut.* smear (a ship) with pitch, tar, etc. as a defence against wet. [OF *peier* f. L *picare* f. *pix picis* PITCH[2]]

payable /ˈpeɪəb(ə)l/ *adj.* **1** that must be paid; due (*payable in April*). **2** that may be paid. **3** (of a mine etc.) profitable.

payback /ˈpeɪbæk/ *n.* **1** a financial return; a reward. **2** the profit from an investment etc., esp. one equal to the initial outlay. □ **payback period** the length of time required for an investment to pay for itself in terms of profits or savings.

PAYE *abbr. Brit.* pay-as-you-earn.

payee /peɪˈiː/ n. a person to whom money is paid or is to be paid.

payload /ˈpeɪləʊd/ n. **1** the part of an aircraft's load from which revenue is derived. **2 a** the explosive warhead carried by an aircraft or rocket. **b** the instruments etc. carried by a spaceship. **3** the goods carried by a road vehicle.

paymaster /ˈpeɪˌmɑːstə(r)/ n. **1** an official who pays troops, workers, etc. **2** a person, organization, etc., to whom another owes duty or loyalty because of payment given. **3** (in full **Paymaster General**) *Brit.* the minister at the head of the Treasury department responsible for payments.

payment /ˈpeɪmənt/ n. **1** the act or an instance of paying. **2** an amount paid. **3** reward, recompense. [ME f. OF *paiement* (as PAY¹)]

paynim /ˈpeɪnɪm/ n. *archaic* **1** a pagan. **2** a non-Christian, esp. a Muslim. [ME f. OF *pai(e)nime* f. eccl.L *paganismus* heathenism (as PAGAN)]

payola /peɪˈəʊlə/ n. esp. *US* **1** a bribe offered in return for unofficial promotion of a product etc. in the media. **2** the practice of such bribery. [PAY¹ + -*ola* as in *Victrola*, make of gramophone]

payroll /ˈpeɪrəʊl/ n. a list of employees receiving regular pay.

paysage /peɪˈzɑːʒ/ n. **1** a rural scene; a landscape. **2** landscape painting. □ **paysagist** /ˈpeɪzaːʒɪst/ n. [F f. *pays* country: see PEASANT]

Pays Basque /peɪ bask/ the French name for the BASQUE COUNTRY.

Pays de la Loire /ˌpeɪ də læ ˈlwaː(r)/ a region of western France, on the Bay of Biscay, centred on the Loire valley.

Paz /pæz/, Octavio (b.1914), Mexican poet and essayist. His poems are noted for their preoccupation with Aztec mythology, as in *Sun Stone* (1957). He is also known for his essays, particularly *The Labyrinth of Solitude* (1950), a critique of Mexican culture, and *Postscript* (1970), a response to Mexico's brutal suppression of student demonstrators in 1968. Paz was awarded the Nobel Prize for literature in 1990.

Pb *symb. Chem.* the element lead. [L *plumbum*]

PBX *abbr.* private branch exchange (private telephone switchboard).

PC *abbr.* **1** (in the UK) police constable. **2** (in the UK) Privy Counsellor. **3** personal computer. **4** political correctness, politically correct.

p.c. *abbr.* **1** per cent. **2** postcard.

PCB *abbr.* **1** *Computing* printed circuit board. **2** *Chem.* polychlorinated biphenyl, any of a class of toxic aromatic compounds containing two benzene molecules in which hydrogens have been replaced by chlorine atoms, formed as waste in industrial processes.

P-Celtic see BRYTHONIC.

PCM *abbr. Electronics* pulse code modulation.

PCMCIA *abbr. Computing* Personal Computer Memory Card International Association, denoting a standard specification for memory cards and interfaces used in small portable computers.

PCP *abbr.* **1** *Pharm.* = PHENCYCLIDINE. **2** *Med.* pneumocystis carinii pneumonia, a fatal lung infection esp. of immunodeficient patients.

pct. *abbr. US* per cent.

PD *abbr. US* Police Department.

Pd *symb. Chem.* the element palladium.

pd. *abbr.* paid.

p.d.q. *abbr. colloq.* pretty damn quick.

PDT *abbr.* Pacific Daylight Time, one hour ahead of Pacific Standard Time.

PE *abbr.* physical education.

p/e *abbr.* price/earnings (ratio).

pea /piː/ n. **1 a** a hardy leguminous climbing plant, *Pisum sativum*, with round green seeds growing in pods and used for food. **b** its seed. **2** any similar leguminous plant (*sweet pea; chick pea*). □ **pea-brain** *colloq.* a stupid or dim-witted person. **pea-green** bright green. **pea-souper** *Brit. colloq.* a thick yellowish fog. [back-form. f. PEASE (taken as pl.: cf. CHERRY)]

peace /piːs/ n. **1 a** quiet; tranquillity (*needs peace to work well*). **b** mental calm; serenity (*peace of mind*). **2 a** (often *attrib.*) freedom from, or the cessation of, war (*peace talks*). **b** (esp. **Peace**) a treaty of peace between states etc. at war. **3** freedom from civil disorder. **4** *Eccl.* a ritual liturgical greeting. □ **at peace 1** in a state of friendliness. **2** serene. **3** *euphem.* dead. **hold one's peace** keep silence. **keep the peace** prevent, or refrain from, strife. **make one's peace** (often foll. by *with*) re-establish

friendly relations. **make peace** bring about peace; reconcile. **the peace** (or **queen's** or **king's peace**) peace existing within a realm; civil order. **peace dividend** money saved by the reduction of expenditure on weapons and defence, esp. after the relaxation of tension between Western and Soviet bloc countries in 1990. **peace-offering 1** a propitiatory or conciliatory gift. **2** *Bibl.* an offering presented as a thanksgiving to God. **peace-pipe** a tobacco-pipe as a token of peace among North American Indians. **peace studies** analysis of international relations, defence policies, the role of the military, etc., as a school or college subject. [ME f. AF *pes*, OF *pais* f. L *pax pacis*]

peaceable /ˈpiːsəb(ə)l/ adj. **1** disposed to peace; unwarlike. **2** free from disturbance; peaceful. □ **peaceableness** n. **peaceably** adv. [ME f. OF *peisible, plaisible* f. LL *placibilis* pleasing f. L *placere* please]

Peace Corps *US* an organization sending young people to work as volunteers in developing countries.

peaceful /ˈpiːsfʊl/ adj. **1** characterized by peace; tranquil. **2** not violating or infringing peace (*peaceful coexistence*). **3** belonging to a state of peace. □ **peacefully** adv. **peacefulness** n.

peacemaker /ˈpiːsˌmeɪkə(r)/ n. a person who brings about peace. □ **peacemaking** n. & adj.

peace movement n. a broad movement opposed to preparations for war, particularly any of those organizations in Britain and western Europe attempting since the 1950s to bring about a reduction in or elimination of nuclear weapons. In Britain the movement was particularly active in the late 1970s when opposing plans to site US cruise missiles on British soil; the Campaign for Nuclear Disarmament held huge demonstrations in London, while women protesters set up camps at US bases such as that at Greenham Common in Berkshire. Since the ending of the cold war the movement has been less prominent.

peacenik /ˈpiːsnɪk/ n. often *derog.* a pacifist. [PEACE + -NIK]

peacetime /ˈpiːstaɪm/ n. a period when a country is not at war.

peach¹ /piːtʃ/ n. **1 a** a round juicy stone-fruit with downy cream or yellow skin flushed with red. **b** the rosaceous tree, *Prunus persica*, bearing it. **2** the yellowish-pink colour of a peach. **3** *colloq.* **a** a person or thing of superlative quality. **b** often *offens.* an attractive young woman. □ **peach-bloom** an oriental porcelain-glaze of reddish-pink, usu. with green markings. **peach-blow 1** a delicate purplish-pink colour. **2** = *peach-bloom*. **peaches and cream** (of a complexion) creamy skin with downy pink cheeks. **peach Melba** see MELBA². □ **peachy** adj. (**peachier, peachiest**). **peachiness** n. [ME f. OF *peche, pesche,* f. med.L *persica* f. L *persicum* (*malum*), lit. 'Persian apple']

peach² /piːtʃ/ v. intr. (usu. foll. by *against, on*) *colloq.* turn informer; inform. **2** tr. archaic inform against. [ME f. *appeach* f. AF *enpecher*, OF *empechier* IMPEACH]

pea-chick /ˈpiːtʃɪk/ n. a young peafowl. [formed as PEACOCK + CHICK¹]

Peacock /ˈpiːkɒk/, Thomas Love (1785–1866), English novelist and poet. He is chiefly remembered for his prose satires, including *Nightmare Abbey* (1818) and *Crotchet Castle* (1831), lampooning the romantic poets.

peacock /ˈpiːkɒk/ n. **1** a male peafowl, having brilliant plumage and a tail (with eyelike markings) that can be expanded erect in display like a fan. **2** an ostentatious strutting person. □ **peacock blue** the lustrous greenish-blue of a peacock's neck. **peacock butterfly** a butterfly, *Inachis io*, with eyelike markings on its wings. [ME *pecock* f. OE *pēa* f. L *pavo* + COCK¹]

peafowl /ˈpiːfaʊl/ n. a large Asian pheasant of the genus *Pavo*, of which the male is the peacock and the female is the peahen; esp. the common *P. cristatus*.

peahen /ˈpiːhen/ n. a female peafowl.

pea-jacket /ˈpiːˌdʒækɪt/ n. a sailor's short double-breasted overcoat of coarse woollen cloth. [prob. f. Du. *pijjakker* f. *pij* coat of coarse cloth + *jekker* jacket: assim. to JACKET]

peak¹ /piːk/ n., v, & adj. ● n. **1** a projecting usu. pointed part, esp.: **a** the pointed top of a mountain. **b** a mountain with a peak. **c** a stiff brim at the front of a cap. **d** a pointed beard. **e** *Naut.* the narrow part of a ship's hold at the bow or stern (*forepeak; after-peak*). **f** *Naut.* the upper outer corner of a sail extended by a gaff. **2 a** the highest point in a curve (*on the peak of the wave*). **b** the time of greatest success (in a career etc.). **c** the highest point on a graph etc. ● v.intr. reach the highest value, quality, etc. (*output peaked in September*). ● attrib.adj. of or at the highest value, quality, frequency, rate, level, etc. (*peak shopping times*). □ **peak hour** the time of the most intense traffic etc. **peak-load** the maximum of electric power demand etc. □ **peaked** adj. **peaky** adj. **peakiness** n. [prob. back-form. f. *peaked* var. of dial. *picked* pointed (PICK²)]

peak² /piːk/ v.intr. **1** waste away. **2** (as **peaked** adj.) esp. *US* sharp-featured; pinched, sickly-looking. [16th c.: orig. unkn.]

Peak District a limestone plateau in Derbyshire, at the southern end of the Pennines. It rises to 636 m (2,088 ft) at Kinder Scout. A large part of the area was designated a national park in 1951.

Peake /piːk/, Mervyn (Laurence) (1911–68), British novelist, poet, and artist, born in China. He is principally remembered for the trilogy comprising *Titus Groan* (1946), *Gormenghast* (1950), and *Titus Alone* (1959), set in the surreal world of Gormenghast Castle. Peake was also a notable book illustrator.

peaky /'piːkɪ/ adj. (**peakier, peakiest**) **1** sickly; puny. **2** white-faced. □ **peakish** adj.

peal¹ /piːl/ n. & v. ● n. **1 a** the loud ringing of a bell or bells, esp. a series of changes. **b** a set of bells. **2** a loud repeated sound, esp. of thunder, laughter, etc. ● v. **1** intr. sound forth in a peal. **2** tr. utter sonorously. **3** tr. ring (bells) in peals. [ME *pele* f. *apele* APPEAL]

peal² /piːl/ n. a salmon grilse. [16th c.: orig. unkn.]

pean¹ /piːn/ n. Heraldry fur represented as sable spotted with or. [16th c.: orig. unkn.]

pean² *US* var. of PAEAN.

peanut /'piːnʌt/ n. **1** a leguminous plant, *Arachis hypogaea*, native to South America, bearing pods that ripen underground and contain seeds used as food and yielding oil. Also called *earth-nut, groundnut*. **2** the seed of this plant. Also called *monkey-nut*. **3** (in *pl.*) colloq. a paltry or trivial thing or amount, esp. of money. □ **peanut butter** a paste of ground roasted peanuts.

pear /peə(r)/ n. **1** a yellowish or brownish-green fleshy fruit, tapering towards the stalk. **2** a rosaceous tree of the genus *Pyrus* bearing it, esp. *P. communis*. □ **pear-drop** a small sweet with the shape and flavour of a pear. [OE *pere, peru* ult. f. L *pirum*]

pearl¹ /pɜːl/ n. & v. ● n. **1 a** (often *attrib.*) a usu. white or bluish-grey hard mass formed within the shell of a pearl-oyster or other bivalve mollusc, highly prized as a gem for its lustre (*pearl necklace*). **b** an imitation of this. **c** (in *pl.*) a necklace of pearls. **d** = *mother-of-pearl*. **2** a precious thing; the finest example. **3** anything resembling a pearl, e.g. a dewdrop, tear, etc. ● v. **1** tr. *poet.* **a** sprinkle with pearly drops. **b** make pearly in colour etc. **2** tr. reduce (barley etc.) to small rounded grains. **3** intr. fish for pearl-oysters. **4** intr. *poet.* form pearl-like drops. □ **cast pearls before swine** offer a treasure to a person unable to appreciate it. **pearl ash** commercial potassium carbonate. **pearl barley** barley reduced to small round grains by grinding. **pearl bulb** a translucent electric light bulb. **pearl button** a button made of mother-of-pearl or an imitation of it. **pearl-diver** a person who dives for pearl-oysters. **pearl millet** a tall cereal, *Pennisetum typhoides*. **pearl onion** a very small onion used in pickles. **pearl-oyster** a marine bivalve mollusc of the genus *Pinctada*, bearing pearls. □ **pearler** n. [ME f. OF *perle* prob. f. L *perna* leg (applied to leg-of-mutton-shaped bivalve)]

pearl² /pɜːl/ n. *Brit.* = PICOT. [var. of PURL¹]

pearled /pɜːld/ adj. **1** adorned with pearls. **2** formed into pearl-like drops or grains. **3** pearl-coloured.

pearlescent /pɜː'les(ə)nt/ adj. having or producing the appearance of mother-of-pearl.

Pearl Harbor a harbour on the island of Oahu, in Hawaii, the site of a major American naval base, where a surprise attack on 7 December 1941 by Japanese carrier-borne aircraft inflicted heavy damage and brought the US into the Second World War. [transl. Hawaiian *Wai Momi*, lit. 'pearl waters']

pearlite var. of PERLITE.

pearlized /'pɜːlaɪzd/ adj. treated so as to resemble mother-of-pearl.

Pearl River a river of southern China, flowing from Guangzhou (Canton) southwards to the South China Sea and forming part of the delta of the Xi river. Its lower reaches widen to form the Pearl River estuary, the inlet between Hong Kong and Macao.

pearlware /'pɜːlweə(r)/ n. a fine white glazed earthenware.

pearlwort /'pɜːlwɜːt/ n. a small herbaceous plant of the genus *Sagina*, of the pink family, inhabiting rocky and sandy areas.

pearly /'pɜːlɪ/ adj. & n. ● adj. (**pearlier, pearliest**) **1** resembling a pearl; lustrous. **2** containing pearls or mother-of-pearl. **3** adorned with pearls. ● n. (pl. **-ies**) (in pl.) *Brit.* **1** pearly kings and queens. **2** a pearly king's or queen's clothes or pearl buttons. **3** *sl.* teeth. □ **pearly king**

(or **queen**) *Brit.* a London costermonger (or his wife) wearing clothes covered with pearl buttons. **pearly nautilus** see NAUTILUS 1. □ **pearliness** n.

Pearly Gates n. colloq. the gates of heaven.

pearmain /'peəmeɪn, 'pɜːmeɪn/ n. a variety of apple with firm white flesh. [ME, = warden pear, f. OF *parmain, permain*, prob. ult. f. L *parmensis* of Parma]

Pearson¹ /'pɪəs(ə)n/, Karl (1857–1936), English mathematician, and the principal founder of 20th-century statistics. He realized from Francis Galton's work on the measurement of human variation that such data are amenable to mathematical treatment. The fields of heredity and evolution were the first to receive statistical analysis. Pearson defined the concept of standard deviation, and devised the chi-square test. He wrote numerous articles, and was the principal editor of the journal *Biometrika* for many years.

Pearson² /'pɪəs(ə)n/, Lester Bowles (1897–1972), Canadian diplomat and Liberal statesman, Prime Minister 1963–8. As Secretary of State for External Affairs (1948–57), he headed the Canadian delegation to the United Nations, served as chairman of NATO (1951), and acted as a mediator in the resolution of the Suez crisis (1956), for which he received the Nobel Peace Prize in 1957. Pearson became leader of the Liberal Party in 1958; he resigned as Prime Minister and Liberal Party leader in 1968.

peart /pɜːt/ adj. *US* lively; cheerful. [var. of PERT]

Peary /'pɪərɪ/, Robert Edwin (1856–1920), American explorer. He made eight Arctic voyages before becoming the first person to reach the North Pole, on 6 April 1909.

Peary Land a mountainous region on the Arctic coast of northern Greenland. It is named after Robert Peary, who explored it in 1892 and 1900.

peasant /'pez(ə)nt/ n. **1** a worker on the land, a farm labourer or small farmer, esp. a member of an agricultural class dependent on subsistence farming. **2** *derog.* a boor, a lout; a person of low social status. □ **peasantry** n. (pl. **-ies**). **peasanty** adj. [ME f. AF *paisant*, OF *paisent*, earlier *paisence* f. *pais* country ult. f. L *pagus* canton]

Peasants' Revolt an uprising in England (1381), when widespread unrest, caused by poor economic conditions and repressive legislation, culminated in revolt among the peasant and artisan classes, particularly in Kent and Essex. The rebels marched on London, occupying the city and executing unpopular ministers, but after the death of their leader, Wat Tyler, they were persuaded to disperse by the young king Richard II, who granted some of their demands. Afterwards the government went back on its promises and rapidly re-established control.

pease /piːz/ n.pl. *archaic* peas. □ **pease-pudding** boiled split peas (served esp. with boiled ham). [OE *pise* pea, pl. *pisan*, f. LL *pisa* f. L *pisum* f. Gk *pison*: cf. PEA]

peashooter /'piː.ʃuːtə(r)/ n. a small tube for blowing dried peas through as a toy.

peat /piːt/ n. **1** vegetable matter partly decomposed in wet acid conditions to form a brown soil-like deposit, used for fuel, in horticulture, etc. **2** a cut piece of this. □ **peaty** adj. [ME f. AL *peta*, perh. f. Celt.]

peatbog /'piːtbɒg/ n. a bog composed of peat.

peatmoss /'piːtmɒs/ n. **1** a peatbog. **2** = SPHAGNUM.

peau-de-soie /ˌpəʊdə'swʌ/ n. a smooth finely ribbed satiny fabric of silk or rayon. [F, = skin of silk]

pebble /'peb(ə)l/ n. **1** a small smooth stone worn by the action of water. **2 a** a type of colourless transparent rock-crystal used for spectacles. **b** a lens of this. **c** (*attrib.*) colloq. (of a spectacle-lens) very thick and convex. **3** an agate or other gem, esp. when found as a pebble in a stream etc. **4** esp. *Austral. sl.* a high-spirited person or animal, esp. one hard to control. □ **not the only pebble on the beach** (esp. of a person) easily replaced. **pebble-dash** mortar with pebbles in it used as a coating for external walls. □ **pebbly** adj. [OE *papel-stān* pebble-stone, *pyppelrīpig* pebble-stream, of unkn. orig.]

p.e.c. abbr. photoelectric cell.

pecan /'piːkən/ n. **1** a pinkish-brown smooth nut with an edible kernel. **2** a hickory, *Carya illinoensis*, of the southern US, producing this. [earlier *paccan*, of Algonquian orig.]

peccable /'pekəb(ə)l/ adj. *formal* liable to sin. □ **peccability** /ˌpekə'bɪlɪtɪ/ n. [F, f. med.L *peccabilis* f. *peccare* sin]

peccadillo /ˌpekəˈdɪləʊ/ n. (pl. **-oes** or **-os**) a trifling offence; a venial sin. [Sp. *pecadillo*, dimin. of *pecado* sin f. L (as PECCANT)]

peccant /ˈpekənt/ adj. formal **1** sinning. **2** inducing disease; morbid. □ **peccancy** n. [F *peccant* or L *peccare* sin]

peccary /ˈpekərɪ/ n. (pl. **-ies**) a wild piglike mammal of the family Tayassuidae, native to Central and South America; esp. the *collared peccary* (*Tayassu tajacu*), which occurs as far north as Texas. [Carib *pakira*]

peccavi /peˈkɑːvɪ/ int. & n. ● int. expressing guilt. ● n. (pl. **peccavis**) a confession of guilt. [L, = I have sinned]

pêche Melba /peʃ ˈmelbə/ n. = peach Melba (see MELBA²). [F]

Pechenga /ˈpetʃɪŋgə/ a region of NW Russia, lying west of Murmansk on the border with Finland. Formerly part of Finland, it was ceded to the Soviet Union in 1940. It was known by its Finnish name, Petsamo, from 1920 until 1944.

Pechora /pɪˈtʃɔːrə/ a river of northern Russia, which rises in the Urals and flows some 1,800 km (1,125 miles) north and east to the Barents Sea.

Peck /pek/, (Eldred) Gregory (b.1916), American actor. He made his screen début in 1944. His many films range from the thriller *Spellbound* (1945) to the western *The Big Country* (1958). Peck won an Oscar for his role as the lawyer Atticus in the literary adaptation *To Kill a Mockingbird* (1962). More recent films include *The Omen* (1976) and *The Old Gringo* (1989).

peck¹ /pek/ v. & n. ● v.tr. **1** strike or bite (something) with a beak. **2** kiss (esp. a person's cheek) hastily or perfunctorily. **3 a** make (a hole) by pecking. **b** (foll. by *out, off*) remove or pluck out by pecking. **4** (also *absol.*) eat (food) listlessly; nibble at. **5** mark with short strokes. **6** (usu. foll. by *up, down*) break with a pick etc. **7** (often foll. by *away, out*) type at a typewriter etc. ● n. **1 a** a stroke or bite with a beak. **b** a mark made by this. **2** a hasty or perfunctory kiss. **3** archaic sl. food. □ **peck at 1** eat (food) listlessly; nibble. **2** carp at; nag. **3** strike (a thing) repeatedly with a beak. **pecking** (or **peck**) **order** a social hierarchy, orig. as observed among hens. [ME prob. f. MLG *pekken*, of unkn. orig.]

peck² /pek/ n. **1** a measure of capacity for dry goods, equal to (in the UK) 2 gallons or (in the US) 8 quarts. **2** a vessel used to contain this amount. □ **a peck of** a large number or amount of (troubles, dirt, etc.). [ME f. AF *pek*, of unkn. orig.]

pecker /ˈpekə(r)/ n. **1** a bird that pecks (*woodpecker*). **2** esp. N. Amer. coarse sl. the penis. □ **keep your pecker up** Brit. colloq. remain cheerful.

peckish /ˈpekɪʃ/ adj. colloq. **1** hungry. **2** US irritable.

pecorino /ˌpekəˈriːnəʊ/ n. (pl. **-os**) an Italian cheese made from ewes' milk. [It. f. *pecorino* (adj.) of ewes f. *pecora* sheep]

Pécs /peɪtʃ/ an industrial city in SW Hungary; pop. (1993) 171,560. Formerly the capital of the southern part of the Roman province of Pannonia, it was made a bishopric by the first king of Hungary, St Stephen, in 1009. It was occupied by the Turks between 1543 and 1686, and developed rapidly as a coal-mining centre in the 19th and 20th centuries.

pecten /ˈpektɪn/ n. (pl. **pectens** or **pectines** /-ˌniːz/) Zool. **1** a comblike structure of various kinds in animal bodies. **2** a bivalve mollusc of the genus *Pecten* (cf. SCALLOP n. 1). □ **pectinate** /-nət/ adj. **pectinated** /-ˌneɪtɪd/ adj. **pectination** /ˌpektɪˈneɪʃ(ə)n/ n. (all in sense 1). [L *pecten pectinis* comb]

pectin /ˈpektɪn/ n. Biochem. any of a class of soluble gelatinous polysaccharides found in ripe fruits etc. and used as setting agents in jams and jellies. □ **pectic** adj. [Gk *pēktos* congealed f. *pēgnumi* make solid]

pectoral /ˈpektərəl/ adj. & n. ● adj. **1** of or relating to the breast or chest; thoracic (*pectoral fin*; *pectoral muscle*). **2** worn on the chest (*pectoral cross*). ● n. **1** (esp. in pl.) a pectoral muscle. **2** a pectoral fin. **3** an ornamental breastplate, esp. of a Jewish high priest. [ME f. OF f. L *pectorale* (n.), *pectoralis* (adj.) f. *pectus pectoris* breast, chest]

pectose /ˈpektəʊz, -təʊs/ n. Biochem. an insoluble polysaccharide derivative found in unripe fruits and converted into pectin by ripening, heating, etc. [*pectic* (see PECTIN) + -OSE²]

peculate /ˈpekjʊleɪt/ v.tr. & intr. embezzle (money). □ **peculator** n. **peculation** /ˌpekjʊˈleɪʃ(ə)n/ n. [L *peculari* rel. to PECULIAR]

peculiar /pɪˈkjuːlɪə(r)/ adj. & n. ● adj. **1** strange; odd; unusual (*a peculiar flavour*; *is a little peculiar*). **2 a** (usu. foll. by *to*) belonging exclusively (*a fashion peculiar to the time*). **b** belonging to the individual (*in their own peculiar way*). **3** particular; special (*a point of peculiar interest*). ● n. **1** archaic a peculiar property, privilege, etc. **2** Eccl. a parish or church exempt

from the jurisdiction of the diocese in which it lies. [ME f. L *peculiaris* of private property f. *peculium* f. *pecu* cattle]

peculiarity /pɪˌkjuːlɪˈærɪtɪ/ n. (pl. **-ies**) **1 a** idiosyncrasy; unusualness; oddity. **b** an instance of this. **2** a characteristic or habit (*meanness is his peculiarity*). **3** the state of being peculiar.

peculiarly /pɪˈkjuːlɪəlɪ/ adv. **1** more than usually; especially (*peculiarly annoying*). **2** oddly. **3** as regards oneself alone; individually (*does not affect him peculiarly*).

pecuniary /pɪˈkjuːnɪərɪ/ adj. **1** of, concerning, or consisting of, money (*pecuniary aid*; *pecuniary considerations*). **2** (of an offence) entailing a money penalty or fine. □ **pecuniarily** adv. [L *pecuniarius* f. *pecunia* money f. *pecu* cattle]

pedagogue /ˈpedəˌgɒg/ n. archaic or derog. a schoolmaster; a teacher, esp. a strict or pedantic one. □ **pedagogism** n. (also **pedagoguism**). **pedagogic** /ˌpedəˈgɒdʒɪk, -ˈgɒgɪk/ adj. **pedagogical** adj. **pedagogically** adv. [ME f. L *paedagogus* f. Gk *paidagōgos* f. *pais paidos* boy + *agōgos* guide]

pedagogy /ˈpedəˌgɒdʒɪ, -ˌgɒgɪ/ n. the science of teaching. □ **pedagogics** /ˌpedəˈgɒdʒɪks, -ˈgɒgɪks/ n. [F *pédagogie* f. Gk *paidagōgia* (as PEDAGOGUE)]

pedal¹ /ˈped(ə)l/ n. & v. ● n. **1** any of several types of foot-operated levers or controls for mechanisms, esp.: **a** either of a pair of levers for transmitting power to a bicycle or tricycle wheel etc. **b** any of the foot-operated controls in a motor vehicle. **c** any of the foot-operated keys of an organ used for playing notes, or for drawing out several stops at once etc. **d** each of the foot-levers on a piano etc. for making the tone fuller or softer. **e** each of the foot-levers on a harp for altering the pitch of the strings. **2** Mus. a note sustained in one part, usu. the bass, through successive harmonies, some of which are independent of it. ● v. (**pedalled, pedalling**; US **pedaled, pedaling**) **1** intr. operate a cycle, organ, etc. by using the pedals. **2** tr. work (a bicycle etc.) with the pedals. □ **pedal cycle** a bicycle. **pedal steel** (**guitar**) a type of electric guitar, usu. mounted on a stand, in which a characteristic glissando effect is produced by sliding a metal bar along the strings as they are plucked. [F *pédale* f. It. *pedale* f. L (as PEDAL²)]

pedal² /ˈped(ə)l, ˈpiːd-/ adj. Zool. of the foot or feet (esp. of a mollusc). [L *pedalis* f. *pes pedis* foot]

pedalo /ˈpedəˌləʊ/ n. (pl. **-os**) a pedal-operated pleasure-boat.

pedant /ˈped(ə)nt/ n. **1** a person who insists on strict adherence to formal rules or literal meaning at the expense of a wider view. **2** a person who rates academic learning or technical knowledge above everything. **3** a person who is obsessed by a theory; a doctrinaire. □ **pedantry** n. (pl. **-ies**). **pedantic** /pɪˈdæntɪk/ adj. **pedantically** adv. [F *pédant* f. It. *pedante*: app. formed as PEDAGOGUE]

pedate /ˈpedeɪt/ adj. **1** Zool. having feet. **2** Bot. (of a leaf) having divisions like toes or a bird's claws. [L *pedatus* f. *pes pedis* foot]

peddle /ˈped(ə)l/ v. **1** tr. **a** sell (goods), esp. in small quantities, as a pedlar. **b** advocate or promote (ideas, a philosophy, a way of life, etc.). **2** tr. sell (drugs) illegally. **3** intr. engage in selling, esp. as a pedlar. [back-form. f. PEDLAR]

peddler /ˈpedlə(r)/ n. **1** a person who sells drugs illegally. **2** US var. of PEDLAR.

pederast /ˈpedəˌræst/ n. (also **paederast** /ˈpiːd-, ˈped-/) a man who performs pederasty.

pederasty /ˈpedəˌræstɪ/ n. (also **paederasty** /ˈpiːd-, ˈped-/) anal intercourse between a man and a boy. [mod.L *paederastia* f. Gk *paiderastia* f. *pais paidos* boy + *erastēs* lover]

pedestal /ˈpedɪst(ə)l/ n. & v. ● n. **1** Archit. the part of a column below the base, comprising the plinth and the dado if present. **2** the stone etc. base of a statue etc. **3** either of the two supports of a knee-hole desk or table, usu. containing drawers. **4 a** an upright support of a machine or apparatus. **b** the column supporting a wash-basin. **c** a lavatory pan; the base of this. ● v.tr. (**pedestalled, pedestalling**; US **pedestaled, pedestaling**) set or support on a pedestal. □ **pedestal table** a table with a single central support. **put** (or **set**) **on a pedestal** admire disproportionately, idolize. [F *piédestal* f. It. *piedestallo* f. *piè* foot f. L *pes pedis* + *di* of + *stallo* STALL¹]

pedestrian /pɪˈdestrɪən/ n. & adj. ● n. **1** (often attrib.) a person who is walking, esp. in a town (*pedestrian crossing*). **2** a person who walks competitively. ● adj. prosaic; dull; uninspired. □ **pedestrian crossing** Brit. a specified part of a road where pedestrians have right of way to cross. **pedestrian precinct** an area of a town restricted to pedestrians. □ **pedestrianism** n. **pedestrianize** v.tr. & intr. (also **-ise**).

pedestrianization /-ˌdestrɪənaɪˈzeɪʃ(ə)n/ n. [F pédestre or L pedester -tris]

pediatrics US var. of PAEDIATRICS.

pedicab /ˈpedɪˌkæb/ n. a pedal-operated rickshaw.

pedicel /ˈpedɪs(ə)l/ n. **1** Bot. a small stalk, esp. each of those that bear the individual flowers in an inflorescence (cf. PEDUNCLE 1). **2** Zool. a small stalklike structure in an animal (cf. PEDICLE 2). □ **pedicellate** /-lət/ adj. [mod.L pedicellus dimin. of L pediculus PEDICLE]

pedicle /ˈpedɪk(ə)l/ n. **1** Med. part of a graft, esp. a skin graft, left temporarily attached to its original site. **2** Anat. & Zool. a small stalklike structure (cf. PEDICEL 2). □ **pediculated** /pɪˈdɪkjʊˌleɪtɪd/ adj. [L pediculus dimin. of pes pedis foot]

pedicular /pɪˈdɪkjʊlə(r)/ adj. (also **pediculous** /-ləs/) infested with lice. □ **pediculosis** /-ˌdɪkjʊˈləʊsɪs/ n. [L pedicularis, -losus f. pediculus louse]

pedicure /ˈpedɪˌkjʊə(r)/ n. & v. ● n. **1** the care or treatment of the feet, esp. of the toenails. **2** a person practising this, esp. professionally. ● v.tr. treat (the feet) by removing corns etc. [F pédicure f. L pes pedis foot + curare: see CURE]

pedigree /ˈpedɪˌɡriː/ n. **1** (often attrib.) a recorded line of descent of a person or esp. a pure-bred domestic or pet animal. **2** the derivation of a word. **3** a genealogical table. **4** the history of a person, thing, idea, etc. □ **pedigreed** adj. [ME pedegru etc. f. AF f. OF pie de grue (unrecorded) crane's foot, a mark denoting succession in pedigrees]

pediment /ˈpedɪmənt/ n. **1** Archit. **a** the triangular front part of a building in classical style, surmounting esp. a portico of columns. **b** a similar part of a building in baroque or mannerist style, irrespective of shape. **2** Geol. a broad flattish rock surface at the foot of a mountain slope. □ **pedimented** /-ˌmentɪd/ adj. **pedimental** /ˌpedɪˈment(ə)l/ adj. [earlier pedament, periment, perh. corrupt. of PYRAMID]

pedlar /ˈpedlə(r)/ n. (US **peddler**) **1** a travelling seller of small items esp. carried in a pack etc. **2** (usu. foll. by of) a retailer of gossip etc. □ **pedlary** n. [ME pedlere alt. of pedder f. ped pannier, of unkn. orig.]

pedo- US var. of PAEDO-.

pedology /pɪˈdɒlədʒɪ/ n. the scientific study of soil, esp. its formation, nature, and classification. □ **pedologist** n. **pedological** /ˌpedəˈlɒdʒɪk(ə)l/ adj. [f. Gk pedon ground]

pedometer /pɪˈdɒmɪtə(r)/ n. an instrument for estimating the distance travelled on foot by recording the number of steps taken. [F pédomètre f. L pes pedis foot]

peduncle /pɪˈdʌŋk(ə)l/ n. **1** Bot. the stalk of a flower, fruit, or cluster, esp. a main stalk bearing a solitary flower or subordinate stalks (cf. PEDICEL 1). **2 a** Zool. a stalklike part of an animal's body, e.g. the narrow fleshy base of a fish's fin. **b** the stalk by which a barnacle etc. is attached to the substrate. □ **peduncular** /-kjʊlə(r)/ adj. [mod.L pedunculus f. L pes pedis foot: see -UNCLE]

pedunculate /pɪˈdʌŋkjʊlət/ adj. Bot. & Zool. having a peduncle. □ **pedunculate oak** an oak tree, Quercus robur, having clusters of acorns borne on long stalks, found commonly on neutral soils (as in southern and eastern Britain).

pee /piː/ v. & n. colloq. ● v. (**pees**, **peed**) **1** intr. urinate. **2** tr. pass (urine, blood, etc.) from the bladder. ● n. **1** an act of urination. **2** urine. [initial letter of PISS]

Peeblesshire /ˈpiːb(ə)lzˌʃɪə(r)/ a former county of southern Scotland. It became a part of Borders region in 1975.

peek /piːk/ v. & n. ● v.intr. (usu. foll. by in, out, at) look quickly or slyly; peep. ● n. a quick or sly look. [ME pike, pyke, of unkn. orig.]

peekaboo /ˈpiːkəˌbuː/ adj. & n. ● adj. **1** (of a garment etc.) transparent or having a pattern of holes which reveal the skin below. **2** (of a hairstyle) concealing one eye with a fringe or wave. ● n. a game of hiding and suddenly reappearing, played with a young child (also as int.). [PEEK + BOO]

Peel /piːl/, Sir Robert (1788–1850), British Conservative statesman, Prime Minister 1834–5 and 1841–6. During his second term as Home Secretary (1828–30), Peel established the Metropolitan Police (and gave his name to the nicknames bobby and peeler). As leader of the new Conservative Party, he affirmed his belief in moderate electoral reform in the Tamworth Manifesto (1834). His repeal of the Corn Laws in 1846, however, split the Conservatives and forced his resignation. In the last years of his career he came to support the Whig policies of free trade.

peel¹ /piːl/ v. & n. ● v. **1 a** tr. strip the skin, rind, bark, wrapping, etc. from (a fruit, vegetable, tree, etc.). **b** (usu. foll. by off) strip (skin, peel, wrapping, etc.) from a fruit etc. **2** intr. **a** (of a tree, an animal's or

person's body, a painted surface, etc.) become bare of bark, skin, paint, etc. **b** (often foll. by off) (of bark, a person's skin, paint, etc.) flake off. **3** intr. (often foll. by off) colloq. (of a person) strip for exercise etc. **4** tr. Croquet send (another player's ball) through the hoops. ● n. the outer covering of a fruit, vegetable, prawn, etc.; rind. □ **peel off 1** veer away and detach oneself from a group of marchers, a formation of aircraft, etc. **2** colloq. strip off one's clothes. □ **peeler** n. (in sense 1 of v.). [earlier pill, pele (orig. = plunder) f. ME pilien etc. f. OE pilian (unrecorded) f. L pilare f. pilus hair]

peel² /piːl/ n. a shovel, esp. a baker's shovel for bringing loaves etc. into or out of an oven. [ME & OF pele f. L pala, rel. to pangere fix]

peel³ /piːl/ n. (also **pele**) hist. a small square tower built in the 16th century in the border counties of England and Scotland for defence against raids. [ME pel stake, palisade, f. AF & OF pel f. L palus stake: cf. PALE²]

peeler /ˈpiːlə(r)/ n. Brit. archaic sl. or dial. a policeman. [PEEL]

peeling /ˈpiːlɪŋ/ n. a strip of the outer skin of a vegetable, fruit, etc. (potato peelings).

Peelite /ˈpiːlaɪt/ n. hist. a Conservative supporting Sir Robert Peel, esp. with reference to the repeal of the Corn Laws (1846).

peen /piːn/ n. & v. ● n. the wedge-shaped or thin or curved end of a hammer-head (opp. FACE n. 5a). ● v.tr. **1** hammer with a peen. **2** treat (sheet metal) with a stream of metal shot in order to shape it. [17th c.: also pane, app. f. F panne f. Du. pen f. L pinna point]

Peenemunde /ˌpeɪnəˈmʊndə/ a village in NE Germany, on a small island just off the Baltic coast. During the Second World War it was the chief site of German rocket research and testing.

peep¹ /piːp/ v. & n. ● v.intr. **1** (usu. foll. by at, in, out, into) look through a narrow opening; look furtively. **2** (usu. foll. by out) **a** (of daylight, a flower beginning to bloom, etc.) come slowly into view; emerge. **b** (of a quality etc.) show itself unconsciously. ● n. **1** a furtive or peering glance. **2** the first appearance (at peep of day). □ **peep-bo** = PEEKABOO n. **peep-hole** a small hole that may be looked through. **peeping Tom** a furtive voyeur (from the name of a tailor said to have peeped at Lady Godiva when she rode naked through Coventry). **peep-of-day boys** hist. a Protestant organization in Ireland (1784–95) which searched opponents' houses at daybreak for arms. **peep-sight** the aperture backsight of some rifles. **peep-toe** (or **-toed**) (of a shoe) leaving the toes partly bare. [ME: cf. PEEK, PEER¹]

peep² /piːp/ v. & n. ● v.intr. make a shrill feeble sound as of young birds, mice, etc.; squeak; chirp. ● n. **1** such a sound, a cheep. **2** the slightest sound or utterance, esp. of protest etc. [imit.: cf. CHEEP]

peeper /ˈpiːpə(r)/ n. **1** a person who peeps. **2** colloq. an eye. **3** US sl. a private detective.

peep-show /ˈpiːpʃəʊ/ n. a device, usually in the form of a box, with a small eyepiece, inside which are arranged the receding elements of a perspective view. Originally a scientific toy for the educated rich, the peep-show became a public entertainment, a fairground sideshow, and a children's plaything.

peepul /ˈpiːp(ə)l/ n. (also **pipal** /ˈpaɪp-/) = BO TREE. [Hindi pīpal f. Skr. pippala]

peer¹ /pɪə(r)/ v.intr. **1** (usu. foll. by into, at, etc.) look keenly or with difficulty (peered into the fog). **2** peep out. **3** archaic come into view, appear. [var. of pire, LG pīren; perh. partly f. APPEAR]

peer² /pɪə(r)/ n. & v. ● n. **1 a** (fem. **peeress** /-rɪs/) a member of one of the degrees of the nobility in Britain or Ireland, i.e. a duke, marquis, earl, viscount, or baron. (See note below.) **b** a noble of any country. **2** a person who is equal in ability, standing, age, rank, or value (tried by a jury of his peers). ● v.intr. & tr. (usu. foll. by with) rank or cause to rank equally. □ **peer group** a group of people of the same age, status, interests, etc. **peer of the realm** (or **of the United Kingdom**) any of the class of peers whose adult members may all sit in the House of Lords. □ **peerless** adj. (in sense 2 of n.) [ME f. AF & OF pe(e)r, perer f. LL pariare f. L par equal]

▪ In the British peerage, earldoms and baronetcies were the first to be conferred; dukes were created from 1337, marquises from the end of the 14th century, and viscounts from 1440. Such peerages are hereditary, although since 1958 there have also been non-hereditary life peerages. The main privileges of peers nowadays are exemption from jury service and a seat in the House of Lords; they are debarred from election to the House of Commons.

peerage /ˈpɪərɪdʒ/ n. **1** peers as a class; the nobility. **2** the rank of peer or peeress (was given a life peerage). **3** a book containing a list of peers with their genealogy etc.

peeve /piːv/ v. & n. ● v.tr. (usu. as **peeved** adj.) annoy; vex; irritate. ● n. **1** a cause of annoyance. **2** vexation. [back-form. f. PEEVISH]

peevish /ˈpiːvɪʃ/ adj. querulous; irritable. □ **peevishly** adv. **peevishness** n. [ME, = foolish, mad, spiteful, etc., of unkn. orig.]

peewee /ˈpiːwiː/ n. **1** Sc. a lapwing. **2** Austral. the magpie-lark, Grallina cyanoleuca. [imit.: cf. PEWEE]

peewit /ˈpiːwɪt/ n. (also **pewit**) **1** a lapwing. **2** its cry. [imit.]

peg /peg/ n. & v. ● n. **1 a** a usu. cylindrical pin or bolt of wood, metal, etc., often tapered at one end, and used for holding esp. two things together. **b** such a peg attached to a wall etc. and used for hanging garments etc. on. **c** a peg driven into the ground and attached to a rope for holding up a tent. **d** a bung for stoppering a cask etc. **e** each of several pegs used to tighten or loosen the strings of a violin etc. **f** a small peg, matchstick, etc. stuck into holes in a board for calculating the scores at cribbage. **2** Brit. = clothes-peg. **3** Brit. a measure of spirits or wine. **4** a place allotted to a competitor to fish etc. from (usu. marked by a numbered peg). ● v.tr. (**pegged**, **pegging**) **1** (usu. foll. by down, in, out, etc.) fix (a thing) with a peg. **2** Econ. **a** stabilize (prices, wages, exchange rates, etc.). **b** prevent the price of (stock etc.) from falling or rising by freely buying or selling at a given price. **3** mark (the score) with pegs on a cribbage-board. **4** (usu. foll. by at) throw (a missile etc.). □ **off the peg** (of clothes) ready-made. **peg away** (often foll. by at) work consistently and esp. for a long period. **peg down** restrict (a person etc.) to rules, a commitment, etc. **peg-leg** colloq. **1** an artificial leg. **2** a person with an artificial leg. **peg on** = peg away. **peg out 1** sl. die. **2** score the winning point at cribbage. **3** Croquet hit the peg with the ball as the final stroke in a game. **4** mark the boundaries of (land etc.). **a peg to hang an idea** etc. **on** a suitable occasion or pretext etc. for it. **a square peg in a round hole** a misfit. **take a person down a peg or two** humble a person. [ME, prob. of LG or Du. orig.: cf. MDu. pegge, Du. dial. peg, LG pigge]

Pegasus /ˈpegəsəs/ **1** Gk Mythol. a winged horse which sprang from the blood of Medusa when Perseus cut off her head. Pegasus was ridden by Perseus in his rescue of Andromeda, and by Bellerophon when he fought the Chimera. The spring Hippocrene arose from a blow of his hoof. **2** Astron. a large northern constellation, said to represent a winged horse. The three brightest stars, together with one star of Andromeda, form the prominent 'Square of Pegasus'.

pegboard /ˈpegbɔːd/ n. a board having a regular pattern of small holes for pegs, used for commercial displays, games, etc.

pegmatite /ˈpegmətaɪt/ n. Geol. a coarsely crystalline type of granite. [Gk pēgma -atos thing joined together f. pēgnumi fasten]

pegtop /ˈpegtɒp/ n. & adj. ● n. a pear-shaped spinning-top with a metal pin or peg forming the point, spun by the rapid uncoiling of a string wound round it. ● adj. (of a garment) wide at the top and narrow at the bottom.

Pegu /peˈguː/ a city and river port of southern Burma, on the Pegu river north-east of Rangoon; pop. (1983) 150,400. Founded in 825 as the capital of the Mon kingdom, it is a centre of Buddhist culture.

Pehlevi var. of PAHLAVI[1].

PEI abbr. Prince Edward Island.

Pei /peɪ/, I(eoh) M(ing) (b.1917), American architect, born in China. His name is associated with monumental public buildings, in which simple geometric forms are placed in dramatic juxtaposition. Major works include the John F. Kennedy Memorial Library at Harvard University (1964), the east wing of the National Gallery of Art, Washington, DC (1971–8), and the controversial glass and steel pyramid in the forecourt of the Louvre, Paris (1989).

Peigan /ˈpiːgən/ n. & adj. (also **Piegan**) ● n. (pl. same or **Peigans**) a member of a people of the Blackfoot confederacy, inhabiting the Rocky Mountain region of Alberta and Montana. ● adj. of or relating to this people. [Blackfoot Piikániwa]

peignoir /ˈpeɪnwɑː(r)/ n. a woman's loose dressing-gown. [F f. peigner to comb]

Peirce /pɪəs/, Charles Sanders (1839–1914), American philosopher and logician. One of the founders of American pragmatism, he proposed a theory of meaning in which the meaning of a belief or an idea is to be understood by the actions, uses, and habits to which it gives rise. Logic was central to Peirce's philosophic concerns; his work on logical atomism and the distinction between arbitrary and non-arbitrary signs was highly influential in the development of modern semantics and semiotics. He also pioneered the logic of relations, in which he argued that induction is an indispensable correlative of deduction,

and, as a formal logician, discovered the quantifier shortly after Gottlob Frege.

Peisistratus see PISISTRATUS.

pejorative /prˈdʒɒrətɪv, ˈpiːdʒərə-/ adj. & n. ● adj. (of a word, an expression, etc.) depreciatory. ● n. a depreciatory word. □ **pejoratively** adv. [F péjoratif -ive f. LL pejorare make worse (pejor)]

pekan /ˈpekən/ n. = FISHER 1b. [Canad. F f. Amer. Indian name]

peke /piːk/ n. colloq. a Pekingese dog. [abbr.]

Peking see BEIJING.

Pekingese /ˌpiːkɪˈniːz/ n. & adj. (also **Pekinese**) ● n. (pl. same) **1** a short-legged breed of dog with long hair and a snub nose, originally brought to Europe from the Summer Palace at Beijing (Peking) in 1860. **2** a citizen of Beijing (Peking). **3** the form of the Chinese language used in Beijing. ● adj. of or concerning Beijing or its language or citizens.

Peking man n. a late form of the fossil hominid Homo erectus, originally named as Sinanthropus pekinensis from remains found in 1926 in caves at Zhoukoudian near Beijing (Peking). Almost all of the original fossils were lost during the Second World War, but later excavations have produced large quantities of cranial fossils and abundant evidence of the activities of these hominids, including the use of controlled fire and the manufacture and use of stone tools. The cave deposits in which the fossils were found date from c.500,000–230,000 BP (the Middle Pleistocene period). (See also JAVA MAN.)

pekoe /ˈpiːkəʊ/ n. a superior kind of black tea. [Chin. dial. pek-ho f. pek white + ho down, leaves being picked young with down on them]

pelage /ˈpelɪdʒ/ n. the fur, hair, wool, etc. of a mammal. [F f. OF pel (mod. poil) hair]

pelagian /prˈleɪdʒɪən/ adj. & n. ● adj. inhabiting the open sea. ● n. an inhabitant of the open sea. [L pelagius f. Gk pelagios of the sea (pelagos)]

pelagic /prˈlædʒɪk/ adj. **1** of or performed on the open sea (pelagic whaling). **2** (of marine life) belonging to the upper layers of the open sea (cf. DEMERSAL). [L pelagicus f. Gk pelagikos (as PELAGIAN)]

Pelagius /prˈleɪdʒɪəs/ (c.360–c.420), British or Irish monk. He denied the doctrines of original sin and predestination, defending innate human goodness and free will. His beliefs were opposed by St Augustine of Hippo and condemned as heretical by the Synod of Carthage in about 418. □ **Pelagian** adj. & n. **Pelagianism** n.

pelargonium /ˌpeləˈgəʊnɪəm/ n. a plant of the genus Pelargonium, with red, pink, or white flowers and often fragrant leaves. See also GERANIUM. [mod.L f. Gk pelargos stork: cf. GERANIUM]

Pelasgian /prˈlæzgɪən/ n. & adj. ● n. a member of an ancient people inhabiting the coasts and islands of the eastern Mediterranean, especially the Aegean Sea, before the arrival of Greek-speaking peoples in the Bronze Age. ● adj. of or relating to the Pelasgians. [L Pelasgius f. Gk Pelasgios]

Pelé /ˈpeleɪ/ (born Edson Arantes do Nascimento) (b.1940), Brazilian footballer. He played for Brazil at the age of 17, scoring twice in his country's victory in the World Cup Final of 1958. In all he appeared 111 times for Brazil, scoring 97 goals, including one in the 1970 World Cup Final victory. Regarded as one of the greatest footballers of all time, he ended his career with New York Cosmos (1975–7) and is credited with over 1,200 goals in first-class soccer.

pele var. of PEEL[3].

Pelée, Mount /pəˈleɪ/ a volcano on the island of Martinique, in the West Indies. Its eruption in 1902 destroyed the town of St Pierre, which was at that time the island's capital, killing its population of some 30,000.

Peleus /ˈpiːlɪəs/ Gk Mythol. a king of Phthia in Thessaly, who is the subject of a number of legends. He was given as wife the sea-nymph Thetis; their child was Achilles.

pelf /pelf/ n. derog. or joc. money; wealth. [ME f. ONF f. OF pelfre, peufre spoils, of unkn. orig.: cf. PILFER]

Pelham /ˈpeləm/, Henry (1696–1754), British Whig statesman, Prime Minister 1743–54. After serving in Sir Robert Walpole's Cabinet from 1721 onwards, he replaced him as Premier, and introduced a period of peace and prosperity by bringing to an end the War of the Austrian Succession (1740–8). His term also saw the adoption of the Gregorian calendar.

pelham /ˈpeləm/ n. a horse's bit combining a curb and a snaffle. [the surname Pelham]

pelican /ˈpelɪkən/ n. a large gregarious waterbird of the family Pelecanidae, with a large bill below which hangs a pouch for holding fish. □ **pelican crossing** (in the UK) a pedestrian crossing with traffic lights operated by pedestrians. [OE *pellican* & OF *pelican* f. LL *pelicanus* f. Gk *pelekan* prob. f. *pelekus* axe, with ref. to its bill]

Pelion /'pi:lɪən/ a wooded mountain in Greece, near the coast of SE Thessaly, rising to 1,548 m (5,079 ft). It was held in Greek mythology to be the home of the centaurs, and the giants were said to have piled Mounts Olympus and Ossa on its summit in their attempt to reach heaven and destroy the gods.

pelisse /pɪ'liːs/ *n. hist.* **1** a woman's cloak with armholes or sleeves, reaching to the ankles. **2** a fur-lined cloak, esp. as part of a hussar's uniform. [F f. med.L *pellicia* (*vestis*) (garment) of fur f. *pellis* skin]

pelite /'piːlaɪt/ *n. Geol.* a rock composed of claylike sediment. [Gk *pēlos* clay, mud]

pellagra /pɪ'lægrə, -'leɪgrə/ *n. Med.* a disease caused by deficiency of nicotinic acid, characterized by cracking of the skin and often resulting in insanity. □ **pellagrous** *adj.* [It. f. *pelle* skin, after PODAGRA]

pellet /'pelɪt/ *n. & v.* ● *n.* **1** a small compressed ball of paper, bread, etc. **2** a pill. **3 a** a small mass of bones, feathers, etc. regurgitated by a bird of prey. **b** a small hard piece of animal, usu. rodent, excreta. **4 a** a piece of small shot. **b** an imitation bullet for a toy gun. ● *v.tr.* (**pelleted**, **pelleting**) **1** make into a pellet or pellets. **2** hit with (esp. paper) pellets. □ **pelletize** *v.tr.* (also **-ise**). [ME f. OF *pelote* f. L *pila* ball]

Pelletier /pe'letɪˌeɪ/, Pierre-Joseph (1788–1842), French chemist. He specialized in plant products, and began with a study of gum resins and pigments. He is best known as the founder of alkaloid chemistry, having isolated a number of alkaloids for the first time with his friend Joseph-Bienaimé Caventou (1795–1877). Pelletier and Caventou also isolated the green pigment of leaves and gave it the name *chlorophyll*.

pellicle /'pelɪk(ə)l/ *n.* a thin skin, membrane, or film. □ **pellicular** /pɪ'lɪkjʊlə(r)/ *adj.* [F *pellicule* f. L *pellicula*, dimin. of *pellis* skin]

pellitory /'pelɪtərɪ/ *n.* **1** (in full **pellitory of Spain**) a North African daisy-like composite plant, *Anacyclus pyrethrum*, with a pungent-flavoured root, used as a local irritant etc. **2** (in full **pellitory of the wall**) a low bushy plant, *Parietaria judaica* (nettle family), with greenish flowers growing on or at the foot of walls. [(sense 1) alt. f. ME f. OF *peletre*, *peretre* f. L *pyrethrum* f. Gk *purethron* feverfew: (sense 2) ult. f. OF *paritaire* f. LL *parietaria* f. L *paries -etis* wall]

pell-mell /pel'mel/ *adv., adj., & n.* ● *adv.* **1** headlong, recklessly (*rushed pell-mell out of the room*). **2** in disorder or confusion (*stuffed the papers together pell-mell*). ● *adj.* confused, tumultuous. ● *n.* confusion; a mixture. [F *pêle-mêle*, OF *pesle mesle*, *mesle pesle*, etc., redupl. of *mesle* f. *mesler* mix]

pellucid /pɪ'luːsɪd, -'ljuːsɪd/ *adj.* **1** (of water, light, etc.) transparent, clear. **2** (of style, speech, sound, etc.) not confused; clear. **3** mentally clear. □ **pellucidly** *adv.* **pellucidity** /ˌpeluː'sɪdɪtɪ, ˌpeljuː-/ *n.* [L *pellucidus* f. *perlucere* (as PER-, *lucere* shine)]

Pelmanism /'pelməˌnɪz(ə)m/ *n.* **1** a system of memory-training orig. devised by the Pelman Institute for the Scientific Development of Mind, Memory, and Personality in London. **2** a card-game based on this. □ **Pelmanize** *v.tr.* (also **-ise**). [Christopher Louis *Pelman*, founder of the Institute in 1899]

pelmet /'pelmɪt/ *n.* a narrow border of cloth, wood, etc. above esp. a window, concealing the curtain rail. [prob. f. F PALMETTE]

Peloponnese, the /ˌpeləpə'niːz/ (also **Peloponnesus** /-'niːsəs/, Greek **Pelopónnisos** /ˌpɛlɔ'ponisɔs/) the mountainous southern peninsula of Greece, connected to central Greece by the Isthmus of Corinth. Its Greek name means 'island of Pelops'.

Peloponnesian War /ˌpeləpə'niːʃ(ə)n/ the war of 431–404 BC fought between Athens and Sparta with their respective allies, occasioned largely by Spartan opposition to the Athenian empire (see DELIAN LEAGUE). It ended in the total defeat of Athens and the transfer, for a brief period, of the leadership of Greece to Sparta.

Pelops /'piːlɒps/ *Gk Mythol.* son of Tantalus, brother of Niobe, and father of Atreus. He was killed by his father and served up as food to the gods, but only one shoulder was eaten, and he was restored to life with an ivory shoulder replacing the one that was missing.

pelorus /pɪ'lɔːrəs/ *n.* a sighting device like a ship's compass for taking bearings. [perh. f. *Pelorus*, reputed name of Hannibal's pilot]

pelota /pɪ'lɒtə, -'ləʊtə/ *n.* **1** a Basque or Spanish game played in a walled court with a ball and basket-like rackets attached to the hand. **2** the ball used in this. [Sp., = ball, augment. of *pella* f. L *pila*]

pelt[1] /pelt/ *v. & n.* ● *v.* **1** *tr.* (usu. foll. by *with*) **a** hurl many small missiles at. **b** strike repeatedly with missiles. **c** assail (a person etc.) with insults, abuse, etc. **2** *intr.* (usu. foll. by *down*) (of rain etc.) fall quickly and torrentially. **3** *intr.* run fast. **4** *intr.* (often foll. by *at*) fire repeatedly. ● *n.*

the act or an instance of pelting. □ **at full pelt** as fast as possible. [16th c.: orig. unkn.]

pelt[2] /pelt/ *n.* **1** the undressed skin of a fur-bearing mammal. **2** the skin of a sheep, goat, etc. with short wool, or stripped ready for tanning. **3** *joc.* the human skin. □ **peltry** *n.* [ME f. obs. *pellet* skin, dimin. of *pel* f. AF *pell*, OF *pel*, or back-form. f. *peltry*, AF *pelterie*, OF *peleterie* f. *peletier* furrier, ult. f. L *pellis* skin]

pelta /'peltə/ *n.* (*pl.* **peltae** /-tiː/) **1** *hist.* a small light shield used by the ancient Greeks, Romans, etc. **2** *Bot.* a shieldlike structure, esp. of a lichen. □ **peltate** /-teɪt/ *adj.* [L f. Gk *peltē*]

pelvic /'pelvɪk/ *adj. Anat.* of or relating to the pelvis or the organs it encloses (e.g. the uterus) (*pelvic fin, pelvic infection*). □ **pelvic girdle** the bony or cartilaginous structure in vertebrates to which the posterior limbs are attached.

pelvis /'pelvɪs/ *n.* (*pl.* **pelvises** or **pelves** /-viːz/) *Anat.* **1** a basin-shaped cavity at the lower end of the torso of most vertebrates, formed from the innominate bone with the sacrum and other vertebrae. **2** the basin-like cavity of the kidney. [L, = basin]

Pemba /'pembə/ **1** a seaport in northern Mozambique, on the Indian Ocean; pop. (1980) 41,200. **2** an island off the coast of Tanzania, in the western Indian Ocean north of Zanzibar. It is noted for the production of cloves.

Pembroke /'pembrʊk/ a port in SW Wales; pop. (1981) 15,600. It was a Norman stronghold from the 11th century.

Pembrokeshire /'pembrʊkˌʃɪə(r)/ a former county of SW Wales. It became a part of Dyfed in 1974.

Pembs. *abbr.* Pembrokeshire.

pemmican /'pemɪkən/ *n.* **1** a cake of dried pounded meat mixed with melted fat, orig. made by North American Indians. **2** beef so treated and flavoured with currants etc. for use by Arctic travellers etc. [Cree *pimecan* f. *pime* fat]

pemphigus /'pemfɪgəs/ *n. Med.* the formation of watery blisters or eruptions on the skin. □ **pemphigoid** *adj.* **pemphigous** *adj.* [mod.L f. Gk *pemphix -igos* bubble]

PEN /pen/ *abbr.* International Association of Poets, Playwrights, Editors, Essayists, and Novelists.

Pen. *abbr.* Peninsula.

pen[1] /pen/ *n. & v.* ● *n.* **1** an instrument for writing or drawing with ink, orig. consisting of a shaft with a sharpened quill or metal nib, now more widely applied. **2 a** (usu. prec. by *the*) the occupation of writing. **b** a style of writing. **3** *Zool.* the internal feather-shaped cartilaginous shell of a squid. ● *v.tr.* (**penned**, **penning**) **1** write. **2** compose and write. □ **pen and ink** *n.* **1** the instruments of writing. **2** writing. **pen-and-ink** *adj.* drawn or written with ink. **pen-feather** a quill-feather of a bird's wing. **pen-friend** a friend communicated with by letter only. **pen-light** a small electric torch shaped like a fountain-pen. **pen-name** a literary pseudonym. **pen-pal** *colloq.* = *pen-friend*. **pen-pusher** *colloq. derog.* a clerical worker. **pen-pushing** *colloq. derog.* clerical work. **put pen to paper** begin writing. [ME f. OF *penne* f. L *penna* feather]

pen[2] /pen/ *n. & v.* ● *n.* **1** a small enclosure for cows, sheep, poultry, etc. **2** a place of confinement. **3** an enclosure for sheltering submarines. **4** a Jamaican farm or plantation. ● *v.tr.* (**penned**, **penning**) (often foll. by *in*, *up*) enclose or shut up, esp. in a pen. [OE *penn*, of unkn. orig.]

pen[3] /pen/ *n.* a female swan. [16th c.: orig. unkn.]

pen[4] /pen/ *n. US sl.* = PENITENTIARY *n.* 1. [abbr.]

penal /'piːn(ə)l/ *adj.* **1 a** of or concerning punishment or its infliction (*a penal sentence; a penal colony*). **b** (of an offence) punishable, esp. by law. **2** extremely severe (*penal taxation*). □ **penal servitude** *hist.* imprisonment with compulsory labour. □ **penally** *adv.* [ME f. OF *penal* or L *poenalis* f. *poena* PAIN]

penalize /'piːnəˌlaɪz/ *v.tr.* (also **-ise**) **1** subject (a person) to a penalty or comparative disadvantage. **2** make or declare (an action) penal. □ **penalization** /ˌpiːnəlaɪ'zeɪʃ(ə)n/ *n.*

Penal Laws *n.pl.* various statutes passed in Britain and Ireland during the 16th and 17th centuries that imposed restrictions on Roman Catholics. People participating in Catholic services could be fined and imprisoned, while Catholics were banned from voting, holding public office, owning land, and teaching. Hardly enforced by the 18th century, the laws were repealed by various Acts 1791–1926. (See also CATHOLIC EMANCIPATION, TEST ACTS.)

penalty /'pen(ə)ltɪ/ *n.* (*pl.* **-ies**) **1 a** a punishment, esp. a fine, for a breach of law, contract, etc. **b** a fine paid. **2** a disadvantage, loss, etc., esp. as a result of one's own actions (*paid the penalty for his carelessness*). **3 a** a disadvantage imposed on a competitor or side in a game etc. for a breach of the rules etc. **b** (*attrib.*) awarded against a side incurring a penalty (*penalty kick*; *penalty goal*). **4** *Bridge* etc. points gained by opponents when a contract is not fulfilled. □ **penalty area** *Football* the ground in front of the goal in which a foul by defenders involves the award of a penalty kick. **penalty box** *Ice Hockey* an area reserved for penalized players and some officials. **the penalty of** a disadvantage resulting from (a quality etc.). **penalty rate** *Austral.* an increased rate of pay for overtime. **penalty spot** *Football* see SPOT *n.* 2c. **under** (or **on**) **penalty of** under the threat of (dismissal etc.). [AF *penalte* (unrecorded), F *pénalité* f. med.L *penalitas* (as PENAL)]

penance /ˈpenəns/ *n. & v.* ● *n.* **1** an act of self-punishment as reparation for guilt. **2 a** (in the RC and Orthodox Church) a sacrament including confession of and absolution for a sin. **b** a penalty imposed esp. by a priest, or undertaken voluntarily, for a sin. ● *v.tr.* impose a penance on. □ **do penance** perform a penance. [ME f. OF f. L *paenitentia* (as PENITENT)]

Penang /pɪˈnaŋ/ (also **Pinang**) **1** an island of Malaysia, situated off the west coast of the Malay Peninsula. In 1786 it was ceded to the East India Company as a British colony by the sultan of Kedah. Known as Prince of Wales Island until 1867, it united with Malacca and Singapore in a union of 1826, which in 1867 became the British colony called the Straits Settlements. It joined the federation of Malaya in 1948. **2** a state of Malaysia, consisting of this island and a coastal strip on the mainland; capital, George Town (on Penang island). The mainland strip was united with the island in 1798 as part of the British colony. **3** see GEORGE TOWN 2.

penannular /penˈænjʊlə(r)/ *adj.* almost ringlike. [L *paene* almost + ANNULAR]

penates /pɪˈnɑːtiːz, -teɪz/ *n.pl.* *Rom. Hist.* gods, esp. of the storeroom, worshipped by households in ancient Rome in close conjunction with Vesta and the lares. [L f. *penus* provision of food]

pence *pl.* of PENNY.

penchant /ˈpɒnʃɒn, ˈpentʃənt/ *n.* an inclination or liking (*has a penchant for old films*). [F, pres. part. of *pencher* incline]

pencil /ˈpensɪl/ *n. & v.* ● *n.* **1** (often *attrib.*) **a** an instrument for writing or drawing, usu. consisting of a thin rod of graphite etc. enclosed in a wooden cylinder (*a pencil sketch*). **b** a similar instrument with a metal or plastic cover and retractable lead. **c** a cosmetic in pencil form. **2** (*attrib.*) resembling a pencil in shape (*pencil skirt*). **3** *Optics* a set of rays meeting at a point. **4** *Geom.* a figure formed by a set of straight lines meeting at a point. ● *v.tr.* (**pencilled**, **pencilling**; US **penciled**, **penciling**) **1** tint or mark with or as if with a pencil. **2** (usu. foll. by *in*) **a** write, esp. tentatively or provisionally (*have pencilled in the 29th for our meeting*). **b** (esp. as **pencilled** *adj.*) fill (an area) with soft pencil strokes (*pencilled in her eyebrows*). □ **pencil-case** a container for pencils etc. **pencil-pusher** *colloq. derog.* a clerical worker. **pencil-pushing** *colloq. derog.* clerical work. **pencil-sharpener** a device for sharpening a pencil by rotating it against a cutting edge. □ **penciller** *n.* [ME f. OF *pincel* ult. f. L *penicillum* paintbrush, dimin. of *peniculus* brush, dimin. of *penis* tail]

pendant /ˈpendənt/ *n.* **1** a hanging jewel etc., esp. one attached to a necklace, bracelet, etc. **2** a light fitting, ornament, etc., hanging from a ceiling. **3** *Naut.* **a** a short rope hanging from the head of a mast etc., used for attaching tackles. **b** = PENNANT 1. **4** the shank and ring of a pocket-watch by which it is suspended. **5** (also /ˈpɒndɒn/) (usu. foll. by *to*) a match, companion, parallel, complement, etc. [ME f. OF f. *pendre* hang f. L *pendere*]

pendent /ˈpendənt/ *adj.* (also **pendant**) **1 a** hanging. **b** overhanging. **2** undecided; pending. **3** *Gram.* (esp. of a sentence) incomplete; not having a finite verb (*pendent nominative*). □ **pendency** *n.* [ME (as PENDANT)]

pendente lite /penˌdenti ˈlaɪti/ *adv. Law* during the progress of a suit. [L]

pendentive /penˈdentɪv/ *n. Archit.* a curved triangle of vaulting formed by the intersection of a dome with its supporting arches. [F *pendentif* -*ive* (adj.) (as PENDANT)]

Penderecki /ˌpendəˈretskɪ/, Krzysztof (b.1933), Polish composer. His music frequently uses unorthodox effects, including sounds drawn from extra-musical sources and note clusters, as in his *Threnody for the Victims of Hiroshima* (1960) for fifty-two strings. Penderecki's many religious works include *Stabat Mater* (1962), a fusion of conventional and avant-garde elements, and the *Polish Requiem* (1980–4).

pending /ˈpendɪŋ/ *adj. & prep.* ● *predic.adj.* **1** awaiting decision or settlement, undecided (*a settlement was pending*). **2** about to come into existence (*patent pending*). ● *prep.* **1** during (*pending these negotiations*). **2** until (*pending his return*). □ **pending-tray** a tray for documents, letters, etc., awaiting attention. [after F *pendant* (see PENDENT)]

pendragon /penˈdrægən/ *n. hist.* an ancient British or Welsh prince (often as a title). [Welsh, = chief war-leader, f. *pen* head + *dragon* standard]

penduline /ˈpendjʊˌlaɪn/ *adj.* **1** (of a nest) suspended. **2** (of a bird) of a kind that builds such a nest. [F (as PENDULOUS)]

pendulous /ˈpendjʊləs/ *adj.* **1** hanging down; drooping (*the dog's pendulous ears*; *her pendulous breasts*). **2** swinging; oscillating. □ **pendulously** *adv.* [L *pendulus* f. *pendere* hang]

pendulum /ˈpendjʊləm/ *n.* a weight suspended so as to swing or oscillate freely, esp. a rod with a weighted end regulating the mechanism of a clock. (*See note below.*) □ **swing of the pendulum** the tendency of public opinion to oscillate between extremes, esp. between political parties. [L neut. adj. (as PENDULOUS)]

▪ The principle of the pendulum, that for small amplitudes the time of oscillation depends only on its length, was discovered by Galileo *c.*1602. Christiaan Huygens, in 1657, was the first to apply this to regulating clocks, and the pendulum clock was the most accurate timekeeping device until the invention of the quartz clock in the 20th century. A freely suspended pendulum resists changes in its plane of oscillation, a fact employed by Jean Foucault in 1851 to demonstrate the earth's rotation.

Penelope /pɪˈneləpɪ/ *Gk Mythol.* the wife of Odysseus, who was beset by suitors when her husband did not return after the fall of Troy. She put them off by saying that she would marry only when she had finished the piece of weaving on which she was engaged, and every night unravelling the work she had done during the day. (See ODYSSEY.)

peneplain /ˈpiːnɪˌpleɪn/ *n. Geol.* a fairly flat area of land produced by erosion. [L *paene* almost + PLAIN[1]]

penetralia /ˌpenɪˈtreɪlɪə/ *n.pl.* **1** innermost shrines or recesses. **2** secret or hidden parts; mysteries. [L, neut. pl. of *penetralis* interior (as PENETRATE)]

penetrate /ˈpenɪˌtreɪt/ *v.* **1** *tr.* **a** find access into or through, esp. forcibly. **b** (usu. foll. by *with*) imbue (a person or thing) with; permeate. **2** *tr.* see into, find out, or discern (a person's mind, the truth, a meaning, etc.). **3** *tr.* see through (darkness, fog, etc.) (*could not penetrate the gloom*). **4** *intr.* be absorbed by the mind (*my hint did not penetrate*). **5** *tr.* (as **penetrating** *adj.*) **a** having or suggesting sensitivity or insight (*a penetrating remark*). **b** (of a voice etc.) easily heard through or above other sounds; piercing. **c** (of a smell) sharp, pungent. **6** *tr.* (of a man) put the penis into the vagina of (a woman). **7** *intr.* (usu. foll. by *into, through, to*) make a way. □ **penetrant** *adj. & n.* **penetratingly** *adv.* **penetrator** *n.* **penetrative** /-trətɪv/ *adj.* **penetrable** /-trəb(ə)l/ *adj.* **penetrability** /ˌpenɪtrəˈbɪlɪt/ *n.* **penetration** /-ˈtreɪʃ(ə)n/ *n.* [L *penetrare* place or enter within f. *penitus* interior]

penguin /ˈpeŋgwɪn/ *n.* a flightless seabird of the family Spheniscidae, of the southern oceans, with black upperparts and white underparts. Penguins walk clumsily on land with an upright posture, but able to swim rapidly under water by means of the flipper-like wings. [16th c., orig. = great auk: orig. unkn.]

penholder /ˈpenˌhəʊldə(r)/ *n.* the esp. wooden shaft of a pen with a metal nib.

penicillate /ˈpenɪsɪlət, ˌpenɪˈsɪlɪt/ *adj. Biol.* **1** having or forming a small tuft or tufts. **2** marked with streaks as of a pencil or brush. [L *penicillum*: see PENCIL]

penicillin /ˌpenɪˈsɪlɪn/ *n.* an antibiotic produced naturally by moulds of the genus *Penicillium*, or prepared synthetically, and able to prevent the growth of certain disease-causing bacteria. The first antibiotic to be used therapeutically, penicillin was discovered in 1928 by Sir Alexander Fleming and characterized and introduced by Howard Florey and Sir Ernst Chain during the Second World War. [f. mod.L *Penicillium* f. L *penicillum*: see PENCIL]

penile /ˈpiːnaɪl/ *adj.* of or concerning the penis. [mod.L *penilis*]

penillion *pl.* of PENILLION.

peninsula /pɪˈnɪnsjʊlə/ *n.* a piece of land almost surrounded by water or projecting far into a sea or lake etc. □ **peninsular** *adj.* [L *paeninsula* f. *paene* almost + *insula* island]

Peninsular War the campaign waged on the Iberian peninsula between the French and the British, the latter assisted by Spanish and Portuguese forces, from 1808 to 1814 during the Napoleonic Wars. Although an early British expedition was forced to evacuate the peninsula in 1809 after its Spanish allies had been defeated, a second expedition, led by Wellington, finally drove the French back over the Pyrenees in early 1814 after a long and bloody campaign.

penis /'pi:nɪs/ n. (pl. **penises** or **penes** /-ni:z/) **1** the male organ of copulation and (in mammals) urination. **2** the male copulatory organ in lower vertebrates. [L, = tail, penis]

penitent /'penɪt(ə)nt/ adj. & n. ● adj. regretting and wishing to atone for sins etc.; repentant. ● n. **1** a repentant sinner. **2** a person doing penance under the direction of a confessor. **3** (in pl.) various RC orders associated for mutual penitence, the giving of religious aid to criminals, etc. □ **penitence** n. **penitently** adv. [ME f. OF f. L paenitens f. paenitere repent]

penitential /ˌpenɪ'ten∫(ə)l/ adj. of or concerning penitence or penance. □ **penitential psalms** seven psalms (6, 32, 38, 51, 102, 130, 143) expressing penitence. □ **penitentially** adv. [OF penitencial f. LL paenitentialis f. paenitentia penitence (as PENITENT)]

penitentiary /ˌpenɪ'ten∫ərɪ/ n. & adj. ● n. (pl. **-ies**) **1** US a reformatory prison. **2** an office in the papal court deciding questions of penance, dispensations, etc. ● adj. **1** of or concerning penance. **2** of or concerning reformatory treatment. **3** US (of an offence) making a culprit liable to a prison sentence. [ME f. med.L paenitentiarius (adj. & n.) (as PENITENT)]

penknife /'pennaɪf/ n. (pl. **penknives**) a small folding knife, esp. for carrying in a pocket.

penman /'penmən/ n. (pl. **-men**) **1** a person who writes by hand with a specified skill (a good penman). **2** an author. □ **penmanship** n.

Penn /pen/, William (1644–1718), English Quaker, founder of Pennsylvania. He was imprisoned in the Tower of London in 1668 for writing in defence of Quaker practices. Acquitted in 1670, he was granted a charter to land in North America by Charles II (1682), using it to found the colony of Pennsylvania as a sanctuary for Quakers and other Nonconformists in the same year. Penn also co-founded the city of Philadelphia.

Penn. abbr. (also **Penna.**) Pennsylvania.

pennant /'penənt/ n. **1** Naut. a tapering flag, esp. that flown at the masthead of a vessel in commission. **2** = PENDANT 3a. **3** = PENNON. **4** US a flag denoting a sports championship etc. [blend of PENDANT and PENNON]

penne /'penɪ/ n.pl. pasta in the form of short tubes cut diagonally at the ends. [It., pl. of penna feather, quill, pen]

penni /'penɪ/ n. (pl. **penniä** /-nɪˌɑː/) a monetary unit of Finland, equal to one-hundredth of a markka. [Finnish]

penniless /'penɪlɪs/ adj. having no money; destitute. □ **pennilessly** adv. **pennilessness** n.

pennill /'penɪl/ n. (pl. **penillion** /pe'nɪljən/) (usu. in pl.) an improvised stanza sung to a harp accompaniment at an eisteddfod etc. [Welsh f. pen head]

Pennine Hills /'penaɪn/ (also **Pennine Chain, Pennines**) a range of hills in northern England, extending from the Scottish border southwards to the Peak District in Derbyshire. Its highest peak is Cross Fell in Cumbria, which rises to 893 m (2,930 ft). The hills are sometimes described as the 'Backbone of England'.

pennon /'penən/ n. **1** a long narrow flag, triangular or swallow-tailed, esp. as the military ensign of lancer regiments. **2** Naut. a long pointed streamer on a ship. **3** a flag. □ **pennoned** adj. [ME f. OF f. L penna feather]

penn'orth var. of PENNYWORTH.

Pennsylvania /ˌpensɪl'veɪnɪə/ a state of the north-eastern US; pop. (1990) 11,881,640; capital, Harrisburg. Founded in 1682 by William Penn and named after his father, Admiral Sir William Penn (1621–70), it became one of the original thirteen states of the Union in 1787.

Pennsylvania Dutch n. **1** a dialect of High German spoken by descendants of 17th–18th-century German and Swiss immigrants to Pennsylvania. **2** (as pl.) these settlers or their descendants.

Pennsylvanian /ˌpensɪl'veɪnɪən/ n. & adj. ● n. **1** a native or inhabitant of Pennsylvania. **2** (prec. by the) esp. US Geol. the upper Carboniferous period or system. ● adj. **1** of or relating to Pennsylvania. **2** esp. US Geol. of or relating to the upper Carboniferous period or system.

penny /'penɪ/ n. (pl. for separate coins **-ies**, for a sum of money **pence** /pens/) **1** (abbr. **p**) a British bronze coin and monetary unit equal to one-hundredth of a pound. (See note below.) **2** (abbr. **d**) hist. a former British coin and monetary unit equal to one-two-hundred-and-fortieth of a pound. (See note below.) **3** N. Amer. colloq. a one-cent coin. **4** Bibl. a denarius. □ **in for a penny, in for a pound** an exhortation to

total commitment to an undertaking. **like a bad penny** continually returning when unwanted. **pennies from heaven** unexpected benefits. **penny-a-liner** a hack writer. **penny black** the first adhesive postage stamp (1840, value one penny). **penny cress** a cruciferous plant, Thlaspi arvense, with flat round pods. **penny dreadful** Brit. a cheap sensational comic or story-book. **the penny drops** colloq. one begins to understand at last. **penny farthing** Brit. an early type of bicycle with one large and one small wheel. (See BICYCLE.) **a penny for your thoughts** a request to a thoughtful person to confide in the speaker. **penny-in-the-slot** (of a machine) activated by a coin pushed into a slot. **penny-pincher** a niggardly person. **penny-pinching** n. meanness. ● adj. mean. **penny whistle** a tin pipe with six holes giving different notes. **penny wise** too careful in saving small amounts. **penny wise and pound foolish** mean in small expenditures but wasteful of large amounts. **a pretty penny** a large sum of money. **two a penny** almost worthless though easily obtained. [OE penig, penning f. Gmc, perh. rel. to PAWN[2]]

• The word penny has been used at least since the 8th century for an English coin. Originally of silver, the penny began to be minted from copper in 1797, and from bronze in 1860; coining of silver pennies for general circulation ceased with the reign of Charles II, although a small number have since been regularly coined as Maundy money. With decimalization in 1971 the traditional penny was replaced by a penny equal to one-hundredth of a pound and known for a time as the new penny.

-penny /pənɪ/ comb. form Brit. forming attributive adjectives meaning 'costing … pence' (esp. in pre-decimal currency) (fivepenny).

penny post n. a system of carrying letters at a standard charge of one penny each instead of at varying rates according to distance, esp. that established in the UK in 1840 at the instigation of Sir Rowland Hill, who also introduced the postage stamp.

pennyroyal /ˌpenɪ'rɔɪəl/ n. **1** a small creeping mint, Mentha pulegium, formerly much used medicinally. **2** US a similar aromatic labiate plant, Hedeoma pulegioides. [app. f. earlier puliol(e) ryall f. AF puliol, OF pouliol ult. f. L pulegium + real ROYAL]

pennyweight /'penɪˌweɪt/ n. a unit of weight, 24 grains or one-twentieth of an ounce troy.

pennywort /'penɪˌwɜːt/ n. **1** (in full **wall pennywort**) = NAVELWORT. **2** (in full **marsh** or **water pennywort**) a small creeping marsh plant, Hydrocotyle vulgaris, with rounded leaves. [ME, f. PENNY + WORT]

pennyworth /'penɪˌwɜːθ/ n. (also **penn'orth** /'penəθ/) **1** as much as can be bought for a penny. **2** a bargain of a specified kind (a bad pennyworth). □ **not a pennyworth** not the least bit.

penology /piː'nɒlədʒɪ/ n. the study of the punishment of crime and of prison management. □ **penologist** n. **penological** /ˌpiːnə'lɒdʒɪk(ə)l/ adj. [L poena penalty + -LOGY]

pensée /'pɒnseɪ/ n. a thought or reflection put into literary form; an aphorism. [F]

pensile /'pensaɪl/ adj. **1** hanging down; pendulous. **2** (of a bird etc.) building a pensile nest. [L pensilis f. pendere pens- hang]

pension[1] /'pen∫(ə)n/ n. & v. ● n. **1 a** a regular payment made by a government to people above a specified age, to widows, or to the disabled. It was introduced in Germany by Bismarck in the late 19th century and in Britain by Lloyd George in 1908. (See also WELFARE STATE.) **b** a similar payment made by an employer etc. after the retirement of an employee. **2 a** a regular payment made to a scientist, artist, etc. for services to the state, or to fund work. **b** any pension paid esp. by a government on charitable grounds. ● v.tr. **1** grant a pension to. **2** bribe with a pension. □ **pension off 1** dismiss with a pension. **2** cease to employ or use. □ **pensionless** adj. [ME f. OF f. L pensio -onis payment f. pendere pens- pay]

pension[2] /'pɒnsjɒn/ n. a boarding-house in France or Continental Europe etc., providing full or half board at a fixed rate. □ **en pension** /ɒn/ as a boarder. [F: see PENSION[1]]

pensionable /'pen∫ənəb(ə)l/ adj. **1** entitled to a pension. **2** (of a service, job, etc.) entitling an employee to a pension. □ **pensionability** /ˌpen∫ənə'bɪlɪtɪ/ n.

pensionary /'pen∫ənərɪ/ adj. & n. ● adj. of or concerning a pension. ● n. (pl. **-ies**) **1** a pensioner. **2** a creature; a hireling. [med.L pensionarius (as PENSION[1])]

pensioner /'pen∫ənə(r)/ n. a recipient of a pension, esp. the retirement pension. [ME f. AF pensionner, OF pensionnier (as PENSION[1])]

pensive /'pensɪv/ adj. **1** deep in thought. **2** sorrowfully thoughtful.

□ **pensively** *adv.* **pensiveness** *n.* [ME f. OF *pensif, -ive* f. *penser* think f. L *pensare* frequent. of *pendere* pens- weigh]

penstemon var. of PENTSTEMON.

penstock /'penstɒk/ *n.* **1** a sluice; a floodgate. **2** *US* a channel for conveying water to a water-wheel. [PEN² in sense 'mill-dam' + STOCK]

pent /pent/ *adj.* (often foll. by *in, up*) closely confined; shut in (*pent-up feelings*). [past part. of *pend* var. of PEN² *v.*]

penta- /'pentə/ *comb. form* **1** five. **2** *Chem.* (forming the names of compounds) containing five atoms or groups of a specified kind (*pentachloride; pentoxide*). [Gk f. *pente* five]

pentachord /'pentə,kɔːd/ *n.* **1** a musical instrument with five strings. **2** a series of five musical notes.

pentacle /'pentək(ə)l/ *n.* a pentagram or similar figure used as a mystic or magical symbol. [med.L *pentaculum* (as PENTA-)]

pentad /'pentæd/ *n.* **1** the number five. **2** a group of five. [Gk *pentas -ados* f. *pente* five]

pentadactyl /,pentə'dæktɪl/ *adj. Zool.* having five digits on each foot. This is the basic arrangement found in higher vertebrates. □ **pentadactyly** *n.*

pentagon /'pentəgən/ *n.* a plane figure with five sides and angles. □ **pentagonal** /pen'tægən(ə)l/ *adj.* [F *pentagone* or f. LL *pentagonus* f. Gk *pentagōnon* (as PENTA-, -GON)]

Pentagon, the the headquarters of the US Department of Defense, near Washington, DC. Built in 1941–3 in the form of five concentric pentagons, it covers 13.8 hectares (34 acres) and is one of the world's largest office buildings. The name is used allusively for the US military leadership.

pentagram /'pentə,græm/ *n.* a five-pointed star formed by extending the sides of a pentagon both ways until they intersect, used as a mystic symbol. [Gk *pentagrammon* (as PENTA-, -GRAM)]

pentagynous /pen'tædʒɪnəs/ *adj. Bot.* having five pistils.

pentahedron /,pentə'hiːdrən/ *n.* a solid figure with five faces. □ **pentahedral** *adj.*

pentamerous /pen'tæmərəs/ *adj.* **1** *Bot.* having five parts in a flower-whorl. **2** *Zool.* having five joints or parts.

pentameter /pen'tæmɪtə(r)/ *n. Prosody* **1** a verse of five feet, e.g. English iambic verse of ten syllables. **2** a form of Gk or Latin dactylic verse composed of two halves each of two feet and a long syllable, used in elegiac verse. [L f. Gk *pentametros* (as PENTA-, -METER)]

pentandrous /pen'tændrəs/ *adj. Bot.* having five stamens.

pentane /'penteɪn/ *n. Chem.* a hydrocarbon of the alkane series (chem. formula: C_5H_{12}). [Gk *pente* five + ALKANE]

pentangle /'pen,tæŋg(ə)l/ *n.* = PENTAGRAM. [ME perh. f. med.L *pentaculum* PENTACLE, assim. to L *angulus* ANGLE¹]

pentanoic acid /,pentə'nəʊɪk/ *n. Chem.* a colourless liquid carboxylic acid used in making perfumes. [PENTANE]

pentaprism /'pentə,prɪz(ə)m/ *n.* a five-sided prism with two silvered surfaces used in a viewfinder to obtain a constant deviation of all rays of light through 90°.

Pentateuch /'pentə,tjuːk/ *the* first five books of the Bible (Genesis, Exodus, Leviticus, Numbers, Deuteronomy), called the Torah by Jews. Traditionally ascribed to Moses, it is now held by scholars to have been compiled from documents dating from the 9th to the 5th centuries BC and incorporating material from oral traditions of varying dates. □ **Pentateuchal** /,pentə'tjuːk(ə)l/ *adj.* [eccl.L *pentateuchus* f. eccl.Gk *pentateukhos* (as PENTA-, *teukhos* implement, book)]

pentathlon /pen'tæθlən/ *n.* an athletic contest in which each competitor takes part in the same prescribed five different events. It was a feature of the Olympic Games in ancient Greece. The modern men's pentathlon consists of fencing, shooting, swimming, riding, and cross-country running; the women's pentathlon consists of sprinting, hurdling, long jump, high jump, and putting the shot. □ **pentathlete** /-'tæθliːt/ *n.* [Gk f. *pente* five + *athlon* contest]

pentatonic /,pentə'tɒnɪk/ *adj. Mus.* **1** consisting of five notes. **2** relating to such a scale.

pentavalent /,pentə'veɪlənt/ *adj. Chem.* having a valency of five; quinquevalent.

Pentecost /'pentɪ,kɒst/ *n.* **1 a** Whit Sunday. **b** a festival celebrating the descent of the Holy Spirit on Whit Sunday. **2 a** the Jewish harvest festival, on the fiftieth day after the second day of Passover (Lev. 23:15–16). **b** a synagogue ceremony on the anniversary of the giving of the Torah on Mount Sinai. [OE *pentecosten* & OF *pentecoste*, f. eccl.L *pentecoste* f. Gk *pentēkostē (hēmera)* fiftieth (day)]

Pentecostal /,pentɪ'kɒst(ə)l/ *adj. & n.* ● *adj.* (also **pentecostal**) **1** of or relating to Pentecost. **2** of or designating a Christian sect which, or individual who, is frequently fundamentalist in outlook and expresses religious feelings by clapping, shouting, dancing, etc. (*See note below.*) ● *n.* a member of a Pentecostal sect; an adherent of Pentecostalism. □ **Pentecostalism** *n.* **Pentecostalist** *adj. & n.*

▪ The Pentecostal religious movement began in the early 20th century among Christians inspired by the description of the coming of the Holy Spirit at Pentecost in Acts 2:1–4. Emphasis is on the corporate element in worship (often involving great spontaneity) and 'speaking in tongues' (generally unintelligible utterances, arising from the intensity of emotion and religious experience), with prophecy, healing, and exorcism. Pentecostalism first attracted worldwide attention after revivalist meetings at Los Angeles in 1906.

Penthesilea /,penθesɪ'liːə/ *Gk Mythol.* the queen of the Amazons, who came to the help of Troy after the death of Hector and was killed by Achilles.

penthouse /'penthaʊs/ *n.* **1** a house or flat on the roof or the top floor of a tall building. **2** a sloping roof, esp. of an outhouse built on to another building. **3** an awning, a canopy. [ME *pentis* f. OF *apentis, -dis,* f. med.L *appendicium,* in LL = appendage, f. L (as APPEND): infl. by HOUSE]

pentimento /,pentɪ'mentəʊ/ *n.* (*pl.* **pentimenti** /-tɪ/) the phenomenon of earlier painting showing through a layer or layers of paint on a canvas. [It., = repentance]

Pentland Firth /'pentlənd/ a channel separating the Orkney Islands from the northern tip of mainland Scotland. It links the North Sea with the Atlantic.

pentobarbitone /,pentə'bɑːbɪ,təʊn/ *n.* (*US* **pentobarbital** /-,tæl/) a narcotic and sedative barbiturate drug formerly used to relieve insomnia. [PENTA-, BARBITONE, *barbital*]

pentode /'pentəʊd/ *n. Electronics* a thermionic valve having five electrodes. [Gk *pente* five + *hodos* way]

pentose /'pentəʊz, -təʊs/ *n. Biochem.* any monosaccharide containing five carbon atoms, including ribose. [PENTA- + -OSE²]

Pentothal /'pentəʊ,θæl/ *n. propr.* = THIOPENTONE.

pent-roof /'pentruːf/ *n.* a roof sloping in one direction only. [PENTHOUSE + ROOF]

pentstemon /pent'stiːmən, 'pentstəmən/ *n.* (also **penstemon** /pen'stiːmən, 'penstəmən/) an American herbaceous plant of the genus *Penstemon,* with showy flowers and five stamens, one of which is sterile. [mod.L, irreg. f. PENTA- + Gk *stēmōn* warp, used for 'stamen']

pentyl /'pentaɪl, -tɪl/ *n.* = AMYL. [PENTANE + -YL]

penult /pɪ'nʌlt, 'piːnʌlt, 'pen-/ *n. & adj.* ● *n.* the last but one (esp. syllable). ● *adj.* last but one. [abbr. of L *paenultimus* (see PENULTIMATE) or of PENULTIMATE]

penultimate /pɪ'nʌltɪmət/ *adj. & n.* ● *adj.* last but one. ● *n.* **1** the last but one. **2** the last syllable but one. [L *paenultimus* f. *paene* almost + *ultimus* last, after *ultimate*]

penumbra /pɪ'nʌmbrə/ *n.* (*pl.* **penumbrae** /-briː/ or **penumbras**) **1 a** the partially shaded outer region of the shadow cast by an opaque object, esp. (*Astron.*) the area on the earth or moon experiencing the partial phase of an eclipse (cf. UMBRA 1). **b** *Astron.* the less dark outer part of a sunspot. **2 a** a partial shadow. **b** an area of obscurity or uncertainty. □ **penumbral** *adj.* [mod.L f. L *paene* almost + UMBRA shadow]

penurious /pɪ'njʊərɪəs/ *adj.* **1** poor; destitute. **2** stingy; grudging. **3** scanty. □ **penuriously** *adv.* **penuriousness** *n.* [med.L *penuriosus* (as PENURY)]

penury /'penjʊrɪ/ *n.* (*pl.* **-ies**) **1** destitution; poverty. **2** a lack; scarcity. [ME f. L *penuria,* perh. rel. to *paene* almost]

Penza /'pjenzə/ a city in south central Russia; pop. (1990) 548,000. Situated on the River Sura, a tributary of the Volga, it is an industrial and transportation centre.

Penzance /pen'zæns/ a resort town in SW England, on the south coast of Cornwall near Land's End; pop. (1981) 19,600.

peon /'piːən/ *n.* **1 a** a Spanish-American day labourer or farm-worker. **b** a poor or destitute South American. **2** (in India) an office messenger, attendant, or orderly. **3** a bullfighter's assistant. **4** *hist.* a worker held in servitude in the southern US. □ **peonage** *n.* [Port. *peão* & Sp. *peon* f. med.L *pedo -onis* walker f. L *pes pedis* foot: cf. PAWN¹]

peony /'piːənɪ/ *n.* (also **paeony**) (*pl.* **-ies**) a herbaceous plant of the genus *Paeonia*, with large globular red, pink, or white flowers, often double in cultivated varieties. [OE *peonie* f. L *peonia* f. Gk *paiōnia* f. *Paiōn*, physician of the gods]

people /ˈpiːp(ə)l/ *n. & v.* ● *n.* **1** (usu. as *pl.*) **a** persons composing a community, tribe, race, nation, etc. (*the English people; a warlike people; the peoples of the Commonwealth*). **b** a group of persons of a usu. specified kind (*the chosen people; these people here; right-thinking people*). **2** (prec. by *the*; treated as *pl.*) **a** the mass of people in a country etc. not having special rank or position. **b** these considered as an electorate (*the people will reject it*). **3** parents or other relatives (*my people disapprove*). **4 a** subjects, armed followers, a retinue, etc. **b** a congregation of a parish priest etc. **5** persons in general (*people do not like rudeness*). ● *v.tr.* (usu. foll. by *with*) **1** fill with people, animals, etc.; populate. **2** (esp. as **peopled** *adj.*) inhabit; occupy; fill (*thickly peopled*). □ **people's democracy** a political system, formerly esp. in eastern Europe, with power regarded as invested in the people. [ME f. AF *poeple, people*, OF *pople, peuple*, f. L *populus*]

People's Liberation Army (abbr. **PLA**) the armed forces of the People's Republic of China, including all its land, sea, and air forces. The PLA traces its origins to an unsuccessful uprising by Communist-led troops against pro-Nationalist forces in Kiangsi province on 1 August 1927, a date celebrated annually as its anniversary.

Peoples of the Sea see SEA PEOPLES.

People's Republic of China the official name (since 1949) of CHINA.

Peoria /piːˈɔːrɪə/ a river port and industrial city in central Illinois, on the Illinois river; pop. (1990) 113,500. The city, which developed around a fort built by the French in 1680, was named after the American Indians who occupied the area when the French arrived.

PEP *abbr. Brit.* **1** Political and Economic Planning. **2** /pep/ Personal Equity Plan.

pep /pep/ *n. & v. colloq.* ● *n.* vigour; go; spirit. ● *v.tr.* (**pepped, pepping**) (usu. foll. by *up*) fill with vigour. □ **pep pill** a pill containing a stimulant drug. **pep talk** a usu. short talk intended to enthuse, encourage, etc. [abbr. of PEPPER]

peperino /ˌpepəˈriːnəʊ/ *n.* a light porous (esp. brown) volcanic rock formed of small grains of sand, cinders, etc. [It. f. *pepere* pepper]

peperoni var. of PEPPERONI.

peplum /ˈpepləm/ *n.* **1** a short flounce etc. at waist level, esp. of a blouse or jacket over a skirt. **2** *Gk Antiq.* a woman's outer garment. [L f. Gk *peplos*]

pepo /ˈpiːpəʊ/ *n.* (*pl.* **-os**) any fleshy fruit of the melon or cucumber type, with numerous seeds and surrounded by a hard skin. [L, = pumpkin, f. Gk *pepōn* abbr. of *pepōn sikuos* ripe gourd]

pepper /ˈpepə(r)/ *n. & v.* ● *n.* **1 a** a hot aromatic condiment from the dried berries of certain plants used whole or ground. **b** a climbing vine of the genus *Piper*, esp. *P. nigrum*, yielding these berries. **2** anything hot or pungent. **3 a** = CAPSICUM. **b** = CAYENNE. ● *v.tr.* **1** sprinkle or treat with or as if with pepper. **2 a** pelt with missiles. **b** hurl abuse etc. at. **3** punish severely. □ **black pepper** the unripe ground or whole berries of *Piper nigrum*, used as a condiment. **green pepper** the unripe fruit of a capsicum. **pepper-mill** a device for grinding pepper by hand. **pepper-pot 1** a small container with a perforated lid for sprinkling pepper. **2** a West Indian dish of meat etc. stewed with cayenne pepper. **red** (or **yellow**) **pepper** the ripe fruit of a capsicum. **sweet pepper** = CAPSICUM. **white pepper** the ripe or husked ground or whole berries of *Piper nigrum*, used as a condiment. [OE *piper, pipor* f. L *piper* f. Gk *peperi* f. Skr. *pippalī-* berry, peppercorn]

pepperbox /ˈpepəˌbɒks/ *n.* = pepper-pot 1.

peppercorn /ˈpepəˌkɔːn/ *n.* **1** the dried berry of *Piper nigrum* as a condiment. **2** (in full **peppercorn rent**) a nominal rent.

peppermint /ˈpepəˌmɪnt/ *n.* **1 a** a mint plant, *Mentha piperita*, grown for the strong-flavoured oil obtained from its leaves. **b** the oil from this. **2** a sweet flavoured with peppermint. **3** *Austral.* a eucalyptus yielding oil with a similar flavour. □ **pepperminty** *adj.*

pepperoni /ˌpepəˈrəʊni/ *n.* (also **peperoni**) beef and pork sausage seasoned with pepper. [It. *peperone* chilli]

pepperwort /ˈpepəˌwɜːt/ *n.* a peppery-tasting cruciferous plant of the genus *Lepidium*, esp. garden cress.

peppery /ˈpepəri/ *adj.* **1** of, like, or containing much, pepper. **2** hot-tempered. **3** pungent; stinging. □ **pepperiness** *n.*

peppy /ˈpepi/ *adj.* (**peppier, peppiest**) *colloq.* vigorous, energetic, bouncy. □ **peppily** *adv.* **peppiness** *n.*

pepsin /ˈpepsɪn/ *n. Biochem.* an enzyme contained in the gastric juice, which hydrolyses proteins under acid conditions. [G f. Gk *pepsis* digestion]

peptic /ˈpeptɪk/ *adj.* concerning or promoting digestion in the stomach. □ **peptic glands** glands secreting gastric juice. **peptic ulcer** an ulcer in the stomach or duodenum, caused by the action of pepsin and stomach acid. [Gk *peptikos* able to digest (as PEPTONE)]

peptide /ˈpeptaɪd/ *n. Biochem.* a compound consisting of two or more amino-acid molecules linked in sequence, the carboxyl group of each acid being bonded to the amino group of the next (cf. POLYPEPTIDE). [G *Peptid*, back-form. (as POLYPEPTIDE)]

peptone /ˈpeptəʊn/ *n. Biochem.* a protein fragment formed by hydrolysis in the process of digestion. □ **peptonize** /-təˌnaɪz/ *v.tr.* (also **-ise**). [G *Pepton* f. Gk *peptos*, neut. *pepton* cooked]

Pepys /piːps/, Samuel (1633–1703), English diarist and naval administrator. He is particularly remembered for his *Diary* (1660–9), which remains an important document of contemporary events such as the Great Plague (1665–6), the Fire of London (1666), and the sailing of the Dutch fleet up the Thames (1665–7). The *Diary* was written in code and was first deciphered in 1825. Pepys became secretary of the Admiralty in 1672 but was deprived of his post in 1679 and committed to the Tower for his alleged complicity in the Popish Plot, being reappointed in 1684.

per /pɜː(r), pə(r)/ *prep.* **1** for each; for every (*two sweets per child; five miles per hour*). **2** by means of; by; through (*per post; per rail*). **3** (in full **as per**) in accordance with (*as per instructions*). **4** *Heraldry* in the direction of. □ **as per usual** as usual. [L]

per- /pɜː(r), pə(r)/ *prefix* **1** forming verbs, nouns, and adjectives meaning: **a** through; all over (*perforate; perforation; pervade*). **b** completely; very (*perfervid; perturb*). **c** to destruction; to the bad (*perdition; pervert*). **2** *Chem.* having the maximum of some element in combination, esp.: **a** in the names of binary compounds in *-ide* (*peroxide*). **b** in the names of oxides, acids, etc. in *-ic* (*perchloric; permanganic*). **c** in the names of salts of these acids (*perchlorate; permanganate*). [L *per-* (as PER)]

peradventure /ˌpərədˈventʃə(r), ˌper-/ *adv. & n. archaic* or *joc.* ● *adv.* perhaps. ● *n.* uncertainty; chance; conjecture; doubt (esp. *beyond* or *without peradventure*). [ME f. OF *per* or *par auenture* by chance (as PER, ADVENTURE)]

Perak /ˈpeərə, peˈræk/ a state of Malaysia, on the west side of the Malay Peninsula; capital, Ipoh. It is a major tin-mining centre.

perambulate /pəˈræmbjʊˌleɪt/ *v.* **1** *tr.* walk through, over, or about (streets, the country, etc.). **2** *intr.* walk from place to place. **3** *tr.* **a** travel through and inspect (territory). **b** formally establish the boundaries of (a parish etc.) by walking round them. □ **perambulatory** /-lətəri/ *adj.* **perambulation** /-ˌræmbjʊˈleɪʃ(ə)n/ *n.* [L *perambulare perambulat-* (as PER-, *ambulare* walk)]

perambulator /pəˈræmbjʊˌleɪtə(r)/ *n. Brit. formal* = PRAM[1]. [PERAMBULATE]

per annum /pər ˈænəm/ *adv.* for each year. [L]

percale /pəˈkeɪl/ *n.* a closely woven cotton fabric like calico. [F, of uncert. orig.]

per capita /pə ˈkæpɪtə/ *adv. & adj.* (also **per caput** /ˈkæpʊt/) for each person. [L, = by heads]

perceive /pəˈsiːv/ *v.tr.* **1** apprehend, esp. through the sight; observe. **2** (usu. foll. by *that, how*, etc. + clause) apprehend with the mind; understand. **3** regard mentally in a specified manner (*perceives the universe as infinite*). □ **perceivable** *adj.* **perceiver** *n.* [ME f. OF *perçoivre*, f. L *percipere* (as PER-, *capere* take)]

per cent /pə ˈsent/ *adv. & n.* (*US* **percent**) ● *adv.* in every hundred. ● *n.* **1** percentage (symbol: %). **2** one part in every hundred (*half a per cent*). **3** (in *pl.*) *Brit.* public securities yielding interest of so much per cent (*three per cents*). [PER + CENT]

percentage /pəˈsentɪdʒ/ *n.* **1** a rate or proportion per cent. **2** a proportion. **3** *colloq.* personal benefit or advantage.

percentile /pəˈsentaɪl/ *n. Statistics* one of 99 values of a variable dividing a population into 100 equal groups as regards the value of that variable.

percept /ˈpɜːsept/ *n. Philos.* **1** an object of perception. **2** a mental concept resulting from perceiving, esp. by sight. [L *perceptum* perceived (thing), neut. past part. of *percipere* PERCEIVE, after *concept*]

perceptible /pəˈseptɪb(ə)l/ *adj.* capable of being perceived by the senses or intellect. □ **perceptibly** *adv.* **perceptibility** /-ˌseptɪˈbɪlɪti/ *n.* [OF *perceptible* or LL *perceptibilis* f. L (as PERCEIVE)]

perception /pə'sepʃ(ə)n/ n. **1 a** the faculty of perceiving. **b** an instance of this. **2** (often foll. by *of*) **a** the intuitive recognition of a truth, aesthetic quality, etc. **b** an instance of this (*a sudden perception of the true position*). **3** an interpretation or impression based on one's understanding of something. **4** *Philos.* the ability of the mind to refer sensory information to an external object as its cause. □ **perceptional** adj. **perceptual** /-'septjʊəl/ adj. **perceptually** adv. [ME f. L perceptio (as PERCEIVE)]

perceptive /pə'septɪv/ adj. **1** capable of perceiving. **2** sensitive; discerning; observant (*a perceptive remark*). □ **perceptively** adv. **perceptiveness** n. **perceptivity** /ˌpɜːsep'tɪvɪtɪ/ n. [med.L perceptivus (as PERCEIVE)]

Perceval[1] /'pɜːsɪv(ə)l/ (also **Parsifal** /'pɑːsɪf(ə)l/) a legendary figure dating back to ancient times, found in French, German, and English poetry from the late 12th century onwards. He is the father of Lohengrin and the hero of a number of legends, some of which are associated with the Holy Grail. (See CHRÉTIEN DE TROYES.)

Perceval[2] /'pɜːsɪv(ə)l/, Spencer (1762–1812), British Tory statesman, Prime Minister 1809–12. He was shot dead in the lobby of the House of Commons by a bankrupt merchant who blamed the government for his insolvency.

perch[1] /pɜːtʃ/ n. & v. ● n. **1** a usu. horizontal bar, branch, etc. used by a bird to rest on. **2** a usu. high or precarious place for a person or thing to rest on. **3** *hist.* a measure of length, esp. for land, of $5\frac{1}{2}$ yards. Also called *rod, pole.* ● v.intr. & tr. (usu. foll. by *on*) settle or rest, or cause to settle or rest on or as if on a perch etc. (*the bird perched on a branch; a town perched on a hill*). □ **knock a person off his or her perch 1** vanquish, destroy. **2** make less confident or secure. **square perch** *hist.* $30\frac{1}{4}$ sq. yards. [ME f. OF *perche, percher* f. L *pertica* pole]

perch[2] /pɜːtʃ/ n. (pl. same or **perches**) a spiny-finned freshwater fish of the genus *Perca*; esp. the edible *P. fluviatilis*, of Europe. [ME f. OF *perche* f. L *perca* f. Gk *perkē*]

perchance /pə'tʃɑːns/ adv. archaic or poet. **1** by chance. **2** possibly; maybe. [ME f. AF *par chance* f. *par* by, CHANCE]

percher /'pɜːtʃə(r)/ n. any bird with feet adapted for perching; a passerine.

percheron /'pɜːʃəˌrɒn/ n. a powerful breed of cart-horse. [F, orig. bred in le *Perche*, a district of northern France]

perchlorate /pə'klɔːreɪt/ n. Chem. a salt or ester of perchloric acid.

perchloric acid /pə'klɔːrɪk/ n. Chem. a fuming toxic liquid containing heptavalent chlorine, used as a strong oxidizing agent. [PER- + CHLORINE]

percipient /pə'sɪpɪənt/ adj. & n. ● adj. **1** able to perceive; conscious. **2** discerning; observant. ● n. a person who perceives, esp. something outside the range of the senses. □ **percipience** n. **percipiently** adv. [L (as PERCEIVE)]

percolate /'pɜːkəˌleɪt/ v. **1** intr. (often foll. by *through*) **a** (of liquid etc.) filter or ooze gradually (esp. through a porous surface). **b** (of an idea etc.) permeate gradually. **2** tr. prepare (coffee) by repeatedly passing boiling water through ground beans. **3** tr. ooze through; permeate. **4** tr. strain (a liquid, powder, etc.) through a fine mesh etc. □ **percolation** /ˌpɜːkə'leɪʃ(ə)n/ n. [L percolare (as PER-, colare strain f. colum strainer)]

percolator /'pɜːkəˌleɪtə(r)/ n. a machine for making coffee by circulating boiling water through ground beans.

per contra /pɜː 'kɒntrə/ adv. on the opposite side (of an account, assessment, etc.); on the other hand. [It.]

percuss /pə'kʌs/ v.tr. Med. tap (a part of the body) gently with a finger or an instrument as part of a diagnosis. [L percutere percuss- strike (as PER-, cutere = quatere shake)]

percussion /pə'kʌʃ(ə)n/ n. **1** Mus. **a** (often attrib.) the playing of music by striking instruments with sticks etc. (*a percussion band*). **b** the section of such instruments in an orchestra, i.e. drums and cymbals with possible extras such as glockenspiel, xylophone, gongs, bells, rattles, etc. (*asked the percussion to stay behind*). **2** Med. the act or an instance of percussing. **3** the forcible striking of one esp. solid body against another. □ **percussion cap** a small amount of explosive powder contained in metal or paper and exploded by striking, used esp. in toy guns and formerly in some firearms. □ **percussionist** n. **percussive** /-'kʌsɪv/ adj. **percussively** adv. **percussiveness** n. [F percussion or L percussio (as PERCUSS)]

percutaneous /ˌpɜːkjʊ'teɪnɪəs/ adj. esp. Med. made or done through the skin. [L per cutem through the skin]

Percy /'pɜːsɪ/, Sir Henry (known as 'Hotspur' and 'Harry Hotspur') (1364–1403), English soldier. Son of the 1st Earl of Northumberland

(1342–1408), he was killed at the battle of Shrewsbury during his father's revolt against Henry IV.

per diem /pɜː 'diːem, 'daɪem/ adv., adj., & n. ● adv. & adj. for each day. ● n. an allowance or payment for each day. [L]

perdition /pə'dɪʃ(ə)n/ n. eternal death; damnation. [ME f. OF perdiciun or eccl.L perditio f. L perdere destroy (as PER-, dere dit- = dare give)]

perdurable /pə'djʊərəb(ə)l/ adj. formal permanent; eternal; durable. □ **perdurably** adv. **perdurability** /-ˌdjʊərə'bɪlɪtɪ/ n. [ME f. OF f. LL perdurabilis (as PER-, DURABLE)]

père /peə(r)/ n. (added to a surname to distinguish a father from a son of the same name) the father, senior (cf. FILS). [F, = father]

Père David's deer /peə 'deɪvɪdz/ n. a large slender-antlered deer, *Elaphurus davidianus*, native to China but now extinct in the wild. [after Father Armand *David*, Fr. missionary (1826–1900)]

peregrinate /'perɪgrɪˌneɪt/ v.intr. archaic or joc. travel; journey, esp. extensively or at leisure. □ **peregrinator** n. **peregrination** /ˌperɪgrɪ'neɪʃ(ə)n/ n. [L peregrinari (as PEREGRINE)]

peregrine /'perɪgrɪn/ n. & adj. ● n. (in full **peregrine falcon**) a strongly built and widespread falcon, *Falco peregrinus*, breeding esp. on coastal cliffs and much used for falconry. Also called (*N. Amer.*) *duck-hawk*. ● adj. archaic imported from abroad; foreign; outlandish. [L peregrinus f. peregre abroad f. per through + ager field]

Perelman /'perəlmən/, S(idney) J(oseph) (1904–79), American humorist and writer. In the early 1930s he worked in Hollywood as a scriptwriter, notably on some of the Marx Brothers' films. From 1934 his name is linked with the *New Yorker* magazine, for whom he wrote most of his short stories and sketches.

peremptory /pə'remptərɪ, 'perɪmp-/ adj. **1** (of a statement or command) admitting no denial or refusal. **2** (of a person, a person's manner, etc.) dogmatic; imperious; dictatorial. **3** Law not open to appeal or challenge; final. **4** absolutely fixed; essential. □ **peremptory challenge** Law a defendant's objection to a proposed juror, made without needing to give a reason. □ **peremptorily** adv. **peremptoriness** n. [AF peremptorie, OF peremptoire f. L peremptorius deadly, decisive, f. perimere perempt- destroy, cut off (as PER-, emere take, buy)]

perennial /pə'renɪəl/ adj. & n. ● adj. **1** lasting through a year or several years. **2** (of a plant) lasting several years. **3** lasting a long time or for ever. **4** (of a stream) flowing through all seasons of the year. ● n. a perennial plant (cf. ANNUAL n. 2) (*a herbaceous perennial*). □ **perennially** adv. **perenniality** /-ˌrenɪ'ælɪtɪ/ n. [L perennis (as PER-, annus year)]

Peres /'perez/, Shimon (Polish name Szymon Perski) (b.1923), Israeli statesman, Prime Minister 1984–6 and since 1995. Born in Poland, he emigrated to Palestine in 1934. Labour Party leader since 1977, Peres became head of a coalition government with the Likud Party in 1984, later serving as deputy to Yitzhak Shamir. In 1990 he was sacked for supporting American proposals for an Israeli–Palestinian peace conference. He was appointed Foreign Minister under Yitzhak Rabin in 1992 and played a major role in negotiating the PLO–Israeli peace accord (1993). Peres shared the 1994 Nobel Peace Prize with Rabin and Yasser Arafat. He became Prime Minister again following Rabin's assassination in 1995.

perestroika /ˌperɪ'strɔɪkə/ n. (in the former USSR) the policy or practice of restructuring or reforming the economic and political system, first proposed by Leonid Brezhnev in 1979 and actively promoted under Mikhail Gorbachev 1985–91. Perestroika (often mentioned in conjunction with glasnost, or openness) originally referred to increased automation and labour efficiency, but under Gorbachev came to entail greater awareness of economic markets and the ending of central planning. [Russ., = restructuring]

Pérez de Cuéllar /ˌperez də 'kweːjɑː(r)/, Javier (b.1920), Peruvian diplomat. He served as Secretary-General of the United Nations from 1982 to 1991, and played a key role in the diplomatic aftermath of the Falklands War (1982) and in ending the Iran–Iraq War (1980–8). His efforts to avert the Gulf War in 1990 raised his international standing, as did his part in negotiating the release of Western hostages held in the Middle East.

perfect adj., v., & n. ● adj. /'pɜːfɪkt/ **1** complete; not deficient. **2 a** faultless (*a perfect diamond*). **b** blameless in morals or behaviour. **3 a** very satisfactory (*a perfect evening*). **b** (often foll. by *for*) most appropriate, suitable. **4** exact; precise (*a perfect circle*). **5** entire; unqualified (*a perfect stranger*). **6** Math. (of a number) equal to the sum of its divisors. **7** Gram. (of a tense) denoting a completed action or event in the past, formed in English with *have* or *has* and the past participle, as in *they have eaten*. **8** Mus. (of pitch) absolute. **9** Bot. **a** (of a flower) having all four types of whorl. **b** (of a fungus) in the stage where the sexual spores are formed. **10** (often foll. by *in*) thoroughly trained or skilled (*is perfect in geometry*). ● *v.tr.* /pəˈfekt/ **1** make perfect; improve. **2** carry through; complete. **3** *Printing* complete (a sheet) by printing the other side. ● *n.* /ˈpɜːfɪkt/ *Gram.* the perfect tense. □ **perfect binding** a form of bookbinding in which the leaves are attached to the spine by gluing rather than sewing. **perfect interval** *Mus.* a fourth or fifth as it would occur in a major or minor scale starting on the lower note of the interval, or octave. **perfect pitch** = *absolute pitch* 1. □ **perfecter** /pəˈfektə(r)/ *n.* **perfectible** /-ˈfektɪb(ə)l/ *adj.* **perfectibility** /-ˌfektɪˈbɪlɪtɪ/ *n.* **perfectness** /ˈpɜːfɪktnɪs/ *n.* [ME and OF *parfit, perfet* f. L *perfectus* past part. of *perficere* complete (as PER-, *facere* do)]

perfecta /pəˈfektə/ *n.* *US* a form of betting in which the first two places in a race must be predicted in the correct order. [Amer. Sp. *quiniela perfecta* perfect quinella]

perfection /pəˈfekʃ(ə)n/ *n.* **1** the act or process of making perfect. **2** the state of being perfect; faultlessness, excellence. **3** a perfect person, thing, or example. **4** an accomplishment. **5** full development; completion. □ **to perfection** exactly; completely. [ME f. OF f. L *perfectio -onis* (as PERFECT)]

perfectionism /pəˈfekʃəˌnɪz(ə)m/ *n.* **1** the uncompromising pursuit of excellence. **2** *Philos.* the belief that religious or moral perfection is attainable. □ **perfectionist** *n. & adj.* [PERFECTION]

perfective /pəˈfektɪv/ *adj. & n. Gram.* ● *adj.* (of an aspect of a verb etc.) expressing the completion of an action (opp. IMPERFECTIVE). ● *n.* the perfective aspect or form of a verb. [med.L *perfectivus* (as PERFECT)]

perfectly /ˈpɜːfɪktlɪ/ *adv.* **1** completely; absolutely (*I understand you perfectly*). **2** quite, completely (*is perfectly capable of doing it*). **3** in a perfect way. **4** very (*you know perfectly well*).

perfecto /pəˈfektəʊ/ *n.* (*pl.* **-os**) orig. *US* a large thick cigar pointed at each end. [Sp., = perfect]

perfervid /pəˈfɜːvɪd/ *adj. literary* very fervid. □ **perfervidly** *adv.* **perfervidness** *n.* [mod.L *perfervidus* (as PER-, FERVID)]

perfidy /ˈpɜːfɪdɪ/ *n.* breach of faith; treachery. □ **perfidious** /pəˈfɪdɪəs/ *adj.* **perfidiously** *adv.* [L *perfidia* f. *perfidus* treacherous (as PER-, *fides* f. *fides* faith)]

perfoliate /pəˈfəʊlɪət/ *adj.* (of a plant) having the stalk apparently passing through the leaf. [mod.L *perfoliatus* (as PER-, FOLIATE)]

perforate *v. & adj.* ● *v.* /ˈpɜːfəˌreɪt/ **1** *tr.* make a hole or holes through; pierce. **2** *tr.* make a row of small holes in (paper etc.) so that a part may be torn off easily. **3** *tr.* make an opening into; pass into or extend through. **4** *intr.* (usu. foll. by *into, through*, etc.) penetrate. ● *adj.* /ˈpɜːfərət/ perforated. □ **perforative** /-rətɪv/ *adj.* **perforator** /-ˌreɪtə(r)/ *n.* **perforation** /ˌpɜːfəˈreɪʃ(ə)n/ *n.* [L *perforare* (as PER-, *forare* pierce)]

perforce /pəˈfɔːs/ *adv. archaic* unavoidably; necessarily. [ME f. OF *par force* by FORCE[1]]

perforin /ˈpɜːfərɪn/ *n. Biochem.* a protein released by killer cells, destroying targeted cells by perforating the cell membrane. [PERFORATE + -IN]

perform /pəˈfɔːm/ *v.* **1** *tr.* (also *absol.*) carry into effect; be the agent of; do (a command, promise, task, etc.). **2** *tr.* (also *absol.*) go through, execute (a public function, play, piece of music, etc.). **3** *intr.* act in a play; play music, sing, etc. (*likes performing*). **4** *intr.* (of a trained animal) execute tricks etc. at a public show. **5** *intr.* operate, function. **6** *intr.* *Finance* (of an investment) yield a return, esp. a profit. **7** *intr.* *sl.* have sexual intercourse (esp. satisfactorily). □ **performatory** *adj. & n.* (*pl.* **-ies**). **performer** *n.* **performing** *adj.* **performable** *adj.* **performability** /-ˌfɔːməˈbɪlɪtɪ/ *n.* [ME f. AF *parfourmer* f. OF *parfournir* (assim. to *forme* FORM) f. *par* PER- + *fournir* FURNISH]

performance /pəˈfɔːməns/ *n.* **1** (usu. foll. by *of*) **a** the act or process of performing or carrying out. **b** the execution or fulfilment (of a duty etc.). **2** a staging or production (of a drama, piece of music, etc.) (*the afternoon performance*). **3** a person's achievement under test conditions etc. (*put up a good performance*). **4** *colloq.* a fuss; a scene; a public exhibition (*made such a performance about leaving*). **5 a** the capabilities of a machine, esp. a car or aircraft. **b** (*attrib.*) of high capability (*a performance car*). **6** *Finance* the return on an investment, esp. in stocks and shares etc. □ **performance art** a form of visual art in which the activity of the artist forms a central feature.

performative /pəˈfɔːmətɪv/ *adj. & n.* ● *adj.* **1** of or relating to performance. **2** *Linguistics* denoting an utterance that effects an action by being spoken or written (e.g. *I bet, I apologize*). ● *n. Linguistics* a performative utterance.

performing arts *n.pl.* the arts, such as drama, music, and dance, that require performance for their realization.

perfume /ˈpɜːfjuːm/ *n. & v.* ● *n.* **1** a sweet smell. **2** fluid containing the essence of flowers etc.; scent. ● *v.tr.* (also /pəˈfjuːm/) (usu. as **perfumed** *adj.*) impart a sweet scent to; impregnate with a sweet smell. □ **perfumy** *adj.* [F *parfum, parfumer* f. obs. It. *parfumare, perfumare* (as PER-, *fumare* smoke, FUME): orig. of smoke from a burning substance]

perfumer /pəˈfjuːmə(r)/ *n.* a maker or seller of perfumes. □ **perfumery** *n.* (*pl.* **-ies**)

perfunctory /pəˈfʌŋktərɪ/ *adj.* **1 a** done merely for the sake of getting through a duty. **b** done in a cursory or careless manner. **2** superficial; mechanical. □ **perfunctorily** *adv.* **perfunctoriness** *n.* [LL *perfunctorius* careless f. L *perfungi perfunct-* (as PER-, *fungi* perform)]

perfuse /pəˈfjuːz/ *v.tr.* **1** (often foll. by *with*) **a** besprinkle (with water etc.). **b** cover or suffuse (with radiance etc.). **2** pour or diffuse (water etc.) through or over. **3** *Med.* cause a fluid to pass through (an organ etc.). □ **perfusion** /-ˈfjuːʒ(ə)n/ *n.* **perfusive** /-ˈfjuːsɪv/ *adj.* [L *perfundere perfus-* (as PER-, *fundere* pour)]

Pergamum /ˈpɜːɡəməm/ a city in ancient Mysia, in western Asia Minor, situated to the north of Izmir on a rocky hill close to the Aegean coast. The capital in the 3rd and 2nd centuries BC of the Attalid dynasty, it was one of the greatest and most beautiful of the Hellenistic cities. It was famed for its cultural institutions, especially its library, which was second only to that at Alexandria. The city, and the extensive kingdom of Pergamum, later became a province of Rome. The modern town of Bergama lies close to the ruins of the ancient city. □ **Pergamene** /-ˌmiːn/ *adj. & n.*

pergana var. of PARGANA.

pergola /ˈpɜːɡələ/ *n.* an arbour or covered walk, formed of growing plants trained over trellis-work. [It. f. L *pergula* projecting roof f. *pergere* proceed]

pergunnah var. of PARGANA.

perhaps /pəˈhæps/ *adv.* **1** it may be; possibly (*perhaps it is lost*). **2** introducing a polite request (*perhaps you would open the window?*). [PER + HAP]

peri /ˈpɪərɪ/ *n.* (*pl.* **peris**) **1** (in Persian mythology) a fairy; a good (orig. evil) genius. **2** a beautiful or graceful being. [Pers. *parī*]

peri- /ˈperɪ/ *prefix* **1** round, about. **2** *Astron.* the point nearest to (*perigee; perihelion*). [Gk *peri* around, about]

perianth /ˈperɪˌænθ/ *n. Bot.* the outer part of a flower. [F *périanthe* f. mod.L *perianthium* (as PERI- + Gk *anthos* flower)]

periapt /ˈperɪˌæpt/ *n.* a thing worn as a charm; an amulet. [F *périapte* f. Gk *periapton* f. *haptō* fasten]

pericardium /ˌperɪˈkɑːdɪəm/ *n.* (*pl.* **pericardia** /-dɪə/) *Anat.* the membranous sac enclosing the heart. □ **pericardial** *adj.* **pericardiac** /-dɪˌæk/ *adj.* **pericarditis** /-kɑːˈdaɪtɪs/ *n.* [mod.L f. Gk *perikardion* (as PERI- + *kardia* heart)]

pericarp /ˈperɪˌkɑːp/ *n. Bot.* the part of a fruit formed from the wall of the ripened ovary. [F *péricarpe* f. Gk *perikarpion* pod, shell (as PERI-, *karpos* fruit)]

perichondrium /ˌperɪˈkɒndrɪəm/ *n. Anat.* the membrane enveloping cartilage tissue (except at the joints). [PERI- + Gk *khondros* cartilage]

periclase /ˈperɪˌkleɪs/ *n.* a pale mineral consisting of magnesia. [mod.L *periclasia*, erron. f. Gk *peri* exceedingly + *klasis* breaking, from its perfect cleavage]

Pericles /ˈperɪˌkliːz/ (*c*.495–429 BC), Athenian statesman and general. A champion of Athenian democracy, he pursued an imperialist policy and masterminded Athenian strategy in the Peloponnesian War. He commissioned the building of the Parthenon in 447 and promoted the culture of Athens in a golden age that produced such figures as Aeschylus, Socrates, and Phidias.

periclinal /ˌperɪˈklaɪn(ə)l/ *adj. Geol.* (of a mound etc.) sloping down in all directions from a central point. [Gk *periklinēs* sloping on all sides (as PERI-, CLINE)]

pericope /pəˈrɪkəpɪ/ *n.* a short passage or paragraph, esp. a portion of Scripture read in public worship. [LL f. Gk *perikopē* (as PERI-, *kopē* cutting f. *koptō* cut)]

pericranium /ˌperɪˈkreɪnɪəm/ *n. Anat.* the membrane enveloping the skull. [mod.L f. Gk (as PERI-, *kranion* skull)]

peridot /ˈperɪˌdɒt/ *n.* a green variety of olivine, used esp. as a semiprecious stone. [ME f. OF *peritot*, of unkn. orig.]

perigee /ˈperɪˌdʒiː/ *n.* the point in a body's orbit at which it is nearest the earth (opp. APOGEE). □ **perigean** /ˌperɪˈdʒɪən/ adj. [F périgée f. mod.L f. Gk perigeion round the earth (as PERI-, gē earth)]

periglacial /ˌperɪˈɡleɪʃ(ə)l, -ˈɡleɪsɪəl/ adj. of or relating to a region adjoining a glacier.

Périgord /ˈperɪˌɡɔː(r)/ an area of SW France, in the south-western Massif Central. A former countship, it became a part of Navarre in 1470, becoming united with France in 1670. Until the French Revolution it was a part of Guyenne.

perigynous /pəˈrɪdʒɪnəs/ adj. Bot. (of stamens) situated around the pistil or ovary. [mod.L perigynus (as PERI-, -GYNOUS)]

perihelion /ˌperɪˈhiːlɪən/ n. (pl. **perihelia** /-lɪə/) Astron. the point of closest approach to the sun by an orbiting planet, comet, etc. (opp. APHELION). Such a position can generally be calculated by the laws of classical mechanics, but the systematic change in the perihelion position of Mercury could be explained only by Einstein's general theory of relativity, for which it constituted an important test. [Graecized f. mod.L perihelium (as PERI-, Gk hēlios sun)]

peril /ˈperɪl/ n. & v. ● n. serious and immediate danger. ● v.tr. (**perilled**, **perilling**; US **periled**, **periling**) threaten; endanger. □ **at one's peril** at one's own risk. **in peril of** with great risk to (in peril of your life). **peril point** US Econ. a critical threshold or limit. [ME f. OF f. L peric(u)lum]

perilous /ˈperɪləs/ adj. **1** full of risk; dangerous; hazardous. **2** exposed to imminent risk of destruction etc. □ **perilously** adv. **perilousness** n. [ME f. OF perillous f. L periculosus f. periculum: see PERIL]

perilune /ˈperɪˌluːn, -ˌljuːn/ n. Astron. the point at which a spacecraft in lunar orbit is nearest to the moon's centre. [PERI- + L luna moon, after perigee]

perilymph /ˈperɪˌlɪmf/ n. Anat. the fluid in the labyrinth of the ear.

perimeter /pəˈrɪmɪtə(r)/ n. **1 a** the circumference or outline of a closed figure. **b** the length of this. **2 a** the outer boundary of an enclosed area. **b** a defended boundary. **3** an instrument for measuring a field of vision. □ **perimetric** /ˌperɪˈmetrɪk/ adj. [F périmètre or f. L perimetrus f. Gk perimetros (as PERI-, metros f. metron measure)]

perinatal /ˌperɪˈneɪt(ə)l/ adj. of or relating to the time immediately before and after birth.

perineum /ˌperɪˈniːəm/ n. Anat. the region of the body between the anus and the scrotum or vulva. □ **perineal** /-ˈniːəl/ adj. [LL f. Gk perinaion]

period /ˈpɪərɪəd/ n. & adj. ● n. **1** a length or portion of time (showers and bright periods). **2** a distinct portion of history, a person's life, etc. (the Georgian period; Picasso's Blue Period). **3** Geol. a time forming part of a geological era (the Quaternary period). **4 a** an interval between recurrences of an astronomical or other phenomenon. **b** the time taken by a planet etc. to rotate about its axis. **c** the time taken by a planet or satellite to make one circuit of its orbit. **5** the time allowed for a lesson in school. **6** an occurrence of menstruation. **7 a** a complete sentence, esp. one consisting of several clauses. **b** (in pl.) rhetorical language. **8** esp. N. Amer. **a** = full stop (see FULL[1]). **b** used at the end of a sentence etc. to indicate finality, absoluteness, etc. (we want the best, period). **9** Math. **a** a set of figures marked off in a large number to assist in reading. **b** a set of figures repeated in a recurring decimal. **c** the smallest interval over which a function takes the same value. **10** Chem. a sequence of elements between two noble gases forming a row in the periodic table. ● adj. belonging to or characteristic of some past period (period furniture). □ **period piece** an object or work whose main interest lies in its historical etc. associations. [ME f. OF periode f. L periodus f. Gk periodos (as PERI-, odos = hodos way)]

periodate /pəˈraɪəˌdeɪt/ n. Chem. a salt or ester of periodic acid.

periodic /ˌpɪərɪˈɒdɪk/ adj. **1** appearing or occurring at esp. regular intervals. **2** of or concerning the period of a celestial body (periodic motion). **3** (of diction etc.) expressed in periods (PERIOD n. 7a). □ **periodic decimal** Math. a set of figures repeated in a recurring decimal. **periodic function** Math. a function returning to the same value at regular intervals. □ **periodicity** /-rɪəˈdɪsɪtɪ/ n. [F périodique or L periodicus f. Gk periodikos (as PERIOD)]

periodic acid /ˌpɜːraɪˈɒdɪk/ n. Chem. a hygroscopic solid acid containing heptavalent iodine. [PER- + IODINE]

periodical /ˌpɪərɪˈɒdɪk(ə)l/ n. & adj. ● n. a newspaper, magazine, etc. issued at regular intervals, usu. monthly or weekly. ● adj. **1** published at regular intervals. **2** periodic, occasional. □ **periodically** adv.

periodic table n. a table of the chemical elements arranged in order of increasing atomic number, usually in rows, such that elements with similar chemical properties form vertical groups (a consequence of analogous arrangements of electrons in their atoms). The modern

periodic table was devised by Dmitri Mendeleev in 1869, although John Newlands was actually the first to propose some form of periodic arrangement. Mendeleev's table proved to be a powerful predictive and interpretative tool, although it was only in the 20th century that the principles of atomic structure underlying the periodic table were discovered.

periodization /ˌpɪərɪədaɪˈzeɪʃ(ə)n/ n. (also **-isation**) the division of history into periods.

periodontics /ˌperɪəˈdɒntɪks/ n.pl. (treated as sing.) the branch of dentistry concerned with the structures surrounding and supporting the teeth. □ **periodontal** adj. **periodontist** n. [PERI- + Gk odous odont-tooth]

periodontology /ˌperɪədɒnˈtɒlədʒɪ/ n. = PERIODONTICS.

perioperative /ˌperɪˈɒpərətɪv/ adj. Med. occurring or performed around the time of an operation.

periosteum /ˌperɪˈɒstɪəm/ n. (pl. **periostea** /-tɪə/) Anat. a membrane enveloping the bones where no cartilage is present. □ **periosteal** /-ˈɒstɪəl/ adj. **periostitis** /-ɒˈstaɪtɪs/ n. [mod.L f. Gk periosteon (as PERI-, osteon bone)]

peripatetic /ˌperɪpəˈtetɪk/ adj. & n. ● adj. **1** (of a teacher) working in more than one school or college etc. **2** going from place to place; itinerant. **3** (**Peripatetic**) Aristotelian (from Aristotle's habit of walking in the Lyceum while teaching). ● n. a peripatetic person, esp. a teacher. □ **peripatetically** adv. **peripateticism** /-tɪˌsɪz(ə)m/ n. [ME f. OF peripatetique or L peripateticus f. Gk peripatētikos f. peripateō (as PERI-, pateō walk)]

peripeteia /ˌperɪpɪˈtaɪə, -ˈtiːə/ n. a sudden change of fortune in a drama or in life. [Gk (as PERI-, pet- f. piptō fall)]

peripheral /pəˈrɪfərəl/ adj. & n. ● adj. **1** of minor importance; marginal. **2** of the periphery; on the fringe. **3** Anat. near the surface of the body, with special reference to the circulation and nervous system. **4** (of equipment) used with a computer etc. but not an integral part of it. ● n. Computing a peripheral device or piece of equipment. □ **peripheral nervous system** see NERVOUS SYSTEM. □ **peripherally** adv. **peripherality** /-ˌrɪfəˈrælɪtɪ/ n.

periphery /pəˈrɪfərɪ/ n. (pl. **-ies**) **1** the boundary of an area or surface. **2** an outer or surrounding region (built on the periphery of the old town). [LL peripheria f. Gk periphereia circumference (as PERI-, phereia f. phero bear)]

periphrasis /pəˈrɪfrəsɪs/ n. (pl. **periphrases** /-ˌsiːz/) **1** a roundabout way of speaking; circumlocution. **2** a roundabout phrase. [L f. Gk f. periphrazō (as PERI-, phrazō declare)]

periphrastic /ˌperɪˈfræstɪk/ adj. **1** of or involving periphrasis. **2** Gram. (of a case, tense, etc.) formed by combination of words rather than by inflection (e.g. did go, of the people rather than went, the people's). □ **periphrastically** adv. [Gk periphrastikos (as PERIPHRASIS)]

peripteral /pəˈrɪptərəl/ adj. Archit. (of a temple) surrounded by a single row of columns. [Gk peripteron (as PERI-, pteron wing)]

periscope /ˈperɪˌskəʊp/ n. an apparatus with a tube and mirrors or prisms, by which an observer in a trench, submerged submarine, or at the rear of a crowd etc., can see things otherwise out of sight.

periscopic /ˌperɪˈskɒpɪk/ adj. of a periscope. □ **periscopic lens** a lens allowing distinct vision over a wide angle. □ **periscopically** adv.

perish /ˈperɪʃ/ v. **1** intr. be destroyed; suffer death or ruin. **2 a** intr. (esp. of rubber, a rubber object, etc.) lose its normal qualities; deteriorate, rot. **b** tr. cause to rot or deteriorate. **3** tr. (in passive) suffer from cold or exposure (we were perished standing outside). □ **perish the thought** an exclamation of horror against an unwelcome idea. □ **perishless** adj. [ME f. OF perir f. L perire pass away (as PER-, ire go)]

perishable /ˈperɪʃəb(ə)l/ adj. & n. ● adj. liable to perish; subject to decay. ● n. (esp. in pl.) a thing, esp. a foodstuff, subject to speedy decay. □ **perishableness** n. **perishability** /ˌperɪʃəˈbɪlɪtɪ/ n.

perisher /ˈperɪʃə(r)/ n. Brit. sl. an annoying person.

perishing /ˈperɪʃɪŋ/ adj. & adv. colloq. ● adj. **1** confounded. **2** freezing cold, extremely chilly. ● adv. confoundedly. □ **perishingly** adv.

perisperm /ˈperɪˌspɜːm/ n. Bot. a mass of nutritive material outside the embryo-sac in some seeds. [PERI- + Gk sperma seed]

perissodactyl /pəˌrɪsəˈdæktɪl/ adj. & n. Zool. ● adj. of or relating to the order Perissodactyla, which comprises ungulate mammals with one main central toe, or a single toe, on each foot, including horses, rhinoceroses, and tapirs. ● n. an animal of this order. [mod.L f. Gk perissos uneven + daktulos finger, toe]

peristalsis /ˌperɪˈstælsɪs/ n. Physiol. an involuntary muscular wavelike movement by which the contents of the alimentary canal etc. are propelled along. □ **peristaltic** adj. **peristaltically** adv. [mod.L f. Gk peristellō wrap around (as PERI-, stellō place)]

peristome /'perɪˌstəʊm/ n. **1** Bot. a fringe of small toothlike projections around the mouth of a capsule in mosses and certain fungi. **2** Zool. the parts surrounding the mouth of various invertebrates. [mod.L peristoma f. PERI- + Gk stoma mouth]

peristyle /'perɪˌstaɪl/ n. Archit. a row of columns surrounding a temple, court, cloister, etc.; a space surrounded by columns. [F péristyle f. L peristylum f. Gk peristulon (as PERI-, stulos pillar)]

peritoneum /ˌperɪtə'niːəm/ n. (pl. **peritoneums** or **peritonea** /-'niːə/) Anat. the serous membrane lining the cavity of the abdomen. □ **peritoneal** /-'niːəl/ adj. [LL f. Gk peritonaion (as PERI-, tonaion f. -tonos stretched)]

peritonitis /ˌperɪtə'naɪtɪs/ n. Med. an inflammatory disease of the peritoneum.

periwig /'perɪˌwɪg/ n. esp. hist. a wig. □ **periwigged** adj. [alt. of PERUKE, with -wi- for F -u- sound]

periwinkle[1] /'perɪˌwɪŋk(ə)l/ n. **1** an evergreen trailing plant of the genus Vinca, with blue or white flowers. **2** a related tropical shrub, Catharanthus roseus, native to Madagascar. [ME f. AF pervenke, OF pervenche f. LL pervinca, assim. to PERIWINKLE[2]]

periwinkle[2] /'perɪˌwɪŋk(ə)l/ n. = WINKLE. [16th c.: orig. unkn.]

perjure /'pɜːdʒə(r)/ v.refl. Law **1** wilfully tell an untruth when on oath. **2** (as **perjured** adj.) guilty of or involving perjury. □ **perjurer** n. [ME f. OF parjurer f. L perjurare (as PER-, jurare swear)]

perjury /'pɜːdʒərɪ/ n. (pl. **-ies**) Law **1** a breach of an oath, esp. the act of wilfully telling an untruth when on oath. **2** the practice of this. □ **perjurious** /pə'dʒʊərɪəs/ adj. [ME f. AF perjurie f. OF parjurie f. L perjurium (as PERJURE)]

perk[1] /pɜːk/ v. & adj. ● v.tr. (often foll. by up) raise (one's head etc.) briskly. ● adj. perky; pert. □ **perk up 1** recover confidence, courage, life, or zest. **2** restore confidence or courage or liveliness in (esp. another person). **3** smarten up. [ME, perh. f. var. of PERCH[1]]

perk[2] /pɜːk/ n. Brit. colloq. a perquisite. [abbr.]

perk[3] /pɜːk/ v. colloq. **1** intr. (of coffee) percolate, make a bubbling sound in the percolator. **2** tr. percolate (coffee). [abbr. of PERCOLATE]

Perkin /'pɜːkɪn/, Sir William Henry (1838–1907), English chemist and pioneer of the synthetic organic chemical industry. At the age of 18 he prepared the first synthetic dyestuff, mauve, which is made from aniline. The discovery was made by accident when he was trying to synthesize the drug quinine. He and his father then set up a factory to make mauve, which was used for textiles and postage stamps, and other synthetic dyes.

perky /'pɜːkɪ/ adj. (**perkier**, **perkiest**) **1** self-assertive; saucy; pert. **2** lively; cheerful. □ **perkily** adv. **perkiness** n.

Perlis /'pɜːlɪs/ the smallest state of Malaysia and the most northerly of those on the Malay Peninsula; capital, Kangar.

perlite /'pɜːlaɪt/ n. (also **pearlite**) a glassy type of vermiculite, expandable to a solid form by heating, used for insulation etc. [F f. perle pearl]

Perm /pɜːm, Russian pjermj/ an industrial city in Russia, in the western foothills of the Ural Mountains; pop. (1990) 1,094,000. It was known from 1940 to 1957 as Molotov.

perm[1] /pɜːm/ n. & v. ● n. a permanent wave. ● v.tr. give a permanent wave to (a person or a person's hair). [abbr.]

perm[2] /pɜːm/ n. & v. colloq. ● n. a permutation. ● v.tr. make a permutation of. [abbr.]

permaculture /'pɜːməˌkʌltʃə(r)/ n. Ecol. the development of agricultural ecosystems intended to be complete and self-sustaining. [PERMANENT + CULTURE]

permafrost /'pɜːməˌfrɒst/ n. subsoil which remains below freezing-point throughout the year, as in polar regions. [PERMANENT + FROST]

permalloy /'pɜːməˌlɔɪ, pɜːm'ælɔɪ/ n. an alloy of nickel and iron that is easily magnetized and demagnetized. [PERMEABLE + ALLOY]

permanent /'pɜːmənənt/ adj. lasting, or intended to last or function, indefinitely. □ **permanent hardness** hardness of water which is not removed by boiling. **permanent magnet** a magnet retaining its magnetic properties in the absence of an inducing field or current. **permanent set 1** the irreversible deformation of a substance after being subjected to stress. **2** the amount of this. **permanent tooth** a tooth succeeding a milk tooth in a mammal, and lasting most of the mammal's life. **permanent wave** an artificial wave in the hair,

intended to last for some time. **permanent way** Brit. the finished roadbed of a railway. □ **permanence** n. **permanency** n. **permanentize** v.tr. (also **-ise**). **permanently** adv. [ME f. OF permanent or L permanere (as PER-, manere remain)]

Permanent Secretary n. Brit. a senior grade in the Civil Service, usu. a permanent adviser to a minister.

Permanent Under-Secretary n. Brit. **1** a senior permanent adviser to a Secretary of State. **2** a senior civil servant below the rank of Permanent Secretary, usu. the head of a division within a department of state.

permanganate /pɜː'mæŋgəˌneɪt, -nət/ n. Chem. a salt of permanganic acid, containing the anion MnO_4^-; esp. potassium permanganate. (Cf. MANGANATE.)

permanganic acid /ˌpɜːmæŋ'gænɪk/ n. Chem. a strong acid (chem. formula: $HMnO_4$), known only as a purple solution. [PER- + MANGANIC: see MANGANESE]

permeability /ˌpɜːmɪə'bɪlɪtɪ/ n. **1** the state or quality of being permeable. **2** Physics a quantity measuring the influence of a substance on the magnetic flux in the region it occupies.

permeable /'pɜːmɪəb(ə)l/ adj. capable of being permeated. [L permeabilis (as PERMEATE)]

permeate /'pɜːmɪˌeɪt/ v. **1** tr. penetrate throughout; pervade; saturate. **2** intr. (usu. foll. by through, among, etc.) diffuse itself. □ **permeance** n. **permeant** adj. **permeator** n. **permeation** /ˌpɜːmɪ'eɪʃ(ə)n/ n. [L permeare permeat- (as PER-, meare pass, go)]

Permian /'pɜːmɪən/ adj. & n. Geol. ● adj. of or relating to the final period of the Palaeozoic era. ● n. this period or the corresponding geological system. [PERM]

■ The Permian lasted from about 286 to 248 million years ago, between the Carboniferous and Triassic periods. The climate was hot and dry in many parts of the world during this period, which saw the extinction of many marine animals, including trilobites, and the proliferation of reptiles.

per mille /pə 'mɪlɪ/ adv. (also **per mil** /mɪl/) in every thousand. [L]

permissible /pə'mɪsɪb(ə)l/ adj. allowable. □ **permissibly** adv. **permissibility** /-ˌmɪsɪ'bɪlɪtɪ/ n. [ME f. F or f. med.L permissibilis (as PERMIT)]

permission /pə'mɪʃ(ə)n/ n. (often foll. by to + infin.) consent; authorization. [ME f. OF or f. L permissio (as PERMIT)]

permissive /pə'mɪsɪv/ adj. **1** tolerant; liberal, esp. in sexual matters (the permissive society). **2** giving permission. □ **permissive legislation** legislation giving powers but not enjoining their use. □ **permissively** adv. **permissiveness** n. [ME f. OF (-if -ive) or med.L permissivus (as PERMIT)]

permit v. & n. ● v. /pə'mɪt/ (**permitted**, **permitting**) **1** tr. give permission or consent to; authorize (permit me to say). **2 a** tr. allow as possible; give an opportunity to (permit the traffic to flow again). **b** tr. give an opportunity to (circumstances permitting). **3** intr. (foll. by of) admit; allow for. ● n. /'pɜːmɪt/ **1 a** a document giving permission to act in a specified way (was granted a work permit). **b** a document etc. which allows entry into a specified zone. **2** formal permission. □ **permittee** /ˌpɜːmɪ'tiː/ n. **permitter** /pə'mɪtə(r)/ n. [L permittere (as PER-, mittere miss-let go)]

permittivity /ˌpɜːmɪ'tɪvɪtɪ/ n. Electr. a quantity measuring the ability of a substance to store electrical energy in an electric field.

permutate /'pɜːmjʊˌteɪt/ v.tr. change the order or arrangement of. [as PERMUTE, or back-form. f. PERMUTATION]

permutation /ˌpɜːmjʊ'teɪʃ(ə)n/ n. **1 a** an ordered arrangement or grouping of a set of numbers, items, etc. **b** any one of the range of possible groupings. **2** any combination or selection of a specified number of things from a larger group, esp. Brit. matches in a football pool. □ **permutational** adj. [ME f. OF or f. L permutatio (as PERMUTE)]

permute /pə'mjuːt/ v.tr. alter the sequence or arrangement of. [ME f. L permutare (as PER-, mutare change)]

Permutit /'pɜːmjʊtɪt/ n. propr. an artificial zeolite used as an ion exchanger, esp. for the softening of water. [G f. L permutare PERMUTE]

Pernambuco /ˌpɜːnæm'buːkuː/ **1** a state of eastern Brazil, on the Atlantic coast; capital, Recife. **2** the former name for RECIFE.

pernicious /pə'nɪʃəs/ adj. destructive; ruinous; fatal. □ **pernicious anaemia** see ANAEMIA. □ **perniciously** adv. **perniciousness** n. [L perniciosus f. pernicies ruin f. nex necis death]

pernickety /pə'nɪkɪtɪ/ adj. **1** fastidious. **2** precise or over-precise. **3** requiring tact or careful handling. [19th-c. Sc.: orig. unkn.]

pernoctate /pəˈnɒkteɪt/ v.intr. formal pass or spend the night. □ **pernoctation** /ˌpɜːnɒkˈteɪʃ(ə)n/ n. [LL pernoctatio f. L pernoctare pernoctat- (as PER-, noctare f. nox noctis night)]

Pernod /ˈpɜːnəʊ/ n. propr. a clear yellow-green aniseed-flavoured aperitif. [Pernod Fils, the manufacturing firm]

Perón[1] /peˈrɒn/, Eva (full name María Eva Duarte de Perón; known as 'Evita') (1919–52), Argentinian politician. After pursuing a successful career as a radio actress in the 1930s and 1940s, she married Juan Perón and became de facto Minister of Health and of Labour. Idolized by the poor, she organized female workers, secured the vote for women, and earmarked substantial government funds for social welfare. She was nominated for the vice-presidency in 1951, but was forced by the army to withdraw. She died the following year from cancer.

Perón[2] /peˈrɒn/, Juan Domingo (1895–1974), Argentinian soldier and statesman, President 1946–55 and 1973–4. He participated in the military coup organized by pro-fascist army officers in 1943, and was elected President in 1946, when he assumed dictatorial powers. He won popular support with his programme of social reform, but, after the death of his second wife, Evita, the faltering economy and his conflict with the Roman Catholic Church led to his removal and exile in 1955. Following a resurgence by the Peronist Party in the early 1970s, Perón returned to power in 1973, but died in office. □ **Peronism** /ˈperəˌnɪz(ə)m/ n. **Peronist** adj. & n.

peroneal /ˌperəˈniːəl/ adj. Anat. relating to or near the fibula. [mod.L peronaeus peroneal muscle f. perone fibula f. Gk peronē pin, fibula]

perorate /ˈperəˌreɪt/ v.intr. **1** sum up and conclude a speech. **2** speak at length. [L perorare perorat- (as PER-, orare speak)]

peroration /ˌperəˈreɪʃ(ə)n/ n. the concluding part of a speech, forcefully summing up what has been said.

peroxidase /pəˈrɒksɪˌdeɪz, -ˌdeɪs/ n. Biochem. any of a class of enzymes which catalyze the oxidation of a substrate by hydrogen peroxide.

peroxide /pəˈrɒksaɪd/ n. & v. ● n. Chem. **1 a** = hydrogen peroxide. **b** (often attrib.) a solution of hydrogen peroxide used to bleach the hair or as an antiseptic. **2** a compound with a molecule containing two oxygen atoms bonded together. ● v.tr. bleach (the hair) with peroxide. [PER- + OXIDE]

Perpendicular /ˌpɜːpənˈdɪkjʊlə(r)/ adj. Archit. of the third stage of English Gothic (15th–16th centuries), marked by vertical tracery in large windows.

perpendicular /ˌpɜːpənˈdɪkjʊlə(r)/ adj. & n. ● adj. **1 a** at right angles to the plane of the horizon. **b** (usu. foll. by to) Geom. at right angles to a given line, plane, or surface. **2** upright, vertical. **3** (of a slope etc.) very steep. **4** joc. in a standing position. ● n. **1** a perpendicular line. **2** a plumb-rule or a similar instrument. **3** (prec. by the) a perpendicular line or direction (is out of the perpendicular). □ **perpendicularly** adv. **perpendicularity** /-ˌdɪkjʊˈlærɪtɪ/ n. [ME f. L perpendicularis f. perpendiculum plumb-line f. PER- + pendere hang]

perpetrate /ˈpɜːpɪˌtreɪt/ v.tr. commit or perform (a crime, blunder, or anything outrageous). □ **perpetrator** n. **perpetration** /ˌpɜːpɪˈtreɪʃ(ə)n/ n. [L perpetrare perpetrat- (as PER-, patrare effect)]

perpetual /pəˈpetjʊəl/ adj. **1** eternal; lasting for ever or indefinitely. **2** continuous, uninterrupted. **3** frequent, much repeated (perpetual interruptions). **4** permanent during life (perpetual secretary). □ **perpetual calendar** a calendar which can be adjusted to show any combination of day, month, and year. **perpetual check** Chess the position of play when a draw is obtained by repeated checking of the king. **perpetual motion** the motion of a hypothetical machine which once set in motion would run for ever unless subject to an external force or to wear. □ **perpetualism** n. **perpetually** adv. [ME f. OF perpetuel f. L perpetualis f. perpetuus f. perpes -etis continuous]

perpetuate /pəˈpetjʊˌeɪt/ v.tr. **1** make perpetual. **2** preserve from oblivion. □ **perpetuance** n. **perpetuator** n. **perpetuation** /-ˌpetjʊˈeɪʃ(ə)n/ n. [L perpetuare (as PERPETUAL)]

perpetuity /ˌpɜːpɪˈtjuːɪtɪ/ n. (pl. **-ies**) **1** the state or quality of being perpetual. **2** a perpetual annuity. **3** a perpetual possession or position. □ **in** (or **to** or **for**) **perpetuity** for ever. [ME f. OF perpetuité f. L perpetuitas -tatis (as PERPETUAL)]

perpetuum mobile /pəˌpetjʊəm ˈməʊbɪlɪ/ n. **1** = perpetual motion. **2** Mus. = MOTO PERPETUO. [L perpetuus continuous + mobilis movable, after PRIMUM MOBILE]

Perpignan /ˈpɜːpɪˌnjɒn/ a city in southern France, in the north-eastern foothills of the Pyrenees, close to the border with Spain; pop. (1990) 108,050. A former fortress town, it was the capital of the old province of Roussillon.

perplex /pəˈpleks/ v.tr. **1** puzzle, bewilder, or disconcert (a person, a person's mind, etc.). **2** complicate or confuse (a matter). **3** (as **perplexed** adj.) archaic entangled, intertwined. □ **perplexing** adj. **perplexingly** adv. **perplexedly** /-sɪdlɪ/ adv. [back-form. f. perplexed f. obs. perplex (adj.) f. OF perplexe or L perplexus (as PER-, plexus past part. of plectere plait)]

perplexity /pəˈpleksɪtɪ/ n. (pl. **-ies**) **1** bewilderment; the state of being perplexed. **2** a thing which perplexes. **3** the state of being complicated. **4** archaic an entangled state. [ME f. OF perplexité or LL perplexitas (as PERPLEX)]

per pro. /pɜː ˈprəʊ/ abbr. through the agency of (used in signatures). ¶ The correct sequence is A per pro. B, where B is signing on behalf of A. [L per procurationem]

perquisite /ˈpɜːkwɪzɪt/ n. **1** an extra profit or allowance additional to a main income etc. **2** a customary extra right or privilege. **3** an incidental benefit attached to employment etc. **4** a thing which has served its primary use and to which a subordinate or employee has a customary right. [ME f. med.L perquisitum f. L perquirere search diligently for (as PER-, quaerere seek)]

Perrault /peˈrəʊ/, Charles (1628–1703), French writer. He is remembered for his Mother Goose Tales (1697), containing such fairy tales as 'Sleeping Beauty', 'Little Red Riding Hood', 'Puss in Boots', 'Bluebeard', and 'Cinderella'. They were translated into English by Robert Samber in 1729.

Perrier /ˈperɪˌeɪ/ n. propr. an effervescent natural mineral water. [the name of a spring at Vergèze, France, its source]

Perrin /peˈræn/, Jean Baptiste (1870–1942), French physical chemist. He proved that cathode rays are negatively charged, and went on to investigate Brownian motion. The latter studies led to a number of mathematical proofs and determinations, and it was accepted that Perrin had provided the definitive proof of the existence of atoms. He was awarded the Nobel Prize for physics in 1926.

perron /ˈperən/ n. an exterior staircase leading up to a main entrance to a church or other (usu. large) building. [ME f. OF ult. f. L petra stone]

Perry /ˈperɪ/, Frederick John ('Fred') (1909–95), British-born American tennis player. He began his career as a table-tennis player, winning the world singles championship in 1929. In tennis, his record of winning three consecutive singles titles at Wimbledon (1934–6) was unequalled until the success of Björn Borg. Perry has subsequently pursued a career as a commentator for radio and television. In 1950 he founded a successful sportswear company.

perry /ˈperɪ/ n. (pl. **-ies**) Brit. a drink like cider, made from the fermented juice of pears. [ME pereye etc. f. OF peré, ult. f. L pirum pear]

per se /pɜː ˈseɪ/ adv. by or in itself; intrinsically. [L]

persecute /ˈpɜːsɪˌkjuːt/ v.tr. **1** subject (a person etc.) to hostility or ill-treatment, esp. on the grounds of political or religious belief. **2** harass; worry. **3** (often foll. by with) bombard (a person) with questions etc. □ **persecutor** n. **persecutory** adj. [ME f. OF persecuter back-form. f. persecuteur persecutor f. LL persecutor f. L persequi (as PER-; sequi secut- follow, pursue)]

persecution /ˌpɜːsɪˈkjuːʃ(ə)n/ n. the act or an instance of persecuting; the state of being persecuted. □ **persecution complex** (or **mania**) an irrational obsessive fear that others are scheming against one.

Perseids /ˈpɜːsɪˌɪdz/ n.pl. Astron. an annual meteor shower with a radiant in the constellation of Perseus, reaching a peak about 12 August. The shower is apparently derived from the comet Swift-Tuttle. [PERSEUS]

Persephone /pəˈsefənɪ/ Gk Mythol. a goddess, the daughter of Zeus and Demeter, called Proserpina by the Romans. She was carried off by Hades and made queen of the underworld. Demeter, vainly seeking her, refused to let the earth produce its fruits until her daughter was restored to her, but because Persephone had eaten some pomegranate-seeds in the other world, she was obliged to spend part of every year there. Her story symbolizes the return of spring and the life and growth of corn.

Persepolis /pəˈsepəlɪs/ a city in ancient Persia, situated to the north-east of Shiraz. It was founded in the late 6th century BC by Darius I as the ceremonial capital of Persia under the Achaemenid dynasty. It was partially destroyed in 330 BC by Alexander the Great, and though it survived as the capital of the Seleucids it began to decline after this date. The city's impressive ruins include functional and ceremonial buildings and cuneiform inscriptions in Old Persian.

Perseus /ˈpɜːsɪəs, -sjuːs/ **1** Gk Mythol. the son of Zeus and Danae, a hero celebrated for many accomplishments. Riding the winged horse

Pegasus, he cut off the head of the gorgon Medusa and gave it to Athene. He also rescued and married Andromeda, and became king of Tiryns in Greece. **2** *Astron.* a large northern constellation which includes a dense part of the Milky Way. It contains several star clusters and the variable star Algol.

perseverance /ˌpɜːsɪˈvɪərəns/ *n.* **1** the steadfast pursuit of an objective. **2** (often foll. by *in*) constant persistence (in a belief etc.). [ME f. OF f. L *perseverantia* (as PERSEVERE)]

perseverate /pəˈsevəˌreɪt/ *v.intr.* **1** continue action etc. for an unusually or excessively long time. **2** *Psychol.* tend to prolong or repeat a response after the original stimulus has ceased. □ **perseveration** /-ˌsevəˈreɪʃ(ə)n/ *n.* [L *perseverare* (as PERSEVERE)]

persevere /ˌpɜːsɪˈvɪə(r)/ *v.intr.* (often foll. by *in*, *at*, *with*) continue steadfastly or determinedly; persist. [ME f. OF *perseverer* f. L *perseverare* persist f. *perseverus* very strict (as PER-, *severus* severe)]

Pershing /ˈpɜːʃɪŋ/ *n.* (in full **Pershing missile**) a type of American short range surface-to-surface ballistic missile. [John Joseph *Pershing*, Amer. general (1860–1948)]

Persia /ˈpɜːʃə, ˈpɜːʒə/ a country of SW Asia, now known as Iran. The ancient kingdom of Persia, corresponding to the modern district of Fars in SW Iran, became in the 6th century BC the domain of the Achaemenid dynasty. It was extended under Cyrus the Great into a powerful empire, which included Media, Lydia, and Babylonia, and eventually all of western Asia, Egypt, and parts of eastern Europe. The empire, defied by the Greeks in the Persian Wars of the 5th century BC, was eventually overthrown by Alexander the Great in 330 BC. The country was subsequently ruled by a succession of dynasties until it was conquered by the Muslim Arabs between AD 633 and 651. Taken by the Mongols in the 13th century, Persia was ruled by the Kajar dynasty from 1794 until 1925, when Reza Khan Pahlavi became shah. It was renamed Iran in 1935. (See also IRAN.)

Persian /ˈpɜːʃ(ə)n, ˈpɜːʒ(ə)n/ *n.* & *adj.* ● *n.* **1 a** a native or inhabitant of ancient Persia or modern Iran. **b** a person of Persian descent. **2** the language of ancient Persia or modern Iran. (*See note below.*) **3** (in full **Persian cat**) a breed of cat with long silky hair and a thick tail. ● *adj.* of or relating to Persia or its people or language. □ **Persian carpet** (or **rug**) a carpet or rug of a traditional pattern made in Persia. **Persian lamb** = KARAKUL 2. [ME f. OF *persien* f. med.L]

▪ The Persian language, spoken by over 30 million people in Iran, by about 5 million in Afghanistan (as Dari), and by another 2.2 million in Tajikistan (as Tajik), belongs to the Iranian language group. It is attested from the 6th century BC, when Old Persian was the language of the Persian empire, which at one time spread from the Mediterranean to India. Old Persian was written in cuneiform, but in the 2nd century BC the Persians created their own alphabet (Pahlavi), which remained in use until the Islamic conquest in the 7th century; since then Persian has been written in the Arabic script. The modern language is also called Farsi.

Persian Gulf (also called *Arabian Gulf*; informally *the Gulf*) an arm of the Arabian Sea, to which it is connected by the Strait of Hormuz and the Gulf of Oman. It extends north-westwards between the Arabian peninsula and the coast of SW Iran.

Persian Wars the wars fought between Greece and Persia in the 5th century BC, in which the Persians sought to extend their territory over the Greek world. The wars began in 490 BC when Darius I sent an expedition to punish the Greeks for having supported the Ionian cities in their unsuccessful revolt against Persian rule; the Persians were defeated by a small force of Athenians at Marathon. Ten years later Darius' son Xerxes I attempted an invasion with a land and sea force, crossing the Hellespont with a bridge of boats. He won a land-battle at Thermopylae and devastated Attica, but Persian forces were defeated on land at Plataea and in a sea battle at Salamis (480 BC), and retreated. Intermittent war continued in various areas until peace was signed in 449 BC.

persiennes /ˌpɜːsɪˈenz/ *n.pl.* window shutters, or outside blinds, with louvres. [F, fem. pl. of obs. *persien* Persian]

persiflage /ˈpɜːsɪˌflɑːʒ/ *n.* light raillery, banter. [F *persifler* banter, formed as PER- + *siffler* whistle]

persimmon /pɜːˈsɪmən/ *n.* **1** an evergreen tree of the genus *Diospyros*, bearing edible orange tomato-like fruits. **2** the fruit of this (cf. SHARON FRUIT). [corrupt. of an Algonquian word]

persist /pəˈsɪst/ *v.intr.* **1** (often foll. by *in*) continue firmly or obstinately (in an opinion or a course of action) esp. despite obstacles, remonstrance, etc. **2** (of an institution, custom, phenomenon, etc.) continue in existence; survive. [L *persistere* (as PER-, *sistere* stand)]

persistent /pəˈsɪstənt/ *adj.* **1** continuing obstinately; persisting. **2** enduring. **3** constantly repeated (*persistent nagging*). **4** *Biol.* (of horns, leaves, etc.) remaining instead of falling off in the normal manner. □ **persistence** *n.* **persistency** *n.* **persistently** *adv.*

person /ˈpɜːs(ə)n/ *n.* **1** an individual human being (*a cheerful and forthright person*). **2** the living body of a human being (*hidden about your person*). **3** *Gram.* any of three classes of personal pronouns, verb-forms, etc.: the person speaking (**first person**); the person spoken to (**second person**); the person spoken of (**third person**). **4** (in *comb.*) used to replace *-man* in offices open to either sex (*salesperson*). **5** (in Christianity) God as Father, Son, or Holy Ghost (*three persons in one God*). **6** *euphem.* the genitals (*expose one's person*). **7** a character in a play or story. **8** an individual characterized by a preference or liking for a specified thing (*not a party person*). □ **in one's own person** oneself; as oneself. **in person** physically present. **person-to-person 1** between individuals. **2** (of a phone call) booked through the operator to a specified person. [ME f. OF *persone* f. L *persona* actor's mask, character in a play, human being]

persona /pɜːˈsəʊnə/ *n.* (*pl.* **personae** /-niː/) **1** *Psychol.* an aspect of the personality as shown to or perceived by others (opp. ANIMA 1). **2** (in literary criticism) an author's assumed character in his or her writing. □ ***persona grata*** /ˈɡrɑːtə/ (*pl.* ***personae gratae*** /ˈɡrɑːtiː/) a person, esp. a diplomat, acceptable to certain others. ***persona non grata*** /nɒn, nəʊn/ (*pl.* ***personae non gratae*** /ˈɡrɑːtiː/) an unacceptable or unwelcome person. [L (as PERSON)]

personable /ˈpɜːsənəb(ə)l/ *adj.* pleasing in appearance and behaviour. □ **personableness** *n.* **personably** *adv.*

personage /ˈpɜːsənɪdʒ/ *n.* **1** a person, esp. of rank or importance. **2** a character in a play etc. [ME f. PERSON + -AGE, infl. by med.L *personagium* effigy & F *personnage*]

personal /ˈpɜːsən(ə)l/ *adj.* **1** one's own; individual; private. **2** done or made in person (*made a personal appearance; my personal attention*). **3** directed to or concerning an individual (*a personal letter*). **4 a** referring (esp. in a hostile way) to an individual's private life or concerns (*making personal remarks; no need to be personal*). **b** close, intimate (*a personal friend*). **5** of the body and clothing (*personal hygiene; personal appearance*). **6** existing as a person, not as an abstraction or thing (*a personal God*). **7** *Gram.* of or denoting one of the three persons (*personal pronoun*). □ **personal column** the part of a newspaper devoted to private advertisements or messages. **personal computer** a computer designed for use by a single individual, esp. in an office or business environment. **personal equation 1** the allowance for an individual person's time of reaction in making observations, esp. in astronomy. **2** a bias or prejudice. **personal equity plan** (abbr. **PEP**) a scheme for limited personal investment in shares, unit trusts, etc. **personal identification number** (abbr. **PIN**) a number allocated to an individual, serving as a password esp. for a cash dispenser, computer, etc. **personal organizer 1** a loose-leaf notebook with sections for various kinds of information, including a diary etc. **2** a hand-held microcomputer serving the same purpose. **personal pronoun** a pronoun replacing the subject, object, etc., of a clause etc., e.g. *I, we, you, them, us*. **personal property** (or **estate**) *Law* all one's property except land and those interests in land that pass to one's heirs (cf. REAL¹ *adj.* 3). **personal service** individual service given to a customer. **personal space** *Sociol.* the area around an individual where encroachment by others causes anxiety or uneasiness. **personal stereo** a small portable audio cassette player, often with radio, or compact disc player, used with lightweight headphones. **personal touch** a way of treating a matter characteristic of or designed for an individual. [ME f. OF f. L *personalis* (as PERSON)]

personality /ˌpɜːsəˈnælɪtɪ/ *n.* (*pl.* **-ies**) **1 a** the distinctive character or qualities of a person, often as distinct from others (*an attractive personality*). **b** socially attractive qualities (*was clever but had no personality*). **2** a famous person; a celebrity (*a TV personality*). **3** a person who stands out from others by virtue of his or her character (*is a real personality*). **4** personal existence or identity; the condition of being a person. **5** (usu. in *pl.*) personal remarks. □ **have personality** have a lively character or noteworthy qualities. **personality cult** the extreme adulation of an individual, esp. a politician. [ME f. OF *personalité* f. LL *personalitas -tatis* (as PERSONAL)]

personalize /ˈpɜːsənəˌlaɪz/ *v.tr.* (also **-ise**) **1** make personal, esp. by marking with one's name etc. **2** personify. □ **personalization** /ˌpɜːsənəlaɪˈzeɪʃ(ə)n/ *n.*

personally /ˈpɜːsənəlɪ/ *adv.* **1** in person (*see to it personally*). **2** for one's own part (*speaking personally*). **3** in the form of a person (*a God existing* *personally*). **4** in a personal manner (*took the criticism personally*). **5** as a person; on a personal level.

personalty /'pɜːsənltɪ/ *n.* (*pl.* **-ies**) *Law* one's personal property or estate (opp. REALTY). [AF *personalté* (as PERSONAL)]

personate /'pɜːsə‚neɪt/ *v.tr.* **1** play the part of (a character in a drama etc.; another type of person). **2** pretend to be (another person), esp. for fraudulent purposes; impersonate. □ **personator** *n.* **personation** /‚pɜːsə'neɪʃ(ə)n/ *n.* [LL *personare personat-* (as PERSON)]

personhood /'pɜːs(ə)n‚hʊd/ *n.* the quality or condition of being an individual person.

personification /pə‚sɒnɪfɪ'keɪʃ(ə)n/ *n.* **1** the act of personifying. **2** (foll. by *of*) a person or thing viewed as a striking example of (a quality etc.) (*the personification of ugliness*).

personify /pə'sɒnɪ‚faɪ/ *v.tr.* (**-ies**, **-ied**) **1** attribute a personal nature to (an abstraction or thing). **2** symbolize (a quality etc.) by a figure in human form. **3** (usu. as **personified** *adj.*) embody (a quality) in one's own person; exemplify typically (*has always been kindness personified*). □ **personifier** *n.* [F *personnifier* (as PERSON)]

personnel /‚pɜːsə'nel/ *n.* **1** a body of employees, persons involved in a public undertaking, armed forces, etc. **2** (in full **personnel department**) the part of an organization concerned with the appointment, training, and welfare of employees. □ **personnel carrier** an armoured vehicle for transporting troops etc. [F, orig. adj. = personal]

perspective /pə'spektɪv/ *n. & adj.* ● *n.* **1 a** the art of drawing solid objects on a two-dimensional surface so as to give the right impression of relative positions, size, etc. (*See note below.*) **b** a picture drawn in this way. **2** the apparent relation between visible objects as to position, distance, etc. **3** a mental view of the relative importance of things (*keep the right perspective*). **4** a geographical or imaginary prospect. ● *adj.* of or in perspective. □ **in** (or **out of**) **perspective 1** drawn or viewed according (or not according) to the rules of perspective. **2** correctly (or incorrectly) regarded in terms of relative importance. □ **perspectively** *adv.* **perspectival** /‚pɜːspek'taɪv(ə)l/ *adj.* [ME f. med.L *perspectiva* (*ars* art) f. *perspicere perspect-* (as PER-, *specere spect-* look)]

▪ Many of the principles of perspective were known to the Greeks and Romans but later lost. The mathematical laws of perspective were first demonstrated by the architect Brunelleschi in the early 15th century. Shortly afterwards (1436) his fellow architect Alberti devised a perspective construction for the special use of painters. Perspective utilizes such optical effects as the apparent convergence of parallel lines as they recede from the spectator so as to create a pictorial representation of the same kind of spatial relationships that we see in the real world, with objects appearing to diminish in size the further away they are. Masaccio and Uccello were among the first artists to put these theories into practice.

perspex /'pɜːspeks/ *n. propr.* (also **Perspex**) a tough light transparent acrylic thermoplastic used instead of glass. [L *perspicere* look through (as PER-, *specere* look)]

perspicacious /‚pɜːspɪ'keɪʃəs/ *adj.* having mental penetration or discernment. □ **perspicaciously** *adv.* **perspicaciousness** *n.* **perspicacity** /-'kæsɪtɪ/ *n.* [L *perspicax -acis* (as PERSPEX)]

perspicuous /pə'spɪkjʊəs/ *adj.* **1** easily understood; clearly expressed. **2** (of a person) expressing things clearly. □ **perspicuously** *adv.* **perspicuousness** *n.* **perspicuity** /‚pɜːspɪ'kjuːɪtɪ/ *n.* [ME, = transparent f. L *perspicuus* (as PERSPECTIVE)]

perspiration /‚pɜːspɪ'reɪʃ(ə)n/ *n.* **1** sweat. **2** sweating. □ **perspiratory** /pə'spɪrətərɪ, pə'spaɪər-/ *adj.* [F (as PERSPIRE)]

perspire /pə'spaɪə(r)/ *v.* **1** *intr.* sweat or exude perspiration, esp. as the result of heat, exercise, anxiety, etc. **2** *tr.* sweat or exude (fluid etc.). [F *perspirer* f. L *perspirare* (as PER-, *spirare* breathe)]

persuade /pə'sweɪd/ *v.tr. & refl.* **1** (often foll. by *of*, or *that* + clause) cause (another person or oneself) to believe; convince (*persuaded them that it would be helpful; tried to persuade me of its value*). **2 a** (often foll. by *to* + infin.) induce (another person or oneself) (*persuaded us to join them; managed to persuade them at last*). **b** (foll. by *away from, down to,* etc.) lure, attract, entice, etc. (*persuaded them away from the pub*). □ **persuadable** *adj.* **persuadability** /-‚sweɪdə'bɪlɪtɪ/ *n.* **persuasible** /-'sweɪzɪb(ə)l/ *adj.* [L *persuadere* (as PER-, *suadere suas-* advise)]

persuader /pə'sweɪdə(r)/ *n.* **1** a person who persuades. **2** *sl.* a gun or other weapon.

persuasion /pə'sweɪʒ(ə)n/ *n.* **1** persuading (*yielded to persuasion*). **2** persuasiveness (*use all your persuasion*). **3** a belief or conviction (*my private persuasion*). **4** a religious belief, or the group or sect holding it

(*of a different persuasion*). **5** *colloq.* or *joc.* any group or party (*the male persuasion*). [ME f. L *persuasio* (as PERSUADE)]

persuasive /pə'sweɪsɪv/ *adj.* able to persuade. □ **persuasively** *adv.* **persuasiveness** *n.* [F *persuasif -ive* or med.L *persuasivus*, (as PERSUADE)]

PERT /pɜːt/ *abbr.* programme evaluation and review technique.

pert /pɜːt/ *adj.* **1** saucy or impudent, esp. in speech or conduct. **2** (of clothes etc.) neat and suggestive of jauntiness. **3** = PEART. □ **pertly** *adv.* **pertness** *n.* [ME f. OF *apert* f. L *apertus* past part. of *aperire* open & f. OF *aspert* f. L *expertus* EXPERT]

pertain /pə'teɪn/ *v.intr.* **1** (foll. by *to*) **a** relate or have reference to. **b** belong to as a part or appendage or accessory. **2** (usu. foll. by *to*) be appropriate to. [ME f. OF *partenir* f. L *pertinere* (as PER-, *tenere* hold)]

Perth /pɜːθ/ **1** a town in eastern Scotland, at the head of the Tay estuary; pop. (1981) 43,000. It was the capital of Scotland from 1210 until 1452. **2** the capital of the state of Western Australia, on the Indian Ocean; pop. (1991) 1,018,700 (including the port of Fremantle). Founded by the British in 1829, it was named after Perth in Scotland. It developed rapidly after the discovery in 1890 of gold in the region and the opening in 1897 of the harbour at Fremantle.

Perthes, Jacques Boucher de, see BOUCHER DE PERTHES.

Perthshire /'pɜːθʃɪə(r)/ a former county of central Scotland. It became a part of Tayside region in 1975.

pertinacious /‚pɜːtɪ'neɪʃəs/ *adj.* stubborn; persistent; obstinate (in a course of action etc.). □ **pertinaciously** *adv.* **pertinaciousness** *n.* **pertinacity** /-'næsɪtɪ/ *n.* [L *pertinax* (as PER-, *tenax* tenacious)]

pertinent /'pɜːtɪnənt/ *adj.* **1** (often foll. by *to*) relevant to the matter in hand; apposite. **2** to the point. □ **pertinence** *n.* **pertinency** *n.* **pertinently** *adv.* [ME f. OF *pertinent* or L *pertinere* (as PERTAIN)]

perturb /pə'tɜːb/ *v.tr.* **1** throw into confusion or disorder. **2** disturb mentally; agitate. **3** *Physics & Math.* subject (a physical system, or a set of equations, or its solution) to a perturbation. □ **perturbable** *adj.* **perturbingly** *adv.* **perturbative** /pə'tɜːbətɪv, 'pɜːtə‚beɪt-/ *adj.* [ME f. OF *pertourber* f. L (as PER-, *turbare* disturb)]

perturbation /‚pɜːtə'beɪʃ(ə)n/ *n.* **1** the act or an instance of perturbing; the state of being perturbed. **2** a cause of disturbance or agitation. **3** *Physics* a slight alteration of a physical system, e.g. of the electrons in an atom, caused by a secondary influence. **4** *Astron.* a minor deviation in the course of a celestial body, caused by the attraction of a neighbouring body.

pertussis /pə'tʌsɪs/ *n. Med.* whooping cough. [mod.L f. PER- + L *tussis* cough]

Peru /pə'ruː/ a country in South America on the Pacific coast, traversed throughout its length by the Andes; pop. (est. 1991) 22,135,000; official languages, Spanish and Quechua; capital, Lima. The centre of the Inca empire, Peru was conquered by the Spanish conquistador Pizarro in 1532. The Spanish established the viceroyalty of Peru, which initially extended over most of Spanish South America but from the late 18th century included only present-day Peru, Chile, and part of Bolivia. Peru was liberated by Simón Bolívar and José de San Martín in 1820–4, and a republic established. It lost territory in the south in a war with Chile (1879–83) and also had border disputes with Colombia and Ecuador in the 1930s and 1940s. Peru has been troubled by revolutionary guerrilla and terrorist activity in recent years.

Perugia /pə'ruːdʒə/ a city in central Italy, the capital of Umbria; pop. (1990) 150,580. Founded by the Etruscans, it was occupied by the Romans from 310 BC. Taken by the Lombards in the late 6th century, it was contested over the succeeding centuries by powerful local families. It flourished in the 15th century as a centre of the Umbrian school of painting. A papal possession from 1540, it became a part of united Italy in 1860.

peruke /pə'ruːk/ *n. hist.* a wig. [F *perruque* f. It. *perrucca parrucca,* of unkn. orig.]

peruse /pə'ruːz/ *v.tr.* **1** (also *absol.*) read or study, esp. thoroughly or carefully. **2** examine (a person's face etc.) carefully. □ **perusal** *n.* **peruser** *n.* [ME, orig. = use up, prob. f. AL f. Rmc (as PER-, USE)]

Peruvian /pə'ruːvɪən/ *n. & adj.* ● *n.* **1** a native or national of Peru. **2** a person of Peruvian descent. ● *adj.* of or relating to Peru. □ **Peruvian bark** the bark of the cinchona tree. [mod.L *Peruvia* Peru]

perv /pɜːv/ *n. & v.* (also **perve**) *sl.* ● *n.* **1** a sexual pervert. **2** *Austral.* an erotic gaze. ● *v.intr.* **1** act like a sexual pervert. **2** (foll. by *at, on*) *Austral.* gaze with erotic interest. [abbr.]

pervade /pə'veɪd/ *v.tr.* **1** spread throughout, permeate. **2** (of influences etc.) become widespread among or in. **3** be rife among or through. □ **pervasion** /-'veɪʒ(ə)n/ n. [L pervadere (as PER-, vadere vas- go)]

pervasive /pə'veɪsɪv/ adj. **1** pervading. **2** able to pervade. □ **pervasively** adv. **pervasiveness** n.

perve var. of PERV.

perverse /pə'vɜːs/ adj. **1** (of a person or action) deliberately or stubbornly departing from what is reasonable or required. **2** persistent in error. **3** wayward; intractable; peevish. **4** perverted; wicked. **5** (of a verdict etc.) against the weight of evidence or the judge's direction. □ **perversely** adv. **perverseness** n. **perversity** n. (pl. **-ies**). [ME f. OF pervers perverse f. L perversus (as PERVERT)]

perversion /pə'vɜːʃ(ə)n/ n. **1** an act of perverting; the state of being perverted. **2** a perverted form of an act or thing. **3 a** a preference for an abnormal form of sexual activity. **b** such an activity. [ME f. L perversio (as PERVERT)]

pervert v. & n. ● v.tr. /pə'vɜːt/ **1** turn (a person or thing) aside from its proper use or nature. **2** misapply or misconstrue (words etc.). **3** lead astray (a person, a person's mind, etc.) from right opinion or conduct, or esp. religious belief. **4** (as **perverted** adj.) showing perversion. ● n. /'pɜːvɜːt/ **1** a perverted person. **2** a person showing sexual perversion. □ **perversive** /pə'vɜːsɪv/ adj. **pervertedly** /-'vɜːtɪdlɪ/ adv. **perverter** /-'vɜːtə(r)/ n. [ME f. OF pervertir or f. L pervertere (as PER-, vertere vers- turn): cf. CONVERT]

pervious /'pɜːvɪəs/ adj. **1** permeable. **2** (usu. foll. by to) **a** affording passage. **b** accessible (to reason etc.). □ **perviousness** n. [L pervius (as PER-, vius f. via way)]

Pesach /'peɪsɑːx/ n. the Passover festival. [Heb. Pesaḥ]

peseta /pə'seɪtə/ n. **1** the basic monetary unit of Spain, equal to 100 centimos. **2** hist. a silver coin equal to one peseta. [Sp., dimin. of pesa weight f. L pensa pl. of pensum: see POISE¹]

pesewa /pɪ'seɪwə/ n. a monetary unit of Ghana, equal to one-hundredth of a cedi. [Fanti, = PENNY]

Peshawar /pə'ʃɑːwə(r)/ n. the capital of North-West Frontier Province, in Pakistan; pop. (1981) 555,000. Mentioned in early Sanskrit literature, it is one of Pakistan's oldest cities. Under Sikh rule from 1834, it was occupied by the British between 1849 and 1947. Situated near the Khyber Pass on the border with Afghanistan, it is of strategic and military importance.

Peshitta /pə'ʃiːtə/ the ancient Syriac version of the Bible, used in Syriac-speaking Christian countries from the early 5th century and still the official Bible of the Syrian Christian Churches. [Syriac, = simple, plain]

peshmerga /peʃ'mɜːgə/ n. (pl. same or **peshmergas**) a member of a Kurdish nationalist guerrilla organization. [Kurdish f. pêsh in front of + merg death]

pesky /'peskɪ/ adj. (**peskier**, **peskiest**) esp. US colloq. troublesome; confounded; annoying. □ **peskily** adv. **peskiness** n. [18th c.: perh. f. PEST]

peso /'peɪsəʊ/ n. (pl. **-os**) the basic monetary unit of several Latin American countries and of the Philippines, equal to 100 centésimos in Uruguay and 100 centavos elsewhere. [Sp., = weight, f. L pensum: see POISE¹]

pessary /'pesərɪ/ n. (pl. **-ies**) Med. **1** a device worn in the vagina to support the uterus or as a contraceptive. **2** a vaginal suppository. [ME f. LL pessarium, pessulum f. pessum, pessus f. Gk pessos oval stone]

pessimism /'pesɪˌmɪz(ə)m/ n. **1** a tendency to take the worst view or expect the worst outcome. **2** Philos. a belief that this world is as bad as it could be or that all things tend to evil. Opp. OPTIMISM. □ **pessimist** n. **pessimistic** /ˌpesɪ'mɪstɪk/ adj. **pessimistically** adv. [L pessimus worst, after OPTIMISM]

pest /pest/ n. **1** a troublesome or annoying person or thing; a nuisance. **2** a destructive animal, esp. an insect which attacks crops, livestock, etc. **3** archaic a pestilence; a plague. □ **pest-house** hist. a hospital for sufferers from plague or other infectious disease. [F peste or L pestis plague]

Pestalozzi /ˌpestə'lɒtsɪ/, Johann Heinrich (1746–1827), Swiss educational reformer. He pioneered education for poor children and had a major impact on the development of primary education. Influenced by Jean-Jacques Rousseau, Pestalozzi believed in the morally improving effects of a rural environment, and held that education should allow for individual differences in ability and pace of development, as well as acknowledge the role of family life and the importance of moral teaching. His theory and method are set out in *How Gertrude Teaches Her Children* (1801). Pestalozzi's work is

commemorated in the International Children's Villages named after him; the first, for war orphans, was established at Trogen in Switzerland in 1946.

pester /'pestə(r)/ v.tr. trouble or annoy, esp. with frequent or persistent requests. □ **pesterer** n. [prob. f. impester f. F empestrer encumber: infl. by PEST]

pesticide /'pestɪˌsaɪd/ n. a substance used for destroying insects or other organisms harmful to cultivated plants or to animals. □ **pesticidal** /ˌpestɪ'saɪd(ə)l/ adj.

pestiferous /pe'stɪfərəs/ adj. **1** noxious; pestilent. **2** harmful; pernicious; bearing moral contagion. [L pestifer, -ferus (as PEST)]

pestilence /'pestɪləns/ n. **1** a fatal epidemic disease, esp. bubonic plague. **2** something evil or destructive. [ME f. OF f. L pestilentia (as PESTILENT)]

pestilent /'pestɪlənt/ adj. **1** destructive to life, deadly. **2** harmful or morally destructive. **3** colloq. troublesome; annoying. □ **pestilently** adv. [L pestilens, pestilentus f. pestis plague]

pestilential /ˌpestɪ'lenʃ(ə)l/ adj. **1** of or relating to pestilence. **2** dangerous; troublesome; pestilent. □ **pestilentially** adv. [ME f. med.L pestilentialis f. L pestilentia (as PESTILENT)]

pestle /'pes(ə)l/ n. & v. ● n. **1** a club-shaped instrument for pounding substances in a mortar. **2** an appliance for pounding etc. ● v. **1** tr. pound with a pestle or in a similar manner. **2** intr. use a pestle. [ME f. OF pestel f. L pistillum f. pinsare pist- to pound]

pesto /'pestəʊ/ n. an Italian sauce of crushed basil leaves, pine nuts, garlic, parmesan cheese, and olive oil, usually served with pasta. [It. f. pestare pound, crush]

pestology /pe'stɒlədʒɪ/ n. the scientific study of pests (esp. harmful insects) and of methods of dealing with them. □ **pestologist** n. **pestological** /ˌpestə'lɒdʒɪk(ə)l/ adj.

PET abbr. **1** Med. positron emission tomography, a form of tomography used esp. for brain scans. **2** polyethylene terephthalate, a plastic used in recyclable packaging.

Pet. abbr. Peter (New Testament).

pet¹ /pet/ n., adj., & v. ● n. **1** a domestic or tamed animal kept for pleasure or companionship. **2** a darling, a favourite (often as a term of endearment). ● attrib.adj. **1** kept as a pet (pet lamb). **2** of or for pet animals (pet food). **3** often joc. favourite or particular (my pet hate is filling in forms). **4** expressing fondness or familiarity (pet name). ● v.tr. (**petted**, **petting**) **1** treat as a pet. **2** (also absol.) fondle, esp. erotically. □ **petter** n. [16th-c. Sc. & N. Engl. dial.: orig. unkn.]

pet² /pet/ n. a feeling of petty resentment or ill-humour (esp. be in a pet). [16th c.: orig. unkn.]

peta- /'petə/ comb. form denoting a factor of 10^{15}. [perh. f. PENTA-]

Pétain /peɪ'tæn/, (Henri) Philippe (Omer) (1856–1951), French general and statesman, head of state 1940–2. He became a national hero in the First World War for halting the German advance at Verdun (1916) and later became Commander-in-Chief of French forces (1917). In the Second World War he concluded an armistice with Nazi Germany after the collapse of French forces in 1940 and established the French government at Vichy (effectively a puppet regime for the Third Reich) until German occupation in 1942. After the war Pétain received a death sentence for collaboration, but this was commuted to life imprisonment.

petal /'pet(ə)l/ n. each of the parts of the corolla of a flower. □ **petalled** adj. (also in comb.). **petal-like** adj. **petaloid** adj. **petaline** /-ˌlaɪn, -lɪn/ adj. [mod.L petalum, in LL metal plate f. Gk petalon leaf f. petalos outspread]

petard /pɪ'tɑːd/ n. hist. **1** a small bomb used to blast down a door etc. **2** a kind of firework or cracker. □ **hoist with one's own petard** adversely affected oneself by one's schemes against others. [F pétard f. péter break wind]

petasus /'petəsəs/ n. an ancient Greek hat with a low crown and broad brim, esp. (in Greek mythology) as worn by Hermes. [L f. Gk petasos]

petaurist /pɪ'tɔːrɪst/ n. any flying squirrel of the genus Petaurista, native to eastern Asia. [Gk petauristēs performer on a springboard (petauron)]

Pete /piːt/ n. □ **for Pete's sake** see SAKE¹. [abbr. of the name Peter]

petechia /pɪ'tiːkɪə/ n. (pl. **petechiae** /-kɪˌiː/) Med. a small red or purple spot as a result of bleeding into the skin. □ **petechial** adj. [mod.L f. It. petecchia a freckle or spot on one's face]

peter¹ /'piːtə(r)/ v. & n. ● v.intr. **1** (foll. by out) (orig. of a vein of ore etc.) diminish, come to an end. **2** Bridge play an echo. ● n. Bridge an echo. [19th c.: orig. unkn.]

peter[2] /ˈpiːtə(r)/ *n. sl.* **1** a prison cell. **2** a safe. [perh. f. the name *Peter*]

Peter I /ˈpiːtə(r)/ (known as Peter the Great) (1672–1725), tsar of Russia 1682–1725. After the death of his half-brother Ivan in 1689, Peter I assumed sole authority and launched a policy of expansion along the Baltic coast. Modernizing his armed forces he waged the Great Northern War (1700–21) against Charles XII of Sweden, and went on to annex Estonia and Latvia, as well as parts of Finland, following the defeat of the Swedish monarch. Peter I's introduction of extensive government and administration reforms were instrumental in transforming Russia into a significant European power. In 1703 he made St Petersburg his capital.

Peter, St 1 (born Simon) an Apostle. 'Peter' (from *petros* = stone) is the Greek form of the name given him by Jesus, signifying the rock on which he would establish his church. He is regarded by Roman Catholics as the founder and first bishop of the Church at Rome, where he is said to have been martyred in about AD 67. He is often represented as the keeper of the door of heaven; his attribute is a set of keys. Feast day, 29 June. **2** either of the two epistles in the New Testament ascribed to St Peter.

Peterborough /ˈpiːtəbərə/ an industrial city in east central England, in Cambridgeshire; pop. (1991) 148,800. An old city with a 12th-century cathedral, it has been developed as a planned urban centre since the late sixties.

Peterloo /ˌpiːtəˈluː/ (also **Peterloo massacre**) an attack by Manchester yeomanry on 16 August 1819 against a large but peaceable crowd. Sent to arrest the speaker at a rally of supporters of political reform in St Peter's Field, Manchester, the local yeomanry charged the crowd, killing 11 civilians and injuring more than 500. The event was named Peterloo in ironical reference to the Battle of Waterloo.

peterman /ˈpiːtəmən/ *n.* (*pl.* **-men**) *sl.* a safe-breaker.

Peter Pan the hero of J. M. Barrie's play of the same name (1904), a boy with magical powers who never grew up.

Peter Principle *n. joc.* the principle that members of a hierarchy are promoted until they reach the level at which they are no longer competent. [Laurence Johnston *Peter*, Amer. educationist and author (1919–90), its propounder]

petersham /ˈpiːtəʃəm/ *n.* thick corded silk ribbon used for stiffening in dressmaking etc. [Viscount *Petersham*, Engl. army officer (1790–1851)]

Peterson /ˈpiːtəs(ə)n/, Oscar (Emmanuel) (b.1925), Canadian jazz pianist and composer. He toured with the American impresario Norman Granz (b.1918) from 1949, becoming internationally famous in the 1960s, when he often appeared with Ella Fitzgerald. During this period he usually led a trio with a bass and guitar. In the 1970s he often played the piano solo, recording the album *My Favourite Instrument* (1973).

Peter's Pence 1 *Brit. hist.* an annual tribute of a penny from every householder having land of a certain value, paid to the papal see at Rome from Anglo-Saxon times until discontinued in 1534 after Henry VIII's break with Rome. **2** a voluntary payment by Roman Catholics to the papal treasury, made since 1860.

Peters projection /ˈpiːtəz/ *n.* (also **Peters' projection**) a world map projection in which areas are shown in correct proportion at the expense of distorted shape, using a rectangular decimal grid to replace latitude and longitude. It was devised in 1973 to be a fairer representation of equatorial (i.e. mainly developing) countries, whose area is under-represented by the usual projections such as Mercator's. [Arno *Peters*, Ger. historian (b.1916)]

Peter the Hermit (c.1050–1115), French monk. His preaching on the First Crusade was a rallying cry for thousands of peasants throughout Europe to journey to the Holy Land; most were massacred by the Turks in Asia Minor. Peter later became prior of an Augustinian monastery in Flanders.

pethidine /ˈpeθɪˌdiːn/ *n.* a synthetic soluble analgesic drug used esp. in childbirth. [perh. f. PIPERIDINE (from which the drug is derived) + ETHYL]

petiole /ˈpetɪˌəʊl/ *n. Bot.* the slender stalk joining a leaf to a stem. □ **petiolate** /-tɪələt, -ˌleɪt/ *adj.* **petiolar** /ˌpetɪˈəʊlə(r)/ *adj.* [F *pétiole* f. L *petiolus* little foot, stalk]

Petipa /ˌpətɪˈpɑː/, Marius (Ivanovich) (1818–1910), French ballet-dancer and choreographer, resident in Russia from 1847. He became principal dancer for the Russian Imperial Ballet in St Petersburg in 1847 and first ballet master in 1869. Petipa choreographed more than fifty ballets, collaborating closely with Tchaikovsky on the premières of *Sleeping Beauty* (1890) and *The Nutcracker* (1892), works which have had an important influence on modern classical ballet in Russia.

petit /ˈpetɪ/ *adj.* esp. *Law* petty; small; of lesser importance. □ **petit jury** = *petty jury*. [ME f. OF, = small, f. Rmc, perh. imit. of child's speech]

petit bourgeois /ˌpetɪ ˈbʊəʒwʌ/ *n.* (*pl.* **petits bourgeois** *pronunc.* same) a member of the lower middle classes. [F]

petite /pəˈtiːt/ *adj.* (of a woman) of small and dainty build. □ **petite bourgeoisie** the lower middle classes. [F, fem. of PETIT]

petit four /ˌpetɪ ˈfɔː(r)/ *n.* (*pl.* **petits fours** /ˌpetɪ ˈfɔːz/) a very small fancy cake, biscuit, or sweet. [F, = little oven]

petition /pɪˈtɪʃ(ə)n/ *n. & v.* ● *n.* **1** a supplication or request. **2** a formal written request, esp. one signed by many people, appealing to authority in some cause. **3** *Law* an application to a court for a writ etc. ● *v.* **1** *tr.* make or address a petition to (*petition your MP*). **2** *intr.* (often foll. by *for, to*) appeal earnestly or humbly. □ **petitionable** *adj.* **petitionary** *adj.* **petitioner** *n.* [ME f. OF f. L *petitio -onis*]

Petition of Right *n.* **1** *hist.* a parliamentary declaration of rights and liberties of the people assented to by Charles I in 1628. **2** *Law* a common-law remedy against the crown for the recovery of property.

petitio principii /pɪˌtɪʃɪˌəʊ prɪnˈsɪpɪˌaɪ, -ˈkɪpɪˌaɪ/ *n. Logic* a fallacy in which a conclusion is taken for granted in the premiss; begging the question. [L, = assuming a principle: see PETITION]

petit-maître /ˌpetɪˈmetrə/ *n.* **1** an artist, musician, etc. of minor importance. **2** a dandy or coxcomb. [F, = little master]

petit mal /ˌpetɪ ˈmæl/ *n.* a mild form of epilepsy with only momentary loss of consciousness; an epileptic fit of this kind (cf. GRAND MAL). [F, = little sickness]

petit point /ˌpetɪ ˈpɔɪnt/ *n.* **1** embroidery on canvas using small stitches. **2** tent-stitch. [F, = little point]

petits pois /ˌpetɪ ˈpwʌ/ *n.pl.* small green peas. [F]

Petra /ˈpetrə/ an ancient city of SW Asia, in present-day Jordan. It was the capital of the Nabataeans from 312 BC until 63 BC, when they became subject to Rome. The city, which lies in a hollow surrounded by cliffs, is accessible only through narrow gorges. Its extensive ruins include temples and tombs hewn from the rose-red sandstone cliffs.

Petrarch /ˈpetrɑːk/ (Italian name Francesco Petrarca) (1304–74), Italian poet. His reputation is chiefly based on his lyrical poetry, in particular the *Canzoniere* (c.1351–3), a sonnet sequence in praise of a woman he calls Laura; this was to be a major source of inspiration for the English sonnet writers such as Thomas Wyatt and Philip Sidney. Petrarch was also an important figure in the rediscovery of classical antiquity, together with his friend Boccaccio, initiating the revived study of Greek and Latin literature and writing most of his works in Latin. In 1341 Petrarch was crowned Poet Laureate in Rome.

Petrarchan /pɪˈtrɑːkən/ *adj.* denoting a sonnet of the kind used by Petrarch, with an octave rhyming *abbaabba* and a sestet usu. rhyming *cdcdcd* or *cdecde*. (See also SONNET.)

petrel /ˈpetrəl/ *n.* a seabird of the family Procellariidae, Pelecanoididae, or Hydrobatidae, mainly with black and white plumage and a hooked bill, usu. found flying far from land. [17th c. (also *pitteral*), of uncert. orig.: later assoc. with St Peter (Matt. 14:30)]

Petri dish /ˈpetrɪ, ˈpiːt-/ *n.* a shallow covered dish used for the culture of bacteria etc. [Julius Richard *Petri*, Ger. bacteriologist (1852–1921)]

Petrie /ˈpiːtrɪ/, Sir (William Matthew) Flinders (1853–1942), English archaeologist and Egyptologist. After fieldwork at Stonehenge in the 1870s he began excavating the Great Pyramid at Giza in 1880, pioneering the use of mathematical calculation and precise measurement in field archaeology. In his excavations in Egypt and Palestine, Petrie also became the first to establish the system of sequence dating, now standard archaeological practice, by which sites are excavated layer by layer and historical chronology determined by the dating of artefacts found *in situ*.

petrifaction /ˌpetrɪˈfækʃ(ə)n/ *n.* **1** the process of fossilization whereby organic matter is turned into a stony substance. **2** a petrified substance or mass. **3** a state of extreme fear or terror. [PETRIFY after *stupefaction*]

petrify /ˈpetrɪˌfaɪ/ *v.* (**-ies, -ied**) **1** *tr.* (also as **petrified** *adj.*) paralyse with fear, astonishment, etc. **2** *tr.* change (organic matter) into a stony substance. **3** *intr.* become like stone. **4** *tr.* deprive (the mind, a doctrine, etc.) of vitality; deaden. [F *pétrifier* f. med.L *petrificare* f. L *petra* rock f. Gk]

petro- /ˈpetrəʊ/ *comb. form* **1** rock. **2** petroleum (*petrochemistry*). [Gk *petros* stone or *petra* rock]

petrochemical /ˌpetrəʊˈkemɪk(ə)l/ n. & adj. ● n. a substance industrially obtained from petroleum or natural gas. ● adj. of or relating to petrochemistry or petrochemicals.

petrochemistry /ˌpetrəʊˈkemɪstrɪ/ n. **1** the chemistry of rocks. **2** the chemistry of petroleum.

petrodollar /ˈpetrəʊˌdɒlə(r)/ n. a notional unit of currency earned by a country exporting petroleum.

petroglyph /ˈpetrəʊˌglɪf/ n. a rock-carving, esp. a prehistoric one. [PETRO- + Gk glyphē carving]

Petrograd /ˈpetrəˌgræd/ a former name (1914–24) for ST PETERSBURG.

petrography /peˈtrɒgrəfɪ/ n. the scientific description of the composition and formation of rocks. □ **petrographer** n. **petrographic** /ˌpetrəˈgræfɪk/ adj. **petrographical** adj.

petrol /ˈpetrəl/ n. Brit. **1** refined petroleum used as a fuel in motor vehicles, aircraft, etc. **2** (attrib.) concerned with the supply of petrol (petrol pump; petrol station). □ **petrol bomb** a simple bomb made of a petrol-filled bottle and a wick. [F pétrole f. med.L petroleum: see PETROLEUM]

petrolatum /ˌpetrəˈleɪtəm/ n. US petroleum jelly. [mod.L f. PETROL + -atum]

petroleum /pɪˈtrəʊlɪəm/ n. a hydrocarbon oil found in the upper strata of the earth, refined for use as a fuel for heating and in internal-combustion engines, for lighting, dry-cleaning, etc. □ **petroleum ether** a volatile liquid distilled from petroleum, consisting of a mixture of hydrocarbons. **petroleum jelly** a translucent solid mixture of hydrocarbons used as a lubricant, ointment, etc. [med.L f. L petra rock f. Gk + L oleum oil]

petrolic /pɪˈtrɒlɪk/ adj. of or relating to petrol or petroleum.

petrology /pɪˈtrɒlədʒɪ/ n. the study of the origin, structure, composition, etc., of rocks. □ **petrologist** n. **petrologic** /ˌpetrəˈlɒdʒɪk/ adj. **petrological** adj.

Petronius /pɪˈtrəʊnɪəs/, Gaius (known as Petronius Arbiter) (d. AD 66), Roman writer. Petronius is generally accepted as the author of the Satyricon, a work in prose and verse satirizing the excesses of Roman society. Only fragments of the Satyricon survive, most notably that recounting a tastelessly extravagant banquet held by Trimalchio, a character bearing some resemblance to Nero. According to Tacitus, Petronius was 'arbiter of taste' at Nero's court. Petronius committed suicide after being accused of treason by Nero.

Petropavlovsk /ˌpetrəˈpævlɒfsk/ (in full **Petropavlovsk-Kamchatsky** /ˌpetrəˌpævˌlɒfskkæmˈtʃætskɪ/) a Russian fishing port and naval base on the east coast of the Kamchatka peninsula in eastern Siberia; pop. (1990) 245,000.

petrous /ˈpetrəs/ adj. **1** Anat. denoting the hard part of the temporal bone protecting the inner ear. **2** Geol. of, like, or relating to rock. [L petrosus f. L petra rock f. Gk]

Petrozavodsk /ˌpetrəzæˈvɒdsk/ a city in NW Russia, on Lake Onega, capital of the republic of Karelia; pop. (1990) 252,000.

Petsamo /ˈpetsəˌməʊ/ the former name (1920–44) for PECHENGA.

petticoat /ˈpetɪˌkəʊt/ n. **1** a woman's or girl's undergarment consisting of a skirt or a skirt and bodice. **2** sl. a woman or girl. **3** (attrib.) often derog. feminine; associated with women (petticoat pedantry). □ **petticoated** adj. **petticoatless** adj. [ME f. petty coat]

pettifog /ˈpetɪˌfɒg/ v.intr. (**pettifogged**, **pettifogging**) **1** practise legal deception or trickery. **2** quibble or wrangle about petty points. [back-form. f. PETTIFOGGER]

pettifogger /ˈpetɪˌfɒgə(r)/ n. **1** a rascally lawyer; an inferior legal practitioner. **2** a petty practitioner in any activity. □ **pettifoggery** n. **pettifogging** adj. [PETTY + fogger underhand dealer, prob. f. Fugger family of merchants in AUGSBURG in the 15th–16th c.]

pettish /ˈpetɪʃ/ adj. peevish, petulant; easily put out. □ **pettishly** adv. **pettishness** n. [PET² + -ISH¹]

petty /ˈpetɪ/ adj. (**pettier**, **pettiest**) **1** unimportant; trivial. **2** mean, small-minded; contemptible. **3** minor; inferior; on a small scale (petty princes). **4** Law (of a crime) of lesser importance (petty sessions) (cf. COMMON adj. 9, GRAND adj. 8). □ **petty bourgeois** = PETIT BOURGEOIS. **petty bourgeoisie** = petite bourgeoisie. **petty cash** money from or for small items of receipt or expenditure. **petty jury** a jury of twelve persons who try the final issue of fact in civil or criminal cases and pronounce a verdict. **petty officer** a naval NCO. **petty treason** see TREASON. □ **pettily** adv. **pettiness** n. [ME pety, var. of PETIT]

petulant /ˈpetjʊlənt/ adj. peevishly impatient or irritable.

□ **petulance** n. **petulantly** adv. [F pétulant f. L petulans -antis f. petere seek]

petunia /pɪˈtjuːnɪə/ n. **1** a solanaceous plant of the genus Petunia, with white, purple, red, etc., funnel-shaped flowers. **2** a dark violet or purple colour. [mod.L f. F petun f. Guarani petỹ tobacco]

petuntse /pɪˈtʊntsɪ, -ˈtʌntsɪ/ n. a white variable feldspathic mineral used for making porcelain. [Chin. baidunzi f. bai white + dun stone + suffix -zi]

Pevsner /ˈpevznə(r)/, Antoine (1886–1962), Russian-born French sculptor and painter. He was a founder of Russian constructivism, together with his brother, Naum Gabo. In 1920 the theoretical basis of the movement was put forward in their Realistic Manifesto; this advanced the notion of incorporating time and movement in sculpture. Pevsner settled in Paris in 1923, becoming a French citizen in 1930. His first sculptures were mainly in plastic, and he later worked in welded metal.

pew /pjuː/ n. & v. ● n. **1** (in a church) a long bench with a back; an enclosed compartment. **2** Brit. colloq. a seat (esp. take a pew). ● v.tr. furnish with pews. □ **pewless** adj. [ME pywe, puwe f. OF puye balcony f. L podia pl. of PODIUM]

pewee /ˈpiːwiː/ n. a North American tyrant flycatcher of the genus Contopus. [imit.: cf. PEEWEE]

pewit var. of PEEWIT.

pewter /ˈpjuːtə(r)/ n. **1** a grey alloy of tin with copper and antimony (formerly, tin and lead). **2** utensils made of this. **3** sl. a tankard etc. as a prize. □ **pewterer** n. [ME f. OF peutre, peualtre f. Rmc, of unkn. orig.]

peyote /peɪˈəʊtɪ/ n. **1** a small cactus, Lophophora williamsii, of Mexico and Texas, having no spines and button-like tops when dried. Also called mescal. **2** a hallucinogenic drug containing mescaline prepared from this. [Amer. Sp. f. Nahuatl peyotl]

Pf. abbr. pfennig.

Pfc. abbr. US Private First Class.

pfennig /ˈpfenɪg, ˈfen-/ n. a German monetary unit, equal to one-hundredth of a mark. [G, rel. to PENNY]

PG abbr. **1** (of films) classified as suitable for children subject to parental guidance. **2** paying guest.

PGA abbr. Professional Golfers' Association (of America).

pH /piːˈeɪtʃ/ n. Chem. a logarithm of the reciprocal of the hydrogen-ion concentration in moles per litre of a solution, giving a measure of its acidity or alkalinity. It is measured on a scale of 1–14, with neutrality at pH7, increasing acidity below 7, and increasing alkalinity above 7. [G, f. Potenz power + H (symbol for hydrogen)]

Phaeacian /fiːˈeɪʃ(ə)n/ n. an inhabitant of Scheria (Corfu) in the story of Odysseus, whose people were noted for their hedonism. [f. L Phaeacia, Gk. Phaiakia]

Phaedra /ˈfiːdrə/ Gk Mythol. the wife of Theseus. She conceived a passion for her stepson Hippolytus, who rejected her, whereupon she hanged herself, leaving behind a letter which accused him of raping her. Theseus would not believe his son's protestations of innocence and banished him.

Phaethon /ˈfeɪəθən/ Gk Mythol. the son of Helios the sun-god. He asked to drive his father's solar chariot for a day, but could not control the immortal horses and the chariot plunged too near the earth until Zeus, to save the earth from destruction, killed Phaethon with a thunderbolt.

phaeton /ˈfeɪt(ə)n/ n. **1** a light open four-wheeled carriage, usu. drawn by a pair of horses. **2** US a touring-car. [F phaéton f. L Phaethon f. Gk Phaethōn, PHAETHON]

phage /feɪdʒ, fɑːʒ/ n. = BACTERIOPHAGE. [abbr.]

phagocyte /ˈfægəˌsaɪt/ n. a type of cell capable of engulfing and absorbing foreign matter, esp. a leucocyte ingesting bacteria in the body. □ **phagocytic** /ˌfægəˈsɪtɪk/ adj. [Gk phag- eat + -CYTE]

phagocytosis /ˌfægəsaɪˈtəʊsɪs/ n. the ingestion of bacteria etc. by phagocytes. □ **phagocytose** /-ˈsaɪtəʊz/ v.tr. **phagocytize** /ˈfægəsɪˌtaɪz/ v.tr. (also **-ise**).

-phagous /fəgəs/ comb. form that eats (as specified) (ichthyophagous). [L -phagus f. Gk -phagos f. phagein eat]

-phagy /fədʒɪ/ comb. form the eating of (specified food) (ichthyophagy). [Gk -phagia (as -PHAGOUS)]

phalange /ˈfælændʒ/ n. **1** Anat. = PHALANX 3. **2** (**Phalange**) a right-wing activist Maronite party in Lebanon, founded in 1936 by Pierre Gemayel (cf. FALANGE). □ **Phalangist** n. [F f. L phalanx: see PHALANX]

phalangeal /fə'lændʒɪəl/ adj. Anat. of or relating to a phalanx.

phalanger /fə'lændʒə(r)/ n. an arboreal Australasian marsupial of the family Phalangeridae, including cuscuses and brush-tailed possums. (See also *flying phalanger*.) [F f. Gk *phalaggion* spider's web, f. the webbed toes of its hind feet]

phalanx /'fælæŋks/ n. (pl. **phalanxes** or **phalanges** /fə'lændʒiːz/) **1** Gk Antiq. a line of battle, esp. a body of Macedonian infantry drawn up in close order. **2** a set of people etc. forming a compact mass, or banded for a common purpose. **3** Anat. a bone of the finger or toe. **4** Bot. a bundle of stamens united by filaments. [L f. Gk *phalanx -ggos*]

phalarope /'fælə,rəʊp/ n. a small swimming sandpiper of the genus *Phalaropus*, with a straight bill and lobed feet, notable for its reversal of the sexual roles. [F f. mod.L *Phalaropus*, irreg. f. Gk *phalaris* coot + *pous podos* foot]

phalli pl. of PHALLUS.

phallic /'fælɪk/ adj. **1** of, relating to, or resembling a phallus. **2** Psychol. denoting the stage of male sexual development characterized by preoccupation with the genitals. □ **phallically** adv. [F *phallique* & Gk *phallikos* (as PHALLUS)]

phallocentric /,fæləʊ'sentrɪk/ adj. centred on the phallus or on male attitudes. □ **phallocentricity** /-sen'trɪsɪtɪ/ n. **phallocentrism** /-'sentrɪz(ə)m/ n.

phallus /'fæləs/ n. (pl. **phalli** /-laɪ/ or **phalluses**) **1** the (esp. erect) penis. **2** an image of this as a symbol of generative power in nature. □ **phallism** n. **phallicism** /-lɪ,sɪz(ə)m/ n. [LL f. Gk *phallos*]

phanariot /fə'nærɪət/ n. hist. a member of a class of Greek officials in Constantinople under the Ottoman Empire. [mod.Gk *phanariōtēs* f. *Phanar* the part of the city where they lived f. Gk *phanarion* lighthouse (on the Golden Horn)]

phanerogam /'fænərə,gæm/ n. Bot. a plant that has stamens and pistils, a flowering plant (cf. CRYPTOGAM). □ **phanerogamic** /,fænərə'gæmɪk/ adj. **phanerogamous** /-'rɒgəməs/ adj. [F *phanérogame* f. Gk *phaneros* visible + *gamos* marriage]

phantasize archaic var. of FANTASIZE.

phantasm /'fæntæz(ə)m/ n. **1** an illusion, a phantom. **2** (usu. foll. by *of*) an illusory likeness. **3** a supposed vision of an absent (living or dead) person. □ **phantasmal** /fæn'tæzm(ə)l/ adj. **phantasmic** adj. [ME f. OF *fantasme* f. L f. Gk *phantasma* f. *phantazō* make visible f. *phainō* show]

phantasmagoria /,fæntæzmə'gɔːrɪə/ n. **1** a shifting series of real or imaginary figures as seen in a dream. **2** an optical device for rapidly varying the size of images on a screen. □ **phantasmagoric** /-'gɒrɪk/ adj. **phantasmagorical** adj. [prob. f. F *fantasmagorie* (as PHANTASM + fanciful ending)]

phantast var. of FANTAST.

phantasy archaic var. of FANTASY.

phantom /'fæntəm/ n. & adj. ● n. **1** a ghost; an apparition; a spectre. **2** a form without substance or reality; a mental illusion. **3** Med. a model of the whole or part of the body used to practise or demonstrate operative or therapeutic methods. ● adj. merely apparent; illusory. □ **phantom circuit** an arrangement of telegraph or other electrical wires equivalent to an extra circuit. **phantom limb** a continuing sensation of the presence of a limb which has been amputated. **phantom pregnancy** Med. the symptoms of pregnancy in a person not actually pregnant. [ME f. OF *fantosme* ult. f. Gk *phantasma* (as PHANTASM)]

Pharaoh /'feərəʊ/ n. **1** the ruler of ancient Egypt. **2** the title of this ruler. □ **Pharaoh's ant** a small reddish ant, *Monomorium pharaonis*, native to warm regions and a pest of heated buildings elsewhere. **Pharaoh's serpent** an indoor firework burning and uncoiling in serpentine form. □ **Pharaonic** /,feəreɪ'ɒnɪk, feə'rɒn-/ adj. [OE f. eccl.L *Pharao* f. Gk *Pharaō* f. Egypt. *pr-ʿo* great house]

Pharisee /'færɪ,siː/ n. **1** a member of an ancient Jewish sect. (See note below.) **2** a person of the spirit or disposition attributed to the Pharisees in the New Testament; a self-righteous person; a hypocrite. □ **Pharisaism** /-seɪ,ɪz(ə)m/ n. **Pharisaic** /,færɪ'seɪɪk/ adj. **Pharisaical** adj. [OE *fariseus* f. OF *pharise* f. eccl.L *pharisaeus* f. Gk *Pharisaios* f. Aram. *pʾrišayyā* pl. f. Heb. *pārûš* separated]

● The Pharisees are mentioned only by Josephus and in the New Testament, where they are presented as having pretensions to superior sanctity. Unlike the Sadducees, who tried to apply Mosaic law strictly, the Pharisees allowed some freedom of interpretation. Although in the Gospels they are represented as the chief opponents of Christ they seem to have been less hostile than the Sadducees to the nascent Church, with which they shared belief in the Resurrection.

pharmaceutical /,fɑːmə'sjuːtɪk(ə)l/ adj. & n. ● adj. **1** of or engaged in pharmacy. **2** of the use or sale of medicinal drugs. ● n. a medicinal drug. □ **pharmaceutically** adv. **pharmaceutics** n. [LL *pharmaceuticus* f. Gk *pharmakeutikos* f. *pharmakeutēs* druggist f. *pharmakon* drug]

pharmacist /'fɑːməsɪst/ n. a person qualified to prepare and dispense drugs.

pharmacognosy /,fɑːmə'kɒgnəsɪ/ n. the science of drugs, esp. relating to medicinal products in their natural or unprepared state. [Gk *pharmakon* drug + *gnōsis* knowledge]

pharmacology /,fɑːmə'kɒlədʒɪ/ n. the branch of medicine that deals with the uses, effects, and modes of action of drugs. □ **pharmacologist** n. **pharmacological** /-kə'lɒdʒɪk(ə)l/ adj. **pharmacologically** adv. [mod.L *pharmacologia* f. Gk *pharmakon* drug]

pharmacopoeia /,fɑːməkə'piːə/ n. **1** a book, esp. one officially published, containing a list of drugs with directions for use. **2** a stock of drugs. □ **pharmacopoeial** adj. [mod.L f. Gk *pharmakopoiia* f. *pharmakopoios* drug-maker (as PHARMACOLOGY + *-poios* making)]

pharmacy /'fɑːməsɪ/ n. (pl. **-ies**) **1** the preparation and the (esp. medicinal) dispensing of drugs. **2** a pharmacist's shop, a dispensary. [ME f. OF *farmacie* f. med.L *pharmacia* f. Gk *pharmakeia* practice of the druggist f. *pharmakeus* f. *pharmakon* drug]

Pharos /'feərɒs/ a lighthouse, one of the earliest known, erected by Ptolemy II (308–246 BC) in *c*.280 BC on the island of Pharos, off the coast of Alexandria. Often considered one of the Seven Wonders of the World, it is said to have been over 130 m (440 ft) high and to have been visible from 67 km (42 miles) away. It was finally destroyed in 1375.

pharos /'feərɒs/ n. a lighthouse or a beacon to guide sailors. [L f. Gk *Pharos* PHAROS]

pharyngo- /fə'rɪŋgəʊ/ comb. form denoting the pharynx.

pharyngotomy /,færɪŋ'gɒtəmɪ/ n. (pl. **-ies**) Med. an incision into the pharynx.

pharynx /'færɪŋks/ n. (pl. **pharynges** /fə'rɪndʒiːz/) **1** Anat. & Zool. a membrane-lined cavity behind the mouth and nose, connecting them to the oesophagus. **2** Zool. (in invertebrates) part of the alimentary canal immediately posterior to the mouth cavity. □ **pharyngal** /fə'rɪŋg(ə)l/ adj. **pharyngeal** /,færɪn'dʒiːəl/ adj. **pharyngitis** /-'dʒaɪtɪs/ n. [mod.L f. Gk *pharugx -ggos*]

phase /feɪz/ n. & v. ● n. **1** a distinct period or stage in a process of change or development. **2** each of the aspects of the moon or a planet, according to the amount of its illumination, esp. the new moon, the first quarter, the last quarter, and the full moon. **3** Physics a stage in a periodically recurring sequence, esp. of alternating electric currents or light vibrations. **4** a difficult or unhappy period, esp. in adolescence (*just going through a phase*). **5** a genetic or seasonal variety of an animal's coloration etc. **6** Chem. a distinct and homogeneous form of matter separated by its surface from other forms. ● v.tr. carry out (a programme etc.) in phases or stages. □ **in phase** having the same phase at the same time. **out of phase** not in phase. **phase in** (or **out**) bring gradually into (or out of) use. **phase rule** Chem. a rule relating numbers of phases, constituents, and degrees of freedom. **three-phase** (of an electric generator, motor, etc.) designed to supply or use simultaneously three separate alternating currents of the same voltage, but with phases differing by a third of a period. □ **phasic** adj. [F *phase* & f. earlier *phasis* f. Gk *phasis* appearance f. *phainō phan-* show]

Phasmida /'fæzmɪdə/ n.pl. Zool. an order of insects comprising the stick insects and leaf insects, which have very long bodies that look like twigs or leaves. □ **phasmid** n. & adj. [mod.L f. Gk *phasma* spectre]

phatic /'fætɪk/ adj. (of speech etc.) used to convey general sociability rather than to communicate a specific meaning, e.g. 'nice morning, isn't it?' [Gk *phatos* spoken f. *phēmi phan-* speak]

Ph.D. abbr. Doctor of Philosophy. [L *philosophiae doctor*]

pheasant /'fez(ə)nt/ n. a large long-tailed game bird of the family Phasianidae, orig. from Asia; esp. the common *Phasianus colchicus*. □ **pheasantry** n. (pl. **-ies**). [ME f. AF *fesaunt* f. OF *faisan* f. L *phasianus* f. Gk *phasianos* (bird) of the river *Phasis* in Asia Minor]

Pheidippides /faɪ'dɪpɪ,diːz/ (5th century BC), Athenian messenger. He was sent to Sparta to ask for help after the Persian landing at Marathon in 490 and is said to have covered the 250 km (150 miles) in two days on foot.

phenacetin /fɪ'næsɪtɪn/ n. an acetyl derivative of phenol used to treat fever etc. [PHENO- + ACETYL + -IN]

phencyclidine /fen'saɪklɪˌdiːn/ n. (abbr. **PCP**) a compound derived from piperidine, used as a veterinary anaesthetic and a hallucinogenic drug. [PHENO- + CYCLO- + PIPERIDINE]

pheno- /'fiːnəʊ/ comb. form **1** Chem. derived from benzene (phenol; phenyl). **2** showing (phenocryst). [Gk phainō shine (with ref. to substances used for illumination), show]

phenobarbitone /ˌfiːnəʊ'bɑːbɪˌtəʊn/ n. (US **phenobarbital** /-t(ə)l/) a narcotic and sedative barbiturate drug used esp. to treat epilepsy.

phenocryst /'fiːnəˌkrɪst/ n. a large or conspicuous crystal in porphyritic rock. [F phénocryste (as PHENO-, CRYSTAL)]

phenol /'fiːnɒl/ n. Chem. **1** a white hygroscopic crystalline solid (chem. formula: C_6H_5OH), with acidic properties, used in dilute form as an antiseptic and disinfectant. Also called carbolic. **2** any hydroxyl derivative of an aromatic hydrocarbon. □ **phenolic** /fɪ'nɒlɪk/ adj. [F phénole f. phène benzene + -OL]

phenolphthalein /ˌfiːnɒl'θeɪliːn/ n. Chem. a white crystalline solid used in solution as an acid-base indicator and medicinally as a laxative. It is colourless below pH8, changing to red at about pH10 and above. [PHENOL + phthal f. NAPHTHALENE + -IN]

phenomena pl. of PHENOMENON.

phenomenal /fɪ'nɒmɪn(ə)l/ adj. **1** of the nature of a phenomenon. **2** extraordinary, remarkable, prodigious. **3** perceptible by, or perceptible only to, the senses. □ **phenomenalize** v.tr. (also **-ise**). **phenomenally** adv.

phenomenalism /fɪ'nɒmɪnəˌlɪz(ə)m/ n. Philos. the doctrine that knowledge about the external world is confined to what is presented to the senses; the doctrine that phenomena are the only objects of knowledge. □ **phenomenalist** n. **phenomenalistic** /-ˌnɒmɪnə'lɪstɪk/ adj.

phenomenology /fɪˌnɒmɪ'nɒlədʒɪ/ n. Philos. **1** the science of phenomena as distinct from that of being (ontology). **2** the philosophical movement that concentrates on the study of consciousness and its immediate objects, associated particularly with Edmund Husserl and influential on the work of Martin Heidegger and other existentialists. □ **phenomenological** /-nə'lɒdʒɪk(ə)l/ adj. **phenomenologically** adv.

phenomenon /fɪ'nɒmɪnən/ n. (pl. **phenomena** /-nə/) **1** a fact or occurrence that appears or is perceived, esp. one of which the cause is in question. **2** a remarkable person or thing. **3** Philos. the object of a person's perception; what the senses or the mind notice. [LL f. Gk phainomenon neut. pres. part. of phainomai appear f. phainō show]

phenotype /'fiːnəʊˌtaɪp/ n. Biol. a set of observable characteristics of an individual or group, resulting from the interaction of its genotype with its environment. □ **phenotypic** /ˌfiːnəʊ'tɪpɪk/ adj. **phenotypical** adj. **phenotypically** adv. [G Phaenotypus (as PHENO-, TYPE)]

phenyl /'fiːnaɪl, 'fenɪl/ n. Chem. the radical $-C_6H_5$, derived from benzene by removal of a hydrogen atom. [PHENO- + -YL]

phenylalanine /ˌfiːnaɪl'læləˌniːn, ˌfenɪ-/ n. Biochem. an amino acid widely distributed in plant proteins and essential in the human diet. [PHENYL + ALANINE]

phenylketonuria /ˌfiːnaɪlˌkiːtə'njʊərɪə, ˌfenɪl-/ n. Med. an inherited inability to metabolize phenylalanine, ultimately leading to mental handicap if untreated. [PHENYL + KETONE + -URIA]

pheromone /'ferəˌməʊn/ n. a chemical substance secreted and released by an animal for detection and response by another usu. of the same species. □ **pheromonal** /ˌferə'məʊn(ə)l/ adj. [Gk pherō convey + HORMONE]

phew /fjuː/ int. an expression of impatience, discomfort, relief, astonishment, or disgust. [imit. of puffing]

phi /faɪ/ n. the twenty-first letter of the Greek alphabet (Φ, φ). [Gk]

phial /'faɪəl/ n. a small glass bottle, esp. for liquid medicine. [ME f. OF fiole f. L phiola phiala f. Gk phialē, a broad flat vessel: cf. VIAL]

Phi Beta Kappa /ˌfaɪ biːtə 'kæpə/ n. **1** the oldest American college fraternity, an honorary society to which distinguished undergraduate (and occasionally graduate) students may be elected, named from the initial letters of its Greek motto philosophia biou kubernētēs, 'philosophy the guide to life'. **2** a member of this society.

Phidias /'fɪdɪˌæs, 'faɪd-/ (5th century BC), Athenian sculptor. In about 447 he was appointed by Pericles to plan and supervise public building on the Acropolis in Athens. His own contributions to the project included a colossal gold-and-ivory statue of Athene Parthenos for the Parthenon (c.438), which has not survived, and the Elgin Marbles. He

is also noted for his vast statue of Zeus at Olympia (c.430), which was one of the Seven Wonders of the World.

Phil. abbr. **1** Philadelphia. **2** Philharmonic. **3** Epistle to the Philippians (New Testament). **4** Philosophy.

phil- var. of PHILO-.

-phil var. of -PHILE.

philabeg var. of FILIBEG.

Philadelphia /ˌfɪlə'delfɪə/ the chief city of Pennsylvania, on the Delaware river; pop. (1990) 1,585,580. First settled by Swedes in the 1640s, it was established as a Quaker colony by William Penn and others in 1681. It was the site in 1776 of the signing of the Declaration of Independence and in 1787 of the adoption of the Constitution of the United States. The US capital from 1790 to 1800, it is now the second largest city on the east coast. [Gk philadelphia brotherly love]

philadelphus /ˌfɪlə'delfəs/ n. a highly scented deciduous flowering shrub of the genus Philadelphus, esp. the mock orange. [mod.L f. Gk philadelphon]

philander /fɪ'lændə(r)/ v.intr. (often foll. by with) flirt or have casual affairs with women; womanize. □ **philanderer** n. [philander (n.) used in Gk literature as the proper name of a lover, f. Gk philandros fond of men f. anēr male person: see PHIL-]

philanthrope /'fɪlənˌθrəʊp/ n. a philanthropist. [Gk philanthrōpos (as PHIL-, anthrōpos human being)]

philanthropic /ˌfɪlən'θrɒpɪk/ adj. loving one's fellow men; benevolent. □ **philanthropically** adv. [F philanthropique (as PHILANTHROPE)]

philanthropy /fɪ'lænθrəpɪ/ n. **1** a love of humankind. **2** practical benevolence, esp. charity on a large scale. □ **philanthropism** n. **philanthropist** n. **philanthropize** v.tr. & intr. (also **-ise**). [LL philanthropia f. Gk philanthrōpia (as PHILANTHROPE)]

philately /fɪ'lætəlɪ/ n. the collection and study of postage stamps. □ **philatelist** n. **philatelic** /ˌfɪlə'telɪk/ adj. **philatelically** adv. [F philatélie f. Gk PHILO- + atelēs free of charge, ateleia exemption from payment f. a- not + telos toll, tax: Gk ateles was taken as 'postage stamp']

Philby /'fɪlbɪ/, Harold Adrian Russell ('Kim') (1912–88), British Foreign Office official and spy. While chief liaison officer at the British Embassy in Washington, DC (1949–51), he was suspected of being a Soviet agent and interrogated, but, in the absence of firm evidence against him, Philby was merely asked to resign. He defected to the USSR in 1963 and in the same year was officially revealed to have spied for the Soviets from 1933. He became a Soviet citizen in 1963 and was appointed a general in the KGB.

-phile /faɪl/ comb. form (also **-phil** /fɪl/) forming nouns and adjectives denoting fondness for what is specified (bibliophile; Francophile). [Gk philos dear, loving]

Philem. abbr. Epistle to Philemon (New Testament).

Philemon /fɪ'liːmən/ Gk Mythol. a good old countryman living with his wife Baucis in Phrygia who offered hospitality to Zeus and Hermes when the two gods came to earth, without revealing their identities, to test people's piety. Philemon and Baucis were subsequently saved from a flood which covered the district.

Philemon, Epistle to a book of the New Testament, an epistle of St Paul to a well-to-do Christian living probably at Colossae in Phrygia.

philharmonic /ˌfɪlhɑː'mɒnɪk, ˌfɪlə'mɒn-/ adj. **1** fond of music. **2** used characteristically in the names of orchestras, choirs, etc. (Royal Philharmonic Orchestra). [F philharmonique f. It. filarmonico (as PHIL-, HARMONIC)]

philhellene /ˌfɪlhe'liːn, fɪl'heliːn/ n. (often attrib.) **1** a lover of Greece and Greek culture. **2** hist. a supporter of the cause of Greek independence. □ **philhellenic** /ˌfɪlhe'liːnɪk/ adj. **philhellenism** /fɪl'helɪˌnɪz(ə)m/ n. **philhellenist** n. [Gk philellēn (as PHIL-, HELLENE)]

-philia /'fɪlɪə/ comb. form **1** denoting (esp. abnormal) fondness or love for what is specified (necrophilia). **2** denoting undue inclination (haemophilia). □ **-philiac** /-lɪˌæk/ comb. form forming nouns and adjectives. **-philic** comb. form forming adjectives. **-philous** comb. form forming adjectives. [Gk f. philos loving]

Philip[1] /'fɪlɪp/ the name of five kings of ancient Macedonia, notably:

Philip II (known as Philip II of Macedon) (382–336 BC), father of Alexander the Great, reigned 359–336. He unified and expanded ancient Macedonia, as well as carrying out a number of army reforms, such as the introduction of the phalanx formation. His victory over Athens and Thebes at the battle of Chaeronea in 338 established his hegemony over Greece. He was assassinated as he was about to lead an expedition against Persia.

Philip V (238–179 BC), reigned 221–179. His expansionist policies led to a series of confrontations with Rome, culminating in his defeat in Thessaly in 197 and his resultant loss of control over Greece.

Philip[2] /ˈfɪlɪp/ the name of six kings of France, notably:

Philip II (known as Philip Augustus) (1165–1223), son of Louis VII, reigned 1180–1223. His reign was marked by a dramatic expansion of Capetian influence, at the expense of the English Plantagenet empire in France. After mounting a series of military campaigns against the English kings Henry II, Richard I, and John, Philip succeeded in regaining Normandy (1204), Anjou (1204), and most of Poitou (1204–5). Towards the end of his reign, after success in the crusade (1209–31) against the Albigenses, he also managed to add fresh territories in the south to his kingdom.

Philip IV (known as Philip the Fair) (1268–1314), son of Philip III, reigned 1285–1314. He continued the Capetian policy of extending French dominions, waging wars of expansion with England (1294–1303) and Flanders (1302–5). His reign, however, was dominated by his struggle with the papacy; in 1303 he imprisoned Pope Boniface VIII (c.1228–1303), and, in 1305, his influence secured the appointment of the French Clement V (c.1260–1314) as pope. Philip's domination of the papacy was further consolidated when Clement moved the papal seat to Avignon (1309), where it remained until 1377. Philip also persuaded Clement to dissolve the powerful and wealthy order of the Knights Templars; its leaders were executed and its property divided between the Crown and the Knights Hospitallers.

Philip VI (known as Philip of Valois) (1293–1350), reigned 1328–50. The founder of the Valois dynasty, Philip came to the throne on the death of his cousin Charles IV (1294–1328), whose only child was a girl and barred from ruling by Salic Law. His claim was disputed by Edward III of England, who could trace a claim through his mother Isabella of France. War between the two countries, which was to develop into the Hundred Years War, ensued. Philip was defeated by Edward at the Battle of Crécy (1346).

Philip[3] /ˈfɪlɪp/ the name of five kings of Spain, notably:

Philip I (known as Philip the Handsome) (1478–1506), reigned 1504–6. Son of the Holy Roman emperor Maximilian I of Habsburg (1459–1519), Philip married the infanta Joanna, daughter of Ferdinand of Aragon and Isabella of Castile, in 1496. After Isabella's death, he ruled Castile jointly with Joanna, establishing the Habsburgs as the ruling dynasty in Spain.

Philip II (1527–98), son of Charles I, reigned 1556–98. Philip married the second of his four wives, Mary I of England, in 1554, and came to the throne following his father's abdication two years later. His reign came to be dominated by an anti-Protestant crusade which exhausted the Spanish economy. He failed to suppress revolt in the Netherlands (1567–79), and although he conquered Portugal in 1580, his war against England also proved a failure, an attempted Spanish invasion being thwarted by the defeat of the Armada in 1588.

Philip V (1683–1746), grandson of Louis XIV, reigned 1700–24 and 1724–46. The selection of Philip, a Bourbon, as successor to Charles II, and Louis XIV's insistence that Philip remain an heir to the French throne, gave rise to the threat of the union of the French and Spanish thrones and led to the War of the Spanish Succession (1701–14). Internationally recognized as king of Spain by the Peace of Utrecht (1713–14), Philip reigned until 1724, when he abdicated in favour of his son Louis I (1707–24), but returned to the throne following Louis's death in the same year.

Philip, Prince, Duke of Edinburgh (b.1921), husband of Elizabeth II. The son of Prince Andrew of Greece and Denmark (1882–1944), he married Princess Elizabeth in 1947; on the eve of his marriage he was created Duke of Edinburgh. He served in the Royal Navy until Elizabeth's accession in 1952.

Philip, St 1 an Apostle. He is commemorated with St James the Less on 1 May. **2** (known as St Philip the Evangelist) one of seven deacons appointed to superintend the secular business of the Church at Jerusalem (Acts 6:5–6). Feast day, 6 June.

Philip II of Macedon /ˈmæsɪˌdɒn, -d(ə)n/, Philip II of Macedonia (see PHILIP[1]).

Philip Augustus /ɔːˈɡʌstəs/, Philip II of France (see PHILIP[2]).

Philip of Valois /ˈvælwʌ/, Philip VI of France (see PHILIP[2]).

Philippi /ˈfɪlɪˌpaɪ, fɪˈlɪpaɪ/ (Greek **Filippoi** /ˈfilipi/) a city in ancient Macedonia, the scene in 42 BC of two battles in which Mark Antony and Octavian defeated Brutus and Cassius. The ruins lie close to the Aegean coast in NE Greece, near the port of Kaválla (ancient Neapolis).

Philippians, Epistle to the /fɪˈlɪpɪənz/ a book of the New Testament, an epistle of St Paul to the Church at Philippi in Macedonia.

philippic /fɪˈlɪpɪk/ n. a bitter verbal attack or denunciation. [L philippicus f. Gk philippikos the name of Demosthenes' speeches against Philip II of Macedon and of Cicero's against Mark Antony]

Philippine /ˈfɪlɪˌpiːn/ adj. of or relating to the Philippines or its people; Filipino. [Philip II of Spain (see PHILIP[3])]

Philippines /ˈfɪlɪˌpiːnz/ a country in SE Asia consisting of an archipelago of over 7,000 islands separated from the Asian mainland by the South China Sea; pop. (1990) 60,684,890; official languages Pilipino and English; capital, Manila. The main islands are Luzon, Mindanao, Mindoro, Leyte, Samar, Negros, and Panay. The Portuguese navigator Magellan visited the islands in 1521 and was killed there. Conquered by Spain in 1565, and named after the king's son, the islands were ceded to the US following the Spanish-American War in 1898. Occupied by the Japanese between 1941 and 1944, the Philippines achieved full independence as a republic in 1946 and has continued to maintain close links with the US. From 1965 to 1986 the country was under the increasingly dictatorial rule of President Ferdinand Marcos (1917–89). He was driven from power by a popular revolt and replaced by Corazón Aquino (b.1933), President 1986–92, the widow of an assassinated opposition leader; she introduced democratic reforms.

Philippopolis /ˌfɪlɪˈpɒpəlɪs/ the ancient Greek name for PLOVDIV.

Philip the Fair, Philip IV of France (see PHILIP[2]).

Philip the Handsome, Philip I of Spain (see PHILIP[3]).

Philistine /ˈfɪlɪˌstaɪn/ n. & adj. ● n. a member of a non-Semitic people opposing the Israelites in ancient Palestine. (See note below.) ● adj. of or relating to the Philistines or Gk Philistinos = Palaistinos f. Heb. pᵊlištî]

■ The Philistines, from whom the country of Palestine took its name, were one of the Sea Peoples who, according to the Bible, came from Crete and settled the southern coastal plain of Canaan in the 12th century BC. They gained control of the land and sea routes and acquired a monopoly of metal technology, but, after repeated conflicts with the Israelites, were defeated by David c.1000 BC and subsequently declined into obscurity after losing control of the sea trade to the Phoenicians.

philistine /ˈfɪlɪˌstaɪn/ n. & adj. ● n. a person who is hostile or indifferent to culture, or whose interests are material or commonplace. ● adj. having such characteristics. □ **philistinism** /-stɪˌnɪz(ə)m/ n. [PHILISTINE]

Phillips /ˈfɪlɪps/ n. (usu. attrib.) propr. denoting a screw with a cross-shaped slot for turning, or a corresponding screwdriver. [Henry F. Phillips, the original Amer. manufacturer (fl. 1935)]

Phillips curve n. a supposed inverse relationship between the level of unemployment and the rate of inflation. [Alban William Housego Phillips, NZ economist (1914–75)]

phillumenist /fɪˈljuːmənɪst, fɪˈluː-/ n. a collector of matchbox labels. □ **phillumeny** n. [PHIL- + L lumen light]

Philly /ˈfɪlɪ/ n. US sl. Philadelphia. [abbr.]

philo- /ˈfɪləʊ/ comb. form (also **phil-** before a vowel or h) denoting a liking for what is specified.

philodendron /ˌfɪləˈdendrən/ n. (pl. **philodendrons** or **philodendra** /-drə/) a tropical American climbing plant of the genus Philodendron, with ornamental foliage. [PHILO- + Gk dendron tree]

philogynist /fɪˈlɒdʒɪnɪst/ n. a person who likes or admires women. [PHILO- + Gk gunē woman]

Philo Judaeus /ˌfaɪləʊ dʒuːˈdiːəs/ (also known as Philo of Alexandria) (c.15 BC–AD c.50), Jewish philosopher of Alexandria. His numerous works (written in Greek) trace many links between Jewish Scripture and Greek philosophy. He is particularly known for his commentaries on the Pentateuch, which he interpreted allegorically in the light of Platonic and Aristotelian philosophy.

philology /fɪˈlɒlədʒɪ/ n. **1** the science of language, esp. in its historical and comparative aspects. **2** the love of learning and literature. □ **philologize** n. **philologize** (also **-ise**). **philologian** /ˌfɪləˈləʊdʒən/ n. **philological** /-ˈlɒdʒɪk(ə)l/ adj. **philologically** adv. [F philologie f. L philologia love of learning f. Gk (as PHILO-, -LOGY)]

Philomel /ˈfɪləˌmel/ (also **Philomela** /ˌfɪləˈmiːlə/) Gk Mythol. the daughter of Pandion, king of Athens. She was turned into a swallow and her sister Procne into a nightingale (or, in Latin versions, into a nightingale with Procne the swallow) when they were being pursued by the cruel Tereus, who had married Procne and raped Philomel. The name is sometimes used in poetry for the nightingale. [earlier philomene f. med.L philomena f. L philomela nightingale f. Gk philomēla]

philoprogenitive /ˌfɪləʊprəʊˈdʒenɪtɪv/ adj. **1** prolific. **2** loving one's offspring.

philosopher /fɪˈlɒsəfə(r)/ n. **1** a person engaged or learned in philosophy or a branch of it. **2** a person who shows philosophic calmness in trying circumstances. □ **philosophers'** (or **philosopher's**) **stone** the supreme object of alchemy, a substance supposed to change other metals into gold or silver. [ME f. AF philosofre var. of OF, philosophe f. L philosophus f. Gk philosophos (as PHILO-, sophos wise)]

philosophical /ˌfɪləˈsɒfɪk(ə)l/ adj. (also **philosophic**) **1** of or according to philosophy. **2** skilled in or devoted to philosophy or learning; learned (philosophical society). **3** wise; serene; temperate. **4** calm in adverse circumstances. □ **philosophically** adv. [LL philosophicus f. L philosophia (as PHILOSOPHY)]

philosophize /fɪˈlɒsəˌfaɪz/ v. (also **-ise**) **1** intr. reason like a philosopher. **2** intr. moralize. **3** intr. speculate; theorize. **4** tr. render philosophic. □ **philosophizer** n. [app. f. F philosopher]

philosophy /fɪˈlɒsəfɪ/ n. (pl. **-ies**) **1** the use of reason and argument in seeking truth and knowledge of reality, esp. of the causes and nature of things and of the principles governing existence, the material universe, perception of physical phenomena, and human behaviour. **2 a** a particular system or set of beliefs reached by this. **b** a personal rule of life. **3** advanced learning in general (doctor of philosophy). **4** serenity; calmness; conduct governed by a particular philosophy. [ME f. OF filosofie f. L philosophia wisdom f. Gk (as PHILO-, sophos wise)]

philtre /ˈfɪltə(r)/ n. (US **philter**) a drink supposed to excite sexual love in the drinker. [F philtre f. L philtrum f. Gk philtron f. phileō to love]

-phily /fɪlɪ/ comb. form = -PHILIA.

phimosis /faɪˈməʊsɪs/ n. Med. a constriction of the foreskin, making it difficult to retract. □ **phimotic** /-ˈmɒtɪk/ adj. [mod.L f. Gk, = muzzling]

Phintias /ˈfɪntɪˌæs/ see DAMON.

phiz /fɪz/ n. (also **phizog** /ˈfɪzɒg/) Brit. colloq. **1** the face. **2** the expression on a face. [abbr. of phiznomy = PHYSIOGNOMY]

phlebitis /flɪˈbaɪtɪs/ n. Med. inflammation of the walls of a vein. □ **phlebitic** /-ˈbɪtɪk/ adj. [mod.L f. Gk f. phleps phlebos vein]

phlebotomy /flɪˈbɒtəmɪ/ n. Med. **1** the surgical opening or puncture of a vein. **2** esp. hist. blood-letting as a medical treatment. □ **phlebotomist** n. **phlebotomize** v.tr. (also **-ise**). [ME f. OF flebothomi f. LL phlebotomia f. Gk f. phleps phlebos vein + -TOMY]

phlegm /flem/ n. **1** the thick viscous substance secreted by the mucous membranes of the respiratory passages, discharged by coughing. **2** phlegmatic character or behaviour. **3** hist. one of the four humours, characterized as cool and moist and supposed to cause stolid calm or apathy. □ **phlegmy** adj. [ME & OF fleume f. LL phlegma f. Gk phlegma -atos inflammation f. phlegō burn]

phlegmatic /fleg'mætɪk/ adj. stolidly calm; unexcitable, unemotional. □ **phlegmatically** adv.

phloem /ˈfləʊem/ n. Bot. the tissue conducting food material in plants (cf. XYLEM). [Gk phloos bark]

phlogiston /flə'dʒɪstən, -'gɪstən/ n. Chem. hist. a substance supposed by 18th-century chemists to exist in all combustible bodies, and to be released in combustion. Since Joseph Priestley held this theory, he called oxygen, which he had discovered, 'dephlogisticated air'. Antoine Lavoisier, however, demonstrated the true nature of oxygen and discredited the phlogiston theory. [mod.L f. Gk phlogizō set on fire f. phlox phlogos flame]

phlox /flɒks/ n. a North American plant of the genus Phlox, with scented clusters of esp. white, blue, or red flowers, with many cultivated varieties. [L f. Gk phlox, the name of a plant (lit. 'flame')]

Phnom Penh /nɒm 'pen/ the capital of Cambodia, a port at the junction of the Mekong and Tonlé Sap rivers; pop. (est. 1990) 800,000. Founded by the Khmers in the 14th century, it became the capital of a Khmer kingdom in the mid-15th century. It was controlled between 1975 and 1979 by the Khmer Rouge, who forced its population of 2.5 million to leave the city and resettle in the country. The city was repopulated after the arrival of the Vietnamese in 1979.

-phobe /fəʊb/ comb. form forming nouns and adjectives denoting a person having a fear or dislike of what is specified (xenophobe). [F f. L -phobus f. Gk -phobos f. phobos fear]

phobia /ˈfəʊbɪə/ n. an abnormal or morbid fear or aversion. □ **phobic** adj. & n. [-PHOBIA used as a separate word]

-phobia /ˈfəʊbɪə/ comb. form forming abstract nouns denoting a fear or dislike of what is specified (agoraphobia; xenophobia). □ **-phobic** comb. form forming adjectives. [L f. Gk]

Phobos /ˈfəʊbɒs/ **1** Gk Mythol. one of the sons of Ares. **2** Astron. the inner of the two small satellites of Mars, discovered in 1877. It is 27 km long and 22 km across and is heavily cratered.

Phoebe /ˈfiːbɪ/ **1** Gk Mythol. a Titaness, daughter of Uranus (Heaven) and Gaia (Earth). She became the mother of Leto and thus the grandmother of Apollo and Artemis. In the later Greek writers her name was often used for Selene (Moon). **2** Astron. satellite IX of Saturn, the furthest from the planet, discovered in 1898 (average diameter 220 km). It has a very dark patchy surface, and an eccentric retrograde orbit which takes it nearly 13 million km from Saturn. [Gk Phoibē bright one]

phoebe /ˈfiːbɪ/ n. a North American tyrant flycatcher of the genus Sayornis. [imit.: infl. by the name]

Phoebus /ˈfiːbəs/ Gk Mythol. an epithet of Apollo, used in contexts where the god was identified with the sun. [Gk Phoibos bright one]

Phoenicia /fəˈniːʃə/ an ancient country on the shores of the eastern Mediterranean, corresponding to modern Lebanon and the coastal plains of Syria. It consisted of a number of city-states, including Tyre and Sidon, and was a flourishing centre of Mediterranean trade and colonization during the early part of the first millennium BC. (See also PHOENICIAN.)

Phoenician /fəˈnɪʃɪən, -ˈniːʃ(ə)n/ n. & adj. ● n. **1** a member of a Semitic people of ancient Phoenicia or of its colonies. (See note below.) **2** the Semitic language of the Phoenicians. ● adj. of or relating to Phoenicia or its people, language, or colonies. [ME f. OF phenicien f. L PHOENICIA]

■ The Phoenicians were a people of unknown origin but probably descended from the Canaanites of the 2nd millennium BC, who occupied the coastal plain of what is now Lebanon and Syria in the early 1st millennium BC and derived their prosperity from trade and manufacturing industries in textiles, glass, metalware, carved ivory, wood, and jewellery. Their trading contacts extended throughout Asia, and reached westwards as far as Africa (where they founded Carthage), Spain, and possibly Britain. The Phoenicians continued to thrive under Assyrian and then Persian suzerainty until 332 BC, when the capital Tyre was sacked and the country incorporated in the Greek world by Alexander the Great. The Phoenicians invented an alphabet which was borrowed by the Greeks and thence passed down into Western cultural tradition.

Phoenix /ˈfiːnɪks/ the state capital of Arizona; pop. (1990) 983,400. Its dry climate makes it a popular winter resort.

phoenix /ˈfiːnɪks/ n. **1** a mythical bird, the only one of its kind, that after living for five or six centuries in the Arabian Desert, burnt itself on a funeral pyre and rose from the ashes with renewed youth to live through another cycle. **2** a unique person or thing. [OE & OF fenix f. L phoenix f. Gk phoinix Phoenician, purple, phoenix]

Phoenix Islands a group of eight islands lying just south of the equator in the western Pacific. They form a part of Kiribati.

pholas /ˈfəʊləs/ n. Zool. a piddock, esp. of the genus Pholas. [mod.L f. Gk phōlas that lurks in a hole (phōleos)]

phon /fɒn/ n. a unit of the perceived loudness of sounds. [Gk phōnē sound]

phonate /ˈfəʊneɪt, fəʊˈneɪt/ v.intr. utter a vocal sound. □ **phonation** /fəʊˈneɪʃ(ə)n/ n. **phonatory** /ˈfəʊnətərɪ/ adj. [Gk phōnē voice]

phone[1] /fəʊn/ n. & v.tr. & intr. = TELEPHONE. □ **phone book** = telephone directory. **phone-in** a broadcast programme during which the listeners or viewers telephone the studio etc. and participate. [abbr.]

phone[2] /fəʊn/ n. Phonet. a simple vowel or consonant sound. [formed as PHONEME]

-phone /fəʊn/ comb. form forming nouns and adjectives meaning: **1** an instrument using or connected with sound (telephone; xylophone). **2** a person who uses a specified language (anglophone). [Gk phōnē voice, sound]

phonecard /ˈfəʊnkɑːd/ n. a card containing prepaid units for use with a cardphone.

phoneme /ˈfəʊniːm/ n. Phonet. any of the units of sound in a specified language that distinguish one word from another (e.g. p, b, d, t as in English pad, pat, bad, bat). □ **phonemic** /fəˈniːmɪk/ adj. **phonemics** n. [F phonème f. Gk phōnēma sound, speech f. phōneō speak]

phonetic /fəˈnetɪk/ adj. **1** representing vocal sounds. **2** (of a system of spelling etc.) having a direct correspondence between symbols and sounds. **3** of or relating to phonetics. □ **phonetically** adv.

phoneticism /-tɪˌsɪz(ə)m/ *n.* **phoneticist** *n.* **phoneticize** *v.tr.* (also **-ise**). [mod.L *phoneticus* f. Gk *phōnētikos* f. *phōneō* speak]

phonetics /fəˈnetɪks/ *n.pl.* (usu. treated as *sing.*) **1** vocal sounds and their classification. **2** the branch of linguistics that deals with these. □ **phonetician** /ˌfəʊnɪˈtɪʃ(ə)n/ *n.*

phonetist /ˈfəʊnɪtɪst/ *n.* **1** a person skilled in phonetics. **2** an advocate of phonetic spelling.

phoney /ˈfəʊnɪ/ *adj. & n.* (also **phony**) *colloq.* ● *adj.* (**phonier, phoniest**) **1** sham; counterfeit. **2** fictitious; fraudulent. ● *n.* (*pl.* **-eys** or **-ies**) a phoney person or thing. □ **phonily** *adv.* **phoniness** *n.* [20th c.: orig. unkn.]

phoney war the period of comparative inaction at the beginning of the Second World War between the German invasions of Poland (September 1939) and Norway (April 1940).

phonic /ˈfɒnɪk, ˈfəʊn-/ *adj. & n.* ● *adj.* of sound; acoustic; of vocal sounds. ● *n.* (in *pl.*) a method of teaching reading based on sounds. □ **phonically** *adv.* [Gk *phōnē* voice]

phono- /ˈfəʊnəʊ/ *comb. form* denoting sound. [Gk *phōnē* voice, sound]

phonogram /ˈfəʊnəˌɡræm/ *n.* a symbol representing a spoken sound.

phonograph /ˈfəʊnəˌɡrɑːf/ *n.* **1** *Brit.* an early form of gramophone using cylinders and able to record as well as reproduce sound. **2** *N. Amer.* a gramophone.

phonography /fəˈnɒɡrəfɪ/ *n.* **1** writing in esp. shorthand symbols, corresponding to the sounds of speech. **2** the recording of sounds by phonograph. □ **phonographic** /ˌfəʊnəˈɡræfɪk/ *adj.*

phonology /fəˈnɒlədʒɪ/ *n.* the study of sounds in a language. □ **phonologist** *n.* **phonological** /ˌfəʊnəˈlɒdʒɪk(ə)l, ˌfɒn-/ *adj.* **phonologically** *adv.*

phonon /ˈfəʊnɒn/ *n. Physics* a quantum of sound or elastic vibrations. [Gk *phōnē* sound, after PHOTON]

phony var. of PHONEY.

phooey /ˈfuːɪ/ *int. colloq.* an expression of disgust or disbelief. [imit.]

-phore /fɔː(r)/ *comb. form* forming nouns meaning 'bearer' (*ctenophore*; *semaphore*). □ **-phorous** /fərəs/ *comb. form* forming adjectives. [mod.L f. Gk *-phoros -phoron* bearing, bearer f. *pherō* bear]

phoresy /fɒˈriːsɪ, ˈfɒrəsɪ/ *n. Biol.* an association in which one organism is carried by another, without being a parasite. □ **phoretic** /fɒˈretɪk/ *adj.* [F *phorésie* f. Gk *phorēsis* being carried]

phormium /ˈfɔːmɪəm/ *n.* a fibre-yielding New Zealand plant of the genus *Phormium*, of the agave family, esp. New Zealand flax. [mod.L f. Gk *phormion* a species of plant]

phosgene /ˈfɒzdʒiːn/ *n.* a colourless poisonous gas, carbonyl chloride (chem. formula: $COCl_2$), formerly used in warfare. [Gk *phōs* light + -GEN, with ref. to its orig. production by the action of sunlight on chlorine and carbon monoxide]

phosphatase /ˈfɒsfəˌteɪz, -ˌteɪs/ *n. Biochem.* any enzyme that catalyses the synthesis or hydrolysis of an organic phosphate.

phosphate /ˈfɒsfeɪt/ *n.* **1 a** *Chem.* a salt or ester of phosphoric acid. **b** (usu. in *pl.*) rock or fertilizer consisting mainly of calcium phosphate. **2** an effervescent drink containing a small amount of phosphate. □ **phosphatic** /fɒsˈfætɪk/ *adj.* [F f. *phosphore* PHOSPHORUS]

phosphene /ˈfɒsfiːn/ *n.* the sensation of rings of light produced by pressure on the eyeball due to irritation of the retina. [irreg. f. Gk *phōs* light + *phainō* show]

phosphide /ˈfɒsfaɪd/ *n. Chem.* a binary compound of phosphorus with another element or group.

phosphine /ˈfɒsfiːn/ *n. Chem.* a colourless ill-smelling gas, phosphorus trihydride (chem. formula: PH_3) □ **phosphinic** /fɒsˈfɪnɪk/ *adj.* [PHOSPHO- + -INE⁴, after *amine*]

phosphite /ˈfɒsfaɪt/ *n. Chem.* a salt or ester of phosphorous acid. [F (as PHOSPHO-)]

phospho- /ˈfɒsfəʊ/ *comb. form* denoting phosphorus. [abbr.]

phospholipid /ˌfɒsfəˈlɪpɪd/ *n. Biochem.* any lipid consisting of a phosphate group and one or more fatty acids, including the main structural lipids of cell membranes.

phosphor /ˈfɒsfə(r)/ *n.* **1** = PHOSPHORUS. **2** a synthetic fluorescent or phosphorescent substance, esp. used in cathode-ray tubes. □ **phosphor bronze** a tough hard bronze alloy containing a small amount of phosphorus, used esp. for bearings. [G f. L *phosphorus* PHOSPHORUS]

phosphorate /ˈfɒsfəˌreɪt/ *v.tr.* combine or impregnate with phosphorus.

phosphorescence /ˌfɒsfəˈres(ə)ns/ *n.* **1** radiation similar to fluorescence but detectable after excitation ceases. **2** the emission of light without combustion or perceptible heat. □ **phosphoresce** *v.intr.* **phosphorescent** *adj.*

phosphoric /fɒsˈfɒrɪk/ *adj.* **1** *Chem.* containing phosphorus, especially in its higher valency (5). **2** phosphorescent. □ **phosphoric acid** a crystalline acid (chem. formula: H_3PO_4), forming syrupy solutions with water, and having many commercial uses, e.g. in fertilizer and soap manufacture and food processing. [F *phosphorique* (as PHOSPHORUS)]

phosphorite /ˈfɒsfəˌraɪt/ *n. Mineral.* a non-crystalline form of apatite.

phosphorous /ˈfɒsfərəs/ *adj.* **1** *Chem.* of or containing phosphorus, especially in its lower valency (3). **2** phosphorescent. □ **phosphorous acid** a crystalline dibasic acid (chem. formula: H_3PO_3). [PHOSPHORUS + -OUS]

phosphorus /ˈfɒsfərəs/ *n.* a non-metallic chemical element (atomic number 15; symbol **P**). Phosphorus was first isolated by the German alchemist Hennig Brand (fl. 1670) in the 17th century. It exists in a number of allotropic forms; the most familiar, known as *white phosphorus*, is a yellowish waxy solid which ignites spontaneously in air and glows in the dark; *red phosphorus*, which is less reactive, is used in making matches. Phosphorus is abundant in the earth's crust in the form of phosphates. It is essential to living organisms; organic phosphates are involved in storing energy in cells, and calcium phosphate is a major component of bone. The main commercial use of phosphorus compounds is in fertilizers. [L, = morning star, f. Gk *phōsphoros* f. *phōs* light + *-phoros* -bringing]

phosphorylate /fɒsˈfɒrɪˌleɪt/ *v.tr. Chem.* introduce a phosphate group into (an organic molecule etc.). □ **phosphorylation** /-ˌfɒrɪˈleɪʃ(ə)n/ *n.*

phossy jaw /ˈfɒsɪ/ *n. colloq. hist.* gangrene of the jawbone caused by phosphorus poisoning, esp. as a former occupational disease in the match industry. [abbr.]

phot /fɒt, fəʊt/ *n.* a unit of illumination equal to one lumen per square centimetre. [Gk *phōs phōtos* light]

photic /ˈfəʊtɪk/ *adj.* **1** of or relating to light. **2** (of ocean layers) reached by sunlight.

photism /ˈfəʊtɪz(ə)m/ *n.* a hallucinatory sensation or vision of light. [Gk *phōtismos* f. *phōtizō* shine f. *phōs phōtos* light]

Photius /ˈfəʊtɪəs/ (*c.*820–*c.*891), Byzantine scholar and patriarch of Constantinople. His most important work is the *Bibliotheca*, a critical account of 280 earlier prose works and an invaluable source of information about many works now lost.

photo /ˈfəʊtəʊ/ *n. & v.* ● *n.* (*pl.* **-os**) = PHOTOGRAPH *n.* ● *v.tr.* (**-oes, -oed**) = PHOTOGRAPH *v.* □ **photo-call** an occasion on which theatrical performers, famous personalities, etc., pose for photographers by arrangement. **photo finish** a close finish of a race or contest, esp. one where the winner is distinguishable only from a photograph. **photo opportunity** (or *colloq.* **op**) = photo-call. [abbr.]

photo- /ˈfəʊtəʊ/ *comb. form* denoting: **1** light (*photosensitive*). **2** photography (*photocomposition*). [Gk *phōs phōtos* light, or as abbr. of PHOTOGRAPH]

photobiology /ˌfəʊtəʊbaɪˈɒlədʒɪ/ *n.* the study of the effects of light on living organisms.

photocell /ˈfəʊtəʊˌsel/ *n.* = photoelectric cell.

photochemical /ˌfəʊtəʊˈkemɪk(ə)l/ *adj.* of or relating to the chemical action of light. □ **photochemical smog** a condition of the atmosphere caused by the action of sunlight on atmospheric pollutants, resulting in haze and high levels of ozone and nitrogen oxide. □ **photochemically** *adv.*

photochemistry /ˌfəʊtəʊˈkemɪstrɪ/ *n.* the study of the chemical effects of light.

photochromic /ˌfəʊtəʊˈkrəʊmɪk/ *adj.* changing colour or shade reversibly in light of a particular frequency or intensity (*photochromic lens*). [PHOTO- + Gk *khrōma* colour]

photocomposition /ˌfəʊtəʊˌkɒmpəˈzɪʃ(ə)n/ *n.* = FILMSETTING.

photoconductivity /ˌfəʊtəʊˌkɒndʌkˈtɪvɪtɪ/ *n.* conductivity due to the action of light. □ **photoconductive** /-kənˈdʌktɪv/ *adj.* **photoconductor** *n.*

photocopier /ˈfəʊtəʊˌkɒpɪə(r)/ *n.* a machine for producing photocopies.

photocopy /ˈfəʊtəʊˌkɒpɪ/ *n. & v.* ● *n.* (*pl.* **-ies**) a photographic copy of printed or written material produced by a process involving the action of light on a specially prepared surface. ● *v.tr.* (**-ies, -ied**) make a photocopy of. □ **photocopiable** *adj.*

photodegradable /ˌfəʊtəʊdɪˈɡreɪdəb(ə)l/ *adj.* capable of being decomposed by the action of light, esp. sunlight.

photodiode /ˌfəʊtəʊˈdaɪəʊd/ *n. Electronics* a semiconductor diode responding electrically to illumination.

photoelectric /ˌfəʊtəʊɪˈlektrɪk/ *adj.* marked by or using emissions of electrons from substances exposed to light. □ **photoelectric cell** a device using this effect to generate current. □ **photoelectricity** /-ˌɪlekˈtrɪsɪtɪ, -ˌel-/ *n.*

photoelectron /ˌfəʊtəʊɪˈlektrɒn/ *n. Physics* an electron emitted from an atom by interaction with a photon, esp. one emitted from a solid surface by the action of light.

photoemission /ˌfəʊtəʊɪˈmɪʃ(ə)n/ *n. Physics* the emission of electrons from a surface by the action of incident light. □ **photoemitter** *n.*

photoengraving /ˌfəʊtəʊɪnˈɡreɪvɪŋ/ *n.* **1** a process by which an image is photographically transferred to a plate for relief printing, the non-printing areas being etched. **2** a plate or print made in this way.

photofit /ˈfəʊtəʊˌfɪt/ *n.* a reconstructed picture of a person (esp. one sought by the police) made from composite photographs of facial features (cf. IDENTIKIT).

photogenic /ˌfəʊtəˈdʒenɪk, -ˈdʒiːnɪk/ *adj.* **1** (esp. of a person) having an appearance that looks pleasing in photographs. **2** *Biol.* producing or emitting light. □ **photogenically** *adv.*

photogram /ˈfəʊtəʊˌɡræm/ *n.* **1** a picture produced with photographic materials but without a camera. **2** *archaic* a photograph.

photogrammetry /ˌfəʊtəʊˈɡræmɪtrɪ/ *n.* the use of photography for surveying. □ **photogrammetrist** *n.*

photograph /ˈfəʊtəˌɡrɑːf/ *n. & v.* ● *n.* a picture formed by means of the chemical action of light or other radiation on sensitive film. (See PHOTOGRAPHY.) ● *v.* **1** *tr.* (also *absol.*) take a photograph of (a person etc.). **2** *intr.* appear (well etc.) when in a photograph. □ **photographable** *adj.* **photographer** /fəˈtɒɡrəfə(r)/ *n.*

photographic /ˌfəʊtəˈɡræfɪk/ *adj.* **1** of, used in, or produced by photography. **2** having the accuracy of a photograph (*photographic likeness*). □ **photographically** *adv.*

photography /fəˈtɒɡrəfɪ/ *n.* **1** the art or process of taking photographs. **2** the business of producing and printing photographs. (*See note below.*)

▪ The world's first photograph was taken in 1826 by a Frenchman, Joseph-Nicéphore Niépce (1765–1833), who produced an image on a pewter plate, using bitumen that hardened slowly in bright light. Modern photography, however, is based on the property of silver compounds of decomposing to metallic silver when exposed to light. The first practical photographic method was the daguerreotype, developed by Louis-Jacques-Mandé Daguerre and first published in 1839; this produced positive images directly on silvered copper plates. The first negative-positive process, from which modern photography has evolved, was devised by Fox Talbot. He made the first photographic negative in 1835, using coated paper placed inside a camera obscura. His methods allowed an unlimited number of prints to be made from one negative; he later discovered that a brief exposure produced a latent image which could be made visible by chemical development. Modern photography essentially uses a similar process, with the light-sensitive salts held in an emulsion mounted usually on transparent roll film. Colour film, using multiple layers of emulsion, was introduced by the Eastman Kodak Company in 1935. (See also CAMERA.)

Photography has become important in science and industry, in aerial reconnaissance, and in medicine. Interest in it as an art form is as old as the process itself. Photographic societies were soon formed, techniques being explored at first with the idea of giving to the medium the manipulative latitude of painting, then (by *c.*1890) for their own sake. New freedoms in the other arts, and especially in painting, were paralleled in photography as surrealists experimented with dark-room techniques, and Max Ernst and others produced photomontages, until realism again became popular in the 1930s. An immense range of photographic materials, equipment, and skills now makes photography as versatile a medium as painting; famous names of photography include Julia Margaret Cameron, Robert Doisneau, Cecil Beaton, Diane Arbus, and David Bailey.

photogravure /ˌfəʊtəʊɡrəˈvjʊə(r)/ *n.* **1** an image produced from a photographic negative transferred to a metal plate and etched in. **2** this process. [F (as PHOTO-, *gravure* engraving)]

photojournalism /ˌfəʊtəʊˈdʒɜːnəˌlɪz(ə)m/ *n.* the art or practice of relating news chiefly by photographs. □ **photojournalist** *n.*

photolithography /ˌfəʊtəʊlɪˈθɒɡrəfɪ/ *n.* (also **photolitho** /-ˈlaɪθəʊ/) lithography using plates made photographically. □ **photolithographer** *n.* **photolithographic** /-ˌlɪθəˈɡræfɪk/ *adj.* **photolithographically** *adv.*

photolysis /fəʊˈtɒlɪsɪs/ *n. Chem.* decomposition or dissociation of molecules by the action of light. □ **photolyse** /ˈfəʊtəˌlaɪz/ *v.tr. & intr.* **photolytic** /ˌfəʊtəˈlɪtɪk/ *adj.*

photometer /fəʊˈtɒmɪtə(r)/ *n.* an instrument for measuring light. □ **photometry** *n.* **photometric** /ˌfəʊtəʊˈmetrɪk/ *adj.*

photomicrograph /ˌfəʊtəʊˈmaɪkrəˌɡrɑːf/ *n.* a photograph of an image produced by a microscope. □ **photomicrography** /-maɪˈkrɒɡrəfɪ/ *n.*

photomontage /ˌfəʊtəʊmɒnˈtɑːʒ/ *n.* **1** the technique of producing a montage (see MONTAGE 2) using photographs. **2** a composite picture produced in this way.

photomultiplier /ˌfəʊtəʊˈmʌltɪˌplaɪə(r)/ *n.* an instrument containing a photocell and a series of electrodes, used to detect and amplify the light from very faint sources.

photon /ˈfəʊtɒn/ *n. Physics* a quantum of electromagnetic radiation energy (e.g. light), proportional to the frequency of radiation. [Gk *phōs phōtos* light, after *electron*]

photonovel /ˈfəʊtəʊˌnɒv(ə)l/ *n.* a novel told in a series of photographs with superimposed speech bubbles.

photo-offset /ˌfəʊtəʊˈɒfset/ *n.* offset printing with plates made photographically.

photoperiod /ˌfəʊtəʊˈpɪərɪəd/ *n. Biol.* the period of daily illumination which an organism receives. □ **photoperiodic** /-ˌpɪərɪˈɒdɪk/ *adj.*

photoperiodism /ˌfəʊtəʊˈpɪərɪəˌdɪz(ə)m/ *n. Biol.* the response of an organism to changes in the lengths of the daily periods of light.

photophobia /ˌfəʊtəʊˈfəʊbɪə/ *n.* an abnormal fear of or aversion to light. □ **photophobic** *adj.*

photorealism /ˌfəʊtəʊˈrɪəlɪz(ə)m/ *n.* detailed and unidealized representation in art, characteristically of the banal, everyday, or sordid aspects of life.

photoreceptor /ˌfəʊtəʊrɪˈseptə(r)/ *n. Biol.* a structure of a living organism which responds to incident light, esp. a cell in which light is converted to a nervous or other signal.

photosensitive /ˌfəʊtəʊˈsensɪtɪv/ *adj.* reacting chemically, electrically, etc., to light. □ **photosensitivity** /-ˌsensɪˈtɪvɪtɪ/ *n.*

photosetting /ˈfəʊtəʊˌsetɪŋ/ *n.* = FILMSETTING. □ **photoset** *v.tr.* (*past* and *past part.* **-set**). **photosetter** *n.*

photosphere /ˈfəʊtəʊˌsfɪə(r)/ *n. Astron.* the luminous envelope of a star from which its light and heat radiate. □ **photospheric** /ˌfəʊtəʊˈsferɪk/ *adj.*

Photostat /ˈfəʊtəʊˌstæt/ *n. & v.* ● *n. propr.* **1** a type of machine for making photocopies. **2** a copy made by this means. ● *v.tr.* (**photostat**) (**-statted**, **-statting**) make a Photostat of. □ **photostatic** /ˌfəʊtəʊˈstætɪk/ *adj.*

photosynthesis /ˌfəʊtəʊˈsɪnθɪsɪs/ *n.* the process by which the energy of sunlight is used by green plants and some unicellular organisms to synthesize carbohydrates from carbon dioxide and water. (*See note below.*) □ **photosynthesize** *v.tr. & intr.* (also **-ise**). **photosynthetic** /-sɪnˈθetɪk/ *adj.* **photosynthetically** *adv.*

▪ Photosynthesis in plants is usually carried out using the green pigment chlorophyll, which has the ability to absorb the radiant energy of the sun and convert it via a complicated metabolic pathway into chemical energy in the form of sugars. It is the only process by which organisms can manufacture carbohydrates, so that almost all life on earth is ultimately dependent on green plants. During the process of photosynthesis carbon dioxide and water are used up and oxygen is generated: higher plants on land and algae in the sea therefore have a major beneficial effect on the constitution of the atmosphere. Some algae and bacteria have red, blue, or brown photosynthetic pigments.

phototransistor /ˌfəʊtəʊtrænˈzɪstə(r), -trɑːnˈzɪstə(r)/ *n. Electronics* a transistor that responds to incident light by generating and amplifying an electric current.

phototropism /ˌfəʊtəʊˈtrəʊpɪz(ə)m, fəˈtɒtrəˌpɪz(ə)m/ *n.* the tendency of a plant etc. to bend or turn towards or away from a source of light. □ **phototropic** /-ˈtrəʊpɪk, -ˈtrɒpɪk/ *adj.*

phototypesetting /ˌfəʊtəʊˈtaɪpˌsetɪŋ/ *n.* = FILMSETTING.

photovoltaic /ˌfəʊtəʊvɒlˈteɪk/ *adj.* relating to the production of electric current at the junction of two substances exposed to light.

phrasal /'freɪz(ə)l/ adj. Gram. consisting of a phrase. □ **phrasal verb** an idiomatic phrase consisting of a verb and an adverb (e.g. *break down*), a verb and a preposition (e.g. *see to*), or a combination of both (e.g. *look down on*).

phrase /freɪz/ n. & v. ● n. **1** a group of words forming a conceptual unit, but not a sentence. **2** an idiomatic or short pithy expression. **3** a manner or mode of expression (*a nice turn of phrase*). **4** Mus. a group of notes forming a distinct unit within a larger piece. ● v.tr. **1** express in words (*phrased the reply badly*). **2** (esp. when reading aloud or speaking) divide (sentences etc.) into units so as to convey the meaning of the whole. **3** Mus. divide (music) into phrases, esp. in performance. □ **phrase book** a book for tourists etc. listing useful expressions with their equivalent in a foreign language. □ **phrasing** n. [earlier *phrasis* f. L f. Gk f. *phrazō* declare, tell]

phraseogram /'freɪzɪəˌɡræm/ n. a written symbol representing a phrase, esp. in shorthand.

phraseology /ˌfreɪzɪ'ɒlədʒɪ/ n. (pl. **-ies**) **1** a choice or arrangement of words. **2** a mode of expression. □ **phraseological** /-zɪə'lɒdʒɪk(ə)l/ adj. [mod.L *phraseologia* f. Gk *phraseōn* genitive pl. of *phrasis* PHRASE]

phreatic /frɪ'ætɪk/ adj. Geol. **1** (of water) situated underground in the zone of saturation; of or relating to groundwater. **2** (of a volcanic eruption or explosion) caused by the heating and expansion of underground water. [Gk *phrear phreatos* well]

phrenic /'frenɪk/ adj. Anat. of or relating to the diaphragm. [F *phrénique* f. Gk *phrēn phrenos* diaphragm, mind]

phrenology /frɪ'nɒlədʒɪ/ n. hist. the study of the shape and size of the cranium as a supposed indication of character and mental faculties. □ **phrenologist** n. **phrenological** /ˌfrenə'lɒdʒɪk(ə)l/ adj.

Phrygia /'frɪdʒɪə/ an ancient region of west central Asia Minor, to the south of Bithynia. Centred on the city of Gordium, it dominated Asia Minor after the decline of the Hittites in the 12th century BC, reaching the peak of its power in the 8th century under King Midas. Conquered by the Cimmerians in about 676 BC, it was eventually absorbed into the kingdom of Lydia in the 6th century BC.

Phrygian /'frɪdʒɪən/ n. & adj. ● n. **1** a native or inhabitant of ancient Phrygia. **2** the language of this people. ● adj. of or relating to Phrygia or its people or language. □ **Phrygian bonnet** (or **cap**) an ancient conical cap with the top bent forwards, now identified with the cap of liberty. **Phrygian mode** Mus. the mode represented by the natural diatonic scale E–E.

phthalic acid /'fθælɪk, 'θæl-/ n. Chem. one of three isomeric dicarboxylic acids derived from benzene. □ **phthalate** /-leɪt/ n. [abbr. of NAPHTHALIC: see NAPHTHALENE]

phthisis /'fθaɪsɪs, 'θaɪ-/ n. Med. any progressive wasting disease, esp. pulmonary tuberculosis. □ **phthisic** adj. **phthisical** adj. [L f. Gk f. *phthinō* to decay]

Phuket /puː'ket/ **1** an island of Thailand, situated at the head of the Strait of Malacca off the west coast of the Malay Peninsula. **2** a port at the south end of Phuket island, a major resort centre and outlet to the Indian Ocean.

phut /fʌt/ n. a dull abrupt sound as of an impact or explosion. □ **go phut** colloq. (esp. of a scheme or plan) collapse, break down. [perh. f. Hindi *phaṭnā* to burst]

phycology /faɪ'kɒlədʒɪ/ n. the study of algae. □ **phycologist** n. **phycological** /ˌfaɪkə'lɒdʒɪk(ə)l/ adj. [Gk *phukos* seaweed + -LOGY]

phycomycete /ˌfaɪkəʊ'maɪsiːt/ n. an often parasitic fungus which typically forms a non-septate mycelium. [Gk *phukos* seaweed + pl. of Gk *mukēs -ētos* fungus]

phyla pl. of PHYLUM.

phylactery /fɪ'læktərɪ/ n. (pl. **-ies**) **1** a small leather box containing Hebrew texts on vellum, worn by Jewish men at morning prayer as a reminder to keep the law. **2** an amulet; a charm. **3** a usu. ostentatious religious observance. **4** a fringe; a border. [ME f. OF f. LL *phylacterium* f. Gk *phulaktērion* amulet f. *phulassō* guard]

phyletic /faɪ'letɪk/ adj. Biol. of or relating to the development of a species or other group. [Gk *phuletikos* f. *phuletēs* tribesman f. *phulē* tribe]

phyllo- /'fɪləʊ/ comb. form leaf. [Gk *phullo-* f. *phullon* leaf]

phyllode /'fɪləʊd/ n. Bot. a flattened leaf-stalk resembling a leaf. [mod.L *phyllodium* f. Gk *phullōdēs* leaflike (as PHYLLO-)]

phyllophagous /fɪ'lɒfəɡəs/ adj. feeding on leaves.

phylloquinone /ˌfɪləʊ'kwɪnəʊn, -kwɪ'nəʊn/ n. one of the K vitamins, found in cabbage, spinach, and other leafy green vegetables, and essential for the blood-clotting process. Also called *vitamin K₁*.

phyllotaxis /ˌfɪləʊ'tæksɪs/ n. (also **phyllotaxy** /-'tæksɪ/) Bot. the arrangement of leaves on an axis or stem. □ **phyllotactic** adj.

phylloxera /ˌfɪlɒk'sɪərə, fɪ'lɒksə-/ n. a homopteran plant bug of (or formerly of) the genus *Phylloxera*, esp. *Daktulosphaira vitifoliae*, a pest of vines. [mod.L f. Gk *phullon* leaf + *xēros* dry]

phylogenesis /ˌfaɪləʊ'dʒenɪsɪs/ n. (also **phylogeny** /faɪ'lɒdʒɪnɪ/) Biol. the evolutionary development and diversification of groups of organisms, or particular features of organisms. □ **phylogenetic** /-dʒɪ'netɪk/ adj. **phylogenic** /-'dʒenɪk/ adj. [Gk *phulon. phutē* tribe, race + -O- + GENESIS]

phylum /'faɪləm/ n. (pl. **phyla** /-lə/) **1** Biol. a taxonomic rank below kingdom comprising a class or classes and subordinate taxa. **2** Linguistics a group of languages related to each other less closely than those of a family. [mod.L f. Gk *phulon* race]

physalis /'faɪsəlɪs/ n. a solanaceous plant of the genus *Physalis*, bearing fruit surrounded by lantern-like calyxes, e.g. Cape gooseberry and Chinese lantern. [Gk *physallis* bladder, with ref. to the inflated calyx]

physic /'fɪzɪk/ n. & v. archaic ● n. **1** a medicine (*a dose of physic*). **2** the art of healing. **3** the medical profession. ● v.tr. (**physicked, physicking**) dose with physic. □ **physic garden** a garden for cultivating medicinal herbs etc. [ME f. OF *fisique* medicine f. L *physica* f. Gk *physikē* (*epistēmē*) (knowledge) of nature]

physical /'fɪzɪk(ə)l/ adj. & n. ● adj. **1** of or concerning the body (*physical exercise; physical education*). **2** of matter; material (*both mental and physical force*). **3 a** of, or according to, the laws of nature (*a physical impossibility*). **b** belonging to physics (*physical science*). ● n. (in full **physical examination**) a medical examination to determine physical fitness. □ **physical chemistry** the application of physics to the study of chemical behaviour. **physical geography** geography dealing with natural features. **physical jerks** colloq. physical exercises. **physical science** the sciences used in the study of inanimate natural objects, e.g. physics, chemistry, astronomy, etc. **physical training** exercises promoting bodily fitness and strength. □ **physically** adv. **physicalness** n. **physicality** /ˌfɪzɪ'kælɪtɪ/ n. [ME f. med.L *physicalis* f. L *physica* (as PHYSIC)]

physician /fɪ'zɪʃ(ə)n/ n. **1 a** a person legally qualified to practise medicine, esp. non-surgical medicine; a specialist in medical diagnosis and treatment. **b** any medical practitioner. **2** a healer (*work is the best physician*). [ME f. OF *fisicien* (as PHYSIC)]

physicist /'fɪzɪsɪst/ n. a person skilled or qualified in physics.

physico- /'fɪzɪkəʊ/ comb. form **1** physical (and). **2** of physics (and). [Gk *phusikos* (as PHYSIC)]

physico-chemical /ˌfɪzɪkəʊ'kemɪk(ə)l/ adj. relating to physics and chemistry or to physical chemistry.

physics /'fɪzɪks/ n. the science dealing with the properties and interactions of matter and energy. (See note below.) [pl. of *physic* physical (thing), after L *physica*, Gk *phusika* natural things f. *phusis* nature]

■ The subject-matter of physics has undergone marked changes throughout the centuries. For Aristotle it included the study of all natural phenomena, but excluded quantitative subjects such as mechanics, music, and optics which were regarded as dealing with artefacts. By the 17th century astronomy, mechanics, and optics were being treated mathematically, but a qualitative experimental physics remained, dealing especially with so-called 'imponderable fluids' such as heat, electricity, and magnetism. During the 18th century physics concentrated increasingly on the more fundamental and general properties of matter, other phenomena being dealt with under the disciplines of chemistry, biology, etc. All branches of physics were treated mathematically during the 19th century and by 1900 the structure of classical physics, describing the properties of matter on the macroscopic scale and at relatively low velocities, was largely complete. Since that time, however, the theories of modern physics, and in particular relativity, quantum theory, and the physics of atomic and subatomic particles, have transformed our overall understanding of the universe.

physio /'fɪzɪəʊ/ n. (pl. **-os**) colloq. a physiotherapist. [abbr.]

physio- /'fɪzɪəʊ/ comb. form nature; what is natural. [Gk *phusis* nature]

physiocracy /ˌfɪzɪ'ɒkrəsɪ/ n. (pl. **-ies**) government according to the natural order, esp. as advocated by the French physiocrats; a society based on this. [F *physiocratie* (as PHYSIO-, -CRACY)]

physiocrat /'fɪzɪəˌkræt/ n. a member of an 18th-century group of French economists founded by Louis XV's physician François Quesnay (1694–1774) and regarded as constituting the first scientific school of economics. In opposition to the theories of mercantilism, the physiocrats held that agriculture, rather than manufacturing or trade, was the source of all wealth and that agricultural products should be highly priced. Advocating adherence to physiocracy, a supposed natural order of social institutions, they stressed the necessity of free trade and coined the term *laissez-faire*. □ **physiocratic** /ˌfɪzɪə'krætɪk/ *adj.* [f. as PHYSIOCRACY]

physiognomy /ˌfɪzɪ'ɒnəmɪ/ *n.* (*pl.* **-ies**) **1 a** the cast or form of a person's features, expression, body, etc. **b** the art of supposedly judging character from facial characteristics etc. **2** the external features of a landscape etc. **3** a characteristic, esp. moral, aspect. □ **physiognomist** *n.* **physiognomic** /-zɪə'nɒmɪk/ *adj.* **physiognomical** *adj.* **physiognomically** *adv.* [ME *fisnomie* etc. f. OF *phisonomie* f. med.L *phisonomia* f. Gk *phusiognōmonia* judging of a man's nature (by his features) (as PHYSIO-, *gnōmōn* judge)]

physiography /ˌfɪzɪ'ɒɡrəfɪ/ *n.* the description of nature, of natural phenomena, or of a class of objects; physical geography. □ **physiographer** *n.* **physiographic** /-zɪə'ɡræfɪk/ *adj.* **physiographical** *adj.* **physiographically** *adv.* [F *physiographie* (as PHYSIO-, -GRAPHY)]

physiological /ˌfɪzɪə'lɒdʒɪk(ə)l/ *adj.* (also **physiologic**) of or concerning physiology. □ **physiological saline** a solution, usu. mixed with salt, that is isotonic with body fluids. □ **physiologically** *adv.*

physiology /ˌfɪzɪ'ɒlədʒɪ/ *n.* **1** the science of the functions of living organisms and their parts. **2** these functions. □ **physiologist** *n.* [F *physiologie* or L *physiologia* f. Gk *phusiologia* (as PHYSIO-, -LOGY)]

physiotherapy /ˌfɪzɪəʊ'θerəpɪ/ *n.* the treatment of disease, injury, deformity, etc., by physical methods including manipulation, massage, infrared heat treatment, remedial exercise, etc., rather than by drugs. □ **physiotherapist** *n.*

physique /fɪ'ziːk/ *n.* the bodily structure, development, and organization of an individual (*an undernourished physique*). [F, orig. adj. (as PHYSIC)]

-phyte /faɪt/ *comb. form* forming nouns denoting a vegetable or plantlike organism (*saprophyte*; *zoophyte*). □ **-phytic** /'fɪtɪk/ *comb. form* forming adjectives. [Gk *phuton* plant f. *phuō* come into being]

phyto- /'faɪtəʊ/ *comb. form* denoting a plant.

phytochemistry /ˌfaɪtəʊ'kemɪstrɪ/ *n.* the chemistry of plant products. □ **phytochemical** *adj.* **phytochemist** *n.*

phytochrome /'faɪtəʊˌkrəʊm/ *n. Biochem.* a blue-green pigment found in many plants, and regulating various developmental processes according to the nature and timing of the light it absorbs. [PHYTO- + Gk *khrōma* colour]

phytogenesis /ˌfaɪtəʊ'dʒenɪsɪs/ *n.* (also **phytogeny** /faɪ'tɒdʒɪnɪ/) the science of the origin or evolution of plants.

phytogeography /ˌfaɪtəʊdʒɪ'ɒɡrəfɪ/ *n.* the geographical distribution of plants.

phytopathology /ˌfaɪtəʊpə'θɒlədʒɪ/ *n.* the study of plant diseases.

phytophagous /faɪ'tɒfəɡəs/ *adj.* feeding on plants.

phytoplankton /ˌfaɪtəʊ'plæŋktən/ *n.* plankton consisting of plants.

phytotomy /faɪ'tɒtəmɪ/ *n.* the dissection of plants.

phytotoxic /ˌfaɪtəʊ'tɒksɪk/ *adj.* poisonous to plants.

phytotoxin /ˌfaɪtəʊ'tɒksɪn/ *n.* **1** any toxin derived from a plant. **2** a substance poisonous or injurious to plants, esp. one produced by a parasite.

pi¹ /paɪ/ *n.* **1** the sixteenth letter of the Greek alphabet (Π, π). **2** (as π) the symbol of the ratio of the circumference of a circle to its diameter (approx. 3.14159). □ **pi-meson** = PION. [Gk: sense 2 f. Gk *periphereia* circumference]

pi² /paɪ/ *adj. Brit. sl.* pious. □ **pi jaw** a long moralizing lecture or reprimand. [abbr.]

pi³ *US* var. of PIE³.

piacular /paɪ'ækjʊlə(r)/ *adj. formal* **1** expiatory. **2** needing expiation. [L *piacularis* f. *piaculum* expiation f. *piare* appease]

Piaf /'piːæf/, Edith (born Edith Giovanna Gassion) (1915–63), French singer. She acquired her name in 1935, when a cabaret impresario called her *la môme piaf* (= little sparrow), referring to her small size. She became known as a cabaret and music-hall singer in the late 1930s, touring Europe and America in the 1940s. She is especially remembered for her defiant and nostalgic songs, some of which she wrote herself, including 'La Vie en rose'. Other songs include 'Je ne regrette rien'.

piaffe /pɪ'æf/ *v.intr.* (of a horse etc.) move as in a trot, but slower. [F *piaffer* to strut]

piaffer /pɪ'æfə(r)/ *n.* the action of piaffing.

Piaget /pɪ'æʒeɪ/, Jean (1897–1980), Swiss psychologist. Piaget's work provided the single biggest impact on the study of the development of thought processes. His central thesis is that children initially lack intellectual and logical abilities, which they acquire through experience and interaction with the world around them. They then proceed through a series of fixed stages of cognitive development, each being a prerequisite for the next.

pia mater /ˌpaɪə 'meɪtə(r)/ *n. Anat.* the delicate innermost membrane enveloping the brain and spinal cord (see MENINX). [med.L, = tender mother, transl. Arab. *al-'umm al-raḳīḳa*: cf. DURA MATER]

piani *pl.* of PIANO².

pianism /'piːəˌnɪz(ə)m/ *n.* **1** the art or technique of piano-playing. **2** the skill or style of a composer of piano music. □ **pianistic** /ˌpiːə'nɪstɪk/ *adj.* **pianistically** *adv.*

pianissimo /pɪə'nɪsɪˌməʊ/ *adj., adv., & n. Mus.* ● *adj.* performed very softly. ● *adv.* very softly. ● *n.* (*pl.* **-os** or **pianissimi** /-mɪ/) a passage to be performed very softly. [It., superl. of PIANO²]

pianist /'pɪənɪst/ *n.* the player of a piano. [F *pianiste* (as PIANO¹)]

piano¹ /pɪ'ænəʊ/ *n.* (*pl.* **-os**) **1** a large keyboard musical instrument with a wooden case enclosing a soundboard and metal strings, which are struck by hammers when the keys are depressed, and whose vibration is stopped by dampers when the keys are released and can be regulated for length and volume by two or three pedals. (*See note below.*) **2** a smaller instrument producing similar sounds electronically. □ **piano-accordion** an accordion with the melody played on a small vertical keyboard like that of a piano. **piano organ** a mechanical piano constructed like a barrel-organ. **piano-player 1** a pianist. **2** a contrivance for playing a piano automatically. [It., abbr. of PIANOFORTE]

▪ The piano was developed in Italy in the early 18th century. It superseded the clavichord and harpsichord, having a louder tone than the former and a greater expressive range than the latter. The earliest models were of harpsichord shape, but by about 1860 this was largely replaced by the upright piano. The upright piano, with strings running perpendicularly rather than horizontally, is today more common than the larger, less compact, grand piano, in which the strings run horizontally.

piano² /'pjɑːnəʊ/ *adj., adv., & n.* ● *adj.* **1** *Mus.* performed softly. **2** subdued. ● *adv.* **1** *Mus.* softly. **2** in a subdued manner. ● *n.* (*pl.* **-os** or **piani** /-nɪ/) *Mus.* a passage to be performed softly. [It. f. L *planus* flat, (of sound) soft]

pianoforte /pɪˌænəʊ'fɔːtɪ/ *n. Mus. formal* or *archaic* a piano. [It., earlier *piano e forte* soft and loud, expressing its gradation of tone]

Pianola /pɪə'nəʊlə/ *n.* **1** *propr.* a kind of automatic piano; a player-piano. **2** (**pianola**) *Bridge* an easy hand needing no skill. **3** (**pianola**) an easy task. [app. dimin. of PIANO¹]

piano nobile /ˌpjɑːnəʊ 'nəʊbɪˌleɪ/ *n. Archit.* the main storey of a large house. [It., = noble floor]

piassava /ˌpiːə'sɑːvə/ *n.* **1** a stout fibre obtained from the leaf-stalks of various American and African palm trees. **2** any of these trees. [Port. f. Tupi *piaçába*]

piastre /pɪ'æstə(r)/ *n.* (*US* **piaster**) a small coin and monetary unit of several Middle Eastern countries. [F *piastre* f. It. *piastra* (*d'argento*) plate (of silver), formed as PLASTER]

Piauí /pjaʊ'iː/ a state of NE Brazil, on the Atlantic coast; capital, Teresina.

piazza /pɪ'ætsə/ *n.* **1** a public square or market-place esp. in an Italian town. **2** *US* the veranda of a house. [It., formed as PLACE]

pibroch /'piːbrɒx, -brɒk/ *n.* a series of variations on a theme for the bagpipes, esp. of a martial or funerary character. [Gael. *piobaireachd* art of piping f. *piobair* piper f. *piob* f. Engl. PIPE]

pic /pɪk/ *n. colloq.* a picture, esp. a cinema film. [abbr.]

pica¹ /'paɪkə/ *n. Printing* **1** a unit of type-size (¹/₆ inch). **2** a size of letters in typewriting (10 per inch). [AL *pica* 15th-c. book of rules about church feasts, perh. formed as PIE²]

pica² /'paɪkə/ *n. Med.* the eating of substances other than normal food, such as earth or stones. [mod.L or med.L, = magpie]

picador /'pɪkəˌdɔː(r)/ *n.* a person mounted on horseback who goads the bull with a lance in a bullfight. [Sp. f. *picar* prick]

Picardy /'pɪkədɪ/ (French **Picardie** /pikardi/) a region and former province of northern France, centred on the city of Amiens. It was the scene of heavy fighting in the First World War.

picaresque /ˌpɪkəˈresk/ adj. (of a style of fiction) dealing with the episodic adventures of rogues etc. [F f. Sp. picaresco f. pícaro rogue]

picaroon /ˌpɪkəˈruːn/ n. archaic **1 a** a rogue. **b** a thief. **2 a** a pirate. **b** a pirate ship. [Sp. picarón (as PICARESQUE)]

Picasso /pɪˈkæsəʊ/, Pablo (1881–1973), Spanish painter, sculptor, and graphic artist, resident in France from 1904. His prolific inventiveness and technical versatility assured his position as the dominant figure in avant-garde art in the first half of the 20th century. The paintings of Picasso's Blue Period (1901–4) used melancholy blue tones to depict social outsiders. These gave way to his Rose Period (1905–6), in which circus performers were represented in pinks and greys. Les Demoiselles d'Avignon (1907), with its rejection of naturalism and focus on the analysis of form, signalled the emergence of cubism, which he developed with Georges Braque and others from 1908 to 1914. The 1920s and 1930s saw the evolution of a neoclassical figurative style, designs for Diaghilev's Ballets Russes, and the evolution of semi-surrealist paintings using increasingly violent imagery, notably The Three Dancers (1935) and Guernica (1937), his response to the destruction of the Basque capital by German bombers.

picayune /ˌpɪkəˈjuːn/ n. & adj. N. Amer. ● n. **1** a small coin of little value, esp. a 5-cent piece. **2** colloq. an insignificant person or thing. ● adj. colloq. mean; contemptible; petty. [F picaillon Piedmontese coin, cash, f. Prov. picaioun, of unkn. orig.]

Piccadilly /ˌpɪkəˈdɪlɪ/ a street in central London, extending from Hyde Park eastwards to Piccadilly Circus. It is noted for its fashionable shops, hotels, and restaurants. The name was originally applied to a house, Pickadilly Hall, thought to be named from the obsolete word 'piccadil' (= a decorative border inserted on the edge of an article of dress), either because piccadils were manufactured there or because houses built there were on the outskirts of the developed area in the 16th century.

piccalilli /ˌpɪkəˈlɪlɪ/ n. (pl. **piccalillis**) a pickle of chopped vegetables, mustard, and hot spices. [18th c.: perh. f. PICKLE + CHILLI]

piccaninny /ˌpɪkəˈnɪnɪ/ n. & adj. (US **pickaninny**) ● n. (pl. **-ies**) often offens. a small black or Australian Aboriginal child. ● adj. archaic very small. [West Indian Negro f. Sp. pequeño or Port. pequeno little]

piccolo /ˈpɪkələʊ/ n. (pl. **-os**) **1** a small flute sounding an octave higher than the ordinary one. **2** a player of this instrument. [It., = small (flute)]

pichiciago /ˌpɪtʃɪsɪˈeɪgəʊ/ n. (pl. **-os**) a fairy armadillo, esp. Chlamyphorus truncatus. [Sp. pichiciego perh. f. Guarani pichey armadillo + Sp. ciego blind f. L caecus]

pick[1] /pɪk/ v. & n. ● v.tr. **1** (also absol.) choose carefully from a number of alternatives (picked the pink one; picked a team; picked the right moment to intervene). **2** detach or pluck (a flower, fruit, etc.) from a stem, tree, etc. **3 a** probe (the teeth, nose, ears, a pimple, etc.) with the finger, an instrument, etc. to remove unwanted matter. **b** clear (a bone, carcass, etc.) of scraps of meat etc. **4** (also absol.) (of a person) eat (food, a meal, etc.) in small bits; nibble without appetite. **5** (also absol.) esp. N. Amer. pluck the strings of (a banjo etc.). **6** remove stalks etc. from (esp. soft fruit) before cooking. **7 a** select (a route or path) carefully over difficult terrain on foot. **b** place (one's steps etc.) carefully. **8** pull apart (pick oakum). **9** (of a bird) take up (grains etc.) in the beak. ● n. **1** the act or an instance of picking. **2 a** a selection or choice. **b** the right to select (had first pick of the prizes). **3** (usu. foll. by of) the best (the pick of the bunch). □ **pick and choose** select carefully or fastidiously. **pick at 1** eat (food) without interest; nibble. **2** = pick on 1. **pick a person's brains** extract ideas, information, etc., from a person for one's own use. **pick holes** (or **a hole**) **in 1** make holes in (material etc.) by plucking, poking, etc. **2** find fault with (an idea etc.). **pick a lock** open a lock with an instrument other than the proper key, esp. with intent to steal. **pick-me-up 1** a tonic for the nerves etc. **2** a good experience, good news, etc. that cheers. **pick off 1** pluck (leaves etc.) off. **2** shoot (people etc.) one by one without haste. **3** eliminate (opposition etc.) singly. **pick on 1** find fault with; nag at. **2** select. **pick out 1** take from a larger number (picked him out from the others). **2** distinguish from surrounding objects or at a distance (can just pick out the church spire). **3** play (a tune) by ear on the piano etc. **4** (often foll. by in, with) **a** highlight (a painting etc.) with touches of another colour. **b** accentuate (decoration, a painting, etc.) with a contrasting colour (picked out the handles in red). **5** make out (the meaning of a passage etc.). **pick over** select the best from. **pick a person's pockets** steal the contents of a person's pockets. **pick a quarrel** start an argument or a fight deliberately. **pick to pieces** = take to pieces (see PIECE). **pick up 1** grasp and raise (from the ground etc.) (picked up his hat). **2** gain or acquire by

chance or without effort (picked up a cold; picked up French easily). **3 a** fetch (a person, animal, or thing) left in another person's charge. **b** stop for and take along with one, esp. in a vehicle (pick me up on the corner). **4** make the acquaintance of (a person) casually, esp. as a sexual overture. **5** (of one's health, the weather, share prices, etc.) recover, prosper, improve. **6** (of a motor engine etc.) recover speed; accelerate. **7** (of the police etc.) take into charge; arrest. **8** detect by scrutiny or with a telescope, searchlight, radio, etc. (picked up most of the mistakes; picked up a distress signal). **9 a** (often foll. by with) form or renew a friendship. **b** resume, take up anew (pick up where we left off). **10** (esp. in phr. **pick up the tab**) accept the responsibility of paying (a bill etc.). **11** (refl.) raise (oneself etc.) after a fall etc. **12** raise (the feet etc.) clear of the ground. **13** Golf pick up one's ball, esp. when conceding a hole. **pick-up 1** (in full **pick-up truck**) a small open motor truck. **2** a device that produces an electrical signal in response to some other kind of signal or charge, esp.: **a** the part of a record-player carrying the stylus. **b** a device on a musical instrument which converts sound vibrations into electrical signals for amplification. **3** sl. a person met casually, esp. for sexual purposes. **4** the act of picking up, esp. of giving a person a lift. **5** an increase in, or recovery of, speed or prosperity. **6** Mus. a series of introductory notes leading into the opening part of a tune. **7** Fishing a semicircular loop of metal for guiding a fishing-line back on to the spool as it is reeled in. **pick-your-own** (usu. attrib.) (of commercially grown fruit and vegetables) dug or picked by the customer at the place of production. **take one's pick** make a choice. □ **pickable** adj. [ME, earlier pike, of unkn. orig.]

pick[2] /pɪk/ n. & v. ● n. **1** a long-handled tool having a usu. curved iron bar pointed at one or both ends, used for breaking up hard ground, masonry, etc. **2** colloq. a plectrum. **3** any instrument for picking, such as a toothpick. ● v.tr. **1** break the surface of (the ground etc.) with or as if with a pick. **2** make (holes etc.) in this way. [ME, app. var. of PIKE[2]]

pickaback var. of PIGGYBACK.

pickaninny US var. of PICCANINNY.

pickaxe /ˈpɪkæks/ n. & v. (US **pickax**) ● n. = PICK[2] n. 1. ● v. **1** tr. break (the ground etc.) with a pickaxe. **2** intr. work with a pickaxe. [ME pikois f. OF picois, rel. to PIKE[2]: assim. to AXE]

pickelhaube /ˈpɪk(ə)lˌhaʊbə/ n. hist. a German soldier's spiked helmet. [G]

picker /ˈpɪkə(r)/ n. **1** a person or thing that picks. **2** (often in comb.) a person who gathers or collects (hop-picker; rag-picker).

pickerel /ˈpɪkərəl/ n. (pl. same or **pickerels**) a young pike. [ME, dimin. of PIKE[1]]

Pickering /ˈpɪkərɪŋ/, William Hayward (b.1910), New Zealand-born American engineer. He spent most of his career at the California Institute of Technology at Pasadena, becoming director of the Jet Propulsion Laboratory there in 1954. Pickering carried out early work on the telemetry, guidance, and communications systems of rockets, and went on to develop America's first satellite, Explorer I, which was launched in 1958. Several unmanned probes to the moon and planets were launched by the JPL during his directorship, notably the Ranger, Surveyor, and Mariner missions.

picket /ˈpɪkɪt/ n. & v. ● n. **1** a person or group of people outside a place of work, intending to persuade esp. workers not to enter during a strike etc. **2** a pointed stake or peg driven into the ground to form a fence or palisade, to tether a horse, etc. **3** (also **picquet**, **piquet**) Mil. **a** a small body of troops or a single soldier sent out to watch for the enemy, held in readiness, etc. **b** a party of sentries. **c** an outpost. **d** a camp-guard on police duty in a garrison town etc. ● v. (**picketed**, **picketing**) **1** tr. & intr. station or act as a picket. **b** tr. beset or guard (a factory, workers, etc.) with a picket or pickets. **2** tr. secure with stakes. **3** tr. tether (an animal). □ **picket line** a boundary established by workers on strike, esp. at the entrance to the place of work, which others are asked not to cross. □ **picketer** n. [F piquet pointed stake f. piquer prick, f. pic PICK[2]]

Pickford /ˈpɪkfəd/, Mary (born Gladys Mary Smith) (1893–1979), Canadian-born American actress. She was a star of silent films, usually playing the innocent young heroine, as in Rebecca of Sunnybrook Farm (1917) and Pollyanna (1920). In 1919 she co-founded the film production company United Artists; she was married to one of the other founders, Douglas Fairbanks, between 1919 and 1936.

pickings /ˈpɪkɪŋz/ n.pl. **1** perquisites; pilferings (rich pickings). **2** remaining scraps; gleanings.

pickle /ˈpɪk(ə)l/ n. & v. ● n. **1 a** (often in pl.) food, esp. vegetables, preserved in brine, vinegar, mustard, etc. and used as a relish. **b** the brine, vinegar, etc. in which food is preserved. **2** colloq. a plight (a fine *pickle we are in!*). **3** *Brit. colloq.* a mischievous child. **4** an acid solution for cleaning metal etc. ● *v.tr.* **1** preserve in pickle. **2** treat with pickle. **3** (as **pickled** *adj.*) *sl.* drunk. [ME *pekille, pykyl*, f. MDu., MLG *pekel*, of unkn. orig.]

pickler /ˈpɪklə(r)/ *n.* **1** a person who pickles vegetables etc. **2** a vegetable suitable for pickling.

picklock /ˈpɪklɒk/ *n.* **1** a person who picks locks. **2** an instrument for this.

pickpocket /ˈpɪkˌpɒkɪt/ *n.* a person who steals from the pockets of others.

Pickwickian /pɪkˈwɪkɪən/ *adj.* **1** of or like Mr Pickwick in Dickens's *Pickwick Papers*, esp. in being jovial, plump, etc. **2** (of words or their sense) misunderstood or misused, esp. to avoid offence.

picky /ˈpɪkɪ/ *adj.* (**pickier, pickiest**) *colloq.* excessively fastidious; choosy. □ **pickiness** *n.*

picnic /ˈpɪknɪk/ *n. & v.* ● *n.* **1** an outing or excursion taking a packed meal to be eaten out of doors. **2** any meal eaten out of doors or without preparation, tables, chairs, etc. **3** (usu. with *neg.*) *colloq.* something agreeable or easily accomplished etc. (*it was no picnic organizing the meeting*). ● *v.intr.* (**picnicked, picnicking**) take part in a picnic. □ **picnicker** *n.* **picnicky** *adj. colloq.* [F *pique-nique*, of unkn. orig.]

pico- /ˈpaɪkəʊ, ˈpiːk-/ *comb. form* denoting a factor of 10^{-12} (*picometre, picosecond*). [Sp. *pico* beak, peak, little bit]

Pico da Neblina /ˌpiːkuː dɑː neˈbliːnə/ a mountain in NW Brazil, close to the border with Venezuela. Rising to 3,014 m (9,888 ft), it is the highest peak in Brazil.

Pico de Orizaba /ˌpiːkəʊ deɪ ˌɒrɪˈzɑːbə/ the Spanish name for CITLALTÉPETL.

picot /ˈpiːkəʊ/ *n.* a small loop of twisted thread in a lace edging etc. [F, dimin. of *pic* peak, point]

picotee /ˌpɪkəˈtiː/ *n.* a type of carnation of which the flowers have a light ground and dark-edged petals. [F *picoté -ée* past part. of *picoter* prick (as PICOT)]

picquet var. of PICKET *n.* 3.

picric acid /ˈpɪkrɪk/ *n. Chem.* a very bitter yellow compound used in dyeing and surgery and in explosives. □ **picrate** /-reɪt/ *n.* [Gk *pikros* bitter]

Pict /pɪkt/ *n.* a member of an ancient people, of disputed but probably pre-Celtic origin, who formerly inhabited parts of northern Britain. They first appear AD *c*.300, in Roman writings which apply the term *Picti* to the hostile tribes occupying the area north of the Antonine Wall. According to chroniclers the Pictish kingdom was united with that of the southern Scots under Kenneth MacAlpin (Kenneth I) in about 844, and the name of the Picts as a distinct people gradually disappeared. □ **Pictish** *adj.* [ME f. LL *Picti* perh. f. *pingere pict-* paint, tattoo, or perh. assim. to native name]

pictograph /ˈpɪktəˌɡrɑːf/ *n.* (also **pictogram** /-ˌɡræm/) **1 a** a pictorial symbol for a word or phrase. (*See note below*.) **b** an ancient record consisting of these. **2** a pictorial representation of statistics etc. on a chart, graph, etc. □ **pictographic** /ˌpɪktəˈɡræfɪk/ *adj.* **pictography** /pɪkˈtɒɡrəfɪ/ *n.* [L *pingere pict-* paint]

▪ Pictographs constitute the oldest form of writing known, examples having been discovered in Egypt and Mesopotamia from before 3000 BC; they were used widely in Minoan Crete, Meso-America, China, and many other places around the world, with no evidence of a common origin. The cuneiform and hieroglyphic systems developed from pictographs, but acquired additional symbols signifying sounds. A modern application of pictographs is in road signs and similar notices. (See also WRITING.)

pictorial /pɪkˈtɔːrɪəl/ *adj. & n.* ● *adj.* **1** of or expressed in a picture or pictures. **2** illustrated. **3** picturesque. ● *n.* a journal, postage stamp, etc., with a picture or pictures as the main feature. □ **pictorially** *adv.* [LL *pictorius* f. L *pictor* painter (as PICTURE)]

picture /ˈpɪktʃə(r)/ *n. & v.* ● *n.* **1 a** (often *attrib.*) a painting, drawing, photograph, etc., esp. as a work of art (*picture frame*). **b** a portrait, esp. a photograph, of a person (*does not like to have her picture taken*). **c** a beautiful object (*her hat is a picture*). **2 a** a total visual or mental impression produced; a scene (*the picture looks bleak*). **b** a written or spoken description (*drew a vivid picture of moral decay*). **3 a** a film. **b** (in *pl.*; prec. by *the*) *Brit.* a showing of films at a cinema (*went to the pictures*). **c** (in *pl.*) films in general. **4** an image on a television screen. **5 a** esp. *iron.* a person or thing exemplifying something (*he was the picture of innocence*). **b** a person or thing resembling another closely (*the picture of her aunt*). **c** *iron.* a striking expression, pose, etc.; a comic or striking

sight (*her face was a picture*). ● *v.tr.* **1** represent in a picture. **2** (also *refl.*; often foll. by *to*) imagine, esp. visually or vividly (*pictured it to herself*). **3** describe graphically. □ **get the picture** *colloq.* grasp the tendency or drift of circumstances, information, etc. **in the picture** fully informed or noticed. **out of the picture** uninvolved, inactive; irrelevant. **picture-book** a book containing many illustrations. **picture-card** a court-card. **picture-gallery** a place containing an exhibition or collection of pictures. **picture-goer** a person who frequents the cinema. **picture hat** a woman's wide-brimmed highly decorated hat as in pictures by Reynolds and Gainsborough. **picture-moulding 1** woodwork etc. used for framing pictures. **2** a rail on a wall used for hanging pictures from. **picture palace** *Brit. archaic* a cinema. **picture postcard** a postcard with a picture on one side. **picture theatre** see THEATRE 1b. **picture window** a very large window consisting of one pane of glass, usu. facing an attractive view. **picture-writing** a mode of recording events etc. by pictorial symbols as in early hieroglyphics etc. [ME f. L *pictura* f. *pingere pict-* paint]

picturesque /ˌpɪktʃəˈresk/ *adj.* **1** (of landscape etc.) beautiful or striking, as in a picture. **2** (of language etc.) strikingly graphic; vivid. □ **picturesquely** *adv.* **picturesqueness** *n.* [F *pittoresque* f. It. *pittoresco* f. *pittore* painter f. L (as PICTORIAL): assim. to PICTURE]

piddle /ˈpɪd(ə)l/ *v. & n.* ● *v.intr.* **1** *colloq.* urinate (used esp. to or by children). **2** work or act in a trifling way. **3** (as **piddling** *adj.*) *colloq.* trivial; trifling. ● *n. colloq.* (used esp. to or by children) **1** an act of urinating. **2** urine. □ **piddler** *n.* [prob. f. PISS + PUDDLE: senses 2 & 3 of *v.* perh. f. PEDDLE]

piddock /ˈpɪdək/ *n.* a rock-boring bivalve mollusc of the family Pholadidae, with a shell much shorter than the body, used for bait; esp. the common *Pholas dactylus*. [18th c.: orig. unkn.]

pidgin /ˈpɪdʒɪn/ *n.* a simplified language containing vocabulary from two or more languages, used for communication between people not having a common language. □ **pidgin English** a pidgin in which the chief language is English, used orig. between Chinese and Europeans. [corrupt. of *business*]

pi-dog var. of PYE-DOG.

pie[1] /paɪ/ *n.* **1** a baked dish of meat, fish, fruit, etc., usu. with a top and base of pastry. **2** anything resembling a pie in form (*a mud pie*). □ **easy as pie** *colloq.* very easy. **pie chart** a circle divided into sectors to represent relative quantities. **pie-eater** *Austral. sl.* a person of little account. **pie-eyed** *sl.* drunk. **pie in the sky** *colloq.* an unrealistic prospect of future happiness after present suffering; a misleading promise. [ME, perh. = PIE[2] f. miscellaneous contents compared to objects collected by a magpie]

pie[2] /paɪ/ *n. archaic* **1** a magpie. **2** a pied animal. [ME f. OF f. L *pica*]

pie[3] /paɪ/ *n. & v.* (*US* **pi**) ● *n.* **1** a confused mass of printers' type. **2** chaos. ● *v.tr.* (**pieing**) muddle up (type). [perh. transl. of F PÂTÉ = PIE[1]]

pie[4] /paɪ/ *n. hist.* a former monetary unit of India equal to one-twelfth of an anna. [Hind. etc. *pāˈī* f. Skr. *pad, padī* quarter]

piebald /ˈpaɪbɔːld/ *adj. & n.* ● *adj.* **1** (usu. of an animal, esp. a horse) having irregular patches of two colours, esp. black and white. **2** motley; mongrel. ● *n.* a piebald animal, esp. a horse.

piece /piːs/ *n. & v.* ● *n.* **1 a** (often foll. by *of*) one of the distinct portions forming part of or broken off from a larger object; a bit; a part (*a piece of string*). **b** each of the parts of which a set or category is composed (*a five-piece band; a piece of furniture*). **2** a coin of specified value (*50p piece*). **3 a** a usu. short literary or musical composition or a picture. **b** a theatrical play. **4** an item, instance, or example (*a piece of impudence; a piece of news*). **5 a** any of the objects used to make moves in board-games. **b** a chessman (strictly, other than a pawn). **6** a definite quantity in which a thing is sold. **7** (often foll. by *of*) an enclosed portion (of land etc.). **8** *sl. derog.* a woman. **9** (foll. by *of*) *colloq.* a share in, involvement in; a financial share or investment in (*has a piece of the new production*). **10** esp. *N. Amer. sl.* a portable firearm; a handgun. ● *v.tr.* **1** (usu. foll. by *together*) form into a whole; put together; join (*finally pieced his story together*). **2** (usu. foll. by *out*) **a** eke out. **b** form (a theory etc.) by combining parts etc. **3** (usu. foll. by *up*) patch. **4** join (threads) in spinning. □ **break to pieces** break into fragments. **by the piece** (paid) according to the quantity of work done. **go to pieces** collapse emotionally; suffer a breakdown. **in one piece 1** unbroken. **2** unharmed. **in pieces** broken. **of a piece** (often foll. by *with*) uniform, consistent, in keeping. **piece-goods** fabrics, esp. Lancashire cottons, woven in standard lengths. **a piece of the action** *colloq.* a share of the profits; a share in the excitement. **piece of arse** (*US* **ass**) *coarse sl. derog.* a woman regarded as an object of sexual gratification. **a piece of cake** see CAKE. **piece of eight** *hist.* a Spanish dollar, equivalent to 8 reals. **a piece of one's mind** a sharp rebuke or lecture. **piece of water** a small lake etc. **piece of work 1** a thing made by working. **2** a person of a specified kind (*a nasty piece of work*). **piece-rates** a rate paid according to the amount produced. **say one's piece** give one's opinion or make a prepared statement. **take to pieces 1** break up or dismantle. **2** criticize harshly. □ **piecer** *n.* (in sense 4 of *v.*). [ME f. AF *pece*, OF *piece* f. Rmc, prob. of Gaulish orig.]

pièce de résistance /ˌpjes də reɪˈzɪstɒns/ *n.* (*pl.* **pièces de résistance** *pronunc.* same) **1** the most important or remarkable item. **2** the most substantial dish at a meal. [F]

piecemeal /ˈpiːsmiːl/ *adv.* & *adj.* ● *adv.* piece by piece; gradually. ● *adj.* partial; gradual; unsystematic. [ME f. PIECE + -*meal* f. OE *mǣlum* (instrumental dative pl. of *mǣl* MEAL¹)]

piecework /ˈpiːswɜːk/ *n.* work paid for by the amount produced.

piecrust /ˈpaɪkrʌst/ *n.* the baked pastry crust of a pie. □ **piecrust table** a table with an indented edge like a piecrust.

pied /paɪd/ *adj.* particoloured. □ **pied wagtail** a black and white race of the white wagtail, *Motacilla alba*, found mainly in the British Isles. [ME f. PIE²]

pied-à-terre /ˌpjeɪdæˈteə(r)/ *n.* (*pl.* **pieds-à-terre** *pronunc.* same) a usu. small flat, house, etc. kept for occasional use. [F, lit. 'foot to earth']

Piedmont /ˈpiːdmɒnt/ (Italian **Piemonte** /pjeˈmonte/) a region of NW Italy, in the foothills of the Alps; capital, Turin. Dominated by Savoy from 1400, it became a part of the kingdom of Sardinia in 1720. It was the centre of the movement for a united Italy in the 19th century. [It. *piemonte* mountain foot]

piedmont /ˈpiːdmɒnt/ *n.* a gentle slope leading from the foot of mountains to a region of flat land. [as PIEDMONT]

pie-dog var. of PYE-DOG.

Pied Piper the hero of *The Pied Piper of Hamelin*, a poem by Robert Browning (1842), based on an old German legend. The piper, dressed in parti-coloured costume, rid the town of Hamelin (Hameln) in Brunswick of rats by enticing them away with his music, and when refused the promised payment he lured away all the children except two, one of whom was blind, the other lame. The event was long regarded as historical, and has been linked to both the Children's Crusade of 1212 and events in Hamelin in 1284.

pieman /ˈpaɪmən/ *n.* (*pl.* **-men**) a pie seller.

Piemonte see PIEDMONT.

pier /pɪə(r)/ *n.* **1 a** a structure of iron or wood raised on piles and leading out to sea, a lake, etc., used as a promenade and landing-stage, and often with entertainment arcades etc. **b** a breakwater; a mole. **2 a** a support of an arch or of the span of a bridge; a pillar. **b** solid masonry between windows etc. **3** a long narrow structure projecting from the main body of an airport terminal, along which passengers walk to and from their aircraft. □ **pier-glass** a large mirror, used orig. to fill wall-space between windows. [ME *per* f. AL *pera*, of unkn. orig.]

Pierce /pɪəs/, Franklin (1804–69), American Democratic statesman, 14th President of the US 1853–7. His presidency saw the rise of divisions within the country over slavery and the encouragement of settlement in the north-west.

pierce /pɪəs/ *v.* **1** *tr.* **a** (of a sharp instrument etc.) penetrate the surface of. **b** (often foll. by *with*) prick with a sharp instrument, esp. to make a hole in. **c** make a hole, opening, or tunnel into or through (something); bore through. **d** make (a hole etc.) (*pierced a hole in the belt*). **e** (of cold, grief, etc.) affect keenly or sharply. **f** (of a light, glance, sound, etc.) penetrate keenly or sharply. **2** (as **piercing** *adj.*) (of a glance, intuition, high noise, bright light, etc.) keen, sharp, or unpleasantly penetrating. **3** *tr.* force a way etc. through or into (something); break through or into (*pierced the German line*). **4** *intr.* usu. foll. by *through*, *into*) penetrate. □ **piercer** *n.* **piercingly** *adv.* [ME f. OF *percer* f. L *pertundere* bore through (as PER-, *tundere* tus- thrust)]

Piero della Francesca /ˌpjeərəʊ ˌdelə frænˈtʃeskə/ (1416–92), Italian painter. He worked in a number of cities, including Florence and Arezzo, and was influenced by the work of Masaccio and Uccello. He used perspective, proportion, and geometrical relationships to create ordered and harmonious pictures in which the figures appear to inhabit real space. He is best known as a fresco painter, and among his major works is a fresco cycle in Arezzo depicting the story of the True Cross (begun 1452). After a long period of neglect his work was regarded more favourably in the 20th century.

Pierre /pɪə(r)/ the state capital of South Dakota, situated on the Missouri river; pop. (1990) 12,900.

Pierrot /ˈpɪərəʊ, ˈpjeər-/ a stock character in the French and English theatres, with whitened face and loose white costume. Originally a robust country bumpkin, he was transformed by a French mime artist Jean-Gaspard Deburau (1796–1846) in the 1820s into an ever-hopeful but always disappointed lover, a guise under which the character became well known in London. Pierrot was ousted from the harlequinade by the English clown, but was revived towards the end of the 19th century. [F, dimin. of *Pierre* Peter]

pietà /ˌpɪeˈtɑː/ *n.* a picture or sculpture of the Virgin Mary holding the dead body of Jesus on her lap or in her arms. [It. f. L (as PIETY)]

pietas /ˈpaɪətɑːs/ *n.* respect due to an ancestor, a forerunner, etc. [L: see PIETY]

Pietermaritzburg /ˌpiːtəˈmærɪtsˌbɜːɡ/ a city in eastern South Africa, the capital of KwaZulu/Natal; pop. (1985) 192,420. Founded by the Boers in 1839, it was made the capital of Natal when the former province was annexed by the British in 1843.

Pietersburg /ˈpiːtəzˌbɜːɡ/ a town in northern South Africa, the capital of the province of Northern Transvaal; pop. (1985) 29,000.

pietism /ˈpaɪətɪz(ə)m/ *n.* **1 a** pious sentiment. **b** an exaggerated or affected piety. **2** (esp. as **Pietism**) *hist.* a movement originating at Frankfurt *c.*1670 for the revitalizing of orthodox Lutheranism, with devotional circles for prayer, Bible study, etc. Pietism influenced similar movements elsewhere, including that of John Wesley. □ **pietist** *n.* **pietistic** /ˌpaɪəˈtɪstɪk/ *adj.* **pietistical** *adj.* [G *Pietismus* (as PIETY)]

piety /ˈpaɪətɪ/ *n.* (*pl.* **-ies**) **1** the quality of being pious. **2** a pious act. [ME f. OF *pieté* f. L *pietas -tatis* dutifulness (as PIOUS)]

piezoelectricity /paɪˌiːzəʊˌɪlekˈtrɪsɪtɪ, -ˌelekˈtrɪsɪtɪ/ *n. Physics* electric polarization in a substance resulting from the application of mechanical stress, esp. in certain crystals. (*See note below.*) □ **piezoelectric** /-ɪˈlektrɪk/ *adj.* **piezoelectrically** *adv.* [Gk *piezō* press + ELECTRICITY]

▪ Piezoelectricity was discovered by Pierre Curie and his brother in 1880. The converse effect, in which a voltage applied to such a material causes a mechanical deformation, also occurs. These effects are of great practical importance because piezoelectric substances are able to convert mechanical signals (such as sound waves) into electrical signals, and vice versa. They are therefore widely used in microphones, gramophone pick-ups, earphones, etc. They can also be made to resonate very accurately at a fixed frequency, and as such form the basis of accurate time-keeping devices such as quartz clocks.

piezometer /ˌpaɪˈzɒmɪtə(r)/ *n.* an instrument for measuring the magnitude or direction of pressure.

piffle /ˈpɪf(ə)l/ *n.* & *v. colloq.* ● *n.* nonsense; empty speech. ● *v.intr.* talk or act feebly; trifle. □ **piffler** *n.* [imit.]

piffling /ˈpɪflɪŋ/ *adj. colloq.* trivial; worthless.

pig /pɪɡ/ *n.* & *v.* ● *n.* **1 a** an omnivorous hoofed bristly mammal of the family Suidae, with a flat snout for rooting in the soil; esp. a domesticated kind, *Sus domesticus*. (*See note below.*) **b** *US* a young pig; a piglet. **c** (often in *comb.*) any similar animal (*guinea pig*). **2** the flesh of esp. a young or sucking pig as food (*roast pig*). **3** *colloq.* **a** a greedy, dirty, obstinate, sulky, or annoying person. **b** an unpleasant, awkward, or difficult thing, task, etc. **4** an oblong mass of metal (esp. iron or lead) from a smelting-furnace. **5** *sl. derog.* a policeman. **6** *Sc.* & *dial.* an earthenware hot-water bottle. ● *v.* (**pigged, pigging**) **1** *tr.* (also *absol.*) (of a sow) bring forth (piglets). **2** *tr. colloq.* eat (food) greedily. **3** *intr.* herd together or behave like pigs. □ **bleed like a pig** (or **stuck pig**) bleed copiously. **buy a pig in a poke** buy, accept, etc. something without knowing its value or esp. seeing it. **in pig** (of a sow) pregnant. **in a pig's eye** *colloq.* certainly not. **make a pig of oneself** overeat. **make a pig's ear of** *colloq.* make a mess of; bungle. **pig in the middle** a person who is placed in an awkward situation between two others (after a ball game for three with one in the middle). **pig-iron** crude iron from a smelting-furnace, often cast in pigs (PIG *n.* 4). **pig it** live in a disorderly, untidy, or filthy fashion. **pig-jump** *Austral. sl. n.* a jump made by a horse from all four legs. ● *v.intr.* (of a horse) jump in this manner. **pig Latin** a made-up jargon. **pig-meat** *Brit.* pork, ham, or bacon. **pig out** (often foll. by *on*) esp. *N. Amer. sl.* make a pig of oneself; overeat. **pigs might fly** *iron.* an expression of disbelief. **pig-sticker** a long sharp knife. **pig's wash** = PIGSWILL. □ **piggish** *adj.* **piggishly** *adv.* **piggishness** *n.* **piglet** *n.* **piglike** *adj.* **pigling** *n.* [ME *pigge* f. OE *pigga* (unrecorded)]

▪ The farmyard pig is derived from the Eurasian wild boar, *Sus scrofa*, and was probably domesticated in the Middle East over 8,000 years ago. Many varieties have been bred in recent centuries, and the world's most popular breed is the Yorkshire or Large White, developed in England in the 18th century. There are seven other species of wild pig in the Old World, but the so-called 'wild pigs' of

North America and Australasia are actually feral domestic animals. The piglike peccaries of the New World belong to a different family.

pigeon[1] /ˈpɪdʒɪn/ n. **1** an often grey and white bird of the family Columbidae, usu. larger than a dove; esp. a domesticated form of the rock dove, *Columba livia*, bred for showing, racing, and carrying messages. **2** *sl.* a person easily swindled; a simpleton. □ **pigeon-breast** (or **-chest**) a deformed human chest with a projecting breastbone. **pigeon-breasted** (or **-chested**) having a pigeon-breast. **pigeon-fancier** a person who keeps and breeds fancy pigeons. **pigeon-fancying** this pursuit. **pigeon-hawk** *N. Amer.* = MERLIN. **pigeon-hearted** cowardly. **pigeon-hole** n. **1** each of a set of compartments in a cabinet or on a wall for papers, letters, etc. **2** a small recess for a pigeon to nest in. ● *v.tr.* **1** assign (a person or thing) to a preconceived category. **2** put (a matter) aside for future consideration or to forget it. **3** deposit (a document) in a pigeon-hole. **pigeon pair** *Brit.* **1** boy and girl twins. **2** a boy and girl as sole children. **pigeon's milk 1** a secretion from the oesophagus with which pigeons feed their young. **2** an imaginary article for which children are sent on a fool's errand. **pigeon-toed** (of a person) having the toes turned inwards. □ **pigeonry** n. (pl. **-ies**). [ME f. OF *pijon* f. LL *pipio -onis* (imit.)]

pigeon[2] /ˈpɪdʒɪn/ n. **1** = PIDGIN. **2** *colloq.* a particular concern, job, or business (*that's not my pigeon*).

piggery /ˈpɪɡərɪ/ n. (pl. **-ies**) **1** a pig-breeding farm etc. **2** = PIGSTY 1. **3** piggishness.

Piggott /ˈpɪɡət/, Lester (Keith) (b.1935), English jockey. He was champion jockey nine times between 1960 and 1971 and again in 1981 and 1982; he won the Derby a record nine times. He made a comeback as a jockey in 1990 after a period of imprisonment for tax irregularities (1987–8).

piggy /ˈpɪɡɪ/ n. & adj. ● n. (also **piggie**) (pl. **piggies**) *colloq.* **1** a little pig. **2 a** a child's word for a pig. **b** a child's word for a toe. **3** *Brit.* the game of tipcat. ● adj. (**piggier**, **piggiest**) **1** like a pig. **2** (of features etc.) like those of a pig (*little piggy eyes*). □ **piggy bank** a pig-shaped money box. **piggy in the middle** = *pig in the middle*.

piggyback /ˈpɪɡɪˌbæk/ n. & adv. (also **pickaback** /ˈpɪkəˌbæk/) ● n. a ride on the back and shoulders of another person. ● adv. **1** on the back and shoulders of another person. **2** on the back or top of a larger object. [16th c.: orig. unkn.]

pigheaded /pɪɡˈhedɪd/ adj. obstinate. □ **pigheadedly** adv. **pigheadedness** n.

Pig Island *Austral. & NZ sl.* New Zealand.

pigment /ˈpɪɡmənt/ n. & v. ● n. **1** colouring-matter used as paint or dye, usu. as an insoluble suspension. **2** the natural colouring-matter of animal or plant tissue, e.g. chlorophyll, haemoglobin. ● *v.tr.* colour with or as if with pigment. □ **pigmentary** adj. **pigmental** /pɪɡˈment(ə)l/ adj. [ME f. L *pigmentum* f. *pingere* paint]

pigmentation /ˌpɪɡmənˈteɪʃ(ə)n/ n. **1** the natural colouring of plants, animals, etc. **2** the excessive colouring of tissue by the deposition of pigment.

pigmy var. of PYGMY.

pignut /ˈpɪɡnʌt/ n. = earth-nut.

pigpen /ˈpɪɡpen/ n. *US* = PIGSTY 1.

pigskin /ˈpɪɡskɪn/ n. **1** the hide of a pig. **2** leather made from this. **3** *N. Amer.* a football.

pigsticking /ˈpɪɡˌstɪkɪŋ/ n. **1** the hunting of wild boar with a spear on horseback. **2** the butchering of pigs.

pigsty /ˈpɪɡstaɪ/ n. (pl. **-ies**) **1** a pen or enclosure for a pig or pigs. **2** a filthy house, room, etc.

pigswill /ˈpɪɡswɪl/ n. kitchen refuse and scraps fed to pigs.

pigtail /ˈpɪɡteɪl/ n. **1** a plait of hair hanging from the back of the head, or either of a pair at the sides. **2** a thin twist of tobacco. □ **pigtailed** adj.

pigwash /ˈpɪɡwɒʃ/ n. = PIGSWILL.

pigweed /ˈpɪɡwiːd/ n. a plant used for fodder, esp. fat hen or a weedy amaranth.

pika /ˈpaɪkə/ n. a small mammal of the genus *Ochotona*, related to rabbits and hares, but with short rounded ears, short legs, and a very small tail. Also called *mouse hare*. [Tungus *piika*]

pike[1] /paɪk/ n. (pl. same) **1** a large predatory freshwater fish of the family Esocidae, with a long narrow snout and sharp teeth; esp. the common *Esox lucius*. **2** any similar fish, e.g. garpike. □ **pike-perch** a pikelike perch of the genus *Lucioperca* or *Stizostedion*. [ME, = PIKE[2] (because of its pointed jaw)]

pike[2] /paɪk/ n. & v. ● n. **1** *hist.* an infantry weapon with a pointed steel or iron head on a long wooden shaft. **2** *N. Engl.* the peaked top of a hill, esp. in names of hills in the Lake District. ● *v.tr. hist.* thrust through or kill with a pike. □ **pike on** *colloq.* withdraw timidly from. [OE *pīc* point, prick: sense 2 perh. f. ON]

pike[3] /paɪk/ n. **1** a toll-gate; a toll. **2** a turnpike road. [abbr. of TURNPIKE]

pike[4] /paɪk/ n. a jackknife position in diving or gymnastics. [20th c.: orig. unkn.]

pikelet /ˈpaɪklɪt/ n. *N. Engl.* a thin kind of crumpet. [Welsh (*bara*) *pyglyd* pitchy (bread)]

pikeman /ˈpaɪkmən/ n. (pl. **-men**) the keeper of a turnpike.

piker /ˈpaɪkə(r)/ n. a cautious, timid, or mean person.

pikestaff /ˈpaɪkstɑːf/ n. **1** *hist.* the wooden shaft of a pike. **2** a walking-stick with a metal point. □ **plain as a pikestaff** quite plain or obvious (orig. *packstaff*, a smooth staff used by a pedlar).

Pik Pobedy /ˌpiːk pəˈbjedɪ/ a mountain in eastern Kyrgyzstan, situated close to the border with China. Rising to a height of 7,439 m (24,406 ft), it is the highest peak in the Tien Shan range. [Russ., = Victory Peak]

pilaster /pɪˈlæstə(r)/ n. *Archit.* a rectangular column, esp. one projecting from a wall. □ **pilastered** adj. [F *pilastre* f. It. *pilastro* f. med.L *pilastrum* f. L *pila* pillar]

Pilate /ˈpaɪlət/, Pontius (died AD *c.*36), Roman procurator of Judaea *c.*26–*c.*36. Little is known of his life; he is chiefly remembered for presiding at the trial of Jesus Christ and sentencing him to death by crucifixion, as recorded in the New Testament. Pilate was later recalled to Rome to stand trial on charges of cruelty, having ordered a massacre of the Samaritans in 36. According to one tradition, he subsequently committed suicide.

pilau /pɪˈlaʊ/ n. (also **pilaff**, **pilaf** /-ˈlæf/, **pilaw** /-ˈlɔː/) a Middle Eastern or Indian dish of spiced rice or wheat with meat, fish, vegetables, etc. [Turk. *pilâv*]

pilchard /ˈpɪltʃəd/ n. a small marine fish, *Sardinia pilchardus*, of the herring family (see SARDINE[1]). [16th-c. *pilcher* etc.: orig. unkn.]

pile[1] /paɪl/ n. & v. ● n. **1** a heap of things laid or gathered upon one another (*a pile of leaves*). **2 a** a large imposing building (*a stately pile*). **b** a large group of tall buildings. **3** *colloq.* **a** a large quantity. **b** a large amount of money; a fortune (*made his pile*). **4 a** a series of plates of dissimilar metals laid one on another alternately to produce an electric current. **b** = *atomic pile*. **5** a funeral pyre. ● v. **1** *tr.* **a** (often foll. by *up*, *on*) heap up (*piled the plates on the table*). **b** (foll. by *with*) load (*piled the bed with coats*). **2** *intr.* (usu. foll. by *in*, *into*, *on*, *out of*, etc.) crowd hurriedly or tightly (*all piled into the car; piled out of the restaurant*). □ **pile arms** *hist.* place (usu. four) rifles with their butts on the ground and the muzzles together. **pile it on** *colloq.* exaggerate. **pile on the agony** *colloq.* exaggerate for effect or to gain sympathy etc. **pile up 1** accumulate; heap up. **2** *colloq.* run (a ship) aground or cause (a vehicle etc.) to crash. **pile-up** n. *colloq.* a multiple crash of road vehicles. [ME f. OF f. L *pila* pillar, pier, mole]

pile[2] /paɪl/ n. & v. ● n. **1** a heavy beam driven vertically into the bed of a river, soft ground, etc., to support the foundations of a superstructure. **2** a pointed stake or post. **3** *Heraldry* a wedge-shaped device. ● *v.tr.* **1** provide with piles. **2** drive (piles) into the ground etc. □ **pile-driver** a machine for driving piles into the ground. **pile-dwelling** a dwelling built on piles, esp. in a lake. [OE *pīl* f. L *pilum* javelin]

pile[3] /paɪl/ n. **1** the soft projecting surface on velvet, plush, etc., or esp. on a carpet; nap. **2** soft hair or down, or the wool of a sheep. [ME prob. f. AF *pyle*, *peile*, OF *poil* f. L *pilus* hair]

pileated /ˈpaɪlɪˌeɪtɪd/ adj. (of a bird, fungus, etc.) having a cap or pileus. □ **pileated woodpecker** a large red-capped North American woodpecker, *Dryocopus pileatus*. [L *pileatus* f. PILEUS]

piles /paɪlz/ n.pl. haemorrhoids. [ME prob. f. L *pila* ball, f. the globular form of external piles]

pileus /ˈpaɪlɪəs/ n. (pl. **pilei** /-lɪˌaɪ/) the caplike part of a mushroom or other fungus. □ **pileate** /-lɪət/ adj. [L, = felt cap]

pilewort /ˈpaɪlwɜːt/ n. the lesser celandine. [PILES, f. its reputed efficacy against piles]

pilfer /ˈpɪlfə(r)/ *v.tr.* (also *absol.*) steal (objects) esp. in small quantities. □ **pilferage** n. **pilferer** n. [ME f. AF & OF *pelfrer* pillage, of unkn. orig.: assoc. with archaic *pill* plunder: cf. PELF]

pilgrim /ˈpɪlɡrɪm/ n. & v. ● n. **1** a person who journeys to a sacred place for religious reasons. **2** a person regarded as journeying through life etc. **3** a traveller. ● *v.intr.* (**pilgrimed**, **pilgriming**) wander like a pilgrim. □ **pilgrimize** *v.intr.* (also **-ise**). [ME *pilegrim* f. Prov. *pelegrin* f. L *peregrinus* stranger: see PEREGRINE]

pilgrimage /ˈpɪlgrɪmɪdʒ/ *n. & v.* ● *n.* **1** a pilgrim's journey (*go on a pilgrimage*). **2** life viewed as a journey. **3** any journey taken for nostalgic or sentimental reasons. ● *v.intr.* go on a pilgrimage. [ME f. Prov. *pilgrinatge* (as PILGRIM)]

Pilgrim Fathers the pioneers of British colonization of North America, a group of 102 people who sailed in the *Mayflower* and founded a settlement at Plymouth, Massachusetts, in 1620. The expedition was initiated by a group of English Puritans fleeing religious persecution.

Pilipino /ˌpɪlɪˈpiːnəʊ/ *n. & adj.* ● *n.* the national language of the Philippines, a standardized form of Tagalog widely used as a second language. ● *adj.* of or relating to Pilipino. [Tagalog f. Sp. *Filipino*]

pill /pɪl/ *n.* **1 a** a ball or disc etc. of solid medicine for swallowing whole. **b** (usu. prec. by *the*; often **the Pill**) *colloq.* a contraceptive pill. **2** an unpleasant or painful necessity; a humiliation (*a bitter pill*; *must swallow the pill*). **3** *colloq.* or *joc.* a ball, e.g. a football. **4** *sl.* a bore. □ **pill-popper** *colloq.* a person who takes pills freely; a drug addict. **pill-pusher** *colloq.* a drug pusher. **sugar** (or **sweeten**) **the pill** make an unpleasant necessity acceptable. [MDu., MLG *pille* prob. f. L *pilula* dimin. of *pila* ball]

pillage /ˈpɪlɪdʒ/ *v. & n.* ● *v.tr.* (also *absol.*) plunder; sack (a place or a person). ● *n.* **1** the act or an instance of pillaging, esp. in war. **2** *hist.* goods plundered. □ **pillager** *n.* [ME f. OF f. *piller* plunder]

pillar /ˈpɪlə(r)/ *n.* **1 a** a usu. slender vertical structure of wood, metal, or esp. stone used as a support for a roof etc. **b** a similar structure used for ornament. **c** a post supporting a structure. **2** a person regarded as a mainstay or support (*a pillar of the faith*; *a pillar of strength*). **3** a vertical column of air, water, rock, etc. (*pillar of flame*). **4** a solid mass of coal etc. left to support the roof of a mine. □ **from pillar to post** (driven etc.) from one place to another; to and fro. □ **pillared** *adj.* [ME & AF *piler*, OF *pilier* ult. f. L *pila* pillar]

pillar-box /ˈpɪləˌbɒks/ *n.* (in the UK) a public postbox shaped like a pillar and traditionally painted red. The first pillar-boxes were set up on the initiative of the novelist Anthony Trollope when he worked for the General Post Office in the mid-19th century. □ **pillar-box red** a bright red colour, as of pillar-boxes.

pillaret /ˌpɪləˈret/ *n.* a small pillar.

Pillars of Hercules the two promontories known in ancient times as Calpe and Abyla and now known as the Rock of Gibraltar and Mount Acho in Ceuta. Situated opposite one another at the eastern end of the Strait of Gibraltar, they were held by legend to have been parted by the arm of Hercules and were regarded as marking the limit of the known world.

pillbox /ˈpɪlbɒks/ *n.* **1** a small shallow cylindrical box for holding pills. **2** a hat of a similar shape. **3** *Mil.* a small partly underground enclosed concrete fort used as an outpost.

pillion /ˈpɪljən/ *n.* **1** seating for a passenger behind a motorcyclist. **2** *hist.* **a** a woman's light saddle. **b** a cushion attached to the back of a saddle for a usu. female passenger. □ **ride pillion** travel seated behind a motorcyclist etc. [Gael. *pillean*, *pillin* dimin. of *pell* cushion f. L *pellis* skin]

pilliwinks /ˈpɪlɪˌwɪŋks/ *n. hist.* an instrument of torture used for squeezing the fingers. [ME *pyrwykes*, *pyrewinkes*, of unkn. orig.]

pillock /ˈpɪlək/ *n. Brit. sl.* a stupid person; a fool. [16th c., = penis (var. of *pillicock*): 20th c. in sense defined]

pillory /ˈpɪlərɪ/ *n. & v.* ● *n.* (*pl.* **-ies**) *hist.* a wooden framework with holes for the head and hands, into which offenders were formerly locked for exposure to public ridicule or assault. ● *v.tr.* (**-ies**, **-ied**) **1** expose (a person) to ridicule or public contempt. **2** *hist.* put in the pillory. [ME f. AL *pillorium* f. OF *pilori* etc.: prob. f. Prov. *espilori* of uncert. orig.]

pillow /ˈpɪləʊ/ *n. & v.* ● *n.* **1 a** a usu. oblong support for the head, esp. in bed, with a cloth cover stuffed with feathers, flock, foam rubber, etc. **b** any pillow-shaped block or support. **2** = lace-pillow. ● *v.tr.* **1** rest (the head etc.) on or as if on a pillow (*pillowed his head on his arms*). **2** serve as a pillow for (*moss pillowed her head*). □ **pillow-fight** a mock fight with pillows, esp. by children. **pillow-lace** lace made on a lace-pillow. **pillow lava** lava forming rounded masses. **pillow talk** romantic or intimate conversation in bed. □ **pillowy** *adj.* [OE *pyle*, *pylu*, ult. f. L *pulvinus* cushion]

pillowcase /ˈpɪləʊˌkeɪs/ *n.* a washable cotton etc. cover for a pillow.

pillowslip /ˈpɪləʊˌslɪp/ *n.* = PILLOWCASE.

pillule var. of PILULE.

pillwort /ˈpɪlwɜːt/ *n.* an aquatic plant of the genus *Pilularia*, related to ferns, with small globular spore-producing bracts; esp. the European *P. globulifera*.

pilose /ˈpaɪləʊz, -ləʊs/ *adj.* (also **pilous** /-ləs/) covered with hair. □ **pilosity** /paɪˈlɒsɪtɪ/ *n.* [L *pilosus* f. *pilus* hair]

pilot /ˈpaɪlət/ *n. & v.* ● *n.* **1** a person who operates the flying controls of an aircraft. **2** a person qualified to take charge of a ship entering or leaving harbour. **3** (usu. *attrib.*) an experimental undertaking or test, esp. in advance of a larger one (*a pilot project*). **4** a guide; a leader. **5** *archaic* a steersman. ● *v.tr.* (**piloted**, **piloting**) **1** act as a pilot on (a ship) or of (an aircraft). **2** conduct, lead, or initiate as a pilot (*piloted the new scheme*). □ **pilot balloon** a small balloon used to track air currents etc. **pilot-bird** a rare dark brown Australian warbler, *Pycnoptilus floccosus*, with a distinctive loud cry. **pilot chute** a small parachute used to bring the main one into operation. **pilot-cloth** thick blue woollen cloth for seamen's coats etc. **pilot-fish** a small fish, *Naucrates ductor*, said to act as a pilot leading a shark to food. **pilot-house** = WHEEL-HOUSE 1. **pilot-jacket** = PEA-JACKET. **pilot-light 1** a small gas burner kept alight to light another. **2** an electric indicator light or control light. **pilot officer** *Brit.* the lowest commissioned rank in the RAF. **pilot whale** a small whale of the genus *Globicephalus*, of temperate or subtropical waters. □ **pilotage** *n.* **pilotless** *adj.* [F *pilote* f. med.L *pilotus*, *pedot(t)a* f. Gk *pēdon* oar]

Pilsen /ˈpɪls(ə)n/ (Czech **Plzeň** /plzenj/) an industrial city in the western part of the Czech Republic; pop. (1991) 173,130. It is noted for the production of lager.

Pilsner /ˈpɪlznə(r), ˈpɪlsnə(r)/ *n.* (also **Pilsener**) a lager beer brewed or like that brewed at Pilsen in the Czech Republic.

Piltdown man /ˈpɪltdaʊn/ *n.* a fraudulent fossil composed of a human cranium and an ape jaw that was presented in 1912 as a genuine hominid (*Eoanthropus dawsoni*) of great antiquity. The remains were allegedly discovered in a gravel-pit on Piltdown Common near Lewes, East Sussex, in association with early Pleistocene fossil fauna and stone and bone artefacts. The 'discoverer' was Charles Dawson (1864–1916), an amateur palaeontologist and antiquarian. Piltdown man was shown to be a fraud in 1953, after thorough scientific tests, although the find had always aroused suspicions. The perpetrators of the hoax, who presumably included Dawson, have never been identified.

pilule /ˈpɪljuːl/ *n.* (also **pillule**) a small pill. □ **pilular** *adj.* **pilulous** *adj.* [F f. L *pilula*: see PILL]

pimento /pɪˈmentəʊ/ *n.* (*pl.* **-os**) **1** a small tropical tree, *Pimenta dioica*, native to Jamaica. **2** the unripe dried berries of this, crushed to make the spice allspice. **3** = PIMIENTO. [Sp. *pimiento* (as PIMIENTO)]

pimiento /ˌpɪmɪˈentəʊ/ *n.* (*pl.* **-os**) a capsicum, esp. a red pepper. [Sp. f. L *pigmentum* PIGMENT, in med.L = spice]

pimp /pɪmp/ *n. & v.* ● *n.* a man who lives off the earnings of a prostitute or a brothel; a pander; a ponce. ● *v.intr.* act as a pimp. [17th c.: orig. unkn.]

pimpernel /ˈpɪmpəˌnel/ *n.* a small plant of the genus *Anagallis*, esp. the scarlet pimpernel. [ME f. OF *pimpernelle*, *piprenelle* ult. f. L *piper* PEPPER]

pimping /ˈpɪmpɪŋ/ *adj.* **1** small or mean. **2** sickly. [17th c.: orig. unkn.]

pimple /ˈpɪmp(ə)l/ *n.* **1** a small hard inflamed spot on the skin. **2** anything resembling a pimple, esp. in relative size. □ **pimpled** *adj.* **pimply** *adj.* [ME nasalized f. OE *piplian* break out in pustules]

PIN /pɪn/ *abbr.* personal identification number.

pin /pɪn/ *n. & v.* ● *n.* **1 a** a small thin pointed piece of esp. steel wire with a round or flattened head used (esp. in sewing) for holding things in place, attaching one thing to another, etc. **b** any of several types of pin (*drawing-pin*; *safety pin*; *hairpin*). **c** a small brooch (*diamond pin*). **d** a badge fastened with a pin. **2** a peg of wood or metal for various purposes, e.g. a wooden skittle in bowling. **3** something of small value (*don't care a pin*; *for two pins I'd resign*). **4** (in *pl.*) *colloq.* legs (*quick on his pins*). **5** *Med.* a steel rod used to join the ends of fractured bones while they heal. **6** *Chess* a position in which a piece is pinned to another. **7** *Golf* a stick with a flag placed in a hole to mark its position. **8** *Mus.* a peg round which one string of a musical instrument is fastened. **9** a small beer cask that holds half a firkin ($4\frac{1}{2}$ imperial gallons or about 20.5 litres). ● *v.tr.* (**pinned**, **pinning**) **1 a** (often foll. by *to*, *up*, *together*) fasten with a pin or pins (*pinned up the hem*; *pinned the papers together*). **b** transfix with a pin, lance, etc. **2** (usu. foll. by *on*) fix (blame, responsibility, etc.) on a person etc. (*pinned the blame on his friend*). **3** (often foll. by *against*, *on*, etc.) seize and hold fast. **4** *Chess* prevent (an opposing piece) from moving except by exposing a more valuable piece to capture. □ **on pins and needles** in an agitated state of suspense. **pin down 1** (often foll. by *to*) bind (a person etc.) to a promise, arrangement, etc. **2** force (a person) to declare his or her intentions. **3** restrict the actions or movement of (an enemy etc.). **4** specify (a thing) precisely (*could not pin down his unease to a particular cause*). **5** hold (a person etc.) down by force. **pin-down** (often *attrib.*) the action or policy of putting children in care into solitary confinement for long periods. **pin one's faith** (or **hopes** etc.) **on** rely implicitly on. **pin-feather** *Zool.* an ungrown feather. **pin-high** *Golf* (of a ball) at the same distance ahead as the pin. **pin-money 1** *hist.* an allowance to a woman for dress etc. from her husband. **2** a very small sum of money, esp. for spending on inessentials (*only works for pin-money*). **pins and needles** a tingling sensation in a limb recovering from numbness. **pin-table** a table used in playing pinball. **pin-tuck** a very narrow ornamental tuck. **pin-up 1** a photograph of a popular or sexually attractive person, designed to be hung on the wall. **2** a person shown in such a photograph. **pin-wheel** a small Catherine wheel. **split pin** a metal cotter pin passed through a hole and held in place by its gaping split end. [OE *pinn* f. L *pinna* point etc., assoc. with *penna* PEN¹]

pina colada /ˌpiːnə kəˈlɑːdə/ *n.* a drink made from pineapple juice, rum, and coconut. [Sp., lit. 'strained pineapple']

pinafore /ˈpɪnəˌfɔː(r)/ *n.* esp. *Brit.* **1 a** an apron, esp. with a bib. **b** a woman's sleeveless wraparound washable covering for the clothes, tied at the back. **2** (in full **pinafore dress**) a collarless sleeveless dress usu. worn over a blouse or jumper. [PIN + AFORE (because orig. pinned on the front of a dress)]

Pinang see PENANG.

pinaster /paɪˈnæstə(r)/ *n.* = *cluster pine*. [L, = wild pine f. *pinus* pine + -ASTER]

Pinatubo, Mount /ˌpɪnəˈtuːbəʊ/ a volcano on the island of Luzon, in the Philippines. It erupted in 1991, killing more than 300 people and destroying the homes of more than 200,000.

pinball /ˈpɪnbɔːl/ *n.* a game in which small metal balls are shot across a board and score points by striking pins with lights etc.

pince-nez /ˈpænsneɪ/ *n.* (*pl.* same) a pair of eyeglasses with a nose-clip instead of earpieces. [F, lit. 'pinch-nose']

pincers /ˈpɪnsəz/ *n.pl.* **1** (also **pair of pincers**) a gripping-tool resembling scissors but with blunt usu. concave jaws to hold a nail etc. for extraction. **2** the front claws of lobsters and some other crustaceans. □ **pincer movement** *Mil.* a movement by two wings of an army converging on the enemy. [ME *pinsers*, *pinsours* f. AF f. OF *pincier* PINCH]

pincette /pænˈset/ *n.* small pincers; tweezers. [F]

pinch /pɪntʃ/ *v.* & *n.* ● *v.* **1** *tr.* **a** grip (esp. the skin of part of the body or of another person) tightly, esp. between finger and thumb (*pinched my finger in the door; stop pinching me*). **b** (often *absol.*) (of a shoe, garment, etc.) constrict (the flesh) painfully. **2** *tr.* (of cold, hunger, etc.) grip (a person) painfully (*she was pinched with cold*). **3** *tr. colloq.* **a** steal; take without permission. **b** arrest (a person) (*pinched him for loitering*). **4** (as **pinched** *adj.*) (of the features) drawn, as with cold, hunger, worry, etc. **5 a** *tr.* (usu. foll. by *in*, *of*, *for*, etc.) stint (a person). **b** *intr.* be niggardly with money, food, etc. **6** *tr.* (usu. foll. by *out*, *back*, *down*) remove (leaves, buds, etc.) to encourage bushy growth. **7** *intr. Naut.* sail very close to the wind. ● *n.* **1** the act or an instance of pinching etc. the flesh. **2** an amount that can be taken up with fingers and thumb (*a pinch of salt*). **3** the stress or pain caused by poverty, cold, hunger, etc. **4** *sl.* **a** an arrest. **b** a theft. □ **at** (or **in**) **a pinch** in an emergency; if necessary. **feel the pinch** experience the effects of poverty. **pinch-hitter 1** a baseball player who bats instead of another in an emergency. **2** *US* a person acting as a substitute. [ME f. AF & ONF *pinchier* (unrecorded), OF *pincier*, ult. f. L *pungere punct-* prick]

pinchbeck /ˈpɪntʃbek/ *n.* & *adj.* ● *n.* an alloy of copper and zinc resembling gold and used in cheap jewellery etc. ● *adj.* **1** counterfeit; sham. **2** cheap; tawdry. [Christopher *Pinchbeck*, Engl. watchmaker (c.1670–1732)]

pinchpenny /ˈpɪntʃˌpenɪ/ *n.* (*pl.* -**ies**) (also *attrib.*) a miserly person.

pincushion /ˈpɪnˌkʊʃ(ə)n/ *n.* a small cushion for holding pins.

Pindar /ˈpɪndə(r)/ (c.518–c.438 BC), Greek lyric poet. His surviving works include four books of odes (the *Epinikia*) celebrating victories won in athletic contests at Olympia and elsewhere. The odes are often in the form of choral hymns, written in an elevated style and imbued with religious significance. □ **Pindaric** /pɪnˈdærɪk/ *adj.*

Pindus Mountains /ˈpɪndəs/ (Greek **Píndhos** /ˈpɪnðɒs/) a range of

mountains in west central Greece, stretching from the border with Albania southwards to the Gulf of Corinth. The highest peak is Mount Smolikas, which rises to 2,637 m (8,136 ft).

pine¹ /paɪn/ *n.* **1** (in full **pine tree**) an evergreen coniferous tree of the genus *Pinus*, native to northern temperate regions, with needle-shaped leaves growing in clusters. **2** the soft timber of this, often used to make furniture. Also called *deal* (DEAL²). **3** (*attrib.*) made of pine. **4** = PINEAPPLE. □ **pine cone** the cone-shaped fruit of a pine tree. **pine marten** a North Eurasian marten, *Martes martes*, having a dark brown coat with a yellowish throat. **pine nut** the edible seed of various pine trees. □ **pinery** /-nərɪ/ *n.* (*pl.* -**ies**) [ME f. OE *pīn* & OF *pin* f. L *pinus*]

pine² /paɪn/ *v.intr.* **1** (often foll. by *away*) decline or waste away, esp. from grief, disease, etc. **2** (usu. foll. by *for*, *after*, or *to* + infin.) long eagerly; yearn. [OE *pīnian*, rel. to obs. *pine* punishment, f. Gmc f. med.L *pena*, L *poena*]

pineal /ˈpɪnɪəl, paɪˈniːəl/ *adj. Anat.* shaped like a pine cone. □ **pineal body** (or **gland**) a pea-sized conical mass of tissue behind the third ventricle of the brain, secreting a hormone-like substance in some mammals, and sensitive to light in some lower vertebrates. [F *pinéal* f. L *pinea* pine cone: see PINE¹]

pineapple /ˈpaɪnˌæp(ə)l/ *n.* **1** a tropical plant, *Ananas comosus*, with a spiral of sword-shaped leaves and a thick stem bearing a large fruit developed from many flowers. **2** the fruit of this, consisting of yellow flesh surrounded by a tough segmented skin and topped with a tuft of stiff leaves. □ **the rough end of the pineapple** *Austral. colloq.* a raw deal. [PINE¹, from the fruit's resemblance to a pine cone]

Pinero /pɪˈnɪərəʊ/, Sir Arthur Wing (1855–1934), English dramatist and actor. He began writing for the stage in 1877, creating a series of comedies and farces, such as *Dandy Dick* (1887). From 1889 he embarked on a number of serious plays dealing with social issues, especially the double standards of morality for men and women (e.g. *The Second Mrs Tanqueray*, 1893).

pinetum /paɪˈniːtəm/ *n.* (*pl.* **pineta** /-tə/) a plantation of pine trees or other conifers for scientific or ornamental purposes. [L f. *pinus* pine]

pinfold /ˈpɪnfəʊld/ *n.* & *v.* esp. *hist.* ● *n.* a pound for stray cattle etc. ● *v.tr.* confine (cattle) in a pinfold. [OE *pundfald* (as POUND³, FOLD²)]

ping /pɪŋ/ *n.* & *v.* ● *n.* a single short high ringing sound. ● *v.intr.* & *tr.* make or cause to make a ping. [imit.]

pinger /ˈpɪŋə(r)/ *n.* **1** a device that transmits pings at short intervals for purposes of detection or measurement etc. **2** a device to ring a bell.

pingo /ˈpɪŋgəʊ/ *n.* (*pl.* -**os**) *Geol.* a dome-shaped mound found in permafrost areas. [Eskimo]

ping-pong /ˈpɪŋpɒŋ/ *n.* = TABLE TENNIS. [imit. f. the sound of a bat striking a ball]

pinguid /ˈpɪŋgwɪd/ *adj. formal* or *joc.* fat, oily, or greasy. [L *pinguis* fat]

pinhead /ˈpɪnhed/ *n.* **1** the flattened head of a pin. **2** a very small thing. **3** *colloq.* a stupid or foolish person.

pinheaded /pɪnˈhedɪd/ *adj. colloq.* stupid, foolish. □ **pinheadedness** *n.*

pinhole /ˈpɪnhəʊl/ *n.* **1** a hole made by a pin. **2** a hole into which a peg fits. □ **pinhole camera** a camera with a pinhole aperture and no lens.

pinion¹ /ˈpɪnjən/ *n.* & *v.* ● *n.* **1** the outer part of a bird's wing, usu. including the flight feathers. **2** *poet.* a wing; a flight-feather. ● *v.tr.* **1** cut off the pinion of (a wing or bird) to prevent flight. **2 a** bind the arms of (a person). **b** (often foll. by *to*) bind (the arms, a person, etc.) esp. to a thing. [ME f. OF *pignon* ult. f. L *pinna*: see PIN]

pinion² /ˈpɪnjən/ *n.* **1** a small cog-wheel engaging with a larger one. **2** a cogged spindle engaging with a wheel. [F *pignon* alt. f. obs. *pignol* f. L *pinea* pine-cone (as PINE¹)]

pink¹ /pɪŋk/ *n.* & *adj.* ● *n.* **1** a pale red colour (*decorated in pink*). **2 a** a plant of the genus *Dianthus*, with sweet-smelling white, pink, crimson, etc. flowers; esp. a cultivated variety of *D. plumarius*. **b** the flower of this plant. **3** (prec. by *the*) the most perfect condition etc. (*the pink of elegance*). **4** (also **hunting pink**) **a** a fox-hunter's red coat. **b** the cloth for this. **c** a fox-hunter. **5** *colloq.* often *derog.* a person with socialist tendencies. ● *adj.* **1** (often in *comb.*) of a pale red colour of any of various shades (*rose-pink*; *salmon-pink*). **2** *colloq.* often *derog.* tending to socialism. **3** *colloq.* of or relating to homosexuals. □ **in the pink** *colloq.* in very good health. **pink-collar** (usu. *attrib.*) (of a profession etc.) traditionally associated with women (cf. *white-collar*, *blue-collar* (see BLUE¹)). **pink disease** a disease of young children with pink discoloration of the extremities. **pink elephants** *colloq.* hallucinations experienced by a drunk or delirious person. **pink-eye 1** a contagious fever in horses.

2 conjunctivitis in humans and some livestock. **pink gin** gin flavoured with angostura bitters. **pink grapefruit** a grapefruit with a pink sweet pulp. □ **pinkish** *adj.* **pinkly** *adv.* **pinkness** *n.* **pinky** *adj.* [perh. f. dial. *pink-eyed* having small eyes]

pink[2] /pɪŋk/ *v.tr.* **1** pierce slightly with a sword etc. **2** cut a scalloped or zigzag edge on. **3** (often foll. by *out*) ornament (leather etc.) with perforations. **4** esp. *literary* adorn; deck. □ **pinking shears** (or **scissors**) a dressmaker's serrated shears for cutting a zigzag edge. [ME, perh. f. LG or Du.: cf. LG *pinken* strike, peck]

pink[3] /pɪŋk/ *v.intr.* (of a vehicle engine) emit a series of high-pitched explosive sounds caused by faulty combustion. [imit.]

pink[4] /pɪŋk/ *n. hist.* a sailing-ship with a narrow stern, orig. a small and flat-bottomed one. [ME f. MDu. *pin(c)ke*, of unkn. orig.]

pink[5] /pɪŋk/ *n.* a yellowish lake pigment made by combining vegetable colouring matter with a white base (*brown pink*; *French pink*). [17th c.: orig. unkn.]

pink[6] /pɪŋk/ *n. Brit.* **1** a young salmon. **2** *dial.* a minnow. [15th c. *penk*, of unkn. orig.]

Pinkerton /ˈpɪŋkət(ə)n/, Allan (1819–84), Scottish-born American detective. He emigrated to the US in 1842, and in 1850 he established the first American private detective agency (in Chicago), becoming famous after solving a series of train robberies. In the early years of the American Civil War (1861–2) he served as chief of the secret service for the Union side. His agency was later involved in anti-trade union activity, particularly in the coal industry (1877).

Pink Floyd /flɔɪd/ an English rock group formed *c*.1965. The group became known at first for quirky, psychedelic, often controversial material such as the single 'Arnold Layne' (1967). After the departure in 1968 of the lead singer Syd Barrett (b.1946), however, they developed a more sombre style featuring lengthy instrumental passages, as, for example, on the albums *Dark Side of the Moon* (1973) and *The Wall* (1979).

pinkie[1] /ˈpɪŋkɪ/ *n. esp. US & Sc.* the little finger. [cf. dial. *pink* small, half-shut (eye)]

pinkie[2] /ˈpɪŋkɪ/ *n.* **1** esp. *Austral. sl.* cheap red wine. **2** *black sl.* a white person.

Pinkster /ˈpɪŋkstə(r)/ *n. US* Whitsuntide. □ **pinkster flower** a pink azalea, *Rhododendron periclymenoides*. [Du., = Pentecost]

pinna /ˈpɪnə/ *n.* (pl. **pinnae** /-niː/ or **pinnas**) **1** *Anat.* the auricle; the external part of the ear. **2** *Bot.* a primary division of a pinnate leaf. **3** *Zool.* a fin or finlike structure, feather, wing, etc. [L, = *penna* feather, wing, fin]

pinnace /ˈpɪnɪs/ *n. Naut.* a warship's or other ship's small boat, usu. motor-driven, orig. schooner-rigged or eight-oared. [F *pinnace, pinasse* ult. f. L *pinus* PINE[1]]

pinnacle /ˈpɪnək(ə)l/ *n. & v.* ● *n.* **1** the culmination or climax (of endeavour, success, etc.). **2** a natural peak. **3** a small ornamental turret usu. ending in a pyramid or cone, crowning a buttress, roof, etc. ● *v.tr.* **1** set on or as if on a pinnacle. **2** form the pinnacle of. **3** provide with pinnacles. [ME *pinacle* f. OF *pin(n)acle* f. LL *pinnaculum* f. *pinna* wing, point (as PIN, -CULE)]

pinnae *pl.* of PINNA.

pinnate /ˈpɪneɪt/ *adj.* **1** *Bot.* (of a compound leaf) having leaflets arranged on either side of the stem, usu. in pairs opposite each other. **2** *Zool.* having branches, tentacles, etc., on each side of an axis. □ **pinnated** *adj.* **pinnately** *adv.* **pinnation** /pɪˈneɪʃ(ə)n/ *n.* [L *pinnatus* feathered (as PINNA)]

pinni- /ˈpɪnɪ/ *comb. form* wing, fin. [L *pinna*]

pinniped /ˈpɪnɪped/ *adj. & n. Zool.* ● *adj.* denoting any aquatic mammal with limbs ending in flippers, e.g. a seal. ● *n.* a pinniped mammal. [L *pinna* fin + *pes ped-* foot]

pinnule /ˈpɪnjuːl/ *n.* **1** *Bot.* the secondary division of a pinnate leaf. **2** *Zool.* a part or organ like a small wing or fin. □ **pinnular** *adj.* [L *pinnula* dimin. of *pinna* fin, wing]

pinny /ˈpɪnɪ/ *n.* (pl. **-ies**) *colloq.* a pinafore. [abbr.]

Pinochet /ˈpiːnəʃeɪ/, Augusto (full name Augusto Pinochet Ugarte) (b.1915), Chilean general and statesman, President 1974–90. He became Commander-in-Chief of Chile's armed forces in 1973 and in the same year masterminded the military coup which overthrew President Allende. He imposed a repressive military dictatorship until forced to call elections (December 1989), giving way to a democratically elected President in 1990.

pinochle /ˈpiːnɒk(ə)l/ *n. N. Amer.* **1** a card-game with a double pack of

48 cards (nine to ace only). **2** the combination of queen of spades and jack of diamonds in this game. [19th c.: orig. unkn.]

pinocytosis /ˌpiːnəʊsaɪˈtəʊsɪs, ˌpɪnəʊ-, ˌpaɪnəʊ-/ *n. Biol.* the ingestion of droplets of liquid by cells, held in small membrane-bound vesicles. □ **pinocytotic** /-ˈtɒtɪk/ *adj.* [Gk *pinein* to drink + -*cytosis* after PHAGOCYTOSIS]

pinole /pɪˈnəʊlɪ/ *n. US* flour made from parched cornflour, esp. mixed with sweet flour made of mesquite beans, sugar, etc. [Amer. Sp. f. Aztec *pinolli*]

piñon /piːˈnjɒn/ *n.* **1** a small North American pine; esp. *Pinus edulis*, bearing edible seeds. **2** the seed of this, a type of pine nut. [Sp. f. L *pinea* pine cone]

Pinot /ˈpiːnəʊ/ *n.* **1** a variety of black (*Pinot Noir*) or white (*Pinot Blanc*) grape used in making wine. **2** a vine on which either of these varieties grows. **3** a wine made from Pinot Noir or Pinot Blanc grapes. [F]

pinpoint /ˈpɪnpɔɪnt/ *n. & v.* ● *n.* **1** the point of a pin. **2** something very small or sharp. **3** (*attrib.*) **a** very small. **b** precise, accurate. ● *v.tr.* locate with precision (*pinpointed the target*).

pinprick /ˈpɪnprɪk/ *n.* **1** a prick caused by a pin. **2** a trifling irritation.

pinstripe /ˈpɪnstraɪp/ *n.* **1** (often *attrib.*) a very narrow stripe in (esp. worsted or serge) cloth (*pinstripe suit*). **2** (in *sing.* or *pl.*) a pinstripe suit (*came wearing his pinstripes*). □ **pinstriped** *adj.*

pint /paɪnt/ *n.* **1** a unit of liquid or dry capacity equal to one-eighth of a gallon: **a** (in full **imperial pint**) (in Britain) 20 fluid oz, 34.66 cu. in., or 0.565 litre. **b** (in full **US pint**) (in the US) 28.87 cu. in. or 0.473 litre (for liquid measure), or 33.60 cu. in. or 0.551 litre (for dry measure). **2** *Brit.* **a** *colloq.* a pint of beer. **b** a pint of a liquid, esp. milk. **3** *Brit.* a measure of shellfish, being the amount containable in a pint mug (*bought a pint of whelks*). □ **pint-pot** a pot, esp. of pewter, holding one pint, esp. of beer. **pint-sized** *colloq.* very small, esp. of a person. [ME f. OF *pinte*, of unkn. orig.]

pinta /ˈpaɪntə/ *n. Brit. colloq.* a pint of milk. [corrupt. of *pint of*]

pintail /ˈpɪnteɪl/ *n.* a bird with a pointed tail; esp. the migratory duck *Anas acuta*, with two long pointed feathers in the tail.

Pinter /ˈpɪntə(r)/, Harold (b.1930), English dramatist, actor, and director. His plays are associated with the Theatre of the Absurd and are often marked by a sense of brooding menace; they include *The Birthday Party* (1958), *The Caretaker* (1960), and *Party Time* (1991). He has also written screenplays, including the film version of John Fowles's *The French Lieutenant's Woman* (1981).

pintle /ˈpɪnt(ə)l/ *n.* a pin or bolt, esp. one on which some other part turns. [OE *pintel* penis, of unkn. orig.: cf. OFris. etc. *pint*]

pinto /ˈpɪntəʊ/ *adj. & n. N. Amer.* ● *adj.* piebald. ● *n.* (pl. **-os**) a piebald horse. □ **pinto bean** a kidney bean with mottled seeds, cultivated in America. [Sp., = mottled, ult. f. L *pictus* past part. of *pingere* paint]

pinworm /ˈpɪnwɜːm/ *n.* **1** a small nematode worm; esp. the parasitic *Enterobius vermicularis*, of which the female has a pointed tail (cf. THREADWORM). **2** an insect larva that leaves small holes in fruit, wood, etc.

piny /ˈpaɪnɪ/ *adj.* of, like, or full of pines (*a piny smell*).

Pinyin /pɪnˈjɪn/ *n.* a system of romanized spelling for transliterating Chinese, adopted by the People's Republic of China in 1958 (e.g. *Mao Tse-tung* now transliterated as *Mao Zedong*). [Chin. *pīn-yīn*, lit. 'spell sound']

piolet /pjəʊˈleɪ/ *n.* a two-headed ice-axe for mountaineering. [F]

pion /ˈpaɪɒn/ *n. Physics* a meson having a mass approximately 270 times that of an electron. Also called *pi-meson* (see PI[1]). □ **pionic** /paɪˈɒnɪk/ *adj.* [PI[1] (the letter used as a symbol for the particle) + -ON]

Pioneer /ˌpaɪəˈnɪə(r)/ a series of American space probes launched between 1958 and 1973. Pioneers 5 to 9 were placed in orbits around the sun, and Pioneers 10 and 11 provided the first clear pictures of Jupiter and Saturn (1973–79).

pioneer /ˌpaɪəˈnɪə(r)/ *n. & v.* ● *n.* **1** an initiator of a new enterprise, an inventor, etc. **2** an explorer or settler; a colonist. **3** *Mil.* a member of an infantry group preparing roads, terrain, etc. for the main body of troops. ● *v.* **1 a** *tr.* initiate or originate (an enterprise etc.). **b** *intr.* act or prepare the way as a pioneer. **2** *tr. Mil.* open up (a road etc.) as a pioneer. **3** *tr.* go before, lead, or conduct (another person or persons). [F *pionnier* foot-soldier, pioneer, OF *paonier, peon(n)ier* (as PEON)]

pious /ˈpaɪəs/ *adj.* **1** devout; religious. **2** hypocritically virtuous; sanctimonious. **3** dutiful. □ **pious fraud** a deception intended to benefit those deceived, esp. religiously. □ **piously** *adv.* **piousness** *n.* [L *pius* dutiful, pious]

pip[1] /pɪp/ *n. & v.* ● *n.* the seed of an apple, pear, orange, grape, etc. ● *v.tr.* (**pipped, pipping**) remove the pips from (fruit etc.). □ **pipless** *adj.* [abbr. of PIPPIN]

pip[2] /pɪp/ *n.* *Brit.* a short high-pitched sound, usu. mechanically produced, esp. as a radio time signal. [imit.]

pip[3] /pɪp/ *n.* **1** any of the spots on a playing card, dice, or domino. **2** *Brit.* a star (1–3 according to rank) on the shoulder of an army officer's uniform. **3** a single blossom of a clustered head of flowers. **4** a diamond-shaped segment of the surface of a pineapple. **5** an image of an object on a radar screen. [16th c. *peep*, of unkn. orig.]

pip[4] /pɪp/ *n.* **1** a disease of poultry etc. causing thick mucus in the throat and white scale on the tongue. **2** *colloq.* a fit of disgust or bad temper (esp. *give one the pip*). [ME f. MDu. *pippe*, MLG *pip* prob. ult. f. corrupt. of L *pituita* slime]

pip[5] /pɪp/ *v.tr.* (**pipped, pipping**) *Brit. colloq.* **1** hit with a shot. **2** defeat. **3** blackball. □ **pip at** (or **to**) **the post** defeat at the last moment. **pip out** die. [PIP[2] or PIP[1]]

pipa /ˈpiːpə/ *n.* = SURINAM TOAD. [prob. f. Galibi]

pipal var. of PEEPUL.

pipe /paɪp/ *n. & v.* ● *n.* **1** a tube of metal, plastic, wood, etc. used to convey water, oil, gas, etc. **2** (also **tobacco-pipe**) **a** a narrow wooden or clay etc. tube with a bowl at one end containing burning tobacco, the smoke from which is drawn into the mouth. **b** the quantity of tobacco held by this (*smoked a pipe*). **3** *Mus.* **a** a wind instrument consisting of a single tube. **b** any of the tubes by which sound is produced in an organ. **c** (in *pl.*) = BAGPIPE. **d** (in *pl.*) a set of pipes joined together, e.g. pan-pipes. **4** a tubal organ, vessel, etc. in an animal's body. **5** a high note or song, esp. of a bird. **6** a cylindrical vein of ore. **7** a cavity in cast metal. **8 a** a boatswain's whistle. **b** the sounding of this. **9** a cask for wine, esp. as a measure of two hogsheads, usu. equivalent to 105 gallons (about 477 litres). **10** *archaic* the voice, esp. in singing. ● *v.tr.* **1** (also *absol.*) play (a tune etc.) on a pipe or pipes. **2 a** convey (oil, water, gas, etc.) by pipes. **b** provide with pipes. **3** transmit (music, a radio programme, etc.) by wire or cable. **4** (usu. foll. by *up, on, to*, etc.) *Naut.* **a** summon (a crew) to a meal, work, etc. **b** signal the arrival of (an officer etc.) on board. **5** utter in a shrill voice; whistle. **6 a** arrange (icing, cream, etc.) in decorative lines or twists on a cake etc. **b** ornament (a cake etc.) with piping. **7** trim (a dress etc.) with piping. **8** lead or bring (a person etc.) by the sound of a pipe. **9** propagate (pinks etc.) by taking cuttings at the joint of a stem. □ **pipe away** *Naut.* give a signal for (a boat) to start. **pipe-cleaner** a piece of flexible covered wire for cleaning a tobacco-pipe. **piped music** pre-recorded background music played through loudspeakers in a public place. **pipe down 1** *colloq.* be quiet or less insistent. **2** *Naut.* dismiss from duty. **pipe-fish** a long slender fish of the family Syngnathidae, with an elongated snout. **pipe-light** a spill for lighting a pipe. **pipe major** an NCO commanding regimental pipers. **pipe-organ** *Mus.* an organ using pipes instead of or as well as reeds. **pipe-rack** a rack for holding tobacco-pipes. **pipe-rolls** *hist.* the annual records of the British Exchequer from the 12th–19th centuries (prob. because the documents were rolled in pipe form). **pipe-stem** the shaft of a tobacco-pipe. **pipe-stone** a hard red clay used by North American Indians for tobacco-pipes. **pipe up** begin to play, sing, speak, etc. **put that in your pipe and smoke it** *colloq.* a challenge to another to accept something frank or unwelcome. □ **pipeful** *n.* (*pl.* **-fuls**). **pipeless** *adj.* **pipy** *adj.* [OE *pīpe, pīpian* & OF *piper* f. Gmc ult. f. L *pipare* peep, chirp]

pipeclay /ˈpaɪpkleɪ/ *n. & v.* ● *n.* a fine white clay used for tobacco-pipes, whitening leather, etc. ● *v.tr.* **1** whiten (leather etc.) with this. **2** put in order.

pipedream /ˈpaɪpdriːm/ *n.* an unattainable or fanciful hope or scheme. [orig. as experienced when smoking an opium pipe]

pipeline /ˈpaɪplaɪn/ *n.* **1** a long, usu. underground, pipe for conveying esp. oil. **2** a channel supplying goods, information, etc. □ **in the pipeline** awaiting completion or processing.

pip emma /pɪp ˈemə/ *adv. & n. Brit. colloq.* = P.M. [formerly signallers' names for letters PM]

Piper /ˈpaɪpə(r)/, John (1903–92), English painter and decorative designer. His early paintings were abstract, but in the 1930s he turned to a romantic naturalism. During the Second World War he was one of the artists commissioned to depict the results of air raids on Britain. He is best known for watercolours and aquatints of buildings (such as those depicting Windsor Castle, 1941–2) and stained glass for Coventry and Llandaff cathedrals.

piper /ˈpaɪpə(r)/ *n.* **1** a bagpipe-player. **2** a person who plays a pipe, esp. an itinerant musician. [OE *pīpere* (as PIPE)]

Piper Alpha /ˌpaɪpər ˈælfə/ an oil platform in the North Sea off the coast of Scotland, which in July 1988 was destroyed by an explosion with the loss of 167 lives.

piperazine /pɪˈperəˌziːn/ *n. Chem.* a heterocyclic organic compound used as an anthelmintic, insecticide, etc. [PIPERIDINE + AZINE]

piperidine /pɪˈperɪˌdiːn/ *n. Chem.* a peppery-smelling liquid formed by the reduction of pyridine. [L *piper* pepper + -IDE + -INE[4]]

pipette /pɪˈpet/ *n. & v.* ● *n.* a slender tube for transferring or measuring small quantities of liquids esp. in chemistry. ● *v.tr.* transfer or measure (a liquid) using a pipette. [F, dimin. of *pipe* PIPE]

piping /ˈpaɪpɪŋ/ *n. & adj.* ● *n.* **1** the act or an instance of piping, esp. whistling or singing. **2** a thin pipelike fold used to edge hems or frills on clothing, seams on upholstery, etc. **3** ornamental lines of icing, cream, potato, etc. on a cake or other dish. **4** lengths of pipe, or a system of pipes, esp. in domestic use. ● *adj.* (of a noise) high; whistling. □ **piping hot** very or suitably hot (esp. as required of food, water, etc.).

pipistrelle /ˌpɪpɪˈstrel/ *n.* a small bat of the genus *Pipistrellus*, native to temperate regions; esp. *P. pipistrellus*, the commonest European bat. [F f. It. *pipistrello, vip-*, f. L *vespertilio* bat f. *vesper* evening]

pipit /ˈpɪpɪt/ *n.* a mainly ground-dwelling songbird of the genus *Anthus* (family Motacillidae), having brown-streaked plumage, e.g. the meadow pipit. [prob. imit.]

pipkin /ˈpɪpkɪn/ *n.* a small earthenware pot or pan. [16th c.: orig. unkn.]

pippin /ˈpɪpɪn/ *n.* **1 a** an apple grown from seed. **b** a red and yellow dessert apple. **2** *colloq.* an excellent person or thing; a beauty. [ME f. OF *pepin*, of unkn. orig.]

pipsqueak /ˈpɪpskwiːk/ *n. colloq.* an insignificant or contemptible person or thing. [imit.]

piquant /ˈpiːkənt, -kɒnt/ *adj.* **1** agreeably pungent, sharp, or appetizing. **2** pleasantly stimulating, or disquieting, to the mind. □ **piquancy** *n.* **piquantly** *adv.* [F, pres. part. of *piquer* (as PIQUE[1])]

pique[1] /piːk/ *v. & n.* ● *v.tr.* (**piques, piqued, piquing**) **1** wound the pride of, irritate. **2** arouse (curiosity, interest, etc.). **3** (*refl.*; usu. foll. by *on*) pride or congratulate oneself. ● *n.* ill-feeling; enmity; resentment (*in a fit of pique*). [F *piquer* prick, irritate, f. Rmc]

pique[2] /piːk/ *n. & v.* ● *n.* the winning of 30 points on cards and play in piquet before one's opponent scores anything. ● *v.* (**piques, piqued, piquing**) **1** *tr.* score a pique against. **2** *intr.* score a pique. [F *pic*, of unkn. orig.]

piqué /ˈpiːkeɪ/ *n.* a stiff ribbed cotton or other fabric. [F, past part. of *piquer*: see PIQUE[1]]

piquet[1] /pɪˈket/ *n.* a game for two players with a pack of thirty-two cards (seven up to ace only). [F, of unkn. orig.]

piquet[2] var. of PICKET *n.* 3.

piracy /ˈpaɪərəsɪ/ *n.* (*pl.* **-ies**) **1** the practice or an act of robbery of ships at sea. **2** a similar practice or act in other forms, esp. hijacking. **3** unauthorized reproduction or use of something, as a book, recording, computer program, etc., esp. when in contravention of patent or copyright. [med.L *piratia* f. Gk *pirateia* (as PIRATE)]

Piraeus /paɪˈriːəs/ (Greek **Piraiévs** or **Piraiéus** /ˌpɪrɛˈɛfs/) the chief port of Athens, situated on the Saronic Gulf 8 km (5 miles) SW of the city; pop. (1981) 196,400. Used as a port by the ancient Athenians, it was connected to the city by the 'long walls', two parallel walls built in the 5th century BC. It was destroyed by the Roman general Sulla in 86 BC. Extensive development in the 19th century led to its modern status as the principal seaport of Greece.

piragua /pɪˈrægwə/ *n.* **1** a long narrow canoe made from a single tree-trunk. **2** a two-masted sailing barge. [Sp. f. Carib, = dug-out]

Pirandello /ˌpɪrənˈdeləʊ/, Luigi (1867–1936), Italian dramatist and novelist. His plays challenged the conventions of naturalism and had a significant influence on the development of European drama, anticipating the anti-illusionist theatre of Brecht. Of his ten plays the best known include *Six Characters in Search of an Author* (1921) and *Henry IV* (1922). Among his novels are *The Outcast* (1901) and *The Late Mattia Pascal* (1904). He was awarded the Nobel Prize for literature in 1934.

Piranesi /ˌpɪrəˈneɪzɪ/, Giovanni Battista (1720–78), Italian engraver. His interest in classical Roman architecture is reflected in his prints, in which he relied on atypical viewpoints and dramatic chiaroscuro to aggrandize its power and scale. His *Prisons* (1745–61) extended this imagery into the realms of fantasy, producing a nightmare vision of claustrophobic space and endless dimensions that prefigured later romantic concerns.

piranha /pɪˈrɑːnə, -ˈrɑːnjə/ n. (also **piraya** /-ˈrɑːjə/) a predatory South American freshwater fish, esp. of the genus *Serrasalmus*, noted for its voracity and sharp teeth. [Port. f. Tupi, var. of *piraya* scissors]

pirate /ˈpaɪərət/ n. & v. ● n. **1 a** a person who commits piracy. **b** a ship used by pirates. **2** a person who infringes another's copyright or other business rights; a plagiarist. **3** (often *attrib.*) a person, organization, etc., that broadcasts without official authorization (*pirate radio station*). ● v.tr. **1** appropriate or reproduce (the work or ideas etc. of another) without permission, for one's own benefit. **2** plunder. □ **piratic** /ˌpaɪəˈrætɪk/ adj. **piratical** adj. **piratically** adv. [ME f. L *pirata* f. Gk *peiratēs* f. *peiraō* attempt, assault]

piraya var. of PIRANHA.

piripiri /ˈpɪrɪˌpɪrɪ/ n. (pl. **piripiris**) NZ a small prostrate rosaceous plant of the genus *Acaena*, native to New Zealand and having prickly burs. [Maori]

pirogue /pɪˈrəʊg/ n. = PIRAGUA. [F, prob. f. Galibi]

pirouette /ˌpɪrʊˈet/ n. & v. ● n. a dancer's spin on one foot or the point of the toe. ● v.intr. perform a pirouette. [F, = spinning-top]

Pisa /ˈpiːzə/ a city in northern Italy, in Tuscany, on the River Arno; pop. (1990) 101,500. Formerly situated on the coast, it was an important Etruscan town and a naval base in Roman times, becoming a powerful maritime city-state in the Middle Ages. It now lies about 10 km (6 miles) inland, as a result of the silting of the Arno. The city is noted for the 'Leaning Tower of Pisa', a circular bell-tower which leans about 5 m (17 ft) from the perpendicular over its height of 55 m (181 ft), part of this inclination dating from its construction at the end of the 12th century.

pis aller /ˌpiːz æ'leɪ/ n. a course of action followed as a last resort. [F f. *pis* worse + *aller* go]

Pisan, Christine de, see DE PISAN.

Pisano[1] /pɪˈsɑːnəʊ/, Andrea (c.1290–c.1348) and Nino, his son (died c.1368), Italian sculptors. Andrea is notable as the creator of the earliest pair of bronze doors for the baptistery at Florence (completed 1336). Nino was one of the earliest to specialize in free-standing life-size figures.

Pisano[2] /pɪˈsɑːnəʊ/, Nicola (c.1220–c.1278) and Giovanni, his son (c.1250–c.1314), Italian sculptors. Nicola's work departed from medieval conventions and signalled a revival of interest in classical sculpture. He brought human expression and dramatic power to his works, the best known of which are the pulpits in the baptistery at Pisa (c.1255–60) and in Siena cathedral (1265–8). Giovanni's works carried this process further and foreshadow the sculptural renaissance which followed. His works include a pulpit in the church of Santa Andrea in Pistoia (completed 1301), and the richly decorated façade of Siena cathedral (completed in 1284).

piscary /ˈpɪskərɪ/ n. □ **common of piscary** the right of fishing in another's water in common with the owner and others. [ME f. med.L *piscaria* neut. pl. of L *piscarius* f. *piscis* fish]

piscatorial /ˌpɪskəˈtɔːrɪəl/ adj. = PISCATORY 1. □ **piscatorially** adv.

piscatory /ˈpɪskətərɪ/ adj. formal **1** of or concerning fishermen or fishing. **2** very enthusiastic about fishing. [L *piscatorius* f. *piscator* fisherman f. *piscis* fish]

Pisces /ˈpaɪsiːz, ˈpɪskiːz/ n. **1** Astron. a large constellation (the Fish or Fishes), said to represent a pair of fishes tied together by their tails. The First Point of Aries is currently within it, but it contains no bright stars. **2** Astrol. **a** the twelfth sign of the zodiac, which the sun enters about 20 Feb. **b** a person born when the sun is in this sign. □ **Piscean** /ˈpaɪsɪən, ˈpɪskɪ-/ n. & adj. [ME f. L, pl. of *piscis* fish]

pisciculture /ˈpɪsɪˌkʌltʃə(r)/ n. the artificial rearing of fish. □ **piscicultural** /ˌpɪsɪˈkʌltʃərəl/ adj. **pisciculturist** n. [L *piscis* fish, after *agriculture* etc.]

piscina /pɪˈsiːnə, -ˈsaɪmə/ n. (pl. **piscinae** /-niː/ or **piscinas**) **1** a stone basin near the altar in RC and pre-Reformation churches for draining water used in the Mass. **2** a fish-pond. **3** hist. a Roman bathing-pond. [L f. *piscis* fish]

piscine /ˈpɪsaɪn/ adj. of or concerning fish. [L *piscis* fish]

piscivorous /pɪˈsɪvərəs/ adj. fish-eating. [L *piscis* fish + -VOROUS]

pish /pɪʃ/ int. & n. ● int. an expression of contempt, impatience, or disgust. ● n. nonsense, rubbish. [imit.]

Pishpek /pɪʃˈpek/ the former name (until 1926) for BISHKEK.

Pisidia /paɪˈsɪdɪə/ an ancient region of Asia Minor, between Pamphylia and Phrygia. Traversed by the Taurus Mountains, the region maintained its independence until, on the death of King Amyntas in 25 BC, it was incorporated into the Roman province of Galatia. □ **Pisidian** adj. & n.

pisiform /ˈpɪsɪˌfɔːm/ adj. pea-shaped. □ **pisiform bone** Anat. a small bone in the wrist in the upper row of the carpus. [mod.L *pisiformis* f. *pisum* pea]

Pisistratus /paɪˈsɪstrətəs/ (also **Peisistratus**) (c.600–c.527 BC), tyrant of Athens. He seized power in 561 and after twice being expelled ruled continuously from 546 until his death. As ruler, he reduced aristocratic power in rural Attica and promoted the financial prosperity and cultural pre-eminence of Athens.

pismire /ˈpɪsˌmaɪə(r)/ n. dial. an ant. [ME f. PISS (from smell of anthill) + obs. *mire* ant]

piss /pɪs/ v. & n. coarse sl. ● v. **1** intr. urinate. **2** tr. a discharge (blood etc.) when urinating. **b** wet with urine. **3** tr. (as **pissed** adj.) Brit. drunk. **4** refl. **a** wet one's clothing with urine. **b** be very frightened, amused, or excited. ● n. **1** urine. **2** an act of urinating. □ **piss about** fool or mess about. **piss artist 1** a drunkard. **2** a person who fools about. **3** a glib person. **piss down** rain heavily. **piss in the wind** do something to no effect or against one's own interests. **piss off** Brit. **1** go away. **2** (often as **pissed off** adj.) annoy; depress. **piss on** show utter contempt for, esp. by humiliating; defeat heavily. **piss-pot** a chamber-pot. **piss-take** a parody. **piss-taker** a person who mocks. **piss-up** a drinking spree. **take the piss** (often foll. by *out of*) mock; deride. [ME f. OF *pisser* (imit.)]

Pissarro /pɪˈsɑːrəʊ/, Camille (1830–1903), French painter and graphic artist. Born in the West Indies, he moved to France in the 1850s and studied in Paris with Monet. At first influenced by Corot, he later began to paint out of doors and to develop a spontaneous style typical of other impressionist painters; he participated in all eight of the impressionist exhibitions. He also experimented with pointillism in the 1880s as well as encouraging Cézanne and Gauguin in their attempts to depart from an impressionist style of painting.

pissoir /ˈpiːswɑː(r)/ n. a public urinal. [F]

pistachio /pɪˈstɑːʃɪˌəʊ/ n. (pl. **-os**) **1** an evergreen tree, *Pistacia vera*, bearing small brownish-green flowers and ovoid reddish fruit. **2** (in full **pistachio nut**) the edible pale-green seed of this. **3** a pale green colour. [It. *pistaccio* and Sp. *pistacho* f. L *pistacium* f. Gk *pistakion* f. Pers. *pistah*]

piste /piːst/ n. a ski-run of compacted snow. [F, = racetrack]

pistil /ˈpɪstɪl/ n. Bot. the female organs of a flower, comprising the stigma, style, and ovary. □ **pistillary** adj. **pistilline** /-ˌlaɪn/ adj. **pistilliferous** /ˌpɪstɪˈlɪfərəs/ adj. [F *pistile* or L *pistillum* PESTLE]

pistillate /ˈpɪstɪlət/ adj. Bot. **1** having pistils. **2** having pistils but no stamens.

pistol /ˈpɪst(ə)l/ n. & v. ● n. **1** a small hand-held firearm. **2** anything of a similar shape. ● v.tr. (**pistolled, pistolling**; US **pistoled, pistoling**) shoot with a pistol. □ **hold a pistol to a person's head** coerce a person by threats. **pistol-grip** a handle shaped like a pistol-butt. **pistol-shot 1** the range of a pistol. **2** a shot fired from a pistol. **pistol-whip** (**-whipped, -whipping**) beat with a pistol. [obs. F f. G *Pistole* f. Czech *píšt'al*]

pistole /pɪˈstəʊl/ n. hist. a foreign (esp. Spanish) gold coin. [F *pistole* abbr. of *pistolet*, of uncert. orig.]

pistoleer /ˌpɪstəˈlɪə(r)/ n. a soldier armed with a pistol.

piston /ˈpɪst(ə)n/ n. **1** a disc or short cylinder fitting closely within a tube in which it moves up and down against a liquid or gas, used in an internal-combustion engine to receive motion from the expansion of gas, or in a pump to impart motion to a fluid. **2** a sliding valve in a trumpet etc. □ **piston-ring** a ring on a piston sealing the gap between the piston and the cylinder wall. **piston-rod** a rod or crankshaft attached to a piston to drive a wheel or to impart motion. [F f. It. *pistone* var. of *pestone* augment. of *pestello* PESTLE]

pit[1] /pɪt/ n. & v. ● n. **1 a** a usu. large deep hole in the ground. **b** a hole made in digging for industrial purposes, esp. for coal (*chalk pit*; *gravel pit*). **c** a covered hole as a trap for esp. wild animals. **2** an indentation left after smallpox, acne, etc. **b** a hollow in a plant or animal body or on any surface. **3** Brit. Theatr. **a** an orchestra pit (see ORCHESTRA 2a). **b** usu. hist. seating at the back of the stalls. **c** the people in the pit. **4 a** (**the pit** or **bottomless pit**) hell. **b** (**the pits**) sl. a wretched or the worst imaginable place, situation, person, etc. **5 a** an area at the side of a track where racing cars are serviced and refuelled. **b** a sunken area in a workshop floor for access to a car's underside. **6** N. Amer. the part of the floor of an exchange allotted to special trading (*wheat-pit*). **7** = COCKPIT. **8** *Brit. sl.* a bed. ● *v.* (**pitted, pitting**) **1** *tr.* (usu. foll. by *against*) **a** set (one's wits, strength, etc.) in opposition or rivalry. **b** set (a cock, dog, etc.) to fight, orig. in a pit, against another. **2** *tr.* (usu. as **pitted** *adj.*) make pits, esp. scars, in. **3** *intr.* (of the flesh etc.) retain the impression of a finger etc. when touched. **4** *tr.* put (esp. vegetables for storage) into a pit. □ **dig a pit for** try to ensnare. **pit bull terrier** a small heavily built American breed of dog noted for its ferocity. **pit-head 1** the top of a mineshaft. **2** the area surrounding this. **pit of the stomach 1** the floor of the stomach. **2** the depression below the bottom of the breastbone. **pit pony** *hist.* a pony kept underground for haulage in coal mines. **pit-prop** a balk of wood used to support the roof of a coal mine. **pit-saw** a large saw for use in a saw-pit. **pit viper** a viper of the subfamily Crotalinae, native to America and Asia, with a sensory pit on the head; e.g a rattlesnake. [OE *pytt* ult. f. L *puteus* well]

pit² /pɪt/ *n. & v.* esp. *US* ● *n.* the stone of a fruit; a pip. ● *v.tr.* (**pitted, pitting**) remove pits from (fruit). [perh. Du., rel. to PITH]

pita var. of PITTA¹.

pit-a-pat /ˈpɪtəˌpæt/ *adv. & n.* (also **pitter-patter** /ˈpɪtəˌpætə(r)/) ● *adv.* **1** with a sound like quick light steps. **2** with a faltering sound (*heart went pit-a-pat*). ● *n.* such a sound. [imit.]

Pitcairn Islands /ˈpɪtkeən/ a British dependency comprising a group of volcanic islands in the South Pacific, east of French Polynesia. The colony's only settlement is Adamstown, on Pitcairn Island, the chief island of the group; pop. (1991) 61. Pitcairn Island was discovered in 1767 by a British naval officer, Philip Carteret (d.1796), and named after the midshipman who first sighted it. It remained uninhabited until settled in 1790 by mutineers from HMS *Bounty* and their Tahitian companions, some of whose descendants still live there.

pitch¹ /pɪtʃ/ *v. & n.* ● *v.* **1** *tr.* (also *absol.*) erect and fix (a tent, camp, etc.). **2** *tr.* **a** throw; fling. **b** (in games) throw (a flat object) towards a mark. **3** *tr.* fix or plant (a thing) in a definite position. **4** *tr.* express in a particular style or at a particular level (*pitched his argument at the most basic level*). **5** *intr.* (often foll. by *against, into*, etc.) fall heavily, esp. headlong. **6** *intr.* (of a ship etc.) plunge forwards in a longitudinal direction (cf. ROLL *v.* 8a). **7** *tr. Mus.* set at a particular pitch. **8** *intr.* (of a roof etc.) slope downwards. **9** *intr.* (often foll. by *about*) move with a vigorous jogging motion, as in a train, carriage, etc. **10** *Cricket* **a** *tr.* cause (a bowled ball) to strike the ground at a specified point etc. **b** *intr.* (of a bowled ball) strike the ground. **11** *tr. colloq.* tell (a yarn or a tale). **12** *tr. Golf* play (a ball) with a pitch shot. **13** *tr.* pave (a road) with stones. ● *n.* **1 a** the area of play in a field game. **b** *Cricket* the area between the creases. **2** height, degree, intensity, etc. (*the pitch of despair; nerves were strung to a pitch*). **3 a** the steepness of a slope, esp. of a roof, stratum, etc. **b** the degree of such a pitch. **4** *Mus.* **a** that quality of a sound which is governed by the rate of vibrations producing it; the degree of highness or lowness of a tone. **b** = CONCERT PITCH 1. **5** the pitching motion of a ship etc. **6** *Cricket* the act or mode of delivery in bowling, or the spot where the ball bounces. **7** *colloq.* a salesman's advertising or selling approach. **8** *Brit.* a place where a street vendor sells wares, has a stall, etc. **9** (also **pitch shot**) *Golf* a high approach shot with a short run. **10** *Mech.* **a** the distance between successive corresponding points or lines, e.g. between the teeth of a cog-wheel, the convolutions of a screw-thread, etc. **b** the inclination of the blades of a propeller or rotor, as determining the forward distance moved in one revolution while exerting no thrust on the medium. **11** the height to which a falcon etc. soars before swooping on its prey. **12** the delivery of a baseball by a pitcher. □ **pitch-and-toss** a gambling game in which coins are pitched at a mark and then tossed. **pitched battle 1** *Mil.* a battle planned beforehand and fought on chosen ground. **2** a vigorous argument etc. **pitched roof** a sloping roof. **pitch in** *colloq.* **1** set to work vigorously. **2** assist, cooperate. **pitch into** *colloq.* **1** attack forcibly with blows, words, etc. **2** assail (food, work, etc.) vigorously. **pitch on** (or **upon**) happen to select. **pitch-pipe** *Mus.* a small pipe blown to set the pitch for singing or tuning. **pitch up** *Cricket* bowl (a ball) to bounce near the batsman. **pitch wickets** *Cricket* fix the stumps in the ground and place the bails. [ME *pic(c)he*, perh. f. OE *picc(e)an* (unrecorded: cf. *picung* stigmata)]

pitch² /pɪtʃ/ *n. & v.* ● *n.* **1** a sticky resinous black or dark brown substance obtained by distilling tar or turpentine, semi-liquid when hot, hard when cold, and used for caulking the seams of ships etc. **2** any bituminous substance, e.g. asphalt. ● *v.tr.* cover, coat, or smear with pitch. □ **pitch-black** (or **-dark**) very or completely dark. **pitch pine** a pine tree; esp. *Pinus rigida* or *P. palustris* with notably resinous wood, of North America. [OE *pic* f. Gmc f. L *pix picis*]

pitchblende /ˈpɪtʃblend/ *n.* the usual form of the mineral uraninite,

occurring in pitchlike masses and yielding radium. [G *Pechblende* (as PITCH², BLENDE)]

pitcher¹ /ˈpɪtʃə(r)/ *n.* **1** a large usu. earthenware jug with a lip and a handle, for holding liquids. **2** the pitcher-like modified leaf of a pitcher-plant. **3** (in *pl.*) broken pottery crushed and reused. □ **pitcher-plant** a plant of the family Nepenthaceae or Sarraceniaceae with pitcher-like structures containing liquid to trap insects etc. □ **pitcherful** *n.* (*pl.* **-fuls**). [ME f. OF *pichier, pechier*, f. Frank.]

pitcher² /ˈpɪtʃə(r)/ *n.* **1** a person or thing that pitches. **2** *Baseball* a player who delivers the ball to the batter. **3** a stone used for paving.

pitchfork /ˈpɪtʃfɔːk/ *n. & v.* ● *n.* a long-handled two-pronged fork for pitching hay etc. ● *v.tr.* **1** throw with or as if with a pitchfork. **2** (usu. foll. by *into*) thrust (a person) forcibly into a position, office, etc. [in ME *pickfork*, prob. f. PICK² + FORK, assoc. with PITCH¹]

pitchstone /ˈpɪtʃstəʊn/ *n.* a dull vitreous rock resembling pitch.

pitchy /ˈpɪtʃɪ/ *adj.* (**pitchier, pitchiest**) of, like, or dark as pitch.

piteous /ˈpɪtɪəs/ *adj.* deserving or causing pity; wretched. □ **piteously** *adv.* **piteousness** *n.* [ME *pito(u)s* etc. f. AF *pitous*, OF *pitos* f. Rmc (as PIETY)]

pitfall /ˈpɪtfɔːl/ *n.* **1** an unsuspected snare, danger, or drawback. **2** a covered pit for trapping animals etc.

pith /pɪθ/ *n. & v.* ● *n.* **1** spongy white tissue lining the rind of an orange, lemon, etc. **2** the essential part; the quintessence (*came to the pith of his argument*). **3** *Bot.* the spongy cellular tissue in the stems and branches of dicotyledonous plants. **4 a** physical strength; vigour. **b** force; energy. **5** importance, weight. **6** *archaic* spinal cord. ● *v.tr.* **1** remove the pith or marrow from. **2** slaughter or immobilize (an animal) by severing the spinal cord. □ **pith helmet** a lightweight sun-helmet made from the dried pith of the sola etc. □ **pithless** *adj.* [OE *pitha* f. WG]

Pithecanthropus /ˌpɪθɪˈkænθrəpəs/ *n.* a genus name formerly applied to some fossil hominids of the species *Homo erectus*, named from remains found in Java in 1891. (See JAVA MAN.) [mod.L f. Gk *pithēkos* ape + *anthrōpos* man]

pithos /ˈpɪθɒs/ *n.* (*pl.* **pithoi** /-θɔɪ/) *Archaeol.* a large storage jar. [Gk]

pithy /ˈpɪθɪ/ *adj.* (**pithier, pithiest**) **1** (of style, speech, etc.) condensed, terse, and forcible. **2** of, like, or containing much pith. □ **pithily** *adv.* **pithiness** *n.*

pitiable /ˈpɪtɪəb(ə)l/ *adj.* **1** deserving or causing pity. **2** contemptible. □ **pitiableness** *n.* **pitiably** *adv.* [ME f. OF *piteable, pitoiable* (as PITY)]

pitiful /ˈpɪtɪˌfʊl/ *adj.* **1** causing pity. **2** contemptible. **3** *archaic* compassionate. □ **pitifully** *adv.* **pitifulness** *n.*

pitiless /ˈpɪtɪlɪs/ *adj.* showing no pity (*the pitiless heat of the desert*). □ **pitilessly** *adv.* **pitilessness** *n.*

Pitman /ˈpɪtmən/, Sir Isaac (1813–97), English inventor of a shorthand system. Inspired by the phonetic shorthand system designed in 1786 by Samuel Taylor (1749–1811), Pitman devised his own system, published as *Stenographic Sound Hand* (1837). Pitman shorthand, first adopted (in the US) in 1852, is still widely used in the UK and elsewhere.

pitman /ˈpɪtmən/ *n.* **1** (*pl.* **-men**) a collier. **2** *US* (*pl.* **-mans**) a connecting rod in machinery.

piton /ˈpiːtɒn/ *n.* a peg or spike driven into a rock or crack to support a climber or a rope. [F, = eye-bolt]

Pitons, the /ˈpiːtɒnz/ two conical mountains in St Lucia in the West Indies. Reaching a height of 798 m (2,618 ft) and 750 m (2,461 ft), they rise up out of the Caribbean Sea just off the SW coast of the island.

Pitot tube /ˈpiːtəʊ/ *n.* a device used to measure the relative speed of a fluid, consisting essentially of a right-angled tube with an open end facing the fluid flow. The velocity is measured from the pressure difference between this tube and a parallel tube that registers the static pressure. The most familiar application of the Pitot tube is in the determination of the air speed of an aircraft. [Henri *Pitot*, Fr. physicist (1695–1771)]

pitpan /ˈpɪtpæn/ *n.* a Central American boat made from a tree-trunk. [Miskito]

Pitt¹ /pɪt/, William (known as Pitt the Elder), 1st Earl of Chatham (1708–78), British Whig statesman. He became Secretary of State (effectively Prime Minister) in 1756 and headed coalition governments 1756–61 and 1766–8. He brought the Seven Years War to an end in 1763 by using a successful maritime strategy to defeat France. He also masterminded the conquest of French possessions overseas, particularly in Canada and India.

Pitt² /pɪt/, William (known as Pitt the Younger) (1759–1806), British statesman, Prime Minister 1783–1801 and 1804–6. The son of Pitt the

Elder, he became Prime Minister at the age of 24, the youngest ever to hold this office. He restored the authority of Parliament, introduced financial reforms, reduced the enormous national debt he had inherited, and reformed the administration of India. With Britain's entry into war against France (1793), Pitt became almost entirely occupied with the conduct of the war and with uniting European opposition to France. Having secured the Union of Great Britain and Ireland in 1800, he resigned in 1801 over the issue of Catholic Emancipation (which George III refused to accept). He returned as Premier in 1804 after hostilities with France had been resumed, and died in office.

pitta[1] /'pɪtə/ n. (also **pita**) a flat hollow unleavened bread which can be split and filled with salad etc. [mod.Gk, = a cake]

pitta[2] /'pɪtə/ n. a brightly coloured passerine bird of the Old World genus *Pitta* and family Pittidae, with a strong bill and short tail. [Telugu *piṭṭa* (young) bird]

pittance /'pɪt(ə)ns/ n. **1** a scanty or meagre allowance, remuneration, etc. (*paid him a mere pittance*). **2** a small number or amount. **3** *hist.* a pious bequest to a religious house for extra food etc. [ME f. OF *pitance* f. med.L *pi(e)tantia* f. L *pietas* PITY]

pitter-patter var. of PIT-A-PAT.

Pitti /'pɪti/ an art gallery and museum in Florence, housed in the Pitti Palace (built 1440–c.1549). It contains about 500 masterpieces from the Medici collections, a profusion of art treasures including Gobelin tapestries, and a rich collection of plate, goldsmiths' work, ivories, enamels, etc.

Pitt Island see CHATHAM ISLANDS.

pittosporum /pɪ'tɒspərəm/ n. an evergreen shrub of the genus *Pittosporum*, chiefly native to Australasia, with small often fragrant flowers. [mod.L f. Gk *pitta* PITCH[2] + *sporos* seed]

Pitt-Rivers /pɪt'rɪvəz/, Augustus Henry Lane Fox (1827–1900), English archaeologist and anthropologist. In 1882 he retired from the army and began a series of excavations of the prehistoric, Roman, and Saxon sites on his Wiltshire estate. His scientific approach and emphasis on the importance of everyday objects greatly influenced the development of modern archaeological techniques. He also carried out pioneering work in establishing typological sequences of artefacts from different cultures. He donated his collection of weapons and artefacts to found the ethnological museum in Oxford which bears his name.

Pittsburgh /'pɪtsbɜːg/ an industrial city in SW Pennsylvania, at the junction of the Allegheny and Monongahela rivers; pop. (1990) 369,880. The city, originally named Fort Pitt after the British statesman William Pitt the Elder, was founded in 1758 on the site of a former French settlement. A coal-mining and steel-producing town for many years, it is now a major centre of high technology.

pituitary /pɪ'tjuːɪtəri/ n. & adj. ● n. (also **pituitary gland** or **body**) (*pl.* **-ies**) a pea-sized ductless gland at the base of the brain which secretes various hormones essential for growth and other bodily functions. (*See note below.*) ● adj. of or relating to this gland. [L *pituitarius* secreting phlegm f. *pituita* phlegm]

- The anterior lobe of the pituitary secretes various hormones stimulating body growth, regulating the activity of the gonads and the adrenal and thyroid glands, and stimulating the growth and activity of the mammary glands. Its posterior lobe, consisting mainly of nervous tissue, releases two hormones, oxytocin and vasopressin.

pity /'pɪti/ n. & v. ● n. (*pl.* **-ies**) **1** sorrow and compassion aroused by another's condition (*felt pity for the child*). **2** something to be regretted; grounds for regret (*what a pity!*; *the pity of it is that he didn't mean it*). ● v.tr. (**-ies, -ied**) feel (often contemptuous) pity for (*they are to be pitied*; *I pity you if you think that*). □ **for pity's sake** an exclamation of urgent supplication, anger, etc. **more's the pity** so much the worse. **take pity on** feel or act compassionately towards. □ **pitying** adj. **pityingly** adv. [ME f. OF *pité* f. L *pietas* (as PIETY)]

pityriasis /ˌpɪtɪ'raɪəsɪs/ n. Med. any of a group of skin diseases characterized by the shedding of fine flaky scales. [mod.L f. Gk *pituriasis* f. *pituron* bran]

più /pju:/ adv. Mus. more (*più piano*). [It.]

Pius XII /'paɪəs/ (born Eugenio Pacelli) (1876–1958), pope 1939–58. He upheld the neutrality of the Roman Catholic Church during the Second World War, maintaining diplomatic relations with both Allied and Axis governments. After the war there was criticism of his failure to condemn Nazi atrocities and of his apparent ambivalence towards anti-Semitism. Pius XII took steps to counter the rise of Communism in postwar Italy, threatening to excommunicate its supporters.

pivot /'pɪvət/ n. & v. ● n. **1** a short shaft or pin on which something turns or oscillates. **2** a crucial or essential person, point, etc., in a scheme or enterprise. **3** *Mil.* the man or men about whom a body of troops wheels. ● v. (**pivoted, pivoting**) **1** intr. turn on or as if on a pivot. **2** intr. (foll. by on) hinge on; depend on. **3** tr. provide with or attach by a pivot. □ **pivotal** adj. **pivotable** adj. **pivotability** /ˌpɪvətə'bɪlɪti/ n. [F, of uncert. orig.]

pix[1] /pɪks/ n.pl. colloq. pictures, esp. photographs. [abbr.: cf. PIC]

pix[2] var. of PYX.

pixel /'pɪks(ə)l/ n. Electronics any of the minute areas of uniform illumination of which an image on a display screen is composed. [abbr. of *picture element*: cf. PIX[1]]

pixie /'pɪksi/ n. (also **pixy**) (*pl.* **-ies**) a being like a fairy; an elf. □ **pixie hat** (or **hood**) a child's hat with a pointed crown. [17th c.: orig. unkn.]

pixilated /'pɪksɪˌleɪtɪd/ adj. (also **pixillated**) **1** bewildered; crazy. **2** drunk. [var. of *pixie-led* (as PIXIE, LED)]

Pizan, Christine de, see DE PISAN.

Pizarro /pɪ'zɑːrəʊ/, Francisco (c.1478–1541), Spanish conquistador. In 1531 he set out from Panama to conquer the Inca empire in Peru. Crossing the mountains, he defeated the Incas and in 1533 executed their emperor Atahualpa (b.1502), setting up an Inca puppet monarchy at Cuzco and building his own capital at Lima (1535). He was assassinated in Lima by supporters of his rival Diego de Almagro (1475–1538).

pizazz /pɪ'zæz/ n. (also **pizzazz**, **pzazz** etc.) sl. verve, energy, liveliness, sparkle.

pizza /'piːtsə/ n. a flat round base of dough with a topping of tomatoes, cheese, onions, etc. [It., = pie]

pizzeria /ˌpiːtsə'riːə/ n. a place where pizzas are made or sold; a pizza restaurant. [It. (as PIZZA)]

pizzicato /ˌpɪtsɪ'kɑːtəʊ/ adv., adj., & n. Mus. ● adv. plucking the strings of a violin etc. with the finger. ● adj. (of a note, passage, etc.) performed pizzicato. ● n. (*pl.* **pizzicatos** or **pizzicati** /-tɪ/) a note, passage, etc. played pizzicato. [It., past part. of *pizzicare* twitch f. *pizzare* f. *pizza* edge]

pizzle /'pɪz(ə)l/ n. the penis of an animal, esp. a bull, formerly used as a whip. [LG *pesel*, dimin. of MLG *pēse*, MDu. *pēze*]

pk. abbr. **1** park. **2** peak. **3** peck(s).

pl. abbr. **1** plural. **2** place. **3** plate. **4** esp. Mil. platoon.

PLA abbr. **1** (in the UK) Port of London Authority. **2** see PEOPLE'S LIBERATION ARMY.

placable /'plækəb(ə)l/ adj. easily placated; mild; forgiving. □ **placably** adv. **placability** /ˌplækə'bɪlɪti/ n. [ME f. OF *placable* or L *placabilis* f. *placare* appease]

placard /'plækɑːd/ n. & v. ● n. a printed or handwritten poster esp. for advertising. ● v.tr. **1** set up placards on (a wall etc.). **2** advertise by placards. **3** display (a poster etc.) as a placard. [ME f. OF *placquart* f. *plaquier* to plaster f. MDu. *placken*]

placate /plə'keɪt/ v.tr. pacify; conciliate. □ **placatingly** adv. **placation** /-'keɪʃ(ə)n/ n. **placatory** /'plækətəri, plə'keɪt-/ adj. [L *placare placat-*]

place /pleɪs/ n. & v. ● n. **1 a** a particular portion of space. **b** a portion of space occupied by a person or thing (*it has changed its place*). **c** a proper or natural position (*he is out of his place*; *take your places*). **d** situation, circumstances (*put yourself in my place*). **2** a city, town, village, etc. (*was born in this place*). **3** a residence; a dwelling (*has a place in the country*; *come round to my place*). **4 a** a group of houses in a town etc., esp. a square. **b** a country house with its surroundings. **5** a person's rank or status (*know their place*; *a place in history*). **6** a space, esp. a seat, for a person (*two places in the coach*). **7** a building or area for a specific purpose (*place of worship*; *bathing-place*). **8 a** a point reached in a book etc. (*lost my place*). **b** a passage in a book. **9** a particular spot on a surface, esp. of the skin (*a sore place on his wrist*). **10 a** employment or office, esp. government employment (*lost his place at the Ministry*). **b** the duties or entitlements of office etc. (*is his place to hire staff*). **11** a position as a member of a team, a student in a college, etc. **12** Brit. any of the first three or sometimes four positions in a race, esp. other than the winner (*backed it for a place*). **13** the position of a figure in a series indicated in decimal or similar notation (*calculated to 50 decimal places*). ● v.tr. **1** put (a thing etc.) in a particular place or state; arrange. **2** identify, classify, or remember correctly (*cannot place him*). **3** assign to a particular place; locate. **4 a** appoint (a person, esp. a member of the clergy) to a post. **b** find a situation, living, etc. for. **c** (usu. foll. by with) consign to a person's care etc. (*placed her with her aunt*). **5** assign rank, importance, or worth to (*place him among the best teachers*). **6 a** dispose of (goods) to a customer. **b** make (an order for goods etc.). **7** (often foll. by *in*, *on*, etc.) have (confidence etc.). **8** invest (money). **9** *Brit.* state the position of (any of the first three or sometimes four runners) in a race. **10** (as **placed** *adj.*) **a** *Brit.* among the first three or sometimes four in a race. **b** *US* second in a race. **11** *Football* get (a goal) by a place-kick. □ **all over the place** in disorder; chaotic. **give place to 1** make room for. **2** yield precedence to. **3** be succeeded by. **go places** *colloq.* be successful. **in place** in the right position; suitable. **in place of** in exchange for; instead of. **in places** at some places or in some parts, but not others. **keep a person in his** or **her place** suppress a person's pretensions. **out of place 1** in the wrong position. **2** unsuitable. **place-bet 1** *Brit.* a bet on a horse to come first, second, third, or sometimes fourth in a race. **2** *US* a bet on a horse to come second. **place-brick** an imperfectly burnt brick from the windward side of the kiln. **place card** a card marking a person's place at a table etc. **place in the sun** a favourable situation, position, etc. **place-kick** *Football* a kick made when the ball is previously placed on the ground. **place-mat** a small mat on a table underneath a person's plate. **place-name** the name of a town, village, hill, field, lake, etc. **place-setting** a set of plates, cutlery, etc. for one person at a meal. **put oneself in another's place** imagine oneself in another's position. **put a person in his** or **her place** deflate or humiliate a person. **take place** occur. **take one's place** go to one's correct position, be seated, etc. **take the place of** be substituted for; replace. □ **placeless** *adj.* **placement** *n.* [ME f. OF f. L *platea* f. Gk *plateia* (*hodos*) broad (way)]

placebo /pləˈsiːbəʊ/ *n.* (*pl.* **-os**) **1 a** *Med.* a pill, medicine, etc. prescribed more for psychological reasons than for any physiological effect. **b** a blank sample used as a control in testing new drugs etc. **2** something that is said or done to calm or humour a person but does not address the cause of his or her anxiety. **3** *RC Ch.* the opening antiphon of the vespers for the dead. □ **placebo effect** a beneficial (or adverse) effect produced by a placebo and not due to any property of the placebo itself. [L, = I shall be acceptable or pleasing f. *placere* please, first word of Ps. 114:9]

placenta /pləˈsentə/ *n.* (*pl.* **placentae** /-tiː/ or **placentas**) **1** a flattened circular organ that develops in the uterus of pregnant eutherian mammals and serves to nourish the foetus, to which it is attached by the umbilical cord. (*See note below.*) **2** (in flowers) part of the ovary wall carrying the ovules. [L f. Gk *plakous -ountos* flat cake f. the root of *plax plakos* flat plate]

▪ The placenta consists of vascular tissue in which oxygen and nutrients can pass from the mother's blood into that of the foetus, and waste products can pass in the reverse direction. The placenta is expelled from the uterus at the birth of the foetus, when it is often called the *afterbirth*. Marsupials and monotremes do not develop placentas.

placental /pləˈsent(ə)l/ *adj.* & *n.* ● *adj.* **1** of or relating to a placenta. **2** *Zool.* = EUTHERIAN *adj.* ● *n. Zool.* = EUTHERIAN *n.*

placer /ˈpleɪsə(r), ˈplæs-/ *n.* a deposit of sand, gravel, etc., in the bed of a stream etc., containing valuable minerals in particles. [Amer. Sp., rel. to *placel* sandbank f. *plaza* PLACE]

placet /ˈpleɪset/ *n.* an affirmative vote in a church or university assembly. [L, = it pleases]

placid /ˈplæsɪd/ *adj.* **1** (of a person) not easily aroused or disturbed; peaceful. **2** mild; calm; serene. □ **placidly** *adv.* **placidness** *n.* **placidity** /pləˈsɪdɪtɪ/ *n.* [F *placide* or L *placidus* f. *placere* please]

placket /ˈplækɪt/ *n.* **1** an opening or slit in a garment, for fastenings or access to a pocket. **2** the flap of fabric under this. [var. of PLACARD]

placoid /ˈplækɔɪd/ *adj.* & *n. Zool.* ● *adj.* **1** (of a fish-scale) toothlike, with a flat base and a spiny backward projection, characteristic of sharks and rays (cf. CTENOID). **2** (of a fish) covered with these scales. ● *n.* a placoid fish, e.g. a shark. [Gk *plax plakos* flat plate]

plafond /pləˈfɒn/ *n.* **1 a** an ornately decorated ceiling. **b** such decoration. **2** an early form of contract bridge. [F f. *plat* flat + *fond* bottom]

plagal /ˈpleɪɡ(ə)l/ *adj. Mus.* (of a church mode) having sounds between the dominant and its octave (cf. AUTHENTIC). □ **plagal cadence** (or **close**) a cadence in which the chord of the subdominant immediately precedes that of the tonic. [med.L *plagalis* f. *plaga* plagal mode f. L *plagius* f. med. Gk *plagios* (in classical Gk = oblique) f. Gk *plagos* side]

plage /plɑːʒ/ *n.* **1** *Astron.* an unusually bright region on the sun. **2** a sea beach, esp. at a fashionable resort. [F, = beach]

plagiarism /ˈpleɪdʒəˌrɪz(ə)m/ *n.* **1** the act or an instance of

plagiarizing. **2** something plagiarized. □ **plagiarist** *n.* **plagiaristic** /ˌpleɪdʒəˈrɪstɪk/ *adj.*

plagiarize /ˈpleɪdʒəˌraɪz/ *v.tr.* (also **-ise**) (also *absol.*) **1** take and use (the thoughts, writings, inventions, etc. of another person) as one's own. **2** pass off the thoughts etc. of (another person) as one's own. □ **plagiarizer** *n.* [L *plagiarius* kidnapper f. *plagium* a kidnapping f. Gk *plagion*]

plagio- /ˈpleɪdʒɪəʊ/ *comb. form* oblique. [Gk *plagios* oblique f. *plagos* side]

plagioclase /ˈpleɪdʒɪəʊˌkleɪz, -ˌkleɪs/ *n.* a series of feldspar minerals forming glassy crystals. [PLAGIO- + Gk *klasis* cleavage]

plague /pleɪɡ/ *n.* & *v.* ● *n.* **1 a** (often prec. by *the*) a deadly contagious disease spreading rapidly over a wide area; often = BUBONIC PLAGUE. **b** (**the Plague**) = GREAT PLAGUE. **2** (foll. by *of*) an unusual infestation of a pest etc. (*a plague of frogs*). **3 a** great trouble. **b** an affliction, esp. as regarded as divine punishment. **4** *colloq.* a nuisance. **5** (in *int.*) *joc.* or *archaic* a curse etc. (*a plague on it!*). ● *v.tr.* (**plagues, plagued, plaguing**) **1** afflict, torment (*plagued by war*). **2** *colloq.* pester or harass continually. □ **plaguesome** *adj.* [ME f. L *plaga* stroke, wound prob. f. Gk *plaga*, *plēgē*]

plaice /pleɪs/ *n.* (*pl.* same) **1** a European flatfish, *Pleuronectes platessa*, having a brown back with orange spots and a white underside, much used for food. **2** (in full **American plaice**) a North Atlantic flatfish, *Hippoglossoides platessoides*. [ME f. OF *plaïz* f. LL *platessa* app. f. Gk *platus* broad]

plaid /plæd/ *n.* **1** (often *attrib.*) chequered or tartan, esp. woollen, twilled cloth (*a plaid skirt*). **2** a long piece of plaid worn over the shoulder as part of Scottish Highland costume. □ **plaided** *adj.* [Gael. *plaide*, of unkn. orig.]

Plaid Cymru /plaɪd ˈkʌmrɪ/ the Welsh Nationalist party, founded in 1925 and dedicated to seeking autonomy for Wales. It won its first parliamentary seat in 1966, and since 1974 has maintained a small number of representatives in Parliament. [Welsh, = party of Wales]

plain[1] /pleɪn/ *adj.*, *adv.*, & *n.* ● *adj.* **1** clear; evident (*is plain to see*). **2** readily understood; simple (*in plain words*). **3 a** (of food, sewing, decoration, etc.) uncomplicated; not elaborate; unembellished; simple. **b** without a decorative pattern. **4** (esp. of a woman or girl) ugly. **5** outspoken; straightforward. **6** (of manners, dress, etc.) unsophisticated; homely (*a plain man*). **7** (of drawings etc.) not coloured (*penny plain, twopence coloured*). **8** not in code. **9** (of a knitting stitch) made by putting the needle through the front of the stitch from left to right (opp. PURL[1] *adj.*). ● *adv.* **1** clearly; unequivocally (*to speak plain, I don't approve*). **2** simply (*that is plain stupid*). ● *n.* a level tract of esp. treeless country. □ **as plain as day** obvious. **be plain with** speak bluntly to. **plain card** neither a trump nor a court-card. **plain chocolate** dark chocolate without added milk. **plain clothes** ordinary clothes worn esp. as a disguise by police officers etc. **plain-clothes** (*attrib.*) wearing plain clothes. **plain cook** a person competent in plain English cooking. **plain dealing** candour; straightforwardness. **plain sailing 1** sailing a straightforward course. **2** an uncomplicated situation or course of action. **plain service** *Eccl.* a church service without music. **plain-spoken** outspoken; blunt. **plain suit** a suit that is not trumps. **plain text** a text not in cipher or code. **plain time** time not paid for at overtime rates. **plain weaving** weaving with the weft alternately over and under the warp. □ **plainly** *adv.* **plainness** /ˈpleɪnnɪs/ *n.* [ME f. OF *plain* (adj. & n.) f. L *planus* (adj.), *planum* (n.)]

plain[2] /pleɪn/ *v.intr. archaic* or *poet.* **1** mourn. **2** complain. **3** make a plaintive sound. [ME f. OF *plaindre* (stem *plaign-*) f. L *plangere planct-* lament]

plainchant /ˈpleɪntʃɑːnt/ *n.* = PLAINSONG.

Plains Indian *n.* a member of any of various North American Indian peoples who formerly inhabited the Great Plains area. Although a few of the Plains Indian peoples were sedentary farmers, most, including the Blackfoot, Cheyenne, and Comanche, were nomadic buffalo hunters, who gathered in tribes during the summer and dispersed into family groups in the winter. They hunted on foot until they acquired horses from the Spanish in the early 18th century. The introduction of the horse also led to other peoples, such as the Sioux and the Cree, moving into the Plains area.

plainsman /ˈpleɪnzmən/ *n.* (*pl.* **-men**) a person who lives on a plain, esp. in North America.

Plains of Abraham a plateau beside the city of Quebec, overlooking the St Lawrence River. It was the scene in 1759 of a decisive battle in which the British army under General Wolfe, having scaled the heights above the city under cover of darkness, surprised and defeated the

French. The battle led to British control over Canada, but both Wolfe and the French commander Montcalm died of their wounds.

plainsong /ˈpleɪnsɒŋ/ n. traditional church music in medieval modes and in free rhythm depending on accentuation of the words, sung in unison, with a single line of vocal melody to words taken from the liturgy. (See also GREGORIAN CHANT.)

plaint /pleɪnt/ n. **1** Brit. Law an accusation; a charge. **2** literary or archaic a complaint; a lamentation. [ME f. OF plainte fem. past part. of plaindre, and OF plaint f. L planctus (as PLAIN²)]

plaintiff /ˈpleɪntɪf/ n. Law a person who brings a case against another into court (opp. DEFENDANT). [ME f. OF plaintif (adj.) (as PLAINTIVE)]

plaintive /ˈpleɪntɪv/ adj. **1** expressing sorrow; mournful. **2** mournful-sounding. □ **plaintively** adv. **plaintiveness** n. [ME f. OF (-if, -ive) f. plainte (as PLAINT)]

plait /plæt/ n. & v. ● n. **1** a length of hair, straw, etc., in three or more interlaced strands. **2** = PLEAT. ● v.tr. **1** form (hair etc.) into a plait. **2** make (a belt, mat, etc.) by plaiting. [ME f. OF pleit fold ult. f. L plicare fold]

plan /plæn/ n. & v. ● n. **1 a** a formulated and esp. detailed method by which a thing is to be done; a design or scheme. **b** an intention or proposed proceeding (my plan was to distract them; plan of campaign). **2 a** a drawing or diagram made by projection on a horizontal plane, esp. showing a building or one floor of a building (cf. ELEVATION 3). **3 a** fairly large-scale map of a town or district. **4 a** a table etc. indicating times, places, etc. of intended proceedings. **b** a diagram of an arrangement (prepared the seating plan). **5** an imaginary plane perpendicular to the line of vision and containing the objects shown in a picture. ● v. (**planned**, **planning**) **1** tr. (often foll. by that + clause or to + infin.) arrange (a procedure etc.) beforehand; form a plan (planned to catch the evening ferry). **2** tr. design (a building, new town, etc.). **b** make a plan of (an existing building, an area, etc.). **3** tr. (as **planned** adj.) in accordance with a plan (his planned arrival; planned parenthood). **4** intr. make plans. □ **planned economy** (esp. in (former) Communist countries) an economy in which prices, incomes, etc., are determined centrally by government rather than through the operation of the free market, and in which industrial production is governed by an overall national plan (also called command economy). **planning permission** Brit. formal permission for building development etc., esp. from a local authority. **plan on** aim at doing or having; intend. □ **planning** n. [F f. earlier plant, f. It. pianta plan of building: cf. PLANT]

planar /ˈpleɪnə(r)/ adj. Math. of, relating to, or in the form of a plane.

planarian /pləˈneərɪən/ n. Zool. a turbellarian flatworm of the division Tricladida, usu. living in fresh water. [mod.L Planaria genus name, fem. of L planarius lying flat]

planchet /ˈplɑːntʃɪt/ n. a plain metal disc from which a coin is made. [dimin. of planch slab of metal f. OF planche: see PLANK]

planchette /plɑːnˈʃet/ n. a small usu. heart-shaped board on castors with a pencil, said to trace letters etc. at spiritualist seances without conscious direction when one or more persons rest their fingers lightly on the board. [F, dimin. of planche PLANK]

Planck /plæŋk/, Max (Karl Ernst Ludwig) (1858–1947), German theoretical physicist, who founded the quantum theory. He published fundamental papers on thermodynamics before taking up the problem of black-body radiation. In 1900 he announced his radiation law, according to which electromagnetic radiation from heated bodies was not emitted as a continuous flow but was made up of discrete units or quanta of energy, the size of which involved a fundamental physical constant (Planck's constant). The quantum concept was immediately used to explain atomic structure and the photoelectric effect. Planck was awarded the 1918 Nobel Prize for physics. (See also QUANTUM.)

Planck's constant n. (also **Planck constant**) a fundamental constant, equal to the energy of quanta of electromagnetic radiation divided by its frequency, with a value of 6.626×10^{-34} joules. [PLANCK]

plane¹ /pleɪn/ n., adj., & v. ● n. **1 a** a flat surface on which a straight line joining any two points on it would wholly lie. **b** an imaginary flat surface through or joining etc. material objects. **2** a level surface. **3** colloq. = AEROPLANE. **4** a flat surface producing lift by the action of air or water over and under it (usu. in comb.: hydroplane). **5** (often foll. by of) a level of attainment, thought, knowledge, etc. **6** a flat thin object such as a tabletop. ● adj. **1** (of a surface etc.) perfectly level. **2** (of an angle, figure, etc.) lying in a plane. ● v.intr. **1** (often foll. by down) travel or glide in an aeroplane. **2** (of a speedboat etc.) skim over water. **3** soar. □ **plane chart** a chart on which meridians and parallels of latitude are represented by equidistant straight lines, used in plane

sailing. **plane polarization** a process restricting the vibrations of electromagnetic radiation, esp. light, to one direction. **plane sailing 1** the practice of determining a ship's position on the theory that it is moving on a plane. **2** = plain sailing (see PLAIN¹). **plane-table** a surveying instrument used for direct plotting in the field, with a circular drawing-board and pivoted alidade. [L planum flat surface, neut. of planus PLAIN¹ (differentiated f. PLAIN¹ in 17th c.): adj. after F plan, plane]

plane² /pleɪn/ n. & v. ● n. **1** a tool consisting of a wooden or metal block with a projecting steel blade, used to smooth a wooden surface by paring shavings from it. **2** a similar tool for smoothing metal. ● v.tr. **1** smooth (wood, metal, etc.) with a plane. **2** (often foll. by away, down) pare (irregularities) with a plane. **3** archaic level (plane the way). [ME f. OF var. of plaine f. LL plana f. L planus PLAIN¹]

plane³ /pleɪn/ n. (in full **plane tree**) a tall tree of the genus Platanus, with maple-like leaves and bark which peels in uneven patches. (See also LONDON PLANE.) [ME f. OF f. L platanus f. Gk platanos f. platus broad]

planet /ˈplænɪt/ n. **1 a** a celestial body moving in an elliptical orbit round a star. (See note below.) **b** the earth. **2** esp. Astrol. & hist. a celestial body distinguished from the fixed stars by having an apparent motion of its own (including the moon and sun), esp. with reference to its supposed influence on people and events. □ **planetology** /ˌplænɪˈtɒlədʒɪ/ n. [ME f. OF planete f. LL planeta, planetes f. Gk planētēs wanderer, planet f. planaomai wander]

▪ The nine planets of the solar system are either gas giants or smaller rocky bodies. The former comprise Jupiter, Saturn, Uranus, and Neptune, all of which have dense atmospheres of hydrogen-rich gases subject to violent winds and eddies. They are orbited both by retinues of satellites (some as large as the other planets), and by concentric rings composed of small particles. Of the rocky planets — Mercury, Venus, Earth, Mars, and Pluto — only Earth and Venus have substantial atmospheres. Only Earth has significant amounts of surface water today, and it is the only planet known to support life. The minor planets or asteroids orbit mainly between the orbits of Mars and Jupiter. Planetary systems are probably a common phenomenon in the universe, though there is as yet no unequivocal evidence for them: a periodic wobble in the motion of certain stars suggests accompanying planets, and discs of matter around stars such as Vega may be planetary systems in the making. (See also SOLAR SYSTEM.)

planetarium /ˌplænɪˈteərɪəm/ n. (pl. **planetariums** or **planetaria** /-rɪə/) **1** a domed building in which images of stars, planets, constellations, etc. are projected for public entertainment or education. **2** the device used for such projection. **3** = ORRERY. [mod.L (as PLANET)]

planetary /ˈplænɪtərɪ/ adj. **1** of or like planets (planetary influence). **2** terrestrial; mundane. **3** wandering; erratic. □ **planetary nebula** Astron. a ring-shaped nebula formed by an expanding shell of gas round an ageing star. [LL planetarius (as PLANET)]

planetesimal /ˌplænɪˈtesɪm(ə)l/ n. Astron. any of a vast number of minute planets or planetary bodies. □ **planetesimal hypothesis** the theory that planets were formed by the accretion of planetesimals in a cold state. [PLANET, after infinitesimal]

planetoid /ˈplænɪˌtɔɪd/ n. = ASTEROID.

plangent /ˈplændʒənt/ adj. **1** (of a sound) loud and reverberating. **2** (of a sound) plaintive; sad. □ **plangency** n. [L plangere plangent- lament]

planimeter /pləˈnɪmɪtə(r)/ n. an instrument for mechanically measuring the area of a plane figure. □ **planimetry** n. **planimetric** /ˌplænɪˈmetrɪk/ adj. **planimetrical** adj. [F planimètre f. L planus level]

planish /ˈplænɪʃ/ v.tr. flatten (sheet metal, coining-metal, etc.) with a smooth-faced hammer or between rollers. □ **planisher** n. [ME f. OF planir smooth f. plain PLANE¹ adj.]

planisphere /ˈplænɪˌsfɪə(r)/ n. a map formed by the projection of a sphere or part of a sphere on a plane, esp. an adjustable circular star map that shows the appearance of the heavens at a specific time and place. □ **planispheric** /ˌplænɪˈsferɪk/ adj. [ME f. med.L planisphaerium (as PLANE¹, SPHERE): infl. by F planisphère]

plank /plæŋk/ n. & v. ● n. **1** a long flat piece of timber used esp. in building, flooring, etc. **2** an item of a political or other programme (cf. PLATFORM 6). ● v.tr. **1** provide, cover, or floor, with planks. **2** (usu. foll. by down; also absol.) esp. US colloq. **a** put (a thing, person, etc.) down roughly or violently. **b** pay (money) on the spot or abruptly (planked down £5). □ **plank bed** a bed of boards without a mattress, esp. in prison. **walk the plank** hist. (of a pirate's captive etc.) be made to walk blindfold along a plank over the side of a ship to one's death in the sea. [ME f. ONF *planke*, OF *planche* f. LL *planca* board f. *plancus* flat-footed]

planking /'plæŋkɪŋ/ *n.* planks as flooring etc.

plankton /'plæŋktən/ *n. Biol.* the small and microscopic organisms drifting or floating in the sea or fresh water (cf. BENTHOS, NEKTON). □ **planktonic** /plæŋk'tɒnɪk/ *adj.* [G f. Gk *plagktos* wandering f. *plazomai* wander]

planner /'plænə(r)/ *n.* **1** a person who controls or plans the development of new towns, designs buildings, etc. **2** a person who makes plans. **3** a list, table, etc., with information helpful in planning.

plano- /'pleɪməʊ/ *comb. form* level, flat. [L *planus* flat]

planoconcave /ˌpleɪnəʊ'kɒŋkeɪv/ *adj.* (of a lens etc.) with one surface plane and the other concave.

planoconvex /ˌpleɪnəʊ'kɒnveks/ *adj.* (of a lens etc.) with one surface plane and the other convex.

planographic /ˌpleɪnə'græfɪk/ *adj.* relating to or produced by a process in which printing is done from a plane surface. □ **planography** /plə'nɒgrəfɪ/ *n.*

planometer /plə'nɒmɪtə(r)/ *n.* a flat plate used as a gauge for plane surfaces in metalwork.

plant /plɑːnt/ *n. & v.* ● *n.* **1 a** a living organism which usu. contains chlorophyll, and lacks specialized sense-organs and the power of voluntary movement. (*See note below.*) **b** a small organism of this kind, as distinguished from a shrub or tree. **2 a** machinery, fixtures, etc., used in industrial processes. **b** a factory. **3 a** *colloq.* something, esp. incriminating or compromising, positioned or concealed so as to be discovered later. **b** *sl.* a spy or detective; hidden police officers. ● *v.tr.* **1** place (a seed, bulb, or growing thing) in the ground so that it may take root and flourish. **2** (often foll. by *in, on,* etc.) **a** put or fix in position. **b** *refl.* take up a position (*planted myself by the door*). **c** place (a bomb) in a building etc. **3** deposit (young fish, spawn, oysters, etc.) in a river or lake. **4** station (a person etc.), esp. as a spy or source of information. **5** cause (an idea etc.) to be established esp. in another person's mind. **6** deliver (a blow, kiss, etc.) with a deliberate aim. **7** *colloq.* position or conceal (something incriminating or compromising) for later discovery. **8 a** settle or people (a colony etc.). **b** found or establish (a city, community, etc.). **9** bury. □ **plant-louse** a small homopteran bug that infests plants, esp. an aphid. **plant out** transfer (a plant) from a pot or frame to the open ground; set out (seedlings) at intervals. □ **plantable** *adj.* **plantlet** /'plɑːntlɪt/ *n.* **plantlike** *adj.* [OE *plante* & F *plante* f. L *planta* sprout, slip, cutting]

▪ The possession of chlorophyll enables plants to synthesize organic molecules with the aid of sunlight, enabling them to live on inorganic substances. The plant kingdom (Plantae) was formerly held to include fungi and bacteria as well as green plants. Now, however, these two groups, along with unicellular algae, are frequently assigned to separate kingdoms, and the plant kingdom is restricted to the multicellular green plants. The main groups of these are the multicellular algae, mosses, ferns, gymnosperms, and angiosperms (flowering plants).

Plantagenet /plæn'tædʒɪnɪt/ the English royal dynasty which held the throne from the accession of Henry II (1154) to the death of Richard III (1485). Plantagenet originated as the nickname of Geoffrey IV, Count of Anjou (1113–51), father of Henry II of England, probably deriving from the sprig of broom worn as a distinctive mark. In the 15th century the line divided into two branches, the House of Lancaster and the House of York (see WARS OF THE ROSES). [L *planta* plant, *genista* broom]

plantain[1] /'plæntɪn/ *n.* a low-growing plant of the genus *Plantago*, with usu. broad flat leaves forming a rosette on the ground, and seeds used as food for birds and as a mild laxative. □ **plantain lily** = HOSTA. [ME f. OF f. L *plantago -ginis* f. *planta* sole of the foot (from its broad prostrate leaves)]

plantain[2] /'plæntɪn/ *n.* **1** a banana-like plant, *Musa paradisiaca*, widely grown for its fruit. **2** the starchy fruit of this, containing less sugar than a dessert banana, and chiefly used in cooking. [earlier *platan* f. Sp. *plá(n)tano* plane tree, prob. assim. f. Galibi *palatana* etc.]

plantar /'plæntə(r)/ *adj.* of or relating to the sole of the foot. [L *plantaris* f. *planta* sole]

plantation /plæn'teɪʃ(ə)n, plɑː-/ *n.* **1** an estate on which cotton, tobacco, etc. is cultivated, esp. in former colonies, formerly by slave labour. **2** an area planted with trees etc. **3** *hist.* a colony; colonization. □ **plantation song** a song of the kind formerly sung by blacks on American plantations. [ME f. OF *plantation* or L *plantatio* (as PLANT)]

Plantation of Ireland the government-sponsored settlement of

English and later Scottish families in Ireland in the 16th–17th centuries. (See IRELAND.)

planter /'plɑːntə(r)/ *n.* **1** a person who cultivates the soil. **2** the manager or occupier of a coffee, cotton, tobacco, etc. plantation. **3** a large container for decorative plants. **4** a machine for planting seeds etc. (*potato-planter*).

plantigrade /'plæntɪˌgreɪd/ *adj. & n. Zool.* ● *adj.* (of an animal) walking on the soles of its feet. ● *n.* a plantigrade animal, e.g. humans or bears (cf. DIGITIGRADE). [F f. mod.L *plantigradus* f. L *planta* sole + *-gradus* -walking]

plaque /plæk, plɑːk/ *n.* **1** an ornamental tablet of metal, porcelain, etc., esp. affixed to a building in commemoration. **2** a deposit on teeth where bacteria proliferate. **3** *Med.* **a** a patch or eruption of skin etc. as a result of damage. **b** a fibrous lesion in atherosclerosis. **4** a small badge of rank in an honorary order. □ **plaquette** /plæ'ket/ *n.* [F f. Du. *plak* tablet f. *plakken* stick]

plash[1] /plæʃ/ *n. & v.* ● *n.* **1** a splash; a plunge. **2 a** a marshy pool. **b** a puddle. ● *v.* **1** *tr. & intr.* splash. **2** *tr.* strike the surface of (water). □ **plashy** *adj.* [OE *plæsc*, prob. imit.]

plash[2] /plæʃ/ *v.tr.* **1** bend down and interweave (branches, twigs, etc.) to form a hedge. **2** make or renew (a hedge) in this way. [ME f. OF *pla(i)ssier* ult. f. L *plectere* plait: cf. PLEACH]

plasma /'plæzmə/ *n.* (also **plasm** /-z(ə)m/) **1 a** the colourless fluid part of blood, lymph, or milk, in which corpuscles or fat-globules are suspended. **b** this taken from donated blood for administering in transfusions. **2** = PROTOPLASM. **3** *Physics* **a** an electrically conducting gas of positive ions and free negative electrons, often electrically neutral overall, formed especially at very high temperatures. (*See note below.*) **b** any analogous collection of mobile charged particles. **4** a green variety of quartz used in mosaic and for other decorative purposes. □ **plasmic** *adj.* **plasmatic** /plæz'mætɪk/ *adj.* [LL, = mould f. Gk *plasma -atos* f. *plassō* to shape]

▪ Examples of hot plasmas are the ionized gases found in the sun and other stars, and in artificial nuclear fusion reactions, glowing electrical discharges, and the gaseous interior of working fluorescent lamps. Such plasmas behave differently from ordinary gases, notably in being affected by magnetic fields. For this reason they are sometimes regarded as constituting a fourth state of matter alongside solids, liquids, and gases. Examples of plasmas in the broader sense include molten salts and the mobile electrons within metals. The study of plasmas is very important for developing controlled nuclear fusion reactions, in which the fuel becomes a plasma hot enough to vaporize any container; the development of magnetic containment systems is seen as the potential solution to the key problem of how to contain the reacting materials.

plasmid /'plæzmɪd/ *n. Biol.* a genetic structure in a cell that can replicate independently of the chromosomes, esp. a circular strand of DNA in a bacterium or protozoan. [PLASMA + -ID²]

plasmodesma /ˌplæzmə'dezmə/ *n.* (*pl.* **plasmodesmata** /-mətə/) *Biol.* a narrow thread of membrane-bound cytoplasm that passes through cell walls and affords communication between plant cells. [PLASMA + Gk *desma* bond, fetter]

plasmodium /plæz'məʊdɪəm/ *n.* (*pl.* **plasmodia** /-dɪə/) *Biol.* **1** a parasitic protozoan of the genus *Plasmodium*, including several causing malaria in humans. **2** a form within the life cycle of various microorganisms including slime moulds, usu. consisting of a mass of naked protoplasm containing many nuclei. □ **plasmodial** *adj.* [mod.L f. PLASMA + -odium: see -ODE¹]

plasmolyse /'plæzməˌlaɪz/ *v.intr. & tr.* (*US* **plasmolyze**) undergo or subject to plasmolysis.

plasmolysis /plæz'mɒlɪsɪs/ *n.* contraction of the protoplast of a plant cell as a result of loss of water from the cell. [mod.L (as PLASMA, -LYSIS)]

Plassey /'plæsɪ/ a village in NE India, in West Bengal, north-west of Calcutta. It was the scene in 1757 of a battle in which a small British army under Robert Clive defeated the much larger forces of the nawab of Bengal (Siraj-ud-Dawlah, *c.*1729–57), a victory won partly because Clive had previously bribed some of the Indian generals. The victory established British supremacy in Bengal.

plaster /'plɑːstə(r)/ *n. & v.* ● *n.* **1** a soft pliable mixture esp. of lime putty with sand or Portland cement etc. for spreading on walls, ceilings, etc., to form a smooth hard surface when dried. **2** *Brit.* = *sticking-plaster* (see STICK²). **3** *hist.* a curative or protective substance spread on a bandage etc. and applied to the body (*mustard plaster*). ● *v.tr.* **1** cover (a wall etc.) with plaster or a similar substance. **2** (often foll. by *with*) coat thickly or to excess; bedaub (*plastered the bread with jam; the wall was plastered with slogans*). **3** stick or apply (a thing) thickly like plaster (*plastered glue all over it*). **4** (often foll. by *down*) make (esp. hair) smooth with water, cream, etc.; fix flat. **5** (as **plastered** adj.) sl. drunk. **6** apply a medical plaster or plaster cast to. **7** sl. bomb or shell heavily. □ **plaster cast 1** a bandage stiffened with plaster of Paris and applied to a broken limb etc. **2** a statue or mould made of plaster. **plaster of Paris** fine white plaster made of gypsum and used for making plaster casts etc. **plaster saint** iron. a person regarded as being without moral faults or human frailty. □ **plasterer** n. **plastery** adj. [ME f. OE & OF *plastre* or F *plastrer* f. med.L *plastrum* f. L *emplastrum* f. Gk *emplastron*]

plasterboard /ˈplɑːstəˌbɔːd/ n. two boards with a filling of plaster used to form or line the inner walls of houses etc.

plastic /ˈplæstɪk/ n. & adj. ● n. **1** a synthetic polymeric substance that can be moulded into a required shape when soft and then set into a rigid or slightly elastic form. (*See note below.*) **2** (attrib.) made of plastic (*plastic bag*); made of cheap materials. **3** = *plastic money*. ● adj. **1 a** capable of being moulded; pliant; supple. **b** susceptible, impressionable. **2** moulding or giving form to clay, wax, etc. **3** Biol. exhibiting an adaptability to environmental changes. **4** (esp. in philosophy) formative, creative. □ **plastic arts** art forms involving modelling or moulding, e.g. sculpture and ceramics, or art involving the representation of solid objects with three-dimensional effects. **plastic bomb** a bomb containing plastic explosive. **plastic explosive** a putty-like explosive capable of being moulded by hand. **plastic money** colloq. a credit card, charge card, or other plastic card that can be used in place of money. **plastic surgeon** a qualified practitioner of plastic surgery. **plastic surgery** the process of reconstructing or repairing parts of the body by the transfer of tissue, either in the treatment of injury or for cosmetic reasons. □ **plastically** adv. **plasticky** adj. **plasticity** /plæˈstɪsɪtɪ/ n. [F *plastique* or L *plasticus* f. Gk *plastikos* f. *plassō* mould]

▪ Plastics consist of long chain-like organic molecules (polymers) intertwined with one another. The first plastics, such as celluloid (developed in 1869), used chemically modified forms of cellulose or other natural polymers. In the 20th century, however, polymers made artificially from small molecules became available, leading to the great diversity of plastics available today. The structure of plastics allows them to be softened during manufacture, and thus they can be moulded into many different forms. Other advantages of plastics include toughness, insulating ability, resistance to chemical and biological attack (although this leads to environmental problems), and ability to be given any desired colour. Among the major successes of the plastics industry are celluloid, bakelite (the first heat-proof plastic), polythene, and nylon.

Plasticine /ˈplæstɪˌsiːn/ n. propr. a soft plastic material used, esp. by children, for modelling. [PLASTIC + -INE⁴]

plasticize /ˈplæstɪˌsaɪz/ v.tr. (also **-ise**) make plastic or mouldable. □ **plasticizer** n. **plasticization** /ˌplæstɪsaɪˈzeɪʃ(ə)n/ n.

plastid /ˈplæstɪd/ n. Bot. a small organelle in the cytoplasm of a plant cell, containing pigment or food. [G f. Gk *plastos* shaped]

plastron /ˈplæstrən/ n. **1 a** a fencer's leather-covered breastplate. **b** a lancer's breast-covering of facings-cloth. **2 a** an ornamental front on a woman's bodice. **b** a man's starched shirt-front. **3 a** the ventral part of the shell of a tortoise or turtle. **b** the corresponding part in other animals. **4** hist. a steel breastplate. □ **plastral** adj. [F f. It. *piastrone* augment. of *piastra* breastplate, f. L *emplastrum* PLASTER]

plat¹ /plæt/ n. US **1** a plot of land. **2** a plan of an area of land. [16th c.: collateral form of PLOT]

plat² /plæt/ n. & v. ● n. = PLAIT n. 1. ● v.tr. (**platted, platting**) = PLAIT v.

Plataea, Battle of /pləˈtiːə/ a battle in 479 BC, during the Persian Wars, in which the Persian forces were defeated by the Greeks near the city of Plataea in Boeotia, central Greece.

platan /ˈplæt(ə)n/ n. = PLANE³. [ME f. L *platanus*: see PLANE³]

plat du jour /ˌplɑː dju ˈʒʊə(r)/ n. (pl. ***plats du jour*** pronunc. same) a dish specially featured on a day's menu. [F, = dish of the day]

plate /pleɪt/ n. & v. ● n. **1 a** a shallow vessel, usu. circular and of earthenware or china, from which food is eaten or served. **b** the contents of this (*ate a plate of sandwiches*). **2** a similar vessel usu. of metal or wood, used esp. for making a collection in a church etc. **3** US a main course of a meal, served on one plate. **4** Austral. & NZ a contribution of cakes, sandwiches, etc., to a social gathering. **5 a** (collect.) utensils of silver, gold, or other metal. **b** (collect.) objects of plated metal. **c** = PLATING 1. **6** a piece of metal with a name or inscription for affixing to a door, container, etc. **7** an illustration on

special paper in a book (cf. *book-plate*). **8** a thin sheet of metal, glass, etc., coated with a sensitive film for photography. **9** a flat thin usu. rigid sheet of metal etc. with an even surface and uniform thickness, often as part of a mechanism. **10 a** a smooth piece of metal etc. for engraving. **b** an impression made from this. **11 a** a silver or gold cup as a prize for a horse-race etc. **b** a race with this as a prize. **12 a** a thin piece of plastic material, moulded to the shape of the mouth and gums, to which artificial teeth or another orthodontic appliance are attached. **b** colloq. a complete denture or orthodontic appliance. **13** Geol. each of the several rigid pieces of the earth's lithosphere which together form the whole of the earth's surface and whose relative motion and interactions are responsible for many geomorphological and geophysical phenomena. (See also PLATE TECTONICS.) **14** Biol. a thin flat organic structure or formation. **15** a light shoe for a racehorse. **16** a stereotype, electrotype, or plastic cast of a page of composed movable types, or a metal or plastic copy of filmset matter, from which sheets are printed. **17** Baseball a flat piece of whitened rubber marking the station of a batter or pitcher. **18** US the anode of a thermionic valve. **19** a horizontal timber laid along the top of a wall to support the ends of joists or rafters (*window-plate*). ● v.tr. **1** apply a thin coat esp. of silver, gold, or tin to (another metal). **2** cover (esp. a ship) with plates of metal, esp. for protection. **3** make a plate (of type etc.) for printing. □ **on a plate** colloq. available with little trouble to the recipient. **on one's plate** for one to deal with or consider. **plate armour** armour of metal plates, for a man, ship, etc. **plate glass** thick fine quality glass used for shop windows etc., originally cast in plates, rolled through heated rollers, then ground and polished, until about 1959 when the float process was introduced. **plate-mark** a hallmark. **plate-rack** Brit. a rack in which plates are placed to drain. **plate tracery** Archit. tracery with perforations in otherwise continuous stone. □ **plateful** n. (pl. **-fuls**). **plateless** adj. **plater** n. [ME f. OF f. med.L *plata* plate armour f. *platus* (adj.) ult. f. Gk *platus* flat]

Plate, River /pleɪt/ (called in Spanish *Río de la Plata*) a wide estuary on the Atlantic coast of South America at the border between Argentina and Uruguay, formed by the confluence of the rivers Paraná and Uruguay. The cities of Buenos Aires and Montevideo lie on its shores. Its name refers to the export from the region of silver (Sp. *plata*) in the Spanish colonial period. In 1939 it was the scene of a naval battle in which the British defeated the Germans.

plateau /ˈplætəʊ/ n. & v. ● n. (pl. **plateaux** /-təʊz/ or **plateaus**) **1** an area of fairly level high ground. **2** a state of little variation after an increase. ● v.intr. (**plateaus, plateaued**) (often foll. by *out*) reach a level or stable state after an increase. [F f. OF *platel* dimin. of *plat* flat surface]

platelayer /ˈpleɪtˌleɪə(r)/ n. Brit. a person employed in fixing and repairing railway rails.

platelet /ˈpleɪtlɪt/ n. a small colourless disc-shaped cell fragment found in large numbers in blood and involved in clotting. Also called *thrombocyte*. (See also BLOOD.)

platen /ˈplæt(ə)n/ n. **1** a plate in a printing-press which presses the paper against the type. **2** a cylindrical roller in a typewriter against which the paper is held. [OF *platine* a flat piece f. *plat* flat]

plateresque /ˌplætəˈrɛsk/ adj. richly ornamented in a style suggesting silverware. [Sp. *plateresco* f. *platero* silversmith f. *plata* silver]

plate tectonics n. Geol. a theory of the earth's surface based on the motion and interaction of rigid lithospheric plates (PLATE n. 13). The plates, some of which are large enough to include a whole continent, lie on top of the more plastic region of the earth's mantle and move slowly relative to each other under the influence of convection currents. The boundaries between adjacent plates are regions of seismic, volcanic, and tectonic activity. Plate tectonics, which since the 1960s has revolutionized geology, provides explanations for the phenomenon of continental drift and the distribution of earthquakes, mid-ocean ridges, deep-sea trenches, and mountain chains. (See also CONTINENTAL DRIFT.)

platform /ˈplætfɔːm/ n. **1** a raised level surface; a natural or artificial terrace. **2** a raised surface from which a speaker addresses an audience. **3** Brit. a raised elongated structure along the side of a track in a railway station. **4** the floor area at the entrance to a bus. **5 a** a thick sole of a shoe. **b** a shoe with such a sole. **6** the declared policy of a political party, organization, etc. (cf. PLANK n. 2). **7** = oil platform. □ **platform ticket** a ticket allowing a non-traveller access to a station platform. [F *plateforme* ground-plan f. *plate* flat + *forme* FORM]

Plath /plæθ/, Sylvia (1932–63), American poet. She married Ted Hughes in 1956 and first published her poetry in 1960. Her life was marked by periods of severe depression and her work is notable for its controlled and intense treatment of extreme and painful states of mind. In 1963 she committed suicide; it was only after the posthumous publication of *Ariel* (1965) that she gained wide recognition. She also wrote a novel, *The Bell Jar* (1963).

plating /'pleɪtɪŋ/ *n.* **1** a coating of gold, silver, etc. **2** racing for plates.

platinic /plə'tɪnɪk/ *adj.* of or containing (esp. tetravalent) platinum.

platinize /'plætɪˌnaɪz/ *v.tr.* (also **-ise**) coat with platinum. □ **platinization** /ˌplætɪnaɪ'zeɪʃ(ə)n/ *n.*

platinoid /'plætɪˌnɔɪd/ *n.* an alloy of copper, zinc, nickel, and tungsten.

platinum /'plætɪnəm/ *n.* a silvery-white metallic chemical element (atomic number 78; symbol **Pt**. (*See note below.*) □ **platinum black** platinum in powder form like lampblack. **platinum blonde** (or **blond**) *adj.* silvery-blond. ● *n.* a person with silvery-blond hair. **platinum disc** a framed platinum disc awarded to a recording artist or group for sales of a recording of at least one million. **platinum metals** *Chem.* the six metals platinum, palladium, ruthenium, osmium, rhodium, and iridium, which have similar physical and chemical properties and tend to occur together in nature. [earlier *platina* f. Sp., dimin. of *plata* silver]

- Platinum was first encountered by the Spanish in America in the 16th century. A transition metal, it occurs in nature in the uncombined state and in ores, always with other metals. It has a high melting-point and is resistant to chemical attack. The pure metal and its alloys are used in jewellery, electrical contacts, and laboratory equipment, and it is an important industrial catalyst.

platitude /'plætɪˌtjuːd/ *n.* **1** a trite or commonplace remark, esp. one solemnly delivered. **2** the use of platitudes; dullness, insipidity. □ **platitudinize** /ˌplætɪ'tjuːdɪˌnaɪz/ *v.intr.* (also **-ise**). **platitudinous** /-'tjuːdɪnəs/ *adj.* [F f. *plat* flat, after *certitude, multitudinous,* etc.]

Plato /'pleɪtəʊ/ (*c.*429–*c.*347 BC), Greek philosopher. He was a disciple of Socrates and the teacher of Aristotle, and he founded the Academy in Athens. His system of thought had a profound influence on Christian theology and Western philosophy. His philosophical writings, which cover metaphysics, politics, and ethics, are presented in the form of dialogues, with Socrates as the principal speaker; they include the *Symposium* and the *Phaedo*. An integral part of his thought is the theory of 'ideas' or 'forms', in which abstract entities or *universals* are contrasted with their objects or *particulars* in the material world. Plato's political theories appear in the *Republic*, in which he explored the nature and structure of a just society. He proposed a political system based on the division of the population into three classes, determined by education rather than birth or wealth: rulers, police and armed forces, and civilians.

Platonic /plə'tɒnɪk/ *adj.* **1** of or associated with Plato or his ideas. **2** (**platonic**) (of love or friendship) purely spiritual, not sexual. **3** (**platonic**) confined to words or theory; not leading to action; harmless. □ **Platonic solid** (or **body**) any of the five regular solids (tetrahedron, cube, octahedron, dodecahedron, icosahedron). □ **Platonically** *adv.* [L *Platonicus* f. Gk *Platōnikos* f. *Platōn* PLATO]

Platonism /'pleɪtəˌnɪz(ə)m/ *n.* **1** the philosophy of Plato or his followers. (See PLATO.) **2 a** = NEOPLATONISM. **b** any of various other revivals of Platonic doctrines or related ideas, especially Cambridge Platonism (17th century), which centred on Cambridge and attempted to reconcile Christianity with humanism and science. **3** *Math.* the theory that numbers exist independently of the physical world and the symbols used to represent them. □ **Platonist** *n.*

platoon /plə'tuːn/ *n.* **1** *Mil.* a subdivision of a company, a tactical unit commanded by a lieutenant and usu. divided into three sections. **2** a group of persons acting together. [F *peloton* small ball, dimin. of *pelote*: see PELLET, -OON]

Plattdeutsch /'plætdɔɪtʃ/ *n. & adj.* = LOW GERMAN. [G f. Du. *Platduitsch*, f. *plat* flat, low + *Duitsch* German]

platteland /'plɑːtəˌlɑːnt/ *n.* *S. Afr.* remote country districts. □ **plattelander** /-ˌlɑːndə(r)/ *n.* [Afrik., = flat land]

platter /'plætə(r)/ *n.* **1** a large flat dish or plate, esp. for food. **2** *sl.* a gramophone record. **3** the rotating metal disc of a record-player turntable. **4** *Computing* a hard disk. □ **on a platter** = on a plate (see PLATE). [ME & AF *plater* f. AF *plat* PLATE]

platy- /'plætɪ/ *comb. form* broad, flat. [Gk *platu-* f. *platus* broad, flat]

platyhelminth /ˌplætɪ'hɛlmɪnθ/ *n.* *Zool.* = FLATWORM.

platypus /'plætɪpəs/ *n.* (*pl.* **platypuses**) (in full **duck-billed platypus**) a semiaquatic Australian egg-laying mammal, *Ornithorhynchus anatinus*, having a leathery ducklike bill, brown fur,

and webbed feet bearing (in the male) poisonous spurs. Also called *duckbill*.

platyrrhine /'plætɪˌraɪn/ *adj. & n.* *Zool.* ● *adj.* (of primates) having the nostrils widely spaced and directed sideways, characteristic of New World monkeys (cf. CATARRHINE). ● *n.* such an animal. (See PRIMATE.) [PLATY- + Gk *rhis rhin-* nose]

plaudit /'plɔːdɪt/ *n.* (usu. in *pl.*) **1** a round of applause. **2** an emphatic expression of approval. [shortened f. L *plaudite* applaud, imper. pl. of *plaudere plaus-* applaud, said by Roman actors at the end of a play]

plausible /'plɔːzɪb(ə)l/ *adj.* **1** (of an argument, statement, etc.) seeming reasonable or probable. **2** (of a person) persuasive but deceptive. □ **plausibly** *adv.* **plausibility** /ˌplɔːzɪ'bɪlɪtɪ/ *n.* [L *plausibilis* (as PLAUDIT)]

Plautus /'plɔːtəs/, Titus Maccius (*c.*250–184 BC), Roman comic dramatist. His plays, of which twenty-one survive, are modelled on the New Comedy of Greek dramatists such as Menander, but with a few important differences. Fantasy and imagination are more important than realism in the development of the plots, for example; his stock characters, which follow Greek types, are often larger than life and their language is correspondingly exuberant.

play /pleɪ/ *v. & n.* ● *v.* **1** *intr.* (often foll. by *with*) occupy or amuse oneself pleasantly with some recreation, game, exercise, etc. **2** *intr.* (foll. by *with*) act light-heartedly or flippantly (with feelings etc.). **3 a** *tr.* perform on or be able to perform on (a musical instrument). **b** *tr.* perform (a piece of music etc.). **c** *tr.* cause (a record, record-player, etc.) to produce sounds. **d** *intr.* (of a record, record-player, etc.) produce sounds. **4 a** *intr.* (foll. by *in*) perform a role in (a drama etc.). **b** *tr.* perform (a drama or role) on stage, or in a film or broadcast. **c** *tr.* give a dramatic performance at (a particular theatre or place). **5** *tr.* act in real life the part of (*play truant; play the fool*). **6** *tr.* (foll. by *on*) perform (a trick or joke etc.) on (a person). **7** *tr.* (foll. by *for*) regard (a person) as (something specified) (*played me for a fool*). **8** *intr. colloq.* participate, cooperate; do what is wanted (*they won't play*). **9** *intr.* gamble. **10** *tr.* gamble on. **11** *tr.* **a** take part in (a game or recreation). **b** compete with (another player or team) in a game. **c** occupy (a specified position) in a team for a game. **d** (foll. by *in, on, at,* etc.) assign (a player) to a position. **12** *tr.* move (a piece) or display (a playing card) in one's turn in a game. **13** *tr.* (also *absol.*) strike (a ball etc.) or execute (a stroke) in a game. **14** *tr.* move about in a lively or unrestrained manner. **15** *intr.* (often foll. by *on*) **a** touch gently. **b** emit light, water, etc. (*fountains gently playing*). **16** *tr.* allow (a fish) to exhaust itself pulling against a line. **17** *intr.* (often foll. by *at*) **a** engage in a half-hearted way (in an activity). **b** pretend to be. **18** *intr.* (of a cricket ground etc.) be conducive to play as specified (*the pitch is playing fast*). **19** *intr. colloq.* act or behave (as specified) (*play fair*). **20** *tr.* (foll. by *in, out,* etc.) accompany (a person) with music (*were played out with bagpipes*). ● *n.* **1** recreation, amusement, esp. as the spontaneous activity of children and young animals. **2 a** the playing of a game. **b** the action or manner of this. **c** the status of the ball etc. in a game as being available to be played according to the rules (*in play; out of play*). **3** a dramatic piece for the stage etc. **4** activity or operation (*are in full play; brought into play*). **5** a freedom of movement. **b** space or scope for this. **6** brisk, light, or fitful movement. **7** gambling. **8** an action or manoeuvre, esp. in or as in a game. □ **at play** engaged in recreation. **in play** for amusement; not seriously. **make play** act effectively. **make a play for** *colloq.* make a conspicuous attempt to acquire. **make play with** use ostentatiously. **play about** (or **around**) **1** behave irresponsibly. **2** philander. **play along** pretend to cooperate. **play back** play (sounds recently recorded), esp. to monitor recording quality etc. **play-back** *n.* a playing back of a sound or sounds. **play ball** see BALL[1]. **play by ear 1** perform (music) without having seen a score of it. **2** (also **play it by ear**) proceed instinctively or step by step according to results and circumstances. **play one's cards close to one's chest** see CHEST. **play one's cards right** (or **well**) make good use of opportunities; act shrewdly. **play down** minimize the importance of. **play ducks and drakes with** see DUCK[1]. **played out** exhausted of energy or usefulness. **play false** act, or treat a (person), deceitfully or treacherously. **play fast and loose** act unreliably; ignore one's obligations. **play the field** see FIELD. **play for time** seek to gain time by delaying. **play the game** see GAME[1]. **play God** see GOD. **play havoc with** see HAVOC. **play hell with** see HELL. **play hookey** see HOOKEY. **play into a person's hands** act so as unwittingly to give a person an advantage. **play it cool** *colloq.* **1** affect indifference. **2** be relaxed or unemotional. **play the man** = *be a man* (see MAN). **play the market** speculate in stocks etc. **play off** (usu. foll. by *against*) **1** oppose (one person against another), esp. for one's own advantage. **2** play an extra match to decide a draw or tie. **play-off** *n.* a match played to decide a draw or tie. **play on 1** continue to play. **2** take advantage of (a person's feelings etc.). **play oneself in** become accustomed to the prevailing conditions in a game etc. **play on words** a pun. **play-pen** a portable enclosure for young children to play in. **play possum** see POSSUM. **play safe** (or **for safety**) avoid risks. **play-suit** a garment for a young child. **play to the gallery** see GALLERY. **play up 1** behave mischievously. **2** cause trouble; be irritating (*my rheumatism is playing up again*). **3** obstruct or annoy in this way (*played the teacher up*). **4** put all one's energy into a game. **5** make the most of; emphasize. **play up to** flatter, esp. to win favour. **play with fire** take foolish risks. □ **playable** *adj.* **playability** /ˌpleɪə'bɪlɪtɪ/ *n.* [OE *plega* (n.), *pleg(i)an* (v.), orig. = (to) exercise]

playa /'plɑːjə/ *n.* a flat dried-up area, esp. a desert basin from which water evaporates quickly. [Sp., = beach, f. LL *plagia*]

play-act /'pleɪækt/ *v.* **1** *intr.* act in a play. **2** *intr.* behave affectedly or insincerely. **3** *tr.* act (a scene, part, etc.). □ **play-acting** *n.* **play-actor** *n.*

playbill /'pleɪbɪl/ *n.* **1** a poster announcing a theatrical performance. **2** *US* a theatre programme.

playboy /'pleɪbɔɪ/ *n.* an irresponsible pleasure-seeking man, esp. a wealthy one.

Player /'pleɪə(r)/, Gary (b.1936), South African golfer. He has won numerous championships including the British Open (1959; 1968; 1974), the Masters (1961; 1974; 1978), the PGA (1962; 1972), and the US Open (1965).

player /'pleɪə(r)/ *n.* **1 a** a person taking part in a sport or game. **b** a gambler. **2** a person playing a musical instrument. **3** a person who plays a part on the stage; an actor. **4** any device for playing records, compact discs, cassettes, etc. □ **player-piano** a piano fitted with an apparatus enabling it to be played automatically. [OE *plegere* (as PLAY)]

Playfair /'pleɪfeə(r)/, John (1748–1819), Scottish mathematician and geologist. A friend of James Hutton, he is chiefly remembered for his *Illustrations of the Huttonian Theory of the Earth* (1802), which presented Hutton's views on geology – and some of his own – in a concise and readable form, enabling them to reach a far wider audience than Hutton's own writings.

playfellow /'pleɪˌfeləʊ/ *n.* a playmate.

playful /'pleɪfʊl/ *adj.* **1** fond of or inclined to play. **2** done in fun; humorous, jocular. □ **playfully** *adv.* **playfulness** *n.*

playgoer /'pleɪˌɡəʊə(r)/ *n.* a person who goes often to the theatre.

playground /'pleɪɡraʊnd/ *n.* an outdoor area for children to play on.

playgroup /'pleɪɡruːp/ *n.* a group of preschool children who play regularly together at a particular place under supervision.

playhouse /'pleɪhaʊs/ *n.* **1** a theatre. **2** a toy house for children to play in.

playing card /'pleɪɪŋ ˌkɑːd/ *n.* each of a set of usu. fifty-two oblong pieces of card or other material with an identical pattern on one side and different values represented by numbers and symbols on the other, divided into four suits, used to play various games. The origin of playing cards is uncertain. They were not known to Graeco-Roman antiquity, and though it has been argued that they reached Europe from the Far East, European and Asian cards are so different that there may not be a common origin. They have been in use in Europe since the 14th century and in the East for at least as long. From the 14th century the various suit systems were becoming established in different parts of Europe. The standard pack now contains fifty-two cards divided into four suits of thirteen cards each. Each suit has an ace, nine cards numbered 2 to 10, and three picture or 'court cards' (jack or knave, queen, and king); the pack used in Britain also includes a joker. The signs of the four suits reached England from France in about the 15th century; since then no major change has been made in the composition of the pack, and figures on the court cards are still represented in the costume of Henry VII's reign. (See also TAROT.)

playing-field /'pleɪɪŋfiːld/ *n.* a field used for outdoor team games. □ **level playing-field** a political, commercial, etc. situation in which none of the participants are favoured or handicapped.

playlet /'pleɪlɪt/ *n.* a short play or dramatic piece.

playmate /'pleɪmeɪt/ *n.* **1** a child's companion in play. **2** a lover.

playschool /'pleɪskuːl/ *n.* a nursery for preschool children.

plaything /'pleɪθɪŋ/ *n.* **1** a toy or other thing to play with. **2** a person treated as a toy.

playtime /'pleɪtaɪm/ *n.* time for play or recreation.

playwright /'pleɪraɪt/ *n.* a person who writes plays.

plaza /'plɑːzə/ *n.* a market-place or open square (esp. in a Spanish town). [Sp., = place]

plc *abbr.* (also **PLC**) (in the UK) Public Limited Company.

plea /pliː/ *n.* **1** an earnest appeal or entreaty. **2** *Law* a formal statement by or on behalf of a defendant. **3** an argument or excuse. □ **plea bargaining** esp. *US* an arrangement between prosecutor and defendant whereby the defendant pleads guilty to a lesser charge in the expectation of leniency. [ME & AF *ple*, *plai*, OF *plait*, *plaid* agreement, discussion f. L *placitum* a decree, neut. past part. of *placere* to please]

pleach /pliːtʃ/ *v.tr.* entwine or interlace (esp. branches to form a hedge). [ME *pleche* f. OF (as PLASH²)]

plead /pliːd/ *v.* (past and past part. **pleaded** or esp. *N. Amer., Sc.,* & *dial.* **pled** /pled/) **1** *intr.* (foll. by *with*) make an earnest appeal to. **2** *intr.* *Law* address a lawcourt as an advocate on behalf of a party. **3** *tr.* maintain (a cause) esp. in a lawcourt. **4** *tr.* *Law* declare to be one's state as regards guilt in or responsibility for a crime (*plead guilty; plead insanity*). **5** *tr.* offer or allege as an excuse (*pleaded forgetfulness*). **6** *intr.* make an appeal or entreaty. □ **pleadable** *adj.* **pleader** *n.* **pleadingly** *adv.* [ME f. AF *pleder*, OF *plaidier* (as PLEA)]

pleading /'pliːdɪŋ/ *n.* (usu. in *pl.*) a formal statement of the cause of an action or defence.

pleasance /'plez(ə)ns/ *n.* a secluded enclosure or part of a garden, esp. one attached to a large house. [ME f. OF *plaisance* (as PLEASANT)]

pleasant /'plez(ə)nt/ *adj.* (**pleasanter, pleasantest**) pleasing to the mind, feelings, or senses. □ **pleasantly** *adv.* **pleasantness** *n.* [ME f. OF *plaisant* (as PLEASE)]

pleasantry /'plez(ə)ntrɪ/ *n.* (*pl.* **-ies**) **1** (usu. in *pl.*) a courteous or polite remark, esp. made in casual conversation. **2** (esp. in *pl.*) an amusing remark. **3** jocularity. [F *plaisanterie* (as PLEASANT)]

please /pliːz/ *v.* **1** *tr.* (also *absol.*) be agreeable to; make glad; give pleasure to (*the gift will please them; anxious to please*). **2** *tr.* (in *passive*) **a** (foll. by *to* + infin.) be glad or willing to (*am pleased to help*). **b** (often foll. by *about, at, with*) derive pleasure or satisfaction (from). **3** *tr.* (*prec.* by *it* as subject; usu. foll. by *to* + infin.) be the inclination or wish of (*it did not please them to attend*). **4** *intr.* think fit; have the will or desire (*take as many as you please*). **5** *tr.* (short for **may it please you**) used in polite requests (*come in, please*). □ **as pleased as Punch** see PUNCH. **if you please** if you are willing, esp. *iron.* to indicate unreasonableness (*then, if you please, we had to pay*). **please oneself** do as one likes. □ **pleased** *adj.* **pleasing** *adj.* **pleasingly** *adv.* [ME *plaise* f. OF *plaisir* f. L *placere*]

pleasurable /'pleʒərəb(ə)l/ *adj.* causing pleasure; agreeable. □ **pleasurableness** *n.* **pleasurably** *adv.* [PLEASURE + -ABLE, after *comfortable*]

pleasure /'pleʒə(r)/ *n.* & *v.* ● *n.* **1** a feeling of satisfaction or joy. **2** enjoyment. **3** a source of pleasure or gratification (*painting was my chief pleasure; it is a pleasure to talk to them*). **4** *formal* a person's will or desire (*what is your pleasure?*). **5** sensual gratification or enjoyment (*a life of pleasure*). **6** (*attrib.*) done or used for pleasure (*pleasure boat; pleasure trip*). ● *v.* **1** *tr.* give (esp. sexual) pleasure to. **2** *intr.* (often foll. by *in*) take pleasure. □ **pleasure principle** *Psychol.* the theory that the basic motivation in human life is to gain pleasure and avoid pain. **take pleasure in** like doing. **with pleasure** gladly. [ME & OF *plesir, plaisir* PLEASE, used as a noun]

pleat /pliːt/ *n.* & *v.* ● *n.* a fold or crease, esp. a flattened fold in cloth doubled upon itself. ● *v.tr.* make a pleat or pleats in. [ME, var. of PLAIT]

pleb /pleb/ *n.* *colloq.* usu. *derog.* = PLEBEIAN *n.* 2. □ **plebby** *adj.* [abbr. of PLEBEIAN]

plebeian /plɪ'biːən/ *n.* & *adj.* ● *n.* **1** a commoner, esp. in ancient Rome (cf. PATRICIAN). **2** a working-class person, esp. an uncultured one. ● *adj.* **1** of low birth; of the common people. **2** uncultured. **3** coarse, ignoble. □ **plebeianism** *n.* [L *plebeius* f. *plebs plebis* the common people]

plebiscite /'plebɪsɪt, -ˌsaɪt/ *n.* **1** the direct vote of all the electors of a state etc. on an important public question, e.g. a change in the constitution. **2** the public expression of a community's opinion, with or without binding force. **3** *Rom. Hist.* a law enacted by the plebeians' assembly. □ **plebiscitary** /plə'bɪsɪtərɪ/ *adj.* [F *plébiscite* f. L *plebiscitum* f. *plebs plebis* the common people + *scitum* decree f. *sciscere* vote for]

plectrum /'plektrəm/ *n.* (*pl.* **plectrums** or **plectra** /-trə/) **1** a thin flat piece of plastic or horn etc. held in the hand and used to pluck a string, esp. of a guitar. **2** the corresponding mechanical part of a harpsichord etc. [L f. Gk *plēktron* f. *plēssō* strike]

pled see PLEAD.

pledge /pledʒ/ *n.* & *v.* ● *n.* **1** a solemn promise or undertaking. **2** a thing given as security for the fulfilment of a contract, the payment of a debt, etc., and liable to forfeiture in the event of failure. **3** a thing put in pawn. **4 a** the promise of a donation to charity. **b** such a donation. **5** a thing given as a token of love, favour, or something to come. **6** the drinking of a person's health; a toast. **7** a solemn undertaking to abstain from alcohol (*sign the pledge*). **8** the state of being pledged (*goods lying in pledge*). ● *v.tr.* **1 a** deposit as security. **b** pawn. **2** promise solemnly by the pledge of (one's honour, word, etc.). **3** (often *refl.*) bind by a solemn promise. **4** drink to the health of. □ **pledge one's troth** see TROTH. □ **pledgeable** *adj.* **pledger** *n.* **pledgor** /pleˈdʒɔː(r)/ *n.* [ME *plege* f. OF *plege* f. LL *plebium* f. *plebire* assure]

pledgee /pleˈdʒiː/ *n.* a person to whom a pledge is given.

pledget /ˈpledʒɪt/ *n.* a small wad of lint etc. [16th c.: orig. unkn.]

pleiad /ˈplaɪəd/ *n.* a brilliant group of (usu. seven) persons or things. [PLEIADES]

Pleiades /ˈplaɪəˌdiːz/ **1** *Gk Mythol.* the seven daughters of the Titan Atlas and the Oceanid Pleione, who were pursued by the hunter Orion until Zeus changed them into a cluster of stars. **2** *Astron.* an open cluster of stars in the constellation of Taurus, also known as 'the Seven Sisters'. They are the best-known such cluster in the sky, containing six (or more) stars visible to the naked eye. There are actually some five hundred members, which were formed very recently in stellar terms. [ME f. L *Pleias* f. Gk *Pleias -ados*]

pleiotropy /plaɪˈɒtrəpɪ/ *n.* *Biol.* (of a gene) the production of two or more unrelated effects in the phenotype. □ **pleiotropic** /ˌplaɪəˈtrəʊpɪk, -ˈtrɒpɪk/ *adj.* **pleiotropism** /ˌplaɪəˈtrəʊpɪz(ə)m/ *n.* [Gk *pleiōn* more + *tropē* turning]

Pleistocene /ˈplaɪstəˌsiːn/ *adj. & n.* *Geol.* of, relating to, or denoting the first epoch of the Quaternary period, between the Pliocene and the Holocene (Recent). The Pleistocene lasted from about 2,000,000 to 10,000 years ago, and is notable for a succession of ice ages. It saw the evolution of modern humankind, and towards the end of the epoch many animal species became extinct. [Gk *pleistos* most + *kainos* new]

plenary /ˈpliːnərɪ/ *adj.* **1** entire, unqualified, absolute (*plenary indulgence*). **2** (of an assembly) to be attended by all members. [LL *plenarius* f. *plenus* full]

plenipotentiary /ˌplenɪpəˈtenʃərɪ/ *n. & adj.* ● *n.* (*pl.* **-ies**) a person (esp. a diplomat) invested with the full power of independent action. ● *adj.* **1** having this power. **2** (of power) absolute. [med.L *plenipotentiarius* f. *plenus* full + *potentia* power]

plenitude /ˈplenɪˌtjuːd/ *n.* *literary* **1** fullness, completeness. **2** abundance. [ME f. OF f. LL *plenitudo* f. *plenus* full]

plenteous /ˈplentɪəs/ *adj.* *poet.* plentiful. □ **plenteously** *adv.* **plenteousness** *n.* [ME f. OF *plentivous* f. *plentif -ive* f. *plenté* PLENTY: cf. *bounteous*]

plentiful /ˈplentɪˌfʊl/ *adj.* abundant, copious. □ **plentifully** *adv.* **plentifulness** *n.*

plenty /ˈplentɪ/ *n., adj., & adv.* ● *n.* **1** (often foll. by *of*) a great or sufficient quantity or number (*we have plenty; plenty of time*). **2** abundance (*in great plenty*). ● *adj.* *colloq.* existing in an ample quantity. ● *adv.* *colloq.* fully, entirely (*it is plenty large enough*). [ME *plenteth*, *plente* f. OF *plentet* f. L *plenitas -tatis* f. *plenus* full]

plenum /ˈpliːnəm/ *n.* **1** a full assembly of people or a committee etc. **2** *Physics* space filled with matter. [L, neut. of *plenus* full]

pleochroic /ˌpliːəˈkrəʊɪk/ *adj.* *Crystallog. & Mineral.* showing different colours when viewed in different directions. □ **pleochroism** /pliːˈɒkrəʊˌɪz(ə)m/ *n.* [Gk *pleiōn* more + *-khroos* f. *khrōs* colour]

pleomorphism /ˌpliːəˈmɔːfɪz(ə)m/ *n.* the occurrence of more than one distinct form, esp. (*Biol.*) at different stages in the life cycle. □ **pleomorphic** *adj.* [Gk *pleiōn* more + *morphē* form]

pleonasm /ˈpliːəˌnæz(ə)m/ *n.* the use of more words than are needed to give the sense (e.g. *see with one's eyes*). □ **pleonastic** /ˌpliːəˈnæstɪk/ *adj.* **pleonastically** *adv.* [LL *pleonasmus* f. Gk *pleonasmos* f. *pleonazō* be superfluous]

plesiosaur /ˈpliːzɪəˌsɔː(r)/ *n.* (also **plesiosaurus** /ˌpliːzɪəˈsɔːrəs/) an extinct marine Mesozoic reptile of the suborder Plesiosauria, with a broad body, short tail, and large paddle-like limbs; esp. one of the superfamily Plesiosauroidea, with a small head and long flexible neck, e.g. *Elasmosaurus* (cf. PLIOSAUR). [mod.L f. Gk *plēsios* near + *sauros* lizard]

plessor var. of PLEXOR.

plethora /ˈpleθərə/ *n.* **1** an oversupply, glut, or excess. **2** *Med.* **a** an excess of red cells in the blood. **b** an excess of any body fluid. □ **plethoric** /plɪˈθɒrɪk/ *adj.* **plethorically** *adv.* [LL f. Gk *plēthōrē* f. *plēthō* be full]

pleura[1] /ˈplʊərə/ *n.* (*pl.* **pleurae** /-riː/) **1** *Anat.* each of a pair of serous membranes lining the thorax and enveloping the lungs in mammals. **2** *Zool.* lateral extensions of the body-wall in arthropods. □ **pleural** *adj.* [med.L f. Gk, = side of the body, rib]

pleura[2] *pl.* of PLEURON.

pleurisy /ˈplʊərɪsɪ/ *n.* *Med.* inflammation of the pleurae, marked by pain in the chest or side, fever, etc. □ **pleuritic** /plʊəˈrɪtɪk/ *adj.* [ME f. OF *pleurisie* f. LL *pleurisis* alt. f. L *pleuritis* f. Gk (as PLEURA[1])]

pleuro- /ˈplʊərəʊ/ *comb. form* **1** denoting the pleura. **2** denoting the side.

pleuron /ˈplʊərɒn/ *n.* (*pl.* **pleura** /-rə/) = PLEURA[1] 2. [Gk, = side of the body, rib]

pleuropneumonia /ˌplʊərəʊnjuːˈməʊnɪə/ *n.* *Med.* pneumonia complicated with pleurisy.

Pléven /ˈplev(ə)n/ an industrial town in northern Bulgaria, north-east of Sofia; pop. (1990) 168,000. An important fortress town and trading centre of the Ottoman Empire, it was taken from the Turks by the Russians in the Russo-Turkish War of 1877, after a siege of 143 days.

Plexiglas /ˈpleksɪˌɡlɑːs/ *n. propr.* = PERSPEX. [formed as PLEXOR + GLASS]

plexor /ˈpleksə(r)/ *n.* (also **plessor** /ˈples-/) *Med.* a small hammer used to test reflexes and in percussing. [irreg. f. Gk *plēxis* percussion + -OR[1]]

plexus /ˈpleksəs/ *n.* (*pl.* same or **plexuses**) **1** *Anat.* a network of nerves or vessels in an animal body (*gastric plexus*). **2** any network or weblike formation. □ **plexiform** *adj.* [L f. *plectere plex-* plait]

pliable /ˈplaɪəb(ə)l/ *adj.* **1** bending easily; supple. **2** yielding, compliant. □ **pliably** *adv.* **pliableness** *n.* **pliability** /ˌplaɪəˈbɪlɪtɪ/ *n.* [F f. *plier* bend: see PLY[1]]

pliant /ˈplaɪənt/ *adj.* = PLIABLE 1. □ **pliancy** *n.* **pliantly** *adv.* [ME f. OF (as PLIABLE)]

plicate /ˈplaɪkeɪt/ *adj.* *Biol. & Geol.* folded, crumpled, corrugated. □ **plicated** /ˈplaɪkeɪtɪd, plɪˈkeɪt-/ *adj.* [L *plicatus* past part. of *plicare* fold]

plication /plɪˈkeɪʃ(ə)n/ *n.* **1** the act of folding. **2** a fold; a folded condition. [ME f. med.L *plicatio* or L *plicare* fold, after *complication*]

plié /ˈpliːeɪ/ *n.* *Ballet* a bending of the knees with the feet on the ground. [F, past part. of *plier* bend: see PLY[1]]

pliers /ˈplaɪəz/ *n.pl.* pincers with parallel flat usu. serrated surfaces for holding small objects, bending wire, etc. [(dial.) *ply* bend (as PLIABLE)]

plight[1] /plaɪt/ *n.* a condition or state, esp. an unfortunate one. [ME & AF *plit* = OF *pleit* fold: see PLAIT: *-gh-* by confusion with PLIGHT[2]]

plight[2] /plaɪt/ *v. & n.* *archaic* ● *v.tr.* **1** pledge or promise solemnly (one's faith, loyalty, etc.). **2** (foll. by *to*) engage, esp. in marriage. ● *n.* an engagement or act of pledging. □ **plight one's troth** see TROTH. [orig. as noun, f. OE *pliht* danger f. Gmc]

plimsoll /ˈplɪms(ə)l/ *n.* (also **plimsole**) *Brit.* a rubber-soled canvas sports shoe. [prob. from the resemblance of the side of the sole to a PLIMSOLL LINE]

Plimsoll line *n.* (also **Plimsoll mark**) a marking on a ship's side showing the limit of legal submersion under various conditions. It is named after Samuel Plimsoll (1824–98), the English politician whose agitation in the 1870s resulted in the Merchant Shipping Act of 1876, putting an end to the practice of sending to sea overloaded and heavily insured old ships from which the owners made a profit if they sank.

plinth /plɪnθ/ *n.* **1** *Archit.* the lower square slab at the base of a column. **2** a base supporting a vase or statue etc. [F *plinthe* or L *plinthus* f. Gk *plinthos* tile, brick, squared stone]

Pliny[1] /ˈplɪnɪ/ (known as Pliny the Elder; Latin name Gaius Plinius Secundus) (23–79), Roman statesman and scholar. He combined a busy life in public affairs with prodigious activity in reading and writing. His *Natural History* (77) is a vast encyclopedia of the natural and human worlds, and is one of the earliest known works of its kind. He died while observing the eruption of Vesuvius in 79, an event which was described by his nephew Pliny the Younger.

Pliny[2] /ˈplɪnɪ/ (known as Pliny the Younger; Latin name Gaius Plinius Caecilius Secundus) (*c.*61–*c.*112), Roman senator and writer. He was the nephew of Pliny the Elder. His books of letters deal with both public and private affairs; they include a description of the eruption of Vesuvius in 79 which destroyed the town of Pompeii and in which his uncle died. The letters also include Pliny's correspondence with the Emperor Trajan, which contains one of the earliest descriptions of non-Christian attitudes towards Christians.

Pliocene /ˈplaɪəˌsiːn/ *adj. & n.* *Geol.* of, relating to, or denoting the last epoch of the Tertiary period, between the Miocene and the Pleistocene. The Pliocene lasted from about 5.1 to 2 million years ago, and was a time when temperatures were falling and many mammals that had flourished earlier in the Tertiary were becoming extinct. [Gk *pleiōn* more + *kainos* new]

pliosaur /'plaɪəˌsɔː(r)/ *n.* (also **pliosaurus** /ˌplaɪə'sɔːrəs/) a plesiosaur of the superfamily Pliosauroidea, with a short neck, large head, and massive toothed jaws. [mod.L f. Gk *pleiōn* more (like a) + *sauros* lizard]

plissé /'pliːseɪ/ *adj. & n.* ● *adj.* (of cloth etc.) treated so as to cause permanent puckering. ● *n.* material treated in this way. [F, past part. of *plisser* pleat]

PLO see PALESTINE LIBERATION ORGANIZATION.

plod /plɒd/ *v. & n.* ● *v.* (**plodded, plodding**) **1** *intr.* (often foll. by *along, on,* etc.) walk doggedly or laboriously; trudge. **2** *intr.* (often foll. by *at*) work slowly and steadily. **3** *tr.* tread or make (one's way) laboriously. ● *n.* the act or a spell of plodding. □ **plodder** *n.* **ploddingly** *adv.* [16th c.: prob. imit.]

-ploid /plɔɪd/ *comb. form Biol.* forming adjectives denoting the number of sets of chromosomes in a cell (*diploid; polyploid*). [after HAPLOID]

ploidy /'plɔɪdɪ/ *n. Biol.* the number of sets of chromosomes in a cell. [after DIPLOIDY, POLYPLOIDY, etc.]

Ploieşti /plɔɪ'eʃt/ an oil-refining city in central Romania, north of Bucharest; pop. (1989) 247,500.

plonk[1] /plɒŋk/ *v. & n.* ● *vtr.* **1** set down hurriedly or clumsily. **2** (usu. foll. by *down*) set down firmly. ● *n.* **1** an act of plonking. **2** a heavy thud. [imit.]

plonk[2] /plɒŋk/ *n. colloq.* cheap or inferior wine. [orig. Austral.: prob. corrupt. of *blanc* in F *vin blanc* white wine]

plonker /'plɒŋkə(r)/ *n. sl.* **1** the penis. **2** a stupid person. [PLONK[1]]

plonko /'plɒŋkəʊ/ *n.* (*pl.* **-os**) *Austral. sl.* an excessive drinker of cheap wine; an alcoholic.

plop /plɒp/ *n., v., & adv.* ● *n.* **1** a sound as of a smooth object dropping into water without a splash. **2** an act of falling with this sound. ● *v.* (**plopped, plopping**) *intr. & tr.* fall or drop with a plop. ● *adv.* with a plop. [19th c.: imit.]

plosion /'pləʊʒ(ə)n/ *n. Phonet.* the sudden release of air in the pronunciation of a stop consonant. [EXPLOSION]

plosive /'pləʊsɪv/ *adj. & n. Phonet.* ● *adj.* denoting a consonant that is produced by stopping the airflow using the lips, teeth, or palate, followed by a sudden release of air. ● *n.* a plosive sound. [EXPLOSIVE]

plot /plɒt/ *n. & v.* ● *n.* **1** a defined and usu. small piece of ground. **2** the interrelationship of the main events in a play, novel, film, etc. **3** a conspiracy or secret plan, esp. to achieve an unlawful end. **4** a plan, map, or chart. **5** a graph showing the relation between two variables. ● *vtr.* (**plotted, plotting**) **1** make a plan or map of (an existing object, a place or thing to be laid out, constructed, etc.). **2** (also *absol.*) plan or contrive secretly (a crime, conspiracy, etc.). **3** mark (a point or course etc.) on a chart or diagram. **4 a** mark out or allocate (points) on a graph. **b** make (a curve etc.) by marking out a number of points. **5** devise or plan the plot of (a play, novel, film, etc.). □ **plotless** *adj.* **plotlessness** *n.* **plotter** *n.* [OE and f. OF *complot* secret plan: both of unkn. orig.]

Plotinus /plə'taɪnəs/ (c.205–70), philosopher, probably of Roman descent. The founder and leading exponent of Neoplatonism, he studied in Alexandria and later Persia before finally settling in Rome in 244 and setting up a school of philosophy. His writings were published after his death by his pupil Porphyry. (See NEOPLATONISM.)

plough /plaʊ/ *n. & v.* (*N. Amer.* **plow**) ● *n.* **1** an implement with a cutting blade fixed in a frame drawn by a tractor or by animals, for cutting furrows in the soil and turning it up. (*See note below.*) **2** an implement resembling this and having a comparable function (*snowplough*). **3** ploughed land. **4** (**the Plough**) *Astron.* the seven brightest stars in the constellation of Ursa Major. Also called *Charles's Wain, Big Dipper*. ● *v.* **1** *tr.* (also *absol.*) turn up (the earth) with a plough, esp. before sowing. **2** *tr.* (foll. by *out, up, down,* etc.) turn or extract (roots, weeds, etc.) with a plough. **3** *tr.* furrow or scratch (a surface) as if with a plough. **4** *tr.* produce (a furrow or line) in this way. **5** *intr.* (foll. by *through*) advance laboriously, esp. through work, a book, etc. **6** *intr.* (foll. by *through, into*) move like a plough violently. **7** *intr. & tr. Brit. colloq.* fail in an examination. □ **plough back 1** plough (grass etc.) into the soil to enrich it. **2** reinvest (profits) in the business producing them. **put one's hand to the plough** undertake a task (Luke 9:62). □ **ploughable** *adj.* **plougher** *n.* [OE *plōh* f. ON *plógr* f. Gmc]

▪ One of the earliest agricultural tools, the plough is used to loosen and aerate the soil, to bury stubble and weeds, and to expose fresh soil to weathering so that it can be harrowed into a good seed-bed.

The essential parts of a plough are a coulter (a vertical knife which cuts a thin vertical slice), a share (which cuts a horizontal slice below the earth), and a mould-board or breast, which turns the horizontal slice cut and thus exposes a furrow which is subsequently filled by the next slice of turned earth.

ploughman /'plaʊmən/ *n.* (*N. Amer.* **plowman**) (*pl.* **-men**) a person who uses a plough. □ **ploughman's lunch** a meal of bread and cheese with pickle or salad. **ploughman's spikenard** a fragrant composite plant, *Inula conyzae*, with yellow rayless flower-heads.

Plough Monday the first Monday after the Epiphany.

ploughshare /'plaʊʃeə(r)/ *n.* (*N. Amer.* **plowshare**) the cutting blade of a plough.

Plovdiv /'plɒvdɪf/ an industrial and commercial city in southern Bulgaria; pop. (1990) 379,080. A city in ancient Macedonia from 341 BC, when it was conquered by Philip II of Macedon, it became a part of Roman Thrace in AD 46 and was taken by the Turks in 1364. Known to the Greeks as Philippopolis and to the Romans as Trimontium, it assumed its present name after the First World War.

plover /'plʌvə(r)/ *n.* a short-billed gregarious bird of the family Charadriidae, esp. of the genus *Charadius*, often found by water. [ME & AF f. OF *plo(u)vier* ult. f. L *pluvia* rain]

plow *N. Amer.* var. of PLOUGH.

ploy /plɔɪ/ *n.* a stratagem; a cunning manoeuvre to gain an advantage. [orig. Sc., 18th c.: orig. unkn.]

PLP *abbr.* (in the UK) Parliamentary Labour Party.

PLR *abbr.* (in the UK) Public Lending Right.

pluck /plʌk/ *v. & n.* ● *v.* **1** *tr.* (often foll. by *out, off,* etc.) remove by picking or pulling out or away. **2** *tr.* strip (a bird) of feathers. **3** *tr.* pull at, twitch. **4** *intr.* (foll. by *at*) tug or snatch at. **5** *tr.* sound (the string of a musical instrument) with the finger or plectrum etc. **6** *tr.* plunder. **7** *tr.* swindle. ● *n.* **1** courage, spirit. **2** an act of plucking; a twitch. **3** the heart, liver, and lungs of an animal as food. □ **pluck up** summon up (one's courage, spirits, etc.). □ **plucker** *n.* **pluckless** *adj.* [OE *ploccian, pluccian,* f. Gmc]

plucky /'plʌkɪ/ *adj.* (**pluckier, pluckiest**) brave, spirited. □ **pluckily** *adv.* **pluckiness** *n.*

plug /plʌg/ *n. & v.* ● *n.* **1** a piece of solid material fitting tightly into a hole, used to fill a gap or cavity or act as a wedge or stopper. **2 a** a device of metal pins in an insulated casing fitting into holes in a socket for making an electrical connection, esp. between an appliance and the mains. **b** *colloq.* an electric socket. **3** = *spark-plug* (see SPARK[1]). **4** *colloq.* a piece of (often free) publicity for an idea, product, etc. **5** a mass of solidified lava filling the neck of a volcano. **6** a cake or stick of tobacco; a piece of this for chewing. **7** = *fire-plug.* ● *v.* (**plugged, plugging**) **1** *tr.* (often foll. by *up*) stop (a hole etc.) with a plug. **2** *tr. sl.* shoot or hit (a person etc.). **3** *tr. colloq.* seek to popularize (an idea, product, etc.) by constant recommendation. **4** *intr.* (often foll. by *at*) *colloq.* work steadily away (at). □ **plug in** connect electrically by inserting a plug in a socket. **plug-in** *adj.* able to be connected by means of a plug. **plug-ugly** *US sl. n.* (*pl.* **-ies**) a thug or ruffian. ● *adj.* villainous-looking. □ **plugger** *n.* [MDu. & MLG *plugge,* of unkn. orig.]

plugola /plʌg'əʊlə/ *n.* esp. *US colloq.* **1** a bribe offered in return for incidental or surreptitious promotion of a person or product, esp. on radio or television. **2** the practice of such bribery. [PLUG + -*ola,* prob. after PAYOLA]

plum /plʌm/ *n.* **1 a** an oval fleshy fruit, usu. purple or reddish when ripe, with sweet pulp and a flattish pointed stone. **b** a deciduous rosaceous tree of the genus *Prunus,* esp. *P. domestica,* bearing this. **c** the wood of such a tree. **2** a reddish-purple colour. **3** a dried grape or raisin used in cooking. **4** *colloq.* the best of a collection; something especially prized (often *attrib.: a plum job*). □ **plum cake** a cake containing raisins, currants, etc. **plum duff** a plain flour pudding with raisins or currants. **a plum in one's mouth** a rich-sounding voice or affected accent. **plum pudding** a rich boiled suet pudding with raisins, currants, spices, etc. [OE *plūme* f. med.L *pruna* f. L *prunum*]

plumage /'pluːmɪdʒ/ *n.* a bird's feathers. □ **plumaged** *adj.* (usu. in comb.). [ME f. OF (as PLUME)]

plumassier /ˌpluːmæ'sɪə(r)/ *n.* a person who trades or works in ornamental feathers. [F f. *plumasse* augment. of *plume* PLUME]

plumb[1] /plʌm/ *n., adv., adj., & v.* ● *n.* a ball of lead or other heavy material, esp. one attached to the end of a line for finding the depth of water or determining the vertical on an upright surface. ● *adv.* **1** exactly (*plumb in the centre*). **2** vertically. **3** *N. Amer. sl.* quite, utterly (*plumb crazy*). ● *adj.* **1** vertical. **2** downright, sheer (*plumb nonsense*). **3** *Cricket* (of the wicket) level, true. ● *vtr.* **1 a** measure the depth of (water) with a plumb.

b determine (a depth). **2** test (an upright surface) to determine the vertical. **3** reach or experience in extremes (*plumb the depths of fear*). **4** learn in detail the facts about (a matter). □ **out of plumb** not vertical. **plumb-line** a line with a plumb attached. **plumb-rule** a mason's plumb-line attached to a board. [ME, prob. ult. f. L *plumbum* lead, assim. to OF *plomb* lead]

plumb[2] /plʌm/ v. **1** tr. provide (a building or room etc.) with plumbing. **2** tr. (often foll. by *in*) fit as part of a plumbing system. **3** intr. work as a plumber. [back-form. f. PLUMBER]

plumbago /plʌmˈbeɪgəʊ/ n. (pl. **-os**) **1** = GRAPHITE. **2** a plant of the genus *Plumbago*, with spikes of grey or blue flowers. Also called *leadwort*. [L f. *plumbum* LEAD[2]]

plumbeous /ˈplʌmbɪəs/ adj. **1** of or like lead. **2** lead-glazed. [L *plumbeus* f. *plumbum* LEAD[2]]

plumber /ˈplʌmə(r)/ n. a person who fits and repairs the apparatus of a water-supply, heating, etc. [ME *plummer* etc. f. OF *plommier* f. L *plumbarius* f. *plumbum* LEAD[2]]

plumbic /ˈplʌmbɪk/ adj. **1** Chem. containing lead esp. in its tetravalent form. **2** Med. due to the presence of lead. [L *plumbum* lead]

plumbing /ˈplʌmɪŋ/ n. **1** the system or apparatus of water-supply, heating, etc., in a building. **2** the work of a plumber. **3** colloq. lavatory installations.

plumbism /ˈplʌmbɪz(ə)m/ n. Med. = lead poisoning.

plumbless /ˈplʌmlɪs/ adj. (of a depth of water etc.) that cannot be plumbed.

plumbous /ˈplʌmbəs/ n. Chem. containing lead in its divalent form.

plume /pluːm/ n. & v. ● n. **1** a feather, esp. a large one used for ornament. **2** an ornament of feathers etc. attached to a helmet or hat or worn in the hair. **3** something resembling this (*a plume of smoke*). **4** Zool. a feather-like part or formation. **5** Geol. a column of hotter magma rising in the earth's mantle, believed to cause volcanic activity away from the margins of plates. ● v. **1** tr. decorate or provide with a plume or plumes. **2** refl. (foll. by *on, upon*) pride (oneself on esp. something trivial). **3** tr. (of a bird) preen (itself or its feathers). □ **plume moth** a small moth of the family Pterophoridae, with narrow divided feathery wings and long legs. □ **plumeless** adj. **plumelike** adj. **plumery** n. [ME f. OF f. L *pluma* down]

plummet /ˈplʌmɪt/ n. & v. ● n. **1** a plumb or plumb-line. **2** a sounding-line. **3** a weight attached to a fishing-line to keep the float upright. ● v.intr. (**plummeted, plummeting**) fall or plunge rapidly. [ME f. OF *plommet* dimin. (as PLUMB[1])]

plummy /ˈplʌmɪ/ adj. (**plummier, plummiest**) **1** full of or rich in plums. **2** colloq. (of a voice) sounding affectedly rich or deep in tone. **b** snobbish. **3** colloq. good, desirable.

plumose /ˈpluːməʊz, -məʊs/ adj. **1** feathered. **2** feather-like. [L *plumosus* (as PLUME)]

plump[1] /plʌmp/ adj. & v. ● adj. (esp. of a person or animal or part of the body) having a full rounded shape; fleshy; filled out. ● v.tr. & intr. (often foll. by *up, out*) make or become plump; fatten. □ **plumpish** adj. **plumply** adv. **plumpness** n. **plumpy** adj. [ME *plompe* f. MDu. *plomp* blunt, MLG *plump, plomp* shapeless etc.]

plump[2] /plʌmp/ v., n., adv., & adj. ● v. **1** intr. & tr. (often foll. by *down*) drop or fall abruptly (*plumped down on the chair; plumped it on the floor*). **2** intr. (foll. by *for*) decide definitely in favour of (one of two or more possibilities). **3** tr. (often foll. by *out*) utter abruptly; blurt out. ● n. an abrupt plunge; a heavy fall. ● adv. colloq. **1** with a sudden or heavy fall. **2** directly, bluntly (*I told him plump*). ● adj. colloq. direct, unqualified (*answered with a plump 'no'*). [ME f. MLG *plumpen*, MDu. *plompen*: orig. imit.]

plumule /ˈpluːmjuːl/ n. **1** Bot. the rudimentary shoot or stem of an embryo plant. **2** Zool. a down feather on a young bird. □ **plumular** adj. (in sense 1). **plumulaceous** /ˌpluːmjʊˈleɪʃəs/ adj. (in sense 2). [F *plumule* or L *plumula*, dimin. (as PLUME)]

plumy /ˈpluːmɪ/ adj. (**plumier, plumiest**) **1** plumelike, feathery. **2** adorned with plumes.

plunder /ˈplʌndə(r)/ v. & n. ● v.tr. **1** rob (a place or person) or steal (goods), esp. systematically or in war. **2** steal from (another's writings etc.). ● n. **1** the violent or dishonest acquisition of property. **2** property acquired by plundering. **3** colloq. profit, gain. □ **plunderer** n. [LG *plündern* lit. 'rob of household goods' f. MHG *plunder* clothing etc.]

plunge /plʌndʒ/ v. & n. ● v. **1** (usu. foll. by *in, into*) **a** tr. thrust forcefully or abruptly. **b** intr. dive; propel oneself forcibly. **c** intr. & tr. enter or cause to enter a certain condition or embark on a certain course abruptly or

impetuously (*they plunged into a lively discussion; the room was plunged into darkness*). **2** tr. immerse completely. **3** intr. **a** move suddenly and dramatically downward. **b** (foll. by *down, into*, etc.) move with a rush (*plunged down the stairs*). **c** diminish rapidly (*share prices have plunged*). **4** intr. (of a horse) start violently forward. **5** intr. (of a ship) pitch. **6** intr. colloq. gamble heavily; run into debt. ● n. a plunging action or movement; a dive. □ **plunging** (or **plunge**) **neckline** a low-cut neckline. **take the plunge** colloq. commit oneself to a (usu. risky) course of action. [ME f. OF *plungier* ult. f. L *plumbum* plummet]

plunger /ˈplʌndʒə(r)/ n. **1** a part of a mechanism that works with a plunging or thrusting movement. **2** a rubber cup on a handle for clearing blocked pipes by a plunging and sucking action. **3** colloq. a reckless gambler.

plunk /plʌŋk/ n. & v. ● n. **1** the sound made by the sharply plucked string of a stringed instrument. **2** US a heavy blow. **3** US = PLONK[1] n. ● v. **1** intr. & tr. sound or cause to sound with a plunk. **2** tr. US hit abruptly. **3** tr. US = PLONK[1] v. [imit.]

pluperfect /pluːˈpɜːfɪkt/ adj. & n. Gram. ● adj. (of a tense) denoting an action completed prior to some past point of time specified or implied, formed in English by *had* and the past participle, as: *he had gone by then*. ● n. the pluperfect tense. [mod.L *plusperfectum* f. L *plus quam perfectum* more than perfect]

plural /ˈplʊərəl/ adj. & n. ● adj. **1** more than one in number. **2** Gram. (of a word or form) denoting more than one, or (in languages with dual number) more than two. ● n. Gram. **1** a plural word or form. **2** the plural number. □ **plurally** adv. [ME f. OF *plurel* f. L *pluralis* f. *plus pluris* more]

pluralism /ˈplʊərəˌlɪz(ə)m/ n. **1** the holding of more than one office, esp. an ecclesiastical office or benefice, at a time. **2 a** a political theory advocating autonomy and retention of identity for individual bodies rather than the development of centralized state power. **b** a form of society in which the members of minority groups maintain their independent cultural traditions. **3** Philos. **a** a system that recognizes more than one ultimate principle or kind of being (opp. MONISM 2). **b** (in moral philosophy) the theory that there is more than one value and that they cannot be reduced one to another. □ **pluralist** n. **pluralistic** /ˌplʊərəˈlɪstɪk/ adj. **pluralistically** adv.

plurality /plʊəˈrælɪtɪ/ n. (pl. **-ies**) **1** the state of being plural. **2** = PLURALISM 1. **3** a large or the greater number. **4** US a majority that is not absolute. [ME f. OF *pluralité* f. LL *pluralitas* (as PLURAL)]

pluralize /ˈplʊərəˌlaɪz/ v. (also **-ise**) **1** tr. & intr. make or become plural. **2** tr. express in the plural. **3** intr. hold more than one ecclesiastical office or benefice.

pluri- /ˈplʊərɪ/ comb. form several. [L *plus pluris* more, *plures* several]

plurry /ˈplʌrɪ/ adj. & adv. Austral. & NZ sl. bloody, damn. [alt. of BLOODY]

plus /plʌs/ prep., adj., n., & conj. ● prep. **1** Math. with the addition of (*3 plus 4 equals 7*) (symbol: +). **2** (of temperature) above zero (*plus 2° C*). **3** colloq. with; having gained; newly possessing (*returned plus a new car*). ● adj. **1** (after a number) at least (*fifteen plus*). **2** (after a grade etc.) rather better than (*beta plus*). **3** Math. positive. **4** having a positive electrical charge. **5** (attrib.) additional, extra (*plus business*). ● n. **1** = plus sign. **2** Math. an additional or positive quantity. **3** an advantage (*experience is a definite plus*). ● conj. colloq. disp. also; and furthermore (*they arrived late, plus they were hungry*). □ **plus sign** the symbol +, indicating addition or a positive value. [L, = more]

plus-fours /plʌsˈfɔːz/ n.pl. men's long wide knickerbockers formerly commonly worn for golf etc. They were so named because the overhang at the knee required an extra four inches of material.

plush /plʌʃ/ n. & adj. ● n. cloth of silk or cotton etc., with a long soft nap. ● adj. **1** made of plush. **2** plushy. □ **plushly** adv. **plushness** n. [obs. F *pluche* contr. f. *peluche* f. OF *peluchier* f. It. *peluzzo* dimin. of *pelo* f. L *pilus* hair]

plushy /ˈplʌʃɪ/ adj. (**plushier, plushiest**) colloq. stylish, luxurious. □ **plushiness** n.

Plutarch /ˈpluːtɑːk/ (Latin name Lucius Mestrius Plutarchus) (c.46–c.120), Greek biographer and philosopher. He is chiefly known for his *Parallel Lives*, a collection of biographies of prominent Greeks and Romans in which the moral character of his subjects is illustrated by a series of anecdotes. The work was an important source for Shakespeare's Roman plays, and was used as a model by biographers such as Izaak Walton.

plutarchy /ˈpluːtɑːkɪ/ n. (pl. **-ies**) plutocracy. [Gk *ploutos* wealth + -*arkhia* -rule]

Pluto /ˈpluːtəʊ/ **1** Gk Mythol. the god of the underworld (cf. HADES).

2 *Astron.* the ninth planet from the sun in the solar system, orbiting at an average distance of 5,900 million km from the sun. (*See note below.*) [L f. Gk *Ploutōn*]

■ Pluto was discovered in 1930 by Clyde Tombaugh, following analysis of perturbations in the orbit of Uranus and Neptune, but it is now known to be too small a body (diameter about 2,250 km) to have produced these effects. Its orbit is more steeply inclined and more eccentric than that of other planets, so that at perihelion Pluto is closer to the sun than Neptune (as in 1979–99). It was discovered in 1978 to have a single satellite, Charon, which is so large that the pair should properly be regarded as a binary system.

plutocracy /plu:'tɒkrəsɪ/ *n.* (*pl.* -**ies**) **1** a government by the wealthy. **b** a state governed in this way. **2** a wealthy élite or ruling class. □ **plutocratic** /ˌpluːtəˈkrætɪk/ *adj.* **plutocratically** *adv.* [Gk *ploutokratia* f. *ploutos* wealth + -CRACY]

plutocrat /'pluːtəˌkræt/ *n. derog. or joc.* **1** a member of a plutocracy or wealthy élite. **2** a wealthy and influential person.

pluton /'pluːt(ə)n/ *n. Geol.* a body of plutonic rock. [back-form. f. PLUTONIC]

Plutonian /pluːˈtəʊnɪən/ *adj.* **1** infernal. **2** of the infernal regions. [L *Plutonius* f. Gk *Ploutōnios* (as PLUTO)]

plutonic /pluːˈtɒnɪk/ *adj.* **1** *Geol.* **a** (of rock) formed as igneous rock by solidification below the surface of the earth. **b** (**Plutonic**) attributing most geological phenomena to the action of internal heat (opp. *Neptunian*). (See also PLUTONIST.) **2** (**Plutonic**) = PLUTONIAN. [formed as PLUTONIST]

Plutonist /'pluːtənɪst/ *n. Geol. hist.* an advocate of the Plutonic theory, which held (correctly) that rocks such as granite are of igneous origin; also called *Vulcanist*. The Scottish geologist James Hutton was the main originator of the theory, which had prevailed over the rival Neptunian theory by the early 19th century. (See also NEPTUNIST.) □ **Plutonism** *n.*

plutonium /pluːˈtəʊnɪəm/ *n.* a dense silvery radioactive chemical element (atomic number 94; symbol **Pu**). A member of the actinide series, plutonium was obtained by Glenn Seaborg and his colleagues in 1940 by bombarding uranium with deuterons. It only occurs in trace amounts in nature. Plutonium was soon discovered to undergo fission, and the weapon exploded at Nagasaki in 1945 was a plutonium bomb. The use of plutonium as a nuclear fuel is economically important because it can be manufactured in nuclear reactors from the commonest to non-fissile isotope of uranium, uranium-238. [PLUTO, after *uranium*, *neptunium*, as the next planet beyond Neptune]

pluvial /'pluːvɪəl/ *adj. & n.* ● *adj.* **1** of rain; rainy. **2** *Geol.* **a** (of a geological period, esp. in the Pleistocene epoch) characterized by relatively high average rainfall. **b** caused by rain. ● *n. Geol.* a pluvial period. □ **pluvious** *adj.* (in sense 1). [L *pluvialis* f. *pluvia* rain]

pluviometer /ˌpluːvɪˈɒmɪtə(r)/ *n.* a rain-gauge. □ **pluviometric** /-vɪəˈmetrɪk/ *adj.* **pluviometrical** *adj.* **pluviometrically** *adv.* [L *pluvia* rain + -METER]

ply[1] /plaɪ/ *n.* (*pl.* -**ies**) **1** a thickness or layer of certain materials, esp. wood or cloth (*three-ply*). **2** a strand of yarn or rope etc. [ME f. F *pli* f. *plier, pleier* f. L *plicare* fold]

ply[2] /plaɪ/ *v.* (-**ies**, -**ied**) **1** *tr.* use or wield vigorously (a tool, weapon, etc.). **2** *tr.* work steadily at (one's business or trade). **3** *tr.* (foll. by *with*) **a** supply (a person) continuously (with food, drink, etc.). **b** approach repeatedly (with questions, demands, etc.). **4 a** *intr.* (often foll. by *between*) (of a vehicle etc.) travel regularly (to and fro between two points). **b** *tr.* work (a route) in this way. **5** *intr.* (of a taxi-driver, boatman, etc.) attend regularly for custom (*ply for trade*). **6** *intr.* sail to windward. [ME *plye*, f. APPLY]

Plymouth /'plɪməθ/ **1** a port and naval base in SW England, on the Devon coast; pop. (1991) 238,800. It was from there that Sir Francis Drake set sail in 1588 against the Spanish Armada and it was the scene in 1620 of the departure in the *Mayflower* of the Pilgrim Fathers to North America. **2** A shipping forecast area covering the English Channel roughly between the meridians of the Scilly Isles in the west and Start Point in the east. **3** a town in SE Massachusetts, on the Atlantic coast; pop. (1986) 40,290. The site in 1620 of the landing of the Pilgrim Fathers, it was the earliest permanent European settlement in New England. **4** the capital of the island of Montserrat in the West Indies; pop. (1985) 3,500.

Plymouth Brethren *n.pl.* a fundamentalist Christian Protestant denomination with no formal creed and no official order of ministers, named after its first centre, established in 1830 by John Nelson Darby

(1800–82) at Plymouth in Devon. Its teaching combines elements of Calvinism and Pietism, with emphasis on an expected millennium. Of austere outlook, members renounce many secular occupations, allowing only those compatible with New Testament standards. As a result of doctrinal and other differences, a split in 1849 resulted in the formation of the Exclusive Brethren and the Open Brethren.

Plymouth Rock a granite boulder at Plymouth, Massachusetts, on to which the Pilgrim Fathers are said to have stepped from the *Mayflower*.

plywood /'plaɪwʊd/ *n.* a strong thin board consisting of two or more layers glued and pressed together with the direction of the grain alternating.

Plzeň see PILSEN.

PM *abbr.* **1** Prime Minister. **2** post-mortem. **3** Provost Marshal.

Pm *symb. Chem.* the element promethium.

p.m. *abbr.* after noon. [L *post meridiem*]

PMG *abbr.* **1** Paymaster General. **2** Postmaster General.

PMS *abbr.* premenstrual syndrome.

PMT *abbr.* premenstrual tension.

PNdB *abbr.* perceived noise decibel(s).

pneumatic /njuːˈmætɪk/ *adj.* **1** of or relating to air or wind. **2** containing or operated by compressed air. **3** connected with or containing air cavities esp. in the bones of birds or in fish. □ **pneumatic drill** a drill driven by compressed air, for breaking up a hard surface. **pneumatic trough** a shallow container used in laboratories to collect gases in jars over the surface of water or mercury. □ **pneumatically** *adv.* **pneumaticity** /ˌnjuːməˈtɪsɪtɪ/ *n.* [F *pneumatique* or L *pneumaticus* f. Gk *pneumatikos* f. *pneuma* wind f. *pneō* breathe]

pneumatics /njuːˈmætɪks/ *n.pl.* (treated as *sing.*) the science of the mechanical properties of gases.

pneumatic tyre *n.* a tyre inflated with air. (See TYRE.)

pneumato- /ˈnjuːmətəʊ/ *comb. form* denoting: **1** air. **2** breath. **3** spirit. [Gk f. *pneuma* (as PNEUMATIC)]

pneumatology /ˌnjuːməˈtɒlədʒɪ/ *n.* **1** the branch of theology concerned with the Holy Ghost and other spiritual concepts. **2** *archaic* psychology. □ **pneumatological** /-təˈlɒdʒɪk(ə)l/ *adj.*

pneumatophore /ˈnjuːmətəˌfɔː(r), njuːˈmætə-/ *n.* **1** *Zool.* the air-filled float of some colonial hydrozoa, such as the Portuguese man-of-war. **2** *Bot.* an aerial root specialized for gaseous exchange found in mangroves etc. growing in swampy areas.

pneumo- /ˈnjuːməʊ/ *comb. form* denoting the lungs. [abbr. of *pneumono-* f. Gk *pneumōn* lung]

pneumoconiosis /ˌnjuːməʊˌkəʊnɪˈəʊsɪs/ *n. Med.* a lung disease caused by inhalation of dust or small particles. [PNEUMO- + Gk *konis* dust]

pneumogastric /ˌnjuːməʊˈɡæstrɪk/ *adj.* of or relating to the lungs and stomach.

pneumonectomy /ˌnjuːməˈnektəmɪ/ *n.* (*pl.* -**ies**) the surgical removal of a lung or part of a lung.

pneumonia /njuːˈməʊnɪə/ *n.* a bacterial inflammation of one lung (*single pneumonia*) or both lungs (*double pneumonia*) causing the air sacs to fill with pus and become solid. □ **pneumonic** /-ˈmɒnɪk/ *adj.* [L f. Gk f. *pneumōn* lung]

pneumonitis /ˌnjuːməˈnaɪtɪs/ *n. Med.* inflammation of the lungs usu. caused by a virus.

pneumothorax /ˌnjuːməʊˈθɔːræks/ *n. Med.* the presence of air or gas in the cavity between the lungs and the chest wall.

PNG *abbr.* Papua New Guinea.

PO *abbr.* **1** Post Office. **2** postal order. **3** Petty Officer. **4** Pilot Officer.

Po[1] *symb. Chem.* the element polonium.

Po[2] /pəʊ/ a river in northern Italy. Italy's longest river, it rises in the Alps near the border with France and flows 668 km (415 miles) eastwards to the Adriatic.

po /pəʊ/ *n.* (*pl.* **pos**) *Brit. colloq.* a chamber-pot.

POA *abbr.* (in the UK) Prison Officers' Association.

poach[1] /pəʊtʃ/ *v.tr.* **1** cook (an egg) without its shell in or over boiling water. **2** cook (fish etc.) by simmering in a small amount of liquid. □ **poacher** *n.* [ME f. OF *pochier* f. *poche* POKE[2]]

poach[2] /pəʊtʃ/ *v.* **1** *tr.* (also *absol.*) catch (game or fish) illegally. **2** *intr.* (often foll. by *on*) trespass or encroach (on another's property, ideas, etc.). **3** *tr.* appropriate illicitly or unfairly (a person, thing, idea, etc.). **4** *tr.* Tennis etc. take (a shot) in one's partner's portion of the court. **5 a** *tr.*

trample or cut up (turf) with hooves. **b** *intr.* (of land) become sodden by being trampled. □ **poacher** *n.* [earlier *poche*, perh. f. F *pocher* put in a pocket (as POACH[1])]

Pocahontas /ˌpɒkəˈhɒntəs/ (c.1595–1617), American Indian princess, daughter of Powhatan (d.1618), an Algonquian chief in Virginia. According to the story of an English colonist, Captain John Smith (1580–1631), she rescued him from death at the hands of her father. In 1613 she was seized as a hostage by the English, and she later married a colonist, John Rolfe (1585–1622). In 1616 she and her husband visited England, where she died.

pochard /ˈpəʊtʃəd/ *n.* (*pl.* same or **pochards**) a duck of the genus *Aythya*; esp. *A. ferina*, the male of which has a bright reddish-brown head and neck and a grey breast. [16th c.: orig. unkn.]

pochette /pɒˈʃet/ *n.* a woman's envelope-shaped handbag. [F, dimin. of *poche* pocket: see POKE[2]]

pock /pɒk/ *n.* (also **pock-mark**) **1** a small pus-filled spot on the skin, esp. caused by chickenpox or smallpox. **2** a mark resembling this. □ **pock-marked** bearing marks resembling or left by such spots. □ **pocky** *adj.* [OE *poc* f. Gmc]

pocket /ˈpɒkɪt/ *n. & v.* ● *n.* **1** a small bag sewn into or on clothing, for carrying small articles. **2** a pouchlike compartment in a suitcase, car door, etc. **3** one's financial resources (*it is beyond my pocket*). **4** an isolated group or area (*a few pockets of resistance remain*). **5** a cavity in rock, usu. filled with ore (esp. gold) or water. **6** a pouch at the corner or on the side of a billiard- or snooker-table into which balls are driven. **7** = *air pocket*. **8** (*attrib.*) **a** of a suitable size and shape for carrying in a pocket. **b** smaller than the usual size. ● *v.tr.* (**pocketed, pocketing**) **1** put into one's pocket. **2** appropriate, esp. dishonestly. **3** confine as in a pocket. **4** submit to (an injury or affront). **5** conceal or suppress (one's feelings). **6** *Billiards* etc. drive (a ball) into a pocket. □ **in pocket 1** having gained in a transaction. **2** (of money) available. **in a person's pocket 1** under a person's control. **2** close to or intimate with a person. **out of pocket** having lost in a transaction. **out-of-pocket expenses** the actual outlay of cash incurred. **pocket battleship** *hist.* a warship armoured and equipped like, but smaller than, a battleship. **pocket gopher** = GOPHER[1]. **pocket knife** a knife with a folding blade or blades, for carrying in the pocket. **pocket money 1** money for minor expenses. **2** *Brit.* an allowance of money made to a child. **put one's hand in one's pocket** spend or provide money. □ **pocketable** *adj.* **pocketless** *adj.* **pockety** *adj.* (in sense 5 of *n.*). [ME f. AF *poket(e)* dimin. of *poke* POKE[2]]

pocketbook /ˈpɒkɪtˌbʊk/ *n.* **1** a notebook. **2** a booklike case for papers or money carried in a pocket. **3** *US* a purse or handbag. **4** *N. Amer.* a paperback or other small book.

pocket borough *n.* in the early 19th century, a British borough where the population was usually small and elections could be controlled by a single wealthy private person or family. Such boroughs were abolished by the Reform Acts of 1832 and 1867.

pocketful /ˈpɒkɪtˌfʊl/ *n.* (*pl.* **-fuls**) as much as a pocket will hold.

poco /ˈpəʊkəʊ/ *adv. Mus.* a little; rather (*poco adagio*). [It.]

pod[1] /pɒd/ *n. & v.* ● *n.* **1** a long seed-vessel esp. of a leguminous plant, e.g. a pea. **2** the cocoon of a silkworm. **3** the case surrounding locust eggs. **4** a narrow-necked eel-net. **5** a compartment suspended under an aircraft for equipment etc. ● *v.* (**podded, podding**) **1** *intr.* bear or form pods. **2** *tr.* remove (peas etc.) from pods. □ **in pod** *colloq.* pregnant. [back-form. f. dial. *podware*, *podder* field crops, of unkn. orig.]

pod[2] /pɒd/ *n.* a small herd or school of marine mammals, esp. whales. [19th c., originally US: orig. unkn.]

podagra /pəˈdægrə, ˈpɒdəgrə/ *n. Med.* gout of the foot, esp. the big toe. □ **podagral** *adj.* **podagric** *adj.* **podagrous** *adj.* [L f. Gk *pous podos* foot + *agra* seizure]

poddy /ˈpɒdɪ/ *adj., n., & v. colloq.* ● *adj.* **1** corpulent, obese. **2** *Austral.* (of a calf, lamb, etc.) fed by hand. ● *n.* (*pl.* **-ies**) *Austral.* **1** an unbranded calf. **2** a calf fed by hand. ● *v.tr.* (**-ies, -ied**) *Austral.* feed (a young animal) by hand. [POD[1]]

Podgorica /pɒdˈgɔːrɪtsə/ the capital of Montenegro; pop. (1981) 132,400. It was under Turkish rule from 1474 until 1878. Between 1946 and 1993 it was named Titograd in honour of Marshal Tito.

podgy /ˈpɒdʒɪ/ *adj.* (**podgier, podgiest**) **1** (of a person) short and fat. **2** (of a face etc.) plump, fleshy. □ **podginess** *n.* [19th c.: f. *podge* a short fat person]

podiatry /pəˈdaɪətrɪ/ *n.* esp. *N. Amer.* = CHIROPODY. □ **podiatrist** *n.* [Gk *pous podos* foot + *iatros* physician]

podium /ˈpəʊdɪəm/ *n.* (*pl.* **podiums** or **podia** /-dɪə/) **1** a continuous projecting base or pedestal round a room or house etc. **2** a raised platform round the arena of an amphitheatre. **3** a platform or rostrum. [L f. Gk *podion* dimin. of *pous pod-* foot]

Podolsk /pəˈdɒlsk/ an industrial city in Russia, south of Moscow; pop. (1990) 209,000.

podzol /ˈpɒdzɒl/ *n.* (also **podsol** /ˈpɒdsɒl/) an acidic infertile soil with minerals leached from its surface layers into a lower stratum. □ **podzolize** *v.tr. & intr.* (also **-ise**). [Russ. f. *pod* under, *zola* ashes]

Poe /pəʊ/, Edgar Allan (1809–49), American short-story writer, poet, and critic. He spent most of his life in poverty and ill health. His fiction and poetry are Gothic in style and characterized by their exploration of the macabre, the fantastic, and the grotesque. His most famous short stories include the Gothic romance 'The Fall of the House of Usher' (which appeared in *Tales of the Grotesque and Arabesque*, 1840) and 'The Pit and the Pendulum' (1843), while his poems include 'The Raven' (1845) and 'Annabel Lee' (1849). His story 'The Murders in the Rue Morgue' (1841) is often regarded as the first detective story in English literature. His critical writings include 'The Poetic Principle' (1850), which anticipated many of the concerns of the Aesthetic Movement.

poem /ˈpəʊɪm/ *n.* **1** a metrical composition, usu. concerned with feeling or imaginative description. **2** an elevated composition in verse or prose. **3** something with poetic qualities (*a poem in stone*). [F *poème* or L *poema* f. Gk *poēma* = *poiēma* f. *poieō* make]

poesy /ˈpəʊɪzɪ, -ɪsɪ/ *n. archaic* **1** poetry. **2** the art or composition of poetry. [ME f. OF *poesie* ult. f. L *poesis* f. Gk *poēsis* = *poiēsis* making, poetry (as POEM)]

poet /ˈpəʊɪt/ *n.* (*fem.* **poetess** /-tɪs/) **1** a writer of poems. **2** a person possessing high powers of imagination or expression etc. □ **Poets' Corner** part of Westminster Abbey where several poets are buried or commemorated. [ME f. OF *poete* f. L *poeta* f. Gk *poētēs* = *poiētēs* maker, poet (as POEM)]

poetaster /ˌpəʊɪˈtæstə(r)/ *n.* a paltry or inferior poet. [mod.L (as POET): see -ASTER]

poetic /pəʊˈetɪk/ *adj.* (also **poetical** /-k(ə)l/) **1 a** of or like poetry or poets. **b** written in verse. **2** elevated or sublime in expression. □ **poetic justice** well-deserved unforeseen retribution or reward. **poetic licence** a writer's or artist's transgression of established rules for effect. □ **poetically** *adv.* [F *poétique* f. L *poeticus* f. Gk *poētikos* (as POET)]

poeticize /pəʊˈetɪˌsaɪz/ *v.tr.* (also **-ise**) make (a theme) poetic.

poetics /pəʊˈetɪks/ *n.* **1** the art of writing poetry. **2** the study of poetry and its techniques.

poetize /ˈpəʊɪˌtaɪz/ *v.* (also **-ise**) **1** *intr.* play the poet. **2** *intr.* compose poetry. **3** *tr.* treat poetically. **4** *tr.* celebrate in poetry. [F *poétiser* (as POET)]

Poet Laureate *n.* an eminent poet appointed as a member of the British royal household. The first Poet Laureate in the modern sense was Ben Jonson, but the title became established with the appointment of John Dryden in 1668; Chaucer, Skelton, Spenser, and many others had previously received royal patronage. The Poet Laureate was formerly expected to write poems for state occasions, but since Victorian times the post has carried no specific duties. Wordsworth, Tennyson, and Betjeman are among later holders of the title, which since 1984 has been held by Ted Hughes. In 1985 the US government created a position of Poet Laureate, of which the first holder was Robert Penn Warren in 1986. [see LAUREATE]

poetry /ˈpəʊɪtrɪ/ *n.* **1** the art or work of a poet. **2** poems collectively. **3** a poetic or tenderly pleasing quality. **4** anything compared to poetry. [ME f. med.L *poetria* f. L *poeta* POET, prob. after *geometry*]

po-faced /ˈpəʊˈfeɪst/ *adj.* **1** solemn-faced, humourless. **2** smug. [20th c.: perh. f. PO, infl. by *poker-faced*]

pogo /ˈpəʊgəʊ/ *n.* (also **pogo stick**) (*pl.* **-os**) a toy consisting of a spring-loaded stick with rests for the feet, for springing about on. [20th c.: orig. uncert.]

pogrom /ˈpɒgrəm, -rɒm/ *n.* an organized massacre (orig. of Jews in Russia). [Russ., = devastation f. *gromit'* destroy]

Po Hai see Bo HAI.

poignant /ˈpɔɪnjənt/ *adj.* **1** painfully sharp to the emotions or senses; deeply moving. **2** arousing sympathy. **3** sharp or pungent in taste or smell. **4** pleasantly piquant. **5** *archaic* (of words etc.) sharp, severe. □ **poignance** *n.* **poignancy** *n.* **poignantly** *adv.* [ME f. OF, pres. part. of *poindre* prick f. L *pungere*]

poikilotherm /ˈpɔɪkɪləˌθɜːm/ *n. Zool.* an organism that regulates its body temperature by behavioural means, such as basking or burrowing; a cold-blooded organism (cf. HOMEOTHERM).

☐ **poikilothermy** *n.* **poikilothermal** /ˌpɔɪkɪlə'θɜːm(ə)l/ *adj.* **poikilothermia** *n.* **poikilothermic** *adj.* [Gk *poikilos* multicoloured, changeable + *thermē* heat]

poilu /'pwʌljuː/ *n. hist.* a French private soldier, esp. as a nickname. [F, lit. 'hairy' f. *poil* hair]

Poincaré /ˌpwʌnkæ'reɪ/, Jules-Henri (1854–1912), French mathematician and philosopher of science. Poincaré made far-reaching contributions to pure and applied mathematics. He worked extensively on differential equations which allowed him to transform celestial mechanics, and was one of the pioneers of algebraic topology. By 1900 he was proposing a relativistic philosophy, suggesting that it implied the absolute velocity of light, which nothing could exceed.

poinciana /ˌpɔɪnsɪ'ɑːnə/ *n.* a tropical leguminous tree of the genus *Caesalpinia* or *Delonix*, with bright showy red flowers. [mod.L *Poinciana* former genus name, f. M. de *Poinci*, 17th-c. Fr. governor in the West Indies + *-ana* fem. suffix]

poind /pɔɪnd/ *v. & n. Sc.* ● *v.tr.* distrain upon; impound. ● *n.* **1** an act of poinding. **2** an animal or chattel poinded. [ME f. OE *pyndan* impound]

poinsettia /pɔɪn'setɪə/ *n.* a tropical shrub, *Euphorbia pulcherrima*, with large showy scarlet bracts surrounding small greenish flowers. [mod.L *Poinsettia* former genus name, f. Joel Roberts *Poinsett*, Amer. diplomat (1779–1851)]

point /pɔɪnt/ *n. & v.* ● *n.* **1** the sharp or tapered end of a tool, weapon, pencil, etc. **2** a tip or extreme end. **3** that which in geometry has position but not magnitude, e.g. the intersection of two lines. **4** a particular place or position (*Bombay and points east; point of contact*). **5 a** a precise or particular moment (*at the point of death*). **b** the critical or decisive moment (*when it came to the point, he refused*). **6** a very small mark on a surface. **7 a** a dot or other punctuation mark, esp. = *full point* (see FULL[1]). **b** a dot or small stroke used in Semitic languages to indicate vowels or distinguish consonants. **8** = *decimal point*. **9** a stage or degree in progress or increase (*abrupt to the point of rudeness; at that point we gave up*). **10** a level of temperature at which a change of state occurs (*freezing-point*). **11** a single item; a detail or particular (*we differ on these points; it is a point of principle*). **12 a** a unit of scoring in games or of measuring value etc. **b** an advantage or success in less quantifiable contexts such as an argument or discussion. **c** a unit of weight (2 mg) for diamonds. **d** a unit (of varying value) in quoting the price of stocks etc. **13 a** (usu. prec. by *the*) the significant or essential thing; what is actually intended or under discussion (*that was the point of the question*). **b** (usu. with *neg.* or *interrog.*; often foll. by *in*) sense or purpose; advantage or value (*saw no point in staying*). **c** (usu. prec. by *the*) a salient feature of a story, joke, remark, etc. (*don't see the point*). **14** a distinctive feature or characteristic (*it has its points; tact is not his good point*). **15** pungency, effectiveness (*their comments lacked point*). **16** each of thirty-two directions marked at equal distances round a compass. **b** the corresponding direction towards the horizon. **17** (usu. in *pl.*) *Brit.* a junction of two railway lines, with a pair of linked tapering rails that can be moved laterally to allow a train to pass from one line to the other. **18** *Brit.* = *power point*. **19** (usu. in *pl.*) each of a set of electrical contacts in the distributor of a motor vehicle. **20** *Cricket* **a** a fielder on the off side near the batsman. **b** this position. **21** the tip of the toe in ballet. **22** a promontory. **23** the prong of a deer's antler. **24** (usu. in *pl.*) the extremities of a dog, horse, etc. **25** *Printing* a unit of measurement for type bodies (in the UK and US 0.351 mm, in Europe 0.376 mm). **26** *Hunting* a spot to which a straight run is made. **b** such a run. **27** *Heraldry* any of nine particular positions on a shield used for specifying the position of charges etc. **28** *Boxing* the tip of the chin as a spot for a knockout blow. **29** *Mil.* a small leading party of an advanced guard. **30** *hist.* a tagged lace for lacing a bodice, attaching a hose to a doublet, etc. **31** *Naut.* a short piece of cord at the lower edge of a sail for tying up a reef. **32** the act or position of a dog in pointing. ● *v.* **1** (usu. foll. by *to, at*) **a** *tr.* direct or aim (a finger, weapon, etc.). **b** *intr.* direct attention in a certain direction (*pointed to the house across the road*). **2** *intr.* (foll. by *at, towards*) **a** aim or be directed to. **b** tend towards. **3** *intr.* (foll. by *to*) indicate; be evidence of (*it all points to murder*). **4** *tr.* give point or force to (words or actions). **5** *tr.* fill in or repair the joints of (brickwork) with smoothly finished mortar or cement. **6** *tr.* **a** punctuate. **b** insert points in (written Hebrew etc.). **c** mark (Psalms etc.) with signs for chanting. **7** *tr.* sharpen (a pencil, tool, etc.). **8** *tr.* (also *absol.*) (of a dog) indicate the presence of (game) by acting as pointer. ☐ **at all points** in every part or respect. **at the point of** (often foll. by verbal noun) on the verge of; about to do (the action specified). **beside the point** irrelevant or irrelevantly. **case in point** an instance that is relevant or (prec. by *the*) under consideration. **have a point** be correct or effective in one's contention. **in point** apposite, relevant. **in point**

of fact see FACT. **make** (or **prove**) **a** (or **one's**) **point** establish a proposition; prove one's contention. **make a point of** (often foll. by verbal noun) insist on; treat or regard as essential. **nine points** nine tenths, i.e. nearly the whole (esp. *possession is nine points of the law*). **on** (or **upon**) **the point of** (foll. by verbal noun) about to do (the action specified). **point-duty** the positioning of a police officer or traffic warden at a crossroad or other point to control traffic. **point lace** thread lace made wholly with a needle. **point of honour** an action or circumstance that affects one's reputation or conscience. **point of no return** a point in a journey or enterprise at which it becomes essential or more practical to continue to the end. **point of order** a query in a debate etc. as to whether correct procedure is being followed. **point-of-sale** (usu. *attrib.*) denoting publicity etc. associated with the place at which goods are retailed. **point of view 1** a position from which a thing is viewed. **2** a particular way of considering a matter. **point out** (often foll. by *that* + clause) indicate, show; draw attention to. **point-to-point** a steeplechase over a marked course for horses used regularly in hunting. **point up** emphasize; show as important. **score points off** get the better of in an argument etc. **take a person's point** concede that a person has made a valid contention. **to the point** relevant or relevantly. **up to a point** to some extent but not completely. **win on points** *Boxing* win by scoring more points, not by a knockout. [ME f. OF *point*, *pointer* f. L *punctum* f. *pungere* punct-prick]

point-blank /pɔɪnt'blæŋk/ *adj. & adv.* ● *adj.* **1 a** (of a shot) aimed or fired horizontally at a range very close to the target. **b** (of a distance or range) very close. **2** (of a remark, question, etc.) blunt, direct. ● *adv.* **1** at very close range. **2** directly, bluntly. [prob. f. POINT + BLANK = white spot in the centre of a target]

Pointe-à-Pitre /ˌpwæntæ'piːtrə/ the chief port and commercial capital of the French island of Guadeloupe in the West Indies; pop. (1988) 25,310.

pointed /'pɔɪntɪd/ *adj.* **1** sharpened or tapering to a point. **2** (of a remark etc.) having point; penetrating, cutting. **3** emphasized; made evident. ☐ **pointedly** *adv.* **pointedness** *n.*

Pointe-Noire /pwænt'nwɑː(r)/ the chief seaport of the Congo, an oil terminal on the Atlantic coast; pop. (1990) 387,770.

pointer /'pɔɪntə(r)/ *n.* **1** a thing that points, e.g. the index hand of a gauge etc. **2** a rod for pointing to features on a map, chart, etc. **3** *colloq.* a hint, clue, or indication. **4** a large breed of dog, often used as a gun dog, with the habit of responding to the scent of game by standing rigid looking towards it.

Pointers /'pɔɪntəz/ *n.pl. Astron.* **1** two stars of the Plough or Big Dipper in Ursa Major, a line through which points nearly to the Pole Star. **2** two stars in the Southern Cross, a line through which points nearly to the south celestial pole.

pointillism /'pwæntɪˌlɪz(ə)m/ *n.* a technique and style of painting, associated with neo-impressionism and developed by Seurat. The technique, derived from a scientific concern with form and colour, involved the application of small dots of unmixed colour, which, though separate on the canvas, would blend when viewed from a distance; the aim was to produce a greater degree of luminosity and brilliance of colour. Although *pointillism* is the better-known term, Seurat himself preferred the term *divisionism*. ☐ **pointillist** *n. & adj.* **pointillistic** /ˌpwæntɪ'lɪstɪk/ *adj.* [F *pointillisme* f. *pointiller* mark with dots]

pointing /'pɔɪntɪŋ/ *n.* **1** cement or mortar filling the joints of brickwork. **2** facing produced by this. **3** the process of producing this.

pointless /'pɔɪntlɪs/ *adj.* **1** without a point. **2** lacking force, purpose, or meaning. **3** (in games) without a point scored. ☐ **pointlessly** *adv.* **pointlessness** *n.*

pointsman /'pɔɪntsmən/ *n.* (*pl.* **-men**) *Brit.* **1** a person in charge of railway points. **2** a policeman or traffic warden on point-duty.

pointy /'pɔɪntɪ/ *adj.* (**pointier**, **pointiest**) having a noticeably sharp end; pointed.

Poirot /'pwʌrəʊ/, Hercule, a fictional Belgian private detective in the crime stories of Agatha Christie.

poise[1] /pɔɪz/ *n. & v.* ● *n.* **1** composure or self-possession of manner. **2** equilibrium; a stable state. **3** carriage (of the head etc.). ● *v.* **1** *tr.* balance; hold suspended or supported. **2** *tr.* carry (one's head etc. in a specified way). **3** *intr.* be balanced; hover in the air etc. [ME f. OF *pois*, *peis*, *peser* ult. f. L *pensum* weight f. *pendere* pens- weigh]

poise[2] /pɔɪz/ *n. Physics* a unit of dynamic viscosity, such that a tangential force of one dyne per square centimetre causes a velocity change one centimetre per second between two parallel planes separated by one centimetre in a liquid. [Jean Léonard Marie *Poiseuille*, Fr. physician (1799–1869)]

poised /pɔɪzd/ *adj.* **1** composed, self-assured. **2** (often foll. by *for*, or *to* + infin.) ready for action.

poisha /'pɔɪʃə/ *n.* (*pl.* same) a monetary unit of Bangladesh, equal to one-hundredth of a taka. [Bengali, alt. of PAISA]

poison /'pɔɪz(ə)n/ *n. & v.* ● *n.* **1** a substance that when introduced into or absorbed by a living organism causes death or injury, esp. one that kills by rapid action even in a small quantity. **2** *colloq.* a harmful influence or principle etc. **3** *Physics & Chem.* a substance that interferes with the normal progress of a nuclear reaction, chain reaction, catalytic reaction, etc. ● *v.tr.* **1** administer poison to (a person or animal). **2** kill or injure with or infect with poison. **3** infect (air, water, etc.) with poison. **4** (esp. as **poisoned** *adj.*) treat (a weapon) with poison. **5** corrupt or pervert (a person or mind). **6** spoil or destroy (a person's pleasure etc.). **7** render (land etc.) foul and unfit for its purpose by a noxious application etc. □ **poisoned chalice** an assignment, honour, etc., likely to prove a source of problems to the recipient. **poison gas** = GAS *n.* 4. **poison ivy** a North American climbing plant, *Rhus radicans*, secreting an irritant oil from its leaves. **poison oak** either of two North American shrubs, *Rhus toxicodendron* and *R. diversilobia*, related to poison ivy and having similar properties. **poison-pen letter** an anonymous libellous or abusive letter. □ **poisoner** *n.* **poisonous** *adj.* **poisonously** *adv.* [ME f. OF *poison*, *poisonner* (as POTION)]

Poisson /'pwʌsɒn/, Siméon-Denis (1781–1840), French mathematical physicist. Early in his career he began applying the integration of differential equations to problems in physics, an approach which he used for many years with great effect. He added to the work of Laplace and Lagrange on planetary motions, and went on to study electrostatics, heat, elasticity, and magnetism. Perhaps his major contributions were in probability theory, in which he greatly improved Laplace's work and developed several concepts that are now named after him.

Poisson distribution *n.* *Statistics* a discrete frequency distribution which gives the probability of events occurring in a fixed time. [POISSON]

Poitiers /'pwʌtɪˌeɪ/ a city in west central France, the chief town of Poitou-Charentes region and capital of the former province of Poitou; pop. (1990) 82,500. It was the site in AD 507 of the defeat of the Visigoths by Clovis and in 732 of Charles Martel's victory over the invading Muslims. In 1356 the city fell to the English forces of Edward, the Black Prince, but was reclaimed by the French some thirteen years later.

Poitou /'pwʌtu:/ a former province of west central France, now united with Charente to form the region of Poitou-Charentes. Formerly part of Aquitaine, it was held by the French and English in succession until it was finally united with France at the end of the Hundred Years War.

Poitou-Charentes /ˌpwʌtu:ʃæ'rɒnt/ a region of western France, on the Bay of Biscay, centred on Poitiers.

poke[1] /pəʊk/ *v. & n.* ● *v.* **1** (foll. by *in*, *up*, *down*, etc.) **a** *tr.* thrust or push with the hand, point of a stick, etc. **b** *intr.* be thrust forward. **2** *intr.* (foll. by *at* etc.) make thrusts with a stick etc. **3** *tr.* thrust the end of a finger etc. against. **4** *tr.* (foll. by *in*) produce (a hole etc. in a thing) by poking. **5** *tr.* thrust forward, esp. obtrusively. **6** *tr.* stir (a fire) with a poker. **7** *intr.* **a** (often foll. by *about*, *around*) move or act desultorily; potter. **b** (foll. by *about*, *into*) pry; search casually. **8** *tr. coarse sl.* have sexual intercourse with. **9** *tr.* (foll. by *up*) *colloq.* confine (esp. oneself) in a poky place. ● *n.* **1** the act or an instance of poking. **2** a thrust or nudge. **3** a device fastened on cattle etc. to prevent them breaking through fences. **4 a** a projecting brim or front of a woman's bonnet or hat. **b** (in full **poke-bonnet**) a bonnet having this. □ **poke fun at** ridicule, tease. **poke one's nose into** see NOSE. [ME f. MDu. and MLG *poken*, of unkn. orig.]

poke[2] /pəʊk/ *n. dial.* a bag or sack. □ **buy a pig in a poke** see PIG. [ME f. ONF *poke*, *poque* = OF *poche*: cf. POUCH]

poker[1] /'pəʊkə(r)/ *n.* a metal rod with a handle, for stirring an open fire. □ **poker-work 1** the technique of burning designs on white wood etc. with a heated metal rod. **2** a design made in this way.

poker[2] /'pəʊkə(r)/ *n.* a card-game for two or more people in which players bet on the value of the five-card hand dealt to them. (*See note below*.) □ **poker-dice** dice with card designs from ace to nine instead of spots. **poker-face 1** the impassive countenance appropriate to a poker-player. **2** a person with this. **poker-faced** having a poker-face. [19th c.: orig. unkn.: cf. G *pochen* to brag, *Pochspiel* bragging game]

▪ Each player in poker aims to win the pool, either by having the best hand at the showdown or by inducing the other players to concede

without showing their hands, sometimes by means of bluff. Poker is particularly popular in the US, where it developed in the 19th century from European gambling games.

pokeweed /'pəʊkwi:d/ *n.* (in full **Virginian pokeweed**) a tall succulent North American plant, *Phytolacca americana*, with spikes of cream flowers, and purple berries that yield emetics and purgatives. [*poke*, Amer. Indian word + WEED]

pokey /'pəʊkɪ/ *n.* (usu. prec. by *the*) esp. *US sl.* prison. [perh. f. POKY]

poky /'pəʊkɪ/ *adj.* (**pokier**, **pokiest**) (of a room etc.) small and cramped. □ **pokily** *adv.* **pokiness** *n.* [POKE[1] (in colloq. sense 'confine') + -Y[1]]

polack /'pəʊlæk/ *n. N. Amer. sl. offens.* a person of Polish origin. [F *Polaque* and G *Polack* f. Pol. *Polak*]

Poland /'pəʊlənd/ (called in Polish *Polska*) a country in central Europe with a coastline on the Baltic Sea; pop. (1990) 38,183,160; official language, Polish; capital, Warsaw. First united as a nation in the 11th century, Poland became a dominant power in the region in the 16th century. Thereafter it suffered severely from the rise of Russian, Swedish, Prussian, and Austrian power, losing territory and eventually its independence in three partitions between 1772 and 1795. Poland regained full independence (as a republic) after the First World War. Its invasion by German forces in 1939 precipitated the Second World War, in which it was at first divided between Germany and the USSR and from which it eventually emerged, with its boundaries displaced westwards, as a Communist state under Soviet domination. In the 1980s, the rise of the independent trade union movement Solidarity eventually led to the end of Communist rule (1989) and the introduction of democratic and free-market reforms under President Lech Wałęsa.

Polanski /pə'lænskɪ/, Roman (b.1933), French film director of Polish descent. Born in France, he grew up in Poland, where he pursued a career as an actor from the age of 14 and directed the film *Knife in the Water* (1962), which established his international reputation. He subsequently worked in Hollywood, having success with films such as *Rosemary's Baby* (1968) and *Chinatown* (1974). He left the US for France in 1977 under threat of prosecution for drug and sex offences. His later films, including *Tess* (1979) and *Frantic* (1988), have not had the success of his earlier titles. His second wife, Hollywood actress Sharon Tate (1943–69), was one of the victims of a multiple murder by followers of the cult leader Charles Manson (b.1934).

polar /'pəʊlə(r)/ *adj.* **1 a** of or near a pole of the earth or a celestial body, or of the celestial sphere. **b** (of a species or variety) living in the north polar region. **2** having magnetic polarity. **3 a** (of a molecule) having a positive charge at one end and a negative charge at the other. **b** (of a compound) having electric charges. **4** *Geom.* of or relating to a pole. **5** directly opposite in character or tendency. **6** *colloq.* (esp. of weather) very cold. □ **polar axis 1** an axis passing through the poles of an object. **2** *Astron.* the axis of an equatorial mount aligned to the celestial pole, about which a telescope is turned to alter right ascension or to keep an object in view as the earth rotates. **polar bear** a large white bear, *Thalarctos maritimus*, of the Arctic regions. **polar body** *Biol.* a small cell produced from an oocyte during the formation of an ovum, which does not develop further. **polar coordinates** a system by which a point can be located with reference to two angles. **polar curve** a curve related in a particular way to a given curve and to a fixed point called a *pole*. **polar distance** the angular distance of a point on a sphere from the nearest pole. **polar star** = *pole star* (see POLE[2]). □ **polarly** *adv.* [F *polaire* or mod.L *polaris* (as POLE[2])]

polari- /'pəʊlərɪ/ *comb. form* polar. [mod.L *polaris* (as POLAR)]

polarimeter /ˌpəʊlə'rɪmɪtə(r)/ *n.* an instrument used to measure the polarization of light or the effect of a substance on the rotation of the plane of polarized light. □ **polarimetry** *n.* **polarimetric** /-rɪ'metrɪk/ *adj.*

Polaris /pə'lɑ:rɪs/ **1** *Astron.* the North Star or Pole Star, located within one degree of the celestial north pole, in the constellation of Ursa Minor. It is a double star, the bright component being a cepheid variable. **2** a US type of submarine-launched ballistic missile, designed to carry nuclear warheads, formerly in service with the US and British navies. [as POLAR]

polariscope /pə'lærɪˌskəʊp/ *n.* = POLARIMETER. □ **polariscopic** /-ˌlærɪ'skɒpɪk/ *adj.*

polarity /pə'lærɪtɪ/ *n.* (*pl.* **-ies**) **1** the tendency of a lodestone, magnetized bar, etc., to point with its extremities to the magnetic poles of the earth. **2** the condition of having two poles with contrary qualities. **3** the state of having two opposite tendencies, opinions, etc.

4 the electrical condition of a body (positive or negative). **5** a magnetic attraction towards an object or person.

polarize /ˈpəʊləˌraɪz/ v. (also **-ise**) **1** tr. restrict the vibrations of (a transverse wave, esp. light) to one direction. **2** tr. give magnetic or electric polarity to (a substance or body). **3** tr. reduce the voltage of (an electric cell) by the action of electrolysis products. **4** tr. & intr. divide into two groups of opposing opinion etc. □ **polarizer** n. **polarizable** adj. **polarization** /ˌpəʊləraɪˈzeɪʃ(ə)n/ n.

polarography /ˌpəʊləˈrɒgrəfɪ/ n. Chem. chemical analysis of a substance by electrolysing it at successively higher voltages and measuring the resulting current. □ **polarographic** /-rəˈgræfɪk/ adj.

Polaroid /ˈpəʊləˌrɔɪd/ n. propr. **1** material in thin plastic sheets that produces a high degree of plane polarization in light passing through it. **2** a type of camera with internal processing that produces a finished print rapidly after each exposure. **3** (in pl.) sunglasses with lenses made from Polaroid. [POLARI- + -OID]

polder /ˈpəʊldə(r)/ n. a piece of low-lying land reclaimed from the sea or a river, esp. in the Netherlands. [MDu. polre, Du. polder]

Pole /pəʊl/ n. **1** a native or national of Poland. **2** a person of Polish descent. [G f. Pol. Polanie, lit. 'field-dwellers' f. pole field]

pole[1] /pəʊl/ n. & v. ● n. **1** a long slender rounded piece of wood or metal, esp. with the end placed in the ground as a support etc. **2** a wooden shaft fitted to the front of a vehicle and attached to the yokes or collars of the draught animals. **3** Athletics the long slender flexible rod of wood, fibreglass, etc. used by a competitor in pole-vaulting. **4** esp. US a simple fishing rod. **5** hist. = PERCH[1] 3. ● v.tr. **1** provide with poles. **2** (usu. foll. by off) push off (a punt etc.) with a pole. □ **pole position** the most favourable position at the start of a motor-race (orig. next to the inside boundary-fence). **pole-vault** (or **-jump**) n. the athletic sport of vaulting over a high bar with the aid of a long flexible pole held in the hands and giving extra spring. ● v.intr. take part in this sport. **pole-vaulter** a person who pole-vaults. **under bare poles** Naut. with no sail set. **up the pole** sl. **1** crazy, eccentric. **2** in difficulty. [OE pāl ult. f. L palus stake]

pole[2] /pəʊl/ n. **1** (in full **north pole, south pole**) **a** each of the two points in the celestial sphere about which the stars appear to revolve. **b** (as a geographical location **North Pole, South Pole**) each of the extremities of the axis of rotation of the earth or another body. (See note below.) **c** see magnetic pole. **2** each of the two opposite points on the surface of a magnet at which magnetic forces are strongest. **3** either of two terminals (positive and negative) of an electric cell or battery etc. **4** either of two opposed principles or ideas. **5** Geom. either of two points in which the axis of a circle cuts the surface of a sphere. **6** a fixed point to which others are referred. **7** Biol. an extremity of the main axis of any spherical or oval organ. □ **be poles apart** differ greatly, esp. in nature or opinion. **pole star 1** Astron. = POLARIS 1. **2 a** a thing or principle serving as a guide. **b** a centre of attraction. □ **poleward** adj. **polewards** adj. & adv. [ME f. L polus f. Gk polos pivot, axis, sky]

▪ The earth's North Pole is situated in the Arctic Ocean and is permanently covered with drifting pack ice. It was first reached by the American Robert Peary in 1909. The South Pole, located in the centre of the Antarctic continent, was reached by the Norwegian Roald Amundsen in 1911. The US established the permanently staffed Amundsen–Scott scientific station at the South Pole in the 1970s; there is no permanent station at the North Pole. The earth's magnetic poles are neither fixed nor symmetrically located with reference to the geographical poles. They are currently at approximately 78°N, 104°W (in the Canadian Arctic islands) and 65°S, 139° E (off the Adélie Coast of Antarctica).

poleaxe /ˈpəʊlæks/ n. & v. (US **poleax**) ● n. **1** hist. a battleaxe. **2** a butcher's cleaver. ● v.tr. hit or kill with or as if with a poleaxe. [ME pol(l)ax, -ex f. MDu. pol(l)aex, MLG pol(l)exe (as POLL[1], AXE)]

polecat /ˈpəʊlkæt/ n. **1** a small flesh-eating mammal of the weasel family; esp. Mustela putorius, native to Europe, with dark brown fur and a fetid smell. (See also FERRET.) **2** US a skunk. [pole (unexplained) + CAT[1]]

polemic /pəˈlemɪk/ n. & adj. ● n. **1** a controversial discussion. **2** Polit. a verbal or written attack, esp. on a political opponent. ● adj. (also **polemical**) involving dispute; controversial. □ **polemically** adv. **polemicist** /-mɪsɪst/ n. **polemicize** /-mɪˌsaɪz/ v.intr. (also **-ise**). **polemize** /ˈpɒlɪˌmaɪz/ v.intr. (also **-ise**). [med.L polemicus f. Gk polemikos f. polemos war]

polemics /pəˈlemɪks/ n.pl. the art or practice of controversial discussion.

polenta /pəˈlentə/ n. porridge made of maize meal etc. [It. f. L, = pearl barley]

police /pəˈliːs/ n. & v. ● n. **1** (usu. prec. by the; usu. treated as pl.) the civil force of a state, responsible for maintaining public order. (See note below.) **2** (treated as pl.) the members of a police force (several hundred police). **3** (usu. treated as pl.) a force with similar functions of enforcing regulations (military police; railway police). ● v.tr. **1** control (a country or area) by means of police. **2** provide with police. **3** keep order in; control; monitor. □ **police constable** see CONSTABLE 1b. **police dog** a dog, esp. an Alsatian, used in police work. **police officer** a policeman or policewoman. **police state** a totalitarian state controlled by political police supervising the citizens' activities. **police station** the office of a local police force. [F f. med.L politia POLICY[1]]

▪ Compared with many countries (where the role of the military was not always sharply distinguished from that of the police), Britain was fairly late in establishing a regular force, the first being that organized in London in 1829 by Sir Robert Peel. From medieval times a petty or parish constable was appointed to preserve the peace in his area and to execute the orders of Justices of the Peace; in towns, the 'watch' patrolled and guarded the streets. In London in the 18th century some justices organized bodies of constables, of which the Bow Street Runners (established in 1749) were one. The modern British police force is not a national one; a number of forces exist for defined areas, maintained by local police authorities, the Home Office being responsible for the London Metropolitan Police. Unlike most other police forces, the British police are not armed except in special circumstances.

policeman /pəˈliːsmən/ n. (pl. **-men**; fem. **policewoman**, pl. **-women**) a member of a police force.

policy[1] /ˈpɒlɪsɪ/ n. (pl. **-ies**) **1** a course or principle of action adopted or proposed by a government, party, business, or individual etc. **2** prudent conduct; sagacity. [ME f. OF policie f. L politia f. Gk politeia citizenship f. politēs citizen f. polis city]

policy[2] /ˈpɒlɪsɪ/ n. (pl. **-ies**) **1** a contract of insurance. **2** a document containing this. [F police bill of lading, contract of insurance, f. Prov. poliss(i)a prob. f. med.L apodissa, apodixa, f. L apodixis f. Gk apodeixis evidence, proof (as APO-, deiknumi show)]

policyholder /ˈpɒlɪsɪˌhəʊldə(r)/ n. a person or body holding an insurance policy.

polio /ˈpəʊlɪəʊ/ n. = POLIOMYELITIS. [abbr.]

poliomyelitis /ˌpəʊlɪəʊˌmaɪəˈlaɪtɪs/ n. Med. an infectious viral disease that affects the central nervous system and which can cause temporary or permanent paralysis. [mod.L f. Gk polios grey + muelos marrow]

polis /ˈpəʊlɪs, ˈpɒl-/ n. Sc. & Ir. **1** the police. **2** a police officer. [repr. regional pronunc.]

Polish /ˈpəʊlɪʃ/ adj. & n. ● adj. **1** of or relating to Poland. **2** of the Poles or their language. ● n. the language of Poland, belonging to the Slavonic language group and written in the Roman alphabet. □ **Polish notation** Math. a system of formula notation without brackets and punctuation, often used to represent the order in which operations are performed in computers etc. **reverse Polish notation** the usual form of Polish notation, with operators following rather than preceding their operands. [POLE + -ISH[1]]

polish /ˈpɒlɪʃ/ v. & n. ● v. **1** tr. & intr. make or become smooth or glossy by rubbing. **2** (esp. as **polished** adj.) refine or improve; add finishing touches to. ● n. **1** a substance used for polishing. **2** smoothness or glossiness produced by friction. **3** the act or an instance of polishing. **4** refinement or elegance of manner, conduct, etc. □ **polish off 1** finish (esp. food) quickly. **2** colloq. kill; murder. **polish up** revise or improve (a skill etc.). □ **polishable** adj. **polisher** n. [ME f. OF polir f. L polire polit-]

Polish Corridor a former region of Poland, which extended northwards to the Baltic coast and separated East Prussia from the rest of Germany. A part of Polish Pomerania in the 18th century, the area had since then been subject to German colonization. It was granted to Poland after the First World War to ensure Polish access to the coast. Its annexation by Germany in 1939, with the German occupation of the rest of Poland, precipitated the Second World War. After the war the area was restored to Poland.

Politburo /ˈpɒlɪtˌbjʊərəʊ/ n. (pl. **-os**) the principal policy-making committee of a Communist party, esp. in the former USSR. Founded in 1917, the Politburo of the USSR, a subcommittee of the Central Committee, consisted of about twelve voting members and six non-voting candidate members. From 1952 to 1966 it was known as the Presidium. [Russ. politbyuro f. politicheskoe byuro political bureau]

polite /pə'laɪt/ adj. (**politer**, **politest**) **1** having good manners; courteous. **2** cultivated, cultured. **3** refined, elegant (*polite letters*). □ **politely** adv. **politeness** n. [L politus (as POLISH)]

politesse /ˌpɒlɪ'tes/ n. formal politeness. [F f. It. politezza, pulitezza f. pulito polite]

politic /'pɒlɪtɪk/ adj. & v. ● adj. **1** (of an action) judicious, expedient. **2** (of a person:) **a** prudent, sagacious. **b** scheming, sly. **3** political (now only in body politic). ● v.intr. (**politicked**, **politicking**) engage in politics. □ **politicly** adv. [ME f. OF politique f. L politicus f. Gk politikos f. politēs citizen f. polis city]

political /pə'lɪtɪk(ə)l/ adj. **1 a** of or concerning the state or its government, or public affairs generally. **b** of, relating to, or engaged in politics. **c** belonging to or forming part of a civil administration. **2** having an organized form of society or government. **3** taking or belonging to a side in politics or in controversial matters. **4** relating to or affecting interests of status or authority in an organization rather than matters of principle (*a political decision*). □ **political asylum** see ASYLUM. **political correctness** (or **incorrectness**) conformity (or failure to conform) to politically correct views; the use (or avoidance) of politically correct language etc. **political economist** a student of or expert in political economy. **political economy** the study of the economic aspects of government. **political geography** that dealing with boundaries and the possessions of states. **political prisoner** a person imprisoned for political beliefs or actions. **political science** the study of the state and systems of government. **political scientist** a specialist in political science. [L politicus (as POLITIC)]

politically /pə'lɪtɪkəlɪ/ adv. **1** by political means; in a political manner. **2** as regards politics. □ **politically correct** (or **incorrect**) conforming (or not conforming) to a prevailing body of liberal opinion, esp. in avoiding (or not avoiding) language, behaviour, etc., which might conceivably be regarded as discriminatory or pejorative to racial or cultural minorities or as reflecting undesirable implicit assumptions.

politician /ˌpɒlɪ'tɪʃ(ə)n/ n. **1** a person engaged in or concerned with politics, esp. as a practitioner. **2** a person skilled in politics. **3** US derog. a person with self-interested political concerns.

politicize /pə'lɪtɪˌsaɪz/ v. (also **-ise**) **1** tr. **a** give a political character to. **b** make politically aware. **2** intr. engage in or talk politics. □ **politicization** /-ˌlɪtɪsaɪ'zeɪʃ(ə)n/ n.

politico /pə'lɪtɪˌkəʊ/ n. (pl. **-os**) colloq. a politician or political enthusiast. [Sp. or It. (as POLITIC)]

politico- /pə'lɪtɪˌkəʊ/ comb. form **1** politically. **2** political and (*politicosocial*). [Gk politikos: see POLITIC]

politics /'pɒlɪtɪks/ n.pl. **1** (treated as sing. or pl.) **a** the art and science of government. **b** public life and affairs as involving authority and government. **2** (usu. treated as pl.) **a** a particular set of ideas, principles, or commitments in politics (*what are their politics?*). **b** activities concerned with the acquisition or exercise of authority or government. **c** an organizational process or principle affecting authority, status, etc. (*the politics of the decision*).

polity /'pɒlɪtɪ/ n. (pl. **-ies**) **1** a form or process of civil government or constitution. **2** an organized society; a state as a political entity. [L politia f. Gk politeia f. politēs citizen f. polis city]

Polk /pəʊk/, James Knox (1795–1849), American Democratic statesman, 11th President of the US 1845–9. His term of office resulted in major territorial additions to the US: Texas was admitted to the Union in 1845, and the successful outcome of the conflict with Mexico resulted in the annexation of California and the south-west two years later.

polka /'pɒlkə, 'pəʊl-/ n. & v. ● n. **1** a lively dance of Bohemian origin in duple time. **2** the music for this. ● v.intr. (**polkas**, **polkaed** /-kəd/ or **polka'd**, **polkaing** /-kəɪŋ/) dance the polka. □ **polka dot** a round dot as one of many forming a regular pattern on a textile fabric etc. [F and G f. Czech půlka half-step f. půl half]

poll[1] /pəʊl/ n. & v. ● n. **1 a** the process of voting at an election. **b** the counting of votes at an election. **c** the result of voting. **d** the number of votes recorded (*a heavy poll*). **2** = GALLUP POLL. **3** dial. a human head. **b** the crown or top of the head (*redpoll*). **c** the part of this on which hair grows (*flaxen poll*). **4** a hornless animal, esp. one of a breed of hornless cattle. ● v. **1** tr. **a** take the vote or votes of. **b** (in passive) have one's vote taken. **c** (of a candidate) receive (so many votes). **d** give (a vote). **2** tr. record the opinion of (a person or group) in an opinion poll. **3** intr. give one's vote. **4** tr. cut off the top of (a tree or plant), esp. make a pollard of. **5** tr. (esp. as **polled** adj.) cut the horns off (cattle). **6** tr.

Computing check the status of (a computer system) at intervals. □ **pollster** n. **pollee** /pəʊ'liː/ n. (in sense 2 of n.). [ME, perh. f. LG or Du.]

poll[2] /pɒl/ n. a tame parrot (*Pretty poll!*). □ **poll parrot** a user of conventional or clichéd phrases and arguments. [Poll, a conventional name for a parrot, alt. f. Moll, a familiar form of Mary]

pollack /'pɒlək/ n. (also **pollock**) a European marine food fish, Pollachius pollachius, of the cod family, with a protruding lower jaw. [earlier (Sc.) podlock: orig. unkn.]

Pollaiuolo /ˌpɒlaɪ'wəʊləʊ/, Antonio (c.1432–98) and Piero (1443–96), Italian sculptors, painters, and engravers. Antonio assisted Ghiberti with the doors for the baptistery of Florence (1452) and both brothers worked on the monuments to the popes Sixtus IV and Innocent VIII in St Peter's, Rome; other joint works include the painting the Martyrdom of St Sebastian (1475). Antonio is particularly known for his realistic depiction of the human form, which reflects his studies of anatomy; his works include the engraving Battle of the Naked Men (c.1470) and the bronze statuette Hercules and Antaeus (c.1475).

pollan /'pɒlən/ n. a form of the freshwater whitefish Coregonus albula, found in Irish lakes (cf. VENDACE). [perh. f. Ir. poll deep water]

pollard /'pɒləd/ n. & v. ● n. **1** an animal that has lost or cast its horns; an ox, sheep, or goat of a hornless breed. **2** a tree whose branches have been cut off to encourage the growth of new young branches, esp. a riverside willow. **3 a** the bran sifted from flour. **b** a fine bran containing some flour. ● v.tr. make (a tree) a pollard. [POLL[1] + -ARD]

pollen /'pɒlən/ n. the fine dustlike grains discharged from the male part of a flower, each containing the gametes that fertilize the female ovule. □ **pollen analysis** = PALYNOLOGY. **pollen count** an index of the amount of pollen in the air, published esp. for the benefit of those allergic to it. □ **pollenless** adj. [L pollen pollinis fine flour, dust]

pollex /'pɒleks/ n. (pl. **pollices** /-lɪˌsiːz/) Zool. the innermost digit of a forelimb, usu. the thumb in primates. [L, = thumb or big toe]

pollie var. of POLLY[2].

pollinate /'pɒlɪˌneɪt/ v.tr. (also absol.) sprinkle (a stigma) with pollen, thus fertilizing the flower etc. □ **pollinator** n. **pollination** /ˌpɒlɪ'neɪʃ(ə)n/ n.

polling /'pəʊlɪŋ/ n. the registering or casting of votes. □ **polling-booth** a compartment in which a voter stands to mark the ballot-paper. **polling-day** the day of a local or general election. **polling-station** a building, often a school, where voting takes place during an election.

pollinic /pə'lɪnɪk/ adj. of or relating to pollen.

polliniferous /ˌpɒlɪ'nɪfərəs/ adj. Bot. bearing or producing pollen.

polliwog /'pɒlɪˌwɒg/ n. (also **pollywog**) US dial. a tadpole. [earlier polwigge, polwygle f. POLL[1] + WIGGLE]

Pollock /'pɒlək/, Jackson (1912–56), American painter. His earlier work shows the influence of surrealist painters such as Miró, but he later became a leading figure of abstract expressionism and from 1947 onwards developed the style known as action painting. Fixing the canvas to the floor or wall, he poured, splashed, or dripped paint on it, covering the whole canvas and avoiding any point of emphasis in the picture. He often used sticks, trowels, and knives instead of brushes and occasionally mixed the paint with sand or broken glass. He was killed in a car accident; he had earlier been treated for alcoholism and in 1938 suffered a mental breakdown.

pollock var. of POLLACK.

poll tax n. a fixed tax levied on every adult, without reference to their income or resources. Poll taxes are arithmetically simple, but have often been extremely unpopular because they weigh disproportionately heavily on poorer people. Such taxes were levied in England in 1377, 1379, and 1380; the last of these is generally regarded as having contributed to the 1381 Peasants' Revolt. Poll taxes did not appear again in Britain until the introduction of the community charge in place of rates (for local government expenditure) by the Conservative government in 1990 (1989 in Scotland). This tax proved not only very unpopular but also expensive and difficult to administer; it was replaced in 1993 by the council tax. [POLL[1]]

pollute /pə'luːt/ v.tr. **1** contaminate or defile (the environment). **2** make foul or filthy. **3** destroy the purity or sanctity of. □ **pollutant** adj. & n. **polluter** n. **pollution** /-'luːʃ(ə)n/ n. [ME f. L polluere pollut-]

Pollux /'pɒləks/ **1** Gk Mythol. the twin brother of Castor (see DIOSCURI). Also called Polydeuces. **2** Astron. the brightest star in the constellation of Gemini, close to Castor.

polly[1] /ˈpɒlɪ/ n. (pl. **-ies**) colloq. a bottle or glass of Apollinaris water. [abbr.]

polly[2] /ˈpɒlɪ/ n. (also **pollie**) (pl. **-ies**) Austral. & US a politician. [abbr.]

Pollyanna /ˌpɒlɪˈænə/ n. a cheerful optimist; an excessively cheerful person. □ **Pollyannaish** adj. **Pollyannaism** n. [character in a novel (1913) by Eleanor Hodgman Porter, Amer. author (1868–1920)]

pollywog var. of POLLIWOG.

Polo, Marco, see MARCO POLO.

polo /ˈpəʊləʊ/ n. a four-a-side game resembling hockey, played on horseback with a long-handled mallet (polo-stick). (See note below.) □ **polo-neck 1** a high round turned-over collar. **2** a pullover with this. **polo-stick** a mallet for playing polo. [Balti, = ball]

▪ Polo is a game of eastern origin, first described in Persia in about 600 BC. Having spread over Asia to China and Japan, by the 19th century it survived only in a few mountain areas in the north-western and north-eastern frontiers of India, where it was discovered and adopted by visiting British officers. Polo is a minority sport in Britain and the US, largely because of the expense of acquiring and maintaining a stable of ponies, but since the 1920s has been extremely popular in Argentina.

polonaise /ˌpɒləˈneɪz/ n. & adj. ● n. **1** a dance of Polish origin in triple time. **2** the music for this. **3** hist. a woman's dress consisting of a bodice and a skirt open from the waist downwards to show an underskirt. ● adj. cooked in a Polish style. [F, fem. of polonais Polish f. med.L Polonia Poland]

polonium /pəˈləʊnɪəm/ n. a radioactive metallic chemical element (atomic number 84; symbol **Po**). Polonium was discovered in the mineral pitchblende by Marie and Pierre Curie in 1898. It is very rare, occurring only as a product of radioactive decay of uranium. It can also be created artificially in nuclear reactors, and has some use as an energy source in satellites. [med.L Polonia Poland (Marie Curie's native country)]

Polonnaruwa /ˌpɒləˈnɑːrʊwə/ a town in NE Sri Lanka; pop. (1981) 11,600. Succeeding Anuradhapura in the 8th century as the capital of Ceylon, it became an important Buddhist centre in the 12th century. It was subsequently deserted until a modern town was built there in the 20th century.

polony /pəˈləʊnɪ/ n. (pl. **-ies**) Brit. = BOLOGNA SAUSAGE. [app. corrupt.]

Pol Pot /pɒl ˈpɒt/ (born c.1925), Cambodian Communist statesman, Prime Minister 1976–9. From 1968 he led the Khmer Rouge, becoming Prime Minister soon after its seizure of power in 1975. During his regime the Khmer Rouge embarked on a brutal reconstruction programme in which many thousands of Cambodians were killed. Overthrown in 1979, Pol Pot led the Khmer Rouge in a guerrilla war against the new Vietnamese-backed government until his official retirement from the leadership in 1985.

Polska /ˈpɒlska/ the Polish name for POLAND.

Poltava /pɒlˈtɑːvə/ a city in east central Ukraine; pop. (1990) 317,000. It was besieged unsuccessfully in 1709 by Charles XII's Swedish forces, who were defeated by the Russians under Peter the Great.

poltergeist /ˈpɒltəˌɡaɪst/ n. a noisy ghost, esp. one manifesting itself by moving physical objects. [G f. poltern create a disturbance + Geist GHOST]

Poltoratsk /ˌpɒltəˈrɑːtsk/ a former name (1919–27) for ASHGABAT.

poltroon /pɒlˈtruːn/ n. a spiritless coward. □ **poltroonery** n. [F poltron f. It. poltrone perh. f. poltro sluggard]

poly /ˈpɒlɪ/ n. (pl. **polys**) Brit. colloq. polytechnic. [abbr.]

poly- /ˈpɒlɪ/ comb. form **1** denoting many or much. **2** Chem. denoting the presence of several or many radicals etc. of a particular kind in a molecule, esp. a polymer (polythene, polysaccharide). [Gk polu- f. polus much, polloi many]

polyadelphous /ˌpɒlɪəˈdelfəs/ adj. Bot. having numerous stamens grouped into three or more bundles.

polyamide /ˌpɒlɪˈeɪmaɪd, -ˈæmaɪd/ n. Chem. any of a class of condensation polymers produced from the interaction of an amino group of one molecule and a carboxylic acid group of another, and which includes many synthetic fibres such as nylon.

polyandry /ˈpɒlɪˌændrɪ/ n. **1** polygamy in which a woman has more than one husband. **2** Bot. the state of having numerous stamens. □ **polyandrous** /ˌpɒlɪˈændrəs/ adj. [POLY- + andry f. Gk anēr andros male]

polyanthus /ˌpɒlɪˈænθəs/ n. (pl. **polyanthuses**) a flower cultivated from hybridized primulas. [mod.L, formed as POLY- + Gk anthos flower]

Polybius /pəˈlɪbɪəs/ (c.200–c.118 BC), Greek historian. After an early political career in Greece, he was deported to Rome. His forty books of Histories (only partially extant) chronicled the rise of the Roman Empire from 220 to 146 BC.

polycarbonate /ˌpɒlɪˈkɑːbəˌneɪt/ n. any of a class of polymers in which the units are linked through a carbonate group, mainly used as moulding materials.

Polycarp, St /ˈpɒlɪˌkɑːp/ (c.69–c.155), Greek bishop of Smyrna in Asia Minor. His dates are uncertain but he was probably the leading Christian figure in Smyrna in the mid-2nd century. He was arrested during a pagan festival, refused to recant his faith, and was burnt to death. His followers buried his remains and wrote an account of his martyrdom, which provides one of the oldest such records to survive. Feast day, 23 Feb.

polychaete /ˈpɒlɪˌkiːt/ n. Zool. an aquatic annelid worm of the class Polychaeta, which includes lugworms and ragworms, having numerous bristles on the fleshy lobes of each body segment. □ **polychaetan** /ˌpɒlɪˈkiːt(ə)n/ adj. **polychaetous** adj.

polychlorinated biphenyl /ˌpɒlɪˌklɔːrɪˌneɪtɪd baɪˈfiːnaɪl, -ˈfenɪl/ see PCB 2.

polychromatic /ˌpɒlɪkrəʊˈmætɪk/ adj. **1** many-coloured. **2** Physics (of radiation) containing more than one wavelength. □ **polychromatism** /-ˈkrəʊməˌtɪz(ə)m/ n.

polychrome /ˈpɒlɪˌkrəʊm/ adj. & n. ● adj. painted, printed, or decorated in many colours. ● n. **1** a work of art in several colours, esp. a coloured statue. **2** varied colouring. □ **polychromic** /ˌpɒlɪˈkrəʊmɪk/ adj. **polychromous** adj. [F f. Gk polukhrōmos as POLY-, khrōma colour]

polychromy /ˈpɒlɪˌkrəʊmɪ/ n. the art of painting in several colours, esp. as applied to ancient pottery, architecture, etc. [F polychromie (as POLYCHROME)]

polyclinic /ˈpɒlɪˌklɪnɪk/ n. a clinic devoted to various diseases; a general hospital.

Polyclitus /ˌpɒlɪˈklaɪtəs/ (5th century BC), Greek sculptor. He is known for his statues of idealized male athletes. Two Roman copies of his works survive, the Doryphoros (spear-bearer) and the Diadumenos (youth fastening a band round his head). His other works include a large gold and ivory statue of the goddess Hera.

polycrystalline /ˌpɒlɪˈkrɪstəˌlaɪn/ adj. (of a solid substance) consisting of many crystalline parts at various orientations, e.g. a metal casting.

polycyclic /ˌpɒlɪˈsaɪklɪk/ adj. Chem. having more than one ring of atoms in the molecule.

polydactyl /ˌpɒlɪˈdæktɪl/ adj. Zool. & Med. having more than five digits on each hand or foot. □ **polydactyly** n.

Polydeuces /ˌpɒlɪˈdjuːsiːz/ see POLLUX 1.

polyester /ˌpɒlɪˈestə(r)/ n. any of a group of condensation polymers used to form synthetic fibres such as Terylene or to make resins.

polyethene /ˌpɒlɪˈeθiːn, -ˈiːθiːn/ n. Chem. = POLYTHENE.

polyethylene /ˌpɒlɪˈeθɪliːn/ n. Chem. = POLYTHENE.

polygamous /pəˈlɪɡəməs/ adj. **1** having more than one wife or husband at the same time. **2** having more than one mate. **3** Bot. bearing some flowers with stamens only, some with pistils only, some with both, on the same or different plants. □ **polygamist** n. **polygamously** adv. **polygamy** n. **polygamic** /ˌpɒlɪˈɡæmɪk/ adj. [Gk polugamos (as POLY-, -gamos marrying)]

polygene /ˈpɒlɪˌdʒiːn/ n. Biol. each of a group of independent genes that collectively affect a characteristic.

polygenesis /ˌpɒlɪˈdʒenɪsɪs/ n. Biol. the (usu. postulated) origination of a race or species from several independent stocks. □ **polygenetic** /-dʒɪˈnetɪk/ adj.

polygeny /pəˈlɪdʒənɪ/ n. Biol. the theory that humankind originated from several independent pairs of ancestors. □ **polygenism** n. **polygenist** n.

polyglot /ˈpɒlɪɡlɒt/ adj. & n. ● adj. **1** of many languages. **2** (of a person) speaking or writing several languages. **3** (of a book, esp. the Bible) with the text translated into several languages. ● n. **1** a polyglot person. **2** a polyglot book, esp. a Bible. □ **polyglottism** n. **polyglottal** /ˌpɒlɪˈɡlɒt(ə)l/ adj. **polyglottic** adj. [F polyglotte f. Gk poluglōttos (as POLY-, glōtta tongue)]

polygon /ˈpɒlɪɡən/ n. a plane figure with many (usu. a minimum of three) sides and angles. □ **polygon of forces** a polygon that represents by the length and direction of its sides all the forces acting on a body or point. □ **polygonal** /pəˈlɪɡən(ə)l/ adj. [LL polygonum f. Gk polugōnon (neut. adj.) (as POLY- + -gōnos angled)]

polygonum /pəˈlɪɡənəm/ n. a plant of (or formerly of) the genus

Polygonum, with small bell-shaped flowers, e.g. bistort and knotgrass. Also called *knotweed*. [mod.L f. Gk *polugonon*]

polygraph /'pɒlɪˌgrɑːf/ *n.* a machine designed to detect and record changes in physiological characteristics (e.g. rates of pulse and breathing), used esp. as a lie-detector.

polygyny /pə'lɪdʒɪnɪ/ *n.* polygamy in which a man has more than one wife. □ **polygynous** *adj.* [POLY- + gyny f. Gk *gunē* woman]

polyhedron /ˌpɒlɪ'hiːdrən/ *n.* (*pl.* **polyhedra** /-drə/) a solid figure with many (usu. more than six) faces. □ **polyhedral** *adj.* **polyhedric** *adj.* [Gk *poluedron* neut. of *poluedros* (as POLY-, *hedra* base)]

polyhistor /ˌpɒlɪ'hɪstə(r)/ *n.* = POLYMATH.

Polyhymnia /ˌpɒlɪ'hɪmnɪə/ *Gk & Rom. Mythol.* the Muse of the mimic art. [Gk, = she of the many hymns]

polymath /'pɒlɪˌmæθ/ *n.* **1** a person of much or varied learning. **2** a great scholar. □ **polymathic** /ˌpɒlɪ'mæθɪk/ *adj.* **polymathy** /pə'lɪməθɪ/ *n.* [Gk *polumathēs* (as POLY-, *math-* stem *manthanō* learn)]

polymer /'pɒlɪmə(r)/ *n. Chem.* a compound composed of one or more large molecules that are formed from repeated units of smaller molecules. □ **polymerism** *n.* **polymerize** *v.intr. & tr.* (also **-ise**) **polymerization** /ˌpɒlɪməraɪ'zeɪʃ(ə)n/ *n.* **polymeric** /-'merɪk/ *adj.* [G f. Gk *polumeros* having many parts (as POLY-, *meros* share)]

polymerase /'pɒlɪməˌreɪz, pə'lɪməˌreɪz/ *n. Biochem.* an enzyme which catalyses the formation of a polymer, esp. of DNA or RNA.

polymerous /pə'lɪmərəs/ *adj. Biol.* having many parts.

polymorphism /ˌpɒlɪ'mɔːfɪz(ə)m/ *n.* the occurrence of something in several different forms. □ **polymorphic** *adj.* **polymorphous** *adj.*

Polynesia /ˌpɒlɪ'niːzɪə, -'niːʒə/ a region of the central Pacific, lying to the east of Micronesia and Melanesia and containing the easternmost of the three great groups of Pacific islands, including Hawaii, the Marquesas Islands, Samoa, the Cook Islands, and French Polynesia. [as POLY- + Gk *nēsos* island]

Polynesian /ˌpɒlɪ'niːzɪən, -'niːʒ(ə)n/ *n. & adj.* ● *n.* **1 a** a native or inhabitant of Polynesia. **b** a person of Polynesian descent. **2** a group of Malayo-Polynesian languages including Maori, Hawaiian, and Samoan. ● *adj.* of or relating to Polynesia or its people or languages. [POLY- + Gk *nēsos* island]

polyneuritis /ˌpɒlɪˌnjʊə'raɪtɪs/ *n. Med.* a disorder that affects many of the peripheral nerves. □ **polyneuritic** /-'rɪtɪk/ *adj.*

polynomial /ˌpɒlɪ'nəʊmɪəl/ *n. & adj. Math.* ● *n.* an expression of more than two algebraic terms, esp. the sum of several terms that contain different powers of the same variable(s). ● *adj.* of or being a polynomial. [POLY-, after *binomial*]

polynya /pə'lɪnjə/ *n.* a stretch of open water surrounded by ice, esp. in the Arctic seas. [Russ. f. *pole* field]

polyp /'pɒlɪp/ *n.* **1** *Zool.* an individual sessile coelenterate, esp. in a colony such as a coral. **2** *Med.* a small usu. benign growth protruding from a mucous membrane. [F *polype* (as POLYPUS)]

polypary /'pɒlɪpərɪ/ *n.* (*pl.* **-ies**) *Zool.* the common stem or support of a colony of polyps. [mod.L *polyparium* (as POLYPUS)]

polypeptide /ˌpɒlɪ'peptaɪd/ *n. Biochem.* a peptide formed by the combination of about ten or more amino acids. [G *Polypeptid* (as POLY-, PEPTONE)]

polyphagous /pə'lɪfəgəs/ *adj. Zool.* able to feed on various kinds of food.

polyphase /'pɒlɪˌfeɪz/ *adj. Electr.* (of a device or circuit) designed to supply or use simultaneously several alternating currents of the same voltage but with different phases.

Polyphemus /ˌpɒlɪ'fiːməs/ *Gk Mythol.* a Cyclops who trapped Odysseus and some of his companions in a cave, from which they escaped by putting out his one eye while he slept. In another story Polyphemus loved the sea-nymph Galatea, and in jealousy killed his rival Acis.

polyphone /'pɒlɪˌfəʊn/ *n.* a symbol or letter that represents several different sounds.

polyphonic /ˌpɒlɪ'fɒnɪk/ *adj.* **1** *Mus.* (of vocal music etc.) in two or more relatively independent parts; contrapuntal. **2** (of a letter etc.) representing more than one sound. □ **polyphonically** *adv.* [Gk *poluphōnos* (as POLY-, *phōnē* voice, sound)]

polyphony /pə'lɪfənɪ/ *n.* (*pl.* **-ies**) **1** *Mus.* **a** a polyphonic style in musical composition; counterpoint. **b** a composition written in this style. **2** *Philol.* the symbolization of different vocal sounds by the same letter or character. □ **polyphonous** *adj.*

polyphyletic /ˌpɒlɪfaɪ'letɪk/ *adj. Biol.* (of a group of organisms) derived from more than one common evolutionary ancestor or ancestral group. [POLY- + PHYLETIC]

polypi *pl.* of POLYPUS.

polyploid /'pɒlɪˌplɔɪd/ *n. & adj. Biol.* ● *n.* a nucleus or organism that contains more than two sets of chromosomes. ● *adj.* of or being a polyploid. □ **polyploidy** *n.* [G (as POLY-, -PLOID)]

polypod /'pɒlɪˌpɒd/ *adj. Zool.* having many feet or footlike appendages. [F *polypode* (adj.) f. Gk (as POLYPUS)]

polypody /'pɒlɪˌpəʊdɪ/ *n.* (*pl.* **-ies**) a fern of the genus *Polypodium*, usu. growing on trees, old walls, banks, and rocks. [ME f. L *polypodium* f. Gk *polupodion* (as POLYPUS)]

polypoid /'pɒlɪˌpɔɪd/ *adj. Zool. & Med.* of or like a polyp. □ **polypous** *adj.*

polypropene /ˌpɒlɪ'prəʊpiːn/ *n.* = POLYPROPYLENE.

polypropylene /ˌpɒlɪ'prəʊpɪˌliːn/ *n. Chem.* any polymer of propylene, including thermoplastic materials used for films, fibres, or moulding materials. Also called *polypropene*.

polypus /'pɒlɪpəs/ *n.* (*pl.* **polypi** /-ˌpaɪ/ or **polypuses**) *Med.* = POLYP 2. [ME f. L *polypus* f. Gk *pōlupos, polupous* cuttlefish, polyp (as POLY-, *pous podos* foot)]

polyrhythm /'pɒlɪˌrɪðəm/ *n. Mus.* **1** the use of two or more different rhythms simultaneously. **2** a passage of music using such rhythms.

polysaccharide /ˌpɒlɪ'sækəˌraɪd/ *n. Chem.* any of a group of carbohydrates whose molecules consist of long chains of monosaccharides.

polysemy /ˌpɒlɪ'siːmɪ/ *n. Philol.* the existence of many meanings (of a word etc.). □ **polysemic** *adj.* **polysemous** *adj.* [POLY- + Gk *sēma* sign]

polystyrene /ˌpɒlɪ'staɪəˌriːn/ *n.* a polymer of styrene, esp. a thermoplastic polymer often expanded with a gas to produce a lightweight rigid white substance used for insulation and in packaging.

polysyllabic /ˌpɒlɪsɪ'læbɪk/ *adj.* **1** (of a word) having many syllables. **2** characterized by the use of words of many syllables. □ **polysyllabically** *adv.*

polysyllable /'pɒlɪˌsɪləb(ə)l/ *n.* a polysyllabic word.

polysynthetic /ˌpɒlɪsɪn'θetɪk/ *adj.* **1** *Linguistics* denoting a language characterized by the combination of several or all syntactic elements of a sentence in one word. **2** *Crystallog.* consisting of a series of twin crystals united so as to form a laminated structure. □ **polysynthesis** /ˌpɒlɪ'sɪnθɪsɪs/ *n.*

polytechnic /ˌpɒlɪ'teknɪk/ *n. & adj.* ● *n.* an institution of higher education offering courses in many subjects at degree level or below. (*See note below.*) ● *adj.* dealing with or devoted to various vocational or technical subjects. [F *polytechnique* f. Gk *polutekhnos* (as POLY- + *tekhnē* art)]

▪ The first British Polytechnic Institution (in Regent Street, London) was founded in 1838 by George Cayley and others to provide workers with technical and vocational education. More polytechnics were set up in the late 19th century, and a second generation was instituted in 1966, created largely from existing colleges. These new polytechnics offered degree courses in a wide range of arts and social science subjects in addition to the established scientific and vocational courses, though they continued to differ from universities in that they did not award their own degrees but awarded them through the CNAA. In 1989 they became independent from local education authorities, and from 1992 the CNAA was abolished and the polytechnics were able to call themselves universities.

polytetrafluoroethylene /ˌpɒlɪˌtetrəˌflʊərəʊ'eθɪˌliːn/ *n.* (abbr. **PTFE**) *Chem.* a tough translucent polymer resistant to chemicals and with a low coefficient of friction, used for seals and bearings, to coat non-stick cooking utensils, etc. [POLY- + TETRA- + FLUORO- + ETHYLENE]

polytheism /'pɒlɪθiːˌɪz(ə)m/ *n.* the belief in or worship of more than one god. □ **polytheist** *n.* **polytheistic** /ˌpɒlɪθiː'ɪstɪk/ *adj.* [F *polythéisme* f. Gk *polutheos* of many gods (as POLY-, *theos* god)]

polythene /'pɒlɪˌθiːn/ *n.* a tough light thermoplastic polymer of ethylene, usu. translucent and flexible or opaque and rigid, used for packaging and insulating materials. Also called *polyethylene, polyethene.*

polytonality /ˌpɒlɪtəʊ'nælɪtɪ/ *n. Mus.* the simultaneous use of two or more keys in a composition. □ **polytonal** /-'təʊn(ə)l/ *adj.*

polyunsaturated /ˌpɒlɪʌn'sætʃəˌreɪtɪd, -'sætjʊˌreɪtɪd/ *adj. Chem.* (of a compound, esp. a fat or oil molecule) containing several double or triple bonds and therefore capable of further reaction.

polyurethane /ˌpɒlɪ'jʊərəˌθeɪn/ *n.* any polymer containing the urethane group, used in adhesives, paints, plastics, rubbers, foams, etc.

polyvalent /ˌpɒlɪˈveɪlənt/ adj. Chem. = MULTIVALENT. □ **polyvalency** n.

polyvinyl acetate /ˌpɒlɪˈvaɪnɪl/ n. (abbr. **PVA**) a soft plastic polymer used in paints and adhesives.

polyvinyl chloride /ˌpɒlɪˈvaɪnɪl/ n. (abbr. **PVC**) a tough transparent solid polymer of vinyl chloride, easily coloured and used for a wide variety of products including pipes, flooring, waterproof clothing, etc.

polyzoan /ˌpɒlɪˈzəʊən/ n. = BRYOZOAN.

pom /pɒm/ n. **1** a Pomeranian dog. **2** Austral. & NZ sl. offens. = POMMY. [abbr.]

pomace /ˈpʌmɪs/ n. **1** the mass of crushed apples in cider-making before or after the juice is pressed out. **2** the refuse of fish etc. after the oil has been extracted, generally used as a fertilizer. [ME f. med.L pomacium cider f. L pomum apple]

pomade /pəˈmɑːd/ n. & v. ● n. scented dressing for the hair and the skin of the head. ● v.tr. anoint with pomade. [F pommade f. It. pomata f. med.L f. L pomum apple (from which it was orig. made)]

pomander /pəˈmændə(r)/ n. **1** a ball of mixed aromatic substances placed in a cupboard etc. or (hist.) carried in a box, bag, etc. as a supposed protection against infection. **2** a (usu. spherical) container for this. **3** a spiced orange etc. similarly used. [earlier pom(e)amber f. AF f. OF pome d'embre f. med.L pomum de ambra apple of ambergris]

pomatum /pəˈmɑːtəm/ n. & v.tr. = POMADE. [mod.L f. L pomum apple]

pome /pəʊm/ n. Bot. a firm-fleshed fruit in which the carpels from the central core enclose the seeds, e.g. the apple, pear, and quince. □ **pomiferous** /pəˈmɪfərəs/ adj. [ME f. OF ult. f. poma pl. of L pomum fruit, apple]

pomegranate /ˈpɒmɪˌɡrænɪt, ˈpɒmˌɡræn-/ n. **1 a** an orange-sized fruit with a tough golden-orange outer skin, containing many seeds in a reddish pulp. **b** the tree bearing this fruit, Punica granatum, native to North Africa and western Asia. **2** an ornamental representation of a pomegranate. [ME f. OF pome grenate (as POME, L granatum having many seeds f. granum seed)]

pomelo /ˈpʌmɪˌləʊ/ n. (pl. **-os**) **1** = SHADDOCK. **2** US = GRAPEFRUIT. [19th c.: orig. unkn.]

Pomerania /ˌpɒməˈreɪnɪə/ a region of northern Europe, extending along the south shore of the Baltic Sea between Stralsund in NE Germany and the Vistula in Poland. Inhabited originally by Celtic, Germanic, and Slavonic tribes, it was conquered by Polish princes during the 10th and 11th centuries. The western part was settled by the Germans from the late 12th century. The region was controlled variously by Germany, Poland, the Holy Roman Empire, Prussia, and Sweden, until the larger part was restored to Poland in 1945, the western portion becoming a part of the German state of Mecklenburg-West Pomerania.

Pomeranian /ˌpɒməˈreɪnɪən/ n. a small breed of dog with long silky hair, a pointed muzzle, and pricked ears.

pomfret /ˈpɒmfrɪt/ n. **1** a marine food fish of the genus Pampus, of the Indian and Pacific Oceans. **2** a dark-coloured deep-bodied marine food fish, Brama brama, of the Atlantic Ocean. [app. f. Port. pampo]

pomfret-cake /ˈpʌmfrɪtˌkeɪk, ˈpɒm-/ n. (also **Pontefract-cake** /ˈpɒntɪˌfrækt-/) Brit. a small round flat liquorice sweetmeat orig. made at Pontefract (earlier Pomfret) in Yorkshire.

pomiculture /ˈpɒmɪˌkʌltʃə(r)/ n. fruit-growing. [L pomum fruit + CULTURE]

pommel /ˈpʌm(ə)l, ˈpɒm-/ n. & v. ● n. **1** a knob, esp. at the end of a sword-hilt. **2** the upward projecting front part of a saddle. ● v.tr. (**pommelled**, **pommelling**; US **pommeled**, **pommeling**) = PUMMEL. □ **pommel horse** a vaulting horse fitted with a pair of curved handgrips. [ME f. OF pomel f. Rmc pomellum (unrecorded), dimin. of L pomum fruit, apple]

pommy /ˈpɒmɪ/ n. (also **pommie**) (pl. **-ies**) Austral. & NZ sl. offens. a British person, esp. a recent immigrant. [20th c.: orig. uncert.]

pomology /pəˈmɒlədʒɪ/ n. the science of fruit-growing. □ **pomologist** n. **pomological** /ˌpɒməˈlɒdʒɪk(ə)l/ adj. [L pomum fruit + -LOGY]

pomp /pɒmp/ n. **1** a splendid display; splendour. **2** (often in pl.) vainglory (the pomps and vanities of this wicked world). [ME f. OF pompe f. L pompa f. Gk pompē procession, pomp f. pempō send]

Pompadour /ˈpɒmpəˌdʊə(r)/, Marquise de (title of Jeanne Antoinette Poisson; known as Madame de Pompadour) (1721–64), French noblewoman. In 1744 she became the mistress and lifelong confidante of Louis XV; although she did not remain his only mistress, she retained her influence and place at court. She was a notable patron of the arts and founded the porcelain factory at Sèvres. She became unpopular through her interference with political affairs, particularly through obtaining ministerial appointments for her favourites.

pompadour /ˈpɒmpəˌdʊə(r)/ n. a woman's hairstyle with the hair in a high turned-back roll round the face. [POMPADOUR]

pompano /ˈpɒmpəˌnəʊ/ n. (pl. **-os**) a fish of the family Carangidae or Stromateidae of the Atlantic and Pacific Oceans, used as food. [Sp. pámpano]

Pompeii /pɒmˈpeɪɪ/ an ancient city in western Italy, south-east of Naples. The life of the city came to an abrupt end following an eruption of Mount Vesuvius in 79 AD, as described by Pliny the Younger. The city lay buried for centuries beneath several metres of volcanic ash until excavations of the site began in 1748. The well-preserved remains of the city include not only buildings and mosaics but wall-paintings, furniture, graffiti, and the personal possessions of its inhabitants, providing an unusually vivid insight into the life, art, and architecture of the ancient Roman period.

Pompey /ˈpɒmpɪ/ (known as Pompey the Great; Latin name Gnaeus Pompeius Magnus) (106–48 BC), Roman general and statesman. His greatest achievements were the suppression of the Mediterranean pirates (66), and the defeat of Mithridates in the east (63). He formed the First Triumvirate with Caesar and Crassus in 60, but disagreement with Caesar resulted in civil war. Pompey was defeated at the battle of Pharsalus, after which he fled to Egypt, where he was murdered.

Pompidou /ˈpɒmpɪˌduː/, Georges (Jean Raymond) (1911–74), French statesman, Prime Minister 1962–8 and President 1969–74. In 1944 he became an adviser of de Gaulle; when the latter was returned to power in 1959, Pompidou assisted in the drafting of the new constitution. He was instrumental in ending the conflict in Algeria between French forces and nationalist guerrillas. He resigned after criticism of his handling of the strikes and riots of 1968. Pompidou was elected President the following year after de Gaulle's resignation; he died in office.

Pompidou Centre a modern art gallery, exhibition centre, and concert hall in Paris, designed by the British architect Richard Rogers and the Italian Renzo Piano (b.1937) and opened in 1977. The design features brightly coloured pipes, ducts, elevators, etc., on the outside of the exterior walls, giving the building an industrial appearance; in front of the building there is a large open area for street theatre etc. Also called the Beaubourg.

pom-pom /ˈpɒmpɒm/ n. an automatic quick-firing gun esp. on a ship. [imit.]

pompon /ˈpɒmpɒn/ n. (also **pompom** /-pɒm/) **1** an ornamental ball or bobble made of wool, silk, or ribbons, usu. worn on women's or children's hats or clothing. **2** the round tuft on a soldier's cap, the front of a shako, etc. **3** (often attrib.) a dahlia or chrysanthemum with small tightly clustered petals. [F, of unkn. orig.]

pompous /ˈpɒmpəs/ adj. **1** self-important, affectedly grand or solemn. **2** (of language) pretentious; unduly grand in style. **3** archaic magnificent; splendid. □ **pompously** adv. **pompousness** n. **pomposity** /pɒmˈpɒsɪtɪ/ n. (pl. **-ies**). [ME f. OF pompeux f. LL pomposus (as POMP)]

'pon /pɒn/ prep. archaic = UPON. [abbr.]

ponce /pɒns/ n. & v. Brit. sl. ● n. **1** a man who lives off a prostitute's earnings; a pimp. **2** offens. a homosexual; an effeminate man. ● v.intr. act as a ponce. □ **ponce about** move about effeminately or ineffectually. □ **poncey** adj. (also **poncy**) (in sense 2 of n.). [perh. f. POUNCE[1]]

Ponce de León /ˌpɒns də ˈliːɒn, ˌpɒnseɪ də leɪˈɒn/, Juan (c.1460–1521), Spanish explorer. He accompanied Columbus on his second voyage to the New World in 1493 and later became governor of Puerto Rico (1510–12). He landed on the coast of Florida in 1513, claiming the area for Spain and becoming its governor the following year.

poncho /ˈpɒntʃəʊ/ n. (pl. **-os**) **1** a South American cloak made of a blanket-like piece of cloth with a slit in the middle for the head. **2** a garment in this style. [S. Amer. Sp., f. Araucanian]

pond /pɒnd/ n. & v. ● n. **1** a fairly small body of still water formed naturally or by hollowing or embanking. **2** (prec. by the) joc. the sea esp. the Atlantic Ocean. ● v. **1** tr. hold back, dam up (a stream etc.). **2** intr. form a pond. □ **pond-life** animals (esp. invertebrates) that live in ponds. **pond-skater** a predatory heteropterous bug of the family Gerridae, which runs on water supported by the surface tension. **pond**

snail a freshwater snail inhabiting ponds, esp. one of the genus *Limnaea*. [ME var. of POUND²]

ponder /ˈpɒndə(r)/ v. **1** tr. weigh mentally; think over; consider. **2** intr. (usu. foll. by on, over) think; muse. [ME f. OF ponderer f. L ponderare f. pondus -eris weight]

ponderable /ˈpɒndərəb(ə)l/ adj. literary having appreciable weight or significance. □ **ponderability** /ˌpɒndərəˈbɪlɪtɪ/ n. [LL ponderabilis (as PONDER)]

ponderation /ˌpɒndəˈreɪʃ(ə)n/ n. literary the act or an instance of weighing, balancing, or considering. [L ponderatio (as PONDER)]

ponderosa /ˌpɒndəˈrəʊsə/ n. US **1** a North American pine tree, *Pinus ponderosa*. **2** the red timber of this tree. [mod.L, fem. of L ponderosus: see PONDEROUS]

ponderous /ˈpɒndərəs/ adj. **1** heavy; unwieldy. **2** laborious. **3** (of style etc.) dull; tedious. □ **ponderously** adv. **ponderousness** n. **ponderosity** /ˌpɒndəˈrɒsɪtɪ/ n. [ME f. L ponderosus f. pondus -eris weight]

Pondicherry /ˌpɒndɪˈtʃerɪ/ **1** a Union Territory of SE India, on the Coromandel Coast, formed from several former French territories and incorporated into India in 1954. **2** its capital city; pop. (1991) 202,650.

pondweed /ˈpɒndwiːd/ n. an aquatic plant, esp. of the genus *Potamogeton*, growing in still or running water.

pone¹ /pəʊn/ n. US **1** unleavened maize bread, esp. as made by North American Indians. **2** a fine light bread made with milk, eggs, etc. **3** a cake or loaf of this. [Algonquian, = bread]

pone² /ˈpəʊnɪ/ n. the dealer's opponent in two-handed card-games. [L, 2nd sing. imper. of ponere place]

pong /pɒŋ/ n. & v. Brit. colloq. ● n. an unpleasant smell. ● v.intr. stink. □ **pongy** /ˈpɒŋɪ/ adj. (**pongier**, **pongiest**). [20th c.: orig. unkn.]

pongal /ˈpɒŋg(ə)l/ n. **1** the Tamil New Year festival at which new rice is cooked. **2** a dish of cooked rice. [Tamil poṅkal boiling]

pongee /pɒnˈdʒiː, pʌn-/ n. **1** a soft usu. unbleached type of Chinese silk fabric. **2** an imitation of this in cotton etc. [Chin. běnjī 'own loom', or běnzhī lit. 'home-woven']

pongid /ˈpɒndʒɪd/ n. & adj. Zool. ● n. an ape of the family Pongidae, including gorillas, chimpanzees, and orang-utans. ● adj. of or relating to this family. [mod.L Pongidae f. Pongo the genus name: see PONGO¹]

pongo¹ /ˈpɒŋgəʊ/ n. (pl. **-os**) **1** an orang-utan. **2** Naut. sl. a soldier. [Congolese mpongo, orig. of African apes]

pongo² /ˈpɒŋgəʊ/ n. (pl. **-os**) Austral. & NZ sl. offens. an Englishman. [20th c.: orig. unkn.]

poniard /ˈpɒnjəd/ n. literary a small slim dagger. [F poignard f. OF poignal f. med.L pugnale f. L pugnus fist]

pons /pɒnz/ n. (pl. **pontes** /ˈpɒntiːz/) (in full **pons Varolii** /vəˈrəʊlɪˌaɪ/) Anat. the part of the brain stem that links the medulla oblongata and the thalamus. □ **pons asinorum** /ˌæsɪˈnɔːrəm/ any difficult proposition, orig. a rule of geometry from Euclid ('bridge of asses'). [L, = bridge: Varolii f. C. Varoli, It. anatomist (1543–75)]

pont /pɒnt/ n. S. Afr. a flat-bottomed ferry-boat. [Du.]

Pont du Gard /ˌpɒ̃ djuː ˈgɑː(r)/ an arched structure built by the Romans AD c.14 over the River Gard in southern France as part of an aqueduct carrying water to Nîmes. Three tiers of limestone arches of diminishing span support the covered water-channel at a height of 55 metres (180 ft) above the valley. In the 18th century the lowest tier was widened to form a road bridge, which is still in use.

Ponte, Lorenzo Da, see DA PONTE.

Pontefract-cake var. of POMFRET-CAKE.

pontes pl. of PONS.

Pontianak /ˌpɒntɪˈɑːnæk/ a seaport in Indonesia, on the west coast of Borneo; pop. (1980) 304,770. It is situated on the equator at the delta of the Kapuas river.

pontifex /ˈpɒntɪˌfeks/ n. (pl. **pontifices** /pɒnˈtɪfɪˌsiːz/) **1** = PONTIFF. **2** Rom. Antiq. a member of the principal college of priests in Rome. □ **Pontifex Maximus** /ˈmæksɪməs/ the head of this. [L pontifex -ficis f. pons pontis bridge + -fex f. facere make]

pontiff /ˈpɒntɪf/ n. (in full **sovereign** or **supreme pontiff**) RC Ch. the pope. [F pontife (as PONTIFEX)]

pontifical /pɒnˈtɪfɪk(ə)l/ adj. & n. ● adj. **1** RC Ch. of or befitting a pontiff; papal. **2** pompously dogmatic; with an attitude of infallibility. ● n. RC Ch. **1** an office-book of the Western Church containing rites to be performed by the pope or bishops. **2** (in pl.) the vestments and insignia of a bishop, cardinal, or abbot. □ **pontifical mass** a high mass,

usu. celebrated by a cardinal, bishop, etc. □ **pontifically** adv. [ME f. F pontifical or L pontificalis (as PONTIFEX)]

pontificate v. & n. ● v.intr. /pɒnˈtɪfɪˌkeɪt/ **1 a** play the pontiff; pretend to be infallible. **b** be pompously dogmatic. **2** RC Ch. officiate as bishop, esp. at Mass. ● n. /pɒnˈtɪfɪkət/ **1** the office of pontifex, bishop, or pope. **2** the period of this. [L pontificatus (as PONTIFEX)]

pontifices pl. of PONTIFEX.

Pontine Marshes /ˈpɒntaɪn/ (called in Italian Agro Pontino) an area of reclaimed marshland in western Italy, on the Tyrrhenian coast south of Rome. Inhabited and cultivated in ancient times, it became infested with malaria in the later years of the Roman Republic. In 1928 an extensive scheme to drain the marshes was begun. Several new towns have been built in the region, which is now a productive agricultural area.

pontoon¹ /pɒnˈtuːn/ n. Brit. **1** a card-game in which players try to acquire cards with a face value totalling twenty-one and no more. **2** = NATURAL n. 4a. [prob. alt. of vingt-un, obs. var. of VINGT-ET-UN]

pontoon² /pɒnˈtuːn/ n. & v. ● n. **1** a flat-bottomed boat. **2 a** each of several boats, hollow metal cylinders, etc., used to support a temporary bridge. **b** a bridge so formed; a floating platform. **3** = CAISSON 1, 2. ● v.tr. cross (a river) by means of pontoons. [F ponton f. L ponto -onis f. pons pontis bridge]

Pontormo /pɒnˈtɔːməʊ/, Jacopo da (1494–1557), Italian painter. He was a pupil of Andrea del Sarto and was influenced by the work of Leonardo da Vinci and later by Michelangelo. From about 1518 onwards he began to develop a style characterized by dynamic composition, anatomical exaggeration, and bright colours, which placed him at the forefront of early mannerism. His works include the Deposition (c.1525), an altarpiece in the chapel of Santa Felicità in Florence.

Pontus /ˈpɒntəs/ an ancient region of northern Asia Minor, on the Black Sea coast north of Cappadocia. Established as an independent kingdom by the end of the 4th century BC, it reached its height between 120 and 63 BC under Mithridates VI. At this time it dominated the whole of Asia Minor, but by the end of the 1st century BC it had been defeated by Rome and absorbed into the Roman Empire.

pony /ˈpəʊnɪ/ n. (pl. **-ies**) **1** a horse of any small breed. **2** a small drinking-glass. **3** (in pl.) sl. racehorses. **4** Brit. sl. £25. □ **pony-tail** a hairstyle in which a person's hair is drawn back, tied, and made to hang down like a pony's tail; a bunch of hair formed in this way. **pony-trekking** the activity or pastime of travelling across country on a pony. [perh. f. poulney (unrecorded) f. F poulenet dimin. of poulain foal]

Pony Express a system of mail delivery in the US in 1860–1, over a distance of 2,900 km (1,800 miles) between St Joseph in Missouri and Sacramento in California, by continuous relays of horse-riders. Buffalo Bill (William Cody) was one of its riders.

poo /puː/ int. & n. = POOH.

pooch /puːtʃ/ n. esp. US sl. a dog. [20th c.: orig. unkn.]

poodle /ˈpuːd(ə)l/ n. & v. ● n. **1** a breed of dog with a curly coat that is usually clipped. **2** a lackey or servile follower. ● v.intr. colloq. move or travel in a leisurely manner. [G Pudel(hund) f. LG pud(d)eln splash in water: cf. PUDDLE]

poof¹ /pʊf, puːf/ n. (also **pouf**) Brit. sl. derog. **1** an effeminate man. **2** a male homosexual. □ **poofy** /ˈpʊfɪ/ adj. [19th c.: perh. alt. of PUFF in same sense]

poof² /pʊf, puːf/ int. & n. (also **pouf**) ● int. **1** expressing contemptuous rejection. **2** imitating a short sharp puff of breath etc. ● n. an utterance of this. [imit.; cf. PUFF]

poofter /ˈpʊftə(r), ˈpuːf-/ n. Brit. sl. derog. = POOF¹. [extension of POOF¹]

pooh /puː/ int. & n. (also **poo**) ● int. **1** expressing impatience or contempt. **2** expressing disgust at a bad smell. ● n. sl. (esp. as a child's word) **1** excrement. **2** an act of defecation. [imit.]

Pooh-Bah /ˈpuːˈbɑː/ n. (also **pooh-bah**) **1** a holder of many offices at once. **2** a pompous self-important person. [a character in Gilbert and Sullivan's The Mikado (1885)]

pooh-pooh /puːˈpuː/ v.tr. express contempt for; ridicule; dismiss (an idea etc.) scornfully. [redupl. of POOH]

pooja var. of PUJA.

pooka /ˈpuːkə/ n. Ir. a hobgoblin. [Ir. púca]

pool¹ /puːl/ n. & v. ● n. **1** a small body of still water, usu. of natural formation. **2** a small shallow body of any liquid (pool of blood). **3** = swimming-pool (see SWIM). **4** a deep place in a river. ● v.intr. **1** (of liquid) form into a pool. **2** (of blood) accumulate in parts of the venous system. [OE pōl, MLG, MDu. pōl, OHG pfuol f. WG]

pool[2] /puːl/ *n. & v.* ● *n.* **1 a** (often *attrib.*) a common supply of persons, vehicles, commodities, etc. for sharing by a group of people (*a typing pool*; *a pool car*). **b** a group of persons sharing duties etc. **2 a** the collective amount of players' stakes in gambling etc. **b** a receptacle for this. **3 a** a joint commercial venture, esp. an arrangement between competing parties to fix prices and share business to eliminate competition. **b** the common funding for this. **4 a** *N. Amer.* a game on a billiard-table played with balls numbered one to fifteen, the number of each ball pocketed being added to a player's score. **b** *Brit.* a game on a billiard-table in which each player has a ball of a different colour for use as a cue-ball to pocket the other balls in fixed order, the winner taking all the stakes. **5** a group of contestants who compete against each other in a tournament for the right to advance to the next round. ● *v.tr.* **1** put (resources etc.) into a common fund. **2** share (things) in common. **3** (of transport or organizations etc.) share (traffic, receipts). **4** *Austral. sl.* **a** involve (a person) in a scheme etc., often by deception. **b** implicate, inform on. □ **the pools** *Brit.* = *football pool.* [F *poule* stake, orig. hen: assoc. with POOL¹]

Poole /puːl/ a port and resort town on the south coast of England, in Dorset just west of Bournemouth; pop. (1991) 130,900. Poole Harbour, a large, sheltered, shallow bay, is a noted yachting centre.

poolroom /ˈpuːlruːm, -rʊm/ *n.* **1** *US* a betting shop. **2** a place for playing pool.

poon[1] /puːn/ *n.* a large Indo-Malayan evergreen tree of the genus *Calophyllum.* □ **poon oil** an oil from the seeds of this tree, used in medicine and for lamps. [Sinh. *pūna*]

poon[2] /puːn/ *n. esp. Austral. sl.* a simple or foolish person. [20th c.: orig. unkn.]

Poona /ˈpuːnə/ (also **Pune**) an industrial city in Maharashtra, western India, in the hills south-east of Bombay; pop. (1991) 1,560,000. It was the capital of the Maratha people, apart from a short period of Mogul occupation, from the 17th century until it fell to the British in 1817. It was a military and administrative centre under British rule.

poop[1] /puːp/ *n. & v.* ● *n.* **1** the stern of a ship. **2** (also **poop deck**) the aftermost and highest deck. ● *v.tr.* **1** (of a wave) break over the stern of (a ship). **2** (of a ship) receive (a wave) over the stern. [ME f. OF *pupe, pope* ult. f. L *puppis*]

poop[2] /puːp/ *v.tr.* (esp. as **pooped** *adj.*) *US colloq.* exhaust; tire out. [20th c.: orig. unkn.]

poop[3] /puːp/ *n. esp. N. Amer. sl.* up to date or inside information; the low-down, the facts. [20th c.: orig. unkn.]

poor /pʊə(r), pɔː(r)/ *adj.* **1** lacking adequate money or means to live comfortably. **2 a** (foll. by *in*) deficient in (a possession or quality) (*the poor in spirit*). **b** (of soil, ore, etc.) unproductive. **3 a** scanty, inadequate (*a poor crop*). **b** less good than is usual or expected (*poor visibility*; *is a poor driver*; *in poor health*). **c** paltry; inferior (*poor condition*; *came a poor third*). **4 a** deserving pity or sympathy; unfortunate (*you poor thing*). **b** with reference to a dead person (*as my poor father used to say*). **5** spiritless; despicable (*is a poor creature*). **6** often *iron.* or *joc.* humble; insignificant (*in my poor opinion*). □ **poor-box** a collection-box, esp. in church, for the relief of the poor. **poor man's** an inferior or cheaper substitute for. **poor man's orchid** = SCHIZANTHUS. **poor man's weather-glass** the scarlet pimpernel. **poor-rate** *hist.* a rate or assessment for relief or support of the poor. **poor relation** an inferior or subordinate member of a family or any other group. **poor-spirited** timid; cowardly. **poor white** esp. *US. derog.* a member of a group of white people regarded as socially inferior. **take a poor view of** regard with disfavour or pessimism. [ME & OF *pov(e)re, poure* f. L *pauper*]

Poor Clare *n.* a member of an order of Franciscan nuns founded by St Clare of Assisi *c.*1212.

poorhouse /ˈpʊəhaʊs, ˈpɔːhaʊs/ *n. hist.* = WORKHOUSE 1.

Poor Law *n.* any of various laws relating to the support of the poor in England. In the 16th century legislation placed the responsibility for the relief of the unemployed, infirm, and disabled on the parish, and levied a compulsory rate for the provision of funds; a statute of 1576 also stipulated that work should be organized for the able-bodied poor. By the late 18th century there was a system to supplement the wages of low-paid workers with parish allowances; the Poor Law Amendment Act of 1834 sought to end this type of outdoor relief by ensuring that relief was given either through the harsh regime of workhouses (for the 'deserving poor') or through that of the even harsher houses of correction (for the 'undeserving poor'). The system proved both repressive and, with increasing urbanization during the 19th century, unable to cope with demand. In the early 20th century the Poor Law was dismantled and replaced by schemes of social security.

poorly /ˈpʊəlɪ, ˈpɔːlɪ/ *adv. & adj.* ● *adv.* **1** scantily; defectively. **2** with no great success. **3** meanly; contemptibly. ● *predic.adj.* esp. *Brit.* unwell.

poorness /ˈpʊənɪs, ˈpɔːnɪs/ *n.* **1** the state or condition of being poor; poverty. **2** the lack of some desirable quality or constituent.

POP *abbr.* Post Office Preferred (size of envelopes etc.).

pop[1] /pɒp/ *n., v., & adv.* ● *n.* **1** a sudden sharp explosive sound as of a cork when drawn. **2** *colloq.* an effervescent sweet drink. ● *v.* (**popped**, **popping**) **1** *intr. & tr.* make or cause to make a pop. **2** *intr. & tr.* (foll. by *in, out, up, down,* etc.) go, move, come, or put unexpectedly or in a quick or hasty manner (*pop out to the shops*; *pop in for a visit*; *pop it on your head*). **3 a** *intr. sl.* burst, making a popping sound. **b** *tr.* heat (popcorn etc.) until it pops. **4** *intr.* (often foll. by *at*) *colloq.* fire a gun (at birds etc.). **5** *tr. sl.* pawn. **6** *tr. sl.* take or inject (a drug etc.). **7** *intr.* (often foll. by *up*) (of a cricket-ball) rise sharply off the pitch. ● *adv.* with the sound of a pop (*heard it go pop*). □ **in pop** *Brit. sl.* in pawn. **pop off** *colloq.* **1** die. **2** quietly slip away (cf. sense 2 of *v.*). **pop one's cherry** *sl.* lose one's virginity. **pop one's clogs** *sl.* die. **pop the question** *colloq.* propose marriage. **pop-shop** *Brit. sl.* a pawnbroker's shop. **pop-up 1** (of a toaster etc.) operating so as to move the object (toast when ready etc.) quickly upwards. **2** (of a book, greetings card, etc.) containing three-dimensional figures, illustrations, etc., that rise up when the page is turned. **3** *Computing* (of a menu) able to be superimposed on the screen being worked on and suppressed rapidly. [ME: imit.]

pop[2] /pɒp/ *adj. & n. colloq.* ● *adj.* **1** in a popular or modern style. **2** of, performing, or relating to pop music (*pop concert*; *pop group*). ● *n.* **1** (in full **pop music**) commercial popular music, esp. that since the 1950s. **2** a pop record or song (*top of the pops*). □ **pop culture** commercial culture based on popular taste. **pop festival** a festival at which popular music etc. is performed. [abbr.]

pop[3] /pɒp/ *n. esp. US colloq.* father. [abbr. of POPPA]

pop. *abbr.* population.

popadam var. of POPPADAM.

pop art *n.* a style and movement in art based on popular culture and the mass media, esp. as a critical comment on traditional fine art values. The term is applied specifically to the works of a group of artists who, largely in the mid-1950s and 1960s, used images from comic books, advertisements, consumer products, television, and cinema in order to challenge the notion of a distinction between highbrow and popular in art. The best-known exponents of pop art are Andy Warhol, Roy Lichtenstein, Jasper Johns, and Peter Blake.

popcorn /ˈpɒpkɔːn/ *n.* **1** maize which bursts open when heated. **2** these kernels when popped.

Pope /pəʊp/, Alexander (1688–1744), English poet. He was a major figure of the Augustan period in England, famous for his caustic wit and metrical skill, in particular his use of the heroic couplet. His *Essay on Criticism* (1711), a poem on the art of writing, drew him to the attention of Addison's literary circle; he later associated with Jonathan Swift, John Gay, and others. Among Pope's other major works are the mock-heroic *The Rape of the Lock* (1712; enlarged 1714), the philosophical poem *An Essay on Man* (1733–4), and the *Epistle to Dr Arbuthnot* (1735), a fierce and ironic attack on his critics. He published an edition of Shakespeare's plays (1725) which was criticized for inaccuracies by the scholar Lewis Theobald (1688–1744); Pope retaliated by making Theobald the hero of his satire attacking 'Dulness', *The Dunciad* (1728). Pope also made notable translations of the *Iliad* (1715–20) and the *Odyssey* (1726).

pope[1] /pəʊp/ *n.* **1** (as title usu. **Pope**) the head of the Roman Catholic Church. (*See note below.*) **2** the head of the Coptic Church. **3** = RUFFE. □ **pope's eye** **1** a lymphatic gland surrounded with fat in the middle of a sheep's leg. **2** *Sc.* a cut of steak. □ **popedom** *n.* **popeless** *adj.* [OE f. eccl.L *pāpa* bishop, pope f. eccl.Gk *papas* = Gk *pappas* father: cf. PAPA]

▪ In the Western Church *pope* is today a title given only to the bishop of Rome as head of the Roman Catholic Church, but in the early Church it was applied to bishops generally, becoming confined to the Roman official from about the 9th century; in the Eastern or Orthodox Church priests, and the patriarch of Alexandria, are known by the title (see POPE²). Elected by the College of Cardinals, the pope derives his authority from being regarded as the direct successor of St Peter, the leader of the Apostles and the first bishop of Rome. The pope formerly had great political power, particularly during the Middle Ages, as ruler of the Papal States; today papal sovereignty is restricted to the Vatican City.

pope[2] /pəʊp/ *n.* a parish priest of the Orthodox Church in Russia etc. [Russ. *pop* f. Old Ch. Slav. *popŭ* f. WG f. eccl.Gk (as POPE¹)]

Pope Joan /dʒəʊn/ (according to a legend widely believed in the Middle Ages) a woman in male disguise who (c.1100) became a distinguished scholar and then pope, reigned for more than two years, and died after giving birth to a child during a procession.

popery /ˈpəʊpərɪ/ n. derog. the papal system; the doctrines and ceremony associated with the pope; the Roman Catholic Church.

pop-eyed /ˈpɒpaɪd/ adj. colloq. **1** having bulging eyes. **2** wide-eyed (with surprise etc.).

popgun /ˈpɒpɡʌn/ n. **1** a child's toy gun which shoots a pellet etc. by the compression of air with a piston. **2** derog. an inefficient firearm.

popinjay /ˈpɒpɪnˌdʒeɪ/ n. **1** a fop, a conceited person, a coxcomb. **2 a** archaic a parrot. **b** hist. a figure of a parrot on a pole as a mark to shoot at. [ME f. AF papeiaye, OF papingay etc. f. Sp. papagayo f. Arab. babaġā: assim. to JAY]

popish /ˈpəʊpɪʃ/ adj. derog. Roman Catholic. □ **popishly** adv.

Popish Plot a fictitious Jesuit plot concocted by an English Protestant clergyman, Titus Oates, in 1678. The plot involved a plan to kill Charles II, massacre Protestants, and put the Catholic Duke of York on the English throne. The 'discovery' of the plot led to widespread panic and the execution of about thirty-five Catholics.

poplar /ˈpɒplə(r)/ n. **1** a tree of the genus Populus, often tall and fast-growing. **2** N. Amer. = tulip tree. [ME f. AF popler, OF poplier f. pople f. L populus]

poplin /ˈpɒplɪn/ n. a plain-woven fabric usu. of cotton, with a corded surface. [obs. F papeline perh. f. It. papalina (fem.) PAPAL, f. the papal town Avignon where it was made]

popliteal /pɒpˈlɪtɪəl/ adj. of the hollow at the back of the knee. [mod.L popliteus f. L poples -itis this hollow]

Popocatépetl /ˌpɒpəˈkætəˌpet(ə)l/, -ˌkætəˈpet(ə)l/ a dormant volcano in Mexico, south-east of Mexico City, which rises to 5,452 m (17,700 ft).

poppa /ˈpɒpə/ n. US colloq. father (esp. as a child's word). [var. of PAPA]

poppadam /ˈpɒpədəm/ n. (also **poppadom**, **popadam**) a thin, crisp, spiced bread usu. eaten with curry or other Indian food. [Tamil pappaḍam]

Popper /ˈpɒpə(r)/, Sir Karl Raimund (1902–94), Austrian-born British philosopher. He was originally associated with the Vienna Circle, but was highly critical of the emphasis placed by logical positivism on verification. In The Logic of Scientific Discovery (1934) he posits instead that scientific hypotheses can never be finally confirmed as true and are acceptable only in so far as they manage to survive frequent attempts to falsify them. He is also known for his criticism of the historicist social philosophies of Plato, Hegel, and Marx, as, for example, in The Open Society and its Enemies (1945). He left Vienna on Hitler's rise to power and eventually settled in England, where he was a professor at the London School of Economics (1949–69).

popper /ˈpɒpə(r)/ n. **1** Brit. colloq. a press-stud. **2** a person or thing that pops. **3** colloq. a small vial of amyl nitrite used for inhalation.

poppet /ˈpɒpɪt/ n. **1** Brit. colloq. (esp. as a term of endearment) a small or dainty person. **2** (in full **poppet-head**) the head of a lathe. **3** a small square piece of wood fitted inside the gunwale or washstrake of a boat. **4** esp. hist. a small human figure, used in sorcery or witchcraft. □ **poppet-head** Brit. the frame at the top of a mine-shaft supporting pulleys for the ropes used in hoisting. **poppet-valve** Engin. a mushroom-shaped valve, lifted bodily from its seat rather than hinged. [ME popet(te), ult. f. L pup(p)a: cf. PUPPET]

popping-crease /ˈpɒpɪŋˌkriːs/ n. Cricket a line four feet in front of and parallel to the wicket, within which the batsman must keep the bat or one foot grounded to avoid the risk of being stumped. [POP¹, perh. in obs. sense 'strike']

popple /ˈpɒp(ə)l/ v. & n. ● v.intr. (of water) tumble about, toss to and fro. ● n. the act or an instance of rolling, tossing, or rippling of water. □ **popply** adj. [ME prob. f. MDu. popelen murmur, quiver, of imit. orig.]

poppy /ˈpɒpɪ/ n. (pl. **-ies**) a plant of the family Papaveraceae, with showy flowers, milky sap, and rounded seed capsules; esp. the corn poppy (Papaver rhoeas), with red flowers, and the opium poppy. □ **poppy-head 1** the seed capsule of the poppy. **2** an ornamental top on the end of a church pew. □ **poppied** adj. [OE popig, papæg, etc. f. med.L papauum f. L papaver]

poppycock /ˈpɒpɪˌkɒk/ n. sl. nonsense. [Du. dial. pappekak]

Poppy Day Remembrance Sunday, on which artificial poppies are worn (see FLANDERS POPPY).

popsy /ˈpɒpsɪ/ n. (also **popsie**) (pl. **-ies**) colloq. (usu. as a term of endearment) a young woman. [shortening of POPPET]

populace /ˈpɒpjʊləs/ n. **1** the ordinary people; the common people. **2** derog. the rabble. [F f. It. popolaccio f. popolo people + -accio pejorative suffix]

popular /ˈpɒpjʊlə(r)/ adj. **1** liked or admired by many people or by a specified group (popular teachers; a popular hero). **2 a** of or carried on by the general public (popular meetings). **b** prevalent among the general public (popular discontent). **3** adapted to the understanding, taste, or means of the people (popular science; popular medicine). □ **popularism** n. **popularly** adv. **popularity** /ˌpɒpjʊˈlærɪtɪ/ n. [ME f. AF populer, OF populeir or L popularis f. populus people]

Popular Front n. **1** an international alliance of Communist, radical, and socialist elements formed and gaining some power in the 1930s. In France such an alliance won elections in 1936, under the leadership of Léon Blum. In Spain the Popular Front government was in office 1936–9, and fought the Spanish Civil War against Franco and the Nationalists. In Chile a Popular Front government ruled from 1938 to 1946. **2** (also **popular front**) a political party or coalition representing left-wing elements.

popularize /ˈpɒpjʊləˌraɪz/ v.tr. (also **-ise**) **1** make popular. **2** cause (a person, principle, etc.) to be generally known or liked. **3** present (a technical subject, specialized vocabulary, etc.) in a popular or readily understandable form. □ **popularizer** n. **popularization** /ˌpɒpjʊləraɪˈzeɪʃ(ə)n/ n.

popular music n. music appealing to the popular taste, esp. as opposed to classical, jazz, or folk music. The origins of popular music may be traced back to the 18th century, when hymns, folk-tunes, and songs from ballad operas were first published in sheet-music form. The 19th century saw the popularity in Britain of the music-hall; in the US popular songs included those written by Stephen Foster. In the early 20th century the invention of sound recording assisted the spread of ragtime, early jazz, and the songs of Tin Pan Alley beyond the US, while radio and gramophone popularized country music, big-band jazz, and blues from the late 1940s, and then rock and roll from the mid-1950s. Amplification of singing and frequently of instrumentation was by then usual. In recent years technology has become increasingly sophisticated and popular music is now an important industry, yet its basis is still the short, concise song. In the past decade the ubiquity of rock and pop among white audiences has been challenged by various forms of black music, such as reggae, rap, and house music, and world music. (See also ROCK MUSIC, COUNTRY MUSIC, FOLK MUSIC, JAZZ, etc.)

populate /ˈpɒpjʊˌleɪt/ v.tr. **1** inhabit; form the population of (a town, country, etc.). **2** supply with inhabitants; people (a densely populated district). [med.L populare populat- (as PEOPLE)]

population /ˌpɒpjʊˈleɪʃ(ə)n/ n. **1 a** the inhabitants of a place, country, etc. referred to collectively. **b** any specified group within this (the Irish population of Liverpool). **2** the total number of any of these (a population of 8 million; the seal population). **3** the act or process of supplying with inhabitants (the population of forest areas). **4** Statistics any finite or infinite collection of items under consideration. □ **population explosion** a sudden large increase of population. [LL populatio (as PEOPLE)]

populist /ˈpɒpjʊlɪst/ n. & adj. ● n. **1** a member or adherent of a political party seeking support mainly from the ordinary people. **2** a person who seeks to appeal to or represents the views of ordinary people. ● adj. of or relating to a populist or populists; appealing to a mass audience. □ **populism** n. **populistic** /ˌpɒpjʊˈlɪstɪk/ adj. [L populus people]

populous /ˈpɒpjʊləs/ adj. thickly inhabited. □ **populously** adv. **populousness** n. [ME f. LL populosus (as PEOPLE)]

porbeagle /ˈpɔːˌbiːɡ(ə)l/ n. a large shark, Lamna nasus, having a pointed snout. [18th-c. Cornish dial., perh. f. Cornish porth harbour + bugel shepherd]

porcelain /ˈpɔːsəlɪn/ n. **1** a hard vitrified translucent ceramic material; china. (See also POTTERY.) **2** objects made of this. □ **porcelain clay** kaolin. □ **porcellaneous** /ˌpɔːsəˈleɪnɪəs/ adj. **porcellanous** /pɔːˈselənəs/ adj. [F porcelaine f. It. porcellana cowrie, porcelain f. porcella dimin. of porca sow f. L fem. of porcus pig]

porch /pɔːtʃ/ n. **1** a covered shelter for the entrance of a building. **2** N. Amer. a veranda. **3** (the Porch) = STOA 2. □ **porched** adj. **porchless** adj. [ME f. OF porche f. L porticus (transl. Gk stoa) f. porta passage]

porcine /ˈpɔːsaɪn/ adj. of or like pigs. [F porcin or f. L porcinus f. porcus pig]

porcupine /ˈpɔːkjʊˌpaɪn/ n. **1** a large rodent of the family Hystricidae, native to Africa, Asia, and SE Europe, or the family Erethizontidae, native to America, having defensive spines or quills. **2** (attrib.) denoting any other animal or plant with spines. □ **porcupine fish** a tropical marine fish, *Diodon hystrix*, covered with sharp spines and able to inflate itself into a spherical shape. [ME f. OF *porc espin* f. Prov. *porc espi(n)* ult. f. L *porcus* pig + *spina* thorn]

pore[1] /pɔː(r)/ *n.* esp. *Biol.* a minute opening in a surface through which gases, liquids, or fine solids may pass. [ME f. OF f. L *porus* f. Gk *poros* passage, pore]

pore[2] /pɔː(r)/ *v.intr.* (foll. by *over*) **1** be absorbed in studying (a book etc.). **2** meditate on, think intently about (a subject). [ME *pure* etc. perh. f. OE *purian* (unrecorded): cf. PEER[1]]

porgy /ˈpɔːgɪ/ *n.* (pl. **-ies**) N. Amer. a sparid fish often used as food, found chiefly in coastal West Atlantic waters; esp. one of the genus *Calamus*. [18th c.: orig. uncert.: cf. Sp. & Port. *pargo*]

Pori /ˈpɔːrɪ/ an industrial port in SW Finland, on the Gulf of Bothnia; pop. (1990) 76,360.

poriferan /pəˈrɪfərən/ *n.* & *adj.* Zool. ● *n.* an aquatic invertebrate of the phylum Porifera, which comprises the sponges. ● *adj.* of or relating to this phylum. [mod.L *Porifera* f. L *porus* PORE[1] + *-fer* bearing]

pork /pɔːk/ *n.* the (esp. unsalted) flesh of a pig, used as food. □ **pork-barrel** N. Amer. colloq. government funds as a source of political benefit. **pork-butcher** a person who slaughters pigs for sale, or who sells pork rather than other meats. **pork pie** a pie of minced pork etc. eaten cold. **pork pie hat** a hat with a flat crown and a brim turned up all round. [ME *porc* f. OF *porc* f. L *porcus* pig]

porker /ˈpɔːkə(r)/ *n.* **1** a pig raised for food. **2** a young fattened pig.

porkling /ˈpɔːklɪŋ/ *n.* a young or small pig.

porky[1] /ˈpɔːkɪ/ *adj.* & *n.* ● *adj.* (**porkier, porkiest**) **1** colloq. fleshy, fat. **2** of or like pork. ● *n. rhyming sl.* a lie (short for *porky pie*).

porky[2] /ˈpɔːkɪ/ *n.* (pl. **-ies**) US colloq. a porcupine. [abbr.]

porn /pɔːn/ *n.* & *adj.* (also **porno** /ˈpɔːnəʊ/) colloq. ● *n.* pornography. ● *attrib.adj.* pornographic. [abbr.]

pornography /pɔːˈnɒgrəfɪ/ *n.* **1** the explicit description or exhibition of sexual subjects or activity in literature, films, etc., intended to stimulate erotic rather than aesthetic or emotional feelings. **2** literature etc. characterized by this. □ **pornographer** *n.* **pornographic** /ˌpɔːnəˈgræfɪk/ *adj.* **pornographically** *adv.* [Gk *pornographos* writing of harlots f. *pornē* prostitute + *graphō* write]

porous /ˈpɔːrəs/ *adj.* **1** full of pores. **2** letting through air, water, etc. **3** (of an argument, security system, etc.) leaky, admitting infiltration. □ **porously** *adv.* **porousness** *n.* **porosity** /pɔːˈrɒsɪtɪ/ *n.* [ME f. OF *poreux* f. med.L *porosus* f. L *porus* PORE[1]]

porphyria /pɔːˈfɪrɪə/ *n.* Med. a rare hereditary disease in which there is abnormal metabolism of the blood pigment haemoglobin. There is excretion of porphyrins in the urine, which becomes dark; other symptoms include mental disturbances and extreme sensitivity of the skin to light. [PORPHYRIN + -IA[1]]

porphyrin /ˈpɔːfɪrɪn/ *n.* Biochem. a crystalline pigment with a flat molecule consisting of four heterocyclic nuclei linked in a ring by CH groups. Porphyrins occur widely in living organisms; haem, the molecule responsible for the red colour of haemoglobin, is a porphyrin containing an iron atom bonded in the centre of the ring. Another example is chlorophyll, containing a central magnesium atom. [Gk *porphura* purple + -IN]

Porphyry /ˈpɔːfɪrɪ/ (c.232–303), Neoplatonist philosopher. Born in Tyre, he studied first at Athens. He then moved to Rome, where he became a pupil of Plotinus, whose works he edited after the latter's death. Porphyry's own works include *Against the Christians*, of which only fragments survive.

porphyry /ˈpɔːfɪrɪ/ *n.* (pl. **-ies**) **1** a hard rock quarried in ancient Egypt, composed of crystals of white or red feldspar in a red matrix. **2** Geol. an igneous rock with large crystals scattered in a matrix of much smaller crystals. □ **porphyritic** /ˌpɔːfɪˈrɪtɪk/ *adj.* [ME ult. f. med.L *porphyreum* f. Gk *porphurītēs* f. *porphura* purple]

porpoise /ˈpɔːpəs/ *n.* a small toothed whale of the family Phocaenidae, esp. of the genus *Phocaena*, with a low triangular dorsal fin and a blunt rounded snout. [ME *porpays* etc. f. OF *po(u)rpois* etc. ult. f. L *porcus* pig + *piscis* fish]

porridge /ˈpɒrɪdʒ/ *n.* **1** a dish consisting of oatmeal or another meal or cereal boiled in water or milk. **2** Brit. sl. imprisonment. □ **porridgy** *adj.* [16th c.: alt. of POTTAGE]

porringer /ˈpɒrɪndʒə(r)/ *n.* a small bowl, often with a handle, for soup, stew, etc. [earlier *pottinger* f. OF *potager* f. *potage* (see POTTAGE): -n- as in *messenger* etc.]

Porsche /pɔːʃ/, German /ˈpɔrʃə/, Ferdinand (1875–1952), Austrian car designer. In 1934, with backing from the Nazi government, he designed the Volkswagen (= people's car), a small economical car with a rear engine, developed and produced in great numbers by the Volkswagen company after the Second World War. Porsche's name has since become famous for the high-performance sports and racing cars produced by his company, originally to his designs.

Porsenna /pɔːˈsenə/, Lars (also **Porsena** /ˈpɔːsɪnə/) (6th century BC), a legendary Etruscan chieftain, king of the town of Clusium. He was summoned by Tarquinius Superbus after the latter's overthrow and exile from Rome and as a result laid siege to Rome but was ultimately unsuccessful in capturing the city.

port[1] /pɔːt/ *n.* **1** a harbour. **2** a place of refuge. **3** a town or place possessing a harbour where ships load or unload, or begin or end their voyages, esp. one where customs officers are stationed. □ **port of call** a place where a ship, person, etc. stops on a journey. [OE f. L *portus* & ME prob. f. OF f. L *portus*]

port[2] /pɔːt/ *n.* (in full **port wine**) a strong, sweet, dark red (sometimes tawny or white) fortified wine of Portugal. [f. OPORTO, city in Portugal from which port is shipped]

port[3] /pɔːt/ *n.* & *v.* ● *n.* the left-hand side (looking forward) of a ship, boat, or aircraft (cf. STARBOARD). ● *v.tr.* (also *absol.*) turn (the helm) to port. □ **port tack** see TACK[1] *n.* 4a. **port watch** see WATCH *n.* 3b. [prob. orig. the side turned towards PORT[1]]

port[4] /pɔːt/ *n.* **1 a** an opening in the side of a ship for entrance, loading, etc. **b** a porthole. **2** an aperture for the passage of steam, water, etc. **3** Electr. a socket or aperture in an electronic circuit, esp. in a computer network, where connections can be made with peripheral equipment. **4** an aperture in a wall etc. for a gun to be fired through. **5** esp. Sc. a gate or gateway, esp. of a walled town. [ME & OF *porte* f. L *porta*]

port[5] /pɔːt/ *v.* & *n.* ● *v.tr.* Mil. carry (a rifle, or other weapon) diagonally across and close to the body with the barrel etc. near the left shoulder (esp. *port arms!*). ● *n.* **1** Mil. this position. **2** external deportment; carriage; bearing. [ME f. OF *port* ult. f. L *portare* carry]

port[6] /pɔːt/ *n.* Austral. colloq. **1** a suitcase or travelling bag. **2** a shopping bag etc. [abbr. of PORTMANTEAU]

portable /ˈpɔːtəb(ə)l/ *adj.* & *n.* ● *adj.* **1** easily movable, convenient for carrying (*portable TV; portable computer*). **2** (of a right, privilege, etc.) capable of being transferred or adapted in altered circumstances (*portable pension*). ● *n.* a portable object, e.g. a radio, typewriter, etc. (*decided to buy a portable*). □ **portably** *adv.* **portableness** *n.* **portability** /ˌpɔːtəˈbɪlɪtɪ/ *n.* [ME f. OF *portable* or LL *portabilis* f. L *portare* carry]

portage /ˈpɔːtɪdʒ/ *n.* & *v.* ● *n.* **1** the carrying of boats or goods between two navigable waters. **2** a place at which this is necessary. **3 a** the act or an instance of carrying or transporting. **b** the cost of this. ● *v.tr.* convey (a boat or goods) between navigable waters. [ME f. OF f. *porter*: see PORT[5]]

Portakabin /ˈpɔːtəˌkæbɪn/ *n. propr.* a portable room or building designed for quick assembly. [PORTABLE + CABIN]

portal[1] /ˈpɔːt(ə)l/ *n.* a doorway or gate etc., esp. a large and elaborate one. [ME f. OF f. med.L *portale* (neut. adj.): see PORTAL[2]]

portal[2] /ˈpɔːt(ə)l/ *adj.* Med. **1** of or relating to the transverse fissure of the liver through which blood vessels etc. pass. **2** of or relating to the portal vein. □ **portal vein** a vein conveying blood to the liver from the spleen, stomach, pancreas, and intestines. [mod.L *portalis* f. L *porta* gate]

portamento /ˌpɔːtəˈmentəʊ/ *n.* (pl. **portamenti** /-tɪ/) Mus. **1** the act or an instance of gliding from one note to another in singing, playing the violin, etc. **2** piano-playing in a manner intermediate between legato and staccato. [It., = carrying]

Port Arthur a former name (1898–1905) for LUSHUN.

portative /ˈpɔːtətɪv/ *adj.* **1** serving to carry or support. **2** Mus. hist. (esp. of a small pipe-organ) portable. [ME f. OF *portatif*, app. alt. of *portatil* f. med.L *portatilis* f. L *portare* carry]

Port-au-Prince /ˌpɔːtəʊˈprɪns, French pɔrtoprɛ̃s/ the capital of Haiti, a port on the west coast; pop. (1988) 1,143,630. Founded by the French in 1749, it became capital of the new republic in 1806.

Port Blair /bleə(r)/ a port on the southern tip of South Andaman Island in the Bay of Bengal; pop. (1991) 74,810. It is the capital of the Andaman and Nicobar Islands.

portcullis /pɔːtˈkʌlɪs/ *n.* **1** a strong heavy grating sliding up and down in vertical grooves, lowered to block a gateway in a fortress etc. **2** (**Portcullis**) Heraldry one of the four pursuivants of the English College of Arms, with this as a badge. □ **portcullised** *adj.* [ME f. OF *porte coleïce* sliding door f. *porte* door f. L *porta* + *col(e)ïce* fem. of *couleïs* sliding ult. f. L *colare* filter]

Porte /pɔːt/ *n.* (in full **the Sublime** or **Ottoman Porte**) *hist.* the Ottoman court at Constantinople. [F (*la Sublime Porte* = the exalted gate), transl. Turk. title of the central office of the Ottoman government]

porte-cochère /ˌpɔːtkɒˈʃeə(r)/ *n.* **1** a porch large enough for vehicles to pass through, usu. into a courtyard. **2** *US* a roofed structure extending from the entrance of a building over a place where vehicles stop to discharge passengers. [F f. *porte* PORT⁴ + *cochère* (fem. adj.) f. *coche* COACH]

Port Elizabeth a port in South Africa, on the coast of the province of Eastern Cape; pop. (1985) 652,000. Settled by the British in 1820, it is now a motor-manufacturing city and beach resort.

portend /pɔːˈtend/ *v.tr.* **1** foreshadow as an omen. **2** give warning of. [ME f. L *portendere* portent- f. *por-* PRO-¹ + *tendere* stretch]

portent /ˈpɔːtent, -t(ə)nt/ *n.* **1** an omen, a sign of something to come, esp. something of a momentous or calamitous nature. **2** a prodigy; a marvellous thing. [L *portentum* (as PORTEND)]

portentous /pɔːˈtentəs/ *adj.* **1** like or serving as a portent. **2** pompously solemn. □ **portentously** *adv.*

Porter¹ /ˈpɔːtə(r)/, Cole (1892–1964), American songwriter. He made his name with a series of Broadway musicals during and after the 1930s; these include *Anything Goes* (1934) and *Kiss me, Kate* (1948). He also wrote songs for films, including *Rosalie* (1937) and *High Society* (1956). Among his best-known songs are 'Let's Do It', 'Night and Day', and 'Begin the Beguine'.

Porter² /ˈpɔːtə(r)/, Katherine Anne (1890–1980), American short-story writer and novelist. Her collections of short stories include *Pale Horse, Pale Rider* (1939) and *Collected Short Stories* (1965), for which she won a Pulitzer Prize. Her novel *Ship of Fools* (1962) is an allegorical treatment of a voyage from Mexico to Germany during the period of the rise of Nazism.

Porter³ /ˈpɔːtə(r)/, Peter (Neville Frederick) (b.1929), Australian poet, resident chiefly in England since 1951. His early collections, such as *Poems, Ancient and Modern* (1964), provide a sharply satiric portrait of London in the 1960s. His later work became increasingly meditative, complex, and allusive. Other collections include *English Subtitles* (1981) and *The Automatic Oracle* (1987).

porter¹ /ˈpɔːtə(r)/ *n.* **1 a** a person employed to carry luggage etc., esp. a railway, airport, or hotel employee. **b** a hospital employee who moves equipment, trolleys, etc. **2** a dark brown bitter beer brewed from charred or browned malt (apparently orig. made esp. for porters). **3** *US* a sleeping-car attendant. □ **porterage** *n.* [ME f. OF port(e)our f. med.L *portator -oris* f. *portare* carry]

porter² /ˈpɔːtə(r)/ *n.* *Brit.* a gatekeeper or doorkeeper, esp. of a large building. [ME & AF, OF *portier* f. LL *portarius* f. *porta* door]

porterhouse /ˈpɔːtəˌhaʊs/ *n.* esp. *N. Amer.* **1** *hist.* a place at which porter and other drinks were retailed. **2** a place where steaks, chops, etc. were served. □ **porterhouse steak** a thick steak cut from the thick end of a sirloin.

Port Étienne /eɪˈtjen/ the former name for NOUADHIBOU.

portfire /ˈpɔːtˌfaɪə(r)/ *n.* a device for firing rockets, igniting explosives in mining, etc. [after F *porte-feu* f. *porter* carry + *feu* fire]

portfolio /pɔːtˈfəʊlɪəʊ/ *n.* (pl. **-os**) **1** a case for keeping loose sheets of paper, drawings, etc. **2** a range of investments held by a person, a company, etc. **3** the office of a Minister of State (cf. *Minister without Portfolio*). **4** samples of an artist's work. [It. *portafogli* f. *portare* carry + *foglio* leaf f. L *folium*]

Port-Gentil /ˌpɔːʒɒnˈtiː/ the principal port of Gabon, on the Atlantic coast south of Libreville; pop. (1983) 123,300.

Port Harcourt /ˈhɑːkɔːt/ a port in SE Nigeria, on the Gulf of Guinea at the eastern edge of the Niger delta; pop. (1983) 296,200.

Port Hedland /ˈhedlənd/ a seaport on the NW coast of Western Australia; pop. (est. 1987) 13,600. It is linked by rail with the mineral mines to the south near the town of Newman and is important for the export of iron ore.

porthole /ˈpɔːthəʊl/ *n.* **1** an (esp. glazed) aperture in a ship's or aircraft's side for the admission of light. **2** *hist.* an aperture for pointing a cannon through.

portico /ˈpɔːtɪˌkəʊ/ *n.* (pl. **-oes** or **-os**) a colonnade; a roof supported by columns at regular intervals usu. attached as a porch to a building. [It. f. L *porticus* PORCH]

portière /ˌpɔːtɪˈeə(r)/ *n.* a curtain hung over a door or doorway. [F f. *porte* door f. L *porta*]

Porţile de Fier /porˌtsiːlɛ dɛ ˈfjɛr/ the Romanian name for IRON GATE.

portion /ˈpɔːʃ(ə)n/ *n.* & *v.* ● *n.* **1** a part or share. **2** the amount of food allotted to one person. **3** a specified or limited quantity. **4** one's destiny or lot. **5** a dowry. ● *v.tr.* **1** divide (a thing) into portions. **2** (foll. by *out*) distribute. **3** give a dowry to. **4** (foll. by *to*) assign (a thing) to (a person). □ **portionless** *adj.* (in sense 5 of *n.*). [ME f. OF *porcion* portion f. L *portio -onis*]

Portland /ˈpɔːtlənd/ **1** an industrial port in NW Oregon, on the Willamette river near its confluence with the Columbia river; pop. (1990) 437,320. Founded in 1845, it developed as a supply centre in the gold rushes of the 1860s and 1870s, and as a port for the lumber trade. It is now the largest city in Oregon. **2** a shipping forecast area covering the English Channel roughly between the meridians of Start Point in the west and Poole in the east, taking its name from the Isle of Portland.

Portland, Isle of a rocky limestone peninsula on the south coast of England, in Dorset. Its southernmost tip is known as the Bill of Portland or Portland Bill. The peninsula is quarried for its fine building stone.

Portland cement *n.* a cement manufactured from chalk and clay which when hard resembles Portland stone in colour.

Portland stone *n.* a limestone from the Isle of Portland in Dorset, used in building.

Portland vase a Roman vase dating from around the 1st century AD, of dark blue transparent glass with an engraved figured decoration in white opaque glass. Acquired in the 18th century by the Duchess of Portland, it was later housed in the British Museum, where it is today; it was extensively restored after being damaged by an attacker in 1845.

Portlaoise /pɔːtˈliːʃ/ (also **Portlaoighise**) the county town of Laois in the Republic of Ireland; pop. (est. 1990) 9,500. It is the site of a top-security prison.

Port Louis /ˈluːɪs, ˈluːɪ/ the capital of Mauritius, a port on the NW coast; pop. (1991) 143,000.

portly /ˈpɔːtlɪ/ *adj.* (**portlier**, **portliest**) **1** corpulent; stout. **2** *archaic* of a stately appearance. □ **portliness** *n.* [PORT⁵ (in the sense 'bearing') + -LY¹]

Port Mahon see MAHÓN.

portmanteau /pɔːtˈmæntəʊ/ *n.* (pl. **portmanteaus** /-təʊz/ or **portmanteaux**) a leather trunk for clothes etc., opening into two equal parts. □ **portmanteau word** a word blending the sounds and combining the meanings of two others, e.g. *motel*, *Oxbridge*. [F *portmanteau* f. *porter* carry f. L *portare* + *manteau* MANTLE]

Port Moresby /ˈmɔːzbɪ/ the capital of Papua New Guinea, situated on the south coast of the island of New Guinea, on the Coral Sea; pop. (1990) 193,240.

Porto see OPORTO.

Pôrto Alegre /ˌpɔːtu əˈlegreɪ/ a major port and commercial city in SE Brazil, capital of the state of Rio Grande do Sul; pop. (1990) 1,254,640. Situated on the Lagoa dos Patos, a lagoon separated from the Atlantic by a sandy peninsula, it is accessible to ocean-going ships via the port of Rio Grande.

Port of London Authority the corporate body controlling the London harbour and docks.

Port-of-Spain /ˌpɔːtəvˈspeɪn/ the capital of Trinidad and Tobago, a port on the NW coast of the island of Trinidad; pop. (1988) 58,400.

portolan /ˈpɔːtəˌlæn/ *n.* (also **portolano** /ˌpɔːtəˈlɑːnəʊ/) (pl. **portolans** or **portolanos**) *hist.* a book of sailing directions with charts, descriptions of harbours, etc. [It. *portolano* f. *porto* PORT¹]

Porto Novo /ˌpɔːtəʊ ˈnəʊvəʊ/ the capital of Benin, a port on the Gulf of Guinea close to the border with Nigeria; pop. (1982) 208,260. A Portuguese settlement in the 17th century, it became a centre of the Portuguese slave trade.

Pôrto Velho /ˌpɔːtu ˈveljuː/ a town in western Brazil, capital of the state of Rondônia; pop. (1991) 286,000.

Port Petrovsk /pɪˈtrɒfsk/ a former name (until 1922) for MAKHACHKALA.

Port Pirie /ˈpɪrɪ/ a port on the coast of South Australia, on the Spencer Gulf north of Adelaide; pop. (est. 1987) 15,160. Founded as a wheat-trading port in the 1870s, it has since 1898 been a centre for the smelting and refining of ores mined to the north-west around Broken Hill.

portrait /ˈpɔːtrɪt/ n. **1** a representation of a person or animal, esp. of the face, made by drawing, painting, photography, etc. **2** a verbal picture; a graphic description. **3** a person etc. resembling or typifying another (*is the portrait of his father*). **4** (in graphic design etc.) a format in which the height of an illustration etc. is greater than the width (cf. LANDSCAPE *n.* 3). [F, past part. of OF *portraire* PORTRAY]

portraitist /ˈpɔːtrɪtɪst/ n. a person who takes or paints portraits.

portraiture /ˈpɔːtrɪtʃə(r)/ n. **1** the art of painting or taking portraits. **2** graphic description. **3** a portrait. [ME f. OF (as PORTRAIT)]

portray /pɔːˈtreɪ/ v.tr. **1** represent (an object) by a painting, carving, etc; make a likeness of. **2** represent in words; describe graphically. **3** represent dramatically. □ **portrayable** *adj.* **portrayal** *n.* **portrayer** *n.* [ME f. OF *portraire* f. *por-* = PRO-[1] + *traire* draw f. L *trahere*]

Port Said /saɪd/ a port in Egypt, on the Mediterranean coast at the north end of the Suez Canal; pop. (est. 1986) 382,000. It was founded in 1859 at the start of the construction of the Suez Canal.

Port Salut /ˌpɔː səˈluː/ n. a pale mild type of French cheese. [*Port Salut*, a Trappist monastery in NW France where it was first produced]

Portsmouth /ˈpɔːtsməθ/ a port and naval base on the south coast of England, in Hampshire; pop. (1991) 174,700. The naval dockyard was established there in 1496.

Port Stanley see STANLEY[1].

Port Sudan the chief port of Sudan, on the Red Sea; pop. (1983) 206,700.

Port Sunlight a village in Merseyside, on the south bank of the Mersey. Founded and built in the 1880s by Viscount Leverhulme, it provided model housing for the employees of his 'Sunlight' soap factory.

Portugal /ˈpɔːtjʊg(ə)l, ˈpɔːtʃʊ-/ a country occupying the western part of the Iberian peninsula in SW Europe; pop. (est. 1991) 10,393,000; official language, Portuguese; capital, Lisbon. In Roman times the region formed the province of Lusitania; the country's history was linked with that of Spain until it became an independent kingdom under Alfonso I (*c.*1073–1134) in the 12th century. Dynastic disputes with the Spanish kingdoms to the east resulted in the formation of Portugal's long-standing alliance with England in the 14th century, and in the following two hundred years it emerged as one of the leading European colonial powers. Independence was lost to Philip II of Spain in 1580 and not regained until 1688. Portugal became a republic in 1911, after the expulsion of the monarchy. A long period of dictatorship by Antonio Salazar (Prime Minister 1932–68), and his successor Marcello Caetano (1906–80), was ended in 1974 by a coup by army officers opposed to continuing colonial wars in Angola and Mozambique. This led to Portugal's rapid withdrawal from its African colonies and eventually to democratic reform. Portugal became a member of the EEC in 1986.

Portuguese /ˌpɔːtjʊˈgiːz, ˌpɔːtʃʊ-/ n. & adj. ● n. (pl. same) **1 a** a native or national of Portugal. **b** a person of Portuguese descent. **2** the official language of Portugal and its territories, and of Brazil, where it was taken by 15th-century colonists. It is a Romance language, most closely related to (but clearly distinct from) Spanish, and is spoken by about 155 million people. ● adj. of or relating to Portugal or its people or language. □ **Portuguese man-of-war** a large tropical or subtropical marine hydrozoan of the genus *Physalia*, with a float like a sail and stinging tentacles. [Port. *portuguez* f. med.L *portugalensis*]

Port Vila see VILA.

POS abbr. point-of-sale.

pose[1] /pəʊz/ v. & n. ● v. **1** intr. assume a certain attitude of body, esp. when being photographed or being painted for a portrait. **2** intr. (foll. by *as*) set oneself up as or pretend to be (another person etc.) (*posing as a celebrity*). **3** intr. behave affectedly in order to impress others. **4** tr. put forward or present (a question etc.). **5** tr. place (an artist's model etc.) in a certain attitude or position. ● n. **1** an attitude of body or mind. **2** an attitude or pretence, esp. one assumed for effect (*his generosity is a mere pose*). [F *poser* (v.), *pose* (n.) f. LL *pausare* PAUSE: some senses by confusion with L *ponere* place (cf. COMPOSE)]

pose[2] /pəʊz/ v.tr. puzzle (a person) with a question or problem. [obs. *appose* f. OF *aposer* var. of *oposer* OPPOSE]

Poseidon /pɒˈsaɪd(ə)n/ Gk Mythol. the god of the sea, water, earthquakes, and horses, son of Cronus and Rhea and brother of Zeus, identified by the Romans with Neptune. In classical times he was prominent as the sea-god, often depicted with a trident in his hand.

Posen see POZNAŃ.

poser /ˈpəʊzə(r)/ n. **1** = POSEUR. **2** a puzzling question or problem.

poseur /pəʊˈzɜː(r)/ n. (fem. **poseuse** /-ˈzɜːz/) a person who poses for effect or behaves affectedly. [F f. *poser* POSE[1]]

posh /pɒʃ/ adj. & adv. colloq. ● adj. **1** smart; stylish. **2** of or associated with the upper classes (*spoke with a posh accent*). ● adv. in a stylish or upper-class way (*talk posh*; *act posh*). □ **posh up** smarten up. □ **poshly** adv. **poshness** n. [20th c.: perh. f. sl. *posh* a dandy: there is no evidence to support the popular derivation of this word from the initials of *port out starboard home* (referring to the more comfortable accommodation on ships travelling between England and the East)]

posit /ˈpɒzɪt/ v. & n. ● v.tr. (**posited**, **positing**) **1** assume as a fact, postulate. **2** put in place or position. ● n. Philos. a statement which is made on the assumption that it will prove valid. [L *ponere posit-* place]

position /pəˈzɪʃ(ə)n/ n. & v. ● n. **1** a place occupied by a person or thing. **2** the way in which a thing or its parts are placed or arranged (*sitting in an uncomfortable position*). **3** the proper place (*in position*). **4** the state of being advantageously placed (*jockeying for position*). **5** a person's mental attitude; a way of looking at a question (*changed their position on nuclear disarmament*). **6** a person's situation in relation to others (*puts one in an awkward position*). **7** rank or status; high social standing. **8** paid employment; a job. **9** a place where troops etc. are posted for strategical purposes (*the position was stormed*). **10** the configuration of chessmen etc. during a game. **11** a specific pose in ballet etc. (*hold first position*). **12** Logic **a** a proposition. **b** a statement of a proposition. ● v.tr. place in position. □ **in a position to** enabled by circumstances, resources, information, etc. to (do, state, etc.). **position paper** (in business etc.) a written report of attitude or intentions. **position vector** Math. a vector which determines the position of a point. □ **positional** adj. **positionally** adv. **positioner** n. [ME f. OF *position* or L *positio -onis* (as POSIT)]

positive /ˈpɒzɪtɪv/ adj. & n. ● adj. **1** formally or explicitly stated; definite, unquestionable (*positive proof*). **2** (of a person) convinced, confident, or assured in his or her opinion (*positive that I was not there*). **3 a** absolute; not relative. **b** Gram. (of an adjective or adverb) expressing a simple quality without comparison (cf. COMPARATIVE adj. 4, SUPERLATIVE adj. 2). **4** colloq. downright; complete (*it would be a positive miracle*). **5 a** constructive; directional (*positive criticism*; *positive thinking*). **b** favourable; optimistic (*positive reaction*; *positive outlook*). **6** marked by the presence rather than absence of qualities or Med. symptoms (*the test was positive*). **7** esp. Philos. dealing only with matters of fact; practical (cf. POSITIVISM 1). **8** tending in a direction naturally or arbitrarily taken as that of increase or progress (*clockwise rotation is positive*). **9** greater than zero (*positive and negative integers*) (opp. NEGATIVE adj. 5). **10** of, containing, or producing electric charge having the same polarity as that electrode of a voltaic cell from which the current is held to flow (and towards which the actual flow of electrons occurs) (opp. NEGATIVE adj. 6a). (See also ELECTRICITY.) **11** (of a photographic image) showing lights and shades or colours true to the original (opp. NEGATIVE). ● n. a positive adjective, photograph, quantity, etc. □ **positive discrimination** the practice of making distinctions in favour of groups considered to be underprivileged. **positive feedback 1** a constructive response to an experiment, questionnaire, etc. **2** Electronics the return of part of an output signal to the input, tending to increase the amplification etc. **3** esp. Biol. feedback that tends to amplify or enhance the process giving rise to it. **positive geotropism** see GEOTROPISM. **positive pole** the north-seeking pole. **positive ray** Physics a canal ray. **positive sign** = plus sign. **positive vetting** Brit. an exhaustive inquiry into the background and character of a candidate for a post in the Civil Service that involves access to secret material. □ **positively** adv. **positiveness** n. **positivity** /ˌpɒzɪˈtɪvɪtɪ/ n. [ME f. OF *positif -ive* or L *positivus* (as POSIT)]

positivism /ˈpɒzɪtɪˌvɪz(ə)m/ n. **1** the theory (held by Francis Bacon and David Hume amongst others, including Auguste Comte) that every rationally justifiable assertion can be scientifically verified or is capable of logical or mathematical proof, and that philosophy can do no more than attest to the logical and exact use of language through which such observation or verification can be expressed. **2** = LOGICAL POSITIVISM. **3** the theory that laws are to be understood as social rules, valid because they are enacted by authority or derive logically from existing decisions, and that ideal or moral considerations (e.g. that a rule is unjust) should not limit the scope or operation of the law. □ **positivist** n. & adj. **positivistic** /ˌpɒzɪtɪˈvɪstɪk/ adj. **positivistically** adv. [F *positivisme* (as POSITIVE)]

positron /ˈpɒzɪˌtrɒn/ n. Physics a subatomic particle having the same mass as an electron and a numerically equal but positive charge. □ **positronic** /ˌpɒzɪˈtrɒnɪk/ *adj.* [POSITIVE + -TRON]

posology /pəˈsɒlədʒɪ/ *n.* the study of the dosage of medicines. □ **posological** /ˌpɒsəˈlɒdʒɪk(ə)l/ *adj.* [F *posologie* f. Gk *posos* how much]

posse /ˈpɒsɪ/ *n.* **1 a** a strong force or company or assemblage. **b** *esp. US sl.* a gang, esp. a criminal gang. **c** *colloq.* usu. *derog.* a band of persons acting or going about together (*a posse of photographers*). **2** (in full **posse comitatus** /ˌkɒmɪˈtaːtəs, -ˈteɪtəs/) **a** a body of constables, law-enforcers, etc. **b** *esp. US* a body of men summoned by a sheriff etc. to enforce the law. [med.L, = power f. L *posse* be able: *comitatus* = of the county]

possess /pəˈzes/ *v.tr.* **1** hold as property; own. **2** have as a faculty, quality, etc. (*they possess a special value for us*). **3 a** (of a demon etc.) occupy; have power over (a person etc.) (*possessed by the devil*). **b** (of an emotion, infatuation, etc.) dominate, be an obsession of (*possessed by fear*). **4** (also *refl.*; foll. by *in*) maintain (oneself, one's soul, etc.) in a specified state (*possess oneself in patience*). **5** have sexual intercourse with (esp. a woman). □ **be possessed of** own, have. **possess oneself of** take or get for one's own. **what possessed you?** an expression of incredulity. □ **possessor** *n.* **possessory** *adj.* [OF *possesser* f. L *possidere* *possess-* f. *potis* able + *sedere* sit]

possession /pəˈzeʃ(ə)n/ *n.* **1** the act or state of possessing or being possessed. **2 a** the thing possessed. **b** a foreign territory subject to a state or ruler. **3** the act or state of actual holding or occupancy. **4** *Law* power or control similar to lawful ownership but which may exist separately from it (*prosecuted for possession of narcotic drugs*). **5** (in *pl.*) property, wealth, subject territory, etc. **6** *Football* etc. temporary control of the ball by a particular player. □ **in possession 1** (of a person) possessing. **2** (of a thing). **in possession of 1** having in one's possession. **2** maintaining control over (*in possession of one's wits*). **in the possession of** held or owned by. **possession order** an order made by a court directing that possession of a property be given to the owner. **take possession** (often foll. by *of*) become the owner or possessor (of a thing). □ **possessionless** *adj.* [ME f. OF *possession* or L *possessio -onis* (as POSSESS)]

possessive /pəˈzesɪv/ *adj. & n.* ● *adj.* **1** showing a desire to possess or retain what one already owns. **2** showing jealous and domineering tendencies towards another person. **3** *Gram.* indicating possession. ● *n.* (in full **possessive case**) *Gram.* the case of nouns and pronouns expressing possession. □ **possessive pronoun** each of the pronouns indicating possession (*my, your, his, their,* etc.) or the corresponding absolute forms (*mine, yours, his, theirs,* etc.). □ **possessively** *adv.* **possessiveness** *n.* [L *possessivus* (as POSSESS), transl. Gk *ktētikē* (*ptōsis* case)]

posset /ˈpɒsɪt/ *n. hist.* a drink made of hot milk curdled with ale, wine, etc., often flavoured with spices, formerly used as a remedy for colds etc. [ME *poshote*: orig. unkn.]

possibility /ˌpɒsɪˈbɪlɪtɪ/ *n.* (*pl.* **-ies**) **1** the state or fact of being possible, or an occurrence of this (*outside the range of possibility; saw no possibility of going away*). **2** a thing that may exist or happen (*there are three possibilities*). **3** (usu. in *pl.*) the capability of being used, improved, etc.; the potential of an object or situation (esp. *have possibilities*). [ME f. OF *possibilité* or LL *possibilitas -tatis* (as POSSIBLE)]

possible /ˈpɒsɪb(ə)l/ *adj. & n.* ● *adj.* **1** capable of existing or happening; that may be managed, achieved, etc. (*came as early as possible; did as much as possible*). **2** that is likely to happen etc. (*few thought their victory possible*). **3** acceptable; potential (*a possible way of doing it*). ● *n.* **1** a possible candidate, member of a team, etc. **2** (prec. by *the*) whatever is likely, manageable, etc. **3** the highest possible score, esp. in shooting etc. [ME f. OF *possible* or L *possibilis* f. *posse* be able]

possibly /ˈpɒsɪblɪ/ *adv.* **1** perhaps. **2** in accordance with possibility (*cannot possibly refuse*).

possum /ˈpɒsəm/ *n.* **1** *colloq.* = OPOSSUM 1. **2** *Austral.* a phalanger or other tree-living marsupial. □ **play possum 1** pretend to be asleep or unconscious when threatened. **2** feign ignorance. [abbr.]

post[1] /pəʊst/ *n. & v.* ● *n.* **1** a long stout piece of timber or metal set upright in the ground etc.: **a** to support something, esp. in building. **b** to mark a position, boundary, etc. **c** to carry notices. **2** a pole etc. marking the start or finish of a race. ● *v.tr.* **1** (often foll. by *up*) **a** attach (a paper etc.) in a prominent place; stick up (*post no bills*). **b** announce or advertise by placard or in a published text. **2** publish the name of (a ship etc.) as overdue or missing. **3** placard (a wall etc.) with bills etc. **4** *N. Amer.* achieve (a score in a game etc.). □ **post-mill** a windmill pivoted on a post and turning to catch the wind. [OE f. L *postis*: in ME also f. OF etc.]

post[2] /pəʊst/ *n., v., & adv.* ● *n.* **1** *Brit.* the official conveyance of parcels, letters, etc. (*send it by post*). (*See note below.*) **2** *Brit.* a single collection, dispatch, or delivery of these; the letters etc. dispatched (*has the post arrived yet?*). **3** *Brit.* a place where letters etc. are dealt with; a post office or postbox (*take it to the post*). **4** *hist.* **a** one of a series of couriers who carried mail on horseback between fixed stages. **b** a letter-carrier; a mail cart. ● *v.* **1** *tr.* put (a letter etc.) in the post. **2** *tr.* (esp. as **posted** *adj.*) supply a person with information (*keep me posted*). **3** *tr.* **a** enter (an item) in a ledger. **b** (often foll. by *up*) complete (a ledger) in this way. **c** carry (an entry) from an auxiliary book to a more formal one, or from one account to another. **4** *intr.* **a** travel with haste, hurry. **b** *hist.* travel with relays of horses. ● *adv.* express; with haste. □ **post-chaise** *hist.* a travelling carriage hired from stage to stage or drawn by horses hired in this manner. **post exchange** *US Mil.* a shop at a military camp etc. **post-free** *Brit.* carried by post free of charge or with the postage prepaid. **post-haste** with great speed. **post-horn** *hist.* a valveless horn formerly used to announce the arrival of the post. **post office 1** (**Post Office**) the public department or corporation responsible for postal services and (in some countries) telecommunications. **2 a** a room or building where postal business is carried on. **b** *US* = *postman's knock*. **post-office box** a private box or pigeon-hole at a post office, in which mail for an individual or firm is put and kept until collected. **post-paid** on which postage has been prepaid. **post room** the department of a company that deals with incoming and outgoing mail. **post-town** a town with a post office, esp. one that is not a sub-office of another. [F *poste* (fem.) f. It. *posta* ult. f. L *ponere posit-* place]

▪ The term *post* was applied from the beginning of the 16th century to men with horses stationed or posted at suitable distances along main routes, the duty of each being to take or despatch to the next stage the monarch's mail, and subsequently also other people's letters, as well as to provide horses for use in this. Corresponding terms in French and Italian are used by Marco Polo of the stations, 40 kilometres (25 miles) apart, at which the messengers of the emperor of China changed horses. In 18th-century England stagecoaches carried the mail, succeeded in the mid-19th century by the railways. In 1840 the 'penny post' was introduced, and similar developments in other countries led to the establishment of a Postal Union in 1874 which stimulated the development of international mail services. The first regular airmail service (London–Paris) was introduced in 1919. Since the 1960s there have been improvements in sorting in Britain and elsewhere with the introduction of postcodes and electronic technology. (See also PENNY POST, STAMP.)

post[3] /pəʊst/ *n. & v.* ● *n.* **1** a place where a soldier is stationed or which he patrols. **2** a place of duty. **3 a** a position taken up by a body of soldiers. **b** a force occupying this. **c** a fort. **4** a situation, paid employment. **5** = *trading post.* **6** *Naut. hist.* a commission as an officer in command of a vessel of twenty guns or more. ● *v.tr.* **1** place or station (soldiers, an employee, etc.). **2** appoint to a post or command. □ **first post** *Brit.* the earliest of several bugle-calls giving notice of the hour of retiring at night. **last post** *Brit.* the final such bugle-call, also blown at military funerals etc. [F *poste* (masc.) f. It. *posto* f. Rmc *postum* (unrecorded) f. L *ponere posit-* place]

post- /pəʊst/ *prefix* after in time or order. [from or after L *post* (adv. & prep.)]

postage /ˈpəʊstɪdʒ/ *n.* the amount charged for sending a letter etc. by post, usu. prepaid in the form of a stamp (*£25 including postage and packing*). □ **postage meter** *N. Amer.* a franking-machine. **postage stamp** an official stamp affixed to or imprinted on a letter etc. indicating the amount of postage paid (see STAMP *n.* 3).

postal /ˈpəʊst(ə)l/ *adj. & n.* ● *adj.* **1** of the post. **2** by post (*postal vote*). ● *n. US* a postcard. □ **postal card** *US* = POSTCARD. **postal code** = POSTCODE. **postal meter** a franking-machine. **postal note** *Austral. & NZ* = *postal order.* **postal order** a money order issued by the Post Office, payable to a specified person. □ **postally** *adv.* [F (*poste* POST[2])]

Postal Union a union of the governments of various countries for the regulation of international postage.

postbag /ˈpəʊstbæg/ *n. Brit.* = MAILBAG.

postbox /ˈpəʊstbɒks/ *n. Brit.* a public box in which mail is posted.

postcard /ˈpəʊstkaːd/ *n.* a card, often with a photograph or picture on one side, for sending a short message by post without an envelope.

post-classical /pəʊstˈklæsɪk(ə)l/ *adj.* (esp. of Greek and Roman literature) later than the classical period.

postcode /ˈpəʊstkəʊd/ *n.* a group of figures or letters and figures which are added to a postal address to assist sorting.

post-coital /ˌpəʊst'kəʊɪt(ə)l/ adj. occurring or existing after sexual intercourse. □ **post-coitally** adv.

post-colonial /ˌpəʊstkə'ləʊnɪəl/ adj. occurring or existing after the end of colonial rule.

postdate v. & n. ● v.tr. /pəʊst'deɪt/ affix or assign a date later than the actual one to (a document, event, etc.). ● n. /'pəʊstdeɪt/ such a date.

post-doctoral /pəʊst'dɒktərəl/ adj. of or relating to research undertaken after the completion of doctoral research.

post-entry /pəʊst'entrɪ/ n. (pl. **-ies**) a late or subsequent entry, esp. in a race or in bookkeeping.

poster /'pəʊstə(r)/ n. **1** a placard in a public place. **2** a large printed picture. **3** a billposter. □ **poster paint** a gummy opaque paint.

poste restante /pəʊst 'restɒnt, re'stɒnt/ n. **1** a direction on a letter to indicate that it should be kept at a specified post office until collected by the addressee. **2** the service department in a post office keeping such letters. [F, = letter(s) remaining]

posterior /pɒ'stɪərɪə(r)/ adj. & n. ● adj. **1** later; coming after in series, order, or time. **2** situated at the back. ● n. the buttocks. □ **posteriorly** adv. **posteriority** /-ˌstɪərɪ'ɒrɪtɪ/ n. [L, compar. of posterus following f. post after]

posterity /pɒ'sterɪtɪ/ n. **1** all succeeding generations. **2** the descendants of a person. [ME f. OF posterité f. L posteritas -tatis f. posterus: see POSTERIOR]

postern /'pɒstən, 'pəʊst-/ n. **1** a back door. **2** a side way or entrance. [ME f. OF posterne, posterle, f. LL posterula dimin. of posterus: see POSTERIOR]

post-feminist /pəʊst'femɪnɪst/ adj. & n. ● adj. of or relating to ideas, attitudes, etc. which ignore or reject the feminist ideas of the 1960s and subsequent decades. ● n. a person holding such ideas or attitudes. □ **post-feminism** n.

postfix n. & v. ● n. /'pəʊstfɪks/ a suffix. ● v.tr. /pəʊst'fɪks/ append (letters) at the end of a word.

postglacial /pəʊst'gleɪʃ(ə)l, -'gleɪsɪəl/ adj. & n. ● adj. formed or occurring after a glacial period. ● n. a postglacial period or deposit.

postgraduate /pəʊst'grædjʊət/ adj. & n. ● adj. **1** (of a course of study) carried on after taking a first degree. **2** of or relating to students following this course of study (postgraduate accommodation). ● n. a postgraduate student.

posthumous /'pɒstjʊməs/ adj. **1** occurring after death. **2** (of a child) born after the death of its father. **3** (of a book etc.) published after the author's death. □ **posthumously** adv. [L postumus last (superl. f. post after): in LL posth- by assoc. with humus ground]

postiche /pɒ'stiːʃ/ n. a coil of false hair, worn as an adornment. [F, = false, f. It. posticcio]

postie /'pəʊstɪ/ n. colloq. a postman or postwoman. [abbr.]

postil /'pɒstɪl/ n. hist. **1** a marginal note or comment, esp. on a text of Scripture. **2** a commentary. [ME f. OF postille f. med.L postilla, of uncert. orig.]

postilion /pɒ'stɪljən/ n. (also **postillion**) the rider on the near (left-hand side) horse drawing a coach etc. when there is no coachman. [F postillon f. It. postiglione post-boy f. posta POST²]

post-impressionism /ˌpəʊstɪm'preʃəˌnɪz(ə)m/ n. the work or style of a group of late 19th and early 20th-century painters including Van Gogh, Gauguin, and Cézanne, a term applied retrospectively after an exhibition in London in 1910. These painters moved away from the naturalism of impressionism, tending towards a concern with form and its analysis or simplification (as Cézanne, for example), and also towards a more symbolic and expressionistic content (as Van Gogh). □ **post-impressionist** n. **post-impressionistic** /-ˌpreʃə'nɪstɪk/ adj.

post-industrial /ˌpəʊstɪn'dʌstrɪəl/ adj. relating to or characteristic of a society or economy which no longer relies on heavy industry.

postliminy /pəʊst'lɪmɪnɪ/ n. **1** (in international law) the restoration to their former status of persons and things taken in war. **2** (in Roman law) the right of a banished person or captive to resume civic privileges on return from exile. [L postliminium (as POST-, limen liminis threshold)]

postlude /'pəʊstluːd, -ljuːd/ n. Mus. a concluding voluntary. [POST-, after PRELUDE]

postman /'pəʊstmən/ n. (pl. **-men**; fem. **postwoman**, pl. **-women**) a person who is employed to deliver and collect letters etc. □ **postman's knock** Brit. a game in which imaginary letters are delivered in exchange for kisses.

postmark /'pəʊstmɑːk/ n. & v. ● n. an official mark stamped on a letter, esp. one giving the place, date, etc. of dispatch or arrival, and serving to cancel the stamp. ● v.tr. mark (an envelope etc.) with this.

postmaster /'pəʊstˌmɑːstə(r)/ n. a man in charge of a post office. □ **postmaster general** the head of a country's postal service. ¶ The office was abolished in the UK in 1969.

post-millennial /ˌpəʊstmɪ'lenɪəl/ adj. following the millennium.

post-millennialism /ˌpəʊstmɪ'lenɪəˌlɪz(ə)m/ n. the doctrine that a second Advent will follow the millennium. □ **post-millennialist** n.

postmistress /'pəʊstˌmɪstrɪs/ n. a woman in charge of a post office.

post-modern /pəʊst'mɒd(ə)n/ adj. (in literature, architecture, the arts, etc.) of or relating to post-modernism.

post-modernism /pəʊst'mɒdəˌnɪz(ə)m/ n. a late 20th-century style and concept in the arts, architecture, and criticism, which represents a departure from modernism and has at its heart a general distrust of grand theories and ideologies as well as a problematical relationship with any notion of 'art'. Typical features include a deliberate mixing of different artistic styles and media, the self-conscious use of earlier styles and conventions, and often the incorporation of images relating to the consumerism and mass communication of late 20th-century post-industrial society. Post-modernist architecture was pioneered by Robert Venturi; the AT&T skyscraper in New York (completed in 1984) is a prime example of the style. Influential literary critics include Jean Baudrillard (b.1929) and Jean-François Lyotard (b.1924). □ **post-modernist** adj. & n.

post-mortem /pəʊst'mɔːtəm/ n., adv., & adj. ● n. **1** (in full **post-mortem examination**) an examination made after death, esp. to determine its cause. **2** colloq. a discussion analysing the course and result of a game, election, etc. ● adv. & adj. after death. [L]

postnatal /pəʊst'neɪt(ə)l/ adj. characteristic of or relating to the period after childbirth. □ **postnatal depression** depression suffered by a mother in the period after a birth.

post-nuptial /pəʊst'nʌpʃ(ə)l/ adj. after marriage.

post-obit /pəʊst'əʊbɪt/ n. & adj. ● n. a bond given to a lender by a borrower securing a sum for payment on the death of another person from whom the borrower expects to inherit. ● adj. taking effect after death. [L post obitum f. post after + obitus decease f. obire die]

postoperative /pəʊst'ɒpərətɪv/ adj. relating to or occurring in a period after a surgical operation.

post-partum /pəʊst'pɑːtəm/ adj. following parturition.

postpone /pəʊst'pəʊn, pə'spəʊn/ v.tr. cause or arrange (an event etc.) to take place at a later time. □ **postponable** adj. **postponement** n. **postponer** n. [L postponere (as POST-, ponere posit- place)]

postposition /ˌpəʊstpə'zɪʃ(ə)n/ n. **1** a word or particle, esp. an enclitic, placed after the word it modifies, e.g. -ward in homeward and at in the books we looked at. **2** the use of a postposition. □ **postpositional** adj. & n. **postpositive** /-'pɒzɪtɪv/ adj. & n. **postpositively** adv. [LL postpositio (as POSTPONE)]

postprandial /pəʊst'prændɪəl/ adj. formal or joc. after dinner or lunch. [POST- + L prandium a meal]

postscript /'pəʊstskrɪpt/ n. **1** an additional paragraph or remark, usu. at the end of a letter after the signature and introduced by 'PS'. **2** any additional information, action, etc. [L postscriptum neut. past part. of postscribere (as POST-, scribere write)]

post-structuralism /pəʊst'strʌktʃərəˌlɪz(ə)m/ n. a school of thought representing an extension and critique of structuralism, which emerged in French intellectual life in the late 1960s and early 1970s. It embraced various standpoints, including the philosophical deconstruction of Jacques Derrida and the later work of Roland Barthes, the psychoanalytic theories of Jacques Lacan and Julia Kristeva (b.1941), the historical critiques of Michel Foucault, and the writings in culture and politics of figures such as Jean-François Lyotard (b.1924) and Jean Baudrillard (b.1929). It departed from the claims to objectivity and comprehensiveness made by structuralism and emphasized instead plurality and deferral of meaning, rejecting the fixed binary oppositions of structuralism and the validity of authorial authority. After initially being influential in France, it subsequently became important in Britain, the US, and elsewhere in the late 1970s and 1980s. □ **post-structural** adj. **post-structuralist** adj. & n.

post-tax /pəʊst'tæks/ adj. (of income) after the deduction of taxes.

post-traumatic stress disorder /ˌpəʊsttrɔː'mætɪk/ n. (abbr. **PTSD**) (also **post-traumatic stress syndrome**) Med. a condition of mental stress that sometimes follows injury or psychological shock, characterized by withdrawal and anxiety, and a tendency to physical illness.

postulant /'pɒstjʊlənt/ n. a candidate, esp. for admission into a religious order. [F postulant or L postulans -antis (as POSTULATE)]

postulate v. & n. ● v.tr. /'pɒstjʊˌleɪt/ **1** (often foll. by that + clause) assume as a necessary condition, esp. as a basis for reasoning; take for granted. **2** claim. **3** (in ecclesiastical law) nominate or elect to a higher rank. ● n. /'pɒstjʊlət/ **1** a thing postulated. **2** a fundamental prerequisite or condition. **3** Math. an assumption used as a basis for mathematical reasoning. □ **postulation** /ˌpɒstjʊ'leɪʃ(ə)n/ n. [L postulare postulat-demand]

postulator /'pɒstjʊˌleɪtə(r)/ n. **1** a person who postulates. **2** RC Ch. a person who presents a case for canonization or beatification.

posture /'pɒstʃə(r)/ n. & v. ● n. **1** the relative position of parts, esp. of the body (in a reclining posture). **2** carriage or bearing (improved by good posture and balance). **3** a mental or spiritual attitude or condition. **4** the condition or state (of affairs etc.) (in more diplomatic postures). ● v. **1** intr. assume a mental or physical attitude, esp. for effect (inclined to strut and posture). **2** tr. pose (a person). □ **postural** adj. **posturer** n. [F f. It. postura f. L positura f. ponere posit- place]

postwar /pəʊst'wɔː(r)/ adj. occurring or existing after a war (esp. the most recent major war).

posy /'pəʊzɪ/ n. (pl. -ies) **1** a small bunch of flowers. **2** archaic a short motto, line of verse, etc., inscribed within a ring. □ **posy-ring** a ring with this inscription. [alt. f. POESY]

pot[1] /pɒt/ n. & v. ● n. **1 a** a vessel, usu. rounded, of ceramic ware or metal or glass for holding liquids or solids or for cooking in. **b** a particular type of vessel designed for holding coffee, tea, jam, glue, etc. (coffee pot; teapot). **2 a** = FLOWERPOT. **b** = chimney-pot. **c** = lobster-pot. **d** = chamber-pot, POTTY[2]. **3** a drinking-vessel of pewter etc. **4** the contents of a pot (ate a whole pot of jam). **5** the total amount of the bet in a game etc. **6** colloq. a large sum (pots of money). **7** sl. a vessel given as a prize in an athletic contest, esp. a silver cup. **8** = pot-belly. ● v.tr. (**potted**, **potting**) **1** place in a pot. **2** (usu. as **potted** adj.) preserve in a sealed pot (potted shrimps). **3** sit (a young child) on a chamber-pot. **4** pocket (a ball) in billiards etc. **5** shoot at, hit, or kill (an animal) with a pot-shot. **6** seize or secure. **7** (esp. as **potted** adj.) abridge or epitomize (in a potted version; potted wisdom). □ **go to pot** colloq. deteriorate; be ruined. **pot-bellied** having a pot-belly. **pot-belly** (pl. -ies) **1** a protruding stomach. **2** a person with this. **3** a small bulbous stove. **pot-boiler 1** a work of literature or art done merely to make the writer or artist a living. **2** a writer or artist who does this. **pot-bound** (of a plant) having roots which fill the flowerpot, leaving no room to expand. **pot cheese** US cottage cheese. **pot-herb** any herb grown in a kitchen garden. **pot-hook 1** a hook over a hearth for hanging a pot etc. on, or for lifting a hot pot. **2** a curved stroke in handwriting, esp. as made in learning to write. **pot-hunter 1** a person who hunts for game at random. **2** a person who takes part in a contest merely for the sake of the prize. **pot luck** whatever is available. **pot of gold** an imaginary reward; an ideal; a jackpot. **pot pie** a pie of meat etc. or fruit with a crust baked in a pot. **pot plant** a plant grown in a flowerpot. **pot roast** a piece of meat cooked slowly in a covered dish. **pot-roast** v.tr. cook (a piece of meat) in this way. **pot-shot 1** a random shot. **2** a shot aimed at an animal etc. within easy reach. **3** a shot at a game bird etc. merely to provide a meal. **pot-valiant** courageous because of drunkenness. **pot-valour** this type of courage. **put a person's pot on** Austral. & NZ sl. inform on a person. □ **potful** n. (pl. -fuls). [OE pott, corresp. to OFris., MDu., MLG pot, f. pop.L]

pot[2] /pɒt/ n. sl. marijuana. □ **pot-head** a person who smokes this. [prob. f. Mex. Sp. potiguaya]

pot[3] /pɒt/ n. & v. Austral. & NZ ● n. a dropped goal in rugby. ● v.tr. (**potted**, **potting**) score (a dropped goal). [perh. f. pot-shot]

potable /'pəʊtəb(ə)l/ adj. drinkable. □ **potability** /ˌpəʊtə'bɪlɪtɪ/ n. [F potable or LL potabilis f. L potare drink]

potage /pɒ'tɑːʒ/ n. thick soup. [F (as POTTAGE)]

potamic /pə'tæmɪk/ adj. of rivers. □ **potamology** /ˌpɒtə'mɒlədʒɪ/ n. [Gk potamos river]

potash /'pɒtæʃ/ n. an alkaline potassium compound, usu. potassium carbonate or hydroxide. [17th-c. pot-ashes f. Du. pot-asschen (as POT[1], ASH[1]): orig. obtained by leaching vegetable ashes and evaporating the solution in iron pots]

potassium /pə'tæsɪəm/ n. a soft silver-white reactive metallic chemical element (atomic number 19; symbol **K**). (See note below.) □ **potassium-argon dating** a method of dating rocks from the relative proportions of radioactive potassium-40 and its decay product, argon-40. **potassium chloride** a white crystalline solid (chem. formula: KCl) used as a fertilizer and in photographic processing. **potassium cyanide** a highly toxic solid (chem. formula: KCN) that releases poisonous hydrogen cyanide gas when hydrolysed. **potassium iodide** a white crystalline solid (chem. formula: KI) used as an additive in table salt to prevent iodine deficiency. **potassium permanganate** a purple crystalline solid (chem. formula: $KMnO_4$) that is used in solution as an oxidizing agent and disinfectant. [POTASH]

▪ Potassium was first isolated by Sir Humphry Davy in 1807, and is a member of the alkali-metal group. Its compounds occur widely in minerals and in sea water. The metal itself has few uses, but potassium compounds are widely used, for example in the manufacture of fertilizers, soaps, and glass. In the body, potassium ions are involved in the transmission of nerve impulses and are an important constituent of the fluid within cells. The symbol K is from the Latin name kalium.

potation /pə'teɪʃ(ə)n/ n. **1** a drink. **2** the act or an instance of drinking. **3** (usu. in pl.) the act or an instance of tippling. □ **potatory** /'pəʊtətərɪ/ adj. [ME f. OF potation or L potatio f. potare drink]

potato /pə'teɪtəʊ/ n. (pl. -oes) **1** a starchy plant tuber that is cooked and used for food. **2** the solanaceous plant, Solanum tuberosum, bearing this. (See note below.) **3** colloq. a hole in (esp. the heel of) a sock or stocking. □ **potato chip** = CHIP n. 3. **potato crisp** Brit. = CRISP n. 1. [Sp. patata var. of Taino batata]

▪ The potato was first cultivated in the Andes of Peru and Bolivia about 1,800 years ago, and was introduced into Europe c.1570 by the Spaniards. It had become a major crop in Europe by the end of the 18th century, and it is now one of the world's main food crops.

pot-au-feu /ˌpɒtəʊ'fɜː/ n. (pl. same) **1** a French soup of usu. boiled beef and vegetables cooked in a large pot. **2** the broth from this. [F, = pot on the fire]

potch /pɒtʃ/ n. an opal of inferior quality. [19th c.: orig. unkn.]

poteen /pɒ'tiːn/ n. (also **potheen** /pɒ'θiːn, -'tʃiːn/) Ir. alcohol made illicitly, usu. from potatoes. [Ir. poitín dimin. of pota POT[1]]

Potemkin /pə'temkɪn/ a battleship whose crew mutinied in the Russian Revolution of 1905 when in the Black Sea, bombarding Odessa before seeking asylum in Romania. The incident, commemorated in Eisenstein's 1925 film The Battleship Potemkin, persuaded the tsar to agree to a measure of reform.

potent[1] /'pəʊt(ə)nt/ adj. **1** powerful; strong. **2** (of a reason) cogent; forceful. **3** (of a male) capable of sexual erection or orgasm. **4** literary mighty. □ **potence** n. **potency** n. **potently** adv. [L potens -entis pres. part. of posse be able]

potent[2] /'pəʊt(ə)nt/ adj. & n. Heraldry ● adj. **1** with a crutch-head shape. **2** (of a fur) formed by a series of such shapes. ● n. this fur. [ME f. OF potence crutch f. L potentia power (as POTENT[1])]

potentate /'pəʊtənˌteɪt/ n. a monarch or ruler. [ME f. OF potentat or L potentatus dominion (as POTENT[2])]

potential /pə'tenʃ(ə)l/ adj. & n. ● adj. capable of coming into being or action; latent. ● n. **1** the capacity for use or development; possibility (achieved its highest potential). **2** usable resources. **3** Physics the quantity determining the energy of mass in a gravitational field or of charge in an electric field. □ **potential barrier** Physics a region of high potential impeding the movement of particles etc. **potential difference** Physics the difference of electric potential between two points. **potential energy** Physics a body's ability to do work by virtue of its position relative to others, stresses within itself, electric charge, etc. □ **potentialize** v.tr. (also **-ise**). **potentially** adv. **potentiality** /-ˌtenʃɪ'ælɪtɪ/ n. [ME f. OF potencial or LL potentialis f. potentia (as POTENT[1])]

potentiate /pə'tenʃɪˌeɪt/ v.tr. **1** make more powerful, esp. increase the effectiveness of (a drug). **2** make possible. [as POTENT[1] after SUBSTANTIATE]

potentilla /ˌpəʊtən'tɪlə/ n. an ornamental herbaceous plant or shrub of the genus Potentilla, of the rose family, with usu. yellow flowers. (See also CINQUEFOIL.) [med.L, dimin. of L potens POTENT[1]]

potentiometer /pəˌtenʃɪ'ɒmɪtə(r)/ n. an instrument for measuring or adjusting small electrical potentials. □ **potentiometry** n. **potentiometric** /-ʃɪə'metrɪk/ adj.

Potenza /pə'tenzə/ a market town in southern Italy, capital of Basilicata region; pop. (1990) 68,500. Founded by the Romans in the 2nd century BC, it was taken by the Lombards in the 6th century AD.

potheen var. of POTEEN.

pother /ˈpɒðə(r)/ n. & v. literary ● n. a noise; commotion; fuss. ● v. **1** tr. fluster, worry. **2** intr. make a fuss. [16th c.: orig. unkn.]

pothole /ˈpɒthəʊl/ n. & v. ● n. **1** a deep hole or system of caves and underground river beds formed by the erosion of rock esp. by the action of water. **2** a deep hole in the ground or a river bed. **3** a hole in a road surface caused by wear or subsidence. ● v. intr. Brit. explore pot-holes as a leisure activity. □ **potholed** adj. **potholer** n. **potholing** n.

potion /ˈpəʊʃ(ə)n/ n. a liquid medicine, drug, poison, etc. [ME f. OF f. L potio -onis f. potus having drunk]

Potiphar /ˈpɒtɪfə(r)/ (in the Bible) an Egyptian officer whose wife tried to seduce Joseph and then falsely accused him of attempting to rape her (Gen. 39).

potlatch /ˈpɒtlætʃ/ n. (among some North American Indians) a ceremonial feast at which presents are given away or property is destroyed in order to enhance status. [Chinook f. Nootka patlatsh gift]

Potomac /pəˈtəʊmək/ a river of the eastern US, which rises in the Appalachian Mountains in West Virginia and flows about 459 km (285 miles) through Washington, DC, into Chesapeake Bay on the Atlantic coast.

potoroo /ˌpɒtəˈruː/ n. a rat kangaroo, esp. of the genus Potorous, native to Australia and Tasmania. [Aboriginal, prob. f. Dharuk badaru]

Potosí /ˌpɒtəʊˈsiː/ a city in southern Bolivia; pop. (1990) 120,000. It was founded by the Spanish in the 1540s as a silver-mining town and is now a major centre of tin, lead, and copper mining. Situated at an altitude of about 4,205 m (13,780 ft), it is one of the highest cities in the world.

pot-pourri /pəʊˈpʊərɪ, ˌpəʊpʊəˈriː/ n. (pl. **pot-pourris**) **1** a mixture of dried petals and spices used to perfume a room etc. **2** a musical or literary medley. [F, = rotten pot]

potrero /pəˈtreərəʊ/ n. (pl. **-os**) **1** (in the south-western US and South America) a paddock or pasture for horses or cattle. **2** (in the south-western US) a narrow steep-sided plateau. [Sp. f. potro colt, pony]

Potsdam /ˈpɒtsdæm/ a city in eastern Germany, the capital of Brandenburg, situated just south-west of Berlin on the Havel river; pop. (est. 1990) 95,000. The former summer residence of the Prussian royal family, it is the site of the rococo Sans Souci palace built for Frederick II between 1745 and 1747. In 1945 it was the venue for the Potsdam Conference.

Potsdam Conference a meeting held in the summer of 1945 between US, Soviet, and British leaders, which established principles for the Allied occupation of Germany following the end of the Second World War. From this conference an ultimatum was sent to Japan demanding unconditional surrender.

potsherd /ˈpɒtʃɜːd/ n. a broken piece of ceramic material, esp. one found on an archaeological site.

pottage /ˈpɒtɪdʒ/ n. archaic soup, stew. [ME f. OF potage (as POT[1])]

Potter[1] /ˈpɒtə(r)/, Dennis (Christopher George) (1935–94), English television dramatist. He began suffering from a crippling form of psoriasis in the 1960s, after which he wrote his most acclaimed works, the series Pennies from Heaven (1978) and The Singing Detective (1986). Both use popular songs of the 1920s and 1930s to contrast the humdrum or painful realities of everyday life with the imagination's capacity for hope and self-delusion. Other plays include Blue Remembered Hills (1979), in which adults played the parts of children, and the controversial Brimstone and Treacle (1976).

Potter[2] /ˈpɒtə(r)/, (Helen) Beatrix (1866–1943), English writer for children. She is known for her series of animal stories, illustrated with her own delicate watercolours, which began with The Tale of Peter Rabbit (first published privately in 1900).

potter[1] /ˈpɒtə(r)/ v. (US **putter** /ˈpʌt-/) **1** intr. **a** (often foll. by about, around) work or occupy oneself in a desultory but pleasant manner (likes pottering about in the garden). **b** (often foll. by at, in) dabble in a subject or occupation. **2** intr. go slowly, dawdle, loiter (pottered up to the pub). **3** tr. (foll. by away) fritter away (one's time etc.). □ **potterer** n. [frequent. of dial. pote push f. OE potian]

potter[2] /ˈpɒtə(r)/ n. a maker of ceramic vessels. [OE pottere (as POT[1])]

Potteries, the /ˈpɒtərɪz/ a district in Staffordshire, including Stoke-on-Trent, where the English pottery industry is centred.

potter's field n. a burial place for paupers, strangers, etc. (after Matt. 27:7).

potter's wheel n. a horizontal revolving disc to carry clay for making pots. At first such a wheel was turned by hand, later by a foot-operated

wheel. The wheel probably came into use during the 4th millennium BC, spreading from Mesopotamia to Egypt and India and reaching Crete and China by about 2000 BC, Europe at varying dates, and southern Britain by the mid-1st century BC. It was unknown in America until the arrival of European conquerors and settlers.

pottery /ˈpɒtərɪ/ n. (pl. **-ies**) **1** vessels etc. made of fired clay. (See note below.) **2** a potter's work. **3** a potter's workshop. [ME f. OF poterie f. potier POTTER[2]]

▪ The shaping and baking of clay vessels is among the oldest and most widely practised of all the crafts: earthenware pottery dating from about 9,000 years ago has been found on the Anatolian plateau of Turkey. At first it was shaped entirely by hand, and many centuries passed before the introduction of the potter's wheel. The brilliant achievements of the Chinese and their neighbours, especially in the field of glazed stoneware and porcelain, were almost unknown in Europe until the opening up of the direct sea route in the 16th century, though these wares were not without influence on the Islamic pottery of the Near East. In medieval times pottery was little used at table but by the 18th century there was an immense demand for table and ornamental wares of all kinds, and new factories appeared in many countries. Notable were those at Sèvres in France and Meissen in Germany, while Josiah Wedgwood's establishment of a factory in Staffordshire led to that area becoming the pottery centre of Britain. The Industrial Revolution saw technical and practical innovations which culminated in the change from craft to industry, although the craft of the individual potter has recently undergone something of a revival. The main divisions in pottery are often given as earthenware, porcelain, and stoneware.

potting shed n. a building in which plants are potted and tools etc. are stored.

pottle /ˈpɒt(ə)l/ n. **1** a small punnet or carton for strawberries etc. **2** archaic **a** a measure for liquids; a half gallon. **b** a pot etc. containing this. [ME f. OF potel (as POT[1])]

potto /ˈpɒtəʊ/ n. (pl. **-os**) a small slow-climbing primate, Perodicticus potto, native to West Africa. [perh. f. Guinea dial.]

Pott's fracture /pɒts/ n. a fracture of the lower end of the fibula, usu. with dislocation of the ankle. [Sir Percivall Pott, Engl. surgeon (1713–88)]

potty[1] /ˈpɒtɪ/ adj. (**pottier, pottiest**) Brit. sl. **1** foolish or crazy. **2** insignificant, trivial (esp. potty little). □ **pottiness** n. [19th c.: orig. unkn.]

potty[2] /ˈpɒtɪ/ n. (pl. **-ies**) a chamber-pot, esp. for a child.

pouch /paʊtʃ/ n. & v. ● n. **1** a small bag or detachable outside pocket. **2** a baggy area of skin underneath the eyes etc. **3 a** a pocket-like receptacle in which marsupials carry their young during lactation. **b** any similar structure in an animal, e.g. in the cheeks of a rodent. **4** a soldier's leather ammunition bag. **5** a lockable bag for mail or dispatches. **6** Bot. a baglike cavity, esp. the seed-vessel, in a plant. ● v.tr. **1** put or make into a pouch. **2** take possession of; pocket. **3** make (part of a dress etc.) hang like a pouch. □ **pouched** adj. **pouchy** adj. [ME f. ONF pouche: cf. POKE[2]]

pouf[1] var. of POOF[1].

pouf[2] var. of POOF[2].

pouffe /puːf/ n. (also **pouf**) a large firm cushion used as a low seat or footstool. [F pouf; ult. imit.]

poulard /ˈpuːlɑːd/ n. a domestic hen that has been spayed and fattened for eating. [F poularde f. poule hen]

Poulenc /ˈpuːlæŋk/, Francis (Jean Marcel) (1899–1963), French composer. He was a member of the group Les Six. The influence of Satie and Cocteau can be seen particularly in his adoption of the idioms of popular music such as jazz. His work is also characterized by a lyricism seen especially in his many songs and in such instrumental works as the sonatas for flute (1957) and oboe (1962). He also wrote a series of lyrical and contrapuntal sacred choral pieces, while his works for the theatre include the opera Dialogues des Carmélites (1957) and the ballet Les Biches (1923).

poult[1] /pəʊlt/ n. a young domestic fowl, turkey, pheasant, etc. [ME, contr. f. PULLET]

poult[2] /puːlt/ n. (in full **poult-de-soie** /dəˈswʌ/) a fine corded silk or taffeta, usu. coloured. [F, of unkn. orig.]

poulterer /ˈpəʊltərə(r)/ n. a dealer in poultry and usu. game. [ME poulter f. OF pouletier (as PULLET)]

poultice /ˈpəʊltɪs/ n. & v. ● n. a soft medicated and usu. heated mass applied to the body and kept in place with muslin etc., for relieving soreness and inflammation. ● *v.tr.* apply a poultice to. [orig. *pultes* (pl.) f. L *puls pultis* pottage, pap, etc.]

poultry /ˈpəʊltrɪ/ *n.* domesticated fowls (ducks, geese, turkeys, chickens, etc.), esp. as a source of food. [ME f. OF *pouletrie* (as POULTERER)]

pounce[1] /paʊns/ *v. & n.* ● *v.intr.* **1** spring or swoop, esp. as in capturing prey. **2** (often foll. by *on*, *upon*) **a** make a sudden attack. **b** seize eagerly upon an answer, remark, etc. (*pounced on what we said*). ● *n.* **1** the act or an instance of pouncing. **2** the claw or talon of a bird of prey. □ **pouncer** *n.* [perh. f. PUNCHEON[1]]

pounce[2] /paʊns/ *n. & v.* ● *n.* **1** a fine powder formerly used to prevent ink from spreading on unglazed paper. **2** powdered charcoal etc. dusted over a perforated pattern to transfer the design to the object beneath. ● *v.tr.* **1** dust with pounce. **2** transfer (a design etc.) by use of pounce. **3** smooth (paper etc.) with pounce or pumice. □ **pouncer** *n.* [F *ponce*, *poncer* f. L *pumex* PUMICE]

pouncet-box /ˈpaʊnsɪtˌbɒks/ *n. archaic* a small box with a perforated lid for perfumes etc. [16th c.: perh. orig. erron. f. *pounced* (= perforated) *box*]

Pound /paʊnd/, Ezra (Weston Loomis) (1885–1972), American poet and critic. He came to Europe in 1908 and co-founded the imagist movement in London; collections of poetry from this period include *Ripostes* (1912). He also co-edited the magazine of the vorticist movement, *Blast*, from 1914 to 1915. After this he gradually moved away from imagism and began to develop a highly eclectic poetic voice, drawing on a vast range of classical and other references, which ensured his reputation as one of the foremost modernist poets. Work from this later period includes *Hugh Selwyn Mauberley* (1920) and the long (unfinished) series of *Cantos* (1917–70). In 1925 he settled in Italy; he was charged with treason in 1945 following his pro-Fascist radio broadcasts during the Second World War, but adjudged insane and committed to a mental institution until 1958.

pound[1] /paʊnd/ *n.* **1** a unit of weight equal to 16 oz avoirdupois (0.4536 kg), or 12 oz troy (0.3732 kg). **2** (in full **pound sterling**) (*pl.* same or **pounds**) the chief monetary unit of the UK and several other countries. □ **pound cake** a rich cake containing a pound (or equal weights) of each chief ingredient. **pound coin** (or **note**) a coin (or note) worth one pound sterling. **pound of flesh** any legitimate but crippling or morally offensive demand (with allusion to Shylock's demand for a pound of Antonio's flesh in Shakespeare's *Merchant of Venice*). **pound Scots** *hist.* 1s. 8d. **pound sign** the sign £, representing a pound. [OE *pund* ult. f. L *pondo* Roman pound weight of 12 ounces]

pound[2] /paʊnd/ *v. & n.* ● *v.* **1** *tr.* **a** crush or beat with repeated heavy blows. **b** thump or pummel, esp. with the fists. **c** grind to a powder or pulp. **2** *intr.* (foll. by *at*, *on*) deliver heavy blows or gunfire. **3** *intr.* (foll. by *along* etc.) make one's way heavily or clumsily. **4** *intr.* (of the heart) beat heavily. ● *n.* a heavy blow or thump; the sound of this. □ **pound into** instil (an attitude, behaviour, etc.) forcefully (*pounded into me*). **pound out 1** produce with or as if with heavy blows. **2** remove (an attitude, behaviour, etc.) forcefully (*pounded out of him*). □ **pounder** *n.* [OE *pūnian*, rel. to Du. *puin*, LG *pün* rubbish]

pound[3] /paʊnd/ *n. & v.* ● *n.* **1** an enclosure where stray animals or officially removed vehicles are kept until redeemed. **2** a place of confinement. ● *v.tr.* enclose (cattle etc.) in a pound. □ **pound lock** a lock with two gates to confine water and often a side reservoir to maintain the water level. [ME f. OE *pund-* in *pundfald*: see PINFOLD]

poundage /ˈpaʊndɪdʒ/ *n.* **1** a commission or fee of so much per pound sterling or weight. **2** a percentage of the total earnings of a business, paid as wages. **3** a person's weight, esp. that which is regarded as excess.

poundal /ˈpaʊnd(ə)l/ *n. Physics* a unit of force equal to the force required to give a mass of one pound an acceleration of one foot per second per second. [POUND[1] + *-al* perh. after *quintal*]

pounder /ˈpaʊndə(r)/ *n.* (usu. in comb.) **1** a thing or person weighing a specified number of pounds (*a five-pounder*). **2** a gun carrying a shell of a specified number of pounds. **3** a thing worth, or a person possessing, so many pounds sterling.

pour /pɔː(r)/ *v.* **1** *intr. & tr.* (usu. foll. by *down*, *out*, *over*, etc.) flow or cause to flow esp. downwards in a stream or shower. **2** *tr.* dispense (a drink, e.g. tea) by pouring. **3** *intr.* (of rain, or prec. by *it* as subject) fall heavily. **4** *intr.* (usu. foll. by *in*, *out*, etc.) come or go in profusion or rapid succession (*the crowd poured out*; *letters poured in*). **5** *tr.* discharge or send freely (*poured forth arrows*). **6** *tr.* (often foll. by *out*) utter at length or in a rush (*poured out their story*). □ **it never rains but it pours** misfortunes rarely come singly. **pour cold water on** see COLD. **pour oil on the**

waters (or **on troubled waters**) calm a disagreement or disturbance, esp. with conciliatory words. □ **pourable** *adj.* **pourer** *n.* [ME: orig. unkn.]

pourboire /ˈpʊəbwɑː(r)/ *n.* a gratuity or tip. [F, = *pour boire* (money) for drinking]

Poussin /puˈsɛ̃/, Nicolas (1594–1665), French painter. He is regarded as the chief representative of French classicism in art and a master of the grand manner and was extremely influential in the development of French art. From 1624 he lived mostly in Rome and was influenced by the work of Italian painters, particularly Titian and Raphael. In the 1630s and 1640s he developed in his painting a harmony and sense of order suffused with a rich colour sense. His subject-matter included biblical scenes (*The Adoration of the Golden Calf*, c.1635), classical mythology (*Et in Arcadia Ego*, c.1655), and historical landscapes.

poussin /ˈpuːsæn/ *n.* a young chicken bred for eating. [F]

pout[1] /paʊt/ *v. & n.* ● *v.* **1** *intr.* **a** push the lips forward as an expression of displeasure or sulking. **b** (of the lips) be pushed forward. **2** *tr.* push (the lips) forward in pouting. ● *n.* **1** such an action or expression. **2** (**the pouts**) a fit of sulking. □ **pouter** *n.* **poutingly** *adv.* **pouty** *adj.* [ME, perh. f. OE *putian* (unrecorded) be inflated: cf. POUT[2]]

pout[2] /paʊt/ *n.* **1** = BIB[1] 3. **2** = EELPOUT. [OE *-puta* in *ælepūta* eelpout, f. WG]

pouter /ˈpaʊtə(r)/ *n.* **1** a person who pouts. **2** a kind of pigeon able to inflate its crop considerably.

poverty /ˈpɒvətɪ/ *n.* **1** the state of being poor; want of the necessities of life. **2** (often foll. by *of*, *in*) scarcity or lack. **3** inferiority, poorness, meanness. **4** *Eccl.* renunciation of the right to individual ownership of property. □ **poverty line** the minimum income level needed to secure the necessities of life. **poverty-stricken** extremely poor. **poverty trap** a situation in which an increase of income incurs a loss of state benefits, making real improvement impossible. [ME f. OF *poverte*, *poverté* f. L *paupertas -tatis* f. *pauper* poor]

POW *abbr.* prisoner of war.

pow /paʊ/ *int.* expressing the sound of a blow or explosion. [imit.]

powder /ˈpaʊdə(r)/ *n. & v.* ● *n.* **1** a substance in the form of fine dry particles. **2** a medicine or cosmetic in this form. **3** = GUNPOWDER. ● *v.tr.* **1 a** apply powder to. **b** sprinkle or decorate with or as with powder. **2** (esp. as **powdered** *adj.*) reduce to a fine powder (*powdered milk*). □ **keep one's powder dry** be cautious and alert. **powder blue** pale blue. **powder-flask** *hist.* a small case for carrying gunpowder. **powder-keg 1** a barrel of gunpowder. **2** a dangerous or volatile situation. **powder metallurgy** the production of metals as fine powders which are then sintered into compact objects. **powder-monkey** *hist.* a boy employed on board ship to carry powder to the guns. **powder-puff** a soft pad for applying powder to the skin, esp. the face. **powder-room** a women's cloakroom or lavatory in a public building. **powder snow** loose dry snow on a ski-run etc. **take a powder** *sl.* depart quickly. □ **powdery** *adj.* [ME f. OF *poudre* f. L *pulvis pulveris* dust]

Powell[1] /ˈpəʊəl/, Anthony (Dymoke) (b.1905), English novelist. He is best known for his sequence of twelve novels, *A Dance to the Music of Time*, beginning with *A Question of Upbringing* (1951) and ending with *Hearing Secret Harmonies* (1975). These novels are a satirical and panoramic portrayal of the fortunes of the English upper middle classes between the First and Second World Wars.

Powell[2] /ˈpəʊəl/, (John) Enoch (b.1912), British politician. He was a classical scholar before the Second World War, joining the Conservative Party in 1946. After serving as Minister of Health (1960–3), he attracted public attention in 1968 with his frank condemnation of multiracial immigration into Britain; as a result he was dismissed from the shadow Cabinet. Powell also opposed British entry into the Common Market, resigning from the Conservative Party in 1974 on this issue. He later served as an Ulster Unionist MP (1974–87).

power /ˈpaʊə(r)/ *n. & v.* ● *n.* **1** the ability to do or act (*will do all in my power*; *has the power to change colour*). **2** a particular faculty of body or mind (*lost the power of speech*; *powers of persuasion*). **3 a** government, influence, or authority. **b** political or social ascendancy or control (*the party in power*; *black power*). **4** authorization; delegated authority (*power of attorney*; *police powers*). **5** (often foll. by *over*) personal ascendancy. **6** an influential person, group, or organization (*the press is a power in the land*). **7 a** military strength. **b** a state having international influence, esp. based on military strength (*the leading powers*). **8** vigour, energy. **9** an active property or function (*has a high heating power*). **10** *colloq.* a large number or amount (*has done me a power of good*). **11** the capacity for exerting mechanical force or doing work (*horsepower*).

12 mechanical or electrical energy as distinct from hand-labour (often *attrib.: power tools; power steering*). **13 a** a public supply of (esp. electrical) energy. **b** a particular source or form of energy (*hydroelectric power*). **14** a mechanical force applied e.g. by means of a lever. **15** *Physics* the rate of energy output. **16** the product obtained when a number is multiplied by itself a certain number of times (*2 to the power of 3 = 8*). **17** the magnifying capacity of a lens. **18 a** a deity. **b** (in *pl.*) the sixth order of the ninefold celestial hierarchy (see ORDER *n.* 19). ● *v.tr.* **1** supply with mechanical or electrical energy. **2** (foll. by *up, down*) increase or decrease the power supplied to (a device); switch on or off. □ **in the power of** under the control of. **more power to your elbow!** an expression of encouragement or approval. **power behind the throne** a person who asserts authority or influence without having formal status. **power block** a group of nations constituting an international political force. **power cut** a temporary withdrawal or failure of an electric power supply. **power-dive** *n.* a steep dive of an aircraft with the engines providing thrust. ● *v.intr.* perform a power-dive. **power line** a conductor supplying electrical power, esp. one supported by pylons or poles. **power of attorney** see ATTORNEY. **power pack 1** a unit for supplying power. **2** the equipment for converting an alternating current (from the mains) to a direct current at a different (usu. lower) voltage. **power play 1** tactics involving the concentration of players at a particular point. **2** similar tactics in business, politics, etc., involving a concentration of resources, effort, etc. **power point** *Brit.* a socket in a wall etc. for connecting an electrical device to the mains. **power politics** political action based on power or influence. **power-sharing** a policy agreed between parties or within a coalition to share responsibility for decision-making and political action. **power station** a building where electrical power is generated for distribution. **the powers that be** those in authority (after Rom. 13:1). **power stroke** the stroke of an internal-combustion engine, in which the piston is moved downward by the expansion of gases. □ **powered** *adj.* (also in *comb.*). [ME & AF *poer* etc., OF *poeir* ult. f. L *posse* be able]

powerboat /ˈpaʊəˌbəʊt/ *n.* a powerful motor boat.

powerful /ˈpaʊəˌfʊl/ *adj.* **1** having much power or strength. **2** politically or socially influential. **3** having a strong emotional effect, impressive. □ **powerfully** *adv.* **powerfulness** *n.*

powerhouse /ˈpaʊəˌhaʊs/ *n.* **1** = power station. **2** a person or thing of great energy.

powerless /ˈpaʊəlɪs/ *adj.* **1** without power or strength. **2** (often foll. by *to* + infin.) wholly unable (*powerless to help*). □ **powerlessly** *adv.* **powerlessness** *n.*

powerplant /ˈpaʊəˌplɑːnt/ *n.* an apparatus or an installation which provides power for industry, a machine, etc.

powwow /ˈpaʊwaʊ/ *n. & v.* ● *n.* a conference or meeting for discussion (orig. among North American Indians). ● *v.intr.* hold a powwow. [Algonquian *powah, powwaw* magician (lit. 'he dreams')]

Powys /ˈpəʊɪs, ˈpaʊ-/ **1** a county of east central Wales, on the border with England, formed in 1974 from the former counties of Montgomeryshire, Radnorshire, and most of Breconshire; administrative centre, Llandrindod Wells. **2** a former Welsh kingdom. At its most powerful in the early 12th century, Powys was divided in 1160 into two principalities. It was finally conquered by the English (1284) after the death of the Welsh Prince Llewelyn in 1282.

pox /pɒks/ *n.* **1** any viral disease producing a rash of pimples that become pus-filled and leave pock-marks on healing. **2** *colloq.* = SYPHILIS. **3** a plant disease that causes pocklike spots. □ **a pox on** *archaic* an exclamation of anger or impatience with (a person). [alt. spelling of *pocks* pl. of POCK]

poxy /ˈpɒksɪ/ *adj.* (**poxier, poxiest**) **1** infected by pox. **2** *sl.* of poor quality; worthless.

Poznań /ˈpɒznænj/ (German **Posen** /ˈpoːz(ə)n/) a city in NW Poland; pop. (1990) 590,100. An area of German colonization since the 13th century, it was annexed to Prussia in 1793, remaining under German control, apart from a short period of Russian rule between 1807 and 1815, until the First World War. It was overrun by the Germans again in 1939. Severely damaged during the Second World War, it was rebuilt and is now a major industrial and commercial city.

Pozsony /ˈpoʒɒn/ the Hungarian name for BRATISLAVA, capital of Hungary 1526–1784.

pozzolana /ˌpɒtsəˈlɑːnə/ *n.* (also **puzzolana** /ˌpʊts-/) a volcanic ash used for mortar or hydraulic cement. [It., f. *pozz(u)olano* (adj.) of *Pozzuoli*, a town near Naples]

pp *abbr. Mus.* pianissimo.

pp. *abbr.* pages.

p.p. *abbr.* (also **pp**) *per pro.*

PPARC *abbr.* (in the UK) Particle Physics and Astronomy Research Council.

PPE *abbr. Brit.* philosophy, politics, and economics (as a degree course at Oxford University).

p.p.m. *abbr.* parts per million.

PPP see PAKISTAN PEOPLE'S PARTY.

PPS *abbr.* **1** *Brit.* Parliamentary Private Secretary. **2** additional postscript.

PR *abbr.* **1** public relations. **2** see PROPORTIONAL REPRESENTATION. **3** *US* Puerto Rico.

Pr *symb. Chem.* the element praseodymium.

pr. *abbr.* pair.

PRA *abbr.* (in the UK) President of the Royal Academy.

praam var. of PRAM².

practicable /ˈpræktɪkəb(ə)l/ *adj.* **1** that can be done or used. **2** possible in practice. □ **practicably** *adv.* **practicableness** *n.* **practicability** /ˌpræktɪkəˈbɪlɪtɪ/ *n.* [F *praticable* f. *pratiquer* put into practice (as PRACTICAL)]

practical /ˈpræktɪk(ə)l/ *adj. & n.* ● *adj.* **1** of or concerned with practice or use rather than theory. **2** suited to use or action; designed mainly to fulfil a function (*practical shoes*). **3** (of a person) inclined to action rather than speculation; able to make things function well. **b** skilled at manual tasks. **4 a** that is such in effect, though not nominally (*for all practical purposes*). **b** virtual (*in practical control*). **5** feasible, realistic; concerned with what is actually possible (*practical politics; practical solutions*). ● *n.* a practical examination or lesson. □ **practical joke** a humorous trick played on a person. **practical joker** a person who plays practical jokes. □ **practicalness** *n.* **practicality** /ˌpræktɪˈkælɪtɪ/ *n.* (pl. **-ies**). [earlier *practic* f. obs. F *practique* or LL *practicus* f. Gk *praktikos* f. *prassō* do, act]

practically /ˈpræktɪkəlɪ/ *adv.* **1** virtually, almost (*practically nothing*). **2** in a practical way.

practice /ˈpræktɪs/ *n. & v.* ● *n.* **1** habitual action or performance (*the practice of teaching; makes a practice of saving*). **2** a habit or custom (*has been my regular practice*). **3 a** repeated exercise in an activity requiring the development of skill (*to sing well needs much practice*). **b** a session of this (*they had a short practice*). **4** action or execution as opposed to theory. **5** the professional work or business of a doctor, lawyer, etc. (*has a practice in town*). **6** an established method of legal procedure. **7** procedure generally, esp. of a specified kind (*bad practice*). ● *v.tr. & intr. US* var. of PRACTISE. □ **in practice 1** when actually applied; in reality. **2** skilful because of recent exercise in a particular pursuit. **out of practice** lacking a former skill from lack of recent practice. **put into practice** actually apply (an idea, method, etc.). [ME f. PRACTISE, after *advice, device*]

practician /prækˈtɪʃ(ə)n/ *n.* a worker; a practitioner. [obs. F *practicien* f. *practique* f. med.L *practica* f. Gk *praktikē* fem. of *praktikos*: see PRACTICAL]

practise /ˈpræktɪs/ *v.* (*US* **practice**) **1** *tr.* perform habitually; carry out in action (*practise the same method; practise what you preach*). **2** *tr.* & (foll. by *in, on*) *intr.* do repeatedly as an exercise to improve a skill; exercise oneself in or on (an activity requiring skill) (*had to practise in the art of speaking; practise your reading*). **3** *tr.* (as **practised** *adj.*) experienced, expert (*a practised liar; with a practised hand*). **4** *tr.* **a** pursue or be engaged in (a profession, religion, etc.). **b** (as **practising** *adj.*) currently active or engaged in (a profession or activity) (*a practising Christian; a practising lawyer*). **5** (foll. by *on, upon*) take advantage of; impose upon. **6** *intr. archaic* scheme, contrive (*when first we practise to deceive*). □ **practiser** *n.* [ME f. OF *pra(c)tiser* or med.L *practizare* alt. f. *practicare* (as PRACTICAL)]

practitioner /prækˈtɪʃənə(r)/ *n.* a person practising a profession, esp. medicine (*general practitioner*). [obs. *practitian* = PRACTICIAN]

prad /præd/ *n.* esp. *Austral. sl.* a horse. [by metathesis f. Du. *paard* f. LL *paraveredus*; see PALFREY]

Prado /ˈprɑːdəʊ/ the Spanish national art gallery, in Madrid. Established in 1818, it houses the greatest collection in the world of Spanish masters — Velázquez, El Greco, Zurbarán, Ribera, Murillo, Goya — as well as important examples of Flemish and Venetian art.

prae- /priː/ *prefix* = PRE- (esp. in words regarded as Latin or relating to Roman antiquity). [L: see PRE-]

praecipe /ˈpriːsɪpɪ/ *n.* **1** a writ demanding action or an explanation of non-action. **2** an order requesting a writ. [L (the first word of the writ), imper. of *praecipere* enjoin: see PRECEPT]

praecocial var. of PRECOCIAL.

praemunire /ˌpriːmjuːˈnɪərɪ, -mjʊˈnaɪərɪ/ n. hist. a writ charging a sheriff to summon a person accused of asserting or maintaining papal jurisdiction in England. [med.L, = forewarn, for L praemonere (as PRAE-, monere warn): the words praemunire facias that you warn (a person to appear) occur in the writ]

praenomen /priːˈnəʊmen/ n. an ancient Roman's first or personal name (e.g. Marcus Tullius Cicero). [L f. prae before + nomen name]

praepostor /priːˈpɒstə(r)/ n. (also **prepostor**) Brit. (at some public schools) a prefect or monitor. [praepositor alt. f. L praepositus past part. of praeponere set over (as PRAE-, ponere posit- place)]

Praesepe /praɪˈsiːpɪ/ Astron. a large open cluster of stars in the constellation of Cancer. Also called the Beehive. [L., = manger, enclosure, hive]

Praesidium var. of PRESIDIUM.

praetor /ˈpriːtə(r), -tɔː(r)/ n. (US **pretor**) Rom. Hist. either of two ancient Roman magistrates ranking below consul. □ **praetorship** n.
praetorial /priːˈtɔːrɪəl/ adj. [ME f. F préteur or L praetor (perh. as PRAE-, ire it- go)]

praetorian /priːˈtɔːrɪən/ adj. & n. (US **pretorian**) Rom. Hist. ● adj. of or having the powers of a praetor. ● n. a man of praetorian rank. □ **praetorian guard** the bodyguard of the Roman emperor. [ME f. L praetorianus (as PRAETOR)]

pragmatic /præɡˈmætɪk/ adj. 1 dealing with matters with regard to their practical requirements or consequences. 2 treating the facts of history with reference to their practical lessons. 3 hist. of or relating to the affairs of a state. 4 (also **pragmatical**) a concerning pragmatism. b meddlesome. c dogmatic. □ **pragmatically** adv. **pragmaticality** /-ˌmætɪˈkælɪtɪ/ n. [LL pragmaticus f. Gk pragmatikos f. pragma -matos deed]

pragmatics /præɡˈmætɪks/ n.pl. (usu. treated as sing.) the branch of linguistics dealing with language in use.

Pragmatic Sanction n. an imperial or royal ordinance issued as a fundamental law, esp. regarding a question of royal succession. The term refers specifically to a document drafted by the Emperor Charles VI after the birth of his daughter Maria Theresa in 1717 making provision for her to succeed to all his territories should he die without a son. It was accepted by Austria, Hungary, and the Austrian Netherlands in 1720–3, but the campaign to have it recognized by other states dominated the diplomatic scene in Europe for two decades afterwards. The opposition to its acceptance led to the War of the Austrian Succession on Charles's death in 1740.

pragmatism /ˈpræɡməˌtɪz(ə)m/ n. 1 a pragmatic attitude or procedure. 2 a philosophy, expounded notably by C. S. Peirce and William James, that evaluates assertions solely by their practical consequences and bearing on human interests. □ **pragmatist** n.
pragmatistic /ˌpræɡməˈtɪstɪk/ adj. [Gk pragma: see PRAGMATIC]

pragmatize /ˈpræɡməˌtaɪz/ v.tr. (also **-ise**) 1 represent as real. 2 rationalize (a myth).

Prague /prɑːɡ/ (Czech **Praha** /ˈpraha/) the capital of the Czech Republic, in the north-east on the River Vltava; pop. (1991) 1,212,000. The Bohemian king and Holy Roman emperor Charles IV made it the capital of Bohemia in the 14th century. It was the scene of religious conflict in the early 15th century between the followers of John Huss and the Catholic Church, and again in the 17th century when, in response to the oppression of the ruling Catholic Habsburgs, the Protestant citizens threw Catholic officials from the windows of Hradčany Castle. This event, known as the Defenestration of Prague (1618), contributed to the outbreak of the Thirty Years War. Prague was the capital of Czechoslovakia from 1918 until the partition of the country at the beginning of 1993.

Prague Spring the attempted democratization of Czech political life in 1968. The movement was crushed when Soviet troops moved into Czechoslovakia in August of that year.

Praha see PRAGUE.

prahu var. of PROA.

Praia /ˈpraɪə/ the capital of the Cape Verde Islands, a port on the island of São Tiago; pop. (1990) 62,000.

prairie /ˈpreərɪ/ n. a large area of usu. treeless grassland esp. in North America. □ **prairie chicken** (or **hen**) a North American grouse of the genus Tympanuchus. **prairie dog** a North American rodent of the genus Cynomys, living gregariously in burrows and having a barking call. **prairie oyster** a seasoned raw egg, often served in spirits and swallowed in one as a cure for a hangover. **prairie schooner** US a covered wagon used by the 19th-century pioneers in crossing the North

American prairies. **prairie wolf** = COYOTE. [F f. OF praerie ult. f. L pratum meadow]

praise /preɪz/ v. & n. ● v.tr. 1 express warm approval or admiration of. 2 glorify (God) in words. ● n. the act or an instance of praising; commendation (won high praise; were loud in their praises). □ **praise be!** an exclamation of pious gratitude. **sing the praises of** commend (a person) highly. □ **praiseful** adj. **praiser** n. [ME f. OF preisier price, prize, praise, f. LL pretiare f. L pretium price: cf. PRIZE¹]

praiseworthy /ˈpreɪzˌwɜːðɪ/ adj. worthy of praise; commendable. □ **praiseworthily** adv. **praiseworthiness** n.

Prakrit /ˈprɑːkrɪt/ n. any of the (esp. ancient or medieval) vernacular dialects of North and Central India existing alongside or derived from Sanskrit. [Skr. prākṛta natural, original: cf. SANSKRIT]

praline /ˈprɑːliːn/ n. a smooth sweet substance made by boiling nuts in sugar and used esp. as a filling for chocolates. [F f. Marshal de Plessis-Praslin, Fr. soldier (1598–1675), whose cook invented it]

pralltriller /ˈprɑːlˌtrɪlə(r)/ n. a musical ornament consisting of one rapid alternation of the written note with the note immediately above it. [G f. prallen rebound + Triller TRILL]

pram¹ /præm/ n. Brit. a four-wheeled carriage for a baby, pushed by a person on foot. [abbr. of PERAMBULATOR]

pram² /prɑːm/ n. (also **praam**) 1 a flat-bottomed gunboat or Baltic cargo-boat. 2 a ship's dinghy. [MDu. prame, praem, MLG prām(e), perh. f. Czech prám]

prana /ˈprɑːnə/ n. 1 Hinduism breath as a life-giving force. 2 the breath; breathing. [Skr.]

prance /prɑːns/ v. & n. ● v.intr. 1 (of a horse) raise the forelegs and spring from the hind legs. 2 (often foll. by about) walk or behave in an elated or arrogant manner. ● n. 1 the act of prancing. 2 a prancing movement. □ **prancer** n. [ME: orig. unkn.]

prandial /ˈprændɪəl/ adj. formal or joc. of or relating to dinner or lunch. [L prandium meal]

Prandtl /ˈprænt(ə)l/, Ludwig (1875–1953), German physicist. Prandtl is remembered for his studies of both aerodynamics and hydrodynamics. He established the existence of the boundary layer, and made important studies on streamlining. The design of an efficient shape, weight, and mass for aircraft and ships owes much to his work.

prang /præŋ/ v. & n. Brit. sl. ● v.tr. 1 crash or damage (an aircraft or vehicle). 2 bomb (a target) successfully. ● n. the act or an instance of pranging. [imit.]

prank /præŋk/ n. a practical joke; a piece of mischief. □ **prankish** adj. **pranksome** adj. [16th c.: orig. unkn.]

prankster /ˈpræŋkstə(r)/ n. a person fond of playing pranks.

prase /preɪz/ n. a translucent green type of quartz. [F f. L prasius f. Gk prasios (adj.) leek-green f. prason leek]

praseodymium /ˌpreɪzɪəˈdɪmɪəm/ n. a soft silvery-white metallic chemical element (atomic number 59; symbol **Pr**). Praseodymium was identified by the Austrian chemist Auer von Welsbach in 1885, when he showed that the supposed rare-earth element didymium was in fact a mixture (see also NEODYMIUM). Praseodymium is a member of the lanthanide series. The metal is a component of misch metal and other alloys and its compounds are used for colouring glass and ceramics. [G Praseodym f. Gk prasios (see PRASE) from its green salts, + G Didym DIDYMIUM]

prat /præt/ n. sl. 1 Brit. a silly or foolish person. 2 the buttocks. [16th-c. cant (in sense 2): orig. unkn.]

prate /preɪt/ v. & n. ● v. 1 chatter; talk too much. 2 intr. talk foolishly or irrelevantly. 3 tr. tell or say in a prating manner. ● n. prating; idle talk. □ **prater** n. **prating** adj. [ME f. MDu., MLG praten, prob. imit.]

pratfall /ˈprætfɔːl/ n. US sl. 1 a fall on the buttocks. 2 a humiliating failure.

pratie /ˈpreɪtɪ/ n. esp. Ir. a potato. [corrupt.]

pratincole /ˈprætɪŋˌkəʊl/ n. an insectivorous fork-tailed plover-like bird of the genus Glareola, living near water and having a swallow-like flight. [mod.L pratincola f. L pratum meadow + incola inhabitant]

pratique /præˈtiːk/ n. a licence to have dealings with a port, granted to a ship after quarantine or on showing a clean bill of health. [F, = practice, intercourse, f. It. pratica f. med.L practica: see PRACTICIAN]

Prato /ˈprɑːtəʊ/ a city in northern Italy, north-west of Florence; pop. (1990) 166,690.

prattle /ˈpræt(ə)l/ v. & n. ● v.intr. & tr. chatter or say in a childish or inconsequential way. ● n. 1 childish chatter. 2 inconsequential talk. □ **prattler** n. **prattling** adj. [MLG pratelen (as PRATE)]

prau var. of PROA.

Pravda /ˈprɑːvdə/ a Russian daily newspaper, founded in 1912 as an underground publication and from 1918 the official organ of the Soviet Communist Party. Banned twice under President Yeltsin, the paper is now regarded as being broadly representative of the views of Communists in Russia. [Russ., = truth]

prawn /prɔːn/ n. & v. ● n. a marine crustacean resembling a shrimp, but usu. larger and with two pairs of pincers. ● v.intr. fish for prawns. □ **come the raw prawn** see RAW. [ME pra(y)ne, of unkn. orig.]

praxis /ˈpræksɪs/ n. **1** accepted practice or custom. **2** the practising of an art or skill. [med.L f. Gk, = doing, f. prassō do]

Praxiteles /prækˈsɪtəˌliːz/ (mid-4th century BC), Athenian sculptor. Although only one of his works, *Hermes Carrying the Infant Dionysus*, survives, he is regarded as one of the greatest Greek sculptors of his day. Other examples of his work survive in Roman copies or are known from their descriptions by writers; they include a statue of Aphrodite, which represents the first important female nude in sculpture.

pray /preɪ/ v. (often foll. by for or to + infin. or that + clause) **1** intr. (often foll. by to) say prayers (to God etc.); make devout supplication. **2 a** tr. entreat, beseech. **b** tr. & intr. ask earnestly (prayed to be released). **3** tr. (as imper.) archaic & formal please (pray tell me). □ **praying mantis** see MANTIS. [ME f. OF preier f. LL precare f. L precari entreat]

prayer[1] /preə(r)/ n. **1 a** a solemn request or thanksgiving to God or an object of worship (say a prayer). **b** a formula or form of words used in praying (the Lord's prayer). **c** the act of praying (be at prayer). **d** a religious service consisting largely of prayers (morning prayers). **2 a** an entreaty to a person. **b** a thing entreated or prayed for. □ **not have a prayer** N. Amer. colloq. have no chance (of success etc.). **prayer-book** a book containing the forms of prayer in regular use, esp. the Book of Common Prayer. **prayer-mat** a small carpet used by Muslims when praying. **prayer-wheel** a revolving cylindrical box inscribed with or containing prayers, used esp. by Tibetan Buddhists. □ **prayerless** adj. [ME f. OF preiere ult. f. L precarius obtained by entreaty f. prex precis prayer]

prayer[2] /preɪə(r)/ n. a person who prays.

prayerful /ˈpreəfʊl/ adj. **1** (of a person) given to praying; devout. **2** (of speech, actions, etc.) characterized by or expressive of prayer. □ **prayerfully** adv. **prayerfulness** n.

Prayer of Manasses /məˈnæsiːz/ a book of the Apocrypha consisting of a penitential prayer put into the mouth of Manasseh, king of Judah. His life and reign are described at 2 Kings 21:1–18.

pre- /priː/ prefix before (in time, place, order, degree, or importance). [from or after L prae- f. prae (adv. & prep.)]

preach /priːtʃ/ v. **1 a** intr. deliver a sermon or religious address. **b** tr. deliver (a sermon); proclaim or expound (the Gospel etc.). **2** intr. give moral advice in an obtrusive way. **3** tr. advocate or inculcate (a quality or practice etc.). □ **preach to the converted** commend an opinion to those who already assent to it. □ **preachable** adj. [ME f. OF prechier f. L praedicare proclaim, in eccl.L preach (as PRAE-, dicare declare)]

preacher /ˈpriːtʃə(r)/ n. a person who preaches, esp. a minister of religion. [ME f. AF prech(o)ur, OF prech(e)or f. eccl.L praedicator (as PREACH)]

preachify /ˈpriːtʃɪˌfaɪ/ v.intr. (**-ies**, **-ied**) colloq. preach or moralize tediously.

preachment /ˈpriːtʃmənt/ n. usu. derog. preaching, sermonizing.

preachy /ˈpriːtʃɪ/ adj. (**preachier**, **preachiest**) colloq. inclined to preach or moralize. □ **preachiness** n.

preadolescent /ˌpriːædəˈlesənt/ adj. & n. ● adj. **1** (of a child) having nearly reached adolescence. **2** of or relating to the two or three years preceding adolescence. ● n. a preadolescent child. □ **preadolescence** n.

preamble /priːˈæmb(ə)l/ n. **1** a preliminary statement or introduction. **2** the introductory part of a statute or deed etc. □ **preambular** /-bjʊlə(r)/ adj. [ME f. OF preambule f. med.L praeambulum f. LL praeambulus (adj.) going before (as PRE-, AMBLE)]

pre-amp /ˈpriːæmp/ n. = PREAMPLIFIER. [abbr.]

preamplifier /priːˈæmplɪˌfaɪə(r)/ n. an electronic device that amplifies a very weak signal (e.g. from a microphone or pick-up) and transmits it to a main amplifier. □ **preamplified** adj.

prearrange /ˌpriːəˈreɪndʒ/ v.tr. arrange beforehand. □ **prearranged** adj. **prearrangement** n.

preatomic /ˌpriːəˈtɒmɪk/ adj. existing or occurring before the use of atomic energy.

Preb. abbr. Prebendary.

prebend /ˈpreb(ə)nd/ n. **1** = PREBENDARY 1. **2** hist. the stipend of a canon or member of chapter. **3** hist. a portion of land or tithe from which this is drawn. [ME f. OF prebende f. LL praebenda pension, neut.pl. gerundive of L praebere grant f. prae forth + habere hold]

prebendary /ˈprebəndərɪ/ n. (pl. **-ies**) **1** an honorary canon. **2** hist. the holder of a prebend. □ **prebendaryship** n. [ME f. med.L praebendarius (as PREBEND)]

Precambrian /priːˈkæmbrɪən/ adj. & n. Geol. of, relating to, or denoting the earliest geological era, preceding the Palaeozoic. The Precambrian includes the whole of the earth's history from its origin about 4,600 million years ago to the beginning of the Cambrian period about 590 million years ago. Fossils of animals with hard skeletons are absent from Precambrian rocks, and the era was once thought devoid of organic life, but it is now known that a variety of organisms did exist during that time. The oldest known Precambrian rocks are about 3,800 million years old. (cf. ARCHAEAN, PROTEROZOIC)

precancerous /priːˈkænsərəs/ adj. Med. tending to precede the development of cancer.

precarious /prɪˈkeərɪəs/ adj. **1** uncertain; dependent on chance (makes a precarious living). **2** insecure, perilous (precarious health). □ **precariously** adv. **precariousness** n. [L precarius: see PRAYER[1]]

precast /priːˈkɑːst/ adj. (of concrete) cast in its final shape before positioning.

precative /ˈprekətɪv/ adj. (of a word or form) expressing a wish or request. [LL precativus f. precari pray]

precaution /prɪˈkɔːʃ(ə)n/ n. **1** an action taken beforehand to avoid risk or ensure a good result. **2** (in pl.) colloq. the use of contraceptives. **3** caution exercised beforehand; prudent foresight. □ **precautionary** adj. [F précaution f. LL praecautio -onis f. L praecavere (as PRAE-, cavere caut- beware of)]

precede /prɪˈsiːd/ v.tr. **1 a** (often as **preceding** adj.) come or go before in time, order, importance, etc. (preceding generations; the paragraph which precedes this one). **b** walk etc. in front of (preceded by our guide). **2** (foll. by by) cause to be preceded (must precede this measure by milder ones). [OF preceder f. L praecedere (as PRAE-, cedere cess- go)]

precedence /ˈpresɪd(ə)ns/ n. (also **precedency** /-dənsɪ/) **1** priority in time, order, or importance, etc. **2** the right of preceding others on formal occasions. □ **take precedence** (often foll. by over, of) have priority (over).

precedent n. & adj. ● n. /ˈpresɪd(ə)nt/ a previous case or legal decision etc. taken as a guide for subsequent cases or as a justification. ● adj. /prɪˈsiːd(ə)nt, ˈpresɪd-/ preceding in time, order, importance, etc. □ **precedently** /prɪˈsiːdəntlɪ, ˈpresɪd-/ adv. [ME f. OF (n. & adj.) (as PRECEDE)]

precedented /ˈpresɪˌdentɪd/ adj. having or supported by a precedent.

precent /prɪˈsent/ v. **1** intr. act as a precentor. **2** tr. lead the singing of (a psalm etc.). [back-form. f. PRECENTOR]

precentor /prɪˈsentə(r)/ n. **1** a person who leads the singing or (in a synagogue) the prayers of a congregation. **2** a minor canon who administers the musical life of a cathedral. □ **precentorship** n. [F précenteur or L praecentor f. praecinere (as PRAE-, canere sing)]

precept /ˈpriːsept/ n. **1** a command; a rule of conduct. **2 a** moral instruction (example is better than precept). **b** a general or proverbial rule; a maxim. **3 a** a writ or warrant. **b** Brit. an order for collection or payment of money under a local rate. □ **preceptive** /prɪˈseptɪv/ adj. [ME f. L praeceptum neut. past part. of praecipere praecept- warn, instruct (as PRAE-, capere take)]

preceptor /prɪˈseptə(r)/ n. (fem. **preceptress** /-trɪs/) a teacher or instructor. □ **preceptorship** n. **preceptorial** /ˌpriːsepˈtɔːrɪəl/ adj. [L praeceptor (as PRECEPT)]

precession /prɪˈseʃ(ə)n/ n. the slow movement of the axis of a spinning body around another axis, as seen in the circle slowly traced out by the pole of a spinning gyroscope. □ **precession of the equinoxes** Astron. **1** the slow retrograde motion of equinoctial points along the ecliptic. **2** the resulting earlier occurrence of equinoxes in each successive sidereal year. (See note below.) □ **precessional** adj. [LL praecessio (as PRECEDE)]

▪ As the earth rotates about its axis it responds to the gravitational attraction of the sun upon its equatorial bulge, so that its axis of rotation describes a circle in the sky. The points where the celestial equator and the ecliptic intersect define the sun's position at the equinoxes, which travel through the constellations of the zodiac once in a period of about 26,000 years. The precession of the equinoxes was discovered by Hipparchus in *c*.125 BC, when the vernal equinox was in Aries (see ARIES, FIRST POINT OF).

pre-Christian /priːˈkrɪstɪən, -ˈkrɪstʃən/ *adj.* before Christ or the advent of Christianity.

precinct /ˈpriːsɪŋkt/ *n.* **1** an enclosed or clearly defined area, e.g. around a cathedral, college, etc. **2** a specially designated area in a town, esp. with the exclusion of traffic (*shopping precinct*). **3** (in *pl.*) **a** the surrounding area or environs. **b** the boundaries. **4** *US* **a** a subdivision of a county, city, etc., for police or electoral purposes. **b** (in *pl.*) a neighbourhood. [ME f. med.L *praecinctum* neut. past part. of *praecingere* encircle (as PRAE-, *cingere* gird)]

preciosity /ˌprɛʃɪˈɒsɪtɪ/ *n.* overrefinement in art or language, esp. in the choice of words. [OF *préciosité* f. L *pretiositas* f. *pretiosus* (as PRECIOUS)]

precious /ˈprɛʃəs/ *adj. & adv.* ● *adj.* **1** of great value or worth. **2** beloved; much prized (*precious memories*). **3** affectedly refined, esp. in language or manner. **4** *colloq.* often *iron.* **a** considerable (*a precious lot you know about it*). **b** expressing contempt or disdain (*you can keep your precious flowers*). ● *adv. colloq.* extremely, very (*tried precious hard*; *had precious little left*). □ **precious metals** gold, silver, and platinum. **precious stone** a piece of mineral having great value esp. as used in jewellery. □ **preciously** *adv.* **preciousness** *n.* [ME f. OF *precios* f. L *pretiosus* f. *pretium* price]

precipice /ˈprɛsɪpɪs/ *n.* **1** a vertical or steep face of a rock, cliff, mountain, etc. **2** a dangerous situation. [F *précipice* or L *praecipitium* falling headlong, precipice (as PRECIPITOUS)]

precipitant /prɪˈsɪpɪt(ə)nt/ *adj. & n.* ● *adj.* = PRECIPITATE *adj.* ● *n. Chem.* a substance that causes another substance to precipitate. □ **precipitance** *n.* **precipitancy** *n.* [obs. F *précipitant* pres. part. of *précipiter* (as PRECIPITATE)]

precipitate *v., adj., & n.* ● *v.tr.* /prɪˈsɪpɪˌteɪt/ **1** hasten the occurrence of; cause to occur prematurely. **2** (foll. by *into*) send rapidly into a certain state or condition (*were precipitated into war*). **3** throw down headlong. **4** *Chem.* cause (a substance) to be deposited in solid form from a solution. **5** *Physics* **a** cause (dust etc.) to be deposited from the air on a surface. **b** condense (vapour) into drops and so deposit it. ● *adj.* /prɪˈsɪpɪtət/ **1** headlong; violently hurried (*precipitate departure*). **2** (of a person or act) hasty, rash, inconsiderate. ● *n.* /prɪˈsɪpɪtət/ **1** *Chem.* a substance precipitated from a solution. **2** *Physics* moisture condensed from vapour by cooling and depositing, e.g. rain or dew. □ **precipitately** /-tətlɪ/ *adv.* **precipitateness** /-tətnɪs/ *n.* **precipitator** /-ˌteɪtə(r)/ *n.* **precipitable** /-təb(ə)l/ *adj.* **precipitability** /-ˌsɪpɪtəˈbɪlɪtɪ/ *n.* [L *praecipitare praecipitat-* f. *praeceps praecipitis* headlong (as PRAE-, *caput* head)]

precipitation /prɪˌsɪpɪˈteɪʃ(ə)n/ *n.* **1** the act of precipitating or the process of being precipitated. **2** rash haste. **3 a** rain or snow etc. falling to the ground. **b** a quantity of this. [F *précipitation* or L *praecipitatio* (as PRECIPITATE)]

precipitous /prɪˈsɪpɪtəs/ *adj.* **1 a** of or like a precipice. **b** dangerously steep. **2** = PRECIPITATE *adj.* □ **precipitously** *adv.* **precipitousness** *n.* [obs. F *précipiteux* f. L *praeceps* (as PRECIPITATE)]

précis /ˈpreɪsiː/ *n. & v.* ● *n.* (*pl.* same /-siːz/) a summary or abstract, esp. of a text or speech. ● *v.tr.* (**précises** /-siːz/; **précised** /-siːd/; **précising** /-siːɪŋ/) make a précis of. [F, = PRECISE (as n.)]

precise /prɪˈsaɪs/ *adj.* **1 a** accurately expressed. **b** (of an instrument etc.) definite, exact. **2 a** punctilious; scrupulous in being exact, observing rules, etc. **b** often *derog.* rigid; fastidious. **3** identical, exact (*at that precise moment*). □ **preciseness** *n.* [F *précis -ise* f. L *praecidere praecis-* cut short (as PRAE-, *caedere* cut)]

precisely /prɪˈsaɪslɪ/ *adv.* **1** in a precise manner; exactly. **2** (as a reply) quite so; as you say.

precisian /prɪˈsɪʒ(ə)n/ *n.* a person who is rigidly precise or punctilious, esp. in religious observance. □ **precisianism** *n.*

precision /prɪˈsɪʒ(ə)n/ *n.* **1** the condition of being precise; accuracy. **2** the degree of refinement in measurement etc. **3** (*attrib.*) marked by or adapted for precision (*precision instruments*; *precision timing*). □ **precisionism** *n.* **precisionist** *n. & adj.* [F *précision* or L *praecisio* (as PRECISE)]

preclassical /priːˈklæsɪk(ə)l/ *adj.* before a period regarded as classical, esp. in music and literature.

preclinical /priːˈklɪnɪk(ə)l/ *adj.* **1** of or relating to the first, chiefly theoretical, stage of a medical education. **2** (of a stage in a disease) before symptoms can be identified.

preclude /prɪˈkluːd/ *v.tr.* **1** (foll. by *from*) prevent, exclude (*precluded from taking part*). **2** make impossible; remove (*so as to preclude all doubt*). □ **preclusion** /-ˈkluːʒ(ə)n/ *n.* **preclusive** /-ˈkluːsɪv/ *adj.* [L *praecludere praeclus-* (as PRAE-, *claudere* shut)]

precocial /prɪˈkəʊʃ(ə)l/ *adj. & n.* (also **praecocial**) ● *adj.* (of a bird) having young that can feed themselves as soon as they are hatched. ● *n.* a precocial bird (cf. ALTRICIAL). [L *praecox -cocis* (as PRECOCIOUS)]

precocious /prɪˈkəʊʃəs/ *adj.* **1** often *derog.* (of a person, esp. a child) prematurely developed in some faculty or characteristic. **2** (of an action etc.) indicating such development. **3** (of a plant) flowering or fruiting early. □ **precociously** *adv.* **precociousness** *n.* **precocity** /-ˈkɒsɪtɪ/ *n.* [L *praecox -cocis* f. *praecoquere* ripen fully (as PRAE-, *coquere* cook)]

precognition /ˌpriːkɒɡˈnɪʃ(ə)n/ *n.* **1** (supposed) foreknowledge, esp. of a supernatural kind. **2** *Sc.* the preliminary examination of witnesses etc., esp. to decide whether there is ground for a trial. □ **precognitive** /-ˈkɒɡnɪtɪv/ *adj.* [LL *praecognitio* (as PRE-, COGNITION)]

precoital /priːˈkəʊɪt(ə)l/ *adj.* preceding sexual intercourse. □ **precoitally** *adv.*

pre-Columbian /ˌpriːkəˈlʌmbɪən/ *adj.* of or relating to the period of history in the Americas before the arrival in America of Columbus in 1492.

preconceive /ˌpriːkənˈsiːv/ *v.tr.* (esp. as **preconceived** *adj.*) form (an idea or opinion etc.) beforehand; anticipate in thought.

preconception /ˌpriːkənˈsepʃ(ə)n/ *n.* **1** a preconceived idea. **2** a prejudice.

preconcert /ˌpriːkənˈsɜːt/ *v.tr.* arrange or organize beforehand.

precondition /ˌpriːkənˈdɪʃ(ə)n/ *n. & v.* ● *n.* a prior condition, that must be fulfilled before other things can be done. ● *v.tr.* bring into a required condition beforehand.

preconize /ˈpriːkəˌnaɪz/ *v.tr.* (also **-ise**) **1** proclaim or commend publicly. **2** summon by name. **3** *RC Ch.* (of the pope) approve publicly the appointment of (a bishop). □ **preconization** /ˌpriːkənaɪˈzeɪʃ(ə)n/ *n.* [ME f. med.L *praeconizare* f. L *praeco -onis* herald]

preconscious /priːˈkɒnʃəs/ *adj. & n. Psychol.* ● *adj.* **1** preceding consciousness. **2** of or associated with a part of the mind below the level of immediate conscious awareness, from which memories and emotions can be recalled. ● *n.* this part of the mind. □ **preconsciousness** *n.*

precook /priːˈkʊk/ *v.tr.* cook in advance.

precool /priːˈkuːl/ *v.tr.* cool in advance.

precordial /priːˈkɔːdɪəl/ *adj.* in front of or about the heart.

precostal /priːˈkɒst(ə)l/ *adj.* in front of the ribs.

precursor /prɪˈkɜːsə(r)/ *n.* **1 a** a forerunner. **b** a person who precedes in office etc. **2** a harbinger. **3** a substance from which another is formed by decay or chemical reaction etc. [L *praecursor* f. *praecurrere praecurs-* (as PRAE-, *currere* run)]

precursory /prɪˈkɜːsərɪ/ *adj.* (also **precursive** /-sɪv/) **1** preliminary, introductory. **2** (foll. by *of*) serving as a harbinger of. [L *praecursorius* (as PRECURSOR)]

precut /priːˈkʌt/ *v.tr.* (*past* and *past part.* **-cut**) cut in advance.

predacious /prɪˈdeɪʃəs/ *adj.* (also **predaceous**) **1** (of an animal) predatory. **2** relating to such animals (*predacious instincts*). □ **predaciousness** *n.* **predacity** /-ˈdæsɪtɪ/ *n.* [L *praeda* booty: cf. *audacious*]

predate[1] /priːˈdeɪt/ *v.tr.* exist or occur at a date earlier than.

predate[2] /prɪˈdeɪt/ *v.tr.* & (foll. by *on, upon*) *intr.* act as a predator of; prey on. [back-form. f. PREDATOR, PREDATION]

predation /prɪˈdeɪʃ(ə)n/ *n.* **1** (usu. in *pl.*) = DEPREDATION. **2** *Zool.* the natural preying of one animal on others. [L *praedatio -onis* taking of booty f. L *praeda* booty]

predator /ˈpredətə(r)/ *n.* **1** an animal naturally preying on others. **2** a person, state, etc., whose behaviour is rapacious or exploitative. [L *praedator* plunderer f. *praedari* seize as plunder f. *praeda* booty (as PREDACIOUS)]

predatory /ˈpredətərɪ/ *adj.* **1** (of an animal) preying naturally upon others. **2** ruthlessly acquisitive, exploitative; aggressive in business etc. □ **predatorily** *adv.* **predatoriness** *n.* [L *praedatorius* (as PREDATOR)]

predecease /ˌpriːdɪˈsiːs/ *v. & n.* ● *v.tr.* die earlier than (another person). ● *n.* a death preceding that of another.

predecessor /ˈpriːdɪˌsesə(r)/ *n.* **1** a former holder of an office or position with respect to a later holder (*my immediate predecessor*). **2** an ancestor. **3** a thing to which another has succeeded (*the new plan will share the fate of its predecessor*). [ME f. OF *predecesseur* f. LL *praedecessor* (as PRAE-, *decessor* retiring officer, as DECEASE)]

pre-decimal /priːˈdesɪm(ə)l/ *adj.* of or relating to a time before the introduction of a decimal system, esp. of coinage. □ **pre-decimalization** /-ˌdesɪməlaɪˈzeɪʃ(ə)n/ *n.*

predella /prɪˈdelə/ *n.* **1** an altar-step, or raised shelf at the back of an altar. **2** a painting or sculpture on this, or any picture forming an appendage to a larger one esp. beneath an altarpiece. [It., = stool]

predestinarian /ˌpriːdestɪˈneərɪən/ *n. & adj.* ● *n.* a person who believes in predestination. ● *adj.* of or relating to predestination.

predestinate *v. & adj.* ● *v.tr.* /priːˈdestɪˌneɪt/ = PREDESTINE. ● *adj.* /priːˈdestɪnət/ predestined. [ME f. eccl.L *praedestinare praedestinat-* (as PRAE-, *destinare* establish)]

predestination /ˌpriːdestɪˈneɪʃ(ə)n/ *n.* (in Christian theology) the divine foreordaining of all that will happen, esp. with regard to the salvation of some and not others. [ME f. eccl.L *praedestinatio* (as PREDESTINATE)]

predestine /priːˈdestɪn/ *v.tr.* **1** determine beforehand. **2** ordain in advance by divine will or as if by fate. [ME f. OF *predestiner* or eccl.L *praedestinare* PREDESTINATE *v.*]

predetermine /ˌpriːdɪˈtɜːmɪn/ *v.tr.* **1** determine or decree beforehand. **2** predestine. □ **predeterminable** *adj.* **predeterminate** *adj.* **predetermination** /-ˌtɜːmɪˈneɪʃ(ə)n/ *n.* [LL *praedeterminare* (as PRAE-, DETERMINE)]

predial /ˈpriːdɪəl/ *adj. & n. hist.* ● *adj.* **1 a** of land or farms. **b** rural, agrarian. **c** (of a slave, tenant, etc.) attached to farms or the land. **2** (of a tithe) consisting of agricultural produce. ● *n.* a predial slave. [med.L *praedialis* f. L *praedium* farm]

predicable /ˈpredɪkəb(ə)l/ *adj. & n.* ● *adj.* that may be predicated or affirmed. ● *n.* **1** a predicable thing. **2** (in *pl.*) Logic the five classes to which predicates belong: genus, species, difference, property, and accident. □ **predicability** /ˌpredɪkəˈbɪlɪtɪ/ *n.* [med.L *praedicabilis* that may be affirmed (as PREDICATE)]

predicament /prɪˈdɪkəmənt/ *n.* **1** a difficult, unpleasant, or embarrassing situation. **2** *Philos.* a category in (esp. Aristotelian) logic. [ME (in sense 2) f. LL *praedicamentum* thing predicated: see PREDICATE]

predicant /ˈpredɪkənt/ *adj. & n.* ● *adj. hist.* (of a religious order, esp. the Dominicans) engaged in preaching. ● *n.* **1** *hist.* a predicant person, esp. a Dominican friar. **2** *S. Afr.* = PREDIKANT. [L *praedicans* part. of *praedicare* (as PREDICATE)]

predicate *v. & n.* ● *v.tr.* /ˈpredɪˌkeɪt/ **1** assert (something) about the subject of a proposition. **2** (foll. by *on*) found or base (a statement etc.) on. ● *n.* /ˈpredɪkət/ **1** *Gram.* what is said about the subject of a sentence etc. (e.g. *went home* in *John went home*). **2** *Logic* **a** what is predicated. **b** what is affirmed or denied of the subject by means of the copula (e.g. *mortal* in *all men are mortal*). □ **predication** /ˌpredɪˈkeɪʃ(ə)n/ *n.* [L *praedicare praedicat-* proclaim (as PRAE-, *dicare* declare)]

predicative /prɪˈdɪkətɪv/ *adj.* **1** *Gram.* (of an adjective or noun) forming or contained in the predicate, as *old* in *the dog is old* (but not in *the old dog*) and *house* in *there is a large house* (opp. ATTRIBUTIVE). **2** that predicates. □ **predicatively** *adv.* [L *praedicativus* (as PREDICATE)]

predict /prɪˈdɪkt/ *v.tr.* (often foll. by *that* + clause) make a statement about the future; foretell, prophesy. □ **predictive** *adj.* **predictively** *adv.* **predictor** *n.* [L *praedicere praedict-* (as PRAE-, *dicere* say)]

predictable /prɪˈdɪktəb(ə)l/ *adj.* **1** that can be predicted or is to be expected. **2** (of a person) behaving in a way that is easy to predict. □ **predictably** *adv.* **predictability** /-ˌdɪktəˈbɪlɪtɪ/ *n.*

prediction /prɪˈdɪkʃ(ə)n/ *n.* **1** the art of predicting or the process of being predicted. **2** a thing predicted; a forecast. [L *praedictio -onis* (as PREDICT)]

predigest /ˌpriːdaɪˈdʒest, -dɪˈdʒest/ *v.tr.* **1** render (food) easily digestible before being eaten. **2** make (reading matter) easier to read or understand. □ **predigestion** /-ˈdʒestʃən/ *n.*

predikant /ˌpreɪdɪˈkɑːnt, ˈpredɪˌkænt/ *n. S. Afr.* a minister of the Dutch Reformed Church. [Du. (as PREDICANT)]

predilection /ˌpriːdɪˈlekʃ(ə)n/ *n.* (often foll. by *for*) a preference or special liking. [F *prédilection* ult. f. L *praediligere praedilect-* prefer (as PRAE-, *diligere* select): see DILIGENT]

predispose /ˌpriːdɪˈspəʊz/ *v.tr.* **1** influence favourably in advance. **2** (foll. by *to*, or *to* + infin.) render liable or inclined beforehand. □ **predisposition** /-ˌdɪspəˈzɪʃ(ə)n/ *n.*

prednisone /ˈprednɪˌzəʊn/ *n.* a synthetic drug similar to cortisone,

used to relieve rheumatic and allergic conditions and to treat leukaemia. [perh. f. *pregnant* + *diene* + *cortisone*]

predominant /prɪˈdɒmɪnənt/ *adj.* **1** predominating. **2** being the strongest or main element. □ **predominance** *n.* **predominantly** *adv.*

predominate /prɪˈdɒmɪˌneɪt/ *v.intr.* **1** (foll. by *over*) have or exert control. **2** be superior. **3** be the strongest or main element; preponderate (*a garden in which dahlias predominate*). [med.L *praedominari* (as PRAE-, DOMINATE)]

predominately /prɪˈdɒmɪnətlɪ/ *adv.* = PREDOMINANTLY (see PREDOMINANT). [rare *predominate* (adj.) = PREDOMINANT]

predoom /priːˈduːm/ *v.tr.* doom beforehand.

predorsal /priːˈdɔːs(ə)l/ *adj.* in front of the dorsal region.

predynastic /ˌpriːdɪˈnæstɪk/ *adj.* of or relating to a period before the normally recognized dynasties (esp. of ancient Egypt).

pre-echo /priːˈekəʊ/ *n.* (*pl.* **-oes**) **1** a faint copy heard just before an actual sound in a recording, caused by the accidental transfer of signals. **2** a foreshadowing.

pre-eclampsia /ˌpriːɪˈklæmpsɪə/ *n. Med.* a condition of pregnancy characterized by high blood pressure and other symptoms associated with eclampsia. □ **pre-eclamptic** *adj. & n.*

pre-elect /ˌpriːɪˈlekt/ *v.tr.* elect beforehand.

pre-election /ˌpriːɪˈlekʃ(ə)n/ *n. & adj.* ● *n.* an election held beforehand. ● *attrib.adj.* done, given, or existing before an election (*pre-election meeting*).

pre-embryo /priːˈembrɪəʊ/ *n. Med.* a human embryo in the first fourteen days after fertilization. □ **pre-embryonic** /-ˌembrɪˈɒnɪk/ *adj.*

pre-eminent /priːˈemɪnənt/ *adj.* **1** excelling others. **2** outstanding; distinguished in some quality. **3** principal, leading; predominant. □ **pre-eminence** *n.* **pre-eminently** *adv.* [ME f. L *praeeminens* (as PRAE-, EMINENT)]

pre-empt /priːˈempt/ *v.* **1** *tr.* **a** forestall. **b** acquire or appropriate in advance. **2** *tr.* prevent (an attack) by disabling the enemy. **3** *tr.* obtain by pre-emption. **4** *tr. US* take for oneself (esp. public land) so as to have the right of pre-emption. **5** *intr. Bridge* make a pre-emptive bid. □ **pre-emptor** *n.* **pre-emptory** *adj.* [back-form. f. PRE-EMPTION]

pre-emption /priːˈempʃ(ə)n/ *n.* **1 a** the purchase or appropriation by one person or party before the opportunity is offered to others. **b** *N. Amer. & Austral. hist.* the right to purchase (esp. public land) in this way. **c** a piece of land so obtained. **2** prior appropriation or acquisition. **3** *Mil.* the action or strategy of making a pre-emptive attack. [med.L *praeemptio* (as PRAE-, *emere empt-* buy)]

pre-emptive /priːˈemptɪv/ *adj.* **1** pre-empting; serving to pre-empt. **2** (of military action) intended to prevent attack by disabling the enemy (*a pre-emptive strike*). **3** *Bridge* (of a bid) intended to be high enough to discourage further bidding.

preen /priːn/ *v.tr. & refl.* **1** (of a bird) tidy (the feathers or itself) with its beak. **2** (of a person) smarten or admire (oneself, one's hair, clothes, etc.). **3** (often foll. by *on*) congratulate or pride (oneself). □ **preen gland** a gland situated at the base of a bird's tail and producing oil used in preening. □ **preener** *n.* [ME, app. var. of earlier *prune* (perh. rel. to PRUNE[2]): assoc. with Sc. & dial. *preen* pierce, pin]

pre-engage /ˌpriːɪnˈɡeɪdʒ/ *v.tr.* engage beforehand. □ **pre-engagement** *n.*

pre-establish /ˌpriːɪˈstæblɪʃ/ *v.tr.* establish beforehand.

pre-exist /ˌpriːɪɡˈzɪst/ *v.intr. & tr.* **1** *intr.* exist at an earlier time. **2** *tr.* exist earlier than. □ **pre-existence** *n.* **pre-existent** *adj.*

pref. *abbr.* **1** prefix. **2** preface. **3 a** preference. **b** preferred.

prefab /ˈpriːfæb/ *n. Brit. colloq.* a prefabricated building. [abbr.]

prefabricate /priːˈfæbrɪˌkeɪt/ *v.tr.* **1** manufacture sections of (a building etc.) prior to their assembly on a site. **2** produce in an artificially standardized way. □ **prefabrication** /-ˌfæbrɪˈkeɪʃ(ə)n/ *n.*

preface /ˈprefɪs/ *n. & v.* ● *n.* **1** an introduction to a book stating its subject, scope, etc. **2** the preliminary part of a speech. **3** *Eccl.* the introduction to the central part of the Eucharistic service. ● *v.tr.* **1** (foll. by *with*) introduce or begin (a speech or event) (*prefaced my remarks with a warning*). **2** provide (a book etc.) with a preface. **3** (of an event etc.) lead up to (another). □ **prefatorial** /ˌprefəˈtɔːrɪəl/ *adj.* **prefatory** /ˈprefətərɪ/ *adj.* [ME f. OF f. med.L *praefatia* for L *praefatio* f. *praefari* (as PRAE-, *fari* speak)]

prefect /ˈpriːfekt/ *n.* **1 a** a chief officer, magistrate, or governor. **b** the chief administrative officer of a department in France. **2** esp. *Brit.* a senior pupil in a school etc. assigned various duties and authorized to enforce discipline etc. **3** *Rom. Hist.* a senior magistrate or military commander. □ **prefectoral** /prɪˈfektərəl/ adj. **prefectorial** /ˌpriːfekˈtɔːrɪəl/ adj. [ME f. OF f. L praefectus past part. of praeficere set in authority over (as PRAE-, facere make)]

prefecture /ˈpriːfektjʊə(r)/ n. **1** a district under the government of a prefect. **2 a** a prefect's office or tenure. **b** the official residence of a prefect. □ **prefectural** /prɪˈfektjʊərəl/ adj. [F préfecture or L praefectura (as PREFECT)]

prefer /prɪˈfɜː(r)/ v.tr. (**preferred, preferring**) **1** (often foll. by to, or to + infin.) choose rather; like better (would prefer to stay; prefers coffee to tea). **2** submit (information, an accusation, etc.) for consideration. **3** promote or advance (a person). □ **preferred shares** (or **stock**) = preference shares (or stock). [ME f. OF preferer f. L praeferre (as PRAE-, ferre latbear)]

preferable /ˈprefərəb(ə)l, disp. prɪˈfɜːr-/ adj. **1** to be preferred. **2** more desirable. □ **preferably** adv.

preference /ˈprefərəns/ n. **1** the act or an instance of preferring or being preferred. **2** a thing preferred. **3 a** the favouring of one person etc. before others. **b** Commerce the favouring of one country by admitting its products at a lower import duty. **4** Law a prior right, esp. to the payment of debts. □ **in preference to** as a thing preferred over (another). **preference shares** (or **stock**) Brit. shares or stock whose entitlement to dividend takes priority over that of ordinary shares. [F préférence f. med.L praeferentia (as PREFER)]

preferential /ˌprefəˈren∫(ə)l/ adj. **1** of or involving preference (preferential treatment). **2** giving or receiving a favour. **3** Commerce (of a tariff etc.) favouring particular countries. **4** (of voting) in which the voter puts candidates in order of preference. □ **preferentially** adv. [as PREFERENCE, after differential]

preferment /prɪˈfɜːmənt/ n. promotion to office.

prefigure /priːˈfɪɡə(r)/ v.tr. **1** represent beforehand by a figure or type. **2** imagine beforehand. □ **prefigurement** n. **prefiguration** /-ˌfɪɡəˈreɪ∫(ə)n/ n. **prefigurative** /-ˈfɪɡərətɪv/ adj. [ME f. eccl.L praefigurare (as PRAE-, FIGURE)]

prefix /ˈpriːfɪks/ n. & v. ● n. **1** a verbal element placed at the beginning of a word to adjust or qualify its meaning (e.g. ex-, non-, re-) or (in some languages) as an inflectional formative. **2** a title placed before a name (e.g. Mr). ● v.tr. (often foll. by to) **1** add as an introduction. **2** join (a word or element) as a prefix. □ **prefixation** /ˌpriːfɪkˈseɪ∫(ə)n/ n. **prefixion** /priːˈfɪk∫(ə)n/ n. [earlier as verb: ME f. OF prefixer (as PRE-, FIX): (n.) f. L praefixum]

preflight /ˈpriːflaɪt/ attrib.adj. occurring or provided before an aircraft flight.

preform /priːˈfɔːm/ v.tr. form beforehand. □ **preformation** /ˌpriːfɔːˈmeɪ∫(ə)n/ n.

preformative /priːˈfɔːmətɪv/ adj. & n. ● adj. **1** forming beforehand. **2** prefixed as the formative element of a word. ● n. a preformative syllable or letter.

prefrontal /priːˈfrʌnt(ə)l/ adj. **1** in front of the frontal bone of the skull. **2** in the forepart of the frontal lobe of the brain.

preglacial /priːˈɡleɪ∫(ə)l, -ˈɡleɪsɪəl/ adj. before a glacial period.

pregnable /ˈpreɡnəb(ə)l/ adj. able to be captured etc.; not impregnable. [ME f. OF prenable takable: see IMPREGNABLE[1]]

pregnancy /ˈpreɡnənsɪ/ n. (pl. **-ies**) the condition or an instance of being pregnant.

pregnant /ˈpreɡnənt/ adj. **1** (of a woman or female animal) having a child or young developing in the uterus. **2** full of meaning; significant or suggestive (a pregnant pause). **3** (esp. of a person's mind) imaginative, inventive. **4** (foll. by with) plentifully provided (pregnant with danger). □ **pregnant construction** Gram. one in which more is implied than the words express (e.g. not have a chance implying of success etc.). □ **pregnantly** adv. (in sense 2). [ME f. F prégnant or L praegnans -antis, earlier praegnas (prob. as PRAE-, (g)nasci be born)]

preheat /priːˈhiːt/ v.tr. heat beforehand.

prehensile /prɪˈhensaɪl/ adj. Zool. (of a tail or limb) capable of grasping. □ **prehensility** /ˌpriːhenˈsɪlɪtɪ/ n. [F préhensile f. L prehendere prehens- (as PRE-, hendere grasp)]

prehension /prɪˈhen∫(ə)n/ n. **1** grasping, seizing. **2** mental apprehension. [L prehensio (as PREHENSILE)]

prehistoric /ˌpriːhɪˈstɒrɪk/ adj. **1** of or relating to the period before written records. (See PREHISTORY.) **2** colloq. utterly out of date. □ **prehistorically** adv. **prehistorian** /-ˈstɔːrɪən/ n. [F préhistorique (as PRE-, HISTORIC)]

prehistory /priːˈhɪstərɪ/ n. the period before written records. The prehistoric era is conventionally divided into Stone Age, Bronze Age, and Iron Age periods, on the basis of the material used for weapons and tools. Originally devised by the Danish museum curator Christian Thomsen (1788–1865), the system was later elaborated and refined, and was confirmed (at least for European areas) by stratification of finds; however, it is neither a guide to absolute dates nor an essential evolutionary sequence, since not all its stages are represented in all parts of the world, and there is often a considerable time-lag between the first appearance of metal artefacts and a fully developed metal-working technology in an area. Until its formulation, however, there was no framework into which archaeological discoveries could be fitted, and it remains a convenient terminology.

prehuman /priːˈhjuːmən/ adj. existing before the time of humans.

pre-ignition /ˌpriːɪɡˈnɪ∫(ə)n/ n. the premature firing of the explosive mixture in an internal-combustion engine.

prejudge /priːˈdʒʌdʒ/ v.tr. **1** form a premature judgement on (a person, issue, etc.). **2** pass judgement on (a person) before a trial or proper enquiry. □ **prejudgement** n. **prejudication** /-ˌdʒuːdɪˈkeɪ∫(ə)n/ n.

prejudice /ˈpredʒʊdɪs/ n. & v. ● n. **1 a** a preconceived opinion. **b** (usu. foll. by against, in favour of) bias or partiality. **c** intolerance of or discrimination against a person or group, esp. on account of race, religion, or gender; bigotry (racial prejudice). **2** harm or injury that results or may result from some action or judgement (to the prejudice of). ● v.tr. **1** impair the validity or force of (a right, claim, statement, etc.). **2 a** cause (a person) to have a prejudice. **b** (as **prejudiced** adj.) not impartial; bigoted. □ **without prejudice** (often foll. by to) without detriment (to any existing right or claim). [ME f. OF prejudice f. L praejudicium (as PRAE-, judicium judgement)]

prejudicial /ˌpredʒʊˈdɪ∫(ə)l/ adj. causing prejudice; detrimental. □ **prejudicially** adv. [ME f. OF prejudiciel (as PREJUDICE)]

prelacy /ˈpreləsɪ/ n. (pl. **-ies**) **1** church government by prelates. **2** (prec. by the) prelates collectively. **3** the office or rank of prelate. [ME f. AF prelacie f. med.L prelatia (as PRELATE)]

prelapsarian /ˌpriːlæpˈseərɪən/ adj. before the Fall of Man.

prelate /ˈprelət/ n. **1** a high ecclesiastical dignitary, e.g. a bishop. **2** hist. an abbot or prior. □ **prelatic** /prɪˈlætɪk/ adj. **prelatical** adj. [ME f. OF prelat f. med.L praelatus past part.: see PREFER]

prelature /ˈprelətjʊə(r)/ n. **1** the office of prelate. **2** (prec. by the) prelates collectively. [F prélature f. med.L praelatura (as PRELATE)]

prelim /ˈpriːlɪm, prɪˈlɪm/ n. colloq. **1** a preliminary examination, esp. at a university. **2** (in pl.) the pages preceding the text of a book. [abbr.]

preliminary /prɪˈlɪmɪnərɪ/ adj., n., & adv. ● adj. introductory, preparatory. ● n. (pl. **-ies**) (usu. in pl.) **1** a preliminary action or arrangement (dispense with the preliminaries). **2** a preliminary trial or contest. **3** (in pl.) = PRELIM 2. ● adv. (foll. by to) preparatory to; in advance of (was completed preliminary to the main event). □ **preliminarily** adv. [mod.L praeliminaris or F préliminaire (as PRE-, L limen liminis threshold)]

preliterate /priːˈlɪtərət/ adj. of or relating to a society or culture that has not developed the use of writing.

prelude /ˈpreljuːd/ n. & v. ● n. (often foll. by to) **1** an action, event, or situation serving as an introduction. **2** the introductory part of a poem etc. **3 a** an introductory piece of music, often preceding a fugue or forming the first piece of a suite or beginning an act of an opera. **b** a short piece of music of a similar type, esp. for the piano. ● v.tr. **1** serve as a prelude to. **2** introduce with a prelude. □ **preludial** /prɪˈljuːdɪəl/ adj. [F prélude or med.L praeludium f. L praeludere praelus- (as PRAE-, ludere play)]

premarital /priːˈmærɪt(ə)l/ adj. existing or (esp. of sexual relations) occurring before marriage. □ **premaritally** adv.

premature /ˈpremətjʊə(r), ˌpreməˈtjʊə(r)/ adj. **1 a** occurring or done before the usual or proper time; too early (a premature decision). **b** too hasty (must not be premature). **2** (of a baby) born (esp. three or more weeks) before the end of the full term of gestation. □ **prematurely** adv. **prematureness** n. **prematurity** /ˌpreməˈtjʊərɪtɪ/ n. [L praematurus very early (as PRAE-, MATURE)]

premaxillary /ˌpriːmækˈsɪlərɪ/ adj. Anat. & Zool. in front of the upper jaw.

premed /priːˈmed/ n. colloq. **1** = PREMEDICATION. **2** a premedical course or student. [abbr.]

premedical /priːˈmedɪk(ə)l/ adj. of or relating to study in preparation for a course in medicine.

premedication /ˌpriːmedɪˈkeɪ∫(ə)n/ n. medication to prepare for an operation or other treatment.

premeditate /priːˈmedɪˌteɪt/ v.tr. (often as **premeditated** adj.) think out or plan (an action) beforehand (*premeditated murder*). □ **premeditation** /-ˌmedɪˈteɪʃ(ə)n/ n. [L *praemeditari* (as PRAE-, MEDITATE)]

premenstrual /priːˈmenstrʊəl/ adj. of, occurring, or experienced, before menstruation (*premenstrual tension*). □ **premenstrual syndrome** any of a complex of symptoms (including tension, fluid retention, etc.) experienced by some women in the days immediately preceding menstruation. □ **premenstrually** adv.

premier /ˈpremɪə(r)/ n. & adj. ● n. (usu. **Premier**) a Prime Minister or other head of government. ● adj. **1** first in importance, order, or time. **2** of earliest creation (*premier earl*). □ **premiership** n. [ME f. OF = first, f. L (as PRIMARY)]

première /ˈpremɪˌeə(r)/ n., adj., & v. ● n. the first performance or showing of a play or film. ● adj. = PREMIER adj. 1 ● v.tr. give a première of. [F, fem. of *premier* (adj.) (as PREMIER)]

premillennial /ˌpriːmɪˈlenɪəl/ adj. existing or occurring before the millennium, esp. with reference to the supposed second coming of Christ. □ **premillennialism** n. **premillennialist** n.

premise n. & v. ● n. /ˈpremɪs/ **1** Logic = PREMISS. **2** (in pl.) **a** a house or building with its grounds and appurtenances. **b** Law houses, lands, or tenements previously specified in a document etc. ● v. /prɪˈmaɪz, ˈpremɪs/ **1** tr. say or write by way of introduction. **2** tr. & intr. assert or assume as a premiss. □ **on the premises** in the building etc. concerned. [ME f. OF *premisse* f. med.L *praemissa* (*propositio*) (proposition) set in front f. L *praemittere praemiss-* (as PRAE-, *mittere* send)]

premiss /ˈpremɪs/ n. Logic a previous statement from which another is inferred. [var. of PREMISE]

premium /ˈpriːmɪəm/ n. **1** an amount to be paid for a contract of insurance. **2 a** a sum added to interest, wages, etc.; a bonus. **b** a sum added to ordinary charges. **3** a reward or prize. **4** (attrib.) (of a commodity) of best quality and therefore more expensive. **5** an item offered free or cheaply as an incentive to buy, sample, or subscribe to something. □ **at a premium 1** above the usual or nominal price. **2** scarce and in demand. **put a premium on 1** provide or act as an incentive to. **2** attach special value to. [L *praemium* booty, reward (as PRAE-, *emere* buy, take)]

Premium Bond n. (in the UK) a government security without interest but with a number included in a regular draw for cash prizes, issued since 1956. Also called *Savings Bond*.

premolar /priːˈməʊlə(r)/ adj. & n. ● adj. in front of a molar tooth. ● n. (in an adult human) each of eight teeth situated in pairs between each of the four canine teeth and each first molar.

premonition /ˌpreməˈnɪʃ(ə)n, ˌpriːm-/ n. a forewarning; a presentiment. □ **premonitor** /prɪˈmɒnɪtə(r)/ n. **premonitory** adj. [F *prémonition* or LL *praemonitio* f. L *praemonere praemonit-* (as PRAE-, *monere* warn)]

Premonstratensian /prɪˌmɒnstrəˈtensɪən/ adj. & n. hist. ● adj. of or relating to an order of regular canons founded at Prémontré in France in 1120, or of the corresponding order of nuns. ● n. a member of either of these orders. [med.L *Praemonstratensis* f. *Praemonstratus* the abbey of Prémontré (lit. 'foreshown')]

premorse /prɪˈmɔːs/ adj. Bot. & Zool. with the end abruptly terminated. [L *praemordere praemors-* bite off (as PRAE-, *mordere* bite)]

prenatal /priːˈneɪt(ə)l/ adj. of or concerning the period before birth. □ **prenatally** adv.

prentice /ˈprentɪs/ n. & v. archaic ● n. = APPRENTICE. ● v.tr. (as **prenticed** adj.) apprenticed. □ **prentice hand** an inexperienced hand. □ **prenticeship** n. [ME f. APPRENTICE]

preoccupation /priːˌɒkjʊˈpeɪʃ(ə)n/ n. **1** the state of being preoccupied. **2** a thing that engrosses the mind. [F *préoccupation* or L *praeoccupatio* (as PREOCCUPY)]

preoccupy /priːˈɒkjʊˌpaɪ/ v.tr. (-ies, -ied) **1** (of a thought etc.) dominate or engross the mind of (a person) to the exclusion of other thoughts. **2** (as **preoccupied** adj.) otherwise engrossed; mentally distracted. **3** occupy beforehand. [PRE- + OCCUPY, after L *praeoccupare* seize beforehand]

preocular /priːˈɒkjʊlə(r)/ adj. Zool. in front of the eye.

preordain /ˌpriːɔːˈdeɪn/ v.tr. ordain or determine beforehand.

prep /prep/ n. colloq. **1** Brit. **a** school work done outside lessons, esp. in an independent school. **b** a period when this is done. **2** US a student in a preparatory school. □ **prep school** = *preparatory school*. [abbr. of PREPARATION, PREPARATORY]

prep. abbr. preposition.

prepack /priːˈpæk/ v.tr. (also **pre-package** /-ˈpækɪdʒ/) pack (goods) on the site of production or before retail.

prepaid past and past part. of PREPAY.

preparation /ˌprepəˈreɪʃ(ə)n/ n. **1** the act or an instance of preparing; the process of being prepared. **2** (often in pl.) something done to make ready. **3** a specially prepared substance, esp. a food or medicine. **4** work done by school pupils to prepare for a lesson. **5** Mus. the sounding of the discordant note in a chord in the preceding chord where it is not discordant, lessening the effect of the discord. [ME f. OF f. L *praeparatio -onis* (as PREPARE)]

preparative /prɪˈpærətɪv/ adj. & n. ● adj. preparatory. ● n. **1** Mil. & Naut. a signal on a drum, bugle, etc., as an order to make ready. **2** a preparatory act. □ **preparatively** adv. [ME f. OF *preparatif -ive* f. med.L *praeparativus* (as PREPARE)]

preparatory /prɪˈpærətərɪ/ adj. & adv. ● adj. (often foll. by *to*) serving to prepare; introductory. ● adv. (often foll. by *to*) in a preparatory manner (*was packing preparatory to departure*). □ **preparatory school 1** (in Britain) a junior school in which pupils are prepared for public school. **2** (in the US) a usu. private school in which pupils are prepared for college or university. □ **preparatorily** adv. [ME f. LL *praeparatorius* (as PREPARE)]

prepare /prɪˈpeə(r)/ v. **1** tr. make or get ready for use, consideration, etc. **2** tr. make ready or assemble (food, a meal, etc.) for eating. **3 a** tr. make (a person or oneself) ready or disposed in some way (*prepares students for university*; *prepared them for a shock*). **b** intr. put oneself or things in readiness, get ready (*prepare to jump*). **4** tr. make (a chemical product etc.) by a regular process; manufacture. **5** tr. Mus. lead up to (a discord). □ **be prepared** (often foll. by *for*, or to + infin.) be ready; be disposed or willing. □ **preparer** n. [ME f. F *préparer* or L *praeparare* (as PRAE-, *parare* make ready)]

preparedness /prɪˈpeərɪdnɪs/ n. a state of readiness, esp. for war.

prepay /priːˈpeɪ/ v.tr. (past and past part. **prepaid** /-ˈpeɪd/) **1** pay (a charge) in advance. **2** pay postage on (a letter or parcel etc.) before posting. □ **prepayable** adj. **prepayment** n.

prepense /prɪˈpens/ adj. (usu. placed after noun) esp. Law deliberate, intentional (*malice prepense*). □ **prepensely** adv. [earlier *prepensed* past part. of obs. *prepense* (v.) alt. f. earlier *purpense* f. AF & OF *purpenser* (as PUR-, *penser*): see PENSIVE]

preplan /priːˈplæn/ v.tr. (**preplanned**, **preplanning**) plan in advance.

preponderant /prɪˈpɒndərənt/ adj. surpassing in influence, power, number, or importance; predominant. □ **preponderance** n. **preponderantly** adv.

preponderate /prɪˈpɒndəˌreɪt/ v.intr. (often foll. by *over*) **1 a** be greater in influence, quantity, or number. **b** predominate. **2 a** be of greater importance. **b** weigh more. [L *praeponderare* (as PRAE-, PONDER)]

preposition /ˌprepəˈzɪʃ(ə)n/ n. Gram. a word governing (and usu. preceding) a noun or pronoun and expressing a relation to another word or element, as in: 'the man *on* the platform', 'came *after* dinner', 'what did you do it *for*?'. □ **prepositional** adj. **prepositionally** adv. [ME f. L *praepositio* f. *praeponere praeposit-* (as PRAE-, *ponere* place)]

prepositive /priːˈpɒzɪtɪv/ adj. Gram. (of a word, particle, etc.) that should be placed before or prefixed. [LL *praepositivus* (as PREPOSITION)]

prepossess /ˌpriːpəˈzes/ v.tr. **1** (usu. in passive) (of an idea, feeling, etc.) take possession of (a person); imbue. **2 a** prejudice (usu. favourably and spontaneously). **b** (as **prepossessing** adj.) attractive, appealing. □ **prepossession** /-ˈzeʃ(ə)n/ n.

preposterous /prɪˈpɒstərəs/ adj. **1** utterly absurd; outrageous. **2** contrary to nature, reason, or common sense. □ **preposterously** adv. **preposterousness** n. [L *praeposterus* reversed, absurd (as PRAE-, *posterus* coming after)]

prepostor var. of PRAEPOSTOR.

prepotent /prɪˈpəʊt(ə)nt/ adj. **1** greater than others in power, influence, etc. **2** Biol. **a** having a greater power of fertilization. **b** dominant in transmitting hereditary qualities. □ **prepotence** n. **prepotency** n. [ME f. L *praepotens -entis*, part. of *praeposse* (as PRAE-, *posse* be able)]

preppy /ˈprepɪ/ n. & adj. N. Amer. colloq. ● n. (pl. **-ies**) a person attending an expensive private school or who looks like such a person (with short hair, blazer, etc.). ● adj. (**preppier**, **preppiest**) **1** like a preppy. **2** neat and fashionable. [*prep school* + -Y²]

preprandial /priːˈprændɪəl/ adj. formal or joc. before dinner or lunch. [PRE- + L *prandium* a meal]

pre-preference /pri:ˈprefərəns/ adj. Brit. (of shares, claims, etc.) ranking before preference shares etc.

preprint /ˈpri:prɪnt/ n. a printed document issued in advance of general publication.

preprocessor /pri:ˈprəʊsesə(r)/ n. a computer program that modifies data to conform with the input requirements of another program.

prepubescence /ˌpri:pjuˈbes(ə)ns/ n. the time, esp. the last two or three years, before puberty. □ **prepubescent** adj.

prepublication /ˌpri:pʌblɪˈkeɪʃ(ə)n/ adj. & n. ● attrib.adj. produced or occurring before publication. ● n. publication in advance or beforehand.

prepuce /ˈpri:pju:s/ n. **1** = FORESKIN. **2** the fold of skin surrounding the clitoris. □ **preputial** /pri:ˈpju:ʃ(ə)l/ adj. [ME f. L praeputium]

prequel /ˈpri:kwəl/ n. a story, film, etc., whose events or concerns precede those of an existing work. [PRE- + SEQUEL]

Pre-Raphaelite /pri:ˈræfəˌlaɪt/ n. & adj. ● n. any of a group of English 19th-century artists, who consciously set out to emulate the work of Italian artists from before the time of Raphael. (See note below.) ● adj. **1** of or relating to the Pre-Raphaelites. **2** (**pre-Raphaelite**) (esp. of a woman) resembling the models painted by the later Pre-Raphaelites, esp. in having wavy auburn hair, pale complexion, and fey demeanour. □ **Pre-Raphaelite Brotherhood** the chosen name of the Pre-Raphaelites. (See note below.) □ **Pre-Raphaelitism** n.

▪ The Pre-Raphaelite Brotherhood was founded in 1848 by seven young English artists and writers, the major figures being Holman Hunt, Sir John Millais, and Dante Gabriel Rossetti. Abhorring the slickness and sentimentality of much Victorian art, they aimed to reintroduce into their work the simplicity and sincerity of early Italian painting. Stylistically, their work is characterized by bright colours, strong boundary lines, and a meticulous attention to detail, while their subjects are often biblical or literary. Much criticized by the art Establishment, although defended by the English critic John Ruskin, the movement began to dissipate in the early 1850s. The name, however, continued to be used in association with the markedly different later pictures of Rossetti and the work of Sir Edward Burne-Jones and William Morris, which typically depict scenes from classical mythology or medieval romance in a dreamy style which has given rise to the association of the term with a kind of romantic escapism.

pre-record /ˌpri:rɪˈkɔ:d/ v.tr. (esp. as **pre-recorded** adj.) record (esp. material for broadcasting) in advance.

prerequisite /pri:ˈrekwɪzɪt/ adj. & n. ● adj. required as a precondition. ● n. a prerequisite thing.

prerogative /prɪˈrɒgətɪv/ n. **1** a right or privilege exclusive to an individual or class. **2** (in full **royal prerogative**) Brit. the right of the sovereign, theoretically subject to no restriction. [ME f. OF prerogative or L praerogativa privilege (orig. to vote first) f. praerogativus asked first (as PRAE-, rogare ask)]

Pres. abbr. President.

presage /ˈpresɪdʒ/ n. & v. ● n. **1** an omen or portent. **2** a presentiment or foreboding. ● v.tr. (also /prɪˈseɪdʒ/) **1** portend, foreshadow. **2** give warning of (an event etc.) by natural means. **3** (of a person) predict or have a presentiment of. □ **presager** n. **presageful** /prɪˈseɪdʒfʊl/ adj. [ME f. F présage, présager f. L praesagium f. praesagire forebode (as PRAE-, sagire perceive keenly)]

presbyopia /ˌprezbɪˈəʊpɪə/ n. long-sightedness caused by loss of elasticity of the eye lens, occurring esp. in middle and old age. □ **presbyopic** /-ˈɒpɪk/ adj. [mod.L f. Gk presbus old man + ōps ōpos eye]

presbyter /ˈprezbɪtə(r)/ n. **1** an elder in the early Christian Church. **2** (in the Episcopal Church) a minister of the second order; a priest. **3** (in the Presbyterian Church) an elder. □ **presbytership** n. **presbyteral** /prezˈbɪtərəl/ adj. **presbyterate** /-ˈbɪtərət/ n. **presbyterial** /ˌprezbɪˈtɪərɪəl/ adj. [eccl.L f. Gk presbuteros elder, compar. of presbus old]

Presbyterian /ˌprezbɪˈtɪərɪən/ adj. & n. ● adj. (of a Church) governed by elders all of equal rank, esp. with reference to the national Church of Scotland. (See also PRESBYTERIANISM.) ● n. **1** a member of a Presbyterian Church. **2** an adherent of the Presbyterian system. [eccl.L presbyterium (as PRESBYTERY)]

Presbyterianism /ˌprezbɪˈtɪərɪəˌnɪz(ə)m/ n. a form of Protestant church government by presbyters or elders. The system was first put forward during the Reformation by John Calvin, who in 1541 established the first Presbyterian government in Geneva. He believed his system to be based on the Bible and early Church; Christ alone was the head of the Church, and all other members were equal under him.

Today Presbyterianism is followed by Reformed Churches in Europe, North America, and elsewhere, notably by the Church of Scotland, the only national Presbyterian Church. Government is by elected lay elders of equal rank as well as by ministers, in courts at local (congregational), regional, and national levels.

presbytery /ˈprezbɪtərɪ/ n. (pl. **-ies**) **1** the eastern part of a chancel beyond the choir; the sanctuary. **2 a** a body of presbyters, esp. a court next above a Kirk-session. **b** a district represented by this. **3** the house of a Roman Catholic priest. [ME f. OF presbiterie f. eccl.L f. Gk presbuterion (as PRESBYTER)]

preschool /pri:ˈsku:l/ adj. of or relating to the time before a child is old enough to go to school. □ **preschooler** n.

prescient /ˈpresɪənt/ adj. having foreknowledge or foresight. □ **prescience** n. **presciently** adv. [L praescire praescient- know beforehand (as PRAE-, scire know)]

prescind /prɪˈsɪnd/ v. **1** tr. (foll. by from) cut off (a part from a whole), esp. prematurely or abruptly. **2** intr. (foll. by from) leave out of consideration. [L praescindere (as PRAE-, scindere cut)]

prescribe /prɪˈskraɪb/ v. **1** tr. **a** advise the use of (a medicine etc.), esp. by an authorized prescription. **b** recommend, esp. as a benefit (prescribed a change of scenery). **2** tr. lay down or impose authoritatively. **3** intr. (foll. by to, for) assert a prescriptive right or claim. □ **prescriber** n. [L praescribere praescript- direct in writing (as PRAE-, scribere write)]

prescript /ˈpri:skrɪpt/ n. an ordinance, law, or command. [L praescriptum neut. past part.: see PRESCRIBE]

prescription /prɪˈskrɪpʃ(ə)n/ n. **1** the act or an instance of prescribing. **2 a** a doctor's (usu. written) instruction for the composition and use of a medicine. **b** a medicine prescribed. **3** (in full **positive prescription**) uninterrupted use or possession from time immemorial or for the period fixed by law as giving a title or right. **4 a** an ancient custom viewed as authoritative. **b** a claim founded on long use. □ **negative prescription** the time-limit within which an action or claim can be raised. [ME f. OF f. L praescriptio -onis (as PRESCRIBE)]

prescriptive /prɪˈskrɪptɪv/ adj. **1** prescribing. **2** Linguistics concerned with or laying down rules of usage (opp. DESCRIPTIVE 3). **3** based on prescription (prescriptive right). **4** prescribed by custom. □ **prescriptively** adv. **prescriptiveness** n. **prescriptivism** n. **prescriptivist** n. & adj. [LL praescriptivus (as PRESCRIBE)]

preselect /ˌpri:sɪˈlekt/ v.tr. select in advance. □ **preselection** /-ˈlekʃ(ə)n/ n.

preselective /ˌpri:sɪˈlektɪv/ adj. that can be selected or set in advance.

preselector /ˌpri:sɪˈlektə(r)/ n. a device for selecting a mechanical or electrical operation in advance of its execution, e.g. of a gear-change in a motor vehicle.

presence /ˈprez(ə)ns/ n. **1 a** the state or condition of being present (your presence is requested). **b** existence; location (the presence of a hospital nearby). **2** a place where a person is (was admitted to their presence). **3 a** a person's appearance or bearing, esp. when imposing (an august presence). **b** a person's force of personality (esp. have presence). **4** a person or thing that is present (the royal presence; there was a presence in the room). **5** representation for reasons of political influence (maintained a presence). □ **in the presence of** in front of; observed by. **presence chamber** a room in which a monarch or other distinguished person receives visitors. **presence of mind** calmness and self-command in sudden difficulty etc. [ME f. OF f. L praesentia (as PRESENT[1])]

present[1] /ˈprez(ə)nt/ adj. & n. ● adj. **1** (usu. predic.) being in the place in question (was present at the trial). **2 a** now existing, occurring, or being such (the present Duke; during the present season). **b** now being considered or discussed etc. (in the present case). **3** Gram. expressing an action etc. now going on or habitually performed (present participle; present tense). ● n. (prec. by the) **1** the time now passing (no time like the present). **2** Gram. the present tense. □ **at present** now. **by these presents** Law by this document (know all men by these presents). **for the present 1** just now. **2** as far as the present is concerned. **present company excepted** excluding those who are here now. **present-day** attrib.adj. of this time; modern. [ME f. OF f. L praesens -entis part. of praeesse be at hand (as PRAE-, esse be)]

present[2] /prɪˈzent/ v. & n. ● v.tr. **1** introduce, offer, or exhibit, esp. for public attention or consideration. **2 a** (with a thing as object, usu. foll. by to) offer, give, or award as a gift (to a person), esp. formally or ceremonially. **b** (with a person as object, foll. by with) make available to; cause to have (presented them with a new car; that presents us with a problem). **3 a** (of a company, producer, etc.) put (a form of entertainment) before the public. **b** (of a performer, compère, etc.) introduce or put before an audience. **4** introduce (a person) formally (*may I present my fiancé?; was presented at court*). **5** offer, give (compliments etc.) (*may I present my card; present my regards to your family*). **6 a** (of a circumstance) reveal (some quality etc.) (*this presents some difficulty*). **b** exhibit (an appearance etc.) (*presented a rough exterior*). **7** (of an idea etc.) offer or suggest itself. **8** deliver (a cheque, bill, etc.) for acceptance or payment. **9 a** (usu. foll. by *at*) aim (a weapon). **b** hold out (a weapon) in a position for aiming. **10** (*refl.* or *absol.*) *Med.* **a** (of an illness etc.) show itself, be manifest. **b** (of a patient) come forward for or undergo initial medical examination. **11** (*absol.*) *Med.* (of a part of a foetus) be directed toward the cervix at the time of delivery. **12** (foll. by *to*) *Law* bring formally under notice, submit (an offence, complaint, etc.). **13** (foll. by *to*) *Eccl.* recommend (a clergyman) to a bishop for institution to a benefice. ● *n.* the position of presenting arms in salute. □ **present arms** hold a rifle etc. vertically in front of the body as a salute. **present oneself 1** appear. **2** come forward for examination etc. □ **presenter** *n.* (esp. in sense 3b of *v.*). [ME f. OF *presenter* f. L *praesentare* (as PRESENT[1])]

present[3] /ˈprez(ə)nt/ *n.* a gift; a thing given or presented. □ **make a present of** give as a gift. [ME f. OF (as PRESENT[1]), orig. in phr. *mettre une chose en present à quelqu'un* put a thing into the presence of a person]

presentable /prɪˈzentəb(ə)l/ *adj.* **1** of good appearance; fit to be presented to other people. **2** fit for presentation. □ **presentably** *adv.* **presentableness** *n.* **presentability** /-ˌzentəˈbɪlɪtɪ/ *n.*

presentation /ˌprezənˈteɪʃ(ə)n/ *n.* **1 a** the act or an instance of presenting; the process of being presented. **b** a thing presented. **2** the manner or quality of presenting. **3** a demonstration or display of materials, information, etc.; a lecture. **4** an exhibition or theatrical performance. **5** a formal introduction. **6** the position of the foetus in relation to the cervix at the time of delivery. □ **presentational** *adj.* **presentationally** *adv.* [ME f. OF f. LL *praesentatio -onis* (as PRESENT[2])]

presentationism /ˌprezənˈteɪʃəˌnɪz(ə)m/ *n. Philos.* the doctrine that in perception the mind has immediate cognition of the object. □ **presentationist** *n.*

presentative /prɪˈzentətɪv/ *adj.* **1** *Philos.* subject to direct cognition. **2** *hist.* (of a benefice) to which a patron has the right of presentation. [prob. f. med.L (as PRESENTATION)]

presentee /ˌprezənˈtiː/ *n.* **1** the recipient of a present. **2** a person presented. [ME f. AF (as PRESENT[2])]

presentient /prɪˈsenʃ(ə)nt, prɪˈzen-/ *adj.* (often foll. by *of*) having a presentiment. [L *praesentiens* (as PRAE-, SENTIENT)]

presentiment /prɪˈzentɪmənt/ *n.* a vague expectation; a foreboding (esp. of misfortune). [obs. F *présentiment* (as PRE-, SENTIMENT)]

presently /ˈprez(ə)ntlɪ/ *adv.* **1** soon; after a short time. **2** esp. *N. Amer.* & *Sc.* at the present time; now.

presentment /prɪˈzentmənt/ *n.* the act of presenting information, esp. a statement on oath by a jury of a fact known to them. [ME f. OF *presentement* (as PRESENT[2])]

preservation /ˌprezəˈveɪʃ(ə)n/ *n.* **1** the act of preserving or process of being preserved. **2** a state of being well or badly preserved (*in an excellent state of preservation*). [ME f. OF f. med.L *praeservatio -onis* (as PRESERVE)]

preservationist /ˌprezəˈveɪʃənɪst/ *n.* a supporter or advocate of preservation, esp. of antiquities and historic buildings.

preservative /prɪˈzɜːvətɪv/ *n.* & *adj.* ● *n.* a substance for preserving perishable foodstuffs, wood, etc. ● *adj.* tending to preserve. [ME f. OF *preservatif -ive* f. med.L *praeservativus -um* (as PRESERVE)]

preserve /prɪˈzɜːv/ *v.* & *n.* ● *v.tr.* **1 a** keep safe or free from harm, decay, etc. **b** keep alive (a name, memory, etc.). **2** maintain (a thing) in its existing state. **3** retain (a quality or condition). **4 a** treat or refrigerate (food) to prevent decomposition or fermentation. **b** prepare (fruit) by boiling it with sugar, for long-term storage. **5** keep (game, a river, etc.) undisturbed for private use. ● *n.* (in *sing.* or *pl.*) **1** preserved fruit; jam. **2** a place where game or fish etc. are preserved. **3** a sphere or area of activity regarded as a person's own. □ **well-preserved 1** carefully stored; remaining in good condition. **2** (of an elderly person) showing little sign of ageing. □ **preservable** *adj.* **preserver** *n.* [ME f. OF *preserver* f. LL *praeservare* (as PRAE-, *servare* keep)]

pre-set /priːˈset/ *v.tr.* (**-setting**; *past* and *past part.* **-set**) **1** set or fix (a device) in advance of its operation. **2** settle or decide beforehand.

preshrunk /priːˈʃrʌŋk/ *adj.* (of a fabric or garment) treated so that it shrinks during manufacture and not in use.

preside /prɪˈzaɪd/ *v.intr.* **1** (often foll. by *at*, *over*) be in a position of authority, esp. as the chairperson or president of a meeting.

2 a exercise control or authority. **b** (foll. by *at*) *colloq.* play an instrument in company (*presided at the piano*). [F *présider* f. L *praesidere* (as PRAE-, *sedere* sit)]

presidency /ˈprezɪdənsɪ/ *n.* (*pl.* **-ies**) **1** the office of president. **2** the period of this. [Sp. & Port. *presidencia*, It. *presidenza* f. med.L *praesidentia* (as PRESIDE)]

president /ˈprezɪd(ə)nt/ *n.* **1** the elected head of a republican state. **2** the head of a society or council etc. **3** the head of certain colleges. **4** *N. Amer.* **a** the head of a university. **b** the head of a company, etc. **5** a person in charge of a meeting, council, etc. □ **presidentship** *n.* **presidential** /ˌprezɪˈden(ʃ)(ə)l/ *adj.* **presidentially** *adv.* [ME f. OF f. L (as PRESIDE)]

Presidential Medal of Freedom *n.* (in the US) a medal constituting the highest award that can be given to a civilian in peacetime.

Presidium /prɪˈsɪdɪəm, prɪˈzɪd-/ *n.* (also **Praesidium**) a standing executive committee in a Communist country, esp. that of the Supreme Soviet in the former Soviet Union, which functioned as the ultimate legislative authority when the Soviet itself was not sitting. From 1952 to 1966 the Soviet Politburo was also known as the Presidium. [Russ. *prezidium* f. L *praesidium* protection etc. (as PRESIDE)]

Presley /ˈprezlɪ/, Elvis (Aaron) (1935–77), American pop singer. He was the dominant personality of early rock and roll, known particularly for the vigour and frank sexuality of his style. He first gained fame in 1956 with the success of such records as 'Heartbreak Hotel' and 'Blue Suede Shoes', attracting a worldwide following. After making a number of films during the 1960s, he resumed his personal appearances in the 1970s, mostly in Las Vegas. He lived much of his life in his Memphis mansion Graceland. He died from a drug overdose.

Presocratic /ˌpriːsəˈkrætɪk/ *adj.* & *n.* ● *adj.* of or relating to the philosophy or philosophers of the time before Socrates. (*See note below.*) ● *n.* a Presocratic philosopher.

▪ Prominent in the Greek world in the 5th and 6th centuries BC, the Presocratics were speculative thinkers who attempted to find rational rather than supernatural explanations for natural phenomena. Among Presocratic philosophers were Parmenides, Anaxagoras, Empedocles, and Heraclitus; of their work only fragments and later quotations and references survive.

press[1] /pres/ *v.* & *n.* ● *v.* **1** *tr.* apply steady force to (a thing in contact) (*press a switch; pressed the two surfaces together*). **2** *tr.* **a** compress or apply pressure to a thing to flatten, shape, or smooth it, as by ironing (*got the curtains pressed*). **b** squeeze (a fruit etc.) to extract its juice. **c** manufacture (a gramophone record etc.) by moulding under pressure. **3** *tr.* (foll. by *out of*, *from*, etc.) squeeze (juice etc.). **4** *tr.* embrace or caress by squeezing (*pressed my hand*). **5** *intr.* (foll. by *on*, *against*, etc.) exert pressure. **6** *intr.* be urgent; demand immediate action (*time was pressing*). **7** *intr.* (foll. by *for*) make an insistent demand. **8** *intr.* (foll. by *up*, *round*, etc.) form a crowd. **9** *intr.* (foll. by *on*, *forward*, etc.) hasten insistently. **10** *tr.* (often in *passive*) (of an enemy etc.) bear heavily on. **11** *tr.* (often foll. by *for*, or *to* + infin.) urge or entreat (*pressed me to stay; pressed me for an answer*). **12** *tr.* (foll. by *on*, *upon*) **a** put forward or urge (an opinion, claim, or course of action). **b** insist on the acceptance of (an offer, a gift, etc.). **13** *tr.* insist on (*did not press the point*). **14** *intr.* (foll. by *on*) produce a strong mental or moral impression; oppress; weigh heavily. **15** *intr. Golf* try too hard for a long shot etc. and so strike the ball imperfectly. ● *n.* **1** the act or an instance of pressing (*give it a slight press*). **2 a** a device for compressing, flattening, shaping, extracting juice, etc. (*trouser press; flower press; wine press*). **b** a frame for preserving the shape of a racket when not in use. **c** a machine that applies pressure to a workpiece by means of a tool, in order to punch shapes, bend it, etc. **3** = *printing-press*. **4** (prec. by *the*) **a** the art or practice of printing. **b** newspapers, journalists, etc., generally or collectively (*read it in the press; pursued by the press*). **5** a notice or piece of publicity in newspapers etc. (*got a good press*). **6** (**Press**) **a** a printing house or establishment. **b** a publishing company (*Athlone Press*). **7 a** crowding. **b** a crowd (of people etc.). **8** the pressure of affairs. **9** esp. *Ir.* & *Sc.* a large usu. shelved cupboard for clothes, books, etc., esp. in a recess. □ **at** (or **in**) **press** (or **the press**) being printed. **be pressed for** have barely enough (time etc.). **go** (or **send**) **to press** go or send to be printed. **press agent** a person employed to attend to advertising and press publicity. **press-box** a reporters' enclosure esp. at a sports event. **press the button 1** set machinery in motion. **2** *colloq.* initiate an action or train of events. **press-button** = *push-button*. **press conference** an interview given to journalists to make an announcement or answer questions. **press the flesh** esp. *US colloq.* shake hands. **press gallery** a gallery for reporters esp. in a legislative assembly. **press-on** (of a material) that can be pressed or ironed on to something. **press release** an official statement issued to newspapers for information. **press-stud** a small fastening device engaged by pressing its two halves together. **press-up** an exercise in which the prone body is raised from the legs or trunk upwards by pressing down on the hands to straighten the arms. [ME f. OF *presser, presse* f. L *pressare* frequent. of *premere* press-]

press[2] /pres/ *v. & n.* ● *v.tr.* **1** *hist.* force to serve in the army or navy. **2** bring into use as a makeshift (*was pressed into service*). ● *n. hist.* compulsory enlistment esp. in the navy. [alt. f. obs. *prest* (v. & n.) f. OF *prest* loan, advance pay f. *prester* f. L *praestare* furnish (as PRAE-, *stare* stand)]

Pressburg /'presburk/ the German name for BRATISLAVA.

press-gang /'presgæŋ/ *n. & v.* ● *n.* **1** *hist.* a body of men employed to press men into service in the army or navy. **2** any group using similar coercive methods. ● *v.tr.* force into service.

pressie var. of PREZZIE.

pressing /'presɪŋ/ *adj. & n.* ● *adj.* **1** urgent (*pressing business*). **2 a** urging strongly (*a pressing invitation*). **b** persistent, importunate (*since you are so pressing*). ● *n.* **1** a thing made by pressing, esp. a gramophone record. **2** a series of these made at one time. **3** the act or an instance of pressing a thing, esp. a gramophone record, grapes, etc. (*all at one pressing*). □ **pressingly** *adv.*

pressman /'presmən/ *n.* (*pl.* **-men**) **1** a journalist. **2** an operator of a printing-press.

pressmark /'presmɑːk/ *n.* a library shelf-mark showing the location of a book etc.

pressure /'preʃə(r)/ *n. & v.* ● *n.* **1 a** the exertion of continuous force on or against a body by another in contact with it. **b** the force exerted. **c** the amount of this (expressed by the force on a unit area) (*atmospheric pressure*). **2** urgency; the need to meet a deadline etc. (*work under pressure*). **3** affliction or difficulty (*under financial pressure*). **4** constraining influence (*if pressure is brought to bear*). ● *v.tr.* **1** apply pressure to. **2 a** coerce. **b** (often foll. by *into*) persuade (*was pressured into attending*). □ **pressure gauge** a gauge showing the pressure of steam etc. **pressure group** a group or association formed to promote a particular interest or cause by influencing public policy. **pressure point 1** a point where an artery can be pressed against a bone to inhibit bleeding. **2** a point on the skin sensitive to pressure. **3** a target for political pressure or influence. **pressure suit** an inflatable suit for flying at a high altitude. [ME f. L *pressura* (as PRESS[1])]

pressure-cooker /'preʃə‚kukə(r)/ *n.* an airtight pan for cooking quickly under steam pressure. □ **pressure-cook** *v.tr.*

pressurize /'preʃə‚raɪz/ *v.tr.* (also **-ise**) **1** (esp. as **pressurized** *adj.*) maintain normal atmospheric pressure in (an aircraft cabin etc.) at a high altitude. **2** raise to a high pressure. **3** pressure (a person). □ **pressurized-water reactor** a nuclear reactor in which the coolant is water at high pressure. □ **pressurization** /‚preʃəraɪ'zeɪʃ(ə)n/ *n.*

Prestel /'prestel/ *n. propr.* (in the UK) a computerized visual information system operated by British Telecom. [blend of PRESS[1] + TELECOMMUNICATION]

Prester John /'prestə(r)/ a legendary medieval Christian king of Asia, said to have defeated the Muslims and to be destined to bring help to the Holy Land. The legend spread in Europe in the mid-12th century. He was later identified with a real king of Ethiopia; another theory identifies him with a Chinese prince who defeated the sultan of Persia in 1141. [ME f. OF *prestre Jehan*, med.L *presbyter Johannes* priest John]

prestidigitator /‚prestɪ'dɪdʒɪ‚teɪtə(r)/ *n. formal* a conjuror. □ **prestidigitation** /-‚dɪdʒɪ'teɪʃ(ə)n/ *n.* [F *prestidigitateur* f. *preste* nimble (as PRESTO) + L *digitus* finger]

prestige /pre'stiːʒ/ *n.* **1** respect, reputation, or influence derived from achievements, power, associations, etc. **2** (*attrib.*) having or conferring prestige. □ **prestigeful** *adj.* [F, = illusion, glamour, f. LL *praestigium* (as PRESTIGIOUS)]

prestigious /pre'stɪdʒəs/ *adj.* having or showing prestige. □ **prestigiously** *adv.* **prestigiousness** *n.* [orig. = deceptive, f. L *praestigiosus* f. *praestigiae* juggler's tricks]

prestissimo /pre'stɪsɪ‚məʊ/ *adv. & n. Mus.* ● *adv.* in a very quick tempo. ● *n.* (*pl.* **-os**) a movement or passage played in this way. [It., superl. (as PRESTO)]

presto /'prestəʊ/ *adv., adj., n., & int.* ● *adv. & adj. Mus.* in quick tempo. ● *n.* (*pl.* **-os**) *Mus.* a presto passage or movement. ● *int.* = hey presto! (see HEY[1]). [It. f. LL *praestus* f. L *praesto* ready]

Preston /'prestən/ a city in NW England, the administrative centre of Lancashire, on the River Ribble; pop. (1991) 126,200. A spinning and

weaving centre since the 15th century, it was the site in the 18th century of the first English cotton mills.

Prestonpans, Battle of /‚prestən'pænz/ a battle in 1745 near the town of Prestonpans just east of Edinburgh, the first major engagement of the Jacobite uprising of 1745–6. A small Hanoverian army was routed by an equally small Jacobite army, leaving the way clear for the Young Pretender's (Charles Edward Stuart's) subsequent invasion of England.

prestressed /priː'strest/ *adj.* strengthened by stressing in advance, esp. of concrete by means of stretched rods or wires put in during manufacture.

Prestwick /'prestwɪk/ a town to the south of Glasgow in Strathclyde region, SW Scotland, the site of an international airport; pop. (1989) 14,052.

presumably /prɪ'zjuːməblɪ/ *adv.* as may reasonably be presumed.

presume /prɪ'zjuːm/ *v.* **1** *tr.* (often foll. by *that* + clause) suppose to be true; take for granted. **2** *tr.* (often foll. by *to* + infin.) **a** take the liberty; be impudent enough (*presumed to question their authority*). **b** dare, venture (*may I presume to ask?*). **3** *intr.* be presumptuous; take liberties. **4** *intr.* (foll. by *on, upon*) take advantage of or make unscrupulous use of (a person's good nature etc.). □ **presumable** *adj.* **presumedly** /-mɪdlɪ/ *adv.* [ME f. OF *presumer* f. L *praesumere praesumpt-* anticipate, venture (as PRAE-, *sumere* take)]

presuming /prɪ'zjuːmɪŋ/ *adj.* presumptuous. □ **presumingly** *adv.* **presumingness** *n.*

presumption /prɪ'zʌmpʃ(ə)n/ *n.* **1** arrogance; presumptuous behaviour. **2 a** the act of presuming a thing to be true. **b** a thing that is or may be presumed to be true; a belief based on reasonable evidence. **3** a ground for presuming (*a strong presumption against their being guilty*). **4** *Law* an inference from known facts. [ME f. OF *presumpcion* f. L *praesumptio -onis* (as PRESUME)]

presumptive /prɪ'zʌmptɪv/ *adj.* **1** based on presumption or inference. **2** giving reasonable grounds for presumption (*presumptive evidence*). □ **presumptively** *adv.* [F *présomptif -ive* f. LL *praesumptivus* (as PRESUME)]

presumptuous /prɪ'zʌmptjʊəs/ *adj.* unduly or overbearingly confident and presuming. □ **presumptuously** *adv.* **presumptuousness** *n.* [ME f. OF *presumptueux* f. LL *praesumptuosus, -tiosus* (as PRESUME)]

presuppose /‚priːsə'pəʊz/ *v.tr.* (often foll. by *that* + clause) **1** assume beforehand. **2** require as a precondition; imply. [ME f. OF *presupposer*, after med.L *praesupponere* (as PRE-, SUPPOSE)]

presupposition /‚priːsʌpə'zɪʃ(ə)n/ *n.* **1** the act or an instance of presupposing. **2** a thing assumed beforehand as the basis of argument etc. [med.L *praesuppositio* (as PRAE-, *supponere* as SUPPOSE)]

pre-tax /priː'tæks/ *adj.* (of income or profits) before the deduction of taxes.

pre-teen /priː'tiːn/ *adj. & n.* ● *adj.* of or relating to children just under the age of 13. ● *n.* a pre-teen child.

pretence /prɪ'tens/ *n.* (*US* **pretense**) **1** pretending, make-believe. **2 a** a pretext or excuse (*on the slightest pretence*). **b** a false show of intentions or motives (*under the pretence of friendship*; *under false pretences*). **3** (foll. by *to*) a claim, esp. a false or ambitious one (*has no pretence to any great talent*). **4 a** affectation, display. **b** pretentiousness, ostentation (*stripped of all pretence*). [ME f. AF *pretense* ult. f. med.L *pretensus* pretended (as PRETEND)]

pretend /prɪ'tend/ *v. & adj.* ● *v.* **1** *tr.* (usu. foll. by *to* + infin., or *that* + clause) claim or assert falsely so as to deceive (*pretended to know the answer*; *pretended that they were foreigners*). **2 a** *tr.* imagine to oneself in play (*pretended to be monsters*; *pretended it was night*). **b** *absol.* make pretence, esp. in imagination or play; make believe (*they're just pretending*). **3** *tr.* **a** profess, esp. falsely or extravagantly (*does not pretend to be a scholar*). **b** (as **pretended** *adj.*) falsely claim to be such (*a pretended friend*). **4** *intr.* (foll. by *to*) lay claim to (a right or title etc.). **b** profess to have (a quality etc.). **5** *tr.* (foll. by *to*) aspire or presume; venture (*I cannot pretend to guess*). ● *adj. colloq.* pretended; in pretence (*pretend money*). [ME f. F *prétendre* or f. L (as PRAE-, *tendere tent-*, later *tens-* stretch)]

pretender /prɪ'tendə(r)/ *n.* **1** a person who claims a throne or title etc. **2** a person who pretends.

pretense *US* var. of PRETENCE.

pretension /prɪ'tenʃ(ə)n/ *n.* **1** (often foll. by *to*) **a** an assertion of a claim. **b** a justifiable claim (*has no pretensions to the name*; *has some pretensions to be included*). **2** pretentiousness. [med.L *praetensio, -tio* (as PRETEND)]

pretentious /prɪ'tenʃəs/ *adj.* **1** making an excessive claim to great merit or importance. **2** ostentatious. □ **pretentiously** *adv.* **pretentiousness** *n.* [F *prétentieux* (as PRETENSION)]

preter- /'priːtə(r)/ *comb. form* more than. [L *praeter* (adv. & prep.) = past, beyond]

preterite /'pretərɪt/ *adj. & n.* (US **preterit**) *Gram.* ● *adj.* expressing a past action or state. ● *n.* a preterite tense or form. [ME f. OF *preterite* or L *praeteritus* past part. of *praeterire* pass (as PRETER-, *ire it-* go)]

preterm /priː'tɜːm/ *adj. & adv.* born or occurring prematurely.

pretermit /ˌpriːtə'mɪt/ *v.tr.* (**pretermitted, pretermitting**) *formal* **1** omit to mention (a fact etc.). **2** omit to do or perform; neglect. **3** leave off (a custom or continuous action) for a time. □ **pretermission** /-'mɪʃ(ə)n/ *n.* [L *praetermittere* (as PRETER-, *mittere miss-* let go)]

preternatural /ˌpriːtə'nætʃrəl/ *adj.* outside the ordinary course of nature; supernatural. □ **preternaturalism** *n.* **preternaturally** *adv.*

pretext /'priːtekst/ *n.* **1** an ostensible or alleged reason or intention. **2** an excuse offered. □ **on** (or **under**) **the pretext** (foll. by *of*, or *that* + clause) professing as one's object or intention. [L *praetextus* outward display f. *praetexere praetext-* (as PRAE-, *texere* weave)]

pretor US var. of PRAETOR.

Pretoria /prɪ'tɔːrɪə/ the administrative capital of South Africa; pop. (1985) 443,000. It was founded in 1855 by Marthinus Wessel Pretorius (1819–1901), the first President of the South African Republic, and named after his father Andries. It was taken by the British under Frederick Roberts in 1900, during the Second Boer War.

pretorian US var. of PRAETORIAN.

Pretoria-Witwatersrand-Vereeniging (abbr. **PWV**) a province of north-eastern South Africa, formerly part of Transvaal; capital, Johannesburg.

prettify /'prɪtɪˌfaɪ/ *v.tr.* (**-ies, -ied**) make (a thing or person) pretty esp. in an affected way. □ **prettifier** *n.* **prettification** /ˌprɪtɪfɪ'keɪʃ(ə)n/ *n.*

pretty /'prɪtɪ/ *adj., adv., n., & v.* ● *adj.* (**prettier, prettiest**) **1** attractive in a delicate way without being truly beautiful or handsome; pleasing to the eye or ear (*a pretty child; a pretty dress; a pretty tune*). **2** fine or good of its kind (esp. *a pretty wit*). **3** (of a quantity or amount) considerable, great (*a pretty penny; a pretty packet*). ● *adv. colloq.* fairly, moderately; considerably (*am pretty well; find it pretty difficult*). ● *n.* (*pl.* **-ies**) a pretty person (esp. as a form of address to a child). ● *v.tr.* (**-ies, -ied**) (often foll. by *up*) make pretty or attractive. □ **pretty much** (or **nearly** or **well**) *colloq.* very nearly; very nearly. **pretty-pretty** too pretty. **sitting pretty** *colloq.* in a favourable or advantageous position. □ **prettily** *adv.* **prettiness** *n.* **prettyish** *adj.* **prettyism** *n.* [OE *prættig* f. WG]

pretzel /'prets(ə)l/ *n.* (also **bretzel** /'bret-/) a crisp knot-shaped or stick-shaped salted biscuit. [G]

prevail /prɪ'veɪl/ *v.intr.* **1** (often foll. by *against, over*) be victorious or gain mastery. **2** be the more usual or predominant. **3** exist or occur in general use or experience; be current. **4** (foll. by *on, upon*) persuade. **5** (as **prevailing** *adj.*) predominant; generally current or accepted (*prevailing opinion*). □ **prevailing wind** the wind that most frequently occurs at a place. □ **prevailingly** *adv.* [ME f. L *praevalere* (as PRAE-, *valere* have power), infl. by AVAIL]

prevalent /'prevələnt/ *adj.* **1** generally existing or occurring. **2** predominant. □ **prevalence** *n.* **prevalently** *adv.* [as PREVAIL]

prevaricate /prɪ'værɪˌkeɪt/ *v.intr.* **1** speak or act evasively or misleadingly. **2** quibble, equivocate. ¶ Often confused with *procrastinate.* □ **prevaricator** *n.* **prevarication** /-ˌværɪ'keɪʃ(ə)n/ *n.* [L *praevaricari* walk crookedly, practise collusion, in eccl.L transgress (as PRAE-, *varicari* straddle f. *varus* bent, knock-kneed)]

prevenient /prɪ'viːnɪənt/ *adj. formal* preceding something else. [L *praeveniens* pres. part of *praevenire* (as PREVENT)]

prevent /prɪ'vent/ *v.tr.* **1** (often foll. by *from* + verbal noun) stop from happening or doing something; hinder; make impossible (*the weather prevented me from going*). **2** *archaic* go or arrive before, precede. □ **preventer** *n.* **preventable** *adj.* (also **preventible**). **preventability** /-ˌventə'bɪlɪtɪ/ *n.* (also **preventibility**). **prevention** /-'venʃ(ə)n/ *n.* [ME = anticipate, f. L *praevenire praevent-* come before, hinder (as PRAE-, *venire* come)]

preventative /prɪ'ventətɪv/ *adj. & n.* = PREVENTIVE. □ **preventatively** *adv.*

preventive /prɪ'ventɪv/ *adj. & n.* ● *adj.* serving to prevent, esp. preventing disease, breakdown, etc. (*preventive medicine; preventive maintenance*). ● *n.* a preventive agent, measure, drug, etc. □ **preventive detention** the imprisonment of a criminal for corrective training etc. □ **preventively** *adv.*

preview /'priːvjuː/ *n. & v.* ● *n.* **1** the act of seeing in advance. **2 a** the showing of a film, play, exhibition, etc., before the official opening. **b** (*N. Amer.* **prevue**) a film trailer. ● *v.tr.* see or show in advance.

Previn /'previn/, André (George) (b.1929), German-born American conductor, pianist, and composer. He is most famous as a conductor, notably with the London Symphony Orchestra (1968–79), the Pittsburgh Symphony Orchestra (1976–86), and the Royal Philharmonic Orchestra (1987–91). He has also composed musicals, film scores, and orchestral and chamber works, and is a noted jazz and classical pianist.

previous /'priːvɪəs/ *adj. & adv.* ● *adj.* **1** coming before in time or order. **2** done or acting hastily. ● *adv.* (foll. by *to*) before (*had called previous to writing*). □ **previous question** *Parl.* a motion concerning the vote on a main question. □ **previously** *adv.* **previousness** *n.* [L *praevius* (as PRAE-, *via* way)]

previse /prɪ'vaɪz/ *v.tr. literary* foresee or forecast (an event etc.). □ **prevision** /-'vɪʒ(ə)n/ *n.* **previsional** *adj.* [L *praevidere praevis-* (as PRAE-, *videre* see)]

Prévost d'Exiles /ˌpreɪvəʊ deg'ziːl/, Antoine-François (known as Abbé Prévost) (1696–1763), French novelist. He became a Benedictine monk in 1721 and was ordained as a priest five years later. He is remembered for his novel *Manon Lescaut* (1731), the story of a mutually destructive passion between a nobleman and a *demi-mondaine*, which inspired operas by Jules Massenet (1842–1912) and Puccini.

prevue *N. Amer.* = PREVIEW *n.* 2b.

pre-war /priː'wɔː(r)/ *adj.* existing or occurring before a war (esp. the most recent major war).

prex /preks/ *n.* (also **prexy** /'preksɪ/) US *sl.* a president (esp. of a college). [abbr.]

prey /preɪ/ *n. & v.* ● *n.* **1** an animal that is hunted or killed by another for food. **2** (often foll. by *to*) a person or thing that is influenced by or vulnerable to (something undesirable) (*became a prey to morbid fears*). **3** *Bibl.* or *archaic* plunder, booty, etc. ● *v.intr.* (foll. by *on, upon*) **1** seek or take as prey. **2** make a victim of. **3** (of a disease, emotion, etc.) exert a harmful influence (*fear preyed on his mind*). □ **beast** (or **bird**) **of prey** an animal (or bird) which hunts animals for food. □ **preyer** *n.* [ME f. OF *preie* f. L *praeda* booty]

Prez, Josquin des, see DES PREZ.

prezzie /'prezɪ/ *n.* (also **pressie**) *colloq.* a present or gift. [abbr.]

Priam /'praɪæm/ *Gk Mythol.* the king of Troy at the time of its destruction by the Greeks under Agamemnon. The father of Paris and Hector and husband of Hecuba, he was slain by Neoptolemus, son of Achilles.

priapic /praɪ'æpɪk/ *adj.* phallic. [PRIAPUS (as PRIAPISM) + -IC]

priapism /'praɪəˌpɪz(ə)m/ *n.* **1** lewdness, licentiousness. **2** *Med.* persistent erection of the penis. [F *priapisme* f. LL *priapismus* f. Gk *priapismos* f. *priapizō* be lewd f. PRIAPUS]

Priapus /praɪ'eɪpəs/ *Gk Mythol.* a god of fertility, whose cult spread to Greece (and, later, Italy) from Turkey after Alexander's conquests. He was represented as a distorted human figure with enormous genitals. He was also a god of gardens and the patron of seafarers and shepherds.

Pribilof Islands /'prɪbɪˌlɒf/ a group of four islands in the Bering Sea, off the coast of SW Alaska. First visited in 1786 by the Russian explorer Gavriil Loginovich Pribylov (d.1796), they came into US possession after the purchase of Alaska in 1867.

Price /praɪs/, Vincent (1911–93), American actor. His career in horror films began with *House of Wax* (1953). He is best known for his performances in a series of films of the 1960s based on stories by Edgar Allan Poe; these included *The Fall of the House of Usher* (1960) and *The Pit and the Pendulum* (1961).

price /praɪs/ *n. & v.* ● *n.* **1 a** the amount of money or goods for which a thing is bought or sold. **b** value or worth (*a pearl of great price; beyond price*). **2** what is or must be given, done, sacrificed, etc., to obtain or achieve something. **3** the odds in betting (*starting price*). **4** a sum of money offered or given as a reward, esp. for the capture or killing of a person. ● *v.tr.* **1** fix or find the price of (a thing for sale). **2** estimate the value of. □ **above** (or **beyond** or **without**) **price** so valuable that no price can be stated. **at any price** no matter what the cost, sacrifice, etc. (*peace at any price*). **at a price** at a high cost. **price-fixing** the maintaining of prices at a certain level by agreement between competing sellers. **price-list** a list of current prices of items on sale. **price on a person's head** a reward for a person's capture or death. **price oneself out of the market** lose to one's competitors by charging more than customers are willing to pay. **price-ring** a group of traders acting illegally to control certain prices. **price tag 1** the label on an item showing its price. **2** the cost of an enterprise or undertaking. **price war** fierce competition among traders cutting prices. **set a price on** declare the price of. **what price** …? (often foll. by verbal noun) *colloq.* **1** what is the chance of …? (*what price your finishing the course?*). **2** *iron.* the expected or much boasted … proves disappointing (*what price your friendship now?*). □ **priced** *adj.* (also in *comb.*). **pricer** *n.* [(n.) ME f. OF *pris* f. L *pretium*: (v.) var. of *prise* = PRIZE[1]]

priceless /ˈpraɪslɪs/ *adj.* **1** invaluable; beyond price. **2** *colloq.* very amusing or absurd. □ **pricelessly** *adv.* **pricelessness** *n.*

pricey /ˈpraɪsɪ/ *adj.* (also **pricy**) (**pricier**, **priciest**) *colloq.* expensive. □ **priciness** *n.*

prick /prɪk/ *v. & n.* ● *v.* **1** *tr.* pierce slightly; make a small hole in. **2** *tr.* (foll. by *off*, *out*) mark (esp. a pattern) with small holes or dots. **3** *tr.* trouble mentally (*my conscience is pricking me*). **4** *intr.* feel a pricking sensation. **5** *intr.* (foll. by *at*, *into*, etc.) make a thrust as if to prick. **6** *tr.* (foll. by *in*, *off*, *out*) plant (seedlings etc.) in small holes pricked in the earth. **7** *tr. Brit. archaic* mark off (a name in a list, esp. to select a sheriff) by pricking. **8** *tr. archaic* spur or urge on (a horse etc.). ● *n.* **1** the act of or an instance of pricking. **2** a small hole or mark made by pricking. **3** a pain caused as by pricking. **4** a mental pain (*felt the pricks of conscience*). **5** *coarse sl.* **a** the penis. **b** *derog.* an objectionable man. **6** *archaic* a goad for oxen. □ **kick against the pricks** persist in futile resistance. **prick up one's ears 1** (of a dog etc.) make the ears erect when on the alert. **2** (of a person) become suddenly attentive. □ **pricker** *n.* [OE *prician* (v.), *pricca* (n.)]

pricket /ˈprɪkɪt/ *n.* **1** *Brit.* a male fallow deer in its second year, having straight unbranched horns. **2** a spike for holding a candle. [ME f. AL *prikettus -um*, dimin. of PRICK]

prickle /ˈprɪk(ə)l/ *n. & v.* ● *n.* **1 a** a small thorn. **b** *Bot.* a thornlike process developed from the epidermis in a plant. **2** a hard-pointed spine of a hedgehog etc. **3** a prickling sensation. ● *v.* **1** *tr. & intr.* affect or be affected with a sensation as of pricking. **2** *intr.* react defensively or aggressively to a situation. [OE *pricel* PRICK: (v.) also dimin. of PRICK]

prickly /ˈprɪklɪ/ *adj.* (**pricklier**, **prickliest**) **1** (esp. in the names of plants and animals) having prickles. **2 a** (of a person) ready to take offence. **b** (of a topic, argument, etc.) full of contentious or complicated points; thorny. **3** tingling. □ **prickly heat** an itchy inflammation of the skin, causing a tingling sensation and common in hot countries. **prickly pear 1** an opuntia cactus native to arid regions of America, bearing large pear-shaped prickly fruits. **2** its fruit. **prickly poppy** a tropical poppy, *Argemone mexicana*, with prickly leaves and yellow flowers. □ **prickliness** *n.*

pricy var. of PRICEY.

pride /praɪd/ *n. & v.* ● *n.* **1 a** a feeling of elation or satisfaction at achievements or qualities or possessions etc. that do one credit. **b** an object of this feeling. **c** the foremost or best of a group. **2** a high or overbearing opinion of one's worth or importance. **3** (in full **proper pride**) a consciousness of what befits one's position; self-respect. **4** a group or company (of animals, esp. lions). **5** esp. *literary* the best condition; the prime. ● *v.refl.* (foll. by *on*, *upon*) be proud of. □ **my, his,** etc. **pride and joy** a thing of which one is very proud. **pride of the morning** a mist or shower at sunrise, supposedly indicating a fine day to come. **pride of place** the most important or prominent position. **take pride** (or **a pride**) **in 1** be proud of. **2** maintain in good condition or appearance. □ **prideful** *adj.* **pridefully** *adv.* **prideless** *adj.* [OE *prȳtu*, *prȳte*, *prȳde* f. *prūd* PROUD]

Pride's Purge /praɪdz/ *n.* (in English history) the exclusion or arrest of about 140 members of parliament by soldiers under the command of Colonel Thomas Pride (d.1658) when, in December 1648, members of the English army removed those members likely to vote against a trial of the captive Charles I. Following the purge, the remaining members, known as the Rump Parliament, voted for the trial which resulted in Charles's execution.

prie-dieu /priːˈdjɜː/ *n.* (*pl.* **prie-dieux** *pronunc.* same) a kneeling-desk for prayer. [F, = pray God]

priest /priːst/ *n. & v.* ● *n.* **1** an ordained minister of the Roman Catholic or Orthodox Church, or of the Anglican Church (above a deacon and below a bishop), authorized to perform certain rites and administer certain sacraments. **2** an official minister of a non-Christian religion. ● *v.tr.* make (a person) a priest; ordain. □ **priest's hole** *hist.* a hiding-place for a Roman Catholic priest during times of religious persecution. □ **priestless** *adj.* **priestlike** *adj.* **priestling** *n.* [OE *prēost*, ult. f. eccl.L *presbyter*: see PRESBYTER]

priestcraft /ˈpriːstkrɑːft/ *n.* usu. *derog.* the work and influence of priests.

priestess /ˈpriːstɪs/ *n.* a female priest of a non-Christian religion.

priesthood /ˈpriːsthʊd/ *n.* (usu. prec. by *the*) **1** the office or position of priest. **2** priests in general.

Priestley[1] /ˈpriːstlɪ/, J(ohn) B(oynton) (1894–1984), English novelist, dramatist, and critic. His first major success came with the picaresque novel *The Good Companions* (1929); this was followed by many other novels, including the more sombre *Angel Pavement* (1930). His plays include *Time and the Conways* (1937) and the mystery drama *An Inspector Calls* (1947). During and after the Second World War he was a popular radio broadcaster on current affairs.

Priestley[2] /ˈpriːstlɪ/, Joseph (1733–1804), English scientist and theologian. Priestley was the author of about 150 books, mostly theological or educational. His chief work was on the chemistry of gases, a number of which he managed to isolate, including ammonia, sulphur dioxide, nitrous oxide, and nitrogen dioxide. Priestley's most significant discovery was of 'dephlogisticated air' (oxygen) in 1774; he demonstrated that it was important to animal life, and that plants give off this gas in sunlight. In his theological writings he maintained a Unitarian position. His support of the French Revolution provoked so much hostility that he settled in America in 1794. (See also PHLOGISTON.)

priestly /ˈpriːstlɪ/ *adj.* of or associated with priests. □ **priestliness** *n.* [OE *prēostlic* (as PRIEST)]

prig /prɪg/ *n.* a self-righteously correct or moralistic person. □ **priggery** *n.* **priggish** *adj.* **priggishly** *adv.* **priggishness** *n.* [16th-c. cant, = tinker: orig. unkn.]

prim /prɪm/ *adj. & v.* ● *adj.* (**primmer, primmest**) **1** (of a person or manner) stiffly formal and precise. **2** (of a woman or girl) demure. **3** prudish. ● *v.tr.* (**primmed, primming**) **1** form (the face, lips, etc.) into a prim expression. **2** make prim. □ **primly** *adv.* **primness** *n.* [17th c.: prob. orig. cant f. OF *prin prime* excellent f. L *primus* first]

prima ballerina /ˌpriːmə ˌbæləˈriːnə/ *n.* (*pl.* **prima ballerinas**) the chief female dancer in a ballet or ballet company. [It.]

primacy /ˈpraɪməsɪ/ *n.* (*pl.* **-ies**) **1** pre-eminence. **2** the office of a primate. [ME f. OF *primatie* or med.L *primatia* (as PRIMATE)]

prima donna /ˌpriːmə ˈdɒnə/ *n.* (*pl.* **prima donnas**) **1** the chief female singer in an opera or opera company. **2** a temperamentally self-important person. □ **prima donna-ish** *adj.* [It.]

primaeval var. of PRIMEVAL.

prima facie /ˌpraɪmə ˈfeɪʃiː/ *adv. & adj.* ● *adv.* at first sight; from a first impression (*seems prima facie to be guilty*). ● *adj.* (of evidence) based on the first impression (*can see a prima facie reason for it*). [ME f. L, fem. ablat. of *primus* first, *facies* FACE]

primal /ˈpraɪm(ə)l/ *adj.* **1** primitive, primeval. **2** chief, fundamental. □ **primally** *adv.* [med.L *primalis* f. L *primus* first]

primary /ˈpraɪmərɪ/ *adj. & n.* ● *adj.* **1 a** of the first importance; chief (*that is our primary concern*). **b** fundamental, basic. **2** earliest, original; first in a series. **3** of the first rank in a series; not derived (*the primary meaning of a word*). **4** (of a battery or cell) generating electricity by irreversible chemical reaction. **5** (of education) for young children, esp. below the age of 11. **6** (**Primary**) *Geol.* of the earliest formation or lowest strata; esp. = PALAEOZOIC *adj.* **7** *Biol.* belonging to the first stage of development. **8** (of an industry or source of production) concerned with obtaining or using raw materials. **9** *Gram.* (of a tense in Latin and Greek) present, future, perfect, or future perfect (cf. HISTORIC 2). ● *n.* (*pl.* **-ies**) **1** a thing that is primary. **2** (in full **primary election**) (in the US) a preliminary election to appoint delegates to a party conference or to select the candidates for a principal (esp. presidential) election. **3** *Astron.* **a** the body orbited by a satellite etc. **b** = *primary planet.* **4** (**Primary**) *Geol.* the Primary period. **5** = *primary feather.* **6** = *primary coil.* □ **primary coil** a coil to which current is supplied in a transformer. **primary colour** any of the colours from a mixture of which all the other colours can be produced, i.e. red, green, and blue, or for pigments, red, blue, and yellow (see also COLOUR). **primary feather** a large flight-feather of a bird's wing. **primary industry** industry (such as mining, forestry, agriculture, etc.) that provides raw materials for conversion into commodities and products for the consumer. **primary planet** a planet that directly orbits the sun (cf. *secondary planet*). **primary school** a school where young children are taught, esp. below the age of 11. □ **primarily** /ˈpraɪmərɪlɪ, praɪˈmeərɪlɪ, -ˈmærɪlɪ/ *adv.* [ME f. L *primarius* f. *primus* first]

primate /ˈpraɪmeɪt/ *n.* **1** a mammal of the order Primates, which includes lemurs, tarsiers, monkeys, apes, and humans. (*See note below.*) **2** an archbishop. □ **primatology** /-məˈtɒlədʒɪ/ *n.* (in sense 1).

primatial /praɪˈmeɪʃ(ə)l/ *adj.* [ME f. OF *primat* f. L *primas -atis* (adj.) of the first rank f. *primus* first, in med.L = primate]

▪ The primates evolved from tree-living shrewlike ancestors in the Cretaceous period. They are generally characterized by having a large brain, good binocular vision, and opposable thumbs. There are two suborders, of which the primitive Prosimii (prosimians) appeared in the Eocene, and the more advanced Anthropoidea (monkeys, apes, and humans) in the Oligocene. The latter are themselves divided into the platyrrhines (New World monkeys), which have widely spaced nostrils that face sideways, and the catarrhines (Old World monkeys, apes, and humans), which have the nostrils close together and facing downwards. (See also MONKEY.)

Primate of All England the Archbishop of Canterbury (see CANTERBURY, ARCHBISHOP OF).

Primate of England the Archbishop of York (see YORK, ARCHBISHOP OF).

primavera /ˌpriːməˈveərə/ *n.* **1** a Central American tree, *Cybistax donnellsmithii*, bearing yellow blooms. **2** the hard light-coloured timber from this. [Sp., = spring (the season) f. L *primus* first + *ver* SPRING]

prime[1] /praɪm/ *adj. & n.* ● *adj.* **1** chief, most important (*the prime agent; the prime motive*). **2** (esp. of cattle and provisions) first-rate, excellent. **3** primary, fundamental. **4** *Math.* **a** (of a number) divisible only by itself and unity (e.g. 2, 3, 5, 7, 11). **b** (of numbers) having no common factor but unity. ● *n.* **1** the state of the highest perfection of something (*in the prime of life*). **2** (prec. by *the*; foll. by *of*) the best part. **3** the beginning or first age of anything. **4** *Eccl.* **a** the second canonical hour of prayer, appointed for the first hour of the day (i.e. 6 a.m.). **b** the office of this. **c** *archaic* this time. **5** a prime number. **6** *Printing* a symbol (′) added to a letter etc. as a distinguishing mark, or to a figure as a symbol for minutes or feet. **7** the first of eight parrying positions in fencing. □ **prime cost** the direct cost of a commodity in terms of materials, labour, etc. **prime meridian 1** the meridian from which longitude is reckoned, esp. that passing through Greenwich. **2** the corresponding line on a map. **prime mover 1** an initial natural or mechanical source of motive power. **2** the author of a fruitful idea. **prime rate** the lowest rate at which money can be borrowed commercially. **prime time** the time at which a radio or television audience is expected to be at its highest. **prime vertical** the great circle of the heavens passing through the zenith and the east and west points of the horizon. □ **primeness** *n.* [(n.) OE *prīm* f. L *prima* (*hora*) first (hour), & MF f. OF *prime*: (adj.) ME f. OF f. L *primus* first]

prime[2] /praɪm/ *v.tr.* **1** prepare (a thing) for use or action. **2** prepare (a gun) for firing or (an explosive) for detonation. **3 a** pour a liquid into (a pump) to prepare it for working. **b** inject petrol into (the cylinder or carburettor of an internal-combustion engine). **4** prepare (wood etc.) for painting by applying a substance that prevents paint from being absorbed. **5** equip (a person) with information etc. **6** ply (a person) with food or drink in preparation for something. [16th c.: orig. unkn.]

Prime Minister *n.* (abbr. **PM**) the head of the executive branch of government in most countries with a parliamentary system. In Britain Robert Walpole is generally regarded as having been the first Prime Minister in the modern sense, although the term was then merely descriptive and unofficial, or even derogatory. It was little used in the later part of the 18th century, 'Premier' or 'First Minister' being used instead, but by the middle of the 19th century it had become common in informal use and began to creep into official use from 1878. In 1905 it was fully recognized and the precedence of the Prime Minister was defined by King Edward VII. In current use, the terms *Premier* and *Prime Minister* refer to the same office in Britain, but in Canada and Australia the government of a province or state is headed by a Premier, that of the federal government by a Prime Minister. In countries such as France, where the President has an executive function, the Prime Minister is in a subordinate position.

primer[1] /ˈpraɪmə(r)/ *n.* **1** a substance used to prime wood etc. **2** a cap, cylinder, etc., used to ignite the powder of a cartridge etc.

primer[2] /ˈpraɪmə(r)/ *n.* **1** an elementary textbook for teaching children to read. **2** an introductory book. [ME f. AF f. med.L *primarius -arium* f. L *primus* first]

primeval /praɪˈmiːv(ə)l/ *adj.* (also **primaeval**) **1** of or relating to the first age of the world. **2** ancient, primitive. □ **primeval soup** see SOUP *n.* 4. □ **primevally** *adv.* [L *primaevus* f. *primus* first + *aevum* age]

primigravida /ˌpriːmɪˈɡrævɪdə, ˌpraɪm-/ *n.* (*pl.* **primigravidae** /-ˌdiː/) a woman who is pregnant for the first time. [mod.L fem. f. L *primus* first + *gravidus* pregnant: see GRAVID]

priming[1] /ˈpraɪmɪŋ/ *n.* **1** a mixture used by painters for a preparatory

coat. **2** a preparation of sugar added to beer. **3 a** gunpowder placed in the pan of a firearm. **b** a train of powder connecting the fuse with the charge in blasting etc.

priming[2] /ˈpraɪmɪŋ/ *n.* an acceleration of the tides taking place from the neap to the spring tides. [*prime* (v.) f. PRIME[1] + -ING[1]]

primipara /praɪˈmɪpərə/ *n.* (*pl.* **primiparae** /-ˌriː/) a woman who is bearing a child for the first time. □ **primiparous** *adj.* [mod.L fem. f. *primus* first + *-parus* f. *parere* bring forth]

primitive /ˈprɪmɪtɪv/ *adj. & n.* ● *adj.* **1** early, ancient; at an early stage of civilization (*primitive man*). **2** undeveloped, crude, simple (*primitive methods*). **3** original, primary. **4** *Gram. & Philol.* (of words or language) radical; not derivative. **5** *Math.* (of a line, figure, etc.) from which another is derived, from which some construction begins, etc. **6** (of a colour) primary. **7 a** of the pre-Renaissance period of western European art. **b** (of art) straightforward or naive in style; suggesting the artist's lack or rejection of formal training. **c** (of an artist) using a primitive or naive style. **8** *Geol.* of the earliest period. **9** *Biol.* appearing in the earliest or a very early stage of growth or evolution. ● *n.* **1 a** a painter of the pre-Renaissance period. **b** a modern imitator of such. **c** a painter with a primitive or naive style. **d** a picture by such a painter. **2** a primitive word, line, etc. □ **primitively** *adv.* **primitiveness** *n.* [ME f. OF *primitif -ive* or L *primitivus* first of its kind f. *primitus* in the first place f. *primus* first]

Primitive Church *n.* the Christian Church in its earliest times.

primitivism /ˈprɪmɪtɪˌvɪz(ə)m/ *n.* **1** primitive behaviour. **2** belief in the superiority of what is primitive. **3** the practice of primitive art. □ **primitivist** *n. & adj.*

primo /ˈpriːməʊ/ *n.* (*pl.* **-os**) *Mus.* the leading or upper part in a duet etc.

Primo de Rivera /ˌpriːməʊ deɪ rɪˈveərə/, Miguel (1870–1930), Spanish general and statesman, head of state 1923–30. He came to power after leading a military coup in 1923, when he assumed dictatorial powers with the consent of Alfonso XIII. The decline of the economy contributed to his forced resignation in 1930. His son, José Antonio Primo de Rivera (1903–36), founded the Falange in 1933 and was executed by Republicans in the Spanish Civil War.

primogenitor /ˌpraɪməʊˈdʒenɪtə(r)/ *n.* **1** the earliest ancestor of a people etc. **2** an ancestor. [var. of *progenitor*, after PRIMOGENITURE]

primogeniture /ˌpraɪməʊˈdʒenɪtʃə(r)/ *n.* **1** the fact or condition of being the first-born child. **2** (in full **right of primogeniture**) the right of succession belonging to the first-born, esp. the feudal rule by which the whole real estate of an intestate passes to the eldest son. □ **primogenital** /-ˈdʒenɪt(ə)l/ *adj.* **primogenitary** *adj.* [med.L *primogenitura* f. L *primo* first + *genitura* f. *gignere genit-* beget]

primordial /praɪˈmɔːdɪəl/ *adj.* **1** existing at or from the beginning, primeval. **2** original, fundamental. □ **primordial soup** see SOUP *n.* 4. □ **primordially** *adv.* **primordiality** /-ˌmɔːdɪˈælɪtɪ/ *n.* [ME f. LL *primordialis* (as PRIMORDIUM)]

primordium /praɪˈmɔːdɪəm/ *n.* (*pl.* **primordia** /-dɪə/) *Biol.* an organ or tissue in the early stages of development. [L, neut. of *primordius* original f. *primus* first + *ordiri* begin]

Primorsky Krai /priːˈmɔːskɪ ˈkraɪ/ an administrative territory in the far south-east of Siberian Russia, between the Sea of Japan and the Chinese border; pop. (1990) 2,281,000; capital, Vladivostok.

primp /prɪmp/ *v.tr.* **1** make (the hair, one's clothes, etc.) tidy. **2** *refl.* make (oneself) smart. [dial. var. of PRIM]

primrose /ˈprɪmrəʊz/ *n.* **1 a** a primula, esp. *Primula vulgaris*, bearing solitary pale yellow flowers. **b** the flower of this. **2** a pale yellow colour. □ **primrose path** the pursuit of pleasure, esp. with disastrous consequences (with ref. to Shakespeare's *Hamlet* I. iii. 50). [ME *primerose*, corresp. to OF *primerose* and med.L *prima rosa*, lit. 'first rose': reason for the name unkn.]

Primrose Day the anniversary of the death of Disraeli (19 Apr. 1881).

Primrose League a political association, formed in memory of Disraeli (whose favourite flower was reputedly the primrose) in 1883, to promote and sustain conservative principles in Britain. It is still in existence today.

primula /ˈprɪmjʊlə/ *n.* a herbaceous plant of the genus *Primula*, bearing primrose-like flowers in a wide variety of colours during the spring. The genus includes primroses, cowslips, and polyanthuses. [med.L, fem. of *primulus* dimin. of *primus* first]

primum mobile /ˌpraɪmʊm ˈməʊbɪlɪ/ *n.* **1** the central or most important source of motion or action. **2** *Astron.* in the medieval version of the Ptolemaic system, an outer sphere supposed to move round the earth in twenty-four hours carrying the inner spheres with it. [med.L, = first moving thing]

Primus /ˈpraɪməs/ n. propr. a brand of portable stove burning vaporized oil for cooking etc. [L (as PRIMUS)]

primus /ˈpraɪməs/ n. the presiding bishop of the Scottish Episcopal Church. [L, = first]

primus inter pares /ˌpriːməs ˌɪntə ˈpɑːriːz/ n. a first among equals; the senior or representative member of a group. [L]

prince /prɪns/ n. (as a title usu. **Prince**) **1** a male member of a royal family other than a reigning king. **2** (in full **prince of the blood**) a son or grandson of a British monarch. **3** a ruler of a small state, actually or nominally subject to a king or emperor. **4** (as an English rendering of foreign titles) a noble usu. ranking next below a duke. **5** (as a courtesy title in some connections) a duke, marquess, or earl. **6** (often foll. by of) the chief or greatest (*the prince of novelists*). □ **prince's feather** a tall plant, *Amaranthus hypochondriacus*, with feathery spikes of small red flowers. **prince's metal** a brasslike alloy of copper and zinc. □ **princedom** n. **princelike** adj. **princeship** n. [ME f. OF f. L *princeps principis* first, chief, sovereign f. *primus* first + *capere* take]

Prince Albert, Prince Charles, etc. see ALBERT, PRINCE; CHARLES, PRINCE, etc.

Prince Charming a fairy-tale hero. The name is a partial translation of the French *Roi Charmant*, the hero of *L'Oiseau bleu* (*The Blue Bird*, 1697) by the Comtesse d'Aulnoy (c.1650–1705). In English it first appears as that of the hero in *King Charming* or *Prince Charming* by James Robinson Planché (1796–1880), and was later adopted for the hero of various fairy-tale pantomimes.

prince consort (a title conferred on) the husband of a reigning female sovereign who is himself a prince. The title was given to Prince Albert, husband of Queen Victoria, to avoid the word *king* as Albert was not reigning.

Prince Edward Island an island in the Gulf of St Lawrence, in eastern Canada, the country's smallest province; capital, Charlottetown. Explored by Jacques Cartier in 1534 and colonized by the French, it was ceded to the British in 1763. It became a province of Canada in 1873.

princeling /ˈprɪnslɪŋ/ n. **1** a young prince. **2** the ruler of a small principality.

princely /ˈprɪnslɪ/ adj. (**princelier, princeliest**) **1 a** of or worthy of a prince. **b** held by a prince. **2 a** sumptuous, generous, splendid. **b** (of a sum of money) substantial. □ **princeliness** n.

Prince of Darkness Satan, the Devil.

Prince of Peace Jesus.

Prince of the Asturias the title given to the eldest son of the king of Spain.

Prince of Wales (a title conferred on) the nominal ruler of Wales, from 1301 the heir apparent to the English or British throne. The first Englishman to be given the title, which had formerly been held by several medieval Welsh rulers, was the future Edward II. (See also CHARLES, PRINCE.)

Prince of Wales Island 1 an island in the Canadian Arctic, in the Northwest Territories to the east of Victoria Island. **2** a former name for the island of Penang (see PENANG 1).

Prince Regent n. a prince who acts as regent, esp. George (afterwards IV) as regent 1811–20.

Prince Royal n. the eldest son of a reigning monarch.

Prince Rupert's drops n.pl. pear-shaped bubbles of glass with a long tail, made by dropping melted glass into water. They have the property, due to internal strain, of disintegrating explosively when the tail is broken off or the surface scratched.

Prince Rupert's Land see RUPERT'S LAND.

Prince Rupert's metal n. a gold-coloured alloy of about three parts copper and one part zinc.

Princes in the Tower the young sons of Edward IV, namely Edward, Prince of Wales (b.1470), and Richard, Duke of York (b.1472), supposedly murdered in the Tower of London in or shortly after 1483. In 1483 Edward reigned briefly as Edward V on the death of his father but was not crowned; he and his brother were taken to the Tower of London by their uncle (the future Richard III). Richard was appointed Protector and the princes disappeared soon afterwards. They are generally assumed to have been murdered, but whether at the instigation of Richard III (as Tudor propagandists claimed) or of another is not known; two skeletons discovered in 1674 are thought to have been those of the princes.

princess /prɪnˈses/ n. (as a title usu. **Princess** /ˈprɪnses/) **1** the wife of a prince. **2** a female member of a royal family other than a reigning queen. **3** (in full **princess of the blood**) a daughter or granddaughter of a British monarch. **4** a pre-eminent woman or thing personified as a woman. [ME f. OF *princesse* (as PRINCE)]

Princess Anne, Princess Diana, etc. see ANNE, PRINCESS; DIANA, PRINCESS, etc.

Princess Regent n. **1** a princess who acts as regent. **2** the wife of a Prince Regent.

Princess Royal n. the eldest daughter of a reigning monarch, esp. as a title conferred by the British monarch. (See also ANNE, PRINCESS.)

Princeton University /ˈprɪnstən/ a university at Princeton in New Jersey, one of the most prestigious in the US. It was founded in 1746.

principal /ˈprɪnsɪp(ə)l/ adj. & n. ● adj. **1** (usu. attrib.) first in rank or importance; chief (*the principal town of the district*). **2** main, leading (*a principal cause of my success*). **3** (of money) constituting the original sum invested or lent. ● n. **1** a head, ruler, or superior. **2** the head of some schools, colleges, and universities. **3** the leading performer in a concert, play, etc. **4** a capital sum as distinguished from interest or income. **5** a person for whom another acts as agent etc. **6** (in the UK) a civil servant of the grade below secretary. **7** the person actually responsible for a crime. **8** a person for whom another is surety. **9** hist. each of the combatants in a duel. **10 a** a main rafter supporting purlins. **b** a main girder. **11** an organ stop sounding an octave above the diapason. **12** Mus. the leading player in each section of an orchestra. □ **principal boy** the leading male part in a pantomime, usually played by a woman. **principal girl** the leading female part in a pantomime. **principal clause** Gram. a clause to which another clause is subordinate. **principal in the first degree** a person directly responsible for a crime as its actual perpetrator. **principal in the second degree** a person directly responsible for a crime as aiding in its perpetration. **principal parts** Gram. the parts of a verb from which all other parts can be deduced. □ **principalship** n. [ME f. OF f. L *principalis* first, original (as PRINCE)]

principality /ˌprɪnsɪˈpælɪtɪ/ n. (pl. **-ies**) **1** a state ruled by a prince. **2** the government of a prince. **3** (in pl.) the fifth order of the ninefold celestial hierarchy (see ORDER n. 19). **4** (**the Principality**) Brit. Wales. [ME f. OF *principalité* f. LL *principalitas -tatis* (as PRINCIPAL)]

principally /ˈprɪnsɪp(ə)lɪ/ adv. for the most part; chiefly.

principate /ˈprɪnsɪpət/ n. **1** Rom. Hist. **a** the rule of the early emperors. **b** the period of this. [ME f. OF *principat* or L *principatus* first place]

principle /ˈprɪnsɪp(ə)l/ n. **1** a fundamental truth or law as the basis of reasoning or action (*arguing from first principles; moral principles*). **2 a** a personal code of conduct (*a person of high principle*). **b** (in pl.) such rules of conduct (*has no principles*). **3** a general law in physics etc. (*the uncertainty principle*). **4** a law of nature forming the basis for the construction or working of a machine etc. **5** a fundamental source; a primary element (*held water to be the first principle of all things*). **6** Chem. a constituent of a substance, esp. one giving rise to some quality, etc. □ **in principle** as regards fundamentals but not necessarily in detail. **on principle** on the basis of a moral attitude (*I refuse on principle*). [ME f. OF *principe* f. L *principium* source, (in pl.) foundations (as PRINCE)]

principled /ˈprɪnsɪp(ə)ld/ adj. based on or having (esp. praiseworthy) principles of behaviour.

prink /prɪŋk/ v. **1** tr. (usu. refl.) **a** make (oneself etc.) smart. **b** (foll. by up) smarten (oneself) up. **c** (of a bird) preen. **2** intr. dress oneself up. [16th c.: prob. f. *prank* dress, adorn, rel. to MLG *prank* pomp, Du. *pronk* finery]

print /prɪnt/ n. & v. ● n. **1** an indentation or mark on a surface left by the pressure of a thing in contact with it (*fingerprint; footprint*). **2 a** printed lettering or writing (*large print*). **b** words in printed form. **c** a printed publication, esp. a newspaper. **d** the quantity of a book etc. printed at one time. **e** the state of being printed. **3** a picture or design printed from a block or plate. **4** Photog. a picture produced on paper from a negative. **5** a printed cotton fabric. ● v.tr. **1 a** produce or reproduce (a book, picture, etc.) by applying inked types, blocks, or plates, to paper, vellum, etc. **b** (of an author, publisher, or editor) cause (a book or manuscript etc.) to be produced or reproduced in this way. **2** express or publish in print. **3 a** (often foll. by on, in) impress or stamp (a mark or figure on a surface). **b** (often foll. by with) impress or stamp (a soft surface, e.g. of butter or wax, with a seal, die, etc.). **4** (often absol.) write (words or letters) without joining, in imitation of typography. **5** (often foll. by off, out) Photog. produce (a picture) by the transmission of light through a negative. **6** (usu. foll. by out) (of a computer etc.) produce output in printed form. **7** mark (a textile fabric) with a decorative design in colours. **8** (foll. by *on*) impress (an idea, scene, etc. on the mind or memory). **9** transfer (a coloured or plain design) from paper etc. to the unglazed or glazed surface of ceramic ware. □ **appear in print** have one's work published. **in print 1** (of a book etc.) available from the publisher. **2** in printed form. **out of print** no longer available from the publisher. **printed circuit** an electric circuit with thin strips of conductor on a flat insulating sheet, usu. made by a process like printing. □ **printless** *adj.* (in sense 1 of *n.*). **printable** *adj.* **printability** /ˌprɪntəˈbɪlɪtɪ/ *n.* [ME f. OF *priente, preinte*, fem. past part. of *preindre* press f. L *premere*]

printer /ˈprɪntə(r)/ *n.* **1** a person who prints books, magazines, advertising matter, etc. **2** the owner of a printing business. **3** a device that prints, esp. as part of a computer system. □ **printer's devil** an errand-boy in a printer's office. **printer's mark** a device used as a printer's trademark. **printer's pie** = PIE³ *n.*

printery /ˈprɪntərɪ/ *n.* (*pl.* **-ies**) *US* a printer's office or works.

printhead /ˈprɪnthed/ *n.* the component in a printer (see PRINTER 3) that assembles and prints the characters on the paper.

printing /ˈprɪntɪŋ/ *n.* **1** the production of printed books etc. (*See note below.*) **2** a single impression of a book. **3** printed letters or writing imitating them. □ **printing-press** a machine for printing from types or plates etc.

▪ Printing involves transferring an image, or group of images, from an original master to a receptive substrate such as paper, board, or cloth, repeating the process any number of times. It is most commonly achieved by inking a suitable form of the master image (i.e. printers' type, a rubber stamp, or litho plate) and transferring the ink to the substrate by flat or rolling pressure. The process originated in China in about the 8th century AD and spread to Europe in the 15th century. A major advance was Johann Gutenberg's invention in about 1450 of movable type, by which each letter is cast separately, allowing words and spaces to be formed into lines and pages, which in turn form relief master images for inking and impressing on to paper. This method has been largely superseded by photographic image-forming techniques and faster rotary printing. Film is now the most widely used master-image material, through which the printing surface is selectively sensitized and etched, ready for the modern rotary printing-press. Multicolour images are formed by superimposing one printed colour upon another in register. An increasingly common method of printing is that whereby master images are held by computer in digital form, and then put on to a suitable substrate by plotting, xerographic, ink-jet, or laser techniques.

printmaker /ˈprɪntˌmeɪkə(r)/ *n.* a person who makes prints. □ **printmaking** *n.*

printout /ˈprɪntaʊt/ *n. Computing* **1** output in printed form. **2** an instance of this.

printworks /ˈprɪntwɜːks/ *n.pl.* a factory where fabrics etc. are printed.

prion¹ /ˈpraɪən/ *n.* a small saw-billed petrel of the genus *Pachyptila*, found in southern seas. [mod.L *Prion*, former genus name, f. Gk *prīon* saw]

prion² /ˈpriːɒn/ *n. Med.* a protein particle associated with, and believed to be the cause of, encephalopathies such as scrapie, BSE, kuru, and Creutzfeldt–Jakob disease (cf. VIRINO). [rearrangement of initial letters of *proteinaceous infectious particle*]

prior /ˈpraɪə(r)/ *adj., adv., & n.* ● *adj.* **1** earlier. **2** (often foll. by *to*) coming before in time, order, or importance. ● *adv.* (foll. by *to*) before (*decided prior to their arrival*). ● *n.* (*fem.* **prioress**) **1** the superior officer of a religious house or order. **2** (in an abbey) the officer next under the abbot or abbess. □ **priorship** *n.* **priorate** /-rət/ *n.* [L, = former, elder, compar. of OL *pri* = L *prae* before]

priority /praɪˈɒrɪtɪ/ *n.* (*pl.* **-ies**) **1** the fact or condition of being earlier or antecedent. **2** precedence in rank etc. **3** an interest having prior claim to consideration. □ **prioritize** *v.tr.* (also **-ise**). **prioritization** /-ˌprʌɪtaɪˈzeɪʃ(ə)n/ *n.* [ME f. OF *priorité* f. med.L *prioritas -tatis* f. L *prior* (as PRIOR)]

priory /ˈpraɪərɪ/ *n.* (*pl.* **-ies**) a monastery governed by a prior or a nunnery governed by a prioress. [ME f. AF *priorie*, med.L *prioria* (as PRIOR)]

Pripyat /ˈpriːpjət/ (also **Pripet** /-pət/) a river of NW Ukraine and southern Belarus, which rises in Ukraine near the border with Poland and flows some 710 km (440 miles) eastwards through the Pripyat Marshes to join the River Dnieper north of Kiev.

Priscian /ˈprɪʃən/ (full name Priscianus Caesariensis) (6th century

AD), Byzantine grammarian. He taught Latin in Constantinople and his *Grammatical Institutions* became one of the standard Latin grammatical works in the Middle Ages.

prise /praɪz/ *v. & n.* (also **prize**) ● *v.tr.* (usu. foll. by adv. or complement) force open or out by leverage (*prised up the lid; prised the box open*). ● *n.* leverage, purchase. [ME & OF *prise* levering instrument (as PRIZE¹)]

prism /ˈprɪz(ə)m/ *n.* **1** a solid geometric figure whose two ends are similar, equal, and parallel rectilinear figures, and whose sides are parallelograms. **2** a transparent body in this form, usu. triangular with refracting surfaces at an acute angle with each other, which separates white light into a spectrum of colours. □ **prismal** /ˈprɪzm(ə)l/ *adj.* [LL *prisma* f. Gk *prisma prismatos* thing sawn f. *prizō* to saw]

prismatic /prɪzˈmætɪk/ *adj.* **1** of, like, or using a prism. **2 a** (of colours) distributed by or as if by a transparent prism. **b** (of light) displayed in the form of a spectrum. □ **prismatically** *adv.* [F *prismatique* f. Gk *prisma* (as PRISM)]

prismoid /ˈprɪzmɔɪd/ *n.* a body like a prism, with similar but unequal parallel polygonal ends. □ **prismoidal** /prɪzˈmɔɪd(ə)l/ *adj.*

prison /ˈprɪz(ə)n/ *n. & v.* ● *n.* **1** a place in which a person is kept in captivity, esp. legally while awaiting trial or for punishment; a jail. (*See note below.*) **2** custody, confinement (*in prison*). ● *v.tr. poet.* (**prisoned**, **prisoning**) put in prison. □ **prison-breaking** escape from prison. **prison camp** a camp for prisoners of war or of state. [ME f. OF *prisun*, *-on* f. L *prensio -onis* f. *prehensio* f. *prehendere prehens-* lay hold of]

▪ In medieval times and earlier imprisonment was not a mode of punishment but a means of holding offenders awaiting trial or execution; it was used also as a means of extorting money, through the holding of wealthy people to ransom. In Britain, the modern idea of prisons grew from the houses of correction used in the 16th century. By the mid-16th century such places (known as *bridewells*) were established in every county under the local justices, and came to be used for imprisoning petty offenders (those guilty of serious offences were generally punished by death, mutilation, and, later, by transportation); similar practices developed on the Continent. Insanitary surroundings, oppression, cruelty, and lack of supervision made prison conditions notoriously bad, and towards the end of the 18th century John Howard began a vigorous crusade for reform. Solitary confinement was at this time regarded as the ideal form of imprisonment, in the belief that the opportunity for contemplation it allowed would encourage penitence, and remained in favour until the late 19th century. In 1878 the government assumed control of all prisons, retaining it until the privatization of some institutions in the 1990s. By the 1960s and 1970s alternative methods of dealing with offenders, including suspended sentences, parole, and community service, were being introduced.

prisoner /ˈprɪznə(r)/ *n.* **1** a person kept in prison. **2** (in full **prisoner at the bar**) a person in custody on a criminal charge and on trial. **3** a person or thing confined by illness, another's grasp, etc. **4** (in full **prisoner of war**) a person who has been captured in war. □ **prisoner of conscience** see CONSCIENCE. **prisoner of state** (or **state prisoner**) a person confined for political reasons. **prisoner's base** a chasing game played, particularly in the street or playground, by two groups of children each occupying a distinct base or home. **take prisoner** seize and hold as a prisoner. [ME f. AF *prisoner*, OF *prisonier* (as PRISON)]

prissy /ˈprɪsɪ/ *adj.* (**prissier**, **prissiest**) prim, prudish. □ **prissily** *adv.* **prissiness** *n.* [perh. f. PRIM + SISSY]

Priština /ˈpriːʃtɪnə/ a city in southern Serbia, the capital of the autonomous province of Kosovo; pop. (1981) 210,000. The capital of medieval Serbia, it was taken by the Turks in 1389, remaining under Turkish control until 1912.

pristine /ˈprɪstiːn, -staɪn/ *adj.* **1** in its original condition; unspoilt. **2** spotless; fresh as if new. **3** ancient, primitive. [L *pristinus* former]

Pritchett /ˈprɪtʃɪt/, Sir V(ictor) S(awdon) (b.1900), English writer and critic. He is chiefly remembered as a writer of short stories; collections include *The Spanish Virgin and Other Stories* (1930). Among his critical works are *The Living Novel* (1946) and *Lasting Impressions* (1990). He is also noted for his novels and for two volumes of autobiography, *A Cab at the Door* (1968) and *Midnight Oil* (1971).

prithee /ˈprɪðiː/ *int. archaic* pray, please. [= *I pray thee*]

privacy /ˈprɪvəsɪ, ˈpraɪv-/ *n.* **1 a** the state of being private and undisturbed. **b** a person's right to this. **2** freedom from intrusion or public attention. **3** avoidance of publicity.

private /ˈpraɪvət/ *adj. & n.* ● *adj.* **1** belonging to an individual; one's own; personal (*private property*). **2** confidential; not to be disclosed to others (*private talks*). **3** kept or removed from public knowledge or observation. **4 a** not open to the public. **b** for an individual's exclusive use (*private room*). **5** (of a place) secluded; affording privacy. **6** (of a person) not holding public office or an official position. **7** (of education or medical treatment) conducted outside the state system, at the individual's expense. **8** (of a person) retiring; reserved; unsociable. ● *n.* **1** a private soldier. **2** (in *pl.*) *colloq.* the genitals. □ **in private** privately; in private company or life. **private bill** a parliamentary bill affecting an individual or corporation only. **private company** *Brit.* a company with restricted membership and no issue of shares. **private detective** a detective engaged privately, outside an official police force. **private enterprise 1** a business or businesses not under state control. **2** individual initiative. **private eye** *colloq.* a private detective. **private first class** *US* a soldier ranking above an ordinary private but below officers. **private hotel** a hotel not obliged to take all comers. **private house** the dwelling-house of a private person, as distinct from a shop, office, or public building. **private law** a law relating to individual persons and private property. **private life** life as a private person, not as an official, public performer, etc. **private means** income from investments etc., apart from earned income. **private member** a member of a legislative body not holding a government office. **private member's bill** a bill introduced by a private member, not part of government legislation. **private parts** the genitals. **private patient** *Brit.* a patient treated by a doctor other than under the National Health Service. **private practice** *Brit.* medical practice that is not part of the National Health Service. **private press** a printing establishment operated by a private person or group not primarily for profit and usu. on a small scale. **private school 1** *Brit.* a school supported wholly by the payment of fees. **2** *US* a school not supported mainly by the state. **private secretary** a secretary dealing with the personal and confidential concerns of a businessman or businesswoman. **private sector** the part of the economy not under direct state control. **private soldier** an ordinary soldier other than the officers (and *US* other than recruits). **private view** the viewing of an exhibition (esp. of paintings) before it is open to the public. **private war 1** a feud between persons or families disregarding the law of murder etc. **2** hostilities against members of another state without the sanction of one's own government. **private wrong** an offence against an individual but not against society as a whole. □ **privately** *adv.* [ME f. L *privatus*, orig. past part. of *privare* deprive]

privateer /ˌpraɪvəˈtɪə(r)/ *n.* **1** an armed vessel owned and officered by private individuals holding a government commission and authorized to use it against a hostile nation, especially in the capture of merchant shipping. **2 a** a commander of such a vessel. **b** (in *pl.*) the crew of such a vessel. □ **privateering** *n.* [PRIVATE, after *volunteer*]

privateersman /ˌpraɪvəˈtɪəzmən/ *n.* (*pl.* **-men**) = PRIVATEER 2.

privation /praɪˈveɪʃ(ə)n/ *n.* **1** lack of the comforts or necessities of life (*suffered many privations*). **2** (often foll. by *of*) loss or absence (of a quality). [ME f. L *privatio* (as PRIVATE)]

privative /ˈprɪvətɪv/ *adj.* **1** consisting in or marked by the loss or removal or absence of some quality or attribute. **2** (of a term) denoting the privation or absence of a quality etc. **3** *Gram.* (of a particle etc.) expressing privation, as Gk *a-* = 'not'. [F *privatif -ive* or L *privativus* (as PRIVATION)]

privatize /ˈpraɪvəˌtaɪz/ *v.tr.* (also **-ise**) make private, esp. assign (a business etc.) to private as distinct from state control or ownership; denationalize. □ **privatizer** *n.* **privatization** /ˌpraɪvətaɪˈzeɪʃ(ə)n/ *n.*

privet /ˈprɪvɪt/ *n.* a semi-evergreen shrub of the genus *Ligustrum*, bearing small white flowers and black berries, and much used for hedges. [16th c.: orig. unkn.]

privilege /ˈprɪvɪlɪdʒ/ *n. & v.* ● *n.* **1 a** a right, advantage, or immunity belonging to a person, class, or office. **b** the freedom of members of a legislative assembly when speaking at its meetings. **2** a special benefit or honour (*it is a privilege to meet you*). **3** a monopoly or patent granted to an individual, corporation, etc. **4** *US Stock Exch.* an option. ● *v.tr.* **1** invest with a privilege. **2** (foll. by *to* + infin.) allow (a person) as a privilege (to do something). **3** (often foll. by *from*) exempt (a person from a liability etc.). [ME f. OF *privilege* f. L *privilegium* bill or law affecting an individual, f. *privus* private + *lex legis* law]

privileged /ˈprɪvɪlɪdʒd/ *adj.* **1 a** invested with or enjoying a certain privilege or privileges; honoured, favoured. **b** exempt from standard regulations or procedures. **c** powerful, affluent. **2** (of information, etc.) confidential, restricted.

privity /ˈprɪvɪtɪ/ *n.* (*pl.* **-ies**) **1** *Law* a relation between two parties that is recognized by law, e.g. that of blood, lease, or service. **2** (often foll. by *to*) the state of being privy (to plans etc.). [ME f. OF *priveté* f. med.L *privitas -tatis* f. L *privus* private]

privy /ˈprɪvɪ/ *adj. & n.* ● *adj.* **1** (foll. by *to*) sharing in the secret of (a person's plans etc.). **2** *archaic* hidden, secret. ● *n.* (*pl.* **-ies**) **1** *US* or *archaic* a lavatory. **2** *Law* a person having a part or interest in any action, matter, or thing. □ **privy purse** *Brit.* **1** an allowance from the public revenue for the monarch's private expenses. **2** the keeper of this. **privy seal** (in the UK) a seal formerly affixed to documents that are afterwards to pass the Great Seal or that do not require it. □ **privily** *adv.* [ME f. OF *privé* f. L *privatus* PRIVATE]

Privy Council *n.* a body of advisers appointed by a sovereign or Governor-General. In Britain, the Privy Council originated in the council of the Norman kings. This took two forms: a large council of the realm (which grew into the Parliament), and a select body of officials who met regularly with the sovereign to carry on everyday government, known from the 14th century as the Privy (= private) Council. In the 18th century the importance of the Cabinet, a smaller group drawn from the Privy Council, increased and the full Privy Council's functions became chiefly formal, except in certain judicial activities. It now consists of about 300 members, chosen by the sovereign from those who hold or have held high political, legal, or ecclesiastical office in the UK or Commonwealth. It is summoned as a body only to sign the proclamation of the accession of a new sovereign, and when a sovereign announces an intention to marry. □ **Privy Counsellor** (or **Councillor**) a private adviser, esp. a member of a Privy Council.

Prix Goncourt /ˌpriː ɡɒnˈkʊə(r)/ *n.* an award given annually for a work of French literature (see GONCOURT).

prize[1] /praɪz/ *n. & v.* ● *n.* **1** something that can be won in a competition or lottery etc. **2** a reward given as a symbol of victory or superiority. **3** something striven for or worth striving for (*missed all the great prizes of life*). **4** (*attrib.*) **a** to which a prize is awarded (*a prize bull; a prize poem*). **b** supremely excellent or outstanding of its kind. ● *v.tr.* value highly (*a much prized possession*). □ **prize-giving** an award of prizes, esp. formally at a school etc. **prize-money** money offered as a prize. **prize-ring 1** an enclosed area (now usu. a square) for prizefighting. **2** the practice of prizefighting. [(n.) ME, var. of PRICE: (v.) ME f. OF *pris-* stem of *preisier* PRAISE]

prize[2] /praɪz/ *n. & v.* ● *n.* **1** a ship or property captured in naval warfare. **2** a find or windfall. ● *v.tr.* make a prize of. □ **prize-court** a department of an admiralty court concerned with prizes. [ME f. OF *prise* taking, booty, fem. past part. of *prendre* f. L *prehendere prehens-* seize: later identified with PRIZE[1]]

prize[3] var. of PRISE.

prizefight /ˈpraɪzfaɪt/ *n.* a boxing-match fought for prize-money. □ **prizefighter** *n.*

prizeman /ˈpraɪzmən/ *n.* (*pl.* **-men**) a winner of a prize, esp. a specified academic one.

prizewinner /ˈpraɪzˌwɪnə(r)/ *n.* a winner of a prize. □ **prizewinning** *adj.*

PRO *abbr.* **1** Public Record Office. **2** public relations officer.

pro[1] /prəʊ/ *n. & adj. colloq.* ● *n.* (*pl.* **-os**) **1** a professional. **2** a prostitute. ● *adj.* professional. □ **pro-am** involving professionals and amateurs. [abbr.]

pro[2] /prəʊ/ *adj., n., & prep.* ● *adj.* (of an argument or reason) for; in favour. ● *n.* (*pl.* **-os**) a reason or argument for or in favour. ● *prep.* in favour of. □ **pros and cons** reasons or considerations for and against a proposition etc. [L, = for, on behalf of]

pro-[1] /prəʊ/ *prefix* **1** favouring or supporting (*pro-government*). **2** acting as a substitute or deputy for (*proconsul*). **3** forwards (*produce*). **4** forwards and downwards (*prostrate*). **5** onwards (*proceed; progress*). **6** in front of (*protect*). [L *pro* in front (of), for, on behalf of, instead of, on account of]

pro-[2] /prəʊ/ *prefix* before in time, place, order, etc. (*problem; proboscis; prophet*). [Gk *pro* before]

proa /ˈprəʊə/ *n.* (also **prau** /praʊ/, **prahu** /ˈprɑːuː/) a Malay boat, esp. with a large triangular sail and a canoe-like outrigger. [Malay *prāū, prāhū*]

proactive /prəʊˈæktɪv/ *adj.* **1** (of a person, policy, etc.) creating or controlling a situation by taking the initiative. **2** of or relating to mental conditioning or a habit etc. which has been learned. □ **proactively** *adv.* **proaction** /-ˈækʃ(ə)n/ *n.* **proactivity** /ˌprəʊækˈtɪvɪtɪ/ *n.* [PRO-[2], after REACTIVE]

probabilistic /ˌprɒbəbə'lɪstɪk/ adj. relating to probability; involving chance variation.

probability /ˌprɒbə'bɪlɪti/ n. (pl. **-ies**) **1** the state or condition of being probable. **2** the likelihood of something happening. **3** a probable or most probable event (the probability is that they will come). **4** Math. the extent to which an event is likely to occur, measured by the ratio of the favourable cases to the whole number of cases possible. □ **in all probability** most probably. [F probabilité or L probabilitas (as PROBABLE)]

probable /'prɒbəb(ə)l/ adj. & n. ● adj. (often foll. by that + clause) that may be expected to happen or prove true; likely (the probable explanation; it is probable that they forgot). ● n. a probable candidate, member of a team, etc. □ **probably** adv. [ME f. OF f. L probabilis f. probare prove]

proband /'prəʊbænd/ n. a person forming the starting-point for the genetic study of a family etc. [L probandus, gerundive of probare test]

probang /'prəʊbæŋ/ n. Surgery a strip of flexible material with a sponge etc. at the end, used to remove a foreign body from the throat or apply a medication to it. [17th c. (named provang by its inventor): orig. unkn., perh. alt. after probe]

probate n. & v. ● n. /'prəʊbeɪt, -bət/ **1** the official proving of a will. **2** a verified copy of a will with a certificate as handed to the executors. ● v.tr. /'prəʊbeɪt/ N. Amer. establish the validity of (a will). [ME f. L probatum neut. past part. of probare PROVE]

probation /prə'beɪʃ(ə)n/ n. **1** Law a system of supervising and monitoring the behaviour of (esp. young) offenders, as an alternative to prison. **2** a process or period of testing the character or abilities of a person in a certain role, esp. of a new employee. **3** a moral trial or discipline. □ **on probation** undergoing probation, esp. legal supervision. **probation officer** an official supervising offenders on probation. □ **probational** adj. **probationary** adj. [ME f. OF probation or L probatio (as PROVE)]

probationer /prə'beɪʃənə(r)/ n. **1** a person on probation, e.g. a newly appointed nurse, teacher, etc. **2** an offender on probation. □ **probationership** n.

probative /'prəʊbətɪv/ adj. affording proof; evidential. [L probativus (as PROVE)]

probe /prəʊb/ n. & v. ● n. **1** a penetrating investigation. **2** any small device, esp. an electrode, for measuring, testing, etc. **3** a blunt-ended surgical instrument usu. of metal for exploring a wound etc. **4** (in full **space probe**) an unmanned exploratory spacecraft transmitting information about its environment. ● v. **1** tr. examine or enquire into closely. **2** tr. explore (a wound or part of the body) with a probe. **3** tr. penetrate with or as with a sharp instrument, esp. in order to explore. **4** intr. make an investigation with or as with a probe (the detective probed into her past life). □ **probeable** adj. **prober** n. **probingly** adv. [LL proba proof, in med.L = examination. f. L probare test]

probit /'prɒbɪt/ n. Statistics a unit of probability based on deviation from the mean of a standard distribution. [probability unit]

probity /'prəʊbɪti, 'prɒb-/ n. uprightness, honesty. [F probité or L probitas f. probus good]

problem /'prɒbləm/ n. **1** a doubtful or difficult matter requiring a solution (how to prevent it is a problem; the problem of ventilation). **2** something hard to understand or accomplish or deal with. **3** (attrib.) causing problems; difficult to deal with (problem child). **4 a** Physics & Math. an inquiry starting from given conditions to investigate or demonstrate a fact, result, or law (cf. THEOREM 1). **b** Geom. a proposition in which something has to be constructed. **5 a** (in various games, esp. chess) an arrangement of men, cards, etc., in which the solver has to achieve a specified result. **b** a puzzle or question for solution. □ **that's your** (or **his** etc.) **problem** said to disclaim responsibility or connection. [ME f. OF probleme or L problema f. Gk problēma -matos f. proballō (as PRO-2, ballō throw)]

problematic /ˌprɒblə'mætɪk/ adj. (also **problematical** /-k(ə)l/) **1** attended by difficulty. **2** doubtful or questionable. **3** Logic enunciating or supporting what is possible but not necessarily true. □ **problematically** adv. [F problématique or LL problematicus f. Gk problēmatikos (as PROBLEM)]

proboscidean /ˌprəʊbə'sɪdɪən/ adj. & n. (also **proboscidian**) Zool. ● adj. **1** having a proboscis. **2** of or like a proboscis. **3** of or relating to the mammalian order Proboscidea, which includes elephants and related extinct animals. ● n. a mammal of this order. [mod.L Proboscidea (as PROBOSCIS)]

proboscis /prəʊ'bɒsɪs/ n. **1** the long flexible trunk or snout of some mammals, e.g. an elephant or tapir. **2** the elongated mouthparts of some insects, used for sucking liquids or piercing. **3** the sucking organ in some worms. **4** joc. the human nose. □ **proboscis monkey** a monkey, Nasalis larvatus, native to Borneo, the male of which has a large pendulous nose. □ **proboscidiferous** /-ˌbɒsɪ'dɪfərəs/ adj. **proboscidiform** /ˌprəʊbə'sɪdɪˌfɔːm/ adj. [L proboscis -cidis f. Gk proboskis f. proboskō (as PRO-2, boskō feed)]

procaine /'prəʊkeɪn/ n. (also **procain**) a synthetic compound used as a local anaesthetic, esp. in dentistry. [PRO-1 + COCAINE]

procaryote var. of PROKARYOTE.

procedure /prə'siːdjə(r), -'siːdʒə(r)/ n. **1** a way of proceeding, esp. a mode of conducting business or a legal action. **2** a mode of performing a task. **3** a series of actions conducted in a certain order or manner. **4** a proceeding. **5** Computing = SUBROUTINE. □ **procedural** adj. **procedurally** adv. [F procédure (as PROCEED)]

proceed /prə'siːd/ v.intr. **1** (often foll. by to) go forward or on further; make one's way. **2** (often foll. by with, or to + infin.) continue; go on with an activity (proceeded with their work; proceeded to tell the whole story). **3** (of an action) be carried on or continued (the case will now proceed). **4** adopt a course of action (how shall we proceed?). **5** go on to say. **6** (foll. by against) start a lawsuit (against a person). **7** (often foll. by from) come forth or originate (shouts proceeded from the bedroom). **8** (foll. by to) Brit. advance to a higher rank, university degree, etc. [ME f. OF proceder f. L procedere process- (as PRO-1, cedere go)]

proceeding /prə'siːdɪŋ/ n. **1** an action or piece of conduct (a high-handed proceeding). **2** (in pl.) (in full **legal proceedings**) an action at law; a lawsuit. **3** (in pl.) a published report of discussions or a conference. **4** (in pl.) business, actions, or events in progress (the proceedings were enlivened by a dog running on to the pitch).

proceeds /'prəʊsiːdz/ n.pl. money produced by a transaction or other undertaking. [pl. of obs. proceed (n.) f. PROCEED]

process[1] /'prəʊses/ n. & v. ● n. **1** a course of action or procedure, esp. a series of stages in manufacture or some other operation. **2** the progress or course of something (in process of construction). **3** a natural or involuntary operation or series of changes (the process of growing old). **4** an action at law; a summons or writ. **5** Anat., Zool., & Bot. a natural appendage or outgrowth on an organism. ● v.tr. **1** handle or deal with by a particular process. **2** treat (food, esp. to prevent decay) (processed cheese). **3** Computing operate on (data) by means of a program. □ **in process** going on, being done. **in process of time** as time goes on. **process server** a sheriff's officer who serves writs. □ **processable** adj. [ME f. OF proces f. L processus (as PROCEED)]

process[2] /prə'ses/ v.intr. walk in procession. [back-form. f. PROCESSION]

procession /prə'seʃ(ə)n/ n. **1** a number of people or vehicles etc. moving forward in orderly succession, esp. at a ceremony, demonstration, or festivity. **2** the movement of such a group (go in procession). **3** a regular succession of things; a sequence. **4** a race in which no competitor is able to overtake another. **5** (in Christian theology) the emanation of the Holy Spirit. □ **processionist** n. [ME f. OF f. L processio -onis (as PROCEED)]

processional /prə'seʃən(ə)l/ adj. & n. ● adj. **1** of or relating to processions. **2** used, carried, or sung in processions. ● n. Eccl. an office-book of processional hymns etc. [med.L processionalis (adj.), -ale (n.) (as PROCESSION)]

processor /'prəʊsesə(r)/ n. a machine or device that processes things, esp.: **1** Computing = central processor. **2** = food processor.

procès-verbal /ˌprəʊsevɜː'baːl/ n. (pl. **procès-verbaux** /-'bəʊ/) a written report of proceedings; minutes. [F]

pro-choice /prəʊ'tʃɔɪs/ adj. & n. ● adj. advocating a woman's legal right to choose whether to have an abortion. ● n. a pro-choice policy.

prochronism /'prəʊkrəˌnɪz(ə)m/ n. the action of referring an event etc. to an earlier date than the true one. [PRO-2 + Gk khronos time]

proclaim /prə'kleɪm/ v.tr. **1** (often foll. by that + clause) announce or declare publicly or officially. **2** declare (a person) to be (a king, traitor, etc.). **3** reveal as being (an accent that proclaims you a Scot). □ **proclaimer** n. **proclamatory** /-'klæmətəri/ adj. **proclamation** /ˌprɒklə'meɪʃ(ə)n/ n. [ME proclame f. L proclamare cry out (as PRO-1, CLAIM)]

proclitic /prə'klɪtɪk/ adj. & n. Gram. ● adj. (of a monosyllable) closely attached in pronunciation to a following word and having itself no accent. ● n. such a word, e.g. at in at home. □ **proclitically** adv. [mod.L procliticus f. Gk proklinō lean forward, after LL encliticus: see ENCLITIC]

proclivity /prə'klɪvɪti/ n. (pl. **-ies**) a tendency or inclination. [L proclivitas f. proclivis inclined (as PRO-1, clivus slope)]

Procne /'prɒkni/ Gk Mythol. the sister of Philomel.

proconsul /prəʊ'kɒns(ə)l/ n. **1** Rom. Hist. a governor of a province, in the later republic usu. an ex-consul. **2** a governor of a modern colony etc. **3** a deputy consul. □ **proconsulship** n. **proconsular** /-sjʊlə(r)/ adj. **proconsulate** /-sjʊlət/ n. [ME f. L, earlier *pro consule* (one acting) for the consul]

Procopius /prə'kəʊpɪəs/ (c.500–c.562), Byzantine historian, born in Caesarea in Palestine. He accompanied Justinian's general Belisarius (c.505–65) on his campaigns between 527 and 540. His principal works are the *History of the Wars of Justinian* and *On Justinian's Buildings*. The authenticity of another work, the *Secret History*, has often been doubted but is now generally accepted; it is a virulent attack on Justinian, his policy, and his officials, and also contains comments on the dubious morals of the empress Theodora.

procrastinate /prəʊ'kræstɪ,neɪt/ v.intr. defer action, esp. without good reason; be dilatory. ¶ Often confused with *prevaricate*. □ **procrastinator** n. **procrastinative** /-nətɪv/ adj. **procrastinatory** /-nətərɪ/ adj. **procrastination** /-,kræstɪ'neɪʃ(ə)n/ n. [L *procrastinare procrastinat-* (as PRO-¹, *crastinus* of tomorrow f. *cras* tomorrow)]

procreate /'prəʊkrɪ,eɪt/ v.tr. (often absol.) bring (offspring) into existence by the natural process of reproduction. □ **procreative** adj. **procreant** /-krɪənt/ adj. **procreation** /,prəʊkrɪ'eɪʃ(ə)n/ n. **procreator** n. [L *procreare procreat-* (as PRO-¹, *creare* create)]

Procrustean /prəʊ'krʌstɪən/ adj. seeking to enforce uniformity by forceful or ruthless methods. [PROCRUSTES]

Procrustes /prəʊ'krʌstiːz/ Gk Mythol. a robber who forced travellers to lie on a bed and made them fit it by stretching their limbs or cutting off the appropriate length of leg. Theseus killed him in like manner. [Gk *Prokroustēs*, lit. 'stretcher', f. *prokrouō* beat out]

proctology /prɒk'tɒlədʒɪ/ n. the branch of medicine concerned with the anus and rectum. □ **proctologist** n. **proctological** /,prɒktə'lɒdʒɪk(ə)l/ adj. [Gk *prōktos* anus + -LOGY]

proctor /'prɒktə(r)/ n. **1** Brit. an officer (usu. one of two) at certain universities, appointed annually and having mainly disciplinary functions. **2** US a supervisor of students in an examination etc. **3** Law a person managing causes in a court (now chiefly ecclesiastical) that administers civil or canon law. **4** a representative of the clergy in the Church of England convocation. □ **proctorship** n. **proctorial** /prɒk'tɔːrɪəl/ adj. [ME, syncopation of PROCURATOR]

proctoscope /'prɒktə,skəʊp/ n. a medical instrument for inspecting the rectum. [Gk *prōktos* anus + -SCOPE]

procumbent /prə'kʌmb(ə)nt/ adj. **1** lying on the face; prostrate. **2** Bot. growing along the ground. [L *procumbere* fall forwards (as PRO-¹, *cumbere* lay oneself)]

procuration /,prɒkjʊə'reɪʃ(ə)n/ n. **1** formal the action of procuring, obtaining, or bringing about. **2** the function or an authorized action of an attorney. [ME f. OF *procuration* or L *procuratio* (as PROCURE)]

procurator /'prɒkjʊə,reɪtə(r)/ n. **1** an agent or proxy, esp. one who has power of attorney. **2** Rom. Hist. a treasury officer in an imperial province. □ **procurator fiscal** an officer of a sheriff's court in Scotland, acting as public prosecutor of a district and with other duties similar to those of a coroner. □ **procuratorship** n. **procuratorial** /,prɒkjʊərə'tɔːrɪəl/ adj. [ME f. OF *procurateur* or L *procurator* administrator, finance-agent (as PROCURE)]

procure /prə'kjʊə(r)/ v.tr. **1** obtain, esp. by care or effort; acquire (*managed to procure a copy*). **2** bring about (*procured their dismissal*). **3** (also absol.) obtain (women) for prostitution. □ **procurable** adj. **procural** n. **procurement** n. [ME f. OF *procurer* f. L *procurare* take care of, manage (as PRO-¹, *curare* see to)]

procurer /prə'kjʊərə(r)/ n. (fem. **procuress** /-rɪs/) a person who obtains women for prostitution. [ME f. AF *procurour*, OF *procureur* f. L *procurator*: see PROCURATOR]

Procyon /'prəʊsɪɒn/ Astron. the eighth brightest star in the sky, and the brightest in the constellation of Canis Minor. [Gk, = before the dog (because it rises before the dog-star)]

prod /prɒd/ v. & n. ● v. (**prodded, prodding**) **1** tr. poke with the finger or a pointed object. **2** tr. stimulate or goad to action. **3** intr. (foll. by at) make a prodding motion. ● n. **1** a poke or thrust. **2** a stimulus to action. **3** a pointed instrument. □ **prodder** n. [16th c.: perh. imit.]

prodigal /'prɒdɪg(ə)l/ adj. & n. ● adj. **1** recklessly wasteful. **2** (foll. by of) lavish. ● n. **1** a prodigal person. **2** (in full **prodigal son**) a repentant wastrel, returned wanderer, etc. (after Luke 15:11–32). □ **prodigally** adv. **prodigality** /,prɒdɪ'gælɪtɪ/ n. [med.L *prodigalis* f. L *prodigus* lavish]

prodigious /prə'dɪdʒəs/ adj. **1** marvellous or amazing. **2** enormous. **3** abnormal. □ **prodigiously** adv. **prodigiousness** n. [L *prodigiosus* (as PRODIGY)]

prodigy /'prɒdɪdʒɪ/ n. (pl. **-ies**) **1** a person endowed with exceptional qualities or abilities, esp. a precocious child. **2** a marvellous thing, esp. one out of the ordinary course of nature. **3** (foll. by of) a wonderful example (of a quality). [L *prodigium* portent]

prodrome /'prəʊdrəʊm/ n. **1** a preliminary book or treatise. **2** Med. a premonitory symptom. □ **prodromal** /prə'drəʊm(ə)l/ adj. **prodromic** /-'drɒmɪk/ adj. [F f. mod.L f. Gk *prodromos* precursor (as PRO-², *dromos* running)]

produce v. & n. ● v.tr. /prə'djuːs/ **1** bring forward for consideration, inspection, or use (*will produce evidence*). **2** manufacture (goods) from raw materials etc. **3** bear or yield (offspring, fruit, a harvest, etc.). **4** bring into existence. **5** cause or bring about (a reaction, sensation, etc.). **6** Geom. extend or continue (a line). **7 a** bring (a play, book, etc.) before the public. **b** supervise the making of (a film, broadcast, record, etc.). ● n. /'prɒdjuːs/ **1 a** what is produced, esp. agricultural and natural products collectively (*dairy produce*). **b** an amount of this. **2** (often foll. by of) a result (of labour, efforts, etc.). **3** a yield, esp. in the assay of ore. □ **producible** /prə'djuːsɪb(ə)l/ adj. **producibility** /-,djuːsɪ'bɪlɪtɪ/ n. [ME f. L *producere* (as PRO-¹, *ducere duct-* lead)]

producer /prə'djuːsə(r)/ n. **1 a** Econ. a person who produces goods or commodities. **b** a person who or thing which produces something or someone. **2** a person who produces a play, film, record, etc. □ **producer gas** a low-grade fuel, consisting mainly of carbon monoxide and nitrogen, formed by passing air, or air and steam, through red-hot carbon.

product /'prɒdʌkt/ n. **1** a thing or substance produced by natural process or manufacture. **2** a result (*the product of their labours*). **3** Math. a quantity obtained by multiplying quantities together. [ME f. L *productum*, neut. past part. of *producere* PRODUCE]

production /prə'dʌkʃ(ə)n/ n. **1** the act or an instance of producing; the process of being produced. **2** the process of being manufactured, esp. in large quantities (*go into production*). **3** a total yield. **4** a thing produced, esp. a literary or artistic work, a film, play, etc. □ **production line** a systematized sequence of mechanical or manual operations involved in producing a commodity. □ **productional** adj. [ME f. OF f. L *productio -onis* (as PRODUCT)]

productive /prə'dʌktɪv/ adj. **1** of or engaged in the production of goods. **2 a** producing much (*productive soil; a productive writer*). **b** (of the mind) inventive, creative. **3** Econ. producing commodities of exchangeable value (*productive labour*). **4** (foll. by of) producing or giving rise to (*productive of great annoyance*). **5** Philol. (of a word-element) frequently used in forming new words. □ **productively** adv. **productiveness** n. [F *productif -ive* or LL *productivus* (as PRODUCT)]

productivity /,prɒdʌk'tɪvɪtɪ/ n. **1** the capacity to produce; the state of being productive. **2** the effectiveness of productive effort, esp. in industry. **3** production per unit of effort.

proem /'prəʊɪm/ n. **1** a preface or preamble to a book or speech. **2** a beginning or prelude. □ **proemial** /prəʊ'iːmɪəl/ adj. [ME f. OF *proeme* or L *prooemium* f. Gk *prooimion* prelude (as PRO-², *oimē* song)]

Prof. abbr. Professor.

prof /prɒf/ n. colloq. a professor. [abbr.]

profane /prə'feɪn/ adj. & v. ● adj. **1** not belonging to what is sacred or biblical; secular. **2 a** irreverent, blasphemous. **b** vulgar, obscene. **3** (of a rite etc.) heathen. **4** not initiated into religious rites or any esoteric knowledge. ● v.tr. **1** treat (a sacred thing) with irreverence or disregard. **2** violate or pollute. □ **profanely** adv. **profaneness** n. **profaner** n. **profanation** /,prɒfə'neɪʃ(ə)n/ n. [ME *prophane* f. OF *prophane* or med.L *prophanus* f. L *profanus* before (i.e. outside) the temple, not sacred (as PRO-¹, *fanum* temple)]

profanity /prə'fænɪtɪ/ n. (pl. **-ies**) **1** a profane act. **2 a** profane language; blasphemy. **b** an oath, a swear-word. [LL *profanitas* (as PROFANE)]

profess /prə'fes/ v. **1** tr. claim openly to have (a quality or feeling). **2** tr. (foll. by to + infin.) pretend. **3** tr. (often foll. by that + clause; also refl.) declare (*profess ignorance; professed herself satisfied*). **4** tr. affirm one's faith in or allegiance to. **5** tr. receive into a religious order under vows. **6** tr. have as one's profession or business. **7 a** tr. teach (a subject) as a professor. **b** intr. perform the duties of a professor. [ME f. L *profiteri profess-* declare publicly (as PRO-¹, *fateri* confess)]

professed /prə'fest/ adj. **1** self-acknowledged (*a professed Christian*). **2** alleged, ostensible. **3** claiming to be duly qualified. **4** (of a monk or nun) having taken the vows of a religious order. □ **professedly** /-sɪdlɪ/ adv. (in senses 1, 2).

profession /prə'feʃ(ə)n/ n. **1** a vocation or calling, esp. one that involves some branch of advanced learning or science (*the medical* _profession_). **2** a body of people engaged in a profession. **3** a declaration or avowal. **4** a declaration of belief in a religion. **5 a** the declaration or vows made on entering a religious order. **b** the ceremony or fact of being professed in a religious order. □ **the oldest profession** _colloq._ or _joc._ prostitution. □ **professionless** _adj._ [ME f. OF f. L _professio -onis_ (as PROFESS)]

professional /prə'feʃən(ə)l/ _adj. & n._ ● _adj._ **1** of or belonging to or connected with a profession. **2 a** having or showing the skill of a professional; competent. **b** worthy of a professional (_professional conduct_). **3** engaged in a specified activity as one's main paid occupation (cf. AMATEUR) (_a professional boxer_). **4** _derog._ engaged in a specified activity regarded with disfavour, esp. habitually (_a professional agitator_). ● _n._ a professional person. □ **professional foul** a deliberate foul in football etc., esp. to prevent an opponent from scoring. □ **professionally** _adv._

professionalism /prə'feʃənə‚lɪz(ə)m/ _n._ the qualities or typical features of a profession or of professionals, esp. competence, skill, etc. □ **professionalize** _v.tr._ (also **-ise**).

professor /prə'fesə(r)/ _n._ **1 a** (often as a title) a university academic of the highest rank; the holder of a university chair. **b** _US_ a university teacher. **2** a person who professes a religion. □ **professorship** _n._ **professorate** /-rət/ _n._ **professorial** /‚prɒfɪ'sɔːrɪəl/ _adj._ **professorially** _adv._ **professoriate** /-'sɔːrɪət/ _n._ [ME f. OF _professeur_ or L _professor_ (as PROFESS)]

proffer /'prɒfə(r)/ _v. & n._ ● _v.tr._ (esp. as **proffered** _adj._) offer (a gift, services, a hand, etc.). ● _n._ _literary_ an offer or proposal. [ME f. AF & OF _proffrir_ (as PRO-¹, _offrir_ OFFER)]

proficient /prə'fɪʃ(ə)nt/ _adj. & n._ ● _adj._ (often foll. by _in_, _at_) adept, expert. ● _n._ a person who is proficient. □ **proficiency** _n._ **proficiently** _adv._ [L _proficiens proficient-_ (as PROFIT)]

profile /'prəʊfaɪl/ _n. & v._ ● _n._ **1 a** an outline (esp. of a human face) as seen from one side. **b** a representation of this. **2 a** a short biographical or character sketch. **b** a report, esp. one written by a teacher on a pupil's academic and social progress. **3** _Statistics_ a representation by a graph or chart of information (esp. on certain characteristics) recorded in a quantified form. **4** a characteristic personal manner or attitude. **5** a vertical cross-section of a structure. **6** a flat outline piece of scenery on stage. ● _v.tr._ **1** represent in profile. **2** give a profile to. **3** write a profile on. □ **in profile** as seen from one side. **keep a low profile** remain inconspicuous. □ **profiler** _n._ **profilist** /-fɪlɪst/ _n._ [obs. It. _profilo_, _profilare_ (as PRO-¹, _filare_ spin f. L _filare_ f. _filum_ thread)]

profit /'prɒfɪt/ _n. & v._ ● _n._ **1** an advantage or benefit. **2** financial gain; excess of returns over outlay. ● _v._ (**profited**, **profiting**) **1** _tr._ (also _absol._) be beneficial to. **2** _intr._ obtain an advantage or benefit (_profited by the experience_). **3** _intr._ make a profit. □ **at a profit** with financial gain. **profit and loss account** an account in which gains are credited and losses debited so as to show the net profit or loss at any time. **profit margin** the profit remaining in a business after costs have been deducted. **profit-sharing** the sharing of profits esp. between employer and employees. **profit-taking** the sale of shares etc. at a time when profit will accrue. □ **profitless** _adj._ [ME f. OF f. L _profectus_ progress, profit f. _proficere profect-_ advance (as PRO-¹, _facere_ do)]

profitable /'prɒfɪtəb(ə)l/ _adj._ **1** yielding profit; lucrative. **2** beneficial; useful. □ **profitably** _adv._ **profitableness** _n._ **profitability** /‚prɒfɪtə'bɪlɪtɪ/ _n._ [ME f. OF (as PROFIT)]

profiteer /‚prɒfɪ'tɪə(r)/ _v. & n._ ● _v.intr._ make or seek to make excessive profits, esp. illegally or in black market conditions. ● _n._ a person who profiteers.

profiterole /prə'fɪtə‚rəʊl/ _n._ a small hollow case of choux pastry usu. filled with cream and covered with chocolate sauce. [F, dimin. of _profit_ PROFIT]

profligate /'prɒflɪgət/ _adj. & n._ ● _adj._ **1** licentious; dissolute. **2** recklessly extravagant. ● _n._ a profligate person. □ **profligacy** _n._ **profligately** _adv._ [L _profligatus_ dissolute, past part. of _profligare_ overthrow, ruin (as PRO-¹, _fligere_ strike down)]

pro forma /prəʊ 'fɔːmə/ _adv., adj., & n._ ● _adv. & adj._ as or being a matter of form. ● _n._ (in full **pro forma invoice**) an invoice sent in advance of goods supplied. [L]

profound /prə'faʊnd/ _adj. & n._ ● _adj._ (**profounder**, **profoundest**) **1 a** having or showing great knowledge or insight. **b** demanding deep study or thought. **2** (of a state or quality) deep, intense, unqualified (_a profound sleep; profound indifference_). **3** at or extending to a great depth (_profound crevasses_). **4** (of a sigh) deep-drawn. **5** (of a disease) deep-seated. ● _n._ (prec. by _the_) _poet._ the vast depth (of the ocean, soul, etc.). □ **profoundly** _adv._ **profoundness** _n._ **profundity** /-'fʌndɪtɪ/ _n._ (pl. **-ies**).

[ME f. AF & OF _profund_, _profond_ f. L _profundus_ deep (as PRO-¹, _fundus_ bottom)]

Profumo /prə'fjuːməʊ/, John (Dennis) (b.1915), British Conservative politician. In 1960 he was appointed Secretary of State for War under Harold Macmillan. Three years later news broke of his relationship with the mistress of a Soviet diplomat, Christine Keeler (b.1942), raising fears of a security breach and precipitating his resignation.

profuse /prə'fjuːs/ _adj._ **1** (often foll. by _in_) lavish; extravagant (_was profuse in her generosity_). **2** (of a thing) exuberantly plentiful; abundant (_profuse bleeding; a profuse variety_). □ **profusely** _adv._ **profuseness** _n._ **profusion** /-'fjuːʒ(ə)n/ _n._ [ME f. L _profusus_ past part. of _profundere profus-_ (as PRO-¹, _fundere fus-_ pour)]

progenitive /prəʊ'dʒenɪtɪv/ _adj._ capable of or connected with the production of offspring.

progenitor /prəʊ'dʒenɪtə(r)/ _n._ **1** the ancestor of a person, animal, or plant. **2** a political or intellectual predecessor. **3** the origin of a copy. □ **progenitorship** _n._ **progenitorial** /-‚dʒenɪ'tɔːrɪəl/ _adj._ [ME f. OF _progeniteur_ f. L _progenitor -oris_ f. _progignere progenit-_ (as PRO-¹, _gignere_ beget)]

progeniture /prəʊ'dʒenɪ‚tjʊə(r)/ _n._ **1** the act or an instance of procreation. **2** young, offspring.

progeny /'prɒdʒɪnɪ/ _n._ **1** the offspring of a person or other organism. **2** a descendant or descendants. **3** an outcome or issue. [ME f. OF _progenie_ f. L _progenies_ f. _progignere_ (as PROGENITOR)]

progesterone /prəʊ'dʒestə‚rəʊn/ _n._ a steroid hormone released by the corpus luteum which stimulates the preparation of the uterus for pregnancy (see also PROGESTOGEN). [_progest_in (as PRO-², GESTATION) + lute_osterone_ f. CORPUS LUTEUM + STEROL]

progestogen /prəʊ'dʒestədʒɪn/ _n._ **1** any of a group of steroid hormones (including progesterone) that maintain pregnancy and prevent further ovulation during it. **2** a similar hormone produced synthetically.

proglottid /prəʊ'glɒtɪd/ _n._ (also **proglottis** /-tɪs/, _pl._ **proglottides** /-‚diːz/) _Zool._ each segment in the strobila of a tapeworm that contains a complete reproductive system. [f. Gk _proglōssid-_, _-glōssis_ (as PRO-², _glōssis_ f. _glōssa_, _glōtta_ tongue), from its shape]

prognathous /prɒg'neɪθəs, 'prɒgnəθ-/ _adj._ **1** having a projecting jaw. **2** (of a jaw) projecting. □ **prognathic** /-'næθɪk/ _adj._ **prognathism** /'prɒgnə‚θɪz(ə)m/ _n._ [PRO-² + Gk _gnathos_ jaw]

prognosis /prɒg'nəʊsɪs/ _n._ (pl. **prognoses** /-siːz/) **1** a forecast; a prognostication. **2** a forecast of the course of a disease. [LL f. Gk _prognōsis_ (as PRO-², _gignōskō_ know)]

prognostic /prɒg'nɒstɪk/ _n. & adj._ ● _n._ **1** (often foll. by _of_) an advance indication or omen, esp. of the course of a disease etc. **2** a prediction; a forecast. ● _adj._ foretelling; predictive (_prognostic of a good result_). □ **prognostically** _adv._ [ME f. OF _pronostique_ f. L _prognosticum_ f. Gk _prognōstikon_ neut. of _prognōstikos_ (as PROGNOSIS)]

prognosticate /prɒg'nɒstɪ‚keɪt/ _v.tr._ **1** (often foll. by _that_ + clause) foretell; foresee; prophesy. **2** (of a thing) betoken; indicate (future events etc.). □ **prognosticator** _n._ **prognosticable** /-kəb(ə)l/ _adj._ **prognosticative** /-kətɪv/ _adj._ **prognosticatory** /-kətərɪ/ _adj._ **prognostication** /-‚nɒstɪ'keɪʃ(ə)n/ _n._ [med.L _prognosticare_ (as PROGNOSTIC)]

programme /'prəʊgræm/ _n. & v._ (_US_ **program**) ● _n._ **1** a usu. printed list of a series of events, performers, etc. at a public function etc. **2** a radio or television broadcast. **3** a plan of future events (_the programme is dinner and an early night_). **4** a course or series of studies, lectures, etc.; a syllabus. **5** (usu. **program**) a series of coded instructions to control the operation of a computer or other machine. ● _v.tr._ (**programmed**, **programming**; _US_ **programed**, **programing**) **1** make a programme or definite plan of. **2 a** (usu. **program**) provide (a computer etc.) with coded instructions for the performance of a particular task. **b** train to behave in a predetermined way. □ **programme music** a piece of music intended to tell a story, evoke images, etc. □ **programmer** _n._ **programmable** _adj._ **programmability** /‚prəʊgræmə'bɪlɪtɪ/ _n._ **programmatic** /-grə'mætɪk/ _adj._ **programmatically** _adv._ [LL _programma_ f. Gk _programma -atos_ f. _prographō_ write publicly (as PRO-², _graphō_ write): spelling after F _programme_]

progress _n. & v._ ● _n._ /'prəʊgres/ **1** forward or onward movement towards a destination. **2** advance or development towards completion, betterment, etc.; improvement (_has made little progress this term; the progress of civilization_). **3** _Brit. archaic_ a state journey or official tour, esp. by royalty. ● _v._ /prə'gres/ **1** _intr._ move or be moved forward or onward; continue (_the argument is progressing_). **2** _intr._ advance or develop towards completion, improvement, etc. (*science progresses*). **3** *tr.* cause (work etc.) to make regular progress. □ **in progress** in the course of developing; going on. **progress-chaser** a person employed to check the regular progress of manufacturing work. **progress report** an account of progress made. [ME f. L *progressus* f. *progredi* (as PRO-[1], *gradi* walk: (v.) readopted f. US after becoming obs. in Brit. use in the 17th c.]

progression /prə'greʃ(ə)n/ *n.* **1** the act or an instance of progressing (*a mode of progression*). **2** a succession; a series. **3** *Math.* **a** = *arithmetic progression*. **b** = *geometric progression*. **c** = *harmonic progression*. **4** *Mus.* passing from one note or chord to another. □ **progressional** *adj.* [ME f. OF *progression* or L *progressio* (as PROGRESS)]

progressionist /prə'greʃənɪst/ *n.* **1** an advocate of or believer in esp. political or social progress. **2** a person who believes in the theory of gradual progression to higher forms of life.

progressive /prə'gresɪv/ *adj.* & *n.* ● *adj.* **1** moving forward (*progressive motion*). **2** proceeding step by step; cumulative (*progressive drug use*). **3 a** (of a political party, government, etc.) favouring or implementing rapid progress or social reform. **b** modern; efficient (*this is a progressive company*). **4** (of disease, violence, etc.) increasing in severity or extent. **5** (of taxation) at rates increasing with the sum taxed. **6** (of a card-game, dance, etc.) with periodic changes of partners. **7** *Gram.* (of an aspect) expressing an action in progress, e.g. *am writing, was writing*. **8** (of education) informal and without strict discipline, stressing individual needs. ● *n.* (also **Progressive**) an advocate of progressive political policies. □ **progressively** *adv.* **progressiveness** *n.* **progressivism** *n.* **progressivist** *n.* & *adj.* [F *progressif -ive* or med.L *progressivus* (as PROGRESS)]

Progressive Conservative Party a Canadian political party advocating free trade and holding moderate views on social policies. Founded in the mid-19th century but operating under its present name since 1942, the party was in power 1984–93 under Brian Mulroney.

pro hac vice /ˌprəʊ hɑːk ˈvaɪsɪ/ *adv.* for this occasion (only). [L]

prohibit /prə'hɪbɪt/ *v.tr.* (**prohibited, prohibiting**) (often foll. by *from* + verbal noun) **1** formally forbid, esp. by authority. **2** prevent; make impossible (*his accident prohibits him from playing football*). □ **prohibited degrees** degrees of blood relationship within which marriage is forbidden. □ **prohibiter** *n.* **prohibitor** *n.* **prohibitory** *adj.* [ME f. L *prohibere* (as PRO-[1], *habere* hold)]

Prohibition /ˌprəʊɪ'bɪʃ(ə)n/ the forbidding of the manufacture and sale of alcoholic drink as established in the US from 1920, after a long campaign, by the 18th Amendment to the Constitution. It led to widespread bootlegging of illicit liquor by organized gangs, and was repealed in 1933 by the 21st Amendment.

prohibition /ˌprəʊɪ'bɪʃ(ə)n/ *n.* **1** the act or an instance of forbidding; a state of being forbidden. **2** *Law* **a** an edict or order that forbids. **b** a writ from a superior court forbidding an inferior court from proceeding in a suit deemed to be beyond its cognizance. **3** the prevention by law of the manufacture and sale of alcohol (see PROHIBITION). □ **prohibitionary** *adj.* **prohibitionist** *n.* [ME f. OF *prohibition* or L *prohibitio* (as PROHIBIT)]

prohibitive /prə'hɪbɪtɪv/ *adj.* **1** prohibiting. **2** (of prices, taxes, etc.) so high as to prevent purchase, use, abuse, etc. (*published at a prohibitive price*). □ **prohibitively** *adv.* **prohibitiveness** *n.* [F *prohibitif -ive* or L *prohibitivus* (as PROHIBIT)]

project *n.* & *v.* ● *n.* /'prɒdʒekt/ **1** a plan; a scheme. **2** a planned undertaking. **3** a usu. long-term task undertaken by a student to be submitted for assessment. ● *v.* /prə'dʒekt/ **1** *tr.* plan or contrive (a course of action, scheme, etc.). **2** *intr.* protrude; jut out. **3** *tr.* throw; cast; impel (*projected the stone into the water*). **4** *tr.* extrapolate (results etc.) to a future time; forecast (*I project that we shall produce 2 million next year*). **5** *tr.* cause (light, shadow, images, etc.) to fall on a surface, screen, etc. **6** *tr.* cause (a sound, esp. the voice) to be heard at a distance. **7** *tr.* (often *refl.* or *absol.*) express or promote (oneself or a positive image) forcefully or effectively. **8** *tr.* *Geom.* **a** draw straight lines from a centre or parallel lines through every point of (a given figure) to produce a corresponding figure on a surface or a line by intersecting it. **b** draw (such lines). **c** produce (such a corresponding figure). **9** *tr.* make a projection of (the earth, sky, etc.). **10** *tr.* *Psychol.* **a** (also *absol.*) attribute (an emotion etc.) to an external object or person, esp. unconsciously. **b** (*refl.*) project (oneself) into another's feelings, the future, etc. [ME f. L *projectum* neut. past part. of *projicere* (as PRO-[1], *jacere* throw)]

projectile /prə'dʒektaɪl/ *n.* & *adj.* ● *n.* **1** a missile, esp. fired by a rocket. **2** a bullet, shell, etc. fired from a gun. **3** any object thrown as a weapon. ● *adj.* **1** capable of being projected by force, esp. from a gun. **2** projecting or impelling. [mod.L *projectilis* (adj.), *-ile* (n.) (as PROJECT)]

projection /prə'dʒekʃ(ə)n/ *n.* **1** the act or an instance of projecting; the process of being projected. **2** a thing that projects or obtrudes. **3** the presentation of an image etc. on a surface or screen. **4 a** a forecast or estimate based on present trends (*a projection of next year's profits*). **b** this process. **5 a** a mental image or preoccupation viewed as an objective reality. **b** the unconscious transfer of one's own impressions or feelings to external objects or persons. **6** *Geom.* the act or an instance of projecting a figure. **7** the representation on a plane surface of any part of the surface of the earth or a celestial sphere (*Mercator projection*). □ **projectionist** *n.* (in sense 3). [L *projectio* (as PROJECT)]

projective /prə'dʒektɪv/ *adj.* **1** *Geom.* **a** relating to or derived by projection. **b** (of a property of a figure) unchanged by projection. **2** *Psychol.* mentally projecting or projected (*a projective imagination*). □ **projective geometry** the study of the projective properties of geometric figures. □ **projectively** *adv.*

projector /prə'dʒektə(r)/ *n.* **1 a** an apparatus containing a source of light and a system of lenses for projecting slides or film on to a screen. **b** an apparatus for projecting rays of light. **2** a person who forms or promotes a project. **3** *archaic* a promoter of speculative companies.

prokaryote /prəʊ'kærɪət/ *n.* (also **procaryote**) *Biol.* a single-celled organism which has neither a distinct nucleus with a membrane nor other specialized organelles. This includes only the bacteria and blue-green algae (cf. EUKARYOTE). □ **prokaryotic** /-ˌkærɪ'ɒtɪk/ *adj.* [PRO-[2] + KARYO- + *-ote* as in ZYGOTE]

Prokofiev /prə'kɒfɪˌef/, Sergei (Sergeevich) (1891–1953), Russian composer. By the age of 13 he had already written operas, sonatas, and piano pieces. In 1918 Prokofiev emigrated to the US; he lived there and in Paris before returning to the Soviet Union in 1933. Notable works include seven symphonies, the operas *The Love for Three Oranges* (1919) and *War and Peace* (1941–3), the *Lieutenant Kijé* suite (1934), and the ballet music for *Romeo and Juliet* (1935–6). He also wrote *Peter and the Wolf* (1936), a young person's guide to the orchestra in the form of a fairy tale.

Prokopevsk /prə'kɒpjefsk/ a coal-mining city in southern Russia, in the Kuznetsk Basin industrial region to the south of Kemerovo; pop. (1990) 274,000.

prolactin /prəʊ'læktɪn/ *n.* *Biochem.* a hormone released from the anterior pituitary gland that stimulates milk production after childbirth. [PRO-[1] + LACTATION]

prolapse /'prəʊlæps/ *n.* & *v.* *Med.* ● *n.* (also **prolapsus** /prəʊ'læpsəs/) **1** the forward or downward displacement of a part or organ. **2** the prolapsed part or organ, esp. the womb or rectum. ● *v.intr.* undergo prolapse. [L *prolabi prolaps-* (as PRO-[1], *labi* slip)]

prolate /'prəʊleɪt/ *adj.* **1** *Geom.* (of a spheroid) lengthened in the direction of a polar diameter (cf. OBLATE[2]). **2** growing or extending in width. **3** widely spread. **4** *Gram.* = PROLATIVE. □ **prolately** *adv.* [L *prolatus* past part. of *proferre* prolong (as PRO-[1], *ferre* carry)]

prolative /prə'leɪtɪv/ *adj.* *Gram.* serving to continue or complete a predication, e.g. *go* (prolative infinitive) in *you may go*.

prole /prəʊl/ *adj.* & *n.* *derog. colloq.* ● *adj.* proletarian. ● *n.* a proletarian. [abbr.]

proleg /'prəʊleg/ *n.* *Zool.* a fleshy abdominal limb of a caterpillar or similar insect larva. [PRO-[1] + LEG]

prolegomenon /ˌprəʊlɪ'gɒmɪnən/ *n.* (*pl.* **prolegomena** /-nə/) (usu. in *pl.*) an introduction or preface to a book etc., esp. when critical or discursive. □ **prolegomenary** *adj.* **prolegomenous** *adj.* [L f. Gk, neut. passive pres. part. of *prolegō* (as PRO-[2], *legō* say)]

prolepsis /prəʊ'lepsɪs, -'liːpsɪs/ *n.* (*pl.* **prolepses** /-siːz/) **1** the anticipation and answering of possible objections in rhetorical speech. **2** anticipation. **3** the representation of a thing as existing before it actually does or did so, as in *he was a dead man when he entered*. **4** *Gram.* the anticipatory use of adjectives, as in *paint the town red*. □ **proleptic** *adj.* [LL f. Gk *prolēpsis* f. *prolambanō* anticipate (as PRO-[2], *lambanō* take)]

proletarian /ˌprəʊlɪ'teərɪən/ *adj.* & *n.* ● *adj.* of or concerning the proletariat. ● *n.* a member of the proletariat. □ **proletarianism** *n.* **proletarianize** *v.tr.* (also **-ise**). [L *proletarius* a person who served the state not with property but with offspring (*proles*)]

proletariat /ˌprəʊlɪ'teərɪət/ *n.* (also **proletariate**) **1 a** *Econ.* wage-earners collectively, esp. those without capital and dependent on selling their labour. **b** esp. *derog.* the lowest class of the community, esp. when considered as uncultured. **2** *Rom. Hist.* the lowest class of citizens. [F *prolétariat* (as PROLETARIAN)]

pro-life /prəʊˈlaɪf/ *adj.* in favour of preserving life, esp. in opposing abortion.

proliferate /prəˈlɪfəˌreɪt/ *v.* **1** *intr.* reproduce; increase rapidly in numbers; grow by multiplication. **2** *tr.* produce (cells etc.) rapidly. □ **proliferative** /-rətɪv/ *adj.* **proliferation** /-ˌlɪfəˈreɪʃ(ə)n/ *n.* [backform. f. *proliferation* f. F *proliférer* (as PROLIFEROUS)]

proliferous /prəˈlɪfərəs/ *adj.* **1** (of a plant) producing many leaf or flower buds; growing luxuriantly. **2** growing or multiplying by budding. **3** spreading by proliferation. [L *proles* offspring + -FEROUS]

prolific /prəˈlɪfɪk/ *adj.* **1** producing many offspring or much output. **2** (often foll. by *of*) abundantly productive. **3** (often foll. by *in*) abounding, copious. □ **prolifically** *adv.* **prolificness** *n.* [med.L *prolificus* (as PROLIFEROUS)]

proline /ˈprəʊliːn/ *n. Biochem.* an amino acid with a cyclic molecule, present in many proteins, esp. collagen. [contr. of PYRROLIDINE (with similar molecular structure)]

prolix /ˈprəʊlɪks, prəˈlɪks/ *adj.* (of speech, writing, etc.) lengthy; tedious. □ **prolixly** *adv.* **prolixity** /prəˈlɪksɪtɪ/ *n.* [ME f. OF *prolixe* or L *prolixus* poured forth, extended (as PRO-[1], *liquere* be liquid)]

prolocutor /prəʊˈlɒkjʊtə(r)/ *n.* **1** *Eccl.* the chairperson esp. of the lower house of convocation of either province of the Church of England. **2** a spokesman. □ **prolocutorship** *n.* [ME f. L f. *proloqui prolocut-* (as PRO-[1], *loqui* speak)]

Prolog /ˈprəʊlɒg/ *n. Computing* a high-level programming language first devised for artificial intelligence applications. [*programming* (see PROGRAMME *v.*) + LOGIC]

prologize /ˈprəʊləˌgaɪz/ *v.intr.* (also **prologuize**, **-ise**) write or speak a prologue. [med.L *prologizare* f. Gk *prologizō* speak prologue (as PROLOGUE)]

prologue /ˈprəʊlɒg/ *n. & v.* ● *n.* **1 a** a preliminary speech, poem, etc., esp. introducing a play (cf. EPILOGUE). **b** the actor speaking the prologue. **2** (usu. foll. by *to*) any act or event serving as an introduction. ● *v.tr.* (**prologues**, **prologued**, **prologuing**) introduce with or provide with a prologue. [ME *prolog* f. OF *prologue* f. L *prologus* f. Gk *prologos* (as PRO-[2], *logos* speech)]

prolong /prəˈlɒŋ/ *v.tr.* **1** extend (an action, condition, etc.) in time or space. **2** lengthen the pronunciation of (a syllable etc.). **3** (as **prolonged** *adj.*) lengthy, esp. tediously so. □ **prolonger** *n.* **prolongedly** /-ˈlɒŋɪdlɪ/ *adv.* **prolongation** /ˌprəʊlɒŋˈgeɪʃ(ə)n/ *n.* [ME f. OF *prolonger* & f. LL *prolongare* (as PRO-[1], *longus* long)]

prolusion /prəˈluːʒ(ə)n, -ˈljuːʒ(ə)n/ *n. formal* **1** a preliminary essay or article. **2** a first attempt. □ **prolusory** /-ˈluːzərɪ, -ˈljuː-/ *adj.* [L *prolusio* f. *proludere prolus-* practise beforehand (as PRO-[1], *ludere lus-* play)]

prom /prɒm/ *n. colloq.* **1** *Brit.* = PROMENADE *n.* 1a. **2** *Brit.* a promenade concert, esp. (**Prom**) any of those supported by the BBC (see PROMENADE CONCERT). **3** *US* = PROMENADE *n.* 3. [abbr.]

promenade /ˌprɒməˈnɑːd, ˈprɒməˌnɑːd/ *n. & v.* ● *n.* **1 a** *Brit.* a paved public walk along the sea front at a resort. **b** any paved public walk. **2** a walk, or sometimes a ride or drive, taken esp. for display, social interaction, etc. **3** *US* a school or university ball or dance. **4** a march of dancers in country dancing etc. ● *v.* **1** *intr.* make a promenade. **2** *tr.* lead (a person etc.) about a place esp. for display. **3** *tr.* make a promenade through (a place). □ **promenade deck** an upper deck on a passenger ship where passengers may promenade. [F f. *se promener* walk, refl. of *promener* take for a walk]

promenade concert *n.* a concert of usu. classical music at which all or part of the auditorium's floor space is without seating so that the audience stands or sits on the floor or can move about. The most famous series of such concerts is the annual BBC Promenade Concerts ('the Proms'), instituted by Sir Henry Wood in 1895 and held since the Second World War chiefly in the Albert Hall in London. The last night of each year's season has tended to become a patriotic celebration, with audience participation.

promenader /ˌprɒməˈnɑːdə(r)/ *n.* **1** a person who promenades. **2** *Brit.* a person who attends a promenade concert, esp. regularly.

promethazine /prəʊˈmeθəˌziːn/ *n.* an antihistamine drug used to treat allergies, motion sickness, etc. [PROPYL + di*methylamine* + phenothi*azine*]

Promethean /prəˈmiːθɪən/ *adj.* daring or inventive like Prometheus.

Prometheus /prəˈmiːθɪəs/ *Gk Mythol.* a demigod, one of the Titans, who was worshipped by craftsmen. When Zeus hid fire away from man Prometheus stole it by trickery and returned it to earth. As punishment Zeus chained him to a rock where an eagle fed each day on his liver,

which (since he was immortal) grew again each night; he was rescued by Hercules.

promethium /prəˈmiːθɪəm/ *n.* a radioactive metallic chemical element (atomic number 61; symbol **Pm**). Promethium was first produced artificially in a nuclear reactor by US scientists in the 1940s. A member of the lanthanide series, it does occur in nature in traces as a product of uranium fission. Its radiation has been used to power miniature batteries. [PROMETHEUS]

prominence /ˈprɒmɪnəns/ *n.* **1** the state of being prominent. **2** a prominent thing, esp. a jutting outcrop, mountain, etc. **3** *Astron.* a stream of incandescent gas projecting above the sun's chromosphere. [obs.F f. L *prominentia* jutting out (as PROMINENT)]

prominent /ˈprɒmɪnənt/ *adj. & n.* ● *adj.* **1** jutting out; projecting. **2** conspicuous. **3** distinguished; important. ● *n.* (in full **prominent moth**) a moth of the family Notodontidae, with tufted forewings and larvae with humped backs. □ **prominency** *n.* **prominently** *adv.* [L *prominere* jut out: cf. EMINENT]

promiscuous /prəˈmɪskjʊəs/ *adj.* **1 a** (of a person) having frequent and diverse sexual relationships, esp. transient ones. **b** (of sexual relationships) of this kind. **2** of mixed and indiscriminate composition or kinds; indiscriminate (*promiscuous hospitality*). **3** *colloq.* carelessly irregular; casual. □ **promiscuously** *adv.* **promiscuousness** *n.* **promiscuity** /ˌprɒmɪˈskjuːɪtɪ/ *n.* [L *promiscuus* (as PRO-[1], *miscere* mix)]

promise /ˈprɒmɪs/ *n. & v.* ● *n.* **1** an assurance that one will or will not undertake a certain action, behaviour, etc. (*a promise of help*; *gave a promise to be generous*). **2** a sign or signs of future achievements, good results, etc. (*a writer of great promise*). ● *v.tr.* **1** (usu. foll. by *to* + infin., or *that* + clause; also *absol.*) make (a person) a promise, esp. to do, give, or procure (a thing) (*I promise you a fair hearing*; *they promise not to be late*; *promised that he would be there*; *cannot positively promise*). **2 a** afford expectations of (*the discussions promise future problems*; *promises to be a good cook*). **b** (foll. by *to* + infin.) seem likely to (*is promising to rain*). **3** *colloq.* assure, confirm (*I promise you, it will not be easy*). **4** (usu. in *passive*) *archaic* betroth (*she is promised to another*). □ **the promised land 1** *Bibl.* Canaan (Gen. 12:7 etc.). **2** any desired place or situation, esp. heaven. **promise oneself** look forward to (a pleasant time etc.). **promise well** (or **ill** etc.) hold out good (or bad etc.) prospects. □ **promiser** *n.* **promisee** /ˌprɒmɪˈsiː/ *n.* esp. *Law.* **promisor** /-ˈsɔː(r)/ *n.* esp. *Law.* [ME f. L *promissum* neut. past part. of *promittere* put forth, promise (as PRO-[1], *mittere* send)]

promising /ˈprɒmɪsɪŋ/ *adj.* likely to turn out well; hopeful; full of promise (*a promising start*). □ **promisingly** *adv.*

promissory /ˈprɒmɪsərɪ/ *adj.* **1** conveying or implying a promise. **2** (often foll. by *of*) full of promise. □ **promissory note** a signed document containing a written promise to pay a stated sum to a specified person or the bearer at a specified date or on demand. [med.L *promissorius* f. L *promissor* (as PROMISE)]

promo /ˈprəʊməʊ/ *n. & adj. colloq.* ● *n.* (*pl.* **-os**) **1** publicity, advertising. **2** a trailer for a television programme. ● *adj.* promotional. [abbr.]

promontory /ˈprɒməntərɪ/ *n.* (*pl.* **-ies**) **1** a point of high land jutting out into the sea etc.; a headland. **2** *Anat.* a prominence or protuberance in the body. [med.L *promontorium* alt. (after *mons montis* mountain) f. L *promunturium* (perh. f. PRO-[1], *mons*)]

promote /prəˈməʊt/ *v.tr.* **1** (often foll. by *to*) advance or raise (a person) to a higher office, rank, etc. (*was promoted to captain*). **2** transfer (a sports team) to a higher division of a league etc. **3** help forward; encourage; support actively (a cause, process, desired result, etc.) (*promoted women's suffrage*). **4** publicize and sell (a product). **5** attempt to ensure the passing of (a private Act of Parliament). **6** *Chess* raise (a pawn) to the rank of queen etc. when it reaches the opponent's end of the board. □ **promotive** *adj.* **promotable** *adj.* **promotability** /-ˌməʊtəˈbɪlɪtɪ/ *n.* **promotion** /-ˈməʊʃ(ə)n/ *n.* **promotional** *adj.* [ME f. L *promovere promot-* (as PRO-[1], *movere* move)]

promoter /prəˈməʊtə(r)/ *n.* **1** a person who promotes. **2** a person who finances, organizes, etc. a sporting event, theatrical production, etc. **3** (in full **company promoter**) a person who promotes the formation of a joint-stock company. **4** *Chem.* an additive that increases the activity of a catalyst. [earlier *promotour* f. AF f. med.L *promotor* (as PROMOTE)]

prompt /prɒmpt/ *adj., adv., v., & n.* ● *adj.* **1** acting with alacrity; ready. **b** made, done, etc. readily or at once (*a prompt reply*). **2 a** (of a payment) made forthwith. **b** (of goods) for immediate delivery and payment. ● *adv.* punctually (*at six o'clock prompt*). ● *v.tr.* **1** (usu. foll. by *to*, or *to* + infin.) incite; urge (*prompted them to action*). **2 a** (also *absol.*) supply a forgotten word, sentence, etc., to (an actor, reciter, etc.). **b** assist (a hesitating speaker) with a suggestion. **3** give rise to; inspire (a feeling, thought, action, etc.). ● *n.* **1 a** an act of prompting. **b** a thing said to help the memory of an actor etc. **c** = PROMPTER 2. **d** *Computing* an indication or sign on a VDU screen to show that the system is waiting for input. **2** the time-limit for the payment of an account, stated on a prompt note. □ **prompt-book** a copy of a play for a prompter's use. **prompt-box** a box in front of the footlights beneath the stage where the prompter sits. **prompt-note** a note sent to a customer as a reminder of payment due. **prompt side** the side of the stage where the prompter sits, usu. to the actor's left. □ **prompting** *n.* **promptitude** *n.* **promptly** *adv.* **promptness** *n.* [ME f. OF *prompt* or L *promptus* past part. of *promere prompt-* produce (as PRO-¹, *emere* take)]

prompter /ˈprɒmptə(r)/ *n.* **1** a person who prompts. **2** *Theatr.* a person seated out of sight of the audience who prompts the actors.

promulgate /ˈprɒməlˌgeɪt/ *v.tr.* **1** make known to the public; disseminate; promote (a cause etc.). **2** proclaim (a decree, news, etc.). □ **promulgator** *n.* **promulgation** /ˌprɒməlˈgeɪʃ(ə)n/ *n.* [L *promulgare* (as PRO-¹, *mulgere* milk, cause to come forth)]

promulge /prəʊˈmʌldʒ/ *v.tr. archaic* = PROMULGATE. [PROMULGATE]

pronaos /prəʊˈneɪɒs/ *n.* (*pl.* **pronaoi** /-ˈneɪɔɪ/) *Gk Antiq.* the space in front of the body of a temple, enclosed by a portico and projecting side walls. [L f. Gk *pronaos* hall of a temple (as PRO-², NAOS)]

pronate /prəʊˈneɪt/ *v.tr.* put (the hand, forearm, etc.) into a prone position (with the palm etc. downwards) (cf. SUPINATE). □ **pronation** /-ˈneɪʃ(ə)n/ *n.* [back-form. f. *pronation* (as PRONE)]

pronator /prəʊˈneɪtə(r)/ *n. Anat.* a muscle producing or assisting in pronation.

prone /prəʊn/ *adj.* **1 a** lying face downwards (cf. SUPINE *adj.* 1). **b** lying flat; prostrate. **c** having the front part downwards, esp. the palm of the hand. **2** (usu. foll. by *to*, or *to* + infin.) disposed or liable, esp. to a bad action, condition, etc. (*is prone to bite his nails*). **3** (usu. in *comb.*) more than usually likely to suffer (*accident-prone*). **4** *archaic* with a downward slope or direction. □ **pronely** *adv.* **proneness** *n.* [ME f. L *pronus* f. *pro* forwards]

proneur /prɒˈnɜː(r)/ *n.* a person who extols; a flatterer. [F *prôneur* f. *prôner* eulogize f. *prône* place in church where addresses were delivered]

prong /prɒŋ/ *n. & v.* ● *n.* each of two or more projecting pointed parts at the end of a fork etc. ● *v.tr.* **1** pierce or stab with a fork. **2** turn up (soil) with a fork. □ **prong-horned antelope** a pronghorn. **three-pronged attack** an attack on three separate points at once. □ **pronged** *adj.* (also in *comb.*). [ME (also *prang*), perh. rel. to MLG *prange* pinching instrument]

pronghorn /ˈprɒŋhɔːn/ *n.* an antelope-like ruminant, *Antilocapra americana*, native to North America, the male of which has bony horns with forward-pointing prongs.

pronominal /prəʊˈnɒmɪn(ə)l/ *adj.* of, relating to, or serving as a pronoun. □ **pronominalize** *v.tr.* (also **-ise**). **pronominally** *adv.* [LL *pronominalis* f. L *pronomen* (as PRO-¹, *nomen, nominis* noun)]

pronoun /ˈprəʊnaʊn/ *n.* a word used instead of and to indicate a noun already mentioned or known, esp. to avoid repetition (e.g. *we, their, this, ourselves*). [PRO-¹, + NOUN, after F *pronom*, L *pronomen* (as PRO-¹, *nomen* name)]

pronounce /prəˈnaʊns/ *v.* **1** *tr.* (also *absol.*) utter or speak (words, sounds, etc.) in a certain way. **2** *tr.* **a** utter or deliver (a judgement, sentence, curse, etc.) formally or solemnly. **b** proclaim or announce officially (*I pronounce you man and wife*). **3** *tr.* state or declare, as being one's opinion (*the apples were pronounced excellent*). **4** *intr.* (usu. foll. by *on, for, against, in favour of*) pass judgement; give one's opinion (*pronounced for the defendant*). □ **pronounceable** *adj.* **pronouncement** *n.* **pronouncer** *n.* **pronounceability** /-ˌnaʊnsəˈbɪlɪtɪ/ *n.* [ME f. OF *pronuncier* f. L *pronuntiare* (as PRO-¹, *nuntiare* announce f. *nuntius* messenger)]

pronounced /prəˈnaʊnst/ *adj.* **1** (of a word, sound, etc.) uttered. **2** strongly marked; decided (*a pronounced flavour; a pronounced limp*). □ **pronouncedly** /-sɪdlɪ/ *adv.*

pronto /ˈprɒntəʊ/ *adv. colloq.* promptly, quickly. [Sp. f. L (as PROMPT)]

pronunciation /prəˌnʌnsɪˈeɪʃ(ə)n/ *n.* **1** the way in which a word is pronounced, esp. with reference to a standard. **2** the act or an instance of pronouncing. **3** a person's way of pronouncing words etc. [ME f. OF *prononciation* or L *pronuntiatio* (as PRONOUNCE)]

proof /pruːf/ *n., adj., & v.* ● *n.* **1** facts, evidence, argument, etc. establishing or helping to establish a fact (*proof of their honesty; no proof that he was there*). **2** *Law* the spoken or written evidence in a trial. **3** a demonstration or act of proving (*not capable of proof; in proof of my assertion*). **4** a test or trial (*put them to the proof; the proof of the pudding is in the eating*). **5** the standard of strength of distilled alcoholic liquors. **6** *Printing* a trial impression taken from type or film, used for making corrections before final printing. **7** the stages in the resolution of a mathematical or philosophical problem. **8** each of a limited number of impressions from an engraved plate before the ordinary issue is printed and usu. (in full **proof before letters**) before an inscription or signature is added. **9** a photographic print made for selection etc. **10** *Sc. Law* a trial before a judge instead of by a jury. ● *adj.* **1** impervious to penetration, ill effects, etc. (*proof against the severest weather; his soul is proof against corruption*). **2** (in *comb.*) able to withstand damage or destruction by a specified agent (*soundproof; childproof*). **3** being of proof alcoholic strength. **4** (of armour) of tried strength. ● *v.tr.* **1** make (something) proof, esp. make (fabric) waterproof. **2** make a proof of (a printed work, engraving, etc.). □ **above proof** (of alcohol) having a stronger than standard strength. **proof-plane** a small flat conductor on an insulating handle for measuring the electrification of a body. **proof positive** absolutely certain proof. **proof-sheet** a sheet of printer's proof. **proof spirit** a mixture of alcohol and water having proof strength. □ **proofless** *adj.* [ME *prōf prōve*, earlier *prēf* etc. f. OF *proeve, prueve* f. LL *proba* f. L *probare* (see PROVE; adj. and sometimes v. formed app. by ellipsis f. phr. *of proof* = proved to be impenetrable]

proofread /ˈpruːfriːd/ *v.tr.* (*past* and *past part.* **-read** /-red/) read (printer's proofs) and mark any errors. □ **proofreader** *n.* **proofreading** *n.*

prop¹ /prɒp/ *n. & v.* ● *n.* **1** a rigid support, esp. one not an integral part of the thing supported. **2** a person who supplies support, assistance, comfort, etc. **3** (in full **prop forward**) *Rugby* a forward at either end of the front row of a scrum. **4** esp. *Austral.* a horse's action of propping. ● *v.* (**propped, propping**) **1** *tr.* (often foll. by *against, up, etc.*) support with or as if with a prop (*propped him against the wall; propped it up with a brick*). **2** *intr.* esp. *Austral.* (of a horse etc.) come to a dead stop with the forelegs rigid. [ME prob. f. MDu. *proppe*: cf. MLG, MDu. *proppen* (v.)]

prop² /prɒp/ *n. Theatr. colloq.* **1** = PROPERTY 3. **2** (in *pl.*) a property man or mistress. [abbr.]

prop³ /prɒp/ *n. colloq.* an aircraft propeller. □ **prop-jet** a turboprop. [abbr.]

prop. *abbr.* **1** proprietor. **2** proposition.

propaedeutic /ˌprəʊpɪˈdjuːtɪk/ *adj. & n.* ● *adj.* serving as an introduction to higher study; introductory. ● *n.* (esp. in *pl.*) preliminary learning; a propaedeutic subject, study, etc. □ **propaedeutical** *adj.* [PRO-² + Gk *paideutikos* of teaching, after Gk *propaideuō* teach beforehand]

propaganda /ˌprɒpəˈgændə/ *n.* **1 a** an organized programme of publicity, selected information, etc., used to propagate a doctrine, practice, etc. **b** usu. *derog.* the information, doctrines, etc., propagated in this way, esp. regarded as misleading or dishonest. **2** (**Propaganda**) *RC Ch.* a committee of cardinals responsible for foreign missions. [It. f. mod.L *congregatio de propaganda fide* congregation for propagation of the faith]

propagandist /ˌprɒpəˈgændɪst/ *n. & adj.* ● *n.* a member or agent of a propaganda organization; a person who spreads propaganda. ● *adj.* consisting of or spreading propaganda. □ **propagandism** *n.* **propagandize** *v.intr. & tr.* (also **-ise**). **propagandistic** /-gænˈdɪstɪk/ *adj.* **propagandistically** *adv.*

propagate /ˈprɒpəˌgeɪt/ **1** *tr.* **a** breed specimens of (a plant, animal, etc.) by natural processes from the parent stock. **b** (*refl.* or *absol.*) (of a plant, animal, etc.) reproduce itself. **2 a** *tr.* disseminate; spread (a statement, belief, theory, etc.). **b** *intr.* grow more widespread or numerous; spread. **3** *tr.* hand down (a quality etc.) from one generation to another. **4** *tr.* extend the operation of; transmit (a vibration, earthquake, etc.). □ **propagative** *adj.* **propagation** /ˌprɒpəˈgeɪʃ(ə)n/ *n.* [L *propagare propagat-* multiply plants from layers, f. *propago* (as PRO-¹, *pangere* fix, layer)]

propagator /ˈprɒpəˌgeɪtə(r)/ *n.* **1** a person or thing that propagates. **2** a small box that can be heated, used for germinating seeds or raising seedlings.

propane /ˈprəʊpeɪn/ *n.* a gaseous hydrocarbon of the alkane series (chem. formula: C_3H_8) used as bottled fuel. □ **propanoate** *n.* [PROPIONIC ACID + -ANE²]

propanoic acid /ˌprəʊpəˈnəʊɪk/ *n. Chem.* = PROPIONIC ACID. [PROPANE + -IC]

propanone /ˈprəʊpəˌnəʊn/ *n. Chem.* = ACETONE. [PROPANE + -ONE]

propel /prəˈpel/ *v.tr.* (**propelled, propelling**) **1** drive or push forward. **2** urge on; encourage. □ **propelling pencil** a pencil with a replaceable lead moved upward by twisting the outer case. [ME, = expel, f. L *propellere* (as PRO-¹, *pellere puls-* drive)]

propellant /prə'pelənt/ *n. & adj.* ● *n.* **1** a thing that propels. **2** an inert compressed fluid in which the active contents of an aerosol are dispersed. **3** an explosive that fires bullets etc. from a firearm. **4** a substance used as a reagent in a rocket engine etc. to provide thrust. ● *adj.* = PROPELLENT.

propellent /prə'pelənt/ *adj.* propelling; capable of driving or pushing forward.

propeller /prə'pelə(r)/ *n.* **1** a person or thing that propels. **2** a revolving shaft with blades, esp. for propelling a ship or aircraft (cf. SCREW *n.* 6). □ **propeller shaft** a shaft transmitting power from an engine to a propeller or to the driven wheels of a motor vehicle. **propeller turbine** a turbo-propeller.

propene /'prəʊpi:n/ *n. Chem.* = PROPYLENE.

propensity /prə'pensɪtɪ/ *n.* (*pl.* **-ies**) an inclination or tendency (*has a propensity for wandering*). [*propense* f. L *propensus* inclined, past part. of *propendere* (as PRO-¹, *pendere* hang)]

proper /'prɒpə(r)/ *adj., adv., & n.* ● *adj.* **1 a** accurate, correct (*in the proper sense of the word; gave him the proper amount*). **b** fit, suitable, right (*at the proper time; do it the proper way*). **2** decent; respectable, esp. excessively so (*not quite proper*). **3** (usu. foll. by *to*) belonging or relating exclusively or distinctively; particular, special (*with the respect proper to them*). **4** (usu. placed after noun) strictly so called; real; genuine (*this is the crypt, not the cathedral proper*). **5** *colloq.* thorough; complete (*had a proper row about it*). **6** (usu. placed after noun) in the natural, not conventional, colours (*a peacock proper*). **7** *archaic* (of a person) handsome; comely. **8** (usu. with possessive pronoun) *archaic* own (*with my proper eyes*). ● *adv. Brit. dial.* or *colloq.* **1** completely; very (*felt proper daft*). **2** (with reference to speech) in a genteel manner (*learn to talk proper*). ● *n. Eccl.* the part of a service that varies with the season or feast. □ **proper fraction** a fraction that is less than unity, with the numerator less than the denominator. **proper motion** *Astron.* the part of the apparent motion of a fixed star etc. that is due to its actual movement in space relative to the sun. **proper name** (or **noun**) *Gram.* a name used for an individual person, place, animal, country, title, etc., and spelt with a capital letter, e.g. Jane, London, Everest. **proper psalms** (or **lessons** etc.) psalms or lessons etc. appointed for a particular day. □ **properness** *n.* [ME f. OF *propre* f. L *proprius* one's own, special]

properly /'prɒpəlɪ/ *adv.* **1** fittingly; suitably (*do it properly*). **2** accurately; correctly (*properly speaking*). **3** rightly (*he very properly refused*). **4** with decency; respectably (*behave properly*). **5** *colloq.* thoroughly (*they were properly ashamed*).

propertied /'prɒpətɪd/ *adj.* having property, esp. land.

Propertius /prə'pɜːʃəs/, Sextus (*c.*50–*c.*16 BC), Roman poet. His four books of elegies are largely concerned with his love affair with a woman whom he called Cynthia, though the later poems also deal with mythological and historical themes.

property /'prɒpətɪ/ *n.* (*pl.* **-ies**) **1 a** something owned; a possession, esp. a house, land, etc. **b** *Law* the right to possession, use, etc. **c** possessions collectively, esp. real estate (*has money in property*). **2** an attribute, quality, or characteristic (*has the property of dissolving grease*). **3** a movable object used on a theatre stage, in a film, etc. **4** *Logic* a quality common to a whole class but not necessary to distinguish it from others. □ **common property** a thing known by most people. **property man** (or **mistress**) a man (or woman) in charge of theatrical properties. **property qualification** a qualification for office, or for the exercise of a right, based on the possession of property. **property tax** a tax levied directly on property. [ME through AF f. OF *propriété* f. L *proprietas -tatis* (as PROPER)]

prophase /'prəʊfeɪz/ *n. Biol.* the phase in cell division in which chromosomes contract and each becomes visible as two chromatids. [PRO-² + PHASE]

prophecy /'prɒfɪsɪ/ *n.* (*pl.* **-ies**) **1 a** a prophetic utterance, esp. biblical. **b** a prediction of future events (*a prophecy of massive inflation*). **2** the faculty, function, or practice of prophesying (*the gift of prophecy*). [ME f. OF *profecie* f. LL *prophetia* f. Gk *prophēteia* (as PROPHET)]

prophesy /'prɒfɪsaɪ/ *v.* (**-ies, -ied**) **1** *tr.* (usu. foll. by *that, who,* etc.) foretell (an event etc.). **2** *intr.* speak as a prophet; foretell future events. **3** *intr. archaic* expound the Scriptures. □ **prophesier** /-,saɪə(r)/ *n.* [ME f. OF *profecier* (as PROPHECY)]

prophet /'prɒfɪt/ *n.* (*fem.* **prophetess** /-tɪs/) **1** a person regarded as a teacher or interpreter of the will of God. **2** any of the prophetical writers in the Old Testament (see OLD TESTAMENT). **3 a** a person who foretells events. **b** a person who advocates and speaks innovatively for a cause (*a prophet of the new order*). **4** (**the Prophet**) **a** Muhammad. **b** Joseph Smith, founder of the Mormons, or one of his successors. **5** *colloq.* a tipster. □ **major prophet** each of the prophets (Isaiah, Jeremiah, and Ezekiel) for whom the longer prophetic books of the Bible are named and whose prophecies they record. **minor prophet** any of the prophets from Hosea to Malachi, for whom the twelve shorter prophetic books of the Bible are named and whose prophecies they record. □ **prophethood** *n.* **prophetism** *n.* [ME f. OF *prophete* f. L *propheta, prophetes* f. Gk *prophētēs* spokesman (as PRO-², *phētēs* speaker f. *phēmi* speak)]

prophetic /prə'fetɪk/ *adj.* **1** (often foll. by *of*) containing a prediction; predicting. **2** of or concerning a prophet. □ **prophetical** *adj.* **prophetically** *adv.* [F *prophétique* or LL *propheticus* f. Gk *prophētikos* (as PROPHET)]

Prophets, the 1 in Judaism, one of the three canonical divisions of the Hebrew Bible, including the books of Joshua, Judges, 1 & 2 Samuel, 1 & 2 Kings, Isaiah, Jeremiah, Ezekiel, Daniel, and the twelve minor prophets, Hosea to Malachi. **2** in the Christian Church, the prophetic books of Isaiah, Jeremiah, Ezekiel, Daniel, and the minor prophets. (See also OLD TESTAMENT.)

prophylactic /,prɒfɪ'læktɪk/ *adj. & n.* ● *adj.* tending to prevent disease. ● *n.* **1** a preventive medicine or course of action. **2** esp. *N. Amer.* a condom. [F *prophylactique* f. Gk *prophulaktikos* f. *prophulassō* (as PRO-², *phulassō* guard)]

prophylaxis /,prɒfɪ'læksɪs/ *n.* preventive treatment against disease. [mod.L f. PRO-² + Gk *phulaxis* act of guarding]

propinquity /prə'pɪŋkwɪtɪ/ *n.* **1** nearness in space; proximity. **2** close kinship. **3** similarity. [ME f. OF *propinquité* or L *propinquitas* f. *propinquus* near f. *prope* near to]

propionic acid /,prəʊpɪ'ɒnɪk/ *n. Chem.* a colourless sharp-smelling liquid carboxylic acid (chem. formula: C_2H_5COOH), used for inhibiting the growth of mould in bread Also called *propanoic acid*. □ **propionate** /'prəʊpɪə,neɪt/ *n.* [F *propionique*, formed as PRO-² + Gk *pïōn* fat, as being the first member of the fatty acid series to form fats]

propitiate /prə'pɪʃɪeɪt/ *v.tr.* appease (an offended person etc.). □ **propitiator** *n.* [L *propitiare* (as PROPITIOUS)]

propitiation /prə,pɪʃɪ'eɪʃ(ə)n/ *n.* **1** appeasement. **2** *Bibl.* atonement, esp. Christ's. **3** *archaic* a gift etc. meant to propitiate. [ME f. LL *propitiatio* (as PROPITIATE)]

propitiatory /prə'pɪʃɪətərɪ/ *adj.* serving or intended to propitiate (*a propitiatory smile*). □ **propitiatorily** *adv.* [ME f. LL *propitiatorius* (as PROPITIATE)]

propitious /prə'pɪʃəs/ *adj.* **1** (of an omen etc.) favourable. **2** (often foll. by *for, to*) (of the weather, an occasion, etc.) suitable. **3** well-disposed (*the fates were propitious*). □ **propitiously** *adv.* **propitiousness** *n.* [ME f. OF *propicieus* or L *propitius*]

propolis /'prɒpəlɪs/ *n.* a red or brown resinous substance collected by bees from buds for use in constructing hives. [L f. Gk, lit. 'suburb' f. PRO-² + *polis* city]

proponent /prə'pəʊnənt/ *n. & adj.* ● *n.* a person advocating a motion, theory, or proposal. ● *adj.* proposing or advocating a theory etc. [L *proponere* (as PROPOUND)]

Propontis /prə'pɒntɪs/ the ancient name for the Sea of Marmara (see MARMARA, SEA OF).

proportion /prə'pɔːʃ(ə)n/ *n. & v.* ● *n.* **1 a** a comparative part or share (*a large proportion of the profits*). **b** a comparative ratio (*the proportion of births to deaths*). **2** the correct or pleasing relation of things or parts of a thing (*the house has fine proportions; exaggerated out of all proportion*). **3** (in *pl.*) dimensions; size (*large proportions*). **4** *Math.* **a** an equality of ratios between two pairs of quantities, e.g. 3:5 and 9:15. **b** a set of such quantities. **c** *Math.* = *rule of three* (see also *direct proportion, inverse proportion*). ● *v.tr.* (usu. foll. by *to*) make (a thing etc.) proportionate (*must proportion the punishment to the crime*). □ **in proportion 1** by the same factor. **2** without exaggerating (importance etc.) (*must get the facts in proportion*). □ **proportioned** *adj.* (also in *comb.*). **proportionless** *adj.* **proportionment** *n.* [ME f. OF *proportion* or L *proportio* (as PRO-¹, PORTION)]

proportionable /prə'pɔːʃənəb(ə)l/ *adj.* = PROPORTIONAL. □ **proportionably** *adv.*

proportional /prə'pɔːʃən(ə)l/ *adj. & n.* ● *adj.* in due proportion; comparable (*a proportional increase in the expense; resentment proportional to his injuries*). ● *n. Math.* each of the terms of a proportion. □ **proportionally** *adv.* **proportionality** /-,pɔːʃə'nælɪtɪ/ *n.*

proportionalist /prəˈpɔːʃənəlɪst/ *n.* an advocate of proportional representation.

proportional representation *n.* (abbr. **PR**) an electoral system in which parties gain seats in proportion to the number of votes cast for them. Whereas the 'first past the post' system, as used in Britain and North America, may assign authority to represent a whole constituency or country to a party that has won less than half of the votes cast, proportional representation gives minority parties a measure of representation commensurate with their support, and thus in this respect may be said to be more democratic. On the other hand, its use often means that no party wins a clear majority of seats, which can give substantial power to minority groups and lead to coalitions and to weak, unstable government. There are several kinds of electoral system based on PR, the most common being the single transferable vote system and voting on the basis of a party list. PR is in use throughout much of Europe.

proportionate /prəˈpɔːʃənət/ *adj.* proportional, appropriate. □ **proportionately** *adv.*

proposal /prəˈpəʊz(ə)l/ *n.* **1 a** the act or an instance of proposing something. **b** a course of action etc. so proposed (*the proposal was never carried out*). **2** an offer of marriage.

propose /prəˈpəʊz/ *v.* **1** *tr.* (also *absol.*) put forward for consideration or as a plan. **2** *tr.* (usu. foll. by *to* + infin., or verbal noun) intend; purpose (*propose to open a restaurant*). **3** *intr.* (usu. foll. by *to*) offer oneself in marriage. **4** *tr.* nominate (a person) as a member of a society, for an office, etc. **5** *tr.* offer (a person's health, a person, etc.) as a subject for a toast. □ **proposer** *n.* [ME f. OF *proposer* f. L *proponere* (as PROPOUND)]

proposition /ˌprɒpəˈzɪʃ(ə)n/ *n. & v.* ● *n.* **1** a statement or assertion. **2** a scheme proposed; a proposal. **3** *Logic* a statement consisting of subject and predicate that is subject to proof or disproof. **4** *colloq.* a problem, opponent, prospect, etc. that is to be dealt with (*a difficult proposition*). **5** *Math.* a formal statement of a theorem or problem, often including the demonstration. **6 a** an enterprise etc. with regard to its likelihood of commercial etc. success. **b** a person regarded similarly. **7** *colloq.* a sexual proposal. ● *v.tr. colloq.* make a proposal (esp. of sexual intercourse) to (*was notorious for propositioning his female students*). □ **not a proposition** unlikely to succeed. □ **propositional** *adj.* [ME f. OF *proposition* or L *propositio* (as PROPOUND)]

propound /prəˈpaʊnd/ *v.tr.* **1** offer for consideration; propose. **2** *Law* produce (a will etc.) before the proper authority so as to establish its legality. □ **propounder** *n.* [earlier *propoune*, *propone* f. L *proponere* (as PRO-[1], *ponere posit*- place): cf. *compound*, *expound*]

proprietary /prəˈpraɪətərɪ/ *adj.* **1 a** of, holding, or concerning property (*the proprietary classes*). **b** of or relating to a proprietor (*proprietary rights*). **2** held in private ownership. **3** (of a product, esp. a drug or medicine) marked under and protected by a registered trade name. □ **proprietary name** (or **term**) a name of a product etc. registered by its owner as a trademark and not usable by another without permission. [LL *proprietarius* (as PROPERTY)]

proprietor /prəˈpraɪətə(r)/ *n.* (*fem.* **proprietress** /-trɪs/) **1** a holder of property. **2** the owner of a business etc., esp. of a hotel. □ **proprietorship** *n.* **proprietorial** /-ˌpraɪəˈtɔːrɪəl/ *adj.* **proprietorially** *adv.*

propriety /prəˈpraɪətɪ/ *n.* (*pl.* **-ies**) **1** fitness; rightness (*doubt the propriety of refusing him*). **2** correctness of behaviour or morals (*highest standards of propriety*). **3** (in *pl.*) the details or rules of correct conduct (*must observe the proprieties*). [ME, = ownership, peculiarity f. OF *propriété* PROPERTY]

proprioceptive /ˌprəʊprɪəˈseptɪv/ *adj. Biol.* relating to stimuli produced and perceived within an organism, esp. relating to the position and movement of the body. [L *proprius* own + RECEPTIVE]

proptosis /prɒpˈtəʊsɪs/ *n. Med.* protrusion or displacement, esp. of an eye. [LL f. Gk *proptōsis* (as PRO-[2], *piptō* fall)]

propulsion /prəˈpʌlʃ(ə)n/ *n.* **1** the act or an instance of driving or pushing forward. **2** an impelling influence; motive power (*jet propulsion*). □ **propulsive** /-ˈpʌlsɪv/ *adj.* [med.L *propulsio* f. L *propellere* (as PROPEL)]

propulsor /prəˈpʌlsə(r)/ *n.* a ducted propeller which can be swivelled to give forward, upward, or downward flight to an airship. [as PROPULSION]

propyl /ˈprəʊpaɪl, -pɪl/ *n.* (usu. *attrib.*) *Chem.* the radical –C$_3$H$_7$, derived from propane by removal of a hydrogen atom.

propyla *pl.* of PROPYLON.

propylaeum /ˌprɒpɪˈliːəm/ *n.* (*pl.* **propylaea** /-ˈliːə/) **1** the entrance to a temple. **2** (**the Propylaeum**) the entrance to the Acropolis at Athens. [L f. Gk *propulaion* (as PRO-[2], *pulē* gate)]

propylene /ˈprəʊpɪˌliːn/ *n. Chem.* a gaseous hydrocarbon of the alkene series (chem. formula: C$_3$H$_6$) used in the manufacture of chemicals.

propylon /ˈprɒpɪˌlɒn/ *n.* (*pl.* **propylons** or **propyla** /-lə/) = PROPYLAEUM. [L f. Gk *propulon* (as PRO-[2], *pulē* gate)]

pro rata /prəʊ ˈrɑːtə, ˈreɪtə/ *adj. & adv.* ● *adj.* proportional. ● *adv.* proportionally. [L, = according to the rate]

prorate /prəʊˈreɪt/ *v.tr.* allocate or distribute *pro rata*. □ **proration** /-ˈreɪʃ(ə)n/ *n.*

prorogue /prəˈrəʊg/ *v.* (**prorogues**, **prorogued**, **proroguing**) **1** *tr.* discontinue the meetings of (a parliament etc.) without dissolving it. **2** *intr.* (of a parliament etc.) be prorogued. □ **prorogation** /ˌprəʊrəˈgeɪʃ(ə)n/ *n.* [ME *proroge* f. OF *proroger*, *-guer* f. L *prorogare* prolong (as PRO-[1], *rogare* ask)]

pros- /prɒs/ *prefix* **1** to, towards. **2** in addition. [Gk f. *pros* (prep.)]

prosaic /prəʊˈzeɪk/ *adj.* **1** like prose, lacking poetic beauty. **2** unromantic; dull; commonplace (*took a prosaic view of life*). □ **prosaically** *adv.* **prosaicness** *n.* [F *prosaïque* or LL *prosaicus* (as PROSE)]

prosaist /ˈprəʊzeɪɪst/ *n.* **1** a prose-writer. **2** (also /prəʊˈzeɪɪst/) a prosaic person. □ **prosaism** /ˈprəʊzeɪˌɪz(ə)m, prəʊˈzeɪ-/ *n.* [F *prosaïste* f. L *prosa* PROSE]

proscenium /prəˈsiːnɪəm/ *n.* (*pl.* **prosceniums** or **proscenia** /-nɪə/) **1** (often *attrib.*) the part of the stage in front of the drop or curtain, usu. with the enclosing arch (esp. *proscenium arch*). **2** the stage of an ancient theatre. [L f. Gk *proskēnion* (as PRO-[2], *skēnē* stage)]

prosciutto /prəˈʃuːtəʊ/ *n.* Italian ham, esp. cured and eaten as an hors-d'œuvre. [It.]

proscribe /prəʊˈskraɪb/ *v.tr.* **1** banish, exile (*proscribed from the club*). **2** put (a person) outside the protection of the law. **3** reject or denounce (a practice etc.) as dangerous etc. □ **proscription** /-ˈskrɪpʃ(ə)n/ *n.* **proscriptive** /-ˈskrɪptɪv/ *adj.* [L *proscribere* (as PRO-[1], *scribere script*-write)]

prose /prəʊz/ *n. & v.* ● *n.* **1** the ordinary form of the written or spoken language (opp. POETRY, VERSE) (also *attrib.*: *Milton's prose works*). **2** a passage of prose, esp. for translation into a foreign language. **3** dull or commonplace speech, writing, etc. **4** a plain matter-of-fact quality (*the prose of existence*). **5** *Eccl.* = SEQUENCE *n.* 8. ● *v.* **1** *intr.* (usu. foll. by *about*, *away*, etc.) talk tediously (*was prosing away about his dog*). **2** *tr.* turn (a poem etc.) into prose. □ **prose idyll** a short description in prose of a picturesque, esp. rustic, incident, character, etc. **prose poem** (or **poetry**) a piece of imaginative poetic writing in prose. □ **proser** *n.* [ME f. OF f. L *prosa* (*oratio*) straightforward (discourse), fem. of *prosus*, earlier *prorsus* direct]

prosector /prəˈsektə(r)/ *n.* esp. *hist.* a person who dissects dead bodies in preparation for an anatomical lecture etc. [LL = anatomist, f. *prosecare prosect*- (as PRO-[1], *secare* cut), perh. after F *prosecteur*]

prosecute /ˈprɒsɪˌkjuːt/ *v.tr.* **1** (also *absol.*) **a** institute legal proceedings against (a person). **b** institute a prosecution with reference to (a claim, crime, etc.). **2** follow up, pursue (an inquiry, studies, etc.). **3** carry on (a trade, pursuit, etc.). □ **prosecutable** *adj.* [ME f. L *prosequi prosecut*- (as PRO-[1], *sequi* follow)]

prosecution /ˌprɒsɪˈkjuːʃ(ə)n/ *n.* **1 a** the institution and carrying on of a criminal charge in a court. **b** the carrying on of legal proceedings against a person. **c** the prosecuting party in a court case (*a witness for the prosecution*). **2** the act or an instance of prosecuting (*met her in the prosecution of his hobby*). [OF *prosecution* or LL *prosecutio* (as PROSECUTE)]

prosecutor /ˈprɒsɪˌkjuːtə(r)/ *n.* (*fem.* **prosecutrix** /-trɪks/) a person who prosecutes, esp. in a criminal court. □ **prosecutorial** /ˌprɒsɪkjuːˈtɔːrɪəl/ *adj.*

proselyte /ˈprɒsɪˌlaɪt/ *n. & v.* ● *n.* **1** a person converted, esp. recently, from one opinion, creed, party, etc., to another. **2** a Gentile convert to Judaism. ● *v.tr. US* = PROSELYTIZE. □ **proselytism** /-lɪˌtɪz(ə)m/ *n.* [ME f. LL *proselytus* f. Gk *prosēluthos* stranger, convert (as PROS-, stem *ēluth*- of *erkhomai* come)]

proselytize /ˈprɒsɪlɪˌtaɪz/ *v.tr.* (also **-ise**) (also *absol.*) convert (a person or people) from one belief etc. to another, esp. habitually. □ **proselytizer** *n.*

prosencephalon /ˌprɒsɛnˈsefəˌlɒn, -ˈkefəˌlɒn/ *n. Anat.* = FOREBRAIN. [Gk *prosō* forwards + ENCEPHALON]

prosenchyma /prɒsˈsɛŋkɪmə/ *n. Bot.* a plant tissue of elongated cells with interpenetrating tapering ends, occurring esp. in vascular tissue.

□ **prosenchymatous** /ˌprɒsɛŋˈkɪmətəs, -ˈkaɪmətəs/ adj. [Gk pros toward + egkhuma infusion, after parenchyma]

Proserpina /prəˈsɜːpɪnə/ (also **Proserpine** /-pɪnɪ/) Rom. Mythol. the Roman name for Persephone.

prosify /ˈprəʊzɪˌfaɪ/ v. (**-ies, -ied**) **1** tr. turn into prose. **2** tr. make prosaic. **3** intr. write prose.

prosimian /prəʊˈsɪmɪən/ n. & adj. Zool. ● n. a primitive primate of the suborder Prosimii, which includes the lemurs, lorises, galagos, and tarsiers. (See PRIMATE.) ● adj. of or relating to this suborder. [PRO-2 + SIMIAN]

prosit /ˈprəʊzɪt/ int. an expression used in drinking a person's health etc. [G f. L, = may it benefit]

prosody /ˈprɒsədɪ/ n. **1** the theory and practice of versification; the laws of metre. **2** the study of speech rhythms. □ **prosodist** n. **prosodic** /prəˈsɒdɪk/ adj. [ME f. L prosodia accent f. Gk prosōidia (as PROS-, ODE)]

prosopography /ˌprɒsəˈpɒɡrəfɪ/ n. (pl. **-ies**) **1** a description of a person's appearance, personality, social and family connections, career, etc. **2** the study of such descriptions, esp. in Roman history. □ **prosopographer** n. **prosopographic** /-pəˈɡræfɪk/ adj. **prosopographical** adj. [mod.L prosopographia f. Gk prosōpon face, person]

prosopopoeia /ˌprɒsəpəˈpiːə/ n. the rhetorical introduction of a pretended speaker or the personification of an abstract thing. [L f. Gk prosōpopoiia f. prosōpon person + poieō make]

prospect n. & v. ● n. /ˈprɒspekt/ **1 a** (often in pl.) an expectation, esp. of success in a career etc. (his prospects were brilliant; offers a gloomy prospect; no prospect of success). **b** something one has to look forward to (don't relish the prospect of meeting him). **2** an extensive view of landscape etc. (a striking prospect). **3** a mental picture (a new prospect in his mind). **4** a possible or probable customer, subscriber, etc. **5 a** a place likely to yield mineral deposits. **b** a sample of ore for testing. **c** the resulting yield. ● v. /prəˈspekt/ **1** intr. (usu. foll. by for) **a** explore a region for gold etc. **b** look out for or search for something. **2** tr. **a** explore (a region) for gold etc. **b** work (a mine) experimentally. **c** (of a mine) promise (a specified yield). □ **in prospect 1** in sight, within view. **2** within the range of expectation, likely. **prospect well** (or **ill** etc.) (of a mine) promise well (or ill etc.). □ **prospectless** /ˈprɒspektlɪs/ adj. **prospector** /prəˈspektə(r)/ n. [ME f. L prospectus: see PROSPECTUS]

prospective /prəˈspektɪv/ adj. **1** concerned with or applying to the future (implies a prospective obligation) (cf. RETROSPECTIVE adj. 1). **2** some day to be; expected; future (prospective bridegroom). □ **prospectively** adv. **prospectiveness** n. [obs. F prospectif -ive or LL prospectivus (as PROSPECTUS)]

prospectus /prəˈspektəs/ n. (pl. **prospectuses**) a printed document advertising or describing a school, commercial enterprise, forthcoming book, etc. [L, = prospect f. prospicere (as PRO-1, specere look)]

prosper /ˈprɒspə(r)/ v. **1** intr. succeed; thrive (nothing he touches prospers). **2** tr. make successful (heaven prosper him). [ME f. OF prosperer or L prosperare (as PROSPEROUS)]

prosperity /prɒˈsperɪtɪ/ n. a state of being prosperous; wealth or success.

prosperous /ˈprɒspərəs/ adj. **1** successful; rich (a prosperous merchant). **2** flourishing; thriving (a prosperous enterprise). **3** auspicious (a prosperous wind). □ **prosperously** adv. **prosperousness** n. [ME f. obs. F prospereus f. L prosper(us)]

Prost /prɒst/, Alain (b.1955), French motor-racing driver. He was the first Frenchman to win the Formula One world championship (1985); he won the championship again in 1986, 1989, and 1993, after which he retired from racing. Since 1987 Prost has held the record for the most Grand Prix victories.

prostaglandin /ˌprɒstəˈɡlændɪn/ n. Biochem. any of a group of cyclic fatty acids which occur in many mammalian tissues. Their effects resemble those of hormones, and act esp. to cause contraction of smooth muscle. [G f. PROSTATE + GLAND1 + -IN]

prostate /ˈprɒsteɪt/ n. (in full **prostate gland**) a gland surrounding the neck of the bladder in male mammals and releasing a fluid forming part of the semen. □ **prostatic** /prɒˈstætɪk/ adj. [F f. mod.L prostata f. Gk prostatēs one that stands before (as PRO-2, statos standing)]

prosthesis /ˈprɒsθɪsɪs, prɒsˈθiːsɪs/ n. (pl. **prostheses** /-ˌsiːz/) **1 a** an artificial part supplied to remedy a deficiency, e.g. a false breast, leg, tooth, etc. **b** the branch of surgery supplying and fitting prostheses. **2** Gram. the addition of a letter or syllable at the beginning of a word,

e.g. be- in beloved. □ **prosthetic** /prɒsˈθetɪk/ adj. **prosthetically** adv. [LL f. Gk prosthesis f. prostithēmi (as PROS-, tithēmi place)]

prosthetics /prɒsˈθetɪks/ n.pl. (usu. treated as sing.) = PROSTHESIS 1b.

prostitute /ˈprɒstɪˌtjuːt/ n. & v. ● n. **1 a** a woman who engages in sexual activity for payment. **b** (usu. **male prostitute**) a man or boy who engages in sexual activity, esp. with homosexual men, for payment. **2** a person who debases himself or herself for personal gain. ● v.tr. **1** (esp. refl.) make a prostitute of (esp. oneself). **2 a** misuse (one's talents, skills, etc.) for money. **b** offer (oneself, one's honour, etc.) for unworthy ends, esp. for money. □ **prostitutor** n. **prostitution** /ˌprɒstɪˈtjuːʃ(ə)n/ n. [L prostituere prostitut- offer for sale (as PRO-1, statuere set up, place)]

prostrate adj. & v. ● adj. /ˈprɒstreɪt/ **1 a** lying face downwards, esp. in submission. **b** lying horizontally. **2** overcome, esp. by grief, exhaustion, etc. (prostrate with self-pity). **3** Bot. growing along the ground. ● v.tr. /prɒˈstreɪt/ **1** lay (a person etc.) flat on the ground. **2** (refl.) throw (oneself) down in submission etc. **3** (of fatigue, illness, etc.) overcome; reduce to extreme physical weakness. □ **prostration** /prɒˈstreɪʃ(ə)n/ n. [ME f. L prostratus past part. of prosternere (as PRO-1, sternere strat- lay flat)]

prostyle /ˈprəʊstaɪl/ n. & adj. Archit. ● n. a portico with not more than four columns. ● adj. (of a building) having such a portico. [L prostylos having pillars in front (as PRO-2, STYLE)]

prosy /ˈprəʊzɪ/ adj. (**prosier, prosiest**) tedious; commonplace; dull (prosy talk). □ **prosily** adv. **prosiness** n.

Prot. abbr. **1** Protectorate. **2** Protestant.

protactinium /ˌprəʊtækˈtɪnɪəm/ n. a radioactive metallic chemical element (atomic number 91; symbol **Pa**). A member of the actinide series, protactinium was discovered by Otto Hahn and Lise Meitner in 1917. It is a rare element, occurring as a product of the natural decay of uranium. It is so named because one of its isotopes decays to form actinium. [f. PROTO- + ACTINIUM]

protagonist /prəʊˈtæɡənɪst/ n. **1** the chief person in a drama, story, etc. **2** the leading person in a contest etc.; a principal performer. **3** (usu. foll. by of, for) disp. an advocate or champion of a cause, course of action, etc. (a protagonist of women's rights). [Gk prōtagōnistēs (as PROTO-, agōnistēs actor)]

protamine /ˈprəʊtəˌmiːn/ n. Biochem. any of a group of proteins which occur combined with nucleic acids, esp. in fish sperm. [PROTO- + AMINE]

protasis /ˈprɒtəsɪs/ n. (pl. **protases** /-ˌsiːz/) the clause expressing the condition in a conditional sentence. □ **protatic** /prɒˈtætɪk/ adj. [L, f. Gk protasis proposition (as PRO-2, teinō stretch)]

protea /ˈprəʊtɪə/ n. a shrub of the genus Protea, native to southern Africa, with conelike flower-heads. [mod.L f. PROTEUS, with ref. to the many species]

protean /ˈprəʊtɪən, prəʊˈtiːən/ adj. **1** variable, taking many forms. **2** (of an artist, writer, etc.) versatile. [f. PROTEUS]

protease /ˈprəʊtɪˌeɪs/ n. Biochem. an enzyme able to hydrolyse proteins and peptides by proteolysis. [PROTEIN + -ASE]

protect /prəˈtekt/ v.tr. **1** (often foll. by from, against) keep (a person, thing, etc.) safe; defend; guard (goggles protected her eyes from dust; guards protected the queen). **2** Econ. shield (home industry) from competition by imposing import duties on foreign goods. **3** Brit. provide funds to meet (a bill, draft, etc.). **4** provide (machinery etc.) with appliances to prevent injury from it. [L protegere protect- (as PRO-1, tegere cover)]

protection /prəˈtekʃ(ə)n/ n. **1 a** the act or an instance of protecting. **b** the state of being protected; defence (affords protection against the weather). **c** a thing, person, or animal that provides protection (bought a dog as protection). **2** (also **protectionism** /-ˌnɪz(ə)m/) Econ. the theory or practice of protecting home industries. **3** colloq. a immunity from molestation obtained by payment to gangsters etc. under threat of violence. **b** (in full **protection money**) the money so paid, esp. on a regular basis. **4** = safe conduct 1. **5** archaic the keeping of a woman as a mistress. □ **protectionist** n. [ME f. OF protection or LL protectio (as PROTECT)]

protective /prəˈtektɪv/ adj. & n. ● adj. **1 a** protecting; intended or intending to protect. **b** (of a person) tending to protect in a possessive way. **2** (of food) protecting against deficiency diseases. ● n. something that protects, esp. a condom. □ **protective clothing** clothing worn to shield the body from dangerous substances or a hostile environment. **protective colouring** colouring disguising or camouflaging a plant or animal. **protective custody** the detention of a person for his or her own protection. □ **protectively** adv. **protectiveness** n.

protector /prəˈtektə(r)/ n. (fem. **protectress** /-trɪs/) **1 a** a person who protects. **b** a guardian or patron. **2** (often in *comb.*) a thing or device that protects (*chest-protector*). **3** (usu. **Protector**) *hist.* a regent in charge of a kingdom during the minority, absence, etc. of the sovereign. **4** (**Protector**) (in full **Lord Protector of the Commonwealth**) *hist.* the title taken by Oliver Cromwell during his government of Britain in 1653–8 and passed on to his son Richard Cromwell 1658–9. (See also PROTECTORATE 3b.) □ **protectoral** *adj.* **protectorship** *n.* [ME f. OF *protecteur* f. LL *protector* (as PROTECT)]

protectorate /prə'tektərət/ *n.* **1** a state that is controlled and protected by another. **2** such a relation of one state to another. **3** (usu. **Protectorate**) **a** the office of the Protector of a kingdom or state. **b** the period of this, esp. (**Protectorate**) in England and Wales under the Cromwells, 1653–9. (*See note below.*)

▪ Oliver Cromwell was appointed Lord Protector in December 1653 at the behest of the army, and retained the position until his death in September 1658. Although the Protectorate achieved considerable success in foreign wars, it depended almost entirely on its leader's personality at home and was continually threatened by the unstable relationship between the Protector, Parliament, and the army. After the elder Cromwell's death his son Richard proved incapable of holding the regime together, and its subsequent collapse led to the restoration of Charles II.

protégé /'prɒtɪˌʒeɪ, 'prəʊt-/ *n.* (*fem.* **protégée** *pronunc.* same) a person under the protection, patronage, tutelage, etc. of another. [F, past part. of *protéger* f. L *protegere* PROTECT]

protein /'prəʊtiːn/ *n.* **1** any of a class of nitrogenous organic compounds forming an essential part of living organisms and having large molecules consisting of one or more chains of amino acids linked together. (*See note below.*) **2** food etc. consisting mainly of such compounds. □ **proteinaceous** /ˌprəʊtiː'neɪʃəs/ *adj.* **proteinic** /prəʊ'tiːnɪk/ *adj.* **proteinous** /-'tiːnəs, -'tiːnəs/ *adj.* [F *protéine*, G *Protein* f. Gk *prōteios* primary]

▪ Proteins have many different functions. Some form strong fibres and make up the structural component of body tissues such as muscle, collagen, hair, nails, etc. Others with more compact molecules include most enzymes, antibodies, and receptors in cell membranes. Protein molecules are built up from about twenty amino acids which are arranged in different orders in one or more long polypeptide chains. Protein in the diet is broken down into amino acids, from which the body makes the proteins it requires. The information needed for the synthesis of particular proteins in a living cell comes from the DNA in the cell's nucleus.

pro tem /prəʊ 'tem/ *adj. & adv. colloq.* = PRO TEMPORE. [abbr.]

pro tempore /prəʊ 'tempərɪ/ *adj. & adv.* for the time being. [L]

proteolysis /ˌprəʊtɪ'ɒlɪsɪs/ *n. Biochem.* the splitting of proteins or peptides by the action of enzymes esp. during the process of digestion. □ **proteolytic** /-tɪə'lɪtɪk/ *adj.* [f. PROTEIN + -LYSIS]

Proterozoic /ˌprɒtərəʊ'zəʊɪk/ *adj. & n. Geol.* of, relating to, or denoting the later part of the Precambrian era, characterized by the oldest forms of life (cf. ARCHAEAN). [Gk *proteros* former + *zōē* life, *zōos* living]

protest *n. & v.* ● *n.* /'prəʊtest/ **1** a statement of dissent or disapproval; a remonstrance (*made a protest*). **2** (often *attrib.*) a usu. public demonstration of objection to government etc. policy (*marched in protest*; *protest demonstration*). **3** a solemn declaration. **4** *Law* a written declaration, usu. by a notary public, that a bill has been presented and payment or acceptance refused. ● *v.* /prə'test/ **1** *intr.* (usu. foll. by *against, at, about,* etc.) make a protest against an action, proposal, etc. **2** *tr.* (often foll. by *that* + clause; also *absol.*) affirm (one's innocence etc.) solemnly, esp. in reply to an accusation etc. **3** *tr. Law* write or obtain a protest in regard to (a bill). **4** *tr. N. Amer.* object to (a decision etc.). □ **under protest** unwillingly. □ **protester** /prə'testə(r)/ *n.* **protestor** *n.* **protestingly** /-'testɪŋlɪ/ *adv.* [ME f. OF *protest* (n.), *protester* (v.), f. L *protestari* (as PRO-[1], *testari* assert f. *testis* witness)]

Protestant /'prɒtɪstənt/ *n. & adj.* ● *n.* **1** a member or follower of any of the Western Christian Churches that are separate from the Roman Catholic Church in accordance with the principles of the Reformation. (*See note below.*) **2** (**protestant**) (also /prə'testənt/) a protesting person. ● *adj.* **1** of or relating to any of the Protestant Churches or their members etc. **2** (**protestant**) (also /prə'testənt/) protesting. □ **Protestant ethic** (or **work ethic**) the principle that hard work leads to success and moral improvement, attributed by Max Weber to the teachings of Calvin. □ **Protestantism** *n.* **Protestantize** *v.tr. & intr.* (also **-ise**). [mod.L *protestans*, part. of L *protestari* (see PROTEST)]

▪ Protestants are so called after the declaration (*protestatio*) of Martin Luther and his supporters dissenting from the decision of the Diet of Spires (1529), which reaffirmed the edict of the Diet of Worms against the Reformation. All Protestants rejected the authority of the papacy, both religious and political, and found authority in the text of the Scriptures, made available to all in vernacular translation. The authority of the clergy and the sacramental system was weakened in varying degrees, and the way opened for religious individualism. In the 16th century the name was generally taken in Germany by the Lutherans, while the Swiss and French called themselves 'Reformed'. In England the use has varied with time and circumstances. In the 17th century 'Protestant' was generally accepted and used by members of the established Church, and sometimes excluded Presbyterians, Quakers, and Nonconformists; it was primarily opposed to 'papist'. It is now usually opposed to 'Roman Catholic' or 'Catholic'. In broader terms, the Protestant Churches encompass the Anglican Communion and the Baptists, Lutherans, Methodists, Pentecostalists, and Presbyterians, and have a total membership of about 300 million.

protestation /ˌprɒtɪ'steɪʃ(ə)n/ *n.* **1** a strong affirmation. **2** a protest. [ME f. OF *protestation* or LL *protestatio* (as PROTESTANT)]

Proteus /'prəʊtɪəs/ **1** *Gk Mythol.* a minor sea-god who had the power of prophecy but who would assume different shapes to avoid answering questions. His name is sometimes used to mean a changing or inconstant person or thing. **2** *Astron.* a satellite of Neptune, the sixth closest to the planet, discovered by the Voyager 2 space probe in 1989. With a diameter of 400 km it is the second largest of Neptune's moons.

proteus /'prəʊtɪəs/ *n.* a Gram-negative bacterium of the genus *Proteus*, found in the intestines of animals and in the soil. **2** = OLM.

prothalamium /ˌprəʊθə'leɪmɪəm/ *n.* (also **prothalamion** /-mɪən/) (*pl.* **prothalamia** /-mɪə/) a song or poem to celebrate a forthcoming wedding. [title of a poem by Spenser, after *epithalamium*]

prothallium /prəʊ'θælɪəm/ *n.* (*pl.* **prothallia** /-lɪə/) *Bot.* = PROTHALLUS. [mod.L f. PRO-[2] + Gk *thallion* dimin. of *thallos*: see PROTHALLUS]

prothallus /prəʊ'θæləs/ *n.* (*pl.* **prothalli** /-laɪ/) *Bot.* the gametophyte of certain plants, esp. of a fern. [mod.L f. PRO-[2] + Gk *thallos* green shoot]

prothesis /'prɒθɪsɪs/ *n.* (*pl.* **protheses** /-ˌsiːz/) **1** *Eccl.* **a** the placing of the Eucharistic elements on the credence table. **b** a credence table. **c** the part of a church where this stands. **2** *Gram.* = PROSTHESIS 2. □ **prothetic** /prə'θetɪk/ *adj.* [Gk f. *protithēmi* (as PRO-[2], *tithēmi* place)]

prothonotary var. of PROTONOTARY.

protist /'prəʊtɪst/ *n. Biol.* a primitive organism of the kingdom Protista, with both plant and animal characteristics. (*See note below.*) □ **protistology** /ˌprəʊtɪ'stɒlədʒɪ/ *n.* [mod.L *Protista* f. Gk *prōtista* neut. pl. superl. f. *prōtos* first]

▪ Protista is now usually held to comprise unicellular eukaryotes that typically possess a flagellum, a grouping that includes only the protozoans and simple algae and fungi. However, the bacteria and other algae and fungi have sometimes also been included.

protium /'prəʊtɪəm/ *n.* the ordinary isotope of hydrogen as distinct from heavy hydrogen (cf. DEUTERIUM, TRITIUM). [f. PROTO- + -IUM]

proto- /'prəʊtəʊ/ *comb. form* **1** original, primitive (*proto-Germanic*; *proto-Slavic*). **2** first, original (*protomartyr*; *protophyte*). [Gk *prōto-* f. *prōtos* first]

protocol /'prəʊtəˌkɒl/ *n. & v.* ● *n.* **1 a** official, esp. diplomatic, formality and etiquette observed on state occasions etc. **b** the rules, formalities, etc. of any procedure, group, etc. **2** the original draft of a diplomatic document, esp. of the terms of a treaty agreed to in conference and signed by the parties. **3** a formal statement of a transaction. **4** the official formulae at the beginning and end of a charter, papal bull, etc. **5** *US* a record of experimental observations etc. **6** a set of rules governing the exchange or transmission of data electronically between devices. ● *v.* (**protocolled, protocolling**) **1** *intr.* draw up a protocol or protocols. **2** *tr.* record in a protocol. [orig. Sc. *prothocoll* f. OF *prothocole* f. med.L *protocollum* f. Gk *protokollon* flyleaf (as PROTO-, *kolla* glue)]

Proto-Indo-European /ˌprəʊtəʊˌɪndəʊˌjʊərə'pɪən/ *n. & adj.* ● *n.* the language from which all Indo-European languages are believed to derive. (See INDO-EUROPEAN.) ● *adj.* of or relating to this language.

protomartyr /ˌprəʊtəʊ'mɑːtə(r)/ *n.* the first martyr in any cause, esp. the first Christian martyr St Stephen.

proton /'prəʊtɒn/ *n. Physics* a stable subatomic particle with a positive electric charge, occurring in all atomic nuclei. (*See note below.*) □ **protonic** /prə'tɒnɪk/ *adj.* [Gk, neut. of *prōtos* first: see -ON]

▪ Protons were identified by the German physicist Wilhelm Wien (1864–1928) in 1898 and by J. J. Thomson in 1910. The charge on the proton is equal and opposite to that on the electron, although the mass of the proton is 1,836 times greater. The atoms of each chemical element have a characteristic number of protons in the nucleus; this is known as the atomic number. The common isotope of hydrogen has a nucleus consisting of a single proton. The transfer of protons between molecules is an important chemical reaction; such reactions in water are the basis of acid-base chemistry and of metabolism in living systems. High-velocity protons are used in particle accelerators to probe the structure of atomic nuclei. (See also PROUT.)

protonotary /ˌprəʊtəˈnəʊtəri, prəˈtɒnətə-/ n. (also **prothonotary** /ˌprəʊθəˈnəʊtəri, prəˈθɒnətə-/) (pl. **-ies**) a chief clerk in some law courts, orig. in the Byzantine court. □ **prothonotary warbler** a North American wood warbler, *Protonotaria citrea*, with a bright yellow head and breast, green back, and blue-grey wings. **Protonotary Apostolic** (or **Apostolical**) a member of the college of prelates who register papal acts, direct the canonization of saints, etc. [med.L *protonotarius* f. late Gk *protonotarios* (as PROTO-, NOTARY)]

protopectin /ˌprəʊtəˈpektɪn/ n. = PECTOSE.

protophyte /ˈprəʊtəˌfaɪt/ n. Bot. a unicellular plant bearing gametes.

protoplasm /ˈprəʊtəˌplæz(ə)m/ n. Biol. the material comprising the living part of a cell. (See note below.) □ **protoplasmal** /ˌprəʊtəˈplæzm(ə)l/ adj. **protoplasmatic** /-plæzˈmætɪk/ adj. **protoplasmic** /-ˈplæzmɪk/ adj. [Gk *protoplasma* (as PROTO-, PLASMA)]

▪ Protoplasm consists of the nucleus and other specialized structures together with the complex, translucent, membrane-bound material (cytoplasm) in which they are embedded. In the 19th century protoplasm was thought of as a basic, essentially homogeneous, form of living matter. With the recognition, especially after the invention of the electron microscope, of the complex structural organization within cells, this concept of protoplasm as a distinct substance was gradually abandoned.

protoplast /ˈprəʊtəˌplæst/ n. Biol. the protoplasm of one cell, esp. a living plant or bacterial cell whose cell wall has been removed. □ **protoplastic** /ˌprəʊtəˈplæstɪk/ adj. [F *protoplaste* or LL *protoplastus* f. Gk *protoplastos* (as PROTO-, *plassō* mould)]

prototherian /ˌprəʊtəˈθɪəriən/ n. & adj. Zool. ● n. an egg-laying mammal of the subclass Prototheria, which comprises the monotremes. ● adj. of or relating to this subclass. [PROTO- + Gk *thēr* wild beast]

prototype /ˈprəʊtəˌtaɪp/ n. 1 an original thing or person of which or whom copies, imitations, improved forms, representations, etc. are made. 2 a trial model or preliminary version of a vehicle, machine, etc. 3 a thing or person representative of a type; an exemplar. □ **prototypal** /ˌprəʊtəˈtaɪp(ə)l/ adj. **prototypic** /-ˈtɪpɪk/ adj. **prototypical** **prototypically** adv. [F *prototype* or LL *prototypus* f. Gk *prototupos* (as PROTO-, TYPE)]

protozoan /ˌprəʊtəˈzəʊən/ n. & adj. Zool. ● n. (also **protozoon** /-ˈzəʊɒn/) (pl. **protozoa** /-ˈzəʊə/ or **protozoans**) a microscopic unicellular eukaryotic organism of the phylum or subkingdom Protozoa, which includes amoebas and ciliates. (See note below.) ● adj. (also **protozoic** /-ˈzəʊɪk/) of or relating to this group of organisms. □ **protozoal** adj. [mod.L (as PROTO-, Gk *zōion* animal)]

▪ First observed by Antoni van Leeuwenhoek in the 17th century, protozoans are ubiquitous in aquatic and damp habitats. There are also numerous parasitic species, among them the trypanosomes which cause sleeping sickness, and the malaria parasite. Protozoans display a great variety of form, and among the most important groups are the flagellates, ciliates, and amoebas. Some protozoans, esp. foraminifers, bear shells, the residue of which forms a significant proportion of some limestone rocks. The protozoans are now usually assigned to the kingdom Protista, along with the unicellular algae and fungi.

protract /prəˈtrækt/ v.tr. 1 **a** prolong or lengthen in space or esp. time (*protracted their stay for some weeks*). **b** (as **protracted** adj.) of excessive length or duration (*a protracted illness*). 2 draw (a plan of ground etc.) to scale. □ **protractedly** adv. **protractedness** n. [L *protrahere* protract- (as PRO-¹, *trahere* draw)]

protractile /prəˈtræktaɪl/ adj. (of a part of the body etc.) capable of being protruded or extended.

protraction /prəˈtrækʃ(ə)n/ n. 1 the act or an instance of protracting; the state of being protracted. 2 a drawing to scale. 3 the action of a protractor muscle. [F *protraction* or LL *protractio* (as PROTRACT)]

protractor /prəˈtræktə(r)/ n. 1 an instrument for measuring angles, usu. in the form of a graduated semicircle. 2 Anat. a muscle serving to extend a limb etc.

protrude /prəˈtruːd/ v. 1 intr. extend beyond or above a surface; project. 2 tr. thrust or cause to thrust out. □ **protrudent** adj. **protrusible** /-ˈtruːsɪb(ə)l/ adj. **protrusive** adj. **protrusion** /-ˈtruːʒ(ə)n/ n. [L *protrudere* (as PRO-¹, *trudere* trus- thrust)]

protrusile /prəˈtruːsaɪl/ adj. (of a limb etc.) capable of being thrust forward. [PRO-¹ + *extrusile*: see EXTRUDE]

protuberant /prəˈtjuːbərənt/ adj. bulging out; prominent (*protuberant eyes*). □ **protuberance** n. [LL *protuberare* (as PRO-¹, *tuber* bump)]

proud /praʊd/ adj. 1 feeling greatly honoured or pleased (*am proud to know him; proud of his friendship*). 2 **a** (often foll. by *of*) valuing oneself, one's possessions, etc. highly, or esp. too highly; haughty; arrogant (*proud of his ancient name*). **b** (often in *comb.*) having a proper pride; satisfied (*house-proud; proud of a job well done*). 3 **a** (of an occasion etc.) justly arousing pride (*a proud day for us; a proud sight*). **b** (of an action etc.) showing pride (*a proud wave of the hand*). 4 (of a thing) imposing; splendid. 5 slightly projecting from a surface etc. (*the nail stood proud of the plank*). 6 (of flesh) overgrown round a healing wound. 7 (of water) swollen in flood. □ **do proud** *colloq.* 1 treat (a person) with lavish generosity or honour (*they did us proud on our anniversary*). 2 (*refl.*) act honourably or worthily. **proud-hearted** haughty; arrogant. □ **proudly** adv. **proudness** n. [OE *prūt, prūd* f. OF *prud, prod* oblique case of *pruz* etc. valiant, ult. f. LL *prode* f. L *prodesse* be of value (as PRO-¹, *esse* be)]

Proudhon /ˈpruːdɒn/, Pierre Joseph (1809–65), French social philosopher and journalist. His criticism of Napoleon III (Louis-Napoleon) and the Second Republic led to his imprisonment 1849–52; he later spent a period (1858–62) in exile in Belgium. His writings exercised considerable influence on the development of anarchism and socialism in Europe. He is chiefly remembered for his pamphlet *What is Property?* (1840), arguing that property, in the sense of the exploitation of one person's labour by another, is theft. His theories were developed by his disciple Bakunin.

Proust[1] /pruːst/, Joseph Louis (1754–1826), French analytical chemist. He is remembered mainly for proposing the law of constant proportions, demonstrating that any pure sample of a chemical compound (such as an oxide of a metal) always contains the same elements in fixed proportions. Berzelius later established the connection between this and John Dalton's atomic theory, giving Proust full credit, though some exceptions to the law were found many years later.

Proust[2] /pruːst/, Marcel (1871–1922), French novelist, essayist, and critic. Although he moved in fashionable Paris society during the 1890s, he was severely incapacitated by asthma and became a virtual recluse after his mother's death in 1905. He devoted the remainder of his life to writing his novel *À la recherche du temps perdu* (published in seven sections between 1913 and 1927). Influenced by the philosophy of Henri Bergson, the work traces the life of the narrator from childhood to middle age; its central theme is the recovery of the lost past and the releasing of its creative energies through the stimulation of unconscious memory.

Prout /praʊt/, William (1785–1850), English chemist and biochemist. He was trained as a physician, and carried out analyses of urine, gastric juices, and foodstuffs. In theoretical chemistry he developed the hypothesis that hydrogen is the primary substance from which all other elements are formed, and if the atomic weight of hydrogen is regarded as unity the weights of all other elements are exact multiples of it. Although this hypothesis was later found to be incorrect, it stimulated research in atomic theory, and in modern particle physics the hydrogen nucleus (proton) is indeed considered a fundamental particle.

Prov. abbr. 1 (in the Bible) Proverbs. 2 Province. 3 Provençal.

prove /pruːv/ v. (past part. **proved** or **proven** /ˈpruːv(ə)n, ˈprəʊv-/) 1 tr. (often foll. by *that* + clause) demonstrate the truth of by evidence or argument. 2 intr. **a** (usu. foll. by *to* + infin.) be found (*it proved to be untrue*). **b** emerge incontrovertibly as (*will prove the winner*). 3 tr. Math. test the accuracy of (a calculation). 4 tr. establish the genuineness and validity of (a will). 5 intr. (of dough) rise in bread-making. 6 tr. = PROOF v. 7 tr. subject (a gun etc.) to a testing process. 8 tr. archaic test the qualities of; try. □ **not proven** (in Scottish Law) a verdict that there is insufficient evidence to establish guilt or innocence. **prove oneself** show one's abilities, courage, etc. □ **provable** adj. **provability** /ˌpruːvəˈbɪlɪti/ n. [ME f. OF *prover* f. L *probare* test, approve, demonstrate f. *probus* good]

provenance /ˈprɒvɪnəns/ n. 1 the place of origin or earliest known history, esp. of a work of art, manuscript, etc. **2** origin. [F f. *provenir* f. L *provenire* (as PRO-¹, *venire* come)]

Provençal /ˌprɒvɒnˈsɑːl/ *adj. & n.* ● *adj.* of or concerning the language, inhabitants, landscape, etc., of Provence. ● *n.* **1** a native of Provence. **2** the language of Provence. (*See note below.*) [F (as PROVINCIAL f. *provincia*; see PROVENCE)]

▪ Provençal is a Romance language closely related to French, Italian, and Catalan. Strictly speaking, it is just one dialect within *langue d'oc* (or Occitan), although it is sometimes used as a synonym for it. In the 12th–14th centuries it was the language of the troubadours and cultured speakers of southern France, but the subsequent spread of the northern dialects of French led to its gradual decline in prestige and importance. Its use as a literary language was revived in the 19th century, and a spoken form, now regarded as a dialect, is still widely used in Provence.

Provence /prɒˈvɒns/ a former province of SE France, on the Mediterranean coast east of the Rhône. Settled by the Greeks in the 6th century BC, the area around Marseilles became, in the first century BC, part of the Roman colony of Gaul. It was united with France under Louis XI in 1481 and is now part of the region of Provence–Alpes–Côte d'Azur. [L *provincia* province, as L colloq. name for southern Gaul, which was the first Roman province to be established outside Italy]

Provence–Alpes–Côte d'Azur /prɒˌvɒns ælp ˌkəʊt dæˈzjʊə(r)/ a mountainous region of SE France, on the border with Italy and including the French Riviera.

provender /ˈprɒvɪndə(r)/ *n.* **1** animal fodder. **2** *joc.* food for human beings. [ME f. OF *provendre, provende* ult. f. L *praebenda* (see PREBEND)]

provenience /prəˈviːnɪəns/ *n. US* = PROVENANCE. [L *provenire* f. *venire* come]

proverb /ˈprɒvɜːb/ *n.* **1** a short pithy saying in general use, held to embody a general truth. **2** a person or thing that is notorious (*he is a proverb for inaccuracy*). [ME f. OF *proverbe* or L *proverbium* (as PRO-¹, *verbum* word)]

proverbial /prəˈvɜːbɪəl/ *adj.* **1** (esp. of a specific characteristic etc.) as well known as a proverb; notorious (*his proverbial honesty*). **2** of or referred to in a proverb (*the proverbial ill wind*). □ **proverbially** *adv.* **proverbiality** /-ˌvɜːbɪˈælɪtɪ/ *n.* [ME f. L *proverbialis* (as PROVERB)]

Proverbs (also **Book of Proverbs**) a book of the Bible containing maxims attributed mainly to Solomon.

provide /prəˈvaɪd/ *v.* **1** *tr.* supply; furnish (*provided them with food; provided food for them; provided a chance for escape*). **2** *intr.* **a** (usu. foll. by *for, against*) make due preparation (*provided for any eventuality; provided against invasion*). **b** (usu. foll. by *for*) prepare for the maintenance of a person etc. **3** *tr.* (also *refl.*) equip with necessities (*they had to provide themselves*). **4** *tr.* (usu. foll. by *that*) stipulate in a will, statute, etc. **5** *tr.* (usu. foll. by *to*) *Eccl. hist.* **a** appoint (an incumbent) to a benefice. **b** (of the pope) appoint (a successor) to a benefice not yet vacant. [ME f. L *providere* (as PRO-¹, *videre vis-* see)]

provided /prəˈvaɪdɪd/ *adj. & conj.* ● *adj.* supplied, furnished. ● *conj.* (often foll. by *that*) on the condition or understanding (that).

Providence /ˈprɒvɪd(ə)ns/ the state capital of Rhode Island, a port on the Atlantic coast; pop. (1990) 160,730. It was founded in 1636 as a haven for religious dissenters by Roger Williams (1604–83), who had been banished from the colony at Plymouth, Massachusetts. It developed in the 18th century as a major port for trade with the West Indies.

providence /ˈprɒvɪd(ə)ns/ *n.* **1** the protective care of God or nature. **2** (**Providence**) God in this aspect. **3** timely care or preparation; foresight; thrift. □ **special providence** a particular instance of God's providence. [ME f. OF *providence* or L *providentia* (as PROVIDE)]

provident /ˈprɒvɪd(ə)nt/ *adj.* having or showing foresight; thrifty. □ **Provident Society** *Brit.* = FRIENDLY SOCIETY. □ **providently** *adv.* [ME f. L (as PROVIDE)]

providential /ˌprɒvɪˈdenʃ(ə)l/ *adj.* **1** of or by divine foresight or interposition. **2** opportune, lucky. □ **providentially** *adv.* [PROVIDENCE + -IAL, after *evidential* etc.]

provider /prəˈvaɪdə(r)/ *n.* **1** a person or thing that provides. **2** the breadwinner of a family etc.

providing /prəˈvaɪdɪŋ/ *conj.* = PROVIDED *conj.*

province /ˈprɒvɪns/ *n.* **1** a principal administrative division of a country etc. **2 a** (**the provinces**) the whole of a country outside the capital, esp. regarded as uncultured, unsophisticated, etc. **b** *Brit.* (**the Province**) Northern Ireland. **3** a sphere of action; business (*outside my province as a teacher*). **4** a branch of learning etc. (*in the province of*

aesthetics). **5** *Eccl.* a district under an archbishop or a metropolitan. **6** *Rom. Hist.* a territory outside Italy under a Roman governor. [ME f. OF f. L *provincia* charge, province]

provincial /prəˈvɪnʃ(ə)l/ *adj. & n.* ● *adj.* **1 a** of or concerning a province. **b** of or concerning the provinces. **2** unsophisticated or uncultured in manner, speech, opinion, etc. ● *n.* **1** an inhabitant of a province or the provinces. **2** an unsophisticated or uncultured person. **3** *Eccl.* the head or chief of a province or of a religious order in a province. □ **provincialize** *v.tr.* (also **-ise**). **provincially** *adv.* **provinciality** /-ˌvɪnʃɪˈælɪtɪ/ *n.* [ME f. OF f. L *provincialis* (as PROVINCE)]

provincialism /prəˈvɪnʃəˌlɪz(ə)m/ *n.* **1** provincial manners, fashion, mode of thought, etc., esp. regarded as restricting or narrow. **2** a word or phrase peculiar to a provincial region. **3** concern for one's local area rather than one's country. □ **provincialist** *n.*

provision /prəˈvɪʒ(ə)n/ *n. & v.* ● *n.* **1 a** the act or an instance of providing (*made no provision for his future*). **b** something provided (*a provision of bread*). **2** (in *pl.*) food, drink, etc., esp. for an expedition. **3 a** a legal or formal statement providing for something. **b** a clause of this. **4** *Eccl. hist.* an appointment to a benefice not yet vacant (cf. PROVIDE 5). ● *v.tr.* supply (an expedition etc.) with provisions. □ **provisioner** *n.* **provisionless** *adj.* **provisionment** *n.* [ME f. OF f. L *provisio -onis* (as PROVIDE)]

provisional /prəˈvɪʒən(ə)l/ *adj. & n.* ● *adj.* providing for immediate needs only; temporary. ● *n.* (**Provisional**) a member of the Provisional IRA. □ **provisionally** *adv.* **provisionalness** *n.* **provisionality** /-ˌvɪʒəˈnælɪtɪ/ *n.*

Provisional IRA see IRISH REPUBLICAN ARMY.

proviso /prəˈvaɪzəʊ/ *n.* (*pl.* **-os**) **1** a stipulation. **2** a clause of stipulation or limitation in a document. [L, neut. ablat. past part. of *providere* PROVIDE, in med.L phr. *proviso quod* it being provided that]

provisor /prəˈvaɪzə(r)/ *n. Eccl.* **1** a deputy of a bishop or archbishop. **2** *hist.* the holder of a provision (see PROVISION *n.* 4). [ME f. AF *provisour* f. L *provisor -oris* (as PROVIDE)]

provisory /prəˈvaɪzərɪ/ *adj.* **1** conditional; having a proviso. **2** making provision (*provisory care*). □ **provisorily** *adv.* [F *provisoire* or med.L *provisorius* (as PROVISOR)]

Provo /ˈprəʊvəʊ/ *n.* (*pl.* **-os**) *colloq.* a member of the Provisional IRA. [abbr.]

provocation /ˌprɒvəˈkeɪʃ(ə)n/ *n.* **1** the act or an instance of provoking; a state of being provoked (*did it under severe provocation*). **2** a cause of annoyance. **3** *Law* an action, insult, etc. held to be likely to provoke physical retaliation. [ME f. OF *provocation* or L *provocatio* (as PROVOKE)]

provocative /prəˈvɒkətɪv/ *adj. & n.* ● *adj.* **1** (usu. foll. by *of*) tending to provoke, esp. anger or sexual desire. **2** intentionally annoying. ● *n.* a provocative thing. □ **provocatively** *adv.* **provocativeness** *n.* [ME f. obs. F *provocative* f. LL *provocativus* (as PROVOKE)]

provoke /prəˈvəʊk/ *v.tr.* **1 a** (often foll. by *to*, or *to* + infin.) rouse or incite (*provoked him to fury*). **b** (often as **provoking** *adj.*) annoy, irritate; exasperate. **2** call forth; instigate (indignation, an inquiry, a storm, etc.). **3** (usu. foll. by *into* + verbal noun) irritate or stimulate (a person). **4** tempt; allure. **5** cause, give rise to (*will provoke fermentation*). □ **provokable** *adj.* **provokingly** *adv.* [ME f. OF *provoquer* f. L *provocare* (as PRO-¹, *vocare* call)]

provost /ˈprɒvəst/ *n.* **1** *Brit.* the head of some colleges esp. at Oxford or Cambridge. **2** *Eccl.* **a** the head of a chapter in a cathedral. **b** *hist.* the head of a religious community. **3** *Sc.* the head of a municipal corporation or burgh. **4** the Protestant minister of the principal church of a town etc. in Germany etc. **5** *US* a high administrative officer in a university. **6** /prəˈvəʊ, ˈprəʊvəʊ/ = *provost marshal.* □ **provost guard** /prəˈvəʊ, ˈprəʊvəʊ/ *US* a body of soldiers under a provost marshal. **provost marshal** /prəˈvəʊ, ˈprəʊvəʊ/ **1** the head of military police in camp or on active service. **2** the master-at-arms of a ship in which a court-martial is held. □ **provostship** *n.* [ME f. OE *profost* & AF *provost, prevost* f. med.L *propositus* for *praepositus*: see PRAEPOSTOR]

prow /praʊ/ *n.* **1** the fore-part or bow of a ship adjoining the stem. **2** a pointed or projecting front part. [F *proue* f. Prov. *proa* or It. dial. *prua* f. L *prora* f. Gk *prōira*]

prowess /ˈpraʊɪs/ *n.* **1** skill; expertise. **2** valour; gallantry. [ME f. OF *proesce* f. *prou* valiant]

prowl /praʊl/ *v. & n.* ● *v.* **1** *tr.* roam (a place) in search or as if in search of prey, plunder, etc. **2** *intr.* (often foll. by *about, around*) move about like a hunter. ● *n.* the act or an instance of prowling. □ **on the prowl** moving about secretly or rapaciously. **prowl car** *US* a police squad car. □ **prowler** *n.* [ME *prolle*, of unkn. orig.]

prox. *abbr.* proximo.

prox. acc. *abbr. proxime accessit.*

proxemics /prɒkˈsiːmɪks/ *n.* (in sociology) the study of socially conditioned spatial factors in ordinary human relations. [PROXIMITY + -emics: cf. *phonemics*]

Proxima Centauri /ˌprɒksɪmə senˈtɔːraɪ/ *Astron.* a faint red dwarf star associated with the bright binary star Alpha Centauri. It is the closest known star to the solar system (distance 4.24 light-years). [L, = nearest (star) of CENTAURUS]

proximal /ˈprɒksɪm(ə)l/ *adj. Anat.* situated nearer to the centre of the body or towards the point of attachment (opp. DISTAL). □ **proximally** *adv.* [L *proximus* nearest]

proximate /ˈprɒksɪmət/ *adj.* **1** nearest or next before or after (in place, order, time, causation, thought process, etc.). **2** approximate. □ **proximately** *adv.* [L *proximatus* past part. of *proximare* draw near (as PROXIMAL)]

proxime accessit /ˌprɒksɪmɪ əkˈsiːsɪt/ *n.* **1** second place in an examination etc. **2** a person gaining this. [L, = came very near]

proximity /prɒkˈsɪmɪti/ *n.* nearness in space, time, etc. (*sat in close proximity to them*). □ **proximity fuse** an electronic detonator causing a projectile to explode within a predetermined distance of a target. **proximity of blood** kinship. [ME f. F *proximité* or L *proximitas* (as PROXIMAL)]

proximo /ˈprɒksɪˌməʊ/ *adj. Commerce* of next month (*the third proximo*). [L *proximo mense* in the next month]

proxy /ˈprɒksi/ *n.* (*pl.* **-ies**) (also *attrib.*) **1** the authorization given to a substitute or deputy (*a proxy vote*; *was married by proxy*). **2** a person authorized to act as a substitute etc. **3 a** a document giving the power to act as a proxy, esp. in voting. **b** a vote given by this. [ME f. obs. *procuracy* f. med.L *procuratia* (as PROCURATION)]

PRS *abbr.* **1** (in the UK) President of the Royal Society. **2** Performing Right Society.

prude /pruːd/ *n.* a person having or affecting an attitude of extreme propriety or modesty esp. in sexual matters. □ **prudish** *adj.* **prudishly** *adv.* **prudishness** *n.* **prudery** /-dəri/ *n.* (*pl.* **-ies**). [F, back form. f. *prudefemme* fem. of *prud'homme* good man and true f. *prou* worthy]

prudent /ˈpruːd(ə)nt/ *adj.* **1** (of a person or conduct) careful to avoid undesired consequences; circumspect. **2** discreet. □ **prudence** *n.* **prudently** *adv.* [ME f. OF *prudent* or L *prudens* = *providens* PROVIDENT]

prudential /pruːˈdenʃ(ə)l/ *adj. & n.* ● *adj.* of, involving, or marked by prudence (*prudential motives*). ● *n.* (in *pl.*) **1** prudential considerations or matters. **2** *US* minor administrative or financial matters. □ **prudentialism** *n.* **prudentialist** *n.* **prudentially** *adv.* [PRUDENT + -IAL, after *evidential* etc.]

Prudhoe Bay /ˈpruːdəʊ/ an inlet of the Arctic Ocean on the north coast of Alaska. It is a major centre of Alaskan oil production.

pruinose /ˈpruːɪˌnəʊz, -ˌnəʊs/ *adj.* esp. *Bot.* covered with white powdery granules; frosted in appearance. [L *pruinosus* f. *pruina* hoar-frost]

prune¹ /pruːn/ *n.* **1** a plum preserved by drying, with a black, wrinkled appearance. **2** *colloq.* a silly or disliked person. [ME f. OF ult. f. L *prunum* f. Gk *prou(m)non* plum]

prune² /pruːn/ *v.tr.* **1 a** (often foll. by *down*) trim (a tree etc.) by cutting away dead or overgrown branches etc. **b** (usu. foll. by *off*, *away*) lop (branches etc.) from a tree. **2** reduce (costs etc.) (*must try to prune expenses*). **3 a** (often foll. by *of*) clear (a book etc.) of superfluities. **b** remove (superfluities). □ **pruning-hook** a long-handled hooked cutting tool used for pruning. □ **pruner** *n.* [ME *prouyne* f. OF *pro(o)ignier* ult. f. L *rotundus* ROUND]

prunella¹ /pruːˈnelə/ *n.* a small labiate plant of the genus *Prunella*, bearing pink, purple, or white flower spikes; esp. *P. vulgaris*, formerly thought to cure quinsy. (Also called *self-heal*.) [mod.L, = quinsy: earlier *brunella* dimin. of med.L *brunus* brown]

prunella² /pruːˈnelə/ *n.* a strong silk or worsted fabric used formerly for barristers' gowns, the uppers of women's shoes, etc. [perh. f. F *prunelle*, of uncert. orig.]

prurient /ˈprʊərɪənt/ *adj.* **1** having an unhealthy obsession with sexual matters. **2** encouraging such an obsession. □ **prurience** *n.* **pruriency** *n.* **pruriently** *adv.* [L *prurire* itch, be wanton]

prurigo /prʊəˈraɪgəʊ/ *n.* a skin disease marked by severe itching. □ **pruriginous** /-ˈrɪdʒɪnəs/ *adj.* [L *prurigo -ginis* f. *prurire* to itch]

pruritus /prʊəˈraɪtəs/ *n.* severe itching of the skin. □ **pruritic** /-ˈrɪtɪk/ *adj.* [L, = itching (as PRURIGO)]

Prussia /ˈprʌʃə/ a former kingdom of Germany, which grew from a small country on the south-eastern shores of the Baltic to an extensive domain covering much of modern NE Germany and Poland. Its nucleus was a forested area east of the Vistula, originally inhabited by a Baltic people and taken in the 13th century by the Teutonic Knights. In the 16th century Prussia became a duchy of the Hohenzollerns, passing in 1618 to the electors of Brandenburg. The kingdom of Prussia, proclaimed in 1701, with its capital at Berlin, grew in the 18th century under Frederick the Great to become a major European power. After victory in the Franco-Prussian War of 1870–1, Prussia under Wilhelm I became the centre of the new German Empire created by Bismarck. With Germany's defeat in the First World War, the Prussian monarchy was abolished and Prussia's supremacy came to an end. (See also EAST PRUSSIA.)

Prussian /ˈprʌʃ(ə)n/ *adj. & n.* ● *adj.* of or relating to Prussia. ● *n.* a native of Prussia. □ **Old Prussian** the ancient Baltic language spoken in Prussia until the 17th century. **Prussian blue** a deep blue pigment, ferric ferrocyanide, used in painting and dyeing.

prussic /ˈprʌsɪk/ *adj.* of or obtained from Prussian blue. □ **prussic acid** hydrocyanic acid. □ **prussate** /-seɪt/ *n.* [F *prussique* f. *Prusse* Prussia]

Prut /pruːt/ (also **Pruth**) a river of SE Europe, which rises in the Carpathian Mountains in southern Ukraine and flows south-east for 850 km (530 miles), joining the Danube near Galaţi in Romania. For much of its course it forms the border between Romania and Moldova.

pry¹ /praɪ/ *v.intr.* (**pries**, **pried**) **1** (usu. foll. by *into*) inquire impertinently (into a person's private affairs etc.). **2** (usu. foll. by *into*, *about*, etc.) look or peer inquisitively. □ **prying** *adj.* **pryingly** *adv.* [ME *prie*, of unkn. orig.]

pry² /praɪ/ *v.tr.* (**pries**, **pried**) *N. Amer.* (often foll. by *out of*, *open*, etc.) = PRISE. [PRISE taken as *pries* 3rd sing. pres.]

PS *abbr.* **1** Police Sergeant. **2** postscript. **3** private secretary. **4** prompt side.

Ps. *abbr.* (*pl.* **Pss.**) (in the Bible) Psalm, Psalms.

psalm /sɑːm/ *n.* **1** (also **Psalm**) any of the sacred songs contained in the Book of Psalms, esp. when set for metrical chanting in a service. **2** a sacred song or hymn. □ **psalm-book** a book containing psalms, esp. with metrical settings for worship. □ **psalmic** *adj.* [OE (p)sealm f. LL *psalmus* f. Gk *psalmos* song sung to a harp f. *psallō* pluck]

psalmist /ˈsɑːmɪst/ *n.* **1** the author or composer of a psalm. **2** (**the Psalmist**) David or the author of (part of) the Book of Psalms. [LL *psalmista* (as PSALM)]

psalmody /ˈsɑːmədɪ, ˈsælm-/ *n.* **1** the practice or art of singing psalms, hymns, etc., esp. in public worship. **2 a** the arrangement of psalms for singing. **b** the psalms so arranged. □ **psalmodist** *n.* **psalmodize** *v.intr.* (also **-ise**). **psalmodic** /sɑːˈmɒdɪk, sælˈmɒd-/ *adj.* [ME f. LL *psalmodia* f. Gk *psalmōidia* singing to a harp (as PSALM, ōidē song)]

Psalms (also **Book of Psalms**) the book of the Bible containing psalms, used in both Jewish and Christian worship and traditionally ascribed to King David.

psalter /ˈsɔːltə(r), ˈsɒl-/ *n.* (also **Psalter**) **1 a** the Book of Psalms. **b** a version of this (*the English Psalter*; *Prayer-Book Psalter*). **2** a copy of the Psalms, esp. for liturgical use. [ME f. AF *sauter*, OF *sautier*, & OE (p)*saltere* f. LL *psalterium* f. Gk *psaltērion* stringed instrument (*psallō* pluck), in eccl.L Book of Psalms]

psalterium /sɔːlˈtɪərɪəm, sɒl-/ *n.* = OMASUM. [L (see PSALTER): named from the many folds of tissue inside it]

psaltery /ˈsɔːltərɪ, ˈsɒl-/ *n.* (*pl.* **-ies**) an ancient and medieval instrument resembling a dulcimer but played by plucking the strings with the fingers or a plectrum. [ME f. OF *sauterie* etc. f. L (as PSALTER)]

PSBR *abbr. Brit.* public sector borrowing requirement.

psephology /seˈfɒlədʒɪ, siːˈfɒl-/ *n.* the statistical study of elections, voting, etc. □ **psephologist** *n.* **psephological** /ˌsefəˈlɒdʒɪk(ə)l, ˌsiːf-/ *adj.* **psephologically** *adv.* [Gk *psēphos* pebble, vote + -LOGY]

pseud /sjuːd/ *adj. & n. colloq.* ● *adj.* intellectually or socially pretentious; not genuine. ● *n.* such a person; a poseur. [abbr. of PSEUDO]

pseud- var. of PSEUDO-.

pseudepigrapha /ˌsjuːdɪˈpɪgrəfə/ *n.pl.* **1** Jewish writings ascribed to various biblical patriarchs and prophets etc. but composed *c.*200 BC–AD 200. **2** spurious writings. □ **pseudepigraphal** *adj.* **pseudepigraphic** /-depɪˈgræfɪk/ *adj.* **pseudepigraphical** *adj.* [neut. pl. of Gk *pseudepigraphos* with false title (as PSEUDO-, EPIGRAPH)]

pseudo /ˈsjuːdəʊ/ *adj. & n.* ● *adj.* **1** sham; spurious. **2** insincere. ● *n.* (*pl.* **-os**) a pretentious or insincere person. [see PSEUDO-]

pseudo- /'sjuːdəʊ/ *comb. form* (also **pseud-** before a vowel) **1** supposed or purporting to be but not really so; false; not genuine (*pseudo-intellectual*; *pseudepigrapha*). **2** resembling or imitating (*pseudo-language*; *pseudo-science*). [Gk f. *pseudēs* false, *pseudos* falsehood]

pseudocarp /'sjuːdəʊˌkɑːp/ *n. Bot.* a fruit formed from parts other than the ovary, e.g. the strawberry or fig. [PSEUDO- + Gk *karpos* fruit]

pseudomorph /'sjuːdəˌmɔːf/ *n.* **1** a crystal etc. consisting of one mineral with the form proper to another. **2** a false form. □ **pseudomorphic** /ˌsjuːdəˈmɔːfɪk/ *adj.* **pseudomorphism** *n.* **pseudomorphous** *adj.* [PSEUDO- + Gk *morphē* form]

pseudonym /'sjuːdənɪm/ *n.* a fictitious name, esp. one assumed by an author. [F *pseudonyme* f. Gk *pseudōnymos* (as PSEUDO-, *-ōnumos* f. *onoma* name)]

pseudonymous /sjuːˈdɒnɪməs/ *adj.* writing or written under a false name. □ **pseudonymously** *adv.* **pseudonymity** /ˌsjuːdəˈnɪmɪtɪ/ *n.*

pseudopod /'sjuːdəʊˌpɒd/ *n. Biol.* = PSEUDOPODIUM. [mod.L (as PSEUDOPODIUM)]

pseudopodium /ˌsjuːdəʊˈpəʊdɪəm/ *n.* (*pl.* **pseudopodia** /-dɪə/) *Biol.* (in amoeboid cells) a temporary protrusion of the cell surface for movement, feeding, etc. [mod.L (as PSEUDO-, PODIUM)]

pseudo-science /'sjuːdəʊˌsaɪəns/ *n.* a pretended or spurious science; a collection of beliefs, mistakenly regarded as based on scientific method. □ **pseudo-scientific** /ˌsjuːdəʊˌsaɪənˈtɪfɪk/ *adj.*

pshaw /pʃɔː, ʃɔː/ *int. archaic* an expression of contempt or impatience. [imit.]

psi /psaɪ/ *n.* **1** the twenty-third letter of the Greek alphabet (Ψ, ψ). **2** supposed parapsychological faculties, phenomena, etc. regarded collectively. [Gk]

p.s.i. *abbr.* pounds per square inch.

psilocybin /ˌsɪləˈsaɪbɪn/ *n.* a hallucinogenic alkaloid found in toadstools of the genus *Psilocybe*. [*Psilocybe* f. Gk *psilos* bald + *kubē* head]

psilosis /saɪˈləʊsɪs/ *n.* = SPRUE². [Gk *psilōsis* f. *psilos* bare]

psittacine /'sɪtəˌsaɪn/ *adj.* of or relating to parrots; parrot-like. [L *psittacinus* f. *psittacus* f. Gk *psittakos* parrot]

psittacosis /ˌsɪtəˈkəʊsɪs/ *n. Med.* a contagious disease of birds caused by chlamydiae, transmissible (esp. from parrots) to human beings as a form of pneumonia. [f. L *psittacus* (as PSITTACINE) + -OSIS]

psoas /'səʊəs/ *n.* either of two muscles used in flexing the hip joint. [Gk, accus. pl. of *psoa*, taken as sing.]

psoriasis /səˈraɪəsɪs/ *n.* a skin disease marked by red scaly patches. □ **psoriatic** /ˌsɔːrɪˈætɪk/ *adj.* [mod.L f. Gk *psōriasis* f. *psōriaō* have an itch f. *psōra* itch]

psst /pst/ *int.* (also **pst**) a whispered exclamation seeking to attract a person's attention surreptitiously. [imit.]

PST *abbr.* Pacific Standard Time.

PSV *abbr. Brit.* public service vehicle.

psych /saɪk/ *v. colloq.* **1** *tr.* (usu. foll. by *up*; often *refl.*) prepare (oneself or another person) mentally for an ordeal etc. **2** *tr.* **a** (usu. foll. by *out*) analyse (a person's motivation etc.) for one's own advantage (*can't psych him out*). **b** subject to psychoanalysis. **3** *tr.* (often foll. by *out*) influence a person psychologically, esp. negatively; intimidate, frighten. **4** *intr. Bridge* make a psychic bid. □ **psych out** break down mentally; become confused or deranged. [abbr.]

Psyche /'saɪkɪ/ *Gk Mythol.* a Hellenistic personification of the soul as female, or sometimes as a butterfly. The allegory of Psyche's love for Cupid is told in *The Golden Ass* by Apuleius. [as PSYCHE]

psyche /'saɪkɪ/ *n.* **1** the soul; the spirit. **2** the mind. [L f. Gk *psukhē* breath, life, soul]

psychedelia /ˌsaɪkɪˈdiːlɪə/ *n.pl.* **1** psychedelic articles, esp. posters, paintings, etc. **2** psychedelic drugs.

psychedelic /ˌsaɪkɪˈdelɪk/ *adj. & n.* ● *adj.* **1 a** expanding the mind's awareness etc., esp. through the use of hallucinogenic drugs. **b** (of an experience) hallucinatory; bizarre. **c** (of a drug) producing hallucinations. **2** *colloq.* **a** producing an effect resembling that of a psychedelic drug; having vivid colours or designs etc. **b** (of colours, patterns, etc.) bright, bold and often abstract. ● *n.* a hallucinogenic drug. □ **psychedelically** *adv.* [irreg. f. Gk (as PSYCHE, *dēlos* clear, manifest)]

psychiatry /saɪˈkaɪətrɪ/ *n.* the study and treatment of mental disease. □ **psychiatrist** *n.* **psychiatric** /ˌsaɪkɪˈætrɪk/ *adj.* **psychiatrical** *adj.* **psychiatrically** *adv.* [as PSYCHE + *iatreia* healing f. *iatros* healer]

psychic /'saɪkɪk/ *adj. & n.* ● *adj.* **1 a** (of a person) considered to have occult powers, such as telepathy, clairvoyance, etc. **b** (of a faculty, phenomenon, etc.) inexplicable by natural laws. **2** of the soul or mind. **3** *Bridge* (of a bid) that deliberately misrepresents the bidder's hand. ● *n.* **1** a person considered to have psychic powers; a medium. **2** *Bridge* a psychic bid. **3** (in *pl.*) the study of psychic phenomena. □ **psychicism** /-ɪˌsɪz(ə)m/ *n.* **psychicist** /-ɪˌsɪst/ *n.* [Gk *psukhikos* (as PSYCHE)]

psychical /'saɪkɪk(ə)l/ *adj.* **1** concerning psychic phenomena or faculties (*psychical research*). **2** of the soul or mind. □ **psychically** *adv.*

psycho /'saɪkəʊ/ *n. & adj. colloq.* ● *n.* (*pl.* **-os**) a psychopath. ● *adj.* psychopathic. [abbr.]

psycho- /'saɪkəʊ/ *comb. form* relating to the mind or psychology. [Gk *psukho-* (as PSYCHE)]

psychoactive /ˌsaɪkəʊˈæktɪv/ *adj.* affecting the mind.

psychoanalysis /ˌsaɪkəʊəˈnælɪsɪs/ *n.* **1** therapy for disorders of personality or behaviour in which the patient is helped to investigate his or her unconscious feelings and to recognise repressed fears and conflicts. **2** the practice of such therapy or the theory of personality and psychical life, originated by Sigmund Freud, on which it is based. □ **psychoanalyse** /-kəʊˈænəˌlaɪz/ *v.tr.* **psychoanalyst** /-ˈænəlɪst/ *n.* **psychoanalytic** /-ˌænəˈlɪtɪk/ *adj.* **psychoanalytical** *adj.* **psychoanalytically** *adv.*

psychobabble /'saɪkəʊˌbæb(ə)l/ *n. US colloq. derog.* jargon used in popular psychology.

psychodrama /'saɪkəʊˌdrɑːmə/ *n.* **1** a form of psychotherapy in which patients act out events from their past. **2** a play or film etc. in which psychological elements are the main interest.

psychodynamics /ˌsaɪkəʊdaɪˈnæmɪks/ *n.pl.* (treated as *sing.*) the study of the activity of and the interrelation between the various parts of an individual's personality or psyche. □ **psychodynamic** *adj.* **psychodynamically** *adv.*

psychogenesis /ˌsaɪkəʊˈdʒenɪsɪs/ *n.* the study of the origin and development of the mind.

psychokinesis /ˌsaɪkəʊkɪˈniːsɪs, -kaɪˈniːsɪs/ *n.* the movement of objects supposedly by mental effort without the action of physical forces.

psycholinguistics /ˌsaɪkəʊlɪŋˈgwɪstɪks/ *n.pl.* (treated as *sing.*) the study of the psychological aspects of language and language-learning. □ **psycholinguistic** *adj.* **psycholinguist** /-ˈlɪŋgwɪst/ *n.*

psychological /ˌsaɪkəˈlɒdʒɪk(ə)l/ *adj.* **1** of, relating to, or arising in the mind. **2** of or relating to psychology. **3** *colloq.* (of an ailment etc.) having a basis in the mind; imaginary (*her cold is psychological*). □ **psychological block** a mental inability or inhibition caused by emotional factors. **psychological moment** the most appropriate time for achieving a particular effect or purpose. **psychological warfare** a campaign directed at reducing an opponent's morale. □ **psychologically** *adv.*

psychology /saɪˈkɒlədʒɪ/ *n.* (*pl.* **-ies**) **1** the scientific study of the human mind and its functions, esp. those affecting behaviour in a given context. **2** a treatise on or theory of this. **3 a** the mental characteristics or attitude of a person or group. **b** the mental factors governing a situation or activity (*the psychology of crime*). □ **psychologist** *n.* **psychologize** *v.tr. & intr.* (also **-ise**). [mod.L *psychologia* (as PSYCHO-, -LOGY)]

psychometrics /ˌsaɪkəʊˈmetrɪks/ *n.pl.* (treated as *sing.*) the science of measuring mental capacities and processes.

psychometry /saɪˈkɒmɪtrɪ/ *n.* **1** the supposed divination of facts about events, people, etc., from inanimate objects associated with them. **2** the measurement of mental abilities. □ **psychometrist** *n.* **psychometric** /ˌsaɪkəˈmetrɪk/ *adj.* **psychometrically** *adv.*

psychomotor /'saɪkəʊˌməʊtə(r)/ *adj.* of or relating to the origination of movement in conscious mental activity.

psychoneurosis /ˌsaɪkəʊnjʊəˈrəʊsɪs/ *n.* neurosis, esp. with the indirect expression of emotions. □ **psychoneurotic** /-ˈrɒtɪk/ *adj.*

psychopath /'saɪkəˌpæθ/ *n.* **1** a person suffering from chronic mental disorder esp. with abnormal or violent social behaviour. **2** a mentally or emotionally unstable person. □ **psychopathic** /ˌsaɪkəˈpæθɪk/ *adj.* **psychopathically** *adv.*

psychopathology /ˌsaɪkəʊpəˈθɒlədʒɪ/ *n.* **1** the scientific study of mental disorders. **2** a mentally or behaviourally disordered state. □ **psychopathological** /-ˌpæθəˈlɒdʒɪk(ə)l/ *adj.*

psychopathy /saɪˈkɒpəθɪ/ *n.* psychopathic or psychologically abnormal behaviour.

psychophysics /ˌsaɪkəʊˈfɪzɪks/ n. the science of the relation between the mind and the body. □ **psychophysical** adj.

psychophysiology /ˌsaɪkəʊˌfɪzɪˈɒlədʒɪ/ n. the branch of physiology dealing with mental phenomena. □ **psychophysiological** /-zɪəˈlɒdʒɪk(ə)l/ adj.

psychosexual /ˌsaɪkəʊˈseksjʊəl, -ˈsekʃʊəl/ adj. of or involving the psychological aspects of the sexual impulse. □ **psychosexually** adv.

psychosis /saɪˈkəʊsɪs/ n. (pl. **psychoses** /-siːz/) a severe mental derangement, esp. when resulting in delusions and loss of contact with external reality. [Gk psukhōsis f. psukhoō give life to (as PSYCHE)]

psychosocial /ˌsaɪkəʊˈsəʊʃ(ə)l/ adj. of or involving the influence of social factors or human interactive behaviour. □ **psychosocially** adv.

psychosomatic /ˌsaɪkəʊsəˈmætɪk/ adj. **1** (of an illness etc.) caused or aggravated by mental conflict, stress, etc. **2** of the mind and body together. □ **psychosomatically** adv.

psychosurgery /ˌsaɪkəʊˈsɜːdʒərɪ/ n. brain surgery as a means of treating mental disorder. □ **psychosurgical** adj.

psychotherapy /ˌsaɪkəʊˈθerəpɪ/ n. the treatment of mental disorder by psychological means. □ **psychotherapist** n. **psychotherapeutic** /-ˌθerəˈpjuːtɪk/ adj.

psychotic /saɪˈkɒtɪk/ adj. & n. ● adj. of or characterized by a psychosis. ● n. a person suffering from a psychosis. □ **psychotically** adv.

psychotropic /ˌsaɪkəʊˈtrəʊpɪk, -ˈtrɒpɪk/ n. (of a drug) acting on the mind.

psychrometer /saɪˈkrɒmɪtə(r)/ n. a thermometer consisting of a dry bulb and a wet bulb for measuring atmospheric humidity. [Gk psukhros cold + -METER]

PT abbr. physical training.

Pt symb. Chem. the element platinum.

pt. abbr. **1** part. **2** pint. **3** point. **4** port.

PTA abbr. **1** parent-teacher association. **2** Passenger Transport Authority.

Ptah /tɑː/ Egyptian Mythol. an ancient deity of Memphis, creator of the universe, god of artisans, and husband of Sekhmet. He became one of the chief deities of Egypt, and was identified by the Greeks with Hephaestus.

ptarmigan /ˈtɑːmɪɡən/ n. a grouselike northern game bird of the genus Lagopus, with feathered legs and feet and plumage that usually changes to white in winter; esp. L. mutus, found on mountains and in the Arctic (also called (N. Amer.) rock ptarmigan). [Gael. tàrmachan: p- after Gk words in pt-]

PT boat n. US a motor torpedo-boat. [Patrol Torpedo]

Pte. abbr. Private (soldier).

pteranodon /teˈrænəˌdɒn/ n. a large toothless pterosaur of the genus Pteranodon, with a bony crest on the head and lacking a tail. [mod.L f. PTERO- + -AN + Gk odous odontos tooth]

pteridology /ˌterɪˈdɒlədʒɪ/ n. the study of ferns. □ **pteridologist** n. **pteridological** /-dəˈlɒdʒɪk(ə)l/ adj. [Gk pteris -idos fern + -LOGY]

pteridophyte /ˈterɪdəˌfaɪt/ n. Bot. a flowerless vascular plant of the division Pteridophyta, which includes ferns, clubmosses, and horsetails. [f. Gk pteris -idos fern + phuton plant]

ptero- /ˈterəʊ/ comb. form wing. [Gk pteron wing]

pterodactyl /ˌterəˈdæktɪl/ n. a pterosaur, esp. one of the genus Pterodactylus, with teeth and a long slender head, neck, and tail. [f. PTERO- + Gk daktulos finger]

pteropod /ˈterəˌpɒd/ n. Zool. a swimming marine gastropod of the class Pteropoda, with the middle part of its foot expanded into a pair of winglike lobes. [f. PTERO- + Gk pous podos foot]

pterosaur /ˈterəˌsɔː(r)/ n. an extinct Mesozoic flying reptile of the order Pterosauria, with long membranous wings supported by an elongated fourth digit, e.g. pterodactyls and pteranodons. [f. PTERO- + Gk saura lizard]

pteroylglutamic acid /ˌterəʊˌaɪlɡluˈtæmɪk/ n. Biochem. = FOLIC ACID. □ **pteroylglutamate** /-ˈɡluːtəˌmeɪt/ n. [pteroic acid (a precursor: ult. f. Gk pteron wing, with ref. to insect pigments) + -YL + GLUTAMIC ACID]

pterygoid process /ˈterɪˌɡɔɪd/ n. Anat. each of a pair of processes from the sphenoid bone in the skull. [Gk pterux -ugos wing]

PTFE abbr. polytetrafluoroethylene.

ptisan /ˈtɪz(ə)n, tɪˈzæn/ n. a nourishing drink, esp. barley water. [ME & OF tizanne etc. f. L ptisana f. Gk ptisanē peeled barley]

PTO abbr. please turn over.

Ptolemaic /ˌtɒlɪˈmeɪɪk/ adj. **1** of or relating to Ptolemy or his theories. (See PTOLEMY².) **2** of or relating to the Ptolemies, Macedonian rulers of Egypt from the death of Alexander the Great (323 BC) to the death of Cleopatra (30 BC). (See PTOLEMY¹.) □ **Ptolemaic system** hist. the theory that the earth is the stationary centre of the universe, with the planets moving in epicyclic orbits within surrounding concentric spheres (cf. Copernican system). [L Ptolemaeus f. Gk Ptolemaios]

Ptolemy¹ /ˈtɒlɪmɪ/ the name of all the Macedonian kings of Egypt, a dynasty founded by Ptolemy, the close friend and general of Alexander the Great, who took charge of Egypt after the latter's death and declared himself king (Ptolemy I) in 304 BC. The dynasty ended with the death of Cleopatra in 30 BC. Under the Ptolemies their capital, Alexandria, became a leading commercial and cultural centre of the Greek world. (See also HELLENISTIC.)

Ptolemy² /ˈtɒlɪmɪ/ (2nd century) Greek astronomer and geographer. His major work, known by its Arabic title (Almagest), was a textbook of astronomy based on the geocentric system of Hipparchus. Ptolemy's teachings had enormous influence on medieval thought, the geocentric view of the cosmos being adopted as Christian doctrine until the late Renaissance. The Almagest included detailed tables of lunar and solar motion with eclipse predictions, and a catalogue giving the positions and magnitudes of 1,022 stars. Ptolemy's Geography, giving lists of places with their longitudes and latitudes, was also a standard work for centuries, despite its inaccuracies.

ptomaine /ˈtəʊmeɪn/ n. Biochem. any of a class of amine compounds, some toxic, formed in putrefying animal and vegetable matter. □ **ptomaine poisoning** archaic food poisoning. [F ptomaïne f. It. ptomaina irreg. f. Gk ptōma corpse]

ptosis /ˈtəʊsɪs/ n. Med. a drooping of the upper eyelid due to paralysis etc. □ **ptotic** /ˈtɒtɪk/ adj. [Gk ptōsis f. piptō fall]

PTSD abbr. Med. post-traumatic stress disorder.

Pty. abbr. Austral., NZ, & S. Afr. proprietary.

ptyalin /ˈtaɪəlɪn/ n. Biochem. an amylase found in the saliva of humans and some other animals. [Gk ptualon spittle]

Pu symb. Chem. the element plutonium.

pub /pʌb/ n. colloq. **1** Brit. a public house. **2** Austral. a hotel. □ **pub-crawl** Brit. colloq. a drinking tour of several pubs. [abbr.]

puberty /ˈpjuːbətɪ/ n. the period during which adolescents reach sexual maturity and become capable of reproduction. □ **age of puberty** the age at which puberty begins, in law usu. 14 in boys and 12 in girls. □ **pubertal** adj. [ME f. F puberté or L pubertas f. puber adult]

pubes¹ /ˈpjuːbiːz, pjuːbz/ n. (pl. same) the lower part of the abdomen at the front of the pelvis, covered with hair from puberty. [L]

pubes² pl. of PUBIS.

pubescence /pjuːˈbes(ə)ns/ n. **1** the time when puberty begins. **2** Bot. soft down on the leaves and stems of plants. **3** Zool. soft down on various parts of animals, esp. insects. □ **pubescent** adj. [F pubescence or med.L pubescentia f. L pubescere reach puberty]

pubic /ˈpjuːbɪk/ adj. of or relating to the pubes or pubis.

pubis /ˈpjuːbɪs/ n. (pl. **pubes** /-biːz/) either of a pair of bones forming the two sides of the pelvis. [L os pubis bone of the pubes]

public /ˈpʌblɪk/ adj. & n. ● adj. **1** of or concerning the people as a whole (a public holiday; the public interest). **2** open to or shared by all the people (public baths; public library; public meeting). **3** done or existing openly (made his views public; a public protest). **4** (of a service, funds, etc.) provided by or concerning local or central government (public money; public records; public expenditure). **5 a** of or involved in the affairs of the community, esp. in government or entertainment (a distinguished public career; public figures). **b** of or relating to a person in his or her capacity as a public figure (had a likeable public face). **6** Brit. of, for, or acting for, a university (public examination). ● n. **1** (as sing. or pl.) the community in general, or members of the community. **2** a section of the community having a particular interest or in some special connection (the reading public; my public demands my loyalty). **3** Brit. colloq. **a** = public bar. **b** = public house. □ **go public 1** become a public company. **2** make one's intentions plain; come out into the open. **in public** openly, publicly. **in the public domain** belonging to the public as a whole, esp. not subject to copyright. **in the public eye** famous or notorious. **make public** publicize, make known; publish. **public act** an act of legislation affecting the public as a whole. **public-address system** loudspeakers, microphones, amplifiers, etc., used in addressing large audiences. **public bar** Brit. the least expensive bar in a public house. **public bill** a bill of legislation affecting the public as a whole. **public company** Brit. a company that sells shares to all buyers on the open market. **public enemy** a notorious wanted criminal. **public figure** a famous person. **public health** the provision of adequate sanitation, drainage, etc. by government. **public house 1** *Brit.* an inn providing alcoholic drinks for consumption on the premises. **2** an inn. **public law 1** the law of relations between individuals and the state. **2** = *public act.* **public lending right** the right of authors to payment when their books etc. are lent by public libraries. **public libel** see LIBEL. **public nuisance 1** an illegal act against the public generally. **2** *colloq.* an obnoxious person. **public opinion** views, esp. moral, prevalent among the general public. **public ownership** state ownership of the means of production, distribution, and exchange. **public prosecutor** a law officer conducting criminal proceedings on behalf of the state or in the public interest. **public relations** the professional maintenance of a favourable public image, esp. by a company, famous person, etc. **public relations officer** a person employed by a company etc. to promote a favourable public image. **public sector** that part of an economy, industry, etc., that is controlled by the state. **public servant** a state official. **public spirit** a willingness to engage in community action. **public-spirited** having a public spirit. **public-spiritedly** in a public-spirited manner. **public-spiritedness** the quality of being public-spirited. **public transport** buses, trains, etc., charging set fares and running on fixed routes, esp. when state-owned. **public utility** an organization supplying water, gas, etc. to the community. **public works** building operations etc. done by or for the state on behalf of the community. **public wrong** an offence against society as a whole. □ **publicly** *adv.* [ME f. OF *public* or L *publicus* f. *pubes* adult]

publican /ˈpʌblɪkən/ *n.* **1 a** *Brit.* the keeper of a public house. **b** *Austral.* the keeper of a hotel. **2** *Rom. Hist. & Bibl.* a tax-collector or tax-farmer. [ME f. OF *publican* f. L *publicanus* f. *publicum* public revenue (as PUBLIC)]

publication /ˌpʌblɪˈkeɪʃ(ə)n/ *n.* **1 a** the preparation and issuing of a book, newspaper, engraving, music, etc. to the public. **b** a book etc. so issued. **2** the act or an instance of making something publicly known. [ME f. OF f. L *publicatio -onis* (as PUBLISH)]

publicist /ˈpʌblɪsɪst/ *n.* **1** a publicity agent or public relations officer. **2** a journalist, esp. concerned with current affairs. **3** *archaic* a writer or other person skilled in international law. □ **publicistic** /ˌpʌblɪˈsɪstɪk/ *adj.* [F *publiciste* f. L (*jus*) *publicum* public law]

publicity /pʌbˈlɪsɪtɪ/ *n.* **1 a** the professional exploitation of a product, company, person, etc., by advertising or popularizing. **b** material or information used for this. **2** public exposure; notoriety. □ **publicity agent** a person employed to produce or heighten public exposure. [F *publicité* (as PUBLIC)]

publicize /ˈpʌblɪˌsaɪz/ *v.tr.* (also **-ise**) advertise; make publicly known.

Public Record Office *n.* an institution keeping official archives, esp. birth, marriage, and death certificates, for public inspection.

public school *n.* **1** *Brit.* a private fee-paying secondary school, esp. for boarders. (*See note below.*) **2** *N. Amer., Austral., & Sc.* etc. a non-fee-paying school.

▪ In England, public schools had their origin in some of the grammar schools of the Tudor period or earlier, carried on under some kind of public management or control, which gained sufficient reputation to attract boys living far away as residential pupils; they were contrasted with 'private schools', which were run for the profit of their proprietors, and with education at home under a tutor. In the 19th century there was a surge of expansion, as well-to-do members of the middle class sought for their sons the education that they had not had themselves, and of reform: it was during this period that the traditional public school ethos, of classical education, public service, and character-building sports, arose. Public schools are now rather more similar in nature to state schools, although often having a wider curriculum and smaller classes. Among the foremost are Winchester (founded in 1382), Eton (1440), Shrewsbury (1552), Westminster (1560), Rugby (1567), Harrow (1571), and Charterhouse (1611); Benenden (1923) and Roedean (1885) are exclusively for girls.

publish /ˈpʌblɪʃ/ *v.tr.* **1** (also *absol.*) (of an author, publisher, etc.) prepare and issue (a book, newspaper, engraving, etc.) for public sale. **2** make generally known. **3** announce (an edict etc.) formally; read (marriage banns). **4** *Law* communicate (a libel etc.) to a third party. **5** (esp. as **published** *adj.*) publish the works of (a particular writer or composer) (*she was already a published poet at the age of 20*). □ **publishable** *adj.* [ME *puplise* etc. f. OF *puplier, publier* f. L *publicare* (as PUBLIC)]

publisher /ˈpʌblɪʃə(r)/ *n.* **1** a person or esp. a company that produces and distributes copies of a book, newspaper, etc. for sale. **2** *N. Amer.* a newspaper proprietor. **3** a person or thing that publishes.

publishing /ˈpʌblɪʃɪŋ/ *n.* the production and issuing of books, magazines, newspapers, and other material for sale. Before the invention of printing, texts were often disseminated by being read aloud to an audience. In the early days of printing the printer was also publisher and bookseller, but during the 15th and 16th centuries these functions began to separate and eventually develop into separate trades. Technological developments in the printing industry in the 19th century dramatically increased output and reduced prices. In the 20th century the introduction of paperbacks, pioneered in the Penguin series in 1935, tapped a wider market by offering books at an unprecedentedly low price, and after the Second World War social and economic changes, and the spread of higher and further education, resulted in an increased demand for books of all kinds. The advent of computers has changed many aspects of publishing in a technological revolution as important as the invention of printing: more titles are now published than ever before, with consequently fiercer competition; specialist books and magazines are produced to cater for every interest, sometimes by means of desktop publishing; and reference books and other works are increasingly available in electronic form, on disk rather than paper.

Puccini /pʊˈtʃiːnɪ/, Giacomo (1858–1924), Italian composer. He established his reputation with his third opera, *Manon Lescaut* (1893). Puccini's sense of the dramatic, gift for melody, and skilful use of the orchestra have ensured that his works remain among the most popular in the repertoire. Several operas followed, including *La Bohème* (1896), *Tosca* (1900), *Madama Butterfly* (1904), and *Turandot*, which was completed by a pupil after his death and produced in 1926.

puce /pjuːs/ *adj. & n.* dark red or purple-brown. [F, = flea(-colour) f. L *pulex -icis*]

puck[1] /pʌk/ *n.* **1 a** a sprite or goblin. **b** (**Puck**) = ROBIN GOODFELLOW. **2** a mischievous child. □ **puckish** *adj.* **puckishly** *adv.* **puckishness** *n.* **pucklike** *adj.* [OE *pūca*: cf. Welsh *pwca*, Ir. *púca*]

puck[2] /pʌk/ *n.* a rubber disc used like a ball in ice hockey. [19th c.: orig. unkn.]

pucka var. of PUKKA.

pucker /ˈpʌkə(r)/ *v. & n.* ● *v.tr. & intr.* (often foll. by *up*) gather or cause to gather into wrinkles, folds, or bulges (*puckered her eyebrows; this seam is puckered up*). ● *n.* such a wrinkle, bulge, fold, etc. □ **puckery** *adj.* [prob. frequent., formed as POKE[2], POCKET (cf. PURSE)]

pud /pʊd/ *n. colloq.* = PUDDING. [abbr.]

pudding /ˈpʊdɪŋ/ *n.* **1 a** a sweet cooked dish often containing fruit (*plum pudding; rice pudding*). **b** a savoury dish containing flour, suet, etc. (*Yorkshire pudding; steak and kidney pudding*). **c** *Brit.* the sweet course of a meal. **d** the intestines of a pig etc. stuffed with oatmeal, spices, blood, etc. (*black pudding*). **2** *colloq.* a fat, dumpy, or stupid person. **3** (*Naut.* **puddening** /ˈpʊd(ə)nɪŋ/) a pad or tow binding to prevent chafing etc. □ **in the pudding club** *sl.* pregnant. **pudding-cloth** a cloth used for tying up some puddings for boiling. **pudding face** *colloq.* a large fat face. **pudding-head** *colloq.* a stupid person. **pudding-stone** a conglomerate rock consisting of rounded pebbles in a siliceous matrix. □ **puddingy** /-dɪŋɪ/ *adj.* [ME *poding* f. OF *boudin* black pudding ult. f. L *botellus* sausage: see BOWEL]

puddle /ˈpʌd(ə)l/ *n. & v.* ● *n.* **1** a small pool, esp. of rainwater on a road etc. **2** clay and sand mixed with water and used as a watertight covering for embankments etc. **3** a circular patch of disturbed water made by the blade of an oar at each stroke. ● *v.* **1** *tr.* **a** knead (clay and sand) into puddle. **b** line (a canal etc.) with puddle. **2** *intr.* make puddle from clay etc. **3** *tr.* (usu. as **puddling** *n.* or **puddled** *adj.*) stir (molten iron) to produce wrought iron by expelling carbon. (See also CORT.) **4** *intr.* **a** dabble or wallow in mud or shallow water. **b** busy oneself in an untidy way. **5** *tr.* make (water etc.) muddy. **6** *tr.* work (mixed water and clay) to separate gold or opal. □ **puddler** *n.* **puddly** *adj.* [ME *podel, puddel,* dimin. of OE *pudd* ditch]

pudency /ˈpjuːd(ə)nsɪ/ *n. literary* modesty; shame. [LL *pudentia* (as PUDENDUM)]

pudendum /pjuːˈdendəm/ *n.* (*pl.* **pudenda** /-də/) (usu. in *pl.*) the genitals, esp. of a woman. □ **pudendal** *adj.* **pudic** /ˈpjuːdɪk/ *adj.* [L *pudenda* (*membra* parts), neut. pl. of gerundive of *pudere* be ashamed]

pudgy /ˈpʌdʒɪ/ *adj.* (**pudgier, pudgiest**) *colloq.* (esp. of a person) plump, thickset. □ **pudge** *n.* **pudgily** *adv.* **pudginess** *n.* [cf. PODGY]

Puebla /ˈpweblə/ **1** a state of south central Mexico. **2** (in full **Puebla de Zaragoza** /deɪ ˌsærəˈɡɒsə/) its capital city; pop. (1990) 1,054,920.

Founded by the Spanish in 1532, it lies at the edge of the central Mexican plateau at an altitude of 2,150 m (7,055 ft).

Pueblo /'pwebləʊ/ *n. & adj.* ● *n.* (*pl.* **-os**) a member of any of various American Indian peoples, including the Hopi, occupying pueblo settlements chiefly in New Mexico and Arizona. Their prehistoric period is known as the Anasazi culture. ● *adj.* of or relating to these peoples. [PUEBLO]

pueblo /'pwebləʊ/ *n.* (*pl.* **-os**) a town or village in Latin America or the south-western US, esp. an American Indian settlement. [Sp., = people, f. L *populus*]

puerile /'pjʊəraɪl/ *adj.* **1** trivial, childish, immature. **2** of or like a child. □ **puerilely** *adv.* **puerility** /pjʊə'rɪlɪti/ *n.* (*pl.* **-ies**). [F *puéril* or L *puerilis* f. *puer* boy]

puerperal /pju:'ɜ:pərəl/ *adj.* of or caused by childbirth. □ **puerperal fever** fever following childbirth and caused by uterine infection. [L *puerperus* f. *puer* child + *-parus* bearing]

Puerto Cortés /ˌpwɜ:təʊ kɔː'tez/ a port in NW Honduras, on the Caribbean coast at the mouth of the Ulua river; pop. (1986) 40,000.

Puerto Limón see LIMÓN.

Puerto Plata /ˌpwɜ:təʊ 'plɑːtə/ a resort town in the Dominican Republic, on the north coast; pop. (1986) 96,500.

Puerto Rico /ˌpwɜːtəʊ 'ri:kəʊ/ an island of the Greater Antilles in the West Indies; pop. (est. 1990) 3,522,040; official languages, Spanish and English; capital, San Juan. Visited in 1493 by Christopher Columbus, the island was one of the earliest Spanish settlements in the New World. It was ceded to the US in 1898 after the Spanish-American War, and in 1952 it became a commonwealth in voluntary association with the US, with full powers of local government. □ **Puerto Rican** *adj. & n.*

Puerto Rico Trench an ocean trench extending in an east–west direction to the north of Puerto Rico and the Leeward Islands. It reaches a depth of 9,220 m (28,397 ft).

puff /pʌf/ *n. & v.* ● *n.* **1 a** a short quick light blast of breath or wind. **b** the sound of this; a similar sound. **c** a small quantity of vapour, smoke, etc., emitted in one blast; an inhalation or exhalation from a cigarette, pipe, etc. (*went up in a puff of smoke; took a puff from his cigarette*). **2** a cake etc. containing jam, cream, etc., and made of light esp. puff pastry. **3** a gathered mass of material in a dress etc. (*puff sleeve*). **4** a rolled protuberant mass of hair. **5 a** an extravagantly enthusiastic review of a book etc., esp. in a newspaper. **b** an advertisement for goods etc., esp. in a newspaper. **6** = powder-puff. **7** *N. Amer.* an eiderdown. **8** *colloq.* one's life (*in all my puff*). ● *v.* **1** *intr.* emit a puff of air or breath; blow with short blasts. **2** *intr.* (usu. foll. by *away, out*, etc.) (of a person smoking, a steam engine, etc.) emit or move with puffs (*puffing away at his cigar; a train puffed out of the station*). **3** *tr.* (usu. in *passive*; often foll. by *out*) put out of breath (*arrived somewhat puffed; completely puffed him out*). **4** *intr.* breathe hard; pant. **5** *tr.* utter pantingly ('*No more,*' he puffed). **6** *intr. & tr.* (usu. foll. by *up, out*) become or cause to become inflated; swell (*his eye was inflamed and puffed up; puffed up the balloon*). **7** *tr.* (usu. foll. by *out, up, away*) blow or emit (dust, smoke, a light object, etc.) with a puff. **8** *tr.* smoke (a pipe etc.) in puffs. **9** *tr.* (usu. as **puffed up** *adj.*) elate; make proud or boastful. **10** *tr.* advertise or promote (goods, a book, etc.) with exaggerated or false praise. □ **puff-adder** a large venomous African viper, *Bitis arietans*, which inflates the upper part of its body and hisses when alarmed. **puff and blow** = sense 4 of *v.* **puff-ball 1** a fungus with a ball-shaped spore case. **2** a short full skirt gathered around the hemline to produce a soft puffy shape. **3** a powder-puff or an object resembling one. **puff pastry** light flaky pastry. **puff-puff** *Brit.* a childish word for a steam-engine or train. **puff up** = sense 9 of *v.* [ME *puf*, *puffe*, perh. f. OE, imit. of the sound of breath]

puffer /'pʌfə(r)/ *n.* **1** a person or thing that puffs. **2** = puff-puff. □ **puffer fish** = globe-fish. □ **puffery** *n.*

puffin /'pʌfɪn/ *n.* a small hole-nesting auk of the genus *Fratercula* or *Lunda*, of the North Atlantic and North Pacific, having a large head with a deep brightly coloured bill; esp. the *Atlantic puffin* (*F. arctica*). [ME (orig. denoting a shearwater), app. f. PUFF + -ING³, with ref. to that bird's fat nestlings]

puffy /'pʌfi/ *adj.* (**puffier, puffiest**) **1** swollen, esp. of the face etc. **2** fat. **3** gusty. **4** short-winded; puffed out. □ **puffily** *adv.* **puffiness** *n.*

pug¹ /pʌg/ *n.* **1** (in full **pug-dog**) a dwarf breed of dog like a bulldog with a broad flat nose and deeply wrinkled face. **2** a fox. **3** a small geometrid moth of *Eupithecia* or a related genus. □ **pug-nose** a short squat or snub nose. **pug-nosed** having such a nose. □ **puggish** *adj.* **puggy** *adj.* [16th c.: perh. f. LG or Du.]

pug² /pʌg/ *n. & v.* ● *n.* loam or clay mixed and prepared for making bricks, pottery, etc. ● *v.tr.* (**pugged, pugging**) **1** prepare (clay) thus. **2** pack (esp. the space under the floor to deaden sound) with pug, sawdust, etc. □ **pug-mill** a mill for preparing pug. □ **pugging** *n.* [19th c.: orig. unkn.]

pug³ /pʌg/ *n. sl.* a boxer. [abbr. of PUGILIST]

pug⁴ /pʌg/ *n. & v.* ● *n.* the footprint of an animal. ● *v.tr.* (**pugged, pugging**) track by pugs. [Hindi *pag* footprint]

Puget Sound /'pju:dʒɪt/ an inlet of the Pacific on the coast of Washington State in the US. It was explored by George Vancouver in 1792 and named after his aide Peter Puget. Linked to the ocean by the Strait of Juan de Fuca, it is overlooked by the city of Seattle, which is situated on its eastern shore.

puggaree /'pʌgəri/ *n.* **1** an Indian turban. **2** a thin muslin scarf tied round a sun-helmet etc. and shielding the neck. [Hindi *pagṛī* turban]

pugilist /'pju:dʒɪlɪst/ *n.* a boxer, esp. a professional. □ **pugilism** *n.* **pugilistic** /ˌpju:dʒɪ'lɪstɪk/ *adj.* [L *pugil* boxer]

Pugin /'pju:dʒɪn/, Augustus Welby Northmore (1812–52), English architect, theorist, and designer. He converted to Roman Catholicism in 1835 and became the main champion of the Gothic revival; he believed that the Gothic style was the only proper architectural style because of its origins in medieval Christian society. Among his chief contributions to architecture and design is his work on the external detail and internal fittings for the Houses of Parliament designed by Sir Charles Barry. Pugin's views, set out in works such as *Contrasts* (1836), influenced John Ruskin and ultimately the Arts and Crafts Movement.

Puglia see APULIA.

pugnacious /pʌg'neɪʃəs/ *adj.* quarrelsome; disposed to fight. □ **pugnaciously** *adv.* **pugnaciousness** *n.* **pugnacity** /-'næsɪti/ *n.* [L *pugnax -acis* f. *pugnare* fight f. *pugnus* fist]

puisne /'pju:ni/ *adj. Law* denoting a judge of a superior court inferior in rank to chief justices. [OF f. *puis* f. L *postea* afterwards + *né* born f. L *natus*: cf. PUNY]

puissance /'pju:ɪs(ə)ns, 'pwɪs-/ *n.* **1** (also /pwi:'sɒns/) a test of a horse's ability to jump large obstacles in showjumping. **2** *archaic* great power, might, or influence. [ME (in sense 2) f. OF (as PUISSANT)]

puissant /'pju:ɪs(ə)nt, 'pwɪs-/ *adj. literary* or *archaic* having great power or influence; mighty. □ **puissantly** *adv.* [ME f. OF f. L *posse* be able: cf. POTENT¹]

puja /'pu:dʒə/ *n.* (also **pooja**) a Hindu rite of worship; an offering. [Skr.]

puke /pju:k/ *v. & n. sl.* ● *v.tr. & intr.* vomit. ● *n.* vomit. □ **pukey** *adj.* [16th c.: prob. imit.]

pukeko /'pu:keˌkəʊ/ *n.* (*pl.* **-os**) *NZ* a purple gallinule. [Maori]

pukka /'pʌkə/ *adj.* (also **pukkah, pucka**) *Anglo-Ind.* **1** genuine. **2** of good quality; reliable (*did a pukka job*). **3** of full weight. [Hindi *pakkā* cooked, ripe, substantial]

pul /pu:l/ *n.* (*pl.* **puls** or **puli** /-li/) a monetary unit of Afghanistan, equal to one-hundredth of an afghani. [Pashto f. Pers. *pūl* copper coin]

pula /'pu:lə/ *n.* the basic monetary unit of Botswana, equal to 100 thebe. [Setswana, = rain]

Pulau Seribu /ˌpu:laʊ 'serɪˌbu:/ the Indonesian name for the Thousand Islands (see THOUSAND ISLANDS 2).

pulchritude /'pʌlkrɪˌtju:d/ *n. literary* beauty. □ **pulchritudinous** /ˌpʌlkrɪ'tju:dɪnəs/ *adj.* [ME f. L *pulchritudo -dinis* f. *pulcher -chri* beautiful]

pule /pju:l/ *v.intr. literary* cry querulously or weakly; whine, whimper. [16th c.: prob. imit.: cf. F *piauler*]

Pulitzer /'pʊlɪtsə(r)/, Joseph (1847–1911), Hungarian-born American newspaper proprietor and editor. A pioneer of campaigning popular journalism, he owned a number of newspapers, including the *New York World*. Through his journalism he aimed to remedy abuses and reform social and economic inequalities by the exposure of striking instances and by the vigorous expression of popular opinion. He made provisions in his will for the establishment of the annual Pulitzer Prizes.

Pulitzer Prize *n.* any of a group of money prizes established under Joseph Pulitzer's will and offered annually to American citizens for work in music, journalism, American history and biography, poetry, drama, and fiction.

pull /pʊl/ *v. & n.* ● *v.* **1** *tr.* exert force on (a thing) tending to move it to oneself or the origin of the force (*stop pulling my hair*). **2** *tr.* cause to move in this way (*pulled it nearer; pulled me into the room*). **3** *intr.* exert a pulling force (*the horse pulls well; the engine will not pull*). **4** *tr.* extract (a cork or tooth) by pulling. **5** *tr.* damage (a muscle etc.) by abnormal strain. **6 a** *tr.* move (a boat) by pulling on the oars. **b** *intr.* (of a boat etc.) be caused to move, esp. in a specified direction. **7** *intr.* (often foll. by *up*) proceed with effort (up a hill etc.). **8** *tr.* (foll. by *on*) bring out (a weapon) for use against (a person). **9 a** *tr.* check the speed of (a horse), esp. so as to make it lose the race. **b** *intr.* (of a horse) strain against the bit. **10** *tr.* attract or secure (custom or support). **11** *tr.* draw (liquor) from a barrel etc. **12** *intr.* (foll. by *at*) tear or pluck at. **13** *intr.* (often foll. by *on*, *at*) inhale deeply; draw or suck (on a pipe etc.). **14** *tr.* (often foll. by *up*) remove (a plant) by the root. **15** *tr.* **a** *Cricket* hit (the ball) round to the leg side from the off with a pulling movement of the bat. **b** *Golf* strike (the ball) widely to the left. **16** *tr.* print (a proof etc.). **17** *tr. colloq.* achieve or accomplish (esp. something illicit). **18** *tr. sl.* pick up or attract (a person) for sexual purposes. ● *n.* **1** the act of pulling. **2** the force exerted by this. **3** a means of exerting influence; an advantage. **4** something that attracts or draws attention. **5** a deep draught of liquor. **6** a prolonged effort, e.g. in going up a hill. **7** a handle etc. for applying a pull. **8** a spell of rowing. **9** a printer's rough proof. **10** *Cricket & Golf* a stroke which pulls the ball. **11** a suck at a cigarette. □ **pull about 1** treat roughly. **2** pull from side to side. **pull apart** (or **to pieces**) = *take to pieces* (see PIECE). **pull away** withdraw, move away. **pull back** retreat or cause to retreat. **pull-back** *n.* **1** a retarding influence. **2** a withdrawal of troops. **pull down 1** demolish (esp. a building). **2** humiliate. **3** *colloq.* earn (a sum of money) as wages etc. **pull a face** assume a distinctive or specified (e.g. sad or angry) expression. **pull a fast one** see FAST¹. **pull in 1** (of a bus, train, etc.) arrive to take passengers. **2** (of a vehicle) move to the side of or off the road. **3** earn or acquire. **4** *colloq.* arrest. **pull-in** *n. Brit.* a roadside café or other stopping-place. **pull a person's leg** deceive a person playfully. **pull off 1** remove by pulling. **2** succeed in achieving or winning. **pull oneself together** recover control of oneself. **pull the other one** *colloq.* expressing disbelief (with ref. to *pull a person's leg*). **pull out 1** take out by pulling. **2** depart. **3** withdraw from an undertaking. **4** (of a bus, train, etc.) leave with its passengers. **5** (of a vehicle) move out from the side of the road, or from its normal position to overtake. **pull-out** *n.* **1** something that can be pulled out, esp. a section of a magazine. **2** the act of pulling out; a withdrawal, esp. from military involvement. **pull over** (of a vehicle) pull in. **pull the plug** (often foll. by *on*) *colloq.* put an end to an enterprise etc.; destroy; cut off (supplies etc.). **pull one's punches** avoid using one's full force. **pull rank** take unfair advantage of one's seniority. **pull round** (or **through**) recover or cause to recover from an illness. **pull strings** exert (esp. clandestine) influence. **pull the strings** be the real actuator of what another does. **pull together** work in harmony. **pull up 1** stop or cause to stop moving. **2** pull out of the ground. **3** reprimand. **4** check oneself. **pull one's weight** do one's fair share of work. **pull wires** esp. *US* = *pull strings*. □ **puller** *n.* [OE (ā)*pullian*, perh. rel. to LG *pūlen*, MDu. *polen* to shell]

pullet /ˈpʊlɪt/ *n.* a young hen, esp. one less than one year old. [ME f. OF *poulet* dimin. of *poule* ult. fem. of L *pullus* chicken]

pulley /ˈpʊlɪ/ *n. & v.* ● *n.* (*pl.* **-eys**) **1** a grooved wheel or set of wheels for a cord etc. to pass over, set in a block and used for changing the direction of a force. **2** a wheel or drum fixed on a shaft and turned by a belt, used esp. to increase speed or power. ● *v.tr.* (**-eys**, **-eyed**) **1** hoist or work with a pulley. **2** provide with a pulley. [ME f. OF *polie* prob. ult. f. med. Gk *polidion* (unrecorded) pivot, dimin. of *polos* POLE²]

Pullman /ˈpʊlmən/ *n.* (*pl.* **Pullmans**) **1** a railway carriage or motor coach affording special comfort. **2** a sleeping-car. **3** a train consisting of Pullman carriages. [George M. *Pullman*, Amer. designer (1831–97)]

pullover /ˈpʊlˌəʊvə(r)/ *n.* a knitted garment put on over the head and covering the top half of the body.

pullulate /ˈpʌljʊˌleɪt/ *v.intr.* **1** (of a seed, shoot, etc.) bud, sprout, germinate. **2** (esp. of an animal) swarm, throng; breed prolifically. **3** develop; spring up; come to life. **4** (foll. by *with*) abound. □ **pullulant** *adj.* **pullulation** /ˌpʌljʊˈleɪʃ(ə)n/ *n.* [L *pullulare* sprout f. *pullulus* dimin. of *pullus* young of an animal]

pulmonary /ˈpʌlmənərɪ/ *adj.* **1** of or relating to the lungs. **2** having lungs or lunglike organs. **3** affected with or susceptible to lung disease. □ **pulmonary artery** the artery conveying blood from the heart to the lungs. **pulmonary tuberculosis** the common form of tuberculosis, affecting the lungs and caused by inhaling the tubercle bacillus. **pulmonary vein** the vein carrying oxygenated blood from the lungs to the heart. □ **pulmonate** /-nət/ *adj.* [L *pulmonarius* f. *pulmo -onis* lung]

pulmonic /pʌlˈmɒnɪk/ *adj.* = PULMONARY 1. [F *pulmonique* or f. mod.L *pulmonicus* f. L *pulmo* (as PULMONARY)]

pulp /pʌlp/ *n. & v.* ● *n.* **1** the soft fleshy part of fruit etc. **2** any soft thick wet mass. **3** a soft shapeless mass derived from rags, wood, etc., used in paper-making. **4** (often *attrib.*) poor quality (often sensational) writing orig. printed on rough paper (*pulp fiction*). **5** vascular tissue filling the interior cavity and root canals of a tooth. **6** *Mining* pulverized ore mixed with water. ● *v.* **1** *tr.* reduce to pulp. **2** *tr.* withdraw (a publication) from the market, usu. recycling the paper. **3** *tr.* remove pulp from. **4** *intr.* become pulp. □ **pulper** *n.* **pulpy** *adj.* **pulpiness** *n.* [L *pulpa*]

pulpit /ˈpʊlpɪt/ *n.* **1** a raised enclosed platform in a church etc. from which the preacher delivers a sermon. **2** (*prec. by the*) preachers or preaching collectively. [ME f. L *pulpitum* scaffold, platform]

pulpwood /ˈpʌlpwʊd/ *n.* timber suitable for making pulp.

pulque /ˈpʊlkeɪ/ *n.* a Mexican fermented drink made from the sap of the maguey. □ **pulque brandy** a strong intoxicant made from pulque. [17th c.: Amer. Sp., of unkn. orig.]

pulsar /ˈpʌlsɑː(r)/ *n. Astron.* a celestial object, believed to be a rapidly rotating neutron star, emitting pulses of radio waves and other electromagnetic radiation with great regularity. Pulsars were first discovered in 1967; the pulses are believed to be due to the interaction of the magnetic field of the neutron star with surrounding material, accelerating electrons which radiate energy in a beam. As this beam sweeps past the observer (up to one thousand times a second in some cases), the characteristic pulsations in signal strength are noted. [*pulsating star*, after *quasar*]

pulsate /pʌlˈseɪt, ˈpʌlseɪt/ *v.intr.* **1** expand and contract rhythmically; throb. **2** vibrate, quiver, thrill. □ **pulsator** *n.* **pulsation** /pʌlˈseɪʃ(ə)n/ *n.* **pulsatory** /ˈpʌlsətərɪ/ *adj.* [L *pulsare* frequent. of *pellere puls-* drive, beat]

pulsatile /ˈpʌlsəˌtaɪl/ *adj.* **1** of or having the property of pulsation. **2** (of a musical instrument) played by percussion. [med.L *pulsatilis* (as PULSATE)]

pulsatilla /ˌpʌlsəˈtɪlə/ *n.* a ranunculaceous plant of the genus *Pulsatilla*, esp. the pasque-flower. [mod.L dimin. of *pulsata* fem. past part. (as PULSATE), because it quivers in the wind]

pulse¹ /pʌls/ *n. & v.* ● *n.* **1 a** a rhythmical throbbing of the arteries as blood is propelled through them, esp. as felt in the wrists, temples, etc. **b** each successive beat of the arteries or heart. **2** a throb or thrill of life or emotion. **3** a general feeling or opinion (*tried to read the pulse of the nation*). **4** a single vibration of sound, electric current, light, etc., esp. as a signal. **5** a musical beat. **6** any regular or recurrent rhythm, e.g. of the stroke of oars. ● *v.intr.* **1** pulsate. **2** (foll. by *out*, *in*, etc.) transmit etc. by rhythmical beats. □ **pulse code** coding information in pulses. **pulse code modulation** a pulse modulation technique of representing a signal by a sequence of binary codes. **pulse modulation** a type of modulation in which pulses are varied to represent a signal. □ **pulseless** *adj.* [ME f. OF *pous* f. L *pulsus* f. *pellere puls-* drive, beat]

pulse² /pʌls/ *n.* (as *sing.* or *pl.*) **1** the edible seeds of leguminous plants, e.g. chick peas, lentils, beans, etc. **2** the plant or plants producing this. [ME f. OF *pols* f. L *puls pultis* porridge of meal etc.]

pulsimeter /pʌlˈsɪmɪtə(r)/ *n.* an instrument for measuring the rate or force of a pulse.

Pulu /ˈpuːluː/ Tiglath-pileser III, king of Assyria (see TIGLATH-PILESER).

pulverize /ˈpʌlvəˌraɪz/ *v.* (also **-ise**) **1** *tr.* reduce to fine particles. **2** *tr. & intr.* crumble to dust. **3** *colloq. tr.* **a** demolish. **b** defeat utterly. □ **pulverizable** *adj.* **pulverizator** *n.* **pulverizer** *n.* **pulverization** /ˌpʌlvəraɪˈzeɪʃ(ə)n/ *n.* [ME f. LL *pulverizare* f. *pulvis pulveris* dust]

pulverulent /pʌlˈverʊlənt/ *adj.* **1** consisting of fine particles; powdery. **2** likely to crumble. [L *pulverulentus* (as PULVERIZE)]

puma /ˈpjuːmə/ *n.* a large American feline, *Felis concolor*, with a usu. tawny or greyish coat. Also called *cougar*, *panther*, *mountain lion*. [Sp. f. Quechua]

pumice /ˈpʌmɪs/ *n. & v.* ● *n.* (in full **pumice-stone**) **1** a light porous volcanic rock often used as an abrasive in cleaning or polishing substances. **2** a piece of this used for removing hard skin etc. ● *v.tr.* rub or clean with a pumice. □ **pumiceous** /pjuːˈmɪʃəs/ *adj.* [ME f. OF *pomis* f. L *pumex pumicis* (dial. *pom-*): cf. POUNCE²]

pummel /ˈpʌm(ə)l/ *v.tr.* (**pummelled, pummelling**; *US* **pummeled, pummeling**) strike repeatedly esp. with the fist. [alt. f. POMMEL]

pump¹ /pʌmp/ *n. & v.* ● *n.* **1** a machine, usu. with rotary action or the reciprocal action of a piston, for raising or moving liquids, compressing gases, inflating tyres, etc. **2** an instance of pumping; a stroke of a pump. ● *v.* **1** *tr.* (often foll. by *in*, *out*, *into*, *up*, etc.) raise or remove (liquid, gas, etc.) with a pump. **2** *tr.* (often foll. by *up*) fill (a tyre etc.) with air.

3 *tr.* **a** remove (water etc.) with a pump. **b** (foll. by *out*) remove liquid from (a place, well, etc.) with a pump. **4** *intr.* work a pump. **5** *tr.* (often foll. by *out*) cause to move, pour forth, etc., as if by pumping. **6** *tr.* question (a person) persistently to obtain information. **7** *tr.* **a** move vigorously up and down. **b** shake (a person's hand) effusively. **8** *tr.* (usu. foll. by *up*) arouse, excite. □ **pump-action** (of a repeating firearm) activated by a horizontally operating slide action. **pump-brake** the handle of a pump, esp. with a transverse bar for several people to work at. **pump-handle** *colloq.* shake (a person's hand) effusively. **pump iron** *colloq.* exercise with weights. **pump-priming 1** the introduction of fluid etc. into a pump to prepare it for working. **2** esp. *US* the stimulation of commerce etc. by investment. **pump room 1** a room where fuel pumps etc. are stored or controlled. **2** a room at a spa etc. where medicinal water is dispensed. [ME *pumpe, pompe* (orig. Naut.): prob. imit.]

pump² /pʌmp/ *n.* **1** a plimsoll. **2** a light shoe for dancing etc. **3** *N. Amer.* a court shoe. [16th c.: orig. unkn.]

pumpernickel /ˈpʌmpəˌnɪk(ə)l, ˈpʊm-/ *n.* German wholemeal rye bread. [G, earlier = lout, bumpkin, of uncert. orig.]

pumpkin /ˈpʌmpkɪn/ *n.* **1** a plant of the genus *Cucurbita*, with tendrils and large lobed leaves, esp. *C. maxima*. **2** the large rounded yellow fruit of this plant, with a thick rind and edible flesh. [alt. f. earlier *pompon, pumpion* f. obs. F *po(m)pon* f. L *pepo -onis* f. Gk *pepōn* large melon: see PEPO]

pun¹ /pʌn/ *n. & v.* ● *n.* the humorous use of a word to suggest different meanings, or of words of the same sound and different meanings. ● *v.intr.* (**punned, punning**) (foll. by *on*) make a pun or puns with (words). □ **punningly** *adv.* [17th c.: perh. f. obs. *pundigrion*, a fanciful formation]

pun² /pʌn/ *v.tr.* (**punned, punning**) *Brit.* consolidate (earth or rubble) by pounding or ramming. □ **punner** *n.* [dial. var. of POUND²]

puna /ˈpuːnə/ *n.* **1** a high plateau in the Peruvian Andes. **2** = *mountain sickness*. [Quechua, in sense 1]

Punch /pʌntʃ/ *n.* (also called *Punchinello*) a grotesque hook-nosed hump-backed buffoon, the chief male character of the Punch and Judy show. Punch is the English stock character derived ultimately from Italian *commedia dell'arte*. □ **as pleased** (or **proud**) **as Punch** showing great pleasure or pride. [PUNCHINELLO]

punch¹ /pʌntʃ/ *v. & n.* ● *v.tr.* **1** strike bluntly, esp. with a closed fist. **2** prod or poke with a blunt object. **3 a** pierce a hole in (metal, paper, a ticket, etc.) as or with a punch. **b** pierce (a hole) by punching. **4** *N. Amer.* drive (cattle) by prodding with a stick etc. ● *n.* **1** a blow with a fist. **2** the ability to deliver this. **3** *colloq.* vigour, momentum; effective force. □ **punch-bag** (or **punching bag**) a suspended stuffed bag used as a punchball. **punch card** = PUNCHED CARD. **punch-drunk** stupefied from or as though from a series of heavy blows. **punch-line** words giving the point of a joke or story. **punch-up** *Brit. colloq.* a fist-fight; a brawl. □ **puncher** *n.* [ME, var. of POUNCE¹]

punch² /pʌntʃ/ *n.* **1** a device or machine for punching holes in a material (e.g. paper, leather, metal, plaster). **2** a tool or machine for impressing a design or stamping a die on a material. [perh. an abbr. of PUNCHEON¹, or f. PUNCH¹]

punch³ /pʌntʃ/ *n.* a drink of wine or spirits mixed with water, fruit juices, spices, etc., and usu. served hot. □ **punch-bowl 1** a bowl in which punch is mixed. **2** a deep round hollow in a hill. [17th c.: orig. unkn.]

punch⁴ /pʌntʃ/ *n.* (in full **Suffolk punch**) a short-legged thickset draught horse. [prob. as PUNCH]

Punch and Judy /ˈdʒuːdɪ/ an English puppet-show of uncertain origin, probably introduced from the Continent in the 17th century. It is presented on the miniature stage of a tall collapsible booth traditionally covered with striped canvas. Punch is on the manipulator's right hand, remaining on stage all the time, while the left hand provides a series of characters — baby, wife (Judy), priest, doctor, policeman, hangman — for him to nag, beat, and finally kill. His live dog, Toby, sits on the ledge of the booth. (See also PUNCH.)

punchball /ˈpʌntʃbɔːl/ *n.* **1** a stuffed or inflated ball suspended or mounted on a stand, for a boxer to practise with. **2** *US* a ball game in which a rubber ball is punched with the fist or head.

punched card *n.* (also **punch card**) a card perforated according to a specified code, used in recording and analysing data and for conveying instructions to a machine. Such cards (introduced by the American engineer Herman Hollerith) were first used in the US census of 1890, and thereafter found widespread use in business and later for scientific

and technical purposes. Early computers were often fed with information from punched cards of the Hollerith type.

puncheon¹ /ˈpʌntʃən/ *n.* **1** a short post, esp. one supporting a roof in a coal mine. **2** = PUNCH². [ME f. OF *poinson, po(i)nchon*, ult. f. L *pungere punct-* prick]

puncheon² /ˈpʌntʃən/ *n. hist.* a large cask for liquids etc. holding from 72 to 120 gallons. [ME f. OF *poinson, po(i)nchon*, of unkn. orig. (prob. not the same as in PUNCHEON¹)]

Punchinello /ˌpʌntʃɪˈneləʊ/ *n.* (*pl.* **-os**) **1** see PUNCH. **2** a short stout person of comical appearance. [f. Neapolitan dial. *polecenella*, It. *Pulcinella*, perh. dimin. of It. *pollecena* young turkey-cock (with hooked beak which the nose of Punch's mask resembles) f. *pulcino* chicken ult. f. L *pullus*]

punchy /ˈpʌntʃɪ/ *adj.* (**punchier, punchiest**) having punch or vigour; forceful. □ **punchily** *adv.* **punchiness** *n.*

puncta *pl.* of PUNCTUM.

punctate /ˈpʌŋkteɪt/ *adj. Biol.* marked or studded with points, dots, spots, etc. □ **punctation** /pʌŋkˈteɪʃ(ə)n/ *n.* [L *punctum* (as POINT)]

punctilio /pʌŋkˈtɪlɪˌəʊ/ *n.* (*pl.* **-os**) **1** a delicate point of ceremony or honour. **2** the etiquette of such points. **3** petty formality. [It. *puntiglio* & Sp. *puntillo* dimin. of *punto* POINT]

punctilious /pʌŋkˈtɪlɪəs/ *adj.* **1** attentive to formality or etiquette. **2** precise in behaviour. □ **punctiliously** *adv.* **punctiliousness** *n.* [F *pointilleux* f. *pointillé* f. It. (as PUNCTILIO)]

punctual /ˈpʌŋktʃʊəl, -tjʊəl/ *adj.* observing the appointed time; neither early nor late. □ **punctually** *adv.* **punctuality** /ˌpʌŋktʃʊˈælɪtɪ, ˌpʌŋktjʊ-/ *n.* [ME f. med.L *punctualis* f. L *punctum* POINT]

punctuate /ˈpʌŋktʃʊˌeɪt, ˈpʌŋktjʊ-/ *v.tr.* **1** insert punctuation marks in. **2** interrupt at intervals (*punctuated his tale with heavy sighs*). [med.L *punctuare punctuat-* (as PUNCTUAL)]

punctuated equilibrium *n. Biol.* evolutionary development marked by isolated episodes of rapid speciation between long periods of little or no change.

punctuation /ˌpʌŋktʃʊˈeɪʃ(ə)n, ˌpʌŋktjʊ-/ *n.* **1** the system or arrangement of marks used to punctuate a written passage. **2** the practice or skill of punctuating. [med.L *punctuatio* (as PUNCTUATE)]

punctuation mark *n.* a mark (e.g. a full stop, comma, dash, apostrophe, bracket, or exclamation mark) used in writing to separate sentences and phrases etc. and to clarify meaning.

punctum /ˈpʌŋktəm/ *n.* (*pl.* **puncta** /-tə/) *Biol.* a speck, dot, spot of colour, etc., or an elevation or depression on a surface. [L, = POINT]

puncture /ˈpʌŋktʃə(r)/ *n. & v.* ● *n.* **1** a prick or pricking, esp. the accidental piercing of a pneumatic tyre. **2** a hole made in this way. ● *v.* **1** *tr.* make a puncture in. **2** *intr.* undergo puncture. **3** *tr.* prick or pierce. **4** *tr.* cause (hopes, confidence, etc.) to collapse; dash, deflate. [ME f. L *punctura* f. *pungere punct-* prick]

pundit /ˈpʌndɪt/ *n.* **1** (also **pandit**) a Hindu learned in Sanskrit and in the philosophy, religion, and jurisprudence of India. **2** often *iron.* a learned expert or teacher. □ **punditry** *n.* [Hind. *paṇḍit* f. Skr. *paṇḍita* learned]

Pune see POONA.

pungent /ˈpʌndʒ(ə)nt/ *adj.* **1** having a sharp or strong taste or smell, esp. so as to produce a pricking sensation. **2** (of remarks) penetrating, biting, caustic. **3** mentally stimulating. **4** *Biol.* having a sharp point. □ **pungency** *n.* **pungently** *adv.* [L *pungent-* pres. part. of *pungere* prick]

Punic /ˈpjuːnɪk/ *adj. & n.* ● *adj.* of or relating to ancient Carthage or its language. ● *n.* the language of Carthage, related to Phoenician. □ **Punic faith** treachery. [L *Punicus, Poenicus* f. *Poenus* f. Gk *Phoinix* Phoenician]

Punic Wars three wars between Rome and Carthage, which led to the unquestioned dominance of Rome in the western Mediterranean, a position of power which was not endangered for centuries afterwards. In the first Punic War (264–241 BC), Rome secured Sicily from Carthage and established herself as a naval power; in the second (218–201 BC), the defeat of Hannibal (largely through the generalship of Fabius Cunctator and Scipio Africanus) put an end to Carthage's position as a Mediterranean power; the third (149–146 BC) ended in the total destruction of the city of Carthage.

punish /ˈpʌnɪʃ/ *v.tr.* **1** cause (an offender) to suffer for an offence. **2** inflict a penalty for (an offence). **3** *colloq.* inflict severe blows on (an opponent). **4 a** tax severely; subject to severe treatment. **b** abuse or treat improperly. □ **punishable** *adj.* **punisher** *n.* **punishing** *adj.* (in sense 4a). **punishingly** adv. [ME f. OF punir f. L punire = poenire f. poena penalty]

punishment /ˈpʌnɪʃmənt/ n. **1** the act or an instance of punishing; the condition of being punished. **2** the loss or suffering inflicted in this. **3** colloq. severe treatment or suffering. [ME f. AF & OF punissement f. punir]

punitive /ˈpjuːnɪtɪv/ adj. (also **punitory** /-təri/) **1** inflicting or intended to inflict punishment. **2** (of taxation etc.) extremely severe. □ **punitive damages** Law = vindictive damages. □ **punitively** adv. [F punitif -ive or med.L punitivus (as PUNISHMENT)]

Punjab /pʌnˈdʒɑːb, ˈpʌndʒɑːb/ (also **the Punjab**) a region of NW India and Pakistan, a wide fertile plain traversed by the Indus and the five tributaries which gave the region its name (Hindustani panj five, āb waters). Under Muslim influence from the 11th century, the region became a centre of Sikhism in the 15th century and, after the capture of Lahore in 1799 by Ranjit Singh, a powerful Sikh kingdom. It was annexed by the British in 1849 and became a part of British India. In the partition of 1947 it was divided between Pakistan and India. The province of Punjab in Pakistan is centred on the former state capital, Lahore. The state of Punjab in India was divided in 1966 into the two states of Punjab and Haryana, both having Chandigarh as their capital.

Punjabi /pʌnˈdʒɑːbi/ n. & adj. (also **Panjabi** /pʌnˈdʒɑːbi, pæn-/) ● n. (pl. **Punjabis**) **1** a native or inhabitant of Punjab. **2** (usu. **Panjabi**) the Indic language of Punjab. ● adj. of or relating to Punjab or the language Panjabi. [Hindi pañjābī]

punk /pʌŋk/ n. & adj. ● n. **1 a** a worthless person or thing (often as a general term of abuse). **b** nonsense. **2 a** (in full **punk rock**) a loud, fast-paced, aggressively performed form of rock music, first popular in the late 1970s. **b** the subculture or style associated with punk rock, including, for example, spiked hair and the wearing of safety pins, leather, and deliberately ripped clothing. **c** (in full **punk rocker**) a devotee of punk rock or its associated style or subculture. **3** N. Amer. a hoodlum or ruffian. **4** US a passive male homosexual. **5** US an inexperienced person; a novice. **6** soft crumbly wood that has been attacked by fungus, used as tinder. ● adj. **1** worthless, poor in quality. **2** of or relating to punk or punk rock. **3** US (of wood) rotten, decayed. □ **punky** adj. [18th c.: orig. unkn.: cf. SPUNK]

punkah /ˈpʌŋkə/ n. **1** (in India) a fan usu. made from the leaf of the palmyra. **2** a large swinging cloth fan on a frame worked by a cord or electrically. □ **punkah-wallah** a person who works a punkah. [Hindi pankhā fan f. Skr. pakṣaka f. pakṣa wing]

punnet /ˈpʌnɪt/ n. Brit. a small light basket or container for fruit or vegetables. [19th c.: perh. dimin. of dial. pun POUND¹]

punster /ˈpʌnstə(r)/ n. a person who makes puns, esp. habitually.

punt¹ /pʌnt/ n. & v. ● n. a long narrow flat-bottomed boat, square at both ends, used mainly on rivers and propelled using a long pole. Formerly used for transporting goods and cattle, punts are now used for pleasure and sometimes for racing. ● v. **1** tr. propel (a punt) with a pole. **2** intr. & tr. travel or convey in a punt. □ **punter** n. [ME f. MLG punte, punto & MDu. ponte ferry-boat f. L ponto Gaulish transport vessel]

punt² /pʌnt/ v. & n. ● v.tr. kick (a ball) after it has dropped from the hands and before it reaches the ground. ● n. such a kick. □ **punter** n. [prob. f. dial. punt push forcibly: cf. BUNT³]

punt³ /pʌnt/ v. & n. ● v.intr. **1** (in some card-games) lay a stake against the bank. **2** Brit. colloq. **a** bet on a horse etc. **b** speculate in shares etc. ● n. **1** a bet. **2** a point in faro. **3** a person who plays against the bank in faro. □ **punter** f. ponte player against the bank f. Sp. punto POINT]

punt⁴ /pʊnt/ n. the chief monetary unit of the Republic of Ireland; the Irish pound. [Ir., = pound]

Punta Arenas /ˌpʊntə əˈreɪnəs/ a port in southern Chile, on the Strait of Magellan; pop. (est. 1987) 111,720.

punter /ˈpʌntə(r)/ n. **1** a person who gambles or lays a bet. **2 a** colloq. a customer or client; a member of an audience. **b** colloq. a participant in any activity; a person. **c** sl. a prostitute's client. **3** a point in faro.

puny /ˈpjuːnɪ/ adj. (**punier**, **puniest**) **1** undersized. **2** weak, feeble. **3** petty. □ **punily** adv. **puniness** n. [phonetic spelling of PUISNE]

pup /pʌp/ n. & v. ● n. **1** a young dog. **2** a young wolf, rat, seal, etc. **3** Brit. an unpleasant or arrogant young man. ● v.tr. (**pupped**, **pupping**) (also absol.) (of a bitch etc.) bring forth (young). □ **in pup** (of a bitch) pregnant. **sell a person a pup** swindle a person, esp. by selling something worthless. [back-form. f. PUPPY as if a dimin. in -Y²]

pupa /ˈpjuːpə/ n. (pl. **pupae** /-piː/) the inactive immature form of an insect, being the resting stage between larva and adult, e.g. a chrysalis. □ **pupal** adj. [mod.L f. L pupa girl, doll]

pupate /pjuːˈpeɪt/ v.intr. become a pupa. □ **pupation** /-ˈpeɪʃ(ə)n/ n.

pupil¹ /ˈpjuːpɪl, -p(ə)l/ n. **1** a person who is taught by another, esp. a schoolchild or student in relation to a teacher. **2** Law a trainee barrister. □ **pupillary** adj. (also **pupilary**). [ME, orig. = orphan, ward f. OF pupille or L pupillus, -illa, dimin. of pupus boy, pupa girl]

pupil² /ˈpjuːpɪl, -p(ə)l/ n. the dark circular opening in the centre of the iris of the eye, varying in size to regulate the passage of light to the retina. □ **pupilar** adj. (also **pupilar**). **pupillary** adj. (also **pupilary**). [OF pupille or L pupilla, dimin. of pūpa doll (as PUPIL¹): so called from the tiny reflected images visible in the eye]

pupillage /ˈpjuːpɪlɪdʒ/ n. (also **pupilage**) **1** the condition of being a pupil or student. **2** Law apprenticeship to a member of the Bar, qualifying a barrister to practise independently.

pupiparous /pjuːˈpɪpərəs/ adj. Zool. (of certain parasitic flies) bringing forth young which immediately pupate. [mod.L pupipara neut. pl. of pupiparus (as PUPA, parere bring forth)]

puppet /ˈpʌpɪt/ n. **1** a small figure representing a human being or animal and moved by various means as entertainment. (See note below.) **2** a person whose actions are controlled by another. □ **puppet state** a country that is nominally independent but actually under the control of another power. □ **puppetry** n. [later form of POPPET]

■ Puppetry as a form of theatre is probably as old as the theatre itself. There are glove or hand puppets, puppets worked by rods, marionettes operated by strings, and the flat puppets used in the shadow theatre of Java, Bali, and Thailand. Famous hand puppets include the English Punch, the French Guignol, and the Italian Pulcinella. In Europe puppets were at their most popular in the 17th and 18th centuries, when special theatres were built for often satirical shows; Haydn wrote an operetta, Dido (c.1776), for puppets. Since then puppetry in the West has been associated more with children's entertainment, although television has revived the use of puppets in satire. The popularity of puppets is now highest in the Far East and eastern Europe.

puppeteer /ˌpʌpɪˈtɪə(r)/ n. a person who works puppets.

puppy /ˈpʌpɪ/ n. (pl. **-ies**) **1** a young dog. **2** colloq. a conceited or arrogant young man. □ **puppy-fat** temporary fatness of a child or adolescent. **puppy love** = calf-love (see CALF¹). □ **puppyhood** n. **puppyish** adj. [ME perh. f. OF po(u)pee doll, plaything, toy f. Rmc (as POPPET)]

pur- /pɜː(r), pə(r)/ prefix = PRO-¹ (purchase; pursue). [AF f. OF por-, pur-, pour- f. L por-, pro-]

Purana /pʊˈrɑːnə/ n. any of a class of Sanskrit sacred writings on Hindu mythology, folklore, etc., of varying date and origin. The most ancient dates from the 4th century AD. □ **Puranic** adj. [Skr. purāṇa ancient legend, ancient, f. purā formerly]

Purbeck marble /ˈpɜːbek/ n. (also **Purbeck stone**) a hard usu. polished limestone from Purbeck in Dorset, used in pillars, effigies, etc.

purblind /ˈpɜːblaɪnd/ adj. **1** partly blind; dim-sighted. **2** obtuse, dim-witted. □ **purblindness** n. [ME pur(e) blind f. PURE orig. in sense 'utterly', with assim. to PUR-]

Purcell /pɜːˈsel, ˈpɜːs(ə)l/, Henry (1659–95), English composer. He enjoyed royal patronage and was organist for Westminster Abbey (1679–95) and the Chapel Royal (1682–95). He composed many choral odes and songs for royal occasions as well as sacred anthems for the Chapel Royal. His main interest was in music for the theatre; he composed the first English opera, Dido and Aeneas (1689), moving away from the tradition of the masque, breaking new dramatic ground and accommodating a wide emotional range. He also composed the incidental music for many plays, while his instrumental music includes a series of Fantasias for the viol (1680).

purchase /ˈpɜːtʃɪs/ v. & n. ● v.tr. **1** acquire by payment; buy. **2** obtain or achieve at some cost. **3** Naut. haul up (an anchor etc.) by means of a pulley, lever, etc. ● n. **1** the act or an instance of buying. **2** something bought. **3** Law the acquisition of property by one's personal action and not by inheritance. **4 a** a firm hold on a thing to move it or to prevent it from slipping; leverage. **b** a device or tackle for moving heavy objects. **5** the annual rent or return from land. □ **purchase tax** Brit. hist. a tax on goods bought, levied at higher rates for non-essential or luxury goods. □ **purchasable** adj. **purchaser** n. [ME f. AF purchacer, OF pourchacier seek to obtain (as PUR-, CHASE¹)]

purdah /ˈpɜːdə/ n. Ind. **1** a system in certain Muslim and Hindu societies of screening women from strangers by means of a veil or curtain. **2** a curtain in a house, used for this purpose. [Urdu & Pers. pardah veil, curtain]

pure /pjʊə(r)/ adj. **1** unmixed, unadulterated (*pure white; pure alcohol*). **2** of unmixed origin or descent (*pure-blooded*). **3** chaste. **4** morally or sexually undefiled; not corrupt. **5** conforming absolutely to a standard of quality; faultless. **6** guiltless. **7** sincere. **8** mere, simple, nothing but, sheer (*it was pure malice*). **9** (of a sound) not discordant, perfectly in tune. **10** (of a subject of study) dealing with abstract concepts and not practical application. **11 a** (of a vowel) not joined with another in a diphthong. **b** (of a consonant) not accompanied by another. □ **pure mathematics** see MATHEMATICS. **pure science** a science depending on deductions from demonstrated truths (e.g. mathematics or logic), or one studied without practical applications. □ **pureness** n. [ME f. OF *pur* pure f. L *purus*]

purée /'pjʊəreɪ/ n. & v. ● n. a pulp of vegetables or fruit etc. reduced to a smooth cream. ● v.tr. (**purées, puréed**) make a purée of. [F]

purely /'pjʊəlɪ/ adv. **1** in a pure manner. **2** merely, solely, exclusively.

purfle /'pɜː:f(ə)l/ n. & v. ● n. **1** an ornamental border, esp. on a violin etc. **2** archaic the ornamental or embroidered edge of a garment. ● v.tr. **1** decorate with a purfle. **2** (often foll. by *with*) ornament (the edge of a building). **3** beautify. □ **purfling** n. [ME f. OF *porfil, porfiler* ult. f. L *filum* thread]

purgation /pɜː:'geɪʃ(ə)n/ n. **1** purification. **2** purging of the bowels. **3** spiritual cleansing, esp. (*RC Ch.*) of a soul in purgatory. **4** hist. the freeing of oneself from accusation or suspicion by an oath or ordeal. [ME f. OF *purgation* or L *purgatio* (as PURGE)]

purgative /'pɜː:gətɪv/ adj. & n. ● adj. **1** serving to purify. **2** strongly laxative. ● n. **1** a purgative thing. **2** a laxative. [ME f. OF *purgatif -ive* or LL *purgativus* (as PURGE)]

purgatory /'pɜː:gətərɪ/ n. & adj. ● n. (pl. **-ies**) **1** the condition or supposed place of spiritual cleansing, esp. (*RC Ch.*) of those who die in the grace of God but have to expiate venial sins etc. **2** a place or state of temporary suffering or expiation. ● adj. purifying. □ **purgatorial** /,pɜː:gə'tɔ:rɪəl/ adj. [ME f. AF *purgatorie*, OF *-oire* f. med.L *purgatorium*, neut. of LL *purgatorius* (as PURGE)]

purge /pɜː:dʒ/ v. & n. ● v.tr. **1** (often foll. by *of, from*) make physically or spiritually clean. **2** remove by a cleansing process. **3 a** rid (an organization, party, etc.) of persons regarded as undesirable. **b** remove (a person regarded as undesirable) from an organization, party, etc., often violently or by force. **4 a** empty (the bowels). **b** administer a laxative to. **5** Law atone for or wipe out (an offence, esp. contempt of court). ● n. **1 a** the act or an instance of purging. **b** the removal, often in a forcible or violent manner, of people regarded as undesirable from an organization, party, etc. **2** a purgative. □ **purger** n. [ME f. OF *purg(i)er* f. L *purgare* purify f. *purus* pure]

purify /'pjʊərɪfaɪ/ v.tr. (**-ies, -ied**) **1** (often foll. by *of, from*) cleanse or make pure. **2** make ceremonially clean. **3** remove extraneous elements from. □ **purifier** n. **purificatory** /-fɪ,keɪtərɪ/ adj. **purification** /,pjʊərɪfɪ'keɪʃ(ə)n/ n. [ME f. OF *purifier* f. L *purificare* (as PURE)]

Purim /'pjʊərɪm, puː'riːm/ n. a Jewish spring festival commemorating the defeat of Haman's plot to massacre the Jews (Esth. 9). [Heb., pl. of *pūr*, perh. = LOT n. 2]

purine /'pjʊəriːn/ n. **1** Chem. a cyclic organic nitrogenous base forming uric acid on oxidation. **2** any of a group of compounds with a similar structure, including the nucleotide constituents adenine and guanine. [G *Purin* L *purus* pure + *uricum* uric acid + *-in* -INE[4]]

Purism /'pjʊərɪz(ə)m/ n. an early 20th-century artistic movement emphasizing purity of geometric form. It arose out of a rejection of cubism and was characterized by a return to the representation of recognizable objects. Founded by the French architect Le Corbusier and painter Amédée Ozenfant (1886–1966), the movement had a significant influence on modern architecture and design.

purist /'pjʊərɪst/ n. **1** an advocate of scrupulous purity, esp. in language. **2** (**Purist**) an adherent of Purism in art. □ **purism** n. **puristic** /pjʊə'rɪstɪk/ adj. [F *puriste* f. *pur* PURE]

Puritan /'pjʊərɪt(ə)n/ n. & adj. ● n. a member of a group of English Protestants of the late 16th and 17th centuries. (*See note below.*) ● adj. of or relating to the Puritans. □ **Puritanism** n. [LL *puritas* (as PURITY) after earlier *Catharan* (see CATHAR)]

▪ The Puritans were Protestants who, dissatisfied with the elements of Catholicism retained by the Elizabethan religious settlement, sought a further purification of the Church from supposedly unscriptural forms. At first they tried to rid the Church of ornaments, vestments, organs, etc.; from 1570 the more extreme attacked the institution of episcopacy itself, wishing to substitute government by Church elders (Presbyterianism). Oppressed under James I and

Charles I, in particular by Archbishop Laud, many (such as the Pilgrim Fathers) emigrated to the Netherlands and America. The Civil War of the 1640s led to the temporary pre-eminence of Puritanism. Soon, however, the movement fragmented into sects, and the term *Puritan* began to be less used; after the Restoration such people tended to be called Dissenters or Nonconformists.

puritan /'pjʊərɪt(ə)n/ n. & adj. ● n. **1** a purist member of any party. **2** a person practising or affecting extreme strictness in religion or morals. ● adj. scrupulous and austere in religion or morals. □ **puritanism** n. [PURITAN]

puritanical /,pjʊərɪ'tænɪk(ə)l/ adj. often derog. practising or affecting strict religious or moral behaviour. □ **puritanically** adv.

purity /'pjʊərɪtɪ/ n. **1** pureness, cleanness. **2** freedom from physical contamination or moral pollution. [ME f. OF *pureté*, with assim. to LL *puritas -tatis* f. L *purus* pure]

purl[1] /pɜː:l/ adj., n., & v. ● adj. (of a knitting stitch) made by putting the needle through the front of the stitch from right to left (opp. PLAIN[1] adj. 9). ● n. **1** a cord of twisted gold or silver wire for bordering. **2** a chain of minute loops; a picot. **3** the ornamental edges of lace, ribbon, etc. ● v.tr. (also *absol.*) knit with a purl stitch. [orig. *pyrle, pirle* f. Sc. *pirl* twist: the knitting sense may be f. a different word]

purl[2] /pɜː:l/ v. & n. ● v.intr. (of a brook etc.) flow with a swirling motion and babbling sound. ● n. this motion or sound. [16th c.: prob. imit.: cf. Norw. *purla* bubble up]

purler /'pɜː:lə(r)/ n. Brit. colloq. a headlong fall. [*purl* upset, rel. to PURL[1]]

purlieu /'pɜː:ljuː/ n. (pl. **purlieus**) **1** a person's bounds or limits. **2** a person's usual haunts. **3** Brit. hist. a tract on the border of a forest, esp. one earlier included in it and still partly subject to forest laws. **4** (in pl.) the outskirts; an outlying region. [ME *purlew*, prob. alt. after F *lieu* place f. AF *purale(e)*, OF *pourallee* a going round to settle the boundaries f. po(u)raler traverse]

purlin /'pɜː:lɪn/ n. a horizontal beam along the length of a roof, resting on principals and supporting the common rafters or boards. [ME: orig. uncert.]

purloin /pə'lɔɪn/ v.tr. formal or joc. steal, pilfer. □ **purloiner** n. [ME f. AF *purloigner* put away, do away with (as PUR-, *loign* far f. L *longe*)]

purple /'pɜː:p(ə)l/ n., adj., & v. ● n. **1** a colour intermediate between red and blue. **2** (in full **Tyrian purple**) a crimson dye obtained from some molluscs. **3** a purple robe, esp. as the dress of an emperor or senior magistrate. **4** the scarlet official dress of a cardinal. **5** (prec. by *the*) a position of rank, authority, or privilege. ● adj. of a purple colour. ● v.tr. & intr. make or become purple. □ **born in the purple 1** born into a reigning family. **2** belonging to the most privileged class. **purple emperor** a large Eurasian butterfly, *Apatura iris*, the male of which has mainly black wings with purple iridescence. **purple gallinule** a rail of the genus *Porphyrio*, of the Old World, or *Porphyrula*, of America, having mainly bluish iridescent plumage and a red bill. **purple heart** Brit. colloq. a heart-shaped stimulant tablet, esp. of amphetamine. **purple passage** an ornate or elaborate passage in a literary composition. **purple patch** colloq. a period of success or good fortune in sport etc. □ **purpleness** n. **purplish** adj. **purply** adj. [OE, alt. f. *purpure purpuran* f. L *purpura* (as PURPURA)]

Purple Heart n. (in the US) a decoration for those wounded in action.

purport v. & n. ● v.tr. /pə'pɔ:t/ **1** profess; be intended to seem (*purports to be the royal seal*). **2** (often foll. by *that* + clause) (of a document or speech) have as its meaning; state. ● n. /'pɜː:pɔ:t/ **1** the ostensible meaning of something. **2** the sense or tenor of a document or statement). □ **purportedly** /pə'pɔ:tɪdlɪ/ adv. [ME f. AF & OF *purport, porport* f. *purporter* f. med.L *proportare* (as PRO-[1], *portare* carry)]

purpose /'pɜː:pəs/ n. & v. ● n. **1** an object to be attained; a thing intended. **2** the intention to act. **3** resolution, determination. **4** the reason for which something is done or made. ● v.tr. have as one's purpose; design, intend. □ **on purpose** intentionally. **purpose-built** (or **-made**) built or made for a specific purpose. **to no purpose** with no result or effect. **to the purpose 1** relevant. **2** useful. [ME f. OF *porpos, purpos* f. L *proponere* (as PROPOUND)]

purposeful /'pɜː:pəsfʊl/ adj. **1** having or indicating purpose. **2** intentional. **3** resolute. □ **purposefully** adv. **purposefulness** n.

purposeless /'pɜː:pəslɪs/ adj. having no aim or plan. □ **purposelessly** adv. **purposelessness** n.

purposely /'pɜː:pəslɪ/ adv. on purpose; intentionally.

purposive /'pɜː:pəsɪv/ adj. **1** having or serving a purpose. **2** done with a purpose. **3** (of a person or conduct) having purpose or resolution; purposeful. □ **purposively** adv. **purposiveness** n.

purpura /ˈpɜːpjʊrə/ n. Med. a skin rash of purple spots caused by internal bleeding from small blood vessels. □ **purpuric** /pɜːˈpjʊərɪk/ adj. [L f. Gk porphura purple]

purpure /ˈpɜːpjʊə(r)/ n. & adj. Heraldry purple. [OE purpure & OF purpre f. L purpura (as PURPURA)]

purpurin /ˈpɜːpjʊrɪn/ n. a red colouring-matter occurring naturally in madder roots, or manufactured synthetically.

purr /pɜː(r)/ v. & n. ● v. 1 intr. (of a cat) make a low vibratory sound expressing contentment. 2 intr. (of machinery etc.) make a similar sound. 3 intr. (of a person) make contented sounds, express pleasure. 4 tr. utter or express (words or contentment) in this way. ● n. a purring sound. [imit.]

purse /pɜːs/ n. & v. ● n. 1 a small pouch of leather etc. for carrying money on the person. 2 N. Amer. a handbag. 3 a receptacle resembling a purse in form or purpose. 4 money, funds. 5 a sum collected as a present or given as a prize in a contest. ● v. 1 tr. (often foll. by up) pucker or contract (the lips). 2 intr. become contracted and wrinkled. □ **hold the purse-strings** have control of expenditure. **the public purse** the national treasury. [OE purs f. med.L bursa, byrsa purse f. Gk bursa hide, leather]

purser /ˈpɜːsə(r)/ n. an officer on a ship who keeps the accounts, esp. the head steward in a passenger vessel.

purslane /ˈpɜːslɪn, -leɪn/ n. a low succulent plant of the genus Portulaca; esp. P. oleracea, with green or golden leaves, used as a herb and salad vegetable. [ME f. OF porcelaine (cf. PORCELAIN) alt. f. L porcil(l)aca, portulaca]

pursuance /pəˈsjuːəns/ n. (foll. by of) the carrying out or observance (of a plan, idea, etc.).

pursuant /pəˈsjuːənt/ adj. & adv. ● adj. pursuing. ● adv. (foll. by to) conforming to or in accordance with. □ **pursuantly** adv. [ME, = prosecuting, f. OF po(u)rsuiant part. of po(u)rsu(iv)ir (as PURSUE): assim. to AF pursuer and PURSUE]

pursue /pəˈsjuː/ v. (**pursues, pursued, pursuing**) 1 tr. follow with intent to overtake or capture or do harm to. 2 tr. continue or proceed along (a route or course of action). 3 tr. follow or engage in (study or other activity). 4 tr. proceed in compliance with (a plan etc.). 5 tr. seek after, aim at. 6 tr. continue to investigate or discuss (a topic). 7 tr. seek the attention or acquaintance of (a person) persistently. 8 tr. (of misfortune etc.) persistently assail. 9 tr. persistently attend, stick to. 10 intr. go in pursuit. □ **pursuable** adj. **pursuer** n. [ME f. AF pursiwer, -suer = OF porsivre etc. ult. f. L prosequi follow after]

pursuit /pəˈsjuːt/ n. 1 the act or an instance of pursuing. 2 an occupation or activity pursued. □ **in pursuit of** pursuing. [ME f. OF poursuite (as PUR-, SUIT)]

pursuivant /ˈpɜːsɪv(ə)nt/ n. 1 (**Pursuivant**) Brit. an officer of the College of Arms ranking below a herald. 2 archaic a follower or attendant. [ME f. OF pursivant pres. part. of pursivre (as PURSUE)]

pursy /ˈpɜːsɪ/ adj. 1 short-winded; puffy. 2 corpulent. □ **pursiness** n. [ME, earlier pursive f. AF porsif f. OF polsif f. polser breathe with difficulty f. L pulsare (as PULSATE)]

purulent /ˈpjʊərʊlənt/ adj. 1 consisting of or containing pus. 2 discharging pus. □ **purulence** n. **purulency** n. **purulently** adv. [F purulent or L purulentus (as PUS)]

purvey /pəˈveɪ/ v. 1 tr. provide or supply (articles of food) as one's business. 2 intr. (often foll. by for) **a** make provision. **b** act as supplier. □ **purveyor** n. [ME f. AF purveier, OF porveir f. L providere PROVIDE]

purveyance /pəˈveɪəns/ n. 1 the act of purveying. 2 Brit. hist. the right of the sovereign to provisions etc. at a fixed price. [ME f. OF porveance f. L providentia PROVIDENCE]

purview /ˈpɜːvjuː/ n. 1 the scope or range of a document, scheme, etc. 2 the range of physical or mental vision. [ME f. AF purveü, OF porveü past part. of porveir (as PURVEY)]

pus /pʌs/ n. a thick yellowish or greenish liquid produced from infected tissue, consisting of dead bacteria and leucocytes with tissue debris and serum. [L pus puris]

Pusan /puːˈsæn/ an industrial city and seaport on the SE coast of South Korea; pop. (1990) 3,797,570.

Pusey /ˈpjuːzɪ/, Edward Bouverie (1800–82), English theologian. In 1833, while professor of Hebrew at Oxford, he founded the Oxford Movement together with John Henry Newman and John Keble; he became leader of the Movement after the withdrawal of Newman (1841). His many writings include a series of Tracts for the Times and a

statement of his doctrinal views, The Doctrine of the Real Presence (1856–7).

push /pʊʃ/ v. & n. ● v. 1 tr. exert a force on (a thing) to move it away from oneself or from the origin of the force. 2 tr. cause to move in this direction. 3 intr. exert such a force (do not push against the door). 4 tr. press, depress (push the button for service). 5 intr. & tr. **a** thrust forward or upward. **b** project or cause to project (pushes out new roots; the cape pushes out into the sea). 6 intr. move forward by force or persistence. 7 tr. make (one's way) by pushing. 8 intr. exert oneself, esp. to surpass others. 9 tr. (often foll. by to, into, or to + infin.) urge or impel. 10 tr. tax the abilities or tolerance of; press (a person) hard. 11 tr. pursue (a claim etc.). 12 tr. promote the use or sale or adoption of, e.g. by advertising. 13 intr. (foll. by for) demand persistently (pushed hard for reform). 14 tr. colloq. sell (a drug) illegally. ● n. 1 the act or an instance of pushing; a shove or thrust. 2 the force exerted in this. 3 a vigorous effort. 4 a military attack in force. 5 enterprise, determination to succeed. 6 the use of influence to advance a person. 7 the pressure of affairs. 8 a crisis. 9 (prec. by the) colloq. dismissal, esp. from employment. 10 Austral. sl. a group of people with a common interest; a clique. □ **be pushed for** colloq. have very little of (esp. time). **get the push** colloq. be dismissed or sent away. **give a person the push** colloq. dismiss or send away a person. **push about** colloq. = push around. **push along** (often in imper.) colloq. depart, leave. **push around** colloq. 1 move (a person) roughly from place to place. 2 bully. **push-bike** Brit. colloq. a bicycle worked by pedals. **push-button 1** a button to be pushed esp. to operate an electrical device. 2 (attrib.) operated in this way. **push one's luck 1** take undue risks. 2 act presumptuously. **push off 1** push with an oar etc. to get a boat out into a river etc. 2 (often in imper.) colloq. go away. **push-pull 1** operated by pushing and pulling. 2 Electr. consisting of two valves etc. operated alternately. **push-start** n. the starting of a motor vehicle by pushing it to turn the engine. ● v.tr. start (a vehicle) in this way. **push through** get (a scheme, proposal, etc.) completed or accepted quickly. **push-up** = press-up. [ME f. OF pousser, pou(l)ser f. L pulsare (as PULSATE)]

pushcart /ˈpʊʃkɑːt/ n. a handcart or barrow.

pushchair /ˈpʊʃtʃeə(r)/ n. Brit. a folding chair on wheels, in which a child can be pushed along.

pusher /ˈpʊʃə(r)/ n. 1 colloq. an illegal seller of drugs. 2 colloq. a pushing or pushy person. 3 a child's utensil for pushing food on to a spoon etc.

pushful /ˈpʊʃfʊl/ adj. pushy; arrogantly self-assertive. □ **pushfully** adv.

pushing /ˈpʊʃɪŋ/ adj. 1 pushy; aggressively ambitious. 2 colloq. having nearly reached (a specified age).

Pushkin /ˈpʊʃkɪn/, Aleksandr (Sergeevich) (1799–1837), Russian poet, novelist, and dramatist. His revolutionary beliefs and atheistic writings led to his dismissal from the civil service and eventual internal exile; he was rehabilitated in 1826 after the accession of Nicholas I. A leading figure in Russian literature, he wrote prolifically in many genres; his first success was the romantic narrative poem Ruslan and Ludmilla (1820). Other notable works include the verse novel Eugene Onegin (1833), and the blank-verse historical drama Boris Godunov (1831). He was fatally wounded in a duel with his wife's admirer.

pushover /ˈpʊʃˌəʊvə(r)/ n. colloq. 1 something that is easily done. 2 a person who can easily be overcome, persuaded, etc.

pushrod /ˈpʊʃrɒd/ n. a rod operated by cams, that opens and closes the valves in an internal-combustion engine.

Pushtu /ˈpʌʃtuː/ n. & adj. = PASHTO. [Pers. puštū]

pushy /ˈpʊʃɪ/ adj. (**pushier, pushiest**) colloq. 1 excessively or unpleasantly self-assertive. 2 selfishly determined to succeed. □ **pushily** adv. **pushiness** n.

pusillanimous /ˌpjuːsɪˈlænɪməs/ adj. lacking courage; timid. □ **pusillanimously** adv. **pusillanimity** /-ləˈnɪmɪtɪ/ n. [eccl.L pusillanimis f. pusillus very small + animus mind]

Puskas /ˈpʊʃkəs/, Ferenc (b.1927), Hungarian footballer. He came to prominence in the celebrated Hungarian national team of the early 1950s. In 1956 he left Hungary to play for Real Madrid, scoring four goals in their 1960 European Cup Final victory and a hat trick in the corresponding 1962 final, in which Real Madrid lost.

puss /pʊs/ n. colloq. 1 a cat (esp. as a form of address). 2 a playful or coquettish girl. 3 a hare. □ **puss moth** a large European moth, Cerura vinula, with a very fluffy body. [prob. f. MLG pūs, Du. poes, of unkn. orig.]

pussy /ˈpʊsɪ/ n. (pl. **-ies**) 1 (also **pussy-cat**) colloq. a cat. 2 coarse sl. **a** the female genitals. **b** women considered sexually. □ **pussy willow** a willow with soft fluffy catkins appearing before the leaves.

pussyfoot /'pʊsɪˌfʊt/ v.intr. **1** move stealthily or warily. **2** act cautiously or noncommittally. □ **pussyfooter** n.

pustulate v. & adj. ● v.tr. & intr. /'pʌstjʊˌleɪt/ form into pustules. ● adj. /'pʌstjʊlət/ of or relating to a pustule or pustules. □ **pustulation** /ˌpʌstjʊ'leɪʃ(ə)n/ n. [LL pustulare f. pustula: see PUSTULE]

pustule /'pʌstjuːl/ n. a pimple containing pus. □ **pustular** adj. **pustulous** adj. [ME f. OF pustule or L pustula]

put[1] /pʊt/ v. & n. ● v. (**putting**; past and past part. **put**) **1** tr. move to or cause to be in a specified place or position (put it in your pocket; put the children to bed; put your signature here). **2** tr. bring into a specified condition, relation, or state (puts me in great difficulty; an accident put the car out of action). **3** tr. **a** (often foll. by on) impose or assign (put a tax on beer; where do you put the blame?). **b** (foll. by on, to) impose or enforce the existence of (put a veto on it; put a stop to it). **4** tr. **a** cause (a person) to go or be, habitually or temporarily (put them at their ease; put them on the right track). **b** refl. imagine (oneself) in a specified situation (put yourself in my shoes). **5** tr. (foll. by for) substitute (one thing for another). **6** tr. express (a thought or idea) in a specified way (to put it mildly). **7** tr. (foll. by at) estimate (an amount etc. at a specified amount) (put the cost at £50). **8** tr. (foll. by into) express or translate in (words, or another language). **9** tr. (foll. by into) invest (money in an asset, e.g. land). **10** tr. (foll. by on) stake (money) on (a horse etc.). **11** tr. (foll. by to) apply or devote to a use or purpose (put it to good use). **12** tr. (foll. by to) submit for consideration or attention (let me put it to you another way; shall now put it to a vote). **13** tr. (foll. by to) subject (a person) to (death, suffering, etc.). **14** tr. throw (esp. a shot or weight) as an athletic sport or exercise. **15** tr. (foll. by to) couple (an animal) with (another of the opposite sex) for breeding. **16** intr. (foll. by back, off, out, etc.) (of a ship etc.) proceed or follow a course in a specified direction. **17** intr. (foll. by in, out of) US (of a river) flow in a specified direction. ● n. **1** a throw of the shot or weight. **2** Stock Exch. the option of selling stock at a fixed price at a given date. □ **not know where to put oneself** feel deeply embarrassed. **put about 1** spread (information, rumour, etc.). **2** Naut. turn round; put (a ship) on the opposite tack. **3** trouble, distress. **put across 1** make acceptable or effective. **2** express in an understandable way. **put aside 1** = put by. **2** set aside, ignore. **put away 1** put (a thing) back in the place where it is normally kept. **2** lay (money etc.) aside for future use. **3 a** confine or imprison. **b** commit to a home or mental institution. **4** colloq. consume (food and drink), esp. in large quantities. **5** put (an old or sick animal) to death. **put back 1** restore to its proper or former place. **2** change (a planned event) to a later date or time. **3** move back the hands of (a clock or watch). **4** check the advance of. **put a bold** etc. **face on it** see FACE. **put the boot in** see BOOT[1]. **put by** lay (money etc.) aside for future use. **put down 1** suppress by force or authority. **2** colloq. snub or humiliate. **3** record or enter in writing. **4** enter the name of (a person) on a list, esp. as a member or subscriber. **5** (foll. by as, for) account or reckon. **6** (foll. by to) attribute (put it down to bad planning). **7** put (an old or sick animal) to death. **8** preserve or store (eggs etc.) for future use. **9** pay (a specified sum) as a deposit. **10** put (a baby) to bed. **11** land (an aircraft). **12** stop to let (passengers) get off. **put-down** n. colloq. a snub or humiliating criticism. **put an end to** see END. **put one's foot down** see FOOT. **put one's foot in it** see FOOT. **put forth 1** (of a plant) send out (buds or leaves). **2** formal submit or put into circulation. **put forward 1** suggest or propose. **2** advance the hands of (a clock or watch). **3** (often refl.) put into a prominent position; draw attention to. **put in 1** enter or submit (a claim etc.). **b** (foll. by for) submit a claim for (a specified thing). **2** (foll. by for) be a candidate for (an appointment, election, etc.). **3** spend (time). **4** perform (a spell of work) as part of a whole. **5** interpose (a remark, blow, etc.). **6** insert as an addition. **put in an appearance** see APPEARANCE. **put a person in mind of** see MIND. **put it across** colloq. get the better of, deceive (a person). **put it to a person** (often foll. by that + clause) challenge a person to deny. **put one's mind to** see MIND. **put off 1 a** postpone. **b** postpone an engagement with (a person). **2** (often foll. by with) evade (a person) with an excuse etc. **3** hinder or dissuade. **4** offend, disconcert; cause (a person) to lose interest in something. **put on 1** clothe oneself with. **2** cause (an electrical device, light, etc.) to function. **3** cause (esp. transport) to be available; provide. **4** stage (a play, show, etc.). **5** advance the hands of (a clock or watch). **6 a** pretend to be affected by (an emotion). **b** assume, take on (a character or appearance). **c** (**put it on**) exaggerate one's feelings etc. **7** increase one's weight by (a specified amount). **8** send (a cricketer) on to bowl. **9** (foll. by to) make aware of or put in touch with (put us on to their new accountant). **10** colloq. tease, play a trick on. **put-on** n. colloq. a deception or hoax. **put one across** (or **over**) (foll. by on) colloq. get the better of; trick. **put out 1 a** (often as **put out** adj.)

disconcert or annoy. **b** (often refl.) inconvenience (don't put yourself out). **2** extinguish (a fire or light). **3** cause (a batsman or side) to be out. **4** dislocate (a joint). **5** exert (strength etc.). **6** lend (money) at interest. **7** allocate (work) to be done off the premises. **8** blind (a person's eyes). **9** issue, publish. **put out of its** etc. **misery** see MISERY. **put over 1** make acceptable or effective. **2** express in an understandable way. **3** US postpone. **4** US achieve by deceit. **put a sock in it** see SOCK[1]. **put store by** see STORE. **put through 1** carry out or complete (a task or transaction). **2** (often foll. by to) connect (a person) by telephone to another subscriber. **put to flight** see FLIGHT[2]. **put together 1** assemble (a whole) from parts. **2** combine (parts) to form a whole. **put under** render unconscious by anaesthetic etc. **put up 1** build or erect. **2** raise (a price etc.). **3** take or provide accommodation for (friends put me up for the night). **4** engage in (a fight, struggle, etc.) as a form of resistance. **5** present (a proposal). **6 a** present oneself for election. **b** propose for election. **7** provide (money) as a backer in an enterprise. **8** display (a notice). **9** publish (banns). **10** offer for sale or competition. **11** cause (game) to rise from cover. **12** put (a sword) back in its sheath. **put-up** adj. (usu. in phr. **put-up job**) fraudulently presented or devised. **put upon** colloq. make unfair or excessive demands on; take advantage of (a person). **put a person up to** inform or instruct a person about. **2** (usu. foll. by verbal noun) instigate a person in (put them up to stealing the money). **put up with** endure, tolerate; submit to. **put the wind up** see WIND[1]. **put a person wise** see WISE[1]. **put words into a person's mouth** see MOUTH. □ **putter** n. [ME f. an unrecorded OE form putian, f. unkn. orig.]

put[2] var. of PUTT.

putative /'pjuːtətɪv/ adj. reputed, supposed (his putative father). □ **putatively** adv. [ME f. OF putatif -ive or LL putativus f. L putare think]

putlog /'pʌtlɒg/ n. (also **putlock** /-lɒk/) a short horizontal timber projecting from a wall, on which scaffold floorboards rest. [17th c.: orig. uncert.]

put-put /'pʌtpʌt/ n. & v. ● n. the rapid intermittent sound of a small petrol engine. ● v.intr. (**put-putted, put-putting**) make this sound. [imit.]

putrefy /'pjuːtrɪˌfaɪ/ v. (**-ies, -ied**) **1** intr. & tr. become or make putrid; go bad. **2** intr. fester, suppurate. **3** intr. become morally corrupt. □ **putrefacient** /ˌpjuːtrɪ'feɪʃ(ə)nt/ adj. **putrefaction** /-'fækʃ(ə)n/ n. **putrefactive** /-'fæktɪv/ adj. [ME f. L putrefacere f. puter putris rotten]

putrescent /pjuː'tres(ə)nt/ adj. **1** in the process of rotting. **2** of or accompanying this process. □ **putrescence** n. [L putrescere inceptive of putrere (as PUTRID)]

putrid /'pjuːtrɪd/ adj. **1** decomposed, rotten. **2** foul, noxious. **3** corrupt. **4** sl. of poor quality; contemptible; very unpleasant. □ **putridly** adv. **putridness** n. **putridity** /pjuː'trɪdɪtɪ/ n. [L putridus f. putrere to rot f. puter putris rotten]

putsch /pʊtʃ/ n. an attempt at political revolution; a violent uprising. [Swiss G, = thrust, blow]

putt /pʌt/ v. & n. (also **put**) ● v.tr. (**putted, putting**) strike (a golf ball) gently so that it rolls across the green into or nearer to the hole. ● n. a putting stroke. □ **putting-green** (in golf) the smooth area of grass round a hole. [differentiated f. PUT[1]]

puttee /'pʌtɪ/ n. **1** a long strip of cloth wound spirally round the leg from ankle to knee for protection and support. **2** US a leather legging. [Hindi paṭṭī band, bandage]

putter[1] /'pʌtə(r)/ n. **1** a golf club used in putting. **2** a golfer who putts.

putter[2] /'pʌtə(r)/ n. & v. = PUT-PUT. [imit.]

putter[3] US var. of POTTER[1].

putto /'pʊtəʊ/ n. (pl. **putti** /-tɪ/) a representation of a naked child (esp. a cherub or a cupid) in (esp. Renaissance) art. [It., = boy, f. L putus]

putty /'pʌtɪ/ n. & v. ● n. (pl. **-ies**) **1** a cement made from whiting and raw linseed oil, used for fixing panes of glass, filling holes in woodwork, etc. **2** a fine white mortar of lime and water, used in pointing brickwork, etc. **3** a polishing powder usu. made from tin oxide, used in jewellery work. ● v.tr. (**-ies, -ied**) cover, fix, join, or fill up with putty. □ **be putty in a person's hands** be subservient to, or easily influenced by, a person. [F potée, lit. 'potful']

puy /pwiː/ n. a small extinct volcanic cone, esp. in the Auvergne, France. [F, = hill, f. L podium: see PODIUM]

puzzle /'pʌz(ə)l/ n. & v. ● n. **1** a difficult or confusing problem; an enigma. **2** a problem or toy designed to test knowledge or ingenuity (crossword puzzle; jigsaw puzzle). ● v. **1** tr. confound or disconcert mentally. **2** intr. (usu. foll. by over etc.) be perplexed (about). **3** tr. (usu. as **puzzling** adj.) require much thought to comprehend (a puzzling situation). **4** tr.

(foll. by *out*) solve or understand by hard thought. □ **puzzlement** *n.* **puzzlingly** *adv.* [16th c.: orig. unkn.]

puzzler /ˈpʌzlə(r)/ *n.* a difficult question or problem.

puzzolana var. of POZZOLANA.

PVA *abbr.* polyvinyl acetate.

PVC *abbr.* polyvinyl chloride.

PVS *abbr. Med.* persistent vegetative state.

Pvt. *abbr.* **1** private. **2** *US* private soldier.

PW *abbr.* policewoman.

p.w. *abbr.* per week.

PWA *abbr.* person with Aids.

PWR *abbr.* pressurized-water reactor.

PWV *abbr.* Pretoria–Witwatersrand–Vereeniging.

PX *abbr. US* post exchange.

pya /pjɑː/ *n.* a monetary unit of Burma (Myanmar), equal to one-hundredth of a kyat. [Burmese]

pyaemia /paɪˈiːmɪə/ *n.* (*US* **pyemia**) *Med.* blood-poisoning caused by the spread of pus-forming bacteria in the bloodstream from a source of infection. □ **pyaemic** *adj.* [mod.L f. Gk *puon* pus + *haima* blood]

pycnic var. of PYKNIC.

pye-dog /ˈpaɪdɒg/ *n.* (also **pie-dog, pi-dog**) a vagrant mongrel, esp. in Asia. [Anglo-Ind. *pye, paë,* Hindi *pāhī* outsider + DOG]

pyelitis /ˌpaɪəˈlaɪtɪs/ *n. Med.* inflammation of the renal pelvis. [Gk *puelos* trough, basin + -ITIS]

pyemia *US* var. of PYAEMIA.

Pygmalion[1] /pɪgˈmeɪlɪən/ *Gk Mythol.* a king of Cyprus who fashioned an ivory statue of a beautiful woman and loved it so deeply that in answer to his prayer Aphrodite gave it life. The woman (at some point named Galatea) bore him a daughter, Paphos.

Pygmalion[2] /pɪgˈmeɪlɪən/ a legendary king of Tyre, brother of Elissa (Dido), whose husband he killed in the hope of obtaining his fortune.

pygmy /ˈpɪgmɪ/ *n.* (also **pigmy**) (*pl.* **-ies**) **1** a member of any of several peoples of very short stature. (*See note below.*) **2** a very small person, animal, or thing. **3** an insignificant person. **4** (*attrib.*) **a** of or relating to pygmies. **b** (of a person, animal, etc.) dwarf. □ **pygmaean** /pɪgˈmiːən/ *adj.* **pygmean** *adj.* [ME f. L *pygmaeus* f. Gk *pugmaios* dwarf f. *pugmē* the length from elbow to knuckles, fist]

■ Peoples known as pygmies inhabit areas of equatorial Africa and of SE Asia. They are typically dark-skinned, nomadic hunter-gatherers having an average male height of not greater than 150 cm (4 ft 11 in.), e.g. the Twa of Burundi, Rwanda, and Zaire. The term *Negrito* is also used in reference to such populations of SE Asia.

pyjamas /pəˈdʒɑːməz/ *n.pl.* (*US* **pajamas**) **1** a suit of loose trousers and jacket for sleeping in. **2** loose trousers tied round the waist, worn by both sexes in some Asian countries. **3** (**pyjama**) (*attrib.*) designating parts of a suit of pyjamas (*pyjama jacket; pyjama trousers*). [Urdu pā(ē)jāma f. Pers. *pae,* pay leg + Hindi *jāma* clothing]

pyknic /ˈpɪknɪk/ *adj.* & *n.* (also **pycnic**) *Anthropol.* ● *adj.* characterized by a thick neck, large abdomen, and relatively short limbs. ● *n.* a person of this bodily type. [Gk *puknos* thick]

pylon /ˈpaɪlən/ *n.* **1** a tall structure erected as a support (esp. for electric power cables) or boundary or decoration. **2** a gateway or gate-tower, esp. the monumental gateway to an Egyptian temple, usually formed by two truncated pyramidal towers connected by a lower architectural member containing the gate. **3** a structure marking a path for aircraft. **4** a structure on an aircraft's wing supporting an engine, weapon, etc. [Gk *pulōn* f. *pulē* gate]

pylorus /paɪˈlɔːrəs/ *n.* (*pl.* **pylori** /-raɪ/) *Anat.* the opening from the stomach into the duodenum. □ **pyloric** /-ˈlɒrɪk/ *adj.* [LL f. Gk *pulōros, pulouros* gatekeeper f. *pulē* gate + *ouros* warder]

Pyongyang /pjɒŋˈjæŋ/ the capital of North Korea; pop. (est. 1984) 2,639,450. The oldest city on the Korean peninsula, it was first mentioned in records of 108 BC. It fell to the Japanese in the late 16th century and was devastated by the Manchus in the early 17th century. It developed as an industrial city during the years of Japanese occupation, from 1910 to 1945.

pyorrhoea /ˌpaɪəˈrɪə/ *n.* (*US* **pyorrhea**) *Med.* **1** a disease of periodontal tissue causing shrinkage of the gums and loosening of the teeth. **2** a discharge of pus. [Gk *puo-* f. *puon* pus + *rhoia* flux f. *rheō* flow]

pyracantha /ˌpaɪrəˈkænθə/ *n.* an ornamental evergreen thorny shrub of the genus *Pyracantha,* having white flowers and bright red or yellow berries. Also called *firethorn.* [L f. Gk *purakantha*]

pyralid /ˈpaɪrælɪd, -ˈreɪlɪd/ *n.* & *adj.* ● *n.* a small moth of the family Pyralidae. ● *adj.* of or relating to this family. [mod.L f. Gk *puralis* a mythical fly said to live in fire]

pyramid /ˈpɪrəmɪd/ *n.* **1** a monumental structure, usu. of stone, with a square base and sloping sides meeting centrally at an apex, esp. an ancient Egyptian royal tomb. (*See note below.*) **2** a polyhedron or solid figure of this type with a base of three or more sides. **3** a pyramid-shaped heap or pile of things. **4** (in *pl.*) a game played on a billiard-table with (usu. 15) coloured balls and a cue-ball. □ **pyramid selling** a system of selling goods in which agency rights are sold to an increasing number of distributors at successively lower levels. □ **pyramidal** /pɪˈræmɪd(ə)l/ *adj.* **pyramidally** *adv.* **pyramidic** /ˌpɪrəˈmɪdɪk/ *adj.* **pyramidical** *adj.* **pyramidically** *adv.* [ME f. L *pyramis* f. Gk *puramis -idos*]

■ The pyramid is the characteristic form of tomb built for Egyptian pharaohs from the 3rd dynasty (*c.*2649 BC) until *c.*1640 BC. There are two principal types: the step pyramid, consisting of several stepped levels rising to a flat top, and the true pyramid, which developed from this and was introduced in the 4th dynasty (*c.*2575 BC, e.g. at Giza). The pyramid was the focal point of a vast funerary complex, including a temple at its side, linked by a causeway to a lower temple near the cultivated land and flood-waters of the river. The exact building procedure, and the mathematical calculations involved, have not been fully explained. The pyramids are imposing in their austere simplicity; those at Giza were one of the Seven Wonders of the World. Monuments of similar shape are associated with the civilizations of Meso-America and South America *c.*1200 BC–AD 750. They were built by the Aztecs and Mayas as centres of worship, usually as part of a complex of courtyards, platforms, and temples.

Pyramus /ˈpɪrəməs/ *Rom. Mythol.* a Babylonian youth, lover of Thisbe. Forbidden to marry by their parents, who were neighbours, the lovers conversed through a chink in a wall and agreed to meet at a tomb outside the city. There, Thisbe was frightened away by a lioness coming from its kill, and Pyramus, seeing her bloodstained cloak and supposing her dead, stabbed himself. Thisbe, finding his body when she returned, threw herself upon his sword. Their blood stained a mulberry-tree, whose fruit has ever since been black when ripe, in sign of mourning for them. The story, known from Ovid, also features in Shakespeare's *A Midsummer Night's Dream.*

pyre /ˈpaɪə(r)/ *n.* a heap of combustible material, esp. a funeral pile for burning a corpse. [L *pyra* f. Gk *pura* f. *pur* fire]

Pyrenees /ˌpɪrəˈniːz/ a range of mountains extending along the border between France and Spain from the Atlantic coast to the Mediterranean. Its highest peak is the Pico de Aneto in northern Spain, which rises to a height of 3,404 m (11,168 ft). □ **Pyrenean** /-ˈniːən/ *adj.*

pyrethrin /paɪˈriːθrɪn/ *n. Chem.* any of a class of compounds found in pyrethrum flowers and used in the manufacture of insecticides.

pyrethroid /paɪˈriːθrɔɪd/ *n. Chem.* any of a group of substances similar to pyrethrins in structure and properties.

pyrethrum /paɪˈriːθrəm/ *n.* **1** an aromatic composite plant of the genus *Tanacetum* (formerly *Pyrethrum*); esp. *T. coccineum,* with brightly coloured flowers. **2** an insecticide made from the dried flowers of these plants, esp. *Tanacetum cinerariifolium.* [L f. Gk *purethron* feverfew]

pyretic /paɪˈretɪk, pɪ-/ *adj.* of, for, or producing fever. [mod.L *pyreticus* f. Gk *puretos* fever]

Pyrex /ˈpaɪəreks/ *n. propr.* a hard heat-resistant type of glass, often used for ovenware. [invented word]

pyrexia /paɪˈreksɪə, pɪ-/ *n. Med.* = FEVER. □ **pyrexial** *adj.* **pyrexic** *adj.* **pyrexical** *adj.* [mod.L f. Gk *purexis* f. *puressō* be feverish f. *pur* fire]

pyridine /ˈpɪrɪˌdiːn/ *n. Chem.* a colourless volatile odorous liquid (chem. formula: C_5H_5N), originally obtained from coal tar, used as a solvent and in chemical manufacture. [Gk *pur* fire + -ID[4] + -INE[4]]

pyridoxine /ˌpɪrɪˈdɒksiːn, -sɪn/ *n.* a vitamin of the B complex found in yeast, and important in the body's use of unsaturated fatty acids. Also called *vitamin B₆.* [PYRIDINE + OX- + -INE[4]]

pyrimidine /paɪˈrɪmɪˌdiːn/ *n.* **1** *Chem.* a cyclic organic nitrogenous base (chem. formula: $C_4H_4N_2$). **2** any of a group of compounds with similar structure, including the nucleotide constituents uracil, thymine, and cytosine. [G *Pyrimidin* f. *Pyridin* (as PYRIDINE, IMIDE)]

pyrite /ˈpaɪraɪt/ *n.* = PYRITES. [F *pyrite* or L (as PYRITES)]

pyrites /paɪˈraɪtiːz/ *n.* (in full **iron pyrites**) a yellow lustrous form of iron disulphide. □ **pyritic** /-ˈrɪtɪk/ *adj.* **pyritiferous** /ˌpaɪərɪˈtɪfərəs/ *adj.* **pyritize** /ˈpaɪərɪˌtaɪz/ *v.tr.* (also **-ise**). **pyritous** /-təs/ *adj.* [L f. Gk *puritēs* of fire (*pur*)]

pyro /ˈpaɪərəʊ/ n. colloq. = PYROGALLIC ACID.

pyro- /ˈpaɪərəʊ/ comb. form **1** denoting fire. **2** Chem. denoting a new substance formed from another by elimination of water (pyrophosphate). **3** Mineral. denoting a mineral etc. showing some property or change under the action of heat, or having a fiery red or yellow colour. [Gk puro- f. pur fire]

pyroclastic /ˌpaɪərəʊˈklæstɪk/ adj. & n. Geol. ● adj. of or formed from fragments of rock from a volcanic eruption. ● n. (in pl.) pyroclastic rocks or rock fragments. □ **pyroclastic flow** a dense, destructive mass of very hot ash, lava fragments, and gases ejected explosively from a volcano and often flowing at high speed. □ **pyroclast** /ˈpaɪərəʊˌklæst/ n.

pyroelectric /ˌpaɪərəʊɪˈlektrɪk/ adj. having the property of becoming electrically charged when heated. □ **pyroelectricity** /-ˌɪlekˈtrɪsɪtɪ, -ˌel-/ n.

pyrogallic acid /ˌpaɪərəʊˈgælɪk/ n. Chem. a weak acid used as a developer in photography, etc.

pyrogallol /ˌpaɪərəʊˈgælɒl/ n. = PYROGALLIC ACID.

pyrogenic /ˌpaɪərəʊˈdʒenɪk/ adj. (also **pyrogenous** /paɪˈrɒdʒɪnəs/) **1 a** producing heat, esp. in the body. **b** producing fever. **2** produced by combustion or volcanic processes.

pyrography /paɪˈrɒgrəfɪ/ n. = poker-work (see POKER¹).

pyrolyse /ˈpaɪərəˌlaɪz/ v.tr. (US **pyrolyze**) decompose by pyrolysis. [PYROLYSIS after analyse]

pyrolysis /paɪˈrɒlɪsɪs/ n. chemical decomposition brought about by heat. □ **pyrolytic** /ˌpaɪərəˈlɪtɪk/ adj.

pyromania /ˌpaɪərəʊˈmeɪnɪə/ n. an obsessive desire to set fire to things. □ **pyromaniac** /-nɪˌæk/ n.

pyrometer /paɪˈrɒmɪtə(r)/ n. an instrument for measuring high temperatures, esp. in furnaces and kilns. □ **pyrometry** n. **pyrometric** /ˌpaɪərəˈmetrɪk/ adj. **pyrometrically** adv.

pyrope /ˈpaɪərəʊp/ n. a deep red variety of garnet. [ME f. OF pirope f. L pyropus f. Gk purōpos gold-bronze, lit. 'fiery-eyed', f. pur fire + ōps eye]

pyrophoric /ˌpaɪərəʊˈfɒrɪk/ adj. (of a substance) liable to ignite spontaneously on exposure to air. [mod.L pyrophorus f. Gk purophoros fire-bearing f. pur fire + pherō bear]

pyrosis /paɪˈrəʊsɪs/ n. Med. a burning sensation in the lower part of the chest, combined with the return of gastric acid to the mouth. [mod.L f. Gk purōsis f. puroō set on fire f. pur fire]

pyrotechnic /ˌpaɪərəʊˈteknɪk/ adj. **1** of or relating to fireworks. **2** (of wit etc.) brilliant or sensational. □ **pyrotechnical** adj. **pyrotechnist** n. **pyrotechny** /ˈpaɪərəʊˌteknɪ/ n. [PYRO- + Gk tekhnē art]

pyrotechnics /ˌpaɪərəʊˈteknɪks/ n.pl. **1** the art of making fireworks. **2** a display of fireworks. **3** any brilliant display.

pyroxene /paɪˈrɒksiːn/ n. any of a group of minerals commonly found as components of igneous rocks, composed of silicates of calcium, magnesium, and iron. [PYRO- + Gk xenos stranger (because supposed to be alien to igneous rocks)]

pyroxylin /paɪˈrɒksɪlɪn/ n. a form of nitrocellulose, soluble in ether and alcohol, used as a basis for lacquers, artificial leather, etc. [F pyroxyline (as PYRO-, Gk xulon wood)]

Pyrrha /ˈpɪrə/ Gk Mythol. the wife of Deucalion.

pyrrhic¹ /ˈpɪrɪk/ adj. (of a victory) won at too great a cost to be of use to the victor. [PYRRHUS]

pyrrhic² /ˈpɪrɪk/ n. & adj. ● n. a metrical foot of two short or unaccented syllables. ● adj. written in or based on pyrrhics. [L pyrrhichius f. Gk purrhikhios (pous) pyrrhic (foot)]

Pyrrho /ˈpɪrəʊ/ (c.365–c.270 BC), Greek philosopher. Regarded as the founder of scepticism, he established the Pyrrhonic school of philosophy at Elis. He held that certainty of knowledge is impossible and that true happiness must therefore come from suspending judgement.

Pyrrhonism /ˈpɪrəˌnɪz(ə)m/ n. **1** the philosophy of Pyrrho of Elis, maintaining that certainty of knowledge is unattainable. **2** scepticism; philosophic doubt. □ **Pyrrhonist** n. **Pyrrhonic** /pɪˈrɒnɪk/ adj. [Gk Purrhōn PYRRHO]

Pyrrhus /ˈpɪrəs/ (c.318–272 BC), king of Epirus c.307–272. After invading Italy in 280, he defeated the Romans at Asculum in 279, but sustained heavy losses; the term pyrrhic victory is named in allusion to this.

pyrrole /ˈpɪrəʊl/ n. Chem. a sweet-smelling weakly basic liquid (chem. formula: C_4H_4NH), present in coal tar. Its molecule contains a five-membered aromatic ring which is found in many biological molecules, e.g. porphyrins. [f. Gk purrhos reddish + -OLE]

pyrrolidine /pɪˈrɒlɪˌdiːn/ n. Chem. a pungent liquid (chem. formula: C_4H_8NH), made by reduction of pyrrole. [PYRROLE + -IDE + -INE⁴]

pyruvic acid /paɪˈruːvɪk/ n. Biochem. a weak acid (chem. formula: $CH_3COCOOH$) occurring as an intermediate in many metabolic pathways, esp. the Krebs cycle. □ **pyruvate** /-veɪt/ n. [as PYRO- + L uva grape]

Pythagoras /paɪˈθægərəs/ (known as Pythagoras of Samos) (c.560–480 BC), Greek philosopher. Pythagoras is said to have discovered the numerical ratios determining the intervals of the musical scale, leading to his attempt at interpreting the entire physical world in terms of numbers, and founding their systematic (and mystical) study. In astronomy, his analysis of the courses of the sun, moon, and stars into circular motions was not set aside until the 17th century. Pythagoras also founded a secret religious, political, and scientific sect in Italy: the Pythagoreans held that the soul is condemned to a cycle of reincarnation, from which it may escape by attaining a state of purity. □ **Pythagorean** /-ˌθægəˈriːən/ adj. & n.

Pythagoras' theorem n. the geometrical theorem, said to have been discovered by Pythagoras, that the square on the hypotenuse of a right-angled triangle is equal to the sum of the squares on the other two sides.

Pythia /ˈpɪθɪə/ the priestess of Apollo at Delphi in ancient Greece (see DELPHI). [as PYTHIAN]

Pythian /ˈpɪθɪən/ adj. & n. ● adj. of or relating to Delphi or its ancient oracle of Apollo. ● n. the god Apollo or his priestess at Delphi. □ **Pythian games** the games celebrated by the ancient Greeks every four years at Delphi. [L Pythius f. Gk Puthios f. Puthō, an older name of Delphi]

Pythias /ˈpɪθɪˌæs/ see DAMON.

python /ˈpaɪθən/ n. a large non-venomous snake of the family Pythonidae, found mainly in the Old World tropics. Pythons kill their prey by compressing and asphyxiating it. □ **pythonic** /paɪˈθɒnɪk/ adj. [L f. Gk Puthōn a huge serpent or monster killed by Apollo]

pythoness /ˈpaɪθənɪs/ n. **1** the Pythian priestess. **2** a witch. [ME f. OF phitonise f. med.L phitonissa f. LL pythonissa fem. of pythō f. Gk puthōn soothsaying demon: cf. PYTHON]

pyuria /paɪˈjʊərɪə/ n. Med. the presence of pus in the urine. [Gk puon pus + -URIA]

pyx /pɪks/ n. (also **pix**) **1** Eccl. the vessel in which the consecrated bread of the Eucharist is kept. **2** (in the UK) a box at the Royal Mint in which specimen gold and silver coins are deposited. □ **trial of the pyx** an annual test of specimen coins at the Royal Mint by a group of members of the Goldsmith's Company. [ME f. L (as PYXIS)]

pyxidium /pɪkˈsɪdɪəm/ n. (pl. **pyxidia** /-dɪə/) Bot. a seed-capsule with a top that comes off like the lid of a box. [mod.L f. Gk puxidion, dimin. of puxis: see PYXIS]

pyxis /ˈpɪksɪs/ n. (pl. **pyxides** /-sɪˌdiːz/) **1** a small box or casket. **2** = PYXIDIUM. [ME f. L f. Gk puxis f. puxos BOX³]

pzazz var. of PIZAZZ.

Qq

Q¹ /kjuː/ n. (also **q**) (pl. **Qs** or **Q's**) the seventeenth letter of the alphabet.

Q² abbr. (also **Q.**) **1** Queen, Queen's. **2** question. **3** (in Christian theology) used to denote the hypothetical source of the passages shared by the gospels of Matthew and Luke, but not found in Mark. [sense 3 probably from German *Quelle* source]

Qabis see GABÈS.

Qaddafi see GADDAFI.

Qafsah see GAFSA.

Qantas /ˈkwɒntəs/ n. the international airline of Australia. [abbr. of Queensland *and* Northern Territory Aerial Service]

Qaraghandy /ˈkærəˌɡændɪ/ (Russian **Karaganda** /ˌkaraɡanˈdaː/) an industrial city in eastern Kazakhstan, at the centre of a major coal-mining region; pop. (1990) 613,000.

QARANC abbr. Queen Alexandra's Royal Army Nursing Corps.

Qatar /kæˈtɑː(r), ˈkʌtə(r)/ a sheikhdom occupying a peninsula on the west coast of the Persian Gulf; pop. (est. 1991) 402,000; official language, Arabic; capital, Doha. The country was a British protectorate from 1916 until 1971, when it became a sovereign independent state. Oil is the chief source of revenue. □ **Qatari** adj. & n.

Qattara Depression /kəˈtɑːrə/ an extensive, low-lying, and largely impassable area of desert in NE Africa, to the west of Cairo, that falls to 133 m (436 ft) below sea level.

QB abbr. Queen's Bench.

QC abbr. Law Queen's Counsel.

QCD abbr. quantum chromodynamics.

Q-Celtic see GOIDELIC.

QED abbr. quod erat demonstrandum.

Q fever /kjuː/ n. a mild febrile disease caused by rickettsiae. [Q = query]

qibla var. of KIBLAH.

Qin /tʃɪn/ (also **Ch'in**) a dynasty that ruled China 221–206 BC and was the first to establish rule over a united China. The construction of the Great Wall of China was begun during this period.

Qing /tʃɪŋ/ (also **Ch'ing**) a dynasty established by the Manchus that ruled China 1644–1912. Its overthrow in 1912 by Sun Yat-sen and his supporters ended imperial rule in China.

Qingdao /tʃɪŋˈdaʊ/ a port in eastern China, in Shandong province on the Yellow Sea coast; pop. (1990) 2,040,000.

Qinghai /tʃɪŋˈhaɪ/ (also **Tsinghai**) a mountainous province in north central China; capital, Xining.

Qiqihar /ˌtʃɪtʃɪˈhɑː(r)/ a port on the River Nen, in Heilongjiang province, NE China; pop. (1990) 1,370,000.

Qld. abbr. Queensland.

QM abbr. quartermaster.

QMG abbr. Quartermaster General.

QMS abbr. Quartermaster Sergeant.

Qom /kʊm/ (also **Qum, Kum**) a city in central Iran; pop. (1986) 551,000. It is a holy city and a place of pilgrimage for Shiite Muslims.

QPM abbr. (in the UK) Queen's Police Medal.

qr. abbr. quarter(s).

Q-ship /ˈkjuːʃɪp/ n. hist. an armed and disguised merchant ship used as a decoy or to destroy submarines. [Q = query]

QSO abbr. quasi-stellar object, quasar.

qt. abbr. quart(s).

q.t. n. colloq. quiet (esp. *on the q.t.*). [abbr.]

qu. abbr. **1** query. **2** question.

Qua var. of KWA.

qua /kwɑː/ conj. in the capacity of; as being (*Napoleon qua general*). [L, ablat. fem. sing. of *qui* who (rel. pron.)]

quack¹ /kwæk/ n. & v. ● n. the harsh sound made by ducks. ● v.intr. **1** utter this sound. **2** colloq. talk loudly and foolishly. [imit.: cf. Du. *kwakken*, G *quacken* croak, quack]

quack² /kwæk/ n. **1 a** an unqualified practiser of medicine. **b** (attrib.) of or characteristic of unskilled medical practice (*quack cure*). **2 a** a charlatan. **b** (attrib.) of or characteristic of a charlatan; fraudulent, sham. **3** sl. a doctor or medical officer. □ **quackery** n. **quackish** adj. [abbr. of *quacksalver* f. Du. (prob. f. obs. *quacken* prattle + *salf* SALVE¹)]

quad¹ /kwɒd/ n. colloq. a quadrangle. [abbr.]

quad² /kwɒd/ n. colloq. = QUADRUPLET 1. [abbr.]

quad³ /kwɒd/ n. Printing a piece of blank metal type used in spacing. [abbr. of earlier QUADRAT]

quad⁴ /kwɒd/ n. & adj. ● n. quadraphony. ● adj. quadraphonic. [abbr.]

quadragenarian /ˌkwɒdrədʒɪˈneərɪən/ n. & adj. ● n. a person from 40 to 49 years old. ● adj. of this age. [LL *quadragenarius* f. *quadrageni* distributive of *quadraginta* forty]

Quadragesima /ˌkwɒdrəˈdʒesɪmə/ n. the first Sunday in Lent. [LL, fem. of L *quadragesimus* fortieth f. *quadraginta* forty, Lent having 40 days]

quadragesimal /ˌkwɒdrəˈdʒesɪm(ə)l/ adj. **1** (of a fast, esp. in Lent) lasting forty days. **2** Lenten.

quadrangle /ˈkwɒdˌræŋɡ(ə)l/ n. **1** a four-sided plane figure, esp. a square or rectangle. **2 a** a four-sided court, esp. enclosed by buildings, as in some colleges. **b** such a court with the buildings round it. □ **quadrangular** /kwɒˈdræŋɡjʊlə(r)/ adj. [ME f. OF f. LL *quadrangulum* square, neut. of *quadrangulus* (as QUADRI-, ANGLE¹)]

quadrant /ˈkwɒdrənt/ n. **1** a quarter of a circle's circumference. **2** a plane figure enclosed by two radii of a circle at right angles and the arc cut off by them. **3** a quarter of a sphere or spherical body. **4** each of four parts of a plane divided by two lines at right angles. **5 a** a thing, esp. a graduated strip of metal, shaped like a quarter-circle. **b** an instrument graduated (esp. through an arc of 90°) for taking angular measurements. □ **quadrantal** /kwɒˈdrænt(ə)l/ adj. [ME f. L *quadrans -antis* quarter f. *quattuor* four]

Quadrantids /kwɒˈdræntɪdz/ n.pl. Astron. an annual meteor shower with a radiant in the constellation of Boötes, reaching a peak about 3 Jan. [f. L *Quadrans Muralis* the Mural Quadrant, a former constellation]

quadraphonic /ˌkwɒdrəˈfɒnɪk/ adj. (also **quadrophonic**) (of sound reproduction) using four transmission channels. □ **quadraphonically** adv. **quadraphonics** n.pl. **quadraphony** /kwɒˈdrɒfənɪ/ n. [QUADRI- + STEREOPHONIC]

quadrat /ˈkwɒdrət/ n. Ecol. **1** a small square area marked out for study. **2** a portable frame or grid used to mark out such an area. [var. of QUADRATE]

quadrate adj., n., & v. ● adj. /ˈkwɒdrət/ esp. Anat. & Zool. square or rectangular (*quadrate bone; quadrate muscle*). ● n. /ˈkwɒdrət/ **1** a quadrate bone or muscle. **2** a rectangular object. ● v. /kwɒˈdreɪt/ **1** tr. make square. **2** *intr. & tr.* (often foll. by *with*) conform or make conform. [ME f. L *quadrare quadrat-* make square f. *quattuor* four]

quadratic /kwɒˈdrætɪk/ *adj. & n. Math.* ● *adj.* **1** involving the second and no higher power of an unknown quantity or variable (*quadratic equation*). **2** square. ● *n.* **1** a quadratic equation. **2** (in *pl.*) the branch of algebra dealing with these. [F *quadratique* or mod.L *quadraticus* (as QUADRATE)]

quadrature /ˈkwɒdrətʃə(r)/ *n.* **1** *Math.* the process of constructing a square with an area equal to that of a figure bounded by a curve, e.g. a circle. **2** *Astron.* the position of two celestial objects when they appear 90° apart as viewed from earth; esp. a point at which the moon or a planet is 90° from the sun as viewed from earth. [F *quadrature* or L *quadratura* (as QUADRATE)]

quadrennial /kwɒˈdrɛnɪəl/ *adj.* **1** lasting four years. **2** recurring every four years. □ **quadrennially** *adv.* [as QUADRENNIUM]

quadrennium /kwɒˈdrɛnɪəm/ *n.* (*pl.* **quadrenniums** or **quadrennia** /-nɪə/) a period of four years. [L *quadriennium* (as QUADRI-, *annus* year)]

quadri- /ˈkwɒdrɪ/ *comb. form* denoting four. [L f. *quattuor* four]

quadric /ˈkwɒdrɪk/ *adj. & n. Geom.* ● *adj.* (of a surface) described by an equation of the second degree. ● *n.* a quadric surface. [L *quadra* square]

quadriceps /ˈkwɒdrɪˌsɛps/ *n. Anat.* a four-headed muscle at the front of the thigh. [mod.L (as QUADRI-, BICEPS)]

quadrifid /ˈkwɒdrɪfɪd/ *adj. Bot.* having four divisions or lobes. [L *quadrifidus* (as QUADRI-, *findere fid-* cleave)]

quadrilateral /ˌkwɒdrɪˈlætərəl/ *adj. & n.* ● *adj.* having four sides. ● *n.* a four-sided figure. [LL *quadrilaterus* (as QUADRI-, *latus lateris* side)]

quadrille[1] /kwɒˈdrɪl/ *n.* **1** a square dance containing usu. five figures. **2** the music for this. [F f. Sp. *cuadrilla* troop, company f. *cuadra* square or It. *quadriglia* f. *quadra* square]

quadrille[2] /kwɒˈdrɪl/ *n.* a card-game for four players with forty cards (i.e. an ordinary pack without the 8s, 9s, and 10s), fashionable in the 18th century. [F, perh. f. Sp. *cuartillo* f. *cuarto* fourth, assim. to QUADRILLE[1]]

quadrillion /kwɒˈdrɪljən/ *n.* (*pl.* same or **quadrillions**) a thousand raised to the fifth (or formerly, esp. *Brit.*, the eighth) power (10^{15} and 10^{24} respectively). [F (as QUADRI-, MILLION)]

quadrinomial /ˌkwɒdrɪˈnəʊmɪəl/ *n. & adj. Math.* ● *n.* an expression of four algebraic terms. ● *adj.* of or being a quadrinomial. [QUADRI- + Gk *nomos* part, portion]

quadripartite /ˌkwɒdrɪˈpɑːtaɪt/ *adj.* **1** consisting of four parts. **2** shared by or involving four parties.

quadriplegia /ˌkwɒdrɪˈpliːdʒə/ *n. Med.* paralysis of all four limbs. □ **quadriplegic** /-dʒɪk/ *adj. & n.* [mod.L (as QUADRI-, Gk *plēgē* blow, strike)]

quadrivalent /ˌkwɒdrɪˈveɪlənt/ *adj. Chem.* = TETRAVALENT.

quadrivium /kwɒˈdrɪvɪəm/ *n. hist.* a medieval university course of arithmetic, geometry, astronomy, and music. [L, = the place where four roads meet (as QUADRI-, *via* road)]

quadroon /kwɒˈdruːn/ *n.* the offspring of a white person and a mulatto; a person of one quarter black ancestry. [Sp. *cuarterón* f. *cuarto* fourth, assim. to QUADRI-]

quadrophonic var. of QUADRAPHONIC.

quadrumanous /kwɒˈdruːmənəs/ *adj.* (of primates other than humans) four-handed, i.e. with opposable digits on all four limbs. [mod.L *quadrumana* neut. pl. of *quadrumanus* (as QUADRI-, L *manus* hand)]

quadruped /ˈkwɒdrʊˌpɛd/ *n. & adj.* ● *n.* a four-footed animal, esp. a four-footed mammal. ● *adj.* four-footed. □ **quadrupedal** /kwɒˈdruːpɪd(ə)l/ *adj.* [F *quadrupède* or L *quadrupes -pedis* f. *quadru-* var. of QUADRI- + L *pes ped-* foot]

quadruple /ˈkwɒdrʊp(ə)l, kwɒˈdruː-/ *adj., n., & v.* ● *adj.* **1** fourfold. **2 a** having four parts. **b** involving four participants. **3** being four times as many or as much. **4** (of time in music) having four beats in a bar. ● *n.* a fourfold number or amount. ● *v.tr. & intr.* multiply by four; increase fourfold. □ **quadruply** /-plɪ/ *adv.* [F f. L *quadruplus* (as QUADRI-, *-plus* as in *duplus* DUPLE)]

Quadruple Alliance any of several alliances involving four powers. They include that of 1718, when Austria joined Britain, the Netherlands, and France against Spain (which had seized Sicily and Sardinia); that of 1813 (renewed in 1815) when Britain, Russia, Austria, and Prussia united to defeat Napoleon and to maintain the international order established in Europe at the end of the Napoleonic Wars; and that of 1834 when Britain, France, Spain, and Portugal

united in their support of the claims of the regent Maria Cristiana (1806–78) to the Spanish throne and of Maria da Glória (1819–53) to that of Portugal.

quadruplet /ˈkwɒdrʊplɪt, kwɒˈdruːp-/ *n.* **1** each of four children born at one birth. **2** a set of four similar or associated things. **3** *Mus.* a group of four notes to be performed in the time of three. [QUADRUPLE, after *triplet*]

quadruplicate *adj. & v.* ● *adj.* /kwɒˈdruːplɪkət/ **1** fourfold. **2** of which four copies are made. ● *v.tr.* /kwɒˈdruːplɪˌkeɪt/ **1** multiply by four. **2** make four identical copies of. □ **in quadruplicate** in four identical copies. □ **quadruplication** /-ˌdruːplɪˈkeɪʃ(ə)n/ *n.* [L *quadruplicare* f. *quadruplex -plicis* fourfold: cf. QUADRUPED, DUPLEX]

quadruplicity /ˌkwɒdrʊˈplɪsɪtɪ/ *n.* the state of being fourfold. [L *quadruplex -plicis* (see QUADRUPLICATE), after *duplicity*]

quaestor /ˈkwiːstɔː(r)/ *n.* either of two ancient Roman magistrates with mainly financial responsibilities. □ **quaestorial** /kwɪˈstɔːrɪəl/ *adj.* [ME f. L f. *quaerere quaesit-* seek]

quaff /kwɒf, kwɑːf/ *v. literary* **1** *tr. & intr.* drink deeply. **2** *tr.* drain (a cup etc.) in long draughts. □ **quaffable** *adj.* **quaffer** *n.* [16th c.: perh. imit.]

quag /kwæg, kwɒg/ *n.* a marshy or boggy place. □ **quaggy** *adj.* [rel. to dial. *quag* (v.) = shake: prob. imit.]

quagga /ˈkwægə/ *n.* an extinct zebra formerly native to southern Africa, which was yellowish-brown with stripes only on the head, neck, and foreparts. It was exterminated in 1883; recent research has shown that it was probably simply a variety of the common zebra, *Equus burchelli*. [perh. imit. of its call, or f. Nguni *iqwara*]

quagmire /ˈkwæg̩maɪə(r), ˈkwɒg-/ *n.* **1** a soft boggy or marshy area that gives way underfoot. **2** a hazardous or awkward situation. [QUAG + MIRE]

quahog /ˈkwɔːhɒg, ˈkəʊhɒg/ *n.* (also **quahaug**) *US* the edible round clam, *Venus mercenaria*, of the Atlantic coast of North America. [Narragansett]

quaich /kweɪx/ *n.* (also **quaigh**) *Sc.* a kind of drinking-cup, usu. of wood and with two handles. [Gael. *cuach* cup, prob. f. L *caucus*]

Quai d'Orsay /ˌkeɪ dɔːˈseɪ/ **1** a riverside street on the left bank of the Seine in Paris. **2** the French ministry of foreign affairs, which has its headquarters in this street.

quail[1] /kweɪl/ *n.* (*pl.* same or **quails**) a small short-tailed game bird of the genus *Coturnix*, related to the partridges, esp. the migratory *C. coturnix*. [ME f. OF *quaille* f. med.L *coacula* (prob. imit.)]

quail[2] /kweɪl/ *v.intr.* flinch; be apprehensive with fear. [ME, of unkn. orig.]

quaint /kweɪnt/ *adj.* **1** piquantly or attractively unfamiliar or old-fashioned. **2** daintily odd. □ **quaintly** *adv.* **quaintness** *n.* [earlier senses 'wise, cunning': ME f. OF *cointe* f. L *cognitus* past part. of *cognoscere* ascertain]

quake /kweɪk/ *v. & n.* ● *v.intr.* **1** shake, tremble. **2** rock to and fro. **3** (of a person) shake or shudder (*was quaking with fear*). ● *n.* **1** *colloq.* an earthquake. **2** an act of quaking. □ **quaking-grass** a grass of the genus *Briza*, having compressed spikelets on slender stalks that tremble in the wind, esp. *B. media* (also called *dodder-grass*). □ **quaky** *adj.* (**quakier**, **quakiest**). [OE *cwacian*]

Quaker /ˈkweɪkə(r)/ *n. & adj.* ● *n.* a member of the Society of Friends, a Christian body founded by George Fox c.1650, noted for pacifism and social and educational work. (*See note below.*) ● *adj.* of or relating to Quakers. □ **Quakerish** *adj.* **Quakerism** *n.*

▪ Central to the Quakers' belief is the doctrine of the 'Inner Light', or sense of Christ's direct working in the soul; this has led them to reject all set forms and trappings of worship. They have no priests or ministers, and their meetings begin in silence until some member feels stirred to speak. They were formerly much persecuted, but refused to meet in secret and marked themselves out by plainness of dress and by distinctive forms of speech; many early Quakers emigrated to found new communities, for example, the new colony of Pennsylvania, established in 1682. In the 20th century, refusal of military service as combatants and of oaths has brought Quakers into conflict with the authorities. Originally derogatory, the name may have arisen from Fox's direction to his opponents to 'quake at the word of the Lord', or from the fits allegedly experienced by worshippers.

qualification /ˌkwɒlɪfɪˈkeɪʃ(ə)n/ *n.* **1** the act or instance of qualifying. **2** (often in *pl.*) **a** a quality, skill, or accomplishment fitting a person for a position or purpose. **b** an award gained on successful completion of a course of education or training (*left school without any qualifications*). **3 a** a circumstance, condition, etc., that modifies or limits (*the statement had many qualifications*). **b** a thing that detracts from completeness or absoluteness (*their relief had one qualification*). **4** a condition that must be fulfilled before a right can be acquired etc. **5** an attribution of a quality (*the qualification of our policy as opportunist is unfair*). □ **qualificatory** /ˈkwɒlɪfɪˌkeɪtərɪ/ *adj.* [F *qualification* or med.L *qualificatio* (as QUALIFY)]

qualify /ˈkwɒlɪˌfaɪ/ *v.* (**-ies, -ied**) **1** *tr.* make competent or fit for a position or purpose. **2** *tr.* make legally entitled. **3** *intr.* (usu. foll. by *for* or *as*) (of a person) satisfy the conditions or requirements for (a position, award, competition, etc.). **4** *tr.* add reservations to; modify or make less absolute (a statement or assertion). **5** *tr. Gram.* (of a word, esp. an adjective) attribute a quality to another word, esp. a noun. **6** *tr.* moderate, mitigate; make less severe or extreme. **7** *tr.* alter the strength or flavour of. **8** *tr.* (foll. by *as*) attribute a specified quality to, describe as (*the idea was qualified as absurd*). **9** *tr.* (as **qualifying** *adj.*) serving to determine those that qualify (*qualifying examination*). **10** (as **qualified** *adj.*) **a** having the qualifications necessary for a particular office or function. **b** dependent on other factors; not definite (*a qualified 'yes'*). □ **qualifiable** *adj.* **qualifier** *n.* [F *qualifier* f. med.L *qualificare* f. L *qualis* such as]

qualitative /ˈkwɒlɪtətɪv, -ˌteɪtɪv/ *adj.* concerned with or depending on quality (*led to a qualitative change in society*). □ **qualitative analysis** *Chem.* detection of the constituents, as elements, functional groups, etc., present in a substance (cf. *quantitative analysis*). □ **qualitatively** *adv.* [LL *qualitativus* (as QUALITY)]

quality /ˈkwɒlɪtɪ/ *n.* (pl. **-ies**) **1** the degree of excellence of a thing (*of good quality; poor in quality*). **2 a** general excellence (*their work has quality*). **b** (*attrib.*) of high quality (*a quality product*). **3** a distinctive attribute or faculty; a characteristic trait. **4** the relative nature or kind or character of a thing (*is made in three qualities*). **5** the distinctive timbre of a voice or sound. **6** *archaic* high social standing (*people of quality*). **7** *Logic* the property of a proposition's being affirmative or negative. □ **quality circle** a group of employees who meet to consider ways of resolving problems and improving production in their organization. **quality control** a system of maintaining standards in manufactured products by testing a sample of the output against the specification. [ME f. OF *qualité* f. L *qualitas -tatis* f. *qualis* of what kind]

qualm /kwɑːm, kwɔːm/ *n.* **1** a misgiving; an uneasy doubt esp. about one's own conduct. **2** a scruple of conscience. **3** a momentary faint or sick feeling. □ **qualmish** *adj.* [16th c.: orig. uncert.]

quandary /ˈkwɒndərɪ/ *n.* (pl. **-ies**) **1** a state of perplexity. **2** a difficult situation; a practical dilemma. [16th c.: orig. uncert.]

quango /ˈkwæŋɡəʊ/ *n.* (pl. **-os**) a semi-public body with financial support from and senior appointments made by the government. [abbr. of *quasi* (or *quasi-autonomous*) *non-government(al) organization*]

Quant /kwɒnt/, Mary (b.1934), English fashion designer. She was a principal creator of the '1960s look', launching the miniskirt in 1966 and promoting bold colours and geometric designs. She was one of the first to design for the ready-to-wear market, and her styles, created especially for the young, did much to make London a leading fashion centre at the time. Since the 1970s she has concentrated on marketing a range of cosmetics.

quant /kwɒnt/ *n. & v.* ● *n. Brit.* a punting-pole with a prong at the bottom to prevent it sinking into the mud, as used by Norfolk bargemen etc. ● *v.tr.* (also *absol.*) propel (a boat) with a quant. [15th c.: perh. f. L *contus* f. Gk *kontos* boat-pole]

quanta *pl.* of QUANTUM.

quantal /ˈkwɒnt(ə)l/ *adj.* **1** composed of discrete units; varying in steps, not continuously. **2** of or relating to a quantum or quantum theory. □ **quantally** *adv.* [L *quantus* how much]

quantic /ˈkwɒntɪk/ *n. Math.* a rational integral homogeneous function of two or more variables.

quantifier /ˈkwɒntɪˌfaɪə(r)/ *n.* **1** a person or thing that quantifies something. **2** *Logic & Gram.* an expression (e.g. *all, some*) that indicates the scope of a term to which it is attached; a determiner indicative of quantity. [f. QUANTIFY + -ER[1]]

quantify /ˈkwɒntɪˌfaɪ/ *v.tr.* (**-ies, -ied**) **1** determine the quantity of. **2** measure or express as a quantity. **3** *Logic* define the application of (a term or proposition) by the use of *all, some*, etc., e.g. 'for all *x* if *x* is A then *x* is B'. □ **quantifiable** *adj.* **quantifiability** /ˌkwɒntɪˌfaɪəˈbɪlɪtɪ/ *n.* **quantification** /-fɪˈkeɪʃ(ə)n/ *n.* [med.L *quantificare* (as QUANTAL)]

quantitative /ˈkwɒntɪtətɪv, -ˌteɪtɪv/ *adj.* **1 a** concerned with quantity. **b** measured or measurable by quantity. **2** of or based on the quantity of syllables. □ **quantitative analysis** *Chem.* measurement of the amounts of the constituents of a substance (cf. *qualitative analysis*). □ **quantitatively** *adv.* [med.L *quantitativus* (as QUANTITY)]

quantitive /ˈkwɒntɪtɪv/ *adj.* = QUANTITATIVE. □ **quantitively** *adv.*

quantity /ˈkwɒntɪtɪ/ *n.* (pl. **-ies**) **1** the property of things that is measurable. **2** the size or extent or weight or amount or number. **3** a specified or considerable portion or number or amount (*buys in quantity; the quantity of heat in a body*). **4** (in *pl.*) large amounts or numbers; an abundance (*quantities of food; is found in quantities on the shore*). **5** the length or shortness of vowel sounds or syllables. **6** *Math.* **a** a value, component, etc. that may be expressed in numbers. **b** the figure or symbol representing this. □ **quantity mark** a mark put over a vowel etc. to indicate its length. **quantity surveyor** a person who measures and prices building work. **quantity theory** the hypothesis that prices vary according to changes in the monetary supply. [ME f. OF *quantité* f. L *quantitas -tatis* f. *quantus* how much]

quantize /ˈkwɒntaɪz/ *v.tr.* (also **-ise**) **1** form into quanta. **2** apply quantum theory to. □ **quantization** /ˌkwɒntaɪˈzeɪʃ(ə)n/ *n.*

quantum /ˈkwɒntəm/ *n.* (pl. **quanta** /-tə/) **1** *Physics* **a** a discrete quantity of energy proportional in magnitude to the frequency of the radiation it represents. **b** an analogous discrete amount of any other physical quantity. (*See note below.*) **2 a** a required or allowed amount. **b** a share or portion. □ **quantum chromodynamics** see CHROMODYNAMICS. **quantum jump** (or **leap**) **1** a sudden large increase or advance. **2** *Physics* an abrupt transition in an atom or molecule from one quantum state to another. **quantum-mechanical** *adj.* of or relating to quantum mechanics. **quantum mechanics** a mathematical form of quantum theory dealing with the motion and interaction of (esp. subatomic) particles and incorporating the concept that these particles can also be regarded as waves. **quantum number** a number expressing the value of some property of a particle occurring in quanta. **quantum state** an energy state of an atom, molecule, etc. described by a particular set of quantum numbers. **quantum theory** the body of theory based on the existence of quanta of energy. [L, neut. of *quantus* how much]

▪ In 1900 Max Planck accounted for certain puzzling characteristics of the electromagnetic radiation given off by hot bodies by postulating that energy did not form a continuous spectrum but was emitted only in fixed amounts or 'quanta' (see also PLANCK). The absorption and emission of discrete quanta of energy by atoms and molecules resulted in the characteristic features visible in the spectra of substances. However, this radical idea also had a more revolutionary effect on understanding of the physical world. Einstein suggested that all electromagnetic radiation, including light, could be regarded as consisting of particles or quanta of energy (photons). In 1913 Niels Bohr proposed a model of the atom that incorporated quantum theory, by supposing that electrons circle around a central nucleus in orbits corresponding to particular amounts of energy. In 1924 Louis-Victor de Broglie conjectured that, just as light waves possess some attributes of particles, perhaps particles such as electrons could have some of the properties of waves. This idea was incorporated by Schrödinger, Heisenberg, and others into a more sophisticated theory of atomic structure called quantum mechanics, in which Bohr's orbits are replaced by more diffuse orbitals, reflecting the dual properties of the electron as a wave and a particle. The principles of relativity theory were incorporated into quantum mechanics by Paul Dirac in 1928. Since then, despite its paradoxical and counter-intuitive aspects, quantum theory has been extremely successful in describing many physical phenomena, including the properties of the atomic nucleus and subatomic particles, and of electromagnetic fields.

quaquaversal /ˌkweɪkwəˈvɜːs(ə)l/ *adj. Geol.* pointing in every direction. [LL *quaquaversus* f. *quaqua* wheresoever + *versus* towards]

quarantine /ˈkwɒrənˌtiːn/ *n. & v.* ● *n.* **1** isolation imposed on persons or animals that have arrived from elsewhere or been exposed to, and might spread, infectious or contagious disease. **2** the period of this isolation. ● *v.tr.* impose such isolation on, put in quarantine. [It. *quarantina* forty days f. *quaranta* forty]

quark[1] /kwɑːk/ *n. Physics* a hypothetical particle carrying a fractional electric charge, postulated as a building block of other subatomic particles. The hadrons are thought to be built up from combinations of quarks; three kinds of quark were originally proposed but this number has risen to account for properties such as strangeness and charm. Although quarks have not been experimentally observed (it may be theoretically impossible for them to occur in the free state), many predictions of quark theory have been corroborated by experiments. [coined by Murray Gell-Mann (1964), from phrase 'Three quarks for Muster Mark!' in Joyce's *Finnegans Wake*]

quark[2] /kwɑːk/ *n.* a type of low-fat curd cheese. [G]

quarrel[1] /ˈkwɒrəl/ *n. & v.* ● *n.* **1** a violent contention or altercation between individuals or with others. **2** a rupture of friendly relations. **3** an occasion of complaint against a person, a person's actions, etc. ● *v.intr.* (**quarrelled, quarrelling**; *US* **quarreled, quarreling**) **1** (often foll. by *with*) take exception; find fault. **2** fall out; have a dispute; break off friendly relations. □ **quarreller** *n.* [ME f. OF *querele* f. L *querel(l)a* complaint f. *queri* complain]

quarrel[2] /ˈkwɒrəl/ *n. hist.* a short heavy square-headed arrow or bolt used in a crossbow or arbalest. [ME f. OF *quar(r)el* ult. f. LL *quadrus* square]

quarrelsome /ˈkwɒrəlsəm/ *adj.* given to or characterized by quarrelling. □ **quarrelsomely** *adv.* **quarrelsomeness** *n.*

quarrian /ˈkwɒrɪən/ *n.* (also **quarrion**) *Austral.* a cockatiel. [Wiradhuri *guwarraying*]

quarry[1] /ˈkwɒrɪ/ *n. & v.* ● *n.* (pl. **-ies**) **1** an excavation made by taking stone etc. for building etc. from its bed. **2** a place from which stone etc. may be extracted. **3** a source of information, knowledge, etc. ● *v.* (**-ies, -ied**) **1** *tr.* extract (stone) from a quarry. **2** *tr.* extract (facts etc.) laboriously from books etc. **3** *intr.* laboriously search documents etc. [ME f. med.L *quare(r)ia* f. OF *quarriere* f. L *quadrum* square]

quarry[2] /ˈkwɒrɪ/ *n.* (pl. **-ies**) **1** the object of pursuit by a bird of prey, hounds, hunters, etc. **2** an intended victim or prey. [ME f. AF f. OF *cuiree, couree* (assim. to *cuir* leather and *curer* disembowel) ult. f. L *cor* heart: orig. = parts of deer placed on hide and given to hounds]

quarry[3] /ˈkwɒrɪ/ *n.* (pl. **-ies**) **1** a diamond-shaped pane of glass as used in lattice windows. **2** (in full **quarry tile**) an unglazed floor-tile. [a later form of QUARREL[2] in the same sense]

quarryman /ˈkwɒrɪmən/ *n.* (pl. **-men**) a worker in a quarry.

quart /kwɔːt/ *n.* **1** a liquid measure equal to a quarter of a gallon; two pints. **2** a vessel containing this amount. **3** *US* a dry measure, equivalent to one-thirty-second of a bushel (1.1 litre). **4** /kɑːt/ (also **quarte** or **carte**) the fourth of eight parrying positions in fencing. □ **a quart into a pint pot 1** a large amount etc. fitted into a small space. **2** something difficult or impossible to achieve. [ME f. OF *quarte* f. L *quarta* fem. of *quartus* fourth f. *quattuor* four]

quartan /ˈkwɔːt(ə)n/ *adj.* (of a fever etc.) recurring every fourth day. [ME f. OF *quartaine* f. L (*febris* fever) *quartana* f. *quartus* fourth]

quarte var. of QUART 4.

quarter /ˈkwɔːtə(r)/ *n. & v.* ● *n.* **1** each of four equal parts into which a thing is or might be divided. **2** a period of three months, usu. for which payments become due on the quarter day. **3** a point of time 15 minutes before or after any hour. **4** a school or *US* university term. **5 a** 25 *US* or Canadian cents. **b** a coin of this denomination. **6** a part of a town, esp. as occupied by a particular class or group (*residential quarter*). **7 a** a point of the compass. **b** a region at such a point. **8** the direction, district, or source of supply etc. (*help from any quarter; came from all quarters*). **9** (in *pl.*) **a** lodgings; an abode. **b** *Mil.* the living accommodation of troops etc. **10 a** one fourth of a lunar month. **b** the moon's position when halfway between new and full moon (*first quarter*) or between full and new moon (*last quarter*). **11 a** each of the four parts into which an animal's or bird's carcass is divided, each including a leg or wing. **b** (in *pl.*) *hist.* the four parts into which a traitor etc. was cut after execution. **c** (in *pl.*) = HINDQUARTERS. **12** mercy offered or granted to an enemy in battle etc. on condition of surrender. **13 a** one-fourth of a pound weight. **b** *Brit.* a grain measure equivalent to 8 bushels. **c** one-fourth of a hundredweight (28 lb or *US* 25 lb). **14 a** each of four divisions on a shield. **b** a charge occupying this, placed in chief. **15** either side of a ship abaft the beam. **16** (in American and Australian football) each of four equal periods into which a match is divided. ● *v.tr.* **1** divide into quarters. **2** *hist.* divide (the body of an executed person) in this way. **3 a** put (troops etc.) into quarters. **b** station or lodge in a specified place. **4** (foll. by *on*) impose (a person) on another as a lodger. **5** cut (a log) into quarters, and these into planks so as to show the grain well. **6** (of a dog, hunting bird of prey, etc.) range or traverse (the ground) in every direction. **7** *Heraldry* **a** place or bear (charges or coats of arms) on the four quarters of a shield's surface. **b** add (another's coat) to one's hereditary arms. **c** (foll. by *with*) place in alternate quarters with. **d** divide (a shield) into four or more parts by vertical and horizontal lines. □ **quarter-binding** the type of bookbinding in which the spine is bound in one material (usu. leather) and the sides in another. **quarter day** one of four days on which quarterly payments are due, tenancies begin and end, etc. **quarter-**

final a match or round preceding the semifinal. **quarter-hour 1** a period of 15 minutes. **2** = sense 3 of *n.* **quarter-light** *Brit.* a window in the side of a motor vehicle, closed carriage, etc. other than the main door-window. **quarter-line** *Rugby* a space enclosed by a line across the ground 22 metres from the goal-line. **quarter note** esp. *N. Amer. Mus.* a crotchet. **quarter-plate 1** a photographic plate or film 8.3 × 10.8 cm. **2** a photograph reproduced from it. **quarter sessions** *hist.* (in the UK) a court of limited criminal and civil jurisdiction and of appeal, usu. held quarterly. **quarter-tone** *Mus.* half a semitone. [ME f. AF *quarter*, OF *quartier* f. L *quartarius* fourth part (of a measure) f. *quartus* fourth]

quarterage /ˈkwɔːtərɪdʒ/ *n.* **1** a quarterly payment. **2** a quarter's wages, allowance, pension, etc.

quarterback /ˈkwɔːtəˌbæk/ *n.* a player in American football who directs attacking play.

quarterdeck /ˈkwɔːtəˌdek/ *n.* **1** part of a ship's upper deck near the stern, usu. reserved for officers. **2** the officers of a ship or the navy.

quartering /ˈkwɔːtərɪŋ/ *n.* **1** (in *pl.*) the coats of arms marshalled on a shield to denote the alliances of a family with the heiresses of others. **2** the provision of quarters for soldiers. **3** the act or an instance of dividing, esp. into four equal parts. **4** timber sawn into lengths, used for high-quality floor-boards etc.

quarterly /ˈkwɔːtəlɪ/ *adj., adv., & n.* ● *adj.* **1** produced or occurring once every quarter of a year. **2** (of a shield) quartered. ● *adv.* **1** once every quarter of a year. **2** in the four, or in two diagonally opposite, quarters of a shield. ● *n.* (pl. **-ies**) a quarterly review or magazine.

quartermaster /ˈkwɔːtəˌmɑːstə(r)/ *n.* **1** a regimental officer in charge of quartering, rations, etc. **2** a naval petty officer in charge of steering, signals, etc. □ **quartermaster sergeant** a sergeant assisting an army quartermaster.

Quartermaster General *n.* the head of the army department in charge of quartering etc.

quartern /ˈkwɔːt(ə)n/ *n. Brit. archaic* a quarter of a pint. □ **quartern loaf** a four-pound loaf. [ME, = quarter f. AF *quartrun*, OF *quart(e)ron* f. QUART fourth or *quartier* QUARTER]

quarterstaff /ˈkwɔːtəˌstɑːf/ *n. hist.* a stout pole 6–8 feet long, formerly used as a weapon.

quartet /kwɔːˈtet/ *n.* (also **quartette**) **1** *Mus.* **a** a composition for four voices or instruments. **b** the performers of such a piece. **2** any group of four. [F *quartette* f. It. *quartetto* f. *quarto* fourth f. L *quartus*]

quartic /ˈkwɔːtɪk/ *adj. & n. Math.* ● *adj.* involving the fourth and no higher power of an unknown quantity or variable. ● *n.* a quartic equation. [L *quartus* fourth]

quartile /ˈkwɔːtaɪl/ *adj. & n.* ● *adj. Astrol.* relating to the aspect of two celestial bodies 90° apart. ● *n.* **1** a quartile aspect. **2** *Statistics* each of three values of a variable dividing a population into four equal groups as regards the value of that variable. [med.L *quartilis* f. L *quartus* fourth]

quarto /ˈkwɔːtəʊ/ *n.* (pl. **-os**) (abbr. **4to**) *Printing* **1** the size of book or paper given by folding a (usu. specified) sheet of paper twice, commonly 10 in. × 8 in. **2** a book consisting of sheets folded in this way. □ **quarto paper** paper folded in this way and cut into sheets. [L (*in*) *quarto* (in) the fourth (of a sheet), ablat. of *quartus* fourth]

quartz /kwɔːts/ *n.* a mineral form of silica that crystallizes as hexagonal prisms. □ **quartz clock** a clock operated by vibrations of an electrically driven quartz crystal. **quartz lamp** a quartz tube containing mercury vapour and used as a light source. [G *Quarz* f. West Slavonic *kwardy*]

quartzite /ˈkwɔːtsaɪt/ *n.* a metamorphic rock consisting mainly of quartz.

quasar /ˈkweɪzɑː(r), ˈkweɪsɑː(r)/ *n. Astron.* an apparently starlike source of light visible in large telescopes, often associated with intense radio emission. The spectra of quasars show large red shifts, suggesting that they are as far away as the most remote galaxies. Quasars must therefore be very massive and emit exceptionally large amounts of energy, the origin of which is not yet understood. It has been suggested that quasars contain massive black holes and may represent a stage in the evolution of some galaxies. [*quasi-stellar object*]

quash /kwɒʃ/ *v.tr.* **1** annul; reject as not valid, esp. by a legal procedure. **2** suppress; crush (a rebellion etc.). [ME f. OF *quasser, casser* annul f. LL *cassare* f. *cassus* null, void or f. L *cassare* frequent. of *quatere* shake]

quasi /ˈkweɪzaɪ, ˈkwɑːzɪ/ *adv.* that is to say; as it were. [L, = as if, almost]

quasi- /ˈkweɪzaɪ, ˈkwɑːzɪ/ *comb. form* **1** seemingly; apparently but not really (*quasi-scientific*; *quasi-stellar*). **2** being partly or almost (*quasi-independent*). [L *quasi* as if, almost]

Quasimodo[1] /ˌkwɒzɪˈmɑʊdəʊ/ the name of the hunchback in Victor Hugo's novel *Notre-Dame de Paris* (1831).

Quasimodo[2] /kwaːˈzɪmɒˌdəʊ/, Salvatore (1901–68), Italian poet. His early work was influenced by French symbolism. Major collections include *Water and Land* (1930) and *And It's Suddenly Evening* (1942). His later work is more extrovert and concerned with political and social issues. He was awarded the Nobel Prize for literature in 1959.

quassia /ˈkwɒʃə/ *n.* **1** an evergreen tree, *Quassia amara*, native to South America. **2** the wood, bark, or root of this tree, yielding a bitter medicinal tonic and insecticide. [G. *Quassi*, 18th-c. Suriname slave, who discovered its medicinal properties]

quatercentenary /ˌkwætəsənˈtiːnərɪ/ *n.* & *adj.* ● *n.* (*pl.* **-ies**) **1** a four-hundredth anniversary. **2** a festival marking this. ● *adj.* of this anniversary. [L *quater* four times + CENTENARY]

quaternary /kwəˈtɜːnərɪ/ *adj.* & *n.* ● *adj.* **1** having four parts. **2** (**Quaternary**) *Geol.* of or relating to the most recent period in the Cenozoic era, comprising the Pleistocene and Holocene (Recent) epochs and beginning about 2 million years ago. **3** *Chem.* (of an ammonium compound) containing four atoms or groups (other than hydrogen atoms) bonded to a nitrogen atom. ● *n.* (*pl.* **-ies**) **1** a set of four things. **2** (**Quaternary**) *Geol.* the Quaternary period or geological system. [ME f. L *quaternarius* f. *quaterni* distributive of *quattuor* four]

quaternion /kwəˈtɜːnɪən/ *n.* **1** a set of four. **2** *Math.* a complex number of the form $w + xi + yj + zk$, where w, x, y, z are real numbers and i, j, k are imaginary units that satisfy certain conditions. [ME f. LL *quaternio -onis* (as QUATERNARY)]

quatorzain /ˈkætəˌzeɪn/ *n.* a fourteen-line poem; an irregular sonnet. [F *quatorzaine* f. *quatorze* fourteen f. L *quattuordecim*]

quatorze /kəˈtɔːz/ *n.* a set of four aces, kings, queens, or jacks, in one hand at piquet, scoring fourteen. [F: see QUATORZAIN]

quatrain /ˈkwɒtreɪn/ *n.* a stanza of four lines, usu. with alternate rhymes. [F f. *quatre* four f. L *quattuor*]

quatrefoil /ˈkætrəˌfɔɪl/ *n.* a four-pointed or four-leafed figure, esp. as an ornament in architectural tracery, resembling a flower or clover leaf. [ME f. AF f. *quatre* four: see FOIL[2]]

quattrocento /ˌkwætrəʊˈtʃentəʊ/ *n.* the style of Italian art of the 15th century. □ **quattrocentist** *n.* [It., = 400 used with ref. to the years 1400–99]

quaver /ˈkweɪvə(r)/ *v.* & *n.* ● *v.* **1** *intr.* **a** (esp. of a voice or musical sound) vibrate, shake, tremble. **b** use trills or shakes in singing. **2** *tr.* **a** sing (a note or song) with quavering. **b** (often foll. by *out*) say in a trembling voice. ● *n.* **1** *Mus.* a note having the time value of an eighth of a semibreve or half a crotchet and represented by a large dot with a hooked stem. Also called *eighth note*. **2** a trill in singing. **3** a tremble in speech. □ **quaveringly** *adv.* **quavery** *adj.* [ME f. *quave*, perh. f. OE *cwafian* (unrecorded: cf. *cwacian* QUAKE)]

quay /kiː/ *n.* a solid stationary artificial landing-place lying alongside or projecting into water for loading and unloading ships. □ **quayage** *n.* [ME *key(e)*, *kay* f. OF *kay* f. Gaulish *caio* f. OCelt.]

quayside /ˈkiːsaɪd/ *n.* the land forming or near a quay.

Que. *abbr.* Quebec.

quean /kwiːn/ *n.* *archaic* an impudent or ill-behaved girl or woman. [OE *cwene* woman: cf. QUEEN]

queasy /ˈkwiːzɪ/ *adj.* (**-ier**, **-iest**) **1 a** (of a person) feeling nausea. **b** (of a person's stomach) easily upset, weak of digestion. **2** (of the conscience etc.) overscrupulous, tender. **3** (of a feeling, thought, etc.) uncomfortable, uneasy. □ **queasily** *adv.* **queasiness** *n.* [ME *queysy*, *coisy* perh. f. AF & OF, rel. to OF *coisir* hurt]

Quebec /kwɪˈbek/ (French **Québec** /kebɛk/) **1** a heavily forested province in eastern Canada; pop. (1991) 6,845,700. Originally inhabited by the Algonquin and Cree peoples, it was settled by the French in 1608, ceded to the British in 1763, and became one of the original four provinces in the Dominion of Canada in 1867. The majority of its residents are French-speaking and its culture remains predominantly French. It has a certain amount of political independence from the rest of Canada, and is a focal point of the French-Canadian nationalist movement, which advocates independence for Quebec. **2** (also **Quebec City**) its capital city, a port on the St Lawrence River; pop. (1991) 574,400. Founded on the site of a Huron village by Samuel de Champlain in 1608, it is Canada's oldest city. It was a centre of the struggle between the French and the British for control of colonial North America and was captured from the French by a British force in

1759 after the battle of the Plains of Abraham. It became capital of Lower Canada (later Quebec) in 1791. □ **Quebecker** *n.* (also **Quebecer**).

Quebecois /ˌkeɪbeˈkwʌ/ *n.* & *adj.* ● *n.* (*pl.* same) **1** a French-speaking inhabitant or native of Quebec. **2** the French spoken in Quebec. ● *adj.* of or relating to Quebec or the Quebecois. [F *Québecois*]

Quechua /ˈketʃwə/ *n.* (also **Quichua** /ˈkɪtʃ-/) (*pl.* same) **1** a member of a native people of Peru and neighbouring parts of Bolivia, Chile, Colombia, and Ecuador. **2** the language or group of related languages spoken by this people. It is one of the official languages of Peru. □ **Quechuan** *adj.* & *n.* [Sp. f. Quechua, = plunderer, despoiler]

Queen[1] /kwiːn/, Ellery (pseudonym of Frederic Dannay, 1905–82, and Manfred Lee, 1905–71), American writers of detective fiction. Their many detective novels, featuring the detective also called Ellery Queen, include *The French Powder Mystery* (1930). They went on to found and edit *Ellery Queen's Mystery Magazine* (1941).

Queen[2] /kwiːn/ a British rock group formed in 1971, featuring flamboyant vocalist Freddie Mercury (born Frederick Bulsara, 1946–91) and guitarist Brian May (b.1947). Queen initially played heavy rock but soon added dramatic, almost operatic, elements, as exemplified by the hugely successful 'Bohemian Rhapsody' (1975). The group was effectively disbanded after Freddie Mercury died of Aids in 1991.

queen /kwiːn/ *n.* & *v.* ● *n.* **1** (as a title usu. **Queen**) **a** a female sovereign etc., esp. the hereditary ruler of an independent state. **2** (in full **queen consort**) a king's wife. **3** a woman, country, or thing pre-eminent or supreme in a specified area or of its kind (*tennis queen*; *the queen of roses*). **4** the fertile female among ants, bees, etc. **5** the most powerful piece in chess. **6** a court card with a picture of a queen. **7** *sl.* a male homosexual, esp. an effeminate one. **8 a** an honoured female, e.g. the Virgin Mary (*queen of heaven*). **b** an ancient goddess (*Venus, queen of love*). **9** a belle or mock sovereign on some occasion (*beauty queen*; *queen of the May*). **10** a person's sweetheart, wife, or mistress. **11** (**the Queen**) (in the UK) the national anthem when there is a female sovereign. ● *v.tr.* **1** make (a woman) queen. **2** *Chess* convert (a pawn) into a queen when it reaches the opponent's side of the board. □ **queen bee 1** the fertile female in a hive. **2** the chief or controlling woman in an organization or social group. **queen-cake** a small soft cake often with raisins etc. **queen dowager** the widow of a king. **queen it** play the queen. **queen mother** the dowager who is mother of the sovereign (see also ELIZABETH, THE QUEEN MOTHER). **queen of the meadows** meadowsweet. **queen of puddings** a pudding made with bread, jam, and meringue. **queen-post** either of two upright timbers between the tie-beam and principal rafters of a roof-truss. **queen's bishop, knight**, etc. *Chess* (of pieces which exist in pairs) the piece starting on the queen's side of the board. **queen's bounty** see BOUNTY. **Queen's colour** see COLOUR. **Queen's evidence** see EVIDENCE. **Queen's highway** see HIGHWAY. **queen-size** (or **-sized**) of an extra-large size, usu. smaller than king-size. **queen's pawn** *Chess* the pawn in front of the queen at the beginning of a game. **Queen's speech** see SPEECH. **queen's-ware** cream-coloured Wedgwood. □ **queendom** *n.* **queenhood** *n.* **queenless** *adj.* **queenlike** *adj.* **queenship** *n.* [OE *cwēn* f. Gmc; cf. QUEAN]

Queen Anne *attrib.adj.* of the style of furniture, architecture, etc. characteristic of the reign of Queen Anne (1702–14), or more loosely the late 17th and early 18th centuries. Having a simple, proportioned style, the furniture was also noted for cabriole legs and the use of walnut. The architecture was characterized by a Dutch-influenced use of red brick in simple, basically rectangular designs which prefigured the Georgian style, and which was revived in the late 19th century.

Queen Anne's Bounty *n.* duties called 'first fruits and tenths', payable originally to the pope but made payable to the Crown by Henry VIII, and directed by Queen Anne in 1704 to be used to provide for the augmentation of livings of the poorer clergy.

Queen Anne's lace *n.* = cow-parsley.

Queen Charlotte Islands a group of more than 150 islands off the west coast of Canada, in British Columbia, noted for their timber and fishing resources.

queenie /ˈkwiːnɪ/ *n.* *sl.* = QUEEN *n.* 7.

queenly /ˈkwiːnlɪ/ *adj.* (**queenlier**, **queenliest**) **1** fit for or appropriate to a queen. **2** majestic; queenlike. □ **queenliness** *n.*

Queen Maud Land /mɔːd/ a part of Antarctica bordering the Atlantic Ocean, claimed since 1939 by Norway. It is named after Queen Maud of Norway (1869–1938).

Queens /kwiːnz/ a borough of New York City, at the western end of Long Island; pop. (1990) 1,951,600.

Queen's Award n. (in the UK) any of several annual awards given to firms for achievements in exporting goods or services or in advancing technology. The original award, the Queen's Award to Industry, instituted in 1965, was replaced in 1976 by the Queen's Award for Export Achievement and the Queen's Award for Technological Achievement, to which was added in 1992 the Queen's Award for Environmental Achievement.

Queen's Bench (in the UK) a division of the High Court of Justice.

Queensberry Rules /ˈkwiːnzbəri/ n.pl. a code of rules governing boxing in the UK drawn up in 1867 under the direction of the 8th Marquess of Queensberry (1844–1900); the standard rules of modern boxing.

Queen's Champion see CHAMPION OF ENGLAND.

Queen's Counsel n. (in the UK) a counsel to the Crown, taking precedence over barristers.

Queen's County the former name for LAOIS.

Queen's English see ENGLISH.

Queen's Guide see GUIDE.

Queensland /ˈkwiːnzlənd/ a state comprising the north-eastern part of Australia; pop. (est. 1990) 2,921,700; capital, Brisbane. Originally established in 1824 as a penal settlement, Queensland was constituted a separate colony in 1859, having previously formed part of New South Wales, and was federated with the other states of Australia in 1901. □ **Queenslander** n.

Queen's Messenger n. a courier in the British diplomatic service.

Queen's Proctor n. (in the UK) an official who has the right to intervene in probate, divorce, and nullity cases when collusion or the suppression of facts is alleged.

Queen's Scout see SCOUT[1].

queer /kwɪə(r)/ adj., n., & v. ● adj. **1** strange; odd; eccentric. **2** shady; suspect; of questionable character. **3 a** slightly ill; giddy; faint. **b** Brit. sl. drunk. **4** derog. sl. (esp. of a man) homosexual. **5** colloq. (of a person or behaviour) crazy; unbalanced; slightly mad. ● n. derog. sl. a homosexual. ● v.tr. sl. spoil; put out of order. □ **in Queer Street** sl. in a difficulty, in debt or trouble or disrepute. **queer a person's pitch** spoil a person's chances, esp. secretly or maliciously. □ **queerish** adj. **queerly** adv. **queerness** n. [perh. f. G quer oblique (as THWART)]

quell /kwel/ v.tr. **1 a** crush or put down (a rebellion etc.). **b** reduce (rebels etc.) to submission. **2** suppress or alleviate (fear, anger, etc.). □ **queller** n. (also in comb.). [OE cwellan kill f. Gmc]

quench /kwentʃ/ v.tr. **1** satisfy (thirst) by drinking. **2** extinguish (a fire or light etc.). **3** cool, esp. with water (heat, a heated thing). **4** cool (esp. hot metal) in cold water, air, oil, etc. **5 a** stifle or suppress (desire etc.). **b** Physics & Electronics inhibit or prevent (oscillation, luminescence, etc.) by counteractive means. **6** sl. reduce (an opponent) to silence. □ **quenchable** adj. **quencher** n. **quenchless** adj. [ME f. OE -cwencan causative f. -cwincan be extinguished]

quenelle /kəˈnel/ n. a seasoned ball or roll of pounded fish or meat. [F, of unkn. orig.]

Quercia, Jacopo della, see DELLA QUERCIA.

Querétaro /keˈretəˌrəʊ/ **1** a state of central Mexico. **2** its capital city; pop. (1990) 454,050. In 1847 it was the scene of the signing of the treaty ending the US–Mexican war.

querist /ˈkwɪərɪst/ n. literary a person who asks questions; a questioner. [L quaerere ask]

quern /kwɜːn/ n. **1** a hand-mill for grinding corn, usu. consisting of two circular stones, the upper of which is rubbed to and fro, or rotated, on the lower one. **2** a small hand-mill for grinding pepper etc. □ **quern-stone** either of the two stones forming a quern. [OE cweorn(e) f. Gmc]

querulous /ˈkwerʊləs/ adj. complaining, peevish. □ **querulously** adv. **querulousness** n. [LL querulosus or L querulus f. queri complain]

query /ˈkwɪərɪ/ n. & v. ● n. (pl. **-ies**) **1** a question, esp. expressing doubt or objection. **2** a question mark, or the word query spoken or written to question accuracy or as a mark of interrogation. ● v. (**-ies, -ied**) **1** tr. (often foll. by whether, if, etc. + clause) ask or inquire. **2** tr. call (a thing) in question in speech or writing. **3** tr. dispute the accuracy of. **4** intr. put a question. [anglicized form of quaere f. L quaerere ask, after INQUIRY]

quest /kwest/ n. & v. ● n. **1** a search or the act of seeking. **2** the thing sought, esp. the object of a medieval knight's pursuit. ● v. **1** intr. (often foll. by about) **a** (often foll. by after, for) go about in search of something. **b** (of a dog etc.) search about for game. **2** tr. poet. search for, seek out. □ **in quest of** seeking. □ **quester** n. **questingly** adv. [ME f. OF queste, quester ult. f. L quaerere quaest- seek]

question /ˈkwestʃən/ n. & v. ● n. **1** a sentence worded or expressed so as to seek information. **2 a** doubt about or objection to a thing's truth, credibility, advisability, etc. (allowed it without question; is there any question as to its validity?). **b** the raising of such doubt etc. **3** a matter to be discussed or decided or voted on. **4** a problem requiring an answer or solution. **5** (foll. by of) a matter or concern depending on conditions (it's a question of money). ● v.tr. **1** ask questions of; interrogate. **2** subject (a person) to examination. **3** throw doubt upon; raise objections to. **4** seek information from the study of (phenomena, facts). □ **be a question of time** be certain to happen sooner or later. **beyond all question** undoubtedly. **call in** (or **into**) **question** make a matter of dispute; query. **come into question** be discussed; become of practical importance. **in question 1** that is being discussed or referred to (the person in question). **2** in dispute (that was never in question). **is not the question** is irrelevant. **no question of** no possibility of (there's no question of my giving in). **out of the question** too impracticable etc. to be worth discussing; impossible. **put the question** require supporters and opponents of a proposal to record their votes, divide a meeting. **question mark 1** a punctuation mark (?) indicating a question. **2** a cause for doubt or uncertainty (there's still a question mark over the plans). **question-master** Brit. a person who presides over a quiz game etc. **question time** a period during parliamentary proceedings when MPs may question ministers. **without question** = beyond all question above. □ **questioner** n. **questioningly** adv. **questionless** adj. [ME f. AF questiun, OF question, questionner f. L quaestio -onis f. quaerere quaest- seek]

questionable /ˈkwestʃənəb(ə)l/ adj. **1** doubtful as regards truth or quality. **2** not clearly in accordance with honesty, honour, wisdom, etc. □ **questionably** adv. **questionableness** n. **questionability** /ˌkwestʃənəˈbɪlɪtɪ/ n.

questionnaire /ˌkwestʃəˈneə(r)/ n. **1** a formulated series of questions, esp. for statistical study. **2** a document containing these. [F f. questionner QUESTION + -aire -ARY[1]]

Quetta /ˈkwetə/ a city in western Pakistan, the capital of Baluchistan province; pop. (est. 1991) 350,000. The city was severely damaged by an earthquake in 1935.

quetzal /ˈkets(ə)l/ n. **1** a brilliantly coloured trogon of the genus Pharomachrus, of Central and South America; esp. the green and red P. mocinno, the male of which has very long tail coverts and was regarded as sacred in ancient Meso-America. **2** the chief monetary unit of Guatemala. [Sp. f. Aztec f. quetzalli the bird's tail feather]

Quetzalcóatl /ˌkets(ə)lkəʊˈɑːt(ə)l/ the plumed serpent god of the Toltec and Aztec civilizations. Traditionally the god of the morning and evening star, he later became known as the patron of priests, inventor of books and of the calendar, and as the symbol of death and resurrection. His worship involved human sacrifice. Legend said that he would return in another age, and when Montezuma, last king of the Aztecs, received news of the landing of Cortés and his men in 1519, he thought that Quetzalcóatl had returned.

queue /kjuː/ n. & v. esp. Brit. ● n. **1** a line or sequence of persons, vehicles, etc., awaiting their turn to be attended to or to proceed. **2** a pigtail or plait of hair. ● v.intr. (**queues, queued, queuing** or **queueing**) (often foll. by up) (of persons etc.) form a queue; take one's place in a queue. □ **queue-jump** Brit. push forward out of turn in a queue. [F f. L cauda tail]

Quezon City /ˈkeɪzɒn/ a city on the island of Luzon in the northern Philippines; pop. (1990) 1,667,000. Forming part of a conurbation with Manila, Quezon City was established in 1940 and named after Manuel Luis Quezon (1878–1944), the first President of the republic. From 1948 to 1976 it was the capital of the Philippines.

Qufu /tʃuːˈfuː/ a small town in Shandong province in eastern China, where Confucius was born in 551 BC and lived for much of his life.

quibble /ˈkwɪb(ə)l/ n. & v. ● n. **1** a petty objection; a trivial point of criticism. **2** a play on words; a pun. **3** an evasion; an insubstantial argument which relies on an ambiguity etc. ● v.intr. use quibbles. □ **quibbler** n. **quibbling** adj. **quibblingly** adv. [dimin. of obs. quib prob. f. L quibus dative & ablat. pl. of qui who (familiar from use in legal documents)]

quiche /kiːʃ/ n. an open flan or tart with a savoury filling. [F]

Quichua var. of QUECHUA.

quick /kwɪk/ adj., adv., & n. ● adj. **1** taking only a short time (a quick worker; a quick visit). **2 a** arriving after a short time, prompt (quick action; quick results). **b** (of an action, occurrence, etc.) sudden; hasty; abrupt. **3** with only a short interval (in quick succession). **4** lively, intelligent. **5 a** acute, alert (has a quick ear). **b** agile, nimble; energetic. **6** (of a temper) easily roused. **7** *archaic* living, alive (*the quick and the dead*). ● *adv.* **1** quickly, at a rapid rate. **2** (as *int.*) come, go, etc., quickly. ● *n.* **1** the soft flesh below the nails, or the skin, or a sore. **2** the seat of feeling or emotion (*cut to the quick*). □ **be quick** act quickly. **quick-fire 1** (of repartee etc.) rapid. **2** firing shots in quick succession. **quick fix** a hasty remedial measure not taking account of long term consequences. **quick-freeze** freeze (food) rapidly so as to preserve its natural qualities. **quick march** *Mil.* **1** a march in quick time. **2** the command to begin this. **quick one** *colloq.* an alcoholic drink taken quickly. **quick-tempered** quick to lose one's temper; irascible. **quick step** *Mil.* a step used in quick time (cf. QUICKSTEP). **quick time** *Mil.* marching at about 120 paces per minute. **quick trick** *Bridge* **1** a trick in the first two rounds of a suit, with no trumping. **2** a card that should win this. **quick with child** *archaic* at a stage of pregnancy when movements of the foetus have been felt. □ **quickly** *adv.* **quickness** *n.* [OE *cwic(u)* alive f. Gmc]

quicken /ˈkwɪkən/ *v.* **1** *tr.* & *intr.* make or become quicker; accelerate. **2** *tr.* give life or vigour to; rouse; animate; stimulate. **3** *intr.* **a** (of a woman) reach a stage in pregnancy when movements of the foetus can be felt. **b** (of a foetus) begin to show signs of life. **4** *tr. archaic* kindle; make (a fire) burn brighter. **5** *intr.* come to life.

quickie /ˈkwɪkɪ/ *n. colloq.* **1** a thing done or made quickly or hastily. **2** an alcoholic drink taken quickly.

quicklime /ˈkwɪklaɪm/ *n.* = LIME¹ *n.* 1.

quicksand /ˈkwɪksænd/ *n.* **1 a** loose wet sand that sucks in anything placed or falling into it. **b** a bed of this. **2** a treacherous thing or situation.

quickset /ˈkwɪkset/ *adj.* & *n.* ● *adj.* (of a hedge) formed of slips of plants, esp. hawthorn set in the ground to grow. ● *n.* **1** such slips. **2** a hedge formed in this way.

quicksilver /ˈkwɪkˌsɪlvə(r)/ *n.* & *v.* ● *n.* **1** liquid mercury. **2** mobility of temperament or mood. ● *v.tr.* coat (a mirror-glass) with an amalgam of tin.

quickstep /ˈkwɪkstep/ *n.* & *v.* ● *n.* a fast foxtrot (cf. *quick step*). ● *v.intr.* (**-stepped**, **-stepping**) dance the quickstep.

quickthorn /ˈkwɪkθɔːn/ *n.* the hawthorn.

quick-witted /kwɪkˈwɪtɪd/ *adj.* quick to grasp a situation, make repartee, etc. □ **quick-wittedness** *n.*

quid¹ /kwɪd/ *n.* (*pl.* same) *Brit. sl.* one pound sterling. □ **not the full quid** *Austral. sl.* mentally deficient. **quids in** *sl.* in a position of profit. [prob. f. *quid* the nature of a thing f. L *quid* what, something]

quid² /kwɪd/ *n.* a lump of tobacco for chewing. [dial. var. of CUD]

quiddity /ˈkwɪdɪtɪ/ *n.* (*pl.* **-ies**) **1** *Philos.* the essence of a person or thing; what makes a thing what it is. **2** a quibble; a trivial objection. [med.L *quidditas* f. L *quid* what]

quidnunc /ˈkwɪdnʌŋk/ *n. archaic* a newsmonger, a person given to gossip. [L *quid* what + *nunc* now]

quid pro quo /ˌkwɪd prəʊ ˈkwəʊ/ *n.* (*pl.* **quid pro quos**) **1** a thing given as compensation. **2** return made (for a gift, favour, etc.). [L, = something for something]

quiescent /kwɪˈes(ə)nt/ *adj.* **1** motionless, inert. **2** silent, dormant. □ **quiescence** *n.* **quiescency** *n.* **quiescently** *adv.* [L *quiescere* f. *quies* QUIET]

quiet /ˈkwaɪət/ *adj., n.,* & *v.* ● *adj.* (**quieter, quietest**) **1** with little or no sound or motion. **2 a** of gentle or peaceful disposition. **b** shy; reticent; reserved. **3** (of a colour, piece of clothing, etc.) unobtrusive; not showy. **4** not overt; private; disguised (*quiet resentment*). **5** undisturbed; uninterrupted; free or far from vigorous action (*a quiet time for prayer*). **6** informal; simple (*just a quiet wedding*). **7** enjoyed in quiet (*a quiet smoke*). **8** tranquil; not anxious or remorseful. ● *n.* **1** silence; stillness. **2** an undisturbed state; tranquillity. **3** a state of being free from urgent tasks or agitation (*a period of quiet*). **4** a peaceful state of affairs (*could do with some quiet*). ● *v.* **1** *tr.* soothe, make quiet. **2** *intr.* (often foll. by *down*) become quiet or calm. □ **be quiet** (esp. in *imper.*) cease talking etc. **keep quiet 1** refrain from making a noise. **2** (often foll. by *about*) suppress or refrain from disclosing information etc. **on the quiet** *colloq.* unobtrusively; secretly. □ **quietly** *adv.* **quietness** *n.* [ME f. AF *quiete* f. OF *quiet(e), quieté* f. L *quietus* past part. of *quiescere*: see QUIESCENT]

quieten /ˈkwaɪət(ə)n/ *v.tr.* & *intr.* (often foll. by *down*) *Brit.* make or become quiet.

quietism /ˈkwaɪəˌtɪz(ə)m/ *n.* **1** a passive attitude towards life, with devotional contemplation and abandonment of the will, as a form of religious mysticism. **2** the principle of non-resistance. □ **quietist** *n.* & *adj.* **quietistic** /ˌkwaɪəˈtɪstɪk/ *adj.* [It. *quietismo* (as QUIET)]

quietude /ˈkwaɪəˌtjuːd/ *n.* a state of quiet.

quietus /kwaɪˈiːtəs/ *n.* (*pl.* **quietuses**) **1** a discharge or release from life; death; that which brings death. **2** something which quiets or represses; a sedative. [med.L *quietus est* 'he is quit', used as a form of receipt (cf. QUIT)]

quiff /kwɪf/ *n. esp. Brit.* **1** a man's tuft of hair, brushed upward over the forehead. **2** a curl plastered down on the forehead. [20th c.: orig. unkn.]

quill /kwɪl/ *n.* & *v.* ● *n.* **1** (in full **quill-feather**) a large feather in a bird's wing or tail. **2** the hollow stem of this. **3** (in full **quill pen**) a pen made of a quill. **4** (usu. in *pl.*) the spines of a porcupine. **5** a musical pipe made of a hollow stem. ● *v.tr.* form into cylindrical quill-like folds; goffer. □ **quill-coverts** the feathers covering the base of quill-feathers. [ME prob. f. (M)LG *quiele*]

quilling /ˈkwɪlɪŋ/ *n.* the art or craft of paper filigree. [QUILL]

quilt¹ /kwɪlt/ *n.* & *v.* ● *n.* **1** a bed-covering made of padding enclosed between layers of cloth etc. and kept in place by cross lines of stitching. **2** a bedspread of similar design (*patchwork quilt*). ● *v.tr.* **1** cover or line with padded material. **2** make or join together (pieces of cloth with padding between) after the manner of a quilt. **3** sew up (a coin, letter, etc.) between two layers of a garment etc. **4** compile (a literary work) out of extracts or borrowed ideas. □ **quilter** *n.* **quilting** *n.* [ME f. OF *coilte, cuilte* f. L *culcita* mattress, cushion]

quilt² /kwɪlt/ *v.tr. Austral. sl.* thrash; clout. [perh. f. QUILT¹]

quim /kwɪm/ *n. coarse sl.* the female genitals. [18th c.: orig. unkn.]

quin /kwɪn/ *n. esp. Brit. colloq.* a quintuplet. [abbr.]

quinacrine /ˈkwɪnəˌkriːn, -krɪn/ *n.* an anti-malarial drug derived from acridine. [*quinine + acridine*]

quinary /ˈkwaɪnərɪ/ *adj.* **1** of the number five. **2** having five parts. [L *quinarius* f. *quini* distributive of *quinque* five]

quinate /ˈkwaɪneɪt/ *adj. Bot.* (of a leaf) having five leaflets. [L *quini* (as QUINARY)]

quince /kwɪns/ *n.* **1** a hard acid pear-shaped fruit used as a preserve or flavouring. **2** a rosaceous shrub or small tree of the genus *Cydonia*, esp. *C. oblonga*, bearing this fruit (see also *Japanese quince*). [ME, orig. collect. pl. of obs. *quoyn, coyn,* f. OF *cooin* f. L *cotoneum* var. of *cydoneum* (apple) of *Cydonia* in Crete]

quincentenary /ˌkwɪnsenˈtiːnərɪ/ *n.* & *adj.* ● *n.* (*pl.* **-ies**) **1** a five-hundredth anniversary. **2** a festival marking this. ● *adj.* of this anniversary. □ **quincentennial** /-ˈtenɪəl/ *adj.* & *n.* [irreg. f. L *quinque* five + CENTENARY]

Quincey, Thomas De, see DE QUINCEY.

quincunx /ˈkwɪnkʌŋks/ *n.* **1** five objects set so that four are at the corners of a square or rectangle and the fifth is at its centre, e.g. the five on dice or cards. **2** this arrangement, esp. in planting trees. □ **quincuncial** /kwɪnˈkʌnʃ(ə)l/ *adj.* **quincuncially** *adv.* [L, = five-twelfths f. *quinque* five, *uncia* twelfth]

Quine /kwaɪn/, Willard Van Orman (b.1908), American philosopher and logician. A radical critic of modern empiricism, Quine took issue with the philosophy of language proposed by Rudolf Carnap, arguing that 'no statement is immune from revision' and that even the principles of logic themselves can be questioned and replaced. In *Word and Object* (1961), he held that there is no such thing as satisfactory translation. He also developed the work on the foundations of mathematics begun by Frege and Russell, specializing in the theory of sets, published in *Set Theory and its Logic* (1963).

quinella /kwɪˈnelə/ *n.* a form of betting in which the better must select the first two place-winners in a race, not necessarily in the correct order. [Amer. Sp. *quiniela*]

quinine /ˈkwɪniːn, kwɪˈniːn/ *n.* **1** an alkaloid found esp. in cinchona bark. **2** a toxic bitter drug containing this, formerly used as a tonic and esp. as a remedy for malaria. [*quina* cinchona bark f. Sp. *quina* f. Quechua *kina* bark]

quinol /ˈkwɪnɒl/ *n.* = HYDROQUINONE.

quinoline /ˈkwɪnəˌliːn/ *n. Chem.* an oily amine (chem. formula: C_9H_7N) obtained from coal tar or by synthesis and used in the preparation of drugs etc.

quinone /ˈkwɪnəʊn, kwɪˈnəʊn/ *n. Chem.* **1** = BENZOQUINONE. **2** any compound with the same ring structure as benzoquinone. [as QUININE + -ONE]

quinquagenarian /ˌkwɪŋkwədʒɪˈneərɪən/ *n.* & *adj.* ● *n.* a person from 50 to 59 years old. ● *adj.* of or relating to this age. [L *quinquagenarius* f. *quinquageni* distributive of *quinquaginta* fifty]

Quinquagesima /ˌkwɪŋkwəˈdʒesɪmə/ n. (in full **Quinquagesima Sunday**) the Sunday before the beginning of Lent. [med.L, fem. of L *quinquagesimus* fiftieth f. *quinquaginta* fifty, after QUADRAGESIMA]

quinque- /ˈkwɪŋkwɪ/ comb. form five. [L f. *quinque* five]

quinquennial /kwɪnˈkwenɪəl/ adj. **1** lasting five years. **2** recurring every five years. □ **quinquennially** adv. [L *quinquennis* (as QUINQUENNIUM)]

quinquennium /kwɪnˈkwenɪəm/ n. (pl. **quinquenniums** or **quinquennia** /-nɪə/) a period of five years. [L f. *quinque* five + *annus* year]

quinquereme /ˈkwɪŋkwɪˌriːm/ n. an ancient Roman galley probably having five oarsmen to each bank of oars. [L *quinqueremis* (as QUINQUE-, *remus* oar)]

quinquevalent /ˌkwɪŋkwɪˈveɪlənt/ adj. Chem. = PENTAVALENT.

quinsy /ˈkwɪnzɪ/ n. an inflammation of the throat, esp. an abscess in the region of the tonsils. □ **quinsied** adj. [ME f. OF *quinencie* f. med.L *quinancia* f. Gk *kunagkhē* f. *kun-* dog + *agkhō* throttle]

quint /kwɪnt/ n. **1** a sequence of five cards in the same suit in piquet etc. **2** N. Amer. a quintuplet. □ **quint major** a quint headed by an ace. [F *quinte* f. L *quinta* fem. of *quintus* fifth f. *quinque* five]

quintain /ˈkwɪntɪn/ n. hist. **1** a post set up as a mark in tilting, and often provided with a sandbag to swing round and strike an unsuccessful tilter. **2** the medieval military exercise of tilting at such a mark. [ME f. OF *quintaine* perh. ult. f. L *quintana* camp market f. *quintus* (*manipulus*) fifth (maniple)]

quintal /ˈkwɪnt(ə)l/ n. **1** a weight of about 100 lb. **2** a hundredweight (112 lb). **3** a weight of 100 kg. [ME f. OF *quintal*, med.L *quintale* f. Arab. *ḳinṭār*]

quintan /ˈkwɪntən/ adj. (of a fever etc.) recurring every fifth day. [L *quintana* f. *quintus* fifth]

Quintana Roo /kiːnˌtɑːnə ˈrəʊ/ a state of SE Mexico, on the Yucatán Peninsula; capital, Chetumal.

quinte /kænt, kwɪnt/ n. the fifth of eight parrying positions in fencing. [F: see QUINT]

quintessence /kwɪnˈtes(ə)ns/ n. **1** (usu. foll. by *of*) the purest and most perfect, or most typical, form, manifestation, or embodiment of some quality or class. **2** the most essential part of a substance; a refined extract. **3** (in ancient philosophy) a fifth substance (beside the four elements) forming celestial bodies and pervading all things. □ **quintessential** /ˌkwɪntɪˈsenʃ(ə)l/ adj. **quintessentially** adv. [ME (in sense 3) f. F f. med.L *quinta essentia* fifth ESSENCE]

quintet /kwɪnˈtet/ n. (also **quintette**) **1** Mus. **a** a composition for five voices or instruments. **b** the performers of such a piece. **2** any group of five. [F *quintette* f. It. *quintetto* f. *quinto* fifth f. L *quintus*]

Quintilian /kwɪnˈtɪlɪən/ (Latin name Marcus Fabius Quintilianus) (AD c.35–c.96), Roman rhetorician. A famous teacher, he is best known for his *Education of an Orator*, a comprehensive treatment of the art of rhetoric and the training of an orator; the work was highly influential in the Middle Ages and the Renaissance.

quintillion /kwɪnˈtɪljən/ n. (pl. same or **quintillions**) a thousand raised to the sixth (or formerly, esp. Brit., the tenth) power (10^{18} and 10^{30} respectively). □ **quintillionth** adj. & n. [L *quintus* fifth + MILLION]

quintuple /ˈkwɪntjʊp(ə)l, kwɪnˈtjuːp-/ adj., n., & v. ● adj. **1** fivefold; consisting of five parts. **2** involving five parties. **3** (of time in music) having five beats in a bar. ● n. a fivefold number or amount. ● v.tr. & intr. multiply by five; increase fivefold. □ **quintuply** /-plɪ/ adv. [F *quintuple* f. L *quintus* fifth, after QUADRUPLE]

quintuplet /ˈkwɪntjʊplɪt, kwɪnˈtjuːp-/ n. **1** each of five children born at one birth. **2** a set of five things working together. **3** Mus. a group of five notes to be performed in the time of three or four. [QUINTUPLE, after QUADRUPLET, TRIPLET]

quintuplicate adj. & v. ● adj. /kwɪnˈtjuːplɪkət/ **1** fivefold. **2** of which five copies are made. ● v.tr. & intr. /kwɪnˈtjuːplɪˌkeɪt/ multiply by five. □ **in quintuplicate 1** in five identical copies. **2** in groups of five. [F *quintuple* f. L *quintus* fifth, after QUADRUPLICATE]

quip /kwɪp/ n. & v. ● n. **1** a clever saying; an epigram; a sarcastic remark etc. **2** a quibble; an equivocation. ● v.intr. (**quipped**, **quipping**) make quips. □ **quipster** n. [abbr. of obs. *quippy* perh. f. L *quippe* forsooth]

quipu /ˈkiːpuː, ˈkwiː-/ n. the ancient Peruvian method of recording information by variously knotting threads of various colours. [Quechua, = knot]

quire /ˈkwaɪə(r)/ n. **1** four sheets of paper etc. folded to form eight leaves, as often in medieval manuscripts. **2** any collection of leaves one within another in a manuscript or book. **3** 25 (formerly 24) sheets of paper; one-twentieth of a ream. □ **in quires** unbound; in sheets. [ME f. OF *qua(i)er* ult. f. L *quaterni* set of four (as QUATERNARY)]

quirk /kwɜːk/ n. **1** a peculiarity of behaviour. **2** a trick of fate; a freak. **3** a flourish in writing. **4** (often attrib.) Archit. a hollow in a moulding. □ **quirkish** adj. **quirky** adj. (**quirkier**, **quirkiest**). **quirkily** adv. **quirkiness** n. [16th c.: orig. unkn.]

quirt /kwɜːt/ n. & v. ● n. a short-handled riding-whip with a braided leather lash. ● v.tr. strike with this. [Sp. *cuerda* CORD]

quisling /ˈkwɪzlɪŋ/ n. **1** a person cooperating with an occupying enemy; a collaborator or fifth-columnist. **2** a traitor. □ **quislingite** adj. & n. [Vidkun *Quisling*, Norwegian Army officer and Nazi collaborator (1887–1945)]

quit /kwɪt/ v. & adj. ● v.tr. (**quitting**; past and past part. **quitted** or **quit**) **1** (also absol.) give up; let go; abandon (a task etc.). **2** N. Amer. cease; stop (*quit grumbling*). **3 a** leave or depart from (a place, person, etc.). **b** (absol.) (of a tenant) leave occupied premises (esp. *notice to quit*). **4** (refl.) acquit; behave (*quit oneself well*). ● predic.adj. (foll. by *of*) rid (*glad to be quit of the problem*). □ **quit hold of** loose. [ME f. OF *quitte*, *quitter* f. med.L *quittus* f. L *quietus* QUIET]

quitch /kwɪtʃ/ n. (in full **quitch-grass**) = COUCH^2. [OE *cwice*, perh. rel. to QUICK]

quite /kwaɪt/ adv. **1** completely; entirely; wholly; to the utmost extent; in the fullest sense (*I quite agree*). **2** somewhat; rather; to some extent (*she's quite nice*). **3** (often foll. by *so*) said to indicate agreement. **4** absolutely; definitely; very much. □ **quite a** a remarkable or outstanding (person or thing) (*it was quite an event*). **quite another** (or **other**) very different (*that's quite another matter*). **quite a few** colloq. a fairly large number of. **quite some** a large amount of (*quite some time*). **quite something** a remarkable thing. [ME f. obs. *quite* (adj.) = QUIT]

Quito /ˈkiːtəʊ/ the capital of Ecuador; pop. (1990) 1,387,890. It is situated in the Andes just south of the equator, at an altitude of 2,850 m (9,350 ft). An ancient pre-Columbian settlement, Quito was captured by the Spanish in 1533.

quits /kwɪts/ predic.adj. on even terms by retaliation or repayment (*then we'll be quits*). □ **call it** (or **cry**) **quits** acknowledge that things are now even; agree not to proceed further in a quarrel etc. [perh. colloq. abbr. of med.L *quittus*: see QUIT]

quittance /ˈkwɪt(ə)ns/ n. archaic or poet. **1** (foll. by *from*) a release. **2** an acknowledgement of payment; a receipt. [ME f. OF *quitance* f. *quiter* QUIT]

quitter /ˈkwɪtə(r)/ n. colloq. **1** a person who gives up easily. **2** a shirker.

quiver^1 /ˈkwɪvə(r)/ v. & n. ● v. **1** intr. tremble or vibrate with a slight rapid motion, esp.: **a** (usu. foll. by *with*) as the result of emotion (*quiver with anger*). **b** (usu. foll. by *in*) as the result of air currents etc. (*quiver in the breeze*). **2** tr. (of a bird, esp. a skylark) make (its wings) quiver. ● n. a quivering motion or sound. □ **quiveringly** adv. **quivery** adj. [ME f. obs. *quiver* nimble: cf. QUAVER]

quiver^2 /ˈkwɪvə(r)/ n. an archer's portable holder for arrows. □ **have an arrow** (or **shaft**) **left in one's quiver** not be resourceless. [ME f. OF *quivre* f. WG (cf. OE *cocor*)]

quiverful /ˈkwɪvəˌfʊl/ n. (pl. **-fuls**) **1** as much as a quiver can hold. **2** many children of one parent (Ps. 127:5). [QUIVER^2]

qui vive /kiː ˈviːv/ n. □ **on the qui vive** on the alert; watching for something to happen. [F, = lit. '(long) live who?', i.e. on whose side are you?, as a sentry's challenge]

Quixote see DON QUIXOTE.

quixotic /kwɪkˈsɒtɪk/ adj. **1** extravagantly and romantically chivalrous. **2** visionary; pursuing lofty but unattainable ideals. **3** derog. ridiculously impractical; preposterous; foolhardy. □ **quixotically** adv. **quixotism** /ˈkwɪksəˌtɪz(ə)m/ n. **quixotry** /-trɪ/ n. [DON QUIXOTE]

quiz^1 /kwɪz/ n. & v. ● n. (pl. **quizzes**) **1** a test of knowledge, esp. between individuals or teams as a form of entertainment. **2** an interrogation, examination, or questionnaire. ● v.tr. (**quizzed**, **quizzing**) examine by questioning. □ **quiz-master** a person who presides over a quiz. [19th-c. dial.: orig. unkn.]

quiz^2 /kwɪz/ v. & n. archaic ● v.tr. (**quizzed**, **quizzing**) **1** look curiously at; observe the ways or oddities of; survey through an eyeglass. **2** make sport of; regard with a mocking air. ● n. (pl. **quizzes**) **1** a hoax, a thing done to burlesque or expose another's oddities. **2 a** an odd or eccentric person; a person of ridiculous appearance. **b** a person given to quizzing. □ **quizzer** n. [18th c.: orig. unkn.]

quizzical /ˈkwɪzɪk(ə)l/ adj. **1** expressing or done with mild or amused perplexity. **2** strange; comical. □ **quizzically** *adv.* **quizzicalness** *n.* **quizzicality** /ˌkwɪzɪˈkælɪtɪ/ *n.* [QUIZ²]

Qum see QOM.

Qumran /kʊmˈrɑːn/ a region on the western shore of the Dead Sea. The Dead Sea scrolls were found (1947–56) in caves at nearby Khirbet Qumran, the site of an ancient Jewish settlement.

quod /kwɒd/ *n. Brit. sl.* prison. [17th c.: orig. unkn.]

quod erat demonstrandum /kwɒd ˌeræt ˌdemənˈstrændʊm/ (abbr. **QED**) (esp. at the conclusion of a proof etc.) which was the thing to be proved. [L]

quodlibet /ˈkwɒdlɪˌbet/ *n.* **1** *hist.* **a** a topic for philosophical or theological discussion. **b** an exercise on this. **2** a light-hearted medley of well-known tunes. □ **quodlibetarian** /ˌkwɒdlɪbəˈteərɪən/ *n.* [ME f. L f. *quod* what + *libet* it pleases one]

quod vide /kwɒd ˈviːdeɪ/ (abbr. **q.v.**) which see (in cross-references etc.). [L]

quoin /kɔɪn/ *n. & v.* ● *n.* **1** an external angle of a building. **2** a stone or brick forming an angle; a cornerstone. **3** a wedge used for locking type in a forme. **4** a wedge for raising the level of a gun, keeping the barrel from rolling, etc. ● *v.tr.* secure or raise with quoins. □ **quoining** *n.* [var. of COIN]

quoit /kɔɪt/ *n. & v.* ● *n.* **1** a heavy flattish sharp-edged iron ring thrown to encircle an iron peg or to land as near as possible to the peg. **2** (in *pl.*) a game consisting of aiming and throwing these. **3** a ring of rope, rubber, etc. for use in a similar game. **4 a** the flat stone of a dolmen. **b** the dolmen itself. ● *v.tr.* fling like a quoit. [ME: orig. unkn.]

quokka /ˈkwɒkə/ *n.* a small Australian short-tailed wallaby, *Setonix brachyurus.* [Aboriginal name]

quondam /ˈkwɒndæm/ *attrib.adj.* that once was; sometime; former. [L (*adv.*), = formerly]

Quonset /ˈkwɒnsɪt/ *n. US propr.* a prefabricated metal building with a semicylindrical corrugated roof. [*Quonset* Point, Rhode Island, where first made]

quorate /ˈkwɔːrət, -reɪt/ *adj. Brit.* (of a meeting) attended by a quorum. [QUORUM]

Quorn /kwɔːn/ *n. propr.* a textured vegetable protein made from edible fungus, used as a meat substitute. [*Quorn*, fox-hunt held near Leicester, where manufactured]

quorum /ˈkwɔːrəm/ *n.* the fixed minimum number of members that must be present to make the proceedings of an assembly or society valid. [L, = of whom (we wish that you be two, three, etc.), in the wording of commissions]

quota /ˈkwəʊtə/ *n.* (*pl.* **quotas**) **1** the share that an individual person or company is bound to contribute to or entitled to receive from a total. **2** a quantity of goods etc. which under official controls must be manufactured, exported, imported, etc. **3** the number of yearly immigrants allowed to enter a country, students allowed to enrol for a course, etc. [med.L *quota* (*pars*) how great (a part), fem. of *quotus* f. *quot* how many]

quotable /ˈkwəʊtəb(ə)l/ *adj.* worth, or suitable for, quoting. □ **quotability** /ˌkwəʊtəˈbɪlɪtɪ/ *n.*

quotation /kwəʊˈteɪʃ(ə)n/ *n.* **1** the act or an instance of quoting or being quoted. **2** a passage or remark quoted. **3** a contractor's estimate. **4** *Stock Exch.* an amount stated as the current price of stocks or commodities. **5** *Mus.* a short passage or tune taken from one piece of music to another. □ **quotation mark** each of a set of punctuation marks, single (' ') or double (" "), used to mark the beginning and end of a quoted passage, a book title, etc., or words regarded as slang or jargon. [med.L *quotatio* (as QUOTE)]

quote /kwəʊt/ *v. & n.* ● *v.tr.* **1** cite or appeal to (an author, book, etc.) in confirmation of some view. **2** repeat a statement by (another person) or copy out a passage from (*don't quote me*). **3** (often *absol.*) **a** repeat or copy out (a passage) usu. with an indication that it is borrowed. **b** (foll. by *from*) cite (an author, book, etc.). **4** (foll. by *as*) cite (an author etc.) as proof, evidence, etc. **5 a** enclose (words) in quotation marks. **b** (as *int.*) (in dictation, reading aloud, etc.) indicate the presence of opening quotation marks (*he said, quote, 'I shall stay'*). **6 a** state (the price) of a job to a person (*they quoted me £600*). **b** (often foll. by *at*) state the price of (a commodity, bet, etc.) (*quoted at 200 to 1*). **7** *Stock Exch.* regularly list the price of. ● *n. colloq.* **1** a passage quoted. **2** a price quoted. **3** (usu. in *pl.*) quotation marks. [ME, earlier 'mark with numbers', f. med.L *quotare* f. *quot* how many, or as QUOTA]

quoth /kwəʊθ/ *v.tr.* (only in 1st and 3rd person) *archaic* said. [OE *cwæth* past of *cwethan* say f. Gmc]

quotidian /kwɒˈtɪdɪən/ *adj. & n.* ● *adj.* **1** daily, of every day. **2** commonplace, trivial. ● *n.* (in full **quotidian fever**) a fever recurring every day. [ME f. OF *cotidien* & L *cotidianus* f. *cotidie* daily]

quotient /ˈkwəʊʃ(ə)nt/ *n.* a result obtained by dividing one quantity by another. [ME f. L *quotiens* how many times f. *quot* how many, by confusion with -ENT]

Qur'an var. of KORAN.

q.v. *abbr. quod vide.*

Qwaqwa /ˈkwækwə/ (also **QwaQwa**) a former homeland established in South Africa for the South Sotho people, situated in the Drakensberg Mountains in Orange Free State. (See also HOMELAND.)

qwerty /ˈkwɜːtɪ/ *attrib.adj.* denoting the standard keyboard on English-language typewriters, word processors, etc., with q, w, e, r, t, and y as the first keys on the top row of letters.

qy. *abbr.* query.

Rr

R[1] /ɑ:(r)/ *n.* (also **r**) (*pl.* **Rs** or **R's**) the eighteenth letter of the alphabet. □ **the r months** the months with r in their names (September to April) as the season for oysters. **the three Rs** see THREE.

R[2] *abbr.* (also **R.**) **1** *Regina* (Elizabeth R). **2** *Rex.* **3** River. **4** (also ®) registered as a trademark. **5** *Chess* rook. **6** right. **7** rand. **8** Réaumur. **9** radius. **10** reverse (gear).

R[3] *symb.* **1** roentgen. **2** electrical resistance. **3** (in chemical formulae) an organic radical or group.

r. *abbr.* (also **r**) **1** right. **2** recto. **3** run(s). **4** radius.

RA *abbr.* **1 a** (in the UK) Royal Academy. **b** (in the UK) Royal Academician. **2** (in the UK) Royal Artillery. **3** right ascension.

Ra[1] /rɑ:/ (also **Re** /reɪ/) *Egyptian Mythol.* the sun-god, the supreme Egyptian deity, worshipped as the creator of all life and often portrayed with a falcon's head bearing the solar disc. He appears travelling in his ship with other gods, crossing the sky by day and journeying through the underworld of the dead at night. From earliest times he was associated with the pharaoh.

Ra[2] *symb. Chem.* the element radium.

RAAF *abbr.* Royal Australian Air Force.

Rabat /rə'bæt/ the capital of Morocco, an industrial port on the Atlantic Coast; pop. (1982) 518,620. Founded as a military fort in the 12th century by the Almohads, its name derives from Arabic *Ribat el-Fath* 'fort of victory'.

Rabaul /rə'baʊl/ the chief town and port of the island of New Britain, Papua New Guinea; pop. (1990) 17,020.

rabbet /'ræbɪt/ *n. & v.* ● *n.* a step-shaped channel etc. cut along the edge or face or projecting angle of a length of wood etc., usu. to receive the edge or tongue of another piece. Also called *rebate.* ● *v.tr.* (**rabbeted, rabbeting**) **1** join or fix with a rabbet. **2** make a rabbet in. □ **rabbet plane** a plane for cutting a groove along an edge. [ME f. OF *rab(b)at* abatement, recess f. *rabattre* REBATE[1]]

rabbi /'ræbaɪ/ *n.* (*pl.* **rabbis**) **1** a Jewish scholar or teacher, esp. of the law. **2** a person appointed as a Jewish religious leader. □ **Chief Rabbi** the religious head of the Jewish communities in Britain. □ **rabbinate** /-bɪnət/ [ME & OE f. eccl.L f. Gk *rhabbi* f. Heb. *rabbi* my master f. *rab* master + pronominal suffix]

rabbinical /rə'bɪnɪk(ə)l/ *adj.* of or relating to rabbis, or to Jewish law or teaching. □ **rabbinically** *adv.*

rabbit /'ræbɪt/ *n. & v.* ● *n.* **1 a** a burrowing gregarious plant-eating mammal of the family Leporidae, with long ears, long hind legs, and a short tail, esp. the widespread *Oryctolagus cuniculus.* (See note below.) **b** *US* a hare. **c** the fur of the rabbit. **2** *Brit. colloq.* a poor performer in any sport or game. ● *v.intr.* (**rabbited, rabbiting**) **1** hunt rabbits. **2** (often foll. by *on, away*) *Brit. colloq.* talk excessively or pointlessly; chatter (*rabbiting on about his holiday*). □ **rabbit punch** a short chop with the edge of the hand to the nape of the neck. **rabbit's foot** the foot of a rabbit, carried to bring luck. **rabbit warren** an area in which rabbits have their burrows, or are kept for meat etc. □ **rabbity** *adj.* [ME perh. f. OF: cf. F dial. *rabotte*, Walloon *robète*, Flem. *robbe*]

▪ The rabbits and hares constitute the family Leporidae, with about forty-six living species, and are placed with the pikas in the order Lagomorpha. They were formerly considered to be rodents, but they differ from these in having additional small incisor teeth. The common rabbit, *Oryctolagus cuniculus*, is native to Iberia and NW Africa, but has been introduced to many other parts of the world, where it has frequently become a pest. It has been developed into various breeds for keeping as a pet or show animal, or for meat.

rabble[1] /'ræb(ə)l/ *n.* **1** a disorderly crowd; a mob. **2** a contemptible or inferior set of people. **3** (prec. by *the*) the lower or disorderly classes of the populace. □ **rabble-rouser** a person who stirs up the rabble or a crowd of people in agitation for social or political change. **rabble-rousing** *adj.* tending to arouse the emotions of a crowd. ● *n.* the act or process of doing this. [ME: orig. uncert.]

rabble[2] /'ræb(ə)l/ *n.* an iron bar with a bent end for stirring molten metal etc. [F *râble* f. med.L *rotabulum*, L *rutabulum* fire-shovel f. *ruere* rut-rake up]

Rabelais /'ræbə,leɪ/, François (*c.*1494–1553), French satirist. He spent a period in his early life as a Franciscan monk, although his works were later condemned by the Church; he also later worked as a physician. He is remembered for his sequence of allegorical works parodying medieval learning and literature, attacking asceticism, and affirming humanist values. These are marked by coarse humour and an imaginative and exuberant use of language, and include *Pantagruel* (*c.*1532) and *Gargantua* (1534). The English translations of his work that appeared from 1653 were to exert a marked influence on writers such as Swift and Sterne.

Rabelaisian /,ræbə'leɪzɪən/ *adj. & n.* ● *adj.* **1** of or like Rabelais or his writings. **2** marked by exuberant imagination and language, coarse humour, and satire. ● *n.* an admirer or student of Rabelais.

rabid /'ræbɪd, 'reɪb-/ *adj.* **1** furious, violent (*rabid hate*). **2** unreasoning; headstrong; fanatical (*a rabid anarchist*). **3** (esp. of a dog) affected with rabies; mad. **4** of or connected with rabies. □ **rabidly** *adv.* **rabidness** *n.* [L *rabidus* f. *rabere* rave]

rabies /'reɪbiːz/ *n.* a fatal viral disease of dogs and other mammals, usually transmitted by saliva in the bite of an infected animal. It produces paralysis or a vicious excitability, and in humans causes an encephalitis with convulsions and with throat spasm on swallowing (cf. HYDROPHOBIA). Treatment with vaccine and antiserum is now usually effective, even after infection. The disease is endemic among wild animals in many areas, and vigorous action is taken to prevent it from reaching areas currently free of it. [L f. *rabere* rave]

Rabin /rə'biːn/, Yitzhak (1922–95), Israeli statesman and military leader, Prime Minister 1974–7 and 1992–5. As Chief of Staff (1964–8), he led Israel's armed forces to victory during the Six Day War of 1967. During his first term as Prime Minister, he supported Henry Kissinger's moves to bring peace to the Middle East and negotiated Israel's partial withdrawal from Sinai with President Sadat of Egypt. Rabin resigned as both Prime Minister and Labour Party leader in 1977. He was re-elected to the premiership in 1992 and in 1993 he negotiated a PLO–Israeli peace accord with Yasser Arafat, which allowed for limited Palestinian autonomy in the West Bank and Gaza Strip. In 1994 he was awarded the Nobel Peace Prize, together with Arafat and Shimon Peres. He was assassinated in 1995.

RAC *abbr.* **1** (in the UK) Royal Automobile Club. **2** (in the UK) Royal Armoured Corps.

raccoon /rə'kuːn/ *n.* (also **racoon**) **1** a furry nocturnal flesh-eating American mammal of the genus *Procyon*; esp. the common *P. lotor*, with greyish-brown fur, a ringed bushy tail, and a dark stripe across the face. **2** the fur of the raccoon. [Algonquian dial.]

race[1] /reɪs/ *n. & v.* ● *n.* **1** a contest of speed between runners, horses, vehicles, ships, etc. **2** (in *pl.*) a series of these for horses, dogs, etc., at a fixed time on a regular course. **3** a contest between persons to be first to achieve something. **4 a** a strong or rapid current flowing through a narrow channel in the sea or a river (*a tide race*). **b** the channel of a stream etc. (*a mill-race*). **5** either of two grooved rings in a ball-bearing or roller bearing. **6** *Austral.* a fenced passageway for drafting sheep etc. **7** a passageway along which football players etc. run to enter the field. **8** (in weaving) the channel along which the shuttle moves. **9** *archaic* **a** the course of the sun or moon. **b** the course of life (*has run his race*). ● *v.* **1** *intr.* take part in a race. **2** *tr.* have a race with. **3** *tr.* try to surpass in speed. **4** *intr.* (foll. by *with*) compete in speed with. **5** *tr.* cause (a horse, car, etc.) to race. **6 a** *intr.* move swiftly; go at full or (of an engine, propeller, the pulse, etc.) excessive speed. **b** *tr.* cause (a person or thing) to do this (*raced the bill through the House*). **7** *intr.* (usu. as **racing** *adj.*) follow or take part in horse-racing (*a racing man*). □ **not in the race** *Austral. sl.* having no chance. **race meeting** a sporting event consisting of a sequence of horse-races at one place. **racing car** a motor car built for racing on a prepared track. [ME, = running, f. ON *rás*]

race² /reɪs/ *n.* **1** any of the major divisions of humankind, having in common distinct physical features or ethnic background. (*See note below.*) **2** a tribe, nation, etc., regarded as being of a distinct ethnic stock. **3** the fact or concept of division into races (*discrimination based on race*). **4** a genus, species, breed, or variety of animals, plants, or micro-organisms. **5** a group of persons, animals, or plants connected by common descent. **6** any great division of living creatures (*the feathered race; the four-footed race*). **7** descent; kindred (*of noble race; separate in language and race*). **8** a class of persons etc. with some common feature (*the race of poets*). □ **race relations** relations between members of different races usu. in the same country. **race riot** an outbreak of violence due to racial antagonism. [F f. It. *razza*, of unkn. orig.]

▪ Although ideas of race are centuries old, it was not until the 19th century, with the growth of interest in ethnology and physical anthropology, that attempts to systematize racial divisions were made. It was then generally believed that most individuals could be assigned unambiguously to a racial group, and that such groupings could account not only for discernible physical characteristics but also for cultural and social differences between geographically distinct peoples. Although this notion of race as a rigid construct has now been abandoned, some of the terms used in the classification (such as Caucasoid, Mongoloid, and Negroid), outlined by early 19th-century anthropologists such as Blumenbach, are still in use. Ideas of supposed racial superiority and social Darwinism put forward by writers such as Gobineau and Nietzsche reached their culmination in Nazi ideology of the 1930s and gave pseudo-scientific justification to policies and attitudes of discrimination, segregation, exploitation, and slavery. Such theories of race, in particular those asserting a link between racial type and intelligence, are now discredited; scientifically it is accepted as obvious that there are subdivisions of the human species, but it is also clear that genetic variation between individuals of the same race can be as great as that between members of different races, and that racial characteristics which might once have been distinctive have, with few exceptions, been complicated and obscured by long histories of interbreeding.

race³ /reɪs/ *n.* a ginger root. [OF *rais, raiz* f. L *radix radicis* root]

racecard /ˈreɪskɑːd/ *n.* a programme of races.

racecourse /ˈreɪskɔːs/ *n.* a ground or track for horse-racing.

racegoer /ˈreɪsˌɡəʊə(r)/ *n.* a person who frequents horse-races.

racehorse /ˈreɪshɔːs/ *n.* a horse bred or kept for racing.

racemate /ˈræsɪˌmeɪt/ *n. Chem.* a racemic mixture.

raceme /rəˈsiːm/ *n. Bot.* a flower-cluster with the separate flowers attached by short equal stalks at equal distances along a central stem (cf. CYME). [L *racemus* grape-bunch]

racemic /rəˈsiːmɪk, -ˈsemɪk/ *adj. Chem.* composed of equal numbers of dextrorotatory and laevorotatory molecules of a compound. □ **racemize** /ˈræsɪˌmaɪz/ *v.tr. & intr.* (also **-ise**). [RACEME + -IC, orig. of tartaric acid in grape juice]

racemose /ˈræsɪˌməʊz, -ˌməʊs/ *adj.* **1** *Bot.* in the form of a raceme. **2** *Anat.* (of a gland etc.) clustered. [L *racemosus* (as RACEME)]

racer /ˈreɪsə(r)/ *n.* **1** a horse, yacht, bicycle, etc., of a kind used for racing. **2** a circular horizontal rail along which the traversing-platform of a heavy gun moves. **3** a person or thing that races.

racetrack /ˈreɪstræk/ *n.* **1** = RACECOURSE. **2** a track for motor-racing.

raceway /ˈreɪsweɪ/ *n.* **1** a track or channel along which something runs, esp.: **a** a channel for water. **b** a groove in which ball-bearings

run. **c** a pipe or tubing enclosing electrical wires. **2** esp. *US* **a** a track on which trotting-races are held. **b** a racecourse.

rachis /ˈreɪkɪs/ *n.* (pl. **rachides** /ˈrækɪˌdiːz, ˈreɪk-/) **1** *Bot.* **a** a stem of grass etc. bearing flower-stalks at short intervals. **b** the axis of a compound leaf or frond. **2** *Anat.* the vertebral column or the cord from which it develops. **3** *Zool.* a feather-shaft, esp. the part bearing the barbs. □ **rachidial** /rəˈkɪdɪəl/ *adj.* [mod.L f. Gk *rhakhis* spine: the pl. *-ides* is erron.]

rachitis /rəˈkaɪtɪs/ *n.* rickets. □ **rachitic** /-ˈkɪtɪk/ *adj.* [mod.L f. Gk *rhakhitis* (as RACHIS)]

Rachmaninov /rækˈmænɪˌnɒf/, Sergei (Vasilevich) (1873–1943), Russian composer and pianist, resident in the US from 1917. Rachmaninov belongs to the Russian romantic tradition, and was particularly influenced by Tchaikovsky. He was a celebrated pianist and is primarily known for his compositions for piano, in particular the Prelude in C sharp minor (1892), the Second Piano Concerto (1901), and the *Rhapsody on a Theme of Paganini* (1934) for piano and orchestra. He also wrote three other piano concertos, as well as three symphonies and three operas.

Rachmanism /ˈrækməˌnɪz(ə)m/ *n.* exploitation or intimidation of a tenant by an unscrupulous landlord. It is named after a London landlord, Peter Rachman (1919–62), whose practices became notorious in the early 1960s.

racial /ˈreɪʃ(ə)l/ *adj.* **1** of or concerning race (*racial diversities; racial minority*). **2** on the grounds of or connected with difference in race (*racial discrimination; racial tension*). □ **racially** *adv.*

racialism /ˈreɪʃəˌlɪz(ə)m/ *n.* = RACISM 1. □ **racialist** *n. & adj.*

Racine /ræˈsiːn/, Jean (1639–99), French dramatist. The principal tragedian of the French classical period, he drew on many different sources, including Greek and Roman literature (*Andromaque*, 1667; *Iphigénie*, 1674; *Phèdre*, 1677) and the Bible (*Athalie*, 1691). Central to the majority of his tragedies is a perception of the blind folly of human passion, continually enslaved to the pursuit of its object and destined always to be unsatisfied. Like his contemporary Corneille he wrote within the constraints of the rules governing tragic composition (see TRAGEDY).

racism /ˈreɪsɪz(ə)m/ *n.* **1 a** a belief in the superiority of a particular race; prejudice based on this. **b** antagonism towards other races, esp. as a result of this. **2** the theory that human abilities etc. are determined by race. □ **racist** *n. & adj.*

rack¹ /ræk/ *n. & v.* ● *n.* **1 a** a framework, usu. with rails, bars, hooks, etc., for holding or storing things. **b** a frame for holding animal fodder. **2** a cogged or toothed bar or rail engaging with a wheel or pinion etc., or using pegs to adjust the position of something. **3 a** *hist.* an instrument of torture stretching the victim's joints by the turning of rollers to which the wrists and ankles were tied. **b** a cause of suffering or anguish. ● *v.tr.* (also **wrack**) **1** (of disease or pain) inflict suffering on. **2** *hist.* torture (a person) on the rack. **3** place in or on a rack. **4** shake violently. **5** injure by straining. **6** oppress (tenants) by exacting excessive rent. **7** exhaust (the land) by excessive use. □ **on the rack** suffering intense distress or strain. **rack one's brains** make a great mental effort (*racked my brains for something to say*). **rack-railway** a railway with a cogged rail between the bearing rails, esp. used for ascending steep slopes. **rack-rent** *n.* **1** a high rent, annually equalling the full value of the property to which it relates. **2** an extortionate rent. ● *v.tr.* exact this from (a tenant) or for (land). **rack-renter** a tenant paying or a landlord exacting an extortionate rent. **rack up** N. *Amer.* accumulate or achieve (a score etc.). **rack-wheel** a cog-wheel. [ME *rakke* f. MDu., MLG *rak, rek*, prob. f. *recken* stretch]

rack² /ræk/ *n.* destruction (esp. *rack and ruin*). [var. of WRACK¹, WRECK]

rack³ /ræk/ *n.* a joint of lamb etc. including the front ribs. [perh. f. RACK¹]

rack⁴ /ræk/ *v.tr.* (often foll. by *off*) draw off (wine, beer, etc.) from the lees. [ME f. Prov. *arracar* f. *raca* stems and husks of grapes, dregs]

rack⁵ /ræk/ *n. & v.* ● *n.* driving clouds. ● *v.intr.* (of clouds) be driven before the wind. [ME, prob. of Scand. orig.: cf. Norw. and Sw. dial. *rak* wreckage etc. f. *reka* drive]

rack⁶ /ræk/ *n. & v.* ● *n.* a horse's gait between a trot and a canter. ● *v.intr.* progress in this way.

racket¹ /ˈrækɪt/ *n.* (also **racquet**) **1** a bat with a round or oval frame strung with catgut, nylon, etc., used in tennis, squash, etc. **2** a snow shoe resembling a racket. □ **racket-tail** a South American hummingbird, *Loddigesia mirabilis*, with a racket-shaped tail. [F *racquette* f. It. *racchetta* f. Arab. *rāḥa* palm of the hand]

racket² /ˈrækɪt/ n. **1 a** a disturbance; an uproar; a din. **b** social excitement; gaiety. **2** colloq. **a** a scheme for obtaining money or attaining other ends by fraudulent and often violent means. **b** a dodge; a sly game. **3** colloq. an activity; a way of life; a line of business (*starting up a new racket*). □ **rackety** adj. [16th c.: perh. imit.]

racketeer /ˌrækɪˈtɪə(r)/ n. a person who operates a dishonest business. □ **racketeering** n.

rackets /ˈrækɪts/ n.pl. (treated as *sing.*) a ball game for two or four persons played with rackets in a plain four-walled court. Rackets is the forerunner of many racket and ball games, notably squash. Now a minority sport, it is distinguished from squash in particular by the use of a solid, harder ball. The game as it is played today developed in England in the 19th century, but its origins may be traced to medieval times; until the first indoor courts were built in the mid-19th century it was played in open courts in the backyards of inns and taverns.

racon /ˈreɪkɒn/ n. esp. *US* a radar beacon that can be identified and located by its response to a radar signal from a ship etc. [radar + beacon]

raconteur /ˌrækɒnˈtɜː(r)/ n. (fem. **raconteuse** /-ˈtɜːz/) a teller of anecdotes. [F f. *raconter* relate, RECOUNT]

racoon var. of RACCOON.

racquet var. of RACKET¹.

racy /ˈreɪsɪ/ adj. (**racier, raciest**) **1** lively and vigorous in style. **2** risqué, suggestive. **3** having characteristic qualities in a high degree (*a racy flavour*). □ **racily** adv. **raciness** n. [RACE² + -Y¹]

rad¹ /ræd/ n. (pl. same) radian. [abbr.]

rad² /ræd/ n. *sl.* a political radical. [abbr.]

rad³ /ræd/ n. *Physics* a unit of absorbed dose of ionizing radiation, corresponding to the absorption of 0.01 joule per kilogram of absorbing material. [radiation *absorbed* dose]

rad⁴ /ræd/ adj. esp. *US sl.* fine, excellent, outstanding. [perh. abbr. RADICAL]

RADA /ˈrɑːdə/ abbr. (in the UK) Royal Academy of Dramatic Art.

radar /ˈreɪdɑː(r)/ n. **1** a system for determining the direction, range, or presence of (usu. moving) objects, by sending out pulses of high-frequency radio waves and detecting the returning echo. (*See note below*.) **2** the apparatus used for this. □ **radar trap** the use of radar to detect vehicles exceeding a speed limit. [radio detection and ranging]

▪ The principle of radar was established in 1886, when Heinrich Hertz showed that radio waves could be reflected from solid objects. However, it remained chiefly theoretical until 1922, when Marconi suggested that radio echoes could be used to detect ships in bad visibility, and the idea was tested experimentally in the US. Radar systems for air and sea navigation and for detecting enemy aircraft and ships were developed in Britain, France, Germany, and the US in the 1930s, and used by both sides in the Second World War. Many systems now use *secondary radar*, in which the target automatically transmits a return signal when it receives a scanning signal.

RADC abbr. (in the UK) Royal Army Dental Corps.

Radcliffe /ˈrædklɪf/ (Mrs Ann (1764–1823), English novelist. She was a leading exponent of the Gothic novel and was influential on the work of other writers, including Byron, Shelley, and Charlotte Brontë. Her five novels include *The Mysteries of Udolpho* (1794) and *The Italian* (1797).

raddle /ˈræd(ə)l/ n. & v. ● n. red ochre (often used to mark sheep). ● v.tr. **1** colour with raddle or too much rouge. **2** (as **raddled** adj.) worn out; untidy, unkempt. [var. of RUDDLE]

Radha /ˈrɑːdɑː/ *Hinduism* the favourite mistress of the god Krishna, and an incarnation of Lakshmi. In devotional religion she represents the longing of the human soul for God. [Skr., = prosperity]

Radhakrishnan /ˌrɑːdəˈkrɪʃnən/ (Sir Sarvepalli (1888–1975), Indian philosopher and statesman, President 1962–7. A teacher of philosophy at Mysore, Calcutta, and Oxford universities, he introduced some of the main ideas of classical Indian philosophy to the West. Major works include *Indian Philosophy* (1923–7) and *Eastern Religions and Western Thought* (1939). Radhakrishnan was Indian ambassador to the Soviet Union (1949–52) before returning to India in 1952 to become Vice-President under Nehru; he was elected President ten years later.

radial /ˈreɪdɪəl/ adj. & n. ● adj. **1** of, concerning, or in rays. **2 a** arranged like rays or radii; having the position or direction of a radius. **b** having spokes or radiating lines. **c** acting or moving along lines diverging from a centre. **3** *Anat.* relating to the radius (*radial artery*). **4** (in full **radial-ply**) (of a vehicle tyre) having the core fabric layers arranged radially at right angles to the circumference and the tread strengthened. ● n. **1** *Anat.* the radial nerve or artery. **2** a radial-ply tyre.

□ **radial engine** an engine having cylinders arranged along radii. **radial keratotomy** *Med.* see KERATOTOMY. **radial symmetry** symmetry occurring about any number of lines or planes passing through the centre of an organism etc. **radial velocity** esp. *Astron.* the speed of motion along a radial line, esp. between a star etc. and an observer. □ **radially** adv. [med.L *radialis* (as RADIUS)]

radian /ˈreɪdɪən/ n. *Geom.* the SI unit of angle (symbol **rad**), equal to the angle subtended at the centre of a circle by an arc equal in length to the radius. One radian equals about 57.3°. [RADIUS + -AN]

radiant /ˈreɪdɪənt/ adj. & n. ● adj. **1** emitting rays of light. **2** (of eyes or looks) beaming with joy or hope or love. **3** (of beauty) splendid or dazzling. **4** (of light) issuing in rays. **5** operating radially. **6** extending radially; radiating. ● n. **1** the point or object from which light or heat radiates, esp. in an electric or gas heater. **2** *Astron.* a radiant point. □ **radiant heat** heat transmitted by radiation, not by conduction or convection. **radiant heater** a heater that works by this method. **radiant point 1** a point from which rays or radii proceed. **2** *Astron.* the apparent focal point of a meteor shower. □ **radiance** n. **radiancy** n. **radiantly** adv. [ME f. L *radiare* (as RADIUS)]

radiate v. & adj. ● v. /ˈreɪdɪˌeɪt/ **1** intr. **a** emit rays of light, heat, or other electromagnetic waves. **b** (of light or heat) be emitted in rays. **2** tr. emit (light, heat, or sound) from a centre. **3** tr. transmit or demonstrate (life, love, joy, etc.) (*radiates happiness*). **4** intr. & tr. diverge or cause to diverge or spread from a centre. **5** tr. (as **radiated** adj.) with parts arranged in rays. ● adj. /ˈreɪdɪət/ having divergent rays or parts radially arranged. □ **radiately** /-dɪətlɪ/ adv. **radiative** /-dɪətɪv/ adj. [L *radiare radiat-* (as RADIUS)]

radiation /ˌreɪdɪˈeɪʃ(ə)n/ n. **1** the act or an instance of radiating; the process of being radiated. **2** *Physics* **a** the emission of energy as electromagnetic waves or as moving particles. **b** the energy transmitted in this way, esp. invisibly. **3** (in full **radiation therapy**) treatment of cancer and other diseases using radiation, such as X-rays or ultraviolet light. □ **radiation chemistry** the study of the chemical effects of radiation on matter. **radiation sickness** sickness caused by exposure to radiation, such as X-rays or gamma rays. □ **radiational** adj. **radiationally** adv. [L *radiatio* (as RADIATE)]

radiator /ˈreɪdɪˌeɪtə(r)/ n. **1** a person or thing that radiates. **2 a** a device for heating a room etc., consisting of a metal case through which hot water or steam circulates. **b** a usu. portable oil or electric heater resembling this. **3** an engine-cooling device in a motor vehicle or aircraft with a large surface for cooling circulating water. □ **radiator grille** a grille at the front of a motor vehicle allowing air to circulate to the radiator.

radical /ˈrædɪk(ə)l/ adj. & n. ● adj. **1** of the root or roots; fundamental (*a radical error*). **2** far-reaching; thorough; going to the root (*radical change*). **3 a** advocating thorough reform; holding extreme political views; left-wing, revolutionary. **b** (of a measure etc.) advanced by or according to principles of this kind. **4** forming the basis; primary (*the radical idea*). **5** *Math.* of the root of a number or quantity. **6** (of surgery etc.) seeking to ensure the removal of all diseased tissue. **7** of the roots of words. **8** *Mus.* belonging to the root of a chord. **9** *Bot.* of, or springing direct from, the root. **10** *hist.* belonging to an extreme section of the Liberal Party. **11** *US hist.* seeking extreme anti-South action at the time of the Civil War. **12** *sl.* excellent, outstanding, cool. ● n. **1** a person holding radical views or belonging to a radical party. **2** *Chem.* **a** a free radical. **b** an element or atom or a group of these normally forming part of a compound and remaining unaltered during the compound's ordinary chemical changes. **3 a** the root of a word. **b** any of the basic set of around 214 Chinese characters from which more complex ones are mainly derived. **4** a fundamental principle; a basis. **5** *Math.* **a** a quantity forming or expressed as the root of another. **b** a radical sign. □ **radical chic** the fashionable affectation of radical left-wing views or of dress, lifestyle, etc., associated with such views. **radical sign** √, ∛, etc., indicating the square, cube, etc., root of the number following. □ **radicalism** n. **radically** adv. **radicalness** n. **radicalize** v.tr. & intr. (also **-ise**). **radicalization** /ˌrædɪkəlarˈzeɪʃ(ə)n/ n. [ME f. LL *radicalis* f. L *radix radicis* root]

radicchio /rəˈdiːkɪˌəʊ/ n. (pl. **-os**) a variety of chicory with dark red-coloured leaves. [It., = chicory]

radices pl. of RADIX.

radicle /ˈrædɪk(ə)l/ n. **1** the part of a plant embryo that develops into the primary root; a rootlet. **2** a rootlike subdivision of a nerve or vein. □ **radicular** /rəˈdɪkjʊlə(r)/ adj. [L *radicula* (as RADIX)]

radii pl. of RADIUS.

radio /ˈreɪdɪəʊ/ n. & v. ● n. (pl. **-os**) **1** (often *attrib.*) **a** the transmission and reception of sound messages etc. by electromagnetic waves of radio frequency, without a connecting wire (cf. WIRELESS). (*See note below.*) **b** an apparatus for receiving, broadcasting, or transmitting radio signals. **c** a message sent or received by radio. **2 a** sound broadcasting in general (*prefers the radio*). **b** a broadcasting station or channel (*Radio One*). ● v. (**-oes, -oed**) **1** tr. a send (a message) by radio. **b** send a message to (a person) by radio. **2** intr. communicate or broadcast by radio. □ **radio cab** (or **car**) a cab or car equipped with a two-way radio. **radio fix** the position of an aircraft, ship, etc., found by radio. **radio frequency** the frequency band of telecommunication, ranging from 10^4–10^{11} or 10^{12} Hz. **radio galaxy** *Astron.* a galaxy emitting radiation in the radio-frequency range of the electromagnetic spectrum. **radio ham** see HAM n. 4. [short for *radio-telegraphy* etc.]

■ Marconi in 1901 was the first to use invisible electromagnetic waves, described theoretically by Maxwell and experimentally confirmed by Hertz, for the transmission of signals. Marconi's early transmissions were really telegraphic signals; the first transmissions of sound messages were by R. A. Fessenden in 1906. By 1921 scheduled radio programmes were being broadcast in the US; the BBC was transmitting in the UK by 1923, and by 1925 there were 600 stations worldwide. As the number of users has increased, higher frequencies (including microwaves) have been exploited. Two methods are in regular use for carrying information by radio: amplitude modulation, first used by Fessenden, and frequency modulation, developed by W. H. Armstrong in the 1930s.

radio- /ˈreɪdɪəʊ/ comb. form **1** denoting radio or broadcasting. **2 a** connected with radioactivity. **b** denoting artificially prepared radioisotopes of elements (*radio-caesium*). **3** connected with rays or radiation. **4** *Anat.* belonging to the radius in conjunction with some other part (*radio-carpal*). [RADIUS + -O- or f. RADIO]

radioactive /ˌreɪdɪəʊˈæktɪv/ adj. of or exhibiting radioactivity. □ **radioactively** adv.

radioactivity /ˌreɪdɪəʊækˈtɪvɪtɪ/ n. **1** the spontaneous disintegration of atomic nuclei, with the emission of usu. penetrating radiation or particles. (*See note below.*) **2** radioactive substances, or the radiation emitted by these.

■ Radioactivity was discovered in 1896 by Becquerel, who found that uranium salt crystals affected a photographic emulsion. Studies by Rutherford and others showed that there are three main types of radiation emitted by radioactive substances: alpha particles (helium nuclei), beta particles (fast-moving electrons), and gamma rays (high-energy electromagnetic radiation). Naturally occurring radioactive elements such as uranium and thorium are a major source of heat in the earth's interior and of background radiation on the surface. Radioactivity is also released in nuclear explosions and in the waste products of nuclear reactors, and severe damage can be caused to living tissue by dosages which significantly exceed the natural background radiation. (See also HALF-LIFE.)

radio-assay /ˌreɪdɪəʊəˈseɪ/ n. an analysis of a substance based on radiation from a sample.

radio astronomy n. the branch of astronomy concerned with radio emissions from celestial objects. Such emission originates within the Galaxy from ionized gas, supernova remnants, and objects such as pulsars, and is also emitted by remote radio galaxies, quasars, and black holes, where poorly understood processes release large amounts of energy. The discovery in 1965 of a uniform background of radiation pervading the universe at microwave wavelengths is now believed to be a direct observation of the remnant of radiation from the big bang. (See RADIO TELESCOPE.)

radiobiology /ˌreɪdɪəʊbaɪˈɒlədʒɪ/ n. the biology concerned with the effects of radiation on organisms and the application in biology of radiological techniques. □ **radiobiologist** n. **radiobiological** /-ˌbaɪəˈlɒdʒɪk(ə)l/ adj. **radiobiologically** adv.

radiocarbon /ˌreɪdɪəʊˈkɑːb(ə)n/ n. a radioactive isotope of carbon with an atomic mass of 14. Radiocarbon has been used since the late 1940s in a technique of assigning absolute dates to ancient organic material. All living things absorb carbon, either from the carbon dioxide in the atmosphere or by eating plants etc. that contain it. Once they are dead, the proportion of carbon-14 in their remains falls at a steady rate (the half-life of carbon-14 is about 5,568 years), and by measuring its concentration the approximate date of death of the specimen can be calculated. The amount of carbon-14 in the atmosphere, however, has not remained constant throughout time, and radiocarbon dates require adjustment, e.g. by calibration against the precise dates derived from dendrochronology.

radiochemistry /ˌreɪdɪəʊˈkemɪstrɪ/ n. the chemistry of radioactive materials. □ **radiochemical** adj. **radiochemist** n.

radio-controlled /ˌreɪdɪəʊkənˈtrəʊld/ adj. (of a model aircraft etc.) controlled from a distance by radio.

radio-element /ˌreɪdɪəʊˈelɪmənt/ n. a natural or artificial radioactive element or isotope.

radiogenic /ˌreɪdɪəʊˈdʒenɪk/ adj. **1** produced by radioactivity. **2** suitable for broadcasting by radio. □ **radiogenically** adv.

radio-goniometer /ˌreɪdɪəʊ ˌgəʊnɪˈɒmɪtə(r)/ n. an instrument for finding direction using radio waves.

radiogram /ˈreɪdɪəʊˌɡræm/ n. **1** *Brit.* a combined radio and record-player. **2** a picture obtained by X-rays, gamma rays, etc. **3** a radio-telegram. [RADIO- + -GRAM, GRAMOPHONE]

radiograph /ˈreɪdɪəʊˌɡrɑːf/ n. & v. ● n. **1** an instrument recording the intensity of radiation. **2** = RADIOGRAM 2. ● v.tr. obtain a picture of by X-ray, gamma ray, etc. □ **radiographer** /ˌreɪdɪˈɒɡrəfə(r)/ n. **radiography** n. **radiographic** /-dɪəˈɡræfɪk/ adj. **radiographically** adv.

radioimmunology /ˌreɪdɪəʊˌɪmjʊˈnɒlədʒɪ/ n. the application of radiological techniques in immunology.

radioisotope /ˌreɪdɪəʊˈaɪsəˌtəʊp/ n. a radioactive isotope. □ **radioisotopic** /-ˌaɪsəˈtɒpɪk/ adj. **radioisotopically** adv.

radiolarian /ˌreɪdɪəʊˈleərɪən/ n. *Zool.* a marine protozoan of the superclass Actinopoda, having a siliceous skeleton with projecting spines and radiating rodlike pseudopodia. [mod.L *Radiolaria* former order name, f. L *radiolus* dimin. of RADIUS]

radiology /ˌreɪdɪˈɒlədʒɪ/ n. the scientific study of X-rays and other high-energy radiation, esp. as used in medicine. □ **radiologist** n. **radiologic** /-dɪəˈlɒdʒɪk/ adj. **radiological** adj.

radiometer /ˌreɪdɪˈɒmɪtə(r)/ n. an instrument for measuring the intensity or force of radiation, invented by William Crookes in 1875. □ **radiometry** n.

radiometric /ˌreɪdɪəʊˈmetrɪk/ adj. of or relating to the measurement of radioactivity. □ **radiometric dating** a method of dating geological specimens by determining the relative proportions of the isotopes of a radioactive element present in a sample.

radionics /ˌreɪdɪˈɒnɪks/ n.pl. (usu. treated as *sing.*) the study and interpretation of radiation believed to be emitted from substances, esp. as a form of diagnosis. [RADIO- + -*onics*, after ELECTRONICS]

radionuclide /ˌreɪdɪəʊˈnjuːklaɪd/ n. a radioactive nuclide.

radiopaque /ˌreɪdɪəʊˈpeɪk/ adj. (also **radio-opaque** /ˌreɪdɪəʊəʊˈpeɪk/) opaque to X-rays or similar radiation. □ **radiopacity** /-ˈpæsɪtɪ/ n. [RADIO- + OPAQUE]

radiophonic /ˌreɪdɪəʊˈfɒnɪk/ adj. of or relating to synthetic sound, esp. music, produced electronically.

radioscopy /ˌreɪdɪˈɒskəpɪ/ n. the examination by X-rays etc. of objects opaque to light. □ **radioscopic** /-dɪəˈskɒpɪk/ adj.

radiosonde /ˈreɪdɪəʊˌsɒnd/ n. a miniature radio transmitter broadcasting information about pressure, temperature, etc., from various levels of the atmosphere, carried esp. by balloon. [RADIO- + G *Sonde* probe]

radio-telegram /ˌreɪdɪəʊˈtelɪˌɡræm/ n. a telegram sent by radio, usu. from a ship to land.

radio-telegraphy /ˌreɪdɪəʊtɪˈleɡrəfɪ/ n. telegraphy using radio transmission. □ **radio-telegraph** /-ˈtelɪˌɡrɑːf/ n.

radio-telephony /ˌreɪdɪəʊtɪˈlefənɪ/ n. telephony using radio transmission. □ **radio-telephone** /-ˈtelɪˌfəʊn/ n. **radio-telephonic** /-ˌtelɪˈfɒnɪk/ adj.

radio telescope n. an instrument used to detect radio emissions from the sky, whether from natural celestial objects or from artificial satellites. Most familiar are large dish antennae such as the steerable reflector at Jodrell Bank, near Manchester, UK, but telescopes may also be constructed as linear arrays of aerials. Greater resolution is obtained by electronically linking signals from widely separated antennae, which then operate as an interferometer. Radio signals from outside the earth's atmosphere were first detected in 1932 by Karl Jansky (1905–50), and five years later his fellow American Grote Reber (b.1911) built the first parabolic reflector to collect and focus radio waves.

radiotelex /ˌreɪdɪəʊˈteleks/ n. a telex sent usu. from a ship to land.

radiotherapy /ˌreɪdɪəʊˈθerəpɪ/ n. the treatment of cancer and other diseases by X-rays or other forms of radiation. □ **radiotherapist** n. **radiotherapeutic** /-ˌθerəˈpjuːtɪk/ adj.

radish /'rædɪʃ/ n. **1** a cruciferous plant, *Raphanus sativus*, with a fleshy pungent root. **2** this root, eaten esp. raw in salads etc. [OE *rædic* f. L *radix radicis* root]

radium /'reɪdɪəm/ n. a radioactive metallic chemical element (atomic number 88; symbol **Ra**). (*See note below.*) □ **radium emanation** = RADON. **radium therapy** the treatment of disease by the use of gamma radiation from radium. [L *radius* ray]

▪ Radium was discovered in the mineral pitchblende in 1898 by Pierre and Marie Curie. It is the most reactive member of the alkaline earth metals. It is rare in nature, occurring chiefly as an impurity in uranium ores. Because of its radioactive properties it was formerly used extensively in the treatment of tumours and in luminous materials, but it has now been largely replaced by other substances for these purposes.

radius /'reɪdɪəs/ n. & v. ● n. (pl. **radii** /-dɪˌaɪ/ or **radiuses**) **1** *Math.* **a** a straight line from the centre to the circumference of a circle or sphere. **b** a radial line from the focus to any point of a curve. **c** the length of the radius of a circle etc. **2** a usu. specified distance from a centre in all directions (*within a radius of 20 miles; has a large radius of action*). **3 a** the thicker and shorter of the two bones in the human forearm (cf. ULNA). **b** the corresponding bone in a vertebrate's foreleg or a bird's wing. **4** any of the five arm-like structures of a starfish. **5 a** any of a set of lines diverging from a point like the radii of a circle. **b** an object of this kind, e.g. a spoke. **6 a** the outer rim of a composite flower-head, e.g. a daisy. **b** a radiating branch of an umbel. ● *v.tr.* give a rounded form to (an edge etc.). □ **radius vector** *Math.* a variable line drawn from a fixed point to an orbit or other curve, or to any point as an indication of the latter's position. [L, = staff, spoke, ray]

radix /'reɪdɪks/ n. (pl. **radices** /'rædɪˌsiːz, 'reɪd-/) **1** *Math.* a number or symbol used as the basis of a numeration scale (e.g. ten in the decimal system). **2** (usu. foll. by of) a source or origin. [L, = root]

Radnorshire /'rædnəˌʃɪə(r)/ a former county of eastern Wales. It became part of Powys in 1974.

Radom /'rɑːdɒm/ an industrial city in central Poland; pop. (1990) 228,490.

radome /'reɪdəʊm/ n. a dome or other structure, transparent to radio waves, protecting radar equipment, esp. on the outer surface of an aircraft. [radar + dome]

radon /'reɪdɒn/ n. a gaseous radioactive chemical element (atomic number 86; symbol **Rn**). One of the noble gases, radon was discovered by the Curies and others as a product of the radioactive decay of radium. It is used as a source of alpha particles in radiotherapy; the half-life of the most stable isotope is only 3.825 days. Small quantities of the gas are also emitted naturally from the ground in granite-rich areas. [RADIUM after argon etc.]

radula /'rædjʊlə/ n. (pl. **radulae** /-ˌliː/) a filelike structure in molluscs for scraping off food particles and drawing them into the mouth. □ **radular** adj. [L, = scraper f. *radere* scrape]

Raeburn /'reɪbɜːn/, Sir Henry (1756–1823), Scottish portrait painter. Influenced by Sir Joshua Reynolds, he became the leading Scottish portraitist of his day, depicting the local intelligentsia and Highland chieftains in a bold and distinctive style. Raeburn was famed for painting directly on to the canvas, dispensing with preliminary drawings. Major works include *The Reverend Robert Walker Skating* (c.1784) and *The MacNab* (1803–13).

RAF /ˌɑːreɪ'ef, colloq. ræf/ see ROYAL AIR FORCE.

Rafferty's rules /'ræfətɪz/ n. Austral. & NZ colloq. no rules at all, esp. in boxing. [prob. corrupt. of *refractory*]

raffia /'ræfɪə/ n. **1** a palm tree, *Raphia ruffia*, native to Madagascar, having very long leaves. **2** the fibre from its leaves, used for making hats, baskets, etc., and for tying plants etc. [Malagasy]

raffinate /'ræfɪˌneɪt/ n. *Chem.* a refined liquid oil produced by solvent extraction of impurities. [F *raffiner* + -ATE[1]]

raffish /'ræfɪʃ/ adj. **1** disreputable, esp. in an attractive manner; rakish. **2** tawdry. □ **raffishly** adv. **raffishness** n. [as RAFT[2] + -ISH[1]]

raffle[1] /'ræf(ə)l/ n. & v. ● n. a fund-raising lottery with goods as prizes. ● *v.tr.* (often foll. by off) dispose of by means of a raffle. [ME, a kind of dice-game, f. OF *raf(f)le*, of unkn. orig.]

raffle[2] /'ræf(ə)l/ n. **1** rubbish; refuse. **2** lumber; debris. [ME, perh. f. OF *ne rifle, ne rafle* nothing at all]

Raffles[1] /'ræf(ə)lz/ A. J., a debonair cricket-loving gentleman burglar, hero of novels (1899 onwards) by E. W. Hornung.

Raffles[2] /'ræf(ə)lz/, Sir (Thomas) Stamford (1781–1826), British colonial administrator. Born in Jamaica, he joined the East India Company in

1795, becoming Lieutenant General of Java in 1811. He later served as Lieutenant General of Sumatra (1818–23), during which time he persuaded the company to purchase the undeveloped island of Singapore (1819) and undertook much of the preliminary work for transforming it into an international port and centre of commerce. He was also a keen botanical collector.

Rafsanjani /ˌræfsæn'dʒɑːnɪ/, Ali Akbar Hashemi (b.1934), Iranian statesman and religious leader, President since 1989. A supporter and former pupil of Ayatollah Khomeini, in 1978 he helped organize the mass demonstrations that led to the shah's overthrow the following year. In 1988 he helped to bring an end to the Iran–Iraq War, having persuaded Khomeini to accept the UN's peace terms. When Khomeini died in 1989 Rafsanjani emerged from the ensuing power struggle as Iran's leader. He has sought to improve Iran's relations with the West and kept his country neutral during the Gulf War of 1991.

raft[1] /rɑːft/ n. & v. ● n. **1** a flat floating structure of timber or other materials for conveying persons or things. **2** a lifeboat or small (often inflatable) boat, esp. for use in emergencies. **3** a floating accumulation of trees, ice, etc. ● *v.* **1** *tr.* transport as or on a raft. **2** *tr.* cross (water) on a raft. **3** *tr.* form into a raft. **4** *intr.* (often foll. by *across*) work a raft (across water etc.). [ME f. ON *raptr* RAFTER[1]]

raft[2] /rɑːft/ n. **1** a large collection. **2** (foll. by of) a crowd. [*raff* rubbish, perh. of Scand. orig.]

rafter[1] /'rɑːftə(r)/ n. each of the sloping beams forming the framework of a roof. □ **raftered** adj. [OE *ræfter*, rel. to RAFT[1]]

rafter[2] /'rɑːftə(r)/ n. **1** a person who rafts timber. **2** a person who travels by raft.

raftsman /'rɑːftsmən/ n. (pl. **-men**) a worker on a raft.

rag[1] /ræg/ n. **1 a** a torn, frayed, or worn piece of woven material. **b** one of the irregular scraps to which cloth etc. is reduced by wear and tear. **2 a** (in pl.) old or worn clothes. **b** (usu. in pl.) colloq. a garment of any kind. **3** (collect.) scraps of cloth used as material for paper, stuffing, etc. **4** derog. **a** a newspaper. **b** a flag, handkerchief, curtain, etc. **5** (usu. with neg.) the smallest scrap of cloth etc. (*not a rag to cover him*). **6** an odd scrap; an irregular piece. **7** a jagged projection, esp. on metal. □ **in rags 1** much torn. **2** in old torn clothes. **rag-and-bone man** *Brit.* an itinerant dealer in old clothes, furniture, etc. **rag-bag 1** a bag in which scraps of fabric etc. are kept for use. **2** a miscellaneous collection. **3** *sl.* a sloppily dressed woman. **rag bolt** a bolt with barbs to keep it tight when it has been driven in. **rag book** a children's book made of untearable cloth. **rag doll** a stuffed doll made of cloth. **rag paper** paper made from rags. **rag-picker** a collector and seller of rags. **rags to riches** poverty to affluence. **rag trade** colloq. the business of designing, making, and selling clothes. [ME, prob. backform. f. RAGGED]

rag[2] /ræg/ n. & v. ● n. Brit. **1** a fund-raising programme of stunts, parades, and entertainment organized by students. **2** colloq. a prank. **3 a** a rowdy celebration. **b** a noisy disorderly scene. ● *v.* (**ragged, ragging**) **1** *tr.* tease; torment; play rough jokes on. **2** *tr.* scold; reprove severely. **3** *intr.* Brit. engage in rough play; be noisy and riotous. [18th c.: orig. unkn.: cf. BALLYRAG]

rag[3] /ræg/ n. **1** a large coarse roofing-slate. **2** a hard coarse sedimentary stone that breaks into thick slabs. [ME: orig. unkn., but assoc. with RAG[1]]

rag[4] /ræg/ n. Mus. a ragtime composition or tune. [perh. f. RAGGED: see RAGTIME]

raga /'rɑːgə/ n. (also **rag** /rɑːg/) Ind. Mus. **1** a pattern of notes used as a basis for improvisation. **2** a piece using a particular raga. [Skr., = colour, musical tone]

ragamuffin /'rægəˌmʌfɪn/ n. a person in ragged dirty clothes, esp. a child. [prob. based on RAG[1]: cf. 14th-c. *ragamoffyn* the name of a demon]

rage /reɪdʒ/ n. & v. ● n. **1** fierce or violent anger. **2** a fit of this (*flew into a rage*). **3** the violent action of a natural force (*the rage of the storm*). **4** (foll. by for) **a** a vehement desire or passion. **b** a widespread temporary enthusiasm or fashion. **5** poet. poetic, prophetic, or martial enthusiasm or ardour. **6** esp. Austral. & NZ colloq. a lively party. ● *v.intr.* **1** be full of anger. **2** (often foll. by at, against) speak furiously or madly; rave. **3 a** (of wind, battle, fever, etc.) be violent; be at its height; continue unchecked. **b** (as **raging** adj.) extreme, very painful (*raging thirst; a raging headache*). **4** Austral. & NZ colloq. seek enjoyment; have a good time, revel. □ **all the rage** very popular, fashionable. □ **rager** n. (esp. in sense 4 of v.). [ME f. OF *rager* ult. f. L RABIES]

ragee /'rɑːgiː/ n. (also **raggee**) a coarse type of millet, *Eleusine coracana*, forming a staple food in parts of India etc. [Hindi *rāgī*]

ragga /ˈrægə/ n. a style of popular music combining elements of reggae and hip-hop. [RAGAMUFFIN, from the style of clothing worn by its followers]

ragged /ˈrægɪd/ adj. **1 a** (of clothes etc.) torn; frayed. **b** (of a place) dilapidated. **2** rough; shaggy; hanging in tufts. **3** (of a person) in ragged clothes. **4** with a broken or jagged outline or surface. **5** Printing (of a right margin) unjustified and so uneven. **6** faulty; imperfect. **7 a** lacking finish, smoothness, or uniformity (ragged rhymes). **b** (of a sound) harsh, discordant. **8** exhausted (esp. be run ragged). □ **ragged robin** a pink-flowered campion, Lychnis flos-cuculi, with tattered petals (also called cuckoo flower). □ **raggedly** adv. **raggedness** n. **raggedy** adj. [ME f. ON roggvathr tufted]

raggee var. of RAGEE.

raggle-taggle /ˈrægəlˌtæg(ə)l/ adj. (also **wraggle-taggle**) ragged; rambling, straggling. [app. fanciful var. of RAGTAG]

raglan /ˈræglən/ n. (often attrib.) an overcoat without shoulder seams, the sleeves running up to the neck. □ **raglan sleeve** a sleeve of this kind. [Lord Raglan, Brit. commander (1788–1855)]

ragman /ˈrægmən/ n. a person who collects or deals in rags, old clothes, etc.

Ragnarök /ˈrægnəˌrɒk/ Scand. Mythol. the final battle between the gods and the powers of evil, the Scandinavian equivalent of the Götterdämmerung. [transl. Icel. ragna rökr twilight of the gods, altered from the original ragna rök history or judgement of the gods]

ragout /ræˈguː/ n. & v. ● n. a dish of meat in small pieces stewed with vegetables and highly seasoned. ● v.tr. cook (food) in this way. [F ragoût f. ragoûter revive the taste of]

ragstone /ˈrægstəʊn/ n. = RAG³ 2.

ragtag /ˈrægtæg/ n. (in full **ragtag and bobtail**) derog. the rabble or common people. [earlier tag-rag, tag and rag, f. RAG¹ + TAG¹]

ragtime /ˈrægtaɪm/ n. & adj. ● n. a style of early popular music characterized by a syncopated melodic line and regularly accented accompaniment, evolved by American black musicians (such as Scott Joplin) in the 1890s and played esp. on the piano. ● adj. **1** of or resembling ragtime. **2** sl. disorderly, disreputable, inferior (a ragtime army). [prob. f. RAG⁴]

raguly /ˈrægjʊlɪ/ adj. Heraldry like a row of sawn-off branches. [perh. f. RAGGED after nebuly]

Ragusa /raˈguːza/ the Italian name (until 1918) for DUBROVNIK. It is the probable source of the word argosy, referring to the large and richly freighted merchant ships of Ragusa in the 16th century.

ragweed /ˈrægwiːd/ n. **1** = RAGWORT. **2** N. Amer. a greyish composite plant of the genus Ambrosia, with allergenic pollen.

ragworm /ˈrægwɜːm/ n. a carnivorous marine polychaete worm of the family Nereidae; esp. Nereis diversicolor, often used for bait (cf. NEREID). [RAG¹]

ragwort /ˈrægwɜːt/ n. a yellow-flowered ragged-leaved composite plant of the genus Senecio; esp. S. jacobaea, a common weed of pastures that is poisonous to cattle and horses.

rah /rɑː/ int. esp. N. Amer. colloq. an expression of encouragement, approval, etc. [shortening of HURRAH]

Rahman see ABDUL RAHMAN, MUJIBUR RAHMAN.

rah-rah /ˈrɑːrɑː/ n. & adj. US sl. ● n. a shout of support and encouragement as for a college team. ● adj. characteristic of college students; marked by great enthusiasm or excitement. □ **rah-rah skirt** a short flounced skirt similar to those worn by cheer-leaders in the US.

rai /raɪ/ n. a style of popular music which fuses Arabic and Algerian folk elements with Western styles. [Algerian or Moroccan Arab.]

raid /reɪd/ n. & v. ● n. **1** a rapid surprise attack, esp.: **a** by troops, aircraft, etc., in warfare. **b** to commit a crime or do harm. **2** a surprise attack by police etc. to arrest suspected persons or seize illicit goods. **3** Stock Exch. an attempt to lower prices by the concerted selling of shares. **4** (foll. by on, upon) a forceful or insistent attempt to make a person or thing provide something. ● v.tr. **1** make a raid on (a person, place, or thing). **2** plunder, deplete. □ **raider** n. [ME, Sc. form of OE rād ROAD¹]

rail¹ /reɪl/ n. & v. ● n. **1** a level or sloping bar or series of bars: **a** used to hang things on. **b** running along the top of a set of banisters. **c** forming part of a fence or barrier as protection against contact, falling over, etc. **2** a steel bar or continuous line of bars laid on the ground, usu. as one of a pair forming a railway track. **3** (often attrib.) a railway (send it by rail; rail fares). **4** (in pl.) the inside boundary fence of a racecourse. **5** a horizontal piece in the frame of a panelled door etc. (cf. STILE²). ● v.tr. **1** furnish with a rail or rails. **2** (usu. foll. by in, off) enclose with

rails (a small space was railed off). **3** convey (goods) by rail. □ **off the rails** disorganized; out of control; deranged. **over the rails** over the side of a ship. **rail fence** esp. US a fence made of posts and rails. **rail gun** an electromagnetic projectile launcher used esp. as an anti-missile weapon. □ **railage** n. **railless** /ˈreɪllɪs/ adj. [ME f. OF reille iron rod f. L regula RULE]

rail² /reɪl/ v.intr. (often foll. by at, against) complain using abusive language; rant. □ **railer** n. **railing** n. & adj. [ME f. F railler f. Prov. ralhar jest, ult. f. L rugire bellow]

rail³ /reɪl/ n. a bird of the family Rallidae, which includes the coot and corncrake; esp. one of the genus Rallus, inhabiting marshes (water rail). [ME f. ONF raille f. Rmc, perh. imit.]

railcar /ˈreɪlkɑː(r)/ n. a railway vehicle consisting of a single powered coach.

railcard /ˈreɪlkɑːd/ n. Brit. a pass entitling the holder to reduced rail fares.

railhead /ˈreɪlhed/ n. **1** the furthest point reached by a railway under construction. **2** the point on a railway at which road transport of goods begins.

railing /ˈreɪlɪŋ/ n. **1** (usu. in pl.) a fence or barrier made of rails. **2** the material for these.

raillery /ˈreɪlərɪ/ n. (pl. **-ies**) **1** good-humoured ridicule; rallying. **2** an instance of this. [F raillerie (as RAIL²)]

railman /ˈreɪlmən/ n. (pl. **-men**) = RAILWAYMAN.

railroad /ˈreɪlrəʊd/ n. & v. ● n. esp. US = RAILWAY. ● v.tr. **1** (often foll. by to, into, through, etc.) rush or coerce (a person or thing) (railroaded me into going too). **2** send (a person) to prison by means of false evidence.

railway /ˈreɪlweɪ/ n. **1** a track or set of tracks made of usu. steel rails on which trains run. (See note below.) **2** such a system worked by a single company (Great Western Railway). **3** the organization and personnel required for its working. **4** a similar set of tracks for other vehicles etc. □ **railway-yard** an area where rolling-stock is kept and made up into trains.

■ Tracks of wooden rails were used in Europe from the 16th century for horse-drawn loads in mines and collieries, iron rails appearing in England in 1738. A breakthrough came in 1804 in Britain with the construction of the first steam locomotive by Richard Trevithick, and the first railway company, the Stockton and Darlington, opened in England in 1825. Following the success of George Stephenson's Rocket in 1830, the railways expanded rapidly, particularly in Britain and the US, and their development played a large part in the growing industrial power of the West in the 19th century. In the second half of the 20th century steam locomotives have mainly given way to diesel and electric traction; many busy routes (including all underground railways) are now electrified, using either overhead wires or an insulated third rail. Today, speeds of 225 kilometres an hour (140 m.p.h.) are routinely reached on some electrified main lines. Steam power persists in some countries and is often the focus of an active preservation movement in countries where it is no longer in ordinary use, such as the UK and US. The expansion of road transport (and later, especially in the US, air transport) has led to a decline in the importance of rail transport and the closure of many lines, though increasing road congestion has recently led some countries to consider the re-expansion of their railways.

railwayman /ˈreɪlweɪmən/ n. (pl. **-men**) a railway employee.

raiment /ˈreɪmənt/ n. archaic clothing. [ME f. obs. arrayment (as ARRAY)]

rain /reɪn/ n. & v. ● n. **1 a** the condensed moisture of the atmosphere falling visibly in separate drops. **b** the fall of such drops. **2** (in pl.) **a** (prec. by the) the rainy season in tropical countries. **b** rainfalls. **3 a** a falling liquid or solid particles or objects. **b** the rainlike descent of these. **c** a large or overwhelming quantity (a rain of congratulations). ● v. **1** intr. (prec. by it as subject) rain falls (it is raining; if it rains). **2 a** intr. fall in showers or like rain (tears rained down their cheeks; blows rain upon him). **b** tr. (prec. by it as subject) send in large quantities (it rained blood; it is raining invitations). **3** tr. send down like rain; lavishly bestow (rained benefits on us; rained blows upon him). **4** intr. (of the sky, the clouds, etc.) send down rain. □ **rain cats and dogs** see CAT¹. **rain check** esp. N. Amer. **1** a ticket given for later use when a sporting fixture or other outdoor event is interrupted or postponed by rain. **2** a promise that an offer will be maintained though deferred. **rain-cloud** a cloud bringing rain. **rain dance** a dance performed by a tribal group in the hope of summoning rain. **rain-gauge** an instrument measuring rainfall. **rain-making** the action of attempting to increase rainfall by artificial means. **rain off** (or N. Amer. **out**) (esp. in passive) cause (an event etc.) to be terminated or cancelled because of rain. **rain or shine** whether it rains or not. **rain-shadow** a region shielded from rain by mountains etc. **rain-wash 1** loose material carried away by rain. **2** the movement of this. **rain-worm** the common earthworm. □ **rainless** adj. [OE regn, rēn, regnian f. Gmc]

rainbird /'reɪnbɜːd/ n. a bird said to foretell rain by its cry, esp. the green woodpecker.

rainbow /'reɪnbəʊ/ n. & adj. ● n. **1** an arch of colours (conventionally red, orange, yellow, green, blue, indigo, violet) formed in the sky (or across a cataract etc.) opposite the sun by reflection, twofold refraction, and dispersion of the sun's rays in falling rain or in spray or mist. **2** a similar effect formed by the moon's rays. **3** a wide variety of related things (a rainbow of colours; a rainbow of political opinion). ● adj. many-coloured. □ **rainbow lorikeet** a small brightly coloured Polynesian parrot, Trichoglossus haematodus. **rainbow trout** a large trout, Salmo gairdneri, orig. of the Pacific coast of North America. **secondary rainbow** an additional arch with the colours in reverse order formed inside or outside a rainbow by twofold reflection and twofold refraction. [OE regnboga (as RAIN, BOW[1])]

Rainbow Bridge a bridge of natural rock, the world's largest natural bridge, situated in southern Utah, just north of the border with Arizona. Its span is 86 m (278 ft).

rainbow coalition n. (esp. in the US) a loose coalition or alliance of several different left-of-centre political groups, representing social, ethnic, and other minorities. The phrase came to wide public attention when used in connection with Jesse Jackson's attempt to be nominated as the Democratic Party's presidential candidate in 1984.

raincoat /'reɪnkəʊt/ n. a waterproof or water-resistant coat.

raindrop /'reɪndrɒp/ n. a single drop of rain. [OE regndropa]

rainfall /'reɪnfɔːl/ n. **1** a fall of rain. **2** the quantity of rain falling within a given area in a given time.

rainforest /'reɪnfɒrɪst/ n. luxuriant forest in an area of heavy rainfall and little seasonality. Rainforest is typical of many tropical regions but temperate rainforest occurs in moist coastal areas, e.g. in the north-western US, western Canada, and eastern Australia. Tropical evergreen rainforest has been stable for millions of years and is the richest of all vegetation types in terms of the numbers of plant and animal species; it has a high proportion of lianas, epiphytes, and very tall trees. Rainforests are important in removing excess carbon dioxide from the atmosphere and as a reservoir of water. However, because almost all the nutrients are held in the vegetation rather than the soil, removal of the plant cover quickly leads to impoverished soils that are easily eroded. There is currently much concern over the rate at which such forests are being irreversibly destroyed (see DEFORESTATION).

Rainier, Mount /rə'nɪə(r), 'reɪnɪə(r)/ a volcanic peak in the south-west of Washington State in the US. Rising to a height of 4,395 m (14,410 ft), it is the highest peak in the Cascade Range.

rainproof /'reɪnpruːf/ adj. (esp. of a building, garment, etc.) resistant to rainwater.

rainstorm /'reɪnstɔːm/ n. a storm with heavy rain.

rainwater /'reɪn,wɔːtə(r)/ n. water obtained from collected rain, as distinct from a well etc.

rainy /'reɪnɪ/ adj. (**rainier, rainiest**) **1** (of weather, a climate, day, region, etc.) in or on which rain is falling or much rain usually falls. **2** (of cloud, wind, etc.) laden with or bringing rain. □ **rainy day** a time of special need in the future. □ **rainily** adv. **raininess** n. [OE rēnig (as RAIN)]

Raipur /raɪ'pʊə(r)/ a city in central India, in Madhya Pradesh; pop. (1991) 438,000.

raise /reɪz/ v. & n. ● v.tr. **1** put or take into a higher position. **2** (often foll. by up) cause to rise or stand up or be vertical; set upright. **3** increase the amount or value or strength of (raised their prices). **4** (often foll. by up) construct or build up. **5** levy or collect or bring together (raise money; raise an army). **6** cause to be heard or considered (raise a shout; raise an objection). **7** set going or bring into being; arouse (raise a protest; raise hopes). **8** rouse from sleep or death, or from a lair. **9** bring up; educate. **10** breed or grow (raise one's own vegetables). **11** promote to a higher rank. **12** (foll. by to) Math. multiply a quantity to a specified power. **13** cause (bread) to rise, e.g. with yeast. **14** Cards bet more than (another player). **b** increase (a stake). **c** Bridge make a bid contracting for more tricks in the same suit as (one's partner); increase (a bid) in this way. **15** abandon or force an enemy to abandon (a siege or blockade). **16** remove (a barrier or embargo). **17** cause (a ghost etc.) to appear (opp. LAY[1] 6b). **18** colloq. find (a person etc. wanted). **19** establish contact

with (a person etc.) by radio or telephone. **20** (usu. as **raised** adj.) cause (pastry etc.) to stand without support (a raised pie). **21** Naut. come in sight of (land, a ship, etc.). **22** make a nap on (cloth). **23** extract from the earth. ● n. **1** Cards an increase in a stake or bid (cf. sense 14 of v.). **2** esp. N. Amer. an increase in salary. □ **raise Cain** see CAIN. **raised beach** Geol. a former beach lying above the level of the present shoreline owing to earth movement or a fall in sea level. **raise the devil** colloq. make a disturbance. **raise a dust 1** cause turmoil. **2** obscure the truth. **raise one's eyebrows** see EYEBROW. **raise one's eyes** see EYE. **raise from the dead** restore to life. **raise one's glass to** drink the health of. **raise one's hand to** make as if to strike (a person). **raise one's hat** (often foll. by to) remove it momentarily as a gesture of courtesy or respect. **raise hell** colloq. make a disturbance. **raise a laugh** cause others to laugh. **raise a person's spirits** give him or her new courage or cheerfulness. **raise one's voice** speak, esp. louder. **raise the wind** Brit. procure money for a purpose. □ **raisable** adj. [ME f. ON reisa, rel. to REAR[2]]

raisin /'reɪz(ə)n/ n. a partially dried grape. □ **raisiny** adj. [ME f. OF ult. f. L racemus grape-bunch]

raison d'être /,reɪzɒn 'detrə/ n. (pl. **raisons d'être** pronunc. same) a purpose or reason that accounts for or justifies or originally caused a thing's existence. [F, = reason for being]

raita /raː'iːtə, 'raɪtə/ n. an Indian side dish consisting of chopped cucumber (or other vegetables) and spices in yoghurt. [Hind. rāytā]

Raj /raːdʒ/ n. (prec. by the) hist. **1** British sovereignty in India. **2** the period of this. [Hindi rāj reign]

raja /'raːdʒə/ n. (also **rajah**) hist. **1** an Indian king or prince. **2** a petty dignitary or noble in India. **3** a Malay or Javanese chief. [Skr. rājan king f. rāj to reign, rel. to L rex king]

Rajasthan /,raːdʒə'staːn/ a state in western India, on the Pakistani border; capital, Jaipur. It was formed as the Union of Rajasthan in 1948 from the former region of Rajputana. In 1956 additional territory was added and its name became simply Rajasthan. The western part of the state consists largely of the Thar Desert and is sparsely populated. □ **Rajasthani** adj. & n.

Rajasthan Canal a former name for the INDIRA GANDHI CANAL.

raja yoga n. a form of yoga intended to achieve control over the mind and emotions. (See YOGA.) [Skr. f. rājan king + YOGA]

Rajkot /raːdʒ'kəʊt/ a city in Gujarat, western India; pop. (1991) 556,000.

Rajneesh /raːdʒ'niːʃ/, Bhagwan Shree (born Chandra Mohan Jain; known as 'the Bhagwan' from a Sanskrit word meaning 'lord') (1931–90), Indian guru. After founding an ashram in Poona, India, he moved to the US in 1981 and founded a commune in Oregon. He became notorious for his doctrine of communal therapy, in particular for his preaching of salvation through free love. Rajneesh was deported in 1985 for immigration violations and the Oregon commune broke up shortly afterwards.

Rajput /'raːdʒpʊt/ n. (also **Rajpoot**) a member of a Hindu soldier caste claiming Kshatriya descent. [Hindi rājpūt f. Skr. rājan king + putra son]

Rajputana /,raːdʒpʊ'taːnə/ an ancient region of India consisting of a collection of princely states ruled by dynasties, which came to power between the 9th and 16th centuries. Following independence from Britain in 1947, they united to form the state of Rajasthan, parts also being incorporated into Gujarat and Madhya Pradesh.

Rajshahi /raːdʒ'ʃaːhɪ/ a port on the Ganges river in western Bangladesh; pop. (1991) 324,530.

rake[1] /reɪk/ n. & v. ● n. **1 a** an implement consisting of a pole with a crossbar toothed like a comb at the end, or with several tines held together by a crosspiece, for drawing together hay etc. or smoothing loose soil or gravel. **b** a wheeled implement for the same purpose. **2** a similar implement used for other purposes, e.g. by a croupier drawing in money at a gaming-table. ● v. **1** tr. (usu. foll. by out, together, up, etc.) collect or gather or remove with or as with a rake. **2** tr. make tidy or smooth with a rake (raked it level). **3** intr. use a rake. **4** tr. & intr. search with or as with a rake, search thoroughly, ransack. **5** tr. **a** direct gunfire along (a line) from end to end. **b** sweep with the eyes. **c** (of a window etc.) have a commanding view of. **6** tr. scratch or scrape. □ **rake in** colloq. amass (profits etc.). **rake it in** colloq. make much money. **rake-off** colloq. a commission or share, esp. in a disreputable deal. **rake up** (or **over**) revive the memory of (past quarrels, grievances, etc.). □ **raker** n. [OE raca, racu f. Gmc, partly f. ON raka scrape, rake]

rake[2] /reɪk/ n. a dissolute man of fashion. □ **rake's progress** a progressive deterioration, esp. through self-indulgence (the title of a series of engravings by Hogarth 1735). [short for archaic *rakehell* in the same sense]

rake[3] /reɪk/ v. & n. ● v. **1** tr. & intr. set or be set at a sloping angle. **2** intr. **a** (of a mast or funnel) incline from the perpendicular towards the stern. **b** (of a ship or its bow or stern) project at the upper part of the bow or stern beyond the keel. ● n. **1** a raking position or build. **2** the amount by which a thing rakes. **3** the slope of the stage or the auditorium in a theatre. **4** the slope of the back of a seat etc. **5** the angle of the edge or face of a cutting tool. [17th c.: prob. rel. to G *ragen* project, of unkn. orig.]

rake[4] /reɪk/ n. a series of wagons or carriages on a railway or of wagons or trucks in a mine or factory. [ME f. ON *rák* stripe, streak f. alt. of *rek-* to drive]

raki /rəˈkiː, ˈrækɪ/ n. (pl. **rakis**) any of various spirits made in eastern Europe and the Middle East. [Turk. *raqi*]

rakish[1] /ˈreɪkɪʃ/ adj. of or like a rake (see RAKE[2]); dashing, jaunty. □ **rakishly** adv. **rakishness** n.

rakish[2] /ˈreɪkɪʃ/ adj. (of a ship) smart and fast-looking, seemingly built for speed and therefore open to suspicion of piracy. [RAKE[3], assoc. with RAKE[2]]

Rákosi /ˈraːkɒʃi/, Mátyás (1892–1971), Hungarian Communist statesman, First Secretary of the Hungarian Socialist Workers' Party 1945–56 and Prime Minister 1952–3 and 1955–6. After the Communist seizure of power at the end of the Second World War he did much to establish a firmly Stalinist regime. Ousted as Premier by the more liberal Imre Nagy in 1953, Rákosi retained his party leadership and briefly returned to power in 1955, but in 1956 was dismissed, fleeing to the USSR during the Hungarian uprising later that year.

raku /ˈraːkuː/ n. a kind of Japanese lead-glazed earthenware, primarily for use in the tea ceremony. [Jap., lit. 'enjoyment']

rale /raːl/ n. an abnormal rattling sound heard in the auscultation of unhealthy lungs. [F f. *râler* to rattle]

Raleigh[1] /ˈraːlɪ, ˈrɔːlɪ/ the state capital of North Carolina; pop. (1990) 207,950.

Raleigh[2] /ˈraːlɪ, ˈrɔːlɪ/, Sir Walter (also **Ralegh**) (c.1552–1618), English explorer, courtier, and writer. A favourite of Elizabeth I, he organized several voyages of exploration and colonization to the Americas, including an unsuccessful attempt to settle Virginia (1584–9) and a journey up the Orinoco river in search of gold (1595); from his travels he brought back potato and tobacco plants to England. Raleigh was imprisoned in 1603 by James I on a charge of conspiracy, but released in 1616 to lead a second expedition up the Orinoco in search of the fabled land of El Dorado. He returned empty-handed after a clash with some Spanish settlers, and was subsequently executed on the original charge.

rall. abbr. Mus. rallentando.

rallentando /ˌrælənˈtændəʊ/ adv., adj., & n. Mus. ● adv. & adj. with a gradual decrease of speed. ● n. (pl. **-os** or **rallentandi** /-dɪ/) a passage to be performed in this way. [It.]

ralli car /ˈrælɪ/ n. (also **ralli cart**) hist. a light two-wheeled horse-drawn vehicle for four persons. [*Ralli*, name of the first purchaser 1885]

rally[1] /ˈrælɪ/ v. & n. ● v. (**-ies, -ied**) **1** tr. & intr. (often foll. by *round, behind, to*) bring or come together as support or for concentrated action. **2** tr. & intr. bring or come together again after a rout or dispersion. **3 a** tr. revive (courage etc.) by an effort of will. **b** tr. rouse (a person or animal) to fresh energy. **c** intr. pull oneself together. **4** intr. recover after illness or prostration or fear, regain health or consciousness, revive. **5** intr. (of share prices etc.) increase after a fall. ● n. (pl. **-ies**) **1** an act of reassembling forces or renewing conflict; a reunion for fresh effort. **2** a recovery of energy after or in the middle of exhaustion or illness. **3** a mass meeting of supporters or persons having a common interest. **4** a competition for motor vehicles, usu. over public roads. (See MOTOR-RACING.) **5** Tennis etc. an extended exchange of strokes between players. □ **rally-cross** a form of motor-racing over roads and cross-country. □ **rallier** n. [F *rallier* (as RE-, ALLY[1])]

rally[2] /ˈrælɪ/ v.tr. (**-ies, -ied**) subject to good-humoured ridicule. [F *railler*: see RAIL[2]]

RAM abbr. **1** /ræm/ Computing random-access memory. **2** (in the UK) Royal Academy of Music.

ram /ræm/ n. & v. ● n. **1** an uncastrated male sheep, a tup. **2** (**the Ram**) the zodiacal sign or constellation Aries. **3** hist. **a** = *battering-ram* (see BATTER[1]). **b** a projection at the bow of a warship, for piercing the sides of other ships. **c** a battleship with such a beak. **4** the falling weight of a

pile-driving machine. **5 a** a hydraulic water-raising or lifting machine. **b** the piston of a hydrostatic press. **c** the plunger of a force-pump. **6** Austral. sl. an accomplice in petty crime. ● v.tr. (**rammed, ramming**) **1** force or squeeze into place by pressure. **2** (usu. foll. by *down, in, into*) beat down or drive in by heavy blows. **3** (of a ship, vehicle, etc.) strike violently, crash against. **4** (foll. by *against, at, on, into*) dash or violently impel. □ **ram home** stress forcefully (an argument, lesson, etc.). **ram-raid** an instance of ram-raiding. **ram-raider** a person who engages in ram-raiding. **ram-raiding** a form of robbery in which the front of a shop is rammed using a car and the shop is looted by the occupants of the car. **ram's-horn snail** a herbivorous freshwater snail of the family Planorbidae, having a flat spiral shell. □ **rammer** n. [OE *ram(m)*, perh. rel. to ON *rammr* strong]

Rama /ˈraːmə/ Hinduism the hero of the Ramayana, husband of Sita. He is the Hindu model of the ideal man, the seventh incarnation of Vishnu, and is widely venerated, by some sects as the supreme god.

Ramadan /ˈræməˌdæn, ˌræməˈdaːn/ n. (also **Ramadhan**) the ninth month of the Muslim year, during which strict fasting is observed from sunrise to sunset. [Arab. *ramaḍān* f. *ramaḍa* be hot; reason for name uncert.]

Ramakrishna /ˌraːməˈkrɪʃnə/ (born Gadadhar Chatterjee) (1836–86), Indian yogi and mystic. In his teachings he condemned lust, money, and the caste system, and preached that all religions leading to the attainment of mystical experience are equally good and true. His doctrines were spread widely in the US and Europe by his disciple Vivekananda.

ramal /ˈreɪm(ə)l/ adj. Bot. of or proceeding from a branch. [L *ramus* branch]

Raman /ˈraːmən/, Sir Chandrasekhara Venkata (1888–1970), Indian physicist. His discovered the optical effect that is named after him, which was one of the most important proofs of the quantum theory of light. He also studied vibrations and sound, and the theory of musical instruments. In optics he went on to investigate the properties of crystals and minerals and the physiology of colour vision. Raman was awarded the 1930 Nobel Prize for physics..

Raman effect n. Physics a change of wavelength exhibited by some of the light etc. scattered in a medium (cf. RAYLEIGH SCATTERING). The effect is specific to the molecules which cause it, and so can be used in spectroscopic analysis. [RAMAN]

Ramanujan /raːˈmaːnʊdʒən/, Srinivasa Aaiyangar (1887–1920), Indian mathematician. Largely self-taught, he was a mathematical genius who produced a number of original discoveries in number theory and power series. Collaborating with G. H. Hardy (1877–1947) in Cambridge, he made what is probably his most important contribution — a theorem concerning the partition of numbers into a sum of smaller integers. He was elected to the Royal Society in 1918.

Ramayana /raːˈmaːjənə/ n. one of the two great Sanskrit epics of the Hindus (the other is the Mahabharata), composed c.300 BC. It describes how Rama, aided by his brother and the monkey Hanuman, rescued his wife Sita from the clutches of Ravana, the ten-headed demon king of Lanka. [Skr., = exploits of Rama]

Rambert /ˈrɒmbeə(r)/, Dame Marie (born Cyvia Rambam) (1888–1982), British ballet-dancer, teacher, and director, born in Poland. In 1913 she joined Diaghilev's Ballets Russes as a teacher of eurhythmics, moving to London in 1917. Rambert later formed the Ballet Club (1930), which became known as the Ballet Rambert (1935). For over fifty years the company, under her direction, promoted new British ballets and young choreographers and dancers such as her pupil Frederick Ashton.

ramble /ˈræmb(ə)l/ v. & n. ● v.intr. **1** walk for pleasure, with or without a definite route. **2** wander in discourse, talk or write disconnectedly. ● n. a walk taken for pleasure. [prob. f. MDu. *rammelen* (of an animal) wander about in sexual excitement, frequent. of *rammen* copulate with, rel. to RAM]

rambler /ˈræmblə(r)/ n. **1** a person who rambles. **2** a straggling or climbing rose (crimson rambler).

rambling /ˈræmblɪŋ/ adj. **1** peripatetic, wandering. **2** disconnected, desultory, incoherent. **3** (of a house, street, etc.) irregularly arranged. **4** (of a plant) straggling, climbing. □ **ramblingly** adv.

Rambo /ˈræmbəʊ/ the hero of David Morrell's novel First Blood (1972), a Vietnam War veteran characterized as macho, self-sufficient, and bent on violent retribution, popularized in the films First Blood (1982) and Rambo: First Blood Part II (1985).

rambunctious /ræmˈbʌŋkʃəs/ adj. N. Amer. colloq. **1** uncontrollably exuberant. **2** unruly. □ **rambunctiously** *adv.* **rambunctiousness** *n.* [19th c.: orig. unkn.]

rambutan /ræmˈbuːt(ə)n/ *n.* **1** a red plum-sized prickly fruit. **2** a Malaysian tree, *Nephelium lappaceum*, that bears this. [Malay *rambūtan* f. *rambut* hair, in allusion to its spines]

RAMC *abbr.* (in the UK) Royal Army Medical Corps.

Rameau /rɑːˈməʊ/, Jean-Philippe (1683–1764), French composer, musical theorist, and organist. In 1722 he published his influential *Treatise on Harmony*. He is best known for his four volumes of harpsichord pieces (1706–41); noted for their bold harmonies and textural diversity, these consist largely of genre pieces with descriptive titles, such as 'La Poule'. Rameau also wrote many operas, including *Castor and Pollux* (1737).

ramekin /ˈræmɪkɪn/ *n.* **1** (in full **ramekin case** or **dish**) a small dish for baking and serving an individual portion of food. **2** food served in such a dish, esp. a small quantity of cheese baked with breadcrumbs, eggs, etc. [F *ramequin*, of LG or Du. orig.]

ramen /ˈrɑːmen/ *n.pl.* quick-cooking noodles, usu. served in a broth with meat and vegetables. [Jap., f. Chin. *la* pull + *mian* noodle]

Rameses see RAMSES.

ramie /ˈræmɪ/ *n.* **1** a tall East Asian plant, *Boehmeria nivea*, of the nettle family. **2** a strong fibre obtained from this, woven into cloth. [Malay *rāmī*]

ramification /ˌræmɪfɪˈkeɪʃ(ə)n/ *n.* **1** the act or an instance of ramifying; the state of being ramified. **2** a subdivision of a complex structure or process comparable to a tree's branches. **3** a consequence, esp. when complex or unwelcome. [F f. *ramifier*: see RAMIFY]

ramify /ˈræmɪfaɪ/ *v.* (**-ies**, **-ied**) **1** *intr.* form branches or subdivisions or offshoots, branch out. **2** *tr.* (usu. in *passive*) cause to branch out; arrange in a branching manner. [F *ramifier* f. med.L *ramificare* f. L *ramus* branch]

Ramillies, Battle of /ˈræmɪlɪz, French ramiji/ a battle in the War of the Spanish Succession which took place in 1706 near the village of Ramillies, north of Namur, central Belgium. The British army under General Marlborough defeated the French.

ramin /ˈræˈmiːn/ *n.* **1** a Malaysian tree of the genus *Gonystylus*, esp. *G. bancanus*. **2** the light-coloured hardwood obtained from this tree. [Malay]

ramjet /ˈræmdʒet/ *n.* a type of jet engine in which the air drawn in for combustion is compressed solely by the forward motion of the aircraft.

rammer see RAM.

rammies /ˈræmɪz/ *n.pl. Austral. & S. Afr. sl.* trousers. [20th c.: orig. uncert.]

rammy /ˈræmɪ/ *n.* (pl. **-ies**) *Sc. sl.* a brawl, a fight (esp. between gangs); a quarrel. [perh. f. Sc. *rammle* row, uproar, var. of RAMBLE]

Ramón y Cajal /rəˌmɒn iː kəˈhaːl/, Santiago (1852–1934), Spanish physician and histologist. He is best known for his research on nerve cells and the brain, and was a founder of the science of neurology. He identified the neuron as the fundamental unit of the nervous system, but argued (incorrectly) that the axons end only in the brain and do not join up with other axons or neurons. Ramón y Cajal shared a Nobel Prize with Camillo Golgi in 1906.

ramose /ˈræməʊz, ˈreɪm-, -əʊs/ *adj.* branched; branching. [L *ramosus* f. *ramus* branch]

ramp[1] /ræmp/ *n. & v.* ● *n.* **1** a slope or inclined plane, esp. for joining two levels of ground, floor, etc. **2** movable stairs for entering or leaving an aircraft. **3** an upward bend in a stair-rail. **4** *Brit.* a transverse ridge in a road to control the speed of vehicles. ● *v.* **1** *tr.* furnish or build with a ramp. **2** *intr.* **a** assume or be in a threatening posture. **b** (often foll. by *about*) storm, rage, rush. **c** *Heraldry* be rampant. **3** *intr. Archit.* (of a wall) ascend or descend to a different level. [ME (as verb in heraldic sense) f. F *rampe* f. OF *ramper* creep, crawl]

ramp[2] /ræmp/ *n. & v. Brit. sl.* ● *n.* a swindle or racket, esp. one conducted by the levying of exorbitant prices. ● *v.* **1** *intr.* engage in a ramp. **2** *tr.* subject (a person etc.) to a ramp. [16th c.: orig. unkn.]

rampage /ræmˈpeɪdʒ/ *v. & n.* ● *v.intr.* **1** (often foll. by *about*) rush wildly or violently about. **2** rage, storm. ● *n.* /often ˈræmpeɪdʒ/ wild or violent behaviour. □ **on the rampage** rampaging. □ **rampageous** *adj.* **rampager** *n.* [18th c., perh. f. RAMP[1]]

rampant /ˈræmp(ə)nt/ *adj.* **1** (placed after noun) *Heraldry* (of an animal) standing on its left hind foot with its forepaws in the air (*lion rampant*). **2** unchecked, flourishing excessively (*rampant violence*). **3** violent or extravagant in action or opinion (*rampant theorists*). **4** rank, luxuriant. □ **rampancy** *n.* **rampantly** *adv.* [ME f. OF, part. of *ramper*: see RAMP[1]]

rampart /ˈræmpɑːt/ *n. & v.* ● *n.* **1 a** a defensive wall with a broad top and usu. a stone parapet. **b** a walkway on top of such a wall. **2** a defence or protection. ● *v.tr.* fortify or protect with or as with a rampart. [F *rempart*, *rempar* f. *remparer* fortify f. *emparer* take possession of, ult. f. L *ante* before + *parare* prepare]

rampion /ˈræmpɪən/ *n.* **1** a bellflower, *Campanula rapunculus*, with white tuberous roots used as a salad. **2** a plant of the related genus *Phyteuma*, with clusters of hornlike buds and flowers. [ult. f. med.L *rapuncium*, *rapontium*, prob. f. L *rapum* RAPE[2]]

ramrod /ˈræmrɒd/ *n.* **1** a rod for ramming down the charge of a muzzle-loading firearm. **2** a thing that is very straight or rigid.

Ramsay[1] /ˈræmzɪ/, Allan (1713–84), Scottish portrait painter. From the late 1730s he was based in London, where he became much in demand as a portraitist in the 1750s. His style is noted for its French rococo grace and sensitivity, particularly in his portraits of women; major works of this period include *The Artist's Wife* (1755). In 1767 he was appointed painter to George III.

Ramsay[2] /ˈræmzɪ/, Sir William (1852–1916), Scottish chemist, discoverer of the noble gases. Investigating the reason why laboratory-prepared nitrogen was less dense than that isolated from air, he decided that the latter must be contaminated by a heavier gas. He went on to discover five chemically inert gases — argon, helium, and (with the help of M. W. Travers, 1872–1961) neon, krypton, and xenon — and determined their atomic weights and places in the periodic table. In 1910, with Frederick Soddy and Sir Robert Whytlaw-Gray (1877–1958), he identified the last noble gas, radon. He was awarded the Nobel Prize for chemistry in 1904.

Ramses /ˈræmsiːz/ (also **Rameses** /ˈræmɪˌsiːz/) the name of eleven Egyptian pharaohs, notably.

Ramses II (known as Ramses the Great) (died *c.*1225 BC), reigned *c.*1292–*c.*1225 BC. The third pharaoh of the 19th dynasty, he is famed for the vast monuments and statues that he built, including the two rock temples at Abu Simbel. He launched a major offensive against the Hittites, leading his troops in person and winning a military victory at the Hittite stronghold of Kadesh, but failing to capture the city; the battle is celebrated in a long poem and carvings on temple walls in Egypt and Nubia.

Ramses III (died *c.*1167 BC), reigned *c.*1198–*c.*1167 BC. The second pharaoh of the 20th dynasty, he fought decisive battles against the Libyans and the Sea Peoples, who attempted invasions. After his death the power of Egypt declined steadily.

ramshackle /ˈræmˌʃæk(ə)l/ *adj.* (usu. of a house or vehicle) tumbledown, rickety. [earlier *ramshackled* past part. of obs. *ransackle* RANSACK]

ramsons /ˈræmz(ə)nz/ *n.* (usu. treated as *sing.*) **1** a broad-leaved woodland garlic, *Allium ursinum*, with elongate pungent-smelling bulbous roots. **2** the root of this, eaten as a relish. [OE *hramsan* pl. of *hramsa* wild garlic, later taken as sing.]

RAN *abbr.* Royal Australian Navy.

ran past of RUN.

ranch /rɑːntʃ/ *n. & v.* ● *n.* **1 a** a cattle-breeding establishment, esp. in the US and Canada. **b** a farm where other animals are bred (*mink ranch*). **2** *N. Amer.* a single-storey or split-level house. ● *v.intr.* farm on a ranch. [Sp. *rancho* group of persons eating together]

rancher /ˈrɑːntʃə(r)/ *n.* **1** a person who farms on a ranch. **2** *N. Amer.* a modern single-storey house.

ranchero /rɑːnˈtʃeərəʊ/ *n.* (pl. **-os**) a person who farms or works on a ranch, esp. in Mexico. [Sp. (as RANCH)]

Ranchi /ˈrɑːntʃɪ/ a city in Bihar, NE India; pop. (1991) 598,000.

rancid /ˈrænsɪd/ *adj.* smelling or tasting like rank stale fat. □ **rancidness** *n.* **rancidity** /rænˈsɪdɪtɪ/ *n.* [L *rancidus* stinking]

rancour /ˈræŋkə(r)/ *n.* (US **rancor**) inveterate bitterness, malignant hate, spitefulness. □ **rancorous** *adj.* **rancorously** *adv.* [ME f. OF f. LL *rancor -oris* (as RANCID)]

rand[1] /rænd, rɑːnt/ *n.* **1** the chief monetary unit of South Africa and Namibia. **2** *S. Afr.* a ridge of high ground on either side of a river. [Afrik., = edge, rel. to RAND[2]: sense 1 f. the gold-field district of the Rand (see WITWATERSRAND, THE]

rand[2] /rænd/ *n.* a levelling-strip of leather between the heel and sides of a shoe or boot. [OE f. Gmc]

Rand, the /rænd, rɑːnt/ = WITWATERSRAND, THE.

R & B *abbr.* (also **R. & B.**) rhythm and blues.

R & D *abbr.* (also **R. & D.**) research and development.

Randers /ˈrɑːnəz/ a port of Denmark, on the Randers Fjord on the east coast of the Jutland peninsula; pop. (1990) 61,020.

random /ˈrændəm/ *adj.* **1** made, done, etc., without method or conscious choice (*random selection*). **2** *Statistics* with equal chances for each item. **b** given by a random process. **3** (of masonry) with stones of irregular size and shape. □ **at random** without aim or purpose or principle. **random-access** *Computing* (of a memory or file) having all parts directly accessible, so that it need not be read sequentially. **random error** *Statistics* an error in measurement caused by factors which vary from one measurement to another. □ **randomly** *adv.* **randomness** *n.* **randomize** *v.tr.* (also **-ise**). **randomization** /ˌrændəmaɪˈzeɪʃ(ə)n/ *n.* [ME f. OF *randon* great speed f. *randir* gallop]

R and R *abbr.* (also **R. and R., R & R**) **1** rescue and resuscitation. **2** rest and recreation. **3** rock and roll.

Randstad /ˈrɔːnstɑːt/ a conurbation in the north-west of the Netherlands that stretches in a horseshoe shape from Dordrecht and Rotterdam round to Utrecht and Amersfoort via The Hague, Leiden, Haarlem, and Amsterdam. The majority of the people of the Netherlands live in this area.

randy /ˈrændɪ/ *adj.* (**randier, randiest**) **1** lustful; eager for sexual gratification. **2** *Sc.* loud-tongued, boisterous, lusty. □ **randily** *adv.* **randiness** *n.* [perh. f. obs. *rand* f. obs. Du. *randen, ranten* RANT]

ranee /ˈrɑːniː/ *n.* (also **rani**) *hist.* a raja's wife or widow; a Hindu queen. [Hindi *rānī* = Skr. *rājñī* fem. of *rājan* king]

rang *past of* RING[1].

rangatira /ˌræŋəˈtɪərə/ *n.* NZ a Maori chief or noble. [Maori]

range /reɪndʒ/ *n. & v.* ● *n.* **1 a** the region between limits of variation, esp. as representing a scope of effective operation (*a voice of astonishing range; the whole range of politics*). **b** such limits. **c** a limited scale or series (*the range of the thermometer readings is about 10 degrees*). **d** a series representing variety or choice; a selection. **2** the area included in or concerned with something. **3 a** the distance attainable by a gun or projectile (*the enemy are out of range*). **b** the distance between a gun or projectile and its objective. **4** a row, series, line, or tier, esp. of mountains or buildings. **5 a** an open or enclosed area with targets for shooting. **b** a testing-ground for military equipment. **6 a** a large cooking stove of which the burners and oven(s) are kept continually hot. **b** *N. Amer.* an electric or gas cooker. **7** the area over which a thing, esp. a plant or animal, is distributed (*gives the ranges of all species*). **8** the distance that can be covered by a vehicle or aircraft without refuelling. **9** the distance between a camera and the subject to be photographed. **10** the extent of time covered by a forecast etc. **11 a** a large area of open land for grazing or hunting. **b** a tract over which one wanders. **12** lie, direction (*the range of the strata is east and west*). ● *v.* **1** *intr.* a reach; lie spread out; extend; be found or occur over a specified district; vary between limits (*ages ranging from twenty to sixty*). **b** run in a line (*ranges north and south*). **2** *tr.* (usu. in *passive* or *refl.*) place or arrange in a row or ranks or in a specified situation or order or company (*ranged their troops; ranged themselves with the majority party; trees ranged in ascending order of height*). **3** *intr.* rove, wander (*ranged through the woods; his thoughts range over past, present, and future*). **4** *tr.* traverse in all directions (*ranging the woods*). **5** *Printing* **a** *tr. Brit.* make (type) lie flush at the ends of successive lines. **b** *intr.* (of type) lie flush. **6** *intr.* **a** (often foll. by *with*) be level. **b** (foll. by *with, among*) rank; find one's right place (*ranges with the great writers*). **7** *intr.* **a** (of a gun) send a projectile over a specified distance (*ranges over a mile*). **b** (of a projectile) cover a specified distance. **c** obtain the range of a target by adjustment after firing past it or short of it. □ **ranging-pole** (or **-rod**) *Surveying* a pole or rod for setting a straight line. [ME f. OF *range* row, rank f. *ranger* f. *rang* RANK[1]]

rangé /ˈrɒnʒeɪ/ *adj.* (*fem.* **rangée** *pronunc.* same) domesticated, orderly, settled. [F]

rangefinder /ˈreɪndʒˌfaɪndə(r)/ *n.* an instrument for estimating the distance of an object, esp. one to be shot at or photographed.

Ranger /ˈreɪndʒə(r)/ a series of nine American moon probes launched between 1961 and 1965. The last three took many photographs before crashing into the moon.

ranger /ˈreɪndʒə(r)/ *n.* **1** a keeper of a royal or national park, or of a forest. **2** a member of a body of armed men, esp.: **a** a mounted soldier. **b** US a commando. **3** (**Ranger**) a senior Guide (see GUIDES ASSOCIATION). **4** a wanderer. □ **rangership** *n.*

Rangoon /ræŋˈguːn/ (Burmese **Yangon** /jæŋˈɡɒn/) the capital of Burma (Myanmar), a port in the Irrawaddy delta; pop. (1983) 2,458,710. For centuries a Buddhist religious centre, it is the site of the Shwe Dagon Pagoda, built over 2,500 years ago to house eight sacred hairs of the Buddha. The city was named Yangon — 'the end of strife' — in

1755, when King Alaungpaya captured the riverside village of Dagon. The modern city was established by the British in the mid-19th century and became capital in 1886.

rangy /ˈreɪndʒɪ/ *adj.* (**rangier, rangiest**) **1** (of a person) tall and slim. **2** *Austral.* hilly, mountainous.

rani var. of RANEE.

Ranjit Singh /ˌrʌndʒɪt ˈsɪŋ/ (known as the 'Lion of the Punjab') (1780–1839), Indian maharaja, founder of the Sikh state of Punjab. After succeeding his father as a Sikh ruler at the age of 12, he seized Lahore from the Afghans in 1799 and proclaimed himself maharaja of Punjab in 1801. He proceeded to make the state the most powerful in India, securing the holy city of Amritsar (1802) and expanding his control north-west with the capture of Peshawar (1818) and Kashmir (1819). At the end of the Sikh Wars which followed his death most of his territory was annexed by Britain.

Ranjitsinhji Vibhaji /ˌrʌndʒɪtˌsɪndʒɪ vɪˈbɑːdʒɪ/, Kumar Shri, Maharaja Jam Sahib of Navanagar (1872–1933), Indian cricketer and statesman. He made his cricketing début for Sussex in 1895, going on to score a total of 72 centuries as a batsman for Sussex and England (when he was popularly known as 'Ranji'). In 1907 he succeeded his cousin as maharaja of the state of Navanagar and promoted a number of modernization schemes to improve the state's infrastructure. He was knighted in 1917.

Rank /ræŋk/, J(oseph) Arthur, 1st Baron (1888–1972), English industrialist and film executive, founder of the Rank Organization. In the 1930s he became interested in films when, as a Methodist Sunday school teacher, he realized that they could be an ideal medium for spreading the Gospel. He founded the film production and distribution company known as the Rank Organization (1941). Under his chairmanship, it went on to own or control the leading British studios and cinema chains in the 1940s and 1950s.

rank[1] /ræŋk/ *n. & v.* ● *n.* **1 a** a position in a hierarchy, a grade of advancement. **b** a distinct social class, a grade of dignity or achievement (*people of all ranks; in the top rank of performers*). **c** high social position (*persons of rank*). **d** a place in a scale. **2** a row or line. **3 a** single line of soldiers drawn up abreast. **4** *Brit.* a place where taxis stand to await customers. **5** order, array. **6** *Chess* a row of squares across the board (cf. FILE[2] *n.* 3). ● *v.* **1** *intr.* have rank or place (*ranks next to the king*). **2** *tr.* classify, give a certain grade to. **3** *tr.* arrange (esp. soldiers) in a rank or ranks. **4** *US* **a** *tr.* take precedence over (a person) in respect to rank. **b** *intr.* have the senior position among the members of a hierarchy etc. □ **break rank** (or **ranks**) fail to remain in line; fail to maintain solidarity. **close ranks** maintain solidarity. **keep rank** remain in line. **other ranks** soldiers other than commissioned officers. **rank and fashion** high society. **rank and file** ordinary undistinguished people (orig. = *the ranks*). **rank correlation** *Statistics* correlation between ways of ranking the members of a set. **the ranks** the common soldiers, i.e. privates and corporals. **rise from the ranks 1** (of a private or a non-commissioned officer) receive a commission. **2** (of a self-made man or woman) advance by one's own exertions. [OF *ranc, renc,* f. Gmc, rel. to RING[1]]

rank[2] /ræŋk/ *adj.* **1** too luxuriant; choked with or apt to produce weeds or excessive foliage. **2 a** a foul-smelling, offensive. **b** loathsome, indecent, corrupt. **3** flagrant, virulent, gross, complete, unmistakable, strongly marked (*rank outsider*). □ **rankly** *adv.* **rankness** *n.* [OE *ranc* f. Gmc]

ranker /ˈræŋkə(r)/ *n.* **1** a soldier in the ranks. **2** a commissioned officer who has been in the ranks.

ranking /ˈræŋkɪŋ/ *n. & adj.* ● *n.* ordering by rank; classification. ● *adj.* US having a high rank or position.

rankle /ˈræŋk(ə)l/ *v.intr.* **1** (of envy, disappointment, etc., or their cause) cause persistent annoyance or resentment. **2** *archaic* (of a wound, sore, etc.) fester, continue to be painful. [ME (in sense 2) f. OF *rancler* f. *rancle, draoncle* festering sore f. med.L *dranculus, dracunculus* dimin. of *draco* serpent; cf. DRACO[1]]

Rann of Kutch see KUTCH, RANN OF.

ransack /ˈrænsæk/ *v.tr.* **1** pillage or plunder (a house, country, etc.). **2** thoroughly search (a place, a receptacle, a person's pockets, one's conscience, etc.). □ **ransacker** *n.* [ME f. ON *rannsaka* f. *rann* house + *-saka* f. *sœkja* seek]

ransom /ˈrænsəm/ *n. & v.* ● *n.* **1** a sum of money or other payment demanded or paid for the release of a prisoner. **2** the liberation of a prisoner in return for this. ● *v.tr.* **1** buy the freedom or restoration of; redeem. **2** hold to ransom. **3** release for a ransom. [ME f. OF *ransoun(er)* f. L *redemptio -onis* REDEMPTION]

Ransome /ˈrænsəm/, Arthur (Michell) (1884–1967), English novelist. Ransome is best known for his children's classics, such as *Swallows and Amazons* (1930) and *Great Northern?* (1947), which depict the imaginative world of children while reflecting a keen interest in sailing, fishing, and the countryside. Before writing these books, however, Ransome covered the Russian Revolution as a journalist and published a successful collection of Russian legends and fairy stories.

rant /rænt/ v. & n. ● v. **1** intr. use bombastic language. **2** tr. & intr. declaim, recite theatrically. **3** tr. & intr. preach noisily. **4** intr. (often foll. by *about*, *on*) speak vehemently or intemperately. ● n. **1** a piece of ranting, a tirade. **2** empty turgid talk. □ **rant and rave** express anger noisily and forcefully. □ **ranter** n. **rantingly** adv. [Du. *ranten* rave]

Ranter /ˈræntə(r)/ n. a member of an antinomian Christian sect in England during the mid-17th century which denied the authority of scripture and clergy. In the 19th century the word was applied to members of certain Nonconformist, in particular Methodist, groups.

ranunculaceous /rəˌnʌŋkjʊˈleɪʃəs/ adj. Bot. of or relating to the family Ranunculaceae of flowering plants, including buttercups, clematis, delphiniums, etc.

ranunculus /rəˈnʌŋkjʊləs/ n. (pl. **ranunculuses** or **ranunculi** /-ˌlaɪ/) a ranunculaceous plant of the genus *Ranunculus*, usu. having bowl-shaped flowers with many stamens and carpels, including buttercups and crowfoots. [L, orig. dimin. of *rana* frog]

Rao /raʊ/, P(amulaparti) V(enkata) Narasimha (b.1921), Indian statesman, Prime Minister since 1991.

RAOC abbr. (in the UK) Royal Army Ordnance Corps.

rap[1] /ræp/ n. & v. ● n. **1** a smart slight blow. **2** a knock, a sharp tapping sound. **3** sl. blame, censure, or punishment. **4** sl. a conversation. **5 a** (in full **rap music**) a style of black popular music with a pronounced beat and words rhythmically recited rather than sung. **b** a piece of music or a rhythmic recital in this style. ● v. (**rapped**, **rapping**) **1** tr. strike smartly. **2** intr. knock; make a sharp tapping sound (*rapped on the table*). **3** tr. criticize adversely. **4** intr. a sl. talk. **b** talk or sing in the style of rap music. □ **beat the rap** N. Amer. escape punishment. **rap on** (or **over**) **the knuckles** n. a reprimand or reproof. ● v. reprimand, reprove. **rap out 1** utter (an oath, order, pun, etc.) abruptly or on the spur of the moment. **2** express or reproduce (a rhythm, signal, etc.) by raps. **take the rap** suffer the consequences. □ **rapper** n. [ME, prob. imit.]

rap[2] /ræp/ n. a small amount, the least bit (*don't care a rap*). [Ir. *ropaire* Irish counterfeit coin]

rapacious /rəˈpeɪʃəs/ adj. grasping, extortionate, predatory. □ **rapaciously** adv. **rapaciousness** n. **rapacity** /-ˈpæsɪtɪ/ n. [L *rapax -acis* f. *rapere* snatch]

RAPC abbr. (in the UK) Royal Army Pay Corps.

rape[1] /reɪp/ n. & v. ● n. **1 a** the act of forcing a woman to have sexual intercourse against her will. **b** forcible sodomy. **2** (often foll. by *of*) violent assault, forcible interference, violation. **3** poet. carrying off (esp. of a woman) by force. **4** an instance of rape. ● v.tr. **1** commit rape on (a person, usu. a woman). **2** violate, assault, pillage. **3** poet. carry off by force. [ME f. AF *rap(er)* f. L *rapere* seize]

rape[2] /reɪp/ n. (also **oilseed rape**) a yellow-flowered cruciferous plant, *Brassica napus*, grown as food for livestock and for its seed, from which oil is made. Also called *colza*, *cole*. □ **rape-cake** rapeseed pressed into a flat shape after the extraction of oil and used as manure or food for livestock. **rape-oil** an oil made from rapeseed and used as a lubricant and in foodstuffs. [ME f. L *rapum*, *rapa* turnip]

rape[3] /reɪp/ n. hist. any of the six ancient divisions of Sussex. [OE, var. of *rāp* ROPE, with ref. to the fencing-off of land]

rape[4] /reɪp/ n. **1** the refuse of grapes after wine-making, used in making vinegar. **2** a vessel used in vinegar-making. [F *râpe*, med.L *raspa*]

rapeseed /ˈreɪpsiːd/ n. the seed of the rape plant. [RAPE[2] + SEED]

Raphael[1] /ˈræfeɪəl/ (in the Bible) one of the seven archangels in the apocryphal Book of Enoch. He is said to have 'healed' the earth when it was defiled by the sins of the fallen angels. [Heb., = God has healed]

Raphael[2] /ˈræfeɪəl/ (Italian name Raffaello Sanzio) (1483–1520), Italian painter and architect. He was a leading figure of the High Renaissance in Italy. In 1505 he went to Florence, where he painted a series of small madonnas distinguished by a serenity of expression. On moving to Rome he was commissioned to paint the frescos in one of the papal rooms in the Vatican (1509). At this time he worked on further madonnas, including his best-known altarpiece the *Sistine Madonna* (c.1513), in which Mother and Child appear among the clouds, simultaneously human and divine, in a significant departure from the naturalism characteristic of previous paintings of the Madonna. He was also an important architect and was put in charge of the work on St Peter's Basilica in Rome (1514).

raphide /ˈreɪfaɪd/ n. (pl. **raphides** /ˈreɪfɪˌdiːz/) Bot. a needle-like crystal of calcium oxalate formed within the tissues of a plant. [back-form. f. *raphides* pl. of *raphis* f. Gk *rhaphis -idos* needle]

rapid /ˈræpɪd/ adj. & n. ● adj. (**rapider**, **rapidest**) **1** quick, swift. **2** acting or completed in a short time. **3** (of a slope) descending steeply. **4** Photog. fast. ● n. (usu. in pl.) a steep descent in a river bed, with a swift current. □ **rapid eye-movement** a type of jerky movement of the eyes during periods of dreaming. **rapid-fire** (attrib.) fired, asked, etc., in quick succession. **rapid transit** (attrib.) denoting high-speed urban transport of passengers. □ **rapidly** adv. **rapidness** n. **rapidity** /rəˈpɪdɪtɪ/ n. [L *rapidus* f. *rapere* seize]

rapier /ˈreɪpɪə(r)/ n. **1** a light slender sword used for thrusting. **2** (attrib.) quick and sharp (*rapier wit*). [prob. f. Du. *rapier* or LG *rappir*, f. F *rapière*, of unkn. orig.]

rapine /ˈræpaɪn, -pɪn/ n. rhet. plundering, robbery. [ME f. OF or f. L *rapina* f. *rapere* seize]

rapist /ˈreɪpɪst/ n. a person who commits rape.

rapparee /ˌræpəˈriː/ n. hist. a 17th-century Irish irregular soldier or freebooter. [Ir. *rapaire* short pike]

rappee /ræˈpiː/ n. a coarse kind of snuff. [F (*tabac*) *râpé* rasped (tobacco)]

rappel /ræˈpel/ n. & v.intr. (**rappelled**, **rappelling**) = ABSEIL. [F, = recall, f. *rappeler* (as RE-, APPEAL)]

rapport /ræˈpɔː/ n. **1** relationship or communication, esp. when useful and harmonious (*in rapport with*; *establish a rapport*). **2** Spiritualism communication through a medium. [F f. *rapporter* (as RE-, AP-[1], *porter* f. L *portare* carry)]

rapporteur /ˌræpɔːˈtɜː(r)/ n. a person who prepares an account of the proceedings of a committee etc. for a higher body. [F (as RAPPORT)]

rapprochement /ræˈprɒʃmɒ̃/ n. the resumption of harmonious relations, esp. between states. [F f. *rapprocher* (as RE-, APPROACH)]

rapscallion /ræpˈskæljən/ n. archaic or joc. a rascal, scamp, or rogue. [earlier *rascallion*, perh. f. RASCAL]

rapt /ræpt/ adj. **1** fully absorbed or intent, enraptured (*listen with rapt attention*). **2** carried away with joyous feeling or lofty thought. **3** poet. carried away bodily. **4** Austral. colloq. overjoyed, delighted. □ **raptly** adv. **raptness** n. [ME f. L *raptus* past part. of *rapere* seize]

raptor /ˈræptə(r)/ n. a bird of prey, e.g. an eagle, falcon, owl, etc. [L, = ravisher, plunderer f. *rapere rapt-* seize]

raptorial /ræpˈtɔːrɪəl/ adj. & n. ● adj. (of a bird or animal) adapted for seizing prey; predatory. ● n. **1** = RAPTOR. **2** a predatory animal. [L *raptor*: see RAPTOR]

rapture /ˈræptʃə(r)/ n. **1 a** ecstatic delight, mental transport. **b** (in pl.) great pleasure or enthusiasm or the expression of it. **2** archaic the act of transporting a person from one place to another. □ **go into** (or **be in**) **raptures** be enthusiastic; talk enthusiastically. □ **rapturous** adj. **rapturously** adv. **rapturousness** n. [obs. F *rapture* or med.L *raptura* (as RAPT)]

rara avis /ˌreərə ˈeɪvɪs, ˌrɑːrə ˈævɪs/ n. (pl. **rarae aves** /ˌreəriː ˈeɪviːz, ˌrɑːriː ˈæviːz/) a rarity; a kind of person or thing rarely encountered. [L, = rare bird]

rare[1] /reə(r)/ adj. (**rarer**, **rarest**) **1** seldom done or found or occurring; uncommon, unusual, few and far between. **2** exceptionally good (*had a rare time*). **3** of less than the usual density, with only loosely packed substance (*the rare atmosphere of the mountain tops*). □ **rare bird** = RARA AVIS. **rare gas** = NOBLE GAS. □ **rareness** n. [ME f. L *rarus*]

rare[2] /reə(r)/ adj. (**rarer**, **rarest**) (of meat) underdone. [var. of obs. *rear* half-cooked (of eggs), f. OE *hrēr*]

rarebit /ˈreəbɪt/ n. = Welsh rabbit. [RARE[1] + BIT[1]]

rare earth n. (also **rare-earth element** or **metal**) Chem. any of a group of seventeen chemically similar metallic elements, including scandium, yttrium, and the lanthanides. Originally the term referred to their oxides, and it is now sometimes applied to the lanthanides alone. They are not in fact especially rare, but they tend to occur together in nature and are difficult to separate from one another. It took over a century for all of them to be identified after the oxide of the first, yttrium, was isolated in 1794.

raree-show /ˈreəriˌʃəʊ/ n. **1** a show or spectacle. **2** a show carried about in a box; a peep-show. [app. = *rare show* as pronounced by Savoyard showmen]

rarefy /ˈreərɪˌfaɪ/ v. (**-ies**, **-ied**) (esp. as **rarefied** adj.) **1** tr. & intr. make or become less dense or solid (*rarefied air*). **2** tr. purify or refine (a person's nature etc.). **3** tr. **a** make (an idea etc.) subtle. **b** (as **rarefied** adj.) refined, subtle; elevated; exalted; select. □ **rarefaction** /ˌreərɪˈfæk(ə)n/ n. **rarefactive** /-ˈfæktɪv/ adj. **rarefication** /-fɪˈkeɪʃ(ə)n/ n. [ME f. OF *rarefier* or med.L *rarificare* f. L *rarefacere* f. *rarus* rare + *facere* make]

rarely /ˈreəlɪ/ adv. **1** seldom; not often. **2** in an unusual degree; exceptionally. **3** exceptionally well.

raring /ˈreərɪŋ/ adj. (foll. by *to* + infin.) *colloq.* enthusiastic, eager (*raring to go*). [part. of *rare*, dial. var. of ROAR or REAR²]

rarity /ˈreərɪtɪ/ n. (pl. **-ies**) **1** rareness. **2** an uncommon thing, esp. one valued for being rare. [F *rareté* or L *raritas* (as RARE¹)]

Rarotonga /ˌrærəˈtɒŋgə/ a mountainous island in the South Pacific, the chief island of the Cook Islands. Its chief town, Avarua, is the capital of the Cook Islands.

Rarotongan /ˌrærəˈtɒŋgən/ n. & adj. ● n. **1** a native or inhabitant of Rarotonga. **2** the Polynesian language of Rarotonga. ● adj. of or relating to Rarotonga or its people or language.

Ras al Khaimah /ˌrɑːs æl ˈkaɪmə/ **1** one of the seven member states of the United Arab Emirates; pop. (1985) 116,470. It joined the United Arab Emirates in 1972, after the British withdrawal from the Persian Gulf. **2** its capital, a port on the Gulf; pop. (1980) 42,000.

rascal /ˈrɑːsk(ə)l/ n. often *joc.* a dishonest or mischievous person, esp. a child. □ **rascally** adj. **rascality** /rɑːˈskælɪtɪ/ n. (pl. **-ies**). [ME f. OF *rascaille* rabble, prob. ult. f. L *radere ras-* scrape]

rase var. of RAZE.

rash¹ /ræʃ/ adj. reckless, impetuous, hasty; acting or done without due consideration. □ **rashly** adv. **rashness** n. [ME, prob. f. OE *ræsc* (unrecorded) f. Gmc]

rash² /ræʃ/ n. **1** an eruption of the skin in spots or patches. **2** (usu. foll. by *of*) a sudden widespread phenomenon, esp. of something unwelcome (*a rash of strikes*). [18th c.: prob. rel. to OF *ra(s)che* eruptive sores, = It. *raschia* itch]

rasher /ˈræʃə(r)/ n. a thin slice of bacon or ham. [16th c.: orig. unkn.]

rasp /rɑːsp/ n. & v. ● n. **1** a coarse kind of file having separate teeth. **2** a rough grating sound. ● v. **1** tr. **a** scrape with a rasp. **b** scrape roughly. **c** (foll. by *off, away*) remove by scraping. **2 a** intr. make a grating sound. **b** tr. say gratingly or hoarsely. **3** tr. grate upon (a person or a person's feelings), irritate. □ **raspingly** adv. **raspy** adj. [ME f. OF *raspe(r)* ult. f. WG]

raspberry /ˈrɑːzbərɪ/ n. (pl. **-ies**) **1 a** a bramble, *Rubus idaeus*, having usu. red berries consisting of numerous drupels on a conical receptacle. **b** this berry. **2** any of various red colours. **3** *colloq.* **a** a sound made with the lips expressing dislike, derision, or disapproval (orig. *raspberry tart*, rhyming sl. = *fart*). **b** a show of strong disapproval (*got a raspberry from the audience*). □ **raspberry-cane** a raspberry plant. [16th-c. *rasp* (now dial.) f. obs. *raspis*, of unkn. orig., + BERRY]

rasper /ˈrɑːspə(r)/ n. **1** a person or thing that rasps. **2** *Hunting* a high difficult fence.

Rasputin /ræˈspjuːtɪn/, Grigori (Efimovich) (1871–1916), Russian monk. He came to exert great influence over Tsar Nicholas II and his family during the First World War by claiming miraculous powers to heal the heir to the throne, who suffered from haemophilia. His appropriation of ecclesiastical, political, and military powers, combined with a reputation for debauchery, steadily discredited the imperial family. Rasputin was eventually assassinated by a group loyal to the tsar.

Rasta /ˈræstə/ n. & adj. = RASTAFARIAN. [abbr.]

Rastafari /ˌræstəˈfɑːrɪ/ adj. & n. ● adj. of or relating to Rastafarians or Rastafarianism (see RASTAFARIAN). ● n. (pl. same or **Rastafaris**) a Rastafarian. [*Ras Tafari*; see RASTAFARIAN]

Rastafarian /ˌræstəˈfeərɪən/ n. & adj. ● n. a member of the Rastafari sect, a religious movement of Jamaican origin, largely inspired by the thinking of Marcus Garvey in the 1930s. (*See note below.*) ● adj. of or relating to the Rastafari sect. □ **Rastafarianism** n. [*Ras Tafari*, the name by which Haile Selassie was known 1916–30]

▪ Rastafarianism combines elements of Christianity, African religion, and Afro-Caribbean culture. Its adherents believe that blacks are the chosen people, that Emperor Haile Selassie of Ethiopia was the Messiah, and that black people will eventually return to their African homeland. Rastafarians have distinctive codes of behaviour and dress, including the wearing of dreadlocks and the smoking of cannabis; they also reject Western medicine and follow a diet that excludes pork, shellfish, and milk.

raster /ˈræstə(r)/ n. a pattern of scanning lines for a cathode-ray tube picture. [G, = screen, f. L *rastrum* rake f. *radere ras-* scrape]

rasterize /ˈræstəˌraɪz/ v.tr. (also **-ise**) *Computing* convert (a digitized image) into a form that can be displayed on a cathode-ray tube or printed out. □ **rasterizer** n. **rasterization** /ˌræstəraɪˈzeɪʃ(ə)n/ n.

Rastyapino /ræˈstjɑːpɪˌnəʊ/ a former name (1919–29) for DZERZHINSK.

rat /ræt/ n. & v. ● n. **1 a** a medium-sized rodent of the mouse family, esp. of the genus *Rattus*. (See note below.) **b** any similar rodent (*muskrat; water-rat*). **2** a deserter from a party, cause, difficult situation, etc.; a turncoat (from the superstition that rats desert a sinking ship). **3** *colloq.* an unpleasant person. **4** a worker who refuses to join a strike, or who blacklegs. **5** (in pl.) *sl.* an exclamation of contempt, annoyance, etc. ● v.intr. (**ratted, ratting**) **1** (of a person or dog) hunt or kill rats. **2** *colloq.* desert a cause, party, etc. **3** (foll. by *on*) *colloq.* **a** betray; let down. **b** inform on. **4** (as **ratted** adj.) *sl.* drunk. □ **rat-arsed** *sl.* drunk. **rat-catcher** a person who rids buildings of rats etc. **rat kangaroo** a small ratlike Australian marsupial of the family Potoroidae, having kangaroo-like hind limbs for jumping. **rat race** a fiercely competitive struggle for position, power, etc. **rat's tail** a thing shaped like a rat's tail, e.g. a tapering cylindrical file. **rat-tail 1** the grenadier fish. **2** a horse with a hairless tail. **3** such a tail. **rat-tail** (or **-tailed**) **spoon** a spoon with a tail-like moulding from the handle to the back of the bowl. [OE *ræt* & OF *rat*]

▪ There are several hundred species of rat, generally only distinguished from mice in being somewhat larger. Some Old World rats of the genus *Rattus* have become cosmopolitan, especially the brown rat *R. norvegicus*, also known as the common, Norwegian, or sewer rat, and the black rat *R. rattus*, also known as the ship, house, or roof rat. Both of these can constitute a health hazard as a result of the diseases that they or their parasites transmit, notably bubonic plague.

rata /ˈrɑːtə/ n. a large New Zealand tree or woody climber of the genus *Metrosideros*; esp. *M. robusta*, with crimson flowers and hard red wood. [Maori]

ratable var. of RATEABLE.

ratafia /ˌrætəˈfiːə/ n. **1** a liqueur flavoured with almonds or kernels of peach, apricot, or cherry. **2** a kind of biscuit similarly flavoured. [F, perh. rel. to TAFIA]

ratan var. of RATTAN.

Ratana /ˈrɑːtənə/, Tahupotiki Wiremu (1873–1939), Maori political and religious leader. A Methodist farmer, he founded the Ratana Church (1920), an interdenominational movement whose aim was to unite Maoris of all tribes. Its doctrine of faith-healing and many unorthodox rituals led to a rift with other Christian denominations in 1925. Politically Ratana struggled for Maori rights by pressing for full implementation of the Treaty of Waitangi.

rataplan /ˌrætəˈplæn/ n. & v. ● n. a drumming sound. ● v. (**rataplanned, rataplanning**) **1** tr. play (a tune) on or as on a drum. **2** intr. make a rataplan. [F: imit.]

ratatat (also **rat-a-tat**) var. of RAT-TAT.

ratatouille /ˌrætəˈtuːi, -ˈtwiː/ n. a dish made of fried and stewed onions, courgettes, tomatoes, aubergines, and peppers. [F dial.]

ratbag /ˈrætbæg/ n. *sl.* an unpleasant or disgusting person.

ratch /rætʃ/ n. **1** a ratchet. **2** a ratchet-wheel. [perh. f. G *Ratsche*: cf. RATCHET]

ratchet /ˈrætʃɪt/ n. & v. ● n. **1** a set of teeth on the edge of a bar or wheel in which a device engages to ensure motion in one direction only. **2** (in full **ratchet-wheel**) a wheel with a rim so toothed. ● v. (**ratcheted, ratcheting**) **1** tr. provide with a ratchet. **b** make into a ratchet. **2** tr. & intr. move as under the control of a ratchet. [F *rochet* blunt lance-head, bobbin, ratchet, etc., prob. ult. f. Gmc]

rate¹ /reɪt/ n. & v. ● n. **1** a stated numerical proportion between two sets of things (the second usu. expressed as unity), esp. as a measure of amount or degree (*moving at a rate of 50 miles per hour*) or as the basis of calculating an amount or value (*rate of taxation*). **2** a fixed or appropriate charge or cost or value; a measure of this (*postal rates; the rate for the job*). **3** rapidity of movement or change (*travelling at a great rate; prices increasing at a dreadful rate*). **4** class or rank (*first-rate*). **5** *Brit.* **a** a tax levied by local authorities on commercial properties (and *hist.* dwellings) at so much per pound of the assessed value of buildings and land owned or leased. ¶ Now replaced by the *council tax* for dwellings. **b** (in pl.) the amount payable by this. ● v. **1** tr. **a** estimate the worth or value of (*I do not rate him very highly; how do you rate your chances of winning the race?*). **b** assign a fixed value to (a coin or metal) in relation to a monetary standard. **c** assign a value to (work, the power of a machine, etc.). **2** *tr.* consider; regard as (*I rate them among my benefactors*). **3** *intr.* (foll. by *as*) rank or be rated. **4** *tr. Brit. hist.* **a** subject to the payment of a local rate. **b** value for the purpose of assessing rates. **5** *tr.* be worthy of, deserve. **6** *tr. Naut.* place in a specified class (cf. RATING[1]). □ **at any rate** in any case, whatever happens. **at this** (or **that**) **rate** if this example is typical or this assumption is true. **rate-capping** *Brit. hist.* the imposition of an upper limit on the rate leviable by a local authority. [ME f. OF f. med.L *rata* f. L *pro rata parte* or *portione* according to the proportional share f. *ratus* past part. of *reri* reckon]

rate[2] /reɪt/ *v.tr.* scold angrily. [ME: orig. unkn.]

rate[3] var. of RET.

rateable /ˈreɪtəb(ə)l/ *adj.* (also **ratable**) **1** *Brit. esp. hist.* liable to payment of local rates. **2** able to be rated or estimated. □ **rateable value** *Brit.* the value at which a house etc. is assessed for payment of rates. □ **rateably** *adv.* **rateability** /ˌreɪtəˈbɪlɪtɪ/ *n.*

ratel /ˈreɪt(ə)l, ˈrɑːt-/ *n.* a nocturnal flesh-eating burrowing mammal, *Mellivora capensis*, native to Africa and southern Asia, with powerful claws. Also called *honey-badger*. [Afrik., of unkn. orig.]

ratepayer /ˈreɪtˌpeɪə(r)/ *n. Brit. esp. hist.* a person liable to pay local rates.

ratfink /ˈrætfɪŋk/ *n. esp. US sl.* = FINK.

rathe /reɪð/ *adj. poet.* coming, blooming, etc., early in the year or day. □ **rathe-ripe 1** ripening early. **2** precocious. [OE *hræth*, *hræd* f. Gmc]

rather /ˈrɑːðə(r)/ *adv.* **1** (often foll. by *than*) by preference; for choice (*would rather not go*; *would rather stay than go*). **2** (usu. foll. by *than*) more truly; as a more likely alternative (*is stupid rather than honest*). **3** more precisely (*a book, or rather, a pamphlet*). **4** slightly; to some extent; somewhat (*became rather drunk*; *I rather think you know him*). **5** /rɑːˈðɜː(r)/ *Brit.* (as an emphatic response) indeed, assuredly (*Did you like it? — Rather!*). □ **had rather** would rather. [ME f. OE *hrathor*, compar. of *hræthe* (adv.) f. *hræth* (adj.): see RATHE]

Rathlin Island /ˈræθlɪn/ an island in the North Channel, off the north coast of Ireland.

rathskeller /ˈrɑːtsˌkelə(r)/ *n. US* a beer-hall or restaurant in a basement. [G, = (restaurant in) town-hall cellar]

ratify /ˈrætɪˌfaɪ/ *v.tr.* (**-ies, -ied**) confirm or accept (an agreement made in one's name) by formal consent, signature, etc. □ **ratifier** *n.* **ratifiable** *adj.* **ratification** /ˌrætɪfɪˈkeɪʃ(ə)n/ *n.* [ME f. OF *ratifier* f. med.L *ratificare* (as RATE[1])]

rating[1] /ˈreɪtɪŋ/ *n.* **1** the act or an instance of placing in a rank or class or assigning a value to. **2** the estimated standing of a person as regards credit etc. **3** *Naut.* **a** *Brit.* a non-commissioned sailor. **b** a person's position or class on a ship's books. **4** *Brit. hist.* an amount fixed as a local rate. **5** (usu. in *pl.*) the relative popularity of a broadcast programme as determined by the estimated size of the audience. **6** *Naut.* any of the classes into which racing yachts are distributed by tonnage.

rating[2] /ˈreɪtɪŋ/ *n.* an angry reprimand.

ratio /ˈreɪʃɪəʊ/ *n.* (*pl.* **-os**) the quantitative relation between two similar magnitudes determined by the number of times one contains the other integrally or fractionally (*in the ratio of three to two*; *the ratios 1:5 and 20:100 are the same*). [L (as RATE[1])]

ratiocinate /ˌrætɪˈɒsɪˌneɪt, ˌræʃɪ-/ *v.intr. literary* go through logical processes, reason, esp. using syllogisms. □ **ratiocinator** *n.* **ratiocinative** /-nətɪv/ *adj.* **ratiocination** /-ˌɒsɪˈneɪʃ(ə)n/ *n.* [L *ratiocinari* (as RATIO)]

ration /ˈræʃ(ə)n/ *n. & v.* ● *n.* **1** a fixed official allowance of food, clothing, etc., in a time of shortage. **2** (foll. by *of*) a single portion of provisions, fuel, clothing, etc. **3** (usu. in *pl.*) a fixed daily allowance of food, esp. in the armed forces (and formerly of forage for animals). **4** (in *pl.*) provisions. ● *v.tr.* **1** limit (persons or provisions) to a fixed ration. **2** (usu. foll. by *out*) share out (food etc.) in fixed quantities. □ **given out with the rations** *Mil. sl.* awarded without regard to merit. **ration book** (or **card**) a document entitling the holder to a ration. [F f. It. *razione* or Sp. *ración* f. L *ratio -onis* reckoning, RATIO]

rational /ˈræʃ(ə)n(ə)l/ *adj.* **1** of or based on reasoning or reason. **2** sensible, sane, moderate; not foolish or absurd or extreme. **3** endowed with reason, reasoning. **4** rejecting what is unreasonable or cannot be tested by reason in religion or custom. **5** *Math.* (of a quantity or ratio) expressible as a ratio of whole numbers. □ **rational dress** *hist.* a style of dress adopted by some women in the late 19th century, including bloomers or knickerbockers. **rational horizon** see HORIZON 1c. □ **rationally** *adv.* **rationality** /ˌræʃəˈnælɪtɪ/ *n.* [ME f. L *rationalis* (as RATION)]

rationale /ˌræʃəˈnɑːl/ *n.* **1** (often foll. by *for*) the fundamental reason or logical basis of anything. **2** a reasoned exposition; a statement of reasons. [mod.L, neut. of L *rationalis*: see RATIONAL]

rationalism /ˈræʃ(ə)nəˌlɪz(ə)m/ *n.* **1** *Philos.* the theory that reason rather than sense-experience is the foundation of certainty in knowledge. Expounded notably by Descartes, Spinoza, and Leibniz, the theory emphasizes a priori rather than empirical concepts (opp. EMPIRICISM). **2** *Theol.* the practice of treating reason as the ultimate authority in religion. **3** *Philos.* a belief in reason rather than religion as the guiding principle in life. **4** the practice of using reason or reasoning as a basis for action or thought. □ **rationalist** *n.* **rationalistic** /ˌræʃ(ə)nəˈlɪstɪk/ *adj.* **rationalistically** *adv.*

rationalize /ˈræʃ(ə)nəˌlaɪz/ *v.* (also **-ise**) **1 a** *tr.* offer or subconsciously adopt a plausible but specious explanation of (one's behaviour or attitude). **b** *intr.* explain one's behaviour or attitude in this way. **2** *tr.* make logical and consistent. **3** *tr.* make (a business etc.) more efficient by reorganizing it to reduce or eliminate waste of labour, time, or materials. **4** *tr.* (often foll. by *away*) explain or explain away rationally. **5** *tr. Math.* clear of surds. **6** *intr.* be or act as a rationalist. □ **rationalizer** *n.* **rationalization** /ˌræʃ(ə)nəlaɪˈzeɪʃ(ə)n/ *n.*

ratite /ˈrætaɪt/ *adj. & n.* ● *adj.* (of a bird) having a keelless breastbone, and unable to fly (opp. CARINATE). ● *n.* a flightless bird, e.g. an ostrich, emu, cassowary, etc. [L *ratis* raft]

ratline /ˈrætlɪn/ *n.* (also **ratlin**) (usu. in *pl.*) any of the small lines fastened across a sailing-ship's shrouds like ladder-rungs. [ME: orig. unkn.]

ratoon /rəˈtuːn/ *n. & v.* ● *n.* a new shoot springing from a root of sugar cane etc. after cropping. ● *v.intr.* send up ratoons. [Sp. *retoño* sprout]

ratsbane /ˈrætsbeɪn/ *n. literary* rat-poison.

rattan /rəˈtæn/ *n.* (also **ratan**) **1** a Malaysian climbing palm, esp. of the genus *Calamus*, with long thin jointed pliable stems. **2** a piece of rattan stem used as a walking-stick etc. [earlier *rot(t)ang* f. Malay *rōtan* prob. f. *raut* pare]

rat-tat /ˈrætˈtæt/ *n.* (also **rat-tat-tat** /ˌrættætˈtæt/, **ratatat**, **rat-a-tat** /ˌrætəˈtæt/) a rapping sound, esp. of a knocker. [imit.]

ratter /ˈrætə(r)/ *n.* **1** a dog or other animal that hunts rats. **2** *sl.* a person who betrays a cause, party, friend, etc.

Rattigan /ˈrætɪɡən/, Sir Terence (Mervyn) (1911–77), English dramatist. His plays include *The Winslow Boy* (1946), concerning a father's fight to clear the name of his accused son, *The Browning Version* (1948), about a repressed and unpopular schoolmaster, and *Ross* (1960), based on the life of T. E. Lawrence. He also wrote screenplays for several films, including *The Yellow Rolls Royce* (1965).

Rattle /ˈræt(ə)l/, Sir Simon (Denis) (b.1955), English conductor. He made his reputation as principal conductor with the City of Birmingham Symphony Orchestra, a post which he held from 1980 until 1991, when he became the orchestra's music director. He is noted particularly for his interpretation of works by early 20th-century composers such as Mahler and as a champion of new music.

rattle /ˈræt(ə)l/ *v. & n.* ● *v.* **1 a** *intr.* give out a rapid succession of short sharp hard sounds, usu. through being shaken or knocking against something. **b** *tr.* make (a chair, window, crockery, etc.) do this. **c** *intr.* cause such sounds by shaking something (*rattled at the door*). **2 a** *intr.* move with a rattling noise. **b** *intr.* drive a vehicle or ride or run briskly. **c** *tr.* cause to move quickly (*the bill was rattled through Parliament*). **3 a** *tr.* (usu. foll. by *off*) say or recite rapidly. **b** *intr.* (usu. foll. by *on*) talk in a lively thoughtless way. **4** *tr. colloq.* disconcert, alarm, fluster, make nervous, frighten. ● *n.* **1** a rattling sound. **2** an instrument or plaything made to rattle, esp. in order to amuse babies or to give an alarm. **3** the set of horny rings in a rattlesnake's tail. **4** a plant with seeds that rattle in their cases when ripe (*red rattle*; *yellow rattle*). **5** uproar, bustle, noisy gaiety, racket. **6 a** a noisy flow of words. **b** empty chatter, trivial talk. **7** *archaic* a lively or thoughtless incessant talker. □ **rattle the sabre** threaten war. □ **rattly** *adj.* [ME, prob. f. MDu. & LG *ratelen* (imit.)]

rattlebox /ˈræt(ə)lˌbɒks/ *n.* **1** a rattle consisting of a box with objects inside. **2** a rickety old vehicle etc.

rattler /ˈrætlə(r)/ *n.* **1** a thing that rattles, esp. an old or rickety vehicle. **2** *colloq.* a rattlesnake. **3** *sl.* a remarkably good specimen of anything.

rattlesnake /ˈræt(ə)lˌsneɪk/ *n.* a venomous American pit viper of the genus *Crotalus* or *Sistrurus*, with loose horny rings in the tail which rattle when shaken.

rattletrap /ˈræt(ə)lˌtræp/ *n. & adj. colloq.* ● *n.* a rickety old vehicle etc. ● *adj.* rickety.

rattling /'rætlɪŋ/ adj. & adv. ● adj. **1** that rattles. **2** brisk, vigorous (a rattling pace). ● adv. remarkably (a rattling good story).

ratty /'ræti/ adj. **1** relating to or infested with rats. **2** colloq. irritable or angry. **3** colloq. **a** wretched, nasty. **b** unkempt; seedy, dirty. □ **rattily** adv. **rattiness** n.

raucous /'rɔːkəs/ adj. harsh-sounding, loud and hoarse. □ **raucously** adv. **raucousness** n. [L raucus]

raunchy /'rɔːntʃi/ adj. (**raunchier, raunchiest**) colloq. **1** coarse, earthy, boisterous; sexually explicit or provocative. **2** esp. US slovenly, grubby. □ **raunchily** adv. **raunchiness** n. [20th c.: orig. unkn.]

ravage /'rævɪdʒ/ v. & n. ● v.tr. & intr. devastate, plunder. ● n. **1** the act or an instance of ravaging; devastation, damage. **2** (usu. in pl.; foll. by of) destructive effect (survived the ravages of winter). □ **ravager** n. [F ravage(r) alt. f. ravine rush of water]

rave[1] /reɪv/ v. & n. ● v. **1** intr. talk wildly or furiously in or as in delirium. **2** intr. (usu. foll. by about, of, over) speak with rapturous admiration; go into raptures. **3** tr. bring into a specified state by raving (raved himself hoarse). **4** tr. utter with ravings (raved their grief). **5** intr. (of the sea, wind, etc.). howl, roar. **6** tr. & intr. colloq. enjoy oneself freely (esp. rave it up). ● n. **1** (usu. attrib.) colloq. a highly enthusiastic review of a film, play, etc. (a rave review). **2** sl. **a** an infatuation. **b** a temporary fashion or craze. **3 a** (also **rave-up**) colloq. a lively party. **b** a large often illicit party or event organized for dancing to fast electronic music. **c** music associated with such events. **4** the sound of the wind etc. raving. [ME, prob. f. ONF raver, rel. to (M)LG reven be senseless, rave]

rave[2] /reɪv/ n. **1** a rail of a cart. **2** (in pl.) a permanent or removable framework added to the sides of a cart to increase its capacity. [var. of dial. rathe (15th c., of unkn. orig.)]

Ravel /ræ'vel/, Maurice (Joseph) (1875–1937), French composer. His early music was influenced by impressionism and the piano music of Liszt, but his mature works have a distinctive tone colour as well as an ironic flavour derived from the use of unresolved dissonances. Major works include the ballet Daphnis and Chloë (1912), staged by Diaghilev's Ballets Russes, the opera L'Enfant et les sortilèges (1925), and the orchestral work Boléro (1928).

ravel /'ræv(ə)l/ v. & n. ● v. (**ravelled, ravelling**; US **raveled, raveling**) **1** tr. & intr. entangle or become entangled or knotted. **2** tr. confuse or complicate (a question or problem). **3** intr. fray out. **4** tr. (often foll. by out) disentangle, unravel, distinguish the separate threads or subdivisions of. ● n. **1** a tangle or knot. **2** a complication. **3** a frayed or loose end. [prob. f. Du. ravelen tangle, fray out, unweave]

ravelin /'rævlɪn/ n. hist. an outwork of fortifications, with two faces forming a salient angle. [F f. obs. It. ravellino, of unkn. orig.]

ravelling /'rævəlɪŋ/ n. a thread from fabric which is frayed or unravelled.

raven[1] /'reɪv(ə)n/ n. & adj. ● n. a large crow of the genus Corvus, feeding chiefly on carrion etc. and having a hoarse cry; esp. C. corax, with glossy all-black plumage. ● adj. glossy black (raven tresses). [OE hræfn f. Gmc]

raven[2] /'ræv(ə)n/ v. **1** intr. **a** plunder. **b** (foll. by after) seek prey or booty. **c** (foll. by about) go plundering. **d** prowl for prey (ravening beast). **2 a** tr. devour voraciously. **b** intr. (esp. as **ravening** adj.) (sometimes foll. by for) have a ravenous appetite. **c** intr. (often foll. by on) feed voraciously. [OF raviner ravage ult. f. L rapina RAPINE]

Ravenna /rə'venə/ a city near the Adriatic coast in NE central Italy; pop. (1991) 136,720. An important centre in Roman times, Ravenna became the capital of the Western Roman Empire in 402 and then of the Ostrogothic kingdom of Italy, afterwards serving as capital of Byzantine Italy. It became an independent republic in the 13th century and then a papal possession in 1509, remaining in papal hands until 1859. In 1861 it became part of the kingdom of Italy. It is noted for its ancient mosaics dating from the early Christian period.

ravenous /'rævənəs/ adj. **1** very hungry, famished. **2** (of hunger, eagerness, etc., or of an animal) voracious. **3** rapacious. □ **ravenously** adv. **ravenousness** n. [ME f. OF ravineus (as RAVEN[2])]

raver /'reɪvə(r)/ n. **1** colloq. an uninhibited pleasure-loving person. **2** a person who attends a rave. **3** a person who raves; a madman or madwoman.

Ravi /'rɑːvɪ/ a river in the north of the Indian subcontinent, one of the headwaters of the Indus, which rises in the Himalayas in Himachel Pradesh, NW India, and flows for 725 km (450 miles) generally south-westwards into Pakistan, where it empties into the Chenab river just north of Multan. It is one of the five rivers that gave Punjab its name.

ravin /'rævɪn/ n. poet. or rhet. **1** robbery, plundering. **2** the seizing and devouring of prey. **3** prey. □ **beast of ravin** a beast of prey. [ME f. OF ravine f. L rapina RAPINE]

ravine /rə'viːn/ n. a deep narrow gorge or cleft. □ **ravined** adj. [F (as RAVIN)]

raving /'reɪvɪŋ/ n., adj., & adv. ● n. (usu. in pl.) wild or delirious talk. ● adj. delirious, frenzied. ● adj. & adv. colloq. as an intensive (a raving beauty; raving mad).

ravioli /ˌrævɪ'əʊlɪ/ n.pl. small pasta envelopes containing minced meat etc. [It.]

ravish /'rævɪʃ/ v.tr. **1** rape. **2** enrapture; fill with delight. **3** archaic **a** carry off (a person or thing) by force. **b** (of death, circumstances, etc.) take from life or from sight. □ **ravisher** n. **ravishment** n. [ME f. OF ravir ult. f. L rapere seize]

ravishing /'rævɪʃɪŋ/ adj. entrancing, delightful; very beautiful. □ **ravishingly** adv.

raw /rɔː/ adj. & n. ● adj. **1** (of food) uncooked. **2** in the natural state; not processed or manufactured (raw sewage). **3** (of alcoholic spirit) undiluted. **4** (of statistics etc.) not analysed or processed. **5** (of a person) inexperienced, untrained; new to an activity (raw recruits). **6 a** stripped of skin; having the flesh exposed. **b** sensitive to the touch from having the flesh exposed. **c** sensitive to emotional pain etc. **7** (of the atmosphere, day, etc.) chilly and damp. **8 a** crude in artistic quality; lacking finish. **b** unmitigated; brutal. **9** (of the edge of cloth) without hem or selvedge. **10** (of silk) as reeled from cocoons. **11** (of grain) unmalted. ● n. a raw place on a person's or horse's body. □ **come the raw prawn** Austral. sl. attempt to deceive. **in the raw 1** in its natural state without mitigation (life in the raw). **2** naked. **raw-boned** gaunt and bony. **raw deal** harsh or unfair treatment. **raw material** that from which the process of manufacture makes products. **raw sienna** a brownish-yellow ferruginous earth used as a pigment. **raw umber** umber in its natural state, dark yellow in colour. **touch on the raw** upset (a person) on a sensitive matter. □ **rawish** adj. **rawly** adv. **rawness** n. [OE hrēaw f. Gmc]

Rawalpindi /rɔː'pɪndɪ, ˌrɑː'wəl-/ a city in Punjab province, northern Pakistan, in the foothills of the Himalayas; pop. (est. 1991) 955,000. A former military station, constructed on the site of an ancient village, it was the interim capital of Pakistan, 1959–67, during the construction of Islamabad.

rawhide /'rɔːhaɪd/ n. **1** untanned hide. **2** a rope or whip of this.

Rawlplug /'rɔːlplʌg/ n. propr. a thin cylinder of fibre or plastic for holding a screw or nail in masonry. [Rawlings, name of the engineers who introduced it]

Rawls /rɔːlz/, John (b.1921), American philosopher. He is the author of A Theory of Justice (1972), which invokes the philosophical concept of social contract and attacks the utilitarian doctrine of subjugating individual needs to the more pressing claims of the general good, arguing for principles to be formulated that guarantee basic liberties.

Ray[1] /reɪ/, John (1627–1705), English naturalist. His principal interest was botany, and his major work was the three-volume Historia Plantarum (1686–1704). He toured Europe with F. Willughby (1635–72) in search of specimens of flora and fauna. Ray was the first to classify flowering plants into monocotyledons and dicotyledons, he established the species as the basic taxonomic unit, and his systematic scheme was not improved upon until that of Linnaeus. The Ray Society of London is named in his honour.

Ray[2] /reɪ/, Man (born Emmanuel Rudnitsky) (1890–1976), American photographer, painter, and film-maker. He helped to found the New York Dada movement, before moving to Paris in 1921 and becoming a leading figure in the European Dada and surrealist movements. Ray pioneered the photogram or 'rayograph', placing objects on sensitized paper and exposing them to light; he later applied the technique to film-making. He is perhaps best known for his photograph the Violin d'Ingres (1924), which achieved the effect of making the back of a female nude resemble a violin.

Ray[3] /raɪ/, Satyajit (1921–92), Indian film director. His first film, Pather Panchali (1955), won a prize at Cannes and brought Indian films to the attention of Western audiences. Filmed in neo-realist style and set in his native Bengal, it formed part of a trilogy completed by Aparijito (1956) and Apur Sansar (1959). His other films include Kanchenjunga (1962), for which he also wrote the music, and The Home and the World (1984).

ray[1] /reɪ/ n. & v. ● n. **1** a single line or narrow beam of light from a small or distant source. **2** a straight line in which radiation travels to a given point. **3** (in pl.) radiation of a specified type (gamma rays; X-rays). **4** a trace or beginning of an enlightening or cheering influence (*a ray of hope*). **5 a** any of a set of radiating lines or parts or things. **b** any of a set of straight lines passing through one point. **6** the marginal portion of a composite flower, e.g. a daisy. **7 a** a radial division of a starfish. **b** each of a set of bones etc. supporting a fish's fin. ● *v.* **1** *intr.* (foll. by *forth, out*) (of light, thought, emotion, etc.) issue in or as if in rays. **2** *intr.* & *tr.* radiate. □ **ray gun** (esp. in science fiction) a gun causing injury or damage by the emission of rays. □ **rayed** *adj.* **rayless** *adj.* **raylet** *n.* [ME f. OF *rai* f. L *radius*: see RADIUS]

ray² /reɪ/ *n.* a broad flat cartilaginous fish of the order Batiformes, with winglike pectoral fins and a long slender tail, e.g. a skate. [ME f. OF *raie* f. L *raia*]

ray³ /reɪ/ *n.* (also **re**) *Mus.* **1** (in tonic sol-fa) the second note of a major scale. **2** the note D in the fixed-doh system. [ME *re* f. L *resonare*: see GAMUT]

Rayleigh /ˈreɪlɪ/, John William Strutt, 3rd Baron (1842–1919), English physicist. He published a major work on acoustics, *The Theory of Sound*, and carried out pioneering work on atmospheric airglow and black-body radiation. He was director of the Cavendish Laboratory after Maxwell, his researches including the establishment of electrical units of resistance, current, and electromotive force. He worked with William Ramsay from 1894, and their accurate measurement of the constituents of the atmosphere led to the discovery of argon and other inert gases. In 1904 Rayleigh was awarded the Nobel Prize for physics.

Rayleigh scattering *n. Physics* the scattering of light etc. by particles in a medium, without change in wavelength (cf. RAMAN EFFECT). It accounts, for example, for the blue colour of the sky, since blue light is scattered slightly more efficiently than red. [RAYLEIGH]

rayon /ˈreɪɒn/ *n.* any of various textile fibres or fabrics made from viscose. [arbitrarily f. RAY¹]

raze /reɪz/ *v.tr.* (also **rase**) **1** completely destroy; tear down (esp. *raze to the ground*). **2** erase; scratch out (esp. in abstract senses). [ME *rase* = wound slightly f. OF *raser* shave close ult. f. L *radere ras-* scrape]

razoo /rɑːˈzuː/ *n.* (also **brass razoo**) *Austral.* & *NZ sl.* an imaginary coin of trivial value; a very small sum of money (*I hadn't a brass razoo*). [20th c.: orig. unkn.]

razor /ˈreɪzə(r)/ *n.* & *v.* ● *n.* an instrument with a sharp blade used in cutting hair, esp. from the skin. ● *v.tr.* **1** use a razor on. **2** shave; cut down close. □ **razor-back** an animal with a sharp ridged back, esp. a rorqual or a semi-wild hog of the southern US. **razor-blade** a blade used in a razor, esp. a flat piece of metal with a sharp edge or edges used in a safety razor. **razor-cut** a haircut made with a razor. **razor-** (or **razor's**) **edge 1** a keen edge. **2** a sharp mountain-ridge. **3** a critical situation (*found themselves on a razor-edge*). **4** a sharp line of division. **razor-shell** (or **-fish**) a bivalve mollusc of the superfamily Solenacea, with a shell like the handle of a cutthroat razor. [ME f. OF *rasor* (as RAZE)]

razorbill /ˈreɪzəˌbɪl/ *n.* a black and white auk, *Alca torda*, of the North Atlantic, with a deep bill that is said to resemble a cutthroat razor.

razz /ræz/ *n.* & *v.* *US sl.* ● *n.* = RASPBERRY 3. ● *v.tr.* tease, ridicule. [*razzberry*, corrupt. of RASPBERRY]

razzle-dazzle /ˈræz(ə)lˌdæz(ə)l/ *n.* (also **razzle**) *sl.* **1 a** glamorous excitement; bustle. **b** a spree. **2** extravagant publicity. [redupl. of DAZZLE]

razzmatazz /ˌræzməˈtæz/ *n.* (also **razzamatazz** /ˌræzəmə-/) *colloq.* **1** = RAZZLE-DAZZLE. **2** insincere actions; humbug. [prob. alt. f. RAZZLE-DAZZLE]

Rb *symb. Chem.* the element rubidium.

RC *abbr.* **1** Roman Catholic. **2** Red Cross. **3** reinforced concrete.

RCA *abbr.* **1** (in the UK) Royal College of Art. **2** (in the US) Radio Corporation of America.

RCAF *abbr. hist.* Royal Canadian Air Force.

RCM *abbr.* (in the UK) Royal College of Music.

RCMP *abbr.* Royal Canadian Mounted Police.

RCN *abbr.* **1** (in the UK) Royal College of Nursing. **2** *hist.* Royal Canadian Navy.

RCP *abbr.* (in the UK) Royal College of Physicians.

RCS *abbr.* (in the UK): **1** Royal College of Science. **2** Royal College of Surgeons. **3** Royal Corps of Signals.

RCVS *abbr.* (in the UK) Royal College of Veterinary Surgeons.

RD *abbr.* **1** refer to drawer. **2** (in the UK) Royal Naval Reserve Decoration.

Rd. *abbr.* Road (in names).

RDC *abbr. Brit. hist.* Rural District Council.

RDF *abbr.* radio direction-finder.

RDX *abbr.* a powerful high explosive. [Research Department explosive]

RE *abbr.* **1** (in the UK) Royal Engineers. **2** religious education.

Re¹ var. of RA¹.

Re² *symb. Chem.* the element rhenium.

re¹ /reɪ, riː/ *prep.* **1** in the matter of (as the first word in a heading, esp. of a legal document). **2** about, concerning. [L, ablat. of *res* thing]

re² var. of RAY³.

re- /riː, rɪ, re/ *prefix* **1** attachable to almost any verb or its derivative, meaning: **a** once more; afresh, anew (*readjust*; *renumber*). **b** back; with return to a previous state (*reassemble*; *reverse*). ¶ A hyphen is normally used when the word begins with *e* (*re-enact*), or to distinguish the compound from a more familiar one-word form (*re-form* = form again). **2** (also **red-** before a vowel, as in *redolent*) in verbs and verbal derivatives denoting: **a** in return; mutually (*react*; *resemble*). **b** opposition (*repel*; *resist*). **c** behind or after (*relic*; *remain*). **d** retirement or secrecy (*recluse*; *reticence*). **e** off, away, down (*recede*; *relegate*; *repress*). **f** frequentative or intensive force (*redouble*; *refine*; *resplendent*). **g** negative force (*recant*; *reveal*). [L *re-*, *red-*, again, back, etc.]

're /ə(r)/ *abbr.* (usu. after pronouns) are (*they're*, *we're*).

reabsorb /ˌriːəbˈsɔːb, -ˈzɔːb/ *v.tr.* absorb again. □ **reabsorption** /-ˈsɔːpʃ(ə)n, -ˈzɔːpʃ(ə)n/ *n.*

reaccept /ˌriːəkˈsept/ *v.tr.* accept again. □ **reacceptance** *n.*

reaccustom /ˌriːəˈkʌstəm/ *v.tr.* accustom again.

reach /riːtʃ/ *v.* & *n.* ● *v.* **1** *intr.* & *tr.* (often foll. by *out*) stretch out; extend. **2** *intr.* stretch out a limb, the hand, etc.; make a reaching motion or effort. **3** *intr.* (often foll. by *for*) make a motion or effort to touch or get hold of, or to attain (*reached for his pipe*). **4** *tr.* get as far as; arrive at (*reached Lincoln at lunchtime*; *your letter reached me today*). **5** *tr.* get to or attain (a specified point) on a scale (*the temperature reached 90°*; *the number of applications reached 100*). **6** *intr.* (foll. by *to*) be adequate for (*my income will not reach to it*). **7** *tr.* succeed in achieving; attain (*have reached agreement*). **8** *tr.* make contact with the hand etc., or by telephone etc. (*was out all day and could not be reached*). **9** *tr.* (of a broadcast, broadcasting station, etc.) be received by. **10** *tr.* succeed in influencing or having the required effect on (*could not manage to reach their audience*). **11** *tr.* hand, pass (*reach me that book*). **12** *tr.* take with an outstretched hand. **13** *intr.* *Naut.* sail with the wind abeam or abaft the beam. ● *n.* **1** the extent to which a hand etc. can be reached out, influence exerted, motion carried out, or mental powers used. **2** an act of reaching out. **3** a continuous extent, esp. a stretch of river between two bends, or the part of a canal between locks. **4** *Naut.* a distance traversed in reaching. **5** the number of people who watch a specified television channel or listen to a specified radio station at any time during a specified period. □ **out of reach** not able to be reached or attained. **reach-me-down** ready-made; second-hand; inferior. □ **reachable** *adj.* [OE *ræcan* f. WG]

reacquaint /ˌriːəˈkweɪnt/ *v.tr.* & *refl.* (usu. foll. by *with*) make (a person or oneself) acquainted again. □ **reacquaintance** *n.*

reacquire /ˌriːəˈkwaɪə(r)/ *v.tr.* acquire anew. □ **reacquisition** /ˌriːˌækwɪˈzɪʃ(ə)n/ *n.*

react /rɪˈækt/ *v.* **1** *intr.* (foll. by *to*) respond to a stimulus; undergo a change or show behaviour due to some influence (*how did they react to the news?*). **2** *intr.* (often foll. by *against*) be actuated by repulsion to; tend in a reverse or contrary direction. **3** *intr.* (often foll. by *upon*) produce a reciprocal or responsive effect; act upon the agent (*they react upon each other*). **4** *intr.* (foll. by *with*) *Chem.* & *Physics* (of a substance or particle) be the cause of activity or interaction with another (*nitrous oxide reacts with the metal*). **5** *tr.* (foll. by *with*) *Chem.* cause (a substance) to react with another. **6** *intr.* *Mil.* make a counter-attack. **7** *intr.* *Stock Exch.* (of shares) fall after rising. [RE- + ACT or med.L *reagere react-* (as RE-, L *agere* do, act)]

reactance /rɪˈæktəns/ *n.* *Electr.* a component of impedance in an AC circuit, due to capacitance or inductance or both.

reactant /rɪˈæktənt/ *n.* *Chem.* a substance that takes part in, and undergoes change during a reaction.

reaction /rɪˈækʃ(ə)n/ *n.* **1** the act or an instance of reacting; a responsive or reciprocal action. **2 a** a responsive feeling (*what was your reaction to the news?*). **b** an immediate or first impression. **3** the occurrence of a (physical or emotional) condition after a period of its opposite. **4 a** a bodily response to an external stimulus. **b** an adverse response to a drug. **5** a tendency to oppose change or to advocate return to a former system, esp. in politics. **6** the interaction of substances undergoing chemical change. **7** propulsion by emitting a jet of particles etc. in the direction opposite to that of the intended motion. □ **reactionist** n. & adj. [REACT + -ION or med.L reactio (as RE-, ACTION)]

reactionary /rɪˈækʃənərɪ/ adj. & n. ● adj. tending to oppose (esp. political) change and advocate return to a former system. ● n. (pl. **-ies**) a reactionary person.

reactivate /rɪˈæktɪˌveɪt/ v.tr. restore to a state of activity; bring into action again. □ **reactivation** /-ˌæktɪˈveɪʃ(ə)n/ n.

reactive /rɪˈæktɪv/ adj. **1** showing reaction. **2** reacting rather than taking the initiative. **3** having a tendency to react chemically. **4** of or relating to reactance. □ **reactivity** /ˌrɪækˈtɪvɪtɪ/ n.

reactor /rɪˈæktə(r)/ n. **1** a person or thing that reacts. **2** (in full **nuclear reactor**) an apparatus or structure in which a controlled nuclear chain reaction releases energy. **3** Electr. a component used to provide reactance, esp. an inductor. **4** an apparatus for the chemical reaction of substances. **5** Med. a person who has a reaction to a drug etc.

read /riːd/ v. & n. ● v. (past and past part. **read** /red/) **1** tr. (also absol.) reproduce mentally or (often foll. by aloud, out, off, etc.) vocally the written or printed words of (a book, author, etc.) by following the symbols with the eyes or by feeling embossed symbols with the fingers. **2** tr. convert or be able to convert into the intended words or meaning (written or other symbols or the things expressed in this way). **3** tr. interpret mentally. **4** tr. deduce or declare an (esp. accurate) interpretation of (read the expression on my face). **5** tr. (often foll. by that + clause) find (a thing) recorded or stated in print etc. (I read somewhere that you are leaving). **6** tr. interpret (a statement or action) in a certain sense (my silence is not to be read as consent). **7** tr. (often foll. by into) assume as intended or deducible from a writer's words; find (implications) (you read too much into my letter). **8** tr. bring into a specified state by reading (read myself to sleep). **9** tr. a (of a meter or other recording instrument) show (a specified figure etc.) (the thermometer reads 20°). **b** inspect and record elsewhere the figure shown on such an instrument (read the meter). **10** intr. convey meaning in a specified manner when read (it reads persuasively). **11** intr. sound or affect a hearer or reader as specified when read (the book reads like a parody). **12 a** tr. study by reading (esp. a subject at university). **b** intr. carry out a course of study by reading (is reading for the Bar). **13** tr. (as **read** /red/ adj.) versed in a subject (esp. literature) by reading (a well-read person; was widely read in law). **14** tr. **a** (of a computer) copy or transfer (data). **b** (foll. by in, out) enter or extract (data) in an electronic storage device. **15** receive and understand the words of (a person) by radio or telephone (do you read me?). **16** tr. **a** understand or interpret (a person) by hearing words or seeing signs, gestures, etc. **b** interpret (cards, a person's hand, etc.) as a fortune-teller. **c** interpret (the sky) as an astrologer or meteorologist. **17** tr. Printing check the correctness of and emend (a proof). **18** tr. **a** (of a text) have at a particular place (reads 'battery' not 'buttery'). **b** substitute (a word etc.) for an incorrect one (for 'illitterate' read 'illiterate'). ● n. **1** a spell of reading. **2** colloq. a book etc. as regards its readability (is a really good read). □ **read between the lines** look for or find hidden meaning (in a document etc.). **read-in** the entry of data in an electronic storage device. **read a person like a book** clearly understand a person's motives etc. **read-only memory** Computing a memory read at high speed but not capable of being changed by program instructions. **read out 1** read aloud. **2** US expel from a political party etc. **read-out** information retrieved from a computer. **read up** (also foll. by on) make a special study of (a subject). **read-write** Computing capable of reading existing data and accepting alterations or further input (cf. read-only memory). [OE rǣdan advise, consider, discern f. Gmc]

readable /ˈriːdəb(ə)l/ adj. **1** able to be read; legible. **2** interesting or pleasant to read. □ **readably** adj. **readableness** n. **readability** /ˌriːdəˈbɪlɪtɪ/ n.

readapt /ˌriːəˈdæpt/ v.intr. & tr. become or cause to become adapted anew. □ **readaptation** /riːˌædæpˈteɪʃ(ə)n/ n.

readdress /ˌriːəˈdres/ v.tr. **1** change the address of (a letter or parcel). **2** address (a problem etc.) anew. **3** speak or write to anew.

Reade /riːd/, Charles (1814–84), English novelist and dramatist. He is remembered for his historical romance The Cloister and the Hearth (1861); set in the 15th century, it relates the adventures of Gerard, father of Erasmus.

reader /ˈriːdə(r)/ n. **1** a person who reads or is reading. **2** a book of extracts for learning, esp. a language. **3** a device for producing an image that can be read from microfilm etc. **4** (also **Reader**) Brit. a university lecturer of the highest grade below professor. **5** a publisher's employee who reports on submitted manuscripts. **6** a printer's proof-

corrector. **7** a person appointed to read aloud, esp. parts of a service in a church. **8** a person entitled to use a particular library. □ **readerly** adj. [OE (as READ)]

readership /ˈriːdəˌʃɪp/ n. **1** the readers of a newspaper etc. **2** the number or extent of these. **3** (also **Readership**) Brit. the position of Reader.

readily /ˈredɪlɪ/ adv. **1** without showing reluctance; willingly. **2 a** without difficulty. **b** without delay.

Reading /ˈredɪŋ/ a town in southern England, the county town of Berkshire, on the River Kennet near its junction with the Thames; pop. (1991) 122,600. It is the site of the ruins of a 12th-century Cluniac abbey, in which Henry I is buried.

reading /ˈriːdɪŋ/ n. **1 a** the act or an instance of reading or perusing (the reading of the will). **b** matter to be read (have plenty of reading with me). **c** the specified quality of such matter (it made exciting reading). **2** (in comb.) used for reading (reading-lamp; reading-room). **3** literary knowledge (a person of wide reading). **4** an entertainment at which a play, poems, etc., are read (poetry reading). **5** a figure etc. shown by a meter or other recording instrument. **6** an interpretation or view taken (what is your reading of the facts?). **7** an interpretation made (of drama, music, etc.). **8** each of the successive occasions on which a bill must be presented to a legislature for acceptance (see also first reading, second reading, third reading). **9** the version of a text, or the particular wording, conjectured or given by an editor etc. □ **reading age** reading ability expressed as the age for which the same ability is calculated as average (has a reading age of 8). [OE (as READ)]

readjust /ˌriːəˈdʒʌst/ v.tr. adjust again or to a former state. □ **readjustment** n.

readmit /ˌriːədˈmɪt/ v.tr. (**readmitted, readmitting**) admit again. □ **readmission** /-ˈmɪʃ(ə)n/ n.

readopt /ˌriːəˈdɒpt/ v.tr. adopt again. □ **readoption** /-ˈdɒpʃ(ə)n/ n.

ready /ˈredɪ/ adj., adv., n., & v. ● adj. (**readier, readiest**) (usu. predic.) **1** with preparations complete (dinner is ready). **2** in a fit or appropriate state (are you ready to go?). **3** willing, inclined, or resolved (he is always ready to complain; I am ready for anything; a ready accomplice). **4** within reach; easily secured (a ready source of income). **5** fit for immediate use (was ready to hand). **6** immediate, unqualified (found ready acceptance). **7** prompt, quick, facile (is always ready with excuses; has a ready wit). **8** (foll. by to + infin.) about to do something (a bud just ready to burst). **9** provided beforehand. ● adv. **1** beforehand. **2** so as not to require doing when the time comes for use (the cases are ready packed). ● n. (pl. **-ies**) sl. **1** (prec. by the) = ready money. **2** (in pl.) bank notes. ● v.tr. (**-ies, -ied**) make ready; prepare. □ **at the ready** ready for action. **make ready** prepare. **ready-mix** (or **-mixed**) (of concrete, paint, food, etc.) having some or all of the constituents already mixed together. **ready money 1** actual coin or notes. **2** payment on the spot. **ready reckoner** a book or table listing standard numerical calculations as used esp. in commerce. **ready, steady** (or **get set**), **go** the usual formula for starting a race. □ **readiness** n. [ME rædi(g), re(a)di, f. OE rǣde f. Gmc]

ready-made /ˌredɪˈmeɪd/ adj. & n. ● adj. (also **ready-to-wear**) (esp. of clothes) made in a standard size, not to measure. ● n. a type of art form consisting of mass-produced articles selected by the artist and displayed as works of art. (See note below.)

▪ The ready-made was invented (and named) in 1912 by Marcel Duchamp, and intended as a Dada anti-art statement. Duchamp's first ready-made was a bicycle wheel placed on a kitchen stool; other examples include a bottle rack, and a urinal entitled Fountain and signed R. Mutt (1917).

reaffirm /ˌriːəˈfɜːm/ v.tr. affirm again. □ **reaffirmation** /riːˌæfəˈmeɪʃ(ə)n/ n.

reafforest /ˌriːəˈfɒrɪst/ v.tr. replant (former forest land) with trees. □ **reafforestation** /-ˌfɒrɪˈsteɪʃ(ə)n/ n.

Reagan /ˈreɪɡən/, Ronald (Wilson) (b.1911), American Republican statesman, 40th President of the US 1981–9. He was a Hollywood actor before entering politics and becoming governor of California (1966–74). In 1981, at the age of 69, he became the oldest-ever President of the US. During his presidency military expenditure was increased, the Strategic Defense Initiative was launched, taxes and spending on social services were reduced, and the national budget deficit rose to record levels. His interventionist policy in Central America led to the Irangate scandal of 1987. Reagan held several summit meetings with President Gorbachev towards the end of his term of office, resulting in an intermediate nuclear forces non-proliferation treaty signed in 1987.

reagency /riːˈeɪdʒənsɪ/ n. reactive power or operation.

reagent /riːˈeɪdʒənt/ n. *Chem.* **1** a substance used to cause a reaction, esp. to detect another substance. **2** a reactive substance or force. [RE- + AGENT: cf. REACT]

real[1] /rɪːl/ adj. & adv. ● adj. **1** actually existing as a thing or occurring in fact. **2** genuine; rightly so called; not artificial or merely apparent. **3** *Law* consisting of or relating to immovable property such as land or houses (*real estate*) (cf. *personal property*). **4** appraised by purchasing power; adjusted for changes in the value of money (*real value*; *income in real terms*). **5** *Philos.* having an absolute and necessary and not merely contingent existence. **6** *Math.* (of a quantity) having no imaginary part (see IMAGINARY 2). **7** *Optics* (of an image etc.) such that light actually passes through it. ● adv. *Sc.* & *N. Amer. colloq.* really, very. □ **for real** *colloq.* as a serious or actual concern; in earnest. **real ale** beer regarded as brewed in a traditional way, with secondary fermentation in the cask. **real life** that lived by actual people, as distinct from fiction, drama, etc. **real live** (*attrib.*) often *joc.* actual; not pretended or simulated (*a real live burglar*). **the real McCoy** see McCOY. **real money** current coin or notes; cash. **the real thing** **1** genuine, not illusory, counterfeit, or inferior. **2** true love, not infatuation or flirtation. **real time** the actual time during which a process or event occurs. **real-time** (*attrib.*) *Computing* **1** (of a system) in which input data is processed within milliseconds so that it is available virtually immediately as feedback to the process from which it is coming, e.g. in a missile-guidance or an airline booking system. **2** (of information, an image, etc.) responding virtually immediately to changes in the state of affairs it reflects (*real-time stock-exchange prices*). □ **realness** n. [AF = OF *reel*, LL *realis* f. L *res* thing]

real[2] /reɪˈaːl/ n. **1** *hist.* a former coin and monetary unit of various Spanish-speaking countries. **2** the basic monetary unit in Brazil, introduced in 1994. [Sp. & Port., noun use of *real* (adj.) (as ROYAL)]

realgar /rɪˈælgə(r)/ n. a mineral of arsenic sulphide used as a pigment and in fireworks. Also called *red arsenic*. [ME f. med.L f. Arab. *rahj al-ḡār* dust of the cave]

realign /ˌriːəˈlaɪn/ v.tr. **1** align again. **2** regroup in politics etc. □ **realignment** n.

realism /ˈrɪəlɪz(ə)m/ n. (usu. opp. IDEALISM) **1** the practice of regarding things in their true nature and dealing with them as they are; practical views and policy. **2** (also **Realism**) in art, fidelity of representation, truth to nature, and insistence on details; the showing of life as it is without glossing over what is ugly or painful. (*See note below.*) **3** *Philos.* **a** the theory that abstract entities or universals have an objective existence (cf. NOMINALISM, CONCEPTUALISM). (*See note below.*) **b** belief that matter as an object of perception has real existence and is not dependent on the existence of mental states or minds (cf. IDEALISM). □ **realist** n. & adj.

▪ While realism in art is often used in the same contexts as naturalism, implying a concern to depict or describe accurately and objectively, it also suggests a deliberate rejection of conventionally beautiful or appropriate subjects in favour of sincerity and a focus on simple and unidealized treatment of contemporary life. Specifically, the term is applied to a late 19th-century movement in French painting and literature represented by Gustave Courbet in the former and Balzac, Stendhal, and Flaubert in the latter; in this context a deliberate concern with social issues is often implied, in which case the term *social realism* is sometimes used. (See also NEO-REALISM, SOCIALIST REALISM.)

▪ The theory that universals have their own reality is sometimes called *Platonic realism* because it was first outlined by Plato's doctrine of 'forms' or ideas. It is sometimes justified by the observation that people automatically recognize ordinary objects as being members of a more generalized class; according to the theory these classes or universals exist separately in another world.

realistic /rɪəˈlɪstɪk/ adj. **1** regarding things as they are; following a policy of realism. **2** based on facts rather than ideals. □ **realistically** adv.

reality /rɪˈælɪtɪ/ n. (pl. **-ies**) **1** what is real or existent or underlies appearances. **2** (foll. by *of*) the real nature of (a thing). **3** real existence; the state of being real. **4** resemblance to an original (*the model was impressive in its reality*). □ **in reality** in fact. [med.L *realitas* or F *réalité* (as REAL[1])]

realize /ˈrɪəlaɪz/ v.tr. (also **-ise**) **1** (often foll. by *that* + clause) (also *absol.*) be fully aware of; conceive as real. **2** (also *absol.*) understand clearly. **3** present as real; make realistic; give apparent reality to (*the story was powerfully realized on stage*). **4 a** convert into actuality; achieve (*realized a childhood dream*). **b** *refl.* develop one's own faculties, abilities, etc.

5 a convert into money. **b** acquire (profit). **c** be sold for (a specified price). **6** *Mus.* reconstruct (a part) in full from a figured bass. □ **realizer** n. **realizable** adj. **realizability** /ˌrɪəlaɪzəˈbɪlɪtɪ/ n. **realization** /-ˈzeɪʃ(ə)n/ n.

reallocate /riːˈæləˌkeɪt/ v.tr. allocate again or differently. □ **reallocation** /-ˌæləˈkeɪʃ(ə)n/ n.

reallot /ˌriːəˈlɒt/ v.tr. (**reallotted**, **reallotting**) allot again or differently. □ **reallotment** n.

really /ˈrɪəlɪ/ adv. **1** in reality; in fact. **2** positively, assuredly (*really useful*). **3** (as a strong affirmative) indeed, I assure you. **4** an expression of mild protest or surprise. **5** (in *interrog.*) (expressing disbelief) is that so? (*They're musicians. — Really?*).

realm /relm/ n. **1** *formal* esp. *Law* a kingdom. **2** a sphere or domain (*the realm of imagination*). [ME f. OF *realme*, *reaume*, f. L *regimen -minis* (see REGIMEN): infl. by OF *reiel* ROYAL]

Realpolitik /reɪˈaːlpɒlɪˌtiːk/ n. politics based on realities and material needs, rather than on morals or ideals. [G]

real tennis n. a game played by two or four persons in a walled court, the forerunner of modern lawn tennis. A net divides the court into equal but dissimilar halves: the service side, from which service is always delivered, and the hazard side, on which service is received. A small solid ball is struck with rackets over the net, or rebounds from the side walls. The scoring system is similar to that used in lawn tennis. A similar game was played in monastery cloisters in the 11th century, at first with a bare (or gloved) hand. The game's popularity declined partly because of the need for special courts, and partly because of the rise of lawn tennis and other forms of racket games; today it is very much a minority sport. [REAL[1]]

realtor /ˈriːəltə(r)/ n. *N. Amer.* a real-estate agent, esp. (**Realtor**) a member of the National Association of Realtors.

realty /ˈriːəltɪ/ n. *Law* real estate (opp. PERSONALTY).

ream[1] /riːm/ n. **1** twenty quires or 500 (formerly 480) sheets of paper (or a larger number, to allow for waste). **2** (in *pl.*) a large quantity of paper or writing (*wrote reams about it*). [ME *rēm*, *rīm* f. OF *raime* etc., ult. f. Arab. *rīzma* bundle]

ream[2] /riːm/ v.tr. **1** widen (a hole in metal etc.) with a borer. **2** turn over the edge of (a cartridge case etc.). **3** *Naut.* open (a seam) for caulking. **4** *US* squeeze the juice from (fruit). □ **reamer** n. [19th c.: orig. uncert.]

reanimate /riːˈænɪˌmeɪt/ v.tr. **1** restore to life. **2** restore to activity or liveliness. □ **reanimation** /-ˌænɪˈmeɪʃ(ə)n/ n.

reap /riːp/ v.tr. **1** cut or gather (a crop, esp. grain) as a harvest. **2** harvest the crop of (a field etc.). **3** receive as the consequence of one's own or others' actions. [OE *rīpan*, *reopan*, of unkn. orig.]

reaper /ˈriːpə(r)/ n. **1** a person who reaps. **2** a machine for cutting grain (and now also, binding the sheaves).

Reaper, the (also **the Grim Reaper**) death personified.

reappear /ˌriːəˈpɪə(r)/ v.intr. appear again or as previously. □ **reappearance** n.

reapply /ˌriːəˈplaɪ/ v.tr. & intr. (**-ies**, **-ied**) apply again, esp. submit a further application (for a position etc.). □ **reapplication** /ˌriːˌæplɪˈkeɪʃ(ə)n/ n.

reappoint /ˌriːəˈpɔɪnt/ v.tr. appoint again to a position previously held. □ **reappointment** n.

reapportion /ˌriːəˈpɔːʃ(ə)n/ v.tr. apportion again or differently. □ **reapportionment** n.

reappraise /ˌriːəˈpreɪz/ v.tr. make a fresh appraisal of, esp. in the light of new facts; reassess. □ **reappraisal** n.

rear[1] /rɪə(r)/ n. & adj. ● n. **1** the back part of anything. **2** the space behind, or position at the back of, anything (*a large house with a terrace at the rear*). **3** the hindmost part of an army or fleet. **4** *colloq.* the buttocks. ● adj. at the back. □ **bring up the rear** come last. **in the rear** behind; at the back. **rear admiral** a naval officer ranking below vice admiral. **rear end** the back section of a motor vehicle. **rear-end** v.tr. esp. *US colloq.* crash into the back of (a motor vehicle). **rear-lamp** (or **-light**) a usu. red light at the rear of a vehicle. **rear sight** the sight nearest to the stock on a firearm. **rear-view mirror** a mirror fixed inside the windscreen of a motor vehicle enabling the driver to see traffic etc. behind. **take in the rear** *Mil.* attack from behind. [prob. f. (*in the*) REARWARD or REARGUARD]

rear[2] /rɪə(r)/ v. **1** *tr.* **a** bring up and educate (children). **b** breed and care for (animals). **c** cultivate (crops). **2** *intr.* (of a horse etc.) raise itself on its hind legs. **3** *tr.* **a** set upright. **b** build. **c** hold upwards (*rear one's head*). **4** *intr.* extend to a great height. □ **rearer** n. [OE *rǣran* f. Gmc]

rearguard /ˈrɪəgɑːd/ n. **1** a body of troops detached to protect the rear, esp. in retreats. **2** a defensive or conservative element in an organization etc. □ **rearguard action 1** Mil. an engagement undertaken by a rearguard. **2** a defensive stand in argument etc., esp. when losing. [OF rereguarde (as RETRO-, GUARD)]

rearm /riːˈɑːm/ v.tr. (also absol.) arm again, esp. with improved weapons. □ **rearmament** n.

rearmost /ˈrɪəməʊst/ adj. furthest back.

rearrange /ˌriːəˈreɪndʒ/ v.tr. arrange again in a different way. □ **rearrangement** n.

rearrest /ˌriːəˈrest/ v. & n. ● v.tr. arrest again. ● n. an instance of rearresting or being rearrested.

rearward /ˈrɪəwəd/ n., adj., & adv. ● n. rear, esp. in prepositional phrases (to the rearward of; in the rearward). ● adj. to the rear. ● adv. (also **rearwards** /-wədz/) towards the rear. [AF rerewarde = REARGUARD]

reascend /ˌriːəˈsend/ v.tr. & intr. ascend again or to a former position. □ **reascension** /-ˈsenʃ(ə)n/ n.

reason /ˈriːz(ə)n/ n. & v. ● n. **1** a motive, cause, or justification (has good reasons for doing this; there is no reason to be angry). **2** a fact adduced or serving as this (I can give you my reasons). **3** the intellectual faculty by which conclusions are drawn from premisses. **4** sanity (has lost his reason). **5** Logic a premiss of a syllogism, esp. a minor premiss when given after the conclusion. **6** a faculty transcending the understanding and providing a priori principles; intuition. **7** sense; sensible conduct; what is right or practical or practicable; moderation. ● v. **1** intr. form or try to reach conclusions by connected thought. **2** intr. (foll. by with) use an argument (with a person) by way of persuasion. **3** tr. (foll. by that + clause) conclude or assert in argument. **4** tr. (foll. by why, whether, what + clause) discuss; ask oneself. **5** tr. (foll. by into, out of) persuade or move by argument (I reasoned them out of their fears). **6** tr. (foll. by out) think or work out (consequences etc.). **7** tr. (often as **reasoned** adj.) express in logical or argumentative form. **8** tr. embody reason in (an amendment etc.). □ **by reason of** owing to. **in** (or **within**) **reason** within the bounds of sense or moderation. **it stands to reason** (often foll. by that + clause) it is evident or logical. **listen to reason** be persuaded to act sensibly. **see reason** acknowledge the force of an argument. **with reason** justifiably. □ **reasoner** n. **reasoning** n. **reasonless** adj. [ME f. OF reisun, res(o)un, raisoner, ult. f. L ratio -onis f. reri rat- consider]

reasonable /ˈriːz(ə)nəb(ə)l/ adj. **1** having sound judgement; moderate; ready to listen to reason. **2** in accordance with reason; not absurd. **3 a** within the limits of reason; not greatly less or more than might be expected. **b** inexpensive; not extortionate. **c** tolerable, fair. **4** archaic endowed with the faculty of reason. □ **reasonableness** n. **reasonably** adv. [ME f. OF raisonable (as REASON) after L rationalis]

reassemble /ˌriːəˈsemb(ə)l/ v.intr. & tr. assemble again or into a former state. □ **reassembly** n.

reassert /ˌriːəˈsɜːt/ v.tr. assert again. □ **reassertion** /-ˈsɜːʃ(ə)n/ n.

reassess /ˌriːəˈses/ v.tr. assess again, esp. differently. □ **reassessment** n.

reassign /ˌriːəˈsaɪn/ v.tr. assign again or differently; appoint (employees) to a new position. □ **reassignment** n.

reassume /ˌriːəˈsjuːm/ v.tr. take on oneself or undertake again. □ **reassumption** /-ˈsʌmpʃ(ə)n/ n.

reassure /ˌriːəˈʃʊə(r)/ v.tr. **1** restore confidence to; dispel the apprehensions of. **2** confirm in an opinion or impression. □ **reassurance** n. **reassuring** adj. **reassuringly** adv.

reattach /ˌriːəˈtætʃ/ v.tr. attach again or in a former position. □ **reattachment** n.

reattain /ˌriːəˈteɪn/ v.tr. attain again. □ **reattainment** n.

reattempt /ˌriːəˈtempt/ v.tr. attempt again, esp. after failure.

Réaumur /ˈreɪəˌmjʊə(r)/, René Antoine Ferchault de (1683–1757), French naturalist. He compiled a list of France's arts, industries, and professions, and, as a consequence, suggested improvements in several manufacturing processes. He is chiefly remembered for his thermometer scale, now obsolete, which set the melting-point of ice at 0° and the boiling-point of water at 80°. Réaumur also carried out pioneering work on insects and other invertebrates.

reave /riːv/ v. (past and past part. **reft** /reft/) archaic **1** tr. **a** (foll. by of) forcibly deprive of. **b** (foll. by away, from) take by force or carry off. **2** intr. make raids; plunder; = REIVE. [OE rēafian f. Gmc: cf. ROB]

reawaken /ˌriːəˈweɪkən/ v.tr. & intr. awaken again.

rebadge /riːˈbædʒ/ v.tr. relaunch (a product etc.) under a different name, logo, etc.

rebarbative /rɪˈbɑːbətɪv/ adj. literary repellent, unattractive. [F rébarbatif -ive f. barbe beard]

rebate[1] /ˈriːbeɪt/ n. & v. ● n. **1** a partial refund of money paid. **2** a deduction from a sum to be paid; a discount. ● v.tr. (also /rɪˈbeɪt/) pay back as a rebate. □ **rebatable** /rɪˈbeɪtəb(ə)l/ adj. [earlier = diminish: ME f. OF rabattre (as RE-, ABATE)]

rebate[2] /ˈriːbeɪt, ˈræbɪt/ n. & v.tr. = RABBET. [respelling of RABBET, after REBATE[1]]

rebec /ˈriːbek/ n. (also **rebeck**) Mus. a medieval usu. three-stringed instrument played with a bow. [F rebec var. of OF rebebe rubebe f. Arab. rabāb]

rebel n., adj., & v. ● n. /ˈreb(ə)l/ **1** a person who fights against, resists, or refuses allegiance to the established government. **2** a person or thing that resists authority or control. ● adj. /ˈreb(ə)l/ (attrib.) **1** rebellious. **2** of or concerning rebels. **3** in rebellion. ● v.intr. /rɪˈbel/ (**rebelled**, **rebelling**; US **rebeled**, **rebeling**) (usu. foll. by against) **1** act as a rebel; revolt. **2** feel or display repugnance. [ME f. OF rebelle, rebeller f. L rebellis (as RE-, bellum war)]

rebellion /rɪˈbeljən/ n. **1** open resistance to authority, esp. organized armed resistance to an established government. **2** an instance of this. [ME f. OF f. L rebellio -onis (as REBEL)]

rebellious /rɪˈbeljəs/ adj. **1** tending to rebel, insubordinate. **2** in rebellion. **3** defying lawful authority. **4** (of a thing) unmanageable, refractory. □ **rebelliously** adv. **rebelliousness** n. [ME f. REBELLION + -OUS or f. earlier rebellous + -IOUS]

rebid v. & n. ● v. /riːˈbɪd/ (**rebidding**; past and past part. **rebid**) **1** intr. bid again. **2** tr. Cards bid (a suit) again at a higher level. ● n. /ˈriːbɪd/ **1** the act of rebidding. **2** a bid made in this way.

rebind /riːˈbaɪnd/ v.tr. (past and past part. **rebound** /-ˈbaʊnd/) bind (esp. a book) again or differently.

rebirth /riːˈbɜːθ/ n. **1** a new incarnation. **2** spiritual enlightenment. **3** a revival (the rebirth of learning). □ **reborn** /-ˈbɔːn/ adj.

rebirthing /riːˈbɜːθɪŋ/ n. a treatment for neurosis involving controlled breathing intended to simulate the trauma of being born.

reboot /riːˈbuːt/ v. & n. Computing ● v.tr. (often absol.) boot up (a system) again. ● n. an act or instance of rebooting.

rebore v. & n. ● v.tr. /riːˈbɔː(r)/ make a new boring in, esp. widen the bore of (the cylinder in an internal-combustion engine). ● n. /ˈriːbɔː(r)/ **1** the process of doing this. **2** a rebored engine.

rebound[1] v. & n. ● v.intr. /rɪˈbaʊnd/ **1** spring back after action or impact. **2** (foll. by on, upon) (of an action) have an adverse effect upon (the doer). ● n. /ˈriːbaʊnd/ **1** the act or an instance of rebounding; recoil. **2** a reaction after a strong emotion. □ **on the rebound** while still recovering from an emotional shock, esp. rejection by a lover. [ME f. OF rebonder, rebondir (as RE-, BOUND[1])]

rebound[2] past and past part. of REBIND.

rebroadcast /riːˈbrɔːdkɑːst/ v. & n. ● v.tr. (past **rebroadcast** or **rebroadcasted**; past part. **rebroadcast**) broadcast again. ● n. a repeat broadcast.

rebuff /rɪˈbʌf/ n. & v. ● n. **1** a rejection of a person who makes advances, proffers help or sympathy, shows interest or curiosity, makes a request, etc. **2** a repulse; a snub. ● v.tr. give a rebuff to. [obs. F rebuffe(r) f. It. ribuffo, ribuffare, rabbuffo, rabbuffare (as RE-, buffo puff)]

rebuild /riːˈbɪld/ v.tr. (past and past part. **rebuilt** /-ˈbɪlt/) build again or differently.

rebuke /rɪˈbjuːk/ v. & n. ● v.tr. reprove sharply; subject to protest or censure. ● n. **1** the act of rebuking. **2** the process of being rebuked. **3** a reproof. □ **rebuker** n. **rebukingly** adv. [ME f. AF & ONF rebuker (as RE-, OF buchier beat, orig. cut down wood f. busche log]

rebury /riːˈberɪ/ v.tr. (**-ies**, **-ied**) bury again. □ **reburial** n.

rebus /ˈriːbəs/ n. (pl. **rebuses**) **1** an enigmatic representation of a word (esp. a name), by pictures etc. suggesting its parts. **2** Heraldry a device suggesting the name of its bearer. [F rébus f. L rebus, ablat. pl. of res thing]

rebut /rɪˈbʌt/ v.tr. (**rebutted**, **rebutting**) **1** refute or disprove (evidence or a charge). **2** force or turn back; check. □ **rebutment** n. **rebuttable** adj. **rebuttal** n. [ME f. AF rebuter, OF rebo(u)ter (as RE-, BUTT[1])]

rebutter /rɪˈbʌtə(r)/ n. **1** a refutation. **2** Law a defendant's reply to the plaintiff's surrejoinder. [AF rebuter (as REBUT)]

recalcitrant /rɪˈkælsɪtrənt/ adj. & n. ● adj. **1** obstinately disobedient. **2** objecting to restraint. ● n. a recalcitrant person. □ **recalcitrance** n. **recalcitrantly** adv. [L recalcitrare (as RE-, calcitrare kick out with the heels f. calx calcis heel)]

recalculate /ri:ˈkælkjʊˌleɪt/ v.tr. calculate again. □ **recalculation** /-ˌkælkjʊˈleɪʃ(ə)n/ n.

recalesce /ˌri:kəˈles/ v.intr. grow hot again (esp. of iron allowed to cool from white heat, whose temperature rises at a certain point for a short time). □ **recalescence** n. [L recalescere (as RE-, calescere grow hot)]

recall /rɪˈkɔːl/ v. & n. • v.tr. **1** summon to return from a place or from a different occupation, inattention, a digression, etc. **2** recollect, remember. **3** bring back to memory; serve as a reminder of. **4** revoke or annul (an action or decision). **5** cancel or suspend the appointment of (an official sent overseas etc.). **6** revive, resuscitate. **7** take back (a gift). • n. (also /ˈriːkɔːl/) **1** the act or an instance of recalling, esp. a summons to come back. **2** the act of remembering. **3** the ability to remember. **4** the possibility of recalling, esp. in the sense of revoking (beyond recall). **5** US removal of an elected official from office. □ **recallable** adj.

recant /rɪˈkænt/ v. **1** tr. withdraw and renounce (a former belief or statement) as erroneous or heretical. **2** intr. disavow a former opinion, esp. with a public confession of error. □ **recanter** n. **recantation** /ˌriːkænˈteɪʃ(ə)n/ n. [L recantare revoke (as RE-, cantare sing, chant)]

recap /ˈriːkæp/ v. & n. colloq. • v.tr. & intr. (**recapped**, **recapping**) recapitulate. • n. recapitulation. [abbr.]

recapitalize /riːˈkæpɪtəˌlaɪz/ v.tr. (also **-ise**) capitalize (shares etc.) again. □ **recapitalization** /-ˌkæpɪtəlaɪˈzeɪʃ(ə)n/ n.

recapitulate /ˌriːkəˈpɪtjʊˌleɪt/ v.tr. **1** go briefly through again; summarize. **2** go over the main points or headings of. □ **recapitulative** /-lətɪv/ adj. **recapitulatory** adj. [L recapitulare (as RE-, capitulum CHAPTER)]

recapitulation /ˌriːkəˌpɪtjʊˈleɪʃ(ə)n/ n. **1** the act or an instance of recapitulating. **2** Biol. the appearance during embryonic development of successive forms somewhat resembling those of the organism's evolutionary predecessors. **3** Mus. part of a movement, esp. in sonata form, in which themes from the exposition are restated. [ME f. OF recapitulation or LL recapitulatio (as RECAPITULATE)]

recapture /riːˈkæptʃə(r)/ v. & n. • v.tr. **1** capture again; recover by capture. **2** re-experience (a past emotion etc.). • n. the act or an instance of recapturing.

recast v. & n. • v.tr. /riːˈkɑːst/ (past and past part. **recast**) **1** put into a new form. **2** improve the arrangement of. **3** change the cast of (a play etc.). • n. /ˈriːkɑːst/ **1** the act or an instance of recasting. **2** a recast form.

recce /ˈrekɪ/ n. & v. colloq. • n. a reconnaissance. • v.tr. & intr. (**recced** /-kɪd/, **recceing** /-kɪŋ/) reconnoitre. [abbr.]

recd. abbr. received.

recede /rɪˈsiːd/ v.intr. **1** go or shrink back or further off. **2** be left at an increasing distance by an observer's motion. **3** slope backwards (a receding chin). **4** decline in force or value. **5** (foll. by from) withdraw from (an engagement, opinion, etc.). **6** (of a man's hair) cease to grow at the front, sides, etc. [ME f. L recedere (as RE-, cedere cess- go)]

re-cede /riːˈsiːd/ v.tr. cede back to a former owner.

receipt /rɪˈsiːt/ n. & v. • n. **1** the act or an instance of receiving or being received into one's possession (will pay on receipt of the goods). **2** a written acknowledgement of this, esp. of the payment of money. **3** (usu. in pl.) an amount of money etc. received. **4** archaic a recipe. • v.tr. place a written or printed receipt on (a bill). □ **in receipt of** having received. [ME receit(e) f. AF & ONF receite, OF recoite, recete f. med.L recepta fem. past part. of L recipere RECEIVE: -p- inserted after L]

receive /rɪˈsiːv/ v.tr. **1** take or accept (something offered or given) into one's hands or possession. **2** acquire; be provided with or given (have received no news; will receive a small fee). **3** accept delivery of (something sent). **4** have conferred or inflicted on one (received many honours; received a heavy blow on the head). **5** a stand the force or weight of. **b** bear up against; encounter with opposition. **6** consent to hear (a confession or oath) or consider (a petition). **7** (also absol.) accept or have dealings with (stolen property), knowing of the theft. **8** admit; consent or prove able to hold; provide accommodation for (received many visitors). **9** (of a receptacle) be able to hold (a specified amount or contents). **10** greet or welcome, esp. in a specified manner (how did they receive your offer?). **11** react to (news, a play, etc.) in a specified way (novel was warmly received). **12** entertain as a guest etc. **13** admit (a person) to membership of a society, organization, etc. **14** be marked more or less permanently with (an impression etc.). **15** convert (broadcast signals) into sound or pictures. **16** Tennis be the player to whom the server serves (the ball). **17** (often as **received** adj.) give credit to; accept as authoritative or true (received opinion). **18** Eccl. eat or drink (the Eucharistic bread and

wine). □ **be at** (or **on**) **the receiving end** colloq. bear the brunt of something unpleasant. **received pronunciation** (or **Received Standard**) the form of spoken English based on educated speech in southern England. □ **receivable** adj. [ME f. OF receivre, reçoivre f. L recipere recept- (as RE-, capere take)]

receiver /rɪˈsiːvə(r)/ n. **1** a person or thing that receives. **2** the part of a machine or instrument that receives sound, signals, etc. (esp. the part of a telephone that contains the earpiece). **3** (in full **official receiver**) a person appointed by a court to administer the property of a bankrupt or insane person, or property under litigation. **4** a radio or television receiving apparatus. **5** a person who receives stolen goods. **6** Chem. a vessel for collecting the products of distillation, chromatography, etc. **7** Amer. Football an offensive player eligible to catch a pass.

receivership /rɪˈsiːvəˌʃɪp/ n. **1** the office of official receiver. **2** the state of being dealt with by a receiver (esp. in receivership).

receiving-order /rɪˈsiːvɪŋˌɔːdə(r)/ n. Brit. an order of a court authorizing a receiver (see RECEIVER 3) to act.

recension /rɪˈsenʃ(ə)n/ n. **1** the revision of a text. **2** a particular form or version of a text resulting from such revision. [L recensio f. recensere revise (as RE-, censere review)]

recent /ˈriːs(ə)nt/ adj. & n. • adj. **1** not long past; that happened, appeared, began to exist, or existed lately or just before the present time. **2** not long established; lately begun; modern. **3** (**Recent**) Geol. = HOLOCENE. • n. (**Recent**) Geol. = HOLOCENE. □ **recency** n. **recently** adv. **recentness** n. [L recens recentis or F récent]

receptacle /rɪˈseptək(ə)l/ n. **1** a containing vessel, place, or space. **2** Bot. **a** the common base of floral organs. **b** the part of a leaf or thallus in some algae where the reproductive organs are situated. **3** Anat. & Zool. an organ or space which receives a secretion, sperm, eggs, etc. [ME f. OF receptacle or L receptaculum (as RECEPTION)]

reception /rɪˈsepʃ(ə)n/ n. **1** the act or an instance of receiving or the process of being received, esp. of a person into a place or group. **2** the manner in which a person or thing is received (got a cool reception). **3** a social occasion for receiving guests, esp. after a wedding. **4** a formal or ceremonious welcome. **5** a place where guests or clients etc. report on arrival at a hotel, office, etc. **6 a** the receiving of broadcast signals. **b** the quality of this (we have excellent reception). □ **reception order** an order authorizing the entry of a patient into a mental hospital. **reception room** a room available or suitable for receiving company or visitors. [ME f. OF reception or L receptio (as RECEIVE)]

receptionist /rɪˈsepʃənɪst/ n. a person employed in a hotel, office, etc., to receive guests, clients, etc.

receptive /rɪˈseptɪv/ adj. **1** able or quick to receive impressions or ideas. **2** concerned with receiving stimuli etc. □ **receptively** adv. **receptiveness** n. **receptivity** /ˌriːsepˈtɪvɪtɪ/ n. [F réceptif -ive or med.L receptivus (as RECEPTION)]

receptor /rɪˈseptə(r)/ n. (often attrib.) Biol. **1** an organ or cell able to respond to an external stimulus such as light, heat, or a drug, and transmit a signal to a sensory nerve. **2** a region in a tissue or a site on the surface of a cell which specifically recognizes and responds to a neurotransmitter, hormone, or other substance. [OF receptour or L receptor (as RECEPTIVE)]

recess /rɪˈses, ˈriːses/ n. & v. • n. **1** a space set back in a wall; a niche. **2** (often in pl.) a remote or secret place (the innermost recesses). **3** a temporary cessation from work, esp. of Parliament, or N. Amer. of a lawcourt or during a school day. **4** Anat. a fold or indentation in an organ. **5** Geog. a receding part of a mountain chain etc. • v. **1** tr. make a recess in. **2** tr. place in a recess; set back. **3** N. Amer. & Austral. **a** intr. take a recess; adjourn. **b** tr. order a temporary cessation from the work of (a court etc.). [L recessus (as RECEDE)]

recession /rɪˈseʃ(ə)n/ n. **1** a temporary decline in economic activity or prosperity. **2** a receding or withdrawal from a place or point. **3** a receding part of an object; a recess. □ **recessionary** adj. [L recessio (as RECESS)]

recessional /rɪˈseʃən(ə)l/ adj. & n. Eccl. • adj. sung while the clergy and choir withdraw after a service. • n. a recessional hymn.

recessive /rɪˈsesɪv/ adj. **1** tending to recede. **2** Phonet. (of an accent) falling near the beginning of a word. **3** (of an inherited characteristic) appearing in offspring only when not masked by a dominant characteristic inherited from one parent. □ **recessively** adv. **recessiveness** n. [RECESS after excessive]

Rechabite /ˈrekəˌbaɪt/ n. **1** (in the Bible) a member of an Israelite family, descended from Rechab, who refused to drink wine or live in houses (Jer. 35). **2** a teetotaller, esp. a member of the Independent Order of Rechabites, a benefit society founded in 1835.

recharge v. & n. ● v. /riːˈtʃɑːdʒ/ **1** tr. charge again. **2** tr. reload. **3** intr. (of a battery etc.) be recharged. ● n. /ˈriːtʃɑːdʒ/ **1** an act or the action of recharging. **2** a renewed charge in battle. □ **rechargeable** /riːˈtʃɑːdʒ(ə)l/ adj.

réchauffé /reɪˈʃəʊfeɪ/ n. **1** a warmed-up dish. **2** a rehash. [F past part. of réchauffer (as RE-, CHAFE)]

recheck v. & n. ● v.tr. & intr. /riːˈtʃek/ check again. ● n. /ˈriːtʃek/ a second or further check or inspection.

recherché /rəˈʃeəʃeɪ/ adj. **1** carefully sought out; rare or exotic. **2** far-fetched, obscure. [F, past part. of rechercher (as RE-, chercher seek)]

rechristen /riːˈkrɪs(ə)n/ v.tr. **1** christen again. **2** give a new name to.

recidivist /rɪˈsɪdɪvɪst/ n. a person who relapses into crime. □ **recidivism** n. **recidivistic** /-ˌsɪdɪˈvɪstɪk/ adj. [F récidiviste f. récidiver f. med.L recidivare f. L recidivus f. recidere (as RE-, cadere fall)]

Recife /rəˈsiːfə/ a port on the Atlantic coast of NE Brazil, capital of the state of Pernambuco; pop. (1990) 1,335,680. It was formerly known as Pernambuco.

recipe /ˈresɪpɪ/ n. **1** a statement of the ingredients and procedure required for preparing something, esp. a dish in cookery. **2** an expedient; a device for achieving something. **3** a medical prescription. [2nd sing. imper. (as used in prescriptions) of L recipere take, RECEIVE]

recipient /rɪˈsɪpɪənt/ n. & adj. ● n. a person who receives something. ● adj. **1** receiving. **2** receptive. □ **recipiency** n. [F récipient f. It. recipiente or L recipiens f. recipere RECEIVE]

reciprocal /rɪˈsɪprək(ə)l/ adj. & n. ● adj. **1** in return (offered a reciprocal greeting). **2** mutual (their feelings are reciprocal). **3** Gram. (of a pronoun) expressing mutual action or relation (as in each other). **4** inversely correspondent; complementary (natural kindness matched by a reciprocal severity). ● n. Math. an expression or function so related to another that their product is unity ($\frac{1}{2}$ is the reciprocal of 2). □ **reciprocally** adv. **reciprocality** /-ˌsɪprəˈkælɪtɪ/ n. [L reciprocus ult. f. re- back + pro forward]

reciprocate /rɪˈsɪprəˌkeɪt/ v. **1** tr. return or requite (affection etc.). **2** intr. (foll. by with) offer or give something in return (reciprocated with an invitation to lunch). **3** tr. give and receive mutually; interchange. **4 a** intr. (of a part of a machine) move backwards and forwards. **b** tr. cause to do this. □ **reciprocating engine** an engine using a piston or pistons moving up and down in cylinders. □ **reciprocator** n. **reciprocation** /-ˌsɪprəˈkeɪʃ(ə)n/ n. [L reciprocare reciprocat- (as RECIPROCAL)]

reciprocity /ˌresɪˈprɒsɪtɪ/ n. **1** the condition of being reciprocal. **2** mutual action. **3** give and take, esp. the interchange of privileges between countries and organizations. [F réciprocité f. réciproque f. L reciprocus (as RECIPROCATE)]

recirculate /riːˈsɜːkjʊˌleɪt/ v.tr. & intr. circulate again, esp. make available for reuse. □ **recirculation** /-ˌsɜːkjʊˈleɪʃ(ə)n/ n.

recital /rɪˈsaɪt(ə)l/ n. **1** the act or an instance of reciting or being recited. **2** the performance of a programme of music by a solo instrumentalist or singer or by a small group. **3** (foll. by of) a detailed account of (connected things or facts); a narrative. **4** Law the part of a legal document that states the facts. □ **recitalist** n.

recitation /ˌresɪˈteɪʃ(ə)n/ n. **1** the act or an instance of reciting. **2** a thing recited. [OF recitation or L recitatio (as RECITE)]

recitative /ˌresɪtəˈtiːv/ n. **1** musical declamation of the kind usual in the narrative and dialogue parts of opera and oratorio. **2** the words or part given in this form. [It. recitativo (as RECITE)]

recite /rɪˈsaɪt/ v. **1** tr. repeat aloud or declaim (a poem or passage) from memory, esp. before an audience. **2** intr. give a recitation. **3** tr. mention in order; enumerate. □ **reciter** n. [ME f. OF reciter or L recitare (as RE-, CITE)]

reck /rek/ v. archaic or poet. (only in neg. or interrog.) **1** tr. (foll. by of) pay heed to; take account of; care about. **2** tr. pay heed to. **3** intr. (usu. prec. by it) be of importance (it recks little). [OE reccan, rel. to OHG ruohhen]

reckless /ˈreklɪs/ adj. disregarding the consequences or danger etc.; lacking caution; rash. □ **recklessly** adv. **recklessness** n. [OE recceléas (as RECK)]

reckon /ˈrekən/ v. **1** tr. count or compute by calculation. **2** tr. (foll. by in) count in or include in computation. **3** tr. (often foll. by as or to be) consider or regard (reckon him wise; reckon them to be beyond hope). **4** tr. **a** (foll. by that + clause) conclude after calculation; be of the considered opinion. **b** (foll. by to + infin.) colloq. expect (reckons to finish by Friday). **5** intr. **a** make calculations; add up an account or sum. **b** (foll. by with)

settle accounts with. **6** intr. (foll. by on, upon) rely on, count on, or base plans on. **7** intr. (foll. by with or without) take (or fail to take) into account. □ **reckon up 1** count up; find the total of. **2** settle accounts. **to be reckoned with** of considerable importance; not to be ignored. [OE (ge)recenian f. WG]

reckoner /ˈrekənə(r)/ n. = ready reckoner.

reckoning /ˈrekənɪŋ/ n. **1** the act or an instance of counting or calculating. **2** a consideration or opinion. **3 a** the settlement of an account. **b** an account. □ **day of reckoning** the time when something must be atoned for or avenged.

reclaim /rɪˈkleɪm/ v. & n. ● v.tr. **1** seek the return of (one's property). **2** claim in return or as a rebate etc. **3** bring under cultivation, esp. from a state of being under water. **4** a win back or away from vice or error or a waste condition; reform. **b** tame, civilize. ● n. the act or an instance of reclaiming; the process of being reclaimed. □ **reclaimable** adj. **reclaimer** n. **reclamation** /ˌrekləˈmeɪʃ(ə)n/ n. [ME f. OF reclamer reclaim- f. L reclamare cry out against (as RE-, clamare shout)]

reclassify /riːˈklæsɪˌfaɪ/ v.tr. (**-ies, -ied**) classify again or differently. □ **reclassification** /-ˌklæsɪfɪˈkeɪʃ(ə)n/ n.

reclinate /ˈreklɪˌneɪt/ adj. Bot. bending downwards. [L reclinatus, past part. of reclinare (as RECLINE)]

recline /rɪˈklaɪn/ v. **1** intr. assume or be in a horizontal or leaning position, esp. in resting. **2** tr. cause to recline or move from the vertical. □ **reclinable** adj. [ME f. OE recliner or L reclinare bend back, recline (as RE-, clinare bend)]

recliner /rɪˈklaɪnə(r)/ n. **1** a comfortable chair for reclining in. **2** a person who reclines.

reclothe /riːˈkləʊð/ v.tr. clothe again or differently.

recluse /rɪˈkluːs/ n. & adj. ● n. a person given to or living in seclusion or isolation, esp. as a religious discipline; a hermit. ● adj. favouring seclusion; solitary. □ **reclusive** adj. **reclusion** /-ˈkluːʒ(ə)n/ n. [ME f. OF reclus recluse past part. of reclure f. L recludere reclus- (as RE-, claudere shut)]

recognition /ˌrekəgˈnɪʃ(ə)n/ n. the act or an instance of recognizing or being recognized. □ **recognitory** /rɪˈkɒgnɪtərɪ/ adj. [L recognitio (as RECOGNIZE)]

recognizance /rɪˈkɒgnɪz(ə)ns/ n. (also **recognisance**) Law **1** a bond by which a person undertakes before a court or magistrate to observe some condition, e.g. to appear when summoned. **2** a sum pledged as surety for this. [ME f. OF recon(n)issance (as RE-, COGNIZANCE)]

recognizant /rɪˈkɒgnɪz(ə)nt/ adj. (also **recognisant**) (usu. foll. by of) **1** showing recognition (of a favour etc.). **2** conscious or showing consciousness (of something).

recognize /ˈrekəgˌnaɪz/ v.tr. (also **-ise**) **1** identify (a person or thing) as already known; know again. **2** realize or discover the nature of. **3** (foll. by that) realize or admit. **4** acknowledge the existence, validity, character, or claims of. **5** show appreciation of; reward. **6** (foll. by as, for) treat or acknowledge. **7** (of a chairperson etc.) allow (a person) to speak in a debate etc. □ **recognizer** n. **recognizable** adj. **recognizably** adv. **recognizability** /ˌrekəgˌnaɪzəˈbɪlɪtɪ/ n. [OF recon(n)iss- stem of reconnaistre f. L recognoscere recognit- (as RE-, cognoscere learn)]

recoil /rɪˈkɔɪl/ v. & n. ● v.intr. **1** suddenly move or spring back in fear, horror, or disgust. **2** shrink mentally in this way. **3** rebound after an impact. **4** (foll. by on, upon) have an adverse reactive effect on (the originator). **5** (of a gun) be driven backwards by its discharge. **6** retreat under an enemy's attack. **7** Physics (of an atom etc.) move backwards by the conservation of momentum on emission of a particle. ● n. (also /ˈriːkɔɪl/) **1** the act or an instance of recoiling. **2** the sensation of recoiling. [ME f. OF reculer (as RE-, L culus buttocks)]

recollect /ˌrekəˈlekt/ v.tr. **1** remember. **2** succeed in remembering; call to mind. [L recolligere recollect- (as RE-, COLLECT¹)]

re-collect /ˌriːkəˈlekt/ v.tr. **1** collect again. **2** (refl.) recover control of (oneself).

recollection /ˌrekəˈlekʃ(ə)n/ n. **1** the act or power of recollecting. **2** a thing recollected. **3 a** a person's memory (to the best of my recollection). **b** the time over which memory extends (happened within my recollection). □ **recollective** /-ˈlektɪv/ adj. [F recollection or med.L recollectio (as RECOLLECT)]

recolonize /riːˈkɒləˌnaɪz/ v.tr. (also **-ise**) colonize again. □ **recolonization** /-ˌkɒlənaɪˈzeɪʃ(ə)n/ n.

recolour /riːˈkʌlə(r)/ v.tr. (US **recolor**) colour again or differently.

recombinant /riːˈkɒmbɪnənt/ adj. & n. Biol. ● adj. (of a gene etc.) formed by recombination. ● *n.* a recombinant organism or cell. □ **recombinant DNA** DNA that has been recombined using constituents from different sources.

recombination /ˌriːˌkɒmbɪˈneɪʃ(ə)n/ *n. Biol.* the rearrangement, esp. by crossing over in chromosomes, of genes to form a different combination from that in the parents.

recombine /ˌriːkəmˈbaɪn/ *v.tr. & intr.* combine again or differently.

recommence /ˌriːkəˈmens/ *v.tr. & intr.* commence or begin again. □ **recommencement** *n.*

recommend /ˌrekəˈmend/ *v.tr.* **1 a** suggest as fit for some purpose or use. **b** suggest (a person) as suitable for a particular position. **2** (often foll. by *that* + clause or *to* + infin.) advise as a course of action etc. (*I recommend that you stay where you are*). **3** (of qualities, conduct, etc.) make acceptable or desirable. **4** (foll. by *to*) commend or entrust (to a person or a person's care). □ **recommender** *n.* **recommendable** *adj.* **recommendatory** /-dətəri/ *adj.* **recommendation** /ˌrekəmenˈdeɪʃ(ə)n/ *n.* [ME (in sense 4) f. med.L *recommendare* (as RE-, COMMEND)]

recommit /ˌriːkəˈmɪt/ *v.tr.* (**recommitted, recommitting**) **1** commit again. **2** return (a bill etc.) to a committee for further consideration. □ **recommitment** *n.* **recommittal** *n.*

recompense /ˈrekəmˌpens/ *v. & n.* ● *v.tr.* **1** make amends to (a person) or for (a loss etc.). **2** requite; reward or punish (a person or action). ● *n.* **1** a reward; requital. **2** retribution; satisfaction given for an injury. [ME f. OF *recompense(r)* f. LL *recompensare* (as RE-, COMPENSATE)]

recompose /ˌriːkəmˈpəʊz/ *v.tr.* compose again or differently.

reconcile /ˈrekənˌsaɪl/ *v.tr.* **1** make friendly again after an estrangement. **2** (usu. in *refl.* or *passive*; foll. by *to*) make acquiescent or contentedly submissive to (something disagreeable or unwelcome) (*was reconciled to failure*). **3** settle (a quarrel etc.). **4 a** harmonize; make compatible. **b** show the compatibility of by argument or in practice (*cannot reconcile your views with the facts*). □ **reconcilement** *n.* **reconciler** *n.* **reconcilable** *adj.* **reconcilability** /ˌrekənˌsaɪləˈbɪlɪti/ *n.* **reconciliation** /-ˌsɪliˈeɪʃ(ə)n/ *n.* **reconciliatory** /-ˈsɪliətəri/ *adj.* [ME f. OF *reconcilier* or L *reconciliare* (as RE-, *conciliare* CONCILIATE)]

recondite /ˈrekənˌdaɪt, rɪˈkɒndaɪt/ *adj.* **1** (of a subject or knowledge) abstruse; out of the way; little known. **2** (of an author or style) dealing in abstruse knowledge or allusions; obscure. □ **reconditely** *adv.* **reconditeness** *n.* [L *reconditus* (as RE-, *conditus* past part. of *condere* hide)]

recondition /ˌriːkənˈdɪʃ(ə)n/ *v.tr.* **1** overhaul, refit, renovate. **2** make usable again.

reconfigure /ˌriːkənˈfɪɡə(r)/ *v.tr.* configure again or differently. □ **reconfiguration** /-ˌfɪɡəˈreɪʃ(ə)n/ *n.*

reconfirm /ˌriːkənˈfɜːm/ *v.tr.* confirm, establish, or ratify anew. □ **reconfirmation** /riːˌkɒnfəˈmeɪʃ(ə)n/ *n.*

reconnaissance /rɪˈkɒnɪs(ə)ns/ *n.* **1** a survey of a region, esp. a military examination to locate an enemy or ascertain strategic features. **2** a preliminary survey or inspection. [F (earlier -*oissance*) f. stem of *reconnaître* (as RECONNOITRE)]

reconnect /ˌriːkəˈnekt/ *v.tr.* connect again. □ **reconnection** /-ˈnekʃ(ə)n/ *n.*

reconnoitre /ˌrekəˈnɔɪtə(r)/ *v. & n.* (*US* **reconnoiter**) ● *v.* **1** *tr.* make a reconnaissance of (an area, enemy position, etc.). **2** *intr.* make a reconnaissance. ● *n.* a reconnaissance. [obs. F *reconnoître* f. L *recognoscere* RECOGNIZE]

reconquer /riːˈkɒŋkə(r)/ *v.tr.* conquer again. □ **reconquest** /-ˈkɒŋkwest/ *n.*

reconsider /ˌriːkənˈsɪdə(r)/ *v.tr. & intr.* consider again, esp. for a possible change of decision. □ **reconsideration** /-ˌsɪdəˈreɪʃ(ə)n/ *n.*

reconsign /ˌriːkənˈsaɪn/ *v.tr.* consign again or differently. □ **reconsignment** *n.*

reconsolidate /ˌriːkənˈsɒlɪˌdeɪt/ *v.tr. & intr.* consolidate again. □ **reconsolidation** /-ˌsɒlɪˈdeɪʃ(ə)n/ *n.*

reconstitute /riːˈkɒnstɪˌtjuːt/ *v.tr.* **1** build up again from parts; reconstruct. **2** reorganize. **3** restore the previous constitution of (dried food etc.) by adding water. □ **reconstitution** /-ˌkɒnstɪˈtjuːʃ(ə)n/ *n.*

reconstruct /ˌriːkənˈstrʌkt/ *v.tr.* **1** build or form again. **2 a** form a mental or visual impression of (past events) by assembling the evidence for them. **b** re-enact (a crime). **3** reorganize. □ **reconstructable** *adj.* (also **reconstructible**). **reconstructive** *adj.* **reconstructor** *n.*

reconstruction /ˌriːkənˈstrʌkʃ(ə)n/ *n.* **1** the act or a mode of constructing. **2** a thing reconstructed.

Reconstruction, the in US history, the period (1865–77) following the Civil War, during which the southern states of the Confederacy were controlled by Federal government, and social legislation, including the granting of new rights to blacks, was introduced. There was strong white opposition to the new measures (it was during this period that the Ku Klux Klan was first organized); the period ended when a new Republican government returned power to white southern leaders, who then introduced a policy of racial segregation.

reconvene /ˌriːkənˈviːn/ *v.tr. & intr.* convene again, esp. (of a meeting etc.) after a pause in proceedings.

reconvert /ˌriːkənˈvɜːt/ *v.tr.* convert back to a former state. □ **reconversion** /-ˈvɜːʃ(ə)n/ *n.*

record *n. & v.* ● *n.* /ˈrekɔːd/ **1 a** a piece of evidence or information constituting an (esp. official) account of something that has occurred, been said, etc. **b** a document preserving this. **2** the state of being set down or preserved in writing or some other permanent form (*is a matter of record*). **3 a** (also **gramophone record**) a thin plastic disc carrying recorded sound in grooves on each surface, for reproduction by a record-player. (See SOUND RECORDING.) **b** a trace made on this or some other medium, e.g. magnetic tape. **4 a** an official report of the proceedings and judgement in a court of justice. **b** a copy of the pleadings etc. constituting a case to be decided by a court (see also *court of record*). **5 a** the sum of facts known about a person's past (*has an honourable record of service*). **b** (in full **criminal record**) a list of a person's previous criminal convictions; a history of being convicted of crime (*has a record*). **6** the best performance (esp. in sport) or most remarkable event of its kind on record (often *attrib.: a record attempt*). **7** an object serving as a memorial of a person or thing; a portrait. **8** *Computing* a number of related items of information which are handled as a unit. ● *v.tr.* /rɪˈkɔːd/ **1** set down in writing or some other permanent form for later reference, esp. as an official record. **2** convert (sound, a broadcast, etc.) into permanent form for later reproduction. **3** establish or constitute a historical or other record of. □ **break** (or **beat**) **the record** outdo all previous performances etc. **for the record** as an official statement etc. **go on record** state one's opinion or judgement openly or officially, so that it is recorded. **a matter of record** a thing established as a fact by being recorded. **off the record** as an unofficial or confidential statement etc. **on record** officially recorded; publicly known. **put** (or **get** or **set** etc.) **the record straight** correct a misapprehension. **recorded delivery** a Post Office service in which the dispatch and receipt of a letter or parcel are recorded. **recording angel** an angel said to register each person's good and bad actions. **record-player** an apparatus for reproducing sound from gramophone records (see SOUND RECORDING). □ **recordable** /rɪˈkɔːdəb(ə)l/ *adj.* [ME f. OF *record* remembrance, *recorder* record, f. L *recordari* remember (as RE-, *cor cordis* heart)]

recorder /rɪˈkɔːdə(r)/ *n.* **1** an apparatus for recording, esp. a tape recorder. **2 a** a keeper of records. **b** a person who makes an official record. **3** (usu. **Recorder**) (in England and Wales) **a** a barrister or solicitor of at least ten years' standing, appointed to serve as a part-time judge. **b** *hist.* a judge in certain courts. **4** *Mus.* a woodwind instrument like a flute but blown through the end and having a more hollow tone. □ **recordership** *n.* (in sense 3). [ME f. AF *recordour*, OF *recordeur* f. RECORD (in obs. sense 'practise a tune')]

recording /rɪˈkɔːdɪŋ/ *n.* **1** the process by which audio or video signals are recorded for later reproduction. **2** material or a programme recorded. □ **recording artist** a musician or singer who records performances for reproduction and sale under contract to a record company.

recordist /rɪˈkɔːdɪst/ *n.* a person who records sound.

recount /rɪˈkaʊnt/ *v.tr.* **1** narrate. **2** tell in detail. [ONF & AF *reconter* (as RE-, COUNT[1])]

re-count *v. & n.* ● *v.tr.* /riːˈkaʊnt/ count again. ● *n.* /ˈriːkaʊnt/ a re-counting, esp. of votes in an election.

recoup /rɪˈkuːp/ *v.tr.* **1** recover or regain (a loss). **2** compensate or reimburse for a loss. **3** *Law* deduct or keep back (part of a sum due). □ **recoup oneself** recover a loss. □ **recoupable** *adj.* **recoupment** *n.* [F *recouper* (as RE-, *couper* cut)]

recourse /rɪˈkɔːs/ *n.* **1** resorting to a possible source of help. **2** a person or thing resorted to. □ **have recourse to** turn to (a person or thing) for help. **without recourse** a formula used by the endorser of a bill etc. to disclaim responsibility for payment. [ME f. OF *recours* f. L *recursus* (as RE-, COURSE)]

recover /rɪˈkʌvə(r)/ *v. & n.* ● *v.* **1** *tr.* regain possession or use or control of, reclaim. **2** *intr.* return to health or consciousness or to a normal state or position (*have recovered from my illness; the country never recovered from the war*). **3** *tr.* obtain or secure (compensation etc.) by legal process. **4** *tr.* retrieve or make up for (a loss, setback, etc.). **5** *refl.* regain composure or consciousness or control of one's limbs. **6** *tr.* retrieve (reusable substances) from industrial waste. ● *n.* the recovery of a normal position in fencing etc. □ **recoverer** *n.* **recoverable** *adj.* **recoverability** /-ˌkʌvərə'bɪlɪtɪ/ *n.* [ME f. AF *recoverer*, OF *recovrer* f. L *recuperare* RECUPERATE]

re-cover /riː'kʌvə(r)/ *v.tr.* **1** cover again. **2** provide (a chair etc.) with a new cover.

recovery /rɪ'kʌvərɪ/ *n.* (*pl.* **-ies**) **1** the act or an instance of recovering; the process of being recovered. **2** *Golf* a stroke bringing the ball out of a bunker etc. [ME f. AF *recoverie*, OF *reco(u)vree* (as RECOVER)]

recreant /'rekrɪənt/ *adj. & n. literary* ● *adj.* **1** craven, cowardly. **2** apostate. ● *n.* **1** a coward. **2** an apostate. □ **recreancy** *n.* **recreantly** *adv.* [ME f. OF, part. of *recroire* f. med.L (*se*) *recredere* yield in trial by combat (as RE-, *credere* entrust)]

re-create /ˌriːkrɪ'eɪt/ *v.tr.* create or produce over again. □ **re-creation** /-'eɪʃ(ə)n/ *n.*

recreation /ˌrekrɪ'eɪʃ(ə)n/ *n.* **1** the process or means of refreshing or entertaining oneself. **2** a pleasurable activity. □ **recreation-ground** public land for games etc. □ **recreational** *adj.* **recreationally** *adv.*

recreative /'rekrɪˌeɪtɪv/ *adj.* [ME f. OF f. L *recreatio -onis* f. *recreare* create again, renew]

recriminate /rɪ'krɪmɪˌneɪt/ *v.intr.* make mutual or counter accusations. □ **recriminative** /-nətɪv/ *adj.* **recriminatory** *adj.* **recrimination** /-ˌkrɪmɪ'neɪʃ(ə)n/ *n.* [med.L *recriminare* (as RE-, *criminare* accuse f. *crimen* CRIME)]

recross /riː'krɒs/ *v.tr. & intr.* cross or pass over again.

recrudesce /ˌriːkruː'des, ˌrek-/ *v.intr.* (of a disease or difficulty etc.) break out again, esp. after a dormant period. □ **recrudescence** *n.* **recrudescent** *adj.* [back-form. f. *recrudescent, -ence* f. L *recrudescere* (as RE-, *crudus* raw)]

recruit /rɪ'kruːt/ *n. & v.* ● *n.* **1** a serviceman or servicewoman newly enlisted and not yet fully trained. **2** a new member of a society or organization. **3** a beginner. ● *v.* **1** *tr.* enlist (a person) as a recruit. **2** *tr.* form (an army etc.) by enlisting recruits. **3** *intr.* get or seek recruits. **4** *tr.* replenish or reinvigorate (numbers, strength, etc.). □ **recruitable** *adj.* **recruiter** *n.* **recruitment** *n.* [earlier = reinforcement, f. obs. F dial. *recrute* ult. f. F *recroître* increase again f. L *recrescere*]

recrystallize /riː'krɪstəˌlaɪz/ *v.tr. & intr.* (also **-ise**) crystallize again. □ **recrystallization** /-ˌkrɪstəlaɪ'zeɪʃ(ə)n/ *n.*

recta *pl.* of RECTUM.

rectal /'rektəl/ *adj.* of or by means of the rectum. □ **rectally** *adv.*

rectangle /'rekˌtæŋg(ə)l/ *n.* a plane figure with four straight sides and four right angles, esp. one with the adjacent sides unequal. [F *rectangle* or med.L *rectangulum* f. LL *rectiangulum* f. L *rectus* straight + *angulus* ANGLE[1]]

rectangular /rek'tæŋɡjʊlə(r)/ *adj.* **1 a** shaped like a rectangle. **b** having the base or sides or section shaped like a rectangle. **2 a** placed at right angles. **b** having parts or lines placed at right angles. □ **rectangular coordinates** coordinates measured along axes at right angles. **rectangular hyperbola** a hyperbola with rectangular asymptotes. □ **rectangularly** *adv.* **rectangularity** /-ˌtæŋɡjʊ'lærɪtɪ/ *n.*

recti *pl.* of RECTUS.

rectifier /'rektɪˌfaɪə(r)/ *n.* **1** a person or thing that rectifies. **2** *Electr.* an electrical device that allows a current to flow preferentially in one direction by converting an alternating current into a direct one.

rectify /'rektɪˌfaɪ/ *v.tr.* (**-ies, -ied**) **1** adjust or make right; correct, amend. **2** purify or refine, esp. by repeated distillation. **3** find a straight line equal in length to (a curve). **4** convert (alternating current) to direct current. □ **rectified spirit** *Chem.* an azeotropic mixture of alcohol (95.6 %) and water produced by distillation. □ **rectifiable** *adj.* **rectification** /ˌrektɪfɪ'keɪʃ(ə)n/ *n.* [ME f. OF *rectifier* f. med.L *rectificare* f. L *rectus* right]

rectilinear /ˌrektɪ'lɪnɪə(r)/ *adj.* (also **rectilineal** /-nɪəl/) **1** bounded or characterized by straight lines. **2** in or forming a straight line. □ **rectilinearly** *adv.* **rectilinearity** /-ˌlɪnɪ'ærɪtɪ/ *n.* [LL *rectilineus* f. L *rectus* straight + *linea* LINE[1]]

rectitude /'rektɪˌtjuːd/ *n.* **1** moral uprightness. **2** righteousness. **3** correctness. [ME f. OF *rectitude* or LL *rectitudo* f. L *rectus* right]

recto /'rektəʊ/ *n.* (*pl.* **-os**) **1** the right-hand page of an open book. **2** the front of a printed leaf of paper or manuscript (opp. VERSO 1b). [L *recto* (*folio*) on the right (leaf)]

rector /'rektə(r)/ *n.* **1** (in the Church of England) the incumbent of a parish where all tithes formerly passed to the incumbent (cf. VICAR 1a). **2** *RC Ch.* a priest in charge of a church or religious institution. **3 a** the head of some schools, universities, and colleges. **b** (in Scotland) an elected representative of students on a university's governing body. □ **rectorship** *n.* **rectorate** /-rət/ *n.* **rectorial** /rek'tɔːrɪəl/ *adj.* [ME f. OF *rectour* or L *rector* ruler f. *regere rect-* rule]

rectory /'rektərɪ/ *n.* (*pl.* **-ies**) **1** a rector's house. **2** (in the Church of England) a rector's benefice. [AF & OF *rectorie* or med.L *rectoria* (as RECTOR)]

rectrix /'rektrɪks/ *n.* (*pl.* **rectrices** /-trɪˌsiːz/) a bird's strong tail feather directing flight. [L, fem. of *rector* ruler: see RECTOR]

rectum /'rektəm/ *n.* (*pl.* **rectums** or **recta** /-tə/) the final section of the large intestine, terminating at the anus. [L *rectum* (*intestinum*) straight (intestine)]

rectus /'rektəs/ *n.* (*pl.* **recti** /-taɪ/) *Anat.* a straight muscle. [L, = straight]

recumbent /rɪ'kʌmb(ə)nt/ *adj.* lying down; reclining. □ **recumbency** *n.* **recumbently** *adv.* [L *recumbere* recline (as RE-, *cumbere* lie)]

recuperate /rɪ'kuːpəˌreɪt/ *v.* **1** *intr.* recover from illness, exhaustion, loss, etc. **2** *tr.* regain (health, something lost, etc.). □ **recuperator** *n.* **recuperable** /-rəb(ə)l/ *adj.* **recuperative** *adj.* **recuperation** /-ˌkuːpə'reɪʃ(ə)n/ *n.* [L *recuperare recuperat-* recover]

recur /rɪ'kɜː(r)/ *v.intr.* (**recurred, recurring**) **1** occur again; be repeated. **2** (of a thought, idea, etc.) come back to one's mind. **3** (foll. by *to*) go back in thought or speech. □ **recurring decimal** a decimal fraction in which the same figures are repeated indefinitely. [L *recurrere recurs-* (as RE-, *currere* run)]

recurrent /rɪ'kʌrənt/ *adj.* **1** recurring; happening repeatedly. **2** (of a nerve, vein, branch, etc.) turning back so as to reverse direction. □ **recurrence** *n.* **recurrently** *adv.*

recursion /rɪ'kɜːʃ(ə)n/ *n. Math. & Linguistics* **1** the application or use of a recursive procedure or definition. **2** a recursive definition. □ **recursion formula** *Math.* an expression giving successive terms of a series etc. [LL *recursio* (as RECUR)]

recursive /rɪ'kɜːsɪv/ *adj.* **1** characterized by recurrence or repetition. **2 a** *Math. & Linguistics* relating to or involving the repeated application of a rule, definition, or procedure to successive results. **b** *Computing* relating to or involving a program or routine, a part of which requires the application of the whole. □ **recursively** *adv.*

recurve /rɪ'kɜːv/ *v.tr. & intr.* bend backwards. □ **recurvate** /-veɪt/ *adj.* **recurvature** /-vətʃə(r)/ *n.* [L *recurvare recurvat-* (as RE-, *curvare* bend)]

recusant /'rekjʊz(ə)nt/ *n. & adj.* ● *n.* a person who refuses submission to an authority or compliance with a regulation, esp. *hist.* one who refused to attend services of the Church of England. ● *adj.* of or being a recusant. □ **recusance** *n.* **recusancy** *n.* [L *recusare* refuse]

recycle /riː'saɪk(ə)l/ *v.tr.* return (material) to a previous stage of a cyclic process, esp. convert (waste) to reusable material. □ **recyclable** *adj.*

red /red/ *adj. & n.* ● *adj.* (**redder, reddest**) **1** of or near the colour seen at the least-refracted end of the visible spectrum, of shades ranging from that of blood to pink or deep orange. **2** flushed in the face with shame, anger, etc. **3** (of the eyes) bloodshot or red-rimmed with weeping. **4** (of the hair) reddish-brown, orange, tawny. **5** involving or having to do with bloodshed, burning, violence, or revolution. **6** *colloq.* communist or socialist. **7** (**Red**) *hist.* Soviet or (formerly) Russian (*the Red Army*). **8** (of wine) made from dark grapes and coloured by their skins. ● *n.* **1** a red colour or pigment. **2** red clothes or material (*dressed in red*). **3** *colloq.* a communist or socialist. **4 a** a red ball, piece, etc., in a game or sport. **b** the player using such pieces. **5** the debit side of an account (*in the red*). **6** a red light. □ **red admiral** a butterfly, *Vanessa atalanta*, having black wings with red and white markings. **red-back** a venomous Australian spider of the genus *Latrodectus*. **red bark** a red kind of cinchona. **red biddy** *colloq.* a mixture of cheap wine and methylated spirits. **red blood cell** (or **corpuscle**) = ERYTHROCYTE. **red-blooded** virile, vigorous. **red-bloodedness** vigour, spirit. **red card** *Football* a card shown by the referee to a player being sent off the field. **the red carpet** privileged treatment of an eminent visitor. **red cedar** a large American juniper, *Juniperus virginiana*. **red cell** (or **corpuscle**) = ERYTHROCYTE. **red cent** *N. Amer.* the smallest (orig. copper) coin; a trivial sum. **red coral** see CORAL *n.* 1b. **red cross 1** St George's cross, the national emblem of England. **2** the Christian side in the Crusades. **red deer** a large Eurasian deer, *Cervus elaphus*, with a rich red-brown summer coat turning dull brown in winter. **red**

duster *Brit. colloq.* = *red ensign* (see ENSIGN). **red dwarf** *Astron.* an old relatively cool star of very small size. **red ensign** see ENSIGN. **red-eye 1** = RUDD. **2** *US sl.* cheap whisky. **3** a red reflection from a person's retina, seen in a flash photograph taken with the flash gun too near the camera lens. **red-faced** embarrassed, ashamed. **red flag 1** a symbol of danger. **2** the symbol of socialist revolution. **red fox** the common fox of Eurasia and North America, *Vulpes vulpes*, usu. having a reddish or fawn coat. **red grouse** a race of the willow grouse, *Lagopus lagopus*, native to Britain and familiar as a game bird. **red gum 1** a teething-rash in children. **2 a** a reddish resin. **b** a kind of eucalyptus yielding this. **red-handed** in or just after the act of committing a crime, doing wrong, etc. **red hat 1** a cardinal's hat. **2** the symbol of a cardinal's office. **red-headed 1** (of a person) having red hair. **2** (of birds etc.) having a red head. **red heat 1** the temperature or state of something so hot as to emit red light. **2** great excitement. **red herring 1** dried smoked herring. **2** a misleading clue or distraction (so called from the practice of using the scent of red herring in training hounds). **red-hot 1** heated until red. **2** highly exciting. **3** (of news) fresh; completely new. **4** intensely excited. **5** enraged. **red-hot poker** a cultivated kniphofia (plant), esp. *Kniphofia uvaria*, with the upper flowers in the spike red and the lower ones yellow. **red lead** a red form of lead oxide used as a pigment. **Red Leicester** see LEICESTER[3]. **red-letter day** a day that is pleasantly noteworthy or memorable (orig. a festival marked in red on the calendar). **red light 1** a red light used as a signal to stop on a road, railway, etc. **2** a warning or refusal. **red-light district** an urban district containing many brothels. **red man** = RED INDIAN. **red meat** meat that is red when raw (e.g. beef or lamb). **red mullet** a mullet of the family Mullidae, esp. *Mullus surmuletus* of Europe, valued as food. **red pepper 1** cayenne pepper. **2** the ripe fruit of the capsicum plant, *Capsicum annuum*. **red rag** something that excites a person's rage (so called because red is supposed to provoke bulls). **red rattle** a pink-flowered lousewort, *Pedicularis palustris*, found on marshy ground. **red roan** see ROAN[1]. **red rose** the emblem of Lancashire or the Lancastrians. **red shift** *Astron.* the displacement of spectral lines towards longer wavelengths (the red end of the spectrum) in radiation from distant galaxies etc., interpreted as a Doppler shift which is proportional to the velocity of recession and thus to distance. **red snapper** an edible marine fish of the family Lutjanidae, esp. *Lutjanus campechinus* of the NW Atlantic. **red spider mite** see *spider mite*. **red squirrel** the common Eurasian squirrel, *Sciurus vulgaris*, with usu. reddish fur and tufted ears, now scarce in much of lowland Britain. **red tape** excessive bureaucracy or adherence to formalities, esp. in public business. **red tide** a discoloration of the sea caused by an outbreak of toxic red dinoflagellates, esp. of the genus *Gonyaulax*. □ **reddish** *adj.* **reddy** *adj.* **redly** *adv.* **redness** *n.* [OE *rēad* f. Gmc]

redact /rɪ'dækt/ *v.tr.* put into literary form; edit for publication. □ **redactor** *n.* [L *redigere redact-* (as RE-, *agere* bring)]

redaction /rɪ'dækʃ(ə)n/ *n.* **1** preparation for publication. **2** revision, editing, rearrangement. **3** a new edition. □ **redactional** *adj.* [F *rédaction* f. LL *redactio* (as REDACT)]

redan /rɪ'dæn/ *n.* a fieldwork with two faces forming a salient angle. [F f. *redent* notching (as RE-, *dent* tooth)]

Red Army 1 *hist.* originally, the army of the Bolsheviks, the Workers' and Peasants' Red Army; later, the army of the Soviet Union, formed after the Revolution of 1917. The name was officially dropped in 1946. **2** the army of China or some other (esp. Communist) countries. **3** a left-wing extremist terrorist organization in Japan.

Red Army Faction a left-wing terrorist group in former West Germany, active from 1968 onwards. It was originally led by Andreas Baader (1943–77) and Ulrike Meinhof (1934–76), after whom it is often called the Baader-Meinhof Group.

redbreast /'redbrest/ *n. colloq.* a robin.

redbrick /'redbrɪk/ *adj.* (of a British university) founded in the late 19th or early 20th century.

Red Brigades an extreme left-wing terrorist organization based in Italy, which from the early 1970s was responsible for carrying out kidnappings, murders, and acts of sabotage. A former Prime Minister of Italy, Aldo Moro, was killed by the Red Brigades in 1978.

redbud /'redbʌd/ *n.* an American leguminous tree of the genus *Cercis*, with pale pink flowers.

redcap /'redkæp/ *n.* **1** *Brit.* a member of the military police. **2** *N. Amer.* a railway porter.

redcoat /'redkəʊt/ *n.* **1** *hist.* a British soldier (so called from the scarlet uniform of most regiments). **2** (in the UK) a steward at a Butlin's holiday camp.

Red Crescent the name used by national branches in Muslim countries of the International Movement of the Red Cross and the Red Crescent, adopted in 1906. (See RED CROSS.)

Red Cross the International Movement of the Red Cross and the Red Crescent (formerly the *International Red Cross*), an international humanitarian organization originally established to treat the sick and wounded in war and later also aiming to help those suffering the effects of large-scale natural disasters. It was set up at the instigation of the Swiss philanthropist Henri Dunant (1828–1910) according to the Geneva Convention of 1864, and its headquarters are at Geneva. In non-Muslim countries its emblem is a red cross on a white background, while in Muslim countries (where it is known as the *Red Crescent*) the emblem is a red crescent. The international organization operates through national societies; the British Red Cross Society was incorporated in 1908.

redcurrant /'red,kʌrənt/ *n.* **1** a widely cultivated shrub, *Ribes rubrum*. **2** the small red edible berry of this plant.

redd[1] /red/ *v.tr.* (*past and past part.* **redd**) *dial.* **1** clear up. **2** arrange, tidy, compose, settle. [ME: cf. MLG, MDu. *redden*]

redd[2] /red/ *n.* a hollow in a river bed made by a trout or salmon to spawn in. [17th c.: orig. unkn.]

redden /'red(ə)n/ *v.* **1** *tr. & intr.* make or become red. **2** *intr.* blush.

Redditch /'redɪtʃ/ an industrial town in west central England, in Hereford and Worcester; pop. (1991) 76,900.

reddle /'red(ə)l/ *n.* red ochre; ruddle. [var. of RUDDLE]

rede /riːd/ *n. & v. archaic* ● *n.* advice, counsel. ● *v.tr.* **1** advise. **2** read (a riddle or dream). [OE *rǣd* f. Gmc, rel. to READ (of which the verb is a ME var. retained for archaic senses)]

redecorate /riː'dekə,reɪt/ *v.tr.* decorate again or differently. □ **redecoration** /-,dekə'reɪʃ(ə)n/ *n.*

redeem /rɪ'diːm/ *v.tr.* **1** buy back; recover by expenditure of effort or by a stipulated payment. **2** make a single payment to discharge (a regular charge or obligation). **3** convert (tokens or bonds etc.) into goods or cash. **4** (of God or Christ) deliver from sin and damnation. **5** make up for; be a compensating factor in (*has one redeeming feature*). **6** (foll. by *from*) save from (a defect). **7** *refl.* save (oneself) from blame. **8** purchase the freedom of (a person). **9** save (a person's life) by ransom. **10** save or rescue or reclaim. **11** fulfil (a promise). □ **redeemable** *adj.* [ME f. OF *redimer* or L *redimere redempt-* (as RE-, *emere* buy)]

redeemer /rɪ'diːmə(r)/ *n.* a person who redeems. □ **the Redeemer** Christ.

redefine /,riːdɪ'faɪn/ *v.tr.* define again or differently. □ **redefinition** /riː,defɪ'nɪʃ(ə)n/ *n.*

redemption /rɪ'dempʃ(ə)n/ *n.* **1** the act or an instance of redeeming; the process of being redeemed. **2** (in Christian theology) humankind's deliverance from sin and damnation. **3** a thing that redeems. □ **redemptive** /-'demptɪv/ *adj.* [ME f. OF f. L *redemptio* (as REDEEM)]

redeploy /,riːdɪ'plɔɪ/ *v.tr.* send (troops, workers, etc.) to a new place or task. □ **redeployment** *n.*

redesign /,riːdɪ'zaɪn/ *v.tr.* design again or differently.

redetermine /,riːdɪ'tɜːmɪn/ *v.tr.* determine again or differently. □ **redetermination** /-,tɜːmɪ'neɪʃ(ə)n/ *n.*

redevelop /,riːdɪ'veləp/ *v.tr.* develop anew (esp. an urban area, with new buildings). □ **redeveloper** *n.* **redevelopment** *n.*

redfish /'redfɪʃ/ *n.* **1** a male salmon in the spawning season. **2** a rose-fish.

Red Flag, the a socialist song with words written in 1889 by James Connell (1852–1929), secretary to the Workmen's Legal Friendly Society, and sung to the tune of the German song 'O Tannenbaum'. It is the anthem of Britain's Labour Party.

Redford /'redfəd/, (Charles) Robert (b.1936), American film actor and director. He made his name playing opposite Paul Newman in *Butch Cassidy and the Sundance Kid* (1969), co-starring again with him in *The Sting* (1973). Other notable films include *The Great Gatsby* (1974), *All the President's Men* (1976), and *Out of Africa* (1986). Redford won an Oscar as the director of *Ordinary People* (1980).

red giant *n. Astron.* a very large star of high luminosity and low surface temperature. Red giants are thought to be in a late stage of evolution when no hydrogen remains in the core to fuel nuclear fusion, but reactions involving hydrogen may continue in a spherical shell. The radius may exceed 150 million km, so that the entire orbit of earth about the sun could be fitted inside the star's tenuous envelope. The surface temperature may be as low as 3,000°C, but its great size renders such a star many times more luminous than the sun. Typical examples are Betelgeuse in Orion and Aldebaran in Taurus.

Redgrave /ˈrɛdgreɪv/ a family of English actors. Sir Michael (Scudamore) (1908–85) was a well-known stage actor, who played numerous Shakespearian roles as well as appearing in other plays, notably in the title role of *Uncle Vanya* (1963). He also starred in films such as *The Browning Version* (1951) and *The Importance of Being Earnest* (1952). His elder daughter Vanessa (b.1937) has had a successful career in the theatre and cinema: her films include *Mary Queen of Scots* (1972), *Julia* (1976), for which she won an Oscar, and *Howard's End* (1992). His son Corin (b.1939) is also an actor, as is his younger daughter Lynn (b.1944), who has made a number of stage and screen performances, and is best known for such films as *Georgy Girl* (1966). Vanessa's two daughters Joely (b.1958) and Natasha Richardson (b.1963) are both actresses.

Red Guard *n.* any of various radical or socialist groups. The term is used in particular to refer to an organized detachment of workers during the Russian Revolution of 1917, or, more importantly, to a militant youth movement in China (1966–76), who carried out attacks on intellectuals and other disfavoured groups as part of Mao Zedong's Cultural Revolution.

redhead /ˈrɛdhɛd/ *n.* a person with red hair.

redial /riːˈdaɪəl/ *v.tr. & intr.* (**redialled**, **redialling**; *US* **redialed**, **redialing**) dial again.

redid past of REDO.

rediffusion /ˌriːdɪˈfjuːʒ(ə)n/ *n.* the relaying of broadcast programmes esp. by cable from a central receiver.

Red Indian *n. offens.* an American Indian or Native American.

redingote /ˈrɛdɪŋgəʊt/ *n.* a woman's long coat with a cutaway front or a contrasting piece on the front. [F f. Engl. *riding-coat*]

redintegrate /rɪˈdɪntɪgreɪt/ *v.tr.* **1** restore to wholeness or unity. **2** renew or re-establish in a united or perfect state. □ **redintegrative** *adj.* **redintegration** /-ˌdɪntɪˈgreɪʃ(ə)n/ *n.* [ME f. L *redintegrare* (as RE-, INTEGRATE)]

redirect /ˌriːdaɪˈrɛkt, -dɪˈrɛkt/ *v.tr.* direct again, esp. change the address of (a letter). □ **redirection** /-ˈrɛkʃ(ə)n/ *n.*

rediscover /ˌriːdɪˈskʌvə(r)/ *v.tr.* discover again. □ **rediscovery** *n.* (pl. **-ies**).

redissolve /ˌriːdɪˈzɒlv/ *v.tr. & intr.* dissolve again. □ **redissolution** /riːˌdɪsəˈluːʃ(ə)n/ *n.*

redistribute /ˌriːdɪˈstrɪbjuːt, *disp.* riːˈdɪs-/ *v.tr.* distribute again or differently. □ **redistributive** *adj.* **redistribution** /riːˌdɪstrɪˈbjuːʃ(ə)n/ *n.*

redivide /ˌriːdɪˈvaɪd/ *v.tr.* divide again or differently. □ **redivision** /-ˈvɪʒ(ə)n/ *n.*

redivivus /ˌrɛdɪˈviːvəs/ *adj.* (placed after noun) come back to life. [L as RE-, *vivus* living)]

Redmond /ˈrɛdmənd/, John (Edward) (1856–1918), Irish politician. He succeeded Charles Parnell as leader of the Irish Nationalist Party in the House of Commons (1891–1918). The Home Rule Bill of 1912 was introduced with his support, although it was never implemented because of the First World War.

redneck /ˈrɛdnɛk/ *n. N. Amer. often derog.* a working-class white in the southern US, esp. a politically conservative one.

redo /riːˈduː/ *v.tr.* (3rd sing. present **redoes** /-ˈdʌz/; past **redid** /-ˈdɪd/; past part. **redone** /-ˈdʌn/) **1** do again or differently. **2** redecorate.

redolent /ˈrɛdələnt/ *adj.* **1** (foll. by *of, with*) strongly reminiscent or suggestive or mentally associated. **2** fragrant. **3** having a strong smell; odorous. □ **redolence** *n.* **redolently** *adv.* [ME f. OF *redolent* or L *redolere* (as RE-, *olere* smell)]

Redon /rəˈdɒn/, Odilon (1840–1916), French painter and graphic artist. He was a leading exponent of symbolism, and an important forerunner of the surrealists, especially in his early work, which chiefly consisted of charcoal drawings of fantastic, often nightmarish, subjects. He began to use colour from about 1890 onwards, becoming known for his richly coloured pastels depicting flowers, mythological subjects, and portraits.

redouble /riːˈdʌb(ə)l/ *v. & n.* ● *v.* **1** *tr. & intr.* make or grow greater or more intense or numerous; intensify, increase. **2** *intr. Bridge* double again a bid already doubled by an opponent. ● *n. Bridge* the redoubling of a bid. [F *redoubler* (as RE-, DOUBLE)]

redoubt /rɪˈdaʊt/ *n. Mil.* an outwork or fieldwork usu. square or polygonal and without flanking defences. [F *redoute* f. obs. It. *ridotta* f. med.L *reductus* refuge f. past part. of L *reducere* withdraw (see REDUCE): *-b-* after DOUBT (cf. REDOUBTABLE)]

redoubtable /rɪˈdaʊtəb(ə)l/ *adj.* formidable, esp. as an opponent. □ **redoubtably** *adv.* [ME f. OF *redoutable* f. *redouter* fear (as RE-, DOUBT)]

redound /rɪˈdaʊnd/ *v.intr.* **1** (foll. by *to*) (of an action etc.) make a great contribution to (one's credit or advantage etc.). **2** (foll. by *upon, on*) come as the final result to; come back or recoil upon. [ME, orig. = overflow, f. OF *redonder* f. L *redundare* surge (as RE-, *unda* wave)]

redox /ˈrɛdɒks, ˈriːd-/ *n.* (often *attrib.*) *Chem.* oxidation and reduction. [reduction + oxidation]

redpoll /ˈrɛdpəʊl/ *n.* **1** a finch, *Acanthis flammea*, resembling a linnet but with a red forehead. **2** a breed of red-haired polled cattle. [RED + POLL¹]

redraft /riːˈdrɑːft/ *v.tr.* draft (a writing or document) again.

redraw /riːˈdrɔː/ *v.* (past **redrew** /-ˈdruː/; past part. **redrawn** /-ˈdrɔːn/) draw again or differently.

redress /rɪˈdrɛs/ *v. & n.* ● *v.tr.* **1** remedy or rectify (a wrong or grievance etc.). **2** readjust; set straight again. ● *n.* **1** reparation for a wrong. **2** (foll. by *of*) the act or process of redressing (a grievance etc.). □ **redress the balance** restore equality. □ **redressable** *adj.* **redressal** *n.* **redresser** *n.* (also **redressor**). [ME f. OF *redresse(r)*, *redrecier* (as RE-, DRESS)]

re-dress /riːˈdrɛs/ *v.tr. & intr.* dress again or differently.

Red River 1 (called in Chinese *Yuan Jiang*, Vietnamese *Song Hong*) a river in SE Asia, which rises in southern China and flows 1,175 km (730 miles) generally south-eastwards through northern Vietnam to the Gulf of Tonkin north of Haiphong. **2** a river in the southern US, a tributary of the Mississippi, which rises in northern Texas and flows 1,966 km (1,222 miles) generally south-eastwards, forming part of the border between Texas and Oklahoma, and enters the Mississippi in Louisiana, north of Baton Rouge. It is also known as *Red River of the South*. **3** a river in the northern US and Canada, which rises in North Dakota and flows 877 km (545 miles) northwards, forming for most of its length the border between North Dakota and Minnesota, before entering Canada and emptying into Lake Winnipeg. It is also known as *Red River of the North*.

Red Sea a long, narrow landlocked sea separating Africa from the Arabian peninsula. It is linked to the Indian Ocean in the south by the Gulf of Aden and to the Mediterranean in the north by the Suez Canal.

redshank /ˈrɛdʃæŋk/ *n.* either of two sandpipers, the *common redshank* (*Tringa totanus*) and the *spotted redshank* (*T. erythropus*), with bright red legs.

redskin /ˈrɛdskɪn/ *n. colloq. offens.* an American Indian or Native American.

Red Square a large square in Moscow next to the Kremlin. In existence since the late 15th century, the square was the scene under Communism of great parades celebrating May Day and the October Revolution.

Red Star *n.* the emblem of the former Soviet Union and some other Communist countries.

redstart /ˈrɛdstɑːt/ *n.* **1** a red-tailed Eurasian songbird of the genus *Phoenicurus*, of the thrush family. **2** an American warbler of the family Parulidae, with orange-red markings. [RED + OE *steort* tail]

reduce /rɪˈdjuːs/ *v.* **1** *tr. & intr.* make or become smaller or less. **2** *tr.* (foll. by *to*) bring by force or necessity (to some undesirable state or action) (*reduced them to tears; were reduced to begging*). **3** *tr.* convert to another (esp. simpler) form (*reduced it to a powder*). **4** *tr.* convert (a fraction) to the form with the lowest terms. **5** *tr.* (foll. by *to*) bring or simplify or adapt by classification or analysis (*the dispute may be reduced to three issues*). **6** *tr.* make lower in status or rank. **7** *tr.* lower the price of. **8** *intr.* lessen one's weight or size. **9** *tr.* weaken (*is in a very reduced state*). **10** *tr.* impoverish. **11** *tr.* subdue; bring back to obedience. **12** *Chem. intr. & tr.* **a** combine or cause to combine with hydrogen. **b** undergo or cause to undergo addition of electrons (opp. OXIDIZE 3). **13** *tr. Chem.* convert (oxide etc.) to metal. **14** *tr.* **a** (in surgery) restore (a dislocated etc. part) to its proper position. **b** remedy (a dislocation etc.) in this way. **15** *tr. Photog.* make (a negative or print) less dense. **16** *tr. Cookery* boil so as to concentrate (a liquid, sauce, etc.). □ **reduced circumstances** poverty after relative prosperity. **reduce to the ranks** demote (an NCO) to the rank of private. **reducing agent** *Chem.* a substance that brings about reduction by becoming oxidized and losing electrons. □ **reducer** *n.* **reducible** *adj.* **reducibility** /-ˌdjuːsɪˈbɪlɪti/ *n.* [ME in sense 'restore to original or proper position', f. L *reducere reduct-* (as RE-, *ducere* bring)]

reductio ad absurdum /rɪ,dʌktɪəʊ æd æb'zɜːdəm/ *n.* a method of proving the falsity of a premiss by showing that the logical consequence is absurd; an instance of this. [L, = reduction to the absurd]

reduction /rɪ'dʌkʃ(ə)n/ *n.* **1** the act or an instance of reducing; the process of being reduced. **2** an amount by which prices etc. are reduced. **3** a reduced copy of a picture etc. **4** an arrangement of an orchestral score for piano etc. □ **reductive** /-'dʌktɪv/ *adj.* [ME f. OF *reduction* or L *reductio*]

reductionism /rɪ'dʌkʃə,nɪz(ə)m/ *n.* **1** the tendency to or principle of analysing complex things into simple constituents. **2** often *derog.* the doctrine that a system can be fully understood in terms of its isolated parts, or an idea in terms of simple concepts (cf. HOLISM 1). □ **reductionist** *n.* **reductionistic** /-,dʌkʃə'nɪstɪk/ *adj.*

redundant /rɪ'dʌndənt/ *adj.* **1** superfluous; not needed. **2** that can be omitted without any loss of significance. **3** (of a person) no longer needed at work and therefore unemployed. **4** *Engin. & Computing* (of a component) not needed but included in case of failure in another component. □ **redundancy** *n.* (*pl.* **-ies**). **redundantly** *adv.* [L *redundare redundant-* (as REDOUND)]

reduplicate /rɪ'djuːplɪ,keɪt/ *v.tr.* **1** make double. **2** repeat. **3** repeat (a letter or syllable or word) exactly or with a slight change (e.g. hurly-burly, see-saw). □ **reduplicative** /-kətɪv/ *adj.* **reduplication** /-,djuːplɪ'keɪʃ(ə)n/ *n.* [LL *reduplicare* (as RE-, DUPLICATE)]

redwater /'red,wɔːtə(r)/ *n.* (in full **redwater fever**) a blood disease of cattle characterized by the passing of red or blackish urine, caused esp. by a tick-borne protozoan parasite of the genus *Babesia*.

redwing /'redwɪŋ/ *n.* a migratory northern European thrush, *Turdus iliacus*, with red underwings showing in flight.

redwood /'redwʊd/ *n.* **1** a sequoia; the *Sequoia sempervirens*, native to the coast ranges of Oregon and northern California, which has red wood and is the tallest known tree. **2** any tree yielding red wood. □ **dawn redwood** see DAWN.

reebok var. of RHEBOK.

re-echo /riː'ekəʊ/ *v.intr. & tr.* (**-oes, -oed**) **1** echo. **2** echo repeatedly; resound.

Reed[1] /riːd/, Sir Carol (1906–76), English film director. He made a succession of celebrated films in the postwar years, including *Odd Man Out* (1947) and *The Third Man* (1949), starring Orson Welles. Among his notable later films are the *Outcast of the Islands* (1952) and the musical *Oliver!* (1968), for which he won an Oscar.

Reed[2] /riːd/, Walter (1851–1902), American physician. He worked mainly in the US Army Medical Corps, and finally headed the Yellow Fever Board (1900–01), based in Cuba. His group proved that the disease was transmitted by the mosquito *Aedes aegypti*, and then showed that the agent responsible was a virus — the first to be recognized as the cause of a human disease. The mosquito's breeding places were successfully attacked, and a vaccine was developed some years later.

reed /riːd/ *n. & v.* ● *n.* **1 a** a water or marsh plant with a tall straight stem and long narrow leaves, esp. one of the genus *Phragmites* or *Arundo*. **b** a stalk of this. **2** (*collect.*) reeds growing in a mass or used as material esp. for thatching. **3** *Brit.* wheat-straw prepared for thatching. **4** a pipe of reed or straw. **5 a** the vibrating part of the mouthpiece of some wind instruments, e.g. the oboe and clarinet, made of reed or other material and producing the sound. **b** (esp. in *pl.*) a reed instrument. **6** a weaver's comblike implement for separating the threads of the warp and correctly positioning the weft. **7** (in *pl.*) *Archit.* a set of semicylindrical adjacent mouldings like reeds laid together. ● *v.tr.* **1** thatch with reed. **2** make (straw) into reed. **3** fit (a musical instrument) with a reed. **4** decorate with a moulding of reeds. □ **reed bunting** a brown Eurasian bunting, *Emberiza schoeniclus*, frequenting reed-beds and similar habitats. **reed-mace** a tall reedlike water-plant, *Typha latifolia*, with straplike leaves and a velvety dark brown flower spike (cf. BULRUSH 1, 2). **reed-organ** a harmonium etc. with the sound produced by metal reeds. **reed-pipe 1** a wind instrument with sound produced by a reed. **2** an organ-pipe with a reed. **reed-stop** a reeded organ-stop. **reed-warbler** an Old World warbler of the genus *Acrocephalus*, with plain brown plumage, inhabiting reed-beds; esp. *A. scirpaceus*. [OE *hrēod* f. WG]

reedbuck /'riːdbʌk/ *n.* an African antelope of the genus *Redunca*, with whistling calls and high bouncing jumps; esp. *R. arundinum* of southern and East Africa.

reeded /'riːdɪd/ *adj. Mus.* (of an instrument) having a vibrating reed.

reeding /'riːdɪŋ/ *n. Archit.* a small semicylindrical moulding or ornamentation (cf. REED *n.* 7).

re-edit /riː'edɪt/ *v.tr.* (**-edited, -editing**) edit again or differently. □ **re-edition** /,riːɪ'dɪʃ(ə)n/ *n.*

reedling /'riːdlɪŋ/ *n.* a bearded tit.

re-educate /riː'edjʊ,keɪt/ *v.tr.* educate again, esp. to change a person's views or beliefs. □ **re-education** /-,edjʊ'keɪʃ(ə)n/ *n.*

reedy /'riːdɪ/ *adj.* (**reedier, reediest**) **1** full of reeds. **2** like a reed, esp. in weakness or slenderness. **3** (of a voice) like a reed instrument in tone; not full, weak and thin. □ **reediness** *n.*

reef[1] /riːf/ *n.* **1** a ridge of rock or coral etc. at or near the surface of the sea. **2 a** a lode of ore. **b** the bedrock surrounding this. [earlier *riff(e)* f. MDu., MLG *rif, ref*, f. ON *rif* RIB]

reef[2] /riːf/ *n. & v. Naut.* ● *n.* each of several strips across a sail, for taking it in or rolling it up to reduce the surface area in a high wind. ● *v.tr.* **1** take in a reef or reefs of (a sail). **2** shorten (a topmast or a bowsprit). □ **reefing-jacket** a thick close-fitting double-breasted jacket. **reef-knot** a double knot made symmetrically to hold securely and cast off easily. **reef-point** each of several short pieces of rope attached to a sail to secure it when reefed. [ME *riff, refe* f. Du. *reef, rif* f. ON *rif* RIB, in the same sense: cf. REEF[1]]

reefer /'riːfə(r)/ *n.* **1** *sl.* a marijuana cigarette. **2** = *reefing-jacket* (see REEF[2]). **3** *Naut.* **a** a person who reefs. **b** *colloq.* a midshipman. [REEF[2] (in sense 1, = a thing rolled) + -ER[1]]

reek /riːk/ *v. & n.* ● *v.intr.* (often foll. by *of*) **1** smell strongly and unpleasantly. **2** have unpleasant or suspicious associations (*this reeks of corruption*). **3** give off smoke or fumes. ● *n.* **1** a foul or stale smell. **2** esp. *Sc.* smoke. **3** vapour; a visible exhalation (esp. from a chimney). □ **reeky** *adj.* [OE *rēocan* (v.), *rēc* (n.), f. Gmc]

reel /riːl/ *n. & v.* ● *n.* **1** a cylindrical device on which thread, silk, yarn, paper, film, wire, etc., are wound. **2** a quantity of thread etc. wound on a reel. **3** a device for winding and unwinding a line as required, esp. in fishing. **4** a revolving part in various machines. **5 a** a lively folk or Scottish dance, of two or more couples facing each other. **b** a piece of music for this, usu. in duple time. ● *v.* **1** *tr.* wind (thread, a fishing-line, etc.) on a reel. **2** *tr.* (foll. by *in, up*) draw (fish etc.) in or up by the use of a reel. **3** *intr.* stand or walk or run unsteadily. **4** *intr.* be shaken mentally or physically. **5** *intr.* rock from side to side, or swing violently. **6** *intr.* dance a reel. □ **reel off** say or recite very rapidly and without apparent effort. □ **reeler** *n.* [OE *hrēol*, of unkn. orig.]

re-elect /,riːɪ'lekt/ *v.tr.* elect again, esp. to a further term of office. □ **re-election** /-'lekʃ(ə)n/ *n.*

re-eligible /riː'elɪdʒɪb(ə)l/ *adj.* eligible for re-election to a further term of office.

re-embark /,riːɪm'bɑːk/ *v.intr. & tr.* go or put on board ship again. □ **re-embarkation** /riː,embɑː'keɪʃ(ə)n/ *n.*

re-emerge /,riːɪ'mɜːdʒ/ *v.intr.* emerge again; come back out. □ **re-emergence** *n.* **re-emergent** *adj.*

re-emphasize /riː'emfə,saɪz/ *v.tr.* place renewed emphasis on. □ **re-emphasis** /-əsɪs/ *n.*

re-employ /,riːɪm'plɔɪ/ *v.tr.* employ again. □ **re-employment** *n.*

re-enact /,riːɪn'ækt/ *v.tr.* act out (a past event). □ **re-enactment** *n.*

re-enlist /,riːɪn'lɪst/ *v.intr.* enlist again, esp. in the armed services. □ **re-enlister** *n.*

re-enter /riː'entə(r)/ *v.tr. & intr.* enter again; go back in. □ **re-entrance** /-'entrəns/ *n.*

re-entrant /riː'entrənt/ *adj. & n.* ● *adj.* **1** (of an angle, esp. in fortification) pointing inwards (opp. SALIENT *adj.* 2). **2** *Geom.* reflex. ● *n.* a re-entrant angle.

re-entry /riː'entrɪ/ *n.* (*pl.* **-ies**) **1** the act of entering again, esp. (of a spacecraft, missile, etc.) re-entering the earth's atmosphere. **2** *Law* an act of retaking or repossession.

re-equip /,riːɪ'kwɪp/ *v.tr. & intr.* (**-equipped, -equipping**) provide or be provided with new equipment.

re-erect /,riːɪ'rekt/ *v.tr.* erect again. □ **re-erection** /-'rekʃ(ə)n/ *n.*

re-establish /,riːɪ'stæblɪʃ/ *v.tr.* establish again or anew. □ **re-establishment** *n.*

re-evaluate /,riːɪ'væljʊ,eɪt/ *v.tr.* evaluate again or differently. □ **re-evaluation** /-,væljʊ'eɪʃ(ə)n/ *n.*

reeve[1] /riːv/ *n.* **1** *hist.* **a** the chief magistrate of a town or district. **b** an official supervising a landowner's estate. **c** any of various minor local officials. **2** *Can.* the president of a village or town council. [OE (*ge*)*rēfa*, *girǣfa*]

reeve[2] /riːv/ *v.tr.* (*past* **rove** /rəʊv/ *or* **reeved**) *Naut.* **1** (usu. foll. by *through*) thread (a rope or rod etc.) through a ring or other aperture. **2** pass a rope through (a block etc.). **3** fasten (a rope or block) in this way. [prob. f. Du. *rēven* REEF[2]]

reeve[3] /riːv/ *n.* a female ruff (see RUFF[1] 4). [17th c.: orig. unkn.]

re-examine /ˌriːɪgˈzæmɪn/ *v.tr.* examine again or further (esp. a witness after cross-examination). □ **re-examination** /-ˌzæmɪˈneɪʃ(ə)n/ *n.*

re-export *v. & n.* ● *v.tr.* /ˌriːɪkˈspɔːt, riːˈekspɔːt/ export again (esp. imported goods after further processing or manufacture). ● *n.* /riːˈekspɔːt/ **1** the process of re-exporting. **2** something re-exported. □ **re-exporter** *n.* **re-exportation** /ˌriːekspɔːˈteɪʃ(ə)n/ *n.*

ref /ref/ *n. colloq.* a referee in sports. [abbr.]

ref. *abbr.* **1** reference. **2** refer to.

reface /riːˈfeɪs/ *v.tr.* put a new facing on (a building).

refashion /riːˈfæʃ(ə)n/ *v.tr.* fashion again or differently.

refection /rɪˈfekʃ(ə)n/ *n. literary* **1** refreshment by food or drink (*we took refection*). **2** a light meal. [ME f. OF f. L *refectio -onis* f. *reficere* (as REFECTORY)]

refectory /rɪˈfektərɪ, ˈrefɪk-/ *n.* (*pl.* **-ies**) a room used for communal meals, esp. in a monastery or college. □ **refectory table** a long narrow table. [LL *refectorium* f. L *reficere* refresh (as REFECTORY make)]

refer /rɪˈfɜː(r)/ *v.* (**referred**, **referring**) (usu. foll. by *to*) **1** *tr.* trace or ascribe (a person or thing as a cause or source) (*referred their success to their popularity*). **2** *tr.* consider as belonging (to a certain date or place or class). **3** *tr.* send on or direct (a person, or a question for decision) (*the matter was referred to arbitration; referred him to her previous answer*). **4** *intr.* make an appeal or have recourse to (some authority or source of information) (*referred to his notes*). **5** *tr.* send or direct (a person) to a medical specialist etc. **6** *tr.* (foll. by *back to*) send (a proposal etc.) back to (a lower body, court, etc.). **7** *intr.* (foll. by *to*) (of a person speaking) make an allusion or direct the hearer's attention (*decided not to refer to our other problems*). **8** *intr.* (foll. by *to*) (of a statement etc.) have a particular relation or be directed to (*this paragraph refers to the events of last year*). **9** *tr.* (foll. by *to*) interpret (a statement) as being directed to (a particular context etc.). **10** *tr.* fail (a candidate in an examination). □ **referred pain** pain felt in a part of the body other than its actual source. **refer to drawer** a banker's note suspending payment of a cheque. □ **referrer** *n.* **referable** /rɪˈfɜːrəb(ə)l, ˈrefər-/ *adj.* [ME f. OF *referer* f. L *referre* carry back (as RE-, *ferre* bring)]

referee /ˌrefəˈriː/ *n. & v.* ● *n.* **1** an umpire esp. in football or boxing. **2** a person whose opinion or judgement is sought in some connection, or who is referred to for a decision in a dispute etc. **3** a person willing to testify to the character of an applicant for employment etc. ● *v.* (**referees**, **refereed**) **1** *intr.* act as referee. **2** *tr.* be the referee of (a game etc.).

reference /ˈrefərəns/ *n. & v.* ● *n.* **1** the referring of a matter for decision or settlement or consideration to some authority. **2** the scope given to this authority. **3** (foll. by *to*) **a** a relation or respect or correspondence (*success seems to have little reference to merit*). **b** an allusion (*made no reference to our problems*). **c** a direction to a book etc. (or a passage in it) where information may be found. **d** a book or passage so cited. **4 a** the act of looking up a passage etc. or looking in a book for information. **b** the act of referring to a person etc. for information. **5 a** a written testimonial supporting an applicant for employment etc. **b** a person giving this. ● *v.tr.* provide (a book etc.) with references to authorities. □ **reference book** a book intended to be consulted for information on individual matters rather than read continuously. **reference library** a library in which the books are for consultation not loan. **with** (or **in**) **reference to** regarding; as regards; about. **without reference to** not taking account of. □ **referential** /ˌrefəˈrenʃ(ə)l/ *adj.*

referendum /ˌrefəˈrendəm/ *n.* (*pl.* **referendums** or **referenda** /-də/) **1** the process of referring a political question to the electorate for a direct decision by general vote. **2** a vote taken by referendum. [L, gerund or neut. gerundive of *referre*: see REFER]

referent /ˈrefərənt/ *n.* the idea or thing that a word etc. symbolizes. [L *referens* (as REFERENDUM)]

referral /rɪˈfɜːrəl/ *n.* the referring of an individual to an expert or specialist for advice, esp. the directing of a patient by a GP to a medical specialist.

refill *v. & n.* ● *v.tr.* /riːˈfɪl/ **1** fill again. **2** provide a new filling for. ● *n.* /ˈriːfɪl/ **1** a new filling. **2** the material for this. □ **refillable** /riːˈfɪləb(ə)l/ *adj.*

refine /rɪˈfaɪn/ *v.* **1** *tr.* free from impurities or defects; purify, clarify. **2** *tr. & intr.* make or become more polished or elegant or cultured. **3** *tr. & intr.* make or become more subtle or delicate in thought, feelings, etc. □ **refinable** *adj.* [RE- + FINE[1] *v.*]

refined /rɪˈfaɪnd/ *adj.* **1** characterized by polish or elegance or subtlety. **2** purified; clarified.

refinement /rɪˈfaɪnmənt/ *n.* **1** the act of refining or the process of being refined. **2** fineness of feeling or taste. **3** polish or elegance in behaviour or manner. **4** an added development or improvement (*a car with several refinements*). **5** a piece of subtle reasoning. **6** a fine distinction. **7** a subtle or ingenious example or display (*all the refinements of reasoning*). [REFINE + -MENT, after F *raffinement*]

refiner /rɪˈfaɪnə(r)/ *n.* a person or company whose business is to refine crude oil, metal, sugar, etc.

refinery /rɪˈfaɪnərɪ/ *n.* (*pl.* **-ies**) a place where oil etc. is refined.

refit *v. & n.* ● *v.tr. & intr.* /riːˈfɪt/ (**refitted**, **refitting**) make or become fit or serviceable again (esp. of a ship undergoing renewal and repairs). ● *n.* /ˈriːfɪt/ the act or an instance of refitting; the process of being refitted. □ **refitment** /-ˈfɪtmənt/ *n.*

reflag /riːˈflæg/ *v.tr.* (**reflagged**, **reflagging**) change the national registration of (a ship).

reflate /riːˈfleɪt/ *v.tr.* cause reflation of (a currency or economy etc.). [RE- after *inflate, deflate*]

reflation /riːˈfleɪʃ(ə)n/ *n.* the inflation of a financial system to restore its previous condition after deflation. □ **reflationary** *adj.* [RE- after *inflation, deflation*]

reflect /rɪˈflekt/ *v.* **1** *tr.* **a** (of a surface or body) throw back (heat, light, sound, etc.). **b** cause to rebound (*reflected light*). **2** *tr.* (of a mirror) show an image of; reproduce to the eye or mind. **3** *tr.* correspond in appearance or effect to; have as a cause or source (*their behaviour reflects a wish to succeed*). **4** *tr.* **a** (of an action, result, etc.) show or bring (credit, discredit, etc.). **b** (*absol.*; usu. foll. by *on, upon*) bring discredit on. **5 a** *intr.* (often foll. by *on, upon*) meditate on; think about. **b** *tr.* (foll. by *that, how,* etc. + clause) consider; remind oneself. **6** *intr.* (usu. foll. by *upon, on*) make disparaging remarks. □ **reflecting telescope** = REFLECTOR 2a. [ME f. OF *reflecter* or L *reflectere* (as RE-, *flectere flex-* bend)]

reflectance /rɪˈflektəns/ *n. Physics* a measure of the proportion of light or other radiation falling on a surface which is then reflected or scattered.

reflection /rɪˈflekʃ(ə)n/ *n.* (also **reflexion**) **1** the act or an instance of reflecting; the process of being reflected. **2 a** reflected light, heat, or colour. **b** a reflected image. **3** meditation; reconsideration (*on reflection*). **4** (often foll. by *on*) discredit or a thing bringing discredit. **5** (often foll. by *on, upon*) an idea arising in the mind; a comment or apophthegm. **6** (usu. foll. by *of*) a consequence; evidence (*a reflection of how she feels*). □ **angle of reflection** *Physics* the angle made by a reflected ray with a perpendicular to the reflecting surface. [ME f. OF *reflexion* or LL *reflexio* (as REFLECT), with assim. to *reflect*]

reflective /rɪˈflektɪv/ *adj.* **1** (of a surface etc.) giving a reflection or image. **2** (of mental faculties) concerned in reflection or thought. **3** (of a person or mood etc.) thoughtful; given to meditation. □ **reflectively** *adv.* **reflectiveness** *n.*

reflectivity /ˌriːflekˈtɪvɪtɪ/ *n. Physics* the property of reflecting light or radiation, esp. reflectance as measured independently of the thickness of a material.

reflector /rɪˈflektə(r)/ *n.* **1** a piece of glass or metal etc. for reflecting light in a required direction, e.g. a red one on the back of a motor vehicle or bicycle. **2 a** a telescope using a concave mirror to collect light (cf. REFRACTOR 2). **b** the mirror itself.

reflet /rəˈfleɪ/ *n.* lustre or iridescence, esp. on pottery. [F f. It. *riflesso* reflection, REFLEX]

reflex /ˈriːfleks/ *adj. & n.* ● *adj.* **1** (of an action) independent of the will, as an automatic response to the stimulation of a nerve (e.g. a sneeze). **2** (of an angle) exceeding 180°. **3** bent backwards. **4** (of light) reflected. **5** (of a thought etc.) introspective; directed back upon itself or its own operations. **6** (of an effect or influence) reactive; coming back upon its author or source. ● *n.* **1** a reflex action. **2** a sign or secondary manifestation (*law is a reflex of public opinion*). **3** reflected light or a reflected image. **4** a word formed by development from an earlier stage of a language. □ **reflex arc** *Anat.* the sequence of nerves involved in a reflex action. **reflex camera** a camera with a ground-glass focusing screen on which the image is formed by a combination of lens and mirror, enabling the scene to be correctly composed and focused. □ **reflexly** *adv.* [L *reflexus* (as REFLECT)]

reflexible /rɪˈfleksɪb(ə)l/ adj. capable of being reflected. □ **reflexibility** /-ˌfleksɪˈbɪlɪtɪ/ n.

reflexion var. of REFLECTION.

reflexive /rɪˈfleksɪv/ adj. & n. Gram. ● adj. **1** (of a word or form) referring back to the subject of a sentence (esp. of a pronoun, e.g. *myself*). **2** (of a verb) having a reflexive pronoun as its object (as in *to wash oneself*). ● n. a reflexive word or form, esp. a pronoun. □ **reflexively** adv. **reflexiveness** n. **reflexivity** /ˌriːflekˈsɪvɪtɪ/ n.

reflexology /ˌriːflekˈsɒlədʒɪ/ n. **1** a system of massage through reflex points on the feet, hands, and head, used to relieve tension and treat illness. **2** Psychol. the scientific study of reflexes. □ **reflexologist** n.

refloat /riːˈfləʊt/ v.tr. set (a stranded ship) afloat again.

refluent /ˈreflʊənt/ adj. flowing back (*refluent tide*). □ **refluence** n. [ME f. L *refluere* (as RE-, *fluere* flow)]

reflux /ˈriːflʌks/ n. & v. ● n. **1** a backward flow. **2** Chem. a method of boiling a liquid so that any vapour is liquefied and returned to the boiler. ● v.tr. & intr. Chem. boil or be boiled under reflux.

refocus /riːˈfəʊkəs/ v.tr. (**refocused**, **refocusing** or **refocussed**, **refocussing**) focus again or adjust the focus of.

reforest /riːˈfɒrɪst/ v.tr. = REAFFOREST. □ **reforestation** /-ˌfɒrɪˈsteɪʃ(ə)n/ n.

reforge /riːˈfɔːdʒ/ v.tr. forge again or differently.

reform /rɪˈfɔːm/ v. & n. ● v. **1** tr. & intr. make or become better by the removal of faults and errors. **2** tr. abolish or cure (an abuse or malpractice). **3** tr. US correct (a legal document). **4** tr. Chem. convert (a straight-chain hydrocarbon) by catalytic reaction to a branched-chain form for use as petrol. **5** (as **Reformed** adj.) of or relating to a Reformed Church. ● n. **1** the removal of faults or abuses, esp. of a moral or political or social kind. **2** an improvement made or suggested. □ **reform school** an institution to which young offenders are sent to be reformed. □ **reformable** adj. [ME f. OF *reformer* or L *reformare* (as RE-, FORM)]

re-form /riːˈfɔːm/ v.tr. & intr. form again.

Reform Act n. an act framed to amend the system of parliamentary representation, especially either of those introduced in Britain in 1832 and 1867. The first Reform Act (1832) disenfranchised various rotten boroughs and lowered the property qualification and so widened the electorate by about 50 per cent to include most of the male members of the upper middle class. The second (1867) doubled the electorate (to 2 million) by again lowering the property qualification. There was also a third act, in 1884, which again extended the franchise, increasing the electorate to about 5 million men.

reformat /riːˈfɔːmæt/ v.tr. (**reformatted**, **reformatting**) format again or differently.

reformation /ˌrefəˈmeɪʃ(ə)n/ n. the act of reforming or process of being reformed, esp. a radical change for the better in political or religious or social affairs. [ME f. OF *reformation* or L *reformatio* (as REFORM)]

re-formation /ˌriːfɔːˈmeɪʃ(ə)n/ n. the process or an instance of forming or being formed again.

Reformation, the the 16th-century movement to reform the doctrine and practices of the Roman Catholic Church, which resulted in the establishment of the Protestant Churches. The roots of the Reformation go back to the 14th century with groups such as the Lollards and the Hussites attacking the Church for its wealth and hierarchical structure, and stressing the need to return to the simplicity and asceticism of the Scriptures; the movement was also fuelled by the work of humanists such as Erasmus. The Reformation is usually thought of as beginning in 1517 in Wittenberg when Martin Luther issued ninety-five theses criticizing the sale of indulgences, and launched an attack on papal infallibility, the celibacy of the priesthood, the doctrine of transubstantiation, and corruption within religious orders. Luther confirmed his commitment to reform at the Diet of Worms in 1521. In Denmark, Norway, Sweden, Saxony, Hesse, and Brandenburg, supporters broke away and established Protestant Churches, while in Switzerland a separate movement was led by Zwingli and later Calvin. Calvinism spread to France, Germany, the Netherlands, Scotland, and the US. (See also PROTESTANT.) □ **Reformational** adj.

reformative /rɪˈfɔːmətɪv/ adj. tending or intended to produce reform. [OF *reformatif* -*ive* or med.L *reformativus* (as REFORM)]

reformatory /rɪˈfɔːmətərɪ/ n. & adj. ● n. (pl. -**ies**) N. Amer. & hist. = reform school. ● adj. reformative.

Reformed Church n. any of the Protestant Churches which have accepted the principles of the Reformation, now especially those following Calvinist rather than Lutheran doctrines and characterized by Presbyterian government and simple services. Many countries have their own Reformed Churches; these include the Church of Scotland, the United Reformed Church in England and Wales, and the Netherlands Reformed Church.

reformer /rɪˈfɔːmə(r)/ n. a person who advocates or brings about (esp. political or social) reform.

reformism /rɪˈfɔːmɪz(ə)m/ n. a policy of reform rather than abolition or revolution. □ **reformist** n.

Reform Jew n. an adherent of Reform Judaism.

Reform Judaism n. a movement, initiated in Germany by the philosopher Moses Mendelssohn (1729–86), which has reformed or abandoned aspects of Orthodox Jewish worship and ritual in an attempt to adapt to modern changes in social, political, and cultural life.

reformulate /rɪˈfɔːmjʊˌleɪt/ v.tr. formulate again or differently. □ **reformulation** /-ˌfɔːmjʊˈleɪʃ(ə)n/ n.

refract /rɪˈfrækt/ v.tr. **1** (of water, air, glass, etc.) deflect (a ray of light etc.) at a certain angle when it enters obliquely from another medium. **2** determine the refractive condition of (the eye). □ **refracting telescope** = REFRACTOR 2. [L *refringere refract*- (as RE-, *frangere* break)]

refraction /rɪˈfrækʃ(ə)n/ n. the process by which or the extent to which light is refracted. □ **angle of refraction** the angle made by a refracted ray with the perpendicular to the refracting surface. [F *réfraction* or LL *refractio* (as REFRACT)]

refractive /rɪˈfræktɪv/ adj. of or involving refraction. □ **refractive index** the ratio of the velocity of light in a vacuum to its velocity in a specified medium.

refractometer /ˌriːfrækˈtɒmɪtə(r)/ n. an instrument for measuring a refractive index. □ **refractometry** n. **refractometric** /-təˈmetrɪk/ adj.

refractor /rɪˈfræktə(r)/ n. **1** a refracting medium or lens. **2** a telescope using an objective lens to collect light (cf. REFLECTOR 2a).

refractory /rɪˈfræktərɪ/ adj. & n. ● adj. **1** stubborn, unmanageable, rebellious. **2 a** (of a wound, disease, etc.) not yielding to treatment. **b** (of a person etc.) resistant to infection. **3** (of a substance) hard to fuse or work. ● n. (pl. -**ies**) a substance especially resistant to heat, corrosion, etc. □ **refractorily** adv. **refractoriness** n. [alt. of obs. *refractary* f. L *refractarius* (as REFRACT)]

refrain[1] /rɪˈfreɪn/ v.intr. (foll. by *from*) avoid doing (an action); forbear, desist (*refrain from smoking*). □ **refrainment** n. [ME f. OF *refrener* f. L *refrenare* (as RE-, *frenum* bridle)]

refrain[2] /rɪˈfreɪn/ n. **1** a recurring phrase or number of lines, esp. at the ends of stanzas. **2** the music accompanying this. [ME f. OF *refrain* (earlier *refrait*) ult. f. L *refringere* (as RE-, *frangere* break), because the refrain 'broke' the sequence]

refrangible /rɪˈfrændʒɪb(ə)l/ adj. that can be refracted. □ **refrangibility** /-ˌfrændʒɪˈbɪlɪtɪ/ n. [mod.L *refrangibilis* f. *refrangere* = L *refringere*: see REFRACT]

refreeze /riːˈfriːz/ v.tr. & intr. (past **refroze** /-ˈfrəʊz/; past part. **refrozen** /-ˈfrəʊz(ə)n/) freeze again.

refresh /rɪˈfreʃ/ v.tr. **1 a** (of food, rest, amusement, etc.) give fresh spirit or vigour to. **b** (esp. refl.) revive with food, rest, etc. (*refreshed myself with a short sleep*). **2** revive or stimulate (the memory), esp. by consulting the source of one's information. **3** make cool. **4** restore to a certain condition, esp. by provision of fresh supplies, equipment, etc.; replenish. [ME f. OF *refreschi(e)r* f. *fres fresche* FRESH]

refresher /rɪˈfreʃə(r)/ n. **1** something that refreshes, esp. a drink. **2** Law an extra fee payable to counsel in a prolonged case. □ **refresher course** a course reviewing or updating previous studies.

refreshing /rɪˈfreʃɪŋ/ adj. **1** serving to refresh. **2** welcome or stimulating because sincere or untypical (*refreshing innocence*). □ **refreshingly** adv.

refreshment /rɪˈfreʃmənt/ n. **1** the act of refreshing or the process of being refreshed in mind or body. **2** (usu. in pl.) food or drink that refreshes. **3** something that refreshes or stimulates the mind. [ME f. OF *refreschement* (as REFRESH)]

refrigerant /rɪˈfrɪdʒərənt/ n. & adj. ● n. **1** a substance used for refrigeration. **2** Med. a substance that cools or allays fever. ● adj. cooling. [F *réfrigérant* or L *refrigerant*- (as REFRIGERATE)]

refrigerate /rɪˈfrɪdʒəˌreɪt/ v. **1** tr. & intr. make or become cool or cold.

2 *tr.* subject (food etc.) to cold in order to freeze or preserve it. □ **refrigerative** /-rətɪv/ *adj.* **refrigeration** /-ˌfrɪdʒəˈreɪʃ(ə)n/ *n.* [L *refrigerare* (as RE-, *frigus frigoris* cold)]

refrigerator /rɪˈfrɪdʒəˌreɪtə(r)/ *n.* an appliance or room in which food etc. is refrigerated. Ice houses, pits, or cellars were known in ancient Mesopotamia, Greece, and Rome, using ice collected in winter, and the principle continued in use up to the 19th century. By the 1880s refrigerating machines were being used in ships, transporting meat successfully on long sea journeys. The first mechanically operated domestic refrigerator was developed *c.*1880 from the cooling apparatus used by the brewing industry, powered by a small steam pump; electric refrigerators appeared in the 1920s. A modern refrigerator makes use of the cooling effect produced when a liquid is made to evaporate: the liquid (often a fluorine compound) is pumped through a valve that causes it to expand and become a vapour, which is then made to condense back to a liquid and give up its heat outside the refrigerator.

refrigeratory /rɪˈfrɪdʒərətərɪ/ *adj. & n.* ● *adj.* serving to cool. ● *n.* (*pl.* **-ies**) *hist.* a cold-water vessel attached to a still for condensing vapour. [mod.L *refrigeratorium* (n.), L *refrigeratorius* (adj.) (as REFRIGERATE)]

refringent /rɪˈfrɪndʒənt/ *adj. Physics* refracting. □ **refringence** *n.* **refringency** *n.* [L *refringere*: see REFRACT]

refroze past of REFREEZE.

refrozen past part. of REFREEZE.

reft past and past part. of REAVE.

refuel /riːˈfjuːəl/ *v.* (**refuelled**, **refuelling**; US **refueled**, **refueling**) **1** *intr.* replenish a fuel supply. **2** *tr.* supply with more fuel.

refuge /ˈrefjuːdʒ/ *n.* **1** a shelter from pursuit or danger or trouble. **2** a person or place etc. offering this. **3 a** a person, thing, or course resorted to in difficulties. **b** a pretext, an excuse. **4** a traffic island. [ME f. OF f. L *refugium* (as RE-, *fugere* flee)]

refugee /ˌrefjʊˈdʒiː/ *n.* a person taking refuge, esp. in a foreign country from war or persecution or natural disaster. [F *réfugié* past part. of (*se*) *réfugier* (as REFUGE)]

refugium /rɪˈfjuːdʒɪəm/ *n.* (*pl.* **refugia** /-dʒɪə/) *Biol.* an area in which a population of organisms can survive through a period of unfavourable conditions, esp. glaciation. [L, = place of refuge]

refulgent /rɪˈfʌldʒənt/ *adj. literary* shining; gloriously bright. □ **refulgence** *n.* **refulgently** *adv.* [L *refulgere* (as RE-, *fulgere* shine)]

refund *v. & n.* ● *v.* /rɪˈfʌnd/ *tr.* (also *absol.*) **1** pay back (money or expenses). **2** reimburse (a person). ● *n.* /ˈriːfʌnd/ **1** an act of refunding. **2** a sum refunded; a repayment. □ **refundable** /rɪˈfʌndəb(ə)l/ *adj.* [ME in sense 'pour back', f. OF *refonder* or L *refundere* (as RE-, *fundere* pour), later assoc. with FUND]

re-fund /riːˈfʌnd/ *v.tr.* fund (a debt etc.) again.

refurbish /riːˈfɜːbɪʃ/ *v.tr.* **1** brighten up. **2** restore and redecorate. □ **refurbishment** *n.*

refurnish /riːˈfɜːnɪʃ/ *v.tr.* furnish again or differently.

refusal /rɪˈfjuːz(ə)l/ *n.* **1** the act or an instance of refusing; the state of being refused. **2** (in full **first refusal**) the right or privilege of deciding to take or leave a thing before it is offered to others.

refuse[1] /rɪˈfjuːz/ *v.* **1** *tr.* withhold acceptance of or consent to (*refuse an offer*; *refuse orders*). **2** *tr.* (often foll. by to + infin.) indicate unwillingness (*I refuse to go*; *the car refuses to start*; *I refuse!*). **3** *tr.* (often with double object) not grant (a request) made by (a person) (*refused me a day off*; *I could not refuse them*). **4** *tr.* (also *absol.*) (of a horse) be unwilling to jump (a fence etc.). □ **refuser** *n.* [ME f. OF *refuser* prob. ult. f. L *recusare* (see RECUSANT) after REFUTE]

refuse[2] /ˈrefjuːs/ *n.* items rejected as worthless; waste. [ME, perh. f. OF *refusé* past part. (as REFUSE[1])]

re-fuse /riːˈfjuːz/ *v.tr.* fuse again; provide with a new fuse.

refusenik /rɪˈfjuːznɪk/ *n.* **1** *hist.* a Jew from the former Soviet Union who was refused permission to emigrate to Israel. **2** a person who refuses to follow orders or obey the law, esp. as a protest. [REFUSE[1] + -NIK]

refute /rɪˈfjuːt/ *v.tr.* **1** prove the falsity or error of (a statement etc. or the person advancing it). **2** rebut or repel by argument. **3** *disp.* deny or contradict (without argument). ¶ Often confused in this sense with *repudiate*. □ **refutable** *adj.* **refutal** *n.* **refuter** *n.* **refutation** /ˌrefjʊˈteɪʃ(ə)n/ *n.* [L *refutare* (as RE-: cf. CONFUTE)]

reg /redʒ/ *n. colloq.* = *registration mark*. [abbr.]

regain /rɪˈɡeɪn/ *v.tr.* obtain possession or use of after loss (*regain consciousness*). [F *regagner* (as RE-, GAIN)]

regal /ˈriːɡ(ə)l/ *adj.* **1** royal; of or by a monarch or monarchs. **2** fit for a monarch; magnificent. □ **regally** *adv.* [ME f. OF *regal* or L *regalis* f. *rex regis* king]

regale /rɪˈɡeɪl/ *v.tr.* **1** entertain lavishly with feasting. **2** (foll. by *with*) entertain or divert with (talk etc.). **3** (of beauty, flowers, etc.) give delight to. □ **regalement** *n.* [F *régaler* f. OF *gale* pleasure]

regalia /rɪˈɡeɪlɪə/ *n.pl.* **1** the insignia of royalty used at coronations. **2** the insignia of an order or of civic dignity. **3** any distinctive or elaborate clothes, accoutrements, etc.; trappings, finery. [med.L, = royal privileges, f. L neut. pl. of *regalis* REGAL]

regalism /ˈriːɡəˌlɪz(ə)m/ *n.* the doctrine of a sovereign's ecclesiastical supremacy.

regality /rɪˈɡælɪtɪ/ *n.* (*pl.* **-ies**) **1** the state of being a king or queen. **2** an attribute of sovereign power. **3** a royal privilege. [ME f. OF *regalité* or med.L *regalitas* (as REGAL)]

regard /rɪˈɡɑːd/ *v. & n.* ● *v.tr.* **1** gaze on steadily (usu. in a specified way) (*regarded them suspiciously*). **2** give heed to; take into account; let one's course be affected by. **3** look upon or contemplate mentally in a specified way (*I regard them kindly*; *I regard it as an insult*). **4** (of a thing) have relation to; have some connection with. ● *n.* **1** a gaze; a steady or significant look. **2** (foll. by *to*, *for*) attention or care. **3** (foll. by *for*) esteem; kindly feeling; respectful opinion. **4 a** a respect; a point attended to (*in this regard*). **b** (usu. foll. by *to*) reference; connection, relevance. **5** (in *pl.*) an expression of friendliness in a letter etc.; compliments (*sent my best regards*). □ **as regards** about, concerning; in respect of. **in** (or **with**) **regard to** as concerns; in respect of. [ME f. OF *regard* f. *regarder* (as RE-, *garder* GUARD)]

regardant /rɪˈɡɑːd(ə)nt/ *adj. Heraldry* looking backwards. [AF & OF (as REGARD)]

regardful /rɪˈɡɑːdfʊl/ *adj.* (foll. by *of*) mindful of; paying attention to.

regarding /rɪˈɡɑːdɪŋ/ *prep.* about, concerning; in respect of.

regardless /rɪˈɡɑːdlɪs/ *adj. & adv.* ● *adj.* (foll. by *of*) without regard or consideration for; heedless (*openly regardless of the expense*). ● *adv.* without paying attention (*carried on regardless*). □ **regardlessly** *adv.* **regardlessness** *n.*

regather /riːˈɡæðə(r)/ *v.tr. & intr.* **1** gather or collect again. **2** meet again.

regatta /rɪˈɡætə/ *n.* a sporting event consisting of a series of boat or yacht races. [It. (Venetian)]

regd. *abbr.* registered.

regelate /ˈriːdʒɪˌleɪt, ˈredʒ-/ *v.intr.* freeze again (esp. of pieces of ice etc. frozen together after temporary thawing of the surfaces). □ **regelation** /ˌriːdʒɪˈleɪʃ(ə)n, ˌredʒ-/ *n.* [RE- + L *gelare* freeze]

regency /ˈriːdʒənsɪ/ *n. & adj.* ● *n.* (*pl.* **-ies**) **1** the office of regent. **2** a commission acting as regent. **3 a** the period of office of a regent or regency commission. **b** (**Regency**) a period during which a regent governs, esp. the period of 1811–20 in Britain when George, Prince of Wales, acted as regent, or 1715–23 in France with Philip, Duke of Orleans, as regent. ● *adj.* (**Regency**) characteristic of the architecture, clothing, furniture, etc. of the Regency of 1811–20 in England or, more widely, of the late 18th and early 19th centuries. (*See note below.*) [ME f. med.L *regentia* (as REGENT)]

▪ Regency style was contemporary with the Empire style and shares many of its features: elaborate and ornate, it is generally neoclassical, with a generous borrowing of Greek and Egyptian motifs. The architecture is exemplified by John Nash's exotic Royal Pavilion in Brighton, and by his elegant curving crescent at Regent's Park, London. (See also EMPIRE STYLE.)

regenerate *v. & adj.* ● *v.* /rɪˈdʒenəˌreɪt/ **1** *tr. & intr.* bring or come into renewed existence; generate again. **2** *tr.* improve the moral condition of. **3** *tr.* impart new and more vigorous life to (a person or institution etc.). **4** *intr.* reform oneself. **5** *tr.* invest with a new and higher spiritual nature. **6** *intr. & tr. Biol.* regrow or cause (new tissue) to regrow to replace lost or injured tissue. **7** *tr. & intr. Chem.* restore or be restored to an initial state of reaction or process. ● *adj.* /rɪˈdʒenərət/ **1** spiritually born again. **2** reformed. □ **regenerative** /-rətɪv/ *adj.* **regeneratively** *adv.* **regenerator** /-ˌreɪtə(r)/ *n.* **regeneration** /-ˌdʒenəˈreɪʃ(ə)n/ *n.* [L *regenerare* (as RE-, GENERATE)]

regent /ˈriːdʒənt/ *n. & adj.* ● *n.* **1** a person appointed to administer a state because the monarch is a minor or is absent or incapacitated. **2** *US* a member of the governing body of a state university. ● *adj.* (placed after noun) acting as regent (*Prince Regent*). □ **regent bird** an Australian bowerbird, *Sericulus chrysocephalus*, the male of which has gold and black plumage. [ME f. OF *regent* or L *regere* rule]

regerminate /riːˈdʒɜːmɪˌneɪt/ *v.tr. & intr.* germinate again. □ **regermination** /-ˌdʒɜːmɪˈneɪʃ(ə)n/ *n.*

reggae /'regeɪ/ n. a kind of popular music of Jamaican origin, characterized by a strongly accentuated subsidiary beat and often a prominent bass. Reggae evolved in the late 1960s from ska and other local variations on calypso and rhythm and blues, and became widely known in the 1970s through the work of Bob Marley; its lyrics are much influenced by Rastafarian ideas. [20th c.: orig. unkn.; perh. rel. to Jamaican English *rege-rege* quarrel, row]

Reggio di Calabria /,redʒɪˌəʊ diː kəˈlæbrɪə/ a port at the southern tip of the 'toe' of Italy, on the Strait of Messina, capital of Calabria region; pop. (1991) 183,440. The original settlement, named Rhegion in Greek (Latin Rhegium), was founded about 720 BC by Greek colonists, falling to the Romans in 270 BC. From the 13th century it was included in the kingdom of Sicily and Naples.

regicide /'redʒɪˌsaɪd/ n. **1** a person who kills or takes part in killing a king, esp. any of those involved in the trial and execution of Charles I of England, Scotland, and Ireland or the execution of Louis XVI of France. **2** the act of killing a king. □ **regicidal** /,redʒɪˈsaɪd(ə)l/ adj. [L rex regis king + -CIDE]

regild /riːˈgɪld/ v.tr. gild again, esp. to renew faded or worn gilding.

regime /reɪˈʒiːm/ n. (also **régime**) **1 a** a method or system of government. **b** derog. a particular government. **2** a prevailing order or system of things. **3** the conditions under which a scientific or industrial process occurs. **4** = REGIMEN 1. [F régime (as REGIMEN)]

regimen /'redʒɪˌmen/ n. **1** esp. Med. a prescribed course of exercise, way of life, and diet. **2** archaic a system of government. [L f. regere rule]

regiment n. & v. ● n. /'redʒɪmənt/ **1 a** a permanent unit of an army usu. commanded by a colonel and divided into several companies or troops or batteries and often into two battalions. **b** an operational unit of artillery etc. **2** (usu. foll. by of) a large array or number. **3** archaic rule, government. ● v.tr. /'redʒɪˌment/ **1** organize (esp. oppressively) in groups or according to a system. **2** form into a regiment or regiments. □ **regimented** /-,mentɪd/ adj. **regimentation** /,redʒɪmenˈteɪʃ(ə)n/ n. [ME (in sense 3) f. OF f. LL regimentum (as REGIMEN)]

regimental /,redʒɪˈment(ə)l/ adj. & n. ● adj. of or relating to a regiment. ● n. (in pl.) military uniform, esp. of a particular regiment. □ **regimentally** adv.

Regina[1] /rɪˈdʒaɪnə/ the capital of Saskatchewan, situated in the centre of the wheat-growing plains of south central Canada; pop. (1991) 179,180. Named in 1882 in honour of Queen Victoria (its name means 'queen' in Latin), it was the administrative headquarters of the Northwest Territories until 1905.

Regina[2] /rɪˈdʒaɪnə/ n. the reigning queen (following a name or in the titles of lawsuits, e.g. Regina v. Jones the Crown versus Jones). [L, = queen f. rex regis king]

Regiomontanus /,redʒɪəʊmɒnˈtɑːnəs/, Johannes (born Johannes Müller) (1436–76), German astronomer and mathematician. He was probably the most important astronomer of the 15th century, and worked in Venice, Buda (Hungary), Nuremberg, and finally Rome. Regiomontanus completed a translation of Ptolemy's Mathematical Syntaxis, with revisions and comments, and wrote four monumental works on mathematics (especially trigonometry) and astronomy.

region /'riːdʒən/ n. **1** an area of land, or division of the earth's surface, having definable boundaries or characteristics (a mountainous region; the region between London and the coast). **2** an administrative district esp. in Scotland. **3** a part of the body round or near some organ etc. (the lumbar region). **4** a sphere or realm (the region of metaphysics). **5 a** a separate part of the world or universe. **b** a layer of the atmosphere or the sea according to its height or depth. □ **in the region of** approximately. □ **regional** adj. **regionalism** n. **regionalist** n. & adj. **regionalize** v.tr. (also **-ise**). **regionally** adv. [ME f. OF f. L regio -onis direction, district f. regere direct]

regisseur /,reɪʒɪˈsɜː(r)/ n. the director of a theatrical production, esp. a ballet. [F régisseur stage-manager]

register /'redʒɪstə(r)/ n. & v. ● n. **1** an official list e.g. of births, marriages, and deaths, of shipping, of professionally qualified persons, or of qualified voters in a constituency. **2** a book in which items are recorded for reference. **3** a device recording speed, force, etc. **4** (in electronic devices) a location in a store of data, used for a specific purpose and with quick access time. **5 a** the compass of a voice or instrument. **b** a part of this compass (lower register). **6** an adjustable plate for widening or narrowing an opening and regulating a draught, esp. in a fire-grate. **7 a** a set of organ pipes. **b** a sliding device controlling this. **8** = cash register (see CASH[1]). **9** Linguistics each of several forms of a language (colloquial, formal, literary, etc.) usually used in particular circumstances. **10** Printing the exact correspondence of the position of printed matter on the two sides of a leaf. **11** Printing & Photog. the correspondence of the position of colour-components in a printed positive. ● v. **1** tr. set down (a name, fact, etc.) formally; record in writing. **2** tr. make a mental note of; notice. **3** tr. enter or cause to be entered in a particular register. **4** tr. entrust (a letter etc.) to a post office for transmission by registered post. **5** intr. & refl. put one's name on a register, esp. as an eligible voter or as a guest in a register kept by a hotel etc. **6** tr. (of an instrument) record automatically; indicate. **7 a** tr. express (an emotion) facially or by gesture (registered surprise). **b** intr. (of an emotion) show in a person's face or gestures. **8** intr. make an impression on a person's mind (did not register at all). **9** intr. & tr. Printing correspond or cause to correspond exactly in position. **10** tr. make known formally or publicly; cause (an opinion, grievance, etc.) to be recorded or noted (I wish to register my disapproval). □ **registered nurse** a nurse with a state certificate of competence. **registered post** a postal procedure with special precautions for safety and for compensation in case of loss. **register office** Brit. a state office where civil marriages are conducted and births, marriages, and deaths are recorded with the issue of certificates. ¶ The name in official use, and generally preferred to registry office. □ **registrable** adj. [ME & OF registre, registre or med.L regestrum, registrum, alt. of regestum f. LL regesta things recorded (as RE-, L gerere gest- carry)]

registrar /,redʒɪˈstrɑː(r)/ n. **1** an official responsible for keeping a register or official records. **2** the chief administrative officer in a university. **3** a middle-ranking hospital doctor undergoing training as a specialist. **4** (in the UK) the judicial and administrative officer of the High Court etc. □ **registrarship** n. [med.L registrarius f. registrum REGISTER]

Registrar General n. a government official responsible for holding a population census.

registrary /'redʒɪstrəri/ n. (pl. **-ies**) the registrar of Cambridge University.

registration /,redʒɪˈstreɪʃ(ə)n/ n. the act or an instance of registering; the process of being registered. □ **registration mark** (or **number**) a combination of letters and figures identifying a motor vehicle etc. [obs. F régistration or med.L registratio (as REGISTRAR)]

registry /'redʒɪstri/ n. (pl. **-ies**) **1** a place or office where registers or records are kept. **2** registration. □ **registry office** = register office. [obs. registery f. med.L registerium (as REGISTER)]

Regius professor /'riːdʒɪəs/ n. (in the UK) the holder of a chair founded by a sovereign (esp. one at Oxford or Cambridge instituted by Henry VIII) or filled by Crown appointment. [L, = royal, f. rex regis king]

reglaze /riːˈgleɪz/ v.tr. glaze (a window etc.) again.

reglet /'reglɪt/ n. **1** Archit. a narrow strip separating mouldings. **2** Printing a thin strip of wood or metal separating type. [F réglet dimin. of règle (as RULE)]

regnal /'regn(ə)l/ adj. of a reign. □ **regnal year** a year reckoned from the date or anniversary of a sovereign's accession. [AL regnalis (as REIGN)]

regnant /'regnənt/ adj. **1** reigning (queen regnant). **2** (of things, qualities, etc.) predominant, prevalent. [L regnare REIGN]

regolith /'regəlɪθ/ n. Geol. unconsolidated solid material covering the bedrock of a planet. [erron. f. Gk rhēgos rug, blanket + -LITH]

regorge /rɪˈgɔːdʒ/ v. **1** tr. bring up or expel again after swallowing. **2** intr. gush or flow back from a pit, channel, etc. [F regorger or RE- + GORGE]

regrade /riːˈgreɪd/ v.tr. grade again or differently.

regress v. & n. ● v. /rɪˈgres/ **1** intr. move backwards, esp. (in abstract senses) return to a former state. **2** intr. & tr. Psychol. return or cause to return mentally to a former stage of life, esp. through hypnosis or mental illness. ● n. /'riːgres/ **1** the act or an instance of going back. **2** reasoning from effect to cause. [ME (as n.) f. L regressus f. regredi regress- (as RE-, gradi step)]

regression /rɪˈgreʃ(ə)n/ n. **1** a backward movement, esp. a return to a former state. **2** a relapse or reversion. **3** Psychol. a mental return to an earlier stage of life, esp. through hypnosis or mental illness. **4** Statistics a measure of the relation between the mean value of one variable (e.g. output) and corresponding values of other variables (e.g. time and cost). [L regressio (as REGRESS)]

regressive /rɪˈgresɪv/ adj. **1** regressing; characterized by regression. **2** (of a tax) proportionally greater on lower incomes. □ **regressively** adv. **regressiveness** n.

regret /rɪˈgret/ v. & n. ● v.tr. (**regretted, regretting**) **1** (often foll. by *that* + clause) feel or express sorrow or repentance or distress over (an action or loss etc.) (*I regret that I forgot; regretted your absence*). **2** (often foll. by *to* + infin. or *that* + clause) acknowledge with sorrow or remorse (*I regret to say that you are wrong; regretted he would not be attending*). ● *n.* **1** a feeling of sorrow, repentance, disappointment, etc., over an action or loss etc. **2** (often in *pl.*) an (esp. polite or formal) expression of disappointment or sorrow at an occurrence, inability to comply, etc. (*refused with many regrets; heard with regret of her death*). □ **give** (or **send**) **one's regrets** formally decline an invitation. [ME f. OF *regreter* bewail]

regretful /rɪˈgretfʊl/ *adj.* feeling or showing regret. □ **regretfully** *adv.* **regretfulness** *n.*

regrettable /rɪˈgretəb(ə)l/ *adj.* (of events or conduct) undesirable, unwelcome; deserving censure. □ **regrettably** *adv.*

regroup /riːˈgruːp/ *v.tr.* & *intr.* group or arrange again or differently. □ **regroupment** *n.*

regrow /riːˈgrəʊ/ *v.intr.* & *tr.* (*past* **-grew** /-ˈgruː/, *past part.* **-grown** /-ˈgrəʊn/) grow again, esp. after an interval. □ **regrowth** *n.*

Regt. *abbr.* Regiment.

regulable /ˈregjʊləb(ə)l/ *adj.* able to be regulated.

regular /ˈregjʊlə(r)/ *adj.* & *n.* ● *adj.* **1** conforming to a rule or principle; systematic. **2 a** (of a structure or arrangement) harmonious, symmetrical (*regular features*). **b** (of a surface, line, etc.) smooth, level, uniform. **3** acting or done or recurring uniformly or calculably in time or manner; habitual, constant, orderly. **4** conforming to a standard of etiquette or procedure; correct; according to convention. **5** properly constituted or qualified; not defective or amateur; pursuing an occupation as one's main pursuit (*cooks as well as a regular cook; has no regular profession*). **6** *Gram.* (of a noun, verb, etc.) following the normal type of inflection. **7** *colloq.* complete, thorough, absolute (*a regular hero*). **8** *Geom.* **a** (of a figure) having all sides and all angles equal. **b** (of a solid) bounded by a number of equal figures. **9** (placed before or after noun) *Eccl.* **a** bound by religious rule. **b** belonging to a religious or monastic order (*canon regular*). **10** (of forces or troops etc.) relating to or constituting a permanent professional body (*regular soldiers; regular police force*). **11** (of a person) defecating or menstruating at predictable times. **12** *Bot.* (of a flower) having radial symmetry. **13** *US colloq.* likeable; normal; reliable (esp. as *regular guy*). ● *n.* **1** a regular soldier. **2** *colloq.* a regular customer, visitor, etc. **3** *Eccl.* one of the regular clergy. □ **keep regular hours** do the same thing, esp. going to bed and getting up, at the same time each day. □ **regularly** *adv.* **regularize** *v.tr.* (also **-ise**). **regularization** /ˌregjʊlaraɪˈzeɪʃ(ə)n/ *n.* **regularity** /-ˈlærɪtɪ/ *n.* [ME *reguler, regular* f. OF *reguler* f. L *regularis* f. *regula* RULE]

regulate /ˈregjʊˌleɪt/ *v.tr.* **1** control by rule. **2** subject to restrictions. **3** adapt to requirements. **4** alter the speed of (a machine or clock) so that it may work accurately. □ **regulator** *n.* **regulative** /-lətɪv/ *adj.* **regulatory** *adj.* [LL *regulare regulat-* f. L *regula* RULE]

regulation /ˌregjʊˈleɪʃ(ə)n/ *n.* **1** the act or an instance of regulating; the process of being regulated. **2** a prescribed rule; an authoritative direction. **3** (*attrib.*) **a** in accordance with regulations; of the correct type etc. (*the regulation speed; a regulation tie*). **b** *colloq.* usual (*the regulation soup*).

regulo /ˈregjʊˌləʊ/ *n.* (usu. foll. by a numeral) each of the numbers of a scale denoting temperature in a gas oven (*cook at regulo 6*). [*Regulo*, propr. term for a thermostatic gas oven control]

Regulus /ˈregjʊləs/ *Astron.* the brightest star in the constellation of Leo. It is a triple system of which the primary is a hot dwarf star. [L, = little king]

regulus /ˈregjʊləs/ *n.* (*pl.* **reguluses** or **reguli** /-ˌlaɪ/) *Chem.* **1** the purer or metallic part of a mineral that separates by sinking on reduction. **2** an impure metallic product formed during the smelting of various ores. □ **reguline** /-ˌlaɪn/ *adj.* [L, dimin. of *rex regis* king: orig. of a metallic form of antimony, so called because of its readiness to combine with gold]

regurgitate /rɪˈgɜːdʒɪˌteɪt/ *v.* **1** *tr.* bring (swallowed food) up again to the mouth. **2** *tr.* cast or pour out again (*required by the exam to regurgitate facts*). **3** *intr.* be brought up again; gush back. □ **regurgitation** /-ˌgɜːdʒɪˈteɪʃ(ə)n/ *n.* [med.L *regurgitare* (as RE-, L *gurges gurgitis* whirlpool)]

rehab /ˈriːhæb/ *n. colloq.* rehabilitation. [abbr.]

rehabilitate /ˌriːəˈbɪlɪˌteɪt/ *v.tr.* **1** restore to effectiveness or normal life by training etc., esp. after imprisonment or illness. **2** restore to former privileges or reputation or a proper condition. □ **rehabilitative**

/-ˌtətɪv/ *adj.* **rehabilitation** /-ˌbɪlɪˈteɪʃ(ə)n/ *n.* [med.L *rehabilitare* (as RE-, HABILITATE)]

rehandle /riːˈhænd(ə)l/ *v.tr.* **1** handle again. **2** give a new form or arrangement to.

rehang /riːˈhæŋ/ *v.tr.* (*past* and *past part.* **rehung** /-ˈhʌŋ/) hang (esp. a picture or a curtain) again or differently.

rehash *v.* & *n.* ● *v.tr.* /riːˈhæʃ/ put (old material) into a new form without significant change or improvement. ● *n.* /ˈriːhæʃ/ **1** material rehashed. **2** the act or an instance of rehashing.

rehear /riːˈhɪə(r)/ *v.tr.* (*past* and *past part.* **reheard** /-ˈhɜːd/) hear again.

rehearsal /rɪˈhɜːs(ə)l/ *n.* **1** the act or an instance of rehearsing. **2** a trial performance or practice of a play, recital, etc.

rehearse /rɪˈhɜːs/ *v.* **1** *tr.* practise (a play, recital, etc.) for later public performance. **2** *intr.* hold a rehearsal. **3** train (a person) by rehearsal. **4** *tr.* recite or say over. **5** *tr.* give a list of; enumerate. □ **rehearser** *n.* [ME f. AF *rehearser*, OF *reherc(i)er*, perh. formed as RE- + *hercer* to harrow f. *herse* harrow: see HEARSE]

reheat *v.* & *n.* ● *v.tr.* /riːˈhiːt/ heat again. ● *n.* /ˈriːhiːt/ the process of using the hot exhaust to burn extra fuel in a jet engine and produce extra power. □ **reheater** /-ˈhiːtə(r)/ *n.*

reheel /riːˈhiːl/ *v.tr.* fit (a shoe etc.) with a new heel.

Rehoboam /ˌriːəˈbəʊəm/, son of Solomon, king of ancient Israel *c.*930–*c.*915 BC. His reign witnessed the secession of the northern tribes and their establishment of a new kingdom under Jeroboam, leaving Rehoboam as the first king of Judah (1 Kings 11–14).

rehoboam /ˌriːəˈbəʊəm/ *n.* a wine bottle of about six times the standard size. [REHOBOAM]

rehouse /riːˈhaʊz/ *v.tr.* provide with new housing.

rehung *past* and *past part.* of REHANG.

rehydrate /ˌriːhaɪˈdreɪt/ *v.* **1** *intr.* absorb water again after dehydration. **2** *tr.* add water to (esp. food) again to restore to a palatable state. □ **rehydratable** *adj.* **rehydration** /-ˈdreɪʃ(ə)n/ *n.*

Reich[1] /raɪx/ *n.* a former German state or commonwealth, especially the Third Reich. ¶ Of *First*, *Second*, and *Third Reich*, only *Third Reich* is standard historical terminology. □ **First Reich** the Holy Roman Empire, 962–1806. **Second Reich** the German Empire, 1871–1918. **Third Reich** the Nazi regime, 1933–45. [G, = kingdom, realm, state]

Reich[2] /raɪx/, Steve (b.1936), American composer. He established himself as a leading minimalist in the mid-1960s; his work is influenced by his study of drumming as well as by Balinese and West African music, and he uses both traditional and electronic instruments. His musical style is based on the repetition of short phrases within a simple harmonic field. Major works include *Drumming* (1971), for percussion and two voices, and *The Desert Music* (1984), for chorus and orchestra.

Reichstag /ˈraɪxstɑːg/ *n.* **1** the diet or parliament of the North German Confederation (1867–71), of the German Empire (1871–1918), and of post-imperial Germany until 1945. **2** the building in Berlin in which this parliament met, badly damaged by fire on the Nazi accession to power in 1933. [G]

reify /ˈriːɪˌfaɪ/ *v.tr.* (**-ies**, **-ied**) convert (a person, abstraction, etc.) into a thing; materialize. □ **reification** /ˌriːɪfɪˈkeɪʃ(ə)n/ *n.* **reificatory** /-ˈkeɪtərɪ/ *adj.* [L *res* thing + -FY]

reign /reɪn/ *v.* & *n.* ● *v.intr.* **1** hold royal office; be king or queen. **2** have power or predominance; prevail; hold sway (*confusion reigns*). **3** (as **reigning** *adj.*) (of a winner, champion, etc.) currently holding the title etc. ● *n.* **1** sovereignty, rule. **2** the period during which a sovereign rules. [ME f. OF *reigne* kingdom f. L *regnare* f. *rex regis* king]

reignite /ˌriːɪgˈnaɪt/ *v.tr.* & *intr.* ignite again.

Reign of Terror see TERROR, THE.

Reilly var. of RILEY[1].

reimburse /ˌriːɪmˈbɜːs/ *v.tr.* **1** repay (a person who has expended money). **2** repay (a person's expenses). □ **reimbursable** *adj.* **reimbursement** *n.* **reimburser** *n.* [RE- + obs. *imburse* put in a purse f. med.L *imbursare* (as IM-, PURSE)]

reimport *v.* & *n.* ● *v.tr.* /ˌriːɪmˈpɔːt, riːˈɪmpɔːt/ import (goods processed from exported materials). ● *n.* /riːˈɪmpɔːt/ **1** the act or an instance of reimporting. **2** a reimported item. □ **reimportation** /ˌriːɪmpɔːˈteɪʃ(ə)n/ *n.*

reimpose /ˌriːɪmˈpəʊz/ *v.tr.* impose again, esp. after a lapse. □ **reimposition** /-pəˈzɪʃ(ə)n/ *n.*

Reims /riːmz, French rɛ̃s/ (also **Rheims**) a city of northern France, chief town of Champagne-Ardenne region; pop. (1990) 185,160. Its name derives from the Remi, a Gallic tribe of the region who were conquered by the Romans. It was the traditional coronation place of most French kings and is noted for its fine 13th-century Gothic cathedral.

rein /reɪn/ n. & v. ● n. (in sing. or pl.) **1** a long narrow strap with each end attached to the bit, used to guide or check a horse etc. in riding or driving. **2** a similar device used to restrain a young child. **3** (a means of) control or guidance; a curb, a restraint. ● v.tr. **1** check or manage with reins. **2** (foll. by up, back) pull up or back with reins. **3** (foll. by in) hold in with reins; restrain. **4** govern, restrain, control. □ **draw rein 1** stop one's horse. **2** pull up. **3** abandon an effort. **give free rein to** remove constraints from; allow full scope to. **keep a tight rein on** allow little freedom to. □ **reinless** adj. [ME f. OF rene, reigne, earlier resne, ult. f. L retinere RETAIN]

reincarnation /ˌriːɪnkɑːˈneɪʃ(ə)n/ n. (in some beliefs) the rebirth of a soul in a new body. □ **reincarnate** /-ˈkɑːnət/ adj. **reincarnate** /riːˈɪnkɑː, neɪt, ˌriːɪnˈkɑːn-/ v.tr.

reincorporate /ˌriːɪnˈkɔːpəˌreɪt/ v.tr. incorporate afresh. □ **reincorporation** /-ˌkɔːpəˈreɪʃ(ə)n/ n.

reindeer /ˈreɪndɪə(r)/ n. (pl. same or **reindeers**) a subarctic deer, Rangifer tarandus, both sexes of which have large antlers. It has been domesticated in northern Eurasia for drawing sledges and as a source of milk, flesh, and hide. Also called (N. Amer.) caribou. □ **reindeer moss** a clump-forming Arctic lichen, Cladonia rangiferina, the chief winter food of reindeer. [ME f. ON hreindýri f. hreinn reindeer + dýr DEER]

reinfect /ˌriːɪnˈfekt/ v.tr. infect again. □ **reinfection** /-ˈfekʃ(ə)n/ n.

reinforce /ˌriːɪnˈfɔːs/ v.tr. strengthen or support, esp. with additional personnel or material or by an increase of numbers or quantity or size etc. □ **reinforced concrete** concrete with metal bars or wire etc. embedded to increase its tensile strength. □ **reinforcer** n. [earlier renforce f. F renforcer]

reinforcement /ˌriːɪnˈfɔːsmənt/ n. **1** the act or an instance of reinforcing; the process of being reinforced. **2** a thing that reinforces. **3** (in pl.) reinforcing personnel or equipment etc.

Reinhardt[1] /ˈraɪnhɑːt/, Django (born Jean Baptiste Reinhardt) (1910–53), Belgian jazz guitarist. He became famous in Paris in the 1930s for his original improvisational style, blending swing with influences from his gypsy background. In 1934, together with violinist Stephane Grappelli, he formed the Quintette du Hot Club de France and went on to make many recordings with the group until they disbanded in 1939. Reinhardt also toured the US with Duke Ellington in 1946.

Reinhardt[2] /ˈraɪnhɑːt/, Max (born Max Goldmann) (1873–1943), Austrian director and impresario. He dominated the theatre in Berlin during the first two decades of the 20th century with his large-scale productions of such works as Sophocles' Oedipus Rex (1910) and Vollmöller's The Miracle (1911). Reinhardt also helped establish the Salzburg Festival, with Richard Strauss and Hugo von Hofmannsthal, in 1920.

reinsert /ˌriːɪnˈsɜːt/ v.tr. insert again. □ **reinsertion** /-ˈsɜːʃ(ə)n/ n.

reinstate /ˌriːɪnˈsteɪt/ v.tr. **1** replace in a former position. **2** restore (a person etc.) to former privileges. □ **reinstatement** n.

reinsure /ˌriːɪnˈʃʊə(r)/ v.tr. & intr. insure again (esp. of an insurer securing the risk by transferring some or all of it to another insurer). □ **reinsurance** n. **reinsurer** n.

reintegrate /riːˈɪntɪˌɡreɪt/ v.tr. **1** = REDINTEGRATE. **2** integrate back into society. □ **reintegration** /-ˌɪntɪˈɡreɪʃ(ə)n/ n.

reinter /ˌriːɪnˈtɜː(r)/ v.tr. inter (a corpse) again. □ **reinterment** n.

reinterpret /ˌriːɪnˈtɜːprɪt/ v.tr. (**reinterpreted**, **reinterpreting**) interpret again or differently. □ **reinterpretation** /-ˌtɜːprɪˈteɪʃ(ə)n/ n.

reintroduce /ˌriːɪntrəˈdjuːs/ v.tr. introduce again. □ **reintroduction** /-ˈdʌkʃ(ə)n/ n.

reinvest /ˌriːɪnˈvest/ v.tr. invest again (esp. money in other property etc.). □ **reinvestment** n.

reinvigorate /ˌriːɪnˈvɪɡəˌreɪt/ v.tr. impart fresh vigour to. □ **reinvigoration** /-ˌvɪɡəˈreɪʃ(ə)n/ n.

reissue /riːˈɪʃuː, -ˈɪsjuː/ v. & n. ● v.tr. (**reissues, reissued, reissuing**) issue again or in a different form. ● n. a new issue, esp. of a previously published book.

reiterate /riːˈɪtəˌreɪt/ v.tr. say or do again or repeatedly. □ **reiterative** /-rətɪv/ adj. **reiteration** /-ˌɪtəˈreɪʃ(ə)n/ n. [L reiterare (as RE-, ITERATE)]

Reith /riːθ/, John (Charles Walsham), 1st Baron (1889–1971), Scottish administrator and politician, first general manager (1922–7) and first director-general (1927–38) of the BBC. He played a major part in the growth and developing ethos of the BBC, refusing to treat broadcasting simply as a means of entertainment and championing its moral and intellectual role in the community. Reith later served in various Cabinet posts during the Second World War. In 1948 the BBC established the Reith Lectures, broadcast annually, in his honour.

reive /riːv/ v.intr. esp. Sc. make raids; plunder. □ **reiver** n. [var. of REAVE]

reject v. & n. ● v.tr. /rɪˈdʒekt/ **1** put aside or send back as not to be used or done or complied with etc. **2** refuse to accept or believe in. **3** rebuff or snub (a person). **4** (of a body or digestive system) cast up again; vomit, evacuate. **5** Med. show an immune response to (a transplanted organ or tissue) so that it fails to survive. ● n. /ˈriːdʒekt/ a thing or person rejected as unfit or below standard. □ **rejectable** /rɪˈdʒektəb(ə)l/ adj. **rejecter** n. (also **rejector**). **rejective** adj. **rejection** /-ˈdʒekʃ(ə)n/ n. [ME f. L rejicere reject- (as RE-, jacere throw)]

rejig /riːˈdʒɪɡ/ v.tr. (**rejigged, rejigging**) **1** re-equip (a factory etc.) for a new kind of work. **2** rearrange.

rejoice /rɪˈdʒɔɪs/ v. **1** intr. feel great joy. **2** intr. (foll. by that + clause or to + infin.) be glad. **3** intr. (foll. by in, at) take delight. **4** intr. celebrate some event. **5** tr. cause joy to. □ **rejoicer** n. **rejoicingly** adv. [ME f. OF rejoir rejoiss- (as RE-, JOY)]

rejoin[1] /riːˈdʒɔɪn/ v. **1** tr. & intr. join together again; reunite. **2** tr. join (a companion etc.) again.

rejoin[2] /rɪˈdʒɔɪn/ v. **1** tr. say in answer, retort. **2** intr. Law reply to a charge or pleading in a lawsuit. [ME f. OF rejoindre rejoign- (as RE-, JOIN)]

rejoinder /rɪˈdʒɔɪndə(r)/ n. **1** what is said in reply. **2** a retort. **3** Law a reply by rejoining. [AF rejoinder (unrecorded: as REJOIN[2])]

rejuvenate /rɪˈdʒuːvɪˌneɪt/ v.tr. make young or as if young again. □ **rejuvenator** n. **rejuvenation** /-ˌdʒuːvɪˈneɪʃ(ə)n/ n. [RE- + L juvenis young]

rejuvenesce /rɪˌdʒuːvɪˈnes/ v. **1** intr. become young again; become restored to vitality. **2** Biol. **a** intr. (of cells) change to a more active form. **b** tr. change (cells) into a more active form. □ **rejuvenescence** n. **rejuvenescent** adj. [LL rejuvenescere (as RE-, L juvenis young)]

rekindle /riːˈkɪnd(ə)l/ v.tr. & intr. kindle again.

-rel /rəl/ suffix with diminutive or derogatory force (cockerel; scoundrel). [from or after OF -erel(le)]

relabel /riːˈleɪb(ə)l/ v.tr. (**relabelled, relabelling**; US **relabeled, relabeling**) label (esp. a commodity) again or differently.

relapse /rɪˈlæps/ v. & n. ● v.intr. (usu. foll. by into) fall back or sink again (into a worse state after an improvement). ● n. (also /ˈriːlæps/) the act or an instance of relapsing, esp. a deterioration in a patient's condition after a partial recovery. □ **relapsing fever** an infectious disease characterized by recurrent fever, caused by spirochaetes of the genus Borselia. □ **relapser** n. [L relabi relaps- (as RE-, labi slip)]

relate /rɪˈleɪt/ v. **1** tr. narrate or recount (incidents, a story, etc.). **2** tr. (in passive.) (foll. by to) be connected by blood or marriage. **3** tr. (usu. foll. by to, with) bring into relation (with one another); establish a connection between (cannot relate your opinion to my own experience, a related problem). **4** intr. (foll. by to) have reference to; concern (see only what relates to themselves). **5** intr. (foll. by to) **a** bring oneself into relation to; associate with. **b** feel emotionally or sympathetically involved or connected; respond (they relate well to one another). □ **relatable** adj. [L referre relat- bring back: see REFER]

related /rɪˈleɪtɪd/ adj. **1** connected by blood or marriage. **2** having (mutual) relation; associated, connected. □ **relatedness** n.

relater /rɪˈleɪtə(r)/ n. (also **relator**) a person who relates something, esp. a story; a narrator.

relation /rɪˈleɪʃ(ə)n/ n. **1 a** the way in which one person or thing is related to another. **b** the existence or effect of a connection, correspondence, contrast, or feeling prevailing between persons or things, esp. when qualified in some way (bears no relation to the facts; enjoyed good relations for many years). **2** a relative; a kinsman or kinswoman. **3** (in pl.) **a** (foll. by with) dealings (with others). **b** sexual intercourse. **4** = RELATIONSHIP. **5 a** narration (his relation of the events). **b** a narrative. **6** Law the laying of information. □ **in relation to** as regards. [ME f. OF relation or L relatio (as RELATE)]

relational /rɪˈleɪʃən(ə)l/ adj. **1** of, belonging to, or characterized by relation. **2** having relation. □ **relational database** Computing a database structured to recognize the relation of stored items of information.

relationship /rɪˈleɪʃ(ə)n, ʃɪp/ n. **1** the fact or state of being related. **2 a** a connection or association (enjoyed a good working relationship).

b an emotional (esp. sexual) association between two people. **3** a condition or character due to being related. **4** kinship.

relative /'relətɪv/ *adj. & n.* ● *adj.* **1** considered or having significance in relation to something else (*relative velocity*). **2** (also foll. by *to*) existing or quantifiable only in terms of individual perception or consideration; not absolute or independent (*truth is relative to your perspective; it's all relative, though, isn't it?*). **3** (foll. by *to*) proportionate to (something else); in proportion to (*growth is relative to input; low energy content relative to its bulk*). **4** (foll. by *to*) in comparison with; compared with (*wage levels were low relative to the rest of the South-East*). **5 a** comparative; compared one with another (*their relative advantages*). **b** (foll. by *to*) in relation to (*move slowly relative to each other*). **6** (usu. foll. by *to*) having reference; relating, relevant (*the facts relative to the issue; need more relative proof*). **7** having mutual relations; corresponding in some way; related to each other. **8** *Gram.* **a** (of a word, esp. a pronoun) referring to an expressed or implied antecedent and attaching a subordinate clause to it, e.g. *which, who*. **b** (of a clause) attached to an antecedent by a relative word. **9** *Mus.* (of major and minor keys) having the same key signature. **10** (of a service rank) corresponding in grade to another in a different service. ● *n.* **1** a person connected by blood or marriage. **2** a species related to another by common origin (*the apes, man's closest relatives*). **3** *Gram.* a relative word, esp. a pronoun. **4** *Philos.* a relative thing or term. □ **relative atomic mass** the ratio of the average mass of one atom of an element to one twelfth of the mass of an atom of carbon-12 (also called *atomic weight*). **relative density** *Chem.* the ratio of the density of a substance to the density of a standard, usu. water for a liquid or solid, and air for a gas. **relative humidity** see HUMIDITY. **relative molecular mass** the ratio of the average mass of one molecule of an element or compound to one twelfth of the mass of an atom of carbon-12 (also called *molecular weight*). □ **relatively** *adv.* **relativeness** *n.* **relatival** /ˌrelə'taɪv(ə)l/ *adj.* (in sense 3 of *n.*). [ME f. OF *relatif -ive* or LL *relativus* having reference or relation (as RELATE)]

relativism /'relətɪˌvɪz(ə)m/ *n.* the doctrine that knowledge, morality, etc., are relative rather than absolute. □ **relativist** *n.*

relativistic /ˌrelətɪ'vɪstɪk/ *adj. Physics* (of phenomena etc.) accurately described only by the theory of relativity. □ **relativistically** *adv.*

relativity /ˌrelə'tɪvɪtɪ/ *n.* **1** the fact or state of being relative. **2** *Physics* the dependence of observations on the relative motion of the observer and the observed object; the branch of physics that deals with the description of space and time allowing for this. (*See note below.*)

▪ The concept of relativity arose from late 19th-century efforts to produce a coherent theory of electromagnetism. The chief architect was Einstein, who published his *special theory of relativity* in 1905. Einstein rejected the idea of a stationary 'ether' pervading space, and abandoned notions of absolute space and time as a common framework of reference for all bodies in the universe, instead placing great importance on distinguishing systematically between the viewpoint or framework of the observer and that of the object or process being observed. Fundamental principles of the theory are that the measured velocity of light is the same for all observers, whatever their mutual velocities, and that the mathematical form of the laws of physics which apply to moving objects or systems is independent of the motion of the framework of the observer, provided the latter motion is uniform. Among the theory's consequences (consistent with experimental results) are the following: nothing can go faster than the speed of light in a vacuum; the mass of a body increases and its length (in the direction of motion) shortens as its speed increases; the time interval between two events occurring in a moving body appears greater to a stationary observer; and mass and energy are equivalent and interconvertible. Einstein's *general theory of relativity*, published in 1915, extended this to accelerated motion and gravitation, which was treated as a curvature of the space–time continuum. It predicted that light rays would be deflected, and shifted in wavelength, when passing through a substantial gravitational field, effects which have been experimentally confirmed.

relator /rɪ'leɪtə(r)/ *n.* **1** var. of RELATER. **2** *Law* a person who makes a relation (see RELATION 6). [L (as RELATE)]

relaunch *v. & n.* ● *v.tr.* /riː'lɔ:ntʃ/ launch again. ● *n.* /'riː'lɔ:ntʃ/ a renewed launch, esp. of a business or new product.

relax /rɪ'læks/ *v.* **1 a** *tr. & intr.* (of the body, a muscle, etc.) make or become less stiff or rigid (*his frown relaxed into a smile*). **b** *tr. & intr.* make or become loose or slack; diminish in force or tension (*relaxed my grip*). **c** *tr. & intr.* (also as *int.*) make or become less tense or anxious. **2** *tr. & intr.* make or become less formal or strict (*rules were relaxed*). **3** *tr.* reduce or abate

(one's attention, efforts, etc.). **4** *intr.* cease work or effort. **5** *tr.* (as **relaxed** *adj.*) at ease; unperturbed. □ **relaxer** *n.* **relaxedly** /-sɪdlɪ/ *adv.* **relaxedness** *n.* [ME f. L *relaxare* (as RE-, LAX)]

relaxant /rɪ'læks(ə)nt/ *n. & adj.* ● *n.* a drug etc. that relaxes and reduces tension. ● *adj.* causing relaxation.

relaxation /ˌriːlæk'seɪʃ(ə)n/ *n.* **1** the act of relaxing or state of being relaxed. **2** recreation or rest, esp. after a period of work. **3** a partial remission or relaxing of a penalty, duty, etc. **4** a lessening of severity, precision, etc. **5** *Physics* the restoration of equilibrium following disturbance. [L *relaxatio* (as RELAX)]

relay /'riːleɪ/ *n. & v.* ● *n.* **1** a fresh set of people or horses substituted for tired ones. **2** a gang of workers, supply of material, etc., deployed on the same basis (*operated in relays*). **3** = relay race. **4** a device activating changes in an electric circuit etc. in response to other changes affecting itself. **5 a** a device to receive, reinforce, and transmit a telegraph message, broadcast programme, etc. **b** a relayed message or transmission. ● *v.tr.* (also /rɪ'leɪ/) **1** receive (a message, broadcast, etc.) and transmit it to others. **2 a** arrange in relays. **b** provide with or replace by relays. □ **relay race** a race between teams of which each member in turn covers part of the distance. [ME f. OF *relai* (n.), *relayer* (v.) (as RE-, *laier* ult. f. L *laxare*): cf. RELAX]

re-lay /riː'leɪ/ *v.tr.* (*past and past part.* **re-laid** /-'leɪd/) lay again or differently.

relearn /riː'lɜːn/ *v.tr.* learn again.

release /rɪ'liːs/ *v. & n.* ● *v.tr.* **1** (often foll. by *from*) set free; liberate, unfasten. **2** allow to move from a fixed position. **3 a** make (information, a recording, etc.) publicly or generally available. **b** issue (a film etc.) for general exhibition. **4** *Law* **a** remit (a debt). **b** surrender (a right). **c** make over (property or money) to another. ● *n.* **1** deliverance or liberation from a restriction, duty, or difficulty. **2** a handle or catch that releases part of a mechanism. **3** a document or item of information made available for publication (*press release*). **4 a** a film or record etc. that is released. **b** the act or an instance of releasing or the process of being released in this way. **5** *Law* **a** the act of releasing (property, money, or a right) to another. **b** a document effecting this. □ **releasable** *adj.* **releaser** *n.* **releasor** *n.* (in sense 4 of *v.*). **releasee** /ˌriːliː'siː/ *n.* (in sense 4 of *v.*). [ME f. OF *reles* (n.), *relesser* (v.), *relaiss(i)er* f. L *relaxare*: see RELAX]

relegate /'relɪˌgeɪt/ *v.tr.* **1** consign or dismiss to an inferior or less important position; demote. **2** transfer (a sports team) to a lower division of a league etc. **3** banish or send into exile. **4** (foll. by *to*) **a** transfer (a matter) for decision or implementation. **b** refer (a person) for information. □ **relegable** /-gəb(ə)l/ *adj.* **relegation** /ˌrelɪ'geɪʃ(ə)n/ *n.* [L *relegare relegat-* (as RE-, *legare* send)]

relent /rɪ'lent/ *v.intr.* **1** abandon a harsh intention. **2** yield to compassion. **3** relax one's severity; become less stern. [ME f. med.L *relentare* (unrecorded), formed as RE- + L *lentare* bend f. *lentus* flexible]

relentless /rɪ'lentlɪs/ *adj.* **1** unrelenting; insistent and uncompromising. **2** continuous; oppressively constant (*the pressure was relentless*). □ **relentlessly** *adv.* **relentlessness** *n.*

re-let *v. & n.* ● *v.tr.* /riː'let/ (**-letting**; *past and past part.* **-let**) let (a property) for a further period or to a new tenant. ● *n.* /'riː'let/ a re-let property.

relevant /'relɪv(ə)nt/ *adj.* (often foll. by *to*) bearing on or having reference to the matter in hand. □ **relevance** *n.* **relevancy** *n.* **relevantly** *adv.* [med.L *relevans*, part. of L *relevare* RELIEVE]

reliable /rɪ'laɪəb(ə)l/ *adj.* **1** that may be relied on. **2** of sound and consistent character or quality. □ **reliably** *adv.* **reliableness** *n.* **reliability** /-ˌlaɪə'bɪlɪtɪ/ *n.*

reliance /rɪ'laɪəns/ *n.* **1** (foll. by *in, on*) trust, confidence (*put full reliance in you*). **2** a thing relied upon. □ **reliant** *adj.*

relic /'relɪk/ *n.* **1** an object interesting because of its age or association. **2** a part of a deceased holy person's body or belongings kept as an object of reverence. **3** a surviving custom or belief etc. from a past age. **4** a memento or souvenir. **5** (in *pl.*) what has survived destruction or wasting or use. **6** (in *pl.*) the dead body or remains of a person. [ME *relike, relique*, etc. f. OF *relique* f. L *reliquiae*: see RELIQUIAE]

relict /'relɪkt/ *n.* **1 a** a geological or other object surviving in its primitive form. **b** an animal or plant known to have existed in the same form in previous geological ages. **2** (foll. by *of*) *archaic* a widow. [L *relinquere relict-* leave behind (as RE-, *linquere* leave): sense 2 f. OF *relicte* f. L *relicta*]

relief /rɪ'liːf/ *n.* **1 a** the alleviation of or deliverance from pain, distress, anxiety, etc. **b** the feeling accompanying such deliverance. **2** a feature etc. that diversifies monotony or relaxes tension. **3** assistance (esp.

financial) given to those in special need or difficulty (*rent relief*). **4 a** the replacing of a person or persons on duty by another or others. **b** a person or persons replacing others in this way. **5** (usu. *attrib.*) a thing supplementing another in some service, esp. an extra vehicle providing public transport at peak times. **6 a** a method of moulding or carving or stamping in which the design stands out from the surface, with projections proportioned and more (**high relief**) or less (**low relief**) closely approximating to those of the objects depicted (cf. ROUND *n.* 9). **b** a piece of sculpture etc. in relief. **c** a representation of relief given by an arrangement of line or colour or shading. **7** vividness, distinctness (*brings the facts out in sharp relief*). **8** (foll. by *of*) the reinforcement (esp. the raising of a siege) of a place. **9** esp. *Law* the redress of a hardship or grievance. □ **relief map 1** a map indicating hills and valleys by shading etc. rather than by contour lines alone. **2** a map-model showing elevations and depressions, usu. on an exaggerated relative scale. **relief printing** = LETTERPRESS 2. **relief road** a road taking traffic around a congested (esp. urban) area. [ME f. AF *relef*, OF *relief* (in sense 6 F *relief* f. It. *rilievo*) f. *relever*: see RELIEVE]

relieve /rɪˈliːv/ *v.tr.* **1** bring or provide aid or assistance to. **2** alleviate or reduce (pain, suffering, etc.). **3** mitigate the tedium or monotony of. **4** bring military support for (a besieged place). **5** release (a person) from a duty by acting as or providing a substitute. **6** (foll. by *of*) take (a burden or responsibility) away from (a person). **7** bring into relief; cause to appear solid or detached. □ **relieve one's feelings** use strong language or vigorous behaviour when annoyed. **relieve oneself** urinate or defecate. □ **relievable** *adj.* **reliever** *n.* [ME f. OF *relever* f. L *relevare* (as RE-, *levis* light)]

relieved /rɪˈliːvd/ *predic.adj.* freed from anxiety or distress (*am very relieved to hear it*). □ **relievedly** /-vɪdlɪ/ *adv.*

relievo /rɪˈliːvəʊ, -ˈljeɪvəʊ/ *n.* (also **rilievo**) (*pl.* **-os**) = RELIEF 6. [It. *rilievo* RELIEF 6]

relight /riːˈlaɪt/ *v.tr.* (*past* and *past part.* **relighted** or **relit** /-ˈlɪt/) light (a fire etc.) again.

religio- /rɪˈlɪɡɪəʊ, -ˈlɪdʒɪəʊ/ *comb. form* **1** religion. **2** religious.

religion /rɪˈlɪdʒən/ *n.* **1** the belief in a superhuman controlling power, esp. in a personal God or gods entitled to obedience and worship. **2** the expression of this in worship. **3** a particular system of faith and worship. **4** life under monastic vows (*the way of religion*). **5** a thing that one is devoted to (*football is their religion*). □ **freedom of religion** the right to follow whatever religion one chooses. □ **religionless** *adj.* [ME f. AF *religiun*, OF *religion* f. L *religio -onis* obligation, bond, reverence]

religionism /rɪˈlɪdʒəˌnɪz(ə)m/ *n.* excessive religious zeal. □ **religionist** *n.* & *adj.*

religiose /rɪˈlɪdʒɪəʊs/ *adj.* excessively religious. [L *religiosus* (as RELIGIOUS)]

religiosity /rɪˌlɪdʒɪˈɒsɪtɪ/ *n.* the condition of being religious or religiose. [ME f. L *religiositas* (as RELIGIOUS)]

religious /rɪˈlɪdʒəs/ *adj.* & *n.* ● *adj.* **1** devoted to religion; pious, devout. **2** of or concerned with religion. **3** of or belonging to a monastic order. **4** scrupulous, conscientious (*a religious attention to detail*). ● *n.* (*pl.* same) a person bound by monastic vows. □ **religiously** *adv.* **religiousness** *n.* [ME f. AF *religius*, OF *religious* f. L *religiosus* (as RELIGION)]

reline /riːˈlaɪn/ *v.tr.* renew the lining of (a garment etc.).

relinquish /rɪˈlɪŋkwɪʃ/ *v.tr.* **1** surrender or resign (a right or possession). **2** give up or cease from (a habit, plan, belief, etc.). **3** relax hold of (an object held). □ **relinquishment** *n.* [ME f. OF *relinquir* f. L *relinquere* (as RE-, *linquere* leave)]

reliquary /ˈrelɪkwərɪ/ *n.* (*pl.* **-ies**) a receptacle for (esp. religious) relics. [F *reliquaire* (as RELIC)]

reliquiae /rɪˈlɪkwɪˌiː/ *n.pl.* **1** remains. **2** *Geol.* fossil remains of animals or plants. [L f. *reliquus* remaining, formed as RE- + *linquere* liq- leave]

relish /ˈrelɪʃ/ *n.* & *v.* ● *n.* **1** (often foll. by *for*) **a** a great liking or enjoyment. **b** keen or pleasurable longing (*had no relish for travelling*). **2 a** an appetizing flavour. **b** an attractive quality (*fishing loses its relish in winter*). **3** a condiment eaten with plainer food to add flavour, esp. a piquant sauce, pickle, etc. **4** (foll. by *of*) a distinctive taste or tinge. ● *v.tr.* **1 a** get pleasure out of; enjoy greatly. **b** look forward to, anticipate with pleasure (*did not relish what lay before her*). **2** add relish to. □ **relishable** *adj.* [alt. (with assim. to -ISH[2]) of obs. *reles* f. OF *reles*, *relais* remainder f. *relaisser*: see RELEASE]

relive /riːˈlɪv/ *v.tr.* live (an experience etc.) over again, esp. in the imagination.

reload /riːˈləʊd/ *v.tr.* (also *absol.*) load (esp. a gun) again.

relocate /ˌriːləʊˈkeɪt/ *v.* **1** *tr.* locate in a new place. **2** *tr.* & *intr.* move to a new place (esp. to live or work). □ **relocation** /-ˈkeɪʃ(ə)n/ *n.*

reluctant /rɪˈlʌkt(ə)nt/ *adj.* (often foll. by *to* + infin.) unwilling or disinclined (*most reluctant to agree*). □ **reluctance** *n.* **reluctantly** *adv.* [L *reluctari* (as RE-, *luctari* struggle)]

rely /rɪˈlaɪ/ *v.intr.* (**-ies**, **-ied**) (foll. by *on*, *upon*) **1** depend on with confidence or assurance (*am relying on your judgement*). **2** be dependent on (*relies on her for everything*). [ME (earlier senses 'rally, be a vassal of') f. OF *relier* bind together f. L *religare* (as RE-, *ligare* bind)]

REM *abbr.* rapid eye-movement.

rem /rem/ *n.* (*pl.* same) a unit of effective absorbed dose of ionizing radiation in human tissue, equivalent to one roentgen of X-rays. [*roentgen equivalent man*]

remade *past* and *past part.* of REMAKE.

remain /rɪˈmeɪn/ *v.intr.* **1 a** be left over after others or other parts have been removed or used or dealt with. **b** (of a period of time) be still to elapse. **2** be in the same place or condition during further time; continue to exist or stay; be left behind (*remained at home*). **3** (foll. by complement) continue to be (*remained calm; remains President*). **4** (as **remaining** *adj.*) left behind; not having been used or dealt with. [ME f. OF *remain-* stressed stem of *remanoir* or f. OF *remaindre* ult. f. L *remanere* (as RE-, *manere* stay)]

remainder /rɪˈmeɪndə(r)/ *n.* & *v.* ● *n.* **1** a part remaining or left over. **2** remaining persons or things. **3** a number left after division or subtraction. **4** the copies of a book left unsold when demand has fallen. **5** *Law* an interest in an estate that becomes effective in possession only when a prior interest (devised at the same time) ends. ● *v.tr.* dispose of (a remainder of books) at a reduced price. [ME (in sense 5) f. AF, = OF *remaindre*: see REMAIN]

remains /rɪˈmeɪnz/ *n.pl.* **1** what remains after other parts have been removed or used etc. **2** relics of antiquity, esp. of buildings (*Roman remains*). **3** a person's body after death. **4** an author's (esp. unpublished) works left after death.

remake *v.* & *n.* ● *v.tr.* /riːˈmeɪk/ (*past* and *past part.* **remade** /-ˈmeɪd/) make again or differently. ● *n.* /ˈriːmeɪk/ a thing that has been remade, esp. a cinema film.

reman /riːˈmæn/ *v.tr.* (**remanned**, **remanning**) **1** equip (a fleet etc.) with new personnel. **2** *poet.* make courageous again.

remand /rɪˈmɑːnd/ *v.* & *n.* ● *v.tr.* return (a prisoner) to custody, esp. to allow further inquiries to be made. ● *n.* a recommittal to custody. □ **on remand** in custody pending trial. **remand centre** (in the UK) an institution to which accused persons are remanded pending trial. [ME f. LL *remandare* (as RE-, *mandare* commit)]

remanent /ˈremənənt/ *adj.* **1** remaining, residual. **2** (of magnetism) remaining after the magnetizing field has been removed. □ **remanence** *n.* [ME f. L *remanere* REMAIN]

remark /rɪˈmɑːk/ *v.* & *n.* ● *v.* **1** *tr.* (often foll. by *that* + clause) **a** say by way of comment. **b** take notice of; regard with attention. **2** *intr.* (usu. foll. by *on*, *upon*) make a comment. ● *n.* **1** a written or spoken comment; anything said. **2 a** the act of noticing or observing (*worthy of remark*). **b** the act of commenting (*let it pass without remark*). [F *remarque*, *remarquer* (as RE-, MARK[1])]

remarkable /rɪˈmɑːkəb(ə)l/ *adj.* **1** worth notice; exceptional, extraordinary. **2** striking, conspicuous. □ **remarkableness** *n.* **remarkably** *adv.* [F *remarquable* (as REMARK)]

remarry /riːˈmærɪ/ *v.intr.* & *tr.* (**-ies**, **-ied**) marry again. □ **remarriage** /-ˈmærɪdʒ/ *n.*

remaster /riːˈmɑːstə(r)/ *v.tr.* make a new master of (a recording), esp. to improve the sound quality.

rematch /ˈriːmætʃ/ *n.* a return match or game.

Rembrandt /ˈrembrænt/ (full name Rembrandt Harmensz van Rijn) (1606–69), Dutch painter. After working at first in his native Leiden, he moved to Amsterdam, where he made his name as a portrait painter with the *Anatomy Lesson of Dr Tulp* (1632), a strongly lit group portrait in the manner of Caravaggio. With his most celebrated painting, the *Night Watch* (1642), he used chiaroscuro to give his subjects a more spiritual and introspective quality, a departure which was to transform the Dutch portrait tradition. Rembrandt is especially identified with the series of over sixty self-portraits painted from 1629 to 1669, a sustained exercise in self-analysis. His prolific output also included many religious, genre, and landscape paintings, drawings, and etchings. Although his name remained well known after his death, it was not until the romantic period that Rembrandt was recognized as a supreme artist.

REME /ˈriːmiː/ *abbr.* (in the UK) Royal Electrical and Mechanical Engineers.

remeasure /riːˈmeʒə(r)/ *v.tr.* measure again. □ **remeasurement** *n.*

remedial /rɪˈmiːdɪəl/ *adj.* **1** affording or intended as a remedy (*remedial therapy*). **2** (of teaching) for children with learning difficulties. □ **remedially** *adv.* [LL *remedialis* f. L *remedium* (as REMEDY)]

remedy /ˈremɪdɪ/ *n. & v.* ● *n.* (*pl.* **-ies**) (often foll. by *for, against*) **1** a medicine or treatment (for a disease etc.). **2** a means of counteracting or removing anything undesirable. **3** redress; legal or other reparation. **4** the margin within which coins as minted may differ from the standard fineness and weight. ● *v.tr.* (**-ies, -ied**) **1** rectify; make good. **2** heal, cure (a person, diseased part, etc.) □ **remediable** /rɪˈmiːdɪəb(ə)l/ *adj.* [ME f. AF *remedie*, OF *remede* or L *remedium* (as RE-, *mederi* heal)]

remember /rɪˈmembə(r)/ *v.tr.* **1** keep in the memory; not forget. **2 a** (also *absol.*) bring back into one's thoughts, call to mind (knowledge or experience etc.). **b** (often foll. by *to* + infin. or *that* + clause) have in mind (a duty, commitment, etc.) (*will you remember to lock the door?*). **3** think of or acknowledge (a person) in some connection, esp. in making a gift etc. **4** (foll. by *to*) convey greetings from (one person) to (another) (*remember me to your mother*). **5** mention (in prayer). □ **remember oneself** recover one's manners or intentions after a lapse. □ **rememberer** *n.* [ME f. OF *remembrer* f. LL *rememorari* (as RE-, *memor* mindful)]

remembrance /rɪˈmembrəns/ *n.* **1** the act of remembering or process of being remembered. **2** a memory or recollection. **3** a keepsake or souvenir. **4** (in *pl.*) greetings conveyed through a third person. [ME f. OF (as REMEMBER)]

Remembrance Day 1 = REMEMBRANCE SUNDAY. **2** *hist.* Armistice Day.

Remembrance Sunday (in the UK) the Sunday nearest 11 Nov., commemorating those killed in the First and Second World Wars and in later conflicts.

remex /ˈriːmeks/ *n.* (*pl.* **remiges** /ˈremɪˌdʒiːz/) a primary or secondary feather in a bird's wing. [L, = rower, f. *remus* oar]

remind /rɪˈmaɪnd/ *v.tr.* **1** (foll. by *of*) cause (a person) to think of. **2** (usu. foll. by *to* + infin. or *that* + clause) cause (a person) to remember a commitment etc. (*remind them to pay their subscriptions*).

reminder /rɪˈmaɪndə(r)/ *n.* **1 a** a thing that reminds, esp. a letter or bill. **b** a means of reminding; an aid to the memory. **2** (often foll. by *of*) a memento or souvenir.

remindful /rɪˈmaɪndfʊl/ *adj.* (often foll. by *of*) acting as a reminder; reviving the memory.

reminisce /ˌremɪˈnɪs/ *v.intr.* (often foll. by *about*) indulge in reminiscence. □ **reminiscer** *n.* [back-form. f. REMINISCENCE]

reminiscence /ˌremɪˈnɪs(ə)ns/ *n.* **1** the act of remembering things past; the recovery of knowledge by mental effort. **2 a** a past fact or experience that is remembered. **b** the process of narrating this. **3** (in *pl.*) a collection in literary form of incidents and experiences that a person remembers. **4** *Philos.* (esp. in Platonism) the theory of the recovery of things known to the soul in previous existences. **5** a characteristic of one thing reminding or suggestive of another. □ **reminiscential** /-nɪˈsenʃ(ə)l/ *adj.* [LL *reminiscentia* f. L *reminisci* remember]

reminiscent /ˌremɪˈnɪs(ə)nt/ *adj.* **1** (foll. by *of*) tending to remind one of or suggest. **2** concerned with reminiscence. **3** (of a person) given to reminiscing. □ **reminiscently** *adv.*

remise /rɪˈmiːz, -ˈmaɪz/ *v. & n.* ● *v.intr.* **1** *Law* surrender or make over (a right or property). **2** *Fencing* make a remise. ● *n. Fencing* a second thrust made after the first has failed. [F f. *remis, remise* past part. of *remettre* put back: cf. REMIT]

remiss /rɪˈmɪs/ *adj.* careless of duty; lax, negligent. □ **remissly** *adv.* **remissness** *n.* [ME f. L *remissus* past part. of *remittere* slacken: see REMIT]

remissible /rɪˈmɪsɪb(ə)l/ *adj.* that may be remitted. [F *rémissible* or LL *remissibilis* (as REMIT)]

remission /rɪˈmɪʃ(ə)n/ *n.* **1** the reduction of a prison sentence on account of good behaviour. **2** the remitting of a debt or penalty etc. **3** a diminution of force, effect, or degree (esp. of disease or pain); the temporary disappearance of symptoms. **4** (often foll. by *of*) forgiveness (of sins etc.). □ **remissive** /-ˈmɪsɪv/ *adj.* [ME f. OF *remission* or L *remissio* (as REMIT)]

remit *v. & n.* ● *v.* /rɪˈmɪt/ (**remitted, remitting**) **1** *tr.* cancel or refrain from exacting or inflicting (a debt or punishment etc.). **2** *intr. & tr.* abate or slacken; cease or cease from partly or entirely. **3** *tr.* send (money etc.) in payment. **4** *tr.* cause to be conveyed by post. **5** *tr.* **a** (foll. by *to*) refer (a matter for decision etc.) to some authority. **b** *Law* send back (a case) to a lower court. **6** *tr.* **a** (often foll. by *to*) postpone or defer. **b** (foll. by *in, into*) send or put back into a previous state. **7** *tr.* (in Christian theology) (usu. of God) pardon (sins etc.). ● *n.* /ˈriːmɪt, rɪˈmɪt/ **1** the terms of reference of a committee etc. **2** an item remitted for consideration. □ **remittable** /rɪˈmɪtəb(ə)l/ *adj.* **remittal** *n.* **remitter** *n.* **remittee** /ˌrɪmɪˈtiː/ *n.* [ME f. L *remittere remiss-* (as RE-, *mittere* send)]

remittance /rɪˈmɪt(ə)ns/ *n.* **1** money sent, esp. by post, for goods or services or as an allowance. **2** the act of sending money. □ **remittance man** *hist.* an emigrant living on remittances from home.

remittent /rɪˈmɪt(ə)nt/ *adj.* (of a fever) that abates at intervals. [L *remittere* (as REMIT)]

remix *v. & n.* ● *v.tr.* /riːˈmɪks/ mix again, esp. create a new version of (a sound recording) by combining the separate tracks in a different way. ● *n.* /ˈriːmɪks/ a version of a sound recording created by remixing.

remnant /ˈremnənt/ *n.* **1** a small remaining quantity. **2** a piece of cloth etc. left when the greater part has been used or sold. **3** (foll. by *of*) a surviving trace (*a remnant of empire*). [ME (earlier *remenant*) f. OF *remenant* f. *remenoir* REMAIN]

remodel /riːˈmɒd(ə)l/ *v.tr.* (**remodelled, remodelling;** *US* **remodeled, remodeling**) **1** model again or differently. **2** reconstruct.

remodify /riːˈmɒdɪˌfaɪ/ *v.tr.* (**-ies, -ied**) modify again. □ **remodification** /-ˌmɒdɪfɪˈkeɪʃ(ə)n/ *n.*

remold *US* var. of REMOULD.

remonetize /riːˈmʌnɪˌtaɪz/ *v.tr.* (also **-ise**) restore (a metal etc.) to its former position as legal tender. □ **remonetization** /-ˌmʌnɪtaɪˈzeɪʃ(ə)n/ *n.*

remonstrance /rɪˈmɒnstrəns/ *n.* **1** the act or an instance of remonstrating. **2** an expostulation or protest. [ME f. obs. F *remonstrance* or med.L *remonstrantia* (as REMONSTRATE)]

Remonstrance, the a document drawn up in 1610 by the Arminians of the Dutch Reformed Church, presenting the differences between their doctrines and those of the strict Calvinists.

Remonstrant /rɪˈmɒnstrənt/ *n.* a member of the Arminian party in the Dutch Reformed Church, so called from the Remonstrance of 1610.

remonstrate /ˈremənˌstreɪt/ *v.* **1** *intr.* (foll. by *with*) make a protest; argue forcibly (*remonstrated with them over the delays*). **2** *tr.* (often foll. by *that* + clause) urge protestingly. □ **remonstrator** *n.* **remonstrant** /rɪˈmɒnstrənt/ *adj.* **remonstrative** *adj.* **remonstration** /ˌremənˈstreɪʃ(ə)n/ *n.* [med.L *remonstrare* (as RE-, *monstrare* show)]

remontant /rɪˈmɒntənt/ *adj. & n.* ● *adj.* (esp. of a rose) blooming more than once a year. ● *n.* a remontant rose etc. [F f. *remonter* REMOUNT]

remora /ˈremərə/ *n.* a slender marine fish of the family Echeneidae, which attaches itself by a modified sucker-like dorsal fin to larger fish and to ships. [L, = hindrance (as RE-, *mora* delay, from the former belief that the fish slowed ships down)]

remorse /rɪˈmɔːs/ *n.* **1** deep regret for a wrong committed. **2** compunction; a compassionate reluctance to inflict pain (esp. in *without remorse*). [ME f. OF *remors* f. med.L *remorsus* f. L *remordere remors-vex* (as RE-, *mordere* bite)]

remorseful /rɪˈmɔːsfʊl/ *adj.* filled with repentance. □ **remorsefully** *adv.*

remorseless /rɪˈmɔːslɪs/ *adj.* **1** without compassion or compunction. **2** relentless; unabating. □ **remorselessly** *adv.* **remorselessness** *n.*

remortgage /riːˈmɔːgɪdʒ/ *v. & n.* ● *v.tr.* (also *absol.*) mortgage again; revise the terms of an existing mortgage on (a property). ● *n.* a different or altered mortgage.

remote /rɪˈməʊt/ *adj.* (**remoter, remotest**) **1** far away in place or time. **2** out of the way; situated away from the main centres of population, society, etc. **3** distantly related (*a remote ancestor*). **4** slight, faint (esp. in *not the remotest chance, idea*, etc.). **5** (of a person) aloof; not friendly. **6** (foll. by *from*) widely different; separate by nature (*ideas remote from the subject*). □ **remote control 1** control of a machine or apparatus from a distance by means of signals transmitted from a radio or electronic device. **2** such a device. **remote-controlled** (of a machine etc.) operated by remote control. **remote sensing** the scanning of the earth or another planet by a satellite or spacecraft in order to obtain information about it. □ **remotely** *adv.* **remoteness** *n.* [ME f. L *remotus* (as REMOVE)]

remould v. & n. (US **remold**) ● v.tr. /riːˈməʊld/ **1** mould again; refashion. **2** re-form the tread of (a tyre). ● n. /ˈriːməʊld/ a remoulded tyre.

remount v. & n. ● v. /riːˈmaʊnt/ **1 a** tr. mount (a horse etc.) again. **b** intr. get on horseback again. **2** tr. get on to or ascend (a ladder, hill, etc.) again. **3** tr. provide (a person) with a fresh horse etc. **4** tr. put (a picture etc.) on a fresh mount. ● n. /ˈriːmaʊnt/ **1** a fresh horse for a rider. **2** a supply of fresh horses for a regiment.

removal /rɪˈmuːv(ə)l/ n. **1** the act or an instance of removing; the process of being removed. **2** the transfer of furniture and other contents on moving house. **3 a** dismissal from an office or post; deposition. **b** (an act of) murder.

remove /rɪˈmuːv/ v. & n. ● v. **1** tr. take off or away from the place or position occupied; detach (*remove the top carefully*). **2** tr. **a** move or take to another place; change the situation of (*will you remove the tea things?*). **b** get rid of; eliminate (*will remove all doubts*). **3** tr. cause to be no longer present or available; take away (*all privileges were removed*). **4** tr. (often foll. by *from*) dismiss (from office). **5** tr. colloq. kill, assassinate. **6** tr. (in passive; foll. by *from*) distant or remote in condition (*the country is not far removed from anarchy*). **7** tr. (as **removed** adj.) (esp. of cousins) separated by a specified number of steps of descent (*a first cousin twice removed = a grandchild of a first cousin*). **8** formal **a** intr. (usu. foll. by *from*, *to*) change one's home or place of residence. **b** tr. conduct the removal of. ● n. **1 a** degree of remoteness; a distance. **2** a stage in a gradation; a degree (*is several removes from what I expected*). **3** (in the UK) a form or division in some schools. □ **remover** n. (esp. in sense 8b of v.). **removable** adj. **removability** /-ˌmuːvəˈbɪlɪtɪ/ n. [ME f. OF *removeir* f. L *removere* *remot-* (as RE-, *movere* move)]

remunerate /rɪˈmjuːnəˌreɪt/ v.tr. **1** reward; pay for services rendered. **2** serve as or provide recompense for (toil etc.) or to (a person). □ **remunerative** /-rətɪv/ adj. **remuneratory** adj. **remuneration** /-ˌmjuːnəˈreɪʃ(ə)n/ n. [L *remunerari* (as RE-, *munus muneris* gift)]

Remus /ˈriːməs/ *Rom. Mythol.* the twin brother of Romulus.

Renaissance /rɪˈneɪs(ə)ns, -sɒns/ n. **1** the revival of art and learning under the influence of classical models which began in Italy in the late Middle Ages. (*See note below.*) **2** the period of this. **3** the culture and style of art, architecture, etc., developed during this era. **4** (**renaissance**) any similar revival. [F *renaissance* (as RE-, F *naissance* birth f. L *nascentia* or F *naître naiss*- be born f. Rmc: cf. NASCENT)]

▪ The Renaissance is generally regarded as beginning in Florence in the early 14th century, where humanist ideas were arising and there was a revival of interest in classical antiquity with its associations of naturalism and fidelity to nature. Important early figures are the writers Petrarch, Dante, and Boccaccio and the painter Giotto. Classical techniques and styles were studied in Rome by the sculptor Donatello as well as by the architects Bramante and Brunelleschi, who worked on the theory of perspective, which was developed in the innovative frescos and paintings of Masaccio. The period from the end of the 15th century has become known as the High Renaissance, when Venice and Rome began to share Florence's importance and Botticelli, Cellini, Raphael, and particularly Leonardo da Vinci and Michelangelo were active. The latter two are the foremost examples of the 'Renaissance man', the ideal figure of the period, versed in all aspects of art and learning. The ideas of the Renaissance did not begin to spread beyond Italy until the early 16th century, but were then important throughout western Europe for the next hundred years, influencing, among others, Dürer, Erasmus, Montaigne, Cervantes, and Shakespeare. From the art of the Italian High Renaissance developed mannerism.

renal /ˈriːn(ə)l/ adj. of or concerning the kidneys. □ **renal dialysis** see DIALYSIS 2. [F *rénal* f. LL *renalis* f. L *renes* kidneys]

rename /riːˈneɪm/ v.tr. name again; give a new name to.

Renamo /rəˈnɑːməʊ/ a right-wing movement that fought a guerrilla war against the Frelimo government in Mozambique from shortly after Mozambican independence in 1975, originally with support from Rhodesia (Zimbabwe) and South Africa. A peace agreement was signed in 1992. [Port., acronym f. *Resistance Nationale de Mozambique*]

Renan /rəˈnɒn/, (Joseph) Ernest (1823–92), French historian, theologian, and philosopher. A major figure in 19th-century French theology and philosophy, he provoked a controversy with the publication of his *Vie de Jésus* (1863), which rejected the supernatural element in the life of Jesus. His belief that the future of the world lay in the progress of science found expression in *L'Avenir de la science* (1890).

renascence /rɪˈnæs(ə)ns/ n. **1** rebirth; renewal. **2** = RENAISSANCE. [RENASCENT]

renascent /rɪˈnæs(ə)nt/ adj. springing up anew; being reborn or renewed. [L *renasci* (as RE-, *nasci* be born)]

Renault[1] /ˈrenəʊ/, Louis (1877–1944), French engineer and motor manufacturer. Together with his brothers he established the original Renault company in 1898 and became known for designing and manufacturing a series of racing cars. In 1918 the company produced its first tank; Renault subsequently expanded the firm's range to incorporate industrial and agricultural machinery, as well as further military technology. In 1944 Renault was imprisoned, accused of collaborating with the Germans; he died before the trial began. His company was nationalized in 1945 and subsequently became one of France's leading manufacturers of motor cars.

Renault[2] /ˈrenəʊ/, Mary (pseudonym of Mary Challans) (1905–83), British novelist, resident in South Africa from 1948. Her reputation is based on her historical novels set in ancient Greece and Asia Minor. They include two trilogies, one recalling the legend of Theseus and the other the story of Alexander the Great (*Fire from Heaven*, 1970; *The Persian Boy*, 1972; *Funeral Games*, 1981).

rencontre /renˈkɒntə(r)/ n. archaic = RENCOUNTER. [F (as RENCOUNTER)]

rencounter /renˈkaʊntə(r)/ n. & v. ● n. **1** an encounter; a chance meeting. **2** a battle, skirmish, or duel. ● v.tr. encounter; meet by chance. [F *rencontre(r)* (as RE-, ENCOUNTER)]

rend /rend/ v. (*past* and *past part.* **rent** /rent/) *archaic* or *rhet.* **1** tr. (foll. by *off*, *from*, *away*, etc.; also *absol.*) tear or wrench forcibly. **2** tr. & intr. split or divide in pieces or into factions (*a country rent by civil war*). **3** tr. cause emotional pain to (the heart etc.). □ **rend the air** sound piercingly. **rend one's garments** (or **hair**) display extreme grief or rage. [OE *rendan*, rel. to MLG *rende*]

Rendell /ˈrend(ə)l/, Ruth (Barbara) (b.1930), English writer of detective fiction and thrillers. She is known as the creator of Chief Inspector Wexford, who appears in a series of detective novels starting with *From Doon with Death* (1964). Rendell is also noted for her psychological crime novels, including *A Judgement in Stone* (1977) and — under the pseudonym of Barbara Vine — *A Dark-Adapted Eye* (1986).

render /ˈrendə(r)/ v.tr. **1** cause to be or become; make (*rendered us helpless*). **2** give or pay (money, service, etc.), esp. in return or as a thing due (*render thanks; rendered good for evil*). **3** (often foll. by *to*) **a** give (assistance) (*rendered aid to the injured man*). **b** show (obedience etc.). **c** do (a service etc.). **4 a** submit; send in; present (an account, reason, etc.). **b** *Law* (of a judge or jury) deliver formally (a judgement or verdict). **5 a** represent or portray artistically, musically, etc. **b** act (a role); represent (a character, idea, etc.) (*the dramatist's conception was well rendered*). **c** *Mus.* perform; execute. **6** translate (*rendered the poem into French*). **7** (often foll. by *down*) melt down (fat etc.), esp. to clarify; extract by melting. **8** cover (stone or brick) with a coat of plaster. **9** formal **a** give back; hand over; deliver, give up, surrender (*render unto Caesar the things that are Caesar's*). **b** show (obedience). □ **render-set** v.tr. (-**setting**; *past* and *past part.* **-set**) plaster (a wall etc.) with two coats. ● n. a plastering of two coats. ● adj. of two coats. □ **renderer** n. [ME f. OF *rendre* ult. f. L *reddere reddit*- (as RE-, *dare* give)]

rendering /ˈrendərɪŋ/ n. **1 a** the act or an instance of performing music, drama, etc.; an interpretation or performance (*an excellent rendering of the part*). **b** a translation. **2 a** the act or an instance of plastering stone, brick, etc. **b** this coating. **3** formal the act or an instance of giving, yielding, or surrendering.

rendezvous /ˈrɒndɪˌvuː, -deɪˌvuː/ n. & v. ● n. (*pl.* same /-ˌvuːz/) **1** an agreed or regular meeting-place. **2** a meeting by arrangement. **3** a place appointed for assembling troops, ships, etc. **4** a prearranged meeting between spacecraft in space. ● v.intr. (**rendezvouses** /-ˌvuːz/; **rendezvoused** /-ˌvuːd/; **rendezvousing** /-ˌvuːɪŋ/) meet at a rendezvous. [F *rendez-vous* present yourselves f. *rendre*: see RENDER]

rendition /renˈdɪʃ(ə)n/ n. (often foll. by *of*) **1** an interpretation or rendering of a dramatic role, piece of music, etc. **2** a visual representation. [obs. F f. *rendre* RENDER]

rendzina /rendˈziːnə/ n. a fertile lime-rich soil with dark humus above a pale soft calcareous layer, typical of grassland on chalk or limestone. [Russ. f. Pol. *rędzina*]

renegade /ˈrenɪˌgeɪd/ n., adj., & v. ● n. **1** a person who deserts a party or principles. **2** an outlaw or rebel. **3** archaic an apostate; a person who abandons one religion for another. ● adj. traitorous, heretical; rebellious. ● v.intr. be a renegade. [Sp. *renegado* f. med.L *renegatus* (as RE-, L *negare* deny)]

renegado /ˌrenɪˈgeɪdəʊ/ n. (*pl.* -**oes**) archaic = RENEGADE. [Sp. (as RENEGADE)]

renege /rɪˈniːg, -ˈneg, -ˈneɪg/ v. (also **renegue**) **1** intr. **a** go back on one's word; change one's mind; recant. **b** (foll. by on) go back on (a promise or undertaking or contract). **2** tr. deny, renounce, abandon (a person, faith, etc.). **3** intr. Cards revoke. □ **reneger** n. **reneguer** n. [med.L *renegare* (as RE-, L *negare* deny)]

renegotiate /ˌriːnɪˈɡəʊʃɪˌeɪt, -ˈɡəʊsɪˌeɪt/ v.tr. (also absol.) negotiate again or on different terms. □ **renegotiable** /-ˈɡəʊʃəb(ə)l, -ˈɡəʊsɪ-/ adj. **renegotiation** /-ˌɡəʊʃɪˈeɪʃ(ə)n, -ˌɡəʊsɪ-/ n.

renew /rɪˈnjuː/ v.tr. **1** revive; regenerate; make new again; restore to the original state. **2** reinforce; resupply; replace. **3** repeat or re-establish, resume after an interruption (*renewed our acquaintance; a renewed attack*). **4** get, begin, make, say, give, etc., anew. **5** (also absol.) grant or be granted a continuation of or continued validity of (a licence, subscription, lease, etc.). **6** recover (one's youth, strength, etc.). □ **renewal** n. **renewer** n.

renewable /rɪˈnjuːəb(ə)l/ adj. & n. ● adj. **1** able to be renewed. **2** (of a fuel or source of energy) not depleted by utilization. (*See note below.*) ● n. a renewable fuel or source of energy. □ **renewability** /-ˌnjuːəˈbɪlɪtɪ/ n.

▪ As non-renewable natural resources such as fossil and nuclear fuels are being depleted, increasing efforts are being made to make more use of renewable resources. There are various ways of generating electricity or obtaining heat from natural sources of energy, depending on local climate and topography. Hydroelectric power now accounts for nearly a fifth of the world's electricity. However, other schemes — wind generators, solar and geothermal energy, and tidal power — are still of only local importance, and wave power has yet to be economically proven. Renewable natural materials such as wood, vegetable matter, and manure can be used directly for heating and cooking, or they can be processed into solid, liquid, or gaseous fuels such as biogas, sometimes yielding fertilizers as a by-product.

Renfrewshire /ˈrenfruːˌʃɪə(r)/ a former county of west central Scotland, on the Firth of Clyde. Since 1975 it has been part of Strathclyde region.

reniform /ˈriːnɪˌfɔːm, ˈren-/ adj. esp. Med. kidney-shaped. [L *ren* kidney + -FORM]

Rennes /ren/ an industrial city in NW France; pop. (1990) 203,530. It was established as the capital of a Celtic tribe, the Redones, from whom it derives its name, later becoming the capital of the ancient kingdom of Brittany.

rennet /ˈrenɪt/ n. **1** curdled milk containing rennin found in the stomach of an unweaned calf, used in curdling milk for cheese, junket, etc. **2** a preparation containing rennin made from the stomach-membrane of a calf or from certain fungi, used for the same purpose. [ME, prob. f. an OE form *rynet* (unrecorded), rel. to RUN]

Rennie /ˈrenɪ/, John (1761–1821), Scottish civil engineer. He is best known as the designer of the London and East India Docks (built c.1800), the Inchcape Rock lighthouse (1807–c.1811), and Waterloo Bridge, Southwark Bridge, and London Bridge (1811–31).

rennin /ˈrenɪn/ n. Biochem. an enzyme secreted into the stomach of unweaned mammals causing the clotting of milk. [RENNET + -IN]

Reno /ˈriːnəʊ/ a city in western Nevada; pop. (1990) 133,850. It is noted as a gambling resort and for its liberal laws enabling quick marriages and divorces.

Renoir[1] /rəˈnwɑː(r), ˈrenwɑː(r)/, Jean (1894–1979), French film director, son of Auguste Renoir. His fame is based chiefly on the films he made in France in the 1930s, including *La Grande illusion* (1937), concerning prisoners of war in the First World War, and *La Règle du jeu* (1939), a black comedy about a weekend shooting-party. After spending the Second World War in the US, he returned to Europe, where he had an important influence on the *nouvelle vague* film directors of the 1960s.

Renoir[2] /rəˈnwɑː(r), ˈrenwɑː(r)/, (Pierre) Auguste (1841–1919), French painter. One of the early impressionists, he developed a style characterized by light, fresh colours and indistinct, subtle outlines. In his later work he concentrated on the human, especially female, form. His best-known paintings include *Le Moulin de la galette* (1876), *Les Grandes baigneuses* (1884–7), and *The Judgement of Paris* (c.1914).

renominate /riːˈnɒmɪˌneɪt/ v.tr. nominate for a further term of office. □ **renomination** /-ˌnɒmɪˈneɪʃ(ə)n/ n.

renounce /rɪˈnaʊns/ v. **1** tr. consent formally to abandon; surrender; give up (a claim, right, possession, etc.). **2** tr. repudiate; refuse to recognize any longer (*renouncing their father's authority*). **3** tr. **a** decline further association or disclaim relationship with (*renounced my former friends*). **b** withdraw from; discontinue; forsake. **4** intr. Law refuse or resign a right or position esp. as an heir or trustee. **5** intr. Cards follow

with a card of another suit when having no card of the suit led (cf. REVOKE v. 2). □ **renounce the world** abandon society or material affairs. □ **renounceable** adj. **renouncement** n. **renouncer** n. [ME f. OF *renoncer* f. L *renuntiare* (as RE-, *nuntiare* announce)]

renovate /ˈrenəˌveɪt/ v.tr. **1** restore to good condition; repair. **2** make new again. □ **renovative** adj. **renovator** n. **renovation** /ˌrenəˈveɪʃ(ə)n/ n. [L *renovare* (as RE-, *novus* new)]

renown /rɪˈnaʊn/ n. fame; high distinction; celebrity (*a city of great renown*). [ME f. AF *ren(o)un*, OF *renon*, *renom* f. *renomer* make famous (as RE-, L *nominare* NOMINATE)]

renowned /rɪˈnaʊnd/ adj. famous; celebrated.

rent[1] /rent/ n. & v. ● n. **1** a tenant's periodical payment to an owner or landlord for the use of land or premises. **2** payment for the use of a service, equipment, etc. ● v. **1** tr. (often foll. by from) take, occupy, or use at a rent (*rented a cottage from the local farmer*). **2** tr. (often foll. by out) let or hire (a thing) for rent. **3** intr. (foll. by at) be let or hired out at a specified rate (*the land rents at £100 per month*). □ **for rent** N. Amer. available to be rented. **rent-a-** (in comb.) often joc. denoting availability for hire (*rent-a-van; rent-a-crowd*). **rent-boy** a young male prostitute. **rent-free** with exemption from rent. **rent-roll** the register of a landlord's lands etc. with the rents due from them; the sum of one's income from rent. [ME f. OF *rente* f. Rmc (as RENDER)]

rent[2] /rent/ n. **1** a large tear in a garment etc. **2** an opening in clouds etc. **3** a cleft, fissure, or gorge. [obs. *rent* var. of REND]

rent[3] /rent/ past and past part. of REND.

rentable /ˈrentəb(ə)l/ adj. **1** available or suitable for renting. **2** giving an adequate ratio of profit to capital. □ **rentability** /ˌrentəˈbɪlɪtɪ/ n.

rental /ˈrent(ə)l/ n. **1** the amount paid or received as rent. **2** the act of renting. **3** N. Amer. a rented house etc. □ **rental library** US a library which rents books for a fee. [ME f. AF *rental* or AL *rentale* (as RENT[1])]

renter /ˈrentə(r)/ n. **1** a person who rents. **2** Cinematog. (in the UK) a person who distributes cinema films. **3** sl. a male prostitute.

rentier /ˈrɒntɪˌeɪ/ n. a person living on dividends from property, investments, etc. [F f. *rente* dividend]

renumber /riːˈnʌmbə(r)/ v.tr. change the number or numbers given or allocated to.

renunciation /rɪˌnʌnsɪˈeɪʃ(ə)n/ n. **1** the act or an instance of renouncing or giving up. **2** self-denial. **3** a document expressing renunciation. □ **renunciant** /rɪˈnʌnsɪənt/ n. & adj. **renunciative** /-sɪətɪv/ adj. **renunciatory** adj. [ME f. OF *renonciation* or LL *renuntiatio* (as RENOUNCE)]

renvoi /ˈrenvɔɪ, ˈrɒnvwʌ/ n. Law the act or an instance of referring a case, dispute, etc., to a different jurisdiction. [F f. *renvoyer* send back]

reoccupy /riːˈɒkjʊˌpaɪ/ v.tr. (-ies, -ied) occupy again. □ **reoccupation** /-ˌɒkjʊˈpeɪʃ(ə)n/ n.

reoccur /ˌriːəˈkɜː(r)/ v.intr. (**reoccurred**, **reoccurring**) occur again or habitually. □ **reoccurrence** /-ˈkʌrəns/ n.

reoffend /ˌriːəˈfend/ v.intr. (of a convicted criminal) offend again.

reopen /riːˈəʊp(ə)n/ v.tr. & intr. open again.

reorder /riːˈɔːdə(r)/ v. & n. ● v.tr. order again. ● n. a renewed or repeated order for goods.

reorganize /riːˈɔːɡəˌnaɪz/ v.tr. (also **-ise**) organize differently. □ **reorganizer** n. **reorganization** /-ˌɔːɡənaɪˈzeɪʃ(ə)n/ n.

reorient /riːˈɔːrɪˌent, -ˈɒrɪˌent/ v.tr. **1** give a new direction to (ideas etc.); redirect (a thing). **2** help (a person) find his or her bearings again. **3** change the outlook of (a person). **4** (refl., often foll. by to) adjust oneself to or come to terms with something.

reorientate /riːˈɔːrɪenˌteɪt, -ˈɒrɪenˌteɪt/ v.tr. = REORIENT. □ **reorientation** /-ˌɔːrɪenˈteɪʃ(ə)n, -ˌɒr-/ n.

Rep. abbr. **1** (in the US) a Representative in Congress. **2** a Republican.

rep[1] /rep/ n. & v. colloq. ● n. a representative, esp. a commercial traveller. ● v.intr. (**repped**, **repping**) act as a representative for a company, product, etc. [abbr.]

rep[2] /rep/ n. colloq. **1** repertory. **2** a repertory theatre or company. [abbr.]

rep[3] /rep/ n. (also **repp**) a textile fabric with a corded surface, used in curtains and upholstery. [F *reps*, of unkn. orig.]

rep[4] /rep/ n. US sl. reputation. [abbr.]

repack /riːˈpæk/ v.tr. pack again.

repackage /riːˈpækɪdʒ/ v.tr. **1** package again or differently. **2** present in a new form. □ **repackaging** n.

repaginate /riːˈpædʒɪˌneɪt/ v.tr. paginate again; renumber the pages of. □ **repagination** /-ˌpædʒɪˈneɪʃ(ə)n/ n.

repaid *past* and *past part.* of REPAY.

repaint *v. & n.* ● *v.tr.* /riː'peɪnt/ **1** paint again or differently. **2** restore the paint or colouring of. ● *n.* /'riːpeɪnt/ **1** the act of repainting. **2** a repainted thing, esp. a golf ball.

repair[1] /rɪ'peə(r)/ *v. & n.* ● *v.tr.* **1** restore to good condition after damage or wear. **2** renovate or mend by replacing or fixing parts or by compensating for loss or exhaustion. **3** set right or make amends for (loss, wrong, error, etc.). ● *n.* **1** the act or an instance of restoring to sound condition (*in need of repair; closed during repair*). **2** the result of this (*the repair is hardly visible*). **3** good or relative condition for working or using (*must be kept in repair; in good repair*). □ **repairable** *adj.* **repairer** *n.* [ME f. OF *reparer* f. L *reparare* (as RE-, *parare* make ready)]

repair[2] /rɪ'peə(r)/ *v. & n.* ● *v.intr.* (foll. by *to*) resort; have recourse; go often or in great numbers or for a specific purpose (*repaired to Spain*). ● *n. archaic* **1** resort (*have repair to*). **2** a place of frequent resort. **3** popularity (*a place of great repair*). [ME f. OF *repaire(r)* f. LL *repatriare* REPATRIATE]

repairman /rɪ'peəmən/ *n.* (*pl.* **-men**) a man who repairs machinery etc.

repand /rɪ'pænd/ *adj. Bot.* with an undulating margin; wavy. [L *repandus* (as RE-, *pandus* bent)]

repaper /riː'peɪpə(r)/ *v.tr.* paper (a wall etc.) again.

reparable /'repərəb(ə)l/ *adj.* (of a loss etc.) that can be made good. □ **reparably** *adv.* **reparability** /,repərə'bɪlɪtɪ/ *n.* [F f. L *reparabilis* (as REPAIR[1])]

reparation /,repə'reɪʃ(ə)n/ *n.* **1** the act or an instance of making amends. **2 a** compensation. **b** (esp. in *pl.*) compensation for war damage paid by the defeated state. **3** the act or an instance of repairing or being repaired. □ **reparative** /'repərətɪv, rɪ'pærə-/ *adj.* [ME f. OF f. LL *reparatio -onis* (as REPAIR[1])]

repartee /,repɑː'tiː/ *n.* **1** the practice or faculty of making witty retorts; sharpness or wit in quick reply. **2 a** a witty retort. **b** witty retorts collectively. [F *repartie* fem. past part. of *repartir* start again, reply promptly (as RE-, *partir* PART)]

repartition /,riːpɑː'tɪʃ(ə)n/ *v.tr.* partition again.

repass /riː'pɑːs/ *v.tr. & intr.* pass again, esp. on the way back. [ME f. OF *repasser*]

repast /rɪ'pɑːst/ *n. formal* **1** a meal, esp. of a specified kind (*a light repast*). **2** food and drink supplied for or eaten at a meal. [ME f. OF *repaistre* f. LL *repascere repast-* feed]

repat /'riːpæt/ *n. colloq.* **1** a repatriate. **2** repatriation. [abbr.]

repatriate /riː'pætrɪ,eɪt/ *v. & n.* ● *v.* **1** *tr.* restore (a person) to his or her native land. **2** *intr.* return to one's own native land. ● *n.* a person who has been repatriated. □ **repatriation** /-,pætrɪ'eɪʃ(ə)n/ *n.* [LL *repatriare* (as RE-, L *patria* native land)]

repay /riː'peɪ/ *v.* (*past* and *past part.* **repaid** /-'peɪd/) **1** *tr.* pay back (money). **2** *tr.* return (a blow, visit, etc.). **3** *tr.* make repayment to (a person). **4** *tr.* make return for; requite (a service, action, etc.) (*must repay their kindness; the book repays close study*). **5** *tr.* (often foll. by *for*) give in recompense. **6** *intr.* make repayment. □ **repayable** *adj.* **repayment** *n.* [OF *repaier* (as RE-, PAY[1])]

repeal /rɪ'piːl/ *v. & n.* ● *v.tr.* revoke, rescind, or annul (a law, Act of Parliament, etc.). ● *n.* the act or an instance of repealing. □ **repealable** *adj.* [ME f. AF *repeler*, OF *rapeler* (as RE-, APPEAL)]

repeat /rɪ'piːt/ *v. & n.* ● *v.* **1** *tr.* say or do over again. **2** *tr.* recite, rehearse, or reproduce (something from memory) (*repeated a poem*). **3** *tr.* say or report (something heard). **4** *tr.* imitate (an action etc.). **5 a** *intr.* recur; appear again, perhaps several times (*a repeating pattern*). **b** *refl.* recur in the same or a similar form (*history repeats itself*). **6** *tr.* used for emphasis (*am not, repeat not, going*). **7** *intr.* (of food) be tasted intermittently for some time after being swallowed as a result of belching or indigestion. **8** *intr.* (of a watch etc.) strike the last quarter etc. over again when required. **9** *intr.* (of a firearm) fire several shots without reloading. **10** *intr. US* illegally vote more than once in an election. ● *n.* **1 a** the act or an instance of repeating. **b** a thing repeated (often *attrib.: repeat prescription*). **2** a repeated broadcast. **3** *Mus.* **a** a passage intended to be repeated. **b** a mark indicating this. **4** a pattern repeated in wallpaper etc. **5** *Commerce* **a** a consignment similar to a previous one. **b** an order given for this; a reorder. □ **repeating decimal** a recurring decimal. **repeat oneself** say or do the same thing over again. □ **repeatedly** *adv.* **repeatable** *adj.* **repeatability** /-,piːtə'bɪlɪtɪ/ *n.* [ME f. OF *repeter* f. L *repetere* (as RE-, *petere* seek)]

repeater /rɪ'piːtə(r)/ *n.* **1** a person or thing that repeats. **2** a firearm which fires several shots without reloading. **3** a watch or clock which repeats its last strike when required. **4** a device for the automatic re-

transmission or amplification of an electrically transmitted message. **5** a signal lamp indicating the state of another that is invisible.

repêchage /,repeɪ'ʃɑːʒ/ *n.* (in rowing etc.) an extra contest in which the runners-up in the eliminating heats compete for a place in the final. [F *repêcher* fish out, rescue]

repel /rɪ'pel/ *v.tr.* (**repelled, repelling**) **1** drive back; ward off; repulse (*repel an assailant*). **2** refuse to accept or admit (an argument etc.). **3** be repulsive or distasteful to. **4** resist mixing with or admitting (*oil repels water; surface repels moisture*). **5** (often *absol.*) (of a magnetic pole) push away from itself (*like poles repel*). □ **repeller** *n.* [ME f. L *repellere* (as RE-, *pellere puls-* drive)]

repellent /rɪ'pelənt/ *adj. & n.* ● *adj.* **1** that repels. **2** disgusting, repulsive. ● *n.* a substance that repels esp. insects etc. □ **repellence** *n.* **repellency** *n.* **repellently** *adv.* [L *repellere* (as REPEL)]

repent[1] /rɪ'pent/ *v.* **1** *intr.* (often foll. by *of*) feel deep sorrow about one's actions etc. **2** *tr.* (also *absol.*) wish one had not done, regret (one's wrong, omission, etc.); resolve not to continue (a wrongdoing etc.). **3** *refl.* (often foll. by *of*) *archaic* feel regret or penitence about (*now I repent me*). □ **repentance** *n.* **repentant** *adj.* **repenter** *n.* [ME f. OF *repentir* (as RE-, *pentir* ult. f. L *paenitere*)]

repent[2] /'riːp(ə)nt/ *adj. Bot.* creeping, esp. growing along the ground or just under the surface. [L *repere* creep]

repeople /riː'piːp(ə)l/ *v.tr.* repopulate; increase the population of again.

repercussion /,riːpə'kʌʃ(ə)n/ *n.* **1** (often foll. by *of*) an indirect effect or reaction following an event or action (*consider the repercussions of moving*). **2** the recoil after impact. **3** an echo or reverberation. □ **repercussive** /-'kʌsɪv/ *adj.* [ME f. OF *repercussion* or L *repercussio* (as RE-, PERCUSSION)]

repertoire /'repə,twɑː(r)/ *n.* **1** a stock of pieces etc. that a company or a performer knows or is prepared to give. **2** a stock of regularly performed pieces, regularly used techniques, etc. (*went through his repertoire of excuses*). [F *répertoire* f. LL (as REPERTORY)]

repertory /'repətərɪ/ *n.* (*pl.* **-ies**) **1** = REPERTOIRE. **2** the theatrical performance of various plays for short periods by one company. **3 a** a repertory company. **b** repertory theatres regarded collectively. **4** a store or collection, esp. of information, instances, etc. □ **repertory company** a theatrical company that performs plays from a repertoire. [LL *repertorium* f. L *reperire repert-* find]

repetend /'repɪ,tend/ *n.* **1** the recurring figures of a decimal. **2** the recurring word or phrase; a refrain. [L *repetendum* (as REPEAT)]

répétiteur /rɪ,petɪ'tɜː(r)/ *n.* **1** a tutor or coach of musicians, esp. opera singers. **2** a person who supervises ballet rehearsals etc. [F]

repetition /,repɪ'tɪʃ(ə)n/ *n.* **1 a** the act or an instance of repeating or being repeated. **b** the thing repeated. **2** a copy or replica. **3** a piece to be learned by heart. **4** the ability of a musical instrument to repeat a note quickly. □ **repetitional** *adj.* **repetitionary** *adj.* [F *répétition* or L *repetitio* (as REPEAT)]

repetitious /,repɪ'tɪʃəs/ *adj.* characterized by repetition, esp. when unnecessary or tiresome. □ **repetitiously** *adv.* **repetitiousness** *n.*

repetitive /rɪ'petɪtɪv/ *adj.* characterized by, or consisting of, repetition; monotonous. □ **repetitive strain injury** injury arising from the continued repeated use of particular muscles, esp. during keyboarding etc. □ **repetitively** *adv.* **repetitiveness** *n.*

rephrase /riː'freɪz/ *v.tr.* express in an alternative way.

repine /rɪ'paɪn/ *v.intr.* (often foll. by *at, against*) *literary* fret; be discontented. [RE- + PINE[2], after *repent*]

repique /rɪ'piːk/ *n. & v.* ● *n.* (in piquet) the winning of 30 points on cards alone before beginning to play. ● *v.* (**repiques, repiqued**) **1** *intr.* score repique. **2** *tr.* score repique against (another person). [F *repic* (as RE-, PIQUE[2])]

replace /rɪ'pleɪs/ *v.tr.* **1** put back in place. **2** take the place of; succeed; be substituted for. **3** find or provide a substitute for; renew. **4** (often foll. by *with, by*) fill up the place of. **5** (in *passive*, often foll. by *by*) be succeeded or have one's place filled by another; be superseded. □ **replaceable** *adj.* **replacer** *n.*

replacement /rɪ'pleɪsmənt/ *n.* **1** the act or an instance of replacing or being replaced. **2** a person or thing that takes the place of another.

replan /riː'plæn/ *v.tr.* (**replanned, replanning**) plan again or differently.

replant /riː'plɑːnt/ *v.tr.* **1** transfer (a plant etc.) to a larger pot, a new site, etc. **2** plant (ground) again; provide with new plants.

replay *v. & n.* ● *v.tr.* /riː'pleɪ/ play (a match, recording, etc.) again. ● *n.*

/ˈriːpleɪ/ the act or an instance of replaying a match, a recording, or a recorded incident in a game etc.

replenish /rɪˈplenɪʃ/ v.tr. **1** (often foll. by *with*) fill up again. **2** renew (a supply etc.). **3** (as **replenished** adj.) filled; fully stored or stocked; full. □ **replenisher** n. **replenishment** n. [ME f. OF *replenir* (as RE-, *plenir* f. *plein* full f. L *plenus*)]

replete /rɪˈpliːt/ adj. (often foll. by *with*) **1** filled or well-supplied with. **2** stuffed; gorged; sated. □ **repleteness** n. **repletion** /-ˈpliːʃ(ə)n/ n. [ME f. OF *replet replete* or L *repletus* past part. of *replere* (as RE-, *plere plet*-fill)]

replevin /rɪˈplevɪn/ n. Law **1** the provisional restoration or recovery of distrained goods pending the outcome of trial and judgement. **2** a writ granting this. **3** the action arising from this process. [ME f. AF f. OF *replevir* (as REPLEVY)]

replevy /rɪˈplevɪ/ v.tr. (**-ies, -ied**) Law recover by replevin. [OF *replevir* recover f. Gmc]

replica /ˈreplɪkə/ n. **1** a duplicate of a work made by the original artist. **2 a** a facsimile, an exact copy. **b** (of a person) an exact likeness, a double. **3** a copy or model, esp. on a smaller scale. [It. f. *replicare* REPLY]

replicate v., adj., & n. ● v.tr. /ˈreplɪ̩keɪt/ **1** repeat (an experiment etc.). **2** make a replica of. **3** fold back. ● adj. /ˈreplɪkət/ Bot. folded back on itself. ● n. /ˈreplɪkət/ Mus. a tone one or more octaves above or below the given tone. □ **replicative** /-kətɪv/ adj. **replicable** adj. (in sense 1 of v.). **replicability** /ˌreplɪkəˈbɪlɪtɪ/ n. (in sense 1 of v.). [L *replicare* (as RE-, *plicare* fold)]

replication /ˌreplɪˈkeɪʃ(ə)n/ n. **1** a reply or response, esp. a reply to an answer. **2** Law the plaintiff's reply to the defendant's plea. **3 a** the act or an instance of copying. **b** a copy. **c** the process by which genetic material or a living organism gives rise to a copy of itself. [ME f. OF *replicacion* f. L *replicatio -onis* (as REPLICATE)]

reply /rɪˈplaɪ/ v. & n. ● v. (**-ies, -ied**) **1** intr. (often foll. by *to*) make an answer, respond in word or action. **2** tr. say in answer (*he replied, 'Please yourself'*). ● n. (pl. **-ies**) **1** the act of replying (*what did they say in reply?*). **2** what is replied; a response. **3** Law = REPLICATION 2. □ **reply coupon** a coupon exchangeable for stamps in any country for prepaying the reply to a letter. **reply paid 1** hist. (of a telegram) with the cost of a reply prepaid by the sender. **2** (of an envelope etc.) for which the addressee undertakes to pay postage. □ **replier** n. [ME f. OF *replier* f. L (as REPLICATE)]

repo /ˈriːpəʊ/ n. (pl. **-os**) US sl. repossession (esp. attrib.: *repo man*). [abbr.]

repoint /riːˈpɔɪnt/ v.tr. point (esp. brickwork) again.

repolish /riːˈpɒlɪʃ/ v.tr. polish again.

repopulate /riːˈpɒpjʊ̩leɪt/ v.tr. populate again or increase the population of. □ **repopulation** /-ˌpɒpjʊˈleɪʃ(ə)n/ n.

report /rɪˈpɔːt/ v. & n. ● v. **1** tr. **a** bring back or give an account of. **b** state as fact or news, narrate or describe or repeat, esp. as an eyewitness or hearer etc. **c** relate as spoken by another. **2** tr. make an official or formal statement about. **3** tr. (often foll. by *to*) name or specify (an offender or offence) (*shall report you for insubordination; reported them to the police*). **4** intr. (often foll. by *to*) present oneself as having returned or arrived (*report to the manager on arrival*). **5** tr. (also absol.) take down word for word or summarize or write a description of for publication. **6** intr. make or draw up or send in a report. **7** intr. (foll. by *to*) be responsible to (a superior, supervisor, etc.) (*reports directly to the managing director*). **8** tr. Parl. (of a committee chairman) announce that the committee has dealt with (a bill). **9** intr. (often foll. by *of*) give a report to convey that one has received good, bad, etc., impressions (*reports well of the prospects*). **10** intr. (usu. foll. by *on*) investigate or scrutinize for a journalistic report; act as a reporter. ● n. **1** an account given or opinion formally expressed after investigation or consideration. **2** a description, summary, or reproduction of an event or speech or law case, esp. for newspaper publication or broadcast. **3** common talk; rumour. **4** the way a person or thing is spoken of (*I hear a good report of you*). **5** a periodical statement on (esp. a school pupil's) work, conduct, etc. **6** the sound of an explosion. □ **report back** deliver a report to the person, organization, etc., for whom one acts etc. **reported speech** the speaker's words with the changes of person, tense, etc., usual in reports, e.g. *he said that he would go* (opp. *direct speech*). **report progress** state what has been done so far. **report stage** (in the UK) the debate on a bill in the House of Commons or House of Lords after it is reported. □ **reportable** adj. **reportedly** adv. [ME f. OF *reporter* f. L *reportare* (as RE-, *portare* bring)]

reportage /ˌrepɔːˈtɑːʒ/ n. **1** the describing of events, esp. the reporting of news etc. for the press and for broadcasting. **2** the typical style of this. **3** factual presentation in a book etc. [REPORT, after F]

reporter /rɪˈpɔːtə(r)/ n. **1** a person employed to report news etc. for newspapers or broadcasts. **2** a person who reports.

reportorial /ˌrepɔːˈtɔːrɪəl/ adj. US of or characteristic of newspaper reporters. □ **reportorially** adv. [REPORTER, after *editorial*]

repose[1] /rɪˈpəʊz/ n. & v. ● n. **1** the cessation of activity or excitement or toil. **2** sleep. **3** a peaceful or quiescent state; stillness; tranquillity. **4** Art a restful effect; harmonious combination. **5** composure or ease of manner. ● v. **1** intr. & refl. lie down in rest (*reposed on a sofa*). **2** tr. (often foll. by *on*) lay (one's head etc.) to rest (on a pillow etc.). **3** intr. (often foll. by *in, on*) lie, be lying or laid, esp. in sleep or death. **4** tr. give rest to; refresh with rest. **5** intr. (foll. by *on, upon*) be supported or based on. **6** intr. (foll. by *on*) (of memory etc.) dwell on. □ **reposal** n. **reposeful** adj. **reposefully** adv. **reposefulness** n. [ME f. OF *repos(er)* f. LL *repausare* (as RE-, *pausare* PAUSE)]

repose[2] /rɪˈpəʊz/ v.tr. (foll. by *in*) place (trust etc.) in. □ **reposal** n. [RE- + POSE[1] after L *reponere reposit*-]

reposition /ˌriːpəˈzɪʃ(ə)n/ n. **1** tr. move or place in a different position. **2** intr. adjust or alter one's position.

repository /rɪˈpɒzɪtərɪ/ n. (pl. **-ies**) **1** a place where things are stored or may be found, esp. a warehouse or museum. **2** a receptacle. **3** (often foll. by *of*) **a** a book, person, etc., regarded as a store of information etc. **b** the recipient of confidences or secrets. [obs. F *repositoire* or L *repositorium* (as REPOSE[2])]

repossess /ˌriːpəˈzes/ v.tr. regain possession of (esp. property or goods on which repayment of a debt is in arrears). □ **repossessor** n. **repossession** /-ˈzeʃ(ə)n/ n.

repot /riːˈpɒt/ v.tr. (**repotted, repotting**) put (a plant) in another, esp. larger, pot.

repoussé /rəˈpuːseɪ/ adj. & n. ● adj. hammered into relief from the reverse side. ● n. ornamental metalwork fashioned in this way. [F, past part. of *repousser* (as RE-, *pousser* PUSH)]

repp var. of REP[3].

repped /rept/ adj. having a surface like rep.

repr. abbr. **1** represent, represented, etc. **2** reprint, reprinted.

reprehend /ˌreprɪˈhend/ v.tr. rebuke; blame; find fault with. □ **reprehension** /-ˈhenʃ(ə)n/ n. [ME f. L *reprehendere* (as RE-, *prehendere* seize)]

reprehensible /ˌreprɪˈhensɪb(ə)l/ adj. deserving censure or rebuke; blameworthy. □ **reprehensibly** adv. **reprehensibility** /-ˌhensɪˈbɪlɪtɪ/ n. [LL *reprehensibilis* (as REPREHEND)]

represent /ˌreprɪˈzent/ v.tr. **1** stand for or correspond to (*the comment does not represent all our views*). **2** (often in passive) be a specimen or example of; exemplify (*all types of people were represented in the audience*). **3** act as an embodiment of; symbolize (*the sovereign represents the majesty of the state; numbers are represented by letters*). **4** call up in the mind by description or portrayal or imagination; place a likeness of before the mind or senses. **5** serve or be meant as a likeness of. **6 a** state by way of expostulation or persuasion (*represented the rashness of it*). **b** (foll. by *to*) try to bring (the facts influencing conduct) home to (*represented the risks to his client*). **7 a** (often foll. by *as, to be*) describe or depict as; declare or make out (*represented them as martyrs; not what you represent it to be*). **b** (often refl.; usu. foll. by *as*) portray; assume the guise of, pose as (*represents himself as an honest broker*). **8** (foll. by *that* + clause) allege (*represents a rural constituency*). **9** show, or play the part of, on stage. **10** fill the place of; be a substitute or deputy for; be entitled to act or speak for (*the Queen was represented by the Prince of Wales*). **11** be elected as a member of parliament, a legislature, etc., for (*represents a rural constituency*). □ **representable** adj. **representability** /-ˌzentəˈbɪlɪtɪ/ n. [ME f. OF *representer* or f. L *repraesentare* (as RE-, PRESENT[2])]

representation /ˌreprɪzenˈteɪʃ(ə)n/ n. **1** the act or an instance of representing or being represented. **2** an image, likeness, or reproduction of a thing, e.g. a painting or drawing. **3** (esp. in pl.) a statement made by way of allegation or to convey opinion. [ME f. OF *representation* or L *repraesentatio* (as REPRESENT)]

representational /ˌreprɪzenˈteɪʃən(ə)l/ adj. **1** of representation. **2** (of a work of art or artistic style) seeking to depict things as they appear to the eye. □ **representationalism** n. **representationalist** adj. & n.

representationism /ˌreprɪzenˈteɪʃə̩nɪz(ə)m/ n. the doctrine that perceived objects are only a representation of real external objects. □ **representationist** n.

representative /ˌreprɪˈzentətɪv/ adj. & n. ● adj. **1** typical of a class or category. **2** containing typical specimens of all or many classes (*a representative sample*). **3 a** consisting of elected deputies etc. **b** based on the representation of a nation etc. by such deputies (*representative* government). **4** (foll. by *of*) serving as a portrayal or symbol of (*representative of their attitude to work*). **5** that presents or can present ideas to the mind (*imagination is a representative faculty*). **6** (of art) representational. ● *n.* **1** (foll. by *of*) a sample, specimen, or typical embodiment or analogue of. **2 a** the agent of a person or society. **b** a commercial traveller. **3** a delegate; a substitute. **4** a deputy in a representative assembly. □ **representatively** *adv.* **representativeness** *n.* [ME f. OF *representatif -ive* or med.L *repraesentativus* (as REPRESENT)]

Representatives, House of the lower house of the US Congress and some other legislatures. (See CONGRESS.)

repress /rɪˈpres/ *v.tr.* **1 a** check; restrain; keep under; quell. **b** suppress; prevent from sounding, rioting, or bursting out. **2** *Psychol.* actively exclude (an unwelcome thought) from conscious awareness; suppress into the unconscious. **3** (usu. as **repressed** *adj.*) subject (a person) to the suppression of his or her thoughts or impulses. □ **represser** *n.* **repressible** *adj.* **repressor** *n.* **repression** /-ˈpreʃ(ə)n/ *n.* [ME f. L *reprimere* (as RE-, *premere* PRESS[1])]

repressive /rɪˈpresɪv/ *adj.* that represses; of the nature of or tending to repression. □ **repressively** *adv.* **repressiveness** *n.*

reprice /riːˈpraɪs/ *v.tr.* price again or differently.

reprieve /rɪˈpriːv/ *v. & n.* ● *v.tr.* **1** relieve or rescue from impending punishment, esp. remit, commute, or postpone the execution of (a condemned person). **2** give respite to. ● *n.* **1 a** the act or an instance of reprieving or being reprieved. **b** a warrant for this. **2** respite; a respite or temporary escape. [ME as past part. *repryed* f. AF & OF *repris* past part. of *reprendre* (as RE-, *prendre* f. L *prehendere* take): 16th-c. *-v-* unexpl.]

reprimand /ˈreprɪˌmɑːnd/ *n. & v.* ● *n.* (often foll. by *for*) an official or sharp rebuke (for a fault etc.). ● *v.tr.* administer this to. [F *réprimande(r)* f. Sp. *reprimenda* f. L *reprimenda* neut. pl. gerundive of *reprimere* REPRESS]

reprint *v. & n.* ● *v.tr.* /riːˈprɪnt/ print again. ● *n.* /ˈriːprɪnt/ **1** the act or an instance of reprinting a book etc. **2 a** a book etc. reprinted. **b** an offprint. **3** the quantity reprinted. □ **reprinter** /riːˈprɪntə(r)/ *n.*

reprisal /rɪˈpraɪz(ə)l/ *n.* **1** (an act of) retaliation. **2** *hist.* the forcible seizure of a foreign subject or his or her goods as an act of retaliation. [ME (in sense 2) f. AF *reprisaille* f. med.L *reprisalia* f. *repraehensalia* (as REPREHEND)]

reprise /rɪˈpriːz/ *n.* **1** a repeated passage in music. **2** a repeated item in a musical programme. [F, fem. past part. of *reprendre* (see REPRIEVE)]

repro /ˈriːprəʊ/ *n.* (pl. **-os**) (often *attrib.*) a reproduction or copy. [abbr.]

reproach /rɪˈprəʊtʃ/ *v. & n.* ● *v.tr.* **1** express disapproval to (a person) for a fault etc. **2** scold; rebuke; censure. **3** *archaic* rebuke (an offence). ● *n.* **1** a rebuke or censure (*heaped reproaches on them*). **2** (often foll. by *to*) a thing that brings disgrace or discredit (*their behaviour is a reproach to us all*). **3** a disgraced or discredited state (*live in reproach and ignominy*). **4** (in *pl.*) *RC Ch.* a set of antiphons and responses for Good Friday representing the reproaches of Christ to his people. □ **above** (or **beyond**) **reproach** perfect. □ **reproachable** *adj.* **reproacher** *n.* **reproachingly** *adv.* [ME f. OF *reproche(r)* f. Rmc (as RE-, L *prope* near)]

reproachful /rɪˈprəʊtʃfʊl/ *adj.* full of or expressing reproach. □ **reproachfully** *adv.* **reproachfulness** *n.*

reprobate /ˈreprəˌbeɪt/ *n., adj., & v.* ● *n.* **1** an unprincipled person; a person of highly immoral character. **2** a person who is condemned by God. ● *adj.* **1** immoral. **2** hardened in sin. ● *v.tr.* **1** express or feel disapproval of; censure. **2** (of God) condemn; exclude from salvation. □ **reprobation** /ˌreprəˈbeɪʃ(ə)n/ *n.* [ME f. L *reprobare reprobat-* disapprove (as RE-, *probare* approve)]

reprocess /riːˈprəʊses/ *v.tr.* process again or differently.

reproduce /ˌriːprəˈdjuːs/ *v.* **1** *tr.* produce a copy or representation of. **2** *tr.* cause to be seen or heard etc. again (*tried to reproduce the sound exactly*). **3** *intr.* produce further members of the same species by natural means. **4** *refl.* produce offspring (*reproduced itself several times*). **5** *intr.* give a specified quality or result when copied (*reproduces badly in black and white*). **6** *tr.* *Biol.* form afresh (a lost part etc. of the body). □ **reproducer** *n.* **reproducible** *adj.* **reproducibly** *adv.* **reproducibility** /-ˌdjuːsɪˈbɪlɪtɪ/ *n.*

reproduction /ˌriːprəˈdʌkʃ(ə)n/ *n.* **1** the act or an instance of reproducing. **2** a copy of a work of art, esp. a print or photograph of a painting. **3** (*attrib.*) (of furniture etc.) made in imitation of a certain style or of an earlier period. **4** the quality of reproduced sound. □ **reproductive** /-ˈdʌktɪv/ *adj.* **reproductively** *adv.* **reproductiveness** *n.*

reprogram /riːˈprəʊgræm/ *v.tr.* (also **reprogramme**)

(**reprogrammed**, **reprogramming**; *US* **reprogramed**, **reprograming**) program (esp. a computer) again or differently. □ **reprogramable** *adj.* (also **reprogrammable**).

reprography /rɪˈprɒgrəfɪ/ *n.* the science and practice of copying documents by photography, xerography, etc. □ **reprographer** *n.* **reprographic** /ˌriːprəˈgræfɪk/ *adj.* **reprographically** *adv.* [REPRODUCE + -GRAPHY]

reproof[1] /rɪˈpruːf/ *n.* **1** blame (*a glance of reproof*). **2** a rebuke; words expressing blame. [ME f. OF *reprove* f. *reprover* REPROVE]

reproof[2] /riːˈpruːf/ *v.tr.* **1** render (a coat etc.) waterproof again. **2** make a fresh proof of (printed matter etc.).

reprove /rɪˈpruːv/ *v.tr.* rebuke (a person, a person's conduct, etc.). □ **reprovable** *adj.* **reprover** *n.* **reprovingly** *adv.* [ME f. OF *reprover* f. LL *reprobare* disapprove: see REPROBATE]

reptile /ˈreptaɪl/ *n. & adj.* ● *n.* **1** a cold-blooded scaly animal of the class Reptilia, which includes snakes, lizards, crocodiles, turtles, tortoises, etc. (See note below.) **2** a mean, grovelling, or repulsive person. ● *adj.* **1** (of an animal) creeping. **2** mean, grovelling, repulsive. □ **reptilian** /repˈtɪlɪən/ *adj. & n.* [ME f. LL *reptilis* f. L *repere rept-* crawl]

▪ Reptiles, which first appeared in the Carboniferous period, evolved from amphibians but differ from them in several important respects. They breathe air all their lives, they have a waterproof scaly skin, their eggs have waterproof shells, and the newly hatched young resemble their parents. There are now about 6,000 species, but there were many more in the Mesozoic era (284–65 million years ago), when life on earth was dominated by flying and swimming reptiles as well as by dinosaurs. However, many of these may not fit easily into the traditional concept of a reptile: they may, for example, have been warm-blooded.

Repton /ˈreptən/, Humphry (1752–1818), English landscape gardener. Repton's reconstructions of estates often used regular bedding and straight paths close to the house, but his parks were carefully informal after the model of Capability Brown. Important designs include the park at Cobham in Kent (*c.*1789–*c.*1793) and the house and grounds at Sheringham Hall in Norfolk (1812).

republic /rɪˈpʌblɪk/ *n.* **1** a state in which supreme power is held by the people or their elected representatives or by an elected or nominated president, not by a monarch etc. **2** a society with equality between its members (*the literary republic*). [F *république* f. L *respublica* f. *res* concern + *publicus* PUBLIC]

republican /rɪˈpʌblɪkən/ *adj. & n.* ● *adj.* **1** of or constituted as a republic. **2** characteristic of a republic. **3** (also **Republican**) advocating or supporting republican government or the Republican Party; advocating a united Ireland. ● *n.* **1** a person advocating or supporting republican government. **2** (**Republican**) (in the US) a member or supporter of the Republican Party. **3** (also **Republican**) an advocate of a united Ireland. Largely Catholic, Republicans are opposed by Loyalists or Unionists. □ **republicanism** *n.*

Republican Party one of the two main US political parties (the other being the Democratic Party). It was formed in 1854 in support of the anti-slavery movement preceding the Civil War; Abraham Lincoln was the first of its leaders to become President. Its political stance is more right-wing than that of the Democratic Party, favouring limited central government and tough, interventionist foreign policy; it tends to appeal to white, middle-class voters, traditionally in the north and west.

Republic Day the day on which the foundation of a republic is commemorated; (in India) 26 Jan.

republish /riːˈpʌblɪʃ/ *v.tr.* (also *absol.*) publish again or in a new edition etc. □ **republication** /-ˌpʌblɪˈkeɪʃ(ə)n/ *n.*

repudiate /rɪˈpjuːdɪˌeɪt/ *v.tr.* **1 a** disown; disavow; reject. **b** refuse dealings with. **c** deny. **2** refuse to recognize or obey (authority or a treaty). **3** refuse to discharge (an obligation or debt). **4** (esp. of the ancients or non-Christians) divorce (one's wife). □ **repudiator** *n.* **repudiable** /-dɪəb(ə)l/ *adj.* **repudiation** /-ˌpjuːdɪˈeɪʃ(ə)n/ *n.* [L *repudiare* f. *repudium* divorce]

repugnance /rɪˈpʌgnəns/ *n.* (also **repugnancy**) **1** (usu. foll. by *to, against*) antipathy; aversion. **2** (usu. foll. by *of, between, to, with*) inconsistency or incompatibility of ideas, statements, etc. [ME (in sense 2) f. F *répugnance* or L *repugnantia* f. *repugnare* oppose (as RE-, *pugnare* fight)]

repugnant /rɪˈpʌgnənt/ *adj.* **1** (often foll. by *to*) extremely distasteful. **2** (often foll. by *to*) contradictory. **3** (often foll. by *with*) incompatible.

4 *poet.* refractory; resisting. □ **repugnantly** *adv.* [ME f. F *répugnant* or L (as REPUGNANCE)]

repulse /rɪˈpʌls/ *v. & n.* ● *v.tr.* **1** drive back (an attack or attacking enemy) by force of arms. **2 a** rebuff (friendly advances or their maker). **b** refuse (a request or offer or its maker). **3** be repulsive to, repel. **4** foil in controversy. ● *n.* **1** the act or an instance of repulsing or being repulsed. **2** a rebuff. [L *repellere repuls-* drive back (as REPEL)]

repulsion /rɪˈpʌlʃ(ə)n/ *n.* **1** aversion; disgust. **2** esp. *Physics* the force by which bodies tend to repel each other or increase their mutual distance (opp. ATTRACTION 2). [LL *repulsio* (as REPEL)]

repulsive /rɪˈpʌlsɪv/ *adj.* **1** causing aversion or loathing; loathsome, disgusting. **2** *Physics* exerting repulsion. **3** *archaic* (of behaviour etc.) cold, unsympathetic. □ **repulsively** *adv.* **repulsiveness** *n.* [F *répulsif -ive* or f. REPULSE]

repurchase /riːˈpɜːtʃɪs/ *v. & n.* ● *v.tr.* purchase again. ● *n.* the act or an instance of purchasing again.

repurify /riːˈpjʊərɪˌfaɪ/ *v.tr.* (**-ies, -ied**) purify again. □ **repurification** /-ˌpjʊərɪfɪˈkeɪʃ(ə)n/ *n.*

reputable /ˈrepjʊtəb(ə)l, *disp.* rɪˈpjuːt-/ *adj.* of good repute; respectable. □ **reputably** *adv.* [obs. F or f. med.L *reputabilis* (as REPUTE)]

reputation /ˌrepjʊˈteɪʃ(ə)n/ *n.* **1** what is generally said or believed about a person's or thing's character or standing (*has a reputation for dishonesty*). **2** the state of being well thought of; distinction; respectability (*have my reputation to think of; the artist has gained an international reputation*). **3** (foll. by *of, for* + verbal noun) fame, credit, or notoriety for doing something (*has the reputation of driving hard bargains*). [ME f. L *reputatio* (as REPUTE)]

repute /rɪˈpjuːt/ *n. & v.* ● *n.* reputation (*known by repute*). ● *v.tr.* **1** (as **reputed** *adj.*) (often foll. by *to* + infin.) be generally considered or reckoned (*is reputed to be the best*). **2** (as **reputed** *adj.*) passing as being, but probably not being (*his reputed father*). □ **reputedly** *adv.* [ME f. OF *reputer* or L *reputare* (as RE-, *putare* think)]

request /rɪˈkwest/ *n. & v.* ● *n.* **1** the act or an instance of asking for something; a petition (*came at his request*). **2** a thing asked for. **3** the state of being sought after; demand (*in great request*). **4 a** a letter etc. asking for a particular record etc. to be played on a radio programme, often with a personal message. **b** a record etc. played in response to such a letter etc. ● *v.tr.* **1** ask to be given or allowed or favoured with (*request a hearing; requests your presence*). **2** (foll. by *to* + infin.) ask a person to do something (*requested her to answer*). **3** (foll. by *that* + clause) ask that. □ **by** (or **on**) **request** in response to an expressed wish. **request programme** a radio etc. programme composed of items requested by the audience. **request stop** a stop at which a bus etc. stops only on a person's request. □ **requester** *n.* [ME f. OF *requeste(r)* ult. f. L *requaerere* (as REQUIRE)]

requiem /ˈrekwɪˌem/ *n.* **1** (**Requiem**) (also *attrib.*) *chiefly RC Ch.* a mass for the repose of the souls of the dead. **2** a musical setting for this. **3** a piece of music serving as a memorial for a dead person. [ME f. accus. of L *requies* rest, the initial word of the mass]

requiem shark *n.* a shark of the family Carcharhinidae, which includes some large voracious sharks such as tiger sharks and whaler sharks. [obs. F *requiem*, var. of *requin* shark, infl. by REQUIEM]

requiescat /ˌrekwiˈeskæt/ *n.* a wish or prayer for the repose of a dead person. [L, = may he or she rest (in peace)]

require /rɪˈkwaɪə(r)/ *v.tr.* **1** need; depend on for success or fulfilment (*the work requires much patience*). **2** lay down as an imperative (*did all that was required by law*). **3** command; instruct (a person etc.). **4** order; insist on (an action or measure). **5** (often foll. by *of, from*, or *that* + clause) demand (of or from a person) as a right. **6** wish to have (*is there anything else you require?*). □ **requirer** *n.* **requirement** *n.* [ME f. OF *requere* ult. f. L *requirere* (as RE-, *quaerere* seek)]

requisite /ˈrekwɪzɪt/ *adj. & n.* ● *adj.* required by circumstances; necessary to success etc. ● *n.* (often foll. by *for*) a thing needed (for some purpose). □ **requisitely** *adv.* [ME f. L *requisitus* past part. (as REQUIRE)]

requisition /ˌrekwɪˈzɪʃ(ə)n/ *n. & v.* ● *n.* **1** an official order laying claim to the use of property or materials. **2** a formal written demand that some duty should be performed. **3** being called or put into service. ● *v.tr.* demand the use or supply of, esp. by requisition order. □ **under** (or **in**) **requisition** being used or applied. □ **requisitioner** *n.* **requisitionist** *n.* [F *réquisition* or L *requisitio* (as REQUIRE)]

requite /rɪˈkwaɪt/ *v.tr.* **1** make return for (a service). **2** (often foll. by *with*) reward or avenge (a favour or injury). **3** (often foll. by *for*) make return to (a person). **4** (often foll. by *for, with*) repay with good or evil (*requite like for like; requite hate with love*). □ **requital** *n.* [RE- + *quite* var. of QUIT]

reran *past* of RERUN.

reread /riːˈriːd/ *v. & n.* ● *v.tr.* (*past* and *past part.* **reread** /-ˈred/) read again. ● *n.* an instance of reading again. □ **re-readable** *adj.*

reredos /ˈrɪədɒs/ *n. Eccl.* an ornamental screen covering the wall at the back of an altar. [ME f. AF f. OF *areredos* f. *arere* behind + *dos* back: cf. ARREARS]

re-release /ˌriːrɪˈliːs/ *v. & n.* ● *v.tr.* release (a record, film, etc.) again. ● *n.* a re-released record, film, etc.

re-route /riːˈruːt/ *v.tr.* (**-routeing**) send or carry by a different route.

rerun *v. & n.* ● *v.tr.* /riːˈrʌn/ (**rerunning**; *past* **reran** /-ˈræn/; *past part.* **rerun**) run (a race, film, etc.) again. ● *n.* /ˈriːrʌn/ **1** the act or an instance of rerunning. **2** a film etc. shown again.

resale /riːˈseɪl/ *n.* the sale of a thing previously bought. □ **resale price maintenance** a manufacturer's practice of setting a minimum resale price for goods. □ **resalable** *adj.*

resat *past* and *past part.* of RESIT.

reschedule /riːˈʃedjuːl, -ˈskedjuːl/ *v.tr.* alter the schedule of; replan.

rescind /rɪˈsɪnd/ *v.tr.* abrogate, revoke, cancel. □ **rescindable** *adj.* **rescindment** *n.* **rescission** /-ˈsɪʒ(ə)n/ *n.* [L *rescindere resciss-* (as RE-, *scindere* cut)]

rescript /ˈriːskrɪpt/ *n.* **1** *Rom. Hist.* an emperor's written reply to an appeal for guidance, esp. on a legal point. **2** *RC Ch.* the pope's decision on a question of doctrine or papal law. **3** an official edict or announcement. **4 a** the act or an instance of rewriting. **b** the thing rewritten. [L *rescriptum*, neut. past part. of *rescribere rescript-* (as RE-, *scribere* write)]

rescue /ˈreskjuː/ *v. & n.* ● *v.tr.* (**rescues, rescued, rescuing**) **1** (often foll. by *from*) save or set free or bring away from attack, custody, danger, or harm. **2** *Law* **a** unlawfully liberate (a person). **b** forcibly recover (property). ● *n.* the act or an instance of rescuing or being rescued; deliverance. □ **rescue bid** *Bridge* a bid made to get one's partner out of a difficult situation. □ **rescuable** *adj.* **rescuer** *n.* [ME *rescowe* f. OF *rescoure* f. Rmc, formed as RE- + L *excutere* (as EX-¹, *quatere* shake)]

reseal /riːˈsiːl/ *v.tr.* seal again. □ **resealable** *adj.*

research /rɪˈsɜːtʃ, ˈriːsɜːtʃ/ *n. & v.* ● *n.* **1 a** the systematic investigation into and study of materials, sources, etc., in order to establish facts and reach new conclusions. **b** (usu. in *pl.*) an endeavour to discover new or collate old facts etc. by the scientific study of a subject or by a course of critical investigation. **2** (*attrib.*) engaged in or intended for research (*research assistant*). ● *v.* **1** *tr.* do research into or for. **2** *intr.* make researches. □ **research and development** (in industry etc.) work directed towards the innovation, introduction, and improvement of products and processes. □ **researchable** *adj.* **researcher** *n.* [obs. F *recherche* (as RE-, SEARCH)]

reseat /riːˈsiːt/ *v.tr.* **1** (also *refl.*) seat (oneself, a person, etc.) again. **2** provide with a fresh seat or seats.

resect /rɪˈsekt/ *v.tr. Surgery* **1** cut out part of (a lung etc.). **2** pare down (bone, cartilage, etc.). □ **resection** /-ˈsekʃ(ə)n/ *n.* **resectional** *adj.* **resectionist** *n.* [L *resecare resect-* (as RE-, *secare* cut)]

reseda /ˈresɪdə/ *n.* **1** a plant of the genus *Reseda*, with a spike of small sweet-scented flowers, e.g. mignonette. **2** (also /ˈrezɪdə/) the pale green colour of mignonette flowers. [L, perh. f. imper. of *resedare* assuage, with ref. to its supposed curative powers]

reselect /ˌriːsɪˈlekt/ *v.tr.* select again or differently. □ **reselection** /-ˈlekʃ(ə)n/ *n.*

resell /riːˈsel/ *v.tr.* (*past* and *past part.* **resold** /-ˈsəʊld/) sell (an object etc.) after buying it.

resemblance /rɪˈzembləns/ *n.* (often foll. by *to, between, of*) a likeness or similarity. □ **resemblant** *adj.* [ME f. AF (as RESEMBLE)]

resemble /rɪˈzemb(ə)l/ *v.tr.* be like; have a similarity to, or features in common with, or the same appearance as. □ **resembler** *n.* [ME f. OF *resembler* (as RE-, *sembler* f. L *similare* f. *similis* like)]

resent /rɪˈzent/ *v.tr.* show or feel indignation at; be aggrieved by (a circumstance, action, or person) (*we resent being patronized*). [obs. F *resentir* (as RE-, L *sentire* feel)]

resentful /rɪˈzentfʊl/ *adj.* feeling resentment. □ **resentfully** *adv.* **resentfulness** *n.*

resentment /rɪˈzentmənt/ *n.* (often foll. by *at, of*) indignant or bitter feelings; anger. [It. *risentimento* or F *ressentiment* (as RESENT)]

reserpine /rɪˈsɜːpiːn, -pɪn/ *n.* an alkaloid obtained from plants of the genus *Rauwolfia*, used as a tranquillizer and in the treatment of hypertension. [G *Reserpin* f. mod.L *Rauwolfia* (f. Leonhard *Rauwolf*, Ger. botanist (d.1596)) *serpentina*]

reservation /ˌrezəˈveɪʃ(ə)n/ n. **1** the act or an instance of reserving or being reserved. **2** a booking (of a room, berth, seat, etc.). **3** the thing booked, e.g. a room in a hotel. **4** an express or tacit limitation or exception to an agreement etc. (had reservations about the plan). **5** (in full **central reservation**) Brit. a strip of land between the carriageways of a road. **6** an area of land reserved for occupation by American Indians in the US, Australian Aboriginals, etc. (cf. RESERVE n. 10). **7 a** a right or interest retained in an estate being conveyed. **b** the clause reserving this. **8** Eccl. **a** the practice of retaining for some purpose a portion of the Eucharistic elements (esp. the bread) after celebration. **b** RC Ch. the power of absolution reserved to a superior. **c** RC Ch. the right reserved to the pope of nomination to a vacant benefice. [ME f. OF reservation or LL reservatio (as RESERVE)]

reserve /rɪˈzɜːv/ v. & n. ● v.tr. **1** postpone, put aside, keep back for a later occasion or special use. **2** order to be specially retained or allocated for a particular person or at a particular time. **3** retain or secure, esp. by formal or legal stipulation (reserve the right to). **4** postpone delivery of (judgement etc.) (reserved my comments until the end). ● n. **1** a thing reserved for future use; an extra stock or amount (a great reserve of strength; huge energy reserves). **2** a limitation, qualification, or exception attached to something (accept your offer without reserve). **3 a** self-restraint; reticence; lack of cordiality (difficult to overcome his reserve). **b** (in artistic or literary expression) absence from exaggeration or ill-proportioned effects. **4** a company's profit added to capital. **5** (in sing. or pl.) assets kept readily available as cash or at a central bank, or as gold or foreign exchange (reserve currency). **6** (in sing. or pl.) **a** troops withheld from action to reinforce or protect others. **b** forces in addition to the regular army, navy, air force, etc., but available in an emergency. **7** a member of the military reserve. **8 a** a player who is second choice for a team or who is kept to replace another if required. **b** (in pl.) the second-choice or reserve team. **9** a place reserved for special use, esp. as a habitat for wildlife (game reserve; nature reserve). **10** (in Canada) an area of land set aside for the use of a group of native people (cf. RESERVATION 6). **11** the intentional suppression of the truth (exercised a certain amount of reserve). **12** (in the decoration of ceramics or textiles) an area which still has the original colour of the material or the colour of the background. □ **in reserve** unused and available if required. **reserve grade** Austral. a second-grade team. **reserve price** the lowest acceptable price stipulated for an item sold at an auction. **with all** (or **all proper**) **reserve** without endorsing. □ **reservable** adj. **reserver** n. [ME f. OF reserver f. L reservare (as RE-, servare keep)]

re-serve /riːˈsɜːv/ v.tr. serve again.

reserved /rɪˈzɜːvd/ adj. **1** reticent; slow to reveal emotion or opinions; uncommunicative. **2 a** set apart, destined for some use or fate. **b** (often foll. by for, to) left by fate for; falling first or only to. □ **reserved occupation** an occupation from which a person will not be taken for military service. □ **reservedly** /-vɪdlɪ/ adv. **reservedness** n.

reservist /rɪˈzɜːvɪst/ n. a member of the reserve forces.

reservoir /ˈrezəˌvwɑː(r)/ n. **1** a large natural or artificial lake used as a source of water supply. **2 a** any natural or artificial receptacle, esp. for or of fluid. **b** a place where fluid etc. collects. **3** a part of a machine etc. holding fluid. **4** (usu. foll. by of) a reserve or supply, esp. of information. [F réservoir f. réserver RESERVE]

reset /riːˈset/ v.tr. (**resetting**; past and past part. **reset**) set (a broken bone, gems, a mechanical device, etc.) again or differently. □ **resettable** adj. **resettability** /-ˌsetəˈbɪlɪtɪ/ n.

resettle /riːˈset(ə)l/ v.tr. & intr. settle again in a new or former place. □ **resettlement** n.

reshape /riːˈʃeɪp/ v.tr. shape or form again or differently.

reshuffle /riːˈʃʌf(ə)l/ v. & n. ● v.tr. **1** shuffle (cards) again. **2** interchange the posts of (government ministers etc.). ● n. the act or an instance of reshuffling.

reside /rɪˈzaɪd/ v.intr. **1** (often foll. by at, in, abroad, etc.) (of a person) have one's home, dwell permanently. **2** (of power, a right, etc.) rest or be vested in. **3** (of an incumbent official) be in residence. **4** (foll. by in) (of a quality) be present or inherent in. [ME, prob. back-form. f. RESIDENT infl. by F résider or L residere (as RE-, sedere sit)]

residence /ˈrezɪd(ə)ns/ n. **1** the act or an instance of residing. **2 a** the place where a person resides; an abode. **b** the official house of a government minister etc. **c** a house, esp. a superior or impressive one; a mansion (returned to their London residence). □ **in residence** living or occupying a specified place, esp. for the performance of duties or work. [ME f. OF residence or med.L residentia f. residēre: see RESIDE]

residency /ˈrezɪd(ə)nsɪ/ n. (pl. **-ies**) **1** = RESIDENCE 1, 2a. **2** N. Amer. a period of specialized medical training; the position of a resident. **3** hist.

the official residence of the Governor-General's representative or other government agent at the court of an Indian state; the territory supervised by this official. **4** a musician's regular engagement at a club etc. **5** a group or organization of intelligence agents in a foreign country.

resident /ˈrezɪd(ə)nt/ n. & adj. ● n. **1** (often foll. by of) **a** a permanent inhabitant (of a town or neighbourhood). **b** a bird belonging to a species that does not migrate. **2** a guest in a hotel etc. staying overnight. **3** hist. a British government agent in any semi-independent state, esp. the Governor-General's agent at the court of an Indian state. **4** N. Amer. a medical graduate engaged in specialized practice under supervision in a hospital. **5** an intelligence agent in a foreign country. ● adj. **1** residing; in residence. **2 a** having quarters on the premises of one's work etc. (resident housekeeper; resident doctor). **b** working regularly in a particular place. **3** located in; inherent (powers of feeling are resident in the nerves). **4** (of birds etc.) non-migratory. □ **residentship** n. (in sense 3 of n.). [ME f. OF resident or L: see RESIDE]

residential /ˌrezɪˈdenʃ(ə)l/ adj. **1** suitable for or occupied by private houses (residential area). **2** used as a residence (residential hotel). **3** based on or connected with residence (the residential qualification for voters; a residential course of study). □ **residentially** adv.

residentiary /ˌrezɪˈdenʃərɪ/ adj. & n. ● adj. of, subject to, or requiring, official residence. ● n. (pl. **-ies**) an ecclesiastic who must officially reside in a place. [med.L residentiarius (as RESIDENCE)]

residua pl. of RESIDUUM.

residual /rɪˈzɪdjʊəl/ adj. & n. ● adj. **1** remaining; left as a residue or residuum. **2** Math. resulting from subtraction. **3** (in calculation) still unaccounted for or not eliminated. ● n. **1** a quantity left over or Math. resulting from subtraction. **2** an error in calculation not accounted for or eliminated. □ **residually** adv.

residuary /rɪˈzɪdjʊərɪ/ adj. **1** Law of the residue of an estate (residuary bequest). **2** of or being a residuum; residual; still remaining.

residue /ˈrezɪˌdjuː/ n. **1** what is left over or remains; a remainder; the rest. **2** Law what remains of an estate after the payment of charges, debts, and bequests. **3** esp. Chem. a residuum. [ME f. OF residu f. L residuum: see RESIDUUM]

residuum /rɪˈzɪdjʊəm/ n. (pl. **residua** /-jʊə/) **1** Chem. a substance left after combustion or evaporation. **2** a remainder or residue. [L, neut. of residuus remaining f. residere: see RESIDE]

resign /rɪˈzaɪn/ v. **1** intr. **a** (often foll. by from) give up office, one's employment, etc. (resigned from the Home Office). **b** (often foll. by as) retire; relinquish one's post (resigned as chief executive). **2** tr. (often foll. by to, into) give up (office, one's employment, etc.); surrender; hand over (a right, charge, task, etc.). **3** tr. give up (hope etc.). **4** refl. (usu. foll. by to) **a** reconcile (oneself, one's mind, etc.) to the inevitable (have resigned myself to the idea). **b** surrender (oneself to another's guidance). **5** intr. Chess etc. discontinue play and admit defeat. □ **resigner** n. [ME f. OF resigner f. L resignare unseal, cancel (as RE-, signare sign, seal)]

re-sign /riːˈsaɪn/ v.tr. & intr. sign again.

resignation /ˌrezɪɡˈneɪʃ(ə)n/ n. **1** the act or an instance of resigning, esp. from one's job or office. **2** the document etc. conveying this intention. **3** the state of being resigned; the uncomplaining endurance of a sorrow or difficulty. [ME f. OF f. med.L resignatio (as RESIGN)]

resigned /rɪˈzaɪnd/ adj. (often foll. by to) having resigned oneself; submissive, acquiescent. □ **resignedly** /-nɪdlɪ/ adv. **resignedness** n.

resile /rɪˈzaɪl/ v.intr. **1** (of something stretched or compressed) recoil to resume a former size and shape; spring back. **2** have or show resilience or recuperative power. **3** (usu. foll. by from) withdraw from a course of action. [obs. F resilir or L resilire (as RE-, salire jump)]

resilient /rɪˈzɪlɪənt/ adj. **1** (of a substance etc.) recoiling; springing back; resuming its original shape after bending, stretching, compression, etc. **2** (of a person) readily recovering from shock, depression, etc.; buoyant. □ **resilience** n. **resiliency** n. **resiliently** adv. [L resiliens resilient- (as RESILE)]

resin /ˈrezɪn/ n. & v. ● n. **1** an adhesive inflammable substance insoluble in water, secreted by some plants, and often extracted by incision, esp. from fir and pine (cf. GUM¹ n. 1). **2** (in full **synthetic resin**) a solid or liquid organic compound made by polymerization etc. and used in plastics etc. ● v.tr. (**resined, resining**) rub or treat with resin. □ **resinate** /-nət/ n. **resinate** /-ˌneɪt/ v.tr. **resinoid** adj. & n. **resinous** adj. [ME resyn, rosyn f. L resina & med.L rosina, rosinum]

resist /rɪˈzɪst/ v. & n. ● v. **1** tr. withstand the action or effect of; repel. **2** tr. stop the course or progress of; prevent from reaching, penetrating, etc. **3** tr. abstain from (pleasure, temptation, etc.). **4** tr. strive against;

try to impede; refuse to comply with (*resist arrest*). **5** *intr.* offer opposition; refuse to comply. ● *n.* a protective coating of a resistant substance, applied esp. to parts of calico that are not to take dye or to parts of pottery that are not to take glaze or lustre. □ **cannot** (or **could not** etc.) **resist 1** (foll. by verbal noun) feel obliged or strongly inclined to (*cannot resist teasing me about it*). **2** is certain to be amused, attracted, etc., by (*can't resist children's clothes*). □ **resistant** *adj.* **resister** *n.* **resistible** *adj.* **resistibility** /-ˌzɪstɪˈbɪlɪtɪ/ *n.* [ME f. OF *resister* or L *resistere* (as RE-, *sistere* stop, redupl. of *stare* stand)]

resistance /rɪˈzɪstəns/ *n.* **1** the act or an instance of resisting; refusal to comply. **2** the power of resisting (*showed resistance to wear and tear*). **3** *a Biol.* the ability to withstand adverse conditions. **b** *Med. & Biol.* lack of sensitivity to a drug, insecticide, etc., esp. due to continued exposure or genetic change. **4** the impeding, slowing, or stopping effect exerted by one material thing on another. **5** *Physics* **a** the property of hindering the conduction of electricity, heat, etc. **b** the measure of this in a body (symbol: **R**). **6** a resistor. **7** (in full **resistance movement**) a secret organization resisting authority, esp. in an occupied country (see RESISTANCE, THE). [ME f. F *résistance, résistence* f. LL *resistentia* (as RESIST)]

resistance, the (also **Resistance**) the underground movement formed in France during the Second World War to fight the German occupying forces and the Vichy government. In common with resistance movements elsewhere in occupied Europe, it performed acts of sabotage, supplied the Allies with intelligence, and assisted the escape of Jews and prisoners of war, constituting a considerable hindrance to the Germans. The resistance was composed of various groups, dominated but not centrally controlled by Communists, which were coordinated into the Forces Françaises de l'Intérieur in 1944; in this form the resistance joined with Free French forces in the liberation of Paris and northern France. Also called the *Maquis*.

resistive /rɪˈzɪstɪv/ *adj.* **1** able to resist. **2** *Electr.* of or concerning resistance.

resistivity /ˌrɪzɪˈstɪvɪtɪ/ *n.* *Electr.* a measure of the resisting power of a specified material to the flow of an electric current.

resistless /rɪˈzɪstlɪs/ *adj.* *archaic poet.* **1** irresistible; relentless. **2** unresisting. □ **resistlessly** *adv.*

resistor /rɪˈzɪstə(r)/ *n.* *Electr.* a device having resistance to the passage of an electrical current.

resit *v. & n.* ● *v.tr.* /riːˈsɪt/ (**resitting**; *past* and *past part.* **resat** /-ˈsæt/) sit (an examination) again after failing. ● *n.* /ˈriːsɪt/ **1** the act or an instance of resitting an examination. **2** an examination held specifically to enable candidates to resit.

re-site /riːˈsaɪt/ *v.tr.* place on another site; relocate.

Resnais /rəˈneɪ/, Alain (b.1922), French film director. He was one of the foremost directors of the *nouvelle vague*; his films are noted for their use of experimental techniques to explore memory and time, and often focus on the theme of a loss of touch with humanity. Throughout his career Resnais has collaborated with writers such as Marguerite Duras, notably in *Hiroshima mon amour* (1959), and Alain Robbe-Grillet, in *L'Année dernière à Marienbad* (1961). More recent films include *Mon oncle d'Amérique* (1980) and *L'Amour à mort* (1984).

resold *past* and *past part.* of RESELL.

resoluble /rɪˈzɒljʊb(ə)l/ *adj.* **1** that can be resolved. **2** (foll. by *into*) analysable. [F *résoluble* or L *resolubilis* (as RESOLVE, after *soluble*)]

re-soluble /riːˈsɒljʊb(ə)l/ *adj.* that can be dissolved again.

resolute /ˈrezəˌluːt, -ˌljuːt/ *adj.* (of a person or a person's mind or action) determined; decided; firm of purpose; not vacillating. □ **resolutely** *adv.* **resoluteness** *n.* [L *resolutus* past part. of *resolvere* (see RESOLVE)]

resolution /ˌrezəˈluːʃ(ə)n, -ˈljuːʃ(ə)n/ *n.* **1** a resolute temper or character; boldness and firmness of purpose. **2** a thing resolved on; an intention (*New Year's resolutions*). **3 a** a formal expression of opinion or intention by a legislative body or public meeting. **b** the formulation of this (*passed a resolution*). **4** (usu. foll. by *of*) the act or an instance of solving doubt or a problem or question (*towards a resolution of the difficulty*). **5 a** esp. *Chem.* separation into components. **b** *Mech.* the replacing of a single force etc. by two or more jointly equivalent to it. **6** (foll. by *into*) analysis; conversion into another form. **7** *Mus.* the act or an instance of causing discord to pass into concord. **8** *Physics* etc. **a** the smallest interval measurable by a scientific (esp. optical) instrument; the resolving power. **b** the degree of detail visible in a photographic or television image. **9** *Med.* the disappearance of inflammation etc. without suppuration. **10** *Prosody* the substitution of two short syllables for one long. [ME f. L *resolutio* (as RESOLVE)]

resolutive /ˈrezəˌluːtɪv, -ˌljuːtɪv, rɪˈzɒljʊtɪv/ *adj.* *Med.* having the power or ability to dissolve. □ **resolutive condition** *Law* a condition whose fulfilment terminates a contract etc. [med.L *resolutivus* (as RESOLVE)]

resolve /rɪˈzɒlv/ *v. & n.* ● *v.* **1** *intr.* make up one's mind; decide firmly (*resolve to do better*). **2** *tr.* (of circumstances etc.) cause (a person) to do this (*events resolved him to leave*). **3** *tr.* (foll. by *that* + clause) (of an assembly or meeting) pass a resolution by vote (*the committee resolved that immediate action should be taken*). **4** *intr. & tr.* (often foll. by *into*) separate or cause to separate into constituent parts; disintegrate; analyse; dissolve. **5** *tr.* (of optical or photographic equipment) separate or distinguish between closely adjacent objects. **6** *tr. & intr.* (foll. by *into*) convert or be converted. **7** *tr. & intr.* (foll. by *into*) reduce by mental analysis into. **8** *tr.* solve; explain; clear up; settle (doubt, argument, etc.). **9** *tr. & intr. Mus.* convert or be converted into concord. **10** *tr. Med.* remove (inflammation etc.) without suppuration. **11** *tr. Prosody* replace (a long syllable) by two short syllables. **12** *tr. Mech.* replace (a force etc.) by two or more jointly equivalent to it. ● *n.* **1 a** a firm mental decision or intention; a resolution (*made a resolve not to go*). **b** *US* a formal resolution by a legislative body or public meeting. **2** resoluteness; steadfastness. □ **resolving power** an instrument's ability to distinguish very small or very close objects. □ **resolver** *n.* **resolvable** *adj.* **resolvability** /-ˌzɒlvəˈbɪlɪtɪ/ *n.* [ME f. L *resolvere resolut-* (as RE-, SOLVE)]

resolved /rɪˈzɒlvd/ *adj.* resolute, determined. □ **resolvedly** /-vɪdlɪ/ *adv.* **resolvedness** *n.*

resolvent /rɪˈzɒlv(ə)nt/ *adj. & n.* esp. *Med.* ● *adj.* (of a drug, application, substance, etc.) effecting the resolution of a tumour etc. ● *n.* such a drug etc.

resonance /ˈrezənəns/ *n.* **1** the reinforcement or prolongation of sound by reflection or synchronous vibration. **2** *Mech.* a condition in which an object or system is subjected to an oscillating force having a frequency close to its own natural frequency. **3** *Chem.* the property of a molecule having a structure best represented by two or more forms rather than a single structural formula. **4** *Physics* a short-lived subatomic particle that is an excited state of a more stable particle. [OF f. L *resonantia* echo (as RESONANT)]

resonant /ˈrezənənt/ *adj.* **1** (of sound) echoing, resounding; continuing to sound; reinforced or prolonged by reflection or synchronous vibration. **2** (of a body, room, etc.) tending to reinforce or prolong sounds esp. by synchronous vibration. **3** (often foll. by *with*) (of a place) resounding (*resonant with the sound of bees*). **4** of or relating to resonance. □ **resonantly** *adv.* [F *résonnant* or L *resonare resonant-* (as RE-, *sonare* sound)]

resonate /ˈrezəˌneɪt/ *v.intr.* produce or show resonance; resound. [L *resonare resonat-* (as RESONANT)]

resonator /ˈrezəˌneɪtə(r)/ *n.* *Mus.* **1** an instrument responding to a single note and used for detecting it in combinations. **2** an appliance for giving resonance to sound or other vibrations.

resorb /rɪˈsɔːb, -ˈzɔːb/ *v.tr.* absorb again. □ **resorbence** *n.* **resorbent** *adj.* [L *resorbere resorpt-* (as RE-, *sorbere* absorb)]

resorcin /rɪˈzɔːsɪn/ *n.* = RESORCINOL. [RESIN + ORCIN]

resorcinol /rɪˈzɔːsɪˌnɒl/ *n.* *Chem.* a crystalline organic compound usu. made by synthesis and used in the production of dyes, drugs, resins, etc. [RESORCIN + -OL]

resorption /rɪˈsɔːpʃ(ə)n, -ˈzɔːpʃ(ə)n/ *n.* **1** the act or an instance of resorbing; the state of being resorbed. **2** *Physiol.* the breaking down and absorption of tissue within the body. □ **resorptive** /-tɪv/ *adj.* [RESORB after *absorption*]

resort /rɪˈzɔːt/ *n. & v.* ● *n.* **1** a place frequented esp. for holidays or for a specified purpose or quality (*seaside resort; health resort*). **2 a** a thing to which one has recourse; an expedient or measure (*a taxi was our best resort*). **b** (foll. by *to*) recourse to; use of (*without resort to violence*). **3** a tendency to frequent or be frequented (*places of great resort*). ● *v.intr.* **1** (foll. by *to*) turn to as an expedient (*resorted to threats*). **2** (foll. by *to*) go often or in large numbers to. □ **in the** (or **as a**) **last resort** when all else has failed. □ **resorter** *n.* [ME f. OF *resortir* (as RE-, *sortir* come or go out)]

re-sort /riːˈsɔːt/ *v.tr.* sort again or differently.

resound /rɪˈzaʊnd/ *v.* **1** *intr.* (often foll. by *with*) (of a place) ring or echo (*the hall resounded with laughter*). **2** *intr.* (of a voice, instrument, sound, etc.) produce echoes; go on sounding; fill a place with sound. **3** *intr.* **a** (of fame, a reputation, etc.) be much talked of. **b** (foll. by *through*) produce a sensation (*the incident resounded through Europe*). **4** *tr.* (often foll. by *of*) proclaim or repeat loudly (the praises) of a person or thing (*resounded the praises of Greece*). **5** *tr.* (of a place) re-echo (a sound). [ME f. RE- + SOUND[1] *v.*, after OF *resoner* or L *resonare*: see RESONANT]

resounding /rɪˈzaʊndɪŋ/ *adj.* **1** in senses of RESOUND. **2** unmistakable; emphatic (*was a resounding success*). □ **resoundingly** *adv.*

resource /rɪˈzɔːs, -ˈsɔːs/ *n.* & *v.* ● *n.* **1** an expedient or device (*escape was their only resource*). **2** (usu. in *pl.*) **a** the means available to achieve an end, fulfil a function, etc. **b** a stock or supply that can be drawn on. **c** *N. Amer.* available assets. **3** (in *pl.*) a country's collective wealth or means of defence. **4** a leisure occupation (*reading is a great resource*). **5 a** (often in *pl.*) skill in devising expedients (*a person of great resource*). **b** practical ingenuity; quick wit (*full of resource*). **6** *archaic* the possibility of aid (*lost without resource*). ● *v.tr.* provide with resources. □ **one's own resources** one's own abilities, ingenuity, etc. □ **resourceful** *adj.* **resourcefully** *adv.* **resourcefulness** *n.* **resourceless** *adj.* **resourcelessness** *n.* [F *ressource, ressourse*, fem. past part. of OF dial. *resourdre* (as RE-, L *surgere* rise)]

respect /rɪˈspekt/ *n.* & *v.* ● *n.* **1** deferential esteem felt or shown towards a person or quality. **2 a** (foll. by *of, for*) heed or regard. **b** (foll. by *to*) attention to or consideration of (*without respect to the results*). **3** an aspect, detail, particular, etc. (*correct except in this one respect*). **4** reference, relation (*a morality that has no respect to religion*). **5** (in *pl.*) a person's polite messages or attentions (*give my respects to your mother*). ● *v.tr.* **1** regard with deference, esteem, or honour. **2 a** avoid interfering with, harming, degrading, insulting, injuring, or interrupting. **b** treat with consideration. **c** refrain from offending, corrupting, or tempting (a person, a person's feelings, etc.). □ **in respect of** (or **with respect to**) as concerns; with reference to. **in respect that** because. **pay one's last respects** see PAY[1]. **pay one's respects** see PAY[1]. **with** (or **with all due**) **respect** a mollifying formula preceding an expression of one's disagreement with another's views. □ **respecter** *n.* [ME f. OF *respect* or L *respectus* f. *respicere* (as RE-, *specere* look at) or f. *respectare* frequent. of *respicere*]

respectable /rɪˈspektəb(ə)l/ *adj.* **1** deserving or enjoying respect (*an intellectually respectable hypothesis, a respectable elder statesman*). **2 a** (of people) of good social standing or reputation (*comes from a respectable middle-class family*). **b** characteristic of or associated with people of such status or character (*a respectable neighbourhood; a respectable profession*). **3 a** honest and decent in character or conduct. **b** characterized by (a sense of) convention or propriety; socially acceptable (*respectable behaviour; a respectable publication*). **c** *derog.* highly conventional; prim. **4 a** commendable, meritorious (*an entirely respectable ambition*). **b** comparatively good or competent; passable, tolerable (*a respectable effort; speaks respectable French*). **5** reasonably good in condition or appearance; presentable. **6** appreciable in number, size, amount, etc. (*earns a very respectable salary*). **7** accepted or tolerated on account of prevalence (*materialism has become respectable again*). □ **respectably** *adv.* **respectability** /-ˌspektəˈbɪlɪtɪ/ *n.*

respectful /rɪˈspektfʊl/ *adj.* showing deference (*stood at a respectful distance*). □ **respectfully** *adv.* **respectfulness** *n.*

respecting /rɪˈspektɪŋ/ *prep.* with reference or regard to; concerning.

respective /rɪˈspektɪv/ *adj.* concerning or appropriate to each of several individually; proper to each (*go to your respective places*). [F *respectif -ive* f. med.L *respectivus* (as RESPECT)]

respectively /rɪˈspektɪvlɪ/ *adv.* for each separately or in turn, and in the order mentioned (*she and I gave £10 and £1 respectively*).

respell /riːˈspel/ *v.tr.* (*past* and *past part.* **respelt** /-ˈspelt/ or **respelled**) spell again or differently, esp. phonetically.

Respighi /reˈspiːgiː/, Ottorino (1879–1936), Italian composer. He is best known for his suites the *Fountains of Rome* (1917) and the *Pines of Rome* (1924), based on the poems of Gabriele d'Annunzio and influenced by Rimsky-Korsakov, his former composition teacher, in their orchestration. As well as his many operas, he also arranged Diaghilev's ballet *La Boutique fantasque* (1919) from Rossini's original music.

respirable /ˈrespɪrəb(ə)l, rɪˈspaɪər-/ *adj.* (of air, gas, etc.) able or fit to be breathed. [F *respirable* or LL *respirabilis* (as RESPIRE)]

respirate /ˈrespɪˌreɪt/ *v.tr.* subject to artificial respiration. [back-form. f. RESPIRATION]

respiration /ˌrespɪˈreɪʃ(ə)n/ *n.* **1 a** the act or an instance of breathing. **b** a single inspiration or expiration; a breath. **2** *Biol.* in living organisms, the process involving the production of energy and release of carbon dioxide from the oxidation of complex organic substances. [ME f. F *respiration* or L *respiratio* (as RESPIRE)]

respirator /ˈrespɪˌreɪtə(r)/ *n.* **1** an apparatus worn over the face to prevent poison gas, cold air, dust particles, etc., from being inhaled. **2** *Med.* an apparatus for maintaining artificial respiration.

respire /rɪˈspaɪə(r)/ *v.* **1** *intr.* breathe air. **2** *intr.* inhale and exhale air. **3** *intr.* (of a plant) carry out respiration. **4** *tr.* breathe (air etc.). **5** *intr.* breathe again; take a breath. **6** *intr.* get rest or respite; recover hope or spirit. □ **respiratory** /rɪˈspɪrətərɪ, ˈrespɪˌreɪt-/ *adj.* [ME f. OF *respirer* or f. L *respirare* (as RE-, *spirare* breathe)]

respite /ˈrespaɪt, -pɪt/ *n.* & *v.* ● *n.* **1** an interval of rest or relief. **2** a delay permitted before the discharge of an obligation or the suffering of a penalty. ● *v.tr.* **1** grant respite to; reprieve (a condemned person). **2** postpone the execution or exaction of (a sentence, obligation, etc.). **3** give temporary relief from (pain or care) or to (a sufferer). [ME f. OF *respit* f. L *respectus* RESPECT]

resplendent /rɪˈsplendənt/ *adj.* brilliant, dazzlingly or gloriously bright. □ **resplendence** *n.* **resplendency** *n.* **resplendently** *adv.* [ME f. L *resplendere* (as RE-, *splendere* glitter)]

respond /rɪˈspɒnd/ *v.* & *n.* ● *v.* **1** *intr.* answer, give a reply. **2** *intr.* act or behave in an answering or corresponding manner. **3** *intr.* (usu. foll. by *to*) show sensitiveness to by behaviour or change (*does not respond to kindness*). **4** *intr.* (of a congregation) make answers to a priest etc. **5** *intr. Bridge* make a bid on the basis of a partner's preceding bid. **6** *tr.* say (something) in answer. ● *n.* **1** *Archit.* a half-pillar or half-pier attached to a wall to support an arch, esp. at the end of an arcade. **2** *Eccl.* a responsory; a response to a versicle. □ **respondence** *n.* **respondency** *n.* **responder** *n.* [ME f. OF *respondre* answer ult. f. L *respondere* respons-answer (as RE-, *spondere* pledge)]

respondent /rɪˈspɒndənt/ *n.* & *adj.* ● *n.* **1** a defendant, esp. in an appeal or divorce case. **2** a person who makes an answer or defends an argument etc. ● *adj.* **1** making answer. **2** (foll. by *to*) responsive. **3** in the position of defendant.

response /rɪˈspɒns/ *n.* **1** an answer given in word or act; a reply. **2** a feeling, movement, change, etc., caused by a stimulus or influence. **3** (often in *pl.*) *Eccl.* any part of the liturgy said or sung in answer to the priest; a responsory. **4** *Bridge* a bid made in responding. [ME f. OF *respons(e)* or L *responsum* neut. past part. of *respondere* RESPOND]

responsibility /rɪˌspɒnsɪˈbɪlɪtɪ/ *n.* (*pl.* **-ies**) **1 a** (often foll. by *for, of*) the state or fact of being responsible (*refuses all responsibility for it; will take the responsibility of doing it*). **b** authority; the ability to act independently and make decisions (*a job with more responsibility*). **2** the person or thing for which one is responsible (*the food is my responsibility*). □ **on one's own responsibility** without authorization.

responsible /rɪˈspɒnsɪb(ə)l/ *adj.* **1** (often foll. by *to, for*) liable to be called to account (to a person or for a thing). **2** morally accountable for one's actions; capable of rational conduct. **3** of good credit, position, or repute; respectable; evidently trustworthy. **4** (often foll. by *for*) being the primary cause (*a short circuit was responsible for the power failure*). **5** (of a ruler or government) not autocratic. **6** involving responsibility (*a responsible job*). □ **responsibleness** *n.* **responsibly** *adv.* [obs. F f. L *responder*: see RESPOND]

responsive /rɪˈspɒnsɪv/ *adj.* **1** (often foll. by *to*) responding readily (to some influence). **2** sympathetic; impressionable. **3 a** answering. **b** by way of answer. **4** *Eccl. archaic* (of a liturgy etc.) using responses. □ **responsively** *adv.* **responsiveness** *n.* [F *responsif -ive* or LL *responsivus* (as RESPOND)]

responsorial /ˌrɪspɒnˈsɔːrɪəl/ *adj.* relating to or involving (esp. liturgical) responses.

responsory /rɪˈspɒnsərɪ/ *n.* (*pl.* **-ies**) *Eccl.* an anthem said or sung by a soloist and choir after a lesson. [ME f. LL *responsorium* (as RESPOND)]

respray *v.* & *n.* ● *v.tr.* /riːˈspreɪ/ spray again (esp. to change the colour of the paint on a vehicle). ● *n.* /ˈriːspreɪ/ the act or an instance of respraying.

rest[1] /rest/ *v.* & *n.* ● *v.* **1** *intr.* cease, abstain, or be relieved from exertion, action, movement, or employment; be tranquil. **2** *intr.* be still or asleep, esp. to refresh oneself or recover strength. **3** *tr.* give relief or repose to; allow to rest (*a chair to rest my legs*). **4** *intr.* (foll. by *on, upon, against*) lie on; be supported by; be spread out on; be propped against. **5** *intr.* (foll. by *on, upon*) depend, be based, or rely on. **6** *intr.* (foll. by *on, upon*) (of a look) alight or be steadily directed on. **7** *tr.* (foll. by *on, upon*) place for support or foundation. **8** *intr.* (of a problem or subject) be left without further investigation or discussion (*let the matter rest*). **9** *intr.* **a** lie in death. **b** (foll. by *in*) lie buried in (a churchyard etc.). **10** *tr.* (as **rested** *adj.*) refreshed or reinvigorated by resting. **11** *intr. US* conclude the calling of witnesses in a law case (*the prosecution rests*). **12** *intr.* (of land) lie fallow. **13** *intr.* (foll. by *in*) repose trust in (*am content to rest in God*). ● *n.* **1** repose or sleep, esp. in bed at night (*get a good night's rest*). **2** freedom from or the cessation of exertion, worry, activity, etc. (*give the subject a rest*). **3** a period of resting (*take a 15-minute rest*). **4** a support or prop for holding or steadying something. **5** *Mus.* **a** an interval of silence of a specified duration. **b** the sign denoting this. **6** a place of resting or abiding, esp. a lodging place or shelter provided for sailors, taxi-drivers, etc. **7** a pause in elocution. **8** *Prosody* a caesura in verse. □ **at rest** not moving; not agitated or troubled; dead. **be resting** *Brit. euphem.* (of an actor) be out of work. **rest-baulk** a ridge left unploughed between furrows. **rest one's case** conclude one's argument etc. **rest-cure** a rest usu. of some weeks as a medical treatment. **rest-day 1** a day spent in rest. **2** = *day of rest*. **rest** (or **God rest**) **his** or **her soul** may God grant his (or her) soul repose. **rest-home** a place where old or frail people can be cared for. **rest-house** *Ind.* a house for travellers to rest in. **resting-place** a place provided or used for resting. **rest mass** *Physics* the mass of a body when at rest. **rest on one's laurels** see LAUREL. **rest on one's oars** see OAR. **rest room** esp. *US* a public lavatory in a factory, shop, etc. **rest up** *US* rest oneself thoroughly. **set at rest** settle or relieve (a question, a person's mind, etc.). □ **rester** n. [OE ræst, rest (n.), ræstan, restan (v.)]

rest² /rest/ n. & v. ● n. (prec. by *the*) **1** the remaining part or parts; the others; the remainder of some quantity or number (*finish what you can and leave the rest*). **2** *Brit. Econ.* the reserve fund, esp. of the Bank of England. **3** *hist.* a rally in tennis. ● v.intr. **1** remain in a specified state (*rest assured*). **2** (foll. by *with*) be left in the hands or charge of (*the final arrangements rest with you*). □ **and all the rest** (or **the rest of it**) and all else that might be mentioned; et cetera. **for the rest** as regards anything else. [ME f. OF *reste rester* f. L *restare* (as RE-, *stare* stand)]

restart v. & n. ● v.tr. & intr. /riːˈstɑːt/ begin again. ● n. /ˈriːstɑːt/ a new beginning.

restate /riːˈsteɪt/ v.tr. express again or differently, esp. more clearly or convincingly. □ **restatement** n.

restaurant /ˈrestərɒnt, -ˌrɒŋ/ n. public premises where meals or refreshments may be had. □ **restaurant car** *Brit.* a dining-car on a train. [F f. *restaurer* RESTORE]

restaurateur /ˌrestərəˈtɜː(r)/ n. a restaurant-keeper. ¶ Frequently misspelt *restauranteur*. [F (as RESTAURANT)]

restful /ˈrestfʊl/ adj. **1** favourable to quiet or repose. **2** free from disturbing influences. **3** soothing. □ **restfully** adv. **restfulness** n.

rest-harrow /ˈrestˌhærəʊ/ n. a tough-rooted leguminous plant of the genus *Ononis*, native to Europe and the Mediterranean, with usu. pink flowers. [obs. *rest* (v.) = ARREST (in sense 'stop') + HARROW]

restitution /ˌrestɪˈtjuːʃ(ə)n/ n. **1** (often foll. by *of*) the act or an instance of restoring a thing to its proper owner. **2** reparation for an injury (esp. *make restitution*). **3** (esp. in Christian theology) the restoration of a thing to its original state. **4** the resumption of an original shape or position because of elasticity. □ **restitutive** /ˈrestɪˌtjuːtɪv/ adj. [ME f. OF *restitution* or L *restitutio* f. *restituere restitut-* restore (as RE-, *statuere* establish)]

restive /ˈrestɪv/ adj. **1** fidgety; restless. **2** (of a horse) refusing to advance, stubbornly standing still or moving backwards or sideways; refractory. **3** (of a person) unmanageable; rejecting control. □ **restively** adv. **restiveness** n. [ME f. OF *restif -ive* f. Rmc (as REST²)]

restless /ˈrestlɪs/ adj. **1** finding or affording no rest. **2** uneasy; agitated. **3** constantly in motion, fidgeting, etc. □ **restlessly** adv. **restlessness** n. [OE *restlēas* (as REST¹, -LESS)]

restock /riːˈstɒk/ v.tr. (also *absol.*) stock again or differently.

restoration /ˌrestəˈreɪʃ(ə)n/ n. **1 a** the act or an instance of restoring (a building etc.) or of being restored. **b** = RESTITUTION 1. **2** a model or drawing representing the supposed original form of an extinct animal, ruined building, etc. **3 a** the restoring of a hereditary monarch to the throne; the period immediately following this. (See also RESTORATION, THE.) **b** (often *attrib.*) the literature of the period following the Restoration of Charles II in 1660. (See also RESTORATION COMEDY.) [17th-c. alt. (after RESTORE) of *restauration*, ME f. OF *restauration* or LL *restauratio* (as RESTORE)]

Restoration, the 1 the re-establishment of the Stuart monarchy in England with the return of Charles II to the throne in 1660. After the death of Oliver Cromwell in 1658, his son Richard (1626–1712) proved incapable of maintaining the Protectorate, and, there appearing to be no other viable form of government, a faction led by General Monck organized the king's return from exile. **2** the restoration of the Bourbon monarchy in France in 1814, following the fall of Napoleon. Louis XVIII, who had been king in title since 1795, was recalled from exile and installed on the throne by Talleyrand; he reigned until 1824,

with an interruption of 100 days when Napoleon briefly returned to power in 1815.

Restoration comedy n. a style of drama which flourished in London after the Restoration in 1660. Restoration comedies were generally sexual intrigues marked by wit, cynicism, and licentiousness, often with highly complicated plots featuring such standard characters as foppish noblemen, scheming servants, and sexually rapacious young widows. Principal exponents include William Congreve, William Wycherley, George Farquhar, Sir John Vanbrugh, and Aphra Behn. The genre marked the first appearance of women, most notably Nell Gwynn, on stage in Britain.

restorative /rɪˈstɒrətɪv/ adj. & n. ● adj. tending to restore health or strength. ● n. a restorative medicine, food, etc. (*needs a restorative*). □ **restoratively** adv. [ME var. of obs. *restaurative* f. OF *restauratif -ive* (as RESTORE)]

restore /rɪˈstɔː(r)/ v.tr. **1** bring back or attempt to bring back to the original state by rebuilding, repairing, repainting, emending, etc. **2** bring back to health etc.; cure. **3** give back to the original owner etc.; make restitution of. **4** reinstate; bring back to dignity or right. **5** replace; put back; bring back to a former condition. **6** make a representation of the supposed original state of (a ruin, extinct animal, etc.). **7** reinstate by conjecture (missing words in a text, missing pieces, etc.). □ **restorable** adj. **restorer** n. [ME f. OF *restorer* f. L *restaurare*]

restrain /rɪˈstreɪn/ v.tr. **1** (often *refl.*, usu. foll. by *from*) check or hold in; keep in check or under control or within bounds. **2** repress; keep down. **3** confine; imprison. □ **restrainable** adj. **restrainer** n. [ME f. OF *restrei(g)n-* stem of *restreindre* f. L *restringere restrict-* (as RE-, *stringere* tie)]

re-strain /riːˈstreɪn/ v.tr. strain again.

restrainedly /rɪˈstreɪnɪdlɪ/ adv. with self-restraint.

restraint /rɪˈstreɪnt/ n. **1** the act or an instance of restraining or being restrained. **2** a stoppage; a check; a controlling agency or influence. **3 a** self-control; avoidance of excess or exaggeration. **b** austerity of literary expression. **4** reserve of manner. **5** confinement, esp. because of insanity. **6** something which restrains or holds in check; bondage, shackles. □ **in restraint of** in order to restrain. **restraint of trade** action seeking to interfere with free-market conditions. [ME f. OF *restreinte* fem. past part. of *restreindre*: see RESTRAIN]

restrict /rɪˈstrɪkt/ v.tr. (often foll. by *to, within*) **1** confine, bound, limit (*restricted parking; restricted them to five days a week*). **2** subject to limitation. **3** withhold from general circulation or disclosure. □ **restricted area 1** *Brit.* an area in which there is a special speed limit for vehicles. **2** *US* an area which military personnel are not allowed to enter. □ **restrictedly** adv. **restrictedness** n. [L *restringere*: see RESTRAIN]

restriction /rɪˈstrɪkʃ(ə)n/ n. **1** the act or an instance of restricting; the state of being restricted. **2** a thing that restricts. **3** a limitation placed on action. □ **restriction enzyme** *Biochem.* an enzyme which divides DNA at or near a specific sequence of bases. □ **restrictionist** adj. & n. [ME f. OF *restriction* or L *restrictio* (as RESTRICT)]

restrictive /rɪˈstrɪktɪv/ adj. imposing restrictions. □ **restrictive clause** *Gram.* a relative clause, usu. without surrounding commas. **restrictive practice** *Brit.* an agreement to limit competition or output in industry. □ **restrictively** adv. **restrictiveness** n. [ME f. OF *restrictif -ive* or med.L *restrictivus* (as RESTRICT)]

restring /riːˈstrɪŋ/ v.tr. (*past* and *past part.* **restrung** /-ˈstrʌŋ/) **1** fit (a musical instrument) with new strings. **2** thread (beads etc.) on a new string.

restructure /riːˈstrʌktʃə(r)/ v.tr. give a new structure to; rebuild; rearrange.

restudy /riːˈstʌdɪ/ v.tr. (**-ies, -ied**) study again.

restyle /riːˈstaɪl/ v.tr. **1** reshape; remake in a new style. **2** give a new designation to (a person or thing).

result /rɪˈzʌlt/ n. & v. ● n. **1** a consequence, issue, or outcome of something. **2** a satisfactory outcome; a favourable result (*gets results*). **3** a quantity, formula, etc., obtained by calculation. **4** (in *pl.*) a list of scores or winners etc. in an examination or sporting event. ● v.intr. **1** (often foll. by *from*) arise as the actual consequence or follow as a logical consequence (from conditions, causes, etc.). **2** (often foll. by *in*) have a specified end or outcome (*resulted in a large profit*). □ **without result** in vain; fruitless. □ **resultful** adj. **resultless** adj. [ME f. med.L *resultare* f. L (as RE-, *saltare* frequent. of *salire* jump)]

resultant /rɪˈzʌlt(ə)nt/ adj. & n. ● adj. resulting, esp. as the total outcome of more or less opposed forces. ● *n. Math.* a force etc. equivalent to two or more acting in different directions at the same point.

resume /rɪ'zju:m/ *v. & n.* ● *v.* **1** *tr. & intr.* begin again or continue after an interruption. **2** *tr. & intr.* begin to speak, work, or use again; recommence. **3** *tr.* get back; take back; recover; reoccupy (*resume one's seat*). ● *n.* = RÉSUMÉ. □ **resumable** *adj.* [ME f. OF *resumer* or L *resumere resumpt-* (as RE-, *sumere* take)]

résumé /'rezjʊˌmeɪ/ *n.* **1** a summary. **2** *N. Amer.* a curriculum vitae. [F past part. of *résumer* (as RESUME)]

resumption /rɪ'zʌmpʃ(ə)n/ *n.* the act or an instance of resuming (*ready for the resumption of negotiations*). □ **resumptive** /-'zʌmptɪv/ *adj.* [ME f. OF *resumption* or LL *resumptio* (as RESUME)]

resupinate /rɪ'su:pɪnət, rɪ'sju:-/ *adj. esp. Bot.* (of a leaf etc.) upside down. [L *resupinatus* past part. of *resupinare* bend back: see SUPINE]

resurface /ri:'sɜ:fɪs/ *v.* **1** *tr.* lay a new surface on (a road etc.). **2** *intr.* rise or arise again; turn up again.

resurgent /rɪ'sɜ:dʒ(ə)nt/ *adj.* **1** rising or arising again. **2** tending to rise again. □ **resurgence** *n.* [L *resurgere resurrect-* (as RE-, *surgere* rise)]

resurrect /ˌrezə'rekt/ *v.* **1** *tr.* revive the practice, use, or memory of. **2** *tr.* take from the grave; exhume. **3** *tr.* dig up. **4** *tr. & intr.* raise or rise from the dead. [back-form. f. RESURRECTION]

resurrection /ˌrezə'rekʃ(ə)n/ *n.* **1** the act or an instance of rising from the dead. **2** (**Resurrection**) **a** Christ's rising from the dead. **b** the rising of the dead at the Last Judgement. **3** a revival after disuse, inactivity, or decay. **4** exhumation. **5** the unearthing of a lost or forgotten thing; restoration to vogue or memory. □ **resurrection plant** a plant that appears dead during drought but revives when moistened, esp. a clubmoss of the genus *Selaginella*, or the rose of Jericho. □ **resurrectional** *adj.* [ME f. OF f. LL *resurrectio -onis* (as RESURGENT)]

resurvey *v. & n.* ● *v.tr.* /ˌri:sə'veɪ/ survey again; reconsider. ● *n.* /ri:'sɜ:veɪ/ the act or an instance of resurveying.

resuscitate /rɪ'sʌsɪˌteɪt/ *v.tr. & intr.* **1** revive from unconsciousness or apparent death. **2** return or restore to vogue, vigour, or vividness. □ **resuscitator** *n.* **resuscitative** /-tətɪv, -ˌteɪtɪv/ *adj.* **resuscitation** /-ˌsʌsɪ'teɪʃ(ə)n/ *n.* [L *resuscitare* (as RE-, *suscitare* raise)]

ret /ret/ *v.* (also **rate**) (**retted, retting**) **1** *tr.* soften (flax, hemp, etc.) by soaking or by exposure to moisture. **2** *intr.* (often as **retted** *adj.*) (of hay etc.) be spoilt by wet or rot. [ME, rel. to ROT]

ret. *abbr.* retired; returned.

retable /rɪ'teɪb(ə)l/ *n.* **1** a frame enclosing decorated panels above the back of an altar. **2** a shelf. [F *rétable*, *retable* f. Sp. *retablo* f. med.L *retrotabulum* rear table (as RETRO-, TABLE)]

retail /'ri:teɪl/ *n., adj., adv., & v.* ● *n.* the sale of goods in relatively small quantities to the public, and usu. not for resale (cf. WHOLESALE). ● *adj. & adv.* by retail; at a retail price (*do you buy wholesale or retail?*). ● *v.* (also /rɪ'teɪl/) **1** *tr.* sell (goods) in retail trade. **2** *intr.* (often foll. by *at, of*) (of goods) be sold in this way (esp. for a specified price) (*retails at £4.95*). **3** *tr.* recount; relate details of. □ **retail price index** an index of the variation in the prices of retail goods. □ **retailer** *n.* [ME f. OF *retaille* a piece cut off f. *retaillier* (as RE-, TAIL²)]

retain /rɪ'teɪn/ *v.tr.* **1 a** keep possession of; not lose; continue to have, practise, or recognize. **b** not abolish, discard, or alter. **2** keep in one's memory. **3 a** keep in place; hold fixed. **b** hold (water etc.). **4** secure the services of (a person, esp. a barrister) with a preliminary payment. □ **retaining fee** a fee paid to secure a person, service, etc. **retaining wall** a wall supporting and confining a mass of earth or water. □ **retainable** *adj.* **retainment** *n.* **retainability** /-ˌteɪnə'bɪlɪtɪ/ *n.* [ME f. AF *retei(g)n-* f. stem of OF *retenir* ult. f. L *retinere retent-* (as RE-, *tenere* hold)]

retainer /rɪ'teɪnə(r)/ *n.* **1** a person or thing that retains. **2** *Law* a fee for retaining a barrister etc. **3 a** *hist.* a dependant or follower of a person of rank. **b** *joc.* an old and faithful friend or servant (esp. *old retainer*). **4** *Brit.* a reduced rent paid to retain accommodation during a period of non-occupancy.

retake *v. & n.* ● *v.tr.* /ri:'teɪk/ (*past* **retook** /-'tʊk/; *past part.* **retaken** /-'teɪkən/) **1** take again. **2** recapture. ● *n.* /'ri:teɪk/ **1 a** the act or an instance of retaking. **b** a thing retaken, e.g. an examination. **2 a** the act or an instance of filming a scene or recording music etc. again. **b** the scene or recording obtained in this way.

retaliate /rɪ'tælɪˌeɪt/ *v.* **1** *intr.* repay an injury, insult, etc., in kind; attack in return; make reprisals. **2** *tr.* **a** (usu. foll. by *upon*) cast (an accusation) back upon a person. **b** repay (an injury or insult) in kind. □ **retaliator**

n. **retaliative** /-lɪətɪv/ *adj.* **retaliatory** *adj.* **retaliation** /-ˌtælɪ'eɪʃ(ə)n/ *n.* [L *retaliare* (as RE-, *talis* such)]

retard /rɪ'tɑ:d/ *v. & n.* ● *v.tr.* **1** make slow or late. **2** delay the progress, development, arrival, or accomplishment of. ● *n.* **1** retardation. **2** /'ri:tɑ:d/ *US sl.* usu. *derog.* a person with a mental handicap. □ **in retard** delayed, in the rear. □ **retardant** *adj. & n.* **retardative** *adj.* **retardatory** *adj.* **retarder** *n.* **retardment** *n.* **retardation** /ˌri:tɑ:'deɪʃ(ə)n/ *n.* [F *retarder* f. L *retardare* (as RE-, *tardus* slow)]

retardate /rɪ'tɑ:deɪt/ *adj. & n. N. Amer.* ● *adj.* mentally retarded. ● *n.* a mentally retarded person. [L *retardare*: see RETARD]

retarded /rɪ'tɑ:dɪd/ *adj.* backward in mental or physical development.

retch /retʃ, ri:tʃ/ *v. & n.* ● *v.intr.* make a motion of vomiting, esp. involuntarily and without effect. ● *n.* such a motion or the sound of it. [var. of (now dial.) *reach* f. OE *hræcan* spit, ON *hrækja* f. Gmc, of imit. orig.]

retd. *abbr.* **1** retired. **2** returned.

rete /'ri:tɪ/ *n.* (*pl.* **retia** /'ri:tɪə, 'ri:ʃɪə/) *Anat.* an elaborate network or plexus of blood vessels or nerve cells. [L *rete* net]

reteach /ri:'ti:tʃ/ *v.tr.* (*past and past part.* **retaught** /-'tɔ:t/) teach again or differently.

retell /ri:'tel/ *v.tr.* (*past and past part.* **retold** /-'təʊld/) tell again or in a different version.

retention /rɪ'tenʃ(ə)n/ *n.* **1 a** the act or an instance of retaining; the state of being retained. **b** the ability to retain things experienced or learned; memory. **2** *Med.* the failure to evacuate urine or another secretion. [ME f. OF *retention* or L *retentio* (as RETAIN)]

retentive /rɪ'tentɪv/ *adj.* **1** (often foll. by *of*) tending to retain (moisture etc.). **2** (of memory or a person) not forgetful. **3** *Surgery* (of a ligature etc.) serving to keep something in place. □ **retentively** *adv.* **retentiveness** *n.* [ME f. OF *retentif -ive* or med.L *retentivus* (as RETAIN)]

retexture /ri:'tekstʃə(r)/ *v.tr.* treat (material, a garment, etc.) so as to restore its original texture.

rethink *v. & n.* ● *v.tr.* /ri:'θɪŋk/ (*past and past part.* **rethought** /-'θɔ:t/) think about (something) again, esp. with a view to making changes. ● *n.* /'ri:θɪŋk/ a reassessment; a period of rethinking.

Rethymnon /'reθɪmˌnɒn/ (Greek **Réthimnon** /'rɛθɪmˌnɔn/) a port on the north coast of Crete; pop. (1981) 17,700.

retia *pl.* of RETE.

retiarius /ˌretɪ'ɑ:rɪəs, ˌri:ʃɪ'eər-/ *n.* (*pl.* **retiarii** /-rɪˌaɪ/) *Rom. Hist.* a Roman gladiator using a net to trap his opponent. [L f. *rete* net]

reticence /'retɪs(ə)ns/ *n.* **1** the avoidance of saying all one knows or feels, or of saying more than is necessary; reserve in speech. **2** a disposition to silence; taciturnity. **3** the act or an instance of holding back some fact. □ **reticent** *adj.* **reticently** *adv.* [L *reticentia* f. *reticere* (as RE-, *tacere* be silent)]

reticle /'retɪk(ə)l/ *n.* a network of fine threads or lines in the focal plane of an optical instrument to help accurate observation. [L *reticulum*: see RETICULUM]

reticula *pl.* of RETICULUM.

reticulate *v. & adj.* ● *v.tr. & intr.* /rɪ'tɪkjʊˌleɪt/ **1** divide or be divided in fact or appearance into a network. **2** arrange or be arranged in small squares or with intersecting lines. ● *adj.* /rɪ'tɪkjʊlət/ reticulated. □ **reticulately** /-lətlɪ/ *adv.* **reticulation** /-ˌtɪkjʊ'leɪʃ(ə)n/ *n.* [L *reticulatus* reticulated (as RETICULUM)]

reticule /'retɪˌkju:l/ *n.* **1** = RETICLE. **2** esp. *hist.* a woman's netted or other bag, esp. with a drawstring, carried or worn to serve the purpose of a pocket. [F *réticule* f. L (as RETICULUM)]

reticulum /rɪ'tɪkjʊləm/ *n.* (*pl.* **reticula** /-lə/) **1** esp. *Biol.* a netlike structure; a fine network, esp. of membranes etc. in living organisms. **2** *Zool.* a ruminant's second stomach. □ **reticular** *adj.* **reticulose** /-ˌləʊs/ *adj.* [L, dimin. of *rete* net]

retie /ri:'taɪ/ *v.tr.* (**retying**) tie again.

retiform /'ri:tɪˌfɔ:m, 'ret-/ *adj.* netlike, reticulated. [L *rete* net + -FORM]

retina /'retɪnə/ *n.* (*pl.* **retinas, retinae** /-ˌni:/) a layer at the back of the eyeball sensitive to light, and triggering nerve impulses via the optic nerve to the brain where the visual image is formed. □ **retinal** *adj.* [ME f. med.L f. L *rete* net]

retinitis /ˌretɪ'naɪtɪs/ *n.* inflammation of the retina.

retinol /'retɪˌnɒl/ *n. Biochem.* vitamin A (see VITAMIN). [RETINA + -OL]

retinue /'retɪˌnju:/ *n.* a body of attendants accompanying an important person. [ME f. OF *retenue* fem. past part. of *retenir* RETAIN]

retiral /rɪ'taɪərəl/ *n.* esp. *Sc.* retirement from office etc.

retire /rɪˈtaɪə(r)/ v. **1** intr. leave office or employment, esp. because of age (retire from the army; retire on a pension). **b** tr. cause (a person) to retire from work. **2** intr. withdraw; go away; retreat. **3** intr. seek seclusion or shelter. **4** intr. go to bed. **5** tr. withdraw (troops). **6** intr. & tr. Cricket (of a batsman) voluntarily end or be compelled to suspend one's innings (retired hurt). **7** tr. Econ. withdraw (a bill or note) from circulation or currency. □ **retire from the world** become a recluse. **retire into oneself** become uncommunicative or unsociable. **retiring age** = retirement age. □ **retirer** n. **retiree** /ˌrɪtaɪəˈriː/ n. esp. US. [F retirer (as RE-, tirer draw)]

retired /rɪˈtaɪəd/ adj. **1 a** having retired from employment (a retired teacher). **b** relating to a retired person (received retired pay). **2** withdrawn from society or observation; secluded (lives a retired life).

retirement /rɪˈtaɪəmənt/ n. **1 a** the act or an instance of retiring. **b** the condition of having retired. **2 a** seclusion or privacy. **b** a secluded place. □ **retirement age** the age at which most people normally retire from work. **retirement pension** Brit. a pension paid by the state to retired people above a certain age.

retiring /rɪˈtaɪərɪŋ/ adj. shy, reserved; fond of seclusion. □ **retiringly** adv.

retold past and past part. of RETELL.

retook past of RETAKE.

retool /riːˈtuːl/ v.tr. equip (a factory etc.) with new tools.

retort[1] /rɪˈtɔːt/ n. & v. ● n. **1** an incisive or witty or angry reply. **2** the turning of a charge or argument against its originator. **3** a retaliation. ● v. **1 a** tr. say by way of a retort. **b** intr. make a retort. **2** tr. repay (an insult or attack) in kind. **3** tr. (often foll. by on, upon) return (mischief, a charge, sarcasm, etc.) to its originator. **4** tr. (often foll. by against) make (an argument) tell against its user. **5** tr. (as **retorted** adj.) recurved; twisted or bent backwards. [L retorquere retort- (as RE-, torquere twist)]

retort[2] /rɪˈtɔːt/ n. & v. ● n. **1** a container, usu. of glass, with a long recurved neck used in distilling liquids. **2** a large receptacle or furnace for heating mercury for purification, coal to generate gas, or iron and carbon to make steel. ● v.tr. purify (mercury) by heating in a retort. [F retorte f. med.L retorta fem. past part. of retorquēre: see RETORT[1]]

retortion /rɪˈtɔːʃ(ə)n/ n. **1** the act or an instance of bending back; the condition of being bent back. **2** (in international law) retaliation by a state on the subjects of another. [RETORT[1], perh. after contortion]

retouch v. & n. ● v.tr. /riːˈtʌtʃ/ improve or repair (a composition, picture, photographic negative or print, etc.) by fresh touches or alterations. ● n. /ˈriːtʌtʃ/ the act or an instance of retouching. □ **retoucher** /riːˈtʌtʃə(r)/ n. [prob. f. F retoucher (as RE-, TOUCH)]

retrace /rɪˈtreɪs/ v.tr. **1** go back over (one's steps etc.). **2** trace back to a source or beginning. **3** recall the course of in one's memory. [F retracer (as RE-, TRACE[1])]

retract /rɪˈtrækt/ v. **1** tr. (also absol.) withdraw or revoke (a statement or undertaking). **2 a** tr. & intr. (esp. with ref. to part of the body) draw or be drawn back or in. **b** tr. draw (an undercarriage etc.) into the body of an aircraft. □ **retractable** adj. **retractive** adj. **retraction** /-ˈtrækʃ(ə)n/ n. [L retrahere or (in sense 1) retractare (as RE-, trahere tract- draw)]

retractile /rɪˈtræktaɪl/ adj. capable of being retracted. □ **retractility** /ˌriːtrækˈtɪlɪtɪ/ n. [RETRACT, after contractile]

retractor /rɪˈtræktə(r)/ n. **1** Anat. a muscle used for retracting. **2** a device for retracting.

retrain /riːˈtreɪn/ v.tr. & intr. train again or further, esp. for new work.

retral /ˈriːtrəl, ˈret-/ adj. Biol. hinder, posterior; at the back. [RETRO- + -AL[1]]

retranslate /ˌriːtrænzˈleɪt, ˌriːtrɑːnz-/ v.tr. translate again, esp. back into the original language. □ **retranslation** /-ˈleɪʃ(ə)n/ n.

retransmit /ˌriːtrænzˈmɪt, ˌriːtrɑːnz-/ v.tr. (**retransmitted**, **retransmitting**) transmit (esp. radio signals or broadcast programmes) back again or to a further distance. □ **retransmission** /-ˈmɪʃ(ə)n/ n.

retread v. & n. ● v.tr. /riːˈtred/ **1** (past **retrod** /-ˈtrɒd/; past part. **retrodden** /-ˈtrɒd(ə)n/) tread (a path etc.) again. **2** (past and past part. **retreaded**) put a fresh tread on (a tyre). ● n. /ˈriːtred/ a retreaded tyre.

retreat /rɪˈtriːt/ v. & n. ● v. **1 a** intr. (esp. of military forces) go back, retire; relinquish a position. **b** tr. cause to retreat; move back. **2** intr. (esp. of features) recede; slope back. ● n. **1 a** the act or an instance of retreating. **b** Mil. a signal for this. **2** withdrawal into privacy or security. **3** a place of shelter or seclusion. **4** a period of seclusion for prayer and meditation. **5** Mil. a bugle-call at sunset. **6** an establishment for the

care of the elderly, the mentally ill, etc. [ME f. OF retret (n.), retraiter (v.) f. L retrahere: see RETRACT]

retrench /rɪˈtrentʃ/ v. **1 a** tr. reduce the amount of (costs). **b** intr. cut down expenses; introduce economies. **2** tr. shorten or abridge. □ **retrenchment** n. [obs. F retrencher (as RE-, TRENCH)]

retrial /riːˈtraɪəl/ n. a second or further (judicial) trial.

retribution /ˌretrɪˈbjuːʃ(ə)n/ n. requital usu. for evil done; vengeance. □ **retributive** /rɪˈtrɪbjʊtɪv/ adj. **retributory** adj. [ME f. LL retributio (as RE-, tribuere tribut- assign)]

retrieve /rɪˈtriːv/ v. & n. ● v.tr. **1** regain possession of. **b** recover by investigation or effort of memory. **2 a** restore to knowledge or recall to mind. **b** obtain (information stored in a computer etc.). **3** (of a dog) find and bring in (killed or wounded game etc.). **4** (foll. by from) recover or rescue (esp. from a bad state). **5** restore to a flourishing state; revive. **6** repair or set right (a loss or error etc.) (managed to retrieve the situation). ● n. the possibility of recovery (beyond retrieve). □ **retrieval** n. **retrievable** adj. **retrievability** /-ˌtriːvəˈbɪlɪtɪ/ n. [ME f. OF retroeve-stressed stem of retrover (as RE-, trover find)]

retriever /rɪˈtriːvə(r)/ n. **1** a breed of dog used for retrieving game. **2** a person who retrieves something.

retro /ˈretrəʊ/ adj. & n. ● adj. imitative or characteristic of a style, fashion, etc. from the past. ● n. a retro fashion or style.

retro- /ˈretrəʊ/ comb. form **1** denoting action back or in return (retroact; retroflex). **2** Anat. & Med. denoting location behind. [L retro backwards]

retroact /ˌretrəʊˈækt/ v.intr. **1** operate in a backward direction. **2** have a retrospective effect. **3** react. □ **retroaction** /-ˈækʃ(ə)n/ n.

retroactive /ˌretrəʊˈæktɪv/ adj. (esp. of legislation) having retrospective effect. □ **retroactively** adv. **retroactivity** /-ækˈtɪvɪtɪ/ n.

retrocede /ˌretrəʊˈsiːd/ v. **1** intr. move back; recede. **2** tr. cede back again. □ **retrocedence** n. **retrocedent** adj. **retrocession** /-ˈseʃ(ə)n/ n. **retrocessive** /-ˈsesɪv/ adj. [L retrocedere (as RETRO-, cedere cess- go)]

retrochoir /ˈretrəʊˌkwaɪə(r)/ n. the part of a cathedral or large church behind the high altar. [med.L retrochorus (as RETRO-, CHOIR)]

retrod past of RETREAD.

retrodden past part. of RETREAD.

retrofit /ˈretrəʊˌfɪt/ v.tr. (**-fitted**, **-fitting**) modify (machinery, vehicles, etc.) to incorporate changes and developments introduced after manufacture. [RETROACTIVE + REFIT]

retroflex /ˈretrəˌfleks/ adj. (also **retroflexed**) **1** Anat., Med., & Bot. turned backwards. **2** Phonet. = CACUMINAL. □ **retroflexion** /ˌretrəˈflekʃ(ə)n/ n. [L retroflectere retroflex- (as RETRO-, flectere bend)]

retrogradation /ˌretrəʊɡrəˈdeɪʃ(ə)n/ n. Astron. **1** the apparent temporary reverse motion of an outer planet, from east to west, seen when it is overtaken in its orbit by the earth. **2** the orbiting of a celestial body, esp. a planet's satellite, in a reverse direction to normal. [LL retrogradatio (as RETRO-, GRADATION)]

retrograde /ˈretrəˌɡreɪd/ adj., n., & v. ● adj. **1** directed backwards; retreating. **2** reverting esp. to an inferior state; declining. **3** inverse, reversed (in retrograde order). **4** Astron. in or showing retrogradation. ● n. a degenerate person. ● v.intr. **1** move backwards; recede, retire. **2** decline, revert. **3** Astron. show retrogradation. □ **retrogradely** adv. [ME f. L retrogradus (as RETRO-, gradus step, gradi walk)]

retrogress /ˌretrəˈɡres/ v.intr. **1** go back; move backwards. **2** deteriorate. □ **retrogressive** adj. [RETRO-, after PROGRESS v.]

retrogression /ˌretrəˈɡreʃ(ə)n/ n. **1** backward or reversed movement. **2** a return to a less advanced state; a reversal of development; a decline or deterioration. **3** Astron. = RETROGRADATION. □ **retrogressive** /-ˈɡresɪv/ adj. **retrogressively** adv. [RETRO-, after progression]

retroject /ˈretrəʊˌdʒekt/ v.tr. throw back (usu. opp. PROJECT v.). [RETRO-, after PROJECT v.]

retro-rocket /ˈretrəʊˌrɒkɪt/ n. an auxiliary rocket for slowing down a spacecraft etc., e.g. when re-entering the earth's atmosphere.

retrorse /rɪˈtrɔːs/ adj. Biol. turned back or down. □ **retrorsely** adv. [L retrorsus = retroversus (as RETRO-, versus past part. of vertere turn)]

retrospect /ˈretrəˌspekt/ n. **1** (foll. by to) regard or reference to precedent or authority, or to previous conditions. **2** a survey of past time or events. □ **in retrospect 1** when looked back on. **2** when looking back; with hindsight. [RETRO-, after PROSPECT n.]

retrospection /ˌretrəˈspekʃ(ə)n/ n. **1** the action of looking back, esp. into the past. **2** an indulgence or engagement in retrospect. [prob. f. retrospect (v.) (as RETROSPECT)]

retrospective /ˌretrəˈspektɪv/ adj. & n. ● adj. **1** looking back on or dealing with the past (cf. PROSPECTIVE *adj.* 1). **2** (of an exhibition, recital, etc.) showing an artist's development over his or her lifetime. **3** (of a statute etc.) applying to the past as well as the future; retroactive. **4** (of a view) lying to the rear. ● *n.* a retrospective exhibition, recital, etc. □ **retrospectively** *adv.*

retrosternal /ˌretrəʊˈstɜːn(ə)l/ *adj.* *Anat.* & *Med.* behind the breastbone.

retroussé /rəˈtruːseɪ/ *adj.* (of the nose) turned up at the tip. [F, past part. of *retrousser* tuck up (as RE-, TRUSS)]

retrovert /ˈretrəʊˌvɜːt/ *v.tr.* **1** turn backwards. **2** *Med.* (as **retroverted** *adj.*) (of the womb) having a backward inclination. □ **retroversion** /ˌretrəʊˈvɜːʃ(ə)n/ *n.* [LL *retrovertere* (as RETRO-, *vertere vers-* turn)]

retrovirus /ˈretrəʊˌvaɪərəs/ *n.* *Biol.* any of a group of RNA viruses which insert a DNA copy of their genome into the host cell, e.g. HIV. [mod.L f. initial letters of *reverse transcriptase* + VIRUS]

retry /riːˈtraɪ/ *v.tr.* (**-ies, -ied**) try (a defendant or lawsuit) a second or further time.

retsina /retˈsiːnə/ *n.* a Greek white wine flavoured with resin. [mod.Gk]

retune /riːˈtjuːn/ *v.tr.* **1** tune (a musical instrument) again or differently. **2** tune (a radio etc.) to a different frequency.

returf /riːˈtɜːf/ *v.tr.* provide with new turf.

return /rɪˈtɜːn/ *v.* & *n.* ● *v.* **1** *intr.* come or go back. **2** *tr.* bring or put or send back to the person or place etc. where originally belonging or obtained (*returned the fish to the river; have you returned my scissors?*). **3** *tr.* pay back or reciprocate; give in response (*decided not to return the compliment*). **4** *tr.* yield (a profit). **5** *tr.* say in reply; retort. **6** *tr.* (in cricket or tennis etc.) hit or send (the ball) back after receiving it. **7** *tr.* state or mention or describe officially, esp. in answer to a writ or formal demand. **8** *tr.* (of an electorate) elect as an MP, government, etc. **9** *tr.* *Cards* **a** lead (a suit) previously led or bid by a partner. **b** lead (a suit or card) after taking a trick. **10** *tr.* *Archit.* continue (a wall etc.) in a changed direction, esp. at right angles. ● *n.* **1** the act or an instance of coming or going back. **2 a** the act or an instance of giving or sending or putting or paying back. **b** a thing given or sent back. **3** (in full **return ticket**) esp. *Brit.* a ticket for a journey to a place and back to the starting-point. **4** (in *sing.* or *pl.*) **a** the proceeds or profit of an undertaking. **b** the acquisition of these. **5** a formal report or statement compiled or submitted by order (*an income-tax return*). **6** (in full **return match** or **game**) a second match etc. between the same opponents. **7** *Electr.* a conductor bringing a current back to its source. **8** *Brit.* a sheriff's report on a writ. **9** esp. *Brit.* **a** a person's election as an MP etc. **b** a returning officer's announcement of this. **10** *Archit.* a part receding from the line of the front, e.g. the side of a house or of a window-opening. □ **by return (of post)** by the next available post in the return direction. **in return** as an exchange or reciprocal action. **many happy returns (of the day)** a greeting on a birthday. **return crease** *Cricket* either of two lines joining the popping-crease and bowling-crease at right angles to the bowling-crease and extending beyond it. **returning officer** *Brit.* an official conducting an election in a constituency and announcing the results. **return thanks** express thanks esp. in a grace at meals or in response to a toast or condolence. □ **returnable** *adj.* **returner** *n.* **returnless** *adj.* [ME f. OF *returner* (as RE-, TURN)]

returnee /rɪtɜːˈniː/ *n.* a person who returns home from abroad, esp. after war service.

retuse /rɪˈtjuːs/ *adj.* esp. *Bot.* having a broad end with a central depression. [L *retundere retus-* (as RE-, *tundere* beat)]

retying *pres. part.* of RETIE.

retype /riːˈtaɪp/ *v.tr.* type again, esp. to correct errors.

Reuben /ˈruːbɪn/ (in the Bible) **1** a Hebrew patriarch, eldest son of Jacob and Leah (Gen. 29:32). **2** the tribe of Israel traditionally descended from him.

reunify /riːˈjuːnɪˌfaɪ/ *v.tr.* (**-ies, -ied**) restore (esp. separated territories) to a political unity. □ **reunification** /-ˌjuːnɪfɪˈkeɪʃ(ə)n/ *n.*

Réunion /riːˈjuːnjən, -nɪən, French reynjɔ̃/ a volcanically active, subtropical island in the Indian Ocean east of Madagascar, one of the Mascarene Islands; pop. (1990) 596,700; capital, Saint-Denis. A French possession since 1638, the island became an overseas department of France in 1946 and an administrative region in 1974.

reunion /riːˈjuːnjən, -nɪən/ *n.* **1 a** the act or an instance of reuniting. **b** the condition of being reunited. **2** a social gathering esp. of people formerly associated. [F *réunion* or AL *reunio* f. L *reunire* unite (as RE-, UNION)]

reunite /ˌriːjuːˈnaɪt/ *v.tr.* & *intr.* bring or come back together.

reupholster /ˌriːʌpˈhəʊlstə(r)/ *v.tr.* upholster again or anew. □ **reupholstery** *n.*

reuse *v.* & *n.* ● *v.tr.* /riːˈjuːz/ use again or more than once. ● *n.* /riːˈjuːs/ a second or further use. □ **reusable** /-ˈjuːzəb(ə)l/ *adj.*

Reuter /ˈrɔɪtə(r)/, Paul Julius, Baron von (born Israel Beer Josaphat) (1816–99), German pioneer of telegraphy and news reporting. After establishing a service for sending commercial telegrams in Aachen (1849), he moved his headquarters to London, where he founded the news agency Reuters.

Reuters /ˈrɔɪtəz/ an international news agency founded in London in 1851 by Paul Julius Reuter. The agency pioneered the use of telegraphy, building up a worldwide network of correspondents to produce a service used today by newspapers and radio and television stations in most countries.

reutilize /riːˈjuːtɪˌlaɪz/ *v.tr.* (also **-ise**) utilize again or for a different purpose. □ **reutilization** /-ˌjuːtɪlaɪˈzeɪʃ(ə)n/ *n.*

Rev. *abbr.* **1** Reverend. **2** Revelation (New Testament).

rev /rev/ *n.* & *v.* *colloq.* ● *n.* (in *pl.*) the number of revolutions of an engine per minute (*running at 3,000 revs*). ● *v.* (**revved, revving**) **1** *intr.* (of an engine) revolve; turn over. **2** *tr.* (also *absol.*; often foll. by *up*) cause (an engine) to run quickly. □ **rev counter** = revolution counter. [abbr.]

revaccinate /riːˈvæksɪˌneɪt/ *v.tr.* vaccinate again. □ **revaccination** /-ˌvæksɪˈneɪʃ(ə)n/ *n.*

revalue /riːˈvæljuː/ *v.tr.* (**revalues, revalued, revaluing**) *Econ.* give a different value to, esp. give a higher value to (a currency) in relation to other currencies or gold (opp. DEVALUE 2). □ **revaluation** /-ˌvæljʊˈeɪʃ(ə)n/ *n.*

revamp /riːˈvæmp/ *v.tr.* **1** renovate, revise, improve. **2** patch up. [RE- + VAMP[1]]

revanchism /rɪˈvæntʃɪz(ə)m/ *n.* *Polit.* a policy of seeking to retaliate, esp. to recover lost territory. □ **revanchist** *n.* & *adj.* [F *revanche* (as REVENGE)]

revarnish /riːˈvɑːnɪʃ/ *v.tr.* varnish again.

Revd *abbr.* Reverend.

reveal[1] /rɪˈviːl/ *v.tr.* **1** display or show; allow to appear. **2** (often as **revealing** *adj.*) disclose, divulge, betray (*revealed his plans; a revealing remark*). **3** *tr.* (in *refl.* or *passive*) come to sight or knowledge. **4** *Relig.* (esp. of God) make known by inspiration or supernatural means. □ **revealed religion** a religion based on revelation (opp. *natural religion*). □ **revealable** *adj.* **revealer** *n.* **revealingly** *adv.* [ME f. OF *reveler* or L *revelare* (as RE-, *velum* veil)]

reveal[2] /rɪˈviːl/ *n.* an internal side surface of an opening or recess, esp. of a doorway or window-aperture. [obs. *revale* (v.) lower f. OF *revaler* f. *avaler* (as RE-, VAIL)]

revegetate /riːˈvedʒɪˌteɪt/ *v.tr.* produce a new growth of vegetation on (disturbed or barren ground). □ **revegetation** /-ˌvedʒɪˈteɪʃ(ə)n/ *n.*

reveille /rɪˈvælɪ/ *n.* a military waking-signal sounded in the morning on a bugle or drums etc. [F *réveillez* imper. pl. of *réveiller* awaken (as RE-, *veiller* f. L *vigilare* keep watch)]

revel /ˈrev(ə)l/ *v.* & *n.* ● *v.* (**revelled, revelling**; *US* **reveled, reveling**) **1** *intr.* have a good time. **2** *intr.* (foll. by *in*) take keen delight in. **3** *tr.* (foll. by *away*) throw away (money or time) in revelry. ● *n.* (in *sing.* or *pl.*) the act or an instance of revelling. □ **reveller** *n.* **revelry** *n.* (*pl.* **-ies**). [ME f. OF *reveler* riot f. L *rebellare* REBEL *v.*]

revelation /ˌrevəˈleɪʃ(ə)n/ *n.* **1 a** the act or an instance of revealing, esp. the supposed disclosure of knowledge to humankind by a divine or supernatural agency. **b** knowledge disclosed in this way. **2** a striking disclosure (*it was a revelation to me*). **3** (**Revelation** or *colloq.* **Revelations**) (in full **the Revelation of St John the Divine**) the last book of the New Testament, recounting a divine revelation of the future to St John. □ **revelational** *adj.* [ME f. OF *revelation* or LL *revelatio* (as REVEAL[1])]

revelationist /ˌrevəˈleɪʃənɪst/ *n.* a believer in divine revelation.

revelatory /ˌrevəˈleɪtərɪ/ *adj.* serving to reveal, esp. something significant. [L *revelare*: see REVEAL[1]]

revenant /ˈrevənənt/ *n.* a person who has returned, esp. supposedly from the dead. [F, pres. part. of *revenir*: see REVENUE]

revenge /rɪˈvendʒ/ *n.* & *v.* ● *n.* **1** retaliation for an offence or injury. **2** an act of retaliation. **3** the desire for this; a vindictive feeling. **4** (in games) a chance to win after an earlier defeat. ● *v.* **1** *tr.* (in *refl.* or *passive*; often foll. by *on, upon*) inflict retaliation for an offence. **2** *tr.* take revenge for (an offence). **3** *tr.* avenge (a person). **4** *intr.* take vengeance.

□ **revenger** n. [ME f. OF revenger, revencher f. LL revindicare (as RE-, vindicare lay claim to)]

revengeful /rɪ'vendʒfʊl/ adj. eager for revenge. □ **revengefully** adv. **revengefulness** n.

revenge tragedy n. a style of drama, popular in England during the late sixteenth and seventeenth centuries, in which the basic plot was a quest for vengeance. Typical elements of the genre were scenes of carnage and mutilation, real or feigned insanity, and the appearance of ghosts. Thomas Kyd's *The Spanish Tragedy* (1592) is regarded as the first revenge tragedy; others include *The Revenger's Tragedy* (c.1606) by Cyril Tourneur (c.1576–1626), John Webster's *The Duchess of Malfi* (1623), Shakespeare's *Hamlet*, and Thomas Middleton's *The Changeling* (1622).

revenue /'revə,njuː/ n. **1 a** income, esp. of a large amount, from any source. **b** (in pl.) items constituting this. **2** a state's annual income from which public expenses are met. **3** the department of the civil service collecting this. □ **revenue tax** a tax imposed to raise revenue, rather than to affect trade. [ME f. OF revenu(e) past part. of revenir f. L revenire return (as RE-, venire come)]

reverb /rɪ'vɜːb, 'riːvɜːb/ n. Mus. colloq. **1** reverberation. **2** a device to produce this. [abbr.]

reverberate /rɪ'vɜːbə,reɪt/ v. **1 a** intr. (of sound, light, or heat) be returned or echoed or reflected repeatedly. **b** tr. return (a sound etc.) in this way. **2** intr. **a** (of a story, rumour, etc.) be heard much or repeatedly. **b** (of an event) have continuing effects. □ **reverberating furnace** a furnace constructed to throw heat back on to the substance exposed to it. □ **reverberator** n. **reverberant** adj. **reverberantly** adv. **reverberative** /-rətɪv/ adj. **reverberatory** adj. **reverberation** /-,vɜːbə'reɪʃ(ə)n/ n. [L reverberare (as RE-, verberare lash f. verbera (pl.) scourge)]

Revere /rɪ'vɪə(r)/, Paul (1735–1818), American patriot. He was one of the demonstrators involved in the Boston Tea Party of 1773; two years later he made his famous midnight ride from Boston to Lexington to warn fellow American revolutionaries of the approach of British troops. The journey is immortalized in Longfellow's poem 'Paul Revere's Ride' (1863).

revere /rɪ'vɪə(r)/ v.tr. hold in deep and usu. affectionate or religious respect; venerate. [F révérer or L revereri (as RE-, vereri fear)]

reverence /'revərəns/ n. & v. ● n. **1 a** the act of revering or the state of being revered (hold in reverence; feel reverence for). **b** the capacity for revering (lacks reverence). **2** archaic a gesture showing that one reveres; a bow or curtsy. **3** (**Reverence**) a title used of or to some members of the clergy. ● v.tr. regard or treat with reverence. [ME f. OF f. L reverentia (as REVERE)]

reverend /'revərənd/ adj. & n. ● adj. (esp. as the title of a clergyman) deserving reverence. ● n. colloq. a clergyman. [ME f. OF reverend or L reverendus gerundive of revereri: see REVERE]

Reverend Mother n. the title of the Mother Superior of a convent.

reverent /'revərənt/ adj. feeling or showing reverence. □ **reverently** adv. [ME f. L reverens (as REVERE)]

reverential /,revə'renʃ(ə)l/ n. of the nature of, due to, or characterized by reverence. □ **reverentially** adv. [med.L reverentialis (as REVERE)]

reverie /'revərɪ/ n. **1** a fit of abstracted musing (was lost in a reverie). **2** archaic a fantastic notion or theory; a delusion. **3** Mus. an instrumental piece suggesting a dreamy or musing state. [obs. F resverie f. OF reverie rejoicing, revelry f. rever be delirious, of unkn. orig.]

revers /rɪ'vɪə(r)/ n. (pl. same /-'vɪəz/) **1** the turned-back edge of a garment revealing the undersurface. **2** the material on this surface. [F, = REVERSE]

reverse /rɪ'vɜːs/ v., adj., & n. ● v. **1** tr. turn the other way round or up or inside out. **2** tr. change to the opposite character or effect (reversed the decision). **3** intr. & tr. travel or cause to travel backwards. **4** tr. make (an engine etc.) work in a contrary direction. **5** tr. revoke or annul (a decree, act, etc.). **6** intr. (of a dancer, esp. in a waltz) revolve in the opposite direction. ● adj. **1** placed or turned in an opposite direction or position. **2** opposite or contrary in character or order; inverted. ● n. **1** the opposite or contrary (the reverse is the case; is the reverse of the truth). **2** the contrary of the usual manner. **3** an occurrence of misfortune; a disaster, esp. a defeat in battle (suffered a reverse). **4** reverse gear or motion. **5** the reverse side of something. **6 a** the side of a coin or medal etc. bearing the secondary design. **b** this design. Opp. OBVERSE n. 1. **7** the verso of a leaf of a book etc. □ **reverse arms** hold a rifle with the butt upwards. **reverse the charges** Brit. make the recipient of a telephone call responsible for payment. **reverse engineering** the reproduction of another manufacturer's product following

detailed examination of its construction or composition. **reverse fault** Geol. an oblique fault in which the rock strata above the fault are displaced upwards in relation to those below it. **reverse gear** a gear used to make a vehicle etc. travel backwards. **reversing light** a white light at the rear of a vehicle operated when the vehicle is in reverse gear. **reverse Polish notation** see POLISH. □ **reversal** n. **reversely** adv. **reverser** n. **reversible** adj. **reversibly** adv. **reversibility** /-,vɜːsɪ'bɪlɪtɪ/ n. [ME f. OF revers (n.), reverser (v.), f. L revertere revers- (as RE-, vertere turn)]

reversion /rɪ'vɜːʃ(ə)n/ n. **1** a return to a previous state, habit, etc. **2 a** the legal right (esp. of the original owner, or his or her heirs) to possess or succeed to property on the death of the present possessor. **b** property to which a person has such a right. **3** Biol. a return to ancestral type. **4** a sum payable on a person's death, esp. by way of life insurance. □ **reversional** adj. **reversionary** adj. [ME f. OF reversion or L reversio (as REVERSE)]

revert /rɪ'vɜːt/ v. **1** intr. (foll. by to) return to a former state, practice, opinion, etc. **2** intr. (of property, an office, etc.) return by reversion. **3** intr. fall back into a wild state. **4** tr. turn (one's eyes or steps) back. □ **reverter** n. (in sense 2). [ME f. OF revertir or L revertere (as REVERSE)]

revertible /rɪ'vɜːtɪb(ə)l/ adj. (of property) subject to reversion.

revet /rɪ'vet/ v.tr. (**revetted, revetting**) face (a rampart, wall, etc.) with masonry, esp. in fortification. [F revêtir f. OF revestir f. LL revestire (as RE-, vestire clothe f. vestis)]

revetment /rɪ'vetmənt/ n. a retaining wall or facing of masonry supporting or protecting a rampart etc. [F revêtement (as REVET)]

review /rɪ'vjuː/ n. & v. ● n. **1** a general survey or assessment of a subject or thing. **2** a retrospect or survey of the past. **3** revision or reconsideration (is under review). **4** a display and formal inspection of troops etc. **5** a published account or criticism of a book, play, etc. **6** a periodical publication with critical articles on current events, the arts, etc. **7** a second view. **8** a facility for playing a tape recording during a fast rewind. ● v.tr. **1** survey or look back on. **2** reconsider or revise. **3** hold a review of (troops etc.). **4** write a review of (a book, play, etc.). **5** view again. □ **court of review** a court before which sentences etc. come for revision. □ **reviewable** adj. **reviewal** n. **reviewer** n. [obs. F reveue f. revoir (as RE-, voir see)]

revile /rɪ'vaɪl/ v. **1** tr. abuse; criticize abusively. **2** intr. talk abusively; rail. □ **revilement** n. **reviler** n. **reviling** n. [ME f. OF reviler (as RE-, VILE)]

revise /rɪ'vaɪz/ v. & n. ● v.tr. **1** examine or re-examine and improve or amend (esp. written or printed matter). **2** consider and alter (an opinion etc.). **3** (also absol.) Brit. read again (work learned or done) to improve one's knowledge, esp. for an examination. ● n. Printing a proof-sheet including corrections made in an earlier proof. □ **revisable** adj. **revisal** n. **reviser** n. **revisory** adj. [F réviser look at, or L revisere (as RE-, visere intensive of videre vis- see)]

Revised Standard Version a modern American version of the Bible, published in 1946, 1952, and 1957, and based on the American Standard Version (published in 1901).

Revised Version (abbr. **RV**) a British and American version of the Bible, published in 1881, 1885, and 1895, and based on the Authorized Version.

revision /rɪ'vɪʒ(ə)n/ n. **1** the act or an instance of revising; the process of being revised. **2** a revised edition or form. □ **revisionary** adj. [OF revision or LL revisio (as REVISE)]

revisionism /rɪ'vɪʒə,nɪz(ə)m/ n. often derog. a policy of revision or modification, esp. of Marxism on evolutionary socialist (rather than revolutionary) or pluralist principles. □ **revisionist** n. & adj.

revisit /riː'vɪzɪt/ v.tr. (**revisited, revisiting**) **1** visit again. **2** revise, reinspect, or re-examine.

revitalize /riː'vaɪtə,laɪz/ v.tr. (also **-ise**) imbue with new life and vitality. □ **revitalization** /-,vaɪtəlaɪ'zeɪʃ(ə)n/ n.

revival /rɪ'vaɪv(ə)l/ n. **1** the act or an instance of reviving; the process of being revived. **2** a new production of an old play etc. **3** a revived use of an old practice, custom, etc. **4 a** a reawakening of religious fervour. **b** a series of evangelistic meetings to promote this. **5** restoration to bodily or mental vigour or to life or consciousness.

revivalism /rɪ'vaɪvə,lɪz(ə)m/ n. **1** belief in or the promotion of a revival of religious fervour. **2** a tendency or desire to revive a former custom or practice. □ **revivalist** n. & adj. **revivalistic** /-,vaɪvə'lɪstɪk/ adj.

revive /rɪ'vaɪv/ v.intr. & tr. **1** come or bring back to consciousness or life or strength. **2** come or bring back to existence, use, notice, etc. □ **revivable** adj. [ME f. OF revivre or LL revivere (as RE-, L vivere live)]

reviver /rɪ'vaɪvə(r)/ n. **1** a person or thing that revives. **2** colloq. a stimulating drink. **3** a preparation used for restoring a faded colour, polish, etc.

revivify /rɪˈvɪvɪˌfaɪ/ v.tr. (**-ies, -ied**) restore to animation, activity, vigour, or life. □ **revivification** /-ˌvɪvɪfɪˈkeɪʃ(ə)n/ n. [F *revivifier* or LL *revivificare* (as RE-, VIVIFY)]

revoke /rɪˈvəʊk/ v. & n. ● v. **1** tr. rescind, withdraw, or cancel (a decree or promise etc.). **2** intr. Cards fail to follow suit when able to do so (cf. RENOUNCE 5). ● n. Cards the act of revoking. □ **revoker** n. **revocable** /ˈrevəkəb(ə)l/ adj. **revocability** /ˌrevəkəˈbɪlɪtɪ/ n. **revocation** /ˌrevəˈkeɪʃ(ə)n/ n. **revocatory** /ˈrevəkətərɪ/ adj. [ME f. OF *revoquer* or L *revocare* (as RE-, *vocare* call)]

revolt /rɪˈvəʊlt/ v. & n. ● v. **1** intr. **a** rise in rebellion. **b** (as **revolted** adj.) having revolted. **2** tr. (often in *passive*) affect with loathing; nauseate (*was revolted by the thought of it*). **b** intr. (often foll. by *at, against*) feel strong disgust. ● n. **1** an act of rebelling. **2** a state of insurrection (*in revolt*). **3** a sense of loathing. **4** a mood of protest or defiance. [F *révolter* f. It. *rivoltare* ult. f. L *revolvere* (as REVOLVE)]

revolting /rɪˈvəʊltɪŋ/ adj. disgusting, horrible. □ **revoltingly** adv.

revolute /ˈrevəˌluːt, -ˌljuːt/ adj. Bot. etc. having a rolled-back edge. [L *revolutus* past part. of *revolvere*: see REVOLVE]

revolution /ˌrevəˈluːʃ(ə)n/ n. **1 a** the complete overthrow of an established government or social order by those previously subject to it. **b** (in Marxism) the replacement of one ruling class by another; the historically inevitable transition from one economic system to another which is expected to lead to political change and the triumph of communism. **2** any fundamental change or reversal of conditions. **3** the act or an instance of revolving. **4 a** motion in orbit or a circular course or round an axis or centre; rotation. **b** the single completion of an orbit or rotation. **c** the time taken for this. **5** a cyclic recurrence. □ **revolution counter** a device for indicating the number or rate of revolutions of an engine etc. □ **revolutionism** n. **revolutionist** n. [ME f. OF *revolution* or LL *revolutio* (as REVOLVE)]

revolutionary /ˌrevəˈluːʃənərɪ/ adj. & n. ● adj. **1** involving great and often violent change or innovation. **2** of or causing political revolution. **3** (also **Revolutionary**) of or relating to a particular revolution, esp. the War of American Independence. ● n. (pl. **-ies**) an instigator or supporter of political revolution.

Revolutionary Tribunal a court established in Paris in October 1793 to try political opponents of the French Revolution. There was no right of appeal and from June 1794 the only penalty was death. A principal instrument of the Terror, it lasted until May 1795 and was responsible for ordering more than 2,600 executions, including those of Charlotte Corday and Marie Antoinette, as well as many aristocrats and priests.

revolutionize /ˌrevəˈluːʃəˌnaɪz/ v.tr. (also **-ise**) introduce fundamental change to.

Revolutions of 1848 a series of revolts against monarchical rule in Europe during 1848. They sprang from a shared background of autocratic government, lack of representation for the middle classes, economic grievances, and growing nationalism. Revolution occurred first in France, where socialists and supporters of universal suffrage caused the overthrow of King Louis Philippe. In the German and Italian states there were uprisings and demonstrations demanding that Germany and Italy be united as democratic republics; in Austria rioting caused the flight of the emperor and the Foreign Minister Metternich, and peoples subject to the Habsburg empire, notably the Hungarians, demanded autonomy. All of the revolutions ended in failure and repression, but some of the liberal reforms gained as a result (such as universal male suffrage in France) survived, and nationalist aims in Germany and Italy were soon achieved (see RISORGIMENTO).

revolve /rɪˈvɒlv/ v. **1** intr. & tr. turn or cause to turn round, esp. on an axis; rotate. **2** intr. move in a circular orbit. **3** tr. ponder (a problem etc.) in the mind. **4** intr. (foll. by *around*) have as its chief concern; be centred upon (*his life revolves around his job*). □ **revolve on** depend on. **revolving credit** credit that is automatically renewed as debts are paid off. **revolving door** a door with usu. four partitions turning round a central axis. □ **revolvable** adj. [ME f. L *revolvere* (as RE-, *volvere* roll)]

revolver /rɪˈvɒlvə(r)/ n. a pistol with revolving chambers enabling several shots to be fired without reloading.

revue /rɪˈvjuː/ n. a light theatrical entertainment consisting of a series of short (usu. satirical) sketches, songs, and dances, often dealing with topical issues. In France revues were seen in the 1820s, but it was not until the end of the 19th century that the genre spread to Britain and America. [F, = REVIEW n.]

revulsion /rɪˈvʌlʃ(ə)n/ n. **1** abhorrence; a sense of loathing. **2** a sudden violent change of feeling. **3** a sudden reaction in taste, fortune, trade, etc. **4** Med. counterirritation; the treatment of one disordered organ etc. by acting upon another. [F *revulsion* or L *revulsio* (as RE-, *vellere vuls-* pull)]

revulsive /rɪˈvʌlsɪv/ adj. & n. Med. ● adj. producing revulsion. ● n. a revulsive substance.

reward /rɪˈwɔːd/ n. & v. ● n. **1 a** a return or recompense for service or merit. **b** requital for good or evil; retribution. **2** a sum offered for the detection of a criminal, the restoration of lost property, etc. ● v.tr. give a reward to (a person) or for (a service etc.). □ **rewardless** adj. [ME f. AF, ONF *reward* = OF *reguard* REGARD]

rewarding /rɪˈwɔːdɪŋ/ adj. (of an activity etc.) well worth doing; providing satisfaction. □ **rewardingly** adv.

rewarewa /ˌreɪwəˈreɪwə/ n. a tall red-flowered tree, *Knightia excelsa*, of New Zealand. [Maori]

rewash /riːˈwɒʃ/ v.tr. wash again.

reweigh /riːˈweɪ/ v.tr. weigh again.

rewind /riːˈwaɪnd/ v.tr. (*past* and *past part.* **rewound** /-ˈwaʊnd/) wind (a film or tape etc.) back to the beginning. □ **rewinder** n.

rewire /riːˈwaɪə(r)/ v.tr. provide (a building etc.) with new wiring. □ **rewirable** adj.

reword /riːˈwɜːd/ v.tr. change the wording of.

rework /riːˈwɜːk/ v.tr. revise; refashion, remake. □ **reworking** n.

rewound past and past part. of REWIND.

rewrap /riːˈræp/ v.tr. (**rewrapped, rewrapping**) wrap again or differently.

rewrite v. & n. ● v.tr. /riːˈraɪt/ (*past* **rewrote** /-ˈrəʊt/; *past part.* **rewritten** /-ˈrɪt(ə)n/) write again or differently. ● n. /ˈriːraɪt/ **1** the act or an instance of rewriting. **2** a thing rewritten.

Rex /reks/ n. the reigning king (following a name or in the titles of lawsuits, e.g. *Rex v. Jones* the Crown versus Jones). [L]

Rexine /ˈreksiːn/ n. propr. an artificial leather used in upholstery, bookbinding, etc. [20th c.: orig. unkn.]

Reykjavik /ˈreɪkjəˌvɪk, -ˌviːk/ the capital of Iceland, a port on the west coast; pop. (1990) 97,570. Its name derives from the Icelandic *rejkja* 'smoky' referring to the steam rising from its many hot springs.

Reynard /ˈrenɑːd, ˈreɪn-/ the fox in the *Roman de Renart*, a series of popular satirical fables written in France c.1175–1250. [F *renard*; infl. in spelling by MDu. *Reynaerd, -aert*]

Reynolds[1] /ˈren(ə)ldz/, Albert (b.1933), Irish statesman, Prime Minister 1992–4. He was involved with John Major in drafting the 'Downing Street Declaration' (1993), intended as the basis of a peace initiative in Northern Ireland.

Reynolds[2] /ˈren(ə)ldz/, Sir Joshua (1723–92), English painter. He became the first president of the Royal Academy (1768), and through his professional and social prestige succeeded in raising the status of painting in Britain. He insisted on the intellectual basis of painting and became a respected figure in literary circles. Reynolds sought to raise portraiture to the status of history painting by adapting poses and settings from classical statues and Renaissance paintings (e.g. *Mrs Siddons as the Tragic Muse*, 1784). His theories were presented in the *Discourses* delivered annually at the Royal Academy (1769–90).

Reynolds number n. Physics a quantity indicating the degree of turbulence of flow past an obstacle etc. [Osborne *Reynolds*, Engl. physicist (1842–1912)]

Reza Shah /ˌreɪzə ˈʃɑː/ see PAHLAVI[1].

Rf symb. Chem. the element rutherfordium.

r.f. abbr. radio frequency.

RFA abbr. (in the UK) Royal Fleet Auxiliary.

RFC abbr. **1** Rugby Football Club. **2** hist. Royal Flying Corps.

RGS abbr. Royal Geographical Society.

Rh[1] symb. Chem. the element rhodium.

Rh[2] abbr. Med. rhesus (factor).

r.h. abbr. right hand.

RHA abbr. (in the UK) Royal Horse Artillery.

rhabdomancy /ˈræbdəˌmænsɪ/ n. the use of a divining-rod, esp. for discovering subterranean water or mineral ore. [Gk *rhabdomanteia* f. *rhabdos* rod: see -MANCY]

Rhadamanthine /ˌrædəˈmænθaɪn/ adj. stern and incorruptible in judgement, like Rhadamanthus.

Rhadamanthus /ˌrædəˈmænθəs/ Gk Mythol. the son of Zeus and

Europa, and brother of Minos, who, as a ruler and judge in the underworld, was renowned for his justice.

Rhaeto-Romance /ˌriːtəʊrəʊˈmæns/ *adj.* & *n.* (also **Rhaeto-Romanic** /-ˈmænɪk/) ● *adj.* of or in any of the Romance dialects of SE Switzerland and Tyrol, esp. Romansh and Ladin. ● *n.* any of these dialects. [L *Rhaetus* of Rhaetia in the Alps + ROMANIC]

rhapsode /ˈræpsəʊd/ *n. Gk Hist.* a reciter of epic poems, esp. one whose profession it was to recite the poems of Homer. [Gk *rhapsōidos* f. *rhaptō* stitch + *ōidē* song, ODE]

rhapsodist /ˈræpsədɪst/ *n.* **1** a person who rhapsodizes. **2 a** *Gk Hist.* = RHAPSODE. **b** a reciter of poems.

rhapsodize /ˈræpsəˌdaɪz/ *v.intr.* (also **-ise**) talk or write in a rhapsodic manner.

rhapsody /ˈræpsədɪ/ *n.* (*pl.* **-ies**) **1** an enthusiastic, ecstatic, or extravagant utterance or composition. **2** *Mus.* a piece of music in one extended movement, usu. emotional in character. **3** *Gk Hist.* an epic poem, or part of it, of a length for one recitation. □ **rhapsodic** /ræpˈsɒdɪk/ *adj.* **rhapsodical** *adj.* (in senses 1, 2). [L *rhapsodia* f. Gk *rhapsōidia* (as RHAPSODE)]

rhatany /ˈrætənɪ/ *n.* (*pl.* **-ies**) **1** a South American shrub, *Krameria trianda*, from which an astringent root extract is obtained. **2** the root of this. [mod.L *rhatania* f. Port. *ratanha*, Sp. *ratania*, f. Quechua *rataña*]

Rhea /ˈriːə/ **1** *Gk Mythol.* one of the Titans, wife of Cronus and mother of Zeus, Demeter, Poseidon, Hera, and Hades. Frightened of betrayal by their children, Cronus ate them; Rhea rescued Zeus from this fate by hiding him and giving Cronus a stone wrapped in blankets instead. **2** *Astron.* satellite V of Saturn, the fourteenth closest to the planet, discovered by Giovanni Cassini in 1672 (diameter 1,530 km). It is probably composed mainly of ice and is heavily cratered.

rhea /ˈriːə/ *n.* a large flightless South American bird of the genus *Rhea*, resembling but smaller than the ostrich; esp. the common *R. americana*. [mod.L genus name, f. RHEA]

rhebok /ˈriːbɒk/ *n.* (also **reebok**) a small southern African antelope, *Pelea capreolus*, with a long slender neck and short straight horns. [Du. *reebok* roebuck]

Rheims see REIMS.

Rhein see RHINE.

Rheinland-Pfalz see RHINELAND-PALATINATE.

Rhenish /ˈrɛnɪʃ/ *adj.* & *n.* ● *adj.* of the Rhine and the regions adjoining it. ● *n.* wine from this area. [ME *rynis, rynisch* etc., f. AF *reneis*, OF *r(a)inois* f. L *Rhenanus* f. *Rhenus* RHINE]

rhenium /ˈriːnɪəm/ *n.* a silvery-white metallic chemical element (atomic number 75; symbol **Re**). A transition metal, rhenium was discovered by the German chemists Walter (1893–1960) and Ida Eva Tacke (b.1896) Noddack in 1925. It is very rare, occurring in trace amounts in ores of molybdenum and other metals. The metal is used in some specialized alloys. [L *Rhenus* RHINE]

rheology /riːˈɒlədʒɪ/ *n.* the science dealing with the flow and deformation of matter. □ **rheologist** *n.* **rheological** /ˌriːəˈlɒdʒɪk(ə)l/ *adj.* [Gk *rheos* stream + -LOGY]

rheostat /ˈriːəˌstæt/ *n. Electr.* an instrument used to control a current by varying the resistance. □ **rheostatic** /ˌriːəˈstætɪk/ *adj.* [Gk *rheos* stream + -STAT]

rhesus /ˈriːsəs/ *n.* (in full **rhesus monkey**) a common southern Asian macaque monkey, *Macaca mulatta*, often kept in captivity and as a laboratory animal. □ **rhesus baby** an infant with a haemolytic disorder caused by the incompatibility of its own rhesus-positive blood with its mother's rhesus-negative blood. **rhesus factor** an antigen occurring on the red blood cells of most humans and some other primates (as in the rhesus monkey, in which it was first observed). **rhesus negative** lacking the rhesus factor. **rhesus positive** having the rhesus factor. [mod.L, arbitrary use of L *Rhesus* f. Gk *Rhēsos*, mythical king of Thrace]

rhetor /ˈriːtə(r)/ *n.* **1** an ancient Greek or Roman teacher or professor of rhetoric. **2** usu. *derog.* an orator. [ME f. LL *rethor* f. L *rhetor* f. Gk *rhētōr*]

rhetoric /ˈrɛtərɪk/ *n.* **1** the art of effective or persuasive speaking or writing. **2** language designed to persuade or impress (often with an implication of insincerity or exaggeration etc.). [ME f. OF *rethorique* f. L *rhetorica, -ice* f. Gk *rhētorikē (tekhnē)* (art) of rhetoric (as RHETOR)]

rhetorical /rɪˈtɒrɪk(ə)l/ *adj.* **1 a** expressed with a view to persuasive or impressive effect; artificial or extravagant in language. **b** (of a question) assuming a preferred answer. **2** of the nature of rhetoric. **3 a** of or relating to the art of rhetoric. **b** given to rhetoric; oratorical.

□ **rhetorical question** a question asked not for information but to produce an effect, e.g. *who cares?* for *nobody cares.* □ **rhetorically** *adv.* [ME f. L *rhetoricus* f. Gk *rhētorikos* (as RHETOR)]

rhetorician /ˌrɛtəˈrɪʃ(ə)n/ *n.* **1** an orator. **2** a teacher of rhetoric. **3** a rhetorical speaker or writer. [ME f. OF *rethoricien* (as RHETORICAL)]

rheum /ruːm/ *n. Med.* a watery discharge from a mucous membrane, esp. of the eyes or nose. □ **rheumy** *adj.* [ME f. OF *reume* ult. f. Gk *rheuma -atos* stream f. *rheō* flow]

rheumatic /ruːˈmætɪk/ *adj.* & *n.* ● *adj.* **1** of, relating to, or suffering from rheumatism. **2** producing or produced by rheumatism. ● *n.* a person suffering from rheumatism. □ **rheumatic fever** a non-infectious fever with inflammation and pain in the joints. □ **rheumatically** *adv.* **rheumaticky** *adj. colloq.* [ME f. OF *reumatique* or L *rheumaticus* f. Gk *rheumatikos* (as RHEUM)]

rheumatics /ruːˈmætɪks/ *n.pl.* (treated as *sing.*; often prec. by *the*) *colloq.* rheumatism.

rheumatism /ˈruːməˌtɪz(ə)m/ *n.* any disease marked by inflammation and pain in the joints, muscles, or fibrous tissue, esp. rheumatoid arthritis. [F *rhumatisme* or L *rheumatismus* f. Gk *rheumatismos* f. *rheumatizō* f. *rheuma* stream]

rheumatoid /ˈruːməˌtɔɪd/ *adj.* having the character of rheumatism. □ **rheumatoid arthritis** a chronic progressive disease causing inflammation and stiffening of the joints.

rheumatology /ˌruːməˈtɒlədʒɪ/ *n.* the study of rheumatic diseases. □ **rheumatologist** *n.* **rheumatological** /-təˈlɒdʒɪk(ə)l/ *adj.*

RHG *abbr.* (in the UK) Royal Horse Guards.

rhinal /ˈraɪn(ə)l/ *adj. Anat.* of a nostril or the nose. [Gk *rhis rhin-*: see RHINO-]

Rhine /raɪn/ (German **Rhein** /raɪn/, French **Rhin** /rɛ̃/) a river in western Europe which rises in the Swiss Alps and flows for 1,320 km (820 miles) to the North Sea, first westwards through Lake Constance, forming the German–Swiss border, then turning northwards through Germany, forming the southern part of the German–French border, before flowing westwards again through the Netherlands to empty into the North Sea near Rotterdam. On its course it flows through several major cities including Basle, Mannheim, Mainz, Cologne, and Düsseldorf.

Rhineland /ˈraɪnlænd/ (German **Rheinland** /ˈraɪnlant/) the region of western Germany through which the Rhine flows, especially the part to the west of the river. The area was demilitarized as part of the Versailles Treaty in 1919 but was reoccupied by Hitler in 1936.

Rhineland-Palatinate /ˌraɪnlændpəˈlætɪnət/ (German **Rheinland-Pfalz** /ˌraɪnlantˈpfalts/) a state of western Germany; capital, Mainz.

rhinestone /ˈraɪnstəʊn/ *n.* an imitation diamond. [RHINE + STONE]

rhinitis /raɪˈnaɪtɪs/ *n. Med.* inflammation of the mucous membrane of the nose. [Gk *rhis rhinos* nose]

rhino¹ /ˈraɪnəʊ/ *n.* (*pl.* same or **-os**) *colloq.* a rhinoceros. [abbr.]

rhino² /ˈraɪnəʊ/ *n. Brit. sl.* money. [17th c.: orig. unkn.]

rhino- /ˈraɪnəʊ/ *comb. form Anat.* the nose. [Gk *rhis rhinos* nostril, nose]

rhinoceros /raɪˈnɒsərəs/ *n.* (*pl.* same or **rhinoceroses**) a large heavy hoofed mammal of the family Rhinocerotidae, of Africa and southern Asia, with either one or two horns on the nose and a thick folded hide. □ **rhinoceros beetle** a very large horned scarabaeid beetle of the subfamily Dynastinae. **rhinoceros bird** = ox-pecker. **rhinoceros horn** the horn of a rhinoceros, consisting of a mass of keratinized fibres, alleged to have medicinal or aphrodisiac powers. □ **rhinocerotic** /-nɒsəˈrɒtɪk/ *adj.* [ME f. L f. Gk *rhinokerōs* (as RHINO-, keras horn)]

rhinopharyngeal /ˌraɪnəʊfəˈrɪndʒɪəl/ *adj. Anat.* of or relating to the nose and pharynx.

rhinoplasty /ˈraɪnəʊˌplæstɪ/ *n.* plastic surgery of the nose. □ **rhinoplastic** /ˌraɪnəʊˈplæstɪk/ *adj.*

rhizo- /ˈraɪzəʊ/ *comb. form esp. Bot.* a root. [Gk *rhiza* root]

rhizobium /raɪˈzəʊbɪəm/ *n.* a nitrogen-fixing soil bacterium of the genus *Rhizobium*, found esp. in the root nodules of leguminous plants. [mod.L f. RHIZO- + Gk *bios* life + -IUM]

rhizocarp /ˈraɪzəʊˌkɑːp/ *n. Bot.* a plant with a perennial root but stems that wither. [RHIZO- + Gk *karpos* fruit]

rhizoid /ˈraɪzɔɪd/ *adj.* & *n. Bot.* ● *adj.* rootlike. ● *n.* a root-hair or filament in mosses, ferns, etc.

rhizome /ˈraɪzəʊm/ *n. Bot.* an underground rootlike stem bearing both roots and shoots. [Gk *rhizōma* f. *rhizoō* take root (as RHIZO-)]

rhizopod /ˈraɪzəˌpɒd/ n. Zool. a protozoan of the class Rhizopoda, having extensible pseudopodia, e.g amoeba. [mod.L f. RHIZO- + Gk *pous podos* foot]

rho /rəʊ/ n. the seventeenth letter of the Greek alphabet (*P*, ρ). [Gk]

rhodamine /ˈrəʊdəmɪn, -ˌmiːn/ n. Chem. a usu. red or pink synthetic dye of a group used esp. to colour textiles. [RHODO- + AMINE]

Rhode Island /rəʊd/ a state in the north-eastern US, on the Atlantic coast; pop. (1990) 1,003,460; capital, Providence. Settled from England in the 17th century, it was one of the original thirteen states of the Union (1776) and is the smallest and most densely populated. It is named after an island in Narragansett Bay, which was one of the original settlements.

Rhode Island Red n. an orig. American breed of reddish-black domestic fowl.

Rhodes[1] /rəʊdz/ (Greek **Ródhos** /ˈrɔðɔs/) **1** a Greek island in the SE Aegean, off the Turkish coast, the largest of the Dodecanese and the most easterly island in the Aegean; pop. (1981) 87,800. Rhodes flourished in the late Bronze Age, becoming a significant trading nation and dominating several islands in the Aegean. It came under Byzantine rule in the 5th century. Throughout the 14th and 15th centuries it was a centre for the struggle against Turkish domination, eventually falling under Turkish administration in 1522, which lasted until 1912 when it was ceded to Italy. In 1947, following the Second World War, it was awarded to Greece. **2** its capital, a port on the northernmost tip; pop. (1981) 40,390. It was founded c.408 BC and was the site of the Colossus of Rhodes.

Rhodes[2] /rəʊdz/, Cecil (John) (1853–1902), British-born South African statesman, Prime Minister of Cape Colony 1890–6. He went to South Africa in 1870, where he became a successful diamond prospector, and twenty years later owned 90 per cent of the world's production of diamonds. Entering politics in 1881, he expanded British territory in southern Africa, annexing Bechuanaland (now Botswana) in 1884 and developing Rhodesia from 1889 onwards through the British South Africa Company, which he founded. While Premier, Rhodes was implicated in the Jameson Raid of 1895 and forced to resign. In his will, he established the system of Rhodes Scholarships to allow students from the British Empire (now the Commonwealth), the US, and Germany to study at Oxford University.

Rhodes[3] /rəʊdz/, Wilfred (1877–1973), English cricketer. An all-rounder, he played for Yorkshire (1898–1930) and for England (1899–1926), scoring almost 40,000 runs during this time and taking 4,187 first-class wickets, more than any other player.

Rhodesia /rəʊˈdiːʃə, -ˈdiːʒə/ **1** the former name of a large territory in central southern Africa, divided into Northern Rhodesia (now Zambia) and Southern Rhodesia (now Zimbabwe). The region was developed by and named after Cecil Rhodes, through the British South Africa Company, which administered it until Southern Rhodesia became a self-governing British colony in 1923 and Northern Rhodesia a British protectorate in 1924. From 1953 to 1963 Northern and Southern Rhodesia were united with Nyasaland (now Malawi) to form the Federation of Rhodesia and Nyasaland. **2** the name adopted by Southern Rhodesia when Northern Rhodesia left the Federation in 1963 to become the independent republic of Zambia. (See ZIMBABWE.) □ **Rhodesian** adj. & n.

rhodium /ˈrəʊdɪəm/ n. a hard silvery-white metallic chemical element (atomic number 45; symbol **Rh**). A transition element and one of the platinum metals, rhodium was discovered by William Wollaston in 1804. It is a rare element, usually found associated with platinum. It is chiefly used in alloys with platinum, where it increases hardness; the pure metal is used in electroplating for decorative purposes and to form reflecting surfaces. [Gk *rhodon* rose (the colour of a solution of its salts)]

rhodo- /ˈrəʊdəʊ/ comb. form esp. Mineral. & Chem. rose-coloured. [Gk *rhodon* rose]

rhodochrosite /ˌrəʊdəʊˈkrəʊsaɪt/ n. a mineral form of manganese carbonate occurring in rose-red crystals. [Gk *rhodokhrous* rose-coloured]

rhododendron /ˌrəʊdəˈdendrən/ n. a large evergreen ericaceous shrub of the genus Rhododendron, with clusters of large trumpet-shaped flowers. [L, = oleander, f. Gk (as RHODO-, *dendron* tree)]

Rhodope Mountains /ˈrɒdəpɪ/ a mountain system in the Balkans, SE Europe, on the frontier between Bulgaria and Greece, rising to a height of over 2,000 m (6,600 ft) and including the Rila Mountains in the north-west. In ancient times it formed the boundary between Thrace and Macedonia.

rhodopsin /rəʊˈdɒpsɪn/ n. = visual purple. [Gk *rhodon* rose + *opsis* sight]

rhodora /rəˈdɔːrə/ n. a North American pink-flowered azalea, Rhododendron canadense. [mod.L Rhodora former genus name, f. L plant-name f. Gk *rhodon* rose]

rhomb /rɒm/ n. Geom. = RHOMBUS. □ **rhombic** /ˈrɒmbɪk/ adj. [F *rhombe* or L *rhombus*]

rhombencephalon /ˌrɒmbenˈsefəˌlɒn, -ˈkefəˌlɒn/ n. Anat. = HINDBRAIN. [RHOMB + ENCEPHALON]

rhombi pl. of RHOMBUS.

rhombohedron /ˌrɒmbəˈhiːdrən/ n. (pl. **-hedrons** or **-hedra** /-drə/) **1** a solid bounded by six equal rhombuses. **2** a crystal in this form. □ **rhombohedral** adj. [after *polyhedron* etc.]

rhomboid /ˈrɒmbɔɪd/ adj. & n. Geom. ● adj. (also **rhomboidal** /rɒmˈbɔɪd(ə)l/) having or nearly having the shape of a rhombus. ● n. a quadrilateral of which only the opposite sides and angles are equal. [F *rhomboïde* or LL *rhomboides* f. Gk *rhomboeidēs* (as RHOMB)]

rhomboideus /rɒmˈbɔɪdɪəs/ n. (pl. **rhomboidei** /-dɪˌaɪ/) Anat. a muscle connecting the shoulder-blade to the vertebrae. [mod.L *rhomboideus* RHOMBOID]

rhombus /ˈrɒmbəs/ n. (pl. **rhombuses** or **rhombi** /-baɪ/) Geom. a parallelogram with oblique angles and equal sides. [L f. Gk *rhombos*]

Rhondda /ˈrɒndə, ˈrɒnðə/ an urbanized district of Mid Glamorgan, South Wales, which extends along the valleys of the rivers Rhondda Fawr and Rhondda Fach. It was formerly noted as a coal-mining area.

Rhône /rəʊn/ a river in SW Europe which rises in the Swiss Alps and flows 812 km (505 miles), at first westwards through Lake Geneva into France, then to Lyons, where it turns southwards, passing Avignon, to the Mediterranean west of Marseilles, where it forms a wide delta that includes the Camargue.

Rhône-Alpes /rəʊnˈælp/ a region of SE France, extending from the Rhône valley to the borders with Switzerland and Italy and including much of the former duchy of Savoy.

RHS abbr. **1** Royal Historical Society. **2** Royal Horticultural Society. **3** Royal Humane Society.

rhubarb /ˈruːbɑːb/ n. **1 a** a plant of the genus Rheum, with very large leaves; esp. a hybrid of *R. rhaponticum*, producing long fleshy dark red leaf-stalks used cooked as food. **b** the leaf-stalks of this. (See note below.) **2 a** a root of a Chinese and Tibetan plant of the genus Rheum. **b** a purgative made from this. **3 a** colloq. a murmurous conversation or noise, esp. the repetition of the word *rhubarb* by crowd actors. **b** sl. nonsense; worthless stuff. **4** US sl. a heated dispute. [ME f. OF r(e)ubarbe, shortening of med.L r(h)eubarbarum, alt. (by assoc. with Gk *rhēon* rhubarb) of *rhabarbarum* foreign 'rha', ult. f. Gk *rha* + *barbaros* foreign]

▪ The earliest use of a rhubarb was medicinal; the dried rootstock, principally of the Chinese species *Rheum officinale*, was employed as a purgative. Culinary use of the leaf-stalks of *R. rhaponticum* dates only from the mid-18th century and did not become popular until the introduction of forced rhubarb (of the hybrid form) in the early 19th century. The leaves are poisonous on account of their oxalic acid content.

Rhum /rʌm/ (also **Rum**) an island in the Inner Hebrides, to the south of Skye. In 1957 it was designated a nature reserve.

rhumb /rʌm/ n. Naut. **1** any of the thirty-two points of the compass. **2** the angle between two successive compass-points. **3** (in full **rhumb-line**) **a** a line cutting all meridians at the same angle. **b** the line followed by a ship sailing in a fixed direction. [F *rumb* prob. f. Du. *ruim* room, assoc. with L *rhombus*: see RHOMBUS]

rhumba var. of RUMBA.

rhyme /raɪm/ n. & v. ● n. **1** identity of sound between words or the endings of words, esp. in verse. **2** (in sing. or pl.) verse having rhymes. **3 a** the use of rhyme. **b** a poem having rhymes. **4** a word providing a rhyme. ● v. **1** intr. (of words or lines) produce a rhyme. **b** (foll. by with) act as a rhyme (with another). **2** intr. make or write rhymes; versify. **3** tr. put or make (a story etc.) into rhyme. **4** tr. (foll. by with) treat (a word) as rhyming with another. □ **rhyming slang** slang that replaces words by rhyming words or phrases, e.g. *stairs* by *apples and pears*, often with the rhyming element omitted (as in TITFER). **without rhyme or reason** lacking discernible sense or logic. □ **rhymeless** adj. **rhymer** n. **rhymist** n. [ME *rime* f. OF *rime* f. med.L *rithmus*, *rythmus* f. L f. Gk *rhuthmos* RHYTHM]

rhymester /ˈraɪmstə(r)/ n. a writer of (esp. simple) rhymes.

rhyolite /ˈraɪəˌlaɪt/ n. a fine-grained volcanic rock of granitic composition. [G *Rhyolit* f. Gk *rhuax* lava-stream + *lithos* stone]

Rhys /riːs/, Jean (pseudonym of Ella Gwendolen Rees Williams) (1890–1979), British novelist and short-story writer, born in Dominica. Her novels include *Good Morning, Midnight* (1939) and *Wide Sargasso Sea* (1966); the latter, set in Dominica and Jamaica in the 1830s, recreates Charlotte Brontë's *Jane Eyre* from the point of view of Mrs Rochester, the 'mad woman in the attic'.

rhythm /ˈrɪðəm/ n. **1** a measured flow of words and phrases in verse or prose determined by various relations of long and short or accented and unaccented syllables. **2 a** the aspect of musical composition concerned with periodical accent and the duration of notes. **b** a particular type of pattern formed by this (*samba rhythm*). **3** *Physiol.* movement with a regular succession of strong and weak elements. **4** a pattern of regularly recurring events or actions. **5** a sense of musical rhythm. **6** *Art* a harmonious correlation of parts. □ **rhythm method** birth control by avoiding sexual intercourse when ovulation (which takes place once a month) is likely to occur. **rhythm section** the part of a pop, rock, or jazz band mainly supplying rhythm, usu. consisting of bass, drums, and sometimes piano. □ **rhythmless** adj. [F *rhythme* or L *rhythmus* f. Gk *rhuthmos*, rel. to *rhéō* flow]

rhythm and blues n. (also **rhythm 'n' blues**) a form of US black popular music which arose in the 1940s from blues, with the addition of driving rhythms taken from jazz. It was an immediate precursor of rock and roll.

rhythmic /ˈrɪðmɪk/ adj. (also **rhythmical** /-k(ə)l/) **1** relating to or characterized by rhythm. **2** regularly occurring. □ **rhythmically** adv. [F *rhythmique* or L *rhythmicus* (as RHYTHM)]

rhythmicity /rɪðˈmɪsɪtɪ/ n. **1** rhythmical quality or character. **2** the capacity for maintaining a rhythm.

RI abbr. **1** King and Emperor. **2** Queen and Empress. **3** US Rhode Island (also in official postal use). **4** Royal Institute or Institution. **5** religious instruction. [sense 1 f. L *rex et imperator*: sense 2 f. L *regina et imperatrix*]

ria /ˈriːə/ n. *Geog.* a long narrow inlet formed by the partial submergence of a river valley. [Sp. *ría* estuary]

rial /ˈriːɑːl, ˈraɪəl/ n. **1** the chief monetary unit of Iran and Oman. **2** (as **riyal**) the chief monetary unit of Saudi Arabia, Qatar, and Yemen. [Pers. f. Arab. *riyal* f. Sp. *real* ROYAL]

Rialto /rɪˈæltəʊ/ an island in Venice, containing the old mercantile quarter of medieval Venice. The Rialto Bridge, completed in 1591, crosses the Grand Canal in a single span of 48 metres (157.5 ft) between Rialto and San Marco islands.

rib /rɪb/ n. & v. ● n. **1** *Anat.* each of the curved bones articulated in pairs to the spine and protecting the thoracic cavity and its organs. **2** a joint of meat from this part of an animal. **3** a ridge or long raised piece often of stronger or thicker material across a surface or through a structure serving to support or strengthen it. **4** any of a ship's transverse curved timbers forming the framework of the hull. **5** *Knitting* a combination of plain and purl stitches producing a ribbed somewhat elastic fabric. **6** each of the hinged rods supporting the fabric of an umbrella. **7** a vein of a leaf or an insect's wing. **8** *Aeron.* a structural member in an aerofoil. ● v.tr. (**ribbed, ribbing**) **1** provide with ribs; act as the ribs of. **2** *colloq.* make fun of; tease. **3** mark with ridges. **4** plough with spaces between the furrows. □ **rib vault** *Archit.* a vault supported or apparently supported by transverse arches springing from the walls. □ **ribless** adj. [OE *rib, ribb* f. Gmc]

RIBA abbr. Royal Institute of British Architects.

ribald /ˈrɪb(ə)ld/ adj. & n. ● adj. (of language or its user) coarsely or disrespectfully humorous; scurrilous. ● n. a user of ribald language. [ME (earlier sense 'low-born retainer') f. OF *ribau(l)d* f. *riber* pursue licentious pleasures f. Gmc]

ribaldry /ˈrɪb(ə)ldrɪ/ n. ribald talk or behaviour.

riband /ˈrɪb(ə)nd/ n. a ribbon. [ME f. OF *riban*, prob. f. a Gmc compound of BAND¹]

ribbed /rɪbd/ adj. having ribs or riblike markings.

Ribbentrop /ˈrɪb(ə)nˌtrɒp/, Joachim von (1893–1946), German Nazi politician. A close associate of Hitler, Ribbentrop served as Foreign Minister from 1938 to 1945. During his ministry, he signed the non-aggression pact with the Soviet Union (1939). He was convicted as a war criminal in the Nuremberg trials and hanged.

ribbing /ˈrɪbɪŋ/ n. **1** ribs or a riblike structure. **2** *colloq.* the act or an instance of teasing.

ribbon /ˈrɪb(ə)n/ n. **1 a** a narrow strip or band of fabric, used esp. for trimming or decoration. **b** material in this form. **2** a ribbon of a special colour etc. worn to indicate some honour or membership of a sports team etc. **3** a long narrow strip of anything, e.g. impregnated material

forming the inking agent in a typewriter. **4** (in pl.) ragged strips (*torn to ribbons*). □ **ribbon development** the building of houses along a main road, usu. one leading out of a town or village. **ribbon worm** a nemertean. □ **ribboned** adj. [var. of RIBAND]

ribbonfish /ˈrɪb(ə)nˌfɪʃ/ n. a long slender flattened deep-sea fish, esp. of the family Regalecidae or Trachipteridae. (See also OARFISH.)

ribcage /ˈrɪbkeɪdʒ/ n. *Anat.* the wall of bones formed by the ribs round the chest.

Ribera /rɪˈbeərə/, José (or Jusepe) de (known as 'Lo Spagnoletto' = the little Spaniard) (c.1591–1652), Spanish painter and etcher, resident in Italy from 1616. He is best known for his paintings of religious subjects and for his genre scenes; these are noted for their dramatic chiaroscuro effects and for their realistic depiction of torture and martyrdom. Important works include the *Martyrdom of St Bartholomew* (c.1630).

riboflavin /ˌraɪbəʊˈfleɪvɪn/ n. (also **riboflavine** /-viːn/) a vitamin of the B complex, found in liver, milk, and eggs, essential for energy production. Also called *vitamin B₂*. [RIBOSE + L *flavus* yellow]

ribonucleic acid /ˌraɪbənjuːˈkliːɪk, -ˈkleɪɪk/ n. see RNA. [RIBOSE + NUCLEIC ACID]

ribose /ˈraɪbəʊz, -bəʊs/ n. *Chem.* a sugar found in many nucleosides and in several vitamins and enzymes. [G, alt. f. *Arabinose* a related sugar]

ribosome /ˈraɪbəˌsəʊm/ n. *Biochem.* each of the minute particles consisting of RNA and associated proteins found in the cytoplasm of living cells, concerned with the synthesis of proteins. □ **ribosomal** /ˌraɪbəˈsəʊm(ə)l/ adj. [RIBONUCLEIC ACID + -SOME³]

ribwort /ˈrɪbwɜːt/ n. a kind of plantain (see PLANTAIN¹) with long narrow ribbed leaves.

rice /raɪs/ n. & v. ● n. **1** a swamp grass, *Oryza sativa*, cultivated in standing water, esp. in Asia. (See note below.) **2** the grains of this, used as cereal food. ● v.tr. US sieve (cooked potatoes etc.) into thin strings. □ **rice-bowl** an area producing much rice. **rice-paper** edible paper made from the pith of an oriental tree and used for painting and in cookery. □ **ricer** n. [ME *rys* f. OF *ris* f. It. *riso*, ult. f. Gk *oruza*, of oriental orig.]

▪ Rice provides the staple diet of half the world's population and is second only to wheat in terms of total output. It was first cultivated in India c.3000 BC and the great majority is still grown in eastern and SE Asia, although it has also become an important crop in the southern US, Brazil, and other parts of the New World. Rice seedlings are usually planted in flooded fields or paddies, so that terraces are necessary on hillsides and a reliable source of water is essential. The grain often has a coloured and nutritious skin, and brown rice (which only has the husk removed) is still rich in protein, vitamins, and minerals. However, this is removed by the milling and polishing processes, so that polished white rice is much less nutritious. African rice belongs to the related species *O. glaberrima*, whereas the so-called wild rice is not a true rice at all.

ricercar /ˌriːtʃeəˈkɑː(r)/ n. (also **ricercare** /-ˈkɑːrɪ/) *Mus.* an elaborate contrapuntal instrumental composition in fugal or canonic style, esp. of the 16th–18th centuries. [It., = seek out]

rich /rɪtʃ/ adj. **1** having much wealth. **2** (often foll. by *in, with*) splendid, costly, elaborate (*rich tapestries; rich with lace*). **3** valuable (*rich offerings*). **4** copious, abundant, ample (*a rich harvest; a rich supply of ideas*). **5** (often foll. by *in, with*) (of soil or a region etc.) abounding in natural resources or means of production; fertile (*rich in nutrients; rich with vines*). **6** (of food or diet) containing much fat or spice etc. **7** (of the mixture in an internal-combustion engine) containing a high proportion of fuel. **8** (of colour or sound or smell) mellow and deep, strong and full. **9 a** (of an incident or assertion etc.) highly amusing or ludicrous; outrageous. **b** (of humour) earthy. □ **richen** v.intr. & tr. **richness** n. [OE *rice* f. Gmc f. Celt., rel. to L *rex* king: reinforced in ME f. OF *riche* rich, powerful, of Gmc orig.]

Richard¹ /ˈrɪtʃəd/ the name of three kings of England:

Richard I (known as Richard Coeur de Lion or Richard the Lion-Heart) (1157–99), son of Henry II, reigned 1189–99. Richard's military exploits made him a medieval legend, but meant that he spent most of his reign absent from his kingdom, leading to a growth in the power of the barons. Soon after succeeding his father he left to lead the Third Crusade, defeating Saladin at Arsuf (1191), but failing to capture Jerusalem. He was taken prisoner on his way home in 1192 by Duke Leopold of Austria (1157–94) and subsequently held hostage by the Holy Roman emperor Henry VI (1165–97), only being released in 1194 following the payment of a huge ransom. After later embarking on a campaign against Philip II of France, Richard was fatally wounded during the siege of the castle of Châlus.

Richard II (1367–1400), son of the Black Prince, reigned 1377–99. On his accession as a minor the government was placed in the hands of selected nobles, dominated by his uncle John of Gaunt. During this time Richard helped to put down the Peasants' Revolt, but was soon facing a threat to his power from rebel nobles; in 1389 he asserted his right to rule independently of his protectors and later executed or banished most of his former opponents (1397–8). However, his confiscation of John of Gaunt's estate on the latter's death provoked Henry Bolingbroke's return from exile to overthrow him.

Richard III (1452–85), brother of Edward IV, reigned 1483–5. During the Wars of the Roses he served as a commander in the battle at Tewkesbury (1471), which restored Edward IV to the throne. After his brother's death he served as Protector to his nephew Edward V, who, two months later, was declared illegitimate on dubious grounds and subsequently disappeared. (See PRINCES IN THE TOWER.) As king, Richard ruled with some success for a brief period, before being defeated and killed at Bosworth Field (1485) by Henry Tudor (later Henry VII). Historical opinion on the popular picture of Richard as a hunchbacked cutthroat usurper is still divided; many modern historians argue that he was demonized as part of Tudor propaganda.

Richard² /'rɪtʃəd/, Cliff (born Harry Roger Webb) (b.1940), British pop singer, born in India. Influenced by rock and roll, he formed his own group the Drifters (later called the Shadows) in 1958, recording such songs as 'Living Doll' (1959) and 'Bachelor Boy' (1961) with them. Richard went on to act in several films, mainly musicals such as *Expresso Bongo* (1960) and *Summer Holiday* (1962). He left the Shadows in 1968, and in the 1970s became a born-again Christian; he has since combined a successful pop career with evangelism.

Richard Coeur de Lion /ˌkɜː də 'liːɒn/, Richard I of England (see RICHARD¹).

Richards¹ /'rɪtʃədz/, Sir Gordon (1904–86), English jockey. He was champion jockey twenty-six times between 1925 and 1953.

Richards² /'rɪtʃədz/, Viv (full name Isaac Vivian Alexander Richards) (b.1952), West Indian cricketer. Born in Antigua, he made his début for the West Indies in 1974, and captained the team from 1985 until 1991, a period during which his country dominated international cricket. He scored over 6,000 runs during his test career; his century against England in Antigua in 1986 remains the fastest ever in test cricket (in terms of the number of balls received). He also played county cricket in England for Somerset (1974–86) and Glamorgan (1990–3).

Richardson /'rɪtʃəds(ə)n/, Samuel (1689–1761), English novelist. His first novel *Pamela* (1740–1) was entirely in the form of letters and journals, and was responsible for popularizing the epistolary novel. He experimented further with the genre in *Clarissa Harlowe* (1747–8), which explored moral issues in a detailed social context with psychological intensity.

Richard the Lion-Heart /'laɪən ˌhɑːt/, Richard I of England (see RICHARD¹).

Richelieu /'riːʃljɜː/ Armand Jean du Plessis (1585–1642), French cardinal and statesman. From 1624 to 1642 he was chief minister of Louis XIII, dominating French government. He destroyed the power base of the Huguenots in the late 1620s and set out to undermine the Habsburg empire by supporting the Swedish king Gustavus Adolphus in the Thirty Years War, involving France from 1635. In the same year, Richelieu was also responsible for establishing the Académie française.

riches /'rɪtʃɪz/ *n.pl.* abundant means; valuable possessions. [ME *richesse* f. OF *richeise* f. *riche* RICH, taken as pl.]

richly /'rɪtʃlɪ/ *adv.* **1** in a rich way. **2** fully, thoroughly (*richly deserves success*).

Richmond /'rɪtʃmənd/ **1** a town in northern England, on the River Swale in North Yorkshire; pop. (1981) 7,600. **2** (in full **Richmond-upon-Thames**) a residential borough of Greater London, situated on the Thames. It contains Hampton Court Palace and the Royal Botanic Gardens at Kew. **3** the state capital of Virginia, a port on the James river; pop. (1990) 203,060. During the American Civil War it was the Confederate capital from July 1861 until its capture in 1865.

Richter scale /'rɪktə(r), 'rɪxt-/ *n.* a logarithmic scale of 0 to 10 for representing the strength of an earthquake. [Charles Francis *Richter*, Amer. seismologist (1900–85)]

ricin /'rɪsɪn, 'raɪs-/ *n.* a toxic substance obtained from castor oil beans and causing gastroenteritis, jaundice, and heart failure. [mod.L *Ricinus* genus name f. L *ricinus* castor oil plant]

rick¹ /rɪk/ *n. & v.* ● *n.* a stack of hay, corn, etc., built into a regular shape and usu. thatched. ● *v.tr.* form into a rick or ricks. [OE *hrēac*, of unkn. orig.]

rick² /rɪk/ *n. & v.* (also **wrick**) ● *n.* a slight sprain or strain. ● *v.tr.* sprain or strain slightly. [ME *wricke* f. MLG *wricken* move about, sprain]

rickets /'rɪkɪts/ *n.* (treated as *sing.* or *pl.*) a disease of children with softening of the bones (esp. the spine) and bow-legs, caused by a deficiency of vitamin D. [17th c.: orig. uncert., but assoc. by medical writers with Gk *rhakhitis* RACHITIS]

rickettsia /rɪ'ketsɪə/ *n.* a parasitic micro-organism of the genus *Rickettsia*, causing typhus and other febrile diseases. □ **rickettsial** *adj.* [mod.L f. Howard Taylor *Ricketts*, Amer. pathologist (1871–1910)]

rickety /'rɪkɪtɪ/ *adj.* **1 a** insecure or shaky in construction; likely to collapse. **b** feeble. **2 a** suffering from rickets. **b** resembling or of the nature of rickets. □ **ricketiness** *n.* [RICKETS + -Y¹]

rickey /'rɪkɪ/ *n.* (pl. **-eys**) a drink of spirit (esp. gin), lime juice, etc. [20th c.: prob. f. the surname *Rickey*]

rickrack var. of RICRAC.

rickshaw /'rɪkʃɔː/ *n.* (also **ricksha** /-ʃə/) a light two-wheeled hooded vehicle drawn by one or more persons. [abbr. of *jinricksha, jinrikshaw* f. Jap. *jinrikisha* f. *jin* person + *riki* power + *sha* vehicle]

ricochet /'rɪkəˌʃeɪ, -ˌʃet/ *n. & v.* ● *n.* **1** the action of a projectile, esp. a shell or bullet, in rebounding off a surface. **2** a hit made after this. ● *v.intr.* (**ricocheted** /-ˌʃeɪd/; **ricocheting** /-ˌʃeɪɪŋ/ or **ricochetted** /-ˌʃetɪd/; **ricochetting** /-ˌʃetɪŋ/) (of a projectile) rebound one or more times from a surface. [F, of unkn. orig.]

ricotta /rɪ'kɒtə/ *n.* a soft Italian cheese. [It., = recooked, f. L *recoquere* (as RE-, *coquere* cook)]

ricrac /'rɪkræk/ *n.* (also **rickrack**) a zigzag braided trimming for garments. [redupl. of RACK¹]

RICS *abbr.* Royal Institution of Chartered Surveyors.

rictus /'rɪktəs/ *n.* **1** Anat. & Zool. the expanse or gape of a mouth or beak. **2** a fixed grimace or grin. □ **rictal** *adj.* [L, = open mouth f. *ringi rict-* to gape]

rid /rɪd/ *v.tr.* (**ridding**; past and past part. **rid** or archaic **ridded**) (foll. by *of*) make (a person or place) free of something unwanted. □ **be** (or **get**) **rid of** be freed or relieved of (something unwanted); dispose of. [ME, earlier = 'clear (land etc.)' f. ON *rythja*]

riddance /'rɪd(ə)ns/ *n.* the act of getting rid of something. □ **good riddance** welcome relief from an unwanted person or thing.

ridden past part. of RIDE.

riddle¹ /'rɪd(ə)l/ *n. & v.* ● *n.* **1** a question or statement testing ingenuity in divining its answer or meaning. **2** a puzzling fact or thing or person. ● *v.* **1** *intr.* speak in or propound riddles. **2** *tr.* solve or explain (a riddle). □ **riddler** *n.* [OE *rædels, rædelse* opinion, riddle, rel. to READ]

riddle² /'rɪd(ə)l/ *v. & n.* ● *v.tr.* (usu. foll. by *with*) **1** make many holes in, esp. with gunshot. **2** (in passive) fill; spread through; permeate (*was riddled with errors*). **3** pass through a riddle or sieve. ● *n.* a coarse sieve. [OE *hriddel*, earlier *hrīder*: cf. *hrīdrian* sift]

riddling /'rɪdlɪŋ/ *adj.* expressed in riddles; puzzling. □ **riddlingly** *adv.*

ride /raɪd/ *v. & n.* ● *v.* (past **rode** /rəʊd/; past part. **ridden** /'rɪd(ə)n/) **1** *tr.* travel or be carried on (a bicycle etc.) or esp. N. Amer. in (a vehicle). **2** *intr.* (often foll. by *on, in*) travel or be conveyed (on a bicycle or in a vehicle). **3** *tr.* sit on and control or be carried by (a horse etc.). **4** *intr.* (often foll. by *on*) be carried (on a horse etc.). **5** *tr.* be carried or supported by (*the ship rides the waves*). **6** *tr.* **a** traverse on horseback etc., ride over or through (*ride 50 miles; rode the prairie*). **b** compete or take part in on horseback etc. (*rode a good race*). **7** *intr.* **a** (of a ship) lie at anchor; float buoyantly. **b** (of the moon) seem to float. **8** *intr.* (foll. by *in, on*) rest in or on while moving. **9** *tr.* yield to (a blow) so as to reduce its impact. **10** *tr.* give a ride to; cause to ride (*rode the child on his back*). **11** *tr.* (of a rider) cause (a horse etc.) to move forward (*rode their horses at the fence*). **12** *tr.* **a** (in *passive*; foll. by *by, with*) be oppressed or dominated by; be infested with (*was ridden with guilt*). **b** (as **ridden** *adj.*) infested or afflicted (usu. in *comb.*: *a rat-ridden cellar*). **13** *intr.* (of a thing normally level or even) project or overlap. **14** *tr.* coarse *sl.* have sexual intercourse with. **15** *tr.* US annoy or seek to annoy. ● *n.* **1** an act or period of travel in a vehicle. **2** a spell of riding on a horse, bicycle, person's back, etc. **3** a path (esp. through woods) for riding on. **4** the quality of sensations when riding (*gives a bumpy ride*). **5** a roundabout, roller-coaster, etc., ridden at an amusement park or fairground. □ **let a thing ride** leave it alone; let it take its natural course. **ride again** reappear, esp. unexpectedly and reinvigorated. **ride down** overtake or trample on horseback. **ride for a fall** act recklessly risking defeat or failure. **ride herd on** see HERD.

ride high be elated or successful. **ride out** come safely through (a storm etc., or a danger or difficulty). **ride roughshod over** see ROUGHSHOD. **ride to hounds** see HOUND. **ride up** (of a garment, carpet, etc.) work or move out of its proper position. **take for a ride 1** *colloq.* hoax or deceive. **2** *sl.* abduct in order to murder. □ **ridable** *adj.* [OE *rīdan*]

rider /'raɪdə(r)/ *n.* **1** a person who rides (esp. a horse). **2 a** an additional clause amending or supplementing a document. **b** *Brit. Parl.* an addition or amendment to a bill at its third reading. **c** a corollary. **d** *Brit.* a recommendation etc. added to a judicial verdict. **3** *Math.* a problem arising as a corollary of a theorem etc. **4** a piece in a machine etc. that surmounts or bridges or works on or over others. **5** (in *pl.*) *Naut.* an additional set of timbers or iron plates strengthening a ship's frame. □ **riderless** *adj.* [OE *rīdere* (as RIDE)]

ridge /rɪdʒ/ *n. & v.* ● *n.* **1** the line of the junction of two surfaces sloping upwards towards each other (*the ridge of a roof*). **2** a long narrow hilltop, mountain range, or watershed. **3** any narrow elevation across a surface. **4** *Meteorol.* an elongated region of high barometric pressure. **5** a raised strip of arable land, usu. one of a set separated by furrows. **6** a raised hotbed for melons etc. ● *v.* **1** *tr.* mark with ridges. **2** *tr.* break up (land) into ridges. **3** *tr.* plant (cucumbers etc.) in ridges. **4** *tr. & intr.* gather into ridges. □ **ridge-piece** (or **-tree**) a beam along the ridge of a roof. **ridge-pole 1** the horizontal pole of a long tent. **2** = ridge-piece. **ridge-tile** a tile used in making a roof-ridge. □ **ridgy** *adj.* [OE *hrycg* f. Gmc]

ridgeway /'rɪdʒweɪ/ *n.* a road or track along a ridge.

ridicule /'rɪdɪˌkjuːl/ *n. & v.* ● *n.* derision or mockery. ● *v.tr.* make fun of; subject to ridicule; laugh at. [F or f. L *ridiculum* neut. of *ridiculus* laughable f. *ridere* laugh]

ridiculous /rɪ'dɪkjʊləs/ *adj.* **1** deserving or inviting ridicule. **2** unreasonable, absurd. □ **ridiculously** *adv.* **ridiculousness** *n.* [L *ridiculosus* (as RIDICULE)]

riding[1] /'raɪdɪŋ/ *n.* **1** in senses of RIDE *v.* **2** the practice or skill of riders of horses. **3** = RIDE *n.* 3. □ **riding-light** (or **-lamp**) a light shown by a ship at anchor. **riding-school** an establishment teaching skills in horsemanship.

riding[2] /'raɪdɪŋ/ *n.* **1** each of three former administrative divisions (*East Riding*, *North Riding*, *West Riding*) of Yorkshire. **2** an electoral division of Canada. [OE *thriding* (unrecorded) f. ON *thrithjungr* third part f. *thrithi* THIRD: *th*- was lost owing to the preceding -*t* or -*th* of *east* etc.]

Ridley /'rɪdlɪ/, Nicholas (c.1500–55), English Protestant bishop and martyr. He became one of Thomas Cranmer's chaplains in 1537 and, during the reign of Edward VI, was appointed bishop of Rochester (1547) and then of London (1550). During this period, he emerged as one of the leaders of the Reformation, opposing the Catholic policies of Edward's sister and successor Mary I, for which he was later imprisoned (1553) and burnt at the stake in Oxford.

Riefenstahl /'riːf(ə)nˌʃtɑːl/, Leni (full name Bertha Helene Amalie Riefenstahl) (b.1902), German film-maker and photographer. She is chiefly known for two films which she made during the 1930s; *Triumph of the Will* (1934), a powerful depiction of the 1934 Nuremberg Nazi Party rallies, and *Olympia* (1938), a two-part documentary of the 1936 Berlin Olympic Games. She was not a Nazi Party member and insisted on full control over these films, but outside Germany her work was regarded as Nazi propaganda and her postwar reputation suffered as a result.

Riel /riː'el/, Louis (1844–85), Canadian political leader. He headed the rebellion of the Metis at Red River Settlement (now in Manitoba) in 1869 to protest against the planned transfer of the territorial holdings of the Hudson's Bay Company to Canadian jurisdiction, a move which the Metis feared would result in the loss of some of their land to Anglo-Protestant settlers. Having formed a provisional government with himself at its head, Riel oversaw negotiations for acceptable terms for union with Canada, including the establishment of the province of Manitoba. He was executed for treason after leading a further rebellion of the Metis in the Saskatchewan valley (1884–5).

Riemann /'riːmən/, (Georg Friedrich) Bernhard (1826–66), German mathematician. He studied under Karl Gauss at Göttingen and became professor there. He founded Riemannian geometry, which is of fundamental importance to both mathematics and physics. The *Riemann hypothesis*, about the complex numbers which are roots of a certain transcendental equation, remains one of the unsolved problems of mathematics.

Riemannian geometry /riː'mænɪən/ *n.* a form of differential non-Euclidean geometry, in which space is everywhere positively curved. It provided Einstein with a basis for his general theory of relativity.

Riesling /'riːzlɪŋ, 'riːslɪŋ/ *n.* **1** a kind of dry white wine produced in Germany, Austria, and elsewhere. **2** the variety of grape from which this is produced. [G]

rife /raɪf/ *predic.adj.* **1** of common occurrence; widespread. **2** (foll. by *with*) abounding in; teeming with. □ **rifeness** *n.* [OE *rȳfe* prob. f. ON *rífr* acceptable f. *reifa* enrich, *reifr* cheerful]

riff /rɪf/ *n. & v.* ● *n.* a short repeated phrase in jazz etc. ● *v.intr.* play riffs. [20th c.: abbr. of RIFFLE *n.*]

riffle /'rɪf(ə)l/ *v. & n.* ● *v.* **1** *tr.* **a** turn (pages) in quick succession. **b** shuffle (playing cards) esp. by flexing and combining the two halves of a pack. **2** *intr.* (often foll. by *through*) leaf quickly (through pages). ● *n.* **1** the act or an instance of riffling. **2** (in gold-washing) a groove or slat set in a trough or sluice to catch gold particles. **3** *N. Amer.* **a** a shallow part of a stream where the water flows brokenly. **b** a patch of waves or ripples on water. [perh. var. of RUFFLE]

riff-raff /'rɪfræf/ *n.* (often prec. by *the*) rabble; disreputable or undesirable persons. [ME *riff and raff* f. OF *rif et raf*]

rifle[1] /'raɪf(ə)l/ *n. & v.* ● *n.* **1** a gun with a long rifled barrel, esp. one fired from shoulder-level. **2** (in *pl.*) riflemen. ● *v.tr.* make spiral grooves in (a gun or its barrel or bore) to make a bullet spin. □ **rifle bird** a bird of paradise of the genus *Ptiloris*, with mainly velvety black plumage. **rifle-range** a place for rifle-practice. **rifle-shot 1** the distance coverable by a shot from a rifle. **2** a shot fired with a rifle. [OF *rifler* graze, scratch f. Gmc]

rifle[2] /'raɪf(ə)l/ *v.* **1** *intr.* (foll. by *through*) make a vigorous search. **2** *tr.* **a** ransack (a place etc.) with intent to steal. **b** rob (a person), esp. by searching clothes or pockets. **c** plunder, steal. [ME f. OF *rifler* graze, scratch, plunder, ult. f. Gmc]

rifleman /'raɪf(ə)lmən/ *n.* (*pl.* **-men**) **1** a soldier armed with a rifle. **2** a small yellow and green New Zealand wren, *Acanthisitta chloris*.

rifling /'raɪflɪŋ/ *n.* the arrangement of grooves on the inside of a gun's barrel.

Rif Mountains /rɪf/ (also **Er Rif** /eə/) a mountain range of northern Morocco, running parallel to the Mediterranean for about 290 km (180 miles) eastwards from Tangier. Rising to over 2,250 m (7,000 ft), it forms a westward extension of the Atlas Mountains.

rift /rɪft/ *n. & v.* ● *n.* **1 a** a crack or split in an object. **b** an opening in a cloud etc. **2** a cleft or fissure in earth or rock. **3** a disagreement; a breach in friendly relations. ● *v.tr.* tear or burst apart. □ **rift valley** a steep-sided valley formed by subsidence of the earth's crust between nearly parallel faults. □ **riftless** *adj.* **rifty** *adj.* [ME, of Scand. orig.]

Rift Valley see GREAT RIFT VALLEY.

rig[1] /rɪg/ *v. & n.* ● *v.tr.* (**rigged, rigging**) **1 a** provide (a sailing-ship) with sails, rigging, etc. **b** prepare ready for sailing. **2** (often foll. by *out, up*) fit with clothes or other equipment. **3** (foll. by *up*) set up hastily or as a makeshift. **4** assemble and adjust the parts of (an aircraft). ● *n.* **1** the arrangement of masts, sails, rigging, etc., of a sailing-ship. **2** equipment for a special purpose, e.g. a radio transmitter. **3 a** = *oil rig*. **b** = *drilling rig*. **4** a person's or thing's look as determined by clothing, equipment, etc., esp. uniform. □ **in full rig** *colloq.* smartly or ceremonially dressed. **rig-out** *Brit. colloq.* an outfit of clothes. □ **rigged** *adj.* (also in *comb.*). [ME, perh. of Scand. orig.: cf. Norw. *rigga* bind or wrap up]

rig[2] /rɪg/ *v. & n.* ● *v.tr.* (**rigged, rigging**) manage or conduct fraudulently (*they rigged the election*). ● *n.* **1** a trick or dodge. **2** a way of swindling. □ **rig the market** cause an artificial rise or fall in prices. □ **rigger** *n.* [19th c.: orig. unkn.]

Riga /'riːgə/ a port on the Baltic Sea, capital of Latvia; pop. (1990) 915,000.

rigadoon /ˌrɪgə'duːn/ *n.* **1** a lively dance in duple or quadruple time for two persons. **2** the music for this. [F *rigodon*, *rigaudon*, perh. f. its inventor *Rigaud*]

rigatoni /ˌrɪgə'təʊnɪ/ *n.pl.* pasta in the form of short hollow fluted tubes. [It. f. *rigare* draw a line, make a groove]

Rigel /'raɪdʒəl, 'raɪg(ə)l/ *Astron.* the seventh brightest star in the sky, and the brightest in the constellation of Orion. It is a blue supergiant nearly sixty thousand times as luminous as our sun. [Arab. *rijl* foot (of Orion)]

rigger /'rɪgə(r)/ *n.* **1 a** a person who rigs or who arranges rigging. **b** a person who erects and maintains scaffolding, lifting tackle, etc. **2** = OUTRIGGER 5a. **3** (as second element in *comb.*) a ship rigged in a specified way. **4** a worker on an oil rig.

rigging /'rɪgɪŋ/ *n.* **1** a ship's spars, ropes, etc., supporting and controlling the sails. **2** the ropes and wires supporting the structure of an airship or biplane.

right /raɪt/ *adj., n., v., adv., & int.* ● *adj.* **1** (of conduct etc.) just, morally or socially correct (*it is only right to tell you; I want to do the right thing*). **2** true, correct; not mistaken (*the right time; you were right about the weather*). **3** less wrong or not wrong (*which is the right way to town?*). **4** more or most suitable or preferable (*the right person for the job; along the right lines*). **5** in a sound or normal condition; physically or mentally healthy; satisfactory (*the engine doesn't sound right*). **6 a** on or towards the side of the human body which corresponds to the position of east if one regards oneself as facing north. **b** on or towards that part of an object which is analogous to a person's right side or (with opposite sense) which is nearer to a spectator's right hand. Opp. LEFT[1]. **7** (of a side of fabric etc.) meant for display or use (*turn it right side up*). **8** *colloq.* or *archaic* real; properly so called (*made a right mess of it; a right royal welcome*). **9** (also **Right**) *Polit.* of the Right. ● *n.* **1** that which is morally or socially correct or just; fair treatment (often in *pl.*: *the rights and wrongs of the case*). **2** (often foll. by *to*, or *to* + infin.) a justification or fair claim (*has no right to speak like that*). **3** a thing one may legally or morally claim; the state of being entitled to a privilege or immunity or authority to act (*a right of reply; human rights*). **4** the right-hand part or region or direction. **5** *Boxing* **a** the right hand. **b** a blow with this. **6** (often **Right**) *Polit.* **a** a group or section favouring conservatism (orig. the more conservative section of a continental legislature, seated on the president's right). **b** such conservatives collectively. **7** the side of a stage which is to the right of a person facing the audience. **8** (esp. in marching) the right foot. **9** the right wing of an army. ● *v.tr.* **1** (often *refl.*) restore to a proper or straight or vertical position. **2 a** correct (mistakes etc.); set in order. **b** avenge (a wrong or a wronged person); make reparation for or to. **c** vindicate, justify, rehabilitate. ● *adv.* **1** straight (*go right on*). **2** *colloq.* immediately; without delay (*I'll be right back; do it right now*). **3 a** (foll. by *to*, *round*, *through*, etc.) all the way (*sank right to the bottom; ran right round the block*). **b** (foll. by *off*, *out*, etc.) completely (*came right off its hinges; am right out of butter*). **4** exactly, quite (*right in the middle*). **5** justly, properly, correctly, truly, satisfactorily (*did not act right; not holding it right; if I remember right*). **6** on or to the right side. **7** *archaic* very; to the full (*am right glad to hear it; dined right royally*). ● *int. colloq.* expressing agreement or assent. □ **as right as rain** perfectly sound and healthy. **at right angles** placed to form a right angle. **by right** (or **rights**) if right were done. **do right by** act dutifully towards (a person). **in one's own right** through one's own position or effort etc. **in the right** having justice or truth on one's side. **in one's right mind** sane; competent to think and act. **of** (or **as of**) **right** having legal or moral etc. entitlement. **on the right side of 1** in the favour of (a person etc.). **2** somewhat less than (a specified age). **put** (or **set**) **right 1** restore to order, health, etc. **2** correct the mistaken impression etc. of (a person). **put** (or **set**) **to rights** make correct or well ordered. **right about** (or **about-turn** or **about-face**) **1** (esp. in marching) a right turn continued to face the rear. **2** a reversal of policy. **3** a hasty retreat. **right and left** (or **right, left, and centre**) on all sides. **right angle** an angle of 90°, made by lines meeting with equal angles on either side. **right-angled 1** containing or making a right angle. **2** involving right angles, not oblique. **right arm** one's most reliable helper. **right ascension** see ASCENSION. **right away** (or **off**) immediately. **right bank** the bank of a river on the right facing downstream. **right bower** see BOWER[3]. **right field** *Baseball* the part of the outfield to the right of the batter when facing the pitcher. **right hand 1** = *right-hand man*. **2** the most important position next to a person (*stood at the President's right hand*). **right-hand** *adj.* **1** on or towards the right side of a person or thing (*right-hand drive*). **2** done with the right hand (*right-hand blow*). **3** (of a screw) = RIGHT-HANDED 4b. **right-hand man** an indispensable or chief assistant. **right-minded** (or **-thinking**) having sound views and principles. **right of search** *Naut.* see SEARCH. **right of way 1** a right established by usage to pass over another's ground. **2** a path subject to such a right. **3** the right of one vehicle to proceed before another. **right oh!** (or **ho!**) = RIGHTO. **right-on** *colloq.* having fashionably liberal or progressive views. **right on!** *sl.* an expression of strong approval or encouragement. **a right one** *Brit. colloq.* a silly or foolish person. **right sphere** *Astron.* see SPHERE. **right turn** a turn that brings one's front to face as one's right side did before. **right whale** a baleen whale of the family Balaenidae (regarded by whalers as 'the right whale to hunt'), having a deep jaw with long plates of whalebone, esp. the *black right whale* (*Balaena glacialis*) (see also BOWHEAD). **right you are!** *colloq.* an exclamation of assent. **she's** (or **she'll be**) **right** *Austral. colloq.* that will be all right. **too right** *colloq.* an expression of agreement. **within one's rights** not exceeding one's authority or entitlement.

□ **rightable** *adj.* **righter** *n.* **rightish** *adj.* **rightless** *adj.* **rightlessness** *n.* **rightness** *n.* [OE *riht* (adj.), *rihtan* (v.), *rihte* (adv.)]

Right Bank a district of the city of Paris, situated on the right bank of the River Seine, to the north of the river. The area contains the Champs Élysées and the Louvre.

righten /ˈraɪt(ə)n/ *v.tr.* make right or correct.

righteous /ˈraɪtʃəs/ *adj.* (of a person or conduct) morally right; virtuous, law-abiding. □ **righteously** *adv.* **righteousness** *n.* [OE *rihtwīs* (as RIGHT *n.* + -WISE or RIGHT *adj.* + WISE[2]), assim. to *bounteous* etc.]

rightful /ˈraɪtfʊl/ *adj.* **1 a** (of a person) legitimately entitled to (a position etc.) (*the rightful heir*). **b** (of status or property etc.) that one is entitled to. **2** (of an action etc.) equitable, fair. □ **rightfully** *adv.* **rightfulness** *n.* [OE *rihtful* (as RIGHT *n.*)]

right-handed /raɪtˈhændɪd/ *adj.* **1** using the right hand by preference as more serviceable than the left. **2** (of a tool etc.) made to be used with the right hand. **3** (of a blow) struck with the right hand. **4 a** turning to the right; towards the right. **b** (of a screw) advanced by turning to the right (clockwise). □ **right-handedly** *adv.* **right-handedness** *n.*

right-hander /raɪtˈhændə(r)/ *n.* **1** a right-handed person. **2** a right-handed blow.

Right Honourable *n.* *Brit.* a title given to certain high officials, e.g. Privy Counsellors.

rightism /ˈraɪtɪz(ə)m/ *n.* *Polit.* the principles or policy of the right. □ **rightist** *n. & adj.*

rightly /ˈraɪtlɪ/ *adv.* justly, properly, correctly, justifiably.

rightmost /ˈraɪtməʊst/ *adj.* furthest to the right.

righto /ˈraɪtəʊ, raɪtˈəʊ/ *int.* *Brit. colloq.* expressing agreement or assent.

Right Reverend *n.* the title of a bishop.

rights of man *n.pl.* rights held to be justifiably belonging to any person; human rights. The phrase is associated with the Declaration of the Rights of Man and of the Citizen, adopted by the French National Assembly in 1789 and used as a preface to the French Constitution of 1791. The declaration enshrined the equal rights of all to liberty, equality of opportunity, freedom of speech and religion, representation, safety from injustice, violence, and oppression, and security of private property.

rightward /ˈraɪtwəd/ *adv. & adj.* ● *adv.* (also **rightwards** /-wədz/) towards the right. ● *adj.* going towards or facing the right.

right wing *n. & adj.* ● *n.* **1** the conservative or reactionary section of society, a political party, etc. (See also LEFT WING 1.) **2** the right side of a football etc. team on the field. **3** the right side of an army. ● *adj.* (also **right-wing**) **1** of or relating to the right wing. **2** *Polit.* conservative; opposed to liberal or socialist policies. □ **right-winger** *n.*

rigid /ˈrɪdʒɪd/ *adj. & n.* ● *adj.* **1** not flexible; that cannot be bent (*a rigid frame*). **2** (of a person, conduct, etc.) **a** inflexible, unbending, harsh (*a rigid disciplinarian; rigid economy*). **b** strict, precise, punctilious. ● *n.* a lorry or truck which is not articulated. □ **rigidly** *adv.* **rigidness** *n.* **rigidity** /rɪˈdʒɪdɪtɪ/ *n.* [F *rigide* or L *rigidus* f. *rigere* be stiff]

rigidify /rɪˈdʒɪdɪˌfaɪ/ *v.tr. & intr.* (**-ies, -ied**) make or become rigid.

rigmarole /ˈrɪɡməˌrəʊl/ *n.* **1** a lengthy and complicated procedure. **2 a** a rambling or meaningless account or tale. **b** such talk. [orig. *ragman roll* = a catalogue, of unkn. orig.]

rigor[1] /ˈrɪɡə(r), ˈraɪɡɔ:(r)/ *n.* *Med.* **1** a sudden feeling of cold with shivering accompanied by a rise in temperature, preceding a fever etc. **2** rigidity of the body caused by shock or poisoning etc. [ME f. L f. *rigere* be stiff]

rigor[2] *US* var. of RIGOUR.

rigor mortis /ˌrɪɡə ˈmɔːtɪs/ *n.* stiffening of the body after death. [L, = stiffness of death]

rigorous /ˈrɪɡərəs/ *adj.* **1** characterized by or showing rigour; strict, severe. **2** strictly exact or accurate. **3** (of the weather) cold, severe. □ **rigorously** *adv.* **rigorousness** *n.* [OF *rigorous* or LL *rigorosus* (as RIGOR[1])]

rigour /ˈrɪɡə(r)/ *n.* (*US* **rigor**) **1 a** severity, strictness, harshness. **b** (often in *pl.*) severity of weather or climate; extremity of cold. **c** (in *pl.*) harsh measures or conditions. **2** logical exactitude. **3** strict enforcement of rules etc. (*the utmost rigour of the law*). **4** austerity of life; puritanical discipline. [ME f. OF *rigour* f. L *rigor* (as RIGOR[1])]

Rig-veda /rɪɡˈveɪdə, -ˈviːdə/ *Hinduism* a sacred text containing a collection of hymns in early Sanskrit. It was composed in the 2nd millennium BC, and is the oldest and most important of the four Vedas. [Skr. *ṛgveda* f. *ṛc* praise + *veda* VEDA]

Rijeka /ri:ˈekə/ (called in Italian *Fiume*) a port on the Adriatic coast of Croatia; pop. (1991) 167,900.

Rijksmuseum /ˈraɪksmuˌzeɪəm/ the national art gallery of the Netherlands, in Amsterdam. Established in the late 19th century and developed from the collection of the House of Orange, it now contains the most representative collection of Dutch art in the world.

Riksmål /ˈriːksmɔːl/ *n.* a form of the Norwegian language (see NORWEGIAN). [Norw., f. *rike* realm + *mål* language]

Rila Mountains /ˈriːlə/ a range of mountains in western Bulgaria, forming the westernmost extent of the Rhodope Mountains. It is the highest range in Bulgaria, rising to a height of 2,925 m (9,596 ft) at Mount Musala.

rile /raɪl/ *v.tr.* **1** *colloq.* anger, irritate. **2** *US* make (water) turbulent or muddy. [var. of ROIL]

Riley[1] /ˈraɪlɪ/ *n.* (also **Reilly**) □ **the life of Riley** *colloq.* a carefree existence. [20th c.: orig. unkn.]

Riley[2] /ˈraɪlɪ/, Bridget (Louise) (b.1931), English painter. A leading exponent of op art, she worked with flat patterns of lines, dots, and squares, initially in black and white and later in colour, to create optical illusions of light and movement. Famous paintings include *Movement in Squares* (1961) and *Fall* (1963).

rilievo var. of RELIEVO.

Rilke /ˈrɪlkə/, Rainer Maria (pseudonym of René Karl Wilhelm Josef Maria Rilke) (1875–1926), Austrian poet, born in Bohemia. Two trips to Russia (1899–1900) inspired him to write the *Book of Hours* (1905), written from the perspective of a Russian monk. Rilke's conception of art as a quasi-religious vocation culminated in the hymnic lyrics for which he is best known: the *Duino Elegies* and *Sonnets to Orpheus* (both 1923). In these he sought to define a poet's spiritual role in the face of transience and death.

rill /rɪl/ *n.* **1** a small stream. **2** a shallow channel cut in the surface of soil or rocks by running water. **3** var. of RILLE. [LG *ril, rille*]

rille /rɪl/ *n.* (also **rill**) *Astron.* a cleft or narrow valley on the moon's surface. [G (as RILL)]

rim /rɪm/ *n. & v.* ● *n.* **1 a** a raised edge or border. **b** a margin or verge, esp. of something circular. **2** the part of a pair of spectacles surrounding the lenses. **3** the outer edge of a wheel, on which the tyre is fitted. **4** a boundary line (*the rim of the horizon*). ● *v.tr.* (**rimmed**, **rimming**) **1 a** provide with a rim. **b** be a rim for or to. **2** edge, border. □ **rim-brake** a brake acting on the rim of a wheel. □ **rimless** *adj.* **rimmed** *adj.* (also in *comb.*). [OE *rima* edge: cf. ON *rimi* ridge (the only known cognate)]

Rimbaud /ˈræmbəʊ/, (Jean Nicholas) Arthur (1854–91), French poet. Rimbaud wrote his most famous poem, 'Le Bateau ivre', at the age of 17. In the same year he began a passionate relationship with the poet Paul Verlaine, which partly inspired his collection of symbolist prose poems *Une Saison en enfer* (1873). In this and *Les Illuminations* (c.1872; published 1884), he explored the visionary possibilities of systematically 'disorientating the senses'. He stopped writing at about the age of 20 and spent the rest of his life travelling.

rime[1] /raɪm/ *n. & v.* ● *n.* **1** frost, esp. formed from cloud or fog. **2** *poet.* hoar-frost. ● *v.tr.* cover with rime. [OE *hrīm*]

rime[2] *archaic* var. of RHYME.

Rimini /ˈrɪmɪnɪ/ a port and resort on the Adriatic coast of NE Italy; pop. (1990) 130,900. The original settlement was established by the Romans in 268 BC, at the junction of the ancient Via Flaminia and Via Aemilia; its ancient name was Ariminum. An independent city-state in the 12th century, it came under the control of the Malatesta family in the 13th century and passed to the Papal States in 1509. It was annexed to the kingdom of Italy in 1860. It suffered widespread destruction by Allied bombing during Second World War.

Rimmon /ˈrɪmən/ (in the Bible) a deity worshipped in ancient Damascus (2 Kings 5: 18). □ **bow down in the house of Rimmon** compromise one's convictions.

rimose /ˈraɪməʊz, -məʊs/ *adj.* (also **rimous** /-məs/) esp. *Bot.* full of chinks or fissures. [L *rimosus* f. *rima* chink]

Rimsky-Korsakov /ˌrɪmskɪˈkɔːsəˌkɒf/, Nikolai (Andreevich) (1844–1908), Russian composer. He achieved fame with his orchestral suite *Scheherazade* (1888) and his many operas drawing on Russian and Slavonic folk tales, notably *The Golden Cockerel* (1906–7); the latter was based on Pushkin's poem lampooning autocracy and was banned in Russia until 1909. Rimsky-Korsakov was also a noted orchestrator, completing works by composers such as Borodin and Mussorgsky.

rimu /ˈriːmuː/ *n.* **1** a tall coniferous tree, *Dacrydium cupressinum*, native to New Zealand. **2** its light brown streaked wood, used for furniture and interior fittings. [Maori]

rimy /ˈraɪmɪ/ *adj.* (**rimier, rimiest**) frosty; covered with frost.

rind /raɪnd/ *n. & v.* ● *n.* **1** the tough outer layer or covering of fruit and vegetables, cheese, bacon, etc. **2** the bark of a tree or plant. ● *v.tr.* strip the bark from. □ **rinded** *adj.* (also in *comb.*). **rindless** *adj.* [OE *rind(e)*]

rinderpest /ˈrɪndəˌpest/ *n.* a virulent infectious viral disease of ruminants (esp. cattle). [G f. *Rinder* cattle + *Pest* PEST]

ring[1] /rɪŋ/ *n. & v.* ● *n.* **1** a circular band, usu. of precious metal, worn on a finger as an ornament or a token of marriage or betrothal. **2** a circular band of any material. **3** the rim of a cylindrical or circular object, or a line or band round it. **4** a mark or part having the form of a circular band (*had rings round his eyes; smoke rings*). **5** = annual ring. **6 a** an enclosure for a circus performance, betting at races, the showing of cattle, etc. **b** (prec. by *the*) bookmakers collectively. **c** a roped enclosure for boxing or wrestling. **7 a** a group of people or things arranged in a circle. **b** such an arrangement. **8** a combination of traders, bookmakers, spies, politicians, etc. acting together usu. illicitly for the control of operations or profit. **9** a circular or spiral course. **10** = gas ring. **11** *Astron.* **a** a thin band or disc of particles etc. round a planet. **b** a halo round the moon. **12** *Archaeol.* a circular prehistoric earthwork, usu. consisting of a bank and ditch. **13** *Chem.* a group of atoms each bonded to two others in a closed sequence. **14** *Math.* a set of elements with two binary operations, addition and multiplication, the second being distributive over the first and associative. ● *v.tr.* **1** make or draw a circle round. **2** (often foll. by *round, about, in*) encircle or hem in (game or cattle). **3** put a ring on (a bird etc.) or through the nose of (a pig, bull, etc.). **4** cut (fruit, vegetables, etc.) into rings. □ **ring-binder** a loose-leaf binder with ring-shaped clasps that can be opened to pass through holes in the paper. **ring circuit** an electrical circuit serving a number of power points with one fuse in the supply to the circuit. **ring-dove 1** the woodpigeon. **2** the collared dove. **ringed plover** a small Eurasian plover, *Charadrius hiaticula*, with a black band across the breast. **ring-fence** *n.* **1** a fence completely enclosing an estate, piece of land, etc. **2** an effective or comprehensive barrier or means of segregation. ● *v.* **1** enclose (a piece of land etc.) with a ring-fence. **2** segregate, keep separate. **ring finger** the finger next to the little finger, esp. of the left hand, on which the wedding ring is usu. worn. **ring main 1** an electrical supply serving a series of consumers and returning to the original source, so that each consumer has an alternative path in the event of a failure. **2** = ring circuit. **ring-neck** a ring-necked bird; esp. the common pheasant, *Phasianus colchicus*, with a white neck-ring. **ring-necked** *Zool.* having a band or bands of colour round the neck. **ring ouzel** a thrush, *Turdus torquatus*, the male of which resembles a blackbird with a white crescent across its breast. **ring-pull** (of a tin) having a ring for pulling to break its seal. **ring road** a bypass encircling a town. **ring-tailed 1** (of monkeys, lemurs, raccoons, etc.) having a tail ringed in alternate colours. **2** with the tail curled at the end. **run** (or **make**) **rings round** *colloq.* outclass or outwit (another person). □ **ringed** *adj.* (also in *comb.*). **ringless** *adj.* [OE *hring* f. Gmc]

ring[2] /rɪŋ/ *v. & n.* ● *v.* (*past* **rang** /ræŋ/; *past part.* **rung** /rʌŋ/) **1** *intr.* (often foll. by *out* etc.) give a clear resonant or vibrating sound of or as of a bell (*a shot rang out; a ringing laugh; the telephone rang*). **2** *tr.* make (esp. a bell) ring. **b** (*absol.*) call for service or attention by ringing a bell (*you rang, madam?*). **3** *tr.* (also *absol.*; often foll. by *up*) *Brit.* telephone (*will ring you on Monday; did you ring?*). **4** *intr.* (usu. foll. by *with, to*) (of a place) resound or be permeated with a sound, or an attribute, e.g. fame (*the theatre rang with applause*). **5** *intr.* (of the ears) be filled with a sensation of ringing. **6** *tr.* **a** sound (a peal etc.) on bells. **b** (of a bell) sound (the hour etc.). **7** *tr.* (foll. by *in, out*) usher in or out with bell-ringing (*ring in the May; rang out the Old Year*). **8** *intr.* (of sentiments etc.) convey a specified impression (*words rang hollow*). ● *n.* **1** a ringing sound or tone. **2 a** the act of ringing a bell. **b** the sound caused by this. **3** *colloq.* a telephone call (*give me a ring*). **4** a specified feeling conveyed by an utterance (*had a melancholy ring*). **5** a set of esp. church bells. □ **ring back** make a return telephone call to (a person who has telephoned earlier). **ring a bell** see BELL[1]. **ring the changes (on)** see CHANGE. **ring down** (or **up**) *Theatr.* cause the stage curtain to be lowered or raised. **2** (foll. by *on*) mark the end or the beginning of (an enterprise etc.). **ring in 1** report or make contact by telephone. **2** *Austral. & NZ sl.* substitute fraudulently. **ring in one's ears** (or **heart** etc.) linger in the memory. **ringing tone** a sound heard by a telephone caller when the number dialled is being rung. **ring off** *Brit.* end a telephone call by replacing the receiver. **ring round** telephone several people. **ring true** (or **false**) convey an impression of truth or falsehood. **ring up 1** *Brit.* call by telephone. **2** record (an amount etc.) on a cash register. □ **ringed** *adj.* (also in *comb.*). **ringer** *n.* **ringing** *adj.* **ringingly** *adv.* [OE *hringan*]

ringbark /ˈrɪŋbɑːk/ *v.tr.* cut a ring in the bark of (a tree) to kill it or retard its growth and thereby improve fruit production.

ringbolt /ˈrɪŋbəʊlt/ *n.* a bolt with a ring attached for fitting a rope to etc.

ringer /ˈrɪŋə(r)/ *n. sl.* **1 a** esp. *US* an athlete or horse entered in a competition by fraudulent means, esp. as a substitute. **b** a person's double, esp. an impostor. **2** *Austral.* **a** the fastest shearer in a shed. **b** a stockman or station hand. **3** a person who rings, esp. a bell-ringer. □ **be a ringer** (or **dead ringer**) **for** resemble (a person) exactly. [RING² + -ER¹]

ringhals var. of RINKHALS.

ringleader /ˈrɪŋˌliːdə(r)/ *n.* a leading instigator in an illicit or illegal activity.

ringlet /ˈrɪŋlɪt/ *n.* **1** a curly lock of hair, esp. a long one. **2** a butterfly with brown wings marked with small rings; esp. the common *Aphantopus hyperantus*. □ **ringleted** *adj.* **ringlety** *adj.*

ringmaster /ˈrɪŋˌmɑːstə(r)/ *n.* the person directing a circus performance.

ringside /ˈrɪŋsaɪd/ *n.* (often *attrib.*) **1** the area immediately beside a boxing ring or circus ring etc. **2** an advantageous position from which to observe or monitor something. □ **ringsider** *n.*

ringster /ˈrɪŋstə(r)/ *n.* a person who participates in a political or commercial ring (see RING¹ *n.* 8).

ringtail /ˈrɪŋteɪl/ *n.* **1** a ring-tailed opossum, lemur, or phalanger. **2** a golden eagle up to its third year. **3** a female hen harrier.

ringworm /ˈrɪŋwɜːm/ *n. Med.* a fungal infection of the skin causing circular inflamed patches, esp. on the scalp.

rink /rɪŋk/ *n.* **1** an area of natural or artificial ice for skating or the game of curling etc. **2** an enclosed area for roller-skating. **3** a building containing either of these. **4** *Bowls* a strip of the green used for playing a match. **5** a team in bowls or curling. [ME (orig. Sc.), = jousting-ground: perh. orig. f. OF *renc* RANK¹]

rinkhals /ˈrɪŋkhæls/ *n.* (also **ringhals** /ˈrɪŋhæls/) a large venomous spitting cobra, *Hemachatus hemachatus*, of southern Africa, with one or two white rings across the neck. [Afrik. f. *ring* RING¹ + *hals* neck]

rinse /rɪns/ *v. & n.* • *v.tr.* (often foll. by *through*, *out*) **1** wash with clean water. **2** apply liquid to. **3** wash lightly. **4** put (clothes etc.) through clean water to remove soap or detergent. **5** (foll. by *out*, *away*) clear (impurities) by rinsing. **6** treat (hair) with a rinse. • *n.* **1** the act or an instance of rinsing (*give it a rinse*). **2** a solution for cleansing the mouth. **3** a dye for the temporary tinting of hair (*a blue rinse*). □ **rinser** *n.* [ME f. OF *rincer*, *raincier*, of unkn. orig.]

Rio Branco /ˌriːuː ˈbræŋkuː/ a city in western Brazil, capital of the state of Acre; pop. (1990) 197,000.

Rio de Janeiro /ˌriːəʊ də dʒəˈnɪərəʊ/ **1** a state of eastern Brazil, on the Atlantic coast. **2** (also **Rio**) its capital; pop. (1990) 5,487,350. The chief port of Brazil, it was the country's capital from 1763 until 1960, when it was replaced by Brasilia. Its skyline is dominated by two rocky peaks: Sugar Loaf Mountain to the east and Corcovado, with its statue of Christ, to the south.

Río de la Plata /ˌrriːo de la ˈplata/ the Spanish name for the River Plate (see PLATE, RIVER).

Río de Oro /ˌriːo deɪ ˈɔːrəʊ/ an arid region on the Atlantic coast of NW Africa, forming the southern part of Western Sahara. Occupied by the Spanish in 1884, it was united with Saguia el Hamra in 1958 to form the province of Spanish Sahara (now Western Sahara).

Rio Grande /ˌriːəʊ ˈgrænd, ˈgrændi/ a river of North America which rises in the Rocky Mountains of SW Colorado and flows 3,030 km (1,880 miles) generally south-eastwards to the Gulf of Mexico, forming the US–Mexico frontier from El Paso to the sea.

Rio Grande do Norte /ˌriːuː ˌgrændɪ duː ˈnɔːtɪ/ a state of NE Brazil, on the Atlantic coast; capital, Natal.

Rio Grande do Sul /ˌriːuː ˌgrændɪ duː ˈsʊl/ a state of Brazil, situated on the Atlantic coast at the southern tip of the country, on the border with Uruguay; capital, Pôrto Alegre.

Rio Muni /ˌriːəʊ ˈmuːnɪ/ the part of Equatorial Guinea that lies on the mainland of West Africa. Its chief town is Bata.

Rio Negro /ˌriːəʊ ˈneɪgrəʊ, ˈneg-/ a river of South America, which rises as the Guainia in eastern Colombia and flows for about 2,255

km (1,400 miles) through NW Brazil before joining the Amazon near Manaus.

riot /ˈraɪət/ *n. & v.* • *n.* **1 a** a disturbance of the peace by a crowd; an occurrence of public disorder. **b** (*attrib.*) involved in suppressing riots (*riot police; riot shield*). **2** uncontrolled revelry; noisy behaviour. **3** (foll. by *of*) a lavish display or enjoyment (*a riot of emotion; a riot of colour and sound*). **4** *colloq.* a very amusing thing or person. • *v.intr.* **1** make or engage in a riot. **2** live wantonly; revel. □ **run riot 1** throw off all restraint. **2** (of plants) grow or spread uncontrolled. □ **rioter** *n.* **riotless** *adj.* [ME f. OF *riote*, *rioter*, *rihoter*, of unkn. orig.]

Riot Act an Act passed by the British government in 1715 and repealed in 1967, designed to prevent civil disorder. It was introduced in the wake of the Jacobite rebellion of 1715, which followed George I's accession to the throne. The Act made it a felony for an assembly of more than twelve people to refuse to disperse after being ordered to do so and having been read a specified portion of the Act by lawful authority. □ **read the Riot Act** insist that noise, disobedience, etc., must cease.

riotous /ˈraɪətəs/ *adj.* **1** marked by or involving rioting. **2** characterized by wanton conduct. **3** wildly profuse. □ **riotously** *adv.* **riotousness** *n.* [ME f. OF (as RIOT)]

RIP *abbr.* may he or she or they rest in peace. [L *requiescat* (pl. *requiescant*) *in pace*]

rip¹ /rɪp/ *v. & n.* • *v.* (**ripped**, **ripping**) **1** *tr.* tear or cut (a thing) quickly or forcibly away or apart (*ripped out the lining; ripped the book up*). **2** *tr.* **a** make (a hole etc.) by ripping. **b** make a long tear or cut in. **3** *intr.* come violently apart; split. **4** *intr.* rush along. • *n.* **1** a long tear or cut. **2** an act of ripping. □ **let rip** *colloq.* **1** act or proceed without restraint. **2** speak violently. **3** not check the speed of or interfere with (a person or thing). **rip-cord** a cord for releasing a parachute from its pack. **rip into** attack (a person) verbally. **rip off** *colloq.* defraud, steal. **rip-off** *n. colloq.* **1** a fraud or swindle. **2** financial exploitation. [ME: orig. unkn.]

rip² /rɪp/ *n.* a stretch of rough water in the sea or in a river, caused by the meeting of currents. □ **rip current** (or **tide**) **1** a strong surface current from the shore. **2** a state of conflicting psychological forces. [18th c.: perh. rel. to RIP¹]

rip³ /rɪp/ *n.* **1** a dissolute person. **2** a rascal. **3** a worthless horse. [perh. f. *rep*, abbr. of REPROBATE]

riparian /raɪˈpeərɪən/ *adj. & n.* esp. *Law* • *adj.* of or on a river-bank (*riparian rights*). • *n.* an owner of property on a river-bank. [L *riparius* f. *ripa* bank]

ripe /raɪp/ *adj.* **1** (of grain, fruit, cheese, etc.) ready to be harvested or picked or eaten. **2** mature; fully developed (*ripe in judgement; a ripe beauty*). **3** (of a person's age) advanced. **4** (often foll. by *for*) fit or ready (*when the time is ripe; land ripe for development*). **5** (of the complexion etc.) red and full like ripe fruit. □ **ripely** *adv.* **ripeness** *n.* [OE *rīpe* f. WG]

ripen /ˈraɪp(ə)n/ *v.tr. & intr.* make or become ripe.

ripieno /rɪˈpjeɪnəʊ/ *n.* (pl. **-os** or **ripieni** /-nɪ/) *Mus.* a body of accompanying instruments in baroque concerto music. [It. (as RE-, *pieno* full)]

riposte /rɪˈpɒst/ *n. & v.* • *n.* **1** a quick sharp reply or retort. **2** a quick return thrust in fencing. • *v.intr.* deliver a riposte. [F *ri(s)poste*, *ri(s)poster* f. It. *risposta* RESPONSE]

ripper /ˈrɪpə(r)/ *n.* **1** a person or thing that rips. **2** a murderer who rips or mutilates the victims' bodies.

ripping /ˈrɪpɪŋ/ *adj. Brit. archaic sl.* splendid, excellent (*a ripping good yarn*). □ **rippingly** *adv.*

ripple¹ /ˈrɪp(ə)l/ *n. & v.* • *n.* **1** a ruffling of the water's surface, a small wave or series of waves. **2 a** a gentle lively sound that rises and falls, e.g. of laughter or applause. **b** a brief wave of emotion, excitement, etc. (*the new recruit caused a ripple of interest in the company*). **3** a wavy appearance in hair, material, etc. **4** *Electr.* a slight variation in the strength of a current etc. **5** ice-cream with added syrup giving a coloured ripple effect (*raspberry ripple*). **6** *US* a riffle in a stream. • *v.* **1** *intr.* form ripples; flow in ripples. **b** *tr.* cause to do this. **2** *intr.* show or sound like ripples. □ **ripple mark** a ridge or ridged surface left on sand, mud, or rock by the action of water or wind. □ **ripply** *adj.* **ripplet** /ˈrɪplɪt/ *n.* [17th c.: orig. unkn.]

ripple² /ˈrɪp(ə)l/ *n. & v.* • *n.* a toothed implement used to remove seeds from flax. • *v.tr.* treat with a ripple. [corresp. to MDu. & MLG *repel(en)*, OHG *riffila*, *rifilōn*]

riprap /ˈrɪpræp/ *n. US* a collection of loose stone as a foundation for a structure. [redupl. of RAP¹]

rip-roaring /ˈrɪpˌrɔːrɪŋ/ *adj.* **1** wildly noisy or boisterous. **2** excellent, first-rate. □ **rip-roaringly** *adv.*

ripsaw /ˈrɪpsɔː/ *n.* a coarse saw for sawing wood along the grain.

ripsnorter /ˈrɪpˌsnɔːtə(r)/ *n. colloq.* an energetic, remarkable, or excellent person or thing. □ **ripsnorting** *adj.* **ripsnortingly** *adv.*

Rip Van Winkle /ˌrɪp væn ˈwɪŋk(ə)l/ the hero of a story in Washington Irving's *Sketch Book* (1819–20), who fell asleep in the Catskill Mountains in New York State, and awoke after 20 years to find the world completely changed.

RISC /rɪsk/ *n. Computing* reduced instruction set computer, using processors designed to carry out a limited number of simple instructions very quickly. [acronym]

rise /raɪz/ *v. & n.* ● *v.intr.* (*past* **rose** /rəʊz/; *past part.* **risen** /ˈrɪz(ə)n/) **1** move from a lower position to a higher one; come or go up. **2** grow, project, expand, or incline upwards; become higher. **3** (of the sun, moon, or stars) appear above the horizon. **4 a** get up from lying or sitting or kneeling (*rose to their feet; rose from the table*). **b** get out of bed, esp. in the morning (*do you rise early?*). **5** recover a standing or vertical position; become erect (*rose to my full height*). **6** (of a meeting etc.) cease to sit for business; adjourn (*Parliament rises next week; the court will rise*). **7** reach a higher position or level or amount (*the flood has risen; prices are rising*). **8** develop greater intensity, strength, volume, or pitch (*the colour rose in her cheeks; the wind is rising; their voices rose with excitement*). **9** make progress; reach a higher social position (*rose from the ranks*). **10 a** come to the surface of liquid (*bubbles rose from the bottom; waited for the fish to rise*). **b** (of a person) react to provocation (*rise to the bait*). **11** become or be visible above the surroundings etc., stand prominently (*mountains rose to our right*). **12 a** (of buildings etc.) undergo construction from the foundations (*office blocks were rising all around*). **b** (of a tree etc.) grow to a usu. specified height. **13** come to life again (*rise from the ashes; risen from the dead*). **14** (of dough) swell by the action of yeast etc. **15** (often foll. by *against*, *up*) cease to be quiet or submissive; rebel (*rose against their oppressors*). **16** originate; have as its source (*the river rises in the mountains*). **17** (of wind) start to blow. **18** (of a person's spirits) become cheerful. **19** (of a barometer) show a higher atmospheric pressure. **20** (of a horse) rear (*rose on its hind legs*). **21** (of a bump, blister, etc.) form. **22** (of the stomach) show nausea. ● *n.* **1** an act or manner or amount of rising. **2** an upward slope or hill or movement (*a rise in the road; the house stood on a rise; the rise and fall of the waves*). **3** an increase in sound or pitch. **4 a** an increase in amount, extent, etc. (*a rise in unemployment*). **b** *Brit.* an increase in salary, wages, etc. **5** an increase in status or power. **6** social, commercial, or political advancement; upward progress. **7** the movement of a fish to the surface of the water to take a fly or bait. **8** origin. **9 a** the vertical height of a step, arch, incline, etc. **b** = RISER 2. □ **get** (or **take**) **a rise out of** *colloq.* provoke an emotional reaction from (a person), esp. by teasing. **on the rise** on the increase. **rise above 1** be superior to (petty feelings etc.). **2** show dignity or strength in the face of (difficulty, poor conditions, etc.). **rise and shine** (usu. as *imper.*) *colloq.* get out of bed smartly; wake up. **rise in the world** attain a higher social position. **rise to** develop powers equal to (an occasion). **rise with the sun** (or **lark**) get up early in the morning. [OE *rīsan* f. Gmc]

riser /ˈraɪzə(r)/ *n.* **1** a person who rises esp. from bed (*an early riser*). **2** a vertical section between the treads of a staircase. **3** a vertical pipe for the flow of liquid or gas. **4** a low platform on a stage, in a studio, etc.

rishi /ˈrɪʃiː/ *n.* (*pl.* **rishis**) a Hindu sage or saint. [Skr. *ṛiṣi*]

risible /ˈrɪzɪb(ə)l/ *adj.* **1** laughable, ludicrous. **2** inclined to laugh. **3** *Anat.* relating to laughter (*risible nerves*). □ **risibly** *adv.* **risibility** /ˌrɪzɪˈbɪlɪtɪ/ *n.* [LL *risibilis* f. L *ridere ris-* laugh]

rising /ˈraɪzɪŋ/ *adj. & n.* ● *adj.* **1** going up; getting higher. **2** increasing (*rising costs*). **3** advancing to maturity or high standing (*the rising generation; a rising young lawyer*). **4** approaching a specified age (*the rising fives*). **5** (of ground) sloping upwards. ● *n.* a revolt or insurrection. □ **rising damp** moisture absorbed from the ground into a wall.

risk /rɪsk/ *n. & v.* ● *n.* **1** a chance or possibility of danger, loss, injury, or other adverse consequences (*a health risk; a risk of fire*). **2** a person or thing causing a risk or regarded in relation to risk (*is a poor risk*). ● *v.tr.* **1** expose to risk. **2** accept the chance of (*could not risk getting wet*). **3** venture on. □ **at risk** exposed to danger. **at one's** (**own**) **risk** accepting responsibility, agreeing to make no claims. **at the risk of** with the possibility of (an adverse consequence). **put at risk** expose to danger. **risk capital** money put up for speculative business investment. **risk one's neck** put one's own life in danger. **run a** (or **the**) **risk** (often foll. by *of*) expose oneself to danger or loss etc. **take**

(or **run**) **a risk** chance the possibility of danger etc. [F *risque, risquer* f. It. *risco* danger, *riscare* run into danger]

risky /ˈrɪskɪ/ *adj.* (**riskier, riskiest**) **1** involving risk. **2** = RISQUÉ. □ **riskily** *adv.* **riskiness** *n.*

Risorgimento /rɪˌsɔːdʒɪˈmentəʊ/ the movement for the unification and independence of Italy in the 19th century. Much of Italy had experienced liberal reforms and an end to feudal and ecclesiastical privilege during the Napoleonic Wars, and the restoration of repressive regimes led to revolts in Naples and Piedmont (1821) and Bologna (1831). Italian nationalists began to support Guiseppe Mazzini and the Young Italy movement, and were given impetus in their struggle by Camillo Cavour's newspaper *Il Risorgimento* from 1847. Attempts to drive the Austrians out of Lombardy and Venetia in 1848 were unsuccessful, and uprisings during the Revolutions of 1848 also failed; with French aid, however, the Austrians were driven out of northern Italy by 1859, and the south was won over by Garibaldi. Voting was held and resulted in the acceptance of Victor Emmanuel II as the first king of a united Italy in 1861. [It., = resurrection]

risotto /rɪˈzɒtəʊ/ *n.* (*pl.* **-os**) an Italian dish of rice cooked in stock with meat, onions, etc. [It.]

risqué /ˈrɪskeɪ, rɪˈskeɪ/ *adj.* (of a story etc.) slightly indecent or liable to shock. [F, past part. of *risquer* RISK]

rissole /ˈrɪsəʊl/ *n.* a compressed mixture of meat and spices, coated in breadcrumbs and fried. [F f. OF *ruissole, roussole* ult. f. LL *russeolus* reddish f. L *russus* red]

rit. *abbr. Mus.* ritardando.

ritardando /ˌrɪtɑːˈdændəʊ/ *adv. & n.* (*pl.* **-os** or **ritardandi** /-dɪ/) *Mus.* = RALLENTANDO. [It.]

rite /raɪt/ *n.* **1** a religious or solemn observance or act (*burial rites*). **2** an action or procedure required or usual in this. **3** a body of customary observances characteristic of a Church or a part of it (*the Latin rite*). □ **rite of passage** (often in *pl.*) a ritual or event marking a stage of a person's advance through life, e.g. marriage. □ **riteless** *adj.* [ME f. OF *rit, rite* or L *ritus* (esp. religious) usage]

ritenuto /ˌrɪtəˈnuːtəʊ/ *adv. & n. Mus.* ● *adv.* with immediate reduction of speed. ● *n.* (*pl.* **-os** or **ritenuti** /-tɪ/) a passage played in this way. [It.]

ritornello /ˌrɪtɔːˈneləʊ/ *n.* (*pl.* **-os** or **ritornelli** /-lɪ/) *Mus.* a short instrumental refrain, interlude, etc., in a vocal work. [It., dimin. of *ritorno* RETURN]

ritual /ˈrɪtjʊəl/ *n. & adj.* ● *n.* **1** a prescribed order of performing rites. **2** a procedure regularly followed. ● *adj.* of or done as a ritual or rites (*ritual murder*). □ **ritually** *adv.* **ritualize** *v.tr. & intr.* (also **-ise**). **ritualization** /ˌrɪtjʊəlaɪˈzeɪʃ(ə)n/ *n.* (also **-isation**). [L *ritualis* (as RITE)]

ritualism /ˈrɪtjʊəˌlɪz(ə)m/ *n.* the regular or excessive practice of ritual. □ **ritualist** *n.* **ritualistic** /ˌrɪtjʊəˈlɪstɪk/ *adj.* **ritualistically** *adv.*

ritzy /ˈrɪtsɪ/ *adj.* (**ritzier, ritziest**) *colloq.* **1** high-class, luxurious. **2** ostentatiously smart. □ **ritzily** *adv.* **ritziness** *n.* [*Ritz*, the name of luxury hotels f. César Ritz, Swiss hotel-owner (1850–1918)]

rival /ˈraɪv(ə)l/ *n. & v.* ● *n.* **1** a person competing with another for the same objective. **2** a person or thing that equals another in quality. **3** (*attrib.*) being a rival or rivals (*a rival firm*). ● *v.tr.* (**rivalled, rivalling**; *US* **rivaled, rivaling**) **1** be the rival of or comparable to. **2** seem or claim to be as good as. [L *rivalis,* orig. = using the same stream, f. *rivus* stream]

rivalry /ˈraɪv(ə)lrɪ/ *n.* (*pl.* **-ies**) the state or an instance of being rivals; competition.

rive /raɪv/ *v.* (*past* **rived**; *past part.* **riven** /ˈrɪv(ə)n/) *archaic* or *poet.* **1** *tr.* split or tear apart violently. **2 a** *tr.* split (wood or stone). **b** *intr.* be split. [ME f. ON *rífa*]

river /ˈrɪvə(r)/ *n.* **1** a copious natural stream of water flowing in a channel to the sea or a lake etc. **2** a copious flow (*a river of lava; rivers of blood*). **3** (*attrib.*) (in the names of animals, plants, etc.) living in or associated with rivers. □ **river blindness** a tropical disease of the skin transmitted by the bite of blackflies that breed in rivers and caused by a parasitic nematode worm, the larvae of which can migrate into the eye and cause blindness. **river capture** the diversion of the upper headwaters of a mountain stream into a more powerful one. **sell down the river** *colloq.* betray or let down. □ **rivered** *adj.* (also in *comb.*). **riverless** *adj.* [ME f. AF *river, rivere,* OF *riviere* river or river-bank ult. f. L *riparius* f. *ripa* bank]

Rivera /rɪˈveərə/, Diego (1886–1957), Mexican painter. His monumental frescos of the 1920s and 1930s gave rise to a revival of fresco painting in Latin America and the US. Rivera's largest and most ambitious mural was a history of Mexico for the National Palace in

Mexico City; begun in 1929 and unfinished at his death, it explicitly sought to construct a sense of nationalist and socialist identity.

riverine /ˈrɪvəˌraɪn/ adj. of or on a river or river-bank; riparian.

Riverside /ˈrɪvəˌsaɪd/ a city in southern California, situated in the centre of an orange-growing region; pop. (1990) 226,505.

riverside /ˈrɪvəˌsaɪd/ n. the ground along a river-bank.

rivet /ˈrɪvɪt/ n. & v. ● n. a nail or bolt for holding together metal plates etc., its headless end being beaten out or pressed down when in place. ● v.tr. (**riveted, riveting**) **1 a** join or fasten with rivets. **b** beat out or press down the end of (a nail or bolt). **c** fix; make immovable. **2 a** (foll. by on, upon) direct intently (one's eyes or attention etc.). **b** (esp. as **riveting** adj.) engross (a person or the attention). □ **riveter** n. [ME f. OF f. river clench, of unkn. orig.]

Riviera, the /ˌrɪviˈeərə/ part of the Mediterranean coastal region of southern France and northern Italy, extending from Cannes to La Spezia, famous for its scenic beauty, fertility, and mild climate, and with many fashionable and expensive resorts. [It., = sea-shore]

rivière /riːˈvjeə(r), ˈrɪviˌeə(r)/ n. a gem necklace, esp. of more than one string. [F. = RIVER]

Rivne /ˈrɪvnə/ (Russian **Rovno** /ˈrɒvnə/) an industrial city in western Ukraine north-east of Lviv; pop. (1990) 232,900.

rivulet /ˈrɪvjʊlɪt/ n. a small stream. [obs. riveret f. F, dimin. of rivière RIVER, perh. after It. rivoletto dimin. of rivolo dimin. of rivo f. L rivus stream]

Riyadh /riːˈɑːd/ the capital of Saudi Arabia; pop. (est. 1988) 2,000,000. It is situated on a high plateau in the centre of the country.

riyal var. of RIAL.

RKO a US film production and distribution company founded in 1928, whose trademark was a pylon on a globe. After producing classic films such as King Kong (1933) and Citizen Kane (1941) the company was bought by Howard Hughes in 1948, but ceased film production in 1953. [abbr. Radio-Keith-Orpheum, f. a merging of Radio Corporation of America (RCA) with the Keith and Orpheum cinema chains]

RL abbr. Rugby League.

rly. abbr. railway.

RM abbr. **1** (in the UK) Royal Marines. **2** Resident Magistrate. **3** (in the UK) Royal Mail.

rm. abbr. room.

RMA abbr. Royal Military Academy.

r.m.s. abbr. Math. root-mean-square.

RMT abbr. (in the UK) the National Union of Rail, Maritime, and Transport Workers. It was formed in 1990 by a merger of the NUR and National Union of Seamen.

RN abbr. **1** (in the UK) Royal Navy. **2** (in the UK) Registered Nurse.

Rn symb. Chem. the element radon.

RNA /ˌɑːrenˈeɪ/ abbr. Biol. ribonucleic acid, a nucleic acid yielding ribose on hydrolysis. It is present in all cells, its principal function being to act as a messenger carrying instructions from DNA for controlling the synthesis of proteins (see DNA, mRNA, tRNA). RNA normally occurs as a single strand, with the base uracil replacing the thymine of DNA. In some viruses RNA carries the genetic information.

RNAS abbr. (in the UK) Royal Naval Air Service (or Station).

RNLI see ROYAL NATIONAL LIFEBOAT INSTITUTION.

RNZAF abbr. Royal New Zealand Air Force.

RNZN abbr. Royal New Zealand Navy.

roach[1] /rəʊtʃ/ n. (pl. same) **1** a small Eurasian freshwater fish, Rutilus rutilus, of the carp family. **2** any similar freshwater fish. [ME f. OF roc(h)e, of unkn. orig.]

roach[2] /rəʊtʃ/ n. **1** N. Amer. colloq. a cockroach. **2** sl. the butt of a marijuana cigarette. [abbr.]

roach[3] /rəʊtʃ/ n. Naut. **1 a** a curved part of a fore-and-aft sail extending beyond a straight line between its corners, esp. on the leech side. **b** the breadth of this. **2** an upward curve in the foot of a square sail. [18th c.: orig. unkn.]

road[1] /rəʊd/ n. **1 a** a path or way with a specially prepared surface, used by vehicles, pedestrians, etc. **b** the part of this used by vehicles (don't step in the road). **2 a** one's way or route (our road took us through unexplored territory). **b** a method or means of accomplishing something. **3** an underground passage in a mine. **4** US a railway. **5** (usu. in pl.) a partly sheltered piece of water near the shore in which ships can ride at anchor. □ **any road** dial. = ANYWAY. **by road** using transport along roads. **get out of the** (or **my** etc.) **road** colloq. cease to obstruct a

person. **in the** (or **my** etc.) **road** colloq. obstructing a person or thing. **one for the road** colloq. a final (esp. alcoholic) drink before departure. **on the road** travelling, esp. as a firm's representative, itinerant performer, or vagrant. **road fund** Brit. hist. a fund for the construction and maintenance of roads and bridges. **road fund licence** Brit. a disc displayed on a vehicle certifying payment of road tax. **road-hog** colloq. a reckless or inconsiderate road-user, esp. a motorist. **road-holding** the capacity of a moving vehicle to remain stable when cornering at high speeds etc. **road-house** an inn or club on a major road. **road hump** = sleeping policeman (see SLEEP). **road kill** N. Amer. colloq. an animal killed on a road by a motor vehicle. **road-manager** the organizer and supervisor of a musicians' tour. **road-map** a map showing the roads of a country or area. **road-metal** broken stone used in road-making or for railway ballast. **road sense** a person's capacity for safe behaviour on the road, esp. in traffic. **road show 1 a** a performance given by a touring company, esp. a group of pop musicians. **b** a company giving such performances. **2** a radio or television programme done on location, esp. a series of programmes each from a different venue. **3** a touring political or advertising campaign. **road sign** a sign giving information or instructions to road users. **road tax** a periodic tax payable on road vehicles. **road test** a test of the performance of a vehicle on the road. **road-test** v.tr. test (a vehicle) on the road. **the road to** the way of getting to or achieving (the road to London; the road to ruin). **road train** a large lorry pulling one or more trailers. **rule of the road** the custom or law regulating which side of the road is to be taken by vehicles (also riders or ships) meeting or passing each other. **take the road** set out. □ **roadless** adj. [OE rād f. rīdan RIDE]

road[2] /rəʊd/ v.tr. (also absol.) (of a dog) follow up (a game bird) by the scent of its trail. [19th c.: orig. unkn.]

roadbed /ˈrəʊdbed/ n. **1** the foundation structure of a railway. **2** the material laid down to form a road. **3** US the part of a road on which vehicles travel.

roadblock /ˈrəʊdblɒk/ n. a barrier or barricade on a road, esp. one set up by the authorities to stop and examine traffic.

roadie /ˈrəʊdɪ/ n. colloq. an assistant employed by a touring band of musicians to erect and maintain equipment.

roadman /ˈrəʊdmən/ n. (pl. -men) a man employed to repair or maintain roads.

roadroller /ˈrəʊdˌrəʊlə(r)/ n. a motor vehicle with a heavy roller, used in road-making.

roadrunner /ˈrəʊdˌrʌnə(r)/ n. a fast-running American bird of the genus Geococcyx, of the cuckoo family; esp. Geococcyx californianus, of arid parts of Mexico and the southern US.

roadside /ˈrəʊdsaɪd/ n. the strip of land beside a road.

roadstead /ˈrəʊdsted/ n. = ROAD[1] 5. [ROAD[1] + stead in obs. sense 'place']

roadster /ˈrəʊdstə(r)/ n. **1** an open two-seater motor car. **2** a horse or bicycle for use on the road.

Road Town the capital of the British Virgin Islands, situated on the island of Tortola; pop. (1991) 6,330.

roadway /ˈrəʊdweɪ/ n. **1** a road. **2** = ROAD[1] 1b. **3** the part of a bridge or railway used for traffic.

roadwork /ˈrəʊdwɜːk/ n. **1** (in pl.) the construction or repair of roads, or other work involving digging up a road surface. **2** athletic exercise or training involving running on roads.

roadworthy /ˈrəʊdˌwɜːðɪ/ adj. fit to be used on the road. □ **roadworthiness** n.

roam /rəʊm/ v. & n. ● v. **1** intr. ramble, wander. **2** tr. travel unsystematically over, through, or about. ● n. an act of roaming; a ramble. □ **roamer** n. [ME: orig. unkn.]

roan[1] /rəʊn/ adj. & n. ● adj. (of an animal, esp. a horse or cow) having a coat of which the prevailing colour is thickly interspersed with hairs of another colour, esp. bay or sorrel or chestnut mixed with white or grey. ● n. a roan animal. **blue roan** adj. black mixed with white. ● n. a blue roan animal. **red roan** adj. bay mixed with white or grey. ● n. a red roan animal. **strawberry roan** adj. chestnut mixed with white or grey. ● n. a strawberry roan animal. [OF, of unkn. orig.]

roan[2] /rəʊn/ n. soft sheepskin leather used in bookbinding as a substitute for morocco. [ME, perh. f. Roan, old name of ROUEN]

roar /rɔː(r)/ n. & v. ● n. **1** a loud deep hoarse sound, as made by a lion, a person in pain or rage or excitement, thunder, a loud engine, etc. **2** a loud laugh. ● v. **1** intr. **a** utter or make a roar. **b** utter loud laughter. **c** (of a horse) make a loud noise in breathing as a symptom of disease. **2** intr. travel in a vehicle at high speed, esp. with the engine roaring.

3 *tr.* (often foll. by *out*) say, sing, or utter (words, an oath, etc.) in a loud tone. □ **roarer** *n.* [OE *rārian*, of imit. orig.]

roaring /ˈrɔːrɪŋ/ *adj.* in senses of ROAR *v.* □ **roaring drunk** very drunk and noisy. **roaring success** a great success. **roaring trade** (or **business**) very brisk trade or business. **roaring twenties** the decade of the 1920s (with ref. to its postwar buoyancy). □ **roaringly** *adv.*

Roaring Forties *n.pl.* stormy ocean tracts between lat. 40° and 50° S (cf. FORTIES 1).

roast /rəʊst/ *v., adj., & n.* ● *v.* **1** *tr.* **a** cook (food, esp. meat) in an oven or by exposure to open heat. **b** heat (coffee beans) before grinding. **2** *tr.* heat (the ore of metal) in a furnace. **3** *tr.* **a** expose (a torture victim) to fire or great heat. **b** *tr. & refl.* expose (oneself or part of oneself) to warmth. **4** *tr.* criticize severely, denounce. **5** *intr.* undergo roasting. ● *attrib.adj.* (of meat or a potato, chestnut, etc.) roasted. ● *n.* **1 a** roast meat. **b** a dish of this. **c** a piece of meat for roasting. **2** the process of roasting. **3** *esp. US* a party where roasted food is eaten (*invited to the pig roast*). [ME f. OF *rost, rostir*, f. Gmc]

roaster /ˈrəʊstə(r)/ *n.* **1** a person or thing that roasts. **2 a** an oven or dish for roasting food in. **b** an ore-roasting furnace. **c** a coffee-roasting apparatus. **3** something fit for roasting, e.g. a fowl, a potato, etc.

roasting /ˈrəʊstɪŋ/ *adj. & n.* ● *adj.* very hot and dry. ● *n.* **1** in senses of ROAST *v.* **2** a severe criticism or denunciation.

rob /rɒb/ *v.tr.* (**robbed, robbing**) (often foll. by *of*) **1** take unlawfully from, esp. by force or threat of force (*robbed the safe; robbed her of her jewels*). **2** deprive of what is due or normal (*was robbed of my sleep*). **3** (*absol.*) commit robbery. **4** *colloq.* cheat, swindle. □ **rob Peter to pay Paul** take away from one to give to another, discharge one debt by incurring another. [ME f. OF *rob(b)er* f. Gmc: cf. REAVE]

Robbe-Grillet /rɒbˈɡriːeɪ/, Alain (b.1922), French novelist. He established himself as a leading exponent of the *nouveau roman* in the 1950s; his first novel *The Erasers* (1953) was an early example of the form. Among his later fictional works are *The Voyeur* (1955) and *Jealousy* (1957). His theories on fiction appeared in his collection of essays *Towards a New Novel* (1963). He also wrote screenplays in which he explored the visual potential of his fictional techniques, most notably that for *L'Année dernière à Marienbad* (1961).

Robben Island /ˈrɒb(ə)n/ a small island off the coast of South Africa, near Cape Town. It is the site of a prison which was formerly used for the detention of political prisoners, including Nelson Mandela.

robber /ˈrɒbə(r)/ *n.* a person who commits robbery. □ **robber baron 1** *hist.* a plundering feudal lord. **2** an unscrupulous plutocrat. [ME f. AF & OF (as ROB)]

robbery /ˈrɒbərɪ/ *n.* (*pl.* **-ies**) **1 a** the act or process of robbing, esp. with force or threat of force. **b** an instance of this. **2** excessive financial demand or cost (*set us back £20 — it was sheer robbery*). [ME f. OF *roberie* (as ROB)]

Robbia see DELLA ROBBIA.

Robbins /ˈrɒbɪnz/, Jerome (b.1918), American ballet-dancer and choreographer. He choreographed a long series of successful musicals including *The King and I* (1951), *West Side Story* (1957), and the *Fiddler on the Roof* (1964). Although chiefly inspired by jazz and modern dance, he has also created a number of ballets with music by classical composers.

robe /rəʊb/ *n. & v.* ● *n.* **1** a long loose outer garment. **2** *esp. N. Amer.* a dressing-gown. **3** a baby's outer garment esp. at a christening. **4** (often in *pl.*) a long outer garment worn as an indication of the wearer's rank, office, profession, etc.; a gown or vestment. **5** *N. Amer.* a blanket or wrap of fur. ● *v.* **1** *tr.* clothe (a person) in a robe; dress. **2** *intr.* put on one's robes or vestments. [ME f. OF f. Gmc (as ROB, orig. sense 'booty')]

Robert /ˈrɒbət/ the name of three kings of Scotland:

Robert I (known as Robert the Bruce) (1274–1329), reigned 1306–29. He led the Scottish campaign against Edward I after the death of Sir William Wallace. His subsequent campaign against Edward II culminated in victory at Bannockburn (1314). He then went on to re-establish Scotland as a separate kingdom, negotiating the Treaty of Northampton (1328), which committed the Plantagenets to recognizing his title as king of Scotland and relinquishing their claims to overlordship.

Robert II (1316–90), grandson of Robert the Bruce, reigned 1371–90. He was steward of Scotland from 1326 to 1371, and the first of the Stuart line.

Robert III (born John) (*c.*1337–1406), son of Robert II, reigned 1390–1406. Before ascending the throne, he was involved in an accident in which the kick of a horse made him physically disabled. As a result, his reign was marked by a power struggle amongst members of his family and Scotland was chiefly ruled by his brother Robert, Duke of Albany (*c.*1340–1420).

Roberts /ˈrɒbəts/, Frederick Sleigh, 1st Earl Roberts of Kandahar (1832–1914), British Field Marshal. He won a Victoria Cross in 1858 for his part in suppressing the Indian Mutiny and commanded the British army in its victory at Kandahar (1880), which ended the Second Afghan War (1878–80). As Commander-in-Chief (1899–90) during the Second Boer War, he planned the successful march on the Boer capital of Pretoria (1900).

Robert the Bruce, Robert I of Scotland (see ROBERT).

Robeson /ˈrəʊbs(ə)n/, Paul (Bustill) (1898–1976), American singer and actor. His singing of 'Ol' Man River' in Jerome Kern's musical *Show Boat* (1927) established his international reputation. Noted for his rich and resonant bass voice, he gave many recitals of spirituals. As an actor, Robeson was particularly identified with the title role of *Othello*, which he performed to great acclaim in London (1930) and on Broadway (1943). He was a prominent black activist, and had his passport revoked in 1950 because of his Communist affiliations.

Robespierre /ˈrəʊbzpjeə(r)/, Maximilien François Marie Isidore de (1758–94), French revolutionary. Robespierre was the leader of the radical Jacobins in the National Assembly and, as such, backed the execution of Louis XVI and implemented a successful purge of the moderate Girondists (both 1793). Later the same year he consolidated his power with his election to the Committee of Public Safety and his appointment as president of the National Assembly. Robespierre was guillotined for his role in the Terror, although he objected to the scale of the executions.

Robey /ˈrəʊbɪ/, Sir George (born George Edward Wade) (1869–1954), British comedian and actor. From the 1890s, he performed in music-halls and was billed as the 'Prime Minister of Mirth'. He later appeared in films such as Laurence Olivier's *Henry V* (1944).

robin /ˈrɒbɪn/ *n.* **1** (also **robin redbreast**) a small brown European bird, *Erithacus rubecula*, of the thrush family, with a red throat and breast. **2** *N. Amer.* a red-breasted thrush, *Turdus migratorius*. **3** any bird similar in appearance to the European robin. [ME f. OF, familiar var. of the name *Robert*]

Robin Goodfellow /ˈɡʊdˌfeləʊ/ (also called *Puck*) a mischievous sprite or goblin believed esp. in the 16th and 17th centuries to haunt the English countryside.

Robin Hood a semi-legendary English medieval outlaw, reputed to have robbed the rich and helped the poor. Although generally associated with Sherwood Forest in Nottinghamshire, it seems likely that the real Robin Hood operated further north, in Yorkshire, most probably in the early decades of the 13th century.

robinia /rəˈbɪnɪə/ *n.* a North American leguminous tree or shrub of the genus *Robinia*, e.g. a locust tree or false acacia. [mod.L. f. Jean and Vespasien *Robin*, 17th-c. French royal gardeners]

Robinson[1] /ˈrɒbɪns(ə)n/, Edward G. (born Emanuel Goldenberg) (1893–1972), Romanian-born American actor. After playing the part of Rico Bandello in the gangster film *Little Caesar* (1930), he went on to appear in a string of similar films in the 1930s. He later played a wider range of screen roles, such as the father in Arthur Miller's *All My Sons* (1948). Robinson was also a noted art collector.

Robinson[2] /ˈrɒbɪns(ə)n/, Mary (Terese Winifred) (b.1944), Irish Labour stateswoman, President since 1990. She was called to the bar in 1967 and entered politics in 1969, when she became a member of the Irish Senate. In 1990 she became Ireland's first woman President. She is noted for her platform of religious toleration and for her liberal attitude to abortion, divorce, and homosexuality.

Robinson[3] /ˈrɒbɪns(ə)n/, Sugar Ray (born Walker Smith) (1920–89), American boxer. He was world welterweight champion (1946–51) and seven times middleweight champion (1951, twice; 1955; 1957; 1958–60).

Robinson Crusoe /ˈkruːsəʊ/ the hero of Daniel Defoe's novel *Robinson Crusoe* (1719), who survives a shipwreck and lives on a desert island. His adventures were based on those of Alexander Selkirk, who lived alone on one of the uninhabited Juan Fernandez Islands in the Pacific for five years (1704–9).

roborant /ˈrəʊbərənt, ˈrɒb-/ *adj. & n. Med.* ● *adj.* strengthening. ● *n.* a strengthening drug. [L *roborare* f. *robur -oris* strength]

robot /ˈrəʊbɒt/ *n.* **1** a machine with a human appearance or functioning like a human. **2** a machine capable of carrying out a complex series of actions automatically. (See also ROBOTICS.) **3** a person who works mechanically and efficiently but insensitively. **4** *S. Afr.* an automatic traffic-signal. □ **robotize** v.tr. (also **-ise**). **robotic** /rəʊˈbɒtɪk/ adj. [Czech (in Karel Čapek's play *R.U.R.* (*Rossum's Universal Robots*) 1920), f. *robota* forced labour]

robotics /rəʊˈbɒtɪks/ n.pl. (usu. treated as *sing.*) the study of robots; the art or science of their design, construction, operation, and application. The term was coined by the science-fiction writer Isaac Asimov in 1941. Robots today consist of fixed or mobile mechanical manipulators linked to a computer, and follow a fixed but reprogrammable sequence of instructions. Simple robots are widely used in production engineering, especially where a variety of goods is to be produced with minimum changeover time. The performance of robots is being continually improved, especially by the provision of sensors and of automatic planning and recovery from errors, and they can handle increasingly complex tasks.

Rob Roy /rɒb ˈrɔɪ/ (born Robert Macgregor) (1671–1734), Scottish outlaw. His escapades as a highland cattle thief and opponent of the government's agents on the eve of the Jacobite uprising of 1715 were popularized in Sir Walter Scott's novel of the same name (1817).

Robsart /ˈrɒbsɑːt/, Amy (1532–60), English noblewoman, wife of Robert Dudley, Earl of Leicester. Her mysterious death at a country house near Oxford aroused suspicions that her husband (the favourite of Queen Elizabeth I) had had her killed so that he could be free to marry the queen. Sir Walter Scott's novel *Kenilworth* (1821) follows this version of her fate.

Robson /ˈrɒbs(ə)n/, Dame Flora (1902–84), English actress. She is noted for her screen performances of historical parts such as the Empress Elizabeth in *Catherine the Great* (1934) and Queen Elizabeth I in *Fire Over England* (1937). Her many acclaimed stage roles include Mrs Alving in Ibsen's *Ghosts* (1959).

robust /rəʊˈbʌst/ adj. (**robuster**, **robustest**) **1** (of a person, animal, or thing) strong and sturdy, esp. in physique or construction. **2** (of exercise, discipline, etc.) vigorous, requiring strength. **3** (of intellect or mental attitude) straightforward, not given to nor confused by subtleties. **4** (of a statement, reply, etc.) bold, firm, unyielding. **5** (of wine etc.) full-bodied. □ **robustly** adv. **robustness** n. [F *robuste* or L *robustus* firm and hard f. *robus*, *robur* oak, strength]

robusta /rəʊˈbʌstə/ n. (also **robusta coffee**) **1** a coffee plant of a widely grown African species, *Coffea canephora* (formerly *robusta*). **2** beans or coffee from this (cf. ARABICA). [mod.L f. L *robustus* robust]

ROC abbr. (in the UK) Royal Observer Corps.

roc /rɒk/ n. a gigantic bird of Eastern legend. [Sp. *rocho* ult. f. Arab *ruḵ*]

rocaille /rəʊˈkaɪ/ n. **1** an 18th-century style of ornamentation based on rock and shell motifs. **2** a rococo style. [F f. *roc* (as ROCK[1])]

rocambole /ˈrɒkəmˌbəʊl/ n. an allium, *Allium scorodoprasum*, with a garlic-like bulb used for seasoning. [F f. G *Rockenbolle*]

roche moutonnée /ˌrɒʃ muːˈtɒneɪ, ˌrəʊʃ muːtəˈneɪ/ n. Geol. a bare outcrop of rock shaped by glacial erosion, with one side smooth and gently sloping and the other steep, rough and irregular. [F, = fleecy rock]

Rochester[1] /ˈrɒtʃɪstə(r)/ **1** a town in SE England, on the Medway estuary in Kent; pop. (1981) 24,400. **2** a city in NW New York state, on Lake Ontario; pop. (1990) 231,640.

Rochester[2] /ˈrɒtʃɪstə(r)/, 2nd Earl of (title of John Wilmot) (1647–80), English poet and courtier. Infamous for his dissolute life at the court of Charles II, he wrote many sexually explicit love poems and, with his social and literary verse satires, is regarded as one of the first Augustans. Famous works include his *Satire against Mankind* (1675).

rochet /ˈrɒtʃɪt/ n. Eccl. a vestment resembling a surplice, used chiefly by bishops and abbots. [ME f. OF, dimin. f. Gmc]

rock[1] /rɒk/ n. **1 a** the hard material of the earth's crust, exposed on the surface or underlying the soil. **b** a similar material on other planets. **2** Geol. any natural material, hard or soft (e.g. clay), consisting of one or more minerals. **3 a** a mass of rock projecting and forming a hill, cliff, reef, etc. **b** (**the Rock**) Gibraltar. **4** a large detached stone. **5** US a stone of any size. **6** a firm and dependable support or protection. **7** a source of danger or destruction. **8** Brit. a kind of hard confectionery usu. made in cylindrical peppermint-flavoured sticks. **9** (in pl.) US sl. money. **10** sl. a precious stone, esp. a diamond. **11** sl. a solid form of cocaine. **12** (in pl.) coarse sl. the testicles. □ **get one's rocks off** coarse sl. **1** achieve sexual satisfaction. **2** obtain enjoyment. **on the rocks** colloq. **1** short of money. **2** broken down. **3** (of a drink) served undiluted with ice-cubes. **rock-bed** a base of rock or a rocky bottom. **rock-bottom** adj. (of prices etc.) the very lowest. ● n. the very lowest level. **rock-bound** (of a coast) rocky and inaccessible. **rock-cake** a small

currant cake with a hard uneven surface. **rock-candy** N. Amer. = sense 8 of n. **rock-climber** a person who practises rock-climbing. **rock-climbing** the sport of climbing rock-faces, esp. with the aid of ropes etc. **rock cress** = ARABIS. **rock-crystal** transparent colourless quartz usu. in hexagonal prisms. **rock dove** a wild cliff-nesting dove, *Columba livia*, ancestor of the domestic pigeon. **rock-face** a vertical surface of natural rock. **rock-fish** a rock-frequenting goby, bass, wrasse, catfish, etc. living among rocks. **rock garden** an artificial mound or bank of earth and stones planted with rock-plants etc.; a garden in which rockeries are the chief feature. **rock hyrax** a hyrax of the genus *Procavia*, found on rocky outcrops. **rock-pigeon** = *rock dove*. **rock pipit** a dark-coloured European pipit, *Anthus spinoletta*, frequenting rocky shores. **rock-plant** any plant growing on or among rocks. **rock ptarmigan** N. Amer. see PTARMIGAN. **rock python** a large snake of the family Boidae, esp. the African python *Python sebae*. **rock-rabbit** = *rock hyrax*. **rock rose** a plant of the genus *Cistus*, *Helianthemum*, etc., with rose-like flowers. **rock-salmon 1** Brit. a catfish or dogfish sold as food. **2** US an amberjack. **rock-salt** common salt as a solid mineral. **rock-wool** inorganic material made into matted fibre esp. for insulation or soundproofing. □ **rockless** adj. **rocklike** adj. **rocklet** /ˈrɒklɪt/ n. [ME f. OF *ro(c)que*, *roche*, med.L *rocca*, of unkn. orig.]

rock[2] /rɒk/ v. & n. ● v. **1** tr. move gently to and fro or in or as if in a cradle; set or maintain such motion (*rock him to sleep*; *the ship was rocked by the waves*). **2** intr. be or continue in such motion (*sat rocking in his chair*; *the ship was rocking on the waves*). **3 a** intr. sway from side to side; shake, oscillate, reel (*the house rocks*). **b** tr. cause to do this (*an earthquake rocked the house*). **4** tr. distress, perturb. **5** intr. dance to or play rock music. ● n. **1** a rocking movement (*gave the chair a rock*). **2** a spell of rocking (*had a rock in his chair*). **3 a** = ROCK AND ROLL. **b** = ROCK MUSIC. □ **rock the boat** colloq. disturb the equilibrium of a situation. **rocking-chair** a chair mounted on rockers or springs for gently rocking in. **rocking-horse** a model of a horse on rockers or springs for a child to rock on. **rocking-stone** a poised boulder easily rocked (also called *logan-stone*). **rock-shaft** Mech. a shaft that oscillates about an axis without making complete revolutions. [OE *roccian*, prob. f. Gmc]

rockabilly /ˈrɒkəˌbɪli/ n. a type of popular music originating in the south-eastern US, combining elements of rock and roll and country music. [blend of *rock and roll* and *hillbilly*]

Rockall /ˈrɒkɔːl/ **1** a rocky islet in the North Atlantic, about 400 km (250 miles) north-west of Ireland. It was formally annexed by Britain in 1955 but has since become the subject of territorial dispute between Britain, Denmark, Iceland, and Ireland over mineral, oil, and fishing rights. **2** a shipping forecast area in the NE Atlantic, containing the islet of Rockall near its northern boundary.

rock and roll n. (also **rock 'n' roll**) a type of lively popular music originating in the US in the mid-1950s. Rock and roll was an amalgam of black rhythm and blues and white country music, usually based around a twelve-bar structure and an instrumentation of guitar, double bass, and drums. 'Rock Around the Clock' (1954) by Bill Haley was the first widely known rock and roll record; it caused a sensation after it appeared in the film *Blackboard Jungle*. The emergence of performers such as Elvis Presley and Chuck Berry in the mid and late 1950s and their immense popularity was an indication of the new cultural and economic importance of teenagers. The term is also an alternative name for *rock music*. □ **rock and roller** a devotee of rock and roll.

rockburst /ˈrɒkbɜːst/ n. a sudden rupture or collapse of highly stressed rock in a mine.

Rockefeller /ˈrɒkəˌfelə(r)/, John D(avison) (1839–1937), American industrialist and philanthropist. One of the first to recognize the industrial possibilities of oil, Rockefeller established the Standard Oil Company (1870) and by the end of the decade exercised a virtual monopoly over oil refining in the US. Early in the 20th century he handed over his business interests to his son, John D(avidson) Rockefeller Jr. (1874–1960), and devoted his private fortune to numerous philanthropic projects, such as the establishment of the Rockefeller Foundation (1913). His son's many philanthropic institutions include the Rockefeller Center in New York (1939).

rocker /ˈrɒkə(r)/ n. **1** a person or thing that rocks. **2** a curved bar or similar support, on which something can rock. **3** a rocking-chair. **4** a devotee of rock music, characteristically associated with leather clothing and motorcycles. **5** a skate with a highly curved blade. **6** a switch constructed on a pivot mechanism operating between the 'on' and 'off' positions. **7** any rocking device forming part of a mechanism. □ **off one's rocker** colloq. crazy.

rockery /'rɒkərɪ/ n. (pl. **-ies**) a heaped arrangement of rough stones with soil between them for growing rock-plants on.

rocket[1] /'rɒkɪt/ n. & v. ● n. **1** a cylindrical projectile that can be propelled to a great height or distance by combustion of its contents, used esp. as a firework or signal. (*See note below.*) **2** (in full **rocket engine** or **motor**) an engine using a similar principle but not dependent on air intake for its operation. **3** a missile, spacecraft, etc., powered by such an engine. **4** *Brit. colloq.* a severe reprimand. ● v. (**rocketed, rocketing**) **1** tr. bombard with rockets. **2** intr. **a** move rapidly upwards or away. **b** increase rapidly (*prices rocketed*). [F *roquette* f. It. *rochetto* dimin. of *rocca* distaff, with ref. to its cylindrical shape]

■ The earliest use of rockets was in China, where rockets propelled by gunpowder were in use probably by AD 1232. Their use in war and as fireworks spread quickly to the Middle East and Europe, but they failed to displace guns as the main artillery weapons. At the end of the 19th century a Russian schoolteacher, Konstantin Tsiolkovsky, proposed the use of liquid fuels and pointed out that rockets could be used to travel in space. In the US, Robert Goddard launched the first liquid-fuelled rocket in 1926, and in Germany and Russia experimenters worked on both liquid and solid-fuelled rockets. In Germany, a team led by Wernher von Braun developed the V2 rocket (see V[3]), which first flew in 1942 and was used against the Allies in the Second World War. It was the first long-range ballistic weapon and the first rocket capable of reaching the edge of space. After the war von Braun and other Germans were recruited by the Americans and Soviets, and worked to develop rockets to carry nuclear warheads, launch satellites, and assist in space exploration; this culminated in the American *Saturn V* rocket used to lift the Apollo missions into earth orbit. At the same time solid-fuelled rockets have continued to be developed as artillery and aircraft weapons, such as the French Exocet and American Sidewinder.

rocket[2] /'rɒkɪt/ n. **1** (also **sweet rocket**) a fast-growing plant, esp. of the genus *Hesperis* or *Sisymbrium*. **2** a variety of a cruciferous annual plant, *Eruca vesicaria*, grown for salad. Also called *roquette*, (*N. Amer.*) *arugula*. □ **wall-rocket** a yellow-flowered cruciferous plant, *Diplotaxis muralis*, growing on walls etc. and emitting a foul smell when crushed. **yellow rocket** winter cress. [F *roquette* f. It. *rochetta*, *ruchetta* dimin. of *ruca* f. L *eruca* downy-stemmed plant]

rocketeer /ˌrɒkɪ'tɪə(r)/ n. **1** a discharger of rockets. **2** a rocket expert or enthusiast.

rocketry /'rɒkɪtrɪ/ n. the science or practice of rocket propulsion.

rockfall /'rɒkfɔ:l/ n. **1** a descent of loose rocks. **2** a mass of fallen rock.

Rockhampton /rɒk'hæmptən/ a port on the Fitzroy river, in Queensland, NE Australia; pop. (1991) 55,790. It is the centre of Australia's largest beef-producing area.

rockhopper /'rɒkˌhɒpə(r)/ n. a small penguin, *Eudyptes crestatus*, of the Antarctic and New Zealand, with a crest of feathers on the forehead.

rockling /'rɒklɪŋ/ n. a small marine fish of the cod family, esp. of the genus *Ciliata* or *Rhinomenus*, found in pools among rocks.

rock music n. a form of popular music which evolved from rock and roll and pop music during the period of social change in the mid and late 1960s. Harsher and often self-consciously more serious than what had gone before, rock music was initially characterized by musical experimentation and drug-related or anti-Establishment lyrics, although it has since become an accepted form. Early performers were new groups such as Jefferson Airplane, Pink Floyd, and the Grateful Dead, and established bands like the Byrds and the Beatles; later exponents include Jimi Hendrix, Led Zeppelin and U2. The term is also an alternative name for *rock and roll*.

Rock of Gibraltar see GIBRALTAR.

rocky[1] /'rɒkɪ/ adj. & n. ● adj. (**rockier, rockiest**) **1** of or like rock. **2** full of or abounding in rock or rocks (*a rocky shore*). **3** **a** firm as a rock; determined, steadfast. **b** unfeeling, cold, hard. □ **rockiness** n. [ROCK[1]]

rocky[2] /'rɒkɪ/ adj. (**rockier, rockiest**) *colloq.* unsteady, tottering. □ **rockily** adv. **rockiness** n. [ROCK[2]]

Rocky Mountain goat n. a white goat-antelope, *Oreamnos americanus*, of western North American mountains.

Rocky Mountains (also **Rockies**) the chief mountain system of North America, which extends from the US–Mexico border to the Yukon Territory of northern Canada. It separates the Great Plains from the Pacific coast and forms the Continental Divide. Several peaks rise to over 4,300 m (14,000 ft), the highest being Mount Elbert at 4,399 m (14,431 ft).

Rocky Mountain spotted fever n. see *spotted fever* 3.

rococo /rə'kəʊkəʊ/ adj. & n. ● adj. **1** of or relating to a style of architecture and design coming after baroque. (*See note below.*) **2** (of literature, music, architecture, and the decorative arts) highly ornamented, florid. ● n. the rococo style. [F, joc. alt. f. ROCAILLE]

■ Where baroque is dramatic and large scale, the ornateness of the rococo style, which flourished principally in France, southern Germany, and Austria in the first half of the 18th century, tends to be more delicate, playful, and on a smaller scale; rococo architecture is more often associated with interior decoration. The style is characterized by light colours, especially pale pinks, blues, and greens, scrollwork, and shell motifs; in painting, it is best illustrated by the work of Watteau, Boucher, Tiepolo, and Fragonard.

rod /rɒd/ n. **1** a slender straight bar esp. of wood or metal. **2** this as a symbol of office. **3** **a** a stick or bundle of twigs used in caning or flogging. **b** (prec. by *the*) the use of this; punishment, chastisement. **4** **a** = *fishing-rod*. **b** an angler using a rod. **5** **a** slender straight round stick growing as a shoot on a tree. **b** this when cut. **6** *hist.* (as a measure) a perch or square perch (see PERCH[1] n. 3). **7** *US sl.* = *hot rod*. **8** *US sl.* a pistol or revolver. **9** *Anat.* any of numerous rod-shaped structures in the eye, detecting dim light. □ **make a rod for one's own back** act in a way that will bring one trouble later. □ **rodless** adj. **rodlike** adj. **rodlet** /'rɒdlɪt/ n. [OE *rodd*, prob. rel. to ON *rudda* club]

Roddenberry /'rɒd(ə)nˌberɪ/, Gene (full name Eugene Wesley Roddenberry) (1921–91), American television producer and scriptwriter. He is best-known as the creator of the TV science-fiction drama series *Star Trek*, first broadcast 1966–9. Roddenberry wrote many scripts for the series, which attracted an international cult following. He later worked on feature films and launched a successful follow-up series, *Star Trek: The Next Generation*, in 1987.

rode[1] past of RIDE.

rode[2] /rəʊd/ v.intr. **1** (of wildfowl) fly landwards in the evening. **2** (of woodcock) fly in the evening during the breeding season. [18th c.: orig. unkn.]

rodent /'rəʊd(ə)nt/ n. & adj. ● n. a mammal of the order Rodentia, with strong gnawing incisors and no canine teeth, e.g. rat, mouse, squirrel, beaver, porcupine. (*See note below.*) ● adj. **1** of the order Rodentia. **2** gnawing (esp. *Med.* of slow-growing ulcers). □ **rodent officer** *Brit.* an official dealing with rodent pests. □ **rodential** /rəʊ'denʃ(ə)l/ adj. [L *rodere ros-* gnaw]

■ There are about 1,800 species in the order Rodentia, which represents about 42 per cent of living mammals. They are characterized by a single pair of large constantly growing incisor teeth and a large gap between these and the premolars. They are divided into three suborders, the Sciuromorpha (squirrels, marmots, gophers, beavers, etc.), the Myomorpha (rats, mice, voles, hamsters, etc.), and the Hystricomorpha (porcupines, guinea pigs, capybaras, agoutis, African mole rats, etc.). The rabbits and hares, formerly thought of as rodents, are now placed in the order Lagomorpha because of their double incisors.

rodenticide /rəʊ'dentɪˌsaɪd/ n. a poison used to kill rodents.

rodeo /'rəʊdɪəʊ, rə'deɪəʊ/ n. (pl. **-os**) **1** an exhibition or entertainment involving cowboys' skills in handling animals. (*See note below.*) **2** an exhibition of other skills, e.g. in motorcycling. **3** **a** a round-up of cattle on a ranch for branding etc. **b** an enclosure for this. [Sp. f. *rodear* go round ult. f. L *rotare* ROTATE[1]]

■ Rodeo was born in the southern states of the US after the Civil War, when Texan cowboys, driving their cattle to the north and west in search of new markets, organized events to display their daring and horsemanship. The principal rodeo events include bull riding, bareback riding, calf roping, steer wrestling, and bronco riding.

Rodgers /'rɒdʒəz/, Richard (Charles) (1902–79), American composer. Together with librettist Lorenz Hart (1895–1943), he created musicals such as *On Your Toes* (1936). After Hart's death, Rodgers collaborated with Oscar Hammerstein II on a succession of popular musicals, including *Oklahoma!* (1943), *Carousel* (1945), *South Pacific* (1949), and *The Sound of Music* (1959).

rodham /'rɒdəm/ n. a raised bank in the Fen district of East Anglia, formed on the bed of a dry river-course. [20th c.: orig. uncert.]

Ródhos see RHODES[1].

Rodin /'rəʊdæn/, Auguste (1840–1917), French sculptor. Influenced by Michelangelo, Rodin was chiefly concerned with the human form in his work, and developed a naturalistic style that made him a controversial figure in his day. His first major work, *The Age of Bronze* (1875–6), was considered so lifelike that Rodin was alleged to have taken a cast from a live model. By 1880 he had been publicly

commissioned to create *The Gate of Hell* for the Musée des arts décoratifs; it remained unfinished at his death and its many figures inspired such independent statues as *The Thinker* (1880) and *The Kiss* (1886).

rodomontade /ˌrɒdəmɒnˈteɪd, -ˈtɑːd/ *n., adj., & v.* ● *n.* **1** boastful or bragging talk or behaviour. **2** an instance of this. ● *adj.* boastful or bragging. ● *vintr.* brag, talk boastfully. [F f. obs. It. *rodomontada* f. F *rodomont* & It. *rodomonte* f. the name of a boastful character in Ariosto's epic *Orlando Furioso*]

roe[1] /rəʊ/ *n.* **1** (also **hard roe**) the mass of eggs in a female fish's ovary. **2** (also **soft roe**) the milt of a male fish. □ **roe-stone** oolite. □ **roed** *adj.* (also in *comb.*). [ME *row*(e), *rough*, f. MLG, MDu. *roge*(n), OHG *rogo*, *rogan*, ON *hrogn*]

roe[2] /rəʊ/ *n.* (also **roe-deer**) (*pl.* same or **roes**) a small Eurasian deer, *Capreolus capreolus*, with short pointed antlers. [OE *rā(ha)*]

roebuck /ˈrəʊbʌk/ *n.* a male roe.

Roedean /ˈrəʊdiːn/ an independent boarding-school for girls, on the south coast of England east of Brighton. It was founded in 1885.

Roeg /rəʊg/, Nicholas (Jack) (b.1928), English film director. His work is often unsettling and impressionistic, and uses cutting techniques to create disjointed narratives. His films include *Performance* (1970), *Walkabout* (1970), *Don't Look Now* (1972), *The Man Who Fell to Earth* (1975), and *Castaway* (1986).

roentgen /ˈrʌntjən, ˈrɒntgən/ *n.* (also **röntgen**) a unit of ionizing radiation, the amount producing one electrostatic unit of positive or negative ionic charge in one cubic centimetre of air under standard conditions. □ **roentgen rays** X-rays. [RÖNTGEN]

roentgenography /ˌrʌntjəˈnɒgrəfi, ˌrɒntgə-/ *n.* (also **röntgenography**) photography using X-rays.

roentgenology /ˌrʌntjəˈnɒlədʒi, ˌrɒntgə-/ *n.* (also **röntgenology**) = RADIOLOGY.

Roeselare /ˈruːsəˌlɑːrə/ (French **Roulers** /rulɛrs/) a town in NW Belgium, in the province of West Flanders; pop. (1991) 57,890.

rogation /rəʊˈgeɪʃ(ə)n/ *n.* (usu. in *pl.*) Eccl. solemn prayers consisting of the litany of the saints chanted on the three days before Ascension Day (see also ROGATION DAY). □ **rogational** *adj.* [ME f. L *rogatio* f. *rogare* ask]

Rogation Day *n.* (usu. *pl.*) in the Christian Church, each of the three days preceding Ascension Day, prescribed for prayer and fasting. Rogation Days were originally associated with pre-Christian harvest rituals.

Rogation Sunday the Sunday before Ascension Day.

roger /ˈrɒdʒə(r)/ *int. & v.* ● *int.* **1** your message has been received and understood (used in radio communication etc.). **2** *colloq.* I agree. ● *v. coarse sl.* **1** *intr.* have sexual intercourse. **2** *tr.* have sexual intercourse with (a woman). [the name *Roger*, code for *R*]

Rogers[1] /ˈrɒdʒəz/, Ginger (born Virginia Katherine McMath) (1911–95), American actress and dancer. She is best known for her dancing partnership with Fred Astaire; from 1933 they appeared in a number of film musicals, including *Top Hat* (1935), *Swing Time* (1936), and *Shall We Dance?* (1937). Rogers's solo acting career included the film *Kitty Foyle* (1940), for which she won an Oscar.

Rogers[2] /ˈrɒdʒəz/, Sir Richard (George) (b.1933), British architect, born in Italy. He was a leading exponent of high-tech architecture and founded his own practice, Team 4, in 1963. He gained international recognition in the 1970s for the Pompidou Centre in Paris (1971–7), which he designed in partnership with the Italian architect Renzo Piano (b.1937). Rogers's other major work, the Lloyd's Building in London (1986), followed a similarly original high-tech design and also featured ducts and pipes on the outside of the steel and glass building.

Roget /ˈrɒʒeɪ/, Peter Mark (1779–1869), English scholar. He worked as a physician but is remembered as the compiler of *Roget's Thesaurus of English Words and Phrases*, which he completed after his retirement and was first published in 1852. The work, which has been revised many times since, is important for its innovative classification of words according to underlying concept or meaning.

rogue /rəʊg/ *n. & v.* ● *n.* **1** a dishonest or unprincipled person. **2** *joc.* a mischievous person, esp. a child. **3** (usu. *attrib.*) **a** a wild animal driven away or living apart from the herd and of fierce temper (*rogue elephant*). **b** a stray, irresponsible, or undisciplined person or thing (*rogue trader*). **4** an inferior or defective specimen among many acceptable ones. ● *v.tr.* remove inferior or defective specimens from. □ **rogues' gallery** a collection of photographs of known criminals etc., used for identification of suspects. [16th-c. cant word: orig. unkn.]

roguery /ˈrəʊgəri/ *n.* (*pl.* **-ies**) conduct or an action characteristic of rogues.

roguish /ˈrəʊgiʃ/ *adj.* **1** playfully mischievous. **2** characteristic of rogues. □ **roguishly** *adv.* **roguishness** *n.*

roil /rɔɪl/ *v.tr.* **1** make (a liquid) turbid by agitating it. **2** *US* = RILE 1. [perh. f. OF *ruiler* mix mortar f. LL *regulare* REGULATE]

roister /ˈrɔɪstə(r)/ *v.intr.* (esp. as **roistering** *adj.*) revel noisily; be uproarious. □ **roisterer** *n.* **roistering** *n.* **roisterous** *adj.* [obs. *roister* roisterer f. F *rustre* ruffian var. of *ruste* f. L *rusticus* RUSTIC]

Roland /ˈrəʊlənd/ the most famous of Charlemagne's paladins, hero of the *Chanson de Roland* (12th century) and other (esp. French and Italian) medieval romances. He is said to have become a friend of Oliver, another paladin, after engaging him in single combat in which neither won. Roland was killed in a rearguard action at the Battle of Roncesvalles. □ **a Roland for an Oliver** an effective retort; a well-balanced combat or exchange.

role /rəʊl/ *n.* (also **rôle**) **1** an actor's part in a play, film, etc. **2** a person's or thing's characteristic or expected function (*the role of the tape recorder in language-learning*). □ **role model** a person regarded by others as an example in a particular role. **role-playing** an exercise or game in which participants act the part of another character, used in psychotherapy, language-teaching, etc. [F *rôle* and obs. F *roule, rolle* ROLL *n.*]

roll /rəʊl/ *v. & n.* ● *v.* **1 a** *intr.* move or go in some direction by turning over and over on an axis or by a rotary movement (*the ball rolled under the table; a barrel started rolling*). **b** *tr.* cause to do this (*rolled the barrel into the cellar*). **2** *tr.* make revolve between two surfaces (*rolled the clay between his palms*). **3 a** *intr.* (foll. by *along, by,* etc.) move or advance on or (of time etc.) as if on wheels etc. (*the bus rolled past; the pram rolled off the pavement; the years rolled by*). **b** *tr.* cause to do this (*rolled the tea trolley into the kitchen*). **c** *intr.* (of a person) be conveyed in a vehicle (*the farmer rolled by on his tractor*). **4 a** *tr.* turn over and over on itself to form a more or less cylindrical or spherical shape (*rolled a newspaper*). **b** *tr.* make by forming material into a cylinder or ball (*rolled a cigarette; rolled a huge snowball*). **c** *tr.* accumulate into a mass (*rolled the dough into a ball*). **d** *intr.* (foll. by *into*) make a specified shape of itself (*the hedgehog rolled into a ball*). **5** *tr.* flatten or form by passing a roller etc. over or by passing between rollers (*roll the lawn; roll pastry; roll thin foil*). **6** *intr. & tr.* change or cause to change direction by rotatory movement (*his eyes rolled; he rolled his eyes*). **7** *intr.* **a** wallow, turn about in a fluid or a loose medium (*the dog rolled in the dust*). **b** (of a horse etc.) lie on its back and kick about, esp. so as to dislodge its rider. **8** *intr.* **a** (of a moving ship, aircraft, or vehicle) sway to and fro on an axis parallel to the direction of motion. **b** walk with an unsteady swaying gait (*they rolled out of the pub*). **9 a** *intr.* undulate, show or go with an undulating surface or motion (*rolling hills; rolling mist; the waves roll in*). **b** *tr.* carry or propel with such motion (*the river rolls its waters to the sea*). **10 a** *intr.* (of machinery) start functioning or moving (*the cameras rolled; the train began to roll*). **b** *tr.* cause (machinery) to do this. **11 a** *tr.* display (credits for a film or television programme) moving up the screen. **b** *intr.* (of credits) be displayed in this way. **12** *intr. & tr.* sound or utter with a vibratory or trilling effect (*words rolled off his tongue; thunder rolled in the distance; he rolls his* rs). **13** *N. Amer. sl.* **a** *tr.* overturn (a car etc.). **b** *intr.* (of a car etc.) overturn. **14** *tr. US* throw (dice). **15** *tr. sl.* rob (esp. a helpless victim). ● *n.* **1** a rolling motion or gait; rotation, spin; undulation (*the roll of the hills*). **2 a** a spell of rolling (*a roll in the mud*). **b** a gymnastic exercise in which the body is rolled into a tucked position and turned in a forward or backward circle. **c** (esp. **a roll in the hay**) *colloq.* an act of sexual intercourse or erotic fondling. **3** a continuous rhythmic sound of thunder or a drum. **4** *Aeron.* a complete revolution of an aircraft about its longitudinal axis. **5 a** a cylinder formed by turning flexible material over and over on itself without folding (*a roll of carpet; a roll of wallpaper*). **b** a filled cake or pastry of similar form (*fig roll; sausage roll*). **6 a** a small portion of bread individually baked. **b** this with a specified filling (*ham roll*). **7** a more or less cylindrical or semicylindrical straight or curved mass of something (*rolls of fat; a roll of hair*). **8 a** an official list or register (*the electoral roll*). **b** the total numbers on this (*the schools' rolls have fallen*). **c** a document, esp. an official record, in scroll form. **9** a cylinder or roller, esp. to shape metal in a rolling-mill. **10** *Archit.* **a** a moulding of convex section. **b** a spiral scroll of an Ionic capital. **11** *US & Austral.* money, esp. as banknotes rolled together. □ **be rolling** *colloq.* be very rich. **be rolling in** *colloq.* have plenty of (esp. money). **on a roll** *N. Amer. sl.* experiencing a bout of success or progress; engaged in a period of intense activity. **roll back** *N. Amer.* cause (esp. prices) to decrease. **rollback** *n.* a reduction (esp. in price). **roll bar** an overhead metal bar strengthening the frame of a vehicle (esp. in racing) and protecting the occupants if the vehicle overturns. **roll-call** a process of calling out a list of names to establish who is present. **rolled gold** gold in the form of a thin coating applied to a baser metal by rolling. **rolled into one** combined in one person or thing. **rolled oats** oats that have been husked and crushed. **roll in 1** arrive in great numbers or quantity. **2** wallow, luxuriate in. **rolling barrage** = creeping barrage. **rolling drunk** swaying or staggering from drunkenness. **rolling-mill** a machine or factory for rolling metal into shape. **rolling-pin** a cylinder for rolling out pastry, dough, etc. **rolling-stock 1** the locomotives, carriages, or other vehicles, used on a railway. **2** US the road vehicles of a company. **rolling stone** a person who is unwilling to settle for long in one place. **rolling strike** industrial action through a series of limited strikes by consecutive groups. **roll-neck** (of a garment) having a high loosely turned-over neck. **roll of honour** a list of those honoured, esp. the dead in war. **roll on 1** put on or apply by rolling. **2** (in imper.) colloq. (of a time, in eager expectation) come quickly (roll on Friday!). **roll-on** attrib.adj. (of deodorant etc.) applied by means of a rotating ball in the neck of the container. ● n. a light elastic corset. **roll-on roll-off** (of a passenger ferry, cargo ship, etc.) in which vehicles are driven directly on at the start of the voyage and off at the end of it. **roll out 1** unroll, spread out. **2** launch (a new product). **roll-out** n. **1** the official launch of a new product. **2** the official wheeling out of a new aircraft or spacecraft. **roll over 1** send (a person) sprawling or rolling. **2** Econ. finance the repayment of (maturing stock etc.) by an issue of new stock. **roll-over** n. **1** Econ. the extension or transfer of a debt or other financial relationship. **2** colloq. the overturning of a vehicle etc. **roll-top desk** a desk with a flexible cover sliding in curved grooves. **roll up 1** colloq. arrive in a vehicle; appear on the scene. **2** make into or form a roll. **3** Mil. drive the flank of (an enemy line) back and round so that the line is shortened or surrounded. **roll-up** (or **roll-your-own**) n. a hand-rolled cigarette. **roll up one's sleeves** see SLEEVE. **strike off the rolls** debar (esp. a solicitor) from practising after dishonesty etc. □ **rollable** adj. [ME f. OF rol(l)er, rouler, ro(u)lle f. L rotulus dim. of rota wheel]

Rolland /'rɒlɒn/, Romain (1866–1944), French novelist, dramatist, and essayist. His interest in genius led to biographies of Beethoven (1903), Michelangelo (1905), and Tolstoy (1911), and ultimately to Jean-Christophe (1904–12), a cycle of ten novels about a German composer. These epitomize the literary form known as the roman-fleuve and in their portrayal of the composer's friendship with a Frenchman symbolized Rolland's desire for harmony between nations. He was awarded the Nobel Prize for literature in 1915.

rollaway /'rəʊlə,weɪ/ adj. US (of a bed etc.) that can be removed on wheels or castors.

roller /'rəʊlə(r)/ n. **1 a** a hard revolving cylinder for smoothing the ground, spreading ink or paint, crushing or stamping, rolling up cloth on, hanging a towel on, etc., used alone or as a rotating part of a machine. **b** a cylinder for diminishing friction when moving a heavy object. **2** a small cylinder on which hair is rolled for setting. **3** a long swelling wave. **4** (also **roller bandage**) a long surgical bandage rolled up for convenient application. **5** a kind of tumbler-pigeon. **6 a** a brilliantly plumaged bird of the family Coraciidae, with characteristic tumbling display-flight. **b** a breed of canary with a trilling song. □ **roller bearing** a bearing like a ball-bearing but with small cylinders instead of balls. **roller-coaster** n. a switchback at a fair etc. ● adj. that goes up and down, or changes, suddenly and repeatedly. ● v.intr. (or **roller-coast**) go up and down or change in this way. **roller-skate** each of a pair of skates used in roller-skating, consisting of metal frames with small wheels (or of boots with wheels attached). (See also ROLLER-SKATING.) **roller towel** a towel with the ends joined, hung on a roller.

rollerball /'rəʊlə,bɔːl/ n. a ball-point pen using thinner ink than other ball-points.

Rollerblade /'rəʊlə,bleɪd/ n. propr. a type of roller-skate with the wheels set in one straight line beneath the boot.

roller-skating /'rəʊlə,skeɪtɪŋ/ n. skating on a hard surface other than ice, either as a sport or pastime. Although roller-skating using a primitive form of roller-skate was known in the Netherlands from the 18th century, the first practical four-wheel skate was not patented until 1863 in the US. Its original purpose was to enable ice-skaters to practise when there was no natural ice, but roller-skating became popular as an independent sport, now with competitions similar in many ways to those in ice-skating.

rollick /'rɒlɪk/ v. & n. ● v.intr. (esp. as **rollicking** adj.) be jovial or exuberant, indulge in high spirits, revel. ● n. **1** exuberant gaiety. **2** a spree or escapade. [19th-c., prob. dial.: perh. f. ROMP + FROLIC]

Rolling Stones, the an English rock group formed c.1962, featuring singer Mick Jagger (b.1943) and guitarist Keith Richards (b.1943). Originally a rhythm-and-blues band, the Rolling Stones became successful with a much-imitated rebel image and songs like 'Satisfaction' (1965) and 'Jumping Jack Flash' (1968). In the 1970s they evolved a simple, derivative, yet distinctive style, heard on albums such as Exile on Main Street (1972) and Some Girls (1978), that they retained almost unchanged into the 1990s.

rollmop /'rəʊlmɒp/ n. a rolled uncooked pickled herring fillet. [G Rollmops]

Rolls /rəʊlz/, Charles Stewart (1877–1910), English motoring and aviation pioneer. He was one of the founder members of the Royal Automobile Club (RAC) in 1897 and the Royal Aero Club (1903). In 1906 he and Henry Royce formed the company Rolls-Royce Ltd., with Royce as chief engineer and Rolls as demonstrator-salesman. (See also ROLLS-ROYCE.) Rolls was the first Englishman to fly across the English Channel, and made the first double crossing in 1910 shortly before he was killed in an air crash, the first English casualty of aviation.

Rolls-Royce /rəʊlz'rɔɪs/ a British company producing aircraft engines and luxury motor cars, founded in 1906 by Charles Stewart Rolls and Henry Royce. The company established its reputation with production of cars such as the Silver Ghost model (produced 1906–25), with its distinguishing vertical radiator grill and winged figure on the front of the bonnet; subsequent models included the Silver Shadow. From 1914 Rolls-Royce also produced aircraft engines, the first being the Eagle, used extensively in the First World War; subsequent designs were used in the Spitfires and Hurricanes of the Second World War. After becoming bankrupt in 1971, Rolls-Royce was formed into two separate companies.

roly-poly /,rəʊlɪ'pəʊlɪ/ n. & adj. ● n. (pl. **-ies**) **1** (also **roly-poly pudding**) a pudding made of a sheet of suet pastry covered with jam etc., formed into a roll, and steamed or baked. **2** US a tumbler toy. **3** Austral. a bushy plant, esp. Salsola kali, that breaks off and is rolled by the wind. ● adj. (usu. of a child) podgy, plump. [prob. formed on ROLL]

ROM /rɒm/ n. Computing read-only memory. [abbr.]

Rom /rɒm/ n. (pl. **Roma** /'rɒmə/) a male gypsy. [Romany, = man, husband]

Rom. abbr. Epistle to the Romans (New Testament).

rom. abbr. roman (type).

Roma see ROME.

Romaic /rəʊ'meɪɪk/ n. & adj. ● n. the vernacular language of modern Greece. ● adj. of or relating to this language. [Gk Rhōmaikos Roman (used esp. of the Eastern Empire)]

romaine /rə'meɪn/ n. N. Amer. a cos lettuce. [F, fem. of romain (as ROMAN)]

romaji /'rəʊmədʒɪ/ n. a system of romanized spelling used to transliterate Japanese. [Jap.]

Roman /'rəʊmən/ adj. & n. ● adj. **1 a** of ancient Rome or its territory or people. **b** archaic of its language. **2** of medieval or modern Rome. **3** of papal Rome, esp. = ROMAN CATHOLIC. **4** of a kind ascribed to the early Romans (Roman honesty; Roman virtue). **5** surviving from a period of ancient Roman rule (Roman road). **6** (**roman**) (of type) of a plain upright kind used in ordinary print. **7** (of the alphabet etc.) based on the ancient Roman system with letters A–Z. ● n. **1 a** a citizen of the ancient Roman Republic or Empire. **b** a soldier of the Roman Empire. **2** a citizen of modern Rome. **3** = ROMAN CATHOLIC. **4** (**roman**) roman type. □ **Roman candle** a firework discharging a series of flaming coloured balls and sparks. **Roman holiday** enjoyment derived from others' discomfiture. **Roman law** the law-code developed by the ancient Romans and forming the basis of many modern codes. **Roman nose** one with a high bridge; an aquiline nose. [ME f. OF Romain (n. & adj.) f. L Romanus f. Roma ROME]

roman-à-clef /,rəʊmɒnæ'kleɪ/ n. (pl. **romans-à-clef** pronunc. same) a novel in which real persons or events appear with invented names. [F, = novel with a key]

Roman baths n.pl. a building containing a complex of rooms designed for bathing, relaxing, and socializing, as used in ancient Rome. Such baths have been found throughout the territories of the Roman Empire. They generally followed a similar design, which had become standard by the 1st century BC. On arrival, a bather was covered with oil before taking exercise and then going into the hot room (caldarium), which was warmed by underfloor heating. Next was the hot-air or steam-room (sudatorium), where the bather's body was scraped of sweat and dirt with a curved instrument (a strigil), followed by the warm room (tepidarium) and finally the cold room (frigidarium).

Roman Britain Britain during the period AD 43–410, when most of Britain was part of the Roman Empire. Britain was twice invaded briefly by Julius Caesar 55–54 BC, but was not conquered until the invasion of Claudius in AD 43. The conquest of England as far north as Fosse Way was swiftly accomplished, and the frontier of the Roman province of Britain was eventually established at Hadrian's Wall; the more northerly Antonine Wall was breached and abandoned (c.181). The Roman military presence was reinforced by a large garrison and a network of roads radiating out from London (Londinium), but a wider policy of romanization was also followed: Roman settlers and traders built villas, and Roman towns including York (Eboracum), Lincoln (Lindum Colonia), St Albans (Verulamium), and Colchester (Camulodunum) were established or developed.

Roman Catholic *adj. & n.* ● *adj.* of the Roman Catholic Church. ● *n.* a member of the Roman Catholic Church. □ **Roman Catholicism** *n.*

Roman Catholic Church the branch of the Christian Church which acknowledges the pope as its head, especially as it has developed since the Reformation. It is the largest Church of Western Christianity, dominant particularly in South America and southern Europe. The Church claims an authority derived directly from the Apostles, particularly St Peter, who is regarded as the first pope. Much modern Roman Catholic thought and practice is grounded in scholastic theology and in the Church's response to the Reformation made at the Council of Trent (1545–63). The doctrine of papal infallibility was formally defined in 1870, the culmination of centuries of tension between papal authority and that of the Councils of the Church. Roman Catholicism differs from Protestantism generally in the importance it grants to tradition and ritual, and especially in its doctrine of the Eucharist (transubstantiation), its celibate male priesthood, its emphasis on confession, its doctrine of the seven sacraments, and the veneration of the Virgin Mary and other saints. Its relations with the Orthodox Church are hampered by disagreements concerning the Filioque and the status of Uniat churches. The Roman Catholic Church became less rigid after the Second Vatican Council (1962–5), abandoning a universal Latin liturgy and being influenced by movements such as ecumenism and liberation theology, but its continuing opposition to divorce, abortion, and artificial contraception remains controversial.

Romance /rəʊˈmæns/ *n. & adj.* ● *n.* the group of European languages descended from Latin, including French, Spanish, Portuguese, Italian, Catalan, Provençal, and Romanian. ● *adj.* of or relating to any of the languages descended from Latin. [ME f. OF *romanz, -ans, -ance,* ult. f. L *Romanicus* ROMANIC]

romance /rəʊˈmæns/ *n. & v.* ● *n.* (also *disp.* /ˈrəʊmæns/) **1** an atmosphere or tendency characterized by a sense of remoteness from or idealization of everyday life. **2 a** a prevailing sense of wonder or mystery surrounding the mutual attraction in a love affair. **b** sentimental or idealized love. **c** a love affair. **3 a** a literary genre with romantic love or highly imaginative unrealistic episodes forming the central theme. **b** a work of this genre. **4** a medieval tale, usu. in verse, of some hero of chivalry, of the kind common in the Romance languages. **5 a** exaggeration or picturesque falsehood. **b** an instance of this. **6** *Mus.* a short informal piece. ● *v.* **1** *intr.* exaggerate or distort the truth, esp. fantastically. **2** *tr.* court, woo; court the favour of, esp. by flattery. [ME f. OF *romanz, -ans, -ance,* ult. f. L *Romanicus* ROMANIC]

romancer /rəʊˈmænsə(r)/ *n.* **1** a writer of romances, esp. in the medieval period. **2** a liar who resorts to fantasy.

Roman de la rose /ˌrəʊmɒn də læ ˈrəʊz/ a French poem of the 13th century, an allegorical romance embodying the aristocratic ethic of courtly love. It was composed by two different authors some forty years apart. The poem was extremely influential on the literature of the Middle Ages. [F, = romance of the rose]

Roman Empire the overseas territories under Roman rule from 27 BC, when Augustus became the first emperor, to AD 476, when the last Roman emperor, Romulus Augustulus (reigned 475–6), was deposed. At its greatest extent Roman rule or influence extended from Armenia and Mesopotamia in the east to the Iberian peninsula in the west, and from the Rhine and Danube in the north to Egypt and provinces on the Mediterranean coast of North Africa. The empire was divided after the death of Theodosius I (AD 395) into the Western Empire and the Eastern or Byzantine Empire (centred on Constantinople). Peace was maintained largely by the substantial presence of the Roman army, and a degree of unity was achieved by an extensive network of roads, a single legal system, and a common language (Latin in the West, Greek in the East). Eventually, the sheer extent of the territories led to the collapse of the Western Empire: Rome was sacked by the Visigoths

under Alaric in 410, and the last emperor of the West, Romulus Augustulus, was deposed in 476. The Eastern Empire, which was stronger, lasted until 1453 (see BYZANTINE EMPIRE).

Romanesque /ˌrəʊməˈnesk/ *adj. & n.* ● *adj.* of or relating to a style of architecture which prevailed in Europe *c.*900–1200, although sometimes dated back to the end of the Roman Empire (5th century). (*See note below.*) ● *n.* Romanesque architecture. [F f. *roman* ROMANCE]

▪ Romanesque architecture is characterized by round arches and massive vaulting, and by heavy piers, columns, and walls with small windows. It arose through the growth of monasticism and the revival of the Church after a period of instability; additionally, an increase in the functions of churches necessitated buildings with more chapels and larger choirs. Although disseminated throughout western Europe, the style reached its fullest development in central and northern France; the equivalent style in England is usually called Norman.

roman-fleuve /ˌrəʊmɒnˈflɜːv/ *n.* (*pl.* **romans-fleuves** *pronunc.* same) **1** a novel featuring the leisurely description of the lives of members of a family etc. **2** a sequence of self-contained novels. Prefigured by Honoré de Balzac and Émile Zola, this form was practised particularly by French writers in the first half of the 20th century, notably Romain Rolland. [F, = river novel]

Romania /rəʊˈmeɪnɪə/ (also **Rumania** /ruːˈmeɪ-/) a country in SE Europe with a coastline on the Black Sea; pop. (est. 1991) 23,276,000; official language, Romanian; capital, Bucharest. In Roman times Romania formed the imperial province of Dacia, and in the Middle Ages it consisted of the principalities of Wallachia and Moldavia, which were swallowed up by the Ottoman Empire in the 15th–16th centuries. The two principalities united in 1861 and gained independence in 1878, and although conquered in 1916 by the Central Powers, Romania emerged from the peace settlement with fresh territorial gains in Bessarabia (lost in 1940) and Transylvania. After the Second World War, in which it supported Germany, Romania became a Communist state under Soviet domination. After 1974 the country pursued an increasingly independent course under the virtual dictatorship of Nicolae Ceauşescu. His regime collapsed in violent popular unrest in 1989, and a new constitution, enshrining a democratic, free-market system, was approved by referendum in 1991.

Romanian /rəʊˈmeɪnɪən/ *n. & adj.* (also **Rumanian** /ruːˈmeɪ-/) ● *n.* **1 a** a native or national of Romania. **b** a person of Romanian descent. **2** the official language of Romania. (*See note below.*) ● *adj.* of or relating to Romania or its people or language.

▪ Romanian is the only Romance language spoken in eastern Europe; it developed from the Latin introduced by Trajan when he conquered the area in the 2nd century AD and annexed it to the Roman Empire. Although influenced by Slavonic languages, Romanian has kept its Latin character. It is spoken by over 23 million people in Romania itself and by the majority of the population of Moldova.

Romanic /rəʊˈmænɪk/ *n. & adj.* ● *n.* = ROMANCE *n.* ● *adj.* **1 a** of or relating to Romance languages. **b** using a Romance language. **2** descended from the ancient Romans or inheriting aspects of their social or political life. [L *Romanicus* (as ROMAN)]

Romanism /ˈrəʊmə,nɪz(ə)m/ *n.* Roman Catholicism.

Romanist /ˈrəʊmənɪst/ *n.* **1** a student of Roman history or law or of the Romance languages. **2 a** a supporter of Roman Catholicism. **b** a Roman Catholic. **3** any of several 16th-century Dutch and Flemish painters influenced by Italian Renaissance art. [mod.L *Romanista* (as ROMAN)]

romanize /ˈrəʊmə,naɪz/ *v.tr.* (also **-ise**) **1** make Roman or Roman Catholic in character. **2** put into the Roman alphabet or into roman type. □ **romanization** /ˌrəʊmənaɪˈzeɪʃ(ə)n/ *n.*

Roman numeral *n.* any of the Roman letters representing numbers: I = 1, V = 5, X = 10, L = 50, C = 100, D = 500, M = 1,000. (See also ARABIC NUMERAL.)

Romano /rəʊˈmɑːnəʊ/ *n.* a strong-tasting hard cheese, orig. made in Italy. [It.,= ROMAN]

Romano- /rəʊˈmɑːnəʊ/ *comb. form* Roman; Roman and (*Romano-British*).

Romanov /rəʊˈmɑːnɒf/ a dynasty that ruled in Russia from the accession of Michael Romanov in 1613 until the overthrow of the last tsar, Nicholas II, in 1917.

Roman Republic the ancient Roman state 509 BC–27 BC from the expulsion of the Etruscan monarchs (see TARQUINIUS) until the assumption of power by Augustus (Octavian) (see ROMAN EMPIRE). The republic was dominated by a landed aristocracy, the patricians, who ruled through the advisory Senate and two annually elected chief magistrates or consuls; the plebeians or common people had their own representatives, the tribunes, who in time gained the power of veto over the other magistrates. There was also a considerable slave population, largely made up of enslaved prisoners of war. During the life of the republic Rome came to dominate the rest of Italy and, following the Punic and Macedonian Wars, began to acquire extensive dominions in the Mediterranean and Asia Minor. But the system of government that served a city-state well proved inadequate for control of an empire: a series of ambitious military leaders (Marius, Sulla, Pompey, Julius Caesar, Mark Antony) came to the fore, dissatisfied with the Senate's control of government and each able to count on the support of a devoted soldiery, until civil wars culminated in Caesar's brief dictatorship. This established the principle of personal autocracy, and after Caesar's assassination another round of civil war ended with Octavian's assumption of authority.

Romans, Epistle to the a book of the New Testament, an epistle of St Paul to the Church at Rome.

Romansh /rəʊˈmænʃ, -ˈmɑːnʃ/ n. & adj. (also **Rumansh** /ruːˈmænʃ, -ˈmɑːnʃ/, **Romansch**) ● n. the Rhaeto-Romanic dialects, esp. as spoken in the Swiss canton of Grisons. ● adj. of these dialects. [Romansh *Ruman(t)sch, Roman(t)sch* f. med.L *romanice* (adv.) (as ROMANCE)]

romantic /rəʊˈmæntɪk/ adj. & n. ● adj. **1** of, characterized by, or suggestive of an idealized, sentimental, or fantastic view of reality; remote from experience (*a romantic picture; a romantic setting*). **2** inclined towards or suggestive of romance in love (*a romantic woman; a romantic evening; romantic words*). **3** (of a person) imaginative, visionary, idealistic. **4 a** of style in art, music, etc.) concerned more with feeling and emotion than with form and aesthetic qualities; preferring grandeur or picturesqueness to finish and proportion. (See ROMANTICISM.) **b** (also **Romantic**) of or relating to the 18th and 19th-century romantic movement or style in the European arts (cf. CLASSICAL). (See ROMANTICISM.) **5** (of a project etc.) unpractical, fantastic. ● n. **1** a romantic person. **2** a romanticist. □ **romantically** adv. [*romant* tale of chivalry etc. f. OF f. *romanz* ROMANCE]

romanticism /rəʊˈmæntɪˌsɪz(ə)m/ n. (also **Romanticism**) a movement in the arts and literature which originated in the late 18th century as a reaction against the order and restraint of classicism and neoclassicism, and was a rejection of the rationalism which characterized the Enlightenment. In their place were inspiration, irrationality, subjectivity, and the primacy of the individual. In music, the period embraces much of the 19th century, with composers including Schubert, Schumann, Liszt, Wagner, and (in the opinion of some critics) Beethoven. Writers exemplifying the movement include Wordsworth, Coleridge, Byron, Shelley, and Keats; among romantic painters are such stylistically diverse artists as William Blake, J. M. W. Turner, Delacroix, and Goya. In its implicit idea of an artist as an isolated misunderstood genius the movement has not yet ended.

romanticist /rəʊˈmæntɪsɪst/ n. (also **Romanticist**) a writer or artist of the romantic school.

romanticize /rəʊˈmæntɪˌsaɪz/ v. (also **-ise**) **1** tr. **a** make or render romantic or unreal (*a romanticized account of war*). **b** describe or portray in a romantic fashion. **2** intr. indulge in romantic thoughts or actions. □ **romanticization** /-ˌmæntɪsaɪˈzeɪʃ(ə)n/ n.

Romany /ˈrɒmənɪ, ˈrəʊm-/ n. & adj. ● n. (pl. **-ies**) **1** a gypsy. **2** the language of the gypsies. (See *note below*.) ● adj. of or concerning gypsies or their language. [Romany *Romani* fem. and pl. of *Romano* (adj.) (ROM)]

■ Romany shares features with Sanskrit and is probably of Indic origin, but the nomadic lifestyle of the Romany people and hence the contact with other languages are reflected in regional variation. It is spoken by a dispersed group of about 1 million people.

Romberg /ˈrɒmbɜːɡ/, Sigmund (1887–1951), Hungarian-born American composer. He wrote a succession of popular operettas, including *The Student Prince* (1924), *The Desert Song* (1926), and *New Moon* (1928).

Rome /rəʊm/ (Italian **Roma** /ˈroːma/) the capital of Italy and of Lazio region, situated on the River Tiber about 25 km (16 miles) inland; pop. (1990) 2,791,350. The name is used allusively to refer to the ancient Roman Republic and Empire, and also the Roman Catholic Church, based in the city at the Vatican. According to tradition, the ancient city was founded by Romulus (after whom it is named) in 753 BC on the Palatine Hill. As it grew it spread to the other six hills of Rome: Aventine, Caelian, Capitoline, Esquiline, and Quirinal. Rome was ruled by kings until the expulsion of Tarquinius Superbus in 510 BC led to the establishment of a republic (see ROMAN REPUBLIC). By the mid-2nd

century BC Rome had subdued the whole of Italy and had come to dominate the western Mediterranean and the Hellenistic world in the east, acquiring the first of the overseas possessions that became the empire (see ROMAN EMPIRE). By the time of the Roman Empire's decline and fall the city was overshadowed politically by Constantinople, but emerged as the seat of the papacy and as the spiritual capital of Western Christianity. In the 14th and 15th centuries Rome became a centre of the Renaissance and a great focus of cultural and artistic activity. It remained under papal control, forming part of the Papal States, until 1871, when it was made the capital of a unified Italy. In 1929, following the Lateran Treaty, the Vatican City was granted the status of an independent sovereign state within the city.

Rome, Treaty of a treaty signed at Rome on 25 March 1957, setting up and defining the aims of the European Economic Community. It was signed by six European states: France, West Germany, Italy, Belgium, the Netherlands, and Luxembourg.

Romeo /ˈrəʊmɪəʊ/ n. (pl. **-os**) a passionate male lover; a seducer of women; a womanizer. [the hero of Shakespeare's romantic tragedy *Romeo and Juliet*]

romer /ˈrəʊmə(r)/ n. a small piece of plastic or card marked with scales along two edges meeting at a right angle, or (if transparent) bearing a grid, used for measuring grid references on a map. [Carrol *Romer*, British barrister, its inventor (1883–1951)]

Romish /ˈrəʊmɪʃ/ adj. usu. derog. Roman Catholic.

Rommel /ˈrɒm(ə)l/, Erwin (known as the 'Desert Fox') (1891–1944), German Field Marshal. Rommel was posted to North Africa in 1941 after the collapse of the Italian offensive, and, as commander of the Afrika Korps, he deployed a series of surprise manoeuvres and succeeded in capturing Tobruk (1942). After being defeated by Montgomery at El Alamein (1942), he was ordered home the following year to serve as Inspector of Coastal Defences. He was forced to commit suicide after being implicated in the officers' conspiracy against Hitler in 1944.

Romney /ˈrɒmnɪ, ˈrʌm-/, George (1734–1802), English portrait painter. Based in London from 1762, he rivalled Thomas Gainsborough and Sir Joshua Reynolds for popularity in the late 18th century. From the early 1780s he produced over fifty portraits of Lady Hamilton in historical poses.

romneya /ˈrɒmnɪə, ˈrʌm-/ n. a shrubby poppy of the genus *Romneya*, bearing showy white flowers. [Thomas *Romney* Robinson, Ir. astronomer (1792–1882)]

romp /rɒmp/ v. & n. ● v.intr. **1** play about roughly and energetically. **2** (foll. by *along, past,* etc.) colloq. proceed without effort. ● n. **1** a spell of romping or boisterous play. **2** Sport an easy victory. □ **romp in** (or **home**) colloq. finish as the easy winner. □ **rompy** adj. (**rompier, rompiest**). [perh. var. of RAMP¹]

romper /ˈrɒmpə(r)/ n. (usu. in pl.) (also **romper suit**) a young child's one-piece garment covering legs and trunk.

Romulus /ˈrɒmjʊləs/ Rom. Mythol. the traditional founder of Rome, one of the twin sons of Mars by the Vestal Virgin Rhea Silvia. He and his brother Remus were abandoned at birth in a basket on the River Tiber but were found and suckled by a she-wolf and later brought up by a shepherd family. Their noble birth was discovered, but Remus was killed before the founding of the city, which was named after Romulus.

Roncesvalles, Battle of /ˈrɒnsəˌvæl/ (French **Roncevaux** /rɔ̃svo/) a battle which took place in 778 at a mountain pass in the Pyrenees, near the village of Roncesvalles in northern Spain. The rearguard of Charlemagne's army was attacked by the Basques and massacred; one of the nobles, Roland, was killed, an event much celebrated in medieval literature, notably in the *Chanson de Roland* (in which the attackers are wrongly identified as the Moors).

rondavel /rɒnˈdɑːvel/ n. S. Afr. **1** a round tribal hut usu. with a thatched conical roof. **2** a similar building, esp. as a holiday cottage, or as an outbuilding on a farm etc. [Afrik. *rondawel*]

ronde /rɒnd/ n. **1** a dance in which the dancers move in a circle. **2** a course of talk, activity, etc. [F, fem. of *rond* ROUND adj.]

rondeau /ˈrɒndəʊ/ n. (pl. **rondeaux** /-dəʊz/) a poem of ten or thirteen lines with only two rhymes throughout and with the opening words used twice as a refrain. [F, earlier *rondel*: see RONDEL]

rondel /ˈrɒnd(ə)l/ n. a rondeau, esp. one of special form. [ME f. OF f. *rond* ROUND: cf. ROUNDEL]

rondo /ˈrɒndəʊ/ n. (pl. **-os**) Mus. a form with a recurring leading theme, often found in the final movement of a sonata or concerto etc. [It. f. F *rondeau*: see RONDEAU]

Rondônia /rɒnˈdɒnjə/ a state of NW Brazil, on the border with Bolivia; capital, Pôrto Velho.

rone /rəʊn/ n. Sc. a gutter for carrying off rain from a roof.

ronin /ˈrəʊnɪn/ n. **1** hist. (in feudal Japan) a lordless wandering samurai; an outlaw. **2** a Japanese student retaking a university examination. [Jap.]

Röntgen /ˈrʌntjən, ˈrɒntgən, German ˈrœntg(ə)n/, Wilhelm Conrad (1845–1923), German physicist, the discoverer of X-rays. He was a skilful experimenter and worked on a variety of topics. In 1888 he demonstrated the existence of a magnetic field caused by the motion of electrostatic charges, predicted by James Clerk Maxwell and important for future electrical theory. In 1895 Röntgen observed that a nearby fluorescent screen glowed when a current was passed through a Crookes' vacuum tube. He investigated the properties of the radiation responsible, which he called 'X-rays', and produced the first X-ray photograph (of his wife's hand). He was awarded the first Nobel Prize for physics in 1901.

röntgen var. of ROENTGEN.

röntgenography var. of ROENTGENOGRAPHY.

röntgenology var. of ROENTGENOLOGY.

roo /ruː/ n. (also **'roo**) Austral. colloq. a kangaroo. [abbr.]

rood /ruːd/ n. **1** a crucifix, esp. one raised on a screen or beam at the entrance to a chancel. **2** hist. a measure of land, usu. a quarter of an acre. □ **rood-loft** a gallery on top of a rood-screen. **rood-screen** a wooden or stone carved screen separating the nave from the chancel in a church, found throughout western Europe, dating especially from the 14th–16th centuries. [OE rōd]

roof /ruːf/ n. & v. ● n. (pl. **roofs** or disp. **rooves** /ruːvz/) **1 a** the upper covering of a building, usu. supported by its walls. **b** the top of a covered vehicle. **c** the top inner surface of an oven, refrigerator, etc. **2** the overhead rock in a cave or mine etc. **3** esp. poet. the branches or the sky etc. overhead. **4** (of prices etc.) the upper limit or ceiling. ● v.tr. **1** (often foll. by in, over) cover with or as with a roof. **2** be the roof of. □ **go through the roof** colloq. (of prices etc.) reach extreme or unexpected heights. **hit** (or **go through** or **raise**) **the roof** colloq. become very angry. **roof-garden** a garden on the flat roof of a building. **roof of the mouth** the palate. **a roof over one's head** somewhere to live. **roof-rack** a framework for carrying luggage etc. on the roof of a vehicle. **roof-tree** the ridge-piece of a roof. **under one roof** in the same building. **under a person's roof** in a person's house (esp. with ref. to hospitality). □ **roofed** adj. (also in comb.). **roofless** adj. [OE hrōf]

roofage /ˈruːfɪdʒ/ n. the expanse of a roof or roofs.

roofer /ˈruːfə(r)/ n. a person who constructs or repairs roofs.

roofing /ˈruːfɪŋ/ n. **1** material for constructing a roof. **2** the process of constructing a roof or roofs.

roofscape /ˈruːfskeɪp/ n. a scene or view of roofs.

rooftop /ˈruːftɒp/ n. **1** the outer surface of a roof. **2** (esp. in pl.) the level of a roof. □ **shout it from the rooftops** make a thing embarrassingly public.

rooibos /ˈrɔɪbɒs, ˈruːɪ-/ n. S. Afr. **1** an evergreen leguminous shrub of the genus Aspalathus, of southern Africa, with leaves used to make tea. **2** an African shrub or small tree, Combretum apiculatum, with spikes of scented yellow flowers. [Afrik., = red bush]

rooinek /ˈrɔɪnek, ˈruːɪ-/ n. S. Afr. sl. offens. a British or English-speaking South African. [Afrik., = red-neck]

rook[1] /rʊk/ n. & v. ● n. **1** a black Eurasian bird, Corvus frugilegus, of the crow family, nesting in colonies in tree-tops. **2** sl. a sharper, esp. at dice or cards; a person who lives off inexperienced gamblers etc. ● v.tr. sl. **1** charge (a customer) extortionately. **2** win money from (a person) at cards etc. esp. by swindling. [OE hrōc]

rook[2] /rʊk/ n. a chess piece with its top in the shape of a battlement. [ME f. OF roc(k) ult. f. Arab. rukk, orig. sense uncert.]

rookery /ˈrʊkərɪ/ n. (pl. **-ies**) **1 a** a colony of rooks. **b** a clump of trees having rooks' nests. **2** a colony of seabirds (esp. penguins) or seals.

rookie /ˈrʊkɪ/ n. sl. **1** a new recruit, esp. in the army or police. **2** N. Amer. a new member of a sports team. [corrupt. of recruit, after ROOK[1]]

room /ruːm, rʊm/ n. & v. ● n. **1 a** space that is or might be occupied by something; capaciousness or ability to accommodate contents (it takes up too much room; there is plenty of room; we have no room here for idlers). **b** space in or on (houseroom; shelf-room). **2 a** a part of a building enclosed by walls or partitions, floor and ceiling. **b** (in pl.) a set of these occupied by a person or family; apartments or lodgings. **c** persons present in a

room (the room fell silent). **3** (in comb.) a room or area for a specified purpose (auction-room). **4** (foll. by for, or to + infin.) opportunity or scope (room to improve things; no room for dispute). ● v.intr. esp. N. Amer. (often foll. by with) have a room or rooms; lodge, board. □ **make room** (often foll. by for) clear a space (for a person or thing) by removal of others; make way, yield place. **not** (or **no**) **room to swing a cat** a very confined space. **rooming-house** a lodging house. **room-mate** a person occupying the same room as another. **room service** (in a hotel etc.) service of food or drink taken to a guest's room. □ **-roomed** adj. (in comb.). **roomful** n. (pl. **-fuls**). [OE rūm f. Gmc]

roomer /ˈruːmə(r), ˈrʊm-/ n. N. Amer. a lodger occupying a room or rooms without board.

roomette /ruːˈmet, rʊˈmet/ n. N. Amer. **1** a private single compartment in a sleeping-car. **2** a small bedroom for letting.

roomie /ˈruːmɪ, ˈrʊmɪ/ n. N. Amer. colloq. a room-mate.

roomy /ˈruːmɪ, ˈrʊmɪ/ adj. (**roomier**, **roomiest**) having much room, spacious. □ **roomily** adv. **roominess** n.

Roosevelt[1] /ˈrəʊzə‚velt/, (Anna) Eleanor (1884–1962), American humanitarian and diplomat. She was the niece of Theodore Roosevelt, and married Franklin D. Roosevelt in 1905. She was involved in a wide range of liberal causes, including civil and women's rights. After Roosevelt died in 1945 she became a delegate to the United Nations, and, as chair of the UN Commission on Human Rights, played a major role in drafting the Declaration of Human Rights (1948).

Roosevelt[2] /ˈrəʊzə‚velt/, Franklin D(elano) (known as FDR) (1882–1945), American Democratic statesman, 32nd President of the US 1933–45. Roosevelt's early political career was curtailed by his contraction of polio in 1921; in spite of the disease, he resumed public life in 1928 and received the Democratic presidential nomination in 1932. His New Deal of 1933 helped to lift the US out of the Great Depression, and after the American entry into the Second World War he played an important part in the coordination of the Allied war effort. In 1940 Roosevelt became the first American President to be elected for a third term in office and subsequently secured a fourth term. He was the joint author, with Churchill, of the Atlantic Charter (1941).

Roosevelt[3] /ˈrəʊzə‚velt/, Theodore ('Teddy') (1858–1919), American Republican statesman, 26th President of the US 1901–9. He was elected Vice-President in 1900, succeeding William McKinley in 1901 following the latter's assassination. At home Roosevelt was noted for his antitrust laws, while abroad he successfully engineered the American bid to build the Panama Canal (1904–14) and won the Nobel Peace Prize in 1906 for negotiating the end of the Russo-Japanese War. The teddy bear is named after Roosevelt, with reference to his bear-hunting.

roost[1] /ruːst/ n. & v. ● n. **1** a branch or other support on which a bird perches, esp. a place where birds regularly settle to sleep. **2** a place offering temporary sleeping-accommodation. ● v. **1** intr. **a** (of a bird) settle for rest or sleep. **b** (of a person) stay for the night. **2** tr. provide with a sleeping-place. □ **come home to roost** (of a scheme etc.) recoil unfavourably upon the originator. [OE hrōst]

roost[2] /ruːst/ n. a tidal race in the Orkney and Shetland Islands. [ON röst]

rooster /ˈruːstə(r)/ n. esp. N. Amer., Austral., etc. a domestic cock.

root[1] /ruːt/ n. & v. ● n. **1 a** the part of a plant normally below the ground, attaching it to the earth and conveying nourishment to it from the soil. **b** (in pl.) such a part divided into branches or fibres. **c** the corresponding organ of an epiphyte; the part attaching ivy to its support. **d** the permanent underground stock of a plant. **e** any small plant with a root for transplanting. **2 a** any plant, e.g. a turnip or carrot, with an edible root. **b** (also **root vegetable**) such a root. **3** (in pl.) social, ethnic, or cultural origins, esp. as the reasons for one's long-standing emotional attachment to a place, community, etc. **4 a** the embedded part of a bodily organ or structure, e.g. hair, tooth, nail, etc. **b** the part of a thing attaching it to a greater or more fundamental whole. **c** (in pl.) the base of a mountain etc. **5 a** the basic cause, source, or origin (love of money is the root of all evil; has its roots in the distant past). **b** (attrib.) (of an idea etc.) from which the rest originated. **6** the basis of something, its means of continuance or growth (has its root(s) in selfishness; has no root in the nature of things). **7** the essential substance or nature of something (get to the root of things). **8** Math. **a** a number or quantity that when multiplied by itself a usu. specified number of times gives a specified number or quantity (the cube root of eight is two). **b** a square root. **c** a value of an unknown quantity satisfying a given equation. **9** Philol. any ultimate unanalysable element of language; a basis, not necessarily surviving as a word in itself, on which words are made by the addition of prefixes or suffixes or by other modification. **10** *Mus.* the fundamental note of a chord. **11** *Bibl.* a scion, an offshoot (*there shall be a root of Jesse*). **12** *Austral. & NZ coarse sl.* **a** an act of sexual intercourse. **b** a sexual partner. ● *v.* **1 a** *intr.* take root or grow roots. **b** *tr.* cause to do this (*take care to root the plants firmly*). **2** *tr.* **a** fix firmly; establish (*fear rooted him to the spot*). **b** (as **rooted** *adj.*) firmly established (*her affection was deeply rooted; rooted objection to*). **3** *tr.* (usu. foll. by *out, up*) drag or dig up by the roots. **4** *tr. Austral. & NZ coarse sl.* **a** have sexual intercourse with. **b** exhaust, frustrate. □ **pull up by the roots 1** uproot. **2** eradicate, destroy. **put down roots 1** begin to draw nourishment from the soil. **2** become settled or established. **root and branch** thorough(ly), radical(ly). **root beer** *N. Amer.* an effervescent drink made from an extract of roots. **root canal** *Dentistry* **1** the pulp-filled cavity in the root of a tooth. **2** *US* a procedure to remove infected pulp from such a cavity and fill the cavity with inert material. **root-mean-square** *Math.* the square root of the arithmetic mean of the squares of a set of values. **root out** find and get rid of. **root sign** *Math.* = *radical sign*. **strike at the root** (or **roots**) **of** set about destroying. **strike** (or **take**) **root 1** begin to grow and draw nourishment from the soil. **2** become fixed or established. □ **rootage** *n.* **rootedness** *n.* **rootless** *adj.* **rootlike** *adj.* **rooty** *adj.* **rootlet** /ˈruːtlɪt/ *n.* [OE *rōt* f. ON *rót*, rel. to WORT & L *radix*: see RADIX]

root[2] /ruːt/ *v.* **1 a** *intr.* (of an animal, esp. a pig) turn up the ground with the snout, beak, etc., in search of food. **b** *tr.* (foll. by *up*) turn up (the ground) by rooting. **2 a** *intr.* (foll. by *around, in,* etc.) rummage. **b** *tr.* (foll. by *out* or *up*) find or extract by rummaging. **3** *intr.* (foll. by *for*) esp. *US colloq.* encourage by applause or support. □ **rooter** *n.* (in sense 3). [earlier *wroot* f. OE *wrōtan* & ON *róta*: rel. to OE *wrōt* snout]

rootle /ˈruːt(ə)l/ *v.intr. & tr. Brit.* = ROOT[2] 1, 2. [ROOT[2]]

roots /ruːts/ *adj.* (esp. of music) expressive of a distinctive ethnic or cultural (esp. West Indian) identity; traditional. □ **rootsy** *adj. colloq.* [pl. of ROOT[1]]

rootstock /ˈruːtstɒk/ *n.* **1** a rhizome. **2** a plant into which a graft is inserted. **3** a primary form from which offshoots have arisen.

rooves see ROOF.

rope /rəʊp/ *n. & v.* ● *n.* **1 a** stout cord made by twisting together strands of hemp, sisal, flax, cotton, nylon, wire, or similar material. **b** a piece of this. **c** *US* a lasso. **2** (foll. by *of*) a quantity of onions, ova, or pearls strung together. **3** (in *pl.*, prec. by *the*) **a** the conditions in some sphere of action (*know the ropes; show a person the ropes*). **b** the ropes enclosing a boxing or wrestling-ring or cricket ground. **4** (prec. by *the*) **a** a halter for hanging a person. **b** execution by hanging. ● *v.* **1** *tr.* fasten, secure, or catch with rope. **2** *tr.* (usu. foll. by *off, in*) enclose (a space) with rope. **3** *Mountaineering* **a** *tr.* connect (a party) with a rope; attach (a person) to a rope. **b** (*absol.*) put on a rope. **c** *intr.* (foll. by *down, up*) climb down or up using a rope. □ **give a person plenty of rope** (or **enough rope to hang himself** or **herself**) give a person enough freedom of action to bring about his or her own downfall. **on the rope** *Mountaineering* roped together. **on the ropes 1** *Boxing* forced against the ropes by the opponent's attack. **2** near defeat. **rope in** persuade to take part. **rope into** persuade to take part in (*was roped into doing the washing-up*). **rope-ladder** two long ropes connected by short crosspieces, used as a ladder. **rope-moulding** *Archit.* a moulding cut spirally in imitation of rope-strands. **rope of sand** delusive security. **rope's end** *hist.* a short piece of rope used to flog (formerly, esp. a sailor) with. **rope-walk** *hist.* a long piece of ground where ropes are made. **rope-walker** a performer on a tightrope. **rope-walking** the action of performing on a tightrope. **rope-yard** a rope-making establishment. **rope-yarn 1** material obtained by unpicking rope-strands, or used for making them. **2** a piece of this. **3** a mere trifle. [OE *rāp* f. Gmc]

ropeable /ˈrəʊpəb(ə)l/ *adj.* (also **ropable**) **1** capable of being roped. **2** *Austral. & NZ sl.* angry.

ropemanship /ˈrəʊpmənˌʃɪp/ *n.* skill in rope-walking or climbing with ropes.

ropeway /ˈrəʊpweɪ/ *n.* a cable railway.

roping /ˈrəʊpɪŋ/ *n.* a set or arrangement of ropes.

ropy /ˈrəʊpɪ/ *adj.* (also **ropey**) (**ropier, ropiest**) **1** *Brit. colloq.* **a** poor in quality. **b** unwell. **2** (of wine, bread, etc.) forming viscous or gelatinous threads. **3** like a rope. □ **ropily** *adv.* **ropiness** *n.*

roque /rəʊk/ *n. US* croquet played on a hard court surrounded by a bank. [alt. form of ROQUET]

Roquefort /ˈrɒkfɔː(r)/ *n. propr.* a soft blue cheese made from ewes' milk at Roquefort. It is ripened in limestone caves and has a strong flavour. [*Roquefort*, a village in SW France]

roquet /ˈrəʊkeɪ, -kɪ/ *v. & n. Croquet* ● *v.* (**roqueted** /-keɪd/; **roqueting** /-keɪɪŋ/) **1** *tr.* **a** cause one's ball to strike (another ball). **b** (of a ball) strike (another). **2** *intr.* strike another ball thus. ● *n.* an instance of roqueting. [app. arbitrarily f. CROQUET *v.*, orig. used in the same sense]

roquette /rɒˈket/ *n.* = ROCKET[2] 2. [F: see ROCKET[2]]

Roraima /rɔːˈraɪmə/ **1** a mountain in the Guiana Highlands of South America, situated at the junction of the borders of Venezuela, Brazil, and Guyana. Rising to 2,774 m (9,094 ft), it is the highest peak in the range. **2** a state of northern Brazil, on the borders with Venezuela and Guyana; capital, Boa Vista.

ro-ro /ˈrəʊrəʊ/ *adj.* roll-on roll-off. [abbr.]

rorqual /ˈrɔːkwəl/ *n.* a baleen whale of the family Balaenopteridae, with a pleated throat and small dorsal fin, esp. the fin whale. [F f. Norw. *røyrkval* f. OIcel. *reythr* the specific name + *hvalr* WHALE[1]]

Rorschach test /ˈrɔːʃɑːk, -ʃɑːx/ *n. Psychol.* a personality test in which a standard set of ink-blots is presented to the subject, who is then asked to describe what they suggest or resemble. Also called *ink-blot test*. [Hermann *Rorschach*, Swiss psychiatrist (1884–1922), who devised it]

rort /rɔːt/ *n. Austral. sl.* **1** a trick, a fraud; a dishonest practice. **2** a wild party. [back-form. f. RORTY]

rorty /ˈrɔːtɪ/ *adj.* (**rortier, rortiest**) *Brit. sl.* **1** splendid; boisterous, rowdy (*had a rorty time*). **2** coarse, earthy. [19th c.: orig. unkn.]

Rosa /ˈrəʊzə/, Salvator (1615–73), Italian painter and etcher. His reputation is chiefly based on his landscapes, often peopled with bandits and containing scenes of violence in wild natural settings; their picturesque and 'sublime' qualities were an important influence on the romantic art of the 18th and 19th centuries.

rosace /ˈrəʊzeɪs/ *n.* **1** a rose-window. **2** a rose-shaped ornament or design. [F f. L *rosaceus*: see ROSACEOUS]

rosaceous /rəʊˈzeɪʃəs/ *adj. Bot.* of the large plant family Rosaceae, which includes the rose. [L *rosaceus* f. *rosa* ROSE[1]]

rosaline /ˈrəʊzəˌliːn/ *n.* a variety of fine needlepoint or pillow lace. [prob. F]

rosaniline /rəʊˈzænɪˌliːn, -lɪn, -ˌlaɪn/ *n.* **1 a** *Chem.* an organic base derived from aniline. **b** a red dye obtained from this. **2** fuchsine. [ROSE[1] + ANILINE]

rosarian /rəˈzeərɪən/ *n.* a person who cultivates roses, esp. professionally. [L *rosarium* ROSARY]

Rosario /rəʊˈsɑːrɪəʊ/ an inland port on the Paraná river in east central Argentina; pop. (1991) 1,096,000.

rosarium /rəˈzeərɪəm/ *n.* (*pl.* **rosariums** or **rosaria** /-rɪə/) a rose-garden. [L (as ROSARY)]

rosary /ˈrəʊzərɪ/ *n.* (*pl.* **-ies**) **1** *RC Ch.* **a** a form of devotion in which five (or fifteen) decades of Hail Marys are repeated, each decade preceded by an Our Father and followed by a Glory Be. The practice dates from the 15th century. **b** a string of 55 (or 165) beads for keeping count in this. **c** a book containing this devotion. **2** a similar form of bead-string used in other religions. **3** a rose-garden or rose-bed. [ME f. L *rosarium* rose-garden, neut. of *rosarius* (as ROSE[1])]

Roscius /ˈrɒsɪəs, ˈrɒʃɪ-/ (full name Quintus Roscius Gallus) (d.62 BC), Roman actor. He achieved phenomenal success as a comic actor during his lifetime and later became identified with all that was considered best in acting; many notable English actors from the 16th century onwards were nicknamed in reference to him.

roscoe /ˈrɒskəʊ/ *n. US sl.* a gun, esp. a pistol or revolver. [the name *Roscoe*]

Roscommon /rɒsˈkɒmən/ **1** a county in the north central part of the Republic of Ireland, in the province of Connacht; pop. (1991) 51,880. **2** its county town; pop. (1991) 17,700.

rose[1] /rəʊz/ *n., adj., & v.* ● *n.* **1** a prickly bush or shrub of the genus *Rosa* (family Rosaceae), bearing usu. fragrant flowers generally of a red, pink, yellow, or white colour. **2** this flower. **3** a flowering plant resembling this (*Christmas rose; rock rose*). **4 a** a light crimson colour, pink. **b** (usu. in *pl.*) a rosy complexion (*roses in her cheeks*). **5 a** a representation of the flower in heraldry or decoration (esp. as the national emblem of England). **b** a rose-shaped design, e.g. on a compass card or on the sound-hole of a lute etc. **6** the sprinkling-nozzle of a watering-can or hose. **7** a circular mounting on a ceiling through which the wiring of an electric light passes. **8 a** a rose diamond. **b** a rose-window. **9** (in *pl.*) used in various phrases to express favourable circumstances, ease, success, etc. (*roses all the way; everything's roses*). **10** an excellent person or thing, esp. a beautiful woman (*English rose;*

rose between two thorns). ● *adj.* light crimson, pink. ● *v.tr.* (esp. as **rosed** *adj.*) make (one's face) rosy. □ **rose-apple 1** a tropical tree of the genus *Syzygium*, native to Malaysia, cultivated for its foliage and fragrant fruit. **2** this fruit. **rose-bush** a rose plant. **rose-chafer** a metallic green or copper-coloured beetle, *Cetonia aurata*, feeding on rose flowers etc. **rose-colour** the colour of a pale red rose, warm pink. **rose-coloured 1** of rose-colour. **2** optimistic, sanguine, cheerful (*takes rose-coloured views*). **rose comb** a flat fleshy comb of a fowl. **rose-cut** cut as a rose diamond. **rose diamond** a hemispherical diamond with the curved part cut in triangular facets. **rose-engine** an appendage to a lathe for engraving curved patterns. **rose-fish** a bright red food fish, *Sebastes marinus*, of the North Atlantic. **rose geranium** a pink-flowered sweet-scented pelargonium, *Pelargonium graveolus*. **rose-hip** = HIP[2]. **rose-leaf** (*pl.* **-leaves**) **1** a petal of a rose. **2** a leaf of a rose. **rose madder** a pale pink pigment. **rose-mallow** = HIBISCUS. **rose nail** a nail with a head shaped like a rose diamond. **rose of Jericho** a cruciferous desert plant, *Anastatica hierochuntica*, of North Africa and the Middle East (also called *resurrection plant*). **rose of Sharon 1** a shrubby St John's wort, *Hypericum calycinum*, with dense foliage and golden-yellow flowers (also called *Aaron's beard*). **2** *Bibl.* a flowering plant of unknown identity. **rose-pink** = rose-colour, rose-coloured 1. **rose-point** a point lace with a design of roses. **rose quartz** a translucent pink variety of quartz. **rose-red** *adj.* red like a rose, rose-coloured. ● *n.* this colour. **rose-root** a yellow-flowered stonecrop, *Sedum* (or *Rhodiola*) *rosea*, with roots smelling like a rose when dried or bruised. **rose-tinted** = rose-coloured. **rose-tree** a rose plant, esp. a standard rose. **rose-water** perfume made from roses. **rose-window** a circular window, usu. with roselike or spokelike tracery. **see through rose-coloured** (or **-tinted**) **spectacles** regard (circumstances etc.) with unfounded favour or optimism. **under the rose** in confidence; under pledge of secrecy. □ **roseless** *adj.* **roselike** *adj.* [ME f. OE *rōse* f. L *rosa*]

rose[2] past of RISE.

rosé /ˈrəʊzeɪ/ *n.* any light pink wine, coloured by only brief contact with red grape-skins. [F, = pink]

roseate /ˈrəʊzɪət/ *adj.* **1** rose-coloured. **2** having a partly pink plumage (*roseate spoonbill*; *roseate tern*). [L *roseus* rosy (as ROSE[1])]

Roseau /rəʊˈzəʊ/ the capital of Dominica in the West Indies; pop. (est. 1990) 20,000.

rosebay /ˈrəʊzbeɪ/ *n.* **1 a** the oleander. **b** a North American azalea. **2** (in full **rosebay willowherb**) a tall willowherb, *Chamerion angustifolium*, with showy pink flowers. Also called *fireweed*.

Rosebery /ˈrəʊzbərɪ/, 5th Earl of (title of Archibald Philip Primrose) (1847–1929), British Liberal statesman, Prime Minister 1894–5. He succeeded Gladstone as Premier after the latter's retirement and subsequently alienated Liberal supporters as a result of his imperialist loyalties during the Second Boer War (1899–1902).

rosebowl /ˈrəʊzbəʊl/ *n.* a bowl for displaying cut roses.

rosebud /ˈrəʊzbʌd/ *n.* **1** a bud of a rose. **2** a pretty young woman.

rosella /rəˈzelə/ *n.* **1** a brightly coloured Australian parakeet of the genus *Platycercus*. **2** *Austral.* an easily shorn sheep. [corrupt. of *Rosehill*, NSW, where the bird was first found]

rosemaling /ˈrəʊzəˌmɑːlɪŋ, -ˌmɔːlɪŋ/ *n.* the art of painting wooden furniture etc. with flower motifs. [Norw., = rose painting]

rosemary /ˈrəʊzmərɪ/ *n.* a fragrant evergeen labiate shrub, *Rosmarinus officinalis*, with narrow leaves, used as a culinary herb, in perfumery, etc., and taken as an emblem of remembrance. [ME, earlier *rosmarine* ult. f. L *ros marinus* f. *ros* dew + *marinus* MARINE, with assim. to ROSE[1] and *Mary* name of the Virgin]

roseola /rəʊˈziːələ/ *n. Med.* **1** a rosy rash occurring in measles, typhoid, syphilis, and other diseases. **2** *archaic* = RUBELLA. □ **roseolar** *adj.* **roseolous** *adj.* [mod. var. of RUBEOLA f. L *roseus* rose-coloured]

rosery /ˈrəʊzərɪ/ *n.* (*pl.* **-ies**) a rose-garden.

Roses, Wars of the see WARS OF THE ROSES.

Rose Theatre a theatre in Southwark, London, built in 1587. Many of Shakespeare's plays were performed there, including some (*Henry VI* and *Titus Andronicus*) for the first time. Remains of the theatre, which was demolished *c*.1605, were uncovered in 1989.

Rosetta Stone /rəʊˈzetə/ an inscribed stone found near Rosetta on the western mouth of the Nile by one of Napoleon's officers in 1799. Its text, a decree commemorating the accession of Ptolemy V (reigned 205–180 BC) is written in two languages and three scripts: hieroglyphic and demotic Egyptian, and Greek. The deciphering of hieroglyphs forming Egyptian parts of the inscription by Jean-François Champollion in 1822 led to the interpretation of many other early

records of Egyptian civilization. The stone is now in the British Museum.

rosette /rəʊˈzet/ *n.* **1** a rose-shaped ornament made usu. of ribbon and worn esp. as a supporter's badge, or as an award or the symbol of an award in a competition, esp. by a prizewinning animal. **2 a** *Archit.* a carved or moulded ornament resembling or representing a rose. **b** a rose-window. **3** an object or symbol or arrangement of parts resembling a rose. **4** *Biol.* **a** a roselike cluster of parts. **b** markings resembling a rose. **5** a rose diamond. □ **rosetted** *adj.* [F dimin. of *rose* ROSE[1]]

rosewood /ˈrəʊzwʊd/ *n.* **1** a fragrant close-grained wood used in making furniture. **2** a tree that yields this, esp. a tropical leguminous tree of the genus *Dalbergia*.

Rosh Hashana /ˌrɒʃ həˈʃɑːnə, ˌrəʊʃ hɑːʃɑːˈnɑː/ (also **Rosh Hashanah**) the Jewish New Year, celebrated on the first (and sometimes second) day of the month Tishri (September–October). [Heb., = beginning (lit. 'head') of the year]

Roshi /ˈrəʊʃɪ/ *n.* (*pl.* **Roshis**) the spiritual leader of a community of Zen Buddhist monks. [Jap.]

Rosicrucian /ˌrəʊzɪˈkruːʃ(ə)n/ *n.* a member of a 17th and 18th century occult society, devoting itself to metaphysical and mystical lore, especially that concerning the transmutation of metals, prolongation of life, and power over the elements. An anonymous pamphlet of 1614, telling the story of a mythical knight of the 15th century called Christian Rosenkreuz, is said to have launched the movement. The name is also applied to various similar modern organizations. □ **Rosicrucianism** *n.* [mod.L *rosa crucis* (or *crux*), as Latinization of G *Rosenkreuz*]

rosin /ˈrɒzɪn/ *n.* & *v.* ● *n.* resin, esp. the solid residue after distillation of oil of turpentine from crude turpentine. ● *v.tr.* (**rosined, rosining**) **1** rub (esp. the bow of a violin etc.) with rosin. **2** smear or seal up with rosin. □ **rosiny** *adj.* [ME, alt. f. RESIN]

Rosinante /ˌrɒzɪˈnæntɪ/ the name of Don Quixote's horse. The name is used allusively to refer to a worn-out or inferior horse.

Roskilde /ˈrɒskɪlə/ a port in Denmark, on the island of Zealand; pop. (1990) 49,080. It was the seat of Danish kings from *c*.1020 and the capital of Denmark until 1443.

rosolio /rəˈzəʊlɪˌəʊ/ *n.* (also **rosoglio**) (*pl.* **-os**) a sweet cordial of spirits, sugar, and flavouring. [It., f. mod.L *ros solis* dew of the sun]

RoSPA /ˈrɒspə/ *abbr.* (in the UK) Royal Society for the Prevention of Accidents.

Ross[1] /rɒs/, Diana (b.1944), American pop and soul singer. She made her name as the lead singer of the Supremes, with whom she recorded many hit singles. She left the group in 1969 and became a successful solo artist, recording songs such as 'Remember Me' (1971). She has also appeared in several films, including *Lady Sings the Blues* (1973), for which she received an Oscar for her role as the jazz singer Billie Holiday.

Ross[2] /rɒs/, Sir James Clark (1800–62), British explorer. He discovered the north magnetic pole in 1831, and headed an expedition to the Antarctic from 1839 to 1843, in the course of which he discovered Ross Island, Ross Dependency, and the Ross Sea, all named after him. He was the nephew of Sir John Ross.

Ross[3] /rɒs/, Sir John (1777–1856), British explorer. He led an expedition to Baffin Bay in 1818 and another in search of the North-west Passage between 1829 and 1833, during which he surveyed King William Land, the Boothia Peninsula, and the Gulf of Boothia. He was the uncle of Sir James Clark Ross.

Ross[4] /rɒs/, Sir Ronald (1857–1932), British physician. He worked in the Indian Medical Service and became interested in malaria, and while on a visit to England met Patrick Manson who suggested that it was transmitted by a mosquito. Ross confirmed that the *Anopheles* mosquito was indeed the vector, and went on to elucidate the stages in the malarial parasite's life cycle. He was awarded a Nobel Prize in 1902.

Ross and Cromarty /ˈkrɒmətɪ/ a former county of northern Scotland, stretching from the Moray Firth to the North Minch. Since 1975 it has been part of Highland Region.

Ross Dependency part of Antarctica administered by New Zealand. It was explored in 1841 by Sir James Clark Ross, after whom it is named, and brought within the jurisdiction of New Zealand in 1923. The territory consists of everything lying to the south of latitude 60° south between longitudes 150° and 160° west.

Rossellini /ˌrɒsəˈliːnɪ/, Roberto (1906–77), Italian film director. He is known for his neo-realist films, particularly his quasi-documentary trilogy about the Second World War, filmed using a mainly non-professional cast; this comprises *Open City* (1945), *Paisà* (1946), and *Germany, Year Zero* (1947).

Rossetti[1] /rə'zetɪ/, Christina (Georgina) (1830–94), English poet. She contributed several poems to the Pre-Raphaelite journal *The Germ* in 1850. Influenced by the Oxford Movement, Rossetti wrote much religious poetry reflecting her High Anglican faith, although she also wrote love poetry and children's verse. Marked by technical virtuosity, a sense of melancholy, and recurrent themes of frustrated love and premature resignation, her work includes the verse collection *Goblin Market and Other Poems* (1862). She was the sister of Dante Gabriel Rossetti.

Rossetti[2] /rə'zetɪ/, Dante Gabriel (full name Gabriel Charles Dante Rossetti) (1828–82), English painter and poet. He was a founder member of the Pre-Raphaelite brotherhood (1848), and encouraged the movement to make links between painting and literature, basing many of his paintings on the work of the Italian poet Dante. Rossetti is best known for his dreamy and idealized images of women including *Beata Beatrix* (c.1863) and *The Blessed Damozel* (1871–9); the latter took its subject from his poem of 1850. From 1861 Rossetti was associated with William Morris's firm Morris & Company. He was the brother of Christina Rossetti.

Rossini /rɒ'si:nɪ/, Gioacchino Antonio (1792–1868), Italian composer. He wrote over thirty operas, of which the best known are the comic opera *The Barber of Seville* (1816) and the grand opera *William Tell* (1829). He was one of the creators of the Italian bel canto style of singing, along with Bellini and Donizetti.

Rosslare /rɒs'leə(r)/ a ferry port on the SE coast of the Republic of Ireland, in County Wexford.

Ross Sea a large arm of the Pacific forming a deep indentation in the coast of Antarctica. It was first explored in January 1841 by an expedition led by Sir James Clark Ross, after whom many features of this area are named. At its head is the Ross Ice Shelf, the world's largest body of floating ice, which is approximately the size of France. On the eastern shores of the Ross Sea lies Ross Island, which is the site of Mount Erebus and of Scott Base, an Antarctic station established by New Zealand in 1957.

Rostand /'rɒstɒn/, Edmond (1868–1918), French dramatist and poet. His reputation is chiefly based on his poetic drama *Cyrano de Bergerac* (1897), which romanticized the life of the 17th-century soldier, duellist, and writer Cyrano de Bergerac.

roster /'rɒstə(r), 'rəʊst-/ n. & v. ● n. a list or plan showing turns of duty or leave for individuals or groups, esp. of a military force. ● v.tr. place on a roster. [Du. *rooster* list, orig. gridiron f. *roosten* ROAST, with ref. to its parallel lines]

Rostock /'rɒstɒk/ an industrial port on the Baltic coast of Germany; pop. (1991) 244,000.

Rostov-on-Don /ˌrɒstɒvɒn'dɒn/ (also **Rostov** /'rɒstɒf/) a port and industrial city in SW Russia, on the River Don near its point of entry into the Sea of Azov; pop. (1990) 1,025,000. The city is built around a fortress erected by the Turks in the 18th century.

rostra pl. of ROSTRUM.

rostral /'rɒstrəl/ adj. 1 *Zool. & Bot.* of or on the rostrum. 2 *Anat.* **a** nearer the hypophysial area in the early embryo. **b** nearer the region of the nose and mouth in post-embryonic life. 3 *Archit.* (of a column etc.) adorned with the beaks of ancient war-galleys or with representations of these. □ **rostrally** adv.

rostrated /rɒ'streɪtɪd/ adj. 1 *Zool. & Bot.* having or ending in a rostrum. 2 *Archit.* = ROSTRAL 3. [L *rostratus* (as ROSTRUM)]

rostrum /'rɒstrəm/ n. (pl. **rostra** /-trə/ or **rostrums**) 1 **a** a platform for public speaking. **b** a conductor's platform facing the orchestra. **c** a similar platform for other purposes, e.g. for supporting a film or television camera. 2 *Zool. & Bot.* a beak, stiff snout, or beaklike part, esp. of an insect or arachnid. 3 *Rom. Antiq.* the beak of a war-galley. □ (all in sense 2) **rostrate** /-trət/ adj. **rostriform** /-trɪ,fɔ:m/ adj. **rostriferous** /rɒ'strɪfərəs/ adj. [L, = beak f. *rodere* ros- gnaw: orig. *rostra* (pl., in sense 1a) in the Roman forum adorned with beaks of captured galleys]

rosy /'rəʊzɪ/ adj. (**rosier, rosiest**) 1 coloured like a pink or red rose (*rosy cheeks; rosy dawn*). 2 optimistic, hopeful, cheerful (*a rosy future; a rosy attitude to life*). □ **rosily** adv. **rosiness** n.

rot /rɒt/ v., n., & int. ● v. (**rotted, rotting**) 1 intr. **a** (of animal or vegetable matter) lose its original form by the chemical action of bacteria, fungi, etc.; decay. **b** (foll. by *off, away*) crumble or drop from a stem etc. through

decomposition. 2 intr. **a** (of society, institutions, etc.) gradually perish from lack of vigour or use. **b** (of a prisoner etc.) waste away (*left to rot in prison*) languish. 3 tr. cause to rot, make rotten. 4 tr. *Brit. sl.* tease, abuse, denigrate. 5 intr. *Brit. sl.* joke. ● n. 1 the process or state of rotting. 2 *colloq.* nonsense; an absurd or foolish statement, argument, or proposal. 3 a sudden series of (usu. unaccountable) failures; a rapid decline in standards etc. (*a rot set in; we must try to stop the rot*). 4 (often prec. by *the*) a necrotic liver disease of sheep caused by liver-fluke infestation. ● int. *colloq.* expressing incredulity or ridicule. □ **rot-gut** *sl.* cheap harmful alcoholic liquor. [OE *rotian* (v.): (n.) ME, perh. f. Scand.: cf. Icel., Norw. *rot*]

rota /'rəʊtə/ n. 1 esp. *Brit.* a list of persons acting, or duties to be done, in rotation; a roster. 2 (**Rota**) *RC Ch.* the supreme ecclesiastical and secular court. [L, = wheel]

Rotarian /rəʊ'teərɪən/ n. & adj. ● n. a member of Rotary International. ● adj. of Rotary International. [ROTARY + -AN]

rotary /'rəʊtərɪ/ adj. & n. ● adj. acting by rotation (*rotary drill; rotary pump*). ● n. (pl. **-ies**) 1 a rotary machine. 2 *N. Amer.* a traffic roundabout. □ **rotary-wing** (of an aircraft) deriving lift from rotary aerofoils. [med.L *rotarius* (as ROTA)]

Rotary club n. a local branch of Rotary International.

Rotary International an international society for business and professional men having as its aim the promotion of charitable work. Its name derives from the fact that the first local group, formed at Chicago in 1905, met at each member's premises in rotation.

rotate[1] /rəʊ'teɪt/ v. 1 intr. & tr. move round an axis or centre, revolve. 2 tr. take or arrange in rotation. **b** intr. act or take place in rotation (*the chairmanship will rotate*). □ **rotatable** adj. **rotative** /'rəʊtətɪv/ adj. **rotatory** /'rəʊtətərɪ, rəʊ'teɪtərɪ/ adj. [L *rotare* f. *rota* wheel]

rotate[2] /'rəʊteɪt/ adj. *Bot.* wheel-shaped. [formed as ROTA]

rotation /rəʊ'teɪʃ(ə)n/ n. 1 the act or an instance of rotating or being rotated. 2 a recurrence; a recurrent series or period; a regular succession of various members of a group in office etc. 3 a system of growing different crops in regular order to avoid exhausting the soil. □ **rotational** adj. **rotationally** adv. [L *rotatio*]

rotator /rəʊ'teɪtə(r)/ n. 1 a machine or device for causing something to rotate. 2 *Anat.* a muscle that rotates a limb etc. 3 a revolving apparatus or part. [L (as ROTATE[1])]

Rotavator /'rəʊtə,veɪtə(r)/ n. (also **Rotovator**) *propr.* a machine with a rotating blade for breaking up or tilling the soil. □ **rotavate** v.tr. [ROTARY + CULTIVATOR]

rote /rəʊt/ n. (usu. prec. by *by*) mechanical or habitual repetition (with ref. to acquiring knowledge). [ME: orig. unkn.]

rotenone /'rəʊtə,nəʊn/ n. a toxic crystalline substance obtained from the roots of derris and other plants, used as an insecticide. [Jap. *rotenon* f. *roten* derris]

Roth /rɒθ/, Philip (Milton) (b.1933), American novelist and short-story writer. The complexity and diversity of contemporary American Jewish life is the subject of many of his works, to which he often brings both irony and humour. His best-known novel, *Portnoy's Complaint* (1969), records the intimate, often sexually explicit, confessions of an adolescent boy, Alexander Portnoy, to his psychiatrist.

Rotherham /'rɒðərəm/ an industrial town in northern England, in South Yorkshire; pop. (1991) 247,100.

Rothko /'rɒθkəʊ/, Mark (born Marcus Rothkovich) (1903–70), American painter, born in Latvia. In the late 1940s he became a leading figure in colour-field painting, creating canvases consisting of hazy and apparently floating rectangles of colour, usually arranged vertically and in parallel, with the intention of absorbing the spectator in an act of contemplation. Famous works include his series of nine paintings for the Seagram Building in New York, notably *Black on Maroon* (1958).

Rothschild[1] /'rɒθstʃaɪld, 'rɒθtʃaɪld/ a famous Jewish banking-house, first established in Frankfurt at the end of the 18th century by Meyer Rothschild and eventually spreading its operations all over western Europe.

Rothschild[2] /'rɒθstʃaɪld, 'rɒθtʃaɪld/, Meyer Amschel (1743–1812), German financier. He was the founder of the Rothschild banking-house in Frankfurt and financial adviser to the landgrave of Hesse. By the time of his death, his firm had already conducted significant financial transactions for a number of European governments. He had five sons, all of whom entered banking, setting up branches of the organization across western Europe. Notable among them were Nathan Meyer, Baron de Rothschild (1777–1836), who founded a bank in London (1804) and became a British citizen; Nathan's son, Lionel Nathan, Baron de Rothschild (1808–79), was Britain's first Jewish MP.

rotifer /ˈrəʊtɪfə(r)/ n. Zool. a microscopic aquatic animal of the phylum Rotifera, with a characteristic wheel-like ciliated organ used in swimming and feeding. [mod.L *rotiferus* f. L *rota* wheel + *-fer* bearing]

rotisserie /rəʊˈtɪsəri/ n. **1** a restaurant etc. where meat is roasted or barbecued. **2** a cooking appliance with a rotating spit for roasting and barbecuing meat. [F *rôtisserie* (as ROAST)]

rotogravure /ˌrəʊtəgrəˈvjʊə(r)/ n. **1** a printing system using a rotary press with intaglio cylinders, usu. running at high speed. **2** a sheet etc. printed with this system. [G *Rotogravur* (name of a company) assim. to PHOTOGRAVURE]

rotor /ˈrəʊtə(r)/ n. **1** a rotary part of a machine, esp. in the distributor of an internal-combustion engine. **2** a set of radiating aerofoils round a hub on a helicopter, providing lift when rotated. [irreg. for ROTATOR]

Rotorua /ˌrəʊtəˈruːə/ a city and health resort on North Island, New Zealand, on the south-west shore of Lake Rotorua; pop. (1991) 53,700. It lies at the centre of a region of thermal springs and geysers.

Rotovator var. of ROTAVATOR.

rotten /ˈrɒt(ə)n/ adj. (**rottener, rottenest**) **1** rotting or rotted; falling to pieces or liable to break or tear from age or use. **2 a** morally, socially, or politically corrupt. **b** despicable, contemptible. **3** colloq. **a** disagreeable, unpleasant (*had a rotten time*). **b** (of a plan etc.) ill-advised, unsatisfactory (*a rotten idea*). **c** disagreeably ill (*feel rotten today*). □ **rotten apple** a morally corrupt person in a group, organization, etc. **rotten-stone** decomposed siliceous limestone used as a powder for polishing metals. □□ **rottenly** adv. **rottenness** /-t(ə)nnɪs/ n. [ME f. ON *rotinn*, rel. to ROT, RET]

rotten borough n. hist. a British borough represented in Parliament although no longer having a real constituency. Before the Reform Act of 1832, in which such boroughs were largely disenfranchised, elections in rotten boroughs were rarely contested, the choice of MP often being in the hands of one person or family.

rotter /ˈrɒtə(r)/ n. esp. Brit. colloq. an objectionable, unpleasant, or reprehensible person. [ROT]

Rotterdam /ˈrɒtədæm/ a city in and the principal port of the Netherlands, at the mouth of the River Meuse, 25 km (15 miles) inland from the North Sea; pop. (1991) 582,270. It is one of the world's largest ports and a major oil refinery, with extensive shipbuilding and petrochemical industries.

Rottweiler /ˈrɒtˌvaɪlə(r), -ˌwaɪlə(r)/ n. a large powerful black-and-tan breed of dog with short hair, broad head, and pendent ears. [G f. *Rottweil* in SW Germany]

rotund /rəʊˈtʌnd/ adj. **1 a** circular, round. **b** (of a person) large and plump, podgy. **2** (of speech, literary style, etc.) sonorous, grandiloquent. □□ **rotundity** n. **rotundly** adv. [L *rotundus* f. *rotare* ROTATE[1]]

rotunda /rəʊˈtʌndə/ n. **1** a building with a circular ground-plan, esp. one with a dome. **2** a circular hall or room. [earlier *rotonda* f. It. *rotonda* (*camera*) round (chamber), fem. of *rotondo* round (as ROTUND)]

Rouault /ruːˈəʊ, French rwo/, Georges (Henri) (1871–1958), French painter and engraver. Although he exhibited with the fauves in 1905, he is chiefly associated with expressionism. His best-known paintings are characterized by the use of vivid colours and simplified forms enclosed in thick black outlines, reflecting the influence of his apprenticeship to a stained-glass window-maker (1885–90). A devout Roman Catholic, from the 1930s he turned increasingly towards religious subject-matter; notable among such works is *Christ Mocked by Soldiers* (1932).

rouble /ˈruːb(ə)l/ n. (also **ruble**) the basic monetary unit of Russia, the USSR (hist.), and some other former republics of the USSR, equal to 100 copecks. [F f. Russ. *rubl'*]

roué /ˈruːeɪ/ n. a debauchee; a rake. [F, past part. of *rouer* break on wheel, = one deserving this]

Rouen /ˈruːɒn, French rwɑ̃/ a port on the River Seine in NW France, chief town of Haute-Normandie; pop. (1990) 105,470. Known during Roman times as Rotomagus, Rouen was in English possession from the time of the Norman Conquest (1066) until captured by the French in 1204, becoming the medieval capital of Normandy. It returned briefly to English rule (1419–49) after its capture by Henry V during the Hundred Years War. In 1431 Joan of Arc was tried and burnt at the stake there.

rouge /ruːʒ/ n. & v. ● n. **1** a red powder or cream used for colouring the cheeks. **2** powdered ferric oxide etc. as a polishing agent esp. for metal.

● v. **1** tr. colour with rouge. **2** intr. **a** apply rouge to one's cheeks. **b** become red, blush. □ **rouge-et-noir** /-eɪˈnwɑː(r)/ a gambling game using a table with red and black marks, on which players place stakes. [F, = red, f. L *rubeus*, rel. to RED]

rough /rʌf/ adj., adv., n., & v. ● adj. **1 a** having an uneven or irregular surface, not smooth or level or polished. **b** Tennis applied to the side of a racket from which the twisted gut projects. **2** (of ground, country, etc.) having many bumps, obstacles, etc. **3 a** hairy, shaggy. **b** (of cloth) coarse in texture. **4 a** (of a person or behaviour) not mild or quiet or gentle; boisterous, unrestrained (*rough manners; rough play*). **b** (of language etc.) coarse, indelicate. **c** (of wine etc.) sharp or harsh in taste. **d** (of a sound, the voice) harsh, discordant; gruff, hoarse. **5** (of the sea, weather, etc.) violent, stormy. **6** disorderly, riotous (*a rough part of town*). **7** harsh, insensitive, inconsiderate (*rough words; rough treatment*). **8 a** unpleasant, severe, demanding (*had a rough time*). **b** unfortunate, unreasonable, undeserved (*had rough luck*). **c** (foll. by *on*) hard or unfair towards. **9** lacking finish, elaboration, comfort, etc. (*rough lodgings; a rough welcome*). **10** incomplete, rudimentary (*a rough attempt; a rough makeshift*). **11 a** inexact, approximate, preliminary (*a rough estimate; a rough sketch*). **b** (of stationery etc.) for use in writing rough notes etc. **12** colloq. **a** ill, unwell (*am feeling rough*). **b** depressed, dejected. ● adv. in a rough manner (*the land should be ploughed rough; play rough*). ● n. **1** (usu. prec. by *the*) a hard part or aspect of life; hardship (*take the rough with the smooth*). **2** rough ground (*over rough and smooth*). **3** a rough or violent person (*met a bunch of roughs*). **4** Golf rough ground off the fairway between tee and green. **5** an unfinished or provisional or natural state (*have written it in rough; shaped from the rough*); a rough draft or sketch. **6** (prec. by *the*) the general way or tendency (*is true in the rough*). ● v.tr. **1** (foll. by *up*) ruffle (feathers, hair, etc.) by rubbing in the wrong direction. **2 a** (foll. by *out*) shape or plan roughly. **b** (foll. by *in*) sketch roughly. **3** give the first shaping to (a gun, lens, etc.). □ **bit of rough** sl. a (usu. male) sexual partner whose toughness or lack of sophistication is a source of physical attraction. **rough-and-ready** rough or crude but effective; not elaborate or over-particular. **rough-and-tumble** adj. irregular, scrambling, disorderly. ● n. a haphazard fight; a scuffle. **rough breathing** see BREATHING 2. **rough coat** a first coat of plaster applied to a surface. **rough copy 1** a first or original draft. **2** a copy of a picture etc. showing only the essential features. **rough deal** hard or unfair treatment. **rough diamond 1** an uncut diamond. **2** a person of good nature but rough manners. **rough-dry** (**-dries, -dried**) dry (clothes) without ironing. **the rough edge** (or **side**) **of one's tongue** severe or harsh words. **rough-handle** treat or handle roughly. **rough-hew** (*past part.* **-hewed** or **-hewn**) shape out roughly; give crude form to. **rough-hewn** uncouth, unrefined. **rough hound** a rough-skinned dogfish; esp. the lesser spotted dogfish, *Scyliorhinus canicula*. **rough-house** sl. a disturbance or row; boisterous play. **rough-house** v. sl. **1** tr. handle (a person) roughly. **2** intr. make a disturbance; act violently. **rough it** do without basic comforts. **rough justice 1** treatment that is approximately fair. **2** treatment that is not at all fair. **rough passage 1** a crossing over rough sea. **2** a difficult time or experience. **rough ride** a difficult time or experience. **rough-rider** a person who breaks in or can ride unbroken horses. **rough stuff** colloq. boisterous or violent behaviour. **rough tongue** a habit of rudeness in speaking. **rough trade** sl. a tough or sadistic element among male homosexuals. **rough up** colloq. treat (a person) with violence; attack violently. **rough work 1** preliminary or provisional work. **2** colloq. violence. **3** a task requiring the use of force. **sleep rough** sleep outdoors, or not in a proper bed. □□ **roughness** n. [OE *rūh* f. WG]

roughage /ˈrʌfɪdʒ/ n. **1** fibrous indigestible matter in vegetable foodstuffs which aids the passage of food etc. through the gut. **2** coarse fodder. [ROUGH + -AGE 3]

roughcast /ˈrʌfkɑːst/ n., adj., & v. ● n. plaster of lime and gravel, used on outside walls. ● adj. **1** (of a wall etc.) coated with roughcast. **2** (of a plan etc.) roughly formed, preliminary. ● v.tr. (*past and past part.* **-cast**) **1** coat (a wall) with roughcast. **2** prepare (a plan, essay, etc.) in outline.

roughen /ˈrʌf(ə)n/ v.tr. & intr. make or become rough.

roughie /ˈrʌfi/ n. sl. **1** esp. Sc. & Austral. a rough; a hooligan. **2** Austral. **a** an outsider in a horse-race etc. **b** an unfair or unreasonable act. **3** see ROUGHY.

roughish /ˈrʌfɪʃ/ adj. somewhat rough.

roughly /ˈrʌfli/ adv. **1** in a rough manner. **2** approximately (*roughly 20 people attended*). □ **roughly speaking** in an approximate sense (*it is, roughly speaking, a square*).

roughneck /ˈrʌfnek/ n. colloq. **1** a rough or rowdy person. **2** a worker on an oil rig.

roughshod /ˈrʌfʃɒd/ adj. (of a horse) having shoes with nail-heads projecting to prevent slipping. □ **ride roughshod over** treat inconsiderately or arrogantly.

roughy /ˈrʌfɪ/ n. (also **roughie**) (pl. **-ies**) esp. Austral. either of two rough-scaled fishes, the tommy ruff, *Arripis georgianus*, or a small reef fish, *Trachichthys australis*. [perh. f. ROUGH OR RUFF²]

roulade /ruːˈlɑːd/ n. **1** a dish cooked or served in the shape of a roll, esp. a rolled piece of meat or sponge with a filling. **2** a florid passage of runs etc. in solo vocal music, usu. sung to one syllable. [F f. *rouler* to roll]

rouleau /ˈruːləʊ/ n. (pl. **rouleaux** or **rouleaus** /-ləʊz/) **1** a cylindrical packet of coins. **2** a coil or roll of ribbon etc., esp. as trimming. [F f. *rôle* ROLL n.]

Roulers see ROESELARE.

roulette /ruːˈlet/ n. **1** a gambling game in which a ball is dropped on to a revolving wheel with numbered compartments in the centre of a table, players betting on the number at which the ball comes to rest. **2** Math. a curve generated by a point on a curve rolling on another. **3 a** a revolving toothed wheel used in engraving. **b** a similar wheel for making perforations between postage stamps in a sheet. □ **rouletted** adj. (in sense 3b). [F, dimin. of *rouelle* f. LL *rotella* dimin. of L *rota* wheel]

Roumelia see RUMELIA.

round /raʊnd/ adj., n., adv., prep., & v. ● adj. **1** shaped like or approximately like a circle, sphere, or cylinder; having a convex or circular outline or surface; curved, not angular. **2** done with or involving circular motion. **3 a** entire, continuous, complete (*a round dozen*); fully expressed or developed; all together, not broken or defective or scanty. **b** (of a sum of money) considerable. **4** genuine, candid, outspoken; (of a statement etc.) categorical, unmistakable. **5** (usu. *attrib.*) (of a number) expressed for convenience or as an estimate in fewer significant numerals or with a fraction removed (*spent £297.32, or in round figures £300*). **6 a** (of a style) flowing. **b** (of a voice) not harsh. **7** Phonet. (of a vowel) pronounced with rounded lips. ● n. **1** a round object or form. **2 a** a revolving motion, a circular or recurring course (*the earth in its yearly round*). **b** a regular recurring series of activities or functions (*one's daily round; a continuous round of pleasure*). **c** a recurring succession or series of meetings for discussion etc. (*a new round of talks on disarmament*). **3 a** a fixed route on which things are regularly delivered (*milk round*). **b** (often in *pl.*) a route or sequence by which people or things are regularly supervised or inspected (*a doctor's rounds*). **4** an allowance of something distributed or measured out, esp.: **a** a single provision of drinks etc. to each member of a group. **b** ammunition to fire one shot; the act of firing this. **5 a** a slice across a loaf of bread. **b** a sandwich made from whole slices of bread. **c** a thick disc of beef cut from the haunch as a joint. **6** each of a set or series, a sequence of actions by each member of a group in turn, esp.: **a** one spell of play in a game etc. **b** one stage in a competition. **7** Golf the playing of all the holes in a course once. **8** Archery a fixed number of arrows shot from a fixed distance. **9** (**the round**) a form of sculpture in which the figure stands clear of any ground (cf. RELIEF 6a). **10** Mus. a canon for three or more unaccompanied voices singing at the same pitch or in octaves. **11** (in *pl.*) Mil. **a** a watch that goes round inspecting sentries. **b** a circuit made by this. **12** a rung of a ladder. **13** (foll. by *of*) the circumference, bounds, or extent of (*in all the round of Nature*). ● adv. **1** with circular motion (*wheels go round*). **2** with return to the starting-point or an earlier state (*summer soon comes round*). **3 a** with rotation, or change to an opposite position (*he turned round to look*). **b** with change to an opposite opinion etc. (*they were angry but I soon won them round*). **4** to, at, or affecting all or many points of a circumference or an area or the members of a company etc. (*tea was then handed round; may I look round?*). **5** in every direction from a centre or within a radius (*spread destruction round; everyone for a mile round*). **6** by a circuitous way (*will you jump over or go round?; go a long way round*). **7 a** to a person's house etc. (*ask him round; will be round soon*). **b** to a more prominent or convenient position (*brought the car round*). **8** measuring a (specified distance) in girth. **9** from beginning to end; through the whole time or course (*all the year round*). ● prep. **1** so as to encircle or enclose (*tour round the world; has a blanket round him*). **2** at or to points on the circumference of (*sat round the table*). **3** with successive visits to (*hawks them round the cafés*). **4** in various directions from or with regard to (*towns round Birmingham; shells bursting round them*). **5** having as an axis of revolution or as a central point (*turns round its centre of gravity; write a book round an event*). **6 a** so as to double or pass in a curved course (*go round the corner*). **b** having passed in this way (*be round the corner*). **c** in the position that would result from this (*find them round the corner*). **7** so as to come close from various sides but not into contact. **8** at various places in or

around (*had lots of clocks round the house to always know the time*). ● v. **1 a** tr. give a round shape to. **b** intr. assume a round shape. **2** tr. double or pass round (a corner, cape, etc.). **3** tr. express (a number) in a less exact but more convenient form (also foll. by *down* when the number is decreased and *up* when it is increased). **4** tr. Phonet. pronounce (a vowel) with rounded lips. □ **go the round** (or **rounds**) (of news etc.) be passed on from person to person. **in the round 1** with all features shown; all things considered. **2** Theatr. with the audience round at least three sides of the stage. **3** (of sculpture) with all sides shown; not in relief. **make the round of** go round. **make** (or **go**) **one's rounds** take a customary route for inspection etc. **round about 1** in a ring (about); all round; on all sides (of). **2** with a change to an opposite position. **3** approximately (*cost round about £50*). **round and round** several times round. **round-arm** Cricket (of bowling) with the arm swung horizontally. **round the bend** see BEND¹. **round brackets** brackets of the form (). **round dance 1** a dance in which couples move in circles round the ballroom. **2** a dance in which the dancers form one large circle. **round down** see sense 3 of *v.* **round off 1** bring to a complete or symmetrical or well-ordered state. **2** smooth out; blunt the corners or angles of. **round on a person** make a sudden verbal attack on or unexpected retort to a person. **round out 1** = *round off* 1. **2** provide more detail about. **a round peg in a square hole** = a square peg in a round hole (see PEG). **round robin 1** a petition esp. with signatures written in a circle to conceal the order of writing. **2** N. Amer. a tournament in which each competitor plays in turn against every other. **round-shouldered** with shoulders bent forward so that the back is rounded. **round trip** a trip to one or more places and back again (esp. by a circular route). **round the twist** see TWIST. **round up** collect or bring together, esp. by going round (see also sense 3 of *v.*). **round-up** n. **1** a systematic rounding up of people or things. **2** a summary; a résumé of facts or events. □ **roundish** adj. **roundness** n. [ME f. OF *ro(u)nd-* stem of *ro(o)nt, reont* f. L *rotundus* ROTUND]

roundabout /ˈraʊndəˌbaʊt/ n. & adj. ● n. **1** Brit. a road junction at which traffic moves in one direction round a central island. **2** Brit. **a** a large revolving device in a playground, for children to ride on. **b** a merry-go-round at a fair etc. ● adj. circuitous, circumlocutory, indirect.

roundel /ˈraʊnd(ə)l/ n. **1 a** a small disc, esp. a decorative medallion. **b** Heraldry any of various circular charges. **2** a circular identifying mark painted on military aircraft, esp. the red, white, and blue of the RAF. **3** a poem of eleven lines in three stanzas. [ME f. OF *rondel(le)* (as ROUND)]

roundelay /ˈraʊndɪˌleɪ/ n. a short simple song with a refrain. [F *rondelet* (as RONDEL), with assim. to LAY³ or *virelay*]

rounder /ˈraʊndə(r)/ n. a complete run of a player through all the bases as a unit of scoring in the game of rounders.

rounders /ˈraʊndəz/ n.pl. (often treated as *sing.*) a field game between two teams played with a bat and ball in which players after hitting the ball run round a circle of bases. First recorded in 1744, rounders has never become a serious competitive sport, but is widely played in British schools. Rounders resembles baseball, which is thought to have developed from it.

Roundhead /ˈraʊndhed/ n. hist. a member or supporter of the party opposing the king (Charles I) in the English Civil War (1642–9), so called because of the short-cropped hairstyle of the Puritans, who were an important element in the forces. Also called *Parliamentarian*.

roundhouse /ˈraʊndhaʊs/ n. **1** a repair-shed for railway locomotives, built round a turntable. **2** sl. **a** a blow given with a wide sweep of the arm. **b** Baseball a pitch made with a sweeping sidearm motion. **3** hist. a prison; a place of detention. **4** Naut. a cabin or set of cabins on the after part of the quarterdeck, esp. on a sailing-ship.

roundly /ˈraʊndlɪ/ adv. **1** bluntly, in plain language, severely (*was roundly criticized; told them roundly that he refused*). **2** in a thoroughgoing manner (*go roundly to work*). **3** in a circular way (*swells out roundly*).

roundsman /ˈraʊndzmən/ n. (pl. **-men**) **1** Brit. a trader's employee going round delivering and taking orders. **2** US a police officer in charge of a patrol. **3** Austral. a journalist covering a specified subject (*political roundsman*).

Round Table n. **1** the table at which King Arthur and his knights sat so that none should have precedence, first mentioned in 1155. Its importance lies in the chivalric fellowship which it came to represent. **2** an international charitable association which holds discussions and undertakes community service. **3** (**round table**) an assembly for discussion, esp. at a conference (often *attrib.*: *round-table talks*).

roundworm /ˈraʊndwɜːm/ n. a nematode worm, esp. a parasitic one such as an ascarid.

roup[1] /raʊp/ n. & v. Sc. & N. Engl. ● n. an auction. ● v.tr. sell by auction. [ME 'to shout', f. Scand. orig.]

roup[2] /ruːp/ n. an infectious poultry-disease, esp. of the respiratory tract. □ **roupy** adj. [16th c.: orig. unkn.]

rouse /raʊz/ v. **1 a** tr. (often foll. by from, out of) bring out of sleep, wake. **b** intr. (often foll. by up) cease to sleep, wake up. **2** (often foll. by up) **a** tr. stir up, make active or excited, startle out of inactivity or confidence or carelessness (roused them from their complacency; was roused to protest). **b** intr. become active. **3** tr. provoke to anger (is terrible when roused). **4** tr. evoke (feelings). **5** tr. (usu. foll. by in, out, up) Naut. haul vigorously. **6** tr. startle (game) from a lair or cover. **7** tr. stir (liquid, esp. beer while brewing). □ **rouse oneself** overcome one's indolence. □ **rousable** adj. **rouser** n. [orig. as a hawking and hunting term, so prob. f. AF: orig. unkn.]

rouseabout /ˈraʊzəˌbaʊt/ n. Austral. & NZ an unskilled labourer or odd jobber, esp. on a farm.

rousing /ˈraʊzɪŋ/ adj. **1** exciting, stirring (a rousing cheer; a rousing song). **2** (of a fire) blazing strongly. □ **rousingly** adv.

Rousse see RUSE.

Rousseau[1] /ˈruːsəʊ/, Henri (Julien) (known as 'le Douanier' = customs officer) (1844–1910), French painter. After retiring as a customs official in 1893, he devoted himself fully to painting, although it was only after his death that he was recognized as a notable naive artist. Fantastic dreams and exotic jungle landscapes often form the subjects of his bold and colourful paintings. Famous works include the Sleeping Gypsy (1897) and Tropical Storm with Tiger (1891).

Rousseau[2] /ˈruːsəʊ/, Jean-Jacques (1712–78), French philosopher and writer, born in Switzerland. From 1750 he came to fame with a series of works highly critical of the existing social order; his philosophy is underpinned by a belief in the fundamental goodness of human nature, encapsulated in the concept of the 'noble savage', and the warping effects of civilization. In his novel Émile (1762), Rousseau formulated new educational principles giving the child full scope for individual development in natural surroundings, shielded from the corrupting influences of civilization. His Social Contract (1762) anticipated much of the thinking of the French Revolution (see SOCIAL CONTRACT). Rousseau is also noted for his Confessions (1782), one of the earliest autobiographies.

Rousseau[3] /ˈruːsəʊ/, (Pierre Étienne) Théodore (1812–67), French painter. He was a leading landscapist of the Barbizon School and placed great importance on making preliminary studies for studio paintings out of doors, directly from nature. His works typically depict the scenery and changing light effects of the forest of Fontainebleau, and include Under the Birches, Evening (1842–4).

Roussillon /ˈruːsɪˌɒn/ a former province of southern France, on the border with Spain in the eastern Pyrenees, now part of Languedoc-Roussillon. For much of its history it was part of Spain until, in 1659, it was acquired by Louis XIV for France. Roussillon retains many of its Spanish characteristics and traditions and Catalan is widely spoken.

roust /raʊst/ v.tr. **1** (often foll. by up, out) **a** rouse, stir up. **b** root out. **2** N. Amer. sl. jostle, harass, rough up. □ **roust around** rummage. [perh. alt. of ROUSE]

roustabout /ˈraʊstəˌbaʊt/ n. **1** a labourer on an oil rig. **2** an unskilled or casual labourer. **3** US a dock labourer or deck hand. **4** Austral. = ROUSEABOUT.

rout[1] /raʊt/ n. & v. ● n. **1 a** a disorderly retreat of defeated troops. **b** a heavy defeat. **2 a** an assemblage or company, esp. of revellers or rioters. **b** Law an assemblage of three or more persons who have made a move towards committing an illegal act. **3** riot, tumult, disturbance, clamour, fuss. **4** Brit. archaic a large evening party or reception. ● v.tr. compel (esp. troops) to retreat in disorder; defeat. □ **put to rout** put to flight, defeat utterly. [ME f. AF rute, OF route ult. f. L ruptus broken]

rout[2] /raʊt/ v. **1** intr. & tr. = ROOT[2]. **2** tr. cut a groove, or any pattern not extending to the edges, in (a wooden or metal surface). □ **rout out** force or fetch out of bed or from a house or hiding-place. [var. of ROOT[2]]

route /ruːt/, Mil. also raʊt/ n. & v. ● n. **1** a way or course taken (esp. regularly) in getting from a starting-point to a destination. **2** N. Amer. a round travelled in delivering, selling, or collecting goods. **3** Mil. archaic marching orders. ● v.tr. (**routeing**) send or forward or direct to be sent by a particular route. □ **route man** N. Amer. = ROUNDSMAN 1. **route march** a training-march for troops. [ME f. OF r(o)ute road ult. f. L ruptus broken]

router /ˈraʊtə(r)/ n. a type of plane with two handles used in routing.

routine /ruːˈtiːn/ n., adj., & v. ● n. **1** a regular course or procedure, an unvarying performance of certain acts. **2** a set sequence in a performance, esp. a dance, comedy act, etc. **3** Computing a sequence of instructions for performing a task. ● adj. **1** performed as part of a routine; unvarying, mechanical (routine duties; a routine job shelling peas). **2** of a customary or standard kind. ● v.tr. organize according to a routine. □ **routinely** adv. [F (as ROUTE)]

routinism /ruːˈtiːnɪz(ə)m/ n. the prevalence of routine. □ **routinist** n. & adj.

routinize /ruːˈtiːnaɪz/ v.tr. (also **-ise**) subject to a routine; make into a matter of routine. □ **routinization** /ˌruːtɪnaɪˈzeɪʃ(ə)n/ n.

roux /ruː/ n. (pl. same) a mixture of fat (esp. butter) and flour used in making sauces etc. [F, = browned (butter): see RUSSET]

Rovaniemi /ˈrɒvəˌnɪəmɪ/ the principal town of Finnish Lapland; pop. (1990) 33,500. Nearby is Sodankylä, the site of the Santa Claus Village, where all letters sent to Finland addressed to Santa Claus are dealt with and answered.

rove[1] /rəʊv/ v. & n. ● v. **1** intr. wander without a settled destination, roam, ramble. **2** intr. (of eyes) look in changing directions. **3** tr. wander over or through. ● n. an act of roving (on the rove). □ **rove-beetle** a long-bodied beetle of the large family Staphylinidae, with short wing-cases that leave most of the abdomen visible. **roving commission** authority given to a person or persons conducting an inquiry to travel as may be necessary. **have a roving eye** colloq. be habitually seeking sexual partners. [ME, orig. a term in archery = shoot at a casual mark with the range not determined, perh. f. dial. rave stray, prob. of Scand. orig.]

rove[2] past of REEVE[2].

rove[3] /rəʊv/ n. a sliver of cotton, wool, etc., drawn out and slightly twisted. ● v.tr. form into roves. [18th c.: orig. unkn.]

rove[4] /rəʊv/ n. a small metal plate or ring for a rivet to pass through and be clenched over, esp. in boat-building. [ON ró, with excrescent v]

rover[1] /ˈrəʊvə(r)/ n. **1** a roving person; a wanderer. **2** Croquet **a** a ball that has passed all the hoops but not pegged out. **b** a player whose ball is a rover. **3** Archery **a** a mark chosen at undetermined range. **b** a mark for long-distance shooting. **4** Austral. Rules one of three players making up the ruck, usu. small, fast, and adept at securing possession of the ball. **5** (**Rover**) Brit. hist. a senior Scout. ¶ Now called Venture Scout. (See SCOUT ASSOCIATION.)

rover[2] /ˈrəʊvə(r)/ n. a sea robber, a pirate. [ME f. MLG, MDu. rōver f. rōven rob, rel. to REAVE]

rover[3] /ˈrəʊvə(r)/ n. a person or machine that makes roves of fibre.

Rovno see RIVNE.

row[1] /rəʊ/ n. **1** a number of persons or things in a more or less straight line. **2** a line of seats across a theatre etc. (in the front row). **3** a street with a continuous line of houses along one or each side. **4** a line of plants in a field or garden. **5** a horizontal line of entries in a table etc. □ **a hard row to hoe** a difficult task. **in a row 1** forming a row. **2** in succession (two Sundays in a row). **row-house** N. Amer. a terrace house. [ME raw, row, f. OE f. Gmc]

row[2] /rəʊ/ v. & n. ● v. **1** tr. propel (a boat) with oars. **2** tr. convey (a passenger) in a boat in this way. **3** intr. propel a boat in this way. **4** tr. make (a stroke) or achieve (a rate of striking) in rowing. **5** tr. compete in (a race) by rowing. **6** tr. row a race with. ● n. **1** a spell of rowing. **2** an excursion in a rowing-boat. □ **row-boat** N. Amer. = rowing-boat. **row down** overtake in a rowing, esp. bumping, race. **rowing-boat** Brit. a small boat propelled by oars. **rowing-machine** a device for exercising the muscles used in rowing. **row out** exhaust by rowing (the crew were completely rowed out at the finish). **row over** complete the course of a boat race with little effort, owing to the absence or inferiority of competitors. □ **rower** n. [OE rōwan f. Gmc, rel. to RUDDER, L remus oar]

row[3] /raʊ/ n. & v. ● n. **1** a loud noise or commotion. **2** a fierce quarrel or dispute. **3 a** a severe reprimand. **b** the condition of being reprimanded (shall get into a row). ● v. **1** intr. make or engage in a row. **2** tr. reprimand. □ **make** (or **kick up**) **a row** colloq. **1** raise a noise. **2** make a vigorous protest. [18th-c. sl.: orig. unkn.]

rowan /ˈrəʊən, ˈraʊ-/ n. **1** (in full **rowan tree**) **a** Brit. the mountain ash. **b** US a similar tree, Sorbus americana, native to America. **2** (in full **rowan-berry**) the scarlet berry of either of these trees. [Scand., corresp. to Norw. rogn, raun, Icel. reynir]

rowdy /ˈraʊdɪ/ adj. & n. ● adj. (**rowdier, rowdiest**) noisy and disorderly. ● n. (pl. **-ies**) a rowdy person. □ **rowdily** adv. **rowdiness** n. **rowdyism** n. [19th-c. US, orig. = lawless backwoodsman: orig. unkn.]

Rowe /rəʊ/, Nicholas (1674–1718), English dramatist. He is best known for his tragedies Tamerlane (1701) and The Fair Penitent (1703). The latter, marked by pathos and suffering, provided Mrs Siddons with one of her most celebrated roles.

rowel /ˈraʊəl/ n. & v. ● n. **1** a spiked revolving disc at the end of a spur. **2** *hist.* a circular piece of leather etc. with a hole in the centre inserted between a horse's skin and flesh to discharge an exudate. ● *v.tr.* (**rowelled, rowelling**; US **roweled, roweling**) **1** urge with a rowel. **2** *hist.* insert a rowel in (a horse). [ME f. OF *roel(e)* f. LL *rotella* dimin. of L *rota* wheel]

rowen /ˈraʊən/ n. (in *sing.* or *pl.*) US a second growth of grass, an aftermath. [ME f. OF *regain* (as GAIN)]

rowing /ˈraʊɪŋ/ n. the sport or pastime of propelling a boat by means of oars. The organized modern form of the sport can be traced to the institution of the Doggett's Coat and Badge race in 1715. Racing takes place in light boats (*shells*), between single rowers (*scullers*) with two oars, or between crews of two, four, or eight people with one oar each; crews are often steered by a coxswain. Major competitive events in England are the Boat Race (between teams from Oxford and Cambridge universities) and Henley Royal Regatta.

Rowlandson /ˈraʊlənds(ə)n/, Thomas (1756–1827), English painter, draughtsman, and caricaturist. He is remembered for his many watercolours and drawings satirizing Georgian manners, morals, and occupations. His best-known illustrations feature in a series of books known as *The Tours of Dr Syntax* (1812–21).

rowlock /ˈrɒlək, ˈrʌl-/ n. a device on a boat's gunwale, esp. a pair of thole-pins, serving as a fulcrum for an oar and keeping it in place. [alt. of earlier OARLOCK, after ROW²]

Rowntree /ˈraʊntriː/, a family of English business entrepreneurs and philanthropists. Joseph (1801–59) was a grocer, who established several Quaker schools. His son Henry Isaac Rowntree (1838–83) founded the family cocoa and chocolate manufacturing firm in York; Henry's brother Joseph Rowntree (1836–1925) became a partner in 1869 and subsequently founded three Rowntree trusts (1904) to support research into social welfare and policy. The latter's son B(enjamin) Seebohm Rowntree (1871–1954) conducted surveys of poverty in York (1897–8; 1936).

Rowton house /ˈraʊt(ə)n/ n. *Brit. hist.* a type of cheap lodging-house providing accommodation of a decent standard for poor men. [Lord Rowton, English social reformer (1838–1903)]

Roxburghshire /ˈrɒksbərəˌʃɪə(r)/ a former county of the Scottish Borders. Since 1975 it has been part of Borders region.

royal /ˈrɔɪəl/ adj. & n. ● adj. **1** of or suited to or worthy of a king or queen. **2** in the service or under the patronage of a king or queen. **3** belonging to the king or queen (*the royal hands; the royal anger*). **4** of the family of a king or queen. **5** majestic, stately, splendid. **6** on a great scale, of exceptional size or quality, first-rate (*gave us royal entertainment; in royal spirits; had a royal time*). ● n. **1** *colloq.* a member of the royal family. **2** *Naut.* a royal sail or mast. **3** a royal stag. **4** a size of paper, about 620 x 500 mm (25 x 20 in.). **5** (**the Royals**) the Royal Marines. □ **royal assent** see ASSENT. **royal blue** a deep vivid blue. **royal burgh** *hist.* (in Scotland) a burgh holding a charter from the Crown. **royal duke** see DUKE. **royal family** the family to which a sovereign belongs. **royal fern** a fern, *Osmunda regalis*, with very large spreading fronds. **royal flush** see FLUSH³. **royal icing** a hard white icing made from icing sugar and egg-whites. **royal jelly** a substance secreted by honey-bee workers and fed by them to future queen bees. **royal mast** *Naut.* a mast above a topgallant mast. **royal oak** a sprig of oak worn on 29 May to commemorate the restoration of Charles II (1660), who hid in an oak after the battle of Worcester (1651). **royal plural** the first person plural 'we' used by a single person. **royal prerogative** see PREROGATIVE 2. **royal road to** way of attaining without trouble. **royal sail** *Naut.* a sail above a topgallant sail. **royal stag** a stag with a head of 12 or more points. **royal standard** a banner bearing royal heraldic arms. **royal tennis** = REAL TENNIS. **royal warrant** a warrant authorizing a tradesperson to supply goods to a specified royal person. □ **royally** adv. [ME f. OF *roial* f. L *regalis* REGAL]

Royal Academy of Arts (also **Royal Academy**) an institution established in London under the patronage of George III in 1768, whose purpose was to cultivate the arts of painting, sculpture, and architecture in Britain, and to raise the status of the artist in society. Sir Joshua Reynolds was its first president and he instituted a highly influential series of annual lectures. Although it became increasingly unrepresentative of modernist tendencies (and often at odds with new talent), it still commands considerable influence and prestige; it holds an annual summer exhibition of contemporary work as well as regular historical exhibitions. Its premises are in Burlington House, Piccadilly.

Royal Air Force (abbr. **RAF**) the British air force, formed in 1918 by amalgamation of the Royal Flying Corps (founded 1912) and the Royal Naval Air Service (founded 1914).

Royal Ascot see *Ascot week*.

Royal Ballet a British ballet company, established in 1956 by the incorporation of the Sadler's Wells Ballet, the Sadler's Wells Theatre Company and the Sadler's Wells ballet school. Under the direction of Ninette de Valois and later Frederick Ashton the company and its school were responsible for establishing British ballet and a characteristic lyrical style. The Royal Ballet is based at the Royal Opera House, Covent Garden. (See also SADLER'S WELLS THEATRE.)

Royal British Legion a national association for the charitable support of ex-members of the British armed forces and their immediate dependants, founded in 1921.

Royal Canadian Mounted Police the Canadian police force, founded in 1873 as the North West Mounted Police. Members of the force are popularly called *Mounties*.

Royal Commission n. **1** a commission of inquiry appointed by the Crown at the instance of the government. **2** a committee so appointed.

Royal Engineers the engineering branch of the British army.

Royal Greenwich Observatory the official astronomical institution of Great Britain. It was founded at Greenwich in London in 1675 by Charles II and occupied a building designed by Sir Christopher Wren (see GREENWICH). In 1948 the Observatory headquarters were moved to Herstmonceux Castle in East Sussex, and subsequently transferred to Cambridge in 1990. Because of poor viewing conditions in Britain, the 2.5-metre Isaac Newton telescope and the 4.2-metre William Herschel telescope are sited at La Palma in the Canary Islands.

Royal Institution a British society founded in 1799 for the dissemination of scientific knowledge, based in London. The society organizes educational meetings, exhibitions, lectures, etc., particularly for young people; it also promotes research, particularly in photochemistry and the history of science, and maintains a museum, library, and information service.

royalist /ˈrɔɪəlɪst/ n. **1 a** a supporter of monarchy. **b** (**Royalist**) *hist.* a member or supporter of the side allied to the king (Charles I) in the English Civil War (1642–9). **2** US a reactionary, esp. a reactionary business tycoon. □ **royalism** n.

Royal Leamington Spa see LEAMINGTON SPA.

Royal Marines a British armed service (part of the Royal Navy) trained for service at sea, or on land under specific circumstances. The Royal Marines were founded in 1644, and the first commando units were formed during the Second World War.

Royal Mint the establishment responsible for the manufacture of British coins. Set up in 1810 in London, it moved in 1968 to Llantrisant in South Wales. In addition to British coins, it mints coins on behalf of certain foreign and Commonwealth governments, and also manufactures medals, decorations, and seals. Since 1869 the office of Master Worker and Warden of the Mint has been nominally held by the Chancellor of the Exchequer.

Royal National Lifeboat Institution (abbr. **RNLI**) (in the UK) an organization formed in 1824 to operate an offshore rescue service with lifeboats. It is financed by voluntary contributions and staffed by volunteers.

Royal Navy (abbr. **RN**) the British navy. The Royal Navy traces its origin to the fleet of warships launched during the reign of King Alfred to repel Viking invaders, but the establishment of a regular government-controlled force did not begin until the 16th century. Reforms and expansion during the 17th century built the Royal Navy into the most powerful navy in the world (a position retained until the Second World War), enabling the extension of Britain's colonial power worldwide.

Royal Shakespeare Company (abbr. **RSC**) a British professional theatre company founded in 1961 from the Shakespeare Memorial Company at Stratford-upon-Avon. Its programme always includes plays by Shakespeare, but additionally the company performs a wide variety of plays by other dramatists. It is based at Stratford, where it has two main venues, and has also usually had a London base, now at the Barbican Centre.

Royal Society the oldest and most prestigious scientific society in Britain, formed to promote scientific discussion especially in the physical sciences. It received its charter from Charles II in 1662, having originated privately among the followers of Francis Bacon, and it numbered among its fellows such famous scientists as Robert Boyle and Sir Christopher Wren. Newton was president from 1703 to 1727. Its *Philosophical Transactions*, founded in 1665, is the oldest scientific journal.

Royal Society for the Prevention of Cruelty to Animals (abbr. **RSPCA**) (in the UK) a charitable organization formed in 1824 to safeguard the welfare of animals.

Royal Society for the Protection of Birds (abbr. **RSPB**) (in the UK) a charitable organization founded in 1889 for the conservation of wild birds.

Royal Society of Arts (abbr. **RSA**) an institution established in London in 1754 whose original purpose was to forge a link between art and commerce following a decline in craftsmanship after the onset of the Industrial Revolution. It now exists to encourage the arts, manufacturing, and commerce, and holds examinations for a wide range of vocational and professional qualifications.

Royal Tunbridge Wells see TUNBRIDGE WELLS.

royalty /'rɔɪəltɪ/ n. (pl. **-ies**) **1** the office or dignity or power of a king or queen, sovereignty. **2 a** royal persons. **b** a member of a royal family. **3** a sum paid to a patentee for the use of a patent or to an author etc. for each copy of a book etc. sold or for each public performance of a work. **4 a** a royal right (now esp. over minerals) granted by the sovereign to an individual or corporation. **b** a payment made by a producer of minerals, oil, or natural gas to the owner of the site or of the mineral rights over it. [ME f. OF *roialté* (as ROYAL)]

Royal Victorian Chain n. (in the UK) an order founded by Edward VII in 1902 and conferred by the sovereign on special occasions.

Royal Victorian Order n. (in the UK) an order founded by Queen Victoria in 1896 and conferred usu. for great service rendered to the sovereign.

Royal Worcester see WORCESTER[2].

Royce /rɔɪs/, Sir (Frederick) Henry (1863–1933), English engine designer. He founded the company of Rolls-Royce Ltd. with Charles Stewart Rolls in 1906, previously having established his own successful electrical manufacturing business and designing and building his own car and engine. He became famous as the designer of the Rolls-Royce Silver Ghost motor car and later also became known for his aircraft engines (see ROLLS-ROYCE).

rozzer /'rɒzə(r)/ n. Brit. sl. a policeman. [19th c.: orig. unkn.]

RP abbr. received pronunciation.

RPG abbr. **1** Computing report program generator, a high-level commercial programming language. **2** rocket-propelled grenade. **3** role-playing game.

RPI abbr. retail price index.

r.p.m. abbr. **1** revolutions per minute. **2** resale price maintenance.

RPO abbr. Royal Philharmonic Orchestra.

RR abbr. US **1** railroad. **2** rural route.

RS abbr. **1** see ROYAL SOCIETY. **2** US Received Standard. **3** (in the UK) Royal Scots.

Rs. abbr. rupee(s).

RSA abbr. **1** Republic of South Africa. **2** see ROYAL SOCIETY OF ARTS. **3** Royal Scottish Academy; Royal Scottish Academician.

RSC abbr. **1** see ROYAL SHAKESPEARE COMPANY. **2** (in the UK) Royal Society of Chemistry.

RSFSR abbr. hist. Russian Soviet Federative Socialist Republic (see RUSSIA).

RSI abbr. repetitive strain injury.

RSJ abbr. rolled steel joist.

RSM abbr. Regimental Sergeant-Major.

RSPB see ROYAL SOCIETY FOR THE PROTECTION OF BIRDS.

RSPCA see ROYAL SOCIETY FOR THE PREVENTION OF CRUELTY TO ANIMALS.

RSV abbr. Revised Standard Version (of the Bible).

RSVP abbr. (in an invitation etc.) please answer. [F *répondez s'il vous plaît*]

RT abbr. **1** radio-telegraphy. **2** radio-telephony.

rt. abbr. right.

Rt. Hon. abbr. Brit. Right Honourable.

Rt. Revd. abbr. (also **Rt. Rev.**) Right Reverend.

RU abbr. Rugby Union.

Ru symb. Chem. the element ruthenium.

rub[1] /rʌb/ v. & n. ● v. (**rubbed, rubbing**) **1** tr. move one's hand or another object with firm pressure over the surface of. **2** tr. (usu. foll. by *against*, *in*, *on*, *over*) apply (one's hand etc.) in this way. **3** tr. clean or polish or make dry or bare by rubbing. **4** tr. (often foll. by *over*) apply (polish, ointment, etc.) by rubbing. **5** tr. (foll. by *in*, *into*, *through*) use rubbing to make (a substance) go into or through something. **6** tr. (often foll. by *together*) move or slide (objects) against each other. **7** intr. (foll. by *against*, *on*) move with contact or friction. **8** tr. chafe or make sore by rubbing. **9** intr. (of cloth, skin, etc.) become frayed or worn or sore or bare with friction. **10** tr. reproduce the design of (a sepulchral brass or a stone) by rubbing paper laid on it with heelball or coloured chalk etc. **11** tr. (foll. by *to*) reduce to powder etc. by rubbing. **12** intr. Bowls (of a bowl) be slowed or diverted by the unevenness of the ground. ● n. **1** a spell or an instance of rubbing (*give it a rub*). **2 a** an impediment or difficulty (*there's the rub*). **b** Bowls an inequality of the ground impeding or diverting a bowl; the diversion or hindering of a bowl by this. □ **rub along** colloq. cope or manage without undue difficulty. **rub away** remove by rubbing. **rub down** dry or smooth or clean by rubbing. **rub-down** n. an instance of rubbing down. **rub elbows with** US = rub shoulders with. **rub one's hands** rub one's hands together usu. in sign of keen satisfaction, or for warmth. **rub it in** (or **rub a person's nose in it**) emphasize or repeat an embarrassing fact etc. **rub noses** rub one's nose against another's in greeting. **rub off 1** (usu. foll. by *on*) be transferred by contact, be transmitted (*some of his attitudes have rubbed off on me*). **2** remove by rubbing. **rub of** (or **on**) **the green** Golf an accidental interference with the course or position of a ball. **rub on** colloq. = rub along. **rub out 1** erase with a rubber. **2** esp. N. Amer. sl. kill, eliminate. **rub shoulders with** associate or come into contact with (another person). **rub up 1** polish (a tarnished object). **2** brush up (a subject or one's memory). **3** mix (pigment etc.) into paste by rubbing. **rub-up** n. the act or an instance of rubbing up. **rub (up) the wrong way** irritate or repel as by stroking a cat against the lie of its fur. [ME *rubben*, perh. f. LG *rubben*, of unkn. orig.]

rub[2] /rʌb/ n. = RUBBER[2]. [abbr.]

Rub' al Khali /ˌrʊb æl 'kɑːliː/ a vast desert in the Arabian peninsula, extending from central Saudi Arabia southwards to Yemen and eastwards to the United Arab Emirates and Oman. It is also known as the Great Sandy Desert and the Empty Quarter.

rub-a-dub /'rʌbəˌdʌb/ n. & v. ● n. **1** the rolling sound of a drum. **2** (also **rub-a-dub-dub** /ˌrʌbədʌb'dʌb/) Austral. rhyming sl. a pub. ● v.intr. (**rub-a-dubbed, rub-a-dubbing**) make this sound. [imit.]

rubato /ruːˈbɑːtəʊ/ adj. & n. Mus. ● n. (pl. **-os** or **rubati** /-tiː/) the temporary disregarding of strict tempo. ● adj. performed with a flexible tempo. [It., = robbed]

rubber[1] /'rʌbə(r)/ n. **1** a tough elastic polymeric substance made from the latex of plants or synthetically. **2** esp. Brit. a piece of this or another substance for erasing pencil or ink marks. **3** colloq. a condom. **4** (in pl.) US galoshes. **5** a person who rubs; a masseur or masseuse. **6 a** an implement used for rubbing. **b** part of a machine operating by rubbing. □ **rubber band** a loop of rubber for holding papers etc. together. **rubber plant 1** an evergreen plant, *Ficus elastica*, with dark green shiny leaves, often cultivated as a house plant. **2** (also **rubber tree**) a tropical tree yielding latex, esp. *Hevea brasiliensis*. **rubber solution** a liquid drying to a rubber-like material, used esp. as an adhesive in mending rubber articles. **rubber stamp 1** a device for inking and imprinting on a surface. **2 a** a person who mechanically copies or agrees to others' actions. **b** an indication of such agreement. **rubber-stamp** v.tr. approve automatically without proper consideration. □ **rubbery** adj. **rubberiness** n. [RUB[1] + -ER[1], from its early use to rub out pencil marks]

rubber[2] /'rʌbə(r)/ n. **1** a match of three or five successive games between the same sides or persons at whist, bridge, cricket, lawn tennis, etc. **2** (prec. by the) **a** the act of winning two games in a rubber. **b** a third game when each side has won one. [orig. unkn.: used as a term in bowls from c.1600]

rubberize /'rʌbəˌraɪz/ v.tr. (also **-ise**) treat or coat with rubber.

rubberneck /'rʌbəˌnek/ n. & v. colloq. ● n. a person, esp. a tourist, who stares inquisitively or stupidly. ● v.intr. act in this way.

rubbing /'rʌbɪŋ/ n. **1** in senses of RUB[1] v. **2** an impression or copy made by rubbing (see RUB[1] v. 10).

rubbish /'rʌbɪʃ/ n. & v. ● n. esp. Brit. **1** waste material; debris, refuse, litter. **2** worthless material or articles; junk. **3** (often as int.) absurd ideas or suggestions; nonsense. ● v.tr. colloq. **1** criticize severely. **2** reject as worthless. □ **rubbishy** adj. [ME f. AF *rubbous* etc., perh. f. RUBBLE]

rubbity /'rʌbɪtɪ/ n. (also **rubbity-dub** /ˌrʌbɪtɪ'dʌb/) = RUB-A-DUB n. 2.

rubble /'rʌb(ə)l/ n. **1** waste or rough fragments of stone or brick etc. **2** pieces of undressed stone used, esp. as filling, for walls. **3** Geol. loose angular stones etc. as the covering of some rocks. **4** water-worn stones. □ **rubbly** *adj.* [ME *robyl, rubel,* of uncert. orig.: cf. OF *robe* spoils]

Rubbra /ˈrʌbrə/, (Charles) Edmund (1901–86), English composer and pianist. He wrote eleven symphonies, of which the fifth (1947–8) is the most frequently performed; the ninth, the *Sinfonia Sacra* (1971–2), is in the nature of a choral passion. He also wrote two masses (1945; 1949) and many songs.

rube /ruːb/ *n.* N. Amer. *colloq.* a country bumpkin. [abbr. of the name *Reuben*]

rubella /ruːˈbelə/ *n.* Med. an acute infectious viral disease with a red rash; German measles. [mod.L, neut. pl. of L *rubellus* reddish]

rubellite /ˈruːbəˌlaɪt/ *n.* a red variety of tourmaline. [L *rubellus* reddish]

Rubens /ˈruːbɪnz/, Sir Peter Paul (1577–1640), Flemish painter. The foremost exponent of northern Baroque, he spent a period of time in Italy (1600–8), where he studied the work of artists such as Titian and Raphael, before settling in Antwerp and becoming court portraitist in 1609. He quickly gained fame as a religious painter with altarpieces such as *Descent from the Cross* (1611–14). He built up a prestigious workshop that executed numerous commissions from across Europe ranging from decorative ceilings to landscapes; its more famous assistants included Anthony Van Dyck and Jacob Jordaens. On a visit to England (1629–30) he was knighted by Charles I and executed several commissions for him, including a series of decorative ceilings at the Banqueting Hall in Whitehall. In addition to his portraits, Rubens is perhaps best known for mythological paintings featuring voluptuous female nudes, as in *Venus and Adonis* (c.1635).

rubeola /ruːˈbiːələ/ *n.* Med. measles. [med.L f. L *rubeus* red]

Rubicon /ˈruːbɪkən/ a stream in NE Italy which marked the ancient boundary between Italy and Cisalpine Gaul. By taking his army across it into Italy from his own province in 49 BC, Julius Caesar broke the law forbidding a general to lead an army out of his province, and so committed himself to war against the Senate and Pompey. The ensuing civil war resulted in victory for Caesar after three years. Thus to 'cross the Rubicon' is to commit oneself irrevocably to a course of action.

rubicon /ˈruːbɪkən/ *n.* the act of winning a game in piquet before an opponent has scored 100. [RUBICON]

rubicund /ˈruːbɪˌkʌnd/ *adj.* (of a face, complexion, or person in these respects) ruddy, high-coloured. □ **rubicundity** /ˌruːbɪˈkʌndɪtɪ/ *n.* [F *rubicond* or L *rubicundus* f. *rubere* be red]

rubidium /ruːˈbɪdɪəm/ *n.* a soft silvery reactive metallic chemical element (atomic number 37; symbol **Rb**). Discovered spectroscopically by Robert Bunsen and Gustav Kirchhoff in 1861, rubidium is a member of the alkali-metal group. It is a rare element, occurring in traces in some rocks and minerals; it has few commercial uses. [L *rubidus* red (with ref. to its spectral lines)]

rubiginous /ruːˈbɪdʒɪnəs/ *adj.* formal rust-coloured. [L *rubigo- inis* rust]

Rubik's cube /ˈruːbɪks/ *n.* a puzzle in which the aim is to restore the faces of a composite cube to single colours by rotating layers of constituent smaller cubes. [Erno *Rubik*, its Hungarian inventor (b.1944)]

Rubinstein[1] /ˈruːbɪnˌstaɪn/, Anton (Grigorevich) (1829–94), Russian composer and pianist. In 1862 he founded the St Petersburg Conservatory and was its director 1862–7 and 1887–91; Tchaikovsky was among his pupils. Rubinstein composed symphonies, operas, songs, and piano music, including *Melody in F* (1852). His brother Nikolai (1835–81) was also a pianist and composer; he was prominent in Moscow's musical life and founded the Moscow Conservatory.

Rubinstein[2] /ˈruːbɪnˌstaɪn/, Artur (1888–1982), Polish-born American pianist. He first came to public attention with his Berlin début in 1900 at the age of 12, when he played the Mozart Concerto in A major. Thereafter he toured extensively in Europe as well as the US and made many recordings, including the complete works of Chopin. He became an American citizen in 1946.

Rubinstein[3] /ˈruːbɪnˌstaɪn/, Helena (1882–1965), American beautician and businesswoman. Born in Poland, she trained in medicine there before going to Australia in 1902, where she opened her first beauty salon. Her success enabled her to return to Europe and open salons in London (1908) and Paris (1912), and later to go to the US and open salons in New York (1915) and elsewhere. After the First World War her organization expanded to become an international cosmetics manufacturer and distributor.

ruble var. of ROUBLE.

rubric /ˈruːbrɪk/ *n.* **1** Eccl. a direction for the conduct of divine service inserted in a liturgical book. **2** a heading or passage in red or special

lettering. **3** explanatory words. **4** an established custom. □ **rubrical** *adj.* [ME f. OF *rubrique, rubrice* or L *rubrica* (*terra*) red (earth or ochre) as writing material, rel. to *rubeus* red]

rubricate /ˈruːbrɪˌkeɪt/ *v.tr.* **1** mark with red; print or write in red. **2** provide with rubrics. □ **rubricator** *n.* **rubrication** /ˌruːbrɪˈkeɪʃ(ə)n/ *n.* [L *rubricare* f. *rubrica*: SEE RUBRIC]

ruby /ˈruːbɪ/ *n., adj.,* & *v.* ● *n.* (pl. **-ies**) **1** a rare precious stone consisting of corundum with a colour varying from deep crimson or purple to pale rose. **2** a glowing purple-tinged red colour. ● *adj.* of this colour. ● *v.tr.* (**-ies, -ied**) dye or tinge ruby-colour. □ **ruby glass** glass coloured with oxides of copper, iron, lead, tin, etc. **ruby-tailed wasp** a small parasitic wasp, esp. of the genus *Chrysis*, with a shiny red end to the abdomen. **ruby wedding** the fortieth anniversary of a wedding. [ME f. OF *rubi* f. med.L *rubinus* (*lapis*) red (stone), rel. to L *rubeus* red]

RUC *abbr.* Royal Ulster Constabulary.

ruche /ruːʃ/ *n.* a frill or gathering of lace etc. as a trimming. □ **ruched** *adj.* **ruching** *n.* [F f. med.L *rusca* tree-bark, of Celt. orig.]

ruck[1] /rʌk/ *n.* & *v.* ● *n.* **1** (prec. by *the*) the main body of competitors not likely to overtake the leaders. **2** an undistinguished crowd of persons or things. **3** *Rugby* a loose scrum with the ball on the ground. **4** *Austral. Rules* a group of three players who follow the play without fixed positions. ● *v.intr. Rugby & Austral. Rules* participate in a ruck. [ME, = stack of fuel, heap, rick: app. Scand., = Norw. *ruka* in the same senses]

ruck[2] /rʌk/ *v.* & *n.* ● *v.tr.* & *intr.* (often foll. by *up*) make or become creased or wrinkled. ● *n.* a crease or wrinkle. [ON *hrukka*]

ruckle /ˈrʌk(ə)l/ *v.* & *n.* Brit. = RUCK[2].

rucksack /ˈrʌksæk, ˈrʊk-/ *n.* a bag slung by straps from both shoulders and resting on the back. [G f. *rucken* dial. var. of *Rücken* back + *Sack* SACK[1]]

ruckus /ˈrʌkəs/ *n.* esp. N. Amer. a row or commotion. [cf. RUCTION, RUMPUS]

ruction /ˈrʌkʃ(ə)n/ *n. colloq.* **1** a disturbance or tumult. **2** (in *pl.*) unpleasant arguments or reactions. [19th c.: orig. unkn.]

rudaceous /ruːˈdeɪʃəs/ *adj.* (of rock) composed of fragments of relatively large size. [L *rudus* rubble]

rudbeckia /rʌdˈbekɪə/ *n.* a composite garden plant of the genus *Rudbeckia*, native to North America, with dark-centred yellow or orange flowers. [mod.L f. Olaf *Rudbeck*, Swedish botanist (1660–1740)]

rudd /rʌd/ *n.* (pl. same) a freshwater fish, *Scardinius erythrophthalmus*, resembling a roach and having red fins. [app. rel. to *rud* red colour f. OE *rudu*, rel. to RED]

rudder /ˈrʌdə(r)/ *n.* **1 a** a flat piece hinged vertically to the stern of a ship for steering. **b** a vertical aerofoil pivoted from the tailplane of an aircraft, for controlling its horizontal movement. **2** a guiding principle etc. □ **rudderless** *adj.* [OE *rōther* f. WG *rōthra-* f. the stem of ROW[2]]

ruddle /ˈrʌd(ə)l/ *n.* & *v.* ● *n.* a red ochre, esp. of a kind used for marking sheep. ● *v.tr.* mark or colour with or as with ruddle. [rel. to obs. *rud*: see RUDD]

ruddock /ˈrʌdək/ *n. dial.* the robin redbreast. [OE *rudduc* (as RUDDLE)]

ruddy /ˈrʌdɪ/ *adj.* & *v.* ● *adj.* (**ruddier, ruddiest**) **1 a** (of a person or complexion) freshly or healthily red. **b** (of health, youth, etc.) marked by this. **2** reddish. **3** Brit. colloq. bloody, damnable. ● *v.tr. & intr.* (**-ies, -ied**) make or grow ruddy. □ **ruddy duck** an American duck, *Oxyura jamaicensis*, the male of which has deep red-brown plumage, naturalized in Britain and elsewhere. □ **ruddily** *adv.* **ruddiness** *n.* [OE *rudig* (as RUDD)]

rude /ruːd/ *adj.* **1** (of a person, remark, etc.) impolite or offensive. **2** roughly made or done; lacking subtlety or accuracy (*a rude plough*). **3** uncivilized or uneducated. **4** abrupt, sudden, startling, violent (*a rude awakening; a rude reminder*). **5** indecent, lewd (*a rude joke*). **6** vigorous or hearty (*rude health*). □ **be rude to** speak impolitely to; insult. □ **rudely** *adv.* **rudeness** *n.* **rudish** *adj.* **rudery** /ˈruːdərɪ/ *n.* [ME f. OF f. L *rudis* unwrought]

ruderal /ˈruːdərəl/ *adj.* & *n.* ● *adj.* (of a plant) growing on or in rubbish or rubble. ● *n.* a ruderal plant. [mod.L *ruderalis* f. L *rudera* pl. of *rudus* rubble]

rudiment /ˈruːdɪmənt/ *n.* **1** (in *pl.*) the elements or first principles of a subject. **2** (in *pl.*) an imperfect beginning of something undeveloped or yet to develop. **3** Biol. an undeveloped or immature part or organ, esp. a structure in an embryo or larva which will develop into a limb etc. [F *rudiment* or L *rudimentum* (as RUDE, after *elementum* ELEMENT)]

rudimentary /ˌruːdɪˈmentərɪ/ *adj.* **1** involving basic principles;

fundamental. **2** incompletely developed; vestigial. □ **rudimentarily** adv. **rudimentariness** n.

Rudolf, Lake /ˈruːdɒlf/ the former name (until 1979) for Lake Turkana. (See TURKANA, LAKE.)

Rudra /ˈrʊdrə/ Hinduism **1** (in the Rig-veda) a Vedic minor god, associated with the storm, father of the Maruts. **2** one of the names of Siva.

Rudras see MARUTS.

rue[1] /ruː/ v. & n. ● v.tr. (**rues, rued, rueing** or **ruing**) repent of; bitterly feel the consequences of; wish to be undone or non-existent (esp. rue the day). ● n. archaic **1** repentance; dejection at some occurrence. **2** compassion or pity. [OE hrēow, hrēowan]

rue[2] /ruː/ n. a perennial evergreen shrub, Ruta graveolens, with bitter strong-scented leaves formerly used in medicine. [ME f. OF f. L ruta f. Gk rhutē]

rueful /ˈruːfʊl/ adj. expressing sorrow or regret, genuine or humorously affected. □ **ruefully** adv. **ruefulness** n. [ME, f. RUE[1]]

rufescent /ruːˈfes(ə)nt/ adj. Zool. etc. reddish. □ **rufescence** n. [L rufescere f. rufus reddish]

ruff[1] /rʌf/ n. **1** a projecting starched frill worn round the neck esp. in the 16th century. **2** a projecting or conspicuously coloured ring of feathers or hair round a bird's or animal's neck. **3** a domestic pigeon like a jacobin. **4** (fem. **reeve** /riːv/) a wading bird, Philomachus pugnax, of which the male has a ruff and ear-tufts in the breeding season. □ **rufflike** adj. [perh. f. ruff = ROUGH]

ruff[2] /rʌf/ n. **1** (usu. **ruffe**) a rough-scaled fish, esp. a perchlike freshwater fish, Gymnocephalus cernua, found in European lakes and rivers. **2** (in full **tommy ruff**) esp. Austral. a rough-scaled marine food fish, Arripis georgianus, common in Australian waters and related to the Australian salmon. Also called roughy. [ME, prob. f. ROUGH]

ruff[3] /rʌf/ v. & n. ● v.intr. & tr. trump at cards. ● n. an act of ruffing. [orig. the name of a card-game: f. OF roffle, rouffle, = It. ronfa (perh. alt. of trionfo TRUMP[1])]

ruffe var. of RUFF[2] 1.

ruffian /ˈrʌfɪən/ n. a violent lawless person. □ **ruffianism** n. **ruffianly** adv. [F ruffian f. It. ruffiano, perh. f. dial. rofia scurf]

ruffle /ˈrʌf(ə)l/ v. & n. ● v. **1** tr. disturb the smoothness or tranquillity of. **2** tr. upset the calmness of (a person). **3** tr. gather (lace etc.) into a ruffle. **4** tr. (often foll. by up) (of a bird) erect (its feathers) in anger, display, etc. **5** intr. undergo ruffling. **6** intr. lose smoothness or calmness. ● n. **1** an ornamental gathered or goffered frill of lace etc. worn at the opening of a garment esp. round the wrist, breast, or neck. **2** perturbation, bustle. **3** a rippling effect on water. **4** the ruff of a bird etc. (see RUFF[1] 2). **5** Mil. a vibrating drum-beat. [ME: orig. unkn.]

rufiyaa /ˈruːfiːˌjɑː/ n. (pl. same) the basic monetary unit of the Maldives, equal to 100 laris. [Maldivian]

rufous /ˈruːfəs/ adj. (esp. of animals) reddish-brown. [L rufus red, reddish]

rug /rʌg/ n. **1** a floor-mat of shaggy material or thick pile. **2** a thick woollen coverlet or wrap. □ **pull the rug from under** deprive of support; weaken, unsettle. [prob. f. Scand.: cf. Norw. dial. rugga coverlet, Sw. rugg ruffled hair: rel. to RAG[1]]

Rugby /ˈrʌɡbɪ/ a town in central England, on the River Avon in Warwickshire; pop. (1981) 59,720. Rugby School, where rugby football was developed in the early 19th century, was founded there in 1567.

rugby /ˈrʌɡbɪ/ n. (also **Rugby**) (in full **rugby football**) a form of football, in which points are scored by carrying (and grounding) an oval-shaped ball across the opponents' goal-line (thereby scoring a try) or by kicking it between the two posts and over the crossbar of the opponents' goal. Named after Rugby School in Warwickshire where it was developed (in 1823 or later), it is played chiefly in the UK, France, Australia, and New Zealand. The ball may be carried or kicked forwards, but a thrown pass may be made only to the side or rear. The level of physical contact is high, especially when play is restarted with a scrum. (See also RUGBY LEAGUE, RUGBY UNION.)

Rugby League n. a professional form of rugby played with a team of thirteen. It dates from the breakaway from the Rugby Union of a group of northern English clubs (called the Northern Union) in 1895.

Rugby Union n. a form of rugby played with a team of fifteen. Originally strictly amateur, the game was opened to professionalism in 1995. The name is also given to the game's governing body, formed in 1871.

Rügen /ˈruːɡən/ an island in the Baltic Sea off the north coast of Germany, to which it is linked by a causeway. It forms part of the state of Mecklenburg-West Pomerania.

rugged /ˈrʌɡɪd/ adj. **1** (of ground or terrain) having a rough uneven surface. **2** (of features) strongly marked; irregular in outline. **3 a** unpolished; lacking gentleness or refinement (rugged grandeur). **b** harsh in sound. **c** austere, unbending (rugged honesty). **d** involving hardship (a rugged life). **4** (esp. of a machine) robust, sturdy. □ **ruggedly** adv. **ruggedness** n. [ME, prob. f. Scand.: cf. RUG, and Sw. rugga, roughen]

rugger /ˈrʌɡə(r)/ n. Brit. colloq. rugby.

rugosa /ruːˈɡəʊzə/ n. a Japanese rose, Rosa rugosa, which has dark green wrinkled leaves and deep pink flowers. [L, fem. of rugosus (see RUGOSE) used as specific epithet]

rugose /ˈruːɡəʊz, -ɡəʊs/ adj. esp. Biol. wrinkled, corrugated. □ **rugosely** adv. **rugosity** /ruːˈɡɒsɪtɪ/ n. [L rugosus f. ruga wrinkle]

Ruhr /rʊə(r)/ a region of coal mining and heavy industry in North Rhine-Westphalia, western Germany. It is named after the River Ruhr, which flows through it, meeting the Rhine near Duisburg. The Ruhr was occupied by French troops 1923–4, after Germany defaulted on war reparation payments.

ruin /ˈruːɪn/ n. & v. ● n. **1** a destroyed or wrecked state (after centuries of neglect, the palace fell to ruin). **2 a** a person's or thing's downfall or elimination (the ruin of my hopes). **b** archaic a woman's loss of chastity by seduction or rape; dishonour resulting from this. **3 a** the complete loss of one's property or position (bring to ruin). **b** a person who has suffered ruin. **4** (in sing. or pl.) the remains of a building etc. that has suffered ruin (an old ruin; ancient ruins). **5** a cause of ruin; a destructive thing or influence (will be the ruin of us). ● v. **1** tr. **a** bring to ruin (your extravagance has ruined me). **b** utterly impair or wreck (the rain ruined my hat). **c** archaic seduce and abandon (a woman). **2** tr. (esp. as **ruined** adj.) reduce to ruins. **3** intr. poet. fall headlong or with a crash. □ **in ruins 1** in a state of ruin. **2** completely wrecked (their hopes were in ruins). [ME f. OF ruine f. L ruina f. ruere fall]

ruination /ˌruːɪˈneɪʃ(ə)n/ n. **1** the act of bringing to ruin. **2** the act of ruining or the state of being ruined. [obs. ruinate (as RUIN)]

ruinous /ˈruːɪnəs/ adj. **1** bringing ruin; disastrous (at ruinous expense). **2** in ruins; dilapidated. □ **ruinously** adv. **ruinousness** n. [ME f. L ruinosus (as RUIN)]

Ruisdael /ˈrɪzdɑːl, ˈrɔɪz-, -deɪl/, Jacob van (also **Ruysdael**) (c.1628–82), Dutch landscape painter. Born in Haarlem, he painted the surrounding landscape from the mid-1640s until his move to Amsterdam in 1657, where he spent the rest of his life. His typical subject-matter was forest scenes, seascapes, and cloudscapes, and his work demonstrated the possibilities of investing landscape with subtle intimations of mood. Meindert Hobbema was his most famous pupil, while among those influenced by his work, were Thomas Gainsborough, John Constable, and the Barbizon School.

Ruiz de Alarcón y Mendoza /ruːˈiːz deɪ ˌælaˈkɒn iː menˈdəʊzə/, Juan (1580–1639), Spanish dramatist, born in Mexico City. His most famous play, the moral comedy La Verdad sospechosa, was the basis of Corneille's Le Menteur (1642).

rule /ruːl/ n. & v. ● n. **1** a principle to which an action conforms or is required to conform. **2** a prevailing custom or standard; the normal state of things. **3** government or dominion (under British rule; the rule of law). **4** a graduated straight measure used in carpentry etc.; a ruler. **5** Printing **a** a thin strip of metal for separating headings, columns, etc. **b** a thin line or dash. **6** a code of discipline of a religious order. **7** Law an order made by a judge or court with reference to a particular case only. **8** (**Rules**) Austral. = AUSTRALIAN RULES FOOTBALL. ● v. **1** tr. exercise decisive influence over; keep under control. **2** tr. & (often foll. by over) intr. have sovereign control of (rules over a vast kingdom). **3** tr. (often foll. by that + clause) pronounce authoritatively (was ruled out of order). **4** tr. **a** make parallel lines across (paper). **b** make (a straight line) with a ruler etc. **5** intr. (of prices or goods etc. in regard to price or quality etc.) have a specified general level; be for the most part (the market ruled high). **6** tr. (in passive; foll. by by) consent to follow (advice etc.); be guided by. □ **as a rule** usually; more often than not. **by rule** in a regulation manner; mechanically. **rule of the road** see ROAD[1]. **rule of three** Math. a method of finding a number in the same ratio to one given as exists between two others given. **rule of thumb** a rule for general guidance, based on experience or practice rather than theory. **rule out** exclude; pronounce irrelevant or ineligible. **rule the roost** (or **roast**) be in control. **run the rule over** examine cursorily for correctness or adequacy. □ **ruleless** adj. [ME f. OF reule, reuler f. LL regulare f. L regula straight stick]

Rule 43 n. (in the UK) a prison regulation whereby offenders can be isolated or segregated for their own protection.

ruler /ˈruːlə(r)/ n. **1** a person exercising government or dominion. **2** a straight usu. graduated strip or cylinder of wood, metal, etc., used to draw lines or measure distance. □ **rulership** n.

ruling /ˈruːlɪŋ/ n. & adj. ● n. an authoritative decision or announcement. ● adj. dominant; prevailing; currently in force (ruling prices). □ **ruling passion** a motive that habitually directs one's actions.

Rum see RHUM.

rum[1] /rʌm/ n. **1** a spirit distilled from sugar-cane residues or molasses. **2** N. Amer. intoxicating liquor. □ **rum baba** see BABA. **rum butter** a rich sweet hard sauce made with rum, butter, and sugar. [17th c.: perh. abbr. of contemporary forms rumbullion, rumbustion, of unkn. orig.]

rum[2] /rʌm/ adj. (**rummer**, **rummest**) Brit. colloq. **1** odd, strange, queer. **2** difficult, dangerous. □ **rum go** (or **do** or **start**) a surprising occurrence or unforeseen turn of affairs. □ **rumly** adv. **rumness** n. [16th-c. cant, orig. = fine, spirited, perh. var. of ROM[1]]

Rumania, Rumanian see ROMANIA, ROMANIAN.

Rumansh var. of ROMANSH.

rumba /ˈrʌmbə, ˈrum-/ n. & v. (also **rhumba**) ● n. **1** an Afro-Cuban dance. **2 a** a ballroom dance imitative of this. **b** the music for it. ● v.intr. (**rumbas, rumbaed** /-bəd/ or **rumba'd, rumbaing** /-bəɪŋ/) dance the rumba. [Amer. Sp.]

rumble /ˈrʌmb(ə)l/ v. & n. ● v. **1** intr. make a continuous deep resonant sound as of distant thunder. **2** intr. (foll. by along, by, past, etc.) (of a person or vehicle) move with a rumbling noise. **3** tr. (often foll. by out) utter or say with a rumbling sound. **4** tr. Brit. sl. find out about (esp. something illicit), discover the misbehaviour of (a person). ● n. **1** a rumbling sound. **2** N. Amer. sl. a street-fight between gangs. □ **rumble seat** N. Amer. an uncovered folding seat in the rear of a motor car. □ **rumbler** n. [ME romble, prob. f. MDu. rommelen, rummelen (imit.)]

rumbustious /rʌmˈbʌstʃəs/ adj. boisterous, noisy, uproarious. □ **rumbustiously** adv. **rumbustiousness** n. [prob. var. of robustious boisterous, ROBUST]

Rumelia /ruːˈmiːlɪə/ (also **Roumelia**) the territories in Europe which formerly belonged to the Ottoman Empire, including Macedonia, Thrace, and Albania. Its name is derived from the Turkish word Rumeli 'land of the Romans'.

rumen /ˈruːmen/ n. (pl. **rumens** or **rumina** /-mɪnə/) the first stomach of a ruminant, in which food, esp. cellulose, is partly digested by bacteria. [L rumen ruminis throat]

ruminant /ˈruːmɪnənt/ n. & adj. ● n. a hoofed animal that chews the cud. (See note below.) ● adj. **1** of or belonging to ruminants. **2** contemplative; given to or engaged in meditation. [L ruminari ruminant- (as RUMEN)]

▪ The ruminants are all even-toed ungulates (Artiodactyla). They have a complex stomach consisting of several chambers, including a rumen from which the cud is regurgitated for chewing. Included in the suborder Ruminantia are the deer, giraffes, antelopes, cattle, sheep, and related families. The camels and llamas are also ruminants, but are usually classified separately.

ruminate /ˈruːmɪˌneɪt/ v. **1** tr. & (foll. by over, on, etc.) intr. meditate, ponder. **2** intr. (of ruminants) chew the cud. □ **ruminator** n. **ruminative** /-nətɪv/ adj. **ruminatively** adv. **rumination** /ˌruːmɪˈneɪʃ(ə)n/ n.

rummage /ˈrʌmɪdʒ/ v. & n. ● v. **1** tr. & (foll. by in, through, among) intr. search, esp. untidily and unsystematically. **2** tr. (foll. by out, up) find among other things. **3** tr. (foll. by about) disarrange; make untidy in searching. ● n. **1** an instance of rummaging. **2** things found by rummaging; a miscellaneous accumulation. □ **rummage sale** esp. N. Amer. a jumble sale. □ **rummager** n. [earlier as noun in obs. sense 'arranging of casks etc. in a hold': OF arrumage f. arrumer stow (as AD-, run ship's hold f. MDu. ruim ROOM)]

rummer /ˈrʌmə(r)/ n. a large drinking-glass. [rel. to Du. roemer, LG römer f. roemen praise, boast]

rummy[1] /ˈrʌmɪ/ n. a card-game played usu. with two packs, in which the players try to form sets and sequences of cards. [20th c.: orig. unkn.]

rummy[2] /ˈrʌmɪ/ adj. (**-ier, -iest**) Brit. colloq. = RUM[2].

rumour /ˈruːmə(r)/ n. & v. (US **rumor**) ● n. **1** general talk or hearsay of doubtful accuracy. **2** (often foll. by of, or that + clause) a current but unverified statement or assertion (heard a rumour that you are leaving). ● v.tr. (usu. in passive) report by way of rumour (it is rumoured that you are leaving; you are rumoured to be leaving). [ME f. OF rumur, rumor f. L rumor -oris noise]

rump /rʌmp/ n. **1** the hind part of a mammal, esp. the buttocks. **2 a** a

small or contemptible remnant. **b** (**the Rump**) = RUMP PARLIAMENT. □ **rump steak** a cut of beef from the rump. □ **rumpless** adj. [ME, prob. f. Scand.]

rumple /ˈrʌmp(ə)l/ v.tr. & intr. make or become creased or ruffled. □ **rumply** adj. [obs. rumple (n.) f. MDu. rompel f. rompe wrinkle]

Rump Parliament that part of the Long Parliament which continued to sit after Pride's Purge in 1648, and voted for the trial which resulted in the execution of Charles I. Dissolved by Oliver Cromwell in 1653, the Rump Parliament was briefly reconvened in 1659 but voted its own dissolution early in 1660.

rumpus /ˈrʌmpəs/ n. (pl. **rumpuses**) a disturbance, brawl, row, or uproar. □ **rumpus room** N. Amer., Austral., & NZ a room in the basement of a house for games and play. [18th c.: prob. fanciful]

rumpy-pumpy /ˌrʌmpɪˈpʌmpɪ/ n. joc. sexual intercourse. [fanciful f. RUMP + PUMP[1]]

run /rʌn/ v. & n. ● v. (**running**; past **ran** /ræn/; past part. **run**) **1** intr. go with quick steps on alternate feet, never having both or all feet on the ground at the same time. **2** intr. flee, abscond. **3** intr. go or travel hurriedly, briefly, etc. **4** intr. advance by or as by rolling or on wheels, or smoothly or easily. **b** be in action or operation (left the engine running). **5** intr. be current or operative; have duration (the lease runs for 99 years). **6** intr. (of a bus, train, etc.) travel or be travelling on its route (the train is running late). **7** intr. (of a play, exhibition, etc.) be staged or presented (is now running at the Apollo). **8** intr. extend; have a course or order or tendency (the road runs by the coast; prices are running high). **9 a** intr. compete in a race. **b** intr. finish a race in a specified position. **c** tr. compete in (a race). **10** intr. (often foll. by for) seek election (ran for president). **11 a** intr. (of a liquid etc.) flow, drip profusely. **b** tr. flow with (a specified liquid) (after the massacre, the rivers ran blood). **c** intr. (foll. by with) flow or be wet; drip (his face ran with sweat). **12** tr. **a** cause (water etc.) to flow. **b** fill (a bath) with water. **13** intr. spread or pass rapidly (a shiver ran down my spine). **14** intr. Cricket (of a batsman) run from one wicket to the other in scoring a run. **15** tr. traverse or make one's way through or over (a course, race, or distance). **16** tr. perform (an errand). **17** tr. publish (an article etc.) in a newspaper or magazine. **18 a** tr. cause (a machine or vehicle etc.) to operate. **b** intr. (of a mechanism or component etc.) move or work freely. **19** tr. direct or manage (a business, household, etc.). **20** tr. own and use (a vehicle) regularly. **21** tr. take (a person) for a journey in a vehicle (shall I run you to the shops?). **22** tr. cause to run or go in a specified way (ran the car into a tree). **23** tr. enter (a horse etc.) for a race. **24** tr. smuggle (guns etc.). **25** tr. chase or hunt. **26** tr. allow (an account) to accumulate for a time before paying. **27** intr. Naut. (of a ship etc.) go straight and fast. **28** intr. (of salmon) go up river from the sea. **29** intr. (of a colour in a fabric) spread from the dyed parts. **30 a** intr. (of a thought, the eye, the memory, etc.) pass in a transitory or cursory way (ideas ran through my mind). **b** tr. cause (one's eye) to look cursorily (ran my eye down the page). **c** tr. pass (a hand etc.) rapidly over (ran his fingers down her spine). **31** intr. (of hosiery) ladder. **32** intr. (of a candle) gutter. **33** intr. (of an orifice, esp. the eyes or nose) exude liquid matter. **34** tr. sew (fabric) loosely or hastily with running stitches. **35** tr. turn (cattle etc.) out to graze. ● n. **1** an act or spell of running. **2** a short trip or excursion, esp. for pleasure. **3** a distance travelled. **4** a general tendency of development or movement. **5** a rapid motion. **6** a regular route. **7 a** a continuous or long stretch or spell or course (a metre's run of wiring; had a run of bad luck). **b** a series or sequence, esp. of cards in a specified suit. **8** (often foll. by on) **a** a high general demand (for a commodity, currency, etc.) (a run on the dollar). **b** a sudden demand for repayment by a large number of customers of (a bank). **9** a quantity produced in one period of production (a print run). **10** a general or average type or class (not typical of the general run). **11 a** Cricket a point scored by the batsmen each running to the other's wicket, or an equivalent point awarded for some other reason. **b** Baseball a point scored usu. by the batter returning to the plate after touching the other bases. **12** (foll. by of) free use of or access to (had the run of the house). **13 a** an animal's regular track. **b** an enclosure for domestic animals or fowls. **c** a range of pasture. **14** a ladder in hosiery. **15** Mus. a rapid scale passage. **16** a class or line of goods. **17** a batch or drove of animals born or reared together. **18** a shoal of fish in motion. **19** a trough for water to run in. **20** US a small stream or brook. **21 a** a single journey, esp. by an aircraft. **b** (of an aircraft) a flight on a straight and even course at a constant speed before or while dropping bombs. **c** an offensive military operation. **22** a slope used for skiing or tobogganing. **23** (**the runs**) colloq. an attack of diarrhoea. □ **at a** (or **the**) **run** running. **on the run 1** escaping, running away. **2** hurrying about from place to place. **run about 1** bustle; hurry from one person or place to another. **2** (esp. of children) play or wander without restraint. **run across 1** happen to meet. **2** (foll. by *to*) make a brief journey or a flying visit (to a place). **run after 1** pursue with attentions; seek the society of. **2** give much time to (a pursuit etc.). **3** pursue at a run. **run against** happen to meet. **run along** *colloq.* depart. **run around 1** *Brit.* take from place to place by car etc. **2** deceive or evade repeatedly. **3** (often foll. by *with*) *colloq.* engage in sexual relations (esp. casually or illicitly). **run-around** *n.* (esp. in phr. **give a person the run-around**) deceit or evasion. **run at** attack by charging or rushing. **run away 1** get away by running; flee, abscond. **2** elope. **3** (of a horse) bolt. **run away with 1** carry off (a person, stolen property, etc.). **2** win (a prize) easily. **3** accept (a notion) hastily. **4** (of expense etc.) consume (money etc.). **5** (of a horse) bolt with (a rider, a carriage or its occupants). **6** leave home to have a relationship with. **7** deprive of self-control or common sense (*let his ideas run away with him*). **run a blockade** see BLOCKADE. **run down 1** knock down or collide with. **2** reduce the strength or numbers of (resources). **3** (of an unwound clock etc.) stop. **4** (of a person or a person's health) become feeble from overwork or undernourishment. **5** discover after a search. **6** disparage. **run-down** *n.* **1** a reduction in numbers. **2** a summary or brief analysis. ● *adj.* **1** decayed after prosperity. **2** enfeebled through overwork etc. **run dry** cease to flow, be exhausted. **run one's eye over** see EYE. **run for it** seek safety by fleeing. **a run** (or **a good run**) **for one's money 1** vigorous competition. **2** pleasure or reward derived from an activity. **run foul of 1** collide or become entangled with (another vessel etc.). **2** quarrel with. **run the gauntlet** see GAUNTLET². **run a person hard** (or **close**) press a person severely in a race or competition, or in comparative merit. **run high 1** (of the sea) have a strong current with a high tide. **2** (of feelings) be strong. **run in 1** run (a new engine or vehicle) carefully in the early stages. **2** *colloq.* arrest. **3** (of a combatant) rush to close quarters. **4** incur (a debt). **run-in** *n.* **1** the approach to an action or event. **2** *colloq.* a quarrel. **run in the family** (of a trait) be common in the members of a family. **run into 1** collide with. **2** encounter. **3** reach as many as (a specified figure). **4** fall into (a practice, absurdity, etc.). **5** be continuous or coalesce with. **run into the ground** *colloq.* bring (a person) to exhaustion etc. **run it fine** see FINE¹. **run its course** follow its natural progress; be left to itself. **run low** (or **short**) become depleted, have too little (*our tea ran low; we ran short of sugar*). **run off 1** flee. **2** produce (copies etc.) on a machine. **3** decide (a race or other contest) after a series of heats or in the event of a tie. **4** flow or cause to flow away. **5** write or recite fluently. **6** digress suddenly. **run-off** *n.* **1** an additional competition, election, race, etc., after a tie. **2** an amount of rainfall that is carried off an area by streams and rivers. **3** *NZ* a separate area of land where young animals etc. are kept. **run off at the mouth** *US sl.* talk incessantly. **run off one's feet** very busy. **run-of-the-mill** ordinary, undistinguished. **run on 1** (of written characters) be joined together. **2** continue in operation. **3** elapse. **4** speak volubly. **5** talk incessantly. **6** *Printing* continue on the same line as the preceding matter. **run out 1** come to an end; become used up. **2** (foll. by *of*) exhaust one's stock of. **3** *Cricket* put down the wicket of (a batsman who is running). **4** (of a liquid etc.) escape from a containing receptacle. **5** (of rope) pass out; be paid out. **6** jut out. **7** come out of a contest in a specified position etc. or complete a required score etc. (*they ran out worthy winners*). **8** complete (a race). **9** advance (a gun etc.) so as to project. **10** exhaust oneself by running. **run-out** *n. Cricket* the dismissal of a batsman by being run out. **run out on** *colloq.* desert (a person). **run over 1** overflow; extend beyond. **2** study or repeat quickly. **3** (of a vehicle or its driver) pass over, knock down or crush. **4** touch (the keys of a piano etc.) in quick succession. **5** (often foll. by *to*) go quickly by a brief journey or for a flying visit. **run ragged** exhaust (a person). **run rings round** see RING¹. **run riot** see RIOT. **run a** (or **the**) **risk** see RISK. **run the show** *colloq.* dominate in an undertaking etc. **run a temperature** be feverish. **run through 1** examine or rehearse briefly. **2** peruse. **3** deal successively with. **4** consume (an estate etc.) by reckless or quick spending. **5** pass through by running. **6** pervade. **7** pierce with a sword etc. **8** draw a line through (written words). **run-through** *n.* **1** a rehearsal. **2** a brief survey. **run to 1** have the money or ability for. **2** reach (an amount or number). **3** (of a person) show a tendency to (*runs to fat*). **4 a** be enough for (some expense or undertaking). **b** have the resources or capacity for. **5** fall into (ruin). **run to earth 1** *Hunting* chase to its lair. **2** discover after a long search. **run to meet** anticipate (one's troubles etc.). **run to seed** see SEED. **run up 1** accumulate (a debt etc.) quickly. **2** build or make hurriedly. **3** raise (a flag). **4** grow quickly. **5** rise in price. **6** (foll. by *to*) amount to. **7** force (a rival bidder) to bid higher. **8** add up (a column of figures). **9** (foll. by *to*) go quickly by a brief journey or for a flying visit. **run-up** *n.* **1** (often foll. by *to*) the period preceding an important event. **2** *Golf* a low approach shot. **run up against** meet with (a difficulty or difficulties). **run upon** (of a person's thoughts etc.) be

engrossed by; dwell upon. **run wild** grow or stray unchecked or undisciplined or untrained. □ **runnable** *adj.* [OE *rinnan*]

runabout /ˈrʌnəˌbaʊt/ *n.* a light car, aircraft, or (esp. *US*) motor boat.

runaway /ˈrʌnəˌweɪ/ *n.* **1** a fugitive. **2** an animal or vehicle that is running out of control. **3** (*attrib.*) **a** that is running away or out of control (*runaway inflation; had a runaway success*). **b** done or performed after running away (*a runaway wedding*).

runcible spoon /ˈrʌnsɪb(ə)l/ *n.* a fork curved like a spoon, with three broad prongs, one edged. [nonsense word used by Edward Lear, perh. after *rouncival* large pea]

runcinate /ˈrʌnsɪnət/ *adj. Bot.* (of a leaf) saw-toothed, with lobes pointing towards the base. [mod.L *runcinatus* f. L *runcina* PLANE² (formerly taken to mean saw)]

Runcorn /ˈrʌŋkɔːn/ an industrial town in NW England, on the River Mersey in Cheshire; pop. (est. 1985) 64,600. It was developed as a new town from 1964.

rune /ruːn/ *n.* **1** any of the letters of the earliest Germanic alphabet used by Scandinavians and Anglo-Saxons from about the 3rd century and formed by modifying Roman or Greek characters to suit carving. **2** a similar mark of mysterious or magic significance. **3** a Finnish poem or a division of it. □ **rune-staff 1** a magic wand inscribed with runes. **2** a runic calendar. □ **runic** *adj.* [ON *rún* (only in pl. *rúnar*) magic sign, rel. to OE *rún*]

rung¹ /rʌŋ/ *n.* **1** each of the horizontal supports of a ladder. **2** a strengthening crosspiece in the structure of a chair etc. □ **runged** *adj.* **rungless** *adj.* [OE *hrung*]

rung² *past part.* of RING².

runlet /ˈrʌnlɪt/ *n.* a small stream.

runnel /ˈrʌn(ə)l/ *n.* **1** a brook or rill. **2** a gutter. [later form (assim. to RUN) of *rinel* f. OE *rynel* (as RUN)]

runner /ˈrʌnə(r)/ *n.* **1** a person, horse, etc. that runs, esp. in a race. **2 a** a creeping plant-stem that can take root. **b** a twining plant. **3** a rod or groove or blade on which a thing slides. **4** a sliding ring on a rod etc. **5** a messenger, scout, collector, or agent for a bank etc.; a tout. **6** *hist.* a police officer. **7** a running bird. **8 a** (usu. with a qualifying word) a smuggler. **b** = *blockade-runner.* **9** a revolving millstone. **10** *Naut.* a rope in a single block with one end round a tackle-block and the other having a hook. **11** (in full **runner bean**) *Brit.* a twining bean plant, *Phaseolus coccineus*, with red flowers and long green seed pods. Also called *scarlet runner.* **12** each of the long pieces on the underside of a sledge etc. that forms the contact in sliding. **13** a roller for moving a heavy article. **14** a long narrow ornamental cloth or rug. □ **do a runner** *sl.* leave hastily; abscond. **runner-up** (*pl.* **runners-up** or **runner-ups**) the competitor or team taking second place.

running /ˈrʌnɪŋ/ *n. & adj.* ● *n.* **1 a** the action of a runner. **b** the sport of racing on foot. **2** the way a race etc. proceeds. **3** management, control; operation. ● *adj.* **1** continuing on an essentially continuous basis though changing in detail (*a running battle*). **2** consecutive; one after another (*three days running*). **3** done with a run (*a running jump*). □ **in** (or **out of**) **the running** (of a competitor) with a good (or poor) chance of winning. **make** (or **take up**) **the running** take the lead; set the pace. **running account** a current account. **running back** *Amer. Football* a back whose main function is to run carrying the ball. **running-board** a footboard on either side of a vehicle. **running commentary** an oral description of events as they occur, esp. a broadcast report of a sports contest. **running fire** successive shots from a line of troops etc. **running gear 1** the moving parts of a machine, esp. the wheels, steering, and suspension of a vehicle. **2** the rope and tackle used in handling a boat. **running hand** writing in which the pen etc. is not lifted after each letter. **running head** (or **headline**) a heading printed at the top of a number of consecutive pages of a book etc. **running knot** a knot that slips along the rope etc. and changes the size of a noose. **running light 1** = *navigation light.* **2** each of a small set of lights on a motor vehicle that remain illuminated while the vehicle is running. **running mate** esp. *US* **1** a candidate for a secondary position in an election. **2** a horse entered in a race in order to set the pace for another horse from the same stable which is intended to win. **running repairs** minor or temporary repairs etc. to machinery while in use. **running rope** a rope that is freely movable through a pulley etc. **running sore** a suppurating sore. **running stitch 1** a line of small non-overlapping stitches for gathering etc. **2** one of these stitches. **running water** water flowing in a stream or from a tap etc. **take a running jump** (esp. as *int.*) *colloq.* go away.

runny /'rʌni/ adj. (**runnier**, **runniest**) **1** tending to run or flow. **2** excessively fluid.

Runnymede /'rʌni,mi:d/ a meadow on the south bank of the Thames at Egham near Windsor, Surrey. It is famous for its association with Magna Carta, which was signed by King John in 1215 there or nearby.

runt /rʌnt/ n. **1** a small pig, esp. the smallest in a litter. **2** a weakling; an undersized person. **3** a large domestic pigeon. **4** a small ox or cow, esp. of various Scottish Highland or Welsh breeds. □ **runty** adj. [16th c.: orig. unkn.]

runway /'rʌnweɪ/ n. **1** a specially prepared surface along which aircraft take off and land. **2** a trail to an animals' watering-place. **3** an incline down which logs are slid. **4** a raised gangway in a theatre, fashion show, etc.

Runyon /'rʌnjən/, (Alfred) Damon (1884–1946), American author and journalist. He is best known for his short stories about New York's Broadway and underworld characters, written in a highly individual style with much use of colourful slang. His collections include *Guys and Dolls* (1932) which formed the basis for the musical of the same name (1950).

rupee /ru:'pi:/ n. the basic monetary unit of India, Pakistan, Sri Lanka, Nepal, Mauritius, and the Seychelles. [Hind. *rūpiyah* f. Skr. *rūpya* wrought silver]

Rupert, Prince /'ru:pət/ (1619–82), English Royalist general, son of Frederick V, elector of the Palatinate, and nephew of Charles I. Born in Bohemia, he went to England and joined the Royalist side just before the outbreak of the Civil War in 1642. He made his name in the early years of the war as a leader of cavalry, but after a series of victorious engagements was defeated by Parliamentarian forces at Marston Moor (1644) and Naseby (1645). He later lived chiefly in France until the Restoration (1660), when he returned to England and commanded naval operations against the Dutch (1665–7 and 1672–4). In 1670 Rupert became the first governor of the Hudson's Bay Company in Canada. He was also responsible for the introduction of mezzotint engraving into England.

Rupert's Land (also **Prince Rupert's Land**) a historic region of northern and western Canada, originally granted in 1670 by Charles II to the Hudson's Bay Company and named after Prince Rupert, the first governor of the Company. It comprised the territory in the drainage basin of Hudson Bay, roughly corresponding to what is now Manitoba, Saskatchewan, Yukon, Alberta, and the southern part of the Northwest Territories. It ceased to exist in 1870, when the land was purchased from the Hudson's Bay Company by Canada, although the name is still used for an ecclesiastical diocese.

rupiah /ru:'pi:ə/ n. the basic monetary unit of Indonesia. [as RUPEE]

rupture /'rʌptʃə(r)/ n. & v. ● n. **1** the act or an instance of breaking; a breach. **2** a breach of harmonious relations; a disagreement and parting. **3** *Med.* an abdominal hernia. ● v. **1** tr. break or burst (a cell or membrane etc.). **2** tr. sever (a connection). **3** intr. undergo a rupture. **4** tr. & intr. affect with or suffer a hernia. □ **rupturable** adj. [ME f. OF *rupture* or L *ruptura* f. *rumpere rupt-* break]

rural /'rʊərəl/ adj. **1** in, of, or suggesting the country (opp. URBAN); pastoral or agricultural (*in rural seclusion; a rural constituency*). **2** often derog. characteristic of country people; rustic, plain, simple. □ **rural dean** see DEAN[1] 1b. **rural district** *Brit. hist.* a group of country parishes governed by an elected council. □ **ruralism** n. **ruralist** n. **rurally** adv. **ruralize** v. (also **-ise**) /ˌrʊərəlaɪ'zeɪʃ(ə)n/ n. **rurality** /rʊə'ræliti/ n. [ME f. OF *rural* or LL *ruralis* f. *rus ruris* the country]

Rurik /'rʊərɪk/ n. & adj. (also **Ryurik**) ● n. a member of a dynasty that ruled in Russia from the 9th century until 1598, reputedly founded by a Varangian chief who settled in Novgorod in 862. The Ruriks established themselves as rulers of the principality of Moscow and gradually extended their dominions into the surrounding territory. ● adj. of or relating to the Ruriks.

Ruritania /ˌrʊərɪ'teɪnɪə/ an imaginary kingdom in SE Europe used as a fictional background for the novels of courtly intrigue and romance written by Anthony Hope (1863–1933). □ **Ruritanian** adj. & n. [as RURAL, after *Lusitania*]

rusa /'ru:sə/ n. a large deer, *Cervus timorensis*, native to Indonesia. [mod.L *Rusa* former genus name, f. Malay]

Ruse /'ru:seɪ/ (also **Rousse**) an industrial city and the principal port of Bulgaria, on the Danube; pop. (1990) 209,760. Turkish during the Middle Ages, it was captured by Russia in 1877 and ceded to Bulgaria.

ruse /ru:z/ n. a stratagem or trick. [ME f. OF f. *ruser* drive back, perh. ult. f. L *rursus* backwards: cf. RUSH[1]]

rush[1] /rʌʃ/ v. & n. ● v. **1** intr. go, move, or act precipitately or with great speed. **2** tr. move or transport with great haste (*was rushed to hospital*). **3** intr. (foll. by *at*) **a** move suddenly and quickly towards. **b** begin impetuously. **4** tr. perform or deal with hurriedly (*don't rush your dinner; the bill was rushed through Parliament*). **5** tr. force (a person) to act hastily. **6** tr. attack or capture by sudden assault. **7** tr. sl. overcharge (a customer). **8** tr. US colloq. pay attentions to (a person) with a view to securing acceptance of a proposal. **9** tr. pass (an obstacle) with a rapid dash. **10** intr. flow, fall, spread, or roll impetuously or fast (*felt the blood rush to my face; the river rushes past*). ● n. **1 a** an act of rushing; a violent advance or attack. **b** a sudden flow or flood. **2** a period of great activity; a commotion. **3** (*attrib.*) done with great haste or speed (*a rush job*). **4** a sudden migration of large numbers. **5** a surge of emotion, excitement, etc. **6** (foll. by *on, for*) a sudden strong demand for a commodity. **7** (in *pl.*) the first prints of a film after a period of shooting. **8** *Rugby & Amer. Football* an attempt by one or more players to force the ball through a line of defenders. □ **rush one's fences** act with undue haste. **rush hour** a time each day when traffic is at its heaviest. □ **rusher** n. **rushingly** adv. [ME f. AF *russher*, = OF *ruser, russer*: see RUSE]

rush[2] /rʌʃ/ n. **1 a** a marsh or waterside plant of the family Juncaceae, with naked slender tapering pith-filled stems (properly leaves), formerly used for strewing floors and still used for making chair-bottoms and plaiting baskets etc. **b** a stem of this. **c** (*collect.*) rushes as a material. **2** *archaic* a thing of no value (*not worth a rush*). □ **rush candle** a candle made by dipping the pith of a rush in tallow. □ **rushlike** adv. **rushy** adj. [OE *rysc, rysce*, corresp. to MLG, MHG *rusch*]

Rushdie /'rʌʃdi, 'rʊʃ-/, (Ahmed) Salman (b.1947), Indian-born British novelist. He was educated in England and became a British citizen in 1964. His work is chiefly associated with magic realism; his Booker Prize-winning novel *Midnight's Children* (1981) views the development of India since independence through the eyes of a telepathic child. His later novel *The Satanic Verses* (1988), with its portrayal of a figure that many identified with Muhammad, was regarded by Muslims as blasphemous; in 1989 Ayatollah Khomeini issued a fatwa condemning Rushdie to death and he has since lived in hiding with a permanent police guard.

rushlight /'rʌʃlaɪt/ n. a rush candle.

Rushmore, Mount /'rʌʃmɔ:(r)/ a mountain in the Black Hills of South Dakota, noted for its giant busts of four US Presidents — George Washington, Thomas Jefferson, Abraham Lincoln, and Theodore Roosevelt — carved (1927–41) under the direction of the sculptor Gutzon Borglum (1867–1941).

rusk /rʌsk/ n. a slice of bread rebaked usu. as a light biscuit, esp. as food for babies. [Sp. or Port. *rosca* twist, coil, roll of bread]

Ruskin /'rʌskɪn/, John (1819–1900), English art and social critic. His prolific writings profoundly influenced 19th-century opinion and the development of the Labour movement. He was a champion of the painter J. M. W. Turner (at that time a controversial figure), the Pre-Raphaelite Brotherhood, and of Gothic architecture, which (following Pugin) he saw as a religious expression of medieval piety. *The Stones of Venice* (1851–3), attacking Renaissance art, led on to later attacks on capitalism in his lectures 'The Political Economy of Art' (1857), and on utilitarianism in *Unto This Last* (1860). His *Fors Clavigera* (1871–8) or 'Letters to the Workmen and Labourers of Great Britain' was an attempt to spread his notions of social justice, coupled with aesthetic improvement. His religious and philanthropic instincts also expressed themselves in the founding of the Guild of St George in 1871, a major contribution to the Arts and Crafts Movement.

Russell[1] /'rʌs(ə)l/, Bertrand (Arthur William), 3rd Earl Russell (1872–1970), British philosopher, mathematician, and social reformer. His work on mathematical logic had great influence on symbolic logic and on set theory in mathematics; his major work in this field is *Principia Mathematica* (1910–13), written with A. N. Whitehead (see MATHEMATICAL LOGIC). Although his philosophical views underwent continual development and revision, he wrote several books in the empiricist tradition (e.g. *Our Knowledge of the External World*, 1914) and was a principal proponent of neutral monism (see MONISM) and of logical atomism. During the First World War Russell became widely known as a conscientious objector; he also campaigned for women's suffrage and later took a leading role in CND. He was awarded the Nobel Prize for literature in 1950.

Russell[2] /'rʌs(ə)l/, George William (1867–1935), Irish poet. He met W. B. Yeats in 1886 and became interested in theosophy and mysticism; the first of several volumes of verse (published under the pseudonym AE) appeared in 1894. After the performance of his poetic drama *Deirdre* (1902) Russell became a leading figure in the Irish literary revival. His interests extended to public affairs and he edited *The Irish Homestead* (1905–23) and *The Irish Statesman* (1923–30).

Russell[3] /ˈrʌs(ə)l/, Henry Norris (1877–1957), American astronomer. He worked mainly in astrophysics and spectroscopy, and is best known for his independent discovery of the relationship between stellar magnitude and spectral type (see HERTZSPRUNG–RUSSELL DIAGRAM). He believed that this diagram represented a sequence of stellar evolution, a view no longer accepted. Russell carried out spectroscopic analyses to determine the constituent elements of stars, and he discovered that the sun contained much more hydrogen than had been expected.

Russell[4] /ˈrʌs(ə)l/, John, 1st Earl Russell (1792–1878), British Whig statesman, Prime Minister 1846–52 and 1865–6. As a member of Lord Grey's government (1830–4), he was responsible for introducing the Reform Bill of 1832 into Parliament. He became Prime Minister when Sir Robert Peel was defeated (1846) and later served as Foreign Secretary in Lord Aberdeen's coalition government (1852–4); Russell's second premiership ended with his resignation when his attempt to extend the franchise again in a further Reform Bill was unsuccessful.

russet /ˈrʌsɪt/ *adj. & n.* ● *adj.* **1** reddish-brown. **2** *archaic* rustic, homely, simple. ● *n.* **1** a reddish-brown colour. **2** a kind of rough-skinned russet-coloured apple. **3** *hist.* a coarse homespun reddish-brown or grey cloth used for simple clothing. □ **russety** *adj.* [ME f. AF f. OF *rosset, rousset*, dimin. of *roux* red f. Prov. *ros*, It. *rosso* f. L *russus* red]

Russia /ˈrʌʃə/ (official name **Russian Federation**) a country in northern Asia and eastern Europe; pop. (est. 1991) 148,930,000; official language, Russian; capital, Moscow. The modern state originated from the great expansion of the principality of Muscovy, under the Rurik and Romanov dynasties, into an empire stretching from the Arctic Ocean in the north to the Black Sea in the south and from the Baltic in the west to the Bering Strait in the east. Russia played an increasing role in Europe from the time of Peter the Great in the early 18th century, and pursued imperial ambitions in the East in the second half of the 19th century. Social and economic problems, exacerbated by the First World War, led to the overthrow of the tsar in the Russian Revolution of 1917. As the Russian Soviet Federative Socialist Republic (RSFSR), Russia formed the largest of the constituent republics of the Soviet Union, with more than three-quarters of the area and over half of the population. On the breakup of the Soviet Union and the collapse of Communist control in 1991, Russia emerged as an independent state and a founder member of the Commonwealth of Independent States. A new federal treaty establishing the Russian Federation was signed in 1992 by the majority of the Russian republics and other territories. (See also SOVIET UNION.)

Russia leather *n.* a durable bookbinding leather made from skins impregnated with birch-bark oil.

Russian /ˈrʌʃ(ə)n/ *n. & adj.* ● *n.* **1 a** a native or national of Russia or the former Soviet Union. **b** a person of Russian descent. **2** the Slavonic language of Russia and the official language of the former Soviet Union, spoken by about 155 million people. ● *adj.* **1** of or relating to Russia. **2** of or in Russian. □ **Russian boot** a boot that loosely encloses the calf. **Russian olive** = OLEASTER. **Russian roulette 1** an act of daring in which a person (usu. with others in turn) squeezes the trigger of a revolver held to his or her head with one chamber loaded, having first spun the chamber. **2** a potentially dangerous enterprise. **Russian salad** a salad of mixed diced vegetables with mayonnaise. □ **Russianness** /-ʃ(ə)nnɪs/ *n.* **Russianize** *v.tr.* (also **-ise**). **Russianization** /ˌrʌʃənaɪˈzeɪʃ(ə)n/ *n.* [med.L *Russianus*]

Russian Civil War a conflict fought in Russia (1918–21) after the Revolution, between the Bolshevik Red Army and the counter-revolutionary White Russians. Forces opposed to the Bolsheviks, led by army commanders, began armed resistance at the end of 1917 but were met by the Red Army hastily organized by Trotsky. Despite being aided by British, French, US, and Japanese forces, the White Russians were hampered by disunity and problems with communications; although revolts in Poland, Finland, and the Baltic states were successful, the Bolsheviks were ultimately victorious, and the Union of Soviet Socialist Republics was established.

Russian Federation the official name for RUSSIA.

Russian Orthodox Church the national Church of Russia (see ORTHODOX CHURCH).

Russian Revolution the revolution in the Russian empire in 1917, in which the tsarist regime was overthrown and replaced by Bolshevik rule under Lenin. There were two phases to the Revolution: the first, in March (Old Style, February, whence *February Revolution*), was largely supported by the liberal Mensheviks; it was sparked off by food and

fuel shortages during the First World War and began with strikes and riots in Petrograd (St Petersburg). The tsar was forced to abdicate, and a provisional government was set up. The second phase, in November 1917 (Old Style, October, whence *October Revolution*), was marked by the seizure of power by the Bolsheviks in a coup led by Lenin. After workers' councils or *soviets* took power in major cities, the new Soviet constitution was declared in 1918.

Russian Revolution of 1905 the uprising in Russia in 1905. Popular discontent, fuelled by heavy taxation and the country's defeat in the Russo-Japanese War, led to a peaceful demonstration in St Petersburg, which was fired on by troops. The crew of the battleship *Potemkin* mutinied and a soviet (or workers' council) was formed in St Petersburg, prompting Tsar Nicholas II to make a number of short-lived concessions including the formation of an elected legislative body or Duma.

Russify /ˈrʌsɪˌfaɪ/ *v.tr.* (**-ies**, **-ied**) make Russian in character. □ **Russification** /ˌrʌsɪfɪˈkeɪʃ(ə)n/ *n.*

Russki /ˈrʌskɪ/ *n.* (also **Russky**) (*pl.* **Russkis** or **-ies**) *sl.* often *offens.* a Russian or Soviet. [RUSSIAN after Russ. surnames ending in *-ski*]

Russo- /ˈrʌsəʊ/ *comb. form* Russian; Russian and.

Russo-Japanese War /ˌrʌsəʊˌdʒæpəˈniːz/ a war between the Russian empire and Japan 1904–5, caused by territorial disputes in Manchuria and Korea. Russia suffered a series of humiliating defeats which contributed to the Revolution of 1905, while the peace settlement gave Japan the ascendancy in the disputed region.

Russophile /ˈrʌsəʊˌfaɪl/ *n.* a person who is fond of Russia or the Russians.

Russo-Turkish Wars /ˌrʌsəʊˈtɜːkɪʃ/ a series of wars between Russia and the Ottoman Empire, fought largely in the Balkans, the Crimea, and the Caucasus in the 19th century. The wars accelerated the decline of the Ottoman Empire and stimulated nationalist aspirations throughout the area of conflict; the treaty ending the war of 1877–8 freed the nations of Romania, Serbia, and Bulgaria from Turkish rule.

rust /rʌst/ *n. & v.* ● *n.* **1 a** a reddish or yellowish-brown coating formed on iron or steel by oxidation, esp. as a result of moisture. **b** a similar coating on other metals. **2 a** a plant disease characterized by rust-coloured spots, caused by fungi of the class Urediniomycetes or the genus *Albugo*. **b** the fungus causing this. **3** an impaired state due to disuse or inactivity. **4** a reddish-brown or brownish-red colour. ● *v.* **1** *tr. & intr.* affect or be affected with rust; undergo oxidation. **2** *intr.* (of bracken etc.) become rust-coloured. **3** *intr.* (of a plant) be attacked by rust. **4** *intr.* lose quality or efficiency by disuse or inactivity. □ **rust-belt** (often *attrib.*) esp. *US* a declining major industrial area, esp. the steel-producing region of the American Midwest and north-east. □ **rustless** *adj.* [OE *rūst* f. Gmc]

rustic /ˈrʌstɪk/ *adj. & n.* ● *adj.* **1** having the characteristics of or associations with the country or country life. **2** unsophisticated, simple, unrefined. **3** of rude or country workmanship. **4** made of untrimmed branches or rough timber (*a rustic bench*). **5** (of lettering) freely formed. **6** *Archit.* with rough-hewn or roughened surface or with sunk joints. ● *n.* a person from or living in the country, esp. a simple unsophisticated one. □ **rustically** *adv.* **rusticity** /rʌˈstɪsɪtɪ/ *n.* [ME f. L *rusticus* f. *rus* the country]

rusticate /ˈrʌstɪˌkeɪt/ *v.* **1** *tr.* send down (a student) temporarily from university. **2** *intr.* retire to or live in the country. **3** *tr.* make rural. **4** *tr. Archit.* mark (masonry) with sunk joints or a roughened surface. □ **rustication** /ˌrʌstɪˈkeɪʃ(ə)n/ *n.* [L *rusticari* live in the country (as RUSTIC)]

rustle /ˈrʌs(ə)l/ *v. & n.* ● *v.* **1** *intr. & tr.* make or cause to make a gentle sound as of dry leaves blown in a breeze. **2** *intr.* (often foll. by *along* etc.) move with a rustling sound. **3** *tr.* (also *absol.*) steal (cattle, horses, or sheep). **4** *intr. US colloq.* hustle. ● *n.* a rustling sound or movement. □ **rustle up** *colloq.* produce quickly when needed. □ **rustler** *n.* (esp. in sense 3 of *v.*). [ME *rustel* etc. (imit.): cf. obs. Flem. *ruysselen*, Du. *ritselen*]

rustproof /ˈrʌstpruːf/ *adj. & v.* ● *adj.* (of a metal) not susceptible to corrosion by rust. ● *v.tr.* make rustproof.

rustre /ˈrʌstə(r)/ *n. Heraldry* a lozenge with a round hole. [F]

rusty /ˈrʌstɪ/ *adj.* (**rustier, rustiest**) **1** rusted or affected by rust. **2** stiff with age or disuse. **3** (of knowledge etc.) faded or impaired by neglect (*my French is a bit rusty*). **4** rust-coloured. **5** (of black clothes) discoloured by age. **6 a** of antiquated appearance. **b** antiquated or behind the times. **7** (of a voice) croaking or creaking. □ **rustily** *adv.* **rustiness** *n.* [OE *rūstig* (as RUST)]

rut[1] /rʌt/ *n. & v.* ● *n.* **1** a deep track made by the passage of wheels. **2** an established (esp. tedious) mode of practice or procedure. ● *v.tr.* (**rutted**, **rutting**) mark with ruts. □ **in a rut** following a fixed (esp. tedious or dreary) pattern of behaviour that is difficult to change. □ **rutty** *adj.* [prob. f. OF *rote* (as ROUTE)]

rut[2] /rʌt/ *n. & v.* ● *n.* the periodic sexual excitement of a male deer, goat, sheep, etc. ● *v.intr.* (**rutted**, **rutting**) be affected with rut. □ **ruttish** *adj.* [ME f. OF *rut*, *ruit* f. L *rugitus* f. *rugire* roar]

rutabaga /ˌruːtəˈbeɪɡə/ *n.* a swede. [Sw. dial. *rotabagge*]

Ruth[1] /ruːθ/ a book of the Bible telling the story of Ruth, a Moabite woman, who married her deceased husband's kinsman Boaz. David is descended from her.

Ruth[2] /ruːθ/, Babe (born George Herman Ruth) (1895–1948), American baseball player. He played for the Boston Red Sox (1914–19) and the New York Yankees (1919–35); during his career he set the record for the most home runs (714), which remained unbroken until 1974.

Ruthenia /ruːˈθiːnɪə/ a region of central Europe on the southern slopes of the Carpathian Mountains, now forming the Transcarpathian region of western Ukraine. The region takes its name from the Ruthenes or Russniaks, a Slavonic people who were ancestors of the Ukrainians. Formerly part of the Austro-Hungarian empire, it was divided between Poland, Czechoslovakia, and Romania. Gaining independence for a single day in 1938, it was occupied by Hungary and in 1945 ceded to the Soviet Union, becoming part of Ukraine. □ **Ruthenian** *adj. & n.*

ruthenium /ruːˈθiːnɪəm/ *n.* a hard silvery-white metallic chemical element (atomic number 44; symbol **Ru**). A transition element and one of the platinum metals, ruthenium was isolated by the Russian chemist Karl (Karlovich) Klauss (1796–1864) in 1845. It is rare, occurring chiefly in platinum ores, and is used as a catalyst and in certain alloys. [med.L *Ruthenia* Russia (from its discovery in ores from the Urals)]

Rutherford[1] /ˈrʌðəfəd/, Sir Ernest, 1st Baron Rutherford of Nelson (1871–1937), New Zealand physicist. He is regarded as the founder of nuclear physics, and worked mainly in Britain. He established the nature of alpha and beta particles, and (with Frederick Soddy) proposed the laws of radioactive decay. He later concluded that the positive charge in an atom, and virtually all its mass, is concentrated in a central nucleus, with negatively charged electrons in orbit round it. In 1919 Rutherford announced the first artificial transmutation of matter — he had changed nitrogen atoms into oxygen by bombarding them with alpha particles. He was awarded the Nobel Prize for chemistry in 1908.

Rutherford[2] /ˈrʌðəfəd/, Dame Margaret (1892–1972), English actress. She is chiefly remembered for her roles as a formidable but jovial eccentric; they include Miss Prism in *The Importance of Being Earnest* which she played on stage in 1939 and on film in 1952. Among her other films were *Passport to Pimlico* (1949), several film versions of Agatha Christie novels in which she played Miss Marple, and *The VIPs* (1963), for which she won an Oscar.

rutherfordium /ˌrʌðəˈfɔːdɪəm/ *n.* a name proposed in the US for the chemical element unnilquadium (cf. KURCHATOVIUM). The corresponding symbol is **Rf**. [RUTHERFORD[1]]

ruthless /ˈruːθlɪs/ *adj.* having no pity or compassion. □ **ruthlessly** *adv.* **ruthlessness** *n.* [ME, f. *ruth* compassion f. RUE[1]]

rutile /ˈruːtaɪl/ *n.* a mineral form of titanium dioxide. [F *rutile* or G *Rutil* f. L *rutilus* reddish]

Ruwenzori /ˌruːenˈzɔːri/ a mountain range in central Africa, on the Ugandan–Zaire border between Lake Edward and Lake Albert, rising to 5,110 m (16,765 ft) at Margherita Peak on Mount Stanley. The range is generally thought to be the 'Mountains of the Moon' mentioned by Ptolemy, and as such the supposed source of the Nile.

Ruysdael see RUISDAEL.

RV *abbr.* **1** see REVISED VERSION. **2** *US* recreational vehicle, esp. a motorized caravan.

Rwanda /ruːˈændə/ a landlocked country in central Africa, to the north of Burundi and the south of Uganda; pop. (est. 1991) 7,403,000; official languages, Rwanda (a Bantu language) and French; capital, Kigali. Inhabited largely by Hutu and Tutsi peoples, the area was claimed by Germany from 1890, and after the First World War became part of a Belgian trust territory; as in Burundi, the country was ruled by the Belgians through the minority Tutsi kings. Rwanda became independent as a republic in 1962, shortly after the violent overthrow of the monarchy by the majority Hutu people. In 1990 an army of Tutsi exiles invaded the country from Uganda, and in 1991 a new multi-party system was introduced in response to the rebels' demands for greater democracy. International attention was focused on the country in April 1994 following the death in a plane crash of the President, together with the President of Burundi. Up to 500,000 people, largely Tutsis, were slaughtered by predominantly Hutu supporters of the government, and over a million fled as refugees into Zaire and neighbouring countries. The Tutsi-dominated Rwandan Patriotic Front took power as the new government. □ **Rwandan** *adj. & n.*

Ry. *abbr.* Railway.

-ry /rɪ/ *suffix* = -ERY (*infantry; rivalry*). [shortened f. -ERY, or by analogy]

Ryazan /rɪəˈzɑːn/ an industrial city in European Russia, situated to the south-east of Moscow; pop. (1990) 522,000.

Rybinsk /ˈrɪbɪnsk/ a city in NW Russia, a port on the River Volga; pop. (1990) 252,000. It was formerly known as Shcherbakov (1946–57) and, in honour of the former President of the Soviet Union, Yuri Andropov, as Andropov (1984–9).

Ryder /ˈraɪdə(r)/, Sue, Baroness Ryder of Warsaw and Cavendish (b.1923), English philanthropist. After the Second World War she co-founded an organization to care for former inmates of concentration camps; later known as the Sue Ryder Foundation for the Sick and the Disabled, it expanded to provide homes for the mentally and physically disabled in the UK and elsewhere in Europe. Ryder married the philanthropist Leonard Cheshire in 1959, and is a trustee of the Cheshire Foundation.

Ryder Cup a golf tournament held every two years and played between teams of male professionals from the US and Europe (originally the US and Great Britain). It was first held in 1927; the trophy was donated by Samuel Ryder (1859–1936), an English seed-merchant.

rye /raɪ/ *n.* **1 a** a cereal plant, *Secale cereale*, with spikes bearing florets which yield wheatlike grains. **b** the grain of this used for bread and fodder. **2** (in full **rye whisky**) whisky distilled from fermented rye. [OE *ryge* f. Gmc]

ryegrass /ˈraɪɡrɑːs/ *n.* a forage or lawn grass of the genus *Lolium*, esp. *L. perenne*. [obs. *ray-grass*, of unkn. orig.]

Ryle[1] /raɪl/, Gilbert (1900–76), English philosopher. He was professor of metaphysical philosophy at Oxford (1945–68) and did much to make Oxford a leading centre for philosophical research. He was a prominent figure in the linguistic school of philosophy and held that philosophy should identify 'the sources in linguistic idioms of recurrent misconstructions and absurd theories'. His most famous work, *The Concept of Mind* (1949), is a strong attack on the mind-and-body dualism of Descartes. He was a cousin of the astronomer Sir Martin Ryle.

Ryle[2] /raɪl/, Sir Martin (1918–84), English astronomer. He carried out pioneering work in radio astronomy in the 1950s, when he produced the first detailed sky map of radio sources. His demonstration that remote objects appeared to be different from closer ones helped to establish the big bang as opposed to the steady-state theory of the universe. Ryle was Astronomer Royal 1972–82 and was awarded the Nobel Prize for physics in 1974, the first astronomer to be so honoured. He was a cousin of the philosopher Gilbert Ryle.

ryokan /rɪˈəʊkən/ *n.* a traditional Japanese inn. [Jap.]

ryot /ˈraɪət/ *n.* an Indian peasant. [Urdu *ra'īyat* f. Arab. *ra'īya* flock, subjects f. *ra'ā* to pasture]

Rysy /ˈrɪsɪ/ a peak in the Tatra Mountains rising to a height of 2,499 m (8,197 ft).

Ryukyu Islands /rɪˈuːkjuː/ a chain of islands in the western Pacific, stretching for about 960 km (600 miles) from the southern tip of the island of Kyushu, Japan, to Taiwan. The largest island is Okinawa. Part of China in the 14th century, the archipelago was incorporated into Japan by 1879. They were placed under US military control in 1945 and returned to Japan in 1972.

Ryurik var. of RURIK.

Ss

S[1] /es/ *n.* (also **s**) (*pl.* **Ss** or **S's** /'esɪz/) **1** the nineteenth letter of the alphabet. **2** an S-shaped object or curve.

S[2] *abbr.* (also **S.**) **1** Saint. **2** South, Southern.

S[3] *symb.* **1** *Chem.* the element sulphur. **2** siemens.

s. *abbr.* **1** second(s). **2** shilling(s). **3** singular. **4** son. **5** succeeded. **6** solid. [sense 2 orig. f. L *solidus*: see SOLIDUS]

's /s/ (after a vowel sound or voiced consonant /z/) *abbr.* **1** is, has (*he's; it's; John's; Charles's*). **2** us (*let's*). **3** *colloq.* does (*what's he say?*).

's- /s, z/ *prefix archaic* (esp. in oaths) God's (*'sblood; 'struth*). [abbr.]

-s[1] /s/ (after a vowel sound or voiced consonant, e.g. *ways, bags* /z/) *suffix* denoting the plurals of nouns (cf. -ES[1]). [OE -*as* pl. ending]

-s[2] /s/ (after a vowel sound or voiced consonant, e.g. *ties, begs* /z/) *suffix* forming the 3rd person sing. present of verbs (cf. -ES[2]). [OE dial., prob. f. OE 2nd person sing. present ending -*es*, -*as*]

-s[3] /s/ (after a vowel sound or voiced consonant, e.g. *besides* /z/) *suffix* **1** forming adverbs (*afterwards; besides; mornings*). **2** forming possessive pronouns (*hers; ours*). [formed as -'s[1]]

-s[4] /s/ (after a vowel sound or voiced consonant /z/) *suffix* forming nicknames or pet names (*Fats; ducks*). [after -s[1]]

-s' /s/ (after a vowel sound or voiced consonant /z/) *suffix* denoting the possessive case of plural nouns and sometimes of singular nouns ending in *s* (*the boys' shoes; Charles' book*). [as -'s[1]]

-'s[1] /s/ (after a vowel sound or voiced consonant /z/) *suffix* denoting the possessive case of singular nouns and of plural nouns not ending in -*s* (*John's book; the book's cover; the children's shoes*). [OE genitive sing. ending]

-'s[2] /s/ (after a vowel sound or voiced consonant /z/) *suffix* denoting the plural of a letter or symbol (*S's; 8's*). [as -s[1]]

SA *abbr.* **1** Salvation Army. **2** sex appeal. **3 a** South Africa. **b** South America. **c** South Australia. **4** *hist.* Sturmabteilung (see BROWNSHIRT).

Saadi see SADI.

Saale /'sɑːlə/ a river of east central Germany. Rising in northern Bavaria near the border with the Czech Republic, it flows 425 km (265 miles) north to join the Elbe near Magdeburg.

Saar /sɑː(r)/ (French **Sarre** /sar/) **1** a river of western Europe. Rising in the Vosges mountains in eastern France, it flows 240 km (150 miles) northwards to join the Mosel river in Germany, just east of the border with Luxembourg. **2** the Saarland.

Saarbrücken /sɑː'brʊkən/ an industrial city in western Germany, the capital of Saarland, on the River Saar close to the border with France; pop. (1991) 361,600.

Saarland /'sɑːlænd/ a state of western Germany, on the border with France; capital, Saarbrücken. It is traversed by the River Saar and has rich deposits of coal and iron ore. Between 1684 and 1697 and again between 1792 and 1815 the area belonged to France. After the First World War, until 1935, it was administered by the League of Nations. A plebiscite in 1935 — and a referendum after the Second World War — indicated the desire on the part of the population to be part of Germany. The area became the tenth German state in 1959.

Saba /'sɑːbɑ/ an island in the Netherlands Antilles, in the Caribbean. The smallest island in the group, it is situated to the north-west of St Kitts.

sabadilla /ˌsæbə'dɪlə/ *n.* **1** a Mexican plant, *Schoenocaulon officinale*, of the lily family, with seeds yielding veratrine. **2** a preparation of these seeds, used in medicine and agriculture. [Sp. *cebadilla* dimin. of *cebada* barley]

Sabaean /sə'biːən/ *n.* & *adj.* ● *n.* a member of an ancient Semitic-speaking people. (*See note below.*) ● *adj.* of or relating to the Sabaeans. [L *Sabaeus* f. Gk *Sabaios*, ult. f. Heb. *Sheba* people of Yemen (see SHEBA)]

▪ The Sabaeans inhabited Saba or Sheba in SW Arabia, home of the biblical queen of Sheba. By the 3rd century AD they had established an elaborate system of government and succeeded in uniting southern Arabia into a single state, which was overthrown by the Abyssinians in AD 525.

Sabah /'sɑːbɑ/ a state of Malaysia, comprising the northern part of Borneo and some offshore islands; capital, Kota Kinabalu. A British protectorate from 1888, it gained independence and joined Malaysia in 1963.

Sabaoth /'sæbeɪˌɒθ, sæ'beɪnθ/ *n.pl. Bibl.* heavenly hosts (see HOST[1] 2) (*Lord of Sabaoth*). [ME f. LL f. Gk *Sabaōth* f. Heb. *ṣᵉbāōṯ* pl. of *ṣābā* host (of heaven)]

Sabbatarian /ˌsæbə'teərɪən/ *n.* & *adj.* ● *n.* **1** a strict sabbath-keeping Jew. **2** a Christian who favours observing Sunday strictly as the sabbath. **3** a Christian who observes Saturday as the sabbath. ● *adj.* relating to or holding the tenets of Sabbatarians. □ **Sabbatarianism** *n.* [LL *sabbatarius* f. L *sabbatum*: see SABBATH]

sabbath /'sæbəθ/ *n.* (also **sabbat** /-bət/ in sense 3) **1** (in full **sabbath day**) **a** the last day of the week (Saturday) as a day for religious observance and abstinence from work for Jews and some Christian sects. **b** Sunday as a Christian day of religious observance and abstinence from work. **c** Friday as a Muslim day of religious observance and abstinence from work. **2** a period of rest. **3** (in full **witches' sabbath**) a midnight meeting of witches, allegedly presided over by the Devil and involving satanic rites. [OE *sabat*, L *sabbatum*, & OF *sabbat*, f. Gk *sabbaton* f. Heb. *šabbāṯ* f. *šāḇaṯ* to rest]

sabbatical /sə'bætɪk(ə)l/ *adj.* & *n.* ● *adj.* **1** of or appropriate to the sabbath. **2** (of leave) granted at intervals to a university teacher for study or travel, orig. every seventh year. ● *n.* a period of sabbatical leave. □ **sabbatical year 1** *Bibl.* every seventh year, prescribed by the Mosaic law to be observed as a 'sabbath', during which the land was allowed to rest. **2** a year's sabbatical leave. □ **sabbatically** *adv.* [LL *sabbaticus* f. Gk *sabbatikos* of the sabbath]

Sabellian[1] /sə'belɪən/ *n.* & *adj.* ● *n.* **1** a member of a group of peoples in ancient Italy (including Sabines, Samnites, Campanians, etc.). **2** the Italic language or dialects of these peoples. ● *adj.* of or relating to the Sabellians or their language or dialects. [L *Sabellus*]

Sabellian[2] /sə'belɪən/ *n.* & *adj.* ● *n.* (in Christian theology) a holder of the doctrine of the Roman or African heretic Sabellius (*fl. c.*220), that the Father, Son, and Holy Spirit are merely aspects of one divine Person. ● *adj.* of or relating to the Sabellians.

saber *US var. of* SABRE.

Sabian /'seɪbɪən/ *n.* & *adj.* ● *n.* **1** an adherent of a religious sect mentioned in the Koran and by later Arabian writers. (*See note below.*) **2** a member of a group of star-worshippers in ancient Arabia and Mesopotamia. ● *adj.* of or relating to the Sabians. [f. Arab. *ṣābi'*, prob. as Heb. *sābā* host]

▪ In the Koran, the Sabians are classed with Muslims, Jews, and Christians as believers in the true God. It is not known who the original Sabians were, but because the sect was tolerated by Muslims, some later non-Islamic sects, including the Mandaeans, adopted the name of Sabians in order to escape religious persecution.

sabicu /'sæbɪˌkuː/ n. **1** a leguminous West Indian tree, *Lysiloma latisiliqua*, grown for timber. **2** the mahogany-like wood of this tree. [Cuban Sp. *sabicú*]

Sabine /'sæbaɪn/ n. & adj. ● n. a member of a people of ancient Italy of the area north-east of Rome. (*See note below.*) ● adj. of or relating to the Sabines. [L *Sabinus*]

■ Renowned in antiquity for their frugal and hardy character and their superstitious practices, the Sabines were finally conquered by Rome in 290 BC. The (unhistorical) legend of the Rape of the Sabine Women (said to have been carried off by the Romans at a spectacle to which the Sabines had been invited) reflects the early intermingling of Romans and Sabines; some Roman religious institutions were said to have a Sabine origin.

Sabin vaccine /'seɪbɪn/ n. Med. an oral vaccine giving immunity against poliomyelitis. [Albert Bruce *Sabin*, Amer. microbiologist (1906–93)]

sable[1] /'seɪb(ə)l/ n. **1 a** a small flesh-eating mammal, *Martes zibellina*, of northern Europe and parts of northern Asia, which is related to the martens and has valuable dark brown fur. **b** its skin or fur. **2** a fine paintbrush made of sable fur. [ME f. OF f. med.L *sabelum* f. Slav.]

sable[2] /'seɪb(ə)l/ n. & adj. ● n. **1** esp. poet. black. **2** (in *pl.*) mourning garments. ● adj. **1** (usu. placed after noun) *Heraldry* black. **2** esp. poet. dark, gloomy. □ **sable antelope** a large African antelope, *Hippotragus niger*, with long curved horns, the male of which has a black coat and white belly. □ **sabled** adj. **sably** adv. [ME f. OF (in Heraldry): usu. taken to be identical with SABLE[1], although sable fur is dark brown]

sabot /'sæbəʊ/ n. **1** a kind of simple shoe hollowed out from a block of wood. **2** a wooden-soled shoe. **3** *Austral.* a small snub-nosed yacht. □ **saboted** /-bəʊd/ adj. [F, blend of *savate* shoe + *botte* boot]

sabotage /'sæbəˌtɑːʒ/ n. & v. ● n. deliberate damage to productive capacity, esp. as a political act. ● v.tr. **1** commit sabotage on. **2** destroy, spoil; make useless (*sabotaged my plans*). [F f. *saboter* make a noise with sabots, bungle, wilfully destroy: see SABOT]

saboteur /ˌsæbəˈtɜː(r)/ n. a person who commits sabotage. [F]

sabra /'sɑːbrə/ n. a Jew born in Israel. [mod. Heb. *sābrāh* opuntia fruit]

Sabratha /'sæbrəθə/ (also **Sabrata** /-brətə/) an ancient city on the coast of North Africa, near present-day Tripoli. It was one of three Phoenician colonies established in the region in the 7th century BC (see TRIPOLITANIA).

sabre /'seɪbə(r)/ n. & v. (US **saber**) ● n. **1** a cavalry sword with a curved blade. **2** a cavalry soldier and horse. **3** a light fencing-sword with a tapering blade. ● v.tr. cut down or wound with a sabre. □ **sabre-cut 1** a blow with a sabre. **2** a wound made or a scar left by this. **sabre-rattling** a display or threat of military force. **sabre-toothed cat** (or **tiger**) an extinct cat or catlike mammal with long curved upper canine teeth (also called *sabretooth*); esp. one of the Pleistocene genus *Smilodon* of America. [F, earlier *sable* f. G *Sabel*, *Säbel*, *Schabel* f. Pol. *szabla* or Hungarian *szablya*]

sabretache /'sæbəˌtæʃ/ n. a flat satchel on long straps worn by some cavalry officers from the left of the waist-belt. [F f. G *Säbeltasche* (as SABRE, *Tasche* pocket)]

sabretooth /'seɪb(ə)ˌtuːθ/ n. = sabre-toothed cat.

sabreur /sæˈbrɜː(r)/ n. a user of the sabre, esp. a cavalryman. [F f. *sabrer* SABRE v.]

sabrewing /'seɪbəˌwɪŋ/ n. a South American hummingbird of the genus *Campylopterus*, with long curved wings.

SAC abbr. (in the UK) Senior Aircraftman.

sac /sæk/ n. **1** Bot. & Zool. a baglike cavity, enclosed by a membrane, in an animal or plant. **2** Med. the distended membrane surrounding a hernia, cyst, tumour, etc. [F *sac* or L *saccus* SACK[1]]

saccade /sæˈkɑːd/ n. a brief rapid movement of the eye between fixation points. □ **saccadic** /səˈkædɪk/ adj. [F, = violent pull, f. OF *saquer*, *sachier* pull]

saccate /'sækeɪt/ adj. Bot. **1** dilated into a bag. **2** contained in a sac.

saccharide /'sækəˌraɪd/ n. Chem. = SUGAR n. 2. [mod.L *saccharum* sugar + -IDE]

saccharimeter /ˌsækəˈrɪmɪtə(r)/ n. a polarimeter for measuring the sugar content of a solution. [F *saccharimètre* (as SACCHARIDE)]

saccharin /'sækərɪn/ n. a very sweet substance used as a non-fattening substitute for sugar. [G (as SACCHARIDE) + -IN]

saccharine /'sækəˌriːn/ adj. **1** sugary. **2** of, containing, or like sugar. **3** unpleasantly over-polite, sentimental, etc.

saccharo- /'sækərəʊ/ comb. form sugar; sugar and. [Gk *sakkharon* sugar]

saccharometer /ˌsækəˈrɒmɪtə(r)/ n. a hydrometer for estimating the sugar content of a solution.

saccharose /'sækəˌrəʊz, -ˌrəʊs/ n. Chem. sucrose. [mod.L *saccharum* sugar + -OSE[2]]

sacciform /'sæksɪˌfɔːm/ adj. sac-shaped. [L *saccus* sac + -FORM]

saccule /'sækjuːl/ n. a small sac or cyst. □ **saccular** adj. [L *sacculus* (as SAC)]

sacerdotal /ˌsækəˈdəʊt(ə)l, ˌsæsə-/ adj. **1** of priests or the priestly office; priestly. **2** (of a doctrine etc.) ascribing sacrificial functions and supernatural powers to ordained priests; claiming excessive authority for the priesthood. □ **sacerdotalism** n. **sacerdotalist** n. **sacerdotally** adv. [ME f. OF *sacerdotal* or L *sacerdotalis* f. *sacerdos -dotis* priest]

sachem /'seɪtʃəm/ n. **1** the supreme chief of some North American Indian tribes. **2** N. Amer. a political leader. [Algonquian *sontim*]

sachet /'sæʃeɪ/ n. **1** a small usu. sealed and airtight bag or packet. **2** a small scented bag. **3** a dry perfume for laying among clothes etc. **b** a packet of this. [F, dimin. of *sac* f. L *saccus*]

Sachs /sæks, zæks/, Hans (1494–1576), German poet and dramatist. A shoemaker by trade, he was a renowned member of the Guild of Meistersinger in Nuremberg, as well as the prolific author of verse and some 200 plays. Some of his poetry celebrated Luther and furthered the Protestant cause, while other pieces were comic verse dramas. Forgotten after his death, he was restored to fame in a poem by Goethe, and Wagner made him the hero of his opera *Die Meistersinger von Nürnberg* (1868).

Sachsen see SAXONY.

Sachsen-Anhalt see SAXONY-ANHALT.

sack[1] /sæk/ n. & v. ● n. **1 a** a large strong bag, usu. made of hessian, paper, or plastic, for storing or conveying goods. **b** (usu. foll. by *of*) this with its contents (*a sack of potatoes*). **c** a quantity contained in a sack. **2** (prec. by *the*) colloq. dismissal from employment. **3** (prec. by *the*) N. Amer. sl. bed. **4 a** a woman's short loose dress with a sacklike appearance. **b** archaic or hist. a woman's loose gown, or a silk train attached to the shoulders of this. **5** a man's or woman's loose-hanging coat not shaped to the back. ● v.tr. **1** put into a sack or sacks. **2** colloq. dismiss from employment. □ **hit the sack** colloq. go to bed. **sack race** a race between competitors in sacks up to the waist or neck. □ **sackful** n. (pl. **-fuls**). **sacklike** adj. [OE *sacc* f. L *saccus* f. Gk *sakkos*, of Semitic orig.]

sack[2] /sæk/ v. & n. ● v.tr. **1** plunder and destroy (a captured town etc.). **2** steal valuables from (a place). ● n. the sacking of a captured place. [orig. as noun, f. F *sac* in phr. *mettre à sac* put to sack, f. It. *sacco* SACK[1]]

sack[3] /sæk/ n. hist. a white wine formerly imported into Britain from Spain and the Canaries (*sherry sack*). [16th-c. *wyne seck*, f. F *vin sec* dry wine]

sackbut /'sækbʌt/ n. an early form of trombone. [F *saquebute*, earlier *saqueboute* hook for pulling a person off a horse f. *saquer* pull + *boute* (as BUTT[1])]

sackcloth /'sækklɒθ/ n. **1** a coarse fabric of flax or hemp. **2** clothing made of this, formerly worn as a penance or in mourning (esp. *sackcloth and ashes*).

sacking /'sækɪŋ/ n. material for making sacks; sackcloth.

Sackville-West /ˌsækvɪlˈwest/, Victoria Mary ('Vita') (1892–1962), English novelist and poet. Her works include the long poem *The Land* (1927), notable for its evocation of the English countryside, and the novel *All Passion Spent* (1931). She is also known for the garden which she created at Sissinghurst in Kent and for her friendship with Virginia Woolf; the central character of Woolf's novel *Orlando* (1928) is said to have been based on her.

sacra pl. of SACRUM.

sacral /'seɪkrəl/ adj. **1** Anat. of or relating to the sacrum. **2** Anthropol. of or for sacred rites. [f. SACRUM, or L *sacrum* sacred thing f. *sacer* SACRED]

sacrament /'sækrəmənt/ n. **1** a religious ceremony or act of the Christian Churches regarded as an outward and visible sign of inward and spiritual grace. (*See note below.*) **2** a thing of mysterious and sacred significance; a sacred influence, symbol, etc. **3** (also **Blessed** or **Holy Sacrament**) (prec. by *the*) **a** the Eucharist. **b** the consecrated elements, esp. the bread or Host. **4** an oath or solemn engagement taken. [ME f. OF *sacrement* f. L *sacramentum* solemn oath etc. f. *sacrare* hallow f. *sacer* SACRED, used in Christian L as transl. of Gk *mustērion* MYSTERY[1]]

▪ The term is applied by the Eastern, pre-Reformation Western, and Roman Catholic Churches to the seven rites of baptism, confirmation, the Eucharist, penance, extreme unction, ordination, and matrimony, but restricted by most Protestants to baptism and the Eucharist.

sacramental /ˌsækrəˈment(ə)l/ adj. & n. ● adj. **1** of or of the nature of a sacrament or the sacrament. **2** (of a doctrine etc.) attaching great importance to the sacraments. ● n. an observance analogous to but not reckoned among the sacraments, e.g. the use of holy water or the sign of the cross. □ **sacramentalism** n. **sacramentalist** n. **sacramentally** adv. **sacramentality** /-menˈtælɪtɪ/ n. [ME f. F sacramental or LL sacramentalis (as SACRAMENT)]

Sacramento /ˌsækrəˈmentəʊ/ **1** a river of northern California, which rises near the border with Oregon and flows some 611 km (380 miles) southwards to San Francisco Bay. **2** the state capital of California, situated on the Sacramento river to the north-east of San Francisco; pop. (1990) 369,365. First settled in 1839, it developed rapidly after the gold rush of 1849, becoming the state capital in 1854.

sacrarium /səˈkreərɪəm/ n. (pl. **sacraria** /-rɪə/) **1** the sanctuary of a church. **2** RC Ch. a piscina. **3** Rom. Antiq. a shrine; the room (in a house) containing the penates. [L f. sacer sacri holy]

sacred /ˈseɪkrɪd/ adj. **1 a** (often foll. by to) exclusively dedicated or appropriated (to a god or to some religious purpose). **b** made holy by religious association. **c** connected with religion; used for a religious purpose (sacred music). **2 a** safeguarded or required by religion, reverence, or tradition. **b** sacrosanct. **3** (of writings etc.) embodying the laws or doctrines of a religion. □ **sacred cow** colloq. an idea or institution unreasonably held to be above criticism (with ref. to the Hindus' respect for the cow as a holy animal). **sacred number** a number associated with religious symbolism, e.g. 7. □ **sacredly** adv. **sacredness** n. [ME, past part. of obs. sacre consecrate f. OF sacrer f. L sacrare f. sacer sacri holy]

Sacred College see COLLEGE OF CARDINALS.

Sacred Heart n. RC Ch. the heart of Christ as an object of devotion.

sacrifice /ˈsækrɪˌfaɪs/ n. & v. ● n. **1 a** the act of giving up something valued for the sake of something else more important or worthy. **b** a thing given up in this way. **c** the loss entailed in this. **2 a** the slaughter of an animal or person or the surrender of a possession as an offering to a deity. **b** an animal, person, or thing offered in this way. **3** an act of prayer, thanksgiving, or penitence as propitiation. **4** (in Christian theology) **a** Christ's offering of himself in the Crucifixion. **b** the Eucharist as either a propitiatory offering of the body and blood of Christ or an act of thanksgiving. **5** (in games) a loss incurred deliberately to avoid a greater loss or to obtain a compensating advantage. ● v. **1** tr. give up (a thing) as a sacrifice. **2** tr. (foll. by to) devote or give over to. **3** tr. (also absol.) offer or kill as a sacrifice. [ME f. OF f. L sacrificium f. sacrificus (as SACRED)]

sacrificial /ˌsækrɪˈfɪʃ(ə)l/ adj. of, relating to, or constituting a sacrifice. □ **sacrificial anode** a metal anode that is used up in protecting another metal against electrolytic corrosion (cf. cathodic protection). □ **sacrificially** adv.

sacrilege /ˈsækrɪlɪdʒ/ n. the violation or misuse of what is regarded as sacred. □ **sacrilegious** /ˌsækrɪˈlɪdʒəs/ adj. **sacrilegiously** adv. [ME f. OF f. L sacrilegium f. sacrilegus stealer of sacred things, f. sacer sacri sacred + legere take possession of]

sacring /ˈseɪkrɪŋ/ n. archaic **1** the consecration of the Eucharistic elements. **2** the ordination and consecration of a bishop, sovereign, etc. □ **sacring bell** a bell rung at the elevation of the elements in the Eucharist. [ME f. obs. sacre: see SACRED]

sacristan /ˈsækrɪstən/ n. **1** a person in charge of a sacristy and its contents. **2** archaic the sexton of a parish church. [ME f. med.L sacristanus (as SACRED)]

sacristy /ˈsækrɪstɪ/ n. (pl. **-ies**) a room in a church, where the vestments, sacred vessels, etc., are kept and the celebrant can prepare for a service. [F sacristie or It. sacrestia or med.L sacristia (as SACRED)]

sacro- /ˈseɪkrəʊ, ˈsæk-/ comb. form Anat. denoting the sacrum (sacro-iliac).

sacroiliac /ˌseɪkrəʊˈɪlɪæk/ adj. Anat. relating to the sacrum and the ilium, esp. designating the rigid joint between them at the back of the pelvis.

sacrosanct /ˈseɪkrəʊˌsæŋkt/ adj. (of a person, place, law, etc.) most sacred; inviolable. □ **sacrosanctity** /ˌseɪkrəʊˈsæŋktɪtɪ/ n. [L sacrosanctus f. sacro ablat. of sacrum sacred rite (see SACRED) + sanctus (as SAINT)]

sacrum /ˈseɪkrəm/ n. (pl. **sacra** /-krə/ or **sacrums**) Anat. a triangular bone formed from fused vertebrae and situated between the two hip-bones of the pelvis. [L os sacrum transl. Gk hieron osteon sacred bone (from its sacrificial use)]

SACW abbr. (in the UK) Senior Aircraftwoman.

SAD abbr. Med. seasonal affective disorder.

sad /sæd/ adj. (**sadder, saddest**) **1** unhappy; feeling sorrow or regret. **2** causing or suggesting sorrow (a sad story). **3** regrettable. **4** shameful, deplorable (is in a sad state). **5** (of a colour) dull, neutral-tinted. **6** (of dough etc.) heavy, having failed to rise. **7** sl. contemptible, pathetic, unfashionable. □ **sad-iron** a solid flat-iron. **sad sack** US colloq. a very inept person. □ **saddish** adj. **sadly** adv. **sadness** n. [OE sæd f. Gmc, rel. to L satis]

Sadat /səˈdæt/, (Muhammad) Anwar al- (1918–81), Egyptian statesman, President 1970–81. He broke with the foreign policies of his predecessor President Nasser, for example by dismissing the Soviet military mission to Egypt, removing the ban on political parties, and introducing measures to decentralize Egypt's political structure and diversify the economy. He later worked to achieve peace in the Middle East, visiting Israel (1977), and attending talks with Prime Minister Begin at Camp David in 1978, the year they shared the Nobel Peace Prize. Also in that year he founded the National Democratic Party, with himself as leader. He was assassinated by members of the Islamic Jihad.

Saddam Hussein /səˈdæm/ see HUSSEIN³.

sadden /ˈsæd(ə)n/ v.tr. & intr. make or become sad.

saddle /ˈsæd(ə)l/ n. & v. ● n. **1** a seat of leather etc., usu. raised at the front and rear, fastened on a horse etc. for riding. **2** a seat for the rider of a bicycle or motorcycle. **3** a joint of meat consisting of the two loins. **4** a ridge rising to a summit at each end. **5** the part of a draught-horse's harness to which the shafts are attached. **6** a part of an animal's back resembling a saddle in shape or marking. **7** the rear part of a male fowl's back. **8** a support for a cable or wire on top of a suspension-bridge, pier, or telegraph pole. **9** a fireclay bar for supporting ceramic ware in a kiln. ● v.tr. **1** put a saddle on (a horse etc.). **2 a** (foll. by with) burden (a person) with a task, responsibility, etc. **b** (foll. by on, upon) impose (a burden) on a person. **3** (of a trainer) enter (a horse) for a race. □ **in the saddle 1** mounted. **2** in office or control. **saddle-bag 1** each of a pair of bags laid across a horse etc. behind the saddle. **2** a bag attached beneath the saddle of a bicycle or motorcycle. **saddle-bow** the arched front or rear of a saddle. **saddle-cloth** a cloth laid on a horse's back under the saddle. **saddle-horse** a horse for riding. **saddle-sore** chafed by riding on a saddle. **saddle stitch** a stitch of thread or a wire staple passed through the centre of a magazine or booklet. **saddle-tree 1** the frame of a saddle. **2** a tulip tree (with saddle-shaped leaves). □ **saddleless** adj. [OE sadol, sadul f. Gmc]

saddleback /ˈsæd(ə)l,bæk/ n. **1** Archit. a roof of a tower with two opposite gables. **2** a hill with a concave upper outline. **3** a black pig with a white stripe across the back. **4** a bird with a saddle-like marking; esp. a New Zealand wattlebird, Creadion carunculatus. □ **saddlebacked** adj.

saddler /ˈsædlə(r)/ n. a maker of or dealer in saddles and other equipment for horses.

saddlery /ˈsædlərɪ/ n. (pl. **-ies**) **1** the saddles and other equipment of a saddler. **2** a saddler's business or premises.

Sadducee /ˈsædjʊˌsiː/ n. a member of a Jewish sect or party of the time of Christ that denied the resurrection of the dead, the existence of spirits, and the obligation of oral tradition, emphasizing acceptance of the written Law only. (Cf. PHARISEE.) □ **Sadducean** /ˌsædjʊˈsiːən/ adj. [OE sadducēas f. LL Sadducaeus f. Gk Saddoukaios f. Heb. ṣᵉḏûḳî, prob. = descendant of Zadok (2 Sam. 8:17)]

Sade /sɑːd/, Donatien Alphonse François, Comte de (known as the Marquis de Sade) (1740–1814), French writer and soldier. His career as a cavalry officer was interrupted by prolonged periods of imprisonment for cruelty and debauchery. While in prison he wrote a number of sexually explicit works, which include Les 120 Journées de Sodome (1784), Justine (1791), and La Philosophie dans le boudoir (1795). The word sadism owes its origin to his name, referring to the cruel sexual practices which he described.

sadhu /ˈsɑːduː/ n. (in India) a holy man, sage, or ascetic. [Skr., = holy man]

Sadi /ˈsɑːdɪ/ (also **Saadi**) (born Sheikh Muslih Addin) (c.1213–c.1291), Persian poet. His principal works were the collections known as the Bustan (1257) and the Gulistan (1258); the former is a series of poems on religious themes, while the latter is a mixture of poems, prose, and maxims concerning moral issues.

sadism /'seɪdɪz(ə)m/ n. **1** the condition or state of deriving esp. sexual gratification from inflicting pain or suffering on others (cf. MASOCHISM). **2** colloq. the enjoyment of cruelty to others. □ **sadist** n. **sadistic** /sə'dɪstɪk/ adj. **sadistically** adv. [F sadisme f. SADE]

Sadler's Wells Theatre /ˌsædləz 'welz/ a London theatre opened by Lilian Baylis in 1931 on the site of an earlier theatre, and known for its ballet and opera companies. Although the Sadler's Wells Ballet moved to the Covent Garden Theatre in 1946, a second company, the Sadler's Wells Theatre Company, and a ballet school were set up at the theatre, the three merging as the Royal Ballet in 1956; in 1968 the opera company moved to the Coliseum in London as the English National Opera. The theatre has since housed visiting companies. Named after a medicinal spring discovered in 1683 by a Thomas Sadler and incorporated into a pleasure garden, the original theatre developed in the 18th century from a music room built by Sadler on the site.

sado-masochism /ˌseɪdəʊ'mæsəˌkɪz(ə)m/ n. a combination of sadism and masochism, either in one person, or within a sexual relationship where a partner plays either a sadistic or a masochistic role. □ **sado-masochist** n. **sado-masochistic** /-ˌmæsə'kɪstɪk/ adj.

s.a.e. abbr. stamped addressed envelope.

Safaqis see SFAX.

safari /sə'fɑːrɪ/ n. (pl. **safaris**) **1** a hunting or scientific expedition, esp. in East Africa (go on safari). **2** a sightseeing trip to see African animals in their natural habitat. □ **safari park** an enclosed area where lions etc. are kept in the open and through which visitors may drive. **safari suit** a lightweight suit usu. with short sleeves and four pleated pockets in the jacket. [Swahili f. Arab. safara to travel]

Safavid /'sæfə,vɪd/ n. & adj. ● n. a member of a dynasty which ruled Persia (Iran) 1502–1736. The dynasty installed Shia rather than Sunni Islam as the state religion. ● adj. of or relating to this dynasty.

safe /seɪf/ adj. & n. ● adj. **1 a** free of danger or injury. **b** (often foll. by from) out of or not exposed to danger (safe from their enemies). **2** affording security or not involving danger or risk (put it in a safe place). **3** reliable, certain; that can be reckoned on (a safe catch; a safe method; is safe to win). **4** prevented from escaping or doing harm (have got him safe). **5** (also **safe and sound**) uninjured; with no harm done. **6** cautious and unenterprising; consistently moderate. ● n. **1** a strong lockable cabinet etc. for valuables. **2** = meat safe. □ **on the safe side** with a margin of security against risks. **safe bet** a bet that is certain to succeed. **safe-breaker** (or **-blower** or **-cracker**) a person who breaks open and robs safes. **safe conduct 1** a privilege of immunity from arrest or harm, esp. on a particular occasion. **2** a document securing this. **safe deposit** a building containing strongrooms and safes let separately. **safe haven 1** a place of refuge or security. **2** a protected zone in a country designated for members of a religious or ethnic minority. **safe house** a place of refuge or rendezvous for spies, criminals, etc. **safe keeping** preservation in a safe place. **safe light** Photog. a filtered light for use in a darkroom. **safe period** the time during and near the menstrual period when conception is least likely. **safe seat** a seat in Parliament etc. that is usually won with a large margin by a particular party. **safe sex** sexual activity in which precautions are taken to reduce the risk of spreading sexually transmitted diseases, esp. Aids. □ **safely** adv. **safeness** n. [ME f. AF saf, OF sauf f. L salvus uninjured: (n.) orig. save f. SAVE[1]]

safeguard /'seɪfgɑːd/ n. & v. ● n. **1** a proviso, stipulation, quality, or circumstance, that tends to prevent something undesirable. **2** a safe conduct. ● v.tr. guard or protect (rights etc.) by a precaution or stipulation. [ME f. AF salve garde, OF sauve garde (as SAFE, GUARD)]

safety /'seɪftɪ/ n. (pl. **-ies**) **1** the condition of being safe; freedom from danger or risks. **2** (attrib.) **a** designating any of various devices for preventing injury from machinery (safety bar; safety lock). **b** designating items of protective clothing (safety helmet). **3** Amer. Football an act of advancing the ball in one's own end zone by a down, conceding two points. **b** (in full **safety man**) a defensive back who plays in a deep position. □ **safety belt 1** = seat-belt. **2** a belt or strap worn by a person working at a height to prevent a fall. **safety-catch** a contrivance for locking a gun-trigger or preventing the accidental operation of machinery. **safety curtain** a fireproof curtain that can be lowered to cut off the auditorium in a theatre from the stage. **safety factor** (or **factor of safety**) **1** the ratio of a material's strength to an expected strain. **2** a margin of security against risks. **safety film** a cinematographic film on a slow-burning or non-flammable base.

safety first a motto advising caution. **safety fuse 1** a fuse (see FUSE[2]) containing a slow-burning composition for firing detonators from a distance. **2** Electr. a protective fuse (see FUSE[1]). **safety glass** glass that will not splinter when broken. **safety harness** a system of belts or restraints to hold a person to prevent falling or injury. **safety lamp** a miner's lamp so protected as not to ignite firedamp. **safety match** a match igniting only on a specially prepared surface. **safety net 1** a net placed to catch an acrobat etc. in case of a fall. **2** a safeguard against possible hardship or adversity. **safety razor** a razor with a guard to reduce the risk of cutting the skin. **safety-valve 1** (in a steam boiler) a valve opening automatically to relieve excessive pressure. **2** a means of giving harmless vent to excitement etc. **safety zone** US an area of a road marked off for pedestrians etc. to wait safely. [ME sauvete f. OF sauveté f. med.L salvitas -tatis f. L salvus (as SAFE)]

safety pin n. a pin with a point that is bent back to the head and can be held in a guard to stop the user being pricked and the pin coming out unintentionally. Fasteners made on the same principle, consisting of a single length of metal wire coiled on itself at its middle point so as to form a spring, are known from the Bronze Age (13th century BC) and seem to have been a European invention. The modern type (with a clasp) was reinvented and patented in the US by Walter Ireland Hunt (1796–1859) in 1849.

safflower /'sæflaʊə(r)/ n. **1** an orange-flowered thistle-like composite plant, Carthamus tinctorius, whose seeds yield an edible oil. **2 a** its dried petals. **b** a red dye made from these, used in rouge etc. [Du. saffloer or G Safflor f. OF saffleur f. obs. It. saffiore from Arabic asfar]

saffron /'sæfrən/ n. & adj. ● n. **1** an orange flavouring and food colouring made from the dried stigmas of a crocus, Crocus sativus. **2** the colour of this. **3** = meadow saffron. ● adj. saffron-coloured. □ **saffrony** adj. [ME f. OF safran f. Arab. za'farān]

safranine /'sæfrə,niːn/ n. (also **safranin** /-nɪn/) any of a large group of mainly red dyes used in biological staining etc. [F safranine (as SAFFRON): orig. of dye from saffron]

sag /sæg/ v. & n. ● v.intr. (**sagged**, **sagging**) **1** sink or subside under weight or pressure, esp. unevenly. **2** have a downward bulge or curve in the middle. **3 a** a fall in price. **b** (of a price) fall. **4** (of a ship) drift from its course, esp. to leeward. ● n. **1 a** the amount that a rope etc. sags. **b** the distance from the middle of its curve to a straight line between its supports. **2** a sinking condition; subsidence. **3** a fall in price. **4** Naut. a tendency to leeward. □ **saggy** adj. [ME f. MLG sacken, Du. zakken subside]

saga /'sɑːgə/ n. **1** a long story of heroic achievement, esp. a medieval Icelandic or Norse prose narrative. **2** a series of connected books giving the history of a family etc. **3** a long involved story. [ON, = narrative, rel. to SAW[3]]

sagacious /sə'geɪʃ(ə)s/ adj. **1** mentally penetrating; gifted with discernment; having practical wisdom. **2** acute-minded, shrewd. **3** (of a saying, plan, etc.) showing wisdom. **4** (of an animal) exceptionally intelligent; seeming to reason or deliberate. □ **sagaciously** adv. **sagacity** /-'gæsɪtɪ/ n. [L sagax sagacis]

sagamore /'sægə,mɔː(r)/ n. = SACHEM 1. [Penobscot sagamo]

Sagan[1] /'seɪgən/, Carl (Edward) (b. 1934), American astronomer. He has specialized in studies of the planets Mars and Venus, and in 1968 became director of the Laboratory of Planetary Studies at Cornell, dealing with information from space probes to those planets. Sagan has shown that amino acids can be synthesized in an artificial primordial soup irradiated by ultraviolet light — a possible origin of life on earth. In 1983 he and several other scientists put forward the concept of a nuclear winter as a likely consequence of global nuclear war. He has written several popular science books, and was co-producer and narrator of the television series Cosmos (1980).

Sagan[2] /sæ'gãn/, Françoise (pseudonym of Françoise Quoirez) (b.1935), French novelist, dramatist, and short-story writer. She rose to fame with her first novel Bonjour Tristesse (1954); in this and subsequent novels, she examined the transitory nature of love as experienced in brief liaisons. Other novels include Un Certain sourire (1956) and Aimez-vous Brahms? (1959).

sage[1] /seɪdʒ/ n. **1** an aromatic labiate shrub, Salvia officinalis, with dull greyish-green leaves. **2** its leaves used in cookery. □ **sage and onion** (or **onions**) a stuffing used with poultry, pork, etc. **sage Derby** (or **cheese**) a cheese made with an infusion of sage which flavours and mottles it. **sage-green** the colour of sage leaves. **sage grouse** a large grouse, Centrocercus urophasianus, of western North America, noted for the male's courtship display. **sage tea** a medicinal infusion of sage leaves. □ **sagy** adj. [ME f. OF sauge f. L salvia healing plant f. salvus safe]

sage² /seɪdʒ/ n. & adj. ● n. **1** often iron. a profoundly wise man. **2** any of the ancients traditionally regarded as the wisest of their time. ● adj. **1** profoundly wise, esp. from experience. **2** of or indicating profound wisdom. **3** often iron. wise-looking; solemn-faced. □ **sagely** adv. **sageness** n. **sageship** n. [ME f. OF ult. f. L sapere be wise]

sagebrush /ˈseɪdʒbrʌʃ/ n. **1** an area covered by shrubby aromatic composite plants of the genus Artemisia, esp. A. tridentata, in semi-arid regions of western North America. **2** this plant.

saggar /ˈsægə(r)/ n. (also **sagger**) a protective fireclay box enclosing ceramic ware while it is being fired. [prob. contr. of SAFEGUARD]

sagittal /ˈsædʒɪt(ə)l/ adj. Anat. **1** of or relating to the suture on top of the skull between the parietal bones. **2** in a plane parallel to this, esp. that which divides the body in the midline into right and left halves. [F f. med.L sagittalis f. sagitta arrow]

Sagittarius /ˌsædʒɪˈteərɪəs/ n. **1** Astron. a large constellation (the Archer), said to represent a centaur carrying a bow and arrow. It contains many bright stars, star clusters, and nebulae, and the centre of the Galaxy is situated within it. **2** Astrol. **a** the ninth sign of the zodiac, which the sun enters about 22 Nov. **b** a person born when the sun is in this sign. □ **Sagittarian** adj. & n. [ME f. L, = archer, f. sagitta arrow]

sagittate /ˈsædʒɪˌteɪt/ adj. Bot. & Zool. shaped like an arrowhead.

sago /ˈseɪɡəʊ/ n. (pl. **-os**) **1** a kind of starch, made from the powdered pith of the sago palm and used in puddings etc. **2** (in full **sago palm**) a tropical palm or cycad, esp. Cycas circinalis or Metroxylon sagu, from which sago is made. [Malay sāgū (orig. through Port.)]

saguaro /sæˈɡwɑːrəʊ/ n. (also **sahuaro** /-ˈwɑːrəʊ/) (pl. **-os**) a giant cactus, Carnegiea gigantea, of the SW United States and Mexico. [Mex. Sp.]

Saguia el Hamra /səˌɡiːə el ˈhæmrə/ **1** an intermittent river in the north of Western Sahara. It flows into the Atlantic west of La'youn. **2** the region through which it flows. A territory of Spain from 1934, it united with Río de Oro in 1958 to become a part of the former province of Spanish Sahara.

Saha /ˈsɑːhɑː/, Meghnad (1894–1956), Indian theoretical physicist. He worked on thermal ionization in stars, using both thermodynamics and quantum theory, and laid the foundations for modern astrophysics. He showed that the ionization of metal atoms increases with temperature, leading to a reduction in the absorption lines visible in stellar spectra. Saha devised an equation, now named after him, expressing the relationship between ionization and temperature.

Sahara Desert /səˈhɑːrə/ (also **Sahara**) a vast desert in North Africa, extending from the Atlantic in the west to the Red Sea in the east, and from the Mediterranean and the Atlas Mountains in the north to the Sahel in the south. The largest desert in the world, it covers an area of about 9,065,000 sq. km (3,500,000 sq. miles). In recent years it has been extending southwards into the Sahel. □ **Saharan** adj. [Arab., = desert]

Sahel /səˈhel/ a vast semi-arid region of North Africa, to the south of the Sahara. An area of dry savannah, it forms a transitional zone at the southern limits of the desert and comprises the northern part of the region known as Sudan. □ **Sahelian** /-ˈhiːlɪən/ adj.

sahib /sɑːb, ˈsɑːhɪb/ n. **1** hist. (in India) a form of address, often placed after the name, to European men. **2** colloq. a gentleman (pukka sahib). [Urdu f. Arab. ṣāḥib friend, lord]

Said /sæˈiːd/, Edward W. (b.1935), American critic, born in Palestine. A professor of English and Comparative Literature at the University of Columbia, he came to public notice with Orientalism (1978), a study of Western attitudes towards Eastern culture. In The Question of Palestine (1985), Said defended the Palestinian struggle for political autonomy and has since played an active role in moves to form a Palestinian state. Other works include Culture and Imperialism (1993), a critique of Western culture.

said past and past part. of SAY.

Saida see SIDON.

saiga /ˈsaɪɡə, ˈseɪɡə/ n. an antelope, Saiga tartarica, of the cold central Asian steppes, having a distinctive inflated snout. [Russ.]

Saigon /saɪˈɡɒn/ a city and port on the south coast of Vietnam; pop. (1991) 3,450,000. It was the capital of the French colony established in Vietnam in the 19th century, becoming capital of South Vietnam in the partition of 1954. Officially renamed Ho Chi Minh City after the reunification of Vietnam in 1975, it is now the largest city and chief industrial centre of Vietnam.

sail /seɪl/ n. & v. ● n. **1** a piece of material (orig. canvas, now usu. nylon etc.) extended on rigging to catch the wind and propel a boat or ship. **2** a ship's sails collectively. **3 a** a voyage or excursion in a sailing-ship. **b** a voyage of specified duration. **4** a ship, esp. as discerned from its sails. **5** (collect.) ships in a squadron or company (a fleet of twenty sail). **6** (in pl.) Naut. **a** sl. a maker or repairer of sails. **b** hist. a chief petty officer in charge of rigging. **7** a wind-catching apparatus, usu. a set of boards, attached to the arm of a windmill. **8 a** the dorsal fin of a sailfish. **b** the tentacle of a nautilus. **c** the float of a Portuguese man-of-war. ● v. **1** intr. travel on water by the use of sails or engine-power. **2** tr. **a** navigate (a ship etc.). **b** travel on (a sea). **3** tr. set (a toy boat) afloat. **4** intr. glide or move smoothly or in a stately manner. **5** intr. (often foll. by through) colloq. succeed easily (sailed through the exams). □ **sail-arm** the arm of a windmill. **sail close to** (or **near**) **the wind 1** sail as nearly against the wind as possible. **2** colloq. come close to indecency or dishonesty; risk overstepping the mark. **sail-fluke** = MEGRIM². **sailing-boat** (or **-ship** or **vessel**) a vessel driven by sails. **sailing-master** an officer navigating a ship, esp. Brit. a yacht. **sailing orders** instructions to a captain regarding departure, destination, etc. **sail into** colloq. attack physically or verbally with force. **take in sail 1** furl the sail or sails of a vessel. **2** moderate one's ambitions. **under sail** with sails set. □ **sailable** adj. **sailed** adj. (also in comb.). **sailless** /ˈseɪllɪs/ adj. [OE segel f. Gmc]

sailboard /ˈseɪlbɔːd/ n. a board with a mast and sail, used in windsurfing. □ **sailboarder** n. **sailboarding** n.

sailboat /ˈseɪlbəʊt/ n. US a sailing-boat.

sailcloth /ˈseɪlklɒθ/ n. **1** canvas or other material for sails. **2** a canvas-like dress material.

sailer /ˈseɪlə(r)/ n. a ship of specified sailing-power (a good sailer).

sailfish /ˈseɪlfɪʃ/ n. **1** a large marine game-fish of the genus Istiophorus, with a tall and long dorsal fin and a spearlike snout. **2** a basking shark.

sailor /ˈseɪlə(r)/ n. **1** a member of the crew of a ship or boat, esp. one below the rank of officer. **2** a person considered as liable or not liable to seasickness (a good sailor). □ **sailor hat 1** a straw hat with a straight narrow brim and flat top. **2** a hat with a turned-up brim in imitation of a sailor's, worn by women and children. □ **sailoring** n. **sailorless** adj. **sailorly** adj. [var. of SAILER]

sailplane /ˈseɪlpleɪn/ n. a glider designed for sustained flight.

sainfoin /ˈseɪnfɔɪn, ˈsæn-/ n. a pink-flowered leguminous plant, Onobrychis viciifolia, grown for fodder. [obs. F saintfoin f. mod.L sanum foenum wholesome hay (because of its medicinal properties)]

Sainsbury /ˈseɪnzbəri/, John James (1844–1928), English grocer. He opened his first grocery store in London in 1875. After his death the business was continued by members of his family, developing into the large supermarket chain bearing the Sainsbury name.

saint /seɪnt, before a name usu. sənt/ n. & v. ● n. (abbr. **St** or **S**; pl. **Sts** or **SS**) **1** a holy person, one declared (in the Roman Catholic or Orthodox Church) worthy of veneration, whose intercession may be publicly sought. (See note below.) **2** (**Saint** or **St**) the title of a saint or archangel, hence used in the names of churches etc. (St Paul's), sometimes without reference to a saint (St Cross; St Saviour's), or (often without the apostrophe) in the names of places (St Andrews; St Albans). **3** a very virtuous person; a person of great real or affected holiness. **4** a member of the company of heaven (with all the angels and saints). **5** (Bibl., archaic, and used by Puritans, Mormons, etc.) one of God's chosen people; a member of the Christian Church or a person's own branch of it. ● v.tr. **1** canonize; admit to the calendar of saints. **2** call or regard as a saint. **3** (as **sainted** adj.) sacred; of a saintly life; worthy to be regarded as a saint. □ **my sainted aunt** see AUNT. **saint's day** a Church festival in memory of a saint. □ **saintdom** n. **sainthood** n. **saintlike** adj. **saintling** n. **saintship** n. [ME f. OF seint, saint f. L sanctus holy, past part. of sancire consecrate]

▪ The original Christian ideal of the saint was the martyr, largely replaced, after persecution ceased in the 4th century, by the devout member of a religious order. A saint in this context is a person who is close to God and can intercede with God on behalf of other Christians, and one through whom divine power is therefore manifest. A cult of the saints, focused on their physical remains, has early attestation (e.g. that of Polycarp, 2nd century), and developed rapidly from the 4th century onwards. Procedures for the approval of such veneration (called canonization) were gradually formalized, being eventually vested in the papacy in the West and episcopal synods in the East: performance of miracles and lives of heroic sanctity are the criteria for canonization. At the Reformation the cult of the saints was attacked as blurring the unique status of Christ and as an occasion of religious commercialization. The persecution of Christians in some countries in the 20th century has to an extent reinforced the early association of sanctity with martyrdom.

St Agnes, St Barnabas, etc. see AGNES, ST; BARNABAS, ST; etc.

St Albans /ˈɔːlbənz/ a city in Hertfordshire, in SE England; pop. (1981) 55,700. The city developed around the abbey of St Albans, which was founded in Saxon times on the site of the martyrdom in the 3rd century of St Alban, a Christian Roman from the nearby Roman city of Verulamium. The ruins of ancient Verulamium lie within the modern city of St Albans.

St Andrews /ˈændruːz/ a town in east Scotland, in Fife, on the North Sea; pop. (1981) 11,350. The ecclesiastical capital of Scotland until the Reformation, it is noted today for its university, founded in 1410, and its golf courses.

St Andrew's cross /ˈændruːz/ n. an X-shaped cross (see ANDREW, ST).

St Anthony's fire /ˈæntənɪz/ n. ergotism. [from the medieval belief that praying to St Anthony would effect a cure]

St Anthony's cross /ˈæntənɪz/ n. (also **St Anthony cross** /ˈæntənɪ/) a T-shaped cross.

St Bartholomew's Day Massacre see MASSACRE OF ST BARTHOLOMEW.

St Bernard /ˈbɜːnəd/ n. (in full **St Bernard dog**) a very large breed of dog, orig. kept to rescue travellers by the monks of the St Bernard passes in the Alps.

St Bernard Pass either of two passes across the Alps in southern Europe. The *Great St Bernard Pass*, on the border between SW Switzerland and Italy, rises to 2,469 m (8,100 ft). The *Little St Bernard Pass*, on the French–Italian border south-east of Mont Blanc, rises to 2,188 m (7,178 ft). Both are named after the hospices founded on their summits in the 11th century by the French monk St Bernard.

St Christopher and Nevis, Federation of the official name for ST KITTS AND NEVIS.

St Croix /krɔɪ/ an island in the West Indies, the largest of the US Virgin Islands; chief town, Christiansted. Purchased by Denmark in 1753, it was sold to the US in 1917.

St David's a small city near the coast of Dyfed, SW Wales. Its 12th-century cathedral houses the shrine of St David, the patron saint of Wales.

Saint-Denis /ˌsændəˈniː/ **1** a municipality in France, now a northern suburb of Paris. **2** the capital of the French island of Réunion in the Indian Ocean, a port on the north coast; pop. (1990) 121,670.

Sainte-Beuve /sæntˈbɜːv/, Charles Augustin (1804–69), French critic and writer. He is chiefly known for his contribution to 19th-century literary criticism, in which he concentrated on the influence of social and other factors in the development of authors' characters; his critical essays were published in collected form as *Causeries du lundi* (1851–62) and *Nouveaux lundis* (1863–70). He also wrote a study of Jansenism (*Port-Royal*, 1840–59) and was an early champion of French romanticism.

St Elmo's fire /ˈelməʊz/ n. luminous electrical discharges seen esp. on a ship or aircraft during a storm (cf. CORPOSANT). [*St Elmo*, the patron saint of seafarers]

St Émilion /ˌsænt eɪˈmɪljɒn/ a small town situated to the north of the Dordogne in SW France. It gives its name to a group of Bordeaux wines.

St-Étienne /ˌsænteɪˈtjen/ an industrial city in SE central France, south-west of Lyons; pop. (1990) 201,570.

St Eustatius /juːˈsteɪʃəs/ a small volcanic island in the Caribbean, in the Netherlands Antilles.

St George's /ˈdʒɔːdʒɪz/ the capital of Grenada in the West Indies, a port in the south-west of the island; pop. (1989) 35,740.

St George's Channel a channel between Wales and Ireland, linking the Irish Sea with the Celtic Sea.

St George's cross n. a +-shaped cross, red on a white background.

St Gotthard Pass /ˈɡɒtɑːd/ a mountain pass in the Alps in southern Switzerland, situated at an altitude of 2,108 m (6,916 ft). Beneath it is a railway tunnel (constructed in 1872–80) and a road tunnel (opened in 1980). The pass is named after a former chapel and hospice built there in the 14th century, which was dedicated to St Godehard or Gotthard, an 11th-century bishop of Hildesheim in Germany.

St Helena /hɪˈliːnə/ a solitary island in the South Atlantic, a British dependency; pop. (1988) 5,560; official language, English; capital, Jamestown. The island was discovered by the Portuguese in 1502 on 21 May, feast day of St Helena. It was administered by the East India

Company from 1659 until 1834 when it became a British colony. Ascension, Tristan da Cunha, and Gough Island are dependencies of St Helena. It is famous as the place of Napoleon's exile (1815–21) and death. □ **St Helenian** *adj. & n.*

St Helens /ˈhelənz/ an industrial town in NW England, in Merseyside north-east of Liverpool; pop. (1991) 175,300.

St Helens, Mount an active volcano in SW Washington, in the Cascade Range, rising to 2,560 m (8,312 ft). A dramatic eruption in May 1980 reduced its height by several hundred metres and spread volcanic ash and debris over a large area.

St Helier /ˈheliə(r)/ a market town and resort on the south coast of Jersey; pop. (1981) 25,700. It is the capital of the island and is named after a 6th-century Christian hermit.

St James's Palace /ˈdʒeɪmzɪz/ a royal palace in Pall Mall, London, built by Henry VIII on the site of an earlier leper hospital dedicated to St James the Less. The palace was the chief royal residence in London from 1697 until 1837, when Queen Victoria made Buckingham Palace the monarch's London residence. □ **Court of St James's** the official title of the British court, to which ambassadors from foreign countries are accredited.

St John /dʒɒn/ **1** an island in the West Indies, one of the three principal islands of the US Virgin Islands. **2** (usu. **Saint John**) a city in New Brunswick, eastern Canada, a port on the Bay of Fundy at the mouth of the St John river. First settled by the French in the 1630s, it was taken by the British in 1758. It became a loyalist refuge after the War of American Independence.

St John Ambulance an organization providing first aid, nursing, ambulance, and welfare services. It was founded by the Knights Hospitallers in 1878 as the St John Ambulance Association, becoming the St John Ambulance Brigade in 1888.

St John's /dʒɒnz/ **1** the capital of Antigua and Barbuda, situated on the NW coast of Antigua; pop. (1986) 36,000. **2** the capital of Newfoundland, a port on the SE coast of the island; pop. (est. 1991) 121,030; metropolitan area pop. 171,860. One of the oldest European settlements in North America, it was colonized by the British in 1583.

St John's wort n. a yellow-flowered plant of the genus *Hypericum*; esp. *common St John's wort* (*H. perforatum*).

St Kilda /ˈkɪldə/ a small island group of the Outer Hebrides, situated in the Atlantic 64 km (40 miles) west of Lewis with Harris. In 1930 the last inhabitants left the islands, which had been settled since prehistoric times. The islands are now administered as a nature reserve.

St Kitts and Nevis /kɪts/ (official name *Federation of St Christopher and Nevis*) a country in the West Indies consisting of two adjoining islands (St Kitts and Nevis) of the Leeward Islands; pop. (est. 1991) 44,000; languages, English (official), Creole; capital, Basseterre (on St Kitts). St Kitts was visited in 1493 by Columbus. He named it after his patron saint, St Christopher, but the name was shortened by settlers from England who arrived in 1623. They established the first successful English colony in the West Indies, driving out the Carib population and importing African slaves. Nevis gained its name from the resemblance of the clouds around its peak to snow (Sp. *las nieves* the snows). A union between St Kitts, Nevis, and Anguilla was created in 1967, but Anguilla seceded within three months. St Kitts and Nevis gained self-government in 1967 and became a fully independent member of the Commonwealth in 1983.

Saint Laurent /ˌsæn lɔːˈrɒn/, Yves (Mathieu) (b.1936), French couturier. He was Christian Dior's assistant from 1953 and after Dior's death in 1957 succeeded him as head designer. His fashions at this time reflected youth culture, and included a 'beatnik' look of turtle-neck sweaters and black leather jackets. He opened his own fashion house in 1962; four years later, he launched the first of a worldwide chain of Rive Gauche boutiques to sell ready-to-wear garments. From the 1970s he expanded the business to include perfumes and household fabrics.

St Lawrence River /ˈlɒrəns/ a river of North America, which flows for some 1,200 km (750 miles) from Lake Ontario along the border between Canada and the US to the Gulf of St Lawrence on the Atlantic coast.

St Lawrence Seaway a waterway in North America, which flows for 3,768 km (2,342 miles) through the Great Lakes and along the course of the St Lawrence River to the Atlantic. Consisting of channels connecting the lakes and a number of artificial sections which bypass the rapids in the river, it is open along its entire length to ocean-going vessels. It was inaugurated in 1959.

St Leger /ˈledʒə(r)/ an annual flat horse-race for three-year-old colts and fillies, held in September at Doncaster, South Yorkshire. It was instituted by Colonel Barry St Leger (1737–89) in 1776.

St Louis /'luːɪs/ a city and port in eastern Missouri, on the Mississippi just south of its confluence with the Missouri; pop. (1990) 396,685. Founded as a fur-trading post by the French in the 1760s, it passed to the Spanish, the French again, and finally in 1803 to the US as part of the Louisiana Purchase.

St Lucia /'luːʃə/ a country in the West Indies, one of the Windward Islands; pop. (est. 1991) 152,000; languages, English (official), French Creole; capital, Castries. First encountered by Europeans around 1500, St Lucia was settled by both French and British in the 17th century. Slaves were brought in to work sugar plantations, and the native Caribs died out. Possession of the island was long disputed until France ceded it to Britain in 1814. Since 1979 it has been an independent state within the Commonwealth. □ **St Lucian** adj. & n.

St Luke's summer /luːks/ Brit. a period of fine weather occurring around 18 Oct., the feast day of St Luke.

saintly /'seɪntlɪ/ adj. (**saintlier**, **saintliest**) very holy or virtuous. □ **saintliness** n.

St Malo /sæn 'mɑːləʊ/ a walled town and port on the north coast of Brittany, in NW France; pop. (1990) 49,270.

St Mark's Cathedral /mɑːks/ a church in Venice, its cathedral church since 1807. It was built in the 9th century to house relics of St Mark, and rebuilt in the 11th century. It is lavishly decorated with mosaics (11th–13th centuries) and sculptures.

St Martin /ˌsæn mɑːˈtæn/ (Dutch **Sint Maarten** /sɪnt 'mɑːrtə/) a small island in the Caribbean, one of the Leeward Islands; pop. (1990) 31,722. The southern section of the island is administered by the Dutch, forming part of the Netherlands Antilles; chief town and seat of the island's administration, Philipsburg. The larger northern part of the island is part of the French overseas department of Guadeloupe; chief town, Marigot.

St Martin's summer /'mɑːtɪnz/ Brit. a period of fine mild weather occurring around 11 Nov., the feast day of St Martin.

St Moritz /sæn 'mɒrɪts, məˈrɪts/ a resort and winter-sports centre in SE Switzerland.

St-Nazaire /ˌsænnæˈzeə(r)/ a seaport and industrial town in NW France, on the Atlantic coast at the mouth of the Loire; pop. (1990) 66,090. It is a naval dockyard and a commercial port.

St Nicolas see SINT-NIKLAAS.

St Paul /pɔːl/ the state capital of Minnesota, situated on the Mississippi adjacent to Minneapolis, with which it forms the Twin Cities metropolitan area; pop. (1990) 272,235. First settled in 1838, it prospered as a river-trading centre, becoming state capital in 1858.

saintpaulia /sənt'pɔːlɪə/ n. a plant of the genus Saintpaulia, esp. the African violet. [Baron W. von Saint Paul, Ger. soldier (1860–1910), its discoverer]

St Paul's Cathedral /pɔːlz/ a cathedral on Ludgate Hill, London, designed by Sir Christopher Wren in a neoclassical style and built between 1675 and 1711 to replace a medieval cathedral largely destroyed in the Great Fire.

St Peter's Basilica /'piːtəz/ the Roman Catholic basilica in the Vatican City, Rome, the largest church in Christendom. The present 16th-century building replaced a much older structure erected by Constantine on the supposed site of St Peter's crucifixion. A succession of architects (including Bramante and Raphael) in turn made considerable changes in the design; the dome closely follows a design of Michelangelo. The building was consecrated in 1626.

St Petersburg /'piːtəz,bɜːg/ **1** a city and seaport in NW Russia, situated on the delta of the River Neva, on the eastern shores of the Gulf of Finland; pop. (1990) 5,035,000. Founded in 1703 by Peter the Great, it was the capital of Russia from 1712 until the Russian Revolution, after which Moscow became the capital. It was the scene in February and October 1917 of the events which triggered the Revolution. During the Second World War it was subjected by German and Finnish forces to a siege which lasted for more than two years (1941–4). A city of waterways and bridges, it is a major industrial and cultural centre. It was formerly called Petrograd (1914–24) and Leningrad (1924–91). **2** a resort city in western Florida, on the Gulf of Mexico; pop. (1990) 238,630.

St Pierre and Miquelon /sæn 'pjeə(r), 'miːklɒn/ a group of eight small islands in the North Atlantic, off the south coast of Newfoundland; pop. (1990) 6,390. An overseas territory of France, the islands form the last remaining French possession in North America. The capital and chief settlement is St Pierre.

St Pölten /'pɜːlt(ə)n/ a city in NE Austria, capital of the state of Lower Austria; pop. (1991) 50,025.

Saint-Saëns /'sænsɒn, French sɛ̃sɑ̃s/, (Charles) Camille (1835–1921), French composer, pianist, and organist. He was organist at the church of the Madeleine in Paris (1858–77) and played an important role in the city's musical life. His works include operas (notably Samson et Dalila, 1877) and oratorios, but he is probably now best known for his Third Symphony (1886), the symphonic poem Danse macabre (1874), which was the first orchestral piece to use a xylophone, and the Carnaval des animaux (1886).

Saint-Simon[1] /ˌsænsiːˈmɒn/, Claude-Henri de Rouvroy, Comte de (1760–1825), French social reformer and philosopher. In reaction to the chaos engendered by the French Revolution he developed a new theory of social organization and was later claimed as the founder of French socialism. His central theory was that society should be organized in an industrial order, controlled by leaders of industry, and given spiritual direction by scientists. His works, which greatly influenced figures such as J. S. Mill and Friedrich Engels, include Du système industriel (1821) and Nouveau Christianisme (1825).

Saint-Simon[2] /ˌsænsiːˈmɒn/, Louis de Rouvroy, Duc de (1675–1755), French writer. He is best known for his Mémoires, a detailed record of court life between 1694 and 1723, in the reigns of Louis XIV and XV.

St Sophia /səˈfiːə, -ˈfaɪə/ (also called Hagia Sophia, Santa Sophia) the key monument of Byzantine architecture, originally a church, at Constantinople (now Istanbul). Built by order of the Byzantine emperor Justinian, it was inaugurated in 537. Its chief feature is an enormous dome, supported by piers, arches, and pendentives and pierced by forty windows, which crowns the basilica. In 1453, when the Turks invaded, orders were given for its conversion into a mosque; the mosaics which adorned its interior were covered and partly destroyed, and minarets were added. It was used as a mosque until 1935, when Atatürk declared it a museum.

St Stephens /'stiːv(ə)nz/ a name for the House of Commons, derived from the ancient chapel of St Stephen, Westminster, in which the House used to sit (1537–1834).

St Thomas /'tɒməs/ an island in the Caribbean, the second largest of the US Virgin Islands, situated to the east of Puerto Rico; pop. (1990) 48,170; chief town, Charlotte Amalie. Settled by the Dutch in 1657, it passed nine years later to the Danes, who sold it to the US in 1917. Made prosperous by its sugar production until the 19th century, the island is now a thriving holiday resort.

St Trinian's /'trɪnɪənz/ the name of a girls' school invented by the cartoonist Ronald Searle (b.1920) in 1941 and depicted in many books and films, whose pupils are characterized by unruly behaviour, ungainly appearance, and unattractive school uniform. Searle's daughters attended a St Trinnean's school in Edinburgh.

St-Tropez /ˌsæntrəʊˈpeɪ/ a fishing port and resort on the Mediterranean coast of southern France, south-west of Cannes; pop. (1985) 6,250.

St Valentine's Day Massacre the shooting on 14th Feb. 1929 of seven members of the rival 'Bugsy' Moran's gang by some of Al Capone's men disguised as policemen.

St Vincent, Cape (Portuguese **São Vincente** /ˌsaʊ vɪnˈsent/) a headland in SW Portugal, which forms the south-westernmost tip of the country. It was the site of a sea battle in 1797 in which the British fleet under Admiral John Jervis defeated the Spanish.

St Vincent and the Grenadines /'vɪns(ə)nt, ˌgrenəˌdiːnz/ an island state in the Windward Islands in the West Indies, consisting of the mountainous island of St Vincent and some of the Grenadines; pop. (est. 1991) 108,000; languages, English (official), English-based Creole; capital, Kingstown. The French, Dutch, and British all made attempts at settlements in the 18th century, and it finally fell to British possession in 1783. The colonists introduced African slaves, and the indigenous Caribs were eventually wiped out. The state obtained full independence with a limited form of membership of the Commonwealth in 1979.

St Vitus's dance /'vaɪtəsɪz/ n. = Sydenham's chorea (see CHOREA).

Saipan /saɪˈpæn/ the largest of the islands comprising the Northern Marianas in the western Pacific.

saith /seθ/ archaic 3rd sing. present of SAY.

saithe /seɪθ/ n. a codlike food fish, Pollachius virens, with skin that soils fingers like wet coal. Also called coalfish, coley. [ON seithr]

Sakai /saːˈkaɪ/ an industrial city in Japan, on Osaka Bay just south of the city of Osaka; pop. (1990) 807,860.

sake[1] /seɪk/ n. □ **for Christ's** (or **God's** or **goodness'** or **heaven's** or **Pete's** etc.) **sake** an expression of urgency, impatience, supplication, anger, etc. **for old times' sake** in memory of former times. **for the sake of** (or **for a person's sake**) **1** out of consideration for; in the interest of; because of; owing to (*for my own sake as well as yours*). **2** in order to please, honour, get, or keep (*for the sake of uniformity*). [OE *sacu* contention, charge, fault, sake f. Gmc]

sake[2] /ˈsɑːkɪ/ n. a Japanese alcoholic drink made from rice. [Jap.]

saker /ˈseɪkə(r)/ n. **1** a large Eurasian falcon, *Falco cherrug*, used in falconry. **2** *hist.* an old form of cannon. [ME f. OF *sacre* (in both senses), f. Arab. ṣaḳr]

Sakha, Republic of /ˈsɑːkə/ the official name for YAKUTIA.

Sakhalin /ˌsækəˈliːn/ a large island in the Sea of Okhotsk, situated off the coast of eastern Russia and separated from it by the Tartar Strait; capital, Yuzhno-Sakhalinsk. Contested for centuries by the Russians and the Japanese, from 1905 to 1946 it was divided into the northern part, held by Russia, and the southern part (known to the Japanese as Karafuto), which was occupied by Japan. Since 1946 it has been wholly part of Russia.

Sakharov /ˈsækəˌrɒf/, Andrei (Dmitrievich) (1921–89), Russian nuclear physicist. Having helped to develop the Soviet hydrogen bomb, he campaigned against nuclear proliferation and called for Soviet–American cooperation. He fought courageously for reform and human rights in the USSR, for which he was awarded the Nobel Peace Prize in 1975. His international reputation as a scientist kept him out of jail, but in 1980 he was banished to Gorky (Nizhni Novgorod) and kept under police surveillance. He was freed (1986) in the new spirit of glasnost, and at his death he was honoured in his own country as well as in the West.

Saki /ˈsɑːkɪ/ (pseudonym of Hector Hugh Munro) (1870–1916), British short-story writer, born in Burma. His stories include the satiric, comic, macabre, and supernatural, and frequently depict animals as agents seeking revenge upon humankind; collections include *Reginald* (1904). He was killed in France during the First World War.

saki /ˈsɑːkɪ/ n. (pl. **sakis**) a South American monkey of the genus *Pithecia* or *Chiropotes*, having coarse fur and a long non-prehensile tail. [F f. Tupi *çahy*]

Sakta /ˈʃʌktə/ n. a member of a Hindu sect worshipping the Sakti. [Skr. *śākta* relating to power or to the SAKTI]

Sakti /ˈʃʌktɪ/ n. (also **sakti**) *Hinduism* **1** the female principle, esp. when personified as the wife of a god. For example, Durga is the Sakti of Siva. **2** the goddess as supreme deity (Devi). □ **Saktism** n. [Skr. *śakti* power, divine energy]

sal /sɑːl/ n. a northern Indian tree, *Shorea robusta*, yielding teaklike timber and dammar resin. [Hindi *sāl*]

salaam /səˈlɑːm/ n. & v. ● n. **1** the oriental salutation 'Peace'. **2** an Indian obeisance, with or without the salutation, consisting of a low bow of the head and body with the right palm on the forehead. **3** (in *pl.*) respectful compliments. ● v. **1** *tr.* make a salaam to (a person). **2** *intr.* make a salaam. [Arab. *salām*]

salable var. of SALEABLE.

salacious /səˈleɪʃəs/ adj. **1** lustful; lecherous. **2** (of writings, pictures, talk, etc.) tending to cause sexual desire. □ **salaciously** adv. **salaciousness** n. **salacity** /-ˈlæsɪtɪ/ n. [L *salax salacis* f. *salire* leap]

salad /ˈsæləd/ n. **1** a cold dish of various mixtures of raw or cooked vegetables or herbs, usu. seasoned with oil, vinegar, etc. **2** a vegetable or herb suitable for eating raw. □ **salad cream** creamy salad-dressing. **salad days** a period of youthful inexperience. **salad-dressing** a mixture of oil, vinegar, etc., used with salad. [ME f. OF *salade* f. Prov. *salada* ult. f. L *sal* salt]

salade var. of SALLET.

Saladin /ˈsælədɪn/ (Arabic name Salah-ad-Din Yusuf ibn-Ayyub) (1137–93), sultan of Egypt and Syria 1174–93. He invaded the Holy Land and reconquered Jerusalem from the Christians (1187), and, for a period, resisted the Third Crusade, the leaders of which included Richard the Lion-heart. He was later defeated by Richard at Arsuf (1191) and withdrew to Damascus, where he died. Saladin earned a reputation not only for military skill but also for honesty and chivalry.

Salam /səˈlɑːm/, Abdus (b.1926), Pakistani theoretical physicist. He worked on the interaction of subatomic particles, and independently developed a unified theory to explain electromagnetic interactions

and the weak nuclear force. In 1979 he shared the Nobel Prize for physics, the first Nobel laureate from his country.

Salamanca /ˌsæləˈmæŋkə/ a city in western Spain, in Castilla-León; pop. (1991) 185,990. An early Iberian settlement, it was destroyed by Hannibal in 217 BC and subsequently settled by the Romans. It was occupied by the Moors from the 8th to the 11th centuries. It was the scene in 1812, during the Peninsular War, of a British victory over the French.

salamander /ˈsæləˌmændə(r)/ n. **1** a tailed newtlike amphibian of the order Urodela; esp. a terrestrial one of the genus *Salamandra*, once thought able to endure fire. **2** a mythical lizard-like creature credited with this property. **3** US = GOPHER[1] 1. **4** an elemental spirit living in fire. **5** a red-hot iron used for lighting pipes, gunpowder, etc. **6** a metal plate heated and placed over food to brown it. □ **salamandrian** /ˌsæləˈmændrɪən/ adj. **salamandrine** /-ˈmændrɪn/ adj. **salamandroid** adj. & n. (in sense 1). [ME f. OF *salamandre* f. L *salamandra* f. Gk *salamandra*]

salami /səˈlɑːmɪ/ n. (pl. **salamis**) a highly seasoned orig. Italian sausage often flavoured with garlic. [It., pl. of *salame*, f. LL *salare* (unrecorded) to salt]

Salamis /ˈsæləmɪs/ an island in the Saronic Gulf in Greece, to the west of Athens. The strait between the island and the mainland was the scene in 480 BC of a crushing defeat of the Persian fleet under Xerxes I by the Greeks under Themistocles.

sal ammoniac /ˌsæl əˈməʊnɪˌæk/ n. ammonium chloride, a white crystalline salt. [L *sal ammoniacus* 'salt of Ammon', said to have been made from camels' dung near the Roman temple of Ammon (Amun) in North Africa]

Salang Pass /ˈsɑːlæŋ/ a high-altitude route across the Hindu Kush mountain range in Afghanistan. The Salang Pass and tunnel were built by the Soviet Union during the 1960s in an attempt to improve the supply route from the Soviet frontier to Kabul.

salariat /səˈleərɪət/ n. the salaried class. [F f. *salaire* (see SALARY), after *prolétariat*]

salary /ˈsælərɪ/ n. & v. ● n. (pl. **-ies**) a fixed regular payment, usu. monthly or quarterly, made by an employer to an employee, esp. a professional or white-collar worker (cf. WAGE n. 1). ● v.tr. (**-ies, -ied**) (usu. as **salaried** adj.) pay a salary to. [ME f. AF *salarie*, OF *salaire* f. L *salarium* orig. soldier's salt-money f. *sal* salt]

Salazar /ˌsæləˈzɑː(r)/, Antonio de Oliveira (1889–1970), Portuguese statesman, Prime Minister 1932–68. While Finance Minister (1928–40), he formulated austere fiscal policies to effect Portugal's economic recovery. During his long premiership, he ruled the country as a virtual dictator, firmly suppressing opposition and enacting a new authoritarian constitution along Fascist lines. Salazar maintained Portugal's neutrality throughout the Spanish Civil War and the Second World War.

sale /seɪl/ n. **1** the exchange of a commodity for money etc.; an act or instance of selling. **2** the amount sold (*the sales were enormous*). **3** the rapid disposal of goods at reduced prices for a period esp. at the end of a season etc. **4 a** an event at which goods are sold. **b** a public auction. □ **on** (or **for** or **up for**) **sale** offered for purchase. **sale of work** an event where goods made by parishioners etc. are sold for charity. **sale or return** an arrangement by which a purchaser takes a quantity of goods with the right of returning surplus goods without payment. **sale-ring** a circle of buyers at an auction. **sales clerk** N. Amer. a salesman or saleswoman in a shop. **sales department** etc. the section of a firm concerned with selling as opposed to manufacturing or dispatching goods. **sales engineer** a salesperson with technical knowledge of the goods and their market. **sales resistance** the opposition or apathy of a prospective customer etc. to be overcome by salesmanship. **sales talk** persuasive talk to promote the sale of goods or the acceptance of an idea etc. **sales tax** a tax on sales or on the receipts from sales. [OE *sala* f. ON]

saleable /ˈseɪləb(ə)l/ adj. (also **salable**) fit to be sold; finding purchasers. □ **saleability** /ˌseɪləˈbɪlɪtɪ/ n.

Salem /ˈseɪləm/ **1** the state capital of Oregon, situated on the Willamette river south-west of Portland; pop. (1990) 107,790. Settled by missionaries in 1841, it became state capital in 1859. **2** a city and port in NE Massachusetts, on the Atlantic coast north of Boston; pop. (1990) 38,090. First settled in 1626, it was the scene in 1692 of a notorious series of witchcraft trials. **3** an industrial city in Tamil Nadu in southern India; pop. (1991) 364,000. It is a centre for the production of textiles.

salep /ˈsæləp/ n. a starchy preparation of the dried tubers of various orchids, used in cookery and formerly medicinally. [F f. Turk. *sālep* f. Arab. (*ḳuṣa-'l-*) *ta'lab* fox, fox's testicles]

saleratus /ˌsæləˈreɪtəs/ *n.* US an ingredient of baking powder consisting mainly of potassium or sodium bicarbonate. [mod.L *sal aeratus* aerated salt]

Salerno /səˈleənəʊ/ a port on the west coast of Italy, on the Gulf of Salerno south-east of Naples; pop. (1990) 151,370.

saleroom /ˈseɪlruːm, -rʊm/ *n.* esp. *Brit.* a room in which items are sold at auction.

salesgirl /ˈseɪlzɡɜːl/ *n.* a saleswoman.

Salesian /səˈliːʒ(ə)n/ *adj. & n.* ● *adj.* of or relating to an educational religious order within the Roman Catholic Church, founded near Turin in 1859, named after St Francis of Sales. ● *n.* a member of this order.

saleslady /ˈseɪlzˌleɪdɪ/ *n.* (*pl.* **-ies**) a saleswoman.

salesman /ˈseɪlzmən/ *n.* (*pl.* **-men**; *fem.* **saleswoman**, *pl.* **-women**) **1** a person employed to sell goods in a shop, or as an agent between the producer and retailer. **2** US a commercial traveller.

salesmanship /ˈseɪlzmənˌʃɪp/ *n.* **1** skill in selling. **2** the techniques used in selling.

salesperson /ˈseɪlzˌpɜːs(ə)n/ *n.* a salesman or saleswoman (used as a neutral alternative).

salesroom /ˈseɪlzruːm, -rʊm/ *n.* US = SALEROOM.

Salford /ˈsɔːlfəd/ an industrial city in NW England, in Greater Manchester; pop. (1991) 217,900.

Salian /ˈseɪlɪən/ *adj. & n.* ● *adj.* of or relating to the Salii, a 4th-century Frankish people living near the River IJssel, from which the Merovingians were descended. ● *n.* a member of this people. [LL *Salii*]

Salic /ˈsælɪk, ˈseɪl-/ *adj.* = SALIAN. [F *Salique* or med.L *Salicus* f. *Salii* (as SALIAN)]

salicet /ˈsælɪsɪt/ *n.* an organ stop like a salicional but one octave higher. [as SALICIONAL]

salicin /ˈsælɪsɪn/ *n.* (also **salicine** /-ˌsiːn/) *Chem.* a bitter crystalline glucoside with analgesic properties, obtained from poplar and willow bark. [F *salicine* f. L *salix -icis* willow]

salicional /səˈlɪʃ(ə)n(ə)l/ *n.* an organ stop with a soft reedy tone like that of a willow-pipe. [G f. L *salix* as SALICIN]

Salic law *n. hist.* **1** a Frankish law-book extant in Merovingian and Carolingian times. **2** a law excluding females from dynastic succession, esp. in the French monarchy. Such a law was used in the 14th century by the French to deny Edward III's claim to the French throne (based on descent from his Capetian mother Isabella), and was a direct cause of the Hundred Years War.

salicylic acid /ˌsælɪˈsɪlɪk/ *n. Chem.* a bitter chemical used as a fungicide and in the manufacture of aspirin and dyestuffs. □ **salicylate** /səˈlɪsɪˌleɪt/ *n.* [*salicyl* its radical f. F *salicyle* (as SALICIN)]

salient /ˈseɪlɪənt/ *adj. & n.* ● *adj.* **1** jutting out; prominent; conspicuous, most noticeable. **2** (of an angle, esp. in fortification) pointing outwards (opp. RE-ENTRANT *adj.* 1). **3** *Heraldry* (of a lion etc.) standing on its hind legs with the forepaws raised. **4** *archaic* **a** leaping or dancing. **b** (of water etc.) jetting forth. ● *n.* **1** a salient angle or part of a work in fortification. **2** an outward bulge in a line of military attack or defence. □ **salient point 1** an important or significant point. **2** *archaic* the initial stage, origin, or first beginning. □ **salience** *n.* **saliency** *n.* **saliently** *adv.* [L *salire* leap]

salientian /ˌseɪlɪˈenʃ(ə)n/ *adj. & n. Zool.* = ANURAN. [mod.L *Salientia* (as SALIENT)]

Salieri /ˌsælɪˈeərɪ/, Antonio (1750–1825), Italian composer. His output includes over forty operas, four oratorios, and much church music. He lived in Vienna for many years and taught Beethoven, Schubert, and Liszt. Salieri was hostile to Mozart and a story arose that he poisoned him, though the story is apparently without foundation.

saliferous /səˈlɪfərəs/ *adj. Geol.* (of rock etc.) containing much salt. [L *sal* salt + -FEROUS]

salina /səˈlaɪnə/ *n.* a salt lake or salt-pan. [Sp. f. med.L, = salt pit (as SALINE)]

saline /ˈseɪlaɪn/ *adj. & n.* ● *adj.* **1** (of natural waters, springs, etc.) impregnated with or containing salt or salts. **2** (of food or drink etc.) tasting of salt. **3** of chemical salts. **4** of the nature of a salt. **5** *Med.* containing a salt or salts of alkaline metals or magnesium, esp. sodium chloride. ● *n.* **1** a salt lake, spring, marsh, etc. **2** a salt-pan or salt-works. **3** a saline substance (esp. *physiological saline*). **4** a solution of

salt in water. □ **salinity** /səˈlɪnɪtɪ/ *n.* **salinization** /ˌsælɪnaɪˈzeɪʃ(ə)n/ *n.* **salinometer** /-ˈnɒmɪtə(r)/ *n.* [ME f. L *sal* salt]

Salinger /ˈsælɪndʒə(r)/, J(erome) D(avid) (b.1919), American novelist and short-story writer. He is best known for his colloquial novel of adolescence *The Catcher in the Rye* (1951). His other works include *Franny and Zooey* (1961).

Salisbury[1] /ˈsɔːlzbərɪ/ **1** a city in southern England, in Wiltshire; pop. (1981) 35,000. It is noted for its 13th-century cathedral, whose spire, at 123 m (404 ft), is the highest in England. Its diocese is known as Sarum, an old name for the city. **2** the former name (until 1982) for HARARE.

Salisbury[2] /ˈsɔːlzbərɪ/, Robert Arthur Talbot Gascoigne-Cecil, 3rd Marquess of (1830–1903), British Conservative statesman, Prime Minister 1885–6, 1886–92, and 1895–1902. His main area of concern was foreign affairs; he was a firm defender of British imperial interests and supported the policies which resulted in the Second Boer War (1899–1902).

Salish /ˈseɪlɪʃ/ *n. & adj.* ● *n.* (*pl.* same) **1** a member of a group of American Indian peoples inhabiting the north-western US and the west coast of Canada. **2** the group of related languages spoken by the Salish. ● *adj.* of or relating to the Salish or their languages. [Salish]

saliva /səˈlaɪvə/ *n.* liquid secreted into the mouth by glands to provide lubrication, facilitate chewing and swallowing, and (in some animals) aid digestion. □ **saliva test** a scientific test on a saliva sample. □ **salivary** /səˈlaɪvərɪ, ˈsælɪvərɪ/ *adj.* [ME f. L]

salivate /ˈsælɪˌveɪt/ *v.* **1** *intr.* secrete or discharge saliva esp. in excess or in greedy anticipation. **2** *tr.* produce an unusual secretion of saliva in (a person) usu. with mercury. □ **salivation** /ˌsælɪˈveɪʃ(ə)n/ *n.* [L *salivare* (as SALIVA)]

Salk vaccine /sɔːlk/ *n. Med.* the first vaccine developed against poliomyelitis, named after the American virologist Jonas Edward Salk (1914–95), who developed the vaccine in 1954.

sallee /ˈsælɪ/ *n.* (also **sally**) (*pl.* **-ees** or **-ies**) *Austral.* any of several eucalypts and acacias resembling the willow. [dial. var. of SALLOW[2]]

sallet /ˈsælɪt/ *n.* (also **salade** /səˈlɑːd/) *hist.* a light helmet with an outward-curving rear part. [F *salade* ult. f. L *caelare* engrave f. *caelum* chisel]

sallow[1] /ˈsæləʊ/ *adj. & v.* ● *adj.* (**sallower, sallowest**) (of the skin or complexion, or of a person) of a sickly yellow or pale brown. ● *v.tr. & intr.* make or become sallow. □ **sallowish** *adj.* **sallowness** *n.* [OE *salo* dusky f. Gmc]

sallow[2] /ˈsæləʊ/ *n.* **1** a willow tree, esp. one of a low-growing or shrubby kind. **2** the wood or a shoot of this. □ **sallowy** *adj.* [OE *salh salg-* f. Gmc, rel. to OHG *salaha*, ON *selja*, L *salix*]

Sallust /ˈsæləst/ (Latin name Gaius Sallustius Crispus) (86–35 BC), Roman historian and politician. As a historian he was concerned with the political decline of Rome after the fall of Carthage in 146 BC, to which he accorded a simultaneous moral decline. His chief surviving works deal with the Catiline conspiracy and the Jugurthine War.

Sally /ˈsælɪ/ *n.* (*pl.* **-ies**) *colloq.* **1** (usu. prec. by *the*) the Salvation Army. **2** a member of this. [abbr.]

sally[1] /ˈsælɪ/ *n. & v.* ● *n.* (*pl.* **-ies**) **1** a sudden charge from a fortification upon its besiegers; a sortie. **2** a going forth; an excursion. **3** a witticism; a piece of banter; a lively remark esp. by way of attack upon a person or thing or of a diversion in argument. **4** a sudden start into activity; an outburst. **5** *archaic* an escapade. ● *v.intr.* (**-ies, -ied**) **1** (usu. foll. by *out, forth*) go for a walk, set out on a journey etc. **2** (usu. foll. by *out*) make a military sally. **3** *archaic* issue or come out suddenly. □ **sally-port** an opening in a fortification for making a sally from. [F *saillie* fem. past part. of *saillir* issue f. OF *salir* f. L *salire* leap]

sally[2] /ˈsælɪ/ *n.* (*pl.* **-ies**) **1** the part of a bell-rope prepared with inwoven wool for holding. **2 a** the first movement of a bell when set for ringing. **b** the bell's position when set. □ **sally-hole** the hole through which the bell-rope passes. [perh. f. SALLY[1] in sense 'leaping motion']

sally[3] var. of SALLEE.

Sally Lunn /ˌsælɪ ˈlʌn/ *n. Brit.* a sweet light teacake, properly served hot. [perh. f. the name of a woman selling them at Bath c.1800]

salmagundi /ˌsælməˈɡʌndɪ/ *n.* (*pl.* **salmagundis**) **1** a dish of chopped meat, anchovies, eggs, onions, etc., and seasoning. **2** a general mixture; a miscellaneous collection of articles, subjects, qualities, etc. [F *salmigondis* of unkn. orig.]

salmanazar /ˌsælməˈneɪzə(r)/ *n.* a wine bottle of about twelve times the standard size. [*Shalmaneser V* king of Assyria 726–721 BC (2 Kings 17–18)]

salmi /ˈsælmɪ/ *n.* (*pl.* **salmis**) a ragout or casserole esp. of partly roasted game birds. [F, abbr. formed as SALMAGUNDI]

salmon /ˈsæmən/ *n. & adj.* ● *n.* (*pl.* same or (esp. of types) **salmons**) **1** a migratory game-fish of the family Salmonidae, much prized for its pink flesh. (*See note below.*) **2** a similar but unrelated fish, esp.: **a** (in full **Australian salmon**) *Austral. & NZ* a large green and silver marine fish, *Arripis trutta*. **b** *US* an American sea trout of the genus *Cynoscion* (family Sciaenidae). ● *adj.* salmon-pink. □ **salmon-ladder** (or **-leap**) a series of steps or other arrangement incorporated in a dam to allow salmon to pass upstream. **salmon-pink** the colour of salmon flesh. **salmon trout 1** = *sea trout* 1. **2** a similar fish, e.g. the Australian salmon, the North American lake trout. □ **salmonoid** *adj. & n.* (in sense 1). **salmony** *adj.* [ME f. AF *sa*(*u*)*moun*, OF *saumon* f. L *salmo -onis*]

■ The Atlantic salmon, *Salmo salar*, is closely related to the trout and not always easy to distinguish from it. It matures in the sea, but once adult it returns annually to spawn in fresh water. The five species of Pacific salmon belong to the genus *Oncorhynchus*, and include the sockeye, chinook, and coho. These spawn in the rivers of Alaska and western Canada, but die after spawning. All kinds perform spectacular leaps while attempting to migrate upstream against torrents and waterfalls.

salmonella /ˌsælməˈnelə/ *n.* (*pl.* **salmonellae** /-liː/) *Med.* **1** a bacterium of the genus *Salmonella*, esp. of a serotype causing food poisoning. **2** food poisoning caused by infection with salmonellae. □ **salmonellosis** /-nəˈləʊsɪs/ *n.* [mod.L f. D. E. *Salmon*, Amer. veterinary surgeon (1850–1914)]

salmonid /sælˈmɒnɪd, ˈsælmənɪd/ *adj. & n. Zool.* ● *adj.* of or relating to the family Salmonidae, which includes the salmon and trout. ● *n.* a fish of this family.

Salome /səˈləʊmɪ/ (in the New Testament) the daughter of Herodias, who danced before her stepfather Herod Antipas. Given a choice of reward for her dancing, she asked for the head of St John the Baptist and thus caused him to be beheaded. Her name is given by Josephus; she is mentioned but not named in the Gospels.

salon /ˈsælɒn/ *n.* **1** the reception room of a large, esp. French or continental, house. **2** a room or establishment where a hairdresser, beautician, etc., conducts trade. **3** *hist.* a meeting of eminent people in the reception room of a (esp. Parisian) lady of fashion. □ **salon music** light music for the drawing-room etc. [F: see SALOON]

Salon, the the name given to the annual exhibition of the work of living artists held by the Royal Academy of Painting and Sculpture in Paris, originally in the Salon d'Apollon in the Louvre in 1667. Responsibility for the Salon passed out of the hands of the Academy in 1881; the Academy's reputation for hostility to new artists contributed to the decline of the Salon thereafter.

Salonica see THESSALONÍKI.

saloon /səˈluːn/ *n.* **1 a** a large room or hall, esp. in a hotel or public building. **b** a public room or gallery for a specified purpose (*billiard-saloon*; *shooting-saloon*). **2** (in full **saloon car**) a motor car with a closed body, boot, and no partition behind the driver. **3** a public room on a ship. **4** *US* a drinking-bar. **5** (in full **saloon bar**) *Brit.* the more comfortable bar in a public house. **6** (in full **saloon car**) *Brit.* a luxurious railway carriage serving as a lounge etc. □ **saloon deck** a deck for passengers using the saloon. **saloon-keeper** *US* a publican or bartender. **saloon pistol** (or **rifle**) a pistol or rifle adapted for short-range practice in a shooting-saloon. [F *salon* f. It. *salone* augment. of *sala* hall]

Salop /ˈsæləp/ an alternative name for SHROPSHIRE. It was the official name of the county 1974–80. □ **Salopian** /səˈləʊpɪən/ *adj. & n.* [abbrev. of AF *Salopesberia* Shrewsbury f. ME f. OE *Scrobbesbyrig*]

salpiglossis /ˌsælpɪˈɡlɒsɪs/ *n.* a solanaceous plant of the genus *Salpiglossis*, cultivated for its funnel-shaped flowers. [mod.L, irreg. f. Gk *salpigx* trumpet + *glōssa* tongue]

salping- /ˈsælpɪŋ/ *comb. form Med.* denoting the Fallopian tubes. [Gk *salpigx salpiggos*, lit. 'trumpet']

salpingectomy /ˌsælpɪŋˈdʒektəmɪ/ *n.* (*pl.* **-ies**) the surgical removal of the Fallopian tubes.

salpingitis /ˌsælpɪŋˈdʒaɪtɪs/ *n. Med.* inflammation of the Fallopian tubes.

salsa /ˈsælsə/ *n.* **1** a kind of dance music of Latin American origin, incorporating jazz and rock elements. **2** a dance performed to this music. **3** (esp. in Latin American cookery) a usu. spicy sauce, esp. one served with meat. [Sp. (as SAUCE)]

salsify /ˈsælsɪfɪ, -ˌfaɪ/ *n.* (*pl.* **-ies**) **1** a European composite plant, *Tragopogon porrifolius*, with long cylindrical fleshy roots. Also called *vegetable oyster*. **2** this root used as a vegetable. □ **black salsify** = SCORZONERA. [F *salsifis* f. obs. It. *salsefica*, of unkn. orig.]

SALT /sɔːlt, sɒlt/ *abbr.* Strategic Arms Limitation Talks (or Treaty), a series of negotiations between the US and the Soviet Union aimed at the limitation or reduction of nuclear armaments. The talks were organized from 1968 onwards and held in stages until superseded by the START negotiations in 1983; two treaties, SALT I and II, were signed, in 1972 and 1979.

salt /sɔːlt, sɒlt/ *n., adj., & v.* ● *n.* **1** (also **common salt**) sodium chloride; the substance that gives sea water its characteristic taste, obtained in crystalline form by mining from strata consisting of it or by the evaporation of sea water, and used for seasoning or preserving food, or for other purposes. **2** a chemical compound formed from the reaction of an acid with a base, with all or part of the hydrogen of the acid replaced by a metal or metal-like radical. **3** sting; piquancy; pungency; wit (*added salt to the conversation*). **4** (in *sing.* or *pl.*) a crystalline compound used as a medicine, cosmetic, etc. (*bath salts*; *Epsom salts*; *smelling-salts*). **5** a marsh, esp. one flooded by the tide, often used as a pasture or for collecting water for salt-making. **6** (also **old salt**) *colloq.* an experienced sailor. **7** (in *pl.*) an exceptional rush of sea water up river. ● *adj.* **1** impregnated with, containing, or tasting of salt; cured or preserved or seasoned with salt. **2** (of a plant) growing in the sea or in salt marshes. **3** (of tears etc.) bitter. **4** (of wit) pungent. ● *v.tr.* **1** cure or preserve with salt or brine. **2** season with salt. **3** make (a narrative etc.) piquant. **4** sprinkle (the ground etc.) with salt esp. in order to melt snow etc. **5** treat with a solution of salt or mixture of salts. **6** (as **salted** *adj.*) hardened or proof against diseases etc. caused by the climate or by special conditions. □ **eat salt with** be a guest of. **in salt** sprinkled with salt or immersed in brine as a preservative. **not made of salt** not disconcerted by wet weather. **put salt on the tail of** capture (with ref. to jocular directions given to children for catching a bird). **salt an account** *sl.* set an extremely high or low price for articles. **salt-and-pepper** (of materials etc. and esp. of hair) with light and dark colours mixed together. **salt away** (or **down**) *sl.* put (money etc.) by. **salt the books** *sl.* show receipts as larger than they really have been. **salt-cat** a mass of salt mixed with gravel, urine, etc., esp. used by pigeon-fanciers to attract pigeons and keep them at home. **salt dome** a mass of salt forced up into sedimentary rocks. **salt fish** fish preserved in salt. **salt-glaze** a hard stoneware glaze produced by throwing salt into a hot kiln containing the ware. **salt-grass** *US* grass growing in salt meadows or in alkaline regions. **salt horse** *Naut. sl.* **1** salt beef. **2** a naval officer with general duties. **salt lake** a lake of salt water. **salt-lick 1** a place where animals go to lick salt from the ground. **2** this salt. **salt marsh** = sense 5 of *n.* **salt meadow** a meadow subject to flooding with salt water. **salt-mine** a mine yielding rock-salt. **salt a mine** *sl.* introduce extraneous ore, material, etc., to make the source seem rich. **the salt of the earth** a person or people of great worthiness, reliability, honesty, etc.; those whose qualities are particularly valuable to society (Matt. 5:13). **salt-pan** a vessel, or a depression near the sea, used for getting salt by evaporation. **salt-shaker** *N. Amer.* a container of salt for sprinkling on food. **salt-spoon** a small spoon usu. with a short handle and a roundish deep bowl for taking table salt. **salt water 1** sea water. **2** *sl.* tears. **salt-water** *adj.* of or living in the sea. **salt-well** a bored well yielding brine. **salt-works** a place where salt is produced. **take with a pinch** (or **grain**) **of salt** regard as exaggerated; be incredulous about; believe only part of. **worth one's salt** efficient, capable. □ **saltish** *adj.* **saltless** *adj.* **saltly** *adv.* **saltness** *n.* [OE *s*(*e*)*alt s*(*e*)*altan*, OS, ON, Goth. *salt*, OHG *salz* f. Gmc]

saltarello /ˌsæltəˈreləʊ/ *n.* (*pl.* **-os** or **saltarelli** /-lɪ/) an Italian and Spanish dance for one couple, with sudden skips. [It. *salterello*, Sp. *saltarelo*, rel. to It. *saltare* and Sp. *saltar* leap, dance f. L *saltare* (as SALTATION)]

saltation /sælˈteɪʃ(ə)n/ *n.* **1** the act or an instance of leaping or dancing; a jump. **2** a sudden transition. **3** *Biol.* an abrupt evolutionary change. □ **saltatorial** /ˌsæltəˈtɔːrɪəl/ *adj.* **saltatory** /ˈsæltətərɪ/ *adj.* [L *saltatio* f. *saltare* frequent. of *salire salt-* leap]

saltbush /ˈsɔːltbʊʃ, ˈsɒlt-/ *n.* = ORACHE.

salt-cellar /ˈsɔːltˌselə(r), ˈsɒlt-/ *n.* **1** a vessel holding salt for table use. **2** *colloq.* an unusually deep hollow above the collar-bone, esp. found in women. [SALT + obs. *saler* f. AF f. OF *salier* salt-box f. L (as SALARY), assim. to CELLAR]

salter /'sɔ:ltə(r), 'sɒl-/ n. **1** a manufacturer or dealer in salt. **2** a workman at a salt-works. **3** a person who salts fish etc. **4** = dry-salter. [OE sealtere (as SALT)]

saltern /'sɔ:lt(ə)n, 'sɒl-/ n. **1** a salt-works. **2** a set of pools for the natural evaporation of sea water. [OE sealtærn (as SALT, ærn building)]

saltigrade /'sæltɪˌgreɪd/ adj. & n. Zool. ● adj. (of arthropods) moving by leaping or jumping. ● n. a saltigrade arthropod, e.g. a spider, sand-hopper, etc. [mod.L Saltigradae f. L saltus leap f. salire salt- + -gradus walking]

Saltillo /sæl'ti:jəʊ/ a city in northern Mexico, capital of the state of Coahuila, situated in the Sierra Madre south-west of Monterrey; pop. (1990) 441,000.

salting /'sɔ:ltɪŋ, 'sɒl-/ n. **1** in senses of SALT v. **2** (esp. in pl.) Geol. a salt marsh; a marsh overflowed by the sea.

saltire /'sɔ:ltaɪə(r)/ n. Heraldry an ordinary formed by a bend and a bend sinister crossing like a St Andrew's cross. □ **in saltire** arranged in this way. □ **saltirewise** adv. [ME f. OF sau(l)toir etc. stirrup-cord, stile, saltire, f. med.L saltatorium (as SALTATION)]

Salt Lake City the capital of Utah, situated near the south-eastern shores of the Great Salt Lake; pop. (1990) 159,940. Founded in 1847 by Brigham Young for the Mormon community, the city is the world headquarters of the Church of Latter Day Saints (Mormons).

saltpetre /sɔ:lt'pi:tə(r), sɒlt-/ n. (US **saltpeter**) potassium nitrate, a white crystalline salty substance used in preserving meat and as a constituent of gunpowder. (See also CHILE SALTPETRE.) [ME f. OF salpetre f. med.L salpetra prob. for sal petrae (unrecorded) salt of rock (i.e. found as an incrustation): assim. to SALT]

saltus /'sæltəs/ n. literary a sudden transition; a breach of continuity. [L, = leap]

saltwort /'sɔ:ltwɜ:t, 'sɒlt-/ n. a salt marsh plant of the genus Salsola (goosefoot family), rich in alkali; glasswort.

salty /'sɔ:ltɪ, 'sɒl-/ adj. (**saltier**, **saltiest**) **1** tasting of, containing, or preserved with salt. **2** racy, risqué. □ **saltiness** n.

salubrious /sə'lu:brɪəs, sə'lju:-/ adj. **1** health-giving; healthy. **2** (of surroundings etc.) pleasant; agreeable. □ **salubriously** adv. **salubriousness** n. **salubrity** n. [L salubris f. salus health]

saluki /sə'lu:kɪ/ n. (pl. **salukis**) a tall swift slender breed of dog with a silky coat, large drooping ears, and fringed feet. [Arab. salūkī]

salutary /'sæljʊtərɪ/ adj. **1** producing good effects; beneficial. **2** archaic health-giving. [ME f. F salutaire or L salutaris f. salus -utis health]

salutation /ˌsæljʊ'teɪʃ(ə)n/ n. **1** a sign or expression of greeting or recognition of another's arrival or departure. **2** (usu. in pl.) words spoken or written to enquire about another's health or well-being. □ **salutational** adj. [ME f. OF salutation or L salutatio (as SALUTE)]

salutatory /sə'lu:tətərɪ, sə'lju:-/ adj. & n. ● adj. of salutation. ● n. (pl. **-ies**) US an oration, esp. as given by a member of a graduating class, often the second-ranking member. □ **salutatorian** /-ˌlu:tə'tɔ:rɪən, -ˌlju:-/ n. (in sense of n.). [L salutatorius (as SALUTE)]

salute /sə'lu:t, -'lju:t/ n. & v. ● n. **1** a gesture of respect, homage, or courteous recognition, esp. made to or by a person when arriving or departing. **2 a** Mil. & Naut. a prescribed or specified movement of the hand or weapons or flags as a sign of respect or recognition. **b** (prec. by the) the attitude taken by an individual soldier, sailor, policeman, etc., in saluting. **3** the discharge of a gun or guns as a formal or ceremonial sign of respect or celebration. **4** Fencing the formal performance of certain guards etc. by fencers before engaging. ● v. **1 a** tr. make a salute to. **b** intr. (often foll. by to) perform a salute. **2** tr. greet; make a salutation to. **3** tr. (foll. by with) receive or greet with (a smile etc.). **4** tr. archaic hail as (king etc.). □ **take the salute 1** (of the highest officer present) acknowledge it by gesture as meant for him. **2** receive ceremonial salutes by members of a procession. □ **saluter** n. [ME f. L salutare f. salus -utis health]

Salvador /'sælvəˌdɔ:(r)/ a port on the Atlantic coast of eastern Brazil, capital of the state of Bahia; pop. (1990) 2,075,400. Founded in 1549, it was the capital of the Portuguese colony until 1763, when the seat of government was transferred to Rio de Janeiro. It was formerly known as Bahia.

salvage /'sælvɪdʒ/ n. & v. ● n. **1** the rescue of a ship, its cargo, or other property, from loss at sea, destruction by fire, etc. **2** the property etc. saved in this way. **3 a** the saving and utilization of waste paper, scrap material, etc. **b** the materials salvaged. **4** payment made or due to a person who has saved a ship or its cargo. ● v.tr. **1** save from a wreck, fire, etc. **2** retrieve or preserve (something favourable) in adverse circumstances (tried to salvage some dignity). □ **salvageable** adj. **salvager** n. [F f. med.L salvagium f. L salvare SAVE[1]]

salvation /sæl'veɪʃ(ə)n/ n. **1** the act of saving or being saved; preservation from loss, calamity, etc. **2** (in Christian theology) deliverance from sin and its consequences and admission to heaven, brought about by Christ. **3** a religious conversion. **4** a person or thing that saves (was the salvation of). □ **salvationism** n. **salvationist** n. (both nouns esp. with ref. to the Salvation Army). [ME f. OF sauvacion, salvacion, f. eccl.L salvatio -onis f. salvare SAVE[1], transl. Gk sōtēria]

Salvation Army (abbr. **SA**) an international organization for evangelistic and social work, established in 1865 by William Booth and given its present name in 1878. It is organized on a quasi-military basis under a General, and its members may wear a distinctive uniform on public occasions. The Salvation Army is active in all kinds of social work, including the provision of soup kitchens, workers' hostels, and night shelters. It is also noted for its brass bands, which play during open-air meetings, and for its publication War Cry. Its headquarters are in London.

salve[1] /sælv, sɑ:v/ n. & v. ● n. **1** a healing ointment. **2** (often foll. by for) a thing that is soothing or consoling for wounded feelings, an uneasy conscience, etc. **3** archaic a thing that explains away a discrepancy or palliates a fault. ● v.tr. **1** soothe (pride, self-love, conscience, etc.). **2** archaic anoint (a wound etc.). **3** archaic smooth over, make good, vindicate, harmonize, etc. [OE s(e)alf(e), s(e)alfian f. Gmc; senses 1 and 3 of v. partly f. L salvare SAVE[1]]

salve[2] /sælv/ v.tr. **1** save (a ship or its cargo) from loss at sea. **2** save (property) from fire. □ **salvable** adj. [back-form. f. SALVAGE]

salver /'sælvə(r)/ n. a tray usu. of gold, silver, brass, or electroplate, on which drinks, letters, etc., are offered. [F salve tray for presenting food to the king f. Sp. salva assaying of food f. salvar SAVE[1]: assoc. with platter]

Salve Regina /ˌsælveɪ rə'dʒi:nə/ n. **1** a Roman Catholic hymn or prayer said or sung after compline and after the Divine Office from Trinity Sunday to Advent. **2** the music for this. [f. the opening words salve regina hail (holy) queen]

salvia /'sælvɪə/ n. a labiate plant of the genus Salvia; esp. S. splendens, with scarlet flowers. [L, = SAGE[1]]

Salvo /'sælvəʊ/ n. (pl. **-os**) Austral. sl. a member of the Salvation Army. [abbr.]

salvo[1] /'sælvəʊ/ n. (pl. **-oes** or **-os**) **1** the simultaneous firing of artillery or other guns esp. as a salute, or in a sea-fight. **2** a number of bombs released from aircraft at the same moment. **3** a round of applause. [earlier salve. F f. It. salva salutation (as SAVE[1])]

salvo[2] /'sælvəʊ/ n. (pl. **-os**) **1** a saving clause; a reservation (with an express salvo of their rights). **2** a tacit reservation. **3** a quibbling evasion; a bad excuse. **4** an expedient for saving reputation or soothing pride or conscience. [L, ablat. of salvus SAFE as used in salvo jure without prejudice to the rights of (a person)]

sal volatile /ˌsæl və'lætɪlɪ/ n. ammonium carbonate, esp. in the form of a scented solution in alcohol used as smelling-salts. [mod.L, = volatile salt]

salvor /'sælvə(r)/ n. a person or ship making or assisting in salvage. [SALVE[2]]

Salween /'sælwi:n, sæl'wi:n/ a river of SE Asia, which rises in Tibet and flows for 2,400 km (1,500 miles) south-east and south through Burma to the Gulf of Martaban, an inlet of the Andaman Sea.

Salyut /sæ'lju:t, 'sæljuːt/ a series of seven Soviet manned orbiting space stations, launched between 1971 and 1982. [Russ., = salute]

Salzburg /'sæltsbɜ:g, 'sɔ:lts-/ a city in western Austria, near the border with Germany, the capital of a state of the same name; pop. (1991) 143,970. It is noted for its annual music festivals, one of which is dedicated to the composer Mozart, who was born in the city in 1756.

Salzgitter /'zælts.gɪtə(r)/ an industrial city in Germany, in Lower Saxony south-east of Hanover; pop. (1991) 115,380.

Salzkammergut /'zæltskæmə.guːt/ a resort area of lakes and mountains in the state of Salzburg in western Austria.

SAM abbr. surface-to-air missile.

Sam. abbr. (in the Bible) Samuel.

samadhi /sə'mɑːdɪ/ n. Buddhism & Hinduism **1** a state of concentration induced by meditation. **2** a state into which a perfected holy man is said to pass at his apparent death. [Skr. samādhi contemplation]

Samar /'sɑːmɑː(r)/ an island in the Philippines, situated to the south-east of Luzon. It is the third largest island of the group.

Samara /sə'mɑːrə/ a city and river port in SW central Russia, situated on the Volga at its confluence with the River Samara; pop. (1990) 1,258,000. Founded in 1586 as a fortress to protect trade routes along the Volga, it is now a major industrial city. It was known as Kuibyshev from 1935 to 1991.

samara /'sæmərə, sə'mɑːrə/ n. *Bot.* a winged seed from the sycamore, ash, etc. [mod.L f. L, = elm seed]

Samaria /sə'meəriə/ **1** an ancient city of central Palestine, founded in the 9th century BC as the capital of the northern Hebrew kingdom of Israel. It was taken and resettled in *c.*722 BC by the Assyrians under Sargon II (2 Kings 17, 18). Rebuilt by Herod the Great between 30 and 4 BC, it became a flourishing city in the Roman period, noted for its Hellenistic features. The ancient site is situated in the modern West Bank, north-west of Nablus. **2** the region of ancient Palestine around this city, bounded by Galilee in the north and Judaea in the south.

Samarinda /ˌsæmə'rɪndə/ a city in Indonesia, in eastern Borneo; pop. (1980) 264,700.

Samaritan /sə'mærɪt(ə)n/ n. & adj. ● n. **1** (in full **good Samaritan**) a charitable or helpful person (with ref. to Luke 10:33 etc.). **2** a member of an organization (**the Samaritans**) founded in 1953 by Revd. Chad Varah to enable help and support to be given (especially through a telephone service) to the suicidal and despairing. **3 a** a native or inhabitant of Samaria. **b** an adherent of the Samaritan religious system. (*See note below*) **4** the Aramaic dialect formerly spoken in Samaria. ● adj. of Samaria or the Samaritans. □ **Samaritan Pentateuch** a recension used by Samaritans of which the manuscripts are in archaic characters. □ **Samaritanism** n. [LL *Samaritanus* f. Gk *Samareitēs* f. *Samareia* Samaria]

▪ According to Jewish tradition, the Samaritans were descendants of Mesopotamian peoples who settled in Samaria after the Assyrian conquest in *c.*722 BC; the hostility of the Jews to them was proverbial and forms the background to Christ's parable of the Good Samaritan (Luke 10:33). They adopted a form of Judaism which accepted only the Pentateuch. A few survive, living near Nablus in the West Bank.

samarium /sə'meəriəm/ n. a hard silvery-white metallic chemical element (atomic number 62; symbol **Sm**). A member of the lanthanide series, samarium was discovered by the French chemist Paul-Émile Lecoq de Boisbaudran (1838–1912) in 1879. The pure metal is used as a catalyst and its compounds are used in computer hardware, some types of electrode, and special kinds of glass. [f. *samarskite* the mineral in which its spectrum was first observed, f. *Samarsky* name of a 19th-c. Russian mining official]

Samarkand /ˌsæmɑː'kænd, 'sæməˌkænd/ (also **Samarqand**) a city in eastern Uzbekistan; pop. (1990) 369,900. One of the oldest cities of Asia, it was founded in the 3rd or 4th millennium BC. It was destroyed by Alexander the Great in 329 BC, but later grew to prominence as a prosperous centre of the silk trade, situated on the Silk Road, the caravan route to the east. It was destroyed again by Genghis Khan in 1221 but in the 14th century became the capital of Tamerlane's Mongol empire. By 1700 it was almost deserted, but in 1868 it was developed as a provincial capital of the Russian empire. Between 1924 and 1930 it was the capital of the Uzbek Soviet Socialist Republic.

Samarra /sə'mɑːrə/ a city in Iraq, on the River Tigris north of Baghdad; pop. (est. 1985) 62,000. Its 17th-century mosque is a place of Shiite pilgrimage.

Sama-veda /ˌsɑːmə'veɪdə, -'viːdə/ *Hinduism* one of the four Vedas, a collection of liturgical chants in early Sanskrit used in the Vedic religion by the priest in charge of chanting aloud at the sacrifice. Its material is drawn largely from the Rig-veda. [Skr. *sāmaveda* f. *sāman* chant + *veda* VEDA]

samba /'sæmbə/ n. & v. ● n. **1** a Brazilian dance of African origin. **2** a ballroom dance imitative of this. **3** the music for this. ● v.intr. (**sambas**, **sambaed** /-bəd/ or **samba'd**, **sambaing** /-bəɪŋ/) dance the samba. [Port., of Afr. orig.]

sambar /'sæmbə(r)/ n. (also **sambur**) a large deer, *Cervus unicolor*, native to southern Asia. [Hindi *sā(m)bar*]

Sambo /'sæmbəʊ/ n. (pl. **-os** or **-oes**) **1** *sl. offens.* a black person. **2** (**sambo**) *hist.* a person of mixed race esp. of Negro and Indian or Negro and European blood. [Sp. *zambo* perh. = *zambo* bandy-legged; sense 1 perh. a different word f. Foulah *sambo* uncle]

Sam Browne /sæm 'braʊn/ n. (in full **Sam Browne belt**) a belt with a supporting strap that passes over the right shoulder, worn by commissioned officers of the British army and also by members of various police forces etc. It is named after its inventor, the British military commander Sir Samuel J. Brown (1824–1901).

same /seɪm/ adj., pron., & adv. ● adj. **1** (often prec. by *the*) identical; not different; unchanged (*everyone was looking in the same direction; the same car was used in another crime; saying the same thing over and over*). **2** unvarying, uniform, monotonous (*the same old story*). **3** (usu. prec. by *this, these, that, those*) (of a person or thing) previously alluded to; just mentioned; aforesaid (*this same man was later my husband*). ● pron. (prec. by *the*) **1** the same person or thing (*the others asked for the same*). **2** *Law* or *archaic* the person or thing just mentioned (*detected the youth breaking in and apprehended the same*). ● adv. (usu. prec. by *the*) similarly; in the same way (*we all feel the same; I want to go, the same as you do*). □ **all** (or **just**) **the same 1** emphatically the same. **2** in spite of changed conditions, adverse circumstances, etc. (*but you should offer, all the same*). **at the same time 1** simultaneously. **2** notwithstanding; in spite of circumstances etc. **be all** (or **just**) **the same to** an expression of indifference or impartiality (*it's all the same to me what we do*). **by the same token** see TOKEN. **same here** *colloq.* the same applies to me. **the same to you!** may you do, have, find, etc., the same thing; likewise. **the very same** emphatically the same. □ **sameness** n. [ME f. ON *sami, sama*, with Gmc cognates]

samey /'seɪmɪ/ adj. (**samier, samiest**) *colloq.* lacking in variety; monotonous. □ **sameyness** n.

samfu /'sæmfuː/ n. a suit consisting of a jacket and trousers, worn by Chinese women and sometimes men. [Cantonese]

Samhain /saʊn, 'saʊɪn/ n. *Brit.* 1 November, celebrated by the ancient Celts as a festival marking the beginning of winter and their new year. [Ir.]

Sami /'sɑːmɪ, sɑːm/ n. & adj. ● n.pl. the Lapp people; Lapps. ● adj. of or relating to the Lapps. [Lappish]

Samian /'seɪmɪən/ n. & adj. ● n. **1** a native or inhabitant of Samos. **2** (in full **Samian pottery** or **ware**) a kind of glossy red pottery made chiefly in Gaul in the 1st–4th centuries and common in archaeological sites of this period. Also called *terra sigillata*. ● adj. of Samos. [L *Samius* f. Gk *Samios* Samos]

samisen /'sæmɪsɪn/ n. a long three-stringed Japanese guitar, played with a plectrum. [Jap. f. Chin. *san-hsien* f. *san* three + *hsien* string]

samite /'sæmaɪt, 'seɪm-/ n. *hist.* a rich medieval dress fabric of silk sometimes interwoven with gold. [ME f. OF *samit* f. med.L *examitum* f. med. Gk *hexamiton* f. Gk *hexa-* six + *mitos* thread]

samizdat /'sæmɪzˌdæt/ n. **1** the clandestine copying and distribution of literature, esp. in the former communist countries of eastern Europe. **2** literature so produced. [Russ., = self-publishing house]

Samnite /'sæmnaɪt/ n. & adj. ● n. **1** a member of a people of ancient Italy often at war with republican Rome. **2** the language of this people. ● adj. of this people or their language. [ME f. L *Samnites* (pl.), rel. to *Sabinus* SABINE]

Samoa /sə'məʊə/ a group of islands in Polynesia. The group was divided administratively in 1899 into American Samoa in the east and German Samoa in the west. The latter, mandated to New Zealand in 1919, gained independence in 1962 as Western Samoa. See also AMERICAN SAMOA and WESTERN SAMOA.

Samoan /sə'məʊən/ n. & adj. ● n. **1** a native of Samoa. **2** the language of this people. ● adj. of or relating to Samoa or its people or language.

Samos /'seɪmɒs/ a Greek island in the Aegean, situated close to the coast of western Turkey.

samosa /sə'məʊsə/ n. a triangular pastry fried in ghee or oil, containing spiced vegetables or meat. [Hind.]

samovar /'sæməˌvɑː(r)/ n. a Russian tea urn, with an internal heating device to keep the water at boiling-point. [Russ., = self-boiler]

Samoyed /'sæməˌjed/ n. **1** a member of a people of northern Siberia. **2** the language of this people. **3** (also **samoyed**) a white Arctic breed of dog. [Russ. *samoed*]

Samoyedic /ˌsæmə'jedɪk/ n. & adj. ● n. the language of the Samoyeds. ● adj. of or relating to the Samoyeds.

samp /sæmp/ n. *US* **1** coarsely ground maize. **2** porridge made of this. [Algonquian *nasamp* softened by water]

sampan /'sæmpæn/ n. a small boat usu. with a stern-oar or stern-oars, used in the Far East. [Chin. *san-ban* f. *san* three + *ban* board]

samphire /'sæmˌfaɪə(r)/ n. **1** an umbelliferous maritime rock plant, *Crithmum maritimum*, with aromatic fleshy leaves used in pickles. **2** a glasswort of the genus *Salicornia*. [earlier *samp(i)ere* f. F (*herbe de*) *Saint Pierre* St Peter's herb]

sample /'sɑːmp(ə)l/ n. & v. ● n. **1** (also *attrib.*) a small part or quantity intended to show what the whole is like. **2** a small amount of fabric, food, or other commodity, esp. given to a prospective customer. **3** a specimen, esp. one taken for scientific testing or analysis. **4** an illustrative or typical example. **5** *Electronics* a sound, esp. a piece of music, created by sampling. ● *v.tr.* **1** take or give samples of. **2** try the qualities of. **3** get a representative experience of. **4** (usu. as **sampling** *n.*) *Electronics* digitally encode (a piece of sound) for re-use in a musical composition or recording. □ **sample bag** *Austral.* a bag of advertisers' samples. **sampling frame** *Statistics* a list of the items, people, etc., forming a population from which a sample is taken. [ME f. AF *assample*, OF *essample* EXAMPLE]

sampler¹ /ˈsɑːmplə(r)/ *n.* a piece of embroidery worked in various stitches as a specimen of proficiency (often displayed on a wall etc.). [OF *essamplaire* (as EXEMPLAR)]

sampler² /ˈsɑːmplə(r)/ *n.* **1** a person who samples. **2** an electronic device used for sampling music and sound. **3** *US* a collection of representative items etc.

samsara /somˈsɑːrə/ *n. Hinduism* the endless cycle of death and rebirth to which life in the material world is bound. □ **samsaric** /-ˈsærɪk/ *adj.* [Skr. *saṃsāra* a wandering through]

samskara /səmˈskɑːrə/ *n. Hinduism* **1** a purificatory ceremony or rite marking an event in one's life. **2** a mental impression, instinct, or memory. [Skr. *saṃskāra* a making perfect, preparation]

Samson /ˈsæms(ə)n/ *n.* (in the Bible) an Israelite leader (prob. 11th century BC) famous for his strength (Judges 13–16). He fell in love with Delilah and confided to her that his strength lay in his hair, and she betrayed him to the Philistines. They cut off his hair while he slept and captured and blinded him, but when his hair grew again his strength returned and he pulled down the pillars of a house, destroying himself and a large gathering of Philistines. □ **Samson** (or **Samson's**) **post 1** a strong pillar passing through the hold of a ship or between decks. **2** a post in a whaleboat to which a harpoon rope is attached.

Samuel /ˈsæmjʊəl/ *n.* **1** a Hebrew prophet who rallied the Israelites after their defeat by the Philistines and became their ruler. **2** either of two books of the Bible covering the history of ancient Israel from Samuel's birth to the end of the reign of David.

samurai /ˈsæmʊˌraɪ, ˈsæmjʊ-/ *n.* (*pl.* same) **1** a Japanese army officer. **2** *hist.* a member of the feudal warrior class of Japan which was bound by the code of bushido, emphasizing qualities of loyalty, bravery, and endurance. The samurai dominated Japanese society until the demise of the feudal order in the 19th century. [Jap.]

San /sɑːn/ *n. & adj.* ● *n.* (*pl.* same) **1** a member of the aboriginal Bushmen of southern Africa. **2** the group of Khoisan languages spoken by the San; any of these languages. ● *adj.* of or relating to the San or their languages. [Nama]

san /sæn/ *n.* = SANATORIUM 2. [abbr.]

Sana'a /sæˈnɑː, ˈsɑːnə/ (also **Sanaa**) the capital of Yemen; pop. (1990) 500,000. It is noted for its medina, one of the largest completely preserved walled city centres in the Arab world.

San Andreas fault /ˌsæn ænˈdreɪəs/ a fault line or fracture of the earth's crust extending for some 965 km (600 miles) through the length of California. Seismic activity is common along its course and is ascribed to movement of two sections of the earth's crust — the eastern Pacific plate and the North American plate — which abut against each other in this region (see PLATE TECTONICS). The city of San Francisco lies close to the fault, and such movement caused the devastating earthquake of 1906 and a further convulsion in 1989.

San Antonio /ˌsæn ænˈtəʊnɪˌəʊ/ an industrial city in south central Texas; pop. (1990) 935,930. It is the site of the Alamo mission (see ALAMO, THE).

sanative /ˈsænətɪv/ *adj.* **1** healing; curative. **2** of or tending to physical or moral health. [ME f. OF *sanatif* or LL *sanativus* f. L *sanare* cure]

sanatorium /ˌsænəˈtɔːrɪəm/ *n.* (*pl.* **sanatoriums** or **sanatoria** /-rɪə/) **1** an establishment for the treatment of invalids, esp. of convalescents and the chronically sick. **2** *Brit.* a room or building for sick people in a school etc. [mod.L (as SANATIVE)]

San Carlos de Bariloche /sæn ˌkɑːlɒs deɪ ˌbærɪˈləʊtʃɪ/ a ski resort in the Andes of Argentina, on the shores of Lake Nahuel Huapi; pop. (1980) 48,200.

Sanchi /ˈsɑːntʃɪ/ the site in central India, in Madhya Pradesh, of several well-preserved Buddhist stupas. The largest of these was probably begun by the Emperor Asoka in the 3rd century BC.

Sancho Panza /ˌsæntʃəʊ ˈpænzə/ the squire of Don Quixote, who accompanies the latter on his adventures. He is an uneducated and

credulous peasant but has a store of proverbial wisdom, and is thus a foil to his master.

sanctify /ˈsæŋktɪˌfaɪ/ *v.tr.* (**-ies, -ied**) **1** consecrate; set apart or observe as holy. **2** purify or free from sin. **3** make legitimate or binding by religious sanction; justify; give the colour of morality or innocence to. **4** make productive of or conducive to holiness. □ **sanctifier** *n.* **sanctification** /ˌsæŋktɪfɪˈkeɪʃ(ə)n/ *n.* [ME f. OF *saintifier* f. eccl.L *sanctificare* f. L *sanctus* holy]

sanctimonious /ˌsæŋktɪˈməʊnɪəs/ *adj.* making a show of sanctity or piety. □ **sanctimoniously** *adv.* **sanctimoniousness** *n.* **sanctimony** /ˈsæŋktɪmənɪ/ *n.* [L *sanctimonia* sanctity (as SAINT)]

sanction /ˈsæŋkʃ(ə)n/ *n. & v.* ● *n.* **1** approval or encouragement given to an action etc. by custom or tradition; express permission. **2** confirmation or ratification of a law etc. **3 a** a penalty for disobeying a law or rule, or a reward for obeying it. **b** a clause containing this. **4** a consideration operating to enforce obedience to any rule of conduct. **5** (esp. in *pl.*) military or esp. economic action by a state to coerce another to conform to an international agreement or norms of conduct. **6** *Law hist.* a law or decree. ● *v.tr.* **1** authorize, countenance, or agree to (an action etc.). **2** ratify; attach a penalty or reward to; make binding. □ **sanctionable** *adj.* [F f. L *sanctio -onis* f. *sancire sanct-* make sacred]

sanctitude /ˈsæŋktɪˌtjuːd/ *n. archaic* saintliness. [ME f. L *sanctitudo* (as SAINT)]

sanctity /ˈsæŋktɪtɪ/ *n.* (*pl.* **-ies**) **1** holiness of life; saintliness. **2** sacredness; the state of being hallowed. **3** inviolability. **4** (in *pl.*) sacred obligations, feelings, etc. [ME f. OF *sain(c)tité* or L *sanctitas* (as SAINT)]

sanctuary /ˈsæŋktjʊərɪ/ *n.* (*pl.* **-ies**) **1** a holy place; a church, temple, etc. **2 a** the inmost recess or holiest part of a temple etc. **b** the part of the chancel containing the high altar. **3** a place where birds, wild animals, etc., breed and are protected. **4** a place of refuge, esp. for political refugees. **5 a** immunity from arrest. **b** the right to offer this. **6** *hist.* a sacred place where a fugitive from the law or a debtor was secured by medieval Church law against arrest or violence. □ **take sanctuary** resort to a place of refuge. [ME f. AF *sanctuarie*, OF *sanctuaire* f. L *sanctuarium* (as SAINT)]

sanctum /ˈsæŋktəm/ *n.* (*pl.* **sanctums**) **1** a holy place. **2** *colloq.* a person's private room, study, or den. □ **sanctum sanctorum** /sæŋkˈtɔːrəm/ **1** the holy of holies in the Jewish temple. **2** = sense 2 of *n.* **3** an inner retreat. **4** an esoteric doctrine etc. [L, neut. of *sanctus* holy, past part. of *sancire* consecrate: *sanctorum* genitive pl. in transl. of Heb. *ḳōḏeš haḳḳoḏāšîm* holy of holies]

sanctus /ˈsæŋktəs/ *n.* (also **Sanctus**) *Eccl.* **1** the prayer or hymn (from Isa. 6:3) beginning 'Sanctus, sanctus, sanctus' or 'Holy, holy, holy', forming the conclusion of the Eucharistic preface. **2** the music for this. □ **sanctus bell** a handbell or the bell in the turret of a church at the junction of the nave and the chancel, rung at the sanctus or at the elevation of the Eucharist. [ME f. L, = holy]

Sand /sɒnd/, George (pseudonym of Amandine-Aurore Lucille Dupin, Baronne Dudevant) (1804–76), French novelist. In 1831 she left her husband to lead an independent literary life in Paris. Her earlier romantic novels, including *Lélia* (1833), portray women's struggles against conventional morals; she later wrote a number of pastoral novels (e.g. *La Mare au diable*, 1846). Among her other works are *Elle et lui* (1859), a fictionalized account of her affair with the poet Alfred de Musset (1810–57), and *Un Hiver à Majorque* (1841), describing an episode during her ten-year relationship with Chopin.

sand /sænd/ *n. & v.* ● *n.* **1** a loose granular substance resulting from the wearing down of esp. siliceous rocks and found on the seashore, river beds, deserts, etc. **2** (in *pl.*) grains of sand. **3** (in *pl.*) an expanse or tracts of sand. **4** a light yellow-brown colour like that of sand. **5** (in *pl.*) a sandbank. **6** *US colloq.* firmness of purpose; grit. ● *v.tr.* **1** smooth or polish with sand or sandpaper, or a mechanical sander. **2** sprinkle or overlay with, or bury under, sand. **3** adulterate (sugar etc.) with sand. □ **sand bar** a sandbank at the mouth of a river or (*US*) on the coast. **sand-bath** a vessel of heated sand to provide uniform heating. **sand-bed** a stratum of sand. **sand-cloud** driving sand in a simoom. **sand-crack 1** a fissure in a horse's hoof. **2** a crack in the human foot from walking on hot sand. **3** a crack in brick due to imperfect mixing. **sand dollar** a flat asymmetrical sea urchin of the order Clypeasteroida. **sand-dune** (or **-hill**) a mound or ridge of sand formed by the wind. **sand eel** a burrowing eel-like fish of the family Ammodytidae (also called *launce*). **sand-flea** a jigger or a sand-hopper. **sand-glass** = HOURGLASS. **sand-groper** *Austral.* **1** a gold rush pioneer. **2** *joc.* a Western

Australian. **sand-hill** a dune. **sand-hopper** a small jumping crustacean of the order Amphipoda, burrowing on sandy shores. **sand martin** a swallow-like bird, *Riparia riparia*, nesting in holes in sandy banks etc. (also called (*N. Amer.*) bank swallow). **the sands are running out** the allotted time is nearly at an end. **sand-shoe** a shoe with a canvas, rubber, hemp, etc., sole for use on sand. **sand-skipper** = *sand-hopper*. **sand spurrey** see SPURREY. **sand-yacht** a boat on wheels propelled along a beach by wind. □ **sandlike** *adj.* [OE *sand* f. Gmc]

sandal[1] /'sænd(ə)l/ *n. & v.* ● *n.* **1** a light shoe with an openwork upper or no upper, attached to the foot usu. by straps. **2** a strap for fastening a low shoe, passing over the instep or around the ankle. ● *v.tr.* (**sandalled, sandalling**; *US* **sandaled, sandaling**) **1** (esp. as **sandalled** *adj.*) put sandals on (a person, a person's feet). **2** fasten or provide (a shoe) with a sandal. [ME f. L *sandalium* f. Gk *sandalion* dimin. of *sandalon* wooden shoe, prob. of Asiatic orig.]

sandal[2] /'sænd(ə)l/ *n.* = SANDALWOOD 1. □ **sandal-tree** a tree that yields sandalwood, esp. the white sandalwood. [ME f. med.L *sandalum*, ult. f. Skr. *candana*]

sandalwood /'sænd(ə)l,wʊd/ *n.* **1 a** the scented wood of a tree of the genus *Santalum*; esp. the *white sandalwood* (*S. album*) of India. **b** a perfume or incense derived from this. **2** any other tree whose wood resembles sandalwood. □ **red sandalwood** the red wood from either of two SE Asian leguminous trees, *Adenanthera pavonina* and *Pterocarpus santalinus*, used as timber and to produce a red dye (cf. SANDERS). **sandalwood oil** a yellow aromatic oil made from the white sandalwood.

Sandalwood Island an alternative name for SUMBA.

sandarac /'sændə,ræk/ *n.* (also **sandarach**) **1** the gummy resin of a North African conifer, *Tetraclinis articulata*, used in making varnish. **2** = REALGAR. [L *sandaraca* f. Gk *sandarakē*, of Asiatic orig.]

sandbag /'sændbæg/ *n. & v.* ● *n.* a bag filled with sand for use: **1** (in fortification) for making temporary defences or for the protection of a building etc. against blast and splinters or floodwaters. **2** as ballast esp. for a boat or balloon. **3** as a weapon to inflict a heavy blow without leaving a mark. **4** to stop a draught from a window or door. ● *v.* (**-bagged, -bagging**) **1** *tr.* barricade or defend. **2** *tr.* place sandbags against (a window, chink, etc.). **3** *tr.* fell with a blow from a sandbag. **4** *tr. N. Amer.* coerce by harsh means. **5** *intr. Sport* deliberately underperform in a race or competition to gain an unfair advantage. □ **sandbagger** *n.*

sandbank /'sændbæŋk/ *n.* a deposit of sand forming a shallow place in the sea or a river.

sandblast /'sændblɑːst/ *v. & n.* ● *v.tr.* (usu. as **sandblasted** *adj.*) roughen, treat, or clean with a jet of sand driven by compressed air or steam. ● *n.* this jet. □ **sandblaster** *n.*

sandbox /'sændbɒks/ *n.* **1** *Railways* a box of sand on a locomotive for sprinkling slippery rails. **2** *Golf* a container for sand used in teeing. **3** a sandpit enclosed in a box. **4** *hist.* a device for sprinkling sand to dry ink.

sandboy /'sændbɔɪ/ *n.* □ **happy as a sandboy** extremely happy or carefree. [prob. = a boy hawking sand for sale]

sandcastle /'sænd,kɑːs(ə)l/ *n.* a shape like a castle made in sand, usu. by a child on the seashore.

sander /'sændə(r)/ *n.* a power tool for smoothing surfaces.

sanderling /'sændəlɪŋ/ *n.* a small wading bird, *Calidris alba*, of the sandpiper family. [perh. f. an OE form *sandyrthling* (unrecorded, as SAND + *yrthling* ploughman, also the name of a bird)]

sanders /'sændəz/ *n.* (also **saunders** /'sɔːn-/) the red sandalwood, *Pterocarpus santalinus*. [ME f. OF *sandre* var. of *sandle* SANDAL[2]]

sandfly /'sændflaɪ/ *n.* (*pl.* **-flies**) **1** a biting blackfly of the genus *Simulium*. **2** a tropical biting fly of the genus *Phlebotomus*, which transmits the protozoal disease leishmaniasis.

sandgrouse /'sændgraʊs/ *n.* a seed-eating ground bird of the family Pteroclididae, related to the pigeons and found in arid regions of the Old World.

sandhi /'sændɪ/ *n. Gram.* the process whereby the form of a word changes as a result of its position in an utterance (e.g. the change from *a* to *an* before a vowel). [Skr. *saṃdhi* putting together]

sandhog /'sændhɒg/ *n. US* a person who works under water laying foundations, constructing tunnels, etc.

Sandhurst /'sændhɜːst/ the Royal Military Academy, Sandhurst, a training college, now at Camberley, Surrey, for officers for the British army. It was formed in 1946 from an amalgamation of the Royal Military College at Sandhurst in Berkshire (founded 1799) and the Royal Military Academy at Woolwich, London (founded 1741).

San Diego /,sæn dɪ'eɪgəʊ/ an industrial city and naval port on the Pacific coast of southern California, just north of the border with Mexico; pop. (1990) 1,110,550. It was founded as a mission in 1769.

Sandinista /,sændɪ'niːstə/ *n.* a member of a left-wing Nicaraguan political organization, the Sandinista National Liberation Front (FSLN). Led by Daniel Ortega, the Sandinistas came to power in 1979 after overthrowing the dictator Anastasio Somoza. Opposed during most of their period of rule by the US-backed Contras, the Sandinistas confirmed their position by winning elections in 1984 but were voted out of office in 1990. They took their name from a similar organization founded by the nationalist leader Augusto César Sandino (1893–1934).

sandiver /'sændɪvə(r)/ *n.* liquid scum formed in glass-making. [ME app. f. F *suin de verre* exhalation of glass f. *suer* to sweat]

sandlot /'sændlɒt/ *n. US* a piece of unoccupied sandy land used for children's games.

sandman /'sændmæn/ *n.* the personification of tiredness causing children's eyes to smart towards bedtime.

sand painting *n.* **1** an American Indian ceremonial art form, using coloured sands. Also called *dry painting*. (*See note below.*) **2** an example of this.

▪ Sand painting is usually used in connection with healing ceremonies. The colours and techniques used are traditional, and the greater part of the design follows patterns handed down from memory, each sand painting being destroyed after the ceremony. Sand painting continues to be especially important among Navajo and Pueblo peoples.

sandpaper /'sænd,peɪpə(r)/ *n. & v.* ● *n.* paper with sand or another abrasive stuck to it for smoothing or polishing. ● *v.tr.* smooth with sandpaper.

sandpiper /'sænd,paɪpə(r)/ *n.* a small wading bird of the family Scolopacidae, esp. of the genus *Calidris* or *Tringa*.

sandpit /'sændpɪt/ *n.* a hollow partly filled with sand, usu. for children to play in.

Sandringham House /'sændrɪŋəm/ a country residence of the British royal family, north-east of King's Lynn in Norfolk. The estate was acquired in 1861 by Edward VII, then Prince of Wales.

sandsoap /'sændsəʊp/ *n.* heavy-duty gritty soap.

sandstock /'sændstɒk/ *n.* brick made with sand dusted on the surface.

sandstone /'sændstəʊn/ *n.* **1** any clastic rock containing particles visible to the naked eye. **2** a sedimentary rock of consolidated sand commonly red, yellow, brown, grey, or white.

sandstorm /'sændstɔːm/ *n.* a desert storm of wind with clouds of sand.

sandwich /'sænwɪdʒ, -wɪtʃ/ *n. & v.* ● *n.* **1** an item of food comprising two or more slices of sun. buttered bread with a filling of meat, cheese, etc., between them. **2** a cake of two or more layers with jam or cream between (*bake a sponge sandwich*). ● *v.tr.* **1** put (a thing, statement, etc.) between two of another character. **2** squeeze in between others (*sat sandwiched in the middle*). □ **open sandwich** see OPEN. **sandwich-board** one of two advertisement boards carried by a sandwich-man. **sandwich course** a course of training with alternate periods of practical experience and theoretical instruction. **sandwich-man** (*pl.* **-men**) a man who walks the streets with two sandwich-boards hanging from his shoulders, one in front and one behind. [4th Earl of *Sandwich* (1718–92), an English nobleman said to have eaten food in this form so as not to leave the gaming-table for 24 hours]

Sandwich Islands a former name for HAWAII.

Sandwich tern *n.* a crested moderately large tern, *Sterna sandvicensis*, found in both Old and New Worlds. [*Sandwich*, a town in Kent, England]

sandwort /'sændwɜːt/ *n.* a low-growing plant of the genus *Arenaria* (pink family), usu. bearing small white flowers.

sandy /'sændɪ/ *adj.* (**sandier, sandiest**) **1** having the texture of sand. **2** having much sand. **3 a** (of hair) yellowish-red. **b** (of a person) having sandy hair. □ **sandy blight** *Austral.* conjunctivitis with sandlike grains in the eye. □ **sandiness** *n.* **sandyish** *adj.* [OE *sandig* (as SAND)]

sane /seɪn/ *adj.* **1** of sound mind; not mad. **2** (of views etc.) moderate; sensible. □ **sanely** *adv.* **saneness** *n.* [L *sanus* healthy]

San Francisco /,sæn fræn'sɪskəʊ/ a city and seaport on the coast of California, situated on a peninsula between the Pacific and San Francisco Bay; pop. (1990) 723,960. Founded as a mission by Mexican Jesuits in 1776, it was taken by the US in 1846. The fine natural harbour of San Francisco Bay is entered by a channel known as the Golden Gate, which is spanned by a noted suspension bridge (built 1937). The city suffered severe damage from an earthquake and subsequent fire in 1906, and has been frequently shaken by less severe earthquakes since. □ **San Franciscan** *n. & adj.*

sang *past of* SING.

sangar /'sæŋgə(r)/ *n.* (also **sanga** /-gə/) a stone breastwork round a hollow. [Pashto *sangar*]

sangaree /ˌsæŋgə'ri:/ *n.* a cold drink of wine diluted and spiced. [Sp. *sangría* SANGRIA]

Sanger /'sæŋə(r)/, Margaret (Higgins) (1883–1966), American birth-control campaigner. Her experiences as a nurse from 1912 prompted her two years later to distribute the pamphlet *Family Limitation* in defence of birth control. Legal proceedings were initiated against her for disseminating 'obscene' literature, but these were dropped in 1916. In the same year she founded the first American birth-control clinic in Brooklyn, serving as its president for seven years. She also set up the first World Population Conference in Geneva in 1927 and became the first president of the International Planned Parenthood Federation in 1953.

sang-froid /sɒŋ'frwʌ/ *n.* composure, coolness, etc., in danger or under agitating circumstances. [F, = cold blood]

sangha /'sʌŋgə/ *n.* the Buddhist monastic order, including monks, nuns, and novices. [Skr. *saṃgha* community]

Sango /'sæŋgəʊ/ *n. & adj.* ● *n.* **1** a dialect of Ngbandi. **2** a lingua franca developed from this and other dialects of Ngbandi, one of the official languages of the Central African Republic. ● *adj.* of or relating to Sango or its speakers. [Ngbandi]

sangrail /sæŋ'greɪl/ *n.* = GRAIL. [ME f. OF *saint graal* (as SAINT, GRAIL)]

sangria /sæŋ'gri:ə/ *n.* a Spanish drink of red wine with lemonade, fruit, etc. [Sp. *sangría* bleeding: cf. SANGAREE]

sanguinary /'sæŋgwɪnəri/ *adj.* **1** accompanied by or delighting in bloodshed. **2** bloody; bloodthirsty. **3** (of laws) inflicting death freely. □ **sanguinarily** *adv.* **sanguinariness** *n.* [L *sanguinarius* f. *sanguis -inis* blood]

sanguine /'sæŋgwɪn/ *adj. & n.* ● *adj.* **1** optimistic; confident. **2** (of the complexion) florid; bright; ruddy. **3** *hist.* one of the four humours, characterized by a predominance of blood and supposed to be indicated by a ruddy complexion and a courageous and hopeful amorous disposition. **4** *Heraldry* or *literary* blood-red. **5** *archaic* bloody; bloodthirsty. ● *n.* **1** a blood-red colour. **2** a crayon of chalk coloured red or flesh with iron oxide. □ **sanguinely** *adv.* **sanguineness** *n.* (both in sense 1 of *n.*). [ME f. OF *sanguin -ine* blood-red f. L *sanguineus* (as SANGUINARY)]

sanguineous /sæŋ'gwɪnɪəs/ *adj.* **1** sanguinary. **2** *Med.* of or relating to blood. **3** blood-red. **4** full-blooded; plethoric. [L *sanguineus* (as SANGUINE)]

Sanhedrin /'sænɪdrɪn/ (also **Sanhedrim** /-drɪm/) the highest Jewish council in ancient Jerusalem, with religious, political, and judicial functions. It was responsible for pronouncing the sentence of death on Christ. [late Heb. *sanhedrîn* f. Gk *sunedrion* (as SYN-, *hedra* seat)]

sanicle /'sænɪk(ə)l/ *n.* an umbelliferous plant of the genus *Sanicula*, formerly believed to have healing properties; esp. *wood sanicle* (*S. europaea*). [ME ult. f. med.L *sanicula* perh. f. L *sanus* healthy]

sanify /'sænɪfaɪ/ *v.tr.* (**-ies, -ied**) make healthy; improve the sanitary state of. [L *sanus* healthy]

sanitarium /ˌsænɪ'teərɪəm/ *n.* (pl. **sanitariums** or **sanitaria** /-rɪə/) *US* = SANATORIUM. [pseudo-L f. L *sanitas* health]

sanitary /'sænɪtəri/ *adj.* **1** of the conditions that affect health, esp. with regard to dirt and infection. **2** hygienic; free from or designed to kill germs, infection, etc. □ **sanitary engineer** a person dealing with systems needed to maintain public health. **sanitary protection** sanitary towels and tampons regarded collectively. **sanitary towel** (*N. Amer.* **napkin**) an absorbent pad used during menstruation. **sanitary ware** porcelain for lavatories etc. □ **sanitarily** *adv.* **sanitariness** *n.*

sanitarian /ˌsænɪ'teərɪən/ *n. & adj.* [F *sanitaire* f. L *sanitas*: see SANITY]

sanitation /ˌsænɪ'teɪʃ(ə)n/ *n.* **1** sanitary conditions. **2** the maintenance or improving of these. **3** the disposal of sewage and refuse from houses etc. □ **sanitationist** *n.* **sanitate** /'sænɪteɪt/ *v.tr. & intr.* [irreg. f. SANITARY]

sanitize /'sænɪtaɪz/ *v.tr.* (also **-ise**) **1** make sanitary; disinfect. **2** render (information etc.) more acceptable by removing improper or disturbing material. □ **sanitizer** *n.* **sanitization** /ˌsænɪtaɪ'zeɪʃ(ə)n/ *n.*

sanity /'sænɪti/ *n.* **1 a** the state of being sane. **b** mental health. **2** the tendency to avoid extreme views. [ME f. L *sanitas* (as SANE)]

San Jose /ˌsæn həʊ'zeɪ/ a city in western California, situated to the south of San Francisco Bay; pop. (1990) 782,250. It lies in the Santa Clara valley, known as Silicon Valley, a centre of the electronics industries.

San José /ˌsæn həʊ'zeɪ/ the capital and chief port of Costa Rica; pop. (est. 1984) 284,550. Founded in 1737, it was made the capital of the newly independent state in 1823.

San Juan /ˌsæn 'hwɑːn/ the capital and chief port of Puerto Rico, on the north coast of the island; pop. (1990) 437,745.

sank *past of* SINK.

San Luis Potosí /ˌsæn luːˌiːs ˌpɒtəʊ'siː/ **1** a state of central Mexico. **2** its capital; pop. (1990) 525,820. It is an industrial city and a centre for the surrounding silver mines.

San Marino /ˌsæn mə'riːnəʊ/ a republic forming a small enclave in Italy, near Rimini; pop. (est. 1991) 22,680; official language, Italian; capital, the town of San Marino. It is perhaps Europe's oldest state, claiming to have been independent almost continuously since its foundation in the 4th century, according to tradition by a Dalmatian stonecutter, Marino, who fled there to escape the persecution of Christians under Diocletian.

San Martín /ˌsæn mɑː'tiːn/, José de (1778–1850), Argentinian soldier and statesman. Having assisted in the liberation of his country from Spanish rule (1812–13) he went on to aid Bernardo O'Higgins in the liberation of Chile (1817–18). He was also involved in gaining Peruvian independence, becoming Protector of Peru in 1821; he resigned a year later after differences with the other great liberator Bolívar.

sannyasi /sʌ'njɑːsɪ/ *n.* (also **sanyasi**) (pl. same) a Hindu religious mendicant. [Skr. *saṃnyāsī* f. *saṃnyāsin* laying aside, ascetic, f. *saṃ* together + *ni* down + *as* throw]

San Pedro Sula /ˌsæn ˌpedrəʊ 'suːlə/ a city in northern Honduras, near the Caribbean coast; pop. (1988) 460,600.

sanpro /'sænprəʊ/ *n. colloq.* = sanitary protection.

sans /sænz, sɒn/ *prep. archaic* or *joc.* without. [ME f. OF *san(z), sen(s)* ult. f. L *sine*, infl. by L *absentia* in the absence of]

San Salvador /ˌsæn 'sælvəˌdɔː(r)/ the capital of El Salvador; pop. (1989) 1,522,000. Founded by the Spanish in the 1520s, it became the country's capital in 1839. It has been subject to several destructive earthquakes.

sansculotte /ˌsænzkjʊ'lɒt, ˌsɒnkjʊ-/ *n.* **1** *hist.* a lower-class Parisian republican in the French Revolution. **2** an extreme republican or revolutionary. □ **sansculottism** *n.* [F, lit. 'without knee-breeches']

San Sebastián /ˌsæn sɪ'bæstɪən/ a port and resort in northern Spain, situated on the Bay of Biscay close to the border with France; pop. (1991) 174,220.

sanserif /sæn'serɪf/ *n. & adj.* (also **sans-serif**) *Printing* ● *n.* a form of type without serifs. ● *adj.* without serifs. [app. f. SANS + SERIF]

Sanskrit /'sænskrɪt/ *n. & adj.* ● *n.* an ancient language of the Indian subcontinent, belonging to the Indic group of the Indo-European family of languages. (*See note below.*) ● *adj.* of or in this language. □ **Sanskritist** *n.* **Sanskritic** /sæn'skrɪtɪk/ *adj.* [Skr. *saṃskṛta* composed, elaborated, f. *saṃ* together, *kṛ* make, *-ta* past part. ending]

■ Now written mainly in the Devanagari script, Sanskrit is the sacred language of Hinduism. The earliest form (*Vedic Sanskrit*), spoken for about a thousand years from *c.*1800 BC, was the language in which the Vedas and Upanishads were written. Classical Sanskrit, the language of the great Hindu epics the Mahabharata and the Ramayana, flourished *c.*500 BC–AD 1000; since then it has continued in use as a language of learning and religion, although gradually eclipsed by English and the modern languages (e.g. Hindi, Bengali, Gujarati) to which, as a spoken language, Sanskrit gave rise. Sanskrit remains a lingua franca among Hindu scholars, and is one of the languages recognized for official use in India.

Sansovino /ˌsænsə'viːnəʊ/, Jacopo Tatti (1486–1570), Italian sculptor and architect. He was city architect of Venice from 1529, where his buildings include the Palazzo Corner (1533) and St Mark's Library (begun 1536), all of which show the influence of his early training in Rome and the development of classical architectural style for contemporary use. His sculpture includes the colossal statues *Mars* and *Neptune* (1554–6) for the staircase of the Doges' Palace.

Santa Ana /ˌsæntə 'ænə/ **1** a city in El Salvador, situated close to the border with Guatemala; pop. (1989) 239,000. **2** a volcano in El Salvador, situated south-west of the city of Santa Ana. It rises to a height of 2,381 m (7,730 ft). **3** a city in southern California, south-east of Los Angeles;

pop. (1990) 293,740. Lying to the east of the city are the Santa Ana Mountains. The region gives its name to the hot dry winds which blow from the mountains across the coastal plain of southern California.

Santa Barbara /ˌsæntə ˈbɑːbrə/ a resort city in California, on the Pacific coast north-west of Los Angeles; pop. (1990) 85,570.

Santa Catarina /ˌsæntə ˌkætəˈriːnə/ a state of southern Brazil, on the Atlantic coast; capital, Florianópolis.

Santa Claus /ˈsæntə ˌklɔːz/ (also colloq. **Santa**) an imaginary person said to bring children presents on the night before Christmas; Father Christmas. His name is an American corruption of a Dutch dialect form of St Nicholas. (See FATHER CHRISTMAS and NICHOLAS, ST.)

Santa Cruz /ˌsæntə ˈkruːz/ **1** a city in the central region of Bolivia; pop. (1990) 696,000. **2** (in full **Santa Cruz de Tenerife**) a port and the chief city of the island of Tenerife, in the Canary Islands; pop. (1991) 191,970.

Santa Fe /ˌsæntə ˈfeɪ/ (also **Santa Fé**) **1** the state capital of New Mexico; pop. (1990) 55,860. It was founded as a mission by the Spanish in 1610. From 1821 until the arrival of the railway in 1880 it was the terminus of the stagecoach route from Independence, Missouri, known as the Santa Fe Trail. Taken by US forces in 1846 during the Mexican War, it became the capital of New Mexico in 1912. **2** a city in northern Argentina, on the Salado river near its confluence with the Paraná; pop. (1991) 395,000.

Santa Monica /ˌsæntə ˈmɒnɪkə/ a resort city on the coast of SW California, situated on the west side of the Los Angeles conurbation; pop. (1990) 86,905.

Santander /ˌsæntænˈdeə(r)/ a port in northern Spain, on the Bay of Biscay, capital of Cantabria; pop. (1991) 194,220.

Santa Sophia see ST SOPHIA.

Santiago /ˌsæntɪˈɑːgəʊ/ the capital of Chile, situated to the west of the Andes in the central part of the country; pop. (1991) 5,343,000. Founded in 1541 by the Spanish conquistador Pedro de Valdivia (c.1498–1554), it became the capital of the newly independent republic in 1818.

Santiago de Compostela /ˌsæntɪˌɑːgəʊ deɪ ˌkɒmpɒˈstelə/ a city in NW Spain, capital in Galicia; pop. (1991) 105,530. It is named after St James the Great (Spanish Sant Iago), whose remains, according to Spanish tradition, were brought there after his death. From the 9th century, when the relics were discovered, the city became the centre of a national and Christian movement against the Moors and an important place of pilgrimage.

Santiago de Cuba /ˌsæntɪˌɑːgəʊ deɪ ˈkjuːbə/ a port on the coast of SE Cuba, the second largest city on the island; pop. (est. 1989) 974,100. Founded in 1514, it was moved to its present site in 1522 and was until 1589 the capital of the island.

Santo Domingo /ˌsæntəʊ dəˈmɪŋgəʊ/ the capital of the Dominican Republic, a port on the south coast; pop. (1986) 1,601,000. Founded in 1496 by the brother of Christopher Columbus, it is the oldest European settlement in the Americas. From 1936 to 1961 it was called Ciudad Trujillo.

santolina /ˌsæntəˈliːnə/ n. an aromatic composite shrub of the genus Santolina, with finely divided leaves and small usu. yellow flowers. [mod.L, var. of SANTONICA]

santonica /sænˈtɒnɪkə/ n. **1** a shrubby wormwood plant, Artemisia cina, yielding santonin. **2** the dried flower-heads of this used as an anthelmintic. [L f. Santones an Aquitanian tribe]

santonin /ˈsæntənɪn/ n. a toxic drug extracted from santonica and other plants of the genus Artemisia, used as an anthelmintic. [SANTONICA + -IN]

Santorini /ˌsæntɔːˈriːnɪ/ an alternative name for THERA.

Santos /ˈsæntɒs/ a port on the coast of Brazil, situated just south-east of São Paulo; pop. (1990) 546,630.

sanyasi var. of SANNYASI.

São Francisco /ˌsaʊ frænˈsɪsku:/ a river of eastern Brazil. It rises in Minas Gerais and flows for 3,200 km (1,990 miles) northwards then eastwards, meeting the Atlantic to the north of Aracajú.

São Luís /ˌsaʊ luˈiːs/ a port in NE Brazil, on the Atlantic coast, capital of the state of Maranhão; pop. (1990) 695,000.

Saône /səʊn/ a river of eastern France, which rises in the Vosges mountains and flows 480 km (298 miles) south-west to join the Rhône at Lyons. It has been made navigable for most of its length by canalization and the construction of locks.

São Paulo /ˌsaʊ ˈpaʊlu:/ **1** a state of southern Brazil, on the Atlantic

coast. **2** its capital city; pop. (1990) 9,700,110. It is the largest city in Brazil and second largest in South America.

São Tomé and Príncipe /ˌsaʊ tɒˈmeɪ, ˈprɪnsɪpɪ/ a country consisting of two main islands and several smaller ones in the Gulf of Guinea; pop. (est. 1991) 120,000; languages, Portuguese (official), Portuguese Creole; capital, São Tomé. The islands were settled by Portugal from 1493 and became an overseas province of that country. São Tomé and Príncipe became independent in 1975.

sap¹ /sæp/ n. & v. ● n. **1** the vital juice circulating in plants. **2** vigour; vitality. **3** = SAPWOOD. **4** US sl. a bludgeon (orig. one made from a sapling). ● v.tr. (**sapped, sapping**) **1** drain or dry (wood) of sap. **2** remove the sapwood from (a log). **3** US sl. hit with a sap. □ **sap-green** n. **1** the pigment made from buckthorn berries. **2** the colour of this. ● adj. of this colour. □ **sapful** adj. **sapless** adj. [OE sæp prob. f. Gmc]

sap² /sæp/ n. & v. ● n. **1** a tunnel or trench to conceal assailants' approach to a fortified place; a covered siege-trench. **2** an insidious or slow undermining of a belief, resolution, etc. ● v. (**sapped, sapping**) **1** intr. **a** dig a sap or saps. **b** approach by a sap. **2** tr. undermine; make insecure by removing the foundations. **3** tr. weaken or destroy (health, strength, courage, etc.) insidiously. [ult. f. It. zappa spade, spadework, in part through F sappe sap(p)er, prob. of Arab. orig.]

sap³ /sæp/ n. sl. a foolish person. [abbr. of sapskull f. SAP¹ = sapwood + SKULL]

sapanwood var. of SAPPANWOOD.

sapele /səˈpiːlɪ/ n. **1** a large West African hardwood tree of the genus Entandrophragma. **2** the reddish-brown mahogany-like timber of this tree. [W. Afr. name]

sapid /ˈsæpɪd/ adj. literary **1** having (esp. an agreeable) flavour; savoury; palatable; not insipid. **2** literary (of talk, writing, etc.) not vapid or uninteresting. □ **sapidity** /səˈpɪdɪtɪ/ n. [L sapidus f. sapere taste]

sapient /ˈseɪpɪənt/ adj. literary **1** wise. **2** aping wisdom; of fancied sagacity. □ **sapience** n. **sapiently** adv. [ME f. OF sapient or L part. stem of sapere be wise]

sapiential /ˌseɪpɪˈenʃ(ə)l, ˌsæp-/ adj. literary of or relating to wisdom. [ME f. F sapiential or eccl.L sapientialis f. L sapientia wisdom]

Sapir /səˈpɪə(r)/, Edward (1884–1939), German-born American linguistics scholar and anthropologist. One of the founders of American structural linguistics, he carried out important work on American Indian languages and linguistic theory. His book Language (1921) presents his thesis that language should be studied within its social and cultural context.

sapling /ˈsæplɪŋ/ n. **1** a young tree. **2** a youth. **3** a greyhound in its first year.

sapodilla /ˌsæpəˈdɪlə/ n. a large evergreen tropical American tree, Manilkara zapota, with edible fruit and durable wood, and sap from which chicle is obtained. □ **sapodilla plum** the fruit of this tree. [Sp. zapotillo dimin. of zapote f. Aztec tzápotl]

saponaceous /ˌsæpəˈneɪʃəs/ adj. **1** of, like, or containing soap; soapy. **2** joc. unctuous; flattering. [mod.L saponaceus f. L sapo -onis soap]

saponify /səˈpɒnɪˌfaɪ/ v. (**-ies, -ied**) Chem. **1** tr. turn (fat or oil) into soap by reaction with an alkali. **2** tr. convert (an ester) to an acid and alcohol. **3** intr. become saponified. □ **saponifiable** adj. **saponification** /-ˌpɒnɪfɪˈkeɪʃ(ə)n/ n. [F saponifier (as SAPONACEOUS)]

saponin /ˈsæpənɪn/ n. any of a group of plant glycosides, esp. those derived from the bark of the soapbark tree Quillaja saponaria, that foam when shaken with water and are used in detergents and fire extinguishers. [F saponine f. L sapo -onis soap]

sapor /ˈseɪpɔː(r)/ n. **1** a quality perceptible by taste, e.g. sweetness. **2** the distinctive taste of a substance. **3** the sensation of taste. [ME f. L sapere taste]

sappanwood /ˈsæp(ə)nˌwʊd/ n. (also **sapanwood**) the heartwood of an Indo-Malayan leguminous tree, Caesalpinia sappan, formerly used as a source of red dye. [Du. sapan f. Malay sapang, of S. Indian orig.]

sapper /ˈsæpə(r)/ n. **1** a person who digs saps. **2** Brit. a soldier of the Royal Engineers (esp. as the official term for a private).

Sapphic /ˈsæfɪk/ adj. & n. ● adj. **1** of or relating to Sappho or her poetry. **2** lesbian. ● n. (in pl.) (**sapphics**) verse in a metre associated with Sappho. [F sa(p)phique f. L Sapphicus f. Gk Sapphikos f. SAPPHO]

sapphire /ˈsæfaɪə(r)/ n. & adj. ● n. **1** a transparent blue precious stone consisting of corundum. **2** precious transparent corundum of any colour. **3** the bright blue of a sapphire. **4** a hummingbird with bright blue plumage, esp. of the genus Hylocharis. ● adj. of sapphire blue.

□ **sapphire wedding** a 45th wedding anniversary. □ **sapphirine** /ˈsæfəˌriːn, -rɪn/ adj. [ME f. OF safir f. L sapphirus f. Gk sappheiros prob. = lapis lazuli]

Sappho /ˈsæfəʊ/ (early 7th century BC) Greek lyric poet. She was renowned in her own day and became the centre of a circle of women and young girls on her native island of Lesbos. The surviving fragments of her poetry, written in her local dialect, are mainly love poems, dealing with subjects such as passion, jealousy, and enmity. Many of her poems express her affection and love for women, and have given rise to her association with female homosexuality, from which the words *lesbian* and *Sapphic* derive.

Sapporo /səˈpɔːrəʊ/ a city in northern Japan, capital of the island of Hokkaido; pop. (1990) 1,672,000.

sappy /ˈsæpɪ/ adj. (**sappier, sappiest**) **1** full of sap. **2** young and vigorous. □ **sappily** adv. **sappiness** n.

sapro- /ˈsæprəʊ/ comb. form Biol. rotten, putrefying. [Gk sapros putrid]

saprogenic /ˌsæprəˈdʒenɪk/ adj. causing or produced by putrefaction.

saprophagous /sæˈprɒfəgəs/ adj. feeding on decaying matter.

saprophile /ˈsæprəˌfaɪl/ n. a bacterium inhabiting putrid matter. □ **saprophilous** /sæˈprɒfɪləs/ adj.

saprophyte /ˈsæprəˌfaɪt/ n. a plant or micro-organism living on dead or decayed organic matter. □ **saprophytic** /ˌsæprəˈfɪtɪk/ adj.

sapsucker /ˈsæpˌsʌkə(r)/ n. a small American woodpecker of the genus *Sphyrapicus*, which pecks holes in trees and visits them for sap and insects.

sapwood /ˈsæpwʊd/ n. the soft outer layers of recently formed wood between the heartwood and the bark.

Saqqara /səˈkɑːrə/ a vast necropolis at the ancient Egyptian city of Memphis, with monuments dating from the early dynastic period (3rd millennium BC) to the Graeco-Roman age. Notable among them is the step pyramid of the 3rd-dynasty pharaoh Djoser (c.2686–c.2613 BC), the earliest type of pyramid and the first known building entirely of stone.

saraband /ˈsærəˌbænd/ n. **1** a stately old Spanish dance. **2** music for this or in its rhythm, usu. in triple time and often with a long note on the second beat of the bar. [F sarabande f. Sp. & It. zarabanda]

Saracen /ˈsærəs(ə)n/ n. & adj. hist. ● n. **1** an Arab or Muslim at the time of the Crusades. **2** a nomad of the Syrian and Arabian Desert. ● adj. of the Saracens. □ **Saracenic** /ˌsærəˈsenɪk/ adj. [ME f. OF sar(r)azin, sar(r)acin f. LL Saracenus f. late Gk Sarakēnos perh. f. Arab. šarkī eastern]

Saracen's head n. the head of a Saracen or Turk as a heraldic charge or inn-sign.

Saragossa /ˌsærəˈgɒsə/ (Spanish **Zaragoza** /ˌθarəˈgoθa/) a city in northern Spain, capital of Aragon, situated on the River Ebro; pop. (1991) 614,400. The ancient settlement there was taken in the 1st century BC by the Romans, who called it Caesaraugusta (from which the modern name is derived). It was a Moorish city from 714 to 1118, when it was captured by Alfonso I of Aragon.

Sarah /ˈseərə/ (in the Bible) the wife of Abraham and mother of Isaac (Gen. 17:15 ff.).

Sarajevo /ˌsærəˈjeɪvəʊ/ the capital of Bosnia–Herzegovina; pop. (est. 1993) 200,000. Taken by the Austro-Hungarians in 1878, it became a centre of Slav opposition to Austrian rule. It was the scene in June 1914 of the assassination by a Bosnian Serb named Gavrilo Princip of Archduke Franz Ferdinand (1863–1914), the heir to the Austrian throne, an event which triggered the outbreak of the First World War. The city suffered severely from the ethnic conflicts that followed the breakup of Yugoslavia in 1991, and was besieged by Bosnian Serb forces in the surrounding mountains from 1992 to 1994.

sarangi /səˈræŋɡɪ/ n. (pl. **sarangis**) an Indian stringed instrument played with a bow. [Hindi sāraṅgī]

Saransk /səˈrænsk/ a city in European Russia, capital of the autonomous republic of Mordvinia, situated to the south of Nizhni Novgorod; pop. (1990) 316,000. It was founded as a fortress town in 1641.

sarape var. of SERAPE.

Saratoga, Battle of /ˌsærəˈtəʊgə/ either of two battles fought in 1777 during the War of American Independence, near the modern city of Saratoga Springs in New York State. The Americans were victorious in both, and in the second battle the British forces, under General Burgoyne, capitulated. The defeat encouraged French support of the Americans and is often regarded as the turning point in the war in favour of the American side.

Saratov /səˈrɑːtɒf/ a city in SW central Russia, situated on the River

Volga north of Volgograd; pop. (1990) 909,000. Founded in 1590 on a nearby site as a fortress to protect trade routes along the Volga, it was moved to its present site following the destruction of the fortress in the 17th century.

Sarawak /səˈrɑːwək/ a state of Malaysia, comprising the north-western part of Borneo; capital, Kuching. From 1841 it was administered by a series of British rajas, the first of whom, James Brooke (1803–68), a former officer of the East India Company, had been awarded the title by the sultan of Brunei in recognition of his help in the suppression of a revolt. A British Crown Colony from 1946, Sarawak became part of the federation of Malaysia in 1963.

sarcasm /ˈsɑːkæz(ə)m/ n. **1** the use of bitter or wounding, esp. ironic, remarks; language consisting of such remarks (suffered from constant sarcasm about his work). **2** such a remark. □ **sarcastic** /sɑːˈkæstɪk/ adj. **sarcastically** adv. [F sarcasme or f. LL sarcasmus f. late Gk sarkasmos f. Gk sarkazō tear flesh, in late Gk gnash the teeth, speak bitterly f. sarx sarkos flesh]

sarcenet var. of SARSENET.

sarcoma /sɑːˈkəʊmə/ n. (pl. **sarcomas** or **sarcomata** /-mətə/) Med. a malignant tumour of connective or other non-epithelial tissue. □ **sarcomatous** adj. **sarcomatosis** /-ˌkəʊməˈtəʊsɪs/ n. [mod.L f. Gk sarkōma f. sarkoō become fleshy f. sarx sarkos flesh]

sarcophagus /sɑːˈkɒfəgəs/ n. (pl. **sarcophagi** /-ˌgaɪ, -ˌdʒaɪ/) a stone coffin, esp. one adorned with a sculpture or inscription. [L f. Gk sarkophagos flesh-consuming (as SARCOMA, -phagos -eating)]

sarcoplasm /ˈsɑːkəˌplæz(ə)m/ n. Anat. the cytoplasm of striated muscle cells. [Gk sarx sarkos flesh + PLASMA]

sarcous /ˈsɑːkəs/ adj. consisting of flesh or muscle. [Gk sarx sarkos flesh]

sard /sɑːd/ n. a yellow or orange-red cornelian. [ME f. F sarde or L sarda = LL sardius f. Gk sardios prob. f. Sardō Sardinia]

Sardanapalus /ˌsɑːdəˈnæpələs/ the name given by ancient Greek historians to the last king of Assyria (died c.626 BC), portrayed as being notorious for his wealth and sensuality. It may not represent a specific historical person. [Gk Sardanapalos]

sardar /ˈsɑːdɑː(r)/ n. (also **sirdar** /ˈsɜːd-/) esp. in the Indian subcontinent: **1** a person of high political or military rank; a leader. **2** a Sikh. [Urdu sardār f. Pers. sar head + dār possessor]

Sardegna see SARDINIA.

sardelle /sɑːˈdel/ n. a sardine, anchovy, or other small fish similarly prepared for eating. [It. sardella dimin. of sarda f. L (as SARDINE¹)]

sardine¹ /sɑːˈdiːn/ n. a young pilchard or similar young or small herring-like marine fish. □ **like sardines** crowded close together (as sardines are in tins). [ME f. OF sardine = It. sardina f. L f. sarda f. Gk, perh. f. Sardō Sardinia]

sardine² /sɑːˈdaɪn/ n. = SARDIUS. [ME f. LL sardinus f. Gk sardinos var. of sardios SARD: see Rev. 4:3]

Sardinia /sɑːˈdɪnɪə/ (Italian **Sardegna** /sarˈdeɲɲa/) a large Italian island in the Mediterranean Sea to the west of Italy; pop. (1990) 1,664,370; capital, Cagliari. Occupied in succession by Phoenicians, Carthaginians, Romans, and Vandals, the island passed in the early 14th century to Aragon. It remained under Spanish influence until it passed to Austria in the early 18th century. Ceded to Savoy in 1720, it was joined with Savoy and Piedmont to form the kingdom of Sardinia. In the mid-19th century the kingdom formed the nucleus of the Risorgimento, becoming part of a unified Italy under Victor Emmanuel II of Sardinia in 1861. The island is separated from the French island of Corsica to the north by the Strait of Bonifacio. □ **Sardinian** adj. & n.

Sardis /ˈsɑːdɪs/ an ancient city of Asia Minor, the capital of Lydia, whose ruins lie near the west coast of modern Turkey, to the north-east of Izmir. It was destroyed by Tamerlane in the 14th century.

sardius /ˈsɑːdɪəs/ n. a red precious stone mentioned in the Bible, e.g. Exodus 28:17, and in classical writings. Also called sardine. [ME f. LL f. Gk sardios SARD]

sardonic /sɑːˈdɒnɪk/ adj. **1** grimly jocular. **2** (of laughter etc.) bitterly mocking or cynical. □ **sardonically** adv. **sardonicism** /-nɪˌsɪz(ə)m/ n. [F sardonique, earlier sardonien f. L sardonius f. Gk sardonios, sardanius Sardinian (with reference to the ancient belief that eating a 'Sardinian plant' could result in convulsive laughter ending in death)]

sardonyx /ˈsɑːdənɪks/ n. onyx in which white layers alternate with sard. [ME f. L f. Gk sardonux (prob. as SARD, ONYX)]

saree var. of SARI.

sargasso /sɑːˈgæsəʊ/ n. (also **sargassum** /-səm/) (pl. **-os** or **-oes** or

sargassa /-sə/) a seaweed of the genus *Sargassum*, with berry-like air-vessels, found floating in island-like masses, esp. in the Sargasso Sea. Also called *gulfweed*. [Port. *sargaço*, of unkn. orig.]

Sargasso Sea a region of the western Atlantic Ocean between the Azores and the West Indies, around latitude 35°N, so called because of the prevalence in it of floating sargasso seaweed. It is the breeding-place of eels from the rivers of Europe and eastern North America, and is known for its usually calm conditions.

sarge /sɑːdʒ/ *n. sl.* sergeant. [abbr.]

Sargent /ˈsɑːdʒənt/, John Singer (1856–1925), American painter. Born in Florence, he travelled and studied widely in Europe in his youth. In the 1870s he painted some impressionist landscapes, but it was in portraiture that he developed the bold brushwork typical of his style, which reflects the influence of Manet, Hals, and Velázquez. He was much in demand in Parisian circles, but following a scandal over the supposed eroticism of *Madame Gautreau* (1884), he moved to London, where he dominated society portraiture for over twenty years. In the First World War he worked as an official war artist.

Sargodha /səˈgəʊdə/ a city in north central Pakistan; pop. (1981) 294,000. Situated near the Jhelum river, it is an agricultural centre and railway junction.

Sargon /ˈsɑːgɒn/ (2334–2279 BC) the semi-legendary founder of the ancient kingdom of Akkad.

Sargon II (d.705 BC), king of Assyria 721–705. He was probably a son of Tiglath-pileser III, and is thought to have been named after the semi-legendary King Sargon. He is famous for his conquest of a number of cities in Syria and Palestine; he also took ten of the tribes of Israel into captivity (see LOST TRIBES).

sari /ˈsɑːriː/ *n.* (also **saree**) (*pl.* **saris** or **sarees**) a length of cotton or silk draped round the body, traditionally worn as a main garment by Indian women. [Hindi *sāṛ(h)ī*]

sarin /ˈsærɪn/ *n. Chem.* an organic phosphorus compound used as a nerve gas. [Ger.]

Sark /sɑːk/ one of the Channel Islands, a small island lying to the east of Guernsey. It is divided by an isthmus into *Great Sark* and *Little Sark*.

sark /sɑːk/ *n. Sc. & N. Engl.* a shirt or chemise. [ME *serk* f. ON *serkr* f. Gmc]

sarking /ˈsɑːkɪŋ/ *n.* boarding between the rafters and the roof. [SARK + -ING¹]

sarky /ˈsɑːkɪ/ *adj.* (**sarkier, sarkiest**) *Brit. sl.* sarcastic. □ **sarkily** *adv.* **sarkiness** *n.* [abbr.]

Sarmatia /sɑːˈmeɪʃə/ an ancient region situated to the north of the Black Sea. Extending originally from the Urals to the Don, by Roman times the region consisted of the area between the Volga and the Vistula, corresponding to modern Poland and SW Russia. It was inhabited by Slavic peoples from ancient times. □ **Sarmatian** *adj. & n.*

sarmentose /ˈsɑːmənˌtəʊz, -ˌtəʊs/ *adj.* (also **sarmentous** /sɑːˈmentəs/) *Bot.* having long thin trailing shoots. [L *sarmentosus* f. *sarmenta* (pl.) twigs, brushwood, f. *sarpere* to prune]

sarnie /ˈsɑːnɪ/ *n. Brit. sl.* a sandwich. [abbr.]

sarong /səˈrɒŋ/ *n.* **1** a Malay and Javanese garment consisting of a long strip of (often striped) cloth worn by both sexes tucked round the waist or under the armpits. **2** a woman's garment resembling this. [Malay, lit. 'sheath']

Saronic Gulf /səˈrɒnɪk/ an inlet of the Aegean Sea on the coast of SE Greece. Athens and the port of Piraeus lie on its northern shores.

saros /ˈsɑːrɒs, ˈseər-/ *n. Astron.* a period of about eighteen years between repetitions of eclipses. [Gk f. Babylonian *šār(u)* 3,600 (years)]

Sarre see SAAR.

sarrusophone /səˈruːsəˌfəʊn/ *n.* a metal wind instrument played with a double reed like an oboe. [W. *Sarrus*, 19th-c. Fr. bandmaster who designed it]

sarsaparilla /ˌsɑːsəpəˈrɪlə/ *n.* **1** a preparation of the dried roots of various plants, esp. smilax, used to flavour some drinks and medicines and formerly as a tonic. **2** any of the plants yielding this. [Sp. *zarzaparilla* f. *zarza* bramble, prob. + dimin. of *parra* vine]

sarsen /ˈsɑːs(ə)n/ *n. Geol.* a silicified sandstone boulder found on the chalk downs of southern England, esp. in Wiltshire where sarsens were used to construct prehistoric stone monuments. [prob. var. of SARACEN]

sarsenet /ˈsɑːsənɪt/ *n.* (also **sarcenet**) a fine soft silk material used esp. for linings. [ME f. AF *sarzinett* perh. dimin. of *sarzin* SARACEN after OF *drap sarrasinois* Saracen cloth]

Sarto /ˈsɑːtəʊ/, Andrea del (born Andrea d'Agnolo) (1486–1531), Italian painter. He worked chiefly in Florence, where among his works are fresco cycles in the church of Santa Annunziata (e.g. *Nativity of the Virgin*, 1514) and the series of grisailles in the cloister of the Scalzi (1511–26) depicting the story of St John the Baptist. His work displays a feeling for tone and harmonies of colour, while the gracefulness of his figures influenced the mannerist style of his pupils Pontormo and Vasari.

sartorial /sɑːˈtɔːrɪəl/ *adj.* **1** of a tailor or tailoring. **2** of men's clothes (*sartorial elegance*). □ **sartorially** *adv.* [L *sartor* tailor f. *sarcire sart-* patch]

sartorius /sɑːˈtɔːrɪəs/ *n. Anat.* a long narrow muscle running across the front of each thigh. [mod.L f. L *sartor* tailor (the muscle being used in adopting a tailor's cross-legged posture)]

Sartre /ˈsɑːtrə/, Jean-Paul (1905–80), French philosopher, novelist, dramatist, and critic. While studying at the Sorbonne in 1929 he began his lifelong association with Simone de Beauvoir; they founded the review *Les Temps modernes* in 1945. A leading exponent of existentialism, he was originally influenced by the work of Heidegger; his later philosophy deals with the social responsibility of freedom, and attempts to synthesize existentialism with Marxist sociology. His works include the treatise *Being and Nothingness* (1943), the novel *Nausée* (1938), the trilogy *Les Chemins de la liberté* (1945–9), and the plays *Les Mouches* (1943) and *Huis clos* (1944). In 1964 he was offered but refused the Nobel Prize for literature.

Sarum /ˈseərəm/ an old name for Salisbury, Wilts., still used as the name of its diocese. (See also OLD SARUM.) □ **Sarum use** the form of liturgy used in the diocese of Salisbury from the 11th century to the Reformation. [med.L, perh. f. abbreviated form of L *Sarisburia* SALISBURY¹]

SAS see SPECIAL AIR SERVICE.

s.a.s.e. *abbr. US* self-addressed stamped envelope.

sash¹ /sæʃ/ *n.* a long strip or loop of cloth worn over one shoulder or round the waist, usu. as part of a uniform or insignia. □ **sashed** *adj.* [earlier *shash* f. Arab. *šāš* muslin, turban]

sash² /sæʃ/ *n.* **1** a frame holding the glass in a window, esp. in a sash-window. **2** the glazed sliding light of a glasshouse or garden frame. □ **sash-cord** a strong cord attaching the sash-weights to a sash. **sash-tool** a glazier's or painter's brush for work on sash-windows. **sash-weight** a weight attached to each end of a sash to balance it at any height. **sash-window** a window with one or two sashes of which one or both can be slid vertically over the other to make an opening. □ **sashed** *adj.* [*sashes* corrupt. of CHASSIS, mistaken for pl.]

sashay /ˈsæʃeɪ/ *v.intr.* esp. *N. Amer. colloq.* walk or move ostentatiously, casually, or diagonally. [corrupt. of CHASSÉ]

sashimi /ˈsæʃɪmɪ/ *n.* a Japanese dish of bite-size pieces of raw fish eaten with soy sauce and horseradish paste. [Jap.]

sasin /ˈsæsɪn/ *n.* = BLACKBUCK. [Nepali]

sasine /ˈseɪzɪn/ *n. Sc. Law* **1** the possession of feudal property. **2** an act or document granting this. [var. of SEISIN]

Sask. *abbr.* Saskatchewan.

Saskatchewan /səˈskætʃɪwən/ **1** a province of central Canada; pop. (1991) 994,000; capital, Regina. It was administered by the Hudson's Bay Company until 1870, when it was acquired by Canada as part of the Northwest Territories. It became a separate province in 1905. **2** a river of Canada. Rising in two headstreams in the Rocky Mountains, it flows eastwards for 596 km (370 miles) to Lake Winnipeg.

Saskatoon /ˌsæskəˈtuːn/ an industrial city in south central Saskatchewan, situated in the Great Plains on the South Saskatchewan river; pop. (1991) 186,060.

sasquatch /ˈsæskwætʃ/ *n.* a supposed yeti-like animal of NW America. [Salish]

sass /sæs/ *n. & v. N. Amer. colloq.* ● *n.* impudence, cheek. ● *v.tr.* be impudent to, cheek. [var. of SAUCE]

sassaby var. of TSESSEBI.

sassafras /ˈsæsəˌfræs/ *n.* **1** a small North American tree, *Sassafras albidum*, with aromatic leaves and bark. **2** a preparation of oil extracted from the leaves or bark of this tree, used medicinally or in perfumery. [Sp. *sasafrás* or Port. *sassafraz*, of unkn. orig.]

Sassanian /sæˈseɪnɪən/ *adj. & n.* (also **Sassanid** /ˈsæsənɪd/) ● *adj.* of or relating to the dynasty ruling the Persian empire from AD 224 until driven from Mesopotamia by the Arabs (637–51). The dynasty took its name from Sasan, the grandfather or father of the first Sassanian, Ardashir (d.242). ● *n.* a member of this dynasty.

Sassenach /ˈsæsəˌnæx, -ˌnæk/ n. & adj. Sc. & Ir. usu. derog. ● n. an English person. ● adj. English. [Gael. Sasunnoch, Ir. Sasanach f. L Saxones Saxons]

Sassoon[1] /səˈsuːn/, Siegfried (Lorraine) (1886–1967), English poet and writer. He is known for his starkly realistic poems written while serving in the First World War, expressing his contempt for war leaders and what he regarded as patriotic cant, as well as compassion for his comrades; collections include *The Old Huntsman* (1917). While in hospital in 1917 he met and gave encouragement to the poet Wilfred Owen; he published Owen's poems in 1920. After the war he also wrote a number of semi-autobiographical novels, including *Memoirs of a Fox-Hunting Man* (1928).

Sassoon[2] /səˈsuːn/, Vidal (b.1928), English hairstylist. After opening a London salon in 1953, he introduced the cut and blow-dry. Ten years later he created a hairstyle that was short at the back and long at the sides; first modelled at a Mary Quant fashion show, it became known as the 'Sassoon Cut'.

sassy /ˈsæsɪ/ adj. (**sassier, sassiest**) esp. N. Amer. colloq. = SAUCY. □ **sassily** adv. **sassiness** n. [var. of SAUCY]

sastrugi /sæˈstruːgɪ/ n.pl. wavelike irregularities on the surface of hard polar snow, caused by winds. [Russ. zastrugi small ridges]

SAT abbr. **1** N. Amer. scholastic aptitude test. **2** (in the UK) standard assessment test.

Sat. abbr. Saturday.

sat past and past part. of SIT.

Satan /ˈseɪt(ə)n/ n. the Devil; Lucifer. (See DEVIL.) [OE f. LL f. Gk f. Heb. śāṭān lit. 'adversary' f. śāṭan oppose, plot against]

satang /ˈsætæŋ/ n. (pl. **satangs** or same) a monetary unit of Thailand, equal to one-hundredth of a baht. [Thai f. Pali sata hundred]

satanic /səˈtænɪk/ adj. **1** of, like, or befitting Satan. **2** diabolical, hellish. □ **satanically** adv.

Satanism /ˈseɪtəˌnɪz(ə)m/ n. **1** the worship of Satan, with a travesty of Christian forms. **2** the pursuit of evil for its own sake. **3** deliberate wickedness. □ **Satanist** n. **Satanize** v.tr. (also **-ise**).

Satanology /ˌseɪtəˈnɒlədʒɪ/ n. **1** beliefs concerning the Devil. **2** a history or collection of these.

satay /ˈsæteɪ/ n. (also **satai, saté**) an Indonesian and Malaysian dish consisting of small pieces of meat grilled on a skewer and usu. served with spiced sauce. [Malayan satai sate, Indonesian sate]

SATB abbr. Mus. soprano, alto, tenor, and bass (as a combination of voices).

satchel /ˈsætʃəl/ n. a small bag usu. of leather and hung from the shoulder with a strap, for carrying books etc. esp. to and from school. [ME f. OF sachel f. L saccellus (as SACK[1])]

sate /seɪt/ v.tr. **1** gratify (desire, or a desirous person) to the full. **2** cloy, surfeit, weary with over-abundance (sated with pleasure). □ **sateless** adj. poet. [prob. f. dial. sade, OE sadian (as SAD), assim. to SATIATE]

sateen /sæˈtiːn/ n. cotton fabric woven like satin with a glossy surface. [satin after velveteen]

satellite /ˈsætəˌlaɪt/ n. & adj. ● n. **1** a celestial body orbiting the earth or another planet. **2** an artificial object placed in orbit round the earth or other celestial body. (See note below.) **3** a follower; a hanger-on. **4** an underling; a member of an important person's staff or retinue. **5** (in full **satellite state**) a small country etc. nominally independent but controlled by or dependent on another. ● adj. **1** transmitted by satellite (satellite communications; satellite television). **2** esp. Computing secondary; dependent; minor (networks of small satellite computers). □ **satellite dish** a concave dish-shaped aerial for receiving broadcasting signals transmitted by satellite. **satellite television** television in which the signal is transmitted via an artificial satellite. **satellite town** a small town economically or otherwise dependent on a nearby larger town. □ **satellitism** n. **satellitic** /ˌsætəˈlɪtɪk/ adj. [F satellite or L satelles satellitis attendant]

▪ The first artificial satellite, Sputnik I, was launched by the USSR on 4 Oct. 1957. However, the idea had been put forward much earlier, as by Jules Verne in *Begum's Fortune* (tr. 1880). The concept of the communications satellite was outlined by Arthur C. Clarke in 1945, twenty years before the first one was launched. Over 5,000 satellites have since been launched into earth orbit and several hundred are still operational. Many of them provide observation or remote sensing of the earth's surface, for military or meteorological purposes, or for research into mineral resources, land-use, etc. Others act as relays for telephone and microwave communications, or for the broadcasting of television and radio, or provide precise coordinates for air, sea, and land navigation. A number of satellites carry instruments for astronomical observation at various electromagnetic wavelengths, unhindered by the earth's atmosphere.

Sati /ˈsʌtiː/ Hinduism the wife of Siva, reborn as Parvati. According to some accounts, she died by throwing herself into the sacred fire, hence the custom of suttee. [Skr., as SUTTEE.]

sati var. of SUTTEE.

satiate /ˈseɪʃɪˌeɪt/ adj. & v. ● adj. archaic satiated. ● v.tr. = SATE. □ **satiable** /-ʃəb(ə)l/ adj. archaic. **satiation** /ˌseɪʃɪˈeɪʃ(ə)n/ n. [L satiatus past part. of satiare f. satis enough]

Satie /ˈsɑːtɪ/, Erik (Alfred Leslie) (1866–1925), French composer. He formed the centre of an irreverent avant-garde artistic set, associated not only with the composers of Les Six, but also with Cocteau, Dadaism, and surrealism. Many of his works are short and have facetious titles: *Three Pieces in the Shape of a Pear* (1903), for example, is in fact a set of six pieces. One of his few large-scale works is the symphonic drama *Socrate* (1918) for four sopranos and chamber orchestra to a libretto based on the writings of Plato. His work influenced Debussy and the development of minimalism.

satiety /səˈtaɪɪtɪ/ n. **1** the state of being glutted or satiated. **2** the feeling of having too much of something. **3** (foll. by of) a cloyed dislike of. □ **to satiety** to an extent beyond what is desired. [obs. F societé f. L satietas -tatis f. satis enough]

satin /ˈsætɪn/ n., adj., & v. ● n. a fabric of silk or various man-made fibres, with a glossy surface on one side produced by a twill weave with the weft-threads almost hidden. ● adj. smooth as satin. ● v.tr. (**satined, satining**) give a glossy surface to (paper). □ **satin finish 1** a polish given to silver etc. with a metallic brush. **2** any effect resembling satin in texture produced on materials in various ways. **satin paper** fine glossy writing paper. **satin spar** a fibrous variety of gypsum. **satin stitch** a long straight embroidery stitch, giving the appearance of satin. **satin white** a white pigment of calcium sulphate and alumina. □ **satinized** adj. (also **-ised**). **satiny** adj. [ME f. OF f. Arab. zaytūnī of Tseutung in China]

satinette /ˌsætɪˈnet/ n. (also **satinet**) a satin-like fabric made partly or wholly of cotton or synthetic fibre.

satinflower /ˈsætɪnˌflaʊə(r)/ n. a plant whose flowers have a satiny sheen, e.g. greater stitchwort, Stellaria holostea.

satinwood /ˈsætɪnˌwʊd/ n. **1 a** (in full **Ceylon** (or **Sri Lanka**) **satinwood**) a tree, Chloroxylon swietenia, native to central and southern India and Sri Lanka. **b** (in full **West Indian** (or **Jamaican**) **satinwood**) a tree, Fagara flava, native to the West Indies, Bermuda, and southern Florida. **2** the yellow glossy timber of either of these trees.

satire /ˈsætaɪə(r)/ n. **1** the use of ridicule, irony, sarcasm, etc., to expose folly or vice or to lampoon an individual. **2** a work or composition in prose or verse using satire. **3** this branch of literature. **4** a thing that brings ridicule upon something else. **5** Rom. Antiq. a poetic medley, esp. a poem ridiculing prevalent vices or follies. [F satire or L satira later form of satura medley]

satiric /səˈtɪrɪk/ adj. **1** of satire or satires. **2** containing satire (wrote a satiric review). **3** writing satire (a satiric poet). [F satirique or LL satiricus (as SATIRE)]

satirical /səˈtɪrɪk(ə)l/ adj. **1** = SATIRIC. **2** given to the use of satire in speech or writing or to cynical observation of others; sarcastic; humorously critical. □ **satirically** adv.

satirist /ˈsætərɪst/ n. **1** a writer of satires. **2** a satirical person.

satirize /ˈsætəˌraɪz/ v.tr. (also **-ise**) **1** assail or ridicule with satire. **2** write a satire upon. **3** describe satirically. □ **satirization** /ˌsætəraɪˈzeɪʃ(ə)n/ n. [F satiriser (as SATIRE)]

satisfaction /ˌsætɪsˈfækʃ(ə)n/ n. **1** the act or an instance of satisfying; the state of being satisfied (heard this with great satisfaction). **2** a thing that satisfies desire or gratifies feeling (is a great satisfaction to me). **3 a** a thing that settles an obligation or pays a debt. **4 a** (foll. by for) atonement; compensation (demanded satisfaction). **b** (in Christian theology) Christ's atonement for the sins of humankind. □ **to one's satisfaction** so that one is satisfied. [ME f. OF f. L satisfactio -onis (as SATISFY)]

satisfactory /ˌsætɪsˈfæktərɪ/ adj. **1** adequate; causing or giving satisfaction (was a satisfactory pupil). **2** satisfying expectations or needs; leaving no room for complaint (a satisfactory result). □ **satisfactorily** adv. **satisfactoriness** n. [F satisfactoire or med.L satisfactorius (as SATISFY)]

satisfy /ˈsætɪsˌfaɪ/ v. (**-ies, -ied**) **1** tr. **a** meet the expectations or desires of; comply with (a demand). **b** be accepted by (a person, his taste) as adequate; be equal to (a preconception etc.). **2** *tr.* put an end to (an appetite or want) by supplying what was required. **3** *tr.* rid (a person) of an appetite or want in a similar way. **4** *intr.* give satisfaction; leave nothing to be desired. **5** *tr.* pay (a debt or creditor). **6** *tr.* adequately meet, fulfil, or comply with (conditions, obligations, etc.) (*has satisfied all the legal conditions*). **7** *tr.* (often foll. by *of, that*) provide with adequate information or proof, convince (*satisfied the others that they were right; satisfy the court of their innocence*). **8** *tr. Math.* (of a quantity) make (an equation) true. **9** *tr.* (in *passive*) **a** (foll. by *with*) contented or pleased with. **b** (foll. by *to*) demand no more than or consider it enough to do. □ **satisfy the examiners** reach the standard required to pass an examination. **satisfy oneself** (often foll. by *that* + clause) be certain in one's own mind. □ **satisfiedly** /-ˌfaɪdlɪ/ *adv.* **satisfying** *adj.* **satisfyingly** *adv.* **satisfiable** *adj.* **satisfiability** /ˌsætɪsˌfaɪəˈbɪlɪtɪ/ *n.* [ME f. OF *satisfier* f. L *satisfacere satisfact-* f. *satis* enough]

satori /səˈtɔːrɪ/ *n.* Buddhism sudden enlightenment. [Jap.]

satrap /ˈsætræp/ *n.* **1** a provincial governor in the ancient Persian empire. **2** a subordinate ruler, colonial governor, etc. [ME f. OF *satrape* or L *satrapa* f. Gk *satrapēs* f. OPers. *xšathra-pāvan* country-protector]

satrapy /ˈsætrəpɪ/ *n.* (*pl.* **-ies**) a province ruled over by a satrap.

Satsuma /ˈsætsʊmə/ a former province of SW Japan. It comprised the major part of the south-western peninsula of Kyushu island, also known as the Satsuma Peninsula. It was ruled as a feudal domain until 1871, when it became the prefecture of Kagoshima, and was noted for the pottery made there from the end of the 16th century.

satsuma *n.* **1** /sætˈsuːmə/ a variety of tangerine orig. grown in Japan. **2** /ˈsætsʊmə/ (**Satsuma**) (in full **Satsuma ware**) cream-coloured Japanese pottery. [SATSUMA]

saturate /ˈsætʃəˌreɪt, ˈsætjʊˌreɪt/ *v.tr.* **1** fill with moisture; soak thoroughly. **2** (often foll. by *with*) fill to capacity. **3** cause (a substance, solution, vapour, metal, or air) to absorb, hold, or combine with the greatest possible amount of another substance, or of moisture, magnetism, electricity, etc. **4** cause (a substance) to combine with the maximum amount of another substance. **5** supply (a market) beyond the point at which the demand for a product is satisfied. **6** (foll. by *with, in*) imbue with or steep in (learning, tradition, prejudice, etc.). **7** overwhelm (enemy defences, a target area, etc.) by concentrated bombing. **8** (as **saturated** *adj.*) **a** (of colour) full; rich; free from an admixture of white. **b** *Chem.* (of fat molecules) containing the greatest number of hydrogen atoms and no double bonds. □ **saturate** /-rət/ *adj. literary.* **saturable** *adj.* **saturant** *n.* & *adj.* [L *saturare* f. *satur* full]

saturation /ˌsætʃəˈreɪʃ(ə)n, ˌsætjʊˈreɪ-/ *n.* the act or an instance of saturating; the state of being saturated. □ **saturation point** the stage beyond which no more can be absorbed or accepted.

Saturday /ˈsætəˌdeɪ, -dɪ/ *n.* & *adv.* ● *n.* the seventh day of the week, following Friday. ● *adv. colloq.* **1** on Saturday. **2** (**Saturdays**) on Saturdays; each Saturday. [OE *Sætern(es) dæg* transl. L *Saturni dies* day of Saturn]

Saturn /ˈsæt(ə)n/ **1** *Rom. Mythol.* an ancient god identified with the Greek Cronus, often regarded as a god of agriculture. His festival in December, Saturnalia, eventually became one of the elements in the traditional celebrations of Christmas. **2 a** *Astron.* the sixth planet from the sun in the solar system, orbiting between Jupiter and Uranus at an average distance of 1,427 million km from the sun. (*See note below.*) **b** *Astrol.* Saturn as a supposed astrological influence on those born under its sign, characterized by coldness and gloominess. **3** a series of American space rockets, of which the very large *Saturn V* was used as the launch vehicle for the Apollo missions of 1969–72. □ **Saturnian** /səˈtɜːnɪən/ *adj.*

■ The planet Saturn is a gas giant with an equatorial diameter of 120,000 km, with a conspicuous ring system extending out to a distance twice as great. It is known from the Voyager encounters that there are several thousand individual rings composed of small icy particles occupying a wide band of orbits, broken by apparent gaps which are actually regions of lower particle density. The planet has a dense hydrogen-rich atmosphere, similar to that of Jupiter, but with more vigorous atmospheric circulation. There are at least eighteen satellites, the largest of which is Titan, and including small shepherd satellites that orbit close to two of the rings.

saturnalia /ˌsætəˈneɪlɪə/ *n.* (*pl.* same or **saturnalias**) **1** (usu. **Saturnalia**) *Rom. Hist.* the festival of Saturn in December, characterized by unrestrained merrymaking for all, the predecessor of Christmas. **2** (as *sing.* or *pl.*) a scene of wild revelry or tumult; an orgy. □ **saturnalian** *adj.* [L, neut. pl. of *Saturnalis* (as SATURN)]

saturnic /səˈtɜːnɪk/ *adj. Med.* affected with lead poisoning. □ **saturnism** /ˈsætəˌnɪz(ə)m/ [SATURN: cf. SATURNINE]

saturniid /səˈtɜːnɪɪd/ *n. Zool.* a large moth of the family Saturniidae, related to the silk moths, e.g. the emperor moth. [mod.L]

saturnine /ˈsætəˌnaɪn/ *adj.* **1 a** of a sluggish gloomy temperament. **b** (of looks etc.) dark and brooding. **2** *archaic* **a** of the metal lead. **b** *Med.* of or affected by lead poisoning. □ **saturninely** *adv.* [ME f. OF *saturnin* f. med.L *Saturninus* of Saturn (identified with lead by the alchemists)]

satyagraha /sʌtˈjɑːɡrəˌhə/ *n. Ind.* **1** *hist.* a policy of passive resistance to British rule advocated by Gandhi. **2** passive resistance as a policy. [Skr. f. *satya* truth + *āgraha* obstinacy]

satyr /ˈsætə(r)/ *n.* **1** *Gk Mythol.* any of a class of lustful, drunken woodland spirits associated with Dionysus. In Greek art they were represented with the tail and ears of a horse, whereas Roman sculptors assimilated the satyr to the Italian faun and gave it the ears, horns, tail, and legs of a goat. **2** a lustful or sensual man. **3** = SATYRID. [ME f. OF *satyre* or L *satyrus* f. Gk *saturos*]

satyriasis /ˌsætɪˈraɪəsɪs/ *n. Med.* excessive sexual desire in men. [LL f. Gk *saturiasis* (as SATYR)]

satyric /səˈtɪrɪk/ *adj. Greek Mythol.* of or relating to satyrs. □ **satyric drama** a kind of ancient Greek comic play with a chorus of satyrs. [L *satyricus* f. Gk *saturikos* (as SATYR)]

satyrid /səˈtɪrɪd/ *n. Zool.* a butterfly of the family Satyridae, often brown with small eyespots on the wings. [mod.L *Satyridae* f. the genus name *Satyrus* (as SATYR)]

sauce /sɔːs/ *n.* & *v.* ● *n.* **1** a liquid or semi-solid accompaniment to food; the liquid constituent of a dish (*mint sauce; tomato sauce; chicken in a lemon sauce*). **2** something adding piquancy or excitement. **3** *colloq.* impudence, impertinence, cheek. **4** *N. Amer.* stewed fruit etc. eaten as dessert or used as a garnish. ● *v.tr.* **1** *colloq.* be impudent to; cheek. **2** *archaic* **a** season with sauce or condiments. **b** add excitement to. □ **sauce-boat** a kind of jug or dish used for serving sauces etc. **sauce for the goose** what is appropriate in one case (by implication appropriate in others). □ **sauceless** *adj.* [ME f. OF ult. f. L *salsus* f. *salere* *sals-* to salt f. *sal* salt]

saucepan /ˈsɔːspən/ *n.* a usu. metal cooking pan, usu. round with a lid and a long handle at the side, used for boiling, stewing, etc., on top of a cooker. □ **saucepanful** *n.* (*pl.* **-fuls**).

saucer /ˈsɔːsə(r)/ *n.* **1** a shallow circular dish used for standing a cup on and to catch drips. **2** any similar dish used to stand a plant pot etc. on. □ **saucerful** *n.* (*pl.* **-fuls**). **saucerless** *adj.* [ME, = condiment-dish, f. OF *saussier(e)* sauce-boat, prob. f. LL *salsarium* (as SAUCE)]

saucy /ˈsɔːsɪ/ *adj.* (**saucier**, **sauciest**) **1** impudent, cheeky. **2** *colloq.* smart-looking (*a saucy hat*). **3** *colloq.* smutty, suggestive. □ **saucily** *adv.* **sauciness** *n.* [earlier sense 'savoury', f. SAUCE]

Saudi /ˈsaʊdɪ/ *n.* & *adj.* ● *n.* (*pl.* **Saudis**) **1 a** a native or national of Saudi Arabia. **b** a person of Saudi descent. **2** a member of the dynasty founded by King Saud. ● *adj.* of or relating to Saudi Arabia or the Saudi dynasty. [Arab. *sa'ūdī* f. A. Ibn-*Sa'ūd*, Arab. king (1880–1953)]

Saudi Arabia a country in SW Asia occupying most of the Arabian peninsula; pop. (est. 1991) 15,431,000; official language, Arabic; capital, Riyadh. The birthplace of Islam in the 7th century, Saudi Arabia emerged from the Arab revolt against the Turks during the First World War to become an independent kingdom in 1932. Since the Second World War the economy has been revolutionized by the exploitation of the area's oil resources, and Saudi Arabia is the largest oil producer in the Middle East. Ruled by the house of Saud along traditional Islamic lines, the country has tended to exercise a conservative influence over Middle Eastern politics. Saudi Arabia was a leading member of the anti-Iraq coalition in the Gulf War of 1991, serving as a base for coalition forces. □ **Saudi Arabian** *adj.* & *n.*

sauerkraut /ˈsaʊəˌkraʊt/ *n.* a German dish of chopped pickled cabbage. [G f. *sauer* SOUR + *Kraut* vegetable]

sauger /ˈsɔːɡə(r)/ *n. US* a small North American pike-perch, *Stizostedion canadense.* [19th c.: orig. unkn.]

Saul /sɔːl/ **1** (in the Bible) the first king of Israel (11th century BC). **2** (also **Saul of Tarsus**) the original name of St Paul.

Sault Sainte Marie /ˌsuː seɪnt məˈriː/ each of two North American river ports which face each other across the falls of the St Mary's river, between Lakes Superior and Huron. The northern port lies in Ontario, Canada; pop. (1991) 72,822. The southern port is in the US state of Michigan; pop. (1990) 14,700. Both settlements were founded by Jacques Marquette in 1668, the town south of the river being ceded to the US by the British at the end of the 18th century. A system of canals serves to bypass the falls on either side of the river.

sauna /'sɔːnə/ n. **1** (orig. in Finland) a dry heat bath in which water is often thrown on to hot coals to produce steam. **2** a building used for this. [Finnish]

saunders var. of SANDERS.

saunter /'sɔːntə(r)/ v. & n. ● v.intr. **1** walk slowly; amble, stroll. **2** proceed without hurry or effort. ● n. **1** a leisurely ramble. **2** a slow gait. □ **saunterer** n. [ME, = muse: orig. unkn.]

saurian /'sɔːrɪən/ adj. of or like a lizard or a dinosaur. [mod.L *Sauria* f. Gk *saura* lizard]

saurischian /sɔːˈrɪʃɪən, ˌsaʊəˈrɪʃ-, -ˈrɪskɪən/ adj. & n. Zool. ● adj. of or relating to the order Saurischia, which comprises dinosaurs having a pelvic structure like that of lizards. (See DINOSAUR.) ● n. a dinosaur of this order. [mod.L f. Gk *sauros* lizard + *iskhion* hip-joint]

sauropod /'sɔːrəˌpɒd, 'saʊərə-/ n. Zool. a very large plant-eating dinosaur of the group Sauropoda, with a long neck, small head, and massive limbs, e.g. *Apatosaurus* and *Brachiosaurus*. [mod.L f. Gk *sauros* lizard + *pous pod-* foot]

saury /'sɔːrɪ/ n. (pl. **-ies**) an elongated long-beaked marine fish, *Scomberesox saurus*, of temperate waters. [perh. f. LL f. Gk *sauros* horse mackerel]

sausage /'sɒsɪdʒ/ n. **1** a short length of minced pork, beef, or other meat seasoned and often mixed with other ingredients, encased in a cylindrical skin, sold to be cooked before eating. **2 a** a minced pork, beef, or other meat seasoned and preserved and encased in a cylindrical skin, sold mainly to be eaten cold in slices or as a spread. **b** a length of this. **3** a sausage-shaped object. □ **not a sausage** *colloq.* nothing at all. **sausage dog** *Brit. colloq.* a dachshund. **sausage machine 1** a sausage-making machine. **2** a relentlessly uniform process. **sausage meat** minced meat used in sausages or as a stuffing etc. **sausage roll** *Brit.* sausage meat enclosed in a pastry roll and baked. [ME f. ONF *saussiche* f. med.L *salsicia* f. L *salsus*: see SAUCE]

Saussure /saʊ'sjʊə(r)/, Ferdinand de (1857–1913), Swiss linguistics scholar. He is one of the founders of modern linguistics and his work is fundamental to the development of structuralism. Departing from traditional diachronic studies of language, he emphasized the importance of a synchronic approach, treating language as a system of mutually dependent and interacting signs. He also made a distinction between *langue* (the total system of language) and *parole* (individual speech acts), and stressed that linguistic study should focus on the former. In his lifetime he published works of fundamental importance for Indo-European studies, but his most influential work, *Cours de linguistique générale* (1916), was compiled from lecture-notes and published posthumously. (See also STRUCTURALISM.)

sauté /'sɔʊteɪ/ adj., n., & v. ● adj. (esp. of potatoes etc.) quickly fried in a little hot fat. ● n. food cooked in this way. ● v.tr. (**sautéd** or **sautéed**) cook in this way. [F, past part. of *sauter* jump]

Sauternes /sɔʊ'tɜːn/ n. a sweet white wine from Sauternes in the Bordeaux region of France.

Sauveterrian /ˌsɔʊvə'terɪən/ adj. & n. Archaeol. of, relating to, or denoting an early mesolithic culture of France and western Europe. [*Sauveterre*-la-Lémance, France, the type-site]

Savage /'sævɪdʒ/, Michael Joseph (1872–1940), New Zealand Labour statesman, Prime Minister 1935–40. He was born in Australia and, after working as a trade union organizer, moved to New Zealand in 1907 and joined the New Zealand Labour Party on its formation in 1916, becoming party leader in 1933. Two years later his party won the general election and Savage became New Zealand's first Labour Prime Minister, a position he held until his death. He introduced many reforms, including social security legislation which he dubbed 'applied Christianity'.

savage /'sævɪdʒ/ adj., n., & v. ● adj. **1** fierce; cruel (*savage persecution; a savage blow*). **2** wild; primitive (*savage tribes; a savage animal*). **3** archaic (of scenery etc.) uncultivated (*a savage scene*). **4** colloq. angry; bad-tempered (*in a savage mood*). **5** Heraldry (of the human figure) naked. ● n. **1** Anthropol. derog. a member of a primitive tribe. **2** a cruel or barbarous person. ● v.tr. **1** (esp. of a dog, wolf, etc.) attack and bite or trample. **2** (of a critic etc.) attack fiercely. □ **savagedom** n. **savagely** adv. **savageness** n. **savagery** n. (pl. **-ies**). [ME f. OF *sauvage* wild f. L *silvaticus* f. *silva* a wood]

Savai'i /sɑː'vaɪ/ (also **Savaii**) a mountainous volcanic island in the SW Pacific, in Western Samoa. It is the largest of the Samoan islands.

Savannah /sə'vænə/ a port in Georgia, just south of the border with

South Carolina, on the Savannah river close to its outlet on the Atlantic; pop. (1990) 137,600.

savannah /sə'vænə/ n. (also **savanna**) a grassy plain in tropical and subtropical regions, with few or no trees. [Sp. *zavana* perh. of Carib orig.]

Savannakhet /ˌsævænə'ket/ (also **Savannaket**) a town in southern Laos, on the Mekong river where it forms the border with Thailand; pop. (est. 1973) 50,700.

savant /'sæv(ə)nt/ n. (fem. **savante** pronunc. same) a learned person, esp. a distinguished scientist etc. [F, part. of *savoir* know (as SAPIENT)]

savate /sə'vɑːt/ n. a form of boxing in which feet and fists are used. [F, orig. a kind of shoe: cf. SABOT]

save¹ /seɪv/ v. & n. ● v. **1** tr. (often foll. by *from*) rescue, preserve, protect, or deliver from danger, harm, discredit, etc. (*saved my life; saved me from drowning*). **2** tr. **a** (often foll. by *up*) keep for future use; reserve; refrain from spending (*saved up £150 for a new bike; likes to save plastic bags*). **b** Computing move a copy of (data) to a storage location. **3** tr. (often refl.) **a** relieve (another person or oneself) from spending (money, time, etc.); prevent exposure to (inconvenience etc.) (*saved myself £50; a word processor saves time*). **b** obviate the need or likelihood of (*soaking saves scrubbing*). **4** tr. preserve from damnation; convert (*saved her soul*). **5** tr. & refl. husband or preserve (one's strength, health, etc.) (*saving himself for the last lap; save your energy*). **6** intr. (often foll. by *up*) store up money for future use. **7** tr. **a** avoid losing (a game, match, etc.). **b** prevent an opponent from scoring (a goal etc.). **c** stop (a ball etc.) from entering the goal. ● n. **1** Football etc. the act of preventing an opponent's scoring etc. **2** Bridge a sacrifice-bid to prevent unnecessary losses. □ **save-all 1** a device to prevent waste. **2** hist. a pan with a spike for burning up candle-ends. **save appearances** present a prosperous, respectable, etc. appearance. **save-as-you-earn** Brit. a method of saving by regular deduction from earnings at source. **save one's breath** not waste time speaking to no effect. **save a person's face** see FACE. **save the situation** (or day) find or provide a solution to difficulty or disaster. **save one's skin** (or **neck** or **bacon**) avoid loss, injury, or death; escape from danger. **save the tide** get in and out (of port etc.) while the tide lasts. □ **savable** adj. (also **saveable**). [ME f. AF *sa(u)ver*, OF *salver, sauver* f. LL *salvare* f. L *salvus* SAFE]

save² /seɪv/ prep. & conj. archaic or poet. ● prep. except; but (*all save him*). ● conj. (often foll. by *for*) unless; but; except (*happy save for one want; is well save that he has a cold*). [ME f. OF *sauf sauve* f. L *salvo, salva*, ablat. sing. of *salvus* SAFE]

saveloy /'sævə,lɔɪ/ n. a seasoned red pork sausage, dried and smoked, and sold ready to eat. [corrupt. of F *cervelas, -at,* f. It. *cervellata* (*cervello* brain)]

saver /'seɪvə(r)/ n. **1** a person who saves esp. money. **2** (often in *comb.*) a device for economical use (of time etc.) (*found the short cut a time-saver*). **3** (often *attrib.*) a cheap, esp. off-peak fare. **4** Horse-racing sl. a hedging bet.

Savery /'seɪvərɪ/, Thomas (known as 'Captain Savery') (c.1650–1715), English engineer. He took out a number of patents, notably one for an engine to raise water 'by the Impellent Force of Fire' (1698). It was described as being suitable for raising water from mines, supplying towns with water, and operating mills. In fact its use of high-pressure steam made it dangerous, and only a few were actually used (at low pressure) in water mills. However, Savery's patent covered the type of engine developed by Thomas Newcomen, who was therefore obliged to join him in its exploitation.

Save the Children Fund (in the UK) a charity operating internationally to aid children suffering the effects of disasters and to look after the longer-term interests of needy children. Founded in 1919, the charity chose Princess Anne as its president in 1971.

savin /'sævɪn/ n. (also **savine**) **1** a bushy juniper, *Juniperus sabina*, usu. spreading horizontally, and yielding oil formerly used in the treatment of amenorrhoea. **2** US = red cedar. [OE f. OF *savine* f. L *sabina* (*herba*) Sabine (herb)]

saving /'seɪvɪŋ/ adj., n., & prep. ● adj. (often in *comb.*) making economical use of (*labour-saving*). ● n. **1** anything that is saved. **2** an economy (*a saving in expenses*). **3** (usu. in *pl.*) money saved. ● prep. **1** with the exception of; except (*all saving that one*). **2** without offence to (*saving your presence*). □ **saving clause** Law a clause containing a stipulation of exemption etc. **saving grace 1** the redeeming grace of God. **2** a redeeming quality or characteristic. **savings account** a deposit account. **savings and loan** US n. a cooperative association which accepts savings at interest and lends money to savers for house or other purchases. ● adj. of or designating such an association. **savings bank** a bank receiving small deposits at interest and returning the profits to the depositors. **savings certificate** *Brit.* an interest-bearing document issued by the government for the benefit of savers. [ME f. SAVE[1]: prep. prob. f. SAVE[2] after *touching*]

Savings Bond see PREMIUM BOND.

saviour /ˈseɪvjə(r)/ *n.* (US **savior**) **1** a person who saves or delivers from danger, harm, etc. (*the saviour of the nation*). **2** (**Saviour**) (prec. by *the*, *our*) Christ. [ME f. OF *sauvëour* f. eccl.L *salvator -oris* (transl. Gk *sōtēr*) f. LL *salvare* SAVE[1]]

savoir faire /ˌsævwɑː ˈfeə(r)/ *n.* the ability to act suitably in any situation; tact. [F, = know how to do]

savoir vivre /ˌsævwɑː ˈviːvrə/ *n.* knowledge of the world and the ways of society; ability to conduct oneself well; sophistication. [F, = know how to live]

Savonarola /ˌsævənəˈrəʊlə/, Girolamo (1452–98), Italian preacher and religious reformer. A Dominican monk and strict ascetic, in 1482 he moved to Florence, where he attracted great attention for his passionate preaching denouncing immorality, vanity, and corruption within the Church, and for his apocalyptic prophecies. He became virtual ruler of Florence (1494–5), but made many enemies, and in 1495 the pope forbade him to preach and summoned him to Rome. His refusal to comply with these orders led to his excommunication in 1497; he was hanged and burned as a schismatic and heretic.

Savonlinna /ˈsɑːvɒnˌlɪnə/ a town in SE Finland; pop. (1990) 28,560. It was founded in 1639 and later became a lakeside resort of the Russian tsars.

savor US var. of SAVOUR.

savory[1] /ˈseɪvərɪ/ *n.* (*pl.* **-ies**) a labiate herb of the genus *Satureja*, used esp. in cookery; esp. the *summer savory* (*S. hortensis*) and the *winter savory* (*S. montana*). [ME *saverey*, perh. f. OE *sætherie* f. L *satureia*]

savory[2] US var. of SAVOURY.

savour /ˈseɪvə(r)/ *n. & v.* (US **savor**) ● *n.* **1** a characteristic taste, flavour, relish, etc. **2** a quality suggestive of or containing a small amount of another. **3** *archaic* a characteristic smell. ● *v.* **1** *tr.* **a** appreciate and enjoy the taste of (food). **b** enjoy or appreciate (an experience etc.). **2** *intr.* (foll. by *of*) **a** suggest by taste, smell, etc. (*savours of mushrooms*). **b** imply or suggest a specified quality (*savours of impertinence*). □ **savourless** *adj.* [ME f. OF f. L *sapor -oris* f. *sapere* to taste]

savoury /ˈseɪvərɪ/ *adj. & n.* (US **savory**) ● *adj.* **1** (of food) salty or piquant, not sweet (*a savoury omelette*). **2** having an appetizing taste or smell. **3** pleasant; acceptable. ● *n.* (*pl.* **-ies**) *Brit.* a savoury dish, esp. one served as an appetizer or at the end of dinner. □ **savourily** *adv.* **savouriness** *n.* [ME f. OF *savouré* past part. (as SAVOUR)]

Savoy /səˈvɔɪ/ an area of SE France bordering on NW Italy, a former duchy ruled by the counts of Savoy from the 11th century, although frequently invaded and fought over by neighbouring states. In 1720 Savoy was joined with Sardinia and Piedmont to form the kingdom of Sardinia. In the mid-19th century the kingdom of Sardinia served as the nucleus for the formation of a unified Italy, but at the time of unification (1861) Savoy itself was ceded to France.

savoy /səˈvɔɪ/ *n.* a hardy variety of cabbage with wrinkled leaves. [SAVOY]

Savoyard /səˈvɔɪɑːd, ˌsævɔɪˈɑːd/ *n. & adj.* ● *n.* a native of Savoy. ● *adj.* of or relating to Savoy or its people. [F f. *Savoie* SAVOY]

Savu Sea /ˈsɑːvuː/ a part of the Indian Ocean which is encircled by the islands of Sumba, Flores, and Timor.

savvy /ˈsævɪ/ *v., n., & adj. sl.* ● *v.intr. & tr.* (**-ies, -ied**) know. ● *n.* knowingness; shrewdness; understanding. ● *adj.* (**savvier, savviest**) US knowing; wise. [orig. black & Pidgin Engl. after Sp. *sabe usted* you know]

saw[1] /sɔː/ *n. & v.* ● *n.* **1 a** a hand tool having a toothed blade used to cut esp. wood with a to-and-fro movement. **b** any of several mechanical power-driven devices with a toothed rotating disc or moving band, for cutting. **2** *Zool.* etc. a serrated organ or part. ● *v.* (*past part.* **sawn** /sɔːn/ or **sawed**) **1** *tr.* **a** cut (wood etc.) with a saw. **b** make (logs etc.) with a saw. **2** *intr.* use a saw. **3 a** *intr.* move to and fro with a motion as of a saw or person sawing (*sawing away on his violin*). **b** *tr.* divide (the air etc.) with gesticulations. □ **saw-billed** (of a bird) having a serrated bill. **saw-doctor** a machine for sharpening the teeth of a saw. **saw-edged** with a jagged edge like a saw. **saw-frame** a frame in which a saw-blade is held taut. **saw-gate** = *saw-frame*. **saw-gin** = *cotton-gin*. **saw-horse** a rack supporting wood for sawing. **sawn-off** (US **sawed-off**) **1** (of a shotgun) having a specially shortened barrel to make it easier to handle and give a wider field of fire. **2** *colloq.* (of a person) short. **saw-pit** a pit in which the lower of two men working a pit-saw stands. **saw-set** a tool for wrenching saw-teeth in alternate directions to

allow the saw to work freely. **saw-wort** a composite plant, *Serratula tinctoria*, yielding a yellow dye from its serrated leaves. □ **sawlike** *adj.* [OE *saga* f. Gmc]

saw[2] *past of* SEE[1].

saw[3] /sɔː/ *n.* a proverb; a maxim (*that's just an old saw*). [OE *sagu* f. Gmc, rel. to SAY: cf. SAGA]

sawbill /ˈsɔːbɪl/ *n.* a merganser.

sawbones /ˈsɔːbəʊnz/ *n. sl.* a doctor or surgeon.

sawbuck /ˈsɔːbʌk/ *n.* US **1** a saw-horse. **2** *sl.* a $10 note.

sawdust /ˈsɔːdʌst/ *n.* powdery particles of wood produced in sawing.

sawfish /ˈsɔːfɪʃ/ *n.* a large cartilaginous marine fish of the family Pristidae, having a long flat snout with toothlike projections along each side.

sawfly /ˈsɔːflaɪ/ *n.* (*pl.* **-flies**) a hymenopterous insect of the superfamily Tenthredinoidea, with a serrated ovipositor, the larvae of which are often injurious to plants.

sawmill /ˈsɔːmɪl/ *n.* a factory in which wood is sawn mechanically into planks or boards.

sawn *past part.* of SAW[1].

sawtooth /ˈsɔːtuːθ/ *adj.* **1** (also **sawtoothed** /-tuːθt/) (esp. of a roof, wave, etc.) shaped like the teeth of a saw with one steep and one slanting side. **2** (of a wave-form) showing a slow linear rise and rapid linear fall.

sawyer /ˈsɔːjə(r)/ *n.* **1** a person who saws timber professionally. **2** US an uprooted tree held fast by one end in a river. **3** NZ a large wingless horned grasshopper whose larvae bore in wood. [ME, earlier *sawer*, f. SAW[1]]

sax[1] /sæks/ *n. colloq.* **1** a saxophone. **2** a saxophone-player. □ **saxist** *n.* [abbr.]

sax[2] /sæks/ *n.* (also **zax** /zæks/) a slater's chopper, with a point for making nail-holes. [OE *seax* knife f. Gmc]

saxatile /ˈsæksəˌtaɪl, -tɪl/ *adj.* *Zool. & Bot.* living or growing on or among rocks. [F *saxatile* or L *saxatilis* f. *saxum* rock]

saxboard /ˈsæksbɔːd/ *n.* the uppermost strake of an open boat. [SAX[2] (cf. ON use of *sax* = gunwale near the prow) + BOARD]

saxe /sæks/ *n.* (in full **saxe blue**) (often *attrib.*) a lightish blue colour with a greyish tinge. [F, = Saxony, the source of a dye of this colour]

Saxe-Coburg-Gotha /ˌsæksˌkəʊbɜːgˈɡəʊtə, -ˈɡəʊθə/ the name of the British royal house 1901–17. The name dates from the accession of Edward VII, whose father Prince Albert, consort of Queen Victoria, was a prince of the German duchy of Saxe-Coburg and Gotha. During the First World War, with anti-German feeling running high, George V changed the family name to Windsor.

saxhorn /ˈsækshɔːn/ *n.* any of a series of different-sized brass wind instruments, each with valves and a funnel-shaped mouthpiece, used mainly in military and brass bands. The saxhorn was evolved by the Belgian instrument-makers Charles Joseph Sax (1791–1865) and his son Adolphe (or Antoine-Joseph) Sax (1814–94) and patented in 1845.

saxicoline /sækˈsɪkəˌlaɪn/ *adj.* (also **saxicolous** /-ləs/) *Biol.* = SAXATILE. [mod.L *saxicolus* f. *saxum* rock + *colere* inhabit]

saxifrage /ˈsæksɪˌfreɪdʒ/ *n.* a rosaceous plant of the genus *Saxifraga*, often growing on rocky or stony ground, and usu. bearing small white, yellow, or red flowers. [ME f. OF *saxifrage* or LL *saxifraga* (*herba*) f. L *saxum* rock + *frangere* break]

Saxon /ˈsæks(ə)n/ *n. & adj.* ● *n.* **1 a** *hist.* a member of a north German tribe, originally inhabitants of the area round the mouth of the Elbe, one branch of which, along with the Angles and the Jutes, conquered and colonized much of southern Britain in the 5th and 6th centuries. **b** (usu. **Old Saxon**) the language of this people. **2** = ANGLO-SAXON *n.* **3** a native of modern Saxony. **4** the Germanic (as opposed to Latin or Romance) elements of English. ● *adj.* **1** *hist.* of or concerning the Saxons. **2** belonging to or originating from the Saxon language or Old English. **3** of or concerning modern Saxony or Saxons. □ **Saxon architecture** the form of Romanesque architecture preceding the Norman in England. **Saxon blue** a solution of indigo in sulphuric acid as a dye. □ **Saxondom** *n.* **Saxonism** *n.* **Saxonist** *n.* **Saxonize** *v.tr. & intr.* (also **-ise**). [ME f. OF f. LL *Saxo -onis* f. Gk *Saxones* (pl.) f. WG: cf. OE *Seaxan, Seaxe* (pl.)]

Saxony /ˈsæksənɪ/ (German **Sachsen** /ˈzaks(ə)n/) **1** a state of eastern Germany, on the upper reaches of the River Elbe; capital, Dresden. Between 1949 and 1990 it was part of the German Democratic Republic. **2** a large region of Germany, including the modern states of Saxony in the south-east, Saxony-Anhalt in the centre, and Lower Saxony in the north-west. It was annexed by the Frankish emperor Charlemagne in the 8th century, emerging as a duchy by the early 10th century and as a kingdom in 919. Conquered by Napoleon in 1806, the kingdom was much reduced after his overthrow in 1815, the north-western part becoming a province of Prussia. The whole of Saxony became part of the German Empire in 1871 and its monarchy was eventually abolished after the First World War.

saxony /ˈsæksənɪ/ n. **1** a fine kind of wool. **2** cloth made from this. [SAXONY f. LL *Saxonia* (as SAXON)]

Saxony-Anhalt /ˌsæksənɪˈænhælt/ (German **Sachsen-Anhalt** /ˌzaks(ə)nˈanhalt/) a state of Germany, on the plains of the Elbe and the Saale rivers; capital, Magdeburg. It corresponds to the former duchy of Anhalt and the central part of the former kingdom of Saxony.

saxophone /ˈsæksəˌfəʊn/ n. a keyed metal wind instrument in several sizes and registers, used esp. in jazz music. It was invented *c*.1840 by Adolphe (or Antoine-Joseph) Sax (1814–94) and patented in 1846. □ **saxophonic** /ˌsæksəˈfɒnɪk/ adj. **saxophonist** /sækˈsɒfənɪst, ˈsæksəˌfəʊnɪst/ n. [Adolphe *Sax* + -PHONE]

say /seɪ/ v. & n. ● v. (3rd sing. present **says** /sez/; past and past part. **said** /sed/) **1** tr. (often foll. by *that* + clause) **a** utter (specified words) in a speaking voice; remark (*said 'Damn!'; said that he was satisfied*). **b** put into words; express (*that was well said; cannot say what I feel*). **2** tr. (often foll. by *that* + clause) **a** state; promise or prophesy (*says that there will be war*). **b** have specified wording; indicate (*says here that he was killed; the clock says ten to six*). **3** tr. (in passive; usu. foll. by *to* + infin.) be asserted or described (*is said to be 93 years old*). **4** tr. (foll. by *to* + infin.) *colloq.* tell a person to do something (*he said to bring the car*). **5** tr. convey (information) (*spoke for an hour but said little*). **6** tr. put forward as an argument or excuse (*much to be said in favour of it; what have you to say for yourself?*). **7** tr. (often *absol.*) form and give an opinion or decision as to (*who did it I cannot say; do say which you prefer*). **8** tr. select, assume, or take as an example or (a specified number etc.) as near enough (*shall we say this one?; paid, say, £20*). **9** tr. **a** speak the words of (prayers, mass, a grace, etc.). **b** repeat (a lesson etc.); recite (*can't say his tables*). **10** tr. Art etc. convey (inner meaning or intention) (*what is the director saying in this film?*). **11** intr. **a** speak; talk. **b** (in *imper.*) poet. tell me (*what is your name, say!*). **12** tr. (**the said**) Law or joc. the previously mentioned (*the said witness*). **13** intr. (as int.) N. Amer. an exclamation of surprise, to attract attention, etc. ● n. **1 a** an opportunity for stating one's opinion etc. (*let him have his say*). **b** a stated opinion. **2** a share or part (in a decision etc.) (*had no say in the matter*). □ **how say you?** Law how do you find? (addressed to the jury requesting its verdict). **I** etc. **cannot** (or **could not**) **say** I etc. do not know. **I'll say** *colloq.* yes indeed. **I say!** Brit. an exclamation expressing surprise, drawing attention, etc. **it is said** the rumour is that. **not to say** and indeed; or possibly even (*his language was rude not to say offensive*). **said he** (or **I** etc.) *colloq.* or *poet.* he etc. said. **say for oneself** say by way of conversation, oratory, etc. **say much** (or **something**) **for** indicate the high quality of. **say no** refuse or disagree. **say out** express fully or candidly. **says I** (or **he** etc.) *colloq.* I, he, etc., said (used in reporting conversation). **say-so 1** the power of decision. **2** mere assertion (*cannot proceed merely on his say-so*). **say something** make a short speech. **says you!** *colloq.* I disagree. **say when** *colloq.* indicate when enough drink or food has been given. **say the word 1** indicate that you agree or give permission. **2** give the order etc. **say yes** agree. **that is to say 1** in other words, more explicitly. **2** or at least. **they say** it is rumoured. **to say nothing of** = *not to mention* (see MENTION). **what do** (or **would**) **you say to** …? would you like …? **when all is said and done** after all, in the long run. **you can say that again!** (or **you said it!**) *colloq.* I agree emphatically. **you don't say so** *colloq.* an expression of amazement or disbelief. □ **sayable** adj. **sayer** n. [OE *secgan* f. Gmc]

SAYE abbr. Brit. save-as-you-earn.

Sayers /ˈseɪəz/, Dorothy L(eigh) (1893–1957), English novelist and dramatist. She is chiefly known for her detective fiction featuring the amateur detective Lord Peter Wimsey; titles include *Murder Must Advertise* (1933) and *The Nine Tailors* (1934). She also wrote religious plays (e.g. *The Devil to Pay*, 1939), which gained her recognition as a Christian polemicist.

saying /ˈseɪɪŋ/ n. **1** the act or an instance of saying. **2** a maxim, proverb, adage, etc. □ **as the saying goes** (or **is**) an expression used in introducing a proverb, cliché, etc. **go without saying** be too well known or obvious to need mention. **there is no saying** it is impossible to know.

Sb symb. Chem. the element antimony. [L *stibium*]

S-bend /ˈesbend/ n. an S-shaped bend in a road, pipe, etc.

SBN abbr. Standard Book Number (cf. ISBN).

SBS see SPECIAL BOAT SERVICE.

S. by E. abbr. South by East.

S. by W. abbr. South by West.

SC abbr. **1** US South Carolina (also in official postal use). **2** special constable.

Sc symb. Chem. the element scandium.

sc. abbr. scilicet.

s.c. abbr. small capitals.

scab /skæb/ n. & v. ● n. **1** the protective crust which forms over a cut, sore, etc. during healing. **2** (often *attrib.*) *colloq. derog.* a person who refuses to strike or join a trade union, or who tries to break a strike by working; a blackleg. **3** the mange or a similar skin disease esp. in animals. **4** a fungous plant-disease causing scablike roughness. **5** a dislikeable person. ● *v.intr.* (**scabbed, scabbing**) **1** act as a scab. **2** (of a wound etc.) form a scab; heal over. □ **scabbed** adj. **scabby** adj. (**scabbier, scabbiest**). **scabbiness** n. **scablike** adj. [ME f. ON *skabbr* (unrecorded), corresp. to OE *sceabb*]

scabbard /ˈskæbəd/ n. **1** esp. hist. a sheath for a sword, bayonet, etc. **2** US a sheath for a revolver etc. □ **scabbard-fish** an elongated marine fish of the family Trichiuridae, shaped like the scabbard of a sword; esp. the edible silvery-white *Lepidopus caudatus*. [ME *sca(u)berc* etc. f. AF prob. f. Frank.]

scabies /ˈskeɪbiːz/ n. a contagious skin disease caused by the itch-mite, marked by severe itching and red papules (cf. ITCH n. 3). [ME f. L f. *scabere* scratch]

scabious /ˈskeɪbɪəs/ n. & adj. ● n. a plant of *Scabiosa, Knautia*, or a related genus, with esp. bluish or lilac pincushion-shaped flowers. ● adj. affected with mange; scabby. [ME f. med.L *scabiosa* (*herba*) formerly regarded as a cure for skin disease: see SCABIES]

scabrous /ˈskeɪbrəs/ adj. **1** having a rough surface; bearing short stiff hairs, scales, etc.; scurfy. **2** (of a subject, situation, etc.) requiring tactful treatment; hard to handle with decency. **3 a** indecent, salacious. **b** behaving licentiously. □ **scabrously** adv. **scabrousness** n. [F *scabreux* or LL *scabrosus* f. L *scaber* rough]

scad /skæd/ n. a fish of the family Carangidae, native to tropical and subtropical seas, usu. having an elongated body and a row of large scales along each side; esp. the horse mackerel, *Trachurus trachurus*. [17th c.: orig. unkn.]

scads /skædz/ n.pl. US colloq. large quantities. [19th c.: orig. unkn.]

Scafell Pike /skɔːˈfel/ a mountain in the Lake District of NW England, in Cumbria. Rising to a height of 978 m (3,210 ft), it is the highest peak in England.

scaffold /ˈskæfəʊld, -f(ə)ld/ n. & v. ● n. **1 a** hist. a raised wooden platform used for the execution of criminals. **b** a similar platform used for drying tobacco etc. **2** = SCAFFOLDING. **3** (prec. by *the*) death by execution. ● *v.tr.* attach scaffolding to (a building). □ **scaffolder** n. [ME f. AF f. OF (e)*schaffaut*, earlier *escadafaut*: cf. CATAFALQUE]

scaffolding /ˈskæfəʊldɪŋ, -f(ə)ldɪŋ/ n. **1 a** a temporary structure formed of poles, planks, etc., erected by workmen and used by them while building or repairing a house etc. **b** materials used for this. **2** a temporary conceptual framework used for constructing theories etc.

scagliola /skæˈljəʊlə/ n. imitation stone or plaster mixed with glue. [It. *scagliuola* dimin. of *scaglia* SCALE[1]]

scalable /ˈskeɪləb(ə)l/ adj. capable of being scaled or climbed. □ **scalability** /ˌskeɪləˈbɪlɪtɪ/ n.

scalar /ˈskeɪlə(r)/ adj. & n. Math. & Physics ● adj. (of a quantity) having only magnitude, not direction. ● n. a scalar quantity (cf. VECTOR n. 1). [L *scalaris* f. *scala* ladder; see SCALE[3]]

scalawag var. of SCALLYWAG.

scald[1] /skɔːld, skɒld/ v. & n. ● *v.tr.* **1** burn (the skin etc.) with hot liquid or steam. **2** heat (esp. milk) to near boiling-point. **3** (usu. foll. by *out*) clean (a pan etc.) by rinsing with boiling water. **4** treat (poultry etc.) with boiling water to remove feathers etc. ● n. **1** a burn etc. caused by scalding. **2** a disease of fruit marked by browning etc., caused esp. by air pollution etc. □ **like a scalded cat** moving unusually fast. □ **scalder** n. [ME f. AF, ONF *escalder*, OF *eschalder* f. LL *excaldare* (as EX-[1], L *calidus* hot)]

scald[2] var. of SKALD.

scalding /ˈskɔːldɪŋ, ˈskɒl-/ adj. **1** very hot; burning. **2** causing an effect like that of scalding (*some scalding and awful truths*).

scale[1] /skeɪl/ n. & v. ● n. **1** each of the small thin bony or horny overlapping plates protecting the skin of fish and reptiles. **2** something resembling a fish-scale, esp.: **a** a pod or husk. **b** a flake of skin; a scab. **c** a rudimentary leaf, feather, or bract. **d** each of the structures covering the wings of butterflies and moths. **e** *Bot.* a layer of a bulb. **3 a** the flaky deposit which forms on iron when heated and hammered or rolled. **b** the thick white deposit formed in a kettle, boiler, etc. by the action of heat on water. **4** plaque formed on teeth. ● *v.* **1** *tr.* remove scale or scales from (fish, nuts, iron, etc.). **2** *tr.* remove plaque from (teeth) by scraping. **3** *intr.* **a** (of skin, metal, etc.) form, come off in, or drop, scales. **b** (usu. foll. by *off*) (of scales) come off. □ **scale-armour** *hist.* armour formed of metal scales attached to leather etc. **scale-board** very thin wood used for the back of a mirror, picture, etc. **scale-bug** = *scale insect.* **scale-fern** a spleenwort, esp. *Asplenium ceterach.* **scale insect** a homopterous bug of the family Coccidae, clinging to plants and secreting a waxy shieldlike scale as a covering. **scale-leaf** a modified leaf resembling a scale. **scale-moss** a type of liverwort with scalelike leaves. **scales fall from a person's eyes** a person is no longer deceived (in allusion to Acts 9:18). **scale-winged** lepidopterous. **scale-work** an overlapping arrangement. □ **scaled** *adj.* (also in *comb.*). **scaleless** *adj.* **scaler** *n.* [ME f. OF *escale* f. Gmc, rel. to SCALE²]

scale² /skeɪl/ *n. & v.* ● *n.* **1 a** (often in *pl.*) a weighing machine or device. **b** (also **scale-pan**) each of the dishes on a simple scale balance. **2** (**the Scales**) the zodiacal sign or constellation Libra. ● *v.tr.* (of something weighed) show (a specified weight) in the scales. □ **pair of scales** a simple balance. **throw into the scale** cause to be a factor in a contest, debate, etc. **tip** (or **turn**) **the scales 1** (usu. foll. by *at*) outweigh the opposite scale-pan (at a specified weight); weigh. **2** (of a motive, circumstance, etc.) be decisive. [ME f. ON *skál* bowl f. Gmc]

scale³ /skeɪl/ *n. & v.* ● *n.* **1** a series of degrees; a graded classification system (*pay fees according to a prescribed scale; high on the social scale; seven points on the Richter scale*). **2 a** (often *attrib.*) *Geog. & Archit.* a ratio of size in a map, model, picture, etc. (*on a scale of one centimetre to the kilometre; a scale model*). **b** relative dimensions or degree (*generosity on a grand scale*). **3** *Mus.* an arrangement of the notes in any system of music in ascending or descending order of pitch (*chromatic scale; major scale*). **4** a set of marks on a line used in measuring, reducing, enlarging, etc. **b** a rule determining the distances between these. **c** a piece of metal, apparatus, etc. on which these are marked. **5** (in full **scale of notation**) *Math.* the ratio between units in a numerical system (*decimal scale*). ● *v.* **1** *tr.* **a** (also *absol.*) climb (a wall, height, etc.) esp. with a ladder. **b** climb (the social scale, heights of ambition, etc.). **2** *tr.* represent in proportional dimensions; reduce to a common scale. **3** *intr.* (of quantities etc.) have a common scale; be commensurable. □ **economies of scale** proportionate savings gained by using larger quantities. **in scale** (of drawing etc.) in proportion to the surroundings etc. **play** (or **sing**) **scales** *Mus.* perform the notes of a scale as an exercise for the fingers or voice. **scale down** make smaller in proportion; reduce in size. **scale up** make larger in proportion; increase in size. **scaling-ladder** *hist.* a ladder used to climb esp. fortress walls, esp. to break a siege. **to scale** with a uniform reduction or enlargement. □ **scaler** *n.* [(n.) ME (= ladder): (v.) ME f. OF *escaler* or med.L *scalare* f. L *scala scandere* climb]

scalene /ˈskeɪliːn/ *adj. & n.* ● *adj.* **1** *Geom.* (of a triangle) having sides unequal in length. **2** *Anat.* of or relating to a scalenus muscle. ● *n.* **1** (in full **scalene muscle**) = SCALENUS. **2** a scalene triangle. □ **scalene cone** (or **cylinder**) a cone (or cylinder) with the axis not perpendicular to the base. [LL *scalenus* f. Gk *skalēnos* unequal, rel. to *skolios* bent]

scalenus /skəˈliːnəs/ *n.* (*pl.* **scaleni** /-naɪ/) (also **scalenus muscle**) a muscle of a group extending from the neck to the first and second ribs. [mod.L: see SCALENE]

Scaliger¹ /ˈskælɪdʒə(r)/, Joseph Justus (1540–1609), French scholar. The son of Julius Caesar Scaliger, he was a leading Renaissance scholar, often regarded as the founder of historical criticism. His edition of Manilius (1579) and his *De Emendatione Temporum* (1583) revolutionized understanding of ancient chronology and gave it a more scientific foundation by comparing and revising the computations of time made by different civilizations, including those of the Babylonians and Egyptians.

Scaliger² /ˈskælɪdʒə(r)/, Julius Caesar (1484–1558), Italian-born French classical scholar and physician. Appointed physician to the bishop of Agen, he settled there and became a French citizen in 1528. Besides polemical works directed against Erasmus (1531, 1536), he wrote a long Latin treatise on poetics, a number of commentaries on botanical works, and a philosophical treatise.

scallawag var. of SCALLYWAG.

scallion /ˈskæljən/ *n.* a shallot or spring onion; any long-necked onion with a small bulb. [ME f. AF *scal(o)un* = OF *escalo(i)gne* ult. f. L *Ascalonia* (*caepa*) (onion) of ASCALON (ASHQELON) in ancient Palestine]

scallop /ˈskæləp, ˈskɒl-/ *n. & v.* (also **scollop** /ˈskɒl-/) ● *n.* **1** a bivalve mollusc of the family Pectinidae, esp. of the genus *Chlamys* or *Pecten*, used as food. **2** (in full **scallop shell**) **a** a single valve from the shell of a scallop, with grooves and ridges radiating from the middle of the hinge and edged with small rounded lobes, often used for cooking or serving food. **b** *hist.* a representation of this shell worn as a pilgrim's badge. **3** (in *pl.*) an ornamental edging cut in material in imitation of the edge of a scallop shell. **4** a small pan or dish shaped like a scallop shell and used for baking or serving food. **5** = ESCALOPE. ● *v.tr.* (**scalloped, scalloping**) **1** ornament with scallops or scalloping; cut or shape in the form of a scallop. **2** cook in a scallop. □ **scalloper** *n.* **scalloping** *n.* (in sense 3 of *n.*). [ME f. OF *escalope* prob. f. Gmc]

scallywag /ˈskælɪˌwæg/ *n.* (also **scalawag, scallawag** /ˈskæləˌwæg/) **1** a scamp; a rascal. **2** (**scalawag**) *US hist. derog.* a white Southerner who supported the Federal programme of legislation during the Reconstruction, or who favoured the Republican Party. [19th-c. US sl.: orig. unkn.]

scalp /skælp/ *n. & v.* ● *n.* **1** the skin covering the top of the head, with the hair etc. attached. **2 a** *hist.* the scalp of an enemy cut or torn away as a trophy by a North American Indian. **b** a trophy or symbol of triumph, conquest, etc. **3** *Sc.* a bare rock projecting above water etc. ● *v.tr.* **1** *hist.* take the scalp of (an enemy). **2** criticize savagely. **3** *US* defeat; humiliate. **4** *N. Amer. colloq.* resell (shares, tickets, etc.) at a high or quick profit. □ **scalpless** *adj.* [ME, prob. of Scand. orig.]

scalpel /ˈskælp(ə)l/ *n.* a surgeon's small sharp knife shaped for holding like a pen. [F *scalpel* or L *scalpellum* dimin. of *scalprum* chisel f. *scalpere* scratch]

scalper /ˈskælpə(r)/ *n.* **1** a person or thing that scalps (esp. in sense 4 of *v.*). **2** (also **scauper** /ˈskɔːpə(r)/, **scorper**) an engraver's tool for hollowing out woodcut or linocut designs. [SCALP + -ER¹: sense 2 also f. L *scalper* cutting tool f. *scalpere* carve]

scaly /ˈskeɪlɪ/ *adj.* (**scalier, scaliest**) covered in or having many scales or flakes. □ **scaly anteater** = PANGOLIN. □ **scaliness** *n.*

scam /skæm/ *n. & v.* ● *n. US sl.* **1** a trick or swindle; a fraud. **2** a story or rumour. ● *v.* (**scammed, scamming**) **1** *intr.* commit fraud. **2** *tr.* swindle. □ **scammer** *n.* [20th c.: orig. unkn.]

scammony /ˈskæmənɪ/ *n.* (*pl.* **-ies**) an Asian plant, *Convolvulus scammonia*, bearing white or pink flowers, the dried roots of which are used as a purgative. [ME f. OF *scamonee*, *escamonie* or L *scammonia* f. Gk *skammōnia*]

scamp¹ /skæmp/ *n. colloq.* a rascal; a rogue. □ **scampish** *adj.* [*scamp* rob on highway, prob. f. MDu. *schampen* decamp f. OF *esc(h)amper* (as EX-¹, L *campus* field)]

scamp² /skæmp/ *v.tr.* do (work etc.) in a perfunctory or inadequate way. [perh. formed as SCAMP¹: cf. SKIMP]

scamper /ˈskæmpə(r)/ *v. & n.* ● *v.intr.* (usu. foll. by *about*, *through*) run and skip impulsively or playfully. ● *n.* the act or an instance of scampering. [prob. formed as SCAMP¹]

scampi /ˈskæmpɪ/ *n.pl.* **1** Norway lobsters. **2** (often treated as *sing.*) a dish of these, usu. fried. [It.]

scan /skæn/ *v. & n.* ● *v.* (**scanned, scanning**) **1** *tr.* look at or over intently or quickly (*scanned the horizon; scanned the page for errors*). **2** *intr.* (of a verse etc.) be metrically correct; be capable of being recited etc. metrically (*this line doesn't scan*). **3** *tr.* **a** examine all parts of (a surface etc.) to detect radioactivity etc. **b** cause (a particular region) to be traversed by a radar etc. beam. **4** *tr.* resolve (a picture) into its elements of light and shade in a prearranged pattern for the purposes esp. of television transmission. **5** *tr.* test the metre of (a line of verse etc.) by reading with the emphasis on its rhythm, or by examining the number of feet etc. **6** *tr.* examine (a patient, a part of the body) with a scanner. ● *n.* **1** the act or an instance of scanning. **2** an image obtained by scanning or with a scanner. □ **scannable** *adj.* [ME f. L *scandere* climb: in LL = scan (verses) (from the raising of one's foot in marking rhythm)]

scandal /ˈskænd(ə)l/ *n.* **1 a** a thing or a person causing general public outrage or indignation. **b** the outrage etc. so caused, esp. as a subject of common talk. **c** malicious gossip or backbiting. **2** *Law* a public affront, esp. an irrelevant abusive statement in court. □ **scandal sheet** *derog.* a newspaper etc. giving prominence to esp. malicious gossip. □ **scandalous** *adj.* **scandalously** *adv.* **scandalousness** *n.*

[ME f. OF *scandale* f. eccl.L *scandalum* f. Gk *skandalon* snare, stumbling-block]

scandalize /ˈskændəˌlaɪz/ *v.tr.* (also **-ise**) offend the moral feelings, sensibilities, etc. of; shock. [ME in sense 'make a scandal of' f. F *scandaliser* or eccl.L *scandalizo* (as SCANDAL)]

scandalmonger /ˈskænd(ə)lˌmʌŋgə(r)/ *n.* a person who spreads malicious scandal.

Scandinavia /ˌskændɪˈneɪvɪə/ **1** a large peninsula in NW Europe, occupied by Norway and Sweden. It is bounded by the Arctic Ocean in the north, the Atlantic in the west, and the Baltic Sea in the south and east. **2** a cultural region consisting of the countries of Norway, Sweden, and Denmark, and sometimes also of Iceland, Finland, and the Faeroe Islands.

Scandinavian /ˌskændɪˈneɪvɪən/ *n. & adj.* ● *n.* **1 a** a native or inhabitant of Scandinavia. **b** a person of Scandinavian descent. **2** the northern branch of the Germanic languages, including Danish, Norwegian, Swedish, and Icelandic, all descended from Old Norse. ● *adj.* of or relating to Scandinavia or its people or languages.

scandium /ˈskændɪəm/ *n.* a soft silvery-white metallic chemical element (atomic number 21; symbol **Sc**). Scandium was discovered in 1879 by the Swedish chemist Lars Frederick Nilson (1840–99), after its existence had been predicted by Mendeleev on the basis of his periodic table. It occurs in company with the rare-earth elements, which it resembles in chemical properties. The metal and its compounds have few commercial uses. [mod.L *Scandia* Scandinavia (source of the minerals containing it)]

scanner /ˈskænə(r)/ *n.* **1** a device for scanning or systematically examining or reading something. **2** *Med.* a machine for measuring the intensity of radiation, ultrasound reflections, etc., from the body as a diagnostic aid. **3** a person who scans or examines critically.

scanning electron microscope *n.* (abbr. **SEM**) an electron microscope in which a specimen is scanned by a beam of electrons that is then reflected rather than transmitted. It is used for examining the three-dimensional structure of surfaces at relatively low magnifications.

scanning tunnelling microscope *n.* (abbr. **STM**) an electron microscope having a probe with a submicroscopic tip that generates electrons by quantum-mechanical tunnelling, giving a resolution at atomic level.

scansion /ˈskænʃ(ə)n/ *n.* **1** the metrical scanning of verse. **2** the way a verse etc. scans. [L *scansio* (LL of metre) f. *scandere scans-* climb]

scant /skænt/ *adj. & v.* ● *adj.* barely sufficient; deficient (*with scant regard for the truth*; *scant of breath*). ● *v.tr. archaic* provide (a supply, material, a person, etc.) grudgingly; skimp; stint. □ **scantly** *adv.* **scantness** *n.* [ME f. ON *skamt* neut. of *skammr* short]

scantling /ˈskæntlɪŋ/ *n.* **1 a** a timber beam of small cross-section. **b** a size to which a stone or timber is to be cut. **2** a set of standard dimensions for parts of a structure, esp. in shipbuilding. **3** (usu. foll. by *of*) *archaic* **a** a specimen or sample. **b** one's necessary supply; a modicum or small amount. [alt. after -LING¹ f. obs. *scantlon* f. OF *escantillon* sample]

scanty /ˈskæntɪ/ *adj.* (**scantier, scantiest**) **1** of small extent or amount. **2** meagre, barely sufficient. □ **scantily** *adv.* **scantiness** *n.* [SCANT]

Scapa Flow /ˈskɑːpə, ˈskæpə/ a strait in the Orkney Islands, Scotland. It was an important British naval base esp. in the First World War; the base was closed in 1957. The entire German High Seas Fleet, which was interned there after its surrender, was scuttled in 1919 as an act of defiance against the terms of the Versailles peace settlement.

scape /skeɪp/ *n.* **1** a long flower-stalk coming directly from the root. **2** the base of an insect's antenna. [L *scapus* f. Gk *skapos*, rel. to SCEPTRE]

-scape /skeɪp/ *comb. form* forming nouns denoting a view or a representation of a view (*moonscape*; *seascape*). [after LANDSCAPE]

scapegoat /ˈskeɪpgəʊt/ *n. & v.* ● *n.* **1** a person bearing the blame for the sins, shortcomings, etc. of others, esp. as an expedient. **2** *Bibl.* a goat sent into the wilderness after the chief priest had symbolically laid the sins of the people upon it (Lev. 16). ● *v.tr.* make a scapegoat of. □ **scapegoater** *n.* [archaic *scape* = escape + GOAT]

scapegrace /ˈskeɪpgreɪs/ *n.* a rascal; a scamp, esp. a young person or child. [*scape* (as SCAPEGOAT) + GRACE, lit. 'a person who escapes the grace of God']

scaphoid /ˈskæfɔɪd/ *adj. & n. Anat.* = NAVICULAR. [mod.L *scaphoides* f. Gk *skaphoeidēs* f. *skaphos* boat]

scapula /ˈskæpjʊlə/ *n.* (pl. **scapulae** /-ˌliː/ or **scapulas**) the shoulder-blade. [LL, sing. of L *scapulae*]

scapular /ˈskæpjʊlə(r)/ *adj. & n.* ● *adj.* of or relating to the shoulder or shoulder-blade. ● *n.* **1 a** a monastic short cloak covering the shoulders. **b** a symbol of affiliation to an ecclesiastical order, consisting of two strips of cloth hanging down the breast and back and joined across the shoulders. **2** a bandage for or over the shoulders. **3** a scapular feather. □ **scapular feather** a feather growing near the insertion of the wing. [(adj.) f. SCAPULA: (n.) f. LL *scapulare* (as SCAPULA)]

scapulary /ˈskæpjʊlərɪ/ *n.* (pl. **-ies**) **1** = SCAPULAR *n.* 1. **2** = SCAPULAR *n.* 3. [ME f. OF *eschapeloyre* f. med.L *scapelorium, scapularium* (as SCAPULA)]

scar¹ /skɑː(r)/ *n. & v.* ● *n.* **1** a usu. permanent mark on the skin left after the healing of a wound, burn, or sore. **2** the lasting effect of grief etc. on a person's character or disposition. **3** a mark left by damage etc. (*the table bore many scars*). **4** a mark left on the stem etc. of a plant by the fall of a leaf etc. ● *v.* (**scarred, scarring**) **1** *tr.* (esp. as **scarred** *adj.*) mark with a scar or scars (*was scarred for life*). **2** *intr.* heal over; form a scar. **3** *tr.* form a scar on. □ **scarless** *adj.* [ME f. OF *eschar(r)e* f. LL *eschara* f. Gk *eskhara* scab]

scar² /skɑː(r)/ *n.* (also **scaur** /skɔː(r)/) a steep craggy outcrop of a mountain or cliff. [ME f. ON *sker* low reef in the sea]

scarab /ˈskærəb/ *n.* **1 a** a large dung-beetle, *Scarabaeus sacer*, regarded as sacred in ancient Egypt. **b** = SCARABAEID. **2** an ancient Egyptian gem cut in the form of a beetle and engraved with symbols on its flat side, used as a signet etc. [L *scarabaeus* f. Gk *skarabeios*]

scarabaeid /ˌskærəˈbiːɪd/ *n. Zool.* a beetle of the family Scarabaeidae, which includes the dung-beetles, chafers, etc. [mod.L *Scarabaeidae* (as SCARAB)]

scaramouch /ˈskærəˌmuːʃ/ *n. archaic* a boastful coward; a braggart. [It. *Scaramuccia* stock character in Italian farce f. *scaramuccia* = SKIRMISH, infl. by F form *Scaramouche*]

Scarborough /ˈskɑːbərə/ a fishing port and resort on the coast of North Yorkshire; pop. (1981) 38,000.

scarce /skeəs/ *adj. & adv.* ● *adj.* **1** (usu. *predic.*) (esp. of food, money, etc.) insufficient for the demand; scanty. **2** hard to find; rare. ● *adv. archaic* or *literary* scarcely. □ **make oneself scarce** *colloq.* disappear discreetly, keep out of the way. □ **scarceness** *n.* [ME f. AF & ONF (*e*)*scars*, OF *eschars* f. L *excerpere*: see EXCERPT]

scarcely /ˈskeəslɪ/ *adv.* **1** hardly; barely; only just (*I scarcely know him*). **2** surely not (*he can scarcely have said so*). **3** a mild or apologetic or ironical substitute for 'not' (*I could scarcely believe my ears*).

scarcity /ˈskeəsɪtɪ/ *n.* (pl. **-ies**) (often foll. by *of*) a lack or inadequacy, esp. of food.

scare /skeə(r)/ *v. & n.* ● *v.* **1** *tr.* frighten, esp. suddenly (*his expression scared us*). **2** *tr.* (as **scared** *adj.*) (usu. foll. by *of*, or *to* + infin.) frightened; terrified (*scared of his own shadow*). **3** *tr.* (usu. foll. by *away, off, up,* etc.) drive away by frightening. **4** *intr.* become scared (*they don't scare easily*). ● *n.* **1** a sudden attack of fright (*gave me a scare*). **2** a general, esp. baseless, fear of war, invasion, epidemic, etc. (*a measles scare*). **3** a financial panic causing share-selling etc. □ **scaredy-cat** /ˈskeədɪˌkæt/ *colloq.* a timid person. **scare up** (or **out**) esp. *N. Amer.* **1** frighten (game etc.) out of cover. **2** *colloq.* manage to find; discover (*see if we can scare up a meal*). □ **scarer** *n.* [ME *skerre* f. ON *skirra* frighten f. *skjarr* timid]

scarecrow /ˈskeəˌkrəʊ/ *n.* **1** an object resembling a human figure, dressed in old clothes and set up in a field to scare birds away. **2** *colloq.* a badly dressed, grotesque-looking, or very thin person. **3** *archaic* an object of baseless fear.

scaremonger /ˈskeəˌmʌŋgə(r)/ *n.* a person who spreads frightening reports or rumours. □ **scaremongering** *n.*

scarf¹ /skɑːf/ *n.* (pl. **scarves** /skɑːvz/ or **scarfs**) a square, triangular, or esp. long narrow strip of material worn round the neck, over the shoulders, or tied round the head (of a woman), for warmth or ornament. □ **scarf-pin** (or **-ring**) *Brit.* an ornamental device for fastening a scarf. **scarf-skin** the outermost layer of the skin constantly scaling off, esp. that at the base of the nails. **scarf-wise** worn diagonally across the body from shoulder to hip. □ **scarfed** *adj.* [prob. alt. of *scarp* (infl. by SCARF²) f. ONF *escarpe* = OF *escherpe* sash]

scarf² /skɑːf/ *v. & n.* ● *v.tr.* **1** join the ends of (pieces of esp. timber, metal, or leather) by bevelling or notching them to fit and then bolting, brazing, or sewing them together. **2** cut the blubber of (a whale). ● *n.* **1** a joint made by scarfing. **2** a cut on a whale made by scarfing. [ME (earlier as noun) prob. f. OF *escarf* (unrecorded) perh. f. ON]

scarifier /ˈskærɪˌfaɪə(r), ˈskeər-/ *n.* **1** a thing or person that scarifies.

2 a machine with prongs for loosening soil without turning it. **3** a spiked road-breaking machine.

scarify[1] /'skærɪˌfaɪ, 'skeər-/ v.tr. (**-ies, -ied**) **1 a** make superficial incisions in. **b** cut off skin from. **2** hurt by severe criticism etc. **3** loosen (soil) with a scarifier. □ **scarification** /ˌskærɪfɪ'keɪʃ(ə)n, ˌskeər-/ n. [ME f. F scarifier f. LL scarificare f. L scarifare f. Gk skariphaomai f. skariphos stylus]

scarify[2] /'skeərɪˌfaɪ/ v.tr. & intr. (**-ies, -ied**) colloq. scare; terrify.

scarious /'skeərɪəs/ adj. Bot. (of a part of a plant etc.) having a dry membranous appearance; thin and brittle. [F scarieux or mod.L scariosus]

scarlatina /ˌskɑːlə'tiːnə/ n. = scarlet fever. [mod.L f. It. scarlattina (febbre fever) dimin. of scarlatto SCARLET]

Scarlatti[1] /skɑː'lætɪ/, (Giuseppe) Domenico (1685–1757), Italian composer. He was a prolific composer of keyboard music, writing more than 550 sonatas for the harpsichord. His work made an important contribution to the development of the sonata form and did much to expand the range of the instrument. He was the son of Alessandro Scarlatti.

Scarlatti[2] /skɑː'lætɪ/, (Pietro) Alessandro (Gaspare) (1660–1725), Italian composer. He was an important and prolific composer of operas, more than seventy of which survive; in them can be found the elements which carried Italian opera through the baroque period and into the classical. He also established the three-part form of the opera overture which was a precursor of the classical symphony. His many other works include cantatas, masses, and oratorios. He was the father of Domenico Scarlatti.

scarlet /'skɑːlɪt/ n. & adj. ● n. **1** a brilliant red colour tinged with orange. **2** clothes or material of this colour (dressed in scarlet). ● adj. of a scarlet colour. □ **scarlet fever** an infectious bacterial fever, affecting esp. children, with a scarlet rash. **scarlet hat** RC Ch. a cardinal's hat as a symbol of rank. **scarlet pimpernel** a small annual plant, Anagallis arvensis, with small esp. scarlet flowers closing in rainy or cloudy weather (also called poor man's weather-glass). **scarlet rash** = ROSEOLA 1. **scarlet runner** = RUNNER 11. **scarlet woman** derog. a notoriously promiscuous woman, a prostitute. [ME f. OF escarlate: ult. orig. unkn.]

Scarlet Pimpernel the name assumed by the hero of a series of novels (including The Scarlet Pimpernel, 1905) by Baroness Orczy. He was a dashing but elusive English nobleman who rescued aristocrats during the French Revolution and smuggled them out of France.

scaroid /'skærɔɪd, 'skeər-/ n. & adj. Zool. ● n. a colourful marine fish of the family Scaridae, native to tropical and temperate seas, including the scarus or parrot-fish. ● adj. of or relating to this family.

scarp /skɑːp/ n. & v. ● n. **1** the inner wall or slope of a ditch in a fortification (cf. COUNTERSCARP). **2** a steep slope. ● v.tr. **1** make (a slope) perpendicular or steep. **2** provide (a ditch) with a steep scarp and counterscarp. **3** (as **scarped** adj.) (of a hillside etc.) steep; precipitous. [It. scarpa]

scarper /'skɑːpə(r)/ v.intr. Brit. sl. run away; escape. [prob. f. It. scappare escape, infl. by rhyming sl. Scapa Flow = go]

scarus /'skeərəs/ n. a fish of the genus Scarus, with brightly coloured scales, and teeth fused to form a parrot-like beak used for eating coral (see also parrot-fish). [L f. Gk skaros]

scarves pl. of SCARF[1].

scary /'skeərɪ/ adj. (**scarier, scariest**) colloq. scaring, frightening. □ **scarily** adv.

scat[1] /skæt/ v. & int. colloq. ● v.intr. (**scatted, scatting**) depart quickly. ● int. go! [perh. abbr. of SCATTER]

scat[2] /skæt/ n. & v. ● n. improvised jazz singing using sounds imitating instruments, instead of words. ● v.intr. (**scatted, scatting**) sing scat. [prob. imit.]

scat[3] /skæt/ n. **1** excrement. **2** the droppings of an animal, esp. a carnivore. [Gk skōr skatos dung]

scathe /skeɪð/ v. & n. ● v.tr. **1** poet. injure esp. by blasting or withering. **2** (as **scathing** adj.) witheringly scornful (scathing look). **3** (with neg.) do the least harm to (shall not be scathed) (cf. UNSCATHED). ● n. (usu. with neg.) archaic harm; injury (without scathe). □ **scatheless** predic.adj. **scathingly** adv. [(v.) ME f. ON skatha = OE sceathian: (n.) OE f. ON skathi = OE sceatha malefactor, injury, f. Gmc]

scatology /skæ'tɒlədʒɪ/ n. **1 a** a morbid interest in excrement. **b** a preoccupation with obscene literature, esp. that concerned with the excretory functions. **c** such literature. **2** the study of fossilized dung.

3 the study of excrement for esp. diagnosis. □ **scatological** /ˌskætə'lɒdʒɪk(ə)l/ adj. [Gk skōr skatos dung + -LOGY]

scatophagous /skæ'tɒfəgəs/ adj. feeding on dung. [as SCATOLOGY + Gk -phagos -eating]

scatter /'skætə(r)/ v. & n. ● v. **1** tr. **a** throw here and there; strew (scattered gravel on the road). **b** cover by scattering (scattered the road with gravel). **2** tr. & intr. **a** move or cause to move in flight etc.; disperse (scattered to safety at the sound). **b** disperse or cause (hopes, clouds, etc.) to disperse. **3** tr. (as **scattered**) not clustered together; wide apart; sporadic (scattered villages). **4** tr. Physics deflect or diffuse (light, particles, etc.). **5 a** intr. (of esp. a shotgun) fire a charge of shot diffusely. **b** tr. fire (a charge) in this way. ● n. (also **scattering**) **1** the act or an instance of scattering. **2** a small amount scattered. **3** the extent of distribution of esp. shot. □ **scatter cushions** (or **rugs**) cushions, rugs, etc., placed here and there for effect. **scatter diagram** (or **plot**) Statistics a diagram in which two variates are plotted for a number of subjects, the pattern of the resulting points revealing any correlation present. **scatter-shot** n. & adj. esp. N. Amer. firing at random. □ **scatterer** n. [ME, prob. var. of SHATTER]

scatterbrain /'skætəˌbreɪn/ n. a person given to silly or disorganized thought with lack of concentration. □ **scatterbrained** adj.

scatty /'skætɪ/ adj. (**scattier, scattiest**) Brit. colloq. scatterbrained; disorganized. □ **scattily** adv. **scattiness** n. [abbr.]

scaup /skɔːp/ n. a diving duck of the genus Aythya, the male of which has a dark head and breast and a white-sided body. [earlier scaup-duck, f. Sc. var. of scalp mussel-bed]

scauper var. of SCALPER 2.

scaur var. of SCAR[2].

scavenge /'skævɪndʒ/ v. & n. ● v. **1** tr. & intr. (usu. foll. by for) **a** search for and collect (discarded items). **b** (of an animal or bird) feed on (carrion). **2** tr. remove unwanted products from (an internal-combustion engine cylinder etc.). ● n. the action or process of scavenging. [back-form. f. SCAVENGER]

scavenger /'skævɪndʒə(r)/ n. **1** a person who seeks and collects discarded items. **2** an animal feeding on carrion, refuse, etc. **3** Brit. archaic a person employed to clean the streets etc. □ **scavengery** n. [ME scavager f. AF scawager f. scawage f. ONF escauwer inspect f. Flem. scauwen, rel. to SHOW: for -n- cf. MESSENGER]

scazon /'skeɪz(ə)n, 'skæz-/ n. Prosody a modification of the iambic trimeter, in which a spondee or trochee takes the place of the final iambus. [L f. Gk skazōn f. skazō limp]

Sc.D. abbr. Doctor of Science. [L scientiae doctor]

SCE abbr. Scottish Certificate of Education.

scena /'ʃeɪnɑː/ n. Mus. **1** a scene or part of an opera. **2** an elaborate dramatic solo usu. including recitative. [It. f. L: see SCENE]

scenario /sɪ'nɑːrɪəʊ, -'neərɪəʊ/ n. (pl. **-os**) **1** an outline of the plot of a play, film, opera, etc., with details of the scenes, situations, etc. **2 a** a postulated sequence of future events. **b** any situation or sequence of events. □ **scenarist** /'siːnərɪst/ n. (in sense 1). [It. (as SCENA)]

scend /send/ n. & v. Naut. ● n. **1** the impulse given by a wave or waves (scend of the sea). **2** a plunge of a vessel. ● v.intr. (of a vessel) plunge or pitch owing to the impulse of a wave. [alt. f. SEND or DESCEND]

scene /siːn/ n. **1** a place in which events in real life, drama, or fiction occur; the locality of an event etc. (the scene was set in India; the scene of the disaster). **2 a** an incident in real life, fiction, etc. (distressing scenes occurred). **b** a description or representation of an incident etc. (scenes of clerical life). **3** a public incident displaying emotion, temper, etc., esp. when embarrassing to others (made a scene in the restaurant). **4 a** a continuous portion of a play in a fixed setting and usu. without a change of personnel; a subdivision of an act. **b** a similar section of a film, book, etc. **5 a** any of the pieces of scenery used in a play. **b** these collectively. **6** a landscape or a view (a desolate scene). **7** colloq. **a** an area of action or interest (not my scene). **b** a way of life; a milieu (well-known on the jazz scene). **8** archaic the stage of a theatre. □ **behind the scenes 1** Theatr. among the actors, scenery, etc. offstage. **2** unknown to the public; secret(ly). **behind-the-scenes** (attrib.) secret, using secret information (a behind-the-scenes investigation). **change of scene** a change of one's surroundings esp. through travel. **come on the scene** arrive. **quit the scene** die; leave. **scene-dock** a space for storing scenery near the stage. **scene-shifter** a person who moves scenery in a theatre. **scene-shifting** this activity. **set the scene 1** describe the location of events. **2** give preliminary information. [L scena f. Gk skēnē tent, stage]

scenery /'siːnərɪ/ n. **1** the general appearance of the natural features of a landscape, esp. when picturesque. **2** *Theatr.* structures (such as painted representations of landscape, rooms, etc.) used on a theatre stage to represent features in the scene of the action. □ **change of scenery** = *change of scene* (see SCENE). [earlier *scenary* f. It. SCENARIO: assim. to -ERY]

scenic /ˈsiːnɪk/ *adj.* **1 a** picturesque; impressive or beautiful (*took the scenic route*). **b** of or concerning natural scenery (*flatness is the main scenic feature*). **2** (of a picture etc.) representing an incident. **3** *Theatr.* of or on the stage (*scenic performances*). □ **scenic railway 1** a miniature railway running through artificial scenery at funfairs etc. **2** = *big dipper* 1. □ **scenically** *adv.* [L *scenicus* f. Gk *skēnikos* of the stage (as SCENE)]

scent /sent/ *n. & v.* ● *n.* **1** a distinctive, esp. pleasant, smell (*the scent of roses*). **2 a** a scent trail left by an animal perceptible to hounds etc. **b** clues etc. that can be followed like a scent trail (*lost the scent in Paris*). **c** the power of detecting or distinguishing smells etc. or of discovering things (*some dogs have little scent; the scent for talent*). **3** *Brit.* = PERFUME 2. **4** a trail laid in a paper-chase. ● *v.* **1** *tr.* **a** discern by scent (*the dog scented game*). **b** perceive, recognize, detect (*scent treachery*). **2** *tr.* make fragrant or foul-smelling. **3** *tr.* (as **scented** *adj.*) having esp. a pleasant smell (*scented soap*). **4** *tr.* apply the sense of smell to (*scented the air*). □ **false scent 1** a scent trail laid to deceive. **2** false clues etc. intended to deflect pursuers. **on the scent** pursuing the correct line of investigation. **put** (or **throw**) **off the scent** deceive by false clues etc. **scent-bag** a bag of aniseed etc. used to lay a trail in drag-hunting. **scent-gland** (or **-organ**) a gland in some animals secreting musk, civet, etc. **scent out** discover by smelling or searching. □ **scentless** *adj.* [ME *sent* f. OF *sentir* perceive, smell, f. L *sentire*; *-c-* (17th c.) unexpl.]

scepsis /ˈskɛpsɪs/ *n.* (US **skepsis**) **1** philosophic doubt. **2** sceptical philosophy. [Gk *skepsis* inquiry, doubt f. *skeptomai* consider]

scepter US var. of SCEPTRE.

sceptic /ˈskɛptɪk/ *n. & adj.* (US **skeptic**) ● *n.* **1** a person inclined to doubt all accepted opinions; a cynic. **2** a person who doubts the truth of Christianity and other religions. **3** *Philos.* a person who accepts the philosophy of scepticism. ● *adj.* = SCEPTICAL. [F *sceptique* or L *scepticus* f. Gk *skeptikos* (as SCEPSIS)]

sceptical /ˈskɛptɪk(ə)l/ *adj.* (US **skeptical**) **1** inclined to question the truth or soundness of accepted ideas, facts, etc.; critical; incredulous. **2** *Philos.* of or accepting scepticism, denying the possibility of knowledge. □ **sceptically** *adv.*

scepticism /ˈskɛptɪˌsɪz(ə)m/ *n.* (US **skepticism**) **1** a sceptical attitude in relation to accepted ideas, facts, etc.; doubting or critical disposition. **2** *Philos.* the doctrine of the sceptics; the opinion that real knowledge is unattainable. (*See note below.*) [f. SCEPTIC + -ISM]

▪ The ancient doctrine of scepticism (also called *Pyrrhonism*) was established by Pyrrho and continued at the Academy in Athens. In modern philosophy scepticism has taken many forms: the most extreme sceptics have doubted whether any knowledge at all of the external world is possible (see SOLIPSISM), Descartes attempted to question his own existence, while others asked whether objects exist when not experienced (Berkeley's idealism), or whether objects exist at all beyond our experiences of them (Hume).

sceptre /ˈsɛptə(r)/ *n.* (US **scepter**) **1** a staff borne esp. at a coronation as a symbol of sovereignty. **2** royal or imperial authority. □ **sceptred** *adj.* [ME f. OF (s)*ceptre* f. L *sceptrum* f. Gk *skēptron* f. *skēptō* lean on]

sch. *abbr.* **1** scholar. **2** school. **3** schooner.

schadenfreude /ˈʃɑːd(ə)nˌfrɔɪdə/ *n.* the malicious enjoyment of another's misfortunes. [G f. *Schaden* harm + *Freude* joy]

schappe /ˈʃæpə/ *n.* fabric or yarn made from waste silk. [G, = waste silk]

schedule /ˈʃɛdjuːl, ˈskɛd-/ *n. & v.* ● *n.* **1 a** a list or plan of intended events, times, etc.; a timetable. **b** a plan of work (*not on my schedule for next week*). **2** a list of rates or prices. **3** a tabulated inventory etc. esp. as an appendix to a document. ● *v.tr.* **1** include in a schedule. **2** make a schedule of. **3** *Brit.* include (a building) in a list for preservation or protection. □ **according to schedule** as planned; on time. **behind schedule** behind time. **on schedule** on time; as planned. **scheduled flight** (or **service** etc.) a public flight, service, etc., according to a regular timetable. **scheduled territories** *hist.* = *sterling area.* □ **scheduler** *n.* [ME f. OF *cedule* f. LL *schedula* slip of paper, dimin. of *scheda* f. Gk *skhedē* papyrus-leaf]

Scheele /ˈʃiːlə, ˈʃeɪlə/, Carl Wilhelm (1742–86), Swedish chemist. He was a keen experimenter, working in difficult and often hazardous conditions; he discovered a number of substances including glycerol and a green gas that was later named chlorine. He is also noted for his

discovery of oxygen in 1773, which he named 'fire air', although he did not publish his findings until after the publication of Joseph Priestley's work in 1774. He also discovered a process resembling pasteurization.

scheelite /ˈʃiːlaɪt/ *n. Mineral.* calcium tungstate in its mineral crystalline form. [SCHEELE]

Scheherazade /ʃəˌherəˈzɑːd, -ˈzɑːdə/ the female narrator of the *Arabian Nights.*

Scheldt /skelt, ʃelt/ (also **Schelde** /ˈskeldə, ˈʃel-/; called in French *Escaut*) a river of northern Europe. Rising in northern France, it flows 432 km (270 miles) through Belgium and the Netherlands to the North Sea.

schema /ˈskiːmə/ *n.* (*pl.* **schemata** /-mətə/ or **schemas**) **1** a synopsis, outline, or diagram. **2** a proposed arrangement. **3** *Logic* a syllogistic figure. **4** (in Kantian philosophy) a conception of what is common to all members of a class; a general type or essential form. [Gk *skhēma -atos* form, figure]

schematic /skɪˈmætɪk/ *adj. & n.* ● *adj.* **1** of or concerning a scheme or schema. **2** representing objects by symbols etc. ● *n.* a schematic diagram, esp. of an electronic circuit. □ **schematically** *adv.*

schematism /ˈskiːməˌtɪz(ə)m/ *n.* a schematic arrangement or presentation. [mod.L *schematismus* f. Gk *skhēmatismos* f. *skhēmat-* (as SCHEMA)]

schematize /ˈskiːməˌtaɪz/ *v.tr.* (also **-ise**) **1** put in a schematic form; arrange. **2** represent by a scheme or schema. □ **schematization** /ˌskiːmətaɪˈzeɪʃ(ə)n/ *n.*

scheme /skiːm/ *n. & v.* ● *n.* **1 a** a systematic plan or arrangement for work, action, etc. **b** a proposed or operational systematic arrangement (*a colour scheme*). **2** an artful or deceitful plot. **3** a timetable, outline, syllabus, etc. ● *v.* **1** *intr.* (often foll. by *for*, or to + infin.) plan esp. secretly or deceitfully; intrigue. **2** *tr.* plan to bring about, esp. artfully or deceitfully (*schemed their downfall*). □ **schemer** *n.* [L *schema* f. Gk (as SCHEMA)]

scheming /ˈskiːmɪŋ/ *adj. & n.* ● *adj.* artful, cunning, or deceitful. ● *n.* plots; intrigues. □ **schemingly** *adv.*

schemozzle var. of SHEMOZZLE.

scherzando /skeəˈtsændəʊ/ *adv., adj., & n. Mus.* ● *adv. & adj.* in a playful manner. ● *n.* (*pl.* **-os** or **scherzandi** /-dɪ/) a passage played in this way. [It., gerund of *scherzare* to jest (as SCHERZO)]

scherzo /ˈskeətsəʊ/ *n.* (*pl.* **-os** or **scherzi** /-tsɪ/) *Mus.* a vigorous, light, or playful composition, usu. as a movement in a symphony, sonata, etc. [It., lit. 'jest']

Schiaparelli[1] /ˌskjæpəˈrelɪ/, Elsa (1896–1973), Italian-born French fashion designer. She settled in Paris and opened her own establishment in the late 1920s. She introduced padded shoulders to her garments in 1932, and the vivid shade now known as 'shocking pink' in 1947. Her fashions also included touches of the surreal, drawing inspiration from artists such as Salvador Dali. She later expanded her interests into ready-to-wear fashions and ranges of perfume and cosmetics.

Schiaparelli[2] /ˌskjæpəˈrelɪ/, Giovanni Virginio (1835–1910), Italian astronomer. He studied the nature of cometary tails, and showed that many meteors are derived from comets and follow similar orbits. He observed Mars in detail, identifying the southern polar ice cap and features which he termed 'seas', 'continents', and 'channels' (*canali*). The last was mistranslated by Percival Lowell as 'canals', beginning a long-running controversy about intelligent life on Mars (see CANAL 3).

Schiele /ˈʃiːlə/, Egon (1890–1918), Austrian painter and draughtsman. He joined the Vienna Academy of Fine Arts in 1906 but left in 1909 because his teacher disapproved of his paintings, which were influenced by Gustav Klimt and art nouveau. From 1910 onwards he evolved a distinctive expressionist style, characterized by an aggressive linear energy and a neurotic intensity, reflecting his interest in the work of Sigmund Freud; his paintings and drawings (often self-portraits) depicted distorted, frequently emaciated bodies with an explicit erotic content, which resulted in him being imprisoned for a month in 1911. His pictures received international acclaim at the exhibition of the Vienna Secession in 1918; Schiele died in the same year of Spanish influenza.

Schiller /ˈʃɪlə(r)/, (Johann Christoph) Friedrich von (1759–1805), German dramatist, poet, historian, and critic. His early work was influenced by the *Sturm und Drang* movement; his mature work established him as a major figure, with Goethe (with whom he formed a long-standing friendship from 1794 until his death), of the Enlightenment in Germany. His first major work was the historical drama in blank verse *Don Carlos* (1787). His other historical plays include the trilogy *Wallenstein* (1800), which drew on his historical studies of the Thirty Years War, *Mary Stuart* (1800), and *William Tell* (1804). These works are concerned with freedom and responsibility, whether political, personal, or moral. Among his best-known poems are 'Ode to Joy', which Beethoven set to music in his Ninth Symphony, and 'The Artists' (both written *c.*1787). His many essays on aesthetics include *On Naive and Reflective Poetry* (1795–6), in which he contrasts his poetry with that of Goethe.

schilling /ˈʃɪlɪŋ/ n. **1** the chief monetary unit of Austria. **2** a coin of this value. [G (as SHILLING)]

Schindler /ˈʃɪndlə(r)/, Oskar (1908–74), German industrialist. In 1940 he established a lucrative enamelware factory in Cracow, Poland, employing Jewish workers from the city's ghetto. After the Nazi evacuation of the ghetto in 1943, Schindler exercised his financial and political influence to protect his employees from being sent to Plaszów, a nearby labour camp. In 1944 he compiled a list of more than 1,200 Jews from Plaszów and his own factory to be relocated at a new armaments factory in Czechoslovakia, thereby saving them from certain death in concentration camps. Schindler's life and role in rescuing Polish Jews are celebrated in the novel *Schindler's Ark* (1982), by Thomas Keneally, and the film *Schindler's List* (1993), directed by Steven Spielberg.

schipperke /ˈskɪpəki, ˈʃɪp-/ n. a small black tailless breed of dog with a ruff of fur round its neck. [Du. dial., = little boatman, f. its use as a watchdog on barges]

schism /ˈsɪz(ə)m, ˈskɪz-/ n. **1 a** the division of a group into opposing sections or parties. **b** any of the sections so formed. **2 a** the separation of a Church into two Churches or the secession of a group owing to doctrinal, disciplinary, etc., differences. **b** the offence of causing or promoting such a separation. [ME f. OF *s(c)isme* f. eccl.L *schisma* f. Gk *skhisma -atos* cleft f. *skhizō* to split]

schismatic /sɪzˈmatɪk, skɪz-/ adj. & n. (also **schismatical**) ● adj. of, concerning, or inclining to schism. ● n. **1** a holder of schismatic opinions. **2** a member of a schismatic faction or a seceded branch of a Church. □ **schismatically** adv. [ME f. OF *scismatique* f. eccl.L *schismaticus* f. eccl.Gk *skhismatikos* (as SCHISM)]

schist /ʃɪst/ n. a foliated metamorphic rock composed of layers of different minerals and splitting into thin irregular plates. □ **schistose** /-təʊz, -təʊs/ adj. [F *schiste* f. L *schistos* f. Gk *skhistos* split (as SCHISM)]

schistosome /ˈʃɪstəˌsəʊm/ n. a parasitic tropical flatworm of the genus *Schistosoma*, carried by freshwater snails and infesting the blood vessels of birds and mammals, causing bilharzia in humans. Also called *blood fluke*. [Gk *skhistos* divided (as SCHISM) + *sōma* body]

schistosomiasis /ˌʃɪstəsəˈmaɪəsɪs/ n. = BILHARZIA 2. [mod.L *Schistosoma* (the genus name, as SCHISTOSOME)]

schizanthus /skɪˈzænθəs/ n. a solanaceous plant of the genus *Schizanthus*, with showy flowers in various colours, and finely divided leaves. Also called *poor man's orchid*. [mod.L f. Gk *skhizō* to split + *anthos* flower]

schizo /ˈskɪtsəʊ/ adj. & n. colloq. ● adj. schizophrenic. ● n. (pl. **-os**) a schizophrenic. [abbr.]

schizocarp /ˈskɪzəˌkɑːp/ n. Bot. any of a group of dry fruits that split into single-seeded parts when ripe. □ **schizocarpic** /ˌskɪtsəˈkɑːpɪk/ adj. **schizocarpous** adj. [Gk *skhizō* to split + *karpos* fruit]

schizoid /ˈskɪtsɔɪd/ adj. & n. ● adj. **1 a** resembling or tending towards schizophrenia but with milder symptoms. **b** relating to or affected by a personality disorder marked by coldness, inability to form social relations, and excessive introspection. **2** = SCHIZOPHRENIC adj. 2. ● n. a schizoid person.

schizophrenia /ˌskɪtsəˈfriːnɪə/ n. a mental disease marked by a breakdown in the relation between thoughts, feelings, and actions, frequently accompanied by delusions and retreat from social life. [mod.L f. Gk *skhizō* to split + *phrēn* mind]

schizophrenic /ˌskɪtsəˈfrɛnɪk, -ˈfriːnɪk/ adj. & n. ● adj. **1** characteristic of or having schizophrenia. **2** characterized by mutually contradictory or inconsistent elements, attitudes, etc. ● n. a person with schizophrenia.

schizothymia /ˌskɪtsəˈθaɪmɪə/ n. Psychol. an introvert condition with a tendency to schizophrenia. □ **schizothymic** adj. [mod.L (as SCHIZOPHRENIA + Gk *thumos* temper)]

schlemiel /ʃləˈmiːl/ n. US colloq. a foolish or unlucky person. [Yiddish *shlumiel*]

schlep /ʃlɛp/ v. & n. (also **schlepp**) colloq. ● v. (**schlepped, schlepping**) **1** tr. carry, drag. **2** intr. go or work tediously or effortly. ● n. esp. US trouble or hard work. [Yiddish *shlepn* f. G *schleppen* drag]

Schleswig /ˈʃlɛsvɪɡ/ a former Danish duchy, situated in the southern part of the Jutland peninsula. Taken by Prussia in 1866, it was incorporated with the neighbouring duchy of Holstein as the province of Schleswig-Holstein. The northern part of Schleswig was returned to Denmark in 1920 after a plebiscite held in accordance with the Treaty of Versailles.

Schleswig-Holstein /ˌʃlɛsvɪɡˈhɒlstaɪn/ a state of NW Germany, occupying the southern part of the Jutland peninsula; capital, Kiel. It comprises the former duchies of Schleswig and Holstein, annexed by Prussia in 1866. The northern part of Schleswig was returned to Denmark by plebiscite in 1920.

Schliemann /ˈʃliːmən/, Heinrich (1822–90), German archaeologist. A former businessman with an amateur interest in archaeology, he was determined to discover the location of the ancient city of Troy, and in 1871 began excavating the mound of Hissarlik on the NE Aegean coast of Turkey. He discovered the remains of a succession of nine cities on the site, identifying the second oldest as Homer's Troy (and romantically naming a hoard of jewellery 'Priam's Treasure'), although he had in fact uncovered a pre-Homeric site. He subsequently undertook significant excavations at Mycenae (1876) and at other sites in mainland Greece.

schlieren /ˈʃliːrən/ n. **1** a visually discernible area or stratum of different density in a transparent medium. **2** Geol. an irregular streak of mineral in igneous rock. [G, pl. of *Schliere* streak]

schlock /ʃlɒk/ n. N. Amer. colloq. inferior goods; trash. [Yiddish *shlak* a blow]

schmaltz /ʃmɔːlts, ʃmælts/ n. esp. US colloq. sentimentality, esp. in music, drama, etc. □ **schmaltzy** adj. (**schmaltzier, schmaltziest**). [Yiddish f. G *Schmalz* dripping, lard]

Schmidt–Cassegrain telescope /ˈʃmɪtˈkæsɪˌɡreɪn/ n. a type of catadioptric telescope, using the correcting plate of a Schmidt telescope together with the secondary mirror and rear focus of a Cassegrain telescope.

Schmidt telescope (or **camera**) /ʃmɪt/ n. a type of catadioptric telescope used solely for wide-angle astronomical photography, with a thin glass plate at the front to correct for spherical aberration. A curved photographic plate is placed at the prime focus inside the telescope. [Bernhard Voldemar *Schmidt*, Ger. inventor (1879–1935), who invented it]

schmooze /ʃmuːz/ v. & n. N. Amer. colloq. ● v.intr. talk idly, chat; gossip. ● n. idle talk, chat; gossip. [Yiddish *schmuesn* (v.), *schmues* (n.) f. Heb.]

schmuck /ʃmʌk/ n. esp. US sl. a foolish or contemptible person. [Yiddish]

schnapps /ʃnæps/ n. any of various strong spirits resembling genever or Dutch gin. [G, = dram of liquor f. LG & Du. *snaps* mouthful (as SNAP)]

schnauzer /ˈʃnaʊtsə(r), ˈʃnaʊzə(r)/ n. a German breed of dog with a close wiry coat and heavy whiskers round the muzzle. [G f. *Schnauze* muzzle, SNOUT]

schnitzel /ˈʃnɪts(ə)l/ n. an escalope of veal. □ **Wiener** /ˈviːnə/ (or **Vienna**) **schnitzel** a breaded, fried, and garnished schnitzel. [G, = slice]

schnook /ʃnʊk/ n. US sl. a dupe; a simpleton. [perh. f. G *Schnucke* a small sheep or Yiddish *shnuk* snout]

schnorkel var. of SNORKEL.

schnorrer /ˈʃnɔːrə(r)/ n. esp. US sl. a beggar or scrounger; a layabout. [Yiddish f. G *Schnurrer*]

Schoenberg /ˈʃɜːnbɜːɡ/, Arnold (1874–1951), Austrian-born American composer and music theorist. His major contribution to modernism is his development of the concepts of atonality and serialism. He introduced atonality into the final movement of his second string quartet (1907–8) and abolished the distinction between consonance and dissonance in his *Three Piano Pieces* (1909). From these experiments he evolved a serial system of composition (see SERIALISM); the third and fourth movements of the *Serenade* (1923) for seven instruments and bass voice are the first clear examples of this technique. He was a professor of music in Vienna and in Berlin until 1933 when, after condemnation of his music by Hitler, he emigrated to the US. He continued to develop his serial techniques and his work influenced composers such as his pupils Berg and Webern.

scholar /ˈskɒlə(r)/ n. **1** a learned person, esp. in language, literature, etc.; an academic. **2** the holder of a scholarship. **3 a** a person with specified academic ability (*is a poor scholar*). **b** a person who learns (*a scholar of life*). **4** *archaic colloq.* a person able to read and write. **5** *archaic* a schoolboy or schoolgirl. □ **scholar's mate** see MATE². □ **scholarly** *adj.* **scholarliness** *n.* [ME f. OE *scol(i)ere* & OF *escol(i)er* f. LL *scholaris* f. L *schola* SCHOOL¹]

scholarship /'skɒləʃɪp/ *n.* **1 a** academic achievement; learning of a high level. **b** the methods and standards characteristic of a good scholar (*shows great scholarship*). **2 a** payment from the funds of a school, university, local government, etc., to maintain a student in full-time education, awarded on the basis of scholarly achievement. **b** an instance of this.

scholastic /skə'læstɪk/ *adj. & n.* ● *adj.* **1** of or concerning universities, schools, education, teachers, etc. **2 a** academic. **b** pedantic; formal (*shows scholastic precision*). **3** *Philos. hist.* of, resembling, or concerning scholasticism, esp. in dealing with logical subtleties. ● *n.* **1** an adherent of scholasticism. **2** *RC Ch.* a member of any of several religious orders, who is between the novitiate and the priesthood. □ **scholastically** *adv.* [L *scholasticus* f. Gk *skholastikos* studious f. *skholazō* be at leisure, formed as SCHOOL¹]

scholasticism /skə'læstɪˌsɪz(ə)m/ *n.* **1** the educational tradition of the medieval 'schools' or universities, especially a method of philosophical and theological speculation which aimed at a better understanding of Christianity. (*See note below.*) **2** narrow or unenlightened insistence on traditional doctrines etc.

▪ The theoretical foundations of scholasticism were laid by St Augustine and Boethius, who attempted to incorporate the thought of Plato and the Neoplatonists into Christian theology. Of decisive importance was the introduction of the works of Aristotle into western Europe in the 13th century. Among the most prominent figures of scholasticism were St Anselm, Abelard, Duns Scotus, and, most importantly, St Thomas Aquinas. The last major scholastic philosopher was William of Occam (d.1349); from his time scholasticism declined, although interest in it revived at the end of the 19th century.

scholiast /'skəʊlɪˌæst/ *n. hist.* an ancient or medieval scholar, esp. a grammarian, who annotated ancient literary texts. □ **scholiastic** /ˌskəʊlɪ'æstɪk/ *adj.* [med.Gk *skholiastēs* f. *skholiazō* write scholia: see SCHOLIUM]

scholium /'skəʊlɪəm/ *n.* (*pl.* **scholia** /-lɪə/) a marginal note or explanatory comment, esp. by an ancient grammarian on a classical text. [mod.L f. Gk *skholion* f. *skholē* disputation: see SCHOOL¹]

school¹ /skuːl/ *n. & v.* ● *n.* **1 a** an institution for educating or giving instruction, esp. *Brit.* for children under 19 years, or *N. Amer.* for any level of instruction including college or university. **b** (*attrib.*) associated with or for use in school (*a school bag; school dinners*). **2 a** the buildings used by such an institution. **b** the pupils, staff, etc. of a school. **c** the time during which teaching is done, or the teaching itself (*no school today*). **3 a** a branch of study with separate examinations at a university; a department or faculty (*the history school*). **b** *Brit.* the hall in which university examinations are held. **c** (in *pl.*) *Brit.* such examinations. **4 a** the disciples, imitators, or followers of a philosopher, artist, etc. (*the school of Epicurus*). **b** a group of artists etc. whose works share distinctive characteristics. **c** a group of people sharing a cause, principle, method, etc. (*school of thought*). **5** *Brit.* a group of gamblers or of persons drinking together (*a poker school*). **6** *colloq.* instructive or disciplinary circumstances, occupation, etc. (*the school of adversity; learned in a hard school*). **7** *hist.* a medieval lecture-room. **8** (usu. foll. by *of*) *Mus.* a handbook or book of instruction (*school of counterpoint*). **9** (in *pl.*; *prec.* by *the*) *hist.* medieval universities, their teachers, disputations, etc. ● *v.tr.* **1** send to school; provide for the education of. **2** (often foll. by *to*) discipline; train; control. **3** (as **schooled** *adj.*) (foll. by *in*) educated or trained (*schooled in humility*). □ **at** (*US* **in**) **school** attending lessons etc. **go to school 1** begin one's education. **2** attend lessons. **leave school** finish one's education. **of the old school** according to former and esp. better tradition (*a gentleman of the old school*). **school age** the age-range in which children normally attend school. **school board** *N. Amer.* or *hist.* a board or authority for local education. **schooldays** the time of being at school, esp. in retrospect. **school-inspector** a government official reporting on the efficiency, teaching standards, etc. of schools. **school-leaver** *Brit.* a child leaving school esp. at the minimum specified age. **school-leaving age** the minimum age at which a schoolchild may leave school. **school-ma'm** (or **-marm**) *US colloq.* a schoolmistress. **school-marmish** *colloq.* prim and fussy. **school-ship** a training-ship. **school-time 1** lesson-time at school or at home. **2** school-days. **school year** = *academic year*. [ME

f. OE *scōl, scolu*, & f. OF *escole* ult. f. L *schola* school f. Gk *skholē* leisure, disputation, philosophy, lecture-place]

school² /skuːl/ *n. & v.* ● *n.* (often foll. by *of*) a shoal of fish, porpoises, whales, etc. ● *v.intr.* form schools. [ME f. MLG, MDu. *schōle* f. WG]

schoolable /'skuːləb(ə)l/ *adj.* liable by age etc. to compulsory education.

schoolboy /'skuːlbɔɪ/ *n.* a boy attending school.

schoolchild /'skuːltʃaɪld/ *n.* (*pl.* **-children**) a child attending school.

schoolfellow /'skuːlˌfeləʊ/ *n.* a past or esp. present member of the same school.

schoolgirl /'skuːlgɜːl/ *n.* a girl attending school.

schoolhouse /'skuːlhaʊs/ *n. Brit.* **1** a building used as a school, esp. in a village. **2** a dwelling-house adjoining a school.

schoolie /'skuːlɪ/ *n. Austral. sl. & dial.* a schoolteacher.

schooling /'skuːlɪŋ/ *n.* **1** education, esp. at school. **2** training or discipline, esp. of an animal.

schoolman /'skuːlmən/ *n.* (*pl.* **-men**) **1** *hist.* a teacher in a medieval European university. **2** *RC Ch. hist.* a theologian seeking to deal with religious doctrines by the rules of Aristotelian logic. **3** *US* a male teacher.

schoolmaster /'skuːlˌmɑːstə(r)/ *n.* a head or assistant male teacher. □ **schoolmasterly** *adj.*

schoolmate /'skuːlmeɪt/ *n.* = SCHOOLFELLOW.

schoolmistress /'skuːlˌmɪstrɪs/ *n.* a head or assistant female teacher.

schoolroom /'skuːlruːm, -rʊm/ *n.* a room used for lessons in a school or esp. in a private house.

schoolteacher /'skuːlˌtiːtʃə(r)/ *n.* a person who teaches in a school.

schooner /'skuːnə(r)/ *n.* **1** a fore-and-aft rigged ship with two or more masts, the foremast being smaller than the other masts. **2 a** *Brit.* a measure or glass for esp. sherry. **b** *US & Austral.* a tall beer-glass. **3** *US hist.* = *prairie schooner*. [18th c.: orig. uncert.]

Schopenhauer /'ʃəʊp(ə)nˌhaʊə(r)/, 'ʃɒp-/, Arthur (1788–1860), German philosopher. His pessimistic philosophy is based on studies of Kant, Plato, and the Hindu Vedas, and is embodied in his principal work *The World as Will and Idea* (1819). According to this, the will (self-consciousness in man and the unconscious forces of nature) is the only reality; the material world is an illusion created by the will. This will is a malignant thing, which deceives us into reproducing and perpetuating life. Asceticism and chastity are the duty of humankind, with a view to terminating the evil; egoism, which manifests itself principally as 'the will to live', must be overcome. His theory of the predominance of the will influenced the development of Freudian psychoanalysis and, via Nietzsche, existentialism.

schorl /ʃɔːl/ *n.* black tourmaline. [G *Schörl*]

schottische /ʃɒ'tiːʃ/ *n.* **1** a kind of slow polka. **2** the music for this. [G *der schottische Tanz* the Scottish dance]

Schreiner /'ʃraɪnə(r)/, Olive (Emilie Albertina) (1855–1920), South African novelist, short-story writer, and feminist. She worked as a governess, in 1881 moving to England, where she published her best-known novel *The Story of an African Farm* (1883), initially under the pseudonym Ralph Iron. The novel was both acclaimed and condemned for its heroine's unconventional approach to religion and marriage. Schreiner returned to South Africa in 1889, later becoming an active supporter of the women's suffrage movement and writing the influential *Women and Labour* (1911).

Schrödinger /'ʃrɜːdɪŋə(r)/, Erwin (1887–1961), Austrian theoretical physicist. In the 1920s he founded the study of wave mechanics, deriving the equation whose roots define the energy levels of atoms. Professor of physics at Berlin from 1927, Schrödinger left Germany after the Nazis came to power in 1933, returning to Austria in 1936 but fleeing again after the *Anschluss* of 1938 and finally settling in Dublin. He wrote a number of general works, including *What is Life?* (1944), which proved influential among scientists of all disciplines. He shared the Nobel Prize for physics in 1933.

Schrödinger equation *n. Physics* a differential equation used in quantum mechanics for the wave function of a particle. [SCHRÖDINGER]

Schubert /'ʃuːbət/, Franz (1797–1828), Austrian composer. While his music is associated with the romantic movement for its lyricism and emotional intensity, it belongs in formal terms to the classical age of Haydn and Mozart. During his brief life he produced more than 600 songs, nine symphonies, fifteen string quartets, and twenty-one piano sonatas, as well as operas and church music. One of his most important contributions to music was as the foremost composer of German lieder; works include the songs 'Gretchen am Spinnrade' (1814) and 'Erlkönig' (1815; 'Erl-King'), in addition to such song cycles as *Die Schöne Müllerin* (1823) and *Winterreise* (1827). Among his other significant works are the String Quintet in C major (1828), the 'Trout' piano quintet, and the Ninth Symphony ('the Great C Major').

Schulz /ʃʊlts/, Charles (b.1922), American cartoonist. He is the creator of the 'Peanuts' comic strip (originally entitled 'Li'l Folks') featuring a range of characters including the boy Charlie Brown and the dog Snoopy. The comic strip was first published in 1950 after Schulz sold the rights to United Features Syndicate, and it later appeared in many publications around the world.

Schumacher /ˈʃuːˌmæxə(r)/, E(rnst) F(riedrich) (1911–77), German economist and conservationist. His reputation is chiefly based on his book *Small is Beautiful: Economics as if People Mattered* (1973), which argues that economic growth is a false god of Western governments and industrialists, and that mass production needs to be replaced by smaller, more energy-efficient enterprises. Schumacher also worked to encourage conservation of natural resources and supported the development of intermediate technology in developing countries.

Schumann /ˈʃuːmən/, Robert (Alexander) (1810–56), German composer. He was a leading romantic composer and is particularly noted for his songs and piano music. He drew much of his inspiration from literature, writing incidental music for Byron's *Manfred* (1849) and setting to music poems by Heinrich Heine and Robert Burns. Notable among his piano pieces are the miniatures *Papillons* (1829–31), *Carnaval* (1834–5), and *Waldscenen* (1848–9), the Fantasy in C major (1836), and the Piano Concerto in A minor (1845). His other works include four symphonies and much chamber music. He spent the last two years of his life in a mental asylum.

schuss /ʃʊs/ *n. & v.* ● *n.* a straight downhill run on skis. ● *v.intr.* make a schuss. [G, lit. 'shot']

Schütz /ʃʊts/, Heinrich (1585–1672), German composer and organist. He is regarded as the first German baroque composer and his work reflects the influence of periods spent in Italy, for example in the settings of *Psalms of David* (1619). He composed much church music and what is thought to have been the first German opera (*Dafne*, 1627; now lost). His three settings of the Passion story represent a turning towards a simple meditative style, eschewing instrumental accompaniment and relying on voices alone.

schwa /ʃwɑː, ʃvɑː/ *n.* (also **sheva** /ʃəˈvɑː/) *Phonet.* **1** the indistinct unstressed vowel sound as in *a* moment ago. **2** the symbol /ə/ representing this in the International Phonetic Alphabet. [G f. Heb. šᵉwā, app. f. šaw emptiness]

Schwaben see SWABIA.

Schwäbisch Gmünd /ˌʃveɪbɪʃ gəˈmʊnt/ a city in SW Germany, situated to the east of Stuttgart; pop. (1983) 56,400.

Schwann /ʃvæn, ʃvɒn/, Theodor Ambrose Hubert (1810–82), German physiologist. He is chiefly remembered for his support of cell theory, showing that animals (as well as plants) are made up of individual cells, and that the egg begins life as a single cell. He also isolated the first animal enzyme, pepsin, recognized that fermentation is caused by processes in the yeast cells, and discovered the cells forming the myelin sheaths of nerve fibres (*Schwann cells*). Schwann became disillusioned following criticism of his work by chemists, emigrated to Belgium at the age of 28, and withdrew from scientific work.

Schwarzkopf /ˈʃvɑːtskɒpf/, Dame (Olga Maria) Elisabeth (Friederike) (b.1915), German operatic soprano. She made her début in Berlin in 1942, and went on to become especially famous for her recitals of German lieder and for her roles in works by Richard Strauss such as *Der Rosenkavalier*.

Schwarzwald /ˈʃvɑːtsvalt/ the German name for the BLACK FOREST.

Schweinfurt /ˈʃvaɪnfʊət/ a city in western Germany; pop. (1991) 54,520. It became part of Bavaria in 1803.

Schweitzer /ˈʃwaɪtsə(r), ˈʃvaɪ-/, Albert (1875–1965), German theologian, musician, and medical missionary, born in Alsace. He decided to devote the first thirty years of his life to learning and music and the remainder to the service of others. His main contribution to theology was his book *The Quest of the Historical Jesus* (1906), which emphasized the importance of understanding Jesus within the context of the Jewish apocalyptic thought of his day. In 1913 he qualified as a doctor and went as a missionary to Lambaréné in French Equatorial Africa (now Gabon), where he established a hospital and lived for most of the rest of his life. He was awarded the Nobel Peace Prize in 1952.

Schwerin /ʃveˈriːn/ a city in NE Germany, capital of Mecklenburg-West Pomerania, situated on the south-western shores of Lake Schwerin; pop. (1991) 125,960.

Schwyz /ʃviːts/ a city in central Switzerland, situated to the east of Lake Lucerne, the capital of a canton of the same name; pop. (1990) 12,530. The canton was one of the three original cantons of the Swiss Confederation, of which it was a leading member and to which it gave its name.

sciagraphy /saɪˈægrəfɪ/ *n.* (also **skiagraphy** /skaɪ-/) the art of shading in drawing etc. □ **sciagram** /ˈsaɪəˌgræm/ *n.* **sciagraph** /-ˌgrɑːf/ *n. & v.tr.* **sciagraphic** /ˌsaɪəˈgræfɪk/ *adj.* [F *sciagraphie* f. L *sciagraphia* f. Gk *skiagraphia* f. *skia* shadow]

sciamachy /saɪˈæməkɪ/ *n.* (also **skiamachy** /skaɪ-/) *formal* **1** fighting with shadows. **2** imaginary or futile combat. [Gk *skiamakhia* (as SCIAGRAPHY, *-makhia* -fighting)]

sciatic /saɪˈætɪk/ *adj.* **1** of the hip. **2** of or affecting the sciatic nerve. **3** suffering from or liable to sciatica. □ **sciatic nerve** the largest nerve in the human body, running from the pelvis to the thigh. □ **sciatically** *adv.* [F *sciatique* f. LL *sciaticus* f. L *ischiadicus* f. Gk *iskhiadikos* subject to sciatica f. *iskhion* hip-joint]

sciatica /saɪˈætɪkə/ *n.* neuralgia of the hip and thigh; a pain in the sciatic nerve. [ME f. LL *sciatica* (*passio*) fem. of *sciaticus*: see SCIATIC]

science /ˈsaɪəns/ *n.* **1** a branch of knowledge conducted on objective principles involving the systematized observation of and experiment with phenomena, esp. concerned with the material and functions of the physical universe. (See also *natural science*.) **2 a** systematic and formulated knowledge, esp. of a specified type or on a specified subject (*political science*). **b** the pursuit or principles of this. **3** an organized body of knowledge on a subject (*the science of philology*). **4** skilful technique rather than strength or natural ability. **5** *archaic* knowledge of any kind. □ **science park** an area devoted to scientific research or the development of science-based industries. [ME f. OF f. L *scientia* f. *scire* know]

science fiction *n.* (abbr. **SF**) a genre of fiction based on imagined future technological or scientific advances, major environmental or social changes, etc., and frequently portraying space or time travel and life on other planets. Science fiction emerged in the late 19th century in the works of writers such as Jules Verne and H. G. Wells, although there are earlier precedents, such as Mary Shelley's *Frankenstein* (1818). Following the First World War there was an upsurge in science-fiction writing in Britain and the US. It was during this period that the very popular science-fiction 'pulp' magazines became established, such as those of the US writer and publisher Hugo Gernsback (1884–1967), who was responsible for popularizing the term 'science fiction'. After the dystopian visions of George Orwell, Aldous Huxley, and Isaac Asimov, the period after the Second World War produced writers such as Ray Bradbury, Kurt Vonnegut, and Philip K. Dick (1928–82) in the US; in Britain in the 1960s writers associated with the magazine *New Worlds*, such as Michael Moorcock (b.1939) and J. G. Ballard (b.1930), introduced apocalyptic themes and political radicalism. Films such as *2001: A Space Odyssey* (1968) and *Star Wars* (1977), together with television series such as *Dr Who* (from 1963) and *Star Trek* (from 1966), have helped to bring popular science fiction to a wider audience. In the 1980s a new genre, called cyberpunk, emerged, bringing writers such as William Gibson (b.1948) to prominence.

Science Museum a national museum of science, technology, and industry in South Kensington, London. It was founded in 1857 as part of the Victoria and Albert Museum, and moved to its present site in 1909.

scienter /saɪˈentə(r)/ *adv. Law* intentionally; knowingly. [L f. *scire* know]

sciential /saɪˈenʃ(ə)l/ *adj.* concerning or having knowledge. [LL *scientialis* (as SCIENCE)]

scientific /ˌsaɪənˈtɪfɪk/ *adj.* **1 a** (of an investigation etc.) according to rules laid down in exact science for performing observations and testing the soundness of conclusions. **b** systematic, accurate. **2** used in, engaged in, or relating to (esp. natural) science (*scientific discoveries*; *scientific terminology*). **3** assisted by expert knowledge. □ **scientifically** *adv.* [F *scientifique* or LL *scientificus* (as SCIENCE)]

scientism /ˈsaɪənˌtɪz(ə)m/ *n.* **1 a** a method or doctrine regarded as characteristic of scientists. **b** the use or practice of this. **2** often *derog.* an excessive belief in or application of scientific method. □ **scientistic** /ˌsaɪənˈtɪstɪk/ *adj.*

scientist /ˈsaɪəntɪst/ *n.* **1** a person with expert knowledge of a (usu. physical or natural) science. **2** a person using scientific methods.

Scientology /ˌsaɪən'tɒlədʒɪ/ n. propr. a religious system whose adherents seek self-knowledge and spiritual fulfilment through graded courses of study and training. Based on the system of Dianetics, scientology was developed in the early 1950s by the American science-fiction writer L. Ron Hubbard (1911–86); the Church of Scientology was founded by Hubbard in 1955. Scientology has aroused controversy through its apparently coercive methods of recruiting and retaining followers, while Hubbard and the Church were accused in the 1980s of misappropriating funds. □ **Scientologist** n. [L scientia knowledge + -LOGY]

sci-fi /ˈsaɪfaɪ/ n. (often attrib.) colloq. science fiction. [abbr.: cf. HI-FI]

scilicet /ˈsaɪlɪˌset, ˈskiːlɪˌket/ adv. to wit; that is to say; namely (introducing a word to be supplied or an explanation of an ambiguity). [ME f. L, = scire licet one is permitted to know]

scilla /ˈsɪlə/ n. a liliaceous plant of the genus Scilla, related to the bluebell, usu. bearing small blue star-shaped or bell-shaped flowers and having long glossy straplike leaves. [L f. Gk skilla]

Scilly Isles /ˈsɪlɪ/ (also **Isles of Scilly**, **Scillies** /ˈsɪlɪz/) a group of about 140 small islands (of which five are inhabited) off the south-western tip of England; pop. (1991) 2,900; capital, Hugh Town (on St Mary's). □ **Scillonian** /sɪˈləʊnɪən/ adj. & n.

scimitar /ˈsɪmɪtə(r)/ n. an oriental curved sword usu. broadening towards the point. [F cimeterre, It. scimitarra, etc., of unkn. orig.]

scintigram /ˈsɪntɪˌɡræm/ n. an image of an internal part of the body, produced by scintigraphy.

scintigraphy /sɪnˈtɪɡrəfɪ/ n. the use of a radioisotope and a scintillation counter to get an image or record of a bodily organ etc. [SCINTILLATION + -GRAPHY]

scintilla /sɪnˈtɪlə/ n. **1** a trace. **2** a spark. [L]

scintillate /ˈsɪntɪˌleɪt/ v.intr. **1** (esp. as **scintillating** adj.) talk cleverly or wittily; be brilliant. **2** sparkle; twinkle; emit sparks. **3** Physics fluoresce momentarily when struck by a charged particle etc. □ **scintillant** adj. **scintillatingly** adv. [L scintillare (as SCINTILLA)]

scintillation /ˌsɪntɪˈleɪʃ(ə)n/ n. **1** the process or state of scintillating. **2** the twinkling of a star. **3** a flash produced in a material by an ionizing particle etc. □ **scintillation counter** a device for detecting and recording scintillation.

scintiscan /ˈsɪntɪˌskæn/ n. an image or other record showing the distribution of radioactive traces in parts of the body, used in the detection and diagnosis of various diseases. [SCINTILLATION + SCAN]

sciolist /ˈsaɪəlɪst/ n. a superficial pretender to knowledge. □ **sciolism** n. **sciolistic** /ˌsaɪəˈlɪstɪk/ adj. [LL sciolus smatterer f. L scire know]

scion /ˈsaɪən/ n. **1** (US cion) a shoot of a plant etc., esp. one cut for grafting or planting. **2** a descendant; a younger member of (esp. a noble) family. [ME f. OF ciun, cion, sion shoot, twig, of unkn. orig.]

Scipio Aemilianus /ˌsɪpɪəʊ iːˌmɪlɪˈɑːnəs/ (full name Publius Cornelius Scipio Aemilianus Africanus Minor) (c.185–129 BC), Roman general and politician. He achieved distinction in the third Punic War, and blockaded and destroyed Carthage in 146. His successful campaign in Spain (133) ended organized resistance in that country. Returning to Rome in triumph, he initiated moves against the reforms introduced by his brother-in-law Tiberius Gracchus. Scipio's sudden death at the height of the crisis gave rise to the rumour that he had been murdered.

Scipio Africanus /ˌsɪpɪəʊ ˌæfrɪˈkɑːnəs/ (full name Publius Cornelius Scipio Africanus Major) (236–c.184 BC), Roman general and politician. His aggressive tactics were successful in concluding the second Punic War, firstly by the defeat of the Carthaginians in Spain in 206 and then by the defeat of Hannibal in Africa in 202; his victories pointed the way to Roman hegemony in the Mediterranean. His son was the adoptive father of Scipio Aemilianus.

scire facias /ˌsaɪərɪ ˈfeɪʃɪˌæs/ n. Law a writ to enforce or annul a judgement, patent, etc. [L, = let (him) know]

scirocco var. of SIROCCO.

scirrhus /ˈsɪrəs, ˈskɪ-/ n. (pl. **scirrhi** /-raɪ/) a carcinoma which is hard to the touch. □ **scirrhous** adj. **scirrhoid** adj. **scirrhosity** /sɪˈrɒsɪtɪ, skɪ-/ n. [mod.L f. Gk skir(r)os f. skiros hard]

scissel /ˈskɪs(ə)l/ n. waste clippings etc. of metal produced during coin manufacture. [F cisaille f. cisailler clip with shears]

scissile /ˈsɪsaɪl/ adj. able to be cut or divided. [L scissilis f. scindere sciss-cut]

scission /ˈsɪʃ(ə)n/ n. **1** the act or an instance of cutting; the state of being cut. **2** a division or split. [ME f. OF scission or LL scissio (as SCISSILE)]

scissor /ˈsɪzə(r)/ v.tr. **1** (usu. foll. by off, up, into, etc.) cut with scissors. **2** (usu. foll. by out) clip out (a newspaper cutting etc.).

scissors /ˈsɪzəz/ n.pl. **1** (also **pair of scissors** sing.) an instrument for cutting fabric, paper, hair, etc., having two pivoted blades with finger and thumb holes in the handles, operating by closing on the material to be cut. (See note below.) **2** (treated as sing.) **a** a method of high jump in which the athlete, approaching the bar side-on, brings the trailing leg up as the leading leg goes down on the other side of the bar. **b** a hold in wrestling in which the opponent's body or esp. head is gripped between the legs. □ **scissor-bill** = SKIMMER 4. **scissor-bird** (or **-tail**) an American tyrant flycatcher, Tyrannus forficatus, with very long outer tail feathers. **scissors-and-paste** = cut-and-paste. □ **scissorwise** adv. [ME sisoures f. OF cisoires f. LL cisoria pl. of cisorium cutting instrument (as CHISEL): assoc. with L scindere sciss-cut]

▪ Scissors with overlapping blades and with a C-shaped spring at the handle end (like tongs) probably date from the Bronze Age; they were used in Europe until the end of the Middle Ages, and the design survived still later for some specific purposes (e.g. sheep-shearing). Scissors pivoted between handle and blade were used in Roman Europe and the Far East 2,000 years ago, coming into domestic use in Europe in the 16th century.

sciurine /ˈsaɪjʊˌraɪn, -rɪn/ adj. **1** of or relating to the family Sciuridae, including squirrels and chipmunks. **2** squirrel-like. □ **sciuroid** adj. [L sciurus f. Gk skiouros squirrel f. skia shadow + oura tail]

sclera /ˈsklɪərə/ n. the white of the eye; a white membrane coating the eyeball. □ **scleral** adj. **sclerotomy** /-ˈrɒtəmɪ/ n. (pl. **-ies**). [mod.L f. fem. of Gk sklēros hard]

sclerenchyma /sklɪəˈreŋkɪmə/ n. Bot. the woody tissue found in a plant, formed from lignified cells and usu. providing support. [mod.L f. Gk sklēros hard + egkhuma infusion, after parenchyma]

sclerite /ˈsklɪəraɪt, ˈskler-/ n. Zool. a component part of an exoskeleton, esp. each of the plates forming the skeleton of an arthropod. [Gk sklēros hard]

scleritis /sklɪəˈraɪtɪs/ n. Med. inflammation of the sclera of the eye.

scleroderma /ˌsklɪərəˈdɜːmə/ n. Med. a chronic hardening of the skin and connective tissue. [Gk sklēros hard + derma skin]

scleroid /ˈsklɪərɔɪd/ adj. Bot. & Zool. having a hard texture; hardened. [Gk sklēros hard]

scleroma /sklɪəˈrəʊmə/ n. (pl. **scleromas** or **scleromata** /-mətə/) Med. an abnormal patch of hardened skin or mucous membrane. [Gk sklēros hard + -OMA]

sclerometer /sklɪəˈrɒmɪtə(r)/ n. an instrument for determining the hardness of materials. [Gk sklēros hard + -METER]

sclerophyll /ˈsklɪərəfɪl/ n. Bot. any woody plant with leathery leaves retaining water. □ **sclerophyllous** /sklɪəˈrɒfɪləs/ adj. [Gk sklēros hard + phullon leaf]

scleroprotein /ˌsklɪərəʊˈprəʊtiːn/ n. Biochem. any insoluble structural protein, e.g. keratin. [Gk sklēros hard + PROTEIN]

sclerosed /ˈsklɪərəʊst, -rəʊzd/ adj. affected by sclerosis.

sclerosis /sklɪəˈrəʊsɪs/ n. **1** an abnormal hardening of body tissue (see also ARTERIOSCLEROSIS, ATHEROSCLEROSIS). **2** (in full **multiple sclerosis**) a chronic disease of the nervous system resulting in symptoms including fatigue, difficulty in walking, and speech defects. **3** Bot. the hardening of a cell-wall with lignified matter. [ME f. med.L f. Gk sklērōsis f. sklēroō harden]

sclerotic /sklɪəˈrɒtɪk/ adj. & n. ● adj. **1** of or having sclerosis. **2** of or relating to the sclera. ● n. = SCLERA. □ **sclerotitis** /ˌsklɪərəˈtaɪtɪs/ n. [med.L sclerotica (as SCLEROSIS)]

sclerous /ˈsklɪərəs/ adj. Physiol. hardened; bony. [Gk sklēros hard]

SCM abbr. (in the UK) **1** State Certified Midwife. **2** Student Christian Movement.

scoff¹ /skɒf/ v. & n. ● v.intr. (usu. foll. by at) speak derisively, esp. of serious subjects; mock; be scornful. ● n. **1** mocking words; a taunt. **2** an object of ridicule. □ **scoffer** n. **scoffingly** adv. [perh. f. Scand.: cf. early mod. Danish skuf, skof jest, mockery]

scoff² /skɒf/ v. & n. colloq. ● v.tr. & intr. eat greedily. ● n. food; a meal. [(n.) f. Afrik. schoff repr. Du. schoft quarter of a day (hence, meal): (v.) orig. var. of dial. scaff, assoc. with the noun]

scold /skəʊld/ v. & n. ● v. **1** tr. rebuke or chide (esp. a child). **2** intr. find fault noisily; complain; rail. ● n. archaic a nagging or grumbling woman. □ **scolder** n. **scolding** n. [ME (earlier as noun), prob. f. ON skáld SKALD]

scolex /ˈskəʊleks/ n. (pl. **scoleces** /skəʊˈliːsiːz/ or **scolices**

/'skəʊlɪˌsiːz/ *Zool.* the anterior end of a larval or adult tapeworm, bearing suckers and hooks for attachment. [mod.L f. Gk *skōlēx* worm]

scoliosis /ˌskɒlɪ'əʊsɪs/ *n.* abnormal lateral curvature of the spine. □ **scoliotic** /-'ɒtɪk/ *adj.* [mod.L f. Gk f. *skolios* bent]

scollop var. of SCALLOP.

scolopendrium /ˌskɒlə'pendrɪəm/ *n.* the hart's tongue (fern), *Phyllitis scolopendrium.* [mod.L f. Gk *skolopendrion* f. *skolopendra* millipede (because of the supposed resemblance)]

scombroid /'skɒmbrɔɪd/ *n. & adj. Zool.* ● *n.* a marine fish of the family Scombridae, which includes mackerels, tunas, and bonitos, or the superfamily Scombroidea. ● *adj.* of or relating to this family or superfamily. □ **scombrid** *n. & adj.* [mod.L f. L *scomber* f. Gk *skombros*]

sconce[1] /skɒns/ *n.* **1** a flat candlestick with a handle. **2** a bracket candlestick to hang on a wall. [ME f. OF *esconse* lantern or med.L *sconsa* f. L *absconsa* fem. past part. of *abscondere* hide: see ABSCOND]

sconce[2] /skɒns/ *n.* **1** a small fort or earthwork usu. defending a ford, pass, etc. **2** *archaic* a shelter or screen. [Du. *schans* brushwood f. MHG *schanze*]

Scone /skuːn/ an ancient Scottish settlement, believed to be on the site of the capital of the Picts, where the kings of medieval Scotland were crowned. It lay near the modern villages of Old Scone and New Scone in Tayside to the north of Perth. □ **Stone of Scone** the stone on which medieval Scottish kings were crowned, brought to England by Edward I and now preserved in the coronation chair in Westminster Abbey (also called *Coronation Stone*).

scone /skɒn, skəʊn/ *n.* a small sweet or savoury cake of flour, fat, and milk, baked quickly in an oven. [orig. Sc., perh. f. MDu. *schoon(broot)*, MLG *schon(brot)* fine (bread)]

scoop /skuːp/ *n. & v.* ● *n.* **1** any of various objects resembling a spoon, esp.: **a** a short-handled deep shovel used for transferring grain, sugar, coal, coins, etc. **b** a large long-handled ladle used for transferring liquids. **c** the excavating part of a digging-machine etc. **d** *Med.* a long-handled spoonlike instrument used for scraping parts of the body etc. **e** an instrument used for serving portions of mashed potato, ice-cream, etc. **2** a quantity taken up by a scoop. **3** a movement of or resembling scooping. **4** a piece of news published by a newspaper etc. in advance of its rivals. **5** a large profit made quickly or by anticipating one's competitors. **6** *Mus.* a singer's exaggerated portamento. **7** a scooped-out hollow etc. ● *v.tr.* **1** (usu. foll. by *out*) hollow out with or as if with a scoop. **2** (usu. foll. by *up*) lift with or as if with a scoop. **3** forestall (a rival newspaper, reporter, etc.) with a scoop. **4** secure a large profit etc.) esp. suddenly. □ **scoop-neck** the rounded low-cut neck of a garment. **scoop-net** a net used for sweeping a river-bottom, or for catching bait. □ **scooper** *n.* **scoopful** *n.* (*pl.* **-fuls**). [ME f. MDu., MLG *schōpe* bucket etc., rel. to SHAPE]

scoot /skuːt/ *v. & n. colloq.* ● *v.intr.* run or dart away, esp. quickly. ● *n.* the act or an instance of scooting. [19th-c. US (earlier *scout*): orig. unkn.]

scooter /'skuːtə(r)/ *n. & v.* ● *n.* **1** a child's toy consisting of a footboard mounted on two wheels and a long steering-handle, propelled by resting one foot on the footboard and pushing the other against the ground. **2** (in full **motor scooter**) a light two-wheeled open motor vehicle with a shieldlike protective front. **3** *N. Amer.* a sailboat able to travel on both water and ice. ● *v.intr.* travel or ride on a scooter. □ **scooterist** *n.*

scopa /'skəʊpə/ *n.* (*pl.* **scopae** /-piː/) *Zool.* a small brushlike tuft of hairs, esp. on the leg of a bee for collecting pollen (cf. SCOPULA). [sing. of L *scopae* = twigs, broom]

scope[1] /skəʊp/ *n.* **1 a** the extent to which it is possible to range; the opportunity for action etc. (*this is beyond the scope of our research*). **b** the sweep or reach of mental activity, observation, or outlook (*an intellect limited in its scope*). **c** space or range for freedom of movement or activity (*doesn't give us much scope*). **2** *Naut.* the length of cable extended when a ship rides at anchor. **3** *archaic* a purpose, end, or intention. [It. *scopo* aim f. Gk *skopos* target f. *skeptomai* look at]

scope[2] /skəʊp/ *n. colloq.* a telescope, microscope, or other device ending in *-scope.* [abbr.]

-scope /skəʊp/ *comb. form* forming nouns denoting: **1** a device looked at or through (*kaleidoscope; telescope*). **2** an instrument for observing or showing (*gyroscope; oscilloscope*). □ **-scopic** /'skɒpɪk/ *comb. form* forming adjectives. [from or after mod.L *-scopium* f. Gk *skopeō* look at]

scopolamine /skə'pɒləmɪn, -ˌmiːn/ *n.* = HYOSCINE. [*Scopolia* genus name of the plants yielding it, f. Giovanni Antonio *Scopoli*, It. naturalist (1723–88) + AMINE]

scopula /'skɒpjʊlə/ *n.* (*pl.* **scopulae** /-ˌliː/) *Zool.* a small dense tuft of hairs, esp. on the leg of a spider or other arthropod (cf. SCOPA). [LL, dimin. of L *scopa*: see SCOPA]

-scopy /skəpɪ/ *comb. form* indicating viewing or observation, usu. with an instrument ending in *-scope* (*microscopy*).

scorbutic /skɔː'bjuːtɪk/ *adj. & n.* ● *adj.* relating to, resembling, or affected with scurvy. ● *n.* a person affected with scurvy. □ **scorbutically** *adv.* [mod.L *scorbuticus* f. med.L *scorbutus* scurvy, perh. f. MLG *schorbūk* f. *schoren* break + *būk* belly]

scorch /skɔːtʃ/ *v. & n.* ● *v.* **1** *tr.* **a** burn the surface of with flame or heat so as to discolour, parch, injure, or hurt. **b** affect with the sensation of burning. **2** *intr.* become discoloured etc. with heat. **3** *tr.* (as **scorching** *adj.*) *colloq.* **a** (of the weather) very hot. **b** (of criticism etc.) stringent; harsh. **4** *intr. colloq.* (of a motorist etc.) go at excessive speed. ● *n.* **1** a mark made by scorching. **2** *colloq.* a spell of fast driving etc. □ **scorchingly** *adv.* [ME, perh. rel. to *skorkle* in the same sense]

scorched earth policy *n.* the burning of crops etc. and the removing or destroying of anything that might be of use to an enemy force occupying a country.

scorcher /'skɔːtʃə(r)/ *n.* **1** a person or thing that scorches. **2** *colloq.* **a** a very hot day. **b** a fine specimen.

score /skɔː(r)/ *n. & v.* ● *n.* **1 a** the number of points, goals, runs, etc. made by a player, side, etc., in some games. **b** the total number of points etc. at the end of a game (*the score was 5–0*). **c** the act of gaining esp. a goal (*a superb score there!*). **2** (*pl.* same or **scores**) twenty or a set of twenty. **3** (in *pl.*) a great many (*scores of people arrived*). **4 a** a reason or motive (*rejected on the score of absurdity*). **b** topic, subject (*no worries on that score*). **5** *Mus.* **a** a usu. printed copy of a composition showing all the vocal and instrumental parts arranged one below the other. **b** the music composed for a film or play, esp. for a musical. **6** *colloq.* **a** a piece of good fortune. **b** the act or an instance of scoring off another person. **7** *colloq.* the state of affairs; the present situation (*asked what the score was*). **8** a notch, line, etc. cut or scratched into a surface. **9 a** an amount due for payment. **b** a running account kept by marks against a customer's name. **10** *Naut.* a groove in a block or dead-eye to hold a rope. ● *v.* **1** *tr.* **a** win or gain (a goal, run, points, etc., or success etc.) (*scored two goals*). **b** count for a score of (points in a game etc.) (*a bull's-eye scores most points*). **c** allot a score to (a competitor etc.). **d** make a record of (a point etc.). **2** *intr.* **a** make a score in a game (*failed to score*). **b** keep the tally of points, runs, etc. in a game. **3** *tr.* mark with notches, incisions, lines, etc.; slash; furrow (*scored his name on the desk*). **4** *intr.* secure an advantage by luck, cunning, etc. (*that is where he scores*). **5** *tr. Mus.* **a** orchestrate (a piece of music). **b** (usu. foll. by *for*) arrange for an instrument or instruments. **c** write the music for (a film, musical, etc.). **d** write out in a score. **6** *tr.* **a** (usu. foll. by *up*) mark (a total owed etc.) in a score (see sense 9b of *n.*). **b** (usu. foll. by *against, to*) enter (an item of debt to a customer). **7** *intr. sl.* **a** obtain drugs illegally. **b** (of a man) make a sexual conquest. **8** *tr.* (usu. foll. by *against, to*) mentally record (an offence etc.). **9** *tr. N. Amer.* criticize (a person) severely. □ **keep score** (or **the score**) register the score as it is made. **know the score** *colloq.* be aware of the essential facts. **no-score draw** a result in football in which neither team scores a goal. **on the score of** for the reason that; because of. **on that score** so far as that is concerned. **score-book** (or **-card** or **-sheet**) a book etc. prepared for entering esp. cricket scores in. **score draw** a result in football in which each team scores the same number of goals. **score off** (or **score points off**) *colloq.* humiliate, esp. verbally in repartee etc. **score out** draw a line through (words etc.). **score under** underline. □ **scorer** *n.* **scoring** *n. Mus.* [(n.) f. OE: sense 5 f. the line or bar drawn through all staves: (v.) partly f. ON *skora* f. ON *skor* notch, tally, twenty, f. Gmc: see SHEAR]

scoreboard /'skɔːbɔːd/ *n.* a large board for publicly displaying the score in a game or match.

scoreline /'skɔːlaɪn/ *n.* the number of points, goals, etc., scored in a match; the score; the result. [originally the line in a newspaper etc. giving the score in a sports contest]

scoria /'skɔːrɪə/ *n.* (*pl.* **scoriae** /-rɪˌiː/) **1** cellular lava, or fragments of it. **2** the slag or dross of metals. □ **scoriaceous** /ˌskɔːrɪ'eɪʃəs/ *adj.* [L f. Gk *skōria* refuse f. *skōr* dung]

scorify /'skɔːrɪˌfaɪ/ *v.tr.* (**-ies, -ied**) **1** reduce to dross. **2** assay (precious metal) by treating a portion of its ore fused with lead and borax. □ **scorifier** *n.* **scorification** /ˌskɔːrɪfɪ'keɪʃ(ə)n/ *n.*

scorn /skɔːn/ *n. & v.* ● *n.* **1** disdain, contempt, derision. **2** an object of contempt etc. (*the scorn of all onlookers*). ● *v.tr.* **1** hold in contempt or disdain. **2** (often foll. by *to* + infin.) abstain from or refuse to do as unworthy (*scorns lying; scorns to lie*). □ **pour scorn on** hold in contempt or disdain. **think scorn of** *archaic* despise. □ **scorner** *n*. [ME f. OF *esc(h)arn(ir)* ult. f. Gmc: cf. OS *skern* MOCKERY]

scornful /'skɔːnfʊl/ *adj*. (often foll. by *of*) full of scorn; contemptuous. □ **scornfully** *adv*. **scornfulness** *n*.

scorper var. of SCALPER 2.

Scorpio /'skɔːpɪˌəʊ/ *n*. **1** *Astron*. = SCORPIUS 1. **2** *Astrol*. **a** the eighth sign of the zodiac, which the sun enters about 23 Oct. **b** a person born when the sun is in this sign. □ **Scorpian** *adj. & n*. [ME f. L (as SCORPION)]

scorpioid /'skɔːpɪˌɔɪd/ *adj. & n*. ● *adj*. **1** *Zool*. of, relating to, or resembling a scorpion; of the scorpion order. **2** *Bot*. (of an inflorescence) curled up at the end, and uncurling as the flowers develop. ● *n*. this type of inflorescence [Gk *scorpioeidēs* (as SCORPION)]

scorpion /'skɔːpɪən/ *n*. **1** an arachnid of the order Scorpiones, with lobster-like pincers and a jointed tail that can be bent over to inflict a venomous sting on prey held in its pincers. **2** (in full **false scorpion**) a similar arachnid of the order Pseudoscorpiones, smaller and without a tail. **3** (**the Scorpion**) the zodiacal sign Scorpio or the constellation Scorpius. **4** *Bibl*. a whip with metal points (with allusion to 1 Kings 12:11). □ **scorpion fish** a marine fish of the family Scorpaenidae, with venomous spines on the head and gills. **scorpion fly** a winged insect of the order Mecoptera, esp. of the genus *Panorpa*, the males of which have a swollen abdomen curved upwards like a scorpion's sting. **scorpion grass** = forget-me-not. [ME f. OF f. L *scorpio -onis* f. *scorpius* f. Gk *skorpios*]

Scorpius /'skɔːpɪəs/ *n*. **1** *Astron*. a large constellation (the Scorpion), said to represent the scorpion which killed Orion. It contains many bright stars, including the red giant Antares, and the binary star Scorpius X-1 (which is the brightest X-ray source in the sky). **2** *Astrol*. = SCORPIO 2.

Scorsese /skɔːˈseɪzɪ/, Martin (b.1942), American film director. He made his directorial début in 1968, but first gained recognition five years later with *Mean Streets*, a realistic study of New York's Italian community, focusing on the plight of a group of friends entangled in a web of crime and violence. The film marked the beginning of Scorsese's long collaboration with the actor Robert De Niro, which continued in such films as *Taxi Driver* (1976), *Raging Bull* (1980), and *GoodFellas* (1990). Other films include the controversial *The Last Temptation of Christ* (1988).

scorzonera /ˌskɔːzəˈnɪərə/ *n*. **1** a composite plant, *Scorzonera hispanica*, with long tapering purple-brown roots. Also called *black salsify*, *viper's grass*. **2** the root of this used as a vegetable. [It. f. *scorzone* venomous snake ult f. med.L *curtio*]

Scot /skɒt/ *n*. **1 a** a native of Scotland. **b** a person of Scottish descent. **2** *hist*. a member of a Gaelic people that migrated from Ireland to Scotland around the late 5th century, and from whom the name Scotland ultimately derives. (See also DALRIADA.) [OE *Scottas* (pl.) f. LL *Scottus*]

scot /skɒt/ *n. hist*. a payment corresponding to a modern tax, rate, etc. □ **pay scot and lot** share the financial burdens of a borough etc. (and so be allowed to vote). **scot-free** unharmed; unpunished; safe. [ME f. ON *skot* & f. OF *escot*, of Gmc orig.: cf. SHOT¹]

Scotch /skɒtʃ/ *adj. & n*. ● *adj*. var. of SCOTTISH or SCOTS. (See note below.) ● *n*. **1** var. of SCOTTISH or SCOTS. (See note below.) **2** Scotch whisky. □ **Scotch broth** a soup made from beef or mutton with pearl barley etc. **Scotch cap** = BONNET *n*. 1b. **Scotch catch** *Mus*. a short note on the beat followed by a long one. **Scotch egg** a hard-boiled egg enclosed in sausage meat and fried. **Scotch fir** = *Scots pine*. **Scotch kale** a variety of kale with purplish leaves. **Scotch mist 1** a thick drizzly mist common in the Highlands. **2** a retort made to a person implying that he or she has imagined or failed to understand something. **Scotch pancake** = drop scone. **Scotch pebble** agate, jasper, cairngorm, etc., found in Scotland. **Scotch pine** = *Scots pine*. **Scotch snap** = *Scotch catch*. **Scotch tape** *propr*. esp. *US* adhesive usu. transparent cellulose or plastic tape. **Scotch terrier** = *Scottish terrier*. **Scotch whisky** whisky distilled in Scotland, esp. from malted barley. ¶ Except for fixed collocations such as those listed above, *Scottish* (less frequently *Scots*) is now the usual adjective, and *Scots* (pl.) the noun. [contr. of SCOTTISH]

scotch¹ /skɒtʃ/ *v. & n*. ● *v.tr*. **1** put an end to; frustrate (*injury scotched his attempt*). **2** *archaic* **a** wound without killing; slightly disable. **b** make incisions in; score. ● *n*. **1** *archaic* a slash. **2** a line on the ground for hopscotch. [ME: orig. unkn.]

scotch² /skɒtʃ/ *n. & v*. ● *n*. a wedge or block placed against a wheel etc. to prevent its slipping. ● *v.tr*. hold back (a wheel, barrel, etc.) with a scotch. [17th c.: perh. = *scatch* stilt f. OF *escache*]

Scotchman /'skɒtʃmən/ *n*. (pl. **-men**; fem. **Scotchwoman**, pl. **-women**) = SCOTSMAN. ¶ *Scotsman* etc. are generally preferred.

scoter /'skəʊtə(r)/ *n*. (pl. same or **scoters**) a large marine duck of the genus *Melanitta*, often mainly black in colour. [17th c., perh. an error for *sooter* (with ref. to its black plumage)]

scotia /'skəʊʃə/ *n*. a concave moulding, esp. at the base of a column. [L f. Gk *skotia* f. *skotos* darkness, with ref. to the shadow produced]

Scoticism var. of SCOTTICISM.

Scoticize var. of SCOTTICIZE.

Scotland /'skɒtlənd/ a country forming the northernmost part of Great Britain and of the United Kingdom; pop. (1991) 4,957,300; capital, Edinburgh. Much of Scotland is mountainous, especially the Highlands in the north and north-west; the lowlands between the Clyde and Forth estuaries form the most populous region. Scotland was sparsely populated until Celtic peoples arrived from the Continent during the Bronze and early Iron Age. It was never fully subjugated by the Romans; Hadrian's Wall marked the Roman frontier except for a period during the 2nd century, when the Antonine Wall was built. An independent country in the Middle Ages, after the unification of various kingdoms of Scots, Picts, and Anglo-Saxons between the 9th and 11th centuries, Scotland successfully resisted English attempts at domination but was finally amalgamated with her southern neighbour as a result of the union of the Crowns in 1603 and of the Parliaments in 1707. The distinctive Celtic society of the Highlands, based on clans, continued to exist until it was destroyed in the aftermath of the Jacobite uprisings of 1715 and 1745-6 and the Highland clearances of the 18th-19th centuries. Scottish Gaelic is still spoken in the more remote parts of the north and west. Scotland's economy has benefited in the 20th century from the discovery of North Sea oil.

Scotland Yard 1 the headquarters of the London Metropolitan Police, situated from 1829 to 1890 in Great Scotland Yard, a short street off Whitehall in London, from 1890 until 1967 in New Scotland Yard on the Thames Embankment, and from 1967 in New Scotland Yard, Westminster. **2** the Criminal Investigation Department of the London Metropolitan Police force.

scotoma /skɒˈtəʊmə/ *n*. (pl. **scotomas** or **scotomata** /-mətə/) *Med*. a partial loss of vision or blind spot in an otherwise normal visual field. [LL f. Gk *skotōma* f. *skotoō* darken f. *skotos* darkness]

Scots /skɒts/ *adj. & n. esp. Sc*. ● *adj*. **1** = SCOTTISH *adj*. **2** in the dialect, accent, etc., of (esp. Lowlands) Scotland. ● *n*. **1** = SCOTTISH *n*. **2** the form of English spoken in (esp. Lowlands) Scotland. □ **Scots pine** a European pine tree, *Pinus sylvestris*, characteristic of old woods in the Scottish Highlands and much planted for timber etc. (also called *Scotch pine* or *fir*). [ME orig. *Scottis*, north. var. of SCOTTISH]

Scotsman /'skɒtsmən/ *n*. (pl. **-men**; fem. **Scotswoman**, pl. **-women**) **1** a native of Scotland. **2** a person of Scottish descent.

Scott¹ /skɒt/, Sir George Gilbert (1811–78), English architect. Influenced by Pugin, he was a prolific Gothic revivalist, designing or restoring many churches, cathedrals, and other buildings. He attracted controversy for his restoration of medieval churches and in 1858 came into conflict with Palmerston over the design for the new Foreign Office, where Scott was ultimately compelled to adopt an Italianate solution; his Albert Memorial in London (1863–72) reflects more accurately his preferred aesthetic. His grandson Sir Giles Gilbert Scott (1880–1960) also worked as a revivalist architect and is best known for the Gothic Anglican cathedral in Liverpool (begun in 1904, completed in 1978).

Scott² /skɒt/, Sir Peter Markham (1909–89), English naturalist and artist. He was particularly interested in wildfowl and their conservation, and in 1946 founded the Wildfowl Trust at Slimbridge in Gloucestershire. He was well known as a wildfowl artist and a presenter of natural history programmes on television, and he became increasingly involved in conservation worldwide. He was also an accomplished yachtsman, and maintained an interest in the search for a monster at Loch Ness. He was the son of the explorer Robert Falcon Scott.

Scott³ /skɒt/, Robert Falcon (1868–1912), English explorer and naval officer. As commander of the ship *Discovery* he led the National Antarctic Expedition (1900–4), surveying the interior of the continent and charting the Ross Sea. On a second expedition (1910–12) Scott and four companions made a journey to the South Pole by sled, arriving there in Jan. 1912 to discover that the Norwegian explorer Amundsen had beaten them to their goal by a month. Scott and his companions died on the journey back to base, and their bodies and diaries were discovered by a search party eight months later. Scott, a national hero, was posthumously knighted. He was the father of the naturalist and artist Sir Peter Scott.

Scott[4] /skɒt/, Sir Walter (1771–1832), Scottish novelist and poet. He established the form of the historical novel in Britain, and was also influential in his treatment of rural themes and use of regional speech. Among his novels are *Waverley* (1814), *Old Mortality* (1816), *Ivanhoe* (1819), and *Kenilworth* (1821). His poetry was influenced by medieval French and Italian poetry, and by contemporary German poets; he collected and imitated old Borders tales and ballads, while among his original works are the romantic narrative poems *The Lay of the Last Minstrel* (1805) and *The Lady of the Lake* (1810).

Scotticism /ˈskɒtɪˌsɪz(ə)m/ *n.* (also **Scoticism**) a Scottish phrase, word, or idiom. [LL *Scot(t)icus*]

Scotticize /ˈskɒtɪˌsaɪz/ *v.* (also **Scoticize, -ise**) **1** *tr.* imbue with or model on Scottish ways etc. **2** *intr.* imitate the Scottish in idiom or habits.

Scottie /ˈskɒtɪ/ *n. colloq.* **1** (also **Scottie dog**) a Scottish terrier. **2** a Scot.

Scottish /ˈskɒtɪʃ/ *adj. & n.* ● *adj.* of or relating to Scotland or its inhabitants. ● *n.* (prec. by *the*; treated as *pl.*) the people of Scotland (see also Scots). □ **Scottish terrier** a small rough-haired short-legged breed of terrier (also called *Scotch terrier*). □ **Scottishness** *n.*

Scottish Highlands see HIGHLAND 2.

Scottish Nationalist *n.* a member or supporter of the Scottish National Party.

Scottish National Party (abbr. **SNP**) a political party formed in 1934 by an amalgamation of the National Party of Scotland and the Scottish Party, which seeks autonomous government for Scotland. It won its first parliamentary seat in 1945, and has since maintained a small group of MPs.

scoundrel /ˈskaʊndrəl/ *n.* an unscrupulous villain; a rogue. □ **scoundreldom** *n.* **scoundrelism** *n.* **scoundrelly** *adj.* [16th c.: orig. unkn.]

scour[1] /ˈskaʊə(r)/ *v. & n.* ● *v.tr.* **1 a** cleanse or brighten by rubbing, esp. with soap, chemicals, sand, etc. **b** (usu. foll. by *away, off,* etc.) remove (esp. rust or dirt) by hard rubbing with soap, chemicals, etc. **2** (of water) clear out (a pipe, channel, etc.) by flushing through. **3** *hist.* purge (the bowels) drastically. ● *n.* **1 a** the act or an instance of scouring. **b** the state of being scoured, esp. by a swift water current (*the scour of the tide*). **2** diarrhoea in farm animals, esp. cattle and pigs. **3** a substance used for scouring. □ **scouring pad** a pad of abrasive material for cleaning kitchenware etc. **scouring powder** an abrasive powder for cleaning kitchenware etc. **scouring-rush** any of several horsetails, esp. *Equisetum hyemale*, with a rough siliceous coating used for polishing wood etc. □ **scourer** *n.* [ME f. MDu., MLG *schüren* f. F *escurer* f. LL *excurare* clean (off) (as EX-[1], CURE)]

scour[2] /ˈskaʊə(r)/ *v.* **1** *tr.* hasten over (an area etc.) searching thoroughly (*scoured the streets for him; scoured the pages of the newspaper*). **2** *intr.* range hastily esp. in search or pursuit. [ME: orig. unkn.]

scourge /skɜːdʒ/ *n. & v.* ● *n.* **1** a whip used for punishment, esp. of people. **2** a person or thing which punishes or oppresses, esp. on a large scale (*the scourge of famine; Genghis Khan, the scourge of Asia*). ● *v.tr.* **1** whip. **2** punish; afflict; oppress. □ **scourger** *n.* [ME f. OF *escorge* (n.), *escorgier* (v.) (ult. as EX-[1], L *corrigia* thong, whip)]

Scouse /skaʊs/ *n. & adj. colloq.* ● *n.* **1** the dialect of Liverpool. **2** (also **Scouser** /ˈskaʊsə(r)/) a native of Liverpool. **3** (**scouse**) = LOBSCOUSE. ● *adj.* of or relating to Liverpool. [abbr. of LOBSCOUSE]

scout[1] /skaʊt/ *n. & v.* ● *n.* **1** a person, esp. a soldier, sent out to get information about the enemy's position, strength, etc. **2** the act of seeking (esp. military) information (*on the scout*). **3** = *talent-scout*. **4** (**Scout**) a member of the Scout Association. **5** a domestic worker at a college, esp. at Oxford University. **6** *colloq.* a person. **7** a ship or aircraft designed for reconnoitring, esp. a small fast aircraft. ● *v.* **1** *intr.* act as a scout. **2** *intr.* (foll. by *about, around*) make a search. **3** *tr.* (often foll. by *out*) *colloq.* explore to get information about (territory etc.). □ **Queen's** (or **King's**) **Scout** a Scout who has reached the highest standard of proficiency. □ **scouter** *n.* **scouting** *n.* [ME f. OF *escouter* listen, earlier *ascolter* ult. f. L *auscultare*]

scout[2] /skaʊt/ *v.tr.* reject (an idea etc.) with scorn. [Scand.: cf. ON *skúta, skúti* taunt.]

Scout Association a worldwide youth organization (originally called the Boy Scouts) founded for boys in 1908 by Lord Baden-Powell with the aim of developing their character by training them in self-

sufficiency and survival techniques in the outdoors. The organization's motto 'Be Prepared' reflects the initials of the founder's name, while much of the original uniform was adopted from the South African constabulary. The word *Boy* was dropped from the Association's title in 1967, and its sections (Wolf Cubs, Scouts, and Rover Scouts) were subsequently known as Cub Scouts (8–11 years), Scouts (11–16 years), and Venture Scouts (16–20 years). By the late 1980s there were Scout organizations in more than a hundred countries. From 1990 girls were admitted as members.

Scouter /ˈskaʊtə(r)/ *n.* an adult leader in the Scout Association.

Scoutmaster /ˈskaʊtˌmɑːstə(r)/ *n.* a person in charge of a group of Scouts.

scow /skaʊ/ *n. esp. US* a flat-bottomed boat used as a lighter etc. [Du. *schouw* ferry-boat]

scowl /skaʊl/ *n. & v.* ● *n.* a severe frown producing a sullen, bad-tempered, or threatening look on a person's face. ● *v.intr.* make a scowl. □ **scowler** *n.* [ME, prob. f. Scand.: cf. Danish *skule* look down or sidelong]

SCPS *abbr.* (in the UK) Society of Civil and Public Servants.

SCR *abbr. Brit.* Senior Common (or Combination) Room.

scr. *abbr.* scruple(s) (of weight).

scrabble /ˈskræb(ə)l/ *v. & n.* ● *v.intr.* (often foll. by *about, at*) scratch or grope to find or collect or hold on to something. ● *n.* **1** an act of scrabbling. **2** (**Scrabble**) *propr.* a game in which players build up words from letter-blocks on a board. [MDu. *schrabbelen* frequent. of *schrabben* SCRAPE]

scrag /skræg/ *n. & v.* ● *n.* **1** (also **scrag-end**) the inferior end of a neck of mutton. **2** a skinny person or animal. **3** *colloq.* a person's neck. ● *v.tr.* (**scragged, scragging**) *sl.* **1** strangle, hang. **2** seize roughly by the neck. **3** handle roughly; beat up. [perh. alt. f. dial. *crag* neck, rel. to MDu. *crāghe*, MLG *krage*]

scraggly /ˈskræglɪ/ *adj.* sparse and irregular.

scraggy /ˈskrægɪ/ *adj.* (**scraggier, scraggiest**) thin and bony. □ **scraggily** *adv.* **scragginess** *n.*

scram /skræm/ *v.intr.* (**scrammed, scramming**) (esp. in *imper.*) *colloq.* go away. [20th c.: perh. f. SCRAMBLE]

scramble /ˈskræmb(ə)l/ *v. & n.* ● *v.* **1** *intr.* make one's way over rough ground, rocks, etc., by clambering, crawling, etc. **2** (foll. by *for, at*) struggle with competitors (*for a thing or share of it*). **3** *intr.* move with difficulty or awkwardly. **4** *tr.* **a** mix together indiscriminately. **b** jumble or muddle. **5** *tr.* cook (eggs) by heating them when broken and well mixed with butter, milk, etc. **6** *tr.* change the speech frequency of (a broadcast transmission or telephone conversation) so as to make it unintelligible without a corresponding decoding device. **7** *intr.* move hastily. **8** *tr. colloq.* execute (an action etc.) awkwardly and inefficiently. **9** *intr.* (of fighter aircraft or their pilots) take off quickly in an emergency or for action. ● *n.* **1** an act of scrambling. **2** a difficult climb or walk. **3** (foll. by *for*) an eager struggle or competition. **4** *Brit.* a motorcycle race over rough ground. **5** an emergency take-off by fighter aircraft. □ **scrambled egg 1** a dish of eggs cooked by scrambling. **2** *joc.* or *colloq.* gold braid on a military officer's cap. [16th c. (imit.): cf. dial. synonyms *scamble, cramble*]

scrambler /ˈskræmblə(r)/ *n.* **1** a device for scrambling telephone conversations. **2** a motorcycle for racing on rough ground. **3** a plant with long slender stems supported by other plants.

scran /skræn/ *n. sl.* **1** food, eatables. **2** remains of food. □ **bad scran** *Ir.* bad luck. [18th c.: orig. unkn.]

Scranton /ˈskræntən/ an industrial city in NE Pennsylvania; pop. (1990) 81,805. The city developed around the steelworks established in 1840 by the Scranton family.

scrap[1] /skræp/ *n. & v.* ● *n.* **1** a small detached piece; a fragment or remnant. **2** rubbish or waste material. **3** an extract or cutting from something written or printed. **4** discarded metal for reprocessing (often *attrib.*: *scrap metal*). **5** (with *neg.*) the smallest piece or amount (*not a scrap of food left*). **6** (in *pl.*) **a** odds and ends. **b** bits of uneaten food. **7** (in *sing.* or *pl.*) a residuum of melted fat or of fish with the oil expressed. ● *v.tr.* (**scrapped, scrapping**) discard as useless. □ **scrap heap 1** a pile of scrap materials. **2** (as **on the scrap heap**) no longer useful, superseded. **scrap merchant** a dealer in scrap. [ME f. ON *skrap*, rel. to *skrapa* SCRAPE]

scrap[2] /skræp/ *n. & v. colloq.* ● *n.* a fight or rough quarrel, esp. a spontaneous one. ● *v.tr.* (**scrapped, scrapping**) (often foll. by *with*) have a scrap. □ **scrapper** *n.* [perh. f. SCRAPE.]

scrapbook /'skræpbʊk/ n. a book of blank pages for sticking cuttings, drawings, etc., in.

scrape /skreɪp/ v. & n. ● v. **1** tr. **a** move a hard or sharp edge across (a surface), esp. to make something smooth. **b** apply (a hard or sharp edge) in this way. **2** tr. (foll. by away, off, etc.) remove (a stain, projection, etc.) by scraping. **3** tr. **a** rub (a surface) harshly against another. **b** scratch or damage by scraping. **4** tr. draw or move with a sound of, or resembling, scraping. **b** intr. emit or produce such a sound. **c** tr. produce such a sound from. **5** intr. (often foll. by along, by, through, etc.) move or pass along while almost touching close or surrounding features, obstacles, etc. (the car scraped through the narrow lane). **6** tr. just manage to achieve (a living, an examination pass, etc.). **7** intr. (often foll. by by, through) **a** barely manage. **b** just pass an examination etc. **8** tr. (foll. by together, up) contrive to bring or provide; amass with difficulty. **9** intr. be economical. **10** intr. draw back a foot in making a clumsy bow. **11** tr. clear (a ship's bottom) of barnacles etc. **12** tr. completely clear (a plate) of food. **13** tr. (foll. by back) draw (the hair) tightly back off the forehead. ● n. **1** the act or sound of scraping. **2** a scraped place (on the skin etc.). **3** a thinly applied layer of butter etc. on bread. **4** the scraping of a foot in bowing. **5** colloq. an awkward predicament, esp. resulting from an escapade. □ **scrape acquaintance with** contrive to get to know (a person). **scrape the barrel** colloq. be reduced to one's last resources. [ME f. ON skrapa or MDu. schrapen]

scraper /'skreɪpə(r)/ n. a device used for scraping, esp. for removing dirt etc. from a surface.

scraperboard /'skreɪpəˌbɔːd/ n. Brit. cardboard or board with a blackened surface which can be scraped off for making white-line drawings.

scrapie /'skreɪpɪ/ n. a disease of sheep, attacking the central nervous system. Affected animals show lack of coordination and often rub against trees etc. for support. It is believed to be caused by either prions or virinos, and to be related to BSE, CJD, and kuru. [20th c.: f. SCRAPE + -IE]

scraping /'skreɪpɪŋ/ n. **1** in senses of SCRAPE v. **2** (esp. in pl.) a fragment produced by this.

scrappy /'skræpɪ/ adj. (**scrappier**, **scrappiest**) **1** consisting of scraps. **2** incomplete; carelessly arranged or put together. □ **scrappily** adv. **scrappiness** n.

scrapyard /'skræpjɑːd/ n. a place where (esp. metal) scrap is collected.

scratch /skrætʃ/ v., n., & adj. ● v. **1** tr. score or mark the surface of with a sharp or pointed object. **2** tr. **a** make a long narrow superficial wound in (the skin). **b** cause (a person or part of the body) to be scratched (scratched himself on the table). **3** tr. (also absol.) scrape without marking, esp. with the hand to relieve itching (scratched his leg). **4** tr. make or form by scratching. **5** tr. scribble; write hurriedly or awkwardly (scratched a quick reply; scratched a large A). **6** tr. (foll. by together, up, etc.) obtain (a thing) by scratching or with difficulty. **7** tr. (foll. by out, off, through) cancel or strike (out) with a pencil etc. **8** tr. (also absol.) withdraw (a competitor, candidate, etc.) from a race or competition. **9** intr. (often foll. by about, around, etc.) **a** scratch the ground etc. in search. **b** look around haphazardly (they were scratching about for evidence). ● n. **1** a mark or wound made by scratching. **2** a sound of scratching. **3** a spell of scratching oneself (have a scratch). **4** colloq. a superficial wound. **5 a** a line from which competitors in a race (esp. those not receiving a handicap) start. **b** Golf a handicap of zero. **6** (in pl.) a disease of horses in which the pastern appears scratched. **7** sl. money. **8** a technique, used esp. in rap music, of stopping a record on a turntable and moving it back and forth to make a scratching sound. ● attrib.adj. **1** collected by chance. **2** collected or made from whatever is available; heterogeneous (a scratch crew). **3** with no handicap given (a scratch race). □ **from scratch 1** from the beginning. **2** without help or advantage. **scratch along** make a living etc. with difficulty. **scratch card** a card used in a competition, having a section coated in a waxy substance which may be scratched away to reveal whether a prize has been won. **scratch one's head** be perplexed. **scratch my back and I will scratch yours 1** do me a favour and I will return it. **2** used in reference to mutual aid or flattery. **scratch pad 1** esp. N. Amer. a pad of paper for scribbling. **2** Computing a small fast memory for the temporary storage of data. **scratch the surface** deal with a matter superficially; make only a superficial search, investigation, etc. (scratch the surface and you'll find they're all corrupt). **up to scratch** up to the required standard. □ **scratcher** n. [ME, prob. f. synonymous ME scrat & cratch, both of uncert. orig.: cf. MLG kratsen, OHG krazzōn]

scratchings /'skrætʃɪŋz/ n.pl. the crisp residue of pork fat left after rendering lard (pork scratchings).

scratchy /'skrætʃɪ/ adj. (**scratchier**, **scratchiest**) **1** tending to make scratches or a scratching noise. **2** (esp. of a garment) tending to cause itchiness. **3** (of a drawing etc.) done in scratches or carelessly. □ **scratchily** adv. **scratchiness** n.

scrawl /skrɔːl/ v. & n. ● v. **1** tr. & intr. write in a hurried untidy way. **2** tr. (foll. by out) cross out by scrawling over. ● n. **1** a piece of hurried writing. **2** a scrawled note. □ **scrawly** adj. [perh. f. obs. scrawl sprawl, alt. of CRAWL]

scrawny /'skrɔːnɪ/ adj. (**scrawnier**, **scrawniest**) lean, scraggy. □ **scrawniness** n. [var. of dial. scranny: cf. archaic scrannel (of sound) weak, feeble]

scream /skriːm/ n. & v. ● n. **1** a loud high-pitched piercing cry expressing fear, pain, extreme fright, etc. **2** a similar sound, e.g. of sirens. **3** colloq. an irresistibly funny occurrence or person. ● v. **1** intr. emit a scream. **2** tr. speak or sing (words etc.) in a screaming tone. **3** intr. move with a shrill sound like a scream. **4** intr. laugh uncontrollably. **5** intr. be blatantly obvious or conspicuous. **6** intr. colloq. turn informer. [OE or MDu.]

screamer /'skriːmə(r)/ n. **1** a person or thing that screams. **2** a South American gooselike bird of the family Anhimidae, frequenting marshland and having a characteristic shrill cry. **3** colloq. a tale that raises screams of laughter. **4** US colloq. a sensational headline.

screamingly /'skriːmɪŋlɪ/ adv. **1** extremely (screamingly funny). **2** blatantly (screamingly obvious).

scree /skriː/ n. (in sing. or pl.) **1** small loose stones. **2** a mountain slope covered with these. [prob. back-form. f. screes (pl.) ult. f. ON skríða landslip, rel. to skrítha glide]

screech /skriːtʃ/ n. & v. ● n. a harsh high-pitched scream. ● v.tr. & intr. utter with or make a screech. □ **screech owl 1** the barn owl, which has a discordant cry. **2** a small American owl of the genus Otus, esp. Otus asio. □ **screecher** n. **screechy** adj. (**screechier**, **screechiest**). [16th-c. var. of ME scritch (imit.)]

screed /skriːd/ n. **1** a long usu. tiresome piece of writing or speech. **2 a** a strip of plaster or other material placed on a surface as a guide to thickness. **b** a levelled layer of material (e.g. cement) applied to a floor or other surface. [ME, prob. var. of SHRED]

screen /skriːn/ n. & v. ● n. **1** a fixed or movable upright partition for separating, concealing, or sheltering from draughts or excessive heat or light. **2** a thing used as a shelter, esp. from observation. **3 a** a measure adopted for concealment. **b** the protection afforded by this (under the screen of night). **4 a** a blank usu. white or silver surface on which a photographic image is projected. **b** (prec. by the) the cinema industry. **5** the surface of a cathode-ray tube or similar electronic device, esp. of a television, VDU, etc., on which images appear. **6** = sight-screen. **7** = WINDSCREEN. **8** a frame with fine wire netting to keep out flies, mosquitoes, etc. **9** Physics a body intercepting light, heat, electric or magnetic induction, etc., in a physical apparatus. **10** Photog. a piece of ground glass in a camera for focusing. **11** a large sieve or riddle, esp. for sorting grain, coal, etc., into sizes. **12** a system of checking for the presence or absence of a disease, ability, attribute, etc. **13** Printing a transparent finely ruled plate or film used in half-tone reproduction. **14** Mil. a body of troops, ships, etc., detached to warn of the presence of an enemy force. ● v.tr. **1** (often foll. by from) **a** afford shelter to; hide partly or completely. **b** protect from detection, censure, etc. **2** (foll. by off) shut off or hide behind a screen. **3 a** show (a film etc.) on a screen. **b** broadcast (a television programme). **4** prevent from causing, or protect from, electrical interference. **5 a** test (a person or group) for the presence or absence of a disease. **b** check on (a person) for the presence or absence of a quality, esp. reliability or loyalty. **6** pass (grain, coal, etc.) through a screen. □ **screen printing** a process like stencilling with ink forced through a prepared sheet of fine material (orig. silk). **screen test** an audition for a part in a cinema film. □ **screenable** adj. **screener** n. [ME f. ONF escren, escran: cf. OHG skrank barrier]

screening /'skriːnɪŋ/ n. **1** in senses of SCREEN v. **2** a showing of a film or television programme. **3 a** = SCREEN n. 12. **b** a test or check for the presence or absence of a disease etc. **4** (in pl.) refuse separated by sifting.

screenplay /'skriːnpleɪ/ n. the script of a film, with acting instructions, scene directions, etc.

screenwriter /'skriːnˌraɪtə(r)/ n. a person who writes a screenplay.

screw /skruː/ n. & v. ● n. **1** a thin cylinder or cone with a spiral ridge or thread running round the outside (male screw) or the inside (female screw). (See note below.) **2** a metal male screw with a slotted head and a sharp point for fastening things, esp. in carpentry, by being rotated to form a thread in wood etc. **3** (in full **screw-bolt**) a metal male screw with a blunt end on which a nut is threaded to bolt things together.

4 a wooden or metal straight screw used to exert pressure. **5** (in *sing.* or *pl.*) an instrument of torture acting in this way. **6** (in full **screw-propeller**) a form of propeller with twisted blades acting like a screw on the water or air. **7** one turn of a screw. **8** (foll. by *of*) *Brit.* a small twisted-up paper (of tobacco etc.). **9** *Brit.* (in billiards etc.) an oblique curling motion of the ball. **10** *sl.* a prison warder. **11** *Brit. sl.* an amount of salary or wages. **12** *coarse sl.* **a** an act of sexual intercourse. **b** a partner in this. **13** *sl.* a mean or miserly person. **14** *sl.* a worn-out horse. ● *v.* **1** *tr.* fasten or tighten with a screw or screws. **2** *tr.* turn (a screw). **3** *intr.* twist or turn round like a screw. **4** *intr.* (of a ball etc.) swerve. **5** *tr.* **a** put psychological etc. pressure on to achieve an end. **b** oppress. **6** *tr.* **a** (foll. by *out of*) extort (consent, money, etc.) from (a person). **b** swindle. **7** *tr.* (also *absol.*) *coarse sl.* have sexual intercourse with. **8** *intr.* (of a rolling ball, or of a person etc.) take a curling course; swerve. **9** *intr.* (often foll. by *up*) make tenser or more efficient. □ **have one's head screwed on the right way** *colloq.* have common sense. **have a screw loose** *colloq.* be slightly crazy. **put the screws on** (also *absol.*) *colloq.* exert pressure on (a person), esp. to extort something or to intimidate. **screw cap** = *screw top.* **screw-coupling** a female screw with threads at both ends for joining lengths of pipes or rods. **screw eye** *n.* a screw with a loop for passing cord etc. through instead of a slotted head. **screw gear** an endless screw with a cog-wheel or pinion. **screw hook** a hook to hang things on, with a screw point for fastening it. **screw-jack** a vehicle jack (see JACK¹ *n.* 1) worked by a screw device. **screw pine** = PANDANUS. **screw-plate** a steel plate with threaded holes for making male screws. **screw-tap** a tool for making female screws. **screw top** (also with hyphen *attrib.*) a cap or lid that can be screwed on to a bottle, jar, etc. **screw up 1** contract or contort (one's face etc.). **2** contract and crush into a tight mass (a piece of paper etc.). **3** summon up (one's courage etc.). **4** *sl.* **a** bungle or mismanage. **b** spoil or ruin (an event, opportunity, etc.). **5** *colloq.* **a** cause to become mentally disturbed. **b** (as **screwed up** *adj.*) mentally unstable, crazy. **screw-up** *n. sl.* a bungle, muddle, or mess. **screw valve** a stopcock opened and shut by a screw. □ **screwable** *adj.* **screwer** *n.* [ME f. OF *escroue* female screw, nut, f. L *scrofa* sow]

■ The principle of the screw was used by the ancient Greeks in a water-raising device (see *Archimedean screw*), and from the 1st century AD for exerting pressure in wine and olive presses. As fasteners, metal screws and nuts appeared in the 16th century, turned with a box-wrench or (sometimes) a pronged device. The metal screw for fixing various materials to wood is described by a German mining engineer of the mid-16th century, but may have been already in use for some time. The screwdriver as a hand-tool appears from *c.*1800, and in this century other kinds of screws turned by Allen keys or Phillips screwdrivers have been developed.

screwball /ˈskruːbɔːl/ *n. & adj. N. Amer. sl.* ● *n.* a crazy or eccentric person. ● *adj.* crazy.

screwdriver /ˈskruːˌdraɪvə(r)/ *n.* **1** a tool with a shaped tip to fit into the head of a screw to turn it. **2** a cocktail made from vodka and orange juice.

screwed /skruːd/ *adj.* **1** twisted. **2** *sl.* **a** ruined; rendered ineffective. **b** drunk.

screwy /ˈskruːɪ/ *adj.* (**screwier**, **screwiest**) *sl.* **1** crazy or eccentric. **2** absurd. □ **screwiness** *n.*

Scriabin /skrɪˈɑːbɪn/, Aleksandr (Nikolaevich) (also **Skryabin**) (1872–1915), Russian composer and pianist. He wrote symphonies, symphonic poems, and numerous pieces for the piano, including sonatas and preludes. Much of his later music reflects his interest in mysticism and theosophy, especially his third symphony *The Divine Poem* (1903) and the symphonic poem *Prometheus: The Poem of Fire* (1909–10), which is scored for orchestra, piano, optional choir, and 'keyboard of light' (projecting colours on to a screen).

scribble¹ /ˈskrɪb(ə)l/ *v. & n.* ● *v.* **1** *tr. & intr.* write carelessly or hurriedly. **2** *intr.* often *derog.* be an author or writer. **3** *intr. & tr.* draw carelessly or meaninglessly. ● *n.* **1** a scrawl. **2** a hasty note etc. **3** careless handwriting. □ **scribbler** *n.* **scribbly** *adj.* [ME f. med.L *scribillare* dimin. of L *scribere* write]

scribble² /ˈskrɪb(ə)l/ *v.tr.* card (wool, cotton, etc.) coarsely. [prob. f. LG: cf. G *schrubben* (in the same sense), frequent. f. LG *schrubben*: see SCRUB¹]

scribe /skraɪb/ *n. & v.* ● *n.* **1** a person who writes out documents, esp. an ancient or medieval copyist of manuscripts. **2** an ancient Jewish record-keeper or, later, professional theologian and jurist. **3** (in full **scribe-awl**) a pointed instrument for making marks on wood, bricks, etc., to guide a saw, or in sign-writing. **4** *N. Amer. colloq.* a writer, esp. a journalist. ● *v.tr.* mark (wood etc.) with a scribe (see sense 3 of *n.*).

□ **scribal** *adj.* **scriber** *n.* [(n.) ME f. L *scriba* f. *scribere* write: (v.) perh. f. DESCRIBE]

scrim /skrɪm/ *n.* open-weave fabric for lining or upholstery etc. [18th c.: orig. unkn.]

scrimmage /ˈskrɪmɪdʒ/ *n. & v.* ● *n.* **1** a rough or confused struggle; a brawl. **2** *Amer. Football* a sequence of play beginning with the placing of the ball on the ground with its longest axis at right angles to the goal-line. ● *v.* **1** *intr.* engage in a scrimmage. **2** *tr. Amer. Football* put (the ball) into a scrimmage. □ **scrimmager** *n.* [var. of SKIRMISH]

scrimp /skrɪmp/ *v.* **1** *intr.* be sparing or parsimonious. **2** *tr.* use sparingly. □ **scrimpy** *adj.* [18th c., orig. Sc.: perh. rel. to SHRIMP]

scrimshank /ˈskrɪmʃæŋk/ *v.intr. Brit. sl. esp. Mil.* shirk duty. □ **scrimshanker** *n.* [19th c.: orig. unkn.]

scrimshaw /ˈskrɪmʃɔː/ *v. & n.* ● *v.tr.* (also *absol.*) adorn (shells, ivory, etc.) with carved or coloured designs (as sailors' pastime at sea). ● *n.* work or a piece of work of this kind. [19th c.: perh. f. a surname]

scrip /skrɪp/ *n.* **1** a provisional certificate of money subscribed to a bank or company etc. entitling the holder to a formal certificate and dividends. **2** (*collect.*) such certificates. **3** an extra share or shares instead of a dividend. [abbr. of *subscription receipt*]

script /skrɪpt/ *n. & v.* ● *n.* **1** handwriting as distinct from print; written characters. **2** type imitating handwriting. **3** an alphabet or system of writing (*the Russian script*). **4** the text of a play, film, or broadcast. **5** an examinee's set of written answers. **6** *Law* an original document as distinct from a copy. ● *v.tr.* write a script for (a film etc.). [ME, = thing written, f. OF *escri*)t f. L *scriptum*, neut. past part. of *scribere* write]

scriptorium /skrɪpˈtɔːrɪəm/ *n.* (*pl.* **scriptoria** /-rɪə/ or **scriptoriums**) a room set apart for writing, esp. in a monastery. □ **scriptorial** *adj.* [med.L (as SCRIPT)]

scriptural /ˈskrɪptʃərəl/ *adj.* **1** of or relating to a scripture, esp. the Bible. **2** having the authority of a scripture. □ **scripturally** *adv.* [LL *scripturalis* f. L *scriptura*: see SCRIPTURE]

scripture /ˈskrɪptʃə(r)/ *n.* **1** (**Scripture** or **the Scriptures**) the sacred writings of the Christians (the Old and New Testaments) or the Jews (the Hebrew Bible). (See OLD TESTAMENT, NEW TESTAMENT, HEBREW BIBLE.) **2** the sacred writings of any religion. **3** a thing regarded as authoritative. [ME f. L *scriptura* (as SCRIPT)]

scriptwriter /ˈskrɪptˌraɪtə(r)/ *n.* a person who writes a script for a film, broadcast, etc. □ **scriptwriting** *n.*

scrivener /ˈskrɪvənə(r)/ *n. hist.* **1** a copyist or drafter of documents. **2** a notary. **3** a broker. **4** a moneylender. [ME f. obs. *scrivein* f. OF *escrivein* ult. f. L (as SCRIBE)]

scrobiculate /skrəˈbɪkjʊlət/ *adj. Bot. & Zool.* pitted, furrowed. [L *scrobiculus* f. *scrobis* trench]

scrod /skrɒd/ *n. N. Amer.* a young cod or haddock, esp. as food. [19th c.: perh. rel. to SHRED]

scrofula /ˈskrɒfjʊlə/ *n. archaic* a disease with glandular swellings, prob. a form of tuberculosis. (See also *king's evil*.) □ **scrofulous** *adj.* [ME f. med.L (sing.) f. LL *scrofulae* (pl.) scrofulous swelling, dimin. of L *scrofa* sow]

scroll /skrəʊl/ *n. & v.* ● *n.* **1** a roll of parchment or paper esp. with writing on it. **2** a book in the ancient roll form. **3** an ornamental design or carving imitating a roll of parchment. ● *v.* **1** *tr.* (often foll. by *down, up*) move (a display on a VDU screen) in order to view new material. **2** *tr.* inscribe in or like a scroll. **3** *intr.* curl up like paper. □ **scroll saw** a saw for cutting along curved lines in ornamental work. [ME *scrowle* alt. f. *rowle* ROLL, perh. after *scrow* (in the same sense), formed as ESCROW]

scrolled /skrəʊld/ *adj.* having a scroll ornament.

scrollwork /ˈskrəʊlwɜːk/ decoration of spiral lines, esp. as cut by a scroll saw.

Scrooge /skruːdʒ/, Ebenezer. A miserly curmudgeon in Charles Dickens's novel *A Christmas Carol* (1843), frequently used as the type of a mean or tight-fisted person.

scrotum /ˈskrəʊtəm/ *n.* (*pl.* **scrota** /-tə/ or **scrotums**) a pouch of skin containing the testicles. □ **scrotal** *adj.* **scrotitis** /skrəʊˈtaɪtɪs/ *n.* [L]

scrounge /skraʊndʒ/ *v. & n. colloq.* ● *v.* **1** *tr.* (also *absol.*) obtain (things) illicitly or by cadging. **2** *intr.* search about to find something at no cost. ● *n.* an act of scrounging. □ **on the scrounge** engaged in scrounging. □ **scrounger** *n.* [var. of dial. *scrunge* steal]

scrub¹ /skrʌb/ *v. & n.* ● *v.* (**scrubbed**, **scrubbing**) **1** *tr.* rub hard so as to clean, esp. with a hard brush. **2** *intr.* use a brush in this way. **3** *intr.*

(often foll. by *up*) (of a surgeon etc.) thoroughly clean the hands and arms by scrubbing, before operating. **4** *tr. colloq.* scrap or cancel (a plan, order, etc.). **5** *tr.* use water to remove impurities from (gas etc.). **6 a** *intr.* (of tyres) scrape on the road surface, esp. when cornering. **b** *tr.* (foll. by *off*) reduce speed by allowing the tyres to scrape the road. ● *n.* the act or an instance of scrubbing; the process of being scrubbed. □ **scrubbing-brush** (*N. Amer.* **scrub-brush**) a hard brush for scrubbing floors. **scrub round** *colloq.* circumvent, avoid. [ME prob. f. MLG, MDu. *schrobben*, *schrubben*]

scrub² /skrʌb/ *n.* **1 a** vegetation consisting mainly of brushwood or stunted forest growth. **b** an area of land covered with this. **2** any animal of inferior breed or physique (often *attrib.*: *scrub horse*). **3** a small or dwarf variety (often *attrib.*: *scrub pine*). **4** *US Sport colloq.* a team or player not of the first class. □ **scrub turkey** a megapode. **scrub typhus** a rickettsial disease transmitted to humans by mites. □ **scrubby** *adj.* [ME, var. of SHRUB¹]

scrubber /ˈskrʌbə(r)/ *n.* **1** an apparatus using water or a solution for purifying gases etc. **2** *sl. derog.* a sexually promiscuous woman.

scrubland /ˈskrʌblænd/ *n.* land consisting of scrub vegetation.

scruff¹ /skrʌf/ *n.* the back of the neck as used to grasp and lift or drag an animal or person by (esp. *scruff of the neck*). [alt. of *scuff*, perh. f. ON *skoft* hair]

scruff² /skrʌf/ *n.* an untidy or scruffy person. [orig. = SCURF, later 'worthless thing', or back-form. f. SCRUFFY]

scruffy /ˈskrʌfɪ/ *adj.* (**scruffier**, **scruffiest**) *colloq.* shabby, slovenly, untidy. □ **scruffily** *adv.* **scruffiness** *n.* [*scruff* var. of SCURF + -Y¹]

scrum /skrʌm/ *n. & v.* ● *n.* **1** *Rugby* an arrangement of the forwards of each team in two opposing groups, each with arms interlocked and heads down, with the ball thrown in between them to restart play. **2** *colloq.* a confused noisy crowd. ● *v.intr.* **1** (often foll. by *down*) *Rugby* form a scrum. **2** *colloq.* jostle, crowd. □ **scrum-half** a half-back who puts the ball into the scrum. [abbr. of SCRUMMAGE]

scrummage /ˈskrʌmɪdʒ/ *n. Rugby* = SCRUM 1. [as SCRIMMAGE]

scrummy /ˈskrʌmɪ/ *adj.* (**-ier**, **-iest**) *colloq.* excellent; enjoyable; delicious. [SCRUMPTIOUS + -Y¹]

scrump /skrʌmp/ *v.tr. Brit. colloq.* steal (fruit) from an orchard or garden. [cf. SCRUMPY]

scrumple /ˈskrʌmp(ə)l/ *v.tr.* crumple, wrinkle. [var. of CRUMPLE]

scrumptious /ˈskrʌmpʃəs/ *adj. colloq.* **1** delicious. **2** pleasing, delightful. □ **scrumptiously** *adv.* **scrumptiousness** *n.* [19th c.: orig. unkn.]

scrumpy /ˈskrʌmpɪ/ *n. Brit. colloq.* rough cider, esp. as made in the West Country of England. [dial. *scrump* small apple]

scrunch /skrʌntʃ/ *v. & n.* ● *v.* **1** *tr. & intr.* (usu. foll. by *up*) make or become crushed or crumpled. **2** *intr. & tr.* make or cause to make a crunching sound. **3** *tr.* style (hair) by squeezing or crushing in the hands to give a tousled look. ● *n.* the act or an instance of scrunching. [var. of CRUNCH]

scruple /ˈskruːp(ə)l/ *n. & v.* ● *n.* **1** (in *sing.* or *pl.*) **a** regard to the morality or propriety of an action. **b** a feeling of doubt or hesitation caused by this. **2** *Brit. hist.* an apothecaries' weight of 20 grains. **3** *archaic* a very small quantity. ● *v.intr.* **1** (foll. by *to* + infin.; usu. with *neg.*) be reluctant because of scruples (*did not scruple to stop their allowance*). **2** feel or be influenced by scruples. [F *scrupule* or L *scrupulus* f. *scrupus* rough pebble, (fig.) anxiety]

scrupulous /ˈskruːpjʊləs/ *adj.* **1** conscientious or thorough even in small matters. **2** careful to avoid doing wrong. **3** punctilious; over-attentive to details. □ **scrupulously** *adv.* **scrupulousness** *n.* **scrupulosity** /ˌskruːpjʊˈlɒsɪtɪ/ *n.* [ME f. F *scrupuleux* or L *scrupulosus* (as SCRUPLE)]

scrutineer /ˌskruːtɪˈnɪə(r)/ *n.* a person who scrutinizes or examines something, esp. the conduct and result of a ballot.

scrutinize /ˈskruːtɪˌnaɪz/ *v.tr.* (also **-ise**) look closely at; examine with close scrutiny. □ **scrutinizer** *n.*

scrutiny /ˈskruːtɪnɪ/ *n.* (pl. **-ies**) **1** a critical gaze. **2** a close investigation or examination of details. **3** an official examination of ballot-papers to check their validity or accuracy of counting. [ME f. L *scrutinium* f. *scrutari* search f. *scruta* rubbish: orig. of rag-collectors]

scry /skraɪ/ *v.intr.* (**-ies**, **-ied**) divine by crystal-gazing. □ **scryer** *n.* [shortening f. DESCRY]

scuba /ˈskuːbə/ *n.* (pl. **scubas**) an aqualung. [acronym f. *self-contained underwater breathing apparatus*]

scuba-diving /ˈskuːbəˌdaɪvɪŋ, ˈskjuː-/ *n.* swimming under water using a scuba, esp. as a sport. □ **scuba-dive** *v.intr.* **scuba-diver** *n.*

scud /skʌd/ *v. & n.* ● *v.intr.* (**scudded**, **scudding**) **1** fly or run straight, fast, and lightly; skim along. **2** *Naut.* run before the wind. ● *n.* **1** a spell of scudding. **2** a scudding motion. **3** vapoury driving clouds. **4** a driving shower; a gust. **5** wind-blown spray. **6** (in full **Scud missile**) a type of long-range surface-to-surface guided missile originally developed in the former Soviet Union. [perh. alt. of SCUT, as if to race like a hare]

scuff /skʌf/ *v. & n.* ● *v.* **1** *tr.* graze or brush against. **2** *tr.* mark or wear down (shoes) in this way. **3** *intr.* walk with dragging feet; shuffle. ● *n.* a mark made by scuffing. [imit.]

scuffle /ˈskʌf(ə)l/ *n. & v.* ● *n.* a confused struggle or disorderly fight at close quarters. ● *v.intr.* engage in a scuffle. [prob. f. Scand.: cf. Sw. *skuffa* to push, rel. to SHOVE]

sculduggery var. of SKULDUGGERY.

scull /skʌl/ *n. & v.* ● *n.* **1** either of a pair of small oars used by a single rower. **2** an oar placed over the stern of a boat to propel it, usu. by a twisting motion. **3** a small boat propelled with a scull or pair of sculls. **4** (in *pl.*) a race between boats with single pairs of oars. ● *v.tr.* propel (a boat) with sculls. [ME: orig. unkn.]

sculler /ˈskʌlə(r)/ *n.* **1** a user of sculls. **2** a boat intended for sculling.

scullery /ˈskʌlərɪ/ *n.* (pl. **-ies**) a small kitchen or room at the back of a house for washing dishes etc. [ME f. AF *squillerie*, OF *escuelerie* f. *escuele* dish f. L *scutella* salver dimin. of *scutra* wooden platter]

scullion /ˈskʌljən/ *n. archaic* **1** a cook's boy. **2** a person who washes dishes etc. [ME: orig. unkn. but perh. infl. by *scullery*]

sculpin /ˈskʌlpɪn/ *n.* a small mainly marine fish of the bullhead family Cottidae (or a related family), found esp. in the cooler waters of the North Pacific and having a large spiny head. [perh. f. obs. *scorpene* f. L *scorpaena* f. Gk *skorpaina* a fish]

sculpt /skʌlpt/ *v.tr. & intr.* (also **sculp**) sculpture. [F *sculpter* f. *sculpteur* SCULPTOR: now regarded as an abbr.]

sculptor /ˈskʌlptə(r)/ *n.* (*fem.* **sculptress** /-trɪs/) an artist who makes sculptures. [L (as SCULPTURE)]

sculpture /ˈskʌlptʃə(r)/ *n. & v.* ● *n.* **1** the art of making forms, often representational, in the round or in relief by chiselling stone, carving wood, modelling clay, casting metal, etc. **2** a work or works of sculpture. **3** *Zool. & Bot.* raised or sunken markings on a shell etc. ● *v.* **1** *tr.* represent in or adorn with sculpture. **2** *intr.* practise sculpture. □ **sculptural** *adj.* **sculpturally** *adv.* **sculpturesque** /ˌskʌlptʃəˈresk/ *adj.* [ME f. L *sculptura* f. *sculpere* *sculpt-* carve]

scum /skʌm/ *n. & v.* ● *n.* **1** a layer of dirt, froth, or impurities etc. forming at the top of liquid, esp. in boiling or fermentation. **2** (foll. by *of*) the most worthless part of something. **3** *colloq.* a worthless person or group. ● *v.* (**scummed**, **scumming**) **1** *tr.* remove scum from; skim. **2** *tr.* be or form a scum on. **3** *intr.* (of a liquid) develop scum. □ **scummy** *adj.* (**scummier**, **scummiest**) *adj.* [ME f. MLG, MDu. *schūm*, OHG *scūm* f. Gmc]

scumbag /ˈskʌmbæg/ *n. sl.* a worthless despicable person.

scumble /ˈskʌmb(ə)l/ *v. & n.* ● *v.tr.* **1** modify (a painting) by applying a thin opaque coat of paint to give a softer or duller effect. **2** modify (a drawing) similarly with light pencilling etc. ● *n.* **1** material used in scumbling. **2** the effect produced by scumbling. [perh. frequent. of SCUM *v.*]

scuncheon /ˈskʌntʃən/ *n.* the inside face of a door-jamb, window-frame, etc. [ME f. OF *escoinson* (as EX-¹, COIN)]

scunge /skʌndʒ/ *n. Austral. & NZ colloq.* **1** dirt, scum. **2** a dirty or disagreeable person. □ **scungy** *adj.* (**scungier**, **scungiest**). [perh. f. Engl. dial. *scrunge* steal: cf. SCROUNGE]

scunner /ˈskʌnə(r)/ *v. & n. Sc.* ● *v.intr. & tr.* feel disgust; nauseate. ● *n.* **1** a strong dislike (esp. *take a scunner at* or *against*). **2** an object of loathing. [14th c.: orig. uncert.]

Scunthorpe /ˈskʌnθɔːp/ an industrial town in NE England, in Humberside; pop. (1991) 60,500. The town grew rapidly in the late 19th century as a centre of the iron and steel industries.

scupper¹ /ˈskʌpə(r)/ *n.* a hole in a ship's side to carry off water from the deck. [ME (perh. f. AF) f. OF *escopir* f. Rmc *skuppire* (unrecorded) to spit: orig. imit.]

scupper² /ˈskʌpə(r)/ *v.tr. Brit. sl.* **1** sink (a ship or its crew). **2** defeat or ruin (a plan etc.). **3** kill. [19th c.: orig. unkn.]

scurf /skɜːf/ *n.* **1** flakes on the surface of the skin, cast off as fresh skin develops below, esp. those of the head; dandruff. **2** any scaly matter on a surface. □ **scurfy** *adj.* [OE, prob. f. ON & earlier OE *sceorf*, rel. to *sceorfan* gnaw, *sceorfian* cut to shreds]

scurrilous /ˈskʌrɪləs/ adj. **1** (of a person or language) grossly or indecently abusive. **2** given to or expressed with low humour. □ **scurrilously** adv. **scurrilousness** n. **scurrility** /skəˈrɪlɪtɪ/ n. (pl. -ies). [F scurrile or L scurrilus f. scurra buffoon]

scurry /ˈskʌrɪ/ v. & n. ● v.intr. (-ies, -ied) run or move hurriedly, esp. with short quick steps; scamper. ● n. (pl. -ies) **1** the act or sound of scurrying. **2** bustle, haste. **3** a flurry of rain or snow. [abbr. of hurry-scurry redupl. of HURRY]

scurvy /ˈskɜːvɪ/ n. & adj. ● n. a disease caused by a deficiency of vitamin C, characterized by swollen bleeding gums and the opening of previously healed wounds, esp. formerly affecting sailors. ● adj. archaic worthless, dishonourable, contemptible. □ **scurvy grass** a cresslike seaside plant of the genus Cochlearia, orig. taken as a cure for scurvy. □ **scurvied** adj. **scurvily** adv. [SCURF + -Y¹: noun sense by assoc. with F scorbut (cf. SCORBUTIC)]

scut /skʌt/ n. a short tail, esp. of a hare, rabbit, or deer. [ME: orig. unkn.: cf. obs. scut short, shorten]

scuta pl. of SCUTUM.

scutage /ˈskjuːtɪdʒ/ n. hist. money paid by a feudal landowner instead of personal service. [ME f. med.L scutagium f. L scutum shield]

Scutari /skuːˈtɑːrɪ/ **1** a former name for Üsküdar near Istanbul, site of a British army hospital in which Florence Nightingale worked during the Crimean War. **2** see SHKODËR.

scutch /skʌtʃ/ v.tr. dress (fibrous material, esp. retted flax) by beating. □ **scutcher** n. [OF escouche, escoucher (dial.), escousser, ult. f. L excutere excuss- (as EX-¹, quatere shake)]

scutcheon /ˈskʌtʃən/ n. **1** = ESCUTCHEON. **2** an ornamented brass etc. plate round or over a keyhole. **3** a plate for a name or inscription. [ME f. ESCUTCHEON]

scute /skjuːt/ n. Zool. = SCUTUM. [L (as SCUTUM)]

scutellum /skjuːˈteləm/ n. (pl. **scutella** /-lə/) Bot. & Zool. a scale, plate, or any shieldlike formation on a plant, insect, bird, etc., esp. one of the horny scales on a bird's foot. □ **scutellate** /ˈskjuːtələt/ adj. **scutellation** /ˌskjuːtəˈleɪʃ(ə)n/ n. [mod.L dimin. of L scutum shield]

scutter /ˈskʌtə(r)/ v. & n. ● v.intr. colloq. scurry. ● n. the act or an instance of scuttering. [perh. alt. of SCUTTLE²]

scuttle¹ /ˈskʌt(ə)l/ n. **1** a receptacle for carrying and holding a small supply of coal. **2** Brit. the part of a motor-car body between the windscreen and the bonnet. [ME f. ON skutill, OHG scuzzila f. L scutella dish]

scuttle² /ˈskʌt(ə)l/ v. & n. ● v.intr. **1** scurry; hurry along. **2** run away; flee from danger or difficulty. ● n. **1** a hurried gait. **2** a precipitate flight or departure. [cf. dial. scuddle frequent. of SCUD]

scuttle³ /ˈskʌt(ə)l/ n. & v. ● n. a hole with a lid in a ship's deck or side. ● v.tr. let water into (a ship) to sink it, esp. by opening the seacocks. [ME, perh. f. obs. F escoutille f. Sp. escotilla hatchway dimin. of escota cutting out cloth]

scuttlebutt /ˈskʌt(ə)lˌbʌt/ n. **1** a water-butt on the deck of a ship, for drinking from. **2** colloq. rumour, gossip.

scutum /ˈskjuːtəm/ n. (pl. **scuta** /-tə/) each of the shieldlike plates or scales forming the bony covering of a crocodile, sturgeon, turtle, armadillo, etc. □ **scutal** adj. **scutate** /-teɪt/ adj. [L, = oblong shield]

scuzzy /ˈskʌzɪ/ adj. (-ier, -iest) sl. **1** dirty, sordid. **2** disgusting, disreputable. [prob. f. DISGUSTING, infl. by scum and fuzz]

Scylla /ˈsɪlə/ Gk Mythol. a female sea monster who devoured sailors when they tried to navigate the narrow channel between her cave and the whirlpool Charybdis. In later legend Scylla was a dangerous rock, located on the Italian side of the Strait of Messina.

scyphozoan /ˌsaɪfəˈzəʊən/ n. & adj. Zool. ● n. a marine coelenterate of the class Scyphozoa, which comprises the jellyfishes. ● adj. of or relating to this class. [mod.L., as SCYPHOZOA + Gk zōion animal]

scyphus /ˈsaɪfəs/ n. (pl. **scyphi** /-faɪ/) **1** Gk Antiq. a large drinking-cup with two handles. **2** Bot. a cup-shaped structure, e.g. the fruiting body of some lichens or the corolla of a daffodil. □ **scyphose** /-fəʊz, -fəʊs/ adj. [mod.L f. Gk skuphos]

scythe /saɪð/ n. & v. ● n. a mowing and reaping implement with a long curved blade swung over the ground by a long pole with two short handles projecting from it. ● v.tr. cut with a scythe. [OE sīthe f. Gmc]

Scythia /ˈsɪðɪə/ an ancient region of SE Europe and Asia. The Scythian empire, which existed between the 8th and 2nd centuries BC, was centred on the northern shores of the Black Sea and extended from southern Russia to the borders of Persia. □ **Scythian** adj. & n.

SD abbr. US South Dakota (in official postal use).

S.Dak. abbr. South Dakota.

SDI see STRATEGIC DEFENSE INITIATIVE.

SDLP see SOCIAL DEMOCRATIC AND LABOUR PARTY.

SDP see SOCIAL DEMOCRATIC PARTY.

SDR abbr. special drawing right (from the International Monetary Fund).

SE abbr. **1** south-east. **2** south-eastern.

Se symb. Chem. the element selenium.

se- /sə, sɪ/ prefix apart, without (seclude; secure). [L f. OL se (prep. & adv.)]

SEA see SINGLE EUROPEAN ACT.

sea /siː/ n. **1** the expanse of salt water that covers most of the earth's surface and surrounds its land masses. **2** any part of this as opposed to land or fresh water. **3** a particular (usu. named) tract of salt water partly or wholly enclosed by land (the North Sea; the Dead Sea). **4** a large inland lake (the Sea of Galilee). **5** the waves of the sea, esp. with reference to their local motion or state (a choppy sea). **6** (foll. by of) a vast quantity or expanse (a sea of troubles; a sea of faces). **7** (attrib.) living or used in, on, or near the sea (often prefixed to the name of a marine animal, plant, etc., having a superficial resemblance to what it is named after) (sea lettuce). □ **at sea 1** in a ship on the sea. **2** (also **all at sea**) perplexed, confused. **by sea** in a ship or ships. **go to sea** become a sailor. **on the sea 1** in a ship at sea. **2** situated on the coast. **put** (or **put out**) **to sea** leave land or port. **sea anchor** a device such as a heavy bag dragged in the water to retard the drifting of a ship. **sea anemone** a sessile marine coelenterate of the order Actiniaria, having a polypoid body bearing a ring of tentacles around the mouth. **sea-angel** an angelfish. **sea-bass** see BASS² 3. **sea bream** see BREAM¹ 2. **sea breeze** a breeze blowing towards the land from the sea, esp. during the day (cf. land breeze). **sea buckthorn** a maritime shrub, Hippophae rhamnoides, with orange berries. **sea change** a notable or unexpected transformation (with ref. to Shakespeare's The Tempest I. ii. 403). **sea-chest** a sailor's storage-chest. **sea coal** archaic mineral coal, as distinct from charcoal etc. **sea cow** a sirenian, esp. a manatee. **sea cucumber** a holothurian (echinoderm), esp. a bêche-de-mer. **sea dog** an old or experienced sailor. **sea-duck** a duck of sea-coasts, esp. an eider, scoter, or other duck of the tribe Mergini or Somateriini. **sea eagle** a fish-eating eagle, esp. of the genus Haliaeetus. **sea-ear** = ORMER. **sea elephant** = elephant seal. **sea fan** a colonial coral of the order Gorgonacea, supported by a fanlike horny skeleton. **sea front** the part of a coastal town etc. directly facing the sea. **sea-girt** literary surrounded by sea. **sea gooseberry** a ctenophore (coelenterate), with an ovoid body bearing numerous cilia. **sea-green** bluish-green (as of the sea). **sea hare** a large sea slug of the order Anaspidea, esp. of the genus Aplysia, with an internal shell and long extensions from its foot. **sea holly** a spiny-leaved maritime composite plant, Eryngium maritimum, with blue flowers. **sea horse 1** a small upright marine fish of the family Syngnathidae, esp. Hippocampus hippocampus, having a body suggestive of the head and neck of a horse. **2** a mythical creature with a horse's head and fish's tail. **sea-island cotton** a fine-quality long-stapled cotton grown on islands off the southern US. **sea lavender** a maritime plant of the genus Limonium, with small brightly coloured funnel-shaped flowers. **sea legs** the ability to keep one's balance and avoid seasickness when at sea. **sea level** the mean level of the sea's surface, used in reckoning the height of hills etc. and as a barometric standard. **sea lily** a sessile echinoderm of the class Crinoidea, with long jointed stalks and feather-like arms for trapping food. **sea lion** a large eared seal of the Pacific, esp. of the genus Zalophus or Otaria. **sea loch** = LOCH 2. **sea mile** a unit of length varying between approx. 2,014 yards (1,842 metres) at the equator and 2,035 yards (1,861 metres) at the poles. **sea mouse** a marine annelid worm of the genus Aphrodite, with a broad iridescent body. **sea onion** = SQUILL 2. **sea otter** a marine otter, Enhydra lutris, found on North Pacific coasts, often floating on its back and using a stone balanced on its stomach to crack open bivalve molluscs. **sea pen** a colonial hydroid of the order Pennatulacea, resembling a quill pen. **sea-perch** a perchlike marine fish, esp. of the family Percichthyidae, Lutjanidae, or Embiotocidae. **sea pink** a maritime plant, Armeria maritima, with bright pink flowers (also called thrift). **sea purse** the egg-case of a skate or shark. **sea room** clear space at sea for a ship to turn or manoeuvre in. **sea salt** salt produced by evaporating sea water. **sea serpent 1** an enormous legendary serpent-like sea monster. **2** = sea snake 1. **sea shell** the shell of a marine mollusc. **sea slug** a shell-less marine gastropod mollusc; esp. one of the order Nudibranchia, with external gills. **sea snail 1** a small slimy fish of the family Cyclopteridae, with a ventral sucker, esp. Liparis liparis of the NE

Atlantic. **2** a spiral-shelled marine gastropod mollusc, e.g. a whelk.

sea snake 1 a venomous tropical snake of the family Hydrophidae, living in the sea. **2** = *sea serpent* 1. **sea-spider** a marine chelicerate arthropod of the class Pycnogonida, with a very small body and long legs. **sea squirt** = ASCIDIAN. **sea trout 1** a large silvery migratory race of the trout, *Salmo trutta* (also called *salmon trout*). **2** *US* an unrelated marine fish of the genus *Cynoscion*, of the drum-fish family. **sea urchin** a marine echinoderm of the class Echinoidea, with a spherical or flattened shell covered in spines. **sea wall** a wall or embankment erected to prevent encroachment by the sea. **sea wasp** an Indo-Pacific jellyfish of the order Cubomedusae, with a dangerous sting. **sea water** water in or taken from the sea. [OE sǣ f. Gmc]

seabed /ˈsiːbed/ *n.* the ground under the sea; the ocean floor.

seabird /ˈsiːbɜːd/ *n.* a bird frequenting the sea or sea-coast.

seaboard /ˈsiːbɔːd/ *n.* **1** the seashore or coastal region. **2** the line of a coast.

Seaborg /ˈsiːbɔːg/, Glenn (Theodore) (b.1912), American nuclear chemist. During 1940–58 Seaborg and his colleagues at the University of California, Berkeley, produced nine of the transuranic elements (plutonium to nobelium) by bombarding uranium and other elements with nuclei in a cyclotron, and he coined the term *actinide* for the elements in this series. Seaborg and his early collaborator Edwin McMillan (1907–91) shared the Nobel Prize for chemistry in 1951, and Seaborg was chairman of the US Atomic Energy Commission 1962–71.

seaborne /ˈsiːbɔːn/ *adj.* transported by sea.

seacock /ˈsiːkɒk/ *n.* a valve below a ship's water-line for letting water in or out.

Sea Dyak *n.* = IBAN n. 1 (see also DYAK).

seafarer /ˈsiːˌfeərə(r)/ *n.* **1** a sailor. **2** a traveller by sea.

seafaring /ˈsiːˌfeərɪŋ/ *adj. & n.* travelling by sea, esp. regularly.

sea-floor spreading /ˈsiːflɔː(r)/ *n. Geol.* the formation of fresh areas of the earth's crust under the oceans, occurring through the upwelling of magma at mid-ocean ridges and its subsequent outward movement on either side. The discovery of sea-floor spreading in the early 1960s, from submarine magnetic surveys, was crucial to the development of plate tectonics.

seafood /ˈsiːfuːd/ *n.* edible sea fish or shellfish.

seagoing /ˈsiːˌɡəʊɪŋ/ *adj.* **1** (of ships) fit for crossing the sea. **2** (of a person) seafaring.

seagull /ˈsiːɡʌl/ *n.* = GULL[1].

seakale /ˈsiːkeɪl/ *n.* a cruciferous maritime plant, *Crambe maritima*, having coarsely toothed leaves and used as a vegetable. □ **seakale beet** = CHARD.

seal[1] /siːl/ *n. & v.* ● *n.* **1** a piece of wax, lead, paper, etc., with a stamped design, attached to a document as a guarantee of authenticity. **2** a similar material attached to a receptacle, envelope, etc., affording security by having to be broken to allow access to the contents. **3** an engraved piece of metal, gemstone, etc., for stamping a design on a seal. **4 a** a substance or device used to close an aperture or act as a fastening. **b** an amount of water standing in the trap of a drain to prevent foul air from rising. **5** an act or gesture or event regarded as a confirmation or guarantee (*gave her seal of approval to the venture*). **6** a significant or prophetic mark (*has the seal of death in his face*). **7** a decorative adhesive stamp. **8** esp. *Eccl.* a vow of secrecy; an obligation to silence. ● *v.tr.* **1** close securely or hermetically. **2** stamp or fasten with a seal. **3** fix a seal to. **4** certify as correct with a seal or stamp. **5** (often foll. by *up*) confine or fasten securely. **6** settle or decide (*their fate is sealed*). **7** (foll. by *off*) put barriers round (an area) to prevent entry and exit, esp. as a security measure. **8** apply a non-porous coating to (a surface) to make it impervious. □ **one's lips are sealed** one is obliged to keep a secret. **sealed-beam** (*attrib.*) designating a vehicle headlamp with a sealed unit consisting of the light source, reflector, and lens. **sealed book** see BOOK. **sealed orders** orders for procedure not to be opened before a specified time. **sealing-wax** a mixture of shellac and rosin with turpentine and pigment, softened by heating and used to make seals. **seal ring** a finger ring with a seal. **seals of office** (in the UK) those held during tenure esp. by the Lord Chancellor or a Secretary of State. **set one's seal to** (or **on**) authorize or confirm. □ **sealable** *adj.* [ME f. AF *seal*, OF *seel* f. L *sigillum* dimin. of *signum* SIGN]

seal[2] /siːl/ *n. & v.* ● *n.* a fish-eating amphibious sea mammal of the order Pinnipedia, with flippers and webbed feet. (*See note below.*) ● *v.intr.* hunt for seals. [OE *seolh seol-* f. Gmc]

▪ The seals are closely related to some land carnivores, and are sometimes classed with them. The earless seals belong to the family Phocidae, many of which live in polar seas. They range in size from the huge elephant seals to the little seal of Lake Baikal, which is unusual in being confined to fresh water. The eared seals belong to the family Otariidae and are able to support the body on the hindlimbs, making them much more agile on land. They include the sea lions and fur seals, and may not be closely related to the earless seals.

sealant /ˈsiːlənt/ *n.* material for sealing, esp. to make something airtight or watertight.

sealer /ˈsiːlə(r)/ *n.* **1 a** a device or substance used for sealing. **b** a person who seals receptacles etc. **2** a ship or person engaged in hunting seals.

sealery /ˈsiːlərɪ/ *n.* (*pl.* **-ies**) a place for hunting seals.

Sea Lord *n.* (in the UK) either of two senior naval officers (**First Sea Lord**, **Second Sea Lord**) serving originally as members of the Admiralty Board (now of the Ministry of Defence).

sealskin /ˈsiːlskɪn/ *n.* **1** the skin or prepared fur of a seal. **2** (often *attrib.*) a garment made from this.

Sealyham /ˈsiːlɪəm/ *n.* (in full **Sealyham terrier**) a wire-haired short-legged breed of terrier. [*Sealyham* in SW Wales, where the dog was first bred]

seam /siːm/ *n. & v.* ● *n.* **1** a line where two edges join, esp. of two pieces of cloth etc. turned back and stitched together, or of boards fitted edge to edge. **2** a fissure between parallel edges. **3** a wrinkle or scar. **4** a stratum of coal etc. ● *v.tr.* **1** join with a seam. **2** (esp. as **seamed** *adj.*) mark or score with or as with a seam. □ **bursting at the seams** full to overflowing. **seam bowler** *Cricket* a bowler who makes the ball deviate by bouncing off its seam. □ **seamer** *n.* **seamless** *adj.* [OE *sēam* f. Gmc]

seaman /ˈsiːmən/ *n.* (*pl.* **-men**) **1** a sailor, esp. one below the rank of officer. **2** a person regarded in terms of skill in navigation (*a poor seaman*). □ **seamanlike** *adj.* **seamanly** *adj.* [OE *sǣman* (as SEA, MAN)]

seamanship /ˈsiːmənʃɪp/ *n.* skill in managing a ship or boat.

seamstress /ˈsemstrɪs/ *n.* (also **sempstress** /ˈsemps-/) a woman who sews, esp. professionally; a dressmaker. [OE *sēamestre* fem. f. *sēamere* tailor, formed as SEAM + -STER + -ESS[1]]

seamy /ˈsiːmɪ/ *adj.* (**seamier**, **seamiest**) **1** marked with or showing seams. **2** unpleasant, sordid, disreputable (esp. *the seamy side*). □ **seaminess** *n.*

Seanad /ˈʃænəð/ *n.* (in full **Seanad Eireann** /ˈeərən/) the upper House of Parliament in the Republic of Ireland, composed of sixty members, of whom eleven are nominated by the Taoiseach and forty-nine are elected by institutions etc. [Ir., = senate (of Ireland)]

seance /ˈseɪɒns/ *n.* (also **séance**) a meeting at which spiritualists attempt to make contact with the dead. [F *séance* f. OF *seoir* f. L *sedere* sit]

Sea of Azov, Sea of Galilee, etc. see AZOV, SEA OF; GALILEE, SEA OF, etc.

Sea Peoples (also **Peoples of the Sea**) groups of invaders who encroached on Egypt and the eastern Mediterranean by land and sea in the late 13th century BC. Their identity is still being debated. The Egyptians were successful in driving them away, but some, including the Philistines, settled in Palestine.

seaplane /ˈsiːpleɪn/ *n.* an aircraft designed to take off from and land and float on water.

seaport /ˈsiːpɔːt/ *n.* a town with a harbour for seagoing ships.

SEAQ *abbr.* Stock Exchange Automated Quotations (computerized access to share information).

seaquake /ˈsiːkweɪk/ *n.* an earthquake under the sea.

sear /sɪə(r)/ *v. & adj.* ● *v.tr.* **1** scorch, esp. with a hot iron; cauterize, brand. **b** (as **searing** *adj.*) scorching, burning (*searing pain*). **2** cause pain or great anguish to. **3** brown (meat) quickly at a high temperature so that it will retain its juices in cooking. **4** make incapable of feeling. **5** *archaic* blast, wither. ● *adj.* (also **sere**) *literary* (esp. of a plant etc.) withered, dried up. [OE *sēar* (adj.), *sēarian* (v.), f. Gmc]

search /sɜːtʃ/ *n. & v.* ● *v.* **1** *tr.* look through or go over thoroughly to find something. **2** *tr.* examine or feel over (a person) to find anything concealed. **3** *tr.* **a** probe or penetrate into. **b** examine or question (one's mind, conscience, etc.) thoroughly. **4** *intr.* (often foll. by *for*) make a search or investigation. **5** *intr.* (as **searching** *adj.*) (of an examination) thorough; leaving no loopholes. **6** *tr.* (foll. by *out*) look probingly for; seek out. ● *n.* **1** an act of searching. **2** an investigation. □ **in search of** trying to find. **right of search** a belligerent's right to stop a neutral vessel and search it for prohibited goods. **search me!** *colloq.* I do not know. **search-party** a group of people organized to look for a lost person or thing. **search warrant** an official authorization to enter and search a building. □ **searchable** *adj.* **searcher** *n.* **searchingly** *adv.* [ME f. AF *sercher*, OF *cerchier* f. LL *circare* go round (as CIRCUS)]

searchlight /'sɜ:tʃlaɪt/ *n.* **1** a powerful outdoor electric light with a concentrated beam that can be turned in any direction. **2** the light or beam from this.

Sears Tower /sɪəz/ a skyscraper in Chicago, the tallest building in the world when it was completed in 1973. It is 443 m (1,454 ft) high and has 110 floors.

seascape /'si:skeɪp/ *n.* a picture or view of the sea.

Sea Scout *n.* a member of the maritime branch of the Scout Association.

seashore /'si:ʃɔ:(r)/ *n.* **1** land close to or bordering on the sea. **2** *Law* the area between high and low water marks.

seasick /'si:sɪk/ *adj.* suffering from sickness or nausea from the motion of a ship at sea. □ **seasickness** *n.*

seaside /'si:saɪd/ *n.* (often *attrib.*) the sea-coast, esp. as a holiday resort.

season /'si:z(ə)n/ *n. & v.* ● *n.* **1** each of the four divisions of the year (spring, summer, autumn, and winter) associated with a type of weather and a stage of vegetation. **2** a time of year characterized by climatic or other features (*the dry season*). **3 a** the time of year when a plant is mature or flowering etc. **b** the time of year when an animal breeds or is hunted. **4** a proper or suitable time. **5** a time when something is plentiful or active or in vogue. **6** (usu. prec. by *the*) = *high season*. **7** the time of year regularly devoted to an activity (*the football season*). **8** the time of year dedicated to social life generally (*went up to London for the season*). **9** a period of indefinite or varying length. **10** *Brit. colloq.* = *season ticket*. ● *v.* **1** *tr.* flavour (food) with salt, herbs, etc. **2** *tr.* enhance with wit, excitement, etc. **3** *tr.* temper or moderate. **4** *tr. & intr.* **a** make or become suitable or in the desired condition, esp. by exposure to the air or weather; mature. **b** (usu. as **seasoned** *adj.*) make or become experienced or accustomed (*seasoned soldiers*). □ **in season 1** (of foodstuff) available in plenty and in good condition. **2** (of an animal) on heat. **3** timely. **season ticket** a ticket entitling the holder to any number of journeys, admittances, etc., in a given period. [ME f. OF *seson* f. L *satio -onis* (in Rmc sense 'seed-time') f. *serere sat-* sow]

seasonable /'si:zənəb(ə)l/ *adj.* **1** suitable to or usual in the season. **2** opportune. **3** meeting the needs of the occasion. □ **seasonableness** *n.* **seasonably** *adv.*

seasonal /'si:zən(ə)l/ *adj.* of, depending on, or varying with the season. □ **seasonal affective disorder** (abbr. **SAD**) a depressive state associated with late autumn and winter and thought to be caused by a lack of light. □ **seasonally** *adv.* **seasonality** /ˌsi:zə'nælɪtɪ/ *n.*

seasoning /'si:z(ə)nɪŋ/ *n.* condiments added to food.

seat /si:t/ *n. & v.* ● *n.* **1** a thing made or used for sitting on; a chair, stool, saddle, etc. **2** the buttocks. **3** the part of the trousers etc. covering the buttocks. **4** the part of a chair etc. on which the sitter's weight directly rests. **5** a place for one person in a theatre, vehicle, etc. **6** the occupation of a seat. **7** esp. *Brit.* **a** the right to occupy a seat, esp. as a Member of the House of Commons. **b** a member's constituency. **8** the part of a machine that supports or guides another part. **9** a site or location of something specified (*a seat of learning; the seat of the emotions*). **10** a country mansion, esp. with large grounds. **11** the manner of sitting on a horse etc. ● *v.tr.* **1** cause to sit. **2 a** provide sitting accommodation for (*the cinema seats 500*). **b** provide with seats. **3** (as **seated** *adj.*) sitting. **4** put or fit in position. □ **be seated** sit down. **by the seat of one's pants** *colloq.* by instinct rather than logic or knowledge. **seat-belt** a belt securing a person in the seat of a motor vehicle or aircraft. **take a** (or **one's**) **seat** sit down. □ **seatless** *adj.* [ME f. ON *sæti* (= OE *gesete* f. Gmc)]

-seater /'si:tə(r)/ *comb. form* denoting a vehicle, sofa, building, etc. with a specified number of seats (*a 16-seater bus*).

seating /'si:tɪŋ/ *n.* **1** seats collectively. **2** sitting accommodation.

SEATO /'si:təʊ/ see SOUTH-EAST ASIA TREATY ORGANIZATION.

Seattle /sɪ'æt(ə)l/ a port and industrial city in the state of Washington, on the eastern shores of Puget Sound; pop. (1990) 516,260. First settled in 1852, it is now the largest city in the north-western US.

seaward /'si:wəd/ *adv., adj., & n.* ● *adv.* (also **seawards** /-wədz/) towards the sea. ● *adj.* going or facing towards the sea. ● *n.* such a direction or position.

seaway /'si:weɪ/ *n.* **1** an inland waterway open to seagoing ships. **2** a ship's progress. **3** a ship's path across the sea.

seaweed /'si:wi:d/ *n.* any large alga growing in the sea or on rocks in the intertidal zone; such plants collectively.

seaworthy /'si:ˌwɜ:ðɪ/ *adj.* (esp. of a ship) fit to put to sea. □ **seaworthiness** *n.*

sebaceous /sɪ'beɪʃəs/ *adj.* fatty; of or relating to tallow or fat. □ **sebaceous gland** (or **follicle** or **duct**) a gland etc. secreting or conveying oily matter to lubricate the skin and hair. [L *sebaceus* f. *sebum* tallow]

Sebastian, St /sɪ'bæstɪən/ (late 3rd century), Roman martyr. According to legend he was a soldier who was shot by archers on the orders of Diocletian, but who recovered, confronted the emperor, and was then clubbed to death. Feast day, 20 Jan.

Sebastopol /sɪ'bæstəp(ə)l, -ˌppl/ (Russian **Sevastopol** /ˌsjivas'tɔpəlj/) a fortress and naval base in Ukraine, near the southern tip of the Crimea; pop. (1990) 361,000. It was the focal point of military operations during the Crimean War, falling eventually to Anglo-French forces in Sept. 1855 after a year-long siege.

Sebat /'si:bæt/ (also **Shebat** /'ʃi:bæt/, **Shevat** /'ʃi:væt/) (in the Jewish calendar) the fifth month of the civil and eleventh of the religious year, usually coinciding with parts of January and February. [Heb. *šĕbaṭ*]

seborrhoea /ˌsebə'rɪə/ *n.* (*US* **seborrhea**) *Med.* excessive discharge of sebum from the sebaceous glands. □ **seborrhoeic** *adj.* [SEBUM after *gonorrhoea* etc.]

sebum /'si:bəm/ *n.* the oily secretion of the sebaceous glands. [mod.L f. L *sebum* grease]

SEC *abbr. US* Securities and Exchange Commission.

Sec. *abbr.* secretary.

sec[1] *abbr.* secant.

sec[2] /sek/ *n. colloq.* (in phrases) a second (of time). [abbr.]

sec[3] /sek/ *adj.* (of wine) dry. [F f. L *siccus*]

sec. *abbr.* second(s).

secant /'si:kənt, 'sek-/ *adj. & n. Math.* ● *adj.* cutting (*secant line*). ● *n.* **1** a line cutting a curve at one or more points. **2** (abbr. **sec**) the ratio of the hypotenuse to the shorter side adjacent to an acute angle (in a right-angled triangle). [F *sécant(e)* f. L *secare secant-* cut]

secateurs /ˌsekə'tɜ:z/ *n.pl.* esp. *Brit.* a pair of pruning clippers for use with one hand. [F *sécateur* cutter, irreg. f. L *secare* cut]

secco /'sekəʊ/ *n.* the technique of painting on dry plaster with pigments mixed in water. [It., = dry, f. L *siccus*]

secede /sɪ'si:d/ *v.intr.* (usu. foll. by *from*) withdraw formally from membership of a political federation or a religious body. □ **seceder** *n.* [L *secedere secess-* (as SE-, *cedere* go)]

secession /sɪ'seʃ(ə)n/ *n.* **1** the act or an instance of seceding. **2** (**Secession**) *US hist.* the withdrawal of eleven southern states from the Union in 1860, leading to the American Civil War. □ **secessional** *adj.* **secessionism** *n.* **secessionist** *n.* [F *sécession* or L *secessio* (as SECEDE)]

Sechuana var. of SETSWANA.

seclude /sɪ'klu:d/ *v.tr.* (also *refl.*) **1** keep (a person or place) retired or away from company. **2** (esp. as **secluded** *adj.*) hide or screen from view. [ME f. L *secludere seclus-* (as SE-, *claudere* shut)]

seclusion /sɪ'klu:ʒ(ə)n/ *n.* **1** a secluded state; retirement, privacy. **2** a secluded place. □ **seclusionist** *n.* **seclusive** /-'klu:sɪv/ *adj.* [med.L *seclusio* (as SECLUDE)]

second[1] /'sekənd/ *n., adj., & v.* ● *n.* **1** the position in a sequence corresponding to that of the number 2 in the sequence 1–2. **2** something occupying this position. **3** the second person etc. in a race or competition. **4** *Mus.* **a** an interval or chord spanning two consecutive notes in the diatonic scale (e.g. C to D). **b** a note separated from another by this interval. **5** = *second gear*. **6** another person or thing in addition to one previously mentioned or considered (*the policeman was then joined by a second*). **7** (in *pl.*) **a** goods of a second or inferior quality. **b** coarse flour, or bread made from it. **8** (in *pl.*) *colloq.* **a** a second helping of food at a meal. **b** the second course of a meal. **9** an attendant assisting a combatant in a duel, boxing-match, etc. **10 a** a place in the second class of an examination. **b** a person having this. ● *adj.* **1** that is the second; next after first. **2** additional, further; other besides one previously mentioned or considered (*ate a second cake*). **3** subordinate in position or importance etc.; inferior. **4** *Mus.* performing a lower or subordinate part (*second violins*). **5** such as to be

comparable to; closely reminiscent of (*a second Callas*). ● *v.tr.*
1 supplement, support; back up. **2** formally support or endorse (a
nomination or resolution etc., or its proposer). □ **at second hand** by
hearsay, not direct observation etc. **in the second place** as a second
consideration etc. **play second fiddle** see FIDDLE. **second advent**
= *second coming*. **second ballot** a deciding ballot between candidates
coming first (without an absolute majority) and second in a previous
ballot. **second-best** *adj.* next after best. ● *n.* a less adequate or
desirable alternative. **second cause** *Logic* a cause that is itself caused.
second chamber the upper house of a bicameral parliament.
second class 1 a set of persons or things grouped together as second-
best. **2** the second-best accommodation in a train, ship, etc. **3** the class
of mail not given priority in handling. **4 a** the second highest division
in a list of examination results. **b** a place in this. **second-class** *adj.*
1 belonging to or travelling by the second class. **2** inferior in quality,
status, etc. (*second-class citizens*). ● *adv.* by second class (*travelled second-
class*). **second coming** (in Christian theology) the future return of
Christ to earth. **second cousin** see COUSIN. **second-degree** *Med.*
denoting burns that cause blistering but not permanent scars.
second floor 1 *Brit.* the floor two levels above the ground floor. **2** *N.
Amer.* the floor above the ground floor. **second gear** the second (and
next to lowest) in a sequence of gears. **second-generation** denoting
the offspring of a first generation, esp. of immigrants. **second-guess**
colloq. **1** anticipate or predict by guesswork. **2** judge or criticize with
hindsight. **second honeymoon** a holiday like a honeymoon, taken
by a couple after some years of marriage. **second in command** the
officer next in rank to the commanding or chief officer. **second
lieutenant** an army officer next below lieutenant or US first
lieutenant. **second name** a surname. **second nature** (often foll.
by *to*) an acquired tendency that has become instinctive (*is second nature
to him*). **second officer** an assistant mate on a merchant ship.
second person *Gram.* see PERSON 3. **second-rate** of mediocre
quality; inferior. **second-rater** a person or thing that is second-rate.
second reading a second presentation of a bill to a legislative
assembly, in the UK to approve its general principles and in the US to
debate committee reports. **second self** a close friend or associate.
second sight the supposed power of being able to perceive future
or distant events. **second-sighted** having the gift of second sight.
second string an alternative course of action, means of livelihood,
etc., invoked if the main one is unsuccessful. **second teeth** the teeth
that replace the milk teeth in a mammal. **second thoughts** a new
opinion or resolution reached after further consideration. **second to
none** surpassed by no other. **second wind 1** recovery of the power
of normal breathing during exercise after initial breathlessness.
2 renewed energy to continue an effort. □ **seconder** *n.* (esp. in sense
2 of *v.*). [ME f. OF f. L *secundus* f. *sequi* follow]

second² /ˈsekənd/ *n.* **1** a sixtieth of a minute of time (symbol: **s** or ″).
(*See note below.*) **2** *colloq.* a very short time (*wait a second*). **3** a sixtieth of
a minute of angular distance (symbol: ″). □ **second-hand** an extra
hand in some watches and clocks, recording seconds. [F f. med.L
secunda (minuta) secondary (minute)]

▪ The second is the SI base unit of time, defined since 1967 as the
duration of 9,192,631,770 periods of the radiation of a certain
transition of the caesium-133 atom. Originally the second was
defined as the fraction ¹⁄₈₆ ₄₀₀ of the mean solar day, but this proved
insufficiently accurate on account of irregularities in the earth's
rotation.

second³ /sɪˈkɒnd/ *v.tr. Brit.* transfer (a military officer or other official
or worker) temporarily to other employment or to another position.
□ **secondment** *n.* [F *en second* in the second rank (of officers)]

Second Adar see ADAR.

secondary /ˈsekəndərɪ/ *adj. & n.* ● *adj.* **1** coming after or next below
what is primary. **2** derived from or depending on or supplementing
what is primary. **3** (of education, a school, etc.) for those who have had
primary education, usu. from 11 to 18 years. **4** *Electr.* **a** (of a cell or
battery) having a reversible chemical reaction and therefore able to
store energy. **b** denoting a device using electromagnetic induction,
esp. a transformer. ● *n.* (*pl.* **-ies**) **1** a secondary thing. **2** a secondary
device or current. □ **secondary colour** the result of mixing two
primary colours. **secondary feather** a feather growing from the
second joint of a bird's wing. **secondary modern** (or **secondary
modern school**) *hist.* a type of secondary school offering a general
education to children not selected for grammar or technical schools.
secondary picketing the picketing of premises of a firm not
otherwise involved in the dispute in question. **secondary planet** a
satellite of a planet (cf. *primary planet*). **secondary sexual**

characteristics those distinctive of one sex but not directly related
to reproduction. □ **secondarily** *adv.* **secondariness** *n.* [ME f. L
secundarius (as SECOND¹)]

Second Boer War see BOER WAR.

seconde /sɪˈkɒnd/ *n. Fencing* the second of eight parrying positions.
[F, fem. of *second* SECOND¹]

Second Empire *hist.* **1** the imperial government in France of
Napoleon III, 1852–70, followed by the Third Republic. **2** this period in
France.

second-hand /ˌsekənd'hænd/ *adj. & adv.* ● *adj.* **1 a** (of goods) having
had a previous owner; not new. **b** (of a shop etc.) where such goods can
be bought. **2** (of information etc.) accepted on another's authority and
not from original investigation. ● *adv.* **1** on a second-hand basis. **2** at
second hand; not directly.

Second International see INTERNATIONAL.

secondly /ˈsekəndlɪ/ *adv.* **1** furthermore; in the second place. **2** as a
second item.

secondo /sɪˈkɒndəʊ/ *n.* (*pl.* **secondi** /-dɪ/) *Mus.* the second or lower
part in a duet etc. [It.]

Second Republic *hist.* **1** the republican regime in France from the
deposition of King Louis Philippe in 1848 to the beginning of the
Second Empire (1852). **2** this period in France.

Second World War (1939–45), a war in which the Axis Powers
(Germany, Italy, and Japan) were defeated by an alliance eventually
including the United Kingdom (and its dominions), the Soviet Union,
and the United States. War began in Europe when Hitler's invasion of
Poland in September 1939 led Great Britain and France to declare
war on Germany. In the following summer France was defeated and
occupied; although British forces had been driven to retreat from
Dunkirk, Britain (with Winston Churchill now as Prime Minister)
remained in the war, defeating the German air force in the Battle of
Britain. But soon Germany had overrun much of Europe, including
France, the Low Countries, Norway, Denmark, Greece, Yugoslavia, and
Poland, and fighting had spread to the Mediterranean and North Africa
after Italy joined the war (1940). In June 1941 Hitler launched an
initially successful offensive against the Soviet Union, although Stalin
had previously signed a non-aggression pact and colluded in the
occupation of Poland and other countries. The war embroiled the US
(already supplying Britain with arms and other aid) and spread to the
Far East and Pacific in December 1941; Japan, in dispute with the US
over trade embargoes, attacked the US fleet at Pearl Harbor and overran
much of SE Asia and the western Pacific, including Indo-China, Burma,
Malaya, Indonesia, and the Philippines. In 1942–3 the Allies achieved
strategically decisive victories over Germany at Stalingrad, Kursk, and
El Alamein, and over Japan at Midway and in the Solomons, and from
this time the Allies were on the offensive. Italy surrendered after the
loss of Sicily (1943), and the Allies launched a full-scale invasion in
Normandy in June 1944. The war in Europe ended when Germany,
invaded from east and west, surrendered (May 1945) shortly after
Hitler's suicide. Japan, facing invasion, surrendered after the US
dropped atom bombs on Hiroshima and Nagasaki in August 1945. An
estimated 55 million people were killed during the war, a much higher
proportion of them civilians than in the First World War. At least
6 million Jews and other members of persecuted minorities were
systematically exterminated by Nazi Germany (see HOLOCAUST, THE).
Other features of the war were the heavy bombing of civilian targets
(particularly by the Allies), the use of unrestricted submarine warfare
by Germany, and the introduction of new technology such as radar,
rockets, jet aircraft, and nuclear weapons. The US and the Soviet
Union emerged from the war as rival superpowers, and the immediate
postwar period saw extensive redrawing of frontiers, especially in
Europe, and the beginning of the cold war.

secrecy /ˈsiːkrɪsɪ/ *n.* **1** the keeping of secrets as a fact, habit, or faculty.
2 a state in which all information is withheld (*was done in great secrecy*).
□ **sworn to secrecy** having promised to keep a secret. [ME f. *secretie*
f. obs. *secre* (adj.) or SECRET *adj.*]

secret /ˈsiːkrɪt/ *adj. & n.* ● *adj.* **1** kept or meant to be kept private,
unknown, or hidden from all or all but a few. **2** acting or operating
secretly. **3** fond of, prone to, or able to preserve secrecy. **4** (of a place)
hidden, completely secluded. ● *n.* **1** a thing kept or meant to be kept
secret. **2** a thing known only to a few. **3** a mystery. **4** a valid but not
commonly known or recognized method of achieving or maintaining
something (*what's their secret?*; *correct breathing is the secret of good health*).
5 *RC Ch.* a prayer concluding the offertory of the mass. □ **in secret**
secretly. **in** (or **in on**) **the secret** among the number of those who

know a secret. **keep a secret** not reveal a secret. **secret agent** a spy acting for a country. **secret ballot** a ballot in which votes are cast in secret. **secret police** a police force operating in secret for political purposes. **secret service** a government department concerned with espionage. **secret society** a society whose members are sworn to secrecy about it. □ **secretly** adv. [ME f. OF f. L secretus (adj.) separate, set apart f. secernere secret- (as SE-, cernere sift)]

secretaire /ˌsekrɪˈteə(r)/ n. an escritoire. [F (as SECRETARY)]

secretariat /ˌsekrɪˈteərɪət/ n. **1** a permanent administrative office or department, esp. a governmental one. **2** its members or premises. **3** the office of secretary. [F secrétariat f. med.L secretariatus (as SECRETARY)]

secretary /ˈsekrɪtərɪ, ˈsekrətrɪ/ n. (pl. -**ies**) **1** a person employed by an individual or in an office etc. to assist with correspondence, keep records, make appointments, etc. **2** an official appointed by a society etc. to conduct its correspondence, keep its records, organize its affairs, etc. **3** (in some organizations) the chief executive. **4** (in the UK) the principal assistant of a government minister, ambassador, etc. **5** (in comb.) the principal official of a political party, trade union, or other organization (First Secretary; General Secretary). □ **secretary bird** a long-legged African bird of prey, Sagittarius serpentarius, feeding on snakes, with a crest likened to a quill pen stuck over a writer's ear. □ **secretaryship** n. **secretarial** /ˌsekrɪˈteərɪəl/ adj. [ME f. LL secretarius (as SECRET)]

Secretary-General /ˌsekrɪtərɪˈdʒenrəl, ˈsekrətrɪ-/ n. the principal administrator of an organization.

Secretary of State 1 (in the UK) the head of a major government department. The title occurs first during the reign of Queen Elizabeth I; it probably indicates the beginning of the development by which the monarch's secretary became a minister invested with governing functions. **2** (in the US) the chief government official responsible for foreign affairs.

secrete[1] /sɪˈkriːt/ v.tr. Biol. (of a cell, organ, etc.) produce by secretion. □ **secretor** n. **secretory** adj. [back-form. f. SECRETION]

secrete[2] /sɪˈkriːt/ v.tr. conceal; put into hiding. [obs. secret (v.) f. SECRET]

Secret Intelligence Service (abbr. **SIS**) see MI6.

secretion /sɪˈkriːʃ(ə)n/ n. **1** Biol. **a** a process by which substances are produced and discharged from a cell, gland, or organ for a function in the organism or for excretion. **b** the secreted substance. **2** the act or an instance of concealing (the secretion of stolen goods). [F sécrétion or L secretio separation (as SECRET)]

secretive /ˈsiːkrɪtɪv/ adj. inclined to make or keep secrets; uncommunicative. □ **secretively** adv. **secretiveness** n. [back-form. f. secretiveness after F secrétivité (as SECRET)]

sect /sekt/ n. **1 a** a body of people subscribing to religious doctrines different from those of others within the same religion; a group deviating from orthodox tradition, often regarded as heretical. **b** usu. derog. a body separated from an established Church; a nonconformist Church. **c** a party or faction in a religious body. **d** a religious denomination. **2** the followers of a particular philosopher or philosophy, or school of thought in politics etc. [ME f. OF secte or L secta f. the stem of sequi secut- follow]

sect. abbr. section.

sectarian /sekˈteərɪən/ adj. & n. ● adj. **1** of or concerning a sect. **2** bigoted or narrow-minded in following the doctrines of one's sect. ● n. **1** a member of a sect. **2** a bigot. □ **sectarianism** n. **sectarianize** v.tr. (also -**ise**). [SECTARY]

sectary /ˈsektərɪ/ n. (pl. -**ies**) a member of a religious or political sect. [med.L sectarius adherent (as SECT)]

section /ˈsekʃ(ə)n/ n. & v. ● n. **1** a part cut off or separated from something. **2** each of the parts into which a thing is divided (actually or conceptually) or divisible or out of which a structure can be fitted together. **3** a distinct group or subdivision of a larger body of people (the wind section of an orchestra). **4** a subdivision of a book, document, statute, etc. **5 a** an area of land. **b** N. Amer. one square mile of land. **c** esp. US a particular district of a town (residential section). **6** a subdivision of an army platoon. **7** esp. Surgery a separation by cutting. **8** Biol. a thin slice of tissue etc., cut off for microscopic examination. **9 a** the cutting of a solid by or along a plane. **b** the resulting figure or the area of this. **10** a representation of the internal structure of something as if cut across along a vertical or horizontal plane. **11** Biol. a group, esp. a subgenus. ● v.tr. **1** arrange in or divide into sections. **2** (esp. as **sectioned** adj.) Brit. cause (a person) to be compulsorily committed to a psychiatric hospital in accordance with a section of a mental health act. **3** Biol. cut (a specimen) into thin slices for microscopic

examination. □ **section-mark** the sign (§) used as a reference mark to indicate the start of a section of a book etc. [F section or L sectio f. secare sect- cut]

sectional /ˈsekʃən(ə)l/ adj. **1** a relating to a section, esp. of a community. **b** partisan. **2** made in sections. **3** local rather than general. □ **sectionalism** n. **sectionalist** n. & adj. **sectionalize** v.tr. (also -**ise**). **sectionally** adv.

sector /ˈsektə(r)/ n. **1** a distinct part or branch of an enterprise, or of society, the economy, etc. **2** Mil. a subdivision of an area for military operations, controlled by one commander or headquarters. **3** the plane figure enclosed by two radii of a circle, ellipse, etc., and the arc between them. **4** a mathematical instrument consisting of two arms hinged at one end and marked with sines, tangents, etc., for making diagrams etc. □ **sectoral** adj. [orig. in math. senses: f. L sector cutter (as SECTION)]

sectorial /sekˈtɔːrɪəl/ adj. **1** of or like a sector or sectors. **2** = CARNASSIAL.

secular /ˈsekjʊlə(r)/ adj. & n. ● adj. **1** concerned with the affairs of this world; not spiritual or sacred. **2** (of education etc.) not concerned with religion or religious belief. **3 a** not ecclesiastical or monastic. **b** (of clergy) not bound by a religious rule. **4** occurring once in an age or century. **5 a** lasting for or occurring over an indefinitely long time. **b** Astron. of or designating slow changes in the motion of the sun or planets. ● n. a secular priest. □ **secularly** adv. **secularism** n. **secularist** n. **secularize** v.tr. (also -**ise**). **secularization** /ˌsekjʊlərˈzeɪʃ(ə)n/ n. **secularity** /ˌsekjʊˈlærɪtɪ/ n. [ME (in senses 1–3 f. OF seculer) f. L saecularis f. saeculum generation, age]

secund /sɪˈkʌnd/ adj. Bot. arranged on one side only (as the flowers of lily of the valley). □ **secundly** adv. [L secundus (as SECOND[1])]

secure /sɪˈkjʊə(r)/ adj. & v. ● adj. **1** untroubled by danger or fear. **2** safe against attack; impregnable. **3** reliable; certain not to fail (the plan is secure). **4** fixed or fastened so as not to give way or get loose or be lost (made the door secure). **5 a** (foll. by of) certain to achieve (secure of victory). **b** (foll. by against, from) safe, protected (secure against attack). **6** (of a prison etc.) difficult to escape from. **7** (of a person) not a risk to security; trustworthy. ● v.tr. **1** make secure or safe; fortify. **2** fasten, close, or confine securely. **3** succeed in obtaining or achieving (have secured front seats). **4** guarantee against loss (a loan secured by property). **5** compress (a blood vessel) to prevent bleeding. □ **secure arms** Mil. hold a rifle with the muzzle downward and the lock in the armpit to guard it from rain. □ **securable** adj. **securely** adv. **securement** n. [L securus (as SE-, cura care)]

Securitate /sɪˌkjʊərɪˈtɑːteɪ/ hist. the internal security force of Romania, set up in 1948 and officially disbanded during the revolution of December 1989. [Romanian, = security]

security /sɪˈkjʊərɪtɪ/ n. (pl. -**ies**) **1** a secure condition or feeling. **2** a thing that guards or guarantees. **3 a** the safety of a state, company, etc., against espionage, theft, or other danger. **b** an organization for ensuring this. **4** a thing deposited or pledged as a guarantee of the fulfilment of an undertaking or the payment of a loan, to be forfeited in case of default. **5** (often in pl.) a certificate attesting credit or the ownership of stock, bonds, etc. □ **on security of** using as a guarantee. **security blanket 1** an official sanction on information in the interest of security. **2** a blanket or other familiar object given as a comfort to a child. **security guard** a person employed to protect the security of buildings, vehicles, etc. **security risk** a person whose presence may threaten security; a situation endangering security. [ME f. OF securité or L securitas (as SECURE)]

Security Council a principal council of the UN, charged with the duty of maintaining security and peace between nations. It consists of fifteen members, of which five (China, France, UK, US, and Russia (formerly the USSR)) are permanent and the rest are elected for two-year terms. Resolutions must receive at least nine votes to be passed, and each of the permanent members has the power of veto.

Security Service see MI5.

sedan /sɪˈdæn/ n. **1** (in full **sedan chair**) an enclosed chair for conveying one person, carried between horizontal poles by two porters, common in the 17th–18th centuries. **2** N. Amer. an enclosed motor car for four or more people. [perh. alt. f. It. dial., ult. f. L sella saddle f. sedere sit]

Sedan, Battle of /sɪˈdæn/ a battle fought in 1870 near the town of Sedan in NE France, a decisive engagement in the Franco-Prussian War. The Prussian army succeeded in surrounding a smaller French army under Napoleon III and forcing it to surrender, opening the way for a Prussian advance on Paris and marking the end of the French Second Empire. During the Second World War Sedan was again the scene of a battle (1940), in which French forces were defeated and which marked the beginning of the German invasion of France.

sedate[1] /sɪˈdeɪt/ adj. tranquil and dignified; equable, serious. □ **sedately** adv. **sedateness** n. [L sedatus past part. of sedare settle f. sedere sit]

sedate[2] /sɪˈdeɪt/ v.tr. put under sedation. [back-form. f. SEDATION]

sedation /sɪˈdeɪʃ(ə)n/ n. a state of rest or sleep esp. produced by a sedative drug. [F sédation or L sedatio (as SEDATE[1])]

sedative /ˈsedətɪv/ n. & adj. ● n. a drug, influence, etc., that tends to calm or soothe. ● adj. calming, soothing; inducing sleep. [ME f. OF sedatif or med.L sedativus (as SEDATE[1])]

sedentary /ˈsedəntərɪ/ adj. **1** sitting (a sedentary posture). **2** (of work etc.) characterized by much sitting and little physical exercise. **3** (of a person) spending much time seated. **4** Zool. **a** inhabiting the same locality throughout life; non-migratory. **b** confined to one spot; sessile. □ **sedentarily** adv. **sedentariness** n. [F sédentaire or L sedentarius f. sedere sit]

Seder /ˈseɪdə(r)/ n. a Jewish ritual service and ceremonial dinner for the first night or first two nights of the Passover. [Heb. sēder order]

sederunt /sɪˈdeərənt/ n. Sc. a sitting of an ecclesiastical assembly or other body. [L, = (the following persons) sat f. sedere sit]

sedge /sedʒ/ n. **1** a grasslike plant of the genus Carex with triangular stems, usu. growing in wet areas. **2** an expanse of this plant. □ **sedge warbler** a small streaked warbler, Acrocephalus schoenobaenus, often found around marshes and in reed-beds. □ **sedgy** adj. [OE secg f. Gmc]

Sedgemoor, Battle of /ˈsedʒmʊə(r), -mɔː(r)/ a battle fought in 1685 on the plain of Sedgemoor in Somerset. The forces of the rebel Duke of Monmouth, who had landed in Dorset as champion of the Protestant cause and pretender to the throne, were decisively defeated by James II's troops.

Sedgwick /ˈsedʒwɪk/, Adam (1785–1873), English geologist. He was based in Cambridge and specialized in the fossil record of rocks from North Wales, assigning the oldest of these to a period that he named the Cambrian. Sedgwick also amassed one of the greatest geological collections.

sedile /sɪˈdaɪlɪ/ n. (pl. **sedilia** /-ˈdɪlɪə/) (usu. in pl.) Eccl. each of usu. three stone seats for priests in the south wall of a chancel, often canopied and decorated. [L, = seat f. sedere sit]

sediment /ˈsedɪmənt/ n. **1** matter that settles to the bottom of a liquid; dregs. **2** Geol. matter that is carried by water or wind and deposited on the surface of the land, and may in time become consolidated into rock. □ **sedimentation** /ˌsedɪmenˈteɪʃ(ə)n/ n. [F sédiment or L sedimentum (as SEDILE)]

sedimentary /ˌsedɪˈmentərɪ/ adj. **1** of or like sediment. **2** Geol. (esp. of rocks) formed from sediment. (See note below.)

▪ Sedimentary rocks form from sand, mud, or silt that has been deposited by water or wind, and are characteristically laid down in strata which are initially horizontal or nearly so. It was by studying sequences of sedimentary rocks and the fossils associated with them that the geological time-scale was established. By volume, sedimentary rocks or their metamorphosed equivalents constitute only 5 per cent of the known crust of the earth, igneous rocks contributing the other 95 per cent. (See also IGNEOUS, METAMORPHIC.)

sedition /sɪˈdɪʃ(ə)n/ n. **1** conduct or speech inciting to rebellion or a breach of public order. **2** agitation against the authority of a state. □ **seditious** adj. **seditiously** adv. [ME f. OF sedition or L seditio f. sed- = SE- + ire it- go]

seduce /sɪˈdjuːs/ v.tr. **1** tempt or entice into sexual activity. **2** lead astray; tempt or beguile into wrongdoing. **3** attract, allure, entice (seduced by the smell of coffee). □ **seducer** n. **seducible** adj. [L seducere seduct- (as SE-, ducere lead)]

seduction /sɪˈdʌkʃ(ə)n/ n. **1** the act or an instance of seducing; the process of being seduced. **2** something that tempts or allures. [F séduction or L seductio (as SEDUCE)]

seductive /sɪˈdʌktɪv/ adj. tending to seduce; alluring, enticing. □ **seductively** adv. **seductiveness** n. [SEDUCTION after inductive etc.]

seductress /sɪˈdʌktrɪs/ n. a female seducer. [obs. seductor male seducer (as SEDUCE)]

sedulous /ˈsedjʊləs/ adj. **1** persevering, diligent, assiduous. **2** (of an action etc.) deliberately and consciously continued; painstaking. □ **sedulously** adv. **sedulousness** n. **sedulity** /sɪˈdjuːlɪtɪ/ n. [L sedulus zealous]

sedum /ˈsiːdəm/ n. a plant of the genus Sedum, with fleshy leaves

and star-shaped yellow, pink, or white flowers, e.g. stonecrop. [L, = houseleek]

see[1] /siː/ v. (past **saw** /sɔː/; past part. **seen** /siːn/) **1** tr. discern by use of the eyes; observe; look at (can you see that spider?; saw him fall over). **2** intr. have or use the power of discerning objects with the eyes (sees best at night). **3** tr. discern mentally; understand (I see what you mean; could not see the joke). **4** tr. watch; be a spectator of (a film, game, etc.). **5** tr. ascertain or establish by inquiry or research or reflection (I will see if the door is open). **6** tr. consider; deduce from observation (I see that you are a brave man). **7** tr. contemplate; foresee mentally (we saw that no good would come of it; can see myself doing this job indefinitely). **8** tr. look for information (usu. in imper. as a direction in or to a book: see page 15). **9** tr. meet or be near and recognize (I saw your mother in town). **10 a** tr. meet socially (sees her sister most weeks). **b** meet regularly as a boyfriend or girlfriend; court (is still seeing that man). **11** tr. give an interview to (the doctor will see you now). **12** tr. visit to consult (went to see the doctor). **13** tr. find out or learn, esp. from a visual source (I see the match has been cancelled). **14 a** tr. reflect; consider further; wait until one knows more (we shall have to see). **b** tr. (foll. by whether or if + clause) consider, decide (on). **15** tr. interpret or have an opinion of (I see things differently now). **16** tr. experience; have presented to one's attention (I never thought I would see this day). **17** tr. recognize as acceptable; foresee (do you see your daughter marrying this man?). **18** tr. observe without interfering (stood by and saw them squander my money). **19** tr. (usu. foll. by in) find attractive (can't think what she sees in him). **20** intr. (usu. foll. by to, or that + infin.) make provision for; ensure; attend to (shall see to your request immediately; see that he gets home safely) (cf. see to it). **21** tr. escort or conduct (to a place etc.) (saw them home). **22** tr. be a witness of (an event etc.) (see the New Year in). **23** tr. supervise (an action etc.) (will stay and see the doors locked). **24** tr. **a** (in gambling, esp. poker) equal (a bet). **b** equal the bet of (a player), esp. to see the player's cards. □ **as far as I can see** to the best of my understanding or belief. **as I see it** in my opinion. **do you see?** do you understand? **has seen better days** has declined from former prosperity, good condition, etc. **I'll be seeing you** colloq. an expression on parting. **I see** I understand (referring to an explanation etc.). **let me see** an appeal for time to think before speaking etc. **see about 1** attend to. **2** consider; look into. **see after 1** take care of. **2** = see about. **see the back of** colloq. be rid of (an unwanted person or thing). **see a person damned first** colloq. refuse categorically and with hostility to do what a person wants. **see eye to eye** see EYE. **see fit** see FIT[1]. **see here!** see LOOK int. **see into** investigate. **see life** gain experience of the world, often by enjoying oneself. **see the light 1** realize one's mistakes etc. **2** suddenly see the way to proceed. **3** undergo religious conversion. **see the light of day** (usu. with neg.) come into existence. **see off 1** be present at the departure of (a person) (saw them off at Heathrow). **2** colloq. ward off, get the better of (managed to see off an investigation into their working methods). **see out 1** accompany out of a building etc. **2** finish (a project etc.) completely. **3** remain awake, alive, etc., until the end of (a period). **4** last longer than; outlive. **see over** inspect; tour and examine. **see reason** see REASON. **see red** get very angry. **see a person right** make sure that a person is rewarded, safe, etc. **see service** see SERVICE[1]. **see stars** colloq. see lights before one's eyes as a result of a blow on the head. **see things** have hallucinations or false imaginings. **see through 1** not be deceived by; detect the true nature of. **2** penetrate visually. **see-through** adj. (esp. of clothing) translucent. **see a person through** support a person during a difficult time; assist financially. **see a thing through** persist with it until it is completed. **see to** = see about. **see to it** (foll. by that + clause) ensure (see to it that I am not disturbed) (cf. sense 20 of v.). **see one's way clear to** feel able or entitled to. **see the world** see WORLD. **see you** (or **see you later**) colloq. an expression on parting. **we shall see** let us await the outcome. **will see about it** a formula for declining to act at once. **will see about it** a formula for declining to act at once. **you see 1** you understand. **2** you will understand when I explain. □ **seeable** adj. [OE sēon f. Gmc]

see[2] /siː/ n. **1** the area under the authority of a bishop or archbishop, a diocese (the see of Norwich). **2** the office or jurisdiction of a bishop or archbishop (fill a vacant see). [ME f. AF se(d) ult. f. L sedes seat f. sedere sit]

seed /siːd/ n. & v. ● n. **1 a** a flowering plant's unit of reproduction (esp. in the form of grain) capable of developing into another such plant. **b** seeds collectively, esp. as collected for sowing (is full of seed; to be kept for seed). **2 a** semen. **b** milt. **3** (foll. by of) the beginning or germ of some growth, development, or consequence (seeds of doubt). **4** archaic offspring, progeny, descendants (the seed of Abraham). **5** Sport a seeded player. **6** a small seedlike container for the application of radium etc. **7** a seed crystal. ● v. **1** tr. **a** place seeds in. **b** sprinkle with or as with seed. **2** intr. sow seeds. **3** intr. produce or drop seed. **4** tr. remove seeds from (fruit etc.). **5** *tr.* place a crystal or crystalline substance in (a solution etc.) to cause crystallization or condensation (esp. in a cloud to produce rain). **6** *tr. Sport* **a** assign to (a strong competitor in a knockout competition) a position in an ordered list so that strong competitors do not meet each other in early rounds (*is seeded seventh*). **b** arrange (the order of play) in this way. **7** *intr.* go to seed. □ **go** (or **run**) **to seed 1** cease flowering as seed develops. **2** become degenerate, unkempt, ineffective, etc. **raise up seed** *archaic* beget children. **seed-bed 1** a bed of fine soil in which to sow seeds. **2** a place of development. **seed-cake** cake containing whole seeds esp. of caraway as flavouring. **seed-coat** the outer integument of a seed. **seed-corn 1** good quality corn kept for seed. **2** assets reused for future profit or benefit. **seed crystal** a crystal used to initiate crystallization. **seed-eater** a bird (esp. a finch) living mainly on seeds. **seed-fish** a fish that is ready to spawn. **seed-head** a flower-head in seed. **seed-leaf** a cotyledon. **seed-lip** a basket for seed used in sowing by hand. **seed money** money allocated to initiate a project. **seed-pearl** a very small pearl. **seed-plot** a place of development. **seed-potato** a potato kept for seed. **seed-time** the sowing season. **seed-vessel** a pericarp. □ **seedless** *adj.* [OE *sǣd* f. Gmc, rel. to sow[1]]

seeder /'siːdə(r)/ *n.* **1** a person or thing that seeds. **2** a machine for sowing seed, esp. a drill. **3** an apparatus for seeding raisins etc. **4** *Brit.* a spawning fish.

seedling /'siːdlɪŋ/ *n.* a young plant, esp. one raised from seed and not from a cutting etc.

seedsman /'siːdzmən/ *n.* (*pl.* **-men**) a dealer in seeds.

seedy /'siːdɪ/ *adj.* (**seedier, seediest**) **1** full of seed. **2** going to seed. **3** shabby-looking, in worn clothes. **4** *colloq.* unwell. □ **seedily** *adv.* **seediness** *n.*

Seeger /'siːgə(r)/, Pete (b.1919), American folk musician and songwriter. In 1949 he formed the folk group the Weavers, with whom he recorded a series of best-selling protest songs. From the early 1950s he followed a solo career and was prominent in the American folk revival. Among his most famous songs are 'If I Had a Hammer' (*c.*1949) and 'Where Have All the Flowers Gone?' (1956). Seeger fell under suspicion during the McCarthy era and was not cleared of all charges until 1962.

seeing /'siːɪŋ/ *conj.* & *n.* ● *conj.* (usu. foll. by *that* + clause) considering that, inasmuch as, because (*seeing that you do not know it yourself*). ● *n.* *Astron.* the quality of observed images as determined by atmospheric conditions.

seek /siːk/ *v.* (*past* and *past part.* **sought** /sɔːt/) **1** *tr.* make a search or inquiry for. **b** *intr.* (foll. by *for, after*) make a search or inquiry. **2** *tr.* **a** try or want to find or get. **b** ask for; request (*sought help from him; seeks my aid*). **3** *tr.* (foll. by *to* + infin.) endeavour or try. **4** *tr.* make for or resort to (a place or person, for advice, health, etc.) (*sought his bed; sought a fortune-teller; sought the shore*). **5** *tr. archaic* aim at, attempt. **6** *intr.* (foll. by *to*) *archaic* resort. □ **seek dead** an order to a retriever to find killed game. **seek out 1** search for and find. **2** single out for companionship, etc. **sought-after** much in demand; generally desired. **to seek** (or **much to seek** or **far to seek**) deficient, lacking, or not yet found (*the reason is not far to seek; an efficient leader is yet to seek*). □ **seeker** *n.* (also in *comb.*). [OE *sēcan* f. Gmc]

seel /siːl/ *v.tr. archaic* close (a person's eyes). [obs. *sile* f. F *ciller, siller,* or med.L *ciliare* f. L *cilium* eyelid]

seem /siːm/ *v.intr.* **1** give the impression or sensation of being (*seems ridiculous; seems certain to win*). **2** (foll. by *to* + infin.) appear or be perceived or ascertained (*he seems to be breathing; they seem to have left*). □ **can't seem to** *colloq.* seem unable to. **do not seem to** *colloq.* somehow do not (*I do not seem to like him*). **it seems** (or **would seem**) (often foll. by *that* + clause) it appears to be true or the fact (in a hesitant, guarded, or ironical statement). [ME f. ON *sœma* honour f. *sœmr* fitting]

seeming[1] /'siːmɪŋ/ *adj.* **1** apparent but perhaps not real (*with seeming sincerity*). **2** apparent only; ostensible. □ **seemingly** *adv.*

seeming[2] /'siːmɪŋ/ *n. literary* **1** appearance, aspect. **2** deceptive appearance.

seemly /'siːmlɪ/ *adj.* (**seemlier, seemliest**) conforming to propriety or good taste; decorous, suitable. □ **seemliness** *n.* [ME f. ON *sœmiligr* (as SEEM)]

seen *past part.* of SEE[1].

See of Rome see HOLY SEE.

seep /siːp/ *v.* & *n.* ● *v.intr.* ooze out; percolate slowly. ● *n. US* a place where petroleum etc. oozes slowly out of the ground. [perh. dial. form of OE *sipian* to soak]

seepage /'siːpɪdʒ/ *n.* **1** the act of seeping. **2** the quantity that seeps out.

seer[1] *n.* **1** /'siːə(r), sɪə(r)/ a person who sees. **2** /sɪə(r)/ a prophet; a person who sees visions; a person of supposed supernatural insight esp. as regards the future. [ME f. SEE[1]]

seer[2] /sɪə(r)/ *n.* in the Indian subcontinent, a varying unit of weight (about one kilogram) or liquid measure (about one litre). [Hindi *ser*]

seersucker /'sɪəˌsʌkə(r)/ *n.* material of linen, cotton, etc., with a puckered surface. [Pers. *šir o šakar,* lit. 'milk and sugar']

see-saw /'siːsɔː/ *n., v., adj.,* & *adv.* ● *n.* **1 a** a device consisting of a long plank balanced on a central support for children to sit on at each end and move up and down by pushing the ground with their feet. **b** a game played on this. **2** an up-and-down or to-and-fro motion. **3** a contest in which the advantage repeatedly changes from one side to the other. ● *v.intr.* **1** play on a see-saw. **2** move up and down as on a see-saw. **3** vacillate in policy, emotion, etc. ● *adj.* & *adv.* with up-and-down or backward-and-forward motion (*see-saw motion*). □ **go see-saw** vacillate or alternate. [redupl. of SAW[1]]

seethe /siːð/ *v.* **1** *intr.* boil, bubble over. **2** *intr.* be very agitated, esp. with anger (*seething with discontent; I was seething inwardly*). **3** *tr.* & *intr. archaic* cook by boiling. □ **seethingly** *adv.* [OE *sēothan* f. Gmc]

segment /'segmənt/ *n.* & *v.* ● *n.* **1** each of several parts into which a thing is or can be divided or marked off. **2** *Geom.* a part of a figure cut off by a line or plane intersecting it, esp.: **a** the part of a circle enclosed between an arc and a chord. **b** the part of a line included between two points. **c** the part of a sphere cut off by any plane not passing through the centre. **3** the smallest distinct part of a spoken utterance. **4** *Zool.* each of the series of similar anatomical units of which the body and its appendages are composed in various animals, visible externally as, e.g., the transverse rings of an earthworm. ● *v.* (usu. /seg'ment/) **1** *intr.* & *tr.* divide into segments. **2** *intr. Biol.* (of a cell) undergo cleavage or divide into many cells. □ **segmentary** /seg'mentərɪ/ *adj.* **segmental** /-'ment(ə)l/ *adj.* **segmentally** *adv.* **segmentalize** *v.tr.* (also **-ise**). **segmentalization** /-ˌmentəlaɪ'zeɪʃ(ə)n/ *n.* **segmentation** /ˌsegmen'teɪʃ(ə)n/ *n.* [L *segmentum* f. *secare* cut]

sego /'siːgəʊ/ *n.* (*pl.* **-os**) (in full **sego lily**) a North American liliaceous plant, *Calochortus nuttallii*, with green and white bell-shaped flowers. [Paiute]

Segovia[1] /sɪ'gəʊvɪə/ a city in north central Spain, north-east of Madrid; pop. (1991) 58,060. It is the site of an aqueduct built by the Romans in the 2nd century AD. Taken by the Moors in the 8th century, it was reclaimed by the king of Castile, Alfonso VI (d.1109), in 1079, its palace becoming a favourite royal residence.

Segovia[2] /sɪ'gəʊvɪə/, Andrés (1893–1987), Spanish guitarist and composer. He was largely responsible for the revival of interest in the classical guitar, elevating it to use as a major concert instrument. He made a large number of transcriptions of classical music, including Bach, to increase the repertoire of the instrument, and also commissioned works from contemporary composers such as Manuel de Falla.

segregate[1] /'segrɪˌgeɪt/ *v.* **1** *tr.* put apart from the rest; isolate. **2** *tr.* enforce racial segregation on (persons) or in (a community etc.). **3** *intr.* separate from a mass and collect together. **4** *intr. Biol.* (of alleles) separate into dominant and recessive groups. □ **segregative** *adj.* **segregable** /-gəb(ə)l/ *adj.* [L *segregare* (as SE-, *grex gregis* flock)]

segregate[2] /'segrɪgət/ *adj.* **1** *Zool.* simple or solitary, not compound. **2** *archaic* set apart, separate. [L *segregatus* past part. (as SEGREGATE[1])]

segregation /ˌsegrɪ'geɪʃ(ə)n/ *n.* **1** enforced separation of racial groups in a community etc. **2** the act or an instance of segregating; the state of being segregated. □ **segregational** *adj.* **segregationist** *n.* & *adj.* [LL *segregatio* (as SEGREGATE[1])]

segue /'segweɪ/ *v.* & *n.* esp. *Mus.* ● *v.intr.* (**segues, segued, seguing**) (usu. foll. by *into*) go on without a pause. ● *n.* an uninterrupted transition from one song or melody to another. [It., = follows]

seguidilla /ˌsegɪ'diːljə/ *n.* **1** a Spanish dance in triple time. **2** the music for this. [Sp. f. *seguida* following f. *seguir* follow]

Sehnsucht /'zeɪnzʊxt/ *n.* yearning, wistful longing. [G]

seicento /seɪ'tʃentəʊ/ *n.* the style of Italian art and literature of the 17th century. □ **seicentist** *n.* **seicentoist** *n.* [It., = 600, used with ref. to the years 1600–99]

seiche /seɪʃ/ *n.* a fluctuation in the water-level of a lake etc., usu. caused by changes in barometric pressure. [Swiss F]

Seidlitz powder /'sedlɪts/ *n.* (US **Seidlitz powders**) a laxative medicine of two powders mixed separately with water and then poured together to effervesce. [named with ref. to the mineral water of *Seidlitz*, a village in Bohemia]

seif /si:f, seɪf/ *n.* (in full **seif dune**) a sand-dune in the form of a long narrow ridge. [Arab. *saif* sword (from its shape)]

seigneur /seɪˈnjɜ:(r)/ *n.* (also **seignior** /ˈseɪnjə(r)/) a feudal lord; the lord of a manor. □ **grand seigneur** /ɡrɒn/ a person of high rank or noble presence. □ **seigneurial** *adj.* **seigniorial** /-ˈnjɔ:rɪəl/ *adj.* [ME f. OF *seigneur, seignor* f. L SENIOR]

seigniorage /ˈseɪnjərɪdʒ/ *n.* (also **seignorage**) **1 a** a profit made by issuing currency, esp. by issuing coins rated above their intrinsic value. **b** *hist.* the Crown's right to a percentage on bullion brought to a mint for coining. **2** *hist.* something claimed by a sovereign or feudal superior as a prerogative. [ME f. OF *seignorage, seigneurage* (as SEIGNEUR)]

seigniory /ˈseɪnjərɪ/ *n.* (*pl.* **-ies**) **1** lordship, sovereign authority. **2** (also **seigneury**) a seigneur's domain. [ME f. OF *seignorie* (as SEIGNEUR)]

Seikan Tunnel /ˈseɪkən/ the world's longest underwater tunnel, linking the Japanese islands of Hokkaido and Honshu under the Tsungaru Strait. Completed in 1988, the tunnel is 51.7 km (32.3 miles) in length.

Seine /seɪn/ a river of northern France. Rising north of Dijon, it flows north-westwards for 761 km (473 miles), through the cities of Troyes and Paris to the English Channel near Le Havre.

seine /seɪn/ *n. & v.* ● *n.* (also **seine-net**) a fishing-net for encircling fish, with floats at the top and weights at the bottom edge, and usu. hauled ashore. ● *v.intr. & tr.* fish or catch with a seine. □ **seiner** *n.* [ME f. OF *saine,* & OE *segne* f. WG f. L *sagena* f. Gk *sagēnē*]

seise var. of SEIZE 9.

seisin /ˈsi:zɪn/ *n.* (also **seizin**) *Law* **1** possession of land by freehold. **2** the act of taking such possession. **3** what is so held. [ME f. AF *sesine,* OF *seisine, saisine* (as SEIZE)]

seismic /ˈsaɪzmɪk/ *adj.* **1** of or relating to an earthquake or earthquakes or similar vibrations. **2** like an earthquake, of enormous effect or proportion. □ **seismic survey** the use of artificially generated seismic waves to explore the structure of underground rocks etc., used in mining and in the oil and gas extraction industries. □ **seismal** *adj.* **seismical** *adj.* **seismically** *adv.* [Gk *seismos* earthquake f. *seiō* shake]

seismicity /saɪzˈmɪsɪtɪ/ *n.* seismic activity, esp. the frequency of earthquakes per unit area in a region.

seismo- /ˈsaɪzməʊ/ *comb. form* earthquake. [Gk *seismos*]

seismogram /ˈsaɪzməˌɡræm/ *n.* a record given by a seismograph.

seismograph /ˈsaɪzməˌɡrɑːf/ *n.* an instrument that records the force, direction, etc., of earthquakes. □ **seismographic** /ˌsaɪzməˈɡræfɪk/ *adj.* **seismographical** *adj.*

seismology /saɪzˈmɒlədʒɪ/ *n.* the scientific study and recording of earthquakes and related phenomena. □ **seismologist** *n.* **seismological** /ˌsaɪzməˈlɒdʒɪk(ə)l/ *adj.* **seismologically** *adv.*

sei whale /seɪ/ *n.* a small blue-grey rorqual, *Balaenoptera borealis*. [Norw. *sejhval*]

seize /si:z/ *v.* **1** *tr.* take hold of forcibly or suddenly. **2** *tr.* take possession of forcibly (*seized the fortress; seized power*). **3** *tr.* **a** take possession of (contraband goods, documents, etc.) by warrant or legal right, confiscate, impound. **b** arrest or apprehend (a person); take prisoner. **4** *tr.* affect suddenly (*panic seized us; was seized by apoplexy; was seized with remorse*). **5** *tr.* take advantage of (an opportunity). **6** *tr.* comprehend quickly or clearly. **7** *intr.* (usu. foll. by *on, upon*) **a** take hold forcibly or suddenly. **b** take advantage eagerly (*seized on a pretext*). **8** *intr.* (usu. foll. by *up*) **a** (of a moving part in a machine) become stuck or jammed from undue heat, friction, etc. **b** (of part of the body etc.) become stiff. **9** *tr.* (also **seise**) (usu. foll. by *of*) *Law* put in possession of. **10** *tr. Naut.* fasten or attach by binding with turns of yarn etc. □ **seized** (or **seised**) **of 1** possessing legally. **2** aware or informed of. □ **seizable** *adj.* **seizer** *n.* [ME f. OF *seizir, saisir* give seisin f. Frank. f. L *sacire* f. Gmc]

seizin var. of SEISIN.

seizing /ˈsi:zɪŋ/ *n. Naut.* a cord or cords used for seizing (see SEIZE 10).

seizure /ˈsi:ʒə(r)/ *n.* **1** the act or an instance of seizing; the state of being seized. **2** a fit or sudden attack of apoplexy etc., a stroke.

sejant /ˈsi:dʒənt/ *adj.* (placed after noun) *Heraldry* (of an animal) sitting upright on its haunches. [properly *seiant* f. OF var. of *seant* sitting f. *seoir* f. L *sedere* sit]

Sekhmet /ˈsekmet/ *Egyptian Mythol.* a ferocious lioness-goddess, counterpart of the gentle cat-goddess Bastet and wife of Ptah at Memphis. Her messengers were fearful creatures who could inflict disease and other scourges upon humankind.

Sekt /zekt/ *n.* a German sparkling white wine. [G]

selachian /sɪˈleɪkɪən/ *n. & adj. Zool.* ● *n.* a cartilaginous fish of the subclass Selachii, including sharks and dogfish. ● *adj.* of or relating to this subclass. [mod.L *Selachii* f. Gk *selakhos* shark]

seladang /səˈlɑːdæŋ/ *n.* a Malayan gaur. [Malay]

selah /ˈsi:lə/ *int.* occurring frequently at the end of a verse in Psalms and three times in the book of Habakkuk, supposed to be a musical direction. [Heb. *selāh*]

Selangor /səˈlæŋə(r)/ a state of Malaysia, on the west coast of the Malay Peninsula; capital, Shah Alam.

Selcraig /ˈselkreɪɡ/ see SELKIRK.

seldom /ˈseldəm/ *adv. & adj.* ● *adv.* rarely, not often. ● *adj.* rare, uncommon. [OE *seldan* f. Gmc]

select /sɪˈlekt/ *v. & adj.* ● *v.tr.* choose, esp. as the best or most suitable. ● *adj.* **1** chosen for excellence or suitability; choice. **2** (of a society etc.) exclusive, cautious in admitting members. □ **select committee** see COMMITTEE. □ **selectable** *adj.* **selectness** *n.* [L *seligere select-* (as SE-, *legere* choose)]

selectee /ˌsɪlekˈti:/ *n. US* a conscript.

selection /sɪˈlekʃ(ə)n/ *n.* **1** the act or an instance of selecting; the state of being selected. **2** a selected person or thing. **3** things from which a choice may be made. **4** *Biol.* the process in which environmental and genetic influences determine which types of organism thrive better than others, regarded as a factor in evolution. □ **selectional** *adj.* **selectionally** *adv.* [L *selectio* (as SELECT)]

selective /sɪˈlektɪv/ *adj.* **1** using or characterized by selection. **2** able to select, esp. (of a radio receiver) able to respond to a chosen frequency without interference from others. **3** (of the memory etc.) selecting what is convenient. □ **selective service** *US hist.* service in the armed forces under conscription. □ **selectively** *adv.* **selectiveness** *n.* **selectivity** /ˌsɪlekˈtɪvɪtɪ, ˌsel-, ˌsi:l-/ *n.*

selector /sɪˈlektə(r)/ *n.* **1** a person who selects, esp. one who selects a representative team in a sport. **2** a device that selects, esp. a device in a vehicle that selects the required gear.

Selene /sɪˈli:nɪ/ *Gk Mythol.* the goddess of the moon, identified with Artemis. According to one story, she fell in love with Endymion and asked Zeus to grant him a wish. Endymion chose immortality and eternal youth, which Zeus granted, but only on condition that Endymion remain forever asleep. In another story, Selene visited Endymion nightly as he lay asleep, and bore him fifty daughters. [Gk *selēnē* moon]

selenic /sɪˈli:nɪk, -ˈlenɪk/ *adj. Chem.* of selenium, esp. in its higher valency. □ **selenic acid** a strong hygroscopic corrosive dibasic acid (chem. formula: H_2SeO_4). □ **selenate** /ˈselɪˌneɪt/ *n.* [SELENIUM + -IC]

selenite /ˈselɪˌnaɪt/ *n.* a form of gypsum occurring as transparent crystals or thin plates. □ **selenitic** /ˌselɪˈnɪtɪk/ *adj.* [L *selenites* f. Gk *selēnitēs lithos* moonstone f. *selēnē* moon]

selenium /sɪˈli:nɪəm/ *n.* a grey crystalline non-metallic chemical element (atomic number 34; symbol **Se**). (*See note below.*) □ **selenium cell** a photoelectric cell using selenium. □ **selenide** /ˈselɪˌnaɪd/ *n.* [Gk *selēnē* moon (as resembling TELLURIUM)]

■ Selenium was discovered by J. J. Berzelius in 1817. It occurs mainly as a minor constituent of metallic sulphide ores. The grey crystalline form is a semiconductor showing strong photoconductivity; other allotropic forms exist. Selenium is used in rectifiers and photoelectric cells, and for colouring glass and ceramics.

seleno- /sɪˈli:nəʊ/ *comb. form* moon. [Gk *selēnē* moon]

selenography /ˌsi:lɪˈnɒɡrəfɪ/ *n.* the study or mapping of the moon. □ **selenographer** *n.* **selenographic** /-nəˈɡræfɪk/ *adj.*

selenology /ˌsi:lɪˈnɒlədʒɪ/ *n.* the scientific study of the moon. □ **selenologist** *n.*

Seleucid /sɪˈluːsɪd/ *n. & adj.* ● *n.* a dynasty founded by Seleucus Nicator, one of the generals of Alexander the Great, ruling over Syria and a great part of western Asia 311–65 BC. Its capital was at Antioch. ● *adj.* of or relating to this dynasty.

self /self/ *n., adj., & v.* ● *n.* (*pl.* **selves** /selvz/) **1** a person's or thing's own individuality or essence (*showed his true self*). **2** a person or thing as the object of introspection or reflexive action (*the consciousness of self*). **3 a** one's own interests or pleasure (*cares for nothing but self*). **b** concentration on these (*self is a bad guide to happiness*). **4** *Commerce or colloq.* myself, yourself, himself, etc. (*cheque drawn to self; ticket admitting* *self and friend).* **5** used in phrases equivalent to *myself, yourself, himself,* etc. (*his very self; your good selves*). **6** (*pl.* **selfs**) a flower of uniform colour, or of the natural wild colour. ● *adj.* **1** of the same colour as the rest or throughout. **2** (of a flower) of the natural wild colour. **3** (of colour) uniform, the same throughout. ● *v.tr.* (usu. in *pass.*) *Bot.* self-fertilize. □ **one's better self** one's nobler impulses. **one's former** (or **old**) **self** oneself as one formerly was. [OE f. Gmc]

self- /self/ *comb. form* expressing reflexive action: **1** of or directed towards oneself or itself (*self-respect; self-cleaning*). **2** by oneself or itself, esp. without external agency (*self-evident*). **3** on, in, for, or relating to oneself or itself (*self-absorbed; self-confident*).

self-abandon /ˌselfəˈbændən/ *n.* (also **self-abandonment**) the abandonment of oneself, esp. to passion or an impulse. □ **self-abandoned** *adj.*

self-abasement /ˌselfəˈbeɪsmənt/ *n.* abasement or humiliation of oneself; cringing.

self-abhorrence /ˌselfəbˈhɒrəns/ *n.* the abhorrence of oneself; self-hatred.

self-abnegation /ˌselfˌæbnɪˈɡeɪʃ(ə)n/ *n.* the abnegation of oneself, one's interests, needs, etc.; self-sacrifice.

self-absorption /ˌselfəbˈsɔːpʃ(ə)n, -ˈzɔːpʃ(ə)n/ *n.* **1** absorption in oneself. **2** *Physics* the absorption, by a body, of radiation emitted within it. □ **self-absorbed** /-ˈsɔːbd, -ˈzɔːbd/ *adj.*

self-abuse /ˌselfəˈbjuːs/ *n.* **1** the reviling or abuse of oneself. **2** masturbation.

self-accusation /ˌselfˌækjʊˈzeɪʃ(ə)n/ *n.* the accusing of oneself. □ **self-accusatory** /ˌselfəˈkjuːzətərɪ/ *adj.*

self-acting /selfˈæktɪŋ/ *adj.* acting without external influence or control; automatic. □ **self-action** /-ˈækʃ(ə)n/ *n.* **self-activity** /-ækˈtɪvɪtɪ/ *n.*

self-addressed /ˌselfəˈdrest/ *adj.* (of an envelope etc.) having one's own address on for return communication.

self-adhesive /ˌselfədˈhiːsɪv/ *adj.* (of an envelope, label, etc.) adhesive, esp. without being moistened.

self-adjusting /ˌselfəˈdʒʌstɪŋ/ *adj.* (of machinery etc.) adjusting itself. □ **self-adjustment** *n.*

self-admiration /selfˌædmɪˈreɪʃ(ə)n/ *n.* the admiration of oneself; pride; conceit.

self-advancement /ˌselfədˈvɑːnsmənt/ *n.* the advancement of oneself.

self-advertisement /ˌselfədˈvɜːtɪsmənt, -tɪzmənt/ *n.* the advertising or promotion of oneself. □ **self-advertiser** /selfˈædvəˌtaɪzə(r)/ *n.*

self-affirmation /selfˌæfəˈmeɪʃ(ə)n/ *n.* *Psychol.* the recognition and assertion of the existence of the conscious self.

self-aggrandizement /ˌselfəˈɡrændɪzmənt/ *n.* (also **-isement**) the act or process of trying to make oneself more powerful or seemingly more important. □ **self-aggrandizing** /-ˈɡrændaɪzɪŋ/ *adj.*

self-analysis /ˌselfəˈnælɪsɪs/ *n.* *Psychol.* the analysis of oneself, one's motives, character, etc. □ **self-analysing** /selfˈænəˌlaɪzɪŋ/ *adj.*

self-appointed /ˌselfəˈpɔɪntɪd/ *adj.* designated so by oneself, not authorized by another (*a self-appointed guardian*).

self-appreciation /ˌselfəˌpriːʃɪˈeɪʃ(ə)n, ˌselfəˌpriːsɪ-/ *n.* high estimation of oneself; conceit.

self-approbation /selfˌæprəˈbeɪʃ(ə)n/ *n.* approval of oneself.

self-approval /ˌselfəˈpruːv(ə)l/ *n.* = SELF-APPROBATION.

self-assembly /ˌselfəˈsemblɪ/ *n.* (often *attrib.*) the construction of furniture etc. from materials sold in kit form.

self-assertion /ˌselfəˈsɜːʃ(ə)n/ *n.* the aggressive promotion of oneself, one's views, etc. □ **self-asserting** /-ˈsɜːtɪŋ/ *adj.* **self-assertive** *adj.* **self-assertiveness** *n.*

self-assurance /ˌselfəˈʃʊərəns/ *n.* confidence in one's own abilities etc. □ **self-assured** /-ˈʃʊəd/ *adj.* **self-assuredly** /-ˈʃʊərɪdlɪ/ *adv.*

self-aware /ˌselfəˈweə(r)/ *adj.* conscious of one's character, feelings, motives, etc. □ **self-awareness** *n.*

self-begotten /ˌselfbɪˈɡɒt(ə)n/ *adj.* produced by oneself or itself; not made externally.

self-betrayal /selfbɪˈtreɪəl/ *n.* **1** the betrayal of oneself. **2** the inadvertent revelation of one's true thoughts etc.

self-binder /selfˈbaɪndə(r)/ *n.* a reaping machine with an automatic mechanism for binding the sheaves.

self-born /selfˈbɔːn/ *adj.* produced by itself or oneself; not made externally.

self-catering /selfˈkeɪtərɪŋ/ *adj. & n.* ● *adj.* (esp. of a holiday or holiday premises) providing rented accommodation with cooking facilities but without food. ● *n.* the activity of catering for oneself in rented temporary or holiday accommodation.

self-censorship /selfˈsensəˌʃɪp/ *n.* the censoring of oneself.

self-centred /selfˈsentəd/ *adj.* preoccupied with one's own personality or affairs; selfish. □ **self-centredly** *adv.* **self-centredness** *n.*

self-certification /ˌselfˌsɜːtɪfɪˈkeɪʃ(ə)n/ *n.* the practice by which an employee declares in writing that an absence from work was due to illness.

self-cleaning /selfˈkliːnɪŋ/ *adj.* (esp. of an oven) cleaning itself or keeping itself clean automatically.

self-closing /selfˈkləʊzɪŋ/ *adj.* (of a door etc.) closing automatically.

self-cocking /selfˈkɒkɪŋ/ *adj.* (of a gun) with the hammer raised by the trigger, not by hand.

self-collected /ˌselfkəˈlektɪd/ *adj.* composed, serene, self-assured.

self-coloured /selfˈkʌləd/ *adj.* **1 a** having the same colour throughout; not patterned (*buttons and belt are self-coloured*). **b** (of material) natural; undyed. **2 a** (of a flower) of uniform colour. **b** having its colour unchanged by cultivation or hybridization.

self-command /ˌselfkəˈmɑːnd/ *n.* = SELF-CONTROL.

self-communion /ˌselfkəˈmjuːnɪən/ *n.* meditation upon one's own character, conduct, etc.

self-conceit /ˌselfkənˈsiːt/ *n.* = SELF-SATISFACTION. □ **self-conceited** *adj.*

self-condemnation /selfˌkɒndemˈneɪʃ(ə)n/ *n.* **1** the blaming of oneself. **2** the inadvertent revelation of one's own sin, crime, etc. □ **self-condemned** /-kənˈdemd/ *adj.*

self-confessed /ˌselfkənˈfest/ *adj.* openly admitting oneself to be (*a self-confessed thief*).

self-confidence /selfˈkɒnfɪd(ə)ns/ *n.* confidence in oneself. □ **self-confident** *adj.* **self-confidently** *adv.*

self-congratulation /ˌselfkənˌɡrætjʊˈleɪʃ(ə)n/ *n.* = SELF-SATISFACTION. □ **self-congratulatory** /-kənˈɡrætʊlətərɪ/ *adj.*

self-conquest /selfˈkɒŋkwest/ *n.* the overcoming of one's worst characteristics etc.

self-conscious /selfˈkɒnʃəs/ *adj.* **1** unduly aware of oneself as an object of observation by others; embarrassed, shy. **2** *Philos.* having knowledge of one's own existence; self-contemplating. □ **self-consciously** *adv.* **self-consciousness** *n.*

self-consistent /ˌselfkənˈsɪstənt/ *adj.* (of parts of the same whole etc.) consistent; not conflicting. □ **self-consistency** *n.*

self-constituted /selfˈkɒnstɪˌtjuːtɪd/ *adj.* (of a person, group, etc.) assuming a function without authorization or right; self-appointed.

self-contained /ˌselfkənˈteɪnd/ *adj.* **1** (of a person) uncommunicative or reserved; independent, self-possessed. **2** *Brit.* (esp. of living accommodation) complete in itself. □ **self-containment** *n.*

self-contempt /ˌselfkənˈtempt/ *n.* contempt for oneself. □ **self-contemptuous** /-ˈtemptjʊəs/ *adj.*

self-content /ˌselfkənˈtent/ *n.* satisfaction with oneself, one's life, achievements, etc. □ **self-contented** *adj.*

self-contradiction /selfˌkɒntrəˈdɪkʃ(ə)n/ *n.* internal inconsistency. □ **self-contradictory** /-ˈdɪktərɪ/ *adj.*

self-control /ˌselfkənˈtrəʊl/ *n.* the power of controlling one's external reactions, emotions, etc.; equanimity. □ **self-controlled** *adj.*

self-convicted /ˌselfkənˈvɪktɪd/ *adj.* = SELF-CONDEMNED (see SELF-CONDEMNATION).

self-correcting /ˌselfkəˈrektɪŋ/ *adj.* correcting itself without external help.

self-created /ˌselfkriːˈeɪtɪd/ *adj.* created by oneself or itself. □ **self-creation** /-ˈeɪʃ(ə)n/ *n.*

self-critical /selfˈkrɪtɪk(ə)l/ *adj.* critical of oneself, one's abilities, etc. □ **self-criticism** /-tɪˌsɪz(ə)m/ *n.*

self-deception /ˌselfdɪˈsepʃ(ə)n/ *n.* deceiving oneself esp. concerning one's true feelings etc. □ **self-deceit** /-ˈsiːt/ *n.* **self-deceiver** /-ˈsiːvə/ *n.* **self-deceiving** *adj.* **self-deceptive** /-ˈseptɪv/ *adj.*

self-defeating /ˌselfdɪˈfiːtɪŋ/ *adj.* (of an attempt, action, etc.) doomed to failure because of internal inconsistencies etc.

self-defence /ˌselfdɪˈfens/ n. **1** defence or protection of oneself (had to hit him in self-defence). **2** aggressive action, speech, etc., intended as defence. □ **the noble art of self-defence** boxing. □ **self-defensive** adj.

self-delight /ˌselfdɪˈlaɪt/ n. delight in oneself or one's existence.

self-delusion /ˌselfdɪˈluːʒ(ə)n, -ˈljuːʒ(ə)n/ n. the act or an instance of deluding oneself.

self-denial /ˌselfdɪˈnaɪəl/ n. the negation of one's interests, needs, or desires, esp. in favour of those of others; self-control, forbearance. □ **self-denying ordinance** hist. (in English history) a resolution (1645) of the Long Parliament depriving Members of Parliament of civil and military office. □ **self-denying** /-ˈnaɪɪŋ/ adj.

self-dependence /ˌselfdɪˈpendəns/ adj. dependence only on oneself or itself; independence. □ **self-dependent** adj.

self-deprecation /ˌselfˌdeprɪˈkeɪʃ(ə)n/ n. the act of disparaging or belittling oneself. □ **self-deprecating** /-ˈdeprɪˌkeɪtɪŋ/ adj. **self-deprecatingly** adv.

self-despair /ˌselfdɪˈspeə(r)/ n. despair with oneself.

self-destroying /ˌselfdɪˈstrɔɪɪŋ/ adj. destroying oneself or itself.

self-destruct /ˌselfdɪˈstrʌkt/ v. & adj. ● v.intr. (of a spacecraft, bomb, etc.) explode or disintegrate automatically, esp. when pre-set to do so. ● attrib.adj. enabling a thing to self-destruct (a self-destruct device).

self-destruction /ˌselfdɪˈstrʌkʃ(ə)n/ n. **1** destruction of oneself or itself; suicide. **2** the process or an act of self-destructing. □ **self-destructive** /-ˈstrʌktɪv/ adj. **self-destructively** adv.

self-determination /ˌselfdɪˌtɜːmɪˈneɪʃ(ə)n/ n. **1** a nation's right to determine its own allegiance, government, etc. **2** the ability to act with free will, as opposed to fatalism etc. □ **self-determined** /-ˈtɜːmɪnd/ adj. **self-determining** adj.

self-development /ˌselfdɪˈveləpmənt/ n. the development of oneself, one's abilities, etc.

self-devotion /ˌselfdɪˈvəʊʃ(ə)n/ n. the devotion of oneself to a person or cause.

self-discipline /selfˈdɪsɪplɪn/ n. the act of or ability to apply oneself, control one's feelings, etc.; self-control. □ **self-disciplined** adj.

self-discovery /ˌselfdɪˈskʌvərɪ/ n. the process of acquiring insight into oneself, one's character, desires, etc.

self-disgust /ˌselfdɪsˈɡʌst/ n. disgust with oneself.

self-doubt /selfˈdaʊt/ n. lack of confidence in oneself, one's abilities, etc.

self-drive /selfˈdraɪv/ adj. (of a hired vehicle) driven by the hirer.

self-educated /selfˈedjʊˌkeɪtɪd/ adj. educated by oneself by reading etc., without formal instruction. □ **self-education** /-ˌedjʊˈkeɪʃ(ə)n/ n.

self-effacing /ˌselfɪˈfeɪsɪŋ/ adj. retiring; modest; timid. □ **self-effacement** n. **self-effacingly** adv.

self-elective /ˌselfɪˈlektɪv/ adj. (of a committee etc.) proceeding esp. by co-opting members etc.

self-employed /ˌselfɪmˈplɔɪd/ adj. working for oneself, as a freelance or owner of a business etc.; not employed by an employer. □ **self-employment** n.

self-esteem /ˌselfɪˈstiːm/ n. a good opinion of oneself; self-confidence.

self-evident /selfˈevɪd(ə)nt/ adj. obvious; without the need for evidence or further explanation. □ **self-evidence** n. **self-evidently** adv.

self-examination /ˌselfɪɡˌzæmɪˈneɪʃ(ə)n/ n. **1** the study of one's own conduct, reasons, etc. **2** the examining of one's body for signs of illness etc.

self-executing /selfˈeksɪˌkjuːtɪŋ/ adj. Law (of a law, legal clause, etc.) not needing legislation etc. to be enforced; automatic.

self-existent /ˌselfɪɡˈzɪstənt/ adj. existing without prior cause; independent.

self-explanatory /ˌselfɪkˈsplænətərɪ/ adj. easily understood; not needing explanation.

self-expression /ˌselfɪkˈspreʃ(ə)n/ n. the expression of one's feelings, thoughts, etc., esp. in writing, painting, music, etc. □ **self-expressive** /-ˈspresɪv/ adj.

self-faced /selfˈfeɪst/ adj. (of stone) unhewn; undressed.

self-feeder /selfˈfiːdə(r)/ n. **1** a furnace, machine, etc., that renews its own fuel or material automatically. **2** a device for supplying food to farm animals automatically. □ **self-feeding** adj.

self-fertile /selfˈfɜːtaɪl/ adj. (of a plant etc.) self-fertilizing. □ **self-fertility** /ˌselffəˈtɪlɪtɪ/ n.

self-fertilization /selfˌfɜːtɪlaɪˈzeɪʃ(ə)n/ n. (also **-isation**) Biol. the fertilization of plants and some invertebrates by their own pollen or sperm. □ **self-fertilized** /-ˈfɜːtɪˌlaɪzd/ adj. **self-fertilizing** adj.

self-financing /selfˈfaɪnænsɪŋ, -fɪˈnænsɪŋ, -faɪˈnænsɪŋ/ adj. that finances itself, esp. (of a project or undertaking) that pays for its own implementation or continuation. □ **self-financed** adj.

self-flattery /selfˈflætərɪ/ n. = SELF-APPRECIATION. □ **self-flattering** adj.

self-forgetful /ˌselffəˈɡetfʊl/ adj. unselfish. □ **self-forgetfulness** n.

self-fulfilling /ˌselffʊlˈfɪlɪŋ/ adj. (of a prophecy, forecast, etc.) bound to come true as a result of actions brought about by its being made.

self-fulfilment /ˌselffʊlˈfɪlmənt/ n. (US **-fillment**) the fulfilment of one's own hopes and ambitions.

self-generating /selfˈdʒenəˌreɪtɪŋ/ adj. generated by itself or oneself, not externally.

self-glorification /selfˌɡlɔːrɪfɪˈkeɪʃ(ə)n/ n. the proclamation of oneself, one's abilities, etc.; boasting.

self-governing /selfˈɡʌvənɪŋ/ adj. **1** having self-government. **2** (in the UK) (of a hospital or school) having opted out of local authority control.

self-government /selfˈɡʌvənmənt/ n. (esp. of a people or a former colony) administration of one's or its own affairs without external interference. □ **self-governed** adj.

self-gratification /selfˌɡrætɪfɪˈkeɪʃ(ə)n/ n. **1** gratification or pleasing of oneself. **2** self-indulgence, dissipation. **3** masturbation. □ **self-gratifying** /-ˈɡrætɪˌfaɪɪŋ/ adj.

self-hate /selfˈheɪt/ n. = SELF-HATRED.

self-hatred /selfˈheɪtrɪd/ n. hatred of oneself, esp. of one's actual self when contrasted with one's imagined self.

self-heal /selfˈhiːl/ n. a small labiate plant, Prunella vulgaris, with purple flowers. [f. its reputed healing properties]

self-help /selfˈhelp/ n. (often attrib.) **1** the theory that individuals should provide for their own support and improvement in society. **2** the act or faculty of providing for or improving oneself.

selfhood /ˈselfhʊd/ n. personality, separate and conscious existence.

self-image /selfˈɪmɪdʒ/ n. one's own idea or picture of oneself, esp. in relation to others.

self-importance /ˌselfɪmˈpɔːt(ə)ns/ n. a high opinion of oneself; pompousness. □ **self-important** adj. **self-importantly** adv.

self-imposed /ˌselfɪmˈpəʊzd/ adj. (of a task or condition etc.) imposed on and by oneself, not externally (self-imposed exile).

self-improvement /ˌselfɪmˈpruːvmənt/ n. the improvement of one's own position or disposition by one's own efforts.

self-induced /ˌselfɪnˈdjuːst/ adj. **1** induced by oneself or itself. **2** Electr. produced by self-induction.

self-inductance /ˌselfɪnˈdʌktəns/ n. Electr. the property of an electric circuit that causes an electromotive force to be generated in it by a change in the current flowing through it (cf. mutual inductance).

self-induction /ˌselfɪnˈdʌkʃ(ə)n/ n. Electr. the production of an electromotive force in a circuit when the current in that circuit is varied. □ **self-inductive** /-ˈdʌktɪv/ adj.

self-indulgent /ˌselfɪnˈdʌldʒənt/ adj. **1** indulging or tending to indulge oneself in pleasure, idleness, etc. **2** (of a work of art etc.) lacking economy and control. □ **self-indulgence** n. **self-indulgently** adv.

self-inflicted /ˌselfɪnˈflɪktɪd/ adj. (esp. of a wound, damage, etc.) inflicted by and on oneself, not externally.

self-interest /selfˈɪntrəst, -trɪst/ n. one's personal interest or advantage. □ **self-interested** adj.

selfish /ˈselfɪʃ/ adj. **1** deficient in consideration for others; concerned chiefly with one's own personal profit or pleasure; actuated by self-interest. **2** (of a motive etc.) appealing to self-interest. □ **selfishly** adv. **selfishness** n.

self-justification /selfˌdʒʌstɪfɪˈkeɪʃ(ə)n/ n. the justification or excusing of oneself, one's actions, etc.

self-knowledge /selfˈnɒlɪdʒ/ n. the understanding of oneself, one's motives, etc.

selfless /ˈselflɪs/ adj. disregarding oneself or one's own interests; unselfish. □ **selflessly** adv. **selflessness** n.

self-loading /selfˈləʊdɪŋ/ adj. (esp. of a gun) loading itself. □ **self-loader** n.

self-locking /selfˈlɒkɪŋ/ adj. locking itself.

self-love /selfˈlʌv/ n. **1** selfishness; self-indulgence. **2** Philos. regard for one's own well-being and happiness.

self-made /selfˈmeɪd/ adj. **1** successful or rich by one's own effort. **2** made by oneself.

self-mastery /selfˈmɑːstəri/ n. = SELF-CONTROL.

selfmate /ˈselfmeɪt/ n. Chess checkmate in which a player forces the opponent to achieve checkmate.

self-mocking /selfˈmɒkɪŋ/ adj. mocking oneself or itself.

self-motion /selfˈməʊʃ(ə)n/ n. motion caused by oneself or itself, not externally. □ **self-moving** /-ˈmuːvɪŋ/ adj.

self-motivated /selfˈməʊtɪˌveɪtɪd/ adj. acting on one's own initiative without external pressure. □ **self-motivation** /-ˌməʊtɪˈveɪʃ(ə)n/ n.

self-murder /selfˈmɜːdə(r)/ n. = SUICIDE n. 1a. □ **self-murderer** n.

self-neglect /ˌselfnɪˈglekt/ n. neglect of oneself.

selfness /ˈselfnɪs/ n. **1** individuality, personality, essence. **2** selfishness or self-regard.

self-opinionated /ˌselfəˈpɪnjəˌneɪtɪd/ adj. **1** stubbornly adhering to one's own opinions. **2** arrogant. □ **self-opinion** n.

self-perpetuating /ˌselfpəˈpetjʊˌeɪtɪŋ/ adj. perpetuating itself or oneself without external agency. □ **self-perpetuation** /-ˌpetjʊˈeɪʃ(ə)n/ n.

self-pity /selfˈpɪti/ n. extreme sorrow for one's own troubles etc. □ **self-pitying** adj. **self-pityingly** adv.

self-pollination /ˌselfˌpɒlɪˈneɪʃ(ə)n/ n. the pollination of a flower by pollen from the same plant. □ **self-pollinated** /-ˈpɒlɪˌneɪtɪd/ adj. **self-pollinating** adj. **self-pollinator** n.

self-portrait /selfˈpɔːtrɪt/ n. a portrait or description of an artist, writer, etc., by himself or herself.

self-possessed /ˌselfpəˈzest/ adj. habitually exercising self-control; composed. □ **self-possession** /-ˈzeʃ(ə)n/ n.

self-praise /selfˈpreɪz/ n. boasting; self-glorification.

self-preservation /selfˌprezəˈveɪʃ(ə)n/ n. **1** the preservation of one's own life, safety, etc. **2** this as a basic instinct of human beings and animals. □ **self-preserving** /ˌselfprəˈzɜːvɪŋ/ adj.

self-proclaimed /ˌselfprəˈkleɪmd/ adj. proclaimed by oneself or itself to be such.

self-propagating /selfˈprɒpəˌgeɪtɪŋ/ adj. (esp. of a plant) able to propagate itself.

self-propelled /ˌselfprəˈpeld/ adj. (esp. of a motor vehicle etc.) moving or able to move without external propulsion. □ **self-propelling** adj.

self-protection /ˌselfprəˈtekʃ(ə)n/ n. protecting oneself or itself. □ **self-protective** /-ˈtektɪv/ adj.

self-raising /selfˈreɪzɪŋ/ adj. Brit. (of flour) having a raising agent already added.

self-realization /selfˌrɪəlaɪˈzeɪʃ(ə)n/ n. (also **-isation**) **1** the development of one's faculties, abilities, etc. **2** this as an ethical principle.

self-recording /ˌselfrɪˈkɔːdɪŋ/ adj. (of a scientific instrument etc.) automatically recording its measurements.

self-referential /selfˌrefəˈrenʃ(ə)l/ adj. of or characterized by reference to oneself or itself.

self-regard /ˌselfrɪˈgɑːd/ n. **1** a proper regard for oneself. **2 a** selfishness. **b** conceit.

self-registering /selfˈredʒɪstərɪŋ/ adj. (of a scientific instrument etc.) automatically registering its measurements.

self-regulating /selfˈregjʊˌleɪtɪŋ/ adj. regulating oneself or itself without intervention. □ **self-regulatory** /-lətəri/ adj. **self-regulation** /-ˌregjʊˈleɪʃ(ə)n/ n.

self-reliance /ˌselfrɪˈlaɪəns/ n. reliance on one's own resources etc.; independence. □ **self-reliant** adj. **self-reliantly** adv.

self-renewal /ˌselfrɪˈnjuːəl/ n. the act or process of renewing oneself or itself.

self-renunciation /ˌselfrɪˌnʌnsɪˈeɪʃ(ə)n/ n. **1** = SELF-SACRIFICE. **2** unselfishness.

self-reproach /ˌselfrɪˈprəʊtʃ/ n. reproach or blame directed at oneself. □ **self-reproachful** adj.

self-respect /ˌselfrɪˈspekt/ n. respect for oneself, a feeling that one is behaving with honour, dignity, etc. □ **self-respecting** adj.

self-restraint /ˌselfrɪˈstreɪnt/ n. = SELF-CONTROL. □ **self-restrained** adj.

self-revealing /ˌselfrɪˈviːlɪŋ/ adj. revealing one's character, motives, etc., esp. inadvertently. □ **self-revelation** /-ˌrevəˈleɪʃ(ə)n/ n.

Selfridge /ˈselfrɪdʒ/, Harry Gordon (1858–1947), American-born British businessman. In 1906 he came to England and began to build the department store in Oxford Street, London, that bears his name; it opened in 1909.

self-righteous /selfˈraɪtʃəs/ adj. excessively conscious of or insistent on one's rectitude, correctness, etc. □ **self-righteously** adv. **self-righteousness** n.

self-righting /selfˈraɪtɪŋ/ adj. (of a boat) righting itself when capsized.

self-rising /selfˈraɪzɪŋ/ adj. US = SELF-RAISING.

self-rule /selfˈruːl/ n. government of a colony, dependent country, etc., by its own people.

self-sacrifice /selfˈsækrɪˌfaɪs/ n. the negation of one's own interests, wishes, etc., in favour of those of others. □ **self-sacrificing** adj.

selfsame /ˈselfseɪm/ attrib.adj. (prec. by the) the very same (the selfsame village).

self-satisfaction /selfˌsætɪsˈfækʃ(ə)n/ n. excessive and unwarranted satisfaction with oneself, one's achievements, etc.; complacency. □ **self-satisfied** /-ˈsætɪsˌfaɪd/ adj.

self-sealing /selfˈsiːlɪŋ/ adj. **1** (of a pneumatic tyre, fuel tank, etc.) automatically able to seal small punctures. **2** (of an envelope) self-adhesive.

self-seeding /selfˈsiːdɪŋ/ adj. (of a plant) propagating itself by seed. □ **self-seed** v.intr. **self-seeder** n.

self-seeking /selfˈsiːkɪŋ/ adj. & n. ● adj. seeking one's own welfare before that of others. ● n. the activity of doing this. □ **self-seeker** n.

self-selection /ˌselfsɪˈlekʃ(ə)n/ n. the act of selecting oneself or itself. □ **self-selecting** /-ˈlektɪŋ/ adj.

self-service /selfˈsɜːvɪs/ adj. & n. ● adj. (often attrib.) **1** (of a shop, restaurant, garage, etc.) where customers serve themselves and pay at a checkout counter etc. **2** (of a machine) serving goods after the insertion of coins. ● n. colloq. a self-service store, garage, etc.

self-serving /selfˈsɜːvɪŋ/ adj. & n. = SELF-SEEKING.

self-slaughter /selfˈslɔːtə(r)/ n. = SUICIDE n. 1a.

self-sown /selfˈsəʊn/ adj. grown from seed scattered naturally.

self-starter /selfˈstɑːtə(r)/ n. **1** an electric appliance for starting a motor vehicle engine without the use of a crank. **2** an ambitious person who needs no external motivation.

self-sterile /selfˈsteraɪl/ adj. Biol. not self-fertilizing. □ **self-sterility** /ˌselfstəˈrɪlɪti/ n.

self-styled /ˈselfstaɪld/ adj. called so by oneself; would-be; pretended (a self-styled artist).

self-sufficient /ˌselfsəˈfɪʃ(ə)nt/ adj. **1 a** needing nothing; independent. **b** (of a person, nation, etc.) able to supply one's needs for a commodity, esp. food, from one's own resources. **2** content with one's own opinion; arrogant. □ **self-sufficiency** n. **self-sufficiently** adv. **self-sufficing** /-ˈfaɪsɪŋ/ adj.

self-suggestion /ˌselfsəˈdʒestʃən/ n. = AUTO-SUGGESTION.

self-supporting /ˌselfsəˈpɔːtɪŋ/ adj. **1** capable of maintaining oneself or itself financially. **2** staying up or standing without external aid. □ **self-support** n.

self-surrender /ˌselfsəˈrendə(r)/ n. the surrender of oneself or one's will etc. to an influence, emotion, or other person.

self-sustaining /ˌselfsəˈsteɪnɪŋ/ adj. sustaining oneself or itself. □ **self-sustained** adj.

self-tanning /selfˈtænɪŋ/ adj. (of a cream or lotion) causing the skin to turn brown as if sun-tanned.

self-taught /selfˈtɔːt/ adj. educated or trained by oneself, not externally.

self-torture /selfˈtɔːtʃə(r)/ n. the inflicting of pain, esp. mental, on oneself.

self-willed /selfˈwɪld/ adj. obstinately pursuing one's own wishes. □ **self-will** n.

self-winding /selfˈwaɪndɪŋ/ adj. (of a watch etc.) having an automatic winding apparatus.

self-worth /selfˈwɜːθ/ n. = SELF-ESTEEM.

Seljuk /'seldʒʊk, sel'dʒuːk/ n. & adj. ● n. a member of the Turkish dynasty which ruled Asia Minor in the 11th–13th centuries, successfully invading the Byzantine Empire and defending the Holy Land against the Crusaders. ● adj. of or relating to this dynasty. □ **Seljukian** /sel'dʒuːkɪən/ adj. & n. [Turk. seljūq (name of their reputed ancestor)]

Selkirk /'selkɜːk/, Alexander (also called Selcraig) (1676–1721), Scottish sailor. While on a privateering expedition in 1704, Selkirk quarrelled with his captain and was put ashore, at his own request, on one of the uninhabited Juan Fernandez Islands in the Pacific, where he remained until he was rescued in 1709. His experiences later formed the basis of Daniel Defoe's novel Robinson Crusoe (1719).

Selkirkshire /'selkɜːkˌʃɪə(r)/ a former county of SE Scotland. It was made a part of Borders region in 1975.

sell /sel/ v. & n. ● v. (past and past part. **sold** /səʊld/) **1** tr. make over or dispose of in exchange for money. **2** tr. keep a stock of for sale or be a dealer in (do you sell candles?). **3 a** intr. (of goods) be purchased (will never sell; these are selling well). **b** tr. (of a publication or recording) attain sales of (a specified number of copies) (the book has sold 10,000 copies). **4** intr. (foll. by at, for) have a specified price (sells at £5). **5** tr. betray for money or other reward (sell one's country). **6** tr. offer dishonourably for money or other reward; make a matter of corrupt bargaining (sell justice; sell one's honour). **7** tr. **a** advertise or publish the merits of. **b** give (a person) information on the value of something, inspire with a desire to buy or acquire or agree to something. **8** tr. cause to be sold (the author's name alone will sell many copies). **9** tr. sl. disappoint by not keeping an engagement etc., by failing in some way, or by trickery (we've been sold). ● n. colloq. **1** a manner of selling (soft sell). **2** a deception or disappointment. □ **sell-by date** the latest recommended date of sale marked on the packaging of esp. perishable food. **sell down the river** see RIVER. **sell the** (or **a**) **dummy** see DUMMY. **selling-point** an advantageous feature. **selling-race** a horse-race after which the winning horse must be auctioned. **sell one's life dear** (or **dearly**) do great injury before being killed. **sell off** sell the remainder of (goods) at reduced prices. **sell-off** n.**1** the privatization of a state company by the sale of shares. **2** a sale, esp. to dispose of property. **3** esp. US Stock Exch. a sale or disposal of bonds, shares, etc., usu. causing a fall in price. **sell oneself 1** promote one's own abilities. **2 a** offer one's services dishonourably for money or other reward. **b** be a prostitute. **sell out 1 a** sell all one's stock-in-trade, one's shares in a company, etc. **b** sell (all or some of one's stock, shares, etc.). **2 a** betray. **b** be treacherous or disloyal. **sell-out** n. **1** a commercial success, esp. the selling of all tickets for a show. **2** a betrayal. **sell the pass** see PASS². **sell a person a pup** see PUP. **sell short** disparage, underestimate. **sell-through** (often attrib.) sale as opposed to rental, esp. of video recordings. **sell up** Brit. **1** sell one's business, house, etc. **2** sell the goods of (a debtor). **sold on** colloq. enthusiastic about. □ **sellable** adj. [OE sellan f. Gmc]

Sellafield /'seləˌfiːld/ the site of a nuclear power station and reprocessing plant on the coast of Cumbria in NW England. It was the scene in 1957 of a fire which caused a serious escape of radioactive material. The site was known as Windscale between 1947 and 1981.

seller /'selə(r)/ n. **1** a person who sells. **2** a commodity that sells well or badly. □ **seller's** (or **sellers'**) **market** an economic position in which goods are scarce and expensive.

Sellers /'seləz/, Peter (1925–80), English comic actor. He made his name in The Goon Show, a British radio series of the 1950s, with Spike Milligan (b.1918) and Harry Secombe (b.1921). Sellers then turned to films, starring in many comedies such as The Lady Killers (1955), I'm All Right Jack (1959), and Dr Strangelove (1964). In the 'Pink Panther' series of films of the 1960s and 1970s, he played the role of the French detective Inspector Clouseau, for which he is best known.

Sellotape /'seləˌteɪp/ n. & v. ● n. propr. adhesive usu. transparent cellulose or plastic tape. Tape of this kind was first developed in the US (where it is called Scotch tape) in 1928. ● v.tr. (**sellotape**) fix with Sellotape. [CELLULOSE + TAPE]

Selous /sə'luː/, Frederick Courteney (1851–1917), English explorer, naturalist, and soldier. He first visited South Africa in 1871 and spent the following ten years in exploration and big-game hunting in south central Africa. From 1890 he was involved in the British South Africa Company, negotiating mineral and land rights for the British (see also RHODES²). The Selous Game Reserve in Tanzania is named after him.

seltzer /'seltsə(r)/ n. (in full **seltzer water**) **1** medicinal mineral water from the region of Nieder-Selters in Germany. **2** an artificial substitute for this; soda water. [G Selterser (adj.) f. Nieder-Selters]

selva /'selvə/ n. (often in pl.) Geog. land covered by dense tropical rainforest, esp. in the Amazon basin. [Sp. or Port. f. L silva wood]

selvedge /'selvɪdʒ/ n. (also **selvage**) **1 a** an edging that prevents cloth from unravelling (either an edge along the warp or a specially woven edging). **b** a border of different material or finish intended to be removed or hidden. **2** Geol. an alteration zone at the edge of a rock mass. **3** the edge-plate of a lock with an opening for the bolt. [ME f. SELF + EDGE, after Du. selfegghe]

selves pl. of SELF.

Selye /'seljeɪ/, Hans Hugo Bruno (1907–82), Austrian-born Canadian physician. After working in several European countries, he moved to Montreal and later became director of the University of Montreal's Institute of Experimental Medicine and Surgery. He showed that environmental stress and anxiety could result in the release of hormones that, over a long period, could produce many of the biochemical and physiological disorders characteristic of the 20th century. Selye's theory had a profound effect on modern medicine, and he became a popularizer of research on stress.

Selznick /'selznɪk/, David O(liver) (1902–65), American film producer. Based in Hollywood from 1926, he produced such films as King Kong (1933) for RKO and Anna Karenina (1935) for MGM. In 1936 he established his own production company, Selznick International, with whom he produced such screen classics as Gone with the Wind (1939) and Rebecca (1940).

SEM abbr. scanning electron microscope.

semanteme /sɪ'mæntiːm/ n. Linguistics a fundamental element expressing an image or idea. [F sémantème (as SEMANTIC)]

semantic /sɪ'mæntɪk/ adj. relating to meaning in language; relating to the connotations of words. □ **semantically** adv. [F sémantique f. Gk sēmantikos significant f. sēmainō signify f. sēma sign]

semantics /sɪ'mæntɪks/ n.pl. (usu. treated as sing.) the branch of linguistics and logic concerned with meaning. Two main branches of semantics may be distinguished: logical semantics is concerned with matters such as sense and reference, presupposition and implication, the scope of quantifiers, modality (necessity and possibility), and the meaning of connectives such as 'and', 'or', and 'if … then'. Lexical semantics is concerned with the analysis of word meanings and the relations between them such as synonymy, antonymy, and hyponymy. □ **semanticist** /-tɪsɪst/ n. **semantician** /ˌsiːmænˈtɪʃ(ə)n/ n.

semaphore /'seməˌfɔː(r)/ n. & v. ● n. **1** Mil. etc. a system of sending messages by holding the arms or two flags in certain positions according to an alphabetic code. **2** a signalling apparatus consisting of a post with a movable arm or arms, lanterns, etc., for use (esp. on railways) by day or night. ● v.intr. & tr. signal or send by semaphore. □ **semaphoric** /ˌseməˈfɒrɪk/ adj. **semaphorically** adv. [F sémaphore, irreg. f. Gk sēma sign + -phoros -PHORE]

Semarang /sə'mɑːræŋ/ a port in Indonesia, on the north coast of Java; pop. (1980) 1,026,600.

semasiology /sɪˌmeɪsɪˈɒlədʒɪ/ n. semantics. □ **semasiological** /-sɪəˈlɒdʒɪk(ə)l/ adj. [G Semasiologie f. Gk sēmasia meaning f. sēmainō signify]

sematic /sɪ'mætɪk/ adj. Zool. (of colouring, markings, etc.) significant; serving to warn off enemies or attract attention. [Gk sēma sēmatos sign]

semblable /'semblab(ə)l/ n. & adj. ● n. a counterpart or equal. ● adj. archaic having the semblance of something, seeming. [ME f. OF (as SEMBLANCE)]

semblance /'sembləns/ n. **1** the outward or superficial appearance of something (put on a semblance of anger). **2** resemblance. [ME f. OF f. sembler f. L similare, simulare SIMULATE]

semé /'semɪ, 'semeɪ/ adj. (also **semée**) Heraldry covered with small bearings of indefinite number (e.g. stars, fleurs-de-lis) arranged all over the field. [F, past part. of semer to sow]

Semei /sə'meɪ/ (also **Semey**) an industrial city and river port in eastern Kazakhstan, on the Irtysh river close to the border with Russia; pop. (1990) 338,800. Founded in the 18th century, it was known as Semipalatinsk until 1991.

semeiology var. of SEMIOLOGY.

semeiotics var. of SEMIOTICS.

Semele /'semɪlɪ/ Gk Mythol. the mother, by Zeus, of Dionysus. She entreated Zeus to come to her in his full majesty and the fire of his thunderbolts killed her but made her child immortal.

sememe /'semiːm, 'siːm-/ n. Linguistics the unit of meaning carried by a morpheme. [as SEMANTIC + -EME]

semen /'si:mən/ n. the reproductive fluid of male animals, containing spermatozoa in suspension. [ME f. L *semen seminis* seed f. *serere* to sow]

semester /sɪ'mestə(r)/ n. a half-year course or term in (esp. German and US) universities. [G f. L *semestris* six-monthly f. *sex* six + *mensis* month]

Semey see SEMEI.

semi /'semɪ/ n. (pl. **semis**) *colloq.* **1** *Brit.* a semi-detached house. **2** *N. Amer. & Austral.* a semi-trailer. [abbr.]

semi- /'semɪ/ *prefix* **1** half (*semicircle*). **2** partly; in some degree or particular (*semi-official; semi-detached*). **3** almost (*a semi-smile*). **4** occurring or appearing twice in a specified period (*semi-annual*). [F, It., etc. or L, corresp. to Gk HEMI-, Skr. *sámi*]

semi-annual /ˌsemɪ'ænjʊəl/ adj. occurring, published, etc., twice a year. □ **semi-annually** *adv.*

semiaquatic /ˌsemɪə'kwætɪk, -'kwɒtɪk/ adj. **1** (of an animal) living partly on land and partly in water. **2** (of a plant) growing in very wet ground.

semi-automatic /ˌsemɪˌɔːtə'mætɪk/ adj. **1** partially automatic. **2** (of a firearm) having a mechanism for continuous loading but not for continuous firing.

semi-basement /ˌsemɪ'beɪsmənt/ n. a storey partly below ground level.

semi-bold /ˌsemɪ'bəʊld/ adj. *Printing* printed in a type darker than normal but not as dark as bold.

semibreve /'semɪˌbriːv/ n. *Mus.* the longest note now in common use, having the time value of two minims or four crotchets, and represented by a ring with no stem. Also called *whole note*.

semicircle /'semɪˌsɜːk(ə)l/ n. **1** half of a circle or of its circumference. **2** a set of objects ranged in, or an object forming, a semicircle. [L *semicirculus* (as SEMI-, CIRCLE)]

semicircular /ˌsemɪ'sɜːkjʊlə(r)/ adj. **1** forming or shaped like a semicircle. **2** arranged as or in a semicircle. □ **semicircular canal** *Anat.* each of three fluid-filled channels in the ear giving information to the brain to help maintain balance. [LL *semicircularis* (as SEMICIRCLE)]

semi-civilized /ˌsemɪ'sɪvɪˌlaɪzd/ adj. partially civilized.

semicolon /ˌsemɪ'kəʊlən, -lɒn/ n. a punctuation mark (;) of intermediate value between a comma and full stop.

semiconducting /ˌsemɪkən'dʌktɪŋ/ adj. having the properties of a semiconductor.

semiconductor /ˌsemɪkən'dʌktə(r)/ n. a solid substance that is a non-conductor when pure or at a low temperature but has a conductivity between that of insulators and that of most metals when containing a suitable impurity or at a higher temperature. Modern electronics now relies principally upon devices made of semiconductors such as silicon and germanium. The importance of semiconductors lies in the fact that, in comparison with metals, their conductivity is much more sensitive to factors such as heat, light, applied voltage, and traces of impurities, meaning that the performance of transistors and other semiconductor devices can be very precisely controlled. The latter is a prerequisite in the creation of electronic circuits that are able to process complicated information, as in a computer.

semi-conscious /ˌsemɪ'kɒnʃəs/ adj. partially conscious.

semicylinder /ˌsemɪ'sɪlɪndə(r)/ n. half of a cylinder cut longitudinally. □ **semicylindrical** /-sɪ'lɪndrɪk(ə)l/ adj.

semidemisemiquaver /'semɪˌdemɪˌsemɪˌkweɪvə(r)/ n. *Mus.* = HEMIDEMISEMIQUAVER. [SEMI- + DEMISEMIQUAVER]

semi-deponent /ˌsemɪdɪ'pəʊnənt/ adj. *Gram.* (of a Latin verb) having active forms in present tenses and passive forms with active sense in perfect tenses.

semi-detached /ˌsemɪdɪ'tætʃt/ adj. & n. ● adj. (of a house) joined to another by a party-wall on one side only. ● n. a semi-detached house.

semidiameter /ˌsemɪdaɪ'æmɪtə(r)/ n. half of a diameter. [LL (as SEMI-, DIAMETER)]

semi-documentary /ˌsemɪˌdɒkjʊ'mentərɪ/ adj. & n. ● adj. (of a film) having a factual background and a fictitious story. ● n. (pl. **-ies**) a semi-documentary film.

semi-dome /'semɪˌdəʊm/ n. **1** a half-dome formed by vertical section. **2** a part of a structure more or less resembling a dome.

semi-double /ˌsemɪ'dʌb(ə)l/ adj. (of a flower) intermediate between single and double in having only the outer stamens converted to petals.

semifinal /ˌsemɪ'faɪn(ə)l/ n. a match or round immediately preceding the final.

semifinalist /ˌsemɪ'faɪnəlɪst/ n. a competitor in a semifinal.

semi-finished /ˌsemɪ'fɪnɪʃt/ adj. prepared for the final stage of manufacture.

semi-fitted /ˌsemɪ'fɪtɪd/ adj. (of a garment) shaped to the body but not closely fitted.

semifluid /ˌsemɪ'fluːɪd/ adj. & n. ● adj. of a consistency between solid and liquid. ● n. a semifluid substance.

semi-independent /ˌsemɪˌɪndɪ'pendənt/ adj. **1 a** partially independent of control or authority. **b** partially self-governing. **2** partially independent of financial support from public funds.

semi-infinite /ˌsemɪ'ɪnfɪnɪt/ adj. *Math.* limited in one direction and stretching to infinity in the other.

semi-invalid /ˌsemɪ'ɪnvəˌliːd, -lɪd/ n. a partially disabled or somewhat infirm person.

semi-liquid /ˌsemɪ'lɪkwɪd/ adj. & n. = SEMIFLUID.

Sémillon /'semɪˌɒn/ n. (also **Semillon**) **1** a variety of white grape used in wine-making. **2** a white wine made from these grapes. [F *dial.*, ult. f. L *semen* seed]

semi-lunar /ˌsemɪ'luːnə(r)/ adj. shaped like a half moon or crescent. □ **semi-lunar bone** a bone of this shape in the carpus. **semi-lunar cartilage** a cartilage of this shape in the knee. **semi-lunar valve** a valve of this shape in the heart. [mod.L *semilunaris* (as SEMI-, LUNAR)]

semi-metal /ˌsemɪ'met(ə)l/ n. a substance with some of the properties of metals; in particular (*Chem.*), an element whose properties are intermediate between those of metals and solid non-metals or semiconductors. Such elements (as arsenic, antimony, and tin) occupy lower positions in Groups IIIA–VIA (13–16) in the periodic table. □ **semi-metallic** /-mɪ'tælɪk/ adj.

semi-monthly /ˌsemɪ'mʌnθlɪ/ adj. & adv. ● adj. occurring, published, etc., twice a month. ● adv. twice a month.

seminal /'semɪn(ə)l/ adj. **1 a** (of ideas etc.) providing the basis for future development. **b** highly original and influential, central to the understanding of a subject. **2** of or relating to semen. **3** of or relating to the seeds of plants. □ **seminal fluid** semen. □ **seminally** *adv.* [ME f. OF *seminal* or L *seminalis* (as SEMEN)]

seminar /'semɪˌnɑː(r)/ n. **1** a small class at a university etc. for discussion and research. **2** a short intensive course of study. **3** a conference of specialists. [G (as SEMINARY)]

seminary /'semɪnərɪ/ n. (pl. **-ies**) **1** a training-college for priests, rabbis, etc. **2** a place of education or development. □ **seminarist** n. [ME f. L *seminarium* seed-plot, neut. of *seminarius* (adj.) (as SEMEN)]

seminiferous /ˌsemɪ'nɪfərəs/ adj. **1** *Bot.* bearing seed. **2** *Anat.* conveying semen. [L *semin-* f. SEMEN + -FEROUS]

Seminole /'semɪˌnəʊl/ n. & adj. ● n. (pl. same or **Seminoles**) **1** a member of any of several groups of peoples of the Creek confederacy who settled in Florida and Oklahoma. **2** the Muskogean language of the Seminoles. ● adj. of or relating to the Seminoles or their language. [Creek f. Amer. Sp. *cimarrón* wild, untamed]

semi-official /ˌsemɪə'fɪʃ(ə)l/ adj. **1** partly official; rather less than official. **2** (of communications to newspapers etc.) made by an official with the stipulation that the source should not be revealed. □ **semi-officially** *adv.*

semiology /ˌsiːmɪ'ɒlədʒɪ, ˌsem-/ n. (also **semeiology**) = SEMIOTICS. □ **semiologist** n. **semiological** /-mɪə'lɒdʒɪk(ə)l/ adj. [Gk *sēmeion* sign f. *sēma* mark]

semi-opaque /ˌsemɪəʊ'peɪk/ adj. not fully transparent.

semiotics /ˌsiːmɪ'ɒtɪks, ˌsem-/ n. (also **semeiotics**) **1** the study of signs and symbols in various fields, esp. language. (*See note below.*) **2** *Med.* symptomatology. □ **semiotic** adj. **semiotical** adj. **semiotically** adv. **semiotician** /-mɪə'tɪʃ(ə)n/ n. [Gk *sēmeiōtikos* of signs (as SEMIOLOGY)]

▪ The term semiotics is used to denote the study of systems of signs and symbols in the broadest sense, as opposed to semantics, which is used, more narrowly, with reference to the meaning of specifically linguistic and logical signs and symbols. Semiotics includes the study of the meanings assigned to non-arbitrary signs such as the weather, animal behaviour, and medical symptoms, as well as cultural systems such as music, food, dress, painting, and language. Much of the terminology of both semiotics and semantics was invented by the American logician C. S. Peirce. Semioticians such as Umberto Eco have also attempted to devise a general theory of semiotics for the analysis of human cultures and behaviour.

Semipalatinsk /ˌsemɪpə'lɑːtɪnsk/ the former name (until 1991) for SEMEI.

semipalmated /ˌsemɪpælˈmeɪtɪd/ adj. Zool. having toes etc. webbed for part of their length. [SEMI- + L palmatus PALMATE]

semi-permanent /ˌsemɪˈpɜːmənənt/ adj. rather less than permanent.

semi-permeable /ˌsemɪˈpɜːmɪəb(ə)l/ adj. (of a membrane etc.) selectively permeable, esp. allowing passage of a solvent but not of large dissolved molecules.

semi-plume /ˈsemɪˌpluːm/ n. a feather with a firm stem and a downy web.

semiprecious /ˌsemɪˈpreʃəs/ adj. (of a gem) less valuable than a precious stone.

semi-pro /ˌsemɪˈprəʊ/ adj. & n. (pl. **-os**) US colloq. = SEMI-PROFESSIONAL.

semi-professional /ˌsemɪprəˈfeʃən(ə)l/ adj. & n. ● adj. **1** receiving payment for an activity but not relying on it for a living. **2** involving semi-professionals. ● n. a semi-professional musician, sportsman, etc.

semiquaver /ˈsemɪˌkweɪvə(r)/ n. Mus. a note having the time value of half a quaver and represented by a large dot with a two-hooked stem. Also called sixteenth note.

Semiramis /sɪˈmɪrəmɪs/ Gk Mythol. the daughter of a Syrian goddess, who later married a king of Assyria and after his death ruled for many years, becoming one of the founders of Babylon. Semiramis is thought to have been based on Sammuramat, a historical figure who ruled 810–805 BC.

semi-rigid /ˌsemɪˈrɪdʒɪd/ adj. (of an airship) having a stiffened keel attached to a flexible gas container.

semi-skilled /ˌsemɪˈskɪld/ adj. (of work or a worker) having or needing some training but less than for a skilled worker.

semi-skimmed /ˌsemɪˈskɪmd/ adj. (of milk) from which some cream has been skimmed.

semi-smile /ˈsemɪˌsmaɪl/ n. an expression that is not quite a smile.

semi-solid /ˌsemɪˈsɒlɪd/ adj. viscous, semifluid.

semi-sweet /ˌsemɪˈswiːt/ adj. (of biscuits etc.) slightly sweetened.

semi-synthetic /ˌsemɪsɪnˈθetɪk/ adj. Chem. (of a substance) that is prepared synthetically but derives from a naturally occurring material.

Semite /ˈsiːmaɪt, ˈsem-/ n. a member of any of the peoples supposed to be descended from Shem, son of Noah (Gen. 10:21 ff.), including esp. the Jews, Arabs, Assyrians, Babylonians, and Phoenicians. □ **Semitism** /ˈsemɪˌtɪz(ə)m/ n. **Semitist** n. **Semitize** v.tr. (also **-ise**). **Semitization** /ˌsemɪtaɪˈzeɪʃ(ə)n/ n. [mod.L Semita f. LL f. Gk Sēm Shem]

Semitic /sɪˈmɪtɪk/ adj. **1** of or relating to the Semites, esp. the Jews. **2** of or relating to a group of languages including Hebrew, Arabic, and Aramaic and certain ancient languages such as Phoenician, Assyrian, and Babylonian, constituting a subgroup of the Afro-Asiatic family. They are characterized by internal vowel changes (ablaut) distinguishing related words, e.g. Arabic salāma safety, salima be unharmed. [mod.L Semiticus (as SEMITE)]

semitone /ˈsemɪˌtəʊn/ n. Mus. the smallest interval used in classical European music; half a tone.

semi-trailer /ˌsemɪˈtreɪlə(r)/ n. a trailer having wheels at the back but supported at the front by a towing vehicle.

semi-transparent /ˌsemɪtrænsˈpærənt, -trɑːns-, -ˈpeərənt/ adj. partially or imperfectly transparent.

semi-tropics /ˌsemɪˈtrɒpɪks/ n.pl. = SUBTROPICS. □ **semi-tropical** adj.

semi-vowel /ˈsemɪˌvaʊəl/ n. **1** a sound intermediate between a vowel and a consonant (e.g. w, y). **2** a letter representing this. [after L semivocalis]

semi-weekly /ˌsemɪˈwiːklɪ/ adj. & adv. ● adj. occurring, published, etc., twice a week. ● adv. twice a week.

Semmelweis /ˈzem(ə)lˌvaɪs/, Ignaz Philipp (born Ignác Fülöp Semmelweis) (1818–65), Hungarian obstetrician, who spent most of his working life in Vienna. He discovered the infectious character of puerperal fever, at the time a major cause of death following childbirth. Semmelweis demonstrated that the infection was transmitted by the hands of doctors who examined patients after working in the dissecting room, and advocated rigorous cleanliness and the use of antiseptics. Despite the spectacular results obtained by Semmelweis, R. K. Virchow and other senior figures opposed his views for many years, and his involvement in events in Vienna during the 1848 revolution hindered his career.

semmit /ˈsemɪt/ n. Sc. an undershirt or vest. [ME: orig. unkn.]

semolina /ˌseməˈliːnə/ n. **1** the hard grains left after the milling of flour, used in puddings etc. and in pasta. **2** a pudding etc. made of this. [It. semolino dimin. of semola bran f. L simila flour]

sempervivum /ˌsempəˈvaɪvəm/ n. a succulent plant of the genus Sempervivum, esp. the houseleek. [mod.L f. L semper always + vivus living]

sempiternal /ˌsempɪˈtɜːn(ə)l/ adj. rhet. eternal, everlasting. □ **sempiternally** adv. **sempiternity** n. [ME f. OF sempiternel f. LL sempiternalis f. L sempiternus f. semper always + aeternus eternal]

semplice /ˈsemplɪˌtʃeɪ, -tʃɪ/ adv. Mus. in a simple style of performance. [It., = SIMPLE]

sempre /ˈsempreɪ, -rɪ/ adv. Mus. throughout, always (sempre forte). [It.]

sempstress var. of SEAMSTRESS.

Semtex /ˈsemteks/ n. propr. a very pliable, odourless plastic explosive. [prob. f. Semtín (a village in the Czech republic near the place of production) + explosive]

SEN see STATE ENROLLED NURSE.

Sen. abbr. **1** Senior. **2** US **a** Senator. **b** Senate.

Senanayake /ˌsenəˈnaɪəkə/, Don Stephen (1884–1952), Sinhalese statesman, Prime Minister of Ceylon 1947–52. In 1919 he co-founded the Ceylon National Congress, and during the 1920s and 1930s held ministerial positions on Ceylon's legislative and state councils. He became Prime Minister in 1947, and the following year presided over Ceylon's achievement of full dominion status within the Commonwealth.

senarius /sɪˈneərɪəs/ n. (pl. **senarii** /-rɪ,aɪ/) Prosody a verse of six feet, esp. an iambic trimeter. [L: see SENARY]

senary /ˈsiːnərɪ, ˈsen-/ adj. of six, by sixes. [L senarius f. seni distributive of sex six]

senate /ˈsenɪt/ n. **1** (usu. **Senate**) a legislative body, esp. the upper and smaller assembly in the US, France, and other countries, in the states of the US, etc. **2** the governing body of a university or (in the US) a college. **3** (usu. **Senate**) Rom. Hist. the state council of the republic and empire sharing legislative power with the popular assemblies, administration with the magistrates, and judicial power with the knights. [ME f. OF senat f. L senatus f. senex old man]

senator /ˈsenətə(r)/ n. **1** a member of a senate. **2** (in Scotland) a Lord of Session. □ **senatorship** n. **senatorial** /ˌsenəˈtɔːrɪəl/ adj. [ME f. OF senateur f. L senator -oris (as SENATE)]

send /send/ v. (past and past part. **sent** /sent/) **1** tr. **a** order or cause to go or be conveyed (send a message to headquarters; sent me a book; sends goods all over the world). **b** propel; cause to move (send a bullet; sent him flying). **c** cause to go or become (send into raptures; send to sleep). **d** dismiss with or without force (sent her away; sent him about his business). **2** intr. send a message or letter (he sent to warn me). **3** tr. (of God, providence, etc.) grant or bestow or inflict; bring about; cause to be (send rain; send a judgement). **4** tr. sl. affect emotionally, put into ecstasy. **5** tr. (often foll. by forth or off) emit, give out (light, heat, odour, etc.); utter or produce (sound); cause (a voice, cry, etc.) to carry, or travel. □ **send away for** send an order to a dealer for (goods). **send down** Brit. **1** rusticate or expel from a university. **2** sentence to imprisonment. **3** Cricket bowl (a ball or an over). **send for 1** summon. **2** order by post. **send in 1** cause to go in. **2** submit (an entry etc.) for a competition etc. **send off 1** get (a letter, parcel, etc.) dispatched. **2** attend the departure of (a person) as a sign of respect etc. **3** Sport (of a referee) order (a player) to leave the field and take no further part in the game. **send-off** n. a demonstration of goodwill etc. at the departure of a person, the start of a project, etc. **send off for** = send away for. **send on** transmit to a further destination or in advance of one's own arrival. **send a person to Coventry** see COVENTRY. **send up 1** cause to go up. **2** transmit to a higher authority. **3** Brit. colloq. satirize or ridicule, esp. by mimicking. **4** US sentence to imprisonment. **send-up** n. Brit. colloq. a satire or parody. **send word** send information. □ **sendable** adj. **sender** n. [OE sendan f. Gmc]

Sendai /senˈdaɪ/ a city in Japan, situated near the NE coast of the island of Honshu; pop. (1990) 918,000. It is the capital of the region of Tohoku.

sendal /ˈsend(ə)l/ n. hist. **1** a thin rich silk material. **2** a garment of this. [ME f. OF cendal, ult. f. Gk sindōn]

Sendero Luminoso /senˌdeərəʊ ˌluːmɪˈnəʊsəʊ/ = SHINING PATH. [Sp., = Shining Path]

Seneca[1] /ˈsenɪkə/, Lucius Annaeus (known as Seneca the Younger) (c.4 BC–AD 65), Roman statesman, philosopher, and dramatist. Born in Spain, he was the son of Seneca the Elder. He was banished to Corsica by Claudius in 41, charged with adultery; in 49 his sentence was repealed and he became tutor to Nero, through the influence of Nero's mother and Claudius' wife, Agrippina. Seneca was a dominant figure in the early years of Nero's reign and was appointed consul in 57; he retired in 62. His subsequent implication in a plot on Nero's life led to his forced suicide. As a philosopher, he expounded the ethics of Stoicism in such works as *Epistulae Morales*. Seneca wrote nine plays, whose lurid violence and use of rhetoric later influenced Elizabethan and Jacobean tragedy.

Seneca[2] /'senɪkə/, Marcus (or Lucius) Annaeus (known as Seneca the Elder) (*c.*55 BC–*c.* AD 39), Roman rhetorician, born in Spain. He was the father of Seneca the Younger. Seneca is best known for his works on rhetoric, only parts of which survive, including *Oratorum Sententiae Divisiones Colores* and *Suasoriae*.

Seneca[3] /'senɪkə/ *n. & adj.* ● *n.* (*pl.* same or **Senecas**) **1** a member of an American Indian people that was one of the five comprising the original Iroquois confederation. **2** the Iroquoian language of this people. ● *adj.* of or relating to this people or their language. [Du. f. Algonquian]

senecio /sɪ'niːʃɪˌəʊ/ *n.* a composite plant of the genus *Senecio*, which includes ragwort, groundsel, and many cultivated species. [L *senecio* old man, groundsel, with ref. to the hairy fruits]

Senegal /ˌsenɪ'gɔːl/ a country on the coast of West Africa; pop. (est. 1991) 7,632,000; languages, French (official), Wolof, and other West African languages; capital, Dakar. Part of the Mali empire in the 14th and 15th centuries, the area was colonized by the French and became part of French West Africa in 1895. Briefly a partner in the Federation of Mali (1959), Senegal withdrew and became a fully independent republic in 1960. The Gambia forms an enclave within Senegal. □ **Senegalese** /-gə'liːz/ *adj. & n.*

Senegambia /ˌsenə'gæmbɪə/ a region of West Africa consisting of the Senegal and Gambia rivers and the area between them. It lies mostly in Senegal and western Mali.

senesce /sɪ'nes/ *v.intr.* grow old. □ **senescence** *n.* **senescent** *adj.* [L *senescere* f. *senex* old]

seneschal /'senɪʃ(ə)l/ *n.* **1** *hist.* the steward or major-domo of a medieval great house. **2** a judge in Sark. [ME f. OF f. med.L *seniscalus* f. Gmc, = old servant]

senhor /se'njɔː(r)/ *n.* a title used of or to a Portuguese or Brazilian man. [Port. f. L *senior*: see SENIOR]

senhora /se'njɔːrə/ *n.* a title used of or to a Portuguese woman or a Brazilian married woman. [Port., fem. of SENHOR]

senhorita /ˌsenjə'riːtə/ *n.* a title used of or to a young Brazilian esp. unmarried woman. [Port., dimin. of SENHORA]

senile /'siːnaɪl/ *adj. & n.* ● *adj.* **1** of or characteristic of old age (*senile apathy*; *senile decay*). **2** having the weaknesses or diseases of old age. ● *n.* a senile person. □ **senile dementia** a severe form of mental deterioration in old age, characterized by loss of memory and disorientation, and most often due to Alzheimer's desease. □ **senility** /sɪ'nɪlɪtɪ/ *n.* [F *sénile* or L *senilis* f. *senex* old man]

senior /'siːnɪə(r)/ *adj. & n.* ● *adj.* **1** (often foll. by *to*) more or most advanced in age or standing. **2** of high or highest position. **3** (placed after a person's name) senior to another of the same name. **4** (of a school) having pupils in an older age-range (esp. over 11). **5** *N. Amer.* of the final year at a university, high school, etc. ● *n.* **1** a person of advanced age or comparatively long service etc. **2** one's elder, or one's superior in length of service, membership, etc. (*is my senior*). **3** a senior student. □ **senior citizen** an elderly person, esp. an old-age pensioner. **senior college** *N. Amer.* a college in which the last two years' work for a bachelor's degree is done. **senior common** (or **combination**) **room** *Brit.* a room for use by senior members of a college. **senior nursing officer** the person in charge of nursing services in a hospital. **senior officer** an officer to whom a junior is responsible. **senior partner** the head of a firm. **senior service** *Brit.* the Royal Navy as opposed to the Army. **senior tutor** *Brit.* a college tutor in charge of the teaching arrangements. □ **seniority** /ˌsiːnɪ'ɒrɪtɪ/ *n.* [ME f. L, = older, older man, compar. of *senex senis* old man, old]

seniti /'senɪtɪ/ *n.* a monetary unit of Tonga, equal to one-hundredth of a pa'anga. [Tongan, f. CENT]

Senna /'senə/, Ayrton (1960–94), Brazilian motor-racing driver. He won the Formula One world championship in 1988, 1990, and 1991. He died from injuries sustained in a crash during the Italian Grand Prix in 1994.

senna /'senə/ *n.* **1** a cassia tree. **2** a laxative prepared from the dried pod of this. [med.L *sena* f. Arab. *sanā*]

Sennacherib /sɪ'nækərɪb/ (d.681 BC) king of Assyria 705–681. The son of Sargon II, he devoted much of his reign to suppressing revolts in various parts of his empire, including Babylon, which he sacked in 689. In 701 he put down a Jewish rebellion, laying siege to Jerusalem but sparing it from destruction (according to 2 Kings 19:35) after an epidemic of illness amongst his forces. He rebuilt and extended the city of Nineveh and made it his capital, and also initiated irrigation schemes and other civil engineering projects.

sennet[1] /'senɪt/ *n. hist.* a signal call on a trumpet or cornet (in the stage directions of Elizabethan plays). [perh. var. of SIGNET]

sennet[2] var. of SINNET.

sennight /'senaɪt/ *n. archaic* a week. [OE *seofon nihta* seven nights]

sennit /'senɪt/ *n.* **1** *hist.* plaited straw, palm leaves, etc., used for making hats. **2** = SINNET. [var. of SINNET]

señor /se'njɔː(r)/ *n.* (*pl.* **señores** /-rez/) a title used of or to a Spanish-speaking man. [Sp. f. L *senior*: see SENIOR]

señora /se'njɔːrə/ *n.* a title used of or to a Spanish-speaking married woman. [Sp., fem. of SEÑOR]

señorita /ˌsenjə'riːtə/ *n.* a title used of or to a young Spanish-speaking esp. unmarried woman. [Sp., dimin. of SEÑORA]

Senr. *abbr.* Senior.

sensate /'senseɪt/ *adj.* perceived by the senses. [LL *sensatus* having senses (as SENSE)]

sensation /sen'seɪʃ(ə)n/ *n.* **1** the consciousness of perceiving or seeming to perceive some state or condition of one's body or its parts or of the senses; an instance of such consciousness (*lost all sensation in my left arm*; *the sensation of falling*; *a burning sensation in his leg*). **2** an awareness or impression (*created the sensation of time passing*; *a sensation of being watched*). **3 a** a stirring of emotions or intense interest esp. among a large group of people (*the news caused a sensation*). **b** a person, event, etc., causing such interest. **c** the sensational use of literary etc. material. [med.L *sensatio* f. L *sensus* SENSE]

sensational /sen'seɪʃən(ə)l/ *adj.* **1** causing or intended to cause great public excitement etc. **2** of or causing sensation. □ **sensationalize** *v.tr.* **sensationally** *adv.*

sensationalism /sen'seɪʃənəˌlɪz(ə)m/ *n.* **1** the use of or interest in the sensational in literature, political agitation, etc. **2** *Philos.* the theory that ideas are derived solely from sensation. Cf. EMPIRICISM. □ **sensationalist** *n. & adj.* **sensationalistic** /-ˌseɪʃənə'lɪstɪk/ *adj.*

sense /sens/ *n. & v.* ● *n.* **1 a** any of the special bodily faculties by which sensation is roused (*has keen senses*; *has a dull sense of smell*). **b** sensitiveness of all or any of these. **2** the ability to perceive or feel or to be conscious of the presence or properties of things. **3** (foll. by *of*) consciousness; intuitive awareness (*sense of having done well*; *sense of one's own importance*). **4** (often foll. by *of*) **a** a quick or accurate appreciation, understanding, or instinct regarding a specified matter (*sense of the ridiculous*; *road sense*; *the moral sense*). **b** the habit of basing one's conduct on such instinct. **5** practical wisdom or judgement, common sense; conformity to these (*has plenty of sense*; *what is the sense of talking like that?*; *has more sense than to do that*). **6 a** a meaning; the way in which a word etc. is to be understood (*the sense of the word is clear*; *I mean that in the literal sense*). **b** intelligibility or coherence or possession of a meaning. **7** the prevailing opinion among a number of people (*sum up the sense of the meeting*). **8** (in *pl.*) a person's sanity or normal state of mind (*taken leave of his senses*). **9** *Math.* etc. **a** a direction of movement. **b** that which distinguishes a pair of entities which differ only in that each is the reverse of the other. ● *v.tr.* **1** perceive by a sense or senses. **2** be vaguely aware of. **3** realize. **4** (of a machine etc.) detect. **5** *US* understand. □ **bring a person to his** or **her senses 1** cure a person of folly. **2** restore a person to consciousness. **come to one's senses 1** regain consciousness. **2** become sensible after acting foolishly. **the five senses** sight, hearing, smell, taste, and touch. **in a** (or **one**) **sense** if the statement is understood in a particular way (*what you say is true in a sense*). **in one's senses** sane. **make sense** be intelligible or practicable. **make sense of** show or find the meaning of. **man** (or **woman**) **of sense** a sagacious man (or woman). **out of one's senses** in or into a state of madness (*is out of her senses*; *frightened him out of his senses*). **sense-datum** (*pl.* **-data**) *Philos.* an element of experience received through the senses. **sense-experience** experience derived from the senses. **sense of direction** the ability to know without guidance the direction in which one is or should be moving. **sense of humour** see HUMOUR. **sense-organ** a bodily organ conveying (esp. external) stimuli to the sensory system. **take leave of one's senses** go mad. **under a sense of wrong** feeling wronged. [ME f. L *sensus* faculty of feeling, thought, meaning, f. *sentire sens-* feel]

senseless /'senslɪs/ *adj.* **1** unconscious. **2** wildly foolish. **3** without meaning or purpose. **4** incapable of sensation. □ **senselessly** *adv.* **senselessness** *n.*

sensibility /ˌsensɪ'bɪlɪtɪ/ *n.* (*pl.* **-ies**) **1** capacity to feel (*little finger lost its sensibility*). **2 a** openness to emotional impressions, susceptibility, sensitiveness (*sensibility to kindness*). **b** an exceptional or excessive degree of this (*sense and sensibility*). **3 a** (in *pl.*) emotional capacities or feelings (*was limited in his sensibilities*). **b** (in *sing.* or *pl.*) a person's moral, emotional, or aesthetic ideas or standards (*offended the sensibilities of believers*). [ME f. LL *sensibilitas* (as SENSIBLE)]

sensible /'sensɪb(ə)l/ *adj.* **1** having or showing wisdom or common sense; reasonable, judicious (*a sensible person; a sensible compromise*). **2 a** perceptible by the senses (*sensible phenomena*). **b** great enough to be perceived; appreciable (*a sensible difference*). **3** (of clothing etc.) practical and functional. **4** (foll. by *of*) aware; not unmindful (*was sensible of his peril*). □ **sensible horizon** see HORIZON 1b. □ **sensibleness** *n.* **sensibly** *adv.* [ME f. OF *sensible* or L *sensibilis* (as SENSE)]

sensitive /'sensɪtɪv/ *adj. & n.* ● *adj.* **1** (often foll. by *to*) very open to or acutely affected by external stimuli or mental impressions; having sensibility. **2** (of a person) easily offended or emotionally hurt. **3** (often foll. by *to*) (of an instrument etc.) responsive to or recording slight changes. **4** (often foll. by *to*) **a** (of photographic materials) prepared so as to respond (esp. rapidly) to the action of light. **b** (of any material) readily affected by or responsive to external action. **5** (of a topic etc.) subject to restriction of discussion to prevent embarrassment, ensure security, etc. **6** (of a market) liable to quick changes of price. ● *n.* a person who is sensitive (esp. to supposed occult influences). □ **sensitive plant 1** a plant whose leaves curve downwards and leaflets fold together when touched, esp. mimosa. **2** a sensitive person. □ **sensitively** *adv.* **sensitiveness** *n.* [ME, = sensory, f. OF *sensitif -ive* or med.L *sensitivus*, irreg. f. L *sentire sens-* feel]

sensitivity /ˌsensɪ'tɪvɪtɪ/ *n.* the quality or degree of being sensitive.

sensitize /'sensɪˌtaɪz/ *v.tr.* (also **-ise**) **1** make sensitive. **2** *Photog.* make sensitive to light. **3** make (an organism etc.) abnormally sensitive to a foreign substance. □ **sensitizer** *n.* **sensitization** /ˌsensɪtaɪ'zeɪʃ(ə)n/ *n.*

sensitometer /ˌsensɪ'tɒmɪtə(r)/ *n. Photog.* a device for measuring sensitivity to light.

sensor /'sensə(r)/ *n.* a device which detects or measures a physical property and records, indicates, or otherwise responds to it. [SENSORY, after MOTOR]

sensorium /sen'sɔːrɪəm/ *n.* (*pl.* **sensoria** /-rɪə/ or **sensoriums**) **1** the seat of sensation, the brain, brain and spinal cord, or grey matter of these. **2** *Biol.* the whole sensory apparatus including the nerve-system. □ **sensorial** *adj.* **sensorially** *adv.* [LL f. L *sentire sens-* feel]

sensory /'sensərɪ/ *adj.* of sensation or the senses. □ **sensorily** *adv.* [as SENSORIUM]

sensual /'sensjʊəl, 'senʃʊəl/ *adj.* **1 a** of or depending on the senses only and not on the intellect or spirit; carnal, fleshly (*sensual pleasures*). **b** given to the pursuit of sensual pleasures or the gratification of the appetites; self-indulgent sexually or in regard to food and drink; voluptuous, licentious. **c** indicative of a sensual nature (*sensual lips*). **2** of sense or sensation, sensory. **3** *Philos.* of, according to, or holding the doctrine of, sensationalism. □ **sensualism** *n.* **sensualist** *n.* **sensualize** *v.tr.* (also **-ise**). **sensually** *adv.* [ME f. LL *sensualis* (as SENSE)]

sensuality /ˌsensjʊ'ælɪtɪ, ˌsenʃʊ-/ *n.* gratification of the senses, self-indulgence. [ME f. F *sensualité* f. LL *sensualitas* (as SENSUAL)]

sensum /'sensəm/ *n.* (*pl.* **sensa** /-sə/) *Philos.* a sense-datum. [mod.L, neut. past part. of L *sentire* feel]

sensuous /'sensjʊəs, 'senʃʊ-/ *adj.* **1** of or derived from or affecting the senses, esp. aesthetically rather than sensually; aesthetically pleasing. **2** readily affected by the senses. □ **sensuously** *adv.* **sensuousness** *n.* [L *sensus* sense]

sent[1] /sent/ *n.* a monetary unit of Estonia, equal to one-hundredth of a kroon. [Estonian = cent]

sent[2] past and past part. of SEND.

sente /'sentɪ/ *n.* (*pl.* **lisente** /lɪ'sentɪ/) a monetary unit of Lesotho, equal to one-hundredth of a loti. [Sesotho]

sentence /'sentəns/ *n. & v.* ● *n.* **1 a** a set of words complete in itself as the expression of a thought, containing or implying a subject and predicate, and conveying a statement, question, exclamation, or command. **b** a piece of writing or speech between two full stops or equivalent pauses, often including several grammatical sentences (e.g. *I went; he came*). **2 a** a decision of a lawcourt, esp. the punishment allotted to a person convicted in a criminal trial. **b** the declaration of this. **3** *Logic* a series of signs or symbols expressing a proposition in an artificial or logical language. ● *v.tr.* **1** declare the sentence of (a convicted criminal etc.). **2** (foll. by *to*) declare (such a person) to be condemned to a specified punishment. □ **under sentence of** having been condemned to (*under sentence of death*). [ME f. OF f. L *sententia* be of opinion f. *sentire* be of opinion]

sentential /sen'tenʃ(ə)l/ *adj. Gram. & Logic* of a sentence. [L *sententialis* (as SENTENCE)]

sententious /sen'tenʃəs/ *adj.* **1** (of a person) fond of pompous moralizing. **2** (of a style) affectedly formal. **3** aphoristic, pithy, given to the use of maxims, affecting a concise impressive style. □ **sententiously** *adv.* **sententiousness** *n.* [L *sententiosus* (as SENTENCE)]

sentient /'senʃ(ə)nt/ *adj.* having the power of perception by the senses. □ **sentience** *n.* **sentiency** *n.* **sentiently** *adv.* [L *sentire* feel]

sentiment /'sentɪmənt/ *n.* **1** a mental feeling (*the sentiment of pity*). **2 a** the sum of what one feels on some subject. **b** a verbal expression of this. **3** the expression of a view or desire esp. as formulated for a toast (*concluded his speech with a sentiment*). **4** an opinion as distinguished from the words meant to convey it (*the sentiment is good though the words are injudicious*). **5** a view or tendency based on or coloured with emotion (*animated by noble sentiments*). **6** such views collectively, esp. as an influence (*sentiment unchecked by reason is a bad guide*). **7** the tendency to be swayed by feeling rather than by reason. **8 a** mawkish tenderness. **b** the display of this. **9** an emotional feeling conveyed in literature or art. [ME f. OF *sentement* f. med.L *sentimentum* f. L *sentire* feel]

sentimental /ˌsentɪ'ment(ə)l/ *adj.* **1** of or characterized by sentiment. **2** showing or affected by emotion rather than reason. **3** appealing to sentiment. □ **sentimental value** the value of a thing to a particular person because of its associations. □ **sentimentalism** *n.* **sentimentalist** *n.* **sentimentally** *adv.* **sentimentalize** *v.intr. & tr.* (also **-ise**). **sentimentalization** /-ˌmentəlaɪ'zeɪʃ(ə)n/ *n.* **sentimentality** /-men'tælɪtɪ/ *n.*

sentinel /'sentɪn(ə)l/ *n. & v.* ● *n.* a sentry or lookout. ● *v.tr.* (**sentinelled**, **sentinelling**; US **sentineled**, **sentineling**) **1** station sentinels at or in. **2** *poet.* keep guard over or in. [F *sentinelle* f. It. *sentinella*, of unkn. orig.]

sentry /'sentrɪ/ *n.* (*pl.* **-ies**) a soldier etc. stationed to keep guard. □ **sentry-box** a cabin intended to shelter a standing sentry. **sentry-go** the duty of pacing up and down as a sentry. [perh. f. obs. *centrinel*, var. of SENTINEL]

Senussi /se'nuːsɪ/ *n.* (*pl.* same) a member of a North African Muslim religious fraternity founded in 1837 by Sidi Muhammad ibn Ali es-Senussi (d.1859).

Seoul /səʊl/ the capital of South Korea, situated in the north-west of the country on the Han river; pop. (1990) 10,627,790. It was the capital of the Korean Yi dynasty from the late 14th century until 1910, when Korea was annexed by the Japanese. Extensively developed under Japanese rule, it became the capital of South Korea after the partition of 1945.

sepal /'sep(ə)l, 'siːp-/ *n. Bot.* each of the divisions or leaves of the calyx. [F *sépale*, mod.L *sepalum*, perh. formed as SEPARATE + PETAL]

separable /'sepərəb(ə)l/ *adj.* **1** able to be separated. **2** *Gram.* (of a prefix, or a verb in respect of a prefix) written as a separate word in some collocations. □ **separably** *adv.* **separableness** *n.* **separability** /ˌsepərə'bɪlɪtɪ/ *n.* [F *séparable* or L *separabilis* (as SEPARATE)]

separate *adj., n., & v.* ● *adj.* /'sepərət/ (often foll. by *from*) forming a unit that is or may be regarded as apart or by itself; physically disconnected, distinct, or individual (*living in separate rooms; the two questions are essentially separate*). ● *n.* /'sepərət/ **1** (in *pl.*) separate articles of clothing suitable for wearing together in various combinations. **2** an offprint. ● *v.* /'sepəˌreɪt/ **1** tr. make separate, sever, disunite. **2** tr. prevent union or contact of. **3** *intr.* go different ways, disperse. **4** *intr.* cease to live together as a married couple. **5** *intr.* (foll. by *from*) secede. **6** tr. **a** divide or sort (milk, ore, fruit, light, etc.) into constituent parts or sizes. **b** (often foll. by *out*) extract or remove (an ingredient, waste product, etc.) by such a process for use or rejection. **7** tr. US discharge, dismiss. □ **separately** /-rətlɪ/ *adv.* **separateness** *n.* **separative** *adj.* **separatory** *adj.* [L *separare separat-* (as SE-, *parare* make ready)]

separation /ˌsepə'reɪʃ(ə)n/ *n.* **1** the act or an instance of separating; the state of being separated. **2** (in full **judicial separation** or **legal separation**) an arrangement by which a husband and wife remain married but live apart. **3** any of three or more monochrome reproductions of a coloured picture which can combine to reproduce the full colour of the original. □ **separation order** an order of court for judicial separation. [ME f. OF f. L *separatio -onis* (as SEPARATE)]

separatist /'sepərətɪst/ n. a person who favours separation, esp. for political or ecclesiastical independence (opp. UNIONIST 2). □ **separatism** n.

separator /'sepə͵reɪtə(r)/ n. a machine for separating, e.g. cream from milk.

Sephardi /sɪ'fɑːdɪ/ n. (pl. **Sephardim** /-dɪm/) a Jew of Spanish or Portuguese descent. Sephardim lived in Spain and Portugal from the early Middle Ages until they were expelled in 1492. Many fled to North Africa and later settled in western Europe, the Balkans, and Macedonia. A large number of Sephardim today live in Israel. They retain their own distinctive dialect of Spanish (Ladino), customs, and rituals, preserving Babylonian Jewish traditions rather than the Palestinian ones of the Ashkenazim. All Jews of the Middle East and North Africa are also frequently known as Sephardim. Cf. ASHKENAZI. □ **Sephardic** adj. [mod.Heb., f. sᵉp̄āraḏ, a country mentioned in Obad. 20 and taken to be Spain]

sepia /'siːpɪə/ n. **1** a dark reddish-brown colour. **2 a** a brown pigment prepared from a black fluid secreted by cuttlefish, used in monochrome drawing and in watercolours. **b** a brown tint used in photography. **3** a drawing done in sepia. **4** the fluid secreted by cuttlefish. [L f. Gk sēpia cuttlefish]

sepoy /'siːpɔɪ/ n. hist. a native Indian soldier under European, esp. British, discipline. [Urdu & Pers. sipāhī soldier f. sipāh army]

Sepoy Mutiny = INDIAN MUTINY.

seppuku /se'puːkuː/ n. hara-kiri. [Jap.]

sepsis /'sepsɪs/ n. **1** the state of being septic. **2** blood-poisoning. [mod.L f. Gk sēpsis f. sēpō make rotten]

Sept. abbr. **1** September. **2** Septuagint.

sept /sept/ n. a clan, esp. in Ireland. [prob. alt. of SECT]

sept- var. of SEPTI-.

septa pl. of SEPTUM.

septal[1] /'septəl/ adj. **1** of a septum or septa. **2** Archaeol. (of a stone or slab) separating compartments in a burial chamber. [SEPTUM]

septal[2] /'septəl/ adj. of a sept or septs.

septate /'septeɪt/ adj. Bot., Zool., & Anat. having a septum or septa; partitioned. □ **septation** /sep'teɪʃ(ə)n/ n.

septavalent var. of SEPTIVALENT.

septcentenary /͵septsen'tiːnərɪ/ n. & adj. ● n. (pl. **-ies**) **1** a seven-hundredth anniversary. **2** a festival marking this. ● adj. of or concerning a septcentenary.

September /sep'tembə(r)/ n. the ninth month of the year. [ME f. L September f. septem seven: orig. the seventh month of the Roman year]

septenarius /͵septɪ'neərɪəs/ n. (pl. **septenarii** /-rɪ͵aɪ/) Prosody a verse of seven feet, esp. a trochaic or iambic tetrameter catalectic. [L f. septeni distributive of septem seven]

septenary /sep'tiːnərɪ, 'septɪn-/ adj. & n. ● adj. of seven, by sevens, on the basis of seven. ● n. (pl. **-ies**) a group or set of seven (esp. years). **2** a septenarius. [L septenarius (as SEPTENARIUS)]

septenate /'septɪ͵neɪt/ adj. Bot. **1** growing in sevens. **2** having seven divisions. [L septeni (as SEPTENARIUS)]

septennial /sep'tenɪəl/ adj. **1** lasting for seven years. **2** recurring every seven years. [LL septennis f. L septem seven + annus year]

septennium /sep'tenɪəm/ n. (pl. **septenniums** or **septennia** /-nɪə/) a period of seven years.

septet /sep'tet/ n. (also **septette**) **1** Mus. **a** a composition for seven performers. **b** the performers of such a composition. **2** any group of seven. [G Septett f. L septem seven]

septfoil /'setfɔɪl/ n. **1** a seven-lobed ornamental figure. **2** archaic tormentil. [LL septifolium after CINQUEFOIL, TREFOIL]

septi- /'septɪ/ comb. form (also **sept-** before a vowel) seven. [L f. septem seven]

septic /'septɪk/ adj. contaminated with bacteria from a festering wound etc., putrefying. □ **septic tank** a usu. underground tank in which the organic matter in sewage is decomposed through bacterial activity, making it liquid enough to drain away. □ **septically** adv. **septicity** /sep'tɪsɪtɪ/ n. [L septicus f. Gk sēptikos f. sēpō make rotten]

septicaemia /͵septɪ'siːmɪə/ n. (US **septicemia**) blood-poisoning. □ **septicaemic** adj. [mod.L f. Gk sēptikos + haima blood]

septillion /sep'tɪljən/ n. (pl. same) a thousand raised to the eighth (or formerly, esp. Brit., the fourteenth) power (10^{24} and 10^{42} respectively). [F f. sept seven, after billion etc.]

septimal /'septɪm(ə)l/ adj. of the number seven. [L septimus seventh f. septem seven]

septime /'septiːm/ n. Fencing the seventh of the eight parrying positions. [L septimus (as SEPTIMAL)]

septivalent /͵septɪ'veɪlənt/ adj. (also **septavalent** /-tə'veɪlənt/) Chem. = HEPTAVALENT.

septuagenarian /͵septjʊədʒɪ'neərɪən/ n. & adj. ● n. a person from 70 to 79 years old. ● adj. of this age. [L septuagenarius f. septuageni distributive of septuaginta seventy]

Septuagesima /͵septjʊə'dʒesɪmə/ n. (in full **Septuagesima Sunday**) the Sunday before Sexagesima. [ME f. L, = seventieth (day), formed as SEPTUAGINT, perh. after QUINQUAGESIMA or with ref. to the period of 70 days from Septuagesima to the Saturday after Easter]

Septuagint /'septjʊə͵dʒɪnt/ a Greek version of the Old Testament (Hebrew Bible) including the Apocrypha, made for the use of Jewish communities in Egypt whose native language was Greek. It derives its name from the tradition that it was the work of about seventy translators who worked in separate cells, each translating the whole, and whose versions were found to be identical, thereby showing the work to be divinely inspired. Internal evidence suggests that the work was produced by a number of translators in the 3rd and 2nd centuries BC. The early Christian Church, whose language was Greek, used the Septuagint as its Bible, and it is still the standard version of the Old Testament in the Greek Orthodox Church. [L septuaginta seventy]

septum /'septəm/ n. (pl. **septa** /-tə/) Anat., Bot., & Zool. a partition, such as that between the nostrils or the chambers of a poppy-fruit or of a shell. [L s(a)eptum f. saepire saept- enclose f. saepes hedge]

septuple /'septjʊp(ə)l, sep'tjuː-/ adj., n., & v. ● adj. **1** sevenfold, having seven parts. **2** being seven times as many or as much. ● n. a sevenfold number or amount. ● v.tr. & intr. multiply by seven. [LL septuplus f. L septem seven]

septuplet /'septjʊplɪt, sep'tjuː-/ n. **1** each of seven children born at one birth. **2** Mus. a group of seven notes to be played in the time of four or six. [as SEPTUPLE, after TRIPLET etc.]

sepulchral /sɪ'pʌlkrəl/ adj. **1** of a tomb or interment (sepulchral mound; sepulchral customs). **2** suggestive of the tomb, funereal, gloomy, dismal (sepulchral look). □ **sepulchrally** adv. [F sépulchral or L sepulchralis (as SEPULCHRE)]

sepulchre /'sepəlkə(r)/ n. & v. (US **sepulcher**) ● n. a tomb esp. cut in rock or built of stone or brick, a burial vault or cave. ● v.tr. **1** lay in a sepulchre. **2** serve as a sepulchre for. □ **whited sepulchre** a hypocrite (with ref. to Matt. 23:27). [ME f. OF f. L sepulc(h)rum f. sepelire sepult-bury]

sepulture /'sepəltʃə(r)/ n. literary the act or an instance of burying or putting in the grave. [ME f. OF f. L sepultura (as SEPULCHRE)]

seq. abbr. (pl. **seqq.**) the following. [L sequens etc.]

sequacious /sɪ'kweɪʃəs/ adj. **1** (of reasoning or a reasoner) not inconsequent, coherent. **2** archaic inclined to follow, lacking independence or originality, servile. □ **sequaciously** adv. **sequacity** /-'kwæsɪtɪ/ n. [L sequax f. sequi follow]

sequel /'siːkwəl/ n. **1** what follows (esp. as a result). **2** a novel, film, etc., that continues the story of an earlier one. □ **in the sequel** as things developed afterwards. [ME f. OF sequelle or L sequel(l)a f. sequi follow]

sequela /sɪ'kwiːlə/ n. (pl. **sequelae** /-liː/) (esp. in pl.) Med. a morbid condition or symptom following a disease. [L f. sequi follow]

sequence /'siːkwəns/ n. & v. ● n. **1** succession, coming after or next. **2** order of succession (shall follow the sequence of events; give the facts in historical sequence). **3** a set of things belonging next to one another on some principle of order; a series without gaps. **4** a part of a film dealing with one scene or topic. **5** a set of poems on one theme. **6** a set of three or more playing cards next to one another in value. **7** Mus. repetition of a phrase or melody at a higher or lower pitch. **8** Eccl. a hymn said or sung after the Gradual or Alleluia that precedes the Gospel. **9** succession without implication of causality (opp. CONSEQUENCE 1). ● v.tr. **1** arrange in a definite order. **2** Biochem. ascertain the sequence of monomers in (esp. a polypeptide or nucleic acid). □ **sequence of tenses** Gram. the dependence of the tense of a subordinate verb on the tense of the principal verb, according to certain rules (e.g. I think you are, thought you were, wrong). [ME f. LL sequentia f. L sequens pres. part. of sequi follow]

sequencer /'siːkwənsə(r)/ n. **1** a programmable device for storing sequences of musical notes, chords, etc., and transmitting them when required to an electronic musical instrument. **2** an apparatus for performing or initiating operations in the correct sequence, esp. one forming part of the control system of a computer. **3** *Biochem.* an apparatus for determining the sequence of monomers in a biological polymer.

sequent /'si:kwənt/ *adj.* **1** following as a sequence or consequence. **2** consecutive. □ **sequently** *adv.* [OF *sequent* or L *sequens* (as SEQUENCE)]

sequential /sɪ'kwenʃ(ə)l/ *adj.* **1** forming a sequence or consequence or sequela. **2** occurring or performed in a particular order (*sequential processing*). □ **sequentially** *adv.* **sequentiality** /-,kwenʃɪ'ælɪtɪ/ *n.* [SEQUENCE, after CONSEQUENTIAL]

sequester /sɪ'kwestə(r)/ *v.tr.* **1** (esp. as **sequestered** *adj.*) seclude, isolate, set apart (*sequester oneself from the world; a sequestered life; a sequestered cottage*). **2** = SEQUESTRATE. **3** *Chem.* bind (a metal ion) so that it cannot react. [ME f. OF *sequestrer* or LL *sequestrare* commit for safe keeping f. L *sequester* trustee]

sequestrate /sɪ'kwestreɪt, 'si:kwɪs-/ *v.tr.* **1** confiscate, appropriate. **2** *Law* take temporary possession of (a debtor's estate etc.). **3** *Eccl.* apply (the income of a benefice) to clearing the incumbent's debts or accumulating a fund for the next incumbent. □ **sequestrable** /-trəb(ə)l/ *adj.* **sequestrator** /'si:kwɪ,streɪtə(r)/ *n.* **sequestration** /,si:kwɪ'streɪʃ(ə)n/ *n.* [LL *sequestrare* (as SEQUESTER)]

sequestrum /sɪ'kwestrəm/ *n.* (*pl.* **sequestra** /-trə/) *Med.* a piece of dead bone or other tissue detached from the surrounding parts. □ **sequestral** *adj.* **sequestrotomy** /,si:kwɪ'strɒtəmɪ/ *n.* (*pl.* **-ies**). [mod.L, neut. of L *sequester* standing apart]

sequin /'si:kwɪn/ *n.* **1** a circular spangle for attaching to clothing as an ornament. **2** *hist.* a Venetian gold coin. □ **sequinned** *adj.* (also **sequined**). [F f. It. *zecchino* f. *zecca* a mint f. Arab. *sikka* a die]

sequoia /sɪ'kwɔɪə/ *n.* a very large conifer of the American genus *Sequoia* or *Sequoiadendron*; esp. (in full **giant sequoia**) *Sequoiadendron giganteum*, which is native to the Sierra Nevada in California and is (by volume) the world's largest tree (cf. REDWOOD 1). [mod.L genus name, f. *Sequoiah*, the name of a Cherokee]

Sequoia National Park a national park in the Sierra Nevada of California, east of Fresno. It was established in 1890 to protect groves of giant sequoia trees, of which the largest, the General Sherman Tree, is thought to be between 3,000 and 4,000 years old.

sera *pl.* of SERUM.

serac /se'ræk/ *n.* a pinnacle or ridge of ice between crevasses on a glacier. [Swiss F *sérac*, orig. the name of a compact white cheese]

seraglio /se'rɑːlɪˌəʊ/ *n.* (*pl.* **-os**) **1** a harem. **2** *hist.* a Turkish palace, esp. that of the sultan with government offices etc. at Constantinople. [It. *serraglio* f. Turk. f. Pers. *sarāy* palace: cf. SERAI]

serai /se'raɪ/ *n.* a caravanserai. [Turk. f. Pers. (as SERAGLIO)]

Seraing /sə'ræŋ/ an industrial town in Belgium, on the River Meuse just south-west of Liège; pop. (1991) 60,840.

Seram Sea see CERAM SEA.

serang /sə'ræŋ/ *n.* *Anglo-Ind.* a native head of a Lascar crew. [Pers. & Urdu *sar-hang* commander f. *sar* + *hang* authority]

serape /se'rɑːpeɪ/ *n.* (also **sarape** /sæ'rɑː-/, **zarape** /zæ'rɑː-/) a shawl or blanket worn as a cloak by Spanish Americans. [Mexican Sp.]

seraph /'seræf/ *n.* (*pl.* **seraphim** /-fɪm/ or **seraphs**) an angelic being, one of the highest order of the ninefold celestial hierarchy gifted esp. with love and associated with light, ardour, and purity (see ORDER *n.* 19). [back-form. f. *seraphim* (cf. CHERUB) (pl.) f. LL f. Gk *seraphim* f. Heb. *s*'*rāpīm*]

seraphic /sə'ræfɪk/ *adj.* **1** of or like the seraphim. **2** ecstatically adoring, fervent, or serene. □ **seraphically** *adv.* [med.L *seraphicus* f. LL (as SERAPH)]

Seraphic Doctor the nickname of St Bonaventura.

Serapis /'serəpɪs, sə'reɪp-/ *Egyptian Mythol.* a god whose cult was developed by Ptolemy I at Memphis as a combination of Apis and Osiris, to unite Greeks and Egyptians in a common worship.

seraskier /,serə'skɪə(r)/ *n.* *hist.* the Turkish Commander-in-Chief and Minister of War. [Turk. f. Pers. *sar'askar* head of army]

Serb /sɜːb/ *n.* & *adj.* ● *n.* **1** a native or national of Serbia. **2** a person of Serbian descent. ● *adj.* = SERBIAN *adj.* [Serbian *Srb*]

Serbia /'sɜːbɪə/ a republic in the Balkans; pop. (1986) 9,660,000; official language, Serbo-Croat; capital, Belgrade. An independent kingdom as early as the 6th century, Serbia was conquered by the Turks in the 14th century, but with the decline of Ottoman power the Serbs pressed for independence, which was regained in 1878. Serbian ambitions to found a South Slav nation state brought rivalry with the Austro-Hungarian empire and contributed to the outbreak of the First World War. Despite early successes against the Austrians, Serbia was occupied by the Central Powers and was absorbed into the new state of Yugoslavia after the end of hostilities. With the secession of four out of the six Yugoslav republics in 1991–2, Serbia struggled to retain the viability of Yugoslavia (of which it remains, with Montenegro, a nominal constituent). It was involved in armed conflict with neighbouring Croatia, the civil war in Bosnia, and the suppression of Albanian nationalism in Kosovo.

Serbian /'sɜːbɪən/ *n.* & *adj.* (also *archaic* **Servian** /'sɜːvɪ-/) ● *n.* **1** the language of the Serbs, a form of Serbo-Croat written in the Cyrillic alphabet. (See also SERBO-CROAT.) **2** = SERB *n.* ● *adj.* of or relating to Serbia, the Serbs, or their language.

Serbo-Croat /,sɜːbəʊ'krəʊæt/ *n.* & *adj.* (also **Serbo-Croatian** /-krəʊ'eɪʃ(ə)n/) ● *n.* the Slavonic language of the Serbs and Croats. (*See note below.*) ● *adj.* of or relating to this language.

▪ Serbo-Croat is generally classed as one language, although it may be written in either of two alphabets. The form spoken by Serbs (*Serbian*), who generally belong to the Orthodox Church, is written in the Cyrillic alphabet, whereas that spoken by the mainly Roman Catholic Croats (*Croat* or *Croatian*) is in the Roman alphabet.

SERC *abbr.* *hist.* (in the UK) Science and Engineering Research Council.

sere[1] /sɪə(r)/ *n.* a catch of a gunlock holding the hammer at half or full cock. [prob. f. OF *serre* lock, bolt, grasp, f. *serrer* (see SERRIED)]

sere[2] /sɪə(r)/ *n.* *Ecol.* a natural succession of plant (or animal) communities, esp. a full series from a particular uncolonized habitat to the appropriate climax vegetation (cf. SUCCESSION 3). [L *serere* join in a SERIES]

sere[3] var. of SEAR *adj.*

serein /sə'ræn/ *n.* a fine rain falling in tropical climates from a cloudless sky. [F f. OF *serain* ult. f. L *serum* evening f. *serus* late]

Seremban /sə'rembən/ the capital of the state of Negri Sembilan in Malaysia, situated in the south-west of the Malay Peninsula; pop. (1980) 136,252.

serenade /,serə'neɪd/ *n.* & *v.* ● *n.* **1** a piece of music sung or played at night, esp. by a lover under his lady's window, or suitable for this. **2** = SERENATA. ● *v.tr.* sing or play a serenade to. □ **serenader** *n.* [F *sérénade* f. It. *serenata* f. *sereno* SERENE]

serenata /,serə'nɑːtə/ *n.* *Mus.* **1** a cantata with a pastoral subject. **2** a simple form of suite for orchestra or wind band. [It. (as SERENADE)]

serendipity /,serən'dɪpɪtɪ/ *n.* the faculty of making happy and unexpected discoveries by accident. □ **serendipitous** *adj.* **serendipitously** *adv.* [coined by Horace Walpole (1754) after *The Three Princes of Serendip* (Sri Lanka), a fairy tale]

serene /sɪ'riːn/ *adj.* & *n.* ● *adj.* (**serener**, **serenest**) **1** placid, tranquil, unperturbed. **2 a** (of the sky, the air, etc.) clear and calm. **b** (of the sea etc.) unruffled. ● *n.* *poet.* a serene expanse of sky, sea, etc. □ **all serene** *Brit. sl.* all right. □ **serenely** *adv.* **sereneness** *n.* [L *serenus*]

Serene Highness *n.* a title used in addressing and referring to members of some European royal families (*His Serene Highness; Their Serene Highnesses; Your Serene Highness*).

Serengeti /,serən'getɪ/ a vast plain in Tanzania, to the west of the Great Rift Valley. In 1951 the Serengeti National Park was created to protect the area's large numbers of wildebeest, zebra, and Thomson's gazelle.

serenity /sɪ'renɪtɪ/ *n.* (*pl.* **-ies**) **1** tranquillity, being serene. **2** (**Serenity**) a title used in addressing and referring to a reigning prince or similar dignitary (*Your Serenity*). [F *sérénité* or L *serenitas* (as SERENE)]

serf /sɜːf/ *n.* (under the feudal system) a labourer who was not free to move from the land on which he worked; a villein. Serfs were allowed to farm part of their lord's estate for their own benefit, but had to work the lord's land without pay for a certain number of days, and give the lord a share of their produce. Though not slaves, they were restricted in their movements and in the disposal of their property, and were inferior in status to free tenants. Serfdom in England lasted until the 14th or 15th century, but it was not abolished in France until 1789 and in Russia and parts of eastern Europe not until the 19th century. □ **serfage** *n.* **serfdom** *n.* **serfhood** *n.* [OF f. L *servus* slave]

serge /sɜːdʒ/ *n.* a durable twilled worsted etc. fabric. [ME f. OF *sarge*, *serge* ult. f. L *serica* (*lana*): see SILK]

sergeant /'sɑːdʒənt/ *n.* **1** a non-commissioned army or air force officer ranking next below warrant-officer. **2** a police officer ranking below (*Brit.*) inspector or (*US*) lieutenant. □ **company sergeant-major** *Mil.*

the highest non-commissioned officer of a company. **sergeant-fish** a marine fish with lateral stripes suggesting a chevron; esp. *Rachycentron canadum*, a large tropical and subtropical game-fish. **sergeant-major** *Mil.* **1** (in full **regimental sergeant-major**) *Brit.* a warrant-officer assisting the adjutant of a regiment or battalion. **2** *US* the highest-ranking non-commissioned officer. □ **sergeancy** *n.* (*pl.* **-ies**). **sergeantship** *n.* [ME f. OF *sergent* f. L *serviens -entis* servant f. *servire* SERVE]

Sergeant Baker *n. Austral.* a large brightly coloured marine fish, *Aulopus purpurrissatus.*

Sergipe /sɜː'ʒiːpɪ/ a state in eastern Brazil, on the Atlantic coast; capital, Aracajú.

Sergius, St /'sɜːdʒɪəs/ (Russian name Svyatoi Sergi Radonezhsky) (1314–92), Russian monastic reformer and mystic. He founded the monastery of the Holy Trinity near Moscow, and thereby re-established monasticism, which had been lost in Russia through the Tartar invasion; altogether he founded forty monasteries. His political influence was also considerable: he stopped four civil wars between Russian princes, and inspired the resistance which saved Russia from the Tartars in 1380. Feast day, 25 Sept.

Sergt. *abbr.* Sergeant.

serial /'sɪərɪəl/ *n. & adj.* ● *n.* **1** a story, play, or film which is published, broadcast, or shown in regular instalments. **2** a periodical. ● *adj.* **1** of or in or forming a series. **2** (of a story etc.) in the form of a serial. **3** *Mus.* using transformations of a fixed series of notes (see SERIES 8). **4** (of a publication) appearing in successive parts published usu. at regular intervals; periodical. □ **serial killer** a person who commits a sequence of murders. **serial number** a number showing the position of an item in a series. **serial rights** the right to publish a story or book as a serial. □ **serially** *adv.* **seriality** /ˌsɪərɪ'ælɪtɪ/ *n.* [SERIES + -AL¹]

serialism /'sɪərɪəˌlɪz(ə)m/ *n.* a compositional technique in which a fixed series of notes, esp. the twelve notes of the chromatic scale, are used to generate the harmonic and melodic basis of a piece and are subject to change only in specific ways. The first fully serial movements appeared in 1923 in works by Schoenberg; other pioneers were Berg, who adapted Schoenberg's ideas to accommodate some tonal elements, and Webern, who extended serialism to the duration as well as the pitch of notes. Other important exponents of serial techniques include Boulez, Stockhausen, Berio, and Milton Babbitt. (See also TWELVE-NOTE.) □ **serialist** *n.*

serialize /'sɪərɪəˌlaɪz/ *v.tr.* (also **-ise**) **1** publish or produce in instalments. **2** arrange in a series. **3** *Mus.* compose according to a serial technique. □ **serialization** /ˌsɪərɪəlaɪ'zeɪʃ(ə)n/ *n.*

seriate *adj. & v.* ● *adj.* /'sɪərɪɪt/ in the form of a series; in orderly sequence. ● *v.tr.* /'sɪərɪˌeɪt/ arrange in a seriate manner. □ **seriation** /ˌsɪərɪ'eɪʃ(ə)n/ *n.*

seriatim /ˌsɪərɪ'eɪtɪm, ˌser-/ *adv.* point by point; taking one subject etc. after another in regular order (*consider seriatim*). [med.L f. L *series*, after LITERATIM etc.]

Seric /'sɪərɪk/ *adj.* archaic Chinese. [L *sericus*; see SILK]

sericeous /sɪ'rɪʃəs/ *adj. Bot. & Zool.* covered with silky hairs. [LL *sericeus* silken]

sericulture /'serɪˌkʌltʃə(r)/ *n.* **1** silkworm-breeding. **2** the production of raw silk. □ **sericultural** /ˌserɪ'kʌltʃərəl/ *adj.* **sericulturist** *n.* [F *sériciculture* f. LL *sericum*: see SILK, CULTURE]

seriema /ˌserɪ'iːmə/ *n.* (also **cariama** /ˌkærɪ'ɑːmə/) a South American bird of the family Cariamidae, having a long neck and legs, and a crest above the bill. [mod.L f. Tupi *siriema* etc. crested]

series /'sɪəriːz, -rɪz/ *n.* (*pl.* same) **1** a number of things of which each is similar to the preceding or in which each successive pair are similarly related; a sequence, succession, order, row, or set. **2** a set of successive games between the same teams. **3** a set of programmes with the same actors etc. or on related subjects but each complete in itself. **4** a set of lectures by the same speaker or on the same subject. **5 a** a set of successive issues of a periodical, of articles on one subject or by one writer, etc., esp. when numbered separately from a preceding or following set (*second series*). **b** a set of independent books in a common format or under a common title or supervised by a common general editor. **6** *Philately* a set of stamps, coins, etc., of different denominations but issued at one time, in one reign, etc. **7** *Geol.* **a** a set of strata with a common characteristic. **b** the rocks deposited during a specific epoch. **8** *Mus.* an arrangement of the twelve notes of the chromatic scale as a basis for serial music. **9** *Electr.* **a** a set of circuits or components arranged so that the current passes through each successively. **b** a set of batteries

etc. having the positive electrode of each connected with the negative electrode of the next. **10** *Chem.* a set of elements with common properties or of compounds related in composition or structure. **11** *Math.* a set of quantities constituting a progression or having the several values determined by a common relation. □ **arithmetical** (or **geometrical**) **series** a series in arithmetical (or geometrical) progression. **in series 1** in ordered succession. **2** *Electr.* (of a set of circuits or components) arranged so that the current passes through each successively. [L, = row, chain f. *serere* join, connect]

serif /'serɪf/ *n.* a slight projection finishing off a stroke of a letter as in T contrasted with T (cf. SANSERIF). □ **seriffed** *adj.* [perh. f. Du. *schreef* dash, line f. Gmc]

serigraphy /sə'rɪgrəfɪ/ *n.* the art or process of printing designs by means of a silk screen. □ **serigrapher** *n.* **serigraph** /'serɪˌgrɑːf/ *n.* [irreg. f. L *sericum* SILK]

serin /'serɪn/ *n.* a small yellow Mediterranean finch of the genus *Serinus*, esp. the wild canary, *S. serinus*. [F, of uncert. orig.]

serine /'seriːn/ *n. Biochem.* a hydrophilic amino acid present in proteins. [L *sericum* silk + -INE⁴]

serinette /ˌserɪ'net/ *n.* a small barrel-organ for teaching cage-birds to sing. [F (as SERIN)]

seringa /sə'rɪŋgə/ *n.* **1** = SYRINGA 1. **2** (in Brazil), the rubber tree, *Hevea brasiliensis*. [F (as SYRINGA)]

serio-comic /ˌsɪərɪəʊ'kɒmɪk/ *adj.* combining the serious and the comic, jocular in intention but simulating seriousness or vice versa. □ **serio-comically** *adv.*

serious /'sɪərɪəs/ *adj.* **1** important, demanding consideration (*this is a serious matter*). **2** thoughtful, earnest, sober, sedate, responsible, not reckless or given to trifling (*has a serious air; a serious young person*). **3 a** not slight or negligible (*a serious injury; a serious offence*). **b** *colloq.* large in size or amount (*serious money*). **4** sincere, in earnest, not ironical or joking (*are you serious?*). **5** (of music and literature) not merely for amusement (opp. LIGHT² 5a). **6** not perfunctory (*serious thought*). **7** not to be trifled with (*a serious opponent*). **8** concerned with religion or ethics (*serious subjects*). □ **seriousness** *n.* [ME f. OF *serieux* or LL *seriosus* f. L *serius*]

seriously /'sɪərɪəslɪ/ *adv.* **1** in a serious manner (esp. introducing a sentence, implying that irony etc. is now to cease). **2** to a serious extent. **3** *colloq.* (as an intensifier) very, really, substantially (*seriously rich*).

serjeant /'sɑːdʒənt/ *n.* **1** (in full **serjeant-at-law**, *pl.* **serjeants-at-law**) *hist.* a barrister of the highest rank. **2** *Brit.* (in official lists) a sergeant in the army. □ **serjeant-at-arms** (*pl.* **serjeants-at-arms**) an official of a court or city or parliament, with ceremonial duties. □ **serjeantship** *n.* [var. of SERGEANT]

sermon /'sɜːmən/ *n.* **1** a spoken or written discourse on a religious or moral subject, esp. a discourse based on a text or passage of Scripture and delivered in a service by way of religious instruction or exhortation. **2** a piece of admonition or reproof, a lecture. **3** a moral reflection suggested by natural objects etc. (*sermons in stones*). [ME f. AF *sermun*, OF *sermon* f. L *sermo -onis* discourse, talk]

sermonette /ˌsɜːmə'net/ *n.* a short sermon.

sermonize /'sɜːməˌnaɪz/ *v.* (also **-ise**) **1** *tr.* deliver a moral lecture to. **2** *intr.* deliver a moral lecture. □ **sermonizer** *n.*

Sermon on the Mount the discourse of Christ recorded in Matt. 5–7, an important collection of Christian ethical teachings that contains the Beatitudes and the Lord's Prayer.

serology /sɪə'rɒlədʒɪ/ *n.* the scientific study of blood sera and their effects. □ **serologist** *n.* **serological** /-rə'lɒdʒɪk(ə)l/ *adj.*

seronegative /ˌsɪərəʊ'negətɪv/ *adj. Med.* giving a negative result in a test of blood serum, e.g. for presence of a virus.

seropositive /ˌsɪərəʊ'pɒzɪtɪv/ *adj. Med.* giving a positive result in a test of blood serum, e.g. for presence of a virus.

serosa /sə'rəʊsə/ *n. Anat.* a serous membrane. [mod.L, fem. of med.L *serosus* SEROUS]

serotine /'serətɪn/ *n.* (in full **serotine bat**) a chestnut-coloured European bat, *Eptesicus serotinus*. [F *sérotine* f. L *serotinus* late, of the evening, f. *serus* late]

serotonin /ˌserə'təʊnɪn/ *n. Biol.* a compound present in blood platelets and serum, which constricts the blood vessels and acts as a neurotransmitter. [SERUM + TONIC + -IN]

serous /'sɪərəs/ *adj.* of or like or producing serum; watery. □ **serous gland** (or **membrane**) a gland or membrane with a serous secretion. □ **serosity** /sɪə'rɒsɪtɪ/ *n.* [F *séreux* or med.L *serosus* (as SERUM)]

serow /ˈserəʊ/ n. either of two thick-coated goat-antelopes, *Capricornus sumatrensis* of mountains in South and East Asia, and *C. crispus* of Taiwan and Japan. [Lepcha (Sikkim) *sā-ro*]

Serpens /ˈsɜːpenz/ *Astron.* a large constellation (the Serpent), said to represent the snake coiled around Ophiuchus. It is divided into two parts by Ophiuchus, *Serpens Caput* (the head) and *Serpens Cauda* (the tail), and contains few bright stars. [L]

serpent /ˈsɜːp(ə)nt/ n. **1** usu. *literary* a snake, esp. of a large kind. **2** *Mus.* an old type of wind instrument. (*See note below.*) **3 a** a sly or treacherous person, esp. one who exploits a position of trust to betray it. **b** (**the Serpent**) *Bibl.* Satan (see Gen. 3, Rev. 20). [ME f. OF f. L *serpens -entis* part. of *serpere* creep]

▪ The serpent was a roughly S-shaped instrument made of wood or sometimes metal and giving a powerful deep note. First introduced towards the end of the 16th century in France, where it was used in church, it became a popular military-band instrument and was used in English church bands until the mid-19th century.

serpentine /ˈsɜːpənˌtaɪn/ *adj., n.,* & *v.* ● *adj.* **1** of or like a serpent. **2** coiling, tortuous, sinuous, meandering (*the serpentine windings of the stream*). **3** cunning, subtle, treacherous. ● *n.* **1** a soft rock mainly of hydrated magnesium silicate, usu. dark green and sometimes mottled or spotted like a serpent's skin, taking a high polish and used as a decorative material. **2** *Skating* a figure of three circles in a line. ● *v.intr.* move sinuously, meander. □ **serpentine verse** a metrical line beginning and ending with the same word. [ME f. OF *serpentin* f. LL *serpentinus* (as SERPENT)]

serpiginous /sɜːˈpɪdʒɪnəs/ *adj.* **1** like a serpent; winding, tortuous. **2** *Med.* (of a skin lesion) having a wavy margin.

SERPS /sɜːps/ *abbr.* (in the UK) State Earnings-Related Pension Scheme.

serpula /ˈsɜːpjʊlə/ n. (*pl.* **serpulae** /-ˌliː/) *Zool.* a marine worm of the family Serpulidae, living in an intricately twisted shell-like tube. [LL, = small serpent, f. L *serpere* creep]

serra /ˈserə/ n. (*pl.* **serrae** /-riː/) a serrated organ, structure, or edge. [L, = saw]

serradilla /ˌserəˈdɪlə/ n. (*pl.* **serradillas**) a Mediterranean clover, *Ornithopus sativus*, grown as fodder. [Port., dimin. of *serrado* serrated]

serranid /səˈrænɪd/ n. & *adj. Zool.* ● *n.* a marine fish of the family Serranidae, which comprises heavy-bodied predatory fishes such as the sea-basses and groupers. ● *adj.* of or relating to this family. [mod.L *Serranus* f. L *serra* saw]

serrate v. & *adj.* ● *v.tr.* /seˈreɪt/ (usu. as **serrated** *adj.*) provide with a sawlike edge. ● *adj.* /ˈsereɪt/ esp. *Anat., Bot.,* & *Zool.* notched like a saw. □ **serration** /seˈreɪʃ(ə)n/ n. [LL *serrare serrat-* f. L *serra* saw]

serried /ˈserɪd/ *adj.* (of ranks of soldiers, rows of trees, etc.) pressed together; without gaps; close. [past part. of *serry* press close prob. f. F *serré* past part. of *serrer* close ult. f. L *sera* lock, or past part. of obs. *serr* f. OF *serrer*]

serrulate /ˈserʊˌleɪt, -lɪt/ *adj.* esp. *Anat., Bot.,* & *Zool.* finely serrate; with a series of small notches. □ **serrulation** /ˌserʊˈleɪʃ(ə)n/ n. [mod.L *serrulatus* f. L *serrula* dimin. of *serra* saw]

serum /ˈsɪərəm/ n. (*pl.* **sera** /-rə/ or **serums**) **1 a** an amber-coloured liquid that separates from a clot when blood coagulates. **b** whey. **2** *Med.* the blood serum of an animal used esp. to provide immunity to a pathogen or toxin by inoculation, or as a diagnostic agent. **3** a watery fluid in animal bodies. □ **serum sickness** a reaction to an injection of serum, characterized by skin eruption, fever, etc. [L, = whey]

serval /ˈsɜːv(ə)l/ n. a tawny black-spotted African feline, *Felis serval*, with long legs, large ears, and a short tail. [F f. Port. *cerval* deerlike f. *cervo* deer f. L *cervus*]

servant /ˈsɜːv(ə)nt/ n. **1** a person who has undertaken (usu. in return for stipulated pay) to carry out the orders of an individual or corporate employer, esp. a person employed in a house on domestic duties or as a personal attendant. **2** a devoted follower, a person willing to serve another (*a servant of Jesus Christ*). □ **your humble servant** *Brit. archaic* a formula preceding a signature or expressing ironical courtesy. **your obedient servant** *Brit.* a formula preceding a signature, now used only in certain formal letters. [ME f. OF (as SERVE)]

serve /sɜːv/ v. & *n.* ● *v.* **1** *tr.* do a service for (a person, community, etc.). **2** *tr.* (also *absol.*) be a servant to. **3** *intr.* carry out duties (*served on six committees*). **4** *intr.* **a** (foll. by *in*) be employed in (an organization, esp. the armed forces, or a place, esp. a foreign country) (*served in the air force*). **b** be a member of the armed forces. **5 a** *tr.* be useful to or serviceable for; meet the needs of; do what is required for (*serve a purpose; one packet serves him for a week*). **b** *intr.* meet requirements; perform a function (*a*

sofa serving as a bed; *if memory serves*). **c** *intr.* (foll. by *to* + infin.) avail, suffice (*his attempt served only to postpone the inevitable; it serves to show the folly of such action*). **6** *tr.* **a** go through a due period of (office, apprenticeship, a prison sentence, etc.). **b** go through (a due period) of imprisonment etc. **7** *tr.* set out or present (food) for those about to eat it (*asparagus served with butter; dinner was then served*). **8** *intr.* (in full **serve at table**) act as a waiter. **9** *tr.* **a** attend to (a customer in a shop). **b** (foll. by *with*) supply with (goods) (*was serving a customer with apples; served the town with gas*). **10** *tr.* treat or act towards (a person) in a specified way (*has served me shamefully; you may serve me as you will*). **11** *tr.* **a** (often foll. by *on*) deliver (a writ etc.) to the person concerned in a legally formal manner (*served a warrant on him*). **b** (foll. by *with*) deliver a writ etc. to (a person) in this way (*served her with a summons*). **12** *tr. Tennis* etc. **a** (also *absol.*) deliver (a ball etc.) to begin or resume play. **b** make (a fault etc.) in doing this. **13** *tr. Mil.* keep (a gun, battery, etc.) firing. **14** *tr.* (of an animal, esp. a stallion etc. hired for the purpose) copulate with (a female). **15** *tr.* distribute (*served the ammunition out; served the rations round*). **16** *tr.* render obedience to (a deity etc.). **17** *Eccl.* **a** *intr.* act as a server. **b** *tr.* act as a server at (a service). **18** *intr.* (of a tide) be suitable for a ship to leave harbour etc. **19** *tr. Naut.* bind (a rope etc.) with thin cord to strengthen it. **20** *tr.* play (a trick etc.) on. ● *n.* **1** *Tennis* etc. **a** the act or an instance of serving. **b** a manner of serving. **c** a person's turn to serve. **2** *Austral. sl.* a reprimand. □ **it will serve** it will be adequate. **serve one's needs** (or **need**) be adequate. **serve out** retaliate on. **serve the purpose of** take the place of, be used as. **serve a person right** be a person's deserved punishment or misfortune. **serve one's time 1** hold office for the normal period. **2** (also **serve time**) undergo imprisonment, apprenticeship, etc. **serve one's** (or **the**) **turn** be adequate. **serve up** offer for acceptance. [ME f. OF *servir* f. L *servire* f. *servus* slave]

server /ˈsɜːvə(r)/ n. **1** a person who serves or attends to the requirements of another. **2** (in tennis etc.) the player who serves the ball. **3** *Eccl.* a person assisting the celebrant at a service, esp. the Eucharist. **4** *Computing* **a** a program which manages shared access to a centralized resource or service in a network. **b** a device on which such a program is run. **5** (in *pl.*) a spoon and fork for serving food, esp. salad.

servery /ˈsɜːvərɪ/ n. (*pl.* **-ies**) a room from which meals etc. are served and in which utensils are kept.

Servian[1] /ˈsɜːvɪən/ *adj.* of or relating to Servius Tullius, the semi-legendary sixth king of ancient Rome (6th century BC). □ **Servian Wall** a wall encircling the ancient city of Rome, said to have been built by Servius Tullius.

Servian[2] *archaic* var. of SERBIAN.

service[1] /ˈsɜːvɪs/ n. & *v.* ● *n.* **1** the act of helping or doing work for another or for a community etc. **2** work done in this way. **3** assistance or benefit given to someone. **4** the provision or system of supplying a public need, e.g. transport, or (often in *pl.*) the supply of water, gas, electricity, telephone, etc. **5 a** the fact or status of being a servant. **b** employment or a position as a servant. **6** a state or period of employment doing work for an individual or organization (*resigned after 15 years' service*). **7 a** a public or Crown department or organization employing officials working for the state (*civil service; secret service*). **b** employment in this. **8** (in *pl.*) the armed forces. **9** (*attrib.*) of the kind issued to the armed forces (*a service revolver*). **10 a** a ceremony of worship according to prescribed forms. **b** a form of liturgy for this. **11 a** the provision of what is necessary for the installation and maintenance of a machine etc. or operation. **b** a periodic routine maintenance of a motor vehicle etc. **12** assistance or advice given to customers after the sale of goods. **13 a** the act or process of serving food, drinks, etc. **b** an extra charge nominally made for this. **14 a** set of dishes, plates, etc., used for serving meals (*a dinner service*). **15** *Tennis* etc. **a** the act or an instance of serving. **b** a person's turn to serve. **c** the manner or quality of serving. **d** (in full **service game**) a game in which a particular player serves. **16** (in *pl.*, treated as *pl.* or (*colloq.*) *sing.*) = *service area*. ● *v.tr.* **1** provide service or services for, esp. maintain. **2** maintain or repair (a car, machine, etc.). **3** pay interest on (a debt). **4** supply with a service. **5 a** (of a male animal) copulate with (a female animal). **b** *coarse sl.* (of a man) have sexual intercourse with (a woman). □ **at a person's service** ready to serve or assist a person. **be of service** be available to assist. **in service 1** employed as a servant. **2** available for use. **on active service** serving in the armed forces in wartime. **out of service** not available for use. **see service 1** have experience of service, esp. in the armed forces. **2** (of a thing) be much used. **service area 1** an area beside a major road for the supply of petrol, refreshments, etc. **2** the area served by a broadcasting station. **service-book** a book of authorized forms of worship of a Church.

service box *Squash* either of two small rectangular areas from which a player serves. **service bus** (or **car**) *Austral. & NZ* a motor coach. **service charge** an additional charge for service in a restaurant, hotel, etc. **service dress** ordinary military etc. uniform. **service flat** a flat in which domestic service and sometimes meals are provided by the management. **service industry** one providing services not goods. **service line** (in tennis etc.) a line marking the limit of the area into which the ball must be served. **service road** a road parallel to a main road, providing access to houses, shops, etc. **service station** an establishment beside a road selling petrol and oil etc. to motorists and often able to carry out maintenance. **take service with** become a servant to. [ME f. OF *service* or L *servitium* f. *servus* slave]

service² /'sɜːvɪs/ *n.* **1** (in full **service tree**) a southern European rosaceous tree, *Sorbus domestica*, with toothed leaves, cream-coloured flowers, and small round or pear-shaped fruit eaten when overripe. **2** (in full **wild service tree**) a related small Eurasian tree, *S. torminalis*, with lobed leaves and bitter fruit. □ **service-berry 1** the fruit of the service tree. **2 a** an American rosaceous shrub of the genus *Amelanchier*. **b** the edible fruit of this. [earlier *serves*, pl. of obs. *serve* f. OE *syrfe* f. Gmc *surbhjōn* ult. f. L *sorbus*]

serviceable /'sɜːvɪsəb(ə)l/ *adj.* **1** useful or usable. **2** able to render service. **3** durable; capable of withstanding difficult conditions. **4** suited for ordinary use rather than ornament. □ **serviceably** *adv.* **serviceableness** *n.* **serviceability** /ˌsɜːvɪsə'bɪlɪtɪ/ *n.* [ME f. OF *servisable* (as SERVICE¹)]

serviceman /'sɜːvɪsmən/ *n.* (pl. **-men**) **1** a man serving in the armed forces. **2** a man providing service or maintenance.

servicewoman /'sɜːvɪsˌwʊmən/ *n.* (pl. **-women**) a woman serving in the armed forces.

serviette /ˌsɜːvɪ'et/ *n.* esp. *Brit.* a napkin for use at table. [ME f. OF f. *servir* SERVE]

servile /'sɜːvaɪl/ *adj.* **1** of or being or like a slave or slaves. **2** slavish, fawning; completely dependent. □ **servilely** *adv.* **servility** /sɜː'vɪlɪtɪ/ *n.* [ME f. L *servilis* f. *servus* slave]

serving /'sɜːvɪŋ/ *n.* a quantity of food served to one person.

servitor /'sɜːvɪtə(r)/ *n.* **1** *archaic* **a** a servant. **b** an attendant. **2** *hist.* an Oxford undergraduate performing menial duties in exchange for assistance from college funds. □ **servitorship** *n.* [ME f. OF f. LL (as SERVE)]

servitude /'sɜːvɪˌtjuːd/ *n.* **1** slavery. **2** subjection (esp. involuntary); bondage. **3** *Law* the subjection of property to an easement. [ME f. OF f. L *servitudo -inis* f. *servus* slave]

servo /'sɜːvəʊ/ *n.* (pl. **-os**) **1** (in full **servo-mechanism**) a powered mechanism producing motion or forces at a higher level of energy than the input level, e.g. in the brakes and steering of large motor vehicles, esp. where feedback is employed to make the control automatic. **2** (in full **servo-motor**) the motive element in a servo-mechanism. **3** (in *comb.*) of or involving a servo-mechanism (*servo-assisted*). [L *servus* slave]

sesame /'sesəmɪ/ *n.* **1** a herbaceous African plant, *Sesamum orientale*, widely grown in the tropics, with seeds used as food and yielding an edible oil. **2** its seeds. □ **open sesame** a means of acquiring or achieving what is normally unattainable (from the magic words used by Ali Baba in the *Arabian Nights*). [L *sesamum* f. Gk *sēsamon*, *sēsamē*]

sesamoid /'sesəˌmɔɪd/ *adj. & n.* esp. *Anat.* ● *adj.* shaped like a sesame seed; nodular (esp. of small independent bones developed in tendons passing over an angular structure such as the kneecap and the navicular bone). ● *n.* a sesamoid bone.

Sesotho /se'suːtuː/ *n. & adj.* ● *n.* a Bantu language spoken by members of the Sotho people, one of the official languages of Lesotho. ● *adj.* of or relating to this language. [Bantu, = language of the Sotho]

sesqui- /'seskwɪ/ *comb. form* **1** denoting one and a half. **2** *Chem.* (of a compound) in which there are three equivalents of a named element or radical to two others (*sesquioxide*). [L (as SEMI-, *-que* and)]

sesquicentenary /ˌseskwɪsen'tiːnərɪ/ *n.* (pl. **-ies**) a one-hundred-and-fiftieth anniversary.

sesquicentennial /ˌseskwɪsen'tenɪəl/ *n. & adj.* ● *n.* a sesquicentenary. ● *adj.* of or relating to a sesquicentenary.

sesquipedalian /ˌseskwɪpɪ'deɪlɪən/ *adj.* **1** (of a word etc.) polysyllabic, long. **2** characterized by long words, long-winded. [L *sesquipedalis* = a foot and a half long, f. SESQUI- + *pes pedis* foot]

sess var. of CESS¹.

sessile /'sesaɪl/ *adj.* **1** *Bot. & Zool.* (of a flower, leaf, eye, etc.) attached directly by its base without a stalk or peduncle. **2 a** (of an animal, esp. a barnacle) fixed in one spot. **b** immobile. □ **sessile oak** = DURMAST. [L *sessilis* f. *sedere sess-* sit]

session /'seʃ(ə)n/ *n.* **1** the process of assembly of a deliberative or judicial body to conduct its business. **2** a single meeting for this purpose. **3** a period during which such meetings are regularly held. **4 a** an academic year. **b** the period during which a school etc. has classes. **5 a** a period devoted to an activity (*poker session*; *recording session*). **b** *colloq.* a period of heavy or sustained drinking. **6** the governing body of a Presbyterian Church. □ **in session** assembled for business; not on vacation. **petty sessions 1** a meeting of two or more magistrates for the summary trial of certain offences. **2** = *quarter sessions*. □ **sessional** *adj.* [ME f. OF *session* or L *sessio -onis* (as SESSILE)]

sesterce /'sestɜːs/ *n.* (also **sestertius** /se'stɜːʃəs, -'stɜːtɪəs/) (pl. **sesterces** /'sestəˌsiːz/ or **sestertii** /se'stɜːʃɪɪ, -'stɜːtɪɪ/) an ancient Roman coin and monetary unit equal to one quarter of a denarius. [L *sestertius* (*nummus* coin) = $2\frac{1}{2}$ f. *semis* half + *tertius* third]

sestet /se'stet/ *n.* **1** the last six lines of a sonnet. **2** a sextet. [It. *sestetto* f. *sesto* f. L *sextus* a sixth]

sestina /se'stiːnə/ *n.* a form of rhymed or unrhymed poem with six stanzas of six lines and a final triplet, all stanzas having the same six words at the line-ends in six different sequences. [It. (as SESTET)]

Set var. of SETH.

set¹ /set/ *v.* (**setting**; *past* and *past part.* **set**) **1** *tr.* put, lay, or stand (a thing) in a certain position or location (*set it on the table*; *set it upright*). **2** *tr.* (foll. by *to*) apply (one thing) to (another) (*set pen to paper*). **3** *tr.* fix ready or in position. **b** dispose suitably for use, action, or display. **4** *tr.* **a** adjust the hands of (a clock or watch) to show the right time. **b** adjust (an alarm clock) to sound at the required time. **5** *tr.* **a** fix, arrange, or mount. **b** insert (a jewel) in a ring, framework, etc. **6** *tr.* make (a device) ready to operate. **7** *tr.* lay (a table) for a meal. **8** *tr.* arrange (the hair) while damp so that it dries in the required style. **9** *tr.* (foll. by *with*) ornament or provide (a surface, esp. a precious item) (*gold set with gems*). **10** *tr.* bring by placing or arranging or other means into a specified state; cause to be (*set things in motion*; *set it on fire*). **11** *intr. & tr.* harden or solidify (*the jelly is set*; *the cement has set*). **12** *intr.* (of the sun, moon, etc.) appear to move towards and below the earth's horizon (as the earth rotates). **13** *tr.* represent (a story, play, scene, etc.) as happening in a certain time or place. **14** *tr.* **a** (foll. by *to* + infin.) cause or instruct (a person) to perform a specified activity (*set them to work*). **b** (foll. by pres. part.) start (a person or thing) doing something (*set him chatting*; *set the ball rolling*). **15** *tr.* present or impose as work to be done or a matter to be dealt with (*set them an essay*). **16** *tr.* exhibit as a type or model (*set a good example*). **17** *tr.* initiate; take the lead in (*set the fashion*; *set the pace*). **18** *tr.* establish (a record etc.). **19** *tr.* determine or decide (*the itinerary is set*). **20** *tr.* appoint or establish (*set them in authority*). **21** *tr.* join, attach, or fasten. **22** *tr.* **a** put parts of (a broken or dislocated bone, limb, etc.) into the correct position for healing. **b** deal with (a fracture or dislocation) in this way. **23** *tr.* (in full **set to music**) provide (words etc.) with music for singing. **24** *tr.* (often foll. by *up*) *Printing* **a** arrange or produce (type or film etc.) as required. **b** arrange the type or film etc. for (a book etc.). **25** *intr.* (of a tide, current, etc.) have a certain motion or direction. **26** *intr.* (of a face) assume a hard expression. **27** *tr.* **a** cause (a hen) to sit on eggs. **b** place (eggs) for a hen to sit on. **28** *tr.* put (a seed, plant, etc.) in the ground to grow. **29** *tr.* give the teeth of (a saw) an alternate outward inclination. **30** *tr.* esp. *US* start (a fire). **31** *intr.* (of eyes etc.) become motionless. **32** *intr.* feel or show a certain tendency (*opinion is setting against it*). **33** *intr.* **a** (of blossom) form into fruit. **b** (of fruit) develop from blossom. **c** (of a tree) develop fruit. **34** *intr.* (in full **set to partner**) (of a dancer) take a position facing one's partner. **35** *intr.* (of a hunting dog) take a rigid attitude indicating the presence of game. **36** *intr. dial. or sl.* sit. □ **set about 1** begin or take steps towards. **2** *colloq.* attack. **set against 1** consider or reckon (a person or thing) as a counterpoise or compensation for (another). **2** cause (a person or persons) to oppose (a person or thing). **set apart** separate, reserve, differentiate. **set aside** see ASIDE. **set-aside** something set aside for a special purpose, esp. land taken out of agricultural production to reduce surpluses etc. **set back 1** place further back in place or time. **2** impede or reverse the progress of. **3** *colloq.* cost (a person) a specified amount. **set by** *archaic* save for future use. **set down 1** record in writing. **2** allow to alight from a vehicle. **3** (foll. by *to*) attribute to. **4** (foll. by *as*) explain or describe to oneself as. **set eyes on** see EYE. **set one's face against** see FACE. **set foot on** (or **in**) see FOOT. **set forth 1** begin a journey. **2** make known; expound. **set forward** begin to advance. **set free** release. **set one's hand to** see HAND. **set one's heart** (or **hopes**) **on** want or hope for eagerly. **set in 1** (of weather, a condition, etc.) begin (and seem likely to continue), become established. **2** insert (esp. a sleeve etc. into a garment). **set little by** consider to be of little value. **set a person's mind at rest** see MIND. **set much by** consider to be of much value. **set off 1** begin a journey. **2** detonate (a bomb etc.). **3** initiate, stimulate. **4** cause (a person) to start laughing, talking, etc. **5** serve as an adornment or foil to; enhance. **6** (foll. by *against*) use as a compensating item. **set-off** *n.* **1** a thing set off against another. **2** a thing of which the amount or effect may be deducted from that of another or opposite tendency. **3** a counterpoise. **4** a counter-claim. **5** a thing that embellishes; an adornment to something. **6** *Printing* = OFFSET 7. **set on** (or **upon**) **1** attack violently. **2** cause or urge to attack. **set out 1** begin a journey. **2** (foll. by *to* + infin.) aim or intend. **3** demonstrate, arrange, or exhibit. **4** mark out. **5** declare. **set sail 1** hoist the sails. **2** begin a voyage. **set the scene** see SCENE. **set store by** (or **on**) see STORE. **set one's teeth 1** clench them. **2** summon one's resolve. **set to** begin doing something vigorously, esp. fighting, arguing, or eating. **set-to** *n.* (*pl.* **-tos**) *colloq.* a fight or argument. **set up 1** place in position or view. **2** organize or start (a business etc.). **3** establish in some capacity. **4** supply the needs of. **5** begin making (a loud sound). **6** cause or make arrangements for (a condition or situation). **7** prepare (a task etc. for another). **8** restore or enhance the health of (a person). **9** establish (a record). **10** propound (a theory). **11** *colloq.* cause (a person) to incriminate himself or herself or to look foolish; frame (a person). **set-up** *n.* **1 a** the way in which something is organized or arranged. **b** an organization or arrangement. **2** *colloq.* a trick or conspiracy, esp. to make an innocent person appear guilty. **set oneself up as** make pretensions to being. [OE *settan* f. Gmc]

set² /set/ *n.* **1** a number of things or persons that belong together or resemble one another or are usually found together. **2** a collection or group. **3** a section of society consorting together or having similar interests etc. **4** a collection of implements, vessels, etc., regarded collectively and needed for a specified purpose (*cricket set; teaset; a set of teeth*). **5** a piece of electric or electronic apparatus, esp. a radio or television receiver. **6** (in tennis, darts, etc.) a group of games counting as a unit towards a match for the player or side that wins a defined number or proportion of the games. **7** *Math. & Logic* a collection of distinct entities, individually specified or satisfying specified conditions, forming a unit. (See also SET THEORY.) **8** a group of pupils or students having the same average ability. **9 a** a slip, shoot, bulb, etc., for planting. **b** a young fruit just set. **10 a** a habitual posture or conformation; the way the head etc. is carried or a dress etc. flows. **b** (also **dead set**) a setter's pointing in the presence of game. **11** the way, drift, or tendency (of a current, public opinion, state of mind, etc.) (*the set of public feeling is against it*). **12** the way in which a machine, device, etc., is set or adjusted. **13** esp. *Austral. & NZ colloq.* a grudge. **14 a** the alternate outward deflection of the teeth of a saw. **b** the amount of this. **15** the last coat of plaster on a wall. **16** *Printing* **a** the amount of spacing in type controlling the distance between letters. **b** the width of a piece of type. **17** a warp or bend or displacement caused by continued pressure or a continued position. **18** a setting, including stage furniture etc., for a play or film etc. **19** a sequence of songs or pieces performed in jazz or popular music. **20** the setting of the hair when damp. **21** (also **sett**) a badger's burrow. **22** (also **sett**) a granite paving-block. **23** a predisposition or expectation influencing a response. **24** a number of people making up a square dance. □ **make a dead set at 1** make a determined attack on. **2** seek to win the affections of. **set point** *Tennis* etc. **1** the state of a game when one side needs only one more point to win the set. **2** this point. [sense 1 (and related senses) f. OF *sette* f. L *secta* SECT: other senses f. SET¹]

set³ /set/ *adj.* **1** in senses of SET¹. **2** prescribed or determined in advance. **3** fixed, unchanging, unmoving. **4** (of a phrase or speech etc.) having invariable or predetermined wording; not extempore. **5** prepared for action. **6** (foll. by *on, upon*) determined to acquire or achieve etc. **7** (of a book etc.) specified for reading in preparation for an examination. **8** (of a meal) served according to a fixed menu. □ **set fair** (of the weather) fine without a sign of breaking. **set phrase** an invariable or usual arrangement of words. **set piece 1** a formal or elaborate arrangement, esp. in art or literature. **2** an organized movement in a team game by which the ball is returned to play, as at a free kick or scrum. **3** a group of fireworks arranged on scaffolding etc. **set screw** a screw for adjusting or clamping parts of a machine. **set scrum** *Rugby* a scrum ordered by the referee. **set square** a right-angled triangular plate for drawing lines, esp. at 90°, 45°, 60°, or 30°. [past part. of SET¹]

seta /'siːtə/ *n.* (*pl.* **setae** /-tiː/) *Bot. & Zool.* a stiff hair or bristle. □ **setaceous** /sɪˈteɪʃəs/ *adj.* [L, = bristle]

setback /'setbæk/ *n.* **1** a reversal or arrest of progress. **2** a relapse.

Seth /seθ/ (also **Set** /set/) *Egyptian Mythol.* an evil god who murdered his brother Osiris and wounded Osiris's son Horus. Seth is represented as having the head of a beast with a long pointed snout.

SETI /'setɪ/ *n.* a series of projects looking for evidence of intelligent life elsewhere in the universe. More than fifty such projects have been undertaken since 1960, mostly in America and generally listening for artificial radio transmissions at a few selected wavelengths. A more recent programme, begun by NASA and independently continued with reduced scope, uses a multichannel spectrum analyser to scan 10 million different frequencies simultaneously. Results have so far been negative. [acronym f. Search for Extraterrestrial Intelligence]

setiferous /sɪˈtɪfərəs/ *adj.* (also **setigerous** /-'tɪdʒərəs/) *Biol.* having bristles. [L SETA bristle, *setiger* bristly + -FEROUS, -GEROUS]

seton /'siːt(ə)n/ *n. Surgery* a skein of cotton etc. passed below the skin and left with the ends protruding to promote drainage etc. [ME f. med.L *seto, seta* silk, app. f. L SETA bristle]

setose /'siːtəʊz, -təʊs/ *adj. Biol.* bristly. [L SETA bristle]

Setswana /seˈtswɑːnə/ *n. & adj.* (also **Sechuana** /-ˈtʃwɑːnə/) ● *n.* the Bantu language of the Tswana. ● *adj.* of or relating to this language. [f. Setswana *se-* language prefix + TSWANA]

sett var. of SET² 21, 22.

settee /seˈtiː/ *n.* a seat (usu. upholstered), with a back and usu. arms, for more than one person. [18th c.: perh. a fanciful var. of SETTLE²]

setter /'setə(r)/ *n.* **1** a large long-haired breed of dog trained to stand rigid when scenting game (see SET¹ 35). **2** a person or thing that sets.

set theory *n.* the branch of mathematics which deals with sets (i.e. things grouped together as forming a unit), without regard to the nature of their individual constituents. It was originated by Georg Cantor (and, less explicitly, Richard Dedekind) in the years 1873–1900. Set theory exists both as a subject in its own right, closely related to mathematical logic, and as a language in which other areas of mathematics may be easily and precisely expressed.

setting /'setɪŋ/ *n.* **1** the position or manner in which a thing is set. **2** the immediate surroundings (of a house etc.). **3** the surroundings of any object regarded as its framework; the environment of a thing. **4** the place and time, scenery, etc., of a story, drama, etc. **5** a frame in which a jewel is set. **6** the music to which words of a poem, song, etc., are set. **7** a set of cutlery and other accessories for one person at a table. **8** the way in which or level at which a machine is set to operate. □ **setting lotion** lotion used to prepare the hair for being set.

settle¹ /'set(ə)l/ *v.* **1** *tr. & intr.* (often foll. by *down*) establish or become established in a more or less permanent abode or way of life. **2** *intr. & tr.* (often foll. by *down*) **a** cease or cause to cease from wandering, disturbance, movement, etc. **b** adopt a regular or secure style of life. **c** (foll. by *to*) apply oneself (to work, an activity, a way of life, etc.) (*settled down to writing letters*). **3 a** *intr.* sit or come down to stay for some time. **b** *tr.* cause to do this. **4** *tr. & intr.* bring to or attain fixity, certainty, composure, or quietness. **5** *tr.* determine or decide or agree upon (*shall we settle a date?*). **6** *tr.* **a** resolve (a dispute etc.). **b** deal with (a matter) finally. **7** *tr.* terminate (a lawsuit) by mutual agreement. **8** *intr.* **a** (foll. by *for*) accept or agree to (esp. an alternative not one's first choice). **b** (foll. by *on*) decide on. **9** *tr.* (also *absol.*) pay (a debt, an account, etc.). **10** *intr.* (as **settled** *adj.*) not likely to change for a time (*settled weather*). **11** *tr.* **a** aid the digestion of (food). **b** remedy the disordered state of (nerves, the stomach, etc.). **12** *tr.* colonize. **b** establish colonists in. **13** *intr.* subside; fall to the bottom or on to a surface (*the foundations have settled; wait till the sediment settles; the dust will settle*). **14** *intr.* (of a ship) begin to sink. **15** *tr.* get rid of the obstruction of (a person) by argument or conflict or killing. □ **settle one's affairs** make any necessary arrangements (e.g. write a will) when death is near. **settle a person's hash** see HASH¹. **settle in** become established in a place. **settle up 1** (also *absol.*) pay (an account, debt, etc.). **2** finally arrange (a matter). **settle with 1** pay all or part of an amount due to (a creditor). **2** get revenge on. **settling day** the fortnightly pay day on the Stock Exchange. □ **settleable** *adj.* [OE *setlan* (as SETTLE²) f. Gmc]

settle² /'set(ə)l/ *n.* a bench with a high back and arms and often with a box fitted below the seat. [OE *setl* place to sit f. Gmc]

settlement /'set(ə)lmənt/ *n.* **1** the act or an instance of settling; the process of being settled. **2 a** the colonization of a region. **b** a place or area occupied by settlers. **c** a small village. **3 a** a political or financial etc. agreement. **b** an arrangement ending a dispute. **4 a** the terms on which property is given to a person. **b** a deed stating these. **c** the amount of property given. **d** = *marriage settlement.* **5** the process of settling an account. **6** subsidence of a wall, house, soil, etc.

Settlement, Act of a statute of 1701 that established the Hanoverian succession to the British throne. It vested the Crown in Sophia of Hanover (granddaughter of James I of England) and her heirs; her son became George I.

settler /ˈsetlə(r)/ *n.* a person who goes to settle in a new country or place; an early colonist.

settlor /ˈsetlə(r)/ *n. Law* a person who makes a settlement esp. of a property.

Setúbal /səˈtuːb(ə)l/ a port and industrial town on the coast of Portugal, south of Lisbon; pop. (1991) 83,550.

Seurat /ˈsʊərʌ/, Georges Pierre (1859–91), French painter. The founder of neo-impressionism, he is chiefly associated with pointillism, which he developed during the 1880s. Among his major paintings using this technique is *Sunday Afternoon on the Island of La Grande Jatte* (1884–6).

Sevastopol see SEBASTOPOL.

seven /ˈsev(ə)n/ *n. & adj.* ● *n.* **1** one more than six, or three less than ten; the sum of four units and three units. **2** a symbol for this (7, vii, VII). **3** a size etc. denoted by seven. **4** a set or team of seven individuals. **5** the time of seven o'clock. **6** a card with seven pips. ● *adj.* that amount to seven. □ **seven-league boots** (in the fairy story of Hop-o'-my-Thumb) boots enabling the wearer to go seven leagues at each stride. **seven year itch** a supposed tendency to infidelity after seven years of marriage. [OE *seofon* f. Gmc]

seven deadly sins in Christian tradition, pride, covetousness, lust, envy, gluttony, anger, and sloth, regarded as the basic human vices. They are listed (with minor variation) by the monk John Cassian (d.435), St Gregory the Great, and St Thomas Aquinas.

sevenfold /ˈsev(ə)nˌfəʊld/ *adj. & adv.* **1** seven times as much or as many. **2** consisting of seven parts.

Seven Sages seven wise Greeks of the 6th century BC, to each of whom a moral saying is attributed. The seven, named in a traditional list found in Plato, are Bias, Chilon, Cleobulus, Periander, Pittacus, Solon, and Thales.

seven seas the oceans of the world: the Arctic, Antarctic, North Pacific, South Pacific, North Atlantic, South Atlantic, and Indian Oceans.

Seven Sisters the Pleiades.

Seven Sleepers in early Christian legend, seven noble Christian youths of Ephesus who fell asleep in a cave while fleeing from the Decian persecution and awoke 187 years later. The legend was translated from the Syriac by Gregory of Tours (6th century) and is also given by other authors; it occurs in the Koran.

seventeen /ˌsev(ə)nˈtiːn/ *n. & adj.* ● *n.* **1** one more than sixteen, or seven more than ten. **2** a symbol for this (17, xvii, XVII). **3** a size etc. denoted by seventeen. ● *adj.* that amount to seventeen. □ **seventeenth** *adj. & n.* [OE *seofontīene*]

seventh /ˈsev(ə)nθ/ *n. & adj.* ● *n.* **1** the position in a sequence corresponding to the number 7 in the sequence 1–7. **2** something occupying this position. **3** one of seven equal parts of a thing. **4** *Mus.* **a** an interval or chord spanning seven consecutive notes in the diatonic scale (e.g. C to B). **b** a note separated from another by this interval. ● *adj.* that is the seventh. □ **seventh heaven 1** the highest of seven heavens in Muslim and some Jewish systems. **2** a state of intense joy. □ **seventhly** *adv.*

Seventh-Day Adventist /ˈsev(ə)nθˌdeɪ/ *n.* a member of a sect of Adventists who originally expected the second coming of Christ in 1844 and still preach that his return is imminent. They are strict Protestants and are notable for observing Saturday as the sabbath. (See ADVENTIST.)

seventy /ˈsev(ə)ntɪ/ *n. & adj.* ● *n. (pl.* **-ies**) **1** the product of seven and ten. **2** a symbol for this (70, lxx, LXX). **3** *(in pl.)* the numbers from 70 to 79, esp. the years of a century or of a person's life. ● *adj.* that amount to seventy. □ **seventy-first, -second**, etc. the ordinal numbers between seventieth and eightieth. **seventy-one, -two**, etc. the cardinal numbers between seventy and eighty. □ **seventieth** *adj. & n.* **seventyfold** *adj. & adv.* [OE *-seofontig*]

Seven Wonders of the World the seven most spectacular man-made structures of the ancient world. Traditionally they comprise (1) the pyramids of Egypt, especially those at Giza; (2) the Hanging Gardens of Babylon; (3) the Mausoleum of Halicarnassus (see MAUSOLEUM); (4) the temple of Artemis at Ephesus in Asia Minor, rebuilt in 356 BC; (5) the Colossus of Rhodes; (6) the huge ivory and gold statue

of Zeus at Olympia in the Peloponnese, made by Phidias c.430 BC; (7) the Pharos of Alexandria (or in some lists, the walls of Babylon). The earliest extant list of these dates from the 2nd century BC.

Seven Years War a war (1756–63) which ranged Britain, Prussia, and Hanover against Austria, France, Russia, Saxony, Sweden, and Spain. Its main issues were the struggle between Britain and France for supremacy overseas, and that between Prussia and Austria for the domination of Germany. After some early setbacks, the British made substantial gains over France abroad, capturing (under James Wolfe) French Canada and (under Robert Clive) undermining French influence in India. On the Continent, the war was most notable for the brilliant campaigns of Frederick the Great of Prussia against converging enemy armies. The war was ended by the Treaties of Paris and Hubertusburg in 1763, leaving Britain the supreme European naval and colonial power and Prussia in an appreciably stronger position than before in central Europe.

sever /ˈsevə(r)/ *v.* **1** *tr. & intr.* (often foll. by *from*) divide, break, or make separate, esp. by cutting. **2** *tr. & intr.* break off or away; separate, part, divide (*severed our friendship*). **3** *tr.* end the employment contract of (a person). □ **severable** *adj.* [ME f. AF *severer*, OF *sevrer* ult. f. L *separare* SEPARATE v.]

several /ˈsevr(ə)l/ *adj. & n.* ● *n.* more than two but not many. ● *adj.* **1** separate or respective; distinct (*all went their several ways*). **2** *Law* applied or regarded separately (opp. JOINT *adj.* 1). □ **joint and several** see JOINT. □ **severally** *adv.* [ME f. AF f. AL *separalis* f. L *separ* SEPARATE *adj.*]

severalty /ˈsevr(ə)ltɪ/ *n.* **1** separateness. **2** the individual or unshared tenure of an estate etc. (esp. *in severalty*). [ME f. AF *severalte* (as SEVERAL)]

severance /ˈsevərəns/ *n.* **1** the act or an instance of severing. **2** a severed state. □ **severance pay** an amount paid to an employee on the early termination of a contract.

severe /sɪˈvɪə(r)/ *adj.* **1** rigorous, strict, and harsh in attitude or treatment (*a severe critic; severe discipline*). **2** serious, critical (*a severe shortage*). **3** vehement or forceful (*a severe storm*). **4** extreme (in an unpleasant quality) (*a severe winter; severe cold*). **5** arduous or exacting; making great demands on energy, skill, etc. (*severe competition*). **6** unadorned; plain in style (*severe dress*). □ **severely** *adv.* **severity** /-ˈverɪtɪ/ *n.* [F *sévère* or L *severus*]

Severn /ˈsev(ə)n/ a river of SW Britain. Rising in central Wales, it flows north-east then south in a broad curve for some 290 km (180 miles) to its wide mouth on the Bristol Channel. The estuary is spanned by a suspension bridge north of Bristol, opened in 1966, and a second is under construction a few miles to the south.

Severnaya Zemlya /ˌseveəˌnaɪə zɪmˈljɑ/ a group of uninhabited islands in the Arctic Ocean off the north coast of Russia, to the north of the Taimyr Peninsula.

Severodvinsk /ˌsevərəˈdvɪnsk/ a port in NW Russia, on the White Sea coast west of Archangel; pop. (1990) 250,000.

Severus /sɪˈvɪərəs/, Septimius (full name Lucius Septimius Severus Pertinax) (146–211), Roman emperor 193–211. He was active in reforms of the imperial administration and of the army, which he recognized as the real basis of imperial power. In 208 he led an army to Britain to suppress a rebellion in the north of the country, and later died at York.

severy /ˈsevərɪ/ *n. (pl.* **-ies**) *Archit.* a space or compartment in a vaulted ceiling. [ME f. OF *civoire* (as CIBORIUM)]

seviche /seˈviːtʃeɪ/ *n.* a South American dish of raw fish or seafood marinaded in lime or lemon juice. [S. Amer. Sp.]

Seville /səˈvɪl/ (Spanish **Sevilla** /seˈβiʎa/) a city in southern Spain, the capital of Andalusia, situated on the Guadalquivir river; pop. (1991) 683,490. It was taken by the Moors in 711 and became a leading cultural centre of Moorish Spain. Reclaimed in 1248 by the Spanish under Ferdinand III, it rapidly grew to prominence as a centre of trade with the colonies of the New World.

Seville orange /ˈsevɪl, səˈvɪl/ *n.* a bitter orange used for marmalade.

Sèvres /ˈseɪvrə/ *n.* a kind of fine porcelain made at a factory in Sèvres, now a suburb of Paris, from the mid-18th century and characterized by elaborate decoration on backgrounds of intense colour. The factory benefited from patronage by Louis XV and, in particular, by his mistress Madame de Pompadour.

sew /səʊ/ *v.tr. (past part.* **sewn** /səʊn/ or **sewed**) **1** (also *absol.*) fasten, join, etc., by making stitches with a needle and thread or a sewing machine. **2** make (a garment etc.) by sewing. **3** (often foll. by *on, in,* etc.) attach by sewing (*shall I sew on your buttons?*). □ **sew up 1** join or enclose by sewing. **2** (esp. in *passive*) *colloq.* bring to a desired conclusion or condition; complete satisfactorily; ensure the favourable outcome of (a thing). □ **sewer** n. [OE si(o)wan]

sewage /'suːɪdʒ, 'sjuː-/ n. waste matter, esp. excrement, conveyed in sewers. □ **sewage farm** (or **works**) a place where sewage is treated, esp. to produce manure.

sewen var. of SEWIN.

sewer /'suːə(r), 'sjuː-/ n. a conduit, usu. underground, for carrying off drainage water and sewage. (*See note below*.) □ **sewer rat** the common brown rat, *Rattus norvegicus*. [ME f. AF sever(e), ONF se(u)wiere channel to carry off the overflow from a fishpond, ult. f. L ex- out of + aqua water]

▪ Sewers were known in the ancient world: the earliest that survives is at Mohenjo-Daro in the Indus valley (c.2500 BC), and there were elaborate domestic arrangements at the Palace of Minos in Crete (c.2000 BC). Many Roman cities had elaborate underground sewers, but after the decline of Roman power these were allowed to deteriorate. In medieval Europe there was no proper drainage system and the water-supply was inadequate and insanitary, with the consequence that disease was rife. When the Industrial Revolution swelled the numbers of people living in towns it became essential to devise a system of waste-disposal; the first underground sewers were constructed at Hamburg in Germany in 1843.

sewerage /'suːərɪdʒ, 'sjuː-/ n. a system of or drainage by sewers.

sewin /'sjuːɪn/ n. (also **sewen**) in Wales, a sea trout. [16th c.: orig. unkn.]

sewing /'səʊɪŋ/ n. material or work to be sewn.

sewing machine n. a machine for sewing or stitching. The earliest patents were taken out in England by Charles Weisenthal in 1755 and Thomas Saint in 1790, and the first chain-stitch sewing machine was invented in 1829 by Barthélemy Thimonnier, a French tailor. In 1846 the American inventor Elias Howe patented a machine producing a lock stitch; this was developed by Isaac Singer into the first commercially successful machine. It used a needle carrying a loop of thread through the fabric from above and interlocking with a second thread running beneath the fabric on a shuttle. Modern machines, while electrically powered and offering a range of stitches and automatic functions, still operate in the same basic way.

sewn past part. of SEW.

sex /seks/ n., adj., & v. ● n. **1** either of the main divisions (male and female) into which living things are placed on the basis of their reproductive functions. **2** the fact of belonging to one of these divisions. **3** males or females collectively. **4** sexual instincts, desires, etc., or their manifestation. **5** colloq. sexual intercourse (*have sex*). ● adj. **1** of or relating to sex (*sex education*). **2** arising from a difference or consciousness of sex (*sex antagonism; sex urge*). ● v.tr. **1** determine the sex of. **2** (as **sexed** adj.) **a** having a sexual appetite (*highly sexed*). **b** having sexual characteristics. □ **sex act** (usu. prec. by the) the (or an) act of sexual intercourse. **sex appeal** sexual attractiveness. **sex change** an apparent change of sex by surgical means and hormone treatment. **sex chromosome** a chromosome concerned in determining the sex of an organism, which in most animals are of two kinds, esp. (in mammals) the X and Y chromosomes. **sex hormone** a hormone affecting sexual development or behaviour. **sex kitten** colloq. a young woman who asserts her sex appeal. **sex life** a person's activity related to sexual instincts. **sex-linked** carried on or by a sex chromosome. **sex maniac** colloq. a person needing or seeking excessive gratification of the sexual instincts. **sex object** a person regarded mainly in terms of sexual attractiveness. **sex-starved** lacking sexual gratification. **sex symbol** a person widely noted for sex appeal. □ **sexer** n. [ME f. OF sexe or L sexus]

sexagenarian /ˌseksədʒɪˈneərɪən/ n. & adj. ● n. a person from 60 to 69 years old. ● adj. of this age. [L sexagenarius f. sexageni distributive of sexaginta sixty]

Sexagesima /ˌseksəˈdʒesɪmə/ n. the Sunday before Quinquagesima. [ME f. eccl.L, = sixtieth (day), prob. named loosely as preceding QUINQUAGESIMA]

sexagesimal /ˌseksəˈdʒesɪm(ə)l/ adj. & n. ● adj. **1** of sixtieths. **2** of sixty. **3** reckoning or reckoned by sixtieths. ● n. (in full **sexagesimal fraction**) a fraction with a denominator equal to a power of 60 as in the divisions of the degree and hour. □ **sexagesimally** adv. [L sexagesimus (as SEXAGESIMA)]

sexavalent var. of SEXIVALENT.

sexcentenary /ˌseksenˈtiːnərɪ/ n. & adj. ● n. (pl. **-ies**) **1** a six-hundredth anniversary. **2** a celebration of this. ● adj. **1** of or relating to a sexcentenary. **2** occurring every six hundred years.

sexennial /sekˈsenɪəl/ adj. **1** lasting six years. **2** recurring every six years. [SEXI- + L annus year]

sexfoil /'seksfɔɪl/ n. a six-lobed ornamental figure. [SEXI-, after CINQUEFOIL, TREFOIL]

sexi- /'seksɪ/ comb. form (also **sex-** before a vowel) six. [L sex six]

sexism /'seksɪz(ə)m/ n. prejudice or discrimination, esp. against women, on the grounds of sex. □ **sexist** adj. & n.

sexivalent /ˌseksɪˈveɪlənt/ adj. (also **sexavalent** /-səˈveɪlənt/) Chem. = HEXAVALENT.

sexless /'sekslɪs/ adj. **1** Biol. neither male nor female. **2** lacking in sexual desire or attractiveness. □ **sexlessly** adv. **sexlessness** n.

sexology /sekˈsɒlədʒɪ/ n. the study of sexual life or relationships, esp. in human beings. □ **sexologist** n. **sexological** /-səˈlɒdʒɪk(ə)l/ adj.

sexpartite /seksˈpɑːtaɪt/ adj. divided into six parts.

sexploitation /ˌseksplɔɪˈteɪʃ(ə)n/ n. colloq. the exploitation of sex, esp. commercially.

sexpot /'sekspɒt/ n. colloq. a sexy person (esp. a woman).

sext /sekst/ n. Eccl. **1** the canonical hour of prayer appointed for the sixth daytime hour (i.e. noon). **2** the office of sext. [ME f. L sexta hora sixth hour f. sextus sixth]

sextant /'sekstənt/ n. an instrument with a graduated arc of 60⁰ used in navigation and surveying for measuring the angular distance of objects by means of mirrors. [L sextans -ntis sixth part f. sextus sixth]

sextet /sekˈstet/ n. (also **sextette**) **1** Mus. a composition for six voices or instruments. **2** the performers of such a piece. **3** any group of six. [alt. of SESTET after L sex six]

sextillion /sekˈstɪljən/ n. (pl. same or **sextillions**) a thousand raised to the seventh (or formerly, esp. Brit., the twelfth) power (10^{21} and 10^{36} respectively) (cf. BILLION). □ **sextillionth** adj. & n. [F f. L sex six, after septillion etc.]

sextodecimo /ˌsekstəʊˈdesɪməʊ/ n. (pl. **-os**) **1** a size of book or page in which each leaf is one-sixteenth that of a printing-sheet. **2** a book or sheet of this size. [L sextus decimus 16th (as QUARTO)]

sexton /'sekstən/ n. a person who looks after a church and churchyard, often acting as bell-ringer and gravedigger. □ **sexton beetle** a black or red and black beetle of the genus *Nicrophorus*, burying carrion to serve as a nidus for its eggs (also called *burying-beetle*). [ME segerstane etc., f. AF, OF segerstein, secrestein f. med.L sacristanus SACRISTAN]

sextuple /'sekstjʊp(ə)l, seksˈtjuː-/ adj., n., & v. ● adj. **1** sixfold. **2** having six parts. **3** being six times as many or much. ● n. a sixfold number or amount. ● v.tr. & intr. multiply by six; increase sixfold. □ **sextuply** /-plɪ/ adv. [med.L sextuplus, irreg. f. L sex six, after LL quintuplus QUINTUPLE]

sextuplet /'sekstjʊplɪt, seksˈtjuː-/ n. **1** each of six children born at one birth. **2** Mus. a group of six notes to be played in the time of four. [SEXTUPLE, after triplet etc.]

sexual /'seksjʊəl, 'sekʃʊəl/ adj. **1** of or relating to sex, or to the sexes or the relations between them. **2** Bot. (of classification) based on the distinction of sexes in plants. **3** Biol. having a sex. □ **sexual intercourse** physical contact between individuals involving sexual stimulation, esp. the insertion of a man's erect penis into a woman's vagina, followed by rhythmic movement and usu. ejaculation of semen; copulation. □ **sexually** adv. **sexuality** /ˌseksjʊˈælɪtɪ, ˌsekʃʊ-/ n. [LL sexualis (as SEX)]

sexy /'seksɪ/ adj. (**sexier**, **sexiest**) **1** sexually attractive or stimulating. **2** sexually aroused. **3** concerned with sex. **4** colloq. (of a project etc.) exciting, appealing, trendy. □ **sexily** adv. **sexiness** n.

Seychelles, the /seɪˈʃelz, -ˈʃel/ (also **Seychelles**) a country consisting of a group of about ninety islands in the Indian Ocean, about 1,000 km (600 miles) NE of Madagascar; pop. (est. 1991) 69,000; languages, French Creole (official), English, French; capital, Victoria. Although known to Arab traders and to the Portuguese, the islands were uninhabited until the mid-18th century, when the French annexed them and introduced settlers and slaves. The Seychelles were captured by Britain during the Napoleonic Wars and administered from Mauritius before becoming a separate colony in 1903 and an independent republic within the Commonwealth in 1976. The islands are noted for their beauty and attract a considerable tourist trade. □ **Seychellois** /ˌseɪʃelˈwɑ/ adj. & n. (pl. same).

Seymour[1] /'siːmɔː(r)/, Jane (c.1509–37), third wife of Henry VIII and mother of Edward VI. She married Henry in 1536 and finally provided the king with the male heir he wanted, although she died twelve days afterwards.

Seymour[2] /'siːmɔː(r)/, Lynn (b.1939), Canadian ballet-dancer. From

1957 she danced for the Royal Ballet in London, performing principal roles in ballets choreographed by Kenneth MacMillan (1929–92) and Frederick Ashton. Her most acclaimed roles came in Ashton's *Five Brahms Waltzes in the Manner of Isadora Duncan* and *A Month in the Country* (both 1976). Seymour later worked as artistic director of the Bavarian State Opera in Munich (1978–80).

sez /sez/ *sl.* says (*sez you*). [phonetic repr.]

SF see SCIENCE FICTION.

sf *abbr. Mus.* sforzando.

SFA *abbr.* Scottish Football Association.

Sfax /sfæks/ (also **Safaqis** /sə'fɑːkɪs/) a port on the east coast of Tunisia; pop. (1984) 231,900. It is a centre for the region's phosphate industry.

sforzando /sfɔːˈtsændəʊ/ *adj., adv., & n.* (also **sforzato** /-ˈtsɑːtəʊ/) *Mus.* ● *adj. & adv.* with sudden emphasis. ● *n.* (*pl.* **-os** or **sforzandi** /-dɪ/) **1** a note or group of notes especially emphasized. **2** an increase in emphasis and loudness. [It., verbal noun and past part. of *sforzare* use force]

sfumato /sfuːˈmɑːtəʊ/ *n. & adj. Art* ● *n.* the technique of allowing tones and colours to shade gradually into one another. ● *adj.* with indistinct outlines. [It., past part. of *sfumare* shade off f. s- = EX-¹ + *fumare* smoke]

sfz *abbr. Mus.* sforzando.

SG *abbr.* **1** *US* senior grade. **2** *Law* Solicitor-General. **3** specific gravity.

sgd. *abbr.* signed.

sgraffito /sgrɑːˈfiːtəʊ/ *n.* (*pl.* **sgraffiti** /-tɪ/) a form of decoration made by scratching through wet plaster on a wall or through slip on ceramic ware, showing a different-coloured undersurface. [It., past part. of *sgraffire* scratch f. s- = EX-¹ + *graffio* scratch]

's-Gravenhage /ˌsxrɑːvənˈhɑːxə/ a Dutch name for The Hague (see HAGUE, THE).

Sgt. *abbr.* Sergeant.

sh *int.* calling for silence. [var. of HUSH]

sh. *abbr. Brit. hist.* shilling(s).

Shaanxi /ʃɑːnˈʃiː/ (also **Shensi** /ʃenˈsiː/) a mountainous province of central China; capital, Xian. It is the site of the earliest settlements of the ancient Chinese civilizations.

Shaba /ˈʃɑːbə/ a copper-mining region of SE Zaire; capital, Lubumbashi. It was known as Katanga until 1972.

Shabaka /ˈʃæbəkə/ (known as Sabacon) (d.698 BC), Egyptian pharaoh of the 25th dynasty, reigned 712–698 BC. He succeeded his brother Piankhi as king of Cush in about 716; four years later he conquered Egypt and founded its 25th dynasty. A conservative ruler, he promoted the cult of Amun and revived the custom of pyramid burial in his own death arrangements.

shabby /ˈʃæbɪ/ *adj.* (**shabbier, shabbiest**) **1** in bad repair or condition; faded and worn, dingy, dilapidated. **2** dressed in old or worn clothes. **3** of poor quality. **4** contemptible, dishonourable (*a shabby trick*). □ **shabbily** *adv.* **shabbiness** *n.* **shabbyish** *adj.* [*shab* scab f. OE *sceabb* f. ON, rel. to SCAB]

shabrack /ˈʃæbræk/ *n. hist.* a cavalry saddle-cloth. [G *Schabracke* of eastern European orig.: cf. Russ. *shabrak*]

shack /ʃæk/ *n. & v.* ● *n.* a roughly built hut or cabin. ● *v.intr.* (foll. by *up*) *sl.* cohabit, esp. as lovers. [perh. f. Mex. *jacal*, Aztec *xacatli* wooden hut]

shackle /ˈʃæk(ə)l/ *n. & v.* ● *n.* **1** a metal loop or link, closed by a bolt, to connect chains etc. **2** a fetter enclosing the ankle or wrist. **3** (usu. in *pl.*) a restraint or impediment. ● *v.tr.* fetter, impede, restrain. □ **shackle-bolt 1** a bolt for closing a shackle. **2** a bolt with a shackle at its end. [OE *sc(e)acul* fetter, corresp. to LG *shäkel* link, coupling, ON *skökull* wagon-pole f. Gmc]

Shackleton /ˈʃæk(ə)lt(ə)n/, Sir Ernest Henry (1874–1922), British explorer. A junior officer on Scott's National Antarctic Expedition (1900–4), he commanded his own expedition in 1909, getting within 155 km (97 miles) of the South Pole (the farthest south anyone had reached at that time). On a second Antarctic expedition (1914–16), Shackleton's ship *Endurance* was crushed in the ice. He and his crew eventually reached an island, from where he and five others set out in an open boat on a 1,300-km (800-mile) voyage to South Georgia to get help. In 1920 he led a fourth expedition to the Antarctic, but died on South Georgia.

shad /ʃæd/ *n.* (*pl.* same or **shads**) a deep-bodied edible marine fish of the genus *Alosa*, of the herring family, spawning in fresh water. [OE *sceadd*, of unkn. orig.]

shaddock /ˈʃædək/ *n.* **1** the largest citrus fruit, with a thick yellow skin and bitter pulp. Also called *pomelo*. **2** the tree, *Citrus grandis*, bearing this. [Capt. *Shaddock*, who introduced it to the West Indies in the 17th c.]

shade /ʃeɪd/ *n. & v.* ● *n.* **1** comparative darkness (and usu. coolness) caused by shelter from direct light and heat. **2** a place or area sheltered from the sun. **3** a darker part of a picture etc. **4** a colour, esp. with regard to its depth or as distinguished from one nearly like it. **5** a slight amount (*am a shade better today*). **6** a translucent cover for a lamp etc. **7** a screen excluding or moderating light. **8** an eye-shield. **9** (in *pl.*) *colloq.* sunglasses. **10** a slightly differing variety (*all shades of opinion*). **11** *literary* **a** a ghost. **b** (in *pl.*) Hades. **12** (in *pl.*; foll. by *of*) suggesting reminiscence or unfavourable comparison (*shades of Dr Johnson!*). ● *v.* **1** *tr.* screen from light. **2** *tr.* cover, moderate, or exclude the light of. **3** *tr.* darken, esp. with parallel pencil lines, to represent shadow etc. **4** *intr. & tr.* (often foll. by *away, off, into*) (cause to) pass or change by degrees; border on. □ **in the shade** in comparative obscurity. **put in** (or **into**) **the shade** eclipse, surpass; appear superior. □ **shadeless** *adj.* [OE *sc(e)adu* f. Gmc]

shading /ˈʃeɪdɪŋ/ *n.* **1** the representation of light and shade, e.g. by pencilled lines, on a map or drawing. **2** the graduation of tones from light to dark to create a sense of depth.

shadoof /ʃəˈduːf/ *n.* a pole with a bucket and counterpoise used esp. in Egypt for raising water. [Egypt. Arab. *šādūf*]

shadow /ˈʃædəʊ/ *n. & v.* ● *n.* **1** shade or a patch of shade. **2** a dark figure projected by a body intercepting rays of light, often regarded as an appendage. **3 a** an inseparable attendant or companion. **b** a person who shadows another at their work (esp. *work shadow*). **4** a person secretly following another. **5** the slightest trace (*not the shadow of a doubt*). **6** a weak or insubstantial remnant or thing (*a shadow of his former self*). **7** (*attrib.*) *Brit.* denoting members of a political party in opposition holding responsibilities parallel to those of the government (*shadow Home Secretary*; *shadow Cabinet*). **8** the shaded part of a picture. **9** a substance used to colour the eyelids. **10** gloom or sadness. ● *v.tr.* **1** cast a shadow over. **2 a** secretly follow and watch the movements of. **b** accompany and observe (a person) at work, for training or to gain understanding of the work. □ **shadow-boxing** boxing against an imaginary opponent as a form of training. □ **shadower** *n.* **shadowless** *adj.* [repr. OE *scead(u)we*, oblique case of *sceadu* SHADE]

shadowgraph /ˈʃædəʊɡrɑːf/ *n.* **1** an image or photograph made by means of X-rays; = RADIOGRAM 2. **2** a picture formed by a shadow cast on a lighted surface. **3** an image formed by light refracted differently by different densities of a fluid.

shadow theatre *n.* a form of puppetry in which flat jointed figures are moved between a strong light and a translucent screen, while the audience, in front of the screen, sees only their shadows. It originated in the Far East, spread to Turkey and Greece, and became popular in Paris during the 18th and 19th centuries. In the streets of London performances known as galanty shows were given until the end of the 19th century, usually in Punch and Judy booths. Shadow theatre survives in its traditional form in Java and Bali.

shadowy /ˈʃædəʊɪ/ *adj.* **1** like or having a shadow. **2** full of shadows. **3** vague, indistinct. **4 a** unreal, imaginary. **b** spectral, ghostly. □ **shadowiness** *n.*

shady /ˈʃeɪdɪ/ *adj.* (**shadier, shadiest**) **1** giving shade. **2** situated in shade. **3** (of a person or behaviour) disreputable; of doubtful honesty. □ **shadily** *adv.* **shadiness** *n.*

shaft /ʃɑːft/ *n. & v.* ● *n.* **1 a** an arrow or spear. **b** the long slender stem of these. **2** a remark intended to hurt or provoke (*a shaft of malice*; *shafts of wit*). **3** (foll. by *of*) **a** a ray (of light). **b** a bolt (of lightning). **4** the stem or handle of a tool, implement, etc. **5** a column, esp. between the base and capital. **6** a long narrow space, usu. vertical, for access to a mine, a lift in a building, for ventilation, etc. **7** a long and narrow part supporting or connecting or driving a part or parts of greater thickness etc. **8** each of the pair of poles between which a horse is harnessed to a vehicle. **9** the central stem of a feather. **10** *Mech.* a large axle or revolving bar transferring force by belts or cogs. **11** *sl.* a penis. **12** *N. Amer. colloq.* harsh or unfair treatment. ● *v.tr.* **1** *N. Amer. colloq.* treat unfairly. **2** *coarse sl.* (of a man) copulate with. [OE *scæft, sceaft* f. Gmc]

Shaftesbury /ˈʃɑːftsbərɪ/, Anthony Ashley Cooper, 7th Earl of (1801–85), English philanthropist and social reformer. He was a dominant figure of the 19th-century social reform movement, inspiring much of the legislation designed to improve conditions for the large working class created as a result of the Industrial Revolution. His reforms included the introduction of the ten-hour working day (1847); he was also actively involved in improving housing and education for the poor.

shaft grave *n.* a type of grave found in late Bronze Age Greece and Crete in which the burial chamber is approached by a vertical shaft sometimes lined with stones and roofed over with beams, as seen in six famous elaborate examples at Mycenae.

shafting /ˈʃɑːftɪŋ/ *n. Mech.* **1** a system of connected shafts for transmitting motion. **2** material from which shafts are cut.

shag[1] /ʃæg/ *n.* **1 a** a rough growth or mass of hair etc. **b** (*attrib.*) (of a carpet) with a long rough pile. **c** (*attrib.*) (of a pile) long and rough. **2** a coarse kind of cut tobacco. **3** a cormorant, esp. *Phalacrocorax aristotelis* of Europe and North Africa, with greenish-black plumage and a curly crest. [OE *sceacga*, rel. to ON *skegg* beard, OE *sceaga* coppice]

shag[2] /ʃæg/ *v. & n. coarse sl.* ● *v.tr.* (**shagged, shagging**) **1** have sexual intercourse with. **2** (usu. in *passive*; often foll. by *out*) exhaust; tire out. ● *n.* an act of sexual intercourse. [18th c.: orig. unkn.]

shaggy /ˈʃægɪ/ *adj.* (**shaggier, shaggiest**) **1** hairy, rough-haired. **2** unkempt. **3** (of the hair) coarse and abundant. **4** *Biol.* having a hairlike covering. □ **shaggy-dog story** a long rambling story amusing only by its being inconsequential. □ **shaggily** *adv.* **shagginess** *n.*

shagreen /ʃæˈgriːn/ *n.* **1** a kind of untanned leather with a rough granulated surface. **2** a sharkskin rough with natural denticles, used for rasping and polishing. [var. of CHAGRIN in the sense 'rough skin']

Shah[1] /ʃɑː/, Karim Al-Hussain, see AGA KHAN.

Shah[2] /ʃɑː/, Reza, see PAHLAVI[2].

shah /ʃɑː/ *n.* a title of the former monarch of Iran. The last shah, Muhammad Reza Pahlavi (see PAHLAVI[2]), was overthrown in 1979 by Ayatollah Khomeini's Islamic revolution. □ **shahdom** *n.* [Pers. *šāh* f. OPers. *kšāyṯiya* king]

Shah Alam /ʃɑː ˈɑːləm/ the capital of the state of Selangor in Malaysia, near the west coast of the Malay Peninsula; pop. (1980) 24,140.

shaikh var. of SHEIKH.

Shaka /ˈʃækə/ (also **Chaka**) (c.1787–1828), Zulu chief. After seizing the Zulu chieftaincy from his half-brother in 1816, he reorganized his forces and waged war against the Nguni clans in SE Africa, subjugating them and forming a Zulu empire in the region. Shaka's military campaigns led to a huge displacement of people and a lengthy spell of clan warfare in the early 1820s. He was subsequently assassinated by his two half-brothers.

shake /ʃeɪk/ *v. & n.* ● *v.* (*past* **shook**; *past part.* **shaken** /ˈʃeɪkn/) **1** *tr. & intr.* move forcefully or quickly up and down or to and fro. **2 a** *intr.* tremble or vibrate markedly. **b** *tr.* cause to do this. **3** *tr.* **a** agitate or shock. **b** *colloq.* upset the composure of. **4** *tr.* weaken or impair; make less convincing or firm or courageous (*shook his confidence*). **5** *intr.* (of a voice, note, etc.) make tremulous or rapidly alternating sounds; trill (*his voice shook with emotion*). **6** *tr.* brandish; make a threatening gesture with (one's fist, a stick, etc.). **7** *intr. colloq.* shake hands (*they shook on the deal*). **8** *tr. esp. US colloq.* = **shake off**. ● *n.* **1** the act or an instance of shaking; the process of being shaken. **2** a jerk or shock. **3** (in *pl.*; prec. by *the*) a fit of or tendency to trembling or shivering. **4** *Mus.* a trill. **5** = *milk shake*. □ **in two shakes** (**of a lamb's** or **dog's tail**) *colloq.* very quickly. **no great shakes** *colloq.* not very good or significant. **shake a person by the hand** = *shake hands*. **shake down 1** settle or cause to fall by shaking. **2** settle down. **3** become established; get into harmony with circumstances, surroundings, etc. **4** *N. Amer. sl.* extort money from. **shake the dust off one's feet** depart indignantly or disdainfully. **shake hands** (often foll. by *with*) clasp right hands at meeting or parting, in reconciliation or congratulation, or over a concluded bargain. **shake one's head** turn one's head from side to side in refusal, denial, disapproval, or concern. **shake in one's shoes** tremble with apprehension. **shake a leg** *colloq.* **1** begin dancing. **2** make a start. **shake off 1** get rid of (something unwanted). **2** manage to evade (a person who is following or pestering one). **shake out 1** empty by shaking. **2** spread or open (a sail, flag, etc.) by shaking. **3** shed (personnel) as a result of reorganization. **shake-out** *n.* an upheaval or reorganization, esp. in a business and involving streamlining, closures, redundancies, etc. **shake up 1** mix (ingredients) by shaking. **2** restore to shape by shaking. **3** disturb or make uncomfortable. **4** rouse from lethargy, apathy, conventionality, etc. **shake-up** *n.* an upheaval or drastic reorganization. □ **shakeable** *adj.* (also **shakable**). [OE *sc(e)acan* f. Gmc]

shakedown /ˈʃeɪkdaʊn/ *n.* **1** a makeshift bed. **2** a period or process of adjustment or change. **3** *US sl.* a swindle; a piece of extortion. **4** (usu.

attrib.) a voyage, flight, etc., to test a new ship, aircraft, etc. and its crew (*shakedown cruise*).

shaken *past part.* of SHAKE.

Shaker /ˈʃeɪkə(r)/ *n.* (*fem.* **Shakeress** /-rɪs/) a member of a US Christian sect, the United Society of Believers in Christ's Second Coming, whose adherents live austerely in celibate mixed communities. Named from the ecstatic whirling and shaking engaged in by its members, the sect developed in England *c.*1750 among a branch of the Quakers and was taken to America in 1774 by Ann Lee (1736–84), who regarded herself as the second incarnation of Christ. The community that was established prospered and spread in New England, and its members gained a reputation as skilled craftsmen, known for the elegant functionalism of their traditional furniture, interest in which has recently revived. Because of its doctrine of celibacy and hence its dependence on recruitment from the wider society, Shakerism has few adherents today. □ **Shakerism** *n.*

shaker /ˈʃeɪkə(r)/ *n.* **1** a person or thing that shakes. **2** a container for shaking together the ingredients of cocktails etc. [ME, f. SHAKE]

Shakespeare /ˈʃeɪkspɪə(r)/, William (also known as 'the Bard (of Avon)') (1564–1616), English dramatist. He was born a merchant's son in Stratford-upon-Avon in Warwickshire and married Anne Hathaway in about 1582. Some time thereafter he went to London, where he pursued a career as an actor, poet, and dramatist. He probably began to write for the stage in the late 1580s; although his plays were widely performed in his lifetime, many were not printed until the First Folio of 1623. His plays are written mostly in blank verse and include comedies (e.g. *A Midsummer Night's Dream* and *As You Like It*); historical plays, including *Richard III* and *Henry V*; the Greek and Roman plays, which include *Julius Caesar* and *Antony and Cleopatra*; the so-called 'problem plays', enigmatic comedies which include *All's Well that Ends Well* and *Measure for Measure*; the great tragedies, *Hamlet*, *Othello*, *King Lear*, and *Macbeth*; and the group of tragicomedies with which he ended his career, such as *The Winter's Tale* and *The Tempest*. He also wrote more than 150 sonnets, published in 1609, as well as narrative poems such as *The Rape of Lucrece* (1594).

Shakespearian /ʃeɪkˈspɪərɪən/ *adj. & n.* (also **Shakespearean**) ● *adj.* **1** of or relating to William Shakespeare. **2** in the style of Shakespeare. ● *n.* a student of Shakespeare's works etc.

Shakhty /ˈʃɑːktɪ/ a coal-mining city in SW Russia, situated in the Donets Basin north-east of Rostov; pop. (1990) 227,000.

shako /ˈʃeɪkəʊ, ˈʃæk-/ *n.* (*pl.* **-os**) a cylindrical peaked military hat with a plume. [F *schako* f. Hungarian *csákó (süveg)* peaked (cap) f. *csák* peak f. G *Zacken* spike]

shakuhachi /ˌʃækuːˈhɑːtʃɪ/ *n.* (*pl.* **shakuhachis**) a Japanese bamboo flute. [Jap. f. *shaku* a measure of length + *hachi* eight (tenths)]

shaky /ˈʃeɪkɪ/ *adj.* (**shakier, shakiest**) **1** unsteady; apt to shake; trembling. **2** unsound, infirm (*a shaky hand*). **3** unreliable, wavering (*a shaky promise*; *got off to a shaky start*). □ **shakily** *adv.* **shakiness** *n.*

shale /ʃeɪl/ *n.* soft finely stratified rock that splits easily, consisting of consolidated mud or clay. □ **shale oil** oil obtained from bituminous shale. □ **shaly** *adj.* [prob. f. G *Schale* f. OE *sc(e)alu* rel. to ON *skál* (see SCALE[2])]

shall /ʃæl, ʃ(ə)l/ *v.aux.* (3rd sing. present **shall**; archaic 2nd sing. present **shalt** /ʃælt/; *past* **should**) (foll. by *infin.* without *to*, or *absol.*; present and past only in use) **1** (in the 1st person) expressing the future tense (*I shall return soon*) or (with *shall* stressed) emphatic intention (*I shall have a party*). **2** (in the 2nd and 3rd persons) expressing a strong assertion or command rather than a wish (cf. WILL[1]) (*you shall not catch me again*; *they shall go to the party*). ¶ For the other persons in senses 1, 2 see WILL[1]. **3** expressing a command or duty (*thou shalt not steal*; *they shall obey*). **4** (in 2nd-person questions) expressing an enquiry, esp. to avoid the form of a request (cf. WILL[1]) (*shall you go to France?*). □ **shall I?** do you want me to? [OE *sceal* f. Gmc]

shallot /ʃəˈlɒt/ *n.* an onion-like plant, *Allium ascalonicum*, with a cluster of small bulbs. [*eschalot* f. F *eschalotte* alt. of OF *eschaloigne*: see SCALLION]

shallow /ˈʃæləʊ/ *adj., n., & v.* ● *adj.* **1** of little depth. **2** superficial, trivial (*a shallow mind*). ● *n.* (often in *pl.*) a shallow place. ● *v.intr. & tr.* become or make shallow. □ **shallowly** *adv.* **shallowness** *n.* [ME, prob. rel. to *schald*, OE *sceald* SHOAL[2]]

Shalmaneser III /ˌʃælmənəˈniːzə(r)/ (d.824 BC), king of Assyria 859–824. Most of his reign was devoted to the expansion of his kingdom and the conquest of neighbouring lands. According to Assyrian records (though it is not mentioned in the Bible) he defeated an alliance of Syrian kings and Ahab, king of Israel, in a battle at Qarqar on the

Orontes in 853 BC. His other military achievements included the invasion of Cilicia and the capture of Tarsus.

shalom /ʃəˈlɒm/ n. & int. a Jewish salutation at meeting or parting. [Heb. *šālôm* peace]

shalt archaic 2nd person sing. of SHALL.

sham /ʃæm/ v., n., & adj. ● v. (**shammed, shamming**) **1** intr. feign, pretend. **2** tr. **a** pretend to be. **b** simulate (*is shamming sleep*). ● n. **1** imposture, pretence. **2** a person or thing pretending or pretended to be what he or she or it is not. ● adj. pretended, counterfeit. □ **shammer** n. [perh. north. dial. var. of SHAME]

shaman /ˈʃæmən/ n. (pl. **shamans**) a person regarded as having direct access to, and influence in, the spirit world, which is usually contacted during a trance and empowers him or her to guide souls, cure illnesses, etc. The shaman is associated today with certain indigenous peoples of northern Asia and North America, but the general pattern of beliefs, rituals, and techniques associated with shamans is found almost universally in primitive cultures at the food-gathering stage of development. □ **shamanism** n. **shamanist** n. & adj. **shamanistic** /ˌʃæməˈnɪstɪk/ adj. [G *Schamane* & Russ. *shaman* f. Tungusian *samán*]

shamateur /ˈʃæmətə(r)/ n. derog. a sports player who makes money from sporting activities though classed as an amateur. □ **shamateurism** n. [SHAM + AMATEUR]

shamble /ˈʃæmb(ə)l/ v. & n. ● v.intr. walk or run with a shuffling or awkward gait. ● n. a shambling gait. [prob. f. dial. *shamble* (adj.) ungainly, perh. f. *shamble legs* with ref. to straddling trestles: see SHAMBLES]

shambles /ˈʃæmb(ə)lz/ n.pl. (usu. treated as sing.) **1** colloq. a mess or muddle (*the room was a shambles*). **2** a butcher's slaughterhouse. **3** a scene of carnage. [pl. of *shamble* stool, stall f. OE *sc(e)amul* f. WG f. L *scamellum* dimin. of *scamnum* bench]

shambolic /ʃæmˈbɒlɪk/ adj. colloq. chaotic, unorganized. [SHAMBLES, prob. after SYMBOLIC]

shame /ʃeɪm/ n. & v. ● n. **1** a feeling of distress or humiliation caused by consciousness of the guilt or folly of oneself or an associate. **2** a capacity for experiencing this feeling, esp. as imposing a restraint on behaviour (*has no sense of shame*). **3** a state of disgrace, discredit, or intense regret. **4 a** a person or thing that brings disgrace etc. **b** a thing or action that is wrong or regrettable. ● v.tr. **1** bring shame on; make ashamed; put to shame. **2** (foll. by *into, out of*) force by shame (*was shamed into confessing*). □ **for shame!** a reproof to a person for not showing shame. **put to shame** disgrace or humiliate by revealing superior qualities etc. **shame on you!** you should be ashamed. **what a shame!** how unfortunate! [OE *sc(e)amu*]

shamefaced /ʃeɪmˈfeɪst/ adj. **1** showing shame. **2** bashful, diffident. □ **shamefacedly** /-sɪdlɪ/ adv. **shamefacedness** n. [16th-c. alt. of *shamefast*, by assim. to FACE]

shameful /ˈʃeɪmfʊl/ adj. **1** that causes or is worthy of shame. **2** disgraceful, scandalous. □ **shamefully** adv. **shamefulness** n. [OE *sc(e)amful* (as SHAME, -FUL)]

shameless /ˈʃeɪmlɪs/ adj. **1** having or showing no sense of shame. **2** impudent. □ **shamelessly** adv. **shamelessness** n. [OE *sc(e)amlēas* (as SHAME, -LESS)]

Shamir /ʃæˈmɪə(r)/, Yitzhak (Polish name Yitzhak Jazernicki) (b.1915), Israeli statesman, Prime Minister 1983–4 and 1986–92. Born in Poland, he emigrated to Palestine in 1935. On Menachem Begin's retirement in 1983, Shamir became Premier, but his Likud party was narrowly defeated in elections a year later. As Prime Minister of a coalition government with Labour, he sacked Shimon Peres in 1990 and formed a new government with a policy of conceding no land to a Palestinian state. Under his leadership, Israel did not retaliate when attacked by Iraqi missiles during the Gulf War of 1991, thereby possibly averting the formation of a pro-Saddam Hussein Arab coalition.

shammy /ˈʃæmɪ/ n. (pl. **-ies**) (in full **shammy leather**) colloq. = CHAMOIS 2. [repr. corrupted pronunc.]

shampoo /ʃæmˈpuː/ n. & v. ● n. **1** liquid or cream used to lather and wash the hair. **2** a similar substance for washing a car or carpet etc. **3** an act or instance of washing with shampoo. ● v.tr. (**shampoos, shampooed**) wash with shampoo. [Hind. *chhāmpo*, imper. of *chhāmpnā* to press]

shamrock /ˈʃæmrɒk/ n. a leguminous plant with trifoliate leaves, esp. *Trifolium minus*, *T. repens*, or *Medicago lupulina*, used as the national emblem of Ireland. [Ir. *seamróg* trefoil, dimin. of *seamar* clover + *og* young]

shamus /ˈʃeɪməs/ n. US sl. a detective. [20th c.: orig. uncert.]

Shandong /ʃænˈdʊŋ/ (also **Shantung** /-ˈtʊŋ/) a coastal province of eastern China; capital, Jinan. It occupies the Shandong Peninsula, separating southern Bo Hai from the Yellow Sea. The region gives its name to shantung silk.

shandy /ˈʃændɪ/ n. (pl. **-ies**) a drink of beer mixed with lemonade or ginger beer. [19th c.: orig. unkn.]

Shang /ʃæŋ/ a dynasty which ruled China during part of the 2nd millennium BC, probably 16th–11th centuries. The period encompassed the invention of Chinese ideographic script and the discovery and development of bronze casting.

Shangaan /ˈʃæŋɡɑːn/ n. & adj. ● n. (pl. same or **Shangaans**) **1 a** a member of a Bantu-speaking people inhabiting parts of southern Mozambique and South Africa. Also called *Tsonga*. **b** a member of either of two Shona peoples of Zimbabwe. **2 a** the language of the Shangaan of Mozambique and South Africa; Tsonga. **b** the Bantu language of the Zimbabwean Shangaan. ● adj. of or relating to the Shangaan or their languages. [Bantu]

Shanghai /ʃæŋˈhaɪ/ a city on the east coast of China, a port on the estuary of the Yangtze; pop. (1990) 7,780,000. Opened for trade from the west in 1842, Shanghai contained until the Second World War areas of British, French, and American settlement. It was the site in 1921 of the founding of the Chinese Communist Party. It is now China's most populous city.

shanghai /ʃæŋˈhaɪ/ v. & n. ● v.tr. (**shanghais, shanghaied** /-ˈhaɪd/, **shanghaiing** /-ˈhaɪɪŋ/) **1** force (a person) to be a sailor on a ship by using drugs or other trickery. **2** colloq. put into detention or an awkward situation by trickery. **3** Austral. & NZ shoot with a catapult. ● n. (pl. **shanghais**) Austral. & NZ a catapult. [SHANGHAI]

Shangri La /ˈʃæŋɡrɪ ˈlɑː/ a Tibetan utopia in James Hilton's novel *Lost Horizon* (1933), frequently used as the type of an earthly paradise or a place of retreat from the worries of modern civilization. [*Shangri* (invented name) + Tibetan *la* mountain pass]

shank /ʃæŋk/ n. **1 a** the leg. **b** the lower part of the leg; the leg from knee to ankle. **c** the shin-bone. **2** the lower part of an animal's foreleg, esp. as a cut of meat. **3** a shaft or stem. **4 a** the long narrow part of a tool etc. joining the handle to the working end. **b** the stem of a key, spoon, anchor, etc. **c** the straight part of a nail or fish-hook. **5** the narrow middle of the sole of a shoe. □ **shanks's mare** (or **pony**) one's own legs as a means of conveyance. □ **shanked** adj. (also in comb.). [OE *sceanca* f. WG]

Shankar /ˈʃæŋkə(r)/, Ravi (b.1920), Indian sitar player and composer. Already an established musician in his own country, from the mid-1950s he embarked on tours of Europe and the US giving sitar recitals, doing much to stimulate contemporary Western interest in Indian music. He founded schools of Indian music in Bombay (1962) and Los Angeles (1967).

Shannon[1] /ˈʃænən/ **1** the longest river of Ireland. It rises in County Leitrim near Lough Allen and flows 390 km (240 miles) south and west to its estuary on the Atlantic. **2** an international airport in the Republic of Ireland, situated on the River Shannon west of Limerick. **3** a shipping forecast area in the NE Atlantic to the south-west of Ireland.

Shannon[2] /ˈʃænən/, Claude Elwood (b.1916), American engineer. He was the pioneer of mathematical communication theory, which has become vital to the design of both communication and electronic equipment (see INFORMATION THEORY). He also investigated digital circuits, and was the first to use the term *bit* to denote a unit of information.

shanny /ˈʃænɪ/ n. (pl. **-ies**) a long-bodied olive-green European blenny (fish), *Blennius pholis*, of intertidal waters. [19th c.: orig. unkn.: cf. 18th-c. *shan*]

Shansi see SHANXI.

shan't /ʃɑːnt/ contr. shall not.

Shantou /ʃænˈtaʊ/ (formerly called *Swatow*) a port in the province of Guangdong in SE China, situated on the South China Sea at the mouth of the Han river; pop. (1990) 860,000. It was designated a treaty port in 1869.

Shantung see SHANDONG.

shantung /ʃænˈtʌŋ/ n. a heavy textured silk fabric, often mixed with rayon or cotton, originally made in Shandong.

shanty[1] /ˈʃæntɪ/ n. (pl. **-ies**) **1** a hut or cabin. **2** a crudely built shack. □ **shanty town** a poor or depressed area of a town, consisting of shanties. [19th c., orig. N. Amer.: perh. f. Canad. F *chantier*]

shanty[2] /ˈʃæntɪ/ n. (also **chanty**) (pl. **-ies**) (in full **sea shanty**) a song with alternating solo and chorus, of a kind orig. sung by sailors while hauling ropes etc. [prob. F *chantez*, imper. pl. of *chanter* sing: see CHANT]

Shanxi /ʃænˈʃiː/ (also **Shansi** /-ˈsiː/) a province of north central China, to the south of Inner Mongolia; capital, Taiyuan.

SHAPE /ʃeɪp/ abbr. Supreme Headquarters, Allied Powers in Europe.

shape /ʃeɪp/ n. & v. ● n. **1** the total effect produced by the outlines of a thing. **2** the external form or appearance of a person or thing. **3** a specific form or guise. **4** a description or sort or way (*not on offer in any shape or form*). **5** a definite or proper arrangement (*must get our ideas into shape*). **6 a** condition, as qualified in some way (*in good shape; in poor shape*). **b** (when unqualified) good condition (*back in shape*). **7** a person or thing as seen, esp. indistinctly or in the imagination (*a shape emerged from the mist*). **8** a mould or pattern. **9** a jelly etc. shaped in a mould. **10** a piece of material, paper, etc., made or cut in a particular form. ● v. **1** tr. give a certain shape or form to; fashion, create. **2** tr. (foll. by *to*) adapt or make conform. **3** intr. give signs of a future shape or development. **4** tr. frame mentally; imagine. **5** intr. assume or develop into a shape. **6** tr. direct (one's life, course, etc.). □ **lick** (or **knock** or **whip**) **into shape** make presentable or efficient. **shape up 1** take a (specified) form. **2** show promise; make good progress. **shape up well** be promising. □ **shapable** adj. (also **shapeable**). **shaped** adj. (also in comb.). **shaper** n. [OE *gesceap* creation f. Gmc]

shapechanger /ˈʃeɪpˌtʃeɪndʒə(r)/ n. (in science fiction etc.) a being capable by supernatural means of assuming different bodily forms. □ **shapechanging** n. & adj.

shapeless /ˈʃeɪplɪs/ adj. lacking definite or attractive shape. □ **shapelessly** adv. **shapelessness** n.

shapely /ˈʃeɪplɪ/ adj. (**shapelier**, **shapeliest**) **1** well formed or proportioned. **2** of elegant or pleasing shape or appearance. □ **shapeliness** n.

Shapley /ˈʃæplɪ/, Harlow (1885–1972), American astronomer. He studied globular star clusters, using cepheid variables within them to determine their distance. He then used their distribution to locate the likely centre of the Galaxy and infer its structure and dimensions. For 31 years Shapley was director of the Harvard Observatory, where he studied the distribution of stars of different spectral types, investigated the Magellanic Clouds, and carried out an extensive survey of galaxies.

shard /ʃɑːd/ n. **1** a broken piece of pottery or glass etc. **2** = POTSHERD. **3** a fragment of volcanic rock. **4** the wing-case of a beetle. [OE *sceard*: sense 4 f. *shard-borne* (Shakespeare) = born in a shard (dial., = cow-dung), wrongly taken as 'borne on shards']

share[1] /ʃeə(r)/ n. & v. ● n. **1** a portion that a person receives from or gives to a common amount. **2 a** a part contributed by an individual to an enterprise or commitment. **b** a part received by an individual from this (*got a large share of the credit*). **3** part-proprietorship of property held by joint owners, esp. any of the equal parts into which a company's capital is divided entitling its owner to a proportion of the profits. ● v. **1** tr. get or have or give a share of. **2** tr. use or benefit from jointly with others. **3** tr. have in common (*I share your opinion*). **4** intr. have a share; be a sharer (*shall I share with you?*). **5** intr. (foll. by *in*) participate. **6** tr. (often foll. by *out*) **a** divide and distribute. **b** give away part of. □ **share and share alike** make an equal division. **share-farmer** Austral. & NZ a tenant farmer who receives a share of the profits from the owner. □ **shareable** adj. (also **sharable**). **sharer** n. [ME f. OE *scearu* division, rel. to SHEAR]

share[2] /ʃeə(r)/ n. = PLOUGHSHARE. [OE *scear, scær* f. Gmc]

sharecropper /ˈʃeəˌkrɒpə(r)/ n. esp. US a tenant farmer who gives a part of each crop as rent. □ **sharecrop** v.tr. & intr. (**-cropped**, **-cropping**).

shareholder /ˈʃeəˌhəʊldə(r)/ n. an owner of shares in a company.

shareware /ˈʃeəweə(r)/ n. Computing software available without a licence and often distributed free of charge for evaluation, after which a small fee is requested for continued use.

sharia /ʃəˈriːə/ n. (also **shariah**) the sacred law of Islam, including the teachings of the Koran and the traditional sayings of Muhammad, prescribing both religious and secular duties and entailing in some cases strictly retributive penalties. The sharia has been supplemented since the earliest times by legislation adapted to the conditions of the day, and since the 19th century has been replaced in some states (e.g. Turkey, Albania) by westernized legal systems, while many other Islamic countries operate mixed systems. Saudi Arabia and Iran maintain the sharia as the law of the land, while elsewhere it is a source of dispute between Islamic fundamentalists and modernists. [Arab. *šarī'a*]

sharif /ʃəˈriːf/ n. (also **shereef**, **sherif**) **1** a descendant of Muhammad

through his daughter Fatima, entitled to wear a green turban or veil. **2** a Muslim leader. [Arab. *šarīf* noble f. *šarafa* be exalted]

Sharjah /ˈʃɑːdʒə/ (Arabic **Ash Shariqah** /ˌæʃ ʃɑːˈriːkə/) **1** one of the seven member states of the United Arab Emirates; pop. (1985) 268,720. **2** its capital city, situated on the Persian Gulf; pop. (1984) 125,000.

shark[1] /ʃɑːk/ n. a large cartilaginous marine fish of the superorder Selachimorpha, with a long streamlined body and prominent dorsal fin, frequently a voracious carnivore. Sharks have a rough scaly skin, and sharp teeth that are continually grown and shed. They vary in size from the small dogfishes to the basking shark and whale shark, which are the largest known fishes and feed only on plankton. [16th c.: orig. unkn.]

shark[2] /ʃɑːk/ n. colloq. a person who unscrupulously exploits or swindles others. [16th c.: orig. perh. f. G *Schurke* worthless rogue: infl. by SHARK[1]]

sharkskin /ˈʃɑːkskɪn/ n. **1** the skin of a shark. **2** a smooth slightly lustrous fabric.

Sharma /ˈʃɑːmə/, Shankar Dayal (b.1918), Indian statesman, President since 1992. A member of the Congress party, Sharma served as Vice-President 1987–92.

Sharon /ˈʃærən/ a fertile coastal plain in Israel, lying between the Mediterranean Sea and the hills of Samaria.

sharon fruit /ˈʃærən/ n. (also **Sharon**) a persimmon, esp. of an orange variety grown in Israel. [SHARON]

Sharp /ʃɑːp/, Cecil (James) (1859–1924), English collector of folk-songs and folk-dances. His work was responsible for a revival of interest in English folk music; from 1904 onwards he published a number of collections of songs and dances, including morris dances. He founded the English Folk Dance Society in 1911 and helped establish the teaching of this music in schools.

sharp /ʃɑːp/ adj., n., adv., & v. ● adj. **1** having an edge or point able to cut or pierce. **2** tapering to a point or edge. **3** abrupt, steep, angular (*a sharp fall; a sharp turn*). **4** well-defined, clean-cut. **5 a** severe or intense (*has a sharp temper*). **b** (of food or its flavour) pungent, acid. **c** keen (*a sharp appetite*). **d** (of a frost) severe, hard. **6** (of a voice or sound) shrill and piercing. **7** (of sand etc.) composed of angular grains. **8** (of words etc.) harsh or acrimonious (*had a sharp tongue*). **9** (of a person) acute; quick to perceive or comprehend. **10** derog. quick to take advantage; artful, unscrupulous, dishonest. **11** vigorous or brisk. **12** Mus. **a** above the normal pitch. **b** (of a key) having a sharp or sharps in the signature. **c** (as C **sharp, F sharp**, etc.) a semitone higher than C, F, etc. **13** colloq. stylish or flashy with regard to dress (*sharp dresser*). ● n. **1** Mus. **a** a note raised a semitone above natural pitch. **b** the sign (#) indicating this. **2** colloq. a swindler or cheat. **3** a fine sewing-needle. ● adv. **1** punctually (*at nine o'clock sharp*). **2** suddenly, abruptly, promptly (*pulled up sharp*). **3** at a sharp angle. **4** Mus. above the true pitch (*sings sharp*). ● v. **1** intr. archaic cheat or swindle at cards etc. **2** tr. US Mus. make sharp. □ **sharp end 1** joc. the bow of a ship. **2** colloq. the scene of direct action or decision. **sharp-eyed** having good sight; observant. **sharp-featured** (of a person) having well-defined facial features. **sharp practice** dishonest or barely honest dealings. **sharp-set 1** set with a sharp edge. **2** hungry. □ **sharply** adv. **sharpness** n. [OE *sc(e)arp* f. Gmc]

shar-pei /ʃɑːˈpeɪ/ n. a Chinese breed of dog with a compact body and loose, deeply wrinkled skin. [Chin. *sha pei* sand fur]

sharpen /ˈʃɑːp(ə)n/ v.tr. & intr. make or become sharp. □ **sharpener** n.

sharper /ˈʃɑːpə(r)/ n. a swindler, esp. at cards.

Sharpeville massacre /ˈʃɑːpvɪl/ the killing of demonstrators at Sharpeville, a black township in South Africa, south of Johannesburg, on 21 Mar. 1960. Security forces fired on a crowd demonstrating against apartheid laws, killing 67 and wounding about 180. Following the massacre, which was greeted with widespread international condemnation, the South African government banned the African National Congress and the Pan-Africanist Congress and withdrew the country from the Commonwealth.

sharpish /ˈʃɑːpɪʃ/ adj. & adv. colloq. ● adj. fairly sharp. ● adv. **1** fairly sharply. **2** quite quickly.

sharpshooter /ˈʃɑːpˌʃuːtə(r)/ n. a skilled marksman. □ **sharpshooting** n. & adj.

sharp-witted /ʃɑːpˈwɪtɪd/ adj. keenly perceptive or intelligent. □ **sharp-wittedly** adv. **sharp-wittedness** n.

shashlik /ˈʃæʃlɪk/ n. (in Asia and eastern Europe) a kebab of mutton and garnishings. [Russ. *shashlyk*, ult. f. Turk. *šiš* spit, skewer: cf. SHISH KEBAB]

Shasta daisy /'ʃæstə/ n. a composite plant, *Leucanthemum maximum*, native to the Pyrenees but widely cultivated, with large daisy-like flowers. [Mount *Shasta* in California]

Shastra /'ʃɑːstrə/ n. (also in *pl.*) Hindu sacred writings. [Skr. *śāstra*]

shat *past* and *past part.* of SHIT.

Shatt al-Arab /ˌʃæt æl 'ærəb/ a river of SW Asia, formed by the confluence of the Tigris and Euphrates rivers and flowing 195 km (120 miles) through SE Iraq to the Persian Gulf. Its lower course forms the border between Iraq and Iran.

shatter /'ʃætə(r)/ v. **1** tr. & intr. break suddenly in pieces. **2** tr. severely damage or utterly destroy (*shattered hopes*). **3** tr. greatly upset or discompose. **4** tr. colloq. (usu. as **shattered** adj.) exhaust. □ **shatterer** n. **shattering** adj. **shatteringly** adv. **shatter-proof** adj. [ME, rel. to SCATTER]

shave /ʃeɪv/ v. & n. ● v.tr. (past part. **shaved** or (as adj.) **shaven** /'ʃeɪv(ə)n/) **1** remove (bristles or hair) from the face etc. with a razor. **2** (also absol.) remove bristles or hair with a razor from the face etc. of (a person) or from (a part of the body). **3** a reduce by a small amount. **b** take (a small amount) away from. **4** cut thin slices from the surface of (wood etc.) to shape it. **5** pass close to without touching; miss narrowly. ● n. **1** an act of shaving or the process of being shaved. **2** a close approach without contact. **3** a narrow miss or escape; = *close shave* (see CLOSE¹). **4** a tool for shaving wood etc. [OE *sc(e)afan* (sense 4 of noun f. OE *sceafa*) f. Gmc]

shaveling /'ʃeɪvlɪŋ/ n. archaic **1** a shaven person. **2** a monk, friar, or priest.

shaven see SHAVE.

shaver /'ʃeɪvə(r)/ n. **1** a person or thing that shaves. **2** an electric razor. **3** colloq. a young lad.

Shavian /'ʃeɪvɪən/ adj. & n. ● adj. of or in the manner of George Bernard Shaw, or his ideas. ● n. an admirer of Shaw. [*Shavius*, Latinized form of SHAW]

shaving /'ʃeɪvɪŋ/ n. **1** a thin strip cut off the surface of wood etc. **2** (attrib.) used in shaving the face (*shaving-cream*).

Shavuoth /ʃəˈvuːəs, ˌʃɑːvʊˈəʊt/ n. (also **Shavuot**) the Jewish Pentecost. [Heb. *šāḇûʿôṯ*, = weeks, with ref. to the weeks between Passover and Pentecost]

Shaw /ʃɔː/, (George) Bernard (1856–1950), Irish dramatist and writer. He moved to London in 1876 and began his literary career as a critic and unsuccessful novelist. His first play was performed in 1892. His best-known plays combine comedy with intellectual debate in challenging conventional morality and thought; they include *Man and Superman* (1903), *Major Barbara* (1907), *Pygmalion* (1913), *Heartbreak House* (1919), and *St Joan* (1923). He wrote lengthy prefaces for most of his plays, in which he expanded his philosophy and ideas. A socialist, he joined the Fabian Society in 1884 and was an active member during its early period, championing many progressive causes, including feminism. He was awarded the Nobel Prize for literature in 1925.

shaw /ʃɔː/ n. esp. Brit. the stalks and leaves of potatoes, turnips, etc. [perh. = SHOW n.]

shawl /ʃɔːl/ n. a piece of fabric, usu. rectangular and often folded into a triangle, worn over the shoulders or head or wrapped round a baby. □ **shawl collar** a rolled collar extended down the front of a garment without lapel notches. □ **shawled** adj. [Urdu etc. f. Pers. *šāl*, prob. f. *Shāliāt* in India]

shawm /ʃɔːm/ n. Mus. a medieval double-reed wind instrument with a sharp penetrating tone. [ME f. OF *chalemie, chalemel, chalemeaus* (pl.), ult. f. L *calamus* f. Gk *kalamos* reed]

Shawnee /ʃɔːˈniː/ n. & adj. ● n. (pl. same or **Shawnees**) **1** a member of an Algonquian people living formerly in the eastern US and now chiefly in Oklahoma. **2** the language of this people. ● adj. of or relating to the Shawnee or their language. [Delaware]

Shcherbakov /ˌʃtʃɜːbəˈkɒf/ a former name (1946–57) for RYBINSK.

shchi /ʃiː, ʃtʃiː/ n. a Russian cabbage soup. [Russ.]

she /ʃiː/ pron. & n. ● pron. (obj. **her**; poss. **her**; pl. **they**) **1** the woman or girl or female animal previously named or in question. **2** a thing regarded as female, e.g. a vehicle or ship. **3** Austral. & NZ colloq. it; the state of affairs (*she'll be right*). ● n. **1** a female; a woman. **2** (in comb.) female (*she-goat*). □ **she-devil** a malicious or spiteful woman. [ME *scæ, sche*, etc., f. OE fem. demonstr. pron. & adj. *sío, séo,* accus. *síe*]

s/he pron. a written representation of 'he or she' used to indicate both sexes.

shea /ʃiː, 'ʃiːə/ n. a West African tree, *Vitellaria paradoxa*, bearing nuts containing a large amount of fat. □ **shea-butter** a butter made from this fat. [Mandingo *si, se, sye*]

sheading /'ʃiːdɪŋ/ n. each of the six administrative divisions of the Isle of Man. [SHED¹ + -ING¹]

sheaf /ʃiːf/ n. & v. ● n. (pl. **sheaves** /ʃiːvz/) a group of things held lengthways together, esp. a bundle of cornstalks tied after reaping, or a collection of papers. ● v.tr. make into sheaves. [OE *scēaf* f. Gmc (as SHOVE)]

shealing var. of SHIELING.

shear /ʃɪə(r)/ v. & n. ● v. (past **sheared**, archaic except Austral. & NZ **shore** /ʃɔː(r)/; past part. **shorn** /ʃɔːn/ or **sheared**) **1** tr. clip the wool off (a sheep etc.). **2** tr. remove or take off by cutting. **3** tr. cut with scissors or shears etc. **4** tr. (foll. by of) a strip bare. **b** deprive. **5** tr. (usu. as **shorn** adj.) cut the hair of a person. **6** tr. & intr. (often foll. by off) distort or be distorted, or break, from a structural strain. ● n. **1** Mech. & Geol. a strain produced by pressure in the structure of a substance, when its layers are laterally shifted in relation to each other. **2** (in pl.) (also **pair of shears** sing.) a large clipping or cutting instrument shaped like scissors for use in gardens etc. □ **shearer** n. [OE *sceran* f. Gmc]

Shearer /'ʃɪərə(r)/, Moira (full name Moira Shearer King) (b.1926), Scottish ballet-dancer and actress. As a ballerina with Sadler's Wells ballet from 1942 she created roles in a number of works by Sir Frederick Ashton. She is perhaps best known for her portrayal of a dedicated ballerina in the film *The Red Shoes* (1948). Her later acting career included roles in plays by Shaw and Chekhov.

shearling /'ʃɪəlɪŋ/ n. **1** a sheep that has been shorn once. **2** wool from a shearling.

shearwater /'ʃɪəˌwɔːtə(r)/ n. a long-winged seabird of the family Procellariidae, related to petrels, often flying low over the surface of the water.

sheath /ʃiːθ/ n. (pl. **sheaths** /ʃiːðz, ʃiːθs/) **1** a close-fitting cover, esp. for the blade of a knife or sword. **2** a condom. **3** Bot., Anat., & Zool. an enclosing case or tissue. **4** the protective covering round an electric cable. **5** a woman's close-fitting dress. □ **sheath knife** a dagger-like knife carried in a sheath. □ **sheathless** adj. [OE *scǣth, scēath*]

sheathe /ʃiːð/ v.tr. **1** put into a sheath. **2** encase; protect with a sheath. [ME f. SHEATH]

sheathing /'ʃiːðɪŋ/ n. a protective casing or covering.

sheave¹ /ʃiːv/ v.tr. make into sheaves.

sheave² /ʃiːv/ n. a grooved wheel in a pulley-block etc., for a rope to run on. [ME f. OE *scife* (unrecorded) f. Gmc]

sheaves pl. of SHEAF.

Sheba /'ʃiːbə/ the biblical name of Saba, an ancient country in SW Arabia famous for its trade in gold and spices. The queen of Sheba visited King Solomon in Jerusalem. The Hebrew word represents the name of the country's inhabitants, the Sabaeans, erroneously assumed by Greek and Roman writers to be a place-name. (See also SABAEAN.)

shebang /ʃɪˈbæŋ/ n. N. Amer. sl. **1** a matter or affair (esp. *the whole shebang*). **2** a shed or hut. [19th c.: orig. unkn.]

Shebat var. of SEBAT.

shebeen /ʃɪˈbiːn/ n. an unlicensed house selling alcoholic liquor, esp. in Ireland. [Anglo-Ir. *síbín* f. *séibe* mugful]

shed¹ /ʃed/ n. **1** a simple one-storeyed building usu. of wood for storage or shelter for animals etc., or as a workshop. **2** a large roofed structure with one side open, for storing or maintaining machinery, vehicles, etc. **3** Austral. & NZ an open-sided building for shearing sheep or milking cattle. [app. var. of SHADE]

shed² /ʃed/ v.tr. (**shedding**; past and past part. **shed**) **1** let or cause to fall off (*trees shed their leaves*). **2** take off (clothes). **3** reduce (an electrical power load) by disconnection etc. **4** cause to fall or flow (*shed blood; shed tears*). **5** disperse, diffuse, radiate (*shed light*). **6** (of a business) reduce its number of (jobs or employees) through redundancy, natural wastage, etc. □ **shed light on** see LIGHT¹. [OE *sc(e)adan* f. Gmc]

she'd /ʃiːd, ʃɪd/ contr. **1** she had. **2** she would.

shedded /'ʃedɪd/ predic.adj. (foll. by at or in) (of a road or rail vehicle) maintained in or based at a shed.

shedder /'ʃedə(r)/ n. **1** a person or thing that sheds. **2** a female salmon after spawning.

shedhand /'ʃedhænd/ n. Austral. & NZ an unskilled assistant in a shearing shed.

Sheela-na-gig /ˌʃiːlənəˈgɪg/ n. a medieval carved stone female figure, sometimes found on churches or castles in Britain and Ireland, shown naked with the legs wide apart and the hands emphasizing the genitals. [Ir. *Síle na gcíoch*, = Julia of the breasts]

sheen /ʃiːn/ *n.* **1** a gloss or lustre on a surface. **2** radiance, brightness. □ **sheeny** *adj.* [obs. *sheen* beautiful, resplendent f. OE *scēne*: sense assim. to SHINE]

Sheene /ʃiːn/, Barry (born Stephen Frank Sheene) (b.1950), English racing motorcyclist. He won the 500 cc. world championship in 1976 and 1977.

sheep /ʃiːp/ *n.* (*pl.* same) **1** a horned ruminant mammal of the genus *Ovis* with a thick woolly coat, esp. a domesticated animal kept in flocks for its wool or meat, and proverbial for its timidity. (*See note below.*) **2** a bashful, defenceless, or easily led person. **3** (usu. in *pl.*) **a** a member of a minister's congregation. **b** a parishioner. □ **separate the sheep from the goats** divide into superior and inferior groups (cf. Matt. 25:33). **sheep-dip 1** a preparation for cleansing sheep of vermin or preserving their wool. **2** the place where sheep are dipped in this. **sheep-run** an extensive sheepwalk, esp. in Australia. **sheep's-bit** a plant, *Jasione montana*, of the bellflower family, with flowers like those of scabious. □ **sheeplike** *adj.* [OE *scēp, scæp, scēap*]

▪ There are six species of sheep in the genus *Ovis*, as well as three others in related genera, and they are close relatives of the goats. The mouflon, *O. orientalis*, is the closest wild representative of the ancestor of the domestic sheep, *O. aries*, which was first domesticated in the Near East in about 7000 BC. There are now more than 200 breeds, including long-wool breeds that were developed mainly in Britain, and fine-wool breeds that are mostly derived from the Spanish merino.

sheepdog /ʃiːpdɒg/ *n.* **1** a dog trained to guard and herd sheep. **2** a dog of various breeds suitable for this.

sheepfold /ʃiːpfəʊld/ *n.* an enclosure for penning sheep.

sheepish /ʃiːpɪʃ/ *adj.* **1** bashful, shy, reticent. **2** embarrassed through shame. □ **sheepishly** *adv.* **sheepishness** *n.*

sheepshank /ʃiːpʃæŋk/ *n.* a knot used to shorten a rope temporarily.

sheepskin /ʃiːpskɪn/ *n.* **1** a garment or rug of sheep's skin with the wool on. **2** leather from a sheep's skin used in bookbinding.

sheepwalk /ʃiːpwɔːk/ *n. Brit.* a tract of land on which sheep are pastured.

sheer[1] /ʃɪə(r)/ *adj. & adv.* ● *adj.* **1** no more or less than; mere, unqualified, absolute (*sheer luck; sheer determination*). **2** (of a cliff or ascent etc.) perpendicular; very steep. **3** (of a textile) very thin; diaphanous. ● *adv.* **1** directly, outright. **2** perpendicularly. □ **sheerly** *adv.* **sheerness** *n.* [ME *schere* prob. f. dial. *shire* pure, clear f. OE *scīr* f. Gmc]

sheer[2] /ʃɪə(r)/ *v. & n.* ● *v.intr.* **1** esp. *Naut.* swerve or change course. **2** (foll. by *away, off*) go away, esp. from a person or topic one dislikes or fears. ● *n. Naut.* a deviation from a course. [perh. f. MLG *scheren* = SHEAR *v.*]

sheer[3] /ʃɪə(r)/ *n.* the upward slope of a ship's lines towards the bow and stern. [prob. f. SHEAR *n.*]

sheerlegs /ʃɪəlegz/ *n.pl.* (treated as *sing.*) a hoisting apparatus made from poles joined at or near the top and separated at the bottom for masting ships, installing engines, etc. [*sheer*, var. of SHEAR *n.* + LEG]

sheet[1] /ʃiːt/ *n. & v.* ● *n.* **1** a large rectangular piece of cotton or other fabric, used esp. in pairs as inner bedclothes. **2 a** a broad usu. thin flat piece of material (e.g. paper or metal). **b** (*attrib.*) made in sheets (*sheet steel*). **3** a wide continuous surface or expanse of water, ice, flame, falling rain, etc. **4** a set of unseparated postage stamps. **5** *derog.* a newspaper, esp. a disreputable one. **6** a complete piece of paper of the size in which it was made, for printing and folding as part of a book. ● *v.* **1** provide or cover with sheets. **2** *tr.* form into sheets. **3** *intr.* (of rain etc.) fall in sheets. □ **sheet lightning** lightning with its brightness diffused by reflection in clouds etc. **sheet metal** metal formed into thin sheets by rolling, hammering, etc. **sheet music 1** printed music, as opposed to performed or recorded music and books about music. **2** music published in single or interleaved sheets, not bound. [OE *scēte, scīete* f. Gmc]

sheet[2] /ʃiːt/ *n.* **1** a rope or chain attached to the lower corner of a sail for securing or controlling it. **2** (in *pl.*) the space at the bow or stern of an open boat. □ **flowing sheets** sheets eased for free movement in the wind. **sheet anchor 1** a second anchor for use in emergencies. **2** a person or thing depended on in the last resort. **sheet bend** a method of temporarily fastening one rope through the loop of another. [ME f. OE *scēata*, ON *skaut* (as SHEET[1])]

sheeting /ʃiːtɪŋ/ *n.* material for making bed linen.

Sheffield /ʃefiːld/ an industrial city in South Yorkshire; pop. (1991) 500,500. Noted for metal-working since medieval times, Sheffield

became famous for the manufacture of cutlery and silverware, and in the 19th century, after the establishment there by Henry Bessemer of a steelworks, for the production of steel.

sheikh /ʃeɪk/ *n.* (also **shaikh, sheik**) **1** a chief or head of an Arab tribe, family, or village. **2** a Muslim leader. □ **sheikhdom** *n.* [ult. f. Arab. *šayk* old man, sheikh, f. *šāka* be or grow old]

sheila /ʃiːlə/ *n. Austral. & NZ sl.* a girl or young woman. [orig. *shaler* (of unkn. orig.): assim. to the name *Sheila*]

shekel /ʃek(ə)l/ *n.* **1** the chief monetary unit of modern Israel. **2** *hist.* a silver coin and unit of weight used in ancient Israel and the Middle East. **3** (in *pl.*) *colloq.* money; riches. [Heb. *šeḳel* f. *šāḳal* weigh]

shelduck /ʃeldʌk/ *n.* (*pl.* same or **shelducks**; *masc.* **sheldrake** /-dreɪk/, *pl.* same or **sheldrakes**) a large bright-plumaged gooselike duck of the genus *Tadorna*, esp. *T. tadorna* of Eurasian and North African coasts. [ME prob. f. dial. *sheld* pied, rel. to MDu. *schillede* variegated, + DUCK[1], DRAKE]

shelf[1] /ʃelf/ *n.* (*pl.* **shelves** /ʃelvz/) **1 a** a thin flat piece of wood or metal etc. projecting from a wall, or as part of a unit, used to support books etc. **b** the flat bottom surface of a recess in a wall etc. used similarly. **2 a** a projecting horizontal ledge in a cliff face etc. **b** a reef or sandbank under water. **c** = *continental shelf.* □ **on the shelf 1** (of a woman) past the age when she might expect to be married. **2** (esp. of a retired person) no longer active or of use. **shelf-life** the amount of time for which a stored item of food etc. remains usable. **shelf-mark** a notation on a book showing its place in a library. **shelf-room** available space on a shelf. □ **shelfful** *n.* (*pl.* **-fuls**). **shelflike** *adj.* **shelved** /ʃelvd/ *adj.* [ME f. (M)LG *schelf*, rel. to OE *scylfe* partition, *scylf* crag]

shelf[2] /ʃelf/ *n. & v. Austral. sl.* ● *n.* an informer. ● *v.tr.* inform upon. [20th c.: orig. uncert.]

shell /ʃel/ *n. & v.* ● *n.* **1 a** the hard calcareous outer case of many molluscs (*cockleshell*). **b** the hard but fragile outer covering of a bird's egg, or the softer covering of a reptile's egg. **c** the usu. hard outer case of a nut, seed, etc. **d** the carapace of a tortoise, turtle, etc. **e** the wing-case or pupa-case of many insects etc. **2 a** an explosive projectile for use in a big gun or mortar. **b** a hollow metal or paper case used as a container for fireworks, explosives, cartridges, etc. **c** *US* a cartridge. **3** a mere semblance or outer form without substance. **4** anything that resembles a shell in being a hollow receptacle or outer case, esp.: **a** a light racing-boat. **b** a hollow pastry case. **c** the metal framework of a vehicle body etc. **d** the walls of an unfinished or gutted building, ship, etc. **e** an inner or roughly made coffin. **f** a building shaped like a conch. **g** the handguard of a sword. **5** *Physics* a group of electrons with almost equal energy in an atom. ● *v.* **1** *tr.* remove the shell or pod from. **2** *tr.* bombard (a town, troops, etc.) with shells. **3** *tr.* provide or cover with a shell or shells. **4** *intr.* (usu. foll. by *off*) (of metal etc.) come off in scales. **5** *intr.* (of a seed etc.) be released from a shell. □ **come out of one's shell** cease to be shy; become communicative. **shell-bit** a gouge-shaped boring bit. **shell company** a company quoted on the Stock Exchange, although not trading, and used to make take-over bids etc. **shell egg** an egg still in its shell, not dried etc. **shell game** *N. Amer.* **1** = THIMBLERIG. **2** a chancy and usu. disreputable proceeding, a confidence trick. **shell-heap** (or **-mound**) *hist.* a kitchen midden. **shell-jacket** an army officer's tight-fitting undress jacket reaching to the waist. **shell-keep** a form of Norman keep built around the top of a mound, usually on the site of an older wooden structure. **shell-lime** fine quality lime produced by burning sea shells. **shell-money** shells used as a medium of exchange, e.g. wampum. **shell out** (also *absol.*) *colloq.* **1** pay (money). **2** hand over (a required sum). **shell-out** *n.* **1** the act of shelling out. **2** a game of snooker etc. played by three or more people. **shell-pink** a delicate pale pink. **shell program** *Computing* a program which provides an interface between the user and the operating system. **shell-shock** psychological disturbance resulting from exposure to battle. **shell-shocked** suffering from shell-shock. **shell suit** a tracksuit with a soft lining and a weatherproof nylon outer 'shell', used for leisure wear. **shell-work** ornamentation consisting of shells cemented on to wood etc. □ **shelled** *adj.* **shell-less** *adj.* **shell-like** *adj.* **shellproof** *adj.* (in sense 2a of *n.*). **shelly** *adj.* [OE *sc(i)ell* f. Gmc: cf. SCALE[1]]

she'll /ʃiːl, ʃɪl/ *contr.* she will; she shall.

shellac /ʃəˈlæk/ *n. & v.* ● *n.* lac resin in thin flakes, used for making varnish (cf. LAC[1]). ● *v.tr.* (**shellacked, shellacking**) **1** varnish with shellac. **2** *US sl.* defeat or thrash soundly. [SHELL + LAC[1], transl. F *laque en écailles* lac in thin plates]

shellback /ʃelbæk/ *n. sl.* an old sailor.

Shelley[1] /ˈʃelɪ/, Mary (Wollstonecraft) (1797–1851), English writer. The daughter of William Godwin and Mary Wollstonecraft, she eloped with the poet Shelley in 1814 and married him in 1816. She is chiefly remembered as the author of the Gothic novel *Frankenstein, or the Modern Prometheus* (1818). Her other works include further novels, short stories (some with science-fiction elements, others Gothic or historical), and an edition of her husband's poems (1830).

Shelley[2] /ˈʃelɪ/, Percy Bysshe (1792–1822), English poet. He was a leading figure of the romantic movement, with radical political views which are often reflected in his work. After the collapse of his first marriage in 1814 he eloped abroad with Mary Godwin and her stepsister, marrying Mary in 1816; they settled permanently in Italy two years later. Major works include the political poems *Queen Mab* (1813) and *The Mask of Anarchy* (1819), *Prometheus Unbound* (1820), a lyrical drama on his aspirations and contradictions as a poet and radical, lyric poetry (e.g. 'Ode to the West Wind', 1820), the essay *The Defence of Poetry* (1821), vindicating the role of poetry in an increasingly industrial society, and *Adonais* (1821), an elegy on the death of Keats. Shelley was drowned in a boating accident.

shellfish /ˈʃelfɪʃ/ n. **1** an aquatic shelled mollusc, e.g. an oyster, winkle, etc. **2** a crustacean, e.g. a crab, shrimp, etc.

Shelta /ˈʃeltə/ n. an ancient secret language used by Irish gypsies and pipers, Irish and Welsh travelling tinkers, etc. It is composed partly of Irish or Gaelic words, mostly disguised by inversion or by arbitrary alteration of initial consonants. A few Shelta words have entered English, perhaps the most common being *bloke*. [19th c.: orig. unkn.]

shelter /ˈʃeltə(r)/ n. & v. ● n. **1** something serving as a shield or protection from danger, bad weather, etc. **2 a** a place of refuge provided esp. for the homeless etc. **b** *N. Amer.* an animal sanctuary. **3** a shielded condition; protection (*took shelter under a tree*). ● v. **1** tr. act or serve as shelter to; protect; conceal; defend (*sheltered them from the storm*; *had a sheltered upbringing*). **2** intr. & refl. find refuge; take cover (*sheltered under a tree*; *sheltered themselves behind the wall*). □ **shelter-belt** a line of trees etc. planted to protect crops from the wind. □ **shelterer** n. **shelterless** adj. [16th c.: perh. f. obs. *sheltron* phalanx f. OE *scieldtruma* (as SHIELD, *truma* troop)]

sheltie /ˈʃeltɪ/ n. (also **shelty**) (pl. **-ies**) a Shetland pony or sheepdog. [prob. repr. ON *Hjalti* Shetlander, as pronounced in Orkney]

shelve[1] /ʃelv/ v.tr. **1 a** abandon or defer (a plan etc.). **b** remove (a person) from active work etc. **2** put (books etc.) on a shelf. **3** fit (a cupboard etc.) with shelves. □ **shelver** n. **shelving** n. [*shelves* pl. of SHELF[1]]

shelve[2] /ʃelv/ v.intr. (of ground etc.) slope in a specified direction (*land shelved away to the horizon*). [perh. f. *shelvy* (adj.) having underwater reefs f. *shelve* (n.) ledge, f. SHELVE[1]]

shelves pl. of SHELF[1].

Shem /ʃem/ (in the Bible) a son of Noah (Gen. 10:21), traditional ancestor of the Semites.

Shema /ʃeˈmɑː/ a Hebrew text forming an important part of Jewish evening and morning prayer and used as a Jewish confession of faith, beginning 'Hear, O Israel, the Lord our God is one Lord'. [Heb., = hear]

shemozzle /ʃɪˈmɒz(ə)l/ n. (also **schemozzle**) sl. **1** a brawl or commotion. **2** a muddle. [Yiddish after mod.Heb. *šel-lō'-mazzāl* of no luck]

Shenandoah /ˌʃenənˈdəʊə/ a river of Virginia. Rising in two headstreams, one on each side of the Blue Ridge Mountains, it flows some 240 km (150 miles) northwards to join the Potomac at Harpers Ferry.

Shenandoah National Park a national park in the Blue Ridge Mountains of northern Virginia, situated to the south-east of the Shenandoah river. It was established in 1935.

shenanigan /ʃɪˈnænɪɡən/ n. (esp. in pl.) colloq. **1** high-spirited behaviour; nonsense. **2** trickery; dubious manoeuvres. [19th c.: orig. unkn.]

Shensi see SHAANXI.

Shenyang /ʃenˈjæŋ/ a city in NE China; pop. (1990) 4,500,000. An important Manchu city between the 17th and early 20th centuries, it is now the capital of the province of Liaoning. It was formerly known as Mukden.

Shenzhen /ʃenˈʒen/ an industrial city in southern China, just north of Hong Kong; pop. (est. 1986) 335,000.

Sheol /ˈʃiːəʊl, ˈʃiːɒl/ (in the Old Testament and Hebrew Bible) the underworld, the abode of the dead. [Heb. šᵉ'ôl]

shepherd /ˈʃepəd/ n. & v. ● n. **1** (fem. **shepherdess** /-dɪs/) a person employed to tend sheep, esp. at pasture. **2** a minister or member of the clergy in relation to his or her congregation. ● v.tr. **1 a** tend (sheep etc.) as a shepherd. **b** guide (followers etc.). **2** marshal or drive (a crowd etc.) like sheep. □ **the Good Shepherd** Christ. **shepherd dog** a sheepdog. **shepherd** (or **shepherding**) **satellite** Astron. a small moon orbiting close to a planetary ring, esp. of Saturn, and whose gravitational field confines the ring within a narrow band. **shepherd's crook** a staff with a hook at one end used by shepherds. **shepherd's needle** a white-flowered umbelliferous plant, *Scandix pecten-veneris*, with spiny fruit, common as a cornfield weed. **shepherd's pie** a dish of minced meat under a layer of mashed potato. **shepherd's plaid 1** a small black and white check pattern. **2** woollen cloth with this pattern. **shepherd's purse** a white-flowered cruciferous plant, *Capsella bursa-pastoris*, with triangular or cordate pods, common as a garden weed. [OE *scēaphierde* (as SHEEP, HERD)]

sherardize /ˈʃerəˌdaɪz/ v.tr. (also **-ise**) coat (iron or steel) with zinc by heating in contact with zinc dust. [*Sherard* Cowper-Coles, Engl. inventor (1867–1936)]

Sheraton /ˈʃerət(ə)n/ n. (often attrib.) a style of furniture introduced in England c.1790, with delicate and graceful forms. [Thomas *Sheraton*, Engl. furniture-maker (1751–1806)]

sherbet /ˈʃɜːbət/ n. **1 a** a flavoured sweet effervescent powder or drink. **b** *N. Amer.* a water-ice. **2** a cooling drink of sweet diluted fruit juices esp. in Arab countries. **3** *Austral. joc.* beer. [Turk. şerbet, Pers. šerbet f. Arab. šarba drink f. šariba to drink: cf. SHRUB[2], SYRUP]

sherd /ʃɜːd/ n. = POTSHERD. [var. of SHARD]

shereef (also **sherif**) var. of SHARIF.

Sheridan /ˈʃerɪd(ə)n/, Richard Brinsley (1751–1816), Irish dramatist and Whig politician. He settled in London in 1773 and became principal director of Drury Lane Theatre in 1776 and sole proprietor in 1779. His plays are comedies of manners and include *The Rivals* (1775) — whose character Mrs Malaprop gave her name to the word *malapropism* — and *The School for Scandal* (1777). He became the friend and supporter of the Whig politician Charles James Fox (1749–1806) and entered Parliament in 1780, where he became a celebrated orator, held senior government posts, and became a friend and adviser of the Prince Regent.

sheriff /ˈʃerɪf/ n. **1** *Brit.* **a** (also **High Sheriff**) (in England and Wales) the chief executive officer of the Crown in a county, responsible (now mainly nominally) for keeping the peace, administering justice, presiding over elections, etc. **b** (in England) an honorary officer elected annually in some towns. **c** (in Scotland) the chief judge of a county or district. **2** *US* an elected officer in a county, responsible for keeping the peace. □ **sheriff court** *Sc.* a county court. **sheriff-depute** *Sc.* the chief judge of a county or district. □ **sheriffalty** n. (pl. **-ies**). **sheriffdom** n. **sheriffhood** n. **sheriffship** n. [OE *scīr-gerēfa* (as SHIRE, REEVE[1])]

Sherlock /ˈʃɜːlɒk/ n. **1** a person who investigates mysteries or shows great perceptiveness. **2** a private detective. [*Sherlock Holmes* (see HOLMES[3])]

Sherman /ˈʃɜːmən/, William Tecumseh (1820–91), American general. He held various commands in the American Civil War from its outset in 1861, and in Mar. 1864 succeeded Ulysses S. Grant as chief Union commander in the west. He set out with 60,000 men on a march through Georgia, during which he crushed Confederate forces and broke civilian morale through his policy of deliberate destruction of the South's sources of supply. In 1869 he was appointed commander of the US army, a post he held until his retirement in 1884.

Sherpa /ˈʃɜːpə/ n. (pl. same or **Sherpas**) a member of a Himalayan people living on the borders of Nepal and Tibet, and skilled in mountaineering. [Tibetan *sharpa* inhabitant of an eastern country]

Sherrington /ˈʃerɪŋtən/, Sir Charles Scott (1857–1952), English physiologist. His researches contributed greatly to understanding of the nervous system, particularly concerning motor pathways, sensory nerves in muscles, and the areas innervated by spinal nerves. He introduced the concept of reflex actions and the reflex arc and was the first to apply the term *synapsis* (later *synapse*) to the junction of two nerve cells. Sherrington shared a Nobel Prize in 1932.

sherry /ˈʃerɪ/ n. (pl. **-ies**) **1** a fortified wine orig. from southern Spain. **2** a glass of this. □ **sherry cobbler** see COBBLER 2. **sherry-glass** a small wineglass used for sherry. [earlier *sherris* f. Sp. (*vino de*) *Xeres* (wine of) JEREZ]

's-Hertogenbosch /ˌsheətəʊxən'bɒs/ a city in the southern Netherlands, the capital of North Brabant; pop. (1991) 92,060.

she's /ʃiːz, ʃɪz/ *contr.* **1** she is. **2** she has.

Shetland Islands /ˈʃetlənd/ (also **Shetland, Shetlands**) a group of about 100 islands off the north coast of Scotland, north-east of the Orkneys, constituting the administrative region of Shetland; pop. (1991) 22,020; chief town, Lerwick. Taken by Norse invaders in the 8th and 9th centuries, the islands became a part of Scotland in 1472 (see also ORKNEY ISLANDS). Noted for the production of knitwear, they have provided in the late 20th century a base for the North Sea oil and gas industries. □ **Shetlander** *n.*

Shetland lace *n.* knitted lace made with Shetland wool or following traditional Shetland openwork designs.

Shetland pony *n.* a pony of a small hardy rough-coated breed.

Shetland sheep *n.* a sheep of a hardy short-tailed breed, kept esp. for its fine wool.

Shetland sheepdog *n.* a small dog of a collie-like breed.

Shetland wool *n.* a fine loosely twisted wool from Shetland sheep.

sheva var. of SCHWA.

Shevardnadze /ˌʃevɑːdˈnɑːdzɪ/, Eduard (Amvrosievich) (b.1928), Soviet statesman and head of state of Georgia since 1992. He became a candidate member of the Soviet Politburo in 1978 and a full member in 1985. In the same year Shevardnadze was appointed Minister of Foreign Affairs under Mikhail Gorbachev, a position he retained until his resignation in 1990. While in office, Shevardnadze supported Gorbachev's commitment to détente and played a key role in arms control negotiations with the West. In 1992 Shevardnadze was elected head of state of his native Georgia, following the toppling of President Zviad Gamsakhurdia (1939–94).

Shevat var. of SEBAT.

shew *archaic* var. of SHOW.

shewbread /ˈʃəʊbred/ *n.* twelve loaves that were displayed in a Jewish temple and renewed each sabbath.

Shia /ˈʃiːə/ *n. & adj.* (also **Shiah**) ● *n.* the smaller of the two main branches of Islam (the other being Sunni, found esp. in Iran. (See note below.) ● *adj.* of or relating to Shia. [Arab. šīʿa party (of Ali)]

▪ The adherents of Shia (*Shiites*) differ from the Sunni Muslims in their understanding of the Sunna and in their rejection of the first three Sunni caliphs and acceptance of Ali, Muhammad's son-in-law and the fourth caliph, as the Prophet's first true successor; Ali and his descendants are believed to form the line of true imams (successors to Muhammad). The Shia branch broke away after the killing of Ali's son Husayn at the battle of Karbala in AD 680. There are many Shia sects, defined according to which imam they believe to be the twelfth and final one, and also according to the nature and extent of the imam's authority. The twelfth imam, who disappeared in the late 9th century, is thought to have been forced into hiding by the repressive political rule of the majority community, and is expected to return and to triumph over injustice (see also MAHDI).

shiatsu /ʃɪˈætsuː/ *n.* (also **shiatsu massage**) a kind of massage originating in Japan and following the same principles as acupuncture, in which the thumbs, palms, etc. are used to apply pressure to certain points of the body. [Jap., = finger pressure]

shibboleth /ˈʃɪbəleθ/ *n.* a custom, doctrine, phrase, etc. distinguishing a particular class or group of people. [ME f. Heb. *šibbōleṯ* ear of corn, used as a test of nationality for its difficult pronunciation (Judg. 12:6)]

shicer /ˈʃaɪsə(r)/ *n. Austral.* **1** *Mining* an unproductive claim or mine. **2** *sl.* **a** a swindler, welsher, or cheat. **b** a worthless thing; a failure. [G *Scheisser* contemptible person]

shicker /ˈʃɪkə(r)/ *adj.* (also **shickered** /ˈʃɪkəd/) *Austral. & NZ sl.* drunk. [Yiddish *shiker* f. Heb. *šikkôr* f. *šākar* be drunk]

shied past and past part. of SHY[1], SHY[2].

shield /ʃiːld/ *n. & v.* ● *n.* **1 a** esp. *hist.* a piece of armour esp. of metal, carried on the arm or in the hand to deflect blows from the head or body. **b** a thing serving to protect (*insurance is a shield against disaster*). **2** a thing resembling a shield, esp.: **a** a trophy in the form of a shield. **b** a protective plate or screen in machinery etc. **c** a shieldlike part of an animal, esp. a shell. **d** a similar part of a plant. **e** *Geol.* a large rigid area of the earth's crust, usu. of Precambrian rock, which has been unaffected by later tectonic processes (*Canadian Shield*). **f** *US* a policeman's shield-shaped badge. **3** *Heraldry* a stylized representation of a shield used for displaying a coat of arms etc. ● *v.tr.* protect or screen, esp. from blame or lawful punishment. □ **shield bug** a broad flat hemipteran bug, esp. of the family Pentatomidae. **shield fern 1** a common fern of the genus *Polystichum*, with shield-shaped indusia. **2** =

BUCKLER 2. **shield volcano** *Geol.* a broad domed volcano with gently sloping sides. □ **shieldless** *adj.* [OE *sc(i)eld* f. Gmc: prob. orig. = board, rel. to SCALE[1]]

shieling /ˈʃiːlɪŋ/ *n.* (also **shealing**) *Sc.* **1** a roughly constructed hut orig. esp. for pastoral use. **2** pasture for cattle. [Sc. *shiel* hut: ME, of unkn. orig.]

shier *compar.* of SHY[1].

shiest *superl.* of SHY[1].

shift /ʃɪft/ *v. & n.* ● *v.* **1** *intr. & tr.* change or move or cause to change or move from one position to another. **2** *tr.* remove, esp. with effort (*washing won't shift the stains*). **3** *sl.* **a** *intr.* hurry (*we'll have to shift!*). **b** *tr.* consume (food or drink) hastily or in bulk. **c** *tr. colloq.* sell (goods), esp. quickly, in large quantities, or dishonestly. **4** *intr.* contrive or manage as best one can. **5** *N. Amer.* **a** *tr.* change (gear) in a vehicle. **b** *intr.* change gear. **6** *intr.* (of cargo) get shaken out of place. **7** *intr. archaic* be evasive or indirect. ● *n.* **1 a** the act or an instance of shifting. **b** the substitution of one thing for another; a rotation. **2 a** a relay of workers (*the night shift*). **b** the time for which they work (*an eight-hour shift*). **3 a** a device, stratagem, or expedient. **b** a dodge, trick, or evasion. **4 a** a woman's straight unwaisted dress. **b** *archaic* a loose-fitting undergarment. **5** *Physics* etc. a displacement of spectral lines (see also *red shift*). **6** (also **sound shift**) a systematic change in pronunciation as a language evolves. **7** a key on a keyboard used to switch between lower and upper case etc. **8** *Bridge* **a** a change of suit in bidding. **b** *US* a change of suit in play. **9** the positioning of successive rows of bricks so that their ends do not coincide. **10** *N. Amer.* **a** a gear lever in a motor vehicle. **b** a mechanism for this. □ **make shift** manage or contrive; get along somehow (*made shift without it*). **shift off** get rid of (responsibility etc.) to another. **shift for oneself** rely on one's own efforts. **shifting cultivation** a form of agriculture in which an area of ground is cleared of vegetation, cultivated for a few years, and then abandoned for a new area. **shift one's ground** take up a new position in an argument etc. **shift off** get rid of (responsibility etc.) to another. □ **shiftable** *adj.* **shifter** *n.* [OE *sciftan* arrange, divide, etc., f. Gmc]

shiftless /ˈʃɪftlɪs/ *adj.* lacking resourcefulness; lazy; inefficient. □ **shiftlessly** *adv.* **shiftlessness** *n.*

shifty /ˈʃɪftɪ/ *adj. colloq.* (**shiftier, shiftiest**) not straightforward; evasive; deceitful. □ **shiftily** *adv.* **shiftiness** *n.*

shigella /ʃɪˈgelə/ *n.* a bacterium of the genus *Shigella*, often found in the gut, esp. one that can cause dysentery. [mod.L f. Kiyoshi *Shiga*, Jap. bacteriologist (1870–1957) + dimin. suffix]

shih-tzu /ˈʃiːˈtsuː/ *n.* a breed of dog with long silky erect hair and short legs. [Chin. *shizi* lion]

shiitake /ʃɪˈtɑːkeɪ, ʃɪˈtɑː-/ *n.* (in full **shiitake mushroom**) an edible mushroom, *Lentinus edodes*, cultivated in Japan and China on logs of oak trees etc. [Jap., f. *shii* a kind of oak + *take* mushroom]

Shiite /ˈʃiːaɪt/ *n. & adj.* ● *n.* an adherent of the Shia branch of Islam. ● *adj.* of or relating to Shia. □ **Shiism** /ˈʃiːɪz(ə)m/ *n.*

Shijiazhuang /ˌʃiːdʒɪəˈʒwæŋ/ a city in NE central China, capital of Hebei province; pop. (1990) 1,320,000.

shikar /ʃɪˈkɑː(r)/ *n. Ind.* hunting. [Urdu f. Pers. *šikār*]

Shikoku /ʃɪˈkəʊkuː/ the smallest of the four main islands of Japan, constituting an administrative region; pop. (1990) 4,195,000; capital, Matsuyama. It is divided from Kyushu to the west and southern Honshu to the north by the Inland Sea.

shiksa /ˈʃɪksə/ *n.* often *offens.* (used by Jews) a Gentile girl or woman. [Yiddish *shikse* f. Heb. *šiqṣâ* f. *sheqeṣ* detested thing + *-â* fem. suffix]

shill /ʃɪl/ *n. N. Amer.* a person employed to decoy or entice others into buying, gambling, etc. [prob. f. earlier *shillaber*, of unkn. orig.]

shillelagh /ʃɪˈleɪlə, -lɪ/ *n.* a thick stick of blackthorn or oak used in Ireland esp. as a weapon. [*Shillelagh* in County Wicklow, Ireland]

shilling /ˈʃɪlɪŋ/ *n.* **1** *hist.* a former British coin and monetary unit equal to one-twentieth of a pound or twelve pence. **2** a monetary unit in Kenya, Tanzania, Uganda, and Somalia. □ **shilling-mark** *hist.* = SOLIDUS. **take the King's** (or **Queen's**) **shilling** *hist.* enlist as a soldier in the British army (formerly a soldier was paid a shilling on enlisting). [OE *scilling*, f. Gmc]

Shillong /ʃɪˈlɒŋ/ a city in the far north-east of India, capital of the state of Meghalaya; pop. (1991) 130,690. It is situated near the northern border of Bangladesh at an altitude of 1,500 m (4,920 ft).

shilly-shally /ˈʃɪlɪˌʃælɪ/ *v., adj., & n.* ● *v.intr.* (**-ies, -ied**) hesitate to act or choose; be undecided; vacillate. ● *adj.* vacillating. ● *n.* indecision; vacillation. □ **shilly-shallyer** *n.* (also **-shallier**). [orig. *shill I, shall I*, redupl. of *shall I?*]

shim /ʃɪm/ *n. & v.* ● *n.* a thin strip of material used in machinery etc. to make parts fit. ● *v.tr.* (**shimmed, shimming**) fit or fill up with a shim. [18th c.: orig. unkn.]

shimmer /ˈʃɪmə(r)/ *v. & n.* ● *v.intr.* shine with a tremulous or faint diffused light. ● *n.* such a light. □ **shimmeringly** *adv.* **shimmery** *adj.* [OE *scymrian* f. Gmc: cf. SHINE]

shimmy /ˈʃɪmɪ/ *n. & v.* ● *n.* (*pl.* **-ies**) **1** *hist.* a kind of ragtime dance in which the whole body is shaken. **2** *archaic colloq.* = CHEMISE. **3** an abnormal vibration of esp. the front wheels of a motor vehicle. ● *v.intr.* (**-ies, -ied**) **1 a** *hist.* dance a shimmy. **b** move in a similar manner. **2** shake or vibrate abnormally. [20th c.: orig. uncert.]

shin /ʃɪn/ *n. & v.* ● *n.* **1** the front of the leg below the knee. **2** a cut of beef from the lower foreleg. ● *v.tr. &* (usu. foll. by *up, down*) *intr.* (**shinned, shinning**) climb quickly by clinging with the arms and legs. □ **shinbone** = TIBIA 1. **shin-pad** (or **-guard**) a protective pad for the shins, worn when playing football etc. [OE *sinu*]

shindig /ˈʃɪndɪg/ *n. colloq.* **1** a festive, esp. noisy, party. **2** = SHINDY 1. [prob. f. SHINDY]

shindy /ˈʃɪndɪ/ *n.* (*pl.* **-ies**) *colloq.* **1** a brawl, disturbance, or noise (*kicked up a shindy*). **2** = SHINDIG 1. [perh. alt. of SHINTY]

shine /ʃaɪn/ *v. & n.* ● *v.* (*past* and *past part.* **shone** /ʃɒn/ or **shined**) **1** *intr.* emit or reflect light; be bright; glow (*the lamp was shining; his face shone with gratitude*). **2** *intr.* (of the sun, a star, etc.) not be obscured by clouds etc.; be visible. **3** *tr.* cause (a lamp etc.) to shine. **4** *tr.* (*past* and *past part.* **shined**) make bright; polish (*shined his shoes*). **5** *intr.* be brilliant in some respect; excel (*does not shine in conversation; is a shining example*). ● *n.* **1** light; brightness, esp. reflected. **2** a high polish; lustre. **3** *US* the act or an instance of shining esp. shoes. □ **shine up to** *US* seek to ingratiate oneself with. **take the shine out of 1** spoil the brilliance or newness of. **2** throw into the shade by surpassing. **take a shine to** *colloq.* take a fancy to; like. □ **shiningly** *adv.* [OE *scīnan* f. Gmc]

shiner /ˈʃaɪnə(r)/ *n.* **1** a thing that shines. **2** *colloq.* a black eye. **3** *US* a small silvery freshwater fish, esp. of the genus *Notropis*. **4** (usu. in *pl.*) *sl.* **a** *archaic* money. **b** a jewel.

shingle[1] /ˈʃɪŋg(ə)l/ *n.* (in *sing.* or *pl.*) small rounded pebbles, esp. on a sea-shore. □ **shingly** *adj.* [16th c.: orig. uncert.]

shingle[2] /ˈʃɪŋg(ə)l/ *n. & v.* ● *n.* **1** a rectangular wooden tile used on roofs, spires, or esp. walls. **2** *archaic* **a** shingled hair. **b** the act of shingling hair. **3** *N. Amer.* a small signboard, esp. of a doctor, lawyer, etc. ● *v.tr.* **1** roof or clad with shingles. **2** *archaic* **a** cut (a woman's hair) short. **b** cut the hair of (a person or head) in this way. [ME app. f. L *scindula*, earlier *scandula*]

shingles /ˈʃɪŋg(ə)lz/ *n.pl.* (usu. treated as *sing.*) an acute painful inflammation of the nerve ganglia, with a skin eruption often forming a girdle around the middle of the body, caused by the same virus that causes chickenpox. Also called *herpes zoster*. [ME f. med.L *cingulus* f. L *cingulum* girdle f. *cingere* gird]

Shining Path a Peruvian Maoist revolutionary movement and terrorist organization, founded in 1970 and led by Abimael Guzmán (b.1934) until his capture and imprisonment in 1992. At first the movement operated in rural areas, but in the 1980s it began to launch terrorist attacks in Peruvian towns and cities. [transl. Sp. SENDERO LUMINOSO]

Shinkansen /ˈʃɪŋkɑːnˌsen/ *n.* (*pl.* same) **1** a Japanese railway system carrying high-speed trains. **2** a train operating on such a system. [Jap., f. *shin* new + *kansen* (railway) main line]

shinny /ˈʃɪnɪ/ *v.intr.* (**-ies, -ied**) (usu. foll. by *up, down*) *N. Amer. colloq.* shin up or down (a tree etc.).

Shinto /ˈʃɪntəʊ/ *n.* an ancient Japanese religion revering ancestors and nature-spirits and emphasizing ritual and standards of behaviour rather than doctrine. It has no official scripture, but there are records of its traditions dating from the early 8th century. Central to Shinto is the belief in sacred power (*kami*) in both animate and inanimate things; in its mythology the sun-goddess Amaterasu was the ancestress of the imperial household. Shinto became closely associated with the state, a position that it held until after the Second World War, when it was disestablished and the Emperor Hirohito was obliged to disavow his claim to divine descent. □ **Shintoism** *n.* **Shintoist** *n.* [Jap. f. Chin. *shen dao* way of the gods]

shinty /ˈʃɪntɪ/ *n.* (*pl.* **-ies**) **1** a Scottish twelve-a-side game similar to hockey, played with a ball and curved sticks but using taller goalposts. It was taken to Scotland by Irish invaders some 1,400 years ago and is derived from the Irish game of hurling; it is now played chiefly in the Scottish Highlands. **2** a stick or ball used in shinty. [earlier *shinny*, app. f. the cry used in the game *shin ye, shin you, shin t' ye*, of unkn. orig.]

shiny /ˈʃaɪnɪ/ *adj.* (**shinier, shiniest**) **1** having a shine; glistening; polished; bright. **2** (of clothing, esp. the seat of trousers etc.) having the nap worn off. □ **shinily** *adv.* **shininess** *n.*

ship /ʃɪp/ *n. & v.* ● *n.* **1 a** a large seagoing vessel. (*See note below.*) **b** a sailing-vessel with a bowsprit and three, four, or five square-rigged masts. **2** *US colloq.* an aircraft. **3** a spaceship. **4** *colloq.* a boat, esp. a racing-boat. ● *v.* (**shipped, shipping**) **1** *tr.* put, take, or send away (goods, passengers, sailors, etc.) on board ship. **2** *tr.* **a** take in (water) over the side of a ship, boat, etc. **b** take (oars) from the rowlocks and lay them inside a boat. **c** fix (a rudder etc.) in its place on a ship etc. **d** step (a mast). **3** *intr.* **a** take ship; embark. **b** (of a sailor) take service on a ship (*shipped for Africa*). **4** *tr.* deliver (goods) to a forwarding agent for conveyance. □ **ship-breaker** a contractor who breaks up old ships. **ship-broker** an agent in shipping goods and insuring ships. **ship-canal** a canal large enough for ships to pass inland. **ship** (or **ship's**) **chandler** see CHANDLER. **ship-fever** typhus. **ship of the desert** the camel. **ship off 1** send or transport by ship. **2** *colloq.* send (a person) away. **ship of the line** *hist.* a sailing warship large enough to take its place in the line of battle. **ship-rigged** square-rigged. **ship's articles** the terms on which seamen take service on a ship. **ship's biscuit** *hist.* a hard coarse kind of biscuit kept and eaten on board ship. **ship's boat** a small boat carried on board a ship. **ship's company** a ship's crew. **ship's corporal** see CORPORAL[1] 2. **ship a sea** be flooded by a wave. **ship's husband** an agent appointed by the owners to see to the provisioning of a ship in port. **ship's papers** documents establishing the ownership, nationality, nature of the cargo, etc., of a ship. **take ship** embark. **when a person's ship comes home** (or **in**) when a person's fortune is made. □ **shipless** *adj.* **shippable** *adj.* [OE *scip, scipian* f. Gmc]

▪ The first boats were probably roughly constructed rafts made of logs or bundles of reeds, followed by canoes made from hollowed-out tree-trunks. Later, stem and stern pieces were fixed to a keel, with ribs to support the sides. Among the most ancient vessels known are a dugout canoe found at Pese in Holland (8th millennium BC), and an oared funeral ship buried beside the tomb of the Egyptian pharaoh Cheops (4th millennium BC). Early ships were propelled by sail, sometimes supplemented by oars. In the mid-19th century some wooden naval vessels were ironclad, eventually giving way to the steel that is now almost universal. Steamships evolved in the 19th century, using reciprocating steam engines and later steam turbines; these in turn have been replaced by turbocharged diesel engines or gas turbines. Modern ships are worked with small crews through extensive use of automatic control systems. It has been customary in the West to think of ships as feminine since at least the 16th century. (See also BOAT.)

-ship /ʃɪp/ *suffix* forming nouns denoting: **1** a quality or condition (*friendship; hardship*). **2** status, office, or honour (*authorship; lordship*). **3** a tenure of office (*chairmanship*). **4** a skill in a certain capacity (*workmanship*). **5** the collective individuals of a group (*membership*). [OE *-scipe* etc. f. Gmc]

shipboard /ˈʃɪpbɔːd/ *n.* (usu. *attrib.*) used or occurring on board a ship (*a shipboard romance*). □ **on shipboard** on board ship.

shipbuilder /ˈʃɪpˌbɪldə(r)/ *n.* a person, company, etc., that constructs ships. □ **shipbuilding** *n.*

ship burial *n. Archaeol.* a burial in a wooden ship under a mound. The practice is found in Scandinavia and parts of the British Isles in the pagan Anglo-Saxon and Viking periods (6th–11th centuries AD).

shiplap /ˈʃɪplæp/ *v. & n.* ● *v.tr.* fit (boards) together for cladding etc. so that each overlaps the one below. ● *n.* such cladding.

shipload /ˈʃɪpləʊd/ *n.* a quantity of goods forming a cargo.

shipmaster /ˈʃɪpˌmɑːstə(r)/ *n.* a ship's captain.

shipmate /ˈʃɪpmeɪt/ *n.* a fellow member of a ship's crew.

shipment /ˈʃɪpmənt/ *n.* **1** an amount of goods shipped; a consignment. **2** the act or an instance of shipping goods etc.

ship-money /ˈʃɪpˌmʌnɪ/ *n. hist.* a tax originally levied on coastal areas in medieval England to provide ships for the navy during times of war. Charles I revived the tax in 1634, but attempted to impose it on the whole country, in peacetime, and without parliamentary consent; the unpopularity of this was a contributory factor in the outbreak of the English Civil War.

shipowner /ˈʃɪpˌəʊnə(r)/ *n.* a person owning a ship or ships or shares in ships.

shipper /ˈʃɪpə(r)/ *n.* a person or company that sends or receives goods esp. by ship. [OE *scipere* (as SHIP)]

shipping /'ʃɪpɪŋ/ n. **1** the act or an instance of shipping goods etc. **2** ships, esp. the ships of a country, port, etc. □ **shipping-agent** a person acting for a ship or ships at a port etc. **shipping-articles** = *ship's articles.* **shipping-bill** *Brit.* a manifest of goods shipped. **shipping forecast area** each of the sea areas covered by British weather forecasts from the Meteorological Office. **shipping-master** *Brit.* an official presiding at the signing of ship's articles, paying off of seamen, etc. **shipping-office** the office of a shipping-agent or -master.

shipshape /'ʃɪpʃeɪp/ adv. & predic.adj. in good order; trim and neat.

shipway /'ʃɪpweɪ/ n. a slope on which a ship is built and down which it slides to be launched.

shipworm /'ʃɪpwɜːm/ n. = TEREDO.

shipwreck /'ʃɪprek/ n. & v. ● n. **1 a** the destruction of a ship by a storm, striking rocks, foundering, etc. **b** a ship so destroyed. **2** (often foll. by *of*) the destruction of hopes, dreams, etc. ● v. **1** tr. inflict shipwreck on (a ship, a person's hopes, etc.). **2** intr. suffer shipwreck.

shipwright /'ʃɪpraɪt/ n. **1** a shipbuilder. **2** a ship's carpenter.

shipyard /'ʃɪpjɑːd/ n. a place where ships are built, repaired, etc.

shiralee /'ʃɪrəˌliː/ n. Austral. a tramp's swag or bundle. [20th c.: orig. unkn.]

Shiraz[1] /'ʃɪəræz, ʃɪ'ræz/ a city in SW central Iran; pop. (1991) 965,000. An important cultural centre since the 4th century BC, the city is noted for the school of miniature painting based there between the 14th and 16th centuries, and for the manufacture of carpets.

Shiraz[2] /'ʃɪəræz, ʃɪ'ræz/ n. **1** a variety of black grape used in wine-making. **2** a vine on which this grape grows. **3** a red wine made from these grapes. [SHIRAZ[1], app. alt. of F *syrah* infl. by the belief that the vine was brought from Iran by the Crusaders]

shire /'ʃaɪə(r)/ n. **1** *Brit.* **1** a county. (*See note below.*) **2** (**the Shires**) **a** a group of English counties with names ending or formerly ending in -shire, extending NE from Hampshire and Devon. **b** the midland counties of England. **c** the fox-hunting district of mainly Leicestershire and Northants. **3** *Austral.* a rural area with its own elected council. □ **shire county** a non-metropolitan county of the UK. **shire-horse** a heavy powerful type of draught-horse with long white hair covering the lower part of the leg, bred chiefly in the midland counties of England. **shire-moot** the judicial assembly of the shire in Old English times. [OE *scīr*, OHG *scīra* care, official charge: orig. unkn.]

▪ In pre-Norman England a shire was an administrative district made up of a number of smaller districts (*hundreds* or *wapentakes*), ruled jointly by an alderman or sheriff. Under Norman rule this division was retained but the term *shire* was replaced by French *counté* (= county). Until local government reorganization in 1974 *shire* and *county* were generally synonymous, indicating the major rural area of local government, but since that date *shire counties* and *metropolitan counties* have been distinguished.

-shire /ʃɪə(r), ʃə(r)/ suffix forming the names of counties (*Derbyshire; Hampshire*).

shirk /ʃɜːk/ v. & n. ● v.tr. (also absol.) shrink from; avoid; get out of (duty, work, responsibility, fighting, etc.). ● n. a person who shirks. □ **shirker** n. [obs. *shirk* (n.) sponger, perh. f. G *Schurke* scoundrel]

shirr /ʃɜː(r)/ n. & v. ● n. **1** two or more rows of esp. elastic gathered threads in a garment etc. forming smocking. **2** elastic webbing. ● v.tr. **1** gather (material) with parallel threads. **2** *US* bake (eggs) without shells. □ **shirring** n. [19th c.: orig. unkn.]

shirt /ʃɜːt/ n. **1** a man's upper-body garment of cotton etc., having a collar, sleeves, and esp. buttons down the front, and often worn under a jacket or sweater. **2** a similar garment worn by a woman; a blouse. **3** = NIGHTSHIRT. □ **keep one's shirt on** colloq. keep one's temper. **put one's shirt on** colloq. bet all one has on; be sure of. **shirt blouse** = sense 2 of n. **shirt-dress** = SHIRTWAISTER. **shirt-front** the breast of a shirt, esp. of a stiffened evening shirt. **the shirt off one's back** colloq. one's last remaining possessions. **shirt-tail** the lower curved part of a shirt below the waist. □ **shirted** adj. **shirting** n. **shirtless** adj. [OE *scyrte*, corresp. to ON *skyrta* (cf. SKIRT) f. Gmc: cf. SHORT]

shirtsleeve /'ʃɜːtsliːv/ n. (usu. in pl.) the sleeve of a shirt. □ **in shirtsleeves** wearing a shirt with no jacket etc. over it.

shirtwaist /'ʃɜːtweɪst/ n. esp. *US* a woman's blouse resembling a shirt.

shirtwaister /'ʃɜːtˌweɪstə(r)/ n. *US* a woman's dress with a bodice like a shirt. [SHIRT, WAIST]

shirty /'ʃɜːtɪ/ adj. (**shirtier, shirtiest**) colloq. angry; annoyed. □ **shirtily** adv. **shirtiness** n.

shish kebab /ˌʃɪʃ kɪ'bæb/ n. a dish of pieces of marinated meat and vegetables cooked and served on skewers. [Turk. *şiş kebabı* f. *şiş* skewer, KEBAB roast meat]

shit /ʃɪt/ v., n., & int. (also **shite** /ʃaɪt/) coarse sl. ● v. (**shitting;** past and past part. **shitted** or **shit** or **shat** /ʃæt/) **1** intr. & tr. expel faeces from the body or cause (faeces etc.) to be expelled. **2** tr. soil (one's clothes, oneself) with faeces. **3** refl. be very frightened. ● n. **1** faeces. **2** an act of defecating. **3** a contemptible or worthless person or thing. **4** rubbish; nonsense. **5** an intoxicating drug, esp. cannabis. ● int. an exclamation of disgust, anger, etc. □ **be in the shit** be in trouble or a difficult or unpleasant situation. **not give a shit** not care at all. **shit-scared** terrified. **up shit creek (without a paddle)** in an unpleasant situation or an awkward predicament. **when the shit hits the fan** at the moment of crisis or in its disastrous aftermath. [OE *scītan* (unrecorded) f. Gmc]

shitty /'ʃɪtɪ/ adj. (**shittier, shittiest**) coarse sl. **1** disgusting, contemptible. **2** covered with excrement.

Shiva var. of SIVA.

Shivaji /ʃɪ'vɑːdʒɪ/ (also **Sivaji**) (1627–80), Indian raja of the Marathas 1674–80. In 1659 he raised a Hindu revolt against Muslim rule in Bijapur, southern India, inflicting a crushing defeat on the army of the sultan of Bijapur. Shivaji was later captured by the Mogul emperor Aurangzeb, but escaped in 1666 and proceeded to expand Maratha territory. He had himself crowned raja in 1674; during his reign he enforced religious toleration throughout the Maratha empire and blocked Mogul expansionism by forming an alliance with the sultans in the south.

shivaree esp. *US* var. of CHARIVARI.

shiver[1] /'ʃɪvə(r)/ v. & n. ● v.intr. **1** tremble with cold, fear, etc. **2** suffer a quick trembling movement of the body; shudder. ● n. **1** a momentary shivering movement. **2** (in pl.; prec. by the) an attack of shivering, esp. from fear or horror (*got the shivers in the dark*). □ **shiverer** n. **shiveringly** adv. **shivery** adj. [ME *chivere*, perh. f. *chavele* chatter (as JOWL[1])]

shiver[2] /'ʃɪvə(r)/ n. & v. ● n. (esp. in pl.) each of the small pieces into which esp. glass is shattered when broken; a splinter. ● v.tr. & intr. break into shivers. □ **shiver my timbers** a reputed piratical curse. [ME *scifre*, rel. to OHG *scivaro* splinter f. Gmc]

shivoo /ʃɪ'vuː/ n. Austral. colloq. a party or celebration.

Shizuoka /ˌʃɪzuː'əʊkə/ a port on the south coast of the island of Honshu in Japan; pop. (1990) 472,200.

Shkodër /'ʃkəʊdə(r)/ (Italian **Scutari** /skuː'tɑːrɪ/) a city in NW Albania, near the border with Montenegro; pop. (1988) 71,000.

shoal[1] /ʃəʊl/ n. & v. ● n. **1** a great number of fish swimming together (cf. SCHOOL[2]). **2** a multitude; a crowd (*shoals of letters*). ● v.intr. (of fish) form shoals. [prob. re-adoption of MDu. *schōle* SCHOOL[2]]

shoal[2] /ʃəʊl/ n., v., & adj. ● n. **1 a** an area of shallow water. **b** a submerged sandbank visible at low water. **2** (esp. in pl.) hidden danger or difficulty. ● v. **1** intr. (of water) get shallower. **2** tr. (of a ship etc.) move into a shallower part of (water). ● adj. (of water) shallow. □ **shoaly** adj. [OE *sceald* f. Gmc, rel. to SHALLOW]

shoat /ʃəʊt/ n. *US* a young pig, esp. newly weaned. [ME: cf. W.Flem. *schote*]

shock[1] /ʃɒk/ n. & v. ● n. **1** a violent collision, impact, tremor, etc. **2** a sudden and disturbing effect on the emotions, physical reactions, etc. (*the news was a great shock*). **3** an acute state of prostration following a wound, pain, etc., esp. when much blood is lost (*in shock*). **4** = *electric shock.* **5** a disturbance causing instability in an organization, monetary system, etc. **6** N. Amer. a shock absorber. ● v. **1** tr. affect with shock; horrify; outrage; disgust; sadden. **b** (absol.) cause shock. **2** tr. (esp. in passive) affect with an electric or pathological shock. **3** intr. experience shock (*I don't shock easily*). **4** intr. archaic collide violently. □ **shock absorber** a device on a vehicle etc. for absorbing shocks, vibrations, etc. **shock-brigade** (or **-workers**) a body of esp. voluntary workers in the former USSR engaged in an especially arduous task. **shock stall** excessive strain produced by air resistance on an aircraft approaching the speed of sound. **shock tactics 1** sudden and violent action. **2** Mil. a massed cavalry charge. **shock therapy** (or **treatment**) Psychol. a method of treating depressive patients by electric shock or drugs inducing coma and convulsions. **shock troops** troops specially trained for assault. **shock wave** a sharp change of pressure in a narrow region travelling through air etc. caused by explosion or by a body moving faster than sound. □ **shockable** adj. **shockability** /ˌʃɒkə'bɪlɪtɪ/ n. [F *choc, choquer*, of unkn. orig.]

shock[2] /ʃɒk/ n. & v. ● n. a group of usu. twelve corn-sheaves stood up with their heads together in a field. ● *v.tr.* arrange (corn) in shocks. [ME, perh. repr. OE *sc(e)oc* (unrecorded)]

shock³ /ʃɒk/ *n.* (usu. foll. by *of*) an unkempt or shaggy mass of hair. [cf. obs. *shock(-dog)*, earlier *shough*, shaggy-haired poodle]

shocker /ˈʃɒkə(r)/ *n. colloq.* **1** a shocking, horrifying, unacceptable, etc. person or thing. **2** *hist.* a sordid or sensational novel etc. **3** a shock absorber.

shocking /ˈʃɒkɪŋ/ *adj. & adv.* ● *adj.* **1** causing indignation or disgust. **2** *colloq.* very bad (*shocking weather*). ● *adv. colloq.* extremely, terribly (*shocking bad manners*). □ **shocking pink** a vibrant shade of pink. □ **shockingly** *adv.* **shockingness** *n.*

Shockley /ˈʃɒklɪ/, William (Bradford) (1910–89), American physicist. He worked mainly at the Bell Telephone Laboratories, where he organized a group to research into solid-state physics. By 1948 they had developed the transistor, which was eventually to replace the thermionic valve, and Shockley shared with his co-workers the Nobel Prize for physics in 1958. He was appointed Professor of Engineering Science at Stanford in 1963, and later became a controversial figure because of his views on a supposed connection between race and intelligence.

shockproof /ˈʃɒkpruːf/ *adj.* resistant to the effects of (esp. physical) shock.

shod past and past part. of SHOE.

shoddy /ˈʃɒdɪ/ *adj. & n.* ● *adj.* (**shoddier, shoddiest**) **1** trashy; shabby; poorly made. **2** counterfeit. ● *n.* (*pl.* **-ies**) **1 a** an inferior cloth made partly from the shredded fibre of old woollen cloth. **b** such fibre. **2** any thing of shoddy quality. □ **shoddily** *adv.* **shoddiness** *n.* [19th c.: orig. dial.]

shoe /ʃuː/ *n. & v.* ● *n.* **1** either of a pair of protective foot-coverings of leather, plastic, etc., having a sturdy sole and, in Britain, not reaching above the ankle. **2** a metal rim nailed to the hoof of a horse etc.; a horseshoe. **3** anything resembling a shoe in shape or use, esp.: **a** a drag for a wheel. **b** = *brake shoe* (see BRAKE¹). **c** a socket. **d** a ferrule, esp. on a sledge-runner. **e** a step (for a mast). **f** a box from which cards are dealt in casinos at baccarat etc. ● *v.tr.* (**shoes, shoeing;** past and past part. **shod** /ʃɒd/) **1** fit (esp. a horse etc.) with a shoe or shoes. **2** protect (the end of a pole etc.) with a metal shoe. **3** (as **shod** *adj.*) (in *comb.*) having shoes etc. of a specified kind (*dry-shod; roughshod*). □ **be in a person's shoes** be in his or her situation, difficulty, etc. **dead men's shoes** property or a position etc. coveted by a prospective successor. **if the shoe fits** N. Amer. = *if the cap fits* (see CAP). **shoe-buckle** a buckle worn as ornament or as a fastening on a shoe. **shoe-leather** leather for shoes, esp. when worn through by walking. **shoe-tree** a shaped block for keeping a shoe in shape when not worn. **where the shoe pinches** where one's difficulty or trouble is. □ **shoeless** *adj.* [OE *scōh, scōg(e)an* f. Gmc]

shoebill /ˈʃuːbɪl/ *n.* (also **shoe-billed stork**) the whale-headed stork.

shoeblack /ˈʃuːblæk/ *n.* a person who cleans the shoes of passers-by for payment.

shoebox /ˈʃuːbɒks/ *n.* **1** a box for packing shoes. **2** *colloq.* a very small space or dwelling.

shoehorn /ˈʃuːhɔːn/ *n. & v.* ● *n.* a curved piece of horn, metal, etc., for easing the heel into a shoe. ● *v.tr.* (usu. foll. by *into*) cram (people or things) into a small space.

shoelace /ˈʃuːleɪs/ *n.* a cord for lacing up shoes.

shoemaker /ˈʃuːˌmeɪkə(r)/ *n.* a maker of boots and shoes. □ **shoemaking** *n.*

Shoemaker–Levy 9, Comet /ˌʃuːmeɪkə ˈliːvɪ/ a comet discovered in March 1993 by the American astronomers Carolyn and Eugene Shoemaker and the Canadian astronomer David Levy. It had been captured by Jupiter's gravitational field, going into an eccentric orbit around the planet and breaking up as a result of passing very close to Jupiter shortly before its discovery. In July 1994 more than twenty separate fragments impacted successively on the planet, causing large explosions in its atmosphere.

shoeshine /ˈʃuːʃaɪn/ *n.* esp. N. Amer. a polish given to shoes.

shoestring /ˈʃuːstrɪŋ/ *n.* **1** a shoelace. **2** *colloq.* a small esp. inadequate amount of money (*living on a shoestring*). **3** (*attrib.*) barely adequate; precarious (*a shoestring majority*).

shofar /ˈʃəʊfə(r)/ *n.* (*pl.* **shofroth** /-frəʊt/) a ram's-horn trumpet used by Jews in religious ceremonies and as an ancient battle-signal. [Heb. *šōpār*, pl. *šōp̄ārōṯ*]

shogun /ˈʃəʊgʊn/ *n.* a hereditary commander-in-chief in feudal Japan.

Because of the military power concentrated in his hands, and the consequent weakness of the nominal head of state (the mikado or emperor), the shogun was generally the real ruler of the country until the last shogunate (the Tokugawa dynasty) was replaced by renewed imperial power in 1867 and feudalism was abolished. [Jap., = general, f. Chin. *jiang jun*]

shogunate /ˈʃəʊgʊnət/ *n.* the office, position, or dignity of a shogun; the period of a shogun's rule.

Sholapur /ˈʃəʊləˌpʊə(r)/ a city in western India, on the Deccan plateau in the state of Maharashtra; pop. (1991) 604,000.

Shona /ˈʃəʊnə/ *n. & adj.* ● *n.* (*pl.* same or **Shonas**) **1** a member of a group of Bantu-speaking peoples inhabiting parts of southern Africa. The Shona comprise over three-quarters of the population of Zimbabwe, and smaller groups live in South Africa, Zambia, and Mozambique. (See also MASHONA.) **2** any of the languages of these peoples. ● *adj.* of or relating to the Shona or their languages. [Bantu]

shone past and past part. of SHINE.

shonky /ˈʃɒŋkɪ/ *adj. & n. Austral. sl.* ● *adj.* (**shonkier, shonkiest**) unreliable, dishonest. ● *n.* (also **shonk**) a person engaged in shady business activities. [perh. Engl. dial. *shonk* smart]

shoo /ʃuː/ *int. & v.* ● *int.* an exclamation used to frighten away birds, children, etc. ● *v.* (**shoos, shooed**) **1** *intr.* utter the word 'shoo'. **2** *tr.* (usu. foll. by *away*) drive (birds etc.) away by shooing. □ **shoo-in** *N. Amer.* something easy or certain to succeed. [imit.]

shook¹ /ʃʊk/ past of SHAKE. ● *predic.adj. colloq.* **1** (foll. by *up*) emotionally or physically disturbed; upset. **2** (foll. by *on*) *Austral. & NZ* keen on; enthusiastic about (*not too shook on the English climate*).

shook² /ʃʊk/ *n.* US a set of staves and headings for a cask, ready for fitting together. [18th c.: orig. unkn.]

shoot /ʃuːt/ *v., n., & int.* ● *v.* (past and past part. **shot**) **1** *tr.* **a** cause (a gun, bow, etc.) to fire. **b** discharge (a bullet, arrow, etc.) from a gun, bow, etc. **c** kill or wound (a person, animal, etc.) with a bullet, arrow, etc. from a gun, bow, etc. **2** *intr.* discharge a gun etc., esp. in a specified way (*shoots well*). **3** *tr.* send out, discharge, propel, etc., esp. violently or swiftly (*shot out the contents; shot a glance at his neighbour*). **4** *intr.* (often foll. by *out, along, forth, etc.*) come or go swiftly or vigorously. **5** *intr.* **a** (of a plant etc.) put forth buds etc. **b** (of a bud etc.) appear. **6** *intr.* **a** hunt game etc. with a gun. **b** (usu. foll. by *over*) shoot game over an estate etc. **7** *tr.* shoot game in or on (coverts, an estate, etc.). **8** *tr.* film or photograph (a scene, film, etc.). **9** *esp. Football* **a** *tr.* take a shot at the goal. **b** *tr.* score (a goal) esp. with a shot. **10** *tr.* (of a boat) sweep swiftly down or under (a bridge, rapids, falls, etc.). **11** *tr.* move (a door-bolt) to fasten or unfasten a door etc. **12** *tr.* let (rubbish, a load, etc.) fall or slide from a container, lorry, etc. **13** *intr.* **a** (usu. foll. by *through, up, etc.*) (of a pain) pass with a stabbing sensation. **b** (of part of the body) be intermittently painful. **14** *intr.* (often foll. by *out*) project abruptly (*the mountain shoots out against the sky*). **15** *tr.* (often foll. by *up*) *sl.* inject esp. oneself with (a drug). **16** *tr.* N. Amer. colloq. play a game (of craps, pool, etc.). **b** throw (a die or dice). **17** *tr.* Golf colloq. make (a specified score) for a round or hole. **18** *tr. colloq.* pass (traffic lights at red). **19** *tr.* plane (the edge of a board) accurately. **20** *intr.* Cricket (of a ball) dart along the ground after pitching. ● *n.* **1** the act or an instance of shooting. **2 a** a young branch or sucker. **b** the new growth of a plant. **3** *Brit.* **a** a hunting party, expedition, etc. **b** land shot over for game. **4** = CHUTE¹. **5** a rapid in a stream. ● *int. colloq.* **1** a demand for a reply, information, etc.; go ahead. **2** *N. Amer. euphem.* an exclamation of disgust, anger, etc. (see SHIT). □ **shoot ahead** come quickly to the front of competitors etc. **shoot one's bolt** see BOLT¹. **shoot down 1** kill (a person) by shooting. **2** cause (an aircraft, its pilot, etc.) to crash by shooting. **3** argue effectively against (a person, argument, etc.). **shoot it out** *sl.* engage in a decisive gun battle. **shoot a line** *sl.* talk pretentiously. **shoot one's mouth off** *sl.* talk too much or indiscreetly. **shoot-out** *colloq.* **1** a decisive gun battle. **2** (in full **penalty shoot-out**) *Football* a tie-breaker decided by each side taking a specified number of penalty shots. **shoot through** *Austral. & NZ sl.* depart; escape; abscond. **shoot up 1** grow rapidly, esp. (of a person) grow taller. **2** rise suddenly. **3** terrorize (a district) by indiscriminate shooting. **4** *sl.* = sense 15 of *v.* **the whole shoot** = *the whole shooting match* (see SHOOTING). □ **shootable** *adj.* [OE *scēotan* f. Gmc: cf. SHEET¹, SHOT¹, SHUT]

shooter /ˈʃuːtə(r)/ *n.* **1** a person or thing that shoots. **2 a** (in *comb.*) a gun or other device for shooting (*peashooter; six-shooter*). **b** *sl.* a pistol etc. **3** a player who shoots or is able to shoot a goal in football, netball, etc. **4** *Cricket* a ball that shoots. **5** a person who throws a die or dice.

shooting /ˈʃuːtɪŋ/ *n. & adj.* ● *n.* **1** the act or an instance of shooting. (See note below.) **2 a** the right of shooting over an area of land. **b** an estate etc. rented to shoot over. ● *adj.* moving, growing, etc. quickly (*a shooting pain in the arm*). □ **shooting-box** *Brit.* a lodge used by sportsmen in the shooting-season. **shooting-brake** (or **-break**) *Brit.* an estate car. **shooting circle** *Hockey* an approximately semicircular area in front of each goal, from within which an attacker may score a goal. **shooting-coat** (or **-jacket**) a coat designed to be worn when shooting game. **shooting-gallery** a place used for shooting at targets with rifles etc. **shooting-iron** esp. *US colloq.* a firearm. **shooting-range** a ground with butts for rifle practice. **shooting star** a meteor that burns up on entering the earth's atmosphere, forming a bright streak in the night sky. **shooting-stick** a walking-stick with a handle that unfolds to form a seat. **shooting war** a war in which there is shooting (opp. COLD WAR, *war of nerves*, etc.). **the whole shooting match** *colloq.* everything.

▪ Early target shooting used either the harquebus or the bow and arrow; the first known shooting club existed in Switzerland from the late 15th century. Use of rifles for target shooting dates from the mid-16th century in central Europe. In Britain, archery remained more popular than shooting with rifles throughout the 17th and 18th centuries and it was not until 1860 that the British National Rifle Association was formed. Shooting was included in the first Olympic Games in 1896. Clay-pigeon shooting, in which a clay disc is propelled into the air as a target for shooting, dates from 1880, although an earlier version of the sport using live pigeons dates from about 1790.

shop /ʃɒp/ *n. & v.* ● *n.* **1** a building, room, etc., for the retail sale of goods or services (*chemist's shop; betting-shop*). **2** *colloq.* an act of going shopping (*our big weekly shop*). **3** a place in which manufacture or repairing is done; a workshop (*engineering-shop*). **4** a profession, trade, business, etc., esp. as a subject of conversation (*talk shop*). **5** *colloq.* an institution, establishment, place of business, etc. ● *v.* (**shopped**, **shopping**) **1** *intr.* **a** go to a shop or shops to buy goods. **b** *US* = *window-shop*. **2** *tr.* esp. *Brit. sl.* inform against (a criminal etc.). □ **all over the shop** *colloq.* **1** in disorder (*scattered all over the shop*). **2** in every place (*looked for it all over the shop*). **3** wildly (*hitting out all over the shop*). **set up shop** establish oneself in business etc. **shop around** look for the best bargain. **shop assistant** *Brit.* a person who serves customers in a shop. **shop-boy** (or **-girl**) an assistant in a shop. **shop-floor 1** the place where manufacturing work is actually done. **2** the workers in a factory etc. as distinct from management. **shop-soiled 1** (of an article) soiled or faded by display in a shop. **2** (of a person, idea, etc.) grubby; tarnished; no longer fresh or new. **shop steward** a person elected by workers in a factory etc. to represent them in dealings with management. **shop-window 1** a display window in a shop. **2** an opportunity for displaying skills, talents, etc. **shop-worn** = *shop-soiled*. □ **shopless** *adj.* **shoppy** *adj.* [ME f. AF & OF *eschoppe* booth f. MLG *schoppe*, OHG *scopf* porch]

shopaholic /ˌʃɒpəˈhɒlɪk/ *n. colloq.* a person addicted to shopping.

shopkeeper /ˈʃɒpˌkiːpə(r)/ *n.* the owner and manager of a shop. □ **shopkeeping** *n.*

shoplifter /ˈʃɒpˌlɪftə(r)/ *n.* a person who steals goods while appearing to shop. □ **shoplifting** *n.*

shopman /ˈʃɒpmən/ *n.* (*pl.* **-men**) **1** *Brit.* a shopkeeper or shopkeeper's assistant. **2** a workman in a repair shop.

shopper /ˈʃɒpə(r)/ *n.* **1** a person who makes purchases in a shop. **2** a shopping bag or trolley. **3** a small-wheeled bicycle with a basket. **4** *sl.* an informer.

shopping /ˈʃɒpɪŋ/ *n.* **1** (often *attrib.*) the purchase of goods etc. (*shopping expedition*). **2** goods purchased (*put the shopping on the table*). □ **shopping centre** an area or complex of shops, with associated facilities.

shopwalker /ˈʃɒpˌwɔːkə(r)/ *n. Brit.* an attendant in a large shop who directs customers, supervises assistants, etc.

shoran /ˈʃɔːræn/ *n.* a system of aircraft navigation using the return of two radar signals by two ground stations. [*short range navigation*]

shore[1] /ʃɔː(r)/ *n.* **1** the land that adjoins the sea or a large body of water. **2** (usu. in *pl.*) a country; a sea-coast (*often visits these shores; on a distant shore*). **3** *Law* land between ordinary high and low water marks. □ **in shore** on the water near or nearer to the shore (cf. INSHORE). **on shore** ashore. **shore-based** operating from a base on shore. **shore leave** *Naut.* **1** permission to go ashore. **2** a period of time ashore. □ **shoreless** *adj.* **shoreward** *adj. & adv.* **shorewards** *adv.* [ME f. MDu., MLG *schōre*, perh. f. the root of SHEAR]

shore[2] /ʃɔː(r)/ *n. & v.* ● *n.* a prop or beam set obliquely against a ship, wall, tree, etc., as a support. ● *v.tr.* (often foll. by *up*) support with or as

if with a shore or shores; hold up. □ **shoring** *n.* [ME f. MDu., MLG *schore* prop, of unkn. orig.]

shore[3] see SHEAR.

shorebird /ˈʃɔːbɜːd/ *n.* a bird that frequents the shore, esp. (*N. Amer.*) a wader of the order Charadriiformes (cf. WADER 1b).

shoreline /ˈʃɔːlaɪn/ *n.* the line along which a stretch of water, esp. a sea or lake, meets the shore.

shoreweed /ˈʃɔːwiːd/ *n.* a small plantain, *Littorella uniflora*, growing at the edge of fresh water or submerged in shallow water.

shorn *past part.* of SHEAR.

short /ʃɔːt/ *adj., adv., n., & v.* ● *adj.* **1 a** measuring little; not long from end to end (*a short distance*). **b** not long in duration (*a short time ago; had a short life*). **c** seeming less than the stated amount (*a few short years of happiness*). **2** of small height; not tall (*a short square tower; was shorter than average*). **3 a** (usu. foll. by *of*, (*colloq.*) *on*) having a partial or total lack; deficient; scanty (*short of spoons; is rather short on sense*). **b** *colloq.* having little money. **c** not far-reaching; acting or being near at hand (*within short range*). **4 a** concise; brief (*kept his speech short*). **b** curt; uncivil (*was short with her*). **5** (of the memory) unable to remember distant events. **6** *Phonet. & Prosody* of a vowel or syllable: **a** having the lesser of the two recognized durations. **b** (of an English vowel) having a sound other than that called long (cf. LONG[1] *adj.* 8). **7 a** (of pastry) crumbling; not holding together. **b** (of clay) having poor plasticity. **8** esp. *Stock Exch.* **a** (of stocks, a stockbroker, crops, etc.) sold or selling when the amount is not in hand, with reliance on getting the deficit in time for delivery. **b** (of a bill of exchange) maturing at an early date. **9** *Cricket* **a** (of a ball) pitching relatively near the bowler. **b** (of a fielding position) relatively near the batsman. **10** (of a drink of spirits) undiluted. ● *adv.* **1** before the natural or expected time or place; abruptly (*pulled up short; cut short the celebrations*). **2** rudely; uncivilly (*spoke to him short*). ● *n.* **1** *colloq.* a short drink, esp. spirits. **2** a short circuit. **3** a short film. **4** *Stock Exch.* **a** a person who sells short. **b** (in *pl.*) short-dated stocks. **5** *Phonet.* **a** a short syllable or vowel. **b** a mark indicating that a vowel is short. **6** (in *pl.*) a mixture of bran and coarse flour. ● *v.tr. & intr.* (often foll. by *out*) short-circuit. □ **be caught** (or **taken**) **short 1** be put at a disadvantage. **2** *colloq.* urgently need to urinate or defecate. **bring up** (or **pull up**) **short** check or pause abruptly. **come short** be inadequate or disappointing. **come short of** fail to reach or amount to. **for short** as a short name (*Tom for short*). **get** (or **have**) **by the short and curlies** *colloq.* be in complete control of (a person). **go short** (often foll. by *of*) not have enough. **in short** to use few words; briefly. **in short order** immediately, rapidly. **in the short run** over a short period of time. **in short supply** scarce. **in the short term** = *in the short run*. **make short work of** accomplish, dispose of, destroy, consume, etc. quickly. **short and sweet** esp. *iron.* brief and pleasant. **short-arm** (of a blow etc.) delivered with the arm not fully extended. **short back and sides** a haircut in which the hair is cut short at the back and the sides. **short change** insufficient money given as change. **short-change** *v.tr.* rob or cheat by giving short change. **short circuit** an electric circuit through small resistance, esp. instead of the resistance of a normal circuit. **short-circuit** *v.* **1** cause a short circuit or a short circuit in. **2** shorten or avoid (a journey, work, etc.) by taking a more direct route etc. **short commons** insufficient food. **short cut 1** a route shortening the distance travelled. **2** a quick way of accomplishing something. **short date** an early date for the maturing of a bill etc. **short-dated** due for early payment or redemption. **short-day** (of a plant) needing the period of light each day to fall below some limit to cause flowering. **short division** *Math.* division in which the quotient is written directly without being worked out in writing. **short drink** a strong alcoholic drink served in small measures. **short-eared owl** a migratory owl, *Asio flammeus*, frequenting open country and often hunting by day. **short for** an abbreviation for ('*Bob' is short for 'Robert'*). **short fuse** *colloq.* a quick temper. **short game** *Golf* approaching and putting. **short-handed** undermanned or understaffed. **short haul 1** the transport of goods over a short distance. **2** a short-term effort. **short head** *Horse-racing* a distance less than the length of a horse's head. **short-head** *v.tr.* beat by a short head. **short hundredweight** see HUNDREDWEIGHT 3. **short line** a line across a squash court behind which both players must stand at service. **short list** *Brit.* a list of selected candidates from which a final choice is made. **short-list** *v.tr. Brit.* put on a short list. **short-lived** ephemeral; not long-lasting. **short mark** = BREVE 2. **short measure** less than the professed amount. **short metre** *Prosody* a hymn stanza of four lines with 6, 6, 8, and 6 syllables. **short notice** a small, esp. insufficient, length of warning time. **short odds** nearly equal stakes or chances in betting. **short of 1** see sense 3a of *adj.* **2** less than (*nothing* *short of a miracle*). **3** distant from; not reaching a designated point (*two miles short of home; fall short of the mark*). **4** without going so far as; except (*did everything short of destroying it*). **short of breath** panting, short-winded. **short on** *colloq.* see sense 3a of *adj.* **short order** *N. Amer.* an order in a restaurant for quickly cooked food. **short-pitched** *Cricket* (of a ball) pitching relatively near the bowler. **short-range 1** having a short range. **2** relating to a fairly immediate future time (*short-range possibilities*). **short rib** = *floating rib.* **short score** *Mus.* a score not giving all parts. **short shrift** see SHRIFT. **short sight** the inability to focus except on comparatively near objects. **short-sleeved** with sleeves not reaching below the elbow. **short-staffed** having insufficient staff. **short story** a story with a fully developed theme but shorter than a novel. **short suit** a suit of less than four cards. **short temper** a tendency to lose one's temper quickly. **short-tempered** quick to lose one's temper; irascible. **short-term** occurring in or relating to a short period of time. **short-termism** concentration on short-term projects etc. for immediate profit at the expense of long-term security. **short time** the condition of working fewer than the regular hours per day or days per week. **short title** an abbreviated form of a title of a book etc. **short ton** see TON[1] 2. **short view** a consideration of the present only, not the future. **short waist 1** a high or shallow waist of a dress. **2** a short upper body. **short wave** a radio wave of frequency greater than 3 MHz. **short weight** weight less than it is alleged to be. **short whist** whist with ten or five points to a game. **short wind** breathing-power that is quickly exhausted. **short-winded 1** having short wind. **2** incapable of sustained effort. □ **shortish** *adj.* **shortness** *n.* [OE *sceort* f. Gmc: cf. SHIRT, SKIRT]

shortage /ˈʃɔːtɪdʒ/ *n.* (often foll. by *of*) a deficiency; an amount lacking (*a shortage of 100 tons*).

shortbread /ˈʃɔːtbred/ *n.* a crisp rich crumbly type of biscuit made with butter, flour, and sugar.

shortcake /ˈʃɔːtkeɪk/ *n.* **1** = SHORTBREAD. **2** a cake made of short pastry and filled with fruit and cream.

shortcoming /ˈʃɔːtˌkʌmɪŋ/ *n.* a failure to come up to a standard; a defect.

shortcrust /ˈʃɔːtkrʌst/ *n.* (in full **shortcrust pastry**) a type of crumbly pastry made with flour and fat.

shorten /ˈʃɔːt(ə)n/ *v.* **1** *intr. & tr.* become or make shorter or short; curtail. **2** *tr. Naut.* reduce the amount of (sail spread). **3** *intr. & tr.* (with reference to gambling odds, prices, etc.) become or make shorter; decrease.

shortening /ˈʃɔːt(ə)nɪŋ/ *n.* fat used for making pastry, esp. for making short pastry.

shortfall /ˈʃɔːtfɔːl/ *n.* a shortage or deficit.

shorthand /ˈʃɔːthænd/ *n.* **1** (often *attrib.*) a method of rapid writing in abbreviations and symbols esp. for taking dictation. (*See note below.*) **2** an abbreviated or symbolic mode of expression. □ **shorthand typist** *Brit.* a typist qualified to take and transcribe in shorthand.

▪ The use of shortened handwriting dates at least from ancient Roman times. In England systems were produced from the 16th century, based at first on the alphabet and then increasingly on phonetics, notably that of Samuel Taylor (1749–1811). The brevity and simplicity of his system attracted the attention of Sir Isaac Pitman, who devised his own method in 1837. Pitman's system is still widely used, though others are also common; in the US, that devised in 1888 by John R. Gregg (1867–1948) is the most popular. Shorthand writing has become less important with the increasing use of sound recording.

shorthorn /ˈʃɔːthɔːn/ *n.* a breed of cattle with short horns.

shortie var. of SHORTY.

shortly /ˈʃɔːtlɪ/ *adv.* **1** (often foll. by *before, after*) before long; soon (*will arrive shortly; arrived shortly after him*). **2** in a few words; briefly. **3** curtly. [OE *scortlice* (as SHORT, -LY[2])]

Short Parliament the first of two parliaments summoned by Charles I in 1640 (the other being the Long Parliament). Due to its insistence on seeking a general redress of grievances against him before granting the money he required, Charles dismissed it after only three weeks.

shorts /ʃɔːts/ *n.pl.* **1** trousers reaching only to the knees or higher. **2** *US* underpants.

short-sighted /ʃɔːtˈsaɪtɪd/ *adj.* **1** having short sight. **2** lacking imagination or foresight. □ **short-sightedly** *adv.* **short-sightedness** *n.*

shortstop /ˈʃɔːtstɒp/ *n.* a baseball fielder between second and third base.

shorty /ˈʃɔːtɪ/ *n.* (also **shortie**) (*pl.* **-ies**) *colloq.* **1** a person shorter than average. **2** a short garment, esp. a nightdress or raincoat.

Shoshone /ʃəˈʃəʊnɪ/ *n. & adj.* ● *n.* (*pl.* same or **Shoshones**) **1** a member of an American Indian people living chiefly in Wyoming, Idaho, and Nevada. **2** the Shoshonean language of this people. ● *adj.* of or relating to the Shoshone or their language. [19th c.: orig. unkn.]

Shoshonean /ʃəˈʃəʊnɪən/ *n. & adj.* ● *n.* a family of American Indian languages that includes Comanche and Shoshone. ● *adj.* of or relating to this family of languages.

Shostakovich /ˌʃɒstəˈkəʊvɪtʃ/, Dmitri (Dmitrievich) (1906–75), Russian composer. He was a prolific composer whose works include fifteen symphonies, operas, and many chamber works, although it is for the symphonies that he is most renowned. Shostakovich developed a highly personal style and, although he experimented with atonality and twelve-note techniques, his music always returned to a basic tonality. The failure of his music to conform to Soviet artistic ideology earned him official condemnation, especially during the Stalinist period. His later work, particularly the last symphonies, written after Stalin's death when the strictures on artistic freedom were less tight, became increasingly sombre and intense.

shot[1] /ʃɒt/ *n.* **1** the act or an instance of firing a gun, cannon, etc. (*several shots were heard*). **2** an attempt to hit by shooting or throwing etc. (*took a shot at him*). **3 a** a single non-explosive missile for a cannon, gun, etc. **b** (*pl.* same or **shots**) a small lead pellet used in quantity in a single charge or cartridge in a shotgun. **c** (as *pl.*) these collectively. **4 a** a photograph. **b** a film sequence photographed continuously by one camera. **5 a** a stroke or a kick in a ball game; a kick etc. of the ball with the aim of scoring a goal. **b** *colloq.* an attempt to guess or do something (*let him have a shot at it*). **6** *colloq.* a person having a specified skill with a gun etc. (*is not a good shot*). **7** a heavy ball thrown by a shot-putter. **8** the launch of a space rocket (*a moonshot*). **9** the range, reach, or distance to or at which a thing will carry or act (*out of earshot*). **10** a remark aimed at a person. **11** *colloq.* **a** a drink of esp. spirits. **b** an injection of a drug, vaccine, etc. (*has had his shots*). □ **like a shot** *colloq.* without hesitation; willingly. **make a bad shot** guess wrong. **not a shot in one's** (or **the**) **locker 1** no money left. **2** not a chance left. **shot across the bows** see BOW[3]. **shot-blasting** the cleaning of metal etc. by the impact of a stream of shot. **shot-firer** a person who fires a blasting-charge in a mine etc. **shot in the arm** *colloq.* **1** a stimulus or an encouragement. **2** an alcoholic drink. **shot in the dark** a mere guess. **shot-put** an athletic contest in which a shot is thrown some distance. **shot-putter** an athlete who puts the shot. **shot-tower** *hist.* a tower in which shot was made from molten lead poured through sieves at the top and falling into water at the bottom. □ **shotproof** *adj.* [OE *scot(e)ot, gesc(e)ot* f. Gmc: cf. SHOOT]

shot[2] /ʃɒt/ *past and past part.* of SHOOT. ● *adj.* **1** (of coloured material) woven so as to show different colours at different angles. **2** *colloq.* **a** exhausted; finished. **b** drunk. **3** (of a board-edge) accurately planed. □ **be** (or **get**) **shot of** *sl.* be (or get) rid of. **shot through** permeated or suffused.

shot[3] /ʃɒt/ *n. colloq.* a reckoning, a bill, esp. at an inn etc. (*paid his shot*). [ME, = SHOT[1]: cf. OE *scēotan* shoot, pay, contribute, and SCOT]

shotgun /ˈʃɒtgʌn/ *n.* a smooth-bore gun for firing small shot at short range. □ **shotgun marriage** (or **wedding**) *colloq.* an enforced or hurried wedding, esp. because of the bride's pregnancy.

shotten herring /ˈʃɒt(ə)n/ *n.* **1** a herring that has spawned. **2** *archaic* a weakened or dispirited person. [ME, archaic past part. of SHOOT]

should /ʃʊd, ʃəd/ *v.aux.* (3rd sing. **should**) *past* of SHALL, used esp.: **1** in reported speech, esp. with the reported element in the 1st person (*I said I should be home by evening*). ¶ Cf. WILL[1], WOULD, now more common in this sense, esp. to avoid implications of sense 2. **2 a** to express a duty, obligation, or likelihood; = OUGHT[1] (*I should tell you; you should have been more careful; they should have arrived by now*). **b** (in the 1st person) to express a tentative suggestion (*I should like to say something*). **3 a** expressing the conditional mood in the 1st person (cf. WOULD) (*I should have been killed if I had gone*). **b** forming a conditional protasis or indefinite clause (*if you should see him; should they arrive, tell them where to go*). **4** expressing purpose = MAY, MIGHT[1] (*in order that we should not worry*).

shoulder /ˈʃəʊldə(r)/ *n. & v.* ● *n.* **1 a** the part of the body at which the arm, foreleg, or wing is attached. **b** (in full **shoulder joint**) the end of the upper arm joining with the collar-bone and blade-bone. **c** either of the two projections below the neck from which the arms hang. **2** the upper foreleg and shoulder blade of a pig, lamb, etc. when butchered. **3** (often in *pl.*) **a** the upper part of the back and arms. **b** this part of the body regarded as capable of bearing a burden or blame, providing comfort, etc. (*needs a shoulder to cry on*). **4** a strip of land next to a metalled road (*pulled over on to the shoulder*). **5** a part of a garment covering the shoulder. **6** a part of anything resembling a shoulder in form or function, as in a bottle, mountain, tool, etc. ● *v.* **1** a *tr.* push with the shoulder; jostle. **b** *intr.* make one's way by jostling (*shouldered through the crowd*). **2** *tr.* take (a burden etc.) on one's shoulders (*shouldered the family's problems*). □ **put** (or **set**) **one's shoulder to the wheel** set to work vigorously. **shoulder arms** hold a rifle with the barrel against the shoulder and the butt in the hand. **shoulder-bag** a woman's handbag that can be hung from the shoulder. **shoulder-belt** a bandolier or other strap passing over one shoulder and under the opposite arm. **shoulder-blade** *Anat.* either of the large flat bones of the upper back; the scapula. **shoulder-high** up to or as high as the shoulders. **shoulder-holster** a gun holster worn in the armpit. **shoulder-knot** a knot of ribbon, metal, lace, etc. worn as part of a ceremonial dress. **shoulder-length** (of hair etc.) reaching to the shoulders. **shoulder loop** *US* the shoulder-strap of an army, air-force, or marines officer. **shoulder mark** *US* the shoulder-strap of a naval officer. **shoulder-note** *Printing* a marginal note at the top of a page. **shoulder-of-mutton sail** = *leg-of-mutton sail.* **shoulder-pad** a pad sewn into a garment to bulk out the shoulder. **shoulder-strap 1** a strip of fabric, leather, etc. suspending a bag or garment from the shoulder. **2** a strip of cloth from shoulder to collar on a military uniform bearing a symbol of rank etc. **3** a similar strip on a raincoat. **shoulder to shoulder 1** side by side. **2** with closed ranks or united effort. □ **shouldered** *adj.* (also in *comb.*). [OE *sculdor* f. WG]

shouldn't /ˈʃʊd(ə)nt/ *contr.* should not.

shout /ʃaʊt/ *v. & n.* ● *v.* **1** *intr.* make a loud cry or vocal sound; speak loudly (*shouted for attention*). **2** *tr.* express loudly; call out (*shouted that the coast was clear*). **3** *tr.* (also *absol.*) *Austral. & NZ colloq.* treat (another person) to drinks etc. ● *n.* **1** a loud cry expressing joy etc. or calling attention. **2** *colloq.* one's turn to order and pay for a round of drinks etc. (*your shout I think*). □ **all over bar** (or **but**) **the shouting** *colloq.* the contest is virtually decided. **shout at** speak loudly to etc. **shout down** reduce to silence by shouting. **shout for** call for by shouting. **shout-up** *colloq.* a noisy argument. □ **shouter** *n.* [ME, perh. rel. to SHOOT: cf. ON *skúta* SCOUT²]

shove /ʃʌv/ *v. & n.* ● *v.* **1** *tr.* (also *absol.*) push vigorously; move by hard or rough pushing (*shoved him out of the way*). **2** *intr.* (usu. foll. by *along*, *past*, *through*, etc.) make one's way by pushing (*shoved through the crowd*). **3** *tr. colloq.* put somewhere (*shoved it in the drawer*). ● *n.* an act of shoving or of prompting a person into action. □ **shove-halfpenny** a form of shovelboard played with coins etc. on a table esp. in licensed premises. **shove off 1** start from the shore in a boat. **2** *sl.* depart; go away (*told him to shove off*). [OE *scúfan* f. Gmc]

shovel /ˈʃʌv(ə)l/ *n. & v.* ● *n.* **1** a a spadelike tool for shifting quantities of coal, earth, etc., esp. having the sides curved upwards. **b** the amount contained in a shovel; a shovelful. **2** a machine or part of a machine having a similar form or function. ● *v.tr.* (**shovelled, shovelling**; *US* **shoveled, shoveling**) **1** shift or clear (coal etc.) with or as if with a shovel. **2** *colloq.* move (esp. food) in large quantities or roughly (*shovelled peas into his mouth*). □ **shovel hat** a broad-brimmed hat esp. worn by some clergymen. □ **shovelful** *n.* (*pl.* **-fuls**). [OE *scofl* f. Gmc (see SHOVE)]

shovelboard /ˈʃʌv(ə)l‚bɔːd/ *n.* a game played esp. on a ship's deck by pushing discs with the hand or with a long-handled shovel over a marked surface. [earlier *shoveboard* f. SHOVE + BOARD]

shovelhead /ˈʃʌv(ə)l‚hed/ *n.* = BONNETHEAD.

shoveller /ˈʃʌvələ(r)/ *n.* (*US* **shoveler**: see also sense 2) **1** a person or thing that shovels. **2** (usu. **shoveler**) a duck of the genus *Anas* with a broad and long bill, esp. *A. clypeata* of Eurasia and North America. [SHOVEL: sense 2 earlier *shovelard* f. -ARD, perh. after *mallard*]

show /ʃəʊ/ *v. & n.* ● *v.* (also *archaic* **shew**) (*past part.* **shown** /ʃəʊn/ or **showed**, *archaic* **shewn** or **shewed**) **1** *intr. & tr.* be, or allow or cause to be, visible; manifest; appear (*the buds are beginning to show; white shows the dirt*). **2** *tr.* (often foll. by *to*) offer, exhibit, or produce (a thing) for scrutiny etc. (*show your tickets please; showed him my poems*). **3** *tr.* (often foll. by *to*, *towards*) demonstrate (kindness, rudeness, etc.) to a person (*showed respect towards him; showed him no mercy*). **4** *intr.* (of feelings etc.) be manifest (*his dislike shows*). **5** *tr.* a point out; prove (*has shown it to be false; showed that he knew the answer*). **b** (usu. foll. by *how to* + infin.) cause (a person) to understand or be capable of doing (*showed them how to knit*). **6** *tr.* (*refl.*) **a** exhibit oneself as being (*showed herself a generous host*). **b** (foll. by *to be*) exhibit oneself to be (*showed herself to be fair*). **7** *tr. & intr.* (with ref. to a film) be presented or cause to be presented. **8** *tr.* exhibit (a picture, animal, flower, etc.) in a show. **9** *tr.* (often foll. by *in*, *out*, *up*,

etc.) conduct or lead (*showed them to their rooms*). **10** *intr. colloq.* = show up 3 (*waited but he didn't show*). **11** *intr. N. Amer.* finish in the first three in a race. ● *n.* **1** the act or an instance of showing; the state of being shown. **2** a a spectacle, display, exhibition, etc. (*a fine show of blossom*). **b** a collection of things etc. shown for public entertainment or in competition (*dog show; flower show*). **3** a a play etc., esp. a musical. **b** a light entertainment programme on television etc. **c** any public entertainment or performance. **4** a an outward appearance, semblance, or display (*made a show of agreeing; a show of strength*). **b** empty appearance; mere display (*did it for show; that's all show*). **5** *colloq.* an undertaking, business, etc. (*sold the whole show*). **6** *colloq.* an opportunity of acting, defending oneself, etc. (*gave him a fair show; made a good show of it*). **7** *Med.* a discharge of blood etc. from the vagina at the onset of childbirth. □ **get the show on the road** *colloq.* get started, begin an undertaking. **give the show** (or **whole show**) **away** demonstrate the inadequacies or reveal the truth. **good** (or **bad** or **poor**) **show! 1** that was well (or badly) done. **2** that was lucky (or unlucky). **nothing to show for** no visible result of (effort etc.). **on show** being exhibited. **show business** *colloq.* the theatrical profession. **show-card** a card used for advertising. **show one's cards** = *show one's hand.* **show cause** *Law* allege with justification. **show a clean pair of heels** *colloq.* retreat speedily or run away from a person. **show one's colours** make one's opinion clear. **show a person the door** dismiss or eject a person. **show one's face** make an appearance; let oneself be seen. **show fight** be persistent or belligerent. **show the flag** see FLAG¹. **show forth** *archaic* exhibit; expound. **show one's hand 1** disclose one's plans. **2** reveal one's cards. **show house** (or **flat** etc.) a furnished and decorated house (or flat etc.) on a new estate shown to prospective buyers. **show in** see sense 9 of *v.* **show a leg** *colloq.* **1** get out of bed. **2** make one's appearance. **show off 1** display to advantage. **2** *colloq.* act pretentiously; display one's wealth, knowledge, etc. **show-off** *n. colloq.* a person who shows off. **show of force** a demonstration that one is prepared to use force. **show of hands** raised hands indicating a vote for or against, usu. without being counted. **show oneself 1** be seen in public. **2** see sense 6 of *v.* **show out** see sense 9 of *v.* **show-piece 1** an item of work presented for exhibition or display. **2** an outstanding example or specimen. **show-place** a house etc. that tourists go to see. **show round** take (a person) to places of interest; act as guide for (a person) in a building etc. **show-stopper** *colloq.* a performance receiving prolonged applause. **show one's teeth** reveal one's strength; be aggressive. **show through 1** be visible through a covering. **2** (of real feelings etc.) be revealed inadvertently. **show trial** a judicial trial designed by the state to terrorize or impress the public, esp. (as in the Soviet Union under Stalin) a prejudged trial of a political dissident. **show up 1** make or be conspicuous or clearly visible. **2** expose (a fraud, impostor, inferiority, etc.). **3** *colloq.* appear; be present; arrive. **4** *colloq.* embarrass or humiliate (*don't show me up by wearing jeans*). **show the way 1** indicate what has to be done etc. by attempting it first. **2** show others which way to go etc. **show the white feather** appear cowardly (see also *white feather*). **show willing** display a willingness to help etc. **show-window** a window for exhibiting goods etc. [ME f. OE *scēawian* f. WG: cf. SHEEN]

showbiz /ˈʃəʊbɪz/ *n. colloq.* = show business.

showboat /ˈʃəʊbəʊt/ *n.* *US* a river steamer on which theatrical performances are given.

showcase /ˈʃəʊkeɪs/ *n. & v.* ● *n.* **1** a glass case used for exhibiting goods etc. **2** a place or medium for presenting something (esp. attractively) to general attention. ● *v.tr.* display in or as if in a showcase.

showdown /ˈʃəʊdaʊn/ *n.* **1** a final test or confrontation; a decisive situation. **2** the laying down face up of the players' cards in poker.

shower /ˈʃaʊə(r)/ *n. & v.* ● *n.* **1** a brief fall of esp. rain, hail, sleet, or snow. **2** a a brisk flurry of arrows, bullets, dust, stones, sparks, etc. **b** a flurry of gifts, letters, honours, praise, etc. **3** (in full **shower-bath**) **a** a cubicle, bath, etc. in which a person stands under a spray of water. **b** the apparatus etc. used for this. **c** the act of bathing in a shower. **4** a group of particles produced by the impact of a cosmic-ray particle in the earth's atmosphere. **5** *N. Amer.* a party for giving presents to a prospective bride etc. (*baby shower*). **6** *Brit. sl.* a contemptible or unpleasant person or group of people. ● *v.* **1** *tr.* **a** discharge (water, missiles, etc.) in a shower. **b** make wet with (or as if with) a shower. **2** *intr.* use a shower-bath. **3** *tr.* (usu. foll. by *on*, *upon*) lavishly bestow (gifts etc.). **4** *intr.* descend or come in a shower (*it showered on and off all day*). □ **showery** *adj.* [OE *scūr* f. Gmc]

showerproof /ˈʃaʊə‚pruːf/ *adj. & v.* ● *adj.* resistant to light rain. ● *v.tr.* make showerproof.

showgirl /ˈʃəʊgɜːl/ n. an actress who sings and dances in musicals, variety shows, etc.

showing /ˈʃəʊɪŋ/ n. **1** the act or an instance of showing. **2** a usu. specified quality of performance (*made a poor showing*). **3** the presentation of a case; evidence (*on present showing it must be true*). [OE *scēawung* (as SHOW)]

showjumping /ˈʃəʊˌdʒʌmpɪŋ/ n. the sport of riding horses over a course of fences and other obstacles within a prescribed arena. Penalty points are given if the horse knocks down (part of) a fence (or otherwise fails to clear the obstacle properly) or refuses a jump. One form of competition (*puissance*) consists mainly of large obstacles and is a test chiefly of the horse's jumping ability, while other courses are constructed to concentrate on speed and agility; however, most courses are a combination of the two, second and subsequent rounds often being run against the clock and acting as a decider between horses jumping equally well. (See also EQUESTRIANISM.) □ **showjump** v.intr. **showjumper** n.

showman /ˈʃəʊmən/ n. (pl. **-men**) **1** the proprietor or manager of a circus etc. **2** a person skilled in self-advertisement or publicity. □ **showmanship** n.

shown past part. of SHOW.

showroom /ˈʃəʊruːm, -rʊm/ n. a room in a factory, office building, etc. used to display goods for sale.

showy /ˈʃəʊɪ/ adj. (**showier**, **showiest**) **1** brilliant; gaudy, esp. vulgarly so. **2** striking. □ **showily** adv. **showiness** n.

shoyu /ˈʃəʊjuː/ n. = SOY 1. [Jap.]

s.h.p. abbr. shaft horsepower.

shrank past of SHRINK.

shrapnel /ˈʃræpn(ə)l/ n. **1** fragments of a bomb etc. thrown out by an explosion. **2** (usu. attrib.) a shell containing bullets or pieces of metal timed to burst short of impact. [Gen. Henry *Shrapnel*, Brit. soldier (1761–1842), inventor of the shell]

shred /ʃred/ n. & v. ● n. **1** a scrap, fragment, or strip of esp. cloth, paper, etc. **2** the least amount, remnant (*not a shred of evidence*). ● v.tr. (**shredded**, **shredding**) tear or cut into shreds. □ **tear to shreds** completely refute (an argument etc.). [OE *scrēad* (unrecorded) piece cut off, *scrēadian* f. WG: see SHROUD]

shredder /ˈʃredə(r)/ n. **1** a machine used to reduce documents to shreds. **2** any device used for shredding.

Shreveport /ˈʃriːvpɔːt/ an industrial city in NW Louisiana, on the Red River near the border with Texas; pop. (1990) 198,525. Founded in 1839, it grew rapidly after the discovery of oil and gas in the region in 1906.

shrew /ʃruː/ n. **1** a small carnivorous mouselike mammal of the family Soricidae, with a long pointed snout and small eyes, esp. one of the genus *Sorex* or *Crocidura*. **2** a bad-tempered or scolding woman. □ **shrewish** adj. (in sense 2). **shrewishly** adv. **shrewishness** n. [OE *scrēawa*, *scrǣwa* shrew-mouse: cf. OHG *scrawaz* dwarf, MHG *schrawaz* etc. devil]

shrewd /ʃruːd/ adj. **1 a** showing astute powers of judgement; clever and judicious (*a shrewd observer; made a shrewd guess*). **b** (of a face etc.) shrewd-looking. **2** archaic **a** (of pain, cold, etc.) sharp, biting. **b** (of a blow, thrust, etc.) severe, hard. **c** mischievous; malicious. □ **shrewdly** adv. **shrewdness** n. [ME, = malignant, f. SHREW in sense 'evil person or thing', or past part. of obs. *shrew* to curse, f. SHREW]

Shrewsbury /ˈʃrəʊzbəri, ˈʃruːz-/ a town in western England, the county town of Shropshire, situated on the River Severn near the border with Wales; pop. (1981) 59,170.

shriek /ʃriːk/ v. & n. ● v. **1** intr. **a** utter a shrill screeching sound or words, esp. in pain or terror. **b** (foll. by of) provide a clear or blatant indication of. **2** tr. **a** utter (sounds or words) by shrieking (*shrieked his name*). **b** indicate clearly or blatantly. ● n. a high-pitched piercing cry or sound; a scream. □ **shriek out** say in shrill tones. **shriek with laughter** laugh uncontrollably. □ **shrieker** n. [imit.: cf. dial. *screak*, ON *skrækja*, and SCREECH]

shrieval /ˈʃriːv(ə)l/ adj. of or relating to a sheriff. [*shrieve* obs. var. of SHERIFF]

shrievalty /ˈʃriːv(ə)ltɪ/ n. (pl. **-ies**) **1** a sheriff's office or jurisdiction. **2** the tenure of this. [as SHRIEVAL + *-alty* as in mayoralty etc.]

shrift /ʃrɪft/ n. archaic **1** confession to a priest. **2** confession and absolution. □ **short shrift 1** curt or dismissive treatment. **2** archaic little time between condemnation and execution or punishment. [OE *scrift* (verbal noun) f. SHRIVE]

shrike /ʃraɪk/ n. **1** a predatory songbird of the family Laniidae, with a strong hooked and toothed bill, often impaling its prey of mice, insects, etc. on thorns. (See also *butcher-bird*.) **2** a similar bird of another family. [perh. rel. to OE *scric* thrush, MLG *schrīk* corncrake (imit.): cf. SHRIEK]

shrill /ʃrɪl/ adj. & v. ● adj. **1** piercing and high-pitched in sound. **2** derog. (esp. of a protester) sharp, unrestrained, unreasoning. ● v. **1** intr. (of a cry etc.) sound shrilly. **2** tr. (of a person etc.) utter or send out (a song, complaint, etc.) shrilly. □ **shrillness** n. **shrilly** /ˈʃrɪlɪ/ adv. [ME, rel. to LG *schrell* sharp in tone or taste f. Gmc]

shrimp /ʃrɪmp/ n. & v. ● n. (pl. same or **shrimps**) **1** a small usu. marine decapod crustacean, grey-green when alive and pink when boiled for eating. **2** colloq. a very small slight person. ● v.intr. go catching shrimps. □ **shrimp plant** an evergreen shrub, *Justicia brandegeana*, bearing small white flowers in clusters of pinkish-brown bracts and popular as a house plant. □ **shrimper** n. [ME, prob. rel. to MLG *schrempen* wrinkle, MHG *schrimpfen* contract, and SCRIMP]

shrine /ʃraɪn/ n. & v. ● n. **1** esp. RC Ch. **a** a chapel, church, altar, etc., sacred to a saint, holy person, relic, etc. **b** the tomb of a saint etc. **c** a casket esp. containing sacred relics; a reliquary. **d** a niche containing a holy statue etc. **2** a place associated with or containing memorabilia of a particular person, event, etc. **3** a Shinto place of worship. ● v.tr. poet. enshrine. [OE *scrīn* f. Gmc f. L *scrinium* case for books etc.]

shrink /ʃrɪŋk/ v. & n. ● v. (past **shrank** /ʃræŋk/; past part. **shrunk** /ʃrʌŋk/ or (esp. as adj.) **shrunken** /ˈʃrʌŋkən/) **1** tr. & intr. make or become smaller; contract, esp. by the action of moisture, heat, or cold. **2** intr. (usu. foll. by from) **a** retire; recoil; flinch; cower (*shrank from her touch*). **b** be averse from doing (*shrinks from meeting them*). **3** (as **shrunken** adj.) (esp. of a face, person, etc.) having grown smaller esp. because of age, illness, etc. ● n. **1** the act or an instance of shrinking; shrinkage. **2** sl. a psychiatrist (from 'head-shrinker'). □ **shrinking violet** an exaggeratedly shy person. **shrink into oneself** become withdrawn. **shrink on** slip (a metal tyre etc.) on while expanded with heat and allow to tighten. **shrink-resistant** (of textiles etc.) resistant to shrinkage when wet etc. **shrink-wrap** (**-wrapped**, **-wrapping**) enclose (an article) in (esp. transparent) film that shrinks tightly on to it. □ **shrinkable** adj. **shrinker** n. **shrinkingly** adv. **shrink-proof** adj. [OE *scrincan*: cf. *skrynka* to wrinkle]

shrinkage /ˈʃrɪŋkɪdʒ/ n. **1 a** the process or fact of shrinking. **b** the degree or amount of shrinking. **2** an allowance made for the reduction in takings due to wastage, theft, etc.

shrive /ʃraɪv/ v.tr. (past **shrove** /ʃrəʊv/; past part. **shriven** /ˈʃrɪv(ə)n/) RC Ch. archaic **1** (of a priest) hear the confession of, assign penance to, and absolve. **2** (refl.) (of a penitent) submit oneself to a priest for confession etc. [OE *scrīfan* impose as penance, WG f. L *scribere* write]

shrivel /ˈʃrɪv(ə)l/ v.tr. & intr. (**shrivelled**, **shrivelling** or US **shriveled**, **shriveling**) contract or wither into a wrinkled, folded, rolled-up, contorted, or dried-up state. [perh. f. ON: cf. Sw. dial. *skryvla* to wrinkle]

shriven past part. of SHRIVE.

Shropshire /ˈʃrɒpʃə(r)/ (also called *Salop*) a county of England, situated on the border with Wales; county town, Shrewsbury.

shroud /ʃraʊd/ n. & v. ● n. **1** a sheetlike garment for wrapping a corpse for burial. **2** anything that conceals like a shroud (*wrapped in a shroud of mystery*). **3** (in pl.) Naut. a set of ropes forming part of the standing rigging and supporting the mast or topmast. ● v.tr. **1** clothe (a body) for burial. **2** cover, conceal, or disguise (*hills shrouded in mist*). □ **shroud-laid** (of a rope) having four strands laid right-handed on a core. □ **shroudless** adj. [OE *scrūd* f. Gmc: see SHRED]

shrove past of SHRIVE.

Shrovetide /ˈʃrəʊvtaɪd/ n. Shrove Tuesday and the two days preceding it when it was formerly customary to be shriven. [ME *shrove* as irreg. past part. of SHRIVE]

Shrove Tuesday /ʃrəʊv/ n. the day before Ash Wednesday.

shrub[1] /ʃrʌb/ n. a woody plant smaller than a tree and having a very short stem with branches near the ground. □ **shrubby** adj. [ME f. OE *scrubb*, *scrybb* shrubbery: cf. NFris. *skrobb* brushwood, WFlem. *schrobbe* vetch, Norw. *skrubba* dwarf cornel, and SCRUB[2]]

shrub[2] /ʃrʌb/ n. a cordial made of sweetened fruit juice and spirits, esp. rum. [Arab. *šurb*, *šarāb* f. *šariba* to drink: cf. SHERBET, SYRUP]

shrubbery /ˈʃrʌbərɪ/ n. (pl. **-ies**) an area planted with shrubs.

shrug /ʃrʌg/ v. & n. ● v. (**shrugged**, **shrugging**) **1** intr. slightly and momentarily raise the shoulders to express indifference, ignorance, helplessness, contempt, etc. **2** tr. **a** raise (the shoulders) in this way. **b** shrug the shoulders to express (indifference etc.) (*shrugged his consent*).

● *n.* the act or an instance of shrugging. □ **shrug off** dismiss as unimportant etc. by or as if by shrugging. [ME: orig. unkn.]

shrunk (also **shrunken**) *past part.* of SHRINK.

shtick /ʃtɪk/ *n. sl.* a theatrical routine, gimmick, etc. [Yiddish f. G *Stück* piece]

shuck /ʃʌk/ *n. & v. US* ● *n.* **1** a husk or pod. **2** the shell of an oyster or clam. **3** (in *pl.*) *colloq.* an expression of contempt or regret or self-deprecation in response to praise. ● *v.tr.* remove the shucks of; shell. □ **shucker** *n.* [17th c.: orig. unkn.]

shudder /ʃʌdə(r)/ *v. & n.* ● *v.intr.* **1** shiver esp. convulsively from fear, cold, repugnance, etc. **2** feel strong repugnance etc. (*shudder to think what might happen*). **3** (of a machine etc.) vibrate or quiver. ● *n.* **1** the act or an instance of shuddering. **2** (in *pl.*; prec. by *the*) *colloq.* a state of shuddering. □ **shudderingly** *adv.* **shuddery** *adj.* [ME *shod(d)er* f. MDu. *schūderen*, MLG *schōderen* f. Gmc]

shuffle /ʃʌf(ə)l/ *v. & n.* ● *v.* **1** *tr. & intr.* move with a scraping, sliding, or dragging motion (*shuffles along*; *shuffling his feet*). **2** *tr.* **a** (also *absol.*) rearrange (a pack of cards) by sliding them over each other quickly. **b** rearrange; intermingle; confuse (*shuffled the documents*). **3** *tr.* (usu. foll. by *on*, *off*, *into*) assume or remove (clothes, a burden, etc.) esp. clumsily or evasively (*shuffled on his clothes*; *shuffled off responsibility*). **4** *intr.* **a** equivocate; prevaricate. **b** continually shift one's position; fidget. **5** *intr.* (foll. by *out of*) escape evasively (*shuffled out of the blame*). ● *n.* **1** a shuffling movement. **2** the act or an instance of shuffling cards. **3** a general change of relative positions. **4** a piece of equivocation; sharp practice. **5** a quick scraping movement of the feet in dancing (see also *double shuffle*). □ **shuffle-board** = SHOVELBOARD. **shuffle the cards** change policy etc. □ **shuffler** *n.* [perh. f. LG *schuffeln* walk clumsily f. Gmc: cf. SHOVE]

shufti /ʃʊftɪ/ *n.* (*pl.* **shuftis**) *Brit. colloq.* a look or glimpse. [Arab. *šaffa* try to see]

Shumen /ʃuːmən/ an industrial city in NE Bulgaria; pop. (1990) 126,350. It is noted for its medieval fortress and its 18th-century mosque, built during the Turkish occupation of Bulgaria.

shun /ʃʌn/ *v.tr.* (**shunned, shunning**) avoid; keep clear of (*shuns human company*). [OE *scunian*, of unkn. orig.]

shunt /ʃʌnt/ *v. & n.* ● *v.* **1** *intr. & tr.* diverge or cause (a train) to be diverted esp. on to a siding. **2** *tr. Electr.* provide (a current) with a shunt. **3** *tr.* **a** postpone or evade. **b** divert (a decision etc.) on to another person etc. ● *n.* **1** the act or an instance of shunting on to a siding. **2** *Electr.* a conductor joining two points of a circuit, through which more or less of a current may be diverted. **3** *Surgery* an alternative path for the circulation of the blood. **4** *sl.* a motor accident, esp. a collision of vehicles travelling one close behind another. □ **shunter** *n.* [ME, perh. f. SHUN]

shush /ʃʊʃ, ʃʌʃ/ *int. & v.* ● *int.* = HUSH *int.* ● *v.* **1** *intr.* **a** call for silence by saying *shush*. **b** be silent (*they shushed at once*). **2** *tr.* make or attempt to make silent. [imit.]

shut /ʃʌt/ *v.* (**shutting**; *past* and *past part.* **shut**) **1** *tr.* **a** move (a door, window, lid, lips, etc.) into position so as to block an aperture (*shut the lid*). **b** close or seal (a room, window, box, eye, mouth, etc.) by moving a door etc. (*shut the box*). **2** *intr.* become or be capable of being closed or sealed (*the door shut with a bang*; *the lid shuts automatically*). **3** *intr. & tr.* become or make (a shop, business, etc.) closed for trade (*the shops shut at five*; *shuts his shop at five*). **4** *tr.* bring (a book, hand, telescope, etc.) into a folded-up or contracted state. **5** *tr.* (usu. foll. by *in*, *out*) keep (a person, sound, etc.) in or out of a room etc. by shutting a door etc. (*shut out the noise*; *shut them in*). **6** *tr.* (usu. foll. by *in*) catch (a finger, dress, etc.) by shutting something on it (*shut her finger in the door*). **7** *tr.* bar access to (a place etc.) (*this entrance is shut*). □ **be** (or **get**) **shut of** *sl.* be (or get) rid of (*were glad to be shut of him*). **shut the door on** refuse to consider; make impossible. **shut down 1** stop (a factory, nuclear reactor, engine, etc.) from operating. **2** (of a factory etc.) stop operating. **3** push or pull (a window-sash etc.) down into a closed position. **shut-down** *n.* the closure of a factory etc. **shut-eye** *colloq.* sleep. **shut one's eyes** (or **ears** or **heart** or **mind**) to pretend not, or refuse, to see (or hear or feel sympathy for or think about). **shut in** (of hills, houses, etc.) encircle, prevent access etc. to or escape from (*were shut in by the sea on three sides*) (see also sense 5). **shut off 1** stop the flow of (water, gas, etc.) by shutting a valve. **2** turn off (a machine, an engine, etc.). **3** separate from society etc. **shut-off** *n.* **1** something used for stopping an operation. **2** a cessation of flow, supply, or activity. **shut out** (see also sense 5 of *v.*) **1** exclude (a person, light, etc.) from a place, situation, etc. **2** screen (landscape etc.) from view. **3** prevent (a possibility etc.). **4** block (a painful memory etc.) from the mind. **5** *N. Amer.* prevent (an opponent)

from scoring. **shut-out** *adj.* **1** *Bridge* (of a bid) pre-emptive. **2** *N. Amer.* (of a game) in which one's opponents do not score. ● *n.* **1** a shut-out bid. **2** *N. Amer.* a shut-out game. **shut to 1** close (a door etc.). **2** (of a door etc.) close as far as it will go. **shut up 1** close all doors and windows of (a house etc.); bolt and bar. **2** imprison (a person). **3** close (a box etc.) securely. **4** *colloq.* reduce to silence by rebuke etc. **5** put (a thing) away in a box etc. **6** (esp. in *imper.*) *colloq.* stop talking. **shut up shop 1** close a business, shop, etc. **2** cease business etc. permanently. **shut your face** (or **head** or **mouth** or **trap**)! *sl.* an impolite request to stop talking. [OE *scyttan* f. WG: cf. SHOOT]

Shute /ʃuːt/, Nevil (pseudonym of Nevil Shute Norway) (1899–1960), English novelist. After the Second World War he settled in Australia, which provides the setting for his later novels; among the best known are *A Town Like Alice* (1950) and *On the Beach* (1957), which depicts a community facing gradual destruction in the aftermath of a nuclear war.

shutter /ʃʌtə(r)/ *n. & v.* ● *n.* **1** a person or thing that shuts. **2 a** each of a pair or set of panels fixed inside or outside a window for security or privacy or to keep the light in or out. **b** a structure of slats on rollers used for the same purpose. **3** a device that exposes the film in a photographic camera. **4** *Mus.* the blind of a swell-box in an organ used for controlling the sound-level. ● *v.tr.* **1** put up the shutters of. **2** provide with shutters. □ **put up the shutters 1** cease business for the day. **2** cease business etc. permanently. □ **shutterless** *adj.*

shuttering /ʃʌtərɪŋ/ *n.* **1** a temporary structure usu. of wood, used to hold concrete during setting. Also called *formwork*. **2** material for making shutters.

shuttle /ʃʌt(ə)l/ *n. & v.* ● *n.* **1 a** a bobbin with two pointed ends used for carrying the weft-thread across between the warp-threads in weaving. **b** a bobbin carrying the lower thread in a sewing machine. **2** a train, bus, etc., going to and fro over a short route continuously. **3** = SHUTTLECOCK. **4** = SPACE SHUTTLE. ● *v.* **1** *intr. & tr.* move or cause to move to and fro like a shuttle. **2** *intr.* travel in a shuttle. □ **shuttle armature** *Electr.* an armature with a single coil wound on an elongated iron bobbin. **shuttle diplomacy** negotiations conducted by a mediator who travels successively to several countries. **shuttle service** a train or bus etc. service operating to and fro over a short route. [OE *scytel* dart f. Gmc: cf. SHOOT]

shuttlecock /ʃʌt(ə)lˌkɒk/ *n.* **1** a cork with a ring of feathers, or a similar device of plastic, hit by the players in badminton and in battledore and shuttlecock. **2** a thing passed repeatedly back and forth. [SHUTTLE + COCK[1], prob. f. the flying motion]

shy[1] /ʃaɪ/ *adj., v., & n.* ● *adj.* (**shyer, shyest**) **1 a** diffident or uneasy in company; timid. **b** (of an animal, bird, etc.) easily startled; timid. **2** (foll. by *of*) avoiding; chary of (*shy of his aunt*; *shy of going to meetings*). **3** (in *comb.*) showing fear of or distaste for (*gun-shy*; *work-shy*). **4** (often foll. by *of, on*) *colloq.* having less; short of (*shy of the price of admission*). ● *v.intr.* (**shies, shied**) **1** (usu. foll. by *at*) (esp. of a horse) start suddenly aside (at an object, noise, etc.) in fright. **2** (usu. foll. by *away from, at*) avoid accepting or becoming involved in (a proposal etc.) in alarm. ● *n.* a sudden startled movement esp. of a horse. □ **shyer** *n.* **shyly** *adv.* **shyness** *n.* [OE *sceoh* f. Gmc]

shy[2] /ʃaɪ/ *v. & n.* ● *v.tr.* (**shies, shied**) (also *absol.*) fling or throw (a stone etc.). ● *n.* (*pl.* **shies**) the act or an instance of shying. □ **have a shy at** *colloq.* **1** try to hit with a stone etc. **2** make an attempt at. **3** jeer at. □ **shyer** *n.* [18th c.: orig. unkn.]

Shylock /ʃaɪlɒk/ a Jewish moneylender in Shakespeare's *Merchant of Venice*, who lends money to Antonio but demands in return a pound of Antonio's own flesh should the debt not be repaid on time.

shyster /ʃaɪstə(r)/ *n.* esp. *N. Amer. colloq.* a person, esp. a lawyer, who uses unscrupulous methods. [19th c.: orig. uncert.]

SI *abbr.* **1** (Order of the) Star of India. **2** Système International (d'Unités). (See INTERNATIONAL SYSTEM OF UNITS.)

Si *symb. Chem.* the element silicon.

si /siː/ *n. Mus.* = TE. [F f. It., perh. f. the initials of *Sancte Iohannes*: see GAMUT]

Siachen Glacier /sɪˈɑːtʃən/ a glacier in the Karakoram mountains in NW India, situated at an altitude of some 5,500 m (17,800 ft). Extending over 70 km (44 miles), it is one of the world's longest glaciers.

sial /ˈsaɪal/ *n. Geol.* **1** the discontinuous upper layer of the earth's crust represented by the continental masses, which are composed of relatively light rocks rich in silica and alumina and may be regarded as floating on a lower crustal layer of sima. **2** the material of which these masses are composed. [SILICON + ALUMINA]

Sialkot /sɪˈɑːlkɒt/ an industrial city in the province of Punjab, in Pakistan; pop. (1981) 296,000.

sialogogue /ˈsaɪələˌɡɒɡ/ n. & adj. ● n. a medicine inducing the flow of saliva. ● adj. inducing such a flow. [F f. Gk *sialon* saliva + *agōgos* leading]

Siam /saɪˈæm/ the former name (until 1939) for THAILAND.

Siam, Gulf of a former name for the Gulf of Thailand (see THAILAND, GULF OF).

siamang /ˈsaɪəˌmæŋ, ˈsiːə-/ n. a large black gibbon, *Hylobates syndactylus*, native to Sumatra and the Malay Peninsula. [Malay]

Siamese /ˌsaɪəˈmiːz/ n. & adj. ● n. (pl. same) **1 a** a native of Siam (now Thailand) in SE Asia. **b** the language of Siam (see THAI). **2** (in full **Siamese cat**) a cream-coloured short-haired breed of cat with brown ears, face, paws, and tail and blue eyes. ● adj. of or relating to Siam, its people, or language. □ **Siamese fighting fish** see *fighting fish* (see FIGHT).

Siamese twins n.pl. **1** identical twins that are physically conjoined at birth. (*See note below.*) **2** any closely associated pair.

▪ The name *Siamese twins* refers to the Siamese men Chang and Eng (1811–74), who were joined by a fleshy band in the region of the waist; despite this, they lived an active life and each fathered several children. The condition ranges from those joined only by the umbilical blood vessels to those having conjoined heads or trunks and sharing some organs. Conjoined twins can sometimes be separated surgically, depending on the degree of fusion.

Sian see XIAN.

sib /sɪb/ n. & adj. ● n. **1** a brother or sister, a sibling. **2** a blood relative. **3** a group of people recognized by an individual as his or her kindred. ● adj. (usu. foll. by to) esp. Sc. related; akin. [OE *sib(b)*]

Sibelius /sɪˈbeɪlɪəs/, Jean (born Johan Julius Christian Sibelius) (1865–1957), Finnish composer. He is best known for the series of seven symphonies spanning the years 1898 to 1924. His affinity for his country's landscape and legends, especially the epic *Kalevala*, expressed themselves in a series of symphonic poems including *The Swan of Tuonela* (1893), *Finlandia* (1899), and *Tapiola* (1925); he also wrote a violin concerto (1903) and more than 100 songs.

Šibenik /ˈʃɪbenɪk/ an industrial city and port in Croatia, on the Adriatic coast; pop. (1991) 41,000.

Siberia /saɪˈbɪərɪə/ a vast region of Russia, extending from the Urals to the Pacific and from the Arctic coast to the northern borders of Kazakhstan, Mongolia, and China. It was occupied by the Russians in successive stages from 1581, when a Cossack expedition ousted the Tartar khanate of Sibir, which gave the region its name. Noted for the severity of its winters, it was traditionally used as a place of exile. After the opening in 1905 of the Trans-Siberian Railway, Siberia became more accessible and attracted settlers from European Russia. It is now a major source of minerals and hydro-electric power.

Siberian /saɪˈbɪərɪən/ n. & adj. ● n. **1** a native of Siberia. **2** a person of Siberian descent. ● adj. **1** of or relating to Siberia. **2** colloq. (esp. of weather) extremely cold.

sibilant /ˈsɪbɪlənt/ adj. & n. ● adj. **1** (of a speech sound) made with a hiss. **2** hissing (a *sibilant whisper*). ● n. a sibilant letter or letter, as s, sh. □ **sibilance** n. **sibilancy** n. [L *sibilare* sibilant- hiss]

sibilate /ˈsɪbɪˌleɪt/ v.tr. & intr. pronounce with or utter a hissing sound. □ **sibilation** /ˌsɪbɪˈleɪʃ(ə)n/ n.

Sibiu /siːˈbjuː/ an industrial city in central Romania; pop. (1990) 188,000. It was founded by German colonists in the 12th century on the site of a former Roman settlement.

sibling /ˈsɪblɪŋ/ n. each of two or more children having one or both parents in common. [SIB + -LING¹]

sibship /ˈsɪbʃɪp/ n. **1** the state of belonging to a sib or the same sib. **2** a group of children having the same two parents.

sibyl /ˈsɪbɪl/ n. **1** a woman who in antiquity acted as the reputed mouthpiece of a god, uttering prophecies and oracles. The most famous was the sibyl of Cumae in south Italy, who guided Aeneas through the underworld in the *Aeneid*. **2** a prophetess, fortune-teller, or witch. [ME f. OF *Sibile* or med.L *Sibilla* f. L *Sibylla* f. Gk *Sibulla*]

sibylline /ˈsɪbɪˌlaɪn/ adj. **1** of or from a sibyl. **2** oracular; prophetic. [L *Sibyllinus* (as SIBYL)]

Sibylline books a collection of oracles belonging to the ancient Roman state and used for guidance by magistrates etc.

sic /sɪk/ adv. (usu. in brackets) used, spelt, etc., as written (confirming, or calling attention to, the form of quoted or copied words). [L, = so, thus]

siccative /ˈsɪkətɪv/ n. & adj. ● n. a substance causing drying, esp. mixed with oil-paint etc. for quick drying. ● adj. having such properties. [LL *siccativus* f. *siccare* to dry]

sice¹ /saɪs/ n. the six on dice. [ME f. OF *sis* f. L *sex* six]

sice² var. of SYCE.

Sichuan /sɪˈtʃwɑːn/ (also **Szechuan, Szechwan** /seˈtʃwɑːn/) a province of west central China; capital, Chengdu.

Sicilia see SICILY.

siciliano /sɪˌsɪlɪˈɑːnəʊ/ n. (also **siciliana** /-nə/) (pl. **-os**) a dance, song, or instrumental piece in 6/8 or 12/8 time, often in a minor key, and evoking a pastoral mood. [It., = Sicilian]

Sicilian Vespers /sɪˈsɪlɪən/ a massacre of French inhabitants of Sicily, which began near Palermo at the time of vespers on Easter Monday in 1282. The ensuing war resulted in the replacement of the unpopular French Angevin dynasty by the Spanish House of Aragon.

Sicily /ˈsɪsɪlɪ/ (Italian **Sicilia** /siˈtʃiːlja/) a large Italian island in the Mediterranean, off the south-western tip of Italy; capital, Palermo. Settled successively by Phoenicians, Greeks, and Carthaginians, it became a Roman province in 241 BC after the first Punic War. It was conquered towards the end of the 11th century by the Normans, who united the island with southern Italy as the kingdom of the Two Sicilies. Conquered by Charles of Anjou in 1266, the island subsequently passed to the House of Aragon. It was reunited with southern Italy in 1442. The kingdom was claimed in 1816 by the Spanish Bourbon king Ferdinand, but was liberated in 1860 by Garibaldi and incorporated into the new state of Italy. The island's highest point is the volcano Mount Etna. □ **Sicilian** /sɪˈsɪlɪən/ adj. & n.

sick¹ /sɪk/ adj., n., & v. ● adj. **1** (often in comb.) esp. Brit. vomiting or tending to vomit (*I think I'm going to be sick; seasick*). **2** ill; affected by illness (*has been in bed sick for a week; a sick man; sick with measles*). **3 a** (often foll. by at) esp. mentally perturbed; disordered (*the product of a sick mind; sick at heart*). **b** (often foll. by for, or in comb.) pining; longing (*sick for a sight of home; lovesick*). **4** (often foll. by of) colloq. **a** disgusted; surfeited (*sick of chocolates*). **b** angry, esp. because of surfeit (*am sick of being teased*). **5** colloq. (of humour etc.) jeering at misfortune, illness, death, etc.; morbid (*sick joke*). **6** (of a ship) needing repair (esp. of a specified kind) (*paint-sick*). ● n. Brit. colloq. vomit. ● v.tr. (usu. foll. by up) Brit. colloq. vomit (*sicked up his dinner*). □ **go sick** report oneself as ill. **look sick** colloq. be unimpressive or embarrassed. **sick at (or to) one's stomach** US vomiting or tending to vomit. **sick bag** a bag provided on an aeroplane or ferry etc. for use by a travel-sick person. **sick-benefit** Brit. = sickness benefit. **sick building syndrome** a high incidence of illness in office workers, attributed to the immediate working surroundings. **sick-call 1** a visit by a doctor to a sick person etc. **2** Mil. a summons for sick men to attend. **sick-flag** a yellow flag indicating disease at a quarantine station or on ship. **sick headache** a migraine headache with vomiting. **sick-leave** leave of absence granted because of illness (*on sick-leave*). **sick-list** a list of the sick, esp. in a regiment, ship, etc. **sick-making** colloq. sickening. **sick nurse** = NURSE¹ n. **sick-pay** pay given to an employee etc. on sick-leave. **take sick** colloq. be taken ill. □ **sickish** adj. [OE *sēoc* f. Gmc]

sick² /sɪk/ v.tr. (usu. in imper.) (esp. to a dog) set upon (a rat etc.). [19th c., dial. var. of SEEK]

sickbay /ˈsɪkbeɪ/ n. **1** part of a ship used as a hospital. **2** any room etc. for sick people.

sickbed /ˈsɪkbed/ n. **1** an invalid's bed. **2** the state of being an invalid.

sicken /ˈsɪkən/ v. **1** tr. affect with loathing or disgust. **2** intr. **a** (often foll. by for) show symptoms of illness (*is sickening for measles*). **b** (often foll. by at, or to + infin.) feel nausea or disgust (*he sickened at the sight*). **3** (as **sickening** adj.) **a** loathsome, disgusting. **b** colloq. very annoying. □ **sickeningly** adv.

sickener /ˈsɪkənə(r)/ n. **1** colloq. something causing nausea, disgust, or severe disappointment. **2** a red toadstool of the genus *Russula*, esp. the poisonous *R. emetica*.

Sickert /ˈsɪkət/, Walter Richard (1860–1942), British painter, of Danish and Anglo-Irish descent. He was a pupil of Whistler and also worked with Degas. His subjects were mainly urban scenes and figure compositions, particularly pictures of the theatre and music-hall and drab domestic interiors, avoiding the conventionally picturesque. His best-known painting, *Ennui* (1913), portrays a stagnant marriage.

sickie /ˈsɪkɪ/ n. Austral. & NZ colloq. a period of sick-leave, usu. taken with insufficient medical reason.

sickle /ˈsɪk(ə)l/ n. **1** a short-handled farming tool with a semicircular blade, used for cutting corn, lopping, or trimming. **2** anything sickle-shaped, esp. the crescent moon. □ **sickle-bill** a bird with a long narrow curved bill, esp. a hummingbird of the genus *Eutoxeres*. **sickle-cell anaemia** *Med.* a severe hereditary form of anaemia, affecting mainly blacks, in which a mutated form of haemoglobin distorts the red blood cells into a crescent shape at low oxygen levels. **sickle-feather** each of the long middle feathers of a cock's tail. [OE *sicol, sicel* f. L *secula* f. *secare* cut]

sickly /'sɪklɪ/ *adj.* (**sicklier, sickliest**) **1 a** of weak health; apt to be ill. **b** (of a person's complexion, look, etc.) languid, faint, or pale, suggesting sickness (*a sickly smile*). **c** (of light or colour) faint, pale, feeble. **2** causing ill health (*a sickly climate*). **3** (of a book etc.) sentimental or mawkish. **4** inducing or connected with nausea (*a sickly taste*). **5** (of a colour etc.) of an unpleasant shade inducing nausea (*a sickly green*). □ **sickliness** *n.* [ME, prob. after ON *sjúkligr* (as SICK[1])]

sickness /'sɪknɪs/ *n.* **1** the state of being ill; disease. **2** a specified disease (*sleeping sickness*). **3** vomiting or a tendency to vomit. □ **sickness benefit** (in the UK) benefit paid by the state for sickness interrupting paid employment. [OE *sēocnesse* (as SICK[1], -NESS)]

sicko /'sɪkəʊ/ *n. N. Amer. sl.* a mentally ill or perverted person.

sickroom /'sɪkruːm, -rʊm/ *n.* **1** a room occupied by a sick person. **2** a room adapted for sick people.

sidalcea /sɪ'dælsɪə/ *n.* a North American plant of the genus *Sidalcea*, of the mallow family, bearing sprays of white, pink, or purple flowers. [mod.L f. *Sida* + *Alcea*, names of related genera]

Siddhartha Gautama /sɪ,dɑːtə 'gaʊtəmə/ see BUDDHA.

Siddons /'sɪd(ə)nz/, Mrs Sarah (née Kemble) (1755–1831), English actress. The sister of John Kemble, she made her first unsuccessful London appearance in 1775; she then toured the provinces and made a successful return to the London stage in 1782, where she became an acclaimed tragic actress, noted particularly for her role as Lady Macbeth. She retained her pre-eminence until her retirement in 1812.

side /saɪd/ *n. & v.* ● *n.* **1 a** each of the more or less flat surfaces bounding an object (*a cube has six sides; this side up*). **b** a more or less vertical inner or outer surface (*the side of a house; a mountainside*). **c** such a vertical lateral surface or plane as distinct from the top or bottom, front or back, or ends (*at the side of the house*). **2 a** the half of a person or animal that is on the right or the left, esp. of the torso (*has a pain in his right side*). **b** the left or right half or a specified part of a thing, area, building, etc. (*put the box on that side*). **c** (often in *comb.*) a position next to a person or thing (*grave-side; seaside; stood at my side*). **d** a specified direction relating to a person or thing (*on the north side of; came from all sides*). **e** half of a butchered carcass (*a side of bacon*). **3 a** either surface of a thing regarded as having two surfaces. **b** the amount of writing needed to fill one side of a sheet of paper (*write three sides*). **4** any of several aspects of a question, character, etc. (*many sides to his character; look on the bright side*). **5 a** either of two sets of opponents in war, politics, games, etc. (*the side that bats first; much to be said on both sides*). **b** a cause or philosophical position etc. regarded as being in conflict with another (*on the side of right*). **6 a** a part or region near the edge and remote from the centre (*at the side of the room*). **b** (*attrib.*) a subordinate, peripheral, or detached part (*a side-road; a side-table*). **7 a** each of the bounding lines of a plane rectilinear figure (*a hexagon has six sides*). **b** either of two quantities stated to be equal in an equation. **8** a position nearer or further than, or right or left of, a given dividing line (*on this side of the Alps; on the other side of the road*). **9** a line of hereditary descent through the father or the mother. **10** (in full **side spin**) *Brit.* a spinning motion given to a billiard-ball etc. by hitting it on one side, not centrally. **11** *Brit. sl.* boastfulness; swagger (*has no side about him*). **12** *Brit. colloq.* a television channel, considered as one of two or more available channels (*what's on the other side?*). ● *v.intr.* (usu. foll. by *with*) take part or be on the same side as a disputant etc. (*sided with his father*). □ **by the side of 1** close to. **2** compared with. **from side to side 1** right across. **2** alternately each way from a central line. **let the side down** fail one's colleagues, esp. by frustrating their efforts or embarrassing them. **on one side 1** not in the main or central position. **2** aside (*took him on one side to explain*). **on the ... side** fairly, somewhat (qualifying an adjective: *on the high side*). **on the side 1** as a sideline; in addition to one's regular work etc. **2** secretly or illicitly. **3** *N. Amer.* as a side dish. **on this side of the grave** in life. **side-arms** swords, bayonets, or pistols. **side-band** a range of frequencies near the carrier frequency of a radio wave, concerned in modulation. **side-bet** a bet between opponents, esp. in card-games, over and above the ordinary stakes. **side-bone** either of the small forked bones under the wings of poultry. **side by side** standing close together, esp. for mutual support. **side-car 1** a small car for a passenger or passengers attached to the side of a motorcycle. **2** a cocktail of orange liqueur, lemon juice, and brandy. **3** a jaunting car. **side chain** *Chem.* a group of atoms attached to the main part of a molecule. **side-chapel** a chapel in the aisle or at the side of a church. **side dish** an extra dish subsidiary to the main course. **side-door 1** a door in or at the side of a building. **2** an indirect means of access. **side-drum** a small double-headed drum in a jazz or military band or in an orchestra (orig. hung at the drummer's side). **side-effect** a secondary, usu. undesirable, effect. **side-glance** a sideways or brief glance. **side-issue** a point that distracts attention from what is important. **side-note** a marginal note. **side-on** *adv.* from the side. ● *adj.* **1** from or towards one side. **2** (of a collision) involving the side of a vehicle. **side-road** a minor or subsidiary road, esp. joining or diverging from a main road. **side-saddle** *n.* a saddle for a woman rider with both feet on the same side of the horse. ● *adv.* sitting in this position on a horse. **side salad** a salad served as a side dish. **side-seat** a seat in a vehicle etc. in which the occupant has his back to the side of the vehicle. **side-slip** *n.* **1** a skid. **2** *Aeron.* a sideways movement, esp. downwards towards the inside of a turn. ● *v.intr.* **1** skid. **2** *Aeron.* move in a side-slip. **side-splitting** causing violent laughter. **side-street** a minor or subsidiary street. **side-stroke 1** a stroke towards or from a side. **2** an incidental action. **3** a swimming stroke in which the swimmer lies on his or her side. **side-swipe** *n.* **1** a glancing blow along the side. **2** an incidental critical remark etc. ● *v.tr.* hit with or as if with a side-swipe. **side-table** a table placed at the side of a room or apart from the main table. **side-trip** a minor excursion during a voyage or trip; a detour. **side valve** a valve in a vehicle engine, operated from the side of the cylinder. **side-view 1** a view obtained sideways. **2** a profile. **side-wheeler** *US* a steamer with paddle-wheels. **side-whiskers** whiskers growing on the cheeks. **side wind 1** wind from the side. **2** an indirect agency or influence. **take sides** support one or other cause etc. □ **sideless** *adj.* [OE *sīde* f. Gmc]

sidearm /'saɪdɑːm/ *adj. & adv.* ● *adj.* esp. *Baseball* performed or delivered with a swing of the arm extended sideways. ● *adv.* in a sidearm manner.

sideboard /'saɪdbɔːd/ *n.* a table or esp. a flat-topped cupboard at the side of a dining-room for supporting and containing dishes, table linen, decanters, etc.

sideboards /'saɪdbɔːdz/ *n.pl. Brit. colloq.* hair grown by a man down the sides of his face; side-whiskers.

sideburns /'saɪdbɜːnz/ *n.pl.* = SIDEBOARDS. [*burnsides* (US) f. Gen. Ambrose *Burnside*, Amer. soldier (1824–81), who wore them]

sided /'saɪdɪd/ *adj.* **1** having sides. **2** (in *comb.*) having a specified side or number of sides (*flat-sided; three-sided*). □ **-sidedly** *adv.* **sidedness** *n.* (also in *comb.*).

sidehill /'saɪdhɪl/ *n. US* a hillside.

sidekick /'saɪdkɪk/ *n. colloq.* a close associate.

sidelight /'saɪdlaɪt/ *n.* **1** a light from the side. **2** incidental information etc. **3** *Brit.* a light at the side of the front of a motor vehicle to warn of its presence. **4** *Naut.* the red port or green starboard light on a ship under way.

sideline /'saɪdlaɪn/ *n. & v.* ● *n.* **1** an activity done in addition to one's main work etc. **2** (usu. in *pl.*) **a** a line bounding the side of a hockey-pitch, tennis-court, etc. **b** the space next to these where spectators etc. sit. ● *v.tr. US* remove (a player) from a team through injury, suspension, etc. □ **on** (or **from**) **the sidelines** in (or from) a position removed from the main action.

sidelong /'saɪdlɒŋ/ *adj. & adv.* ● *adj.* inclining to one side; oblique (*a sidelong glance*). ● *adv.* obliquely (*moved sidelong*). [*sideling* (as SIDE, -LING[2]): see -LONG]

sidereal /saɪ'dɪərɪəl/ *adj. Astron.* of or concerning the constellations or fixed stars. □ **sidereal clock** a clock showing sidereal time. **sidereal day** the time between successive meridional transits of a star or (esp.) of the First Point of Aries, about four minutes shorter than the solar day. **sidereal time** time measured by the apparent diurnal motion of the stars. **sidereal year** a year longer than the solar year by 20 minutes 23 seconds because of precession. [L *sidereus* f. *sidus sideris* star]

siderite /'sɪdə,raɪt/ *n.* **1** a mineral form of ferrous carbonate. **2** a meteorite consisting mainly of nickel and iron. [Gk *sidēros* iron]

sidero- /'sɪdərəʊ/ *comb. form* **1** of or relating to iron. **2** of or relating to the stars. [sense 1 f. Gk *sidēros* iron; sense 2 f. L *sider- sidus* star]

siderostat /'sɪdərə,stæt/ *n. Astron.* an instrument used for keeping the image of a celestial object in a fixed position. [f. SIDERO- 2, after *heliostat*]

sideshow /ˈsaɪdʃəʊ/ n. **1** a small show or stall in an exhibition, fair, etc. **2** a minor incident or issue.

sidesman /ˈsaɪdzmən/ n. (pl. **-men**) a churchwarden's assistant, who shows worshippers to their seats, takes the collection, etc.

sidestep /ˈsaɪdstep/ n. & v. ● n. a step taken sideways. ● v.tr. (**-stepped**, **-stepping**) **1** esp. Football avoid (esp. a tackle) by stepping sideways. **2** evade. □ **sidestepper** n.

sidetrack /ˈsaɪdtræk/ n. & v. ● n. a railway siding. ● v.tr. **1** turn into a siding; shunt. **2 a** postpone, evade, or divert treatment or consideration of. **b** divert (a person) from considering etc.

sidewalk /ˈsaɪdwɔːk/ n. N. Amer. a pedestrian path at the side of a road; a pavement.

sideward /ˈsaɪdwəd/ adj. & adv. ● adj. = SIDEWAYS adj. ● adv. (also **sidewards** /-wədz/) = SIDEWAYS adv.

sideways /ˈsaɪdweɪz/ adv. & adj. ● adv. **1** to or from a side (*moved sideways*). **2** with one side facing forward (*sat sideways on the bus*). ● adj. **1** to or from a side (*a sideways movement*). **2** unconventional, unorthodox (*a sideways look at recent events*). □ **sidewise** adv. & adj.

sidewinder[1] /ˈsaɪdˌwaɪndə(r)/ n. **1** a North American desert rattlesnake, *Crotalus cerastes*, that moves sideways by throwing its body into S-shaped curves. **2** (**Sidewinder**) a type of American air-to-air heat-seeking missile.

sidewinder[2] /ˈsaɪdˌwaɪndə(r)/ n. N. Amer. a sideways blow.

Sidi bel Abbès /ˌsiːdɪ bel əˈbes/ a town in northern Algeria, situated to the south of Oran; pop. (1989) 186,000. Established as a French military outpost in 1843, it was the headquarters of the French Foreign Legion until Algeria's independence in 1962.

siding /ˈsaɪdɪŋ/ n. **1** a short track at the side of and opening on to a railway line, used for shunting trains. **2** N. Amer. cladding material for the outside of a building.

sidle /ˈsaɪd(ə)l/ v. & n. ● v.intr. (usu. foll. by *along*, *up*) walk in a timid, furtive, stealthy, or cringing manner. ● n. the act or an instance of sidling. [back-form. f. *sideling*, SIDELONG]

Sidney /ˈsɪdnɪ/, Sir Philip (1554–86), English poet and soldier. Generally considered to represent the apotheosis of the Elizabethan courtier, he was a leading poet and patron of poets, including Edmund Spenser. His best-known work is *Arcadia* (published posthumously in 1590), a prose romance including poems and pastoral eclogues in a wide variety of verse forms.

Sidon /ˈsaɪd(ə)n/ (Arabic **Saida** /ˈsaɪdə/) a city in Lebanon, on the Mediterranean coast south of Beirut; pop. (1988) 38,000. Founded in the third millennium BC, it was a Phoenician seaport and city-state.

Sidra, Gulf of /ˈsɪdrə/ (also **Gulf of Sirte** /ˈsɜːtɪ/) a broad inlet of the Mediterranean on the coast of Libya, between the towns of Benghazi and Misratah.

SIDS abbr. sudden infant death syndrome; = *cot-death* (see COT[1]).

Siebengebirge /ˈziːb(ə)ŋɡəˌbɪərɡə/ a range of hills in western Germany, on the right bank of the Rhine south-east of Bonn.

siege /siːdʒ/ n. **1 a** a military operation in which an attacking force seeks to compel the surrender of a fortified place by surrounding it and cutting off supplies etc. **b** a similar operation by police etc. to force the surrender of an armed person. **c** the period during which a siege lasts. **2** a persistent attack or campaign of persuasion. □ **lay siege to** esp. Mil. conduct the siege of. **raise the siege of** abandon or cause the abandonment of an attempt to take (a place) by siege. **siege-gun** hist. a heavy gun used in sieges. **siege-train** artillery and other equipment for a siege, with vehicles etc. [ME f. OF *sege* seat f. *assegier* BESIEGE]

Siegfried /ˈsiːɡfriːd/ the hero of the first part of the Nibelungenlied. A prince of the Netherlands Siegfried obtains a treasure by killing the dragon Fafner. He marries Kriemhild, and helps Gunther to win Brunhild before being treacherously slain by Hagen.

Siegfried Line 1 the line of defence constructed by the Germans along the western frontier of Germany before the Second World War. **2** see HINDENBURG LINE.

Siemens /ˈziːmənz/, Ernst Werner von (1816–92), German electrical engineer. He developed electroplating and an electric generator which used an electromagnet, and set up a factory which manufactured telegraph systems and electric cables and pioneered electrical traction. His brother Karl Wilhelm (Sir Charles William, 1823–83) moved to England, where he developed the open-hearth steel furnace and designed the cable-laying steamship *Faraday*, and also designed the electric railway at Portrush in Northern Ireland. A third brother Friedrich (1826–1904) worked both with Werner in Germany and with Charles in England; he applied the principles of the open-hearth furnace to glassmaking.

siemens /ˈsiːmənz/ n. Electr. the SI unit of conductance (symbol: **S**), equal to the conductance of a body whose resistance is one ohm (cf. MHO). [SIEMENS]

Siena /sɪˈenə/ a city in west central Italy, in Tuscany; pop. (1990) 57,745. It is noted for the school of art which flourished there in the 13th and 14th centuries. □ **Sienese** /sɪəˈniːz/ adj. & n.

sienna /sɪˈenə/ n. **1** a kind of ferruginous earth used as a pigment in paint. **2** its colour of yellowish-brown (*raw sienna*) or reddish-brown (*burnt sienna*). [It. (*terra di*) *Sienna* (earth of) SIENA]

sierra /sɪˈeərə/ n. a long jagged mountain chain, esp. in Spain, Spanish America, or the western US. [Sp. f. L *serra* saw]

Sierra Leone /sɪˌeərə lɪˈəʊn/ a country on the coast of West Africa; pop. (est. 1991) 4,239,000; languages, English (official), English Creole, Temne, and other West African languages; capital, Freetown. An area of British influence from the late 18th century, the district around Freetown on the coast became a colony in 1807, serving as a centre for operations against slave-traders. The large inland territory was not declared a protectorate until 1896. Sierra Leone achieved independence within the Commonwealth in 1961. The government was toppled by a military coup in 1992. □ **Sierra Leonean** adj. & n.

Sierra Madre /sɪˌeərə ˈmɑːdreɪ/ a mountain system in Mexico, extending from the border with the US in the north to the southern border with Guatemala. It is divided into the *Sierra Madre Occidental* in the west, the *Sierra Madre Oriental* in the east, and the *Sierra Madre del Sur* in the south. The highest peak is Citlaltépetl, in the Sierra Madre Oriental south-east of Mexico City, which rises to a height of 5,699 m (18,697 ft).

Sierra Nevada /sɪˌeərə nəˈvɑːdə/ **1** a mountain range in southern Spain, in Andalusia, south-east of Granada. Its highest peak is Mulhacén, which rises to 3,482 m (11,424 ft). **2** a mountain range in eastern California. Rising sharply from the Great Basin in the east, it descends more gently to California's Central Valley in the west. Its highest peak is Mount Whitney on the edge of Sequoia National Park, which rises to 4,418 m (14,495 ft).

siesta /sɪˈestə/ n. an afternoon sleep or rest esp. in hot countries. [Sp. f. L *sexta* (*hora*) sixth hour]

sieve /sɪv/ n. & v. ● n. a utensil having a perforated or meshed bottom for separating solids or coarse material from liquids or fine particles, or for reducing a soft solid to a fine pulp. ● v.tr. **1** put through or sift with a sieve. **2** examine (evidence etc.) to select or separate. □ **head like a sieve** colloq. a memory that retains little. □ **sievelike** adj. [OE *sife* f. WG]

sievert /ˈsiːvət/ n. an SI unit of dose equivalent of ionizing radiation, equal to one joule per kilogram and equivalent to 100 rem. [Rolf M. *Sievert*, Swedish radiologist (1896–1966)]

sifaka /sɪˈfɑːkə/ n. a long-tailed arboreal lemur of the genus *Propithecus*, native to Madagascar, with white to blackish-brown fur. [Malagasy]

siffleur /siːˈflɜː(r)/ n. (fem. **siffleuse** /-ˈflɜːz/) a professional whistler. [F f. *siffler* whistle]

sift /sɪft/ v. **1** tr. sieve (material), esp. to separate finer and coarser parts. **2** tr. (usu. foll. by *from*, *out*) separate (finer or coarser parts) from material. **3** tr. sprinkle (esp. sugar) from a perforated container. **4** tr. examine (evidence, facts, etc.) in order to assess authenticity etc. **5** intr. (of snow, light, etc.) fall as if from a sieve. □ **sift through** examine by sifting. □ **sifter** n. (also in comb.). [OE *siftan* f. WG]

Sig. abbr. Signor.

sigh /saɪ/ v. & n. ● v. **1** intr. emit a long deep audible breath expressive of sadness, weariness, longing, relief, etc. **2** intr. (foll. by *for*) yearn for (a lost person or thing). **3** tr. utter or express with sighs (*'Never!' he sighed*). **4** intr. (of the wind etc.) make a sound like sighing. ● n. **1** the act or an instance of sighing. **2** a sound made in sighing (*a sigh of relief*). [ME *sihen* etc., prob. back-form. f. *sihte* past of *sihen* f. OE *sican*]

sight /saɪt/ n. & v. ● n. **1 a** the faculty of seeing with the eyes (*lost his sight*). **b** the act or an instance of seeing; the state of being seen. **2** a thing seen; a display, show, or spectacle (*not a pretty sight*; *a beautiful sight*). **3** a way of looking at or considering a thing (*in my sight he can do no wrong*). **4** a range of space within which a person etc. can see or an object be seen (*he's out of sight*; *they are just coming into sight*). **5** (usu. in pl.) noteworthy features of a town, area, etc. (*went to see the sights*). **6 a** a device on a gun or optical instrument used for assisting the precise aim or observation. **b** the aim or observation so gained (*got a sight of him*). **7** *colloq.* a person or thing having a ridiculous, repulsive, or dishevelled appearance (*looked a perfect sight*). **8** *colloq.* a great quantity (*will cost a sight of money; is a sight better than he was*). ● *v.tr.* **1** get sight of, esp. by approaching (*they sighted land*). **2** observe the presence of (esp. aircraft, animals, etc.) (*sighted buffalo*). **3** take observations of (a star etc.) with an instrument. **4 a** provide (a gun, quadrant, etc.) with sights. **b** adjust the sight of (a gun etc.). **c** aim (a gun etc.) with sights. □ **at first sight** on first glimpse or impression. **at** (or **on**) **sight** as soon as a person or a thing has been seen (*plays music at sight; liked him on sight*). **catch** (or **lose**) **sight of** begin (or cease) to see or be aware of. **get a sight of** manage to see; glimpse. **have lost sight of** no longer know the whereabouts of. **in sight 1** visible. **2** near at hand (*rescue is in sight*). **in** (or **within**) **sight of** so as to see or be seen from. **lower one's sights** become less ambitious. **out of my sight!** go at once! **out of sight 1** not visible. **2** *sl.* excellent; delightful. **out of sight out of mind** we forget the absent. **put out of sight** hide, ignore. **set one's sights** on aim at (*set her sights on a directorship*). **sight for the gods** (or **sight for sore eyes**) a welcome person or thing, esp. a visitor. **sight-glass** a transparent device for observing the interior of apparatus etc. **sighting shot** an experimental shot to guide gunners in adjusting their sights. **sight-line** a hypothetical line from a person's eye to what is seen. **sight-read** (*past* and *past part.* **-read** /-red/) read and perform (music) at sight. **sight-reader** a person who sight-reads music. **sight-screen** *Cricket* a large usu. white movable screen placed near the boundary in line with the wicket to help the batsman see the ball. **sight-sing** sing (music) at sight. **sight unseen** without previous inspection. □ **sighter** *n.* [OE (*ge*)*sihth*]

sighted /'saɪtɪd/ *adj.* **1** capable of seeing; not blind. **2** (in *comb.*) having a specified kind of sight (*long-sighted*).

sightless /'saɪtlɪs/ *adj.* **1** blind. **2** *poet.* invisible. □ **sightlessly** *adv.* **sightlessness** *n.*

sightly /'saɪtlɪ/ *adj.* pleasing to the sight; not unsightly. □ **sightliness** *n.*

sightseer /'saɪtˌsiːə(r), -sɪə(r)/ *n.* a person who visits places of interest; a tourist. □ **sightsee** *v.intr.* & *tr.* **sightseeing** *n.*

sightworthy /'saɪtˌwɜːðɪ/ *adj.* worth seeing.

sigillate /'sɪdʒɪlət/ *adj.* **1** (of pottery) having impressed patterns. **2** *Bot.* having seal-like marks. [L *sigillatus* f. *sigillum* seal dimin. of *signum* sign]

SIGINT /'sɪɡɪnt/ *n.* (also **Sigint**) intelligence gathered from the monitoring, interception, and interpretation of radio signals etc. [abbr. of *signals* intelligence]

siglum /'sɪɡləm/ *n.* (*pl.* **sigla** /-lə/) a letter (esp. an initial) or other symbol to denote a word in a book, esp. to refer to a particular text. [LL *sigla* (pl.), perh. f. *singula* neut. pl. of *singulus* single]

sigma /'sɪɡmə/ *n.* the eighteenth letter of the Greek alphabet (Σ, σ, or, when final, ς). [L f. Gk]

sigmate /'sɪɡmət, -meɪt/ *adj.* **1** sigma-shaped. **2** S-shaped.

sigmoid /'sɪɡmɔɪd/ *adj.* & *n.* ● *adj.* **1** curved like the uncial sigma (Ϲ); crescent-shaped. **2** S-shaped. ● *n.* (in full **sigmoid flexure**) *Anat.* the curved part of the intestine between the colon and the rectum. [Gk *sigmoeidēs* (as SIGMA)]

sign /saɪn/ *n.* & *v.* ● *n.* **1 a** a thing indicating or suggesting a quality or state etc.; a thing perceived as indicating a future state or occurrence (*violence is a sign of weakness; shows all the signs of decay*). **b** a miracle evidencing supernatural power; a portent (*did signs and wonders*). **2 a** a mark, symbol, or device used to represent something or to distinguish the thing on which it is put (*marked the jar with a sign*). **b** a technical symbol used in algebra, music, etc. (*a minus sign; a repeat sign*). **3 a** a gesture or action used to convey information, an order, request, etc. (*gave him a sign to leave; conversed by signs*). **b** a gesture used in a system of sign language. **c** = *sign language*. **4** a publicly displayed board etc. giving information; a signboard or signpost. **5** an objective indication of a disease (cf. SYMPTOM 1), usu. specified (*Babinski's sign*). **6** a password (*advanced and gave the sign*). **7** each of the twelve divisions of the zodiac, named from the constellations formerly situated in them (*the sign of Cancer*). **8** *US* the trail of a wild animal. **9** *Math.* etc. the positiveness or negativeness of a quantity. ● *v.* **1** *tr.* (also *absol.*) write (one's name, initials, etc.) on a document etc. indicating that one has authorized it. **b** write one's name etc. on (a document) as authorization. **2** *intr.* & *tr.* **a** communicate by gesture (*signed to me to come; signed their assent*). **b** communicate in a sign language. **3** *tr.* & *intr.* engage or be engaged by signing a contract etc. (see also *sign on, sign up*). **4** *tr.* mark with a sign (esp. with the sign of the cross in baptism). □ **make no sign** seem unconscious; not protest. **sign and countersign** secret words etc.

used as passwords. **sign away** convey (one's right, property, etc.) by signing a deed etc. **sign for** acknowledge receipt of by signing. **sign in 1** sign a register on arrival in a hotel etc. **2** authorize the admittance of (a person) by signing a register. **sign language** a system of communication by visual gestures, used esp. by the deaf. **sign of the cross** a Christian sign made in blessing or prayer, by tracing a cross from the forehead to the chest and to each shoulder, or in the air. **sign off 1** end work, broadcasting, a letter, etc., esp. by writing or speaking one's name. **2 a** end a period of employment, contract, etc. **b** end the period of employment or contract of (a person). **3** *Brit.* register to stop receiving unemployment benefit after finding work. **4** *Bridge* indicate by a conventional bid that one is seeking to end the bidding. **sign-off** *n. Bridge* such a bid. **sign of the times** a portent etc. showing a likely trend. **sign on 1** agree to a contract, employment, etc. **2** begin work, broadcasting, etc., esp. by writing or announcing one's name. **3** employ (a person). **4** *Brit.* register as unemployed. **sign out 1** sign a register on leaving a hotel etc. **2** authorize the release or record the departure of (a person or thing) by signing a register. **sign-painter** (or **-writer**) a person who paints signboards etc. **sign up 1** engage or employ (a person). **2** enlist in the armed forces. **3 a** commit (another person or oneself) by signing etc. (*signed you up for dinner*). **b** enrol (*signed up for evening classes*). □ **signable** *adj.* **signer** *n.* [ME f. OF *signe, signer* f. L *signum, signare*]

Signac /siː'njæk/, Paul (1863–1935), French neo-impressionist painter. He was an ardent disciple of Seurat, although his own technique was freer than Seurat's and was characterized by the use of small dashes and patches of pure colour rather than dots; his subject-matter included landscapes, seascapes, and city scenes. He also wrote *D'Eugène Delacroix aux néo-impressionisme* (1899), a manifesto defending the movement.

signal[1] /'sɪɡn(ə)l/ *n.* & *v.* ● *n.* **1 a** a usu. prearranged sign conveying information, guidance, etc., esp. at a distance (*waved as a signal to begin*). **b** a message made up of such signs (*signals made with flags*). **2** an immediate occasion for or cause of movement, action, etc. (*the uprising was a signal for repression*). **3** *Electr.* **a** an electrical impulse or impulses or radio waves transmitted or received. **b** a sequence of these. **4** a light, semaphore, etc., on a railway giving instructions or warnings to train-drivers etc. **5** *Bridge* a prearranged mode of bidding or play to convey information to one's partner. ● *v.* (**signalled**, **signalling**; *US* **signaled**, **signaling**) **1** *intr.* make signals. **2** *tr.* **a** (often foll. by *to* + *infin.*) make signals to; direct. **b** transmit (an order, information, etc.) by signal; announce (*signalled her agreement; signalled that the town had been taken*). □ **signal-book** a list of signals arranged for sending esp. naval and military messages. **signal-box** *Brit.* a building beside a railway track from which signals, points, etc. are controlled. **signal of distress** esp. *Naut.* an appeal for help, esp. from a ship by firing guns. **signal-to-noise ratio** the ratio of the strength of an electrical or other signal carrying information to that of unwanted interference, usu. expressed in decibels. **signal-tower** *US* = signal-box. □ **signaller** *n.* [ME f. OF f. Rmc & med.L *signale* neut. of LL *signalis* f. L *signum* SIGN]

signal[2] /'sɪɡn(ə)l/ *adj.* remarkably good or bad; noteworthy (*a signal victory*). □ **signally** *adv.* [F *signalé* f. It. past part. *segnalato* distinguished f. *segnale* SIGNAL[1]]

signalize /'sɪɡnəˌlaɪz/ *v.tr.* (also **-ise**) **1** make noteworthy or remarkable. **2** lend distinction or lustre to. **3** indicate.

signalman /'sɪɡn(ə)lmən/ *n.* (*pl.* **-men**) **1** a railway employee responsible for operating signals and points. **2** a person who displays or receives naval etc. signals.

signary /'sɪɡnərɪ/ *n.* (*pl.* **-ies**) a list of signs constituting the syllabic or alphabetic symbols of a language. [L *signum* SIGN + -ARY[1], after *syllabary*]

signatory /'sɪɡnətərɪ/ *n.* & *adj.* ● *n.* (*pl.* **-ies**) a party or esp. a state that has signed an agreement or esp. a treaty. ● *adj.* having signed such an agreement etc. [L *signatorius* f. *signare signat-* mark]

signature /'sɪɡnətʃə(r)/ *n.* **1 a** a person's name, initials, or mark used in signing a letter, document, etc. **b** the act of signing a document etc. **2** a distinctive pattern or characteristic by which something can be identified. **3** *Mus.* **a** = key signature. **b** = time signature. **4** *Printing* **a** a letter or figure placed at the foot of one or more pages of each sheet of a book as a guide for binding. **b** such a sheet after folding. **5** *US* directions given to a patient as part of a medical prescription. □ **doctrine of signatures** *hist.* the belief that the form or colouring of a medicinal plant resembled that of the organ or disease which it could be used to treat. **signature tune** esp. *Brit.* a distinctive tune used to introduce a particular programme or performer on television or radio. [med.L *signatura* (LL = marking of sheep), as SIGNATORY]

signboard /ˈsaɪnbɔːd/ n. a board with a name or symbol etc. displayed outside a shop or hotel etc.

signet /ˈsɪgnɪt/ n. **1** a seal used instead of or with a signature as authentication. **2** (prec. by *the*) the royal seal formerly used for special purposes in England and Scotland, and in Scotland later as the seal of the Court of Session. □ **signet-ring** a ring with a seal set in it. [ME f. OF *signet* or med.L *signetum* (as SIGN)]

significance /sɪgˈnɪfɪkəns/ n. **1** importance; noteworthiness (*his opinion is of no significance*). **2** a concealed or real meaning (*what is the significance of his statement?*). **3** the state of being significant. **4** *Statistics* the extent to which a result deviates from that expected on the basis of the null hypothesis. [OF *significance* or L *significantia* (as SIGNIFY)]

significant /sɪgˈnɪfɪkənt/ adj. **1** having a meaning; indicative. **2** having an unstated or secret meaning; suggestive (*refused it with a significant gesture*). **3** noteworthy; important; consequential (*a significant figure in history*). **4** *Statistics* of or relating to significance, departing from the null hypothesis. □ **significant figure** *Math.* a digit conveying information about a number containing it, and not a zero used simply to fill vacant space at the beginning or end (*results expressed to three significant figures*). □ **significantly** adv. [L *significare*: see SIGNIFY]

signification /ˌsɪgnɪfɪˈkeɪʃ(ə)n/ n. **1** the act of signifying. **2** (usu. foll. by *of*) exact meaning or sense, esp. of a word or phrase. [ME f. OF f. L *significatio -onis* (as SIGNIFY)]

significative /sɪgˈnɪfɪkətɪv/ adj. **1** (esp. of a symbol etc.) signifying. **2** having a meaning. **3** (usu. foll. by *of*) serving as a sign or evidence. [ME f. OF *significatif -ive*, or LL *significativus* (as SIGNIFY)]

signified /ˈsɪgnɪˌfaɪd/ n. esp. *Ling.* a meaning or idea, as distinct from its expression by a physical medium (as a sound, symbol, etc.) (cf. SIGNIFIER).

signifier /ˈsɪgnɪˌfaɪə(r)/ n. esp. *Ling.* a physical medium (as a sound, symbol, etc.) expressing meaning, as distinct from the meaning expressed (cf. SIGNIFIED).

signify /ˈsɪgnɪˌfaɪ/ v. (**-ies, -ied**) **1** tr. be a sign or indication of (*a yawn signifies boredom*). **2** tr. mean; have as its meaning (*'Dr' signifies 'doctor'*). **3** tr. communicate; make known (*signified their agreement*). **4** intr. be of importance; matter (*it signifies little*). [ME f. OF *signifier* f. L *significare* (as SIGN)]

signing /ˈsaɪnɪŋ/ n. a person who has recently signed a contract, esp. to join a professional sports team.

signor /ˈsiːnjɔː(r)/ n. (pl. **signori** /siːˈnjɔːrɪ/) **1** a title or form of address used of or to an Italian-speaking man, corresponding to Mr or sir. **2** an Italian man. [It. f. L *senior*: see SENIOR]

signora /siːˈnjɔːrə/ n. **1** a title or form of address used of or to an Italian-speaking married woman, corresponding to Mrs or madam. **2** a married Italian woman. [It., fem. of SIGNOR]

signorina /ˌsiːnjəˈriːnə/ n. **1** a title or form of address used of or to an Italian-speaking unmarried woman. **2** an Italian unmarried woman. [It., dimin. of SIGNORA]

signory /ˈsiːnjərɪ/ n. (pl. **-ies**) **1** = SEIGNIORY. **2** *hist.* the governing body of a medieval Italian republic. [ME f. OF *s(e)ignorie* (as SEIGNEUR)]

signpost /ˈsaɪnpəʊst/ n. & v. ● n. **1** a sign indicating the direction to and sometimes also the distance from various places, esp. a post with arms at a road junction. **2** a means of guidance; an indication. ● v.tr. **1** provide with a signpost or signposts. **2** indicate (a course of action, direction, etc.).

Sigurd /ˈsɪgʊəd/ (in Norse legend) the Norse equivalent of Siegfried, husband of Gudrun.

Sihanouk /ˈsɪənʊk/, Norodom (b.1922), Cambodian king 1941–55 and since 1993, Prime Minister 1955–60, and head of state 1960–70 and 1975–6. Two years after Cambodian independence (1953), Sihanouk abdicated in order to become Premier, passing the throne to his father Prince Norodom Suramarit (d.1960). On his father's death, Prince Sihanouk proclaimed himself head of state, a position he retained until a US-backed military coup ten years later. Sihanouk was reinstated by the Khmer Rouge in 1975, only to be removed the following year. After serving as President of the government-in-exile (1982–9), he was appointed head of state by the provisional government and subsequently crowned for the second time (1993).

sika /ˈsiːkə/ n. a forest-dwelling deer, *Cervus nippon*, native to Japan and SE Asia and naturalized in Britain and elsewhere. [Jap. *shika*]

Sikh /siːk, sɪk/ n. & adj. ● n. an adherent of Sikhism. There are over 10 million Sikhs, most of whom live in Punjab. ● adj. of or relating to the Sikhs or Sikhism. [Punjabi, = disciple, f. Skr. *śiṣya*]

Sikhism /ˈsiːkɪz(ə)m, ˈsɪk-/ a monotheistic religion founded in Punjab in the 15th century by Guru Nanak. It combines elements of Hinduism and Islam, accepting the Hindu concepts of karma and reincarnation but rejecting the caste system, and has one sacred scripture, the Adi Granth. The tenth and last of its gurus, Gobind Singh (1666–1708), prescribed the distinctive outward forms (the so-called five Ks) — long hair (to be covered by a turban) and uncut beard (*kesh*), comb (*kangha*), short sword (*kirpan*), steel bangle (*kara*), and short trousers for horse-riding (*kaccha*); Sikh men take the last name Singh (lion) and women Kaur (princess).

Sikh Wars a series of wars between the Sikhs and the British in 1845 and 1848–9, culminating in the British annexation of Punjab.

Siking /ˌʃiːˈkɪŋ/ a former name for XIAN.

Sikkim /ˈsɪkɪm/ a state of NE India, in the eastern Himalayas between Bhutan and Nepal, on the border with Tibet; capital, Gangtok. It was an independent kingdom until taken by the British in 1861. After British rule it became an Indian protectorate, becoming a state of India in 1975. □ **Sikkimese** /ˌsɪkɪˈmiːz/ adj. & n.

Sikorsky /sɪˈkɔːskɪ/, Igor (Ivanovich) (1889–1972), Russian-born American aircraft designer. He studied aeronautics in Paris before returning to Russia to build the first large four-engined aircraft, the Grand, in 1913. After experimenting unsuccessfully with helicopters he emigrated to New York, where he established the Sikorsky Aero Engineering Co. and produced many famous amphibious aircraft and flying boats. Sikorsky again turned his attention to helicopters, personally flying the prototype of the first mass-produced helicopter in 1939, and was closely associated with their subsequent development.

Siksika /ˈsɪksɪkə/ n.pl. the northernmost of the three peoples forming the Blackfoot confederacy. [Blackfoot, f. *siksi-* black + *-ka* foot]

silage /ˈsaɪlɪdʒ/ n. & v. ● n. **1** storage in a silo. **2** green fodder that has been stored and fermented in a silo for use as food for cattle etc. ● v.tr. put into a silo. [alt. of ENSILAGE after *silo*]

Silbury Hill /ˈsɪlbərɪ/ a neolithic monument near Avebury in Wiltshire, a flat-topped conical mound more than 40 m (130 ft) high, which is the largest man-made prehistoric mound in Europe. Excavations have failed to reveal its purpose.

Silchester /ˈsɪltʃɪstə(r)/ a village in Hampshire, situated to the south-west of Reading. It is the site of an important town of pre-Roman and Roman Britain, known to the Romans as Calleva Atrebatum.

sild /sɪlt/ n. a small immature herring, esp. one caught in northern European seas. [Danish & Norw.]

silence /ˈsaɪləns/ n. & v. ● n. **1** absence of sound. **2** abstinence from speech or noise. **3** the avoidance of mentioning a thing, betraying a secret, etc. **4** oblivion; the state of not being mentioned. ● v.tr. make silent, esp. by coercion or superior argument. □ **in silence** without speech or other sound. **reduce** (or **put**) **to silence** refute in argument. [ME f. OF f. L *silentium* (as SILENT)]

silencer /ˈsaɪlənsə(r)/ n. a device for reducing the noise emitted by the exhaust of a motor vehicle, a gun, etc.

silent /ˈsaɪlənt/ adj. **1** not speaking; not uttering or making or accompanied by any sound. **2** (of a letter) written but not pronounced, e.g. *b* in *doubt*. **3** (of a film) without a synchronized soundtrack. **4** (of a person) taciturn; speaking little. **5** (of an agreement) unspoken, unrecorded. **6** saying or recording nothing on some subject (*the records are silent on the incident*). **7** (of spirits) unflavoured. □ **silent majority** those of moderate opinions who rarely assert them. **silent partner** *N. Amer.* = *sleeping partner* (see SLEEP). □ **silently** adv. [L *silere silent-* be silent]

Silenus /saɪˈliːnəs/ *Gk Mythol.* an aged woodland deity, one of the sileni (see SILENUS), who was entrusted with the education of Dionysus. He is depicted either as dignified, inspired, and musical, or as an old drunkard.

silenus /saɪˈliːnəs/ n. (pl. **sileni** /-naɪ/) *Gk Mythol.* any of a class of woodland spirits, usually shown in art as old and with horse-ears, similar to the satyrs. [L f. Gk *seilēnos*]

Silesia /saɪˈliːzɪə, -ˈliːʒə/ a region of central Europe, centred on the upper Oder valley, now largely in SW Poland. It was partitioned at various times between the states of Prussia, Austria–Hungary, Poland, and Czechoslovakia. □ **Silesian** adj. & n.

silex /ˈsaɪleks/ n. silica, esp. quartz or flint. [L (as SILICA)]

silhouette /ˌsɪluːˈet/ n. & v. ● n. **1** a representation of a person or thing showing the outline only, usu. done in solid black on white or cut from paper. (*See note below.*) **2** the dark shadow or outline of a person or thing against a lighter background. ● v.tr. represent or (usu. in *passive*) show in silhouette. □ **in silhouette** seen or placed in outline. [Étienne de *Silhouette* (see below)]

▪ Silhouette painting is found in Egyptian and Greek art, but its main vogue was from the mid-18th to the mid-19th century, when such portraiture was popularized by neoclassical taste, until the introduction of photography relegated it to a position of curiosity value. The word comes from the name of the French author and politician Étienne de Silhouette (1709–67), an amateur maker of paper cut-outs.

silica /'sɪlɪkə/ n. silicon dioxide, a hard white refractory solid occurring as quartz etc. and as a principal constituent of sandstone and other rocks. □ **silica gel** hydrated silica in a hard granular hygroscopic form used as a desiccant. □ **siliceous** /sɪ'lɪʃəs/ adj. (also **silicious**). [L *silex -icis* flint, after *alumina* etc.]

silicate /'sɪlɪˌkeɪt, -kət/ n. Chem. **1** a salt or ester of silicic acid. **2** any of the many minerals etc. consisting of silica combined with metal oxides, forming a major component of the rocks of the earth's crust.

silicic /sɪ'lɪsɪk/ adj. Chem. of silicon or silica. □ **silicic acid** an oxyacid of silica, esp. a weakly acidic colloidal hydrated form of silica. [SILICA + -IC]

silicify /sɪ'lɪsɪˌfaɪ/ v.t. (**-ies, -ied**) convert into or impregnate with silica. □ **silicification** /sɪˌlɪsɪfɪ'keɪʃ(ə)n/ n.

silicon /'sɪlɪkən/ n. a non-metallic chemical element (atomic number 14; symbol **Si**). (*See note below.*) □ **silicon carbide** = CARBORUNDUM. **silicon chip** a silicon microchip. [L *silex -icis* flint (after *carbon, boron*), alt. of earlier *silicium*]

▪ Silicon was first isolated and described as an element by Berzelius in 1823. After oxygen, it is the most abundant element in the earth's crust; most rocks consist primarily of silica or silicates. Pure silicon can exist in a shiny dark grey crystalline form or as an amorphous powder. The element is used in some alloys but its great importance is in the electronics industry. As a semiconductor, it has largely replaced the chemically similar element germanium, used for the first transistors. Silicon compounds have for centuries been widely used: glass, pottery, and bricks are largely composed of silicate minerals.

silicone /'sɪlɪˌkəʊn/ n. a synthetic organic polymer containing silicon and oxygen, used in polishes, paints, lubricants, rubber, etc. Silicones, developed in the 1940s, are resistant to chemical attack and relatively insensitive to temperature changes; they have a chemical structure based on chains of alternating silicon and oxygen atoms. [SILICON + -ONE]

Silicon Valley n. an area with a high concentration of electronics industries, esp. the Santa Clara valley south-east of San Francisco.

silicosis /ˌsɪlɪ'kəʊsɪs/ n. lung fibrosis caused by the inhalation of dust containing silica. □ **silicotic** /-'kɒtɪk/ adj.

siliqua /'sɪlɪkwə/ n. (also **silique** /sɪ'liːk/) (pl. **siliquae** /'sɪlɪˌkwiː/ or **siliques** /sɪ'liːks/) Bot. the long narrow seed-pod of a cruciferous plant. □ **siliquose** /'sɪlɪˌkwəʊs, -ˌkwəʊs/ adj. **siliquous** adj. [L, = pod]

silk /sɪlk/ n. **1** a fine strong soft lustrous fibre produced by silkworms in making cocoons. **2** a similar fibre spun by some spiders etc. **3 a** thread or cloth made from silk fibre. **b** a fine soft thread (*embroidery silk*). **4** (in pl.) kinds of silk cloth or garments made from it, esp. as worn by a jockey in a horse-owner's colours. **5** (in the UK) Queen's (or King's) Counsel, as having the right to wear a silk gown. **6** (attrib.) made of silk (*silk blouse*). **7** the silky styles of the female maize-flower. □ **silk cotton** kapok or a similar substance. **silk-fowl** a breed of fowl with a silky plumage. **silk-gland** a gland secreting the substance produced as silk. **silk hat** a tall cylindrical hat covered with silk plush. **silk moth** a large moth of the family Bombycidae or Saturniidae, the caterpillar of which spins a protective silken cocoon; esp. the adult of the silkworm. **silk-screen printing** = screen printing. **take silk** (in the UK) become a Queen's (or King's) Counsel. □ **silklike** adj. [OE *sioloc, seoloc* (cf. ON *silki*) f. LL *sericum* neut. of L *sericus* f. *Seres* f. Gk *Sēres* an oriental people]

silken /'sɪlkən/ adj. **1** made of silk. **2** wearing silk. **3** soft or lustrous as silk. **4** (of a person's manner etc.) suave or insinuating. [OE *seolcen* (as SILK)]

Silk Road (also **Silk Route**) an ancient caravan route linking Xian in central China with the eastern Mediterranean. Skirting the northern edge of the Taklimakan Desert and passing through Turkestan, it covered a distance of some 6,400 km (4,000 miles). It was established during the period of Roman rule in Europe, and took its name from the silk which was brought to the west from China. It was also the route by which Christianity spread to the East. A railway

(completed in 1963) follows the Chinese part of the route, from Xian to Urumqi.

silkworm /'sɪlkwɜːm/ n. the caterpillar of the moth *Bombyx mori*, which spins a cocoon of silk and is bred commercially for this purpose. The silkworm had been domesticated in China by 3000 BC, the technology of their culture was fully documented by 1000 BC, and silk became a major trade with the West (see SILK ROAD). Silkworms are fed on the leaves of the white mulberry. Some other kinds of silkworm, especially the tussore, belong to the related family Saturniidae.

silky /'sɪlkɪ/ adj. (**silkier, silkiest**) **1** like silk in smoothness, softness, fineness, or lustre. **2** (of a person's manner etc.) suave, insinuating. □ **silkily** adv. **silkiness** n.

sill /sɪl/ n. (also (esp. Building) **cill**) **1** a shelf or slab of stone, wood, or metal at the foot of a window or doorway. **2** a horizontal timber at the bottom of a dock or lock entrance, against which the gates close. **3** Geol. a tabular sheet of igneous rock intruded between other rocks and parallel with their planar structure. [OE *syll, sylle*]

sillabub var. of SYLLABUB.

sillimanite /'sɪlɪməˌnaɪt/ n. an aluminium silicate occurring in orthorhombic crystals or fibrous masses. [Benjamin *Silliman*, Amer. chemist (1779–1864)]

Sillitoe /'sɪlɪˌtəʊ/, Alan (b.1928), English writer. His fiction is notable for its depiction of working-class provincial life; his first novel *Saturday Night and Sunday Morning* (1958) describes the life of a dissatisfied young Nottingham factory worker, while the title story in *The Loneliness of the Long-Distance Runner* (1959) portrays a rebellious Borstal boy. He has also published volumes of poetry, short stories, and plays.

silly /'sɪlɪ/ adj. & n. ● adj. (**sillier, silliest**) **1** lacking sense; foolish, imprudent, unwise. **2** weak-minded. **3** Cricket (of a fielder or position) very close to the batsman (*silly mid-off*). **4** archaic innocent, simple, helpless. **5** colloq. stunned (as) by a blow (*I was knocked silly*). ● n. (pl. **-ies**) colloq. a foolish person. □ **silly billy** colloq. a foolish person. **the silly season** high summer as the season when newspapers often publish trivial material for lack of important news. □ **sillily** adv. **silliness** n. [later form of ME *sely* (dial. *seely*) happy, repr. OE *sǣlig* (recorded in *unsǣlig* unhappy) f. Gmc]

silo /'saɪləʊ/ n. & v. ● n. (pl. **-os**) **1** a pit or airtight structure in which green crops are pressed and kept for fodder, undergoing fermentation. **2** a pit or tower for the storage of grain, cement, etc. **3** an underground chamber in which a guided missile is kept ready for firing. ● v.tr. (**-oes, -oed**) make silage of. [Sp. f. L *sirus* f. Gk *siros* corn-pit]

Siloam /saɪ'ləʊəm/ (in the New Testament) a spring and pool of water near Jerusalem, where a man born blind was told by Jesus to wash, thereby gaining sight (John 9:7).

silt /sɪlt/ n. & v. ● n. sediment deposited by water in a channel, harbour, etc. ● v.tr. & intr. (often foll. by up) choke or be choked with silt. □ **silty** adj. **siltation** /sɪl'teɪʃ(ə)n/ n. [ME, perh. rel. to Danish, Norw. *sylt*, OLG *sulta*, OHG *sulza* salt marsh, formed as SALT]

siltstone /'sɪltstəʊn/ n. rock of consolidated silt.

Silurian /saɪ'ljʊərɪən, -'lʊərɪən/ adj. & n. Geol. ● adj. of or relating to the third period of the Palaeozoic era. ● n. this period or the corresponding geological system. [L *Silures* a people of ancient SE Wales]

▪ The Silurian lasted from about 438 to 408 million years ago, between the Ordovician and Devonian periods. The first land plants and the first true fish (with jaws) appeared during this period.

silva var. of SYLVA.

silvan var. of SYLVAN.

Silvanus /sɪl'veɪnəs, sɪl'vɑːnəs/ Rom. Mythol. an Italian woodland deity identified with Pan.

silver /'sɪlvə(r)/ n., adj., & v. ● n. **1** a precious shiny greyish-white metallic element (atomic number 47; symbol **Ag**). (*See note below.*) **2** the colour of silver. **3** silver or cupro-nickel coins. **4** esp. Sc. money. **5** silver vessels or implements, esp. cutlery. **6** household cutlery of any material. **7** = silver medal. ● adj. **1** made wholly or chiefly of silver. **2** coloured like silver. ● v. **1** tr. coat or plate with silver. **2** tr. provide (a mirror-glass) with a backing of tin amalgam etc. **3** tr. (of the moon or a white light) give a silvery appearance to. **4 a** tr. turn (the hair) grey or white. **b** intr. (of the hair) turn grey or white. □ **silver age** a period regarded as inferior to a golden age, e.g. that of post-classical Latin literature in the early period of the Roman Empire. **silver band** Brit. a band playing silver-plated instruments. **silver birch** a common birch, *Betula alba*, with silvery bark. **silver fern 1** a New Zealand tree fern, *Cyathea dealbata*. **2** a stylized silver fern leaf as an emblem of New Zealand. **silver fir** a fir of the genus *Abies*, with the undersides of its needles coloured silver. **silver fox 1** an American red fox at a time when its fur is black with white tips. **2** such fur. **silver gilt 1** gilded silver. **2** an imitation gilding of yellow lacquer over silver leaf. **silver-grey** a lustrous grey. **silver jubilee 1** the 25th anniversary of a sovereign's accession. **2** any other 25th anniversary. **silver Latin** see LATIN. **silver-leaf** a fungal disease of fruit trees. **silver lining** a consolation or hopeful feature in misfortune. **silver medal** a silver-coloured medal, usu. awarded as second prize. **silver nitrate** a colourless solid (chem. formula: $AgNO_3$) that is soluble in water, formerly used in photography. **silver paper 1** aluminium or occasionally tin foil. **2** a fine white tissue-paper for wrapping silver. **silver plate** vessels, spoons, etc., of copper etc. plated with silver. **silver salmon** = COHO. **silver sand** a fine pure sand used in gardening. **silver screen** (usu. prec. by *the*) motion pictures collectively. **silver solder** solder containing silver. **silver spoon** a sign of future prosperity. **silver standard** a system by which the value of a currency is defined in terms of silver, for which the currency may be exchanged. **silver thaw** the formation of a glassy coating of ice on the ground or an exposed surface by freezing rain or the refreezing of thawed ice. **silver tongue** eloquence. **silver wedding** the 25th anniversary of a wedding. [OE *seolfor* f. Gmc]

▪ A transition metal, silver is found in nature in the uncombined state as well as in ores. It has been used for jewellery and other ornaments since ancient times, and it is also used in coins, cutlery, the coatings of mirrors, and dental amalgams. A further use is in printed circuits, as silver is a very good electrical conductor. The metal is generally resistant to corrosion, but tarnishes in air through reaction with traces of hydrogen sulphide. Silver salts decompose when exposed to light, depositing metallic silver — a reaction that is the basis of photography. The symbol Ag is from the Latin *argentum*.

silverfish /'sɪlvəˌfɪʃ/ *n.* **1** a silvery bristletail of the order Thysanura; esp. *Lepisma saccharina*, found commonly in houses. **2** a silver-coloured fish, esp. a colourless variety of goldfish.

silvern /'sɪlv(ə)n/ *adj. archaic or poet.* = SILVER *adj.* [OE *seolfren, silfren* (as SILVER)]

silver-point /'sɪlvəˌpɔɪnt/ *n.* the process of drawing with a silver-pointed instrument on paper coated with a special ground of powdered bone or zinc white. Fragments of metal deposited on the paper produce a very delicate fine line that does not smudge and cannot be erased. The process was widely used in the 15th and 16th century in Italy, the Netherlands, and Germany, but is now largely obsolete.

silverside /'sɪlvəˌsaɪd/ *n.* **1** the upper side of a round of beef from the outside of the hind leg. **2** (also **silversides**) a small fish of the mainly marine family Atherinidae, with a silver line along the side.

silversmith /'sɪlvəˌsmɪθ/ *n.* a worker in silver; a manufacturer of silver articles. □ **silversmithing** *n.*

Silverstone /'sɪlvəˌstəʊn/ a motor-racing circuit near Towcester in Northamptonshire, built on a disused airfield after the Second World War.

silverware /'sɪlvəˌweə(r)/ *n.* articles made of or coated with silver.

silverweed /'sɪlvəˌwiːd/ *n.* a low-growing yellow-flowered potentilla, *Potentilla anserina*, with silver-coloured leaves.

silvery /'sɪlvərɪ/ *adj.* **1** like silver in colour or appearance. **2** having a clear gentle ringing sound. **3** (of the hair) white and lustrous. □ **silveriness** *n.*

silviculture /'sɪlvɪˌkʌltʃə(r)/ *n.* (also **sylviculture**) the growing and tending of trees as a branch of forestry. □ **silvicultural** /ˌsɪlvɪˈkʌltʃərəl/ *adj.* **silviculturist** *n.* [F f. L *silva* a wood + F *culture* CULTURE]

sima /'saɪmə/ *n. Geol.* **1** the continuous basal layer of the earth's crust, composed of relatively heavy basic rocks rich in silica and magnesia, which underlies the sial of the continental masses and forms the crust under the oceans, its lower limit being the Mohorovičić discontinuity. **2** the material of which this is composed. [SILICON + MAGNESIUM]

Simbirsk /sɪmˈbɪəsk/ a city in European Russia, a port on the River Volga south-east of Nizhni Novgorod; pop. (1990) 638,000. Between 1924 and 1992 it was called Ulyanovsk, in honour of Lenin (Vladimir Ilich Ulyanov), who was born there in 1870.

Simenon /'siːməˌnɒn/, Georges (Joseph Christian) (1903–89), Belgian-born French novelist. He is best known for his series of detective novels featuring Commissaire Maigret, who was introduced in 1931. Maigret relies on his understanding of the criminal's motives rather than scientific deduction to solve crimes and the novels show considerable insight into human psychology.

Simeon /'sɪmɪən/ **1** a Hebrew patriarch, son of Jacob and Leah (Gen. 29:33). **2** the tribe of Israel traditionally descended from him.

Simeon Stylites, St /ˌsɪmɪən staɪˈlaɪtiːz/ (*c*.390–459), Syrian monk. After living in a monastic community he became the first to practise an extreme form of asceticism which involved living on top of a pillar; this became a site of pilgrimage. (See also STYLITE.)

Simferopol /ˌsɪmfəˈrɒp(ə)l/ a city in the Crimea; pop. (1990) 348,900. An ancient Scythian town and fortress, it was settled by the Tartars in the 16th century, when it was known as Ak-Mechet. It was seized and destroyed in 1736 by the Russians, who founded the modern town on an adjacent site in 1784.

simian /'sɪmɪən/ *adj. & n.* ● *adj.* **1** of or concerning the anthropoid apes. **2** like an ape or monkey (*a simian walk*). ● *n.* an ape or monkey. [L *simia* ape, perh. f. L *simus* f. Gk *simos* flat-nosed]

similar /'sɪmɪlə(r)/ *adj.* **1** like, alike. **2** (often foll. by *to*) having a resemblance. **3** of the same kind, nature, or amount. **4** *Geom.* shaped alike. □ **similarly** *adv.* **similarity** /ˌsɪmɪˈlærɪtɪ/ *n.* (pl. **-ies**). [F *similaire* or med.L *similaris* f. L *similis* like]

simile /'sɪmɪlɪ/ *n.* **1** a figure of speech involving the comparison of one thing with another of a different kind, as an illustration or ornament (e.g. *as brave as a lion*). **2** the use of such comparison. [ME f. L, neut. of *similis* like]

similitude /sɪˈmɪlɪˌtjuːd/ *n.* **1** the likeness, guise, or outward appearance of a thing or person. **2** a comparison or the expression of a comparison. **3** *archaic* a counterpart or facsimile. [ME f. OF *similitude* or L *similitudo* (as SIMILE)]

Simla /'sɪmlə/ a city in NE India, capital of the state of Himachal Pradesh; pop. (1991) 109,860. Situated in the foothills of the Himalayas, it served from 1865 to 1939 as the summer capital of British India, and is now a popular hill resort.

simmer /'sɪmə(r)/ *v. & n.* ● *v.* **1** *intr. & tr.* be or keep bubbling or boiling gently. **2** *intr.* be in a state of suppressed anger or excitement. ● *n.* a simmering condition. □ **simmer down** become calm or less agitated. [alt. of ME (now dial.) *simper*, perh. imit.]

Simnel /'sɪmn(ə)l/, Lambert (*c*.1475–1525), English pretender and rebel. The son of a baker, he was trained by Yorkists to impersonate firstly one of the sons of Edward IV (see PRINCES IN THE TOWER) and subsequently the Earl of Warwick (also imprisoned in the Tower of London), in an attempt to overthrow Henry VII. He was crowned in Dublin in 1487 as Edward VI but captured when the Yorkist uprising was defeated. He was not executed, but given a menial post in the royal household.

simnel cake /'sɪmn(ə)l/ *n. Brit.* a rich fruit cake, usu. with a marzipan layer and decoration, eaten esp. at Easter or during Lent. [ME f. OF *simenel*, ult. f. L *simila* or Gk *semidalis* fine flour]

Simon[1] /'saɪmən/, (Marvin) Neil (b.1927), American dramatist. Most of his plays are wry comedies portraying aspects of middle-class life; they include *Barefoot in the Park* (1963), *The Odd Couple* (1965), and *Brighton Beach Memoirs* (1983). Among his musicals are *Sweet Charity* (1966) and *They're Playing Our Song* (1979).

Simon[2] /'saɪmən/, Paul (b.1942), American singer and songwriter. He formed a folk-rock partnership with his school friend Art Garfunkel (b.1941), which first made its mark with the album *Sounds of Silence* (1966). Further achievements were the soundtrack music to the film *The Graduate* (1968) and the song 'Bridge Over Troubled Water' (1970) from the album of the same name. The duo split up in 1970 and Simon went on to pursue a successful solo career, recording albums such as *Graceland* (1986), which featured many black South African musicians. He has also acted in films, including Woody Allen's *Annie Hall* (1977).

Simon, St (known as Simon the Zealot), an Apostle. According to one tradition he preached and was martyred in Persia along with St Jude. Feast day (with St Jude), 28 Oct.

Simonides /saɪˈmɒnɪˌdiːz/ (*c*.556–468 BC), Greek lyric poet. He wrote for the rulers of Athens, Thessaly, and Syracuse; much of his poetry, which includes elegies, odes, and epigrams, celebrates the heroes of the Persian Wars, and includes verse commemorating those killed at Marathon and Thermopylae.

simon-pure /ˌsaɪmənˈpjʊə(r)/ *adj.* real, genuine. [(*the real*) Simon Pure, a character in Susannah Centlivre's *Bold Stroke for a Wife* (1717)]

simony /'saɪmənɪ, 'sɪm-/ *n.* the buying or selling of ecclesiastical privileges, e.g. pardons or benefices. □ **simoniac** /saɪˈməʊnɪˌæk, sɪ-/ *adj. & n.* **simoniacal** /ˌsaɪməˈnaɪək(ə)l, ˌsɪm-/ *adj.* [ME f. OF *simonie* f. LL *simonia* f. Simon Magus (Acts 8:18), with allusion to his offer of money to the Apostles]

simoom /sɪˈmuːm/ *n.* (also **simoon** /-ˈmuːn/) a hot dry dust-laden wind blowing at intervals esp. in the Arabian Desert. [Arab. *samūm* f. *samma* to poison]

simp /sɪmp/ *n.* US *colloq.* a simpleton. [abbr.]

simpatico /sɪmˈpætɪˌkəʊ/ *adj.* congenial, likeable. [It. & Sp. (as SYMPATHY)]

simper /ˈsɪmpə(r)/ *v. & n.* ● *v.* **1** *intr.* smile in a silly or affected way. **2** *tr.* express by or with simpering. ● *n.* such a smile. □ **simperingly** *adv.* [16th c.: cf. Du. and Scand. *semper, simper*, G *zimp(f)er* elegant, delicate]

simple /ˈsɪmp(ə)l/ *adj. & n.* ● *adj.* (**simpler, simplest**) **1** easily understood or done; presenting no difficulty (*a simple explanation; a simple task*). **2** not complicated or elaborate; without luxury or sophistication. **3** not compound; consisting of or involving only one element or operation etc. **4** absolute, unqualified, straightforward (*the simple truth; a simple majority*). **5** foolish or ignorant; gullible, feeble-minded (*am not so simple as to agree to that*). **6** plain in appearance or manner; unsophisticated, ingenuous, artless. **7** of low rank; humble, insignificant (*simple people*). **8** *Bot.* **a** consisting of one part. **b** (of fruit) formed from one pistil. ● *n. archaic* **1** a herb used medicinally. **2** a medicine made from it. □ **simple eye** an eye of an insect etc. having only one lens (also called *ocellus*) (cf. *compound eye*). **simple fracture** a fracture of the bone only, without a skin wound. **simple harmonic motion** see HARMONIC. **simple interest** interest payable on a capital sum only (cf. *compound interest* (see COMPOUND[1])). **simple interval** *Mus.* an interval of one octave or less. **simple machine** any of the basic mechanical devices for applying a force (e.g. an inclined plane, wedge, or lever). **simple sentence** a sentence with a single subject and predicate. **Simple Simon** a foolish person (from the nursery-rhyme character). **simple time** *Mus.* a time with two, three, or four beats in a bar. □ **simpleness** *n.* [ME f. OF f. L *simplus*]

simple-minded /ˌsɪmp(ə)lˈmaɪndɪd/ *adj.* **1** feeble-minded; stupid, foolish. **2** natural, unsophisticated. □ **simple-mindedly** *adv.* **simple-mindedness** *n.*

simpleton /ˈsɪmp(ə)lt(ə)n/ *n.* a foolish, gullible, or halfwitted person. [SIMPLE after surnames f. place-names in *-ton*]

simplex /ˈsɪmpleks/ *adj. & n.* ● *adj.* **1** simple; not compounded. **2** *Computing* (of a communication system, computer circuit, etc.) allowing transmission of signals in one direction only (opp. DUPLEX). ● *n.* a simple or uncompounded thing, esp. a word. [L, = single, var. of *simplus* simple]

simplicity /sɪmˈplɪsɪtɪ/ *n.* the fact or condition of being simple. □ **be simplicity itself** be extremely easy. [OF *simplicité* or L *simplicitas* (as SIMPLEX)]

simplify /ˈsɪmplɪˌfaɪ/ *v.tr.* (**-ies, -ied**) make simple; make easy or easier to do or understand. □ **simplification** /ˌsɪmplɪfɪˈkeɪʃ(ə)n/ *n.* [F *simplifier* f. med.L *simplificare* (as SIMPLE)]

simplism /ˈsɪmplɪz(ə)m/ *n.* **1** affected simplicity. **2** the unjustifiable simplification of a problem etc.

simplistic /sɪmˈplɪstɪk/ *adj.* **1** excessively or affectedly simple. **2** oversimplified so as to conceal or distort difficulties. □ **simplistically** *adv.*

Simplon /ˈsɪmplɒn/ a pass in the Alps in southern Switzerland, consisting of a road built by Napoleon in 1801–5 at an altitude of 2,028 m (6,591 ft) and a railway tunnel (built in 1922) which links Switzerland and Italy.

simply /ˈsɪmplɪ/ *adv.* **1** in a simple manner. **2** absolutely; without doubt (*simply astonishing*). **3** merely (*was simply trying to please*).

Simpson[1] /ˈsɪmps(ə)n/, Sir James Young (1811–71), Scottish surgeon and obstetrician. He discovered the usefulness of chloroform as an anaesthetic by experimentation on himself and his colleagues shortly after the first use of ether, and he was active in the debate over which of the two was the best agent to use in surgery. Simpson was also a distinguished antiquarian and historian, publishing monographs on archaeology and the history of medicine.

Simpson[2] /ˈsɪmps(ə)n/, Wallis (née Wallis Warfield) (1896–1986), American wife of Edward, Duke of Windsor (Edward VIII). Her relationship with the king caused a national scandal in 1936, especially in view of her impending second divorce, and forced the king's abdication. The couple were married shortly afterwards and she became the Duchess of Windsor. She remained in France after her husband died and lived as a recluse until her death.

Simpson Desert a desert in central Australia, situated between Alice Springs and the Channel Country to the east. It was named in

1929 after A. A. Simpson, who was president of the Royal Geographical Society of Australia at that time.

simulacrum /ˌsɪmjʊˈleɪkrəm/ *n.* (*pl.* **simulacra** /-krə/) **1** an image of something. **2 a** a shadowy likeness; a deceptive substitute. **b** mere pretence. [L (as SIMULATE)]

simulate /ˈsɪmjʊˌleɪt/ *v.tr.* **1 a** pretend to have or feel (an attribute or feeling). **b** pretend to be. **2** imitate or counterfeit. **3 a** imitate the conditions of (a situation etc.), e.g. for training. **b** produce a computer model of (a process). **4** (as **simulated** *adj.*) made to resemble the real thing but not genuinely such (*simulated fur*). **5** (of a word) take or have an altered form suggested by (a word wrongly taken to be its source). □ **simulative** /-lətɪv/ *adj.* **simulation** /ˌsɪmjʊˈleɪʃ(ə)n/ *n.* [L *simulare* f. *similis* like]

simulator /ˈsɪmjʊˌleɪtə(r)/ *n.* **1** a person or thing that simulates. **2** a device designed to simulate the operations of a complex system (e.g. a vehicle or aircraft), used esp. in training.

simulcast /ˈsɪm(ə)lˌkɑːst/ *n.* simultaneous transmission of the same programme on radio and television. [SIMULTANEOUS + BROADCAST]

simultaneous /ˌsɪm(ə)lˈteɪnɪəs/ *adj.* (often foll. by *with*) occurring or operating at the same time. □ **simultaneous equations** equations involving two or more unknowns that are to have the same values in each equation. □ **simultaneously** *adv.* **simultaneousness** *n.* **simultaneity** /-təˈneɪɪtɪ/ *n.* [med.L *simultaneus* f. L *simul* at the same time, prob. after *instantaneus* etc.]

simurg /sɪˈmɜːɡ/ *n.* a monstrous bird of Persian myth, with the power of reasoning and speech. [Pers. *sīmurg* f. Pahlavi *sīn* eagle + *murg* bird]

sin[1] /sɪn/ *n. & v.* ● *n.* **1 a** the breaking of divine or moral law, esp. by a conscious act. **b** such an act. **2** an offence against good taste or propriety etc. ● *v.* (**sinned, sinning**) **1** *intr.* commit a sin. **2** *intr.* (foll. by *against*) offend. **3** *tr. archaic* commit (a sin). □ **as sin** *colloq.* extremely (*ugly as sin*). **for one's sins** *joc.* as a judgement on one for something done. **like sin** *colloq.* vehemently or forcefully. **live in sin** *colloq.* live together without being married. **sin bin** *colloq.* **1** Ice Hockey a penalty box. **2** a place set aside for offenders of various kinds. □ **sinless** *adj.* **sinlessly** *adv.* **sinlessness** *n.* [OE *syn(n)*]

sin[2] /saɪn/ *abbr.* sine.

Sinai /ˈsaɪnaɪ, -nɪˌaɪ/ an arid mountainous peninsula in NE Egypt, extending into the Red Sea between the Gulf of Suez and the Gulf of Aqaba. It was occupied by Israel between 1967 and 1982, when it was fully restored to Egypt following the Camp David agreement. In the south is Mount Sinai, where, according to the Bible, Moses received the Ten Commandments (Exod. 19–34).

Sinaitic /ˌsaɪnɪˈɪtɪk/ *adj.* of or relating to Mount Sinai or the Sinai peninsula.

Sinaloa /ˌsiːnəˈləʊə/ a state on the Pacific coast of Mexico; capital, Culiacán Rosales.

Sinanthropus /sɪˈnænθrəpəs/ *n.* a genus name formerly applied to some fossil hominids of the species *Homo erectus*, named from remains found near Beijing in 1926. (See PEKING MAN.) [mod.L, as SINO- + Gk *anthrōpos* man]

Sinatra /sɪˈnɑːtrə/, Frank (full name Francis Albert Sinatra) (b.1915), American singer and actor. He began his long career as a singer in 1938 performing with big bands on the radio, becoming a solo star in the 1940s with a large teenage following; his many hits include 'Night and Day' and 'My Way'. Among his numerous films are *From Here to Eternity* (1953), for which he won an Oscar.

Sinbad the Sailor /ˈsɪnbæd/ (also **Sindbad** /ˈsɪndbæd/) the hero of one of the tales in the *Arabian Nights*, who relates the fantastic adventures he meets with in his voyages.

since /sɪns/ *prep., conj., & adv.* ● *prep.* throughout, or at a point in, the period between (a specified time, event, etc.) and the time present or being considered (*must have happened since yesterday; has been going on since June; the greatest composer since Beethoven*). ● *conj.* **1** during or in the time after (*what have you been doing since we met?; has not spoken since the dog died*). **2** for the reason that, because; inasmuch as (*since you are drunk I will drive you home*). **3** (*ellipt.*) as being (*a more useful, since better designed, tool*). ● *adv.* **1** from that time or event until now or the time being considered (*have not seen them since; had been healthy ever since; has since been cut down*). **2** ago (*happened many years since*). [ME, reduced form of obs. *sithence* or f. dial. *sin* (f. *sithen*) f. OE *siththon*]

sincere /sɪnˈsɪə(r)/ *adj.* (**sincerer, sincerest**) **1** free from pretence or deceit; the same in reality as in appearance. **2** genuine, honest, frank. □ **sincereness** *n.* **sincerity** /-ˈserɪtɪ/ *n.* [L *sincerus* clean, pure]

sincerely /sɪnˈsɪəlɪ/ adv. in a sincere manner. □ **yours sincerely** a formula for ending a letter, esp. an informal one.

sinciput /ˈsɪnsɪˌpʊt/ n. Anat. the front of the skull from the forehead to the crown. □ **sincipital** /sɪnˈsɪpɪt(ə)l/ adj. [L f. semi- half + caput head]

Sinclair /ˈsɪŋkleə(r)/, Sir Clive (Marles) (b.1940), English electronics engineer and entrepreneur. Sinclair founded a research and development company and launched a range of innovative products including pocket calculators, wrist-watch televisions, and personal computers. A three-wheeled electric car, the C5, powered by a washing-machine motor, failed to achieve commercial success.

Sind /sɪnd/ a province of SE Pakistan, traversed by the lower reaches of the Indus; capital, Karachi. Annexed by Britain in 1843, it was a part of British India until 1947.

Sindebele /ˌsɪndəˈbeɪlɪ, -ˈbiːlɪ/ n. the Bantu language of the Ndebele. [Bantu, f. isi- pref. + NDEBELE]

Sindhi /ˈsɪndɪ/ n. & adj. ● n. **1** a native or inhabitant of Sind. **2** the Indic language of Sind, used also in western India. ● adj. of or relating to Sind or its people or language.

sine /saɪn/ n. Math. **1** the trigonometric function that is equal to the ratio of the side opposite a given angle (in a right-angled triangle) to the hypotenuse. **2** a function of the line drawn from one end of an arc perpendicularly to the radius through the other. □ **sine curve** (or **wave**) a curve representing periodic oscillations of constant amplitude as given by a sine function (also called *sinusoid*). [L sinus curve, fold of a toga, used in med.L as transl. of Arab. *jayb* bosom, sine]

sinecure /ˈsaɪnɪˌkjʊə(r), ˈsɪn-/ n. a position that requires little or no work but usu. yields profit or honour. □ **sinecurism** n. **sinecurist** n. [L sine cura without care]

sine die /ˌsaɪnɪ ˈdaɪɪ, ˌsɪneɪ ˈdiːeɪ/ adv. (of business etc. adjourned indefinitely) with no appointed date. [L, = without day]

sine qua non /ˌsaɪnɪ kwɑː ˈnɒn, ˌsɪneɪ/ n. an indispensable condition or qualification. [L, = without which not]

sinew /ˈsɪnjuː/ n. & v. ● n. **1** tough fibrous tissue uniting muscle to bone; a tendon. **2** (in pl.) muscles; bodily strength; wiriness. **3** (in pl.) that which forms the strength or framework of a plan, city, organization, etc. ● v.tr. poet. serve as the sinews of; sustain; hold together. □ **the sinews of war** money. □ **sinewless** adj. **sinewy** adj. [OE sin(e)we f. Gmc]

sinfonia /ˌsɪnfəˈnɪə, sɪnˈfəʊnɪə/ n. Mus. **1** a symphony. **2** (in baroque music) an orchestral piece used as an introduction to an opera, cantata, or suite. **3** (**Sinfonia**; usu. in names) a small symphony orchestra. [It., = SYMPHONY]

sinfonietta /ˌsɪnfəˈnjetə/ n. Mus. **1** a short or simple symphony. **2** (**Sinfonietta**; usu. in names) a small symphony orchestra. [It., dimin. of SINFONIA]

sinful /ˈsɪnfʊl/ adj. **1** (of a person) committing sin, esp. habitually. **2** (of an act) involving or characterized by sin; highly reprehensible. □ **sinfully** adv. **sinfulness** n. [OE synfull (as SIN[1], -FUL)]

sing /sɪŋ/ v. & n. ● v. (past **sang** /sæŋ/; past part. **sung** /sʌŋ/) **1** intr. utter musical sounds with the voice, esp. words with a set tune. **2** tr. utter or produce by singing (sing another song). **3** intr. (of the wind, a kettle, etc.) make melodious, humming, buzzing, or whistling sounds. **4** intr. (of the ears) be affected as with a buzzing sound. **5** tr. bring to a specified state by singing (sang the child to sleep). **6** tr. (foll. by in, out) usher (esp. the new or old year) in or out with singing. **7** intr. sl. turn informer; confess. **8** intr. archaic compose poetry. **9** tr. & (foll. by of) intr. celebrate in verse. ● n. **1** an act or spell of singing. **2** US a meeting for amateur singing. □ **sing along** sing in accompaniment to a song or piece of music. **sing-along 1** a tune etc. to which one can sing in accompaniment (also attrib.: a sing-along chorus). **2** an occasion of community singing (also attrib.: a sing-along evening). **singing hinny** see HINNY[2]. **singing saw** = musical saw. **sing out** call out loudly; shout. **sing the praises of** see PRAISE. **sing up** sing more loudly. □ **singable** adj. **singer** n. **singingly** adv. [OE singan f. Gmc]

sing. abbr. singular.

Singapore /ˌsɪŋəˈpɔː(r)/ a country in SE Asia consisting of the island of Singapore and about 54 smaller islands; pop. (est. 1991) 3,045,000; official languages, Malay, Chinese, Tamil, and English; capital, Singapore City. Singapore lies off the southern tip of the Malay Peninsula, to which it is linked by a causeway carrying a road and railway. Sir Stamford Raffles established a trading post under the East India Company in 1819, and it was incorporated with Penang and Malacca to form the Straits Settlements in 1826; these came under British colonial rule in 1867. Singapore rapidly grew, by virtue of its large protected harbour, to become the most important commercial centre and naval base in SE Asia. It fell to the Japanese in 1942, and after liberation became first a British Crown Colony in 1946 and then a self-governing state within the Commonwealth in 1959. Federated with Malaysia in 1963, it declared full independence two years later and remains a world trade and financial centre. □ **Singaporean** /-ˈpɔːrɪən/ adj. & n.

singe /sɪndʒ/ v. & n. ● v. (**singeing**) **1** tr. & intr. burn superficially or lightly. **2** tr. burn the bristles or down off (the carcass of a pig or fowl) to prepare it for cooking. **3** tr. burn off the tips of (the hair) in hairdressing. ● n. a superficial burn. □ **singe one's wings** suffer some harm esp. in a risky endeavour. [OE sencgan f. WG]

Singer[1] /ˈsɪŋə(r)/, Isaac Bashevis (1904–91), Polish-born American novelist and short-story writer. His work, written in Yiddish but chiefly known from English translations, blends realistic detail and elements of fantasy, mysticism, and magic to portray the lives of Polish Jews from many periods. Notable titles include the novels The Magician of Lublin (1955) and The Slave (1962) and the short-story collection The Spinoza of Market Street (1961). He was awarded the Nobel Prize for literature in 1978.

Singer[2] /ˈsɪŋə(r)/, Isaac Merrit (1811–75), American inventor. In 1851 he designed and built the first commercially successful sewing machine, which included features already developed by Elias Howe. Although he was successfully sued by Howe for infringement of patent, Singer founded his own company, gained patents for improvements to the machine, and became the world's largest sewing machine manufacturer; his success was partly based on his pioneering use of hire-purchase agreements. (See also SEWING MACHINE.)

Singh /sɪŋ/ n. **1** a title adopted by the warrior castes of northern India. **2** a surname adopted by male Sikhs. [Punjabi siṅgh f. Skr. siṃha lion]

Singhalese var. of SINHALESE.

single /ˈsɪŋg(ə)l/ adj., n., & v. ● adj. **1** one only, not double or multiple. **2** united or undivided. **3 a** designed or suitable for one person (single room). **b** used or done by one person etc. or one set or pair. **4** one by itself; not one of several (a single tree). **5** regarded separately (every single thing). **6** not married; not involved in a relationship, unattached. **7** Brit. (of a ticket) valid for an outward journey only, not for the return. **8** (with neg. or interrog.) even one; not to speak of more (did not see a single person). **9** (of a flower) having only one circle of petals. **10** lonely, unaided. **11** archaic free from duplicity, sincere, consistent, guileless, ingenuous. ● n. **1** a single thing, or item in a series. **2** Brit. a single ticket. **3** a short record of usu. popular music featuring only one piece of music. **4** Cricket a hit for one run. **5** (usu. in pl.) a game with one player on each side. **6** an unmarried person (young singles). **7** US colloq. a one-dollar note. ● v.tr. (foll. by out and often by for, as) select from a group as worthy of special attention, praise, etc. (singled out for praise; singled out as the finest). □ **single acrostic** see ACROSTIC. **single-acting** (of an engine etc.) having pressure applied only to one side of the piston. **single-breasted** (of a jacket etc.) having only one set of buttons and buttonholes, not overlapping. **single combat** a duel. **single cream** thin cream with a relatively low fat-content. **single cut** (of a file) with grooves cut in one direction only, not crossing. **single-decker** esp. Brit. a bus having only one deck. **single entry** a system of bookkeeping in which each transaction is entered in one account only. **single file** n. a line of people or things arranged one behind another. ● adv. one behind another. **single-handed** adv. **1** without help from another. **2** with one hand. ● adj. **1** done etc. single-handed. **2** for one hand. **single-handedly** in a single-handed way. **single-lens reflex** denoting a reflex camera in which a single lens serves the film and the viewfinder. **single-line** with movement of traffic in only one direction at a time. **single parent** a person bringing up a child or children without a partner. **singles bar** a bar for single people seeking company. **single-seater** a vehicle with one seat. **single stick 1** a basket-hilted stick of about a sword's length. **2** one-handed fencing with this. **single-tree** US = SWINGLETREE. □ **singleness** n. **singly** adv. [ME f. OF f. L singulus, rel. to simplus SIMPLE]

Single European Act (abbr. **SEA**) a decree approved by the European Council in Dec. 1985 and coming into force on 1 July 1987. It provided for the establishment of a single European market from 1 Jan. 1993, and gave greater powers to the European Parliament.

single market n. an association of countries trading without restrictions, esp. in the European Community. The European single market came into effect on 1 Jan. 1993.

single-minded /ˌsɪŋg(ə)lˈmaɪndɪd/ adj. having or intent on only one purpose. □ **single-mindedly** adv. **single-mindedness** n.

singlet /ˈsɪŋglɪt/ n. **1** Brit. a garment worn under or instead of a shirt; a vest. **2** a single unresolvable line in a spectrum. [SINGLE + -ET¹, after doublet, the garment being unlined]

singleton /ˈsɪŋg(ə)lt(ə)n/ n. **1** one card only of a suit, esp. as dealt to a player. **2 a** a single person or thing. **b** an only child. **3** a single child or animal born, not a twin etc. [SINGLE, after simpleton]

Sing Sing /ˈsɪŋ sɪŋ/ a New York State prison, built in 1825–8 at Ossining village on the Hudson River and formerly notorious for its severe discipline. It is now called Ossining Correctional Facility.

singsong /ˈsɪŋsɒŋ/ adj., n., & v. ● adj. uttered with a monotonous rhythm or cadence. ● n. **1** a singsong manner. **2** Brit. an informal gathering for singing. ● v.intr. & tr. (past and past part. **singsonged**) speak or recite in a singsong manner.

singular /ˈsɪŋgjʊlə(r)/ adj. & n. ● adj. **1** unique; much beyond the average; extraordinary. **2** eccentric or strange. **3** Gram. (of a word or form) denoting or referring to a single person or thing. **4** Math. possessing unique properties. **5** single, individual. ● n. Gram. **1** a singular word or form. **2** the singular number. □ **singularly** adv. [ME f. OF singuler f. L singularis (as SINGLE)]

singularity /ˌsɪŋgjʊˈlærɪtɪ/ n. (pl. **-ies**) **1** the state or condition of being singular. **2** an odd trait or peculiarity. **3** Physics & Math. a point at which a function takes an infinite value, e.g. in space–time when matter is infinitely dense. [ME f. OF singularité f. LL singularitas (as SINGULAR)]

singularize /ˈsɪŋgjʊləˌraɪz/ v.tr. (also **-ise**) **1** distinguish, individualize. **2** make singular. □ **singularization** /ˌsɪŋgjʊləraɪˈzeɪʃ(ə)n/ n.

sinh /ʃaɪn, samˈeɪtʃ/ abbr. Math. hyperbolic sine. [sine + hyperbolic]

Sinhalese /ˌsɪnhəˈliːz, ˌsɪnə-/ n. & adj. (also **Singhalese**) ● n. (pl. same) **1** a member of a people of northern Indian origin forming the majority of the population in Sri Lanka. **2** the language of the Sinhalese, descended from Sanskrit and spoken by about three-quarters of the population of Sri Lanka. Its alphabet resembles that of the Dravidian languages of southern India. ● adj. of or relating to the Sinhalese or their language. [Skr. Siṅhala Sri Lanka + -ESE]

Sining see XINING.

sinister /ˈsɪnɪstə(r)/ adj. **1** suggestive of evil; looking malignant or villainous. **2** wicked or criminal (a sinister motive). **3** of evil omen. **4** Heraldry of or on the left-hand side of a shield etc. (i.e. to the observer's right). **5** archaic left-hand. □ **sinisterly** adv. **sinisterness** n. [ME f. OF sinistre or L sinister left]

sinistral /ˈsɪnɪstrəl/ adj. & n. ● adj. **1** left-handed. **2** of or on the left. **3** (of a flatfish) with the left side uppermost. **4** (of a spiral shell) with whorls rising to the left and not (as usually) to the right. ● n. a left-handed person. □ **sinistrally** adv. **sinistrality** /ˌsɪnɪˈstrælɪtɪ/ n.

sinistrorse /ˈsɪnɪˌstrɔːs/ adj. rising towards the left, esp. of the spiral stem of a plant. [L sinistrorsus f. sinister left + versus past part. of vertere turn]

sink /sɪŋk/ v. & n. ● v. (past **sank** /sæŋk/ or **sunk** /sʌŋk/; past part. **sunk** or as adj. **sunken**) **1** intr. fall or come slowly downwards. **2** intr. disappear below the horizon (the sun is sinking). **3** intr. **a** go or penetrate below the surface esp. of a liquid. **b** (of a ship) go to the bottom of the sea etc. **4** intr. settle down comfortably (sank into a chair). **5** intr. **a** gradually lose strength or value or quality etc.; decline (my heart sank). **b** (of the voice) descend in pitch or volume. **c** (of a sick person) approach death. **6** tr. send (a ship) to the bottom of the sea etc. **7** tr. cause or allow to sink or penetrate (sank its teeth into my leg). **8** tr. cause the failure of (a plan etc.) or the discomfiture of (a person). **9** tr. dig (a well) or bore (a shaft). **10** tr. engrave (a die) or inlay (a design). **11** tr. **a** invest (money) (sunk a large sum into the business). **b** lose (money) by investment. **12** tr. **a** cause (a ball) to enter a pocket in billiards, a hole at golf, etc. **b** achieve this by (a stroke). **13** tr. overlook or forget; keep in the background (sank their differences). **14** intr. (of a price etc.) become lower. **15** intr. (of a storm or river) subside. **16** intr. (of ground) slope down, or reach a lower level by subsidence. **17** intr. (foll. by on, upon) (of darkness) descend (on a place). **18** tr. lower the level of. **19** tr. (usu. in passive; foll. by in) absorb; hold the attention of (be sunk in thought). ● n. **1** a fixed basin with a water-supply and outflow pipe. **2** a place where foul liquid collects. **3** a place of vice or corruption. **4** a pool or marsh in which a river's water disappears by evaporation or percolation. **5** Physics a body or process used to absorb or dissipate heat. **6** (in full **sink-hole**) Geol. a cavity in limestone etc. into which a stream etc. disappears. □ **sink in 1** penetrate or make its way in. **2** become gradually comprehended (paused to let the words sink in). **sinking feeling** a bodily sensation, esp. in the abdomen, caused by hunger or apprehension. **sinking fund** money set aside for the gradual repayment of a debt. **sink or swim** even at the risk of complete failure (determined to try, sink or swim). **sunk fence** a fence formed by, or along the bottom of, a ditch. □ **sinkable** adj. **sinkage** n. [OE sincan f. Gmc]

sinker /ˈsɪŋkə(r)/ n. **1** a weight used to sink a fishing-line or sounding-line. **2** US a doughnut.

Sinkiang see XINJIANG.

sinner /ˈsɪnə(r)/ n. a person who sins, esp. habitually.

sinnet /ˈsɪnɪt/ n. (also **sennit**) Naut. braided cordage made in flat or round or square form by plaiting together several cords. [17th c.: orig. unkn.]

Sinn Fein /ʃɪn ˈfeɪn/ an Irish movement founded in 1905 by Arthur Griffith, originally aiming at the independence of Ireland and a revival of Irish culture and language and now dedicated to the political unification of Northern Ireland and the Republic of Ireland. Sinn Fein became increasingly committed to Republicanism after the failure of the Home Rule movement and supported the Easter Rising of 1916. Having won a majority of Irish seats in the 1918 general election, its members refused to go to Westminster and set up their own parliament in Ireland in 1919. The Republican section of the party supported Eamon de Valera in his rejection of the Anglo-Irish Treaty, and most of the section's members joined his Fianna Fáil party on its formation in 1926. The remainder of the party began to function as the political wing of the IRA, and in 1969 split like the IRA into Official and Provisional wings. □ **Sinn Feiner** n. [Ir. sinn féin we ourselves]

Sino- /ˈsaɪnəʊ/ comb. form Chinese; Chinese and (Sino-American). [Gk Sinai the Chinese]

Sino-Japanese Wars /ˈsaɪnəʊˌdʒæpəˌniːz/ two wars fought between China and Japan. The first, in 1894–5, caused by rivalry over Korea, was ended by the Treaty of Shimonoseki in Japan's favour. Poor Chinese performance in the war was a factor in the eventual overthrow of the Manchus in 1912. In the second war (1937–45) Japanese expansionism led to trouble in Manchuria in 1931 and to the establishment of a Japanese puppet state (Manchukuo) a year later. Hostilities began in earnest in 1937, but after two years of dramatic Japanese successes degenerated into stalemate. The Japanese position was gradually eroded by Communist guerrillas and finally collapsed at the end of the Second World War.

sinologue /ˈsaɪnəˌlɒg, ˈsɪn-/ n. an expert in sinology. [F, formed as SINO- + Gk -logos speaking]

sinology /saɪˈnɒlədʒɪ, sɪ-/ n. the study of Chinese language, history, customs, etc. □ **sinologist** n. **sinological** /ˌsaɪnəˈlɒdʒɪk(ə)l, ˌsɪn-/ adj.

Sino-Tibetan /ˌsaɪnəʊtɪˈbet(ə)n/ n. a language family which includes Chinese, Burmese, Tibetan, and (according to some scholars) Thai. They are tonal languages, but the exact relationships between them are far from clear.

sinter /ˈsɪntə(r)/ n. & v. ● n. **1** a siliceous or calcareous rock formed by deposition from springs. **2** a substance formed by sintering. ● v.intr. & tr. coalesce or cause to coalesce from powder into solid by heating. [G, cognate with CINDER]

Sint Maarten see ST MARTIN.

Sint-Niklaas /ˌsɪnt niˈklɑːs/ (French **St Nicolas** /sɛ̃ nikɔlɑ/) an industrial town in northern Belgium, south-west of Antwerp; pop. (1991) 68,200. It is noted for its large market square.

Sintra /ˈsiːntrə/ (also **Cintra**) a small town in western Portugal, situated in a mountainous area to the north-west of Lisbon; pop. (1981) 20,000. It was formerly a summer residence of the Portuguese royal family.

sinuate /ˈsɪnjʊət/ adj. esp. Bot. wavy-edged; with distinct inward and outward bends along the edge. [L sinuatus past part. of sinuare bend]

Sinuiju /ˈʃɪnəˌdʒuː/ a city and port in North Korea, situated on the Yalu river near its mouth on the Yellow Sea; pop. (1984) 500,000. It lies across the river from the Chinese city of Dandong.

sinuosity /ˌsɪnjʊˈɒsɪtɪ/ n. (pl. **-ies**) **1** the state of being sinuous. **2** a bend, esp. in a stream or road. [F sinuosité or med.L sinuositas (as SINUOUS)]

sinuous /ˈsɪnjʊəs/ adj. with many curves; tortuous, undulating. □ **sinuously** adv. **sinuousness** n. [F sinueux or L sinuosus (as SINUS)]

sinus /ˈsaɪnəs/ n. **1** a cavity of bone or tissue, esp. in the skull connecting with the nostrils. **2** Med. a fistula, esp. to a deep abscess. **3** Bot. the curve between the lobes of a leaf. [L, = bosom, recess]

sinusitis /ˌsaɪnəˈsaɪtɪs/ n. inflammation of a nasal sinus.

sinusoid /ˈsaɪnəˌsɔɪd/ n. **1** a sine curve. **2** a small irregularly shaped blood vessel, esp. found in the liver. □ **sinusoidal** /ˌsaɪnəˈsɔɪd(ə)l/ adj. [F sinusoïde f. L sinus: see SINUS]

Sion var. of ZION.

-sion /ʃ(ə)n, ʒ(ə)n/ suffix forming nouns (see -ION) from Latin participial stems in -s- (mansion; mission; persuasion).

Sioux /suː/ n. & adj. (also called *Dakota*) ● n. (pl. same) **1** a member of a group of American Indian peoples chiefly inhabiting the upper Mississippi and Missouri river basins. (See note below.) **2** the language of these peoples. ● adj. of or relating to these peoples or their language. □ **Siouan** /ˈsuːən/ adj. & n. [N. Amer. F f. *Nadouessioux* f. Ojibwa (Ottawa dial.) *nātowĕssiwak*: F pl. ending -x replaced Ojibwa pl. ending -ak]

■ Like other nomadic Plains Indians, the Sioux were buffalo hunters. Theirs was a warrior society, in which men gained esteem from performing brave deeds. When incursions from white settlers and prospectors greatly increased after the discovery of gold in their lands, some Sioux resisted violently; a treaty was made with the US government, but was broken by incomers when gold was again discovered, in the Black Hills of Dakota. Under such chiefs as Sitting Bull and Crazy Horse, Sioux resistance led to the massacre of General Custer and his men at Little Bighorn (1876) but was ended by the Battle of Wounded Knee in 1890.

sip /sɪp/ v. & n. ● v.tr. & intr. (**sipped, sipping**) drink in one or more small amounts or by spoonfuls. ● n. **1** a small mouthful of liquid (*a sip of brandy*). **2** the act of taking this. □ **sipper** n. [ME: perh. a modification of SUP¹]

sipe /saɪp/ n. a groove or channel in the tread of a tyre to improve its grip. [dial. *sipe* to ooze f. OE *sīpian*, MLG *sīpen*, of unkn. orig.]

siphon /ˈsaɪf(ə)n/ n. & v. (also **syphon**) ● n. **1** a pipe or tube shaped like an inverted V or U with unequal legs to convey a liquid from a container to a lower level by atmospheric pressure. **2** (in full **siphon-bottle**) a bottle containing carbonated water which is forced out through a tube by the pressure of gas (esp. *soda siphon*). **3** *Zool.* a tubular organ in an aquatic animal, esp. a mollusc, through which water is drawn in or expelled. ● v.tr. & intr. (often foll. by *off*) **1** conduct or flow through a siphon. **2** divert or set aside (funds etc.). □ **siphonage** n. **siphonal** adj. **siphonic** /saɪˈfɒnɪk/ adj. [F *siphon* or L *sipho -onis* f. Gk *siphōn* pipe]

siphonophore /saɪˈfɒnəˌfɔː(r), ˈsaɪfənə-/ n. *Zool.* a colonial pelagic hydrozoan of the order Siphonophora, having both polyps and medusae attached to a float; e.g. the Portuguese man-of-war. [Gk *siphōno-* (as SIPHON, -PHORE)]

sippet /ˈsɪpɪt/ n. **1** a small piece of bread etc. soaked in liquid. **2** a piece of toast or fried bread as a garnish. **3** a fragment. [app. dimin. of SOP]

sir /sɜː(r)/ n. **1** a polite or respectful form of address or mode of reference to a man. **2** (**Sir**) a titular prefix to the forename of a knight or baronet. [ME, reduced form of SIRE]

Siracusa see SYRACUSE 1.

Sirdaryo /ˌsɪədɑːˈjəʊ/ (Russian **Syr Darya, Syrdarya** /ˌsɪ darˈjaˈ/) a river of central Asia. Rising in two headstreams in the Tien Shan mountains in eastern Uzbekistan, it flows for some 2,220 km (1,380 miles) west and north-west through southern Kazakhstan to the Aral Sea.

sire /ˈsaɪə(r)/ n. & v. ● n. **1** the male parent of an animal, esp. a stallion kept for breeding. **2** archaic a respectful form of address, now esp. to a king. **3** archaic poet. a father or male ancestor. ● v.tr. (esp. of an animal) beget. [ME f. OF ult. f. L *senior*: see SENIOR]

siren /ˈsaɪərən/ n. **1 a** a device for making a loud prolonged signal or warning sound, esp. by revolving a perforated disc over a jet of compressed air or steam. **b** the sound made by this. **2** *Gk Mythol.* any of a class of sea creatures who had the power of luring seafarers to destruction on dangerous rocks by their song. (See note below.) **3** a sweet singer. **4 a** a dangerously fascinating woman; a temptress. **b** a tempting pursuit etc. **5** (attrib.) irresistibly tempting. **6** a North American eel-like gilled amphibian of the family Sirenidae, with no hindlimbs and tiny forelimbs. □ **siren suit** a one-piece garment for the whole body, easily put on or taken off, orig. for use in air-raid shelters. [ME f. OF *sereine*, *sirene* f. LL *Sirena* fem. f. L f. Gk *Seirēn*]

■ The sirens are usually depicted as women or as half-woman and half-bird, although their appearance is not described by Homer. In the *Odyssey* Odysseus had himself tied to the mast of his ship in order to hear their song safely, having first ordered his crew to plug their ears with wax; the Argonauts were saved from them because the singing of Orpheus matched that of the sirens.

sirenian /saɪˈriːnɪən/ n. & adj. *Zool.* ● n. a large aquatic plant-eating mammal of the order Sirenia, with flipper-like forelimbs and no hindlimbs, e.g. the manatee and dugong. ● adj. of or relating to this order. [mod.L *Sirenia* (as SIREN)]

Sirius /ˈsɪrɪəs/ *Astron.* the dog-star, the brightest star in the sky, in the constellation of Canis Major. It is conspicuous in the winter sky of the northern hemisphere, apparently following on the heels of the hunter Orion. It was important to the ancient Egyptians (see SOTHIC), as its heliacal rising coincided with the season of flooding of the Nile. It is a binary star with a dim companion, Sirius B or the Pup, which is a white dwarf. [Gk *seirios astēr* scorching star]

sirloin /ˈsɜːlɔɪn/ n. the upper and choicer part of a loin of beef. [OF (as SUR-¹, LOIN)]

sirocco /sɪˈrɒkəʊ/ n. (also **scirocco**) (pl. **-os**) a hot, oppressive, and often dusty or rainy wind blowing from North Africa across the Mediterranean to southern Europe. [F f. It. *scirocco*, ult. f. Arab. *šalūḳ* east wind]

sirrah /ˈsɪrə/ n. archaic = SIR (as a form of address). [prob. f. ME *sīrĕ* SIR]

sirree /sɪˈriː/ int. US colloq. as an emphatic, esp. after *yes* or *no*. [SIR + emphatic suffix]

Sirte, Gulf of see SIDRA, GULF OF.

sirup US var. of SYRUP.

SIS abbr. (in the UK) Secret Intelligence Service (see MI6).

sis /sɪs/ n. colloq. a sister. [abbr.]

sisal /ˈsaɪs(ə)l/ n. **1** a Mexican plant, *Agave sisalana*, with large fleshy leaves. **2** the fibre made from this plant, used for cordage, ropes, etc. [*Sisal*, the port of Yucatán]

siskin /ˈsɪskɪn/ n. a small streaked yellowish-green finch of the genus *Carduelis*; esp. *C. spinus* of northern Eurasia. [MDu. *siseken* dimin., rel. to MLG *sīsek*, MHG *zīse*, *zīsec*, of Slav. origin]

Sisley /ˈsɪslɪ, ˈsɪzlɪ/, Alfred (1839–99), French impressionist painter, of English descent. His development towards impressionism from his early Corot-influenced landscapes was gradual and greatly indebted to Monet. He is chiefly remembered for his paintings of the countryside around Paris in the 1870s, with their concentration on reflecting surfaces and fluid brushwork; like Monet, he also painted the same scenes under different weather conditions.

sissy /ˈsɪsɪ/ n. & adj. (also **cissy**) colloq. ● n. (pl. **-ies**) an effeminate or cowardly person. ● adj. effeminate; cowardly. □ **sissified** adj. **sissiness** n. **sissyish** adj. [SIS + -Y²]

sister /ˈsɪstə(r)/ n. **1** a woman or girl in relation to sons and other daughters of her parents. **2 a** (often as a form of address) a close female friend or associate. **b** a female fellow member of a trade union, class, sect, or the human race. **3** a senior female nurse. **4** a member of a female religious order. **5** (often attrib.) of the same type or design or origin etc. (*sister ship; prose, the younger sister of verse*). □ **sister german** see GERMAN. **sister-in-law** (pl. **sisters-in-law**) **1** the sister of one's wife or husband. **2** the wife of one's brother. **3** the wife of one's brother-in-law. **sister uterine** see UTERINE. □ **sisterless** adj. **sisterly** adj. **sisterliness** n. [ME *sister* (f. ON), *suster* etc. (repr. OE *sweostor* f. Gmc)]

sisterhood /ˈsɪstəˌhʊd/ n. **1** the relationship between sisters. **2 a** a society or association of women, esp. when bound by monastic vows or devoting themselves to religious or charitable work or the feminist cause. **b** its members collectively. **3** community of feeling and mutual support between women.

Sister of Mercy n. a member of an educational or charitable order of women, esp. that founded in Dublin in 1827.

Sistine /ˈsɪstiːn, -taɪn/ adj. of or relating to any of the popes called Sixtus, esp. Sixtus IV (pope 1471–84). [It. *Sistino* f. *Sisto* Sixtus]

Sistine Chapel a chapel in the Vatican, built in the late 15th century by Pope Sixtus IV, containing a painted ceiling and fresco of the Last Judgement by Michelangelo and also frescos by Botticelli, Ghirlandaio, and others. It is used for the principal papal ceremonies and also by the cardinals when meeting for the election of a new pope.

sistrum /ˈsɪstrəm/ n. (pl. **sistra** /-trə/) a jingling metal instrument used by the ancient Egyptians esp. in the worship of Isis. [ME f. L f. Gk *seistron* f. *seiō* shake]

Sisyphean /ˌsɪsɪˈfiːən/ adj. as of Sisyphus; endlessly laborious.

Sisyphus /ˈsɪsɪfəs/ *Gk Mythol.* the son of Aeolus, punished in Hades for his misdeeds in life by being condemned to the eternal task of rolling a large stone to the top of a hill, from which it always rolled down again.

sit /sɪt/ v. & n. ● v. (**sitting**; *past* and *past part.* **sat** /sæt/) **1** intr. adopt or be in a position in which the body is supported more or less upright by the buttocks resting on the ground or a raised seat etc., with the thighs usu. horizontal. **2** tr. cause to sit; place in a sitting position. **3** intr. **a** (of a bird) perch. **b** (of an animal) rest with the hind legs bent and the body close to the ground. **4** *intr.* (of a bird) remain on its nest to hatch its eggs. **5** *intr.* **a** be engaged in an occupation in which the sitting position is usual. **b** (of a committee, legislative body, etc.) be engaged in business. **c** (of an individual) be entitled to hold some office or position (*sat as a magistrate*). **6** *intr.* (usu. foll. by *for*) pose, usu. in a sitting position (for a portrait). **7** *intr.* (foll. by *for*) be a member of parliament for (a constituency). **8** *tr.* & (foll. by *for*) *intr. Brit.* be a candidate for (an examination). **9** *intr.* be in a more or less permanent position or condition (esp. of inactivity or being out of use or out of place). **10** *intr.* (of clothes etc.) fit or hang in a certain way. **11** *tr.* keep or have one's seat on (a horse etc.). **12** *intr.* act as a babysitter. **13** *intr.* (often foll. by *before*) (of an army) take a position outside a city etc. to besiege it. **14** *tr.* = SEAT *v.* 2a. ● *n.* the way clothing etc. sits on a person. □ **be sitting pretty** be comfortably or advantageously placed. **make a person sit up** *colloq.* surprise or interest a person. **sit at a person's feet** be a person's pupil. **sit back** relax one's efforts. **sit by** look on without interfering. **sit down 1** sit after standing. **2** cause to sit. **3** (foll. by *under*) submit tamely to (an insult etc.). **sit-down** *adj.* **1** (of a meal) eaten sitting at a table. **2** (of a protest etc.) in which demonstrators occupy their workplace or sit down on the ground in a public place. ● *n.* **1** a period of sitting. **2** a sit-down protest etc. **sit-down strike** a strike in which workers refuse to leave their place of work. **sit heavy on the stomach** take a long time to be digested. **sit in 1** occupy a place as a protest. **2** (foll. by *for*) take the place of. **3** (foll. by *on*) be present as a guest or observer at (a meeting etc.). **sit-in** *n.* a protest involving sitting in. **sit in judgement** assume the right of judging others; be censorious. **sit loosely on** not be very binding. **sit on 1** be a member of (a committee etc.). **2** hold a session or inquiry concerning. **3** *colloq.* delay action about (*the government has been sitting on the report*). **4** *colloq.* repress or rebuke or snub (*felt rather sat on*). **sit on the fence** see FENCE. **sit on one's hands 1** take no action. **2** refuse to applaud. **sit out 1** take no part in (a dance etc.). **2** stay till the end of (esp. an ordeal). **3** sit outdoors. **4** outstay (other visitors). **sit tight** *colloq.* **1** remain firmly in one's place. **2** not be shaken off or move away or yield to distractions. **sit up 1** rise from a lying to a sitting position. **2** sit firmly upright. **3** go to bed later than the usual time. **4** *colloq.* become interested or aroused etc. **sit-up** *n.* a physical exercise in which a person sits up from a supine position without using the arms for leverage. **sit up and take notice** *colloq.* have one's interest aroused, esp. suddenly. **sit-upon** *colloq.* the buttocks. **sit well** have a good seat in riding. **sit well on** suit or fit. [OE *sittan* f. Gmc]

Sita /ˈsiːtaː/ (in the Ramayana) the wife of Rama. She is the Hindu model of the ideal woman, an incarnation of Lakshmi. [Skr., = furrow]

sitar /ˈsɪtɑː(r), sɪˈtɑː(r)/ *n.* a long-necked Indian lute with movable frets, of Persian origin, one of the most important of Indian musical instruments and now well known also in the West. It now usually has four principal strings (although there were originally three), three drone-strings, and a dozen or more 'sympathetic strings' (made of metal and providing a special background resonance). The strings are plucked with a plectrum. □ **sitarist** *n.* [Pers. & Urdu *sitār*, f. *sih* three + *tār* string.]

sitatunga /ˌsɪtəˈtʌŋɡə/ *n.* a medium-sized brown or greyish antelope, *Tragelaphus spekii*, of swamps in central and East Africa, with splayed hooves and, in the male, spiral horns. [Swahili]

sitcom /ˈsɪtkɒm/ *n. colloq.* a situation comedy. [abbr.]

site /saɪt/ *n.* & *v.* ● *n.* **1** the ground chosen or used for a town or building. **2** a place where some activity is or has been conducted (*camping site; launching site*). ● *v.tr.* **1** locate or place. **2** provide with a site. [ME f. AF *site* or L *situs* local position]

Sitka /ˈsɪtkə/ *n.* (in full **Sitka spruce**) a fast-growing spruce, *Picea sitchensis*, native to North America and yielding timber. [*Sitka* in Alaska]

sitrep /ˈsɪtrep/ *n.* a report on the current military situation in an area. [*situation report*]

sits vac /sɪts ˈvæk/ *n. colloq.* situations vacant. [abbr.]

Sittang /ˈsɪtæŋ/ a river of southern Burma (Myanmar). Rising in the Pegu mountains, it flows some 560 km (350 miles) south into the Bay of Bengal at the Gulf of Martaban.

sitter /ˈsɪtə(r)/ *n.* **1** a person who sits, esp. for a portrait. **2** = BABYSITTER (see BABYSIT). **3** *colloq.* **a** an easy catch or shot. **b** an easy task. **4** a sitting hen.

sitting /ˈsɪtɪŋ/ *n.* & *adj.* ● *n.* **1** a continuous period of being seated, esp. engaged in an activity (*finished the book in one sitting*). **2** a time during which an assembly is engaged in business. **3** a session in which a meal is served (*dinner will be served in two sittings*). **4** *Brit. Law* = TERM 5c. **5** a clutch of eggs. ● *adj.* **1** having sat down. **2** (of an animal or bird) not

running or flying. **3** (of a hen) engaged in hatching. **4** (of an MP) current. □ **sitting duck** (or **target**) *colloq.* a vulnerable person or thing. **sitting pretty** see PRETTY. **sitting-room 1** a room in a house for relaxed sitting in. **2** space enough to accommodate seated persons. **sitting tenant** a tenant already in occupation of premises, esp. when there is a change of owner.

Sitting Bull (Sioux name Tatanka Iyotake) (c.1831–90), Sioux chief. As the main chief of the Sioux peoples from about 1867, he resisted the US government order of 1875 forcibly resettling the Sioux on reservations; when the US army opened hostilities in 1876, Sitting Bull led the Sioux in the fight to retain their lands, which resulted in the massacre of General Custer and his men at Little Bighorn. In 1885 he appeared in Buffalo Bill's Wild West Show, but continued to lead his people; becoming an advocate of the Ghost Dance cult, he was killed in an uprising.

situate *v.* & *adj.* ● *v.tr.* /ˈsɪtjʊˌeɪt/ (usu. in *passive*) **1** put in a certain position or circumstances (*is situated at the top of a hill; how are you situated at the moment?*). **2** establish or indicate the place of; put in a context. ● *adj.* /ˈsɪtjʊət/ *Law* or *archaic* situated. [med.L *situare situat-* f. L *situs* site]

situation /ˌsɪtjʊˈeɪʃ(ə)n/ *n.* **1** a place and its surroundings (*the house stands in a fine situation*). **2** a set of circumstances; a position in which one finds oneself; a state of affairs (*came out of a difficult situation with credit*). **3** an employee's position or job. **4** a critical point or complication in a drama. □ **situation comedy** a comedy series in which the same characters and settings are used in each episode. **situations vacant** (or **wanted**) headings of lists of employment offered and sought. □ **situational** *adj.* [ME f. F *situation* or med.L *situatio* (as SITUATE)]

situationism /ˌsɪtjʊˈeɪʃəˌnɪz(ə)m/ *n.* **1** the theory that human behaviour is determined by (esp. a false representation of) surrounding circumstances rather than by personal attributes or qualities. **2** (also **Situationism**) a revolutionary political theory and movement. (See note below.) □ **situationist** *n.* & *adj.* [SITUATION + -ISM]

▪ Situationism regards modern industrial society as being inevitably oppressive and exploitative, a system in which the individual is no more than a commodity. The movement rejects all conventional politics and demands a comprehensive revolution in relationships, work, and all aspects of everyday life. Influenced by surrealism and Dada, situationism came to public attention through the French magazine *Internationale Situationniste* from 1958, and was an inspiration behind the strikes and student uprisings of May 1968.

Sitwell /ˈsɪtwel/, Dame Edith (Louisa) (1887–1964), English poet and critic. Light-hearted and experimental, her early verse, like that of her brothers Osbert (1892–1969) and Sacheverell (1897–1988), marked a revolt against the prevailing Georgian style of the day. In 1923 she attracted attention with *Façade*, a group of poems in notated rhythm recited to music by William Walton. Her later verse is graver and makes increasing use of Christian symbolism.

sitz-bath /ˈsɪtsbɑːθ/ *n.* a hip-bath. [partial transl. of G *Sitzbad* f. *sitzen* sit + *Bad* bath]

Siva /ˈsiːvə, ˈʃiːvə, ˈʃɪvə/ (also **Shiva** /ˈʃiːvə, ˈʃɪvə/) *Hinduism* one of the major gods, perhaps a later development of the Vedic god Rudra, who forms a triad with Brahma and Vishnu. He is worshipped in many aspects: as fierce destroyer, naked ascetic, lord of the cosmic dance, lord of beasts, and, most commonly, in the form of the phallus (*linga*). In his beneficent aspect he lives in the Himalayas with his wife Parvati and their two sons, Ganesha and Skanda. His mount is the bull Nandi. Typically, Siva is depicted with a third eye in the middle of his forehead, wearing a crescent moon in his matted hair and a necklace of skulls at his throat, entwined with live snakes, and carrying a trident. □ **Sivaism** *n.* **Sivaite** *n.* & *adj.* [Skr. *Śiva*, lit. 'the auspicious one']

Sivaji see SHIVAJI.

Sivan /ˈsiːvɑːn/ *n.* (in the Jewish calendar) the ninth month of the civil and third of the religious year, usually coinciding with parts of May and June. [Heb. *sīwān*]

Siwalik Hills /sɪˈwɑːlɪk/ a range of foothills in the southern Himalayas, extending from NE India across Nepal to Sikkim.

six /sɪks/ *n.* & *adj.* ● *n.* **1** one more than five, or four less than ten; the product of two units and three units. **2** a symbol for this (6, vi, VI). **3** a size etc. denoted by six. **4** a set or team of six individuals. **5** *Cricket* a hit scoring six runs by clearing the boundary without bouncing. **6** the time of six o'clock. **7** a card etc. with six pips. ● *adj.* that amount to six. □ **at sixes and sevens** in confusion or disagreement. **knock**

for six *colloq.* utterly surprise or overcome (a person). **six-gun** = *six-shooter.* **six of one and half a dozen of the other** a situation of little real difference between the alternatives. **six-shooter** a revolver with six chambers. [OE *siex* etc. f. Gmc]

Six, Les see LES SIX.

sixain /'sɪkseɪn/ *n.* a six-line stanza. [F f. *six* six]

Six Counties the Ulster counties of Antrim, Down, Armagh, Londonderry, Tyrone, and Fermanagh, which since 1920 have comprised the province of Northern Ireland.

Six Day War (also **Six Days War**) an Arab–Israeli war, 5–10 June 1967 (known to the Arabs as the *June War*), in which Israel occupied Sinai, the Old City of Jerusalem, the West Bank, and the Golan Heights and defeated an Egyptian, Jordanian, and Syrian alliance.

sixer /'sɪksə(r)/ *n.* **1** the leader of a group of six Brownies or Cubs. **2** *Cricket* a hit for six runs.

sixfold /'sɪksfəʊld/ *adj. & adv.* **1** six times as much or as many. **2** consisting of six parts.

sixpence /'sɪkspəns/ *n. Brit.* **1** the sum of six pence, esp. before decimalization. **2** *hist.* a coin worth six old pence (2$\frac{1}{2}$p). □ **turn on a sixpence** *colloq.* make a sharp turn in a motor vehicle.

sixpenny /'sɪkspənɪ/ *adj. Brit.* costing or worth six pence, esp. before decimalization.

sixte /sɪkst/ *n. Fencing* the sixth of the eight parrying positions. [F f. L *sextus* sixth]

sixteen /sɪks'tiːn/ *n. & adj.* ● *n.* **1** one more than fifteen, or six more than ten. **2** a symbol for this (16, xvi, XVI). **3** a size etc. denoted by sixteen. ● *adj.* that amount to sixteen. □ **sixteenth note** esp. *N. Amer. Mus.* = SEMIQUAVER. □ **sixteenth** *adj. & n.* [OE *siextiene* (as SIX, -TEEN)]

sixteenmo /sɪks'tiːnməʊ/ *n.* (*pl.* **-os**) sextodecimo. [English reading of the symbol 16mo]

sixth /sɪksθ/ *n. & adj.* ● *n.* **1** the position in a sequence corresponding to that of the number 6 in the sequence 1–6. **2** something occupying this position. **3** any of six equal parts of a thing. **4** *Mus.* **a** an interval or chord spanning six consecutive notes in the diatonic scale (e.g. C to A). **b** a note separated from another by this interval. ● *adj.* that is the sixth. □ **sixth form** *Brit.* a form in a secondary school for pupils over 16. **sixth-form college** *Brit.* a college for pupils over 16. **sixth-former** a pupil in the sixth form. **sixth sense 1** a supposed faculty giving intuitive or extrasensory knowledge. **2** such knowledge. □ **sixthly** *adv.* [SIX]

Sixtine /'sɪkstiːn, -taɪn/ *adj.* = SISTINE. [mod.L *Sixtinus* f. *Sixtus*]

sixty /'sɪkstɪ/ *n. & adj.* ● *n.* (*pl.* **-ies**) **1** the product of six and ten. **2** a symbol for this (60, lx, LX). **3** (in *pl.*) the numbers from 60 to 69, esp. the years of a century or of a person's life. **4** a set of sixty persons or things. ● *adj.* that amount to sixty. □ **sixty-first, -second,** etc. the ordinal numbers between sixtieth and seventieth. **sixty-fourmo** /ˌsɪkstɪ'fɔːməʊ/ (*pl.* **-os**) **1** a size of book in which each leaf is one-sixty-fourth of a printing-sheet. **2** a book of this size (after DUODECIMO etc.). **sixty-fourth note** esp. *N. Amer. Mus.* = HEMIDEMISEMIQUAVER. **sixty-four thousand** (or **sixty-four**) **dollar question** a difficult and crucial question (from the top prize in a broadcast quiz show). **sixty-one, -two,** etc. the cardinal numbers between sixty and seventy. □ **sixtieth** *adj. & n.* **sixtyfold** *adj. & adv.* [OE *siextig* (as SIX, -TY²)]

sizable var. of SIZEABLE.

sizar /'saɪzə(r)/ *n.* a student at Cambridge University or at Trinity College, Dublin, receiving an allowance from the college and formerly having certain menial duties. □ **sizarship** *n.* [SIZE¹ = ration]

size¹ /saɪz/ *n. & v.* ● *n.* **1** the relative bigness or extent of a thing, dimensions, magnitude (*is of vast size; size matters less than quality*). **2** each of the classes, usu. numbered, into which things otherwise similar, esp. garments, are divided according to size (*is made in several sizes; takes size 7 in gloves; is three sizes too big*). ● *v.tr.* sort or group in sizes or according to size. □ **of a size** having the same size. **of some size** fairly large. **the size of** as big as. **the size of it** *colloq.* a true account of the matter (*that is the size of it*). **size-stick** a shoemaker's measure for taking the length of a foot. **size up 1** estimate the size of. **2** *colloq.* form a judgement of. **what size?** how big? □ **sized** *adj.* (also in *comb.*). **sizer** *n.* [ME f. OF *sise* f. *assise* ASSIZE, or f. ASSIZE]

size² /saɪz/ *n. & v.* ● *n.* a gelatinous solution used in glazing paper, stiffening textiles, preparing plastered walls for decoration, etc. ● *v.tr.* glaze or stiffen or treat with size. [ME, perh. = SIZE¹]

sizeable /'saɪzəb(ə)l/ *adj.* (also **sizable**) large or fairly large. □ **sizeably** *adv.*

Sizewell /'saɪzwel/ a village on the Suffolk coast, the site of a nuclear power station. The construction of the pressurized-water reactor at Sizewell was the subject of a lengthy public inquiry between 1983 and 1985.

sizzle /'sɪz(ə)l/ *v. & n.* ● *v.intr.* **1** make a sputtering or hissing sound, as when frying. **2** *colloq.* be in a state of great heat or excitement or marked effectiveness; be salacious or risqué. ● *n.* **1** a sizzling sound. **2** *colloq.* a state of great heat or excitement. □ **sizzler** *n.* **sizzling** *adj. & adv.* (*sizzling hot*). [imit.]

SJ *abbr.* Society of Jesus.

SJAA *abbr.* (in the UK) St John Ambulance Association.

SJAB *abbr.* (in the UK) St John Ambulance Brigade.

Sjælland see ZEALAND.

sjambok /'ʃæmbɒk/ *n. & v.* ● *n.* (in South Africa) a long stiff whip, orig. made of rhinoceros hide. ● *v.tr.* flog with a sjambok. [Afrik. f. Malay *samboq, chambok* f. Urdu *chābuk*]

SJC *abbr.* (in the US) Supreme Judicial Court.

ska /skɑː/ *n.* a style of popular music of Jamaican origin, with a fast tempo and strongly accentuated offbeat. Popular in the early and mid-1960s, it was important in the evolution of reggae. [20th c.: orig. uncert.: perh. imit.]

Skagerrak /'skægəˌræk/ a strait separating southern Norway from the NW coast of Denmark.

skald /skɔːld, skɒld/ *n.* (also **scald**) (in ancient Scandinavia) a composer and reciter of poems honouring heroes and their deeds. □ **skaldic** *adj.* [ON *skáld,* of unkn. orig.]

Skanda /'skændə/ *Hinduism* the Hindu war-god, first son of Siva and Parvati and brother of Ganesha. He is depicted as a boy or youth, sometimes with six heads and often with his mount, a peacock.

Skara Brae /ˌskɑːrə 'breɪ, ˌskærə/ a late neolithic (3rd millennium BC) settlement on Mainland in the Orkney Islands. The settlement consists of a group of one-room stone dwellings with built-in stone shelves, chests, and hearths.

skat /skæt/ *n.* a three-handed card-game with bidding. [G f. It. *scarto* discard f. *scartare* discard]

skate¹ /skeɪt/ *n. & v.* ● *n.* **1** each of a pair of steel blades (or of boots with blades attached) for gliding on ice. **2** = *roller-skate.* **3** a device on which a heavy object moves. ● *v.* **1 a** *intr.* move on skates. **b** *tr.* perform (a specified figure) on skates. **2** *intr.* (foll. by *over*) refer fleetingly to, disregard. □ **get one's skates on** *Brit. colloq.* make haste. **skate on thin ice** *colloq.* behave rashly, risk danger, esp. by dealing with a subject needing tactful treatment. **skating-rink 1** a piece of ice artificially made, or a floor used, for skating. **2** a building containing this. □ **skater** *n.* [orig. *scates* (pl.) f. Du. *schaats* (sing.) f. ONF *escace,* OF *eschasse* stilt]

skate² /skeɪt/ *n.* (*pl.* same or **skates**) a ray (fish) of the family Rajidae, with a diamond-shaped body and long thin tail; esp. the large *Raja batis,* used as food. [ME f. ON *skata*]

skate³ /skeɪt/ *n. sl.* a contemptible, mean, or dishonest person (esp. in CHEAPSKATE). [19th c.: orig. uncert.]

skateboard /'skeɪtbɔːd/ *n. & v.* ● *n.* a short narrow board on roller-skate wheels for riding on while standing. Skateboards were introduced in California in the early 1960s and achieved worldwide popularity in the 1970s. ● *v.intr.* ride on a skateboard. □ **skateboarder** *n.*

skean /skiːn, 'skiːən/ *n. hist.* a Gaelic dagger formerly used in Ireland and Scotland. □ **skean-dhu** /-'duː/ *n.* a dagger worn in the stocking as part of Highland costume. [Gael. *sgian* knife, *dubh* black]

sked /sked/ *n. & v. colloq.* ● *n.* = SCHEDULE *n.* ● *v.tr.* (**skedded, skedding**) = SCHEDULE *v.* [abbr.]

skedaddle /skɪ'dæd(ə)l/ *v. & n. colloq.* ● *v.intr.* run away, depart quickly, flee. ● *n.* a hurried departure or flight. [19th c.: orig. unkn.]

skeet /skiːt/ *n.* a shooting sport in which a clay target is thrown from a trap to simulate the flight of a bird. [ON *skjóta* SHOOT]

skeeter¹ /'skiːtə(r)/ *n.* esp. *N. Amer. & Austral. dial. & colloq.* a mosquito. [abbr.]

skeeter² var. of SKITTER.

skeg /skeg/ *n.* **1** a fin underneath the rear of a surfboard. **2** the after part of a vessel's keel or a projection from it. [ON *skeg* beard, perh. via Du. *scheg(ge)*]

skein /skeɪn/ *n.* **1** a loosely coiled bundle of yarn or thread. **2** a flock of wild geese etc. in flight. **3** a tangle or confusion. [ME f. OF *escaigne,* of unkn. orig.]

skeletal /'skelɪt(ə)l/ adj. **1** of, forming, or resembling a skeleton. **2** very thin, emaciated. **3** consisting only of a bare outline; meagre. □ **skeletal muscle** = striated muscle (see STRIATE). □ **skeletally** adv.

skeleton /'skelɪt(ə)n/ n. **1 a** a hard internal or external framework of bones, cartilage, shell, woody fibre, etc., supporting or containing the body of an animal or plant. (See note below.) **b** the dried bones of a human being or other animal fastened together in the same relative positions as in life. **2** the supporting framework or structure or essential part of a thing. **3** a very thin or emaciated person or animal. **4** the remaining part of anything after its life or usefulness is gone. **5** an outline sketch, an epitome or abstract. **6** (attrib.) having only the essential or minimum number of persons, parts, etc. (skeleton plan; skeleton staff). □ **skeleton at the feast** something that spoils one's pleasure; an intrusive worry. **skeleton in the cupboard** (US **closet**) a discreditable or embarrassing fact kept secret. **skeleton key** a key designed to fit many locks by having the interior of the bit hollowed. □ **skeletonize** v.tr. (also **-ise**). [mod.L f. Gk, neut. of skeletos dried-up f. skellō dry up]

▪ Many invertebrates lack an internal skeleton, though a tough external one is seen in the mollusc's shell and the arthropod's exoskeleton. The more primitive fish have a cartilaginous skeleton, and the higher vertebrates one of bone. The human skeleton is conventionally regarded as consisting of two main parts: the skull, spine, and ribs, which support and protect body tissues and organs; and the limb bones and their attachments, which function in conjunction with muscles as levers to provide movement. (See also BONE.)

Skeleton Coast an arid coastal area in Namibia. Comprising the northern part of the Namib desert, it extends from Walvis Bay in the south to the border with Angola. A part of the area was designated a national park in 1971.

Skelton /'skelt(ə)n/, John (c.1460–1529), English poet. He was court poet to Henry VIII, to whom he had acted as tutor. Skelton's principal works include *The Bowge of Courte* (c.1498), a satire on the court of Henry VII, *Magnificence* (1516), a morality play, and *Collyn Cloute* (1522), which contained an attack on Cardinal Wolsey. His characteristic verse consisted of short irregular rhyming lines with rhythms based on colloquial speech, giving rise to the word *Skeltonic* to describe this type of verse or metre.

skep /skep/ n. **1 a** a wooden or wicker basket of any of various forms. **b** the quantity contained in this. **2** a straw or wicker beehive. [ME f. ON skeppa]

skepsis N. Amer. var. of SCEPSIS.

skeptic N. Amer. var. of SCEPTIC.

skeptical N. Amer. var. of SCEPTICAL.

skerrick /'skerɪk/ n. (usu. with neg.) esp. Austral. colloq. the smallest bit (not a skerrick left). [N. Engl. dial.; orig. uncert.]

skerry /'skerɪ/ n. (pl. **-ies**) Sc. a reef or rocky island. [Orkney dial. f. ON sker: cf. SCAR²]

sketch /sketʃ/ n. & v. ● n. **1** a rough, slight, merely outlined, or unfinished drawing or painting, often made to assist in making a more finished picture. **2** a brief account without many details conveying a general idea of something; a rough draft or general outline. **3 a** a very short play, usu. humorous and limited to one scene. **b** an item or scene in a comedy programme. **4** a short descriptive piece of writing. **5** a musical composition of a single movement. **6** colloq. a comical person or thing. ● v. **1** tr. make or give a sketch of. **2** intr. draw sketches esp. of landscape (went out sketching). **3** tr. (often foll. by in, out) indicate briefly or in outline. □ **sketch-book** (or **-block**) a pad of drawing-paper for doing sketches on. **sketch-map** a roughly drawn map with few details. □ **sketcher** n. [Du. schets or G Skizze f. It. schizzo f. schizzare make a sketch ult. f. Gk skhēdios extempore]

sketchy /'sketʃɪ/ adj. (**sketchier, sketchiest**) **1** giving only a slight or rough outline, like a sketch. **2** colloq. unsubstantial or imperfect, esp. through haste. □ **sketchily** adv. **sketchiness** n.

skeuomorph /'skjuːəʊˌmɔːf/ n. Archaeol. **1** an object or feature copying the design of a similar artefact in another material. **2** an ornamental design resulting from the nature of the material used or the method of working it. □ **skeuomorphic** /ˌskjuːəʊˈmɔːfɪk/ adj. [Gk skeuos vessel, implement + morphē form]

skew /skjuː/ adj., n., & v. ● adj. **1** oblique, slanting, set askew. **2** Math. **a** lying in three dimensions (skew curve). **b** (of lines) not coplanar. **c** (of a statistical distribution) not symmetrical. ● n. **1** a slant. **2** Statistics skewness. ● v. **1** tr. make skew. **2** tr. distort. **3** intr. move obliquely. **4** intr.

twist. □ **on the skew** askew. **skew arch** (or **bridge**) an arch (or bridge) with the line of the arch not at right angles to the abutment. **skew chisel** a chisel with an oblique edge. **skew-eyed** Brit. squinting. **skew gear** a gear consisting of two cog-wheels having non-parallel, non-intersecting axes. **skew-whiff** Brit. colloq. askew. □ **skewness** n. [ONF eskiu(w)er (v.) = OF eschuer: see ESCHEW]

skewback /'skjuːbæk/ n. Archit. the sloping face of the abutment on which an extremity of an arch rests.

skewbald /'skjuːbɔːld/ adj. & n. ● adj. (of an animal) with irregular patches of white and another colour (properly not black) (cf. PIEBALD). ● n. a skewbald animal, esp. a horse. [ME skued (orig. uncert.), after PIEBALD]

skewer /'skjuːə(r)/ n. & v. ● n. a long pin designed for holding meat etc. compactly together while cooking. ● v.tr. fasten together or pierce with or as with a skewer. [17th c., var. of dial. skiver: orig. unkn.]

ski /skiː/ n. & v. ● n. (pl. **skis**) **1** each of a pair of long narrow pieces of wood etc., usu. pointed and turned up at the front, fastened under the feet for travelling over snow. (See SKIING.) **2** a similar device under a vehicle or aircraft. **3** = WATER-SKI. **4** (attrib.) for wear when skiing (ski boots). ● v. (**skis, ski'd** or **skied** /skiːd/; **skiing**) **1** intr. travel on skis. **2** tr. ski at (a place). □ **ski-bob** n. a machine like a bicycle with skis instead of wheels. ● v.intr. (**-bobbed, -bobbing**) ride a ski-bob. **ski-bobber** a person who ski-bobs. **ski-jump 1** a steep slope levelling off before a sharp drop to allow a skier to leap through the air. **2** a jump made from this. **ski-jumper** a person who takes part in ski-jumping. **ski-jumping** the sport of leaping off a ski-jump with marks awarded for style and distance attained. **ski-lift** a device for carrying skiers up a slope, usu. on seats hung from an overhead cable. **ski-plane** an aeroplane having its undercarriage fitted with skis for landing on snow or ice. **ski-run** a slope prepared for skiing. □ **skiable** adj. [Norw. f. ON skíth billet, snow-shoe]

skiagraphy var. of SCIAGRAPHY.

skiamachy var. of SCIAMACHY.

Skiathos /skiːˈæθɒs/ (Greek **Skíathos** /'skiaθɒs/) a Greek island in the Aegean Sea, the most westerly of the Northern Sporades group.

skid /skɪd/ v. & n. ● v. (**skidded, skidding**) **1** intr. (of a vehicle, a wheel, or a driver) slide on slippery ground, esp. sideways or obliquely. **2** tr. cause (a vehicle etc.) to skid. **3** intr. slip, slide. **4** intr. colloq. fail or decline or err. **5** tr. support or move or protect or check with a skid. ● n. **1** the act or an instance of skidding. **2** a piece of wood etc. serving as a support, ship's fender, inclined plane, etc. **3** a braking device, esp. a wooden or metal shoe preventing a wheel from revolving or used as a drag. **4** a runner beneath an aircraft for use when landing. □ **hit the skids** colloq. enter a rapid decline or deterioration. **on the skids** colloq. **1** about to be discarded or defeated. **2** ready for launching. **put the skids under** colloq. **1** hasten the downfall or failure of. **2** cause to hasten. **skid-lid** sl. a crash-helmet. **skid-pan** Brit. **1** a slippery surface prepared for vehicle-drivers to practise control of skidding. **2** a braking device. **skid road** N. Amer. **1** a road for hauling logs along. **2** colloq. a part of a town frequented by loggers or vagrants. **skid row** N. Amer. colloq. a part of a town frequented by vagrants, alcoholics, etc. [17th c.: orig. unkn.]

Skidoo /skɪˈduː/ n. & v. (also **skidoo**) propr. ● n. a motorized toboggan. ● v.intr. ride on a Skidoo. [f. SKI]

skidoo /skɪˈduː/ v.intr. (also **skiddoo**) (**-oos, -ooed**) N. Amer. sl. go away; depart. [perh. f. SKEDADDLE]

skier¹ /'skiːə(r)/ n. a person who skis.

skier² var. of SKYER.

skiff /skɪf/ n. a light rowing-boat or sculling-boat. [F esquif f. It. schifo, rel. to SHIP]

skiffle /'skɪf(ə)l/ n. a kind of simple popular music played esp. in the 1950s by small groups characteristically with a rhythmic accompaniment provided by a washboard to a singing guitarist etc. [perh. imit.]

skiing /'skiːɪŋ/ n. the action of travelling over snow on skis, especially as a sport or recreation. Skis have been used in Scandinavia since the 3rd millennium BC for cross-country travel and transport, and ski troops were used in Scandinavia, Poland, and Russia in the 15th–16th centuries. The modern sport and recreation dates only from the mid-19th century, however, when improved skis and bindings made possible more accurate manoeuvring. Competitive skiing falls into two categories: *Nordic* (cross-country racing and jumping) and *Alpine* (downhill or straight racing, and slalom racing between series of markers).

ski-joring /'ski:ˌdʒɔːrɪŋ, ˌʃiːˈjɜːrɪŋ/ n. a winter sport in which a skier is towed by a horse or vehicle. □ **ski-jorer** n. [Norw. *skikjøring* (as SKI, *kjøre* drive)]

skilful /'skɪlfʊl/ adj. (US **skillful**) (often foll. by *at, in*) having or showing skill; practised, expert, ingenious. □ **skilfully** adv. **skilfulness** n.

skill /skɪl/ n. (often foll. by *in*) expertness, practised ability, facility in an action; dexterity or tact. □ **skill-less** adj. (archaic **skilless**). [ME f. ON *skil* distinction]

skilled /skɪld/ adj. **1** (often foll. by *in*) having or showing skill; skilful. **2** (of a worker) highly trained or experienced. **3** (of work) requiring skill or special training.

skillet /'skɪlɪt/ n. **1** Brit. a small metal cooking-pot with a long handle and usu. legs. **2** N. Amer. a frying-pan. [ME, perh. f. OF *escuelete* dimin. of *escuele* platter f. LL *scutella*]

skilly /'skɪlɪ/ n. Brit. **1** a thin broth or soup or gruel (usu. of oatmeal and water flavoured with meat). **2** an insipid beverage; tea or coffee. [abbr. f. *skilligalee*, prob. fanciful]

skim /skɪm/ v. & n. ● v. (**skimmed, skimming**) **1** tr. **a** take scum or cream or a floating layer from the surface of (a liquid). **b** take (cream etc.) from the surface of a liquid. **2** tr. **a** keep touching lightly or nearly touching (a surface) in passing over. **b** deal with or treat (a subject) superficially. **3** intr. **a** (often foll. by *over, along*) go lightly over a surface, glide along in the air. **b** (foll. by *over*) = sense 2b of v. **4 a** tr. read superficially, look over cursorily, gather the salient facts contained in. **b** intr. (usu. foll. by *through*) read or look over cursorily. **5** tr. US sl. conceal or divert (income) to avoid paying tax. ● n. **1** the act or an instance of skimming. **2** a thin covering on a liquid (*skim of ice*). □ **skim the cream off** take the best part of. **skim** (or **skimmed**) **milk** milk from which the cream has been skimmed. [ME, back-form. f. SKIMMER]

skimmer /'skɪmə(r)/ n. **1** a device for skimming liquids. **2** a person who skims. **3** a flat hat, esp. a broad-brimmed straw hat. **4** a mainly tropical long-winged ternlike bird of the genus *Rynchops*, feeding by flying low over water with the tip of its elongated lower bill immersed. Also called *scissor-bill*. **5** a hydroplane, hydrofoil, hovercraft, or other vessel that has little or no displacement at speed. **6** US a sheath-like dress. [ME f. OF *escumoir* f. *escumer* f. *escume* SCUM]

skimmia /'skɪmɪə/ n. an evergreen shrub of the genus *Skimmia*, native to eastern Asia, with red berries. [mod.L f. Jap.]

skimp /skɪmp/ v., adj., & n. ● v. **1** tr. (often foll. by *in*) supply (a person etc.) meagrely with food, money, etc. **2** tr. use a meagre or insufficient amount of, stint (material, expenses, etc.). **3** tr. do hastily or carelessly. **4** intr. be parsimonious. ● adj. scanty. ● n. colloq. a small or scanty thing, esp. a skimpy garment. [18th c.: orig. unkn.: cf. SCRIMP]

skimpy /'skɪmpɪ/ adj. (**skimpier, skimpiest**) meagre; not ample or sufficient. □ **skimpily** adv. **skimpiness** n.

skin /skɪn/ n. & v. ● n. **1** the flexible continuous covering of a human or other animal body. (*See note below.*) **2 a** the skin of a flayed animal with or without the hair etc. **b** a material prepared from skins esp. of smaller animals (cf. HIDE² n. 1). **3** a person's skin with reference to its colour or complexion (*has a fair skin*). **4** an outer layer or covering, esp. the coating of a plant, fruit, or sausage. **5** a film like skin on the surface of a liquid etc. **6** a container for liquid, made of an animal's whole skin. **7 a** the planking or plating of a ship or boat, inside or outside the ribs. **b** the outer covering of any craft or vehicle, esp. an aircraft or spacecraft. **8** Brit. sl. a skinhead. **9** US Cards a game in which each player has one card which he or she bets will not be the first to be matched by a card dealt from the pack. **10** = *gold-beater's skin*. **11** a duplicating stencil. ● v. (**skinned, skinning**) **1** tr. **a** remove the skin from. **b** graze (a part of the body). **2** (often foll. by *over*) **a** tr. cover (a sore etc.) with or as with skin. **b** intr. (of a wound etc.) become covered with new skin. **3** tr. sl. fleece or swindle. □ **be skin and bone** be very thin. **by** (or **with**) **the skin of one's teeth** by a very narrow margin. **change one's skin** undergo an impossible change of character etc. **get under a person's skin** interest or annoy a person intensely. **have a thick** (or **thin**) **skin** be insensitive (or sensitive) to criticism etc. **no skin off one's nose** a matter of indifference or even benefit to one. **skin-deep** (of a wound, or of an emotion, an impression, quality, etc.) superficial, not deep or lasting. **skin-diver** a person who practices skin-diving. **skin-diving** swimming under water without a diving-suit, usu. in deep water with an aqualung and flippers. **skin effect** *Electr.* the tendency of a high-frequency alternating current to flow mainly through the outer layer of a conductor. **skin-flick** sl. an explicitly pornographic film. **skin-food** a cosmetic intended to improve the condition of the skin. **skin friction** friction at the surface

of a solid and a fluid in relative motion. **skin game** US sl. a rigged gambling game; a swindle. **skin-graft 1** the surgical transplanting of skin. **2** a piece of skin transferred in this way. **skin test** a test to determine whether an immune reaction is elicited when a substance is applied to or injected into the skin. **skin-tight** (of a garment) very close-fitting. **to the skin** through all one's clothing (*soaked to the skin*). **with a whole skin** unwounded. □ **skinless** adj. **skinlike** adj. **skinned** adj. (also in comb.). [OE *scin(n)* f. ON *skinn*]

■ Skin protects the body from external injury, excessive heat or cold, fluid loss, and infection. It also acts as a sense-organ, being provided with receptors for temperature, touch, pressure, and pain, and it plays an important part in controlling the temperature of the body and in maintaining the balance of body fluid. The skin's outer layer or epidermis consists of cells that are continually renewed and pushed outwards, being progressively impregnated with keratin and finally dying and sloughing off. The inner layer or dermis is a thick layer of tissue containing blood capillaries and lymph vessels, sensory nerve-endings, sweat glands and their ducts, sebaceous glands, hair follicles, and smooth muscle fibres.

skinflint /'skɪnflɪnt/ n. a miserly person.

skinful /'skɪnfʊl/ n. (pl. **-fuls**) colloq. enough alcoholic liquor to make one drunk.

skinhead /'skɪnhed/ n. **1** Brit. a youth characterized by close-cropped hair and heavy boots, esp. one of an aggressive gang. **2** US a recruit in the Marines.

skink /skɪŋk/ n. a smooth-bodied lizard of the family Scincidae, with the limbs short or even absent. [F *scinc* or L *scincus* f. Gk *skigkos*]

Skinner /'skɪnə(r)/, Burrhus Frederic (1904–90), American psychologist. He promoted the view that the proper aim of psychology should be to predict, and hence be able to control, behaviour. He demonstrated that arbitrary responses in animals could be obtained by using reinforcements — rewards and punishments. He applied similar techniques in both clinical and educational practice, devising one of the first teaching machines, and was involved in the development of programmed learning. Skinner also attempted to account for the nature and development of language as a response to conditioning.

skinner /'skɪnə(r)/ n. **1** a person who skins animals or prepares skins. **2** a dealer in skins, a furrier. **3** Austral. Horse-racing sl. a result very profitable to bookmakers.

skinny /'skɪnɪ/ adj. (**skinnier, skinniest**) **1** thin or emaciated. **2** (of clothing) tight-fitting. **3** made of or like skin. □ **skinny-dipping** esp. N. Amer. colloq. bathing in the nude. □ **skinniness** n.

skint /skɪnt/ adj. Brit. colloq. having no money left, penniless. [= *skinned*, past part. of SKIN]

skip¹ /skɪp/ v. & n. ● v. (**skipped, skipping**) **1** intr. **a** move along lightly, esp. by taking two steps with each foot in turn. **b** jump lightly from the ground, esp. so as to clear a skipping-rope. **c** jump about, gambol, caper, frisk. **2** intr. (often foll. by *from, off, to*) move quickly from one point, subject, or occupation to another; be desultory. **3** tr. (also absol.) omit in dealing with a series or in reading (*skip every tenth row; always skips the small print*). **4** tr. colloq. not participate in. **5** tr. colloq. depart quickly from; leave hurriedly. **6** intr. (often foll. by *out, off*) colloq. make off, disappear. **7** tr. make (a stone) ricochet on the surface of water. ● n. **1** a skipping movement or action. **2** Computing the action of passing over part of a sequence of data or instructions. **3** N. Amer. colloq. a person who defaults or absconds. □ **skip it** sl. **1** abandon a topic etc. **2** make off, disappear. **skipping-rope** (N. Amer. **skip-rope**) a length of rope revolved over the head and under the feet while jumping as a game or exercise. **skip zone** the annular region round a broadcasting station where neither direct nor reflected waves are received. [ME, prob. f. Scand.]

skip² /skɪp/ n. **1** a large container for builders' refuse etc. **2** a cage, bucket, etc., in which men or materials are lowered and raised in mines and quarries. **3** = SKEP. [var. of SKEP]

skip³ /skɪp/ n. & v. ● n. the captain or director of a side at bowls, curling, etc. ● v.tr. (**skipped, skipping**) be the skip of. [abbr. of SKIPPER¹]

skipjack /'skɪpdʒæk/ n. **1** (in full **skipjack tuna**) a small striped Pacific tuna, *Katsuwonus pelamis*, used as food. **2** a click beetle. **3** a kind of sailing-boat used off the east coast of the US. [SKIP¹ + JACK¹]

skipper¹ /'skɪpə(r)/ n. & v. ● n. **1** a sea captain, esp. the master of a small trading or fishing vessel. **2** the captain of an aircraft. **3** the captain of a side in a game or sport. ● v.tr. act as captain of. [ME f. MDu., MLG *schipper* f. *schip* SHIP]

skipper[2] /'skɪpə(r)/ n. **1** a person who skips. **2** a small brownish mothlike butterfly of the family Hesperiidae.

skippet /'skɪpɪt/ n. a small round wooden box to enclose and protect a seal attached to a document. [ME: orig. unkn.]

skirl /skɜ:l/ n. & v. ● n. the shrill sound characteristic of bagpipes. ● v.intr. make a skirl. [prob. Scand.: ult. imit.]

skirmish /'skɜ:mɪʃ/ n. & v. ● n. **1** a piece of irregular or unpremeditated fighting, esp. between small or outlying parts of armies or fleets; a minor engagement. **2** a short argument or contest of wit etc. ● v.intr. engage in a skirmish. □ **skirmisher** n. [ME f. OF eskirmir, escremir f. Frank.]

skirr /skɜ:(r)/ v.intr. move rapidly esp. with a whirring sound. [perh. rel. to SCOUR[1] or SCOUR[2]]

skirret /'skɪrɪt/ n. a perennial umbelliferous plant, Sium sisarum, formerly cultivated in Europe for its edible root. [ME skirwhit(e), perh. formed as SHEER[1], WHITE]

skirt /skɜ:t/ n. & v. ● n. **1** a woman's outer garment hanging from the waist. **2** the part of a dress, coat, etc., that hangs below the waist. **3** a hanging part round the base of a hovercraft to contain or divide the air-cushion. **4** (in sing. or pl.) an edge, border, or extreme part. **5** (also **bit of skirt**) sl. offens. a woman regarded as an object of sexual desire. **6** (in full **skirt of beef** etc.) **a** the diaphragm and other membranes as food. **b** Brit. a cut of meat from the lower flank. **7** a flap of a saddle. **8** a surface that conceals or protects the wheels or underside of a vehicle or aircraft. ● v. **1** tr. go along or round or past the edge of. **2** tr. be situated along. **3** tr. avoid dealing with (an issue etc.). **4** intr. (foll. by along) go along the coast, a wall, etc. □ **skirt-chaser** see CHASER 4. **skirt-dance** a dance with graceful manipulation of a full skirt. □ **skirted** adj. (also in comb.). **skirtless** adj. [ME f. ON skyrta shirt, corresp. to OE scyrte: see SHIRT]

skirting /'skɜ:tɪŋ/ n. (in full **skirting-board**) Brit. a narrow board etc. along the base of the wall of a room next to the floor.

skit[1] /skɪt/ n. (often foll. by on) a light, usu. short, piece of satire or burlesque. [rel. to skit move lightly and rapidly, perh. f. ON (cf. skjóta SHOOT)]

skit[2] /skɪt/ n. colloq. **1** a large number, a crowd. **2** (in pl.) heaps, lots. [20th c.: cf. SCADS]

skite /skaɪt/ v. & n. ● v.intr. Austral. & NZ colloq. boast, brag. ● n. **1** Austral. & NZ colloq. **a** a boaster. **b** boasting; boastfulness. **2** Sc. a drinking-bout; a spree (on the skite). [Sc. & N. Engl. dial., = a person regarded with contempt: cf. BLATHERSKITE]

skitter /'skɪtə(r)/ v.intr. (also **skeeter** /'ski:t-/) **1 a** (usu. foll. by along, across) move lightly or hastily. **b** (usu. foll. by about, off) hurry about, dart off. **2** fish by drawing bait jerkily across the surface of the water. [app. frequent. of dial. skite, perh. formed as SKIT[1]]

skittery /'skɪtəri/ adj. skittish, restless.

skittish /'skɪtɪʃ/ adj. **1** lively, playful. **2** (of a horse etc.) nervous, inclined to shy, fidgety. □ **skittishly** adv. **skittishness** n. [ME, perh. formed as SKIT[1]]

skittle /'skɪt(ə)l/ n. & v. ● n. **1** a pin used in the game of skittles. **2** (in pl.; usu. treated as sing.) **a** a game played with usu. nine wooden pins set up at the end of an alley to be bowled down usu. with wooden balls or a wooden disc (cf. NINEPIN 1). **b** (in full **table skittles**) a game played with similar pins set up on a board to be knocked down by swinging a suspended ball. **c** colloq. chess not played seriously. ● v.tr. (often foll. by out) Cricket get (batsmen) out in rapid succession. [17th c. (also kittle-pins): orig. unkn.]

skive /skaɪv/ v & n. ● v. **1** tr. split or pare (hides, leather, etc.). **2** intr. Brit. colloq. **a** evade a duty, shirk. **b** (often foll. by off) avoid work by absenting oneself, play truant. ● n. colloq. **1** an instance of shirking. **2** an easy option. □ **skiver** n. [ON skífa, rel. to ME schíve slice]

skivvy /'skɪvi/ n. & v. ● n. (pl. **-ies**) **1** colloq. derog. a female domestic servant; a menial or poorly paid worker. **2 a** US & Austral. a thin high-necked long-sleeved garment. **b** (in pl.) N. Amer. underwear of vest and underpants. ● v.intr. (**-ies**, **-ied**) colloq. work as a skivvy. [20th c.: orig. unkn.]

skol /skɒl, skəʊl/ n. (also **skoal**) used as a toast in drinking. [Danish skaal, Sw. skål, f. ON skál bowl]

Skopje /'skɒpjeɪ/ the capital of the republic of Macedonia, situated in the north on the Vardar river; pop. (1989) 563,000. An ancient city founded by the Romans, it was under Turkish control from the late 14th until the 20th centuries. It became the capital of Macedonia in 1945.

Skryabin see SCRIABIN.

skua /'skju:ə/ n. a brownish predatory seabird of the family Stercorariidae, allied to the gulls; esp. (N. Amer.) one of the larger kinds, which belong to the genus Catharacta (cf. JAEGER). They pursue other birds and make them disgorge the fish they have caught. [mod.L f. Faroese skúgvur, ON skúfr]

skulduggery /skʌl'dʌgəri/ n. (also **sculduggery**, **skullduggery**) trickery; unscrupulous behaviour. [earlier sculduddery, orig. Sc., = unchastity (18th c.: orig. unkn.)]

skulk /skʌlk/ v. & n. ● v.intr. **1** move stealthily, lurk, or keep oneself concealed, esp. in a cowardly or sinister way. **2** stay or sneak away in time of danger. **3** shirk duty. ● n. **1** a person who skulks. **2** a company of foxes. □ **skulker** n. [ME f. Scand.: cf. Norw. skulka lurk, Danish skulke, Sw. skolka shirk]

skull /skʌl/ n. **1** the bony case of the brain of a vertebrate. **2 a** the part of the skeleton corresponding to the head. **b** this with the skin and soft internal parts removed. **c** a representation of this. **3** the head as the seat of intelligence. □ **out of one's skull** sl. out of one's mind, crazy; very drunk. **skull and crossbones** a representation of a skull with two thigh-bones crossed below it as an emblem of piracy or death. **skull session** US sl. a discussion or conference. □ **skulled** adj. (also in comb.). [ME scolle: orig. unkn.]

skullcap /'skʌlkæp/ n. **1** a small close-fitting peakless cap. **2** the top part of the skull. **3** a labiate plant of the genus Scutellaria, having a helmet-shaped calyx after flowering.

skunk /skʌŋk/ n. & v. ● n. **1 a** a cat-sized flesh-eating American mammal of the weasel family; esp. the common Mephitis mephitis, with distinctive black and white striped fur and able to spray a foul-smelling liquid from its anal glands in defence. **b** its fur. **2** colloq. a thoroughly contemptible person. ● v.tr. N. Amer. sl. **1** defeat. **2** fail to pay (a bill etc.). □ **skunk-bear** US a wolverine. **skunk-cabbage** a herbaceous North American plant, Lysichiton americanum, of the arum family, with a foul-smelling spathe. [Algonquian segankw, segongw]

sky /skaɪ/ n. & v. ● n. (pl. **skies**) (in sing. or pl.) **1** the region of the atmosphere and outer space seen from the earth. **2** the weather or climate evidenced by this. ● v.tr. (**skies**, **skied**) **1** Cricket etc. hit (a ball) high into the air. **2** hang (a picture) high on a wall. □ **sky-blue** adj. & n. a bright clear blue. **sky-blue pink** an imaginary colour. **sky-clad** sl. naked (esp. in witchcraft). **sky cloth** Theatr. a backcloth painted or coloured to represent the sky. **sky-high** adv. & adj. as if reaching the sky, very high. **the sky is the limit** there is practically no limit. **sky pilot** sl. a clergyman. **sky-rocket** a rocket exploding high in the air. ● v.intr. (**-rocketed**, **-rocketing**) (esp. of prices etc.) rise very steeply or rapidly. **sky-shouting** the sending of messages from an aircraft to the ground by means of a loudspeaker. **sky-sign** an advertisement on the roof of a building. **sky wave** a radio wave reflected from the ionosphere. **sky-writing** a legible pattern of smoke-trails made by an aeroplane esp. for advertising. **to the skies** very highly; without reserve (praised to the skies). **under the open sky** out of doors. □ **skyey** adj. **skyless** adj. [ME ski(es) cloud(s) f. ON ský]

skydiving /'skaɪˌdaɪvɪŋ/ n. the sport of performing acrobatic manoeuvres under free fall with a parachute. □ **skydive** v.intr. **skydiver** n.

Skye /skaɪ/ a mountainous island of the Inner Hebrides, lying just off the west coast of Scotland; chief town, Portree. It is the largest island of the group.

skyer /'skaɪə(r)/ n. (also **skier**) Cricket a high hit.

Skye terrier n. a small long-haired slate-coloured or fawn variety of Scottish terrier.

skyjack /'skaɪdʒæk/ v. & n. sl. ● v.tr. hijack (an aircraft). ● n. an act of skyjacking. □ **skyjacker** n. [SKY + HIJACK]

Skylab /'skaɪlæb/ an American space laboratory launched into earth orbit in 1973. It was used to conduct experiments in conditions of zero gravity as well as for astrophysical studies. Skylab was manned until 1974 and disintegrated in the atmosphere in 1979.

skylark /'skaɪlɑ:k/ n. & v. ● n. a lark, Alauda arvensis of Eurasia and North Africa, that sings while hovering in flight. ● v.intr. play tricks or practical jokes, indulge in horseplay, frolic. [SKY + LARK[1]: (v.) with pun on LARK[2]]

skylight /'skaɪlaɪt/ n. a window set in the plane of a roof or ceiling.

skyline /'skaɪlaɪn/ n. the outline of hills, buildings, etc., defined against the sky; the visible horizon.

skysail /'skaɪseɪl, -s(ə)l/ n. Naut. a light sail above the royal in a square-rigged ship.

skyscape /ˈskaɪskeɪp/ n. **1** a picture chiefly representing the sky. **2** a view of the sky.

skyscraper /ˈskaɪˌskreɪpə(r)/ n. a very tall building with many storeys, especially the type of office building that dominates Manhattan in New York and the centres of other large American cities. First built because of the high cost of land in congested urban areas but subsequently also for their prestige value, skyscrapers were made technically possible by the development of steel-frame construction and the invention of the electric lift. Leroy S. Buffington, a Minneapolis architect, designed the first skyscraper in 1880, but the first one actually erected was the Home Insurance Building in Chicago, a ten-storey structure completed in 1885.

skyward /ˈskaɪwəd/ adv. & adj. ● adv. (also **skywards** /-wədz/) towards the sky. ● adj. moving skyward.

skywatch /ˈskaɪwɒtʃ/ n. the activity of watching the sky for aircraft etc.

skyway /ˈskaɪweɪ/ n. **1** a route used by aircraft. **2** the sky as a medium of transport.

slab /slæb/ n. & v. ● n. **1** a flat broad fairly thick usu. square or rectangular piece of solid material, esp. stone. **2** a large flat piece of cake, chocolate, etc. **3** (of timber) an outer piece sawn from a log. **4** Brit. a mortuary table. ● v.tr. (**slabbed**, **slabbing**) remove slabs from (a log or tree) to prepare it for sawing into planks. [ME: orig. unkn.]

slack¹ /slæk/ adj., n., v., & adv. ● adj. **1** (of rope etc.) not taut. **2** inactive or sluggish. **3** negligent or remiss. **4** (of tide etc.) neither ebbing nor flowing. **5** (of trade or business or a market) with little happening, quiet. **6** loose. **7** Phonet. lax. **8** relaxed, languid. ● n. **1** the slack part of a rope (haul in the slack). **2** a slack time in trade etc. **3** colloq. a spell of inactivity or laziness. **4** (in pl.) full-length loosely cut trousers for informal wear. ● v. **1** a tr. & intr. slacken. **b** tr. loosen (rope etc.). **2** intr. take a rest, be lazy, shirk. **3** tr. slake (lime). ● adv. **1** slackly. **2** slowly or insufficiently (dry slack; bake slack). □ **slack hand** lack of full control in riding or governing. **slack lime** slaked lime. **slack off 1** loosen. **2** lose or cause to lose vigour. **slack rein** = slack hand. **slack suit** US casual clothes of slacks and a jacket or shirt. **slack up** reduce the speed of a train etc. before stopping. **slack water** a time near the turn of the tide, esp. at low tide. **take up the slack** use up a surplus or make up a deficiency; avoid an undesirable lull. □ **slackly** adv. **slackness** n. [OE slæc f. Gmc]

slack² /slæk/ n. coal-dust or small pieces of coal. [ME prob. f. LG or Du.]

slacken /ˈslækən/ v.tr. & intr. make or become slack. □ **slacken off** = slack off (see SLACK¹).

slacker /ˈslækə(r)/ n. a shirker; an indolent person.

slag /slæg/ n. & v. ● n. **1** vitreous refuse left after ore has been smelted, dross separated in a fused state in the reduction of ore, clinkers. **2** volcanic scoria. **3** sl. derog. **a** a prostitute or promiscuous woman. **b** a worthless or insignificant person. ● v. (**slagged**, **slagging**) **1** intr. **a** form slag. **b** cohere into a mass like slag. **2** tr. (often foll. by off) sl. criticize, insult. □ **slag-heap** a hill of refuse from a mine etc. **slag-wool** = mineral wool. □ **slaggy** adj. (**slaggier**, **slaggiest**). [MLG slagge, perh. f. slagen strike, with ref. to fragments formed by hammering]

slain past part. of SLAY¹.

slainte /ˈslɑːndʒə/ int. a Gaelic toast: good health! [Gael. sláinte, lit. 'health']

slake /sleɪk/ v.tr. **1** assuage or satisfy (thirst, revenge, etc.). **2** disintegrate (quicklime) by chemical combination with water. □ **slaked lime** = LIME¹ n. 2. [OE slacian f. slæc SLACK¹]

slalom /ˈslɑːləm/ n. **1** a ski-race down a zigzag course defined by artificial obstacles. **2** an obstacle race in canoes or cars or on skateboards or water-skis. [Norw., lit. 'sloping track']

slam¹ /slæm/ v. & n. ● v. (**slammed**, **slamming**) **1** tr. & intr. shut forcefully and loudly. **2** tr. put down (an object) with a similar sound. **3** intr. move violently (he slammed out of the room). **4** tr. & intr. put or come into sudden action (slam the brakes on). **5** tr. sl. criticize severely. **6** tr. sl. hit. **7** tr. sl. gain an easy victory over. ● n. **1** a sound of or as of a slammed door. **2** the shutting of a door etc. with a loud bang. **3** (usu. prec. by the) N. Amer. sl. prison. □ **slam-dancing** a type of violent uncoordinated dancing to rock music in which the participants deliberately clash with one another. [prob. f. Scand.: cf. ON slam(b)ra]

slam² /slæm/ n. Cards the winning of every trick in a game. (See also GRAND SLAM.) □ **small** (or **little**) **slam** Bridge the winning of twelve tricks. [orig. name of a card-game: perh. f. obs. slampant trickery]

slambang /slæmˈbæŋ/ adv. & adj. ● adv. with the sound of a slam. ● adj. colloq. impressive, exciting, or energetic.

slammer /ˈslæmə(r)/ n. (usu. prec. by the) sl. prison.

slander /ˈslɑːndə(r)/ n. & v. ● n. **1** a malicious, false, and injurious statement spoken about a person. **2** the uttering of such statements; calumny. **3** Law false oral defamation (cf. LIBEL). ● v.tr. utter slander about; defame falsely. □ **slanderer** n. **slanderous** adj. **slanderously** adv. [ME sclaundre f. AF esclaundre, OF esclandre alt. f. escandle f. LL scandalum: see SCANDAL]

slang /slæŋ/ n. & v. ● n. words, phrases, and uses that are regarded as very informal and are often restricted to special contexts or are peculiar to a specified profession, class, etc. (racing slang; schoolboy slang). (See note below.) ● v. **1** tr. use abusive language to. **2** intr. use such language. □ **slanging-match** a prolonged exchange of insults. [18th-c. cant: orig. unkn.]

▪ Slang marks group identity and is often used for secrecy and the exclusion of outsiders; however, it may also be used to shock or attract attention, to be concise, picturesque, or humorous, or to express feelings and attitudes (e.g. of hostility, ridicule, or affection) better than formal words would do. Some words originally regarded as slang (e.g. clever, fun, frisky, mob, which were disliked by Dr Johnson and others in the 18th century) have now passed into standard non-colloquial usage; others (e.g. quid = £1) have remained in the category of slang.

slangy /ˈslæŋɪ/ adj. (**slangier**, **slangiest**) **1** of the character of slang. **2** fond of using slang. □ **slangily** adv. **slanginess** n.

slant /slɑːnt/ v., n., & adj. ● v. **1** intr. slope; diverge from a line; lie or go obliquely to a vertical or horizontal line. **2** tr. cause to do this. **3** tr. (often as **slanted** adj.) present (information) from a particular angle, esp. in a biased or unfair way. ● n. **1** a slope; an oblique position. **2** a way of regarding a thing; a point of view, esp. a biased one. ● adj. sloping, oblique. □ **on a** (or **the**) **slant** aslant. **slant-eyed** having slanting eyes. **slant height** the height of a cone from the vertex to the periphery of the base. [aphetic form of ASLANT: (v.) rel. to ME slent f. ON sletta dash, throw]

slantwise /ˈslɑːntwaɪz/ adv. aslant.

slap /slæp/ v., n., & adv. ● v. (**slapped**, **slapping**) **1** tr. & intr. strike with the palm of the hand or a flat object, or so as to make a similar noise. **2** tr. lay forcefully (slapped the money on the table; slapped a writ on the offender). **3** tr. put hastily or carelessly (slap some paint on the walls). **4** tr. (often foll. by down) colloq. reprimand or snub. ● n. **1** a blow with the palm of the hand or a flat object. **2** a slapping sound. ● adv. **1** with the suddenness or effectiveness or true aim of a blow; suddenly, fully, directly (ran slap into him; hit me slap in the eye). **2** = slap-bang. □ **slap and tickle** Brit. colloq. light-hearted amorous amusement. **slap-bang 1** violently, noisily, headlong. **2** conspicuously, prominently. **slap-happy** colloq. **1** cheerfully casual or flippant. **2** punch-drunk. **slap in the face** a rebuff or affront. **slap on the back** n. congratulations. ● v.tr. congratulate. **slap on the wrist** a mild reprimand or rebuke. ● v. tr. colloq. reprimand. **slap-up** esp. Brit. colloq. excellent, lavish; done regardless of expense (slap-up meal). [LG slapp (imit.)]

slapdash /ˈslæpdæʃ/ adj. & adv. ● adj. hasty and careless. ● adv. in a slapdash manner.

slapjack /ˈslæpdʒæk/ n. N. Amer. a kind of pancake cooked on a griddle (cf. FLAPJACK 2). [SLAP + JACK¹]

slapstick /ˈslæpstɪk/ n. **1** boisterous knockabout comedy. **2** a flexible divided lath used by a clown. [SLAP + STICK¹]

slash /slæʃ/ v. & n. ● v. **1** intr. make a sweeping or random cut or cuts with a knife, sword, whip, etc. **2** tr. make such a cut or cuts at. **3** tr. make a long narrow gash or gashes in. **4** tr. reduce (prices etc.) drastically. **5** tr. censure vigorously. **6** tr. make (one's way) by slashing. **7** tr. **a** lash (a person etc.) with a whip. **b** crack (a whip). ● n. **1 a** a slashing cut or stroke. **b** a wound or slit made by this. **2** an oblique stroke; a solidus. **3** Brit. sl. an act of urinating. **4** N. Amer. debris resulting from the felling or destruction of trees. □ **slash-and-burn 1** (of cultivation) in which vegetation is cut down, allowed to dry, and then burned off before seeds are planted. **2** (of a person, action, etc.) ruthless, destructive. □ **slasher** n. [ME perh. f. OF esclachier break in pieces]

slashed /slæʃt/ adj. (of a sleeve etc.) having slits to show a lining or puffing of other material.

slashing /ˈslæʃɪŋ/ adj. vigorously incisive or effective.

slat /slæt/ n. a thin narrow piece of wood or plastic or metal, esp. used in an overlapping series as in a fence or venetian blind. [ME s(c)lat f. OF esclat splinter etc. f. esclater split f. Rmc]

slate /sleɪt/ n., v., & adj. ● n. **1** a fine-grained grey, green, or bluish-purple metamorphic rock easily split into flat smooth plates. **2** a piece of such a plate used as roofing-material. **3** a piece of such a plate used for writing on, usu. framed in wood. **4** a bluish-grey colour. **5** N. Amer. a list of nominees for office etc. **6** a record, esp. of a person's debts or credits. ● v.tr. **1** cover with slates esp. as roofing. **2** Brit. colloq. criticize severely; scold. **3** N. Amer. make arrangements for (an event etc.). **4** N. Amer. propose or nominate for office etc. ● adj. made of slate. □ **on the slate** Brit. recorded as a debt to be paid. **slate-blue** (or **-black**) a shade of blue (or black) occurring in slate. **slate-colour** a dark bluish or greenish-grey. **slate-coloured** of this colour. **slate-grey** a shade of grey occurring in slate. **slate-pencil** a small rod of soft slate used for writing on slate. **wipe the slate clean** forgive or cancel the record of past offences. □ **slating** n. **slaty** adj. [ME s(c)late f. OF esclate, fem. form of esclat SLAT]

slater /ˈsleɪtə(r)/ n. **1** a person who slates roofs etc. **2** a woodlouse or similar crustacean.

slather /ˈslæðə(r)/ n. & v. ● n. **1** (usu. in pl.) US colloq. a large amount. **2** (often **open slather**) Austral. & NZ sl. unrestricted scope for action. ● v.tr. N. Amer. colloq. **1** spread thickly. **2** squander. [19th c.: orig. unkn.]

slatted /ˈslætɪd/ adj. having slats.

slattern /ˈslæt(ə)n/ n. a slovenly woman. □ **slatternly** adj. **slatternliness** n. [17th c.: rel. to slattering slovenly, f. dial. slatter to spill, slop, waste, frequent. of slat strike]

slaughter /ˈslɔːtə(r)/ n. & v. ● n. **1** the killing of an animal or animals. **2** the killing of many persons or animals at once or continuously; carnage, massacre. ● v.tr. **1** kill (people) in a ruthless manner or on a great scale. **2** kill (animals) esp. in large numbers. **3** colloq. defeat utterly. □ **slaughterer** n. **slaughterous** adj. [ME slahter ult. f. ON slátr butcher's meat, rel. to SLAY[1]]

slaughterhouse /ˈslɔːtəˌhaʊs/ n. **1** a place for the slaughter of animals as food. **2** a place of carnage.

Slav /slɑːv/ n. & adj. ● n. a member of a group of peoples in central and eastern Europe (including the Russians, Poles, Czechs, Bulgarians, Serbs, Croats, and others) speaking Slavonic languages. ● adj. **1** of or relating to the Slavs. **2** Slavonic. □ **Slavism** n. [ME Sclave f. med.L Sclavus, late Gk Sklabos, & f. med.L Slavus]

slave /sleɪv/ n. & v. ● n. **1** a person who is the legal property of another or others and is bound to absolute obedience, a human chattel. **2** a drudge, a person working very hard. **3** (foll. by of, to) a helpless victim of some dominating influence (slave of fashion; slave to duty). **4** a machine, or part of one, directly controlled by another (cf. MASTER n. 13). ● v. **1** intr. (often foll. by at, over) work very hard. **2** tr. (foll. by to) subject (a device) to control by another. □ **slave-bangle** a bangle of gold, glass, etc., worn by a woman usu. above the elbow. **slave-born** born in slavery, born of slave parents. **slave-bracelet** = slave-bangle. **slave-drive** (past **-drove**; past part. **-driven**) work (a person) hard, esp. excessively so. **slave-driver 1** an overseer of slaves at work. **2** a person who works others hard. **slave labour** forced labour. **slave ship** hist. a ship transporting slaves, esp. from Africa. **slave-trade** hist. the procuring, transporting, and selling of human beings, esp. African blacks, as slaves. **slave-trader** hist. a person engaged in the slave-trade. [ME f. OF esclave = med.L sclavus, sclava Slav (captive): see SLAV]

slaver[1] /ˈsleɪvə(r)/ n. hist. a ship or person engaged in the slave-trade.

slaver[2] /ˈslævə(r)/ n. & v. ● n. **1** saliva running from the mouth. **2 a** fulsome or servile flattery. **b** drivel, nonsense. ● v.intr. **1** let saliva run from the mouth, dribble. **2** (foll. by over) show excessive sentimentality over, or desire for. [ME prob. f. LG or Du.: cf. SLOBBER]

slavery /ˈsleɪvərɪ/ n. **1** the condition of a slave. **2** exhausting labour; drudgery. **3** the custom of having slaves. (See note below.)

▪ Slavery was a widespread institution in the classical world, on which the economy was dependent. The island of Delos served as the main slave market in the Aegean; in both the Greek and Roman periods the main source of slaves was the enslavement of prisoners of war. Major slave revolts occurred during the 2nd and 1st centuries BC, but the practice did not begin to diminish until the spread of Christianity and the decline of the Roman Empire in the 3rd to 5th centuries AD.

▪ The transportation of slaves from Africa to the Americas by European traders began on a large scale in the 16th and 17th centuries, and although slavery became illegal in Britain in 1772, it remained an important feature of the economy of the British Empire until the 19th century; the slave-trade was abolished in 1807 and slavery itself throughout the empire in 1833. In the American South slavery was an integral part of the cotton-based economy, and the abolition

campaign waged during the first half of the 19th century eventually led to the American Civil War and to final emancipation. In some parts of the world chattel slavery, the ownership of one person by another, continues to exist.

Slave State n. hist. any of the southern states of the US in which slavery was legal before the Civil War.

slavey /ˈsleɪvɪ/ n. (pl. **-eys**) colloq. a maidservant, esp. a hard-worked one.

Slavic /ˈslɑːvɪk/ adj. & n. = SLAVONIC.

slavish /ˈsleɪvɪʃ/ adj. **1** of, like, or as of slaves. **2** showing no attempt at originality or development. **3** abject, servile, base. □ **slavishly** adv. **slavishness** n.

Slavonic /sləˈvɒnɪk/ adj. & n. (also **Slavic** /ˈslɑːvɪk/) ● adj. **1** of or relating to the branch of Indo-European languages including Russian, Polish, Czech, Bulgarian, and Serbo-Croat. (See note below.) **2** of or relating to the Slavs. ● n. the Slavonic branch of languages. [med.L S(c)lavonicus f. S(c)lavonia country of Slavs f. Sclavus SLAV]

▪ The Slavonic branch constitutes a major division of the Indo-European family. It is made up of three groups: South Slavonic, which includes Serbo-Croat, Slovene, Macedonian, and Bulgarian; West Slavonic, which includes Czech, Slovak, and Polish; and East Slavonic, which includes Russian, Ukrainian, and Belorussian. Important common characteristics of Slavonic languages are that nouns and adjectives are highly inflected (Russian and Polish have as many as seven cases), and that verbs have few tenses but preserve aspect, a distinction between actions thought of as finished or limited in time and those regarded as continuous. The two principal alphabets used are Cyrillic and Roman. The earliest written Slavonic language is Church Slavonic, the language of the Orthodox Church, dating from the 9th century AD. (See also CHURCH SLAVONIC.)

slaw /slɔː/ n. coleslaw. [Du. sla, shortened f. salade SALAD]

slay[1] /sleɪ/ v.tr. (past **slew** /sluː/; past part. **slain** /sleɪn/) **1** literary, joc., or N. Amer. kill. **2** colloq. overwhelm with delight; convulse with laughter. □ **slayer** n. [OE slēan f. Gmc]

slay[2] var. of SLEY.

SLBM abbr. submarine-launched ballistic missile.

SLD see SOCIAL AND LIBERAL DEMOCRATS.

sleaze /sliːz/ n. & v. ● n. **1** colloq. sleaziness. **2** sl. a person of low moral standards, a sleazy person. ● v.intr. sl. move in a sleazy fashion. [back-form. f. SLEAZY]

sleazoid /ˈsliːzɔɪd/ adj. & n. esp. US sl. ● adj. sordid, disgusting; morally low; sleazy. ● n. a morally low or sleazy person. [SLEAZE + -OID]

sleazy /ˈsliːzɪ/ adj. (**sleazier**, **sleaziest**) **1** sordid, disreputable, squalid, tawdry. **2** slatternly. **3** (of textiles etc.) flimsy. □ **sleazily** adv. **sleaziness** n. [17th c.: orig. unkn.]

sled /sled/ n. & v. N. Amer. ● n. a sledge. ● v.intr. (**sledded**, **sledding**) ride on a sledge. [MLG sledde, rel. to SLIDE]

sledge[1] /sledʒ/ n. & v. ● n. **1** a vehicle on runners for conveying loads or passengers esp. over snow, drawn by horses, dogs, or reindeer or pushed or pulled by one or more persons. **2** a toboggan. ● v.intr. & tr. travel or convey by sledge. [MDu. sledse, rel. to SLED]

sledge[2] /sledʒ/ n. = SLEDGEHAMMER 1.

sledgehammer /ˈsledʒˌhæmə(r)/ n. **1** a large heavy hammer used to break stone etc. **2** (attrib.) heavy or powerful (a sledgehammer blow). [OE slecg, rel. to SLAY[1]]

sleek /sliːk/ adj. & v. ● adj. **1** (of hair, fur, or skin, or an animal or person with such hair etc.) smooth and glossy. **2** looking well-fed and comfortable. **3** ingratiating. **4** (of a thing) smooth and polished. ● v.tr. make sleek, esp. by stroking or pressing down. □ **sleekly** adv. **sleekness** n. **sleeky** adj. [later var. of SLICK]

sleep /sliːp/ n. & v. ● n. **1** a condition of body and mind such as that which normally recurs for several hours every night, in which the nervous system is inactive, the eyes closed, the postural muscles relaxed, and consciousness practically suspended. (See note below.) **2** a period of sleep (shall try to get a sleep). **3** a state like sleep, such as rest, quiet, negligence, or death. **4** the prolonged inert condition of hibernating animals. **5** a substance found in the corners of the eyes after sleep. ● v. (past and past part. **slept** /slept/) **1** intr. **a** be in a state of sleep. **b** fall asleep. **2** intr. (foll. by at, in, etc.) spend the night. **3** tr. provide sleeping accommodation for (the house sleeps six). **4** intr. (foll. by with, together) have sexual intercourse, esp. in bed. **5** intr. (foll. by on, over) not decide (a question) until the next day. **6** intr. (foll. by through) fail to be woken by. **7** intr. be inactive or dormant. **8** intr. be dead; lie in the grave.

9 *tr.* **a** (foll. by *off*) remedy by sleeping (*slept off his hangover*). **b** (foll. by *away*) spend in sleeping (*sleep the hours away*). **10** *intr.* (of a top) spin so steadily as to seem motionless. □ **get to sleep** manage to fall asleep. **go to sleep 1** enter a state of sleep. **2** (of a limb) become numbed by pressure. **in one's sleep** while asleep. **last sleep** death. **let sleeping dogs lie** avoid stirring up trouble. **put to sleep 1** anaesthetize. **2** kill (an animal) painlessly. **sleep around** be sexually promiscuous. **sleep in 1** remain asleep later than usual in the morning. **2** sleep by night at one's place of work. **sleeping-bag** a lined or padded bag to sleep in esp. when camping etc. **sleeping-car** (or **-carriage**) a railway coach provided with beds or berths. **sleeping-draught** a drink to induce sleep. **sleeping partner** a partner not sharing in the actual work of a firm. **sleeping-pill** a pill to induce sleep. **sleeping policeman** *Brit.* a ramp etc. in the road intended to cause traffic to reduce speed. **sleeping-suit** a child's one-piece night-garment. **sleep-learning** learning by hearing while asleep. **sleep like a log** (or **top**) sleep soundly. **the sleep of the just** sound sleep. **sleep out** sleep by night out of doors, or not at one's place of work. **sleep-out** *Austral. & NZ* a veranda, porch, or outbuilding providing sleeping accommodation. **sleep over** esp. *N. Amer.* spend the night at a place other than one's own residence. **sleep-over** esp. *N. Amer.* an instance of sleeping over, esp. by a group of teenage friends at another's house. [OE *slēp, slæp* (n.), *slēpan, slǣpan* (v.) f. Gmc]

▪ The capacity for sleep is general among animals, and even invertebrates usually have a period of torpor at some stage in their daily cycle of activity. The pattern of sleep in mammals tends to be related to their feeding habits and dominant senses, but some (e.g. lions, horses, and sheep) show no clear-cut rhythm and may sleep at any time. The physiology of sleep has been studied by analysing the electrical activity of the human brain: the relatively fast electrical rhythm of the waking brain is replaced during sleep by slow waves, with only occasional bursts of faster activity. This activity changes during dreaming, which is associated with rapid movements of the eyes. There is no general agreement about the function of sleep.

sleeper /ˈsliːpə(r)/ *n.* **1** a person or animal that sleeps. **2** *Brit.* a wooden or concrete beam laid horizontally as a support, esp. for railway track. **3 a** a sleeping-car. **b** a berth in this. **4** *Brit.* a ring worn in a pierced ear to keep the hole from closing. **5** a thing that is suddenly successful after being undistinguished. **6** a sleeping-suit. **7** a spy or saboteur etc. who remains inactive while establishing a secure position.

Sleeping Beauty a fairy-tale heroine who slept for 100 years until woken by the kiss of a prince. The story of Sleeping Beauty, as told by Charles Perrault, formed the subject of a ballet by Tchaikovsky.

sleeping sickness *n. Med.* a disease caused by trypanosomes, transmitted by the bite of the tsetse fly and characterized by changes in the central nervous system leading to apathy, coma, and death (cf. TRYPANOSOMIASIS). It is especially prevalent in tropical Africa. The fly favours the dense shade cast by thickly growing shrubs and small trees, as on the banks of watercourses and water-holes: one of the principal control methods is to eliminate such vegetation, especially around villages, fords, and other places where people congregate.

sleepless /ˈsliːplɪs/ *adj.* **1** characterized by lack of sleep (*a sleepless night*). **2** unable to sleep. **3** continually active or moving. □ **sleeplessly** *adv.* **sleeplessness** *n.*

sleepwalk /ˈsliːpwɔːk/ *v.intr.* walk or perform other actions while asleep. □ **sleepwalker** *n.* **sleepwalking** *n.*

sleepy /ˈsliːpɪ/ *adj.* (**sleepier, sleepiest**) **1** drowsy; ready for sleep; about to fall asleep. **2** lacking activity or bustle (*a sleepy little town*). **3** habitually indolent, unobservant, etc. □ **sleepy sickness** encephalitis lethargica, an infection of the brain with drowsiness and sometimes a coma. □ **sleepily** *adv.* **sleepiness** *n.*

sleepyhead /ˈsliːpɪˌhed/ *n.* (esp. as a form of address) a sleepy or inattentive person.

sleet /sliːt/ *n. & v.* ● *n.* **1** a mixture of snow and rain falling together. **2** hail or snow melting as it falls. **3** *US* a thin coating of ice. ● *v.intr.* (prec. by *it* as subject) sleet falls (*it is sleeting; if it sleets*). □ **sleety** *adj.* [ME prob. f. OE: rel. to MLG *slōten* (pl.) hail, MHG *slōz(e)* f. Gmc]

sleeve /sliːv/ *n.* **1** the part of a garment that wholly or partly covers an arm. **2** the cover of a gramophone record. **3** a tube enclosing a rod or smaller tube. **4 a** a wind-sock. **b** a drogue towed by an aircraft. □ **roll up one's sleeves** prepare to fight or work. **sleeve-board** a small ironing-board for pressing sleeves. **sleeve-coupling** a tube for connecting shafts or pipes. **sleeve-link** a cuff-link. **sleeve-note** a descriptive note on a record-sleeve. **sleeve-nut** a long nut with right-

hand and left-hand screw-threads for drawing together pipes or shafts conversely threaded. **sleeve-valve** a valve in the form of a cylinder with a sliding movement. **up one's sleeve** concealed but ready for use, in reserve. □ **sleeved** *adj.* (also in *comb.*). **sleeveless** *adj.* [OE *slēfe, slīefe, slȳf*]

sleeving /ˈsliːvɪŋ/ *n.* tubular covering for electric cable etc.

sleigh /sleɪ/ *n. & v.* ● *n.* a sledge, esp. one for riding on. ● *v.intr.* travel on a sleigh. □ **sleigh-bell** any of a number of tinkling bells attached to the harness of a sleigh-horse etc. [orig. US, f. Du. *slee*, rel. to SLED]

sleight /slaɪt/ *n. archaic* **1** a deceptive trick or device or movement. **2** dexterity. **3** cunning. □ **sleight of hand 1** dexterity esp. in conjuring or fencing. **2** a display of dexterity, esp. a conjuring trick. [ME *slegth* f. ON *slœgth* f. *slœgr* SLY]

slender /ˈslendə(r)/ *adj.* (**slenderer, slenderest**) **1 a** of small girth or breadth (*a slender pillar*). **b** gracefully thin (*a slender waist*). **2** relatively small or scanty; slight, meagre, inadequate (*slender hopes; slender resources*). □ **slender loris** see LORIS. □ **slenderly** *adv.* **slenderness** *n.* [ME: orig. unkn.]

slenderize /ˈslendəˌraɪz/ *v.* (also **-ise**) **1** *tr.* **a** make (a thing) slender. **b** make (one's figure) appear slender. **2** *intr.* make oneself slender; slim.

slept *past* and *past part.* of SLEEP.

sleuth /sluːθ/ *n. & v.* ● *n.* a detective. ● *v.* **1** *intr.* act as a detective. **2** *tr.* investigate. □ **sleuth-hound 1** a bloodhound. **2** *colloq.* a detective, an investigator. [orig. in *sleuth-hound*: ME f. *sleuth* f. ON *slóth* track, trail: cf. SLOT[2]]

slew[1] /sluː/ *v. & n.* (also **slue**) ● *v.tr. & intr.* (often foll. by *round*) turn or swing forcibly or with effort out of the forward or ordinary position. ● *n.* such a change of position. [18th-c. Naut.: orig. unkn.]

slew[2] /sluː/ *n.* esp. *N. Amer. colloq.* a large number or quantity. [Ir. *sluagh*]

slew[3] *past* of SLAY[1].

sley /sleɪ/ *n.* (also **slay**) a weaver's reed. [OE *slege*, rel. to SLAY[1]]

slice /slaɪs/ *n. & v.* ● *n.* **1** a thin broad piece or wedge cut off or out esp. from meat or bread or a cake, pie, or large fruit. **2** a share; a part taken or allotted or gained (*a slice of territory; a slice of the profits*). **3** a kitchen utensil with a broad flat often perforated blade for serving fish, cake, etc. **4** *Golf & Tennis* a slicing stroke. ● *v.* **1** *tr.* (foll. by *up*) cut into slices. **2** *tr.* (foll. by *off*) cut (a piece) off. **3** *intr.* (foll. by *into, through*) cut with or like a knife. **4** *tr.* (also *absol.*) *Sport* strike (the ball) so that it deviates away from the striker; propel (the ball) forward at an angle. **5** *tr.* go through (air etc.) with a cutting motion. □ **slice of life** a realistic representation of everyday experience. □ **sliceable** *adj.* **slicer** *n.* (also in *comb.*). [ME f. OF *esclice, esclicier* splinter f. Frank. *slītjan*, rel. to SLIT]

slick /slɪk/ *adj., n., & v.* ● *adj.* **1 a** (of a person or action) skilful or efficient; dexterous (*gave a slick performance*). **b** superficially or pretentiously smooth and dexterous. **2 a** sleek, smooth. **b** slippery. ● *n.* **1 a** a smooth patch of oil etc., esp. on the sea. **2** *Motor-Racing* a smooth tyre. **3** *US* a glossy magazine. **4** *US sl.* a slick person. ● *v.tr.* **1** make sleek or smart. **2** (usu. foll. by *down*) flatten (one's hair etc.). □ **slickly** *adv.* **slickness** *n.* [ME *slīke(n)*, prob. f. OE: cf. SLEEK]

slicker /ˈslɪkə(r)/ *n. N. Amer.* **1** *colloq.* **a** a plausible rogue. **b** a smart and sophisticated city-dweller (cf. *city slicker*). **2** a raincoat of smooth material.

slide /slaɪd/ *v. & n.* ● *v.* (*past* and *past part.* **slid** /slɪd/) **1 a** *intr.* move along a smooth surface with continuous contact on the same part of the thing moving (cf. ROLL). **b** *tr.* cause to do this (*slide the drawer into place*). **2** *intr.* move quietly; glide; go smoothly along. **3** *intr.* pass gradually or imperceptibly. **4** *intr.* glide over ice on one or both feet without skates (under gravity or with momentum got by running). **5** *intr.* (foll. by *over*) barely touch upon (a delicate subject etc.). **6** *intr. & tr.* (often foll. by *into*) move or cause to move quietly or unobtrusively (*slid his hand into mine*). **7** *intr.* take its own course (*let it slide*). **8** *intr.* decline (*shares slid to a new low*). ● *n.* **1 a** the act or an instance of sliding. **b** a rapid decline. **2** an inclined plane down which children, goods, etc., slide; a chute. **3 a** a track made by or for sliding, esp. on ice. **b** a slope prepared with snow or ice for tobogganing. **4** a part of a machine or instrument that slides, esp. a slide-valve. **5 a** a thing slid into place, esp. a piece of glass holding an object for a microscope. **b** a mounted transparency usu. placed in a projector for viewing on a screen. **6** *Brit.* = hair-slide. **7** a part or parts of a machine on or between which a sliding part works. □ **let things slide** be negligent; allow deterioration. **slide fastener** *US* a zip-fastener. **slide-rule** a ruler with a sliding central strip, graduated logarithmically for making rapid calculations, esp. multiplication and division. **slide-valve** a sliding piece that opens and closes an aperture by sliding across it. **sliding door** a door drawn across an aperture on a slide, not turning on hinges. **sliding keel** *Naut.* a centreboard. **sliding roof** a part of a roof (esp. in a motor car) made able to slide and so form an aperture. **sliding scale** a scale of fees, taxes, wages, etc., that varies as a whole in accordance with variation of some standard. **sliding seat** a seat able to slide to and fro on runners etc., esp. in a racing-boat to adjust the length of a stroke. □ **slidable** *adj.* **slidably** *adv.* **slider** *n.* [OE *slīdan*]

slideway /'slaidwei/ *n.* = SLIDE *n.* 7.

slight /slait/ *adj., v., & n.* ● *adj.* **1 a** inconsiderable; of little significance (*has a slight cold; the damage is very slight*). **b** barely perceptible (*a slight smell of gas*). **c** not much or great or thorough, inadequate, scanty (*a conclusion based on very slight observation; paid him slight attention*). **2** slender, frail-looking (*saw a slight figure approaching; supported by a slight framework*). **3** (in *superl.*, with *neg.* or *interrog.*) any whatever (*paid not the slightest attention*). ● *v.tr.* **1** treat or speak of (a person etc.) as not worth attention, fail in courtesy or respect towards, markedly neglect. **2** *hist.* make militarily useless, raze (a fortification etc.). ● *n.* a marked piece of neglect, a failure to show due respect. □ **not in the slightest** not at all. **put a slight upon** = sense 1 of *v.* □ **slightingly** *adv.* **slightish** *adj.* **slightly** *adv.* **slightness** *n.* [ME *slyght, sleght* f. ON *slēttr* level, smooth f. Gmc]

Sligo /'slaigəʊ/ **1** a county of the Republic of Ireland, in the west in the province of Connacht. **2** its county town, a seaport on Sligo Bay, an inlet of the Atlantic; pop. (1991) 17,300.

slily var. of SLYLY (see SLY).

slim /slim/ *adj., v., & n.* ● *adj.* (**slimmer, slimmest**) **1 a** of small girth or thickness, of long narrow shape. **b** gracefully thin, slenderly built. **c** not fat or overweight. **2** small, insufficient (*a slim chance of success*). **3** clever, artful, crafty, unscrupulous. **4** reduced to an economical or efficient size, level, etc. ● *v.* (**slimmed, slimming**) **1** *intr.* make oneself slimmer by dieting, exercise, etc. **2** *tr.* make slim or slimmer. ● *n.* **1 a** course of slimming. **2** (also **Slim**) (in full **Slim disease**) (esp. in African use) Aids. □ **slimly** *adv.* **slimmer** *n.* **slimming** *n. & adj.* **slimmish** *adj.* **slimness** *n.* [LG or Du. f. Gmc]

slime /slaim/ *n. & v.* ● *n.* thick slippery mud or a substance of similar consistency, e.g. liquid bitumen or a mucus exuded by fish etc. ● *v.tr.* cover with slime. □ **slime mould** a spore-bearing micro-organism secreting slime, esp. a myxomycete. [OE *slīm* f. Gmc, rel. to L *limus* mud, Gk *limnē* marsh]

slimline /'slimlain/ *adj.* **1** of slender design. **2** (of a drink etc.) not fattening.

slimy /'slaimi/ *adj.* (**slimier, slimiest**) **1** of the consistency of slime. **2** covered, smeared with, or full of slime. **3** *colloq.* disgustingly dishonest, meek, or flattering. **4** slippery, hard to hold. □ **slimily** *adv.* **sliminess** *n.*

sling¹ /slɪŋ/ *n. & v.* ● *n.* **1** a strap, belt, etc., used to support or raise a hanging weight, e.g. a rifle, a ship's boat, or goods being transferred. **2** a bandage looped round the neck to support an injured arm. **3** a strap or string used with the hand to give impetus to a small missile, esp. a stone. **4** a pouch or frame supported by a strap round the neck or shoulders for carrying a young child. **5** *Austral. sl.* a tip or bribe. ● *v.tr.* (*past and past part.* **slung** /slʌŋ/) **1** (also *absol.*) hurl (a stone etc.) from a sling. **2** *colloq.* throw. **3** suspend with a sling, allow to swing suspended, arrange so as to be supported from above, hoist or transfer with a sling. □ **sling-back 1** a shoe held in place by a strap above the heel. **2** (in full **sling-back chair**) a chair with a fabric seat suspended from a rigid frame. **sling-bag** a bag with a long strap which may be hung from the shoulder. **sling one's hook** see HOOK. **sling off at** *Austral. & NZ sl.* disparage; mock; make fun of. **slung shot** a metal ball attached by a thong etc. to the wrist and used esp. by criminals as a weapon. [ME, prob. f. ON *slyngva* (v.)]

sling² /slɪŋ/ *n.* a sweetened drink of spirits (esp. gin) and water. [18th c.: orig. unkn.]

slinger /'slɪŋə(r)/ *n.* a person who slings, esp. the user of a sling.

slingshot /'slɪŋʃɒt/ *n.* US a catapult.

slink¹ /slɪŋk/ *v.intr.* (*past and past part.* **slunk** /slʌŋk/) (often foll. by *off, away, by*) move in a stealthy or guilty or sneaking manner. [OE *slincan* crawl]

slink² /slɪŋk/ *v. & n.* ● *v.tr.* (also *absol.*) (of an animal) produce (young) prematurely. ● *n.* **1** an animal, esp. a calf, so born. **2** its flesh. [app. f. SLINK¹]

slinky /'slɪŋki/ *adj.* (**slinkier, slinkiest**) **1** stealthy. **2** (of a garment) close-fitting and flowing, sinuous. **3** gracefully slender. □ **slinkily** *adv.* **slinkiness** *n.*

slip¹ /slɪp/ *v. & n.* ● *v.* (**slipped, slipping**) **1** *intr.* slide unintentionally esp. for a short distance; lose one's footing or balance or place by unintended sliding. **2** *intr.* go or move with a sliding motion (*as the door closes the catch slips into place; slipped into her nightdress*). **3** *intr.* escape restraint or capture by being slippery or hard to hold or by not being grasped (*the eel slipped through his fingers*). **4** *intr.* **a** make one's or its way unobserved or quietly or quickly (*just slip across to the baker's; errors will slip in*). **b** (foll. by *by*) (of time) go by rapidly or unnoticed. **5** *intr.* **a** make a careless or casual mistake. **b** fall below the normal standard, deteriorate, lapse. **6** *tr.* insert or transfer stealthily or casually or with a sliding motion (*slipped a coin into his hand; slipped the papers into his pocket*). **7** *tr.* **a** release from restraint (*slipped the greyhounds from the leash*). **b** detach (an anchor) from a ship. **c** *Brit.* detach (a carriage) from a moving train. **d** release (the clutch of a motor vehicle) for a moment. **e** (of an animal) produce (young) prematurely. **8** *tr. Knitting* move (a stitch) to the other needle without knitting it. **9** *tr.* (foll. by *on, off*) pull (a garment) hastily on or off. **10** *tr.* escape from; give the slip to (*the dog slipped its collar; point slipped my mind*). ● *n.* **1** the act or an instance of slipping. **2** an accidental or slight mistake. **3** a loose covering or garment, esp. a petticoat or pillowcase. **4 a** a reduction in the movement of a pulley etc. due to slipping of the belt. **b** a reduction in the distance travelled by a ship or aircraft arising from the nature of the medium in which its propeller revolves. **5** (in *sing.* or *pl.*) **a** an artificial slope of stone etc. on which boats are landed. **b** an inclined structure on which ships are built or repaired. **6** (in *sing.* or *pl.*) *Cricket* a fielding position close behind the batsman stationed for balls glancing off the bat to the off side (*caught in the slips; caught at slip*); see also *leg slip*. **7** a leash to slip dogs. □ **give a person the slip** escape from or evade him or her. **let slip 1** release accidentally or deliberately, esp. from a leash. **2** miss (an opportunity). **3** utter inadvertently. **let slip the dogs of war** *poet.* open hostilities. **let slip through one's fingers 1** lose hold of. **2** miss the opportunity of having. **slip away** depart without leave-taking etc. **slip-carriage** *Brit.* a railway carriage on an express for detaching at a station where the rest of the train does not stop. **slip-case** a close-fitting case for a book. **slip-coach** *Brit.* = slip-carriage. **slip-cover 1** a detachable cover for a chair, sofa, etc., esp. when out of use; a loose cover. **2** a jacket or slip-case for a book. **slip form** a mould in which a structure of uniform cross-section is cast by filling it with concrete and continually moving and refilling it. **slip-hook** a hook with a contrivance for releasing it readily when necessary. **slip-knot 1** a knot that can be undone by a pull. **2** a running knot. **slip off** depart without leave-taking etc. **slip of the pen** (or **tongue**) a small mistake in which something is written (or said) unintentionally. **slip-on** *adj.* (of shoes or clothes) that can be easily slipped on and off. ● *n.* a slip-on shoe or garment. **slip-over** (of a garment) to be slipped on over the head. **slipped disc** a cartilaginous disc between vertebrae that has become displaced, pinching the spinal nerves and causing lumbar pain. **slip-ring** a ring for sliding contact in a dynamo or electric motor. **slip-road** *Brit.* a road for entering or leaving a motorway etc. **slip-rope** *Naut.* a rope with both ends on board so that casting loose either end frees the ship from its moorings. **slip sheet** *Printing* a sheet of paper placed between newly printed sheets to prevent set-off or smudging. **slip something over on** *colloq.* outwit. **slip-stitch 1** a loose stitch joining layers of fabric and not visible externally. **2** a stitch moved to the other needle without being knitted. ● *v.tr.* sew with slip-stitch. **slip up** *colloq.* make a mistake. **slip-up** *n. colloq.* a mistake, a blunder. **there's many a slip 'twixt cup and lip** nothing is certain till it has happened. [ME prob. f. MLG *slippen*: cf. SLIPPERY]

slip² /slɪp/ *n.* **1 a** a small piece of paper esp. for writing on. **b** a long narrow strip of thin wood, paper, etc. **c** a printer's proof on such paper; a galley proof. **2** a cutting taken from a plant for grafting or planting, a scion. □ **slip of a** small and slim (*a slip of a girl*). [ME, prob. f. MDu., MLG *slippe* cut, strip, etc.]

slip³ /slɪp/ *n.* clay in a creamy mixture with water, used mainly for decorating earthenware. □ **slip casting** the manufacture of ceramic ware by allowing slip to solidify in a mould. **slip-ware** ware decorated with slip. [OE *slipa, slyppe* slime: cf. COWSLIP]

slipover /'slɪpˌəʊvə(r)/ *n.* a pullover, usu. V-necked and without sleeves.

slippage /'slɪpɪdʒ/ *n.* **1** the act or an instance of slipping. **2 a** a decline, esp. in popularity or value. **b** failure to meet a deadline or fulfil a promise; delay.

slipper /'slɪpə(r)/ *n. & v.* ● *n.* **1** a light loose comfortable indoor shoe. **2** a light slip-on shoe for dancing etc. ● *v.tr.* beat or strike with a slipper. □ **slipper bath** *Brit.* a bath shaped like a slipper, with a covered end. □ **slippered** *adj.*

slipperwort /'slɪpəˌwɜːt/ *n.* calceolaria.

slippery /'slɪpərɪ/ *adj.* **1** difficult to hold firmly because of smoothness, wetness, sliminess, or elusive motion. **2** (of a surface) difficult to stand on, causing slips by its smoothness or muddiness. **3** unreliable, unscrupulous, shifty. **4** (of a subject) requiring tactful handling. □ **slippery elm 1** the North American red elm, *Ulmus fulva*. **2** the mucilaginous inner bark of this, used medicinally. **slippery slope** a course leading to disaster. □ **slipperily** *adv.* **slipperiness** *n.* [prob. coined by Coverdale (1535) after Luther's *schlipfferig*, MHG *slipferig* f. *slipfern, slipfen* f. Gmc: partly f. *slipper* slippery (now dial.) f. OE *slipor* f. Gmc]

slippy /'slɪpɪ/ *adj.* (**slippier, slippiest**) *colloq.* slippery. □ **look** (or **be**) **slippy** *Brit.* look sharp; make haste. □ **slippiness** *n.*

slipshod /'slɪpʃɒd/ *adj.* **1** (of speech or writing, a speaker or writer, a method of work, etc.) careless, unsystematic; loose in arrangement. **2** slovenly. **3** having shoes down at heel.

slipstream /'slɪpstriːm/ *n. & v.* ● *n.* **1** a current of air or water driven back by a revolving propeller or a moving vehicle. **2** an assisting force regarded as drawing something along with or behind something else. ● *v.tr.* **1** follow closely behind (another vehicle). **2** pass after travelling in another's slipstream.

slipway /'slɪpweɪ/ *n.* a slip for building ships or landing boats.

slit /slɪt/ *n. & v.* ● *n.* **1** a long straight narrow incision. **2** a long narrow opening comparable to a cut. ● *v.tr.* (**slitting**; *past* and *past part.* **slit**) **1** make a slit in; cut or tear lengthwise. **2** cut into strips. □ **slit-eyed** having long narrow eyes. **slit-pocket** a pocket with a vertical opening giving access to the pocket or to a garment beneath. **slit trench** a narrow trench for a soldier or a weapon. □ **slitter** *n.* [ME *slitte*, rel. to OE *slītan*, f. Gmc]

slither /'slɪðə(r)/ *v. & n.* ● *v.intr.* slide unsteadily; go with an irregular slipping motion. ● *n.* an instance of slithering. □ **slithery** *adj.* [ME var. of *slidder* (now dial.) f. OE *slid(e)rian* frequent. f. *slid-*, weak grade of *slīdan* SLIDE]

slitty /'slɪtɪ/ *adj.* (of the eyes) long and narrow.

Sliven /'sliːv(ə)n/ a commercial city in east central Bulgaria, in the foothills of the Balkan Mountains; pop. (1990) 150,210.

sliver /'slɪvə(r), 'slaɪv-/ *n. & v.* ● *n.* **1** a long thin piece cut or split off. **2** a piece of wood torn from a tree or from timber. **3** a splinter, esp. from an exploded shell. **4** a strip of loose textile fibres after carding. ● *v.tr. & intr.* **1** break off as a sliver. **2** break up into slivers. **3** form into slivers. [ME, rel. to *slive* cleave (now dial.) f. OE]

slivovitz /'slɪvəvɪts/ *n.* a plum brandy made esp. in Romania and the former Yugoslavia. [Serbo-Croat *šljivovica* f. *šljiva* plum]

Sloane[1] /sləʊn/, Sir Hans (1660–1753), English physician and naturalist. He purchased the manor of Chelsea and endowed the Chelsea Physic Garden. His collections, which included a large number of books and manuscripts, were purchased by the nation and placed in Montague House, which afterwards became the British Museum. Sloane's geological and zoological specimens formed the basis of the Natural History Museum in South Kensington.

Sloane[2] /sləʊn/ *n.* (in full **Sloane Ranger**) *Brit. sl.* a fashionable and conventional upper-class young woman, esp. one living in London. □ **Sloaney** *adj.* [*Sloane* Square, London + *Lone Ranger*, a cowboy hero]

slob /slɒb/ *n.* **1** *colloq.* a stupid, careless, coarse, or fat person. **2** *Ir.* muddy land. □ **slobbish** *adj.* [Ir. *slab* mud f. Engl. *slab* ooze, sludge, prob. f. Scand.]

slobber /'slɒbə(r)/ *v. & n.* ● *v.intr.* **1** slaver. **2** (foll. by *over*) show excessive sentiment. ● *n.* saliva running from the mouth; slaver. □ **slobbery** *adj.* [ME, = Du. *slobberen*, of imit. orig.]

sloe /sləʊ/ *n.* **1** = BLACKTHORN 1. **2** the small bluish-black fruit of this, with a sharp sour taste. □ **sloe-eyed 1** having eyes of this colour. **2** slant-eyed. **sloe gin** a liqueur of sloes steeped in gin and sugar. [OE *slā(h)* f. Gmc]

slog /slɒg/ *v. & n.* ● *v.* (**slogged, slogging**) **1** *intr. & tr.* hit hard and usu. wildly, esp. in boxing or cricket. **2** *intr.* (often foll. by *away, on*) walk or work doggedly. ● *n.* **1** a hard random hit. **2** a hard steady work or walking. **b** a spell of this. □ **slogger** *n.* [19th c.: orig. unkn.: cf. SLUG[2]]

slogan /'sləʊgən/ *n.* **1** a short catchy phrase used in advertising etc. **2** a party cry; a watchword or motto. **3** *hist.* a Scottish Highland war-cry. [Gael. *sluagh-ghairm* f. *sluagh* army + *gairm* shout]

sloop /sluːp/ *n.* **1** a small one-masted fore-and-aft-rigged vessel with mainsail and jib. **2** (in full **sloop of war**) *Brit. hist.* a small warship with guns on the upper deck only. □ **sloop-rigged** rigged like a sloop. [Du. *sloep(e)*, of unkn. orig.]

sloosh /sluːʃ/ *n. & v. colloq.* ● *n.* a pouring or pouring sound of water. ● *v.intr.* **1** flow with a rush. **2** make a heavy splashing or rushing noise. [imit.]

sloot /sluːt/ *n.* (also **sluit**) *S. Afr.* a deep gully formed by heavy rain. [Afrik. f. Du. *sloot* ditch]

slop[1] /slɒp/ *v. & n.* ● *v.* (**slopped, slopping**) **1** (often foll. by *over*) **a** *intr.* spill or flow over the edge of a vessel. **b** *tr.* allow to do this. **2** *tr.* make (the floor, clothes, etc.) wet or messy by slopping, spill or splash liquid on. **3** *intr.* (usu. foll. by *over*) gush; be effusive or maudlin. ● *n.* **1** a quantity of liquid spilled or splashed. **2** weakly sentimental language. **3** (in *pl.*) waste liquid, esp. dirty water or the waste contents of kitchen, bedroom, or prison vessels. **4** (in *sing.* or *pl.*) unappetizing weak liquid food. **5** *Naut.* a choppy sea. □ **slop about** move about in a slovenly manner. **slop-basin** *Brit.* a basin for the dregs of cups at table. **slop out** carry slops out (in prison etc.). **slop-pail** a pail for removing bedroom or kitchen slops. [earlier sense 'slush', prob. rel. to *slyppe*: cf. COWSLIP]

slop[2] /slɒp/ *n.* **1** a workman's loose outer garment. **2** (in *pl.*) ready-made or cheap clothing. **3** (in *pl.*) clothes and bedding supplied to sailors in the navy. **4** (in *pl.*) *archaic* wide baggy trousers esp. as worn by sailors. [ME: cf. OE *oferslop* surplice f. Gmc]

slope /sləʊp/ *n. & v.* ● *n.* **1** an inclined position or direction; a state in which one end or side is at a higher level than another; a position in a line neither parallel nor perpendicular to level ground or to a line serving as a standard. **2** a piece of rising or falling ground. **3 a** a difference in level between the two ends or sides of a thing (*a slope of 5 metres*). **b** the rate at which this increases with distance etc. **4** a place for skiing on the side of a hill or mountain. **5** (prec. by *the*) the position of a rifle when sloped. ● *v.* **1** *intr.* have or take a slope; slant esp. up or down; lie or tend obliquely, esp. to ground level. **2** *tr.* place or arrange or make in or at a slope. □ **slope arms** place one's rifle in a sloping position against one's shoulder. **slope off** *colloq.* go away, esp. to evade work etc. [shortening of ASLOPE]

sloppy /'slɒpɪ/ *adj.* (**sloppier, sloppiest**) **1 a** (of the ground) wet with rain; full of puddles. **b** (of food etc.) watery and disagreeable. **c** (of a floor, table, etc.) wet with slops, having water etc. spilt on. **2** unsystematic, careless, not thorough. **3** (of a garment) ill-fitting or untidy; (of a person) wearing such garments. **4** (of sentiment or talk) weakly emotional, maudlin. **5** *colloq.* (of the sea) choppy. □ **sloppily** *adv.* **sloppiness** *n.*

slosh /slɒʃ/ *v. & n.* ● *v.* **1** *intr.* (often foll. by *about*) splash or flounder about, move with a splashing sound. **2** *tr. Brit. sl.* hit esp. heavily. **3** *tr. colloq.* **a** pour (liquid) clumsily. **b** pour liquid on. ● *n.* **1** slush. **2 a** an instance of splashing. **b** the sound of this. **3** *Brit. sl.* a heavy blow. **4** a quantity of liquid. [var. of SLUSH]

sloshed /slɒʃt/ *adj. Brit. sl.* drunk.

sloshy /'slɒʃɪ/ *adj.* (**sloshier, sloshiest**) **1** slushy. **2** sloppy, sentimental.

slot[1] /slɒt/ *n. & v.* ● *n.* **1** a slit or other aperture in a machine etc. for something (esp. a coin) to be inserted. **2** a slit, groove, channel, or long aperture into which something fits or in which something works. **3** an allotted place in an arrangement or scheme, esp. in a broadcasting schedule. ● *v.* (**slotted, slotting**) **1** *tr. & intr.* place or be placed into or as if into a slot. **2** *tr.* provide with a slot or slots. □ **slot-machine** a machine worked by the insertion of a coin, esp.: **1** one for automatic retail of small articles. **2** one allowing a spell of play at a pin-table etc. **3** *US* = *fruit machine*. [ME, = hollow of the breast, f. OF *esclot*, of unkn. orig.]

slot[2] /slɒt/ *n.* **1** the track of a deer etc., esp. as shown by footprints. **2** a deer's foot. [OF *esclot* hoof-print of a horse, prob. f. ON *slóth* trail: cf. SLEUTH]

sloth /sləʊθ/ *n.* **1** laziness or indolence; reluctance to make an effort. **2** a slow-moving edentate arboreal mammal of the family Bradypodidae or Megalonychidae, native to South America, with long limbs and hooked claws for hanging upside down from branches. □ **sloth bear** a black shaggy bear, *Melursus ursinus*, native to India and Sri Lanka, with long curved claws. [ME f. SLOW + -TH[2]]

slothful /'sləʊθfʊl/ *adj.* lazy; characterized by sloth. □ **slothfully** *adv.* **slothfulness** *n.*

slouch /slaʊtʃ/ *v. & n.* ● *v.* **1** *intr.* stand or move or sit in a drooping ungainly fashion. **2** *tr.* bend one side of the brim of (a hat) downwards. **3** *intr.* droop, hang down loosely. ● *n.* **1** a slouching posture or movement, a stoop. **2** a downward bend of a hat-brim. **3** *colloq.* an incompetent or slovenly worker or operator or performance (*he's no slouch*). □ **slouch hat** a hat with a wide flexible brim. □ **slouchy** *adj.* [16th c.: orig. unkn.]

Slough /slaʊ/ a town in Berkshire, to the west of London; pop. (1981) 97,400.

slough[1] /slaʊ/ *n.* a swamp; a miry place; a quagmire. □ **sloughy** *adj.* [OE *slōh, slō(g)g*]

slough[2] /slʌf/ *n. & v.* ● *n.* **1** a part that an animal casts or moults, esp. a snake's cast skin. **2** dead tissue that drops off from living flesh etc. **3** a habit etc. that has been abandoned. ● *v.* **1** *tr.* cast off as a slough. **2** *intr.* (often foll. by *off*) drop off as a slough. **3** *intr.* cast off a slough. **4** *intr.* (often foll. by *away, down*) (of soil, rock, etc.) collapse or slide into a hole or depression. □ **sloughy** *adj.* [ME, perh. rel. to LG *slu(we)* husk]

Slough of Despond /slaʊ/ (in John Bunyan's *The Pilgrim's Progress*) a deep boggy place between the City of Destruction and the gate at the beginning of Christian's journey. The term is sometimes used in literary contexts for a state of hopeless depression.

Slovak /'sləʊvæk/ *n. & adj.* ● *n.* **1 a** a member of a Slavonic people inhabiting Slovakia. **b** a native or national of Slovakia. **2** the language of this people. ● *adj.* of or relating to Slovakia, the Slovaks, or their language. [Slovak *Slovák* f. Slav. base of SLOVENE]

Slovakia /slə'vækɪə, -'vɑːkɪə/ a country in central Europe; pop. (1991) 5,268,935; official language, Slovak; capital, Bratislava. Settled by Slavic tribes in the 5th–6th centuries, Slovakia was dominated by Hungary until it declared independence in 1918 and united with the Czech-speaking areas of Bohemia and Moravia to form Czechoslovakia. The eastern of the two constituent republics of Czechoslovakia, Slovakia became independent on the partition of that country on 1 Jan. 1993. □ **Slovakian** *adj. & n.*

sloven /'slʌv(ə)n/ *n.* a person who is habitually untidy or careless. [ME perh. f. Flem. *sloef* dirty or Du. *slof* careless]

Slovene /'sləʊviːn, slə'viːn/ (also **Slovenian** /slə'viːnɪən/) *n. & adj.* ● *n.* **1 a** a member of a Slavonic people inhabiting Slovenia. **b** a native or national of Slovenia. **2** the language of this people. ● *adj.* of or relating to Slovenia, the Slovenes, or their language. [G *Slowene* f. Slav. *Slov-*, perh. rel. to *slovo* word]

Slovenia /slə'viːnɪə/ a country in SE Europe, formerly a constituent republic of Yugoslavia; pop. (est. 1991) 1,962,000; official language, Slovene; capital, Ljubljana. Occupied by Celtic and Illyrian tribes in ancient times, Slovenia was settled by southern Slavs in the 6th century and later formed part of the Austrian empire. In 1919 the region was ceded to the kingdom of Serbs, Croats, and Slovenes (named Yugoslavia from 1929) and in 1945 it became a constituent republic of Communist-ruled Yugoslavia. Slovenia declared its independence in 1991.

slovenly /'slʌv(ə)nlɪ/ *adj. & adv.* ● *adj.* careless and untidy; unmethodical. ● *adv.* in a slovenly manner. □ **slovenliness** *n.*

slow /sləʊ/ *adj., adv., & v.* ● *adj.* **1 a** taking a relatively long time to do a thing or cover a distance (also foll. by *of: slow of speech*). **b** not quick; acting or moving or done without speed. **2** gradual; obtained over a length of time (*slow growth*). **3** not producing, allowing, or conducive to speed (*in the slow lane*). **4** (of a clock etc.) showing a time earlier than is the case. **5** (of a person) not understanding readily; not learning easily. **6** dull; uninteresting; tedious. **7** slack or sluggish (*business is slow*). **8** (of a fire or oven) giving little heat. **9** *Photog.* **a** (of a film) needing long exposure. **b** (of a lens) having a small aperture. **10 a** reluctant; tardy (*not slow to defend himself*). **b** not hasty or easily moved (*slow to take offence*). **11** (of a cricket-pitch, tennis-court, putting-green, etc.) on which the ball bounces or runs slowly. ● *adv.* **1** at a slow pace; slowly. **2** (in comb.) (*slow-moving traffic*). ● *v.* (usu. foll. by *down, up*) **1** *intr. & tr.* reduce one's speed or the speed of (a vehicle etc.). **2** *intr.* reduce one's pace of life; live or work less actively or intensely. □ **slow and sure** the attitude that haste is risky. **slow but sure** achieving the required result eventually. **slow-down** the action of slowing down; a go-slow. **slow handclap** slow clapping by an audience as a sign of displeasure or boredom. **slow loris** see LORIS. **slow march** the marching time adopted by troops in a funeral procession etc. **slow-match** a slow-burning match for lighting explosives etc. **slow motion 1** the operation or speed of a film using slower projection or more rapid exposure so that actions etc. appear much slower than usual. **2** the simulation of this in real action. **slow neutron** a neutron with low kinetic energy esp. after moderation (cf. *fast neutron* (see FAST[1])). **slow poison** a poison eventually causing death by repeated doses. **slow puncture** a puncture causing only slow deflation of the tyre. **slow**

reactor *Physics* a nuclear reactor using mainly slow neutrons (cf. *fast reactor* (see FAST[1])). **slow virus** a virus or virus-like organism that multiplies slowly in the host organism and has a long incubation period. **slow-witted** stupid. □ **slowish** *adj.* **slowly** *adv.* **slowness** *n.* [OE *slāw* f. Gmc]

slowcoach /'sləʊkəʊtʃ/ *n. Brit.* **1** a slow or lazy person. **2** a dull-witted person. **3** a person behind the times in opinions etc.

slowpoke /'sləʊpəʊk/ *n. N. Amer.* = SLOWCOACH.

slow-worm /'sləʊwɜːm/ *n.* a small European legless lizard, *Anguis fragilis*, giving birth to live young. Also called *blindworm*. [OE *slā-wyrm*: first element of uncert. orig., assim. to SLOW]

SLR *abbr.* **1** *Photog.* single-lens reflex. **2** self-loading rifle.

slub[1] /slʌb/ *n. & adj.* ● *n.* **1** a lump or thick place in yarn or thread. **2** fabric woven from thread etc. with slubs. ● *adj.* (of material etc.) with an irregular appearance caused by uneven thickness of the warp. [19th c.: orig. unkn.]

slub[2] /slʌb/ *n. & v.* ● *n.* wool slightly twisted in preparation for spinning. ● *v.tr.* (**slubbed, slubbing**) twist (wool) in this way. [18th c.: orig. unkn.]

sludge /slʌdʒ/ *n.* **1** thick greasy mud. **2** muddy or slushy sediment. **3** sewage. **4** *Mech.* an accumulation of dirty oil, esp. in the sump of an internal-combustion engine. **5** *Geol.* sea-ice newly formed in small pieces. **6** (usu. *attrib.*) a muddy colour (*sludge green*). □ **sludgy** *adj.* [cf. SLUSH]

slue var. of SLEW[1].

slug[1] /slʌɡ/ *n. & v.* ● *n.* **1** a slow-moving gastropod mollusc, with the shell rudimentary or absent, often destructive to plants. (See also *sea slug*.) **2 a** a bullet esp. of irregular shape. **b** a missile for an airgun. **3** *Printing* **a** a metal bar used in spacing. **b** a line of type in Linotype printing. **4** a tot or drink of liquor etc. **5** a unit of mass, given an acceleration of 1 foot per second per second by a force of 1 lb. **6** a roundish lump of metal. **7** a thick piece or lump of something; a (large) portion. ● *v.tr.* (**slugged, slugging**) drink in large quantities, swig. □ **slug pellet** a pellet containing a substance poisonous to slugs. [ME *slug(e)* sluggard, prob. f. Scand.]

slug[2] /slʌɡ/ *v. & n.* esp. N. Amer. ● *v.tr.* (**slugged, slugging**) strike with a hard blow. ● *n.* a hard blow. □ **slug it out 1** fight it out. **2** endure; stick it out. □ **slugger** *n.* [19th c.: orig. unkn.]

slugabed /'slʌɡəˌbed/ *n. archaic* a lazy person who lies late in bed. [*slug* (v.) (see SLUGGARD) + ABED]

sluggard /'slʌɡəd/ *n.* a lazy sluggish person. □ **sluggardly** *adj.* **sluggardliness** *n.* [ME f. *slug* (v.) be slothful (prob. f. Scand.: cf. SLUG[1]) + -ARD]

sluggish /'slʌɡɪʃ/ *adj.* inert; inactive; slow-moving; torpid; indolent (*a sluggish circulation; a sluggish stream*). □ **sluggishly** *adv.* **sluggishness** *n.* [ME f. SLUG or *slug* (v.): see SLUGGARD]

sluice /sluːs/ *n. & v.* ● *n.* **1** (also **sluice-gate, sluice-valve**) a sliding gate or other contrivance for controlling the volume or flow of water. **2** (also **sluice-way**) an artificial water-channel esp. for washing ore. **3** a place for rinsing. **4** the act or an instance of rinsing. **5** the water above or below or issuing through a floodgate. ● *v.* **1** *tr.* provide or wash with a sluice or sluices. **2** *tr.* rinse, pour or throw water freely upon. **3** *tr.* (foll. by *out, away*) wash out or away with a flow of water. **4** *tr.* flood with water from a sluice. **5** *intr.* (of water) rush out from a sluice, or as if from a sluice. [ME f. OF *escluse* ult. f. L *excludere* EXCLUDE]

sluit var. of SLOOT.

slum /slʌm/ *n. & v.* ● *n.* **1** an overcrowded and squalid back street, district, etc., usu. in a city and inhabited by very poor people. **2** a house or building unfit for human habitation. ● *v.intr.* (**slummed, slumming**) **1** live in slumlike conditions. **2** go about the slums through curiosity, to examine the condition of the inhabitants, or for charitable purposes. □ **slum clearance** the demolition of slums and rehousing of their inhabitants. **slum it** put up with conditions less comfortable than usual. □ **slummy** *adj.* (**slummier, slummiest**). **slumminess** *n.* [19th c.: orig. cant]

slumber /'slʌmbə(r)/ *v. & n. poet. rhet.* ● *v.intr.* **1** sleep, esp. in a specified manner. **2** be idle, drowsy, or inactive. ● *n.* a sleep, esp. of a specified kind (*fell into a fitful slumber*). □ **slumber away** spend (time) in slumber. **slumber-wear** nightclothes. □ **slumberer** *n.* **slumberous** *adj.* **slumbrous** *adj.* [ME *slūmere* etc. f. *slūmen* (v.) or *slūme* (n.) f. OE *slūma*: -*b*- as in *number*]

slump /slʌmp/ *n. & v.* ● *n.* **1** a sudden severe or prolonged fall in prices or values of commodities or securities. **2** a sharp or sudden decline in trade or business, usu. bringing widespread unemployment. **3** a lessening of interest or commitment in a subject or undertaking. ● *v.intr.* **1** undergo a slump; fail; fall in price. **2** sit or fall heavily or limply (*slumped into a chair*). **3** lean or subside. [17th c., orig. 'sink in a bog': imit.]

slung *past* and *past part.* of SLING¹.

slunk *past* and *past part.* of SLINK¹.

slur /slɜː(r)/ *v. & n.* ● *v.* (**slurred**, **slurring**) **1** *tr. & intr.* pronounce or write indistinctly so that the sounds or letters run into one another. **2** *tr. Mus.* **a** perform (a group of two or more notes) legato. **b** mark (notes) with a slur. **3** *tr. archaic* or *US* put a slur on (a person or a person's character); make insinuations against. **4** *tr.* (usu. foll. by *over*) pass over (a fact, fault, etc.) lightly; conceal or minimize. ● *n.* **1** an imputation of wrongdoing; blame; stigma (*a slur on my reputation*). **2** the act or an instance of slurring in pronunciation, singing, or writing. **3** *Mus.* a curved line to show that two or more notes are to be sung to one syllable or played or sung legato. [17th c.: orig. unkn.]

slurp /slɜːp/ *v. & n.* ● *v.tr.* eat or drink noisily. ● *n.* the sound of this; a slurping gulp. [Du. *slurpen*, *slorpen*]

slurry /'slʌrɪ/ *n.* (*pl.* **-ies**) **1** a semi-liquid mixture of fine particles and water; thin mud. **2** thin liquid cement. **3** a fluid form of manure. **4** a residue of water and particles of coal left at pit-head washing plants. [ME, rel. to dial. *slur* thin mud]

slush /slʌʃ/ *n.* **1** watery mud or thawing snow. **2** silly sentiment. □ **slush fund** reserve funding esp. as used for political bribery. [17th c., also *sludge* and *slutch*: orig. unkn.]

slushy /'slʌʃɪ/ *adj.* (**slushier**, **slushiest**) **1** like slush; watery. **2** weakly sentimental; insipid. □ **slushiness** *n.*

slut /slʌt/ *n. derog.* a slovenly or promiscuous woman. □ **sluttish** *adj.* **sluttishness** *n.* [ME: orig. unkn.]

sly /slaɪ/ *adj.* (**slyer**, **slyest**) **1** cunning; crafty; wily. **2 a** (of a person) practising secrecy or stealth. **b** (of an action etc.) done etc. in secret. **3** hypocritical; ironical. **4** knowing; arch; bantering; insinuating. **5** *Austral. & NZ sl.* (esp. of liquor) illicit. □ **on the sly** privately; covertly; without publicity (*smuggled some through on the sly*). **sly dog** *colloq.* a person who is discreet about mistakes or pleasures. □ **slyly** *adv.* (also **slily**). **slyness** *n.* [ME *sleh* etc. f. ON *slœgr* cunning, orig. 'able to strike' f. *slóg*- past stem of *slá* strike: cf. SLEIGHT]

slyboots /'slaɪbuːts/ *n. colloq.* a sly person.

slype /slaɪp/ *n.* a covered way or passage between a cathedral etc. transept and the chapter house or deanery. [perh. = *slipe* a long narrow piece of ground, = SLIP² 1]

SM *abbr.* **1** sadomasochism. **2** Sergeant-Major.

Sm *symb. Chem.* the element samarium.

smack¹ /smæk/ *n., v., & adv.* ● *n.* **1** a sharp slap or blow esp. with the palm of the hand or a flat object. **2** a hard hit at cricket etc. **3** a loud kiss (*gave her a hearty smack*). **4** a loud sharp sound (*heard the smack as it hit the floor*). ● *v.* **1** *tr.* strike with the open hand etc. **2** *tr.* part (one's lips) noisily in eager anticipation or enjoyment of food or another delight. **3** *tr.* crack (a whip). **4** *tr. & intr.* move, hit, etc., with a smack. ● *adv.* **1** with a smack. **2** suddenly; directly; violently (*landed smack on my desk*). **3** exactly (*hit it smack in the centre*). □ **have a smack at** *colloq.* make an attempt, attack, etc., at. **a smack in the eye** (or **face**) *colloq.* a rebuff; a setback. [MDu. *smack(en)* of imit. orig.]

smack² /smæk/ *v. & n.* (foll. by *of*) ● *v.intr.* **1** have a flavour of; taste of (*smacked of garlic*). **2** suggest the presence or effects of (*it smacks of nepotism*). ● *n.* **1** a flavour; a taste that suggests the presence of something. **2** (in a person's character etc.) a barely discernible quality (*just a smack of superciliousness*). **3** (in food etc.) a very small amount (*add a smack of ginger*). [OE *smæc*]

smack³ /smæk/ *n.* a single-masted sailing-boat for coasting or fishing. [Du. *smak* f. earlier *smacke*; orig. unkn.]

smack⁴ /smæk/ *n. sl.* a hard drug, esp. heroin, sold or used illegally. [prob. alt. of Yiddish *schmeck* sniff]

smacker /'smækə(r)/ *n.* **1** *colloq.* a loud kiss. **2** *colloq.* a resounding blow. **3** *sl.* **a** *Brit.* £1. **b** *N. Amer.* $1.

small /smɔːl/ *adj., n., & adv.* ● *adj.* **1** not large or big. **2** slender; thin. **3** not great in importance, amount, number, strength, or power. **4** not much; trifling (*a small token*; *paid small attention*). **5** insignificant; unimportant (*a small matter*; *from small beginnings*). **6** consisting of small particles (*small gravel*; *small shot*). **7** doing something on a small scale (*a small farmer*). **8** socially undistinguished; poor or humble. **9** petty; mean; ungenerous; paltry (*a small spiteful nature*). **10** lacking in imagination (*they have such small minds*). **11** young; not fully grown or developed (*a*

small child). ● *n.* **1** the slenderest part of something (esp. *small of the back*). **2** (in *pl.*) *Brit. colloq.* small items of laundry, esp. underwear. ● *adv.* into small pieces (*chop it small*). □ **feel** (or **look**) **small** be humiliated; appear mean or humiliated. **in a small way** unambitiously; on a small scale. **no small** considerable; a good deal of (*no small excitement about it*). **small arms** portable firearms, esp. rifles, pistols, light machine-guns, sub-machine-guns, etc. **small beer 1** a trifling matter; something unimportant. **2** weak beer. **small-bore** (of a firearm) with a narrow bore, in international and Olympic shooting usu. .22 inch calibre (5.6 millimetre bore). **small capital** a capital letter which is of the same dimensions as the lower-case letters in the same typeface minus ascenders and descenders, as THIS. **small change 1** money in the form of coins as opposed to notes. **2** trivial remarks. **small circle** any circle on a sphere other than a great circle. **small claims court** (in the UK) a local tribunal in which claims for small amounts can be heard and decided quickly and cheaply without legal representation. **small copper** see COPPER¹ *n.* 4. **small craft** a general term for small boats and fishing vessels. **small fry 1** young children or the young of various species. **2** small or insignificant things or people. **small hours** the early hours of the morning after midnight. **small intestine** see INTESTINE. **small letter** (in printed material) a lower-case letter. **small mercy** a minor concession, benefit, etc. (*be grateful for small mercies*). **small potatoes** an insignificant person or thing. **small print 1** printed matter in small type. **2** inconspicuous and usu. unfavourable limitations etc. in a contract. **small profits and quick returns** the policy of a cheap shop etc. relying on large trade. **small-scale** made or occurring in small amounts or to a lesser degree. **small screen** television. **small slam** see SLAM². **small-sword** a light tapering thrusting-sword, esp. *hist.* for duelling. **small talk** light social conversation. **small-time** *colloq.* unimportant or petty. **small-timer** *colloq.* a small-time operator; an insignificant person. **small-town** relating to or characteristic of a small town; unsophisticated; provincial. **small wonder** not very surprising. □ **smallish** *adj.* **smallness** *n.* [OE *smæl* f. Gmc]

smallgoods /'smɔːlɡʊdz/ *n. Austral.* delicatessen meats.

smallholder /'smɔːlˌhəʊldə(r)/ *n. Brit.* a person who farms a smallholding.

smallholding /'smɔːlˌhəʊldɪŋ/ *n. Brit.* an agricultural holding smaller than a farm.

small-minded /smɔːl'maɪndɪd/ *adj.* petty; of rigid opinions or narrow outlook. □ **small-mindedly** *adv.* **small-mindedness** *n.*

smallpox /'smɔːlpɒks/ *n. Med.* an acute contagious viral disease with fever and pustules that usually leave permanent scars, the main devastating disease of the 17th and 18th centuries. It was formerly the practice in the East to infect people with a mild form of the disease in order to give them some immunity, and this was introduced into England by the traveller Lady Mary Wortley Montagu (1689–1762). In 1796 Edward Jenner observed that the mild disease cowpox gave immunity against smallpox, and established the practice of vaccination. Its systematic application resulted in the worldwide eradication of smallpox by 1979. Also called *variola*.

smalt /smɒlt, smɔːlt/ *n.* **1** glass coloured blue with cobalt. **2** a pigment made by pulverizing this. [F f. It. *smalto* f. Gmc, rel. to SMELT¹]

smarm /smɑːm/ *v. & n. colloq.* ● *v.* **1** *tr.* (often foll. by *down*) smooth, plaster down (hair etc.) usu. with cream or oil. **2** *intr.* be ingratiating. ● *n.* obsequiousness. [orig. dial. (also *smalm*), of uncert. orig.]

smarmy /'smɑːmɪ/ *adj.* (**smarmier**, **smarmiest**) *colloq.* ingratiating; flattering; obsequious. □ **smarmily** *adv.* **smarminess** *n.*

smart /smɑːt/ *adj., v., n., & adv.* ● *adj.* **1 a** clever; ingenious; quick-witted (*a smart talker*; *gave a smart answer*). **b** keen in bargaining; quick to take advantage. **c** (of transactions etc.) unscrupulous to the point of dishonesty. **2** well groomed; neat; bright and fresh in appearance (*a smart suit*). **3** in good repair; showing bright colours, new paint, etc. (*a smart red bicycle*). **4** stylish; fashionable; prominent in society (*in all the smart restaurants*; *the smart set*). **5** quick; brisk (*set a smart pace*). **6** painfully severe; sharp; vigorous (*a smart blow*). ● *v.intr.* **1** (of a person or a part of the body) feel or give acute pain or distress (*my eye smarts*; *smarting from the insult*). **2** (of an insult, grievance, etc.) rankle. **3** (foll. by *for*) suffer the consequences of (*you will smart for this*). ● *n.* a bodily or mental sharp pain; a stinging sensation. ● *adv.* smartly; in a smart manner. □ **look smart** make haste. **smart-arse** (or **-ass**) *sl.* = SMART ALEC. **smart card** a plastic card with a built-in microprocessor, enabling access to computerized data, instant transfer of funds, etc. **smart drug** a drug supposedly capable of enhancing cognitive performance. **smart money 1** money paid or exacted as a penalty or compensation. **2** money invested by persons with expert knowledge. □ **smartingly** adv. **smartish** adj. & adv. **smartly** adv. **smartness** n. [OE *smeart, smeortan*]

smart alec /ˈælɪk/ n. (also **aleck**, **alick**) *colloq.* a person displaying ostentatious or smug cleverness. □ **smart-alecky** adj. [SMART + *Alec*, dimin. of the name *Alexander*]

smarten /ˈsmɑːt(ə)n/ v.tr. & intr. (usu. foll. by *up*) make or become smart or smarter.

smarty /ˈsmɑːtɪ/ n. (pl. **-ies**) *colloq.* **1** a know-all; a smugly clever person. **2** a smartly dressed person; a member of a smart set. □ **smarty-boots** (or **-pants**) = SMARTY 1. [SMART + -Y²]

smash /smæʃ/ v., n., & adv. ● v. **1** tr. & intr. (often foll. by *up*) **a** break into pieces; shatter. **b** bring or come to sudden or complete destruction, defeat, or disaster. **2** tr. (foll. by *into*, *through*) (of a vehicle etc.) move with great force and impact. **3** tr. & intr. (foll. by *in*) break in with a crushing blow (*smashed in the window*). **4** tr. (in tennis, squash, etc.) hit (a ball etc.) with great force, esp. downwards (*smashed it back over the net*). **5** intr. *colloq.* (of a business etc.) go bankrupt, come to grief. **6** tr. (as **smashed** adj.) *sl.* intoxicated. ● n. **1** the act or an instance of smashing; a violent fall, collision, or disaster. **2** the sound of this. **3** (in full **smash hit**) a very successful play, song, performer, etc. **4** a stroke in tennis, squash, etc., in which the ball is hit esp. downwards with great force. **5** a violent blow with a fist etc. **6** *colloq.* bankruptcy; a series of commercial failures. **7** a mixture of spirits (usu. brandy) with flavoured water and ice. ● adv. with a smash (*fell smash on the floor*). □ **go to smash** *colloq.* be ruined etc. **smash-and-grab** (of a robbery etc.) in which the thief smashes a shop-window and seizes goods. **smash-up** *colloq.* a violent collision; a complete smash. [18th c., prob. imit. after *smack*, *smite* and *bash*, *mash*, etc.]

smasher /ˈsmæʃə(r)/ n. **1** *colloq.* a very beautiful or pleasing person or thing. **2** a person or thing that smashes.

smashing /ˈsmæʃɪŋ/ adj. *colloq.* superlative; excellent; wonderful; beautiful. □ **smashingly** adv.

smatter /ˈsmætə(r)/ n. (also **smattering**) **1** a slight superficial knowledge of a language or subject. **2** esp. *US colloq.* a small quantity; a scattering. □ **smatterer** n. [ME *smatter* talk ignorantly, prate: orig. unkn.]

smear /smɪə(r)/ v. & n. ● v.tr. **1** daub or mark with a greasy or sticky substance or with something that stains. **2** blot; smudge; obscure the outline of (writing, artwork, etc.). **3** defame the character of; slander; attempt to or succeed in discrediting (a person or his name) publicly. ● n. **1** the act or an instance of smearing. **2** *Med.* **a** material smeared on a microscopic slide etc. for examination. **b** a specimen of this. □ **smear test** = *cervical smear.* □ **smearer** n. **smeary** adj. [OE *smierwan* f. Gmc]

smectic /ˈsmektɪk/ adj. & n. *Physics* ● adj. designating or involving a state of a liquid crystal in which the molecules are oriented in parallel and arranged in well-defined planes (cf. NEMATIC). ● n. a smectic substance. [L *smecticus*, Gk *smēktikos* cleansing (from the soaplike consistency)]

smegma /ˈsmegmə/ n. a sebaceous secretion in the folds of the skin, esp. of the foreskin. □ **smegmatic** /smegˈmætɪk/ adj. [L f. Gk *smēgma -atos* soap f. *smēkhō* cleanse]

smell /smel/ n. & v. ● n. **1** the faculty of perceiving odours or scents (*has a fine sense of smell*). **2** the quality in substances that is perceived by this (*the smell of thyme; this rose has no smell*). **3** an unpleasant odour. **4** the act of inhaling to ascertain smell. ● v. (past and past part. **smelt** /smelt/ or **smelled**) **1** tr. perceive the smell of; examine by smell (*thought I could smell gas*). **2** intr. emit odour. **3** intr. seem by smell to be (*this milk smells sour*). **4** intr. (foll. by *of*) **a** be redolent of (*smells of fish*). **b** be suggestive of (*smells of dishonesty*). **5** intr. have a strong or unpleasant smell. **6** tr. perceive as if by smell; detect, discern, suspect (*smell a bargain; smell blood*). **7** intr. have or use a sense of smell. **8** intr. (foll. by *about*) sniff or search about. **9** intr. (foll. by *at*) inhale the smell of. □ **smelling-bottle** a small bottle of smelling-salts. **smelling-salts** ammonium carbonate mixed with scent to be sniffed as a restorative in faintness etc. **smell out 1** detect by smell; find out by investigation. **2** (of a dog etc.) hunt out by smell. **smell a rat** begin to suspect trickery etc. □ **smellable** adj. **smeller** n. **smell-less** adj. [ME *smel(le)*, prob. f. OE]

smelly /ˈsmelɪ/ adj. (**smellier**, **smelliest**) having a strong or unpleasant smell. □ **smelliness** n.

smelt¹ /smelt/ v.tr. **1** extract metal from (ore) by melting. **2** extract (metal) from ore by melting. □ **smelter** n. **smeltery** n. (pl. **-ies**). [MDu., MLG *smelten*, rel. to MELT]

smelt² past and past part. of SMELL.

smelt³ /smelt/ n. (pl. same or **smelts**) a small silvery fish of the genus *Osmerus*, of a family related to the salmon; esp. *O. eperlanus* of Europe, used as food (also called *sparling*). [OE, of uncert. orig.: cf. SMOLT]

Smersh /smɜːʃ/ the popular name for the Russian counter-espionage organization, originating during the Second World War, responsible for maintaining security within the Soviet armed and intelligence services. [Russ. abbr. of *smert' shpionam*, lit. 'death to spies']

Smetana /ˈsmetənə/, Bedřich (1824–84), Czech composer. Regarded as the founder of Czech music, he was dedicated to the cause of Czech nationalism, as is apparent in his operas (e.g. *The Bartered Bride*, 1866) and in the cycle of symphonic poems *My Country* (1874–9). He also contributed to the cause through his work as conductor of the National Theatre in Prague. He died in an asylum after suffering ten years of deteriorating health due to syphilis, which had left him completely deaf in 1874.

smew /smjuː/ n. a small northern Eurasian merganser, *Mergus albellus*, the male of which has black and white plumage. [17th c., rel. to *smeath*, *smee* = smew, wigeon, etc.]

smidgen /ˈsmɪdʒɪn/ n. (also **smidgin**) *colloq.* a small bit or amount. [perh. f. *smitch* in the same sense: cf. dial. *smitch* wood-smoke]

smilax /ˈsmaɪlæks/ n. **1** a spiny climbing shrub of the genus *Smilax*, the roots of some American species of which yield sarsaparilla. **2** a climbing kind of asparagus, *Myrsiphyllum asparagoides*, used decoratively by florists. [L f. Gk, = bindweed]

smile /smaɪl/ v. & n. ● v. **1** intr. relax the features into a pleased or kind or gently sceptical expression or a forced imitation of these, usu. with the lips parted and the corners of the mouth turned up. **2** tr. express by smiling (*smiled their consent*). **3** tr. give (a smile) of a specified kind (*smiled a sardonic smile*). **4** intr. (foll. by *on*, *upon*) adopt a favourable attitude towards; encourage (*fortune smiled on me*). **5** intr. have a bright or favourable aspect (*the smiling countryside*). **6** tr. (foll. by *away*) drive (a person's anger etc.) away (*smiled their tears away*). **7** intr. (foll. by *at*) **a** ridicule or show indifference to (*smiled at my feeble attempts*). **b** favour; smile on. **8** tr. (foll. by *into*, *out of*) bring (a person) into or out of a specified mood etc. by smiling (*smiled them into agreement*). ● n. **1** the act or an instance of smiling. **2** a smiling expression or aspect. □ **come up smiling** *colloq.* recover from adversity and cheerfully face what is to come. □ **smileless** adj. **smiler** n. **smiley** adj. **smilingly** adv. [ME perh. f. Scand., rel. to SMIRK: cf. OHG *smīlenter*]

smirch /smɜːtʃ/ v. & n. ● v.tr. mark, soil, or smear (a thing, a person's reputation, etc.). ● n. **1** a spot or stain. **2** a blot (on one's character etc.). [ME: orig. unkn.]

smirk /smɜːk/ n. & v. ● n. an affected, conceited, or silly smile. ● v.intr. put on or wear a smirk. □ **smirker** n. **smirkingly** adv. **smirky** adj. **smirkily** adv. [OE *sme(a)rcian*]

smit /smɪt/ archaic past part. of SMITE.

smite /smaɪt/ v. & n. ● v. (past **smote** /sməʊt/; past part. **smitten** /ˈsmɪt(ə)n/) archaic or literary **1** tr. strike or hit. **2** tr. chastise; defeat. **3** tr. (in passive) **a** have a sudden strong effect on (*was smitten by his conscience*). **b** infatuate, fascinate (*was smitten by her beauty*). **4** intr. (foll. by *on*, *upon*) come forcibly or abruptly upon. ● n. a blow or stroke. □ **smiter** n. [OE *smītan* smear f. Gmc]

Smith¹ /smɪθ/, Adam (1723–90), Scottish economist and philosopher. He is regarded by many as the founder of modern economics, and his work marks a significant turning-point in the breakdown of mercantilism and the spread of laissez-faire ideas. Smith retired from academic life to write his *Inquiry into the Nature and Causes of the Wealth of Nations* (1776), establishing theories of labour, distribution, wages, prices, and money, and advocating free trade and minimal state interference in economic matters. His work was highly influential in terms not only of economic but also of political theory in the following century.

Smith² /smɪθ/, Bessie (1894–1937), American blues singer. She became a leading artist in the 1920s and made over 150 recordings, including some with Benny Goodman and Louis Armstrong. She was involved in a car accident and died after being refused admission to a 'whites only' hospital.

Smith³ /smɪθ/, Ian (Douglas) (b.1919), Rhodesian statesman, Prime Minister 1964–79. He founded the white supremacist Rhodesian Front (renamed the Republican Front in 1981) in 1962, becoming Prime Minister and head of the white minority government two years later. In 1965 he issued a unilateral declaration of independence (UDI) after Britain stipulated that it would only grant the country independence if Smith undertook to prepare for black majority rule. He eventually conceded in 1979 and resigned to make way for majority rule; after the country became the independent state of Zimbabwe he remained active in politics, leading the Republican Front until 1987.

Smith[4] /smɪθ/, Joseph (1805–44), American religious leader and founder of the Church of Jesus Christ of Latter-Day Saints (the Mormons). In 1827, according to his own account, he was led by divine revelation to find the sacred texts written by the prophet Mormon, which he later translated and published as the Book of Mormon in 1830. He founded the Mormon Church in the same year and established a large community in Illinois, of which he became mayor. He was murdered by a mob while in prison awaiting trial for conspiracy.

Smith[5] /smɪθ/, Stevie (pseudonym of Florence Margaret Smith) (1902–71), English poet and novelist. Although she first attracted notice with her novel *Novel on Yellow Paper* (1936), she is now mainly remembered for her witty, caustic, and enigmatic verse, often illustrated by her own comic drawings. Collections include *A Good Time was Had By All* (1937), *Not Waving But Drowning* (1957), and the posthumous *Collected Poems* (1975).

Smith[6] /smɪθ/, Sydney (1771–1845), English Anglican churchman, essayist, and wit. He is notable for his witty contributions to the periodical the *Edinburgh Review* and as the author of the *Letters of Peter Plymley* (1807), which defended Catholic Emancipation.

Smith[7] /smɪθ/, William (1769–1839), English land-surveyor and geologist. Despite being self-taught, he became known as the father of English geology. Working initially in the area around Bath, he discovered that rock strata could be distinguished on the basis of their characteristic assemblages of fossils, and that strata in different places could be identified by these. Smith later travelled extensively in Britain, accumulating data which enabled him to produce the first geological map of England and Wales. Many of the names he devised for particular strata are still in use.

smith /smɪθ/ *n. & v.* ● *n.* (esp. in comb.) **1** a worker in metal (*goldsmith*; *tinsmith*). **2** a person who forges iron; a blacksmith. **3** a craftsman (*wordsmith*). ● *v.tr.* make or treat by forging. [OE f. Gmc]

smithereens /ˌsmɪðəˈriːnz/ *n.pl.* (also **smithers** /ˈsmɪðəz/) small fragments (*smash into smithereens*). [19th c.: orig. unkn.]

smithery /ˈsmɪθərɪ/ *n.* (pl. **-ies**) **1** a smith's work. **2** (esp. in naval dockyards) a smithy.

Smithfield /ˈsmɪθfiːld/ a part of London containing the city's principal meat market. Formerly an open area situated just outside the walls of the City of London, it was used as a horse and cattle market, as a fairground, and as a place of execution.

Smithsonian Institution /smɪθˈsəʊnɪən/ a US foundation for education and scientific research in Washington, DC, opened in 1846 and now responsible for administering many museums, art galleries, and other establishments. It originated in a £100,000 bequest in the will of the English chemist and mineralogist James Smithson (1765–1829).

smithy /ˈsmɪðɪ/ *n.* (pl. **-ies**) a blacksmith's workshop; a forge. [ME f. ON *smithja*]

smitten *past part.* of SMITE.

smock /smɒk/ *n. & v.* ● *n.* **1** a loose shirtlike garment with the upper part closely gathered in smocking. **2** (also **smock-frock**) a loose overall, esp. *hist.* a field-labourer's outer linen garment. ● *v.tr.* adorn with smocking. [OE *smoc*, prob. rel. to OE *smūgan* creep, ON *smjúga* put on a garment]

smocking /ˈsmɒkɪŋ/ *n.* an ornamental effect on cloth made by gathering the material tightly into pleats, often with stitches in a honeycomb pattern.

smog /smɒg/ *n.* fog or reduced visibility caused or intensified by atmospheric pollutants, esp. smoke (see also *photochemical smog*). □ **smoggy** *adj.* (**smoggier**, **smoggiest**). [portmanteau word]

smoke /sməʊk/ *n. & v.* ● *n.* **1** a visible suspension of carbon etc. in air, emitted from a burning substance. **2** an act or period of smoking tobacco (*had a quiet smoke*). **3** *colloq.* a cigarette or cigar (*got a smoke?*). **4** (**the Smoke**) *Brit. & Austral. colloq.* a big city, esp. London. ● *v.* **1** *intr.* **a** emit smoke or visible vapour (*smoking ruins*). **b** (of a lamp etc.) burn badly with the emission of smoke. **c** (of a chimney or fire) discharge smoke into the room. **2 a** *intr.* inhale and exhale the smoke of a cigarette etc. **b** *intr.* do this habitually. **c** *tr.* use (a cigarette etc.) in this way. **3** *tr.* cure or darken by the action of smoke (*smoked salmon*). **4** *tr.* spoil the taste of in cooking. **5** *tr.* **a** rid of insects etc. by the action of smoke.

b subdue (insects, esp. bees) in this way. **6** *tr. archaic* make fun of. **7** *tr.* bring (oneself) into a specified state by smoking. □ **go up in smoke** *colloq.* **1** be destroyed by fire. **2** (of a plan etc.) come to nothing. **no smoke without fire** rumours are not entirely baseless. **smoke-ball 1** a puff-ball. **2** a projectile filled with material emitting dense smoke, used to conceal military operations etc. **smoke bomb** a bomb that emits dense smoke on exploding. **smoke-bush** = *smoke-plant*. **smoked glass** glass darkened (as) with smoke. **smoke-dried** cured in smoke. **smoke-ho** *Austral. & NZ colloq.* = SMOKO. **smoke out 1** drive out by means of smoke. **2** drive out of hiding or secrecy etc. **smoke-plant** (or **-tree**) an ornamental Eurasian sumac, *Cotinus coggyria*, with a feathery smokelike inflorescence (cf. *young fustic*). **smoke-ring** smoke from a cigarette etc. exhaled in the shape of a ring. **smoke-room** *Brit.* = SMOKING-ROOM. **smoke-stone** cairngorm. **smoke-tunnel** a form of wind-tunnel using smoke filaments to show the motion of air. □ **smokable** *adj.* (also **smokeable**). [OE *smoca* f. weak grade of the stem of *smēocan* emit smoke]

smokeless /ˈsməʊklɪs/ *adj.* having or producing little or no smoke. □ **smokeless zone** a district in which it is illegal to create smoke and where only smokeless fuel may be used.

smoker /ˈsməʊkə(r)/ *n.* **1** a person or thing that smokes, esp. a person who habitually smokes tobacco. **2** a compartment on a train etc. in which smoking is allowed. **3** esp. *US* an informal social gathering of men. □ **smoker's cough** a persistent cough caused by excessive smoking.

smokescreen /ˈsməʊkskriːn/ *n.* **1** a cloud of smoke diffused to conceal (esp. military) operations. **2** a device or ruse for disguising one's activities.

smokestack /ˈsməʊkstæk/ *n.* **1** a chimney or funnel for discharging the smoke of a locomotive or steamer. **2** a tall chimney.

smoking-jacket /ˈsməʊkɪŋˌdʒækɪt/ *n.* an often ornate jacket of velvet etc., formerly worn by men while smoking.

smoking-room /ˈsməʊkɪŋˌruːm, -ˌrʊm/ *n.* a room in a hotel or house, kept for smoking in.

smoko /ˈsməʊkəʊ/ *n.* (pl. **-os**) *Austral. & NZ colloq.* **1** a stoppage of work for a rest and a smoke. **2** a tea break.

smoky /ˈsməʊkɪ/ *adj.* (**smokier**, **smokiest**) **1** emitting, veiled or filled with, or obscured by smoke (*smoky fire*; *smoky room*). **2** stained with or coloured like smoke (*smoky glass*). **3** having the taste or flavour of smoked food (*smoky bacon*). □ **smokily** *adv.* **smokiness** *n.*

smolder *US* var. of SMOULDER.

Smolensk /sməˈljensk/ a city in western European Russia, on the River Dnieper close to the border with Belarus; pop. (1990) 346,000.

Smollett /ˈsmɒlɪt/, Tobias (George) (1721–71), Scottish novelist. His novels are picaresque tales characterized by fast-moving narrative and humorous caricature; they include *The Adventures of Roderick Random* (1748), *The Adventures of Peregrine Pickle* (1751), and the epistolary work *The Expedition of Humphry Clinker* (1771). Among his other works are *A Complete History of England* (1757–8) and translations of Voltaire and Cervantes.

smolt /sməʊlt/ *n.* a young salmon migrating to the sea for the first time. [ME (orig. Sc. & N. Engl.): orig. unkn.]

smooch /smuːtʃ/ *n. & v. colloq.* ● *n.* **1** *Brit.* a period of slow dancing close together. **2** a spell of kissing and caressing. ● *v.intr.* engage in a smooch. □ **smoocher** *n.* **smoochy** *adj.* (**smoochier**, **smoochiest**). [dial. *smouch* imit.]

smoodge /smuːdʒ/ *v.intr.* (also **smooge**) *Austral. & NZ* **1** behave in a fawning or ingratiating manner. **2** behave amorously. [prob. var. of dial. *smudge* kiss, sidle up to, beg in a sneaking way]

smooth /smuːð/ *adj., v., n., & adv.* ● *adj.* **1** having a relatively even and regular surface; free from perceptible projections, lumps, indentations, and roughness. **2** not wrinkled, pitted, scored, or hairy (*smooth skin*). **3** that can be traversed without check. **4** (of liquids) of even consistency; without lumps (*mix to a smooth paste*). **5** (of the sea etc.) without waves or undulations. **6** (of a journey, passage, progress, etc.) untroubled by difficulties or adverse conditions. **7** having an easy flow or correct rhythm (*smooth breathing*; *a smooth metre*). **8 a** not harsh in sound or taste. **b** (of wine etc.) not astringent. **9** (of a person, his or her manner, etc.) suave, conciliatory, flattering, unruffled, or polite (*a smooth talker*; *he's very smooth*). **10** (of movement etc.) not suddenly varying; not jerky. ● *v.* **1** *tr. & intr.* (often foll. by *out*, *down*) make or become smooth. **2** (often foll. by *out*, *down*, *over*, *away*) **a** *tr.* reduce or get rid of (differences, faults, difficulties, etc.) in fact or appearance. **b** *intr.* (of difficulties etc.) diminish, become less obtrusive (*it will all smooth over*).

3 *tr.* modify (a graph, curve, etc.) so as to lessen irregularities. **4** *tr.* free from impediments or discomfort (*smooth the way; smooth the declining years*). ● *n.* **1** a smoothing touch or stroke (*gave his hair a smooth*). **2** the easy part of life (*take the rough with the smooth*). ● *adv.* smoothly (*the course of true love never did run smooth*). □ **in smooth water** having passed obstacles or difficulties. **smooth-bore** a gun with an unrifled barrel. **smooth-faced** hypocritically friendly. **smooth hound** a small shark of the genus *Mustelus*, found in shallow water. **smoothing-iron** *hist.* a flat-iron. **smoothing-plane** a small plane for finishing the planing of wood. **smooth muscle** muscle without striations, usu. occurring in hollow organs and performing involuntary functions (also called *visceral muscle*). **smooth talk** bland specious language. **smooth-talk** *v.tr.* address or persuade with this. **smooth-tongued** insincerely flattering. □ **smoothable** *adj.* **smoother** *n.* **smoothish** *adj.* **smoothly** *adv.* **smoothness** *n.* [OE *smōth*]

smoothie /'smuːðɪ/ *n. colloq.* **1** a person who is smooth (see SMOOTH *adj.* 9). **2** esp. *US, Austral., & NZ* a thick smooth drink of fresh fruit puréed with milk, yoghurt, or ice cream. [SMOOTH]

smorgasbord /'smɔːɡəsˌbɔːd/ *n.* open sandwiches served with delicacies as hors-d'œuvres or a buffet. [Sw. f. *smör* butter + *gås* goose, lump of butter + *bord* table]

smorzando /smɔːˈtsændəʊ/ *adj., adv., & n. Mus.* ● *adj. & adv.* dying away. ● *n.* (*pl.* **-os** or **smorzandi** /-dɪ/) a smorzando passage. [It., gerund of *smorzare* extinguish]

smote *past* of SMITE.

smother /'smʌðə(r)/ *v. & n.* ● *v.* **1** *tr.* suffocate; stifle; kill by stopping the breath of or excluding air from. **2** *tr.* (foll. by *with*) overwhelm with (kisses, gifts, kindness, etc.) (*smothered with affection*). **3** *tr.* (foll. by *in, with*) cover entirely in or with (*chicken smothered in mayonnaise*). **4** *tr.* extinguish or deaden (a fire or flame) by covering it or heaping it with ashes etc. **5** *intr.* **a** die of suffocation. **b** have difficulty breathing. **6** *tr.* (often foll. by *up*) suppress or conceal; keep from notice or publicity. **7** *tr. US* defeat rapidly or utterly. ● *n.* **1** a cloud of dust or smoke. **2** obscurity caused by this. □ **smothered mate** *Chess* checkmate in which the king, having no vacant square to move to, is checkmated by a knight. [ME *smorther* f. the stem of OE *smorian* suffocate]

smothery /'smʌðərɪ/ *adj.* tending to smother; stifling.

smoulder /'sməʊldə(r)/ *v. & n.* (*US* **smolder**) ● *v.intr.* **1** burn slowly with smoke but without a flame; slowly burn internally or invisibly. **2** (of emotions etc.) exist in a suppressed or concealed state. **3** (of a person) show silent or suppressed anger, hatred, etc. ● *n.* a smouldering or slow-burning fire. [ME, rel. to LG *smöln*, MDu. *smölen*]

smriti /'smrɪtɪ/ *n.* Hindu traditional teachings on religion etc. [Skr. *smṛti* remembrance]

smudge[1] /smʌdʒ/ *n. & v.* ● *n.* **1** a blurred or smeared line or mark; a blot; a smear of dirt. **2** a stain or blot on a person's character etc. ● *v.* **1** *tr.* make a smudge on. **2** *intr.* become smeared or blurred (*smudges easily*). **3** *tr.* smear or blur the lines of (writing, drawing, etc.) (*smudge the outline*). **4** *tr.* defile, sully, stain, or disgrace (a person's name, character, etc.). □ **smudgeless** *adj.* [ME: orig. unkn.]

smudge[2] /smʌdʒ/ *n. N. Amer.* an outdoor fire with dense smoke made to keep off insects, protect plants against frost, etc. □ **smudge-pot** a container holding burning material that produces a smudge. [*smudge* (v.) cure (herring) by smoking (16th c.: orig. unkn.)]

smudgy /'smʌdʒɪ/ *adj.* (**smudgier, smudgiest**) **1** smudged. **2** likely to produce smudges. □ **smudgily** *adv.* **smudginess** *n.*

smug /smʌɡ/ *adj.* (**smugger, smuggest**) self-satisfied; complacent. □ **smugly** *adv.* **smugness** *n.* [16th c., orig. 'neat' f. LG *smuk* pretty]

smuggle /'smʌɡ(ə)l/ *v.tr.* **1** (also *absol.*) import or export (goods) illegally, esp. without payment of customs duties. **2** (foll. by *in, out*) convey secretly. **3** (foll. by *away*) put into concealment. □ **smuggler** *n.* **smuggling** *n.* [17th c. (also *smuckle*) f. LG *smukkeln smuggelen*]

smut /smʌt/ *n. & v.* ● *n.* **1** a small flake of soot etc. **2** a spot or smudge made by this. **3** obscene or lascivious talk, pictures, or stories. **4 a** a fungal disease of cereals in which parts of the ear change to black powder. **b** a fungus of the order Ustilaginales that causes this. ● *v.* (**smutted, smutting**) **1** *tr.* mark with smuts. **2** *tr.* infect (a plant) with smut. **3** *intr.* (of a plant) contract smut. □ **smut-ball** *Agriculture* grain affected by smut. **smut-mill** a machine for freeing grain from smut. □ **smutty** *adj.* (**smuttier, smuttiest**) (esp. in sense 3 of *n.*). **smuttily** *adv.* **smuttiness** *n.* [rel. to LG *smutt*, MHG *smutz(en)* etc.: cf. OE *smitt(ian)* smear, and SMUDGE[1]]

Smuts /smʌts/, Jan Christiaan (1870–1950), South African statesman and soldier, Prime Minister 1919–24 and 1939–48. He led Boer forces during the Second Boer War, but afterwards supported Louis Botha's policy of Anglo-Boer cooperation and was one of the founders of the Union of South Africa. During the First World War he led Allied troops against the Germans in East Africa (1916); he later attended the peace conference at Versailles in 1919 and helped to found the League of Nations. He then succeeded Botha as Prime Minister, a post he held again between 1939 and 1948. After the Second World War Smuts played a leading role in the formation of the United Nations and drafted the preamble to the UN charter.

Smyrna /'smɜːnə/ an ancient city on the west coast of Asia Minor, on the site of modern Izmir in Turkey.

Sn *symb. Chem.* the element tin.

snack /snæk/ *n. & v.* ● *n.* **1** a light, casual, or hurried meal. **2** a small amount of food eaten between meals. **3** *Austral. sl.* something easy to accomplish. ● *v.intr.* (often foll. by *on*) eat a snack. □ **snack bar** a place where snacks are sold. [ME, orig. a snap or bite, f. MDu. *snac(k)* f. *snacken* (v.), var. of *snappen*]

snaffle /'snæf(ə)l/ *n. & v.* ● *n.* (in full **snaffle-bit**) a simple bridle-bit without a curb and usu. with a single rein. ● *v.tr.* **1** put a snaffle on. **2** *colloq.* steal; seize; appropriate. [prob. f. LG or Du.: cf. MLG, MDu. *snavel* beak, mouth]

snafu /snæˈfuː/ *adj. & n. sl.* ● *adj.* in utter confusion or chaos. ● *n.* this state. [acronym for 'situation normal: all fouled (or fucked) up']

snag[1] /snæɡ/ *n. & v.* ● *n.* **1** an unexpected or hidden obstacle or drawback. **2** a jagged or projecting point or broken stump. **3** a tear in material etc. **4** a short tine of an antler. ● *v.tr.* (**snagged, snagging**) **1** catch or tear on a snag. **2** clear (land, a waterway, a tree-trunk, etc.) of snags. **3** *N. Amer.* catch or obtain by quick action. □ **snagged** *adj.* **snaggy** *adj.* [prob. f. Scand.: cf. Norw. dial. *snag(e)* sharp point]

snag[2] /snæɡ/ *n.* (usu. in *pl.*) *Austral. sl.* a sausage. [20th c.: orig. unkn.]

snaggle-tooth /'snæɡ(ə)lˌtuːθ/ *n.* (*pl.* **snaggle-teeth**) an irregular or projecting tooth. □ **snaggle-toothed** *adj.* [SNAG[1] + -LE[2]]

snail /sneɪl/ *n.* **1** any slow-moving gastropod mollusc with a spiral shell able to enclose the whole body. **2** a slow or lazy person; a dawdler. □ **snail's pace** a very slow movement. □ **snail-like** *adj.* [OE *snæg(e)l* f. Gmc]

snake /sneɪk/ *n. & v.* ● *n.* **1 a** an elongated limbless reptile of the suborder Ophidia (or Serpentes), including many kinds with a venomous bite. (*See note below.*) **b** a limbless lizard or amphibian. **2** (also **snake in the grass**) a treacherous person or secret enemy. **3** (prec. by *the*) a system of interconnected exchange rates for the EC currencies. ● *v.intr.* move or twist like a snake. □ **snake bird** = ANHINGA. **snake-charmer** a person appearing to make snakes move by music etc. **snake-pit 1** a pit containing snakes. **2** a scene of vicious behaviour. **snakes and ladders** a game with counters moved along a board with advances up 'ladders' or returns down 'snakes' depicted on the board. **snake's head** a bulbous liliaceous plant, *Fritillaria meleagris*, with hanging bell-shaped chequered flowers (cf. FRITILLARY 1). □ **snakelike** *adj.* [OE *snaca*]

■ There are nearly 2,400 species of snake, belonging to eleven or more families. They evolved from burrowing lizards during the Cretaceous period, and the pythons and other primitive snakes still bear the vestiges of hindlimbs. Unlike lizards, they lack eyelids and shed the skin (including the 'spectacles' over the eyes) in one piece. All snakes are carnivorous and their jaws are capable of considerable extension to assist the swallowing of large prey. Many overpower their prey by constriction, which results in death by asphyxiation. Venomous bites have developed independently in three families, being associated with small fangs at the back of the mouth (some members of Colubridae), larger fixed fangs at the front of the mouth (Elapidae, e.g. cobras and mambas), or large folding fangs (Viperidae, e.g. vipers and rattlesnakes).

snakebite /'sneɪkbaɪt/ *n.* **1** a bite made by a snake. **2** *US sl.* strong alcoholic drink; esp. whisky of inferior quality. **3** *Brit. colloq.* a mixed drink of cider and lager.

Snake River a river of the north-western US. Rising in Yellowstone National Park in Wyoming, it flows for 1,670 km (1,038 miles) through Idaho into the state of Washington, where it joins the Columbia river.

snakeroot /'sneɪkruːt/ *n.* any North American plant with roots reputed to contain an antidote to snake's poison, e.g. Virginia snakeroot (*Aristolochia serpentaria*).

snaky /'sneɪkɪ/ *adj.* **1** of or like a snake. **2** winding; sinuous. **3** showing coldness, ingratitude, venom, or guile. **4** infested with or composed of snakes. **5** *Austral. sl.* angry; irritable. □ **snakily** *adv.* **snakiness** *n.*

snap /snæp/ v., n., adv., & adj. ● v. (**snapped**, **snapping**) **1** intr. & tr. break suddenly or with a snap. **2** intr. & tr. emit or cause to emit a sudden sharp sound or crack. **3** intr. & tr. open or close with a snapping sound (*the bag snapped shut*). **4 a** intr. (often foll. by *at*) speak irritably or spitefully (to a person) (*did not mean to snap at you*). **b** tr. say irritably or spitefully. **5** intr. (often foll. by *at*) (esp. of a dog etc.) make a sudden audible bite. **6** tr. & intr. move quickly (*snap into action*). **7** tr. take a snapshot of. **8** tr. *Amer. Football* put (the ball) into play on the ground by a quick backward movement. ● n. **1** an act or sound of snapping. **2** a crisp biscuit or cake (*brandy snap; ginger snap*). **3** a snapshot. **4** (in full **cold snap**) a sudden brief spell of cold weather. **5** *Brit.* **a** a card-game in which players call 'snap' when two similar cards are exposed. **b** (as *int.*) on noticing the (often unexpected) similarity of two things. **6** crispness of style; fresh vigour or liveliness in action; zest; dash; spring. **7** *N. Amer. sl.* an easy task (*it was a snap*). ● adv. with the sound of a snap (*heard it go snap*). ● adj. done or taken on the spur of the moment, unexpectedly, or without notice (*snap decision*). □ **snap at** accept (bait, a chance, etc.) eagerly (see also senses 4a and 5 of *v.*). **snap bean** *US* a bean grown for its pods, which are broken into pieces and eaten. **snap-bolt** (or **-lock**) a bolt etc. which locks automatically when a door or window closes. **snap-brim** (of a hat) with a brim that can be turned up and down at opposite sides. **snap-fastener** = *press-stud* (see PRESS[1]). **snap one's fingers 1** make a sharp noise by bending the last joint of the finger against the ball of the thumb and suddenly releasing it, esp. in rhythm to music etc. **2** (often foll. by *at*) defy; show contempt for. **snap-hook** (or **-link**) a hook or link with a spring allowing the entrance but barring the escape of a cord, link, etc. **snap off** break off or bite off. **snap off a person's head** address a person angrily or rudely. **snap out** say irritably. **snap out of** *colloq.* get rid of (a mood, habit, etc.) by a sudden effort. **snapping turtle** a large aggressive American freshwater turtle of the family Chelydridae, with a large head and long tail. **snap up 1** accept (an offer, a bargain) quickly or eagerly. **2** pick up or catch hastily or smartly. **3** interrupt (another person) before he or she has finished speaking. □ **snappable** adj. **snappingly** adv. [prob. f. MDu. or MLG *snappen*, partly imit.]

snapdragon /'snæp,dragən/ n. the garden antirrhinum, *Antirrhinum majus*, with an irregular flower which gapes like a mouth when a bee lands on the lower part.

snapper /'snæpə(r)/ n. **1** a person or thing that snaps. **2 a** a large marine fish of the family Lutjanidae, usu. reddish in colour and with a triangular head profile, used as food. **b** *Austral.* a sparid fish, *Chrysophrys auratus*, used as food. **c** *N. Amer.* see BLUEFISH. **3** a snapping turtle. **4** *US* a cracker (as a toy).

snappish /'snæpɪʃ/ adj. **1** (of a person's manner or a remark) curt; ill-tempered; sharp. **2** (of a dog etc.) inclined to snap. □ **snappishly** adv. **snappishness** n.

snappy /'snæpɪ/ adj. (**snappier**, **snappiest**) **1** brisk, full of zest. **2** neat and elegant (*a snappy dresser*). **3** snappish. □ **make it snappy** *colloq.* be quick about it. □ **snappily** adv. **snappiness** n.

snapshot /'snæpʃɒt/ n. a casual photograph taken quickly with a small hand-camera.

snare /sneə(r)/ n. & v. ● n. **1** a trap for catching birds or animals, esp. with a noose of wire or cord. **2** a thing that acts as a temptation. **3** a device for tempting an enemy etc. to expose himself or herself to danger, failure, loss, capture, defeat, etc. **4** (in *sing.* or *pl.*) *Mus.* twisted strings of gut, hide, or wire stretched across the lower head of a side-drum to produce a rattling sound. **5** (in full **snare drum**) a drum fitted with snares. **6** *Surgery* a wire loop for extracting polyps etc. ● v.tr. **1** catch (a bird etc.) in a snare. **2** ensnare; lure or trap (a person) with a snare. □ **snarer** n. (also in *comb.*). [OE *sneare* f. ON *snara*: senses 4 & 5 prob. f. MLG or MDu.]

snark /snɑːk/ n. a fabulous animal, originally the subject of the nonsense poem *The Hunting of the Snark* (1876) by Lewis Carroll.

snarl[1] /snɑːl/ v. & n. ● v. **1** intr. (of a dog) make an angry growl with bared teeth. **2** intr. (of a person) speak cynically; make bad-tempered complaints or criticisms. **3** tr. (often foll. by *out*) **a** utter in a snarling tone. **b** express (discontent etc.) by snarling. ● n. the act or sound of snarling. □ **snarler** n. **snarlingly** adv. **snarly** adj. (**snarlier**, **snarliest**). [earlier *snar* f. (M)LG, MHG *snarren*]

snarl[2] /snɑːl/ v. & n. ● v. **1** tr. (often foll. by *up*) twist; entangle; confuse and hamper the movement of (traffic etc.). **2** intr. (often foll. by *up*) become entangled, congested, or confused. **3** tr. adorn the exterior of (a narrow metal vessel) with raised work. ● n. a knot or tangle. □ **snarling iron** an implement used for snarling metal. **snarl-up**

colloq. a traffic jam; a muddle; a mistake. [ME f. *snare* (n. & v.): sense 3 perh. f. noun in dial. sense 'knot in wood']

snatch /snætʃ/ v. & n. ● v.tr. **1** seize quickly, eagerly, or unexpectedly, esp. with outstretched hands. **2** steal (a wallet, handbag, etc.); kidnap. **3** secure with difficulty (*snatched an hour's rest*). **4** (foll. by *away*, *from*) take away or from, esp. suddenly (*snatched away my hand*). **5** (foll. by *from*) rescue narrowly (*snatched from the jaws of death*). **6** (foll. by *at*) **a** try to seize by stretching or grasping suddenly. **b** take (an offer etc.) eagerly. ● n. **1** an act of snatching (*made a snatch at it*). **2** a fragment of a song or talk etc. (*caught a snatch of their conversation*). **3** esp. *US colloq.* a kidnapping. **4** (in weightlifting) the rapid raising of a weight from the floor to above the head. **5** a short spell of activity etc. □ **in** (or **by**) **snatches** in fits and starts. □ **snatcher** n. (esp. in sense 3 of n.). **snatchy** adj. [ME *snecchen*, *sna(c)che*, perh. rel. to SNACK]

snavel /'snæv(ə)l/ v.tr. (also **snavle**, **snavvle** /-vɪl/) *Austral. sl.* catch; take; steal. [Engl. dial. (as SNAFFLE)]

snazzy /'snæzɪ/ adj. (**snazzier**, **snazziest**) *colloq.* smart or attractive esp. in an ostentatious way. □ **snazzily** adv. **snazziness** n. [20th c.: orig. unkn.]

sneak /sniːk/ v., n., & adj. (*past* and *past part.* **sneaked** or esp. *US* **snuck** /snʌk/) ● v. **1** intr. & tr. (foll. by *in*, *out*, *past*, *away*, etc.) go or convey furtively; slink. **2** tr. steal unobserved; make off with. **3** intr. *Brit. school sl.* tell tales; turn informer. **4** intr. (as **sneaking** adj.) **a** furtive; undisclosed (*have a sneaking affection for him*). **b** persistent in one's mind; nagging (*a sneaking feeling that it is not right*). ● n. **1** a mean-spirited cowardly underhand person. **2** *Brit. school sl.* a tell-tale. ● adj. acting or done without warning; secret (*a sneak attack*). □ **sneak-thief** a thief who steals without breaking in; a pickpocket. □ **sneakingly** adv. [16th c., prob. dial.: perh. rel. to ME *snike*, OE *snican* creep]

sneaker /'sniːkə(r)/ n. esp. *N. Amer.* each of a pair of soft-soled canvas etc. shoes.

sneaky /'sniːkɪ/ adj. (**sneakier**, **sneakiest**) given to or characterized by sneaking; furtive, mean. □ **sneakily** adv. **sneakiness** n.

sneck /snek/ n. & v. *Sc. & N. Engl.* ● n. a latch. ● v.tr. latch (a door etc.); close or fasten with a sneck. [ME, rel. to SNATCH]

Sneek /sneɪk/ a town in the Netherlands, a water-sports centre with a large yachting harbour; pop. (1991) 29,280.

sneer /snɪə(r)/ n. & v. ● n. a derisive smile or remark. ● v. **1** intr. (often foll. by *at*) smile derisively. **2** tr. say sneeringly. **3** intr. (often foll. by *at*) speak derisively, esp. covertly or ironically (*sneered at his attempts*). □ **sneerer** n. **sneering** adj. **sneeringly** adv. [16th c.: orig. unkn.]

sneeze /sniːz/ n. & v. ● n. **1** a sudden involuntary expulsion of air from the nose and mouth caused by irritation of the nostrils. **2** the sound of this. ● v.intr. make a sneeze. □ **not to be sneezed at** *colloq.* not contemptible; considerable; notable. □ **sneezer** n. **sneezy** adj. [ME *snese*, app. alt. of obs. *fnese* f. OE *-fnēsan*, ON *fnýsa* & replacing earlier and less expressive *nese*]

sneezewort /'sniːzwɜːt/ n. a kind of yarrow, *Achillea ptarmica*, whose powdered leaves are used to induce sneezing.

Snell's law /snelz/ n. *Physics* the law that the ratio of the sines of the angles of incidence and refraction of a wave are constant when it passes between two given media. [Willebrord *Snell*, Dutch mathematician (1591–1626)]

snib /snɪb/ v. & n. esp. *Sc. & Ir.* ● v.tr. (**snibbed**, **snibbing**) bolt, fasten, or lock (a door etc.). ● n. a lock, catch, or fastening for a door or window. [19th c.: orig. uncert.]

snick /snɪk/ v. & n. ● v.tr. **1** cut a small notch in. **2** make a small incision in. **3** *Cricket* deflect (the ball) slightly with the bat. ● n. **1** a small notch or cut. **2** *Cricket* a slight deflection of the ball by the bat. [18th c.: prob. f. *snick-a-snee* fight with knives]

snicker /'snɪkə(r)/ v. & n. ● v.intr. **1** = SNIGGER v. **2** whinny, neigh. ● n. **1** = SNIGGER n. **2** a whinny, a neigh. □ **snickeringly** adv. [imit.]

snicket /'snɪkɪt/ n. *dial.* a narrow passage between houses; an alleyway. [19th c.: orig. unkn.]

snide /snaɪd/ adj. & n. ● adj. **1** sneering; slyly derogatory; insinuating. **2** counterfeit; bogus. **3** *US* mean; underhand. ● n. a snide person or remark. □ **snidely** adv. **snideness** n. [19th-c. *colloq.*: orig. unkn.]

sniff /snɪf/ v. & n. ● v. **1** intr. draw up air audibly through the nose to stop it running or to detect a smell or as an expression of contempt. **2** tr. (often foll. by *up*) draw in (a scent, drug, liquid, or air) through the nose. **3** tr. draw in the scent of (food, drink, flowers, etc.) through the nose. ● n. **1** an act or sound of sniffing. **2** the amount of air etc. sniffed up. □ **sniff at 1** try the smell of; show interest in. **2** show contempt for or discontent with. **sniff out** detect; discover by investigation. □ **sniffingly** *adv.* [ME, imit.]

sniffer /'snɪfə(r)/ *n.* **1** a person who sniffs, esp. one who sniffs a drug or toxic substances (often in *comb.*: *glue-sniffer*). **2** *sl.* the nose. **3** *colloq.* any device for detecting gas, radiation, etc. □ **sniffer-dog** *colloq.* a dog trained to sniff out drugs or explosives.

sniffle /'snɪf(ə)l/ *v. & n.* ● *v.intr.* sniff slightly or repeatedly. ● *n.* **1** the act of sniffling. **2** (in *sing.* or *pl.*) a cold in the head causing a running nose and sniffling. □ **sniffler** *n.* **sniffly** *adj.* [imit.: cf. SNIVEL]

sniffy /'snɪfɪ/ *adj.* (**sniffier**, **sniffiest**) *colloq.* **1** inclined to sniff. **2** disdainful; contemptuous. □ **sniffily** *adv.* **sniffiness** *n.*

snifter /'snɪftə(r)/ *n.* **1** *colloq.* a small drink of alcohol. **2** *N. Amer.* a balloon glass for brandy. □ **snifter-valve** a valve in a steam engine to allow air in or out. [dial. *snift* sniff, perh. f. Scand.: imit.]

snig /snɪg/ *v.tr.* (**snigged**, **snigging**) *Austral. & NZ* drag with a jerk. □ **snigging chain** a chain used to move logs. [Engl. dial.]

snigger /'snɪgə(r)/ *n. & v.* ● *n.* a half-suppressed secretive laugh. ● *v.intr.* utter such a laugh. □ **sniggerer** *n.* **sniggeringly** *adv.* [var. of SNICKER]

sniggle /'snɪg(ə)l/ *v.intr.* fish (for eels) by pushing bait into a hole. [ME *snig* small eel, of unkn. orig.]

snip /snɪp/ *v. & n.* ● *v.tr.* (**snipped**, **snipping**) (also *absol.*) cut (cloth, a hole, etc.) with scissors or shears, esp. in small quick strokes. ● *n.* **1** an act of snipping. **2** a piece of material etc. snipped off. **3** *colloq.* **a** something easily achieved. **b** *Brit.* a bargain; something cheaply acquired. **4** (in *pl.*) hand-shears for metal cutting. □ **snip at** make snipping strokes at. □ **snipping** *n.* [LG & Du. *snippen* imit.]

snipe /snaɪp/ *n. & v.* ● *n.* (*pl.* same or **snipes**) a brownish wading bird of the sandpiper family, esp. of the genus *Gallinago*, with a long straight bill and frequenting marshy ground. ● *v.* **1** *intr.* fire shots from hiding, usu. at long range. **2** *tr.* kill by sniping. **3** *intr.* (foll. by *at*) make a sly critical attack. **4** *intr.* go snipe-shooting. □ **snipe eel** a slender eel-like deep-sea fish of the family Nemichthyidae, with a long thin snout. **snipe fish** a marine fish of the family Macrorhamphosidae, with a deep compressed body and long snout. □ **sniper** *n.* [ME, prob. f. Scand.: cf. Icel. *mýrisnípa*, & MDu., MLG *snippe*, OHG *snepfa*]

snippet /'snɪpɪt/ *n.* **1** a small piece cut off. **2** (usu. in *pl.*; often foll. by *of*) **a** a scrap or fragment of information, knowledge, etc. **b** a short extract from a book, newspaper, etc. □ **snippety** *adj.*

snippy /'snɪpɪ/ *adj.* (**snippier**, **snippiest**) *colloq.* fault-finding, snappish, sharp. □ **snippily** *adv.* **snippiness** *n.*

snit /snɪt/ *n. N. Amer.* a rage; a sulk (esp. *in a snit*). [20th c.: orig. unkn.]

snitch /snɪtʃ/ *v. & n. sl.* ● *v.* **1** *tr.* steal. **2** *intr.* (often foll. by *on*) inform on a person. ● *n.* an informer. [17th c.: orig. unkn.]

snivel /'snɪv(ə)l/ *v. & n.* ● *v.intr.* (**snivelled**, **snivelling**; *US* **sniveled**, **sniveling**) **1** weep with sniffling. **2** run at the nose; make a repeated sniffing sound. **3** show weak or tearful sentiment. ● *n.* **1** running mucus. **2** hypocritical talk; cant. □ **sniveller** *n.* **snivelling** *adj.* **snivellingly** *adv.* [ME f. OE *snyflan* (unrecorded) f. *snofl* mucus: cf. SNUFFLE]

snob /snɒb/ *n.* **1 a** a person with an exaggerated respect for social position or wealth and who despises people considered socially inferior. **b** (*attrib.*) related to or characteristic of this attitude. **2** a person who despises others whose (usu. specified) tastes or attainments are considered inferior (*an intellectual snob*; *a wine snob*). □ **snobbery** *n.* (*pl.* **-ies**). **snobbish** *adj.* **snobbishly** *adv.* **snobbishness** *n.* **snobby** *adj.* (**snobbier**, **snobbiest**). [18th c. (now dial.) 'cobbler': orig. unkn.]

SNOBOL /'snəʊbɒl/ *n.* *Computing* a high-level programming language used esp. in manipulating textual data. [partial acronym, f. string-oriented symbolic language, after COBOL]

snoek /snuːk/ *n. S. Afr.* a barracouta. [Afrik. f. Du., = PIKE[1], f. MLG *snōk*, prob. rel. to SNACK]

snog /snɒg/ *v. & n. Brit. sl.* ● *v.intr. & tr.* (**snogged**, **snogging**) engage in kissing and caressing (with). ● *n.* a spell of snogging. □ **snogger** *n.* [20th c.: orig. unkn.]

snood /snuːd/ *n.* **1** an ornamental hairnet usu. worn at the back of the head. **2** a ring of woollen etc. material worn as a hood. **3** a short line attaching a hook to a main line in sea fishing. **4** *hist.* a ribbon or band worn by unmarried women in Scotland to confine their hair. [OE *snōd*]

snook[1] /snuːk/ *n.* a contemptuous gesture with the thumb to the nose and the fingers spread out. □ **cock a snook** (often foll. by *at*) **1** make this gesture. **2** register one's contempt (for a person, establishment, etc.). [19th c.: orig. unkn.]

snook[2] /snuːk/ *n.* a marine fish, *Centropomus undecimalis*, of the Caribbean, used as food. [Du. *snoek*: see SNOEK]

snooker /'snuːkə(r)/ *n. & v.* ● *n.* **1** a game played with cues on a rectangular baize-covered table in which the players use a cue-ball (white) to pocket the other balls (fifteen red and six coloured) in a set order. (*See note below.*) **2** a position in this game in which a direct shot at a permitted ball is impossible. ● *v.tr.* **1** (also *refl.*) subject (oneself or another player) to a snooker. **2** (esp. as **snookered** *adj.*) defeat; thwart. [19th c.: orig. unkn.]

▪ The game of snooker is a combination of pool and pyramids, developed in the 1870s by British army officers in India. In Britain, Joe Davis first brought snooker to a greater pitch of skill and popularity than billiards in the 1920s, but with the advent of a BBC series *Pot Black* and, in the 1960s, with colour television becoming more widespread, snooker developed into one of the most popular sports both for participants and for spectators.

snoop /snuːp/ *v. & n. colloq.* ● *v.intr.* **1** pry into matters one need not be concerned with. **2** (often foll. by *about*, *around*) investigate in order to find out transgressions of the law etc. ● *n.* **1** an act of snooping. **2** a person who snoops; a detective. □ **snooper** *n.* **snoopy** *adj.* [Du. *snœpen* eat on the sly]

snooperscope /'snuːpəˌskəʊp/ *n.* a device which converts infrared radiation into a visible image, esp. used for seeing in the dark.

snoot /snuːt/ *n. colloq.* the nose. [var. of SNOUT]

snooty /'snuːtɪ/ *adj.* (**snootier**, **snootiest**) *colloq.* supercilious; conceited. □ **snootily** *adv.* **snootiness** *n.* [20th c.: orig. unkn.]

snooze /snuːz/ *n. & v. colloq.* ● *n.* a short sleep, esp. in the daytime. ● *v.intr.* take a snooze. □ **snoozer** *n.* **snoozy** *adj.* (**snoozier**, **snooziest**). [18th-c. sl.: orig. unkn.]

snore /snɔː(r)/ *n. & v.* ● *n.* a snorting or grunting sound in breathing during sleep. ● *v.intr.* make this sound. □ **snore away** pass (time) sleeping or snoring. □ **snorer** *n.* **snoringly** *adv.* [ME, prob. imit.: cf. SNORT]

Snorkel /'snɔːk(ə)l/ *n. propr.* a piece of apparatus consisting of a platform which may be elevated and extended, used in fighting fires in tall buildings.

snorkel /'snɔːk(ə)l/ *n. & v.* (also **schnorkel** /'ʃnɔː-/) ● *n.* **1** a breathing-tube for an underwater swimmer. **2** a device for supplying air to a submerged submarine. ● *v.intr.* (**snorkelled**, **snorkelling**; *US* **snorkeled**, **snorkeling**) use a snorkel. □ **snorkeller** *n.* [G *Schnorchel*]

Snorri Sturluson /ˌsnɔːrɪ 'stɜːləs(ə)n/ (1178–1241), Icelandic historian and poet. A leading figure of medieval Icelandic literature, he wrote the *Younger* or *Prose Edda* and the *Heimskringla*, a history of the kings of Norway from mythical times to the year 1177. (See also EDDA.)

snort /snɔːt/ *n. & v.* ● *n.* **1** an explosive sound made by the sudden forcing of breath through the nose, esp. expressing indignation or incredulity. **2** a similar sound made by an engine etc. **3** *colloq.* a small drink of liquor. **4** *sl.* an inhaled dose of a (usu. illegal) powdered drug. ● *v.* **1** *intr.* make a snort. **2** *intr.* (of an engine etc.) make a sound resembling this. **3** *tr.* (also *absol.*) *sl.* inhale (a usu. illegal narcotic drug, esp. cocaine or heroin). **4** *tr.* express (defiance etc.) by snorting. □ **snort out** express (words, emotions, etc.) by snorting. [ME, prob. imit.: cf. SNORE]

snorter /'snɔːtə(r)/ *n. colloq.* **1** something very impressive or difficult. **2** something vigorous or violent.

snot /snɒt/ *n. sl.* **1** nasal mucus. **2** a term of contempt for a person. □ **snot-rag** a handkerchief. [prob. f. MDu., MLG *snotte*, MHG *snuz*, rel. to SNOUT]

snotty /'snɒtɪ/ *adj.* (**snottier**, **snottiest**) *sl.* **1** running or foul with nasal mucus. **2** contemptible. **3** supercilious, conceited. □ **snottily** *adv.* **snottiness** *n.*

snout /snaʊt/ *n.* **1** the projecting nose and mouth of an animal. **2** *derog.* a person's nose. **3** the pointed front of a thing; a nozzle. **4** *Brit. sl.* tobacco or a cigarette. **5** *sl.* an informer. □ **snout-beetle** a weevil. □ **snouted** *adj.* (also in *comb.*). **snouty** *adj.* [ME f. MDu., MLG *snūt*]

Snow /snəʊ/, C(harles) P(ercy), 1st Baron Snow of Leicester (1905–80), English novelist and scientist. The title under which his sequence of eleven novels *Strangers and Brothers* appeared in 1940; the series, which also includes *The Masters* (1951) and *The Corridors of Power* (1964), deals with moral dilemmas and power-struggles in the academic world. He is also known for his lecture *Two Cultures*, delivered in 1959, in which he discussed the deleterious effects of the division between science and the humanities, which provoked an acerbic and controversial response from the critic F. R. Leavis.

snow /snəʊ/ *n. & v.* ● *n.* **1** atmospheric water vapour frozen into ice crystals and falling to earth in light white flakes. **2** a fall of this, or a layer of it on the ground. **3 a** a thing resembling snow in whiteness or texture etc. **b** a frozen vapour resembling snow (*carbon dioxide snow*). **4** a mass of flickering white spots on a television or radar screen, caused by interference or a poor signal. **5** *sl.* cocaine. **6** a dessert or other dish resembling snow. ● *v.* **1** *intr.* (prec. by *it* as subject) snow falls (*it is snowing; if it snows*). **2** *tr.* (foll. by *in, over, up,* etc.) confine or block with large quantities of snow. **3** *tr. & intr.* sprinkle or scatter or fall as or like snow. **4** *intr.* come in large numbers or quantities. **5** *tr. US* deceive or charm with plausible words. □ **be snowed under** be overwhelmed, esp. with work. **snow-blind** temporarily blinded by the glare of light reflected by large expanses of snow. **snow-blindness** this blindness. **snow-blink** the reflection in the sky of snow or ice fields. **snow boot** an overboot of rubber and cloth. **snow-broth** melted or melting snow. **snow bunting** a northern bunting, *Plectrophenax nivalis*, the male of which has mainly white plumage. **snow goose** a white Arctic goose, *Anser caerulescens*, with black-tipped wings. **snow-ice** opaque white ice formed from melted snow. **snow job** esp. *US sl.* an attempt at flattery or deception. **snow-job** *v.tr.* esp. *US sl.* do a snow job on (a person). **snow leopard** a large rare feline, *Panthera uncia*, found in the mountains of central Asia, having a greyish-white coat with black markings (also called *ounce*). **snow owl** *US* = *snowy owl*. **snow partridge** a black and white partridge, *Lerwa lerwa*, found in the Himalayas. **snow-slip** an avalanche. **snow-white** pure white. □ **snowless** *adj.* **snowlike** *adj.* [OE *snāw* f. Gmc]

snowball /ˈsnəʊbɔːl/ *n. & v.* ● *n.* **1** snow pressed together into a ball, esp. for throwing in play. **2** anything that grows or increases rapidly like a snowball rolled on snow. ● *v.* **1** *intr. & tr.* throw or pelt with snowballs. **2** *intr.* increase rapidly. □ **snowball tree** a guelder rose.

snowberry /ˈsnəʊˌbɛrɪ/ *n.* (*pl.* **-ies**) a shrub with white berries; esp. the ornamental *Symphoricarpos albus*, of the honeysuckle family.

snowblower /ˈsnəʊˌbləʊə(r)/ *n.* a machine that clears snow by blowing it to the side of the road etc.

snowbound /ˈsnəʊbaʊnd/ *adj.* prevented by snow from going out or travelling.

snowcap /ˈsnəʊkæp/ *n.* **1** the tip of a mountain when covered with snow. **2** a white-crowned hummingbird, *Microchera albocoronata*, native to Central America. □ **snowcapped** *adj.*

Snowdon /ˈsnəʊd(ə)n/ a mountain in NW Wales. Rising to 1,085 m (3,560 ft), it is the highest mountain in Wales.

Snowdonia /snəʊˈdəʊnɪə/ a massif region in Gwynedd, NW Wales, forming the heart of the Snowdonia National Park. Its highest peak is Snowdon.

snowdrift /ˈsnəʊdrɪft/ *n.* a bank of snow heaped up by the action of the wind.

snowdrop /ˈsnəʊdrɒp/ *n.* a bulbous plant, *Galanthus nivalis*, with white drooping flowers in the early spring.

snowfall /ˈsnəʊfɔːl/ *n.* **1** a fall of snow. **2** *Meteorol.* the amount of snow that falls on one occasion or on a given area within a given time.

snowfield /ˈsnəʊfiːld/ *n.* a permanent wide expanse of snow in mountainous or polar regions.

snowflake /ˈsnəʊfleɪk/ *n.* **1** a flake of snow; esp. a feathery ice crystal, often displaying delicate sixfold symmetry. **2 a** a bulbous plant of the genus *Leucojum*, resembling the snowdrop but with larger flowers. **b** the green-tipped white flower of this plant.

snowline /ˈsnəʊlaɪn/ *n.* the level above which snow never melts entirely.

snowman /ˈsnəʊmæn/ *n.* (*pl.* **-men**) a figure resembling a man, made of compressed snow.

snowmobile /ˈsnəʊməˌbiːl/ *n.* a motor vehicle, esp. with runners or caterpillar tracks, for travelling over snow.

snowplough /ˈsnəʊplaʊ/ *n.* (*US* **snowplow**) **1** a device, or a vehicle equipped with one, for clearing roads of thick snow. **2** *Skiing* a movement involving turning the points of the skis inwards so as to bring the skier to a stop.

snowshoe /ˈsnəʊʃuː/ *n. & v.* ● *n.* a flat device like a racket attached to a boot for walking on snow without sinking in. ● *v.intr.* travel on snowshoes. □ **snowshoe hare** (or **rabbit**) a North American hare, *Lepus americanus*, with large furry hind feet and white fur in winter. □ **snowshoer** /-ˌʃuːə(r)/ *n.*

snowstorm /ˈsnəʊstɔːm/ *n.* a heavy fall of snow, esp. with a high wind.

snowy /ˈsnəʊɪ/ *adj.* (**snowier, snowiest**) **1** of or like snow; pure white.

2 (of the weather etc.) with much snow; covered with snow. □ **snowy owl** a large white owl, *Nyctea scandiaca*, native to the Arctic. □ **snowily** *adv.* **snowiness** *n.*

SNP see SCOTTISH NATIONAL PARTY.

Snr. *abbr.* Senior.

snub /snʌb/ *v., n.,* & *adj.* ● *v.tr.* (**snubbed, snubbing**) **1** rebuff or humiliate with sharp words or a marked lack of cordiality. **2** check the movement of (a boat, horse, etc.), esp. by a rope wound round a post etc. ● *n.* an act of snubbing; a rebuff. ● *adj.* short and blunt in shape. □ **snub nose** a short turned-up nose. **snub-nosed** having a snub nose. □ **snubber** *n.* **snubbingly** *adv.* [ME f. ON *snubba* chide, check the growth of]

snuck esp. *US* past and past part. of SNEAK.

snuff[1] /snʌf/ *n. & v.* ● *n.* the charred part of a candle-wick. ● *v.tr.* trim the snuff from (a candle). □ **snuff it** *Brit. sl.* die. **snuff out 1** extinguish by snuffing. **2** kill; put an end to. [ME *snoffe, snuffe;* orig. unkn.]

snuff[2] /snʌf/ *n. & v.* ● *n.* powdered tobacco or medicine taken by sniffing it up the nostrils. ● *v.intr.* take snuff. □ **snuff-coloured** dark yellowish-brown. **up to snuff** *sl.* **1** *Brit.* knowing; not easily deceived. **2** up to standard. [Du. *snuf* (*tabak* tobacco) f. MDu. *snuffen* snuffle]

snuffbox /ˈsnʌfbɒks/ *n.* a small usu. ornamental box for holding snuff.

snuffer /ˈsnʌfə(r)/ *n.* **1** a small hollow cone with a handle used to extinguish a candle. **2** (in *pl.*) an implement like scissors used to extinguish a candle or trim its wick.

snuffle /ˈsnʌf(ə)l/ *v. & n.* ● *v.* **1** *intr.* make sniffing sounds. **2 a** *intr.* speak nasally, whiningly, or like one with a cold. **b** *tr.* (often foll. by *out*) say in this way. **3** *intr.* breathe noisily as through a partially blocked nose. **4** *intr.* sniff. ● *n.* **1** a snuffling sound or tone. **2** (in *pl.*) a partial blockage of the nose causing snuffling. **3** a sniff. □ **snuffler** *n.* **snuffly** *adj.* [prob. f. LG & Du. *snuffelen* (as SNUFF²): cf. SNIVEL]

snuffy[1] /ˈsnʌfɪ/ *adj.* (**snuffier, snuffiest**) **1** annoyed. **2** irritable. **3** supercilious or contemptuous. [SNUFF + -Y¹]

snuffy[2] /ˈsnʌfɪ/ *adj.* like snuff in colour or substance. [SNUFF² + -Y¹]

snug /snʌg/ *adj. & n.* ● *adj.* (**snugger, snuggest**) **1 a** cosy, comfortable, sheltered; well enclosed or placed or arranged. **b** cosily protected from the weather or cold. **c** close-fitting. **2** (of an income etc.) allowing comfort and comparative ease. ● *n. Brit.* a small room in a pub or inn. □ **snugly** *adv.* **snugness** *n.* [16th c. (orig. Naut.): prob. of LG or Du. orig.]

snuggery /ˈsnʌgərɪ/ *n.* (*pl.* **-ies**) **1** a snug place, esp. a person's private room or den. **2** *Brit.* = SNUG *n.*

snuggle /ˈsnʌg(ə)l/ *v.intr. & tr.* (usu. foll. by *down, up, together*) settle or draw into a warm comfortable position. [SNUG + -LE⁴]

So. *abbr.* South.

so[1] /səʊ/ *adv. & conj.* ● *adv.* **1** (often foll. by *that* + clause) to such an extent, or to the extent implied (*why are you so angry?; do stop complaining so; they were so pleased that they gave us a bonus*). **2** (with *neg.;* often foll. by *as* + clause) to the extent to which … is or does etc., or to the extent implied (*was not so late as I expected; am not so eager as you*). ¶ In positive constructions *as … as …* is used: see AS¹ *adv.* **3** (foll. by *that* or *as* + clause) to the degree or in the manner implied (*so expensive that few can afford it; so small as to be invisible; am not so foolish as to agree to that*). **4** (adding emphasis) to that extent; in that or a similar manner (*I want to leave and so does she; you said it was good, and so it is*). **5** to a great or notable degree (*I am so glad*). **6** (with verbs of state) in the way described (*am not very fond of it but may become so*). **7** (with verb of saying or thinking etc.) as previously mentioned or described (*I think so; so he said; so I should hope*). ● *conj.* (often foll. by *that* + clause) **1** with the result that (*there was none left, so we had to go without*). **2** in order that (*came home early so that I could see you*). **3** and then; as the next step (*so then the car broke down; and so to bed*). **4 a** (introducing a question) then; after that (*so what did you tell them?*). **b** (*absol.*) = so what? □ **and so on** (or **forth**) **1** and others of the same kind. **2** and in other similar ways. **so as** (foll. by *to* + infin.) in order to (*did it so as to get it finished*). **so be it** an expression of acceptance or resignation. **so-called** commonly designated or known as, often incorrectly. **so far** see FAR. **so far as** see FAR. **so far so good** see FAR. **so long!** *colloq.* goodbye till we meet again. **so long as** see *as long as* (see LONG¹). **so much 1** a certain amount (of). **2** a great deal of (*is so much nonsense*). **3** (with *neg.*) a less than; to a lesser extent (*not so much forgotten as ignored*). **b** not even (*didn't give me so much as a penny*). **so much for** that is all that need be done or said about. **so so** *adj.* (usu. *predic.*) indifferent; not very good. ● *adv.* indifferently; only moderately well. **so to say** (or **speak**) an expression of reserve or apology for an exaggeration or neologism etc.

so what? *colloq.* why should that be considered significant? [OE *swā* etc.]

so² var. of SOH.

-so /səʊ/ *comb. form* = -SOEVER.

soak /səʊk/ *v. & n.* ● *v.* **1** *tr. & intr.* make or become thoroughly wet through saturation with or in liquid. **2** *tr.* (of rain etc.) drench. **3** *tr.* (foll. by *in*, *up*) **a** absorb (liquid). **b** acquire (knowledge etc.) copiously. **4** *refl.* (often foll. by *in*) steep (oneself) in a subject of study etc. **5** *intr.* (foll. by *in*, *into*, *through*) (of liquid) make its way or penetrate by saturation. **6** *tr. colloq.* extract money from by an extortionate charge, taxation, etc. (*soak the rich*). **7** *intr. colloq.* drink persistently, booze. **8** *tr.* (as **soaked** *adj.*) very drunk. ● *n.* **1** the act of soaking or the state of being soaked. **2** a drinking-bout. **3** *colloq.* a hard drinker. □ **soakage** *n.* **soaker** *n.* [OE *socian* rel. to *soc* sucking at the breast, *sūcan* SUCK]

soakaway /ˈsəʊkəˌweɪ/ *n.* an arrangement for disposing of waste water by letting it percolate through the soil.

soaking /ˈsəʊkɪŋ/ *adj. & n.* ● *adj.* (in full **soaking wet**) very wet, wet through. ● *n.* the act of soaking; an instance of being soaked.

so-and-so /ˈsəʊəndˌsəʊ/ *n.* (*pl.* **so-and-so's**) **1** a particular person or thing not needing to be specified (*told me to do so-and-so*). **2** *colloq.* a person disliked or regarded with disfavour (*the so-and-so left me behind*).

Soane /səʊn/, Sir John (1753–1837), English architect. After initial training in England and Italy he was appointed architect of the Bank of England in 1788, where he developed a characteristic neoclassical style. By 1810 his style had become more severe, avoiding unnecessary ornament and adopting structural necessity as the basis of design. His collection of pictures is housed in the house he designed for himself in London, the Sir John Soane Museum.

soap /səʊp/ *n. & v.* ● *n.* **1** a cleansing agent that is a compound of fatty acid with soda or potash or (**insoluble soap**) with another metallic oxide, of which the soluble kinds when rubbed in water yield a lather used in washing. **2** *colloq.* = soap opera. ● *v.tr.* **1** apply soap to. **2** scrub or rub with soap. □ **soap flakes** soap in the form of thin flakes, for washing clothes etc. **soap opera** a broadcast drama, usu. serialized in many episodes, dealing with esp. domestic themes (so called because originally sponsored in the US by soap manufacturers). **soap powder** powdered soap esp. with additives. □ **soapless** *adj.* **soaplike** *adj.* [OE *sāpe* f. WG]

soapbark /ˈsəʊpbɑːk/ *n.* **1** a tree bark that contains saponin (which can be used as soap). **2** (in full **soapbark tree**) a tree that yields this bark, esp. *Quillaja saponaria* of Chile.

soapberry /ˈsəʊpˌberɪ/ *n.* (*pl.* **-ies**) a tree or shrub with fruits that contain saponin (which can be used as soap); esp. one of the genus *Sapindus*, of tropical America and Asia.

soapbox /ˈsəʊpbɒks/ *n.* **1** a box for holding soap. **2** a makeshift stand for a public speaker.

soapstone /ˈsəʊpstəʊn/ *n.* steatite.

soapsuds /ˈsəʊpsʌdz/ *n.pl.* = SUDS 1.

soapwort /ˈsəʊpwɜːt/ *n.* a pink-flowered Eurasian plant of the genus *Saponaria*, with juice that forms lather with water; esp. *S. officinalis*, grown in gardens.

soapy /ˈsəʊpɪ/ *adj.* (**soapier**, **soapiest**) **1** of or like soap. **2** containing or smeared with soap. **3** (of a person or manner) unctuous or flattering. □ **soapily** *adv.* **soapiness** *n.*

soar /sɔː(r)/ *v.intr.* **1** fly or rise high. **2** reach a high level or standard (*prices soared*). **3** maintain height in the air without flapping the wings or using power. □ **soarer** *n.* **soaringly** *adv.* [ME f. OF *essorer* ult. f. L (as EX-¹, *aura* breeze)]

S.O.B. *abbr.* esp. *US sl.* = son of a bitch.

sob /sɒb/ *v. & n.* ● *v.* (**sobbed**, **sobbing**) **1** *intr.* draw breath in convulsive gasps when weeping or under mental distress or physical exhaustion; weep in this way. **2** *tr.* (usu. foll. by *out*) utter with sobs. **3** *tr.* bring (oneself) to a specified state by sobbing (*sobbed themselves to sleep*). ● *n.* a convulsive drawing of breath, esp. in weeping. □ **sob story** *colloq.* a story or explanation appealing mainly to the emotions. **sob-stuff** *colloq.* sentimental talk or writing. □ **sobber** *n.* **sobbingly** *adv.* [ME *sobbe* (prob. imit.)]

sober /ˈsəʊbə(r)/ *adj. & v.* ● *adj.* (**soberer**, **soberest**) **1** not under the influence of alcohol. **2** not given to excessive drinking of alcohol. **3** moderate, well-balanced, tranquil, sedate. **4** not fanciful or exaggerated (*the sober truth*). **5** (of a colour etc.) quiet and inconspicuous. ● *v.tr. & intr.* (often foll. by *down*, *up*) make or become sober or less wild, reckless, enthusiastic, visionary, etc. (*a sobering thought*). □ **as sober**

as a judge completely sober. □ **soberingly** *adv.* **soberly** *adv.* [ME f. OF *sobre* f. L *sobrius*]

Sobers /ˈsəʊbəz/, Sir Garfield St Aubrun ('Gary') (b.1936), West Indian cricketer. Born in Barbados, he first played for the West Indies in 1953. Four years later he hit a record test score of 365 not out, which stood until beaten by Brian Lara in 1994. He was captain of the West Indies (1965–72) and of Nottinghamshire (1968–74). During his test career he scored more than 8,000 runs and took 235 wickets, bowling in three different styles; he was also a fine fielder. In 1968 Sobers became the first batsman in first-class cricket to hit all six balls of an over for six.

Sobieski /sɒˈbjeskɪ/, John, see JOHN III.

sobriety /səˈbraɪɪtɪ/ *n.* the state of being sober. [ME f. OF *sobriété* or L *sobrietas* (as SOBER)]

sobriquet /ˈsəʊbrɪˌkeɪ/ *n.* (also **soubriquet** /ˈsuːb-/) **1** a nickname. **2** an assumed name. [F, orig. = 'tap under the chin']

Soc. *abbr.* **1** Socialist. **2** Society.

soca /ˈsəʊkə/ *n.* a kind of popular music originating in Trinidad and combining elements of soul and calypso. [blend of SOUL + CALYPSO]

socage /ˈsɒkɪdʒ/ *n.* (also **soccage**) *hist.* a feudal tenure of land involving payment of rent or other non-military service to a superior. [ME f. AF *socage* f. *soc* f. OE *sōcn* SOKE]

soccer /ˈsɒkə(r)/ *n.* Association football. [ASSOC. + -ER³]

Sochi /ˈsɒtʃɪ/ a port and holiday and health resort in SW Russia, situated in the western foothills of the Caucasus, on the Black Sea coast close to the border with Georgia; pop. (1990) 339,000.

sociable /ˈsəʊʃəb(ə)l/ *adj. & n.* ● *adj.* **1** fitted for or liking the society of other people; ready and willing to talk and act with others. **2** (of a person's manner or behaviour etc.) friendly. **3** (of a meeting etc.) marked by friendliness, not stiff or formal. ● *n.* **1** an open carriage with facing side seats. **2** an S-shaped couch for two occupants partly facing each other. **3** *US* a social. □ **sociably** *adv.* **sociableness** *n.* **sociability** /ˌsəʊʃəˈbɪlɪtɪ/ *n.* [F *sociable* or L *sociabilis* f. *sociare* to unite f. *socius* companion]

social /ˈsəʊʃ(ə)l/ *adj. & n.* ● *adj.* **1** of or relating to society or its organization. **2** concerned with the mutual relations of human beings or of classes of human beings. **3** living in organized communities; unfitted for a solitary life (*man is a social animal*). **4 a** needing companionship; gregarious, interdependent. **b** cooperative; practising the division of labour. **5** existing only as a member of a compound organism. **6** *Zool.* **a** (of insects) living together in organized communities. (*See note below.*) **b** (of animals or birds) breeding or nesting near each other in communities. **7** (of plants) growing thickly together and monopolizing the ground they grow on. ● *n.* **1** a social gathering, esp. one organized by a club, congregation, etc. **2** *colloq.* = social security. □ **social anthropology** the comparative study of peoples through their culture and kinship systems. **social chapter** a social charter forming part of the Maastricht Treaty. **social charter** a document dealing with social policy, esp. workers' rights. **social climber** *derog.* a person anxious to gain a higher social status. **social conscience** a sense of responsibility for or preoccupation with the problems and injustices of society. **social credit** the economic theory that the purchasing power of consumers should be increased either by subsidizing producers so that they can reduce prices or by distributing the profits of industry to consumers. **social democracy** a socialist system achieved by democratic means. **social democrat** a person who advocates social democracy. **social engineering** the application of sociological principles to specific social problems. **social insurance** a system of compulsory contribution to provide state assistance in illness, unemployment, old age, etc. **social mobility** the ability or potential to move between different social levels, fields of employment, etc. **social order** the network of human relationships in society. **social science 1** the scientific study of human society and social relationships. **2** a branch of this (e.g. politics or economics). **social scientist** a student of or expert in the social sciences. **social secretary** a person who makes arrangements for the social activities of a person or organization. **social security** state assistance to those lacking in economic security and welfare, e.g. the aged and the unemployed. **social service** philanthropic activity. **social services** services provided by the state for the community, esp. education, health, and housing. **social studies** the social sciences as a subject. **social war** a war fought between allies. **social work** work of benefit to those in need of help or welfare, esp. done by specially trained personnel. **social worker** a person trained to do social work. □ **socially** *adv.* **sociality** /ˌsəʊʃɪˈælɪtɪ/ *n.* [F *social* or L *socialis* allied f. *socius* friend]

▪ The social insects comprise some Hymenoptera (bees, wasps, and ants), together with the Isoptera (termites). Many of these live in large colonies that have a complex caste structure, with an egg-laying queen, males or drones, infertile workers, and (in some cases) soldiers. Such colonies operate effectively as single organisms, controlled by the pheromones emitted by the queen. Honey bee and ant colonies are usually able to survive the winter, but in the case of most wasps and bumble-bees only the fertilized queen overwinters. (See also BEE.)

Social and Liberal Democrats (abbr. **SLD**) (in the UK) a political party formed in 1988 from a majority of the membership of the Liberal Party and the Social Democratic Party. It was officially renamed in 1989 as the Liberal Democrats.

social contract n. (also **social compact**) an implicit agreement among the members of a society to cooperate for social benefits, e.g. by sacrificing some individual freedom for state protection. Theories of a social contract became popular in the 16th and 17th centuries as a means of explaining the origin of government and the obligations of subjects. Postulating a savage, unregulated original state of nature, theorists such as Thomas Hobbes, John Locke and, later, Jean-Jacques Rousseau held that it would have been in each person's interest to agree to the establishment of government, the terms and form of which would be regulated by the agreement or contract made. Hobbes felt that an absolute government was implied by this contract, but Locke and Rousseau argued for democratic and even revolutionary ideas: if government policy ran counter to the general will, the contract was broken and the subjects had the right to rebel. More recently the idea of the social contract has been revived by the philosopher John Rawls and in reference to unofficial agreements between governments and trade unions.

social Darwinism n. the theory that individuals, groups, and peoples are subject to the same Darwinian laws of natural selection as are plants and animals. Advocated by Herbert Spencer and others in the late 19th and early 20th centuries but now largely discredited, the theory held that superior individuals or groups survived and succeeded while the weaker disappeared, with a consequent benefit to society, and that the existing order must be the natural one. It was used to justify political conservatism, imperialism, and racism and to discourage intervention and reform.

Social Democratic and Labour Party (abbr. **SDLP**) a left-of-centre political party in Northern Ireland, formed in 1970 and supported largely by Catholics. It calls for the establishment of a united Ireland by constitutional means, and rejects the violent tactics of the IRA.

Social Democratic Party (abbr. **SDP**) a UK political party with moderate socialist aims, founded in 1981 by a group of former Labour MPs and disbanded in 1990 after political regroupings. (See also LIBERAL PARTY, GANG OF FOUR 2.)

socialism /ˈsəʊʃəˌlɪz(ə)m/ n. a political and economic theory of social organization which advocates that the means of production, distribution, and exchange should be regulated or owned by the community as a whole; policy or practice based on this theory. Socialism has been used to describe positions as far apart as anarchism, Soviet state Communism, and social democracy; however, it necessarily implies an opposition to the untrammelled workings of the economic market. Although socialist ideas can be traced back to classical times, the word was first used in the early 19th century in reference to pioneering reformers including Saint-Simon in France and Robert Owen in Britain. Marx and Engels attempted to bring precise and scientific organization to theories of socialism, but used the word *communist* to describe their ideal classless state, reserving *socialism* for the transitional state posited as coming between the overthrow of capitalism and the establishment of communism. Since the time of Marx a major source of division among socialists has been the question of whether the transition from capitalism to socialism should be achieved by revolutionary or parliamentary means. Although the Russian Revolution in 1917 put the theories of the revolutionary school into practice, the socialist parties that arose in most European countries from the late 19th century generally tended towards social democracy. Socialism in the West is currently somewhat overshadowed by free-market ideas, but may draw renewed strength from disillusion at the perceived social and environmental costs of capitalism; at a deeper level it has brought concepts such as the welfare state and state intervention in the economy into the political mainstream. (See also COMMUNISM, MARXISM.) ☐ **socialist** n. & adj.

socialistic /ˌsəʊʃəˈlɪstɪk/ adj. **socialistically** adv. [F socialisme (as SOCIAL)]

socialist realism n. the theory of art, literature, and music officially sanctioned by the state in some Communist countries (especially in the Soviet Union under Stalin), by which artistic work was supposed to reflect and promote the ideals of a socialist society. Cf. SOCIAL REALISM.

socialite /ˈsəʊʃəˌlaɪt/ n. a person prominent in fashionable society.

socialize /ˈsəʊʃəˌlaɪz/ v. (also **-ise**) **1** intr. act in a sociable manner, mix socially with others. **2** tr. **a** make social. **b** (usu. in *passive*) imbue (a person) with the values and standards of the social group to which he or she belongs. **3** tr. organize on socialist principles. ☐ **socialized medicine** US often derog. the provision of medical services for all from public funds. ☐ **socialization** /ˌsəʊʃəlaɪˈzeɪʃ(ə)n/ n.

social realism n. in the art of the 19th and 20th centuries, a form of realism which contains explicit or overt social or political comment. An imprecise term, social realism has been applied to some of the paintings of Gustave Courbet and to the work of the Ashcan School. Cf. SOCIALIST REALISM.

society /səˈsaɪətɪ/ n. (pl. **-ies**) **1** the sum of human conditions and activity regarded as a whole functioning interdependently. **2** a social community (all societies must have firm laws). **3 a** a social mode of life. **b** the customs and organization of an ordered community. **4** Ecol. a plant or animal community. **5 a** the socially advantaged or prominent members of a community (society would not approve). **b** this, or a part of it, qualified in some way (is not done in polite society). **6** participation in hospitality; other people's homes or company (avoids society). **7** companionship, company (avoids the society of such people). **8** an association of persons united by a common aim or interest or principle (formed a music society). ☐ **societal** adj. (esp. in sense 1). **societally** adv. [F société f. L societas -tatis f. socius companion]

Society Islands a group of islands in the South Pacific, forming part of French Polynesia. The islands were named by Captain James Cook, who visited them in 1769, in honour of the Royal Society.

Society of Friends the official name for the Quakers (see QUAKER).

Society of Jesus the official name for the Jesuits (see JESUIT).

socio- /ˈsəʊsɪəʊ, ˈsəʊʃɪ-/ comb. form **1** of society (and). **2** of or relating to sociology (and). [L socius companion]

sociobiology /ˌsəʊsɪəʊbaɪˈɒlədʒɪ, ˌsəʊʃɪ-/ n. the scientific study of the biological aspects of social behaviour in animals and in humans. Sociobiology gives special emphasis to social systems considered as ecological adaptations, and controversially attempts a mechanistic explanation of social behaviour in terms of modern biological and evolutionary theory. The theory suggests in particular that the apparent altruism of certain social animals and insects is due to their acting in a way which will best ensure the survival of their genes. This is partly based on the work of W. D. Hamilton (b.1936) on the breeding system of social insects such as ants, in which a worker has more genes in common with its sisters than it would have with its own offspring, and can therefore perpetuate its genes by remaining sterile and raising the queen's offspring. Sociobiology has also been developed and popularized by E. O. Wilson and Richard Dawkins. ☐ **sociobiologist** n. **sociobiological** /-ˌbaɪəˈlɒdʒɪk(ə)l/ adj. **sociobiologically** adv.

sociocultural /ˌsəʊsɪəʊˈkʌltʃərəl, ˌsəʊʃɪ-/ adj. combining social and cultural factors. ☐ **socioculturally** adv.

socio-economic /ˌsəʊsɪəʊˌiːkəˈnɒmɪk, ˌsəʊʃɪ-, -ˌekəˈnɒmɪk/ adj. relating to or concerned with the interaction of social and economic factors. ☐ **socio-economically** adv.

sociolinguistic /ˌsəʊsɪəʊlɪŋˈgwɪstɪk, ˌsəʊʃɪ-/ adj. relating to or concerned with language in its social aspects. ☐ **sociolinguistically** adv. **sociolinguist** /-ˈlɪŋgwɪst/ n.

sociolinguistics /ˌsəʊsɪəʊlɪŋˈgwɪstɪks, ˌsəʊʃɪ-/ n. the study of language in relation to social factors.

sociology /ˌsəʊsɪˈɒlədʒɪ, ˌsəʊʃɪ-/ n. the study of the development, structure, and functioning of human society. Although the word was coined by Auguste Comte c.1838, sociology was founded as a formal discipline by Durkheim, Weber, and Marx in the late 19th century. Sociology asserts that humans act under the influence of history, culture, and the expectations of others. Sociologists are less concerned with individuals than with patterns of behaviour which recur irrespective of the individuals concerned; some sociologists examine the factors, such as class, status, and expected role, which lead to cohesion or conflict in the family or community; others study social problems such as crime, drugs, and domestic violence. Sociological findings are now used extensively by governments and in advertising, public relations, and other businesses. □ **sociologist** n. **sociological** /-sɪə'lɒdʒɪk(ə)l, -ʃɪə-/ adj. **sociologically** adv. [F sociologie (as SOCIO-, -LOGY)]

sociometry /ˌsəʊsɪ'ɒmɪtrɪ, ˌsəʊʃɪ-/ n. the study of relationships within a group of people. □ **sociometrist** n. **sociometric** /-sɪə'metrɪk, -ʃɪə-/ adj. **sociometrically** adv.

socio-political /ˌsəʊsɪəʊpə'lɪtɪk(ə)l, ˌsəʊʃɪ-/ adj. combining social and political factors.

sock[1] /sɒk/ n. (pl. **socks** or colloq. & Commerce **sox** /sɒks/) **1** a short knitted covering for the foot, usu. not reaching the knee. **2** a removable inner sole put into a shoe for warmth etc. **3** an ancient Greek or Roman comic actor's light shoe. **4** comic drama. □ **pull one's socks up** Brit. colloq. make an effort to improve. **put a sock in it** Brit. colloq. be quiet. [OE socc f. L soccus comic actor's shoe, light low-heeled slipper, f. Gk sukkhos]

sock[2] /sɒk/ v. & n. colloq. ● v.tr. hit (esp. a person) forcefully. ● n. **1** a hard blow. **2** US the power to deliver a blow. □ **sock it to** attack or address (a person) vigorously. [c.1700 (cant): orig. unkn.]

socket /'sɒkɪt/ n. & v. ● n. **1** a natural or artificial hollow for something to fit into or stand firm or revolve in. **2** Electr. a device receiving a plug, light bulb, etc., to make a connection. **3** Golf the part of an iron club into which the shaft is fitted. ● v.tr. (**socketed**, **socketing**) **1** place in or fit with a socket. **2** Golf hit (a ball) with the socket of a club. [ME f. AF, dimin. of OF soc ploughshare, prob. of Celt. orig.]

sockeye /'sɒkaɪ/ n. (also **sockeye salmon**) a blue-backed salmon, Oncorhynchus nerka, of Alaska and western Canada. [Salish sukai fish of fishes]

socking /'sɒkɪŋ/ adv. & adj. ● adv. colloq. exceedingly, very (a socking great diamond ring). ● adj. sl. confounded, bloody.

socle /'səʊk(ə)l/ n. Archit. a plain low block or plinth serving as a support for a column, urn, statue, etc., or as the foundation of a wall. [F f. It. zoccolo orig. 'wooden shoe' f. L socculus f. soccus SOCK[1]]

Socotra /sə'kəʊtrə/ an island in the Arabian Sea near the mouth of the Gulf of Aden; capital, Tamridah. It is administered by Yemen.

Socrates /'sɒkrəˌtiːz/ (469–399 BC), Greek philosopher. His interests lay not in the speculation about the natural world engaged in by earlier philosophers but in pursuing questions of ethics. He was the centre of a circle of friends and disciples in Athens and his method of inquiry (the Socratic method) was based on discourse with those around him; his careful questioning was designed to reveal truth and to expose error. Although he wrote nothing himself, he was immensely influential; he is known chiefly through his disciple Plato, who recorded Socrates' dialogues and teachings in, for example, the Symposium and the Phaedo. Charged with introducing strange gods and corrupting the young, Socrates was sentenced to death and condemned to take hemlock, which he did, spurning offers to help him escape into exile.

Socratic /sə'krætɪk/ adj. & n. ● adj. of or relating to the Greek philosopher Socrates or his philosophy, esp. the method associated with him of seeking the truth by a series of questions and answers. ● n. a follower of Socrates. □ **Socratic irony** a pose of ignorance assumed in order to entice others into making statements that can then be challenged. □ **Socratically** adv. [L Socraticus f. Gk Sōkratikos f. Sōkratēs]

sod[1] /sɒd/ n. & v. ● n. **1** turf or a piece of turf. **2** the surface of the ground. ● v.tr. (**sodded, sodding**) cover (the ground) with sods. □ **under the sod** in the grave. [ME f. MDu., MLG sode, of unkn. orig.]

sod[2] /sɒd/ n. & v. esp. Brit. coarse sl. ● n. **1** an unpleasant or awkward person or thing. **2** a person of a specified kind; a fellow (the lucky sod). ● v.tr. (**sodded, sodding**) **1** (often absol. or as int.) an exclamation of annoyance (sod them, I don't care!). **2** (as **sodding** adj.) a general term of contempt. □ **sod off** go away. **Sod's Law** = MURPHY'S LAW. [abbr. of SODOMITE]

soda /'səʊdə/ n. **1** any of several compounds of sodium in common use (washing soda, caustic soda). **2** (in full **soda water**) water made effervescent by impregnation with carbon dioxide under pressure and used alone or with spirits etc. as a drink (orig. made with sodium bicarbonate). **3** esp. US any sweet effervescent soft drink. □ **soda bread** bread leavened with baking soda. **soda fountain 1** a device supplying soda water. **2** a shop or counter equipped with this. **soda lime** a mixture of calcium oxide and sodium hydroxide. [med.L, perh. f. sodanum glasswort (used as a remedy for headaches) f. soda headache f. Arab. ṣudāʿ f. ṣadaʿa split]

sodality /səʊ'dælɪtɪ/ n. (pl. **-ies**) a confraternity or association, esp. a Roman Catholic religious guild or brotherhood. [F sodalité or L sodalitas f. sodalis comrade]

sodden /'sɒd(ə)n/ adj. & v. ● adj. **1** saturated with liquid; soaked through. **2** rendered dazed or dull etc. with drunkenness. **3** (of bread etc.) doughy; heavy and moist. ● v.intr. & tr. become or make sodden. □ **soddenly** adv. **soddenness** n. [archaic past part. of SEETHE]

Soddy /'sɒdɪ/, Frederick (1877–1956), English physicist. He worked with Ernest Rutherford in Canada on radioactive decay and formulated a theory of isotopes, the word isotope being coined by him in 1913. He also assisted William Ramsay in London in the discovery of helium. Soddy wrote on economics, and later concentrated on creating an awareness of the social relevance of science. He was awarded the Nobel Prize for chemistry in 1921.

sodium /'səʊdɪəm/ n. a soft silver-white reactive metallic chemical element (atomic number 11; symbol **Na**). (See note below.) □ **sodium bicarbonate** a white soluble powder (chem. formula: $NaHCO_3$) used in fire extinguishers and effervescent drinks, and as a raising agent in baking (also called baking soda). **sodium carbonate** a white powder (chem. formula: Na_2CO_3) with many commercial applications including the manufacture of soap and glass (also called washing soda). **sodium chloride** a colourless crystalline compound occurring naturally in sea water and halite; common salt (chem. formula: $NaCl$). **sodium hydroxide** a deliquescent, strongly alkaline compound (chem. formula: $NaOH$) (also called caustic soda). **sodium nitrate** a white powdery compound (chem. formula: $NaNO_3$) used mainly in the manufacture of fertilizers. **sodium-vapour lamp** (or **sodium lamp**) a lamp using an electrical discharge in sodium vapour and giving a yellow light. [SODA + -IUM]

▪ Sodium was first isolated by Humphry Davy in 1807, and is a member of the alkali-metal group. It is a common element in the earth's crust, notably as rock salt (sodium chloride). Sodium compounds have many industrial and other uses; the liquid metal is used as a coolant in some types of nuclear reactor. The symbol Na is from the Latin name natrium.

Sodom /'sɒdəm/ a town in ancient Palestine, probably south of the Dead Sea, destroyed by fire from heaven (according to Gen. 19:24), along with Gomorrah, for the wickedness of its inhabitants.

sodomite /'sɒdəˌmaɪt/ n. a person who practises sodomy. [ME f. OF f. LL Sodomita f. Gk Sodomitēs inhabitant of Sodom f. Sodoma SODOM]

sodomy /'sɒdəmɪ/ n. = BUGGERY. □ **sodomize** v.tr. (also **-ise**). [ME f. med.L sodomia f. LL peccatum Sodomiticum sin of Sodom: see SODOM]

Sodor /'səʊdə(r)/ a medieval diocese comprising the Hebrides and the Isle of Man. Formerly belonging to Norway (Norse Sudhr-eyjar = 'southern isles'), the islands passed to Scotland in 1266. The Hebrides were separated in 1334, but Sodor and Man has been the official name for the Anglican diocese of the Isle of Man since 1684.

SOE see SPECIAL OPERATIONS EXECUTIVE.

soever /səʊ'evə(r)/ adv. literary of any kind; to any extent (how great soever it may be).

-soever /səʊ'evə(r)/ comb. form (added to relative pronouns, adverbs, and adjectives) of any kind; to any extent (whatsoever; howsoever).

sofa /'səʊfə/ n. a long upholstered seat with a back and arms, for two or more people. □ **sofa bed** a sofa that can be converted into a temporary bed. [F, ult. f. Arab. ṣuffa]

soffit /'sɒfɪt/ n. the undersurface of an architrave, arch, balcony, etc. [F soffite or It. soffitta, -itto ult. f. L suffixus (as SUFFIX)]

Sofia /'səʊfɪə, sɒ'fiːə/ the capital of Bulgaria; pop. (1990) 1,220,900. An ancient Thracian settlement, it became a province of Rome in the first century AD and grew to prominence as a flourishing Byzantine city. Held by the Turks between the late 14th and late 19th centuries, it became the capital of Bulgaria in 1879.

S. of S. abbr. (in the Bible) Song of Songs.

soft /sɒft/ adj., adv., & n. ● adj. **1** (of a substance, material, etc.) lacking hardness or firmness; yielding to pressure; easily cut. **2** (of cloth etc.) having a smooth surface or texture; not rough or coarse. **3** (of air etc.) mellow, mild, balmy; not noticeably cold or hot. **4** (of water) free from mineral salts and therefore good for lathering. **5** (of a light or colour etc.) not brilliant or glaring. **6** (of a voice or sounds) gentle and pleasing. **7** Phonet. **a** (of a consonant) sibilant or palatal (as c in ice, g in age). **b** voiced or unaspirated. **8** (of an outline etc.) not sharply defined. **9** (of an action or manner etc.) gentle, conciliatory, complimentary; amorous. **10** (of the heart or feelings etc.) compassionate, sympathetic. **11** (of a person) **a** feeble, lenient, silly, sentimental. **b** weak; not robust. **12** colloq. (of a job etc.) easy. **13** (of drugs) mild; not likely to cause addiction. **14** (of radiation) having little penetrating power. **15** (also **soft-core**) (of pornography) suggestive or erotic but not explicit. **16** *Stock Exch.* (of currency, prices, etc.) likely to fall in value. **17** *Polit.* moderate; willing to compromise (*the soft left*). **18** peaceful (*soft slumbers*). **19** *Brit.* (of the weather etc.) rainy or moist or thawing. ● *adv.* softly (*play soft*). ● *n.* a silly weak person. □ **be soft on 1** be lenient towards. **2** be infatuated with. **have a soft spot for** be fond of or affectionate towards (a person). **soft answer** a good-tempered answer to abuse or an accusation. **soft-boiled** (of an egg) lightly boiled leaving the yolk soft or liquid. **soft-centred 1** (of a sweet) having a soft filling or centre. **2** (of a person) soft-hearted, sentimental. **soft coal** bituminous coal. **soft detergent** a biodegradable detergent. **soft drink** a non-alcoholic drink. **soft focus** *Photog.* the slight deliberate blurring of a picture. **soft fruit** *Brit.* small stoneless fruit (strawberry, currant, etc.). **soft furnishings** *Brit.* curtains, rugs, etc. **soft goods** *Brit.* textiles. **soft-headed** feeble-minded. **soft-headedness** feeble-mindedness. **soft in the head** feeble-minded. **soft-land** make a soft landing. **soft landing** a landing by a spacecraft without its suffering major damage. **soft option** the easier alternative. **soft palate** the rear part of the palate. **soft-paste** denoting an 'artificial' porcelain containing glassy materials and fired at a comparatively low temperature. **soft pedal** a pedal on a piano that makes the tone softer. **soft-pedal** *v.tr.* & (often foll. by *on*) *intr.* (**-pedalled, -pedalling**; *US* **-pedaled, -pedaling**) **1** refrain from emphasizing; be restrained (about). **2** play the piano with the soft pedal down. **soft roe** see ROE[1] 2. **soft sell** restrained or subtly persuasive salesmanship. **soft-sell** *v.tr.* (*past* and *past part.* **-sold**) sell by this method. **soft soap 1** a semi-fluid soap, esp. one made with potassium not sodium salts. **2** *colloq.* persuasive flattery. **soft-soap** *v.tr. colloq.* persuade (a person) with flattery. **soft-spoken** speaking with a gentle voice. **soft sugar** granulated or powdered sugar. **soft tack** bread or other good food (opp. *hard tack*). **soft tissues** tissues of the body that are not bony or cartilaginous. **soft-top 1** a motor-vehicle roof that is soft and can be folded back. **2** a vehicle having such a roof. **soft touch** see TOUCH. **soft wicket** a wicket with moist or sodden turf. □ **softish** *adj.* **softness** *n.* [OE *sōfte* agreeable, earlier *sēfte* f. WG]

softa /ˈsɒftə/ *n.* a Muslim student of sacred law and theology. [Turk. f. Pers. *sūḵta* burnt, afire]

softball /ˈsɒftbɔːl/ *n.* a modified form of baseball using a ball like a baseball but softer and larger; the ball used in this sport. Apart from using a different ball, softball differs from baseball chiefly in that it is played on a smaller field, each side has seven rather than nine innings, and bowling is underarm rather than overarm or sidearm. The game evolved in the US, where it is still popular, during the late 19th century from a form of indoor baseball.

soften /ˈsɒf(ə)n/ *v.* **1** *tr.* & *intr.* make or become soft or softer. **2** *tr.* (often foll. by *up*) **a** reduce the strength of (defences) by bombing or some other preliminary attack. **b** reduce the resistance of (a person). □ **softening of the brain** a morbid degeneration of the brain, esp. in old age. □ **softener** *n.*

soft-hearted /sɒftˈhɑːtɪd/ *adj.* tender, compassionate; easily moved. □ **soft-heartedness** *n.*

softie /ˈsɒftɪ/ *n.* (also **softy**) (*pl.* **-ies**) *colloq.* a weak or silly or soft-hearted person.

softly /ˈsɒftlɪ/ *adv.* in a soft, gentle, or quiet manner. □ **softly softly** (of an approach or strategy) cautious; discreet and cunning.

software /ˈsɒftweə(r)/ *n.* the programs and other operating information used by a computer (opp. HARDWARE 3).

softwood /ˈsɒftwʊd/ *n.* the wood of pine, spruce, or other conifers, easily sawn.

SOGAT /ˈsəʊgæt/ *abbr. hist.* (in the UK) Society of Graphical and Allied Trades. (See also GPMU.)

soggy /ˈsɒgɪ/ *adj.* (**soggier, soggiest**) sodden, saturated, dank. □ **soggily** *adv.* **sogginess** *n.* [dial. *sog* a swamp]

Sogne Fiord /ˈsɒnjə/ (Norwegian **Sognafjorden** /ˈsɒŋnəˌfjuːrən/) a fiord on the west coast of Norway. The longest and deepest fiord in the country, it extends inland for some 200 km (125 miles), reaching a maximum depth of 1,308 m (4,291 ft).

soh /səʊ/ *n.* (also **so, sol** /sɒl/) *Mus.* **1** (in tonic sol-fa) the fifth note of a major scale. **2** the note G in the fixed-doh system. [*sol* f. ME *sol* f. L *solve*: see GAMUT]

soi-disant /ˌswʌdiːˈzɒn/ *adj.* self-styled or pretended. [F f. *soi* oneself + *disant* saying]

soigné /ˈswʌnjeɪ/ *adj.* (*fem.* **soignée** *pronunc.* same) carefully finished or

arranged; well groomed. [past part. of F *soigner* take care of f. *soin* care]

soil[1] /sɔɪl/ *n.* **1** the upper layer of earth in which plants grow, consisting of disintegrated rock usu. with an admixture of organic remains (*alluvial soil; rich soil*). **2** ground belonging to a nation; territory (*on British soil*). □ **soil mechanics** the study of the properties of soil as affecting its use in civil engineering. **soil science** pedology. □ **soilless** *adj.* **soily** *adj.* [ME f. AF, perh. f. L *solium* seat, taken in sense of L *solum* ground]

soil[2] /sɔɪl/ *v.* & *n.* ● *v.tr.* **1** make dirty; smear or stain with dirt (*soiled linen*). **2** tarnish, defile; bring discredit to (*would not soil my hands with it*). ● *n.* **1** a dirty mark; a stain, smear, or defilement. **2** filth; refuse matter. □ **soil pipe** the discharge-pipe of a lavatory. [ME f. OF *suiller, soiller*, etc., ult. f. L *sucula* dimin. of *sus* pig]

soil[3] /sɔɪl/ *v.tr.* feed (cattle) on fresh-cut green fodder (orig. for purging). [perh. f. SOIL[2]]

soirée /ˈswʌreɪ/ *n.* an evening party, usu. in a private house, for conversation or music. [F f. *soir* evening]

soixante-neuf /ˌswʌsɒntˈnɜːf/ *n.* sexual activity between two people involving mutual oral stimulation of the genitals; a position enabling this. [F, = sixty-nine, from the position of the couple]

sojourn /ˈsɒdʒən/ *n.* & *v.* ● *n.* a temporary stay. ● *v.intr.* stay temporarily. □ **sojourner** *n.* [ME f. OF *sojorn* etc. f. LL SUB- + *diurnum* day]

Soka Gakkai /ˌsəʊkə ˈgækaɪ/ a political and lay religious organization founded in Japan in 1930 and based on the teachings of the Nichiren Buddhist sect (see NICHIREN). [Jap., f. *sō* to create + *ka* value + *gakkai* (learned) society]

soke /səʊk/ *n. Brit. hist.* **1** a right of local jurisdiction. **2** a district under a particular jurisdiction and administration. [ME f. AL *sōca* f. OE *sōcn* prosecution f. Gmc]

Sol /sɒl/ *Rom. Mythol.* the sun, esp. as a personification. [ME f. L]

sol[1] /sɒl/ *n. Chem.* a liquid suspension of a colloid. [abbr. of SOLUTION]

sol[2] /sɒl/ *n.* (*pl.* **soles** /ˈsɒles/) (also **nuevo sol**) a monetary unit of Peru, which replaced the inti in 1991. [Sp., = sun f. L]

sol[3] var. of SOH.

sola[1] /ˈsəʊlə/ *n.* a pithy-stemmed leguminous swamp plant, *Aeschynomene indica*, of tropical Asia. □ **sola topi** an Indian sun-helmet made from its pith. [Urdu & Bengali *solā*, Hindi *sholā*]

sola[2] *fem.* of SOLUS.

solace /ˈsɒlɪs/ *n.* & *v.* ● *n.* comfort in distress, disappointment, or tedium. ● *v.tr.* give solace to. □ **solace oneself with** find compensation or relief in. [ME f. OF *solas* f. L *solatium* f. *solari* CONSOLE[1]]

solan /ˈsəʊlən/ *n.* (in full **solan goose**) the northern gannet, *Sula bassana*. [prob. f. ON *súla* gannet + *önd, and-* duck]

solanaceous /ˌsɒləˈneɪʃəs/ *adj. Bot.* of or relating to the plant family Solanaceae, which includes potatoes, tomatoes, capsicums, nightshades, tobacco, etc. [mod.L *solanaceae* f. L *sōlānum* nightshade]

solar /ˈsəʊlə(r)/ *adj.* & *n.* ● *adj.* of, relating to, or reckoned by the sun (*solar eclipse; solar time*). ● *n.* **1** a solarium. **2** an upper chamber in a medieval house. □ **solar calendar** a calendar based on the annual cycle of the sun. **solar cell** (or **battery**) a photoelectric device converting solar radiation into electricity. **solar constant** the rate at which energy reaches the earth's surface from the sun, usually taken to be 1,388 watts per square metre. **solar day** the interval between successive meridian transits of the sun at a place. **solar disc** = *sun disc.* **solar eclipse** an eclipse of the sun (see ECLIPSE). **solar energy 1** radiant energy emitted by the sun, esp. as light or heat. **2** = SOLAR POWER. **solar month** one-twelfth of the solar year. **solar myth** a tale ascribing the course or attributes of the sun to a god, hero, etc. **solar panel** a panel that harnesses the energy in the sun's radiation, either to generate electricity using solar cells or to heat water. **solar plexus** a complex of radiating nerves at the pit of the stomach. **solar time** time determined with reference to the position of the sun. **solar wind** the continuous flow of charged particles from the sun into surrounding space. **solar year** the time taken for the earth to travel once round the sun, equal to 365 days, 5 hours, 48 minutes, and 46 seconds. [ME f. L *solaris* f. *sol* sun]

solarium /səˈleərɪəm/ *n.* (*pl.* **solaria** /-rɪə/) a room equipped with sun-lamps or fitted with extensive areas of glass for exposure to the sun. [L, = sundial, sunning-place (as SOLAR)]

solarize /ˈsəʊləˌraɪz/ *v.intr.* & *tr.* (also **-ise**) *Photog.* undergo or cause to undergo change in the relative darkness of parts of an image by long exposure. □ **solarization** /ˌsəʊləraɪˈzeɪʃ(ə)n/ *n.*

solar power *n.* power obtained by harnessing solar energy. Energy from the sun has long been used directly in some processes, such as evaporating sea water to make salt in hot countries. Modern attempts to harness solar radiation use arrays of solar cells or panels to convert it into electricity, a method favoured for powering satellites and space probes. Greater output may be obtained by using steerable mirrors and a parabolic reflector, either to heat a solar furnace or to produce steam to drive a generator, but such methods are too costly for general use. Solar panels may also be used to heat water directly, especially for domestic use, and solar ovens have been used in the tropics.

solar system n. Astron. the collection of nine planets and their moons in orbit round the sun, together with smaller bodies in the form of asteroids, meteoroids, and comets. Most of these objects lie within or close to the plane of the ecliptic, suggesting that the solar system originated in the condensation of a disc of a primordial gaseous nebula, forming a large massive star at the centre of a group of orbiting inert bodies. It remains unexplained why the planets, with mass much less than that of the sun, should contain most of the angular momentum of the system. The postulated Oort cloud extends the limits of the solar system to a distance of one light year from the sun. (See also PLANET.)

solatium /sə'leɪʃɪəm/ n. (pl. **solatia** /-ʃɪə/) a thing given as a compensation or consolation. [L, = SOLACE]

sold past and past part. of SELL.

soldanella /ˌsɒldə'nelə/ n. a dwarf alpine plant of the genus *Soldanella*, of the primrose family, having bell-shaped lilac flowers with fringed petals. [mod.L f. It.]

solder /'səʊldə(r), 'sɒl-/ n. & v. ● n. **1** a fusible alloy used to join less fusible metals or wires etc. **2** a cementing or joining agency. ● v.tr. join with solder. □ **soldering iron** a tool used for applying solder. □ **solderable** n. **solderer** n. [ME f. OF *soudure* f. *souder* f. L *solidare* fasten f. *solidus* SOLID]

soldier /'səʊldʒə(r)/ n. & v. ● n. **1** a person serving in or having served in an army. **2** (in full **common soldier**) a private or NCO in an army. **3** a military commander of specified ability (*a great soldier*). **4** (in full **soldier ant**) a wingless ant or termite with a large head and jaws for fighting in defence of its colony. **5** (in full **soldier beetle**) a carnivorous beetle of the family Cantharidae, with soft reddish wing-cases. **6** colloq. a finger of bread for dipping into a soft-boiled egg. ● v.intr. serve as a soldier (*was off soldiering*). □ **soldier of Christ** an active or proselytizing Christian. **soldier of fortune** an adventurous person ready to take service under any state or person; a mercenary. **soldier on** persevere doggedly. □ **soldierly** adj. **soldiership** n. [ME *souder* etc. f. OF *soudier, soldier* f. *soulde* (soldier's) pay f. L *solidus*: see SOLIDUS]

soldiery /'səʊldʒərɪ/ n. (pl. **-ies**) **1** soldiers, esp. of a specified character. **2** a group of soldiers.

Sole /səʊl/ a shipping forecast area in the NE Atlantic, covering the western approaches to the English Channel.

sole[1] /səʊl/ n. & v. ● n. **1** the undersurface of the foot. **2** the part of a shoe, sock, etc., corresponding to this (esp. excluding the heel). **3** the lower surface or base of an implement, e.g. a plough, golf-club head, etc. **4** the floor of a ship's cabin. ● v.tr. provide (a shoe etc.) with a sole, replace the sole of. □ **sole-plate** the bedplate of an engine etc. □ **-soled** adj. (in comb.). [OF ult. f. L *solea* sandal, sill: cf. OE unrecorded *solu* or *sola* f. *solum* bottom, pavement, sole]

sole[2] /səʊl/ n. a flatfish of the family Soleidae; esp. *Solea solea* of the NE Atlantic, used as food. [ME f. OF f. Prov. *sola* ult. f. L *solea* (as SOLE[1], named from its shape)]

sole[3] /səʊl/ adj. **1** (attrib.) one and only; single, exclusive (*the sole reason; has the sole right*). **2** archaic or Law (esp. of a woman) unmarried. **3** archaic alone, unaccompanied. □ **solely** adv. [ME f. OF *soule* f. L *sola* fem. of *solus* alone]

solecism /'sɒlɪˌsɪz(ə)m/ n. **1** a mistake of grammar or idiom; a blunder in the manner of speaking or writing. **2** a piece of bad manners or incorrect behaviour. □ **solecist** n. **solecistic** /ˌsɒlɪ'sɪstɪk/ adj. [F *solécisme* or L *soloecismus* f. Gk *soloikismos* f. *soloikos* speaking incorrectly]

solemn /'sɒləm/ adj. **1** serious and dignified (*a solemn occasion*). **2** formal; accompanied by ceremony (*a solemn oath*). **3** mysteriously impressive. **4** (of a person) serious or cheerless in manner (*looks rather solemn*). **5** full of importance; weighty (*a solemn warning*). **6** grave, sober, deliberate; slow in movement or action (*a solemn promise; solemn music*). □ **solemn mass** = high mass (see MASS[2]). □ **solemnly** adv. **solemness** /'sɒləmnɪs/ n. [ME f. OF *solemne* f. L *sol(l)emnis* customary, celebrated at a fixed date f. *sollus* entire]

solemnity /sə'lemnɪtɪ/ n. (pl. **-ies**) **1** the state of being solemn; a

solemn character or feeling; solemn behaviour. **2** a rite or celebration; a piece of ceremony. [ME f. OF *solem(p)nité* f. L *sollemnitas -tatis* (as SOLEMN)]

solemnize /'sɒləmˌnaɪz/ v.tr. (also **-ise**) **1** duly perform (a ceremony esp. of marriage). **2** celebrate (a festival etc.). **3** make solemn. □ **solemnization** /ˌsɒləmnaɪ'zeɪʃ(ə)n/ n. [ME f. OF *solem(p)niser* f. med.L *solemnizare* (as SOLEMN)]

Solemn League and Covenant an agreement made in 1643 between the English Parliament and the Scottish Covenanters during the English Civil War, by which the Scots would provide military aid in return for the establishment of a Presbyterian system in England, Scotland, and Ireland. Although the Scottish support proved crucial in the Parliamentary victory, the principal Presbyterian leaders were expelled from Parliament in 1647 and the covenant was never honoured.

solenodon /sə'lenəd(ə)n, sə'liː-/ n. a rare insectivorous mammal of the genus *Solenodon*, confined to the Caribbean islands of Cuba and Hispaniola and resembling a large shrew. [mod.L f. Gk *sōlēn* channel, pipe + *odous odontos* tooth]

solenoid /'səʊləˌnɔɪd, 'sɒl-/ n. a cylindrical coil of wire acting as a magnet when carrying electric current. □ **solenoidal** /ˌsəʊlə'nɔɪd(ə)l, ˌsɒl-/ adj. [F *solénoïde* f. Gk *sōlēn* channel, pipe + -OID]

Solent, the /'səʊlənt/ a channel between the NW coast of the Isle of Wight and the mainland of southern England.

Soleure /sɒlœr/ the French name for SOLOTHURN.

sol-fa /'sɒlfaː/ n. & v. ● n. = SOLMIZATION (cf. TONIC SOL-FA). ● v.tr. (**-fas, -faed** /-faːd/) sing (a tune) with sol-fa syllables. [SOL[3] + FA]

solfatara /ˌsɒlfə'taːrə/ n. a volcanic vent emitting only sulphurous and other vapours. [name of a volcano near Naples, f. It. *solfo* sulphur]

solfeggio /sɒl'fedʒɪˌəʊ/ n. (pl. **solfeggi** /-dʒɪ/) Mus. **1** an exercise in singing using sol-fa syllables. **2** solmization. [It. (as SOL-FA)]

soli pl. of SOLO.

solicit /sə'lɪsɪt/ v. (**solicited, soliciting**) **1** tr. & (foll. by for) intr. ask repeatedly or earnestly for or seek or invite (business etc.). **2** tr. (often foll. by for) make a request or petition to (a person). **3** tr. (also absol.) accost (a person) and offer one's services as a prostitute. □ **solicitation** /-ˌlɪsɪ'teɪʃ(ə)n/ n. [ME f. OF *solliciter* f. L *sollicitare* agitate f. *sollicitus* anxious f. *sollus* entire + *citus* past part., = set in motion]

solicitor /sə'lɪsɪtə(r)/ n. **1** Brit. a member of the legal profession qualified to deal with conveyancing, draw up wills, etc., and to advise clients and instruct barristers. **2** a person who solicits. **3** US a canvasser. **4** US the chief law officer of a city etc. [ME f. OF *solliciteur* (as SOLICIT)]

Solicitor-General /sə'lɪsɪtə'dʒenrəl/ n. (pl. **Solicitors-General**) **1** (in the UK) the Crown law officer below the Attorney-General or (in Scotland) below the Lord Advocate. **2** (in the US) the law officer below the Attorney General.

solicitous /sə'lɪsɪtəs/ adj. **1** (often foll. by of, about, etc.) showing interest or concern. **2** (foll. by to + infin.) eager, anxious. □ **solicitously** adv. **solicitousness** n. [L *sollicitus* (as SOLICIT)]

solicitude /sə'lɪsɪˌtjuːd/ n. **1** the state of being solicitous; solicitous behaviour. **2** anxiety or concern. [ME f. OF *sollicitude* f. L *sollicitudo* (as SOLICITOUS)]

solid /'sɒlɪd/ adj., n., & adv. ● adj. (**solider, solidest**) **1** firm and stable in shape; not liquid or fluid (*solid food; water becomes solid at 0°C*). **2** of such material throughout, not hollow or containing cavities (*a solid sphere*). **3** of the same substance throughout (*solid silver*). **4** of strong material or construction or build, not flimsy or slender etc. **5 a** having three dimensions. **b** concerned with solids (*solid geometry*). **6 a** sound and reliable; genuine (*solid arguments*). **b** staunch and dependable (*a solid Tory*). **7** sound but without any special flair etc. (*a solid piece of work*). **8** financially sound. **9** (of time) uninterrupted, continuous (*spend four solid hours on it*). **10 a** unanimous, undivided (*support has been pretty solid so far*). **b** (foll. by for) united in favour of. **11** (of printing) without spaces between the lines etc. **12** (of a tyre) without a central air space. **13** (foll. by with) US colloq. on good terms. **14** Austral. & NZ colloq. severe, unreasonable. ● n. **1** a solid substance or body. **2** (in pl.) solid food. **3** Geom. a body or magnitude having three dimensions. ● adv. so as to become solid; solidly (*jammed solid; set solid*). □ **solid angle** an angle formed by planes etc. meeting at a point. **solid colour** colour covering the whole of an object, without a pattern etc. **solid-drawn** (of a tube etc.) pressed or drawn out from a solid bar of metal. **solid solution** solid material containing one substance uniformly distributed in another. □ **solidly** adv. **solidness** n. [ME f. OF *solide* f. L *solidus*, rel. to *salvus* safe, *sollus* entire]

Solidarity /ˌsɒlɪˈdærɪtɪ/ an independent trade-union movement in Poland which developed into a mass campaign for political change and inspired popular opposition to Communist regimes in other eastern European countries. It was formed under the leadership of Lech Wałęsa in 1980 following a series of strikes in the shipyards of Gdańsk, but banned in 1981 after the imposition of martial law in Poland. Legalized again in 1989 following further strikes, Solidarity won a majority in the elections of that year. [transl. Polish *Solidarność*]

solidarity /ˌsɒlɪˈdærɪtɪ/ n. **1** unity or agreement of feeling or action, esp. among individuals with a common interest. **2** mutual dependence. [F *solidarité* f. *solidaire* f. *solide* SOLID]

solidi *pl.* of SOLIDUS.

solidify /səˈlɪdɪˌfaɪ/ v.tr. & intr. (**-ies, -ied**) make or become solid. □ **solidifier** n. **solidification** /-ˌlɪdɪfɪˈkeɪʃ(ə)n/ n.

solidity /səˈlɪdɪtɪ/ n. the state of being solid; firmness.

solid state n. *Physics* the state of matter which retains its boundaries without support. (*See note below.*) □ **solid-state** adj. using the electronic properties of solids, especially semiconductors, to replace those of valves.

▪ The solid state is one of the fundamental states in which matter can exist, along with the liquid, gas, and plasma states. The atoms or molecules in a solid occupy fixed positions with respect to each other and cannot move freely; solids oppose deforming forces and solid bodies do not easily penetrate each other. Solids may be broadly classified into crystalline and amorphous. In crystalline solids, such as saltar quartz, the atoms are arranged in an orderly and repeating manner. In amorphous solids, such as amorphous sulphur, this order is absent over large distances (on an interatomic scale); glasses are liquids which have solidified without any crystallization having occurred.

solidus /ˈsɒlɪdəs/ n. (*pl.* **solidi** /-ˌdaɪ/) **1** an oblique stroke (/) used in writing fractions ($^3/_4$), to separate other figures and letters, or to denote alternatives (*and/or*) and ratios (*miles/day*). **2** (in full **solidus curve**) a curve in a graph of the temperature and composition of a mixture, below which the substance is entirely solid. **3** *hist.* a gold coin of the later Roman Empire. [ME (in sense 3) f. L: see SOLID]

solifluction /ˌsəʊlɪˈflʌkʃ(ə)n, ˌsɒl-/ n. the gradual movement of wet soil etc. down a slope. [L *solum* soil + L *fluctio* flowing f. *fluere fluct-* flow]

Solihull /ˈsɒlɪˌhʌl, ˌsəʊlɪˈhʌl/ a town in the Midlands, forming part of the conurbation of Birmingham; pop. (1981) 94,600.

soliloquy /səˈlɪləkwɪ/ n. (*pl.* **-ies**) **1** the act of talking when alone or regardless of any hearers, esp. in drama. **2** part of a play involving this. □ **soliloquist** n. **soliloquize** v.intr. (also **-ise**). [LL *soliloquium* f. L *solus* alone + *loqui* speak]

Soliman see SULEIMAN I.

soliped /ˈsɒlɪˌped/ adj. & n. ● adj. (of an animal) solid-hoofed. ● n. a solid-hoofed animal. [F *solipède* or mod.L *solipes -pedis* f. L *solidipes* f. *solidus* solid + *pes* foot]

solipsism /ˈsɒlɪpˌsɪz(ə)m/ n. **1** *Philos.* the view that the self is all that exists, or is all that can be known. (*See note below.*) **2** self-centredness, selfishness. □ **solipsist** n. **solipsistic** /ˌsɒlɪpˈsɪstɪk/ adj. **solipsistically** adv. [L *solus* alone + *ipse* self]

▪ The most extreme form of scepticism, solipsism is typically based on claims that all we can perceive are our immediate experiences or states of mind, and that these do not supply an adequate basis for knowledge of anything further. A solipsist therefore denies both that there is an external physical world and that there are any other minds.

solitaire /ˈsɒlɪˌteə(r)/ n. **1** a diamond or other gem set by itself. **2** a ring having a single gem. **3** a game for one player played by removing pegs etc. one at a time from a board by jumping others over them until only one is left. **4** *N. Amer.* = PATIENCE 4. **5** a large extinct flightless bird of the dodo family, found formerly on two of the Mascarene Islands. **6** an American thrush of the genus *Myadestes*. [F f. L *solitarius* (as SOLITARY)]

solitary /ˈsɒlɪtərɪ/ adj. & n. ● adj. **1** living alone; not gregarious; without companions; lonely (*a solitary existence*). **2** (of a place) secluded or unfrequented. **3** single or sole (*a solitary instance*). **4** (of an insect) not living in communities. **5** *Biol.* growing singly, not in a cluster. ● n. (*pl.* **-ies**) **1** a recluse or hermit. **2** *colloq.* = solitary confinement. □ **solitary confinement** isolation of a prisoner in a separate cell as a punishment. □ **solitarily** adv. **solitariness** n. [ME f. L *solitarius* f. *solus* alone]

solitude /ˈsɒlɪˌtjuːd/ n. **1** the state of being solitary. **2** a lonely place. [ME f. OF *solitude* or L *solitudo* f. *solus* alone]

solmization /ˌsɒlmɪˈzeɪʃ(ə)n/ n. (also **-isation**) *Mus.* a system of associating each note of a scale with a particular syllable, now usu. *doh ray me fah soh lah te*, with *doh* as C in the fixed-doh system and as the keynote in the movable-doh or tonic sol-fa system. □ **solmizate** /ˈsɒlmɪˌzeɪt/ v.intr. & tr. [F *solmisation* (as SOL³, MI)]

solo /ˈsəʊləʊ/ n., v., & adv. ● n. (*pl.* **-os**) **1** **-os** or **soli** /-lɪ/) **a** a vocal or instrumental piece or passage, or a dance, performed by one person with or without accompaniment. **b** (*attrib.*) performed or performing as a solo (*solo passage; solo violin*). **2 a** an unaccompanied flight by a pilot in an aircraft. **b** anything done by one person unaccompanied. **c** (*attrib.*) unaccompanied, alone. **3** (in full **solo whist**) **a** a card-game like whist in which one player may oppose the others. **b** a declaration or the act of playing to win five tricks at this. ● v. (**-oes, -oed**) **1** intr. perform a solo, esp. a solo flight or a musical solo. **2** tr. perform or achieve as a solo. ● adv. unaccompanied, alone (*flew solo for the first time*). □ **solo stop** an organ stop especially suitable for imitating a solo performance on another instrument. [It. f. L *solus* alone]

soloist /ˈsəʊləʊɪst/ n. a performer of a solo, esp. in music.

Solomon /ˈsɒləmən/, son of David, king of ancient Israel *c.*970–*c.*930 BC. During his reign he extended the kingdom of Israel to the border with Egypt and the Euphrates, and became famous both for his wisdom and for the magnificence of his palaces. In 957 he built the Temple of Jerusalem, with which his name is associated; this, together with his other building schemes and the fortifying of strategic cities, led to a system of levies and enforced labour, and the resulting discontent culminated in the secession of the northern tribes (see ISRAEL¹ 2). In the Bible, Solomon is traditionally associated with the Song of Solomon, Ecclesiastes, and Proverbs; the Wisdom of Solomon in the Apocrypha is also ascribed to him. (See also JUDGEMENT OF SOLOMON.) □ **Solomonic** /ˌsɒləˈmɒnɪk/ adj.

Solomon Islands (also **Solomons**) a country consisting of a group of islands in the SW Pacific, to the east of New Guinea; pop. (est. 1991) 326,000; languages, English (official), Pidgin, local Malayo-Polynesian languages; capital, Honiara. Visited by the Spanish in 1658, the islands were divided between Britain and Germany in the late 19th century; the southern islands became a British protectorate in 1893 while the north remained German until mandated to Australia in 1920. The islands were the scene of heavy fighting between Allied and Japanese forces in 1942–3. With the exception of the northern part of the chain (now part of Papua New Guinea), the Solomons became self-governing in 1976 and fully independent within the Commonwealth two years later. □ **Solomon Islander** n.

Solomon's seal n. **1** a figure like the Star of David. **2** a liliaceous plant of the genus *Polygonatum*, with arching stems and drooping green and white flowers.

Solon /ˈsəʊlɒn/ (*c.*630–*c.*560 BC), Athenian statesman and lawgiver. One of the Seven Sages listed by Plato, he is notable for his economic, constitutional, and legal reforms, begun in about 594. He revised the existing code of laws established by Draco, making them less severe; for example, he abolished the punishment of slavery for debt and freed those already enslaved for this, and reserved the death penalty for murder. His division of the citizens into four classes based on wealth rather than birth with a corresponding division of political responsibility laid the foundations of Athenian democracy.

Solothurn /ˈzəʊləʊˌtʊən/ (called in French *Soleure*) **1** a canton in NW Switzerland, in the Jura mountains. **2** its capital, a town on the River Aare; pop. (1990) 15,430. Occupying a strategic position at the southwestern approach to the Rhine, it became a free imperial city in 1218, acquiring in the 15th century the territories comprising the modern canton. It joined the Swiss Confederation in 1481.

solstice /ˈsɒlstɪs/ n. **1** either of the times when the sun is furthest from the equator at noon, marked by the longest and shortest days. **2** *Astron.* the point in the ecliptic reached by the sun at a solstice. □ **summer solstice 1** the solstice occurring in midsummer at the time of the longest day, about 21 June in the northern hemisphere and about 22 Dec. in the southern hemisphere. **2** *Astron.* the solstice occurring in June. **winter solstice 1** the solstice occurring in midwinter at the time of the shortest day, about 22 Dec. in the northern hemisphere and about 21 June in the southern hemisphere. **2** *Astron.* the solstice occurring in December. □ **solstitial** /sɒlˈstɪʃ(ə)l/ adj. [ME f. OF f. L *solstitium* f. *sol* sun + *sistere stit-* make stand]

solubilize /ˈsɒljʊbɪˌlaɪz/ v.tr. (also **-ise**) make soluble or more soluble. □ **solubilization** /ˌsɒljʊbɪlaɪˈzeɪʃ(ə)n/ n.

soluble /ˈsɒljʊb(ə)l/ adj. **1** that can be dissolved, esp. in water. **2** that can be solved. □ **soluble glass** = water-glass. □ **solubility** /ˌsɒljʊˈbɪlɪtɪ/ n. [ME f. OF f. LL solubilis (as SOLVE)]

solus /ˈsəʊləs/ predic.adj. (fem. **sola** /-lə/) (esp. in a stage direction) alone, unaccompanied. [L]

solute /ˈsɒljuːt/ n. a dissolved substance. [L solutum, neut. of solutus: see SOLVE]

solution /səˈluːʃ(ə)n, -ˈljuːʃ(ə)n/ n. **1 a** the act or a means of solving a problem or difficulty. **b** an explanation, answer, or decision. **2 a** the conversion of a solid or gas into a liquid by mixture with a liquid solvent. **b** a liquid mixture produced by this. **c** the state resulting from this (held in solution). **3** the act of dissolving or the state of being dissolved. **4** the act of separating or breaking. **5** = rubber solution (see RUBBER¹). □ **solution set** Math. the set of all the solutions of an equation or condition. [ME f. OF f. L solutio -onis (as SOLVE)]

Solutrean /səˈluːtrɪən/ adj. & n. (also **Solutrian**) Archaeol. of, relating to, or denoting an upper palaeolithic culture of central and SW France and parts of Iberia, following the Aurignacian and preceding the Magdalenian, dated to c.21,000–18,000 BP. [f. Solutré, the type-site in eastern France]

solvate /ˈsɒlveɪt/ v.intr. & tr. Chem. enter or cause to enter into combination with a solvent. □ **solvation** /sɒlˈveɪʃ(ə)n/ n.

Solvay process /ˈsɒlveɪ/ n. a manufacturing process for obtaining sodium carbonate (washing soda) from limestone, ammonia, and brine. [Ernest Solvay, Belgian chemist, who developed the process (1838–1922)]

solve /sɒlv/ v.tr. find an answer to, or an action or course that removes or effectively deals with (a problem or difficulty). □ **solvable** adj. **solver** n. [ME, = loosen, f. L solvere solut- unfasten, release]

solvent /ˈsɒlv(ə)nt/ adj. & n. ● adj. **1** able to dissolve or form a solution with something. **2** having enough money to meet one's liabilities. ● n. **1** a solvent liquid etc. **2** a dissolving or weakening agent. □ **solvency** n. (in sense 2).

Solway Firth /ˈsɒlweɪ/ an inlet of the Irish Sea, separating Cumbria (England) from Dumfries and Galloway (Scotland).

Solyman see SULEIMAN I.

Solzhenitsyn /ˌsɒlʒəˈnɪtsɪn/, Alexander (Russian name Aleksandr Isaevich Solzhenitsyn) (b.1918), Russian novelist. In 1945 he was imprisoned for eight years in a labour camp for criticizing Stalin and spent another three years in internal exile before being rehabilitated. After his release he was allowed to publish his first novel One Day in the Life of Ivan Denisovich (1962), describing conditions in a labour camp. In 1963, however, he was again in conflict with the authorities and thereafter was unable to have his books published in the Soviet Union. He was deported to West Germany in 1974 following the publication abroad of the first part of his trilogy The Gulag Archipelago in 1973; the first Russian-language edition appeared in 1989. Solzhenitsyn lived in the US until returning to Russia in 1994. He was awarded the Nobel Prize for literature in 1970.

Som. abbr. Somerset.

soma¹ /ˈsəʊmə/ n. (pl. **somata** /-mətə/) **1** the body as distinct from the soul. **2** the body of an organism as distinct from its reproductive cells. [Gk sōma -atos body]

soma² /ˈsəʊmə/ n. **1** an intoxicating drink used in Vedic ritual. **2** a plant yielding this. [Skr. sōma]

Somali /səˈmɑːlɪ/ n. & adj. ● n. **1** (pl. same or **Somalis**) **a** a member of a Hamitic Muslim people of Somalia. **b** a native or national of Somalia. **2** the language of this people, which belongs to the Cushitic branch of the Afro-Asiatic family of languages and is the official language of Somalia. ● adj. of or relating to Somalia, the Somalis, or their language. □ **Somalian** adj. & n. [Afr. name]

Somalia /səˈmɑːlɪə/ a country in the Horn of Africa; pop. (est. 1991) 8,041,000; official languages, Somali and Arabic; capital, Mogadishu. The area of the Horn of Africa was divided between British and Italian spheres of influence in the late 19th century, and the modern republic (which became independent in 1960) is a result of the unification of the former British Somaliland and Italian Somalia. Since independence, Somalia has been involved in border disputes with Kenya and Ethiopia, the latter leading to an intermittent war over the Ogaden. Civil war broke out in Somalia in 1988 and led to the overthrow of the government in 1991; in the same year northern Somalia declared itself independent as the Somaliland Republic. Factional fighting has continued, with the US intervening militarily (1992–4) in an attempt to restore order and ensure the safe passage of famine relief supplies.

Somali Peninsula an alternative name for the HORN OF AFRICA.

somatic /səˈmætɪk/ adj. of or relating to the body, esp. as distinct from the mind. □ **somatic cell** Biol. any cell of a living organism except the reproductive cells. □ **somatically** adv. [Gk sōmatikos (as SOMA¹)]

somato- /ˈsəʊmətəʊ/ comb. form the human body. [Gk sōma -atos body]

somatogenic /ˌsəʊmətəʊˈdʒenɪk/ adj. Biol. originating in the somatic cells.

somatology /ˌsəʊməˈtɒlədʒɪ/ n. the science of living bodies physically considered.

somatotonic /ˌsəʊmətəʊˈtɒnɪk/ adj. having an extrovert and aggressive temperament, thought to be associated esp. with a mesomorphic physique.

somatotrophin /ˌsəʊmətəʊˈtrəʊfɪn/ n. a growth hormone secreted by the pituitary gland. [as SOMATO-, -TROPHIC]

somatotype /ˈsəʊmətəʊˌtaɪp/ n. physique expressed in relation to various extreme types.

sombre /ˈsɒmbə(r)/ adj. (also US **somber**) **1** dark, gloomy (a sombre sky). **2** oppressively solemn or sober. **3** dismal, foreboding (a sombre prospect). □ **sombrely** adv. **sombreness** n. [F sombre f. OF sombre (n.) ult. f. L SUB- + umbra shade]

sombrero /sɒmˈbreərəʊ/ n. (pl. **-os**) a broad-brimmed felt or straw hat worn esp. in Mexico and the south-west US. [Sp. f. sombra shade (as SOMBRE)]

some /sʌm, səm/ adj., pron., & adv. ● adj. **1** an unspecified amount or number of (some water; some apples; some of them). **2** that is unknown or unnamed (will return some day; some fool has locked the door; to some extent). **3** denoting an approximate number (waited some twenty minutes). **4** a considerable amount or number of (went to some trouble). **5** (usu. stressed) **a** at least a small amount of (do have some consideration). **b** to a certain extent (that is some help). **c** colloq. notably such (I call that some story). ● pron. some people or things, some number or amount (I have some already; would you like some more?). ● adv. (colloq. except in **some more**) to some extent (we talked some; do it some more). □ **and then some** sl. and plenty more than that. **some few** see FEW. [OE sum f. Gmc]

-some¹ /səm/ suffix forming adjectives meaning: **1** adapted to; productive of (cuddlesome; fearsome). **2** characterized by being (fulsome; lithesome). **3** apt to (tiresome; meddlesome). [OE -sum]

-some² /səm/ suffix forming nouns from numerals, meaning 'a group of (so many)' (foursome). [OE sum SOME, used after numerals in genitive pl.]

-some³ /səʊm/ comb. form denoting a portion of a body, esp. of a cell (chromosome; ribosome). [Gk sōma body]

somebody /ˈsʌmbədɪ/ pron. & n. ● pron. some person. ● n. (pl. **-ies**) a person of importance (is really somebody now).

someday /ˈsʌmdeɪ/ adv. at some time in the future.

somehow /ˈsʌmhaʊ/ adv. **1** for some reason or other (somehow I never liked them). **2** in some unspecified or unknown way (he somehow dropped behind). **3** no matter how (must get it finished somehow).

someone /ˈsʌmwʌn/ n. & pron. = SOMEBODY.

someplace /ˈsʌmpleɪs/ adv. N. Amer. = SOMEWHERE.

somersault /ˈsʌməˌsɒlt, -ˌsɔːlt/ n. & v. (also **summersault**) ● n. an acrobatic movement in which a person turns head over heels in the air or on the ground and lands on the feet. ● v.intr. perform a somersault. [OF sombresault alt. f. sobresault ult. f. L supra above + saltus leap f. salire to leap]

Somerset /ˈsʌməˌset/ a county of SW England, on the Bristol Channel; county town, Taunton.

something /ˈsʌmθɪŋ/ n., pron., & adv. ● n. & pron. **1 a** some unspecified or unknown thing (have something to tell you; something has happened). **b** (also **something or other**) as a substitute for an unknown or forgotten description (a student of something or other). **2** a known or understood but unexpressed quantity, quality, or extent (there is something about it I do not like; is something of a fool). **3** colloq. an important or notable person or thing (the party was quite something). ● adv. archaic in some degree. □ **or something** or some unspecified alternative possibility (must have run away or something). **see something of** encounter (a person) briefly or occasionally. **something else** **1** something different. **2** colloq. something exceptional. **something like 1** an amount in the region of (left something like a million pounds). **2** somewhat like (shaped something like a cigar). **3** colloq. impressive; a fine specimen of. **something of** to some extent; in some sense (is something of an expert). [OE sum thing (as SOME, THING)]

sometime /ˈsʌmtaɪm/ adv. & adj. ● adv. **1** at some unspecified time. **2** archaic formerly. ● attrib.adj. former (the sometime mayor).

sometimes /ˈsʌmtaɪmz/ adv. at some times; occasionally.

somewhat /ˈsʌmwɒt/ adv., n., & pron. ● adv. to some extent (behaviour that was somewhat strange; answered somewhat hastily). ● n. & pron. archaic something (loses somewhat of its force). □ **more than somewhat** colloq. very (was more than somewhat perplexed).

somewhen /ˈsʌmwen/ adv. colloq. at some time.

somewhere /ˈsʌmweə(r)/ adv. & pron. ● adv. in or to some place. ● pron. some unspecified place. □ **get somewhere** colloq. achieve success. **somewhere about** approximately.

somite /ˈsəʊmaɪt/ n. each body-division of a metamerically segmented animal. □ **somitic** /səʊˈmɪtɪk/ adj. [Gk sōma body + -ITE¹]

Somme /sɒm/ a river of northern France. Rising east of Saint-Quentin, it flows 245 km (153 miles) through Amiens to the English Channel north-east of Dieppe. The area was the scene of heavy fighting in the First World War.

Somme, Battle of the a battle of the First World War between the British and the Germans, on the Western Front in northern France Jul.–Nov. 1916. Following a bombardment of German lines the British advanced from their trenches on foot to face the German machine-guns; although the Germans retreated a few kilometres they took refuge in the fortified Hindenburg Line. More than a million men on both sides were killed or wounded.

sommelier /ˈsɒmə‚ljeɪ/ n. a wine waiter. [F, = butler, f. somme pack (as SUMPTER)]

somnambulism /sɒmˈnæmbjʊ‚lɪz(ə)m/ n. **1** sleepwalking. **2** a condition of the brain inducing this. □ **somnambulant** adj. **somnambulantly** adv. **somnambulist** n. **somnambulistic** /-‚næmbjʊˈlɪstɪk/ adj. **somnambulistically** adv. [L somnus sleep + ambulare walk]

somniferous /sɒmˈnɪfərəs/ adj. inducing sleep; soporific. [L somnifer f. somnium dream]

somnolent /ˈsɒmnələnt/ adj. **1** sleepy, drowsy. **2** inducing drowsiness. **3** Med. in a state between sleeping and waking. □ **somnolence** n. **somnolency** n. **somnolently** adv. [ME f. OF sompnolent or L somnolentus f. somnus sleep]

Somoza /səˈməʊzə/, Anastasio (full surname Somoza García) (1896–1956), Nicaraguan soldier and statesman, President 1937–47 and 1951–6. After becoming Commander-in-Chief of the Nicaraguan army in 1933, he organized a military coup in 1936 and took presidential office the following year. From this time he ruled Nicaragua as a virtual dictator, exiling the majority of his political opponents and building up a fortune from land and business ventures. He was assassinated in 1956, after which his eldest son Luis Somoza Debayle (1922–67) succeeded him as President, serving from 1957 to 1963. Luis's younger brother Anastasio Somoza Debayle (1925–80) became President in 1967, after winning the general election that year. His dictatorial regime was overthrown by the Sandinistas in 1979, and he was assassinated while in exile in Paraguay.

son /sʌn/ n. **1** a boy or man in relation to either or both of his parents. **2 a** a male descendant. **b** (foll. by of) a male member of a family, nation, etc. **3** a person regarded as inheriting an occupation, quality, etc., or associated with a particular attribute (sons of freedom; sons of the soil). **4** (in full **my son**) a form of address esp. to a boy. **5** (**the Son**) (in Christian belief) Jesus Christ, the second person of the Trinity. □ **son-in-law** (pl. **sons-in-law**) the husband of one's daughter. **son of a bitch** US sl. a general term of contempt. **son of a gun** colloq. a jocular or affectionate form of address or reference. □ **sonless** adj. **sonship** n. [OE sunu f. Gmc]

sonant /ˈsəʊnənt/ adj. & n. Phonet. ● adj. (of a speech sound) voiced and syllabic. ● n. a voiced sound forming a syllable; a vowel or any of the consonants l, m, n pronounced as a syllable. □ **sonancy** n. [L sonare sonant- sound]

sonar /ˈsəʊnə(r)/ n. a system for the underwater detection of objects by reflected or emitted sound; an apparatus for this. Hydrophones were used in the 1890s to detect the sound of engines and propellers, but by 1918 a system of transmitting a pulse of sound and using its rebounding echo to detect a stationary submerged craft had been developed. At first known as asdic, it was widely used in the Second World War. Sonar has important civil applications, its simplest form being the echo-sounder used for determining water depth. It is also used for locating fish shoals, and recently developed forms of sonar use sideways scanning to give very detailed images of the seabed and objects lying on it. [sound navigation and ranging, after radar]

sonata /səˈnɑːtə/ n. a musical composition for one or two instruments (one usually being a keyboard instrument), generally in several movements with one (especially the first) or more in what has come to be known as sonata form (see SONATA FORM). The classical sonata, as exemplified in the work of Haydn, Mozart, and Beethoven, is usually a four-movement work, each movement of a standard type and with the movements in a particular order. In earlier use the term indicated only that instruments, rather than voices, were used. [It., = sounded (orig. as distinct from sung): fem. past part. of sonare sound]

sonata form n. Mus. a type of composition in three sections (exposition, development, and recapitulation) in which two themes or subjects are explored according to set key relationships. It forms the basis for much classical music, including the sonata, symphony, and concerto.

sonatina /‚sɒnəˈtiːnə/ n. a simple or short sonata. [It., dimin. of SONATA]

sonde /sɒnd/ n. a device sent up to obtain information about atmospheric conditions, esp. = RADIOSONDE. [F, = sounding(-line)]

Sondheim /ˈsɒndhaɪm/, Stephen (Joshua) (b.1930), American composer and lyricist. He became famous with his lyrics for Leonard Bernstein's West Side Story (1957), and later wrote both words and music for a number of musicals, including A Funny Thing Happened on the Way to the Forum (1962), A Little Night Music (1973), and Sweeney Todd (1979).

sone /səʊn/ n. a unit of subjective loudness, equal to 40 phons. [L sonus sound]

son et lumière /‚sɒn eɪ ˈluːmɪ‚eə(r)/ n. an entertainment by night at a historic monument, building, etc., using lighting effects and recorded sound to give a dramatic narrative of its history. [F, = sound and light]

Song see SUNG.

song /sɒŋ/ n. **1** a musical composition comprising a short poem or other set of words set to music; a set of words meant to be sung. **2** singing or vocal music (burst into song). **3** a musical composition suggestive of a song. **4** the usually repeated musical call of some birds. **5** a short poem in rhymed stanzas. **6** archaic poetry or verse. □ **for a song** colloq. very cheaply. **on song** Brit. colloq. performing exceptionally well. **song and dance** colloq. a fuss or commotion. **song cycle** a set of musically linked songs on a romantic theme. **song sparrow** a North American sparrow-like bird, Melospiza melodia, of the bunting family, with a characteristic musical song. **song thrush** a thrush, Turdus philomelos, of Europe and western Asia, with a speckled breast and a song of repeated phrases. □ **songless** adj. [OE sang f. Gmc (as SING)]

songbird /ˈsɒŋbɜːd/ n. **1** a bird with a musical song. **2** Zool. a perching bird of the group Oscines, possessing a syrinx. (See PASSERINE.)

songbook /ˈsɒŋbʊk/ n. a collection of songs with music.

Song Hong /sɒŋ ˈhɒŋ/ the Vietnamese name for the Red River in SE Asia (see RED RIVER 1).

Song of Solomon (also called the Song of Songs or Canticles) a book of the Bible, an anthology of love poems ascribed to Solomon but dating from a much later period. From an early date Jewish and Christian writers interpreted the book allegorically, in the Talmud as God's dealings with the congregation of Israel and in Christian exegesis as God's relations with the Church or the individual soul.

Song of the Three (Holy Children) a book of the Apocrypha, telling of three Hebrew exiles thrown into a fiery furnace by Nebuchadnezzar.

songsmith /ˈsɒŋsmɪθ/ n. a writer of songs.

songster /ˈsɒŋstə(r)/ n. (fem. **songstress** /-trɪs/) **1** a singer, esp. a fluent and skilful one. **2** a songbird. **3** a poet. **4** US a songbook. [OE sangestre (as SONG, -STER)]

songwriter /ˈsɒŋ‚raɪtə(r)/ n. a writer of songs or the music for them.

sonic /ˈsɒnɪk/ adj. of or relating to or using sound or sound waves. □ **sonic bang** (or **boom**) a loud explosive noise caused by the shock wave from an aircraft when it passes the speed of sound. **sonic barrier** = sound barrier (see SOUND¹). **sonic mine** a mine exploded by the sound of a passing ship. □ **sonically** adv. [L sonus sound]

sonnet /ˈsɒnɪt/ n. & v. ● n. a poem of fourteen lines (usu. pentameters) using any of a number of formal rhyme-schemes, in English usu. having ten syllables per line. (See note below.) ● v. (**sonneted**, **sonneting**) **1** intr. write sonnets. **2** tr. address sonnets to. [F sonnet or It. sonetto dimin. of suono SOUND¹]

■ Originating in Italy, the sonnet was established by Petrarch as a major form of love poetry; introduced to England by Thomas Wyatt and developed by Henry, Earl of Surrey (c.1517–47), it was widely used from the late 16th century. Sir Philip Sidney, Shakespeare, and others wrote sequences of love sonnets, while John Donne and Milton extended the form to encompass religious and political subjects. Also in the late 16th century, Spenser developed a new form of stanza for the sonnet (see SPENSER). The sonnet form was revived in the 19th century, and adopted by such writers as Keats, Wordsworth, Elizabeth Barrett Browning, and Dante Gabriel Rossetti. The rhyme-schemes of the sonnet follow two basic patterns, the Italian or Petrarchan (see PETRARCHAN) and the English or Shakespearian. Shakespeare's sonnets comprise three quatrains and a couplet, rhymed *ababcdcdefefgg*.

sonneteer /ˌsɒnɪˈtɪə(r)/ n. usu. derog. a writer of sonnets.

sonny /ˈsʌnɪ/ n. colloq. a familiar form of address to a young boy.

sonobuoy /ˈsəʊnəˌbɔɪ/ n. a buoy for detecting underwater sounds and transmitting them by radio. [L sonus sound + BUOY]

sonogram /ˈsəʊnəˌɡræm/ n. 1 a graph representing a sound, showing the distribution of energy at different frequencies. 2 esp. Med. a visual image produced from an ultrasound examination. □ **sonograph** n. [L sonus sound + -GRAM]

sonometer /səˈnɒmɪtə(r)/ n. 1 an instrument for measuring the vibration frequency of a string etc. 2 an audiometer. [L sonus sound + -METER]

Sonora /səˈnɔːrə/ a state of NW Mexico, on the Gulf of California; capital, Hermosillo.

Sonora Desert an arid region of North America, comprising SE California and SW Arizona in the US and, in Mexico, much of Baja California and the western part of Sonora.

sonorous /ˈsɒnərəs, səˈnɔːr-/ adj. 1 having a loud, full, or deep sound; resonant. 2 (of a speech, style, etc.) imposing, grand. □ **sonorously** adv. **sonorousness** n. **sonority** /səˈnɒrɪtɪ/ n. (pl. -ies). [L sonorus f. sonor sound]

sonsy /ˈsɒnsɪ/ adj. (also **sonsie**) (**sonsier, sonsiest**) Sc. 1 plump, buxom. 2 of a cheerful disposition. 3 bringing good fortune. [ult. f. Ir. & Gael. sonas good fortune f. sona fortunate]

Sontag /ˈsɒntæɡ/, Susan (b.1933), American writer and critic. She established her reputation as a radical intellectual with a series of essays which were collected in *Against Interpretation* (1966). In the 1970s Sontag made two films and won critical acclaim for her study *On Photography* (1976) and her collection of essays *Illness as Metaphor* (1979); the latter was prompted by her experiences as a cancer patient.

Soochow see SUZHOU.

sook /sʊk/ n. esp. Sc., Austral., & NZ colloq. 1 derog. a timid bashful person; a coward or sissy. 2 a hand-reared calf. [Engl. dial. suck, call-word for a calf]

sool /suːl/ v.tr. Austral. & NZ sl. 1 (of a dog) attack or worry (an animal). 2 (often foll. by on) urge or goad. □ **sooler** n. [var. of 17th-c. (now dial.) sowl seize roughly, of unkn. orig.]

soon /suːn/ adv. 1 after a short interval of time (*shall soon know the result*). 2 relatively early (*must you go so soon?*). 3 (prec. by how) early (with relative rather than distinctive sense) (*how soon will it be ready?*). 4 readily or willingly (in expressing choice or preference: *which would you sooner do?; would as soon stay behind*). □ **as** (or **so**) **soon as** (implying a causal or temporal connection) at the moment that; not later than; as early as (*came as soon as I heard about it; disappears as soon as it's time to pay*). **no sooner ... than** at the very moment that (*we no sooner arrived than the rain stopped*). **sooner or later** at some future time; eventually. □ **soonish** adv. [OE sōna f. WG]

soot /sʊt/ n. & v. ● n. a black carbonaceous substance rising in fine flakes in the smoke of wood, coal, oil, etc., and deposited on the sides of a chimney etc. ● v.tr. cover with soot. [OE sōt f. Gmc]

sooth /suːθ/ n. archaic truth, fact. □ **in sooth** really, truly. [OE sōth (orig. adj., = true) f. Gmc]

soothe /suːð/ v.tr. 1 calm (a person or feelings). 2 soften or mitigate (pain). 3 archaic flatter or humour. □ **soother** n. **soothing** adj. **soothingly** adv. [OE sōthian verify f. sōth true: see SOOTH]

soothsayer /ˈsuːθˌseɪə(r)/ n. a diviner or seer. [ME, = a person who says the truth: see SOOTH]

sooty /ˈsʊtɪ/ adj. (**sootier, sootiest**) 1 covered with or full of soot. 2 (esp. of an animal or bird) black or brownish-black. □ **sootily** adv. **sootiness** n.

sop /sɒp/ n. & v. ● n. 1 a thing given or done to pacify or bribe. 2 a piece of bread etc. dipped in gravy etc. ● v. (**sopped, sopping**) 1 intr. be drenched (*came home sopping*). 2 tr. (foll. by up) absorb (liquid) in a towel etc. 3 tr. wet thoroughly; soak. [OE sopp, corresp. to MLG soppe, OHG sopfa bread and milk, prob. f. a weak grade of the base of OE sūpan: see SUP¹]

sophism /ˈsɒfɪz(ə)m/ n. a false argument, esp. one intended to deceive. [ME f. OF sophime f. L f. Gk sophisma clever device f. sophizomai become wise f. sophos wise]

sophist /ˈsɒfɪst/ n. 1 a member of the last generations of Greek philosophers before Plato, of the period c.450 BC to c.400 BC. (See note below.) 2 a person who reasons with clever but fallacious arguments. □ **sophistic** /səˈfɪstɪk/ adj. **sophistical** adj. **sophistically** adv. [L sophistes f. Gk sophistēs f. sophizomai: see SOPHISM]

■ The sophists were in business offering the equivalent of university education: for a hefty fee they would teach, in particular, rhetoric — how to argue a case in a lawcourt or an assembly. But they were also thinkers with argued views. Perhaps most important was the contrast they emphasized between what exists by nature and what exists only by human convention. This led not only to moral relativism (what is right depends entirely on what society thinks right) but to amoralism (morality is a matter merely of conventions, which the intelligent and strong person will disregard). Such views, together with the idea that they taught how to make the worse case appear the better, brought the sophists into disrepute; they were condemned by Plato, and the popular distrust of sophists is today reflected in the derogatory use of the terms *sophist* and *sophistry*.

sophisticate v., adj., & n. ● v. /səˈfɪstɪˌkeɪt/ 1 tr. make (a person etc.) educated, cultured, or refined. 2 tr. make (equipment or techniques etc.) highly developed or complex. 3 tr. a involve (a subject) in sophistry. b mislead (a person) by sophistry. 4 tr. deprive (a person or thing) of its natural simplicity, make artificial by worldly experience etc. 5 tr. tamper with (a text etc.) for purposes of argument etc. 6 tr. adulterate (wine etc.). 7 intr. use sophistry. ● adj. /səˈfɪstɪkət/ sophisticated. ● n. /səˈfɪstɪkət/ a sophisticated person. □ **sophistication** /-ˌfɪstɪˈkeɪʃ(ə)n/ n. [med.L sophisticare tamper with f. sophisticus (as SOPHISM)]

sophisticated /səˈfɪstɪˌkeɪtɪd/ adj. 1 a (of a person) cultured and refined, worldly, discriminating in taste and judgement. b appealing to sophisticated people or tastes. 2 (of a thing, idea, etc.) highly developed and complex. □ **sophisticatedly** adv.

sophistry /ˈsɒfɪstrɪ/ n. (pl. -ies) 1 the reasoning of sophists; the use of sophisms. 2 a sophism.

Sophocles /ˈsɒfəˌkliːz/ (c.496–406 BC), Greek dramatist. He is one of the trio of major Greek tragedians, with Aeschylus and Euripides. His seven surviving plays are notable for their addition of a third actor to the previous two (in addition to the chorus), thus allowing a greater complexity of plot and fuller depiction of character, and for their examination of the relationship between mortals and the divine order. The plays include *Antigone, Electra*, and *Oedipus Rex* (also called *Oedipus Tyrannus*).

sophomore /ˈsɒfəˌmɔː(r)/ n. N. Amer. a second-year university or high-school student. □ **sophomoric** /ˌsɒfəˈmɒrɪk/ adj. [earlier sophumer f. sophum, obs. var. of SOPHISM]

Sophy /ˈsəʊfɪ/ n. (pl. -ies) hist. a ruler of Persia in the 16th–17th centuries. [Pers. safi surname of the dynasty, f. Arab. safi-ud-din pure of religion, title of the founder's ancestor]

soporific /ˌsɒpəˈrɪfɪk/ adj. & n. ● adj. tending to produce sleep. ● n. a soporific drug or influence. □ **soporiferous** adj. **soporifically** adv. [L sopor sleep + -FIC]

sopping /ˈsɒpɪŋ/ adj. (also **sopping wet**) soaked with liquid; wet through. [pres. part. of SOP v.]

soppy /ˈsɒpɪ/ adj. (**soppier, soppiest**) 1 Brit. colloq. a silly or foolish in a feeble or self-indulgent way. b mawkishly sentimental. 2 (foll. by on) Brit. colloq. foolishly infatuated with. 3 soaked with water. □ **soppily** adv. **soppiness** n. [SOP + -Y¹]

sopranino /ˌsɒprəˈniːnəʊ/ n. (pl. -os) Mus. an instrument higher than soprano, esp. a recorder or saxophone. [It., dimin. of SOPRANO]

soprano /səˈprɑːnəʊ/ n. (pl. -os or **soprani** /-nɪ/) 1 a the highest singing voice. b a female or boy singer with this voice. c a part written for it. 2 a (attrib.) denoting an instrument of a high or the highest pitch in its family. b the player of such an instrument. □ **soprano-clef** an obsolete clef placing middle C on the lowest line of the staff. [It. f. sopra above f. L supra]

Sopwith /ˈsɒpwɪθ/, Sir Thomas (Octave Murdoch) (1888–1989), English aircraft designer. During the First World War he designed a number of planes, including the famous fighter biplane the Sopwith Camel, which were built by his Sopwith Aviation Company (founded 1912). During the Second World War, as chairman of the Hawker Siddeley company, he was responsible for the production of aircraft such as the Hurricane fighter.

sora /'sɔːrə/ n. (in full **sora rail**) a small American crake, *Porzana carolina*, frequenting marshes. [perh. f. Amer. Indian]

sorb /sɔːb/ n. **1** a service tree (see SERVICE² 1). **2** (in full **sorb-apple**) the fruit of this tree. [F *sorbe* or L *sorbus* service tree, *sorbum* service-berry]

sorbefacient /ˌsɔːbɪˈfeɪʃ(ə)nt/ adj. & n. Med. ● adj. causing absorption. ● n. a sorbefacient drug etc. [L *sorbere* suck in + -FACIENT]

sorbet /'sɔːbeɪ, -bɪt/ n. **1** a water-ice. **2** sherbet. [F f. It. *sorbetto* f. Turk. *şerbet* f. Arab. *šarba* to drink: cf. SHERBET]

sorbitol /'sɔːbɪˌtɒl/ n. Chem. a sweet-tasting crystalline alcohol found in some fruit, used in industry and as a food additive. [SORB + -ITE¹ + -OL]

Sorbo /'sɔːbəʊ/ n. Brit. propr. (in full **Sorbo rubber**) a spongy rubber. [ABSORB + -O]

Sorbonne /sɔːˈbɒn/ the seat of the faculties of science and literature of Paris University, originally a theological college founded by Robert de Sorbon, chaplain to Louis IX, c.1257.

sorcerer /'sɔːsərə(r)/ n. (fem. **sorceress** /-rɪs/) a person who claims to use magic powers; a magician or wizard. □ **sorcerous** adj. **sorcery** n. (pl. **-ies**). [obs. sorcer f. OF sorcier ult. f. L sors sortis lot]

sordid /'sɔːdɪd/ adj. **1** dirty or squalid. **2** ignoble, mean, or mercenary. **3** mean or niggardly. **4** dull-coloured. □ **sordidly** adv. **sordidness** n. [F sordide or L sordidus f. sordere be dirty]

sordino /sɔːˈdiːnəʊ/ n. (pl. **sordini** /-nɪ/) Mus. a mute for a bowed or wind instrument. [It. f. sordo mute f. L surdus]

sore /sɔː(r)/ adj., n., & adv. ● adj. **1** (of a part of the body) painful from injury or disease (has a sore arm). **2** (of a person) suffering pain. **3** (often foll. by about, at) aggrieved or vexed. **4** archaic grievous or severe (in sore need). ● n. **1** a sore place on the body, esp. as caused by pressure or friction. **2** a source of distress or annoyance (reopen old sores). ● adv. archaic grievously, severely. □ **sore point** a subject causing distress or annoyance. **sore throat** an inflammation of the lining membrane at the back of the mouth etc. □ **soreness** n. [OE sār (n. & adj.), sāre (adv.), f. Gmc]

sorehead /'sɔːhed/ n. N. Amer. colloq. a touchy or disgruntled person.

sorel /'sɒrəl/ n. Brit. a male fallow deer in its third year. [var. of SORREL²]

sorely /'sɔːlɪ/ adv. **1** extremely, badly (am sorely tempted; sorely in need of repair). **2** severely (am sorely vexed). [OE sārlīce (as SORE, -LY²)]

sorghum /'sɔːgəm/ n. a tropical cereal grass of the genus *Sorghum*; esp. a variety of *S. bicolor*, e.g. durra. [mod.L f. It. sorgo, perh. f. unrecorded Rmc syricum (gramen) Syrian (grass)]

sori pl. of SORUS.

Soroptimist /səˈrɒptɪmɪst/ n. a member of the Soroptimist Club, an international club for professional and business women founded in California in 1921. [L soror sister + OPTIMIST (as OPTIMISM)]

sorority /səˈrɒrɪtɪ/ n. (pl. **-ies**) N. Amer. a female students' society in a university or college. [med.L sororitas or L soror sister, after fraternity]

sorosis /səˈrəʊsɪs/ n. (pl. **soroses** /-siːz/) Bot. a fleshy compound fruit, e.g. a pineapple or mulberry. [mod.L f. Gk sōros heap]

sorption /'sɔːpʃ(ə)n/ n. absorption or adsorption happening jointly or separately. [back-form. f. absorption, adsorption]

sorrel¹ /'sɒrəl/ n. **1** an acid-leaved plant of the dock family; esp. *Rumex acetosa*, used in salads and for flavouring. **2** a plant with similarly acid leaves (wood sorrel). [ME f. OF surele, sorele f. Gmc]

sorrel² /'sɒrəl/ adj. & n. ● adj. of a light reddish-brown colour. ● n. **1** this colour. **2** a sorrel animal, esp. a horse. **3** Brit. a sorel. [ME f. OF sorel f. sor yellowish f. Frank.]

Sorrento /səˈrentəʊ/ a town on the west coast of central Italy, situated on a peninsula separating the Bay of Naples, which it faces, from the Gulf of Salerno; pop. (1990) 17,500.

sorrow /'sɒrəʊ/ n. & v. ● n. **1** mental distress caused by loss or disappointment etc. **2** a cause of sorrow. **3** lamentation. ● v.intr. **1** feel sorrow. **2** mourn. □ **sorrower** n. **sorrowing** adj. [OE sorh, sorg]

sorrowful /'sɒrəʊˌfʊl/ adj. **1** feeling or showing sorrow. **2** distressing, lamentable. □ **sorrowfully** adv. **sorrowfulness** n. [OE sorhful (as SORROW, -FUL)]

sorry /'sɒrɪ/ adj. (**sorrier, sorriest**) **1** (predic.) pained or regretful or penitent (were sorry for what they had done; am sorry that you have to go). **2** (predic.; foll. by for) feeling pity or sympathy for (a person). **3** as an expression of apology. **4** wretched; in a poor state (a sorry sight). □ **sorry for oneself** dejected. □ **sorrily** adv. **sorriness** n. [OE sārig f. WG (as SORE, -Y¹)]

sort /sɔːt/ n. & v. ● n. **1** a group of things etc. with common attributes; a class or kind. **2** (foll. by of) roughly of the kind specified (is some sort of doctor). **3** colloq. a person of a specified character or kind (a good sort). **4** Printing a letter or piece in a fount of type. **5** Computing the arrangement of data in a prescribed sequence. **6** archaic a manner or way. ● v.tr. **1** (often foll. by out, over) arrange systematically or according to type, class, etc. **2** = sort out 3, 4. □ **after a sort** after a fashion. **in some sort** to a certain extent. **of a sort** (or **of sorts**) colloq. not fully deserving the name (a holiday of sorts). **out of sorts 1** slightly unwell. **2** in low spirits; irritable. **sort of** colloq. as it were; to some extent (I sort of expected it). **sort out 1** separate into sorts. **2** select (things of one or more sorts) from a miscellaneous group. **3** disentangle or put into order. **4** resolve (a problem or difficulty). **5** colloq. deal with or reprimand (a person). □ **sortable** adj. **sorter** n. **sorting** n. [ME f. OF sorte ult. f. L sors sortis lot, condition]

sortie /'sɔːtɪ/ n. & v. ● n. **1** a sally, esp. from a besieged garrison. **2** an operational flight by a single military aircraft. ● v.intr. (**sorties, sortied, sortieing**) make a sortie; sally. [F, fem. past part. of sortir go out]

sortilege /'sɔːtɪlɪdʒ/ n. divination by lots. [ME f. OF f. med.L sortilegium sorcery f. L sortilegus sorcerer (as SORT, legere choose)]

sorus /'sɔːrəs/ n. (pl. **sori** /-raɪ/) Bot. a heap or cluster, esp. of spore-cases on the undersurface of a fern-leaf, or in a fungus or lichen. [mod.L f. Gk sōros heap]

SOS /ˌesəʊˈes/ n. (pl. **SOSs**) **1** an international code-signal of extreme distress, used esp. by ships at sea. **2** an urgent appeal for help. **3** Brit. a message broadcast to an untraceable person in an emergency. [chosen as being easily transmitted and recognized in Morse code; by folk etymology an abbr. of save our souls]

Sosnowiec /sɒsˈnɒvjets/ an industrial mining town in SW Poland, west of Cracow; pop. (1990) 259,350.

sostenuto /ˌsɒstəˈnuːtəʊ/ adv., adj., & n. Mus. ● adv. & adj. in a sustained or prolonged manner. ● n. (pl. **-os**) a passage to be played in this way. [It., past part. of sostenere SUSTAIN]

sot /sɒt/ n. & v. ● n. a habitual drunkard. ● v.intr. (**sotted, sotting**) tipple. □ **sottish** adj. [OE sott & OF sot foolish, f. med.L sottus, of unkn. orig.]

soteriology /sɒˌtɪərɪˈɒlədʒɪ/ n. Theol. the doctrine of salvation. [Gk sōtēria salvation + -LOGY]

Sothic /'səʊθɪk, 'sɒθ-/ adj. of or relating to the dog-star, Sirius. □ **Sothic year** the ancient Egyptian year of 365¼ days, fixed by the heliacal rising of Sirius. **Sothic cycle** a cycle (first fixed in AD 139) of 1460 Sothic years, after which the 365-day calendar year gives the same date for this rising. [Gk Sōthis f. the Egypt. name of the dog-star]

Sotho /'suːtuː/ n. & adj. ● n. (pl. same) **1** a member of a Bantu-speaking people living chiefly in Botswana, Lesotho, and northern South Africa. **2** the Bantu languages of the Sotho, especially Sesotho. ● adj. of this people or their languages. [Bantu]

sotto voce /ˌsɒtəʊ ˈvəʊtʃɪ/ adv. in an undertone or aside. [It. sotto under + voce voice]

sou /suː/ n. **1** hist. a former French coin of low value. **2** (usu. with neg.) colloq. a very small amount of money (hasn't a sou). [F, orig. pl. sous f. OF sout f. L SOLIDUS]

soubrette /suːˈbret/ n. **1** a pert maidservant or similar female character in a comedy. **2** an actress taking this part. [F f. Prov. soubreto fem. of soubret coy f. sobrar f. L superare be above]

soubriquet var. of SOBRIQUET.

souchong /'suːʃɒŋ/ n. a fine black kind of China tea. [Chin. xiao small + zhong sort]

souffle /'suːf(ə)l/ n. Med. a low murmur heard in the auscultation of various organs etc. [F f. souffler blow f. L sufflare]

soufflé /'suːfleɪ/ n. & adj. ● n. **1** a light spongy dish usu. made with flavoured egg yolks added to stiffly beaten whites of eggs and baked (cheese soufflé). **2** any of various light sweet or savoury dishes made with beaten egg whites. ● adj. **1** light and frothy or spongy (omelette soufflé). **2** (of ceramics) decorated with small spots. [F past part. (as SOUFFLE)]

Soufrière /ˌsuːfrɪˈeə(r)/ **1** a dormant volcano on the French island of

Guadeloupe in the West Indies. Rising to 1,468 m (4,813 ft), it is the highest peak in the Lesser Antilles. **2** an active volcanic peak on the island of St Vincent in the West Indies. It rises to a height of 1,234 m (4,006 ft). [F, f. *soufre* SULPHUR]

sough /saʊ, sʌf/ v. & n. ● v.intr. make a moaning, whistling, or rushing sound, as of the wind in trees etc. ● n. this sound. [OE *swōgan* resound]

sought past and past part. of SEEK.

souk /suːk/ n. (also **suk**, **sukh**, **suq**) a market-place in Muslim countries. [Arab. *sūk*]

soukous /ˈsuːkuːs/ n. a form of Zairean popular dance music, deriving from the rumba and characterized by a light fast guitar style. [corrupt. of F *secouer* shake]

soul /səʊl/ n. **1** the spiritual or immaterial part of a human being, often regarded as immortal. **2** the moral or emotional or intellectual nature of a person. **3** the personification or pattern of something (*the very soul of discretion*). **4** an individual (*not a soul in sight*). **5 a** a person regarded with familiarity or pity etc. (*the poor soul was utterly confused*). **b** a person regarded as embodying moral or intellectual qualities (*left that to meaner souls*). **6** a person regarded as the animating or essential part of something (*the life and soul of the party*). **7** emotional or intellectual energy or intensity, esp. as revealed in a work of art (*pictures that lack soul*). **8** black American culture or music etc. (see SOUL MUSIC). □ **soul-destroying** (of an activity etc.) deadeningly monotonous. **soul food** the traditional food of American blacks, esp. as eaten in the southern US. **soul mate** a person ideally suited to another. **the soul of honour** a person incapable of dishonourable conduct. **soul-searching** n. the examination of one's emotions and motives. ● adj. characterized by this. **upon my soul** an exclamation of surprise. □ **-souled** adj. (in comb.). [OE *sāwol, sāwel, sāwl*, f. Gmc]

soulful /ˈsəʊlfʊl/ adj. **1** having or expressing or evoking deep feeling. **2** over-emotional. □ **soulfully** adv. **soulfulness** n.

soulless /ˈsəʊllɪs/ adj. **1** lacking sensitivity or noble qualities. **2** having no soul. **3** undistinguished or uninteresting. □ **soullessly** adv. **soullessness** n.

soul music n. (also **soul**) a kind of black American popular music which developed in the early 1950s, combining rhythm and blues with the emotional fervour of gospel music. Soul music is characterized by an emphasis on vocals and on an impassioned improvisatory delivery. It was popularized by record labels such as, in particular, Motown, for which artists such as Marvin Gaye and the Supremes recorded during the 1960s and 1970s; the period is usually regarded as soul's classic era. Many soul performers, for example Aretha Franklin and more recently Whitney Houston (b.1963), began their careers by singing gospel music in church. Other significant performers include James Brown and Otis Redding (1941–67). (See also MOTOWN 2.)

sound[1] /saʊnd/ n. & v. ● n. **1** a sensation caused in the ear by the vibration of the surrounding air or other medium. (*See note below.*) **2 a** vibrations causing this sensation. **b** similar vibrations whether audible or not. **3** what is or may be heard. **4** an idea or impression conveyed by words (*don't like the sound of that*). **5** mere words (*sound and fury*). **6** (in full **musical sound**) sound produced by continuous and regular vibrations (opp. NOISE n. 3). **7** any of a series of articulate utterances (*vowel and consonant sounds*). **8** music, speech, etc., accompanying a film or other visual presentation. **9** (often attrib.) broadcasting by radio as distinct from television. ● v. **1** intr. & tr. emit or cause to emit sound. **2** tr. utter or pronounce (*sound a note of alarm*). **3** intr. convey an impression when heard (*you sound worried*). **4** tr. give an audible signal for (an alarm etc.). **5** tr. test (the lungs etc.) by noting the sound produced. **6** tr. cause to resound; make known (*sound their praises*). □ **sound barrier** the high resistance of air to aircraft etc. moving at speeds near that of sound. **sound bite** a short extract from a recorded interview, chosen for its pungency or appropriateness. **sound check** a test of sound equipment before a musical performance or recording to ensure that the desired sound is being produced. **sound effect** a sound other than speech or music made artificially for use in a play, film, etc. **sound engineer** an engineer dealing with acoustics or with sound equipment for a broadcast, recording, or live musical performance. **sound film** a cinematic film with accompanying recorded sound. **sound-hole** an aperture in the belly of some stringed instruments. **sound off** talk loudly or express one's opinions forcefully. **sound-post** a small prop between the belly and back of some stringed instruments. **sound shift** see SHIFT n. 6. **sound spectrograph** an instrument for analysing sound into its frequency components. **sound wave** a wave of compression and rarefaction, by which sound is propagated in an elastic medium, e.g. air. □ **soundless**

adj. **soundlessly** adv. **soundlessness** n. [ME f. AF *soun*, OF *son* (n.), AF *suner*, OF *soner* (v.) f. L *sonus*]

■ Sound consists of longitudinal waves of pressure passing through solids, liquids, or gases; it cannot travel through a vacuum. The speed of sound is independent of frequency and is constant for a particular medium, although density and temperature are also important: the speed of sound in air is about 344 metres (1,128 ft) per second. In general, sound travels faster through solids than through liquids, and faster through liquids than through gases. The human ear responds to frequencies between about 20 Hz and 20,000 Hz. Frequency is perceived as pitch: if one sound has twice the frequency of another, it is perceived (in musical terms) as being an octave above it. Most musical tones, however, are complex in form, including many higher frequencies or harmonics as well as the fundamental frequency. Frequencies below and above the range of human hearing are called infrasonic and ultrasonic respectively. (See also ULTRASONICS.)

sound[2] /saʊnd/ adj. & adv. ● adj. **1** healthy; not diseased or injured. **2** undamaged; in good condition. **3** (of an opinion or policy etc.) correct, orthodox, well-founded, judicious. **4** financially secure (*a sound investment*). **5** undisturbed; tending to sleep deeply and unbrokenly (*a sound sleep(er)*). **6** severe, hard (*a sound blow*). ● adv. soundly (*sound asleep*). □ **soundly** adv. **soundness** n. [ME *sund*, *isund* f. OE *gesund* f. WG]

sound[3] /saʊnd/ v. & n. ● v.tr. & intr. **1** tr. test the depth or quality of the bottom of (the sea or a river etc.). **2** tr. (often foll. by *out*) inquire (esp. cautiously or discreetly) into the opinions or feelings of (a person). **3** tr. find the depth of water in (a ship's hold). **4** tr. get records of temperature, humidity, pressure, etc. from (the upper atmosphere). **5** tr. examine (a person's bladder etc.) with a probe. **6** intr. (of a whale or fish) dive to the bottom. ● n. a surgeon's probe. □ **sounder** n. [ME f. OF *sonder* ult. f. L SUB- + *unda* wave]

sound[4] /saʊnd/ n. **1 a** a narrow passage of water connecting two seas or a sea with a lake etc. **b** an arm of the sea. **2** a fish's swim-bladder. [OE *sund*, = ON *sund* swimming, strait, f. Gmc (as SWIM)]

Sound, the an alternative name for the ØRESUND.

soundboard /ˈsaʊndbɔːd/ n. a thin sheet of wood over which the strings of a piano etc. pass to increase the sound produced.

soundbox /ˈsaʊndbɒks/ n. the hollow chamber providing resonance and forming the body of a stringed musical instrument.

sounding[1] /ˈsaʊndɪŋ/ n. **1 a** the action or process of measuring the depth of water, now usu. by means of echo. **b** an instance of this (*took a sounding*). **2** (in pl.) **a** a region close to the shore of the right depth for sounding. **b** Naut. measurements taken by sounding. **c** cautious investigation (*made soundings as to his suitability*). **3 a** the determination of any physical property at a depth in the sea or at a height in the atmosphere. **b** an instance of this. □ **sounding-balloon** a balloon used to obtain information about the upper atmosphere. **sounding-line** a line used in sounding the depth of water. **sounding-rod** a rod used in finding the depth of water in a ship's hold (see SOUND[3]).

sounding[2] /ˈsaʊndɪŋ/ adj. **1** giving forth (esp. loud or resonant) sound (*sounding brass*). **2** emptily boastful, resonant, or imposing (*sounding promises*).

sounding-board /ˈsaʊndɪŋˌbɔːd/ n. **1** a canopy over a pulpit etc. to direct sound towards the congregation. **2** = SOUNDBOARD. **3 a** a means of causing opinions etc. to be more widely known (*used his students as a sounding-board*). **b** a person etc. used as a trial audience.

soundproof /ˈsaʊndpruːf/ adj. & v. ● adj. impervious to sound. ● v.tr. make soundproof.

sound recording n. **1** the transcription of sound waves on to a storage medium. (*See note below.*) **2** a permanent record of music or other sounds.

■ The first device able both to record and reproduce sound was Thomas Edison's phonograph (1877), in which the vibrations of a stylus were recorded as indentations on a cylinder. From this was developed the gramophone (1897). Flat gramophone records were originally recorded on a wax master, using an acoustic horn and a stylus to cut a fine spiral groove, and records were played back by a reversal of this process. Recording on disc became electrically controlled in the late 1920s, and from the early 1950s record-players were operated electrically, with the signal from the pick-up amplified electronically and passed to a loudspeaker. The introduction of the long-playing record and stereophonic sound in the 1950s allowed great improvements in the quality of sound reproduction. Magnetic tape, in which sound is recorded as patterns of magnetization in a suitable medium, began to be used to record sound in the 1940s, and with the advent of the audio cassette in the 1960s tapes began to rival records in popularity. Further developments were multi-track and digital recording. The compact disc, which stores audio signals (or other forms of information) optically as digital data, was introduced by the Philips and Sony companies in 1982. Playing a compact disc involves scanning it with a laser beam and converting the information into electrical signals photoelectrically.

soundtrack /'saʊndtræk/ n. **1** the recorded sound element of a film. **2** this recorded on the edge of a film in optical or magnetic form.

soup /suːp/ n. & v. ● n. **1** a usu. savoury, largely liquid dish made by boiling meat, fish, or vegetables etc. in stock or water. **2** US sl. nitroglycerine or gelignite, esp. for safe-breaking. **3** sl. the chemicals in which film is developed. **4** (usu. **primordial** or **primeval soup**) a solution rich in organic compounds in the primitive oceans of the earth, from which life is thought to have originated. **5** colloq. fog; thick cloud. ● v.tr. (usu. foll. by up) colloq. **1** increase the power and efficiency of (an engine). **2** increase the power or impact of (writing, music, etc.). □ **in the soup** colloq. in difficulties. **soup and fish** colloq. men's evening dress, a dinner suit. **soup-kitchen** a place dispensing soup etc. to the poor. **soup-plate** a deep wide-rimmed plate for serving soup. **soup-spoon** a large round-bowled spoon for drinking soup. [F soupe sop, broth, f. LL suppa f. Gmc: cf. SOP, SUP¹]

soupçon /'suːpsɒn/ n. a very small quantity; a dash. [F f. OF sou(s)peçon f. med.L suspectio -onis: see SUSPICION]

soupy /'suːpɪ/ adj. (**soupier, soupiest**) **1** of or resembling soup. **2** colloq. sentimental; mawkish. □ **soupily** adv. **soupiness** n.

sour /'saʊə(r)/ adj., n., & v. ● adj. **1** having an acid taste like lemon or vinegar, esp. because of unripeness (sour apples). **2 a** (of food, esp. milk or bread) bad because of fermentation. **b** smelling or tasting rancid or unpleasant. **3** (of a person, temper, etc.) harsh; morose; bitter. **4** (of a thing) unpleasant; distasteful. **5** (of soil) deficient in lime and usually dank. ● n. **1** N. Amer. a drink with lemon or lime juice (whisky sour). **2** an acid solution used in bleaching etc. ● v.tr. & intr. make or become sour (soured the cream; soured by misfortune). □ **go** (or **turn**) **sour 1** (of food etc.) become sour. **2** turn out badly (the job went sour on him). **3** lose one's keenness. **sour cream** cream deliberately fermented by adding bacteria. **sour grapes** resentful disparagement of something one cannot personally acquire (from the fable of the fox who wanted some grapes but found that they were out of reach and so pretended that they were sour and undesirable anyway). **sour mash** US a brewing or distilling-mash made acid to promote fermentation. □ **sourish** adj. **sourly** adv. **sourness** n. [OE sūr f. Gmc]

source /sɔːs/ n. & v. ● n. **1** a spring or fountain-head from which a stream issues (the sources of the Nile). **2** a place, person, or thing from which something originates (the source of all our troubles). **3** a person or document etc. providing evidence (reliable sources of information; historical source material). **4 a** a body emitting radiation etc. **b** Physics a place from which a fluid or current flows. **c** Electronics a part of a transistor from which carriers flow into the interelectrode channel. ● v.tr. obtain (esp. components) from a specified source. □ **at source** at the point of origin or issue. **source-criticism** the evaluation of different, esp. successive, literary or historical sources. [ME f. OF sors, sourse, past part. of sourdre rise f. L surgere]

sourcebook /'sɔːsbʊk/ n. a collection of documentary sources for the study of a subject.

sourdough /'saʊədəʊ/ n. N. Amer. **1** fermenting dough, esp. that left over from a previous baking, used as leaven. **2** an experienced prospector in Alaska etc., an old-timer. [dial., = leaven]

sourpuss /'saʊəpʊs/ n. colloq. a sour-tempered person. [SOUR + PUSS = face]

soursop /'saʊəsɒp/ n. **1** a West Indian evergreen tree, Annona muricata. **2** the large succulent spiny fruit of this tree.

sous- /suː, suːz/ prefix (in words adopted from French) subordinate, under (sous-chef). [F]

Sousa /'suːzə/, John Philip (1854–1932), American composer and conductor. He became director of the US Marine Band in 1880, and then formed his own band in 1892. His works include more than 100 marches, for example The Stars and Stripes, King Cotton, and Hands Across the Sea. The sousaphone, invented in 1898, was named in his honour.

sousaphone /'suːzəfəʊn/ n. a large brass bass wind instrument encircling the player's body. □ **sousaphonist** n. [SOUSA, after saxophone]

souse /saʊs/ v. & n. ● v. **1** tr. put (gherkins, fish, etc.) in pickle. **2** tr. & intr. plunge into liquid. **3** tr. (as **soused** adj.) colloq. drunk. **4** tr. (usu. foll. by in) soak (a thing) in liquid. **5** tr. (usu. foll. by over) throw (liquid) over a thing. ● n. **1 a** pickle made with salt. **b** US food, esp. a pig's head etc., in pickle. **2** a dip, plunge, or drenching in water. **3** colloq. **a** a drinking-bout. **b** a drunkard. [ME f. OF sous, souz pickle f. OS sultia, OHG sulza brine f. Gmc: cf. SALT]

souslik /'suːslɪk/ n. (also **suslik** /'sʌs-/) a Eurasian ground squirrel of the genus Spermophilus, esp. S. citellus, of SE Europe. [Russ.]

Sousse /suːs/ (also **Susah** /'suːzə/, **Susa**) a port and resort on the east coast of Tunisia; pop. (1984) 83,500. It was founded in the 9th century BC as a Phoenician colony.

soutache /suː'tæʃ/ n. a narrow flat ornamental braid used to trim garments. [F f. Hungarian sujtás]

soutane /suː'tɑːn, -'tæn/ n. RC Ch. a cassock worn by a priest. [F f. It. sottana f. sotto under f. L subtus]

souteneur /ˌsuːtə'nɜː(r)/ n. a pimp. [F, = protector]

souter /'suːtə(r)/ n. Sc. & N. Engl. a shoemaker; a cobbler. [OE sūtere f. L sutor f. suere sew- sew]

souterrain /'suːtəˌreɪn/ n. esp. Archaeol. an underground chamber or passage. [F f. sous under + terre earth]

south /saʊθ/ n., adj., adv., & v. ● n. **1** the point of the horizon 90° clockwise from east. **2** the compass point corresponding to this. **3** the direction in which this lies. **4** (usu. **the South**) **a** the part of the world or a country or a town lying to the south. **b** the southern states of the US, esp. the Confederate states. **5** Bridge a player occupying the position designated 'south'. ● adj. **1** towards, at, near, or facing the south (a south wall; south country). **2** coming from the south (south wind). ● adv. **1** towards, at, or near the south (they travelled south). **2** (foll. by of) further south than. ● v.intr. **1** move towards the south. **2** (of a celestial object) cross the meridian. □ **south by east** (or **west**) between south and south-south-east (or south-south-west). **south pole** see POLE². **south-south-east** the point or direction midway between south and south-east. **south-south-west** the point or direction midway between south and south-west. **south wind** a wind blowing from the south. **to the south** (often foll. by of) in a southerly direction. [OE sūth]

South Africa a country occupying the southernmost part of the continent of Africa; pop. (est. 1991) 36,762,000; official languages, English, Afrikaans, Zulu, Xhosa, and seven other languages; administrative capital, Pretoria; seat of legislature, Cape Town. Settled by the Dutch in the 17th century, the Cape later came under British administration (1806), which set in motion a series of conflicting political and economic developments leading to inland expansion, the subjugation of the native population, and finally war between the British and the Boer (Dutch) settlers at the end of the 19th century (see BOER WAR). The defeated Boer republics of Transvaal and Orange Free State were annexed as British Crown Colonies in 1902, but joined with the colonies of Natal and the Cape to form the self-governing Union of South Africa in 1910. After supporting Britain in both world wars, in 1961 South Africa became a republic and left the Commonwealth. From 1948 it pursued a policy of white minority rule (apartheid), which resulted in conflict with its neighbours and led to international diplomatic isolation. Ten regions were designated as African homelands, four of these (Transkei, Ciskei, Bophuthatswana, and Venda) being designated sovereign states but recognized only by South Africa. A process of constitutional change involving the gradual dismantling of the instruments of apartheid began in 1990 following the release of the African National Congress leader Nelson Mandela. Majority rule was achieved after the country's first democratic elections in April 1994. Conducted peacefully despite fears of violence from white right-wingers and between supporters of the ANC and the Zulu Inkatha party, the elections were won by the ANC, and Nelson Mandela was installed as President at the head of a multiracial, multi-party government. South Africa rejoined the Commonwealth in 1994. (See also AZANIA.) □ **South African** adj. & n.

South America a continent comprising the southern half of the American land mass, connected to North America by the Isthmus of Panama. It includes the Falkland Islands, the Galapagos Islands, and Tierra del Fuego. (See also AMERICA.) □ **South American** adj. & n.

Southampton /saʊθ'hæmptən/ an industrial city and seaport on the south coast of England, in Hampshire; pop. (1991) 194,400. It lies on Southampton Water, an inlet of the English Channel opposite the Isle of Wight.

South Atlantic Ocean see ATLANTIC OCEAN.

South Australia a state comprising the central southern part of Australia; capital, Adelaide. Constituted as a semi-independent colony in 1836, it became a Crown Colony in 1841. It was federated with the other states of Australia in 1901.

southbound /'saʊbaʊnd/ adj. travelling or leading southwards.

South Carolina /ˌkærə'laɪnə/ a state of the US on the Atlantic coast; pop. (1990) 3,486,700; capital, Columbia. Explored by the Spanish, French, and English in the early 17th century, the region was permanently settled by the English from 1663 and named after Charles I. Separated from North Carolina in 1729, South Carolina became one of the original thirteen states of the Union (1788). In 1860 it was the first state to secede from the Union, an action which precipitated the American Civil War.

South China Sea see CHINA SEA.

South Dakota a state in the north central US; pop. (1990) 696,000; capital, Pierre. Acquired partly by the Louisiana Purchase in 1803, it became a part of the former Dakota Territory in 1861. It was the scene of a gold rush in 1874. In 1889 it separated from North Dakota and became the 40th state of the US.

Southdown /'saʊdaʊn/ n. a breed of sheep raised esp. for mutton, orig. on the South Downs of Hampshire and Sussex.

south-east /saʊθ'i:st/ n., adj., & adv. ● n. **1** the point of the horizon midway between south and east. **2** the compass point corresponding to this. **3** the direction in which this lies. **4** (**South-East**) the part of a country or town lying to the south-east. ● adj. of, towards, or coming from the south-east. ● adv. towards, at, or near the south-east. □ **south-easterly** adj. & adv. **south-eastern** adj.

South-East Asia Treaty Organization (abbr. **SEATO**) a defence alliance which existed between 1954 and 1977 for countries of SE Asia and part of the SW Pacific, to further a US policy of containing Communism. Its members were Australia, Britain, France, New Zealand, Pakistan, the Philippines, Thailand, and the US.

southeaster /saʊθ'i:stə(r), Naut. saʊ'i:st-/ n. a south-east wind.

South-East Iceland a shipping forecast area covering part of the NE Atlantic between Iceland and the Faeroes.

south-eastward /saʊθ'i:stwəd/ adj. & adv. (also **south-eastwards** /-wədz/) towards the south-east.

Southend-on-Sea /ˌsaʊθendɒn'si:/ a resort town in Essex, on the Thames estuary east of London; pop. (1991) 153,700.

souther /'saʊðə(r)/ n. a south wind.

southerly /'sʌðəlɪ/ adj., adv., & n. ● adj. & adv. **1** in a southern position or direction. **2** (of a wind) blowing from the south. ● n. (pl. **-ies**) a southerly wind.

southern /'sʌðən/ adj. esp. Geog. **1 a** of or in the south; inhabiting the south. **b** (**Southern**) of or relating to the states in the south of the US. **2** lying or directed towards the south (at the southern end). □ **southern hemisphere** the half of the earth below the equator. **southern lights** the aurora australis. □ **southernmost** adj. [OE sūtherne (as SOUTH, -ERN)]

Southern Alps a mountain range in South Island, New Zealand. Running roughly parallel to the west coast, it extends for almost the entire length of the island. At Mount Cook, its highest peak, it rises to 3,764 m (12,349 ft).

Southern Cone the region of South America comprising the countries of Brazil, Paraguay, Uruguay, and Argentina.

Southern Cross Astron. the constellation of Crux (Australis).

southerner /'sʌðənə(r)/ n. a native or inhabitant of the south.

Southern Ocean an alternative name for the ANTARCTIC OCEAN.

Southern Paiute see PAIUTE.

Southern Rhodesia see ZIMBABWE.

southernwood /'sʌðənˌwʊd/ n. an aromatic shrubby kind of wormwood, Artemisia abrotanum, formerly used medicinally. Also called lad's love.

Southey /'sʌðɪ, 'saʊðɪ/, Robert (1774–1843), English poet. Associated with the Lake Poets, he wrote a number of long narrative poems including Madoc (1805), but is best known for his shorter poems such as the anti-militarist ballad the 'Battle of Blenheim' (1798), and for his biography the Life of Nelson (1813). Southey was made Poet Laureate in 1813.

South Georgia a barren island in the South Atlantic, situated 1,120 km (700 miles) east of the Falkland Islands, of which it is a dependency. First explored in 1775 by Captain James Cook, who named the island after George III, it is the site of a research station maintained by the British Antarctic Survey.

South Glamorgan a county of South Wales, on the Bristol Channel; administrative centre, Cardiff.

southing /'saʊθɪŋ, 'saʊðɪŋ/ n. **1** a southern movement. **2** Naut. the distance travelled or measured southward. **3** Astron. the angular distance of a star etc. south of the celestial equator.

South Island the southernmost and larger of the two main islands of New Zealand, separated from North Island by Cook Strait.

South Korea (official name **Republic of Korea**) a country in the Far East, occupying the southern part of the peninsula of Korea; pop. (est. 1990) 42,793,000; official language, Korean; capital, Seoul. South Korea was formed in 1948, when Korea was partitioned along the 38th parallel; the Korean War (1950–3) has been followed by decades of hostility between North and South Korea, although a series of preliminary accords between the two countries was signed in 1991. The country has been governed mainly by authoritarian regimes, notably that of Park Chung Hee (1961–79). A more democratic constitution was adopted in 1987 after political unrest. An emerging industrial power, South Korea has one of the world's fastest-growing economies. □ **South Korean** adj. & n.

South Orkney Islands a group of uninhabited islands in the South Atlantic, lying to the north-east of the Antarctic Peninsula. Discovered in 1821 by American and British sealers, the islands are now administered as part of the British Antarctic Territory.

South Ossetia an autonomous region of Georgia, situated in the Caucasus on the border with Russia; capital, Tskhinvali. (See also OSSETIA.)

South Pacific Commission an agency established in 1947 to promote the economic and social stability of the islands in the South Pacific, having twenty-seven member governments and administrations.

southpaw /'saʊθpɔ:/ n. & adj. colloq. ● n. a left-handed person, esp. in boxing. ● adj. left-handed.

Southport /'saʊθpɔ:t/ a resort town in NW England, on the Irish Sea coast north of Liverpool; pop. (1981) 90,960.

South Sandwich Islands a group of uninhabited volcanic islands in the South Atlantic, lying 480 km (300 miles) south-east of South Georgia. It is administered from the Falkland Islands.

South Sea (also **South Seas**) hist. the southern Pacific Ocean.

South Sea Bubble a fever of speculation in shares of the South Sea Company (1720). The company had been formed in 1711 to trade with Spanish America. In 1720 it assumed responsibility for the national debt in return for a guaranteed profit, but the speculative boom in this and in ever more implausible projects was quickly followed by the company's failure and a general financial collapse; similar events followed in Paris and Amsterdam. The subsequent inquiry revealed corruption among government ministers, a political crisis saved by Sir Robert Walpole, who transferred the South Sea stock to the Bank of England and the East India Company. A statute was passed severely restricting joint-stock companies for the future.

South Shetland Islands a group of uninhabited islands in the South Atlantic, lying north of the Antarctic Peninsula. Discovered in 1819 by the English navigator William Smith, the islands are now administered as part of the British Antarctic Territory.

South Shields /ʃi:ldz/ a port on the coast of NE England, at the mouth of the Tyne opposite North Shields; pop. (1981) 87,125. Founded in the 13th century, the town was important in the 19th and 20th centuries as a shipbuilding centre.

South Uist see UIST.

South Utsire see UTSIRE.

southward /'saʊθwəd/ adj., adv., & n. ● adj. & adv. (also **southwards** /-wədz/) towards the south. ● n. a southward direction or region.

south-west /saʊθ'west/ n., adj., & adv. ● n. **1** the point of the horizon midway between south and west. **2** the compass point corresponding to this. **3** the direction in which this lies. **4** (**South-West**) the part of a country or town lying to the south-west. ● adj. of, towards, or coming from the south-west. ● adv. towards, at, or near the south-west. □ **south-westerly** adj. & adv. **south-western** adj.

South West Africa the former name for NAMIBIA.

South West Africa People's Organization (abbr. **SWAPO**) a nationalist organization formed in Namibia in 1964–6 to oppose the illegitimate South African rule over the region. Driven from the country, SWAPO began a guerrilla campaign, operating largely from neighbouring Angola; it eventually gained UN recognition, and won elections in 1989.

southwester /saʊθ'westə(r), *Naut.* saʊ'west-/ *n.* a south-west wind.

south-westward /saʊθ'westwəd/ *adj. & adv.* (also **south-westwards** /-wədz/) towards the south-west.

South Yorkshire a metropolitan county of northern England; administrative centre, Barnsley.

Soutine /suː'tiːn/, Chaim (1893–1943), French painter, born in Lithuania. After emigrating to Paris in 1913, he was closely associated with a group of painters that included Chagall and Modigliani. A major exponent of expressionism, Soutine evolved a style distinguished by bright colours, vigorous brushstrokes, and impasto, imbued with a feverish emotional content. During the 1920s he produced pictures of grotesque figures, with twisted faces and deformed bodies. From 1925 he increasingly painted still lifes, including plucked fowl and flayed carcasses.

souvenir /ˌsuːvə'nɪə(r)/ *n. & v.* ● *n.* (often foll. by *of*) a memento of an occasion, place, etc. ● *v.tr. sl.* take as a souvenir; pilfer, steal. [F f. *souvenir* remember f. L *subvenire* occur to the mind (as SUB-, *venire* come)]

souvlaki /suː'vlɑːkɪ/ *n.* (pl. ***souvlakia*** /-kɪə/) a Greek dish of pieces of meat grilled on a skewer. [mod.Gk]

sou'wester /saʊ'westə(r)/ *n.* **1** = SOUTHWESTER. **2** a waterproof hat with a broad flap covering the neck.

sov. /sɒv/ *abbr. Brit.* sovereign.

sovereign /'sɒvrɪn/ *n. & adj.* ● *n.* **1** a supreme ruler, esp. a monarch. **2** *Brit. hist.* a gold coin nominally worth £1. ● *adj.* **1 a** supreme (*sovereign power*). **b** unmitigated (*sovereign contempt*). **2** excellent; effective (*a sovereign remedy*). **3** possessing sovereign power (*a sovereign state*). **4** royal (*our sovereign lord*). □ **the sovereign good** the greatest good, esp. for a state, its people, etc. **sovereign pontiff** see PONTIFF. □ **sovereignly** *adv.* [ME f. OF *so(u)verain* f. L: -*g*- by assoc. with *reign*]

sovereignty /'sɒvrɪntɪ/ *n.* (pl. **-ies**) **1** supremacy. **2** self-government. **3** a self-governing state.

soviet /'səʊvɪət, 'sɒv-/ *n. & adj. hist.* ● *n.* **1** an elected local, district, or national council in the former USSR. **2** (**Soviet**) a citizen of the former USSR. **3** a revolutionary council of workers, peasants, etc., in Russia before 1917. ● *adj.* (usu. **Soviet**) of or concerning the former Soviet Union. □ **Sovietize** *v.tr.* (also **-ise**). **Sovietization** /ˌsəʊvɪətaɪ'zeɪʃ(ə)n, ˌsɒv-/ *n.* [Russ. *sovet* council]

sovietologist /ˌsəʊvɪə'tɒlədʒɪst, ˌsɒv-/ *n.* a person who studies the former Soviet Union, esp. during the cold war.

Soviet Union (called in full **Union of Soviet Socialist Republics**, abbr. **USSR**) a former federation of Communist republics occupying the northern half of Asia and part of eastern Europe; capital, Moscow. Created from the Russian empire as a Communist state in the aftermath of the 1917 Russian Revolution and the Russian Civil War, the Soviet Union was the largest country in the world. It comprised fifteen republics: Russia, Belarus, Ukraine, Georgia, Armenia, Moldova, Azerbaijan, Kazakhstan, Kyrgyzstan, Turkmenistan, Tajikistan, Uzbekistan, and the three Baltic states (annexed in 1940). After the Second World War, in which it was invaded by Germany and sustained possibly over 20 million casualties in bitter fighting mainly on its own soil, the Soviet Union emerged as a superpower in rivalry with the US, leading to the cold war and the polarization of much of the world into Communist and non-Communist blocs. Decades of repression and economic failure eventually led to attempts at liberalization and economic reform under President Mikhail Gorbachev during the 1980s. This resulted, however, in a resurgence of nationalist feeling in the republics, some of which began to secede from the Union. The USSR was formally dissolved in 1991, some of its constituents joining a looser confederation, the Commonwealth of Independent States.

sow[1] /saʊ/ *v.tr.* (*past* **sowed** /saʊd/; *past part.* **sown** /saʊn/ or **sowed**) **1** (also *absol.*) **a** scatter or put (seed) on or in the earth. **b** (often foll. by *with*) plant (a field etc.) with seed. **2** initiate; arouse (*sowed doubt in her mind*). **3** (foll. by *with*) cover thickly with. □ **sow the seed** (or **seeds**) **of** first give rise to; implant (an idea etc.). □ **sower** *n.* **sowing** *n.* [OE *sāwan* f. Gmc]

sow[2] /saʊ/ *n.* **1 a** a female adult pig, esp. after farrowing. **b** a female guinea pig. **c** the female of some other species. **2 a** the main trough through which molten iron runs into side-channels to form pigs. **b** a large block of iron so formed. **3** (in full **sow bug**) esp. *US* a woodlouse. [OE *sugu*]

sowback /'saʊbæk/ *n.* a low ridge of sand etc.

sowbread /'saʊbred/ *n.* a southern European cyclamen, *Cyclamen*

hederifolium, with tuberous roots. [sow[2] (reputedly eaten by wild boars in Sicily)]

Soweto /sə'wetəʊ, -'weɪtəʊ/ a large urban area, consisting of several townships, in South Africa south-west of Johannesburg. It was the scene in June 1976 of demonstrations against the imposition of Afrikaans as the compulsory language in schools, which resulted in violent police activity and the death of hundreds of people, many of them children. □ **Sowetan** *n. & adj.* [South West Township]

sown *past part.* of SOW[1].

sowthistle /'saʊˌθɪs(ə)l/ *n.* a yellow-flowered composite plant of the genus *Sonchus*, with soft thistle-like leaves and milky juice.

sox *colloq.* or *Commerce pl.* of SOCK[1].

soy /sɔɪ/ *n.* **1** (also **soy sauce**) a sauce made in Japan and China with fermented soya beans. **2** (in full **soy bean**) = SOYA 1. [Jap. *shō-yu* f. Chin. *shi-you* f. *shi* salted beans + *you* oil]

soya /'sɔɪə/ *n.* **1 a** a leguminous plant, *Glycine max*, orig. of SE Asia, cultivated for its seeds. **b** (in full **soya bean**) the seed of this (or an extract of it), used as a replacement for animal protein in certain foods and for flour, oil, tofu, soy sauce, etc. **2** (also **soya sauce**) = SOY 1. [Du. *soja* f. Malay *soi* (as SOY)]

Soyinka /ʃɔɪ'ɪŋkə/, Wole (b.1934), Nigerian dramatist, novelist, and critic. He made his name in 1959 with the play *The Lion and the Jewel*. His writing often uses satire and humour to explore the contrast between traditional and modern society in Africa, and combines elements of both African and European aesthetics. His works include the novel *The Interpreters* (1965), the play *Kongi's Harvest* (1964), and the collection of poems and other writings *The Man Died* (1972), a record of his time spent serving a sentence as a political prisoner 1967–9. In 1986 Soyinka became the first African to receive the Nobel Prize for literature.

Soyuz /sɔɪ'juːz, 'sɔɪjʊz/ a series of manned Soviet orbiting spacecraft, used to investigate the operation of orbiting space stations. [Russ., = union]

sozzled /'sɒz(ə)ld/ *adj. colloq.* very drunk. [past part. of dial. *sozzle* mix sloppily (prob. imit.)]

SP *abbr.* starting price.

Spa /spɑː/ a small town in eastern Belgium, south-east of Liège; pop. (1991) 10,140. It has been celebrated since medieval times for the curative properties of its mineral springs.

spa /spɑː/ *n.* **1** a curative mineral spring. **2** a place or resort with this. [SPA]

space /speɪs/ *n. & v.* ● *n.* **1 a** a continuous unlimited area or expanse which may or may not contain objects etc. **b** an interval between one, two, or three-dimensional points or objects (*a space of 10 metres*). **c** an empty area; room (*clear a space in the corner*; *occupies too much space*). **d** any of a limited number of places for a person or thing (*no spaces left at the table*). **2** a large unoccupied region (*the wide open spaces*). **3 a** (also **outer space**) the physical universe beyond the earth's atmosphere. **b** the largely empty regions of the universe between the planets and stars, consisting of a near-vacuum with very small amounts of dust and gas. **4** an interval of time (*in the space of an hour*). **5** the amount of paper used in writing etc. (*hadn't the space to discuss it*). **6 a** a blank between printed, typed, or written words etc. **b** *Printing* a piece of metal providing this. **7** *Mus.* each of the blanks between the lines of a staff. **8** freedom to think, be oneself, etc. (*need my own space*). ● *v.tr.* **1** set or arrange at intervals. **2** put spaces between (esp. words, letters, lines, etc., in printing, typing, or writing). **3** (as **spaced** *adj.*) (often foll. by *out*) *sl.* in a state of euphoria or disorientation, esp. from taking drugs. □ **space age** the era when exploration of space has become possible. **space-bar** a long key in a typewriter (or on a similar keyboard) for making a space between words etc. **space blanket** a light metal-coated sheet designed to retain heat. **space flight 1** a journey through space. **2** = *space travel*. **space out** spread out with more or wider spaces or intervals between. **space probe** = PROBE *n.* 4. **space rocket** a rocket travelling through space, or used to launch a spacecraft. **space-saving** occupying little space. **space station** an artificial satellite used as a base for operations in space. **space-time** (or **space–time continuum**) the fusion of the concepts of space and time, esp. as a four-dimensional continuum. **space travel** travel through outer space. **space traveller** a traveller in outer space; an astronaut. **space vehicle** = SPACECRAFT. **space walk** a physical activity by an astronaut in space outside a spacecraft. □ **spacer** *n.* **spacing** *n.* (esp. in sense 2 of *v.*). [ME f. OF *espace* f. L *spatium*]

spacecraft /'speɪskrɑːft/ *n.* a vehicle for travelling in space. The first spacecraft to be placed in orbit was the Soviet Sputnik I, launched on 4 Oct. 1957. The first manned spacecraft took the Soviet pilot Yuri Gagarin into orbit on 12 Apr. 1961, and on 21 July 1969 two US astronauts, Neil Armstrong and Buzz Aldrin, landed on the moon. The first reusable spacecraft was the American space shuttle, which made its first mission in 1981.

spaceman /'speɪsmæn/ n. (pl. **-men**; fem. **spacewoman**, pl. **-women**) = space traveller.

spaceship /'speɪsʃɪp/ n. a spacecraft, esp. one controlled by its crew.

space shuttle n. a reusable spacecraft for carrying people and cargo into earth orbit and then returning. The first such craft, US Space Shuttle, began flights in 1981 and has been used for all American manned flights since then. It has two solid-fuel rocket boosters that are jettisoned and later recovered, and a large external fuel tank that is discarded. The orbiter later returns and lands as a glider. The space shuttle programme was suspended until Sept. 1988 after the explosion in Jan. 1986 of the shuttle *Challenger*, with the death of the seven astronauts.

spacesuit /'speɪssuːt, -sjuːt/ n. a garment designed to allow an astronaut to survive in space.

spacey /'speɪsɪ/ adj. (also **spacy**) (**spacier**, **spaciest**) **1** large, roomy. **2** esp. N. Amer. sl. out of touch with reality; in a dreamy or euphoric state.

spacial var. of SPATIAL.

spacious /'speɪʃəs/ adj. having ample space; covering a large area; roomy. □ **spaciously** adv. **spaciousness** n. [ME f. OF spacios or L spatiosus (as SPACE)]

spade[1] /speɪd/ n. & v. ● n. **1** a tool used for digging or cutting the ground etc., with a sharp-edged metal blade and a long handle. **2** a tool of a similar shape for various purposes, e.g. for removing the blubber from a whale. **3** anything resembling a spade. ● v.tr. dig over (ground) with a spade. □ **call a spade a spade** speak plainly or bluntly. **spade beard** an oblong-shaped beard. **spade foot** a square spadelike enlargement at the end of a chair-leg. □ **spadeful** n. (pl. **-fuls**). [OE spadu, spada]

spade[2] /speɪd/ n. **1 a** a playing card of a suit denoted by black inverted heart-shaped figures with small stalks. **b** (in pl.) this suit. **2** sl. offens. a black person. □ **in spades** sl. to a high degree, with great force. **spade guinea** hist. a guinea of George III's reign with a spade-shaped shield on the reverse. [It. spade pl. of spada sword f. L spatha f. Gk spathē, rel. to SPADE[1]: assoc. with the shape of a pointed spade]

spadework /'speɪdwɜːk/ n. hard or routine preparatory work.

spadille /spə'dɪl/ n. Cards **1** the ace of spades in ombre and quadrille. **2** the highest trump, esp. the ace of spades. [F f. Sp. espadilla dimin. of espada sword (as SPADE[2])]

spadix /'speɪdɪks/ n. (pl. **spadices** /speɪ'daɪsiːz/) Bot. a spike of flowers closely arranged round a fleshy axis and usu. enclosed in a spathe. □ **spadiceous** /speɪ'dɪʃəs/ adj. [L f. Gk, = palm-branch]

spae /speɪ/ v.intr. & tr. Sc. foretell; prophesy. [ME f. ON spá]

spaewife /'speɪwaɪf/ n. (pl. **-wives**) Sc. a female fortune-teller or witch.

spaghetti /spə'getɪ/ n.pl. pasta made in solid strings, between macaroni and vermicelli in thickness. □ **spaghetti Bolognese** /ˌbɒlə'neɪz/ spaghetti served with a sauce of minced beef, tomato, onion, etc. **spaghetti junction** a multi-level road junction, esp. on a motorway. **spaghetti western** a western film made cheaply in Italy. [It., pl. of dimin. of spago string]

spahi /'spɑːhiː/ n. hist. **1** a member of the Turkish irregular cavalry. **2** a member of the Algerian cavalry in French service. [Turk. sipāhī formed as SEPOY]

Spain /speɪn/ (Spanish **España** /es'paɲa/) a country in SW Europe, occupying the greater part of the Iberian peninsula; pop. (est. 1991) 39,045,000; languages, Spanish (official), Catalan; capital, Madrid. The Iberian peninsula came under Roman occupation after 206 BC, following a period of Carthaginian domination. The Romans were displaced by the Visigoths in the 5th century, and the Moors (Muslim invaders from Morocco) in turn established control 711–18. In the medieval period Moorish rule fragmented and independent Christian kingdoms were established in the north, of which the most important were Aragon and Castile. In the late 15th century the marriage of Ferdinand of Aragon and Isabella of Castile, and their reconquest of the last Moorish stronghold, Granada, reunited Spain. Under the Habsburg kings of the 16th century Spain became the dominant European power, building up a huge empire in America and elsewhere. Thereafter it declined, suffering as a result of the War of the Spanish

Succession and the Napoleonic Wars and losing most of its overseas empire in the early 19th century. A republic was formed in 1931, which was ended by the Spanish Civil War (1936–9) and the establishment of a Fascist dictatorship under General Franco. Franco's death in 1975 was followed by the re-establishment of a constitutional monarchy and considerable democratization. Spain became a member of the EEC on 1 Jan. 1986.

spake /speɪk/ archaic past of SPEAK.

spall /spɔːl/ n. & v. ● n. a splinter or chip, esp. of rock. ● v.intr. & tr. break up or cause (ore) to break up in preparation for sorting. [ME (also spale): orig. unkn.]

Spallanzani /ˌspælən'zɑːnɪ/, Lazzaro (1729–99), Italian physiologist and biologist. A priest as well as a keen traveller and collector, he is known today for his meticulous experiments on a wide variety of subjects. He explained the circulation of the blood and the digestive system of animals, showed that fertilization can result only from contact between egg and seminal fluid, demonstrated that protozoa do not appear as a result of spontaneous generation, and studied regeneration in invertebrates. Spallanzani also worked on various problems in the physical and earth sciences.

spallation /spɔː'leɪʃ(ə)n/ n. Physics the breakup of a bombarded nucleus into several parts.

spalpeen /spæl'piːn/ n. Ir. **1** a rascal; a villain. **2** a youngster. [Ir. spailpín, of unkn. orig.]

Spam /spæm/ n. propr. a tinned meat product made mainly from ham. [spiced ham]

span[1] /spæn/ n. & v. ● n. **1** the full extent from end to end in space or time (the span of a bridge; the whole span of history). **2** each arch or part of a bridge between piers or supports. **3** the maximum lateral extent of an aeroplane, its wing, a bird's wing, etc. **4 a** the maximum distance between the tips of the thumb and little finger. **b** this as a measurement, equal to 9 inches. **5** a short distance or time (our life is but a span). ● v. (**spanned**, **spanning**) **1** tr. **a** (of a bridge, arch, etc.) stretch from side to side of; extend across (the bridge spanned the river). **b** (of a builder etc.) bridge (a river etc.). **2** tr. extend across (space or a period of time etc.). **3** tr. measure or cover the extent of (a thing) with one's hand with the fingers stretched (spanned a tenth on the piano). **4** intr. US move in distinct stretches like the span-worm. □ **span roof** a roof with two inclined sides (cf. PENTHOUSE 2, lean-to (see LEAN[1])). **span-worm** US = LOOPER 1. [OE span(n) or OF espan]

span[2] /spæn/ n. **1** Naut. a rope with both ends fastened to take purchase in a loop. **2** N. Amer. a matched pair of horses, mules, etc. **3** S. Afr. a team of two or more pairs of oxen. [LG & Du. span f. spannen unite]

span[3] see SPICK AND SPAN.

span[4] /spæn/ archaic past of SPIN.

spandrel /'spændrɪl/ n. Archit. **1** the almost triangular space between one side of the outer curve of an arch, a wall, and the ceiling or framework. **2** the space between the shoulders of adjoining arches and the ceiling or moulding above. □ **spandrel wall** a wall built on the curve of an arch, filling in the spandrel. [perh. f. AF spaund(e)re, or f. espaundre EXPAND]

spang /spæŋ/ adv. US colloq. exactly; completely (spang in the middle). [20th c.: orig. unkn.]

spangle /'spæŋg(ə)l/ n. & v. ● n. **1** a small thin piece of glittering material, esp. used in quantity to ornament a dress etc.; a sequin. **2** a small sparkling object. **3** (in full **spangle gall**) a spongy excrescence on oak-leaves. ● v.tr. (esp. as **spangled** adj.) cover with or as with spangles (star-spangled; spangled costume). □ **spangly** adj. [ME f. spang f. MDu. spange, OHG spanga, ON spöng brooch f. Gmc]

Spaniard /'spænjəd/ n. **1 a** a native or national of Spain. **b** a person of Spanish descent. **2** NZ a spear grass. [ME f. OF Espaignart f. Espaigne Spain]

spaniel /'spænjəl/ n. **1** a breed of dog with a long silky coat and drooping ears. **2** an obsequious or fawning person. [ME f. OF espaigneul Spanish (dog) f. Rmc Hispaniolus (unrecorded) f. Hispania Spain]

Spanish /'spænɪʃ/ adj. & n. ● adj. of or relating to Spain or its people or language. ● n. **1** the language of Spain and Spanish America. (See note below.) **2** (prec. by the; treated as pl.) the people of Spain. □ **Spanish bayonet** a yucca with stiff sharp-pointed leaves, esp. Yucca aloifolia of North America. **Spanish chestnut** = CHESTNUT n. 1b. **Spanish fly** a bright green beetle, Lytta vesicatoria, formerly dried and used for raising blisters, and as a supposed aphrodisiac. **Spanish guitar** the standard six-stringed acoustic guitar, used esp. for classical and folk music. **Spanish ibex** (or **goat**) an ibex, Capra pyrenaica, inhabiting the Pyrenees. **Spanish influenza** a serious form of influenza, esp. that responsible for an epidemic in Europe in 1918–19 that resulted in many deaths. **Spanish mackerel** a large Atlantic game-fish allied to the mackerel, esp. *Scomberomorphus maculatus*. **Spanish moss** a tropical American plant, *Tillandsia usneoides*, an epiphytic bromeliad which grows as silvery festoons on trees. **Spanish omelette** an omelette containing chopped vegetables (esp. potatoes) and often not folded. **Spanish onion** a large mild-flavoured onion. **Spanish windlass** a device for tightening ropes etc. using a stick as a lever. [ME f. *Spain*, with shortening of the first element]

▪ Spanish is the most widely spoken of the Romance languages, with about 265 million people using it as a first language. It is the official language of Spain and of most countries of Central and South America (notable exceptions being Brazil and Guyana); the large number of Hispanic immigrants, especially from Mexico, means that it is now also widely spoken in the southern states of the US. It contains many Arabic words dating from the time when the Moors dominated Spain (8th–15th centuries). Dialects of Spanish spoken in Mexico and Central and South America are derived from Castilian Spanish, the standard spoken and literary form in Spain since the 15th century; the main differences between Castilian and Latin American Spanish are phonological, in particular, the replacement of /θ/ for *c* or *z* in Castilian with /s/ in Latin American.

Spanish America the parts of America colonized in the 16th century by the Spanish, including Central and South America and parts of the West Indies.

Spanish-American War /ˈspænɪʃəˌmɛrɪkən/ a war between Spain and the United States in the Caribbean and the Philippines in 1898, American public opinion having been aroused by Spanish atrocities in Cuba and the destruction of the warship *Maine* in Santiago harbour, the US declared war and destroyed the Spanish fleets in both the Pacific and the West Indies before successfully invading Cuba, Puerto Rico, and the Philippines, all of which Spain gave up by the Treaty of Paris, signed at the end of the year.

Spanish Armada see ARMADA.

Spanish Civil War (1936–9) the conflict between Nationalist forces (including monarchists and members of the Falange Party) and Republicans (including socialists and Communists and Catalan and Basque separatists) in Spain. It began with a military uprising against the leftist, Republican Popular Front government in July 1936; the Nationalists were supported by Fascist Germany and Italy, and the Republicans by the Soviet Union, as well as by an International Brigade of volunteers from Europe and America. In bitter fighting the Nationalists, led by General Franco, gradually gained control of the countryside but failed to capture the capital, Madrid. Bombing of civilians was carried out by German aircraft, including the destruction of the Basque town of Guernica in 1937. Various attempts to bring the war to an end failed until, after periods of prolonged stalemate, Franco finally succeeded in capturing Barcelona and Madrid in early 1939. He established a Fascist dictatorship that lasted until his death in 1975.

Spanish Inquisition see INQUISITION.

Spanish Main *hist.* the northern coast and coastal waters of South America, between the Orinoco river and Panama.

Spanish Sahara the former name (1958–75) for the WESTERN SAHARA.

Spanish Succession, War of the (1701–14) a European war, provoked by the death of the Spanish king Charles II without issue, which marked the end of Louis XIV's attempts to establish French dominance over Europe. The Grand Alliance of Britain, the Netherlands, and the Holy Roman emperor, largely through the victories of the Duke of Marlborough, threw back a French invasion of the Low Countries, and, although the Peace of Utrecht (1713–14) confirmed the accession of a Bourbon king in Spain, prevented Spain and France from being united under one crown.

Spanish Town a town in Jamaica, west of Kingston, the second largest town and a former capital of Jamaica; pop. (1982) 89,100.

spank /spæŋk/ *v. & n.* ● *v.* **1** *tr.* slap esp. on the buttocks with the open hand, a slipper, etc. **2** *intr.* (of a horse etc.) move briskly, esp. between a trot and a gallop. ● *n.* a slap esp. with the open hand on the buttocks. [perh. imit.]

spanker /ˈspæŋkə(r)/ *n.* **1** a person or thing that spanks. **2** *Naut.* a fore-and-aft sail set on the after side of the mizen-mast. **3** a fast horse. **4** *colloq.* a person or thing of notable size or quality.

spanking /ˈspæŋkɪŋ/ *adj., adv., & n.* ● *adj.* **1** (esp. of a horse) moving quickly; lively; brisk (*at a spanking trot*). **2** *colloq.* striking; excellent.

● *adv. colloq.* very, exceedingly (*spanking clean*). ● *n.* the act or an instance of slapping, esp. on the buttocks as a punishment for children.

spanner /ˈspænə(r)/ *n.* **1** *Brit.* an instrument for turning or gripping a nut on a screw etc. (cf. WRENCH). **2** the cross-brace of a bridge etc. □ **a spanner in the works** *Brit. colloq.* a drawback or impediment. [G *spannen* draw tight: see SPAN²]

Spansule /ˈspænsjuːl/ *n. propr. Pharm.* a capsule which when swallowed releases one or more drugs over a set period.

spar¹ /spɑː(r)/ *n.* **1** a stout pole esp. used for the mast, yard, etc., of a ship. **2** the main longitudinal beam of an aeroplane wing. □ **spar-buoy** a buoy made of a spar with one end moored so that the other stands up. **spar-deck** the light upper deck of a vessel. [ME *sparre*, *sperre* f. OF *esparre* or ON *sperra* or direct f. Gmc: cf. MDu., MLG *sparre*, OS, OHG *sparro*]

spar² /spɑː(r)/ *v. & n.* ● *v.intr.* (**sparred**, **sparring**) **1** (often foll. by *at*) make the motions of boxing without landing heavy blows. **2** engage in argument (*they are always sparring*). **3** (of a gamecock) fight with the feet or spurs. ● *n.* **1 a** a sparring motion. **b** a boxing-match. **2** a cock-fight. **3** an argument or dispute. □ **sparring partner 1** a boxer employed to engage in sparring with another as training. **2** a person with whom one enjoys arguing. [ME f. OE *sperran*, *spyrran*, of unkn. orig.: cf. ON *sperrask* kick out]

spar³ /spɑː(r)/ *n.* any crystalline, easily cleavable, and non-lustrous mineral, e.g. calcite or fluorspar. □ **sparry** *adj.* [MLG, rel. to OE *spæren* of plaster, *spærstān* gypsum]

sparable /ˈspærəb(ə)l/ *n.* a headless nail used for the soles and heels of shoes. [contr. of *sparrow-bill*, also used in this sense]

sparaxis /spəˈræksɪs/ *n.* a southern African plant of the genus *Sparaxis* (iris family), with showy flowers and jagged spathes. [mod.L f. Gk, = laceration, f. *sparassō* tear]

spare /speə(r)/ *adj., n., & v.* ● *adj.* **1 a** not required for ordinary use; extra (*have no spare cash*; *spare time*). **b** not wanted or used by others (*a spare seat in the front row*). **c** reserved for emergency or occasional use (*slept in the spare room*). **2** lean; thin. **3** scanty; frugal; not copious (*a spare diet*; *a spare prose style*). ● *n.* **1** *Brit.* a spare part; a duplicate. **2** *Bowling* the knocking-down of all the pins with the first two balls. ● *v.* **1** *tr.* afford to give or do without; dispense with (*cannot spare him just now*; *can spare you a couple*). **2** *tr.* **a** abstain from killing, hurting, wounding, etc. (*spared his feelings*; *spared her life*). **b** abstain from inflicting or causing; relieve from (*spare me this talk*; *spare my blushes*). **3** *tr.* be frugal or grudging of (*no expense spared*). **4** *intr. archaic* be frugal. □ **go spare** *colloq.* **1** *Brit.* become extremely angry or distraught. **2** be unwanted by others. **not spare oneself** exert one's utmost efforts. **spare part** a duplicate part to replace a lost or damaged part of a machine etc. **spare tyre 1** an extra tyre carried in a motor vehicle for emergencies. **2** *Brit. colloq.* a roll of fat round the waist. **to spare** left over; additional (*an hour to spare*). □ **sparely** *adv.* **spareness** *n.* **sparer** *n.* [OE *spær*, *sparian* f. Gmc]

spare-rib /speəˈrɪb/ *n.* (usu. in *pl.*) closely trimmed ribs of esp. pork. [prob. f. MLG *ribbesper* (by transposition of the syllables), and assoc. with SPARE]

sparge /spɑːdʒ/ *v.tr.* moisten by sprinkling, esp. in brewing. □ **sparger** *n.* [app. f. L *spargere* sprinkle]

sparid /ˈspærɪd/ *n. & adj. Zool.* ● *n.* a deep-bodied marine fish of the family Sparidae, with long spiny dorsal fins, e.g. a sea bream or a porgy. ● *adj.* of or relating to this family. [mod.L *Sparidae* f. L *sparus* f. Gk *sparos* sea bream]

sparing /ˈspeərɪŋ/ *adj.* **1** inclined to save; economical. **2** restrained; limited. □ **sparingly** *adv.* **sparingness** *n.*

Spark /spɑːk/, Dame Muriel (b.1918), Scottish novelist. Her novels include *Memento Mori* (1959), a comic and macabre study of old age, and *The Prime of Miss Jean Brodie* (1961), a sardonic portrait of an emancipated Edinburgh schoolmistress and her favourite pupils. In 1954 Spark converted to Roman Catholicism; her awareness of the parodoxes and ironies of the faith informs much of her work, particularly her novel *The Mandelbaum Gate* (1965).

spark¹ /spɑːk/ *n. & v.* ● *n.* **1** a fiery particle thrown off from a fire, or alight in ashes, or produced by a flint, match, etc. **2** (often foll. by *of*) a particle of a quality etc. (*not a spark of life*; *a spark of interest*). **3** *Electr.* **a** a light produced by a sudden disruptive discharge through the air etc. **b** such a discharge serving to ignite the explosive mixture in an internal-combustion engine. **4 a** a flash of wit etc. **b** anything causing interest, excitement, etc. **5** a small bright object or point, e.g. in a gem. **6** (**Sparks**) a nickname for a radio operator or an electrician. ● *v.* **1** *intr.* emit sparks of fire or electricity. **2** *tr.* (often foll. by *off*) stir into activity;

initiate (a process) suddenly. **3** *intr. Electr.* produce sparks at the point where a circuit is interrupted. □ **spark chamber** an apparatus designed to show ionizing particles. **spark-gap** the space between electric terminals where sparks occur. **sparking-plug** *Brit.* = spark-plug. **spark-plug** a device for firing the explosive mixture in an internal-combustion engine. □ **sparkless** *adj.* **sparky** *adj.* [ME f. OE *spærca, spearca*]

spark² /spɑːk/ *n. & v.* ● *n.* (also **bright spark**) **1** a clever, witty, or lively person. **2** *archaic* a dashing young man; a gallant, a beau. ● *v.intr. archaic* play the gallant. □ **sparkish** *adj.* [prob. a figurative use of SPARK¹]

sparkle /'spɑːk(ə)l/ *v. & n.* ● *v.intr.* **1 a** emit or seem to emit sparks; glitter; glisten (*her eyes sparkled*). **b** be witty; scintillate (*sparkling repartee*). **2** (of wine etc.) (esp. as **sparkling** *adj.*) effervesce (cf. STILL¹ *adj.* 4). ● *n.* **1 a** gleam, spark. **2** vivacity, liveliness. □ **sparkly** *adj.* [ME f. SPARK¹ + -LE⁴]

sparkler /'spɑːklə(r)/ *n.* **1** a person or thing that sparkles. **2** a hand-held sparkling firework. **3** *colloq.* a diamond or other gem.

sparling /'spɑːlɪŋ/ *n.* the European smelt, *Osmerus eperlanus*. [ME f. OF *esperlinge*, of Gmc orig.]

sparrow /'spærəʊ/ *n.* **1** a small finchlike Old World bird of the family Ploceidae, with brown and grey plumage, esp. the *house sparrow* (*Passer domesticus*). **2** a bird resembling this in size or colour, esp. a New World bird of the bunting family (*hedge sparrow; song sparrow*). □ **sparrow-grass** *dial.* or *colloq.* asparagus. [OE *spearwa* f. Gmc]

sparrowhawk /'spærəʊˌhɔːk/ *n.* **1** a small Old World hawk of the genus *Accipiter*, preying on small birds; esp. the Eurasian *A. nisus*. **2** *N. Amer.* the American kestrel, *Falco sparverius*.

sparse /spɑːs/ *adj.* thinly dispersed or scattered; not dense (*sparse population; sparse greying hair*). □ **sparsely** *adv.* **sparseness** *n.* **sparsity** *n.* [L *sparsus* past part. of *spargere* scatter]

Sparta /'spɑːtə/ a city in the southern Peloponnese in Greece, capital of the department of Laconia; pop. (1981) 14,390. It was a powerful city-state in the 5th century BC, defeating its rival Athens in the Peloponnesian War to become the leading city of Greece until challenged by Thebes in 371 BC. The ancient Spartans were renowned for the military organization of their state and for their rigorous discipline, courage, and austerity.

Spartacist /'spɑːtəsɪst, -təkɪst/ *n.* a member of the Spartacus League, a German revolutionary socialist group founded in 1916 by Rosa Luxemburg and Karl Liebknecht (1871–1919) with the aims of overthrowing the government and ending the First World War. At the end of 1918 the group became the German Communist Party, organizing an uprising in Berlin that was brutally crushed. [G *Spartakist* f. SPARTACUS, adopted as a pseudonym by Liebknecht]

Spartacus /'spɑːtəkəs/ (died *c*.71 BC), Thracian slave and gladiator. He led a revolt against Rome in 73, increasing his army from some seventy gladiators at the outset to several thousand rebels. He was eventually defeated by Crassus in 71 and crucified.

Spartacus League see SPARTACIST.

Spartan /'spɑːt(ə)n/ *adj. & n.* ● *adj.* **1** of or relating to ancient Sparta. **2 a** possessing the qualities of courage, endurance, stern frugality, etc., associated with ancient Sparta. **b** (of a regime, conditions, etc.) lacking comfort; austere. ● *n.* a citizen of ancient Sparta. [ME f. L *Spartanus* f. Sparta f. Gk Sparta, -tē]

spartina /spɑːˈtiːnə/ *n.* a grass of the genus *Spartina*, with rhizomatous roots and growing in wet or marshy ground. [Gk *spartinē* rope]

spasm /'spæz(ə)m/ *n.* **1** a sudden involuntary muscular contraction. **2** a sudden convulsive movement or emotion etc. (*a spasm of coughing*). **3** (usu. foll. by *of*) *colloq.* a brief spell of an activity. [ME f. OF *spasme* or L *spasmus* f. Gk *spasmos, spasma* f. *spaō* pull]

spasmodic /spæzˈmɒdɪk/ *adj.* **1** of, caused by, or subject to, a spasm or spasms (*a spasmodic jerk; spasmodic asthma*). **2** occurring or done by fits and starts (*spasmodic efforts*). □ **spasmodically** *adv.* [mod.L *spasmodicus* f. Gk *spasmōdēs* (as SPASM)]

spastic /'spæstɪk/ *adj. & n.* ● *adj.* **1** *Med.* suffering from cerebral palsy with spasm of the muscles. **2** *sl. offens.* weak, feeble, incompetent. **3** spasmodic. ● *n.* **1** *Med.* a person suffering from cerebral palsy. **2** *sl. offens.* a stupid or incompetent person. □ **spastically** *adv.* **spasticity** /spæˈstɪsɪtɪ/ *n.* [L *spasticus* f. Gk *spastikos* pulling f. *spaō* pull]

spat¹ past and past part. of SPIT¹.

spat² /spæt/ *n.* **1** (usu. in *pl.*) *hist.* a short cloth gaiter protecting the shoe from mud etc. **2** a cover for the upper part of an aircraft wheel. [abbr. of SPATTERDASH]

spat³ /spæt/ *n. & v. colloq.* ● *n.* a petty quarrel. ● *v.intr.* (**spatted, spatting**) quarrel pettily. [prob. imit.]

spat⁴ /spæt/ *n. & v.* ● *n.* the spawn of shellfish, esp. the oyster. ● *v.* (**spatted, spatting**) **1** *intr.* (of an oyster) spawn. **2** *tr.* shed (spawn). [AF, of unkn. orig.]

spatchcock /'spætʃkɒk/ *n. & v.* ● *n.* a chicken or esp. game bird split open and grilled. ● *v.tr.* **1** treat (poultry) in this way. **2** *colloq.* insert or interpolate (a phrase, sentence, story, etc.) esp. incongruously. [orig. in Irish use, explained as f. *dispatch-cock*, but cf. SPITCHCOCK]

spate /speɪt/ *n.* **1** a river-flood (*the river is in spate*). **2** a large or excessive amount (*a spate of enquiries*). [ME, Sc. & N. Engl.: orig. unkn.]

spathe /speɪð/ *n. Bot.* a large bract or pair of bracts enveloping a spadix or flower-cluster. [L f. Gk *spathē* broad blade etc.]

spathic /'spæθɪk/ *adj.* (of a mineral) like spar (see SPAR³), esp. in cleavage. □ **spathic iron ore** = SIDERITE 1. □ **spathose** /-θəʊs/ *adj.* [*spath* f. G Spath]

spatial /'speɪʃ(ə)l/ *adj.* (also **spacial**) of or concerning space (*spatial extent*). □ **spatialize** *v.tr.* (also **-ise**). **spatially** *adv.* **spatiality** /ˌspeɪʃɪˈælɪtɪ/ *n.* [L *spatium* space]

spatio-temporal /ˌspeɪʃɪəʊˈtempərəl/ *adj. Physics & Philos.* belonging to both space and time or to space-time. □ **spatio-temporally** *adv.* [formed as SPATIAL + TEMPORAL¹]

spatter /'spætə(r)/ *v. & n.* ● *v.* **1** *tr.* **a** (often foll. by *with*) splash (a person etc.) (*spattered him with mud*). **b** scatter or splash (liquid, mud, etc.) here and there. **2** *intr.* (of rain etc.) fall here and there (*glass spattered down*). **3** *tr.* slander (a person's honour etc.). ● *n.* **1** (usu. foll. by *of*) a splash (*a spatter of mud*). **2** a quick pattering sound. [frequent. f. base as in Du., LG *spatten* burst, spout]

spatterdash /'spætəˌdæʃ/ *n.* **1** (usu. in *pl.*) *hist.* a long gaiter or legging to keep the stockings etc. clean, esp. when riding. **2** *US* = ROUGHCAST.

spatula /'spætjʊlə/ *n.* **1** a broad-bladed knife-like implement used for spreading, stirring, mixing (paints), etc. **2** a doctor's instrument for holding the tongue down to examine the throat. [L, var. of *spathula*, dimin. of *spatha* SPATHE]

spatulate /'spætjʊlət/ *adj.* having a broad rounded end. [SPATULA]

spavin /'spævɪn/ *n.* a disease of a horse's hock with a hard bony tumour or excrescence. □ **blood** (or **bog**) **spavin** a distension of the joint by effusion of lymph or fluid. **bone spavin** a deposit of bony substance uniting the bones. □ **spavined** *adj.* [ME f. OF *espavin*, var. of *esparvain* f. Gmc]

spawn /spɔːn/ *v. & n.* ● *v.* **1 a** *tr.* (also *absol.*) (of a fish, frog, mollusc, or crustacean) release or deposit (eggs). **b** *intr.* be produced as eggs or young. **2** *tr.* *derog.* (of people) produce (offspring). **3** *tr.* produce or generate, esp. in large numbers. ● *n.* **1** the eggs of fish, frogs, etc. **2** *derog.* human or other offspring. **3** the mycelium of a fungus, esp. of a cultivated mushroom. □ **spawner** *n.* [ME f. AF *espaundre* shed roe, OF *espandre* EXPAND]

spay /speɪ/ *v.tr.* sterilize (a female animal) by removing the ovaries. [ME f. AF *espeier*, OF *espeer* cut with a sword f. *espee* sword f. L *spatha*: see SPATHE]

SPCK *abbr.* Society for Promoting Christian Knowledge.

speak /spiːk/ *v.* (*past* **spoke** /spəʊk/; *past part.* **spoken**) **1** *intr.* make articulate verbal utterances in an ordinary (not singing) voice. **2** *tr.* **a** utter (words). **b** make known or communicate (one's opinion, the truth, etc.) in this way (*never speaks sense*). **3** *intr.* **a** (foll. by *to, with*) hold a conversation (*spoke to him for an hour; spoke with them about their work*). **b** (foll. by *of*) mention in writing etc. (*speaks of it in his novel*). **c** (foll. by *for*) articulate the feelings of (another person etc.) in speech or writing (*speaks for our generation*). **4** *intr.* (foll. by *to*) **a** address; converse with (a person etc.). **b** speak in confirmation of or with reference to (*spoke to the resolution; can speak to his innocence*). **c** *colloq.* reprove (*spoke to them about their lateness*). **5** *intr.* make a speech before an audience etc. (*spoke for an hour on the topic; has a good speaking voice*). **6** *tr.* use or be able to use (a specified language) (*cannot speak French*). **7** *intr.* (of a gun, a musical instrument, etc.) make a sound. **8** *intr.* (usu. foll. by *to*) *poet.* communicate feeling etc., affect, touch (*the sunset spoke to her*). **9** *intr.* (of a hound) bark. **10** *tr.* hail and hold communication with (a ship). **11** *tr.* *archaic* **a** (of conduct etc.) show (a person) to be (*his conduct speaks him generous*). **b** be evidence of (*the loud laugh speaks the vacant mind*). □ **not** (or **nothing**) **to speak of** not (or nothing) worth mentioning; practically none (or nothing). **speak for itself** need no supporting evidence. **speak for oneself 1** give one's own opinions. **2** not presume to speak for others. **speak one's mind** speak bluntly or frankly. **speak out** speak loudly or freely, give one's opinion. **speak up 1 a** raise one's voice. **b** a request to a person to speak more clearly or loudly. **2** = *speak out*. **speak volumes** (of a fact etc.) be very significant. **speak volumes** (or **well** etc.) **for 1** be abundant evidence of. **2** place in a favourable light. □ **speakable** adj. [OE *sprecan*, later *specan*]

speakeasy /ˈspiːkˌiːzɪ/ n. (pl. **-ies**) US hist. sl. an illicit liquor shop or drinking club during Prohibition.

speaker /ˈspiːkə(r)/ n. **1** a person who speaks, esp. in public. **2** a person who speaks a specified language (esp. in *comb.*: *a French-speaker*). **3** (**Speaker**) the presiding officer in a legislative assembly, esp. the House of Commons. (*See note below.*) **4** a loudspeaker. □ **speakership** n.

▪ The Speaker of the House of Commons is an MP chosen to act as the House's representative and to preside over its debates. The first person formally mentioned as holding such an office was Sir Thomas de Hungerford (d.1398) in 1376–7. Originally a royal nominee, the Speaker has been elected since the late 17th century and, although no longer holding ministerial office or taking part in debate, remains chairperson of the House with the casting vote and the power to censure, suspend, or expel members.

speaking /ˈspiːkɪŋ/ n. & adj. ● n. the act or an instance of uttering words etc. ● adj. **1** that speaks; capable of articulate speech. **2** (of a portrait) lifelike; true to its subject (*a speaking likeness*). **3** (in *comb.*) speaking or capable of speaking a specified foreign language (*French-speaking*). **4** with a reference or from a point of view specified (*roughly speaking*; *professionally speaking*). □ **on speaking terms** (foll. by *with*) **1** slightly acquainted. **2** on friendly terms. **speaking acquaintance 1** a person one knows slightly. **2** this degree of familiarity. **speaking clock** Brit. a telephone service giving the correct time in speech. **speaking-trumpet** hist. an instrument for making the voice carry. **speaking-tube** a tube for conveying the voice from one room, building, etc., to another.

spear /spɪə(r)/ n. & v. ● n. **1** a thrusting or throwing weapon with a pointed usu. steel tip and a long shaft. **2** a pointed or barbed instrument used for catching fish etc. **3** archaic a spearman. **4** a plant shoot, esp. a pointed stem of asparagus or broccoli. ● v.tr. pierce or strike with or as if with a spear (*speared an olive*). □ **spear gun** a gun used to propel a spear in underwater fishing. **spear side** the male line of descent; the male side of a family. [OE *spere*]

spearhead /ˈspɪəhɛd/ n. & v. ● n. **1** the point of a spear. **2** an individual or group chosen to lead a thrust or attack. ● v.tr. act as the spearhead of (an attack etc.).

spearman /ˈspɪəmən/ n. (pl. **-men**) archaic a person, esp. a soldier, who uses a spear.

spearmint /ˈspɪəmɪnt/ n. the common garden mint, *Mentha spicata*, used in cookery. An oil extracted from it is used to flavour sweets and chewing-gum.

spearwort /ˈspɪəwɜːt/ n. an aquatic plant, *Ranunculus lingua*, with thick hollow stems, long narrow spear-shaped leaves, and yellow flowers.

spec[1] /spek/ n. colloq. a commercial speculation or venture. □ **on spec** in the hope of success; as a gamble, on the off chance. [abbr. of SPECULATION]

spec[2] /spek/ n. colloq. a detailed working description; a specification. [abbr. of SPECIFICATION]

special /ˈspɛʃ(ə)l/ adj. & n. ● adj. **1 a** particularly good; exceptional; out of the ordinary (*bought them a special present*; *today is a special day*; *took special trouble*). **b** peculiar; specific; not general (*lacks the special qualities required*; *the word has a special sense*). **2** for a particular purpose (*sent on a special assignment*). **3** in which a person specializes (*statistics is his special field*). **4** of or designating education for children with particular needs, esp. with learning difficulties; also *colloq.* in *special ed.* ● n. a special person or thing, e.g. a special constable, train, examination, edition of a newspaper, dish on a menu, etc. □ **special area** Brit. a district for which special economic provision is made in legislation. **special case 1** a written statement of fact presented by litigants to a court. **2** an exceptional or unusual case. **special constable** Brit. a person sworn in to assist the police on special occasions. **special correspondent** a journalist writing for a newspaper on special events or a special area of interest. **special creation** the creation of matter and living things by God as described in the Bible. **special delivery** a delivery of mail in advance of the regular delivery. **special drawing rights** the right to purchase extra foreign currency from the International Monetary Fund. **special edition** an extra edition of a newspaper including later news than the ordinary edition. **special effects** (in film and television) scenic illusions created by props, camerawork, computer graphics, etc. **special intention** see

INTENTION. **special jury** a jury with members of a particular social standing (cf. *common jury*). **special licence** Brit. a marriage licence allowing immediate marriage without banns, or at an unusual time or place. **special pleading 1** Law pleading with reference to new facts in a case. **2** (in general use) a specious or unfair argument favouring the speaker's point of view. **special verdict** Law a verdict stating the facts as proved but leaving the court to draw conclusions from them. □ **specially** adv. **specialness** n. [ME f. OF *especial* ESPECIAL or L *specialis* (as SPECIES)]

Special Air Service (abbr. **SAS**) (in the UK) a specialist army regiment trained in commando techniques of warfare, formed during the Second World War and used in clandestine operations, frequently against terrorists (cf. SPECIAL BOAT SERVICE).

Special Boat Service (also **Special Boat Section**) (abbr. **SBS**) (in the UK) a nautical counterpart of the land-based SAS, provided by the Royal Marines.

Special Branch (in the UK) a police department dealing with political security.

specialist /ˈspɛʃəlɪst/ n. (usu. foll. by *in*) **1** a person who is trained in a particular branch of a profession, esp. medicine (*a specialist in dermatology*). **2** a person who specially or exclusively studies a subject or a particular branch of a subject. □ **specialism** n. **specialistic** /ˌspɛʃəˈlɪstɪk/ adj.

speciality /ˌspɛʃɪˈælɪtɪ/ n. (pl. **-ies**) **1** a special pursuit, product, operation, etc., to which a company or a person gives special attention. **2** a special feature, characteristic, or skill. [ME f. OF *especialité* or LL *specialitas* (as SPECIAL)]

specialize /ˈspɛʃəlaɪz/ v. (also **-ise**) **1** intr. (often foll. by *in*) **a** be or become a specialist (*specializes in optics*). **b** devote oneself to an area of interest, skill, etc. (*specializes in insulting people*). **2** Biol. **a** tr. (esp. in *passive*) adapt or set apart (an organ etc.) for a particular purpose. **b** intr. (of an organ etc.) become adapted etc. in this way. **3** tr. make specific or individual. **4** tr. modify or limit (an idea, statement, etc.). □ **specialization** /ˌspɛʃəlaɪˈzeɪʃ(ə)n/ n. [F *spécialiser* (as SPECIAL)]

Special Operations Executive (abbr. **SOE**) a secret British military service during the Second World War, set up in 1940 to carry out clandestine operations and coordinate with resistance movements in Europe and later the Far East.

specialty /ˈspɛʃ(ə)ltɪ/ n. (pl. **-ies**) **1** esp. N. Amer. = SPECIALITY. **2** Law an instrument under seal; a sealed contract. [ME f. OF *(e)specialté* (as SPECIAL)]

speciation /ˌspiːsɪˈeɪʃən, ˌspiːʃɪ-/ n. Biol. the formation of new and distinct species in the course of evolution. □ **speciate** /ˈspiːsɪˌeɪt, ˈspiːʃɪ-/ v.intr. [SPECIES + -ATION]

specie /ˈspiːʃiː, -ʃɪ/ n. coin money as opposed to paper money. [L, ablat. of *species* in phrase *in specie*]

species /ˈspiːʃiːz, -ʃiːz, ˈspiːsiːz, -siːz/ n. (pl. same) **1** a class of things having some common characteristics. **2** Biol. a taxonomic rank below a genus, consisting of similar individuals capable of exchanging genes or interbreeding and denoted by a Latin binomial. **3** a kind or sort. **4** Logic a group subordinate to a genus and containing individuals agreeing in some common attribute(s) and called by a common name. **5** Law a form or shape given to materials. **6** Eccl. the visible form of each of the elements of consecrated bread and wine in the Eucharist. [L, = appearance, kind, beauty, f. *specere* look]

speciesism /ˈspiːʃiːzˌɪz(ə)m, ˈspiːʃiːz-, ˈspiːsiːz-, ˈspiːsiːz-/ n. discrimination against or exploitation of certain animal species, based on an assumption of human superiority.

specific /spɪˈsɪfɪk/ adj. & n. ● adj. **1** clearly defined; definite (*has no specific name*; *told me so in specific terms*). **2** relating to a particular subject; peculiar (*a style specific to that*). **3 a** of or concerning a species (*the specific name for a plant*). **b** possessing, or concerned with, the properties that characterize a species (*the specific forms of animals*). **4** (of a duty or a tax) assessed by quantity or amount, not by the value of goods. ● n. **1** archaic a medicine or remedy specifically effective for a disease or part of the body. **2** a specific aspect or factor (*shall we discuss specifics?*). □ **specific cause** the cause of a particular form of a disease. **specific difference** a factor that differentiates a species. **specific disease** a disease caused by one identifiable agent. **specific gravity** = *relative density*. **specific heat capacity** the heat required to raise the temperature of the unit mass of a given substance by a given amount (usu. one degree). **specific medicine** a medicine having a distinct effect in curing a certain disease. **specific performance** Law the performance of a contractual duty, as ordered in cases where damages would not be adequate remedy. □ **specifically** *adv.* **specificness** *n.* **specificity** /ˌspesɪˈfɪsɪtɪ/ *n.* [LL *specificus* (as SPECIES)]

specification /ˌspesɪfɪˈkeɪʃ(ə)n/ *n.* **1** the action or an instance of specifying; the state of being specified. **2** (esp. in *pl.*) a detailed description of the construction, workmanship, materials, etc., of work done or to be done, prepared by an architect, engineer, etc. **3** a detailed description of the construction and performance of an appliance, machine, vehicle, etc. **4** a description by an applicant for a patent of the construction and use of his invention. **5** *Law* the conversion of materials into a new product not held to be the property of the owner of the materials. [med.L *specificatio* (as SPECIFY)]

specify /ˈspesɪˌfaɪ/ *v.tr.* (**-ies, -ied**) **1** (also *absol.*) name or mention expressly (*specified the type he needed*). **2** (usu. foll. by *that* + clause) name as a condition (*specified that he must be paid at once*). **3** include in specifications (*a French window was not specified*). □ **specifiable** *adj.* **specifier** *n.* [ME f. OF *specifier* or LL *specificare* (as SPECIFIC)]

specimen /ˈspesɪmən/ *n.* **1** an individual or part taken as an example of a class or whole, esp. when used for investigation or scientific examination (*specimens of copper ore; a specimen of your handwriting*). **2** *Med.* a sample of urine for testing. **3** *colloq.* usu. *derog.* a person of a specified sort. [L f. *specere* look]

specious /ˈspiːʃəs/ *adj.* **1** superficially plausible but actually wrong (*a specious argument*). **2** misleadingly attractive in appearance. □ **speciously** *adv.* **speciousness** *n.* **speciosity** /ˌspiːʃɪˈɒsɪtɪ/ *n.* [ME, = beautiful, f. L *speciosus* (as SPECIES)]

speck /spek/ *n. & v.* ● *n.* **1** a small spot, dot, or stain. **2** (foll. by *of*) a particle (*speck of dirt*). **3** a rotten spot in fruit. ● *v.tr.* (esp. as **specked** *adj.*) mark with specks. □ **speckless** *adj.* [OE *specca*: cf. SPECKLE]

speckle /ˈspek(ə)l/ *n. & v.* ● *n.* a small spot, mark, or stain, esp. in quantity on the skin, a bird's egg, etc. ● *v.tr.* (esp. as **speckled** *adj.*) mark with speckles or patches. □ **speckled wood 1** a kind of timber with speckled markings. **2** a brown European satyrid butterfly, *Pararge aegeria*, with yellowish spots. [ME f. MDu. *spekkel*]

specs /speks/ *n.pl. colloq.* a pair of spectacles. [abbr.]

spectacle /ˈspektək(ə)l/ *n.* **1** a public show, ceremony, etc. **2** anything attracting public attention (*a charming spectacle; a disgusting spectacle*). □ **make a spectacle of oneself** make oneself an object of ridicule. [ME f. OF f. L *spectaculum* f. *spectare* frequent. of *specere* look]

spectacled /ˈspektək(ə)ld/ *adj.* **1** wearing spectacles. **2** (of an animal) having markings resembling spectacles. □ **spectacled bear** a small South American bear, *Tremarctos ornatus*, with white markings around the eyes. **spectacled cobra** the Asian cobra, *Naja naja*, with a spectacle-like marking on the expanded hood.

spectacles /ˈspektək(ə)lz/ *n.pl.* (also **pair of spectacles** *sing.*) a pair of lenses in a frame resting on the nose and ears, used to correct defective eyesight or protect the eyes. Spectacles seem to have been invented in Europe and in China at about the same time (AD *c.*1300). In Europe they originated in Italy, and for centuries they remained the mark of the scholar, since their main use was for reading and most people were unable to read. An early design shows a type of pince-nez (side-pieces came later); early rims were of horn or leather. With the increase of books and the spread of literacy the spectacle trade expanded in the 16th century, but until the 20th century spectacles were worn chiefly by the elderly and scholarly. (Also called *glasses*.)

spectacular /spekˈtækjʊlə(r)/ *adj. & n.* ● *adj.* **1** of or like a public show; striking, amazing, lavish. **2** strikingly large or obvious (*a spectacular increase in output*). ● *n.* an event intended to be spectacular, esp. a musical film or play. □ **spectacularly** *adv.* [SPECTACLE, after *oracular* etc.]

spectate /spekˈteɪt/ *v.intr.* be a spectator, esp. at a sporting event. [back-form. f. SPECTATOR]

spectator /spekˈteɪtə(r)/ *n.* a person who looks on at a show, game, incident, etc. □ **spectator sport** a sport attracting spectators rather than participants. □ **spectatorial** /ˌspektəˈtɔːrɪəl/ *adj.* [F *spectateur* or L *spectator* f. *spectare*: see SPECTACLE]

Spector /ˈspektə(r)/, Phil (b.1940), American record producer and songwriter. In 1961 he formed a record company and pioneered a 'wall of sound' style, using echo and tape loops. He had a succession of hit recordings in the 1960s with groups such as the Ronettes and the Crystals and worked on the last Beatles album *Let it Be* (1970).

spectra *pl.* of SPECTRUM.

spectral /ˈspektrəl/ *adj.* **1 a** of or relating to spectres or ghosts. **b** ghostlike. **2** of or concerning spectra or the spectrum (*spectral colours*). □ **spectral analysis 1** *Chem.* chemical analysis using a spectroscope.

2 *Physics* analysis of light, sound, etc., into a spectrum. **spectral type** *Astron.* the group in which a star is classified on the basis of its spectrum (see HARVARD CLASSIFICATION). □ **spectrally** *adv.*

spectre /ˈspektə(r)/ *n.* (*US* **specter**) **1** a ghost. **2** a haunting presentiment or preoccupation (*the spectre of war*). **3** (in *comb.*) used in the names of some animals because of their thinness, transparency, etc. (*spectre-insect*). [F *spectre* or L *spectrum*: see SPECTRUM]

spectro- /ˈspektrəʊ/ *comb. form* a spectrum.

spectrochemistry /ˌspektrəʊˈkemɪstrɪ/ *n.* chemistry based on the study of the spectra of substances.

spectrogram /ˈspektrəʊˌgræm/ *n.* a record obtained with a spectrograph.

spectrograph /ˈspektrəʊˌgrɑːf/ *n.* an apparatus for photographing or otherwise recording spectra. □ **spectrographic** /ˌspektrəʊˈgræfɪk/ *adj.* **spectrographically** *adv.* **spectrography** /spekˈtrɒgrəfɪ/ *n.*

spectroheliograph /ˌspektrəʊˈhiːlɪəˌgrɑːf/ *n.* an instrument for taking photographs of the sun in light of one wavelength only.

spectrohelioscope /ˌspektrəʊˈhiːlɪəˌskəʊp/ *n.* a device similar to a spectroheliograph, for visual observation.

spectrometer /spekˈtrɒmɪtə(r)/ *n.* an instrument used for the measurement of observed spectra. □ **spectrometry** *n.* **spectrometric** /ˌspektrəˈmetrɪk/ *adj.* [G *Spektrometer* or F *spectromètre* (as SPECTRO-, -METER)]

spectrophotometer /ˌspektrəʊfəʊˈtɒmɪtə(r)/ *n.* an instrument for measuring the intensity of light in various parts of the spectrum, esp. as transmitted or emitted by a substance or solution at a particular wavelength. □ **spectrophotometry** *n.* **spectrophotometric** /-ˌfəʊtəʊˈmetrɪk/ *adj.*

spectroscope /ˈspektrəˌskəʊp/ *n.* an instrument for producing and recording spectra for examination. □ **spectroscopic** /ˌspektrəˈskɒpɪk/ *adj.* **spectroscopical** *adj.* **spectroscopist** /spekˈtrɒskəpɪst/ *n.* [G *Spektroskop* or F *spectroscope* (as SPECTRO-, -SCOPE)]

spectroscopy /spekˈtrɒskəpɪ/ *n.* the branch of science concerned with the investigation and measurement of spectra produced when matter interacts with or emits electromagnetic radiation. Spectroscopic techniques were pioneered by the German scientists Gustav Kirchhoff and Robert Bunsen after the former's discovery in 1859 that each substance, when heated, has its own characteristic emission spectrum. There are now many kinds of spectroscopy which differ according to the region of the electromagnetic spectrum studied (visible, infrared, X-ray, etc.) and the nature of the physical interaction (absorption, emission, scattering, etc.). All depend on the absorption or emission of quanta of energy by atoms and molecules passing between energy states. Spectroscopy is a very powerful tool for chemical analysis and the investigation of the properties of matter, and several techniques are routinely used in laboratories. (See also QUANTUM, SPECTRUM.)

spectrum /ˈspektrəm/ *n.* (*pl.* **spectra** /-trə/) **1** a band of colours, as seen in a rainbow etc., produced by separation of the components of light by their different degrees of refraction according to wavelength. (*See note below.*) **2** the entire range of wavelengths of electromagnetic radiation. **3 a** an image or distribution of components of any electromagnetic radiation arranged in a progressive series according to wavelength. (*See note below.*) **b** this as characteristic of a body or substance when emitting or absorbing radiation. **4** a similar image or distribution of components of sound, particles, etc., arranged according to frequency, charge, energy, etc. **5** the entire range or a wide range of anything arranged by degree or quality etc. **6** (in full **ocular spectrum**) an after-image. □ **ocular analyser** a device for analysing oscillation, esp. sound, into its separate components. **spectrum analysis** = *spectral analysis*. [L, = image, apparition f. *specere* look]

▪ The visible spectrum is seen when a prism splits the white light of the sun into its constituent colours. It is arranged according to wavelength, ranging continuously from red (the longest wavelength) to violet (the shortest). It was Sir Isaac Newton who first analysed light in this way. This, however, is but a small part of the whole spectrum of electromagnetic radiation, which ranges (with increasing frequency and decreasing wavelength) through radio waves, microwaves, infrared, visible light, ultraviolet, X-rays, and gamma rays.

specula *pl.* of SPECULUM.

specular /ˈspekjʊlə(r)/ *adj.* **1** of or having the nature of a speculum.

2 reflecting. □ **specular iron ore** lustrous haematite. [L *specularis* (as SPECULUM)]

speculate /'spekjʊˌleɪt/ v. **1** intr. (usu. foll. by *on*, *upon*, *about*) form a theory or conjecture, esp. without a firm factual basis; meditate (*speculated on their prospects*). **2** tr. (foll. by *that*, *how*, etc. + clause) conjecture, consider (*speculated how he might achieve it*). **3** intr. **a** invest in stocks etc. in the hope of gain but with the possibility of loss. **b** gamble recklessly. □ **speculator** n. [L *speculari* spy out, observe f. *specula* watch-tower f. *specere* look]

speculation /ˌspekjʊ'leɪʃ(ə)n/ n. **1** the act or an instance of speculating; a theory or conjecture (*made no speculation as to her age; is given to speculation*). **2 a** a speculative investment or enterprise (*bought it as a speculation*). **b** the practice of business speculating. **3** a game in which trump cards are bought or sold. [ME f. OF *speculation* or LL *speculatio* (as SPECULATE)]

speculative /'spekjʊlətɪv/ adj. **1** of, based on, engaged in, or inclined to speculation. **2** (of a business investment) involving the risk of loss (*a speculative builder*). □ **speculatively** adv. **speculativeness** n. [ME f. OF *speculatif -ive* or LL *speculativus* (as SPECULATE)]

speculum /'spekjʊləm/ n. (pl. **specula** /-lə/) **1** Surgery an instrument for dilating the cavities of the human body for inspection. **2** a mirror, usu. of polished metal, esp. (formerly) in a reflecting telescope. **3** a lustrous coloured area on the wing of some birds, esp. ducks. □ **speculum-metal** an alloy of copper and tin used as a mirror, esp. (formerly) in a telescope. [L, = mirror, f. *specere* look]

sped past and past part. of SPEED.

speech /spiːtʃ/ n. **1** the faculty or act of speaking. **2** a usu. formal address or discourse delivered to an audience or assembly. **3** a manner of speaking (*a man of blunt speech*). **4** a remark (*after this speech he was silent*). **5** the language of a nation, region, group, etc. **6** Mus. the act of sounding in an organ-pipe etc. □ **Queen's** (or **King's**) **Speech** a statement including the government's proposed measures read by the sovereign at the opening of Parliament. **speech day** Brit. an annual prize-giving day in many schools, usu. marked by speeches etc. **speech-reading** lip-reading. **speech therapist** a person who practises speech therapy. **speech therapy** treatment to improve defective speech. **speech-writer** a person employed to write speeches for a politician etc. to deliver. □ **speechful** adj. [OE *sprǣc*, later *spēc* f. WG, rel. to SPEAK]

speechify /'spiːtʃɪˌfaɪ/ v.intr. (**-ies**, **-ied**) joc. or derog. make esp. boring or long speeches. □ **speechifier** n. **speechification** /ˌspiːtʃɪfɪ'keɪʃ(ə)n/ n.

speechless /'spiːtʃlɪs/ adj. **1** temporarily unable to speak because of emotion etc. (*speechless with rage*). **2** dumb. □ **speechlessly** adv. **speechlessness** n. [OE *spǣclēas* (as SPEECH, -LESS)]

speed /spiːd/ n. & v. ● n. **1** rapidity of movement (*with all speed; at full speed*). **2** a rate of progress or motion over a distance in time (*attains a high speed*). **3 a** a gear appropriate to a range of speeds of a bicycle. **b** esp. US or archaic such a gear in a motor vehicle. **4** Photog. **a** the sensitivity of film to light. **b** the light-gathering power or f-number of a lens. **c** the duration of an exposure. **5** sl. an amphetamine drug, esp. methamphetamine. **6** archaic success, prosperity (*send me good speed*). ● v. (past and past part. **sped** /sped/) **1** intr. go fast (*sped down the street*). **2** (past and past part. **speeded**) **a** intr. (of a motorist etc.) travel at an illegal or dangerous speed. **b** tr. regulate the speed of (an engine etc.). **c** tr. cause (an engine etc.) to go at a fixed speed. **3** tr. send fast or on its way (*speed an arrow from the bow*). **4** intr. & tr. archaic be or make prosperous or successful (*how have you sped?; God speed you!*). □ **at speed** moving quickly. **speed bump** (or **hump**) a transverse ridge in the road to control the speed of vehicles. **speed limit** the maximum speed at which a road vehicle may legally be driven in a particular area etc. **speed merchant** colloq. a motorist who enjoys driving fast. **speed up** or work at greater speed. **speed-up** n. an increase in the speed or rate of working. □ **speeder** n. [OE *spēd*, *spēdan* f. Gmc]

speedball /'spiːdbɔːl/ n. sl. a mixture of cocaine with heroin or morphine.

speedboat /'spiːdbəʊt/ n. a motor boat designed for high speed.

speedo /'spiːdəʊ/ n. (pl. **-os**) colloq. = SPEEDOMETER. [abbr.]

speedometer /spiː'dɒmɪtə(r)/ n. an instrument on a motor vehicle etc. indicating its speed to the driver. [SPEED + METER[1]]

speedway /'spiːdweɪ/ n. **1 a** motorcycle racing. **b** a stadium or track used for this. **2** N. Amer. a road or track used for fast motor traffic.

speedwell /'spiːdwel/ n. a small herbaceous plant of the genus *Veronica*, of the figwort family, with small usu. blue flowers. [app. f. SPEED + WELL[1]]

speedy /'spiːdɪ/ adj. (**speedier**, **speediest**) **1** moving quickly; rapid. **2** done without delay; prompt (*a speedy answer*). □ **speedily** adv. **speediness** n.

speiss /spaɪs/ n. a compound of arsenic, iron, etc., formed in smelting certain lead ores. [G *Speise* food, amalgam]

Speke /spiːk/, John Hanning (1827–64), English explorer. From 1854 to 1858 he accompanied Sir Richard Burton on expeditions to trace the source of the Nile. They became the first Europeans to discover Lake Tanganyika (1858), after which Speke went on to reach a great lake which he identified as the 'source reservoir' of the Nile; he called it Lake Victoria in honour of the queen.

speleology /ˌspiːlɪ'ɒlədʒɪ, ˌspel-/ n. **1** the scientific study of caves. **2** the exploration of caves. □ **speleologist** n. **speleological** /-lɪə'lɒdʒɪk(ə)l/ adj. [F *spéléologie* f. L *spelaeum* f. Gk *spēlaion* cave]

spell[1] /spel/ v.tr. (past and past part. **spelt** /spelt/ or **spelled**) **1** (also absol.) write or name the letters that form (a word etc.) in correct sequence (*spell 'exaggerate'; cannot spell properly*). **2 a** (of letters) make up or form (a word etc.). **b** (of circumstances, a scheme, etc.) result in; involve (*spell ruin*). □ **spell-check** Computing n. a check of the spelling in a file of text using a spelling checker. ● v.tr. check the spelling in (a text) in this way. **spell-checker** = spelling checker. **spell out 1** make out (words, writing, etc.) letter by letter. **2** explain in detail (*spelled out what the change would mean*). □ **spellable** adj. [ME f. OF *espel(l)er*, f. Frank. (as SPELL[2])]

spell[2] /spel/ n. **1** a form of words used as a magical charm or incantation. **2** an attraction or fascination exercised by a person, activity, quality, etc. □ **under a spell** mastered by or as if by a spell. [OE *spel(l)* f. Gmc]

spell[3] /spel/ n. & v. ● n. **1** a short or fairly short period (*a cold spell in April*). **2** a turn of work (*did a spell of woodwork*). **3** Austral. a period of rest from work. ● v. **1** tr. **a** relieve or take the place of (a person) in work etc. **b** allow to rest briefly. **2** intr. Austral. take a brief rest. [earlier as verb: later form of dial. *spele* take place of f. OE *spelian*, of unkn. orig.]

spell[4] /spel/ n. a splinter of wood etc. [perh. f. obs. *speld*]

spellbind /'spelbaɪnd/ v.tr. (past and past part. **spellbound** /-baʊnd/) **1** bind with or as if with a spell; entrance. **2** (as **spellbound** adj.) entranced, fascinated, esp. by a speaker, activity, quality, etc. □ **spellbinder** n. **spellbindingly** adv.

speller /'spelə(r)/ n. **1** a person who spells esp. in a specified way (*is a poor speller*). **2** a book on spelling.

spellican var. of SPILLIKIN.

spelling /'spelɪŋ/ n. **1** the process or activity of writing or naming the letters of a word etc. **2** the way a word is spelled. **3** the ability to spell (*his spelling is weak*). □ **spelling-bee** a spelling competition. **spelling checker** Computing a program that checks the spelling of words in files of text, usually by comparison with a stored list of words.

spelt[1] past and past part. of SPELL[1].

spelt[2] /spelt/ n. a primitive kind of wheat, *Triticum spelta*. [OE f. OS *spelta* (OHG *spelza*), ME f. MLG, MDu. *spelte*]

spelter /'speltə(r)/ n. commercial crude smelted zinc. [corresp. to OF *espeautre*, MDu. *speauter*, G *Spialter*, rel. to PEWTER]

spelunker /spɪ'lʌŋkə(r)/ n. N. Amer. a person who explores caves, esp. as a hobby. □ **spelunking** n. [obs. *spelunk* cave f. L *spelunca*]

Spence /spens/, Sir Basil (Urwin) (1907–76), British architect, born in India. He designed the new Coventry cathedral (1962), embellished with the works of Jacob Epstein, John Piper, and Graham Sutherland. In the 1950s and 1960s he was much in demand for the designs of new British universities.

spence /spens/ n. archaic a buttery or larder. [ME f. OF *despense* f. L *dispensa* fem. past part. of *dispendere*: see DISPENSE]

Spencer[1] /'spensə(r)/, Herbert (1820–1903), English philosopher and sociologist. He was an early adherent of evolutionary theory, which he set down in his *Principles of Psychology* (1855). Spencer embraced Darwin's theory of natural selection proposed four years later, coined the phrase the 'survival of the fittest' (1864), and advocated social and economic laissez-faire. He later sought to synthesize the natural and social sciences in the *Programme of a System of Synthetic Philosophy* (1862–96).

Spencer[2] /'spensə(r)/, Sir Stanley (1891–1959), English painter. He is best known for his religious and visionary works in the modern setting of his native village of Cookham in Berkshire. Famous works include the painting *Resurrection: Cookham* (1926), the series of military murals for the Sandham Memorial Chapel at Burghclere in Hampshire (1927–32), and the sequence of panels portraying the Clyde shipyards during the Second World War when he was an official war artist.

spencer[1] /'spensə(r)/ *n.* **1** a short close-fitting jacket. **2** a woman's thin usu. woollen under-bodice worn for extra warmth in winter. [prob. f. the 2nd Earl *Spencer*, Engl. politician (1758–1834)]

spencer[2] /'spensə(r)/ *n. Naut.* a trysail. [perh. f. K. *Spencer* (early 19th c.)]

spend /spend/ *v.tr.* (*past* and *past part.* **spent** /spent/) **1** (usu. foll. by *on*) **a** (also *absol.*) pay out (money) in making a purchase etc. (*spent £5 on a new pen*). **b** pay out (money) for a particular person's benefit or for the improvement of a thing (*had to spend £200 on the car*). **2 a** use or consume (time or energy) (*shall spend no more effort; how do you spend your Sundays?*). **b** (also *refl.*) use up; exhaust; wear out (*their ammunition was all spent; his anger was soon spent; spent herself campaigning for justice*). **3** *tr.* (as **spent** *adj.*) having lost its original force or strength; exhausted (*the storm is spent; spent bullets*). □ **spending money** pocket money. **spend a penny** *Brit. colloq.* urinate or defecate (with reference to the coin-operated locks of public lavatories). □ **spendable** *adj.* **spender** *n.* [OE *spendan* f. L *expendere* (see EXPEND): in ME perh. also f. obs. *dispend* f. OF *despendre* expend f. L *dispendere*: see DISPENSE]

Spender /'spendə(r)/, Sir Stephen (1909–95), English poet and critic. His *Poems* (1933) contained both personal and political poems including 'The Pylons', which lent its name to the group of young left-wing poets of the 1930s known as the 'Pylon School'; its members used industrial imagery in their work and included W. H. Auden, C. Day-Lewis, and Louis MacNeice. In his critical work *The Destructive Element* (1935), Spender defended the importance of political subject-matter in literature. He later wrote the autobiography *World Within World* (1951), giving an account of his association with the Communist Party.

spendthrift /'spendθrɪft/ *n. & adj.* ● *n.* an extravagant person; a prodigal. ● *adj.* extravagant; prodigal.

Spengler /'speŋglə(r)/, Oswald (1880–1936), German philosopher. His fame rests on his book *The Decline of the West* (1918–22), in which he argues that civilizations undergo a seasonal cycle of about a thousand years and are subject to growth, flowering, and decay analogous to biological species.

Spenser /'spensə(r)/, Edmund (c.1552–99), English poet. His first major poem was the *Shepheardes Calendar* (1579) in twelve eclogues. He is best known for his allegorical romance the *Faerie Queene* (1590; 1596), celebrating Queen Elizabeth I. The poem is written in the stanza invented by Spenser (later used by Keats, Shelley, and Byron) with eight iambic pentametres and an alexandrine, rhyming *ababbcbcc*. He also wrote the marriage poem *Epithalamion* (1594). □ **Spenserian** /spen'sɪərɪən/ *adj.*

spent *past* and *past part.* of SPEND.

sperm /spɜːm/ *n.* (*pl.* same or **sperms**) **1** = SPERMATOZOON. **2** the male reproductive fluid containing spermatozoa; semen. **3** = *sperm whale*. **4** = SPERMACETI. **5** = *sperm oil*. □ **sperm bank** a supply of semen stored for use in artificial insemination. **sperm count** the number of spermatozoa in one ejaculation or a measured amount of semen. **sperm oil** an oil obtained from the head of a sperm whale, used as a lubricant. **sperm whale** a large toothed whale, *Physeter macrocephalus*, hunted for the spermaceti and sperm oil contained in its massive bulbous head, and for the ambergris found in its intestines (also called *cachalot*). [ME f. LL *sperma* f. Gk *sperma -atos* seed f. *speirō* sow: in *sperm whale* an abbr. of SPERMACETI]

spermaceti /ˌspɜːmə'setɪ/ *n.* a white waxy substance produced by the sperm whale to aid buoyancy, and used in the manufacture of candles, ointments, etc. [ME f. med.L f. LL *sperma* sperm + *ceti* genitive of *cetus* f. Gk *kētos* whale, from the belief that it was whale-spawn]

spermatic /spɜː'mætɪk/ *adj.* of or relating to sperm or semen. □ **spermatic cord** *Anat.* a bundle of nerves, ducts, and blood vessels passing to the testicles. [LL *spermaticus* f. Gk *spermatikos* (as SPERM)]

spermatid /'spɜːmətɪd/ *n. Zool.* an immature male sex cell formed from a spermatocyte, which may develop into a spermatozoon. □ **spermatidal** /ˌspɜːmə'taɪd(ə)l/ *adj.*

spermato- /'spɜːmətəʊ/ *comb. form Biol.* a sperm or seed.

spermatocyte /'spɜːmətəʊˌsaɪt/ *n. Zool.* a cell produced from a spermatogonium and which may divide by meiosis into spermatids.

spermatogenesis /ˌspɜːmətəʊ'dʒenɪsɪs/ *n. Zool.* the production or development of mature spermatozoa. □ **spermatogenic** *adj.*

spermatogonium /ˌspɜːmətəʊ'gəʊnɪəm/ *n.* (*pl.* **spermatogonia**

/-nɪə/) *Zool.* a cell produced at an early stage in the formation of spermatozoa, from which spermatocytes develop. [SPERM + mod.L *gonium* f. Gk *gonos* offspring, seed]

spermatophore /'spɜːmətəʊˌfɔː(r)/ *n. Zool.* an albuminous capsule containing spermatozoa found in various invertebrates. □ **spermatophoric** /ˌspɜːmətəʊ'fɒrɪk/ *adj.*

spermatophyte /'spɜːmətəʊˌfaɪt/ *n. Bot.* a seed-bearing vascular plant of the division Spermatophyta, which comprises the gymnosperms and flowering plants. [mod.L *Spermatophyta* f. Gk *spermato- sperma* seed + *phuton* plant]

spermatozoid /ˌspɜːmətəʊ'zəʊɪd/ *n. Bot.* the mature motile male sex cell of some plants.

spermatozoon /ˌspɜːmətəʊ'zəʊɒn/ *n.* (*pl.* **spermatozoa** /-'zəʊə/) *Zool.* the mature motile sex cell in animals. □ **spermatozoal** /-'zəʊəl/ *adj.* **spermatozoan** /-'zəʊən/ *adj.* [SPERM + Gk *zōion* animal]

spermicide /'spɜːmɪˌsaɪd/ *n.* a substance able to kill spermatozoa. □ **spermicidal** /ˌspɜːmɪ'saɪd(ə)l/ *adj.*

spermo- /'spɜːməʊ/ *comb. form* = SPERMATO-.

spew /spjuː/ *v.* (also **spue**) **1** *tr. & intr.* (often foll. by *up*) vomit. **2** (often foll. by *out*) **a** *tr.* expel (contents) rapidly and forcibly. **b** *intr.* (of contents) be expelled in this way. □ **spewer** *n.* [OE *spīwan*, *spēowan* f. Gmc]

Spey /speɪ/ a river of east central Scotland. Rising in the Grampian Mountains east of the Great Glen, it flows 171 km (108 miles) north-eastwards to the North Sea.

sp. gr. *abbr.* specific gravity.

sphagnum /'sfægnəm/ *n.* (*pl.* **sphagna** /-nə/) (in full **sphagnum moss**) a large moss of the genus *Sphagnum*, growing in waterlogged bogs where it forms peat, and used for lining hanging baskets, as a medium for growing orchids, etc. Also called *bog moss*, *peatmoss*. [mod.L f. Gk *sphagnos* a moss]

sphalerite /'sfælə,raɪt/ *n.* = BLENDE. [Gk *sphaleros* deceptive: cf. BLENDE]

spheno- /'sfiːnəʊ/ *comb. form Anat.* the sphenoid bone. [Gk f. *sphēn* wedge]

sphenoid /'sfiːnɔɪd/ *adj. & n.* ● *adj.* **1** wedge-shaped. **2** *Anat.* of or relating to the sphenoid bone. ● *n.* (in full **sphenoid bone**) a compound bone forming the base of the cranium behind the eyes. □ **sphenoidal** /sfiː'nɔɪd(ə)l/ *adj.* [mod.L *sphenoides* f. Gk *sphēnoeidēs* f. *sphēn* wedge]

sphere /sfɪə(r)/ *n. & v.* ● *n.* **1** a solid figure, or its surface, with every point on its surface equidistant from its centre. **2** an object having this shape; a ball or globe. **3 a** a celestial body. **b** a globe representing the earth. **c** *poet.* the heavens; the sky. **d** the sky perceived as a vault upon or in which celestial bodies are represented as lying. **e** *hist.* each of a series of revolving concentrically arranged spherical shells in which celestial bodies were formerly thought to be set in a fixed relationship. **4 a** a field of action, influence, or existence (*have done much within their own sphere*). **b** a (usu. specified) stratum of society or social class (*moves in quite another sphere*). ● *v.tr. archaic* or *poet.* **1** enclose in or as in a sphere. **2** form into a sphere. □ **music** (or **harmony**) **of the spheres** the natural harmonic tones supposedly produced by the movement of the celestial spheres (see sense 3e of *n.*) or the bodies fixed in them. **oblique** (or **parallel** or **right**) **sphere** the sphere of the apparent heavens at a place where there is an oblique, zero, or right angle between the equator and the horizon. **sphere of influence** the claimed or recognized area of a state's interests, an individual's control, etc. □ **spheral** *adj.* [ME *sper(e)* f. OF *espere* f. LL *sphera*, L f. Gk *sphaira* ball]

-sphere /sfɪə(r)/ *comb. form* **1** having the form of a sphere (*bathysphere*). **2** a region round the earth (*atmosphere*).

spheric /'sfɪərɪk/ *adj.* = SPHERICAL. □ **sphericity** /sfɪə'rɪsɪtɪ/ *n.*

spherical /'sferɪk(ə)l/ *adj.* **1** shaped like a sphere; globular. **2 a** of or relating to the properties of spheres (*spherical geometry*). **b** formed inside or on the surface of a sphere (*spherical triangle*). □ **spherical aberration** a loss of definition in the image arising from the surface geometry of a spherical mirror or lens. **spherical angle** an angle formed by the intersection of two great circles of a sphere. □ **spherically** *adv.* [LL *sphaericus* f. Gk *sphairikos* (as SPHERE)]

spheroid /'sfɪərɔɪd/ *n.* **1** a spherelike but not perfectly spherical body. **2** a solid generated by a half-revolution of an ellipse about its major axis (*prolate spheroid*) or minor axis (*oblate spheroid*). □ **spheroidal** /sfɪə'rɔɪd(ə)l/ *adj.* **spheroidicity** /ˌsfɪərɔɪ'dɪsɪtɪ/ *n.*

spherometer /sfɪə'rɒmɪtə(r)/ *n.* an instrument for finding the radius of a sphere and for the exact measurement of the thickness of small bodies. [F *sphéromètre* (as SPHERE, -METER)]

spherule /ˈsferuːl/ n. a small sphere. □ **spherular** adj. [LL *sphaerula* dimin. of L *sphaera* (as SPHERE)]

spherulite /ˈsferəˌlaɪt/ n. Geol. a vitreous globule as a constituent of volcanic rocks. □ **spherulitic** /ˌsferəˈlɪtɪk/ adj.

sphincter /ˈsfɪŋktə(r)/ n. Anat. a ring of muscle surrounding and serving to guard or close an opening or tube, esp. the anus. □ **sphincteral** adj. **sphincteric** /sfɪŋkˈterɪk/ adj. [L f. Gk *sphigktēr* f. *sphiggō* bind tight]

sphingid /ˈsfɪndʒɪd, ˈsfɪŋɡɪd/ n. & adj. Zool. ● n. a moth of the family Sphingidae, which comprises the hawkmoths. ● adj. of or relating to this family. [mod.L *Sphingidae* f. *Sphinx* genus name: see SPHINX]

sphinx /sfɪŋks/ n. **1** Gk Mythol. a monster with a human head and the body of a lion, regarded by the Greeks as female. (*See note below.*) **2 a** (in ancient Egypt) a stone figure with a couchant lion's body and a man's or animal's head. **b** (**Sphinx**) the colossal figure of this kind at Giza, part of the complex of funerary monuments of the pharaoh Chephren (4th dynasty, 3rd millennium BC). **3** an enigmatic or inscrutable person. **4** US a hawkmoth, esp. of the genus *Sphinx*. **5** a baboon, esp. the mandrill, *Mandrillus sphinx*. [L f. Gk *Sphigx*, app. f. *sphiggō* draw tight]

▪ Originating in Egypt, the myth of the sphinx became known early to Syrians, Phoenicians, and Mycenaean Greeks. It was associated particularly with Thebes, where it propounded a riddle about the three ages of man and killed whoever failed to solve this until Oedipus was successful and the sphinx committed suicide (or was killed by him).

sphygmo- /ˈsfɪɡməʊ/ comb. form Physiol. a pulse or pulsation. [Gk *sphugmo-* f. *sphugmos* pulse f. *sphuzō* to throb]

sphygmogram /ˈsfɪɡməʊˌɡræm/ n. a record produced by a sphygmograph.

sphygmograph /ˈsfɪɡməʊˌɡrɑːf/ n. an instrument for showing the character of a pulse in a series of curves. □ **sphygmographic** /ˌsfɪɡməʊˈɡræfɪk/ adj. **sphygmographically** adv. **sphygmography** /sfɪɡˈmɒɡrəfɪ/ n.

sphygmology /sfɪɡˈmɒlədʒɪ/ n. the scientific study of the pulse. □ **sphygmological** /ˌsfɪɡməˈlɒdʒɪk(ə)l/ adj.

sphygmomanometer /ˌsfɪɡməʊməˈnɒmɪtə(r)/ n. an instrument for measuring blood pressure by means of a manometer and an inflatable cuff. □ **sphygmomanometric** /-ˌmænəˈmetrɪk/ adj.

spic /spɪk/ n. US sl. offens. **1** a Spanish-speaking person from Central or South America or the Caribbean, esp. a Mexican. **2** the Spanish language spoken by such a person. [abbr. of *spiggoty*, of uncert. orig.: perh. alt. of *speak the* in 'no speak the English']

Spica /ˈspiːkə/ Astron. the brightest star in the constellation of Virgo. [L, = ear of wheat (in the hand of the goddess): see SPICA]

spica /ˈspaɪkə/ n. **1** Bot. a spike or spikelike form. **2** Surgery a spiral bandage with reversed turns, suggesting an ear of corn. □ **spicate** /-keɪt/ adj. **spicated** /spaɪˈkeɪtɪd/ adj. [L, = spike, ear of corn, rel. to *spina* SPINE: in sense 2 after Gk *stakhus*]

spiccato /spɪˈkɑːtəʊ/ n., adj., & adv. Mus. ● n. (pl. **-os**) **1** a style of staccato playing on stringed instruments involving bouncing the bow on the strings. **2** a passage in this style. ● adj. performed or to be performed in this style. ● adv. in this style. [It., = detailed, distinct]

spice /spaɪs/ n. & v. ● n. **1** an aromatic or pungent vegetable substance used to flavour food, e.g. cloves, pepper, or mace. **2** spices collectively (*a dealer in spice*). **3 a** an interesting or piquant quality. **b** (foll. by *of*) a slight flavour or suggestion (*a spice of malice*). ● v.tr. **1** flavour with spice. **2** add an interesting or piquant quality to (*a book spiced with humour*). [ME f. OF *espice*(r) f. L *species* specific kind: in LL pl. = merchandise]

spicebush /ˈspaɪsbʊʃ/ n. an aromatic shrub, *Lindera benzoin*, of the laurel family, native to North America.

Spice Islands a former name for the MOLUCCA ISLANDS.

spick and span /spɪk, spæn/ adj. (also **spic and span**) **1** smart and new. **2** neat and clean. [16th-c. *spick and span new*, emphatic extension of ME *span new* f. ON *spán-nýr* f. *spánn* chip + *nýr* new]

spicknel /ˈspɪknəl/ n. = BALDMONEY. [var. of SPIGNEL]

spicule /ˈspɪkjuːl/ n. **1** any small sharp-pointed body. **2** Zool. a small hard calcareous or siliceous body, esp. in the framework of a sponge. **3** Bot. a small or secondary spike. **4** Astron. a spikelike prominence, esp. one appearing as a jet of gas in the sun's corona. □ **spicular** adj. **spiculate** /-lət/ adj. [mod.L *spicula*, *spiculum*, dimins. of SPICA]

spicy /ˈspaɪsɪ/ adj. (**spicier**, **spiciest**) **1** of, flavoured with, or fragrant with spice. **2** piquant, pungent; sensational or improper (*a spicy story*). □ **spicily** adv. **spiciness** n.

spider /ˈspaɪdə(r)/ n. & v. ● n. **1 a** an eight-legged arthropod of the order Araneae, with a round unsegmented body, often spinning webs for the capture of insects as food. (*See note below.*) **b** any similar or related arachnid (*red spider*). **2** any object comparable to a spider, esp. as having numerous or prominent legs or radiating spokes. **3** Brit. a radiating series of elastic ties used to hold a load in place on a vehicle etc. ● v.intr. **1** move in a scuttling manner suggestive of a spider (*fingers spidered across the map*). **2** cause to move or appear in this way. **3** (as **spidering** adj.) spiderlike in form, manner, or movement (*spidering streets*). □ **spider crab** a crab of the family Majidae, with a pear-shaped body and long thin legs. **spider mite** a plant-feeding mite of the family Tetranychidae; esp. the *red spider mite* (*Tetranychus urticae*), which is a serious horticultural pest. **spider monkey** a South American monkey of the genus *Ateles* or *Brachyteles*, with very long limbs and a prehensile tail. **spider plant** a liliaceous southern African plant, *Chlorophytum comosum*, with long narrow striped leaves, often grown as a house plant. □ **spiderish** adj. [OE *spīthra* (as SPIN)]

▪ There are some 30,000 species of spider known, and they are distinguished from other arachnids by their leglike palps and single pair of venomous fangs or chelicerae. The head and thorax are fused to form a cephalothorax, which bears several pairs of eyes. The abdomen is usually unsegmented and bears the spinnerets. These secrete the silk that is used for protection and dispersal of the young, as well as for a variety of different methods of prey capture. Most spiders feed on other arthropods, though some large tropical kinds are able to catch young birds and other vertebrates.

spiderman /ˈspaɪdəˌmæn/ n. (pl. **-men**) Brit. colloq. a person who works at great heights in building construction.

spiderwort /ˈspaɪdəˌwɜːt/ n. a tradescantia, esp. a cultivated form of *Tradescantia virginiana*, having flowers with long hairy stamens.

spidery /ˈspaɪdərɪ/ adj. elongated and thin (*spidery handwriting*).

spiegeleisen /ˈspiːɡ(ə)lˌaɪz(ə)n/ n. an alloy of iron and manganese, used in steel-making. [G f. *Spiegel* mirror + *Eisen* iron]

spiel /ʃpiːl/ n. & v. sl. ● n. a glib speech or story, esp. a salesman's patter. ● v. **1** intr. speak glibly; hold forth. **2** tr. reel off (patter etc.). [G, = play, game]

Spielberg /ˈspiːlbɜːɡ/, Steven (b.1947), American film director and producer. He established a wide popular appeal with films concentrating on sensational and fantastic themes, such as *Jaws* (1975), *Close Encounters of the Third Kind* (1977), and *ET* (1982), which he also produced. He later directed a series of adventure films, notably *Raiders of the Lost Ark* (1981) and *Indiana Jones and the Temple of Doom* (1984). Other films include *Jurassic Park* (1992), which like his earlier film *ET* broke box-office records in the US and Britain, and *Schindler's List* (1993), which won seven Oscars, including that for best director and best picture.

spieler /ˈʃpiːlə(r)/ n. sl. **1** esp. US a person who spiels. **2** Austral. a gambler; a swindler. [G (as SPIEL)]

spiffing /ˈspɪfɪŋ/ adj. Brit. archaic sl. **1** excellent. **2** smart, handsome. [19th c.: orig. unkn.]

spiffy /ˈspɪfɪ/ adj. esp. N. Amer. sl. = SPIFFING. □ **spiffily** adv.

spiflicate /ˈspɪflɪˌkeɪt/ v.tr. (also **spifflicate**) esp. joc. **1** destroy. **2** beat (in a fight etc.). [18th c.: fanciful]

spignel /ˈspɪɡn(ə)l/ n. = BALDMONEY. [perh. f. ME *spigurnel* plant-name, f. med.L *spigurnellus*, of unkn. orig.]

spigot /ˈspɪɡət/ n. **1** a small peg or plug, esp. for insertion into the vent-hole of a cask. **2 a** US a tap. **b** a device for controlling the flow of liquid in a tap. **3** the plain end of a pipe-section fitting into the socket of the next one. [ME, perh. f. Prov. *espigou(n)* f. L *spiculum* dimin. of *spicum* = SPICA]

spike[1] /spaɪk/ n. & v. ● n. **1 a** a sharp point. **b** a pointed piece of metal, esp. the top of an iron railing etc. **2 a** a metal point set into the sole of a running-shoe to prevent slipping. **b** (in pl.) a pair of running-shoes with spikes. **3 a** a pointed metal rod standing on a base and used for filing news items etc. esp. when rejected for publication. **b** a similar spike used for bills etc. **4** a large stout nail esp. as used for railways. **5** sl. a hypodermic needle. **6** Brit. sl. a doss-house. **7** Electronics a pulse of very short duration in which a rapid increase in voltage is followed by a rapid decrease. ● v.tr. **1 a** fasten or provide with spikes. **b** fix on or pierce with spikes. **2** (of a newspaper editor etc.) reject (a story) by filing it on a spike. **3** colloq. **a** lace (a drink) with alcohol, a drug, etc.

b contaminate (a substance) with something added. **4** make useless, put an end to, thwart (an idea etc.). **5** *hist.* plug up the vent of (a gun) with a spike. □ **spike a person's guns** spoil his or her plans. **spike heel** a high tapering heel of a shoe. [ME perh. f. MLG, MDu. *spiker*, rel. to SPOKE¹]

spike² /spaɪk/ *n. Bot.* a flower-cluster formed of many flower-heads attached closely on a long stem. [ME, = ear of corn, f. L SPICA]

spikelet /'spaɪklɪt/ *n.* a small spike; esp. (*Bot.*) the basic unit of a grass flower, with two bracts at the base. [SPIKE² + -LET]

spikenard /'spaɪknɑːd/ *n.* **1 a** *hist.* a costly perfumed ointment, much valued in ancient times. **b** the plant from whose rhizome this was prepared, probably the Himalayan *Nardostachys grandiflora*, of the valerian family. **2** = NARD 1. **3** a plant resembling spikenard in fragrance (*ploughman's spikenard*). [ME ult. f. med.L *spica nardi* (as SPIKE², NARD) after Gk *nardostakhus*]

spiky¹ /'spaɪkɪ/ *adj.* (**spikier, spikiest**) **1** like a spike; having many spikes. **2** *colloq.* easily offended; prickly. □ **spikily** *adv.* **spikiness** *n.*

spiky² /'spaɪkɪ/ *adj. Bot.* having spikes or ears.

spile /spaɪl/ *n. & v.* ● *n.* **1** a wooden peg or spigot. **2** a large timber or pile for driving into the ground. **3** *N. Amer.* a small spout for tapping the sap from a sugar-maple etc. ● *v.tr.* broach (a cask etc.) with a spile in order to draw off liquid. [MDu., MLG, = wooden peg etc.: in sense 'pile' app. alt. of PILE²]

spill¹ /spɪl/ *v. & n.* ● *v.* (*past and past part.* **spilt** /spɪlt/ *or* **spilled**) **1** *intr. & tr.* fall or run or cause (a liquid, powder, etc.) to fall or run out of a vessel, esp. unintentionally. **2 a** *tr.* throw (a person etc.) from a vehicle, saddle, etc. **b** *intr.* (esp. of a crowd) tumble out quickly from a place etc. (*the fans spilled into the street*). **3** *tr. sl.* disclose (information etc.). **4** *tr. Naut.* **a** empty (a sail) of wind. **b** lose (wind) from a sail. ● *n.* **1 a** the act or an instance of spilling or being spilt. **b** a quantity spilt. **2 a** tumble or fall, esp. from a horse etc. (*had a nasty spill*). **3** *Austral.* the vacating of all or several posts of a parliamentary party to allow reorganization. □ **spill the beans** *colloq.* divulge information etc., esp. unintentionally or indiscreetly. **spill blood** be guilty of bloodshed. **spill the blood of** kill or injure (a person). **spill over 1** overflow. **2** (of a surplus population) be forced to move (cf. OVERSPILL). □ **spillage** *n.* **spiller** *n.* [OE *spillan* kill, rel. to OE *spildan* destroy: orig. unkn.]

spill² /spɪl/ *n.* a thin strip of wood, folded or twisted paper, etc., used for lighting a fire, candles, a pipe, etc. [ME, rel. to SPILE]

spillikin /'spɪlɪkɪn/ *n.* (also **spellican** /'spelɪkən/) **1** a splinter of wood, bone, etc. **2** (in *pl.*) a game in which a heap of spillikins is to be removed one at a time without moving the others. [SPILL² + -KIN]

spillover /'spɪlˌəʊvə(r)/ *n.* **1 a** the process or an instance of spilling over. **b** a thing that spills over. **2** a consequence, repercussion, or by-product.

spillway /'spɪlweɪ/ *n.* a passage for surplus water from a dam.

spilt *past and past part.* of SPILL¹.

spilth /spɪlθ/ *n.* **1** material that is spilled. **2** the act or an instance of spilling. **3** an excess or surplus.

spin /spɪn/ *v. & n.* ● *v.* (**spinning**; *past and past part.* **spun** /spʌn/) **1** *intr. & tr.* turn or cause (a person or thing) to turn or whirl round quickly. **2** *tr.* (also *absol.*) **a** draw out and twist (wool, cotton, etc.) into threads. **b** make (yarn) in this way. **c** make a similar type of thread from (a synthetic substance etc.). **3** *tr.* (of a spider, silkworm, etc.) make (a web, gossamer, a cocoon, etc.) by extruding a fine viscous thread. **4** *tr.* tell or write (a story, essay, article, etc.) (*spins a good tale*). **5** *tr.* impart spin to (a ball). **6** *intr.* (of a person's head etc.) be dizzy through excitement, astonishment, etc. **7** *tr.* shape (metal) on a mould in a lathe etc. **8** *intr.* esp. *Cricket* (of a ball) move through the air with spin. **9** *tr.* (spin *adj.*) converted into threads (*spun glass*; *spun gold*; *spun sugar*). **10** *tr.* fish in (a stream, pool, etc.) with a spinner. **11** *tr.* toss (a coin). **12** *tr.* = spin-dry. ● *n.* **1** a spinning motion; a whirl. **2** an aircraft's diving descent combined with rotation. **3** a revolving motion through the air, esp. in a rifle bullet or in a billiard, tennis, or table tennis ball struck aslant. **b** *Cricket* a twisting motion given to the ball in bowling. **4** *colloq.* a brief drive in a motor vehicle, aeroplane, etc., esp. for pleasure. **5** *Physics* the intrinsic angular momentum of a subatomic particle. **6** *Austral. & NZ sl.* a piece of good or bad luck. **7** esp. *US Polit.* bias in information to give a favourable impression. □ **spin bowler** *Cricket* an expert at bowling with spin. **spin doctor** *US* a political pundit who is employed to promote a favourable interpretation of political developments to the media. **spin-drier** a machine for drying wet clothes etc. centrifugally in a revolving drum. **spin-dry** (**-dries, -dried**) dry (clothes etc.) in

this way. **spin off** throw off by centrifugal force in spinning. **spin-off** *n.* an incidental result or results esp. as a side benefit from industrial technology. **spin out 1** prolong (a discussion etc.). **2** make (a story, money, etc.) last as long as possible. **3** spend or consume (time, one's life, etc., by discussion or in an occupation etc.). **4** *Cricket* dismiss (a batsman or side) by spin bowling. **spin a yarn** orig. *Naut.* tell a story. **spun silk** a cheap material made of short-fibred and waste silk. **spun yarn** *Naut.* a line formed of rope-yarns twisted together. [OE *spinnan*]

spina bifida /ˌspaɪnə 'bɪfɪdə/ *n. Med.* a congenital defect of the spine, in which part of the spinal cord and its meninges are exposed through a gap in the backbone. [mod.L (as SPINE, BIFID)]

spinach /'spɪnɪdʒ, -nɪtʃ/ *n.* **1** a garden vegetable, *Spinacia oleracea*, with dark green succulent leaves. **2** the leaves of this plant used as food. □ **spinach beet** a variety of beet, *Beta vulgaris*, with edible leaves like those of spinach. □ **spinachy** *adj.* [prob. MDu. *spinaetse*, *spinag(i)e*, f. OF *espinage*, *espinache* f. med.L *spinac(h)ia* etc. f. Arab. 'isfānāk f. Pers. *ispānāk*: perh. assim. to L *spina* SPINE, with ref. to its prickly seeds]

spinal /'spaɪn(ə)l/ *adj.* of or relating to the spine (*spinal curvature*; *spinal disease*). □ **spinal canal** a cavity through the vertebrae containing the spinal cord. **spinal column** the spine. **spinal cord** a cylindrical structure of the central nervous system enclosed in the spine, connecting all parts of the body with the brain. □ **spinally** *adv.* [LL *spinalis* (as SPINE)]

spindle /'spɪnd(ə)l/ *n. & v.* ● *n.* **1 a** a pin in a spinning-wheel used for twisting and winding the thread. **b** a small bar with tapered ends used for the same purpose in hand-spinning. **c** a pin bearing the bobbin of a spinning-machine. **2** a pin or axis that revolves or on which something revolves. **3** a turned piece of wood used as a banister, chair leg, etc. **4** *Biol.* a spindle-shaped mass of microtubules formed when a cell divides. **5** a varying measure of length for yarn. **6** a slender person or thing. ● *v.intr.* have, or grow into, a long slender form. □ **spindle berry** the fruit of the spindle tree. **spindle-shanked** having long thin legs. **spindle-shanks** a person with such legs. **spindle-shaped** having a circular cross-section and tapering towards each end. **spindle side** = distaff side. **spindle tree** a shrub or small tree of the genus *Euonymus*; esp. the Eurasian *E. europaeus*, with greenish-white flowers, pink fruits that are orange inside, and hard wood used for spindles. [OE *spinel* (as SPIN)]

spindly /'spɪndlɪ/ *adj.* (**spindlier, spindliest**) long or tall and thin; thin and weak.

spindrift /'spɪndrɪft/ *n.* spray blown along the surface of the sea. [Sc. var. of *spoondrift* f. *spoon* run before wind or sea + DRIFT]

spine /spaɪn/ *n.* **1** a series of vertebrae extending from the skull to the small of the back, enclosing the spinal cord and providing support for the thorax and abdomen; the backbone. **2** *Anat., Zool., & Bot.* any hard pointed process or structure. **3** a sharp ridge or projection, esp. of a mountain range or slope. **4** a central feature, main support, or source of strength. **5** the part of a book's jacket or cover that encloses the page-fastening part and usu. faces outwards on a shelf. □ **spine-chiller** a frightening story, film, etc. **spine-chilling** (esp. of a story etc.) frightening. **spine-tingling** thrilling, pleasurably exciting. □ **spined** *adj.* [ME f. OF *espine* or L *spina* thorn, backbone]

spinel /spɪ'nel/ *n.* **1** any of a group of hard crystalline minerals of various colours, consisting chiefly of oxides of magnesium and aluminium. **2** any substance of similar composition or properties. □ **spinel ruby** a deep-red variety of spinel used as a gem. [F *spinelle* f. It. *spinella*, dimin. of *spina*: see SPINE]

spineless /'spaɪnlɪs/ *adj.* **1 a** having no spine; invertebrate. **b** having no spines. **2** (of a person) lacking energy or resolution; weak and purposeless. □ **spinelessly** *adv.* **spinelessness** *n.*

spinet /spɪ'net, 'spɪnɪt/ *n. Mus. hist.* a small harpsichord with oblique strings. [obs. F *espinette* f. It. *spinetta* virginal, spinet, dimin. of *spina* thorn etc. (as SPINE), with ref. to the plucked strings]

spinifex /'spɪnɪˌfeks/ *n.* an Australian grass of the genus *Spinifex*, with coarse, spiny leaves. [mod.L f. L *spina* SPINE + *-fex* maker f. *facere* make]

spinnaker /'spɪnəkə(r)/ *n.* a large triangular sail carried opposite the mainsail of a racing-yacht running before the wind. [fanciful f. *Sphinx*, name of yacht first using it, perh. after *spanker*]

spinner /'spɪnə(r)/ *n.* **1** a person or thing that spins. **2** *Cricket* **a** a spin bowler. **b** a spun ball. **3** a spin-drier. **4 a** a real or artificial fly for esp. trout-fishing. **b** revolving bait. **5** a manufacturer or merchant engaged in (esp. cotton-) spinning. **6** = SPINNERET. **7** *archaic* a spider.

spinneret /ˈspɪnəˌret/ *n.* **1** *Zool.* the spinning-organ in a spider, silkworm, etc. **2** a device for forming filaments of synthetic fibre.

spinney /ˈspɪnɪ/ *n.* (*pl.* **-eys**) *Brit.* a small wood; a thicket. [OF *espinei* f. L *spinetum* thicket f. *spina* thorn]

spinning /ˈspɪnɪŋ/ *n.* the act or an instance of spinning. □ **spinning-machine** a machine that spins fibres continuously. **spinning-top** = TOP².

spinning-jenny /ˌspɪnɪŋˈdʒenɪ/ *n.* an early form of spinning-machine, invented by James Hargreaves and patented in 1770. The spinning-jenny was the first machine to use multiple spindles, eight (or later sixteen) of which could be set in motion by a band from one wheel. Improvements made to the process of weaving meant that weavers were working more quickly and spinners could no longer keep up with them. Hargreaves's machinery was opposed, however, and destroyed by a mob fearful of the threat to their employment. The design of the spinning-jenny was later developed by Richard Arkwright and Samuel Crompton.

spinning mule *n.* a kind of spinning-machine producing yarn on spindles, invented in 1779 by Samuel Crompton, so called because it was a cross between Richard Arkwright's 'water frame' and James Hargreaves's spinning-jenny. It was an improvement upon both, producing yarn of higher quality at greater speed.

spinning-wheel /ˈspɪnɪŋˌwiːl/ *n.* a household machine for spinning yarn or thread with a spindle driven by a wheel operated originally by hand, later by a crank or treadle. The spinning-wheel is thought to have been introduced into Europe from India in the early 14th century.

spinose /ˈspaɪnəʊz, -nəʊs/ *adj.* (also **spinous** /-nəs/) *esp. Bot. & Zool.* having spines.

Spinoza /spɪˈnəʊzə/, Baruch (or Benedict) de (1632–77), Dutch philosopher, of Portuguese-Jewish descent. His unorthodox views led to his expulsion from the Amsterdam synagogue in 1656. Spinoza rejected the Cartesian dualism of spirit and matter in favour of a pantheistic system, seeing God as the single infinite substance, the immanent cause of the universe and not a ruler outside it. His *Ethics* (1677) sought to formulate a metaphysical system that was mathematically deduced from theorems and hypotheses; its rationalist method broke new ground in biblical analysis. Spinoza espoused a determinist political doctrine, arguing that the individual surrenders his or her natural rights to the state in order to obtain security. □ **Spinozism** *n.* **Spinozist** *n.* **Spinozistic** /ˌspɪnəʊˈzɪstɪk/ *adj.*

spinster /ˈspɪnstə(r)/ *n.* **1** an unmarried woman. **2** a woman, esp. elderly, thought unlikely to marry. □ **spinsterhood** *n.* **spinsterish** *adj.* **spinsterishness** *n.* [ME, orig. = woman who spins]

spinthariscope /spɪnˈθærɪˌskəʊp/ *n.* an instrument with a fluorescent screen showing the incidence of alpha particles by flashes. [irreg. f. Gk *spintharis* spark + -SCOPE]

spinule /ˈspaɪnjuːl/ *n.* *Bot. & Zool.* a small spine. □ **spinulose** /-ˌləʊz, -ˌləʊs/ *adj.* **spinulous** *adj.* [L *spinula* dimin. of *spina* SPINE]

spiny /ˈspaɪnɪ/ *adj.* (**spinier, spiniest**) **1** full of spines; prickly. **2** perplexing, troublesome, thorny. □ **spiny anteater** = ECHIDNA. **spiny lobster** a large edible crustacean of the family Palinuridae, with a spiny shell and no large anterior claws, esp. *Palinurus vulgaris* (also called *crawfish*). □ **spininess** *n.*

spiracle /ˈspaɪərək(ə)l/ *n.* (also **spiraculum** /ˌspaɪəˈrækjʊləm/) (*pl.* **spiracles** or **spiracula** /-lə/) *Zool.* an external respiratory opening in insects, whales, and some fish. □ **spiracular** /ˌspaɪəˈrækjʊlə(r)/ *adj.* [L *spiraculum* f. *spirare* breathe]

spiraea /ˌspaɪəˈrɪə/ *n.* (*US* **spirea**) a rosaceous shrub of the genus *Spiraea*, with clusters of small white or pink flowers. [L f. Gk *speiraia* f. *speira* coil]

spiral /ˈspaɪərəl/ *adj., n.,* & *v.* ● *adj.* **1** winding about a centre in an enlarging or decreasing continuous circular motion, either on a flat plane or rising in a cone; coiled. **2** winding continuously along or as if along a cylinder, like the thread of a screw. ● *n.* **1** a plane or three-dimensional spiral curve. **2** a spiral spring. **3** a spiral formation in a shell etc. **4** *Astron.* a spiral galaxy. **5** a progressive rise or fall of prices, wages, etc., each responding to an upward or downward stimulus provided by the other (*a spiral of rising prices and wages*). ● *v.* (**spiralled, spiralling**; *US* **spiraled, spiraling**) **1** *intr.* move in a spiral course, esp. upwards or downwards. **2** *tr.* make spiral. **3** *intr.* esp. *Econ.* (of prices, wages, etc.) rise or fall, esp. rapidly (cf. sense 5 of *n.*). □ **spiral balance** a device for measuring weight by the torsion of a spiral spring. **spiral galaxy** a galaxy in which the matter is concentrated mainly in one

or more spiral arms. **spiral staircase** a staircase rising in a spiral round a central axis. □ **spirally** *adv.* **spirality** /ˌspaɪəˈrælɪtɪ/ *n.* [F *spiral* or med.L *spiralis* (as SPIRE²)]

spirant /ˈspaɪərənt/ *adj.* & *n.* *Phonet.* ● *adj.* (of a consonant) uttered with a continuous expulsion of breath, esp. fricative. ● *n.* such a consonant. [L *spirare* spirant- breathe]

spire¹ /ˈspaɪə(r)/ *n.* & *v.* ● *n.* **1** a tapering conical or pyramidal structure built esp. on a church tower (cf. STEEPLE). **2** the continuation of a tree-trunk above the point where branching begins. **3** any tapering thing, e.g. the spike of a flower. ● *v.tr.* provide with a spire. □ **spiry** /ˈspaɪrɪ/ *adj.* [OE *spīr*]

spire² /ˈspaɪə(r)/ *n.* **1 a** a spiral; a coil. **b** a single twist of this. **2** the upper part of a spiral shell. [F f. L *spira* f. Gk *speira* coil]

spirea *US* var. of SPIRAEA.

spirillum /ˌspaɪəˈrɪləm/ *n.* (*pl.* **spirilla** /-lə/) a bacterium with a rigid spiral structure, esp. one of the genus *Spirillum*. [mod.L, irreg. dimin. of L *spira* SPIRE²]

spirit /ˈspɪrɪt/ *n.* & *v.* ● *n.* **1 a** the animating or life-giving principle in a person or animal. **b** the intelligent non-physical part of a person; the soul. **2 a** a rational or intelligent being without a material body. **b** a supernatural being such as a ghost, fairy, etc. (*haunted by spirits*). **3** a prevailing mental or moral condition or attitude; a mood; a tendency (*public spirit; took it in the wrong spirit*). **4 a** (usu. in *pl.*) strong distilled liquor, e.g. brandy, whisky, gin, rum. **b** a distilled liquid essence (*spirit of hartshorn*). **c** a distilled alcohol (*methylated spirit*). **d** a solution of a volatile principle in alcohol; a tincture. **5 a** a person's mental or moral nature or qualities, usu. specified (*has an unbending spirit*). **b** a person viewed as possessing these (*is an ardent spirit*). **c** (in full **high spirit**) courage, energy, vivacity, dash (*played with spirit; infused him with spirit*). **6** the real meaning as opposed to lip-service or verbal expression (*the spirit of the law*). **7** *archaic* an immaterial principle thought to govern vital phenomena (*animal spirits*). ● *v.tr.* (**spirited, spiriting**) (usu. foll. by *away, off,* etc.) convey rapidly and secretly by or as if by spirits. □ **in** (or **in the**) **spirit** inwardly (*shall be with you in spirit*). **spirit duplicator** a duplicator using an alcoholic solution to reproduce copies from a master sheet. **spirit gum** a quick-drying solution of gum used esp. for attaching false hair. **spirit-lamp** a lamp burning methylated or other volatile spirits instead of oil. **the spirit moves a person** he or she feels inclined (to do something) (orig. in Quaker use). **spirit** (or **spirits**) **of wine** *archaic* purified alcohol. **spirits of salt** *archaic* hydrochloric acid. **spirit up** animate or cheer (a person). [ME f. AF (*e*)*spirit*, OF *esp*(*e*)*rit*, f. L *spiritus* breath, spirit f. *spirare* breathe]

spirited /ˈspɪrɪtɪd/ *adj.* **1** full of spirit; animated, lively, brisk, or courageous (*a spirited attack; a spirited translation*). **2** having a spirit or spirits of a specified kind (*high-spirited; mean-spirited*). □ **spiritedly** *adv.* **spiritedness** *n.*

spiritless /ˈspɪrɪtlɪs/ *adj.* lacking courage, vigour, or vivacity. □ **spiritlessly** *adv.* **spiritlessness** *n.*

spirit-level /ˈspɪrɪtˌlev(ə)l/ *n.* a device consisting of a sealed glass tube nearly filled with alcohol or other liquid and containing an air-bubble whose position is used to test horizontality. Such devices were incorporated in telescopes in the 17th century, but were not used as tools by carpenters and builders until the mid-19th century.

spiritual /ˈspɪrɪtjʊəl/ *adj.* & *n.* ● *adj.* **1** of or concerning the spirit as opposed to matter. **2** concerned with sacred or religious things; holy; divine; inspired (*the spiritual life; spiritual songs*). **3** (of the mind etc.) refined, sensitive; not concerned with the material. **4** (of a relationship etc.) concerned with the soul or spirit etc., not with external reality (*his spiritual home*). ● *n.* (also **Negro spiritual**) a religious song of the kind sung in black Baptist and Pentecostal churches of the southern US, regarded as deriving from a combination of the hymns of early white settlers with traditional African elements such as the pentatonic scale. □ **spiritual courts** ecclesiastical courts. **spiritual healer** a person who claims to bring about healing through spiritual rather than physical means; a faith-healer. **spiritual healing** faith-healing. □ **spiritually** *adv.* **spiritualness** *n.* **spirituality** /ˌspɪrɪtjʊˈælɪtɪ/ *n.* [ME f. OF *spirituel* f. L *spiritualis* (as SPIRIT)]

spiritualism /ˈspɪrɪtjʊəˌlɪz(ə)m/ *n.* **1** the belief that the spirits of the dead can communicate with the living, esp. through mediums; the practice of this. (See note below.) **2** *Philos.* the doctrine that the spirit exists as distinct from matter, or that spirit is the only reality (cf. MATERIALISM 2). □ **spiritualist** *n.* **spiritualistic** /ˌspɪrɪtjʊəˈlɪstɪk/ *adj.*

▪ Belief that spirits of the dead can and do communicate with the living is very ancient and is an element in many religions. Saul clandestinely consulted the woman of Endor (1 Sam. 28) in order to speak with the dead prophet Samuel, but the Jewish prophets disapproved of the practice, and this repugnance was maintained by Christianity. The modern spiritualist movement had its origin in 1848, when three sisters of the Fox family in a small town in New York State caused a sensation by claiming to have communicated with a spirit in their home by a system of rappings. Although attacked by established Churches and by those suspecting charlatanry, spiritualism became immensely popular on both sides of the Atlantic among people wishing to communicate with dead friends or relatives or to know about a future life. Spiritualist Churches and societies were formed, and seances were held at which mediums attempted to communicate with the dead and apparently experienced phenomena including clairvoyance, telepathy, levitation, and the appearance of ectoplasm. Spiritualism is associated particularly with the Victorian period, but was given fresh impetus by the two world wars; it was also one of the few means of exploring the supernatural sanctioned in the USSR.

spiritualize /'spɪrɪtjʊəˌlaɪz/ v.tr. (also **-ise**) **1** make (a person or a person's character, thoughts, etc.) spiritual. **2** attach a spiritual as opposed to a literal meaning to. □ **spiritualization** /ˌspɪrɪtjʊəlaɪˈzeɪʃ(ə)n/ n.

spirituel /ˌspɪrɪtjʊˈel/ adj. (also **spirituelle**) (of the mind) refined and yet spirited; witty. [F spirituel, fem. -elle (as SPIRITUAL)]

spirituous /'spɪrɪtjʊəs/ adj. **1** containing much alcohol. **2** distilled, as whisky, rum, etc. (spirituous liquor). □ **spirituousness** n. [L spiritus spirit, or F spiritueux]

spiro-¹ /'spaɪərəʊ/ comb. form a coil. [L spira, Gk speira coil]

spiro-² /'spaɪərəʊ/ comb. form breath. [irreg. f. L spirare breathe]

spirochaete /'spaɪərəˌkiːt/ n. (US **spirochete**) a flexible spirally twisted bacterium of the order Spirochaetales, esp. one that causes syphilis. [SPIRO-¹ + Gk khaitē long hair]

spirograph /'spaɪərəˌɡrɑːf/ n. an instrument for recording breathing movements. □ **spirographic** /ˌspaɪərəˈɡræfɪk/ adj. **spirographically** adv.

spirogyra /ˌspaɪərəʊˈdʒaɪərə/ n. a green filamentous freshwater alga of the genus Spirogyra, with cells containing spiral bands of chlorophyll. [mod.L f. SPIRO-¹ + Gk guros gura round]

spirometer /ˌspaɪəˈrɒmɪtə(r)/ n. an instrument for measuring the air capacity of the lungs.

spirt var. of SPURT.

spit¹ /spɪt/ v. & n. ● v. (**spitting**; past and past part. **spat** /spæt/ or **spit**) **1** intr. **a** eject saliva from the mouth. **b** do this as a sign of hatred or contempt (spat at him). **2** tr. (usu. foll. by out) **a** eject (saliva, blood, food, etc.) from the mouth (spat the meat out). **b** utter (oaths, threats, etc.) vehemently ('Damn you!' he spat). **3** intr. (of a fire, pen, pan, etc.) send out sparks, ink, hot fat, etc. **4** intr. (of rain) fall lightly (it's only spitting). **5** intr. (esp. of a cat) make a spitting or hissing noise in anger or hostility. ● n. **1** spittle. **2** the act or an instance of spitting. **3** the foamy liquid secretion of some insects used to protect their young (cuckoo-spit). □ **the spit** (or **very spit**) **of** colloq. the exact double of (cf. spitting image). **spit and polish 1** the cleaning and polishing duties of a soldier etc. **2** exaggerated neatness and smartness. **spit chips** Austral. sl. **1** feel extreme thirst. **2** be angry or frustrated. **spit it out** colloq. say what is on one's mind. **spitting cobra** a cobra that can spray venom for a considerable distance; esp. the black-necked cobra, Naja nigricollis, of southern Africa. **spitting distance** a very short distance. **spitting image** (foll. by of) colloq. the exact double of (another person or thing). □ **spitter** n. [OE spittan, of imit. orig.: cf. SPEW]

spit² /spɪt/ n. & v. ● n. **1** a slender rod on which meat is skewered before being roasted on a fire etc.; a skewer. **2 a** a small point of land projecting into the sea. **b** a long narrow underwater bank. ● v.tr. (**spitted**, **spitting**) **1** thrust a spit through (meat etc.). **2** pierce or transfix with a sword etc. □ **spit-roast** cook on a spit. □ **spitty** adj. [OE spitu f. WG]

spit³ /spɪt/ n. (pl. same or **spits**) a spade-depth of earth (dig it two spit deep). [MDu. & MLG, = OE spittan dig with spade, prob. rel. to SPIT²]

spitball /'spɪtbɔːl/ n. & v. ● n. N. Amer. **1** a ball of chewed paper etc. used as a missile. **2** a baseball moistened by the pitcher to impart spin. ● v.intr. throw out suggestions for discussion. □ **spitballer** n.

spitchcock /'spɪtʃkɒk/ n. & v. ● n. an eel split and grilled or fried. ● v.tr. prepare (an eel, fish, bird, etc.) in this way. [16th c.: orig. unkn.: cf. SPATCHCOCK]

spite /spaɪt/ n. & v. ● n. **1** ill will, malice towards a person (did it from spite). **2** a grudge. ● v.tr. thwart, mortify, annoy (does it to spite me). □ **in**

spite of notwithstanding. **in spite of oneself** etc. though one would rather have done otherwise. [ME f. OF despit DESPITE]

spiteful /'spaɪtfʊl/ adj. motivated by spite; malevolent. □ **spitefully** adv. **spitefulness** n.

Spitfire /'spɪtˌfaɪə(r)/ a single-seat, single-engined British fighter aircraft of the Second World War, designed by Reginald Mitchell. Perhaps the most famous aircraft of the war, the Spitfire is remembered in particular for its role in the Battle of Britain along with the Hurricane.

spitfire /'spɪtˌfaɪə(r)/ n. a person of fiery temper.

Spithead /'spɪtˈhed/ a channel between the NE coast of the Isle of Wight and the mainland of southern England. It offers sheltered access to Southampton Water and deep anchorage.

Spitsbergen /'spɪtsˌbɜːɡən/ a Norwegian island in the Svalbard archipelago, in the Arctic Ocean north of Norway; principal settlement, Longyearbyen.

spittle /'spɪt(ə)l/ n. saliva, esp. as ejected from the mouth. □ **spittly** adj. [alt. of ME (now dial.) spattle = OE spātl f. spǣtan to spit, after SPIT¹]

spittoon /spɪˈtuːn/ n. a metal or earthenware pot with esp. a funnel-shaped top, used for spitting into.

Spitz /spɪts/, Mark (Andrew) (b.1950), American swimmer. He won seven gold medals in the 1972 Olympic Games at Munich and set twenty-seven world records for free style and butterfly (1967–72).

spitz /spɪts/ n. a small breed of dog with a pointed muzzle, esp. a Pomeranian. [G Spitz(hund) f. spitz pointed + Hund dog]

spiv /spɪv/ n. Brit. colloq. a man, often characterized by flashy dress, who makes a living by illicit or unscrupulous dealings. □ **spivvish** adj. **spivvy** adj. [20th c.: orig. unkn.]

splake /spleɪk/ n. a hybrid trout of North American lakes. [speckled (trout) + lake (trout)]

splanchnic /'splæŋknɪk/ adj. Anat. & Zool. of or relating to the viscera; intestinal. [mod.L splanchnicus f. Gk splagkhnikos f. splagkhna entrails]

splash /splæʃ/ v. & n. ● v. **1** intr. & tr. spatter or cause (liquid) to spatter in small drops. **2** tr. cause (a person) to be spattered with liquid etc. (splashed them with mud). **3** intr. **a** (of a person) cause liquid to spatter (was splashing about in the bath). **b** (usu. foll. by across, along, etc.) move while spattering liquid etc. (splashed across the carpet in his boots). **c** step, fall, or plunge etc. into a liquid etc. so as to cause a splash (splashed into the sea). **4** tr. display (news) prominently. **5** tr. decorate with scattered colour. **6** tr. spend (money) ostentatiously. ● n. **1** the act or an instance of splashing. **2 a** a quantity of liquid splashed. **b** the resulting noise (heard a splash). **3** a spot of dirt etc. splashed on to a thing. **4** a prominent news feature etc. **5** a daub or patch of colour, esp. on an animal's coat. **6** Brit. colloq. a small quantity of liquid, esp. of soda water etc. to dilute spirits. □ **make a splash** attract much attention, esp. by extravagance. **splash down** (esp. of a spacecraft) alight on the sea. **splash out** colloq. spend money freely. □ **splashy** adj. (**splashier**, **splashiest**). [alt. of PLASH¹]

splashback /'splæʃbæk/ n. a panel behind a sink etc. to protect the wall from splashes.

splashdown /'splæʃdaʊn/ n. the alighting of a spacecraft on the sea.

splat¹ /splæt/ n. a flat piece of thin wood in the centre of a chair-back. [splat (v.) split up, rel. to SPLIT]

splat² /splæt/ n., adv., & v. colloq. ● n. a sharp cracking or slapping sound (hit the wall with a splat). ● adv. with a splat (fell splat on his head). ● v.intr. & tr. (**splatted**, **splatting**) fall or hit with a splat. [abbr. of SPLATTER]

splatter /'splætə(r)/ v. & n. ● v. **1** tr. (often foll. by with) make wet or dirty by splashing. **2** tr. & intr. splash, esp. with a continuous noisy action. **3** tr. (often foll. by over) publicize or spread (news etc.) (the story was splattered over the front pages). ● n. **1** a noisy splashing sound. **2** (attrib.) sl. designating or relating to films involving the depiction of many violent deaths. **3** a rough patch of colour etc., esp. splashed on a surface. [imit.]

splay /spleɪ/ v., n., & adj. ● v. **1** tr. (usu. foll. by out) spread (the elbows, feet, etc.) out. **2** intr. (of an aperture or its sides) diverge in shape or position. **3** tr. construct (a window, doorway, aperture, etc.) so that it diverges or is wider at one side of the wall than the other. ● n. a surface making an oblique angle with another, e.g. the splayed side of a window or embrasure. ● adj. **1** wide and flat. **2** turned outward. □ **splay-foot** a broad flat foot turned outward. **splay-footed** having such feet. [ME f. DISPLAY]

spleen /spliːn/ n. **1** an abdominal organ involved in the production and removal of red blood cells in most vertebrates, forming part of the immune system. **2** lowness of spirits; moroseness, ill temper, spite (from the earlier belief that the spleen was the seat of such feelings) (*a fit of spleen; vented their spleen*). □ **spleenful** *adj.* **spleeny** *adj.* [ME f. OF *esplen* f. L *splen* f. Gk *splēn*]

spleenwort /ˈspliːnwɜːt/ *n.* a small fern of the genus *Asplenium*, formerly used as a remedy for disorders of the spleen.

splen- /spliːn, splɪn, splen/ *comb. form Anat.* the spleen. [Gk (as SPLEEN)]

splendent /ˈsplendənt/ *adj. formal* **1** shining; lustrous. **2** illustrious. [ME f. L *splendere* to shine]

splendid /ˈsplendɪd/ *adj.* **1** magnificent, gorgeous, brilliant, sumptuous (*a splendid palace; a splendid achievement*). **2** dignified; impressive (*a splendid isolation*). **3** excellent; fine (*a splendid chance*). □ **splendidly** *adv.* **splendidness** *n.* [F *splendide* or L *splendidus* (as SPLENDENT)]

splendiferous /splenˈdɪfərəs/ *adj. colloq.* or *joc.* splendid. □ **splendiferously** *adv.* **splendiferousness** *n.* [irreg. f. SPLENDOUR]

splendour /ˈsplendə(r)/ *n.* (US **splendor**) **1** great or dazzling brightness. **2** magnificence; grandeur. [ME f. AF *splendeur* or L *splendor* (as SPLENDENT)]

splenectomy /splɪˈnektəmɪ/ *n.* (*pl.* **-ies**) *Med.* the surgical removal of the spleen.

splenetic /splɪˈnetɪk/ *adj. & n.* ● *adj.* **1** ill-tempered; peevish. **2** of or concerning the spleen. ● *n.* a splenetic person. □ **splenetically** *adv.* [LL *spleneticus* (as SPLEEN)]

splenic /ˈsplenɪk, ˈspliːn-/ *adj.* of or in the spleen. □ **splenic fever** anthrax. □ **splenoid** /ˈspliːnɔɪd/ *adj.* [F *splénique* or L *splenicus* f. Gk *splēnikos* (as SPLEEN)]

splenitis /splɪˈnaɪtɪs/ *n. Med.* inflammation of the spleen.

splenius /ˈspliːnɪəs/ *n.* (*pl.* **splenii** /-nɪˌaɪ/) *Anat.* either section of muscle on each side of the neck and back serving to draw back the head. □ **splenial** *adj.* [mod.L f. Gk *splēnion* bandage]

splenology /splɪˈnɒlədʒɪ/ *n.* the scientific study of the spleen.

splenomegaly /ˌspliːnəˈmegəlɪ/ *n. Med.* a pathological enlargement of the spleen. [SPLEN- + *megaly* (as MEGALO-)]

splenotomy /splɪˈnɒtəmɪ/ *n.* (*pl.* **-ies**) *Med.* a surgical incision into or dissection of the spleen.

splice /splaɪs/ *v. & n.* ● *v.tr.* **1** join the ends of (ropes) by interweaving strands. **2** join (pieces of timber, magnetic tape, film, etc.) in an overlapping position. **3** (esp. as **spliced** *adj.*) *colloq.* join in marriage. ● *n.* a joint consisting of two ropes, pieces of wood, film, etc., made by splicing, e.g. the handle and blade of a cricket bat. □ **splice the main brace** *Naut. hist.* issue an extra tot of rum. □ **splicer** *n.* [prob. f. MDu. *splissen*, of uncert. orig.]

spliff /splɪf/ *n.* (also **splif**) *sl.* a cannabis cigarette. [20th c.: orig. unkn.]

spline /splaɪn/ *n. & v.* ● *n.* **1** a rectangular key fitting into grooves in the hub and shaft of a wheel and allowing longitudinal play. **2** a slat. **3** a flexible wood or rubber strip used esp. in drawing large curves. ● *v.tr.* fit with a spline (sense 1). [orig. E. Anglian dial., perh. rel. to SPLINTER]

splint /splɪnt/ *n. & v.* ● *n.* **1 a** a strip of rigid material used for holding a broken bone etc. when set. **b** a rigid or flexible strip of esp. wood used in basketwork etc. **2** a tumour or bony excrescence on the inside of a horse's leg. **3** a thin strip of wood etc. used to light a fire, pipe, etc. **4** = *splint-bone*. ● *v.tr.* secure (a broken limb etc.) with a splint or splints. □ **splint-bone 1** either of two small bones in a horse's foreleg lying behind and close to the cannon-bone. **2** the human fibula. **splint-coal** hard bituminous laminated coal burning with great heat. [ME *splent(e)* f. MDu. *splinte* or MLG *splinte*, *splente* metal plate or pin, rel. to SPLINTER]

splinter /ˈsplɪntə(r)/ *v. & n.* ● *v.tr. & intr.* break into fragments. ● *n.* a small thin sharp-edged piece broken off from wood, stone, etc. □ **splinter-bar** *Brit.* a crossbar in a horse-drawn vehicle to which traces are attached; a swingletree. **splinter group** (or **party**) a group or party that has broken away from a larger one. **splinter-proof** capable of withstanding splinters e.g. from bursting shells or bombs. □ **splintery** *adj.* [ME f. MDu. (= LG) *splinter*, *splenter*, rel. to SPLINT]

Split /splɪt/ a seaport on the coast of southern Croatia; pop. (1991) 189,300. Founded as a Roman colony in 78 BC, it contains the ruins of the palace of the emperor Diocletian, built in about AD 300.

split /splɪt/ *v. & n.* ● *v.* (**splitting**; *past* and *past part.* **split**) **1** *intr. & tr.* **a** break or cause to break forcibly into parts, esp. into halves or (of a log) with the grain. **b** (often foll. by *up*) divide into parts (*split into groups; split up the money equally*). **2** *tr. & intr.* (often foll. by *off, away*) remove or be removed by breaking, separating, or dividing (*split the top off the bottle; split away from the main group*). **3** *intr. & tr.* **a** (usu. foll. by *up, on, over*, etc.) *colloq.* separate esp. through discord (*split up after ten years; they were split on the question of picketing*). **b** (foll. by *with*) *sl.* quarrel or cease association with (another person etc.). **4** *tr.* cause the fission of (an atom). **5** *intr. & tr. sl.* leave, esp. suddenly. **6** *intr.* (usu. foll. by *on*) *colloq.* betray secrets; inform (*split on them to the police*). **7** *intr.* **a** (as **splitting** *adj.*) (esp. of a headache) very painful; acute. **b** (of the head) suffer great pain from a headache, noise, etc. **8** *intr.* (of a ship) be wrecked. **9** *tr. US colloq.* dilute (whisky etc.) with water. ● *n.* **1** the act or an instance of splitting; the state of being split. **2** a fissure, vent, crack, cleft, etc. **3** a separation into parties; a schism. **4** (in *pl.*) *Brit.* the athletic feat of leaping in the air or sitting down with the legs at right angles to the body in front and behind, or at the sides with the trunk facing forwards. **5** a split osier etc. used for parts of basketwork. **6** each strip of steel, cane, etc., of the reed in a loom. **7** a single thickness of split hide. **8** the turning up of two cards of equal value in faro, so that the stakes are divided. **9 a** half a bottle of mineral water. **b** half a glass of liquor. **10** *sl.* a division of money, esp. the proceeds of crime. □ **split the difference** take the average of two proposed amounts. **split gear** (or **pulley** or **wheel**) a gear etc. made in halves for removal from a shaft. **split hairs** make small and insignificant distinctions. **split infinitive** a phrase consisting of an infinitive with an adverb etc. inserted between *to* and the verb, e.g. *to boldly go*. **split-level** (of a building) having a room or rooms a fraction of a storey higher than other parts. **split mind** = SCHIZOPHRENIA. **split pea** a pea dried and split in half for cooking. **split personality** the alteration or dissociation of personality occurring in some mental illnesses, esp. schizophrenia and hysteria. **split pin** a metal cotter passed through a hole and held by the pressing back of the two ends. **split ring** a small steel ring with two spiral turns, such as a key-ring. **split-screen** a screen on which two or more separate images are displayed. **split second** a very brief moment of time. **split shift** a shift comprising two or more separate periods of duty. **split shot** (or **stroke**) *Croquet* a stroke driving two touching balls in different directions. **split one's sides** be convulsed with laughter. **split the ticket** (or **one's vote**) *US* vote for candidates of more than one party. **split the vote** *Brit.* (of a candidate or minority party) attract votes from another so that both are defeated by a third. □ **splitter** *n.* [orig. Naut. f. MDu. *splitten*, rel. to *spletten*, *splīten*, MHG *splīzen*]

splodge /splɒdʒ/ *n. & v. colloq.* ● *n.* a daub, blot, or smear. ● *v.tr.* make a large, esp. irregular, spot or patch on. □ **splodgy** *adj.* [imit., or alt. of SPLOTCH]

splosh /splɒʃ/ *v. & n. colloq.* ● *v.tr. & intr.* move with a splashing sound. ● *n.* **1** a splashing sound. **2** a splash of water etc. **3** *sl.* money. [imit.]

splotch /splɒtʃ/ *n. & v.tr.* = SPLODGE. □ **splotchy** *adj.* [perh. f. SPOT + obs. *plotch* BLOTCH]

splurge /splɜːdʒ/ *n. & v. colloq.* ● *n.* **1** an ostentatious display or effort. **2** an instance of sudden great extravagance. ● *v.intr. & tr.* **1** (usu. foll. by *on*) spend (effort or esp. large sums of money) (*splurged the lot on a holiday*). **2** splash heavily. [19th-c. US: prob. imit.]

splutter /ˈsplʌtə(r)/ *v. & n.* ● *v.* **1** *intr.* **a** speak in a hurried, vehement, or choking manner. **b** emit particles from the mouth, sparks, hot oil, etc., with spitting sounds. **2** *tr.* **a** speak or utter (words, threats, a language, etc.) rapidly or incoherently. **b** emit (food, sparks, hot oil, etc.) with a spitting sound. ● *n.* spluttering speech or sound. □ **splutterer** *n.* **splutteringly** *adv.* [SPUTTER by assoc. with *splash*]

Spock /spɒk/, Benjamin McLane (known as Dr Spock) (b.1903), American paediatrician and writer. His manual *The Common Sense Book of Baby and Child Care* (1946) challenged traditional ideas of discipline and rigid routine in child rearing in favour of a psychological approach and influenced a generation of parents after the Second World War. He was sent to prison in 1968 for helping draft-dodgers.

Spode /spəʊd/ *n.* a kind of fine pottery or porcelain named after the English potter Josiah Spode (1754–1827), its original maker.

spoil /spɔɪl/ *v. & n.* ● *v.* (*past* and *past part.* **spoilt** /spɔɪlt/ or **spoiled**) **1** *tr.* **a** damage; diminish the value of (*was spoilt by the rain; will spoil all the fun*). **b** reduce a person's enjoyment etc. of (*the news spoiled his dinner*). **2** *tr.* injure the character of (esp. a child, pet, etc.) by excessive indulgence. **3** *intr.* **a** (of food) go bad, decay; become unfit for eating. **b** (usu. in *neg.*) (of a joke, secret, etc.) become stale through long keeping. **4** *tr.* render (a ballot paper) invalid by improper marking. **5** *tr.* (foll. by *of*) *archaic* or *literary* plunder or deprive (a person of a thing) by force or stealth (*spoiled him of all his possessions*). ● *n.* **1** (usu. in *pl.*) **a** plunder taken from an enemy in war, or seized by force. **b** esp. *joc.* profit or advantages gained by succeeding to public office, high position, etc. **2** earth etc. thrown up in excavating, dredging, etc. □ **be**

spoiling for aggressively seek (a fight etc.). **spoils system** *US* the practice of a successful political party replacing existing holders of government or public positions with its own supporters. **spoilt for choice** having so many choices that it is difficult to choose. [ME f. OF *espoillier, espoille* f. L *spoliare* f. *spolium* spoil, plunder, or f. DESPOIL]

spoilage /ˈspɔɪlɪdʒ/ *n.* **1** paper spoilt in printing. **2** the spoiling of food etc. by decay.

spoiler /ˈspɔɪlə(r)/ *n.* **1 a** a person or thing that spoils something. **b** a news story published to spoil the impact of another story published elsewhere. **2 a** a device on an aircraft to retard its speed by interrupting the airflow. **b** a similar device on a vehicle to improve its road-holding at speed. **3** an electronic device for preventing unauthorized copying of sound recordings, using a disruptive signal inaudible on the original.

spoilsman /ˈspɔɪlzmən/ *n.* (*pl.* **-men**) *US esp. Polit.* **1** an advocate of the spoils system. **2** a person who seeks to profit by it.

spoilsport /ˈspɔɪlspɔːt/ *n.* a person who spoils others' pleasure or enjoyment.

spoilt *past* and *past part.* of SPOIL.

Spokane /spəʊˈkeɪn/ a city in eastern Washington, situated on the falls of the Spokane river, near the border with Idaho; pop. (1990) 177,200.

spoke[1] /spəʊk/ *n. & v.* ● *n.* **1** each of the bars running from the hub to the rim of a wheel. **2** a rung of a ladder. **3** each radial handle of the wheel of a ship etc. ● *v.tr.* **1** provide with spokes. **2** obstruct (a wheel etc.) by thrusting a spoke in. □ **put a spoke in a person's wheel** *Brit.* thwart or hinder a person. □ **spoked** *adj.* **spokewise** *adv.* [OE *spáca* f. WG]

spoke[2] *past* of SPEAK.

spoken /ˈspəʊkən/ *past part.* of SPEAK. ● *adj.* (in *comb.*) speaking in a specified way (*smooth-spoken; well-spoken*). □ **spoken for** claimed, requisitioned (*this seat is spoken for*).

spokeshave /ˈspəʊkʃeɪv/ *n.* a blade set transversely between two handles, used for shaping spokes and other esp. curved work where an ordinary plane is not suitable.

spokesman /ˈspəʊksmən/ *n.* (*pl.* **-men**; *fem.* **spokeswoman**, *pl.* **-women**) **1** a person who speaks on behalf of others, esp. in the course of public relations. **2** a person deputed to express the views of a group etc. [irreg. f. SPOKE[2] after *craftsman* etc.]

spokesperson /ˈspəʊks,pɜːs(ə)n/ *n.* (*pl.* **-persons** or **-people**) a spokesman or spokeswoman.

Spoleto /spəˈleɪtəʊ/ a town in Umbria, in central Italy; pop. (1990) 38,030. The site of a settlement from the 7th century BC, it became a province of Rome in 241 BC. Taken by the Lombards in about 570 AD, it was one of Italy's principal cities between the 6th and 8th centuries.

spoliation /ˌspəʊlɪˈeɪʃ(ə)n/ *n.* **1 a** plunder or pillage, esp. of neutral vessels in war. **b** extortion. **2** *Eccl.* the taking of the fruits of a benefice under a pretended title etc. **3** *Law* the destruction, mutilation, or alteration, of a document to prevent its being used as evidence. □ **spoliator** /ˈspəʊlɪ,eɪtə(r)/ *n.* **spoliatory** /-lɪətərɪ/ *adj.* [ME f. L *spoliatio* (as SPOIL)]

spondaic /spɒnˈdeɪk/ *adj. Prosody* **1** of or concerning spondees. **2** (of a hexameter) having a spondee as a fifth foot. [F *spondaïque* or LL *spondaicus* = LL *spondiacus* f. Gk *spondeiakos* (as SPONDEE)]

spondee /ˈspɒndiː/ *n. Prosody* a foot consisting of two long (or stressed) syllables. [ME f. OF *spondee* or L *spondeus* f. Gk *spondeios* (*pous* foot) f. *spondē* libation, as being characteristic of music accompanying libations]

spondulicks /spɒnˈdjuːlɪks/ *n.pl. sl.* money. [19th c.: orig. unkn.]

spondylitis /ˌspɒndɪˈlaɪtɪs/ *n. Med.* inflammation of the vertebrae. [L *spondylus* vertebra f. Gk *spondulos* + -ITIS]

sponge /spʌndʒ/ *n. & v.* ● *n.* **1** a simple sessile aquatic animal of the phylum Porifera, with a porous baglike body and a rigid or elastic internal skeleton. **2 a** the skeleton of a sponge, esp. the soft light elastic absorbent kind used in bathing, cleansing surfaces, etc. **b** a piece of porous rubber or plastic etc. used similarly. **c** a piece of sponge or similar material inserted in the vagina as a contraceptive. **3** a thing of spongelike absorbency or consistency, e.g. a sponge pudding, cake, porous metal, etc. (*lemon sponge*). **4** = SPONGER. **5** *colloq.* a person who drinks heavily. **6** cleansing with or as with a sponge (*had a quick sponge this morning*). ● *v.* **1** *tr.* wipe or cleanse with a sponge. **2** *tr.* (also *absol.*; often foll. by *down, over*) sluice water over (the body, a car, etc.). **3** *tr.* (often foll. by *out, away*) wipe off or efface (writing, a memory, etc.) with or as with a sponge. **4** *tr.* (often foll. by *up*) absorb with or as with

a sponge. **5** *intr.* (often foll. by *on, off*) live as a parasite; be meanly dependent upon (another person). **6** *tr.* obtain (drink etc.) by sponging. **7** *intr.* gather sponges. **8** *tr.* apply paint with a sponge to (walls, furniture, etc.). □ **sponge bag** a waterproof bag for toilet articles. **sponge cake** a very light cake with a spongelike consistency. **sponge cloth** **1** soft, lightly woven cloth with a slightly wrinkled surface. **2** a thin spongy material used for cleaning. **sponge pudding** *Brit.* a steamed or baked pudding of fat, flour, and eggs with a usu. specified flavour. **sponge rubber** liquid rubber latex processed into a spongelike substance. **sponge tree** a spiny tropical acacia, *Acacia farnesiana*, with globose heads of fragrant yellow flowers yielding a perfume (cf. OPOPANAX 3). □ **spongeable** *adj.* **spongelike** *adj.* [OE f. L *spongia* f. Gk *spoggia, spoggos*]

sponger /ˈspʌndʒə(r)/ *n.* a person who contrives to live at another's expense.

spongiform /ˈspʌndʒɪ,fɔːm/ *adj.* spongelike; spongy.

spongy /ˈspʌndʒɪ/ *adj.* (**spongier, spongiest**) **1** like a sponge, esp. in being porous, compressible, elastic, or absorbent. **2** (of metal) finely divided and loosely coherent. □ **spongily** *adv.* **sponginess** *n.*

sponsion /ˈspɒnʃ(ə)n/ *n.* **1** being a surety for another. **2** a pledge or promise made on behalf of the state by an agent not authorized to do so. [L *sponsio* f. *spondere spons-* promise solemnly]

sponson /ˈspɒns(ə)n/ *n.* **1** a projection from the side of a warship or tank to enable a gun to be trained forward and aft. **2** a short subsidiary wing to stabilize a seaplane. **3** a triangular platform supporting the wheel on a paddle-steamer. [19th c.: orig. unkn.]

sponsor /ˈspɒnsə(r)/ *n. & v.* ● *n.* **1** a person who supports an activity done for charity by pledging money in advance. **2 a** a person or organization that promotes or supports an artistic or sporting activity etc. **b** esp. *US* a business organization that promotes a broadcast programme in return for advertising time. **3** an organization lending support to an election candidate. **4** a person who introduces a proposal for legislation. **5 a** a godparent at baptism. **b** esp. *RC Ch.* a person who presents a candidate for confirmation. **6** a person who makes himself or herself responsible for another. ● *v.tr.* be a sponsor for. □ **sponsorship** *n.* **sponsorial** /spɒnˈsɔːrɪəl/ *adj.* [L (as SPONSION)]

spontaneous /spɒnˈteɪnɪəs/ *adj.* **1** acting or done or occurring without external cause. **2** voluntary, without external incitement (*made a spontaneous offer of his services*). **3** *Biol.* (of structural changes in plants and muscular activity esp. in young animals) instinctive, automatic, prompted by no motive. **4** (of bodily movement, literary style, etc.) gracefully natural and unconstrained. **5** (of sudden movement etc.) involuntary, not due to conscious volition. **6** growing naturally without cultivation. □ **spontaneous combustion** the ignition of a mineral or vegetable substance (e.g. a heap of rags soaked with oil, a mass of wet coal) from heat engendered within itself, usu. by rapid oxidation. **spontaneous generation** *hist.* the supposed production of living from non-living matter as inferred from the appearance of life (due in fact to bacteria etc.) in some infusions; abiogenesis. **spontaneous suggestion** suggestion from association of ideas without conscious volition. □ **spontaneously** *adv.* **spontaneousness** *n.* **spontaneity** /ˌspɒntəˈniːɪtɪ, -ˈneɪtɪ/ *n.* [LL *spontaneus* f. *sponte* of one's own accord]

spoof /spuːf/ *n. & v. colloq.* ● *n.* **1** a parody. **2** a hoax or swindle. ● *v.tr.* **1** parody. **2** hoax, swindle. □ **spoofer** *n.* **spoofery** *n.* [invented by Arthur Roberts, English comedian (1852–1933)]

spook /spuːk/ *n. & v.* **1** *colloq.* a ghost. **2** *US sl.* a spy. ● *v. N. Amer. sl.* **1** *tr.* frighten, unnerve, alarm. **2** *intr.* take fright, become alarmed. [Du., = MLG *spōk*, of unkn. orig.]

spooky /ˈspuːkɪ/ *adj.* (**spookier, spookiest**) **1** *colloq.* ghostly, eerie. **2** *N. Amer. sl.* nervous; easily frightened. **3** *US sl.* of spies or espionage. □ **spookily** *adv.* **spookiness** *n.*

spool /spuːl/ *n. & v.* ● *n.* **1 a** a reel for winding magnetic tape, photographic film, etc., on. **b** a reel for winding yarn or *US* thread on. **c** a quantity of tape, yarn, etc., wound on a spool. **2** the revolving cylinder of an angler's reel. ● *v.tr.* wind on a spool. [ME f. OF *espole* or f. MLG *spôle*, MDu. *spoele*, OHG *spuolo*, of unkn. orig.]

spoon /spuːn/ *n. & v.* ● *n.* **1 a** a utensil consisting of an oval or round bowl and a handle for conveying food (esp. liquid) to the mouth, for stirring, etc. **b** a spoonful, esp. of sugar. **c** (in *pl.*) *Mus.* a pair of spoons held in the hand and beaten together rhythmically. **2** a spoon-shaped thing, esp.: **a** (in full **spoon-bait**) a bright revolving piece of metal used as a lure in fishing. **b** an oar with a broad curved blade. **c** a wooden-headed golf club. **3** *colloq.* **a** a silly or demonstratively fond lover. **b** a simpleton. ● *v.* **1** *tr.* (often foll. by *up, out*) take (liquid etc.) with a spoon. **2** *tr.* hit (a ball) feebly upwards. **3** *colloq.* **a** *intr.* behave in an amorous way, esp. foolishly. **b** *tr. archaic* woo in a silly or sentimental way. **4** *intr.* fish with a spoon-bait. □ **born with a silver spoon in one's mouth** born in affluence. **spoon-bread** *US* soft maize bread. □ **spooner** *n.* **spoonful** *n.* (*pl.* **-fuls**) [OE *spōn* chip of wood f. Gmc]

spoonbill /ˈspuːnbɪl/ *n.* **1** a large mainly white wading bird of the family Threskiornithidae, allied to the ibises, having a long bill with a broad flat tip. **2** a shoveler duck.

spoonerism /ˈspuːnəˌrɪz(ə)m/ *n.* a transposition, usu. accidental, of the initial letters etc. of two or more words, e.g. *you have hissed the mystery lectures*. [Revd William Archibald *Spooner*, English scholar (1844–1930), reputed to make such errors in speaking]

spoonfeed /ˈspuːnfiːd/ *v.tr.* (*past and past part.* **-fed** /-fed/) **1** feed (a baby etc.) with a spoon. **2** provide help, information, etc., to (a person) without requiring any effort on the recipient's part. **3** artificially encourage (an industry) by subsidies or import duties.

spoony /ˈspuːnɪ/ *adj. & n. colloq. archaic* ● *adj.* (**spoonier, spooniest**) **1** (often foll. by *on*) sentimental, amorous. **2** foolish, silly. ● *n.* (*pl.* **-ies**) a simpleton. □ **spoonily** *adv.* **spooniness** *n.*

spoor /spʊə(r), spɔː(r)/ *n. & v.* ● *n.* the track or scent of an animal. ● *v.tr. intr.* follow by the spoor. □ **spoorer** *n.* [Afrik. f. MDu. *spo(o)r* f. Gmc]

Sporades /ˈspɒrəˌdiːz/ two groups of Greek islands in the Aegean Sea. The *Northern Sporades*, which lie close to the east coast of mainland Greece, include the islands of Euboea, Skiros, Skiathos, and Skopelos. The *Southern Sporades*, situated off the west coast of Turkey, include Rhodes and the other islands of the Dodecanese.

sporadic /spəˈrædɪk/ *adj.* occurring only here and there or occasionally; separate; scattered. □ **sporadically** *adv.* [med.L *sporadicus* f. Gk *sporadikos* f. *sporas -ados* scattered: cf. *speirō* to sow]

sporangium /spəˈrændʒɪəm/ *n.* (*pl.* **sporangia** /-dʒɪə/) *Bot.* a receptacle in which spores are formed. □ **sporangial** *adj.* [SPORO- + Gk *aggeion* vessel]

spore /spɔː(r)/ *n.* **1** a specialized reproductive cell of many plants and micro-organisms. **2** these collectively. [mod.L *spora* f. Gk *spora* sowing, seed f. *speirō* sow]

sporo- /ˈspɔːrəʊ/ *comb. form Biol.* a spore. [Gk *spora* (as SPORE)]

sporogenesis /ˌspɔːrəˈdʒenɪsɪs/ *n. Biol.* the process of spore formation.

sporogenous /spəˈrɒdʒɪnəs/ *adj. Biol.* producing spores.

sporophyll /ˈspɔːrəˌfɪl/ *n. Bot.* a modified leaf bearing sporangia, esp. in the terminal spike of a clubmoss. [SPORO- + Gk *phullon* leaf]

sporophyte /ˈspɔːrəˌfaɪt/ *n. Bot.* the spore-producing form of a plant that has alternation of generations between this and the sexual form (cf. GAMETOPHYTE). □ **sporophytic** /ˌspɔːrəˈfɪtɪk/ *adj.*

sporozoite /ˌspɔːrəˈzəʊaɪt, ˌspɒr-/ *n. Biol. & Med.* a small motile infective stage in the life cycle of some protozoans (e.g. the malaria parasite). [SPORO- + Gk *zoion* animal+ -ITE[1]]

sporran /ˈspɒrən/ *n.* a pouch, usu. of leather or sealskin covered with fur etc., worn by a Highlander in front of the kilt. [Gael. *sporan* f. med.L *bursa* PURSE]

sport /spɔːt/ *n. & v.* ● *n.* **1 a** a game or competitive activity, esp. an outdoor one involving physical exertion, e.g. cricket, football, racing, hunting. **b** such activities collectively (*the world of sport*). **2** (in *pl.*) *Brit.* **a** a meeting for competing in sports, esp. athletics (*school sports*). **b** athletics. **3** amusement, diversion, fun. **4** *colloq.* **a** a fair or generous person. **b** a person behaving in a specified way, esp. regarding games, rules, etc. (*a bad sport at tennis*). **c** *Austral.* a form of address, esp. between males. **5** *US* a playboy. **5** *Biol.* an animal or plant showing an abnormal or striking variation from the parental type. **6** a plaything or butt (*was the sport of Fortune*). ● *v.* **1** *intr.* **a** divert oneself, take part in a pastime. **b** frolic, gambol. **2** *tr.* wear, exhibit, or produce, esp. ostentatiously (*sported a gold tie-pin*). **3** *intr. Biol.* become or produce a sport. □ **have good sport** be successful in shooting, fishing, etc. **in sport** jestingly. **make sport of** make fun of, ridicule. **the sport of kings** horse-racing (less often war, hunting, or surfing). **sports car** a usu. open, low-built fast car. **sports coat** (or **jacket**) a man's jacket for informal wear. **sports writer** a person who writes (esp. as a journalist) on sports. □ **sporter** *n.* [ME f. DISPORT]

sporting /ˈspɔːtɪŋ/ *adj.* **1** interested in sport (*a sporting man*). **2** sportsmanlike, generous (*a sporting offer*). **3** concerned in sport (*a sporting dog; sporting news*). □ **a sporting chance** some possibility of success. **sporting house** *US* a brothel. □ **sportingly** *adv.*

sportive /ˈspɔːtɪv/ *adj.* playful. □ **sportively** *adv.* **sportiveness** *n.*

sportscast /ˈspɔːtskɑːst/ *n. N. Amer.* a broadcast of a sports event or information about sport. □ **sportscaster** *n.*

sportsman /ˈspɔːtsmən/ *n.* (*pl.* **-men**; *fem.* **sportswoman**, *pl.* **-women**) **1** a person who takes part in much sport, esp. professionally. **2** a person who behaves fairly and generously. □ **sportsmanlike** *adj.* **sportsmanly** *adj.* **sportsmanship** *n.*

sportsperson /ˈspɔːtsˌpɜːs(ə)n/ *n.* (*pl.* **-people**) a sportsman or sportswoman.

sportswear /ˈspɔːtsweə(r)/ *n.* clothes worn for sport or for casual outdoor use.

sporty /ˈspɔːtɪ/ *adj.* (**sportier, sportiest**) *colloq.* **1 a** fond of sport. **b** (esp. of clothes) suitable for wearing for sport or for casual outdoor use. **2** rakish, showy. □ **sportily** *adv.* **sportiness** *n.*

sporule /ˈspɔːruːl/ *n. Biol.* a small spore. □ **sporular** /ˈspɒrjʊlə(r)/ *adj.* [F *sporule* or mod.L *sporula* (as SPORE)]

spot /spɒt/ *n. & v.* ● *n.* **1 a** a small part of the surface of a thing distinguished by colour, texture, etc., usu. round or less elongated than a streak or stripe (*a blue tie with pink spots*). **b** a small mark or stain. **c** a pimple. **d** a small circle or other shape used in various numbers to distinguish faces of dice, playing cards in a suit, etc. **e** a moral blemish or stain (*without a spot on his reputation*). **2 a** a particular place; a definite locality (*dropped it on this precise spot; the spot where William III landed*). **b** a place used for a particular activity (often in *comb.*: *nightspot*). **c** (in full **penalty spot**) *Football* the place from which a penalty kick is taken. **3** a particular part of one's body or aspect of one's character. **4 a** *colloq.* one's esp. regular position in an organization, programme of events, etc. **b** a place or position in a performance or show (*did the spot before the interval*). **5** *Brit.* a *colloq.* a small quantity of anything (*a spot of lunch; a spot of bother*). **b** a drop (*a spot of rain*). **c** *colloq.* a drink. **6** = SPOTLIGHT *n.* **7** *colloq.* an awkward or difficult situation (esp. in *in a* (*tight* etc.) *spot*). **8** (usu. *attrib.*) money paid or goods delivered immediately after a sale (*spot cash; spot silver*). **9** *Billiards* etc. **a** a small round black patch to mark the position where a ball is placed at certain times. **b** (in full **spot-ball**) the white ball distinguished from the other by two black spots. ● *v.* (**spotted, spotting**) **1** *tr.* **a** *colloq.* single out beforehand (the winner of a race etc.). **b** *colloq.* recognize the identity, nationality, etc., of (*spotted him at once as the murderer*). **c** watch for and take note of (trains, talent, etc.). **d** *colloq.* catch sight of. **e** *Mil.* locate (an enemy's position), esp. from the air. **2 a** *tr. & intr.* mark or become marked with spots. **b** *tr.* stain, soil (a person's character etc.). **3** *intr.* make spots, rain slightly (*it was spotting with rain*). **4** *tr. Billiards* place (a ball) on a spot. □ **on the spot 1** at the scene of an action or event. **2** *colloq.* in a position such that response or action is required. **3** without delay or change of place, then and there. **4** (of a person) wide awake, equal to the situation, in good form at a game etc. **put on the spot 1** *colloq.* force to make a difficult decision, answer an awkward question, etc. **2** *US sl.* decide to murder. **running on the spot** raising the feet alternately as in running but without moving forwards. **spot check** a test made on the spot or on a randomly selected subject. **spot height 1** the altitude of a point. **2** a figure on a map showing this. **spot on** *Brit. colloq. adj.* precise; on target. ● *adv.* precisely. **spot weld** a weld made in spot welding. **spot-weld** *v.tr.* join by spot welding. **spot welder** a person or device that spot-welds. **spot welding** welding two surfaces together in a series of discrete points. [ME, perh. f. MDu. *spotte*, LG *spot*, ON *spotti* small piece]

spotless /ˈspɒtlɪs/ *adj.* immaculate; absolutely clean or pure. □ **spotlessly** *adv.* **spotlessness** *n.*

spotlight /ˈspɒtlaɪt/ *n. & v.* ● *n.* **1** a beam of light directed on a small area, esp. on a particular part of a theatre stage or of the road in front of a vehicle. **2** a lamp projecting this. **3** full attention or publicity. ● *v.tr.* (*past and past part.* **-lighted** or **-lit** /-lɪt/) **1** direct a spotlight on. **2** make conspicuous, draw attention to.

spotted /ˈspɒtɪd/ *adj.* marked or decorated with spots. □ **spotted dick** *Brit.* a suet pudding containing currants. **spotted dog** a Dalmatian dog. **spotted fever 1** cerebrospinal meningitis. **2** typhus. **3** (in full **Rocky Mountain spotted fever**) a rickettsial disease transmitted by ticks. □ **spottedness** *n.*

spotter /ˈspɒtə(r)/ *n.* **1** (often in *comb.*) a person who spots people or things (*train-spotter*). **2** an aviator or aircraft employed in locating enemy positions etc.

spotty /ˈspɒtɪ/ *adj.* (**spottier, spottiest**) **1** marked with spots. **2** patchy, irregular. □ **spottily** *adv.* **spottiness** *n.*

spouse /spaʊz, spaʊs/ *n.* a husband or wife. [ME *spūs(e)* f. OF *spus(e)* (masc.), *sp(o)use* (fem.), var. of *espous(e)* f. L *sponsus sponsa* past part. of *spondere* betroth]

spout /spaʊt/ n. & v. ● n. **1 a** a projecting tube or lip through which a liquid etc. is poured from a teapot, kettle, jug, etc., or issues from a fountain, pump, etc. **b** a sloping trough down which a thing may be shot into a receptacle. **c** hist. a lift serving a pawnbroker's storeroom. **2** a jet or column of liquid, grain, etc. **3** (in full **spout-hole**) a whale's blowhole. ● v.tr. & intr. **1** discharge or issue forcibly in a jet. **2** utter (verses etc.) or speak in a declamatory manner, speechify. □ **up the spout** sl. **1** useless, ruined, hopeless. **2** pawned. **3** pregnant. □ **spouter** n. **spoutless** adj. [ME f. MDu. spouten, orig. imit.]

SPQR abbr. **1** hist. the Senate and people of Rome. **2** colloq. small profits and quick returns. [sense 1 f. L Senatus Populusque Romanus]

Spr. abbr. (in the UK) Sapper.

sprag /spræg/ n. **1** a thick piece of wood or similar device used as a brake. **2** a prop in a coal mine. [19th c.: orig. unkn.]

sprain /spreɪn/ v. & n. ● v.tr. wrench (an ankle, wrist, etc.) violently so as to cause pain and swelling but not dislocation. ● n. **1** such a wrench. **2** the resulting inflammation and swelling. [17th c.: orig. unkn.]

spraint /spreɪnt/ n. (also in pl.) otter droppings. [OF espreintes f. espraindre squeeze out, rel. to EXPRESS¹]

sprang past of SPRING.

sprat /spræt/ n. & v. ● n. **1** a small European marine fish, Sprattus sprattus, of the herring family, much used as food. **2** a similar fish, e.g. a sand eel or a young herring. ● v.intr. (**spratted**, **spratting**) fish for sprats. □ **a sprat to catch a mackerel** a small risk to gain much. □ **spratter** n. **spratting** n. [OE sprot]

Spratly Islands /ˈsprætlɪ/ a group of small islands and coral reefs in the South China Sea, between Vietnam and Borneo. Dispersed over a distance of some 965 km (600 miles), the islands are variously claimed by China, Taiwan, Vietnam, the Philippines, and Malaysia.

sprauncy /ˈsprɔːnsɪ/ adj. (**sprauncier**, **praunciest**) Brit. sl. smart or showy. [20th c.: perh. rel. to dial. sprouncey cheerful]

sprawl /sprɔːl/ v. & n. ● v. **1 a** intr. sit or lie or fall with limbs flung out or in an ungainly way. **b** tr. spread (one's limbs) in this way. **2** intr. (of handwriting, a plant, a town, etc.) be of irregular or straggling form. ● n. **1** a sprawling movement or attitude. **2** a straggling group or mass. **3** the straggling expansion of an urban or industrial area. □ **sprawlingly** adv. [OE spreawlian]

spray¹ /spreɪ/ n. & v. ● n. **1** water or other liquid flying in small drops from the force of the wind, the dashing of waves, or the action of an atomizer etc. **2** a liquid preparation to be applied in this form with an atomizer etc., esp. for medical purposes. **3** an instrument or apparatus for such application. ● v.tr. (also absol.) **1** throw (liquid) in the form of spray. **2** sprinkle (an object) with small drops or particles, esp. (a plant) with an insecticide. **3** (absol.) (of a male animal, esp. a cat) mark its environment with the smell of its urine, as an attraction to females. □ **spray-dry** (**-dries**, **-dried**) dry (milk etc.) by spraying into hot air etc. **spray-gun** a gunlike device for spraying paint etc. **spray-paint** paint (a surface) by means of a spray. □ **sprayable** adj. **sprayer** n. [earlier spry, perh. rel. to MDu. spra(e)yen, MHG spræjen sprinkle]

spray² /spreɪ/ n. **1** a sprig of flowers or leaves, or a branch of a tree with branchlets or flowers, esp. a slender or graceful one. **2** a bunch of flowers decoratively arranged. **3** an ornament in a similar form (a spray of diamonds). □ **sprayey** /ˈspreɪɪ/ adj. [ME f. OE spræg (unrecorded)]

spread /spred/ v. & n. ● v. (past and past part. **spread**) **1** tr. (often foll. by out) **a** open or extend the surface of. **b** cause to cover a larger surface (spread butter on bread). **c** display to the eye or the mind (the view was spread out before us). **2** intr. (often foll. by out) have a wide or specified or increasing extent (on every side spread a vast desert; spreading trees). **3** intr. & tr. become or make widely known, felt, etc. (rumours are spreading; spread a little happiness). **4** tr. **a** cover the surface of (spread the wall with paint; a meadow spread with daisies). **b** lay (a table). ● n. **1** the act or an instance of spreading. **2** capability of expanding (has a large spread). **3** diffusion (spread of learning). **4** breadth, compass (arches of equal spread). **5** an aircraft's wing-span. **6** increased bodily girth (middle-aged spread). **7** the difference between two rates, prices, etc. **8** colloq. an elaborate meal. **9** a sweet or savoury paste for spreading on bread etc. **10** a bedspread. **11** printed matter spread across two facing pages or across more than one column. **12** US a ranch with extensive land. □ **spread eagle 1** a representation of an eagle with legs and wings extended as an emblem. **2** hist. a person secured with arms and legs spread out, esp. to be flogged. **spread-eagle** v.tr. (usu. as **spread-eagled** adj.) **1** place (a person) in this position. **2** defeat utterly. **3** spread out. ● adj. US bombastic, esp. noisily patriotic. **spread oneself** be lavish or discursive. **spread one's wings** see WING. □ **spreadable** adj. **spreader** n. [OE -sprædan f. WG]

spreadsheet /ˈspredʃiːt/ n. Computing a program allowing manipulation and flexible retrieval of esp. tabulated numerical data.

Sprechgesang /ˈʃprexɡəˌzaːŋ/ n. Mus. a style of dramatic vocalization between speech and song. [G, lit. 'speech song']

spree /spriː/ n. & v. colloq. ● n. **1** a lively extravagant outing (shopping spree). **2** a bout of fun or drinking etc. ● v.intr. (**sprees**, **spreed**) have a spree. □ **on the spree** engaged in a spree. [19th c.: orig. unkn.]

sprig¹ /sprɪɡ/ n. & v. ● n. **1** a small branch or shoot. **2** an ornament resembling this, esp. on fabric. **3** usu. derog. a youth or young man (a sprig of the nobility). ● v.tr. (**sprigged**, **sprigging**) **1** ornament with sprigs (a dress of sprigged muslin). **2** (usu. as **sprigging** n.) decorate (ceramic ware) with ornaments in applied relief. □ **spriggy** adj. [ME f. or rel. to LG sprick]

sprig² /sprɪɡ/ n. a small tapering headless tack. [ME: orig. unkn.]

sprightly /ˈspraɪtlɪ/ adj. (**sprightlier**, **sprightliest**) vivacious, lively, brisk. □ **sprightliness** n. [spright var. of SPRITE + -LY¹]

spring /sprɪŋ/ v. & n. ● v. (past **sprang** /spræŋ/ or US **sprung** /sprʌŋ/; past part. **sprung**) **1** intr. jump; move rapidly or suddenly (sprang from his seat; sprang through the gap; sprang to their assistance). **2** intr. move rapidly as from a constrained position or by the action of a spring (the branch sprang back; the door sprang to). **3** intr. (usu. foll. by from) originate or arise (springs from an old family; their actions spring from a false conviction). **4** intr. (usu. foll. by up) come into being; appear, esp. suddenly (a breeze sprang up; the belief has sprung up). **5** tr. cause to act suddenly, esp. by means of a spring (spring a trap). **6** tr. (often foll. by on) produce or develop or make known suddenly or unexpectedly (has sprung a new theory; loves to spring surprises). **7** tr. sl. contrive the escape or release of. **8** tr. rouse (game) from earth or covert. **9 a** intr. become warped or split. **b** tr. split, crack (wood or a wooden implement). **10** tr. (usu. as **sprung** adj.) provide (a motor vehicle etc.) with springs. **11 a** tr. colloq. spend (money). **b** intr. (usu. foll. by for) N. Amer. & Austral. sl. pay for a treat. **12** tr. cause (a mine) to explode. ● n. **1** a jump (took a spring; rose with a spring). **2** a backward movement from a constrained position; a recoil, e.g. of a bow. **3** elasticity; ability to spring back strongly (a mattress with plenty of spring). **4** a resilient device usu. of bent or coiled metal used esp. to drive clockwork or for cushioning in furniture or vehicles. **5 a** the season in which vegetation begins to appear, the first season of the year, in the northern hemisphere from March to May and in the southern hemisphere from September to November. **b** Astron. the period from the vernal equinox to the summer solstice. **c** (often foll. by of) the early stage of life etc. **d** = spring tide. **6** a place where water, oil, etc., wells up from the earth; the basin or flow so formed (hot springs; mineral springs). **7** the motive for or occasion of an action, custom, etc. (the springs of human action). **8** sl. an escape or release from prison. **9** the upward curve of a beam etc. from a horizontal line. **10** the splitting or yielding of a plank etc. under strain. □ **spring balance** a balance that measures weight by the tension of a spring. **spring bed** a bed with a spring mattress. **spring chicken 1** a young fowl for eating (orig. available only in spring). **2** (esp. with neg.) a young person (she's no spring chicken). **spring-clean** n. a thorough cleaning of a house or room, esp. in spring. ● v.tr. clean (a house or room) in this way. **spring fever** a restless or lethargic feeling sometimes associated with spring. **spring greens** the leaves of young cabbage plants of a variety that does not develop a heart. **spring a leak** develop a leak (orig. Naut., from timbers springing out of position). **spring-loaded** containing a compressed or stretched spring pressing one part against another. **spring mattress** a mattress containing or consisting of springs. **spring onion** an onion taken from the ground before the bulb has formed, and eaten raw in salad. **spring roll** a Chinese snack consisting of a pancake filled with vegetables etc. and fried. **spring tide** a tide just after new and full moon when there is the greatest difference between high and low water. **spring water** water from a spring, as opposed to river or rain water. □ **springless** adj. **springlet** /ˈsprɪŋlɪt/ n. **springlike** adj. [OE springan f. Gmc]

springboard /ˈsprɪŋbɔːd/ n. **1** a springy board giving impetus in leaping, diving, etc. **2** a source of impetus in any activity. **3** N. Amer. & Austral. a platform inserted in the side of a tree, on which a lumberjack stands to chop at some height from the ground.

springbok /ˈsprɪŋbɒk/ n. **1** a southern African gazelle, Antidorcas marsupialis, with the habit of running with high springing leaps when alarmed. **2** (**Springbok**) a South African, esp. one who has played for South Africa in international sporting competitions. [Afrik. f. Du. springen SPRING + bok antelope]

springe /sprɪndʒ/ n. a noose or snare for catching small game. [ME, rel. to obs. sprenge, and SPRING]

springer /ˈsprɪŋə(r)/ n. **1** a person or thing that springs. **2** a small breed of spaniel originally used to spring game. **3** *Archit.* **a** the part of an arch where the curve begins. **b** the lowest stone of this. **c** the bottom stone of the coping of a gable. **d** a rib of a groined roof or vault. **4** a springbok.

Springfield /ˈsprɪŋfiːld/ **1** the state capital of Illinois; pop. (1990) 105,230. First settled in 1819, it became state capital in 1837. It was the home and burial place of Abraham Lincoln. **2** a city in SW Massachusetts, on the Connecticut river; pop. (1990) 156,980. It was first settled in 1636. **3** a city in SW Missouri, on the northern edge of the Ozark Mountains; pop. (1990) 140,490.

Springsteen /ˈsprɪŋstiːn/, Bruce (b.1949), American rock singer, songwriter, and guitarist. He is noted for his songs about working-class life in the US and for his energetic stage performances. Major albums include *Born to Run* (1975) and *Born in the USA* (1984).

springtail /ˈsprɪŋteɪl/ n. a minute primitive wingless insect of the order Collembola, abundant in the soil, and leaping by means of a springlike posterior organ.

springtide /ˈsprɪŋtaɪd/ n. poet. = SPRINGTIME.

springtime /ˈsprɪŋtaɪm/ n. **1** the season of spring. **2** a time compared to this.

springy /ˈsprɪŋɪ/ adj. (**springier, springiest**) **1** springing back quickly when squeezed or stretched, elastic. **2** (of movements) as of a springy substance. □ **springily** adv. **springiness** n.

sprinkle /ˈsprɪŋk(ə)l/ v. & n. ● v.tr. **1** scatter (liquid, ashes, crumbs, etc.) in small drops or particles. **2** (often foll. by *with*) subject (the ground or an object) to sprinkling with liquid etc. **3** (of liquid etc.) fall on in this way. **4** distribute in small amounts. ● n. (usu. foll. by *of*) **1** a light shower. **2** = SPRINKLING. [ME, perh. f. MDu. *sprenkelen*]

sprinkler /ˈsprɪŋklə(r)/ n. a person or thing that sprinkles, esp. a device for sprinkling water on a lawn or to extinguish fires.

sprinkling /ˈsprɪŋklɪŋ/ n. (usu. foll. by *of*) a small thinly distributed number or amount.

sprint /sprɪnt/ v. & n. ● v. **1** intr. run a short distance at full speed. **2** tr. run (a specified distance) in this way. ● n. **1 a** such a run. **b** *Sport* a running-race over a distance of 400 metres or less. **2** a similar short spell of maximum effort in cycling, swimming, motor-racing, etc. [ON *sprinta* (unrecorded), of unkn. orig.]

sprinter /ˈsprɪntə(r)/ n. **1** an athlete who specializes in short distance races. **2** a vehicle, esp. a train, designed for rapid travel over short distances.

sprit /sprɪt/ n. a small spar reaching diagonally from the mast to the upper outer corner of the sail. [OE *sprēot* pole, rel. to SPROUT]

sprite /spraɪt/ n. an elf, fairy, or goblin. [ME f. *sprit* var. of SPIRIT]

spritsail /ˈsprɪtseɪl, *Naut.* -s(ə)l/ n. **1** a sail extended by a sprit. **2** hist. a sail extended by a yard set under the bowsprit.

spritz /sprɪts/ v. & n. N. Amer. ● v.tr. sprinkle, squirt, or spray. ● n. the act or an instance of spritzing. [G *spritzen* to squirt]

spritzer /ˈsprɪtsə(r)/ n. a mixture of wine and soda water. [G *Spritzer* a splash]

sprocket /ˈsprɒkɪt/ n. **1** each of several teeth on a wheel engaging with links of a chain, e.g. on a bicycle, or with holes in film or tape or paper. **2** (also **sprocket-wheel**) a wheel with sprockets. [16th c.: orig. unkn.]

sprog /sprɒg/ n. sl. a child; a baby. [orig. services' sl., = new recruit: perh. f. obs. *sprag* lively young man]

sprout /spraʊt/ v. & n. ● v. **1** tr. put forth, produce (shoots, hair, etc.) (*has sprouted a moustache*). **2** intr. begin to grow, put forth shoots. **3** intr. spring up, grow to a height. ● n. **1** a shoot of a plant. **2** = BRUSSELS SPROUT. [OE *sprūtan* (unrecorded) f. WG]

spruce[1] /spruːs/ adj. & v. ● adj. neat in dress and appearance; trim, smart. ● v.tr. & intr. (also *refl.*; usu. foll. by *up*) make or become smart. □ **sprucely** adv. **spruceness** n. [perh. f. SPRUCE[2] in obs. sense 'Prussian', in the collocation *spruce (leather) jerkin*]

spruce[2] /spruːs/ n. **1** a coniferous tree of the genus *Picea*, with dense foliage and growing in a distinctive conical shape. **2** the wood of this tree used as timber. □ **spruce beer** a fermented beverage using spruce twigs and needles as flavouring. [alt. of obs. *Pruce* Prussia: cf. PRUSSIAN]

spruce[3] /spruːs/ v. Brit. sl. **1** tr. deceive. **2** intr. lie, practise deception. **3** intr. evade a duty, malinger. □ **sprucer** n. [20th c.: orig. unkn.]

sprue[1] /spruː/ n. **1** a channel through which metal or plastic is poured into a mould. **2** a piece of metal or plastic which has filled a sprue and solidified there. [19th c.: orig. unkn.]

sprue[2] /spruː/ n. a tropical disease with ulceration of the mucous membrane of the mouth and chronic enteritis. [Du. *spruw* THRUSH[2]; cf. Flem. *spruwen* sprinkle]

spruik /spruːk/ v.intr. Austral. & NZ sl. speak in public, esp. as a showman. □ **spruiker** n. [20th c.: orig. unkn.]

spruit /spreɪt/ n. S. Afr. a small watercourse, usu. dry except during the rainy season. [Du., rel. to SPROUT]

sprung see SPRING.

sprung rhythm n. *Prosody* an English poetic metre approximating to speech, each foot having one stressed syllable followed by a varying number of unstressed. The term was invented by Gerard Manley Hopkins (see HOPKINS[3]) to describe his own idiosyncratic poetic metre.

spry /spraɪ/ adj. (**spryer, spryest**) active, lively. □ **spryly** adv. **spryness** n. [18th c., dial. & US: orig. unkn.]

spud /spʌd/ n. & v. ● n. **1** sl. a potato. **2** a small narrow spade for cutting the roots of weeds etc. ● v.tr. (**spudded, spudding**) **1** (foll. by *up, out*) remove (weeds) with a spud. **2** (also *absol.*; often foll. by *in*) make the initial drilling for (an oil well). □ **spud-bashing** Brit. sl. a lengthy spell of peeling potatoes. [ME: orig. unkn.]

spue var. of SPEW.

spumante /spuːˈmæntɪ/ n. an Italian sparkling white wine (cf. ASTI). [It., = 'sparkling']

spume /spjuːm/ n. & v.intr. froth, foam. □ **spumous** adj. **spumy** adj. [ME f. OF (e)*spume* or L *spuma*]

spumoni /spuːˈməʊnɪ/ n. N. Amer. a kind of ice-cream dessert. [It. *spumone* f. *spuma* SPUME]

spun past and past part. of SPIN.

spunk /spʌŋk/ n. **1** touchwood. **2** colloq. courage, mettle, spirit. **3** coarse sl. semen. **4** Austral. sl. a sexually attractive person. [16th c.: orig. unkn.: cf. PUNK]

spunky /ˈspʌŋkɪ/ adj. (**spunkier, spunkiest**) colloq. brave, spirited. □ **spunkily** adv.

spur /spɜː(r)/ n. & v. ● n. **1** a device with a small spike or a spiked wheel worn on a rider's heel for urging a horse forward. **2** a stimulus or incentive. **3** a spur-shaped thing, esp.: **a** a projection from a mountain or mountain range. **b** a branch road or railway. **c** a hard projection on a cock's leg. **d** a steel point fastened to the leg of a gamecock. **e** a climbing-iron. **f** a small support for ceramic ware in a kiln. **4** Bot. **a** a slender hollow projection from part of a flower. **b** a short fruit-bearing shoot. ● v. (**spurred, spurring**) **1** tr. prick (a horse) with spurs. **2 a** (often foll. by *on*) incite (a person) (*spurred him on to greater efforts; spurred her to try again*). **b** stimulate (interest etc.). **3** intr. (often foll. by *on, forward*) ride a horse hard. **4** tr. (esp. as **spurred** adj.) provide (a person, boots, a gamecock) with spurs. □ **on the spur of the moment** on a momentary impulse; impromptu. **put** (or **set**) **spurs to 1** spur (a horse). **2** stimulate (resolution etc.). **spur-gear** = spur-wheel. **spur-of-the-moment** adj. unpremeditated; impromptu. **spur royal** hist. a 15-shilling coin of James I bearing a spurlike sun with rays. **spur-wheel** a cog-wheel with radial teeth. □ **spurless** adj. [OE *spora, spura* f. Gmc, rel. to SPURN]

spurge /spɜːdʒ/ n. a plant of the family Euphorbiaceae, esp. the genus *Euphorbia*, exuding an acrid milky juice formerly used medicinally as a purgative. □ **spurge laurel** a low shrub, *Daphne laureola*, related to mezereon, with glossy leaves and greenish flowers. [ME f. OF *espurge* f. *espurgier* f. L *expurgare* (as EX-[1], PURGE)]

spurious /ˈspjʊərɪəs/ adj. **1** not genuine, not being what it purports to be, not proceeding from the pretended source (*a spurious excuse*). **2** having an outward similarity of form or function only. **3** (of offspring) illegitimate. □ **spuriously** adv. **spuriousness** n. [L *spurius* false]

spurn /spɜːn/ v. & n. ● v. **1** reject with disdain; treat with contempt. **2** repel or thrust back with one's foot. ● n. an act of spurning. □ **spurner** n. [OE *spurnan, spornan*, rel. to SPUR]

spurrey /ˈspʌrɪ/ n. (also **spurry**) (pl. **-eys** or **-ies**) a slender plant of the genus *Spergula* or *Spergularia*, of the pink family; esp. the *corn spurrey* (*Spergula arvensis*), a white-flowered cornfield weed, and the *sand spurrey* (*Spergularia rubra*), a pink-flowered plant of sandy ground. [Du. *spurrie*, prob. rel. to med.L *spergula*]

spurrier /ˈspʌrɪə(r)/ n. a spur-maker.

spurt /spɜːt/ v. & n. ● v. **1** (also **spirt**) **a** intr. gush out in a jet or stream. **b** tr. cause (liquid etc.) to do this. **2** intr. make a sudden effort. ● n. **1** (also

spirt) a sudden gushing out, a jet. **2** a short sudden effort or increase of pace esp. in racing. [16th c.: orig. unkn.]

sputnik /ˈsputnɪk, ˈspʌt-/ n. a series of Soviet artificial satellites, the first of which (launched on 4 Oct. 1957) was the first satellite to be placed in orbit. [Russ., = fellow-traveller]

sputter /ˈspʌtə(r)/ v. & n. ● v. **1** intr. emit spitting sounds, esp. when being heated. **2** intr. (often foll. by at) speak in a hurried or vehement fashion. **3** tr. emit with a spitting sound. **4** tr. speak or utter (words, threats, a language, etc.) rapidly or incoherently. **5** tr. Physics deposit (metal) by using fast ions etc. to eject particles of it from a target. ● n. a sputtering sound, esp. sputtering speech. □ **sputterer** n. [Du. sputteren (imit.)]

sputum /ˈspjuːtəm/ n. (pl. **sputa** /-tə/) **1** saliva, spittle. **2** a mixture of saliva and mucus expectorated from the respiratory tract, usu. a sign of disease. [L, neut. past part. of spuere spit]

spy /spaɪ/ n. & v. ● n. (pl. **spies**) **1** a person who secretly collects and reports information on the activities, movements, etc., of an enemy, competitor, etc. **2** a person who keeps watch on others, esp. furtively. ● v. (**spies, spied**) **1** tr. discern or make out, esp. by careful observation (spied a house in the distance). **2** intr. (often foll. by on) act as a spy, keep a close and secret watch. **3** intr. (often foll. by into) pry. □ **I-spy** a children's game of guessing a visible object from the initial letter of its name. **spy-master** colloq. the head of an organization of spies. **spy out** explore or discover, esp. secretly. □ **spying** n. [ME f. OF espie espying, espier espy f. Gmc]

spyglass /ˈspaɪɡlɑːs/ n. a small telescope.

spyhole /ˈspaɪhəʊl/ n. a peep-hole.

sq. abbr. square.

SQL abbr. Computing structured query language.

Sqn. Ldr. abbr. Squadron Leader.

squab /skwɒb/ n. & adj. ● n. **1** a short fat person. **2** a very young bird, esp. an unfledged pigeon. **3 a** a stuffed cushion. **b** Brit. the padded back or side of a car-seat. **4** a sofa or ottoman. ● adj. short and fat, squat. □ **squab-chick** an unfledged bird. **squab pie 1** pigeon pie. **2** a pie of mutton, pork, onions, and apples. [17th c.: orig. unkn.: cf. obs. quab shapeless thing, Sw. dial. sqvabba fat woman]

squabble /ˈskwɒb(ə)l/ n. & v. ● n. a petty or noisy quarrel. ● v.intr. engage in a squabble. □ **squabbler** n. [prob. imit.: cf. Sw. dial. sqvabbel a dispute]

squabby /ˈskwɒbɪ/ adj. (**squabbier, squabbiest**) short and fat; squat.

squacco /ˈskwækəʊ/ n. (in full **squacco heron**) a small crested buff and white heron, Ardeola ralloides, of southern Europe and parts of Africa. [It. dial. sguacco]

squad /skwɒd/ n. **1** a small group of people sharing a task etc. **2** Mil. a small number of men assembled for drill etc. **3** Sport a group of players forming a team. **4 a** (often in comb.) a specialized unit within a police force (drug squad). **b** = flying squad. **5** a group or class of people of a specified kind (the awkward squad). □ **squad car** a police car having a radio link with headquarters. [F escouade var. of escadre f. It. squadra SQUARE]

squaddie /ˈskwɒdɪ/ n. (also **squaddy**) (pl. **-ies**) Brit. Mil. sl. **1** a recruit. **2** a private.

squadron /ˈskwɒdrən/ n. **1** an organized body of persons. **2** a principal division of a cavalry regiment or armoured formation, consisting of two troops. **3** a detachment of warships employed on a particular duty. **4** a unit of the Royal Air Force with ten to eighteen aircraft. □ **squadron leader** the commander of a squadron of the Royal Air Force, the officer next below wing commander. [It. squadrone (as SQUAD)]

squail /skweɪl/ n. **1** (in pl.) a game with small wooden discs propelled across a table or board. **2** each of these discs. □ **squail-board** a board used in squails. [19th c.: orig. unkn.: cf. dial. kayles skittles]

squalid /ˈskwɒlɪd/ adj. **1** filthy, repulsively dirty. **2** mean or poor in appearance. **3** wretched, sordid. □ **squalidly** adv. **squalidness** n. **squalidity** /skwɒˈlɪdɪtɪ/ n. [L squalidus f. squalere be rough or dirty]

squall /skwɔːl/ n. & v. ● n. **1** a sudden or violent gust or storm of wind, esp. with rain or snow or sleet. **2** a discordant cry; a scream (esp. of a baby). **3** (esp. in pl.) trouble, difficulty. ● v. **1** intr. utter a squall; scream, cry out violently as in fear or pain. **2** tr. utter in a screaming or discordant voice. □ **squall line** Meteorol. a narrow band of high winds along a cold front. □ **squally** adj. [prob. f. SQUEAL after BAWL]

squalor /ˈskwɒlə(r)/ n. the state of being filthy or squalid. [L, as SQUALID]

squama /ˈskweɪmə/ n. (pl. **squamae** /-miː/) Zool. & Bot. **1** a scale on an animal or plant. **2** Anat. a thin scalelike plate of bone. □ **squamate** /-meɪt/ adj. **squamose** /-məʊz, -məʊs/ adj. **squamous** adj. **squamule** /-mjuːl/ n. [L squama]

squander /ˈskwɒndə(r)/ v.tr. **1** spend (money, time, etc.) wastefully. **2** dissipate (a fortune etc.) wastefully. □ **squanderer** n. [16th c.: orig. unkn.]

square /skweə(r)/ n., adj., adv., & v. ● n. **1** an equilateral rectangle. **2 a** an object of this shape or approximately this shape. **b** a small square area on a game-board. **c** a square scarf. **d** an academic cap with a stiff square top; a mortarboard. **3 a** an open (usu. four-sided) area surrounded by buildings, esp. one planted with trees etc. and surrounded by houses. **b** an open area at the meeting of streets. **c** Cricket a closer-cut area at the centre of a ground, any strip of which may be prepared as a wicket. **d** an area within barracks etc. for drill. **e** US a block of buildings bounded by four streets. **4** the product of a number multiplied by itself (81 is the square of 9). **5** an L-shaped or T-shaped instrument for obtaining or testing right angles. **6** sl. a conventional or old-fashioned person, one ignorant of or opposed to current trends. **7** a square arrangement of letters, figures, etc. **8** a body of infantry drawn up in rectangular form. **9** a unit of 100 sq. ft as a measure of flooring etc. **10** N. Amer. a square meal (three squares a day). ● adj. **1** having the shape of a square. **2** having or in the form of a right angle (table with square corners). **3** angular and not round; of square section (has a square jaw). **4** designating a unit of measure equal to the area of a square whose side is one of the unit specified (square metre). **5** (often foll. by with) a level, parallel. **b** on a proper footing; even, quits. **6 a** (usu. foll. by to) at right angles. **b** Cricket on a line through the stumps at right angles to the wicket. **7** having the breadth more nearly equal to the length or height than is usual (a man of square frame). **8** properly arranged; in good order, settled (get things square). **9** (also **all square**) **a** not in debt, with no money owed. **b** having equal scores, esp. Golf having won the same number of holes as one's opponent. **c** (of scores) equal. **10** fair and honest (his dealings are not always quite square). **11** uncompromising, direct, thorough (was met with a square refusal). **12** sl. conventional or old-fashioned, unsophisticated, conservative (cf. sense 6 of n.). **13** Mus. (of rhythm) simple, straightforward. ● adv. **1** squarely (sat square on his seat). **2** fairly, honestly (play square). ● v. **1** tr. make square or rectangular, give a rectangular cross-section to (timber etc.). **2** tr. multiply (a number) by itself (3 squared is 9). **3** tr. & intr. (usu. foll. by to, with) adjust; make or be suitable or consistent; reconcile (the results do not square with your conclusions). **4** tr. mark out in squares. **5** tr. settle or pay (a bill etc.). **6** tr. place (one's shoulders etc.) squarely facing forwards. **7** tr. colloq. **a** pay or bribe. **b** secure the acquiescence etc. of (a person) in this way. **8** tr. (also absol.) make the scores of (a match etc.) all square. **9** intr. assume the attitude of a boxer. **10** tr. Naut. **a** lay (yards) at right angles with the keel making them at the same time horizontal. **b** get (dead-eyes) horizontal. **c** get (ratlines) horizontal and parallel to one another. □ **back to square one** colloq. back to the starting-point with no progress made. **get square with** pay or compound with (a creditor). **on the square** adj. **1** colloq. honest, fair. **2** having membership of the Freemasons. ● adv. colloq. honestly, fairly (can be trusted to act on the square). **out of square** not at right angles. **perfect square** = square number. **square accounts with** see ACCOUNT. **square away** US tidy up. **square-bashing** Brit. Mil. sl. drill on a barrack-square. **square brackets** brackets of the form [ ]. **square-built** of comparatively broad shape. **square the circle 1** construct a square equal in area to a given circle (a problem incapable of a purely geometrical solution). **2** do what is impossible. **square dance** a dance with usu. four couples facing inwards from four sides. **square deal** a fair bargain, fair treatment. **squared paper** paper marked out in squares, esp. for plotting graphs. **square-eyed** joc. affected by or given to excessive viewing of television. **square leg** Cricket a fielding position at some distance on the batsman's leg side and close to a line through the stumps at right angles to the wicket. **square meal** a substantial and satisfying meal. **square measure** measure expressed in square units. **square number** the square of an integer e.g. 1, 4, 9, 16. **square off 1** US assume the attitude of a boxer. **2** Austral. placate or conciliate. **3** mark out in squares. **a square peg in a round hole** see PEG. **square piano** an early type of piano, small and oblong in shape. **square-rigged** (of a vessel) with the principal sails at right angles to the length of the ship and extended by horizontal yards slung to the mast by the middle (cf. fore-and-aft rigged). **square root** the number that multiplied by itself gives a specified number (3 is the square root of

9). **square sail** a four-cornered sail extended on a yard slung to the mast by the middle. **square-shouldered** with broad and not sloping shoulders (cf. *round-shouldered*). **square-toed 1** (of shoes or boots) having square toes. **2** wearing such shoes or boots. **3** formal, prim. **square up** settle an account etc. **square up to 1** move towards (a person) in a fighting attitude. **2** face and tackle (a difficulty etc.) resolutely. **square wave** *Physics* a wave with periodic sudden alternations between only two values of quantity. □ **squarely** *adv.* **squareness** *n.* **squarer** *n.* **squarish** *adj.* [ME f. OF *esquare, esquarré, esquarrer*, ult. f. EX-¹ + L *quadra* square]

Square Mile the City, the financial district of London.

squarrose /ˈskwɒrəʊz, -rəʊs/ *adj. Bot. & Zool.* rough with scalelike projections. [L *squarrosus* scurfy, scabby]

squash¹ /skwɒʃ/ *v. & n.* ● *v.* **1** *tr.* crush or squeeze flat or into pulp. **2** *intr.* (often foll. by *into*) make one's way by squeezing. **3** *tr.* pack tight, crowd. **4** *tr.* **a** silence (a person) with a crushing retort etc. **b** dismiss (a proposal etc.). **c** quash (a rebellion). ● *n.* **1** a crowd; a crowded assembly. **2** a sound of or as of something being squashed, or of a soft body falling. **3** *Brit.* a drink made of crushed fruit etc., diluted with water. **4** (also **squash rackets**) a game played with rackets on a closed four-walled court. (*See note below*.) **5** a squashed thing or mass. □ **squash tennis** *US* a game similar to squash, played with a lawn-tennis ball. □ **squashy** *adj.* (**squashier, squashiest**). **squashily** *adv.* **squashiness** *n.* [alt. of QUASH]

▪ Squash developed from rackets at Harrow school in the early 19th century, but it was not until the formation of the Squash Rackets Association in 1928 that the game became more international. It resembles the older game of rackets in many ways, but differs in its use of a softer ball. It is a game for two players, each of whom hits the ball against the front wall of the court in turn, attempting to win points by making the ball too difficult for the other to return. A fast-moving and strenuous game, squash is now popular in the West as a means of keeping fit. World open championships were first held in 1976.

squash² /skwɒʃ/ *n.* (*pl.* same or **squashes**) **1** a trailing gourd plant of the genus *Cucurbita*, having pumpkin-like fruits, esp. *C. maxima, C. moschata*, and *C. melopepo*. **2** the fruit of this cooked and eaten as a vegetable. □ **squash bug** a North American hemipteran bug, *Anasa tristis*, which is a pest of squashes, pumpkins, and melons. [obs. (i)*squoutersquash* f. Narragansett *asquutasquash* f. *asq* uncooked + *squash* green]

squat /skwɒt/ *v., adj., & n.* ● *v.* (**squatted, squatting**) **1** *intr.* **a** crouch with the hams resting on the backs of the heels. **b** sit on the ground etc. with the knees drawn up and the heels close to or touching the hams. **2** *tr.* put (a person) into a squatting position. **3** *intr. colloq.* sit down. **4 a** *intr.* act as a squatter. **b** *tr.* occupy (a building) as a squatter. **5** *intr.* (of an animal) crouch close to the ground. ● *adj.* (**squatter, squattest**) **1** (of a person etc.) short and thick, dumpy. **2** in a squatting posture. ● *n.* **1** a squatting posture. **2 a** place occupied by a squatter or squatters. **b** being a squatter. **3** *N. Amer. sl.* = DIDDLY-SQUAT. □ **squatly** *adv.* **squatness** *n.* [ME f. OF *esquatir* flatten f. *es-* EX-¹ + *quatir* press down, crouch ult. f. L *coactus* past part. of *cogere* compel: see COGENT]

squatter /ˈskwɒtə(r)/ *n.* **1** a person who takes unauthorized possession of unoccupied premises. **2** *Austral.* **a** a sheep-farmer esp. on a large scale. **b** *hist.* a person who gets the right of pasturage from the government on easy terms. **3** a person who settles on new esp. public land without title. **4** a person who squats.

squaw /skwɔː/ *n.* a North American Indian woman or wife. □ **squaw-man** a white married to a squaw. **squaw winter** (in North America) a brief wintry spell before an Indian summer. [Narragansett *squaws*, Massachusetts Algonquian *squa* woman]

squawk /skwɔːk/ *n. & v.* ● *n.* **1** a loud harsh cry esp. of a bird. **2** a complaint. ● *v.tr. & intr.* utter with or make a squawk. □ **squawk-box** *colloq.* a loudspeaker or intercom. □ **squawker** *n.* [imit.]

squeak /skwiːk/ *n. & v.* ● *n.* **1 a** a short shrill cry as of a mouse. **b** a slight high-pitched sound as of an unoiled hinge. **2** (also **narrow squeak**) a narrow escape, a success barely attained. ● *v.* **1** *intr.* make a squeak. **2** *tr.* utter (words) shrilly. **3** *intr.* (foll. by *by, through*) *colloq.* pass narrowly. **4** *intr. sl.* turn informer. [ME, imit.: cf. SQUEAL, SHRIEK, and Sw. *skväka* croak]

squeaker /ˈskwiːkə(r)/ *n.* **1** a person or thing that squeaks. **2** a young bird, esp. a pigeon.

squeaky /ˈskwiːkɪ/ *adj.* (**squeakier, squeakiest**) making a squeaking sound. □ **squeaky clean 1** completely clean. **2** above criticism; beyond reproach. □ **squeakily** *adv.* **squeakiness** *n.*

squeal /skwiːl/ *n. & v.* ● *n.* a prolonged shrill sound, esp. a cry of a child or a pig. ● *v.* **1** *intr.* make a squeal. **2** *tr.* utter (words) with a squeal. **3** *intr. sl.* turn informer. **4** *intr. sl.* protest loudly or excitedly. □ **squealer** *n.* [ME, imit.]

squeamish /ˈskwiːmɪʃ/ *adj.* **1** easily nauseated or disgusted. **2** fastidious or overscrupulous in questions of propriety, honesty, etc. □ **squeamishly** *adv.* **squeamishness** *n.* [ME var. of *squeamous* (now dial.), f. AF *escoymos*, of unkn. orig.]

squeegee /ˈskwiːdʒiː/ *n. & v.* ● *n.* **1** a rubber-edged implement set on a long handle and used for cleaning windows etc. **2** a small similar instrument or roller used in photography. ● *v.tr.* (**squeegees, squeegeed**) treat with a squeegee. [*squeege*, strengthened form of SQUEEZE]

squeeze /skwiːz/ *v. & n.* ● *v.* **1** *tr.* **a** exert pressure on from opposite or all sides, esp. in order to extract moisture or reduce size. **b** compress with one's hand or between two bodies. **c** reduce the size of or alter the shape of by squeezing. **2** *tr.* (often foll. by *out*) extract (moisture) by squeezing. **3 a** *tr.* force (a person or thing) into or through a small or narrow space. **b** *intr.* make one's way by squeezing. **c** *tr.* make (one's way) by squeezing. **4** *tr.* **a** harass by exactions; extort money etc. from. **b** constrain; bring pressure to bear on. **c** (usu. foll. by *out of*) obtain (money etc.) by extortion, entreaty, etc. **d** *Bridge* subject (a player) to a squeeze. **5** *tr.* press or hold closely as a sign of sympathy, affection, etc. **6** *tr.* (often foll. by *out*) produce with effort (*squeezed out a tear*). ● *n.* **1** an instance of squeezing; the state of being squeezed. **2 a** *Brit.* a close embrace. **b** *esp. US sl.* a man's close female friend, esp. a girlfriend. **3** a crowd or crowded state; a crush. **4** a small quantity produced by squeezing (*a squeeze of lemon*). **5** a sum of money extorted or exacted, esp. an illicit commission. **6** *Econ.* a restriction on borrowing, investment, etc., in a financial crisis. **7** an impression of a coin etc. taken by pressing damp paper, wax, etc., against it. **8** (in full **squeeze play**) a *Bridge* leading winning cards until an opponent is forced to discard an important card. **b** *Baseball* hitting a ball short to the infield to enable a runner on third base to start for home as soon as the ball is pitched. **9** *colloq.* a difficult situation, an emergency. □ **put the squeeze on** *colloq.* coerce or pressure (a person). **squeeze bottle** a flexible container whose contents are extracted by squeezing it. **squeeze-box** *sl.* an accordion or concertina. □ **squeezable** *adj.* **squeezer** *n.* [earlier *squise*, intensive of obs. *queise*, of unkn. orig.]

squelch /skweltʃ/ *v. & n.* ● *v.* **1** *intr.* **a** make a sucking sound as of treading in thick mud. **b** move with a squelching sound. **2** *tr.* **a** disconcert, silence. **b** stamp on, crush flat, put an end to. ● *n.* an instance of squelching. □ **squelcher** *n.* **squelchy** *adj.* [imit.]

squib /skwɪb/ *n. & v.* ● *n.* **1** a small firework burning with a hissing sound and usu. with a final explosion. **2** a short satirical composition, a lampoon. ● *v.* (**squibbed, squibbing**) **1** *tr. Amer. Football* kick (the ball) a comparatively short distance on a kick-off; execute (a kick) in this way. **2** *archaic* **a** *intr.* write lampoons. **b** *tr.* lampoon. [16th c.: orig. unkn.: perh. imit.]

squid /skwɪd/ *n. & v.* ● *n.* **1** an elongated fast-swimming cephalopod mollusc of the order Teuthoidea, with eight arms and two long tentacles; esp. a common edible one of the genus *Loligo*. **2** artificial bait for fish imitating a squid in form. ● *v.intr.* (**squidded, squidding**) fish with squid as bait. [17th c.: orig. unkn.]

squidgy /ˈskwɪdʒɪ/ *adj.* (**squidgier, squidgiest**) *colloq.* squashy, soggy. [imit.]

squiffed /skwɪft/ *adj. sl.* = SQUIFFY.

squiffy /ˈskwɪfɪ/ *adj.* (**squiffier, squiffiest**) esp. *Brit. sl.* slightly drunk. [19th c.: orig. unkn.]

squiggle /ˈskwɪg(ə)l/ *n. & v.* ● *n.* a short curly line, esp. in handwriting or doodling. ● *v.* **1** *tr.* write in a squiggly manner; scrawl. **2** *intr.* wriggle, squirm. □ **squiggly** *adj.* [imit.]

squill /skwɪl/ *n.* **1** a bulbous liliaceous plant of *Scilla* or a related genus, typically with star-shaped blue flowers, esp. the *striped squill* (*Puschkinia autumnalis*). **2 a** a related seashore plant, *Drimia maritima*, with white flowers and a large bulb. Also called *sea onion*. **b** an extract of this bulb, used in cough mixtures and other medicines. **3** a shrimplike crustacean of the genus *Squilla*. [ME f. L *squilla, scilla* f. Gk *skilla*]

squillion /ˈskwɪljən/ *n. colloq.* a very large number (*squillions of stars*). [arbitrarily f. TRILLION etc.]

squinch /skwɪntʃ/ *n. Archit.* a straight or arched structure across an interior angle of a square tower to carry a superstructure, e.g. a dome. [var. of obs. *scunch*, abbr. of SCUNCHEON]

squint /skwɪnt/ *v., n., & adj.* ● *v.* **1** *intr.* have the eyes turned in different directions, have a squint. **2** *intr.* (often foll. by *at*) look obliquely or with half-closed eyes. **2** *tr.* close (one's eyes) quickly, hold (one's eyes) half-shut. ● *n.* **1** = STRABISMUS. **2** a stealthy or sidelong glance. **3** *colloq.* a glance or look (*had a squint at it*). **4** an oblique opening through the wall of a church affording a view of the altar. **5** a leaning or inclination towards a particular object or aim. ● *adj.* **1** squinting. **2** looking different ways. □ **squint-eyed 1** squinting. **2** malignant, ill-willed. □ **squinter** *n.* **squinty** *adj.* [ASQUINT: (adj.) perh. f. *squint-eyed* f. obs. *squint* (adv.) f. ASQUINT]

squire /'skwaɪə(r)/ *n. & v.* ● *n.* **1** a country gentleman, esp. the chief landowner in a country district. **2** *hist.* a knight's attendant. **3** *Brit. colloq.* a jocular form of address to a man. **4** *US* a magistrate or lawyer. **5** *Austral.* = COCKNEY *n.* 2. ● *v.tr.* (of a man) attend upon or escort (a woman). □ **squiredom** *n.* **squirehood** *n.* **squirelet** /'skwaɪəlɪt/ *n.* **squireling** *n.* **squirely** *adj.* **squireship** *n.* [ME f. OF *esquier* ESQUIRE]

squirearch /'skwaɪəˌrɑːk/ *n.* a member of the squirearchy. □ **squirearchical** /ˌskwaɪə'rɑːkɪk(ə)l/ *adj.* (also **squirarchical**). [back-form. f. SQUIREARCHY, after MONARCH]

squirearchy /'skwaɪəˌrɑːkɪ/ *n.* (also **squirarchy**) (*pl.* **-ies**) landowners collectively, esp. as a class having political or social influence; a class or body of squires. [SQUIRE, after HIERARCHY etc.]

squireen /ˌskwaɪə'riːn/ *n. Brit.* the owner of a small landed property esp. in Ireland.

squirl /skwɜːl/ *n. colloq.* a flourish or twirl, esp. in handwriting. [perh. f. SQUIGGLE + TWIRL or WHIRL]

squirm /skwɜːm/ *v. & n.* ● *v.intr.* **1** wriggle, writhe. **2** show or feel embarrassment or discomfiture. ● *n.* a squirming movement. □ **squirmer** *n.* **squirmy** *adj.* (**squirmier**, **squirmiest**). [imit., prob. assoc. with WORM]

squirrel /'skwɪrəl/ *n. & v.* ● *n.* **1** a rodent of the family Sciuridae, often tree-living, with a bushy tail arching over its back (*grey squirrel*; *red squirrel*). **2** the fur of this animal. **3** a person who hoards objects, food, etc. ● *v.* (**squirrelled**, **squirrelling**; *US* **squirreled**, **squirreling**) **1** *tr.* (often foll. by *away*) hoard (objects, food, time, etc.) (*squirrelled it away in the cupboard*). **2** *intr.* (often foll. by *around*) bustle about. □ **squirrel cage 1** a small cage containing a revolving cylinder like a treadmill, on which a captive squirrel may exercise. **2** a form of rotor used in small electric motors, resembling the cylinder of a squirrel cage. **3** a monotonous or repetitive way of life. **squirrel** (or **squirrel-tail**) **grass** a grass, *Hordeum jubatum*, with bushy spikelets. **squirrel monkey** a small yellow-haired monkey, *Saimiri sciureus*, native to South America. [ME f. AF *esquirel*, OF *esquireul*, ult. f. L *sciurus* f. Gk *skiouros* f. *skia* shade + *oura* tail]

squirrelly /'skwɪrəlɪ/ *adj.* **1** like a squirrel. **2 a** inclined to bustle about. **b** (of a person) unpredictable, nervous, demented.

squirt /skwɜːt/ *v. & n.* ● *v.* **1** *tr.* eject (liquid or powder) in a jet as from a syringe. **2** *intr.* (of liquid or powder) be discharged in this way. **3** *tr.* splash with liquid or powder ejected by squirting. ● *n.* **1 a** a jet of water etc. **b** a small quantity produced by squirting. **2 a** a syringe. **b** (in full **squirt-gun**) a kind of toy syringe. **3** *colloq.* an insignificant but presumptuous person. □ **squirter** *n.* [ME, imit.]

squish /skwɪʃ/ *n. & v.* ● *n.* a slight squelching sound. ● *v.* **1** *intr.* move with a squish. **2** *tr. colloq.* squash, squeeze. □ **squishy** *adj.* (**squishier**, **squishiest**). [imit.]

squit /skwɪt/ *n. Brit.* **1** *sl.* a small or insignificant person. **2** *dial.* nonsense. **3** (**the squits**) *dial. & colloq.* diarrhoea. [cf. dial. *squit* insignificant person, and *squit* to squirt]

squitch /skwɪtʃ/ *n.* couch grass. [alt. f. QUITCH]

squiz /skwɪz/ *n. Austral. & NZ sl.* a look or glance. [prob. f. QUIZ²]

SR *abbr. hist.* Southern Railway.

Sr *symb. Chem.* the element strontium.

Sr. *abbr.* **1** Senior. **2** Señor. **3** Signor. **4** *Eccl.* Sister.

sr *abbr.* steradian(s).

Sri Lanka /sriː 'læŋkə, ʃriː/ (formerly called *Ceylon*) an island country off the SE coast of India; pop. (est. 1991) 17,194,000; languages, Sinhalese (official), Tamil; capital, Colombo. A centre of Buddhist culture from the 3rd century BC, the island was ruled by a strong native dynasty from the 12th century but was successively dominated by the Portuguese, Dutch, and British from the 16th century and finally annexed by the British in 1815. A Commonwealth state from 1948, the country became an independent republic in 1972, taking the name of Sri Lanka ('resplendent island'). Conflict between the majority Sinhalese, mainly Buddhist, population and the Tamil minority, mainly Hindu, chiefly concentrated in the north, has erupted in recent years, and since 1981 there has been fighting between government forces and Tamil separatist guerrillas. □ **Sri Lankan** *adj. & n.*

Srinagar /srɪ'nʌgə(r)/ a city in NW India, the summer capital of the state of Jammu and Kashmir, situated on the Jhelum river in the foothills of the Himalayas; pop. (1991) 595,000.

SRN see STATE REGISTERED NURSE.

SRO *abbr.* **1** standing room only. **2** self-regulatory organization. **3** Statutory Rules and Orders.

SS¹ *abbr.* **1** Saints. **2** steamship.

SS² /'es'es/ *hist.* a German special police force, the ruthless élite corps of the Nazi Party. Founded in 1925 by Hitler as a personal bodyguard, the SS provided security forces (including the Gestapo) and administered the concentration camps; a section of the SS served as combat troops alongside but independent of the armed forces. From 1929 until 1945 the SS was headed by Heinrich Himmler. [G *Schutzstaffel* defence squadron]

SSAFA *abbr.* (in the UK) Soldiers', Sailors', and Airmen's Families Association.

SSC *abbr.* **1** (in Scotland) Solicitor(s) in the Supreme Court. **2** Superconducting Super Collider.

SSE *abbr.* south-south-east.

SSP *abbr.* (in the UK) statutory sick pay.

SSR *abbr. hist.* Soviet Socialist Republic.

SSSI *abbr.* (in the UK) Site of Special Scientific Interest, the official designation of a site having notable flora, fauna, or geology and considered to be worthy of legal protection.

SST *abbr.* supersonic transport.

SSW *abbr.* south-south-west.

St *abbr.* **1** Saint. **2** Street. **3** stokes.

st. *abbr.* **1** stone (in weight). **2** *Cricket* stumped by.

-st var. of -EST².

Sta. *abbr.* Station.

stab /stæb/ *v. & n.* ● *v.* (**stabbed**, **stabbing**) **1** *tr.* pierce or wound with a (usu. short) pointed tool or weapon e.g. a knife or dagger. **2** *intr.* (often foll. by *at*) aim a blow with such a weapon. **3** *intr.* cause a sensation like being stabbed (*stabbing pain*). **4** *tr.* hurt or distress (a person, feelings, conscience, etc.). **5** *intr.* (foll. by *at*) aim a blow at a person's reputation, etc. ● *n.* **1 a** an instance of stabbing. **b** a blow or thrust with a knife etc. **2** a wound made in this way. **3** a sharply painful (physical or mental) sensation; a blow inflicted on a person's feelings. **4** *colloq.* an attempt, a try. □ **stab in the back** *n.* a treacherous or slanderous attack. ● *v.tr.* slander or betray. □ **stabber** *n.* [ME: cf. dial. *stob* in sense 1 of *v.*]

Stabat Mater /ˌstɑːbæt 'mɑːtə(r)/ *n.* **1** a Latin hymn on the suffering of the Virgin Mary at the Crucifixion. **2** a musical setting for this. [the opening words, L *Stabat mater dolorosa* 'Stood the mother, full of grief']

stabile /'steɪbaɪl, -bɪl/ *n.* a rigid, free-standing abstract sculpture or structure of wire, sheet metal, etc. [L *stabilis* STABLE¹, after MOBILE]

stability /stə'bɪlɪtɪ/ *n.* the quality or state of being stable. [ME f. OF *stableté* f. L *stabilitas* f. *stabilis* STABLE¹]

stabilize /'steɪbɪˌlaɪz/ *v.tr. & intr.* (also **-ise**) make or become stable. □ **stabilization** /ˌsteɪbɪlaɪ'zeɪʃ(ə)n/ *n.*

stabilizer /'steɪbɪˌlaɪzə(r)/ *n.* (also **-iser**) a device or substance used to keep something stable, esp.: **1** a gyroscopic device to prevent rolling of a ship. **2** *N. Amer.* the horizontal tailplane of an aircraft. **3** (in *pl.*) a pair of small wheels fitted to the rear wheel of a child's bicycle. **4** a substance which prevents the breakdown of emulsions, esp. as a food additive maintaining texture.

stable¹ /'steɪb(ə)l/ *adj.* (**stabler**, **stablest**) **1** firmly fixed or established; not easily adjusted, destroyed, or altered (*a stable structure*; *a stable government*). **2 a** firm, resolute; not wavering or fickle (*a stable and steadfast friend*). **b** (of a person) well-adjusted, sane, sensible. **3** *Chem.* (of a compound) not readily decomposing. **4** *Physics* (of an isotope) not subject to radioactive decay. **5** in a settled medical condition after an injury, operation, etc. □ **stable equilibrium** a state in which a body when disturbed tends to return to equilibrium. □ **stableness** *n.* **stably** *adv.* [ME f. AF *stable*, OF *estable* f. L *stabilis* f. *stare* stand]

stable² /'steɪb(ə)l/ *n. & v.* ● *n.* **1** a building set apart and adapted for keeping horses. **2** an establishment where racehorses are kept and trained. **3** the racehorses of a particular stable. **4** persons, products, etc., having a common origin or affiliation. **5** such an origin or affiliation. ● *v.tr.* put or keep (a horse) in a stable. □ **stable-boy** a boy employed in a stable. **stable-companion** (or **-mate**) **1** a horse of the same stable. **2** a person or product from the same source; a member of the same organization. **stable-girl** a girl employed in a stable. **stable-lad** a person employed in a stable. □ **stableful** n. (pl. **-fuls**). [ME f. OF *estable* f. L *stabulum* f. *stare* stand]

stableman /ˈsteɪb(ə)lmən/ n. (pl. **-men**) a person employed in a stable.

stabling /ˈsteɪblɪŋ/ n. accommodation for horses.

stablish /ˈstæblɪʃ/ v.tr. *archaic* fix firmly; establish; set up. [var. of ESTABLISH]

staccato /stəˈkɑːtəʊ/ adv., adj., & n. esp. *Mus.* ● adv. & adj. with each sound or note sharply detached or separated from the others (cf. LEGATO, TENUTO). ● n. (pl. **-os**) **1** a staccato passage in music etc. **2** staccato delivery or presentation. □ **staccato mark** a dot or stroke above or below a note, indicating that it is to be played staccato. [It., past part. of *staccare* = *distaccare* DETACH]

stack /stæk/ n. & v. ● n. **1** a pile or heap, esp. in orderly arrangement. **2** a circular or rectangular pile of hay, straw, etc., or of grain in sheaf, often with a sloping thatched top, a rick. **3** *colloq.* a large quantity (*a stack of work; has stacks of money*). **4 a** = *chimney-stack*. **b** = SMOKESTACK. **c** a tall factory chimney. **5** a stacked group of aircraft. **6** (also **stack-room**) a part of a library where books are compactly stored, esp. one to which the public does not have direct access. **7** *Brit.* a tall column of rock esp. off the coast of Scotland and the Orkneys. **8** a vertical arrangement of hi-fi or public-address equipment. **9** a pyramidal group of rifles, a pile. **10** *Computing* a set of storage locations which store data in such a way that the most recently stored item is the first to be retrieved. **11** *Brit.* a measure for a pile of wood of 108 cu. ft (3.06 cubic metres). ● v.tr. **1** pile in a stack or stacks. **2** a arrange (cards) secretly for cheating. **b** manipulate (circumstances etc.) to one's advantage. **3** cause (aircraft) to fly round the same point at different levels while waiting to land at an airport. □ **stack arms** *hist.* = *pile arms*. **stack up** N. Amer. *colloq.* present oneself, measure up. **stack-yard** an enclosure for stacks of hay, straw, etc. □ **stackable** adj. **stacker** n. [ME f. ON *stakkr* haystack f. Gmc]

stacte /ˈstækti:/ n. a sweet spice used by the ancient Hebrews in making incense. [ME f. L f. Gk *staktē* f. *stazō* drip]

staddle /ˈstæd(ə)l/ n. a platform or framework supporting a rick etc. □ **staddle-stone** a usu. mushroom-shaped stone supporting a staddle or rick etc., to prevent the entry of rodents. [OE *stathol* base f. Gmc, rel. to STAND]

stadium /ˈsteɪdɪəm/ n. (pl. **stadiums**) **1** an athletic or sports ground with tiers of seats for spectators. **2** (pl. **stadiums** or **stadia** /-dɪə/) *Antiq.* **a** a course for a foot-race or chariot-race. **b** a measure of length, about 185 metres. **3** a stage or period of development etc. [ME f. L f. Gk *stadion*]

stadtholder /ˈstæd‚həʊldə(r), ˈstæt‚həʊl-/ n. (also **stadholder** /ˈstæd‚həʊl-/) *hist.* **1** the chief magistrate of the United Provinces of the Netherlands. **2** the viceroy or governor of a province or town in the Netherlands. □ **stadtholdership** n. [Du. *stadhouder* deputy f. *stad* STEAD + *houder* HOLDER, after med.L *locum tenens*]

Staël, Mme de, see DE STAËL.

staff[1] /stɑːf/ n. & v. ● n. **1 a** a stick or pole for use in walking or climbing or as a weapon. **b** a stick or pole as a sign of office or authority. **c** a person or thing that supports or sustains. **d** a flagstaff. **e** *Surveying* a rod for measuring distances, heights, etc. **f** a token given to an engine-driver on a single-track railway as authority to proceed over a given section of line. **g** a spindle in a watch. **2 a** a body of persons employed in a business etc. (*editorial staff of a newspaper*). **b** those in authority within an organization, esp. the teachers in a school. **c** *Mil.* etc. a body of officers assisting an officer in high command and concerned with an army, regiment, fleet, or air force as a whole (*general staff*). **d** (usu. **Staff**) *Mil.* = *staff sergeant*. **3** (pl. **staffs** or **staves** /steɪvz/) *Mus.* a set of usu. five parallel lines on any one or between any adjacent two of which a note is placed to indicate its pitch. ● v.tr. provide (an institution etc.) with staff. □ **staff college** *Brit. Mil.* etc. a college at which officers are trained for staff duties. **staff notation** *Mus.* notation by means of a staff, esp. as distinct from tonic sol-fa. **staff nurse** *Brit.* a nurse ranking just below a sister. **staff officer** *Mil.* an officer serving on the staff of an army etc. **staff sergeant 1** *Brit.* the senior sergeant of a non-infantry company. **2** *US* a non-commissioned officer ranking just above sergeant. □ **staffed** adj. (also in *comb.*). [OE *stæf* f. Gmc]

staff[2] /stɑːf/ n. a mixture of plaster of Paris, cement, etc., as a temporary building-material. [19th c.: orig. unkn.]

Staffa /ˈstæfə/ a small uninhabited island of the Inner Hebrides, west

of Mull. It is the site of Fingal's Cave and is noted for its basalt columns (see also GIANT'S CAUSEWAY).

staffage /stəˈfɑːʒ/ n. accessory items in a painting, esp. figures or animals in a landscape picture. [G f. *staffieren* decorate, perh. f. OF *estoffer*: see STUFF]

staffer /ˈstɑːfə(r)/ n. *US* a member of a staff, esp. of a newspaper.

Stafford /ˈstæfəd/ an industrial town in central England, to the south of Stoke-on-Trent; pop. (1981) 62,240. An important market town since medieval times, it was the scene of conflict in 1643, during the English Civil War, when its city walls and 11th-century castle were destroyed by the Roundheads.

Staffordshire /ˈstæfəd‚ʃɪə(r)/ a county of central England; county town, Stafford.

Staffs. *abbr.* Staffordshire.

stag /stæg/ n. & v. ● n. **1** an adult male deer, esp. a large one of the genus *Cervus* with a set of antlers. **2** *Brit. Stock Exch.* a person who applies for shares of a new issue with a view to selling at once for a profit. **3** *US* a man who attends a social gathering unaccompanied by a woman (*stag-night*). ● v.tr. (**stagged**, **stagging**) *Brit. Stock Exch.* deal in (shares) as a stag. □ **stag beetle** a large beetle of the family Lucanidae, the male of which has large branched mandibles resembling a stag's antlers. **stag-** (or **stag's-)horn 1** the horn of a stag, esp. as a material for cutlery handles. **2** a fern, esp. of the genus *Platycerium*, having fronds like antlers. **stag-night** (or **-party**) an all-male celebration, esp. in honour of a man about to marry. [ME f. OE *stacga, stagga* (unrecorded): cf. *docga* dog, *frogga* frog, etc., and ON *steggr, steggi* male bird]

stage /steɪdʒ/ n. & v. ● n. **1** a point or period in a process or development (*reached a critical stage; is in the larval stage*). **2 a** a raised floor or platform, esp. one on which plays etc. are performed before an audience. **b** (prec. by *the*) the acting or theatrical profession; the art of writing or presenting plays. **c** the scene of action (*the stage of politics*). **d** = *landing-stage*. **3 a** a regular stopping-place on a route. **b** the distance between two stopping-places. **c** *Brit.* = *fare-stage*. **4** a section of a rocket with a separate engine, jettisoned when its propellant is exhausted. **5** *Geol.* a range of strata forming a subdivision of a series. **6** *Electronics* a single amplifying transistor or valve with the associated equipment. **7** the raised plate on which an object is placed for inspection through a microscope. ● v.tr. **1** present (a play etc.) on stage. **2** arrange the occurrence of (*staged a demonstration; staged a comeback*). □ **go on the stage** become an actor. **hold the stage** dominate a conversation etc. **stage direction** an instruction in the text of a play as to the movement, position, tone, etc., of an actor, or sound effects etc. **stage-diving** the practice of jumping off the stage at a rock-concert etc., to be caught by the crowd below. **stage door** an actors' and workmen's entrance from the street to a theatre behind the stage. **stage effect 1** an effect produced in acting or on the stage. **2** an artificial or theatrical effect produced in real life. **stage fright** nervousness on appearing before an audience. **stage-hand** a person handling scenery etc. during a performance on stage. **stage left** (or **right**) on the left (or right) side of the stage, facing the audience. **stage-manage 1** be the stage-manager of. **2** arrange and control for effect. **stage-management** the job or craft of a stage-manager. **stage-manager** the person responsible for lighting and other mechanical arrangements for a play etc. **stage name** a name assumed for professional purposes by an actor. **stage play** a play performed on stage rather than broadcast etc. **stage rights** exclusive rights to perform a particular play. **stage-struck** filled with an inordinate desire to go on the stage. **stage whisper 1** an aside. **2** a loud whisper meant to be heard by people other than the one addressed. □ **stager** n. **stageable** adj. **stageability** /‚steɪdʒəˈbɪlɪtɪ/ n. [ME f. OF *estage* dwelling ult. f. L *stare* stand]

stagecoach /ˈsteɪdʒkəʊtʃ/ n. *hist.* a large closed horse-drawn coach running regularly by stages between two places. Such coaches were used in England from the mid-17th century and reached their heyday in the early 19th century as roads improved; in the US they were often the only method available for long-distance travel. In the mid-19th century they were superseded by the newly developed railways.

stagecraft /ˈsteɪdʒkrɑːft/ n. skill or experience in writing or staging plays.

stagey var. of STAGY.

stagflation /stægˈfleɪʃ(ə)n/ n. *Econ.* a state of inflation without a corresponding increase of demand and employment. [STAGNATION (as STAGNATE) + INFLATION]

stagger /ˈstægə(r)/ v. & n. ● v. **1 a** *intr.* walk unsteadily, totter. **b** *tr.* cause to totter (*was staggered by the blow*). **2 a** *tr.* shock, confuse; cause to hesitate or waver (*the question staggered them; they were staggered at the suggestion*). **b** *intr.* hesitate; waver in purpose. **3** *tr.* arrange (events, hours of work, etc.) so that they do not coincide. **4** *tr.* arrange (objects) so that they are not in line. **b** set (the spokes of a wheel) to incline alternately to right and left. ● *n.* **1** a tottering movement. **2** an overhanging or slantwise or zigzag arrangement of like parts in a structure etc. □ **staggerer** *n.* [alt. of ME *stacker* (now dial.) f. ON *stakra* frequent. of *staka* push, stagger]

staggering /ˈstægərɪŋ/ *adj.* **1** astonishing, bewildering. **2** that staggers. □ **staggeringly** *adv.*

staggers /ˈstægəz/ *n.* **1** a parasitic or deficiency disease of farm animals marked by staggering or loss of balance. **2** (usu. **the staggers**) *colloq.* giddiness.

staghound /ˈstæghaʊnd/ *n.* a large breed of hound originally used for hunting deer.

staging /ˈsteɪdʒɪŋ/ *n.* **1** the presentation of a play etc. **2 a** a platform or support or scaffolding, esp. temporary. **b** shelves for plants in a greenhouse. □ **staging area** an intermediate assembly point for troops in transit. **staging post** a regular stopping-place, esp. on an air route.

stagnant /ˈstægnənt/ *adj.* **1** (of liquid) motionless, having no current; stale or foul due to this. **2** (of life, action, the mind, business, a person) showing no activity, dull, sluggish. □ **stagnancy** *n.* **stagnantly** *adv.* [L *stagnare stagnant-* f. *stagnum* pool]

stagnate /stægˈneɪt/ *v.intr.* be or become stagnant. □ **stagnation** /-ˈneɪʃ(ə)n/ *n.*

stagy /ˈsteɪdʒɪ/ *adj.* (also **stagey**) (**stagier**, **stagiest**) theatrical, artificial, exaggerated. □ **stagily** *adv.* **staginess** *n.*

staid /steɪd/ *adj.* of quiet and steady character; sedate. □ **staidly** *adv.* **staidness** *n.* [= *stayed*, past part. of STAY¹]

stain /steɪn/ *v.* & *n.* ● *v.* **1** *tr.* & *intr.* discolour or be discoloured by the action of liquid sinking in. **2** *tr.* sully, blemish, spoil, damage (a reputation, character, etc.). **3** *tr.* colour (wood, glass, etc.) by a process other than painting or covering the surface. **4** *tr.* *Biol.* impregnate (a specimen) for microscopic examination with colouring matter that makes the structure visible by being deposited in some parts more than in others. **5** *tr.* print colours on (wallpaper). ● *n.* **1** a discoloration, a spot or mark caused esp. by contact with foreign matter and not easily removed (*a cloth covered with tea-stains*). **2 a** a blot or blemish. **b** damage to a reputation etc. (*a stain on one's character*). **3** a substance used in staining. □ **stainable** *adj.* **stainer** *n.* [ME f. *distain* f. OF *desteindre desteign-* (as DIS-, TINGE)]

stained glass *n.* pieces of glass, either dyed or superficially coloured, set in a framework (usually of lead) to form decorative or pictorial designs. The art began in the service of the Christian Church and is of Byzantine origin, but perhaps its highest achievements are to be seen in the windows of medieval cathedrals of western and northern Europe. The production of stained glass has been revived in the 19th and 20th centuries by artists including, for example, Burne-Jones and Chagall.

Stainer /ˈsteɪnə(r)/, Sir John (1840–1901), English composer. He is remembered for his church music, including hymns, cantatas, and the oratorio *Crucifixion* (1887).

stainless /ˈsteɪnlɪs/ *adj.* **1** (esp. of a reputation) without stains. **2** not liable to stain. □ **stainless steel** an iron alloy containing chromium, resistant to tarnishing and rust; esp. steel containing 11–14 per cent chromium, used for cutlery etc.

stair /steə(r)/ *n.* **1** each of a set of fixed steps in a building (*on the top stair but one*). **2** (usu. in *pl.*) a set of such steps (*passed him on the stairs; down a winding stair*). **3** (in *pl.*) a landing-stage. □ **stair-rod** a rod for securing a carpet in the angle between two steps. [OE *stæger* f. Gmc]

staircase /ˈsteəkeɪs/ *n.* **1** a flight of stairs and the supporting structure. **2** a part of a building containing a staircase.

stairhead /ˈsteəhed/ *n.* a level space at the top of stairs.

stairway /ˈsteəweɪ/ *n.* **1** a flight of stairs, a staircase. **2** the way up this.

stairwell /ˈsteəwel/ *n.* the shaft in which a staircase is built.

staithe /steɪð/ *n.* *Brit.* a wharf, esp. a waterside coal depot equipped for loading vessels. [ME f. ON *stöth* landing-stage f. Gmc, rel. to STAND]

stake¹ /steɪk/ *n.* & *v.* ● *n.* **1** a stout stick or post sharpened at one end and driven into the ground as a support, boundary mark, etc. **2** *hist.* **a** the post to which a person was tied to be burnt alive. **b** (prec. by *the*) death by burning as a punishment (*was condemned to the stake*). **3** a long

vertical rod in basket-making. **4** a metalworker's small anvil fixed on a bench by a pointed prop. ● *v.tr.* **1** fasten, secure, or support with a stake or stakes. **2** (foll. by *off*, *out*) mark off (an area) with stakes. **3** state or establish (a claim). □ **pull** (or **pull up**) **stakes** depart; go to live elsewhere. **stake-boat** a boat anchored to mark the course for a boat race etc. **stake-body** (*pl.* -ies) *US* a body for a lorry etc. having a flat open platform with removable posts along the sides. **stake-net** a fishing-net hung on stakes. **stake out** *colloq.* **1** place under surveillance. **2** place (a person) to maintain surveillance. **stake-out** *n.* *colloq.* a period of surveillance. [OE *staca* f. WG, rel. to STICK²]

stake² /steɪk/ *n.* & *v.* ● *n.* **1** a sum of money etc. wagered on an event, esp. deposited with a stakeholder. **2** (often foll. by *in*) an interest or concern, esp. financial. **3** (in *pl.*) **a** money offered as a prize esp. in a horse-race. **b** such a race (*maiden stakes; trial stakes*). ● *v.tr.* **1 a** wager (*staked £5 on the next race*). **b** risk (*staked everything on convincing him*). **2** *US colloq.* give financial or other support to. □ **at stake 1** risked, to be won or lost (*life itself is at stake*). **2** at issue, in question. □ **staker** *n.* [16th c.: perh. f. STAKE¹]

stakeholder /ˈsteɪkˌhəʊldə(r)/ *n.* an independent party with whom each of those who make a wager deposits the money etc. wagered.

Stakhanovite /stəˈkɑːnəˌvaɪt/ *n.* & *adj.* ● *n.* a worker who is exceptionally hard-working and productive. (*See note below.*) ● *adj.* characteristic of a Stakhanovite; exceptionally hard-working. □ **Stakhanovism** *n.* **Stakhanovist** *n.* & *adj.*

▪ The term is derived from the name of Aleksei Grigorevich Stakhanov (1906–77), a Russian miner who in 1935 produced a phenomenal amount of coal by a combination of new methods and great energy, an achievement publicized by the Soviet authorities in their campaign to increase industrial output.

stalactite /ˈstæləkˌtaɪt, stəˈlæk-/ *n.* a deposit of calcite having the shape of a large icicle, formed by the trickling of water from the roof of a cave, cliff overhang, etc. □ **stalactic** /stəˈlæktɪk/ *adj.* **stalactiform** *adj.* **stalactitic** /ˌstælək'tɪtɪk/ *adj.* [mod.L *stalactites* f. Gk *stalaktos* dripping f. *stalassō* drip]

Stalag /ˈstælæg/ *n.* *hist.* a German prison camp, esp. for non-commissioned officers and privates. [G f. *Stamm* base, main stock, *Lager* camp]

stalagmite /ˈstæləgˌmaɪt/ *n.* a deposit of calcite formed by the dripping of water into the shape of a large inverted icicle rising from the floor of a cave etc., often uniting with a stalactite. □ **stalagmitic** /ˌstæləgˈmɪtɪk/ *adj.* [mod.L *stalagmites* f. Gk *stalagma* a drop f. *stalassō* (as STALACTITE)]

stale¹ /steɪl/ *adj.* & *v.* ● *adj.* (**staler**, **stalest**) **1 a** not fresh, not quite new (*stale bread is best for toast*). **b** musty, insipid, or otherwise the worse for age or use. **2** trite or unoriginal (*a stale joke; stale news*). **3** (of an athlete or other performer) having ability impaired by excessive exertion or practice. **4** *Law* (esp. of a claim) having been left dormant for an unreasonably long time. ● *v.tr.* & *intr.* make or become stale. □ **stalely** *adv.* **staleness** *n.* [ME, prob. f. AF & OF f. *estaler* halt: cf. STALL¹]

stale² /steɪl/ *n.* & *v.* ● *n.* the urine of horses and cattle. ● *v.intr.* (esp. of horses and cattle) urinate. [ME, perh. f. OF *estaler* adopt a position (cf. STALE¹)]

stalemate /ˈsteɪlmeɪt/ *n.* & *v.* ● *n.* **1** *Chess* a position counting as a draw, in which a player is not in check but cannot move except into check. **2** a deadlock or drawn contest. ● *v.tr.* **1** *Chess* bring (a player) to a stalemate. **2** bring to a standstill. [obs. *stale* (f. AF *estale* f. *estaler* be placed: cf. STALE¹) + MATE²]

Stalin¹ /ˈstɑːlɪn/ a former name (1924–61) for DONETSK.

Stalin² /ˈstɑːlɪn/, Joseph (born Iosif Vissarionovich Dzhugashvili) (1879–1953), Soviet statesman, General Secretary of the Communist Party of the USSR 1922–53. Born in Georgia, he joined the Bolsheviks under Lenin in 1903 and co-founded the party's newspaper *Pravda* in 1912, adopting the name 'Stalin' (Russ. = 'man of steel') by 1913; in the same year he was exiled to Siberia until just after the Russian Revolution. Following Lenin's death, he became chairman of the Politburo and secured enough support within the party to eliminate Trotsky, who disagreed with his theory of building socialism in the Soviet Union as a base from which Communism could spread. By 1927 he was the uncontested leader of the party, and in the following year he launched a succession of five-year plans for the industrialization and collectivization of agriculture; as a result of this process, some 10 million peasants are thought to have died, either of famine, or, in the case of those who resisted Stalin's policies, of hard labour or by execution. His large-scale purges of the intelligentsia in the 1930s

along similarly punitive and ruthless lines removed all op-
position, while his direction of the armed forces led to victory over
Hitler 1941–5. After 1945 he played a large part in the restructuring
of postwar Europe and attempted to maintain a firm grip on other
Communist states; he was later denounced by Khrushchev and the
Eastern bloc countries.

Stalinabad /'stɑːlɪnəˌbæd/ a former name (1929–61) for DUSHANBE.

Stalingrad /'stɑːlɪnˌgræd/ a former name (1925–61) for VOLGOGRAD.

Stalingrad, Battle of a long and bitterly fought battle of the Second
World War, in which the German advance into the Soviet Union was
turned back at Stalingrad (now Volgograd) in 1942–3. Grim house-to-
house fighting took place until Jan. 1943, when the Germans
surrendered after suffering more than 300,000 casualties.

Stalinism /'stɑːlɪˌnɪz(ə)m/ n. **1** the policies followed by Stalin in the
government of the former USSR, esp. centralization, totalitarianism,
and the pursuit of socialism. **2** any rigid centralized authoritarian
form of socialism. □ **Stalinist** n. & adj.

Stalino /'stɑːlɪnəʊ/ a former name (1924–61) for DONETSK.

stalk¹ /stɔːk/ n. **1** the main stem of a herbaceous plant. **2** the slender
attachment or support of a leaf, flower, fruit, etc. **3** a similar support
for an organ etc. in an animal. **4** a slender support or linking shaft in
a machine, object, etc., e.g. the stem of a wineglass. **5** the tall chimney
of a factory etc. □ **stalk-eyed** (of crabs, snails, etc.) having the eyes
mounted on stalks. □ **stalked** adj. (also in comb.). **stalkless** adj.
stalklet /'stɔːklɪt/ n. **stalklike** adj. **stalky** adj. [ME stalke, prob. dimin.
of (now dial.) stale rung of a ladder, long handle, f. OE stalu]

stalk² /stɔːk/ v. & n. ● v. **1** a tr. pursue or approach (game, prey, or an
enemy) stealthily. **b** intr. steal up to game under cover. **2** intr. stride, walk
in a stately or haughty manner. **3** tr. formal or rhet. move silently or
threateningly through (a place) (fear stalked the streets). ● n. **1** the
stalking of game. **2** an imposing gait. □ **stalking-horse 1** a horse or
screen behind which a hunter is concealed. **2** a pretext concealing
one's real intentions or actions. **3** a weak political candidate who
forces an election in the hope of a more serious contender coming
forward. □ **stalker** n. (also in comb.). [OE f. Gmc, rel. to STEAL]

stall¹ /stɔːl/ n. & v. ● n. **1** a a trader's stand or booth in a market etc., or
out of doors. **b** a compartment in a building for the sale of goods. **c** a
table in this on which goods are displayed. **2** a a stable or cowhouse.
b a compartment for one animal in this. **3** a a fixed seat in the choir
or chancel of a church, more or less enclosed at the back and sides
and often canopied, esp. one appropriated to a clergyman (canon's stall;
dean's stall). **b** the office or dignity of a canon etc. **4** (usu. in pl.) Brit. each
of a set of seats in a theatre, usu. on the ground floor. **5** a a
compartment for one person in a shower, lavatory, etc. **b** a
compartment for one horse at the start of a race. **6** a the stalling of an
engine or aircraft. **b** the condition resulting from this. **7** a receptacle
or sheath for one object (finger-stall). ● v. **1** a tr. (of a motor vehicle or
its engine) stop because of an overload on the engine or an inadequate
supply of fuel to it. **b** intr. (of an aircraft or its pilot) reach a condition
where the speed is too low to allow effective operation of the controls.
c tr. cause (an engine or vehicle or aircraft) to stall. **2** tr. a put or keep
(cattle etc.) in a stall or stalls esp. for fattening (a stalled ox). **b** furnish
(a stable etc.) with stalls. **3** intr. a (of a horse or cart) stick fast as in mud
or snow. **b** US be snowbound. □ **stall-feed** fatten (cattle) in a stall. [OE
steall f. Gmc, rel. to STAND: partly f. OF estal f. Frank.]

stall² /stɔːl/ v. & n. ● v. **1** intr. play for time when being questioned etc.
2 tr. delay, obstruct, block. ● n. the act or an instance of stalling. □ **stall
off** evade or deceive. [stall pickpocket's confederate, orig. 'decoy' f. AF
estal(e), prob. rel. to STALL¹]

stallage /'stɔːlɪdʒ/ n. Brit. **1** space for a stall or stalls in a market etc.
2 the rent for such a stall. **3** the right to erect such a stall. [ME f. OF
estalage f. estal STALL¹]

stallholder /'stɔːlˌhəʊldə(r)/ n. a person in charge of a stall at a market
etc.

stallion /'stæljən/ n. an uncastrated adult male horse, esp. one kept
for breeding. [ME f. OF estalon ult. f. a Gmc root rel. to STALL¹]

stalwart /'stɔːlwət/ adj. & n. ● adj. **1** strongly built, sturdy. **2** courageous,
resolute, determined (stalwart supporters). ● n. a stalwart person, esp. a
loyal uncompromising partisan. □ **stalwartly** adv. **stalwartness** n.
[Sc. var. of obs. stalworth f. OE stælwierthe f. stæl place, WORTH]

Stamboul /stæm'buːl/ an obsolete name for ISTANBUL.

stamen /'steɪmən/ n. Bot. the male fertilizing organ of a flowering
plant, including the anther containing pollen. □ **staminiferous**
/ˌstæmɪ'nɪfərəs/ adj. [L stamen staminis warp in an upright loom, thread]

stamina /'stæmɪnə/ n. the ability to endure prolonged physical or
mental strain; staying power, power of endurance. [L, pl. of STAMEN in
sense 'warp, threads spun by the Fates']

staminate /'stæmɪnət/ adj. Bot. (of a plant or flower) having stamens,
esp. stamens but not pistils.

stammer /'stæmə(r)/ v. & n. ● v. **1** intr. speak (habitually, or on occasion
from embarrassment etc.) with halting articulation, esp. with pauses
or rapid repetitions of the same syllable. **2** tr. (often foll. by out) utter
(words) in this way (stammered out an excuse). ● n. **1** a tendency to
stammer. **2** an instance of stammering. □ **stammerer** n.
stammeringly adv. [OE stamerian f. WG]

stamp /stæmp/ v. & n. ● v. **1** a tr. bring down (one's foot) heavily on the
ground etc. **b** tr. crush, flatten, or bring into a specified state in this
way (stamped down the earth round the plant). **c** intr. bring down one's foot
heavily; walk with heavy steps. **2** tr. a impress (a pattern, mark, etc.) on
metal, paper, butter, etc., with a die or similar instrument of metal,
wood, rubber, etc. **b** impress (a surface) with a pattern etc. in this way.
3 tr. affix a postage or other stamp to (an envelope or document). **4** tr.
assign a specific character to; characterize; mark out (stamps the story
an invention). **5** tr. crush or pulverize (ore etc.). ● n. **1** an instrument for
stamping a pattern or mark. **2** a a mark or pattern made by this.
b the impression of an official mark required to be made for revenue
purposes on deeds, bills of exchange, etc., as evidence of payment of
tax. **3** a small adhesive piece of paper indicating that a price, fee, or
tax has been paid, esp. a postage stamp. (See note below.) **4** a mark
impressed on or label etc. affixed to a commodity as evidence of quality
etc. **5** a a heavy downward blow with the foot. **b** the sound of this.
6 a a characteristic mark or impress (bears the stamp of genius).
b character, kind (avoid people of that stamp). **7** the block that crushes
ore in a stamp-mill. □ **stamp-collecting** the collecting of postage
stamps as objects of interest or value. **stamp-collector** a person
engaged in stamp-collecting. **stamp-duty** a duty imposed on certain
kinds of legal document. **stamp-hinge** see HINGE. **stamping-
ground** a favourite haunt or place of action. **stamp-machine** a coin-
operated machine for selling postage stamps. **stamp-mill** a mill for
crushing ore etc. **stamp-office** an office for the issue of government
stamps and the receipt of stamp-duty etc. **stamp on 1** impress (an
idea etc.) on (the memory etc.). **2** suppress. **stamp out 1** produce by
cutting out with a die etc. **2** put an end to, crush, destroy. **stamp-
paper 1** paper with the government revenue stamp. **2** the gummed
marginal paper of a sheet of postage stamps. □ **stamper** n. [prob. f.
OE stampian (v.) (unrecorded) f. Gmc: infl. by OF estamper (v.) and F
estampe (n.) also f. Gmc]

▪ The world's first postage stamps were the penny black and twopenny
blue, issued by Great Britain in May 1840 and showing the head of
Queen Victoria in profile. The system was introduced by Sir Rowland
Hill, and subsequently adopted throughout the world. The name
stamp was originally applied to marks stamped or impressed by the
Post Office on letters, to state whether they were 'prepaid', 'unpaid',
'free', etc.; when adhesive stamps were introduced these marks
became known as postmarks.

Stamp Act n. an act concerned with stamp-duty, esp. that imposing
the duty on the American colonies in 1765 and repealed in 1766.

stampede /stæm'piːd/ n. & v. ● n. **1** a sudden flight and scattering of
a number of horses, cattle, etc. **2** a sudden flight or hurried movement
of people due to interest or panic. **3** US the spontaneous and
simultaneous response of many persons to a common impulse. ● v.
1 intr. take part in a stampede. **2** tr. cause to do this. **3** tr. cause to act
hurriedly or unreasoningly. □ **stampeder** n. [Sp. estampida crash,
uproar, ult. f. Gmc, rel. to STAMP]

stance /stɑːns, stæns/ n. **1** an attitude or position of the body esp.
when hitting a ball etc. **2** a standpoint; an attitude of mind. **3** Sc. a
site for a market, taxi rank, etc. [F f. It. stanza: see STANZA]

stanch¹ /stɑːntʃ/ v.tr. (also **staunch** /stɔːntʃ/) **1** restrain the flow of
(esp. blood). **2** restrain the flow from (esp. a wound). [ME f. OF estanchier
f. Rmc]

stanch² var. of STAUNCH¹.

stanchion /'stɑːnʃ(ə)n/ n. & v. ● n. **1** a post or pillar, an upright support,
a vertical strut. **2** an upright bar, pair of bars, or frame, for confining
cattle in a stall. ● v.tr. **1** supply with a stanchion. **2** fasten (cattle) to a
stanchion. [ME f. AF stanchon, OF estanchon f. estance prob. ult. f. L stare
stand]

stand /stænd/ v. & n. ● v. (past and past part. **stood** /stʊd/) **1** intr. have or
take or maintain an upright position, esp. on the feet or a base. **2** intr.
be situated or located (here once stood a village). **3** intr. be of a specified

height (*stands six foot three*). **4** *intr.* be in a specified condition (*stands accused; the thermometer stood at 90°; the matter stands as follows; stood in awe of them*). **5** *tr.* place or set in an upright or specified position (*stood it against the wall*). **6** *intr.* **a** move to and remain in a specified position (*stand aside*). **b** take a specified attitude (*stand aloof*). **7** *intr.* maintain a position; avoid falling or moving or being moved (*the house will stand for another century; stood for hours arguing*). **8** *intr.* assume a stationary position; cease to move (*now stand still*). **9** *intr.* remain valid or unaltered; hold good (*the former conditions must stand*). **10** *intr. Naut.* hold a specified course (*stand in for the shore; you are standing into danger*). **11** *tr.* endure without yielding or complaining; tolerate (*cannot stand the pain; how can you stand him?*). **12** *tr.* provide for another or others at one's own expense (*stood him a drink*). **13** *intr.* (often foll. by *for*) *Brit.* be a candidate (for an office, legislature, or constituency) (*stood for Parliament; stood for Finchley*). **14** *intr.* act in a specified capacity (*stood proxy*). **15** *tr.* undergo (trial). **16** *intr. Cricket* act as umpire. **17** *intr.* (of a dog) point, set. **18** *intr.* (in full **stand at stud**) (of a stallion) be available for breeding. ● *n.* **1** a cessation from motion or progress, a stoppage (*was brought to a stand*). **2 a** a halt made, or a stationary condition assumed, for the purpose of resistance. **b** resistance to attack or compulsion (esp. *make a stand*). **c** *Cricket* a prolonged period at the wicket by two batsmen. **3 a** a position taken up (*took his stand near the door*). **b** an attitude adopted. **4** a rack, set of shelves, table, etc., on or in which things may be placed (*music stand; hatstand*). **5 a** a small open-fronted structure for a trader outdoors or in a market etc. **b** a structure occupied by a participating organization at an exhibition. **6** a standing-place for vehicles (*cab-stand*). **7 a** a raised structure for spectators, performers, etc. to sit or stand on. **b** *US* a witness-box (*take the stand*). **8** *Theatr.* etc. each half made on a tour to give one or more performances. **9** a group of growing plants (*stand of trees; stand of clover*). □ **as it stands 1** in its present condition, unaltered. **2** (also **as things stand**) in the present circumstances. **be at a stand** *archaic* be unable to proceed, be in perplexity. **it stands to reason** see REASON. **stand alone** be unequalled. **stand and deliver!** *hist.* a highwayman's order to hand over valuables etc. **stand at bay** see BAY[5]. **stand back 1** withdraw; take up a position further from the front. **2** withdraw psychologically in order to take an objective view. **stand by 1** stand nearby; look on without interfering (*will not stand by and see him ill-treated*). **2** uphold, support, side with (a person). **3** adhere to, abide by (terms or promises). **4 a** *Naut.* stand ready to take hold of or operate (an anchor etc.). **b** be ready to act or assist. **stand-by** *n.* (*pl.* **-bys**) **1** a person or thing ready if needed in an emergency etc. **2** readiness for duty (*on stand-by*). ● *adj.* **1** ready for immediate use. **2** (of air travel, theatre seats, etc.) not booked in advance but allocated on the basis of earliest availability. **stand camera** a camera for use on a tripod, not hand-held. **stand a chance** see CHANCE. **stand corrected** accept correction. **stand down 1** withdraw from a team, match, election, etc. **2** *Mil.* relax after a state of alert; go off duty. **3** leave a witness-box. **stand easy!** see EASY. **stand for 1** represent, signify, imply (*'US' stands for 'United States'; democracy stands for a great deal more than that*). **2** (often with *neg.*) *colloq.* endure, tolerate, acquiesce in. **3** espouse the cause of. **stand one's ground** see hold one's ground (see GROUND[1]). **stand high** be high in status, price, etc. **stand in** (usu. foll. by *for*) deputize; act in place of another. **stand-in** *n.* a deputy or substitute, esp. for an actor when the latter's acting ability is not needed. **stand in the breach** see BREACH. **stand a person in good stead** see STEAD. **stand in with** be in league with. **stand of arms** *Brit. Mil.* a complete set of weapons for one man. **stand of colours** *Brit. Mil.* a regiment's flags. **stand off 1** move or keep away, keep one's distance. **2** *Brit.* temporarily dispense with the services of (an employee). **stand-off** *n.* **1** *N. Amer.* a deadlock. **2** = *stand-off half*. **stand-off half** *Rugby* a half-back who forms a link between the scrum-half and the three-quarters. **stand on 1** insist on, observe scrupulously (*stand on ceremony; stand on one's dignity*). **2** *Naut.* continue on the same course. **stand on me** *sl.* rely on me; believe me. **stand on one's own feet** (or **own two feet**) be self-reliant or independent. **stand out 1** be prominent or conspicuous or outstanding. **2** (usu. foll. by *against, for*) hold out; persist in opposition or support or endurance. **stand over 1** stand close to (a person) to watch, control, threaten, etc. **2** be postponed, be left for later settlement etc. **stand pat** see PAT[2]. **stand to 1** *Mil.* stand ready for an attack (esp. before dawn or after dark). **2** abide by, adhere to (terms or promises). **3** be likely or certain to (*stands to lose everything*). **4** uphold, support, or side with (a person). **stand treat** bear the expense of entertainment etc. **stand up 1 a** rise to one's feet from a sitting or other position. **b** come to or remain in or place in a standing position. **2** (of an argument etc.) be valid. **3** *colloq.* fail to keep an appointment with. **stand-up** *attrib.adj.* **1** (of a meal) eaten standing. **2** (of a fight) violent, thorough, or fair and square. **3** (of a collar) upright, not turned down. **4** (of a comedian

performing by standing before an audience and telling jokes. **stand up for** support, side with, maintain (a person or cause). **stand upon** = *stand on*. **stand up to 1** meet or face (an opponent) courageously. **2** be resistant to the harmful effects of (wear, use, etc.). **stand well** (usu. foll. by *with*) be on good terms or in good repute. **take one's stand on** base one's argument etc. on, rely on. □ **stander** *n.* [OE *standan* f. Gmc]

standalone /ˌstændəˈləʊn/ *adj.* (of a computer) operating independently of a network or other system.

standard /ˈstændəd/ *n. & adj.* ● *n.* **1** an object or quality or measure serving as a basis or example or principle to which others conform or should conform or by which the accuracy or quality of others is judged (*by present-day standards*). **2 a** the degree of excellence etc. required for a particular purpose (*not up to standard*). **b** average quality (*of a low standard*). **3** the ordinary procedure, or quality or design of a product, without added or novel features. **4** a distinctive flag, esp. the flag of a cavalry regiment as distinct from the *colours* of an infantry regiment. **5 a** an upright support. **b** an upright water or gas pipe. **6 a** a tree or shrub that grows on an erect stem of full height and stands alone without support. **b** a shrub grafted on an upright stem and trained in tree form (*standard rose*). **7** a document specifying nationally or internationally agreed properties for manufactured goods etc. (*British Standard*). **8** a thing recognized as a model for imitation etc. **9** a tune or song of established popularity. **10 a** a system by which the value of a currency is defined in terms of gold or silver or both. **b** the prescribed proportion of the weight of fine metal in gold or silver coins. **11** a measure for timber, equivalent to 165 cu. ft (4.67 cubic metres). **12** *Brit. hist.* a grade of classification in elementary schools. ● *adj.* **1** serving or used as a standard (*a standard size*). **2** of a normal or prescribed quality or size etc. **3** having recognized and permanent value; authoritative (*the standard book on the subject*). **4** (of language) conforming to established educated usage (*Standard English*). □ **multiple standard** a standard of value obtained by averaging the prices of a number of products. **raise a standard** take up arms; rally support (*raised the standard of revolt*). **standard-bearer 1** a soldier who carries a standard. **2** a prominent leader in a cause. **standard deviation** see DEVIATION. **standard lamp** *Brit.* a lamp set on a tall upright with its base standing on the floor. **standard of living** the degree of material comfort available to a person or class or community. **standard time** a uniform time for places in approximately the same longitude, established in a country or region by law or custom. [ME f. AF *estaundart*, OF *estendart* f. *estendre*, as EXTEND: in senses 5 and 6 of *n.* affected by association with STAND]

Standardbred /ˈstændədˌbred/ *n. N. Amer.* a breed of horse able to attain a specified speed, developed esp. for trotting.

standardize /ˈstændəˌdaɪz/ *v.* (also **-ise**) **1** *tr.* cause to conform to a standard. **2** *tr.* determine the properties of by comparison with a standard. **3** *intr.* (foll. by *on*) adopt as one's standard or model. □ **standardizable** *adj.* **standardizer** *n.* **standardization** /ˌstændədaɪˈzeɪʃ(ə)n/ *n.*

standee /stænˈdiː/ *n. colloq.* a person who stands, esp. when all seats are occupied.

standing /ˈstændɪŋ/ *n. & adj.* ● *n.* **1** esteem or repute, esp. high; status, position (*people of high standing; is of no standing*). **2** duration (*a dispute of long standing*). **3** length of service, membership, etc. ● *adj.* **1** that stands, upright. **2 a** established, permanent (*a standing rule*). **b** not made, raised, etc., for the occasion (*a standing army*). **3** (of a jump, start, race, etc.) performed from rest or from a standing position. **4** (of water) stagnant. **5** (of corn) unreaped. **6** (of a stallion) that stands at stud. **7** *Printing* (formerly, of type) not yet distributed after use. □ **all standing 1** *Naut.* without time to lower the sails. **2** taken by surprise. **be in good standing** (often foll. by *with*) be in favour or on good terms. **leave a person standing** make far more rapid progress than he or she. **standing committee** see COMMITTEE. **standing joke** an object of permanent ridicule. **standing order** an instruction to a banker to make regular payments, or to a newsagent etc. for a regular supply of a periodical etc. **standing orders** the rules governing the manner in which all business shall be conducted in a parliament, council, society, etc. **standing ovation** see OVATION. **standing rigging** *Naut.* rigging which is fixed in position. **standing-room** space to stand in. **standing wave** *Physics* the vibration of a system in which some particular points remain fixed while others between them vibrate with the maximum amplitude (cf. *travelling wave*).

standoffish /stændˈɒfɪʃ/ *adj.* cold or distant in manner. □ **standoffishly** *adv.* **standoffishness** *n.*

standout /'stændaʊt/ *n. N. Amer.* (often *attrib.*) a remarkable or outstanding person or thing.

standpipe /'stændpaɪp/ *n.* a vertical pipe extending from a water supply, esp. one connecting a temporary tap to the mains.

standpoint /'stændpɔɪnt/ *n.* **1** the position from which a thing is viewed. **2** a mental attitude.

standstill /'stændstɪl/ *n.* a stoppage; an inability to proceed.

Stanford /'stænfəd/, Sir Charles (Villiers) (1852–1924), British composer, born in Ireland. As professor of composition at the Royal College of Music, London (1883–1924), and professor of music at Cambridge University (1887–1924), he was a highly influential teacher and played a significant role in the revival of English music at the turn of the century; his pupils included Gustav Holst and Ralph Vaughan Williams. A prolific composer in many genres, he is now known mainly through his Anglican church music and numerous choral works.

Stanhope /'stænəp/, Lady Hester Lucy (1776–1839), English traveller. She kept house for her uncle William Pitt the Younger from 1803 to 1806, becoming a distinguished political hostess. Stanhope was granted a pension on Pitt's death and later set out for the Middle East (1810), living in a ruined convent in the Lebanon Mountains four years later. She participated in Middle Eastern politics for several years, but eventually died in poverty after her pension was stopped by Lord Palmerston.

stanhope /'stænəp/ *n.* a light open carriage for one with two or four wheels. [Fitzroy *Stanhope*, Engl. clergyman (1787–1864), for whom the first one was made]

staniel /'stænjəl/ *n.* a kestrel. [OE *stāngella* 'stone-yeller' f. *stān* stone + *gellan* yell]

Stanier /'stænɪə(r)/, Sir William (Arthur) (1876–1965), English railway engineer. After working for the Great Western Railway he became chief mechanical engineer of the London Midland and Scottish Railway in 1932. Stanier was given the task of modernizing the company's entire locomotive stock, and he is chiefly remembered for his new standard locomotive designs of the 1930s – especially large engines of 4-6-0 and pacific type. He was elected to the Royal Society in 1944.

Stanislaus, St /'stænɪsˌlɔːs/ (known as St Stanislaus of Cracow; Polish name Stanisław /sta'niswaf/ (1030–79), patron saint of Poland. He became bishop of Cracow in 1072 and, as such, excommunicated King Boleslaus II (1039–81). According to tradition, Stanislaus was murdered by Boleslaus while taking Mass. Feast day, 11 April (formerly 7 May).

Stanislavsky /ˌstænɪ'læfskɪ/, Konstantin (Sergeevich) (born Konstantin Sergeevich Alekseev) (1863–1938), Russian theatre director and actor. In 1898 he founded the Moscow Art Theatre and became known for his innovative productions of works by Chekhov and Maxim Gorky. He trained his actors to take a psychological approach and use latent powers of self-expression when taking on roles; his theory and technique of acting were later adopted in the US and developed into the system known as method acting.

stank *past of* STINK.

Stanley[1] /'stænlɪ/ (also **Port Stanley**) the chief port and town of the Falkland Islands, situated on the island of East Falkland; pop. (1991) 1,557.

Stanley[2] /'stænlɪ/, Sir Henry Morton (born John Rowlands) (1841–1904), Welsh explorer. As a newspaper correspondent he was sent in 1869 to central Africa to find the Scottish missionary and explorer David Livingstone; two years later he found him on the eastern shore of Lake Tanganyika. After Livingstone's death, Stanley continued his exploration, charting Lake Victoria (1874), tracing the course of the Congo (1874–7), mapping Lake Albert (1889), and becoming the first European to discover Lake Edward (1889). Stanley also helped establish the Congo Free State (now Zaire), with Belgian support, from 1879 to 1885.

Stanley, Mount (known in Zaire as *Mount Ngaliema*) a mountain in the Ruwenzori range in central Africa, on the border between Zaire and Uganda. It has several peaks, of which the highest, Margherita Peak, rising to 5,110 m (16,765 ft), is the third-highest peak in Africa. It is named after Sir Henry Morton Stanley, who, in 1889, was the first European to reach it.

Stanley Cup a North American ice-hockey competition played at the end of each season between leading teams in the National Hockey League. The competition is named after Lord Stanley of Preston (1841–1908), the Governor-General of Canada who donated the trophy in 1893.

Stanleyville /'stænlɪˌvɪl/ the former name (1882–1966) for KISANGANI.

stannary /'stænərɪ/ *n.* (*pl.* **-ies**) (usu. in *pl.* prec. by *the*) *hist.* a tin-mining district in Cornwall and Devon. □ **stannary court** a legal body for the regulation of tin miners in the stannaries. [med.L *stannaria* (pl.) f. LL *stannum* tin]

stannic /'stænɪk/ *adj. Chem.* of or relating to tetravalent tin (cf. STANNOUS). [LL *stannum* tin]

stannous /'stænəs/ *adj. Chem.* of or relating to divalent tin (cf. STANNIC).

Stansted /'stænstɪd/ an airport in Essex, north-east of London. The new international terminal was opened in 1991.

stanza /'stænzə/ *n. Prosody* **1** the basic metrical unit in a poem or verse consisting of a recurring group of lines (often four lines and usu. not more than twelve) which may or may not rhyme. **2** a group of four lines in some Greek and Latin metres. □ **stanza'd** /-zəd/ *adj.* (also **stanzaed**) (also in *comb.*). **stanzaic** /stæn'zeɪɪk/ *adj.* [It., = standing-place, chamber, stance, ult. f. L *stare* stand]

stapelia /stə'piːlɪə/ *n.* a southern African plant of the genus *Stapelia*, with flowers having an unpleasant smell. [mod.L f. Jan Bode von *Stapel*, Dutch botanist (died *c.*1636)]

stapes /'steɪpiːz/ *n.* (*pl.* same) *Anat.* a small stirrup-shaped bone in the middle ear transmitting vibrations from the incus to the inner ear. Also called *stirrup-bone*. [mod.L f. med.L *stapes* stirrup]

staphylococcus /ˌstæfɪlə'kɒkəs/ *n.* (*pl.* **staphylococci** /-'kɒksaɪ, -'kɒkaɪ/) a bacterium of the genus *Staphylococcus*, occurring in grapelike clusters, and sometimes causing pus formation usu. in the skin and mucous membranes of animals. □ **staphylococcal** *adj.* [mod.L f. Gk *staphulē* bunch of grapes + *kokkos* berry]

staple[1] /'steɪp(ə)l/ *n. & v.* ● *n.* a U-shaped metal bar or piece of wire with pointed ends for driving into, securing, or fastening together various materials or for driving through and clenching papers, netting, electric wire, etc. ● *v.tr.* provide or fasten with a staple. □ **staple gun** a hand-held device for driving in staples. □ **stapler** *n.* [OE *stapol* f. Gmc]

staple[2] /'steɪp(ə)l/ *n., adj., & v.* ● *n.* **1** the principal or an important article of commerce (*the staples of British industry*). **2** the chief element or a main component, e.g. of a diet. **3** a raw material. **4** the fibre of cotton or wool etc. as determining its quality (*cotton of fine staple*). ● *adj.* **1** main or principal (*staple commodities*). **2** important as a product or an export. ● *v.tr.* sort or classify (wool etc.) according to fibre. [ME f. OF *estaple* market f. MLG, MDu. *stapel* market (as STAPLE[1])]

star /stɑː(r)/ *n. & v.* ● *n.* **1 a** a remote celestial body appearing as a fixed luminous point in the night sky, being a large incandescent gaseous body like the sun. (*See note below.*) **b** *colloq.* a celestial body orbiting the sun and shining by reflected light, esp. a planet (*evening star*). (See also *shooting star.*) **2** *Astrol.* a celestial body regarded as influencing a person's fortunes etc. (*born under a lucky star*). **3** a thing resembling a star in shape or appearance. **4** a star-shaped mark, esp. a white mark on a horse's forehead. **5** a figure or object with radiating points esp. as the insignia of an order, as a decoration or mark of rank, or showing a category of excellence (*a five-star hotel; was awarded a gold star*). **6 a** a famous or brilliant person; the principal or most prominent performer in a play, film, etc. (*the star of the show*). **b** (*attrib.*) outstanding; particularly brilliant (*star pupil*). **7** (in full **star connection**) *Electr.* a Y-shaped arrangement of three-phase windings. **8** = *star prisoner.* ● *v.* (**starred, starring**) **1 a** *tr.* (of a film etc.) feature as a principal performer. **b** *intr.* (of a performer) be featured in a film etc. **2** (esp. as **starred** *adj.*) **a** mark, set, or adorn with a star or stars. **b** put an asterisk or star beside (a name, an item in a list, etc.). □ **my stars!** *colloq.* an expression of surprise. **star-apple** an edible purple apple-like fruit (with a starlike cross-section) of a tropical evergreen tree, *Chrysophyllum cainito.* **star-crossed** *archaic* ill-fated. **star fruit** = CARAMBOLA 2. **star-gaze 1** gaze at or study the stars. **2** gaze intently. **star-gazer 1** *colloq.* usu. *derog.* or *joc.* an astronomer or astrologer. **2** *Austral. sl.* a horse that turns its head when galloping. **star prisoner** *Brit. sl.* a convict serving a first prison sentence. **star route** *US* a postal delivery route served by private contractors. **star sapphire** a cabochon sapphire reflecting a star-like image due to its regular internal structure. **star shell** an explosive projectile designed to burst in the air and light up the enemy's position. **star-spangled** (esp. of the US national flag) covered or glittering with stars. **star stream** a systematic drift of stars. **star-struck** fascinated or greatly impressed by stars or stardom. **star-studded** containing or covered with many stars, esp. featuring many famous performers. **star turn** the principal item in an entertainment or performance. □ **stardom** *n.* **starless** *adj.* **starlike** *adj.* [OE *steorra* f. Gmc]

▪ True stars were formerly known as the 'fixed stars', to distinguish them from the planets or 'wandering stars'. They are gaseous spheres consisting primarily of hydrogen and helium, there being an equilibrium between the compressional force of gravity and the outward pressure of radiation resulting from internal thermonuclear fusion reactions. Apart from the sun, the nearest star to earth is Proxima Centauri. All others are many light-years away, so that they appear in relatively fixed positions in the sky. Some six thousand stars are visible to the naked eye, but there are actually more than a hundred thousand million in our own Galaxy, while billions of other galaxies are known. A star is classified according to its spectral type, which is related to its mass and stage in its life cycle (see HARVARD CLASSIFICATION and HERTZSPRUNG-RUSSELL DIAGRAM).

Stara Zagora /ˌstɑːrə zəˈɡɔːrə/ a city in east central Bulgaria; pop. (1990) 188,230. It was held by the Turks from 1370 until 1877, when it was destroyed by them during the Russo-Turkish War. It has since been rebuilt as a modern planned city.

starboard /ˈstɑːbəd/ n. & v. Naut. & Aeron. ● n. the right-hand side (looking forward) of a ship, boat, or aircraft (cf. PORT³). ● v.tr. (also absol.) turn (the helm) to starboard. □ **starboard tack** see TACK¹ n. 4a. **starboard watch** see WATCH n. 3b. [OE stēorbord = rudder side (see STEER¹, BOARD), early Teutonic ships being steered with a paddle over the right side]

starburst /ˈstɑːbɜːst/ n. **1** a pattern of radiating lines around a central object, light source, etc.; an explosion or Photog. a lens attachment producing this effect (starburst filter). **2** Astron. a period of intense activity, apparently star formation, in some galaxies.

starch /stɑːtʃ/ n. & v. ● n. **1** an odourless tasteless polysaccharide occurring widely as a store of carbohydrate in plants and obtained chiefly from cereals and potatoes, forming an important constituent of the human diet. **2** a preparation of this for stiffening fabric before ironing. **3** stiffness of manner; formality. ● v.tr. stiffen (clothing) with starch. □ **starch-reduced** (esp. of food) containing less than the normal proportion of starch. □ **starcher** n. [earlier as verb: ME sterche f. OE stercan (unrecorded) stiffen f. Gmc: cf. STARK]

Star Chamber 1 (in full **Court of Star Chamber**) hist. a court which tried civil and criminal cases, especially those affecting Crown interests. (See note below.) **2** any arbitrary or oppressive tribunal. **3** colloq. a committee appointed to arbitrate between the Treasury and the spending departments of the British government.

▪ The court, which developed in the late 15th century from the judicial sittings of the Privy Council, sat in an apartment in the royal palace at Westminster, which was said to have had gilt stars on the ceiling. Under the Tudors and early Stuarts (particularly Charles I) the court became an instrument of tyranny, notorious for its arbitrary and oppressive judgements. It was abolished by Parliament in 1641.

starchy /ˈstɑːtʃɪ/ adj. (**starchier**, **starchiest**) **1 a** of or like starch. **b** containing much starch. **2** (of a person) precise, prim. □ **starchily** adv. **starchiness** n.

stardust /ˈstɑːdʌst/ n. **1** a twinkling mass. **2** an illusory or insubstantial substance. **3** a multitude of stars looking like dust. □ **have stardust in one's eyes** be dreamily romantic or unrealistic.

stare /steə(r)/ v. & n. ● v. **1** intr. (usu. foll. by at) look fixedly with eyes open, esp. as the result of curiosity, surprise, bewilderment, admiration, horror, etc. (sat staring at the door; stared in amazement). **2** intr. (of eyes) be wide open and fixed. **3** intr. be unpleasantly prominent or striking. **4** tr. (foll. by into) reduce (a person) to a specified condition by staring (stared me into silence). ● n. a staring gaze. □ **stare down** (or **out**) outstare. **stare a person in the face** be evident or imminent. □ **starer** n. [OE starian f. Gmc]

starfish /ˈstɑːfɪʃ/ n. an echinoderm of the class Asteroidea, with five or more radiating arms.

stark /stɑːk/ adj. & adv. ● adj. **1** desolate, bare (a stark landscape). **2** sharply evident; brutally simple (in stark contrast, the stark reality). **3** downright, sheer (stark madness). **4** completely naked. **5** archaic strong, stiff, rigid. ● adv. completely, wholly (stark mad; stark naked). □ **starkly** adv. **starkness** n. [OE stearc f. Gmc: stark naked f. earlier start-naked f. obs. start tail: cf. REDSTART]

Stark effect /stɑːk/ n. Physics the splitting of a spectrum line into several components by the application of an electric field. [Johannes Stark, Ger. physicist (1874–1957)]

starkers /ˈstɑːkəz/ adj. Brit. colloq. stark naked.

starlet /ˈstɑːlɪt/ n. **1** a promising young performer, esp. a woman. **2** a little star.

starlight /ˈstɑːlaɪt/ n. **1** the light of the stars (walked home by starlight). **2** (attrib.) = STARLIT (a starlight night).

Starling /ˈstɑːlɪŋ/, Ernest Henry (1866–1927), English physiologist. Studying the digestive system, he demonstrated the existence of peristalsis, and showed that a substance secreted by the pancreas passes via the blood to the duodenal wall, where it stimulates the secretion of digestive juices. He coined the term hormone for such substances, and founded the science of endocrinology. Starling also studied the theory of circulation, the functioning of heart muscle, and fluid exchange at capillary level.

starling¹ /ˈstɑːlɪŋ/ n. **1** a small gregarious partly migratory songbird, Sturnus vulgaris, with speckled blackish-brown lustrous plumage, chiefly inhabiting cultivated areas. **2** any other bird of the family Sturnidae. [OE stærlinc f. stær starling f. Gmc: cf. -LING¹]

starling² /ˈstɑːlɪŋ/ n. piles built around or upstream of a bridge or pier to protect it from floating rubbish etc. [perh. corrupt. of (now dial.) staddling STADDLE]

starlit /ˈstɑːlɪt/ adj. **1** lighted by stars. **2** with stars visible.

Star of Bethlehem n. **1** the bright star which in the biblical account of Jesus' birth led the Magi to Bethlehem (Matt. 2:9). **2** a liliaceous plant of the genus Ornithogalum, with starlike flowers; esp. O. umbellatum, with white flowers striped with green on the outside.

Star of David n. a figure consisting of two interlaced equilateral triangles, used as a symbol of Judaism and of the state of Israel.

starry /ˈstɑːrɪ/ adj. (**starrier**, **starriest**) **1** covered with or full of stars. **2** resembling a star. **3** of or relating to stars in entertainment; full of stars. □ **starry-eyed** colloq. **1** visionary; enthusiastic but impractical. **2** euphoric. □ **starrily** adv. **starriness** n.

Stars and Bars hist. the popular name of the flag of the Confederate states of the US.

Stars and Stripes the popular name of the flag of the US. Originally it contained thirteen alternating red and white stripes and thirteen stars (white on a blue background), representing the thirteen states of the Union. Today it retains the thirteen stripes, but has fifty stars, Hawaii having brought the number of states to fifty in 1959.

starship /ˈstɑːʃɪp/ n. (in science fiction) a large manned spacecraft designed for interstellar travel.

Star-spangled Banner /ˈstɑːˌspæŋɡ(ə)ld/ a song written in 1814 with words composed by Francis Scott Key (1779–1843) and a tune adapted from that of a popular English drinking song, To Anacreon in Heaven. Key was inspired to write the words after the heroic defence of Fort McHenry in Baltimore harbour against the British during the War of 1812. It was officially adopted as the US national anthem in 1931.

START /stɑːt/ abbr. Strategic Arms Reduction Talks, a series of arms-reduction negotiations between the US and the Soviet Union begun in 1983. The talks led to the signing of the Intermediate Nuclear Forces (INF) treaty in 1987 and to the signing of the START treaty in 1991, under which significant reductions in nuclear weapons were made. (See also SALT.)

start /stɑːt/ v. & n. ● v. **1** tr. & intr. begin; commence (started work; started crying; started to shout; the play starts at eight). **2** tr. set (proceedings, an event, etc.) in motion (start the meeting; started a fire). **3** intr. (often foll. by on) make a beginning (started on a new project). **4** intr. (often foll. by after, for) set oneself in motion or action ('wait!' he shouted, and started after her). **5** intr. set out; begin a journey etc. (we start at 6 a.m.). **6** (often foll. by up) **a** intr. (of a machine) begin operating (the car wouldn't start). **b** tr. cause (a machine etc.) to begin operating (tried to start the engine). **7** tr. **a** cause or enable (a person) to make a beginning (with something) (started me in business with £10,000). **b** (foll. by pres. part.) cause (a person) to begin (doing something) (the smoke started me coughing). **c** Brit. colloq. complain or be critical (don't you start). **8** tr. (often foll. by up) found or establish; originate. **9** intr. (foll. by at, with) have as the first of a series of items, e.g. in a meal (we started with soup). **10** tr. give a signal to (competitors) to start in a race. **11** intr. (often foll. by up, from, etc.) make a sudden movement from surprise, pain, etc. (started at the sound of my voice). **12** intr. (foll. by out, up, from, etc.) spring out, up, etc. (started up from the chair). **13** tr. conceive (a baby). **14** tr. rouse (game etc.) from its lair. **15 a** intr. (of a thing) be displaced from its proper position by pressure or shrinkage; give way. **b** tr. cause to do this; displace by pressure etc. **c** tr. (of a wooden ship etc.) undergo the giving way of (a plank etc.). **16** intr. (foll. by out, to, etc.) (of a thing) move or appear suddenly (tears started to his eyes). **17** intr. (foll. by from) (of eyes, usu. with exaggeration) burst forward (from their sockets etc.). **18** tr. pour out (liquor) from a cask. ● n. **1** a beginning of an event, action, journey, etc. (missed the start; an early start tomorrow; made a fresh start). **2** the place from which a race etc. begins. **3** an advantage given at the beginning of a race etc. (*a 15-second start*). **4** an advantageous initial position in life, business, etc. (*a good start in life*). **5** a sudden movement of surprise, pain, etc. (*you gave me a start*). **6** an intermittent or spasmodic effort or movement (esp. *in* or *by fits and starts*). **7** *colloq.* a surprising occurrence (*a queer start; a rum start*). □ **for a start** *colloq.* as a beginning; in the first place. **get the start of** gain an advantage over. **start a hare** see HARE. **start in** *colloq.* **1** begin. **2** (foll. by *on*) US make a beginning on. **start off 1** begin; commence (*started off on a lengthy monologue*). **2** begin to move (*it's time we started off*). **start on** *colloq.* attack; nag; bully. **start out 1** begin a journey. **2** (foll. by *to* + infin.) *colloq.* proceed as intending (to do something). **start over** *N. Amer.* begin again. **start school** attend school for the first time. **start something** *colloq.* cause trouble. **start up** arise; occur. **to start with 1** in the first place; before anything else is considered (*should never have been there to start with*). **2** at the beginning (*had six members to start with*). [OE (orig. in sense 11) f. Gmc]

starter /ˈstɑːtə(r)/ *n.* **1** a person or thing that starts. **2** an esp. automatic device for starting the engine of a motor vehicle etc. **3** a person giving the signal for the start of a race. **4** a horse or competitor starting in a race (*a list of probable starters*). **5** the first course of a meal. **6** the initial action etc. □ **for starters** *colloq.* to start with. **under starter's orders** (of racehorses etc.) in a position to start a race and awaiting the starting-signal.

starting /ˈstɑːtɪŋ/ *n.* in senses of START *v.* □ **starting-block** a shaped rigid block for bracing the feet of a runner at the start of a race. **starting-gate** a movable barrier for securing a fair start in horse-races. **starting-handle** *Brit. Mech.* a crank for starting a motor engine. **starting pistol** a pistol used to give the signal for the start of a race. **starting-point** the point from which a journey, process, argument, etc. begins. **starting post** the post from which competitors start in a race. **starting price** the odds ruling at the start of a horse-race. **starting stall** a compartment for one horse at the start of a race.

startle /ˈstɑːt(ə)l/ *v.tr.* give a shock or surprise to; cause (a person etc.) to start with surprise or sudden alarm. □ **startler** *n.* [OE *steartlian* (as START, -LE⁴)]

startling /ˈstɑːtlɪŋ/ *adj.* **1** surprising. **2** alarming (*startling news*). □ **startlingly** *adv.*

Start Point a headland on the south coast of Devon, to the south-west of Torquay.

starve /stɑːv/ *v.* **1** *intr.* die of hunger; suffer from malnourishment. **2** *tr.* cause to die of hunger or suffer from lack of food. **3** *intr.* suffer from extreme poverty. **4** *intr. colloq.* (esp. as **starved** or **starving** *adjs.*) feel very hungry (*I'm starving*). **5** *intr.* **a** suffer from mental or spiritual want. **b** (foll. by *for*) feel a strong craving for (sympathy, amusement, knowledge, etc.). **6** *tr.* **a** (foll. by *of*) deprive of; keep scantily supplied with (*starved of affection*). **b** cause to suffer from mental or spiritual want. **7** *tr.* **a** (foll. by *into*) compel by starving (*starved into submission*). **b** (foll. by *out*) compel to surrender etc. by starving (*starved them out*). **8** *intr. archaic* or *dial.* perish with or suffer from cold. □ **starvation** /stɑːˈveɪʃ(ə)n/ *n.* [OE *steorfan* die]

starveling /ˈstɑːvlɪŋ/ *n. & adj. archaic* ● *n.* a starving or ill-fed person or animal. ● *adj.* **1** starving. **2** meagre.

Star Wars *n. colloq.* = STRATEGIC DEFENSE INITIATIVE.

starwort /ˈstɑːwɜːt/ *n.* a plant with starlike flowers or leaves; esp. greater stitchwort, *Stellaria holostea*, with small white flowers.

stash /staʃ/ *v. & n. colloq.* ● *v.tr.* (often foll. by *away*) **1** conceal; put in a safe or hidden place. **2** hoard, stow, store. ● *n.* **1** a hiding-place or hide-out. **2** a thing hidden; a cache. [18th c.: orig. unkn.]

Stasi /ˈʃtɑːzɪ/ *hist.* the internal security force of the German Democratic Republic (East Germany). [G, acronym f. *Staatssicherheits(dienst)* state security service]

stasis /ˈsteɪsɪs, ˈstæs-/ *n.* (*pl.* **stases** /-siːz/) **1** inactivity; stagnation; a state of equilibrium. **2** a stoppage of circulation of any of the body fluids. [mod.L f. Gk f. *sta-* STAND]

-stasis /ˈsteɪsɪs, ˈstæs-/ *comb. form* (*pl.* **-stases** /-siːz/) *Physiol.* forming nouns denoting a slowing or stopping (*haemostasis*). □ **-static** /ˈstætɪk/ *comb. form* forming adjectives.

-stat /stæt/ *comb. form* forming nouns with ref. to keeping fixed or stationary (*rheostat*). [Gk *statos* stationary]

state /steɪt/ *n. & v.* ● *n.* **1** the existing condition or position of a person or thing (*in a bad state of repair; in a precarious state of health*). **2** *colloq.* **a** an excited, anxious, or agitated mental condition (esp. *in a state*). **b** an untidy condition. **3** (also **State**) **a** an organized political community

under one government; a commonwealth; a nation. **b** such a community forming part of a federal republic, esp. the United States of America. **c** (**the States**) the US. **4** (also **State**) (*attrib.*) **a** of, for, or concerned with the state (*state documents*). **b** reserved for or done on occasions of ceremony (*state apartments; state visit*). **c** involving ceremony (*state opening of Parliament*). **5** (also **State**) civil government (*Church and state; Secretary of State*). **6 a** pomp, rank, dignity (*as befits their state*). **b** imposing display; ceremony, splendour (*arrive in state*). **7** (**the States**) the legislative body in Jersey, Guernsey, and Alderney. **8** each of two or more variant forms of a single edition of a book. **9 a** an etched or engraved plate at a particular stage of its progress. **b** an impression taken from this. ● *v.tr.* **1** express, esp. fully or clearly, in speech or writing (*have stated my opinion; must state full particulars*). **2** fix, specify (*at stated intervals*). **3** *Law* specify the facts of (a case) for consideration. **4** *Mus.* play (a theme etc.) so as to make it known to the listener. □ **in state** with all due ceremony. **of state** concerning politics or government. **state capitalism** a system of state control and use of capital. **state house** US **1** the building where the legislature of a state meets. **2** *NZ* a private house built at the government's expense. **state of the art 1** the current stage of development of a practical or technological subject. **2** (usu. **state-of-the-art**) (*attrib.*) using the latest techniques or equipment (*state-of-the-art weaponry*). **state of grace** the condition of being free from grave sin. **state of life** rank and occupation. **state of things** (or **affairs** or **play**) the circumstances; the current situation. **state of war** the situation when war has been declared or is in progress. **state prisoner** see PRISONER. **state school** a school managed and funded by the public authorities. **state's evidence** see EVIDENCE. **state socialism** a system of state control of industries and services. **states' rights** US the rights and powers not assumed by the United States but reserved to its individual states. **state trial** prosecution by the state. **state university** US a university managed by the public authorities of a state. □ **statable** *adj.* **statedly** *adv.* **statehood** *n.* [ME: partly f. ESTATE, partly f. L STATUS]

statecraft /ˈsteɪtkrɑːft/ *n.* the art of conducting affairs of state.

State Department (in the US) the department of foreign affairs.

State Enrolled Nurse *n.* (abbr. **SEN**) (in the UK) a nurse enrolled on a state register and having a qualification lower than that of a State Registered Nurse.

stateless /ˈsteɪtlɪs/ *adj.* **1** (of a person) having no nationality or citizenship. **2** without a state or political community. □ **statelessness** *n.*

stately /ˈsteɪtlɪ/ *adj.* (**statelier, stateliest**) dignified; imposing; grand. □ **stately home** *Brit.* a large magnificent house, esp. one open to the public. □ **stateliness** *n.*

statement /ˈsteɪtmənt/ *n.* **1** the act or an instance of stating or being stated; expression in words. **2** a thing stated; a declaration (*that statement is unfounded*). **3** a formal account of facts, esp. to the police or in a court of law (*make a statement*). **4** a record of transactions in a bank account etc. **5** a formal notification of the amount due to a tradesman etc.

Staten Island /ˈstæt(ə)n/ an island borough of New York City, named by early Dutch settlers after the Staten or States General of the Netherlands; pop. (1990) 378,980. Situated in the south-western part of the city, it is separated from New Jersey by the narrow channel known as Arthur Kill.

State of the Union message (also **State of the Union address**) a yearly address delivered in January by the President of the US to Congress, giving the administration's view of the state of the nation and plans for legislation.

stater /ˈsteɪtə(r)/ *n. hist.* an ancient Greek gold or silver coin. [ME f. LL f. Gk *statēr*]

State Registered Nurse *n.* (abbr. **SRN**) (in the UK) a nurse enrolled on a state register and more highly qualified than a State Enrolled Nurse.

stateroom /ˈsteɪtruːm, -rʊm/ *n.* **1** a state apartment in a palace, hotel, etc. **2** a private compartment in a passenger ship or US train.

States General *n.* (also **Estates General**) a legislative body, esp. in the Netherlands from the 15th to 18th centuries or in France before 1789, representing the three estates of the realm (i.e. the clergy, the nobility, and the commons). The States General in France was first convened by King Philip IV in 1302 but was called upon only occasionally until it was revived as a political device against the Huguenots during the French Wars of Religion. It met in 1614 but then not again until 1789, when it was urgently summoned to push through much-needed financial and administrative reforms. The same voting methods as in 1614 were used and as a result the radical Third Estate (the commons) gained control and formed themselves into a National Assembly, helping to precipitate the French Revolution. [F *états généraux*, Du. *staten generaal*]

Stateside /ˈsteɪtsaɪd/ *adj.* esp. *US colloq.* of, in, or towards the United States.

statesman /ˈsteɪtsmən/ *n.* (*pl.* **-men**; *fem.* **stateswoman**, *pl.* **-women**) **1** a person skilled in affairs of state, esp. one taking an active part in politics. **2** a distinguished and capable politician. □ **statesmanlike** *adj.* **statesmanly** *adj.* **statesmanship** *n.* [= *state's man* after F *homme d'état*]

statewide /ˈsteɪtwaɪd/ *adj. US* so as to include or cover a whole state.

static /ˈstætɪk/ *adj. & n.* ● *adj.* **1** stationary; not acting or changing; passive. **2** *Physics* **a** concerned with bodies at rest or forces in equilibrium (opp. DYNAMIC *adj.* 2a). **b** acting as weight but not moving (*static pressure*). **c** of statics. ● *n.* **1** static electricity. **2** atmospherics. **3** esp. *US sl.* aggravation, fuss; criticism. □ **static electricity** electricity not flowing as a current. **static line** a length of cord attached to an aircraft etc. which releases a parachute without the use of a ripcord. [mod.L *staticus* f. Gk *statikos* f. *sta-* stand]

statical /ˈstætɪk(ə)l/ *adj.* = STATIC *adj.* □ **statically** *adv.*

statice /ˈstætɪsɪ/ *n.* **1** sea lavender. **2** sea pink. [L f. Gk, fem. of *statikos* STATIC (with ref. to stanching of blood)]

statics /ˈstætɪks/ *n.pl.* (usu. treated as *sing.*) **1** the science of bodies at rest or of forces in equilibrium (opp. DYNAMICS 1a). **2** = STATIC *n.* [from STATIC *n.* + -s[1]: see -ICS]

station /ˈsteɪʃ(ə)n/ *n. & v.* ● *n.* **1 a** a regular stopping place on a public transport route, esp. one on a railway line with a platform and usu. administrative buildings. **b** these buildings (see also *bus station*, *coach station*). **2** a place or building etc. where a person or thing stands or is placed, esp. habitually or for a definite purpose. **3 a** a designated point or establishment where a particular service or activity is based or organized (*police station*; *polling station*). **b** *US* a subsidiary post office. **4** an establishment involved in radio or television broadcasting. **5 a** a military or naval base, esp. *hist.* in India. **b** the inhabitants of this. **6** position in life; rank or status (*ideas above your station*). **7** *Austral. & NZ* a large sheep or cattle farm. **8** *Bot.* a particular place where an unusual species etc. grows. ● *v.tr.* **1** assign a station to. **2** put in position. □ **station-bill** *Naut.* a list showing the prescribed stations of a ship's crew for various drills or in an emergency. **station break** *N. Amer.* a pause between broadcast programmes for an announcement of the identity of the station transmitting them. **station hand** *Austral.* a worker on a large sheep or cattle farm. **station house** *US* a police station. **station-keeping** the maintenance of one's proper relative position in a moving body of ships etc. **station pointer** *Naut.* a ship's navigational instrument, often a three-armed protractor, for fixing one's place on a chart from the angle in the horizontal plane between two land or sea-marks. **station sergeant** *Brit.* the sergeant in charge of a police station. **station-wagon** an estate car. [ME, = standing, f. OF f. L *statio -onis* f. *stare* stand]

stationary /ˈsteɪʃənərɪ/ *adj.* **1** remaining in one place, not moving (*hit a stationary car*). **2** not meant to be moved; not portable (*stationary troops*; *stationary engine*). **3** not changing in magnitude, number, quality, efficiency, etc. (*stationary temperature*). **4** *Astron.* (of a planet) having no apparent motion in longitude. □ **stationary air** air remaining in the lungs during ordinary respiration. **stationary bicycle** a fixed exercise-machine resembling a bicycle. **stationary point** *Math.* a point on a curve where the gradient is zero. **stationary wave** = *standing wave.* □ **stationariness** *n.* [ME f. L *stationarius* (as STATION)]

stationer /ˈsteɪʃənə(r)/ *n.* a person who sells writing materials etc. [ME, = bookseller (as STATIONARY in med.L sense 'shopkeeper', esp. bookseller, as opposed to pedlar)]

Stationers' Hall *n.* (in the UK) the hall of the Stationers' Company in London, at which a book was formerly registered for purposes of copyright.

stationery /ˈsteɪʃənərɪ/ *n.* writing materials etc. sold by a stationer.

Stationery Office *n.* (in the UK) the government's publishing house which also provides stationery for government offices.

stationmaster /ˈsteɪʃ(ə)nˌmɑːstə(r)/ *n.* the official in charge of a railway station.

Stations of the Cross a series of fourteen pictures or carvings, representing successive incidents (at specific locations) during Jesus'

progress from Pilate's house to his Crucifixion at Calvary, before which devotions are performed in some Churches.

statism /ˈsteɪtɪz(ə)m/ *n.* centralized state administration and control of social and economic affairs.

statist *n.* **1** /ˈsteɪtɪst/ a statistician. **2** /ˈsteɪtɪst/ a supporter of statism. [orig. 'politician' f. It. *statista* (as STATE)]

statistic /stəˈtɪstɪk/ *n. & adj.* ● *n.* a statistical fact or item. ● *adj.* = STATISTICAL. [G *statistisch*, *Statistik* f. *Statist* (as STATIST)]

statistical /stəˈtɪstɪk(ə)l/ *adj.* of or relating to statistics. □ **statistical physics** physics as it is concerned with large numbers of particles to which statistics can be applied. **statistical significance** = SIGNIFICANCE 4. □ **statistically** *adv.*

statistics /stəˈtɪstɪks/ *n.pl.* **1** (usu. treated as *sing.*) the science of collecting and analysing numerical data, esp. in or for large quantities, and usu. inferring proportions in a whole from proportions in a representative sample. **2** any systematic collection or presentation of such facts. □ **statistician** /ˌstætɪˈstɪʃ(ə)n/ *n.*

Statius /ˈsteɪʃəs/, Publius Papinius (AD *c.*45–96), Roman poet. He flourished at the court of Domitian and is best known for the *Silvae*, a miscellany of poems addressed to friends, and the *Thebais*, an epic concerning the bloody quarrel between the sons of Oedipus. His work, which often uses mythological or fantastical images, was much admired in the Middle Ages.

stator /ˈsteɪtə(r)/ *n. Electr.* the stationary part of a machine, esp. of an electric motor or generator. [STATIONARY, after ROTOR]

statoscope /ˈstætəˌskəʊp/ *n.* an aneroid barometer used to show minute variations of pressure, esp. to indicate the altitude of an aircraft. [Gk *statos* fixed f. *sta-* stand + -SCOPE]

statuary /ˈstætjʊərɪ, ˈstætʃʊ-/ *adj. & n.* ● *adj.* of or for statues (*statuary art*). ● *n.* (*pl.* **-ies**) **1** statues collectively. **2** the art of making statues. **3** a sculptor. □ **statuary marble** fine-grained white marble. [L *statuarius* (as STATUE)]

statue /ˈstætjuː, ˈstætʃuː/ *n.* a sculptured, cast, carved, or moulded figure of a person or animal, esp. life-size or larger (cf. STATUETTE). □ **statued** *adj.* [ME f. OF f. L *statua* f. *stare* stand]

Statue of Liberty see LIBERTY, STATUE OF.

statuesque /ˌstætjʊˈesk, ˌstætʃʊ-/ *adj.* like, or having the dignity or beauty of, a statue. □ **statuesquely** *adv.* **statuesqueness** *n.* [STATUE + -ESQUE, after *picturesque*]

statuette /ˌstætjʊˈet, ˌstætʃʊ-/ *n.* a small statue; a statue less than life-size. [F, dimin. of *statue*]

stature /ˈstætʃə(r)/ *n.* **1** the height of a (esp. human) body. **2** a degree of eminence, social standing, or advancement; mental or moral calibre (*recruit someone of his stature*). □ **statured** *adj.* (also in *comb.*). [ME f. OF f. L *statura* f. *stare stat-* stand]

status /ˈsteɪtəs/ *n.* **1** rank, social position, relation to others, relative importance (*not sure of their status in the hierarchy*). **2** a superior social etc. position (*considering your status in the business*). **3** *Law* a person's legal standing which determines his or her rights and duties, e.g. citizen, alien, commoner, civilian, etc. **4** the position of affairs (*let me know if the status changes*). □ **status symbol** a possession etc. taken to indicate a person's high status. [L, = standing f. *stare* stand]

status quo /ˌsteɪtəs ˈkwəʊ/ *n.* the existing state of affairs. [L, = the state in which]

statutable /ˈstætjʊtəb(ə)l, ˈstætʃʊ-/ *adj.* statutory, esp. in amount or value. □ **statutably** *adv.*

statute /ˈstætjuːt, ˈstætʃuːt/ *n.* **1** a written law passed by a legislative body, e.g. an Act of Parliament. **2** a rule of a corporation, founder, etc., intended to be permanent (*against the University Statutes*). **3** divine law (*kept thy statutes*). □ **statute-barred** *Law* (of a case etc.) no longer legally enforceable by reason of the lapse of time. **statute-book 1** a book or books containing the statute law. **2** the body of a country's statutes. **statute law 1** (*collect.*) the body of principles and rules of law laid down in statutes as distinct from rules formulated in practical application (cf. *common law*, *case-law* (see CASE[1])). **2** a statute. **statute mile** see MILE 1. **statute-roll 1** the rolls in the Public Records Office containing the statutes of the Parliament of England. **2** a statute-book. **statutes at large** the statutes as originally enacted, regardless of later modifications. [ME f. OF *statut* f. LL *statutum* neut. past part. of L *statuere* set up f. *status*: see STATUS]

statutory /ˈstætjʊtərɪ, ˈstætʃʊ-/ *adj.* **1** required, permitted, or enacted by statute (*statutory minimum*; *statutory provisions*). **2** (of a criminal offence) carrying a penalty prescribed by statute. □ **statutory rape** US the offence of sexual intercourse with a minor. □ **statutorily** adv.

staunch[1] /stɔ:ntʃ/ adj. (also **stanch** /sta:ntʃ/) **1** trustworthy, loyal (*my staunch friend and supporter*). **2** (of a ship, joint, etc.) strong, watertight, airtight, etc. □ **staunchly** adv. **staunchness** n. [ME f. OF *estanche* fem. of *estanc* f. Rmc: see STANCH[1]]

staunch[2] var. of STANCH[1].

Stavanger /stə'væŋə(r)/ a seaport in SW Norway; pop. (1991) 98,180. One of Norway's oldest towns, it was first settled in the 8th century, and developed as a fishing centre in the late 19th century. It is now an important centre servicing offshore oilfields in the North Sea.

stave /steɪv/ n. & v. ● n. **1** each of the curved pieces of wood forming the sides of a cask, pail, etc. **2** = STAFF[1] n. 3. **3** a stanza or verse. **4** the rung of a ladder. ● v.tr. (*past* and *past part.* **stove** /stəʊv/ or **staved**) **1** break a hole in. **2** crush or knock out of shape. **3** fit or provide (a cask etc.) with staves. □ **stave in** crush by forcing inwards. **stave off** (*past* and *past part.* **staved**) avert or defer (danger or misfortune). **stave rhyme** alliteration, esp. in old Germanic poetry. [ME, back-form. f. *staves*, pl. of STAFF[1]]

staves pl. of STAFF[1] n. 3.

stavesacre /'steɪvzˌeɪkə(r)/ n. a larkspur, *Delphinium staphisagria*, yielding seeds used as an insecticide. [ME f. L *staphisagria* f. Gk *staphis agria* wild raisin]

Stavropol /'stævrəˌpɒl, stæv'rɒp(ə)l/ **1** a krai (administrative territory) in southern Russia, in the northern Caucasus; pop. (1989) 2,855,000. **2** its capital city; pop. (1990) 324,000. **3** the former name (until 1964) for TOGLIATTI.

stay[1] /steɪ/ v. & n. ● v. **1** intr. continue to be in the same place or condition; not depart or change (*stay here until I come back*). **2** intr. **a** (often foll. by *at, in, with*) have temporary residence as a visitor etc. (*stayed with them for Christmas*). **b** Sc. & S. Afr. dwell permanently. **3** archaic or literary **a** tr. stop or check (progress, the inroads of a disease, etc.). **b** intr. (esp. in *imper.*) pause in movement, action, speech, etc. (*Stay! You forget one thing*). **4** tr. postpone (judgement, decision, etc.). **5** tr. assuage (hunger etc.) esp. for a short time. **6** a intr. show endurance. **b** tr. show endurance to the end of (a race etc.). **7** tr. (often foll. by *up*) literary support, prop up (as or with a buttress etc.). **8** intr. (foll. by *for, to*) wait long enough to share or join in an activity etc. (*stay to supper; stay for the film*). ● n. **1** **a** the act or an instance of staying or dwelling in one place. **b** the duration of this (*just a ten-minute stay; a long stay in London*). **2** a suspension or postponement of a sentence, judgement, etc. (*was granted a stay of execution*). **3** archaic or literary a check or restraint (*will endure no stay; a stay upon his activity*). **4** endurance, staying power. **5** a prop or support. **6** (in pl.) hist. a corset esp. with whalebone etc. stiffening, and laced. □ **has come** (or **is here**) **to stay** colloq. must be regarded as permanent. **stay-at-home** adj. remaining habitually at home. ● n. a person who does this. **stay-bar** (or **-rod**) a support used in building or in machinery. **stay the course** pursue a course of action or endure a struggle etc. to the end. **stay one's hand** see HAND. **stay in** remain indoors or at home, esp. in school after hours as a punishment. **staying power** endurance, stamina. **stay-in strike** = *sit-down strike*. **stay the night** remain until the next day. **stay put** colloq. remain where it is placed or where one is. **stay up** not go to bed (until late at night). □ **stayer** n. [AF *estai-* stem of OF *ester* f. L *stare* stand: sense 5 f. OF *estaye(r)* prop, formed as STAY[2]]

stay[2] /steɪ/ n. & v. ● n. **1** **a** Naut. a large rope used to brace a mast, leading from the masthead to another mast or spar, or down to another part of the ship. **b** a guy or rope supporting a flagstaff or other upright pole. **2** a tie-piece in an aircraft etc. ● v.tr. **1** support (a mast etc.) by stays. **2** Naut. put (a sailing-ship) on another tack. □ **be in stays** Naut. (of a sailing-ship) be head to the wind while tacking. **miss stays** Naut. fail to be in stays. [OE *stæg* be firm, f. Gmc]

staysail /'steɪseɪl, -s(ə)l/ n. Naut. a triangular fore-and-aft sail extended on a stay.

STD abbr. **1** subscriber trunk dialling. **2** sexually transmitted disease. **3** Doctor of Sacred Theology. [sense 3 f. L *Sanctae Theologiae Doctor*]

stead /sted/ n. □ **in a person's** (or **thing's**) **stead** as a substitute; instead of him or her or it. **stand a person in good stead** be advantageous or serviceable to him or her. [OE *stede* f. Gmc]

steadfast /'stedfɑ:st/ adj. constant, firm, unwavering. □ **steadfastly** adv. **steadfastness** n. [OE *stedefæst* (as STEAD, FAST[1])]

steading /'stedɪŋ/ n. Sc. & N. Engl. a farmstead.

steady /'stedɪ/ adj., v., adv., int., & n. ● adj. (**steadier**, **steadiest**) **1** firmly fixed or supported or standing or balanced; not tottering, rocking, or wavering. **2** done or operating or happening in a uniform and regular manner (*a steady pace; a steady increase*). **3** **a** constant in mind or conduct; not changeable. **b** persistent. **4** (of a person) serious and dependable in behaviour; of industrious and temperate habits; safe; cautious. **5** regular, established (*a steady girlfriend*). **6** accurately directed; not faltering, controlled (*a steady hand; a steady eye; steady nerves*). **7** (of a ship) on course and upright. ● v.tr. & intr. (**-ies**, **-ied**) make or become steady (*steady the boat*). ● adv. steadily (*hold it steady*). ● int. as a command or warning to take care. ● n. (pl. **-ies**) colloq. a regular boyfriend or girlfriend. □ **go steady** (often foll. by *with*) colloq. have as a regular boyfriend or girlfriend. **steady down** become steady. **steady-going** staid; sober. **steady on!** a call to take care. □ **steadier** n. **steadily** adv. **steadiness** n. [STEAD = place, + -Y[1]]

steady state n. an unvarying condition, especially in a physical process. The term is used specifically of a cosmological theory put forward by Sir James Jeans c.1920, in a revised form by the Austrian-born British theoretical physicist Sir Hermann Bondi (b.1919) and the Austrian-born American astronomer Thomas Gold (b.1920) in 1948, and further developed by Sir Fred Hoyle. This postulates that the universe maintains a constant average density, with more matter continuously created to fill the void left by galaxies that are receding from one another. The theory has now largely been abandoned in favour of the big bang theory and an evolving universe. (See also BIG BANG 1.)

steak /steɪk/ n. **1** a thick slice of meat (esp. beef) or fish, often cut for grilling, frying, etc. **2** beef cut for stewing or braising. □ **steak-house** a restaurant specializing in serving beefsteaks. **steak-knife** a knife with a serrated steel blade for eating steak. [ME f. ON *steik* rel. to *steikja* roast on spit, *stikna* be roasted]

steal /sti:l/ v. & n. ● v. (*past* **stole** /stəʊl/; *past part.* **stolen** /'stəʊlən/) **1** tr. (also *absol.*) **a** take (another person's property) illegally. **b** take (property etc.) without right or permission, esp. in secret with the intention of not returning it. **2** tr. obtain surreptitiously or by surprise (*stole a kiss*). **3** tr. **a** gain insidiously or artfully. **b** (often foll. by *away*) win or get possession of (a person's affections etc.), esp. insidiously (*stole her heart away*). **4** intr. (foll. by *in, out, away, up,* etc.) **a** move, esp. silently or stealthily (*stole out of the room*). **b** (of a sound etc.) become gradually perceptible. **5** tr. **a** (in various sports) gain (a run, the ball, etc.) surreptitiously or by luck. **b** Baseball run to (a base) while the pitcher is in the act of delivery. ● n. **1** US colloq. the act or an instance of stealing or theft. **2** colloq. an unexpectedly easy task or good bargain. □ **steal a march on** get an advantage over by surreptitious means; anticipate. **steal the show** outshine other performers, esp. unexpectedly. **steal a person's thunder 1** use another person's idea, policy, etc., and spoil the effect the originator hoped to achieve by expressing it or acting upon it first. **2** take the limelight or attention from another person. (From a remark of the Engl. dramatist John Dennis (1657–1734) when the stage thunder he had invented for his own play was used for another.) □ **stealer** n. (also in comb.). **stealing** n. [OE *stelan* f. Gmc]

stealth /stelθ/ n. **1** secrecy, a secret procedure. **2** (also **stealth bomber**) a type of military aircraft capable of extremely high speed and designed to evade detection, esp. by enemy radar, while in flight. □ **by stealth** surreptitiously. [ME f. OE (as STEAL, -TH[2])]

stealthy /'stelθɪ/ adj. (**stealthier**, **stealthiest**) **1** (of an action) done with stealth; proceeding imperceptibly. **2** (of a person or thing) acting or moving with stealth. □ **stealthily** adv. **stealthiness** n.

steam /sti:m/ n. & v. ● n. **1** **a** the gas into which water is changed by boiling, used as a source of power by virtue of its expansion of volume. **b** a mist of liquid particles of water produced by the condensation of this gas. **2** any similar vapour. **3** **a** energy or power provided by a steam engine or other machine. **b** colloq. power or energy generally. ● v. **1** tr. **a** cook (food) in steam. **b** soften or make pliable (timber etc.) or otherwise treat with steam. **2** intr. give off steam or other vapour, esp. visibly. **3** intr. **a** move under steam power (*the ship steamed down the river*). **b** (foll. by *ahead, away,* etc.) colloq. proceed or travel fast or with vigour. **4** tr. & intr. (usu. foll. by *up*) cover or become covered with condensed steam. **b** (as **steamed up**) colloq. angry or excited. **5** tr. (foll. by *open* etc.) apply steam to the gum of (a sealed envelope) to get it open. **6** intr. sl. (of a gang) pass rapidly through a public place, robbing bystanders by force of numbers. □ **get up steam 1** generate enough power to work a steam engine. **2** work oneself into an energetic or angry state. **let off steam** relieve one's pent-up feelings or energy. **run out of steam** lose one's impetus or energy. **steam age** the era when trains were drawn by steam locomotives. **steam bath** a room etc. filled with steam for bathing in. **steam boiler** a vessel (in a steam engine etc.) in which water is boiled to generate steam. **steam gauge** a pressure gauge attached to a steam boiler. **steam hammer** a forging-hammer powered by steam. **steam-heat** the warmth given out by steam-heated radiators etc. **steam in** *sl.* start or join a fight. **steam iron** an electric iron that emits steam from its flat surface, to improve its pressing ability. **steam-jacket** a casing for steam round a cylinder, for heating its contents. **steam organ** a fairground pipe-organ driven by a steam engine and played by means of a keyboard or a system of punched cards. **steam power** the force of steam applied to machinery etc. **steam shovel** an excavator powered by steam. **steam-tight** impervious to steam. **steam train** a train driven by a steam engine. **steam tug** a steamer for towing ships etc. **steam turbine** a turbine in which a high-velocity jet of steam rotates a bladed disc or drum. **under one's own steam** without assistance; unaided. [OE *stēam* f. Gmc]

steamboat /ˈstiːmbəʊt/ *n.* a boat propelled by a steam engine, especially a paddle-wheel craft of a type used widely on rivers in the 19th century. Such boats were constructed experimentally in the 1780s, following James Watt's improvements of the steam engine, and pioneered most successfully by Robert Fulton. In the US the most famous steamboats were those on the Mississippi, noted for their ornate fittings and for the risk of fire and other hazards to which they were prone. (See also STEAMSHIP.)

steam engine *n.* **1** an engine in which the successive expansion and rapid condensation of steam forces a piston (or pistons) to move up and down in a cylinder (or cylinders) to produce motive power, which is transmitted to a crank by means of a connecting-rod. (*See note below.*) **2** a railway or road locomotive powered by this.
▪ The power of steam had been demonstrated by Hero of Alexandria in AD 100 but remained unrealized until the work of Thomas Newcomen, James Watt, and others in the 18th century. The steam engine has changed little in its essentials since it was improved by Richard Trevithick in the early 19th century. It made possible the Industrial Revolution, being used first for pumping water from mines and later for driving machinery in mills. It had reached its highest development by c.1900, when steam provided the power for factories, ships, railway locomotives, and traction engines. It has now been largely replaced by the steam turbine and the internal-combustion engine.

steamer /ˈstiːmə(r)/ *n.* **1** a person or thing that steams. **2** a vessel propelled by steam, esp. a ship. **3** a receptacle in which things are steamed, esp. cooked by steam. □ **steamer rug** *US* a travelling-rug.

steamroller /ˈstiːmˌrəʊlə(r)/ *n. & v.* ● *n.* **1** a heavy slow-moving vehicle with a roller, used to flatten new-made roads. **2** a crushing power or force. ● *v.tr.* **1** crush forcibly or indiscriminately. **2** (foll. by *through*) force (a measure etc.) through a legislature by overriding opposition.

steamship /ˈstiːmʃɪp/ *n.* a ship propelled by a steam engine. From the early 19th century steam engines were used to power ships. At first they were used as auxiliary engines on what were essentially sailing-ships, but in 1832 HMS *Rhadamanthus* crossed the Atlantic entirely under steam power. Several steamship lines were established to exploit the profitable Atlantic route. A disadvantage of steamships for long journeys was that they were obliged to carry large amounts of fuel, and sailing-ships for carrying cargo survived into the 20th century. (See also SHIP and STEAMBOAT.)

steamy /ˈstiːmɪ/ *adj.* (**steamier, steamiest**) **1** like or full of steam. **2** *colloq.* erotic, salacious. □ **steamily** *adv.* **steaminess** *n.*

stearic /ˈstɪərɪk, stɪˈær-/ *adj. Chem.* derived from stearin. □ **stearic acid** a solid saturated fatty acid obtained from animal or vegetable fats. □ **stearate** /ˈstɪəreɪt/ *n.* [F *stéarique* f. Gk *stear steatos* tallow]

stearin /ˈstɪərɪn/ *n.* **1** *Chem.* a glyceryl ester of stearic acid, esp. in the form of a white crystalline constituent of tallow etc. **2** a mixture of fatty acids used in candle-making. [F *stéarine* f. Gk *stear steatos* tallow]

steatite /ˈstɪətaɪt/ *n.* a soapstone or other impure form of talc. □ **steatitic** /stɪəˈtɪtɪk/ *adj.* [L *steatitis* f. Gk *steatītēs* f. *stear steatos* tallow]

steatopygia /ˌstɪətəʊˈpɪdʒɪə/ *n.* an excess of fat on the buttocks. □ **steatopygous** /-təʊˈpaɪɡəs, -ˈtɒpɪɡəs/ *adj.* [mod.L (as STEATITE + Gk *pugē* rump)]

steed /stiːd/ *n. archaic* or *poet.* a horse, esp. a fast powerful one. [OE *stēda* stallion, rel. to STUD²]

steel /stiːl/ *n., adj., & v.* ● *n.* **1** a hard strong usu. grey or bluish-grey alloy of iron with carbon and usually other elements, much used as a structural and fabricating material (*carbon steel; stainless steel*). (*See note below.*) **2** hardness of character; strength, firmness (*nerves of steel*). **3 a** a rod of steel, usu. roughened and tapering, on which knives are sharpened. **b** a strip of steel for expanding a skirt or stiffening a corset.

4 (not in *pl.*) *literary* a sword, lance, etc. (*foemen worthy of their steel*). ● *adj.* **1** made of steel. **2** like or having the characteristics of steel. ● *v.tr. & refl.* harden or make resolute (*steeled myself for a shock*). □ **cold steel** cutting or thrusting weapons. **pressed steel** steel moulded under pressure. **steel band** a band of musicians who play (chiefly calypso-style) music on steel drums. **steel-clad** wearing armour. **steel engraving** the process of engraving on or an impression taken from a steel-coated copper plate. **steel wool** an abrasive substance consisting of a mass of fine steel shavings. [OE *stȳle, stēli* f. Gmc, rel. to STAY²]
▪ The name steel is generally applied to iron alloys containing up to 2.2 per cent carbon, with the total non-ferrous content not exceeding three per cent. Additives such as chromium, manganese, silicon, and other elements and special manufacturing and finishing techniques are employed to give desired properties of hardness, strength, ductility, corrosion resistance, etc. The basis of modern steel-making is the removal of most of the non-ferrous elements from molten pig-iron by combining them with oxygen, but early techniques of making steel relied on the absorption of charcoal by heated iron. Originally (c.1200 BC) it was only possible to produce steel surfaces on iron weapons and tools, but by c.200 BC steel-making had started in India. However, steel remained a small-scale specialist product, despite the improvements introduced in England in 1740 by the mechanic and engineer Benjamin Huntsman (1704–76), who produced and refined steel by heating in a crucible. Production of steel in bulk became possible with the introduction of the Bessemer converter and the open-hearth process in the 19th century. These methods, largely obsolete, have now given way to the electric-arc furnace and the basic oxygen process.

Steele /stiːl/, Sir Richard (1672–1729), Irish essayist and dramatist. He founded and wrote for the periodicals the *Tatler* (1709–11) and the *Spectator* (1711–12), the latter in collaboration with Joseph Addison; both had an important influence on the manners, morals, and literature of the time. Steele also launched the short-lived periodical the *Guardian* (1713), to which Addison contributed.

steelhead /ˈstiːlhed/ *n.* a large North American rainbow trout.

steelwork /ˈstiːlwɜːk/ *n.* articles of steel.

steelworks /ˈstiːlwɜːks/ *n.pl.* (usu. treated as *sing.*) a place where steel is manufactured. □ **steelworker** *n.*

steely /ˈstiːlɪ/ *adj.* (**steelier, steeliest**) **1** of, or hard as, steel. **2** inflexibly severe; cold; ruthless (*steely composure; steely-eyed glance*). □ **steeliness** *n.*

steelyard /ˈstiːljɑːd/ *n.* a kind of balance with a short arm to take the item to be weighed and a long graduated arm along which a weight is moved until it balances.

steenbok /ˈsteɪnbɒk, ˈstiːn-/ *n.* a small African antelope, *Raphicerus campestris*. [Du. f. *steen* STONE + *bok* BUCK¹]

steep¹ /stiːp/ *adj. & n.* ● *adj.* **1** sloping sharply; almost perpendicular (*a steep hill; steep stairs*). **2** (of a rise or fall) rapid (*a steep drop in share prices*). **3** (*predic.*) *colloq.* **a** (of a demand, price, etc.) exorbitant; unreasonable (*esp. a bit steep*). **b** (of a story etc.) exaggerated; incredible. ● *n.* a steep slope; a precipice. □ **steepen** *v.intr. & tr.* **steepish** *adj.* **steeply** *adv.* **steepness** *n.* [OE *stēap* f. WG, rel. to STOOP¹]

steep² /stiːp/ *v. & n.* ● *v.tr.* soak or bathe in liquid. ● *n.* **1** the act or process of steeping. **2** the liquid for steeping. □ **steep in 1** pervade or imbue with (*steeped in misery*). **2** make deeply acquainted with (a subject etc.) (*steeped in the classics*). [ME f. OE f. Gmc (as STOUP)]

steeple /ˈstiːp(ə)l/ *n.* a tall tower, esp. one surmounted by a spire, above the roof of a church. □ **steeple-crowned** (of a hat) with a tall pointed crown. □ **steepled** *adj.* [OE *stēpel stȳpel* f. Gmc (as STEEP¹)]

steeplechase /ˈstiːp(ə)lˌtʃeɪs/ *n.* **1** a horse-race across the countryside or on a racecourse with ditches, hedges, etc., to jump. Steeplechasing is believed to have developed from cross-country races between two horses using church steeples as landmarks. (See HORSE-RACING.) **2** a cross-country foot-race. □ **steeplechaser** *n.* **steeplechasing** *n.*

steeplejack /ˈstiːp(ə)lˌdʒæk/ *n.* a person who climbs tall chimneys, steeples, etc., to do repairs etc.

steer¹ /stɪə(r)/ *v. & n.* ● *v.* **1 a** guide (a vehicle, aircraft, etc.) by a wheel etc. **b** guide (a vessel) by a rudder or helm. **2** *intr.* guide a vessel or vehicle in a specified direction (*tried to steer left*). **3** *tr.* direct (one's course). **4** *intr.* direct one's course in a specified direction (*steered for the railway station*). **5** *tr.* guide the movement or trend of (*steered them into the garden; steered the conversation away from that subject*). ● *n. US* steering; guidance. □ **steer clear of** take care to avoid. **steering-column** the shaft or column which connects the steering-wheel, handlebars, etc.

of a vehicle to the rest of the steering-gear. **steering committee** a committee deciding the order of dealing with business, or priorities and the general course of operations. **steering-wheel** a wheel by which a vehicle etc. is steered. □ **steerable** *adj.* **steerer** *n.* **steering** *n.* (esp. in senses 1, 2 of v.). [OE *stieran* f. Gmc]

steer[2] /stɪə(r)/ *n.* = BULLOCK. [OE *stēor* f. Gmc]

steerage /ˈstɪərɪdʒ/ *n.* **1** the act of steering. **2** the effect of the helm on a ship. **3** *archaic* the part of a ship allotted to passengers travelling at the cheapest rate. **4** *hist.* (in a warship) quarters assigned to midshipmen etc. just forward of the wardroom. □ **steerage-way** the amount of headway required by a vessel to enable it to be controlled by the helm.

steersman /ˈstɪəzmən/ *n.* (*pl.* **-men**) a person who steers a vessel.

steeve[1] /stiːv/ *n. & v. Naut.* ● *n.* the angle of the bowsprit in relation to the horizontal. ● *v.* **1** *intr.* (of a bowsprit) make an angle with the horizontal. **2** *tr.* cause (the bowsprit) to do this. [17th c.: orig. unkn.]

steeve[2] /stiːv/ *n. & v. Naut.* ● *n.* a long spar used in stowing cargo. ● *v.tr.* stow with a steeve. [ME f. OF *estiver* or Sp. *estivar* f. L *stipare* pack tight]

stegosaurus /ˌstegəˈsɔːrəs/ *n.* (also **stegosaur** /ˈstegəˌsɔː(r)/) a small-headed plant-eating dinosaur of the suborder Stegosauria, with two rows of large bony plates (or spines) along the back. [mod.L f. Gk *stegē* covering + *sauros* lizard]

Steiermark /ˈʃtaɪərˌmark/ the German name for STYRIA.

Stein /staɪn/, Gertrude (1874–1946), American writer. From 1903 she lived mainly in Paris, where during the 1920s and 1930s her home became a focus for the avant-garde, including writers such as Ernest Hemingway and Ford Madox Ford and artists such as Matisse. In her writing, Stein developed an esoteric stream-of-consciousness style, whose hallmarks include use of repetition and lack of punctuation. Her best-known work is *The Autobiography of Alice B. Toklas* (1933), in which her long-standing American companion Alice B. Toklas (1877–1967) is made the ostensible author of her own memoir.

stein /staɪn/ *n.* a large earthenware mug, esp. for beer. [G, lit. 'stone']

Steinbeck /ˈstaɪnbek/, John (Ernst) (1902–68), American novelist. His work is noted for its sympathetic and realistic portrayal of the migrant agricultural workers of California, as in *Of Mice and Men* (1937) and *The Grapes of Wrath* (1939). His later novels include *East of Eden* (1952). Steinbeck was awarded the Nobel Prize for literature in 1962.

steinbock /ˈstaɪnbɒk/ *n.* **1** an ibex native to the Alps. **2** = STEENBOK. [G f. *Stein* STONE + *Bock* BUCK[1]]

Steiner /ˈstaɪnə(r)/, Rudolf (1861–1925), Austrian philosopher, founder of anthroposophy. He joined Annie Besant's theosophist movement in 1902, but ten years later broke away to found his own Anthroposophical Society. Steiner proposed that the spiritual development of humankind had been stunted by over-attention to the material world and that to reverse this process it was necessary to nurture the faculty of cognition. His society is noted for its contribution to child-centred education, and particularly for its Steiner schools for children with learning difficulties, operating in many parts of the Western world.

Steinway /ˈstaɪnweɪ/, Henry (Engelhard) (born Heinrich Engelhard Steinweg) (1797–1871), German piano-builder, resident in the US from 1849. His name is used to designate pianos manufactured by the firm which he founded in New York in 1853.

stela /ˈstiːlə/ *n.* (*pl.* **stelae** /-liː/) *Archaeol.* an upright slab or pillar usu. with an inscription and sculpture, esp. as a gravestone. [L f. Gk (as STELE)]

stele /stiːl, ˈstiːlɪ/ *n.* **1** *Bot.* the axial cylinder of vascular tissue in the stem and roots of most plants. **2** *Archaeol.* = STELA. □ **stelar** *adj.* [Gk *stēlē* standing block]

Stella /ˈstelə/, Frank (Philip) (b.1936), American painter. In the late 1950s he reacted against the subjectivity of abstract expressionism and became an important figure in minimalism, painting a series of all-black paintings. He later experimented with shaped canvases and cut-out shapes in relief.

stellar /ˈstelə(r)/ *adj.* **1** of or relating to a star or stars. **2** having the quality of a star entertainer or performer; leading, outstanding. □ **stelliform** *adj.* [LL *stellaris* f. L *stella* star]

stellate /ˈstelɪt/ *adj.* (also **stellated** /steˈleɪtɪd/) **1** arranged like a star; radiating. **2** *Bot.* (of leaves) surrounding the stem in a whorl. [L *stellatus* f. *stella* star]

Stellenbosch /ˈstelənˌbɒs/ a university town in SW South Africa, just east of Cape Town; pop. (1985) 43,000. Founded in 1679, it is the second-oldest European settlement in South Africa, after Cape Town.

Steller's sea cow /ˈstelə/ *n.* a very large sirenian, *Hydrodamalis gigas*, found in the Bering Sea area until it was exterminated in the 18th century. [Georg Wilhelm *Steller*, Ger. naturalist (1709–46)]

stellini /steˈliːnɪ/ *n.pl.* pasta in the form of small star shapes. [It., pl. of *stellino* dimin. of *stella* star]

stellular /ˈsteljʊlə(r)/ *adj.* shaped like, or set with, small stars. [LL *stellula* dimin. of L *stella* star]

stem[1] /stem/ *n. & v.* ● *n.* **1** the main body or stalk of a plant or shrub, usu. rising above ground, but occasionally subterranean. **2** the stalk supporting a fruit, flower, or leaf, and attaching it to a larger branch, twig, or stalk. **3** a stem-shaped part of an object, esp.: **a** the slender part of a wineglass between the body and the foot. **b** the tube of a tobacco-pipe. **c** a vertical stroke in a letter or musical note. **d** the winding-shaft of a watch. **4** *Gram.* the root or main part of a noun, verb, etc., to which inflections are added; the part that appears unchanged throughout the cases and derivatives of a noun, persons of a tense, etc. **5** *Naut.* the main upright timber or metal piece at the bow of a ship to which the ship's sides are joined at the fore end (*from stem to stern*). **6** a line of ancestry, branch of a family, etc. (*descended from an ancient stem*). **7** (in full **drill stem**) a rotating rod, cylinder, etc., used in drilling. ● *v.* (**stemmed, stemming**) **1** *intr.* (foll. by *from*) spring or originate from (*stems from a desire to win*). **2** *tr.* remove the stem or stems from (fruit, tobacco, etc.). **3** *tr.* (of a vessel etc.) hold its own or make headway against (the tide etc.). □ **stem cell** *Biol.* an undifferentiated cell from which specialized cells develop. **stem stitch** an embroidery stitch used for narrow stems etc. **stem-winder** *US* a watch wound by turning a head on the end of a stem rather than by a key. □ **stemless** *adj.* **stemlet** /ˈstemlɪt/ *n.* **stemlike** *adj.* **stemmed** *adj.* (also in *comb.*). [OE *stemn, stefn* f. Gmc, rel. to STAND]

stem[2] /stem/ *v. & n.* ● *v.* (**stemmed, stemming**) **1** *tr.* check or stop. **2** *tr.* dam up (a stream etc.). **3** *intr.* slide the tail of one ski or both skis outwards usu. in order to turn or slow down. ● *n.* an act of stemming on skis. □ **stem-turn** a turn on skis made by stemming with one ski. [ON *stemma* f. Gmc: cf. STAMMER]

stemma /ˈstemə/ *n.* (*pl.* **stemmata** /-mətə/) **1** a family tree; a pedigree. **2** the line of descent e.g. of variant texts of a work. **3** *Zool.* a simple eye; a facet of a compound eye. [L f. Gk *stemma* wreath f. *stephō* wreathe]

stemple /ˈstemp(ə)l/ *n.* each of several crossbars in a mineshaft serving as supports or steps. [17th c.: orig. uncert.: cf. MHG *stempfel*]

stemware /ˈstemweə(r)/ *n. US* glasses with stems.

stench /stentʃ/ *n.* an offensive or foul smell. □ **stench trap** a trap in a sewer etc. to prevent the upward passage of gas. [OE *stenc* smell f. Gmc, rel. to STINK]

stencil /ˈstensɪl/ *n. & v.* ● *n.* **1** (in full **stencil-plate**) a thin sheet of plastic, metal, card, etc., in which a pattern or lettering is cut, used to produce a corresponding pattern on the surface beneath it by applying ink, paint, etc. **2** the pattern, lettering, etc., produced by a stencil-plate. **3** a waxed sheet etc. from which a stencil is made by means of a typewriter. ● *v.tr.* (**stencilled, stencilling**; *US* **stenciled, stenciling**) **1** (often foll. by *on*) produce (a pattern) with a stencil. **2** decorate or mark (a surface) in this way. [ME f. OF *estanceler* sparkle, cover with stars, f. *estencele* spark ult. f. L *scintilla*]

Stendhal /ˈstɒndɑːl/ (pseudonym of Marie Henri Beyle) (1783–1842), French novelist. His two best-known novels are *Le Rouge et le noir* (1830), relating the rise and fall of a young man from the provinces in the France of the Restoration (1814), and *La Chartreuse de Parme* (1839), set in a small Italian court in the same period. Both are notable for their psychological realism and political analysis.

Sten gun /sten/ *n.* a type of lightweight sub-machine-gun. [f. Major R. V. Shepherd and H. J. Turpin, Engl. designers, + Enfield, after BREN]

Steno /ˈstiːnəʊ/, Nicolaus (Danish name Niels Steensen) (1638–86), Danish anatomist and geologist. He proposed several ideas that are now regarded as fundamental to geology — that fossils are the petrified remains of living organisms, that many rocks arise from consolidation of sediments, and that such rocks occur in layers in the order in which they were laid down, thereby constituting a record of the geological history of the earth. Steno also recognized the constancy of crystal form in particular minerals. He later became a bishop.

steno /ˈstenəʊ/ *n.* (*pl.* **-os**) *N. Amer. colloq.* a stenographer. [abbr.]

stenography /steˈnɒgrəfɪ/ *n.* shorthand or the art of writing this. □ **stenographer** *n.* **stenographic** /ˌstenəˈgræfɪk/ *adj.* [Gk *stenos* narrow + -GRAPHY]

stenosis /stɪˈnəʊsɪs/ n. Med. the abnormal narrowing of a passage in the body. □ **stenotic** /-ˈnɒtɪk/ adj. [mod.L f. Gk stenōsis narrowing f. stenoō make narrow f. stenos narrow]

stenotype /ˈstenəˌtaɪp/ n. **1** a machine like a typewriter for recording speech in syllables or phonemes. **2** a symbol or the symbols used in this process. □ **stenotypist** n. [STENOGRAPHY + TYPE]

Stentor /ˈstentə(r)/ n. (also **stentor**) a person with a powerful voice. □ **stentorian** /stenˈtɔːrɪən/ adj. [Gk Stentōr, herald in the Trojan War (Homer, Iliad v. 785)]

step /step/ n. & v. ● n. **1 a** the complete movement of one leg in walking or running (took a step forward). **b** the distance covered by this. **c** (in pl.) the course followed by a person in walking etc. **2** a unit of movement in dancing. **3** a measure taken, esp. one of several in a course of action (took steps to prevent it; considered it a wise step). **4 a** a flat-topped structure used singly or as one of a series, for passing from one level to another. **b** the rung of a ladder. **c** a notch cut for a foot in ice-climbing. **d** a platform etc. in a vehicle provided for stepping up or down. **5** a short distance (only a step from my door). **6** the sound or mark made by a foot in walking etc. (heard a step on the stairs). **7** the manner of walking etc. as seen or heard (know her by her step). **8 a** a degree in the scale of promotion, advancement, or precedence. **b** one of a series of fixed points on a payscale etc. **9** (in pl.) (also **pair of steps** sing.) = STEPLADDER. **10** esp. US Mus. a melodic interval of one degree of the scale, i.e. a tone or semitone. **11** Naut. a block, socket, or platform supporting a mast. ● v. (**stepped, stepping**) **1** intr. lift and set down one's foot or alternate feet in walking. **2** intr. come or go in a specified direction by stepping. **3** intr. make progress in a specified way (stepped into a new job). **4** tr. (foll. by off, out) measure (distance) by stepping. **5** tr. perform (a dance). **6** tr. Naut. set up (a mast) in a step. □ **in step 1** stepping in time with music or other marchers. **2** conforming with others. **in a person's steps** following a person's example. **keep step** remain in step. **mind** (or **watch**) **one's step** be careful. **out of step** not in step. **step aerobics** a type of aerobics involving stepping up on to and down from a portable block. **step by step** gradually; cautiously; by stages or degrees. **step-cut** (of a gem) cut in straight facets round the centre. **step down 1** resign from a position etc. **2** Electr. decrease (voltage) by using a transformer. **step in 1** enter a room, house, etc. **2 a** intervene to help or hinder. **b** act as a substitute for an indisposed colleague etc. **step-in** attrib.adj. (of a garment) put on by being stepped into without unfastening. ● n. such a garment. **step it** dance. **step on it** (or **on the gas**) colloq. **1** accelerate a motor vehicle. **2** hurry up. **step out 1** leave a room, house, etc. **2** go out, esp. to socialize. **3** take large steps. **step out of line** behave inappropriately or disobediently. **stepping-stone 1** a raised stone, usu. one of a set in a stream, muddy place, etc., to help in crossing. **2** a means or stage of progress to an end. **step pyramid** see PYRAMID. **step this way** a deferential formula meaning 'follow me'. **step up 1** increase, intensify (must step up production). **2** Electr. increase (voltage) using a transformer. **take a step** (or **steps**) implement a course of action leading to a specific result; proceed. **turn one's steps** go in a specified direction. □ **steplike** adj. **stepped** adj. **stepwise** adv. & adj. [OE stæpe, stepe (n.), stæppan, steppan (v.), f. Gmc]

step- /step/ comb. form denoting a relationship like the one specified but resulting from a parent's remarriage. [OE stēop- orphan-]

Stepanakert /ˌstɪpanəˈkert/ the Russian name for XANKÄNDI.

stepbrother /ˈstepˌbrʌðə(r)/ n. a son of a step-parent by a marriage other than with one's father or mother.

stepchild /ˈsteptʃaɪld/ n. (pl. **-children**) a child of one's husband or wife by a previous marriage. [OE stēopcīld (as STEP-, CHILD)]

stepdaughter /ˈstepˌdɔːtə(r)/ n. a female stepchild. [OE stēopdohtor (as STEP-, DAUGHTER)]

stepfather /ˈstepˌfɑːðə(r)/ n. a male step-parent. [OE stēopfæder (as STEP-, FATHER)]

stephanotis /ˌstefəˈnəʊtɪs/ n. a climbing tropical plant of the genus Stephanotis, cultivated for its fragrant waxy usu. white flowers. [mod.L f. Gk, = fit for a wreath f. stephanos wreath]

Stephen /ˈstiːvən/ (c.1097–1154), grandson of William the Conqueror, king of England 1135–54. Stephen seized the throne of England from Matilda a few months after the death of her father Henry I. Having forced Matilda to flee the kingdom, Stephen was confronted with civil war following her invasion in 1139; although captured at Lincoln (1141) and temporarily deposed, he ultimately forced Matilda to withdraw from England in 1148. However, the year before he died Stephen was obliged to recognize Matilda's son, the future Henry II, as heir to the throne.

Stephen, St[1] (died c.35), Christian martyr. He was one of the original seven deacons in Jerusalem appointed by the Apostles. He incurred the hostility of the Jews and was charged with blasphemy before the Sanhedrin and stoned, so becoming the first Christian martyr. Saul (the future St Paul) was present at his execution. Feast day (in the Western Church) 26 Dec.; (in the Eastern Church) 27 Dec.

Stephen, St[2] (c.977–1038), king and patron saint of Hungary, reigned 1000–38. The first king of Hungary, he united Pannonia and Dacia as one kingdom, and took steps to Christianize the country. Feast day, 2 Sept. or (in Hungary) 20 Aug.

Stephenson /ˈstiːv(ə)ns(ə)n/, George (1781–1848), English engineer, the father of railways. He started as a colliery engineman, applied steam power to the haulage of coal wagons by cable, and built his first locomotive in 1814. He became engineer to the Stockton and Darlington Railway, and in 1825 drove the first train on it using a steam locomotive of his own design. George's son Robert (1803–59) assisted him in the building of engines and of the Liverpool to Manchester railway, for which they built the famous Rocket (1829) — the prototype for all future steam locomotives. Robert became famous also as a bridge designer, notably major bridges at Menai Strait and Conwy in Wales, Berwick and Newcastle in northern England, Montreal in Canada, and in Egypt.

stepladder /ˈstepˌlædə(r)/ n. a short ladder with flat steps and a folding prop, used without being leant against a surface.

stepmother /ˈstepˌmʌðə(r)/ n. a female step-parent. [OE stēopmōdor (as STEP-, MOTHER)]

step-parent /ˈstepˌpeərənt/ n. a mother's or father's later husband or wife.

steppe /step/ n. a level grassy unforested plain, esp. in SE Europe and Siberia. [Russ. step']

stepsister /ˈstepˌsɪstə(r)/ n. a daughter of a step-parent by a marriage other than with one's father or mother.

stepson /ˈstepsʌn/ n. a male stepchild. [OE stēopsunu (as STEP-, SON)]

-ster /stə(r)/ suffix denoting a person engaged in or associated with a particular activity or thing (brewster; gangster; youngster). [OE -estre etc. f. Gmc]

steradian /stəˈreɪdɪən/ n. Geom. the SI unit of solid angle (symbol **sr**), equal to the solid angle subtended at the centre of a sphere by an area of the surface equal to the square of the radius. [Gk stereos solid + RADIAN]

stercoraceous /ˌstɜːkəˈreɪʃəs/ adj. **1** consisting of or resembling dung or faeces. **2** (of an insect) living in dung. [L stercus -oris dung]

stere /stɪə(r)/ n. a unit of volume equal to one cubic metre. [F stère f. Gk stereos solid]

stereo /ˈsterɪəʊ, ˈstɪər-/ n. & adj. ● n. (pl. **-os**) **1 a** a stereophonic record-player, tape recorder, etc. **b** stereophony. **2** = STEREOSCOPE. ● adj. **1** = STEREOPHONIC. **2** stereoscopic. [abbr.]

stereo- /ˈsterɪəʊ, ˈstɪər-/ comb. form solid; having three dimensions. [Gk stereos solid]

stereobate /ˈsterɪəˌbeɪt, ˈstɪər-/ n. Archit. a solid mass of masonry as a foundation for a building. [F stéréobate f. L stereobata f. Gk stereobatēs (as STEREO-, bainō walk)]

stereochemistry /ˌsterɪəʊˈkemɪstrɪ, ˌstɪər-/ n. the branch of chemistry dealing with the three-dimensional arrangement of atoms in molecules.

stereography /ˌsterɪˈɒɡrəfɪ, ˌstɪər-/ n. the art of depicting solid bodies in a plane.

stereoisomer /ˌsterɪəʊˈaɪsəmə(r), ˌstɪər-/ n. Chem. any of two or more compounds differing only in their spatial arrangement of atoms.

stereometry /ˌsterɪˈɒmɪtrɪ, ˌstɪər-/ n. the measurement of solid bodies.

stereophonic /ˌsterɪəʊˈfɒnɪk, ˌstɪər-/ adj. (of sound reproduction) using two or more channels in such a way that the sound has the effect of being distributed and of reaching the listener from more than one direction, thus seeming more realistic. □ **stereophonically** adv. **stereophony** /-rɪˈɒfənɪ/ n.

stereopsis /ˌsterɪˈɒpsɪs, ˌstɪər-/ n. the perception of depth produced by combining the visual images from both eyes; binocular vision. □ **stereoptic** adj. [STEREO- + Gk opsis sight]

stereopticon /ˌsterɪˈɒptɪkən, ˌstɪər-/ n. a projector which combines two images to give a three-dimensional effect, or makes one image dissolve into another. [STEREO- + Gk optikon neut. of optikos OPTIC]

stereoscope /ˈsterɪəˌskəʊp, ˈstɪər-/ n. a device by which two photographs of the same object taken at slightly different angles are viewed together, giving an impression of depth and solidity as in ordinary human vision. □ **stereoscopic** /ˌsterɪəˈskɒpɪk, ˌstɪər-/ adj. **stereoscopically** adv. **stereoscopy** /-rɪˈɒskəpɪ/ n.

stereospecific /ˌsterɪəʊsprˈsɪfɪk, ˌstɪər-/ adj. Chem. of or relating to a particular stereoisomer of a substance. □ **stereospecifically** adv. **stereospecificity** /-ˌspesɪˈfɪsɪtɪ/ n.

stereotaxis /ˌsterɪəʊˈtæksɪs, ˌstɪər-/ n. (also **stereotaxy** /-ˈtæksɪ/) Biol. & Med. surgery involving the accurate positioning of probes etc. inside the brain. □ **stereotactic** /-ˈtæktɪk/ adj. **stereotaxic** adj. [STEREO- + Gk taxis orientation]

stereotype /ˈsterɪəˌtaɪp, ˈstɪər-/ n. & v. ● n. **1 a** a person or thing that conforms to an unjustifiably fixed, usu. standardized, mental picture. **b** such an impression or attitude. **2** a printing-plate cast from a mould of composed type. ● v.tr. **1** (esp. as **stereotyped** adj.) formalize, standardize; cause to conform to a type. **2 a** print from a stereotype. **b** make a stereotype of. □ **stereotypy** n. **stereotypic** /ˌsterɪəˈtɪpɪk, ˌstɪər-/ adj. **stereotypical** adj. **stereotypically** adv. [F stéréotype (adj.) (as STEREO-, TYPE)]

steric /ˈstɪərɪk, ˈster-/ adj. Chem. relating to the spatial arrangement of atoms in a molecule. □ **steric hindrance** the inhibiting of a chemical reaction by the obstruction of reacting atoms. [irreg. f. Gk stereos solid]

sterile /ˈsteraɪl/ adj. **1** not able to produce crop or fruit or (of an animal) young; barren. **2** unfruitful, unproductive (sterile discussions). **3** free from living micro-organisms etc. **4** lacking originality or emotive force; mentally barren. □ **sterilely** adv. **sterility** /stəˈrɪlɪtɪ/ n. [F stérile or L sterilis]

sterilize /ˈsterɪˌlaɪz/ v.tr. (also **-ise**) **1** make sterile. **2** deprive of the power of reproduction. □ **sterilizable** adj. **sterilizer** n. **sterilization** /ˌsterɪlaɪˈzeɪʃ(ə)n/ n.

sterlet /ˈstɜːlɪt/ n. a small sturgeon, Acipenser ruthenus, found in the Caspian Sea area and yielding fine caviare. [Russ. sterlyad']

sterling /ˈstɜːlɪŋ/ adj. & n. ● adj. **1** of or in British money (pound sterling). **2** (of a coin or precious metal) genuine; of standard value or purity. **3** (of a person or qualities etc.) of solid worth; genuine, reliable (sterling work). ● n. **1** British money (paid in sterling). **2** Austral. hist. an English-born Australian person (cf. CURRENCY 4). □ **sterling area** a group of countries with currencies tied to British sterling and holding reserves mainly in sterling. **sterling silver** silver of 92½ per cent purity. □ **sterlingness** n. [prob. f. late OE steorling (unrecorded) f. steorra star + -LING (because some early Norman pennies bore a small star): recorded earlier in OF esterlin]

Sterlitamak /ˌsteəlɪtəˈmɑːk/ an industrial city in southern Russia, situated on the Belaya river to the north of Orenburg; pop. (1990) 250,000.

stern[1] /stɜːn/ adj. severe, grim, strict; enforcing discipline or submission (a stern expression; stern treatment). □ **the sterner sex** men. □ **sternly** adv. **sternness** n. [OE styrne, prob. f. a Gmc root = be rigid]

stern[2] /stɜːn/ n. **1** the rear part of a ship or boat. **2** any rear part. □ **stern foremost** moving backwards. **stern on** with the stern presented. **stern-post** the central upright support at the stern of a ship etc., usu. bearing the rudder. □ **sterned** adj. (also in comb.). **sternmost** adj. **sternward** adj. & adv. **sternwards** adv. [ME prob. f. ON stjórn steering f. stýra STEER[1]]

sternal /ˈstɜːn(ə)l/ adj. Anat. of or relating to the sternum. □ **sternal rib** = true rib.

Sterne /stɜːn/, Laurence (1713-68), Irish novelist. He worked as a clergyman in the north of England before publishing the first two volumes of his best-known work The Life and Opinions of Tristram Shandy in 1759. Seven subsequent volumes appeared between 1761 and 1767. Both praised for its humour and condemned for its indecency at the time, Tristram Shandy parodied the developing conventions of the novel form and used devices – including a distinctive fluid narrative – which anticipated many of the stylistic concerns of modernist and later writers. He suffered from tuberculosis and after 1762 spent much of his time in France and Italy, later writing A Sentimental Journey through France and Italy (1768).

Stern Gang /stɜːn/ the British name for a militant Zionist group that campaigned in Palestine during the 1940s for the creation of a Jewish state. Founded by Avraham Stern (1907-42) as an offshoot of Irgun, the group assassinated several officials, notably the British Minister for the Middle East, Lord Moyne (1880-1944), and Count Bernadotte, the UN mediator for Palestine, in 1948.

sternum /ˈstɜːnəm/ n. (pl. **sternums** or **sterna** /-nə/) Anat. the breastbone. [mod.L f. Gk sternon chest]

sternutation /ˌstɜːnjuˈteɪʃ(ə)n/ n. Med. or joc. a sneeze or attack of sneezing. [L sternutatio f. sternutare frequent. of sternuere sneeze]

sternutator /ˈstɜːnjʊˌteɪtə(r)/ n. Med. a substance, esp. poison gas, that causes nasal irritation, violent coughing, etc. □ **sternutatory** /stɜːˈnjuːtətərɪ/ adj. & n. (pl. **-ies**)

sternway /ˈstɜːnweɪ/ n. Naut. backward motion or impetus of a ship.

sternwheeler /ˈstɜːnˌwiːlə(r)/ n. a steamer propelled by a paddle-wheel positioned at the stern.

steroid /ˈstɪərɔɪd, ˈster-/ n. Biochem. any of a group of organic compounds with a characteristic structure of four rings of carbon atoms, including many hormones, alkaloids, and vitamins. □ **steroidal** /stɪəˈrɔɪd(ə)l, steˈrɔɪ-/ adj. [STEROL + -OID]

sterol /ˈsterɒl/ n. Chem. any of a group of naturally occurring steroid alcohols. [CHOLESTEROL, ERGOSTEROL, etc.]

stertorous /ˈstɜːtərəs/ adj. (of breathing etc.) laboured and noisy; sounding like snoring. □ **stertorously** adv. **stertorousness** n. [stertor, mod.L f. L stertere snore]

stet /stet/ v. (**stetted**, **stetting**) **1** intr. (usu. as an instruction written on a proof-sheet etc.) ignore or cancel the correction or alteration; let the original form stand. **2** tr. write 'stet' against; cancel the correction of. [L, = let it stand, f. stare stand]

stethoscope /ˈsteθəˌskəʊp/ n. Med. an instrument used in listening to the sound of the heart and lungs. (See note below.) □ **stethoscopic** /ˌsteθəˈskɒpɪk/ adj. **stethoscopically** adv. **stethoscopist** /steˈθɒskəpɪst/ n. **stethoscopy** /-ˈθɒskəpɪ/ n. [F stéthoscope f. Gk stēthos breast: see -SCOPE]

▪ In 1816 the French physician René-Théophile-Hyacinthe Laënnec (1781-1826) introduced a perforated wooden cylinder which concentrated the sounds of air flowing in and out of the lungs, and described the sounds which it revealed. The modern form, with flexible tubes connecting the earpieces to a circular piece placed against the chest, was developed later in the 19th century.

stetson /ˈstets(ə)n/ n. a slouch hat with a very wide brim and a high crown. [John Batterson Stetson, Amer. hat-maker (1830-1906)]

Stettin see SZCZECIN.

stevedore /ˈstiːvəˌdɔː(r)/ n. a person employed in loading and unloading ships. [Sp. estivador f. estivar stow a cargo f. L stipare: see STEEVE[2]]

Stevenage /ˈstiːvənɪdʒ/ a town in Hertfordshire; pop. (1981) 74,520. An old town with a 12th-century church, it was designated a planned urban centre in 1946 and was developed as a new town.

stevengraph /ˈstiːv(ə)nˌɡrɑːf/ n. a colourful woven silk picture. [Thomas Stevens, Engl. weaver (1828-88), whose firm made them]

Stevens /ˈstiːv(ə)nz/, Wallace (1879-1955), American poet. He spent most of his working life as a lawyer for an insurance firm, writing poetry privately and mostly in isolation from the literary community, developing an original and colourful style. Collections of his work include Harmonium (1923), Man with the Blue Guitar and Other Poems (1937), and Collected Poems (1954), which won a Pulitzer Prize.

Stevenson /ˈstiːv(ə)ns(ə)n/, Robert Louis (Balfour) (1850-94), Scottish novelist, poet, and travel writer. He suffered from a chronic bronchial condition and spent much of his life abroad, notably in the South Seas. Stevenson made his name with the adventure story Treasure Island (1883). His other works include the novel The Strange Case of Dr Jekyll and Mr Hyde (1886) and a series of Scottish romances including Kidnapped (1886) and The Master of Ballantrae (1889). He is also known for A Child's Garden of Verses, a collection of poetry first published as Penny Whistles in 1885.

stew[1] /stjuː/ v. & n. ● v. **1** tr. & intr. cook slowly in simmering liquid in a closed container. **2** intr. colloq. be oppressed by heat or humidity, esp. in a confined space. **3** intr. colloq. **a** suffer prolonged embarrassment, anxiety, etc. **b** (foll. by over) fret or be anxious. **4** intr. (of tea) become bitter or strong with prolonged brewing. **5** tr. (as **stewed** adj.) colloq. drunk. **6** intr. (often foll. by over) colloq. study hard. ● n. **1** a dish of stewed meat etc. **2** colloq. an agitated or angry state (be in a stew). **3** archaic **a** a hot bath. **b** (in pl.) a brothel. □ **stew in one's own juice** be left to suffer the consequences of one's own actions. [ME f. OF estuve, estuver prob. ult. f. EX-[1] + Gk tuphos smoke, steam]

stew[2] /stjuː/ n. Brit. **1** an artificial oyster-bed. **2** a pond or large tank for keeping fish for eating. [ME f. F estui f. estoier confine ult. f. L studium: see STUDY]

steward /'stju:əd/ n. & v. ● n. **1** a passengers' attendant on a ship or aircraft or train. **2** an official appointed to keep order or supervise arrangements at a meeting or show or demonstration etc. **3** = *shop steward*. **4** a person responsible for supplies of food etc. for a college or club etc. **5** a person employed to manage another's property. **6** *Brit.* the title of several officers of state or the royal household. ● *v.tr.* act as a steward of (*will steward the meeting*). □ **stewardship** n. [OE *stiweard* f. *stig* prob. = house, hall + *weard* WARD]

stewardess /ˌstjuːˈdes, ˈstjuːədɪs/ n. a female steward, esp. on a ship or aircraft.

Stewart[1] var. of STUART[1].

Stewart[2] /'stju:ət/, Jackie (born John Young Stewart) (b.1939), British motor-racing driver. He was three times world champion (1969; 1971; 1973).

Stewart[3] /'stju:ət/, James (Maitland) (b.1908), American actor. He made his screen début in 1935. Famous films include *The Philadelphia Story* (1940), which earned him an Oscar, Alfred Hitchcock's *Rear Window* (1954) and *Vertigo* (1958), and westerns such as *The Man from Laramie* (1955).

Stewart Island an island of New Zealand, situated off the south coast of South Island, from which it is separated by the Foveaux Strait; chief settlement, Oban. It is named after Captain William Stewart, a whaler and sealer, who made a survey of the island in 1809.

stg. *abbr.* sterling.

Sth. *abbr.* South.

sthenic /'sθenɪk/ *adj.* **1** *Med.* strong and athletic in physique. **2** *Psychol.* vigorous and aggressive in personality. [Gk *sthenos* strength, after *asthenic*]

stick[1] /stɪk/ n. **1 a** a short slender branch or length of wood broken or cut from a tree. **b** this trimmed for use as a support or weapon. **2** a thin rod or spike of wood etc. for a particular purpose (*cocktail stick*). **3 a** an implement used to propel the ball in hockey or polo etc. **b** (in *pl.*) the raising of the stick above the shoulder in hockey. **4** a gear lever. **5** a conductor's baton. **6 a** a slender piece of a thing, e.g. celery, dynamite, deodorant, etc. **b** a number of bombs or paratroops released rapidly from aircraft. **7** (often prec. by *the*) punishment, esp. by beating. **8** *colloq.* adverse criticism; censure, reproof (*took a lot of stick*). **9** *colloq.* a piece of wood as part of a house or furniture (*a few sticks of furniture*). **10** *colloq.* a person, esp. one who is dull or unsociable (*a funny old stick*). **11** (in *pl.*; prec. by *the*) *colloq.* remote rural areas. **12** (in *pl.*) *Austral. sl.* goalposts. **13** *Naut. sl.* a mast or spar. □ **stick insect** an elongated usu. wingless insect of the family Phasmidae, with a twiglike body. **stick one's nose into** see NOSE. **up sticks** *colloq.* go to live elsewhere. □ **stickless** *adj.* **sticklike** *adj.* [OE *sticca* f. WG]

stick[2] /stɪk/ v. (*past* and *past part.* **stuck** /stʌk/) **1** *tr.* (foll. by *in, into, through*) insert or thrust (a thing or its point) (*stuck a finger in my eye; stick a pin through it*). **2** *tr.* insert a pointed thing into; stab. **3** *tr.* & *intr.* (foll. by *in, into, on,* etc.) **a** fix or be fixed on a pointed thing. **b** fix or be fixed by or as by a pointed end. **4** *tr.* & *intr.* fix or become or remain fixed by or as by adhesive etc. (*stick a label on it; the label won't stick*). **5** *intr.* endure; make a continued impression (*the scene stuck in my mind; the name stuck*). **6** *intr.* lose or be deprived of the power of motion or action through adhesion or jamming or other impediment. **7 a** *tr.* put in a specified position or place, esp. quickly or haphazardly (*stick it in your pocket*). **b** *intr.* remain in a place (*stuck indoors*). **8** *intr.* (of an accusation etc.) be convincing or regarded as valid (*could not make the charges stick*). **b** *tr.* (foll. by *on*) place the blame for (a thing) on (a person). **9** *tr. colloq.* endure, tolerate (*could not stick it any longer*). **10** *tr.* (foll. by *at*) *colloq.* persevere with. □ **be stuck 1** be unable to progress. **2** be confined in a place (*was stuck in the house*). **be stuck for** be at a loss for or in need of. **be stuck on** *colloq.* be infatuated with. **be stuck with** *colloq.* be unable to get rid of or escape from; be permanently involved with. **get stuck in** (or **into**) *colloq.* begin in earnest. **stick around** *colloq.* linger; remain at the same place. **stick at it** *colloq.* persevere. **stick at nothing** allow nothing, esp. no scruples, to deter one. **stick by** (or **with**) stay loyal or close to. **stick 'em up!** *colloq.* hands up! **stick fast** adhere or become firmly fixed or trapped in a position or place. **stick in one's gizzard** see GIZZARD. **sticking-plaster** an adhesive plaster for wounds etc. **sticking-point** the place where obstacles arise to progress or to an agreement etc. **stick-in-the-mud** *colloq.* an unprogressive or old-fashioned person. **stick in one's throat** be against one's principles. **stick it on** *colloq.* **1** make high charges. **2** tell an exaggerated story. **stick it out** *colloq.* put up with or persevere with a burden etc. to the end. **stick one's chin out** show firmness or fortitude. **stick one's neck out** expose oneself to censure etc. by acting or speaking boldly.

stick out protrude or cause to protrude or project (*stuck his tongue out; stick out your chest*). **stick out for** persist in demanding. **stick out a mile** (or **like a sore thumb**) *colloq.* be very obvious or incongruous. **stick pigs** engage in pigsticking. **stick to 1** remain close to or fixed on or to. **2** remain faithful to. **3** keep to (a subject etc.) (*stick to the point*). **stick to a person's fingers** *colloq.* (of money) be embezzled by a person. **stick together** become or remain united or mutually loyal. **stick to one's guns** see GUN. **stick to it** persevere. **stick to one's last** see LAST[3]. **stick up 1** be or make erect or protruding upwards. **2** fasten to an upright surface. **3** *colloq.* rob or threaten with a gun. **stick-up** n. *colloq.* an armed robbery using a gun. **stick up for** support or defend or champion (a person or cause). **stick up to** be assertive in the face of; offer resistance to. **stick with** remain in touch with or faithful to; persevere with. **stuck-up** *colloq.* affectedly superior and aloof, snobbish. [OE *stician* f. Gmc]

stickability /ˌstɪkəˈbɪlɪtɪ/ n. *colloq.* perseverance; staying power. [STICK[2]]

sticker /'stɪkə(r)/ n. **1** an adhesive label or notice etc. **2** a person or thing that sticks. **3** a persistent person.

stickleback /'stɪk(ə)lˌbæk/ n. a small fish of the family Gasterosteidae, with sharp spines along the back; esp. the common *three-spined stickleback* (*Gasterosteus aculeatus*). [ME f. OE *sticel* thorn, sting + *bæc* BACK]

stickler /'stɪklə(r)/ n. (foll. by *for*) a person who insists on something (*a stickler for accuracy*). [obs. *stickle* be umpire, ME *stightle* control, frequent. of *stight* f. OE *stiht(i)an* set in order]

stickpin /'stɪkpɪn/ n. N. Amer. an ornamental tie-pin.

stickweed /'stɪkwiːd/ n. US = RAGWEED 2.

sticky /'stɪkɪ/ *adj. & n.* ● *adj.* (**stickier, stickiest**) **1** tending or intended to stick or adhere. **2** glutinous, viscous. **3 a** (of the weather) humid. **b** damp with sweat. **4** *colloq.* awkward or uncooperative; intransigent (*was very sticky about giving me leave*). **5** *colloq.* difficult, awkward (*a sticky problem*). ● n. *colloq.* glue. □ **come to a sticky end** die or come to grief in an unpleasant or painful way. **sticky wicket 1** *Cricket* a pitch that has been drying after rain and is difficult for the batsman. **2** *colloq.* difficult or awkward circumstances. □ **stickily** *adv.* **stickiness** n.

stickybeak /'stɪkɪˌbiːk/ n. & v. *Austral. & NZ sl.* ● n. an inquisitive person. ● *v.intr.* pry.

Stieglitz /'stiːɡlɪts/, Alfred (1864–1946), American photographer. He was important in his pioneering work to establish photography as a fine art in the US, which he achieved through his galleries and publications, as well as through his own work. Stieglitz gained an international reputation in the 1890s when he experimented with such innovations as night-time photography. He opened the first of three galleries in New York in 1905; known as '291', the gallery exhibited not only photographs but also modern paintings and sculpture, with work by Picasso, Matisse, and Rodin, as well as by contemporary American painters such as Georgia O'Keeffe, whom Stieglitz married in 1924.

stiff /stɪf/ *adj. & n.* ● *adj.* **1** rigid; not flexible. **2** hard to bend or move or turn etc.; not working freely. **3** hard to cope with; needing strength or effort (*a stiff test; a stiff climb*). **4** severe or strong (*a stiff breeze; a stiff penalty; stiff opposition*). **5** (of a person or manner) formal, constrained; lacking spontaneity. **6** (of a muscle or limb etc., or a person affected by these) aching when used, owing to previous exertion, injury, etc. **7** (of an alcoholic or medicinal drink) strong. **8** (*predic.*) *colloq.* to an extreme degree (*bored stiff; scared stiff*). **9** (foll. by *with*) *colloq.* abounding in (*a place stiff with tourists*). **10** *colloq.* (of a price, demand, etc.) unusually high, excessive. ● n. **1** *colloq.* a corpse. **2** *sl.* a foolish or useless person (*you big stiff*). □ **stiff neck** a rheumatic condition in which the head cannot be turned without pain. **stiff-necked** obstinate or haughty. **stiff upper lip** firmness, fortitude. □ **stiffish** *adj.* **stiffly** *adv.* **stiffness** n. [OE *stif* f. Gmc]

stiffen /'stɪf(ə)n/ *v.tr.* & *intr.* make or become stiff. □ **stiffener** n. **stiffening** n.

stifle[1] /'staɪf(ə)l/ v. **1** *tr.* smother, suppress (*stifled a yawn*). **2** *intr.* & *tr.* experience or cause to experience constraint of breathing (*stifling heat*). **3** *tr.* kill by suffocating. □ **stifler** n. **stifling** *adj. & adv.* **stiflingly** *adv.* [perh. alt. of ME *stuffe, stuffle* f. OF *estouffer*]

stifle[2] /'staɪf(ə)l/ n. (in full **stifle-joint**) a joint in the legs of horses, dogs, etc., equivalent to the knee in humans. □ **stifle-bone** the bone in front of this joint. [ME: orig. unkn.]

stigma /'stɪɡmə/ n. (pl. **stigmas** or esp. in sense 4 **stigmata** /'stɪɡmətə, stɪɡˈmɑːtə/) **1** a mark or sign of disgrace or discredit. **2** (foll. by *of*) a distinguishing mark or characteristic. **3** *Bot.* the part of a pistil that receives the pollen in pollination. **4** (in *pl.*) (in Christian belief) marks corresponding to those left on Christ's body by the nails and spear at his Crucifixion. Attributed to divine favour, such marks are first recorded as occurring on St Francis of Assisi. **5** a mark or spot on the skin or on a butterfly-wing. **6** *Med.* a visible sign or characteristic of a disease. **7** an insect's spiracle. [L f. Gk *stigma -atos* a mark made by a pointed instrument, a brand, a dot: rel. to STICK[1]]

stigmatic /stɪgˈmætɪk/ *adj. & n.* ● *adj.* **1** of or relating to a stigma or stigmas. **2** = ANASTIGMATIC. ● *n. Eccl.* a person bearing stigmata. □ **stigmatically** *adv.*

stigmatist /ˈstɪgmətɪst/ *n. Eccl.* = STIGMATIC *n.*

stigmatize /ˈstɪgməˌtaɪz/ *v.tr.* (also **-ise**) **1** (often foll. by *as*) describe as unworthy or disgraceful. **2** *Eccl.* produce stigmata on. □ **stigmatization** /ˌstɪgmətaɪˈzeɪʃ(ə)n/ *n.* [F *stigmatiser* or med.L *stigmatizo* f. Gk *stigmatizō* (as STIGMA)]

Stijl see DE STIJL.

stilb /stɪlb/ *n.* a unit of luminance equal to one candela per square centimetre. [F f. Gk *stilbō* glitter]

stilbene /ˈstɪlbiːn/ *n. Chem.* an aromatic hydrocarbon forming phosphorescent crystals. [as STILB + -ENE]

stilboestrol /stɪlˈbiːstrɒl/ *n.* (US **stilbestrol**) a powerful synthetic oestrogen derived from stilbene. [STILBENE + OESTRUS]

stile[1] /staɪl/ *n.* an arrangement of steps allowing people but not animals to climb over a fence or wall. [OE *stigel* f. a Gmc root *stig-* (unrecorded) climb]

stile[2] /staɪl/ *n.* a vertical piece in the frame of a panelled door, wainscot, etc. (cf. RAIL[1] *n.* 5). [prob. f. Du. *stijl* pillar, doorpost]

stiletto /stɪˈletəʊ/ *n.* (pl. **-os**) **1** (in full **stiletto heel**) a long tapering heel of a shoe. **b** a shoe with such a heel. **2** a short dagger with a thick blade. **3** a pointed instrument for making eyelets etc. [It., dimin. of *stilo* dagger (as STYLUS)]

still[1] /stɪl/ *adj., n., adv., & v.* ● *adj.* **1** not or hardly moving. **2** with little or no sound; calm and tranquil (*a still evening*). **3** (of sounds) hushed, stilled. **4** (of a drink) not effervescing. ● *n.* **1** deep silence (*in the still of the night*). **2** an ordinary static photograph (as opposed to a motion picture), esp. a single shot from a cinema film. ● *adv.* **1** without moving (*stand still*). **2** even now or at a particular time (*they still did not understand*; *why are you still here?*). **3** nevertheless; all the same. **4** (with *compar.* etc.) even, yet, increasingly (*still greater efforts*; *still another explanation*). ● *v.tr. & intr.* make or become still; quieten. □ **still and all** *colloq.* nevertheless. **still life** (pl. **still lifes**) **1** a painting or drawing of inanimate objects such as fruit or flowers. **2** this genre of painting. **still waters run deep** a quiet manner conceals depths of feeling or knowledge or cunning. □ **stillness** *n.* [OE *stille* (adj. & adv.), *stillan* (v.), f. WG]

still[2] /stɪl/ *n.* an apparatus for distilling spirituous liquors etc. □ **still-room** *Brit.* **1** a room for distilling. **2** a housekeeper's storeroom in a large house. [obs. *still* (v.), ME f. DISTIL]

stillage /ˈstɪlɪdʒ/ *n.* a bench, frame, etc., for keeping articles off the floor while draining, drying, waiting to be packed, etc. [app. f. Du. *stellagie* scaffold f. *stellen* to place + F *-age*]

stillbirth /ˈstɪlbɜːθ/ *n.* the birth of a dead child.

stillborn /ˈstɪlbɔːn/ *adj.* **1** (of a child) born dead. **2** (of an idea, plan, etc.) abortive; not able to succeed.

Stillson /ˈstɪls(ə)n/ *n.* (in full **Stillson wrench**) a large wrench with jaws that tighten as pressure is increased. [Daniel C. *Stillson*, Amer. inventor (1830–99)]

stilly *adv. & adj.* ● *adv.* /ˈstɪllɪ/ in a still manner. ● *adj.* /ˈstɪlɪ/ *poet.* still, quiet. [(adv.) OE *stillīce*: (adj.) f. STILL[1]]

stilt /stɪlt/ *n.* **1** either of a pair of poles with supports for the feet enabling the user to walk at a distance above the ground. **2** each of a set of piles or posts supporting a building etc. **3** a wading bird of the genus *Himantopus*, with very long slender legs. **4** a three-legged support for ceramic ware in a kiln. □ **on stilts 1** supported by stilts. **2** bombastic, stilted. □ **stiltless** *adj.* [ME & LG *stilte* f. Gmc]

stilted /ˈstɪltɪd/ *adj.* **1** (of a literary style etc.) stiff and unnatural; bombastic. **2** standing on stilts. **3** *Archit.* (of an arch) with pieces of upright masonry between the imposts and the springers. □ **stiltedly** *adv.* **stiltedness** *n.*

Stilton /ˈstɪlt(ə)n/ *n. propr.* a kind of strong rich cheese, often with blue veins, originally made in Leicestershire. It takes its name from the village of Stilton (now in Cambridgeshire), where the cheese was formerly sold to travellers at a coaching inn on the Great North Road from London.

stimulant /ˈstɪmjʊlənt/ *adj. & n.* ● *adj.* that stimulates, esp. bodily or mental activity. ● *n.* **1** a stimulant substance, esp. a drug or alcoholic drink. **2** a stimulating influence. [L *stimulare stimulant-* urge, goad]

stimulate /ˈstɪmjʊˌleɪt/ *v.tr.* **1** apply or act as a stimulus to. **2** animate, excite, arouse. **3** be a stimulant to. □ **stimulating** *adj.* **stimulatingly** *adv.* **stimulator** *n.* **stimulative** /-lətɪv/ *adj.* **stimulation** /ˌstɪmjʊˈleɪʃ(ə)n/ *n.*

stimulus /ˈstɪmjʊləs/ *n.* (pl. **stimuli** /-ˌlaɪ/) **1** a thing that rouses to activity or energy. **2** a stimulating or rousing effect. **3** *Physiol.* a thing that evokes a specific functional reaction in an organ or tissue. [L, = goad, spur, incentive]

stimy var. of STYMIE.

sting /stɪŋ/ *n. & v.* ● *n.* **1** a sharp often poisonous wounding organ of an insect, snake, nettle, etc. **2 a** the act of inflicting a wound with this. **b** the wound itself or the pain caused by it. **3** a wounding or painful quality or effect (*the sting of hunger*; *stings of remorse*). **4** pungency, sharpness, vigour (*a sting in the voice*). **5** *sl.* **a** a swindle or robbery. **b** a police undercover operation to trap a criminal. ● *v.* (*past* and *past part.* **stung** /stʌŋ/) **1 a** *tr.* wound or pierce with a sting. **b** *intr.* be able to sting; have a sting. **2** *intr. & tr.* feel or cause to feel a tingling physical or sharp mental pain. **3** *tr.* (foll. by *into*) incite by a strong or painful mental effect (*was stung into replying*). **4** *tr. sl.* swindle or charge exorbitantly. □ **stinging nettle** a nettle, *Urtica dioica*, having stinging hairs. **sting in the tail** unexpected pain or difficulty at the end of something; a twist in a story etc. □ **stingingly** *adv.* **stingless** *adj.* **stinglike** *adj.* [OE *sting* (n.), *stingan* (v.), f. Gmc]

stingaree /ˈstɪŋəˌriː, ˌstɪŋəˈriː/ *n. US & Austral.* = STINGRAY.

stinger /ˈstɪŋə(r)/ *n.* **1** a stinging insect, snake, nettle, etc. **2** a sharp painful blow.

stingray /ˈstɪŋreɪ/ *n.* a flattened cartilaginous fish of the family Dasyatidae or Urolophidae, having a broad diamond-shaped body and a long poisonous serrated spine at the base of its tail.

stingy /ˈstɪndʒɪ/ *adj.* (**stingier**, **stingiest**) niggardly, mean. □ **stingily** *adv.* **stinginess** *n.* [perh. f. dial. *stinge* STING]

stink /stɪŋk/ *v. & n.* ● *v.* (*past* **stank** /stæŋk/ or **stunk** /stʌŋk/; *past part.* **stunk**) **1** *intr.* emit a strong offensive smell. **2** *tr.* (often foll. by *out*) fill (a place) with a stink. **3** *tr.* (foll. by *out* etc.) drive (a person) out etc. by a stink. **4** *intr. colloq.* be or seem very unpleasant, contemptible, or scandalous. **5** *intr.* (foll. by *of*) *colloq.* have plenty of (esp. money). ● *n.* **1** a strong or offensive smell; a stench. **2** *colloq.* a row or fuss (*the affair caused quite a stink*). □ **like stink** *colloq.* intensely; extremely hard or fast etc. (*working like stink*). **stink bomb** a device emitting a stink when exploded. [OE *stincan* ult. f. WG: cf. STENCH]

stinker /ˈstɪŋkə(r)/ *n.* **1** a person or thing that stinks. **2** *sl.* an objectionable person or thing. **3** *sl.* **a** a difficult task. **b** a letter etc. conveying strong disapproval.

stinkhorn /ˈstɪŋkhɔːn/ *n.* a foul-smelling fungus of the order Phallales, esp. *Phallus impudicus*.

stinking /ˈstɪŋkɪŋ/ *adj. & adv.* ● *adj.* **1** that stinks. **2** *colloq.* very objectionable. ● *adv. colloq.* extremely and usu. objectionably (*stinking rich*). □ **stinking badger** a teledu. **stinking hellebore** a European hellebore, *Helleborus foetidus*, with purple-tipped greenish flowers (also called *bear's foot*). **stinking iris** the gladdon. □ **stinkingly** *adv.*

stinko /ˈstɪŋkəʊ/ *adj. sl.* drunk.

stinkpot /ˈstɪŋkpɒt/ *n. sl.* **1** a term of contempt for a person. **2** a vehicle or boat that emits foul exhaust fumes.

stinkweed /ˈstɪŋkwiːd/ *n.* = wall-rocket (see ROCKET[2]).

stinkwood /ˈstɪŋkwʊd/ *n.* a tree with foul-smelling timber; esp. *Ocotea bullata*, of southern Africa.

stinky /ˈstɪŋkɪ/ *adj.* (**stinkier**, **stinkiest**) *colloq.* having a strong or unpleasant smell.

stint /stɪnt/ *v. & n.* ● *v.* **1 a** *tr.* supply (food or aid etc.) in a niggardly amount or grudgingly. **b** *intr.* (foll. by *on*) be grudging or mean about. **2** *tr.* (often *refl.*) supply (a person etc.) in this way. ● *n.* **1** a limitation of supply or effort (*without stint*). **2** a fixed or allotted amount of work (*do one's stint*). **3** a small sandpiper, esp. a dunlin. □ **stinter** *n.* **stintless** *adj.* [OE *styntan* to blunt, dull, f. Gmc, rel. to STUNT[1]]

stipe /staɪp/ *n. Bot. & Zool.* a stalk or stem, esp. the support of a carpel, the stalk of a frond, the stem of a fungus, or an eye-stalk. □ **stipiform** /ˈstaɪpɪˌfɔːm, ˈstɪp-/ *adj.* **stipitate** /ˈstɪpɪˌteɪt/ *adj.* **stipitiform** /staɪˈpɪtɪˌfɔːm, stɪ-/ *adj.* [F f. L *stipes*: see STIPES]

stipel /'staɪp(ə)l/ n. Bot. a secondary stipule at the base of the leaflets of a compound leaf. □ **stipellate** /'staɪpə‚leɪt, staɪ'pelət/ adj. [F stipelle f. mod.L stipella dimin. (as STIPULE)]

stipend /'staɪpend/ n. a fixed regular allowance or salary, esp. one paid to a clergyman. [ME f. OF stipend(i)e or L stipendium f. stips wages + pendere to pay]

stipendiary /staɪ'pendɪərɪ/ adj. & n. ● adj. **1** receiving a stipend. **2** working for pay, not voluntarily. ● n. (pl. **-ies**) a person receiving a stipend. □ **stipendiary magistrate** a paid professional magistrate. [L stipendiarius (as STIPEND)]

stipes /'staɪpiːz/ n. (pl. **stipites** /'stɪpɪ‚tiːz/) Bot. & Zool. = STIPE. [L, = log, tree-trunk]

stipple /'stɪp(ə)l/ v. & n. ● v. **1** tr. & intr. draw or paint or engrave etc. with dots instead of lines. **2** tr. roughen the surface of (paint, cement, etc.). ● n. **1** the process or technique of stippling. **2** the effect of stippling. □ **stippler** n. **stippling** n. [Du. stippelen frequent. of stippen to prick f. stip point]

stipulate[1] /'stɪpjʊ‚leɪt/ v.tr. **1** demand or specify as part of a bargain or agreement. **2** (foll. by for) mention or insist upon as an essential condition. **3** (as **stipulated**) laid down in the terms of an agreement. □ **stipulator** n. **stipulation** /‚stɪpjʊ'leɪʃ(ə)n/ n. [L stipulari]

stipulate[2] /'stɪpjʊlət/ adj. Bot. having stipules. [L stipula (as STIPULE)]

stipule /'stɪpjuːl/ n. Bot. a small leaflike appendage to a leaf, usu. at the base of a leaf-stem. □ **stipular** adj. [F stipule or L stipula straw]

stir[1] /stɜː(r)/ v. & n. ● v. (**stirred**, **stirring**) **1** tr. move a spoon or other implement round and round in (a liquid etc.), esp. to mix the ingredients or constituents. **2 a** tr. cause to move or be disturbed, esp. slightly (a breeze stirred the lake). **b** intr. be or begin to be in motion (not a creature was stirring). **c** refl. rouse (oneself), esp. from a lethargic state. **3** intr. rise from sleep (is still not stirring). **4** intr. (foll. by out of) go out of (esp. one's house). **5** tr. arouse or inspire or excite (the emotions etc., or a person as regards these) (was stirred to anger; it stirred the imagination). **6** esp. Austral. colloq. **a** tr. annoy; tease. **b** intr. cause trouble. **7** intr. colloq. cause trouble between people by gossiping etc. ● n. **1** an act of stirring (give it a good stir). **2** commotion or excitement; public attention (caused quite a stir). **3** the slightest movement (not a stir). □ **stir the blood** inspire enthusiasm etc. **stir a finger** = lift a finger. **stir in** mix (an added ingredient) with a substance by stirring. **stir one's stumps** colloq. **1** begin to move. **2** become active. **stir up 1** mix thoroughly by stirring. **2** incite (trouble etc.) (loved stirring things up). **3** stimulate, excite, arouse (stirred up their curiosity). □ **stirless** adj. [OE styrian f. Gmc]

stir[2] /stɜː(r)/ n. sl. a prison (esp. in stir). □ **stir-crazy** deranged from long imprisonment. [19th c.: orig. unkn.]

stir-fry /'stɜːfraɪ/ v. & n. ● v.tr. (**-ies**, **-ied**) fry rapidly on a high heat while stirring and tossing. ● n. a dish consisting of stir-fried meat, vegetables, etc.

stirk /stɜːk/ n. Brit. dial. a yearling bullock or heifer. [OE stirc, perh. dimin. of stēor STEER[2]: see -OCK]

Stirling[1] /'stɜːlɪŋ/ a town in central Scotland, on the River Forth, administrative centre of Central Region; pop. (1981) 38,800. It is dominated by its 12th-century hilltop castle, a residence of the Stuart kings.

Stirling[2] /'stɜːlɪŋ/, James (1692–1770), Scottish mathematician. He proved Newton's work on cubic curves, thus earning Newton's support. His main work, Methodus Differentialis (1730), was concerned with summation and interpolation. A formula named after him, giving the approximate value of the factorial of a large number, was actually first worked out by the French-born mathematician Abraham De Moivre (1667–1754).

Stirling[3] /'stɜːlɪŋ/, Robert (1790–1878), Scottish engineer and Presbyterian minister. In 1816 he was co-inventor (with his brother) of a type of external-combustion engine using heated air, and both the engine and the heat cycle that it uses are named after him. This engine achieved a modest success in the 1890s but development lapsed until 1938, and it has not achieved commercial success despite postwar efforts using pressurized helium.

stirps /stɜːps/ n. (pl. **stirpes** /-piːz/) **1** Biol. a classificatory group. **2** Law **a** a branch of a family. **b** its progenitor. [L, = stock]

stirrer /'stɜːrə(r)/ n. **1** a thing or a person that stirs. **2** colloq. a troublemaker; an agitator.

stirring /'stɜːrɪŋ/ adj. **1** stimulating, exciting, rousing. **2** actively occupied. □ **stirringly** adv. [OE styrende (as STIR[1])]

stirrup /'stɪrəp/ n. **1** each of a pair of devices attached to each side of a horse's saddle, in the form of a loop with a flat base to support the rider's foot. **2** (attrib.) having the shape of a stirrup. **3** Anat. (in full **stirrup bone**) = STAPES. □ **stirrup-cup** a cup of wine etc. offered to a person about to depart, orig. on horseback. **stirrup-iron** the metal loop of a stirrup. **stirrup-leather** (or **-strap**) the strap attaching a stirrup to a saddle. **stirrup-pump** a hand-operated water-pump with a foot-rest, used to extinguish small fires. [OE stigrāp f. stigan climb (as STILE[1]) + ROPE]

stitch /stɪtʃ/ n. & v. ● n. **1 a** (in sewing or knitting or crocheting etc.) a single pass of a needle or the thread or loop etc. resulting from this. **b** a particular method of sewing or knitting etc. (am learning a new stitch). **2** (usu. in pl.) Surgery each of the loops of material used in sewing up a wound. **3** the least bit of clothing (hadn't a stitch on). **4** an acute pain in the side of the body induced by running etc. ● v.tr. **1** sew; make stitches (in). **2** join or close with stitches. □ **in stitches** colloq. laughing uncontrollably. **a stitch in time** a timely remedy. **stitch up 1** join or mend by sewing or stitching. **2** sl. cause (a person) to be charged with a crime, esp. by informing or manufacturing evidence; cheat. **3** colloq. = sew up 2. □ **stitcher** n. **stitchery** n. **stitchless** adj. [OE stice f. Gmc, rel. to STICK[2]]

stitchwort /'stɪtʃwɜːt/ n. a plant of the genus Stellaria, of the pink family; esp. the greater stitchwort (S. holostea), with starry white flowers and narrow leaves. [STITCH n. 4 (f. its reputed ability to cure a stitch in the side)]

stiver /'staɪvə(r)/ n. the smallest quantity or amount (don't care a stiver). [Du. stuiver a small coin, prob. rel. to STUB]

STM abbr. scanning tunnelling microscope.

stoa /'stəʊə/ n. (pl. **stoas**) **1** Archit. a portico or roofed colonnade, esp. in ancient Greece. **2** (**the Stoa**) the Stoic school of philosophy. [Gk: cf. STOIC]

stoat /stəʊt/ n. a small long-bodied flesh-eating mammal, Mustela erminea, of the weasel family, having reddish-brown upperparts and a black-tipped tail. It often turns mainly white in winter in more northern latitudes, when it is also called ermine. [ME: orig. unkn.]

stochastic /stə'kæstɪk/ adj. **1** determined by a random distribution of probabilities. **2** (of a process) characterized by a sequence of random variables. **3** governed by the laws of probability. □ **stochastically** adv. [Gk stokhastikos f. stokhazomai aim at, guess f. stokhos aim]

stock /stɒk/ n., adj., & v. ● n. **1** a store of goods etc. ready for sale or distribution etc. **2** a supply or quantity of anything for use (lay in winter stocks of fuel; a great stock of information). **3** equipment or raw material for manufacture or trade etc. (rolling-stock; paper stock). **4 a** farm animals or equipment. **b** = FATSTOCK. **5 a** the capital of a business company. **b** shares in this. **6** one's reputation or popularity (his stock is rising). **7 a** money lent to a government at fixed interest. **b** the right to receive such interest. **8** a line of ancestry; family origins (comes of Cornish stock). **9** liquid made by stewing bones, vegetables, fish, etc., as a basis for soup, gravy, sauce, etc. **10** a fragrant-flowered plant of the genus Matthiola or Malcolmia, of the pink family (orig. stock-gillyflower, so-called because it had a stronger stem than the clove gillyflower). **11** a plant into which a graft is inserted. **12** the main trunk of a tree etc. **13** (in pl.) hist. a timber frame with holes for the feet and sometimes the hands and head, in which offenders were locked as a public punishment. **14** US **a** = stock company. **b** the repertory of this. **15 a** a base or support or handle for an implement or machine. **b** the crossbar of an anchor. **16** the butt of a rifle etc. **17 a** = HEADSTOCK. **b** = TAILSTOCK. **18** (in pl.) the supports for a ship during building. **19** a band of material worn round the neck esp. in horse-riding or below a clerical collar. **20** hard solid brick pressed in a mould. ● adj. **1** kept in stock and so regularly available (stock sizes). **2** perpetually repeated; hackneyed, conventional (a stock answer). ● v.tr. **1** have or keep (goods) in stock. **2 a** provide (a shop or a farm etc.) with goods, equipment, or livestock. **b** fill with items needed (shelves well-stocked with books). **3** fit (a gun etc.) with a stock. □ **in stock** available immediately for sale etc. **on the stocks** in construction or preparation. **out of stock** not immediately available for sale. **stock-book** a book showing amounts of goods acquired and disposed of. **stock-car 1** a specially strengthened production car for use in racing in which collision occurs. **2** N. Amer. a railway truck for transporting livestock. **stock company** US a repertory company performing mainly at a particular theatre. **stock dove** a grey Eurasian wild pigeon, Columba oenas, smaller than a wood pigeon and nesting in tree holes. **stock-in-trade 1** goods kept on sale by a retailer, dealer, etc. **2** all the requisites of a trade or profession. **3** a ready supply of characteristic phrases, attitudes, etc. **stock market 1** = STOCK EXCHANGE. **2** transactions on this. **stock-still** without moving. **stock up 1** provide with or get stocks or supplies.

2 (foll. by *with*, *on*) get in or gather a stock of (food, fuel, etc.). **take stock 1** make an inventory of one's stock. **2** (often foll. by *of*) make a review or estimate (of a situation etc.). **3** (foll. by *in*) concern oneself with. □ **stocker** *n.* **stockless** *adj.* [OE *stoc*, *stocc* f. Gmc]

stockade /stɒˈkeɪd/ *n. & v.* ● *n.* **1** a line or enclosure of upright stakes. **2** esp. *N. Amer.* a prison, esp. a military one. ● *v.tr.* fortify with a stockade. [obs. F *estocade*, alt. of *estacade* f. Sp. *estacada*: rel. to STAKE[1]]

stockbreeder /ˈstɒkˌbriːdə(r)/ *n.* a farmer who raises livestock. □ **stockbreeding** *n.*

stockbroker /ˈstɒkˌbrəʊkə(r)/ *n.* = BROKER 2. □ **stockbroker belt** *Brit.* an affluent residential area, esp. near a business centre such as London. □ **stockbrokerage** *n.* **stockbroking** *n.*

stock exchange *n.* (also **Stock Exchange** or **the Exchange**) **1** an association of dealers in stocks and shares, conducting business according to fixed rules. (*See note below.*) **2** a building occupied by such dealers. **3** the level of transactions or prices in a stock market.

▪ The world's most important stock exchanges are those based in London, New York (Wall Street), and Tokyo, although the first stock exchange was established in Antwerp in the 15th century. A London stock exchange was not formed until late in the 18th century; before that dealings had previously taken place among bankers, brokers, and financial houses. Members of the new association met regularly at a coffee-house, and early in the 19th century acquired a building of their own. The New York Stock Exchange had even humbler beginnings, for it started (at the end of the 18th century) as a street market under a spreading tree in Lower Wall Street. That in London (now officially called the International Stock Exchange) formerly made a rigid division of membership between the jobbers or dealers and those who acted as brokers; this was abolished in 1986 (see BIG BANG 2).

stockfish /ˈstɒkfɪʃ/ *n.* cod or a similar fish split and dried in the open air without salt.

Stockhausen /ˈstɒkˌhaʊz(ə)n/, Karlheinz (b.1928), German composer. After studying with Olivier Messiaen (1952) he co-founded the new electronic music studio of West German Radio, and created works there such as *Gesang der Jünglinge* (1956), in which the human voice is combined with electronic sound. Later works include the serialist *Gruppen* (1955–7) for three orchestras, influenced by Anton Webern, and *Momente* (1962). With *Donnerstag* (1980), Stockhausen embarked on his *Licht* cycle of musical ceremonies, meant to be performed on each evening of a week; three further parts of the cycle had been completed by 1991.

stockholder /ˈstɒkˌhəʊldə(r)/ *n.* an owner of stocks or shares. □ **stockholding** *n.*

Stockholm /ˈstɒkhəʊm/ the capital of Sweden, a seaport on the east coast, situated on the mainland and on numerous adjacent islands; pop. (1990) 674,450. Founded in the 13th century, it established trading links with the cities of the Hanseatic League, especially Lübeck. It developed rapidly in the 17th century.

stockinet /ˌstɒkɪˈnet/ *n.* (also **stockinette**) an elastic knitted material. [prob. f. *stocking-net*]

stocking /ˈstɒkɪŋ/ *n.* **1 a** a close-fitting knitted or woven garment covering the foot and part or all of the leg, esp. a woman's semi-transparent leg-covering, usu. of nylon, reaching to the thigh. **b** esp. *US* = SOCK[1]. **2** any close-fitting garment resembling a stocking (*bodystocking*). **3** a differently coloured, usu. white, lower part of the leg of a horse etc. □ **in one's stocking** (or **stockinged**) **feet** without shoes (esp. while being measured). **stocking cap** a knitted usu. conical cap. **stocking-filler** *Brit.* a small present suitable for a Christmas stocking. **stocking-stitch** *Knitting* a stitch of alternate rows of plain and purl, making a plain smooth surface on one side. □ **stockinged** *adj.* (also in *comb.*). **stockingless** *adj.* [STOCK in (now dial.) sense 'stocking' + -ING[1]]

stockist /ˈstɒkɪst/ *n. Brit.* a dealer who stocks goods of a particular type for sale.

stockjobber /ˈstɒkˌdʒɒbə(r)/ *n.* **1** *Brit.* = JOBBER 1. **2** *US* = JOBBER 2b. □ **stockjobbing** *n.*

stocklist /ˈstɒklɪst/ *n. Brit.* a regular publication stating a dealer's stock of goods with current prices etc.

stockman /ˈstɒkmən/ *n.* (*pl.* **-men**) **1 a** *Austral.* a man in charge of livestock. **b** *US* an owner of livestock. **2** *US* a person in charge of a stock of goods in a warehouse etc.

stockpile /ˈstɒkpaɪl/ *n. & v.* ● *n.* an accumulated stock of goods,

materials, weapons, etc., held in reserve. ● *v.tr.* accumulate a stockpile of. □ **stockpiler** *n.*

Stockport /ˈstɒkpɔːt/ an industrial town in Greater Manchester; pop. (est. 1991) 130,000. Granted its charter in 1220, the town developed as a cotton-spinning centre in the 19th century.

stockpot /ˈstɒkpɒt/ *n.* a pot for cooking stock for soup etc.

stockroom /ˈstɒkruːm, -rʊm/ *n.* a room for storing goods in stock.

stocktaking /ˈstɒkˌteɪkɪŋ/ *n.* **1** the process of making an inventory of stock in a shop etc. **2** a review of one's position and resources. □ **stocktake** *n.* **stocktaker** *n.*

Stockton-on-Tees /ˌstɒktənɒnˈtiːz/ an industrial town in NE England, a port on the River Tees near its mouth on the North Sea; pop. (1991) 170,200. Granted its charter in 1310, the town developed after the opening in 1825 of the Stockton and Darlington Railway, the first passenger rail service in the world.

stocky /ˈstɒkɪ/ *adj.* (**stockier**, **stockiest**) (of a person, plant, or animal) short and strongly built; thickset. □ **stockily** *adv.* **stockiness** *n.*

stockyard /ˈstɒkjɑːd/ *n.* an enclosure with pens etc. for the sorting or temporary keeping of cattle.

stodge /stɒdʒ/ *n. & v. colloq.* ● *n.* **1** food, esp. of a thick heavy kind. **2** an unimaginative person or idea. ● *v.tr.* stuff with food etc. [earlier as verb: imit., after *stuff* and *podge*]

stodgy /ˈstɒdʒɪ/ *adj.* (**stodgier**, **stodgiest**) **1** (of food) heavy and indigestible. **2** dull and uninteresting. **3** (of a literary style etc.) turgid and dull. □ **stodgily** *adv.* **stodginess** *n.*

stoep /stuːp/ *n. S. Afr.* a terraced veranda in front of a house. [Du., rel. to STEP]

stogy /ˈstəʊɡɪ/ *n.* (also **stogie**) (*pl.* **-ies**) **1** *N. Amer.* a long narrow roughly made cigar. **2** *US* a rough heavy boot. [orig. *stoga*, short for *Conestoga* in Pennsylvania]

Stoic /ˈstəʊɪk/ *n. & adj.* ● *n.* a member of an ancient Greek school of philosophy (see STOICISM). ● *adj.* of or like the Stoics. [ME f. L *stoicus* f. Gk *stōïkos* f. STOA]

stoic /ˈstəʊɪk/ *n. & adj.* ● *n.* a stoical person. ● *adj.* = STOICAL. [STOIC]

stoical /ˈstəʊɪk(ə)l/ *adj.* having or showing great self-control in adversity. □ **stoically** *adv.*

stoichiometry /ˌstɔɪkɪˈɒmɪtrɪ/ *n.* (also **stoichometry** /stɔɪˈkɒm-/) *Chem.* **1** the fixed, usu. rational numerical relationship between the relative quantities of substances in a reaction or compound. **2** the determination or measurement of these quantities. □ **stoichiometric** /-krəˈmetrɪk/ *adj.* [Gk *stoikheion* element + -METRY]

Stoicism /ˈstəʊɪˌsɪz(ə)m/ *n.* the philosophy of the Stoics. Stoicism was an ancient Greek school of philosophy founded *c.*300 BC by Zeno of Citium and named after the *Stoa Poikilē* (painted colonnade) in Athens in which its founder used to lecture. The school taught that virtue, the highest good, is based on knowledge, and that only the wise are truly virtuous; the wise live in harmony with the divine Reason (also identified with Fate and Providence) that governs nature, and are indifferent to the vicissitudes of fortune and to pleasure and pain (and hence 'stoic' in the popular sense). Stoicism was particularly influential among the Roman upper classes, numbering Seneca and Marcus Aurelius among its followers.

stoicism /ˈstəʊɪˌsɪz(ə)m/ *n.* a stoical attitude. [STOICISM]

stoke /stəʊk/ *v.* (often foll. by *up*) **1 a** *tr.* feed and tend (a fire or furnace etc.). **b** *intr.* act as a stoker. **2** *intr. colloq.* consume food, esp. steadily and in large quantities. [back-form. f. STOKER]

stokehold /ˈstəʊkhəʊld/ *n.* a compartment in a steamship, containing its boilers and furnace.

stokehole /ˈstəʊkhəʊl/ *n.* a space for stokers in front of a furnace.

Stoke-on-Trent /ˌstəʊkɒnˈtrent/ a city in Staffordshire, on the River Trent; pop. (1991) 244,800. It has been the centre of the Staffordshire pottery industries.

Stoker /ˈstəʊkə(r)/, Abraham ('Bram') (1847–1912), Irish novelist and theatre manager. He was secretary and touring manager to the actor Henry Irving (1878–1905), but is chiefly remembered as the author of the vampire story *Dracula* (1897).

stoker /ˈstəʊkə(r)/ *n.* a person who tends the furnace on a steamship. [Du. f. *stoken* stoke f. MDu. *stoken* push, rel. to STICK[1]]

stokes /stəʊks/ *n.* (*pl.* same) *Physics* the cgs unit of kinematic viscosity, corresponding to a dynamic viscosity of 1 poise and a density of 1 gram per cubic centimetre, equivalent to 10^{-4} square metres per second. [Sir George Gabriel *Stokes*, Brit. physicist (1819–1903)]

Stokowski /stɒˈkɒfskɪ/, Leopold (1882–1977), British-born American conductor, of Polish descent. He is best known for arranging and conducting the music for Walt Disney's film *Fantasia* (1940), which sought to bring classical music to cinema audiences by means of cartoons.

STOL *abbr. Aeron.* short take-off and landing.

stole[1] /staʊl/ *n.* **1** a woman's long garment like a scarf, worn over the shoulders. **2** a strip of silk etc. worn similarly as a vestment by a priest. [OE *stol*, *stole* (orig. a long robe) f. L *stola* f. Gk *stolē* equipment, clothing]

stole[2] *past of* STEAL.

stolen *past part. of* STEAL.

stolid /ˈstɒlɪd/ *adj.* **1** lacking or concealing emotion or animation. **2** not easily excited or moved. □ **stolidly** *adv.* **stolidness** *n.* **stolidity** /stəˈlɪdɪtɪ/ *n.* [obs. F *stolide* or L *stolidus*]

stolon /ˈstaʊlən/ *n.* **1** *Bot.* a horizontal stem or branch that takes root at points along its length, forming new plants. **2** *Zool.* a branched stemlike structure in some invertebrates such as corals. □ **stolonate** /-lə-neɪt/ *adj.* **stoloniferous** /ˌstaʊləˈnɪfərəs/ *adj.* [L *stolo -onis*]

stoma /ˈstaʊmə/ *n.* (pl. **stomas** or **stomata** /-mətə/) **1** *Bot.* a minute pore in the epidermis of a leaf. **2 a** *Zool.* a small mouthlike opening in some lower animals. **b** *Surgery* a similar artificial orifice made in the abdominal wall. □ **stomal** *adj.* [mod.L f. Gk *stoma -atos* mouth]

stomach /ˈstʌmək/ *n. & v.* ● *n.* **1 a** the internal organ in which the first part of digestion occurs, being in humans a pear-shaped muscular enlargement of the alimentary canal linking the oesophagus to the small intestine. **b** each of several such organs in certain animals; esp. in ruminants, which have four (cf. RUMEN, RETICULUM 2, OMASUM, ABOMASUM). **2 a** the belly, abdomen, or lower front of the body (*pit of the stomach*). **b** a protuberant belly (*what a stomach he has got!*). **3** (usu. foll. by *for*) **a** an appetite (for food). **b** liking, readiness, or inclination (for controversy, conflict, danger, or an undertaking) (*had no stomach for the fight*). ● *v.tr.* **1** find sufficiently palatable to swallow or keep down. **2** submit to or endure (an affront etc.) (usu. with *neg.*: *cannot stomach it*). □ **muscular stomach** *see* MUSCULAR. **on an empty stomach** not having eaten recently. **on a full stomach** soon after a large meal. **stomach-ache** a pain in the belly or bowels. **stomach-pump** a syringe for forcing liquid etc. into or out of the stomach. **stomach-tube** a tube introduced into the stomach via the gullet for cleansing or emptying it. **stomach upset** (or **upset stomach**) a temporary slight disorder of the digestive system. □ **stomachful** *n.* (pl. **-fuls**). **stomachless** *adj.* [ME *stomak* f. OF *stomaque, estomac* f. L *stomachus* f. Gk *stomakhos* gullet f. *stoma* mouth]

stomacher /ˈstʌməkə(r)/ *n. hist.* **1** a pointed front-piece of a woman's dress covering the breast and pit of the stomach, often jewelled or embroidered. **2** an ornament worn on the front of a bodice. [ME, prob. f. OF *estomachier* (as STOMACH)]

stomachic /stəˈmækɪk/ *adj. & n.* ● *adj.* **1** of or relating to the stomach. **2** promoting the appetite or assisting digestion. ● *n.* a medicine or stimulant for the stomach. [F *stomachique* or L *stomachicus* f. Gk *stomakhikos* (as STOMACH)]

stomata *pl.* of STOMA.

stomatal /ˈstaʊmət(ə)l, ˈstɒm-/ *adj. Bot. & Zool.* of or relating to a stoma or stomata.

stomatitis /ˌstaʊməˈtaɪtɪs/ *n. Med.* inflammation of the mucous membrane of the mouth.

stomatology /ˌstaʊməˈtɒlədʒɪ/ *n.* the scientific study of the mouth or its diseases. □ **stomatologist** *n.* **stomatological** /-təˈlɒdʒɪk(ə)l/ *adj.*

stomp /stɒmp/ *v. & n.* ● *v.intr.* tread or stamp heavily. ● *n.* **1** a heavy tramping gait or walk. **2** a lively jazz dance with heavy stamping. □ **stomper** *n.* [US dial. var. of STAMP]

stone /staʊn/ *n. & v.* ● *n.* **1 a** solid non-metallic mineral matter, of which rock is made. **b** a piece of this, esp. a small piece. **2** *Building* **a** = LIMESTONE. **b** = SANDSTONE. **3** *Mineral.* = precious stone. **4** a stony meteorite, an aerolite. **5** (often in *comb.*) a piece of stone of a definite shape or for a particular purpose (*tombstone; stepping-stone*). **6 a** a thing resembling stone in hardness or form, e.g. the hard case of the kernel in some fruits. **b** (often in *pl.*) *Med.* a calculus (*gallstones*). **7** (pl. same) *Brit.* a unit of weight equal to 14 lb (6.35 kg). **8** (*attrib.*) made of stone. **b** of the colour of stone. ● *v.tr.* **1** pelt with stones. **2** remove the stones from (fruit). **3** face or pave etc. with stone. □ **cast** (or **throw**) **stones** make aspersions on a person's character etc. **cast** (or **throw**) **the first stone** be the first to make an accusation, esp. though guilty oneself. **leave no stone unturned** try all possible means. **stone-coal**

anthracite. **stone-cold** completely cold. **stone-cold sober** completely sober. **stone the crows** *Brit. sl.* an exclamation of surprise or disgust. **stone curlew** a mottled brown and grey plover-like bird of the family Burhinidae (also called *thick-knee*); esp. *Burhinus oedicnemus*, inhabiting open stony country. **stone-dead** completely dead. **stone-deaf** completely deaf. **stone-fruit** a fruit with flesh or pulp enclosing a stone. **stone marten** a southern Eurasian marten, *Martes foina*, having a brown coat with a white throat (also called *beech marten*). **stone parsley** an umbelliferous hedge-plant, *Sison amomum*, with strongly scented seeds. **stone pine** a southern European pine tree, *Pinus pinea*, with branches at the top spreading like an umbrella. **stone-pit** a quarry. **a stone's throw** a short distance. □ **stoned** *adj.* (also in *comb.*). **stoneless** *adj.* **stoner** *n.* [OE *stān* f. Gmc]

Stone Age the prehistoric period when weapons and tools were made out of stone or of organic materials such as bone, wood, or horn (see PREHISTORY). It is subdivided into the palaeolithic, mesolithic, and neolithic periods.

stonechat /ˈstaʊntʃæt/ *n.* a small migratory thrush of the genus *Saxicola*, with a call like two stones being knocked together; esp. *S. torquata*, with an orange breast and dark head.

stone circle *n.* a megalithic monument of a type found mainly in western Europe, consisting of usu. standing stones arranged more or less in a circle. The earliest stone circles date from the neolithic period and are often placed within a henge, as seen at Avebury in England. In the early Bronze Age many hundreds of small circles were constructed in western Britain, often from quite small stones. Stone circles frequently have other stone settings both within them and nearby, sometimes forming complex arrangements. Stonehenge is the most sophisticated circle known, though it is atypical in both style and location. Circles often appear to be aligned astronomically, especially with particular sunrise or sunset positions, and it is generally agreed that they had a ritual function.

stonecrop /ˈstaʊnkrɒp/ *n.* a succulent plant of the genus *Sedum*, usu. having yellow or white flowers and growing amongst rocks or on walls.

stonecutter /ˈstaʊnˌkʌtə(r)/ *n.* a person or machine that cuts or carves stone.

stoned /staʊnd/ *adj. sl.* under the influence of alcohol or drugs.

stonefish /ˈstaʊnfɪʃ/ *n.* (pl. same) a venomous tropical fish, *Synanceia verrucosa*, with poison glands at the base of its erect dorsal spines (cf. DEVILFISH 2).

stonefly /ˈstaʊnflaɪ/ *n.* (pl. **-flies**) a winged insect of the order Plecoptera, with aquatic larvae found under stones.

stoneground /ˈstaʊngraʊnd/ *adj.* (of flour) ground with millstones.

stonehatch /ˈstaʊnhætʃ/ *n.* a ringed plover.

Stonehenge /staʊnˈhendʒ/ a megalithic monument on Salisbury Plain in Wiltshire, England, completed in three main constructional phases between *c.*3000 BC and *c.*1500 BC. It is composed of a concentric bank and ditch around a circle of bluestones (believed to be from the Preseli Hills in South Wales), which in turn surrounds a circle of sarsen stones that were all originally surmounted by stone lintels. Within this inner circle is a horseshoe arrangement of five triliths with the axis aligned on the midsummer sunrise — an orientation that was probably for ritual rather than scientific purposes. It has also been suggested that the monument has several other astronomical alignments, which may have been used to predict eclipses and other phenomena. Stonehenge is popularly associated with the Druids, although this connection is now generally rejected by scholars; the monument has also been attributed to the Phoenicians, Romans, Vikings, and visitors from other worlds.

stonemason /ˈstaʊnˌmeɪs(ə)n/ *n.* a person who cuts, prepares, and builds with stone. □ **stonemasonry** *n.*

stonewall /ˈstaʊnwɔːl/ *v.* **1** *tr. & intr.* obstruct (discussion or investigation) or be obstructive with evasive answers or denials etc. **2** *intr. Cricket* bat with excessive caution. □ **stonewaller** *n.* **stonewalling** *n.*

stoneware /ˈstaʊnweə(r)/ *n.* a kind of dense, impermeable, usu. opaque pottery, made from clay containing a high proportion of silica and partly vitrified during firing. (See also POTTERY.)

stonewashed /ˈstaʊnwɒʃt/ *adj.* (of a garment or fabric, esp. denim) washed with abrasives to produce a worn or faded appearance.

stoneweed /ˈstaʊnwiːd/ *n.* = GROMWELL.

stonework /ˈstaʊnwɜːk/ *n.* **1** masonry. **2** the parts of a building made of stone. □ **stoneworker** *n.*

stonewort /ˈstəʊnwɜːt/ n. **1** = stone parsley. **2** a freshwater alga of the genus *Chara*, with a calcareous deposit on the stem.

stonkered /ˈstɒŋkəd/ adj. esp. Austral. & NZ sl. **1** very drunk. **2** utterly defeated or exhausted. [20th c.: orig. unkn.]

stonking /ˈstɒŋkɪŋ/ adv. & adj. sl. ● adv. extremely, very. ● adj. powerful, considerable. [ult. f. 19th c. *stonk* a marble]

stony /ˈstəʊnɪ/ adj. (**stonier, stoniest**) **1** full of or covered with stones (*stony soil; a stony road*). **2 a** hard, rigid. **b** cold, unfeeling, uncompromising (*a stony stare; a stony silence*). □ **stony-broke** Brit. colloq. entirely without money. **stony coral** see CORAL n. 2. **stony-hearted** unfeeling, obdurate. □ **stonily** adv. **stoniness** n. [OE *stānig* (as STONE)]

stood past and past part. of STAND.

stooge /stuːdʒ/ n. & v. colloq. ● n. **1** a butt or foil, esp. for a comedian. **2** an assistant or subordinate, esp. for routine or unpleasant work. **3** a compliant person; a puppet. ● v.intr. **1** (foll. by *for*) act as a stooge for. **2** (foll. by *about, around*, etc.) move about aimlessly. [20th c.: orig. unkn.]

stook /stuːk, stʊk/ n. & v. ● n. a group of sheaves of grain stood on end in a field. ● v.tr. arrange in stooks. [ME *stouk*, from or rel. to MLG *stūke*]

stool /stuːl/ n. & v. ● n. **1** a seat without a back or arms, usu. for one person and consisting of a wooden slab on three or four short legs. **2 a** = FOOTSTOOL. **b** a low bench for kneeling on. **3 a** the action or an act of discharging faeces. **b** (usu. in *pl.*) faeces; a discharge of faecal matter. **4** the root or stump of a tree or plant from which the shoots spring. **5** US a decoy-bird in hunting. ● v.intr. (of a plant) throw up shoots from the root. □ **fall between two stools** fail from vacillation between two courses etc. **stool-pigeon 1** a person acting as a decoy (orig. a decoy of a pigeon fixed to a stool). **2** a police informer. [OE *stōl* f. Gmc, rel. to STAND]

stoolball /ˈstuːlbɔːl/ n. an old game resembling cricket, still played in some places (especially southern England) but now chiefly by women and girls. Forms of stoolball were played in Elizabethan times, when the ball was hit with the hands rather than a bat and the 'stool' or wicket may have been an ordinary stool.

stoolie /ˈstuːlɪ/ n. N. Amer. sl. a person acting as a stool-pigeon.

stoop[1] /stuːp/ v. & n. ● v. **1** tr. bend (one's head or body) forwards and downwards. **2** intr. carry one's head and shoulders bowed forward. **3** intr. (often foll. by *down*) lower the body by bending forward, sometimes also bending at the knee. **4** intr. (foll. by *to* + infin.) deign or condescend. **5** intr. (foll. by *to*) descend or lower oneself to (some conduct) (*has stooped to crime*). **6** intr. (of a hawk etc.) swoop on its prey. ● n. **1** a stooping posture. **2** the downward swoop of a hawk etc. [OE *stūpian* f. Gmc, rel. to STEEP[1]]

stoop[2] /stuːp/ n. N. Amer. a porch or small veranda or set of steps in front of a house. [Du. *stoep*: see STOEP]

stoop[3] var. of STOUP.

stop /stɒp/ v. & n. ● v. (**stopped, stopping**) **1** tr. **a** put an end to (motion etc.); completely check the progress or motion or operation of. **b** effectively hinder or prevent (*stopped them playing so loudly*). **c** discontinue (an action or sequence of actions) (*stopped playing; stopped my visits*). **2** intr. come to an end; cease (*supplies suddenly stopped*). **3** intr. cease from motion or speaking or action; make a halt or pause (*the car stopped at the lights; he stopped in the middle of a sentence; my watch has stopped*). **4** tr. **a** cause to cease action; defeat. **b** Boxing defeat (an opponent) by a knockout. **5** tr. colloq. receive (a blow etc.). **6** intr. remain; stay for a short time. **7** tr. (often foll. by *up*) block or close up (a hole or leak etc.). **8** tr. not permit or supply as usual; discontinue or withhold (*shall stop their wages*). **9** tr. (in full **stop payment of** or **on**) instruct a bank to withhold payment on (a cheque). **10** tr. Brit. put a filling in (a tooth). **11** tr. Mus. obtain the required pitch from (the string of a violin etc.) by pressing at the appropriate point with the finger. **12** tr. Mus. plug the upper end of (an organ-pipe), giving a note an octave lower. **13** tr. Bridge be able to prevent opponents from taking all the tricks in (a suit). **14** tr. make (a sound) inaudible. **15** tr. Fencing, Boxing, etc. parry (a blow). **16** tr. pinch back (a plant). **17** tr. make (a clock, factory, etc.) cease working. **18** tr. Brit. provide with punctuation. **19** tr. Naut. make fast; stopper (a cable etc.). ● n. **1** the act or an instance of stopping; the state of being stopped (*put a stop to; the vehicle was brought to a stop*). **2** a place designated for a bus or train etc. to stop. **3** a punctuation mark, esp. a full stop. **4** a device for stopping motion at a particular point. **5** Mus. a change of pitch effected by stopping a string. **6** Mus. **a** (in an organ) a row of pipes of one character. **b** a knob etc. operating these. **7** a manner of speech adopted to produce a particular effect. **8** Optics & Photog. = DIAPHRAGM 3. **9 a** the effective diameter of a lens. **b** a device for reducing this. **c** a unit of change of relative aperture or exposure (with a reduction of one stop equivalent to halving it). **10** Phonet. a plosive consonant. **11** (in telegrams etc.) a full stop. **12** Bridge a card or

cards stopping a suit. **13** Naut. a small line used as a lashing. □ **pull all the stops out** exert extreme effort. **put a stop to** cause to end, esp. abruptly. **stop at nothing** be ruthless. **stop by** (also absol.) call at (a place). **stop dead** (or **short**) cease abruptly. **stop down** Photog. reduce the aperture of (a lens) with a diaphragm. **stop-drill** a drill with a shoulder limiting the depth of penetration. **stop one's ears 1** put one's fingers in one's ears to avoid hearing. **2** refuse to listen. **stop a gap** serve to meet a temporary need. **stop-go 1** alternate stopping and restarting of progress. **2** Brit. the alternate restriction and stimulation of economic demand. **stop in** US pay a brief visit. **stop-knob** Mus. a knob controlling an organ stop. **stop lamp** a light on the rear of a vehicle showing when the brakes are applied. **stop light 1** a red traffic light. **2** = stop lamp. **stop a person's mouth** induce a person by bribery or other means to keep silent about something. **stop off** (or **over**) break one's journey. **stop out 1** stay out. **2** cover (part of an area) to prevent printing, etching, etc. **stop payment** declare oneself insolvent. **stop press** Brit. **1** (often attrib.) late news inserted in a newspaper after printing has begun. **2** a column in a newspaper reserved for this. **stop valve** a valve closing a pipe against the passage of liquid. **stop-volley** esp. Tennis a checked volley close to the net, dropping the ball dead on the other side. □ **stopless** adj. **stoppable** adj. [ME f. OE *-stoppian* f. LL *stuppare* STUFF: see ESTOP]

stopbank /ˈstɒpbæŋk/ n. Austral. & NZ an embankment built to prevent river-flooding.

stopcock /ˈstɒpkɒk/ n. an externally operated valve regulating the flow of a liquid or gas through a pipe etc.

stope /stəʊp/ n. a steplike part of a mine where ore etc. is being extracted. [app. rel. to STEP n.]

Stopes /stəʊps/, Marie (Charlotte Carmichael) (1880–1958), Scottish birth-control campaigner. After establishing an academic reputation as a botanist, she published the best seller *Married Love* (1918), a frank treatment of sexuality within marriage. In 1921 she founded the Mothers' Clinic for Birth Control in Holloway, London, so pioneering the establishment of birth-control clinics in Britain. Her study *Contraception: Its Theory, History, and Practice* (1923) was one of the first comprehensive works on the subject.

stopgap /ˈstɒpgæp/ n. (often attrib.) a temporary substitute.

stopoff /ˈstɒpɒf/ n. a break in one's journey.

stopover /ˈstɒpˌəʊvə(r)/ n. = STOPOFF.

stoppage /ˈstɒpɪdʒ/ n. **1** the condition of being blocked or stopped. **2** a stopping (of pay). **3** a stopping or interruption of work in a factory etc.

Stoppard /ˈstɒpɑːd/, Tom (b.1937), British dramatist, born in Czechoslovakia. His best-known plays are comedies, which often deal with metaphysical and ethical questions and are characterized by verbal wit and the use of pastiche. His most famous play is *Rosencrantz and Guildenstern are Dead* (1966), based on the characters in *Hamlet*; other works include *Jumpers* (1972) and *The Real Thing* (1982).

stopper /ˈstɒpə(r)/ n. & v. ● n. **1** a plug for closing a bottle etc. **2** a person or thing that stops something. **3** Naut. a rope or clamp etc. for checking a rope cable or chain cable. ● v.tr. close or secure with a stopper. □ **put a stopper on 1** put an end to (a thing). **2** keep (a person) quiet.

stopping /ˈstɒpɪŋ/ n. Brit. a filling for a tooth.

stopple /ˈstɒp(ə)l/ n. & v. ● n. a stopper or plug, esp. US an earplug. ● v.tr. close with a stopple. [ME: partly f. STOP + -LE[1], partly f. ESTOPPEL]

stopwatch /ˈstɒpwɒtʃ/ n. a watch with a mechanism for recording elapsed time, used to time races etc.

storage /ˈstɔːrɪdʒ/ n. **1 a** the storing of goods etc. **b** a particular method of storing or the space available for it. **2** the cost of storing. **3** the electronic retention of data in a computer etc. □ **storage battery** (or **cell**) a battery (or cell) for storing electricity. **storage heater** Brit. an electric heater accumulating heat outside peak hours for later release.

storax /ˈstɔːræks/ n. **1 a** a fragrant resin, obtained from the tree *Styrax officinalis* and formerly used in perfume. **b** this tree. **2** (in full **Levant** or **liquid storax**) a balsam obtained from the tree *Liquidambar orientalis*. [L f. Gk, var. of STYRAX]

store /stɔː(r)/ n. & v. ● n. **1** a quantity of something kept available for use (*a store of wine; a store of wit*). **2** (in *pl.*) **a** articles for a particular purpose accumulated for use (*naval stores*). **b** a supply of these or the place where they are kept. **3 a** = department store. **b** esp. N. Amer. any retail outlet or shop. **c** (often in *pl.*) a shop selling basic necessities (*general stores*). **4** a warehouse for the temporary keeping of furniture etc.

5 *Computing* a device in a computer for storing retrievable data; a memory. ● *v.tr.* **1** put (furniture etc.) in store. **2** (often foll. by *up*, *away*) accumulate (stores, energy, electricity, etc.) for future use. **3** stock or provide with something useful (*a mind stored with facts*). **4** (of a receptacle) have storage capacity for. **5** *Computing* enter or retain (data) for retrieval. □ **in store 1** kept in readiness. **2** coming in the future. **3** (foll. by *for*) destined or intended. **set** (or **lay** or **put**) **store by** (or **on**) consider important or valuable. **store card** a credit card issued by a store to its customers. □ **storable** *adj.* **storer** *n.* [ME f. obs. *astore* (n. & v.) f. OF *estore*, *estorer* f. L *instaurare* renew: cf. RESTORE]

storefront /ˈstɔːfrʌnt/ *n.* esp. N. Amer. **1** the side of a shop facing the street. **2** a room at the front of a shop.

storehouse /ˈstɔːhaʊs/ *n.* a place where things are stored.

storekeeper /ˈstɔːˌkiːpə(r)/ *n.* **1** a storeman. **2** N. Amer. a shopkeeper.

storeman /ˈstɔːmən/ *n.* (*pl.* **-men**) a person responsible for stored goods.

storeroom /ˈstɔːruːm, -rʊm/ *n.* a room in which items are stored.

storey /ˈstɔːrɪ/ *n.* (also **story**) (*pl.* **-eys** or **-ies**) **1** any of the parts into which a building is divided horizontally; the whole of the rooms etc. having a continuous floor (*a third-storey window; a house of five storeys*). **2** a thing forming a horizontal division. □ **-storeyed** (in *comb.*) (also **-storied**). [ME f. AL *historia* HISTORY (perh. orig. meaning a tier of painted windows or sculpture)]

storiated /ˈstɔːrɪˌeɪtɪd/ *adj.* decorated with historical, legendary, or emblematic designs. □ **storiation** /ˌstɔːrɪˈeɪʃ(ə)n/ *n.* [shortening of HISTORIATED]

storied /ˈstɔːrɪd/ *adj. literary* celebrated in or associated with stories or legends.

stork /stɔːk/ *n.* **1** a large long-legged wading bird of the family Ciconiidae; esp. the *white stork* (*Ciconia ciconia*), with white plumage, black wing-tips, long reddish beak, and red feet, nesting esp. on tall buildings in Europe. **2** this bird as the pretended bringer of babies. □ **stork's-bill** a small plant of the genus *Erodium*, of the cranesbill family. [OE *storc*, prob. rel. to STARK (from its rigid posture)]

storm /stɔːm/ *n. & v.* ● *n.* **1** a violent disturbance of the atmosphere with strong winds and usu. with thunder and rain or snow etc. **2** *Meteorol.* a wind intermediate between gale and hurricane, force 10 or 11 on the Beaufort scale (55–72 m.p.h. or 24.5–32.6 metres per second). **3** a violent disturbance of the established order in human affairs. **4** (foll. by *of*) **a** a violent shower of missiles or blows. **b** an outbreak of applause, indignation, hisses, etc. (*they were greeted by a storm of abuse*). **5 a** a direct assault by troops on a fortified place. **b** the capture of a place by such an assault. ● *v.* **1** *intr.* (often foll. by *at*, *away*) talk violently, rage, bluster. **2** *intr.* (usu. foll. by *in*, *out of*, etc.) move violently or angrily (*stormed out of the meeting*). **3** *tr.* attack or capture by storm. **4** *intr.* (of wind, rain, etc.) rage; be violent. □ **storm-bird** = *storm petrel.* **storm centre 1** the point to which the wind blows spirally inward in a cyclonic storm. **2** a subject etc. upon which agitation or disturbance is concentrated. **storm cloud** a dark heavy cloud from which rain falls or threatens to fall. **storm-cock** a mistle-thrush. **storm-collar** a high coat-collar that can be turned up and fastened. **storm cone** *Brit.* a black cone hoisted by coastguards as a gale warning, the number and arrangement of cones giving information about wind direction etc. **storm-door** an additional outer door for protection in bad weather or winter. **storm-finch** *Brit.* = *storm petrel.* **storm-glass** a sealed tube containing a solution of which the clarity is thought to change when storms approach. **storming-party** a detachment of troops ordered to begin an assault. **storm in a teacup** *Brit.* great excitement over a trivial matter. **storm-lantern** *Brit.* a hurricane lamp. **storm petrel** a very small seabird, *Hydrobates pelagicus*, of the North Atlantic, with blackish plumage and a white rump (see also *stormy petrel*). **storm-sail** *Naut.* a sail of smaller size and stouter canvas than the corresponding one used in ordinary weather. **storm-signal** a device warning of an approaching storm. **storm window** an additional outer sash-window for protection in bad weather or winter. **take by storm 1** capture by direct assault. **2** rapidly captivate (a person, audience, etc.). □ **stormless** *adj.* **stormproof** *adj.* [OE f. Gmc]

stormbound /ˈstɔːmbaʊnd/ *adj.* prevented by storms from leaving port or continuing a voyage.

stormer /ˈstɔːmə(r)/ *n.* **1** *sl.* something of surpassing size or excellence. **2** a person who storms.

Stormont /ˈstɔːmənt/ a suburb of the east side of Belfast. Stormont Castle was, until 1972, the seat of the Parliament of Northern Ireland and is now the headquarters of the Northern Ireland Assembly.

Storm Trooper *n.* **1** a member of a force of shock troops. **2** *hist.* a member of the Brownshirts or Nazi political militia (see BROWNSHIRT).

Storm Troops *n.pl.* **1** *hist.* the troops of the Brownshirts or Nazi political militia (see BROWNSHIRT). **2** = *shock troops.*

stormy /ˈstɔːmɪ/ *adj.* (**stormier, stormiest**) **1** of or affected by storms. **2** (of a wind etc.) violent, raging, vehement. **3** full of angry feeling or outbursts; lively, boisterous (*a stormy meeting*). □ **stormy petrel 1** = *storm petrel.* **2** a person who enjoys conflict or attracts trouble. □ **stormily** *adv.* **storminess** *n.*

Stornoway /ˈstɔːnəˌweɪ/ a port on the east coast of Lewis, in the Outer Hebrides; pop. (1981) 8,640. The administrative centre of the Western Isles, it is noted for the manufacture of Harris tweed.

story[1] /ˈstɔːrɪ/ *n.* (*pl.* **-ies**) **1** an account of imaginary or past events; a narrative, tale, or anecdote. **2** the past course of the life of a person or institution etc. (*my story is a strange one*). **3** = STORYLINE. **4** facts or experiences that deserve narration. **5** *colloq.* a fib or lie. **6** a narrative or descriptive item of news. □ **the old** (or **same old**) **story** the familiar or predictable course of events. **story-book 1** a book of stories for children. **2** (*attrib.*) unreal, romantic (*a story-book ending*). **the story goes** it is said. **to cut** (or **make**) **a long story short** a formula excusing the omission of details. [ME *storie* f. AF *estorie* (OF *estoire*) f. L *historia* (as HISTORY)]

story[2] var. of STOREY.

storyboard /ˈstɔːrɪˌbɔːd/ *n.* a displayed sequence of pictures etc. outlining the plan of a film, television advertisement, etc.

storyline /ˈstɔːrɪˌlaɪn/ *n.* the narrative or plot of a novel or play etc.

storyteller /ˈstɔːrɪˌtelə(r)/ *n.* **1** a person who tells stories. **2** *colloq.* a liar. □ **storytelling** *n. & adj.*

stotinka /stɒˈtɪŋkə/ *n.* (*pl.* **stotinki** /-kɪ/) a monetary unit of Bulgaria, equal to one-hundredth of a lev. [Bulg., = hundredth]

stoup /stuːp/ *n.* (also **stoop**) **1** a holy-water basin. **2** *archaic* a flagon, beaker, or drinking-vessel. [ME f. ON *staup* (= OE *stēap*) f. Gmc, rel. to STEEP[2]]

Stour /ˈstaʊə(r)/ **1** a river of southern England which rises in west Wiltshire and flows south-east to meet the English Channel east of Bournemouth. **2** (also /stʊə(r)/) a river of eastern England which rises south-east of Cambridge and flows south-eastwards to the North Sea. **3** a river of central England which rises west of Wolverhampton and flows south-westwards through Stourbridge and Kidderminster to meet the Severn at Stourport-on-Severn.

stoush /staʊʃ/ *v. & n. Austral. & NZ sl.* ● *v.tr.* **1** hit; fight with. **2** attack verbally. ● *n.* a fight; a beating. [19th c.: orig. uncert.]

stout /staʊt/ *adj. & n.* ● *adj.* **1** rather fat; corpulent; bulky. **2** of considerable thickness or strength (*a stout stick*). **3** brave, resolute, vigorous (*a stout fellow; put up stout resistance*). ● *n.* a strong dark beer brewed with roasted malt or barley. □ **a stout heart** courage, resolve. **stout-hearted** courageous. **stout-heartedly** courageously. **stout-heartedness** courage. □ **stoutish** *adj.* **stoutly** *adv.* **stoutness** *n.* [ME f. AF & dial. OF *stout* f. WG, perh. rel. to STILT]

stove[1] /stəʊv/ *n. & v.* ● *n.* **1** a closed apparatus burning fuel or electricity for heating or cooking. **2** *Brit.* a hothouse. ● *v.tr. Brit.* force or raise (plants) in a hothouse. □ **stove-enamel** a heatproof enamel produced by the treatment of enamelled objects in a stove. **stove-pipe** a pipe conducting smoke and gases from a stove to a chimney. **stove-pipe hat** *colloq.* a tall silk hat. [ME = sweating-room, f. MDu., MLG *stove*, OHG *stuba* f. Gmc, perh. rel. to STEW[1]]

stove[2] past and past part. of STAVE v.

stow /stəʊ/ *v.tr.* **1** pack (goods etc.) tidily and compactly. **2** *Naut.* place (a cargo or provisions) in its proper place and order. **3** fill (a receptacle) with articles compactly arranged. **4** (usu. in *imper.*) *sl.* abstain or cease from (*stow the noise!*). □ **stow away 1** place (a thing) where it will not cause an obstruction. **2** be a stowaway on a ship etc. [ME, f. BESTOW: in Naut. use perh. infl. by Du. *stouwen*]

stowage /ˈstəʊɪdʒ/ *n.* **1** the act or an instance of stowing. **2** a place for this.

stowaway /ˈstəʊəˌweɪ/ *n.* a person who hides on board a ship or aircraft etc. to get free passage.

Stowe /stəʊ/, Harriet (Elizabeth) Beecher (1811–96), American novelist. She won fame with her anti-slavery novel *Uncle Tom's Cabin* (1852), which was successfully serialized 1851–2, and strengthened the contemporary abolitionist cause with its descriptions of the sufferings caused by slavery. Other works include the controversial *Lady Byron*

Vindicated (1870), which charged the poet Byron with incestuous relations with his half-sister.

STP *abbr.* **1** Professor of Sacred Theology. **2** standard temperature and pressure.

str. *abbr.* **1** strait. **2** stroke (of an oar).

strabismus /strəˈbɪzməs/ *n. Med.* the abnormal condition of one or both eyes not correctly aligned in direction; a squint. □ **strabismal** *adj.* **strabismic** *adj.* [mod.L f. Gk *strabismos* f. *strabizō* squint f. *strabos* squinting]

Strabo /ˈstreɪbəʊ/ (*c.*63 BC–AD *c.*23), historian and geographer of Greek descent. His only extant work, *Geographica*, in seventeen volumes, provides a detailed physical and historical geography of the ancient world during the reign of Augustus.

Strachey /ˈstreɪtʃɪ/, (Giles) Lytton (1880–1932), English biographer. A prominent member of the Bloomsbury Group, he achieved recognition with *Eminent Victorians* (1918), which attacked the literary Establishment through satirical biographies of Florence Nightingale, General Gordon, and others. His irreverence and independence were influential in the development of biography. Other works include a biography of Queen Victoria (1921) and *Elizabeth and Essex* (1928).

Strad /stræd/ *n. colloq.* a Stradivarius. [abbr.]

straddle /ˈstræd(ə)l/ *v. & n.* ● *v.* **1** *tr.* **a** sit or stand across (a thing) with the legs wide apart. **b** be situated across or on both sides of (*the town straddles the border*). **2** *intr.* **a** sit or stand in this way. **b** (of the legs) be wide apart. **3** *tr.* part (one's legs) widely. **4** *tr.* drop shots or bombs short of and beyond (a target). **5** *tr.* vacillate between two policies etc. regarding (an issue). ● *n.* **1** the act or an instance of straddling. **2** *Stock Exch.* an option giving the holder the right of either calling for or delivering stock at a fixed price. □ **straddler** *n.* [alt. of *striddle*, back-form. f. *striddlings* astride f. *strid-* = STRIDE]

Stradivarius /ˌstrædɪˈveərɪəs/ *n.* a violin or other stringed instrument made by Antonio Stradivari (*c.*1644–1737), the greatest of a family of violin-makers of Cremona in northern Italy, or his followers. Stradivari produced over 1,100 instruments, of which perhaps 400 are known to exist; many of them are still played. Among the many famous Stradivarius violins are those nicknamed the Betts (1704), now in the Library of Congress in Washington, DC, and the Messie or Messiah (1716), now in the Ashmolean Museum, Oxford. Stradivari's finest instruments are those dating from 1700 onwards. [Latinized f. *Stradivari*]

strafe /strɑːf, streɪf/ *v. & n.* ● *v.tr.* **1** bombard; harass with gunfire, esp. from aircraft. **2** reprimand. **3** abuse. **4** thrash. ● *n.* an act of strafing. [joc. adaptation of G catchword (1914) *Gott strafe England* may God punish England]

straggle /ˈstræg(ə)l/ *v. & n.* ● *v.intr.* **1** lack or lose compactness or tidiness. **2** be or become dispersed or sporadic. **3** trail behind others in a march or race etc. **4** (of a plant, beard, etc.) grow long and loose. ● *n.* a body or group of straggling or scattered persons or things. □ **straggler** *n.* **straggly** *adj.* (**stragglier**, **straggliest**). [ME, perh. rel. to dial. *strake* go, rel. to STRETCH]

straight /streɪt/ *adj., n., & adv.* ● *adj.* **1 a** extending uniformly in the same direction; without a curve or bend etc. **b** *Geom.* (of a line) lying on the shortest path between any two of its points. **2** successive, uninterrupted (*three straight wins*). **3** in proper order or place or condition; duly arranged; level, symmetrical (*is the picture straight?*; *put things straight*). **4** honest, candid; not evasive (*a straight answer*). **5** (of thinking etc.) logical, unemotional. **6** (of drama etc.) serious as opposed to popular or comic; employing the conventional techniques of its art form. **7 a** unmodified. **b** (of a drink) undiluted. **8** *colloq.* **a** (of a person etc.) conventional or respectable. **b** heterosexual. **9** (of an arch) flat-topped. **10** (of a person's back) not bowed. **11** (of the hair) not curly or wavy. **12** (of a knee) not bent. **13** (of the legs) not bandy or knock-kneed. **14** (of a garment) not flared. **15** coming direct from its source. **16** (of an aim, look, blow, or course) going direct to the mark. ● *n.* **1** the straight part of something, esp. the concluding stretch of a racecourse. **2** a straight condition. **3** a sequence of five cards in poker. **4** *colloq.* **a** a conventional person. **b** a heterosexual. ● *adv.* **1** in a straight line; direct; without deviation or hesitation or circumlocution (*came straight from Paris*; *I told them straight*). **2** in the right direction, with a good aim (*shoot straight*). **3** correctly (*can't see straight*). **4** *archaic* at once or immediately. □ **go straight** live an honest life after being a criminal. **the straight and narrow** morally correct behaviour. **straight angle** an angle of 180°. **straight away** at once; immediately. **straight-bred** not cross-bred. **straight chain** *Chem.* a chain of atoms, esp. carbon atoms, that is neither branched nor forms

a ring (often, with hyphen, *attrib.*: *straight-chain hydrocarbon*). **straight-cut** (of tobacco) cut lengthwise into long silky fibres. **straight-edge** a bar with one edge accurately straight, used for testing. **straight-eight 1** an internal-combustion engine with eight cylinders in line. **2** a vehicle having this. **straight eye** the ability to detect deviation from the straight. **straight face** an intentionally expressionless face, esp. avoiding a smile though amused. **straight-faced** having a straight face. **straight fight** a simple contest between two opponents, esp. in an election. **straight flush** see FLUSH³. **straight from the shoulder 1** (of a blow) well delivered. **2** (of a verbal attack) delivered in a frank or direct manner. **straight man** a comedian's stooge. **straight off** (or **out**) *colloq.* without hesitation, deliberation, etc. (*cannot tell you straight off*). **straight-out** esp. *US* **1** uncompromising. **2** straightforward, genuine. **straight razor** *US* a cutthroat razor. **straight-up** *N. Amer. colloq.* **1** truthfully, honestly. **2** (of food, drink, etc.) without admixture or dilution. □ **straightish** *adj.* **straightly** *adv.* **straightness** *n.* [ME, past part. of STRETCH]

straightaway /ˈstreɪtəˌweɪ/ *adv., adj., & n.* ● *adv.* = *straight away*. ● *adj.* esp. *US* (of a course etc.) straight, direct. ● *n.* esp. *N. Amer.* a straight course or section.

straighten /ˈstreɪt(ə)n/ *v.tr. & intr.* **1** (often foll. by *out*) make or become straight. **2** (foll. by *up*) stand erect after bending. □ **straightener** *n.*

straightforward /streɪtˈfɔːwəd/ *adj.* **1** honest or frank. **2** (of a task etc.) uncomplicated. □ **straightforwardly** *adv.* **straightforwardness** *n.*

straightway /ˈstreɪtweɪ/ *adv. archaic* = *straight away*.

strain¹ /streɪn/ *v. & n.* ● *v.* **1** *tr. & intr.* stretch tightly; make or become taut or tense. **2** *tr.* exercise (oneself, one's senses, a thing, etc.) intensely or excessively, press to extremes. **3 a** *intr.* make an intensive effort. **b** *intr.* (foll. by *after*) strive intensely for (*straining after perfection*). **4** *intr.* (foll. by *at*) tug, pull (*the dog strained at the leash*). **5** *intr.* hold out with difficulty under pressure (*straining under the load*). **6** *tr.* **a** distort from the true intention or meaning. **b** apply (authority, laws, etc.) beyond their province or in violation of their true intention. **7** *tr.* overtask or injure by overuse or excessive demands (*strain a muscle*; *strained their loyalty*). **8 a** *tr.* clear (a liquid) of solid matter by passing it through a sieve etc. **b** *tr.* (foll. by *out*) filter (solids) out from a liquid. **c** *intr.* (of a liquid) percolate. **9** *tr.* hug or squeeze tightly. **10** *tr.* use (one's ears, eyes, voice, etc.) to the best of one's power. ● *n.* **1 a** the act or an instance of straining. **b** the force exerted in this. **2** an injury caused by straining a muscle etc. **3 a** a severe demand on physical strength or resources. **b** the exertion needed to meet this (*is suffering from strain*). **4** (in *sing.* or *pl.*) a snatch or spell of music or poetry. **5** a tone or tendency in speech or writing (*more in the same strain*). **6** *Physics* **a** the condition of a body subjected to stress; molecular displacement. **b** a quantity measuring this, equal to the amount of deformation usu. divided by the original dimension. □ **at strain** (or **full strain**) exerted to the utmost. **strain every nerve** make every possible effort. **strain oneself 1** injure oneself by effort. **2** make undue efforts. □ **strainable** *adj.* [ME f. OF *estreindre estreign-* f. L *stringere strict-* draw tight]

strain² /streɪn/ *n.* **1** a breed or stock of animals, plants, etc. **2** a moral tendency as part of a person's character (*a strain of aggression*). [ME, = progeny, f. OE *strēon* (recorded in *ģestrēonan* beget), rel. to L *struere* build]

strained /streɪnd/ *adj.* **1** constrained, forced, artificial. **2** (of a relationship) mutually distrustful or tense. **3** (of an interpretation) involving an unreasonable assumption; far-fetched, laboured.

strainer /ˈstreɪnə(r)/ *n.* a device for straining liquids, vegetables, etc.

strait /streɪt/ *n. & adj.* ● *n.* **1** (in *sing.* or *pl.*) a narrow passage of water connecting two seas or large bodies of water. **2** (usu. in *pl.*) difficulty, trouble, or distress (usu. in *dire* or *desperate straits*). ● *adj. archaic* **1** narrow, limited; confined or confining. **2** strict or rigorous. □ **strait-laced** severely virtuous; morally scrupulous; puritanical. □ **straitly** *adv.* **straitness** *n.* [ME *streit* f. OF *estreit* tight, narrow f. L *strictus* STRICT]

straiten /ˈstreɪt(ə)n/ *v.* **1** *tr.* restrict in range or scope. **2** *tr.* (as **straitened** *adj.*) (esp. of circumstances) characterized by poverty. **3** *tr. & intr. archaic* make or become narrow.

strait-jacket /ˈstreɪtˌdʒækɪt/ *n. & v.* ● *n.* **1** a strong garment with long arms for confining the arms of a violent prisoner, mental patient, etc. **2** restrictive measures. ● *v.tr.* (**-jacketed**, **-jacketing**) **1** restrain with a strait-jacket. **2** severely restrict.

Straits Settlements a former British Crown Colony in SE Asia, centred on the Strait of Malacca. Established in 1867, it comprised Singapore, Penang, and Malacca, and later included Labuan, Christmas Island, and the Cocos Islands. It was disbanded in 1946, when Singapore was made a separate Crown Colony.

strake /streɪk/ *n.* **1** *Naut.* a continuous line of planking or plates from the stem to the stern of a ship. **2** a section of the iron rim of a wheel. [ME: prob. rel. to OE *streccan* STRETCH]

Stralsund /ˈstrɑːlzʊnt/ a town and fishing port in northern Germany, on the Baltic coast opposite the island of Rügen; pop. (1991) 71,620. It was a member of the Hanseatic League.

stramonium /strəˈməʊnɪəm/ n. **1** datura. **2** the dried leaves of this plant used in the treatment of asthma. [mod.L, perh. f. Tartar *turman* horse-medicine]

strand[1] /strænd/ v. & n. ● v. **1** tr. & intr. run aground. **2** tr. (as **stranded** adj.) in difficulties, esp. without money or means of transport. ● n. esp. *literary* the margin of a sea, lake, or river, esp. the foreshore. [OE]

strand[2] /strænd/ n. & v. ● n. **1** each of the threads or wires twisted round each other to make a rope or cable. **2 a** a single thread or strip of fibre. **b** a constituent filament. **3** a lock of hair. **4** an element or strain in any composite whole. ● v.tr. **1** break a strand in (a rope). **2** arrange in strands. [ME: orig. unkn.]

strange /streɪndʒ/ adj. & adv. ● adj. **1** unusual, peculiar, surprising, eccentric, novel. **2 a** (often foll. by to) unfamiliar, alien, foreign (*lost in a strange land; surrounded by strange faces; a taste strange to him*). **b** not one's own (*strange gods*). **3** (foll. by to) (of a person) unaccustomed to; unfamiliar with. **4** not at ease; out of one's element (*not having been invited, I felt strange*). ● adv. colloq. in a strange manner; strangely. □ **feel strange** be unwell. **strange attractor** Math. an equation or fractal set representing a complex pattern of behaviour in a chaotic system. **strange particle** Physics a subatomic particle having a non-zero value for strangeness. **strange to say** it is surprising or unusual (that). □ **strangely** adv. [ME f. OF *estrange* f. L *extraneus* EXTRANEOUS]

strangeness /ˈstreɪndʒnɪs/ n. **1** the state or fact of being strange or unfamiliar etc. **2** Physics a property of certain quarks that is conserved in strong interactions.

stranger /ˈstreɪndʒə(r)/ n. **1** a person who does not know or is not known in a particular place or company. **2** (often foll. by to) a person one does not know (*was a complete stranger to me*). **3** (foll. by to) a person entirely unaccustomed to (a feeling, experience, etc.) (*no stranger to controversy*). **4** Parl. a person who is not a member or official of the House of Commons. [ME f. OF *estrangier* ult. f. L (as STRANGE)]

strangle /ˈstræŋg(ə)l/ v.tr. **1** squeeze the windpipe or neck of, esp. so as to kill. **2** hamper or suppress (a movement, impulse, cry, etc.). □ **strangler** n. [ME f. OF *estrangler* f. L *strangulare* f. Gk *straggalaō* f. *straggalē* halter: cf. *straggos* twisted]

stranglehold /ˈstræŋg(ə)l,həʊld/ n. **1** a wrestling hold that throttles an opponent. **2** a deadly grip. **3** complete and exclusive control.

strangles /ˈstræŋg(ə)lz/ n.pl. (usu. treated as sing.) an infectious streptococcal fever, esp. affecting the respiratory tract, in a horse, ass, etc. [pl. of strangle (n.) f. STRANGLE]

strangulate /ˈstræŋgjʊˌleɪt/ v.tr. **1** Med. prevent circulation through (a vein, intestine, etc.) by compression. **2** Med. remove (a tumour etc.) by binding with a cord. **3** (as **strangulated** adj.) (of a voice) sounding as though the speaker's throat is constricted. □ **strangulated hernia** Med. a hernia in which the protruding part is constricted, preventing circulation. [L *strangulare strangulat-* (as STRANGLE)]

strangulation /ˌstræŋgjʊˈleɪʃ(ə)n/ n. **1** the act of strangling or the state of being strangled. **2** the act of strangulating. [L *strangulatio* (as STRANGULATE)]

strangury /ˈstræŋgjʊərɪ/ n. Med. a condition in which urine is passed painfully and in drops. □ **strangurious** /stræŋˈgjʊərɪəs/ adj. [ME f. L *stranguria* f. Gk *straggouria* f. *stragx -ggos* drop squeezed out + *ouron* urine]

Stranraer /strænˈrɑː(r)/ a port and market town in SW Scotland, in Dumfries and Galloway; pop. (1984) 10,170. It is the terminus of a ferry service from Northern Ireland

strap /stræp/ n. & v. ● n. **1** a strip of leather or other flexible material, often with a buckle or other fastening for holding things together etc. **2** a thing like this for keeping a garment in place. **3** a loop for grasping to steady oneself in a moving vehicle. **4 a** a strip of metal used to secure or connect. **b** a leaf of a hinge. **5** Bot. a tongue-shaped part in a floret. **6** (prec. by the) punishment by beating with a strap. ● v.tr. (**strapped**, **strapping**) **1** (often foll. by down, up, etc.) secure with or bind with a strap. **2** beat with a strap. **3** (esp. as **strapped** adj.) (usu. foll. by for) colloq. subject to a shortage (*strapped for cash*). **4** (often foll. by up) close (a wound) or bind (a part) with adhesive plaster. □ **strap-work** ornamentation imitating plaited straps. □ **strapper** n. **strappy** adj. [dial. form of STROP]

straphanger /ˈstræpˌhæŋə(r)/ n. colloq. a standing passenger in a bus or train. □ **straphang** v.intr.

strapless /ˈstræplɪs/ adj. (of a garment) without straps, esp. shoulder-straps.

strappado /strəˈpɑːdəʊ/ n. (pl. **-os**) hist. a form of torture in which the victim is secured to a rope and made to fall from a height almost to the ground then stopped with a jerk; an application of this; the instrument used. [F (*e*)*strapade* f. It. *strappata* f. *strappare* snatch]

strapping /ˈstræpɪŋ/ adj. (esp. of a young person) large and sturdy.

Strasberg /ˈstræzbɜːg/, Lee (born Israel Strassberg) (1901–82), American actor, director, and drama teacher, born in Austria. As artistic director of the Actors' Studio in New York City (1948–82), he was the leading figure in the development of method acting in the US. Among his pupils were Marlon Brando, James Dean, Jane Fonda, and Dustin Hoffman.

Strasbourg /ˈstræzbɜːg, French strasbur/ a city in NE France, in Alsace, close to the border with Germany; pop. (1990) 255,940. Annexed by Germany in 1870, it was returned to France after the First World War. It is the headquarters of the Council of Europe and of the European Parliament.

strata pl. of STRATUM.

stratagem /ˈstrætədʒəm/ n. **1** a cunning plan or scheme, esp. for deceiving an enemy. **2** trickery. [ME f. F *stratagème* f. L *stratagema* f. Gk *stratēgēma* f. *stratēgeō* be a general (*stratēgos*) f. *stratos* army + *agō* lead]

strategic /strəˈtiːdʒɪk/ adj. **1** of or serving the ends of strategy; useful or important with regard to strategy (*strategic considerations, strategic move*). **2** (of materials) essential in fighting a war. **3** (of bombing or weapons) done or for use against an enemy's home territory as a longer-term military objective (opp. TACTICAL 2). □ **strategical** adj. **strategically** adv. **strategics** n.pl. (usu. treated as sing.). [F *stratégique* f. Gk *stratēgikos* (as STRATAGEM)]

Strategic Arms Limitation Talks see SALT.

Strategic Arms Reduction Talks see START.

Strategic Defense Initiative (abbr. **SDI**) (popularly known as *Star Wars*) a proposed US defence system against potential nuclear attack. Based partly in space, it was intended to protect the US from intercontinental ballistic missiles by intercepting and destroying them before they reached their targets. Development of SDI was launched by President Reagan in the early 1980s, but the project was surrounded by controversy and President Bush allowed many of its component elements to be discontinued. The remains of the project were renamed the *Ballistic Missile Defense Organization* in 1993.

strategy /ˈstrætɪdʒɪ/ n. (pl. **-ies**) **1** the art of war. **2 a** the management of an army or armies in a campaign. **b** the planning and direction of the larger military movements and overall operations of a campaign (cf. TACTICS 1). **c** an instance of this or a plan formed according to it. **3** a plan of action or policy in business or politics etc. (*economic strategy*). □ **strategist** n. [F *stratégie* f. Gk *stratēgia* generalship f. *stratēgos*: see STRATAGEM]

Stratford-upon-Avon /ˌstrætfədəpɒnˈeɪv(ə)n/ a town in Warwickshire, on the River Avon; pop. (1981) 20,100. Famous as the birth and burial place of William Shakespeare, it is the site of the Royal Shakespeare Theatre.

strath /stræθ/ n. Sc. a broad mountain valley. [Gael. *srath*]

Strathclyde /stræθˈklaɪd/ a local government region in west central Scotland; administrative centre, Glasgow.

strathspey /stræθˈspeɪ/ n. **1** a slow Scottish dance. **2** the music for this. [*Strathspey*, valley of the River Spey]

strati pl. of STRATUS.

straticulate /strəˈtɪkjʊlət/ adj. Geol. (of rock-formations) arranged in thin strata. [STRATUM, after *vermiculate* etc.]

stratify /ˈstrætɪˌfaɪ/ v.tr. (**-ies, -ied**) **1** (esp. as **stratified** adj.) **a** arrange in layers. **b** Geol. deposit (sediments etc.) in strata. **c** (of a lake etc.) form two or more layers differing in temperature or density. **2** construct in layers, social grades, etc. □ **stratification** /ˌstrætɪfɪˈkeɪʃ(ə)n/ n. [F *stratifier* (as STRATUM)]

stratigraphy /strəˈtɪgrəfɪ/ n. Geol. & Archaeol. **1** the order and relative position of strata. **2** the study of this as a means of historical interpretation. □ **stratigrapher** n. **stratigraphic** /ˌstrætɪˈgræfɪk/ adj. **stratigraphical** adj. [STRATUM + -GRAPHY]

strato- /ˈstrætəʊ/ comb. form stratus.

stratocirrus /ˌstrætəʊˈsɪrəs/ n. Meteorol. a cloud type resembling cirrostratus but more compact.

stratocracy /strəˈtɒkrəsɪ/ n. (pl. **-ies**) **1** a military government. **2** domination by soldiers. [Gk *stratos* army + -CRACY]

stratocumulus /ˌstrætəʊˈkjuːmjʊləs/ n. *Meteorol.* a cloud type consisting of a low horizontal layer of clumped or broken grey cloud.

stratopause /ˈstrætəʊˌpɔːz/ n. the interface between the stratosphere and the ionosphere.

stratosphere /ˈstrætəˌsfɪə(r)/ n. a layer of atmospheric air above the troposphere extending to about 50 km above the earth's surface, in which the lower part changes little in temperature and the upper part increases in temperature with height (cf. IONOSPHERE). □ **stratospheric** /ˌstrætəˈsferɪk/ adj. [STRATUM + SPHERE after *atmosphere*]

stratum /ˈstrɑːtəm, ˈstreɪt-/ n. (pl. **strata** /-tə/) **1** esp. *Geol.* a layer or set of successive layers of any deposited substance. **2** an atmospheric layer. **3** *Anat. & Biol.* a layer of tissue etc. **4 a** a social grade, class, etc. (*the various strata of society*). **b** *Statistics* each of the groups into which a population is divided in stratified sampling. □ **stratal** adj. [L, = something spread or laid down, neut. past part. of *sternere* strew]

stratus /ˈstrɑːtəs, ˈstreɪt-/ n. (pl. **strati** /-taɪ/) *Meteorol.* a continuous horizontal sheet of cloud. [L, past part. of *sternere*: see STRATUM]

Strauss[1] /straʊs/, Johann (known as Strauss the Elder) (1804–49), Austrian composer. He was a leading composer of waltzes from the 1830s, although probably his best-known work is the *Radetzky March* (1838).

Strauss[2] /straʊs/, Johann (known as Strauss the Younger) (1825–99), Austrian composer. The son of Strauss the Elder, he became known as 'the waltz king', composing many famous waltzes such as *The Blue Danube* (1867) and *Tales from the Vienna Woods* (1868). He is also noted for the operetta *Die Fledermaus* (1874).

Strauss[3] /straʊs/, Richard (1864–1949), German composer. From the mid-1880s he composed a succession of symphonic poems, including *Till Eulenspiegels Lustige Streiche* (1895) and *Also Sprach Zarathustra* (1896). His reputation grew as he turned to opera, exploring polytonality in *Salome* (1905) and *Elektra* (1905). The latter marked the beginning of his collaboration with the librettist Hugo von Hofmannsthal; together they produced such popular operas as *Der Rosenkavalier* (1911). Often regarded as the last of the 19th-century romantic composers, Strauss retained his romanticism to the end of his long career, notably in *Four Last Songs* (1948) for soprano and orchestra.

Stravinsky /strəˈvɪnskɪ/, Igor (Fyodorovich) (1882–1971), Russian-born composer. He made his name as a composer for Diaghilev's Ballets Russes, writing the music for the ballets *The Firebird* (1910) and *The Rite of Spring* (1913); both shocked Paris audiences with their irregular rhythms and frequent dissonances. Stravinsky later developed a neoclassical style, typified by the ballet *Pulcinella* (1920) and, ultimately, the opera *The Rake's Progress* (1948–51), based on William Hogarth's paintings. In the 1950s he experimented with serialism in such works as the cantata *Threni* (1957–8). Resident in the US from the outbreak of the Second World War, Stravinsky became an American citizen in 1945.

straw /strɔː/ n. **1** dry cut stalks of grain for use as fodder or as material for thatching, packing, making hats, etc. **2** a single stalk or piece of straw. **3** a thin hollow paper or plastic tube for sucking drink from a glass etc. **4** an insignificant thing (*not worth a straw*). **5** the pale yellow colour of straw. **6** a straw hat. □ **catch** (or **clutch** or **grasp**) **at a straw** (or **straws**) resort in desperation to an utterly inadequate expedient. **draw the short straw** be chosen by lot, esp. for some disagreeable task. **straw boss** *US* an assistant foreman. **straw-colour** pale yellow. **straw-coloured** of pale yellow. **straw in the wind** a slight hint of future developments. **straw man 1** a human figure made of straw. **2** *US* a weak or invented argument set up simply to be easily refuted. **3** *US* = FRONT n. 8. **straw vote** (or **poll**) an unofficial ballot as a test of opinion. **straw-worm** a caddis-worm. □ **strawy** adj. [OE *strēaw* f. Gmc, rel. to STREW]

strawberry /ˈstrɔːbərɪ/ n. (pl. **-ies**) **1 a** a rosaceous plant of the genus *Fragaria*, esp. of a cultivated variety, with white flowers, trifoliate leaves, and runners. **b** the pulpy red edible fruit of this, having a seed-studded surface. **2** a deep pinkish-red colour. □ **strawberry blonde 1** pinkish-blonde hair. **2** a woman with such hair. **strawberry mark** a soft reddish birthmark. **strawberry pear 1** a West Indian cactus, *Hylocereus undatus*. **2** the fruit of this. **strawberry roan** see ROAN[1]. **strawberry tree** an evergreen ericaceous tree, *Arbutus unedo*, bearing strawberry-like fruit. [OE *strēa(w)berige*, *strēowberige* (as STRAW, BERRY): reason for the name unkn.]

strawboard /ˈstrɔːbɔːd/ n. a type of building board made of straw pulp faced with paper.

stray /streɪ/ v., n., & adj. ● v.intr. **1 a** wander from the right place; become separated from one's companions etc.; go astray. **b** (often foll. by *from*, *off*) digress. **2** deviate morally. **3** (as **strayed** adj.) that has gone astray. ● n. **1** a person or thing that has strayed, esp. a domestic animal. **2** (esp. in pl.) electrical phenomena interfering with radio reception. ● adj. **1** strayed or lost. **2** isolated; found or occurring occasionally (*a stray customer or two*; *hit by a stray bullet*). **3** *Physics* wasted or unwanted (*eliminate stray magnetic fields*). □ **strayer** n. [ME f. AF & OF *estrayer* (v.), AF *strey* (n. & adj.) f. OF *estraié* (as ASTRAY)]

streak /striːk/ n. & v. ● n. **1** a long thin usu. irregular line or band, esp. distinguished by colour (*black with red streaks*; *a streak of light above the horizon*). **2** a strain or element in a person's character (*has a streak of mischief*). **3** a spell or series (*a winning streak*). **4** a line of bacteria etc. placed on a culture medium. ● v. **1** tr. mark with streaks. **2** intr. move very rapidly. **3** intr. colloq. run naked in a public place as a stunt. □ **streak of lightning** a sudden prominent flash of lightning. □ **streaker** n. **streaking** n. [OE *strica* pen-stroke f. Gmc: rel. to STRIKE]

streaky /ˈstriːkɪ/ adj. (**streakier**, **streakiest**) **1** full of streaks. **2** (of bacon) with alternate streaks of fat and lean. □ **streakily** adv. **streakiness** n.

stream /striːm/ n. & v. ● n. **1** a flowing body of water, esp. a small river. **2 a** the flow of a fluid or of a mass of people (*a stream of lava*). **b** (in sing. or pl.) a large quantity of something that flows or moves along. **3** a current or direction in which things are moving or tending (*against the stream*). **4** *Brit.* a group of schoolchildren taught together as being of similar ability for a given age. ● v. **1** intr. flow or move as a stream. **2** intr. run with liquid (*my eyes were streaming*). **3** intr. (of a banner or hair etc.) wave or be blown behind in the wind. **4** tr. emit a stream of (blood etc.). **5** tr. *Brit.* arrange (schoolchildren) in streams. □ **go with the stream** do as others do. **on stream** (of a factory etc.) in operation. **stream-anchor** an anchor intermediate in size between a bower and a kedge, esp. for use in warping. □ **streamless** adj. **streamlet** /ˈstriːmlɪt/ n. [OE *strēam* f. Gmc]

streamer /ˈstriːmə(r)/ n. **1** a long narrow flag. **2** a long narrow strip of ribbon or paper, esp. in a coil that unrolls when thrown. **3** a banner headline. **4** (in pl.) the aurora borealis or australis.

streamline /ˈstriːmlaɪn/ v. & n. ● v.tr. **1** give (a vehicle etc.) the form which presents the least resistance to motion. **2** make (an organization, process, etc.) simple or more efficient or better organized. **3** (as **streamlined** adj.) **a** having a smooth, slender, or elongated form; aerodynamic. **b** having a simplified and more efficient structure or organization. ● n. **1** the natural course of water or air currents. **2** (often attrib.) the shape of an aircraft, car, etc., calculated to cause the least resistance to motion.

stream of consciousness n. **1** *Psychol.* a person's thoughts and conscious reactions to events perceived as a continuous flow. The term was introduced by William James in his *Principles of Psychology* (1890). **2** a literary style in which a character's thoughts, feelings, and reactions are depicted in a continuous flow uninterrupted by objective description or conventional dialogue. James Joyce, Virginia Woolf, and Marcel Proust are among notable early exponents of the style.

Streep /striːp/, Meryl (born Mary Louise Streep) (b.1949), American actress. After her screen début in 1977 she became a leading star in the 1980s. She won an Oscar for her part as a divorcee in *Kramer vs. Kramer* (1980). Her other films include *The French Lieutenant's Woman* (1981), *Sophie's Choice* (1982), for which she won a second Oscar, and *Out of Africa* (1986).

street /striːt/ n. **1 a** a public road in a city, town, or village. **b** this with the houses or other buildings on each side. **2** the persons who live or work on a particular street. □ **in the street 1** in the area outside the houses. **2** (of stock exchange business) done after closing-time. **not in the same street with** colloq. utterly inferior to in ability etc. **on the streets 1** living by prostitution. **2** homeless. **street Arab** often offens. **1** a homeless child. **2** an urchin. **street credibility** (or colloq. **cred**) acceptability among young fashionable urban people. **street cries** *Brit.* the cries of street hawkers. **street door** a main outer house-door opening on the street. **street entertainer** a person who entertains people in the street for money, esp. with music, acting, or juggling. **street jewellery** enamel advertising plates as collectors' items. **streets ahead** (often foll. by of) colloq. much superior (to). **street value** the value of drugs etc. sold illicitly. **take to the streets** gather outdoors in order to protest. **up** (or **right up**) **a person's street** (or **alley**) colloq. **1** within a person's range of interest or knowledge. **2** to a person's liking. □ **streeted** *adj.* (also in *comb.*).

streetward *adj.* & *adv.* [OE *strǣt* f. LL *strāta* (*via*) paved (way), fem. past part. of *sternere* lay down]

streetcar /ˈstriːtkɑː(r)/ *n.* N. Amer. a tram.

streetwalker /ˈstriːtˌwɔːkə(r)/ *n.* a prostitute seeking customers in the street. □ **streetwalking** *n.* & *adj.*

streetwise /ˈstriːtwaɪz/ *adj.* esp. N. Amer. familiar with the ways of modern urban life.

Streisand /ˈstraɪs(ə)nd, -sænd/, Barbra (Joan) (b.1942), American singer, actress, and film director. She became a star in 1964 in the Broadway musical *Funny Girl*, winning an Oscar in 1968 for her performance in the film of the same name. She later played the lead in *A Star is Born* (1976), which she also produced; the film's song 'Evergreen', composed by Streisand, won an Oscar. Other starring roles came in films such as *Yentl* (1983), which she also produced and directed.

strelitzia /streˈlɪtsɪə/ *n.* a southern African plant of the genus *Strelitzia*, with showy irregular flowers, each having a long projecting tongue. [mod.L f. Charlotte of Mecklenburg-*Strelitz* (1744–1818), queen of George III]

strength /streŋθ, *disp.* streŋkθ/ *n.* **1** the state of being strong; the degree to which, or respect in which, a person or thing is strong. **2 a** a person or thing affording strength or support. **b** an attribute making for strength of character (*patience is your great strength*). **3** the number of persons present or available. **4** a full complement (*below strength*). □ **from strength** from a strong position. **from strength to strength** with ever-increasing success. **in strength** in large numbers. **on the strength of** relying on; on the basis of. **the strength of** the essence or main features of. □ **strengthless** *adj.* [OE *strengthu* f. Gmc (as STRONG)]

strengthen /ˈstreŋθən, *disp.* ˈstreŋkθən/ *v.tr.* & *intr.* make or become stronger. □ **strengthen a person's hand** (or **hands**) encourage a person to vigorous action. □ **strengthener** *n.*

strenuous /ˈstrenjʊəs/ *adj.* **1** requiring or using great effort. **2** energetic or unrelaxing. □ **strenuously** *adv.* **strenuousness** *n.* [L *strenuus* brisk]

strep /strep/ *n.* colloq. = STREPTOCOCCUS. [abbr.]

streptocarpus /ˌstreptəʊˈkɑːpəs/ *n.* a southern African plant of the genus *Streptocarpus*, with funnel-shaped flowers, often violet or pink, and spirally twisted fruits. [mod.L f. Gk *streptos* twisted + *karpos* fruit]

streptococcus /ˌstreptəˈkɒkəs/ *n.* (*pl.* **streptococci** /-ˈkɒksaɪ, -ˈkɒkaɪ/) a bacterium of the genus *Streptococcus*, usu. found joined together in chains. Streptococci include pathogens causing pneumonia, scarlet fever, etc., and the agents of dental decay and souring of milk. □ **streptococcal** *adj.* [Gk *streptos* twisted f. *strephō* turn + COCCUS]

streptokinase /ˌstreptəʊˈkaɪneɪz/ *n.* Biochem. & Pharm. an enzyme produced by some streptococci and used to treat inflammation and blood clots. [STREPTOCOCCUS + Gk *kinein* move + -ASE]

streptomycin /ˌstreptəˈmaɪsɪn/ *n.* an antibiotic produced by the bacterium *Streptomyces griseus*, effective against many disease-producing bacteria. [as STREPTOCOCCUS + -MYCIN]

stress /stres/ *n.* & *v.* ● *n.* **1 a** pressure or tension exerted on a material object. **b** a quantity measuring this. **2 a** a demand on physical or mental energy. **b** distress caused by this (*suffering from stress*). **3 a** emphasis (*the stress was on the need for success*). **b** accentuation; emphasis laid on a syllable or word. **c** an accent, esp. the principal one in a word (*the stress is on the first syllable*). **4** Mech. force per unit area exerted between contiguous bodies or parts of a body. ● *v.tr.* **1** lay stress on; emphasize. **2** subject to mechanical or physical or mental stress. □ **lay stress on** indicate as important. **stress disease** a disease resulting from continuous mental stress. □ **stressless** *adj.* [ME f. DISTRESS, or partly f. OF *estresse* narrowness, oppression, ult. f. L *strictus* STRICT]

stressed /strest/ *adj.* affected by mental stress. □ **stressed out** colloq. exhausted as a result of stress.

stressful /ˈstresfʊl/ *adj.* causing stress; mentally tiring (*had a stressful day*). □ **stressfully** *adv.* **stressfulness** *n.*

stretch /stretʃ/ *v.* & *n.* ● *v.* **1** *tr.* & *intr.* draw or be drawn or admit of being drawn out into greater length or size. **2** *tr.* & *intr.* make or become taut. **3** *tr.* & *intr.* place or lie at full length or spread out (*with a canopy stretched over them*). **4** *a tr.* extend (an arm, leg, etc.). **b** *intr.* & *refl.* thrust out one's limbs and tighten one's muscles after being relaxed. **5** *intr.* have a specified length or extension; extend (*farmland stretches for many miles*). **6** *tr.* strain or exert extremely or excessively; exaggerate (*stretch the truth*).

7 *tr.* (as **stretched** *adj.*) elongated or extended. ● *n.* **1** a continuous extent or expanse or period (*a stretch of open road*). **2** the act or an instance of stretching; the state of being stretched. **3** (*attrib.*) able to stretch; elastic (*stretch fabric*). **4 a** colloq. a period of imprisonment. **b** a period of service. **5** N. Amer. the straight side of a racetrack. **6** Naut. the distance covered on one tack. **7** (usu. *attrib.*) colloq. an aircraft or motor vehicle modified so as to have extra seating or storage capacity (*stretch limousine*). □ **at full stretch** working to capacity. **at a stretch 1** in one continuous period (*slept for two hours at a stretch*). **2** with much effort. **stretch one's legs** exercise oneself by walking. **stretch marks** marks on the skin resulting from a gain of weight, or on the abdomen after pregnancy. **stretch out 1** extend (a hand or foot etc.). **2** last for a longer period; prolong. **3** make (money etc.) last for a sufficient time. **stretch a point** agree to something not normally allowed. **stretch one's wings** see WING. □ **stretchy** *adj.* **stretchiness** *n.* **stretchable** *adj.* **stretchability** /ˌstretʃəˈbɪlɪti/ *n.* [OE *streccan* f. WG: cf. STRAIGHT]

stretcher /ˈstretʃə(r)/ *n.* & *v.* ● *n.* **1** a framework of two poles with canvas etc. between, for carrying sick, injured, or dead persons in a lying position. **2** a brick or stone laid with its long side along the face of a wall (cf. HEADER 3). **3** a board in a boat against which a rower presses the feet. **4** a rod or bar as a tie between chair-legs etc. **5** a wooden frame over which a canvas is stretched ready for painting. **6** archaic sl. an exaggeration or lie. ● *v.tr.* (often foll. by *off*) convey (a sick or injured person) on a stretcher. □ **stretcher-bearer** a person who helps to carry a stretcher, esp. in war or at a major accident.

stretto /ˈstretəʊ/ *n.* Mus. in quicker time. [It., = narrow]

strew /struː/ *v.tr.* (*past part.* **strewn** /struːn/ or **strewed**) **1** scatter or spread about over a surface. **2** (usu. foll. by *with*) spread (a surface) with scattered things. □ **strewer** *n.* [OE *stre(o)wian*]

'strewth var. of STRUTH.

stria /ˈstraɪə/ *n.* (*pl.* **striae** /ˈstraɪiː/) **1** Anat., Zool., Bot., & Geol. **a** a linear mark on a surface. **b** a slight ridge, furrow, or score. **2** Archit. a fillet between the flutes of a column. [L]

striate *adj.* & *v.* Anat., Zool., Bot., & Geol. ● *adj.* /ˈstraɪət/ (also **striated** /ˈstraɪeɪtɪd/) marked with striae. ● *v.tr.* /ˈstraɪeɪt/ mark with striae. □ **striated muscle** muscle with striations, attached to bones by tendons and under voluntary control (also called *skeletal muscle*). □ **striation** /straɪˈeɪʃ(ə)n/ *n.*

stricken /ˈstrɪkən/ *adj.* **1** affected or overcome with illness or misfortune etc. (*stricken with measles*; *grief-stricken*). **2** (often foll. by *from* etc.) US Law deleted. **3** levelled with a strickle. □ **stricken in years** archaic enfeebled by age. [archaic past part. of STRIKE]

strickle /ˈstrɪk(ə)l/ *n.* **1** a rod used in strike-measure. **2** a whetting tool. [OE *stricel*, rel. to STRIKE]

strict /strɪkt/ *adj.* **1** precisely limited or defined; without exception or deviation (*lives in strict seclusion*). **2 a** (of a person) severe; rigorous in upholding standards of conduct or morality. **b** requiring complete compliance or exact performance; enforced rigidly (*gave strict orders*). □ **strictness** *n.* [L *strictus* past part. of *stringere* tighten]

strictly /ˈstrɪktlɪ/ *adv.* **1** in a strict manner. **2** (also **strictly speaking**) applying words in their strict sense (*he is, strictly, an absconder*). **3** esp. N. Amer. colloq. definitely.

stricture /ˈstrɪktʃə(r)/ *n.* **1** (usu. in *pl.*; often foll. by *on, upon*) a critical or censorious remark. **2** Med. a morbid narrowing of a canal or duct in the body. □ **strictured** *adj.* [ME f. L *strictura* (as STRICT)]

stride /straɪd/ *v.* & *n.* ● *v.* (*past* **strode** /strəʊd/; *past part.* **stridden** /ˈstrɪd(ə)n/) **1** *intr.* & *tr.* walk with long firm steps. **2** *tr.* cross with one step. **3** *tr.* bestride; straddle. ● *n.* **1 a** a single long step. **b** the length of this. **2** a person's gait as determined by the length of stride. **3** (usu. in *pl.*) progress (*has made great strides*). **4** a settled rate of progress (*get into one's stride*; *be thrown out of one's stride*). **5** (in *pl.*) sl. trousers. **6** the distance between the feet parted either laterally or as in walking. □ **take in one's stride 1** clear (an obstacle) without changing one's gait to jump. **2** manage without difficulty. □ **strider** *n.* [OE *strīdan*]

strident /ˈstraɪd(ə)nt/ *adj.* loud and harsh. □ **stridency** *n.* **stridently** *adv.* [L *stridere* *strident-* creak]

stridulate /ˈstrɪdjʊˌleɪt/ *v.intr.* (of insects, esp. the cicada and grasshopper) make a shrill sound by rubbing esp. the legs or wing-cases together. □ **stridulant** *adj.* **stridulation** /ˌstrɪdjʊˈleɪʃ(ə)n/ *n.* [F *striduler* f. L *stridulus* creaking (as STRIDENT)]

strife /straɪf/ *n.* **1** conflict; struggle between opposed persons or things. **2** Austral. colloq. trouble of any kind. [ME f. OF *estrif*: cf. OF *estriver* STRIVE]

strigil /ˈstrɪdʒɪl/ *n.* **1** Gk & Rom. Antiq. an instrument with a curved blade used to scrape sweat and dirt from the skin in a hot-air bath or after exercise. **2** a structure on the leg of an insect used to clean its antennae etc. [L *strigilis* f. *stringere* graze]

strigose /ˈstraɪɡəʊz, -ɡəʊs/ *adj.* **1** (of leaves etc.) having short stiff hairs or scales. **2** (of an insect etc.) streaked, striped, or ridged. [L *striga* swath, furrow]

strike /straɪk/ *v. & n.* ● *v.* (*past* **struck** /strʌk/; *past part.* **struck** or *archaic* **stricken** /ˈstrɪkən/) **1 a** *tr.* subject to an impact. **b** *tr.* deliver (a blow) or inflict a blow on (also with double object: *struck him a blow*). **2** *tr.* come or bring sharply into contact with (*the ship struck a rock*). **3** *tr.* propel or divert with a blow (*struck the ball into the pond*). **4** *intr.* (foll. by *at*) try to hit. **5** *tr.* cause to penetrate (*struck terror into him*). **6** *tr.* ignite (a match) or produce (sparks etc.) by rubbing. **7** *tr.* make (a coin) by stamping. **8** *tr.* produce (a musical note) by striking. **9 a** *tr.* (also *absol.*) (of a clock) indicate the time by the sounding of a chime etc. **b** *intr.* (of time) be indicated in this way. **10** *tr.* **a** attack or affect suddenly (*was struck with sudden terror*). **b** (of a disease) afflict. **11** *tr.* cause to become suddenly (*was struck dumb*). **12** *tr.* reach or achieve (*strike a balance*). **13** *tr.* agree on (a bargain). **14** *tr.* assume (an attitude or pose) suddenly and dramatically. **15** *tr.* **a** discover or come across. **b** find (oil etc.) by drilling. **c** encounter (an unusual thing etc.). **16** *tr.* come to the attention of or appear to (*it strikes me as silly; an idea suddenly struck me*). **17 a** *intr.* (of employees) engage in a strike; cease work as a protest. **b** *tr. N. Amer.* act in this way against (an employer). **18 a** *tr.* lower or take down (a flag or tent etc.). **b** *intr.* signify surrender by striking a flag; surrender. **19** *intr.* take a specified direction (*struck east*). **20** *tr.* (also *absol.*) secure a hook in the mouth of (a fish) by jerking the tackle. **21** *tr.* (of a snake) wound with its fangs. **22** *intr.* (of oysters) attach themselves to a bed. **23 a** *tr.* insert (a cutting of a plant) in soil to take root. **b** *tr.* (also *absol.*) (of a plant or cutting etc.) put forth (roots). **24** *tr.* level (grain etc. or the measure) in strike-measure. **25** *tr.* ascertain (a balance) by deducting credit or debit from the other. **b** arrive at (an average, state of balance) by equalizing all items. **26** *tr.* compose (a jury) esp. by allowing both sides to reject the same number. ● *n.* **1** the act or an instance of striking. **2 a** the organized refusal by employees to work until some grievance is remedied. **b** a similar refusal to participate in some other expected activity. **3 a** a sudden find or success (*a lucky strike*). **b** the discovery of oil etc. **4** an attack, esp. from the air. **5** *Baseball* a batter's unsuccessful attempt to hit a pitched ball, or another event counting equivalently against a batter. **6** the act of knocking down all the pins with the first ball in bowling. **7** horizontal direction in a geological structure. **8** a strickle. **9** infestation of a sheep or cow with flies whose larvae burrow into the skin from open wounds. □ **on strike** taking part in an industrial etc. strike. **strike at the root** (or **roots**) of see ROOT[1]. **strike back 1** strike or attack in return. **2** (of a gas-burner) burn from an internal point before the gas has become mixed with air. **strike down 1** knock down. **2** bring low; afflict (*struck down by a virus*). **strike force 1** a military or police force ready for rapid effective action. **2** the forwards in a football team. **strike home 1** deal an effective blow. **2** have an intended effect (*my words struck home*). **strike in 1** intervene in a conversation etc. **2** (of a disease) attack the interior of the body from the surface. **strike it rich** *colloq.* find a source of abundance or success. **strike a light 1** produce a light by striking a match. **2** (as *int.*) *Brit. colloq.* an expression of surprise, disgust, etc. **strike lucky** (or **strike it lucky**) have a lucky success. **strike-measure** measurement by passing a rod across the top of a heaped container to ensure that it is exactly full. **strike off 1** remove with a stroke. **2** delete (a name etc.) from a list. **3** produce (copies of a document). **strike oil 1** find petroleum by sinking a shaft. **2** attain prosperity or success. **strike out 1** hit out. **2** act vigorously. **3** delete (an item or name etc.). **4** set off or begin (*struck out eastwards*). **5** use the arms and legs in swimming. **6** forge or devise (a plan etc.). **7** *Baseball* **a** dismiss (a batter) by means of three strikes. **b** be dismissed in this way. **strike pay** an allowance paid to strikers by their trade union. **strike-slip fault** *Geol.* a fault in which rock strata are displaced mainly in a horizontal direction, parallel to the line of the fault. **strike through** delete (a word etc.) with a stroke of one's pen. **strike up 1** start (an acquaintance, conversation, etc.) esp. casually. **2** (also *absol.*) begin playing (a tune etc.). **strike upon 1** have (an idea etc.) luckily occur to one. **2** (of light) illuminate. **strike while the iron is hot** act promptly at a good opportunity. **struck on** *colloq.* infatuated with. □ **strikable** *adj.* [OE *strīcan* go, stroke f. WG]

strikebound /ˈstraɪkbaʊnd/ *adj.* immobilized or closed by a strike.

strikebreaker /ˈstraɪkˌbreɪkə(r)/ *n.* a person working or employed in place of others who are on strike. □ **strikebreak** *v.intr.*

striker /ˈstraɪkə(r)/ *n.* **1** a person or thing that strikes. **2** an employee

on strike. **3** *Sport* the player who is to strike, or who is to be the next to strike, the ball. **4** *Football* an attacking player positioned well forward in order to score goals. **5** a device striking the primer in a gun.

striking /ˈstraɪkɪŋ/ *adj. & n.* ● *adj.* **1** impressive; attracting attention. **2** (of a clock) making a chime to indicate the hours etc. ● *n.* the act or an instance of striking. □ **striking-circle** (in hockey) an elongated semicircle in front of the goal, from within which the ball must be hit in order to score. **striking-force** a military body ready to attack at short notice. **within striking distance** near enough to hit or achieve. □ **strikingly** *adv.* **strikingness** *n.*

Strindberg /ˈstrɪndbɜːɡ/, (Johan) August (1849–1912), Swedish dramatist and novelist. Although best known outside Scandinavia as a dramatist and precursor of expressionism in the theatre, Strindberg was also a leading figure in the naturalist movement in literature; his satire *The Red Room* (1879) is regarded as Sweden's first modern novel. His earlier plays, also naturalistic in style, depict a bitter power struggle between the sexes, notably in *The Father* (1887) and *Miss Julie* (1888). His later plays are typically tense, symbolic, psychic dramas; the trilogy *To Damascus* (1898–1904) and, more particularly, *A Dream Play* (1902), introduced expressionist techniques and are of major importance for the development of modern drama.

Strine /straɪn/ *n.* **1** a comic transliteration of Australian speech, e.g. *Emma Chissitt* = 'How much is it?'. **2** (esp. uneducated) Australian English. [= *Australian* in Strine]

string /strɪŋ/ *n. & v.* ● *n.* **1** twine or narrow cord. **2** a piece of this or of similar material used for tying or holding together, pulling, etc. **3** a length of catgut or wire etc. on a musical instrument, producing a note by vibration. **4 a** (in *pl.*) stringed instruments (i.e. violin, viola, cello, double bass) forming a section of an orchestra, group, etc. **b** (*attrib.*) relating to or consisting of stringed instruments (*string quartet*). **5** (in *pl.*) an awkward associated or consequent condition or complication (*the offer has no strings*). **6** a set of things strung together; a series or line of persons or things (*a string of beads; a string of oaths*). **7** a group of racehorses trained at one stable. **8** a tough piece connecting the two halves of a bean-pod etc. **9** a piece of catgut etc. interwoven with others to form the head of a tennis etc. racket. **10** = STRINGBOARD. **11** *Physics* **a** a hypothetical one-dimensional subatomic particle having the dynamical properties of a flexible loop. **b** (in full **cosmic string**) in some cosmological theories, a threadlike concentration of energy hypothesized to exist within the structure of space–time. **12** *Computing* a linear sequence of characters, records, or data. ● *v.* (*past* and *past part.* **strung** /strʌŋ/) **1** *tr.* supply with a string or strings. **2** *tr.* tie with string. **3** *tr.* thread (beads etc.) on a string. **4** *tr.* arrange in or as a string. **5** *tr.* remove the strings from (a bean). **6** *tr.* place a string ready for use on (a bow). **7** *tr.* esp. *N. Amer. colloq.* hoax. **8** *intr.* (of glue etc.) become stringy. **9** *intr. Billiards* make the preliminary strokes that decide which player begins. □ **have two** (or **many**) **strings to one's bow** see BOW[1]. **on a string** under one's control or influence. **string along** *colloq.* **1** deceive, esp. by appearing to comply with (a person). **2** (often foll. by *with*) keep company (with). **string bass** *Mus.* a double bass. **string bean 1** a bean eaten in its fibrous pod, esp. a runner bean or French bean. **2** *colloq.* a tall thin person. **string-course** a raised horizontal band or course of bricks etc. on a building. **string out** extend; prolong (esp. unduly). **string-piece** a long timber supporting and connecting the parts of a framework. **string tie** a very narrow necktie. **string up 1** hang up on strings etc. **2** kill by hanging. **3** (usu. as **strung up** *adj.*) make tense. **string vest** a vest with large meshes. □ **stringless** *adj.* **stringlike** *adj.* [OE *streng* f. Gmc: cf. STRONG]

stringboard /ˈstrɪŋbɔːd/ *n.* a supporting timber or skirting in which the ends of a staircase steps are set.

stringed /strɪŋd/ *adj.* (of musical instruments) having strings (also in *comb.*: *twelve-stringed guitar*).

stringendo /strɪnˈdʒendəʊ/ *adj. & adv. Mus.* with increasing speed. [It. f. *stringere* press: see STRINGENT]

stringent /ˈstrɪndʒənt/ *adj.* **1** (of rules etc.) strict, precise; requiring exact performance; leaving no loophole or discretion. **2** (of a money market etc.) tight; hampered by scarcity; unaccommodating; hard to operate in. □ **stringency** *n.* **stringently** *adv.* [L *stringere* draw tight]

stringer /ˈstrɪŋə(r)/ *n.* **1** a longitudinal structural member in a framework, esp. of a ship or aircraft. **2** *colloq.* a newspaper correspondent not on the regular staff. **3** = STRINGBOARD.

stringhalt /ˈstrɪŋhɒlt, -hɔːlt/ *n.* spasmodic movement of a horse's hind leg.

string quartet *n.* **1** a chamber music ensemble consisting of first and second violins, viola, and cello, first seen in the 18th century. Major composers for the string quartet were Haydn, Mozart, Beethoven, and Schubert. **2** a piece of music for a string quartet.

stringy /ˈstrɪŋɪ/ *adj.* (**stringier, stringiest**) **1** (of food etc.) fibrous, tough. **2** of or like string. **3** (of a person) tall, wiry, and thin. **4** (of a liquid) viscous; forming strings. □ **stringy-bark** *Austral.* a eucalyptus tree with tough fibrous bark, e.g *Eucalyptus obliqua.* □ **stringily** *adv.* **stringiness** *n.*

strip[1] /strɪp/ *v. & n.* ● *v.* (**stripped, stripping**) **1** *tr.* (often foll. by *of*) remove the clothes or covering from (a person or thing). **2** *intr.* (often foll. by *off*) undress oneself. **3** *tr.* (often foll. by *of*) deprive (a person) of property or titles. **4** *tr.* leave bare of accessories or fittings. **5** *tr.* remove bark and branches from (a tree). **6** *tr.* (often foll. by *down*) remove the accessory fittings of or take apart (a machine etc.) to inspect or adjust it. **7** sell off (the assets of a company) for profit. **8** *tr.* milk (a cow) to the last drop. **9** *tr.* remove the old hair from (a dog). **10** *tr.* remove the stems from (tobacco). **11 a** *tr.* tear the thread from (a screw). **b** *intr.* (of a screw) lose its thread. **12** *tr.* tear the teeth from (a gearwheel). **13** *tr.* remove (paint) or remove paint from (a surface) with solvent. **14** *tr.* (often foll. by *from*) pull or tear (a covering or property etc.) off (*stripped the masks from their faces*). **15** *intr.* (of a bullet) issue from a rifled gun without spin owing to a loss of surface. ● *n.* **1** an act of stripping, esp. of undressing in striptease. **2** the identifying outfit worn by the members of a sports team while playing. □ **strip club** a club at which striptease performances are given. **strip mine** *US* a mine worked by removing the material that overlies the ore etc. **strip-search** *n.* a search of a person involving the removal of all clothes. ● *v.tr.* search in this way. [ME f. OE *bestrīepan* plunder f. Gmc]

strip[2] /strɪp/ *n.* **1** a long narrow piece (*a strip of land*). **2** a narrow flat bar of iron or steel. **3** (in full **strip cartoon**) = *comic strip*. □ **strip farming** *hist.* agriculture in which the land was divided into long narrow strips worked by different farmers. **strip light** a tubular fluorescent lamp. **strip mill** a mill in which steel slabs are rolled into strips. **tear a person off a strip** *colloq.* angrily rebuke a person. [ME, from or rel. to MLG *strippe* strap, thong, prob. rel. to STRIPE]

strippagram /ˈstrɪpəˌgræm/ *n.* (also **strippagram**) a novelty telegram or greetings message delivered by a person who performs a striptease for the recipient.

stripe /straɪp/ *n.* **1** a long narrow band or strip differing in colour or texture from the surface on either side of it (*black with a red stripe*). **2** *Mil.* a chevron etc. denoting military rank. **3** *N. Amer.* a category of character, opinion, etc. (*a man of that stripe*). **4** (usu. in *pl.*) *archaic* a blow with a scourge or lash. **5** (in *pl.*, treated as *sing.*) *colloq.* a tiger. [perh. back-form. f. *striped*: cf. MDu., MLG *strīpe*, MHG *strīfe*]

striped /straɪpt/ *adj.* marked with stripes (also in *comb.*: *red-striped*).

stripling /ˈstrɪplɪŋ/ *n.* a youth not yet fully grown. [ME, prob. f. STRIP[2] + -LING[1], in the sense of having a figure not yet filled out]

strippagram var. of STRIPAGRAM.

stripper /ˈstrɪpə(r)/ *n.* **1** a person or thing that strips something. **2** a device or solvent for removing paint etc. **3** a striptease performer.

striptease /ˈstrɪptiːz/ *n. & v.* ● *n.* an entertainment in which the performer gradually undresses before the audience. ● *v.intr.* perform a striptease. □ **stripteaser** *n.*

stripy /ˈstraɪpɪ/ *adj.* (**stripier, stripiest**) striped; having many stripes.

strive /straɪv/ *v.intr.* (*past* **strove** /strəʊv/; *past part.* **striven** /ˈstrɪv(ə)n/) **1** (often foll. by *for*, or *to* + infin.) try hard, make efforts (*strive to succeed*). **2** (often foll. by *with*, *against*) struggle or contend. □ **striver** *n.* [ME f. OF *estriver*, rel. to *estrif* STRIFE]

strobe /strəʊb/ *n. colloq.* **1** a stroboscope. **2** a stroboscopic lamp. [abbr.]

strobila /strəˈbaɪlə/ *n.* (*pl.* **strobilae** /-liː/) *Zool.* **1** a chain of proglottids in a tapeworm. **2** a sessile polyp-like form which divides horizontally to produce jellyfish larvae. [mod.L f. Gk *strobilē* twisted lint-plug f. *strephō* twist]

strobile /ˈstrəʊbaɪl/ *n. Bot.* = STROBILUS. [F *strobile* or LL *strobilus* f. Gk *strobilos* f. *strephō* twist]

strobilus /ˈstrəʊbɪləs/ *n.* (*pl.* **strobili** /-ˌlaɪ/) *Bot.* **1** the cone of a pine etc. **2** a conelike structure, e.g. the flower of a hop, or the aggregate of sporophylls in a clubmoss or horsetail. [LL (as STROBILE)]

stroboscope /ˈstrəʊbəˌskəʊp/ *n.* **1** *Physics* an instrument for determining speeds of rotation etc. by shining a bright light at intervals so that a rotating object appears stationary. **2** a lamp made to flash intermittently, esp. for this purpose. □ **stroboscopic** /ˌstrəʊbəˈskɒpɪk/ *adj.* **stroboscopical** *adj.* **stroboscopically** *adv.* [Gk *strobos* whirling + -SCOPE]

strode *past* of STRIDE.

Stroganoff /ˈstrɒgəˌnɒf/ *adj.* (of meat) cut into strips and cooked in sour-cream sauce (*beef Stroganoff*). [Count Pavel Aleksandrovich *Stroganoff*, Russian diplomat (1772–1817)]

stroke /strəʊk/ *n. & v.* ● *n.* **1** the act or an instance of striking; a blow or hit (*with a single stroke; a stroke of lightning*). **2** a sudden disabling attack or loss of consciousness caused by an interruption in the flow of blood to the brain, esp. through thrombosis; apoplexy. **3 a** an action or movement esp. as one of a series. **b** the time or way in which such movements are done. **c** the slightest such action (*has not done a stroke of work*). **4** the whole of the motion (of a wing, oar, etc.) until the starting-position is regained. **5** (in rowing) the mode or action of moving the oar (*row a fast stroke*). **6** the whole motion (of a piston) in either direction. **7** *Golf* the action of hitting (or hitting at) a ball with a club, as a unit of scoring. **8** a mode of moving the arms and legs in swimming. **9** a method of striking with the bat etc. in games etc. (*played some unorthodox strokes*). **10** a specially successful or skilful effort (*a stroke of diplomacy*). **11 a** a mark made by the movement in one direction of a pen or pencil or paintbrush. **b** a similar mark printed. **12** a detail contributing to the general effect in a description. **13** the sound made by a striking clock. **14** (in full **stroke oar**) the oar or oarsman nearest the stern, setting the time of the stroke. **15** the act or a spell of stroking. ● *v.tr.* **1** pass one's hand gently along the surface of (hair or fur etc.); caress lightly. **2** act as the stroke of (a boat or crew). □ **at a stroke** by a single action. **finishing stroke** a *coup de grâce*; a final and fatal stroke. **off one's stroke** not performing as well as usual. **on the stroke** punctually. **on the stroke of nine** etc. with the clock about to strike nine etc. **stroke a person down** appease a person's anger. **stroke of business** a profitable transaction. **stroke of genius** an original or strikingly successful idea. **stroke of luck** (or **good luck**) an unforeseen opportune occurrence. **stroke play** *Golf* play in which the score is reckoned by counting the number of strokes taken for the round (cf. *match play* (see MATCH[1])). **stroke a person** (or **a person's hair**) **the wrong way** irritate a person. [OE *strācian* f. Gmc, rel. to STRIKE]

stroll /strəʊl/ *v. & n.* ● *v.intr.* **1** saunter or walk in a leisurely way. **2** achieve something easily, without effort. ● *n.* **1** a short leisurely walk (*go for a stroll*). **2** something easily achieved; a walkover. □ **strolling players** actors etc. going from place to place to give performances. [orig. of a vagrant, prob. f. G *strollen, strolchen* f. *Strolch* vagabond, of unkn. orig.]

stroller /ˈstrəʊlə(r)/ *n.* **1** a person who strolls. **2** *US* a pushchair.

stroma /ˈstrəʊmə/ *n.* (*pl.* **stromata** /-mətə/) *Biol.* **1** the framework of an organ or cell. **2** a fungous tissue containing spore-producing bodies. □ **stromatic** /strəʊˈmætɪk/ *adj.* [mod.L f. LL f. Gk *strōma* coverlet]

stromatolite /strəʊˈmætəˌlaɪt/ *n. Biol.* a mound built up of layers of blue-green algae and trapped sediment, found in lagoons in Australasia and fossilized in Precambrian rocks elsewhere. [STROMA + -LITE]

Stromboli /ˈstrɒmbəlɪ, strɒmˈbəʊlɪ/ a volcanic island in the Mediterranean, the most north-easterly of the Lipari Islands. Its volcano has been in a state of continual mild eruption throughout history.

strong /strɒŋ/ *adj. & adv.* ● *adj.* (**stronger** /ˈstrɒŋgə(r)/; **strongest** /ˈstrɒŋgɪst/) **1** having the power of resistance; able to withstand great force or opposition; not easily damaged or overcome (*strong material; strong faith; a strong character*). **2** (of a person's constitution) able to overcome, or not liable to, disease. **3** (of a person's nerves) proof against fright, irritation, etc. **4** (of a patient) restored to health. **5** (of an economy) stable and prosperous; (of a market) having steadily high or rising prices. **6** capable of exerting great force or of doing much; muscular, powerful. **7** forceful or powerful in effect (*a strong wind; a strong protest*). **8** decided or firmly held (*a strong suspicion; strong views*). **9** (of an argument etc.) convincing or striking. **10** powerfully affecting the senses or emotions (*a strong light; strong acting*). **11** powerful in terms of size or numbers or quality (*a strong army*). **12** capable of doing much when united (*a strong combination*). **13 a** formidable; likely to succeed (*a strong candidate*). **b** tending to assert or dominate (*a strong personality*). **14** (of a solution or drink etc.) containing a large proportion of a substance in water or another solvent (*strong tea*). **15** *Chem.* (of an acid or base) fully ionized into cations and anions in aqueous solution. **16** (of a group) having a specified number (*200 strong*). **17** (of a voice) loud or penetrating. **18** (of food or its flavour) pungent. **19** (esp. of a person's breath) ill-smelling. **20** (of a literary style) vivid and terse. **21** (of a measure) drastic. **22** *Gram.* in Germanic languages: **a** (of a verb) forming inflections by change of vowel within the stem rather than by the addition of a suffix (e.g. *swim, swam*). **b** (of a noun or adjective) belonging to a declension in which the stem originally ended otherwise than in *-n* (opp. WEAK 9). **23** having validity or credence (*a strong possibility; a strong chance*). **24** unmistakable, noticeable (*a strong resemblance; a strong accent*). ● *adv.* strongly (*the tide is running strong*). □ **come it strong** *colloq.* go to great lengths; use exaggeration. **come on strong** behave aggressively or assertively. **going strong** continuing action vigorously; in good health or trim. **strong-arm** using force (*strong-arm tactics*). **strong drink** see DRINK. **strong grade** *Philol.* the stressed ablaut-form. **strong interaction** *Physics* interaction between certain subatomic particles that is very strong but is effective only at short distances. **strong language** forceful language; swearing. **strong meat** a doctrine or action acceptable only to vigorous or instructed minds. **strong-minded** having determination. **strong-mindedness** determination. **strong point 1** a thing at which one excels. **2** a specially fortified defensive position. **strong stomach** a stomach not easily affected by nausea. **strong suit 1** a suit at cards in which one can take tricks. **2** a thing at which one excels. □ **strongish** *adj.* **strongly** *adv.* [OE f. Gmc: cf. STRING]

strongbox /'strɒŋbɒks/ *n.* a strongly made small chest for valuables.

stronghold /'strɒŋhəʊld/ *n.* **1** a fortified place. **2** a secure refuge. **3** a centre of support for a cause etc.

strongman /'strɒŋmæn/ *n.* (*pl.* **-men**) **1** a man of great physical strength, esp. a performer of feats of strength at a fair, circus, etc. **2** a forceful leader who exercises firm control over a state, group, etc.

strongroom /'strɒŋruːm, -rʊm/ *n.* a room designed to protect valuables against fire and theft.

strontia /'strɒnʃə/ *n. Chem.* strontium oxide. [*strontian* native strontium carbonate f. Strontian in the Highland Region of Scotland, where it was discovered]

strontium /'strɒntɪəm/ *n.* a soft silver-white metallic chemical element (atomic number 38; symbol **Sr**). Minerals containing strontium were identified in the late 18th century, and the element was isolated by Sir Humphry Davy in 1808. It is a member of the alkaline earth metals. The metal has few uses, but strontium salts are used in fireworks and flares because they give a brilliant red light. The radioactive isotope strontium-90 is a particularly dangerous component of nuclear fallout as it can become concentrated in bones and teeth. [STRONTIA]

strop /strɒp/ *n. & v.* ● *n.* **1** a device, esp. a strip of leather, for sharpening razors. **2** *Naut.* a collar of leather or spliced rope or iron used for handling cargo. ● *v.tr.* (**stropped, stropping**) sharpen on or with a strop. [ME f. MDu., MLG *strop*, OHG *strupf*, WG f. L *stroppus*]

strophanthin /strə'fænθɪn/ *n. Chem.* any of a group of poisonous glycosides obtained from African trees of the genera *Strophanthus* and *Acokanthera* and used as heart stimulants (cf. OUABAIN). [mod.L *strophanthus* f. Gk *strophos* twisted cord + *anthos* flower]

strophe /'strəʊfi/ *n.* **1 a** a turn in dancing made by an ancient Greek chorus. **b** lines recited during this. **c** the first section of an ancient Greek choral ode or of one division of it. **2** *Prosody* a group of lines forming a section of a lyric poem. □ **strophic** *adj.* [Gk *strophē*, lit. 'turning', f. *strephō* turn]

stroppy /'strɒpɪ/ *adj.* (**stroppier, stroppiest**) *Brit. colloq.* bad-tempered; awkward to deal with. □ **stroppily** *adv.* **stroppiness** *n.* [20th c.: perh. abbr. of OBSTREPEROUS]

strove *past* of STRIVE.

strow /strəʊ/ *v.tr.* (*past part.* **strown** /strəʊn/ or **strowed**) *archaic* = STREW. [var. of STREW]

struck *past* and *past part.* of STRIKE.

structural /'strʌktʃərəl/ *adj.* of, concerning, or having a structure. □ **structural engineering** the branch of civil engineering concerned with large modern buildings etc. **structural formula** *Chem.* a formula showing the arrangement of atoms in the molecule of a compound. **structural linguistics** the study of language as a system of interrelated elements, without reference to their historical development. **structural psychology** the study of the arrangement and composition of mental states and conscious experiences. **structural steel** strong mild steel in shapes suited to construction work. □ **structurally** *adv.*

structuralism /'strʌktʃərəlɪz(ə)m/ *n.* a method of analysing and organizing concepts in anthropology, linguistics, psychology, and other cognitive and social sciences in terms of contrasting relations (esp. binary oppositions) among sets of items within conceptual systems. Structuralism has had a profound influence on many disciplines in the 20th century. Its origins can be traced to the Swiss

linguist Ferdinand de Saussure, who formulated a number of fundamental distinctions, such as that between the signifier and the signified and that between syntagmatic association (word in context) and paradigmatic association (words in sets such as the parts of speech). Saussure's ideas were extended by the French anthropologist Claude Lévi-Strauss, who analysed kinship, myths, taboos, and other subjects on the basis of the relational structure of the terms used. These ideas were further developed by a group of French writers and critics, including Roland Barthes, and their followers in many other countries, with the aim of providing a 'grammar' of narrative, fashion, food, and innumerable other aspects of culture. (See also DECONSTRUCTION.) □ **structuralist** *n. & adj.*

structure /'strʌktʃə(r)/ *n. & v.* ● *n.* **1 a** a whole constructed unit, esp. a building. **b** the way in which a building etc. is constructed (*has a flimsy structure*). **2** a set of interconnecting parts of any complex thing; a framework (*the structure of a sentence; a new wages structure*). ● *v.tr.* give structure to; organize; frame. □ **structured** *adj.* (also in *comb.*). **structureless** *adj.* [ME f. OF *structure* or L *structura* f. *struere* struct-build]

strudel /'struːd(ə)l/ *n.* a confection of thin pastry rolled up round a filling and baked (*apple strudel*). [G]

struggle /'strʌg(ə)l/ *v. & n.* ● *v.intr.* **1** make forceful or violent efforts to get free of restraint or constriction. **2** (often foll. by *for*, or *to* + infin.) make violent or determined efforts under difficulties; strive hard (*struggled for supremacy; struggled to get the words out*). **3** (foll. by *with, against*) contend; fight strenuously (*struggled with the disease; struggled against superior numbers*). **4** (foll. by *along, up,* etc.) make one's way with difficulty (*struggled to my feet*). **5** (esp. as **struggling** *adj.*) have difficulty in gaining recognition or a living (*a struggling artist*). ● *n.* **1** the act or a spell of struggling. **2** a hard or confused contest. **3** a determined effort under difficulties. □ **the struggle for existence** (or **life**) the competition between organisms esp. as an element in natural selection, or between persons seeking a livelihood. □ **struggler** *n.* [ME *strugle* frequent. of uncert. orig. (perh. imit.)]

strum /strʌm/ *v. & n.* ● *v.tr.* (**strummed, strumming**) **1** play on (a stringed or keyboard instrument), esp. carelessly or unskilfully. **2** play (a tune etc.) in this way. ● *n.* **1** the sound made by strumming. **2** an instance or spell of strumming. □ **strummer** *n.* [imit.: cf. THRUM¹]

struma /'struːmə/ *n.* (*pl.* **strumae** /-miː/) **1** *Med.* **a** = SCROFULA. **b** = GOITRE. **2** *Bot.* a cushion-like swelling of an organ. □ **strumose** /-məʊz, -məʊs/ *adj.* **strumous** *adj.* [L, = scrofulous tumour]

strumpet /'strʌmpɪt/ *n. archaic* or *joc.* a prostitute. [ME: orig. unkn.]

strung *past* and *past part.* of STRING.

strut /strʌt/ *n. & v.* ● *n.* **1** a bar forming part of a framework and designed to resist compression. **2** a strutting gait. ● *v.* (**strutted, strutting**) **1** *intr.* walk with a pompous or affected stiff erect gait. **2** *tr.* brace with a strut or struts. □ **strutter** *n.* **struttingly** *adv.* [ME 'bulge, swell, strive', earlier *stroute* f. OE *strūtian* be rigid (?)]

'struth /struːθ/ *int.* (also **'strewth**) *colloq.* a mild oath. [*God's truth*]

struthious /'struːθɪəs/ *adj. Zool.* of or like an ostrich. [L *struthio* ostrich]

Struve /'struːvə/, Otto (1897–1963), Russian-born American astronomer. He belonged to the fourth generation of a line of distinguished astronomers that began with the German-born Friedrich Georg Wilhelm Struve (1793–1864). Otto Struve was successively director of four observatories in the US, including the McDonald Observatory in Texas which he was instrumental in founding. He was mainly interested in spectroscopic investigations into the composition, evolution, and rotation of stars, but his most important contribution was his discovery of the presence of ionized hydrogen in interstellar space (1938).

Struwwelpeter /'struːəlˌpiːtə(r)/ a character in a collection of children's stories of the same name by Heinrich Hoffmann (1809–94), with long thick unkempt hair and extremely long fingernails. [G, = shock-headed Peter]

strychnine /'strɪkniːn/ *n.* a vegetable alkaloid obtained from plants of the genus *Strychnos*, esp. nux vomica. It is bitter and highly poisonous, being used as an experimental stimulant and (formerly) as a tonic. □ **strychnic** *adj.* [F f. L *strychnos* f. Gk *strukhnos* a kind of nightshade]

Sts *abbr.* Saints.

Stuart¹ /'stjuːət/ (also **Stewart**) the name of the royal house of Scotland from the accession (1371) of Robert II, one of the hereditary stewards of Scotland, and of Britain from the accession of James VI of

Scotland to the English throne as James I (1603) to the death of Queen Anne (1714).

Stuart[2] /ˈstjuːət/, Charles Edward (known as 'the Young Pretender' or 'Bonnie Prince Charlie') (1720–88), son of James Stuart, pretender to the British throne. He led the Jacobite uprising of 1745–6, gaining the support of the Highlanders, with whom he invaded England and advanced as far as Derby. However, he was driven back to Scotland by the Duke of Cumberland and defeated at the Battle of Culloden (1746). He later died in exile in Rome.

Stuart[3] /ˈstjuːət/, James (Francis Edward) (known as 'the Old Pretender') (1688–1766), son of James II (James VII of Scotland), pretender to the British throne. He arrived in Scotland too late to alter the outcome of the 1715 Jacobite uprising and left the leadership of the 1745–6 uprising to his son Charles Edward Stuart.

Stuart[4] /ˈstjuːət/, John McDouall (1815–66), Scottish explorer. He was a member of Charles Sturt's third expedition to Australia (1844–6), and subsequently crossed Australia from south to north and back again, at his sixth attempt (1860–2).

Stuart[5] /ˈstjuːət/, Mary, see MARY, QUEEN OF SCOTS.

stub /stʌb/ n. & v. ● n. **1** the remnant of a pencil or cigarette etc. after use. **2** the counterfoil of a cheque or receipt etc. **3** a stunted tail etc. **4** the stump of a tree, tooth, etc. **5** (attrib.) going only part of the way through (stub-mortise; stub-tenon). ● v.tr. (**stubbed, stubbing**) **1** strike (one's toe) against something. **2** (usu. foll. by out) extinguish (a lighted cigarette) by pressing the lighted end against something. **3** (foll. by up) grub up by the roots. **4** clear (land) of tree stumps etc. □ **stub-axle** an axle supporting only one wheel of a pair. [OE stub, stubb f. Gmc]

stubble /ˈstʌb(ə)l/ n. **1** the cut stalks of cereal plants left sticking up after the harvest. **2 a** cropped hair or a cropped beard. **b** a short growth of unshaven hair. □ **stubbled** adj. **stubbly** adj. [ME f. AF stuble, OF estuble f. L stupla, stupula var. of stipula straw]

stubborn /ˈstʌb(ə)n/ adj. **1** unreasonably obstinate. **2** unyielding, obdurate, inflexible. **3** refractory, intractable. □ **stubbornly** adv. **stubbornness** n. [ME stiborn, stoburn, etc., of unkn. orig.]

Stubbs[1] /stʌbz/, George (1724–1806), English painter and engraver. Known for the anatomical accuracy of his depictions of animals, he established his reputation with the book Anatomy of the Horse (1766), illustrated with his own engravings. He is particularly noted for his sporting scenes and paintings of horses and lions, for example, the Mares and Foals in a Landscape series (c.1760–70).

Stubbs[2] /stʌbz/, William (1825–1901), English historian and ecclesiastic. He wrote the influential Constitutional History of England (three volumes 1874–8), which charted the history of English institutions from the Germanic invasion to 1485. He was also bishop of Chester (1884–8) and of Oxford (1888–1901).

stubby /ˈstʌbɪ/ adj. & n. ● adj. (**stubbier, stubbiest**) short and thick. ● n. (pl. **-ies**) Austral. colloq. a small squat bottle of beer. □ **stubbily** adv. **stubbiness** n.

stucco /ˈstʌkəʊ/ n. & v. ● n. (pl. **-oes**) plaster or cement used for coating wall surfaces or moulding into architectural decorations. ● v.tr. (**-oes, -oed**) coat with stucco. [It., of Gmc orig.]

stuck past and past part. of STICK[2].

stuck-up /stʌkˈʌp/ see STICK[2].

stud[1] /stʌd/ n. & v. ● n. **1 a** a large-headed nail, boss, or knob, projecting from a surface esp. for ornament. **b** (in full **ear-stud**) a small usu. round ornament worn on the lobe of the ear. **2** a double button, esp. for use with two buttonholes in a shirt-front. **3** a small object projecting slightly from a road-surface as a marker etc. **4** a rivet or crosspiece in each link of a chain-cable. **5 a** a post to which laths are nailed. **b** US the height of a room as indicated by the length of this. ● v.tr. (**studded, studding**) **1** set with or as with studs. **2** (as **studded** adj.) (foll. by with) thickly set or strewn (studded with diamonds). **3** be scattered over or about (a surface). [OE studu, stuthu post, prop, rel. to G stützen to prop]

stud[2] /stʌd/ n. **1 a** a number of horses kept for breeding etc. **b** a place where these are kept. **2** (in full **stud-horse**) a stallion. **3** colloq. a young man: one noted for sexual prowess. **4** (in full **stud poker**) a form of poker with betting after the dealing of successive rounds of cards face up. □ **at stud** (of a male horse) publicly available for breeding on payment of a fee. **stud-book** a book containing the pedigrees of horses. **stud-farm** a place where horses are bred. [OE stōd f. Gmc: rel. to STAND]

studding /ˈstʌdɪŋ/ n. the woodwork of a lath-and-plaster wall.

studding-sail /ˈstʌdɪŋˌseɪl, ˈstʌns(ə)l/ n. Naut. a sail set on a small extra

yard and boom beyond the leech of a square sail in light winds. [16th c.: orig. uncert.: perh. f. MLG, MDu. stōtinge a thrusting]

student /ˈstjuːd(ə)nt/ n. **1** a person who is studying, esp. at university or another place of higher education. **2** (attrib.) studying in order to become (a student nurse). **3** a person of studious habits. **4** Brit. a graduate recipient of a stipend from the foundation of a college, esp. a fellow of Christ Church, Oxford. □ **studentship** n. [ME f. L studere f. studium STUDY]

studio /ˈstjuːdɪəʊ/ n. (pl. **-os**) **1** the workroom of a painter or photographer etc. **2** a place where cinema films or recordings are made or where television or radio programmes are made or produced. □ **studio couch** a couch that can be converted into a bed. **studio flat** a flat containing a room suitable as an artist's studio, or only one main room. [It. f. L (as STUDY)]

studious /ˈstjuːdɪəs/ adj. **1** devoted to or assiduous in study or reading. **2** studied, deliberate, painstaking (with studious care). **3** (foll. by to + infin. or in + verbal noun) showing care or attention. **4** (foll. by of + verbal noun) anxiously desirous. □ **studiously** adv. **studiousness** n. [ME f. L studiosus (as STUDY)]

study /ˈstʌdɪ/ n. & v. ● n. (pl. **-ies**) **1** the devotion of time and attention to acquiring information or knowledge, esp. from books. **2** (in pl.) the pursuit of academic knowledge (continued their studies abroad). **3** a room used for reading, writing, etc. **4** a piece of work, esp. a drawing, done for practice or as an experiment (a study of a head). **5** the portrayal in literature or another art form of an aspect of behaviour or character etc. **6** a musical composition designed to develop a player's skill. **7** a thing worth observing closely (your face was a study). **8** a thing that has been or deserves to be investigated. **9** Theatr. **a** the act of memorizing a role. **b** a person who memorizes a role. **10** archaic a thing to be secured by pains or attention. ● v. (**-ies, -ied**) **1** tr. make a study of; investigate or examine (a subject) (study law). **2** intr. (often foll. by for) apply oneself to study. **3** tr. scrutinize or earnestly contemplate (studied their faces; studying the problem). **4** tr. try to learn (the words of one's role etc.). **5** tr. take pains to achieve (a result) or pay regard to (a subject or principle etc.). **6** tr. (as **studied** adj.) deliberate, intentional, affected (with studied politeness). **7** tr. read (a book) attentively. **8** tr. (foll. by to + infin.) archaic **a** be on the watch. **b** try constantly to manage. □ **in a brown study** in a reverie; absorbed in one's thoughts. **make a study of** investigate carefully. **study group** a group of people meeting from time to time to study a particular subject or topic. □ **studiedly** /-dɪdlɪ/ adv. **studiedness** n. [ME f. OF estudie f. L studium zeal, study]

stuff /stʌf/ n. & v. ● n. **1** the material that a thing is made of; material that may be used for some purpose. **2** a substance or things or belongings of an indeterminate kind or a quality not needing to be specified (there's a lot of stuff about it in the newspapers). **3** a particular knowledge or activity (know one's stuff). **4** woollen fabric (esp. as distinct from silk, cotton, and linen). **5** valueless matter, trash, refuse, nonsense (take that stuff away). **6** (prec. by the) **a** colloq. an available supply of something, esp. drink or drugs. ● v. **1** tr. pack (a receptacle) tightly (stuff a cushion with feathers; a head stuffed with weird notions). **2** tr. (foll. by in, into) force or cram (a thing) (stuffed the socks in the drawer). **3** tr. fill out the skin of (an animal or bird etc.) with material to restore the original shape (a stuffed owl). **4** tr. fill (poultry etc.) with a savoury or sweet mixture, esp. before cooking. **5 a** tr. & refl. fill (a person or oneself) with food. **b** tr. & intr. eat greedily. **6** tr. push, esp. hastily or clumsily (stuffed the note behind the cushion). **7** tr. (usu. in passive; foll. by up) block up (a person's nose etc.). **8** tr. sl. as an expression of contemptuous dismissal) dispose of as unwanted (you can stuff the job). **9** tr. N. Amer. place bogus votes in (a ballot-box). **10** tr. coarse sl. offens. have sexual intercourse with (a woman). □ **bit of stuff** sl. offens. a woman regarded as an object of sexual desire. **do one's stuff** colloq. do what is required or expected of one. **get stuffed** sl. an exclamation of dismissal, contempt, etc. **stuff and nonsense** an exclamation of incredulity or ridicule. **stuffed shirt** colloq. a pompous person. **stuff gown** Brit. a gown worn by a barrister who has not taken silk. **stuff it** sl. an expression of rejection or disdain. **that's the stuff** colloq. that is what is wanted. □ **stuffer** n. (also in comb.). [ME stoffe f. OF estoffe (n.), estoffer (v.) equip, furnish f. Gk stuphō draw together]

stuffing /ˈstʌfɪŋ/ n. **1** padding used to stuff cushions etc. **2** a mixture used to stuff poultry etc., esp. before cooking. □ **knock** (or **take**) **the stuffing out of** colloq. make feeble or weak; defeat. **stuffing-box** a box packed with material, to allow the working of an axle while remaining airtight.

stuffy /ˈstʌfɪ/ adj. (**stuffier, stuffiest**) **1** (of a room or the atmosphere in it) lacking fresh air or ventilation; close. **2** dull or uninteresting. **3** (of a person's nose etc.) stuffed up. **4** (of a person) dull and conventional. □ **stuffily** adv. **stuffiness** n.

stultify /ˈstʌltɪˌfaɪ/ v.tr. (**-ies, -ied**) **1** make ineffective, useless, or futile, esp. as a result of tedious routine (*stultifying boredom*). **2** cause to appear foolish or absurd. **3** negate or neutralize. □ **stultifier** n. **stultification** /ˌstʌltɪfɪˈkeɪʃ(ə)n/ n. [LL *stultificare* f. L *stultus* foolish]

stum /stʌm/ n. & v. ● n. unfermented grape juice; must. ● v.tr. (**stummed, stumming**) **1** prevent from fermenting, or secure (wine) against further fermentation in a cask, by the use of sulphur etc. **2** renew the fermentation of (wine) by adding stum. [Du. *stommen* (v.), *stom* (n.) f. *stom* (adj.) dumb]

stumble /ˈstʌmb(ə)l/ v. & n. ● v. **1** intr. lurch forward or have a partial fall from catching or striking or misplacing one's foot. **2** intr. (often foll. by *along*) walk with repeated stumbles. **3** intr. make a mistake or repeated mistakes in speaking etc. **4** intr. (foll. by *on, upon, across*) find or encounter by chance (*stumbled on a disused well*). ● n. an act of stumbling. □ **stumbling-block** an obstacle or circumstance causing difficulty or hesitation. □ **stumbler** n. **stumblingly** adv. [ME *stumble* (with euphonic *b*) corresp. to Norw. *stumla*: rel. to STAMMER]

stumblebum /ˈstʌmb(ə)lˌbʌm/ n. US colloq. a clumsy or inept person.

stumer /ˈstjuːmə(r)/ n. Brit. sl. **1** a worthless cheque; a counterfeit coin or note. **2** a sham or fraud. **3** a failure. [19th c.: orig. unkn.]

stump /stʌmp/ n. & v. ● n. **1** the projecting remnant of a cut or fallen tree. **2** the similar remnant of anything else (e.g. a branch or limb) cut off or worn down. **3** Cricket each of the three uprights of a wicket. **4** (in pl.) joc. the legs. **5** the stump of a tree, or other place, used by an orator to address a meeting. **6** a cylinder of rolled paper or other material with conical ends for softening pencil-marks and other uses in drawing. ● v. **1** tr. (of a question etc.) be too hard for; puzzle. **2** tr. (as **stumped** adj.) at a loss; baffled. **3** tr. Cricket (esp. of a wicket-keeper) put (a batsman) out by touching the stumps with the ball while the batsman is out of the crease. **4** intr. walk stiffly or noisily as on a wooden leg. **5** tr. (also absol.) US traverse (a district) making political speeches. **6** tr. use a stump on (a drawing, line, etc.). □ **on the stump** colloq. engaged in political speech-making or agitation. **stump up** Brit. colloq. pay or produce (the money required). **up a stump** US in difficulties. [ME *stompe* f. MDu. *stomp*, OHG *stumpf*]

stumper /ˈstʌmpə(r)/ n. colloq. **1** a puzzling question. **2** a wicket-keeper.

stumpy /ˈstʌmpɪ/ adj. (**stumpier, stumpiest**) short and thick. □ **stumpily** adv. **stumpiness** n.

stun /stʌn/ v.tr. (**stunned, stunning**) **1** knock senseless; stupefy. **2** bewilder or shock. **3** (of a sound) deafen temporarily. □ **stun gun** a gun which stuns through an electric shock, ultrasound, etc., without causing serious injury. [ME f. OF *estoner* ASTONISH]

stung past and past part. of STING.

stunk past and past part. of STINK.

stunner /ˈstʌnə(r)/ n. colloq. a stunning person or thing.

stunning /ˈstʌnɪŋ/ adj. colloq. extremely impressive or attractive. □ **stunningly** adv.

stunsail /ˈstʌnseɪl/ Naut. -s(ə)l/ n. (also **stuns'l**) Naut. = STUDDING-SAIL.

stunt[1] /stʌnt/ v.tr. **1** retard the growth or development of. **2** dwarf, cramp. □ **stuntedness** n. [stunt foolish (now dial.), MHG *stunz*, ON *stuttr* short f. Gmc, perh. rel. to STUMP]

stunt[2] /stʌnt/ n. & v. ● n. **1** something unusual done to attract attention. **2** a trick or daring manoeuvre. **3** a display of concentrated energy. ● v.intr. perform stunts, esp. aerobatics. □ **stunt man** a man employed to take an actor's place in performing dangerous stunts. [orig. unkn.: first used in 19th-c. US college athletics]

stupa /ˈstuːpə/ n. a round usu. domed building erected as a Buddhist shrine. [Skr. *stūpa*]

stupe[1] /stjuːp/ n. & v. ● n. a flannel etc. moistened with hot medicated liquid and applied as a fomentation. ● v.tr. treat with this. [ME f. L Gk *stupē* tow]

stupe[2] /stjuːp/ n. sl. a foolish or stupid person.

stupefy /ˈstjuːpɪˌfaɪ/ v.tr. (**-ies, -ied**) **1** make stupid or insensible (*stupefied with drink*). **2** stun with astonishment (*the news was stupefying*). □ **stupefier** n. **stupefying** adj. **stupefyingly** adv. **stupefacient** /ˌstjuːpɪˈfeɪʃ(ə)nt/ adj. & n. **stupefaction** /-ˈfækʃ(ə)n/ n. [F *stupéfier* f. L *stupefacere* f. *stupere* be amazed]

stupendous /stjuːˈpendəs/ adj. amazing or prodigious, esp. in terms of size or degree (*a stupendous achievement*). □ **stupendously** adv. **stupendousness** n. [L *stupendus* gerundive of *stupere* be amazed at]

stupid /ˈstjuːpɪd/ adj. & n. ● adj. (**stupider, stupidest**) **1** unintelligent,

slow-witted, foolish (*a stupid fellow*). **2** typical of stupid persons (*put it in a stupid place*). **3** uninteresting or boring. **4** in a state of stupor or lethargy. **5** obtuse; lacking in sensibility. **6** colloq. a general term of disparagement (*all you do is read your stupid books*). ● n. colloq. a stupid person. □ **stupidly** adv. **stupidity** /stjuːˈpɪdɪtɪ/ n. (pl. **-ies**). [F *stupide* or L *stupidus* (as STUPENDOUS)]

stupor /ˈstjuːpə(r)/ n. a dazed, torpid, or helplessly amazed state. □ **stuporous** adj. [ME f. L (as STUPENDOUS)]

sturdy /ˈstɜːdɪ/ adj. & n. ● adj. (**sturdier, sturdiest**) **1** robust; strongly built. **2** vigorous and determined (*sturdy resistance*). ● n. vertigo in sheep caused by a tapeworm larva encysted in the brain. □ **sturdied** /-dɪd/ adj. (in sense of n.). **sturdily** adv. **sturdiness** n. [ME 'reckless, violent', f. OF *esturdi, estourdi* past part. of *estourdir* stun, daze ult. f. L *ex*-[1] + *turdus* thrush (taken as a type of drunkenness)]

sturgeon /ˈstɜːdʒ(ə)n/ n. a very large mailed sharklike fish of the family Acipenseridae, swimming up river to spawn, used as food and as a source of caviare and isinglass. [ME f. AF *sturgeon*, OF *esturgon* ult. f. Gmc]

Sturmabteilung /ˈʃtʊəmæpˌtaɪlʊŋ/ (abbr. **SA**) hist. the Nazi Brownshirts (see BROWNSHIRT). [G, = storm division]

Sturm und Drang /ˈʃtʊəm ʊnt ˈdræŋ/ n. a literary and artistic movement of the late 18th century in Germany, characterized by the expression of emotional unrest and strong feeling and by a rejection of neoclassical literary norms. Influenced by Jean-Jacques Rousseau, the movement was an early aspect of romanticism; major figures were the young Goethe and Schiller. [G, = storm and stress, the title of a play (1776) by Friedrich von Klinger (1752–1831)]

Sturt /stɜːt/, Charles (1795–1869), English explorer. He led three expeditions into the Australian interior, becoming the first European to discover the Darling River (1828) and the source of the Murray (1830). He wrote about his travels in *Two Expeditions into the Interior of Southern Australia, 1828–31* (1833). Sturt went blind during his third expedition into central Australia (1844–6) and later returned to England.

stutter /ˈstʌtə(r)/ v. & n. ● v. **1** intr. stammer, esp. by involuntarily repeating the first consonants of words. **2** tr. (often foll. by *out*) utter (words) in this way. ● n. **1** the act or habit of stuttering. **2** an instance of stuttering. □ **stutterer** n. **stutteringly** adv. [frequent. of ME (now dial.) *stut* f. Gmc]

Stuttgart /ˈʃtʊtɡɑːt/ an industrial city in western Germany, the capital of Baden-Württemberg, on the Neckar river; pop. (1991) 591,950.

sty[1] /staɪ/ n. & v. ● n. (pl. **sties**) **1** a pen or enclosure for pigs. **2** a filthy room or dwelling. **3** a place of debauchery. ● v.tr. & intr. (**sties, stied**) lodge in a sty. [OE *stī*, prob. = *stig* hall (cf. STEWARD), f. Gmc]

sty[2] /staɪ/ n. (also **stye**) (pl. **sties** or **styes**) an inflamed swelling on the edge of an eyelid. [*styany* (now dial.) = *styan eye* f. OE *stīgend* sty, lit. 'riser' f. *stīgan* rise + EYE, shortened as if = *sty on eye*]

Stygian /ˈstɪdʒɪən/ adj. **1** Gk Mythol. of or relating to the River Styx in the underworld. **2** literary dark, gloomy, indistinct. [L *stygius* f. Gk *stygios* f. *Stux -ugos* Styx f. *stugnos* hateful, gloomy]

style /staɪl/ n. & v. ● n. **1** a kind or sort, esp. in regard to appearance and form (*an elegant style of house*). **2** a manner of writing or speaking or performing (*written in a florid style; started off in fine style*). **3** the distinctive manner of a person or school or period, esp. in relation to painting, architecture, furniture, dress, etc. **4** the correct way of designating a person or thing. **5 a** a superior quality or manner (*do it in style*). **b** = FORM n. 9. **6** a particular make, shape, or pattern (*in all sizes and styles*). **7** a method of reckoning dates (*Old Style; New Style*). **8 a** an ancient writing-implement, a small rod with a pointed end for scratching letters on wax-covered tablets and a blunt end for obliterating them. **b** a thing of a similar shape esp. for engraving, tracing, etc. **9** the gnomon of a sundial. **10** Bot. the narrow extension of the ovary supporting the stigma. **11** Zool. a small slender pointed process or appendage (cf. STYLET 3). **12** (in comb.) = -WISE. ● v.tr. **1** design or make etc. in a particular (esp. fashionable) style. **2** designate in a specified way. □ **styleless** adj. **stylelessness** n. **styler** n. [ME f. OF *stile, style* f. L *stilus*: spelling *style* due to assoc. with Gk *stulos* column]

stylet /ˈstaɪlɪt/ n. **1** a slender pointed instrument; a stiletto. **2** Med. the stiffening wire of a catheter; a probe. **3** Zool. a small style, esp. a piercing mouthpart of an insect. [F *stilet* f. It. STILETTO]

styli pl. of STYLUS.

stylish /ˈstaɪlɪʃ/ adj. **1** fashionable; elegant. **2** having a superior quality, manner, etc. □ **stylishly** adv. **stylishness** n.

stylist /ˈstaɪlɪst/ n. **1 a** a designer of fashionable styles etc. **b** a hairdresser. **2 a** a writer noted for or aspiring to good literary style. **b** (in sport or music) a person who performs with style.

stylistic /staɪˈlɪstɪk/ adj. of or concerning esp. literary style. □ **stylistically** adv. [STYLIST + -IC, after G stilistisch]

stylistics /staɪˈlɪstɪks/ n. the study of literary style.

stylite /ˈstaɪlaɪt/ n. an ancient or medieval ascetic living on top of a pillar, esp. in Syria, Turkey, and Greece in the 5th century AD. The first person known to do this was Simeon Stylites. [eccl.Gk stulitēs f. stulos pillar]

stylize /ˈstaɪlaɪz/ v.tr. (also **-ise**) (esp. as **stylized** adj.) paint, draw, etc. (a subject) in a conventional non-realistic style. □ **stylization** /ˌstaɪlaɪˈzeɪʃ(ə)n/ n. [STYLE + -IZE, after G stilisieren]

stylo /ˈstaɪləʊ/ n. (pl. **-os**) colloq. = STYLOGRAPH. [abbr.]

stylobate /ˈstaɪləˌbeɪt/ n. Archit. a continuous base supporting a row of columns. [L stylobata f. Gk stulobatēs f. stulos pillar, bainō walk]

stylograph /ˈstaɪləˌɡrɑːf/ n. a kind of fountain pen having a point instead of a split nib. □ **stylographic** /ˌstaɪləˈɡræfɪk/ adj. [STYLUS + -GRAPH]

styloid /ˈstaɪlɔɪd/ adj. & n. Anat. & Zool. ● adj. resembling a stylus or pen. ● n. (in full **styloid process**) a spine of bone, esp. that projecting from the base of the temporal bone. [mod.L styloides f. Gk stuloeidēs f. stulos pillar]

stylus /ˈstaɪləs/ n. (pl. **-li** /-laɪ/ or **-luses**) **1 a** a hard, esp. diamond or sapphire, point following a groove in a gramophone record and transmitting the recorded sound for reproduction. **b** a similar point producing such a groove when recording sound. **2** = STYLE n. 8, 9. [erron. spelling of L stilus: cf. STYLE]

stymie /ˈstaɪmɪ/ n. & v. (also **stimy**) ● n. (pl. **-ies**) **1** Golf a situation where an opponent's ball lies between the player and the hole, forming a possible obstruction to play (lay a stymie). **2** a difficult situation. ● v.tr. (**stymies, stymied, stymying** or **stymieing**) **1** obstruct; thwart. **2** Golf block (an opponent, an opponent's ball, or oneself) with a stymie. [19th c.: orig. unkn.]

styptic /ˈstɪptɪk/ adj. & n. ● adj. (of a drug etc.) that checks bleeding. ● n. a styptic drug or substance. □ **styptic pencil** a stick of a styptic substance used to treat small cuts. [ME f. L stypticus f. Gk stuptikos f. stuphō contract]

styrax /ˈstaɪəræks/ n. **1** storax resin. **2** a tree or shrub of the genus Styrax, e.g. the storax-tree. [L f. Gk sturax: cf. STORAX]

styrene /ˈstaɪəriːn/ n. Chem. a liquid hydrocarbon easily polymerized and used in making plastics etc. [STYRAX + -ENE]

Styria /ˈstɪrɪə/ (called in German Steiermark) a mountainous state of SE Austria; capital, Graz.

Styx /stɪks/ Gk Mythol. one of the nine rivers in the underworld, over which Charon ferried the souls of the dead.

suable /ˈsuːəb(ə)l, ˈsjuː-/ adj. capable of being sued. □ **suability** /ˌsuːəˈbɪlɪtɪ, ˌsjuː-/ n.

suasion /ˈsweɪʒ(ə)n/ n. formal persuasion as opposed to force (moral suasion). □ **suasive** /ˈsweɪsɪv/ adj. [ME f. OF suasion or L suasio f. suadere suas- urge]

suave /swɑːv/ adj. **1** (of a person, esp. a man) smooth; polite; sophisticated. **2** (of a wine etc.) bland, smooth. □ **suavely** adv. **suaveness** n. **suavity** n. (pl. **-ies**) [F suave or L suavis agreeable: cf. SWEET]

sub /sʌb/ n. & v. colloq. ● n. **1** a submarine. **2** a subscription. **3** a substitute. **4** a sub-editor. **5** Mil. a subaltern. **6** Brit. an advance or loan against expected income. ● v. (**subbed, subbing**) **1** intr. (usu. foll. by for) act as a substitute for a person. **2** tr. Brit. colloq. lend or advance (a sum) to (a person) against expected income. **3** tr. sub-edit. [abbr.]

sub- /sʌb, səb/ prefix (also **suc-** before c, **suf-** before f, **sug-** before g, **sup-** before p, **sur-** before r, **sus-** before c, p, t) **1** at or to or from a lower position (subordinate; submerge; subtract; subsoil). **2** secondary or inferior in rank or position (subclass; subcommittee; sub-lieutenant; subtotal). **3** somewhat, nearly; more or less (subacid; subarctic; subaquatic). **4** (forming verbs) denoting secondary action (subdivide; sublet). **5** denoting support (subvention). **6** Chem. (of a salt) basic (subacetate). [from or after L sub- f. sub under, close to, towards]

subabdominal /ˌsʌbæbˈdɒmɪn(ə)l/ adj. below the abdomen.

subacid /sʌbˈæsɪd/ adj. moderately acid or tart (subacid fruit; a subacid remark). □ **subacidity** /ˌsʌbəˈsɪdɪtɪ/ n. [L subacidus (as SUB-, ACID)]

subacute /ˌsʌbəˈkjuːt/ adj. Med. (of a condition) between acute and chronic.

subagency /sʌbˈeɪdʒənsɪ/ n. (pl. **-ies**) a secondary or subordinate agency. □ **subagent** n.

subalpine /sʌbˈælpaɪn/ adj. of or situated in the higher slopes of mountains just below the timberline.

subaltern /ˈsʌbəlt(ə)n/ n. & adj. ● n. Brit. Mil. an officer below the rank of captain, esp. a second lieutenant. ● adj. **1** of inferior rank. **2** Logic (of a proposition) particular, not universal. [LL subalternus f. alternus ALTERNATE adj.]

subantarctic /ˌsʌbæntˈɑːktɪk/ adj. of or like regions immediately north of the Antarctic Circle.

sub-aqua /sʌbˈækwə/ adj. of or concerning underwater swimming or diving.

subaquatic /ˌsʌbəˈkwætɪk, -ˈkwɒtɪk/ adj. **1** of more or less aquatic habits or kind. **2** underwater.

subaqueous /sʌbˈeɪkwɪəs/ adj. **1** existing, formed, or taking place under water. **2** lacking in substance or strength; wishy-washy.

subarctic /sʌbˈɑːktɪk/ adj. of or like regions immediately south of the Arctic Circle.

subastral /sʌbˈæstrəl/ adj. terrestrial.

subatomic /ˌsʌbəˈtɒmɪk/ adj. occurring in or smaller than an atom.

subatomic particle n. Physics a particle which is smaller than an atom. The discovery of the electron in 1897 and subsequent discoveries of the photon, the atomic nucleus, the proton, and the neutron showed that atoms themselves are composed of simpler particles and that there are other particles which can exist independently of atoms. Subatomic particles are often grouped into four categories: quanta (including photons), leptons (electrons and neutrinos), mesons, and baryons (protons and neutrons). These may not in themselves be 'elementary', and efforts to study and classify them have led to the postulation of further particles such as quarks and gluons.

subaudition /ˌsʌbɔːˈdɪʃ(ə)n/ n. **1** the act of mentally supplying an omitted word or words in speech. **2** the act or process of understanding the unexpressed; reading between the lines. [LL subauditio f. subaudire understand (as SUB-, AUDITION)]

subaxillary /ˌsʌbækˈsɪlərɪ/ adj. **1** Bot. in or growing beneath the axil. **2** beneath the armpit.

sub-basement /ˈsʌbˌbeɪsmənt/ n. a storey below a basement.

sub-branch /ˈsʌbbrɑːntʃ/ n. a secondary or subordinate branch.

sub-breed /ˈsʌbbriːd/ n. a secondary or inferior breed.

subcategory /ˈsʌbˌkætɪɡərɪ/ n. (pl. **-ies**) a secondary or subordinate category. □ **subcategorize** v.tr. (also **-ise**). **subcategorization** /ˌsʌbkætɪɡəraɪˈzeɪʃ(ə)n/ n.

subcaudal /sʌbˈkɔːd(ə)l/ adj. Anat. & Zool. of or concerning the region under the tail or the back part of the body.

subclass /ˈsʌbklɑːs/ n. **1** a secondary or subordinate class. **2** Biol. a taxonomic category below a class.

sub-clause /ˈsʌbklɔːz/ n. **1** esp. Law a subsidiary section of a clause. **2** Gram. a subordinate clause.

subclavian /sʌbˈkleɪvɪən/ adj. & n. Anat. ● adj. (of an artery etc.) lying or extending under the collar-bone. ● n. such an artery. [mod.L subclavius (as SUB-, clavis key): cf. CLAVICLE]

subclinical /sʌbˈklɪnɪk(ə)l/ adj. Med. (of a disease) not yet presenting definite symptoms.

subcommissioner /ˌsʌbkəˈmɪʃənə(r)/ n. a deputy commissioner.

subcommittee /ˈsʌbkəˌmɪtɪ/ n. a secondary committee.

subconical /sʌbˈkɒnɪk(ə)l/ adj. approximately conical.

subconscious /sʌbˈkɒnʃəs/ adj. & n. ● adj. of or concerning the part of the mind which is not fully conscious but influences actions etc. ● n. this part of the mind. □ **subconsciously** adv. **subconsciousness** n.

subcontinent /ˈsʌbˌkɒntɪnənt/ n. **1** a large land mass, smaller than a continent. **2** a large geographically or politically independent part of a continent. □ **subcontinental** /ˌsʌbkɒntɪˈnent(ə)l/ adj.

subcontract v. & n. ● v. /ˌsʌbkənˈtrækt/ **1** tr. employ a firm etc. to do (work) as part of a larger project. **2** intr. make or carry out a subcontract. ● n. /sʌbˈkɒntrækt/ a secondary contract, esp. to supply materials, labour, etc. □ **subcontractor** /ˌsʌbkənˈtræktə(r)/ n.

subcontrary /sʌbˈkɒntrərɪ/ adj. & n. Logic ● adj. (of a proposition) incapable of being false at the same time as another. ● n. (pl. **-ies**) such a proposition. [LL subcontrarius (as SUB-, CONTRARY), transl. Gk hupenantios]

subcordate /sʌbˈkɔːdeɪt/ adj. approximately heart-shaped.

subcortical /sʌbˈkɔːtɪk(ə)l/ adj. Anat. below the cortex.

subcostal /sʌbˈkɒst(ə)l/ adj. Anat. below the ribs.

subcranial /sʌbˈkreɪnɪəl/ adj. Anat. below the cranium.

subcritical /sʌbˈkrɪtɪk(ə)l/ adj. Physics of less than critical mass etc.

subculture /ˈsʌbˌkʌltʃə(r)/ n. a cultural group within a larger culture, often having beliefs or interests at variance with those of the larger culture. □ **subcultural** /sʌbˈkʌltʃərəl/ adj.

subcutaneous /ˌsʌbkjuːˈteɪnɪəs/ adj. under the skin. □ **subcutaneously** adv.

subdeacon /sʌbˈdiːkən/ n. Eccl. a minister of the order next below a deacon. □ **subdiaconate** /-daɪˈækəˌneɪt, -nət/ n.

subdean /sʌbˈdiːn/ n. an official ranking next below, or acting as a deputy for, a dean. □ **subdeanery** n. (pl. **-ies**). **subdecanal** /-dɪˈkeɪn(ə)l/ adj.

subdelirious /ˌsʌbdɪˈlɪrɪəs/ adj. capable of becoming delirious; mildly delirious. □ **subdelirium** n.

subdivide /ˈsʌbdɪˌvaɪd/ v.tr. & intr. divide again after a first division. [ME f. L subdividere (as SUB-, DIVIDE)]

subdivision /ˈsʌbdɪˌvɪʒ(ə)n/ n. **1** the act or an instance of subdividing. **2** a secondary or subordinate division. **3** N. Amer. & Austral. an area of land divided into plots for sale.

subdominant /sʌbˈdɒmɪnənt/ n. Mus. the fourth note of the diatonic scale of any key.

subduction /səbˈdʌkʃ(ə)n/ n. Geol. the sideways and downward movement of the edge of a plate of the earth's crust into the mantle beneath another plate (subduction zone). □ **subduct** v.tr. & intr. [L subduct- stem of subducere, f. SUB- + ducere lead, bring]

subdue /səbˈdjuː/ v.tr. (**subdues, subdued, subduing**) **1** conquer, subjugate, or tame (an enemy, nature, one's emotions, etc.). **2** (as **subdued** adj.) softened; lacking in intensity; toned down (subdued light; in a subdued mood). □ **subduable** adj. **subdual** n. [ME sodewe f. OF so(u)duire f. L subducere (as SUB-, ducere lead, bring) used with the sense of subdere conquer (as SUB-, -dere put)]

subdural /sʌbˈdjʊərəl/ adj. Anat. & Med. situated or occurring between the dura mater and the arachnoid membrane of the brain and spinal cord. [SUB- + DURA MATER]

sub-editor /sʌbˈedɪtə(r)/ n. **1** an assistant editor. **2** Brit. a person who edits material for printing in a book, newspaper, etc. □ **sub-edit** v.tr. (**-edited, -editing**). **sub-editorial** /-ˌedɪˈtɔːrɪəl/ adj.

suberect /ˌsʌbɪˈrekt/ adj. (of an animal, plant, etc.) almost erect.

subereous /sjuːˈbɪərɪəs/ adj. (also **suberic** /-ˈberɪk/, **suberose** /ˈsjuːbəˌrəʊz, -ˌrəʊs/) Bot. **1** of or concerning cork. **2** corky. [L suber cork, cork-oak]

subfamily /ˈsʌbˌfæmɪlɪ/ n. (pl. **-ies**) **1** Biol. a taxonomic category below a family. **2** any subdivision of a group.

subfloor /ˈsʌbflɔː(r)/ n. a foundation for a floor in a building.

subform /ˈsʌbfɔːm/ n. a subordinate or secondary form.

subfusc /ˈsʌbfʌsk/ adj. & n. ● adj. formal dull; dusky; gloomy. ● n. formal clothing at some universities. [L subfuscus f. fuscus dark brown]

subgenus /sʌbˈdʒiːnəs/ n. (pl. **subgenera** /-ˈdʒenərə/) Biol. a taxonomic category below a genus. □ **subgeneric** /-dʒɪˈnerɪk/ adj.

subglacial /sʌbˈgleɪʃ(ə)l, -ˈgleɪsɪəl/ adj. next to or at the bottom of a glacier.

subgroup /ˈsʌbgruːp/ n. Math. etc. a subset of a group.

subhead /ˈsʌbhed/ n. (also **subheading**) **1** a subordinate heading or title in a chapter, article, etc. **2** a subordinate division in a classification.

subhuman /sʌbˈhjuːmən/ adj. **1** (of an animal) closely related to man. **2** (of behaviour, intelligence, etc.) less than human.

subjacent /sʌbˈdʒeɪs(ə)nt/ adj. underlying; situated below. [L subjacere (as SUB-, jacere lie)]

subject n., adj., adv., & v. ● n. /ˈsʌbdʒɪkt/ **1 a** a matter, theme, etc. to be discussed, described, represented, dealt with, etc. **b** (foll. by for) a person, circumstance, etc., giving rise to specified feeling, action, etc. (a subject for congratulation). **2** a department or field of study (his best subject is geography). **3** Gram. a noun or its equivalent about which a sentence is predicated and with which the verb agrees. **4 a** any person except a monarch living under a monarchy or any other form of government (the ruler and his subjects). **b** any person owing obedience to another. **5** Philos. **a** a thinking or feeling entity; the conscious mind; the ego, esp. as opposed to anything external to the mind. **b** the central substance or core of a thing as opposed to its attributes. **6** Mus. a theme of a fugue or sonata; a leading phrase or motif. **7** a person of specified mental or physical tendencies (a hysterical subject). **8** Logic the part of a proposition about which a statement is made. **9 a** a person or animal undergoing treatment, examination, or experimentation. **b** a dead body for dissection. ● adj. /ˈsʌbdʒɪkt/ (often foll. by to) owing obedience to a government, colonizing power, force, etc.; in subjection. **2** (foll. by to) liable, exposed, or prone to (is subject to infection). **3** (foll. by to) conditional upon; on the assumption of (the arrangement is subject to your approval). ● adv. /ˈsʌbdʒɪkt/ (foll. by to) conditionally upon (subject to your consent, I propose to try again). ● v.tr. /səbˈdʒekt/ **1** (foll. by to) make liable; expose; treat (subjected us to hours of waiting). **2** (usu. foll. by to) subdue (a nation, person, etc.) to one's sway etc. □ **on the subject of** concerning, about. **subject and object** Psychol. the ego or self and the non-ego; consciousness and that of which it is or may be conscious. **subject catalogue** a catalogue, esp. in a library, arranged according to the subjects treated. **subject-heading** a heading in an index collecting references to a subject. **subject-matter** the matter treated of in a book, lawsuit, etc. □ **subjection** /səbˈdʒekʃ(ə)n/ n. **subjectless** /ˈsʌbdʒɪktlɪs/ adj. [ME soget etc. f. OF suget etc. f. L subjectus past part. of subjicere (as SUB-, jacere throw)]

subjective /səbˈdʒektɪv/ adj. & n. ● adj. **1** (of art, literature, written history, a person's views, etc.) proceeding from personal idiosyncrasy or individuality; not impartial or literal. **2** esp. Philos. proceeding from or belonging to the individual consciousness or perception; imaginary, partial, or distorted. **3** Gram. of or concerning the subject. ● n. Gram. the subjective case. □ **subjective case** Gram. the nominative. □ **subjectively** adv. **subjectiveness** n. **subjectivity** /ˌsʌbdʒekˈtɪvɪtɪ/ n. [ME f. L subjectivus (as SUBJECT)]

subjectivism /səbˈdʒektɪˌvɪz(ə)m/ n. Philos. the doctrine that knowledge, morality, perception, etc., is merely subjective and relative and that there is no external or objective truth (opp. OBJECTIVISM 2). □ **subjectivist** n.

subjoin /sʌbˈdʒɔɪn/ v.tr. add or append (an illustration, anecdote, etc.) at the end. [obs. F subjoindre f. L subjungere (as SUB-, jungere junct- join)]

subjoint /ˈsʌbdʒɔɪnt/ n. a secondary joint (in an insect's leg etc.).

sub judice /sʌb ˈdʒuːdɪsɪ, sʊb ˈjuːdɪˌkeɪ/ adj. Law under judicial consideration and therefore prohibited from public discussion elsewhere. [L, = under a judge]

subjugate /ˈsʌbdʒʊˌgeɪt/ v.tr. (often followed by to) bring into subjection; subdue; vanquish. □ **subjugator** n. **subjugable** /-gəb(ə)l/ adj. **subjugation** /ˌsʌbdʒʊˈgeɪʃ(ə)n/ n. [ME f. LL subjugare bring under the yoke (as SUB-, jugum yoke)]

subjunctive /səbˈdʒʌŋktɪv/ adj. & n. Gram. ● adj. (of a mood) denoting what is imagined or wished or possible (e.g. if I were you, God help you, be that as it may). ● n. **1** the subjunctive mood. **2** a verb in this mood. □ **subjunctively** adv. [F subjonctif -ive or LL subjunctivus f. L (as SUBJOIN), transl. Gk hupotaktikos, as being used in subjoined clauses]

subkingdom /ˈsʌbˌkɪŋdəm/ n. Biol. a taxonomic category below a kingdom.

sublease n. & v. ● n. /ˈsʌbliːs/ a lease of a property by a tenant to a subtenant. ● v.tr. /sʌbˈliːs/ lease (a property) to a subtenant.

sublessee /ˌsʌbleˈsiː/ n. a person who holds a sublease.

sublessor /ˌsʌbleˈsɔː(r)/ n. a person who grants a sublease.

sublet n. & v. ● n. /ˈsʌblet/ **1** = SUBLEASE n. **2** colloq. a sublet property. ● v.tr. /sʌbˈlet/ (**-letting**; past and past part. **-let**) = SUBLEASE v.

sub-lieutenant /ˌsʌblefˈtenənt/ n. Brit. a naval officer ranking next below lieutenant.

sublimate v., adj., & n. ● v. /ˈsʌblɪˌmeɪt/ **1** tr. & intr. (also absol.) divert (the energy of a primitive impulse, esp. sexual) into a culturally higher, or socially more acceptable, activity. **2** tr. & intr. Chem. convert (a substance) from the solid state directly to its vapour by heat, and usu. allow it to solidify again. **3** tr. refine; purify; idealize. ● adj. /ˈsʌblɪmət/ **1** Chem. (of a substance) sublimated. **2** purified, refined. ● n. /ˈsʌblɪmət/ Chem. **1** a sublimated substance. **2** = corrosive sublimate. □ **sublimation** /ˌsʌblɪˈmeɪʃ(ə)n/ n. [L sublimare sublimat- SUBLIME v.]

sublime /səˈblaɪm/ adj. & v. ● adj. (**sublimer, sublimest**) **1** of the most exalted, grand, or noble kind; awe-inspiring (sublime genius). **2** (of indifference, impudence, etc.) arrogantly unruffled; extreme (sublime ignorance). ● v. **1** tr. & intr. Chem. = SUBLIMATE v. **2** tr. purify or elevate by or as if by sublimation; make sublime. **3** intr. become pure by or as if by sublimation. □ **sublimely** adv. **sublimity** /-ˈblɪmɪtɪ/ n. [L sublimis (as SUB-, second element perh. rel. to limen threshold, limus oblique)]

Sublime Porte n. see PORTE.

subliminal /sʌbˈlɪmɪn(ə)l/ adj. Psychol. (of a stimulus etc.) below the threshold of sensation or consciousness. □ **subliminal advertising** the use of subliminal images in advertising on television etc. to influence the viewer at an unconscious level. **subliminal self** the part of one's personality outside conscious awareness. □ **subliminally** *adv.* [SUB- + L *limen -inis* threshold]

sublingual /sʌbˈlɪŋgwəl/ *adj.* under the tongue. [SUB- + L *lingua* tongue]

sublittoral /sʌbˈlɪtərəl/ *adj.* **1** (of plants, animals, deposits, etc.) living or found on the seashore just below the low-water mark. **2** of or concerning the seashore.

Sub-Lt. *abbr. Brit.* Sub-Lieutenant.

sublunary /sʌbˈluːnərɪ/ *adj.* **1** beneath the moon. **2** *Astron.* **a** within the moon's orbit. **b** subject to the moon's influence. **3** of this world; earthly. [LL *sublunaris* (as SUB-, LUNAR)]

subluxation /ˌsʌblʌkˈseɪʃ(ə)n/ *n. Med.* partial dislocation. [SUB- + L *luxat-* past part. stem of *luxare* f. *luxus* dislocated]

sub-machine-gun /ˌsʌbməˈʃiːnɡʌn/ *n.* a hand-held lightweight machine-gun.

subman /ˈsʌbmæn/ *n.* (*pl.* **-men**) *derog.* an inferior, brutal, or stupid man.

submarginal /sʌbˈmɑːdʒɪn(ə)l/ *adj.* **1** esp. *Econ.* not reaching minimum requirements. **2** (of land) that cannot be farmed profitably.

submarine /ˌsʌbməˈriːn/ *n. & adj.* ● *n.* a vessel, esp. a warship, capable of operating under water and usu. equipped with torpedoes, missiles, and a periscope. (*See note below.*) ● *adj.* existing, occurring, done, or used under the surface of the sea (*submarine cable*). □ **submariner** /-ˈmærɪnə(r)/ *n.*

■ Classical writers mention attempts to build various submersible craft, but the first authenticated vessel was built by Cornelius Drebbel (1572–1633), a Dutch inventor who manoeuvred his craft below the surface of the River Thames in the 1620s. During the War of American Independence a one-man craft, the *Turtle*, attempted an underwater attack on a British warship. Also of note was Robert Fulton's *Nautilus* (1800), which was capable of submerging to a depth of 7.6 m (25 ft). By the early 20th century the internal-combustion engine, electric motor, and new designs of torpedo enabled the development of an effective warship, and submarines played a significant part in both world wars, especially the Second. Major advances since then have been the introduction of nuclear-powered submarines able to proceed submerged indefinitely at high speed, inertial navigation systems enabling the fixing of position without surfacing, better streamlining allowing much higher speeds, and ballistic missiles capable of being fired while submerged.

submaster /ˈsʌbˌmɑːstə(r)/ *n.* an assistant master or assistant headmaster in a school.

submaxillary /ˌsʌbmækˈsɪlərɪ/ *adj. Anat.* (esp. of a pair of salivary glands) beneath the upper jaw.

submediant /sʌbˈmiːdɪənt/ *n. Mus.* the sixth note of the diatonic scale of any key.

submental /sʌbˈment(ə)l/ *adj. Anat.* under the chin.

submerge /səbˈmɜːdʒ/ *v.* **1** *tr.* **a** place under water; flood; inundate. **b** flood or inundate with work, problems, etc. **2** *intr.* (of a submarine, its crew, a diver, etc.) dive below the surface of water. □ **the submerged tenth** the supposed fraction of the population permanently living in poverty. □ **submergence** *n.* **submergible** *adj.* **submersion** /-ˈmɜːʃ(ə)n/ *n.* [L *submergere* (as SUB-, *mergere* mers- dip)]

submersible /səbˈmɜːsɪb(ə)l/ *n. & adj.* ● *n.* a submarine operating under water for short periods. ● *adj.* capable of being submerged. [*submerse* (v.) = SUBMERGE]

submicroscopic /ˌsʌbmaɪkrəˈskɒpɪk/ *adj.* too small to be seen by an ordinary microscope.

subminiature /sʌbˈmɪnɪtʃə(r)/ *adj.* **1** of greatly reduced size. **2** (of a camera) very small and using 16-mm film.

submission /səbˈmɪʃ(ə)n/ *n.* **1 a** the act or an instance of submitting; the state of being submitted. **b** anything that is submitted. **2** humility, meekness, obedience, submissiveness (*showed great submission of spirit*). **3** *Law* a theory etc. submitted by counsel to a judge or jury. **4** (in wrestling) the surrender of a participant yielding to the pain of a hold. [ME f. OF *submission* or L *submissio* (as SUBMIT)]

submissive /səbˈmɪsɪv/ *adj.* **1** humble; obedient. **2** yielding to power or authority; willing to submit. □ **submissively** *adv.* **submissiveness** *n.* [SUBMISSION after *remissive* etc.]

submit /səbˈmɪt/ *v.* (**submitted, submitting**) **1** (usu. foll. by *to*) **a** *intr.* cease resistance; give way; yield (*had to submit to defeat; will never submit*).

b *refl.* surrender (oneself) to the control of another etc. **2** *tr.* present for consideration or decision. **3** *tr.* (usu. foll. by *to*) subject (a person or thing) to an operation, process, treatment, etc. (*submitted it to the flames*). **4** *tr.* esp. *Law* urge or represent esp. deferentially (*that, I submit, is a misrepresentation*). □ **submitter** *n.* [ME f. L *submittere* (as SUB-, *mittere miss-* send)]

submultiple /sʌbˈmʌltɪp(ə)l/ *n. & adj.* ● *n.* a number that can be divided exactly into a specified number. ● *adj.* being such a number.

subnormal /sʌbˈnɔːm(ə)l/ *adj.* **1** (esp. as regards intelligence) below normal. **2** less than normal. □ **subnormality** /ˌsʌbnɔːˈmælɪtɪ/ *n.*

subnuclear /sʌbˈnjuːklɪə(r)/ *adj. Physics* occurring in or smaller than an atomic nucleus.

subocular /sʌbˈɒkjʊlə(r)/ *adj. Anat. & Zool.* situated below or under the eyes.

suborbital /sʌbˈɔːbɪt(ə)l/ *adj.* **1** *Anat. & Zool.* situated below the orbit of the eye. **2** (of a spaceship etc.) not completing a full orbit of the earth.

suborder /ˈsʌbˌɔːdə(r)/ *n. Biol.* a taxonomic category below an order. □ **subordinal** /sʌbˈɔːdɪn(ə)l/ *adj.*

subordinary /sʌbˈɔːdɪnərɪ, -d(ə)nrɪ/ *n.* (*pl.* **-ies**) *Heraldry* a device or bearing that is common but less so than ordinaries.

subordinate *adj., n., & v.* ● *adj.* /səˈbɔːdɪnət/ (usu. foll. by *to*) of inferior importance or rank; secondary, subservient. ● *n.* /səˈbɔːdɪnət/ a person working under another's control or orders. ● *v.tr.* /səˈbɔːdɪneɪt/ (usu. foll. by *to*) **1** make subordinate; treat or regard as of minor importance. **2** make subservient. □ **subordinate clause** *Gram.* a clause, usu. introduced by a conjunction, that functions like a noun, adjective, or adverb, and qualifies a main clause (e.g. 'when it rang' in 'she answered the phone when it rang'). □ **subordinately** /-nətlɪ/ *adv.* **subordinative** *adj.* **subordination** /-ˌbɔːdɪˈneɪʃ(ə)n/ *n.* [med.L *subordinare, subordinat-* (as SUB-, L *ordinare* ordain)]

suborn /səˈbɔːn/ *v.tr.* induce by bribery etc. to commit perjury or any other unlawful act. □ **suborner** *n.* **subornation** /ˌsʌbɔːˈneɪʃ(ə)n/ *n.* [L *subornare* incite secretly (as SUB-, *ornare* equip)]

suboxide /sʌbˈɒksaɪd/ *n. Chem.* an oxide containing the smallest proportion of oxygen.

subphylum /ˈsʌbˌfaɪləm/ *n.* (*pl.* **subphyla** /-lə/) *Biol.* a taxonomic category below a phylum.

sub-plot /ˈsʌbplɒt/ *n.* a subordinate plot in a play etc.

subpoena /səbˈpiːnə, səˈpiː-/ *n. & v.* ● *n.* a writ ordering a person to attend a lawcourt. ● *v.tr.* (*past* and *past part.* **subpoenaed** /-nəd/ or **subpoena'd**) serve a subpoena on. [ME f. L *sub poena* under penalty (the first words of the writ)]

subregion /ˈsʌbˌriːdʒən/ *n.* a division of a region, esp. with regard to natural life. □ **subregional** /sʌbˈriːdʒən(ə)l/ *adj.*

subreption /səbˈrepʃ(ə)n/ *n. formal* the obtaining of a thing by surprise or misrepresentation. [L *subreptio* purloining f. *subripere* (as SUB-, *rapere* snatch)]

subrogation /ˌsʌbrəˈgeɪʃ(ə)n/ *n. Law* the substitution of one party for another as creditor, with the transfer of rights and duties. □ **subrogate** /ˈsʌbrəˌgeɪt/ *v.tr.* [LL *subrogatio* f. *subrogare* choose as substitute (as SUB-, *rogare* ask)]

sub rosa /sʌb ˈrəʊzə/ *adj. & adv.* (of communication, consultation, etc.) in secrecy or confidence. [L, lit. 'under the rose', as emblem of secrecy]

subroutine /ˈsʌbruːˌtiːn/ *n. Computing* a routine designed to perform a frequently used operation within a program.

sub-Saharan /ˌsʌbsəˈhɑːrən/ *attrib.adj.* from or forming part of the regions of Africa south of the Sahara desert.

subscribe /səbˈskraɪb/ *v.* **1** (usu. foll. by *to, for*) **a** *tr. & intr.* contribute (a specified sum) or make or promise a contribution to a fund, project, charity, etc., esp. regularly. **b** *intr.* enter one's name in a list of contributors to a charity etc. **c** *tr.* raise or guarantee raising (a sum) by so subscribing. **2** *intr.* (usu. foll. by *to*) express one's agreement with an opinion, resolution, etc. (*cannot subscribe to that*). **3** *tr.* **a** write (esp. one's name) at the foot of a document etc. (*subscribed a motto*). **b** write one's name at the foot of, sign (a document, picture, etc.). □ **subscribe for** agree to take a copy or copies of (a book) before publication. **subscribe oneself** sign one's name as. **subscribe to** arrange to receive (a periodical etc.) regularly. [ME f. L *subscribere* (as SUB-, *scribere* script- write)]

subscriber /səbˈskraɪbə(r)/ *n.* **1** a person who subscribes. **2** a person paying for the hire of a telephone line. □ **subscriber trunk dialling** *Brit.* the automatic connection of trunk calls by dialling without the assistance of an operator.

subscript /'sʌbskrɪpt/ adj. & n. ● adj. written or printed below the line, esp. Math. (of a symbol) written below and to the right of another symbol. ● n. a subscript number or symbol. [L subscriptus (as SUBSCRIBE)]

subscription /səbˈskrɪpʃ(ə)n/ n. **1 a** the act or an instance of subscribing. **b** money subscribed. **2** Brit. a fee for the membership of a society etc., esp. paid regularly. **3 a** an agreement to take and pay for usu. a specified number of issues of a newspaper, magazine, etc. **b** the money paid by this. **4** a signature on a document etc. **5** the offer of a reduced price to those ordering a book before publication. □ **subscription concert** etc. each of a series of concerts etc. for which tickets are sold in advance. [ME f. L subscriptio (as SUBSCRIBE)]

subsection /'sʌbˌsekʃ(ə)n/ n. a division of a section.

subsellium /səbˈseliəm/ n. (pl. **subsellia** /-liə/) = MISERICORD 1. [L f. sella seat]

subsequence /'sʌbsɪkwəns/ n. a subsequent incident; a consequence.

sub-sequence /'sʌbˌsiːkwəns/ n. a sequence forming part of a larger one.

subsequent /'sʌbsɪkwənt/ adj. (usu. foll. by to) following a specified event etc. in time, esp. as a consequence. □ **subsequently** adv. [ME f. OF subsequent or L subsequi (as SUB-, sequi follow)]

subserve /səbˈsɜːv/ v.tr. serve as a means in furthering (a purpose, action, etc.). [L subservire (as SUB-, SERVE)]

subservient /səbˈsɜːvɪənt/ adj. **1** cringing; obsequious. **2** (usu. foll. by to) serving as a means; instrumental. **3** (usu. foll. by to) subordinate. □ **subservience** n. **subserviency** n. **subserviently** adv. [L subserviens subservient- (as SUBSERVE)]

subset /'sʌbset/ n. **1** a secondary part of a set. **2** Math. a set all the elements of which are contained in another set.

subshrub /'sʌbʃrʌb/ n. a low-growing or small shrub.

subside /səbˈsaɪd/ v.intr. **1** cease from agitation; become tranquil; abate (excitement subsided). **2** (of water, suspended matter, etc.) sink. **3** (of the ground) cave in; sink. **4** (of a building, ship, etc.) sink lower in the ground or water. **5** (of a swelling etc.) become less. **6** usu. joc. (of a person) sink into a sitting, kneeling, or lying posture. □ **subsidence** /səbˈsaɪd(ə)ns, 'sʌbsɪd-/ n. [L subsidere (as SUB-, sidere settle rel. to sedere sit)]

subsidiarity /səbˌsɪdɪˈærɪtɪ/ n. (pl. **-ies**) **1** the quality of being subsidiary. **2** the principle that a central authority should have a subsidiary function and perform only tasks which cannot be performed effectively at a local level.

subsidiary /səbˈsɪdɪərɪ/ adj. & n. ● adj. **1** serving to assist or supplement; auxiliary. **2** (of a company) controlled by another. **3** (of troops): **a** paid for by subsidy. **b** hired by another nation. ● n. (pl. **-ies**) **1** a subsidiary thing or person; an accessory. **2** a subsidiary company. □ **subsidiarily** adv. [L subsidiarius (as SUBSIDY)]

subsidize /'sʌbsɪˌdaɪz/ v.tr. (also **-ise**) **1** pay a subsidy to. **2** reduce the cost of by subsidy (subsidized lunches). □ **subsidizer** n. **subsidization** /ˌsʌbsɪdaɪˈzeɪʃ(ə)n/ n.

subsidy /'sʌbsɪdɪ/ n. (pl. **-ies**) **1 a** money granted by the state or a public body etc. to keep down the price of commodities etc. (housing subsidy). **b** money granted to a charity or other undertaking held to be in the public interest. **c** any grant or contribution of money. **2** money paid by one state to another in return for military, naval, or other aid. **3** hist. **a** a parliamentary grant to the sovereign for state needs. **b** a tax levied on a particular occasion. [ME f. AF subsidie, OF subside f. L subsidium assistance]

subsist /səbˈsɪst/ v. **1** intr. (often foll. by on) keep oneself alive; be kept alive (subsists on vegetables). **2** intr. remain in being; exist. **3** intr. (foll. by in) be attributable to (its excellence subsists in its freshness). **4** tr. archaic provide sustenance for. □ **subsistent** adj. [L subsistere stand firm (as SUB-, sistere set, stand)]

subsistence /səbˈsɪstəns/ n. **1** the state or an instance of subsisting. **2 a** the means of supporting life; a livelihood. **b** a minimal level of existence or the income providing this (a bare subsistence). □ **subsistence allowance** (or **money**) esp. Brit. an allowance or advance on pay granted esp. as travelling expenses. **subsistence farming** farming which directly supports the farmer's household without producing a significant surplus for trade. **subsistence level** (or **wage**) a standard of living (or wage) providing only the bare necessities of life.

subsoil /'sʌbsɔɪl/ n. soil lying immediately under the surface soil (opp. TOPSOIL).

subsonic /sʌbˈsɒnɪk/ adj. relating to speeds less than that of sound. □ **subsonically** adv.

subspecies /'sʌbˌspiːʃiːz, -ʃiːz, -ˌspiːsɪz, -siːz/ n. (pl. same) Biol. a taxonomic category below a species, usu. a fairly permanent geographically isolated variety. □ **subspecific** /ˌsʌbspɪˈsɪfɪk/ adj.

substance /'sʌbstəns/ n. **1 a** the essential material, esp. solid, forming a thing (the substance was transparent). **b** a particular kind of material having uniform properties (this substance is salt). **2 a** a reality; solidity (ghosts have no substance). **b** seriousness or steadiness of character (there is no substance in him). **3** the theme or subject of esp. a work of art, argument, etc. (prefer the substance to the style). **4** the real meaning or essence of a thing. **5** wealth and possessions (a woman of substance). **6** Philos. the essential nature underlying phenomena, which is subject to changes and accidents. **7** esp. N. Amer. an intoxicating or narcotic chemical or drug, esp. an illegal one (substance abuse). □ **in substance** generally; apart from details. [ME f. OF f. L substantia (as SUB-, stare stand)]

substandard /sʌbˈstændəd/ adj. **1** of less than the required or normal quality or size; inferior. **2** (of language) not conforming to standard usage.

substantial /səbˈstænʃ(ə)l/ adj. **1 a** of real importance, value, or validity (made a substantial contribution). **b** of large size or amount (awarded substantial damages). **2** of solid material or structure; stout (a man of substantial build; a substantial house). **3** commercially successful; wealthy. **4** essential; true in large part (substantial truth). **5** having substance; real. □ **substantially** adv. **substantiality** /-ˌstænʃɪˈælɪtɪ/ n. [ME f. OF substantiel or LL substantialis (as SUBSTANCE)]

substantialism /səbˈstænʃəˌlɪz(ə)m/ n. Philos. the doctrine that behind all phenomena there are substantial or unchanging realities. □ **substantialist** n. & adj.

substantialize /səbˈstænʃəˌlaɪz/ v.tr. & intr. (also **-ise**) invest with or acquire substance or actual existence.

substantiate /səbˈstænʃɪˌeɪt/ v.tr. prove the truth of (a charge, statement, claim, etc.); give good grounds for. □ **substantiation** /-ˌstænʃɪˈeɪʃ(ə)n/ n. [med.L substantiare give substance to (as SUBSTANCE)]

substantive /'sʌbstəntɪv/ adj. & n. ● adj. (also /səbˈstæntɪv/) **1** having separate and independent existence. **2** having a firm or solid basis; important, substantial. **3** Law relating to rights and duties. **4** (of an enactment, motion, resolution, etc.) made in due form as such; not amended. **5** Gram. expressing existence. **6** (of a dye) not needing a mordant. **7** Mil. (of a rank etc.) permanent, not acting or temporary. **8** archaic denoting a substance. ● n. Gram. = NOUN. □ **the substantive verb** the verb 'to be'. □ **substantively** adv. esp. Gram. **substantival** /ˌsʌbstənˈtaɪv(ə)l/ adj. [ME f. OF substantif -ive, or LL substantivus (as SUBSTANCE)]

substation /'sʌbˌsteɪʃ(ə)n/ n. a subordinate station, esp. one reducing the high voltage of electric power transmission to that suitable for supply to consumers.

substituent /sʌbˈstɪtjʊənt/ adj. & n. Chem. ● adj. (of a group of atoms) replacing another atom or group in a compound. ● n. such a group. [L substituere substituent- (as SUBSTITUTE)]

substitute /'sʌbstɪˌtjuːt/ n. & v. ● n. **1 a** (also attrib.) a person or thing acting or serving in place of another. **b** an artificial alternative to a natural substance (butter substitute). **2** Sc. Law a deputy. ● v. **1** intr. & tr. (often foll. by for) act or cause to act as a substitute; put or serve in exchange (substituted for her mother; substituted it for the broken one). **2** tr. (usu. foll. by by, with) replace (a person or thing) with another. **3** tr. Chem. replace (an atom or group in a molecule) with another. □ **substitutive** adj. **substitutable** adj. **substitutability** /ˌsʌbstɪˌtjuːtəˈbɪlɪtɪ/ n. **substitution** /-ˈtjuːʃ(ə)n/ n. **substitutional** adj. **substitutionary** adj. [ME f. L substitutus past part. of substituere (as SUB-, statuere set up)]

substrate /'sʌbstreɪt/ n. **1** = SUBSTRATUM. **2** a surface to be painted, printed, etc., on. **3** Biol. **a** the substance upon which an enzyme acts. **b** the surface or material on which any particular organism grows. [anglicized f. SUBSTRATUM]

substratum /'sʌbˌstrɑːtəm, -ˌstreɪtəm/ n. (pl. **substrata** /-tə/) **1** an underlying layer or substance. **2** a layer of rock or soil beneath the surface. **3** a foundation or basis (there is a substratum of truth in it). [mod.L, past part. of L substernere (as SUB-, sternere strew): cf. STRATUM]

substructure /'sʌbˌstrʌktʃə(r)/ n. an underlying or supporting structure. □ **substructural** /sʌbˈstrʌktʃərəl/ adj.

subsume /səbˈsjuːm/ v.tr. (usu. foll. by under) include (an instance, idea, category, etc.) under another, or in a rule, class, category, etc.

□ **subsumable** adj. **subsumption** /-'sʌmpʃ(ə)n/ n. [med.L subsumere (as SUB-, sumere sumpt- take)]

subtenant /'sʌbˌtenənt/ n. a person who leases a property from a tenant. □ **subtenancy** n.

subtend /sʌb'tend/ v.tr. **1 a** (usu. foll. by at) (of a line, arc, figure, etc.) form (an angle) at a particular point when its extremities are joined at that point. **b** (of an angle or chord) have bounding lines or points that meet or coincide with those of (a line or arc). **2** Bot. (of a bract etc.) extend under so as to embrace or enfold. [L subtendere (as SUB-, tendere stretch)]

subterfuge /'sʌbtəˌfjuːdʒ/ n. **1 a** an attempt to avoid blame or defeat esp. by lying or deceit. **b** a statement etc. resorted to for such a purpose. **2** this as a practice or policy. [F subterfuge or LL subterfugium f. L subterfugere escape secretly f. subter beneath + fugere flee]

subterminal /sʌb'tɜːmɪn(ə)l/ adj. nearly at the end.

subterranean /ˌsʌbtəˈreɪnɪən/ adj. **1** existing, occurring, or done under the earth's surface. **2** secret, underground, concealed. □ **subterraneously** adv. [L subterraneus (as SUB-, terra earth)]

subtext /'sʌbtekst/ n. an underlying often distinct theme in a piece of writing or conversation.

subtilize /'sʌtɪˌlaɪz/ v. (also **-ise**) **1** tr. **a** make subtle. **b** elevate; refine. **2** intr. (usu. foll. by upon) argue or reason subtly. □ **subtilization** /ˌsʌtɪlaɪˈzeɪʃ(ə)n/ n. [F subtiliser or med.L subtilizare (as SUBTLE)]

subtitle /'sʌbˌtaɪt(ə)l/ n. & v. ● n. **1** a secondary or additional title of a book etc. **2** a printed caption at the bottom of a film etc., esp. translating dialogue. ● v.tr. provide with a subtitle or subtitles.

subtle /'sʌt(ə)l/ adj. (**subtler**, **subtlest**) **1** evasive or mysterious; hard to grasp (subtle charm; a subtle distinction). **2** (of scent, colour, etc.) faint, delicate, elusive (subtle perfume). **3 a** capable of making fine distinctions; perceptive; acute (subtle intellect; subtle senses). **b** ingenious; elaborate; clever (a subtle device). **4** archaic crafty, cunning. □ **subtleness** n. **subtly** adv. [ME f. OF sotil f. L subtilis]

subtlety /'sʌt(ə)lti/ n. (pl. **-ies**) **1** something subtle; the quality of being subtle. **2** a fine distinction; a subtle argument. [ME f. OF s(o)utilté f. L subtilitas -tatis (as SUBTLE)]

subtonic /sʌb'tɒnɪk/ n. Mus. the note below the tonic, the seventh note of the diatonic scale of any key.

subtopia /sʌb'təʊpɪə/ n. Brit. derog. unsightly and sprawling suburban development. □ **subtopian** adj. [SUBURB, UTOPIA]

subtotal /'sʌbˌtəʊt(ə)l/ n. the total of one part of a group of figures to be added.

subtract /səb'trækt/ v.tr. (often foll. by from) deduct (a part, quantity, or number) from another. □ **subtracter** n. (cf. SUBTRACTOR).

subtractive adj. **subtraction** /-'trækʃ(ə)n/ n. [L subtrahere subtract- (as SUB-, trahere draw)]

subtractor /səb'træktə(r)/ n. Electronics a circuit or device that produces an output dependent on the difference of two inputs.

subtrahend /'sʌbtrəˌhend/ n. Math. a quantity or number to be subtracted. [L subtrahendus gerundive of subtrahere: see SUBTRACT]

subtropics /sʌb'trɒpɪks/ n.pl. the regions adjacent to or bordering on the tropics. □ **subtropical** adj.

subulate /'sʌbjʊlət/ adj. Bot. & Zool. slender and tapering. [L subula awl]

suburb /'sʌbɜːb/ n. an outlying district of a city, esp. a residential one. [ME f. OF suburbe or L suburbium (as SUB-, urbs urbis city)]

suburban /sə'bɜːb(ə)n/ adj. **1** of or characteristic of suburbs. **2** derog. provincial, narrow-minded, uncultured, or naive. □ **suburbanite** n. **suburbanize** v.tr. (also **-ise**). **suburbanization** /-ˌbɜːbənaɪˈzeɪʃ(ə)n/ n. [L suburbanus (as SUBURB)]

suburbia /sə'bɜːbɪə/ n. often derog. the suburbs, their inhabitants, and their way of life.

subvention /səb'venʃ(ə)n/ n. a grant of money from a government etc.; a subsidy. [ME f. OF f. LL subventio -onis f. L subvenire subvent- assist (as SUB-, venire come)]

subversive /səb'vɜːsɪv/ adj. & n. ● adj. (of a person, organization, activity, etc.) seeking to subvert (esp. a government). ● n. a subversive person; a revolutionary. □ **subversively** adv. **subversiveness** n. **subversion** /-'vɜːʃ(ə)n/ n. [med.L subversivus (as SUBVERT)]

subvert /səb'vɜːt/ v.tr. esp. Polit. overturn, overthrow, or upset (religion, government, the monarchy, morality, etc.). □ **subverter** n. [ME f. OF subvertir or L subvertere (as SUB-, vertere vers- turn)]

subway /'sʌbweɪ/ n. **1 a** a tunnel beneath a road etc. for pedestrians.

b an underground passage for pipes, cables, etc. **2** esp. N. Amer. an underground railway.

subzero /sʌb'zɪərəʊ/ adj. (esp. of temperature) lower than zero.

suc- /sʌk, sək/ prefix assim. form of SUB- before c.

succedaneum /ˌsʌksɪˈdeɪnɪəm/ n. (pl. **succedanea** /-nɪə/) a substitute, esp. for a medicine or drug. □ **succedaneous** adj. [mod.L, neut. of L succedaneus (as SUCCEED)]

succeed /sək'siːd/ v. **1** intr. **a** (often foll. by in) accomplish one's purpose; have success; prosper (succeeded in his ambition). **b** (of a plan etc.) be successful. **2 a** tr. follow in order; come next after (night succeeded day). **b** intr. (foll. by to) come next, be subsequent. **3** intr. (often foll. by to) take one's place as holder of an office, title, property, etc. (succeeded to the throne). **4** tr. take over an office, throne, inheritance, etc. from (succeeded his father; succeeded the manager). □ **nothing succeeds like success** one success leads to others. □ **succeeder** n. [ME f. OF succeder or L succedere (as SUB-, cedere cess- go)]

succentor /sək'sentə(r)/ n. Eccl. a precentor's deputy in some cathedrals. □ **succentorship** n. [LL f. L succinere (as SUB-, canere sing)]

succès de scandale /sʊkˌseɪ də skɒn'dɑːl/ n. **1** a book, play, etc., having great success because of its scandalous nature or associations. **2** success achieved in this way. [F]

success /sək'ses/ n. **1** the accomplishment of an aim; a favourable outcome (their efforts met with success). **2** the attainment of wealth, fame, or position (spoilt by success). **3** a thing or person that turns out well. **4** archaic a usu. specified outcome of an undertaking (ill success). □ **success story 1** a person's rise from poverty to wealth etc. **2** a person whose life has been a success story. [L successus (as SUCCEED)]

successful /sək'sesfʊl/ adj. having or resulting in success; having wealth or status, prosperous. □ **successfully** adv. **successfulness** n.

succession /sək'seʃ(ə)n/ n. **1 a** the process of following in order; succeeding. **b** a series of things or people in succession. **2 a** the right of succeeding to an office, throne, inheritance, etc. **b** the act or process of so succeeding. **c** those having such a right. **3** Ecol. a sequence of changes whereby a plant or animal community successively gives way to another until a climax community is reached (cf. SERE²). **4** Geol. **a** a sequence of fossil forms representing an evolutionary series. **b** a group of strata representing a single chronological sequence. □ **in quick succession** following one another at short intervals. **in succession** one after another, without intervention. **in succession to** as the successor of. **law of succession** the law regulating inheritance. **settle the succession** determine who shall succeed. **succession state** a state resulting from the partition of a previously existing country. □ **successional** adj. [ME f. OF succession or L successio (as SUCCEED)]

Succession, Act of (in English history) each of three Acts of Parliament passed during the reign of Henry VIII. The first (1534) declared Henry's marriage to Catherine of Aragon to be invalid and disqualified their daughter Mary from succeeding to the throne, fixing the succession on any child born to Henry's new wife Anne Boleyn. The second (1536) cancelled this, asserting the rights of Jane Seymour and her issue, while the third (1544) determined the order of succession of Henry's three children, the future Edward VI, Mary I, and Elizabeth I.

successive /sək'sesɪv/ adj. following one after another; running, consecutive. □ **successively** adv. **successiveness** n. [ME f. med.L successivus (as SUCCEED)]

successor /sək'sesə(r)/ n. (often foll. by to) a person or thing that succeeds another. [ME f. OF successour f. L successor (as SUCCEED)]

succinct /sək'sɪŋkt/ adj. briefly expressed; terse, concise. □ **succinctly** adv. **succinctness** n. [ME f. L succinctus past part. of succingere tuck up (as SUB-, cingere gird)]

succinic acid /sʌk'sɪnɪk/ n. Chem. a crystalline dibasic acid derived from amber etc., very important as an intermediate in glucose metabolism. □ **succinate** /'sʌksɪˌneɪt/ n. [F succinique f. L succinum amber]

succor US var. of SUCCOUR.

succory /'sʌkəri/ n. = CHICORY 1. [alt. f. cicorée etc., early forms of CHICORY]

succotash /'sʌkəˌtæʃ/ n. US a dish of green maize and beans boiled together. [Narragansett msiquatash]

Succoth /suː'kəʊt, 'sʌkəθ/ n. the Jewish autumn thanksgiving festival commemorating the sheltering in the wilderness. [Heb. sukkôt pl. of sukkāh thicket, hut]

succour /ˈsʌkə(r)/ *n. & v.* (*US* **succor**) ● *n.* **1** aid; assistance, esp. in time of need. **2** (in *pl.*) *archaic* reinforcements of troops. ● *v.tr.* assist or aid (esp. a person in danger or distress). □ **succourless** *adj.* [ME f. OF *socours* f. med.L *succursus* f. L *succurrere* (as SUB-, *currere* curs- run)]

succubus /ˈsʌkjʊbəs/ *n.* (*pl.* **succubi** /-ˌbaɪ/) a female demon believed to have sexual intercourse with sleeping men. [LL *succuba* prostitute, med.L *succubus* f. *succubare* (as SUB-, *cubare* lie)]

succulent /ˈsʌkjʊlənt/ *adj. & n.* ● *adj.* **1** juicy; palatable. **2** *colloq.* desirable. **3** *Bot.* (of a plant, its leaves, or stems) thick and fleshy. ● *n. Bot.* a succulent plant, esp. a cactus. □ **succulence** *n.* **succulently** *adv.* [L *succulentus* f. *succus* juice]

succumb /səˈkʌm/ *v.intr.* (usu. foll. by *to*) **1** be forced to give way; be overcome (*succumbed to temptation*). **2** be overcome by death (*succumbed to his injuries*). [ME f. OF *succomber* or L *succumbere* (as SUB-, *cumbere* lie)]

succursal /səˈkɜːs(ə)l/ *adj. Eccl.* (of a chapel etc.) subsidiary. [F *succursale* f. med.L *succursus* (as SUCCOUR)]

succussion /səˈkʌʃ(ə)n/ *n.* vigorous shaking, esp. in the preparation of a homeopathic remedy. [L, f. *succutere*, f. SUB- + *quatere* shake]

such /sʌtʃ, sətʃ/ *adj. & pron.* ● *adj.* **1** (often foll. by *as*) of the kind or degree in question or under consideration (*such a person; such people; people such as these*). **2** (usu. foll. by *as* to + infin. or *that* + clause) so great; in such high degree (*not such a fool as to believe them; had such a fright that he fainted*). **3** of a more than normal kind or degree (*we had such an enjoyable evening; such horrid language*). **4** of the kind or degree already indicated, or implied by the context (*there are no such things; such is life*). **5** *Law* or *formal* the aforesaid; of the aforesaid kind. ● *pron.* **1** the thing or action in question or referred to (*such were his words; such was not my intention*). **2 a** *Commerce* or *colloq.* the aforesaid thing or things; it, they, or them (*those without tickets should purchase such*). **b** similar things; suchlike (*brought sandwiches and such*). □ **as such** as being what has been indicated or named (*a stranger is welcomed as such; there is no theatre as such*). **such-and-such** *adj.* of a particular kind but not needing to be specified. ● *n.* a person or thing of this kind. **such-and-such a person** someone; so-and-so. **such as 1** of a kind that; like (*a person such as we all admire*). **2** for example (*insects, such as moths and bees*). **3** those who (*such as don't need help*). **such as it is** despite its shortcomings (*you are welcome to it, such as it is*). **such a one 1** (usu. foll. by *as*) such a person or such a thing. **2** *archaic* some person or thing unspecified. [OE *swilc*, *swylc* f. Gmc: cf. LIKE¹]

suchlike /ˈsʌtʃlaɪk/ *adj. & n.* ● *adj.* of such a kind. ● *n.* things, people, etc., of such a kind.

Suchou see SUZHOU.

Suchow see XUZHOU.

suck /sʌk/ *v. & n.* ● *v.* **1** *tr.* draw (a fluid) into the mouth by contracting the muscles of the lips etc. to make a partial vacuum. **2** *tr.* (also *absol.*) **a** draw milk or other fluid from or through (the breast etc. or a container). **b** extract juice from (a fruit) by sucking. **3** *tr.* **a** draw sustenance, knowledge, or advantage from (a book etc.). **b** imbibe or gain (knowledge, advantage, etc.) as if by sucking. **4** *tr.* roll the tongue round (a sweet, teeth, one's thumb, etc.). **5** *intr.* make a sucking action or sound (*sucking at his pipe*). **6** *intr.* (of a pump etc.) make a gurgling or drawing sound. **7** *tr.* (usu. foll. by *down*, *in*) engulf, smother, or drown in a sucking movement. **8** *intr. US sl.* be or seem very unpleasant or contemptible. ● *n.* **1** the act or an instance of sucking, esp. the breast. **2** the drawing action or sound of a whirlpool etc. **3** (often foll. by *of*) a small draught of liquor. **4** (in *pl.*; esp. as *int.*) *colloq.* **a** an expression of disappointment. **b** an expression of derision or amusement at another's discomfiture. □ **give suck** *archaic* (of a mother, dam, etc.) suckle. **suck dry 1** exhaust the contents of (a bottle, the breast, etc.) by sucking. **2** exhaust (a person's sympathy, resources, etc.) as if by sucking. **suck in 1** absorb. **2** = sense 7 of *v.* involve (a person) in an activity etc., esp. against his or her will. **suck up 1** (often foll. by *to*) *colloq.* behave obsequiously esp. for one's own advantage. **2** absorb. [OE *sūcan*, = L *sugere*]

sucker /ˈsʌkə(r)/ *n. & v.* ● *n.* **1 a** a person or thing that sucks. **b** a sucking-pig, newborn whale, etc. **2** *colloq.* **a** a gullible or easily deceived person. **b** (foll. by *for*) a person especially susceptible to. **c** esp. *US* a person or thing of the kind expressed or implied. **3 a** a rubber cup etc. that adheres to a surface by suction. **b** an organ enabling an organism to cling to a surface by suction. **4** *Bot.* a shoot springing from the rooted part of a stem, from the root at a distance from the main stem, from an axil, or occasionally from a branch. **5** a fish that has a mouth capable of, or seeming to be capable of, adhering by suction. **6 a** the piston of a suction-pump. **b** a pipe through which liquid is drawn by

suction. **7** *N. Amer. colloq.* a lollipop. ● *v.* **1** *tr. Bot.* remove suckers from. **2** *intr. Bot.* produce suckers. **3** *tr.* esp. *N. Amer. sl.* cheat, trick.

sucking /ˈsʌkɪŋ/ *adj.* **1** (of a child, animal, etc.) not yet weaned. **2** *Zool.* unfledged (*sucking dove*). □ **sucking-disc** an organ used for adhering to a surface. **sucking-fish** = REMORA.

suckle /ˈsʌk(ə)l/ *v.* **1** *tr.* **a** feed (young) from the breast or udder. **b** nourish (*suckled his talent*). **2** *intr.* feed by sucking the breast etc. □ **suckler** *n.* [ME, prob. back-form. f. SUCKLING]

Suckling /ˈsʌklɪŋ/, Sir John (1609–42), English poet, dramatist, and Royalist leader. He lived at court from 1632 and was a leader of the Cavaliers during the English Civil War. His poems include 'Ballad upon a Wedding', published in the posthumous collection *Fragmenta Aurea* (1646). According to John Aubrey, Suckling invented the game of cribbage.

suckling /ˈsʌklɪŋ/ *n.* an unweaned child or animal.

Sucre¹ /ˈsuːkreɪ/ the judicial capital and seat of the judiciary of Bolivia; pop. (est. 1988) 105,800. It is situated in the Andes, at an altitude of 2,700 m (8,860 ft). Named Chuquisaca by the Spanish in 1539, the city was renamed in 1825 in honour of Antonio José de Sucre, the first President of Bolivia.

Sucre² /ˈsuːkreɪ/, Antonio José de (1795–1830), Venezuelan revolutionary and statesman, President of Bolivia 1826–8. Sucre served as Simón Bolívar's Chief of Staff, liberating Ecuador (1822), Peru (1824), and Bolivia (1825) from the Spanish. The first President of Bolivia, Sucre resigned following a Peruvian invasion in 1828; he was later assassinated. The Bolivian judicial capital Sucre is named after him.

sucrose /ˈsuːkrəʊz, ˈsjuː-, -krəʊs/ *n. Chem.* sugar; a disaccharide obtained from sugar cane, sugar beet, etc., consisting of a glucose molecule linked to a fructose molecule. [F *sucre* SUGAR]

suction /ˈsʌkʃ(ə)n/ *n.* **1** the act or an instance of sucking. **2 a** the production of a partial vacuum by the removal of air etc. in order to force in liquid etc. or procure adhesion. **b** the force produced by this process (*suction keeps the lid on*). □ **suction-pump** a pump for drawing liquid through a pipe into a chamber emptied by a piston. [LL *suctio* f. L *sugere* suct- SUCK]

suctorial /sʌkˈtɔːrɪəl/ *adj. Zool.* **1** adapted for or capable of sucking. **2** having a sucker for feeding or adhering. □ **suctorian** *n.* [mod.L *suctorius* (as SUCTION)]

Sudan /suːˈdɑːn, -ˈdæn/ (also **the Sudan**) **1** a country in NE Africa south of Egypt, with a coastline on the Red Sea; pop. (est. 1991) 25,855,000; languages, Arabic (official), Dinka, Hausa, and other languages; capital, Khartoum. The north-eastern area was part of ancient Nubia. Under Arab rule from the 13th century, the country was conquered by Egypt in 1820–2. Sudan was separated from its northern neighbour by the Mahdist revolt of 1881–98, and administered after the reconquest of 1898 as an Anglo-Egyptian condominium. It became an independent republic in 1956, but has suffered severely as a result both of drought and of protracted civil war between the Islamic government in the north and separatist forces in the south. In 1989 the country came under military dictatorship. **2** a vast region of North Africa, extending across the width of the continent from the southern edge of the Sahara to the tropical equatorial zone in the south. The northern part of the region, consisting of dry savannah plains, is known as the Sahel. □ **Sudanese** /ˌsuːdəˈniːz/ *adj. & n.* [Arab. *sūdān* (pl. of *sūdā* black), = country of the blacks]

sudarium /suːˈdeərɪəm, sjuː-/ *n.* (*pl.* **sudaria** /-rɪə/) **1** a cloth for wiping the face. **2** *RC Ch.* = VERONICA 2. [L, = napkin f. *sudor* sweat]

sudatorium /ˌsuːdəˈtɔːrɪəm, ˌsjuː-/ *n.* (*pl.* **sudatoria** /-rɪə/) esp. *Rom. Antiq.* **1** a hot-air or steam bath. **2** a room where such a bath is taken. [L, neut. of *sudatorius*: see SUDATORY]

sudatory /ˈsuːdətərɪ, ˈsjuː-/ *adj. & n.* ● *adj.* promoting perspiration. ● *n.* (*pl.* **-ies**) **1** a sudatory drug. **2** = SUDATORIUM. [L *sudatorius* f. *sudare* sweat]

Sudbury /ˈsʌdbərɪ/ a city in SW central Ontario; pop. (1991) 110,670. Situated near extensive sources of nickel, it lies at the centre of Canada's largest mining region.

sudd /sʌd/ *n.* an area of floating vegetation impeding the navigation of the White Nile. [Arab., = obstruction]

sudden /ˈsʌd(ə)n/ *adj. & n.* ● *adj.* occurring or done unexpectedly or without warning; abrupt, hurried, hasty (*a sudden storm; a sudden departure*). ● *n. archaic* a hasty or abrupt occurrence. □ **all of a sudden** unexpectedly; hurriedly; suddenly. **on a sudden** *archaic* suddenly. **sudden death** a means of deciding a tied game etc. consisting of an extra period of play etc. in which the first to concede a point etc. is the loser. **sudden infant death syndrome** *Med.* = cot-death (see COT¹). □ **suddenly** *adv.* **suddenness** /-dənnɪs/ *n.* [ME f. AF *sodein, sudein*, OF *soudain* f. LL *subitanus* f. L *subitaneus* f. *subitus* sudden]

Sudetenland /suːˈdeɪt(ə)nˌlænd/ an area in the north-west part of the Czech Republic, on the border with Germany. Allocated to the new state of Czechoslovakia after the First World War, it became an object of Nazi expansionist policies because of its large German population and was ceded to Germany as a result of the Munich Agreement of Sept. 1938. In 1945 the area was returned to Czechoslovakia.

sudoriferous /ˌsuːdəˈrɪfərəs, ˌsjuː-/ *adj.* (of a gland etc.) secreting sweat. [LL *sudorifer* f. L *sudor* sweat]

sudorific /ˌsuːdəˈrɪfɪk, ˌsjuː-/ *adj. & n.* ● *adj.* (of a drug) causing sweating. ● *n.* a sudorific drug. [mod.L *sudorificus* f. L *sudor* sweat]

Sudra /ˈsuːdrə/ *n.* a member of the lowest of the four great Hindu classes or varnas, the labourer class, with the traditional function of serving the other three varnas. [Skr. *śūdra*]

suds /sʌdz/ *n. & v.* ● *n.pl.* **1** froth of soap and water. **2** *N. Amer. colloq.* beer. ● *v.* **1** *intr.* form suds. **2** *tr.* lather, cover, or wash in soapy water. □ **sudsy** *adj.* [orig. = fen waters etc., of uncert. orig.: cf. MDu., MLG *sudde*, MDu. *sudse* marsh, bog, prob. rel. to SEETHE]

sue /suː, sjuː/ *v.* (**sues, sued, suing**) **1** *tr.* (also *absol.*) *Law* institute legal proceedings against (a person). **2** *tr.* (also *absol.*) entreat (a person). **3** *intr.* (often foll. by *to, for*) *Law* make application to a lawcourt for redress. **4** *intr.* (often foll. by *to, for*) make entreaty to a person for a favour. **5** *tr.* (often foll. by *out*) make a petition in a lawcourt for and obtain (a writ, pardon, etc.). □ **suer** *n.* [ME f. AF *suer, siwer*, etc. f. OF *siu-* etc. stem of *sivre* f. L *sequi* follow]

suede /sweɪd/ *n.* (often *attrib.*) **1** leather, esp. kidskin, with the flesh side rubbed to make a velvety nap. **2** (also **suede-cloth**) a woven fabric resembling suede. [F (*gants de*) *Suède* (gloves of) Sweden]

suet /ˈsuːɪt, ˈsjuː-/ *n.* the hard white fat on the kidneys or loins of oxen, sheep, etc., used to make dough etc. □ **suet pudding** a pudding of suet etc., usu. boiled or steamed. □ **suety** *adj.* [ME f. AF f. OF *seu* f. L *sebum* tallow]

Suetonius /swiːˈtəʊnɪəs/ (full name Gaius Suetonius Tranquillus) (AD c.69–c.150), Roman biographer and historian. His surviving works include *Lives of the Caesars*, covering Julius Caesar and the Roman emperors who followed him, up to Domitian.

Suez, Isthmus of /ˈsuːɪz/ an isthmus between the Mediterranean and the Red Sea, connecting Egypt and Africa to the Sinai peninsula and Asia. The port of Suez lies in the south, at the head of the Gulf of Suez, an arm of the Red Sea. The isthmus is traversed by the Suez Canal.

Suez Canal /ˈsuːɪz/ a shipping canal connecting the Mediterranean at Port Said with the Red Sea. Constructed between 1859 and 1869 by Ferdinand de Lesseps, it is 171 km (106 miles) long, providing the shortest route for sea traffic between Europe and Asia. In 1875 it came under British control, after Britain acquired a majority of its shares at Disraeli's instigation, and after 1888 Britain acted as guarantor of its neutral status. Its nationalization by Egypt in 1956 prompted the Suez crisis. It was closed between 1967 and 1975 as a consequence of the Six Day War.

Suez crisis /ˈsuːɪz/ a conflict following the nationalization of the Suez Canal by President Nasser of Egypt in 1956. Britain and France made a military alliance with Israel to regain control of the canal; Israel launched an assault on Sinai, while Britain and France attacked Egyptian bases and landed troops after Nasser's rejection of an ultimatum. Pressure from the US and USSR and criticism from many other countries, however, forced the withdrawal of forces. Among the consequences of the crisis were a weakening of British and French influence in the Middle East and the resignation of the British Prime Minister Anthony Eden.

suf- /sʌf, səf/ *prefix* assim. form of SUB- before *f*.

suffer /ˈsʌfə(r)/ *v.* **1** *intr.* **a** undergo pain, grief, etc. (*suffers acutely; suffers from neglect*). **b** be damaged, decline (*your reputation will suffer*). **2** *tr.* undergo, experience, or be subjected to (pain, loss, grief, defeat, change, etc.) (*suffered banishment*). **3** *tr.* put up with; tolerate (*does not suffer fools gladly*). **4** *intr. archaic* or *literary* undergo martyrdom. **5** *tr.* (foll. by *to* + infin.) *archaic* allow. □ **sufferable** *adj.* **sufferer** *n.* **suffering** *n.* [ME f. AF *suffrir, soeffrir*, OF *sof(f)rir* f. L *sufferre* (as SUB-, *ferre* bear)]

sufferance /ˈsʌfərəns/ *n.* **1** tacit consent, abstinence from objection. **2** *archaic* submissiveness. □ **on sufferance** with toleration implied by

lack of consent or objection. [ME f. AF, OF *suffraunce* f. LL *sufferentia* (as SUFFER)]

suffice /səˈfaɪs/ *v.* **1** *intr.* (often foll. by *for*, or *to* + infin.) be enough or adequate (*that will suffice for our purpose; suffices to prove it*). **2** *tr.* meet the needs of; satisfy (*six sufficed him*). □ **suffice it to say** I shall content myself with saying. [ME f. OF *suffire* (*suffis-*) f. L *sufficere* (as SUB-, *facere* make)]

sufficiency /səˈfɪʃənsɪ/ *n.* (pl. **-ies**) **1** (often foll. by *of*) an adequate amount or adequate resources. **2** *archaic* being sufficient; ability; efficiency. [LL *sufficientia* (as SUFFICIENT)]

sufficient /səˈfɪʃ(ə)nt/ *adj.* **1** sufficing, adequate, enough (*is sufficient for a family; didn't have sufficient funds*). **2** = SELF-SUFFICIENT. **3** *archaic* competent; of adequate ability, resources, etc. □ **sufficiently** *adv.* [ME f. OF *sufficient* or L *sufficiens* (as SUFFICE)]

suffix /ˈsʌfɪks/ *n. & v.* ● *n.* **1** a verbal element added at the end of a word to form a derivative (e.g. *-ation, -fy, -ing, -itis*). **2** *Math.* = SUBSCRIPT. ● *v.tr.* (also /səˈfɪks/) append, esp. as a suffix. □ **suffixation** /ˌsʌfɪkˈseɪʃ(ə)n/ *n.* [*suffixum, suffixus* past part. of L *suffigere* (as SUB-, *figere fix-* fasten)]

suffocate /ˈsʌfəˌkeɪt/ *v.* **1** *tr.* choke or kill by stopping the breath of, esp. by pressure, fumes, etc. **2** *tr.* (often foll. by *by, with*) produce a choking or breathless sensation in, esp. by excitement, terror, etc. **3** *intr.* be or feel suffocated or breathless. □ **suffocating** *adj.* **suffocatingly** *adv.* **suffocation** /ˌsʌfəˈkeɪʃ(ə)n/ *n.* [L *suffocare* (as SUB-, *fauces* throat)]

Suffolk¹ /ˈsʌfək/ a county of eastern England, on the coast of East Anglia; county town, Ipswich.

Suffolk² /ˈsʌfək/ *n.* a large black-faced hornless breed of sheep. [SUFFOLK¹]

Suffolk punch see PUNCH⁴.

suffragan /ˈsʌfrəgən/ *n.* (in full **suffragan bishop** or **bishop suffragan**) **1** a bishop appointed to help a diocesan bishop in the administration of a diocese. **2** a bishop in relation to his archbishop or metropolitan. □ **suffragan see** the see of a suffragan bishop. □ **suffraganship** *n.* [ME f. AF & OF, repr. med.L *suffraganeus* assistant (bishop) f. L *suffragium* (see SUFFRAGE): orig. of a bishop summoned to vote in synod]

suffrage /ˈsʌfrɪdʒ/ *n.* **1 a** the right of voting in political elections (*full adult suffrage*). **b** a view expressed by voting; a vote (*gave their suffrages for and against*). **c** opinion in support of a proposal etc. **2** (esp. in *pl.*) *Eccl.* **a** a prayer made by a priest in the liturgy. **b** a short prayer made by a congregation esp. in response to a priest. **c** *archaic* an intercessory prayer. [ME f. L *suffragium*, partly through F *suffrage*]

suffragette /ˌsʌfrəˈdʒɛt/ *n. hist.* a woman seeking the right to vote through organized protest. In the UK the suffragettes (officially members of the Women's Suffrage Movement) initiated a campaign of demonstrations and militant action in the early 20th century, after the repeated defeat of women's suffrage bills in Parliament. Under the leadership of the Pankhursts they chained themselves to railings, heckled meetings, refused to pay taxes, and went on hunger strike when imprisoned, eventually winning (in 1918) the vote for women over the age of 30. Ten years later British women were given full equality with men in voting rights. [SUFFRAGE + -ETTE]

suffragist /ˈsʌfrədʒɪst/ *n. esp. hist.* a person who advocates the extension of the suffrage, esp. to women. □ **suffragism** *n.*

suffuse /səˈfjuːz/ *v.tr.* **1** (of colour, moisture, etc.) spread from within to colour or moisten (*a blush suffused her cheeks*). **2** cover with colour etc. □ **suffusion** /-ˈfjuːʒ(ə)n/ *n.* [L *suffundere suffus-* (as SUB-, *fundere* pour)]

Sufi /ˈsuːfɪ/ *n.* (pl. **Sufis**) a Muslim ascetic and mystic; a member of any of several orders of Islamic mystics. (See SUFISM.) □ **Sufic** *adj.* [Arab. *ṣūfī*, perh. f. *ṣūf* wool (from the woollen garment worn)]

Sufism /ˈsuːfɪz(ə)m/ *n.* the mystical system of the Sufis. Sufism is the esoteric dimension of the Islamic faith, the inner way or spiritual path to mystical union with God. Its followers may be ascetics who isolate themselves from society, or (more usually) members of a Sufi order. There are many orders (of which the best known are perhaps the dervish orders), each founded by a devout individual and each having different devotional practices. The movement seems to have begun in the late 7th century as a reaction against the strict formality of orthodox teaching, and reached its peak in the 13th century. Influenced by the beliefs of other faiths, such as Buddhism, but deeply rooted in Islamic spiritualism, Sufism has been responsible for worldwide missionary activity. In the 19th and 20th centuries Sufic orders have sometimes taken on overtly political roles, notably in Libyan resistance to Italian colonial occupation.

sug- /sʌg, səg/ *prefix* assim. form of SUB- before *g*.

sugar /'ʃʊgə(r)/ n., v., & int. ● n. **1** a sweet crystalline substance obtained from various plants, esp. sugar cane and sugar beet, used in cookery, confectionery, brewing, etc.; sucrose. (*See note below.*) **2** Chem. any of a group of soluble usu. sweet-tasting crystalline carbohydrates found esp. in plants, e.g. glucose. **3** esp. US colloq. darling, dear (used as a term of address). **4** sweet words; flattery. **5** anything comparable to sugar encasing a pill in reconciling a person to what is unpalatable. **6** sl. a narcotic drug, esp. heroin or LSD (taken on a lump of sugar). ● v.tr. **1** sweeten with sugar. **2** make (one's words, meaning, etc.) more pleasant or welcome. **3** coat with sugar (*sugared almond*). **4** spread a sugar mixture on (a tree) to catch moths. ● int. euphem. = SHIT int. □ **sugar beet** a variety of the beet, *Beta vulgaris*, from which sugar is extracted. **sugar-candy** see CANDY 1. **sugar cane** a perennial tropical grass of the genus *Saccharum*; esp. *S. officinarum*, with tall stout jointed stems from which sugar is made. **sugar-coated 1** (of food) enclosed in sugar. **2** made superficially attractive. **3** excessively sentimental. **sugar-daddy** (pl. **-ies**) sl. an elderly man who lavishes gifts on a young woman. **sugar-gum** an Australian eucalyptus, *Eucalyptus cladocalyx*, with sweet foliage eaten by cattle. **sugar loaf** a conical moulded mass of sugar. **sugar maple** a North American maple, *Acer saccharum*, from the sap of which maple sugar and maple syrup are made. **sugar of lead** Chem. lead acetate. **sugar pea** a variety of pea eaten whole including the pod; mangetout. **sugar the pill** see PILL. **sugar snap pea** = *sugar pea*. **sugar soap** Brit. an alkaline preparation containing washing soda and soap, used for cleaning paintwork or removing paint. □ **sugarless** adj. [ME f. OF *çukre, sukere* f. It. *zucchero* prob. f. med.L *saccarum* f. Arab. *sūkkar*]

- Cane-sugar was known in India in prehistoric times, reaching Europe from the Roman period as a rare spice from the East. In 1493 sugar cane was taken to the West Indies by Christopher Columbus, and has since been introduced into every tropical country. Sugar beet will grow in colder climates and was long used as a vegetable and cattle-food: in 1747 the German chemist Andreas Marggraf found a way of extracting sugar from it in crystalline form. The development of the beet-sugar industry dates from the period of the Napoleonic Wars, when the British blockade prevented the importing of cane-sugar from the West Indies.

Sugar Loaf Mountain a rocky peak situated to the north-east of Copacabana Beach, in Rio de Janeiro, Brazil. It rises to a height of 390 m (1,296 ft).

sugarplum /'ʃʊgə,plʌm/ n. archaic a small round sweet of flavoured boiled sugar.

sugary /'ʃʊgərɪ/ adj. **1** containing or resembling sugar. **2** excessively sweet or esp. sentimental. **3** falsely sweet or pleasant (*sugary compliments*). □ **sugariness** n.

suggest /sə'dʒest/ v.tr. **1** (often foll. by *that* + clause) propose (a theory, plan, or hypothesis) (*suggested to them that they should wait; suggested a different plan*). **2 a** cause (an idea, memory, association, etc.) to present itself; evoke (*this poem suggests peace*). **b** hint at (*his behaviour suggests guilt*). □ **suggest itself** (of an idea etc.) come into the mind. □ **suggester** n. [L *suggerere suggest-* (as SUB-, *gerere* bring)]

suggestible /sə'dʒestɪb(ə)l/ adj. **1** capable of being suggested. **2** open to suggestion; easily swayed. □ **suggestibility** /-,dʒestɪ'bɪlɪtɪ/ n.

suggestion /sə'dʒestʃən/ n. **1** the act or an instance of suggesting; the state of being suggested. **2** a theory, plan, etc., suggested (*made a helpful suggestion*). **3** a slight trace; a hint (*a suggestion of garlic*). **4** Psychol. **a** the insinuation of a belief etc. into the mind. **b** such a belief etc. [ME f. OF f. L *suggestio -onis* (as SUGGEST)]

suggestive /sə'dʒestɪv/ adj. **1** (usu. foll. by *of*) conveying a suggestion; evocative. **2** (esp. of a remark, joke, etc.) tending to suggest something indecent or improper. □ **suggestively** adv. **suggestiveness** n.

Sui /sweɪ/ a dynasty which ruled in China AD 581–618, unifying the country and preparing the ground for the cultural flowering of the succeeding Tang dynasty.

suicidal /,su:ɪ'saɪd(ə)l, ,sju:-/ adj. **1** inclined to commit suicide. **2** of or concerning suicide. **3** self-destructive; fatally or disastrously rash. □ **suicidally** adv.

suicide /'su:ɪ,saɪd, 'sju:-/ n. & v. ● n. **1 a** the intentional killing of oneself. **b** a person who commits suicide. **2** a self-destructive action or course (*political suicide*). **3** (*attrib.*) Mil. designating a highly dangerous or deliberately suicidal operation etc. (*a suicide mission*). ● v.intr. commit suicide. □ **suicide pact** an agreement between two or more people to commit suicide together. [mod.L *suicida, suicidium* f. L *sui* of oneself]

sui generis /,sju:aɪ 'dʒenərɪs, ,su:ɪ 'gen-/ adj. of its own kind; unique. [L]

sui juris /,sju:aɪ 'dʒʊərɪs, ,su:ɪ 'jʊə-/ adj. Law of age; independent. [L]

suint /swɪnt/ n. the natural grease in sheep's wool. [F f. *suer* sweat]

Suisse see SWITZERLAND.

suit /su:t, sju:t/ n. & v. ● n. **1 a** a set of outer clothes of matching material for men, consisting usu. of a jacket, trousers, and sometimes a waistcoat. **b** a similar set of clothes for women usu. having a skirt instead of trousers. **c** (in full **grey suit**) sl. a person who wears a suit at work, esp. a business executive; a bureaucrat. **d** (esp. in comb.) a set of clothes for a special occasion, occupation, etc. (*play-suit; swimsuit*). **2 a** any of the four sets (spades, hearts, diamonds, clubs) into which a pack of cards is divided. **b** a player's holding in a suit (*his strong suit was clubs*). **c** Bridge one of the suits as proposed trumps in bidding, frequently as opposed to no trumps. **3** (in full **suit at law**) a lawsuit (*criminal suit*). **4 a** a petition esp. to a person in authority. **b** the process of courting a woman (*paid suit to her*). **5** (usu. foll. by *of*) a set of sails, armour, etc. ● v. **1** tr. go well with (a person's figure, features, character, etc.); become. **2** tr. (also absol.) meet the demands or requirements of; satisfy; agree with (*does not suit all tastes; that date will suit*). **3** tr. make fitting or appropriate; accommodate; adapt (*suited his style to his audience*). **4** tr. (as **suited** adj.) appropriate; well-fitted (*not suited to be an engineer*). **5** intr. go well with the appearance etc. of a person; be suitable. □ **suit the action to the word** carry out a promise or threat at once. **suit oneself 1** do as one chooses. **2** find something that satisfies one. [ME f. AF *siute*, OF *si(e)ute* f. fem. past part. of Rmc *sequere* (unrecorded) follow: see SUE]

suitable /'su:təb(ə)l, 'sju:t-/ adj. (usu. foll. by *to, for*) well fitted for the purpose; appropriate. □ **suitably** adv. **suitableness** n. **suitability** /,su:tə'bɪlɪtɪ, ,sju:t-/ n. [SUIT + -ABLE, after *agreeable*]

suitcase /'su:tkeɪs, 'sju:t-/ n. a usu. oblong case for carrying clothes etc., having a handle and a flat hinged lid. □ **suitcaseful** n. (pl. **-fuls**)

suite /swi:t/ n. **1** a set of things belonging together, esp.: **a** a set of rooms in a hotel etc. **b** a sofa, armchairs, etc., of the same design. **2** Mus. a set of instrumental pieces or compositions played as one work. (*See note below.*) **3** a set of people in attendance; a retinue. [F (as SUIT)]

- During the 17th and 18th centuries the suite was one of the most important forms of instrumental music. Originally consisting of a set of dance tunes played in succession, it was superseded in importance by the sonata and the symphony, and in the 19th century the title was given to works of a lighter tone and to assemblages of movements from opera or ballet scores. 20th-century neoclassical composers (e.g. Stravinsky) have revived the term in something like its original meaning.

suiting /'su:tɪŋ, 'sju:-/ n. cloth used for making suits.

suitor /'su:tə(r), 'sju:-/ n. **1** a man seeking to marry a specified woman; a wooer. **2** a plaintiff or petitioner in a lawsuit. **3** a prospective buyer of a business or corporation; the maker of a takeover bid. [ME f. AF *seutor, suitour,* etc., f. L *secutor -oris* f. *sequi secut-* follow]

suk (also **sukh**) var. of SOUK.

Sukarno /su:'kɑ:nəʊ/, Achmad (1901–70), Indonesian statesman, President 1945–67. One of the founders of the Indonesian National Party (1927), he was Indonesian leader during the Japanese occupation (1942–5) and led the struggle for independence, which was formally granted by the Netherlands in 1949. From the mid-1950s his dictatorial tendencies aroused opposition. He was alleged to have taken part in the abortive Communist coup of 1965, after which he steadily lost power to the army, being finally ousted two years later.

Sukhotai /,sʊkə'taɪ/ (also **Sukhothai**) a town in NW central Thailand; pop. (1990) 22,600. It was formerly the capital of an independent state of the same name, which flourished from the mid-13th to the mid-14th centuries. The ruins of the old city lie to the west of the modern town.

sukiyaki /,sʊkɪ'jɑːkɪ/ n. a Japanese dish of sliced meat fried rapidly with vegetables and sauce. [Jap.]

Sukkur /'sʌkə(r)/ a city in SE Pakistan, on the Indus river; pop. (est. 1991) 350,000. Nearby is the Sukkur Barrage (completed in 1932), a dam constructed across the Indus which directs water from the river through irrigation channels to a large area of the Indus valley.

Sulawesi /,sʊlə'weɪsɪ/ a mountainous island in the Greater Sunda group in Indonesia, situated to the east of Borneo; chief town, Ujung Pandang. It is noted as the habitat of numerous endemic species, a fact recorded in the mid-19th century by the naturalist Alfred Wallace. The island was formerly called Celebes.

Sulaymaniyah /,sʊlmə'ni:ə/ (in full **As Sulaymaniyah** /æs/, also **Sulaimaniya**) a town in NE Iraq, in the mountainous region of southern Kurdistan; pop. (est. 1985) 279,400. Founded in 1781, it was an important trade centre on the route between Baghdad and Tabriz. It is the capital of a Kurdish governorate of the same name.

sulcate /ˈsʌlkeɪt/ *adj.* grooved, fluted, channelled. [L *sulcatus*, past part. of *sulcare* furrow (as SULCUS)]

sulcus /ˈsʌlkəs/ *n.* (*pl.* **sulci** /ˈsʌlsaɪ/) *Anat.* a groove or furrow, esp. on the surface of the brain. [L]

Suleiman I /ˈsuːlɪmən, ˌsuːleɪˈmɑːn/ (also **Soliman** /ˈsɒlɪmən/ or **Solyman**) (c.1494–1566), sultan of the Ottoman Empire 1520–66. The Ottoman Empire reached its fullest extent under his rule; his conquests included Belgrade (1521), Rhodes (1522), and Tripoli (1551), in addition to those in Iraq (1534) and Hungary (1562). This and the cultural achievements of the time earned him the nickname in Europe of 'Suleiman the Magnificent'. He was also a noted administrator, known to his subjects as 'Suleiman the Lawgiver'.

sulfa *US* var. of SULPHA.

sulfanilamide *US* var. of SULPHANILAMIDE.

sulfate etc. *US* var. of SULPHATE etc.

sulfur etc. *US* var. of SULPHUR etc.

sulk /sʌlk/ *v. & n.* ● *v.intr.* indulge in a sulk, be sulky. ● *n.* (also in *pl.*, prec. by *the*) a period of sullen esp. resentful silence or aloofness from others (*having a sulk; got the sulks*). □ **sulker** *n.* [perh. back-form. f. SULKY]

sulky /ˈsʌlkɪ/ *adj. & n.* ● *adj.* (**sulkier, sulkiest**) **1** sullen, morose, or silent, esp. from resentment or ill temper. **2** sluggish. ● *n.* (*pl.* **-ies**) a light two-wheeled horse-drawn vehicle for one, esp. used in trotting-races. □ **sulkily** *adv.* **sulkiness** *n.* [perh. f. obs. *sulke* hard to dispose of]

Sulla /ˈsʌlə/ (full name Lucius Cornelius Sulla Felix) (138–78 BC), Roman general and politician. Having come to prominence through military successes in Africa, Sulla became involved in a power struggle with Marius, and in 88 marched on Rome. After a victorious campaign against Mithridates VI Sulla invaded Italy in 83, ruthlessly suppressing his opponents. He was elected dictator in 82, after which he implemented constitutional reforms in favour of the Senate, resigning in 79.

sullage /ˈsʌlɪdʒ/ *n.* filth, refuse, sewage. [perh. f. AF *suillage* f. *souiller* SOIL²]

sullen /ˈsʌlən/ *adj. & n.* ● *adj.* **1** morose, resentful, sulky, unforgiving, unsociable. **2 a** (of a thing) slow-moving. **b** dismal, melancholy (*a sullen sky*). ● *n.* (in *pl.*, usu. prec. by *the*) *archaic* a sullen frame of mind; depression. □ **sullenly** *adv.* **sullenness** /-lənnɪs/ *n.* [16th-c. alt. of ME *solein* f. AF f. *sol* SOLE³]

Sullivan /ˈsʌlɪv(ə)n/, Sir Arthur (Seymour) (1842–1900), English composer. Although he composed much 'serious' music, his fame rests on the fourteen light operas which he wrote in collaboration with the librettist W. S. Gilbert (see GILBERT³), many for Richard D'Oyly Carte's company at the Savoy Theatre.

sully /ˈsʌlɪ/ *v.tr.* (**-ies, -ied**) **1** disgrace or tarnish (a person's reputation or character, a victory, etc.). **2** *poet.* dirty; soil. [perh. f. F *souiller* (as SOIL²)]

sulpha /ˈsʌlfə/ *n.* (*US* **sulfa**) *Pharm.* any drug derived from sulphanilamide (often *attrib.*: *sulpha drug*). [abbr.]

sulphadimidine /ˌsʌlfəˈdɪmɪˌdiːn/ *n.* *Pharm.* a sulphonamide antibiotic used esp. to treat human urinary infections and to control respiratory disease in pigs. [SULPHANILAMIDE + DI-² + PYRIMIDINE]

sulphamic acid /sʌlˈfæmɪk/ *n.* (*US* **sulfamic**) *Chem.* a strong acid used in weed-killer, an amide of sulphuric acid. □ **sulphamate** /ˈsʌlfəˌmeɪt/ *n.* [SULPHUR + AMIDE]

sulphanilamide /ˌsʌlfəˈnɪləˌmaɪd/ *n.* (*US* **sulfanilamide**) *Pharm.* a colourless sulphonamide drug with antibacterial properties. [*sulphanilic* (SULPHUR, ANILINE) + AMIDE]

sulphate /ˈsʌlfeɪt/ *n.* (*US* **sulfate**) *Chem.* a salt or ester of sulphuric acid. [F *sulfate* f. L *sulphur*]

sulphide /ˈsʌlfaɪd/ *n.* (*US* **sulfide**) *Chem.* a binary compound of sulphur.

sulphite /ˈsʌlfaɪt/ *n.* (*US* **sulfite**) *Chem.* a salt or ester of sulphurous acid. [F *sulfite* alt. of *sulfate* SULPHATE]

sulphonamide /sʌlˈfɒnəˌmaɪd/ *n.* (*US* **sulfonamide**) *Pharm.* a substance derived from an amide of a sulphonic acid, able to prevent the multiplication of some pathogenic bacteria. [SULPHONE + AMIDE]

sulphonate /ˈsʌlfəˌneɪt/ *n. & v.* (*US* **sulfonate**) *Chem.* ● *n.* a salt or ester of sulphonic acid. ● *v.tr.* convert into a sulphonate by reaction with sulphuric acid.

sulphone /ˈsʌlfəʊn/ *n.* (*US* **sulfone**) *Chem.* an organic compound containing the SO_2 group united directly to two carbon atoms. □ **sulphonic** /sʌlˈfɒnɪk/ *adj.* [G *Sulfon* (as SULPHUR)]

sulphur /ˈsʌlfə(r)/ *n. & v.* (*US* **sulfur**) ● *n.* **1 a** a yellow non-metallic chemical element (atomic number 16; symbol **S**). Also called *brimstone*. (*See note below.*) **b** (*attrib.*) like or containing sulphur. **2** the material of which hell-fire and lightning were believed to consist. **3** a yellow butterfly of the family Pieridae. **4** a pale greenish-yellow colour. ● *v.tr.* **1** treat with sulphur. **2** fumigate with sulphur. □ **sulphur candle** a candle burnt to produce sulphur dioxide for fumigating. **sulphur dioxide** a colourless pungent gas (chem. formula: SO_2) formed by burning sulphur in air. **sulphur spring** a spring impregnated with sulphur or its compounds. □ **sulphury** *adj.* [ME f. AF *sulf(e)re*, OF *soufre* f. L *sulfur, sulp(h)ur*]

▪ Sulphur occurs uncombined in nature, especially in volcanic and sedimentary deposits, as well as in many minerals and as a constituent of petroleum, and has been known since ancient times. It is normally a bright yellow crystalline solid, but several other allotropic forms can be made. Sulphur is an ingredient of gunpowder, and is used in making matches and as an antiseptic and fungicide. The most important sulphur compound is sulphuric acid, produced in enormous quantities for use in industrial processes. Sulphur is essential to living organisms.

sulphurate /ˈsʌlfjʊˌreɪt/ *v.tr.* (*US* **sulfurate**) impregnate, fumigate, or treat with sulphur, esp. in bleaching. □ **sulphurator** *n.* **sulphuration** /ˌsʌlfjʊˈreɪʃ(ə)n/ *n.*

sulphureous /sʌlˈfjʊərɪəs/ *adj.* (*US* **sulfureous**) **1** of, like, or suggesting sulphur. **2** sulphur-coloured; yellow. [L *sulphureus* f. SULPHUR]

sulphuretted /ˌsʌlfjʊˈretɪd/ *adj.* (*US* **sulfureted**) *archaic* containing sulphur in combination. □ **sulphuretted hydrogen** hydrogen sulphide. [*sulphuret* sulphide f. mod.L *sulphuretum*]

sulphuric /sʌlˈfjʊərɪk/ *adj.* (*US* **sulfuric**) *Chem.* containing hexavalent sulphur (cf. SULPHUROUS 2). □ **sulphuric acid** a dense oily colourless highly acid and corrosive liquid (chem. formula: H_2SO_4) much used in the chemical industry. [F *sulfurique* (as SULPHUR)]

sulphurize /ˈsʌlfjʊˌraɪz/ *v.tr.* (also **-ise**, *US* **sulfurize**) = SULPHURATE. □ **sulphurization** /ˌsʌlfjʊraɪˈzeɪʃ(ə)n/ *n.* [F *sulfuriser* (as SULPHUR)]

sulphurous /ˈsʌlfərəs/ *adj.* (*US* **sulfurous**) **1** relating to or suggestive of sulphur, esp. in colour. **2** *Chem.* containing tetravalent sulphur (cf. SULPHURIC). □ **sulphurous acid** an unstable weak acid used as a reducing and bleaching acid. [L *sulphurosus* f. SULPHUR]

sultan /ˈsʌlt(ə)n/ *n.* **1 a** a Muslim sovereign. **b** (**the Sultan**) *hist.* the sultan of Turkey. **2** a variety of white domestic fowl from Turkey. [F *sultan* or med.L *sultanus* f. Arab. *sulṭān* power, ruler f. *saluṭa* rule]

sultana /sʌlˈtɑːnə/ *n.* **1 a** a seedless raisin used in puddings, cakes, etc. **b** the small pale yellow grape producing this. **2** the mother, wife, concubine, or daughter of a sultan. [It., fem. of *sultano* = SULTAN]

sultanate /ˈsʌltəˌneɪt/ *n.* a territory subject to a sultan; the office, position, or period of rule of a sultan. [SULTAN + -ATE¹]

sultry /ˈsʌltrɪ/ *adj.* (**sultrier, sultriest**) **1** (of the atmosphere or the weather) hot or oppressive; close. **2** (of a person, character, etc.) passionate; sensual. □ **sultrily** *adv.* **sultriness** *n.* [obs. *sulter* SWELTER]

Sulu Sea /ˈsuːluː/ a sea in the Malay Archipelago, encircled by the NE coast of Borneo and the western islands of the Philippines.

sum /sʌm/ *n. & v.* ● *n.* **1** the total amount resulting from the addition of two or more items, facts, ideas, feelings, etc. (*the sum of two and three is five; the sum of their objections is this*). **2** a particular amount of money (*paid a large sum for it*). **3 a** an arithmetical problem (*could not work out the sum*). **b** (esp. *pl.*) *colloq.* arithmetic work, esp. at an elementary level (*was good at sums*). ● *v.tr.* (**summed, summing**) find the sum of. □ **in sum** in brief. **summing-up 1** a review of evidence and a direction given by a judge to a jury. **2** a recapitulation of the main points of an argument, case, etc. **sum total** = sense 1 of *n.* **sum up 1** (esp. of a judge) recapitulate or review the evidence in a case etc. **2** form or express an idea of the character of (a person, situation, etc.). **3** collect into or express as a total or whole. [ME f. OF *summe, somme* f. L *summa* main part, fem. of *summus* highest]

sumac /ˈsuːmæk, ˈʃuː-, ˈsjuː-/ *n.* (also **sumach**) **1** a shrub or small tree of the genus *Rhus* or *Cotinus*; esp. the southern European *R. coriaria*, having reddish cone-shaped fruits used as a spice in cooking. **2** the dried and ground leaves of *R. coriaria*, used in tanning and dyeing. [ME f. OF *sumac* or med.L *sumac(h)* f. Arab. *summāḳ*]

Sumatra /sʊˈmɑːtrə/ a large island of Indonesia, situated to the south-west of the Malay Peninsula, from which it is separated by the Strait of Malacca; chief city, Medan. □ **Sumatran** n. & adj.

Sumba /'sʊmbə/ (also called *Sandalwood Island*) an island of the Lesser Sunda group in Indonesia, lying to the south of the islands of Flores and Sumbawa; chief town, Waingapu.

Sumbawa /sʊm'bɑːwə/ an island in the Lesser Sunda group in Indonesia, situated between Lombok and Flores.

Sumer /'suːmə(r)/ an ancient region of SW Asia in present-day Iraq, comprising the southern part of Mesopotamia. It was the site in the 4th and 3rd millenniums BC of a thriving civilization of city-states. Sumer joined with Akkad in the first half of the 2nd millennium BC to form the kingdom of Babylonia.

Sumerian /suːˈmɪərɪən, sjuː-, -ˈmeərɪən/ adj. & n. ● adj. of or relating to a non-Semitic language, people, and civilization native to Sumer in the 4th millennium BC and possibly earlier. (*See note below.*) ● n. **1** a member of this people. **2** the language of this people.

▪ The Sumerians spoke an agglutinative language, the oldest known written language, whose relationship to any other language is unclear. Theirs is the first historically attested civilization and they are credited with the invention of cuneiform writing, the sexagesimal system of mathematics, and the socio-political institution of the city-state with bureaucracies, legal codes, division of labour, and a money economy. Their art, literature, and theology had a profound cultural and religious influence, which continued long after their demise *c.*2000 BC and was the prototype of Akkadian, Hurrian, Canaanite, Hittite, and eventually biblical literature.

Sumgait see SUMQAYIT.

summa /'sʌmə, 'sʊmə/ n. (pl. **summae** /-miː/) a summary of what is known of a subject. [ME f. L: see SUM]

summa cum laude /kʊm 'laʊdeɪ/ adv. & adj. esp. N. Amer. (of a degree, diploma, etc.) of the highest standard; with the highest distinction. [L, = with highest praise]

summarize /'sʌmərʌɪz/ v.tr. (also **-ise**) make or be a summary of; sum up. □ **summarist** n. **summarizable** adj. **summarizer** n. **summarization** /ˌsʌmərʌɪˈzeɪʃ(ə)n/ n.

summary /'sʌmərɪ/ n. & adj. ● n. (pl. **-ies**) a brief account; an abridgement. ● adj. **1** dispensing with needless details or formalities; brief (*a summary account*). **2** Law (of a trial etc.) without the customary legal formalities (*summary justice*). □ **summary conviction** a conviction made by a judge or magistrates without a jury. **summary jurisdiction** the authority of a court to use summary proceedings and arrive at a judgement. **summary offence** an offence within the scope of a summary court. □ **summarily** adv. **summariness** n. [ME f. L *summarium* f. L *summa* SUM]

summation /sə'meɪʃ(ə)n/ n. **1** the finding of a total or sum; an addition. **2** a summing-up. □ **summational** adj. **summative** /'sʌmətɪv/ adj.

summer¹ /'sʌmə(r)/ n. & v. ● n. **1** the warmest season of the year, in the northern hemisphere from June to August and in the southern hemisphere from December to February. **2** Astron. the period from the summer solstice to the autumnal equinox. **3** the hot weather typical of summer. **4** (often foll. by *of*) the mature stage of life; the height of achievement, powers, etc. **5** (esp. in *pl.*) poet. a year (esp. of a person's age) (*a child of ten summers*). **6** (attrib.) characteristic of or suitable for summer (*summer clothes*). ● v. **1** intr. (usu. foll. by *at, in*) pass the summer. **2** tr. (often foll. by *at, in*) pasture (cattle). □ **summer-house** a light building in a garden etc. used for sitting in fine weather. **summer lightning** sheet lightning without thunder, resulting from a distant storm. **summer pudding** Brit. a pudding of soft summer fruit encased in bread or sponge. **summer school** a course of lectures etc. held during the summer vacation, esp. at a university. **summer solstice** see SOLSTICE. **summer time** time as advanced one hour ahead of standard time for daylight saving in summer, esp. British Summer Time (cf. *daylight time*, SUMMERTIME). **summer-weight** (of clothes) suitable for use in summer, esp. because of their light weight. □ **summerless** adj. **summerly** adv. **summery** adj. [OE *sumor*]

summer² /'sʌmə(r)/ n. (in full **summer-tree**) a horizontal bearing beam, esp. one supporting joists or rafters. [ME f. AF *sumer, somer* packhorse, beam, OF *somier* f. LL *sagmarius* f. *sagma* f. Gk *sagma* pack-saddle]

Summer Palace a palace (now in ruins) of the Chinese emperors near Beijing.

summersault var. of SOMERSAULT.

summertime /'sʌmə,tʌɪm/ n. the season or period of summer (cf. *summer time*).

summit /'sʌmɪt/ n. **1** the highest point, esp. of a mountain; the apex. **2** the highest degree of power, ambition, etc. **3** (in full **summit meeting, talks**, etc.) a discussion, esp. on disarmament etc., between heads of government. □ **summitless** adj. [ME f. OF *somet, som(m)ete* f. *som* top f. L *summum* neut. of *summus*]

summon /'sʌmən/ v.tr. **1** call upon to appear, esp. as a defendant or witness in a lawcourt. **2** (usu. foll. by *to* + infin.) call upon (*summoned her to assist*). **3** call together for a meeting or some other purpose (*summoned the members to attend*). **4** (often foll. by *up* and *to, for*) gather (courage, spirits, resources, etc.) (*summoned up her strength for the task*). □ **summonable** adj. **summoner** n. [ME f. OF *somondre* f. L *summonere* (as SUB-, *monere* warn)]

summons /'sʌmənz/ n. & v. ● n. (pl. **summonses**) **1** an authoritative or urgent call to attend on some occasion or do something. **2 a** a call to appear before a judge or magistrate. **b** the writ containing such a summons. ● v.tr. esp. Law serve with a summons. [ME f. OF *somonce, sumunse* f. L *summonita* fem. past part. of *summonere*: see SUMMON]

summum bonum /ˌsʊməm 'bɒnəm, 'bəʊn-/ n. the highest good, esp. as the end or determining principle in an ethical system. [L]

sumo /'suːməʊ/ n. (pl. **-os**) a style of Japanese heavyweight wrestling; also, a sumo wrestler. Sumo is the national spectator sport of Japan, attracting thousands to watch the championships, which are held six times a year. The huge wrestlers, who follow a special diet to put on weight, try to keep their centre of gravity as low as possible in the ring. A wrestler is defeated by touching the ground with any part of the body other than the soles of the feet or by moving outside a marked area. [Jap.]

sump /sʌmp/ n. **1** a pit, well, hole, etc., in which superfluous liquid collects in mines, machines, etc. **2** a cesspool. [ME, = marsh f. MDu., MLG *sump*, or (mining) G *Sumpf*, rel. to SWAMP]

sumpter /'sʌmptə(r)/ n. archaic **1** a packhorse. **2** any pack-animal (*sumpter-mule*). [ME f. OF *som(m)etier* f. LL f. Gk *sagma -atos* pack-saddle: cf. SUMMER²]

sumptuary /'sʌmptjʊərɪ/ adj. **1** regulating expenditure. **2** (of a law or edict etc.) limiting private expenditure in the interests of the state. [L *sumptuarius* f. *sumptus* cost f. *sumere sumpt-* take]

sumptuous /'sʌmptjʊəs/ adj. rich, lavish, costly (*a sumptuous setting*). □ **sumptuously** adv. **sumptuousness** n. **sumptuosity** /ˌsʌmptʊ'ɒsɪtɪ/ n. [ME f. OF *somptueux* f. L *sumptuosus* (as SUMPTUARY)]

Sumqayit /ˌsʊmkɑː'iːt/ (Russian **Sumgait** /ˌsumɡaˈit/) an industrial city in eastern Azerbaijan, on the Caspian Sea; pop. (1990) 234,600.

Sumy /'suːmɪ/ an industrial city in NE Ukraine, near the border with Russia; pop. (1990) 296,000.

Sun. abbr. Sunday.

sun /sʌn/ n. & v. ● n. **1 a** (also **Sun**) the star round which the earth orbits and from which it receives light and warmth. (*See note below.*) **b** any other star in the universe. **2** the light or warmth received from the sun (*pull down the blinds and keep out the sun*). **3** poet. a day or a year. **4** poet. a person or thing regarded as a source of glory, radiance, etc. ● v. (**sunned, sunning**) **1** refl. bask in the sun. **2** tr. expose to the sun. **3** intr. sun oneself. □ **against the sun** anticlockwise. **beneath** (or **under**) **the sun** anywhere in the world. **in the sun** exposed to the sun's rays. **on which the sun never sets** (of an empire etc.) worldwide. **sun and planet** a system of gearing cog wheels. **sun-baked** dried or hardened or baked from the heat of the sun. **sun-bath** a period of exposing the body to the sun. **sun bear** a small black bear, *Helarctos malayanus*, of SE Asia, with a light-coloured mark on its chest. **sun-blind** Brit. a window awning. **sun-bonnet** a bonnet of cotton etc. covering the neck and shading the face, esp. for children. **sun-bow** a spectrum of colours like a rainbow produced by the sun shining on spray etc. **sun cream** cream for protecting the skin from sunburn and for promoting suntanning. **sun-dance** a dance of North American Indians in honour of the sun. **sun-deck** the upper deck of a steamer. **sun disc** a winged disc, emblematic of the sun-god. **sun-dog** = PARHELION. **sun-dress** a dress without sleeves and with a low neck. **sun-dried** dried by the sun, not by artificial heat. **sunglasses** glasses tinted to protect the eyes from sunlight or glare. **sun-god** the sun worshipped as a deity. **sun-hat** a hat designed to protect the head from the sun. **sun-helmet** a helmet of cork etc. formerly worn by white people in the tropics. **sun in splendour** Heraldry the sun with rays and a human face. **one's sun is set** the time of one's prosperity is over. **sun-kissed** warmed or affected by the sun. **sun-lamp 1** a lamp giving ultraviolet rays for an artificial suntan, therapy, etc.

2 *Cinematog.* a large lamp with a parabolic reflector used in film-making. **sun lounge** a room with large windows, designed to receive sunlight. **sun parlor** *US* = **sun lounge**. **sun-rays 1** sunbeams. **2** ultraviolet rays used therapeutically. **sun-roof** a sliding roof on a car. **sun-stone** a cat's eye gem, esp. feldspar with embedded flecks of haematite etc. **sun-suit** a play-suit, esp. for children, suitable for sunbathing. **sun-up** esp. *N. Amer.* sunrise. **sun visor** a fixed or movable shield at the top of a vehicle windscreen to shield the eyes from the sun. **take** (or **shoot**) **the sun** *Naut.* ascertain the altitude of the sun with a sextant in order to fix the latitude. **with the sun** clockwise. □ **sunless** *adj.* **sunlessness** *n.* **sunlike** *adj.* **sunproof** *adj.* **sunward** *adj. & adv.* **sunwards** *adv.* [OE *sunne, sunna*]

▪ The sun is the central body of the solar system. It provides the light and energy that sustains life on earth and its changing position relative to the earth's axis determines the terrestrial seasons. The sun is a star of a type known as a G2 dwarf, a sphere of hydrogen and helium 1.4 million km in diameter which obtains its energy from nuclear fusion reactions deep within its interior, where the temperature is at about 15 million degrees. The surface is a little under 6,000°C, periodically marked by sunspots where temperatures are 2,000°C cooler and which arise from local intense magnetic fields. Above this region, known as the photosphere, are the chromosphere and corona, regions of much higher temperature. The solar wind, a stream of charged particles from the sun, pervades the solar system to beyond the orbit of Pluto.

sunbathe /ˈsʌnbeɪð/ *v.intr.* bask in the sun, esp. to tan the body. □ **sunbather** *n.*

sunbeam /ˈsʌnbiːm/ *n.* a ray of sunlight.

sunbed /ˈsʌnbed/ *n.* **1** a lightweight, usu. folding chair with a seat long enough to support the legs, used for sunbathing. **2** a bed for lying on under a sun-lamp.

sunbelt /ˈsʌnbelt/ *n.* a strip of territory receiving a high amount of sunshine, esp. **(the Sunbelt)** the region in the southern US stretching from California in the west to Florida in the east.

sunbird /ˈsʌnbɜːd/ *n.* a small bright-plumaged nectar-feeding songbird of the family Nectariniidae, found in tropical and subtropical parts of the Old World, resembling a hummingbird but unable to hover.

sunblock /ˈsʌnblɒk/ *n.* a cream or lotion for protecting the skin from the sun.

sunburn /ˈsʌnbɜːn/ *n. & v.* ● *n.* tanning and inflammation of the skin caused by over-exposure to the sun. ● *v.intr.* **1** suffer from sunburn. **2** (as **sunburnt** or **sunburned** *adj.*) suffering from sunburn; brown or tanned.

sunburst /ˈsʌnbɜːst/ *n.* **1** something resembling the sun and its rays, esp.: **a** an ornament, brooch, etc. **b** a firework. **2** the sun shining suddenly from behind clouds.

sundae /ˈsʌndeɪ, -dɪ/ *n.* a dish of ice-cream with fruit, nuts, syrup, etc. It was apparently originally named in the US in the late 19th century after the day *Sunday*, either because the dish originally included leftover ice-cream unable to be sold on a Sunday and sold cheaply on the Monday, or because it was at first sold only on Sundays, having been devised (according to some accounts) to circumvent Sunday legislation.

Sunda Islands /ˈsʌndə/ a chain of islands in the south-western part of the Malay Archipelago, consisting of two groups: the *Greater Sunda Islands*, which include Sumatra, Java, Borneo, and Sulawesi; and the *Lesser Sunda Islands*, which lie to the east of Java and include Bali, Sumbawa, Flores, Sumba, and Timor.

Sundanese /ˌsʌndəˈniːz/ *n. & adj.* ● *n.* (*pl.* same) **1** a member of a chiefly Muslim people of western Java. **2** the Malayo-Polynesian language of this people. ● *adj.* of or relating to the Sundanese or their language. [Sundanese *Sunda* western Java]

Sundarbans /ˈsʌndəˌbʌnz/ a region of swampland in the Ganges delta, extending from the mouth of the River Hooghly in West Bengal to that of the Tetulia in Bangladesh.

Sunday /ˈsʌndeɪ, -dɪ/ *n. & adv.* ● *n.* **1** the first day of the week, a Christian holiday and day of worship. (*See note below.*) **2** a newspaper published on a Sunday. ● *adv. colloq.* **1** on Sunday. **2** (**Sundays**) on Sundays; each Sunday. □ **Sunday best** *joc.* a person's best clothes, kept for Sunday use. **Sunday letter** = *dominical letter*. **Sunday painter** an amateur painter, esp. one with little training. [OE *sunnandæg*, transl. L *dies solis*, Gk *hēmera hēliou* day of the sun]

▪ Sunday began to replace the Jewish sabbath (Saturday) for Christians in the earliest days of Christianity, chiefly in commemoration of the Resurrection. Its observance as a day of rest and worship began to be

regulated by both ecclesiastical and civil legislation from the 4th century. In the 19th century Sunday was still a day mainly devoted to duties of piety, but in the later 20th century many restrictions on entertainment, shopping, and drinking alcohol which had been in force since medieval times have been lifted, while others remain but arouse controversy.

Sunday school *n.* a school held on Sundays for children, now only for Christian religious instruction but originally also teaching general subjects. Sunday schools derive from a school set up by Robert Raikes (1736–1811) of Gloucester in 1780.

sunder /ˈsʌndə(r)/ *v.tr. & intr. archaic* or *literary* separate, sever. □ **in sunder** apart. [OE *sundrian*, f. *āsundrian* etc.: *in sunder* f. ME f. *o(n)sunder* ASUNDER]

Sunderland /ˈsʌndələnd/ an industrial city in NE England, a port at the mouth of the River Wear; pop. (1991) 286,800.

sundew /ˈsʌndjuː/ *n.* a small bog plant of the family Droseraceae, esp. of the genus *Drosera*, with hairs secreting drops of sticky juice in which small insects are trapped and digested.

sundial /ˈsʌndaɪəl/ *n.* an instrument showing the time by the shadow of a pointer cast by the sun on to a graduated dial. It is probably the most ancient time-measuring instrument. Sundials in which the edge of the pointer or gnomon is parallel to the earth's axis (so that it points to the celestial pole) can show solar time to an accuracy of a minute or two. For centuries sundials were used as a check on the accuracy of the clocks and watches which eventually superseded them, until telegraphic and radio time signals became available.

sundown /ˈsʌndaʊn/ *n.* sunset.

sundowner /ˈsʌnˌdaʊnə(r)/ *n.* **1** *Austral. & NZ colloq.* a tramp, originally one who arrived at a sheep station etc. in the evening (under the pretence of seeking work) so as to obtain food and shelter. **2** *Brit. colloq.* an alcoholic drink taken at sunset.

sundry /ˈsʌndrɪ/ *adj. & n.* ● *adj.* various; several (*sundry items*). ● *n.* (*pl.* **-ies**) **1** (in *pl.*) items or oddments not mentioned individually. **2** *Austral. Cricket* = EXTRA *n.* 5. [OE *syndrig* separate, rel. to SUNDER]

sunfast /ˈsʌnfɑːst/ *adj. US* (of dye) not subject to fading by sunlight.

sunfish /ˈsʌnfɪʃ/ *n.* a fish of rounded form or brilliant appearance; esp. a very large ocean fish, *Mola mola*, with a circular laterally flattened body and very short tail.

sunflower /ˈsʌnˌflaʊə(r)/ *n.* a very tall composite plant of the genus *Helianthus*; esp. *H. annuus*, with very large showy golden-rayed flowers, grown also for its seeds which yield an edible oil.

Sung /sʊŋ/ (also **Song** /sɒŋ/) a dynasty that ruled in China AD 960–1279. It was a period of general prosperity, cultural flowering, and technological advance, marked by the first use of paper money and by advances in printing, firearms, shipbuilding, clock-making, and medicine.

sung *past part.* of SING.

sunk *past* and *past part.* of SINK.

sunken /ˈsʌŋkən/ *adj.* **1** that has been sunk. **2** beneath the surface; submerged. **3** (of the eyes, cheeks, etc.) hollow, depressed. □ **sunken garden** a garden placed below the general level of its surroundings. [past part. of SINK]

Sun King the nickname of Louis XIV of France (see LOUIS[1]).

sunlight /ˈsʌnlaɪt/ *n.* light from the sun.

sunlit /ˈsʌnlɪt/ *adj.* illuminated by sunlight.

sunn /sʌn/ *n.* (in full **sunn hemp**) a hemplike fibre of southern Asia. [Urdu & Hindi *san* f. Skr. *śáṇá* hempen]

Sunna /ˈsʊnə, ˈsʌnə/ the body of traditional customs and practices based on Muhammad's words or acts, accepted (together with the Koran) as authoritative by Muslims and followed particularly by Sunni Muslims. [Arab., = form, way, course, rule]

Sunni /ˈsʊnɪ, ˈsʌnɪ/ *n. & adj.* ● *n.* (*pl.* same or **Sunnis**) **1** the larger of the two main branches of Islam (the other being Shia). (*See note below.*) **2** an adherent of this branch of Islam. ● *adj.* of or relating to Sunni. [Arab., = lawful]

▪ Sunnis constitute the majority community in most Muslim countries other than Iran, where Shiites predominate. Commonly described as orthodox, Sunni Islam differs from Shia in its understanding of the Sunna and in its interpretation of Muhammad's first successor: Sunnis recognize the order of succession of the first four caliphs, whereas the Shiites believe authority begins with Ali, the fourth caliph. There are other differences between the two branches, in particular in community organization and legal practice, but doctrinally Sunni and Shiite Muslims adhere to the same body of tenets.

Sunnite /ˈsʊnaɪt, ˈsʌn-/ n. & adj. ● n. an adherent of the Sunni branch of Islam. ● adj. of or relating to Sunni or Sunnis.

sunny /ˈsʌnɪ/ adj. (**sunnier, sunniest**) **1 a** bright with sunlight. **b** exposed to or warmed by the sun. **2** cheery and bright in temperament. □ **the sunny side 1** the side of a house, street, etc., that gets most sun. **2** the more cheerful aspect of circumstances etc. (always looks on the sunny side). □ **sunnily** adv. **sunniness** n.

sunrise /ˈsʌnraɪz/ n. **1** the sun's rising at dawn. **2** the coloured sky associated with this. **3** the time at which sunrise occurs. □ **sunrise industry** any newly established industry, esp. in electronics and telecommunications, regarded as signalling prosperity.

sunset /ˈsʌnset/ n. **1** the sun's setting in the evening. **2** the coloured sky associated with this. **3** the time at which sunset occurs. **4** the declining period of life.

Sunset Boulevard a road which links the centre of Los Angeles with the Pacific Ocean 48 km (30 miles) to the west. The eastern section of the road between Fairfax Avenue and Beverly Hills is known as Sunset Strip.

sunshade /ˈsʌnʃeɪd/ n. **1** a parasol. **2** an awning.

sunshine /ˈsʌnʃaɪn/ n. **1 a** the light of the sun. **b** an area lit by the sun. **2** fine weather. **3** cheerfulness; joy (brought sunshine into her life). **4** Brit. colloq. a form of address. □ **sunshine roof** = sun-roof. □ **sunshiny** adj.

sunspot /ˈsʌnspɒt/ n. a spot or patch on the sun's surface, appearing dark by contrast with its surroundings. Sunspots are regions of lower surface temperature and are believed to form where loops in the sun's magnetic field intersect the surface; an individual spot may persist for several weeks. The number of sunspots on the solar surface fluctuates according to a regular cycle, with times of maximum sunspot activity recurring every eleven years.

sunstar /ˈsʌnstɑː(r)/ n. a starfish of the genus Solaster, with up to thirteen arms radiating from the body.

sunstroke /ˈsʌnstrəʊk/ n. acute prostration or collapse from the excessive heat of the sun.

suntan /ˈsʌntæn/ n. & v. ● n. the brownish colouring of skin caused by exposure to the sun. ● v.intr. (**-tanned, -tanning**) colour the skin with a suntan.

suntrap /ˈsʌntræp/ n. a place sheltered from the wind and suitable for catching the sunshine.

Sun Yat-sen /ˌsʊn jætˈsen/ (also **Sun Yixian** /ˌsʊn jiːˈʃɑːæn/) (1866–1925), Chinese Kuomintang statesman, provisional President of the Republic of China 1911–12 and President of the Southern Chinese Republic 1923–5. Generally regarded in the West as the father of the modern Chinese state, he spent the period 1895–1911 in exile after an abortive attempt to overthrow the Manchus. During this time he issued an early version of his influential 'Three Principles of the People' (nationalism, democracy, and the people's livelihood) and set up a revolutionary society which became the nucleus of the Kuomintang. He returned to China to play a vital part in the revolution of 1911 in which the Manchu dynasty was overthrown. After being elected provisional President, Sun Yat-sen resigned in 1912 in response to opposition from conservative members of the government and established a secessionist government at Guangzhou. He reorganized the Kuomintang along the lines of the Soviet Communist Party and began a period of uneasy cooperation with the Chinese Communists before dying in office.

Suomi /ˈsuɔmi/ the Finnish name for FINLAND.

sup¹ /sʌp/ v. & n. ● v.tr. (**supped, supping**) **1** take (soup, tea, etc.) by sips or spoonfuls. **2** esp. N. Engl. colloq. drink (alcohol). ● n. a sip of liquid. [OE sūpan]

sup² /sʌp/ v.intr. (**supped, supping**) (usu. foll. by off, on) archaic take supper. [OF super, soper]

sup- /sʌp, səp/ prefix assim. form of SUB- before p.

super /ˈsuːpə(r), ˈsjuː-/ adj. & n. ● adj. **1** (also **super-duper** /-ˈduːpə(r)/) (also as int.) colloq. exceptional; splendid. **2** Commerce of extra quality, superfine. **3** Commerce (of a measure) superficial, in square (not lineal or solid) measure (120 super ft; 120 ft super). ● n. colloq. **1** Theatr. a supernumerary actor. **2** a superintendent. **3** superphosphate. **4** an extra, unwanted, or unimportant person; a supernumerary. **5** Commerce superfine cloth or manufacture. [abbr.]

super- /ˈsuːpə(r), ˈsjuː-/ comb. form forming nouns, adjectives, and verbs,

meaning: **1** above, beyond, or over in place or time or conceptually (superstructure; supernormal; superimpose). **2** to a great or extreme degree (superabundant; supereminent). **3** extra good or large of its kind (supertanker). **4** of a higher kind, esp. in names of classificatory divisions (superclass). [from or after L super- f. super above, beyond]

superable /ˈsuːpərəb(ə)l, ˈsjuː-/ adj. able to be overcome. [L superabilis f. superare overcome]

superabound /ˌsuːpərəˈbaʊnd, ˌsjuː-/ v.intr. be very or too abundant. [LL superabundare (as SUPER-, ABOUND)]

superabundant /ˌsuːpərəˈbʌndənt, ˌsjuː-/ adj. abounding beyond what is normal or right. □ **superabundance** n. **superabundantly** adv. [ME f. LL superabundare: see SUPERABOUND]

superadd /ˌsuːpərˈæd, ˌsjuː-/ v.tr. add over and above. □ **superaddition** /-pərəˈdɪʃ(ə)n/ n. [ME f. L superaddere (as SUPER-, ADD)]

superaltar /ˈsuːpərˌɔːltə(r), ˈsjuː-, -ˌɒltə(r)/ n. Eccl. a portable slab of stone consecrated for use on an unconsecrated altar etc. [ME f. med.L superaltare (as SUPER-, ALTAR)]

superannuate /ˌsuːpərˈænjʊeɪt, ˌsjuː-/ v.tr. **1** retire (a person) with a pension. **2** dismiss or discard as too old for use, work, etc. **3** (as **superannuated** adj.) too old for work or use; obsolete. **4 a** make (a post) pensionable. **b** make pensionable the post of (an employee). □ **superannuable** adj. [back-form. f. superannuated f. med.L superannuatus f. L SUPER- + annus year]

superannuation /ˌsuːpərˌænjʊˈeɪʃ(ə)n, ˌsjuː-/ n. **1** a pension paid to a retired person. **2** a regular payment made towards this by an employed person. **3** the process or an instance of superannuating.

superaqueous /ˌsuːpərˈeɪkwɪəs, ˌsjuː-/ adj. above water.

superb /suːˈpɜːb, sjuː-/ adj. **1** of the most impressive, splendid, grand, or majestic kind (superb courage; a superb specimen). **2** colloq. excellent; fine. □ **superbly** adv. **superbness** n. [F superbe or L superbus proud]

Super Bowl (in American football) a deciding game between the champions of the American Football Conference and the National Football League, played annually in January from 1967 onwards.

supercalender /ˌsuːpəˈkælɪndə(r), ˌsjuː-/ v.tr. give a highly glazed finish to (paper) by extra calendering.

supercargo /ˈsuːpəˌkɑːgəʊ, ˈsjuː-/ n. (pl. **-oes**) an officer in a merchant ship managing sales etc. of cargo. [earlier supracargo f. Sp. sobrecargo f. sobre over + cargo CARGO]

supercede var. of SUPERSEDE.

supercelestial /ˌsuːpəsɪˈlestɪəl, ˌsjuː-/ adj. **1** above the heavens. **2** more than heavenly. [LL supercaelestis (as SUPER-, CELESTIAL)]

supercharge /ˈsuːpətʃɑːdʒ, ˈsjuː-/ v.tr. **1** (usu. foll. by with) charge (the atmosphere etc.) with energy, emotion, etc. **2** use a supercharger on (an internal-combustion engine).

supercharger /ˈsuːpətʃɑːdʒə(r), ˈsjuː-/ n. a device supplying air or fuel to an internal-combustion engine at above normal pressure to increase efficiency.

superciliary /ˌsuːpəˈsɪlɪərɪ, ˌsjuː-/ adj. Anat. of or concerning the eyebrow; over the eye. [L supercilium eyebrow (as SUPER-, cilium eyelid)]

supercilious /ˌsuːpəˈsɪlɪəs, ˌsjuː-/ adj. assuming an air of contemptuous indifference or superiority. □ **superciliously** adv. **superciliousness** n. [L superciliosus (as SUPERCILIARY)]

superclass /ˈsuːpəˌklɑːs, ˈsjuː-/ n. Biol. a taxonomic category between class and phylum.

supercolumnar /ˌsuːpəkəˈlʌmnə(r), ˌsjuː-/ adj. Archit. having one order or set of columns above another. □ **supercolumniation** /-ˌlʌmnɪˈeɪʃ(ə)n/ n.

supercomputer /ˌsuːpəkəmˈpjuːtə(r), ˌsjuː-/ n. a powerful computer capable of dealing with complex problems. □ **supercomputing** n.

superconductivity /ˌsuːpəˌkɒndʌkˈtɪvɪtɪ, ˌsjuː-/ n. Physics the property of having zero electrical resistance, as exhibited in some substances at low absolute temperatures. Discovered in 1911 by H. Kamerlingh Onnes, superconductivity was at first believed to be restricted to some metals at temperatures within a few degrees of absolute zero, and consequently had very limited practical application despite the enormous potential benefits for efficient power transmission and faster electronic circuits. In 1986, however, a class of stable superconductors was discovered (originally at IBM in Zurich) which retained their properties at higher temperatures. These are composite ceramics with sandwich structures, made from mixed oxides of copper, barium, yttrium and other elements; several are now available that display superconductivity above 77 kelvins, the temperature of liquid nitrogen. This makes their use much more practicable, and the first commercial devices appeared in the early 1990s. □ **superconducting** /-kən'dʌktɪŋ/ *adj.* **superconductive** *adj.*

superconductor /ˌsuːpəkən'dʌktə(r), ˌsjuː-/ *n. Physics* a substance having zero electrical resistance. (See also SUPERCONDUCTIVITY.)

superconscious /ˌsuːpə'kɒnʃəs, ˌsjuː-/ *adj.* transcending human consciousness. □ **superconsciously** *adv.* **superconsciousness** *n.*

supercontinent /ˈsuːpəˌkɒntɪnənt, 'sjuː-/ *n. Geol.* a very large land mass, such as Pangaea, thought to have divided to form the present continents in the geological past.

supercool /ˈsuːpəˌkuːl, 'sjuː-/ *v. & adj.* ● *v. Chem.* **1** *tr.* cool (a liquid) below its freezing-point without solidification or crystallization. **2** *intr.* (of a liquid) be cooled in this way. ● *adj. sl.* very cool, relaxed, fine, etc.

supercritical /ˌsuːpə'krɪtɪk(ə)l, ˌsjuː-/ *adj. Physics* of more than critical mass etc.

super-duper *var.* of SUPER *adj.* 1.

superego /ˌsuːpər'iːɡəʊ, ˌsjuː-/ *n.* (*pl.* **-os**) *Psychol.* the part of the mind that acts as a conscience and responds to social rules.

superelevation /ˌsuːpərˌelɪ'veɪʃ(ə)n, ˌsjuː-/ *n.* the amount by which the outer edge of a curve on a road or railway is above the inner edge.

supereminent /ˌsuːpər'emɪnənt, ˌsjuː-/ *adj.* supremely eminent, exalted, or remarkable. □ **supereminence** *n.* **supereminently** *adv.* [L *supereminere* rise above (as SUPER-, EMINENT)]

supererogation /ˌsuːpərˌerə'ɡeɪʃ(ə)n, ˌsjuː-/ *n.* the performance of more than duty requires. □ **works of supererogation** *RC Ch.* actions believed to form a reserve fund of merit that can be drawn on by prayer in favour of sinners. □ **supererogatory** /-pərɪ'rɒɡətərɪ/ *adj.* [LL *supererogatio* f. *supererogare* pay in addition (as SUPER-, *erogare* pay out)]

superexcellent /ˌsuːpər'eksələnt, ˌsjuː-/ *adj.* very or supremely excellent. □ **superexcellence** *n.* **superexcellently** *adv.* [LL *superexcellens* (as SUPER-, EXCELLENT)]

superfamily /ˈsuːpəˌfæmɪlɪ, 'sjuː-/ *n.* (*pl.* **-ies**) *Biol.* a taxonomic category between family and order.

superfatted /ˌsuːpə'fætɪd, ˌsjuː-/ *adj.* (of soap) containing extra fat.

superfecundation /ˌsuːpəˌfiːkən'deɪʃ(ə)n, ˌsjuː-, -ˌfekən'deɪʃ(ə)n/ *n.* = SUPERFETATION 1.

superfetation /ˌsuːpəfiː'teɪʃ(ə)n, ˌsjuː-/ *n.* **1** *Med. & Zool.* a second conception during pregnancy giving rise to embryos of different ages in the uterus. **2** *Bot.* the fertilization of the same ovule by different kinds of pollen. **3** the accretion of one thing on another. [F *superfétation* or f. mod.L *superfetatio* f. L *superfetare* (as SUPER-, *fetus* FOETUS)]

superficial /ˌsuːpə'fɪʃ(ə)l, ˌsjuː-/ *adj.* **1** of or on the surface; lacking depth (*a superficial knowledge; superficial wounds*). **2** swift or cursory (*a superficial examination*). **3** apparent but not real (*a superficial resemblance*). **4** (esp. of a person) having no depth of character or knowledge; trivial; shallow. **5** *Commerce* (of a measure) square (cf. SUPER *adj.* 3). □ **superficially** *adv.* **superficialness** *n.* **superficiality** /-ˌfɪʃɪ'ælɪtɪ/ *n.* (*pl.* **-ies**). [LL *superficialis* f. L (as SUPERFICIES)]

superficies /ˌsuːpə'fɪʃɪˌiːz, ˌsjuː-/ *n.* (*pl.* same) *Geom.* a surface. [L (as SUPER-, *facies* face)]

superfine /ˈsuːpəˌfaɪn, 'sjuː-/ *adj.* **1** *Commerce* of extra quality. **2** pretending great refinement. [med.L *superfinus* (as SUPER-, FINE¹)]

superfluidity /ˌsuːpəflu'ɪdɪtɪ/ *n. Physics* the property of flowing without friction or viscosity, as in liquid helium below about 2.18 kelvins. □ **superfluid** /'suːpə'fluːɪd, 'sjuː-/ *n.*

superfluity /ˌsuːpə'fluːɪtɪ, ˌsjuː-/ *n.* (*pl.* **-ies**) **1** the state of being superfluous. **2** a superfluous amount or thing. [ME f. OF *superfluité* f. LL *superfluitas -tatis* f. L *superfluus*: see SUPERFLUOUS]

superfluous /suː'pɜːfluəs, sjuː-/ *adj.* more than enough, redundant, needless. □ **superfluously** *adv.* **superfluousness** *n.* [ME f. L *superfluus* (as SUPER-, *fluere* to flow)]

supergiant /ˈsuːpəˌdʒaɪənt, 'sjuː-/ *n. Astron.* a star of very great luminosity and size.

superglue /ˈsuːpəˌɡluː, 'sjuː-/ *n.* any of various adhesives with an exceptional bonding capability.

supergrass /ˈsuːpəˌɡrɑːs, 'sjuː-/ *n. colloq.* a police informer who implicates a large number of people.

superheat /ˈsuːpə'hiːt, ˌsjuː-/ *v.tr. Physics* **1** heat (a liquid) above its boiling-point without vaporization. **2** heat (a vapour) above its boiling-point (*superheated steam*). □ **superheater** /'suːpə'hiːtə(r), 'sjuː-/ *n.*

superhero /ˈsuːpəˌhɪərəʊ, 'sjuː-/ *n.* (*pl.* **-oes**) a person or fictional character with extraordinary heroic attributes.

superhet /ˌsuːpə'het, ˌsjuː-/ *n. colloq.* = SUPERHETERODYNE.

superheterodyne /ˌsuːpə'hetərəʊˌdaɪn, ˌsjuː-/ *adj. & n.* ● *adj.* denoting or characteristic of a system of radio reception in which a local variable oscillator is tuned to beat at a constant ultrasonic frequency with carrier-wave frequencies, making it unnecessary to vary the amplifier tuning and securing greater selectivity. ● *n.* a superheterodyne receiver. [SUPERSONIC + HETERODYNE]

superhighway /ˈsuːpəˌhaɪweɪ, 'sjuː-/ *n.* **1** *N. Amer.* a broad main road for fast traffic. **2** (in full **information superhighway**) a means of rapid transfer of information in different digital forms (e.g. video, sound, and graphics) via an extensive electronic network.

superhuman /ˌsuːpə'hjuːmən, ˌsjuː-/ *adj.* **1** beyond normal human capability. **2** higher than humans. □ **superhumanly** *adv.* [LL *superhumanus* (as SUPER-, HUMAN)]

superhumeral /ˌsuːpə'hjuːmərəl, ˌsjuː-/ *n. Eccl.* a vestment worn over the shoulders, e.g. an amice, ephod, or pallium. [LL *superhumerale* (as SUPER-, HUMERAL)]

superimpose /ˌsuːpərɪm'pəʊz, ˌsjuː-/ *v.tr.* (usu. foll. by *on*) lay (a thing) on something else. □ **superimposition** /-ˌɪmpə'zɪʃ(ə)n/ *n.*

superincumbent /ˌsuːpərɪn'kʌmb(ə)nt, ˌsjuː-/ *adj.* lying on something else.

superinduce /ˌsuːpərɪn'djuːs, ˌsjuː-/ *v.tr.* introduce or induce in addition. [L *superinducere* cover over, bring from outside (as SUPER-, INDUCE)]

superintend /ˌsuːpərɪn'tend, ˌsjuː-/ *v.tr. & intr.* be responsible for the management or arrangement of (an activity etc.); supervise and inspect. □ **superintendence** *n.* **superintendency** *n.* [eccl.L *superintendere* (as SUPER-, INTEND), transl. Gk *episkopō*]

superintendent /ˌsuːpərɪn'tendənt, ˌsjuː-/ *n. & adj.* ● *n.* **1 a** a person who superintends. **b** a director of an institution etc. **2 a** (in the UK) a police officer above the rank of inspector. **b** (in the US) the head of a police department. **3** *US* the caretaker of a building. ● *adj.* superintending. [eccl.L *superintendent-* part. stem of *superintendere*: see SUPERINTEND]

superior /suː'pɪərɪə(r), sjuː-/ *adj. & n.* ● *adj.* **1** in a higher position; of higher rank (*a superior officer; a superior court*). **2** above the average in quality etc. (*made of superior leather*). **b** having or showing a high opinion of oneself; supercilious (*had a superior air*). **3** (often foll. by *to*) **a** better or greater in some respect (*superior to its rivals in speed*). **b** above yielding, making concessions, paying attention, etc. (*is superior to bribery; superior to temptation*). **4** further above or out; higher, esp.: **a** *Astron.* (of a planet) having an orbit further from the sun than the earth's, i.e. Mars and the outer planets. **b** *Zool.* (of an insect's wings) folding over others. **c** *Printing* (of figures or letters) placed above the line. **d** *Bot.* (of the calyx) above the ovary. **e** *Bot.* (of the ovary) above the calyx. ● *n.* **1** a person superior to another in rank, character, etc. (*is deferential to his superiors; is his superior in courage*). **2** *Eccl.* the head of a monastery or other religious institution (*Mother Superior; Father Superior*). **3** *Printing* a superior letter or figure. □ **superior numbers** esp. *Mil.* more people, troops, etc. or their strength (*overcome by superior numbers*). □ **superiorly** *adv.* [ME f. OF *superiour* f. L *superior -oris*, compar. of *superus* that is above f. *super* above]

Superior, Lake /suː'pɪərɪə(r), sjuː-/ the largest of the five Great Lakes of North America, on the border between Canada and the US. With an area of 82,350 sq km (31,800 sq miles), it is the largest freshwater lake in the world.

superiority /suːˌpɪərɪ'ɒrɪtɪ, sjuː-/ *n.* the state of being superior. □ **superiority complex** *Psychol.* an undue conviction of one's own superiority to others.

superjacent /ˌsuːpə'dʒeɪs(ə)nt, ˌsjuː-/ *adj.* overlying; superincumbent. [L *superjacere* (as SUPER-, *jacere* lie)]

superlative /suː'pɜːlətɪv, sjuː-/ *adj. & n.* ● *adj.* **1** of the highest quality or degree (*superlative wisdom*). **2** *Gram.* (of an adjective or adverb) expressing the highest or a very high degree of a quality (e.g. *bravest, most fiercely*) (cf. POSITIVE *adj.* 3b, COMPARATIVE *adj.* 4). ● *n.* **1** *Gram.* **a** the superlative expression or form of an adjective or adverb. **b** a word in the superlative. **2** something embodying excellence; the highest form of a thing. **3** (usu. in *pl.*) a hyperbolical or excessively laudatory expression. □ **superlatively** *adv.* **superlativeness** *n.* [ME f. OF *superlatif -ive* f. LL *superlativus* f. L *superlatus* (as SUPER-, *latus* past part. of *ferre* take)]

superluminal /ˌsuːpə'luːmɪn(ə)l/ *adj. Physics* of or having a speed greater than that of light. [SUPER- + L *lumen luminis* a light]

superlunary /ˌsuːpə'luːnərɪ, ˌsjuː-, -'ljuː'nərɪ/ *adj.* **1** situated beyond the moon. **2** belonging to a higher world, celestial. [med.L *superlunaris* (as SUPER-, LUNAR)]

Superman /ˈsuːpəˌmæn, ˈsjuː-/ a US cartoon character having great strength, the ability to fly, and other extraordinary powers. Born on a planet called Krypton, Superman conceals his true nature behind the identity of mild-mannered reporter Clark Kent. Since his first appearance in 1938 in a comic strip by writer Jerry Siegel (1914–96) and artist Joe Shuster (1914–92), Superman has featured in many television and film productions.

superman /ˈsuːpəˌmæn, ˈsjuː-/ n. (pl. **-men**) **1** Philos. = ÜBERMENSCH. **2** colloq. a man of exceptional strength or ability. [SUPER- + MAN, formed by G. B. Shaw after ÜBERMENSCH]

supermarket /ˈsuːpəˌmɑːkɪt, ˈsjuː-/ n. a large self-service store selling foods, household goods, etc.

supermodel /ˈsuːpəˌmɒd(ə)l, ˈsjuː-/ n. a highly successful and internationally famous fashion model.

supermundane /ˌsuːpəˈmʌndeɪn, ˌsjuː-/ adj. superior to earthly things.

supernal /suːˈpɜːn(ə)l, sjuː-/ adj. esp. poet. **1** heavenly; divine. **2** of or concerning the sky. **3** lofty. □ **supernally** adv. [ME f. OF supernal or med.L supernalis f. L supernus f. super above]

supernatant /ˌsuːpəˈneɪt(ə)nt, ˌsjuː-/ adj. & n. esp. Chem. ● adj. (of a liquid) lying above a solid residue after precipitation, centrifugation, etc. ● n. a supernatant liquid. [SUPER- + natant swimming (as NATATION)]

supernatural /ˌsuːpəˈnætʃrəl, ˌsjuː-/ adj. & n. ● adj. attributed to or thought to reveal some force above the laws of nature; magical, occult, mystical. ● n. (prec. by the) supernatural, occult, or magical forces, effects, etc. □ **supernaturalism** n. **supernaturalist** n. **supernaturalize** v.tr. (also **-ise**). **supernaturally** adv. **supernaturalness** n.

supernormal /ˌsuːpəˈnɔːm(ə)l, ˌsjuː-/ adj. beyond what is normal or natural. □ **supernormality** /-nɔːˈmælɪtɪ/ n.

supernova /ˌsuːpəˈnəʊvə, ˌsjuː-/ n. (pl. **-novae** /-viː/ or **-novas**) Astron. a star that suddenly increases very greatly in brightness because of an explosion ejecting most of its mass. Supernovae are believed to be supergiant stars which have exhausted the hydrogen and helium fuel in their interiors. Gravitational contraction then leads to runaway thermonuclear reactions involving heavier elements, catastrophically disrupting the star and ejecting debris at speeds of up to a tenth of the speed of light and temperatures of hundreds of thousands of degrees. Within the resulting shell of material may be left a pulsar or a black hole. Though frequently observed in other galaxies, only three supernovae have been recorded in our own Galaxy: by Chinese astronomers in 1054 (these formed the Crab Nebula), by Tycho Brahe in 1572, and by Kepler in 1604. (Cf. NOVA.)

supernumerary /ˌsuːpəˈnjuːmərərɪ, ˌsjuː-/ adj. & n. ● adj. **1** in excess of the normal number; extra. **2** (of a person) engaged for extra work. **3** (of an actor) appearing on stage but not speaking. ● n. (pl. **-ies**) **1** an extra or unwanted person or thing. **2** a supernumerary actor. **3** a person engaged for extra work. [LL supernumerarius (soldier) added to a legion already complete, f. L super numerum beyond the number]

superorder /ˈsuːpərˌɔːdə(r), ˈsjuː-/ n. Biol. a taxonomic category between order and class. □ **superordinal** /ˌsuːpərˈɔːdɪn(ə)l, ˌsjuː-/ adj.

superordinate /ˌsuːpərˈɔːdɪnət, ˌsjuː-/ adj. & n. ● adj. (usu. foll. by to) of superior importance or rank. ● n. **1** a superordinate person or thing, a superior. **2** Linguistics a linguistic unit operating at a higher level than another, which is subordinate to it; esp. in semantics, a hypernym. [SUPER-, after subordinate]

superphosphate /ˌsuːpəˈfɒsfeɪt, ˌsjuː-/ n. a fertilizer made by treating phosphate rock with sulphuric or phosphoric acid.

superphysical /ˌsuːpəˈfɪzɪk(ə)l, ˌsjuː-/ adj. **1** unexplainable by physical causes; supernatural. **2** beyond what is physical.

superpose /ˌsuːpəˈpəʊz, ˌsjuː-/ v.tr. (usu. foll. by on) esp. Geom. place (a thing or a geometric figure) on or above something else, esp. so as to coincide. □ **superposition** /-pəpəˈzɪʃ(ə)n/ n. [F superposer (as SUPER-, POSE[1])]

superpower /ˈsuːpəˌpaʊə(r), ˈsjuː-/ n. a nation or state having a dominant position in world politics, one with the power to act decisively in pursuit of interests affecting the whole world, esp. the US (and formerly the USSR) since the Second World War.

supersaturate /ˌsuːpəˈsætʃəˌreɪt, ˌsjuː-, -ˈsætjʊˌreɪt/ v.tr. add to (esp. a solution) beyond saturation point. □ **supersaturation** /-ˌsætʃəˈreɪʃ(ə)n, -ˌsætjʊˈreɪʃ(ə)n/ n.

superscribe /ˈsuːpəˌskraɪb, ˈsjuː-/ v.tr. **1** write (an inscription) at the top of or on the outside of a document etc. **2** write an inscription over or on (a thing). □ **superscription** /ˌsuːpəˈskrɪpʃ(ə)n, ˌsjuː-/ n. [L superscribere (as SUPER-, scribere script- write)]

superscript /ˈsuːpəˌskrɪpt, ˈsjuː-/ adj. & n. ● adj. written or printed above the line, esp. Math. (of a symbol) written above and to the right of another. ● n. a superscript number or symbol. [L superscriptus past part. of superscribere: see SUPERSCRIBE]

supersede /ˌsuːpəˈsiːd, ˌsjuː-/ v.tr. (also **supercede**) **1 a** adopt or appoint another person or thing in place of. **b** set aside; cease to employ. **2** (of a person or thing) take the place of. □ **supersedence** n. **supersedure** /-ˈsiːdʒə(r)/ n. **supersession** /-ˈseʃ(ə)n/ n. [OF superseder f. L supersedere be superior to (as SUPER-, sedere sess- sit)]

supersonic /ˌsuːpəˈsɒnɪk, ˌsjuː-/ adj. designating or having a speed greater than that of sound. □ **supersonically** adv.

supersonics /ˌsuːpəˈsɒnɪks, ˌsjuː-/ n.pl. (treated as sing.) = ULTRASONICS.

superstar /ˈsuːpəˌstɑː(r), ˈsjuː-/ n. an extremely famous or renowned actor, film star, musician, etc. □ **superstardom** n.

superstate /ˈsuːpəˌsteɪt, ˈsjuː-/ n. a powerful political state, esp. one formed from a federation of nations.

superstition /ˌsuːpəˈstɪʃ(ə)n, ˌsjuː-/ n. **1** credulity regarding the supernatural. **2** an irrational fear of the unknown or mysterious. **3** misdirected reverence. **4** a practice, opinion, or religion based on these tendencies. **5** a widely held but unjustified idea of the effects or nature of a thing. □ **superstitious** adj. **superstitiously** adv. **superstitiousness** n. [ME f. OF superstition or L superstitio (as SUPER-, stare stat- stand)]

superstore /ˈsuːpəˌstɔː(r), ˈsjuː-/ n. a large supermarket selling a wide range of goods and usu. situated outside a town's main shopping area.

superstratum /ˈsuːpəˌstrɑːtəm, ˈsjuː-, -ˌstreɪtəm/ n. (pl. **-strata** /-tə/) an overlying stratum.

superstring /ˈsuːpəˌstrɪŋ, ˈsjuː-/ n. Physics a subatomic particle in a version of string theory (STRING n. 11a) that incorporates supersymmetry.

superstructure /ˈsuːpəˌstrʌktʃə(r), ˈsjuː-/ n. **1** the part of a building above its foundations. **2** a structure built on top of something else. **3 a** a concept or idea based on others. **b** (in Marxism) the institutions and culture resulting from the economic system (or base) underlying a society. □ **superstructural** adj.

supersubtle /ˌsuːpəˈsʌt(ə)l, ˌsjuː-/ adj. extremely or excessively subtle. □ **supersubtlety** n.

supersymmetry /ˌsuːpəˈsɪmɪtrɪ, ˌsjuː-/ n. Physics a form of mathematical symmetry which relates bosons and fermions. □ **supersymmetric** /-sɪˈmetrɪk/ adj.

supertanker /ˈsuːpəˌtæŋkə(r), ˈsjuː-/ n. a very large tanker ship.

supertax /ˈsuːpəˌtæks, ˈsjuː-/ n. a higher rate of tax on incomes above a certain level, esp. hist. that levied in Britain between 1909 and 1929 (succeeded by surtax).

superterrestrial /ˌsuːpətəˈrestrɪəl, ˌsjuː-/ adj. **1** in or belonging to a region above the earth. **2** celestial.

supertonic /ˌsuːpəˈtɒnɪk, ˌsjuː-/ n. Mus. the note above the tonic, the second note of the diatonic scale of any key.

supervene /ˌsuːpəˈviːn, ˌsjuː-/ v.intr. occur as an interruption in or a change from some state. □ **supervenient** adj. **supervention** /-ˈvenʃ(ə)n/ n. [L supervenire supervent- (as SUPER-, venire come)]

supervise /ˈsuːpəˌvaɪz, ˈsjuː-/ v.tr. **1** superintend, oversee the execution of (a task etc.). **2** oversee the actions or work of (a person). □ **supervisor** n. **supervisory** adj. **supervision** /ˌsuːpəˈvɪʒ(ə)n, ˌsjuː-/ n. [med.L supervidere supervis- (as SUPER-, videre see)]

superwoman /ˈsuːpəˌwʊmən, ˈsjuː-/ n. (pl. **-women**) colloq. a woman of exceptional strength or ability.

supinate /ˈsuːpɪˌneɪt, ˈsjuː-/ v.tr. put (a hand or foreleg etc.) into a supine position (cf. PRONATE). □ **supination** /ˌsuːpɪˈneɪʃ(ə)n, ˌsjuː-/ n. [back-form. f. supination f. L supinatio f. supinare f. supinus: see SUPINE]

supinator /ˈsuːpɪˌneɪtə(r), ˈsjuː-/ n. Anat. a muscle in the forearm effecting supination.

supine /ˈsuːpaɪn, ˈsjuː-/ adj. & n. ● adj. **1** lying face upwards (cf. PRONE 1a). **2** having the front or ventral part upwards; (of the hand) with the palm upwards. **3** inert, indolent; morally or mentally inactive. ● n. a Latin verbal noun used only in the accusative and ablative cases, esp. to denote purpose (e.g. mirabile dictu wonderful to relate). □ **supinely**

adv. **supineness** *n.* [L *supinus*, rel. to *super*: (n.) f. LL *supinum* neut. (reason unkn.)]

supper /ˈsʌpə(r)/ *n.* an evening meal, esp. a light or informal one. □ **sing for one's supper** do something in return for a benefit. □ **supperless** *adj.* [ME f. OF *soper, super*]

supplant /səˈplɑːnt/ *v.tr.* dispossess and take the place of, esp. by underhand means. □ **supplanter** *n.* [ME f. OF *supplanter* or L *supplantare* trip up (as SUB-, *planta* sole)]

supple /ˈsʌp(ə)l/ *adj. & v.* ● *adj.* (**suppler, supplest**) **1** flexible, pliant; easily bent. **2** compliant; avoiding overt resistance; artfully or servilely submissive. ● *v.tr. & intr.* make or become supple. □ **suppleness** *n.* [ME f. OF *souple* ult. f. L *supplex supplicis* submissive]

supplejack /ˈsʌpl̩ˌdʒæk/ *n.* a strong twining tropical shrub; e.g. *Berchemia scandens*, of the southern US. [SUPPLE + JACK[1]]

supplely var. of SUPPLY[2].

supplement *n. & v.* ● *n.* /ˈsʌplɪmənt/ **1** a thing or part added to remedy deficiencies (*dietary supplement*). **2** a part added to a book etc. to provide further information. **3** a separate section, esp. a colour magazine, added to a newspaper or periodical. **4** an additional charge payable for an extra service or facility. **5** *Geom.* the amount by which an angle is less than 180° (cf. COMPLEMENT *n.* 6). ● *v.tr.* (also /ˈsʌplɪˌment/) provide a supplement for. □ **supplemental** /ˌsʌplɪˈment(ə)l/ *adj.* **supplementally** *adv.* **supplementation** /ˌsʌplɪmenˈteɪʃ(ə)n/ *n.* [ME f. L *supplementum* (as SUB-, *plere* fill)]

supplementary /ˌsʌplɪˈmentərɪ/ *adj.* forming or serving as a supplement; additional. □ **supplementary benefit** *hist.* (in the UK) a weekly allowance paid by the state to those not in full-time employment and with an income below a certain level (replaced by *income support*). □ **supplementarily** *adv.*

suppletion /səˈpliːʃ(ə)n/ *n.* the act or an instance of supplementing, esp. *Linguistics* the occurrence of unrelated forms to supply gaps in conjugation (e.g. *went* as the past of *go*). □ **suppletive** /-ˈpliːtɪv/ *adj.* [ME f. OF f. med.L *suppletio -onis* (as SUPPLY[1])]

suppliant /ˈsʌplɪənt/ *adj. & n.* ● *adj.* **1** supplicating. **2** expressing supplication. ● *n.* a supplicating person. □ **suppliantly** *adv.* [ME f. F *supplier* beseech f. L (as SUPPLICATE)]

supplicate /ˈsʌplɪˌkeɪt/ *v. formal* **1** *tr.* petition humbly to (a person) or for (a thing). **2** *intr.* (foll. by *to, for*) make a petition. □ **supplicant** *adj. & n.* **supplicatory** *adj.* **supplication** /ˌsʌplɪˈkeɪʃ(ə)n/ *n.* [ME f. L *supplicare* (as SUB-, *plicare* bend)]

supply[1] /səˈplaɪ/ *v. & n.* ● *v.tr.* (**-ies, -ied**) **1** provide or furnish (a thing needed). **2** (often foll. by *with*) provide (a person etc. with a thing needed). **3** meet or make up for (a deficiency or need etc.). **4** fill (a vacancy, place, etc.) as a substitute. ● *n.* (*pl.* **-ies**) **1** the act or an instance of providing what is needed. **2** a stock, store, amount, etc., of something provided or obtainable (*a large supply of water; the gas-supply*). **3** (in *pl.*) **a** the collected provisions and equipment for an army, expedition, etc. **b** a grant of money by Parliament for the costs of government. **c** a money allowance to a person. **4** (often *attrib.*) a person, esp. a schoolteacher or clergyman, acting as a temporary substitute for another. **5** (*attrib.*) providing supplies or a supply (*supply officer*). □ **in short supply** available in limited quantity. **on supply** (of a schoolteacher etc.) acting as a supply. **supply and demand** *Econ.* quantities available and required as factors regulating the price of commodities. **supply-side** *Econ.* denoting a policy of low taxation and other incentives to produce goods and invest. □ **supplier** *n.* [ME f. OF *so(u)pleer* etc. f. L *supplere* (as SUB-, *plere* fill)]

supply[2] /ˈsʌplɪ/ *adv.* (also **supplely** /ˈsʌpəlɪ/) in a supple manner.

support /səˈpɔːt/ *v. & n.* ● *v.tr.* **1** carry all or part of the weight of. **2** keep from falling or sinking or failing. **3** provide with a home and the necessities of life (*has a family to support*). **4** enable to last out; give strength to; encourage. **5** bear out; tend to substantiate or corroborate (a statement, charge, theory, etc.). **6** give help or countenance to, back up; second, further. **7** speak in favour of (a resolution etc.). **8** be actively interested in (a particular team or sport). **9** (often as **supporting** *adj.*) take a part that is secondary to (a principal actor etc.). **10** assist (a lecturer etc.) by one's presence. **11** endure, tolerate (*can no longer support the noise*). **12** maintain or represent (a part or character) adequately. **13** subscribe to the funds of (an institution). ● *n.* **1** the act or an instance of supporting; the process of being supported. **2** a person or thing that supports. **3** a group or performer providing a secondary act at a pop concert etc. □ **in support of** in order to support. **supporting film** (or **picture** etc.) a less important film in a cinema programme. **support price** a minimum price guaranteed to a farmer for agricultural produce and maintained by subsidy etc.

□ **supportingly** *adv.* **supportless** *adj.* **supportable** *adj.* **supportably** *adv.* **supportability** /-ˌpɔːtəˈbɪlɪtɪ/ *n.* [ME f. OF *supporter* f. L *supportare* (as SUB-, *portare* carry)]

supporter /səˈpɔːtə(r)/ *n.* **1** a person or thing that supports, esp. a person supporting a team or sport. **2** *Heraldry* the representation of an animal etc., usu. one of a pair, holding up or standing beside an escutcheon.

supportive /səˈpɔːtɪv/ *adj.* providing support or encouragement. □ **supportively** *adv.* **supportiveness** *n.*

suppose /səˈpəʊz/ *v.tr.* (often foll. by *that* + clause) **1** assume, esp. in default of knowledge; be inclined to think (*I suppose they will return; what do you suppose he meant?*). **2** take as a possibility or hypothesis (*let us suppose you are right*). **3** (in *imper.*) as a formula of proposal (*suppose we go to the party*). **4** (of a theory or result etc.) require as a condition (*design in creation supposes a creator*). **5** (in *imper.* or *pres. part.* forming a question) in the circumstances that; if (*suppose he won't let you; supposing we stay*). **6** (as **supposed** *adj.*) generally accepted as being so; believed (*his supposed brother; generally supposed to be wealthy*). **7** (in *passive*; foll. by *to* + infin.) **a** be expected or required (*was supposed to write to you*). **b** (with *neg.*) ought not; not be allowed to (*you are not supposed to go in there*). □ **I suppose so** an expression of hesitant agreement. □ **supposable** *adj.* [ME f. OF *supposer* (as SUB-, POSE[1])]

supposedly /səˈpəʊzɪdlɪ/ *adv.* as is generally supposed.

supposition /ˌsʌpəˈzɪʃ(ə)n/ *n.* **1** a fact or idea etc. supposed. **2** the act or an instance of supposing. □ **suppositional** *adj.*

supposititious /ˌsʌpəˈzɪʃəs/ *adj.* hypothetical, assumed. □ **suppositiously** *adv.* **suppositiousness** *n.* [partly f. SUPPOSITITIOUS, partly f. SUPPOSITION + -OUS]

supposititious /səˌpɒzɪˈtɪʃəs/ *adj.* spurious; substituted for the real. □ **supposititiously** *adv.* **supposititiousness** *n.* [L *suppositicius, -icius* f. *supponere supposit-* substitute (as SUB- *ponere* place)]

suppository /səˈpɒzɪtərɪ/ *n.* (*pl.* **-ies**) a medical preparation in the form of a cone, cylinder, etc., to be inserted into the rectum or vagina to dissolve. [ME f. med.L *suppositorium*, neut. of LL *suppositorius* placed underneath (as SUPPOSITITIOUS)]

suppress /səˈpres/ *v.tr.* **1** end the activity or existence of, esp. forcibly. **2** prevent (information, feelings, a reaction, etc.) from being seen, heard, or known (*tried to suppress the report; suppressed a yawn*). **3 a** partly or wholly eliminate (electrical interference etc.). **b** equip (a device) to reduce such interference due to it. **4** *Psychol.* keep out of one's consciousness. □ **suppressible** *adj.* **suppressive** *adj.* **suppressor** *n.* **suppression** /-ˈpreʃ(ə)n/ *n.* [ME f. L *supprimere suppress-* (as SUB-, *premere* press)]

suppressant /səˈpres(ə)nt/ *n.* a suppressing or restraining agent, esp. a drug that suppresses the appetite.

suppurate /ˈsʌpjʊˌreɪt/ *v.intr.* **1** form pus. **2** fester. □ **suppurative** /-rətɪv/ *adj.* **suppuration** /ˌsʌpjʊˈreɪʃ(ə)n/ *n.* [L *suppurare* (as SUB-, *purare* as PUS)]

supra /ˈsuːprə, ˈsjuː-/ *adv.* above or earlier on (in a book etc.). [L, = above]

supra- /ˈsuːprə, ˈsjuː-/ *prefix* **1** above (opp. INFRA- 1). **2** beyond, transcending (*supranational*). [from or after L *supra-* f. *supra* above, beyond, before in time]

supramaxillary /ˌsuːprəmækˈsɪlərɪ, ˌsjuː-/ *adj.* of or relating to the upper jaw.

supramundane /ˌsuːprəˈmʌndeɪn, ˌsjuː-/ *adj.* above or superior to the world.

supranational /ˌsuːprəˈnæʃ(ə)n(ə)l, ˌsjuː-/ *adj.* transcending national limits. □ **supranationalism** *n.* **supranationality** /-ˌnæʃəˈnælɪtɪ/ *n.*

supraorbital /ˌsuːprəˈɔːbɪt(ə)l, ˌsjuː-/ *adj.* situated above the orbit of the eye.

suprarenal /ˌsuːprəˈriːn(ə)l, ˌsjuː-/ *adj.* situated above the kidneys.

supremacist /suːˈpreməsɪst, sjuː-/ *n. & adj.* ● *n.* an advocate of the supremacy of a particular group, esp. determined by race or sex. ● *adj.* relating to or advocating such supremacy. □ **supremacism** *n.*

supremacy /suːˈpreməsɪ, sjuː-/ *n.* the state of being supreme in authority, power, rank, or importance.

Supremacy, Act of (in English history) either of two Acts of Parliament of 1534 and 1559, laying down the position of the sovereigns Henry VIII and Elizabeth I as supreme heads of the Church of England and excluding the authority of the pope. The term is used particularly with reference to the Act of 1534.

suprematism /suːˈpreməˌtɪz(ə)m, sjuː-/ *n.* a Russian art movement initiated and developed by Kazimir Malevich (1878–1935) c.1915 and characterized by a highly abstract style and a concern with spiritual purity. □ **suprematist** n.

supreme /suːˈpriːm, sjuː-/ adj. & n. ● adj. **1** highest in authority or rank. **2** greatest; most important. **3** (of a penalty or sacrifice etc.) involving death. ● n. **1** a rich cream sauce. **2** a dish served in this. □ **supreme pontiff** see PONTIFF. □ **supremely** adv. **supremeness** n. [L supremus, superl. of superus that is above f. super above]

suprême /suːˈprem, sjuː-/ n. = SUPREME n. [F]

Supreme Being a name for God.

Supreme Court n. the highest judicial court in a country or state; (in the US) the highest federal court. The highest court in England and Wales is the Supreme Court of Judicature, formed in 1873 by the amalgamation of several superior courts and divided into the High Court, the Court of Appeal, and (from 1971) the Crown Court.

Supreme Soviet hist. the governing council of the USSR. As the highest legislative authority in the Soviet Union the Supreme Soviet was responsible for electing the Presidium, the supreme authority when the Soviet was not sitting. It was composed of two equal chambers: the Soviet of Union, composed of one delegate for every 300,000 citizens, and the Soviet of Nationalities, elected on a regional basis in the constituent republics and areas.

Supremes, the a US pop and soul vocal group, one of the most successful female groups. Best known as a trio of Diana Ross, Mary Wilson (b.1944), and Florence Ballard (1943–76), the Supremes signed to Motown records in 1960 and produced a sequence of memorable singles, among them 'Baby Love' (1964) and 'You Can't Hurry Love' (1966). Diana Ross left for a solo career in 1969.

supremo /suːˈpriːməʊ, sjuː-/ n. (pl. **-os**) **1** a supreme leader or ruler. **2** a person in overall charge. [Sp., = SUPREME]

Supt. abbr. Superintendent.

sur-¹ /sɜː(r), sə(r)/ prefix = SUPER- (surcharge; surrealism). [OF]

sur-² /sɜː(r), sə(r)/ prefix assim. form of SUB- before r.

sura /ˈsʊərə/ n. (also **surah**) a chapter or section of the Koran. [Arab. sūra]

Surabaya /ˌsʊərəˈbaɪə/ a seaport in Indonesia, on the north coast of Java; pop. (1980) 2,027,900. Formerly the leading port of the Dutch East Indies, it is today Indonesia's principal naval base and its second largest city.

surah /ˈsjʊərə/ n. a soft twilled silk for scarves etc. [F pronunc. of Surat in India, where it was orig. made]

sural /ˈsjʊərəl/ adj. of or relating to the calf of the leg (sural artery). [mod.L suralis f. L sura calf]

Surat /ˈsʊərət/ a city in the state of Gujarat in western India, a port on the Tapti river near its mouth on the Gulf of Cambay; pop. (1991) 1,497,000. Noted for its textiles and gold and silver ware, it was the site of the first trading post of the East India Company, established in 1612.

surcease /sɜːˈsiːs/ n. & v. literary ● n. a cessation. ● v.intr. & tr. cease. [ME f. OF sursis, -ise (cf. AF sursise omission), past part. of OF surseoir refrain, delay f. L (as SUPERSEDE), with assim. to CEASE]

surcharge n. & v. ● n. /ˈsɜːtʃɑːdʒ/ **1** an additional charge or payment. **2** a charge made by assessors as a penalty for false returns of taxable property. **3** a mark printed on a postage stamp changing its value. **4** an additional or excessive load. **5** Brit. an amount in an official account not passed by the auditor and having to be refunded by the person responsible. **6** the showing of an omission in an account for which credit should have been given. ● v.tr. /ˈsɜːtʃɑːdʒ, sɜːˈtʃɑːdʒ/ **1** exact a surcharge from. **2** exact (a sum) as a surcharge. **3** mark (a postage stamp) with a surcharge. **4** overload. **5** fill or saturate to excess. [ME f. OF surcharger (as SUR-¹, CHARGE)]

surcingle /ˈsɜːˌsɪŋg(ə)l/ n. a band round a horse's body, usu. to keep a pack etc. in place. [ME f. OF surcengle (as SUR-¹, cengle girth f. L cingula f. cingere gird)]

surcoat /ˈsɜːkəʊt/ n. **1** hist. a loose robe worn over armour. **2** a similar sleeveless garment worn as part of the insignia of an order of knighthood. **3** hist. an outer coat of rich material. [ME f. OF surcot (as SUR-¹, cot coat)]

surculose /ˈsɜːkjʊˌləʊz, -ˌləʊs/ adj. Bot. producing suckers. [L surculosus f. surculus twig]

surd /sɜːd/ adj. & n. ● adj. **1** Math. (of a number) irrational. **2** Phonet. uttered with the breath and not the voice (e.g. f, k, p, s, t). ● n. **1** Math. a surd number, esp. the root of an integer. **2** Phonet. a surd sound. [L

surdus deaf, mute: sense 1 by mistransl. into L of Gk alogos irrational, speechless, through Arab. jadr aṣamm deaf root]

sure /ʃʊə(r), ʃɔː(r)/ adj., adv., & int. ● adj. **1** having or seeming to have adequate reason for a belief or assertion. **2** (often foll. by of, or that + clause) convinced. **3** (foll. by of) having a certain prospect or confident anticipation or satisfactory knowledge of. **4** reliable or unfailing (there is one sure way to find out). **5** (foll. by to + infin.) certain. **6** undoubtedly true or truthful. ● adv. & int. colloq. certainly. □ **as sure as eggs is eggs** see EGG¹. **as sure as fate** quite certain. **be sure** (in imper. or infin.; foll. by that + clause or to + infin.) take care to; not fail to (be sure to turn the lights out). **for sure** colloq. without doubt. **make sure 1** make or become certain; ensure. **2** (foll. by of) establish the truth or ensure the existence or happening of. **sure enough** colloq. **1** in fact; certainly. **2** with near certainty (they will come sure enough). **sure-fire** colloq. certain to succeed. **sure-footed** never stumbling or making a mistake. **sure-footedly** in a sure-footed way. **sure-footedness** being sure-footed. **sure thing** n. a certainty. ● int. esp. US colloq. certainly. **to be sure 1** it is undeniable or admitted. **2** it must be admitted. □ **sureness** n. [ME f. OF sur sure (earlier seür) f. L securus SECURE]

surely /ˈʃʊəlɪ, ˈʃɔːlɪ/ adv. **1** with certainty (the time approaches slowly but surely). **2** as an appeal to likelihood or reason (surely that can't be right). **3** with safety; securely (the goat plants its feet surely).

surety /ˈʃʊərɪtɪ, ˈʃʊətɪ/ n. (pl. **-ies**) **1** a person who takes responsibility for another's performance of an undertaking, e.g. to appear in court, or payment of a debt. **2 a** money given as a guarantee. **b** a guarantee. **3** certainty. □ **of** (or **for**) **a surety** archaic certainly. **stand surety** become a surety, go bail. □ **suretyship** n. [ME f. OF surté, seürté f. L securitas -tatis SECURITY]

surf /sɜːf/ n. & v. ● n. **1** the swell of the sea breaking on the shore or a reef. **2** the foam produced by this. ● v.intr. **1** go surfing. **2** sl. ride illicitly on the roof or outside of a train. □ **surf-casting** fishing by casting a line into the sea from the shore. **surf-riding** the sport of surfing. □ **surfer** n. **surfy** adj. [app. f. obs. suff, perh. assim. to surge: orig. applied to the Indian coast]

surface /ˈsɜːfɪs/ n. & v. ● n. **1 a** the outside of a material body. **b** the area of this. **2** any of the limits terminating a solid. **3** the upper boundary of a liquid or of the ground etc. **4** the outward aspect of anything; what is apparent on a casual view or consideration (presents a large surface to view; all is quiet on the surface). **5** Geom. a set of points that has length and breadth but no thickness. **6** (attrib.) **a** of or on the surface (surface area). **b** superficial (surface politeness). ● v. **1** tr. give the required surface to (a road, paper, etc.). **2** intr. & tr. rise or bring to the surface. **3** intr. become visible or known. **4** intr. colloq. become conscious; wake up. □ **come to the surface** become perceptible after being hidden. **surface-active** (of a substance, e.g. a detergent) able to affect the wetting properties of a liquid. **surface mail** mail carried over land and by sea, and not by air. **surface noise** extraneous noise in playing a gramophone record, caused by imperfections in the grooves. **surface structure** (in transformational grammar) the representation of grammatical or syntactic elements determining the form of a phrase or sentence (opp. deep structure). **surface tension** the tension of the surface-film of a liquid, tending to minimize its surface area. **surface-to-air** (of a missile) designed to be fired from the ground or at sea at an aircraft etc. **surface-to-surface** (of a missile) designed to be fired from one point on the ground or at sea to another such point. □ **surfaced** adj. (usu. in comb.). **surfacer** n. [F (as SUR-¹, FACE)]

surfactant /sɜːˈfæktənt/ n. a substance which reduces surface tension. [surface-active]

surfboard /ˈsɜːfbɔːd/ n. a long narrow board used in surfing.

surfeit /ˈsɜːfɪt/ n. & v. ● n. **1** an excess esp. in eating or drinking. **2** a feeling of satiety or disgust resulting from this. ● v. (**surfeited**, **surfeiting**) **1** tr. overfeed. **2** intr. overeat. **3** intr. & tr. (foll. by with) be or cause to be wearied through excess. [ME f. OF sorfe(i)t, surfe(i)t (as SUPER-, L facere fact- do)]

surficial /sɜːˈfɪʃ(ə)l/ adj. Geol. of or relating to the earth's surface. □ **surficially** adv. [SURFACE after superficial]

surfing /ˈsɜːfɪŋ/ n. the sport of being carried to the shore on the crest of large waves while standing or lying on a surfboard. Surfing or surf-riding was a pastime for the peoples of the South Sea Islands before European mariners made their historic voyages, and was observed by Captain James Cook in Tahiti in 1777. Body surfing (without a board) was practised by the early Hawaiians. Surfing was developed as a sport in the 1950s and 1960s: the first world championships were held in

1964. It became extremely popular in the 1960s and has remained so, especially in California, Hawaii, and Australia.

surge /sɜːdʒ/ n. & v. ● n. **1** a sudden or impetuous onset (a surge of anger). **2** the swell of the waves at sea. **3** a heavy forward or upward motion. **4** a rapid increase in price, activity, etc., over a short period. **5** a sudden marked increase in voltage of an electric current. ● v.intr. **1** (of waves, the sea, etc.) rise and fall or move heavily forward. **2** (of a crowd etc.) move suddenly and powerfully forwards in large numbers. **3** (of an electric current etc.) increase suddenly. **4** Naut. (of a rope, chain, or windlass) slip back with a jerk. □ **surge chamber** (or **tank**) a chamber designed to neutralize sudden changes of pressure in a flow of liquid. [OF sourdre sourge-, or sorgir f. Catalan, f. L surgere rise]

surgeon /ˈsɜːdʒən/ n. **1** a medical practitioner qualified to practise surgery. **2** a medical officer in a navy or army or military hospital. □ **surgeon fish** a tropical marine fish of the genus Acanthurus, with movable lancet-shaped spines on each side of the tail. **surgeon general** (pl. **surgeons general**) (in the US) the head of a public health service or of an army etc. medical service. **surgeon's knot** a reef-knot with a double twist. [ME f. AF surgien f. OF serurgien (as SURGERY)]

surgery /ˈsɜːdʒərɪ/ n. (pl. **-ies**) **1** the branch of medicine concerned with treatment of injuries or disorders of the body by incision, or manipulation with instruments or with the hands. (See note below.) **2** Brit. **a** a place where a doctor, dentist, etc., treats patients. **b** the occasion of this (the doctor will see you after surgery). **3** Brit. **a** a place where an MP, lawyer, or other professional person gives advice. **b** the occasion of this. [ME f. OF surgerie f. L chirurgia f. Gk kheirourgia handiwork, surgery f. kheir hand + erg- work]

▪ Surgery can be traced back to prehistoric times, when sharpened flints were used for opening abscesses, scarifying the skin, and even trepanning the skull. In the Middle Ages the Church forbade the practice of surgery by its clerics, whose numbers included most educated doctors, so that surgery was left to barber-surgeons and other lowly practitioners. In 1745 the Company of Surgeons was formed, becoming the Royal College of Surgeons in 1800. In the 19th century the great problems of pain and infection began to be overcome (see ANAESTHETIC, ANTISEPTIC). In the 20th century the surgeon's work has been aided by X-rays, safer anaesthetics, prompt replacement of blood and fluid loss, and effective antibiotics. Recent advances include open-heart surgery, cryosurgery, microsurgery, and the use of tomography.

surgical /ˈsɜːdʒɪk(ə)l/ adj. **1** of or relating to or done by surgeons or surgery. **2** resulting from surgery (surgical fever). **3 a** used in surgery. **b** (of a special garment etc.) worn to correct a deformity etc. **4** (esp. of military action) swift and precise. □ **surgical spirit** methylated spirit used in surgery for cleansing etc. □ **surgically** adv. [earlier chirurgical f. chirurgy f. OF sirurgie: see SURGEON]

suricate /ˈsʊərɪˌkeɪt/ n. the grey meerkat, Suricata suricatta. [F, of African origin]

Suriname /ˌsʊərɪˈnæm, -ˈnɑːmə/ (also **Surinam** /-ˈnæm/) a country on the NE coast of South America; pop. (est. 1991) 457,000; languages, Dutch (official), Creoles, Hindi; capital, Paramaribo. Colonized by the Dutch and the English from the 17th century, and known until 1948 as Dutch Guiana, Suriname became fully independent in 1975. The population is descended largely from African slaves and Asian workers brought in to work on sugar plantations. Some blacks have recreated a traditional West African tribal culture in the interior; there is also a small American Indian population. The economy is now chiefly dependent on bauxite. □ **Surinamer** n. **Surinamese** /-nəˈmiːz/ adj. & n.

Suriname toad n. an aquatic South American toad, Pipa pipa, having a flat body with long webbed feet, the female of which carries the eggs and tadpoles in pockets on her back. Also called pipa.

surly /ˈsɜːlɪ/ adj. (**surlier**, **surliest**) bad-tempered and unfriendly; churlish. □ **surlily** adv. **surliness** n. [alt. spelling of obs. sirly haughty f. SIR + -LY[1]]

surmise n. & v. ● n. /səˈmaɪz, ˈsɜːmaɪz/ a conjecture or suspicion about the existence or truth of something. ● v. /səˈmaɪz/ **1** tr. (often foll. by that + clause) infer doubtfully; make a surmise about. **2** tr. suspect the existence of. **3** intr. make a guess. [ME f. AF & OF fem. past part. of surmettre accuse f. LL supermittere supermiss- (as SUPER-, mittere send)]

surmount /səˈmaʊnt/ v.tr. **1** overcome or get over (a difficulty or obstacle). **2** (usu. in passive) cap or crown (peaks surmounted with snow). □ **surmountable** adj. [ME f. OF surmonter (as SUR-[1], MOUNT[1])]

surmullet /sɜːˈmʌlɪt/ n. the red mullet. [F surmulet f. OF sor red + mulet MULLET]

surname /ˈsɜːneɪm/ n. & v. ● n. **1** a hereditary name common to all members of a family, as distinct from a Christian or first name. **2** archaic an additional descriptive or allusive name attached to a person, sometimes becoming hereditary. ● v.tr. **1** give a surname to. **2** give (a person a surname). **3** (as **surnamed** adj.) having as a family name. [ME, alt. of surnoun f. AF (as SUR-[1], NOUN name)]

surpass /səˈpɑːs/ v.tr. **1** outdo, be greater or better than. **2** (as **surpassing** adj.) pre-eminent, matchless (of surpassing intelligence). □ **surpassingly** adv. [F surpasser (as SUR-[1], PASS[1])]

surplice /ˈsɜːplɪs/ n. a loose white linen vestment varying from hip-length to calf-length, worn over a cassock by clergy and choristers at services. □ **surpliced** adj. [ME f. AF surplis, OF sourpelis, f. med.L superpellicium (as SUPER-, pellicia PELISSE)]

surplus /ˈsɜːpləs/ n. & adj. ● n. **1** an amount left over when requirements have been met. **2 a** an excess of revenue over expenditure in a given period, esp. a financial year (opp. DEFICIT). **b** the excess value of a company's assets over the face value of its stock. ● adj. exceeding what is needed or used. □ **surplus value** Econ. the difference between the value of work done and wages paid. [ME f. AF surplus, OF s(o)urplus f. med.L superplus (as SUPER-, + plus more)]

surprise /səˈpraɪz/ n. & v. ● n. **1** an unexpected or astonishing event or circumstance. **2** the emotion caused by this. **3** the act of catching a person etc. unawares, or the process of being caught unawares. **4** (attrib.) unexpected; made or done etc. without warning (a surprise visit). ● v.tr. **1** affect with surprise; turn out contrary to the expectations of (your answer surprised me; I surprised her by arriving early). **2** (usu. in passive; foll. by at) shock, scandalize (I am surprised at you). **3** capture or attack by surprise. **4** come upon (a person) unawares (surprised him taking a biscuit). **5** (foll. by into) startle (a person) by surprise into an action etc. (surprised them into consenting). □ **take by surprise** affect with surprise, esp. by an unexpected encounter or statement. □ **surprisedly** /-zɪdlɪ/ adv. **surprising** adj. **surprisingly** adv. **surprisingness** n. [OF, fem. past part. of surprendre (as SUR-[1], prendre f. L praehendere seize)]

surra /ˈsʊərə, ˈsʌrə/ n. a usu. fatal disease of horses, cattle, etc., in Asia and NE Africa, caused by trypanosomes and transmitted by the bite of horseflies. [Marathi]

surreal /səˈriːl/ adj. **1** having the qualities of surrealism. **2** strange, bizarre. □ **surreally** /-ˈriːllɪ/ adv. **surreality** /ˌsɜːrɪˈælɪtɪ/ n. [back-form. f. SURREALISM etc.]

surrealism /səˈrɪəlɪz(ə)m/ n. a 20th-century avant-garde movement in art and literature aiming to explore and express the subconscious, and to move beyond the accepted conventions of reality by representing in poetry and art the irrational imagery of dreams. Launched in 1924 by a manifesto of André Breton and having a strong political content, the movement grew out of symbolism and Dada and was strongly influenced by Sigmund Freud. In poetry, there was experiment with 'automatic' writing, the setting down of words unfettered by the conscious mind; in the visual arts surrealism is represented both by a fluid abstract style analogous to this (as in the works of André Masson, Jean Arp, and Joan Miró) and by a more 'realistic' style which relied on deliberately ambiguous combinations of recognizable forms, with René Magritte and Salvador Dali creating a disorientating realist imagery often based on dreams, hallucination, and paranoia. The surrealists readily experimented with new media: Max Ernst developed Dadaist collage and invented frottage, while Man Ray and others experimented with photography and photomontage; in film, the movement strongly influenced directors such as Luis Buñuel. □ **surrealist** n. & adj. **surrealistic** /ˌsɜːrɪəˈlɪstɪk/ adj. **surrealistically** adv. [F surréalisme (as SUR-[1], REALISM)]

surrebutter /ˌsʌrɪˈbʌtə(r)/ n. Law the plaintiff's reply to the defendant's rebutter. [SUR-[1] + REBUTTER, after SURREJOINDER]

surrejoinder /ˌsʌrɪˈdʒɔɪndə(r)/ n. Law the plaintiff's reply to the defendant's rejoinder. [SUR-[1] + REJOINDER]

surrender /səˈrendə(r)/ v. & n. ● v. **1** tr. hand over; relinquish possession of, esp. on compulsion or demand; give into another's power or control. **2** intr. **a** accept an enemy's demand for submission. **b** give oneself up; cease from resistance; submit. **3** intr. & refl. (foll. by to) give oneself over to a habit, emotion, influence, etc. **4** tr. give up rights under (a life-insurance policy) in return for a smaller sum received immediately. **5** tr. give up (a lease) before its expiry. **6** tr. abandon (hope etc.). ● n. the act or an instance of surrendering. □ **surrender to bail** duly appear in a lawcourt after release on bail. **surrender value** the amount payable to a person who surrenders a life-insurance policy. [ME f. AF f. OF *surrendre* (as SUR-[1], RENDER)]

surreptitious /ˌsʌrəpˈtɪʃəs/ adj. **1** covert; kept secret. **2** done by stealth; clandestine. □ **surreptitiously** adv. **surreptitiousness** n. [ME f. L *surrepticius -itius* f. *surripere surrept-* (as SUR-[1], *rapere* seize)]

Surrey /ˈsʌrɪ/ a county of SE England; county town, Guildford.

surrey /ˈsʌrɪ/ n. (pl. **surreys**) US a light four-wheeled carriage with two seats facing forwards. [orig. of an adaptation of the *Surrey cart*, orig. made in Surrey]

surrogate /ˈsʌrəgət/ n. **1** a substitute, esp. for a person in a specific role or office. **2** *Brit.* a deputy, esp. of a bishop in granting marriage licences. **3** US a judge in charge of probate, inheritance, and guardianship. □ **surrogate mother 1** a person acting the role of mother. **2** a woman who bears a child on behalf of another woman, either from her own egg fertilized by the other woman's partner, or from the implantation in her womb of a fertilized egg from the other woman. □ **surrogacy** n. **surrogateship** n. [L *surrogatus* past part. of *surrogare* elect as a substitute (as SUR-[1], *rogare* ask)]

surround /səˈraʊnd/ v. & n. ● v.tr. **1** come or be all round; encircle, enclose. **2** (in *passive*; foll. by *by, with*) have on all sides (*the house is surrounded by trees*). ● n. **1** *Brit.* **a** a border or edging, esp. an area between the walls and carpet of a room. **b** a floor-covering for this. **2** an area or substance surrounding something. □ **surrounding** adj. [ME = overflow, f. AF *sur(o)under*, OF *s(o)uronder* f. LL *superundare* (as SUPER-, *undare* flow f. *unda* wave)]

surroundings /səˈraʊndɪŋz/ n.pl. the things in the neighbourhood of, or the conditions affecting, a person or thing.

surtax /ˈsɜːtæks/ n. & v. ● n. **1** an additional tax on something already taxed. **2** a higher rate of tax levied on incomes above a certain level, esp. *hist.* in Britain between 1929 and 1973 (replacing *supertax*). ● v.tr. impose a surtax on. [F *surtaxe* (as SUR-[1], TAX)]

Surtees /ˈsɜːtiːz/, Robert Smith (1805–64), English journalist and novelist. He is remembered for his comic sketches of Mr Jorrocks, the sporting Cockney grocer, collected in *Jorrocks's Jaunts and Jollities* (1838); its style, format, and illustrations by 'Phiz' (H. K. Browne, 1815–82) were to influence Dickens's *Pickwick Papers*. Other famous caricatures, all set against a background of English fox-hunting society, include Mr Soapy Sponge of *Mr Sponge's Sporting Tour* (1849; 1853).

surtitle /ˈsɜːˌtaɪt(ə)l/ n. (esp. in opera) each of a sequence of captions projected above the stage, translating the text being sung.

surtout /sɜːˈtuː, -ˈtuːt/ n. *hist.* a greatcoat or frock-coat. [F f. *sur* over + *tout* everything]

Surtsey /ˈsɜːtsɪ/ a small island to the south of Iceland, formed by a volcanic eruption in 1963.

surveillance /səˈveɪləns, disp. -ˈveɪjəns/ n. close observation, esp. of a suspected person. [F f. *surveiller* (as SUR-[1], *veiller* f. L *vigilare* keep watch)]

survey v. & n. ● v.tr. /səˈveɪ/ **1** take or present a general view of. **2** examine the condition of (a building etc.), esp. on behalf of a prospective buyer or mortgagee. **3** determine the boundaries, extent, ownership, etc., of (a district etc.). ● n. /ˈsɜːveɪ/ **1** a general view or consideration of something. **2 a** the act of surveying property. **b** the result or findings of this, esp. in a written report. **3** an inspection or investigation. **4** a map or plan made by surveying an area. **5** a department carrying out the surveying of land. [ME f. AF *survei(e)r*, OF *so(u)rveeir* (pres. stem *survey-*) f. med.L *supervidere* (as SUPER-, *videre* see)]

Surveyor /səˈveɪə(r)/ a series of unmanned American spacecraft sent to the moon between 1966 and 1968, five of which successfully made soft landings.

surveyor /səˈveɪə(r)/ n. **1** a person who surveys land and buildings, esp. professionally. **2** *Brit.* an official inspector, esp. for measurement and valuation. **3** a person who carries out surveys. □ **surveyorship** n. (esp. in sense 2). [ME f. AF & OF *surve(i)our* (as SURVEY)]

survival /səˈvaɪv(ə)l/ n. **1** the process or an instance of surviving. **2** a person, thing, or practice that has remained from a former time. **3** the practice of coping with harsh or warlike conditions, as a leisure activity or training exercise. □ **survival kit** emergency rations etc., esp. carried by servicemen. **survival of the fittest** the process or result of natural selection.

survivalism /səˈvaɪvəˌlɪz(ə)m/ n. the practising of outdoor survival skills as a sport or hobby. □ **survivalist** adj. & n.

survive /səˈvaɪv/ v. **1** *intr.* continue to live or exist; be still alive or existent. **2** *tr.* live or exist longer than. **3** *tr.* remain alive after going through, or continue to exist in spite of (a danger, accident, etc.). [ME f. AF *survivre*, OF *sourvivre* f. L *supervivere* (as SUPER-, *vivere* live)]

survivor /səˈvaɪvə(r)/ n. **1** a person who survives or has survived. **2** *Law* a joint tenant who has the right to the whole estate on the other's death.

Surya /ˈsʊərɪə/ *Hinduism* the sun-god of later Hindu mythology, originally one of several solar deities in the Vedic religion. [Skr., = sun]

Sus. *abbr.* Susanna (Apocrypha).

sus var. of SUSS.

sus- /sʌs, səs/ *prefix* assim. form of SUB- before *c, p, t*.

Susa /ˈsuːsə/ **1** an ancient city of SW Asia, the capital in the 4th millennium BC of the kingdom of Elam and from the late 6th to the late 4th century BC of the Persian Achaemenid dynasty. **2** see SOUSSE.

Susah see SOUSSE.

Susanna /suːˈzænə/ **1** (in the Apocrypha) a woman of Babylon falsely accused of adultery by two elders but saved by the sagacity of Daniel. **2** the book of the Apocrypha telling the story of Susanna.

susceptibility /səˌseptɪˈbɪlɪtɪ/ n. (pl. **-ies**) **1** the state of being susceptible. **2** (in pl.) a person's sensitive feelings. **3** *Physics* the ratio of magnetization to a magnetizing force.

susceptible /səˈseptɪb(ə)l/ adj. **1** impressionable, sensitive; easily moved by emotion. **2** (*predic.*) **a** (foll. by *to*) likely to be affected by; liable or vulnerable to (*susceptible to pain*). **b** (foll. by *of*) allowing; admitting of (*facts not susceptible of proof*). □ **susceptibly** adv. [LL *susceptibilis* f. L *suscipere suscept-* (as SUB-, *capere* take)]

susceptive /səˈseptɪv/ adj. **1** concerned with the receiving of emotional impressions or ideas. **2** receptive. **3** = SUSCEPTIBLE. [LL *susceptivus* (as SUSCEPTIBLE)]

sushi /ˈsuːʃɪ/ n. a Japanese dish in which various ingredients such as raw fish are added to vinegar-flavoured cold rice and formed into balls or rolls. [Jap.]

suslik var. of SOUSLIK.

suspect v., n., & adj. ● v.tr. /səˈspekt/ **1** have an impression of the existence or presence of (*suspects poisoning*). **2** (foll. by *to be*) believe tentatively, without clear grounds. **3** (foll. by *that* + clause) be inclined to think. **4** (often foll. by *of*) be inclined to accuse mentally; doubt the innocence of (*suspect him of complicity*). **5** doubt the genuineness or truth of. ● n. /ˈsʌspekt/ a suspected person. ● adj. /ˈsʌspekt/ subject to or deserving suspicion or distrust; not sound or trustworthy. [ME f. L *suspicere suspect-* (as SUB-, *specere* look)]

suspend /səˈspend/ v.tr. **1** hang up. **2** keep inoperative or undecided for a time; defer. **3** debar temporarily from a function, office, privilege, etc. **4** (as **suspended** adj.) (of solid particles or a body in a fluid medium) sustained somewhere between top and bottom. □ **suspended animation** a temporary cessation of the vital functions without death. **suspended sentence** a judicial sentence left unenforced subject to good behaviour during a specified period. **suspend payment** (of a company) fail to meet its financial engagements; admit insolvency. □ **suspensible** /-ˈspensɪb(ə)l/ adj. [ME f. OF *suspendre* or L *suspendere suspens-* (as SUB-, *pendere* hang)]

suspender /səˈspendə(r)/ n. **1** an attachment to hold up a stocking or sock by its top. **2** (in pl.) N. Amer. a pair of braces. □ **suspender belt** a woman's undergarment consisting of a belt and elastic suspenders.

suspense /səˈspens/ n. **1** a state of anxious uncertainty or expectation. **2** *Law* a suspension; the temporary cessation of a right etc. □ **keep in suspense** delay informing (a person) of urgent information. **suspense account** an account in which items are entered temporarily before allocation to the right account. □ **suspenseful** adj. [ME f. AF & OF *suspens* f. past part. of L *suspendere* SUSPEND]

suspension /səˈspenʃ(ə)n/ n. **1** the act of suspending or the condition of being suspended. **2** the means by which a vehicle is supported on its axles. **3** a substance consisting of particles suspended in a medium. **4** *Mus.* the prolongation of a note of a chord to form a discord with the following chord. □ **suspension bridge** a bridge with a roadway suspended from cables supported by structures at each end. [F *suspension* or L *suspensio* (as SUSPEND)]

suspensive /səˈspensɪv/ adj. **1** having the power or tendency to suspend or postpone. **2** causing suspense. □ **suspensively** adv. **suspensiveness** n. [F *suspensif -ive* or med.L *suspensivus* (as SUSPEND)]

suspensory /səˈspensərɪ/ adj. (of a ligament, muscle, bandage, etc.) holding an organ etc. suspended. [F *suspensoire* (as SUSPENSION)]

suspicion /səˈspɪʃ(ə)n/ n. **1** the feeling or thought of a person who suspects. **2** the act or an instance of suspecting; the state of being suspected. **3** (foll. by *of*) a slight trace of. □ **above suspicion** too obviously good etc. to be suspected. **under suspicion** suspected. [ME f. AF *suspeciun* (OF *sospeçon*) f. med.L *suspectio -onis* f. L *suspicere* (as SUSPECT): assim. to F *suspicion* & L *suspicio*]

suspicious /sə'spɪʃəs/ *adj.* **1** prone to or feeling suspicion. **2** indicating suspicion (*a suspicious glance*). **3** inviting or justifying suspicion (*a suspicious lack of surprise*). □ **suspiciously** *adv.* **suspiciousness** *n.* [ME f. AF & OF f. L *suspiciosus* (as SUSPICION)]

Susquehanna /ˌsʌskwə'hænə/ a river of the north-eastern US. It has two headstreams, one rising in New York State and one in Pennsylvania, which meet in central Pennsylvania. The river then flows 240 km (150 miles) south to Chesapeake Bay.

suss /sʌs/ *v., n.,* & *adj.* (also **sus**) *Brit. sl.* ● *v.tr.* (**sussed, sussing**) **1** suspect of a crime. **b** (usu. foll. by *out*) investigate, inspect (*go and suss out the restaurants*). **b** work out; grasp, understand, realize (*he had the market sussed*). **3** (as **sussed** *adj.*) well-informed, aware, alert. ● *n.* **1** a suspect. **2** a suspicion; suspicious behaviour. ● *adj.* suspicious, suspect. □ **on suss** on suspicion (of having committed a crime). [abbr. of SUSPECT, SUSPICION]

Sussex[1] /'sʌsɪks/ a former county of southern England. It was divided in 1974 into the counties of East Sussex and West Sussex.

Sussex[2] /'sʌsɪks/ *n.* a speckled or red breed of domestic fowl. [SUSSEX[1]]

sustain /sə'steɪn/ *v.* & *n.* ● *v.tr.* **1** support, bear the weight of, esp. for a long period. **2** give strength to; encourage, support. **3** (of food) give nourishment to. **4** endure, stand; bear up against. **5** undergo or suffer (defeat or injury etc.). **6** (of a court etc.) uphold or decide in favour of (an objection etc.). **7** substantiate or corroborate (a statement or charge). **8 a** maintain or keep (a sound, effort, etc.) going continuously. **b** (as **sustained** *adj.*) maintained continuously over a long period. **9** continue to represent (a part, character, etc.) adequately. ● *n. Mus.* the effect or result of sustaining a note, esp. electronically. □ **sustainedly** /-nɪdlɪ/ *adv.* **sustainer** *n.* **sustainment** *n.* [ME f. AF *sustein-*, OF *so(u)stein-* stressed stem of *so(u)stenir* f. L *sustinere sustent-* (as SUB-, *tenere* hold)]

sustainable /sə'steɪnəb(ə)l/ *adj.* **1** able to be sustained or upheld. **2** (of economic development or the utilization of natural resources) able to be maintained at a particular level without causing damage to the environment or depletion of the resource.

sustenance /'sʌstɪnəns/ *n.* **1 a** nourishment, food. **b** the process of nourishing. **2** a means of support; a livelihood. [ME f. AF *sustenaunce*, OF *so(u)stenance* (as SUSTAIN)]

sustentation /ˌsʌstən'teɪʃ(ə)n/ *n. formal* **1** the support of life. **2** maintenance. [ME f. OF *sustentation* or L *sustentatio* f. *sustentare* frequent. of *sustinere* SUSTAIN]

susurration /ˌsju:sə'reɪʃ(ə)n, ˌsu:-/ *n.* (also **susurrus** /sju:'sʌrəs, su:-/) *literary* a sound of whispering or rustling. [ME f. LL *susurratio* f. L *susurrare*]

Sutherland[1] /'sʌðələnd/, Graham (Vivian) (1903–80), English painter. During the Second World War he was an official war artist, who concentrated on depicting the devastation caused by bombing. Among his portraits are those of Somerset Maugham (1949) and Sir Winston Churchill (1954); the latter was considered unflattering and was destroyed by Churchill's family. His postwar work included the tapestry *Christ in Majesty* (1962), designed for the rebuilt Coventry cathedral.

Sutherland[2] /'sʌðələnd/, Dame Joan (b.1926), Australian operatic soprano. Noted for her dramatic coloratura roles, she is best known for her performance of the title role in Donizetti's *Lucia di Lammermoor* in 1959.

Sutlej /'sʌtlɪdʒ/ a river of northern India and Pakistan. Rising in the Himalayas in SW Tibet, it flows for 1,450 km (900 miles) westwards through the Indian states of Himachal Pradesh and Punjab into Punjab province in Pakistan, where it joins the Chenab river to form the Panjnad, eventually to join the Indus. It is one of the five rivers that gave Punjab its name.

sutler /'sʌtlə(r)/ *n. hist.* a person following an army and selling provisions etc. to the soldiers. [obs. Du. *soeteler* f. *soetelen* befoul, perform mean duties, f. Gmc]

Sutra /'su:trə/ *n.* **1** an aphorism or set of aphorisms in Hindu literature. **2** a narrative part of Buddhist literature. **3** Jainist scripture. [Skr. *sūtra* thread, rule, f. *siv* SEW]

suttee /sʌ'ti:, 'sʌtɪ/ *n.* (also **sati**) (*pl.* **suttees** or **satis**) *hist.* **1** the Hindu practice of a widow immolating herself on her husband's funeral pyre. **2** a widow who undergoes or has undergone this. [Hindi f. Skr. *satī* faithful wife f. *sat* good]

Sutton Coldfield /ˌsʌt(ə)n 'kəʊldfi:ld/ a town in the West Midlands, just north of Birmingham; pop. (1981) 86,494.

Sutton Hoo /ˌsʌt(ə)n 'hu:/ the site in Suffolk of a group of barrows, one of which was found in 1939 to cover the remains of a Saxon ship burial of the 7th century AD. The timbers of the ship had decayed and only their impression was left in the soil, with the iron bolts still in place; in the centre was a magnificent collection of grave goods, including jewellery, decorated weapons, and gold coins, which are now in the British Museum.

suture /'su:tʃə(r)/ *n.* & *v.* ● *n.* **1** *Surgery* **a** the joining of the edges of a wound or incision by stitching. **b** the thread or wire used for this. **2** the seamlike junction of two bones, esp. in the skull. **3** *Bot.* & *Zool.* a similar junction of parts. ● *v.tr. Surgery* stitch up (a wound or incision) with a suture. □ **sutural** *adj.* **sutured** *adj.* [F *suture* or L *sutura* f. *suere* *sut-* sew]

Suva /'su:və/ the capital of Fiji, situated on the SE coast of the island of Viti Levu; pop. (1986) 71,600.

Suwannee /sʊ'wɒnɪ/ (also **Swanee** /'swɒnɪ/) a river of the south-eastern US. Rising in SE Georgia, it flows for some 400 km (250 miles) south-west through northern Florida to the Gulf of Mexico.

suzerain /'su:zərən/ *n.* **1** *hist.* a feudal overlord. **2** a sovereign or state having some control over another state that is internally autonomous. □ **suzerainty** *n.* [F, app. f. *sus* above f. L *su(r)sum* upward, after *souverain* SOVEREIGN]

Suzhou /su:'dʒəʊ/ (also **Suchou, Soochow** /-'tʃaʊ/) a city in eastern China, in the province of Jiangsu, situated west of Shanghai on the Grand Canal; pop. (1990) 840,000. Founded in the 6th century BC, it was the capital of the ancient Wu kingdom and a noted silk-manufacturing city.

Suzman /'su:zmən/, Helen (b.1917), South African politician, of Lithuanian-Jewish descent. In 1953 she became an MP for the opposition United Party, before becoming one of the founders of the anti-apartheid Progressive Party in 1959. In the elections two years later she was the only Progressive candidate to be returned to parliament, and from this time until 1974 she was the sole MP opposed to apartheid. Suzman was awarded the UN Human Rights Award in 1978; she retired in 1989.

s.v. *abbr.* **1** a side valve. **2** (in a reference) under the word or heading given. [sense 2 f. L *sub voce* (or *verbo*)]

Svalbard /'sva:lba:(r)/ a group of islands, including the island of Spitsbergen, in the Arctic Ocean about 640 km (400 miles) north of Norway; pop. (1989) 3,540. Explored by Willem Barents in 1596, the islands came under Norwegian sovereignty in 1925. They contain coal and mineral deposits. The chief settlement (on Spitsbergen) is Longyearbyen.

svelte /svelt/ *adj.* slender, lissom, graceful. [F f. It. *svelto*]

Sven see SWEYN I.

Svengali /sveŋ'gɑ:lɪ/ a character in George du Maurier's novel *Trilby* (1894). He is a musician who trains Trilby's voice and controls her stage singing through hypnotic power; his influence over her is such that when he dies her voice collapses and she loses her eminence. A person who exercises a controlling or mesmeric influence on another is sometimes called a Svengali.

Sverdlovsk /sveəd'lɒfsk/ a former name (1924–91) for YEKATERINBURG.

Sverige /'sværjə/ the Swedish name for SWEDEN.

Svetambara /swe'tɑ:mbərə/ *n.* a member of one of the two principal sects of Jainism (the other is that of the Digambaras), which was formed AD *c*.80 and survives today in parts of India. The sect is characterized by asceticism and the wearing of white clothing. [Skr., = white-clad]

SW *abbr.* **1** south-west. **2** south-western.

swab /swɒb/ *n.* & *v.* (also **swob**) ● *n.* **1** a mop or other absorbent device for cleaning or mopping up. **2 a** an absorbent pad used in surgery. **b** a specimen of a possibly morbid secretion taken with a swab for examination. **3** *sl.* a term of contempt for a person. ● *v.tr.* (**swabbed, swabbing**) **1** clean (a wound, ship's deck, etc.) with a swab. **2** (foll. by *up*) absorb (moisture) with a swab. [back-form. f. *swabber* f. early mod.Du. *zwabber* f. a Gmc base = 'splash, sway']

Swabia /'sweɪbɪə/ (German **Schwaben** /'ʃva:b(ə)n/) a former duchy of medieval Germany. The region is now divided between SW Germany, Switzerland, and France. □ **Swabian** *adj.* & *n.*

swaddle /'swɒd(ə)l/ *v.tr.* swathe (esp. an infant) in garments or bandages etc. □ **swaddling-clothes** narrow bandages formerly wrapped round a newborn child to restrain its movements and quieten it. [ME f. SWATHE + -LE⁴]

swag /swæg/ n. & v. ● n. **1** sl. **a** the booty carried off by burglars etc. **b** illicit gains. **2** an ornamental festoon of flowers etc. **b** a carved etc. representation of this. **c** drapery of similar appearance. **3** Austral. & NZ a traveller's or miner's bundle of personal belongings. ● v. (**swagged**, **swagging**) **1** tr. arrange (a curtain etc.) in swags. **2** intr. **a** hang heavily. **b** sway from side to side. **3** tr. cause to sway or sag. [16th c.: prob. f. Scand.]

swage /sweɪdʒ/ n. & v. ● n. **1** a die or stamp for shaping wrought iron etc. by hammering or pressure. **2** a tool for bending metal etc. ● v.tr. shape with a swage. □ **swage-block** a block with various perforations, grooves, etc., for shaping metal. [F s(o)uage decorative groove, of unkn. orig.]

swagger /ˈswægə(r)/ v., n., & adj. ● v.intr. **1** walk arrogantly or self-importantly. **2** behave arrogantly; be domineering. ● n. **1** a swaggering gait or manner. **2** swaggering behaviour. **3** a dashing or confident air or way of doing something. **4** smartness. ● adj. **1** colloq. smart or fashionable. **2** (of a coat) cut with a loose flare from the shoulders. □ **swagger stick** a short cane carried by a military officer. □ **swaggerer** n. **swaggeringly** adv. [app. f. SWAG v. + -ER⁴]

swagman /ˈswægmæn/ n. (pl. **-men**) Austral. & NZ a tramp carrying a swag (see SWAG n. 3).

Swahili /swəˈhiːlɪ/ n. & adj. ● n. (pl. same) **1** a member of a Bantu-speaking people of Zanzibar and adjacent coasts. **2** the language of this people, a major language of the Bantu family with a vocabulary heavily influenced by Arabic (also called Kiswahili). (See note below.) ● adj. of or relating to the Swahili or their language. [Arab. sawāḥilī pl. of sāḥil coast]

■ Swahili is the most important language in East Africa, spoken also in central and southern regions and expanding rapidly to the west and north. While it is the first language of only a million people it is used as a common language by 20 million who speak different mother tongues. It is an official language of Kenya and Tanzania.

swain /sweɪn/ n. **1** archaic a country youth. **2** poet. a young lover or suitor. [ME swein f. ON sveinn lad = OE swān swineherd, f. Gmc]

swallow¹ /ˈswɒləʊ/ v. & n. ● v. **1** tr. cause or allow (food etc.) to pass down the throat. **2** intr. perform the muscular movement of the oesophagus required to do this. **3** tr. **a** accept meekly; put up with (an affront etc.). **b** accept credulously (an unlikely assertion etc.). **4** tr. repress; resist the expression of (a feeling etc.) (swallow one's pride). **5** tr. articulate (words etc.) indistinctly. **6** tr. (often foll. by up) engulf or absorb; exhaust; cause to disappear. ● n. **1** the act of swallowing. **2** an amount swallowed in one action. □ **swallow-hole** Brit. a sink-hole (see SINK n. 6). □ **swallowable** adj. **swallower** n. [OE swelg (n.), swelgan (v.) f. Gmc]

swallow² /ˈswɒləʊ/ n. a migratory swift-flying insect-eating songbird of the family Hirundinidae, with long pointed wings; esp. Hirundo rustica, with a long forked tail (also called (N. Amer.) barn swallow). □ **one swallow does not make a summer** a warning against a hasty inference from one instance. **swallow-dive** a dive with the arms outspread until close to the water. **swallow-tailed** having a swallowtail. [OE swealwe f. Gmc]

swallowtail /ˈswɒləʊˌteɪl/ n. **1** a deeply forked tail. **2** anything resembling this shape. **3** an animal with a forked tail; esp. a large butterfly of the family Papilionidae, with an elongated extension on each hindwing.

swam past of SWIM.

swami /ˈswɑːmɪ/ n. (pl. **swamis**) a Hindu male religious teacher. [Skr. svāmin, svāmī master, prince]

Swammerdam /ˈswɑːməˌdæm/, Jan (1637–80), Dutch naturalist and microscopist. Qualified in medicine, he preferred to commit himself to research. He worked extensively on insects, describing their anatomy and life history and classifying them into four groups. A pioneer in the use of lenses, Swammerdam was the first to observe red blood cells. He also provided an elegant demonstration of the fact that muscles do not change in volume during motion.

swamp /swɒmp/ n. & v. ● n. a piece of waterlogged ground; a bog or marsh. ● v. **1 a** tr. overwhelm, flood, or soak with water. **b** intr. become swamped. **2** tr. overwhelm or make invisible etc. with an excess or large amount of something. □ **swampy** adj. (**swampier**, **swampiest**). [17th c., = dial. swamp sunk (14th c.), prob. of Gmc orig.]

swampland /ˈswɒmplænd/ n. land consisting of swamps.

Swan /swɒn/, Sir Joseph Wilson (1828–1914), English physicist and chemist. He was a pioneer of electric lighting, devising in 1860 an electric light bulb consisting of a carbon filament inside a glass bulb, and working for nearly twenty years to perfect it. In 1883 he formed a partnership with Thomas Edison to manufacture the bulbs. Swan also devised a dry photographic plate, and bromide paper for the printing of negatives.

swan /swɒn/ n. & v. ● n. **1** a large waterbird, esp. of the genus Cygnus, having a long gracefully curved neck, black webbed feet, and (in most species) all-white plumage. **2** literary a poet. ● v.intr. (**swanned**, **swanning**) (usu. foll. by about, off, etc.) colloq. move or go aimlessly or casually or with a superior air. □ **swan-dive** N. Amer. = swallow-dive (see SWALLOW²). **swan-neck** a curved structure shaped like a swan's neck. **swan-upping** (in England) the annual taking up and marking of Thames swans. □ **swanlike** adj. & adv. [OE f. Gmc]

Swanee see SUWANNEE.

swank /swæŋk/ n., v., & adj. colloq. ● n. ostentation, swagger, bluff. ● v.intr. behave with swank; show off. ● adj. esp. US = SWANKY. [19th c.: orig. uncert.]

swankpot /ˈswæŋkpɒt/ n. Brit. colloq. a person behaving with swank.

swanky /ˈswæŋkɪ/ adj. (**swankier**, **swankiest**) colloq. **1** marked by swank; ostentatiously smart or showy. **2** (of a person) inclined to swank; boastful. □ **swankily** adv. **swankiness** n.

swannery /ˈswɒnərɪ/ n. (pl. **-ies**) a place where swans are bred.

Swan of Avon literary William Shakespeare.

Swan River a river of Western Australia. Rising as the Avon to the south-east of Perth, it flows north and west through Perth to the Indian ocean at Fremantle. Explored in 1697 by the Dutch explorer Willem de Vlamingh, it was the site of the first free European settlement in Western Australia.

swansdown /ˈswɒnzdaʊn/ n. **1** the fine down of a swan, used in trimmings and esp. in powder-puffs. **2** a kind of thick cotton cloth with a soft nap on one side.

Swansea /ˈswɒnzɪ/ a city in South Wales, on the Bristol Channel, the administrative centre of West Glamorgan; pop. (1991) 182,100.

Swanson /ˈswɒns(ə)n/, Gloria (born Gloria May Josephine Svensson) (1899–1983), American actress. She was the most highly paid star of silent films in the 1920s; with her own production company, Swanson Productions, she made such films as Sadie Thompson (1928) and Queen Kelly (1928). Swanson is perhaps now chiefly known for her performance as the fading movie star Norma Desmond in Sunset Boulevard (1950).

swansong /ˈswɒnsɒŋ/ n. **1** a person's last work or act before death or retirement etc. **2** a song like that fabled to be sung by a dying swan.

swap /swɒp/ v. & n. (also **swop**) ● v.tr. & intr. (**swapped**, **swapping**) exchange or barter (one thing for another). ● n. **1** an act of swapping. **2** a thing suitable for swapping. **3** a thing swapped. □ **swapper** n. [ME, orig. = 'hit': prob. imit.]

SWAPO /ˈswɑːpəʊ/ see SOUTH WEST AFRICA PEOPLE'S ORGANIZATION.

Swaraj /swəˈrɑːdʒ/ n. hist. self-government or independence for India. □ **Swarajist** n. [Hind. svarāj f. Skr. svarājya f. sva own + rājya rule: cf. RAJ]

sward /swɔːd/ n. esp. literary **1** an expanse of short grass. **2** turf. □ **swarded** adj. [OE sweard skin]

sware archaic past of SWEAR.

swarf /swɔːf/ n. **1** fine chips or filings of stone, metal, etc. **2** wax etc. removed in cutting a gramophone record. [ON svarf file-dust]

swarm¹ /swɔːm/ n. & v. ● n. **1** a cluster of bees leaving the hive with the queen to establish a new colony. **2** a large number of insects or birds moving in a cluster. **3** a large group of people, esp. moving over or filling a large area. **4** (in pl.; foll. by of) great numbers. **5** Biol. a group of zoospores. ● v.intr. **1** move in or form a swarm. **2** gather or move in large numbers. **3** (foll. by with) (of a place) be overrun, crowded, or infested (was swarming with tourists). [OE swearm f. Gmc]

swarm² /swɔːm/ v.intr. & (foll. by up) tr. climb (a rope or tree etc.), esp. in a rush, by clasping or clinging with the hands and knees etc. [16th c.: orig. unkn.]

swart /swɔːt/ adj. archaic swarthy, dark-hued. [OE sweart f. Gmc]

swarthy /ˈswɔːðɪ/ adj. (**swarthier**, **swarthiest**) dark, dark-complexioned. □ **swarthily** adv. **swarthiness** n. [var. of obs. swarty (as SWART)]

swash¹ /swɒʃ/ v. & n. ● v. **1** intr. (of water etc.) wash about; make the sound of washing or rising and falling. **2** tr. archaic strike violently. **3** intr. archaic swagger. ● n. the motion or sound of swashing water. [imit.]

swash² /swɒʃ/ *adj.* **1** inclined obliquely. **2** (of a letter) having a flourished stroke or strokes. □ **swash-plate** an inclined disc revolving on an axle and giving reciprocating motion to a part in contact with it. [17th c.: orig. unkn.]

swashbuckler /'swɒʃˌbʌklə(r)/ *n.* a swaggering adventurer or blustering ruffian. [SWASH¹ + BUCKLER]

swashbuckling /'swɒʃˌbʌklɪŋ/ *adj. & n.* ● *adj.* characteristic or reminiscent of a swashbuckler; swaggering, blustering, ostentatiously daring. ● *n.* swashbuckling behaviour. □ **swashbuckle** *v.intr.*

swastika /'swɒstɪkə/ *n.* **1** an ancient symbol formed by an equal-armed cross with each arm continued at a right angle. **2** this with clockwise continuations as the symbol of Nazi Germany. [Skr. *svastika* f. *svasti* well-being f. *su* good + *asti* being]

swat /swɒt/ *v. & n.* ● *v.tr.* (**swatted**, **swatting**) **1** crush (a fly etc.) with a sharp blow. **2** hit hard and abruptly. ● *n.* a swatting blow. [17th c. in the sense 'sit down': N. Engl. dial. & US var. of SQUAT]

swatch /swɒtʃ/ *n.* **1** a sample, esp. of cloth or fabric. **2** a collection of samples. [17th c.: orig. unkn.]

swath /swɔːθ/ *n.* (also **swathe** /sweɪð/) (*pl.* **swaths** /swɔːθs, swɔːðz/ or **swathes** /sweɪð/) **1** a ridge of grass or corn etc. lying after being cut. **2** a space left clear after the passage of a mower etc. **3** a broad strip. □ **cut a wide swath** be effective in destruction. [OE *swæth, swathu*]

swathe /sweɪð/ *v. & n.* ● *v.tr.* bind or enclose in bandages or garments etc. ● *n.* a bandage or wrapping. [OE *swathian*]

Swatow /swɒ'taʊ/ the former name for SHANTOU.

swatter /'swɒtə(r)/ *n.* an implement for swatting flies.

sway /sweɪ/ *v. & n.* ● *v.* **1** *intr. & tr.* lean or cause to lean unsteadily in different directions alternately. **2** *intr.* oscillate irregularly; waver. **3** *tr.* **a** control the motion or direction of. **b** have influence or rule over. ● *n.* **1** rule, influence, or government (*hold sway*). **2** a swaying motion or position. **3** *Sc. & N. Engl.* an iron rod in a fireplace used for hanging kettles etc. □ **sway-back** an abnormally hollowed back (esp. of a horse); lordosis. **sway-backed** (esp. of a horse) having a sway-back. [ME: cf. LG *swājen* be blown to and fro, Du. *zwaaien* swing, wave]

Swazi /'swɑːzɪ/ *n. & adj.* ● *n.* (*pl.* same or **Swazis**) **1 a** a member of a people inhabiting Swaziland and parts of Eastern Transvaal in South Africa. **b** a native or national of Swaziland. **2** the Nguni language of this people, an official language of Swaziland. ● *adj.* of or relating to Swaziland, the Swazis, or their language. [*Mswati*, name of a former king of the Swazi]

Swaziland /'swɑːzɪˌlænd/ a small landlocked kingdom in southern Africa, bounded by South Africa and Mozambique; pop. (est. 1991) 825,000; official languages, Swazi and English; capital, Mbabane. The Swazis have occupied the region from the mid-18th century. Swaziland was a South African protectorate from 1894 and came under British rule in 1902 after the Second Boer War. In 1968 it became a fully independent Commonwealth state. Under King Sobhuza II (reigned 1921–82), the constitution was reorganized in the 1970s on traditional tribal lines, the king exercising power with the help of an appointed Cabinet.

swear /sweə(r)/ *v. & n.* ● *v.* (*past* **swore** /swɔː(r)/ or *archaic* **sware** /sweə(r)/; *past part.* **sworn**) **1** *tr.* **a** (often foll. by *to* + infin. or *that* + clause) state or promise solemnly or on oath. **b** take (an oath). **2** *tr. colloq.* say emphatically; insist (*swore he had not seen it*). **3** *tr.* cause to take an oath (*swore them to secrecy*). **4** *intr.* (often foll. by *at*) use profane or indecent language, esp. as an expletive or from anger. **5** *tr.* (often foll. by *against*) make a sworn affirmation of (an offence) (*swear treason against*). **6** *intr.* (foll. by *by*) **a** appeal to as a witness in taking an oath (*swear by Almighty God*). **b** colloq. express great confidence in (*swears by yoga*). **7** *intr.* (foll. by *to*; usu. in *neg.*) admit the certainty of (*could not swear to it*). **8** *intr.* (foll. by *at*) colloq. (of colours etc.) fail to harmonize with. ● *n.* a spell of swearing. □ **swear blind** colloq. affirm emphatically. **swear in** induct into office etc. by administering an oath. **swear off** colloq. promise to abstain from (drink etc.). **swear-word** a profane or indecent word, esp. uttered as an expletive. □ **swearer** *n.* [OE *swerian* f. Gmc, rel. to ANSWER]

sweat /swet/ *n. & v.* ● *n.* **1** moisture exuded through the pores of the skin, esp. from heat or nervousness. **2** a state or period of sweating. **3** colloq. a state of anxiety (*was in a sweat about it*). **4** colloq. a drudgery, effort. **b** a laborious task or undertaking. **5** condensed moisture on a surface. ● *v.* (*past* and *past part.* **sweated** or US **sweat**) **1** *intr.* exude sweat; perspire. **2** *intr.* be terrified, suffering, etc. **3** *intr.* (of a wall etc.) exhibit surface moisture. **4** *intr.* drudge, toil. **5** *tr.* heat (meat or vegetables) slowly in fat or water to extract the juices. **6** *tr.* emit (blood, gum, etc.) like sweat. **7** *tr.* make (a horse, athlete, etc.) sweat by exercise.

8 *tr.* **a** cause to drudge or toil. **b** (as **sweated** *adj.*) (of goods, workers, or labour) produced by or subjected to long hours under poor conditions. **9** *tr.* subject (hides or tobacco) to fermentation in manufacturing. □ **by the sweat of one's brow** by one's own hard work. **no sweat** *sl.* there is no need to worry. **sweat-band** a band of absorbent material inside a hat or round a wrist etc. to soak up sweat. **sweat blood** colloq. **1** work strenuously. **2** be extremely anxious. **sweat gland** *Anat.* a spiral tubular gland below the skin secreting sweat. **sweating-sickness** *hist.* any of various febrile diseases with profuse sweating epidemic in England in the 15th–16th centuries. **sweat it out** colloq. endure a difficult experience to the end. [ME *swet(e)*, alt. (after *swete* v. f. OE *swætan* OHG *sweizzen* roast) of *swote* f. OE *swāt* f. Gmc]

sweater /'swetə(r)/ *n.* **1** a jersey or pullover of a kind worn esp. as an informal garment or before, during, or after exercise. **2** an employer who works employees hard in poor conditions for low pay.

sweatshirt /'swetʃɜːt/ *n.* a loose long-sleeved thick cotton sweater, fleecy on the inside and worn for sports or leisurewear.

sweatshop /'swetʃɒp/ *n.* a workshop where sweated labour is used.

sweatsuit /'swetsuːt, -sjuːt/ *n.* a suit of a sweatshirt and loose trousers, as worn by athletes etc.

sweaty /'swetɪ/ *adj.* (**sweatier**, **sweatiest**) **1** sweating; covered with sweat. **2** causing sweat. □ **sweatily** *adv.* **sweatiness** *n.*

Swede /swiːd/ *n.* a native or national of Sweden; a person of Swedish descent. [MLG & MDu. *Swēde*, prob. f. ON *Svíthjóth* f. *Svíar* Swedes + *thjóth* people]

swede /swiːd/ *n.* **1** a large yellow-fleshed turnip, *Brassica napus*, orig. from Sweden. **2** this used as a vegetable. [SWEDE]

Sweden /'swiːd(ə)n/ (called in Swedish *Sverige*) a country occupying the eastern part of the Scandinavian peninsula; pop. (1990) 8,590,630, official language, Swedish; capital, Stockholm. Its Germanic and Gothic inhabitants took part in the Viking raids of the Middle Ages. Originally united in the 12th century, Sweden formed part of the Union of Kalmar with Denmark and Norway from 1397 until its re-emergence as an independent kingdom under Gustavus Vasa in 1523. Sweden's influence as a European power peaked during the reign of Gustavus Adolphus in the early 17th century and declined following the defeat of Charles XII in the Great Northern War (1700–21). Between 1814 and 1905 it was united with Norway. A constitutional monarchy, Sweden has pursued a policy of non-alignment, and remained neutral in the two world wars. In 1994 Sweden's application to join the European Community was approved by the European Parliament.

Swedenborg /'swiːd(ə)nˌbɔːg/, Emanuel (1688–1772), Swedish scientist, philosopher, and mystic. As a scientist he concerned himself with many subjects, anticipating in his speculative and inventive work later developments such as the nebular theory, crystallography, and flying machines. He became increasingly concerned to show by scientific means the spiritual structure of the universe. However, a series of mystical experiences (1743–5) prompted him to devote the rest of his life to expounding his spiritual beliefs. His doctrines, which blended Christianity with elements of both pantheism and theosophy, were taken up by a group of followers, who founded the New Jerusalem Church in 1787. □ **Swedenborgian** /ˌswiːd(ə)n'bɔːgɪən/ *adj. & n.*

Swedish /'swiːdɪʃ/ *adj. & n.* ● *adj.* of or relating to Sweden or its people or language. ● *n.* the Scandinavian language of Sweden, spoken by some 10 million people in Sweden, Finland, Estonia, and North America.

Sweeney /'swiːnɪ/ *n.* (prec. by *the*) *Brit. sl.* the members of a flying squad. [rhyming sl. f. *Sweeney Todd* (see TODD²)]

sweep /swiːp/ *v. & n.* ● *v.* (*past* and *past part.* **swept** /swept/) **1** *tr.* clean or clear (a room or area etc.) with or as with a broom. **2** *intr.* (often foll. by *up*) clean a room etc. in this way. **3** *tr.* (often foll. by *up*) collect or remove (dirt or litter etc.) by sweeping. **4** *tr.* (foll. by *aside, away*, etc.) **a** push with or as with a broom. **b** dismiss or reject abruptly (*their objections were swept aside*). **5** *tr.* (foll. by *along, down*, etc.) carry or drive along with force. **6** *tr.* (foll. by *off, away*, etc.) remove or clear forcefully. **7** *tr.* traverse swiftly or lightly (*the wind swept the hillside*). **8** *tr.* impart a sweeping motion to (*swept his hand across*). **9** *tr.* swiftly cover or affect (*a new fashion swept the country*). **10** *intr.* **a** glide swiftly; speed along with unchecked motion. **b** go majestically. **11** *intr.* (of geographical features etc.) have continuous extent. **12** *tr.* drag (a river-bottom etc.) to search for something. **13** *tr.* (of artillery etc.) include in the line of fire; cover the whole of. **14** *tr.* propel (a barge etc.) with sweeps. ● *n.* **1** the act or motion or an instance of sweeping. **2** a curve in the road, a sweeping line of a hill, etc. **3** range or scope (*beyond the sweep of the human mind*). **4** = chimney-sweep. **5** a sortie by aircraft. **6** *colloq.* = SWEEPSTAKE. **7** a long oar worked from a barge etc. **8** the sail of a windmill. **9** a long pole mounted as a lever for raising buckets from a well. **10** *Electronics* the movement of a beam across the screen of a cathode-ray tube. **11 a** a survey of an area, esp. the night sky, made in an arc or circle. **b** (in *pl.*) *US* a regular survey of the popularity ratings of local television stations. □ **make a clean sweep of 1** completely abolish or expel. **2** win all the prizes etc. in (a competition etc.). **sweep away 1** abolish swiftly. **2** (usu. in *passive*) powerfully affect, esp. emotionally. **sweep the board 1** win all the money in a gambling-game. **2** win all possible prizes etc. **sweep second-hand** a second hand on a clock or watch, moving on the same dial as the other hands. **sweep under the carpet** see CARPET. **swept-back** (of an aircraft wing) fixed at an acute angle to the fuselage, inclining outwards towards the rear. **swept-up** (of hair) = UPSWEPT 1. **swept-wing** (of an aircraft) having swept-back wings. [ME *swepe* (earlier *swope*) f. OE *swāpan*]

sweepback /ˈswiːpbæk/ *n.* the angle at which the swept-back wing of an aircraft is set back from a position at right angles to the body.

sweeper /ˈswiːpə(r)/ *n.* **1** a person who cleans by sweeping. **2** a device for sweeping carpets etc. **3** *Football* a defensive player usu. playing behind the other defenders across the width of the field.

sweeping /ˈswiːpɪŋ/ *adj. & n.* ● *adj.* **1** wide in range or effect (*sweeping changes*). **2** taking no account of particular cases or exceptions (*a sweeping statement*). ● *n.* (in *pl.*) dirt etc. collected by sweeping. □ **sweepingly** *adv.* **sweepingness** *n.*

sweepstake /ˈswiːpsteɪk/ *n.* **1** a form of gambling on horse-races etc. in which all competitors' stakes are paid to the winners. **2** a race with betting of this kind. **3** a prize or prizes won in a sweepstake.

sweet /swiːt/ *adj. & n.* ● *adj.* **1** having the pleasant taste characteristic of sugar. **2** smelling pleasant like roses or perfume etc.; fragrant. **3** (of sound etc.) melodious or harmonious. **4** a not salt, sour, or bitter. **b** fresh, with flavour unimpaired by rottenness. **c** (of water) fresh and readily drinkable. **5** (of wine) having a sweet taste (opp. DRY *adj.* 2). **6** highly gratifying or attractive. **7** amiable, pleasant (*has a sweet nature*). **8** *colloq.* (of a person or thing) pretty, charming, endearing. **9** (foll. by *on*) *colloq.* fond of; in love with. **10** (esp. *iron.*) one's own; particular, individual (*goes his own sweet way*). ● *n.* **1** *Brit.* a small shaped piece of confectionery usu. made with sugar or sweet chocolate. **2** *Brit.* a sweet dish forming a course of a meal. **3** a sweet part of something; sweetness. **4** (in *pl.*) delights, gratification. **5** (esp. as a form of address) sweetheart etc. □ **she's sweet** *Austral. sl.* all is well. **sweet alyssum** see ALYSSUM 2. **sweet-and-sour** cooked in a sauce containing sugar and vinegar or lemon etc. **sweet basil** see BASIL. **sweet bay** = BAY² 1. **sweet-brier** see BRIER¹. **sweet chestnut** see CHESTNUT *n.* 1b. **sweet cicely** a white-flowered aromatic umbelliferous plant, *Myrrhis odorata.* **sweet corn 1** a kind of maize with kernels having a high sugar content. **2** these kernels, eaten as a vegetable when young. **sweet flag** a waterside plant, *Acorus calamus* (arum family), with leaves like those of the iris, used medicinally and as a flavouring. **sweet-gale** see GALE². **sweet pea** a climbing plant of the genus *Lathyrus*; esp. *L. odoratus*, with fragrant flowers in many colours. **sweet pepper** = CAPSICUM. **sweet potato 1** a tropical climbing plant, *Ipomoea batatas*, with sweet tuberous roots used for food. **2** the root of this. **sweet rocket** see ROCKET² 1. **sweet rush** (or **sedge**) a sedge with a thick creeping aromatic rootstock used in medicine and confectionery. **sweet sultan** a sweet-scented composite plant, *Centaurea moschata* or *C. suaveolens.* **sweet talk** *colloq.* flattery, blandishment. **sweet-talk** *v.tr. colloq.* flatter in order to persuade. **sweet-tempered** amiable. **sweet tooth** a liking for sweet-tasting things. **sweet violet** a sweet-scented violet, *Viola odorata.* **sweet william** a garden pink, *Dianthus barbatus*, with clusters of vivid fragrant flowers. □ **sweetish** *adj.* **sweetly** *adv.* [OE *swēte* f. Gmc]

sweetbread /ˈswiːtbred/ *n.* the pancreas or thymus of an animal, esp. as food.

sweeten /ˈswiːt(ə)n/ *v.* **1** *tr. & intr.* make or become sweet or sweeter. **2** *tr.* make agreeable or less painful. □ **sweeten the pill** see PILL. □ **sweetening** *n.*

sweetener /ˈswiːt(ə)nə(r)/ *n.* **1** a substance used to sweeten food or drink. **2** *colloq.* a bribe or inducement.

sweetheart /ˈswiːthɑːt/ *n.* **1** a lover or darling. **2** a term of endearment (esp. as a form of address). □ **sweetheart agreement** (or **deal**) *colloq.* an industrial agreement reached privately by employers and trade unions in their own interests.

sweetie /ˈswiːtɪ/ *n. colloq.* **1** *Brit.* a sweet. **2** (also **sweetie-pie**) a term

of endearment (esp. as a form of address). **3** a green-skinned hybrid variety of grapefruit, noted for its sweet taste.

sweeting /ˈswiːtɪŋ/ *n.* **1** a sweet-flavoured variety of apple. **2** *archaic* darling.

sweetmeal /ˈswiːtmiːl/ *n.* **1** sweetened wholemeal. **2** a sweetmeal biscuit.

sweetmeat /ˈswiːtmiːt/ *n. archaic* **1** a sweet (see SWEET *n.* 1). **2** a small fancy cake.

sweetness /ˈswiːtnɪs/ *n.* the quality of being sweet; fragrance, melodiousness, etc. □ **sweetness and light** a display of (esp. uncharacteristic) mildness and reason.

sweetshop /ˈswiːtʃɒp/ *n. Brit.* a shop selling sweets as its main item.

sweetsop /ˈswiːtsɒp/ *n.* **1** a tropical American evergreen shrub, *Annona squamosa.* **2** the fruit of this, having a green rind and a sweet pulp.

swell /swel/ *v., n., & adj.* ● *v.* (*past part.* **swollen** /ˈswəʊlən/ or **swelled**) **1** *intr. & tr.* grow or cause to grow bigger or louder or more intense; expand; increase in force or intensity. **2** *intr.* (often foll. by *up*) & *tr.* rise or raise up from the surrounding surface. **3** *intr.* (foll. by *out*) bulge. **4** *intr.* (of the heart as the seat of emotion) feel full of joy, pride, relief, etc. **5** *intr.* (foll. by *with*) be hardly able to restrain (pride etc.). **6** (as **swollen** *adj.*) distended or bulging. ● *n.* **1** an act or the state of swelling. **2** the heaving of the sea with waves that do not break, e.g. after a storm. **3 a** a crescendo. **b** a mechanism in an organ etc. for obtaining a crescendo or diminuendo. **4** *archaic colloq.* a person of distinction or of dashing or fashionable appearance. **5** a protuberant part. ● *adj.* **1** esp. *N. Amer. colloq.* fine, splendid, excellent. **2** *archaic colloq.* smart, fashionable. □ **swell-box** *Mus.* a box in which organ-pipes are enclosed, with a shutter for controlling the sound-level. **swelled** (or **swollen**) **head** *colloq.* conceit. **swell-organ** *Mus.* a section of an organ with pipes in a swell-box. □ **swellish** *adj.* [OE *swellan* f. Gmc]

swelling /ˈswelɪŋ/ *n.* an abnormal protuberance on or in the body.

swelter /ˈsweltə(r)/ *v. & n.* ● *v.intr.* (usu. as **sweltering** *adj.*) be uncomfortably hot. ● *n.* a sweltering atmosphere or condition. □ **swelteringly** *adv.* [base of (now dial.) *swelt* f. OE *sweltan* perish f. Gmc]

swept *past* and *past part.* of SWEEP.

swerve /swɜːv/ *v. & n.* ● *v.intr. & tr.* change or cause to change direction, esp. abruptly. ● *n.* **1** a swerving movement. **2** divergence from a course. □ **swerveless** *adj.* **swerver** *n.* [ME, repr. OE *sweorfan* SCOUR¹]

Sweyn I /sweɪn/ (also **Sven** /sven/; known as Sweyn Forkbeard) (d.1014), king of Denmark *c.*985–1014. From 1003 he launched a series of attacks on England, finally driving the English king Ethelred the Unready to flee to Normandy at the end of 1013. Sweyn then became king of England until his death five weeks later. His son Canute was later king of England, Denmark, and Norway.

SWG *abbr.* standard wire gauge.

Swift /swɪft/, Jonathan (known as Dean Swift) (1667–1745), Irish satirist, poet, and Anglican cleric. He was born in Dublin, a cousin of John Dryden, and divided his life between London and Ireland. His *Journal to Stella* (1710–13) gives a vivid account of life in London, where he was close to Tory ministers. In 1713 he was made Dean of St Patrick's in Dublin, where he wrote his greatest work, *Gulliver's Travels* (1726), a satire on human society and institutions in the form of a fantastic tale of travels in imaginary lands. He also involved himself in Irish affairs and wrote many political pamphlets, such as *A Modest Proposal* (1729), ironically urging that the children of the poor should be fattened to feed the rich. His poems include the *Verses on the Death of Dr Swift* (1739), in which he reviews his life with pathos and humour.

swift /swɪft/ *adj., adv., & n.* ● *adj.* **1** quick, rapid; soon coming or passing. **2** speedy, prompt (*a swift response; was swift to act*). ● *adv.* (*archaic* except in *comb.*) swiftly (*swift-moving*). ● *n.* **1** a swift-flying insect-eating bird of the family Apodidae, with long curved wings and a superficial resemblance to a swallow. **2** (also **swift moth**) a long-winged moth of the family Hepialidae. **3** a revolving frame for winding yarn etc. from. □ **swift fox** a small fox, *Vulpes velox*, of North American prairies. □ **swiftly** *adv.* **swiftness** *n.* [OE, rel. to *swīfan* move in a course]

swiftie /ˈswɪftɪ/ *n. Austral. sl.* **1** a deceptive trick. **2** a person who acts or thinks quickly.

swiftlet /ˈswɪftlɪt/ *n.* a small Asian and Australasian swift of the genus *Collocalia* or *Aerodramus*, nesting communally in caves and building nests largely from saliva.

swig /swɪg/ *v. & n.* ● *v.tr. & intr.* (**swigged, swigging**) drink in large draughts. ● *n.* a large draught or swallow of drink, esp. of alcoholic liquor. □ **swigger** n. [16th c., orig. as noun in obs. sense 'liquor': orig. unkn.]

swill /swɪl/ v. & n. ● v. **1** tr. (often foll. by out) rinse or flush; pour water over or through. **2** tr. & intr. drink greedily. ● n. **1** an act of rinsing. **2** mainly liquid kitchen refuse as pig-food, pigswill. **3** inferior liquor. **4** worthless matter, rubbish. □ **swiller** n. [OE swillan, swilian, of unkn. orig.]

swim /swɪm/ v. & n. ● v. (**swimming**; past **swam** /swæm/; past part. **swum** /swʌm/) **1** intr. propel the body through water by working the arms and legs, or (of a fish) the fins and tail. (See SWIMMING.) **2** tr. **a** traverse (a stretch of water or its distance) by swimming. **b** compete in (a race) by swimming. **c** use (a particular stroke) in swimming. **3** intr. float on or at the surface of a liquid (bubbles swimming on the surface). **4** intr. appear to undulate or reel or whirl. **5** intr. have a dizzy effect or sensation (my head swam). **6** intr. (foll. by in, with) be flooded. ● n. **1** a spell or the act of swimming. **2** a deep pool frequented by fish in a river. □ **in the swim** involved in or acquainted with what is going on. **swim-bladder** a gas-filled sac in fishes used to maintain buoyancy. **swimming-bath** (often in pl.) an artificial indoor pool for swimming, esp. a public one. **swimming-costume** Brit. a garment worn for swimming. **swimming-pool** an artificial indoor or outdoor pool for swimming. □ **swimmable** adj. **swimmer** n. [OE swimman f. Gmc]

swimmeret /ˈswɪməˌret/ n. a swimming-foot in crustaceans.

swimming /ˈswɪmɪŋ/ n. the action of swimming, especially as a sport or recreation. Swimming as a sport, recorded from Japan in 36 BC, today takes place in a pool 50 m (55 yd) long and divided into lanes. Four main strokes are used: breast-stroke, crawl, butterfly, and back-stroke, of which all but the breast-stroke (which evolved in the 16th century) were developed in the 20th century. Races in each of four strokes are usually held at 100 and 200 m, whereas for longer, freestyle races the crawl is generally used. Other races include relays and medleys, in which a different stroke must be used for each stage.

swimmingly /ˈswɪmɪŋlɪ/ adv. with easy and unobstructed progress.

swimsuit /ˈswɪmsuːt, -sjuːt/ n. a one-piece swimming-costume worn by women. □ **swimsuited** adj.

swimwear /ˈswɪmweə(r)/ n. clothing worn for swimming.

Swinburne /ˈswɪnbɜːn/, Algernon Charles (1837–1909), English poet and critic. Associated with Dante Gabriel Rossetti and the Pre-Raphaelites, he came to fame with Atalanta in Calydon (1865), a drama in classical Greek form which was praised for its metrical finesse. In Songs before Sunrise (1871), Swinburne expressed his hatred of authority and his support for Mazzini's struggle for Italian independence. As a critic he contributed to the revival of contemporary interest in Elizabethan and Jacobean drama and produced influential studies of William Blake and the Brontës.

swindle /ˈswɪnd(ə)l/ v. & n. ● v.tr. (often foll. by out of) **1** cheat (a person) of money, possessions, etc. (was swindled out of all his savings). **2** cheat a person of (money etc.) (swindled all his savings out of him). ● n. **1** an act of swindling. **2** a person or thing represented as what it is not. **3** a fraudulent scheme. □ **swindler** n. [back-form. f. swindler f. G Schwindler extravagant maker of schemes, swindler, f. schwindeln be dizzy]

Swindon /ˈswɪndən/ an industrial town in Wiltshire; pop. (est. 1991) 100,000. An old market town, it developed rapidly after the Great Western Railway established its engineering works there in 1841. It expanded again in the 1950s when it was designated to receive overspill population and industry from London.

swine /swaɪn/ n. (pl. same) **1** formal or US a pig. **2** colloq. (pl. **swine** or **swines**) **a** a term of contempt or disgust for a person. **b** a very unpleasant or difficult thing. □ **swine fever** an intestinal viral disease of pigs. **swine vesicular disease** an infectious viral disease of pigs causing mild fever and blisters around the mouth and feet. □ **swinish** adj. (esp. in sense 2). **swinishly** adv. **swinishness** n. [OE swīn f. Gmc]

swineherd /ˈswaɪnhɜːd/ n. a person who tends pigs.

swing /swɪŋ/ v. & n. ● v. (past and past part. **swung** /swʌŋ/) **1** intr. & tr. move or cause to move with a to-and-fro or curving motion, as of an object attached at one end and hanging free at the other. **2** intr. & tr. **a** sway. **b** hang so as to be free to sway. **c** oscillate or cause to oscillate. **3** intr. & tr. revolve or cause to revolve. **4** intr. move by gripping something and leaping etc. (swung from tree to tree). **5** intr. go with a swinging gait (swung out of the room). **6** intr. (foll. by round) move round to the opposite direction. **7** intr. change from one opinion or mood to another. **8 a** intr. (foll. by at) attempt to hit or punch. **b** tr. throw (a punch). **9 a** intr. (also **swing it**) play music with a swing rhythm. **b** tr. play (a tune) with swing. **10** intr. colloq. **a** be lively or up to date; enjoy oneself. **b** be promiscuous. **11** intr. colloq. (of a party etc.) be lively, successful, etc. **12** tr. have a decisive

influence on (esp. voting etc.). **13** tr. colloq. deal with or achieve; manage. **14** intr. colloq. be executed by hanging. **15** Cricket **a** intr. (of the ball) deviate from a straight course in the air. **b** tr. cause (the ball) to do this. ● n. **1** the act or an instance of swinging. **2** the motion of swinging. **3** the extent of swinging. **4** a swinging or smooth gait or rhythm or action. **5 a** a seat slung by ropes or chains etc. for swinging on or in. **b** a spell of swinging on this. **6** an easy but vigorous continued action. **7 a** an easily flowing but vigorous rhythmic feeling in jazz or popular music. **b** dance music characterized by swing, esp. a variety of big-band jazz played in this style by the bands of Glenn Miller, Duke Ellington, Benny Goodman, and others, in the 1930s and early 1940s. **8** a discernible change in opinion, esp. the amount by which votes or points scored etc. change from one side to another. **9** colloq. the regular procedure or course of events (get into the swing of things). □ **swing-boat** a boat-shaped swing at fairs. **swing-bridge** a bridge that can be swung to one side to allow the passage of ships. **swing-door** a door able to open in either direction and close itself when released. **swing the lead** Brit. colloq. malinger; shirk one's duty. **swings and roundabouts** a situation affording no eventual gain or loss (from the phr. lose on the swings what you make on the roundabouts). **swing shift** US a work shift from afternoon to late evening. **swing-wing** an aircraft wing that can move from a right-angled to a swept-back position. **swung dash** a dash (~) with alternate curves. □ **swinger** n. (esp. in sense 10 of v.). [OE swingan to beat f. Gmc]

swinge /swɪndʒ/ v.tr. (**swingeing**) archaic strike hard; beat. [alt. f. ME swenge f. OE swengan shake, shatter, f. Gmc]

swingeing /ˈswɪndʒɪŋ/ adj. esp. Brit. **1** (of a blow) forcible. **2** huge or far-reaching, esp. in severity (swingeing economies). □ **swingeingly** adv.

swinging /ˈswɪŋɪŋ/ adj. **1** (of gait, melody, etc.) vigorously rhythmical. **2** colloq. **a** lively; up to date; excellent. **b** promiscuous. □ **swingingly** adv.

swingle /ˈswɪŋg(ə)l/ n. & v. ● n. **1** a wooden instrument for beating flax and removing the woody parts from it. **2** the swinging part of a flail. ● v.tr. clean (flax) with a swingle. [ME f. MDu. swinghel (as SWING, -LE¹)]

swingletree /ˈswɪŋg(ə)lˌtriː/ n. a crossbar pivoted in the middle, to which the traces are attached in a cart, plough, etc.

swingy /ˈswɪŋɪ/ adj. (**swingier**, **swingiest**) **1** (of music) characterized by swing (see SWING n. 7a). **2** (of a skirt or dress) designed to swing with body movement.

swipe /swaɪp/ v. & n. colloq. ● v. **1** tr. & (often foll. by at) intr. hit hard and recklessly. **2** tr. steal. **3** tr. pass (a credit card etc.) through an electronic device in order to read data magnetically encoded on it. ● n. **1** a reckless hard hit or attempted hit. **2** a casual criticism. □ **swipe card** a credit card etc. which may be read electronically. □ **swiper** n. [perh. var. of SWEEP]

swipple /ˈswɪp(ə)l/ n. the swingle of a flail. [ME, prob. formed as SWEEP + -LE¹]

swirl /swɜːl/ v. & n. ● v.intr. & tr. move or flow or carry along with a whirling motion. ● n. **1** a swirling motion of or in water, air, etc. **2** the act of swirling. **3** a twist or curl, esp. as part of a pattern or design. □ **swirly** adj. [ME (orig. as noun): orig. Sc., perh. of LG or Du. orig.]

swish /swɪʃ/ v., n., & adj. ● v. **1** tr. swing (a scythe or stick etc.) audibly through the air, grass, etc. **2** intr. move with or make a swishing sound. **3** tr. (foll. by off) cut (a flower etc.) in this way. ● n. a swishing action or sound. ● adj. colloq. smart, fashionable. □ **swishy** adj. [imit.]

Swiss /swɪs/ adj. & n. ● adj. of or relating to Switzerland or its people. ● n. (pl. same) **1** a native or national of Switzerland. **2** a person of Swiss descent. □ **Swiss chard** = CHARD. **Swiss cheese plant** a climbing house plant, Monstera deliciosa, with aerial roots and holes in the leaves (as in some Swiss cheeses). **Swiss guards** Swiss mercenary troops employed formerly by sovereigns of France etc. and still at the Vatican. **Swiss roll** a cylindrical cake with a spiral cross-section, made from a flat piece of sponge cake spread with jam etc. and rolled up. [F Suisse f. MHG Swīz]

Swiss Confederation the confederation of cantons forming Switzerland (see SWITZERLAND).

switch /swɪtʃ/ n. & v. ● n. **1** a device for making and breaking the connection in an electric circuit. **2 a** a transfer, change-over, or deviation. **b** an exchange. **3** a slender flexible shoot cut from a tree. **4** a light tapering rod. **5** US a device at the junction of railway tracks for transferring a train from one track to another; = POINT n. 17. **6** a tress of false or detached hair tied at one end, used in hairdressing. **7 a** a computer system which manages the transfer of funds between point-of-sale terminals and institutions. **b** propr. (**Switch**) such a system in the UK. **c** the transfer of funds by such a system. ● v. **1** tr. & absol. (foll.

by *on*, *off*) turn (an electrical device) on or off. **2** *intr.* change or transfer position, subject, etc. **3** *tr.* change or transfer. **4** *tr.* reverse the positions of; exchange (*switched chairs*). **5** *tr.* swing or snatch (a thing) suddenly (*switched it out of my hand*). **6** *tr.* beat or flick with a switch. □ **switch-blade** a pocket knife with the blade released by a spring. **switched-on 1** *colloq.* up to date; aware of what is going on. **2** *sl.* excited; under the influence of drugs. **switch off** *colloq.* cease to pay attention. **switch over** change or exchange. **switch-over** *n.* a change or exchange. □ **switcher** *n.* [earlier *swits*, *switz*, prob. f. LG]

switchback /ˈswɪtʃbæk/ *n.* **1** *Brit.* a railway at a fair etc., in which the train's ascents are effected by the momentum of its previous descents. **2** (often *attrib.*) a railway or road with alternate sharp ascents and descents.

switchboard /ˈswɪtʃbɔːd/ *n.* an apparatus for varying connections between electric circuits, esp. in telephony.

swither /ˈswɪðə(r)/ *v.* & *n.* Sc. ● *v.intr.* hesitate; be uncertain. ● *n.* doubt or uncertainty. [16th c.: orig. unkn.]

Swithin, St /ˈswɪðɪn/ (also **Swithun** /-ðən/) (d.862), English ecclesiastic. He was chaplain to Egbert, king of Wessex, and bishop of Winchester from 852. The tradition that if it rains on St Swithin's day it will do so for the next forty days may have its origin in the heavy rain said to have occurred when his relics were to be transferred to a shrine in Winchester cathedral. Feast day, 15 July.

Switzerland /ˈswɪtsələnd/ (French **Suisse** /sɥis/, German **Schweiz** /ʃvaɪts/, Italian **Svizzera** /ˈzvittsera/; also called by its Latin name *Helvetia*) a landlocked country in central Europe; pop. (1990) 6,673,850; official languages, French, German, Italian, and Romansh; capital, Berne. A mountainous country lying mainly in the Alps and the Jura, Switzerland is divided into French, German, and Italian-speaking areas. The area (occupied by a Celtic people, the Helvetii) was under Roman rule from the 1st century BC until the 5th century AD, and from the 10th century formed part of the Holy Roman Empire. Switzerland emerged as an independent country in the 14th and 15th centuries, when the states or cantons formed a confederation to defeat first their Habsburg overlords and then their Burgundian neighbours. After a period of French domination (1798–1815), the Swiss Confederation's neutrality was guaranteed by the other European powers. Neutral in both world wars, Switzerland has emerged as an international financial centre and as the headquarters of several international organizations such as the Red Cross.

swivel /ˈswɪv(ə)l/ *n.* & *v.* ● *n.* a coupling between two parts enabling one to revolve without turning the other. ● *v.tr.* & *intr.* (**swivelled**, **swivelling**; *US* **swiveled**, **swiveling**) turn on or as on a swivel. □ **swivel chair** a chair with a seat able to be turned horizontally. [ME f. weak grade *swif-* of OE *swīfan* sweep + -LE¹: cf. SWIFT]

swizz /swɪz/ *n.* (also **swiz**) (pl. **swizzes**) *Brit. colloq.* **1** something unfair or disappointing. **2** a swindle. [abbr. of SWIZZLE²]

swizzle¹ /ˈswɪz(ə)l/ *n.* & *v. colloq.* ● *n.* a mixed alcoholic drink esp. of rum or gin and bitters made frothy. ● *v.tr.* stir with a swizzle-stick. □ **swizzle-stick** a stick used for frothing or flattening drinks. [19th c.: orig. unkn.]

swizzle² /ˈswɪz(ə)l/ *n. Brit. colloq.* = SWIZZ. [20th c.: prob. alt. of SWINDLE]

swob var. of SWAB.

swollen *past part.* of SWELL.

swoon /swuːn/ *v.* & *n. literary* ● *v.intr.* faint; fall into a fainting-fit. ● *n.* an occurrence of fainting. [ME *swoune* perh. back-form. f. *swogning* (n.) f. *iswogen* f. OE *geswogen* overcome]

swoop /swuːp/ *v.* & *n.* ● *v.* **1** *intr.* (often foll. by *down*) descend rapidly like a bird of prey. **2** *intr.* (often foll. by *on*) make a sudden attack from a distance. **3** *tr.* (often foll. by *up*) *colloq.* snatch the whole of at one swoop. ● *n.* a swooping or snatching movement or action. □ **at** (or **in**) **one fell swoop** see FELL⁴. [perh. dial. var. of obs. *swōpe* f. OE *swāpan*: see SWEEP]

swoosh /swʊʃ/ *n.* & *v.* ● *n.* the noise of a sudden rush of liquid, air, etc. ● *v.intr.* move with this noise. [imit.]

swop var. of SWAP.

sword /sɔːd/ *n.* **1** a weapon usu. of metal with a long blade and hilt with a handguard, used esp. for thrusting or striking, and also worn as part of ceremonial dress. (*See note below.*) **2** (prec. by *the*) **a** war. **b** military power. □ **put to the sword** kill, esp. in war. **sword-bearer** an official carrying the sovereign's etc. sword on a formal occasion. **sword dance** a dance in which the performers brandish swords or step about swords laid on the ground. **sword grass** a grass, *Scirpus americanus*, with swordlike leaves. **sword knot** a ribbon or

tassel attached to a sword-hilt orig. for securing it to the wrist. **sword lily** = GLADIOLUS. **sword of Damocles** see DAMOCLES. **the sword of justice** judicial authority. **sword of state** a sword borne before the sovereign on state occasions. **sword-swallower** a person ostensibly or actually swallowing sword blades as a public entertainment. □ **swordlike** *adj.* [OE *sw(e)ord* f. Gmc]

▪ The sword evolved in the Bronze Age, when metal-smelting became known, and developed in various forms — as a rapier (for cut-and-thrust and slashing), with a broad straight blade, and with the curved blade which was believed to deal a deeper wound. It became obsolete as an infantry weapon after the development of explosives, but remained in use as a weapon of cavalry units until the early 20th century. Swords are now confined to ceremonial use and are also used in the sport of fencing.

swordbill /ˈsɔːdbɪl/ *n.* a South American hummingbird, *Ensifera ensifera*, with a very long bill.

swordfish /ˈsɔːdfɪʃ/ *n.* a large marine game-fish, *Xiphias gladius*, with an extended swordlike upper jaw.

swordplay /ˈsɔːdpleɪ/ *n.* **1** fencing. **2** repartee; cut-and-thrust argument.

swordsman /ˈsɔːdzmən/ *n.* (*pl.* **-men**) a person of (usu. specified) skill with a sword. □ **swordsmanship** *n.*

swordstick /ˈsɔːdstɪk/ *n.* a hollow walking-stick containing a blade that can be used as a sword.

swordtail /ˈsɔːdteɪl/ *n.* **1** a small tropical freshwater fish, *Xiphophorus helleri*, with a long tail. **2** = horseshoe crab.

swore *past* of SWEAR.

sworn /swɔːn/ **1** *past part.* of SWEAR. **2** *adj.* bound by or as by an oath (*sworn enemies*).

swot /swɒt/ *v.* & *n. Brit. colloq.* ● *v.* (**swotted**, **swotting**) **1** *intr.* study assiduously. **2** *tr.* (often foll. by *up*) study (a subject) hard or hurriedly. ● *n.* **1** a person who swots. **2 a** hard study. **b** a thing that requires this. [dial. var. of SWEAT]

swum *past part.* of SWIM.

swung *past* and *past part.* of SWING.

swy /swaɪ/ *n. Austral.* the game of two-up. [G *zwei* two]

SY *abbr.* steam yacht.

sybarite /ˈsɪbəˌraɪt/ *n.* & *adj.* ● *n.* a person who is self-indulgent or devoted to sensuous luxury. ● *adj.* fond of luxury or sensuousness. □ **sybaritism** *n.* **sybaritic** /ˌsɪbəˈrɪtɪk/ *adj.* **sybaritical** *adj.* **sybaritically** *adv.* [orig. an inhabitant of Sybaris in S. Italy, noted for luxury, f. L *sybarita* f. Gk *subaritēs*]

sycamine /ˈsɪkəˌmaɪn, -mɪn/ *n. Bibl.* the black mulberry tree, *Morus nigra* (see Luke 17:6; in modern versions translated as 'mulberry tree'). [L *sycaminus* f. Gk *sukaminos* mulberry tree f. Heb. *šiḳmāh* sycamore, assim. to Gk *sukon* fig]

sycamore /ˈsɪkəˌmɔː(r)/ *n.* **1** (in full **sycamore maple**) **a** a large Eurasian maple, *Acer pseudoplatanus*, with winged seeds, grown for its shade and timber, and naturalized in Britain. **b** its wood. **2** *N. Amer.* the plane tree or its wood. **3** *Bibl.* a fig tree, *Ficus sycomorus*, growing in Egypt, Syria, etc. [var. of SYCOMORE]

syce /saɪs/ *n.* (also **sice**) *Anglo-Ind.* a groom. [Hind. f. Arab. *sā'is*, *sāyis*]

sycomore /ˈsɪkəˌmɔː(r)/ *n.* = SYCAMORE 3. [ME f. OF *sic(h)amor* f. L *sycomorus* f. Gk *sukomoros* f. *sukon* fig + *moron* mulberry]

syconium /saɪˈkəʊnɪəm/ *n.* (*pl.* **syconia** /-nɪə/) *Bot.* a fleshy hollow receptacle developing into a multiple fruit, as in the fig. [mod.L f. Gk *sukon* fig]

sycophant /ˈsɪkəˌfænt/ *n.* a servile flatterer; a toady. □ **sycophancy** *n.* **sycophantic** /ˌsɪkəˈfæntɪk/ *adj.* **sycophantically** *adv.* [F *sycophante* or L *sycophanta* f. Gk *sukophantēs* informer f. *sukon* fig + *phainō* show: the reason for the name is uncertain, and association with informing against the illegal exportation of figs from ancient Athens (recorded by Plutarch) cannot be substantiated]

sycosis /saɪˈkəʊsɪs/ *n.* a skin disease of the bearded part of the face with inflammation of the hair follicles. [mod.L f. Gk *sukōsis* f. *sukon* fig: orig. of a figlike ulcer]

Sydenham /ˈsɪd(ə)nəm/, Thomas (*c.*1624–89), English physician. He was known as 'the English Hippocrates', because of his contemporary reputation as a physician and his scepticism towards theoretical medicine. He emphasized the healing power of nature, made a study of epidemics, wrote a treatise on gout (from which he suffered), and explained the nature of the type of chorea that is named after him (see CHOREA).

Sydney /'sɪdnɪ/ the capital of New South Wales in SE Australia; pop. (1991) 3,097,950. The site of a penal colony established at Port Jackson in 1788, it was the first British settlement in Australia. Noted for its fine natural harbour, it is the country's largest city and chief port.

Sydney Harbour Bridge an arch bridge crossing Sydney harbour, one of the longest of its kind in the world (its main arch is 503 m (1,652 ft) long). It was opened in 1932.

Sydney Opera House a striking building in Sydney, on a site at the harbour surrounded on three sides by water. Well known internationally for its roof, which is composed of huge white shapes resembling sails or shells, the building was designed by the Danish architect Jørn Utzon (b.1918) and opened in 1973. It contains a concert hall, theatres, and a cinema.

syenite /'saɪə,naɪt/ n. a grey crystalline rock of feldspar and hornblende with or without quartz. □ **syenitic** /,saɪə'nɪtɪk/ adj. [F syénite f. L Syenites (lapis) (stone) of Syene in Egypt]

Syktyvkar /,sɪktɪf'ka:(r)/ a city in NW Russia, capital of the autonomous republic of Komi; pop. (1990) 235,000.

syl- /sɪl/ prefix assim. form of SYN- before l.

syllabary /'sɪləbərɪ/ n. (pl. **-ies**) a list of characters representing syllables and (in some languages or stages of writing) serving the purpose of an alphabet. [mod.L syllabarium (as SYLLABLE)]

syllabi pl. of SYLLABUS.

syllabic /sɪ'læbɪk/ adj. **1** of, relating to, or based on syllables. **2** Prosody based on the number of syllables. **3** (of a symbol) representing a whole syllable. **4** articulated in syllables. □ **syllabically** adv. **syllabicity** /,sɪlə'bɪsɪtɪ/ n. [F syllabique or LL syllabicus f. Gk sullabikos (as SYLLABLE)]

syllabication /sɪ,læbɪ'keɪʃ(ə)n/ n. (also **syllabification** /-,læbɪfɪ'keɪʃ(ə)n/) division into or articulation by syllables. □ **syllabify** /-'læbɪ,faɪ/ v.tr. (**-ies**, **-ied**). [med.L syllabicatio f. syllabicare f. L syllaba: see SYLLABLE]

syllabize /'sɪlə,baɪz/ v.tr. (also **-ise**) divide into or articulate by syllables. [med.L syllabizare f. Gk sullabizō (as SYLLABLE)]

syllable /'sɪləb(ə)l/ n. & v. ● n. **1** a unit of pronunciation uttered without interruption, forming the whole or a part of a word and usu. having one vowel sound often with a consonant or consonants before or after: there are two syllables in water and three in inferno. **2** a character or characters representing a syllable. **3** (usu. with neg.) the least amount of speech or writing (did not utter a syllable). ● v.tr. pronounce by syllables; articulate distinctly. □ **in words of one syllable** expressed plainly or bluntly. □ **syllabled** adj. (also in comb.). [ME f. AF sillable f. OF sillabe f. L syllaba f. Gk sullabē (as SYN-, lambanō take)]

syllabub /'sɪlə,bʌb/ n. (also **sillabub**) a dessert made of cream or milk flavoured, sweetened, and whipped to thicken it. [16th c.: orig. unkn.]

syllabus /'sɪləbəs/ n. (pl. **syllabuses** or **syllabi** /-,baɪ/) **1 a** the programme or outline of a course of study, teaching, etc. **b** a statement of the requirements for a particular examination. **2** RC Ch. a summary of points decided by papal decree regarding heretical doctrines or practices. [mod.L, orig. a misreading of L sittybas accus. pl. of sittyba f. Gk sittuba title-slip or label]

syllepsis /sɪ'lepsɪs/ n. (pl. **syllepses** /-si:z/) a figure of speech in which a word is applied to two others in different senses (e.g. caught the train and a bad cold) or to two others of which it grammatically suits one only (e.g. neither they nor it is working) (cf. ZEUGMA). □ **sylleptic** adj. **sylleptically** adv. [LL f. Gk sullēpsis taking together f. sullambanō: see SYLLABLE]

syllogism /'sɪlə,dʒɪz(ə)m/ n. **1** a form of reasoning in which a conclusion is drawn from two given or assumed propositions (premisses). In a syllogism a common or middle term is present in the two premisses but not in the conclusion, which may be invalid (e.g. all trains are long; some buses are long; therefore some buses are trains: the common term is long). **2** deductive reasoning as distinct from induction. □ **syllogistic** /,sɪlə'dʒɪstɪk/ adj. **syllogistically** adv. [ME f. OF silogisme or L syllogismus f. Gk sullogismos f. sullogizomai (as SYN-, logizomai to reason f. logos reason)]

syllogize /'sɪlə,dʒaɪz/ v. (also **-ise**) **1** intr. use syllogisms. **2** tr. put (facts or an argument) in the form of syllogism. [ME f. OF sillogiser or LL syllogizare f. Gk sullogizomai (as SYLLOGISM)]

sylph /sɪlf/ n. **1** an elemental spirit of the air. **2** a slender graceful woman or girl. **3** a hummingbird of the genus Aglaiocercus or Neolesbia, with a long forked tail. □ **sylphlike** adj. [mod.L sylphes, G Sylphen (pl.), perh. based on L sylvestris of the woods + nympha nymph]

sylva /'sɪlvə/ n. (also **silva**) (pl. **sylvae** /-vi:/ or **sylvas**) **1** the trees of a region, epoch, or environment. **2** a treatise on or a list of such trees. [L silva a wood]

sylvan /'sɪlv(ə)n/ adj. (also **silvan**) esp. poet. **1 a** of the woods. **b** having woods; wooded. **2** rural. [F sylvain (obs. silvain) or L SILVANUS f. silva a wood]

sylviculture var. of SILVICULTURE.

sym- /sɪm/ prefix assim. form of SYN- before b, m, p.

symbiont /'sɪmbɪənt/ n. Biol. an organism living in symbiosis. [Gk sumbiōn -ountos part. of sumbioō live together (as SYMBIOSIS)]

symbiosis /,sɪmbaɪ'əʊsɪs, ,sɪmbɪ-/ n. (pl. **symbioses** /-si:z/) **1 a** a mutually advantageous association or relationship between persons. **b** an instance of this. **2** Biol. **a** an interaction between two different organisms living in close physical association, usu. to the advantage of both (cf. ANTIBIOSIS). **b** an instance of this. □ **symbiotic** /-'ɒtɪk/ adj. **symbiotically** adv. [mod.L f. Gk sumbiōsis a living together f. sumbioō live together, sumbios companion (as SYN-, bios life)]

symbol /'sɪmb(ə)l/ n. & v. ● n. **1** a thing conventionally regarded as typifying, representing, or recalling something, esp. an idea or quality (white is a symbol of purity). **2** a mark or character taken as the conventional sign of some object, idea, function, or process, e.g. the letters standing for the chemical elements or the characters in musical notation. ● v.tr. (**symbolled**, **symbolling**; US **symboled**, **symboling**) symbolize. □ **symbology** /sɪm'bɒlədʒɪ/ n. [ME f. L symbolum f. Gk sumbolon mark, token (as SYN-, ballō throw)]

symbolic /sɪm'bɒlɪk/ adj. (also **symbolical**) **1** of or serving as a symbol. **2** involving the use of symbols or symbolism. □ **symbolic logic** the use of symbols to denote propositions etc. in order to assist reasoning. □ **symbolically** adv. [F symbolique or LL symbolicus f. Gk sumbolikos]

symbolism /'sɪmbə,lɪz(ə)m/ n. **1 a** the use of symbols to represent ideas. **b** symbols collectively. **2** an artistic and poetic movement or style using symbols and indirect suggestion rather than direct description to express ideas, emotions, etc. (See note below.) □ **symbolist** n. **symbolistic** /,sɪmbə'lɪstɪk/ adj.

▪ The French symbolist poets (for example, Mallarmé, Verlaine, and Rimbaud) of the late 19th century reacted against prevailing tendencies of realism and naturalism, laying emphasis on the suggestive or evocative powers of words and the musical properties of language. In the visual arts, the term is widely applied to both French and non-French painters of the fin de siècle, who similarly reacted against the realist insistence on painting from nature, using colour and line as symbolic and expressive tools and dealing in mystical, erotic, or dreamlike subject-matter; Odilon Redon and James Ensor were among its most famous exponents. Symbolism was influential in the development of later theories of abstraction and surrealism in art, while in literature it was an important forerunner of modernism.

symbolize /'sɪmbə,laɪz/ v.tr. (also **-ise**) **1** be a symbol of. **2** represent by means of symbols. □ **symbolization** /,sɪmbəlaɪ'zeɪʃ(ə)n/ n. [F symboliser f. symbole SYMBOL]

symbology /sɪm'bɒlədʒɪ/ n. **1** the study of symbols. **2 a** the use of symbols. **b** symbols collectively.

symmetry /'sɪmɪtrɪ/ n. (pl. **-ies**) **1 a** correct proportion of the parts of a thing; balance, harmony. **b** beauty resulting from this. **2 a** a structure that allows an object to be divided into parts of an equal shape and size and similar position to the point or line or plane of division. **b** the possession of such a structure. **c** approximation to such a structure. **3** the repetition of exactly similar parts facing each other or a centre. **4** Bot. the possession by a flower of sepals and petals and stamens and pistils in the same number or multiples of the same number. □ **symmetrize** v.tr. (also **-ise**). **symmetric** /sɪ'metrɪk/ adj. **symmetrical** adj. **symmetrically** adv. [obs. F symmétrie or L summetria f. Gk (as SYN-, metron measure)]

Symons /'saɪmənz/, Julian (Gustave) (b.1912), English writer of detective fiction. He is an important exponent of psychological crime fiction whose many novels include The Colour of Murder (1957) and The Progress of a Crime (1960).

sympathectomy /,sɪmpə'θektəmɪ/ n. (pl. **-ies**) the surgical removal of a sympathetic ganglion etc.

sympathetic /,sɪmpə'θetɪk/ adj. **1** of, showing, or expressing sympathy. **2** due to sympathy. **3** likeable or capable of evoking sympathy. **4** (of a person) friendly and cooperative. **5** (foll. by to) inclined to favour (a proposal etc.) (was most sympathetic to the idea). **6** (of a landscape etc.) that touches the feelings by association etc. **7** (of a pain etc.) caused by a pain or injury to someone else or in another part of the body. **8** (of a sound, resonance, or string) sounding by a vibration communicated from another vibrating object. **9** (of a nerve or ganglion) belonging to the sympathetic nervous system. □ **sympathetic magic** a type of magic that seeks to achieve an effect by performing an associated action or using an associated thing. **sympathetic nervous system** one of the two divisions of the autonomic nervous system, consisting of nerves from ganglia close to the spinal cord that supply the internal organs, blood vessels, and glands, balancing the action of the parasympathetic nervous system. □ **sympathetically** adv. [SYMPATHY, after pathetic]

sympathize /ˈsɪmpəˌθaɪz/ v.intr. (also **-ise**) (often foll. by with) **1** feel or express sympathy; share a feeling or opinion. **2** agree with a sentiment or opinion. □ **sympathizer** n. [F sympathiser (as SYMPATHY)]

sympathy /ˈsɪmpəθɪ/ n. (pl. **-ies**) **1 a** the state of being simultaneously affected with a feeling similar or corresponding to that of another. **b** the capacity for this. **2** (often foll. by with) **a** the act of sharing in or responding to an emotion, sensation, or condition of another person or thing. **b** the capacity for this. **3** (in sing. or pl.) compassion or commiseration; condolences. **4** (often foll. by for) a favourable attitude; approval. **5** (in sing. or pl.; often foll. by with) agreement (with a person etc.) in opinion or desire. **6** (attrib.) in support of another cause (sympathy strike). □ **in sympathy** (often foll. by with) **1** having or showing or resulting from sympathetic action (working to rule in sympathy). [L sympathia f. Gk sumpatheia (as SYN-, pathēs f. pathos feeling)]

sympatric /sɪmˈpatrɪk/ adj. Biol. occurring within the same geographical area (cf. ALLOPATRIC). [SYM- + Gk patra fatherland]

sympetalous /sɪmˈpetələs/ adj. Bot. having the petals united.

symphonic /sɪmˈfɒnɪk/ adj. (of music) relating to or having the form or character of a symphony. □ **symphonic poem** an extended orchestral piece, usu. in one movement, on a descriptive or rhapsodic theme. □ **symphonically** adv.

symphonist /ˈsɪmfənɪst/ n. a composer of symphonies.

symphony /ˈsɪmfənɪ/ n. (pl. **-ies**) **1** an elaborate composition usu. for full orchestra, and in several movements with one or more in sonata form. (See note below.) **2** an interlude for orchestra alone in a large-scale vocal work. **3** = symphony orchestra. □ **symphony orchestra** a large orchestra suitable for playing symphonies etc. [ME, = harmony of sound, f. OF symphonie f. L symphonia f. Gk sumphōnia (as SYN-, -phōnos f. phōnē sound)]

▪ In the 16th century the term denoted a piece of music for an instrumental ensemble, but since the 18th century it has been applied to a type of orchestral work typically in four movements (in the early years, three): a fast movement, a slow movement, a minuet or scherzo, and a final movement, often in sonata form. Haydn played a significant role in the development of the symphony, as did Mozart, Schubert, and Beethoven. Beethoven's Ninth (or 'Choral') Symphony was the first to use human voice, an innovation taken up notably by Mahler and used in four of his symphonies. After the formidable achievements of Beethoven, the main tradition of the form became associated with Mendelssohn, Schumann, Brahms, Tchaikovsky, and Dvořák. In the 20th century, the form has been further extended by composers including Sibelius, Shostakovich, and Vaughan Williams, often with more or fewer than four movements.

symphyllous /sɪmˈfɪləs/ adj. Bot. having the leaves united. [SYN- + Gk phullon leaf]

symphysis /ˈsɪmfɪsɪs/ n. (pl. **symphyses** /-ˌsiːz/) Anat. & Zool. **1** the process of growing together. **2 a** a union between two bones esp. in the median plane of the body. **b** the place or line of this. □ **symphyseal** /sɪmˈfɪzɪəl/ adj. **symphysial** adj. [mod.L f. Gk sumphusis (as SYN-, phusis growth)]

sympodium /sɪmˈpəʊdɪəm/ n. (pl. **sympodia** /-dɪə/) Bot. the apparent main axis or stem of a vine etc., made up of successive secondary axes. □ **sympodial** adj. [mod.L (as SYN-, Gk pous podos foot)]

symposium /sɪmˈpəʊzɪəm/ n. (pl. **symposia** /-zɪə/) **1 a** a conference or meeting to discuss a particular subject. **b** a collection of essays or papers for this purpose. **2** a philosophical or other friendly discussion. **3** a drinking-party, esp. of the ancient Greeks with conversation etc. after a banquet. [L f. Gk sumposion in sense 3 (as SYN-, -potēs drinker)]

symptom /ˈsɪmptəm/ n. **1** a change in the physical or mental condition of a person, regarded as evidence of a disorder (cf. SIGN n. 5). **2** a sign of the existence of something. [ME synthoma f. med.L sinthoma, & f. LL symptoma f. Gk sumptōma -atos chance, symptom, f. sumpiptō happen (as SYN-, piptō fall)]

symptomatic /ˌsɪmptəˈmatɪk/ adj. serving as a symptom. □ **symptomatically** adv.

symptomatology /ˌsɪmptəˈtɒlədʒɪ/ n. the branch of medicine concerned with the study and interpretation of symptoms.

syn- /sɪn/ prefix with, together, alike. [from or after Gk sun- f. sun with]

synaeresis /sɪˈnɪərɪsɪs/ n. (US **syneresis**) (pl. **synaereses** /-ˌsiːz/) the contraction of two vowels into a diphthong or single vowel. [LL f. Gk sunairesis (as SYN-, hairesis f. haireō take)]

synaesthesia /ˌsiniːsˈθiːzɪə/ n. (US **synesthesia**) **1** Psychol. the production of a mental sense-impression relating to one sense by the stimulation of another sense. **2** a sensation produced in a part of the body by stimulation of another part. □ **synaesthetic** /-ˈθetɪk/ adj. [mod.L f. SYN- after anaesthesia]

synagogue /ˈsɪnəˌɡɒɡ/ n. **1** the building where a Jewish assembly or congregation meets for religious observance and instruction. **2** the assembly itself. □ **synagogal** /ˌsɪnəˈɡɒɡ(ə)l/ adj. **synagogical** /-ˈɡɒdʒɪk(ə)l/ adj. [ME f. OF sinagoge f. LL synagoga f. Gk sunagōgē meeting (as SYN-, agō bring)]

synallagmatic /ˌsɪnəlaɡˈmatɪk/ adj. (of a treaty or contract) imposing reciprocal obligations. [SYN- + Gk allassō exchange]

synapse /ˈsaɪnaps, ˈsɪn-/ n. Anat. a junction of two nerve cells, consisting of a minute gap across which an impulse passes by diffusion of a neurotransmitter. [Gk synapsis (as SYN-, hapsis f. haptō join)]

synapsis /sɪˈnapsɪs/ n. (pl. **synapses** /-siːz/) **1** Anat. = SYNAPSE. **2** Biol. the fusion of chromosome-pairs at the start of meiosis. □ **synaptic** adj. **synaptically** adv.

synarthrosis /ˌsɪnɑːˈθrəʊsɪs/ n. (pl. **synarthroses** /-siːz/) Anat. an immovably fixed bone-joint, e.g. the sutures of the skull. [SYN- + Gk arthrōsis jointing f. arthron joint]

sync /sɪŋk/ n. & v. (also **synch**) colloq. ● n. synchronization. ● v.tr. & intr. synchronize. □ **in** (or **out of**) **sync** (often foll. by with) according or agreeing well (or badly). [abbr.]

syncarp /ˈsɪnkɑːp/ n. Bot. a compound fruit from a flower with several carpels, e.g. a blackberry. [SYN- + Gk karpos fruit]

syncarpous /sɪnˈkɑːpəs/ adj. Bot. (of a flower or fruit) having the carpels united (opp. APOCARPOUS). [SYN- + Gk karpos fruit]

synchondrosis /ˌsɪŋkɒnˈdrəʊsɪs/ n. (pl. **synchondroses** /-siːz/) Anat. an almost immovable bone-joint bound by a layer of cartilage, as in the spinal vertebrae. [SYN- + Gk khondros cartilage]

synchro- /ˈsɪŋkrəʊ/ comb. form synchronized, synchronous.

synchrocyclotron /ˌsɪŋkrəʊˈsaɪkləˌtrɒn/ n. Physics a cyclotron able to achieve higher energies by decreasing the frequency of the accelerating electric field as the particles increase in energy and mass.

synchromesh /ˈsɪŋkrəʊˌmeʃ/ n. & adj. ● n. a system of gear-changing, esp. in motor vehicles, in which the driving and driven gearwheels are made to revolve at the same speed during engagement by means of a set of friction clutches, thereby easing the change. ● adj. relating to or using this system. [abbr. of synchronized mesh]

synchronic /sɪŋˈkrɒnɪk/ adj. describing a subject (esp. a language) as it exists at one point in time (opp. DIACHRONIC). □ **synchronically** adv. [LL synchronus: see SYNCHRONOUS]

synchronicity /ˌsɪŋkrəˈnɪsɪtɪ/ n. **1** the simultaneous occurrence of events which appear meaningfully related but have no discoverable causal connection. **2** synchrony, simultaneity. [f. SYNCHRONIC + -ITY]

synchronism /ˈsɪŋkrəˌnɪz(ə)m/ n. **1** = SYNCHRONY. **2** the process of synchronizing sound and picture in cinematography, television, etc. □ **synchronistic** /ˌsɪŋkrəˈnɪstɪk/ adj. **synchronistically** adv. [Gk sugkhronismos (as SYNCHRONOUS)]

synchronize /ˈsɪŋkrəˌnaɪz/ v. (also **-ise**) **1** intr. (often foll. by with) occur at the same time; be simultaneous. **2** tr. cause to occur at the same time. **3** tr. carry out the synchronism of (a film). **4** tr. ascertain or set forth the correspondence in the date of (events). **5 a** tr. cause (clocks etc.) to show a standard or uniform time. **b** intr. (of clocks etc.) be synchronized. □ **synchronized swimming** a form of swimming in which participants make coordinated leg and arm movements in time to music. □ **synchronizer** n. **synchronization** /ˌsɪŋkrənaɪˈzeɪʃ(ə)n/ n.

synchronous /ˈsɪŋkrənəs/ adj. (often foll. by with) existing or occurring at the same time. □ **synchronous motor** Electr. a motor having a speed exactly proportional to the current frequency. □ **synchronously** adv. [LL synchronus f. Gk sugkhronos (as SYN-, khronos time)]

synchrony /ˈsɪŋkrənɪ/ n. **1** the state of being synchronic or synchronous. **2** the treatment of events etc. as being synchronous. [Gk sugkhronos: see SYNCHRONOUS]

synchrotron /'sɪŋkrə,trɒn/ n. Physics a cyclotron in which the magnetic field strength increases with the energy of the particles to keep their orbital radius constant. □ **synchrotron radiation** polarized radiation emitted by a charged particle spinning in a magnetic field.

syncline /'sɪŋklaɪn/ n. a rock-bed forming a trough. □ **synclinal** /sɪŋ'klaɪn(ə)l/ adj. [synclinal (as SYN-, Gk klinō lean)]

syncopate /'sɪŋkə,peɪt/ v.tr. **1** Mus. displace the beats or accents in (a passage) so that strong beats become weak and vice versa. **2** shorten (a word) by dropping interior sounds or letters, as symbology for symbolology, Gloster for Gloucester. □ **syncopator** n. **syncopation** /ˌsɪŋkə'peɪʃ(ə)n/ n. [LL syncopare swoon (as SYNCOPE)]

syncope /'sɪŋkəpi/ n. **1** Gram. the omission of interior sounds or letters in a word (see SYNCOPATE 2). **2** Med. a temporary loss of consciousness caused by a fall in blood pressure. □ **syncopal** adj. [ME f. LL syncopē f. Gk sugkopē (as SYN-, koptō strike, cut off)]

syncretism /'sɪŋkrə,tɪz(ə)m/ n. **1** Philos. & Theol. attempted union or reconciliation of diverse or opposing tenets or practices; an instance of this. (See note below.) **2** Philol. the merging of different inflectional varieties in the development of a language. □ **syncretist** n. **syncretic** /sɪŋ'krɛtɪk/ adj. **syncretistic** /-krə'tɪstɪk/ adj. [mod.L syncretismus f. Gk sugkrētismos f. sugkrētizō (of two parties) combine against a third f. krēs Cretan (orig. of ancient Cretan communities)]

▪ In a religious context the term has been applied particularly to the principles of the German theologian George Calixtus (1586–1656), who aimed to harmonize the beliefs of Protestant sects and ultimately of all Christians, but many examples are found in the early years of Christianity, notably the heretical movements of Gnosticism and Manichaeism. Neoplatonist philosophy was syncretic in its synthesis of Platonic and other classical ideas with Eastern elements.

syncretize /'sɪŋkrə,taɪz/ v.tr. (also **-ise**) Philos. & Theol. perform syncretism on.

syncytium /sɪn'sɪtɪəm/ n. (pl. **syncytia** /-tɪə/) Biol. a mass of cytoplasm with several nuclei, not divided into separate cells. □ **syncytial** adj. [formed as SYN- + -CYTE + -IUM]

syndactyl /sɪn'dæktɪl/ adj. Zool. & Med. having the digits united, as in webbed feet. □ **syndactyly** n.

syndesis /'sɪndɪsɪs/ n. (pl. **syndeses** /-ˌsiːz/) Biol. = SYNAPSIS 2. [mod.L f. Gk syndesis binding together f. sundeō bind together]

syndesmosis /ˌsɪndez'məʊsɪs/ n. the union and articulation of bones by means of ligaments. [mod.L f. Gk sundesmos binding, fastening + -OSIS]

syndetic /sɪn'dɛtɪk/ adj. Gram. of or using conjunctions. [Gk sundetikos (as SYNDESIS)]

syndic /'sɪndɪk/ n. **1** a government official in various countries. **2** (in the UK) a business agent of certain universities and corporations, esp. (at Cambridge University) a member of a committee of the senate. □ **syndical** adj. [F f. LL syndicus f. Gk sundikos (as SYN-, -dikos f. dikē justice)]

syndicalism /'sɪndɪkə,lɪz(ə)m/ n. a movement aiming to overthrow capitalism and to replace the state with unions of workers, especially through direct action such as a general strike. It was closer in philosophy to anarchism than to socialism, perceiving the state as an inevitable tool of oppression. Influenced by Proudhon and by the French social philosopher Georges Sorel (1847–1922), syndicalism developed in French trade unions during the late 19th century and was at its most vigorous between 1900 and 1914, particularly in France, Italy, Spain, and the US. After the First World War the movement became overshadowed by Communism, although it remained significant in Spain until the end of the Civil War. □ **syndicalist** n. & adj. [F syndicalisme f. syndical (as SYNDIC)]

syndicate n. & v. ● n. /'sɪndɪkət/ **1** a combination of individuals or commercial firms to promote some common interest. **2** an association or agency supplying material simultaneously to a number of newspapers or periodicals. **3** a group of people who combine to buy or rent property, gamble, organize crime, etc. **4** a committee of syndics. ● v.tr. /'sɪndɪ,keɪt/ **1** form into a syndicate. **2** publish (material) through a syndicate. □ **syndication** /ˌsɪndɪ'keɪʃ(ə)n/ n. [F syndicat f. med.L syndicatus f. LL syndicus: see SYNDIC]

syndrome /'sɪndrəʊm/ n. **1** a characteristic combination of opinions, emotions, behaviour, etc. **2** Med. a group of concurrent symptoms of a disease. □ **syndromic** /sɪn'drɒmɪk/ adj. [mod.L f. Gk sundromē (as SYN-, dromē f. dramein to run)]

syne /saɪn/ adv., conj., & prep. Sc. since. [contr. f. ME sithen SINCE]

synecdoche /sɪ'nɛkdəkɪ/ n. a figure of speech in which a part is made to represent the whole or vice versa (e.g. new faces at the meeting; England lost by six wickets). □ **synecdochic** /ˌsɪnɛk'dɒkɪk/ adj. [ME f. L f. Gk sunekdokhē (as SYN-, ekdokhē f. ekdekhomai take up)]

synecology /ˌsɪnɪ'kɒlədʒɪ/ n. the ecological study of plant or animal communities. □ **synecologist** n. **synecological** /sɪn,iːkə'lɒdʒɪk(ə)l/ adj.

syneresis US var. of SYNAERESIS.

synergist /'sɪnədʒɪst/ n. a medicine or a bodily organ (e.g. a muscle) that cooperates with another or others.

synergy /'sɪnədʒɪ/ n. (also **synergism** /-,dʒɪz(ə)m/) **1** the combined effect of drugs, muscles, etc., that exceeds the sum of their individual effects. **2** increased effectiveness or achievement produced by combined action or cooperation. □ **synergic** /sɪ'nɜːdʒɪk/ adj. **synergetic** /ˌsɪnə'dʒɛtɪk/ adj. **synergistic** adj. **synergistically** adv. [Gk sunergos working together (as SYN-, ergon work)]

synesthesia US var. of SYNAESTHESIA.

syngamy /'sɪŋgəmɪ/ n. Biol. the fusion of gametes or nuclei in reproduction. □ **syngamous** adj. [SYN- + Gk gamos marriage]

Synge /sɪŋ/, (Edmund) J(ohn) M(illington) (1871–1909), Irish dramatist. Between 1898 and 1902 he lived with the peasant community on the Aran Islands, an experience that inspired his plays Riders to the Sea (1905) and The Playboy of the Western World (1907). The latter caused outrage and riots at the Abbey Theatre, Dublin, with its explicit language and its implication that Irish peasants would condone a brutal murder.

syngenesis /sɪn'dʒɛnɪsɪs/ n. sexual reproduction from combined male and female elements.

synod /'sɪnəd/ n. **1** a Church council attended by delegated clergy and sometimes laity (see also GENERAL SYNOD). **2** a Presbyterian ecclesiastical court above the presbyteries and subject to the General Assembly. **3** any meeting for debate. [ME f. LL synodus f. Gk sunodos meeting (as SYN-, hodos way)]

synodic /sɪ'nɒdɪk/ adj. Astron. relating to or involving the conjunction of stars, planets, etc. □ **synodic period** the time between the successive conjunctions of a planet with the sun. [LL synodicus f. Gk sunodikos (as SYNOD)]

synodical /sɪ'nɒdɪk(ə)l/ adj. **1** (also **synodal** /'sɪnəd(ə)l/) of, relating to, or constituted as a synod. **2** = SYNODIC.

synoecious /sɪ'niːʃəs/ adj. Bot. having male and female organs in the same flower or receptacle. [SYN- after dioecious etc.]

synonym /'sɪnənɪm/ n. **1** a word or phrase that means exactly or nearly the same as another in the same language (e.g. shut and close). **2** a word denoting the same thing as another but suitable to a different context (e.g. serpent for snake, Hellene for Greek). **3** a word equivalent to another in some but not all senses (e.g. ship and vessel). □ **synonymic** /ˌsɪnə'nɪmɪk/ adj. **synonymity** n. [ME f. L synonymum f. Gk sunōnumon neut. of sunōnumos (as SYN-, onoma name): cf. ANONYMOUS]

synonymous /sɪ'nɒnɪməs/ adj. (often foll. by with) **1** having the same meaning; being a synonym (of). **2** (of a name, idea, etc.) suggestive of or associated with another (excessive drinking regarded as synonymous with violence). □ **synonymously** adv. **synonymousness** n.

synonymy /sɪ'nɒnɪmɪ/ n. (pl. **-ies**) **1** the state of being synonymous. **2** the collocation of synonyms for emphasis (e.g. in any shape or form). **3 a** a system or collection of synonyms. **b** a treatise on synonyms. [LL synonymia f. Gk sunōnumia (as SYNONYM)]

synopsis /sɪ'nɒpsɪs/ n. (pl. **synopses** /-siːz/) **1** a summary or outline. **2** a brief general survey. □ **synopsize** v.tr. (also **-ise**). [LL f. Gk (as SYN-, opsis seeing)]

synoptic /sɪ'nɒptɪk/ adj. & n. ● adj. **1** of, forming, or giving a synopsis. **2** taking or affording a comprehensive mental view. **3** of or relating to the Synoptic Gospels. **4** giving a general view of weather conditions. ● n. **1** a Synoptic Gospel. **2** the writer of a Synoptic Gospel. □ **synoptical** adj. **synoptically** adv. [Gk sunoptikos (as SYNOPSIS)]

Synoptic Gospel n. each of the Gospels of Matthew, Mark, and Luke, which describe events from a similar point of view and have many similarities, whereas the Gospel of John differs greatly.

synoptist /sɪ'nɒptɪst/ n. the writer of a Synoptic Gospel.

synostosis /ˌsɪnɒ'stəʊsɪs/ n. the joining of bones by ankylosis etc. [SYN- + Gk osteon bone + -OSIS]

synovial /saɪ'nəʊvɪəl, sɪ-/ adj. Physiol. denoting or relating to a viscous fluid lubricating joints and tendon sheaths. □ **synovial membrane**

a dense membrane of connective tissue secreting synovial fluid. [mod.L *synovia*, formed prob. arbitrarily by Paracelsus]

synovitis /ˌsaɪnəʊˈvaɪtɪs, ˌsɪn-/ *n. Med.* inflammation of the synovial membrane.

syntactic /sɪnˈtæktɪk/ *adj.* of or according to syntax. □ **syntactical** *adj.* **syntactically** *adv.* [Gk *suntaktikos* (as SYNTAX)]

syntagma /sɪnˈtægmə/ *n.* (*pl.* **syntagmas** or **syntagmata** /-mətə/) **1** a word or phrase forming a syntactic unit. **2** a systematic collection of statements. □ **syntagmic** *adj.* **syntagmatic** /ˌsɪntægˈmætɪk/ *adj.* [LL f. Gk *suntagma* (as SYNTAX)]

syntax /ˈsɪntæks/ *n.* **1** the grammatical arrangement of words, showing their connection and relation. **2** a set of rules for or an analysis of this. [F *syntaxe* or LL *syntaxis* f. Gk *suntaxis* (as SYN-, *taxis* f. *tassō* arrange)]

synth /sɪnθ/ *n. colloq.* = SYNTHESIZER.

synthesis /ˈsɪnθɪsɪs/ *n.* (*pl.* **syntheses** /-ˌsiːz/) **1** the process or result of building up separate elements, esp. ideas, into a connected whole, esp. into a theory or system. **2** a combination or composition. **3** *Chem.* the artificial production of compounds from their constituents as distinct from extraction from plants etc. **4** *Gram.* **a** the process of making compound and derivative words. **b** the tendency in a language to use inflected forms rather than groups of words, prepositions, etc. **5** *Med.* the joining of divided parts in surgery. □ **synthesist** *n.* [L f. Gk *sunthesis* (as SYN-, THESIS)]

synthesize /ˈsɪnθɪsaɪz/ *v.tr.* (also **synthetize** /-ˌtaɪz/, **-ise**) **1** make a synthesis of. **2** combine into a coherent whole.

synthesizer /ˈsɪnθɪsaɪzə(r)/ *n.* an electronic musical instrument, generally operated by a keyboard, producing a wide variety of sounds by generating and combining signals of different frequencies. The development of the synthesizer as a musical instrument dates from the late 1950s, in the US; it has since become an important element in (especially popular) music.

synthetic /sɪnˈθetɪk/ *adj. & n.* ● *adj.* **1** made by chemical synthesis, esp. to imitate a natural product (*synthetic rubber*). **2** (of emotions etc.) affected, insincere. **3** *Logic* (of a proposition) having truth or falsity determinable by recourse to experience (cf. ANALYTIC 3). **4** *Philol.* using combinations of simple words or elements in compounded or complex words (cf. ANALYTICAL 2). ● *n. Chem.* a synthetic substance. □ **synthetic cubism** see CUBISM. **synthetic resin** *Chem.* see RESIN *n.* 2. □ **synthetical** *adj.* **synthetically** *adv.* [F *synthétique* or mod.L *syntheticus* f. Gk *sunthetikos* f. *sunthetos* f. *suntithēmi* (as SYN-, *tithēmi* put)]

syphilis /ˈsɪfɪlɪs/ *n.* a contagious venereal disease caused by the spirochaete *Treponema*, progressing from infection of the genitals via the skin and mucous membrane to the bones, muscles, and brain. □ **syphilize** *v.tr.* (also **-ise**). **syphiloid** *adj.* **syphilitic** /ˌsɪfɪˈlɪtɪk/ *adj.* [mod.L f. title (*Syphilis, sive Morbus Gallicus*) of a Latin poem (1530), f. *Syphilus*, a character in it, the supposed first sufferer from the disease]

syphon var. of SIPHON.

Syracuse 1 /ˈsaɪərəˌkjuːz/ (Italian **Siracusa** /ˌsiraˈkuːza/) a port on the east coast of Sicily; pop. (1990) 125,440. Settled in 734 BC by colonists from Corinth, it became a flourishing centre of Greek culture, especially in the 5th and 4th centuries BC under the rule of Dionysius I and II. It was taken by the Romans at the end of the 3rd century BC. **2** /ˈsɪrəˌkjuːz/ a city in New York State, to the south-east of Lake Ontario; pop. (1990) 163,860. The site of salt springs discovered by the French in 1654, it was an important centre of salt production during the 19th century.

Syr Darya (also **Syrdarya**) see SIRDARYO.

Syria /ˈsɪrɪə/ a country in the Middle East with a coastline on the eastern Mediterranean Sea; pop. (est. 1991) 12,824,000; official language, Arabic; capital, Damascus. In ancient times the name was applied to a much wider area which also encompassed present-day Lebanon, Israel, Jordan, and parts of Iraq and Saudi Arabia. It was the site of various early civilizations, notably that of the Phoenicians, who occupied the coastal plain. Falling successively within the empires of Persia, Macedon, and Rome, it became a centre of Islamic power and civilization from the 7th century and a province of the Ottoman Empire in 1516. After the Turkish defeat in the First World War Syria was mandated to France, becoming independent with the ejection of Vichy troops by the Allies in 1941. From 1958 to 1961 Syria was united with Egypt as the United Arab Republic. Especially since the emergence of Hafiz al-Assad as President (1971), Syria has exercised significant influence in Middle Eastern politics, intervening to enforce a ceasefire

in the Lebanese civil war in 1976. Since then, Syrian forces have remained stationed in parts of Lebanon. □ **Syrian** *adj. & n.*

Syriac /ˈsɪrɪˌæk/ *n. & adj.* ● *n.* the liturgical language of the Maronite and Syrian Catholic Churches, the Syrian Jacobite Church, and the Nestorian Church. (*See note below.*) ● *adj.* of or relating to this language. [L *Syriacus* f. Gk *Suriakos* f. *Suria* Syria]

▪ Syriac is descended from the Aramaic spoken near the city of Edessa (now Urfa) in SE Turkey shortly before the Christian era, and was extensively used in the early Church owing to the active Christian communities in that region. After Greek it was the most important language in the Eastern Roman Empire until the rise of Islam in the 8th century. The Syriac alphabet developed from a form used at Palmyra in Syria.

syringa /sɪˈrɪŋgə/ *n.* **1** the mock orange, *Philadelphus coronarius*. **2** a plant of the genus *Syringa*, esp. the lilac. [mod.L, formed as SYRINX (with ref. to the use of its stems as pipe-stems)]

syringe /sɪˈrɪndʒ/ *n. & v.* ● *n.* **1** *Med.* **a** a tube with a nozzle and piston or bulb for sucking in and ejecting liquid in a fine stream, used in surgery. **b** (in full **hypodermic syringe**) a similar device with a hollow needle for insertion under the skin. **2** any similar device used in gardening, cooking, etc. ● *v.tr.* sluice or spray (the ear, a plant, etc.) with a syringe. [ME f. med.L *syringa* (as SYRINX)]

syrinx /ˈsɪrɪŋks/ *n.* (*pl.* **syrinxes** or **syringes** /sɪˈrɪndʒiːz/) **1** a set of pan-pipes. **2** *Archaeol.* a narrow gallery cut in rock in an ancient Egyptian tomb. **3** the lower larynx or song-organ of birds. □ **syringeal** /sɪˈrɪndʒɪəl/ *adj.* [L *syrinx -ngis* f. Gk *surigx suriggos* pipe, channel]

Syro- /ˈsaɪərəʊ/ *comb. form* Syrian; Syrian and (*Syro-Phoenician*). [Gk *Suro-* f. *Suros* a Syrian]

syrphid /ˈsɜːfɪd/ *adj. & n. Zool.* ● *adj.* of or relating to the dipteran family Syrphidae, which includes the hoverflies. ● *n.* a fly of this family; a hoverfly. [mod.L *Syrphidae* f. *Syrphis* genus name, f. Gk *surphos* gnat]

syrup /ˈsɪrəp/ *n.* (*US* also **sirup**) **1 a** a thick sweet liquid made by dissolving sugar in boiling water, often used for preserving fruit etc. **b** a similar substance of a specified flavour as a drink, medicine, etc. (*rose-hip syrup*). **2** condensed sugar-cane juice; part of this remaining uncrystallized at various stages of refining; molasses, treacle. **3** excessive sweetness of style or manner. □ **syrupy** *adj.* [ME f. OF *sirop* or med.L *siropus* f. Arab. *šarāb* beverage: cf. SHERBET, SHRUB²]

SYSOP /ˈsɪsɒp/ *n. Computing* system operator. [acronym]

syssarcosis /ˌsɪsɑːˈkəʊsɪs/ *n.* (*pl.* **syssarcoses** /-siːz/) *Anat.* a connection between bones formed by intervening muscle. [mod.L f. Gk *sussarkōsis* (as SYN-, *sarx, sarkos* flesh)]

systaltic /sɪˈstæltɪk/ *adj. Physiol.* (esp. of the heart) contracting and dilating rhythmically; pulsatory (cf. SYSTOLE, DIASTOLE). [LL *systalticus* f. Gk *sustaltikos* (as SYN-, *staltos* f. *stellō* put)]

system /ˈsɪstəm/ *n.* **1 a** a complex whole; a set of connected things or parts; an organized body of material or immaterial things. **b** the composition of such a body; arrangement, set-up. **2** a set of devices (e.g. pulleys) functioning together. **3** *Physiol.* **a** a set of organs in the body with a common structure or function (*the digestive system*). **b** the human or animal body as a whole. **4 a** method; considered principles of procedure or classification. **b** classification. **5** orderliness. **6 a** a body of theory or practice relating to or prescribing a particular form of government, religion, etc. **b** (prec. by *the*) the prevailing political or social order, esp. regarded as oppressive and intransigent. **7** a method of choosing one's procedure in gambling etc. **8** *Computing* a group of related hardware units or programs or both, esp. when dedicated to a single application. **9** *Crystallog.* one of seven general types of crystal structure. **10** *Geol.* a major group of geological strata (*the Devonian system*). **11** *Physics* a group of associated bodies moving under mutual gravitation etc. **12** *Mus.* the braced staves of a score. □ **get a thing out of one's system** *colloq.* be rid of a preoccupation or anxiety. **system operator** *Computing* a person who manages the operation of an electronic bulletin board (also called *SYSOP*). **systems analysis** the analysis of a complex process or operation in order to improve its efficiency, esp. by applying a computer system. **systems operator** a person who controls or monitors the operation of complex, esp. electronic, systems. □ **systemless** *adj.* [F *système* or LL *systema* f. Gk *sustēma -atos* (as SYN-, *histēmi* set up)]

systematic /ˌsɪstəˈmætɪk/ *adj.* **1** methodical; done or conceived according to a plan or system. **2** regular, deliberate (*a systematic liar*). □ **systematic theology** a form of theology in which the aim is to arrange religious truths in a self-consistent whole. □ **systematically** *adv.* **systematism** /ˈsɪstəməˌtɪz(ə)m/ *n.* **systematist** *n.* [F *systématique* f. LL *systematicus* f. late Gk *sustēmatikos* (as SYSTEM)]

systematics /ˌsɪstəˈmætɪks/ *n.pl.* (usu. treated as *sing.*) the study or a system of classification; taxonomy.

systematize /ˈsɪstəməˌtaɪz/ *v.tr.* (also **-ise**) **1** arrange systematically; make systematic. **2** devise a system for. □ **systematizer** *n.* **systematization** /ˌsɪstəmətaɪˈzeɪʃ(ə)n/ *n.*

systemic /sɪˈstemɪk, -ˈstiːmɪk/ *adj.* **1** *Physiol.* **a** of or concerning the whole body, not confined to a particular part (*systemic infection*). **b** (of blood circulation) other than pulmonary. **2** (of an insecticide, fungicide, etc.) entering the plant via the roots or shoots and passing through the tissues. □ **systemically** *adv.* [irreg. f. SYSTEM]

systemize /ˈsɪstəˌmaɪz/ *v.tr.* = SYSTEMATIZE. □ **systemizer** *n.* **systemization** /ˌsɪstəmaɪˈzeɪʃ(ə)n/ *n.*

systole /ˈsɪstəlɪ/ *n. Physiol.* the contraction of the heart, when blood is pumped into the arteries (cf. DIASTOLE). □ **systolic** /sɪˈstɒlɪk/ *adj.* [LL f. Gk *sustolē* f. *sustellō* contract (as SYSTALTIC)]

syzygy /ˈsɪzɪdʒɪ/ *n.* (pl. **-ies**) **1** *Astron.* conjunction or opposition, esp. of the moon with the sun. **2** a pair of connected or correlated things. [LL *syzygia* f. Gk *suzugia* f. *suzugos* yoked, paired (as SYN-, *zugon* yoke)]

Szczecin /ˈʃtʃetʃɪn/ (German **Stettin** /ʃteˈtiːn/) a city in NW Poland, a port on the Oder river near the border with Germany; pop. (1990) 413,437.

Szechuan (also **Szechwan**) see SICHUAN.

Szeged /ˈseged/ a city in southern Hungary, a port on the River Tisza near the border with Serbia; pop. (1993) 178,500.

Szilard /ˈsɪlɑːd/, Leo (1898–1964), Hungarian-born American physicist and molecular biologist. Working in Germany before the Second World War, he developed an electromagnetic pump that is now used for refrigerants in nuclear reactors. He fled the Nazis, first to Britain, where he suggested the idea of nuclear chain reactions, and then to the US, where he became a central figure in the Manhattan Project to develop the atom bomb. After the war Szilard turned to experimental and theoretical studies in molecular biology and biochemistry.

Tt

T¹ /tiː/ *n.* (also **t**) (*pl.* **Ts** or **T's**) **1** the twentieth letter of the alphabet. **2** a T-shaped thing (esp. *attrib.*: *T-joint*). □ **to a T** exactly; to a nicety. **2,4,5-T** *abbr.* 2,4,5-trichlorophenoxyacetic acid, a selective herbicide used esp. for controlling brushwood (see also DIOXIN).

T² *symb.* **1** *Chem.* the isotope tritium. **2** tera-. **3** tesla. **4** temperature.

t. *abbr.* **1** ton(s). **2** tonne(s).

't /t/ *pron. contr.* of IT¹ (*'tis*).

-t¹ /t/ *suffix* = -ED² (*crept; sent*).

-t² /t/ *suffix* = -EST² (*shalt*).

TA *abbr.* (in the UK) Territorial Army.

Ta *symb. Chem.* the element tantalum.

ta /taː/ *int. Brit. colloq.* thank you. [childish form]

Taal /taːl/ *n.* (prec. by *the*) the Afrikaans language. [Du., = language, rel. to TALE]

TAB *abbr.* **1** *Med.* typhoid-paratyphoid A and B vaccine. **2** *Austral.* Totalizator Agency Board.

tab¹ /tæb/ *n. & v.* ● *n.* **1 a** a small flap or strip of material attached for grasping, fastening, or hanging up, or for identification. **b** a similar object as part of a garment etc. **2** esp. *N. Amer.* a bill or price (*picked up the tab*). **3** *Brit. Mil.* a marking on the collar distinguishing a staff officer. **4 a** a stage-curtain. **b** a loop for suspending this. ● *v.tr.* (**tabbed**, **tabbing**) provide with a tab or tabs. □ **keep tabs** (or **a tab**) **on** *colloq.* **1** keep account of. **2** have under observation or in check. [prob. f. dial.: cf. TAG¹]

tab² /tæb/ *n.* **1** = TABULATOR 2. **2** = TABULATOR 3. [abbr.]

tab³ /tæb/ *n. sl.* a tablet or pill, esp. one containing an illegal drug. [abbr.]

tabard /'tæbəd/ *n.* **1** a short-sleeved or sleeveless jerkin emblazoned with the arms of the sovereign and forming the official dress of a herald or pursuivant. **2** *hist.* a short surcoat open at the sides and with short sleeves, worn by a knight over armour and emblazoned with armorial bearings. **3** a woman's or girl's sleeveless jerkin open at the sides. [ME f. OF *tabart*, of unkn. orig.]

tabaret /'tæbərɪt/ *n.* an upholstery fabric of alternate satin and plain stripes. [prob. f. TABBY]

Tabasco /tə'bæskəʊ/ a state of SE Mexico, on the Gulf of Mexico; capital, Villahermosa.

tabasco /tə'bæskəʊ/ *n.* **1** a pungent pepper made from the fruit of the chilli, *Capsicum frutescens.* **2** (**Tabasco**) *propr.* a sauce made from this used to flavour food. [TABASCO]

tabbouleh /tə'buːleɪ/ *n.* an Arab vegetable salad made with cracked wheat. [Arab. *tabbūla*]

tabby /'tæbɪ/ *n.* (*pl.* **-ies**) **1** (in full **tabby cat**) **a** a grey or brownish cat mottled or streaked with dark stripes. **b** any domestic cat, esp. female. **2** a kind of watered silk. **3** a plain weave. [F *tabis* (in sense 2) f. Arab. *al-'attabiya* the quarter of Baghdad where tabby was manufactured: connection of other senses uncert.]

tabernacle /'tæbə,næk(ə)l/ *n.* **1** (in the Bible) **a** a fixed or movable habitation, usually of light construction. **b** a tent containing the Ark of the Covenant, used as a portable shrine by the Israelites during their wanderings in the wilderness. **2 a** a meeting-place for worship used by Nonconformists (e.g. Baptists) or by Mormons. **b** *hist.* any of the temporary structures used as churches during the rebuilding after the Fire of London. **3 a** a canopied niche or recess in the wall of a church etc. **b** an ornamental receptacle for the pyx or consecrated elements of the Eucharist. **4** *Naut.* a socket or double post for a hinged mast that can be lowered to pass under low bridges. □ **feast of Tabernacles** = SUCCOTH. □ **tabernacled** *adj.* [ME f. OF *tabernacle* or L *tabernaculum* tent, dimin. of *taberna* hut]

tabes /'teɪbiːz/ *n. Med.* **1** emaciation. **2** locomotor ataxy; a form of neurosyphilis. □ **tabetic** /tə'betɪk/ *adj.* [L, = wasting away]

tabla /'tæblə, 'taːb-/ *n. Ind. Mus.* a pair of small drums played with the hands. [Hind. f. Arab. *ṭabl* drum]

tablature /'tæblətʃə(r)/ *n. Mus.* a system of writing down music to be performed, where diagrams, figures, and letters are used instead of ordinary notation. [F f. It. *tavolatura* f. *tavolare* set to music]

table /'teɪb(ə)l/ *n. & v.* ● *n.* **1 a** a piece of furniture with a flat top and one or more legs, providing a level surface for eating, writing, or working at, playing games on, etc. **2** a flat surface serving a specified purpose (*altar table; bird table*). **3 a** food provided in a household (*keeps a good table*). **b** a group seated at table for dinner etc. **4 a** a set of facts or figures systematically displayed, esp. in columns (*a table of contents*). **b** matter contained in this. **c** = *multiplication table.* **5** a flat surface for working on or for machinery to operate on. **6 a** a slab of wood or stone etc. for bearing an inscription. **b** matter inscribed on this. **7** = TABLELAND. **8** *Archit.* **a** a flat usu. rectangular vertical surface. **b** a horizontal moulding, esp. a cornice. **9 a** a flat surface of a gem. **b** a cut gem with two flat faces. **10** each half or quarter of a folding board for backgammon. **11** (prec. by *the*) *Bridge* the dummy hand. ● *v.tr.* **1** bring forward for discussion or consideration at a meeting. **2** esp. *US* postpone consideration of (a matter). **3** *Naut.* strengthen (a sail) with a wide hem. □ **at table** taking a meal at a table. **lay on the table 1** submit for discussion. **2** esp. *US* postpone indefinitely. **on the table** offered for discussion. **table knife** a knife for use at meals, esp. in eating a main course. **table licence** a licence to serve alcoholic drinks only with meals. **table linen** tablecloths, napkins, etc. **table manners** decorum or correct behaviour while eating at table. **table-mat** a mat for protecting a tabletop from hot dishes etc. **table salt** salt that is ground or easy to grind for use at meals. **table talk** miscellaneous informal talk at table. **table wine** ordinary wine for drinking with a meal. **turn the tables** (often foll. by *on*) reverse one's relations (with), esp. by turning an inferior into a superior position (orig. in backgammon). **under the table** *colloq.* very drunk after a drinking-bout or meal. □ **tableful** *n.* (*pl.* **-fuls**). **tabling** *n.* [ME f. OF f. L *tabula* plank, tablet, list]

tableau /'tæblaʊ/ *n.* (*pl.* **tableaux** /-ləʊz/) **1** a picturesque presentation. **2** = TABLEAU VIVANT. **3** a dramatic or effective situation suddenly brought about. □ **tableau curtains** *Theatr.* a pair of curtains drawn open by a diagonal cord. [F, = picture, dimin. of *table*: see TABLE]

tableau vivant /ˌtæbləʊ 'viːvɒn/ *n.* (*pl.* **tableaux vivants** *pronunc.* same) *Theatr.* a silent and motionless group of people arranged to represent a scene. [F, lit. 'living picture']

tablecloth /'teɪb(ə)l,klɒθ/ *n.* a cloth spread over the top of a table, esp. for meals.

table d'hôte /ˌtaːb(ə)l 'dəʊt/ *n.* a meal consisting of a set menu at a fixed price, esp. in a hotel (cf. À LA CARTE). [F, = host's table]

tableland /'teɪb(ə)l,lænd/ *n.* an extensive elevated region with a level surface; a plateau.

Table Mountain a flat-topped mountain near the south-west tip of

South Africa, overlooking Cape Town and Table Bay, rising to a height of 1,087 m (3,563 ft).

Table of the House (in the UK) the central table in either of the Houses of Parliament.

tablespoon /ˈteɪb(ə)lˌspuːn/ n. **1** a large spoon for serving food. **2** an amount held by this. □ **tablespoonful** n. (pl. **-fuls**).

tablet /ˈtæblɪt/ n. **1** a small measured and compressed amount of a substance, esp. of a medicine or drug. **2** a small flat piece of soap etc. **3** a flat slab of stone or wood, esp. for display or an inscription. **4** Archit. = TABLE 8. **5** N. Amer. a writing-pad. **6** esp. Sc. a type of crumbly fudge. [ME f. OF tablete f. Rmc, dimin. of L tabula TABLE]

table tennis n. an indoor game based on lawn tennis, played by two or four people with small solid bats and a light ball bounced on a table divided by a net. The winner is the first person or pair to reach twenty-one points with a lead of two points or more. Invented in Britain towards the end of the 19th century, table tennis has become very popular elsewhere, especially in China and Korea.

tabletop /ˈteɪb(ə)lˌtɒp/ n. **1** the top or surface of a table. **2** (attrib.) that can be placed or used on a tabletop.

tableware /ˈteɪb(ə)lˌweə(r)/ n. dishes, plates, implements, etc., for use at meals.

tablier /ˈtæblɪˌeɪ/ n. hist. an apron-like part of a woman's dress. [F]

tabloid /ˈtæblɔɪd/ n. **1** a newspaper, usu. popular in style with bold headlines and large photographs, having pages of half the size of those of the average broadsheet. **2** (attrib.) of or in the style of such a newspaper; having mass appeal; sensationalist. **3** anything in a compressed or concentrated form. [orig. the propr. name of a medicine sold in tablets]

taboo /təˈbuː/ n., adj., & v. (also **tabu**) ● n. (pl. **taboos** or **tabus**) **1** a system or the act of setting a person or thing apart as prohibited, sacred, or accursed. **2** a prohibition or restriction on words, forms of behaviour, etc., imposed by social custom. ● adj. **1** avoided or prohibited, esp. by social custom (taboo words). **2** designated as sacred and prohibited. ● v.tr. (**taboos**, **tabooed** or **tabus**, **tabued**) **1** put a thing, practice, etc.) under taboo. **2** exclude or prohibit by authority or social influence. [Tongan tabu]

tabor /ˈteɪbə(r)/ n. hist. a small drum, esp. one used to accompany a pipe. [ME f. OF tabour, tabur: cf. TABLA, Pers. tabīra drum]

tabouret /ˈtæbərɪt/ n. (US **taboret**) a low seat usu. without arms or a back. [F, = stool, dimin. as TABOR]

Tabriz /təˈbriːz/ a city in NW Iran; pop. (1991) 1,089,000. It lies at about 1,367 m (4,485 ft) above sea level at the centre of a volcanic region and has been subject to frequent destructive earthquakes.

tabular /ˈtæbjʊlə(r)/ adj. **1** of or arranged in tables or lists. **2** broad and flat like a table. **3** (of a crystal) having two broad flat faces. **4** formed in thin plates. □ **tabularly** adv. [L tabularis (as TABLE)]

tabula rasa /ˌtæbjʊlə ˈrɑːzə/ n. **1** an erased tablet. **2** the human mind (esp. at birth) viewed as having no innate ideas. [L, = scraped tablet]

tabulate /ˈtæbjʊˌleɪt/ v.tr. arrange (figures or facts) in tabular form. □ **tabulation** /ˌtæbjʊˈleɪʃ(ə)n/ n. [LL tabulare tabulat- f. tabula table]

tabulator /ˈtæbjʊˌleɪtə(r)/ n. **1** a person or thing that tabulates. **2** a device on a typewriter for advancing to a sequence of set positions in tabular work. **3** Computing a machine that produces lists or tables from a data storage medium such as punched cards.

tabun /ˈtɑːbʊn/ n. an organic phosphorus compound used as a nerve gas. [Ger.]

tacamahac /ˈtækəməˌhæk/ n. **1** a resinous gum obtained from certain tropical trees, esp. one of the genus Calophyllum. **2 a** the balsam poplar. **b** the resin of this. [obs. Sp. tacamahaca f. Aztec tecomahiyac]

tac-au-tac /ˈtækəʊˌtæk/ n. Fencing a parry combined with a riposte. [F: imit.]

tacet /ˈtæsɪt, ˈteɪset/ v.intr. Mus. an instruction for a particular voice or instrument to be silent. [L, = is silent]

Taching see DAQING.

tachism /ˈtæʃɪz(ə)m/ n. (also **tachisme**) a chiefly French form of abstract painting popular in the 1940s and 1950s, characterized by dabs of colour arranged randomly with the aim of evoking a subconscious feeling. [F tachisme f. tache stain]

tachistoscope /təˈkɪstəˌskəʊp/ n. an instrument for very brief measured exposure of objects to the eye. □ **tachistoscopic** /-ˌkɪstəˈskɒpɪk/ adj. [Gk takhistos swiftest + -SCOPE]

tacho /ˈtækəʊ/ n. (pl. **-os**) colloq. = TACHOMETER. [abbr.]

tacho- /ˈtækəʊ/ comb. form speed. [Gk takhos speed]

tachograph /ˈtækəˌɡrɑːf/ n. a device used esp. in heavy goods vehicles and coaches etc. for automatically recording speed and travel time.

tachometer /təˈkɒmɪtə(r)/ n. an instrument for measuring the rate of rotation of a shaft and hence the speed or velocity of a vehicle.

tachy- /ˈtæki/ comb. form swift. [Gk takhus swift]

tachycardia /ˌtækɪˈkɑːdɪə/ n. Med. an abnormally rapid heart rate. [TACHY- + Gk kardia heart]

tachygraphy /təˈkɪɡrəfɪ/ n. **1** stenography, esp. that of the ancient Greeks and Romans. **2** the abbreviated medieval writing of Greek and Latin. □ **tachygrapher** n. **tachygraphic** /ˌtækɪˈɡræfɪk/ adj. **tachygraphical** adj.

tachymeter /təˈkɪmɪtə(r)/ n. **1** Surveying an instrument used to locate points rapidly. **2** a speed-indicator.

tachyon /ˈtækjɒn/ n. Physics a hypothetical particle that travels faster than light. [TACHY- + -ON]

tacit /ˈtæsɪt/ adj. understood or implied without being stated (tacit consent). □ **tacitly** adv. [L tacitus silent f. tacere be silent]

taciturn /ˈtæsɪˌtɜːn/ adj. reserved in speech; saying little; uncommunicative. □ **taciturnly** adv. **taciturnity** /ˌtæsɪˈtɜːnɪtɪ/ n. [F taciturne or L taciturnus (as TACIT)]

Tacitus /ˈtæsɪtəs/ (full name Publius, or Gaius, Cornelius Tacitus) (AD c.56–c.120), Roman historian. His major works on the history of the Roman Empire, only partially preserved, are the Annals (covering the years 14–68) and the Histories (69–96). They are written in an elevated and concise style, pervaded by a deep pessimism about the course of Roman history since the end of the Republic.

tack[1] /tæk/ n. & v. ● n. **1** a small sharp broad-headed nail. **2** N. Amer. a drawing-pin. **3** a long stitch used in fastening fabrics etc. lightly or temporarily together. **4** Naut. **a** the direction in which a ship moves as determined by the position of its sails and regarded in terms of the direction of the wind (starboard tack). **b** a temporary change of direction in sailing to take advantage of a side wind etc. **5** a course of action or policy (try another tack). **6** Naut. **a** a rope for securing the corner of some sails. **b** the corner to which this is fastened. **7** a sticky condition of varnish etc. **8** Brit. an extraneous clause appended to a bill in Parliament. ● v. **1** tr. (often foll. by down etc.) fasten with tacks. **2** tr. stitch (pieces of cloth etc.) lightly together. **3** tr. (foll. by to, on) annex (a thing). **4** intr. (often foll. by about) **a** change a ship's course by turning its head to the wind (cf. WEAR[2] + -ON). **b** make a series of tacks. **5** intr. change one's conduct or policy etc. **6** tr. Brit. append (a clause) to a bill. □ **tacker** n. [ME tak etc., of uncert. orig.: cf. biblical tache clasp, link f. OF tache]

tack[2] /tæk/ n. the saddle, bridle, etc., of a horse. [shortened f. TACKLE]

tack[3] /tæk/ n. colloq. cheap or shoddy material; tat; kitsch. [back-form. f. TACKY[2]]

tack[4] /tæk/ n. (esp. in **hard tack**, **soft tack**) food. [16th c.: orig. unkn.]

tacker /ˈtækə(r)/ n. Austral. & dial. a small boy.

tackle /ˈtæk(ə)l/ n. & v. ● n. **1** equipment for a task or sport (fishing-tackle). **2** a mechanism, esp. of ropes, pulley-blocks, hooks, etc., for lifting weights, managing sails, etc. (block and tackle). **3** a windlass with its ropes and hooks. **4** an act of tackling in football etc. **5** Amer. Football **a** the position next to the end of the forward line. **b** the player in this position. ● v.tr. **1** try to deal with (a problem or difficulty). **2** (in football etc.) grapple with or try to overcome (an opponent). **3** initiate discussion with on some esp. disputed subject. **4** obstruct, intercept, or seize and stop (a player running with the ball). **5** secure by means of tackle. □ **tackle-block** a pulley over which a rope runs. **tackle-fall** a rope for applying force to the blocks of a tackle. □ **tackler** n. **tackling** n. [ME, prob. f. MLG takel f. taken lay hold of]

tacky[1] /ˈtækɪ/ adj. (**tackier**, **tackiest**) (of glue or paint etc.) still slightly sticky after application. □ **tackiness** n. [TACK[1] + -Y[1]]

tacky[2] /ˈtækɪ/ adj. (**tackier**, **tackiest**) esp. US colloq. **1** showing poor taste or style. **2** tatty or seedy. □ **tackily** adv. **tackiness** n. [19th c.: orig. unkn.]

taco /ˈtɑːkəʊ/ n. (pl. **-os**) a Mexican dish of meat etc. in a folded or rolled tortilla. [Mex. Sp.]

tact /tækt/ n. **1** adroitness in dealing with others or with difficulties arising from personal feeling. **2** intuitive perception of the right thing to do or say. [F f. L tactus touch, sense of touch f. tangere tact- touch]

tactful /ˈtæktfʊl/ adj. having or showing tact. □ **tactfully** adv. **tactfulness** n.

tactic /'tæktɪk/ *n.* **1** a tactical manoeuvre. **2** = TACTICS 1. [mod.L *tactica* f. Gk *taktikē* (*tekhnē* art): see TACTICS]

tactical /'tæktɪk(ə)l/ *adj.* **1** of, relating to, or constituting tactics (*a tactical retreat*). **2** (of bombing or weapons) done or for use in immediate support of military or naval operations (opp. STRATEGIC 3). **3** adroitly planning or planned. **4** (of voting) aimed at preventing the strongest candidate from winning by supporting the next strongest. □ **tactically** *adv.* [Gk *taktikos* (as TACTICS)]

tactics /'tæktɪks/ *n.pl.* **1** (also treated as *sing.*) the art of disposing armed forces in order of battle and of carrying out manoeuvres in actual contact with an enemy (cf. STRATEGY 2b). **2 a** the plans and means adopted in carrying out a scheme or achieving some end. **b** a skilful device or devices. □ **tactician** /tæk'tɪʃ(ə)n/ *n.* [mod.L *tactica* f. Gk *taktika* neut.pl. f. *taktos* ordered f. *tassō* arrange]

tactile /'tæktaɪl/ *adj.* **1** of or connected with the sense of touch. **2** perceived by touch. **3** tangible. **4** *Art* (in painting) producing or concerning the effect of three-dimensional solidity. □ **tactual** /-tʃʊəl, -tjʊəl/ *adj.* (in senses 1, 2). **tactility** /tæk'tɪlɪtɪ/ *n.* [L *tactilis* f. *tangere* tact-touch]

tactless /'tæktlɪs/ *adj.* having or showing no tact. □ **tactlessly** *adv.* **tactlessness** *n.*

tad /tæd/ *n.* esp. *N. Amer. colloq.* a small amount (often used adverbially: *a tad too salty*). [19th c.: orig. unkn.]

Tadmur /'tædmʊə(r)/ (also **Tadmor** /-mɔ:(r)/) see PALMYRA.

tadpole /'tædpəʊl/ *n.* a larva of an amphibian, esp. a frog, toad, or newt in its aquatic stage and breathing through gills. [ME *taddepolle* (as TOAD, POLL[1] from the size of its head)]

Tadzhik var. of TAJIK.

Tadzhikistan see TAJIKISTAN.

taedium vitae /ˌtaɪdɪəm 'viːtaɪ, ˌtiːdɪəm 'vaɪtiː/ *n.* weariness of life (often as a pathological state, with a tendency to suicide). [L]

Taegu /tæ'guː/ a city in SE South Korea; pop. (1990) 2,228,830. Nearby is the Haeinsa temple, established in AD 802, which contains 80,000 wooden printing-blocks dating from the 13th century, engraved with compilations of Buddhist scriptures.

Taejon /tæ'dʒɒn/ a city in central South Korea; pop. (1990) 1,062,080.

tae kwon do /ˌtaɪ kwɒn 'dəʊ/ *n.* a Korean form of unarmed combat developed chiefly in the mid-20th century, combining elements of karate, ancient Korean martial art, and kung fu. It differs from karate in its wide range of kicking techniques and its emphasis on different methods of breaking objects. [Korean, lit. 'art of hand and foot fighting']

taenia /'tiːnɪə/ *n.* (US **tenia**) (*pl.* **taeniae** /-nɪˌiː/ or **taenias**) **1** *Archit.* a fillet between a Doric architrave and frieze. **2** *Anat.* a flat ribbon-like structure, esp. any of the muscles of the colon. **3** a large tapeworm of the genus *Taenia*; esp. *T. saginata* and *T. soleum*, parasitic in humans. **4** *Gk Antiq.* a fillet or headband. □ **taenioid** *adj.* [L f. Gk *tainia* ribbon]

taffeta /'tæfɪtə/ *n.* a fine lustrous silk or silklike fabric. [ME f. OF *taffetas* or med.L *taffata*, ult. f. Pers. *tāfta* past part. of *tāftan* twist]

taffrail /'tæfreɪl/ *n. Naut.* a rail round a ship's stern. [earlier *tafferel* f. Du. *tafereel* panel, dimin. of *tafel* (as TABLE): assim. to RAIL[1]]

Taffy /'tæfɪ/ *n.* (*pl.* **-ies**) *colloq.* often *offens.* a Welshman. [supposed Welsh pronunc. of *Davy* = *David* (Welsh *Dafydd*)]

taffy /'tæfɪ/ *n.* (*pl.* **-ies**) **1** *N. Amer.* a confection like toffee. **2** *US colloq.* insincere flattery. [19th c.: orig. unkn.]

tafia /'tæfɪə/ *n. West Indies* rum distilled from molasses etc. [18th c.: orig. uncert.]

Taft /tæft/, William Howard (1857–1930), American Republican statesman, 27th President of the US 1909–13. His presidency is remembered for its dollar diplomacy in foreign affairs and for its tariff laws, which were criticized as being too favourable to big business. Taft later served as Chief Justice of the Supreme Court (1921–30).

tag[1] /tæg/ *n. & v.* ● *n.* **1 a** a label, esp. one for tying on an object to show its address, price, etc. **b** *colloq.* an epithet or popular name serving to identify a person or thing. **c** *sl.* the signature or identifying mark of a graffiti artist. **2** a metal or plastic point at the end of a lace etc. to assist insertion. **3** a loop at the back of a boot used in pulling it on. **4** *US* a licence plate of a motor vehicle. **5** an electronic device that can be attached to a person or thing for monitoring purposes, e.g. to track offenders under house arrest or to deter shoplifters. **6** a loose or ragged end of anything. **7** a ragged lock of wool on a sheep. **8** *Theatr.* a closing speech addressed to the audience. **9** a trite quotation or stock phrase. **10 a** the refrain of a song. **b** a musical phrase added to the end of a

piece. **11** an animal's tail, or its tip. ● *v.tr.* (**tagged**, **tagging**) **1** provide with a tag or tags. **2** (often foll. by *on, on to*) join or attach. **3** *colloq.* follow closely or trail behind. **4** *Computing* identify (an item of data) by its type for later retrieval. **5** label radioactively (see LABEL *v.* 3). **6 a** find rhymes for (verses). **b** string (rhymes) together. **7** shear away tags from (sheep). □ **tag along** (often foll. by *with*) go along or accompany passively. **tag end** esp. *US* the last remnant of something. **tag line** *US* **1** = *punch-line*. **2** a slogan. [ME: orig. unkn.]

tag[2] /tæg/ *n. & v.* ● *n.* **1** a children's game in which one chases the rest, and anyone who is caught then becomes the pursuer. **2** *Baseball* the act of tagging a runner. ● *v.tr.* (**tagged**, **tagging**) **1** touch in a game of tag. **2** (often foll. by *out*) *Baseball* put (a runner) out by touching with the ball or with the hand holding the ball. [18th c.: orig. unkn.]

Tagalog /tə'gɑːlɒg/ *n. & adj.* ● *n.* **1** a member of the principal people of the Philippines. **2** the language of this people, which belongs to the Malayo-Polynesian language group although its vocabulary has been heavily influenced by Spanish with some adaptations from Chinese and Arabic. ● *adj.* of or relating to this people or language. [Tagalog f. *taga* native + *ilog* river]

Taganrog /ˌtægən'rɒg/ an industrial port in SW Russia, on the Gulf of Taganrog, an inlet of the Sea of Azov; pop. (1990) 293,000. It was founded in 1698 by Peter the Great as a fortress and naval base.

tagetes /tə'dʒiːtiːz/ *n.* a plant of the genus *Tagetes*, esp. a marigold with bright orange or yellow flowers. [mod.L f. L *Tages* an Etruscan god]

tagliatelle /ˌtæljə'telɪ/ *n.pl.* pasta in the form of narrow ribbons. [It., f. *tagliare* to cut]

tagmeme /'tægmiːm/ *n. Linguistics* **1** the smallest meaningful unit of grammatical form. **2** the correlate of a grammatical function and the class of items which can perform it. □ **tagmemic** /tæg'miːmɪk/ *adj.* [Gk *tagma* arrangement, after *phoneme*]

tagmemics /tæg'miːmɪks/ *n.* the study and description of language in terms of tagmemes, based on the work of K. L. Pike (b.1912), which stresses the functional and structural relations of grammatical units.

Tagore /tə'gɔː(r)/, Rabindranath (1861–1941), Indian writer and philosopher. His poetry pioneered the use of colloquial Bengali instead of the archaic literary idiom then approved for verse; his own translations established his reputation in the West, and he won the Nobel Prize for literature in 1913 for *Gitanjali* (1912), a set of poems modelled on medieval Indian devotional lyrics. He also wrote philosophical plays, novels such as *Gora* (1929), and short fiction which often commented on Indian national and social concerns. He was knighted in 1915, an honour which he renounced after the Amritsar massacre (1919).

Tagus /'teɪgəs/ (Spanish **Tajo** /'taxo/, Portuguese **Tejo** /'teʒu/) a river in SW Europe, the longest river of the Iberian peninsula, which rises in the mountains of eastern Spain and flows over 1,000 km (625 miles) generally westwards into Portugal, where it turns south-westwards, emptying into the Atlantic near Lisbon.

Tahiti /tə'hiːtɪ/ an island in the central South Pacific, one of the Society Islands, forming part of French Polynesia; pop. (1988) 115,820; capital, Papeete. One of the largest islands in the South Pacific, it was claimed for France in 1768 and declared a French colony in 1880. □ **Tahitian** /-'hiːʃ(ə)n/ *adj. & n.*

tahr /tɑː(r)/ *n.* (also **thar**) a goatlike mammal of the genus *Hemitragus*, esp. *H. jemlahicus* of the Himalayas. [local Himalayan name]

tahsil /tɑː'siːl/ *n.* an administrative area in parts of India. [Urdu *taḥsīl* f. Arab., = collection]

Tai'an /taɪ'ɑːn/ a city in NE China, in Shandong province; pop. (1986) 1,370,000.

t'ai chi /taɪ 'tʃiː/ *n.* **1** (also **t'ai chi ch'uan** /ˌtaɪ tʃiː 'tʃwɑːn/) a Chinese martial art and system of callisthenics consisting of sequences of very slow controlled movements. It is believed to have been devised by a Taoist priest in the Sung dynasty. **2** (in Chinese philosophy) the ultimate source and limit of reality, from which spring yin and yang and all of creation. The concept occurs first in the I Ching. [Chin., f. *tài* extreme + *ji* limit]

Taichung /taɪ'tʃʊŋ/ a city in west central Taiwan; pop. (1991) 774,000.

Ta'if /'tɑːɪf/ a city in western Saudi Arabia, situated to the south-east of Mecca in the Asir Mountains; pop. (est. 1986) 204,850. It is the unofficial seat of government of Saudi Arabia during the summer.

Taig /teɪg/ *n. sl. offens.* (in Northern Ireland) a Protestant name for a Catholic. [var. of *Teague*, anglicized spelling of the Irish name *Tadhg*, a nickname for an Irishman]

taiga /ˈtaɪgə/ n. coniferous forest lying between tundra and steppe, esp. in Siberia. [Russ.]

tail[1] /teɪl/ n. & v. ● n. **1** the hindmost part of an animal, esp. when prolonged beyond the rest of the body. **2 a** a thing like a tail in form or position, esp. something extending downwards or outwards at an extremity. **b** the rear end of anything, e.g. of a procession. **c** a long train or line of people, vehicles, etc. **3 a** the rear part of an aeroplane, with the tailplane and rudder, or of a rocket. **b** the rear part of a motor vehicle. **4** the luminous trail of particles following a comet. **5 a** the inferior or weaker part of anything, esp. in a sequence. **b** *Cricket* the end of the batting order, with the weakest batsmen. **6 a** the part of a shirt below the waist. **b** the hanging part of the back of a coat. **7** (in pl.) colloq. **a** a tailcoat. **b** evening dress including this. **8 a** sl. the buttocks. **b** coarse sl. the female genitals; sexual intercourse. **9** (in pl.) the reverse of a coin as a choice when tossing. **10** colloq. a person following or shadowing another. **11** an extra strip attached to the lower end of a kite. **12** the stem of a note in music. **13** the part of a letter (e.g. y) below the line. **14 a** the exposed end of a slate or tile in a roof. **b** the unexposed end of a brick or stone in a wall. **15** the slender backward prolongation of a butterfly's wing. **16** a comparative calm at the end of a gale. **17** a calm stretch following rough water in a stream. ● v. **1** tr. remove the stalks of (fruit). **2** tr. & (foll. by after) intr. colloq. shadow or follow closely. **3** tr. provide with a tail. **4** tr. dock the tail of (a lamb etc.). **5** tr. (often foll. by on to) join (one thing to another). □ **on a person's tail** closely following a person. **tail back** (of traffic) form a tailback. **tail covert** any of the feathers covering the base of a bird's tail feathers. **tail-end 1** the hindmost or lowest or last part. **2** = sense 5 of n. **tail-ender** a person at the tail-end of something, esp. in cricket and athletic races. **tail in** fasten (timber) by one end into a wall etc. **tail-light** (or **-lamp**) esp. N. Amer. a light at the rear of a train, motor vehicle, or bicycle. **tail off** (or **away**) **1** become fewer, smaller, or slighter. **2** fall behind or away in a scattered line. **tail-off** n. a decline or gradual reduction, esp. in demand. **tail-race** the part of a mill-race below the water-wheel. **tail-skid** a support for the tail of an aircraft when on the ground. **tail wind** a wind blowing in the direction of travel of a vehicle or aircraft etc. **with one's tail between one's legs** in a state of dejection or humiliation. **with one's tail up** in good spirits; cheerful. □ **tailed** adj. (also in comb.). **tailless** /ˈteɪllɪs/ adj. [OE tægl, tægel f. Gmc]

tail[2] /teɪl/ n. & adj. Law ● n. limitation of ownership, esp. of an estate limited to a person and that person's heirs. ● adj. so limited (estate tail; fee tail). □ **in tail** under such a limitation. [ME f. OF taille notch, cut, tax, f. taillier cut ult. f. L talea twig]

tailback /ˈteɪlbæk/ n. Brit. a long line of traffic extending back from an obstruction.

tailboard /ˈteɪlbɔːd/ n. a hinged or removable flap at the rear of a lorry etc.

tailcoat /ˈteɪlkəʊt/ n. a man's morning or evening coat with a long skirt divided at the back into tails and cut away in front, worn as part of formal dress.

tailgate /ˈteɪlgeɪt/ n. & v. ● n. **1** esp. US **a** = TAILBOARD. **b** the tail door of an estate car or hatchback. **2** the lower end of a canal lock. ● v. US colloq. **1** intr. drive too closely behind another vehicle. **2** tr. follow (a vehicle) too closely. □ **tailgater** n.

tailie /ˈteɪli/ n. Austral. & NZ (in the game of two-up) a person who bets on the coins falling tail upwards.

tailing /ˈteɪlɪŋ/ n. **1** (in pl.) the refuse or inferior part of grain or ore etc. **2** the part of a beam or projecting brick etc. embedded in a wall.

Tailleferre /taɪˈfeə(r)/, Germaine (1892–1983), French composer and pianist. She was a pupil of Ravel and later became a member of Les Six. Her works include concertos for unusual combinations of instruments, including one for baritone, piano, and orchestra.

tailor /ˈteɪlə(r)/ n. & v. ● n. a maker of clothes, esp. one who makes men's outer garments to measure. ● v. **1** tr. make (clothes) as a tailor. **2** tr. make or adapt for a special purpose. **3** intr. work as or be a tailor. **4** tr. (esp. as **tailored** adj.) make clothes for (he was immaculately tailored). **5** tr. (as **tailored** adj.) = tailor-made. □ **tailor-bird** a small Asian warbler of the genus Orthotomus that stitches leaves together to form a nest. **tailor-made** adj. **1** (of clothing) made to order by a tailor. **2** made or suited for a particular purpose (a job tailor-made for me). ● n. a tailor-made garment. **tailor's chair** a chair without legs for sitting cross-legged like a tailor at work. **tailor's twist** a fine strong silk thread used by tailors. □ **tailoring** n. [ME & AF taillour, OF tailleur cutter, formed as TAIL[2]]

tailored /ˈteɪləd/ adj. (of clothing) well or closely fitted.

tailpiece /ˈteɪlpiːs/ n. **1** an appendage at the rear of anything. **2** the final part of a thing. **3** a decoration in a blank space at the end of a chapter etc. in a book. **4** a piece of wood to which the strings of some musical instruments are attached at their lower ends.

tailpipe /ˈteɪlpaɪp/ n. the rear section of the exhaust pipe of a motor vehicle.

tailplane /ˈteɪlpleɪn/ n. a horizontal aerofoil at the tail of an aircraft.

tailspin /ˈteɪlspɪn/ n. & v. ● n. **1** a spin (see SPIN n. 2) by an aircraft with the tail spiralling. **2** a state of chaos or panic. ● v.intr. (**-spinning**; past and past part. **-spun** /-spʌn/) perform a tailspin.

tailstock /ˈteɪlstɒk/ n. the adjustable part of a lathe holding the fixed spindle.

Taimyr Peninsula /taɪˈmɪə(r)/ (also **Taymyr**) a vast, almost uninhabited peninsula on the north coast of central Russia, extending into the Arctic Ocean and separating the Kara Sea from the Laptev Sea. Its northern tip is the northernmost point of Asia.

Tainan /taɪˈnɑːn/ a city on the SW coast of Taiwan; pop. (1991) 690,000. Settled from mainland China in 1590, it is one of the oldest cities on the island and was its capital from 1684 until 1885, when it was replaced by Taipei. Its original name was Taiwan, the name later given to the whole island.

taint /teɪnt/ n. & v. ● n. **1** a spot or trace of decay, infection, or some bad quality. **2** an unpleasant scent or smell. **3** a corrupt condition or infection. ● v. **1** tr. affect with a taint. **2** tr. (foll. by with) affect slightly. **3** intr. become tainted. □ **taintless** adj. [ME, partly f. OF teint(e) f. L tinctus f. tingere dye, partly f. ATTAINT]

taipan[1] /ˈtaɪpæn/ n. the head of a foreign business in China. [Chin.]

taipan[2] /ˈtaɪpæn/ n. a large venomous Australian snake, Oxyuranus microlepidotus. [Aboriginal]

Taipei /taɪˈpeɪ/ the capital of Taiwan; pop. (1991) 2,718,000. It developed as an industrial city in the 19th century, and became the capital in 1885.

Taiping Rebellion /taɪˈpɪŋ/ a sustained uprising against the Qing dynasty in China 1850–64. The rebellion was led by Hong Xinquan (1814–64), who, claiming to be a son of God, had founded a religious group inspired by elements of Christian theology and proposing egalitarian social policies. An army of a million men and women captured Nanjing in 1853 but failed to take Beijing and suffered defeat at Shanghai at the hands of an army trained by the British general Gordon. The rebellion was finally defeated after the recapture of Nanjing, some 20 million people having been killed, but the Qing dynasty was severely weakened as a result.

Taiwan /taɪˈwɑːn/ (official name **Republic of China**) an island country off the SE coast of China; pop. (1991) 20,400,000; official language, Mandarin Chinese; capital, Taipei. Settled for centuries by the Chinese, the island was sighted by the Portuguese in 1590; they named it Formosa (= beautiful). It was ceded to Japan by China in 1895 but was returned to China after the Second World War. Chiang Kai-shek, leader of the Kuomintang, withdrew there in 1949 with 500,000 nationalist troops, towards the end of the war with the Communist regime of mainland China; Taiwan became the headquarters of the Kuomintang, which has held power continuously since then. Since the 1950s Taiwan has undergone steady economic growth, particularly in its export-oriented industries. In 1971 it lost its seat in the United Nations to the People's Republic of China, which regards Taiwan as one of its provinces. □ **Taiwanese** /ˌtaɪwəˈniːz/ adj. & n.

Taiyuan /ˌtaɪjʊˈɑːn/ a city in northern China, capital of Shanxi province; pop. (1990) 1,900,000.

Tai Yue Shan /ˌtaɪ jʊeɪ ˈʃæn/ the Chinese name for LANTAU.

Ta'iz /tæˈɪz/ a city in SW Yemen; pop. (1987) 178,000. It was the administrative capital of Yemen from 1948 to 1962.

taj /tɑːdʒ/ n. a tall conical cap worn by a dervish. [Arab. tāj]

Tajik /tɑːˈdʒiːk/ n. & adj. (also **Tadzhik**) ● n. (pl. same or **Tajiks**) **1 a** a member of a Muslim people inhabiting Tajikistan and Afghanistan. **b** a native or national of Tajikistan. **2** the Iranian language of this people, a form of Persian. ● adj. of or relating to Tajikistan, the Tajiks, or their language. [Pers.]

Tajikistan /tɑːˌdʒiːkɪˈstɑːn/ (also **Tadzhikistan**) a mountainous republic in central Asia, north of Afghanistan; pop. (est. 1991) 5,412,000; languages, Tajik (official), Russian; capital, Dushanbe. The region was conquered by the Mongols in the 13th century and absorbed into the Russian empire during the 1880s and 1890s. The Tajiks rebelled against Russian authority after the 1917 revolution and were not subjugated until 1921. At first incorporated into Uzbekistan, from 1929 Tajikistan formed a constituent republic of the USSR; it became an independent republic within the Commonwealth of Independent States in 1991.

Taj Mahal /ˌtɑːʒ məˈhɑːl, ˌtɑːˈdʒ/ a mausoleum at Agra in northern India, completed c.1649. The Taj Mahal was built by the Mogul emperor Shah Jahan (1592–1666) in memory of his favourite wife, who had borne him fourteen children. Set in formal gardens, the domed building in white marble is reflected in a pool flanked by cypresses. [perh. corrupt. of Pers. *Mumtaz Mahal*, title of wife of Shah Jahan, f. *mumtāz* chosen one, *mahal* abode]

Tajo see TAGUS.

taka /ˈtɑːkɑː/ n. the basic monetary unit of Bangladesh, equal to 100 poisha. [Bengali]

takahe /ˈtɑːkəhɪ/ n. a rare giant rail, *Porphyrio mantelli*, of New Zealand, with iridescent blue-green plumage and a thick red bill. Also called *notornis*. [Maori]

take /teɪk/ v. & n. ● v. (*past* **took** /tʊk/; *past part.* **taken** /ˈteɪkən/) **1** *tr.* lay hold of; get into one's hands. **2** *tr.* acquire, get possession of, capture, earn, or win. **3** *tr.* get the use of by purchase or formal agreement (*take lodgings*). **4** *tr.* (in a recipe) avail oneself of; use (*take a large mixing bowl*). **5** *tr.* use as a means of transport (*took a taxi*). **6** *tr.* regularly buy or subscribe to (a particular newspaper or periodical etc.). **7** *tr.* obtain after fulfilling the required conditions (*take a degree*). **8** *tr.* occupy (*take a chair*). **9** *tr.* make use of (*take precautions*). **10** *tr.* consume as food or medicine (*took tea; took the pills*). **11** *intr.* **a** be successful or effective (*the inoculation did not take*). **b** (of a plant, seed, etc.) begin to grow. **12** *tr.* **a** require or use up (*will only take a minute; these things take time*). **b** accommodate, have room for (*the lift takes three people*). **13** *tr.* wear (a particular size of garment etc.) (*takes size six*). **14** *tr.* cause to come or go with one; convey (*take the book home; the bus will take you all the way*). **15** *tr.* **a** remove; dispossess a person of (*someone has taken my pen*). **b** destroy, annihilate (*took her own life*). **c** (often foll. by *for*) *sl.* defraud, swindle. **16** *tr.* catch or be infected with (fire or fever etc.). **17** *tr.* **a** experience or be affected by (*take fright; take pleasure*). **b** give play to (*take comfort*). **c** exert (*take courage; take no notice*). **d** exact, get (*take revenge*). **18** *tr.* find out and note (a name and address; a person's temperature etc.) by enquiry or measurement. **19** *tr.* grasp mentally; understand (*I take your point; I took you to mean yes*). **20** *tr.* treat or regard in a specified way (*took the news calmly; took it badly*). **21** *tr.* (foll. by *for* or *to be*) regard as being (*do you take me for an idiot?*). **22** *tr.* **a** accept (*take the offer; take a phone call; takes lodgers*). **b** submit to (*take a joke; take no nonsense; took a risk*). **23** *tr.* choose or assume (*took a different view; took a job; took the initiative; took responsibility*). **24** *tr.* derive (*takes its name from the inventor*). **25** *tr.* (foll. by *from*) subtract (*take 3 from 9*). **26** *tr.* execute, make, or undertake; perform or effect (*take notes; take an oath; take a decision; take a look*). **27** *tr.* occupy or engage oneself in; indulge in; enjoy (*take a rest; take exercise; take a holiday*). **28** *tr.* conduct (*took the school assembly*). **29** *tr.* deal with in a certain way (*took the corner too fast*). **30** *tr.* **a** teach or be taught (a subject). **b** be examined in (a subject). **31** *tr.* make (a photograph) with a camera; photograph (a person or thing). **32** *tr.* use as an instance (*let us take Napoleon*). **33** *tr. Gram.* have or require as part of the appropriate construction (*this verb takes an object*). **34** *tr.* have sexual intercourse with (a woman). **35** *tr.* (in *passive*; foll. by *by*, *with*) be attracted or charmed by. ● n. **1** an amount taken or caught in one session or attempt etc. **2** a scene or sequence of film photographed continuously at one time. **3** esp. *US* takings, esp. money received at a theatre for seats. **4** *Printing* the amount of copy set up at one time. □ **be taken ill** become ill, esp. suddenly. **have what it takes** *colloq.* have the necessary qualities etc. for success. **take account of** see ACCOUNT. **take action** see ACTION. **take advantage of** see ADVANTAGE. **take advice** see ADVICE. **take after** resemble (esp. a parent or ancestor). **take against** begin to dislike, esp. impulsively. **take aim** see AIM. **take apart 1** dismantle. **2** *colloq.* beat or defeat conclusively. **3** *colloq.* criticize severely. **take aside** see ASIDE. **take as read** accept without reading or considering. **take away 1** remove or carry elsewhere. **2** subtract. **3** *Brit.* buy (food etc.) at a shop or restaurant for eating elsewhere. **take-away** *Brit. attrib.adj.* (of food) bought at a shop or restaurant for eating elsewhere. ● n. **1** an establishment selling such food. **2** the food itself (*let's get a take-away*). **take back 1** retract (a statement). **2** convey (a person or thing) to his or her or its original position. **3** carry (a person) in thought to a past time. **4** *Printing* transfer to the previous line. **5 a** return (goods) to a shop. **b** (of a shop) accept such goods. **6** accept (a person) back into one's affections, into employment, etc. **take the biscuit** (or **bun** or **cake**) *colloq.* be the most remarkable. **take a bow** see BOW². **take care of** see CARE. **take a chance** etc. see CHANCE. **take down 1** write down (spoken words).

2 remove (a structure) by dismantling. **3** humiliate. **4** lower (a garment worn below the waist). **take effect** see EFFECT. **take for granted** see GRANT. **take fright** see FRIGHT. **take from** diminish; weaken; detract from. **take heart** be encouraged. **take hold** see HOLD¹. **take home earn. take-home pay** the pay received by an employee after the deduction of tax etc. **take ill** (*US* **sick**) *colloq.* be taken ill. **take in 1** receive as a lodger etc. **2** undertake (work) at home. **3** make (a garment etc.) smaller. **4** understand (*did you take that in?*). **5** cheat (*managed to take them all in*). **6** include or comprise. **7** absorb into the body. **8** *colloq.* visit (a place) on the way to another (*shall we take in Avebury?*). **9** furl (a sail). **10** *Brit.* regularly buy (a newspaper etc.). **take-in** n. a deception. **take in hand 1** undertake; start doing or dealing with. **2** undertake the control or reform of (a person). **take into account** see ACCOUNT. **take it 1** (often foll. by *that* + clause) assume (*I take it that you have finished*). **2** endure a difficulty or hardship in a specified way (*took it badly*). **take it easy** see EASY. **take it from me** (or **take my word for it**) I can assure you. **take it ill** resent it. **take it into one's head** see HEAD. **take it on one** (or **oneself**) (foll. by *to* + infin.) venture or presume. **take it or leave it** (esp. in *imper.*) an expression of indifference or impatience about another's decision after making an offer. **take it out of 1** exhaust the strength of. **2** have revenge on. **take it out on** relieve one's frustration by attacking or treating harshly. **take one's leave of** see LEAVE². **take a lot of** (or **some**) **doing** be hard to do. **take a person's name in vain** see VAIN. **take off 1 a** remove (clothing) from one's or another's body. **b** remove or lead away. **c** withdraw (transport, a show, etc.). **2** deduct (part of an amount). **3** depart, esp. hastily (*took off in a fast car*). **4** *colloq.* mimic humorously. **5** jump from the ground. **6** become airborne. **7** (of a scheme, enterprise, etc.) become successful or popular. **8** have (a period) away from work. **take-off 1** the act of becoming airborne. **2** *colloq.* an act of mimicking. **3** a place from which one jumps. **take oneself off** go away. **take on 1** undertake (work, a responsibility, etc.). **2** engage (an employee). **3** be willing or ready to meet (an adversary in sport, argument, etc., esp. a stronger one). **4** acquire (a new meaning etc.). **5** *colloq.* show strong emotion. **take orders** see ORDER. **take out 1** remove from within a place; extract. **2** escort on an outing. **3** get (a licence or summons etc.) issued. **4** *US* = **take away** 3. **5** *Bridge* remove (a partner or a partner's call) from a suit by bidding a different one or no trumps. **6** *sl.* murder or destroy. **take a person out of himself** or **herself** make a person forget his or her worries. **take over 1** succeed to the management or ownership of. **2** take control (of). **3** *Printing* transfer to the next line. **take part** see PART. **take place** see PLACE. **take a person's point** see POINT. **take shape** assume a distinct form; develop into something definite. **take sides** see SIDE. **take stock** see STOCK. **take the sun** see SUN. **take that!** an exclamation accompanying a blow etc. **take one's time** not hurry. **take to 1** begin or fall into the habit of (*took to smoking*). **2** have recourse to. **3** adapt oneself to. **4** form a liking for. **5** make for (*took to the hills*). **take to heart** see HEART. **take to one's heels** see HEEL¹. **take to pieces** see PIECE. **take the trouble** see TROUBLE. **take up 1** become interested or engaged in (a pursuit, a cause, etc.). **2** adopt as a protégé. **3** occupy (time or space). **4** begin (residence etc.). **5** resume after an interruption. **6** interrupt or question (a speaker). **7** accept (an offer etc.). **8** shorten (a garment). **9** lift up. **10** absorb (*sponges take up water*). **11** take (a person) into a vehicle. **12** pursue (a matter etc.) further. **take a person up on** accept (a person's offer etc.). **take up with** begin to associate with. □ **takable** adj. (also **takeable**). [OE *tacan* f. ON *taka*]

takeover /ˈteɪkˌəʊvə(r)/ n. the assumption of control (esp. of a business); the buying-out of one company by another.

taker /ˈteɪkə(r)/ n. **1** a person who takes a bet. **2** a person who accepts an offer.

takin /ˈtɑːkɪn/ n. a large shaggy horned ruminant, *Budorcas taxicolor*, of Tibet, Bhutan, and Burma. [name in a local language]

taking /ˈteɪkɪŋ/ adj. & n. ● adj. **1** attractive or captivating. **2** catching or infectious. ● n. (in pl.) an amount of money taken in business. □ **takingly** adv. **takingness** n.

Taklimakan Desert /ˌtækləməˈkɑːn/ (also **Takla Makan**) a desert in the Xinjiang autonomous region of NW China, lying between the Kunlun Shan and Tien Shan mountains and forming the greater part of the Tarim Basin.

Takoradi /ˌtɑːkəˈrɑːdɪ/ a seaport in western Ghana, on the Gulf of Guinea; pop. (1984) 615,000. It is part of the joint urban area of Sekondi-Takoradi and is one of the major seaports of West Africa.

tala¹ /ˈtɑːlə/ n. any of the traditional rhythmic patterns of Indian music. [Skr.]

tala² /'tɑːlə/ n. the basic monetary unit of Western Samoa, equal to 100 sene. [Samoan]

Talaing /tə'laɪŋ/ n. & adj. (pl. same or **Talaings**) = MON. [Burmese]

talapoin /'tæləˌpɔɪn/ n. **1** a Buddhist monk or priest. **2** a West African monkey, *Miopithecus talapoin*, which is the smallest in the Old World. [Port. *talapão* f. Mon *tala pói* my lord]

talaria /tə'leərɪə/ n.pl. Rom. Mythol. winged sandals as an attribute of Mercury and other gods and goddesses. [L, neut. pl. of *talaris* f. *talus* ankle]

Talbot /'tɔːlbət, 'tɒl-/, (William Henry) Fox (1800–77), English pioneer of photography. Working at the family seat of Lacock Abbey in Wiltshire, he produced the first photograph on paper in 1835. Five years later he discovered a process for producing a negative from which multiple positive prints could be made, though the independently developed daguerreotype proved to be superior. Apart from patenting a number of other photographic processes and publishing two of the earliest books illustrated with photographs, he also made contributions to mathematics and deciphered cuneiform scripts.

talc /tælk/ n. & v. ● n. **1** talcum powder. **2** any crystalline form of magnesium silicate that occurs in soft flat plates, usu. white or pale green in colour and used as a lubricator etc. ● v.tr. (**talcked, talcking**) powder or treat (a surface) with talc to lubricate or dry it. □ **talcose** /-kəʊz, -kəʊs/ adj. **talcous** adj. **talcy** /-kɪ/ adj. (in sense 1). [F *talc* or med.L *talcum*, f. Arab. *ṭalḳ* f. Pers. *ṭalḳ*]

talcum /'tælkəm/ n. & v. ● n. **1** (in full **talcum powder**) powdered talc for toilet and cosmetic use, usu. perfumed. **2** = TALC 2. ● v.tr. (**talcumed, talcuming**) powder with talcum. [med.L: see TALC]

tale /teɪl/ n. **1** a narrative or story, esp. fictitious and imaginatively treated. **2** a report of an alleged fact, often malicious or in breach of confidence (*all sorts of tales will get about*). **3** archaic or literary a number or total (*the tale is complete*). □ **tale of a tub** an idle fiction. [OE *talu* f. Gmc: cf. TELL¹]

talebearer /'teɪlˌbeərə(r)/ n. a person who maliciously gossips or reveals secrets. □ **talebearing** n. & adj.

talent /'tælənt/ n. **1** a special aptitude or faculty (*a talent for music; has real talent*). **2** high mental ability. **3 a** a person or persons of talent (*is a real talent; plenty of local talent*). **b** colloq. attractive members of the opposite sex. **4** an ancient weight and unit of currency, esp. among the Greeks. □ **talent-scout** (or **-spotter**) a person looking for talented performers, esp. in sport and entertainment. □ **talented** adj. **talentless** adj. [OE *talente* & OF *talent* f. L *talentum* inclination of mind f. Gk *talanton* balance, weight, sum of money]

tales /'teɪliːz/ n. Law **1** a writ for summoning jurors to supply a deficiency. **2** a list of persons who may be summoned. [ME f. L *tales* (*de circumstantibus*) such (of the bystanders), the first words of the writ]

talesman /'teɪliːzmən, 'teɪlz-/ n. (pl. **-men**) Law a person summoned by a *tales*.

taleteller /'teɪlˌtelə(r)/ n. **1** a person who tells stories. **2** a person who spreads malicious reports.

tali pl. of TALUS¹.

talion /'tælɪən/ n. = LEX TALIONIS. [ME f. OF f. L *talio -onis* f. *talis* such]

talipes /'tælɪˌpiːz/ n. Med. = club-foot. [mod.L f. L *talus* ankle + *pes* foot]

talipot /'tælɪˌpɒt/ n. a tall southern Indian palm, *Corypha umbraculifera*, with very large fan-shaped leaves that are used as sunshades etc. [Malayalam *tālipat*, Hindi *tālpāt* f. Skr. *tālapattra* f. *tāla* palm + *pattra* leaf]

talisman /'tælɪzmən/ n. (pl. **talismans**) **1** an object, esp. an inscribed ring or stone, supposed to be endowed with magic powers esp. of averting evil from or bringing good luck to its holder. **2** a charm or amulet; a thing supposed capable of working wonders. □ **talismanic** /ˌtælɪz'mænɪk/ adj. [F & Sp., = It. *talismano*, f. med.Gk *telesmon*, Gk *telesma* completion, religious rite f. *teleō* complete f. *telos* end]

talk /tɔːk/ v. & n. ● v. **1** intr. (often foll. by *to*, *with*) converse or communicate ideas by spoken words. **2** intr. have the power of speech. **3** intr. (foll. by *about*) **a** have as the subject of discussion. **b** (in imper.) colloq. as an emphatic statement (*talk about expense! It cost me £50*). **4** tr. express or utter in words; discuss (*you are talking nonsense; talked cricket all day*). **5** tr. use (a language) in speech (*is talking Spanish*). **6** intr. (foll. by *at*) address pompously. **7** tr. (usu. foll. by *into*, *out of*) bring into a specified condition etc. by talking (*talked himself hoarse; how did you talk them into it?; talked them out of the difficulty*). **8** intr. reveal (esp. secret) information; betray secrets. **9** intr. gossip (*people are beginning to talk*). **10** intr. have influence (*money talks*). **11** intr. communicate by radio. ● n. **1** conversation or talking. **2** a particular mode of speech (*baby-talk*). **3** an informal address

or lecture. **4 a** a rumour or gossip (*there is talk of a merger*). **b** its theme (*their success was the talk of the town*). **c** empty words, verbiage (*mere talk*). **5** (often in pl.) extended discussions or negotiations. **6** empty promises or boasting. □ **know what one is talking about** be expert or authoritative. **now you're talking** colloq. I like what you say, suggest, etc. **talk away 1** consume (time) in talking. **2** carry on talking (*talk away! I'm listening*). **talk back 1** reply defiantly. **2** respond on a two-way radio system. **talk big** colloq. talk boastfully. **talk down** denigrate, belittle. **talk down to** speak patronizingly or condescendingly to. **talk a person down 1** silence a person by greater loudness or persistence. **2** bring (a pilot or aircraft) to landing by radio instructions from the ground. **talk the hind leg off a donkey** talk incessantly. **talk nineteen to the dozen** see DOZEN. **talk of 1** discuss or mention. **2** (often foll. by verbal noun) express some intention of (*talked of moving to London*). **talk of the town** what is being talked about generally. **talk out** Brit. block the course of (a bill in Parliament) by prolonging discussion to the time of adjournment. **talk over** discuss at length. **talk a person over** (or **round**) gain agreement or compliance from a person by talking. **talk shop** talk, esp. tediously or inopportunely, about one's occupation, business, etc. **talk show** = chat show (see CHAT¹). **talk tall** colloq. boast. **talk through** discuss thoroughly. **talk a person through** guide a person in (a task) with continuous instructions. **talk through one's hat** (or **neck**) colloq. **1** exaggerate. **2** bluff. **3** talk wildly or nonsensically. **talk to** reprove or scold (a person). **talk to oneself** soliloquize. **talk turkey** see TURKEY. **talk up** discuss (a subject) in order to arouse interest in it. **you can't** (or **can**) **talk** colloq. a reproof that the person addressed is just as culpable etc. in the matter at issue. □ **talker** n. [ME *talken* frequent. verb f. TALE or TELL¹]

talkathon /'tɔːkəˌθɒn/ n. colloq. a prolonged session of talking or discussion. [TALK + MARATHON]

talkative /'tɔːkətɪv/ adj. fond of or given to talking. □ **talkatively** adv. **talkativeness** n.

talkback /'tɔːkbæk/ n. **1** (often attrib.) a system of two-way communication by loudspeaker. **2** = phone-in (see PHONE¹).

talkie /'tɔːkɪ/ n. esp. US colloq. a film with a soundtrack, as distinct from a silent film. [TALK + -IE, after *movie*]

talking /'tɔːkɪŋ/ adj. & n. ● adj. **1** that talks. **2** having the power of speech (*a talking parrot*). **3** expressive (*talking eyes*). ● n. in senses of TALK v. □ **talking book** a recorded reading of a book, esp. for the blind. **talking film** (or **picture**) a film with a soundtrack. **talking head** colloq. a presenter etc. on television, speaking to the camera and viewed in close-up. **talking of** while we are discussing (*talking of food, what time is lunch?*). **talking-point** a topic for discussion or argument. **talking-shop** derog. an institution regarded as a place of argument rather than action. **talking-to** colloq. a reproof or reprimand (*gave them a good talking-to*).

tall /tɔːl/ adj. & adv. ● adj. **1** of more than average height. **2** of a specified height (*looks about six feet tall*). **3** higher than the surrounding objects (*a tall building*). **4** extravagant or excessive; fanciful, unlikely (*a tall story; tall talk*). ● adv. as if tall; proudly; in a tall or extravagant way (*sit tall*). □ **tall drink** a drink served in a tall glass. **tall hat** = top hat (see TOP¹). **tall order** an exorbitant or unreasonable demand. **tall ship** a sailing-ship with a high mast. □ **tallish** adj. **tallness** n. [ME, repr. OE *getæl* swift, prompt]

tallage /'tælɪdʒ/ n. hist. **1** a form of taxation on towns etc., abolished in the 14th century. **2** a tax on feudal dependants etc. [ME f. OF *taillage* f. *tailler* cut: see TAIL²]

Tallahassee /ˌtælə'hæsɪ/ the state capital of Florida; pop. (1990) 124,770.

tallboy /'tɔːlbɔɪ/ n. a tall chest of drawers sometimes in lower and upper sections or mounted on legs.

Talleyrand /'tælɪˌrænd, French talɛrɑ̃/ (full surname Talleyrand-Périgord), Charles Maurice de (1754–1838), French statesman. Foreign Minister under the Directory from 1797, he was involved in the coup that brought Napoleon to power, and held the same position under the new leader (1799–1807); he then resigned office and engaged in secret negotiations to have Napoleon deposed. Talleyrand became head of the new government after the fall of Napoleon (1814) and recalled Louis XVIII to the throne. He was later instrumental in the overthrow of Charles X and the accession of Louis Philippe (1830).

Tallinn /'tælɪn/ the capital of Estonia, a port on the Gulf of Finland; pop. (est. 1989) 505,000.

Tallis /'tælɪs/, Thomas (c.1505–85), English composer. Organist of the Chapel Royal jointly with William Byrd, he served under Henry VIII,

Edward VI, Mary, and Elizabeth I. In 1575 he and Byrd were given a twenty-one-year monopoly in printing music, and in that year published *Cantiones Sacrae*, a collection of thirty-four of their motets. Tallis is known particularly for his church music, especially the forty-part motet *Spem in Alium*.

tallith /ˈtælɪθ/ *n.* a fringed shawl worn by Jewish men esp. at prayer. [Rabbinical Heb. *ṭallīt* f. *ṭillel* to cover]

tallow /ˈtæləʊ/ *n. & v.* ● *n.* the harder kinds of (esp. animal) fat melted down for use in making candles, soap, etc. ● *v.tr.* grease with tallow. □ **tallow-tree** a tree that yields a substance resembling tallow, esp. *Sapium sebiferum* of China. **vegetable tallow** a vegetable fat used as tallow. □ **tallowish** *adj.* **tallowy** *adj.* [ME *talg, talug,* f. MLG *talg, talch,* of unkn. orig.]

tally /ˈtælɪ/ *n. & v.* ● *n.* (*pl.* **-ies**) **1** the reckoning of a debt or score. **2** a total score or amount. **3 a** a mark registering a fixed number of objects delivered or received. **b** such a number as a unit. **4** *hist.* **a** a piece of wood scored across with notches for the items of an account and then split into halves, each party keeping one. **b** an account kept in this way. **5** a ticket or label for identification. **6** a corresponding thing, counterpart, or duplicate. ● *v.* (**-ies, -ied**) (often foll. by *with*) **1** *intr.* agree or correspond. **2** *tr.* record or reckon by tally. □ **tally clerk** an official who keeps a tally of goods, esp. those loaded or unloaded in docks. **tally sheet** a paper on which a tally is kept. **tally system** a system of sale on short credit or instalments with an account kept by tally. □ **tallier** *n.* [ME f. AF *tallie,* AL *tallia, talia* f. L *talea*: cf. TAIL²]

tally-ho /ˌtælɪˈhəʊ/ *int., n., & v.* ● *int.* a huntsman's cry to the hounds on sighting a fox. ● *n.* (*pl.* **-hos**) an utterance of this. ● *v.* (**-hoes, -hoed**) **1** *intr.* utter a cry of 'tally-ho'. **2** *tr.* indicate (a fox) or urge (hounds) with this cry. [cf. F *taïaut*]

tallyman /ˈtælɪmən/ *n.* (*pl.* **-men**) **1** a person who keeps a tally. **2** a person who sells goods on credit, esp. from door to door.

Talmud /ˈtælmʊd, -məd/ *n.* the body of Jewish civil and ceremonial law and legend comprising the Mishnah and the Gemara, written in a mixture of Hebrew and Aramaic. There are two versions of the Talmud: the Babylonian Talmud (which dates from the 5th century AD but includes earlier material) and the earlier Palestinian or Jerusalem Talmud. □ **Talmudist** *n.* **Talmudic** /tælˈmʊdɪk/ *adj.* **Talmudical** *adj.* [late Heb. *talmūd* instruction f. Heb. *lāmad* learn]

talon /ˈtælən/ *n.* **1** a claw, esp. of a bird of prey. **2** the cards left after the deal in a card-game. **3** the last part of a dividend-coupon sheet, entitling the holder to a new sheet on presentation. **4** the shoulder of a bolt against which the key presses in shooting it in a lock. **5** *Archit.* an ogee moulding. □ **taloned** *adj.* (also in *comb.*). [ME f. OF, = heel, ult. f. L *talus*: see TALUS¹]

talus¹ /ˈteɪləs/ *n.* (*pl.* **tali** /-laɪ/) *Anat.* the ankle-bone supporting the tibia. Also called *astragalus*. [L, = ankle, heel]

talus² /ˈteɪləs/ *n.* (*pl.* **taluses**) **1** the slope of a wall that tapers to the top or rests against a bank. **2** *Geol.* a sloping mass of fragments at the foot of a cliff. [F: orig. unkn.]

tam /tæm/ *n.* a tam-o'-shanter. [abbr.]

tamable var. of TAMEABLE.

tamale /təˈmɑːlɪ/ *n.* a Mexican food of seasoned meat and maize flour steamed or baked in maize husks. [Mex. Sp. *tamal,* pl. *tamales*]

tamandua /təˈmændjʊə, ˌtæmənˈdʊə/ *n.* a small Central and South American arboreal anteater of the genus *Tamandua,* with a prehensile tail used in climbing. [Port. f. Tupi *tamanduà*]

Tamar /ˈteɪmɑː(r)/ a river in SW England which rises in NW Devon and flows 98 km (60 miles) generally southwards, forming the boundary between Devon and Cornwall and emptying into the English Channel through Plymouth Sound.

tamarack /ˈtæməˌræk/ *n.* **1** an American larch, *Larix laricina.* **2** the wood from this. [prob. Algonquian]

tamarillo /ˌtæməˈrɪləʊ/ *n.* (*pl.* **-os**) esp. *Austral. & NZ = tree tomato.* [arbitrary marketing name: cf. TOMATILLO]

tamarin /ˈtæmərɪn/ *n.* a small South American monkey of the genus *Saguinus* or *Leontopithecus,* often eating insects and having a hairy crest and moustache. [F f. Carib]

tamarind /ˈtæmərɪnd/ *n.* **1** a tropical leguminous tree, *Tamarindus indica.* **2** the fruit of this, containing an acid pulp used as food and in making drinks. [med.L *tamarindus* f. Arab. *tamr-hindī* Indian date]

tamarisk /ˈtæmərɪsk/ *n.* an Old World shrub or small tree of the genus *Tamarix,* with slender feathery branches and tiny pink flowers; esp. the

cultivated *T. chinensis,* which thrives by the sea. [ME f. LL *tamariscus,* L *tamarix*]

Tamaulipas /ˌtæmaʊˈliːpæs/ a state of NE Mexico with a coastline on the Gulf of Mexico; capital, Ciudad Victoria.

Tambo /ˈtæmbəʊ/, Oliver (1917–93), South African politician. He joined the African National Congress in 1944, and when the organization was banned by the South African government (1960) he left the country in order to organize activities elsewhere; he returned in 1990 when the ban on the ANC was lifted. During Nelson Mandela's long imprisonment he became acting president of the ANC in 1967 and president in 1977, a position he held until 1991, when he gave it up in favour of the recently released Mandela. Tambo remained as ANC national chairman until his death.

tambour /ˈtæmbʊə(r)/ *n. & v.* ● *n.* **1** a drum. **2 a** a circular frame for holding fabric taut while it is being embroidered. **b** material embroidered in this way. **3** *Archit.* each of a sequence of cylindrical stones forming the shaft of a column. **4** *Archit.* the circular part of various structures. **5** *Archit.* a lobby with a ceiling and folding doors in a church porch etc. to obviate draughts. **6** a sloping buttress or projection in a fives-court etc. ● *v.tr.* (also *absol.*) decorate or embroider on a tambour. [F f. *tabour* TABOR]

tamboura /tæmˈbʊərə/ *n.* *Mus.* an Indian stringed instrument used as a drone. [Arab. *ṭanbūra*]

tambourin /ˈtæmbərɪn/ *n.* **1** a long narrow drum used in Provence. **2 a** a dance accompanied by a tambourin. **b** the music for this. [F, dimin. of TAMBOUR]

tambourine /ˌtæmbəˈriːn/ *n.* a percussion instrument consisting of a hoop with a parchment stretched over one side and jingling discs in slots round the hoop. □ **tambourinist** *n.* [F, dimin. of TAMBOUR]

Tambov /tæmˈbɒf/ an industrial city in SW Russia; pop. (1990) 307,000.

tame /teɪm/ *adj. & v.* ● *adj.* **1** (of an animal) domesticated; not wild or shy. **2** insipid; lacking spirit or interest; dull (*tame acquiescence*). **3** (of a person) amenable or cooperative and available (*tame scientists supported the government's decision*). **4** *US* **a** (of land) cultivated. **b** (of a plant) produced by cultivation. ● *v.tr.* **1** make tame; domesticate; break in. **2** subdue, curb, humble; break the spirit of. □ **tamely** *adv.* **tameness** *n.* **tamer** *n.* (also in *comb.*). [OE *tam* f. Gmc]

tameable /ˈteɪməb(ə)l/ *adj.* (also **tamable**) capable of being tamed. □ **tameableness** *n.* **tameability** /ˌteɪməˈbɪlɪtɪ/ *n.*

Tamerlane /ˈtæməˌleɪn/ (also **Tamburlaine** /ˈtæmbə-/) (born Timur Lenk = 'lame Timur') (1336–1405), Mongol ruler of Samarkand 1369–1405. Leading a force of Mongols and Turks, between about 1364 and 1405 he conquered a large area including Persia, northern India, and Syria and established his capital at Samarkand; he defeated the Ottomans near Ankara in 1402, but died during an invasion of China. He was the ancestor of the Mogul dynasty in India.

Tamil /ˈtæmɪl/ *n. & adj.* ● *n.* **1** a member of a Dravidian people inhabiting southern India and Sri Lanka. **2** the language of the Tamils, belonging to the Dravidian group. Tamil is one of the major languages of southern India, spoken by about 50 million people including some 4 million in Sri Lanka and Malaysia. ● *adj.* of or relating to this people or their language. □ **Tamilian** /tæˈmɪlɪən/ *adj.* [(Port., Du. *Tamul* f.) Tamil *Tamiḷ* = Prakrit *Damiḷa, Daviḷa,* Skr. *Dramiḍa, Draviḍa,* rel. to DRAVIDIAN]

Tamil Nadu /ˌtæmɪl ˈnɑːduː/ a state in the extreme south-east of the Indian peninsula, on the Coromandel Coast, with a largely Tamil-speaking, Hindu population; capital, Madras. Tamil Nadu was formerly an ancient kingdom comprising a much larger area, stretching northwards to Orissa and including the Lakshadweep Islands and part of the Malabar Coast. It was divided when the present state was formed in 1956, parts going to Andhra Pradesh, Kerala, and Karnataka. It was known as Madras until 1968.

Tamil Tiger *n.* (also **Tiger**) a member of the Liberation Tigers for Tamil Eelam, a Sri Lankan guerrilla organization founded in 1972 that seeks the establishment of an independent state (Eelam) in the north-east of the country for the Tamil community.

Tamla Motown /ˈtæmlə/ *propr.* = MOTOWN 2. [*Tamla* US record company]

Tammany /ˈtæmənɪ/ (in the US) **1** a fraternal and benevolent society of New York City, founded in 1789, from an earlier patriotic society. The society was named after an American Indian chief of the late 17th century said to have welcomed William Penn, and regarded as 'patron saint' of Pennsylvania and other northern colonies. **2** a political organization of the Democratic Party, identified with this society and notorious in the 19th century for corruption, maintaining power by the use of bribes etc. It dominated the political life of New York City during the 19th and early 20th centuries before being reduced in power by Franklin D. Roosevelt in the early 1930s.

Tammany Hall *n.* **1** any of the successive buildings used as the headquarters of Tammany. **2** the members of Tammany. **3** a corrupt political organization.

Tammerfors /ˌtamərˈfɔrs/ the Swedish name for TAMPERE.

Tammuz[1] /ˈtæmʊz/ *Mythol.* a Babylonian and Assyrian god, lover of Ishtar, corresponding to the Greek Adonis. He became the personification of the seasonal death and rebirth of crops.

Tammuz[2] var. of THAMMUZ.

tammy /ˈtæmɪ/ *n.* (*pl.* -ies) = TAM-O'-SHANTER.

tam-o'-shanter /ˌtæməˈʃæntə(r)/ *n.* a round woollen or cloth cap of Scottish origin fitting closely round the brows but large and full above. [the hero of Burns's *Tam o' Shanter*]

tamp /tæmp/ *v.tr.* **1** pack (a blast-hole) full of clay etc. to get the full force of an explosion. **2** ram down (road material etc.). □ **tamper** *n.* **tamping** *n.* (in sense 1). [perh. back-form. f. F *tampin* (var. of TAMPION, taken as = *tamping*)]

Tampa /ˈtæmpə/ a port and resort on the west coast of Florida; pop. (1990) 280,015.

tamper /ˈtæmpə(r)/ *v.intr.* (foll. by *with*) **1** meddle or interfere with; make unauthorized changes in. **2** exert a secret or corrupt influence upon; bribe. □ **tamperer** *n.* **tamper-proof** *adj.* [var. of TEMPER.]

Tampere /ˈtæmpəˌreɪ/ (called in Swedish *Tammerfors*) a city in SW Finland; pop. (1990) 172,560.

Tampico /tæmˈpiːkəʊ/ one of Mexico's principal seaports, on the Gulf of Mexico; pop. (1990) 271,640.

tampion /ˈtæmpɪən/ *n.* (also **tompion** /ˈtɒm-/) **1** a wooden stopper for the muzzle of a gun. **2** a plug e.g. for the top of an organ-pipe. [ME f. F *tampon*, nasalized var. of *tapon*, rel. to TAP[1]]

tampon /ˈtæmpɒn/ *n. & v.* ● *n.* a plug of soft material used to stop a wound or absorb secretions, esp. one inserted into the vagina. ● *v.tr.* (**tamponed, tamponing**) plug with a tampon. [F: see TAMPION]

tamponade /ˌtæmpəˈneɪd/ *n.* compression of the heart by an accumulation of fluid in the pericardial sac.

tamponage /ˈtæmpənɪdʒ/ *n.* = TAMPONADE.

tam-tam /ˈtæmtæm/ *n.* a large metal gong. [Hindi: see TOM-TOM]

Tamworth /ˈtæmwɜːθ, -wəθ/ a town in central England, in Staffordshire; pop. (1981) 64,550.

Tamworth Manifesto (in English history) an election speech by Sir Robert Peel in 1834 in his Tamworth constituency, in which he accepted the changes instituted by the Reform Act and expressed his belief in moderate political reform. The manifesto is often held to signal the emergence of the Conservative Party from the old loose grouping of Tory interests.

tan[1] /tæn/ *n., adj., & v.* ● *n.* **1** a brown skin colour resulting from exposure to ultraviolet light. **2** a yellowish-brown colour. **3** bark, esp. of oak, bruised and used to tan hides. **4** (in full **spent tan**) tan from which the tannic acid has been extracted, used for covering roads etc. ● *adj.* yellowish-brown. ● *v.* (**tanned, tanning**) **1** *tr. & intr.* make or become brown by exposure to ultraviolet light. **2** *tr.* convert (raw hide) into leather by soaking in a liquid containing tannic acid or by the use of mineral salts etc. **3** *tr. sl.* beat, thrash. □ **tannable** *adj.* **tanning** *n.* **tannish** *adj.* [OE *tannian*, prob. f. med.L *tanare*, *tannare*, perh. f. Celtic]

tan[2] /tæn/ *abbr.* tangent.

Tana, Lake /ˈtɑːnə/ a lake in northern Ethiopia, the source of the Blue Nile.

tanager /ˈtænədʒə(r)/ *n.* an American songbird of the subfamily Thraupinae, related to the buntings, the male usu. having brightly coloured plumage. [mod.L *tanagra* f. Tupi *tangara*]

Tanagra /ˈtænəgrə/ *n.* a terracotta figurine of a type dating chiefly from the 3rd century BC, many of which have been found at Tanagra in Boeotia, Greece. Carefully modelled and painted, the figurines usually represent elegantly draped young women.

Tananarive /ˌtænənəˈriːv/ the former name (until 1975) for ANTANANARIVO.

tanbark /ˈtænbɑːk/ *n.* the bark of oak and other trees, used to obtain tannin.

tandem /ˈtændəm/ *n. & adv.* ● *n.* **1** a bicycle or tricycle with two or more seats one behind another. **2** a group of two persons or machines etc.

with one behind or following the other. **3** a carriage driven tandem. ● *adv.* with two or more horses harnessed one behind another (*drive tandem*). □ **in tandem 1** one behind another. **2** alongside each other, together. [L, = at length (of time), used punningly]

tandoor /ˈtænduə(r)/ *n.* a clay oven of a type used orig. in northern India and Pakistan. [Hind.]

tandoori /tænˈduərɪ/ *n.* food cooked over charcoal in a tandoor (often *attrib.*: *tandoori chicken*). [Hind.]

Tang /tæŋ/ a dynasty that ruled China AD 618–907, a period noted for territorial conquest and great wealth and regarded as the golden age of Chinese poetry and art. [Chin. *táng*]

tang[1] /tæŋ/ *n.* **1** a strong taste or flavour or smell. **2 a** a characteristic quality. **b** a trace; a slight hint of some quality, ingredient, etc. **3** the projection on the blade of a tool, esp. a knife, by which the blade is held firm in the handle. [ME f. ON *tange* point, tang of a knife]

tang[2] /tæŋ/ *v. & n.* ● *v.tr. & intr.* ring, clang; sound loudly. ● *n.* a tanging sound. [imit.]

Tanga /ˈtæŋgə/ one of the principal ports of Tanzania, situated in the north-east of the country on the Indian Ocean; pop. (1988) 187,630.

tanga /ˈtæŋgə/ *n.* a pair of very skimpy briefs consisting of small panels connected with strings. [Port.]

Tanganyika /ˌtæŋgəˈniːkə, -ˈnjiːkə/ see TANZANIA.

Tanganyika, Lake a lake in East Africa, in the Great Rift Valley. The deepest lake in Africa and the longest freshwater lake in the world, it forms most of the border of Zaire with Tanzania and Burundi.

Tange /ˈtæŋgeɪ/, Kenzo (b.1913), Japanese architect. His work reflects the influence of Le Corbusier and is characterized by the use of modern materials, while retaining a feeling for traditional Japanese architecture. During the 1950s he built a number of civic buildings in Brutalist style, including the Peace Centre at Hiroshima (1955). Later buildings, such as the National Gymnasium in Tokyo (built for the 1964 Olympics), make use of dynamic sweeping curves.

tangelo /ˈtændʒəˌləʊ/ *n.* (*pl.* -os) a hybrid of the tangerine and grapefruit. [TANGERINE + POMELO]

tangent /ˈtændʒənt/ *n. & adj.* ● *n.* **1** a straight line, curve, or surface that meets another curve or curved surface at a point, but if extended does not intersect it at that point. **2** the ratio of the sides (other than the hypotenuse) opposite and adjacent to an angle in a right-angled triangle. ● *adj.* **1** (of a line or surface) that is a tangent. **2** touching. □ **at a tangent** diverging from a previous course of action or thought etc. (*go off at a tangent*). **tangent galvanometer** a galvanometer with a coil through which the current to be measured is passed, its strength being proportional to the tangent of the angle of deflection. □ **tangency** *n.* [L *tangere* tangent- touch]

tangential /tænˈdʒenʃ(ə)l/ *adj.* **1** of or along a tangent. **2** divergent. **3** peripheral. □ **tangentially** *adv.*

tangerine /ˌtændʒəˈriːn/ *n.* **1** a small sweet orange-coloured citrus fruit with a thin skin; a mandarin. **2** a deep orange-yellow colour. [TANGIER]

tangible /ˈtændʒɪb(ə)l/ *adj.* **1** perceptible by touch. **2** definite; clearly intelligible; not elusive or visionary (*tangible proof*). □ **tangibly** *adv.* **tangibleness** *n.* **tangibility** /ˌtændʒɪˈbɪlɪtɪ/ *n.* [F *tangible* or LL *tangibilis* f. *tangere* touch]

Tangier /tænˈdʒɪə(r)/ a seaport on the northern coast of Morocco, on the Strait of Gibraltar commanding the western entrance to the Mediterranean; pop. (1982) 266,300. Tangier had its beginning in the Roman port and town of Tingis, but the present walled city was built in the Middle Ages by the Moors. It was taken by the Portuguese towards the end of the 15th century and given to England as part of the dowry of Princess Catherine of Braganza, when she married Charles II in 1662. Abandoned twenty-two years later to the sultan of Morocco, the port and the surrounding countryside remained under the sultanate until 1904. From then until 1956 (except for five years in the Second World War when it was seized by Spain) the zone was under international control. In 1956 it passed to the newly independent monarchy of Morocco.

tangle[1] /ˈtæŋg(ə)l/ *v. & n.* ● *v.* **1 a** *tr.* intertwine (threads or hairs etc.) in a confused mass; entangle. **b** *intr.* become tangled. **2** *intr.* (foll. by *with*) *colloq.* become involved (esp. in conflict or argument) with (*don't tangle with me*). **3** *tr.* complicate (*a tangled affair*). ● *n.* **1** a confused mass of intertwined threads etc. **2** a confused or complicated state (*be in a tangle; a love tangle*). [ME var. of obs. *tagle*, of uncert. orig.]

tangle[2] /ˈtæŋg(ə)l/ *n.* a coarse seaweed, esp. of the genus *Laminaria*. [prob. f. Norw. *taangel* f. ON *thöngull*]

tangly /ˈtæŋglɪ/ adj. (**tanglier**, **tangliest**) tangled.

tango[1] /ˈtæŋgəʊ/ n. & v. ● n. (pl. **-os**) **1** a ballroom dance of South American origin characterized by slow gliding movements and abrupt pauses. **2** the music for this. ● v.intr. (**-oes**, **-oed**) dance the tango. [Amer. Sp.]

tango[2] /ˈtæŋgəʊ/ n. a tangerine colour. [abbr. after TANGO[1]]

tangram /ˈtæŋgræm/ n. a Chinese puzzle square cut into seven pieces to be combined into various figures. [19th c.: orig. unkn.]

Tangshan /tæŋˈʃæn/ an industrial city in Hebei province, NE China; pop. (1990) 1,500,000. The city had to be rebuilt after a devastating earthquake in 1976, in which approximately 1 million people were killed or injured.

tangy /ˈtæŋɪ/ adj. (**tangier**, **tangiest**) having a strong usu. sharp or spicy taste or smell. □ **tanginess** n.

tanh /θæn, tænʃ, tænˈeɪtʃ/ abbr. hyperbolic tangent.

tanist /ˈtænɪst/ n. hist. the heir apparent to a Celtic chief, usu. his most vigorous adult relation, chosen by election. □ **tanistry** n. [Ir. & Gael. tánaiste heir]

Tanjungkarang /ˌtændʒʊŋkəˈræŋ/ see BANDAR LAMPUNG.

tank /tæŋk/ n. & v. ● n. **1** a large receptacle or storage chamber usu. for liquid or gas. **2** a heavy armoured fighting vehicle carrying guns and moving on a tracked carriage. (See note below.) **3** a container for the fuel supply in a motor vehicle. **4** the part of a steam locomotive tender containing water for the boiler. **5 a** Ind. & Austral. a reservoir. **b** dial. esp. US a pond. ● v. (usu. foll. by up) esp. Brit. **1** tr. fill the tank of (a vehicle etc.) with fuel. **2** intr. & tr. (in passive) colloq. drink heavily; become drunk. □ **tank engine** a steam locomotive carrying fuel and water receptacles in its own frame, not in a tender. **tank-farming** the practice of growing plants in tanks of water without soil. **tank top** a sleeveless close-fitting upper garment with a scoop-neck. □ **tankful** n. (pl. **-fuls**). **tankless** adj. [Gujarati tānkh etc., perh. f. Skr. taḍāga pond]

▪ Early designs for a mechanical armoured vehicle include one by Leonardo da Vinci (1484), but it was not until the development of the internal-combustion engine and the track mechanism that a really successful vehicle was constructed. This was developed in Britain during the First World War, when the name tank was adopted for purposes of secrecy. Due to resistance from senior army officers tanks were not used en masse until Nov. 1917, when they were immediately successful. Their use in the Second World War was one reason why there was no repetition of the static trench warfare of 1914–18. Different types of armoured vehicles have since been designed for different roles, and the modern tank remains very important for its mobility, relatively high ratio of weapon-power to manpower, and moderate protection against the blast and radioactivity from tactical nuclear weapons.

tanka /ˈtæŋkə/ n. a Japanese poem in five lines and thirty-one syllables giving a complete picture of an event or mood. [Jap.]

tankage /ˈtæŋkɪdʒ/ n. **1 a** storage in tanks. **b** a charge made for this. **2** the cubic content of a tank. **3** a kind of fertilizer obtained from refuse bones etc.

tankard /ˈtæŋkəd/ n. **1** a tall beer mug with a handle and sometimes a hinged lid, esp. of silver or pewter. **2** the contents of or an amount held by a tankard (drank a tankard of ale). [ME: orig. unkn.: cf. MDu. tanckaert]

tanker /ˈtæŋkə(r)/ n. a ship, aircraft, or road vehicle for carrying liquids, esp. mineral oils, in bulk.

tanner[1] /ˈtænə(r)/ n. a person who tans hides.

tanner[2] /ˈtænə(r)/ n. Brit. hist. sl. a sixpence. [19th c.: orig. unkn.]

tannery /ˈtænərɪ/ n. (pl. **-ies**) a place where hides are tanned.

Tannhäuser /ˈtænˌhɔɪzə(r)/, (c.1200–c.1270), German poet. In reality a Minnesinger, he became a legendary figure as a knight who visited Venus's grotto and spent seven years in debauchery, then repented and sought absolution from the pope; he is the subject of Wagner's opera Tannhäuser (1845). The real Tannhäuser's surviving works include lyrics and love poetry.

tannic /ˈtænɪk/ adj. of or produced from tan. □ **tannic acid** a complex natural organic compound of a yellowish colour used as a mordant and astringent. □ **tannate** /-neɪt/ n. [F tannique (as TANNIN)]

tannin /ˈtænɪn/ n. any of a group of complex organic compounds found in certain tree-barks and oak-galls, used in leather production and ink manufacture. [F tanin (as TAN[1], -IN)]

tannish see TAN[1].

Tannoy /ˈtænɔɪ/ n. propr. a type of public-address system. [from tantalum alloy (rectifier)]

Tannu-Tuva /ˌtænuːˈtuːvə/ the former name for TUVA.

tanrec var. of TENREC.

Tansen /ˈtænsen/ (c.1500–89), Indian musician and singer. He is regarded as the leading exponent of northern Indian classical music. A native of Gwalior, he became an honoured member of the court of Akbar the Great, and was noted both for his skill as an instrumentalist and as a singer. Many legends arose about his life and musical achievements.

tansy /ˈtænzɪ/ n. (pl. **-ies**) a composite plant of the genus Tanacetum; esp. T. vulgare, with yellow button-like flowers and aromatic leaves, formerly used in medicines and cookery. [ME f. OF tanesie f. med.L athanasia immortality f. Gk]

tantalite /ˈtæntəˌlaɪt/ n. a rare dense black mineral, the principal source of the element tantalum. [G & Sw. tantalit (as TANTALUM)]

tantalize /ˈtæntəˌlaɪz/ v.tr. (also **-ise**) **1** torment or tease by the sight or promise of what is unobtainable. **2** raise and then dash the hopes of; torment with disappointment. □ **tantalizer** n. **tantalizingly** adv. **tantalization** /ˌtæntəlaɪˈzeɪʃ(ə)n/ n. [Gk Tantalos TANTALUS]

tantalum /ˈtæntələm/ n. a hard silver-grey metallic chemical element (atomic number 73, symbol **Ta**). Tantalum was discovered by the Swedish chemist A. G. Ekeberg in 1802 in the mineral now called tantalite. A transition metal, it is rare in nature and usually occurs in association with niobium, which it resembles. Tantalum alloys have some specialized uses, and the metal was formerly used for electric light filaments. [f. TANTALUS (with ref. to its frustrating insolubility in acids)]

Tantalus /ˈtæntələs/ Gk Mythol. a Lydian king, son of Zeus and father of Pelops. For his crimes (which included killing Pelops and offering his flesh to the gods) he was punished in Tartarus by being provided with fruit and water which receded when he reached for them. His name is the origin of the word tantalize.

tantalus /ˈtæntələs/ n. **1** a stand in which spirit-decanters may be locked up but remain visible. **2** an American wood ibis, Mycteria americana. [TANTALUS]

tantamount /ˈtæntəˌmaʊnt/ predic.adj. (foll. by to) equivalent to (was tantamount to a denial). [f. obs. verb f. It. tanto montare amount to so much]

tantivy /tænˈtɪvɪ/ n. & adj. archaic ● n. (pl. **-ies**) **1** a hunting cry. **2** a swift movement; a gallop or rush. ● adj. swift. [17th c.: perh. an imit. of hoof-beats]

tant mieux /tɒn ˈmjɜː/ int. so much the better. [F]

tant pis /tɒn ˈpiː/ int. so much the worse. [F]

tantra /ˈtæntrə/ n. any of a class of Hindu or Buddhist mystical and magical writings, dating from the 7th century or earlier; also, adherence to the doctrines or principles of the tantras, involving mantras, meditation, yoga, and ritual. In Hinduism tantric practice may involve indulgence in normally forbidden taboos and is designed to awaken the energy of Sakti. Tantrism is an important element in Tibetan Buddhism. □ **tantric** adj. **tantrism** n. **tantrist** n. [Skr., = loom, groundwork, doctrine f. tan stretch]

tantrum /ˈtæntrəm/ n. an outburst of bad temper or petulance (threw a tantrum). [18th c.: orig. unkn.]

Tanzania /ˌtænzəˈnɪə/ a country in East Africa with a coastline on the Indian Ocean; pop. (est. 1991) 27,270,000; official languages, Swahili and English; capital, Dodoma. Tanzania consists of a mainland area (the former Tanganyika) and the island of Zanzibar. A German colony (German East Africa) from the late 19th century, Tanganyika became a British mandate after the First World War and a trust territory, administered by Britain, after the Second, before becoming independent within the Commonwealth in 1961. It was named Tanzania after its union with Zanzibar in 1964. □ **Tanzanian** adj. & n.

Tao /taʊ, ˈtɑːəʊ/ n. the metaphysical concept central to Chinese philosophy, the absolute principle underlying the universe, combining within itself the principles of yin and yang. By extension Tao also signifies the way, or code of behaviour, that is in harmony with the natural order. The interpretation of Tao in the Tao-te-Ching developed into the philosophical religion of Taoism. [Chin. dao (right) way]

Taoiseach /ˈtiːʃəx/ n. the Prime Minister of the Republic of Ireland. [Ir., = chief, leader]

Taoism /ˈtaʊɪz(ə)m, ˈtɑːəʊ-/ n. one of two major Chinese religious and philosophical systems (the other is Confucianism), traditionally said to have been founded by Lao-tzu in about the 6th century BC. The central concept and goal is the Tao, and its most important text is the Tao-te-Ching. Like Confucianism, Taoism has both a philosophical and a religious aspect. In contrast with the practical teachings of Confucianism, however, philosophical Taoism emphasizes inner contemplation and mystical union with nature; wisdom, learning, and purposive action should be abandoned in favour of simplicity and *wu-wei* (non-action, or letting things take their natural course). The religious aspect of Taoism developed later, *c*.3rd century AD, incorporating certain Buddhist features and developing a monastic system. In China Taoist religious practice has been suppressed (although it survives in Taiwan), but Taoist thought is still important in Chinese life; it has also attracted interest in the West in the 20th century. □ **Taoist** *n.* & *adj.* **Taoistic** /taʊˈɪstɪk, ˌtɑːəʊ-/ *adj.* [TAO]

Taormina /ˌtɑːɔːˈmiːnə/ a resort town on the east coast of Sicily; pop. (1990) 10,905. It was founded by Greek colonists in the 4th century BC and was an important city under the Romans and the Normans, but declined after the 15th century.

Tao-te-Ching /ˌtaʊtiːˈtʃɪŋ/ (also called *Lao-tzu*) the central Taoist text, ascribed to Lao-tzu, the traditional founder of Taoism. Apparently written as a guide for rulers, it defined the Tao or way and established the philosophical basis of Taoism. [Chin., = the Book of the Way and its Power]

tap¹ /tæp/ *n.* & *v.* ● *n.* **1** a device by which a flow of liquid or gas from a pipe or vessel can be controlled. **2 a** an act of tapping a telephone etc. **b** the device used for this. **3** *Brit.* a taproom. **4** an instrument for cutting the thread of a female screw. ● *v.tr.* (**tapped**, **tapping**) **1 a** provide (a cask) with a tap. **b** let out (a liquid) by means of, or as if by means of, a tap. **2** draw sap from (a tree) by cutting into it. **3 a** obtain information or supplies or resources from. **b** extract or obtain; discover and exploit (*mineral resources to be tapped*; *to tap the skills of young people*). **4** connect a listening device to (a telephone or telegraph line etc.) to listen to a call or transmission. **5** cut a female screw-thread in. □ **on tap 1** ready to be drawn off by tap. **2** *colloq.* ready for immediate use; freely available. **tap root** a tapering root growing vertically downwards. **tap water** water from a piped supply. □ **tapless** *adj.* **tappable** *adj.* [OE *tæppian* (v.), *tæppa* (n.) f. Gmc]

tap² /tæp/ *v.* & *n.* ● *v.* (**tapped**, **tapping**) **1** *intr.* (foll. by *at*, *on*) strike a gentle but audible blow. **2** *tr.* strike lightly (*tapped me on the shoulder*). **3** *tr.* (foll. by *against* etc.) cause (a thing) to strike lightly (*tapped a stick against the window*). **4** *intr.* = TAP-DANCE *v.* (*can you tap?*). **5** *tr.* (often foll. by *out*) make by a tap or taps (*tapped out the rhythm*). ● *n.* **1 a** a light blow; a rap. **b** the sound of this (*heard a tap at the door*). **2 a** tap-dancing (*goes to tap classes*). **b** a piece of metal attached to the toe and heel of a tap-dancer's shoe to make the tapping sound. **3** (in *pl.*, usu. treated as *sing.*) **a** a bugle call for lights to be put out in army quarters. **b** a similar signal at a military funeral. □ **tap-tap** a repeated tap; a series of taps. □ **tapper** *n.* [ME *tappe* (imit.), perh. through F *taper*]

tapa¹ /ˈtɑːpə/ *n.* **1** the bark of a paper-mulberry tree. **2** cloth made from this, used in the Pacific islands. [Polynesian]

tapa² /ˈtæpə/ *n.* (usu. in *pl.*) a savoury snack served in a bar or café providing Spanish food. [Sp., lit. 'cover, lid']

tap-dance /ˈtæpdɑːns/ *n.* & *v.* ● *n.* a dance or form of display dancing performed wearing shoes fitted with metal taps, with rhythmical tapping of the toes and heels. ● *v.intr.* perform a tap-dance. □ **tap-dancer** *n.* **tap-dancing** *n.*

tape /teɪp/ *n.* & *v.* ● *n.* **1** a narrow strip of woven material for tying up, fastening, etc. **2 a** a strip of material stretched across the finishing line of a race. **b** a similar strip for marking off an area or forming a notional barrier. **3** (in full **adhesive tape**) a strip of opaque or transparent paper or plastic etc., esp. coated with adhesive for fastening, sticking, masking, insulating, etc. **4 a** = *magnetic tape*. **b** a tape recording or tape cassette. **5** = *tape-measure*. ● *v.tr.* **1 a** tie up or join etc. with tape. **b** apply tape to. **2** (foll. by *off*) seal or mark off an area or thing with tape. **3** record on magnetic tape. **4** measure with tape. □ **breast the tape** win a race. **have** (or **get**) **a person** (or **thing**) **taped** *Brit. colloq.* understand a person or thing fully. **on tape** recorded on magnetic tape. **tape deck** a piece of equipment for playing cassette tapes, esp. as part of a stereo system. **tape machine** a machine for receiving and recording telegraph messages. **tape-measure** a strip of tape or thin flexible metal marked for measuring lengths. **tape-record** record (sounds) on magnetic tape. **tape recorder** apparatus for recording sounds on magnetic tape and afterwards reproducing them. **tape recording** a recording on magnetic tape. □ **tapeable** *adj.* (esp. in sense 3 of *v.*). **tapeless** *adj.* **tapelike** *adj.* [OE *tæppa*, *tæppe*, of unkn. orig.]

taper /ˈteɪpə(r)/ *n.* & *v.* ● *n.* **1** a wick coated with wax etc. for conveying a flame. **2** a slender candle. ● *v.* (often foll. by *off*) **1** *intr.* & *tr.* diminish or reduce in thickness towards one end. **2** *tr.* & *intr.* make or become gradually less. [OE *tapur*, *-or*, *-er* wax candle, f. L PAPYRUS, whose pith was used for candle-wicks]

tapestry /ˈtæpɪstrɪ/ *n.* (*pl.* **-ies**) **1 a** a thick textile fabric in which coloured weft threads are woven to form pictures or designs. **b** embroidery imitating this, usu. in wools on canvas. **c** a piece of such embroidery. **2** events or circumstances etc. compared with a tapestry in being intricate, interwoven, etc. (*life's rich tapestry*). □ **tapestried** *adj.* [ME, alt. f. *tapissery* f. OF *tapisserie* f. *tapissier* tapestry-worker or *tapisser* to carpet, f. *tapis*: see TAPIS]

tapetum /təˈpiːtəm/ *n.* *Zool.* a light-reflecting part of the choroid membrane in the eyes of certain mammals, e.g. cats. [LL f. L *tapete* carpet]

tapeworm /ˈteɪpwɜːm/ *n.* a parasitic flatworm of the class Cestoda, the adult of which lives in the intestines and has a small head or scolex and a long ribbon-like body composed of many egg-producing segments or proglottids.

taphonomy /tæˈfɒnəmɪ/ *n.* the science concerned with the process of fossilization. □ **taphonomist** *n.* **taphonomic** /-fəˈnɒmɪk/ *adj.* [Gk *taphos* grave + -NOMY]

tapioca /ˌtæpɪˈəʊkə/ *n.* a starchy substance in hard white grains obtained from cassava and used for puddings etc. [Tupi-Guarani *tipioca* f. *tipi* dregs + *og*, *ok* squeeze out]

tapir /ˈteɪpə(r), -pɪə(r)/ *n.* a heavily built hoofed mammal of the family Tapiridae, native to the forests of tropical America and SE Asia, having a short flexible protruding snout used for feeding on vegetation. □ **tapiroid** *adj.* & *n.* [Tupi *tapira*]

tapis /ˈtæpiː/ *n.* (*pl.* same) a covering or tapestry. □ **on the tapis** (of a subject) under consideration or discussion. [ME, a kind of cloth, f. OF *tapiz* f. LL *tapetium* f. Gk *tapētion* dimin. of *tapēs tapētos* tapestry]

tapotement /təˈpəʊtmənt/ *n.* *Med.* rapid and repeated striking of the body as massage treatment. [F f. *tapoter* tap]

tapper see TAP².

tappet /ˈtæpɪt/ *n.* a lever or projecting part used in machinery to give intermittent motion, often in conjunction with a cam. [app. f. TAP² + -ET¹]

taproom /ˈtæpruːm, -rʊm/ *n.* a room in which alcoholic drinks (esp. beer) are available on tap.

tapster /ˈtæpstə(r)/ *n.* a person who draws and serves alcoholic drinks at a bar. [OE *tæppestre* orig. fem. (as TAP¹, -STER)]

tapu /ˈtɑːpuː/ *n.* & *adj.* NZ = TABOO. [Maori]

tar¹ /tɑː(r)/ *n.* & *v.* ● *n.* **1** a dark thick inflammable liquid distilled from wood or coal etc. and used as a preservative of wood and iron, in making roads, as an antiseptic, etc. **2** a similar substance formed in the combustion of tobacco etc. ● *v.tr.* (**tarred**, **tarring**) cover with tar. □ **tar and feather** smear with tar and then cover with feathers as a punishment. **tar-brush** a brush for applying tar. **tarred with the same brush** having the same faults. **tar sand** *Geol.* a deposit of sand impregnated with bitumen. [OE *te(o)ru* f. Gmc, rel. to TREE]

tar² /tɑː(r)/ *n.* *colloq.* a sailor. [abbr. of TARPAULIN]

Tara /ˈtɑːrə/ a hill in County Meath in the Republic of Ireland, site in early times of the residence of the high kings of Ireland and still marked by ancient earthworks.

Tarabulus Al-Gharb see TRIPOLI 1.

Tarabulus Ash-Sham see TRIPOLI 2.

taradiddle /ˈtærəˌdɪd(ə)l/ *n.* (also **tarradiddle**) *colloq.* **1** a petty lie. **2** pretentious nonsense. [18th c.: cf. DIDDLE]

tarakihi /ˌtærəˈkiːhɪ, -ˈkiː/ *n.* (also **terakihi** /ˌterə-/) a silver fish, *Cheilodactylus macropterus*, with a black band behind the head, caught for food off New Zealand coasts. [Maori]

taramasalata /ˌtærəməsəˈlɑːtə/ *n.* (also **tarama** /ˈtærəmə/) a pinkish pâté made from the roe of mullet or other fish with olive oil, seasoning, etc. [mod.Gk *taramas* roe (f. Turk. *tarama*) + *salata* SALAD]

Taranaki /ˌtærəˈnækɪ/ the Maori name for Mount Egmont (see EGMONT, MOUNT).

tarantass /ˌtærənˈtæs/ *n.* a springless four-wheeled Russian vehicle. [Russ. *tarantas*]

tarantella /ˌtærənˈtelə/ *n.* (also **tarantelle** /-ˈtel/) **1** a rapid whirling dance of southern Italy. **2** the music for this. [It., f. TARANTO (because the dance was once thought to be a cure for a tarantula bite): cf. TARANTISM]

tarantism /ˈtærənˌtɪz(ə)m/ n. hist. dancing mania, esp. that originating in southern Italy among those who had (actually or supposedly) been bitten by a tarantula. [mod.L tarantismus, It. tarantismo f. TARANTO]

Taranto /təˈræntəʊ/ a seaport and naval base in Apulia, SE Italy; pop. (1990) 244,030. Founded by the Greeks in the 8th century BC, it came under Roman rule in 272 BC.

tarantula /təˈræntjʊlə/ n. **1** a very large hairy tropical spider of the family Theraphosidae. **2** a large black wolf-spider, Lycosa tarentula, of southern Europe, whose bite was formerly held to cause tarantism. [med.L f. It. tarantola (AS TARANTISM)]

Tarawa /ˈtærəwə, təˈrɑːwə/ an atoll in the South Pacific, one of the Gilbert Islands; pop. (1990) 28,800. The capital of the republic of Kiribati, Bairiki, is located on the atoll.

taraxacum /təˈræksəkəm/ n. **1** a composite plant of the genus Taraxacum, including the dandelions. **2** a tonic etc. prepared from the dried roots of this. [med.L f. Arab. ṭarak̲s̲ak̲ūk f. Pers. talk̲ bitter + chakūk purslane]

tarboosh /tɑːˈbuːʃ/ n. a cap like a fez, sometimes worn by Muslim men as part of a turban. [Egypt. Arab. ṭarbūš, ult. f. Pers. sar-būš head-cover]

Tardenoisian /ˌtɑːdɪˈnɔɪzɪən/ adj. & n. Archaeol. of, relating to, or denoting a late mesolithic culture of western and central Europe. [f. Tardenois, the type-site in NE France]

tardigrade /ˈtɑːdɪˌgreɪd/ n. & adj. Zool. ● n. a minute freshwater animal of the phylum Tardigrada, having a short plump body and four pairs of short legs. Also called water bear. ● adj. of or relating to this phylum. [F tardigrade f. L tardigradus f. tardus slow + gradi walk]

tardy /ˈtɑːdɪ/ adj. (**tardier, tardiest**) **1** slow to act or come or happen. **2** delaying or delayed beyond the right or expected time. □ **tardily** adv. **tardiness** n. [F tardif, tardive ult. f. L tardus slow]

tare¹ /teə(r)/ n. **1** vetch, esp. as corn-weed or fodder. **2** (in pl.) Bibl. an injurious weed resembling corn when young (Matt. 13:24–30). [ME: orig. unkn.]

tare² /teə(r)/ n. **1** an allowance made for the weight of the packing or wrapping around goods. **2** the weight of a motor vehicle without its fuel or load. □ **tare and tret** the arithmetical rule for computing a tare. [ME f. F, = deficiency, tare, f. med.L tara f. Arab. ṭarḥa what is rejected f. ṭaraḥa reject]

targe /tɑːdʒ/ n. archaic = TARGET n. 5. [ME f. OF]

target /ˈtɑːgɪt/ n. & v. ● n. **1** a mark or point fired or aimed at, esp. a round or rectangular object marked with concentric circles. **2** a person or thing aimed at, or exposed to gunfire etc. (they were an easy target). **3** (also attrib.) an objective or result aimed at (our export targets; target date). **4** a person or thing against whom criticism, abuse, etc., is or may be directed. **5** archaic a shield or buckler, esp. a small round one. ● v.tr. (**targeted, targeting**) **1** identify or single out (a person or thing) as an object of attention or attack. **2** aim or direct (missiles targeted on major cities; should target our efforts where needed). □ **targetable** adj. [ME, dimin. of ME and OF targe shield]

Targum /ˈtɑːgəm/ n. any of various ancient Aramaic paraphrases or interpretations of the Hebrew Bible, made from about the 1st century AD when Hebrew was ceasing to be a spoken language. [Chaldee, = interpretation]

tariff /ˈtærɪf/ n. & v. ● n. **1** a table of fixed charges (a hotel tariff). **2 a** a duty on a particular class of imports or exports. **b** a list of duties or customs to be paid. **3** standard charges agreed between insurers etc. ● v.tr. subject (goods) to a tariff. [F tarif f. It. tariffa f. Turk. tarife f. Arab. ta'rīf(a) f. 'arrafa notify]

Tarim /tɑːˈriːm/ a river of NW China, in Xinjiang autonomous region. It rises as the Yarkand in the Kunlun Shan mountains and flows for over 2,000 km (1,250 miles) generally eastwards through the Tarim Basin, the driest region of Eurasia, petering out in the Lop Nor depression. For much of its course the river is unformed, following no clearly defined bed and subject to much evaporation.

tarlatan /ˈtɑːlət(ə)n/ n. a thin stiff open-weave muslin. [F tarlatane, prob. of Ind. orig.]

Tarmac /ˈtɑːmæk/ n. & v. ● n. propr. **1** = TARMACADAM. **2** a surface made of this, e.g. a runway. ● v.tr. (**tarmac**) (**tarmacked, tarmacking**) apply tarmacadam to. [abbr.]

tarmacadam /ˌtɑːməˈkædəm/ n. a material of stone or slag bound with tar, used in paving roads etc. [TAR¹ + MACADAM]

Tarn /tɑːn/ a river of southern France which rises in the Cévennes and

flows 380 km (235 miles) generally south-westwards through deep gorges before meeting the Garonne north-west of Toulouse.

tarn /tɑːn/ n. a small mountain lake. [ME terne, tarne f. ON]

tarnation /tɑːˈneɪʃ(ə)n/ int. esp. US sl. damn, blast. [alt. of DAMNATION, darnation]

tarnish /ˈtɑːnɪʃ/ v. & n. ● v. **1** tr. lessen or destroy the lustre of (metal etc.). **2** tr. impair (one's reputation etc.). **3** intr. (of metal etc.) lose lustre. ● n. **1 a** a loss of lustre. **b** a film of colour formed on an exposed surface of a mineral or metal. **2** a blemish; a stain. □ **tarnishable** adj. [F ternir f. terne dark]

Tarnów /ˈtɑːnʊf/ an industrial city in southern Poland; pop. (1990) 73,740.

taro /ˈtɑːrəʊ/ n. (pl. **-os**) a tropical aroid plant, Colocasia esculenta, with tuberous roots that are used as a staple food in many countries. Also called eddo. [Polynesian]

tarot /ˈtærəʊ/ n. **1 a** a pack of cards having four suits and an additional set of 22 picture-cards, trumps, now used esp. in fortune-telling. (See note below.) **b** (in sing. or pl.) any of several games played with such a pack. **2 a** any of the trump cards. **b** any of the cards from a fortune-telling pack. [F tarot, It. tarocchi, of unkn. orig.]

▪ Tarot cards are thought to go back to the 12th century in Europe, though the earliest surviving set is from 1390. Their origin and symbolism is obscure. The tarot pack consists of 78 cards, of which 56 are arranged in four suits of 14 each, numbered similarly to a 52-card pack except that a knight is included among the court cards. The suits are called cups, swords, money, and batons (clubs). The rest of the pack is composed of a series of pictures symbolizing various aspects of life and having links with astrology and the cabbala: emblematic pictures of Death, the Sun, the Moon, the Devil, etc., are included. Renewed interest in the occult has led to a recent revival in the use of tarot cards for fortune-telling. (See also PLAYING CARD.)

tarp /tɑːp/ n. N. Amer. & Austral. colloq. tarpaulin. [abbr.]

tarpan /ˈtɑːpæn/ n. an extinct northern European primitive wild horse. [Kyrgyz Tartar]

tarpaulin /tɑːˈpɔːlɪn/ n. **1** heavy-duty waterproof cloth esp. of tarred canvas. **2** a sheet or covering of this. **3 a** a sailor's tarred or oilskin hat. **b** archaic a sailor. [prob. f. TAR¹ + PALL¹ + -ING¹]

Tarpeia /tɑːˈpiːə/ Rom. Hist. one of the Vestal Virgins, the daughter of a commander of the Capitol in Rome. According to legend she betrayed the citadel to the Sabines in return for what they wore on their arms, hoping to receive their golden bracelets; however, the Sabines killed her by throwing their shields on to her.

Tarpeian Rock /tɑːˈpiːən/ a cliff in ancient Rome, at the south-western corner of the Capitoline Hill, over which murderers and traitors were hurled. [TARPEIA]

tarpon /ˈtɑːpɒn/ n. **1** a large silvery game-fish, Tarpon atlanticus, of the tropical Atlantic. **2** a similar fish, Megalops cyprinoides, of the Indian and Pacific Oceans. [Du. tarpoen, of unkn. orig.]

Tarquinius /tɑːˈkwɪnɪəs/ (anglicized name Tarquin) the name of two semi-legendary Etruscan kings of ancient Rome:

Tarquinius Priscus (full name Lucius Tarquinius Priscus), reigned c.616–c.578 BC. According to tradition, he was murdered by the sons of the previous king.

Tarquinius Superbus (full name Lucius Tarquinius Superbus; known as Tarquin the Proud), reigned c.534–c.510 BC. Traditionally, he was the son or grandson of Tarquinius Priscus and the seventh and last king of Rome. His reign was noted for its cruelty, and he was ultimately expelled from the city after the rape of a woman called Lucretia by his son; following his expulsion the Republic was founded. He was later engaged in a number of unsuccessful attacks on Rome, assisted by Lars Porsenna.

tarradiddle var. of TARADIDDLE.

tarragon /ˈtærəgən/ n. a bushy aromatic labiate herb, Artemisia dracunculus, with leaves used to flavour salads, stuffings, vinegar, etc. [= med. L tarchon f. med. Gk tarkhōn, perh. through Arab. f. Gk drakōn dragon]

tarras var. of TRASS.

Tarrasa /təˈræsə/ (also **Terrassa**) an industrial city in Catalonia, NE Spain; pop. (1991) 153,520.

tarry¹ /ˈtɑːrɪ/ adj. (**tarrier, tarriest**) of or like or smeared with tar. □ **tarriness** n.

tarry² /ˈtærɪ/ v.intr. (**-ies, -ied**) archaic or literary delay, linger, stay, wait. □ **tarrier** n. [ME: orig. uncert.]

tarsal /'tɑːs(ə)l/ adj. & n. ● adj. of or relating to the bones in the ankle. ● n. a tarsal bone. [TARSUS + -AL¹]

tarsi pl. of TARSUS.

tarsi- /'tɑːsɪ/ comb. form (also **tarso-** /'tɑːsəʊ/) tarsus.

tarsia /'tɑːsɪə/ n. = INTARSIA. [It.]

tarsier /'tɑːsɪə(r)/ n. a small arboreal nocturnal primate of the family Tarsiidae, native to Borneo, the Philippines, etc., with very large eyes and feeding on insects and small vertebrates. [F (as TARSUS), from the structure of its foot]

tarso- comb. form var. of TARSI-.

Tarsus /'tɑːsəs/ an ancient city in southern Turkey, a present-day market-town. It was the capital of Cilicia and the birthplace of St Paul.

tarsus /'tɑːsəs/ n. (pl. **tarsi** /-saɪ/) **1 a** Anat. the group of bones forming the ankle and upper foot. **b** Zool. the shank of a bird's leg. **2** the fifth joint of the leg in insects etc., usu. comprising several small segments together with the terminal claw. **3** Anat. the fibrous connective tissue of the eyelid. [mod.L f. Gk tarsos flat of the foot, rim of the eyelid]

tart¹ /tɑːt/ n. **1** an open pastry case containing jam etc. **2** esp. Brit. a pie with a fruit or sweet filling. □ **tartlet** /'tɑːtlɪt/ n. [ME f. OF tarte = med.L tarta, of unkn. orig.]

tart² /tɑːt/ n. & v. ● n. sl. **1** a prostitute; a promiscuous woman. **2** offens. a girl or woman. ● v. (foll. by up) esp. Brit. colloq. **1** tr. (usu. refl.) smarten (oneself or a thing) up, esp. flashily or gaudily. **2** intr. dress up gaudily. [prob. abbr. of SWEETHEART]

tart³ /tɑːt/ adj. **1** sharp or acid in taste. **2** (of a remark etc.) cutting, bitter. □ **tartly** adv. **tartness** n. [OE teart, of unkn. orig.]

tartan¹ /'tɑːt(ə)n/ n. **1** a pattern of coloured stripes crossing at right angles, esp. the distinctive plaid worn by Scottish Highlanders to denote their clan. **2** woollen etc. cloth woven in this pattern (often attrib.: a tartan scarf). [perh. f. OF tertaine, tiretaine]

tartan² /'tɑːt(ə)n/ n. a lateen-sailed single-masted ship used in the Mediterranean. [F tartane f. It. tartana, perh. f. Arab. ṭarīda]

Tartar /'tɑːtə(r)/ n. & adj. (also **Tatar** except in sense 3 of n.) ● n. **1 a** a member of a group of Turkic peoples inhabiting parts of European and Asiatic Russia, especially parts of Siberia, Crimea, the Caucasus, and districts along the Volga. **b** hist. a member of the combined forces of central Asian peoples, including Mongols and Turks, who under the leadership of Genghis Khan overran and devastated much of Asia and eastern Europe in the early 13th century, and under Tamerlane (14th century) established a large empire in central Asia with its capital at Samarkand. **2** the Turkic language of these peoples. **3** (**tartar**) a violent-tempered or intractable person. ● adj. **1** of or relating to the Tartars or their language. **2** of or relating to central Asia east of the Caspian Sea. □ **Tartarian** /tɑː'teərɪən/ adj. [ME tartre f. OF Tartare or med.L Tartarus]

tartar /'tɑːtə(r)/ n. **1** a hard deposit of saliva, calcium phosphate, etc., that forms on the teeth. **2** a deposit of acid potassium tartrate that forms a hard crust on the inside of a cask during the fermentation of wine. □ **cream of tartar** see CREAM. **tartar emetic** potassium antimony tartrate, a toxic compound used formerly as an emetic and now for the treatment of protozoal diseases in animals and as a mordant in dyeing. □ **tartarize** v.tr. (also **-ise**). [ME f. med.L f. med.Gk tartaron]

tartare /tɑː'tɑː(r)/ adj. (in full **sauce tartare**) = TARTAR SAUCE. [F, = TARTAR]

tartaric /tɑː'tærɪk/ adj. Chem. of or produced from tartar. □ **tartaric acid** a natural carboxylic acid found esp. in unripe grapes, used in baking powders and as a food additive. [F tartarique f. med.L tartarum: see TARTAR]

tartar sauce n. a sauce of mayonnaise and chopped gherkins, capers, etc. [TARTAR]

Tartarus /'tɑːtərəs/ Gk Mythol. **1** a primeval god, offspring of Chaos. **2** a part of the underworld where the wicked suffered punishment for their misdeeds, especially those such as Ixion and Tantalus who had committed some outrage against the gods. □ **Tartarean** /tɑː'teərɪən/ adj. [L f. Gk Tartaros]

Tartary /'tɑːtərɪ/ a historical region of Asia and eastern Europe, especially the high plateau of central Asia and its NW slopes, which formed part of the Tartar empire in the Middle Ages.

tartrate /'tɑːtreɪt/ n. Chem. a salt or ester of tartaric acid. [F (as TARTAR, -ATE¹)]

tartrazine /'tɑːtrəˌziːn/ n. Chem. a brilliant yellow dye derived from tartaric acid and used to colour food, drugs, and cosmetics. [as TARTAR + AZO- + -INE⁴]

tarty /'tɑːtɪ/ adj. colloq. (**tartier, tartiest**) (esp. of a woman) vulgarly provocative, gaudy; promiscuous. □ **tartily** adv. **tartiness** n. [TART² + -Y¹]

Tarzan /'tɑːz(ə)n/ **1** a character in novels by Edgar Rice Burroughs and subsequent films and television series. Tarzan is a white man (Lord Greystoke by birth), orphaned in West Africa in his infancy and reared by apes in the jungle. **2** a man of great agility and powerful physique.

Tas. abbr. Tasmania.

Tashi lama /'tæʃɪ/ n. = PANCHEN LAMA. [f. the Tibetan Buddhist monastery Tashi Lhunpo]

Tashkent /tæʃ'kent/ the capital of Uzbekistan, in the western foothills of the Tien Shan mountains, in the far north-east of the country; pop. (1990) 2,094,000. One of the oldest cities in central Asia, Tashkent was an important centre on the trade route between Europe and the Orient. Under Arab control in the 8th century, it became part of the Mongol empire in the 13th century, and was captured by the Russians in 1865. It was the administrative centre of Turkestan until the republic was divided in 1924. In 1930 it became capital of Uzbekistan, following Samarkand.

task /tɑːsk/ n. & v. ● n. a piece of work to be done or undertaken. ● v.tr. **1** make great demands on (a person's powers etc.). **2** assign a task to. □ **take to task** rebuke, scold. **task force** (or **group**) **1** Mil. an armed force organized for a special operation. **2** a unit specially organized for a task. [ME f. ONF tasque = OF tasche f. med.L tasca, perh. f. taxa f. L taxare TAX]

taskmaster /'tɑːskˌmɑːstə(r)/ n. (fem. **taskmistress** /-ˌmɪstrɪs/) a person who imposes a task or burden, esp. regularly or severely.

TASM abbr. tactical air-to-surface missile.

Tasman /'tæzmən/, Abel (Janszoon) (1603–c.1659), Dutch navigator. In 1642 he was sent by Anthony van Diemen (see TASMANIA) to explore Australian waters; that year he reached Tasmania (which he named Van Diemen's Land) and New Zealand, and in 1643 arrived at Tonga and Fiji. On a second voyage in 1644 he also reached the Gulf of Carpentaria on the north coast of Australia.

Tasmania /tæz'meɪnɪə/ a state of Australia consisting of the mountainous island of Tasmania itself and several smaller islands; pop. (1990) 457,500; capital, Hobart. It is separated from the SE coast of mainland Australia by the Bass Strait. The first European to visit was the Dutch navigator and explorer Abel Tasman, in 1642, who named it Van Diemen's Land after Anthony Van Diemen (1593–1645), Governor-General of the Dutch East Indies; it was renamed Tasmania in 1855. Following the arrival of European settlers, the native Aboriginal inhabitants were displaced and became extinct in 1888.

Tasmanian /tæz'meɪnɪən/ n. & adj. ● n. **1** an inhabitant of Tasmania. **2** a person of Tasmanian descent. ● adj. of or relating to Tasmania. □ **Tasmanian devil** a small nocturnal flesh-eating marsupial, Sarcophilus harrisii, heavily built and of an aggressive disposition, now found only in Tasmania. **Tasmanian wolf** (or **tiger**) = THYLACINE.

Tasman Sea an arm of the South Pacific lying between Australia and New Zealand.

Tass /tæs/ (also **TASS**) the official news agency of the former Soviet Union, founded in 1925 in Leningrad (St Petersburg). In 1992 the agency was renamed ITAR-Tass (Informatsionnoe telegrafnoe agentstvo Rossii Information Telegraph Agency of Russia). [the initials of Russ. Telegrafnoe agentstvo Sovetskogo Soyuza Telegraph Agency of the Soviet Union]

tass /tæs/ n. Sc. **1** a cup or small goblet. **2** a small draught of brandy etc. [ME f. OF tasse f. Arab. ṭāsa basin f. Pers. tast]

tassel¹ /'tæs(ə)l/ n. & v. ● n. **1** a tuft of loosely hanging threads or cords etc. attached for decoration to a cushion, scarf, cap, etc. **2** a tassel-like head of some plants, esp. a flower-head with prominent stamens at the top of a maize stalk. ● v. (**tasselled, tasselling;** US **tasseled, tasseling**) **1** tr. provide with a tassel or tassels. **2** intr. N. Amer. (of maize etc.) form tassels. [ME f. OF tas(s)el clasp, of unkn. orig.]

tassel² /'tæs(ə)l/ n. (also **torsel** /'tɔːs-/) a small piece of stone, wood, etc., supporting the end of a beam or joist. [OF ult. f. L taxillus small die, and tessella: see TESSELLATE]

tassie /'tæsɪ/ n. Sc. a small cup.

taste /teɪst/ n. & v. ● n. **1 a** the sensation characteristic of a soluble substance caused in the mouth and throat by contact with that substance (disliked the taste of garlic). **b** the faculty of perceiving this sensation (was bitter to the taste). **2 a** a small portion of food or drink taken as a sample. **b** a hint or touch of some ingredient or quality. **3** a slight experience (*a taste of success*). **4** (often foll. by *for*) a liking or predilection (*has expensive tastes*; *is not to my taste*). **5 a** aesthetic discernment in art, literature, conduct, etc., esp. of a specified kind (*a person of taste*; *dresses in poor taste*). **b** a style or manner based on this (*a table in the French taste*). ● *v.* **1** *tr.* sample or test the flavour of (food etc.) by taking it into the mouth. **2** *tr.* (also *absol.*) perceive the flavour of (*could taste the lemon*; *cannot taste with a cold*). **3** *tr.* (esp. with *neg.*) eat or drink a small portion of (*had not tasted food for days*). **4** *tr.* have experience of (*had never tasted failure*). **5** *intr.* (often foll. by *of*) have a specified flavour (*tastes bitter*; *tastes of onions*). □ **a bad** (or **bitter** etc.) **taste** a strong feeling of regret or unease after an experience etc. **taste blood** see BLOOD. **taste bud** any of the cells or nerve-endings on the surface of the tongue by which things are tasted. **to taste** in the amount needed for a pleasing result (*add salt and pepper to taste*). □ **tasteable** *adj.* [ME, = touch, taste, f. OF *tast*, *taster* touch, try, taste, ult. perh. f. L *tangere* touch + *gustare* taste]

tasteful /ˈteɪstfʊl/ *adj.* having, or done in, good taste. □ **tastefully** *adv.* **tastefulness** *n.*

tasteless /ˈteɪstlɪs/ *adj.* **1** lacking flavour. **2** having, or done in, bad taste. □ **tastelessly** *adv.* **tastelessness** *n.*

taster /ˈteɪstə(r)/ *n.* **1** a person employed to test food or drink by tasting it, esp. for quality or *hist.* to detect poisoning. **2** a small cup used by a wine-taster. **3** an instrument for extracting a small sample from within a cheese. **4** a sample of something offered. [ME f. AF *tastour*, OF *tasteur* f. *taster*: see TASTE]

tasting /ˈteɪstɪŋ/ *n.* a gathering at which food or drink (esp. wine) is tasted and evaluated.

tasty /ˈteɪstɪ/ *adj.* (**tastier**, **tastiest**) **1** (of food) pleasing in flavour; appetizing. **2** *colloq.* attractive. □ **tastily** *adv.* **tastiness** *n.*

tat[1] /tæt/ *n.* **1 a** a tatty or tasteless clothes; worthless goods. **b** rubbish, junk. **2** a shabby person. [prob. f. Sc. *tatty* shaggy (see TATTY)]

tat[2] /tæt/ *v.* (**tatted**, **tatting**) **1** *intr.* do tatting. **2** *tr.* make by tatting. [19th c.: orig. unkn.]

tat[3] see TIT[2].

ta-ta /tæˈtɑː/ *int. Brit. colloq.* goodbye (said esp. to or by a child). [19th c.: orig. unkn.]

tatami /təˈtɑːmɪ/ *n.* **1** (in full **tatami mat**) a mat made from woven rushes, used as a traditional floor-covering in Japan. **2** a standard unit in room measurement in Japan, approx. 1.83 by 0.91 metres (6 ft by 3 ft). [Jap.]

Tatar var. of TARTAR.

Tatarstan /ˌtɑːtəˈstɑːn/ an autonomous republic in European Russia, in the valley of the River Volga; pop. (1990) 3,658,000; capital, Kazan. In 1552, after a fierce struggle with Ivan the Terrible, the region was absorbed into the Russian empire. In 1920 it was constituted an autonomous republic of the USSR. Seeking greater independence, Tatarstan refused to sign the Russian federal treaty of 1992 (see RUSSIA).

Tate /teɪt/, Nahum (1652–1715), Irish dramatist and poet, resident in London from the 1670s. He wrote a number of plays, chiefly adaptations from earlier writers; he is especially known for his version of Shakespeare's *King Lear*, in which he substituted a happy ending. He also wrote the libretto for Purcell's *Dido and Aeneas* (1689) and (with John Dryden) the second part of *Absalom and Achitophel* (1682). He was appointed Poet Laureate in 1692.

Tate Gallery a national gallery of art at Millbank, London, opened in 1897 and built at the expense of the sugar manufacturer Sir Henry Tate to house examples of British schools neglected by the National Gallery. In the 20th century modern foreign paintings and sculpture (both British and foreign) were added.

tater /ˈteɪtə(r)/ *n.* (also **tatie** /ˈteɪtɪ, ˈtætɪ/, **tato** /ˈteɪtəʊ/) *sl.* = POTATO. [abbr.]

Tati /ˈtætɪ/ Jacques (born Jacques Tatischeff) (1908–82), French film director and actor. Although he made only five full-length films, he became internationally known as a comic actor with his performances as Monsieur Hulot, a character which he introduced in his second film *Monsieur Hulot's Holiday* (1953). Subsequent films featuring the character include the Oscar-winning *Mon oncle* (1958).

tatler *archaic* var. of TATTLER.

Tatra Mountains /ˈtɑːtrə/ (also **Tatras**) a range of mountains in eastern Europe on the Polish–Slovak border, the highest range in the Carpathians, rising to 2,655 m (8,710 ft) at Mount Gerlachovsky.

tats /tæts/ *n.pl.* esp. *Austral. sl.* teeth. [f. *tats* = dice]

tatter /ˈtætə(r)/ *n.* (usu. in *pl.*) a rag; an irregularly torn piece of cloth or paper etc. □ **in tatters 1** torn in many places. **2** (of a negotiation, argument, etc.) ruined, demolished. □ **tattery** *adj.* [ME f. ON *tǫtrar* rags: cf. Icel. *tötur*]

tattered /ˈtætəd/ *adj.* in tatters; ragged.

tattersall /ˈtætəˌsɔːl/ *n.* (in full **tattersall check**) a fabric with a pattern of coloured lines forming squares like a tartan. [Richard *Tattersall* (see TATTERSALLS): from the traditional design of horse blankets]

Tattersalls /ˈtætəˌsɔːlz/ an English firm of horse auctioneers founded in 1776 by the horseman Richard Tattersall (1724–95).

tatting /ˈtætɪŋ/ *n.* **1** a kind of knotted lace made by hand with a small shuttle and used for trimming etc. **2** the process of making this. [19th c.: orig. unkn.]

tattle /ˈtæt(ə)l/ *v. & n.* ● *v.* **1** *intr.* prattle, chatter; gossip idly, speak indiscreetly. **2** *tr.* utter (words) idly, reveal (secrets). ● *n.* gossip; idle or trivial talk. □ **tattle-tale** *N. Amer.* a tell-tale, esp. a child. [ME f. Middle Flemish *tatelen*, *tateren* (imit.)]

tattler /ˈtætlə(r)/ *n.* (also *archaic* **tatler**) a prattler; a gossip.

tattoo[1] /təˈtuː/ *n.* (pl. **tattoos**) **1** *Mil.* an evening drum or bugle signal recalling soldiers to their quarters. **2** an elaboration of this with music and marching, presented as an entertainment. **3** a rhythmic tapping or drumming. [17th-c. *tap-too* f. Du. *taptoe*, lit. 'close the tap' (of the cask)]

tattoo[2] /təˈtuː/ *v. & n.* ● *v.tr.* (**tattoos**, **tattooed**) **1** mark (the skin) with an indelible design by puncturing it and inserting pigment. **2** make (a design) in this way. ● *n.* (pl. **tattoos**) a design made by tattooing. □ **tattooer** *n.* **tattooist** *n.* [Polynesian]

tatty /ˈtætɪ/ *adj.* (**tattier**, **tattiest**) **1** tattered; worn and shabby. **2** inferior. **3** tawdry. □ **tattily** *adv.* **tattiness** *n.* [prob. f. TAT[1]; but cf. Sc. *tatty* shaggy (app. rel. to OE *tættec* rag, TATTER)]

Tatum /ˈteɪtəm/, Arthur ('Art') (1910–56), American jazz pianist. He was born with cataracts in both eyes and as a result was almost completely blind throughout his life. He first became famous in the 1930s as a musician of great technical accomplishment; he performed chiefly in a trio with bass and guitar or as a soloist.

tau /taʊ, tɔː/ *n.* the nineteenth letter of the Greek alphabet (T, τ). □ **tau cross** a T-shaped cross. **tau particle** *Physics* an unstable heavy charged subatomic particle of the lepton class. [ME f. Gk]

taught *past* and *past part.* of TEACH.

taunt /tɔːnt/ *n. & v.* ● *n.* a thing said in order to anger or wound a person. ● *v.tr.* **1** assail with taunts. **2** reproach (a person) contemptuously. □ **taunter** *n.* **tauntingly** *adv.* [16th c., in phr. *taunt for taunt* f. F *tant pour tant* tit for tat, hence a smart rejoinder]

Taunton /ˈtɔːntən/ the county town of Somerset, in SW England; pop. (1981) 48,860.

taupe /təʊp/ *n.* a grey with a tinge of another colour, usu. brown. [F, = MOLE[1]]

Taupo, Lake /ˈtaʊpəʊ/ (called in Maori *Taupomoana*) the largest lake of New Zealand, in the centre of North Island. The town of Taupo is situated on its northern shore.

Tauranga /taʊˈræŋə/ a port on the Bay of Plenty, North Island, New Zealand; pop. (1990) 64,000.

taurine[1] /ˈtɔːriːn, -raɪn/ *adj.* of or like a bull; bullish. [L *taurinus* f. *taurus* bull]

taurine[2] /ˈtɔːriːn/ *n. Biochem.* a sulphur-containing amino acid important in the metabolism of fats. [f. Gk *tauros* bull + -INE[4]]

tauromachy /tɔːˈrɒməkɪ/ *n.* (pl. **-ies**) *archaic* **1** a bullfight. **2** bullfighting. [Gk *tauromakhia* f. *tauros* bull + *makhē* fight]

Taurus /ˈtɔːrəs/ *n.* **1** *Astron.* a constellation (the Bull), said to represent a bull with brazen feet that was tamed by Jason. Its many bright stars include Aldebaran (the bull's eye), and it contains the star clusters of the Hyades and the Pleiades, and the Crab Nebula. **2** *Astrol.* **a** the second sign of the zodiac, which the sun enters about 21 Apr. **b** a person born when the sun is in this sign. □ **Taurean** /ˈtɔːrɪən, tɔːˈriːən/ *adj. & n.* [ME f. L, = bull]

Taurus Mountains a range of mountains in southern Turkey, parallel to the Mediterranean coast. Rising to 3,734 m (12,250 ft) at Mount Aladaë, the range forms the southern edge of the Anatolian plateau.

taut /tɔːt/ *adj.* **1** (of a rope, muscles, etc.) tight; not slack. **2** (of nerves) tense. **3** (of a ship etc.) in good order or condition. □ **tauten** *v.tr. & intr.*

tautly adv. **tautness** n. [ME *touht, togt*, perh. = TOUGH, infl. by *tog-* past part. stem of obs. *tee* (OE *tēon*) pull]

tauto- /'tɔːtəʊ/ comb. form the same. [Gk, f. *tauto*, to *auto* the same]

tautog /tɔːˈtɒg/ n. a fish, *Tautoga onitis*, found off the Atlantic coast of North America, used as food. [Narragansett *tautauog* (pl.)]

tautology /tɔːˈtɒlədʒɪ/ n. (pl. **-ies**) **1** the saying of the same thing twice over in different words, esp. as a fault of style (e.g. *arrived one after the other in succession*). **2** a statement that is necessarily true. □ **tautologist** n. **tautologize** v.intr. (also **-ise**). **tautologous** /-ləgəs/ adj. **tautological** /ˌtɔːtəˈlɒdʒɪk(ə)l/ adj. **tautologically** adv. [LL *tautologia* f. Gk (as TAUTO-, -LOGY)]

tautomer /'tɔːtəˌmə(r)/ n. Chem. a substance that exists as two mutually convertible isomers in equilibrium. □ **tautomeric** /ˌtɔːtəˈmerɪk/ adj. **tautomerism** /tɔːˈtɒməˌrɪz(ə)m/ n. [TAUTO- + -MER]

tautophony /tɔːˈtɒfənɪ/ n. repetition of the same sound. [TAUTO- + Gk *phōnē* sound]

tavern /'tæv(ə)n/ n. esp. *archaic* or *literary* an inn or public house. [ME f. OF *taverne* f. L *taberna* hut, tavern]

taverna /təˈvɜːnə/ n. a Greek restaurant or eating-house. [mod.Gk (as TAVERN)]

TAVR abbr. (in the UK) Territorial and Army Volunteer Reserve. ¶ The name in use 1967–79: now *Territorial Army*.

taw[1] /tɔː/ v.tr. make (hide) into leather without the use of tannin, esp. by soaking in a solution of alum and salt. □ **tawer** n. [OE *tawian* f. Gmc]

taw[2] /tɔː/ n. **1** a large marble. **2** a game of marbles. **3** a line from which players throw marbles. [18th c.: orig. unkn.]

tawdry /'tɔːdrɪ/ adj. & n. ● adj. (**tawdrier, tawdriest**) **1** showy but worthless. **2** over-ornamented, gaudy, vulgar. ● n. cheap or gaudy finery. □ **tawdrily** adv. **tawdriness** n. [earlier as noun: short for *tawdry lace*, orig. *St Audrey's lace* f. *Audrey* = *Etheldrida*, patron saint of ELY]

tawny /'tɔːnɪ/ adj. (**tawnier, tawniest**) of an orange- or yellow-brown colour. □ **tawny eagle** a uniformly brown eagle, *Aquila rapax*, of Africa and Asia. **tawny owl** a usu. reddish-brown European owl, *Strix aluco*, with a distinctive and familiar hoot (also called *brown owl*). □ **tawniness** n. [ME f. AF *tauné* f. OF *tané* f. *tan* TAN[1]]

taws /tɔːz/ n. (also **tawse**) Sc. hist. a thong with a slit end formerly used in schools for punishing children. [app. pl. of obs. *taw* tawed leather, f. TAW[1]]

tax /tæks/ n. & v. ● n. **1** a contribution to state revenue compulsorily levied on individuals, property, or businesses (often foll. by *on*: *a tax on luxury goods*). (See also INCOME TAX.) **2** (usu. foll. by *on, upon*) a strain or heavy demand; an oppressive or burdensome obligation. ● v.tr. **1** impose a tax on (persons or goods etc.). **2** deduct tax from (income etc.). **3** make heavy demands on (a person's powers or resources etc.) (*you really tax my patience*). **4** (foll. by *with*) confront (a person) with a wrongdoing etc. **5** call to account. **6** Law examine and assess (costs etc.). □ **tax avoidance** the arrangement of financial affairs to minimize payment of tax. **tax-deductible** (of expenditure) that may be paid out of income before the deduction of income tax. **tax disc** Brit. a paper disc displayed on the windscreen of a motor vehicle, certifying payment of excise duty. **tax evasion** the illegal non-payment or underpayment of income tax. **tax-free** exempt from taxes. **tax haven** a country etc. where income tax is low. **tax return** a declaration of income for taxation purposes. **tax shelter** a means of organizing business affairs to minimize payment of tax. **tax year** see *financial year*. □ **taxable** adj. **taxer** n. **taxless** adj. [ME f. OF *taxer* f. L *taxare* censure, charge, compute, perh. f. Gk *tassō* fix]

taxa pl. of TAXON.

taxation /tækˈseɪʃ(ə)n/ n. the imposition or payment of tax. [ME f. AF *taxacioun*, OF *taxation* f. L *taxatio -onis* f. *taxare*: see TAX]

taxi /'tæksɪ/ n. & v. ● n. (pl. **taxis**) **1** (in full **taxi-cab**) a motor car licensed to ply for hire and usu. fitted with a taximeter. **2** a boat etc. similarly used. ● v. (**taxis, taxied, taxiing** or **taxying**) **1 a** intr. (of an aircraft or pilot) move along the ground under the machine's own power before take-off or after landing. **b** tr. cause (an aircraft) to taxi. **2** intr. & tr. go or convey in a taxi. □ **taxi dancer** a dancing partner available for hire. **taxi-driver** a driver of a taxi. **taxi rank** (US **stand**) a place where taxis wait to be hired. [abbr. of *taximeter cab*]

taxidermy /'tæksɪˌdɜːmɪ/ n. the art of preparing, stuffing, and mounting the skins of animals with lifelike effect. □ **taxidermist** n. **taxidermal** /ˌtæksɪˈdɜːm(ə)l/ adj. **taxidermic** adj. [Gk *taxis* arrangement + *derma* skin]

taximeter /'tæksɪˌmiːtə(r)/ n. an automatic device fitted to a taxi, recording the distance travelled and the fare payable. [F *taximètre* f. *taxe* tariff, TAX + -METER]

taxis /'tæksɪs/ n. **1** Surgery the restoration of displaced bones or organs by manual pressure. **2** Biol. the directional movement of a cell or organism in response to an external stimulus (cf. KINESIS 2). **3** Gram. order or arrangement of words. [Gk f. *tassō* arrange]

taxman /'tæksmæn/ n. colloq. (pl. **-men**) **1** an inspector or collector of taxes. **2** the personification of the government department dealing with tax.

taxon /'tæks(ə)n/ n. (pl. **taxa** /-sə/) any taxonomic group. [back-form. f. TAXONOMY]

taxonomy /tækˈsɒnəmɪ/ n. **1** the science of classification, esp. of living and extinct organisms. **2** a scheme of classification. □ **taxonomist** n. **taxonomic** /ˌtæksəˈnɒmɪk/ adj. **taxonomical** adj. **taxonomically** adv. [F *taxonomie* (as TAXIS, Gk *-nomia* distribution)]

taxpayer /'tæksˌpeɪə(r)/ n. a person who pays taxes.

Tay /teɪ/ the longest river in Scotland, flowing 192 km (120 miles) eastwards through Loch Tay, entering the North Sea through the Firth of Tay.

Tay, Firth of the estuary of the River Tay, on the North Sea coast of Scotland. It is spanned by the longest railway bridge in Britain, a structure opened in 1888 that has 85 spans and a total length of 3,553 m (11,653 feet), and by a road bridge that was opened nearby in 1966. The first bridge over the Firth of Tay, a cast-iron railway bridge near Dundee, collapsed in a gale in 1879 while a passenger train was crossing it.

tayberry /'teɪbərɪ/ n. (pl. **-ies**) a dark red soft fruit produced by crossing a blackberry and a raspberry. [TAY (near where introduced in 1977)]

Taylor[1] /'teɪlə(r)/, Elizabeth (b.1932), American actress, born in England. She began her career as a child star in films such as *National Velvet* (1944). She went on to star in many films, including *Cat on a Hot Tin Roof* (1958), *Who's Afraid of Virginia Woolf?* (1966), for which she won an Oscar, and *The Mirror Crack'd* (1980). She has been married eight times, including twice to the actor Richard Burton, with whom she starred in a number of films, notably *Cleopatra* (1963).

Taylor[2] /'teɪlə(r)/, Jeremy (1613–67), English Anglican churchman and writer. He was chaplain to Charles I during the English Civil War and lived chiefly in Wales until the Restoration, when he was appointed bishop of Down and Connor (1660). Although a celebrated preacher in his day, he is now remembered chiefly for his devotional writings, especially *The Rule and Exercises of Holy Living* (1650) and *The Rule and Exercises of Holy Dying* (1651).

Taylor[3] /'teɪlə(r)/, Zachary (1784–1850), American Whig statesman, 12th President of the US 1849–50. He became a national hero after his victories in the war with Mexico (1846–8). As President, he came into conflict with Congress over his desire to admit California to the Union as a free state (without slavery). He died in office before the problem could be resolved.

Taymyr Peninsula see TAIMYR PENINSULA.

Tay–Sachs disease /teɪˈsæks/ n. Med. an inherited metabolic disorder in which certain lipids accumulate in the brain, causing spasticity and death in childhood. [Warren *Tay*, English ophthalmologist (1843–1927) + Bernard *Sachs*, Amer. neurologist (1858–1944)]

Tayside /'teɪsaɪd/ a local government region in eastern Scotland; administrative centre, Dundee.

tazza /'tɑːtsə/ n. a saucer-shaped cup, esp. one mounted on a foot. [It.]

TB abbr. **1** tuberculosis. **2** torpedo boat.

Tb symb. Chem. the element terbium.

t.b. abbr. **1** tubercle bacillus. **2** tuberculosis.

T-bar /'tiːbɑː(r)/ n. **1** (in full **T-bar lift**) a type of ski-lift in the form of a series of inverted T-shaped metal bars for towing skiers uphill. **2** (often *attrib.*) a T-shaped fastening on a shoe or sandal.

Tbilisi /təbɪˈliːsɪ/ the capital of Georgia; pop. (1991) 1,267,500. From 1845 until 1936 its name was Tiflis.

T-bone /'tiːbəʊn/ n. a T-shaped bone, esp. in steak from the thin end of a loin.

tbsp. abbr. tablespoonful.

Tc symb. Chem. the element technetium.

TCCB abbr. (in the UK) Test and County Cricket Board.

TCD abbr. Trinity College, Dublin.

TCDD *abbr. Chem.* 2,3,7,8-tetrachlorodibenzo-*para*-dioxin (see DIOXIN).

T-cell /'tiːsel/ *n. Physiol.* = T-LYMPHOCYTE.

Tchaikovsky /tʃaɪˈkɒfski/, Pyotr (Ilich) (1840–93), Russian composer. He is especially known as a composer of the ballets *Swan Lake* (1877), *Sleeping Beauty* (1890), and *The Nutcracker* (1892), the First Piano Concerto (1875), and the overture *1812* (1880); other notable works include the operas *Eugene Onegin* (1879) and *The Queen of Spades* (1890). His music is characterized by melodiousness, depth of expression, and, especially in his later symphonies (including his sixth symphony, the 'Pathétique', 1893), melancholy. His death was officially attributed to cholera, but there is now a theory that he took poison because of a potential scandal arising from an alleged homosexual relationship with a member of the royal family.

TCP *abbr. propr.* a disinfectant and germicide. [trichlorophenylmethyliodasalicyl]

TD *abbr.* **1** (in the UK) Territorial (Officer's) Decoration. **2** see TEACHTA DÁLA.

Te *symb. Chem.* the element tellurium.

te /tiː/ *n.* (also **ti**) *Mus.* **1** (in tonic sol-fa) the seventh note of a major scale. **2** the note B in the fixed-doh system. [earlier *si*: F f. It., perh. f. *Sancte Iohannes*: see GAMUT]

tea /tiː/ *n. & v.* ● *n.* **1 a** (in full **tea plant**) an evergreen shrub or small tree, *Camellia sinensis*, of India, China, etc. **b** its dried leaves. (*See note below.*) **2** a drink made by infusing tea-leaves in hot (at first preferably boiling) water. **3** a similar drink made from the leaves of other plants or from another substance (*camomile tea; beef tea*). **4 a** a light afternoon meal consisting of tea, bread, cakes, etc. **b** *Brit.* a cooked (esp. early) evening meal. ● *v.* (**teaed** or **tea'd** /tiːd/) **1** *intr.* take tea. **2** *tr.* give tea to (a person). □ **tea and sympathy** *colloq.* hospitable behaviour towards a troubled person. **tea bag** a small perforated bag of tea for infusion. **tea-ball** *esp. N. Amer.* a ball of perforated metal to hold tea for infusion. **tea-bread** light or sweet bread for eating at tea. **tea break** *Brit.* a pause in work etc. to drink tea. **tea caddy** a container for tea. **tea ceremony** an elaborate Japanese ritual of serving and drinking tea, as an expression of Zen Buddhist philosophy. **tea chest** a light metal-lined wooden box in which tea is transported. **tea-clipper** *hist.* a clipper or fast sailing-ship formerly used in the tea trade. **tea cloth** = *tea towel*. **tea cosy** a cover to keep a teapot warm. **tea dance** an afternoon tea with dancing. **tea garden** a garden in which afternoon tea is served to the public. **tea lady** a woman employed to make tea in offices etc. **tea-leaf 1** a dried leaf of tea, used to make a drink of tea. **2** (esp. in *pl.*) these after infusion or as dregs. **3** *rhyming sl.* a thief. **tea party** a party at teatime. **tea-planter** a proprietor or cultivator of a tea plantation. **tea rose** a hybrid shrub, *Rosa odorata*, with a scent supposed to resemble that of tea. **tea towel** a towel for drying washed crockery etc. **tea-tree 1** *Austral. & NZ* a shrub or tree of the genus *Leptospermum* or *Melaleuca*, esp. the manuka. **2** an ornamental solanaceous shrub, *Lycium barbarum*, with red berries. **tea trolley** (*US* **wagon**) a small wheeled trolley from which tea is served. [17th-c. *tay*, *tey*, prob. f. Du. *tee* f. Chin. (Amoy dial.) *te*, = Mandarin dial. *cha*]

▪ The tea plant has been cultivated for thousands of years in China, from where it was introduced into Europe in the 17th century. Tea remained extremely scarce and costly until the 19th century, when it was discovered growing also in Assam in NE India. In 1870, when rust destroyed the coffee crop in Sri Lanka, tea was planted there, and later it was introduced into the islands of SE Asia, the Transcaucasian region, and parts of Africa. It grows best in areas of moderate to high rainfall, equable temperatures, and high humidity.

teacake /'tiːkeɪk/ *n. Brit.* a light yeast-based usu. sweet bun eaten at tea, often toasted.

teach /tiːtʃ/ *v.tr.* (*past and past part.* **taught** /tɔːt/) **1 a** give systematic information to (a person) or about (a subject or skill). **b** (*absol.*) practise this professionally. **c** enable (a person) to do something by instruction and training (*taught me to swim; taught me how to dance*). **2 a** advocate as a moral etc. principle (*my parents taught me forgiveness*). **b** communicate, instruct in (*suffering taught me patience*). **3** (foll. by *to* + infin.) **a** induce (a person) by example or punishment to do or not to do a thing (*that will teach you to sit still; that will teach you not to laugh*). **b** *colloq.* make (a person) disinclined to do a thing (*I will teach you to interfere*). □ **teach-in 1** an informal lecture and discussion on a subject of public interest. **2** a series of these. **teach a person a lesson** see LESSON. **teach school** *US* be a teacher in a school. [OE *tæcan* f. a Gmc root = 'show']

teachable /'tiːtʃəb(ə)l/ *adj.* **1** apt at learning. **2** (of a subject) that can be taught. □ **teachableness** *n.* **teachability** /ˌtiːtʃəˈbɪlɪti/ *n.*

teacher /'tiːtʃə(r)/ *n.* a person who teaches, esp. in a school. □ **teacherly** *adj.*

teaching /'tiːtʃɪŋ/ *n.* **1** the profession of a teacher. **2** (often in *pl.*) what is taught; a doctrine. □ **teaching hospital** a hospital where medical students are taught. **teaching machine** any of various devices for giving instruction according to a program that reacts to pupils' responses.

Teachta Dála /ˌtjɒxtə ˈdɔːlə/ *n.* (*pl.* **Teachti** /-tɪ/) (abbr. **TD**) (in the Republic of Ireland) a member of the Dáil or lower House of Parliament. [Ir.]

teacup /'tiːkʌp/ *n.* **1** a cup from which tea is drunk, usu. with a single handle and a matching saucer. **2** an amount held by this, about 150 ml. □ **teacupful** *n.* (*pl.* **-fuls**).

teak /tiːk/ *n.* **1** a large deciduous tree, *Tectona grandis*, native to India and SE Asia. **2** its hard durable timber, much used in shipbuilding and furniture. [Port. *teca* f. Malayalam *tēkka*]

teal /tiːl/ *n.* (*pl.* same) **1** a small freshwater duck of the genus *Anas*, esp. the *green-winged teal* (*A. crecca*). **2** a dark greenish-blue colour. [rel. to MDu. *tēling*, of unkn. orig.]

team /tiːm/ *n. & v.* ● *n.* **1** a set of players forming one side in a game (*a cricket team*). **2** two or more persons working together. **3 a** a set of draught animals. **b** one animal or more in harness with a vehicle. ● *v.* **1** *intr. & tr.* (usu. foll. by *up*) join in a team or in common action (*decided to team up with them*). **2** *tr.* harness (horses etc.) in a team. **3** *tr.* (foll. by *with*) match or coordinate (clothes). □ **team-mate** a fellow member of a team or group. **team spirit** willingness to act as a member of a group rather than as an individual. **team-teaching** teaching by a team of teachers working together. [OE *tēam* offspring f. a Gmc root = 'pull', rel. to TOW[1]]

teamster /'tiːmstə(r)/ *n.* **1** *N. Amer.* a lorry-driver. **2** a driver of a team of animals.

teamwork /'tiːmwɜːk/ *n.* the combined action of a team, group, etc., esp. when effective and efficient.

teapot /'tiːpɒt/ *n.* a pot with a handle, spout, and lid, in which tea is brewed and from which it is poured.

teapoy /'tiːpɔɪ/ *n.* a small three- or four-legged table esp. for tea. [Hindi *tīn*, *tir-* three + Pers. *pāī* foot: sense and spelling infl. by TEA]

tear[1] /teə(r)/ *v. & n.* ● *v.* (*past* **tore** /tɔː(r)/; *past part.* **torn** /tɔːn/) **1** *tr.* (often foll. by *up*) pull apart or to pieces with some force (*tear it in half; tore up the letter*). **2** *tr.* **a** make a hole or rent in by tearing (*have torn my coat*). **b** make (a hole or rent). **3** *tr.* (foll. by *away, off*, etc.) pull violently or with some force (*tore the book away from me; tore off the cover; tore a page out*). **4** *tr.* violently disrupt or divide (*the country was torn by civil war; torn by conflicting emotions*). **5** *intr. colloq.* go or travel hurriedly or impetuously (*tore across the road*). **6** *intr.* undergo tearing (*the curtain tore down the middle*). **7** *intr.* (foll. by *at* etc.) pull violently or with some force. ● *n.* **1** a hole or other damage caused by tearing. **2** a torn part of cloth etc. **3** *sl.* a spree; a drinking-bout. □ **be torn between** have difficulty in choosing between. **tear apart 1** destroy, divide utterly; distress greatly. **2** search (a place) exhaustively. **3** criticize forcefully. **tear one's hair out** behave with extreme desperation or anger. **tear into 1** attack verbally; reprimand. **2** make a vigorous start on (an activity). **tear-off** *attrib.adj.* (of a slip, plastic bag, etc.) designed to be removed by tearing off usu. along a perforated line. **tear oneself away** leave despite a strong desire to stay. **tear sheet** a page that can be removed from a newspaper or magazine etc. for use separately. **tear to shreds** *colloq.* refute or criticize thoroughly. **that's torn it** *Brit. colloq.* that has spoiled things, caused a problem, etc. □ **tearable** *adj.* **tearer** *n.* [OE *teran* f. Gmc]

tear[2] /tɪə(r)/ *n.* **1** a drop of clear salty liquid secreted by glands, that serves to moisten and wash the eye and is shed from it in grief or other strong emotions. **2** a tearlike thing; a drop. □ **in tears** crying; shedding tears. **tear-drop** a single tear. **tear-duct** a passage through which tears pass to the eye or from the eye to the nose. **tear-gas** gas that disables by causing severe irritation to the eyes. **tear-jerker** *colloq.* a story, film, etc., calculated to evoke sadness or sympathy. **without tears** presented so as to be learned or done easily. □ **tearlike** *adj.* [OE *tēar*]

tearaway /'teərəˌweɪ/ *n. Brit.* **1** an impetuous or reckless young person. **2** a hooligan.

tearful /'tɪəfʊl/ *adj.* **1** crying or inclined to cry. **2** causing or accompanied with tears; sad (*a tearful event*). □ **tearfully** *adv.* **tearfulness** *n.*

tearing /ˈteərɪŋ/ *adj.* extreme, overwhelming, violent (*in a tearing hurry*).

tearless /ˈtɪəlɪs/ *adj.* not shedding tears. □ **tearlessly** *adv.* **tearlessness** *n.*

tearoom /ˈtiːruːm, -rʊm/ *n.* a small restaurant or café where tea is served.

tease /tiːz/ *v. & n.* ● *v.tr.* (also *absol.*) **1 a** make fun of (a person or animal) playfully or unkindly or annoyingly. **b** tempt or allure, esp. sexually, while refusing to satisfy the desire aroused. **2** pick (wool, hair, etc.) into separate fibres. **3** dress (cloth) esp. with teasels. ● *n.* **1** *colloq.* a person fond of teasing. **2** an instance of teasing (*it was only a tease*). □ **tease out 1** separate (strands etc.) by disentangling. **2** search out, elicit (information etc.). □ **teasingly** *adv.* [OE *tǣsan* f. WG]

teasel /ˈtiːz(ə)l/ *n. & v.* (also **teazel**, **teazle**) ● *n.* **1** a tall plant of the genus *Dipsacus*, with large prickly heads that were formerly dried and used to raise the nap on woven cloth. **2** a device used as a substitute for teasels. ● *v.tr.* dress (cloth) with teasels. □ **teaseler** *n.* [OE *tǣs(e)l*, = OHG *zeisala* (as TEASE)]

teaser /ˈtiːzə(r)/ *n.* **1** *colloq.* a hard question or task. **2** a teasing person. **3** esp. *US* a short introductory advertisement, trailer for a film, etc.

teaset /ˈtiːset/ *n.* a set of crockery for serving tea.

teashop /ˈtiːʃɒp/ *n.* esp. *Brit.* = TEAROOM.

teaspoon /ˈtiːspuːn/ *n.* **1** a small spoon for stirring tea. **2** an amount held by this, equivalent to about 4 ml. □ **teaspoonful** *n.* (*pl.* **-fuls**).

teat /tiːt/ *n.* **1** a mammary nipple, esp. of an animal. **2** a thing resembling this, esp. a device of rubber etc. for sucking milk from a bottle. [ME f. OF *tete*, prob. of Gmc orig., replacing TIT³]

teatime /ˈtiːtaɪm/ *n.* the time in the afternoon when tea is served.

teazel (also **teazle**) var. of TEASEL.

Tebet /ˈtebet/ *n.* (also **Tevet** /ˈtevet/) (in the Jewish calendar) the fourth month of the civil and tenth of the religious year, usually coinciding with parts of December and January. [Heb. *ṭēbēt*]

tec /tek/ *n. colloq.* a detective. [abbr.]

tech /tek/ *n.* (also **tec**) *colloq.* **1** a technical college. **2** technology. [abbr.]

techie /ˈtekɪ/ *n. colloq.* an expert in or enthusiast for (esp. computing) technology. [TECH + -IE]

technetium /tekˈniːʃəm/ *n.* a radioactive metallic chemical element (atomic number 43; symbol Tc). Technetium was the first element to be created artificially, in 1937, when the Italian scientists Emilio Segrè and Carlo Perrier bombarded molybdenum with deuterons. It is the lightest element without stable isotopes, and is also formed as a product of uranium fission. A transition metal chemically related to rhenium, technetium has some use as a radioactive tracer. [f. Gk *tekhnētos* artificial f. *tekhnē* art]

technic /ˈteknɪk/ *n.* **1** (usu. in *pl.*) **a** technology. **b** technical terms, details, methods, etc. **2** (also /tekˈniːk/) technique. □ **technicist** /-nɪsɪst/ *n.* [L *technicus* f. Gk *tekhnikos* f. *tekhnē* art]

technical /ˈteknɪk(ə)l/ *adj.* **1** of or involving or concerned with the mechanical arts and applied sciences (*technical college*; *a technical education*). **2** of or relating to a particular subject or craft etc. or its techniques (*technical terms*; *technical merit*). **3** (of a book or discourse etc.) using technical language; requiring special knowledge to be understood. **4** due to mechanical failure. **5** legally such; such in strict interpretation (*technical assault*; *lost on a technical point*). □ **technical hitch 1** a temporary breakdown or problem in machinery etc. **2** an unexpected obstacle or snag. **technical knockout** *Boxing* a termination of a fight by the referee on the grounds of a contestant's inability to continue, the opponent being declared the winner. □ **technically** *adv.* **technicalness** *n.*

technicality /ˌteknɪˈkælɪtɪ/ *n.* (*pl.* **-ies**) **1** the state of being technical. **2** a technical expression. **3** a technical point or detail (*was acquitted on a technicality*).

technician /tekˈnɪʃ(ə)n/ *n.* **1** an expert in the practical application of a science. **2** a person skilled in the technique of an art or craft. **3** a person employed to look after technical equipment and do practical work in a laboratory etc.

Technicolor /ˈteknɪˌkʌlə(r)/ *n.* (often *attrib.*) **1** *propr.* a process of colour cinematography using synchronized monochrome films, each of a different colour, to produce a colour print. **2** (usu. **technicolor**) *colloq.* **a** vivid colour. **b** artificial brilliance. □ **technicolored** *adj.* [TECHNICAL + COLOR]

technique /tekˈniːk/ *n.* **1 a** manner of esp. artistic execution or performance in relation to mechanical or formal details. **b** skill or

ability in this area. **2** a means or method of achieving one's purpose, esp. skilfully. [F (as TECHNIC)]

techno /ˈteknəʊ/ *n.* a form of popular dance music using synthesized sounds and having a fast heavy beat. [abbr. TECHNOLOGY]

technobabble /ˈteknəʊˌbæb(ə)l/ *n. colloq.* incomprehensible technical jargon.

technocracy /tekˈnɒkrəsɪ/ *n.* (*pl.* **-ies**) **1** the government or control of society or industry by technical experts. **2** an instance or application of this. [Gk *tekhnē* art + -CRACY]

technocrat /ˈteknəˌkræt/ *n.* an exponent or advocate of technocracy. □ **technocratic** /ˌteknəˈkrætɪk/ *adj.* **technocratically** *adv.*

technological /ˌteknəˈlɒdʒɪk(ə)l/ *adj.* of or using technology. □ **technologically** *adv.*

technology /tekˈnɒlədʒɪ/ *n.* (*pl.* **-ies**) **1** the study or use of the mechanical arts and applied sciences. **2** these subjects collectively. □ **technologist** *n.* [Gk *tekhnologia* systematic treatment f. *tekhnē* art]

technophile /ˈteknəˌfaɪl/ *n. & adj.* ● *n.* an enthusiast about new technology. ● *adj.* **1** of or relating to a technophile. **2** compatible with new technology.

technophobe /ˈteknəˌfəʊb/ *n.* a person who fears or hates using technology. □ **technophobic** /ˌteknəˈfəʊbɪk/ *adj.*

techy var. of TETCHY.

tectonic /tekˈtɒnɪk/ *adj.* **1** of or relating to building or construction. **2** *Geol.* relating to the deformation of the earth's crust or to the structural changes caused by this (see PLATE TECTONICS). □ **tectonically** *adv.* [LL *tectonicus* f. Gk *tektonikos* f. *tektōn -onos* carpenter]

tectonics /tekˈtɒnɪks/ *n.pl.* (usu. treated as *sing.*) **1** *Archit.* the art and process of producing practical and aesthetically pleasing buildings. **2** *Geol.* the study of large-scale structural features (cf. PLATE TECTONICS).

tectorial /tekˈtɔːrɪəl/ *adj. Anat.* **1** forming a covering. **2** (in full **tectorial membrane**) the membrane covering the organ of Corti (see CORTI) in the inner ear. [L *tectorium* a cover (as TECTRIX)]

tectrix /ˈtektrɪks/ *n.* (*pl.* **tectrices** /ˈtektrɪˌsiːz, tekˈtraɪsiːz/) = COVERT *n.* 2. [mod.L f. L *tegere tect-* cover]

Ted /ted/ *n.* (also **ted**) *Brit. colloq.* a Teddy boy. [abbr.]

ted /ted/ *v.tr.* (**tedded**, **tedding**) turn over and spread out (grass, hay, or straw) to dry or for a bedding etc. □ **tedder** *n.* [ME f. ON *tethja* spread manure f. *tad* dung, *toddi* small piece]

teddy /ˈtedɪ/ *n.* (*pl.* **-ies**) **1** (also **Teddy**; in full **teddy bear**) a soft toy bear. Teddy bears are named after the US President Theodore ('Teddy') Roosevelt, whose bear-hunting exploits were much publicized. **2** a woman's all-in-one undergarment combining a camisole with camiknickers. [orig. of sense 2 unkn.]

Teddy boy /ˈtedɪ/ *n. Brit. colloq.* a youth, esp. of the 1950s, affecting an Edwardian style of dress and appearance, esp. a long jacket and drainpipe trousers. [*Teddy*, pet-form of *Edward*]

Te Deum /tiː ˈdiːəm, teɪ ˈdeɪəm/ *n.* **1 a** a hymn beginning *Te Deum laudamus*, 'We praise Thee, O God', sung at matins or on special occasions such as a thanksgiving. **b** the music for this. **2** an expression of thanksgiving or exultation. [L]

tedious /ˈtiːdɪəs/ *adj.* tiresomely long; wearisome. □ **tediously** *adv.* **tediousness** *n.* [ME f. OF *tedieus* or LL *taediosus* (as TEDIUM)]

tedium /ˈtiːdɪəm/ *n.* the state of being tedious; boredom. [L *taedium* f. *taedere* to weary]

tee¹ /tiː/ *n.* = T¹. [phonet. spelling]

tee² /tiː/ *n. & v.* ● *n.* **1** *Golf* **a** a cleared space from which a golf ball is struck at the beginning of play for each hole. **b** a small support of wood or plastic from which a ball is struck at a tee. **2** a mark aimed at in bowls, quoits, curling, etc. ● *v.tr.* (**tees**, **teed**) (often foll. by *up*) *Golf* place (a ball) on a tee ready to strike it. □ **tee off 1** *Golf* play a ball from a tee. **2** *colloq.* start, begin. [earlier (17th-c.) *teaz*, of unkn. orig.: in sense 2 perh. = TEE¹]

tee-hee /tiːˈhiː/ *n. & v.* (also **te-hee**) ● *n.* **1** a titter. **2** a restrained or contemptuous laugh. ● *v.intr.* (**tee-hees**, **tee-heed**) titter or laugh in this way. [imit.]

teem¹ /tiːm/ *v.intr.* **1** be abundant (*fish teem in these waters*). **2** (foll. by *with*) be full of or swarming with (*teeming with fish*; *teeming with ideas*). [OE *tēman* etc. give birth to f. Gmc, rel. to TEAM]

teem² /tiːm/ *v.intr.* (often foll. by *down*) (of water etc.) flow copiously; pour (*it was teeming with rain*). [ME *tēmen* f. ON *tœma* f. *tómr* (adj.) empty]

teen /tiːn/ *adj. & n.* ● *adj.* = TEENAGE. ● *n.* = TEENAGER. [abbr. of TEENAGE, TEENAGER]

-teen /tiːn/ *suffix* forming the names of numerals from 13 to 19. [OE inflected form of TEN]

teenage /ˈtiːneɪdʒ/ *adj.* relating to or characteristic of teenagers. □ **teenaged** *adj.*

teenager /ˈtiːnˌeɪdʒə(r)/ *n.* a person from 13 to 19 years of age.

teens /tiːnz/ *n.pl.* the years of one's age from 13 to 19 (*in one's teens*).

teensy /ˈtiːnzɪ/ *adj.* (**teensier, teensiest**) *colloq.* = TEENY. □ **teensy-weensy** = teeny-weeny.

teeny /ˈtiːnɪ/ *adj.* (**teenier, teeniest**) *colloq.* tiny. □ **teeny-weeny** very tiny. [var. of TINY]

teeny-bopper /ˈtiːnɪˌbɒpə(r)/ *n. colloq.* a young teenager, usu. a girl, who keenly follows the latest fashions in clothes, pop music, etc.

teepee var. of TEPEE.

Tees /tiːz/ a river of NE England which rises in Cumbria and flows 128 km (80 miles) generally south-eastwards to the North Sea at Middlesbrough.

teeshirt var. of T-SHIRT.

Teesside /ˈtiːzsaɪd/ an industrial region in NE England around the lower Tees valley, including Middlesbrough.

teeter /ˈtiːtə(r)/ *v.intr.* **1** totter; stand or move unsteadily. **2** hesitate; be indecisive. □ **teeter on the brink** (or **edge**) be in imminent danger (of disaster etc.). [var. of dial. *titter*]

teeth *pl.* of TOOTH.

teethe /tiːð/ *v.intr.* grow or cut teeth, esp. milk teeth. □ **teething-ring** a small ring for an infant to bite on while teething. **teething troubles** initial difficulties in an enterprise etc., regarded as temporary. □ **teething** *n.*

teetotal /tiːˈtəʊt(ə)l/ *adj.* advocating or characterized by total abstinence from alcoholic drink. □ **teetotalism** *n.* [redupl. of TOTAL, app. first used in a speech in the early 19th c. advocating total abstinence from all alcoholic liquor, as opposed to abstinence from spirits only]

teetotaller /tiːˈtəʊtələ(r)/ *n.* (US **teetotaler**) a person advocating or practising abstinence from alcoholic drink.

teetotum /tiːˈtəʊtəm/ *n.* **1** a spinning-top with four sides lettered to determine whether the spinner has won or lost. **2** any top spun with the fingers. [T (the letter on one side) + L *totum* the whole (stakes), for which T stood]

teff /tef/ *n.* an African cereal, *Eragrostis tef*, grown esp. in Ethiopia. [Amharic *ṭēf*]

TEFL /ˈtef(ə)l/ *abbr.* teaching of English as a foreign language.

Teflon /ˈteflɒn/ *n. propr.* polytetrafluoroethylene, esp. used as a non-stick coating for kitchen utensils. [*tetra-* + *fluor-* + *-on*]

teg /teg/ *n.* a sheep in its second year. [ME *tegge* (recorded in place-names), repr. OE (unrecorded) *tegga* ewe]

Tegucigalpa /teˌɡuːsɪˈɡælpə/ capital of Honduras; pop. (1988) 678,700.

tegument /ˈteɡjʊmənt/ *n. Zool.* integument, esp. of a flatworm. □ **tegumental** /ˌteɡjʊˈment(ə)l/ *adj.* **tegumentary** *adj.* [L *tegumentum* f. *tegere* cover]

te-hee var. of TEE-HEE.

Tehran /teəˈrɑːn/ the capital of Iran, situated in the foothills of the Elburz Mountains; pop. (1986) 6,042,600. It replaced Isfahan as capital of Persia in 1788.

Teilhard de Chardin /ˌteɪɑː də ˈʃɑːdæn/, Pierre (1881–1955), French Jesuit philosopher and palaeontologist. He is best known for his theory, blending science and Christianity, that man is evolving mentally and socially towards a perfect spiritual state. His views were held to be unorthodox by the Roman Catholic Church and his major works (e.g. *The Phenomenon of Man*, 1955) were published posthumously.

Tejo see TAGUS.

Te Kanawa /te ˈkɑːnəwə/, Dame Kiri (Janette) (b.1944), New Zealand operatic soprano, resident in Britain since 1966. She made her début in London in 1970 and since then has sung in the world's leading opera houses, especially in works by Mozart, Richard Strauss, and Verdi.

tektite /ˈtektaɪt/ *n. Geol.* a small black glassy object found lying on the ground or ocean bed, present in large numbers in several areas and believed to result from meteoric impacts. [G *Tektit* f. Gk *tēktos* molten f. *tēkō* melt]

tel. *abbr.* (also **Tel.**) telephone.

telaesthesia /ˌteliːsˈθiːzɪə/ *n.* (US **telesthesia**) *Psychol.* the supposed perception of distant occurrences or objects otherwise than by the recognized senses. □ **telaesthetic** /-ˈθetɪk/ *adj.* [mod.L, formed as TELE- + Gk *aisthēsis* perception]

telamon /ˈteləˌmɒn/ *n.* (*pl.* **telamones** /ˌteləˈməʊniːz/) *Archit.* a male figure used as a pillar to support an entablature. [L *telamones* f. Gk *telamōnes* pl. of *Telamōn*, name of a mythical hero]

Tel Aviv /ˌtel əˈviːv/ (also **Tel Aviv-Jaffa**) a city on the Mediterranean coast of Israel; pop. (1987) 319,500 (with Jaffa). It was founded as a suburb of Jaffa by Russian Jewish immigrants in 1909 and named Tel Aviv a year later.

telco /ˈtelkəʊ/ *n.* US a telecommunications company. [abbr.]

tele- /ˈtelɪ/ *comb. form* **1** at or to a distance (*telekinesis*). **2** forming names of instruments for operating over long distances (*telescope*). **3** television (*telecast*). **4** done by means of the telephone (*telesales*). [Gk *tēle-* f. *tēle* far off: sense 3 f. TELEVISION: sense 4 f. TELEPHONE]

tele-ad /ˈtelɪˌæd/ *n.* an advertisement placed in a newspaper etc. by telephone.

telebanking /ˈtelɪˌbæŋkɪŋ/ *n.* a banking system whereby computerized transactions are effected by telephone.

telecamera /ˈtelɪˌkæmrə, -mərə/ *n.* **1** a television camera. **2** a telephotographic camera.

telecast /ˈtelɪˌkɑːst/ *n. & v.* ● *n.* a television broadcast. ● *v.tr.* transmit by television. □ **telecaster** *n.* [TELE- + BROADCAST]

telecine /ˈtelɪˌsɪnɪ/ *n.* **1** the broadcasting of cinema film on television. **2** equipment for doing this. [TELE- + CINEMA]

telecommunication /ˌtelɪkəˌmjuːnɪˈkeɪʃ(ə)n/ *n.* **1** communication over a distance by cable, telegraph, telephone, or broadcasting. **2** (usu. in *pl.*) the branch of technology concerned with this. [F *télécommunication* (as TELE-, COMMUNICATION)]

telecommute /ˌtelɪkəˈmjuːt/ *v.intr.* work from home by using a computer, fax, etc., to communicate with the employer's base. □ **telecommuter** *n.* **telecommuting** *n.*

telecoms /ˈtelɪˌkɒmz/ *n.* (also **telecomms**) (also *attrib.*) telecommunications (see TELECOMMUNICATION 2). [abbr.]

teleconference /ˈtelɪˌkɒnfərəns/ *n.* a conference with participants in different locations linked by telecommunication devices. □ **teleconferencing** *n.*

telecottage /ˈtelɪˌkɒtɪdʒ/ *n.* a room or building, esp. in a rural area, equipped with computers, faxes, etc., and used as a workplace remote from an employer's base. □ **telecottaging** *n.*

teledu /ˈtelɪˌduː/ *n.* a badger-like mammal, *Mydaus javanensis*, of Java, Sumatra, and Borneo, that secretes a foul-smelling liquid when attacked. [Javanese]

telefacsimile /ˌtelɪfækˈsɪmɪlɪ/ *n.* facsimile transmission (see FACSIMILE *n.* 2).

Telefax /ˈtelɪˌfæks/ *n. propr.* **1** = TELEFACSIMILE. **2** a document etc. sent by facsimile transmission. [abbr.]

telefilm /ˈtelɪˌfɪlm/ *n.* = TELECINE.

telegenic /ˌtelɪˈdʒenɪk/ *adj.* having an appearance or manner that looks pleasing on television. [TELEVISION + *-genic* in PHOTOGENIC]

telegony /tɪˈleɡənɪ/ *n. Biol.* the supposed influence of a previous sire on the offspring of a dam with other sires. □ **telegonic** /ˌtelɪˈɡɒnɪk/ *adj.* [TELE- + Gk *-gonia* begetting]

telegram /ˈtelɪˌɡræm/ *n.* a message sent by telegraph and then usu. delivered in written form. ¶ In UK official use since 1981 only for international messages. [TELE- + -GRAM, after TELEGRAPH]

telegraph /ˈtelɪˌɡrɑːf, -ˌɡræf/ *n. & v.* ● *n.* **1 a** a system of or device for transmitting messages or signals to a distance, esp. by making and breaking an electrical connection. (*See note below.*) **b** (*attrib.*) used in this system (*telegraph wire*). **2** (in full **telegraph board**) a board displaying scores or other information at a match, race meeting, etc. ● *v.* **1** *tr.* send a message by telegraph to. **2** *tr.* send by telegraph. **3** *tr.* give an advance indication of. **4** *intr.* make signals (*telegraphed to me to come up*). □ **telegraph key** a device for making and breaking the electric circuit of a telegraph system. **telegraph plant** a tropical Asian leguminous plant, *Codariocalyx motorius*, whose leaves have a spontaneous jerking motion. **telegraph pole** a pole used to carry telegraph or telephone wires above the ground. □ **telegrapher** /ˈtelɪˌɡrɑːfə(r), tɪˈleɡrəfə(r)/ *n.* [F *télégraphe* (as TELE-, -GRAPH)]

▪ Electric telegraphy began in the 1830s on the new railways. In America, Samuel Morse had made his first working model of a telegraph by 1835, using electromagnets; a similar system was developed by Sir William Cooke and Charles Wheatstone in England, where the first practical telegraph was set up between two stations in London in 1837. The telegraph became the standard means of rapid communication within a district, but skilled operators were needed to send and receive messages in Morse code. The introduction of the teleprinter enabled it to continue alongside the telephone system for the transmission of telegrams etc. Though the telegraph has been superseded by the telephone for domestic purposes, it has survived in the form of the telex, which can use any public telecommunications system.

telegraphese /ˌtelɪgrəˈfiːz/ n. colloq. or joc. an abbreviated style usual in telegrams.

telegraphic /ˌtelɪˈgræfɪk/ adj. **1** of or by telegraphs or telegrams. **2** economically worded. □ **telegraphic address** an abbreviated or other registered address for use in telegrams. □ **telegraphically** adv.

telegraphist /tɪˈlegrəfɪst/ n. a person skilled or employed in telegraphy.

telegraphy /tɪˈlegrəfɪ/ n. the science or practice of using or constructing communication systems for the reproduction of information.

Telegu var. of TELUGU.

telekinesis /ˌtelɪkɪˈniːsɪs, ˌtelɪkaɪ-/ n. Psychol. movement of objects at a distance supposedly by paranormal means. □ **telekinetic** /-ˈnetɪk/ adj. [mod.L (as TELE-, Gk kinēsis motion f. kineō move)]

Telemachus /tɪˈleməkəs/ Gk Mythol. the son of Ulysses and Penelope.

Telemann /ˈteɪləmæn/, Georg Philipp (1681–1767), German composer and organist. His prolific output includes 600 overtures, 44 Passions, 12 complete services, and 40 operas; his work reflects a variety of influences, particularly French composers such as Lully. In his lifetime his reputation was far greater than that of his friend and contemporary J. S. Bach.

telemark /ˈtelɪˌmɑːk/ n. & v. Skiing ● n. a swing turn with one ski advanced and the knee bent, used to change direction or stop short. ● v.intr. perform this turn. [Telemark in Norway]

telemarketing /ˈtelɪˌmɑːkɪtɪŋ/ n. the marketing of goods etc. by means of usu. unsolicited telephone calls. □ **telemarketer** n.

telemessage /ˈtelɪˌmesɪdʒ/ n. a message sent by telephone or telex and delivered in written form. ¶ In UK official use since 1981 for inland messages, replacing telegram.

telemeter /ˈtelɪˌmiːtə(r), tɪˈlemɪtə(r)/ n. & v. ● n. an apparatus for recording the readings of an instrument and transmitting them by radio. ● v. **1** intr. record readings in this way. **2** tr. transmit (readings etc.) to a distant receiving set or station. □ **telemetric** /ˌtelɪˈmetrɪk/ adj. **telemetry** /tɪˈlemɪtrɪ/ n.

teleology /ˌtelɪˈɒlədʒɪ, ˌtiːl-/ n. (pl. -ies) Philos. **1** the explanation of phenomena by the purpose they serve rather than by postulated causes. **2** (in Christian theology) the doctrine of design and purpose in the material world. □ **teleologism** n. **teleologist** n. **teleologic** /-lɪəˈlɒdʒɪk/ adj. **teleological** adj. **teleologically** adv. [mod.L teleologia f. Gk telos teleos end + -LOGY]

teleost /ˈtiːlɪˌɒst/ n. & adj. Zool. ● n. a fish of the subclass Teleostei, comprising the bony fishes, and including most familiar types of fish except for the sharks, rays, sturgeons, and lungfishes. ● adj. of or relating to this subclass. [Gk teleo- complete + osteon bone]

telepath /ˈtelɪˌpæθ/ n. a telepathic person. [back-form. f. TELEPATHY]

telepathy /tɪˈlepəθɪ/ n. the supposed communication of thoughts or ideas otherwise than by the known senses. □ **telepathist** n. **telepathize** v.tr. & intr. (also -ise). **telepathic** /ˌtelɪˈpæθɪk/ adj. **telepathically** adv.

telephone /ˈtelɪˌfəʊn/ n. & v. ● n. **1** an apparatus for transmitting sound (esp. speech) to a distance, esp. by converting acoustic vibrations to electrical signals. (See note below.) **2** a transmitting and receiving instrument used in this. **3** a system of communication using a network of telephones. ● v. **1** tr. speak to (a person) by telephone. **2** tr. send (a message) by telephone. **3** intr. make a telephone call. □ **on the telephone 1** having a telephone. **2** by use of or using the telephone. **over the telephone** by use of or using the telephone. **telephone book** = telephone directory. **telephone box** Brit. = telephone booth. **telephone booth** (or **kiosk**) a public booth or enclosure from which telephone calls can be made. **telephone call** = CALL n. 4. **telephone directory** a book listing telephone subscribers and numbers in a particular area. **telephone exchange** = EXCHANGE n. 3. **telephone number 1** a number assigned to a particular telephone and used in

making connections to it. **2** (often in pl.) colloq. a number with many digits, esp. representing a large sum of money. **telephone operator** esp. US an operator in a telephone exchange. □ **telephoner** n. **telephonic** /ˌtelɪˈfɒnɪk/ adj. **telephonically** adv.

▪ The 'Electrical Speaking Telephone' was invented by Alexander Graham Bell and patented in the US in 1876. The system was quickly adopted, mainly for communication over shorter distances than the telegraph, and by 1887 there were over 100,000 subscribers worldwide. The three basic essentials are a unit with a microphone and earpiece, a system to transmit the signal within acceptable limits of distortion and attenuation, and a (usually automatic) switching system to connect any two telephones. Besides the conventional wires and cables, long distance transmission is facilitated by the use of optical fibres, microwave links, and satellite relays, and digitization permits the accurate transmission of signals. The telephone system is also used to transmit telex, fax, and television signals, and to connect computers via modem links.

telephonist /tɪˈlefənɪst/ n. Brit. an operator in a telephone exchange or at a switchboard.

telephony /tɪˈlefənɪ/ n. the use or a system of telephones.

telephoto /ˈtelɪˌfəʊtəʊ/ n. (pl. -os) (in full **telephoto lens**) a lens with a longer focal length than standard, giving a narrow field of view and a magnified image.

teleport /ˈtelɪˌpɔːt/ v.tr. Psychol. move by telekinesis. □ **teleportation** /ˌtelɪpɔːˈteɪʃ(ə)n/ n. [TELE- + PORT⁴ 3]

telepresence /ˈtelɪˌprez(ə)ns/ n. **1** the use of virtual reality technology, esp. for machine operation or for apparent participation in distant events. **2** a sensation of being elsewhere created in this way.

teleprinter /ˈtelɪˌprɪntə(r)/ n. a device for transmitting telegraph messages as they are keyed, and for printing messages received.

teleprompter /ˈtelɪˌprɒmptə(r)/ n. a device beside a television or cinema camera that slowly unrolls a speaker's script out of sight of the audience (cf. AUTOCUE).

telerecord /ˈtelɪrɪˌkɔːd/ v.tr. record for television broadcasting.

telerecording /ˈtelɪrɪˌkɔːdɪŋ/ n. a recorded television broadcast.

telesales /ˈtelɪˌseɪlz/ n.pl. selling by means of the telephone.

telescope /ˈtelɪˌskəʊp/ n. & v. ● n. **1** an optical instrument using lenses or mirrors or both to make distant objects appear nearer and larger. (See note below.) **2** = RADIO TELESCOPE. ● v. **1** tr. press or drive (sections of a tube, colliding vehicles, etc.) together so that one slides into another like the sections of a folding telescope. **2** intr. close or be driven or be capable of closing in this way. **3** tr. compress so as to occupy less space or time. [It. telescopio or mod.L telescopium (as TELE-, -SCOPE)]

▪ The refracting telescope that is generally used for terrestrial purposes employs a lens to collect the light, and further lenses or prisms to produce an upright magnified image. It was probably invented independently a number of times before Galileo first used one for astronomy in 1609. Astronomical refractors generally dispense with one lens, so as to let more light through at the expense of an inverted image. The telescope not only advanced scientific knowledge but brought consequences for religious and philosophical thought. The reflecting telescope, using a concave mirror to collect the light, was invented in England in the 16th century. The later Gregorian design (1663) was quickly superseded by the Newtonian (1668) and Cassegrain (1672) designs that form the basis for most large optical telescopes today. The compact catadioptric systems use both lenses and mirrors and include the Schmidt, Schmidt–Cassegrain, and Maksutov designs. Several orbiting telescopes, equipped to receive visible, infrared, or ultraviolet light, X-rays, or high-energy microwave radiation, operate without interference from the earth's atmosphere.

telescopic /ˌtelɪˈskɒpɪk/ adj. **1 a** of, relating to, or made with a telescope (telescopic observations). **b** visible only through a telescope (telescopic stars). **2** (esp. of a lens) able to focus on and magnify distant objects. **3** consisting of sections that telescope. □ **telescopic sight** a telescope used for sighting on a rifle etc. □ **telescopically** adv.

telesoftware /ˌtelɪˈsɒftweə(r)/ n. software transmitted or broadcast to receiving terminals.

telesthesia US var. of TELAESTHESIA.

Teletex /ˈtelɪˌteks/ n. propr. an electronic text transmission system.

teletext /ˈtelɪˌtekst/ n. **1** a news and information service, in the form of text and graphics, from a computer source transmitted to televisions with appropriate receivers. **2** (**Teletext Ltd**) Brit. propr. a teletext service provided by a consortium of companies and transmitted on Independent Television (cf. CEEFAX).

telethon /ˈtelɪˌθɒn/ n. an exceptionally long television programme, esp. to raise money for a charity. [TELE- + -thon in MARATHON]

Teletype /ˈtelɪˌtaɪp/ n. & v. ● n. propr. a kind of teleprinter. ● v. (**teletype**) **1** intr. operate a teleprinter. **2** tr. send by means of a teleprinter.

teletypewriter /ˌtelɪˈtaɪpˌraɪtə(r)/ n. esp. US = TELEPRINTER.

televangelist /ˌtelɪˈvændʒəlɪst/ n. esp. US an evangelical preacher who appears regularly on television to promote beliefs and appeal for funds.

televiewer /ˈtelɪˌvjuːə(r)/ v.tr. a person who watches television. □ **televiewing** adj.

televise /ˈtelɪˌvaɪz/ v.tr. transmit by television. □ **televisable** adj. [back-form. f. TELEVISION]

television /ˈtelɪˌvɪʒ(ə)n/ n. **1** a system for reproducing on a screen visual images converted (usu. with sound) into electrical signals and transmitted esp. by radio. (See note below.) **2** (in full **television set**) a device with a screen for receiving these signals. **3** the medium, art form, or occupation of broadcasting on television; the content of television programmes.

 ■ Television was first demonstrated in 1926 by John Logie Baird, who used mechanical picture scanning, but this was displaced by the electronic system developed by Vladimir Zworykin in the 1930s. The broad principle is that of cinematography, reproducing a series of successive images which the human brain registers as a continuous picture because of the persistence of vision. The television camera divides an image into an array of many thousands of picture elements, which it scans several (typically twenty-five) times a second; each element generates an electrical signal proportional to its brightness. These signals can then be transmitted by radio or cable (or stored on videotape) for eventual reconstitution as an image in a cathode-ray tube in a television receiver. In colour television three images, using picture elements in red, blue, and green, are combined to produced a multicoloured picture. The first regular TV broadcasting service was introduced in Germany in 1935; the BBC began broadcasts in the UK in 1936 and NBC in the US in 1939. Television expanded rapidly across the world in postwar years, colour being introduced in the late 1950s and 1960s. Broadcasting by state and commercial networks has been augmented since the 1970s by multinational satellite and cable organizations. (See also CATHODE RAY, VIDEO.)

televisual /ˌtelɪˈvɪʒʊəl, -ˈvɪzjʊəl/ adj. relating to or suitable for television. □ **televisually** adv.

telework /ˈtelɪˌwɜːk/ v.intr. = TELECOMMUTE. □ **teleworker** n. **teleworking** n.

telex /ˈteleks/ n. & v. (also **Telex**) ● n. an international system of telegraphy with printed messages transmitted and received by teleprinters using the public telecommunications network. ● v.tr. send or communicate with by telex. [TELEPRINTER + EXCHANGE]

Telford[1] /ˈtelfəd/ a town in west central England, in Shropshire; pop. (est. 1991) 115,000. It was named after the engineer Thomas Telford and designated a new town in 1963.

Telford[2] /ˈtelfəd/, Thomas (1757–1834), Scottish civil engineer. He built hundreds of miles of roads, especially in Scotland, more than a thousand bridges, and a number of canals. Among his most important achievements are the London to Holyhead road, including the suspension bridge crossing the Menai Strait (1819–26), the Caledonian Canal across Scotland (opened 1822), and the Göta Canal in Sweden. He was the first president of the Institution of Civil Engineers (founded 1818).

Tell /tel/, William. A legendary hero of the liberation of Switzerland from Austrian oppression, who was required to hit with an arrow an apple placed on the head of his son; this he successfully did. The events are placed in the 14th century, but there is no evidence for a historical person of this name, and similar legends are of widespread occurrence.

tell[1] /tel/ v. (past and past part. **told** /təʊld/) **1** tr. relate or narrate in speech or writing; give an account of (tell me a story). **2** tr. make known; express in words; divulge (tell me your name; tell me what you want). **3** tr. reveal or signify to (a person) (your face tells me everything). **4** tr. **a** utter (don't tell lies). **b** warn (I told you so). **5** intr. **a** (often foll. by of, about) divulge information or a description; reveal a secret (I told of the plan; promise you won't tell). **b** (foll. by on) colloq. inform against (a person). **6** tr. (foll. by to + infin.) give (a person) a direction or order (tell them to wait; do as you are told). **7** tr. assure (it's true, I tell you). **8** tr. explain in writing; instruct

(this book tells you how to cook). **9** tr. decide, predict, determine, distinguish (cannot tell what might happen; how do you tell one from the other?). **10** intr. **a** (often foll. by on) produce a noticeable effect (every disappointment tells; the strain was beginning to tell on me). **b** reveal the truth (time will tell). **c** have an influence (the evidence tells against you). **11** tr. (often absol.) count (votes) at a meeting, election, etc. □ **as far as one can tell** judging from the available information. **tell apart** distinguish between (usu. with neg. or interrog.: could not tell them apart). **tell me another** colloq. an expression of incredulity. **tell off 1** colloq. reprimand, scold. **2** count off or detach for duty. **tell a tale** (or **its own tale**) be significant or revealing. **tell tales** report a discreditable fact about another. **tell that to the marines** see MARINE. **tell the time** determine the time from the face of a clock or watch. **that would be telling** colloq. that would be to reveal too much (esp. secret or confidential) information. **there is no telling** it is impossible to know (there's no telling what may happen). **you're telling me** colloq. I agree wholeheartedly. □ **tellable** adj. [OE tellan f. Gmc, rel. to TALE]

tell[2] /tel/ n. Archaeol. an artificial mound in the Middle East etc. formed by the accumulated remains of ancient settlements. [Arab. tall hillock]

Tell el-Amarna see AMARNA, TELL EL-.

Teller /ˈtelə(r)/, Edward (b.1908), Hungarian-born American physicist. After moving to America in the 1930s he worked on the first atomic reactor, later working on the first atom bombs at Los Alamos. Teller studied the feasibility of producing a fusion bomb, and work under his guidance after the Second World War led to the detonation of the first hydrogen bomb in 1952. His own studies on fusion were mainly theoretical and later concerned with its peaceful use, though he was a forceful advocate of the nuclear deterrent.

teller /ˈtelə(r)/ n. **1** a person employed to receive and pay out money in a bank etc. **2** a person who counts (votes). **3** a person who tells esp. stories (a teller of tales). □ **tellership** n.

telling /ˈtelɪŋ/ adj. **1** having a marked effect; striking. **2** significant. □ **tellingly** adv.

telling-off /ˌtelɪŋˈɒf/ n. (pl. **tellings-off**) colloq. a reproof or reprimand.

tell-tale /ˈtelteɪl/ n. **1** a person who reveals (esp. discreditable) information about another's private affairs or behaviour. **2** (attrib.) that reveals or betrays (a tell-tale smile). **3** a device for automatic monitoring or registering of a process etc. **4** a metal sheet extending across the front wall of a squash court, above which the ball must strike the wall.

tellurian /teˈljʊərɪən/ adj. & n. ● adj. of or inhabiting the earth. ● n. an inhabitant of the earth. [L tellus -uris earth]

telluric /teˈljʊərɪk/ adj. **1** of the earth as a planet. **2** of the soil. **3** Chem. of tellurium, esp. in its higher valency. □ **tellurate** /ˈteljʊərət/ n. [L tellus -uris earth: sense 3 f. TELLURIUM]

tellurium /teˈljʊərɪəm/ n. a brittle shiny silvery-white semi-metallic element (atomic number 52; symbol **Te**). (See note below.) □ **telluride** /ˈteljʊəˌraɪd/ n. **tellurous** adj. [L tellus -uris earth, prob. named in contrast to uranium]

 ■ Tellurium was discovered by the Austrian mineralogist Franz Müller (1740–1825) in 1782. Like selenium, which it resembles, tellurium occurs mainly in small amounts in metallic sulphide ores. It is a semiconductor and is used in some electrical devices and in specialized alloys.

telly /ˈtelɪ/ n. (pl. **-ies**) esp. Brit. colloq. **1** television. **2** a television set. [abbr.]

telophase /ˈtiːləˌfeɪz, ˈtel-/ n. Biol. the last stage of cell division, in which the nuclei of the daughter cells are formed. [Gk telos end + PHASE]

telpher /ˈtelfə(r)/ n. a system for transporting goods etc. by electrically driven trucks or cable-cars. [TELE- + -PHORE]

telson /ˈtels(ə)n/ n. Zool. the last segment in the abdomen of crustaceans and arachnids. [Gk, = limit]

Telstar /ˈtelstɑː(r)/ the first of the active communications satellites (i.e. both receiving and retransmitting signals, not merely reflecting signals from the earth). It was launched by the US in 1962 and used in the transmission of television broadcasting and telephone communication.

Telugu /ˈteləˌɡuː/ n. & adj. (also **Telegu**) ● n. (pl. same or **Telugus**) **1** a member of a Dravidian people in SE India. **2** the language of the Telugu, spoken by about 54 million people mainly in Andhra Pradesh. ● adj. of or relating to the Telugu or their language. [Telugu]

temerarious /ˌteməˈreərɪəs/ adj. literary reckless, rash. [L temerarius f. temere rashly]

temerity /tɪ'merɪtɪ/ n. **1** rashness. **2** audacity, impudence. [L temeritas f. temere rashly]

Temesvár see TIMIȘOARA.

Temne /'temnɪ/ n. & adj. ● n. (pl. same or **Temnes**) **1** a member of a people of Sierra Leone. **2** the Niger-Congo language of this people. ● adj. of or relating to this people or their language. [Temne]

temp /temp/ n. & v. colloq. ● n. a temporary employee, esp. a secretary. ● v.intr. work as a temp. [abbr.]

temp.[1] /temp/ abbr. temperature.

temp.[2] /temp/ abbr. in the time of (temp. Henry I). [L tempore ablat. of tempus time]

tempeh /'tempeɪ/ n. an Indonesian dish made by deep-frying fermented soya beans. [Indonesian tempe]

temper /'tempə(r)/ n. & v. ● n. **1** habitual or temporary disposition of mind esp. as regards composure (a person of a placid temper). **2 a** irritation or anger (in a fit of temper). **b** an instance of this (flew into a temper). **3** a tendency to have fits of anger (have a temper). **4** composure or calmness (keep one's temper; lose one's temper). **5** the condition of metal as regards hardness and elasticity. ● v.tr. **1** bring (metal or clay) to a proper hardness or consistency. **2** (often foll. by with) moderate or mitigate (temper justice with mercy). **3** tune or modulate (a piano etc.) so as to distance intervals correctly. □ **in a bad temper** angry, peevish. **in a good temper** in an amiable mood. **out of temper** angry, peevish. **show temper** be petulant. □ **temperable** adj. **temperative** /-rətɪv/ adj. **tempered** adj. **temperedly** adv. **temperer** n. [OE temprian (v.) f. L temperare mingle: infl. by OF temprer, tremper]

tempera /'tempərə/ n. a method of painting using an emulsion of powdered pigment typically held together with egg yolk and water, esp. in fine art on prepared panels. The method was used in Europe from the 12th or early 13th century until the 15th, when it began to give way to oils. [It.: cf. DISTEMPER[1]]

temperament /'temprəmənt/ n. **1** a person's distinct nature and character, esp. as determined by physical constitution and permanently affecting behaviour (a nervous temperament; the artistic temperament). **2** a creative or spirited personality (was full of temperament). **3** Mus. **a** an adjustment of intervals in tuning a piano etc. so as to fit the scale for use in all keys. **b** (**equal temperament**) an adjustment in which the 12 semitones are at equal intervals. [ME f. L temperamentum (as TEMPER)]

temperamental /ˌtemprə'ment(ə)l/ adj. **1** of or having temperament. **2 a** (of a person) liable to erratic or moody behaviour. **b** (of a thing, e.g. a machine) working unpredictably; unreliable. □ **temperamentally** adv.

temperance /'tempərəns/ n. **1** moderation or self-restraint esp. in eating and drinking. **2 a** total or partial abstinence from alcoholic drink. **b** (attrib.) advocating or concerned with abstinence. [ME f. AF temperaunce f. L temperantia (as TEMPER)]

temperance movement n. a movement seeking restrictions on the consumption of alcohol. Groups with such an aim appeared in the early 19th century, first in New England in the US and Northern Ireland in the UK, and spread across Europe and the US; Christian groups (especially Nonconformists), trade unionists, and advocates of women's suffrage were among leading elements in the movement. In the US a lengthy campaign from the second half of the 19th century by groups such as the Anti-Saloon League (founded 1893) led to Prohibition in 1920, while in the UK the Defence of the Realm Act (1916) limited the hours that public houses could open and excluded people under 18.

temperate /'tempərət/ adj. **1** avoiding excess; self-restrained. **2** moderate. **3** (of a region or climate) characterized by mild temperatures. **4** abstemious. □ **temperate zone** the belt of the earth between the frigid and the torrid zones. □ **temperately** adv. **temperateness** n. [ME f. L temperatus past part. of temperare: see TEMPER]

temperature /'temprɪtʃə(r)/ n. **1** the degree or intensity of heat of a body in relation to others, esp. as shown by a thermometer or perceived by touch etc. **2** Med. the degree of internal heat of the body. **3** a body temperature above the normal (have a temperature). **4** the degree of excitement in a discussion etc. □ **take a person's temperature** ascertain a person's body temperature, esp. as a diagnostic aid. **temperature-humidity index** a quantity giving the measure of discomfort due to the combined effects of the temperature and humidity of the air. **temperature inversion** = INVERSION 4. [F température or L temperatura (as TEMPER)]

-tempered /'tempəd/ comb. form having a specified temper or disposition (bad-tempered; hot-tempered). □ **-temperedly** adv. **-temperedness** n.

Tempest /'tempɪst/, Dame Marie (born Mary Susan Etherington) (1864–1942), English actress. Though trained as a singer she made her name in comedy, becoming noted for her playing of elegant middle-aged women; the role of Judith Bliss in Hay Fever (1925) was created for her by Noel Coward.

tempest /'tempɪst/ n. **1** a violent windy storm. **2** violent agitation or tumult. [ME f. OF tempest(e) ult. f. L tempestas season, storm, f. tempus time]

tempestuous /tem'pestjʊəs/ adj. **1** stormy. **2** (of a person, emotion, etc.) turbulent, violent, passionate. □ **tempestuously** adv. **tempestuousness** n. [LL tempestuosus (as TEMPEST)]

tempi pl. of TEMPO.

Templar /'templə(r)/ n. **1** a lawyer or law student with chambers in the Temple, London. **2** hist. a member of the Knights Templars. [ME f. AF templer, OF templier, med.L templarius (as TEMPLE[1])]

template /'templɪt, -pleɪt/ n. (also **templet** /-plɪt/) **1 a** a pattern or gauge, usu. a piece of thin board or metal plate, used as a guide in cutting or drilling metal, stone, wood, etc. **b** a flat card or plastic pattern esp. for cutting cloth for patchwork etc. **2** a timber or plate used to distribute the weight in a wall or under a beam etc. **3** Biochem. the molecular pattern governing the assembly of a protein etc. [orig. templet: prob. f. TEMPLE[3] + -ET[1], alt. after plate]

Temple /'temp(ə)l/, Shirley (b.1928), American child star. In the 1930s she appeared in a succession of films, often adapted from children's classics (e.g. Rebecca of Sunnybrook Farm, 1938). She later became active in Republican politics, represented the US at the United Nations, and served as ambassador in various countries.

temple[1] /'temp(ə)l/ n. **1** a building devoted to the worship, or regarded as the dwelling-place, of a god or gods or other objects of religious reverence. **2** hist. any of two successive religious buildings of the Jews in Jerusalem (see TEMPLE, THE). **3** N. Amer. a synagogue. **4** a place of Christian public worship, esp. a Protestant church in France. **5** a place in which God is regarded as residing, esp. a Christian's person or body. □ **temple block** a percussion instrument consisting of a hollow block of wood which is struck with a stick. [OE temp(e)l, reinforced in ME by OF temple, f. L templum open or consecrated space]

temple[2] /'temp(ə)l/ n. (often in pl.) the flat part of either side of the head between the forehead and the ear. [ME f. OF ult. f. L tempora pl. of tempus temple of the head]

temple[3] /'temp(ə)l/ n. a device in a loom for keeping the cloth stretched. [ME f. OF, orig. the same word as TEMPLE[2]]

Temple, the 1 either of two ancient Jewish temples in Jerusalem. The first, built by Solomon, was completed in 957 BC; it contained the Ark of the Covenant, and became established as the place of sacrifice in Judah. It was destroyed by Nebuchadnezzar's forces in 586 BC and its treasures, including the Ark of the Covenant, were lost. The second temple was completed in 515 BC; rebuilt and enlarged by Herod the Great from 20 BC, it was the Temple standing in Jesus' time. It was destroyed by the Romans in AD 70, during a Jewish revolt; all that remains is the Wailing Wall. **2** a group of buildings in Fleet Street, central London, which stand on land formerly occupied by the headquarters of the Knights Templars. Located there are the Inner and Outer Temple, two of the Inns of Court.

templet var. of TEMPLATE.

tempo /'tempəʊ/ n. (pl. **-os** or **tempi** /-pɪ/) **1** Mus. the speed at which music is or should be played, esp. as characteristic (waltz tempo). **2** the rate of motion or activity (the tempo of the war is quickening). [It. f. L tempus time]

temporal[1] /'tempərəl/ adj. **1** of worldly as opposed to spiritual affairs; of this life; secular. **2** of or relating to time. **3** Gram. relating to or denoting time or tense (temporal conjunction). □ **temporal power** the power of an ecclesiastic, esp. the pope, in temporal matters. □ **temporally** adv. [ME f. OF temporel or f. L temporalis f. tempus -oris time]

temporal[2] /'tempərəl/ adj. Anat. of or situated in the temples of the head (temporal artery; temporal bone). □ **temporal lobe** each of the paired lobes of the brain lying beneath the temples, including areas concerned with the understanding of speech. [ME f. LL temporalis f. tempora the temples (as TEMPLE[2])]

temporality /ˌtempə'ralɪtɪ/ n. (pl. **-ies**) **1** temporariness. **2** (usu. in pl.)

a secular possession, esp. the properties and revenues of a religious corporation or an ecclesiastic. [ME f. LL *temporalitas* (as TEMPORAL[1])]

temporary /'tempərəri/ *adj. & n.* ● *adj.* lasting or meant to last only for a limited time (*temporary buildings*; *temporary relief*). ● *n.* (*pl.* **-ies**) a person employed temporarily (cf. TEMP). □ **temporarily** /'tempərərɪlɪ, *disp.* ˌtempə'reərɪlɪ/ *adv.* **temporariness** *n.* [L *temporarius* f. *tempus -oris* time]

temporize /'tempəˌraɪz/ *v.intr.* (also **-ise**) **1** avoid committing oneself so as to gain time; employ delaying tactics. **2** comply temporarily with the requirements of the occasion, adopt a time-serving policy. □ **temporizer** *n.* **temporization** /ˌtempəraɪ'zeɪʃ(ə)n/ *n.* [F *temporiser* bide one's time f. med. L *temporizare* delay f. *tempus -oris* time]

tempt /tempt/ *v.tr.* **1** entice or incite (a person) to do a wrong or forbidden thing (*tempted him to steal it*). **2** allure, attract. **3** risk provoking (esp. an abstract force or power) (*would be tempting fate to try it*). **4** *archaic* make trial of; try the resolution of (*God did tempt Abraham*). □ **be tempted to** be strongly disposed to (*I am tempted to question this*). □ **temptable** *adj.* **temptability** /ˌtemptə'bɪlɪtɪ/ *n.* [ME f. OF *tenter, tempter* test f. L *temptare* handle, test, try]

temptation /temp'teɪʃ(ə)n/ *n.* **1 a** the act or an instance of tempting; the state of being tempted; incitement esp. to wrongdoing. **b** (**the Temptation**) the tempting of Jesus by the Devil (see Matt. 4). **2** an attractive thing or course of action. **3** *archaic* putting to the test. [ME f. OF *tentacion, temptacion* f. L *temptatio -onis* (as TEMPT)]

tempter /'temptə(r)/ *n.* (*fem.* **temptress** /-trɪs/) **1** a person who tempts. **2** (**the Tempter**) the Devil. [ME f. OF *tempteur* f. eccl.L *temptator -oris* (as TEMPT)]

tempting /'temptɪŋ/ *adj.* **1** attractive, inviting. **2** enticing to evil. □ **temptingly** *adv.*

tempura /'tempʊərə/ *n.* a Japanese dish of fish, shellfish, or vegetables, fried in batter. [Jap.]

ten /ten/ *n. & adj.* ● *n.* **1** one more than nine. **2** a symbol for this (10, x, X). **3** a size etc. denoted by ten. **4** ten o'clock. **5** a card with ten pips. **6** a set of ten. ● *adj.* **1** that amount to ten. **2** (as a round number) several (*ten times as easy*). □ **ten-gallon hat** a cowboy's large broad-brimmed hat. **ten to one** very probably. **ten-week stock** a variety of stock, *Matthiola incana*, said to bloom ten weeks after the sowing of the seed. [OE *tien, tēn* f. Gmc]

ten. *abbr. Mus.* tenuto.

tenable /'tenəb(ə)l/ *adj.* **1** that can be maintained or defended against attack or objection (*a tenable position*; *a tenable theory*). **2** (foll. by *for, by*) (of an office etc.) that can be held for (a specified period) or by (a specified class of person). □ **tenableness** *n.* **tenability** /ˌtenə'bɪlɪtɪ/ *n.* [F f. *tenir* hold f. L *tenere*]

tenace /'tenəs/ *n. Bridge* etc. **1** two cards, one ranking next above, and the other next below, a card held by an opponent. **2** the holding of such cards. [F f. Sp. *tenaza*, lit. 'pincers']

tenacious /tɪ'neɪʃəs/ *adj.* **1** (often foll. by *of*) keeping a firm hold of property, principles, life, etc.; not readily relinquishing. **2** (of memory) retentive. **3** holding fast. **4** strongly cohesive. **5** persistent, resolute. **6** adhesive, sticky. □ **tenaciously** *adv.* **tenaciousness** *n.* **tenacity** /-'næsɪtɪ/ *n.* [L *tenax -acis* f. *tenere* hold]

tenaculum /tɪ'nækjʊləm/ *n.* (*pl.* **tenacula** /-lə/) a surgeon's sharp hook for picking up arteries etc. [L, = holding instrument, f. *tenere* hold]

tenancy /'tenənsɪ/ *n.* (*pl.* **-ies**) **1** the status of a tenant; possession as a tenant. **2** the duration or period of this.

tenant /'tenənt/ *n. & v.* ● *n.* **1** a person who rents land or property from a landlord. **2** (often foll. by *of*) the occupant of a place. **3** *Law* a person holding real property by private ownership. ● *v.tr.* occupy as a tenant. □ **tenant farmer** a person who farms rented land. **tenant right** *Brit.* the right of a tenant to continue a tenancy at the termination of the lease. □ **tenantable** *adj.* **tenantless** *adj.* [ME f. OF, pres. part. of *tenir* hold f. L *tenere*]

tenantry /'tenəntrɪ/ *n.* the tenants of an estate etc.

tench /tentʃ/ *n.* (*pl.* same) a European freshwater fish, *Tinca tinca*, of the carp family. [ME f. OF *tenche* f. LL *tinca*]

Ten Commandments, the (in the Bible) the divine rules of conduct given by God to Moses on Mount Sinai, according to Exod. 20:1–17. The commandments are generally enumerated as: have no other gods; do not make or worship idols; do not take the word of the Lord in vain; keep the sabbath holy; honour one's father and mother; do not kill; do not commit adultery; do not steal; do not give false evidence; do not covet another's property or wife.

tend[1] /tend/ *v.intr.* **1** (usu. foll. by *to*) be apt or inclined (*tends to lose his temper*). **2** serve, conduce. **3** be moving; be directed; hold a course (*tends in our direction*; *tends downwards*; *tends to the same conclusion*). [ME f. OF *tendre* stretch f. L *tendere* tens- or *tent-*]

tend[2] /tend/ *v.* **1** *tr.* take care of, look after (a person esp. an invalid, animals, a machine, etc.). **2** *intr.* (foll. by *on, upon*) wait on. **3** *intr.* (foll. by *to*) esp. *US* give attention to. □ **tendance** *n. archaic.* [ME f. ATTEND]

tendency /'tendənsɪ/ *n.* (*pl.* **-ies**) **1** (often foll. by *to, towards*) a leaning or inclination, a way of tending. **2** a group within a larger political party or movement. [med.L *tendentia* (as TEND[1])]

tendentious /ten'denʃəs/ *adj. derog.* (of writing etc.) calculated to promote a particular cause or viewpoint; having an underlying purpose. □ **tendentiously** *adv.* **tendentiousness** *n.* [as TENDENCY + -OUS]

tender[1] /'tendə(r)/ *adj.* (**tenderer, tenderest**) **1** easily cut or chewed, not tough (*tender steak*). **2** easily touched or wounded, susceptible to pain or grief (*a tender heart*; *a tender conscience*). **3** easily hurt, sensitive (*tender skin*; *a tender place*). **4 a** delicate, fragile (*a tender plant*). **b** gentle, soft (*a tender touch*). **5** loving, affectionate, fond (*tender parents*; *wrote tender verses*). **6** requiring tact or careful handling, ticklish (*a tender subject*). **7** (of age) early, immature (*of tender years*). **8** (usu. foll. by *of*) solicitous, concerned (*tender of his honour*). □ **tender-eyed 1** having gentle eyes. **2** weak-eyed. **tender-hearted** having a tender heart, easily moved by pity etc. **tender-heartedness** being tender-hearted. **tender mercies** *iron.* attention or treatment which is not in the best interests of its recipient. **tender spot** a subject on which a person is touchy. □ **tenderly** *adv.* **tenderness** *n.* [ME f. OF *tendre* f. L *tener*]

tender[2] /'tendə(r)/ *v. & n.* ● *v.* **1** *tr.* **a** offer, present (one's services, apologies, resignation, etc.). **b** offer (money etc.) as payment. **2** *intr.* (often foll. by *for*) make a tender for the supply of a thing or the execution of work. ● *n.* an offer, esp. an offer in writing to execute work or supply goods at a fixed price. □ **legal tender** see LEGAL. **plea of tender** *Law* a plea that the defendant has always been ready to satisfy the plaintiff's claim and now brings the sum into court. **put out to tender** seek tenders in respect of (work etc.). □ **tenderer** *n.* [OF *tendre*: see TEND[1]]

tender[3] /'tendə(r)/ *n.* **1** a person who looks after people or things. **2** a vessel attending a larger one to supply stores, convey passengers or orders, etc. **3** a special truck closely coupled to a steam locomotive to carry fuel, water, etc. [ME f. TEND[2] or f. ATTENDER (as ATTEND)]

tenderfoot /'tendəfʊt/ *n.* (*pl.* **tenderfoots** or **-feet** /-ˌfiːt/) a newcomer or novice, esp. in the bush or in the Scouts or Guides.

tenderize /'tendəˌraɪz/ *v.tr.* (also **-ise**) make tender, esp. make (meat) tender by beating etc. □ **tenderizer** *n.*

tenderloin /'tendəˌlɔɪn/ *n.* **1 a** *Brit.* the middle part of a pork loin. **b** *US* the undercut of a sirloin. **2** *US sl.* a district of a city where vice and corruption are prominent.

tendon /'tendən/ *n. Anat.* **1** a cord or strand of strong fibrous tissue attaching a muscle to a bone etc. **2** (in a quadruped) = HAMSTRING *n.* 2. □ **tendinous** /-dɪnəs/ *adj.* **tendinitis** /ˌtendɪ'naɪtɪs/ *n.* [F *tendon* or med.L *tendo -dinis* f. Gk *tenōn* sinew f. *teinō* stretch]

tendril /'tendrɪl/ *n.* **1** each of the slender leafless shoots by which some climbing plants cling for support. **2** a slender curl of hair etc. [prob. f. obs. F *tendrillon* dimin. of obs. *tendron* young shoot ult. f. L *tener* TENDER[1]]

Tenebrae /'tenəˌbreɪ/ *n.pl.* **1** *RC Ch. hist.* matins and lauds for the last three days of Holy Week, at which candles are successively extinguished. **2** this office set to music. [L, = darkness]

tenebrous /'tenɪbrəs/ *adj. literary* dark, gloomy. [ME f. OF *tenebrus* f. L *tenebrosus* (as TENEBRAE)]

tenement /'tenɪmənt/ *n.* **1** a room or a set of rooms forming a separate residence within a house or block of flats. **2** *US & Sc.* (also **tenement-house**) a house divided into and let in tenements. **3** a dwelling-place. **4 a** a piece of land held by an owner. **b** *Law* any kind of permanent property, e.g. lands or rents, held from a superior. □ **tenemental** /ˌtenɪ'ment(ə)l/ *adj.* **tenementary** *adj.* [ME f. OF f. med.L *tenementum* f. *tenere* hold]

Tenerife /ˌtenə'riːf/ a volcanic island in the Atlantic, the largest of the Canary Islands; pop. (1986) 759,400; capital, Santa Cruz.

tenesmus /tɪ'nezməs/ *n. Med.* a continual inclination to evacuate the bowels or bladder accompanied by painful straining. [med.L f. Gk *teinesmos* straining f. *teinō* stretch]

tenet /'tenɪt, 'tiːnet/ *n.* a doctrine, dogma, or principle held by a group or person. [L, = he etc. holds f. *tenere* hold]

tenfold /ˈtenfəʊld/ *adj.* & *adv.* **1** ten times as much or as many. **2** consisting of ten parts.

Teng Hsiao-p'ing see Deng Xiaoping.

tenia *US* var. of taenia.

Teniers /ˈteniəz/, David (known as David Teniers the Younger) (1610–90), Flemish painter. The son of the painter David Teniers the Elder (1582–1649), he worked chiefly in Antwerp and Brussels, and from 1651 was court painter to successive regents of the Netherlands. His many works include peasant genre scenes in the style of Brouwer, religious subjects, landscapes, and portraits.

Ten Lost Tribes of Israel see Lost Tribes.

Tenn. *abbr.* Tennessee.

Tennant Creek /ˈtenənt/ a mining town between Alice Springs and Darwin in Northern Territory, Australia; pop. (1986) 3,300.

tenné /ˈteni/ *n.* & *adj.* (usu. placed after noun) (also **tenny**) *Heraldry* orange-brown. [obs. F, var. of *tanné* tawny]

tenner /ˈtenə(r)/ *n. colloq.* a ten-pound or ten-dollar note. [ten]

Tennessee /ˌtenɪˈsiː/ **1** a river in the south-eastern US, flowing some 1,400 km (875 miles) in a great loop, generally westwards through Tennessee and Alabama, then northwards to re-enter Tennessee, joining the Ohio river in western Kentucky. The Tennessee Valley Authority provides one of the world's greatest irrigation and hydro-electric power systems. **2** a state in the central south-eastern US; pop. (1990) 4,877,185; capital, Nashville. Ceded by Britain to the US in 1783, it became the 16th state in 1796.

Tennessee Valley Authority (abbr. **TVA**) an independent Federal government agency in the US. Created in 1933 as part of the New Deal proposals to offset unemployment by a programme of public works, its aim was to provide for the development of the whole Tennessee river basin. It was authorized to construct or improve dams that would control flooding and generate cheap hydroelectric power, to check erosion, and to provide afforestation across seven states.

Tenniel /ˈtenɪəl/, Sir John (1820–1914), English illustrator and cartoonist. He is known chiefly for his illustrations for Lewis Carroll's *Alice's Adventures in Wonderland* (1865) and *Through the Looking Glass* (1871). He also worked as a cartoonist for the magazine *Punch* between 1851 and 1901.

tennis /ˈtenɪs/ *n.* **1** (also **lawn tennis**) a game played with rackets by two persons (*singles*) or four (*doubles*) with a soft ball on a usu. outdoor grass or hard court without walls. (*See note below.*) **2** = real tennis. □ **tennis-ball** a ball used in playing tennis. **tennis-court** a court used in playing tennis. **tennis elbow** a sprain caused by or as by playing tennis. **tennis-racket** a racket used in playing tennis. **tennis shoe** a light canvas or leather soft-soled shoe suitable for tennis or general casual wear. [ME *tenetz*, *tenes*, etc., app. f. OF *tenez* 'take, receive', called by the server to an opponent, imper. of *tenir* take]

▪ Tennis developed from the sport of real tennis in the Victorian era, when it was a popular social game frequently played in gardens. The aim of tennis is to win points by hitting the ball over the net (but within the court) in such a way that one's opponent is unable to return it. A game is won by the first player or pair to win four points (scored 15, 30, 40, game), unless both reach 40 (deuce), in which case two consecutive points are needed to take the game. A set is won by the player winning at least six games with a margin of two (a tie-break system sometimes operating); a match generally consists of three or five sets. Although tennis is still a popular social game, there is now a financially valuable professional circuit, in which the famous championships held at Wimbledon are a constituent of the 'grand slam' (the others are the US Open, the French Open, and the Australian Open); international teams also compete in the Davis Cup. (See also real tennis, table tennis.)

tenno /ˈtenəʊ/ *n.* (pl. **-os**) the emperor of Japan viewed as a divinity. [Jap.]

tenny var. of tenné.

Tennyson /ˈtenɪs(ə)n/, Alfred, 1st Baron Tennyson of Aldworth and Freshwater (1809–92), English poet. His first poems, published in the early 1830s, include 'Mariana', 'The Lotos-Eaters', and 'The Lady of Shalott'; a later collection (1842) included 'Morte d'Arthur', the germ of *Idylls of the King* (1859). His reputation was established by *In Memoriam*, a long poem concerned with immortality, change, and evolution, written in memory of his friend Arthur Hallam (1811–33); although begun in about 1833, it was not published until 1850. The same year he was made Poet Laureate; thereafter he enjoyed considerable celebrity and was one of the most popular poets of his day, publishing 'The Charge of the Light Brigade' in 1854 and *Maud* in 1855. His later works include the collection *Tiresias and Other Poems* (1885) and several dramas.

Tennysonian /ˌtenɪˈsəʊnɪən/ *adj.* of or relating to Alfred Tennyson, or his writings.

Tenochtitlán /teˌnɒxtʃtiːˈtlɑːn/ the ancient capital of the Aztec empire, which was founded *c.*1320 on the site of present-day Mexico City. In 1521 the Spanish conquistador Cortés, having deposed the Aztec emperor Montezuma and overthrown his empire, razed Tenochtitlán and established Mexico City.

tenon /ˈtenən/ *n.* & *v.* ● *n.* a projecting piece of wood made for insertion into a corresponding cavity (esp. a mortise) in another piece. ● *v.tr.* **1** cut as a tenon. **2** join by means of a tenon. □ **tenon-saw** a small saw with a strong brass or steel back for fine work. □ **tenoner** *n.* [ME f. F f. *tenir* hold f. L *tenere*]

tenor /ˈtenə(r)/ *n.* **1 a** a singing voice between baritone and alto or counter-tenor, the highest of the ordinary adult male range. **b** a singer with this voice. **c** a part written for it. **2 a** (*attrib.*) denoting an instrument, esp. a viola, recorder, or saxophone, of which the range is roughly that of a tenor voice. **b** a tenor instrument. **c** (in full **tenor bell**) the largest bell of a peal or set. **3** (usu. foll. by *of*) the general purport or drift of a document or speech. **4** (usu. foll. by *of*) a settled or prevailing course or direction, esp. the course of a person's life or habits. **5** *Law* **a** the actual wording of a document. **b** an exact copy. **6** the subject to which a metaphor refers (opp. vehicle 4). □ **tenor clef** *Mus.* a clef placing middle C on the second highest line of the staff. [ME f. AF *tenur*, OF *tenour* f. L *tenor -oris* f. *tenere* hold]

tenorist /ˈtenərɪst/ *n.* a person who sings a tenor part or plays a tenor instrument, esp. the tenor saxophone.

tenosynovitis /ˌtenəʊˌsaɪnəʊˈvaɪtɪs/ *n.* *Med.* inflammation and swelling of a tendon, usu. in the wrist, often caused by repetitive movements such as typing. [Gk *tenōn* tendon + synovitis]

tenotomy /təˈnɒtəmɪ/ *n.* (pl. **-ies**) the surgical cutting of a tendon, esp. as a remedy for a club-foot. [F *ténotomie*, irreg. f. Gk *tenōn -ontos* tendon]

tenpin /ˈtenpɪn/ *n.* **1** a pin used in tenpin bowling. **2** (in pl.) *US* = tenpin bowling. □ **tenpin bowling** a game developed from ninepins in which ten pins are set up at the end of an alley and bowled down with hard rubber balls. (See bowling.)

tenrec /ˈtenrek/ *n.* (also **tanrec** /ˈtæn-/) a hedgehog-like insect-eating mammal of the family Tenrecidae, confined to Madagascar; esp. the tailless tenrec (*Tenrec ecaudatus*). [F *tanrec*, f. Malagasy *tàndraka*]

tense[1] /tens/ *adj.* & *v.* ● *adj.* **1** stretched tight, strained (*tense cord*; *tense muscle*; *tense nerves*; *tense emotion*). **2** causing tenseness (*a tense moment*). **3** *Phonet.* pronounced with the vocal muscles tense. ● *v.tr.* & *intr.* make or become tense. □ **tense up** become tense. □ **tensely** *adv.* **tenseness** *n.* **tensity** *n.* [L *tensus* past part. of *tendere* stretch]

tense[2] /tens/ *n.* *Gram.* **1** a form taken by a verb to indicate the time (also the continuance or completeness) of the action etc. (*present tense*; *imperfect tense*). **2** a set of such forms for the various persons and numbers. □ **tenseless** *adj.* [ME f. OF *tens* f. L *tempus* time]

tensile /ˈtensaɪl/ *adj.* **1** of or relating to tension. **2** capable of being drawn out or stretched. □ **tensile strength** resistance to breaking under tension. □ **tensility** /tenˈsɪlɪtɪ/ *n.* [med.L *tensilis* (as tense[1])]

tensimeter /tenˈsɪmɪtə(r)/ *n.* **1** an instrument for measuring vapour pressure. **2** a manometer. [tension + -meter]

tension /ˈtenʃ(ə)n/ *n.* & *v.* ● *n.* **1** the act or an instance of stretching; the state of being stretched; tenseness. **2** mental strain or excitement. **3** a strained (political, social, etc.) state or relationship. **4** *Mech.* the strained condition resulting from forces acting in opposition to each other. **5** electromagnetic force (*high tension*; *low tension*). ● *v.tr.* subject to tension. □ **tensional** *adj.* **tensionally** *adv.* **tensionless** *adj.* [F *tension* or L *tensio* (as tend[1])]

tenson /ˈtens(ə)n/ *n.* (also **tenzon** /ˈtiːnz(ə)n/) *hist.* **1** a contest in verse-making between troubadours. **2** a piece of verse composed for this. [F *tenson*, = Prov. *tenso* (as tension)]

tensor /ˈtensə(r)/ *n.* **1** *Anat.* a muscle that tightens or stretches a part of the body. **2** *Math.* a generalized form of vector involving an arbitrary number of indices. □ **tensorial** /tenˈsɔːrɪəl/ *adj.* [mod.L (as tend[1])]

tent[1] /tent/ *n.* & *v.* ● *n.* **1** a portable shelter or dwelling of canvas, cloth, etc., supported by a pole or poles and stretched by cords attached to pegs driven into the ground. **2** = oxygen tent. ● *v.* **1** *tr.* cover with or as with a tent. **2** *intr.* **a** encamp in a tent. **b** dwell temporarily. □ **tent-bed** a bed with a tentlike canopy, or for a patient in an oxygen tent. **tent coat** (or **dress**) a coat (or dress) cut very full. **tent-fly** (pl. **-flies**) **1** a flap at the entrance to a tent. **2** a flysheet. **tent-peg** any of the pegs to which the cords of a tent are attached. **tent-pegging** a sport in which a rider tries at full gallop to carry off on the point of a lance a tent-peg fixed in the ground. **tent-stitch 1** a series of parallel diagonal stitches. **2** such a stitch. [ME f. OF *tente* ult. f. L *tendere* stretch: *tent-stitch* may be f. another word]

tent² /tent/ *n.* a deep-red sweet wine chiefly from Spain, used esp. as sacramental wine. [Sp. *tinto* deep-coloured f. L *tinctus* past part.: see TINGE]

tent³ /tent/ *n. Surgery* a piece (esp. a roll) of lint, linen, etc., inserted into a wound or natural opening to keep it open. [ME f. OF *tente* f. *tenter* probe (as TEMPT)]

tentacle /ˈtentək(ə)l/ *n.* **1** *Zool.* a long slender flexible appendage of an (esp. invertebrate) animal, used for feeling, grasping, or moving. **2** a thing used like a tentacle as a feeler etc. **3** *Bot.* a sensitive hair or filament. □ **tentacled** *adj.* (also in *comb.*). **tentacular** /tenˈtækjʊlə(r)/ *adj.* **tentaculate** /-lət/ *adj.* [mod.L *tentaculum* f. L *tentare* = *temptare* (see TEMPT) + *-culum* -CULE]

tentative /ˈtentətɪv/ *adj. & n.* ● *adj.* **1** done by way of trial, experimental. **2** hesitant, not definite (*tentative suggestion*; *tentative acceptance*). ● *n.* an experimental proposal or theory. □ **tentatively** *adv.* **tentativeness** *n.* [med.L *tentativus* (as TENTACLE)]

tented /ˈtentɪd/ *adj.* **1** covered with tents; enclosed in a tent or tents. **2** shaped like a tent.

tenter¹ /ˈtentə(r)/ *n.* **1** a machine for stretching cloth to dry in shape. **2** = TENTERHOOK. [ME ult. f. med.L *tentorium* (as TEND¹)]

tenter² /ˈtentə(r)/ *n. Brit.* **1** a person in charge of something, esp. of machinery in a factory. **2** a worker's unskilled assistant. [*tent* (now Sc.) pay attention, perh. f. *tent* attention f. INTENT or obs. *attent* (as ATTEND)]

tenterhook /ˈtentəhʊk/ *n.* any of the hooks to which cloth is fastened on a tenter. □ **on tenterhooks** in a state of suspense or mental agitation due to uncertainty.

tenth /tenθ/ *n. & adj.* ● *n.* **1** the position in a sequence corresponding to the number 10 in the sequence 1–10. **2** something occupying this position. **3** one of ten equal parts of a thing. **4** *Mus.* **a** an interval or chord spanning an octave and a third in the diatonic scale. **b** a note separated from another by this interval. ● *adj.* that is the tenth. □ **tenth-rate** of extremely poor quality. □ **tenthly** *adv.* [ME *tenthe*, alt. of OE *teogotha*]

tenuis /ˈtenjʊɪs/ *n.* (*pl.* **tenues** /-jʊˌiːz/) *Phonet.* a voiceless stop, e.g. *k*, *p*, *t*. [L, = thin, transl. Gk *psilos* smooth]

tenuity /tɪˈnjuːɪtɪ/ *n.* **1** slenderness. **2** (of a fluid, esp. air) rarity, thinness. [L *tenuitas* (as TENUIS)]

tenuous /ˈtenjʊəs/ *adj.* **1** slight, of little substance (*tenuous connection*). **2** (of a distinction etc.) oversubtle. **3** thin, slender, small. **4** rarefied. □ **tenuously** *adv.* **tenuousness** *n.* [L *tenuis*]

tenure /ˈtenjə(r)/ *n.* **1** *Law* a condition, or form of right or title, under which (esp. real) property is held. **2** (often foll. by *of*) **a** the holding or possession of an office or property. **b** the period of this (*during his tenure of office*). **3** guaranteed permanent employment, esp. as a teacher or lecturer after a probationary period. [ME f. OF f. *tenir* hold f. L *tenere*]

tenured /ˈtenjəd/ *adj.* **1** (of an official position) carrying a guarantee of permanent employment. **2** (of a teacher, lecturer, etc.) having guaranteed tenure of office.

tenurial /teˈnjʊərɪəl/ *adj. Law* of the tenure of land. □ **tenurially** *adv.* [med.L *tenūra* TENURE]

tenuto /təˈnuːtəʊ/ *adv., adj., & n. Mus.* ● *adv. & adj.* (of a note etc.) sustained, given its full time-value (cf. LEGATO, STACCATO). ● *n.* (*pl.* **-os**) a note or chord played tenuto. [It., = held]

Tenzing Norgay /ˌtensɪŋ ˈnɔːɡeɪ/ (1914–86), Sherpa mountaineer. In 1953, as members of the British Expedition, he and Sir Edmund Hillary were the first to reach the summit of Mount Everest.

tenzon var. of TENSON.

teocalli /ˌtiːəˈkælɪ/ *n.* (*pl.* **teocallis**) a temple of the Aztecs or other Mexican peoples, usu. on a truncated pyramid. [Nahuatl f. *teotl* god + *calli* house]

teosinte /ˌtiːəʊˈsɪntɪ/ *n.* a Mexican grass, *Zea mexicana*, grown as fodder. [F, f. Nahuatl *teocintli*]

Teotihuacán /teɪˌəʊtiwəˈkɑːn/ the largest city of pre-Columbian America, situated about 40 km (25 miles) north-east of Mexico City. Built *c.*300 BC, it reached its zenith AD *c.*300–600, when it was the centre of an influential culture which spread throughout Meso-America. By 650 it was declining as a major power and was sacked by the invading

Toltecs *c.*900. Among its monuments are palatial buildings, plazas, and temples, including the Pyramids of the Sun and the Moon and the temple of Quetzalcóatl.

tepal /ˈtep(ə)l/ *n. Bot.* a segment of the outer whorl in a flower that has no differentiation between petals and sepals. [F *tépale*, as blend of PETAL + SEPAL]

tepee /ˈtiːpiː/ *n.* (also **teepee**) a North American Indian conical tent, made of skins, cloth, or canvas on a frame of poles. [Sioux *tīpī*]

tephra /ˈtefrə/ *n. Geol.* fragmented rock etc. ejected by a volcanic eruption. [Gk, = ash]

Tepic /teˈpiːk/ a city in western Mexico, capital of the state of Nayarit; pop. (1990) 238,100.

tepid /ˈtepɪd/ *adj.* **1** slightly warm. **2** unenthusiastic. □ **tepidly** *adv.* **tepidness** *n.* **tepidity** /tɪˈpɪdɪtɪ/ *n.* [L *tepidus* f. *tepere* be lukewarm]

tepidarium /ˌtepɪˈdeərɪəm/ *n.* (*pl.* **tepidariums** or **tepidaria** /-rɪə/) *Rom. Antiq.* in a Roman bath, the warm room between the caldarium and the frigidarium. (See also ROMAN BATHS.) [L (as TEPID)]

tequila /teˈkiːlə/ *n.* a Mexican liquor made from an agave. [*Tequila* in Mexico]

ter- /tɜː(r)/ *comb. form* three; threefold (*tercentenary*; *tervalent*). [L *ter* thrice]

tera- /ˈterə/ *comb. form* denoting a factor of 10^{12}. [Gk *teras* monster]

terai /təˈraɪ/ *n.* (in full **terai hat**) a wide-brimmed felt hat, often with a double crown, worn by travellers etc. in subtropical regions. [*Terai*, belt of marshy jungle between Himalayan foothills and plains, f. Hindi *tarāī* moist (land)]

terakihi var. of TARAKIHI.

terametre /ˈterəˌmiːtə(r)/ *n.* a unit of length equal to 10^{12} metres.

teraph /ˈterəf/ *n.* (*pl.* **teraphim** /-fɪm/, also used as *sing.*) a small image as a domestic deity or oracle of the ancient Israelites. [ME f. LL *theraphim*, Gk *theraphin* f. Heb. *ṯrāpîm*]

terato- /ˈterətəʊ/ *comb. form* monster. [Gk *teras -atos* monster]

teratogen /təˈrætədʒən/ *n. Med.* an agent or factor causing malformation of an embryo. □ **teratogenic** /ˌterətəˈdʒenɪk/ *adj.* **teratogeny** /-ˈtɒdʒənɪ/ *n.*

teratology /ˌterəˈtɒlədʒɪ/ *n.* **1** *Med. & Biol.* the scientific study of congenital abnormalities and abnormal formations. **2** mythology relating to fantastic creatures, monsters, etc. □ **teratologist** *n.* **teratological** /-təˈlɒdʒɪk(ə)l/ *adj.*

teratoma /ˌterəˈtəʊmə/ *n.* (*pl.* **teratomas** or **teratomata** /-mətə/) *Med.* a tumour of heterogeneous tissues, esp. of the gonads.

terbium /ˈtɜːbɪəm/ *n.* a silvery-white metallic chemical element (atomic number 65; symbol **Tb**). A member of the lanthanide series, terbium was discovered in 1843 by the Swedish chemist C. G. Mosander, who separated it from yttrium and erbium. Its main use is in making semiconductors. [f. *Ytterby* in Sweden: see YTTERBIUM]

terce /tɜːs/ *n. Eccl.* **1** the office of the canonical hour of prayer appointed for the third daytime hour (i.e. 9 a.m.). **2** this hour. [var. of TIERCE]

tercel /ˈtɜːs(ə)l/ *n.* (also **tiercel** /ˈtɪəs-/) *Falconry* the male of the hawk, esp. a peregrine or goshawk. [ME f. OF *tercel*, ult. a dimin. of L *tertius* third, perh. from a belief that the third egg of a clutch produced a male bird, or that the male was one-third smaller than the female]

tercentenary /ˌtɜːsenˈtiːnərɪ/ *n. & adj.* ● *n.* (*pl.* **-ies**) **1** a three-hundredth anniversary. **2** a celebration of this. ● *adj.* of this anniversary.

tercentennial /ˌtɜːsenˈtenɪəl/ *adj. & n.* ● *adj.* **1** occurring every three hundred years. **2** lasting three hundred years. ● *n.* a tercentenary.

tercet /ˈtɜːsɪt/ *n.* (also **tiercet** /ˈtɪəs-/) *Prosody* a set or group of three lines rhyming together or connected by rhyme with an adjacent triplet. [F f. It. *terzetto* dimin. of *terzo* third f. L *tertius*]

terebene /ˈterɪˌbiːn/ *n. Pharm.* a mixture of terpenes prepared by treating oil of turpentine with sulphuric acid, used as an expectorant etc. [TEREBINTH + -ENE]

terebinth /ˈterɪbɪnθ/ *n.* a small Mediterranean tree, *Pistacia terebinthus*, yielding resin formerly used as a source of turpentine. [ME f. OF *terebinte* or L *terebinthus* f. Gk *terebinthos*]

terebinthine /ˌterɪˈbɪnθaɪn/ *adj.* **1** of the terebinth. **2** of turpentine. [L *terebinthinus* f. Gk *terebinthinos* (as TEREBINTH)]

teredo /təˈriːdəʊ/ *n.* (*pl.* **-os**) a bivalve mollusc of the genus *Teredo*, esp. *T. navalis*, that bores into wooden ships etc. Also called *shipworm*. [L f. Gk *terēdōn* f. *teirō* rub hard, wear away, bore]

Terence /'terəns/ (Latin name Publius Terentius Afer) (c.190–159 BC), Roman comic dramatist. His six surviving comedies are based on the Greek New Comedy; set in Athens, they use the same stock characters as are found in Plautus, but are marked by a more realistic treatment of character and language, and a greater consistency of plot. Terence's work had an influence on the development of Renaissance and Restoration comedy.

Terengganu see TRENGGANU.

terephthalic acid /ˌterə'fθælɪk, -'θælɪk/ n. Chem. the para-isomer of phthalic acid, used in making plastics and other polymers. □ **terephthalate** /-leɪt/ n. [terebic (f. TEREBINTH) + PHTHALIC ACID]

Teresa, Mother /təˈreɪzə, -ˈriːzə/ (also **Theresa**) (born Agnes Gonxha Bojaxhiu) (b.1910), Roman Catholic nun and missionary, born in what is now Macedonia of Albanian parentage. In 1928 she went to India, where she devoted herself to helping the destitute. In 1948 she became an Indian citizen and founded the Order of Missionaries of Charity, which became noted for its work among the poor and the dying in Calcutta. Her organization now operates in many other parts of the world. Mother Teresa was awarded the Nobel Peace Prize in 1979.

Teresa of Ávila, St /ˈævɪlə/ (1515–82), Spanish Carmelite nun and mystic. She combined vigorous activity as a reformer with mysticism and religious contemplation. Seeking to return the Carmelite Order to its original discipline and observances, she instituted the 'discalced' reform movement, establishing the first of a number of convents in 1562 and encouraging St John of the Cross to found a similar monastic order. Her spiritual writings include *The Way of Perfection* (1583) and *The Interior Castle* (1588). Feast day, 15 Oct.

Teresa of Lisieux, St /liˈzjɜː/ (also **Thérèse** /teˈrez/) (born Marie-Françoise Thérèse Martin) (1873–97), French Carmelite nun. After her death from tuberculosis her cult grew through the publication of her autobiography *L'Histoire d'une âme* (1898), teaching that sanctity can be attained through continual renunciation in small matters, and not only through extreme self-mortification. Feast day, 3 Oct.

Tereshkova /ˌterɪʃˈkəʊvə/, Valentina (Vladimirovna) (b.1937), Russian cosmonaut. In June 1963 she became the first woman in space; her spacecraft returned to earth after three days in orbit.

Teresina /ˌtereˈziːnə/ a river port in NE Brazil, on the Parnaíba river, capital of the state of Piauí; pop. (1990) 591,160.

terete /təˈriːt/ adj. Biol. smooth and rounded; cylindrical. [L teres -etis]

tergal /'tɜːg(ə)l/ adj. Zool. of or relating to the back; dorsal. [L tergum back]

tergiversate /'tɜːdʒɪvəˌseɪt/ v.intr. 1 be apostate; change one's party or principles. 2 equivocate; make conflicting or evasive statements. 3 turn one's back on something. □ **tergiversator** n. **tergiversation** /ˌtɜːdʒɪvəˈseɪʃ(ə)n/ n. [L tergiversari turn one's back f. tergum back + vertere vers- turn]

-teria /'tɪərɪə/ suffix denoting self-service establishments (washeteria). [after CAFETERIA]

teriyaki /ˌterɪˈjɑːkɪ/ n. a Japanese dish of meat or fish marinated in soy sauce and grilled. [Jap., f. teri lustre + yaki grill, fry]

term /tɜːm/ n. & v. ● n. 1 a word used to express a definite concept, esp. in a particular branch of study etc. (a technical term). 2 (in pl.) language used; mode of expression (answered in no uncertain terms). 3 (in pl.) a relation or footing (we are on familiar terms). 4 (in pl.) a conditions or stipulations (cannot accept your terms; do it on your own terms). b charge or price (his terms are £20 a lesson). 5 a a limited period of some state or activity (for a term of five years). b a period over which operations are conducted or results contemplated (in the short term). c a period of some weeks, alternating with holiday or vacation, during which instruction is given in a school, college, or university, or Brit. during which a lawcourt holds sessions. d a period of imprisonment. e a period of tenure. 6 Logic a word or words that may be the subject or predicate of a proposition. 7 Math. a each of the two quantities in a ratio. b each quantity in a series. c a part of an expression joined to the rest by + or – (e.g. a, b, c in a + b – c). 8 the completion of a normal length of pregnancy. 9 an appointed day, esp. a Scottish quarter day. 10 (in full Brit. **term of years** or US **term for years**) Law an interest in land for a fixed period. 11 Archit. = TERMINUS 6. 12 archaic a boundary or limit, esp. of time. ● v.tr. denominate, call; assign a term to (the music termed classical). □ **bring to terms** cause to accept conditions. **come to terms** yield, give way. **come to terms with 1** reconcile oneself to (a difficulty etc.). 2 conclude an agreement with. **in set terms** in definite terms. **in terms** explicitly. **in terms of 1** in the language peculiar to, using as a basis of expression or thought. 2 by way of.

make terms conclude an agreement. **on terms** on terms of friendship or equality. **term paper** N. Amer. an essay or dissertation representative of the work done during a term. **terms of reference** Brit. 1 points referred to an individual or body of persons for decision or report. 2 the scope of an inquiry etc.; a definition of this. **terms of trade** Brit. the ratio between prices paid for imports and those received for exports. **term-time** n. (esp. in phr. **in** or **during term-time**) the period when a school etc. is in session. ● adj. of or relating to this period. □ **termless** adj. **termly** adj. & adv. [ME f. OF terme f. L TERMINUS]

termagant /'tɜːməgənt/ n. & adj. ● n. 1 an overbearing or brawling woman; a virago or shrew. 2 (**Termagant**) hist. an imaginary deity of violent and turbulent character, often appearing in morality plays. ● adj. violent, turbulent, shrewish. [ME Tervagant f. OF Tervagan f. It. Trivigante]

terminable /'tɜːmɪnəb(ə)l/ adj. 1 that may be terminated; not lasting or perpetual. 2 coming to an end after a certain time (terminable annuity). □ **terminableness** n.

terminal /'tɜːmɪn(ə)l/ adj. & n. ● adj. 1 a (of a disease) ending in death, fatal. b (of a patient) in the last stage of a fatal disease. c (of a morbid condition) forming the last stage of a fatal disease. d colloq. ruinous, disastrous, very great (terminal laziness). 2 of or forming a limit or terminus (terminal station). 3 a Zool. etc. ending a series (terminal joints). b Bot. borne at the end of a stem etc. 4 of or done etc. each term (terminal accounts; terminal examinations). ● n. 1 a terminating thing; an extremity. 2 a terminus for trains or long-distance buses. 3 (in full **air terminal**) a departure and arrival building for air passengers at an airport. 4 a point of connection for closing an electric circuit. 5 an apparatus for transmission of messages between a user and a computer, communications system, etc. 6 (in full **terminal figure**) Archit. = TERMINUS 6. 7 an installation where oil is stored at the end of a pipeline or at a port. 8 a patient suffering from a terminal illness. □ **terminal velocity** a velocity of a falling body such that the resistance of the air etc. prevents further increase of speed under gravity. □ **terminally** adv. [L terminalis (as TERMINUS)]

terminate /'tɜːmɪˌneɪt/ v. 1 tr. & intr. bring or come to an end. 2 intr. (foll. by in) (of a word) end in (a specified letter or syllable etc.). 3 tr. Med. end (a pregnancy) before term by artificial means. 4 tr. bound, limit. [L terminare (as TERMINUS)]

termination /ˌtɜːmɪˈneɪʃ(ə)n/ n. 1 the act or an instance of terminating; the state of being terminated. 2 Med. an induced abortion. 3 an ending or result of a specified kind (a happy termination). 4 a word's final syllable or letters or letter esp. as an element in inflection or derivation. □ **put a termination to** (or **bring to a termination**) make an end of. □ **terminational** adj. [ME f. OF termination or L terminatio (as TERMINATE)]

terminator /'tɜːmɪˌneɪtə(r)/ n. 1 a person or thing that terminates. 2 Astron. the dividing line between the light and dark part of a planetary body.

terminer see OYER AND TERMINER.

termini pl. of TERMINUS.

terminism /'tɜːmɪˌnɪz(ə)m/ n. 1 (in Christian theology) the doctrine that everyone has a limited time for repentance. 2 = NOMINALISM. □ **terminist** n. [L]

terminological /ˌtɜːmɪnəˈlɒdʒɪk(ə)l/ adj. relating to terminology. □ **terminological inexactitude** joc. a lie. **terminologically** adv.

terminology /ˌtɜːmɪˈnɒlədʒɪ/ n. (pl. **-ies**) 1 the system of terms used in a particular subject. 2 the science of the proper use of terms. □ **terminologist** n. [G Terminologie f. med.L TERMINUS term]

terminus /'tɜːmɪnəs/ n. (pl. **termini** /-ˌnaɪ/ or **terminuses**) 1 a station at the end of a railway or bus route. 2 a point at the end of a pipeline etc. 3 a final point; a goal. 4 a starting-point. 5 Math. the end-point of a vector etc. 6 Archit. a figure of a human bust or an animal ending in a square pillar from which it appears to spring, orig. as a boundary-marker. □ **terminus ad quem** /æd ˈkwem/ the finishing-point of an argument, policy, period, etc. **terminus ante quem** /æntɪ ˈkwem/ the finishing-point of a period; the latest possible date for something. **terminus a quo** /ɑː ˈkwəʊ/ the starting-point of an argument, policy, period, etc. **terminus post quem** /pəʊst ˈkwem/ the starting-point of a period; the earliest possible date for something. [L, = end, limit, boundary]

termitary /'tɜːmɪtərɪ/ n. (pl. **-ies**) a nest of termites, usu. a large mound of earth.

termite /'tɜːmaɪt/ n. a small chiefly tropical social insect of the order Isoptera, occurring in large colonies with distinct castes of individuals, and very destructive to timber. Also called *white ant*. [LL *termes -mitis*, alt. of L *tarmes* after *terere* rub]

termor /'tɜːmə(r)/ *n. Law* a person who holds lands etc. for a term of years, or for life. [ME f. AF *termer* (as TERM)]

tern[1] /tɜːn/ *n.* any marine bird of the subfamily Sterninae, like a gull but usu. smaller and with a long forked tail. [of Scand. orig.: cf. Danish *terne*, Sw. *tärna* f. ON *therna*]

tern[2] /tɜːn/ *n.* **1** a set of three, esp. three lottery numbers that when drawn together win a large prize. **2** such a prize. [F *terne* f. L *terni* three each]

ternary /'tɜːnərɪ/ *adj.* **1** composed of three parts. **2** *Math.* using three as a base (*ternary scale*). □ **ternary form** *Mus.* the form of a movement in which the first subject is repeated after an interposed second subject in a related key. [ME f. L *ternarius* f. *terni* three each]

ternate /'tɜːneɪt/ *adj.* **1** arranged in threes. **2** *Bot.* (of a leaf): **a** having three leaflets. **b** whorled in threes. □ **ternately** *adv.* [mod.L *ternatus* (as TERNARY)]

terne /tɜːn/ *n.* **1** (in full **terne metal**) a lead alloy containing about 20 per cent tin and some antimony. **2** (in full **terne-plate**) iron or steel plate coated with this. [prob. f. F *terne* dull: cf. TARNISH]

terotechnology /ˌterəʊtek'nɒlədʒɪ, ˌtɪər-/ *n.* the branch of technology and engineering concerned with the installation and maintenance of equipment. [Gk *tēreō* take care of + TECHNOLOGY]

terpene /'tɜːpiːn/ *n. Chem.* any of a large group of unsaturated cyclic hydrocarbons found in the essential oils of plants, esp. conifers and oranges. [*terpentin* obs. var. of TURPENTINE]

Terpsichore /tɜːp'sɪkərɪ/ *Gk & Rom. Mythol.* the Muse of lyric poetry and dance. [Gk, = delighting in dance]

Terpsichorean /ˌtɜːpsɪkə'rɪən/ *adj.* of or relating to dancing.

terra alba /ˌterə 'ælbə/ *n.* a white mineral, esp. pipeclay or pulverized gypsum. [L, = white earth]

terrace /'terəs, -rɪs/ *n. & v.* ● *n.* **1** each of a series of flat areas formed on a slope and used for cultivation. **2** a level paved area next to a house. **3 a** a row of houses on a raised level or along the top or face of a slope. **b** a row of houses built in one block of uniform style. **4 a** a flight of wide shallow steps as for spectators at a sports ground. **b** (in *pl.*) the spectators occupying such steps. **5** *Geol.* a raised beach, or a similar formation beside a river etc. ● *v.tr.* form into or provide with a terrace or terraces. □ **terraced house** *Brit.* = *terrace house.* **terraced roof** a flat roof esp. of an Indian or Eastern house. **terrace house** *Brit.* any of a row of houses joined by party-walls. [OF ult. f. L *terra* earth]

terracotta /ˌterə'kɒtə/ *n.* **1 a** unglazed usu. brownish-red earthenware used chiefly as an ornamental building-material and in modelling. **b** a statuette of this. **2** its colour. [It. *terra cotta* baked earth]

terra firma /ˌterə 'fɜːmə/ *n.* dry land, firm ground. [L, = firm land]

terraform /'terəˌfɔːm/ *v.tr.* (esp. in science fiction) transform (a planet) so as to resemble the earth. [L *terra* earth + FORM]

terrain /te'reɪn/ *n.* **1** a tract of land, esp. as regarded by the physical geographer or the military tactician. **2** a particular area of knowledge; a sphere of influence or action. [F, ult. f. L *terrenum* neut. of *terrenus* TERRENE]

terra incognita /ˌterə ɪn'kɒɡnɪtə, ˌɪŋkɒɡ'niːtə/ *n.* an unknown or unexplored region. [L, = unknown land]

terramara /ˌterə'mɑːrə/ *n.* (*pl.* **terramare** /-rɪ/) = TERRAMARE. [It. dial.: see TERRAMARE]

terramare /ˌterə'mɑːrɪ, -'meə(r)/ *n.* **1** an ammoniacal earthy deposit found in mounds in prehistoric lake-dwellings or settlements in Italy. **2** such a dwelling or settlement. [F f. It. dial. *terra mara* f. *marna* marl]

terrapin /'terəpɪn/ *n.* **1 a** a small freshwater turtle of the family Emydidae; esp. the *European pond terrapin* (*Emys orbicularis*). **b** a small edible turtle, *Malaclemys terrapin*, found in brackish coastal marshes of the eastern US. **2** (**Terrapin**) *propr.* a type of prefabricated one-storey building. [Algonquian]

terrarium /te'reərɪəm/ *n.* (*pl.* **terrariums** or **terraria** /-rɪə/) **1** a vivarium for small land animals. **2** a sealed transparent globe etc. containing growing plants. [mod.L f. L *terra* earth, after AQUARIUM]

terra sigillata /ˌterə ˌsɪdʒɪ'leɪtə, -'lɑːtə/ *n.* **1** *hist.* astringent clay from Lemnos or Samos, formerly used as a medicine and antidote. **2** Samian ware. [med.L, = sealed earth]

Terrassa see TARRASA.

terrazzo /te'rætsəʊ/ *n.* (*pl.* **-os**) a flooring-material of stone chips set in concrete and given a smooth surface. [It., = terrace]

Terre Haute /ˌterə 'həʊt/ a city in western Indiana, on the Wabash river, near the border with Illinois; pop. (1990) 57,480.

terrene /te'riːn/ *adj.* **1** of the earth; earthly, worldly. **2** of earth, earthy. **3** of dry land; terrestrial. [ME f. AF f. L *terrenus* f. *terra* earth]

terreplein /'teəpleɪn/ *n.* a level space where a battery of guns is mounted. [orig. a sloping bank behind a rampart f. F *terre-plein* f. It. *terrapieno* f. *terrapienare* fill with earth f. *terra* earth + *pieno* f. L *plenus* full]

terrestrial /tə'restrɪəl/ *adj. & n.* ● *adj.* **1** of or on or relating to the earth; earthly. **2 a** of or on dry land. **b** *Zool.* living on or in the ground (cf. AQUATIC, ARBOREAL, AERIAL). **c** *Bot.* growing in the soil (cf. AQUATIC, EPIPHYTIC). **3** *Astron.* (of a planet) similar in size or composition to the earth; esp. being one of the four inner planets. **4** of this world, worldly. **5** (of broadcasting) not using satellites. ● *n.* an inhabitant of the earth. □ **terrestrial globe** **1** a globe representing the earth. **2** the earth. **terrestrial magnetism** the magnetic properties of the earth as a whole. **terrestrial telescope** a telescope giving an erect image for observation of terrestrial objects. □ **terrestrially** *adv.* [ME f. L *terrestris* f. *terra* earth]

terret /'terɪt/ *n.* (also **territ**) each of the loops or rings on a harness-pad for the driving-reins to pass through. [ME, var. of *toret* (now dial.) f. OF *to(u)ret* dimin. of TOUR]

terre-verte /teə'veət/ *n.* a soft green earth used as a pigment. [F, = green earth]

terrible /'terɪb(ə)l/ *adj.* **1** *colloq.* very great or bad (*a terrible bore*). **2** *colloq.* very incompetent (*terrible at tennis*). **3** causing terror; fit to cause terror; awful, dreadful, formidable. **4** (*predic.* usu. foll. by *about*) *colloq.* full of remorse, sorry (*I felt terrible about it*). □ **terribleness** *n.* [ME f. F f. L *terribilis* f. *terrere* frighten]

terribly /'terɪblɪ/ *adv.* **1** *colloq.* very, extremely (*he was terribly nice about it*). **2** in a terrible manner.

terricolous /te'rɪkələs/ *adj. Zool. & Bot.* living on or in the earth. [L *terricola* earth-dweller f. *terra* earth + *colere* inhabit]

terrier[1] /'terɪə(r)/ *n.* **1 a** a small dog of various breeds originally used for turning out foxes etc. from their earths. **b** any of these breeds. **2** an eager or tenacious person or animal. **3** (**Terrier**) *Brit. colloq.* a member of the Territorial Army etc. [ME f. OF (*chien*) *terrier* f. med.L *terrarius* f. L *terra* earth]

terrier[2] /'terɪə(r)/ *n. hist.* **1** a book recording the site, boundaries, etc., of the land of private persons or corporations. **2** a rent-roll. **3** a collection of acknowledgements of vassals or tenants of a lordship. [ME f. OF *terrier* (adj.) = med.L *terrarius liber* (as TERRIER[1])]

terrific /tə'rɪfɪk/ *adj.* **1** *colloq.* **a** of great size or intensity. **b** excellent (*did a terrific job*). **c** excessive (*making a terrific noise*). **2** causing terror. □ **terrifically** *adv.* [L *terrificus* f. *terrere* frighten]

terrify /'terɪˌfaɪ/ *v.tr.* (**-ies, -ied**) fill with terror; frighten severely (*terrified them into submission; is terrified of dogs*). □ **terrifier** *n.* **terrifying** *adj.* **terrifyingly** *adv.* [L *terrificare* (as TERRIFIC)]

terrigenous /te'rɪdʒɪnəs/ *adj. Geol.* land-derived, esp. designating a marine deposit consisting of material eroded from the land. [L *terrigenus* earth-born]

terrine /tə'riːn/ *n.* **1** a kind of pâté, esp. a coarse-textured one. **2** an earthenware container, esp. one in which such food is cooked or sold. [orig. form of TUREEN]

territ var. of TERRET.

territorial /ˌterɪ'tɔːrɪəl/ *adj. & n.* ● *adj.* **1** of territory (*territorial possessions*). **2** limited to a district (*the right was strictly territorial*). **3** (of a person or animal etc.) tending to defend an area of territory. **4** (usu. **Territorial**) of any of the Territories of the US or Canada. ● *n.* (**Territorial**) (in the UK) a member of the Territorial Army. □ **territorial waters** the waters under the jurisdiction of a state, esp. the part of the sea within a stated distance of the shore (traditionally three miles from low-water mark). □ **territorially** *adv.* **territorialize** *v.tr.* (also **-ise**). **territorialization** /-ˌtɔːrɪəlaɪ'zeɪʃ(ə)n/ *n.* **territoriality** /-ˌtɔːrɪ'ælɪtɪ/ *n.* [LL *territorialis* (as TERRITORY)]

Territorial Army (in the UK) a volunteer force locally organized to provide a reserve of trained and disciplined manpower for use in an emergency (known as *Territorial and Army Volunteer Reserve 1967–79*).

territory /'terɪtərɪ/ *n.* (*pl.* **-ies**) **1** the extent of the land under the jurisdiction of a ruler, state, city, etc. **2** (also **Territory**) an organized division of a country, esp. one not yet admitted to the full rights of a state. **3** a sphere of action or thought; a province. **4** the area over which a commercial traveller or goods-distributor operates. **5** *Zool.* an area defended by an animal or animals against others of the same species.

6 an area defended by a team or player in a game. **7** a large tract of land. [ME f. L *territorium* f. *terra* land]

terror /'terə(r)/ *n.* **1** extreme fear. **2 a** a person or thing that causes terror. **b** (also **holy terror**) *colloq.* a formidable person; a troublesome person or thing (*the twins are little terrors*). **3** the use of organized intimidation; terrorism. □ **reign of terror** a period of remorseless repression or bloodshed. **terror-stricken** (or **-struck**) affected with terror. [ME f. OF *terrour* f. L *terror -oris* f. *terrere* frighten]

Terror, the (also **Reign of Terror**) the period of the French Revolution between mid-1793 and July 1794 when the ruling Jacobin faction, dominated by Robespierre, ruthlessly executed opponents and anyone else considered a threat to their regime. It ended with the fall and execution of Robespierre, but in its last six weeks more than 1,300 people were guillotined in Paris alone. (See also REVOLUTIONARY TRIBUNAL.)

terrorism /'terə,rɪʒ(ə)m/ *n.* the systematic use of violence and intimidation to coerce a government or community, esp. into acceding to specific political demands.

terrorist /'terərɪst/ *n.* (also *attrib.*) a person who uses or favours violent and intimidating methods of coercing a government or community. □ **terroristic** /,terə'rɪstɪk/ *adj.* **terroristically** *adv.* [F *terroriste* (as TERROR)]

terrorize /'terə,raɪz/ *v.tr.* (also **-ise**) **1** fill with terror. **2** use terrorism against. □ **terrorizer** *n.* **terrorization** /,terəraɪ'zeɪʃ(ə)n/ *n.*

Terry /'terɪ/, Dame (Alice) Ellen (1847–1928), English actress. She was already well known when in 1878 Henry Irving engaged her as his leading lady at the Lyceum Theatre in London. For the next twenty-four years she played in many of his Shakespearian productions, notably in the roles of Desdemona, Portia, and Beatrice. She also acted in plays by George Bernard Shaw, with whom she conducted a long correspondence; the part of Lady Cicely Waynflete in *Captain Brassbound's Conversion* (1905) was among a number of roles Shaw created for her. She was married to the painter George Frederick Watts from 1864 to 1877.

terry /'terɪ/ *n. & adj.* ● *n.* (*pl.* **-ies**) **1** a pile fabric with the loops uncut, used esp. for towels. **2** a baby's nappy made of this fabric. ● *adj.* of this fabric. [18th c.: orig. unkn.]

terse /tɜːs/ *adj.* (**terser**, **tersest**) **1** (of language) brief, concise, to the point. **2** curt, abrupt. □ **tersely** *adv.* **terseness** *n.* [L *tersus* past part. of *tergēre* wipe, polish]

tertian /'tɜːʃ(ə)n/ *adj. Med.* (of a fever) recurring every third day by inclusive counting. [ME (*fever*) *tersiane* f. L (*febris*) *tertiana* (as TERTIARY)]

tertiary /'tɜːʃərɪ/ *adj. & n.* ● *adj.* **1** third in order or rank etc. **2** (**Tertiary**) *Geol.* of or relating to the first period in the Cenozoic era. (*See note below.*) ● *n.* (**Tertiary**) *Geol.* this period or the corresponding geological system. **2** a member of a lay order of certain monastic bodies (e.g. the Franciscan order). □ **tertiary education** education, esp. in a college or university, that follows secondary education. [L *tertiarius* f. *tertius* third]

▪ The Tertiary period, so called because it followed the Mesozoic (formerly also called *Secondary*), lasted from about 65 to 2 million years ago. The climate was generally warm except towards the close, and mammals evolved rapidly, becoming the dominant land vertebrates. The beginning of the Tertiary appears to have been marked by devastating meteorite impacts (see CRETACEOUS).

tertium quid /,tɜːʃɪəm 'kwɪd, ,tɜːtɪəm/ *n.* a third thing, indefinite and undefined, related in some way to two definite or known things, but distinct from both. [L, app. transl. of Gk *triton ti*]

Tertullian /tɜː'tʌlɪən/ (Latin name Quintus Septimius Florens Tertullianus) (*c.*160–*c.*240), early Christian theologian. Born in Carthage after the Roman conquest, he converted to Christianity *c.*195. His writings (in Latin) include Christian apologetics and attacks on pagan idolatry and Gnosticism. He later joined the Montanists, urging asceticism and venerating martyrs.

tervalent /tɜː'veɪlənt/ *adj. Chem.* = TRIVALENT.

Terylene /'terɪ,liːn/ *n. propr.* a synthetic polyester used as a textile fibre. [TEREPHTHALIC ACID + ETHYLENE]

terza rima /,teətsə 'riːmə/ *n. Prosody* an arrangement of (esp. iambic pentameter) triplets rhyming *aba bcb cdc* etc., as in Dante's *The Divine Comedy*. [It., = third rhyme]

terzetto /teə'tsetəʊ, tɜː'tset-/ *n.* (*pl.* **-os** or **terzetti** /-tɪ/) *Mus.* a vocal or instrumental trio. [It.: see TERCET]

TESL /'tes(ə)l/ *abbr.* teaching of English as a second language.

Tesla /'teslə/, Nikola (1856–1943), American electrical engineer and inventor, born in what is now Croatia of Serbian descent. He emigrated to the US in 1884 and worked briefly on motors and direct-current generators with Thomas Edison before joining the Westinghouse company, where he developed the first alternating-current induction motor (1888) and made contributions to long-distance electrical power transmission. Tesla also studied high-frequency current, developing several forms of oscillators and the tesla coil, and developed a wireless guidance system for ships. Although his inventions revolutionized the electrical industry, he died in poverty.

tesla /'teslə/ *n. Physics* the SI unit of magnetic flux density (symbol: **T**), equal to one weber per square metre or 10,000 gauss. [TESLA]

tesla coil *n.* (also **Tesla**) a form of induction coil for producing high-frequency alternating currents, used widely in radios, televisions, and other electronic equipment. [TESLA]

TESOL /'tesɒl/ *abbr.* teaching of English to speakers of other languages.

TESSA /'tesə/ *n.* (also **Tessa**) *Brit.* tax-exempt special savings account, a form of tax-free investment introduced in 1991. [acronym]

tessellate /'tesə,leɪt/ *v.tr.* **1** make from tesserae. **2** *Math.* cover (a plane surface) by repeated use of a single shape. [L *tessellare* f. *tessella* dimin. of TESSERA]

tessellated /'tesə,leɪtɪd/ *adj.* **1** of or resembling mosaic. **2** *Bot. & Zool.* regularly chequered. [L *tessellatus* or It. *tessellato* (as TESSELLATE)]

tessellation /,tesə'leɪʃ(ə)n/ *n.* **1** the act or an instance of tessellating; the state of being tessellated. **2** an arrangement of polygons without gaps or overlapping, esp. in a repeated pattern.

tessera /'tesərə/ *n.* (*pl.* **tesserae** /-,riː/) **1** a small square block used in mosaic. **2** *Gk & Rom. Antiq.* a small square of bone etc. used as a token, ticket, etc. □ **tesseral** *adj.* [L f. Gk, neut. of *tesseres, tessares* four]

Tessin see TICINO.

tessitura /,tesɪ'tʊərə/ *n. Mus.* the range within which most tones of a voice-part fall. [It., = TEXTURE]

test[1] /test/ *n. & v.* ● *n.* **1** a critical examination or trial of a person's or thing's qualities. **2** the means of so examining; a standard for comparison or trial; circumstances suitable for this (*success is not a fair test*). **3** a minor examination, esp. in school (*spelling test*). **4** a test match. **5** a ground of admission or rejection (*is excluded by our test*). **6** *Chem.* a reagent or a procedure employed to reveal the presence of another in a compound. **7** *Brit.* a movable hearth in a reverberating furnace with a cupel used in separating gold or silver from lead. **8** (*attrib.*) done or performed in order to test (*a test run*). ● *v.tr.* **1** put to the test; make trial of (a person or thing or quality). **2** try severely; tax a person's powers of endurance etc. **3** *Chem.* examine by means of a reagent. **4** *Brit.* refine or assay (metal). □ **put to the test** cause to undergo a test. **test bed** equipment for testing aircraft engines before acceptance for general use. **test card** a still television picture transmitted outside normal programme hours and designed for use in judging the quality and position of the image. **test case** *Law* a case setting a precedent for other cases involving the same question of law. **test drive** a drive taken to determine the qualities of a motor vehicle with a view to its regular use. **test-drive** *v.tr.* (*past* **-drove**; *past part.* **-driven**) drive (a vehicle) for this purpose. **test flight** a flight during which the performance of an aircraft is tested. **test-fly** *v.tr.* (**-flies**; *past* **-flew**; *past part.* **-flown**) fly (an aircraft) for this purpose. **test match** a cricket or rugby match between teams of certain countries, usu. each of a series in a tour. **test meal** a meal of specified quantity and composition, eaten to assist tests of gastric secretion. **test out** put (a theory etc.) to a practical test. **test paper 1** a minor examination paper. **2** *Chem.* a paper impregnated with a substance changing colour under known conditions. **test pilot** a pilot who test-flies aircraft. **test-tube** a thin glass tube closed at one end used for chemical tests etc. **test-tube baby** *colloq.* a baby conceived by *in vitro* fertilization. □ **testable** *adj.* **testability** /,testə'bɪlɪtɪ/ *n.* **testee** /te'stiː/ *n.* [ME f. OF f. L *testu(m)* earthen pot, collateral form of *testa* TEST[2]]

test[2] /test/ *n.* the shell or integument of some invertebrates, esp. foraminifers and tunicates. [L *testa* tile, jug, shell, etc.: cf. TEST[1]]

testa /'testə/ *n.* (*pl.* **testae** /-tiː/) *Bot.* a seed-coat. [L (as TEST[2])]

testaceous /te'steɪʃəs/ *adj.* **1** *Biol.* having a hard continuous outer covering. **2** *Bot. & Zool.* of a brick-red colour. [L *testaceus* (as TEST[2])]

Test Acts *n.pl.* various acts that made the holding of public office in Britain conditional on profession of the established religion. Such an act was imposed in Scotland in 1567 and in England in 1673: the effect was the exclusion of Catholics, Nonconformists, and non-Christians from office. In England candidates for public positions had to receive

Anglican communion, acknowledge the monarch as head of the Church of England, and repudiate the doctrine of transubstantiation. This act was repealed in 1828, although laws imposing similar conditions on university entrance remained until 1871. (See also CATHOLIC EMANCIPATION, PENAL LAWS.)

testament /ˈtestəmənt/ n. **1** a will (esp. *last will and testament*). **2** (usu. foll. by *to*) evidence, proof (*is testament to his loyalty*). **3** *Bibl.* **a** a covenant or dispensation. **b** (**Testament**) a division of the Christian Bible (see OLD TESTAMENT, NEW TESTAMENT). **c** (**Testament**) a copy of the New Testament. [ME f. L *testamentum* will (as TESTATE): in early Christian L rendering Gk *diathēkē* covenant]

testamentary /ˌtestəˈmentəri/ adj. of or by or in a will. [L *testamentarius* (as TESTAMENT)]

testate /ˈtesteɪt/ adj. & n. ● adj. having left a valid will at death. ● n. a testate person. □ **testacy** /-təsi/ n. (pl. **-ies**). [L *testatus* past part. of *testari* testify, make a will, f. *testis* witness]

testator /teˈsteɪtə(r)/ n. (fem. **testatrix** /-trɪks/) a person who has made a will, esp. one who dies testate. [ME f. AF *testatour* f. L *testator* (as TESTATE)]

Test-Ban Treaty /ˈtestbæn/ an international agreement not to test nuclear weapons in the atmosphere, in space, or under water, signed in 1963 by the US, the UK, and the USSR. Framed in order to prevent other nations from developing nuclear weapons as well as to protect the environment, the treaty was later signed by more than 100 governments, with the notable exceptions of France and China.

tester[1] /ˈtestə(r)/ n. **1** a person or thing that tests. **2** a sample of a cosmetic etc., allowing customers to try it before purchase.

tester[2] /ˈtestə(r)/ n. a canopy, esp. over a four-poster bed. [ME f. med.L *testerium, testrum, testura*, ult. f. L *testa* tile]

testes pl. of TESTIS.

testicle /ˈtestɪk(ə)l/ n. a male organ that produces spermatozoa etc., esp. one of a pair enclosed in the scrotum behind the penis of a man and most mammals. □ **testicular** /teˈstɪkjʊlə(r)/ adj. [ME f. L *testiculus* dimin. of *testis* witness (of virility)]

testiculate /teˈstɪkjʊlət/ adj. **1** having or shaped like testicles. **2** *Bot.* (esp. of an orchid) having pairs of tubers so shaped. [LL *testiculatus* (as TESTICLE)]

testify /ˈtestɪˌfaɪ/ v. (**-ies**, **-ied**) **1** intr. (of a person or thing) bear witness (*testified to the facts*). **2** intr. *Law* give evidence. **3** tr. affirm or declare (*testified his regret; testified that she had been present*). **4** tr. (of a thing) be evidence of, evince. □ **testifier** n. [ME f. L *testificari* f. *testis* witness]

testimonial /ˌtestɪˈməʊnɪəl/ n. **1** a formal letter etc. testifying to a person's character, conduct, or qualifications. **2** a gift presented to a person (esp. in public) as a mark of esteem, in acknowledgement of services, etc. [ME f. OF *testimoignal* (adj.) f. *tesmoin* or LL *testimonialis* (as TESTIMONY)]

testimony /ˈtestɪməni/ n. (pl. **-ies**) **1** *Law* an oral or written statement under oath or affirmation. **2** declaration or statement of fact. **3** evidence, demonstration (*called him in testimony; produce testimony*). **4** *Bibl.* the Ten Commandments. **5** *archaic* a solemn protest or confession. [ME f. L *testimonium* f. *testis* witness]

testis /ˈtestɪs/ n. (pl. **testes** /-tiːz/) *Anat.* & *Zool.* a testicle. [L, = witness: cf. TESTICLE]

testosterone /teˈstɒstəˌrəʊn/ n. *Biochem.* a steroid hormone that stimulates the development of male secondary sexual characteristics, produced mainly in the testes. [TESTIS + STEROL + -ONE]

testudinal /teˈstjuːdɪn(ə)l/ adj. of or shaped like a tortoise. [as TESTUDO]

testudo /teˈstjuːdəʊ/ n. (pl. **-os** or **testudines** /-dɪˌniːz/) *Rom. Hist.* **1** a screen formed by a body of troops in close array with overlapping shields. **2** a movable screen to protect besieging troops. [L *testudo -dinis*, lit. 'tortoise' (as TEST[2])]

testy /ˈtesti/ adj. (**testier**, **testiest**) irritable, touchy. □ **testily** adv. **testiness** n. [ME f. AF *testif* f. OF *teste* head (as TEST[2])]

tetanic /tɪˈtænɪk/ adj. of or such as occurs in tetanus. □ **tetanically** adv. [L *tetanicus* f. Gk *tetanikos* (as TETANUS)]

tetanus /ˈtetənəs/ n. **1** *Med.* a disease caused by the bacterium *Clostridium tetani*, marked by rigidity and spasms of the voluntary muscles. (See also TRISMUS.) **2** *Physiol.* the prolonged contraction of a muscle caused by rapidly repeated stimuli. □ **tetanize** v.tr. (also **-ise**). **tetanoid** adj. [ME f. L f. Gk *tetanos* muscular spasm f. *teinō* stretch]

tetany /ˈtetəni/ n. *Med.* a disease with intermittent muscular spasms caused by malfunction of the parathyroid glands and a consequent deficiency of calcium. [F *tétanie* (as TETANUS)]

tetchy /ˈtetʃi/ adj. (also **techy**) (**-ier**, **-iest**) peevish, irritable. □ **tetchily** adv. **tetchiness** n. [prob. f. *tecche, tache* blemish, fault f. OF *teche, tache*]

tête-à-tête /ˌteɪtɑːˈteɪt/ n., adv., & adj. ● n. **1** a private conversation or interview usu. between two persons. **2** an S-shaped sofa for two people to sit face to face. ● adv. together in private (*dined tête-à-tête*). ● adj. **1** private, confidential. **2** concerning only two persons. [F, lit. 'head-to-head']

tête-bêche /teɪtˈbeʃ/ adj. (of a postage stamp) printed upside down or sideways relative to another. [F f. *tête* head + *béchevet* double bed-head]

tether /ˈteðə(r)/ n. & v. ● n. **1** a rope etc. by which an animal is tied to confine it to the spot. **2** the extent of one's knowledge, authority, etc.; scope, limit. ● v.tr. tie (an animal) with a tether. □ **at the end of one's tether** having reached the limit of one's patience, resources, abilities, etc. [ME f. ON *tjóthr* f. Gmc]

Tethys /ˈteθɪs/ **1** *Gk Mythol.* a goddess of the sea, daughter of Uranus (Heaven) and Gaia (Earth). **2** *Astron.* satellite III of Saturn, the ninth closest to the planet, discovered by Giovanni Cassini in 1684 (diameter 1,050 km). It is probably composed mainly of ice and is heavily cratered, with one giant crater and a vast trench extending three-quarters of the satellite's circumference. **3** *Geol.* an ocean formerly separating the supercontinents of Gondwana and Laurasia, the forerunner of the present-day Mediterranean.

Tet Offensive /tet/ (in the Vietnam War) an offensive launched in January–February 1968 by the Vietcong and the North Vietnamese army. Timed to coincide with the first day of the Tet (Vietnamese New Year), it was a surprise attack on South Vietnamese cities, notably Saigon. Although repulsed after initial successes, the attack shook US confidence and hastened the withdrawal of its forces.

Tétouan /teɪˈtwɑːn/ a city in northern Morocco; pop. (1982) 199,600.

tetra /ˈtetrə/ n. a small, often brightly coloured tropical fish of the characin family, frequently kept in aquaria. [abbr. of mod.L *Tetragonopterus*, former genus name]

tetra- /ˈtetrə/ comb. form (also **tetr-** before a vowel) **1** four (*tetrapod*). **2** *Chem.* (forming names of compounds) containing four atoms or groups of a specified kind (*tetroxide*). [Gk f. *tettares* four]

tetrachord /ˈtetrəˌkɔːd/ n. *Mus.* **1** a scale-pattern of four notes, the interval between the first and last being a perfect fourth. **2** a musical instrument with four strings.

tetracyclic /ˌtetrəˈsaɪklɪk/ adj. **1** *Bot.* having four circles or whorls. **2** *Chem.* (of a compound) having a molecular structure of four fused hydrocarbon rings.

tetracycline /ˌtetrəˈsaɪkliːn, -klɪn/ n. an antibiotic with a molecule of four rings. [TETRACYCLIC + -INE[4]]

tetrad /ˈtetræd/ n. **1** a group of four. **2** the number four. [Gk *tetras -ados* (as TETRA-)]

tetradactyl /ˌtetrəˈdæktɪl/ adj. *Zool.* having four digits on each foot.

tetraethyl lead /ˌtetrəˈiːθaɪl ˌled, -ˈeθɪl/ n. a liquid added to petrol as an antiknock agent.

tetragon /ˈtetrəˌɡɒn/ n. a plane figure with four angles and four sides. [Gk *tetragōnon* quadrangle (as TETRA-, -GON)]

tetragonal /tɪˈtræɡən(ə)l/ adj. **1** of or like a tetragon. **2** *Crystallog.* (of a crystal) having three axes at right angles, two of them equal. □ **tetragonally** adv.

tetragram /ˈtetrəˌɡræm/ n. a word of four letters.

Tetragrammaton /ˌtetrəˈɡræmətɒn/ n. the Hebrew name of God written in four letters, articulated as *Yahweh* etc. [Gk (as TETRA-, *gramma, -atos* letter)]

tetragynous /tɪˈtrædʒɪnəs/ adj. *Bot.* having four pistils.

tetrahedron /ˌtetrəˈhiːdrən/ n. (pl. **tetrahedra** /-drə/ or **tetrahedrons**) a four-sided solid; a triangular pyramid. □ **tetrahedral** adj. [late Gk *tetraedron* neut. of *tetraedros* four-sided (as TETRA-, -HEDRON)]

tetrahydrocannabinol /ˌtetrəˌhaɪdrəʊˈkænəbɪnɒl, - kəˈnæbɪˌnɒl/ n. (abbr. **THC**) *Chem.* a derivative of cannabinol which is the main active principle in cannabis.

tetralogy /tɪˈtrælədʒɪ/ n. (pl. **-ies**) **1** a group of four related literary or operatic works. **2** *Gk Antiq.* a series of four dramas, three tragic and one satyric.

tetramerous /tɪˈtræmərəs/ adj. having four parts.

tetrameter /tɪˈtræmɪtə(r)/ n. *Prosody* a verse of four measures. [LL *tetrametrus* f. Gk *tetrametros* (as TETRA-, *metron* measure)]

tetrandrous /tɪˈtrændrəs/ *adj. Bot.* having four stamens.

tetraplegia /ˌtetrəˈpliːdʒə/ *n. Med.* = QUADRIPLEGIA. □ **tetraplegic** /-dʒɪk/ *adj. & n.* [mod.L (as TETRA-, Gk *plēgē* blow, strike)]

tetraploid /ˈtetrəˌplɔɪd/ *adj. & n. Biol.* ● *adj.* (of an organism or cell) having four times the haploid set of chromosomes. ● *n.* a tetraploid organism or cell.

tetrapod /ˈtetrəˌpɒd/ *n.* **1** *Zool.* an animal with four feet. **2** a structure supported by four feet radiating from a centre. □ **tetrapodous** /tɪˈtræpədəs/ *adj.* [mod.L *tetrapodus* f. Gk *tetrapous* (as TETRA-, *pous podos* foot)]

tetrapterous /tɪˈtræptərəs/ *adj. Zool.* having four wings. [mod.L *tetrapterus* f. Gk *tetrapteros* (as TETRA-, *pteron* wing)]

tetrarch /ˈtetrɑːk/ *n.* **1** *Rom. Hist.* **a** the governor of a fourth part of a country or province. **b** a subordinate ruler. **2** one of four joint rulers. □ **tetrarchy** *n.* (*pl.* **-ies**). **tetrarchate** /-ˌkeɪt/ *n.* **tetrarchical** /teˈtrɑːkɪk(ə)l/ *adj.* [ME f. LL *tetrarcha* f. L *tetrarches* f. Gk *tetrarkhēs* (as TETRA-, *arkhō* rule)]

tetrastich /ˈtetrəˌstɪk/ *n. Prosody* a group of four lines of verse. [L *tetrastichon* f. Gk (as TETRA-, *stikhon* line)]

tetrastyle /ˈtetrəˌstaɪl/ *n. & adj. Archit.* ● *n.* a building with four pillars esp. forming a portico in front or supporting a ceiling. ● *adj.* (of a building) built in this way. [L *tetrastylos* f. Gk *tetrastulos* (as TETRA-, STYLE)]

tetrasyllable /ˈtetrəˌsɪləb(ə)l/ *n.* a word of four syllables. □ **tetrasyllabic** /ˌtetrəsɪˈlæbɪk/ *adj.*

tetrathlon /tɪˈtræθlən/ *n.* an athletic contest in which each competitor takes part in the same prescribed four events, esp. riding, shooting, swimming, and running. [TETRA- + Gk *athlon* contest, after PENTATHLON]

tetratomic /ˌtetrəˈtɒmɪk/ *adj. Chem.* having four atoms (of a specified kind) in the molecule.

tetravalent /ˌtetrəˈveɪlənt/ *adj. Chem.* having a valency of four; quadrivalent.

tetrode /ˈtetrəʊd/ *n. Electronics* a thermionic valve having four electrodes. [TETRA- + Gk *hodos* way]

tetter /ˈtetə(r)/ *n. archaic* or *dial.* a pustular skin eruption, e.g. eczema. [OE *teter*: cf. OHG *zittaroh*, G dial. *Zitteroch*, Skr. *dadru*]

Teut. *abbr.* Teutonic.

Teuto- /ˈtjuːtəʊ/ *comb. form* = TEUTON.

Teuton /ˈtjuːt(ə)n/ *n.* **1** a member of a Teutonic nation, esp. a German. **2** *hist.* a member of a northern European tribe recorded from the 4th century BC and combining with others to carry out raids on NE and southern France during the Roman period until heavily defeated in 102 BC. [L *Teutones*, *Teutoni*, f. an Indo-European base meaning 'people' or 'country']

Teutonic /tjuːˈtɒnɪk/ *adj. & n.* ● *adj.* **1** relating to or characteristic of the Germanic peoples or their languages. **2** German. ● *n.* the early language usu. called Germanic. □ **Teutonicism** /-nɪˌsɪz(ə)m/ *n.* [F *teutonique* f. L *Teutonicus* (as TEUTON)]

Teutonic Knights a military and religious order of German knights, priests, and lay brothers, originally enrolled *c.*1191 as the Teutonic Knights of St Mary of Jerusalem. They took part in the Crusades from a base in Palestine until expelled from the Holy Land in 1225, whereupon they turned their attentions against Germany's neighbouring non-Christian nations of Prussia, Livonia, etc. Their conquests made them a great sovereign power, but from the 15th century they rapidly declined and were abolished by Napoleon in 1809. The order was re-established in Vienna as an honorary ecclesiastical institution in 1834 and maintains a titular existence.

Tevere see TIBER.

Tevet var. of TEBET.

Tex. *abbr.* Texas.

Texas /ˈteksəs/ a state in the southern US, on the border with Mexico, with a coastline on the Gulf of Mexico; pop. (1990) 16,986,510; capital, Austin. The area was opened up by Spanish explorers (16th–17th centuries) and formed part of Mexico until 1836, when it declared independence and became a republic. It became the 28th state of the US in 1845. □ **Texan** *adj. & n.*

Tex-Mex /ˈteksmeks/ *n. & adj.* ● *n.* **1** a style of cooking incorporating elements of the cuisines of the south-western US (esp. Texas) and Mexico. **2** the variety of Mexican Spanish spoken in Texas. ● *adj.* of or relating to the blend of Texan and Mexican language, culture, or cuisine existing or originating in the south-western US (esp. Texas). [*Texan* + *Mexican*]

text /tekst/ *n.* **1** the main body of a book as distinct from notes, appendices, pictures, etc. **2** the original words of an author or document, esp. as distinct from a paraphrase of or commentary on them. **3** a passage quoted from Scripture, esp. as the subject of a sermon. **4** a subject or theme. **5** (in *pl.*) books prescribed for study. **6** a textbook. **7** *Computing* data in textual form, esp. as stored, processed, or displayed in a word processor etc. **8** (in full **text-hand**) a fine large kind of handwriting esp. for manuscripts. □ **text editor** *Computing* a system or program allowing the user to enter and edit text. **text processing** *Computing* the manipulation of text, esp. transforming it from one format to another. □ **textless** *adj.* [ME f. ONF *tixte*, *texte* f. L *textus* tissue, literary style (in med.L = Gospel) f. L *texere* text- weave]

textbook /ˈtekstbʊk/ *n. & adj.* ● *n.* a book for use in studying, esp. a standard account of a subject. ● *attrib.adj.* **1** exemplary, accurate (cf. COPYBOOK). **2** instructively typical. □ **textbookish** *adj.*

textile /ˈtekstaɪl/ *n. & adj.* ● *n.* **1** a woven or bonded fabric; a cloth. **2** a fibre, filament, or yarn used for weaving into cloth etc. ● *adj.* **1** of or relating to textiles or weaving (*textile industry*). **2** woven (*textile fabrics*). **3** suitable for weaving (*textile materials*). [L *textilis* (as TEXT)]

textual /ˈtekstjʊəl, ˈtekstjʊ-/ *adj.* of, in, or concerning a text (*textual errors*). □ **textual criticism** the process of attempting to ascertain the correct reading of a text. □ **textually** *adv.* [ME f. med.L *textualis* (as TEXT)]

textualist /ˈtekstjʊəlɪst, ˈtekstjʊ-/ *n.* a person who adheres strictly to the letter of a text, esp. that of the Scriptures. □ **textualism** *n.*

textuality /ˌtekstjʊˈælɪti, ˌtekstjʊ-/ *n.* **1** the nature or quality of a text or discourse; the identifying quality of a text. **2** strict adherence to a text; textualism.

texture /ˈtekstʃə(r)/ *n. & v.* ● *n.* **1** the feel or appearance of a surface or substance. **2** the arrangement of threads etc. in textile fabric. **3** the arrangement of small constituent parts. **4** *Art* the representation of the structure and detail of objects. **5** *Mus.* the quality of sound formed by combining parts. **6** the quality of a piece of writing, esp. with reference to imagery, alliteration, etc. **7** distinctive nature or quality resulting from composition (*the texture of her life*). ● *v.tr.* (usu. as **textured** *adj.*) provide with a texture. □ **textural** *adj.* **texturally** *adv.* **textureless** *adj.* [ME f. L *textura* weaving (as TEXT)]

texturize /ˈtekstʃəˌraɪz/ *v.tr.* (also **-ise**) (usu. as **texturized** *adj.*) impart a particular texture to (fabrics or food).

TG *abbr.* **1** transformational-generative (grammar). **2** transformational grammar.

TGV *n.* a type of electrically powered high-speed French passenger train. Introduced in 1981 on the Paris–Lyons route, the TGV is now also operated between Paris, Le Mans, and Tours, with an additional service planned to link Paris with the Channel Tunnel. [F, abbr. of *train à grande vitesse*]

TGWU *abbr.* (in the UK) Transport and General Workers' Union.

Th *symb. Chem.* the element thorium.

Th. *abbr.* Thursday.

-th[1] /θ/ *suffix* (also **-eth** /ɪθ/) forming ordinal and fractional numbers from *four* onwards (*fourth*; *thirtieth*). [OE *-tha*, *-the*, *-otha*, *-othe*]

-th[2] /θ/ *suffix* **1** forming nouns from verbs denoting an action or process (*birth*; *growth*). **2** forming nouns of state from adjectives (*breadth*; *filth*; *length*). [OE *-thu*, *-tho*, *-th*]

-th[3] var. of -ETH[2].

Thackeray /ˈθækəri/, William Makepeace (1811–63), British novelist, born in Calcutta. He worked in London as a journalist and illustrator after leaving Cambridge University without a degree. All of his novels originally appeared in serial form; he established his reputation with *Vanity Fair* (1847–8), a vivid portrayal of early 19th-century society, satirizing upper-middle class pretensions through its central character Becky Sharp. Later novels include *Pendennis* (1848–50), *The History of Henry Esmond* (1852), and *The Virginians* (1857–9). In 1860 Thackeray became the first editor of the *Cornhill Magazine*, in which much of his later work was published.

Thaddaeus /ˈθædɪəs/ an apostle named in St Matthew's Gospel, traditionally identified with St Jude.

Thai /taɪ/ *n. & adj.* ● *n.* (*pl.* same or **Thais**) **1 a** a native or national of Thailand; a member of the largest ethnic group in Thailand. **b** a person of Thai descent. **2** the language of Thailand, a tonal language of the Sino-Tibetan language group spoken at some level by most of the country's population. ● *adj.* of or relating to Thailand or its people or language. [Thai, = free]

Thailand /'taɪlænd/ a kingdom in SE Asia; pop. (1990) 56,303,270; official language, Thai; capital, Bangkok. The country was known as Siam until 1939, when it changed its name to Thailand (literally 'land of the free'). For centuries Thais had filtered into the area and by the 13th century had established a number of principalities. A powerful kingdom emerged in the 14th century and engaged in a series of wars with its neighbour Burma. In the 19th century it lost territory in the east to France and in the south to Britain. Thailand was occupied by the Japanese in the Second World War; it supported the US in the Vietnam War, later experiencing a large influx of refugees from Cambodia, Laos, and Vietnam. Absolute monarchy was abolished in 1932, the king remaining head of state; since then Thailand has often been under military rule.

Thailand, Gulf of an inlet of the South China Sea between the Malay Peninsula to the west and Thailand and Cambodia to the east. It was formerly known as the Gulf of Siam.

thalamus /'θæləməs/ n. (pl. **thalami** /-ˌmaɪ/) **1** Anat. either of two masses of grey matter in the forebrain, serving as relay stations for sensory tracts. **2** Bot. the receptacle of a flower. **3** Gk Antiq. an inner room or women's apartment. □ **thalamic** /θə'læmɪk, 'θæləmɪk/ adj. (in senses 1 and 2). [L f. Gk thalamos]

thalassaemia /ˌθælə'siːmɪə/ n. (US **thalassemia**) Med. a hereditary haemolytic disease caused by faulty haemoglobin synthesis, widespread in various forms in Mediterranean, African, and Asian countries. [Gk thalassa (because first known around the Mediterranean) + -AEMIA]

thalassic /θə'læsɪk/ adj. of the sea or seas, esp. small or inland seas. [F thalassique f. Gk thalassa sea]

thalassotherapy /θəˌlæsəʊ'θerəpɪ/ n. a therapeutic treatment using sea water. [Gk thalassa sea + THERAPY]

thaler /'tɑːlə(r)/ n. hist. a German silver coin. [G T(h)aler: see DOLLAR]

Thales /'θeɪliːz/ (c.624–c.545 BC), Greek philosopher, mathematician, and astronomer, of Miletus. He was one of the Seven Sages listed by Plato and was held by Aristotle to be the founder of physical science; he is also credited with founding geometry. He proposed that water was the primary substance from which all things were derived, and represented the earth as floating on an underlying ocean; his cosmology had Egyptian and Semitic affinities.

Thalia /θə'laɪə/ **1** Gk & Rom. Mythol. the Muse of comedy. **2** Gk Mythol. one of the Graces. [Gk, = rich, plentiful]

thalidomide /θə'lɪdəˌmaɪd/ n. a drug formerly used as a sedative but found in 1961 to cause foetal malformation when taken by a mother early in pregnancy. □ **thalidomide baby** (or **child** etc.) a baby (or child etc.) born deformed from the effects of thalidomide. [phthalimidoglutarimide]

thalli pl. of THALLUS.

thallium /'θælɪəm/ n. a soft silvery-white metallic chemical element (atomic number 81; symbol **Tl**). (See note below.) □ **thallic** adj. **thallous** adj. [formed as THALLUS, from a green line in its spectrum]

■ Thallium was discovered spectroscopically by William Crookes in 1861. It occurs naturally in small amounts in iron pyrites, zinc blende, and other ores. The metal itself has few uses; its compounds have some specialized uses, e.g. in infrared and optical equipment. Thallium compounds are very poisonous and have been used as insecticides and rat poison.

thallophyte /'θæləˌfaɪt/ n. Bot. a plant having a thallus, e.g. alga, fungus, or lichen. [mod.L Thallophyta (as THALLUS) + -PHYTE]

thallus /'θæləs/ n. (pl. **thalli** /-laɪ/) Bot. a plant-body without vascular tissue and not differentiated into root, stem, and leaves. □ **thalloid** adj. [L f. Gk thallos green shoot f. thallō bloom]

thalweg /'tɑːlveg/ n. **1** Geog. a line where opposite slopes meet at the bottom of a valley, river, or lake. **2** Law a boundary between states along the centre of a river etc. [G f. Thal valley + Weg way]

Thames /temz/ **1** a river of southern England, flowing 338 km (210 miles) eastwards from the Cotswolds in Gloucestershire through London to the North Sea. A flood barrier across the river to protect London from high tides was completed in 1982. **2** a shipping forecast area covering the southernmost part of the North Sea, roughly as far north as the latitude of northern Norfolk.

Thammuz /'tæmʊz/ n. (also **Tammuz**) (in the Jewish calendar) the tenth month of the civil and fourth of the religious year, usually coinciding with parts of June and July. [Heb. tammūz]

than /ðən, ðæn/ conj. **1** introducing the second element in a comparison (you are older than he is; you are older than he). ¶ It is also possible to say you are older than him, with than treated as a preposition, esp. in less formal contexts. **2** introducing the second element in a statement of difference (anyone other than me). [OE thanne etc., orig. the same word as THEN]

thanage /'θeɪnɪdʒ/ n. hist. **1** the rank of thane. **2** the land granted to a thane. [ME f. AF thanage (as THANE)]

thanatology /ˌθænə'tɒlədʒɪ/ n. the scientific study of death and its associated phenomena and practices. [Gk thanatos death + -LOGY]

Thanatos /'θænəˌtɒs/ n. (in Freudian psychoanalysis) the urge for destruction or self-destruction. [Gk thanatos death]

thane /θeɪn/ n. hist. **1** in Anglo-Saxon England, a man who held land from a king or other superior by military service, ranking between ordinary freemen and hereditary nobles. **2** a man who held land from a Scottish king and ranked with an earl's son; the chief of a clan. □ **thanedom** n. [OE theg(e)n servant, soldier f. Gmc]

thank /θæŋk/ v. & n. ● v.tr. **1** express gratitude to (thanked him for the present). **2** hold responsible (you can thank yourself for that). ● n. (in pl.) **1** gratitude (expressed his heartfelt thanks). **2** an expression of gratitude (give thanks to heaven). **3** (as a formula) thank you (thanks for your help; thanks very much). □ **give thanks** say grace at a meal. **I will thank you** usu. iron. a polite formula implying reproach (I will thank you to go away). **no** (or **small**) **thanks to** despite. **thank goodness** (or **God** or **heavens** etc.) **1** colloq. an expression of relief or pleasure. **2** an expression of pious gratitude. **thank-offering** an offering made as an act of thanksgiving. **thanks to** as the (good or bad) result of (thanks to my foresight; thanks to your obstinacy). **thank you** a polite formula acknowledging a gift or service or an offer accepted or refused. **thank-you** n. colloq. **1** an instance of expressing thanks. **2** (often attrib.) an action or gift as an expression of thanks (thank-you letter). [OE thancian, thanc f. Gmc, rel. to THINK]

thankful /'θæŋkfʊl/ adj. **1** grateful, pleased. **2** (of words or acts) expressive of thanks. □ **thankfulness** n. [OE thancful (as THANK, -FUL)]

thankfully /'θæŋkfʊlɪ/ adv. **1** in a thankful manner. **2** (qualifying a whole sentence) disp. let us be thankful; fortunately (thankfully, nobody was hurt). [OE thancfullice (as THANKFUL, -LY²)]

thankless /'θæŋklɪs/ adj. **1** not expressing or feeling gratitude. **2** (of a task etc.) giving no pleasure or profit; unappreciated. **3** not deserving thanks. □ **thanklessly** adv. **thanklessness** n.

Thanksgiving /'θæŋks,gɪvɪŋ, θæŋks'gɪvɪŋ/ (also **Thanksgiving Day**) (in North America) a national holiday for giving thanks to God, the fourth Thursday in November in the US and usu. the second Monday in October in Canada. A festival of this kind was first held by the Pilgrim Fathers in 1621 in thankfulness for a successful harvest after a year of hardship. Turkey and pumpkin pie are traditionally eaten.

thanksgiving /'θæŋks,gɪvɪŋ, θæŋks'gɪvɪŋ/ n. **1** the expression of gratitude, esp. to God. **2** a form of words for this.

thar var. of TAHR.

Thar Desert /tɑː(r)/ (also known as Great Indian Desert) a desert region to the east of the River Indus, lying in the Rajasthan and Gujarat states of NW India and the Punjab and Sind regions of SE Pakistan.

that /ðæt/ pron., adj., adv., & conj. ● demonstr. pron. (pl. **those** /ðəʊz/) **1** the person or thing indicated, named, or understood, esp. when observed by the speaker or when familiar to the person addressed (I heard that; who is that in the garden?; I knew all that before; that is not fair). **2** (contrasted with this) the further or less immediate or obvious etc. of two (this bag is much heavier than that). **3** the action, behaviour, or circumstances just observed or mentioned (don't do that again). **4** Brit. (on the telephone etc.) the person spoken to (who is that?). **5** colloq. referring to a strong feeling just mentioned ('Are you glad?' 'I am that'). **6** (esp. in relative constructions) the one, the person, etc., described or specified in some way (those who have cars can take the luggage; those unfit for use; a table like that described above). **7** /ðət/ (pl. **that**) used instead of which or whom to introduce a defining clause, esp. one essential to identification (the book that you sent me; there is nothing here that matters). ¶ As a relative that usually specifies, whereas who or which need not: compare the book that you sent me is lost with the book, which I gave you, is lost. ● demonstr. adj. (pl. **those** /ðəʊz/) **1** designating the person or thing indicated, named, understood, etc. (cf. sense 1 of pron.) (look at that dog; what was that noise?; things were easier in those days). **2** contrasted with this (cf. sense 2 of pron.) (this bag is heavier than that one). **3** expressing strong feeling (shall not easily forget that day). ● adv. **1** to such a degree; so (have done that much; will that go far). **2** Brit. colloq. very (not that good). **3** /ðət/ at which, on which, etc. (at the speed that he was going he could not stop; the day that I first met her). ¶ Often omitted in this sense: the day I first met her.

● *conj.* /ðət except when stressed/ introducing a subordinate clause indicating: **1** a statement or hypothesis (*they say that he is better; there is no doubt that he meant it; the result was that the handle fell off*). **2** a purpose (*we live that we may eat*). **3** a result (*am so sleepy that I cannot keep my eyes open*). **4** a reason or cause (*it is rather that he lacks the time*). **5** a wish (*Oh, that summer were here!*). ¶ Often omitted in senses 1, 3: *they say he is better.* □ **all that** very (*not all that good*). **and all that** (or **and that** *colloq.*) and all or various things associated with or similar to what has been mentioned; and so forth. **like that 1** of that kind (*is fond of books like that*). **2** in that manner, as you are doing, as he has been doing, etc. (*wish they would not talk like that*). **3** *colloq.* without effort (*did the job like that*). **4** of that character (*she would not accept any payment — she is like that*). **that is** (or **that is to say**) a formula introducing or following an explanation of a preceding word or words. **that's** *colloq.* you are (by virtue of present or future obedience etc.) (*that's a good boy*). **that's more like it** an acknowledgement of improvement. **that's right** an expression of approval or *colloq.* assent. **that's that** a formula concluding a narrative or discussion or indicating completion of a task. **that there** *sl.* = sense 1 of *adj.* **that will do** no more is needed or desirable. [OE *thæt*, nom. & accus. sing. neut. of demonstr. pron. & adj. *se, sēo, thæt* f. Gmc; *those* f. OE *thās* pl. of *thes* THIS]

thatch /θætʃ/ *n. & v.* ● *n.* **1** a roof-covering of straw, reeds, palm-leaves, or similar material. **2** *colloq.* the hair of the head. ● *v.tr.* (also *absol.*) cover (a roof or a building) with thatch. □ **thatcher** *n.* [n. late collateral form of *thack* (now dial.) f. OE *thæc*, after v. f. OE *theccan* f. Gmc, assim. to *thack*]

Thatcher /'θætʃə(r)/, Margaret (Hilda), Baroness Thatcher of Kesteven (b.1925), British Conservative stateswoman, Prime Minister 1979–90. She became Conservative Party leader in 1975 and in 1979 was elected the country's first woman Prime Minister; she went on to become the longest-serving British Prime Minister of the 20th century. Her period in office was marked by an emphasis on monetarist policies, privatization of nationalized industries, and trade-union legislation. (See also THATCHERISM.) In international affairs, she was a strong supporter of the policies of President Reagan. She was well known for determination and resolve (she had been dubbed the 'Iron Lady' as early as 1976), especially in her handling of the Falklands War of 1982. She resigned after a leadership challenge and was created a life peer in 1992.

Thatcherism /'θætʃə,rɪz(ə)m/ *n.* the political and economic policies advocated by Margaret Thatcher as UK Prime Minister 1979–90. They included a reliance on the workings of the free market, reduction of income taxation and public spending, privatization of previously nationalized industries, and emphasis on individual responsibility and enterprise in reaction to the collectivism and corporatism of previous decades. □ **Thatcherite** *n.*

thaumatrope /'θɔːmə,trəʊp/ *n. hist.* **1** a disc or card with two different pictures on its two sides, which combine into one by the persistence of visual impressions when the disc is rapidly rotated. **2** a zoetrope. [irreg. f. Gk *thauma* marvel + *-tropos* -turning]

thaumaturge /'θɔːmə,tɜːdʒ/ *n.* a worker of miracles; a wonder-worker. □ **thaumaturgist** *n.* **thaumaturgy** *n.* **thaumaturgic** /,θɔːmə-'tɜːdʒɪk/ *adj.* **thaumaturgical** *adj.* [med.L *thaumaturgus* f. Gk *thaumatourgos* (adj.) f. *thauma -matos* marvel + *-ergos* -working]

thaw /θɔː/ *v. & n.* ● *v.* **1** *intr.* (often foll. by *out*) (of ice or snow or a frozen thing) pass into a liquid or unfrozen state. **2** *intr.* (usu. prec. by *it* as subject) (of the weather) become warm enough to melt ice etc. (*it began to thaw*). **3** *intr.* become warm enough to lose numbness etc. **4** *intr.* become less cold or stiff in manner; become genial. **5** *tr.* (often foll. by *out*) cause to thaw. **6** *tr.* make cordial or animated. ● *n.* **1** the act or an instance of thawing. **2** the warmth of weather that thaws (*a thaw has set in*). **3** *Polit.* a relaxation of control or restriction. □ **thawless** *adj.* [OE *thawian* f. WG; orig. unkn.]

THC *abbr. Chem.* tetrahydrocannabinol.

the /before a vowel ðɪ, before a consonant ðə, when stressed ðiː/ *adj. & adv.* ● *adj.* (called the definite article) **1** denoting one or more persons or things already mentioned, under discussion, implied, or familiar (*gave the man a wave; shall let the matter drop; hurt myself in the arm; went to the theatre*). **2** serving to describe as unique (*the Queen; the Thames*). **3 a** (foll. by defining adj.) which is, who are, etc. (*ignored the embarrassed Mr Smith; Edward the Seventh*). **b** (foll. by adj. used *absol.*) denoting a class described (*from the sublime to the ridiculous*). **4** best known or best entitled to the name (*with the stressed: no relation to the Kipling; this is the book on this subject*). **5** used to indicate a following defining clause or phrase (*the book that you borrowed; the best I can do for you; the bottom of a well*). **6 a** used to indicate that a singular noun represents a species, class,

etc. (*the cat loves comfort; has the novel a future?; plays the harp well*). **b** used with a noun which figuratively represents an occupation, pursuit, etc. (*went on the stage; too fond of the bottle*). **c** (foll. by the name of a unit) a, per (*5p in the pound; £5 the square metre; allow 8 minutes to the mile*). **d** *colloq.* or *archaic* designating a disease, affliction, etc. (*the measles; the toothache; the blues*). **7** (foll. by a unit of time) the present, the current (*man of the moment; questions of the day; book of the month*). **8** *Brit. colloq.* my, our (*the dog; the fridge*). **9** used before the surname of the chief of a Scottish or Irish clan (*the Macnab*). **10** *dial.* (esp. in Wales) used with a noun characterizing the occupation of the person whose name precedes (*Jones the Bread*). ● *adv.* (preceding comparatives in expressions of proportional variation) in or by that (or such a) degree; on that account (*the more the merrier; the more he gets the more he wants*). □ **all the** in the full degree to be expected (*that makes it all the worse*). **so much the** (tautologically) so much, in that degree (*so much the worse for her*). [(adj.) OE, replacing *se, sēo, thæt* (= THAT), f. Gmc: (adv.) f. OE *thȳ, thē*, instrumental case]

theandric /θiːˈændrɪk/ *adj. Theol.* **1** at once human and divine. **2** of the union, or by the joint agency, of the divine and human natures in Christ. [eccl.Gk *theandrikos* f. *theos* god + *anēr andros* man]

theanthropic /,θiːənˈθrɒpɪk/ *adj.* **1** both divine and human. **2** tending to embody deity in human form. [eccl.Gk *theanthrōpos* god-man f. *theos* god + *anthrōpos* human being]

thearchy /'θiːɑːkɪ/ *n.* (pl. **-ies**) **1** government by a god or gods. **2** a system or order of gods (*the Olympian thearchy*). [eccl.Gk *thearkhia* godhead f. *theos* god + *-arkhia* f. *arkhō* rule]

theatre /'θɪətə(r)/ *n.* (US **theater**) **1 a** a building or outdoor area for dramatic performances. **b** (in full **picture theatre**) esp. *N. Amer., Austral., & NZ* a cinema. **2 a** the writing and production of plays. **b** effective material for the stage (*makes good theatre*). **c** action with a dramatic quality; dramatic character or effect. **3** (in full **lecture theatre**) a room or hall for lectures etc. with seats in tiers. **4** *Brit.* an operating theatre. **5 a** a scene or field of action (*the theatre of war*). **b** (*attrib.*) designating weapons intermediate between tactical and strategic (*theatre nuclear missiles*). **6** a natural land-formation in a gradually rising part-circle like ancient Greek and Roman theatres. □ **theatre-goer** a frequenter of theatres. **theatre-going** frequenting theatres. **theatre sister** a nurse supervising the nursing team in an operating theatre. [ME f. OF *t(h)eatre* or f. L *theatrum* f. Gk *theatron* f. *theaomai* behold]

theatre-in-the-round /,θɪətərɪnðəˈraʊnd/ *n.* a form of theatrical presentation in which the audience is seated all round the acting area. One of the earliest forms of theatre, it was probably used for open-air performances, street theatres, and such rustic entertainments as the mummers' play. It was revived in the 20th century — beginning in the Soviet Union in the 1930s — by those who rebelled against the proscenium arch, which they felt to be a barrier between actors and audience.

Theatre of the Absurd *n.* the name given to the works of a group of dramatists, including Samuel Beckett, Eugène Ionesco, and Harold Pinter, who share the belief that human life is without meaning or purpose and that human beings cannot communicate. Such dramatists abandoned conventional dramatic form and coherent dialogue, the futility of existence being conveyed by illogical and meaningless speeches and ultimately by complete silence. The first and perhaps most characteristic play in this style was Beckett's *Waiting for Godot* (1952).

theatric /θɪˈætrɪk/ *adj. & n.* ● *adj.* = THEATRICAL. ● *n.* (in *pl.*) theatrical actions.

theatrical /θɪˈætrɪk(ə)l/ *adj. & n.* ● *adj.* **1** of or for the theatre; of acting or actors. **2** (of a manner, speech, gesture, or person) calculated for effect; showy, artificial, affected. ● *n.* **1** (in *pl.*) dramatic performances (*amateur theatricals*). **2** (in *pl.*) theatrical actions. **3** (usu. in *pl.*) a professional actor or actress. □ **theatricalism** *n.* **theatrically** *adv.* **theatricalize** *v.tr.* (also **-ise**). **theatricalization** /-,ætrɪkəlaɪˈzeɪʃ(ə)n/ *n.* **theatricality** /-,ætrɪˈkælɪtɪ/ *n.* [LL *theatricus* f. Gk *theatrikos* f. *theatron* THEATRE]

thebe /'θeɪbeɪ/ *n.* (pl. same) a monetary unit of Botswana, equal to one-hundredth of a pula. [Setswana, = shield]

Thebes /θiːbz/ **1** the Greek name for an ancient city of Upper Egypt, whose ruins are situated on the Nile about 675 km (420 miles) south of Cairo. It was the capital of ancient Egypt under the 18th dynasty (c.1550–1290 BC). Its monuments (on both banks of the Nile) were the richest in the land, with the town and major temples at Luxor and Karnak on the east bank and the necropolis, with tombs of royalty and nobles, on the west bank. It was already a tourist attraction in the 2nd century AD. **2** (Greek **Thívai** /'θivɛ/) a city in Greece, in Boeotia, north-west of Athens. Traditionally founded by Cadmus and the seat of the legendary king Oedipus, Thebes became a major military power in Greece following the defeat of the Spartans at the battle of Leuctra in 371 BC. It was destroyed by Alexander the Great in 336 BC. □ **Theban** *adj. & n.*

theca /'θi:kə/ *n.* (*pl.* **thecae** /'θi:si:/) **1** *Bot.* a part of a plant serving as a receptacle. **2** *Zool.* a case or sheath enclosing an organ or organism. □ **thecate** /-keit/ *adj.* [L f. Gk *thēkē* case]

thé dansant /tei 'dɒnsɒn/ *n.* (*pl.* *thés dansants* pronunc. same) = tea dance. [F]

thee /ði:/ *pron. objective case of* THOU[1]. [OE]

theft /θeft/ *n.* **1** the act or an instance of stealing. **2** *Law* dishonest appropriation of another's property with intent to deprive him or her of it permanently. [OE *thīefth*, *thēofth*, later *thēoft*, f. Gmc (as THIEF)]

thegn /θein/ *n. hist.* an English thane. [OE: see THANE]

theine /'θi:in, 'θi:i:n/ *n.* = CAFFEINE. [mod.L *thea* tea + -INE[4]]

their /ðeə(r)/ *poss.pron.* (*attrib.*) **1** of or belonging to them or themselves (*their house; their own business*). **2** (**Their**) (in titles) that they are (*Their Majesties*). **3** *disp.* as a third person sing. indefinite meaning 'his or her' (*has anyone lost their purse?*). [ME f. ON *their(r)a* of them, genitive pl. of *sá* THE, THAT]

theirs /ðeəz/ *poss.pron.* the one or ones belonging to or associated with them (*it is theirs; theirs are over here*). □ **of theirs** of or belonging to them (*a friend of theirs*). [ME f. THEIR]

theirselves /ðeə'selvz/ *pron. dial. & colloq.* = THEMSELVES.

theism /'θi:ɪz(ə)m/ *n.* belief in the existence of gods or a god; specifically, belief in a God who is supernaturally revealed to humankind, who created and intervenes in the universe, and who sustains a personal relation to living creatures. Cf. DEISM. □ **theist** *n.* **theistic** /θi:'ɪstɪk/ *adj.* **theistical** *adj.* **theistically** *adv.* [Gk *theos* god + -ISM]

them /ðəm, or, when stressed, ðem/ *pron. & adj.* ● *pron.* **1** objective case of THEY (*I saw them*). **2** they (*it's them again; is older than them*). ¶ Strictly, the pronoun following the verb *to be* should be in the subjective rather than the objective case, i.e. *is older than they* rather than *is older than them*. **3** *archaic* themselves (*they fell and hurt them*). ● *adj. sl.* or *dial.* those (*them bones*). [ME *theim* f. ON: see THEY]

thematic /θi'mætɪk/ *adj.* **1** of or relating to subjects or topics (*thematic database; the arrangement of the anthology is thematic*). **2** *Mus.* of melodic subjects (*thematic treatment*). **3** *Gram.* **a** of or belonging to a theme (*thematic vowel; thematic form*). **b** (of a form of a verb) having a thematic vowel. □ **thematic catalogue** *Mus.* a catalogue giving the opening themes of works as well as their names and other details. □ **thematically** *adv.* [Gk *thematikos* (as THEME)]

theme /θi:m/ *n.* **1** a subject or topic on which a person speaks, writes, or thinks. **2** *Mus.* a prominent or frequently recurring melody or group of notes in a composition. **3** *US* a school exercise, esp. an essay, on a given subject. **4** *Gram.* the stem of a noun or verb; the part to which inflections are added, esp. composed of the root and an added vowel. **5** *hist.* any of the twenty-nine provinces in the Byzantine Empire. □ **theme park** an amusement park organized round a unifying idea. **theme song** (or **tune**) **1** a recurrent melody in a musical play or film. **2** a signature tune. [ME *teme* ult. f. Gk *thema -matos* f. *tithēmi* set, place]

Themis /'θemɪs/ *Gk Mythol.* a goddess, daughter of Uranus (Heaven) and Gaia (Earth), although sometimes also identified with Gaia. In Homer she was the personification of order and justice, who convened the assembly of the gods.

Themistocles /θi'mɪstəˌkli:z/ (*c.*528–462 BC) Athenian statesman. He was instrumental in building up the Athenian fleet, which under his command defeated the Persian fleet at Salamis in 480. In the following years he lost influence, was ostracized in 470, and eventually fled to the Persians in Asia Minor, where he died.

themselves /ðəm'selvz/ *pron.* **1 a** *emphat. form* of THEY or THEM. **b** *refl. form* of THEM; (cf. HERSELF). **2** in their normal state of body or mind (*are quite themselves again*). □ **be themselves** act in their normal, unconstrained manner.

then /ðen/ *adv., adj., & n.* ● *adv.* **1** at that time; at the time in question (*was then too busy; then comes the trouble; the then existing laws*). **2 a** next, afterwards; after that (*then he told me to come in*). **b** and also (*then, there are the children to consider*). **c** after all (*it is a problem, but then that is what we are here for*). **3 a** in that case; therefore; it follows that (*then you should*

have said so*). **b** if what you say is true (*but then why did you take it?*). **c** (implying grudging or impatient concession) if you must have it so (*all right then, have it your own way*). **d** used parenthetically to resume a narrative etc. (*the policeman, then, knocked on the door*). ● *adj.* that or who was such at the time in question (*the then Duke*). ● *n.* that time (*until then*). □ **then and there** immediately and on the spot. [OE *thanne, thonne*, etc., f. Gmc, rel. to THAT, THE]

thenar /'θi:nə(r)/ *n. Anat.* the ball of muscle at the base of the thumb. [earlier = palm of the hand: mod.L f. Gk]

thence /ðens/ *adv.* (also **from thence**) *archaic* or *literary* **1** from that place or source. **2** for that reason. [ME *thannes, thennes* f. *thanne, thenne* f. OE *thanon(e)* etc. f. WG]

thenceforth /ðens'fɔ:θ/ *adv.* (also **from thenceforth**) *archaic* or *literary* from that time onward.

thenceforward /ðens'fɔ:wəd/ *adv. archaic* or *literary* thenceforth.

theo- /'θi:əʊ/ *comb. form* God or gods. [Gk f. *theos* god]

theobromine /θi:ə'brəʊmi:n/ *n. Chem.* a bitter white alkaloid obtained from cacao seeds, related to caffeine. [*Theobroma* cacao genus: mod.L f. Gk *theos* god + *brōma* food, + -INE[4]]

theocentric /θi:ə'sentrɪk/ *adj.* having God as its centre.

theocracy /θi'ɒkrəsi/ *n.* (*pl.* **-ies**) **1** a form of government by God or a god directly or through a priestly order etc. **2** (**the Theocracy**) the commonwealth of Israel from Moses to the election of Saul as king. □ **theocrat** /'θi:əˌkræt/ *n.* **theocratic** /θi:ə'krætɪk/ *adj.* **theocratically** *adv.*

theocrasy /'θi:əˌkreɪsi, θi'ɒkrəsi/ *n.* **1** the mingling of deities into one personality. **2** the union of the soul with God through contemplation (among Neoplatonists etc.). [THEO- + Gk *krasis* mingling]

Theocritus /θi'ɒkrɪtəs/ (*c.*310–*c.*250 BC), Greek poet, born in Sicily. Little is known of his life but he is thought to have lived on the island of Kos and in Alexandria as well as Sicily. He is chiefly known for his bucolic idylls, hexameter poems presenting the song-contests and love-songs of imaginary shepherds. These poems were the model for Virgil's *Eclogues* and for subsequent pastoral poetry.

theodicy /θi'ɒdɪsi/ *n.* (*pl.* **-ies**) **1** the vindication of divine providence in view of the existence of evil. **2** an instance of this. □ **theodicean** /-ˌɒdɪ'si:ən/ *adj.* [THEO- + Gk *dikē* justice]

theodolite /θi'ɒdəˌlaɪt/ *n.* a surveying-instrument for measuring horizontal and vertical angles with a rotating telescope. □ **theodolitic** /-ˌɒdə'lɪtɪk/ *adj.* [16th c. *theodelitus*, of unkn. orig.]

Theodora /θi:ə'dɔ:rə/ (*c.*500–48), Byzantine empress, wife of Justinian. She is reputed (according to Procopius) to have led a dissolute life in her early years. She later became noted for her intellect and learning and, as Justinian's closest adviser, exercised a considerable influence on political affairs and the theological questions of the time.

Theodoric /θi'ɒdərɪk/ (known as Theodoric the Great) (*c.*454–526), king of the Ostrogoths 471–526. He invaded Italy in 488 and completed its conquest in 493, establishing a kingdom with the capital at Ravenna. At its greatest extent his empire included not only the Italian mainland, but Sicily, Dalmatia, and parts of Germany.

Theodosius I /θi:ə'dəʊsɪəs/ (known as Theodosius the Great; full name Flavius Theodosius) (*c.*346–95), Roman emperor 379–95. In 379 he was proclaimed co-emperor by the Emperor Gratian (359–83) and took control of the Eastern Empire. During his reign he brought an end to war with the Visigoths by a treaty in 382. He later defeated two usurpers to the western throne, on which he installed his son in 393. A pious Christian and rigid upholder of the Nicene Creed, in 391 he banned all forms of pagan worship, probably under the influence of St Ambrose.

theogony /θi'ɒgəni/ *n.* (*pl.* **-ies**) **1** the genealogy of the gods. **2** an account of this. [THEO- + Gk *-gonia* begetting]

theologian /θi:ə'ləʊdʒɪən, -dʒən/ *n.* a person trained in theology. [ME f. OF *theologien* (as THEOLOGY)]

theological /θi:ə'lɒdʒɪk(ə)l/ *adj.* of theology. □ **theological virtues** faith, hope, and charity. □ **theologically** *adv.* [med.L *theologicalis* f. L *theologicus* f. Gk *theologikos* (as THEOLOGY)]

theology /θi'ɒlədʒi/ *n.* (*pl.* **-ies**) **1 a** the study of theistic (esp. Christian) religion. **b** a system of theistic (esp. Christian) religion. **c** the rational analysis of a religious faith. **2** a system of theoretical principles, esp. an impractical or rigid ideology. □ **theologist** *n.* **theologize** *v.tr. & intr.* (also **-ise**). [ME f. OF *theologie* f. L *theologia* f. Gk (as THEO-, -LOGY)]

theomachy /θi'ɒməki/ *n.* (*pl.* **-ies**) strife among or against the gods. [THEO- + Gk *makhē* fight]

theophany /θɪˈɒfəni/ n. (pl. **-ies**) a visible manifestation of God or a god to man.

theophoric /θɪəˈfɒrɪk/ adj. bearing the name of a god.

Theophrastus /θɪəˈfræstəs/ (c.370–c.287 BC), Greek philosopher and scientist. He was the pupil and successor of Aristotle, whose method and researches he continued, with a particular emphasis on empirical observation. The few surviving works of Theophrastus include treatises on botany and other scientific subjects, and the *Characters*, a collection of sketches of psychological types, which in post-classical times was the most influential of his works.

theophylline /θɪəˈfɪlɪn, -liːn, θɪˈɒfəlɪn, -ˌliːn/ n. Chem. an alkaloid similar to theobromine, found in tea-leaves. [irreg. f. mod.L *thea* tea + Gk *phullon* leaf + -INE⁴]

theorbo /θɪˈɔːbəʊ/ n. (pl. **-os**) a two-necked musical instrument of the lute class much used in the seventeenth century. □ **theorbist** n. [It. *tiorba*, of unkn. orig.]

theorem /ˈθɪərəm/ n. esp. Math. **1** a general proposition not self-evident but proved by a chain of reasoning; a truth established by means of accepted truths (cf. PROBLEM 4a). **2** a rule in algebra etc., esp. one expressed by symbols or formulae (*binomial theorem*). □ **theorematic** /ˌθɪərəˈmætɪk/ adj. [F *théorème* or LL *theorema* f. Gk *theōrēma* speculation, proposition f. *theōreō* look at]

theoretic /θɪəˈretɪk/ adj. & n. ● adj. = THEORETICAL. ● n. (in sing. or pl.) the theoretical part of a science etc. [LL *theoreticus* f. Gk *theōrētikos* (as THEORY)]

theoretical /θɪəˈretɪk(ə)l/ adj. **1** concerned with knowledge but not with its practical application. **2** based on theory rather than experience or practice. □ **theoretically** adv.

theoretician /ˌθɪərɪˈtɪʃ(ə)n/ n. a person concerned with the theoretical aspects of a subject.

theorist /ˈθɪərɪst/ n. a holder or inventor of a theory or theories.

theorize /ˈθɪəraɪz/ v.intr. (also **-ise**) evolve or indulge in theories. □ **theorizer** n.

theory /ˈθɪəri/ n. (pl. **-ies**) **1** a supposition or system of ideas explaining something, esp. one based on general principles independent of the particular things to be explained (cf. HYPOTHESIS 2) (*atomic theory; theory of evolution*). **2** a speculative (esp. fanciful) view (*one of my pet theories*). **3** the sphere of abstract knowledge or speculative thought (*this is all very well in theory, but how will it work in practice?*). **4** the exposition of the principles of a science etc. (*the theory of music*). **5** Math. a collection of propositions to illustrate the principles of a subject (*probability theory; theory of equations*). [LL *theoria* f. Gk *theōria* f. *theōros* spectator f. *theōreō* look at]

theosophy /θɪˈɒsəfi/ n. (pl. **-ies**) any of various philosophies professing to achieve a knowledge of God by spiritual ecstasy, direct intuition, or special individual relationships, esp. a modern movement following Hindu and Buddhist teachings and seeking universal fellowship. Founded in 1875 as the Theosophical Society by Helena Blavatsky and the American Henry Steel Olcott (1832–1907), the movement taught of the transmigration of souls and the community of humankind irrespective of race or creed; it dealt in complicated systems of psychology and cosmology and denied a personal god. In 1878 Blavatsky and Olcott transferred the headquarters of the society to India, where the movement flourished under the presidency of Annie Besant. The Theosophical Society survives today, albeit in a reduced form. □ **theosopher** n. **theosophist** n. **theosophic** /θɪəˈsɒfɪk/ adj. **theosophical** adj. **theosophically** adv. [med.L *theosophia* f. late Gk *theosophia* f. *theosophos* wise concerning God (as THEO-, *sophos* wise)]

Thera /ˈθɪərə/ (Greek **Thíra** /ˈθiːra/) a Greek island in the southern Cyclades. The island was transformed by a violent volcanic eruption in about 1500 BC; it now consists of the remaining eastern rim of the volcano, with steep cliffs rising above a lagoon in the volcano's caldera, which is encircled by some other small islands which were originally part of the main island. There are important archaeological sites, where remains of an ancient Minoan civilization have been preserved beneath the pumice and volcanic debris. The island is also called Santorini, after St Irene of Thessalonica, who died there in exile in AD 304.

therapeutic /ˌθerəˈpjuːtɪk/ adj. **1** of, for, or contributing to the cure of disease. **2** contributing to general, esp. mental, well-being (*finds walking therapeutic*). □ **therapeutical** adj. **therapeutically** adv. **therapeutist** n. [attrib. use of *therapeutic*, orig. form of THERAPEUTICS]

therapeutics /ˌθerəˈpjuːtɪks/ n.pl. (usu. treated as sing.) the branch of medicine concerned with the treatment of disease and the action of remedial agents. [F *thérapeutique* or LL *therapeutica* (pl.) f. Gk *therapeutika* neut. pl. of *therapeutikos* f. *therapeuō* wait on, cure]

therapsid /θeˈræpsɪd/ n. & adj. ● n. a fossil reptile of the order Therapsida, related to the ancestors of mammals. ● adj. of or relating to this order. [mod.L *Therapsida* f. Gk *thēr* beast + (h)*apsis -idos* arch (referring to the structure of the skull)]

therapy /ˈθerəpi/ n. (pl. **-ies**) **1** the treatment of physical or mental disorders, other than by surgery. **2 a** a particular type of such treatment. **b** psychotherapy. □ **therapist** n. [mod.L *therapia* f. Gk *therapeia* healing]

Theravada /ˌteərəˈvɑːdə/ n. (also **Theravada Buddhism**) one of the two major Buddhist traditions (the other being Mahayana), an orthodox, ancient school that developed from Hinayana Buddhism. It is practised today chiefly in Sri Lanka, Burma (Myanmar), Thailand, Cambodia, and Laos. [Pali *theravāda*, lit. 'doctrine of the elders']

there /ðeə(r)/ adv., n., & int. ● adv. **1** in, at, or to that place or position (*lived there for some years; goes there every day*). **2** at that point (in speech, performance, writing, etc.) (*there he stopped*). **3** in that respect (*I agree with you there*). **4** used for emphasis in calling attention (*you there!; there goes the bell*). **5** used to indicate the fact or existence of something (*there is a house on the corner*). ● n. that place (*lives somewhere near there*). ● int. **1** expressing confirmation, triumph, dismay, etc. (*there! what did I tell you?*). **2** used to soothe a child etc. (*there, there, never mind*). □ **have been there before** colloq. know all about it. **so there** colloq. that is my final decision (whether you like it or not). **there and then** immediately and on the spot. **there it is 1** that is the trouble. **2** nothing can be done about it. **there's** colloq. you are (by virtue of present or future obedience etc.) (*there's a dear*). **there you are** (or **go**) colloq. **1** this is what you wanted etc. **2** expressing confirmation, resignation, etc. [OE *thær, thēr* f. Gmc, rel. to THAT, THE]

thereabouts /ˈðeərəˌbaʊts/ adv. (also **thereabout**) **1** near that place (*ought to be somewhere thereabouts*). **2** near that number, quantity, etc. (*two litres or thereabouts*).

thereafter /ðeərˈɑːftə(r)/ adv. formal after that.

thereanent /ˌðeərəˈnent/ adv. Sc. about that matter.

thereat /ðeərˈæt/ adv. archaic **1** at that place. **2** on that account. **3** after that.

thereby /ðeəˈbaɪ/ adv. by that means, as a result of that. □ **thereby hangs a tale** much could be said about that.

therefor /ðeəˈfɔː(r)/ adv. archaic for that object or purpose.

therefore /ˈðeəfɔː(r)/ adv. for that reason; accordingly, consequently.

therefrom /ðeəˈfrɒm/ adv. archaic from that or it.

therein /ðeərˈɪn/ adv. formal **1** in that place etc. **2** in that respect.

thereinafter /ˌðeərɪnˈɑːftə(r)/ adv. formal later in the same document etc.

thereinbefore /ˌðeərɪnbɪˈfɔː(r)/ adv. formal earlier in the same document etc.

thereinto /ðeərˈɪntu:/ adv. archaic into that place.

thereof /ðeərˈɒv/ adv. formal of that or it.

thereon /ðeərˈɒn/ adv. archaic on that or it (of motion or position).

thereout /ðeərˈaʊt/ adv. archaic out of that, from that source.

Theresa, Mother see TERESA, MOTHER.

Thérèse of Lisieux, St see TERESA OF LISIEUX, ST.

therethrough /ðeəˈθru:/ adv. archaic through that.

thereto /ðeəˈtu:/ adv. archaic **1** to that or it. **2** in addition, to boot.

theretofore /ˌðeətuˈfɔː(r)/ adv. formal before that time.

thereunder /ðeərˈʌndə(r)/ adv. formal **1** in accordance with that. **2** stated below that (in the document etc.).

thereunto /ðeərˈʌntu:/ adv. archaic to that or it.

thereupon /ˌðeərəˈpɒn/ adv. **1** in consequence of that. **2** soon or immediately after that. **3** archaic upon that (of motion or position).

therewith /ðeəˈwɪð/ adv. archaic **1** with that. **2** soon or immediately after that.

therewithal /ˌðeəwɪˈðɔːl/ adv. archaic in addition, besides.

theriac /ˈθɪərɪˌæk/ n. archaic an antidote to the bites of poisonous animals, esp. snakes. [L *theriaca* f. Gk *thēriakē* antidote, fem. of *thēriakos* f. *thēr* wild beast]

therianthropic /ˌθɪərɪænˈθrɒpɪk/ adj. of or worshipping beings represented in combined human and animal forms. [Gk *thērion* dimin. of *thēr* wild beast + *anthrōpos* human being]

theriomorphic /ˌθɪərɪəˈmɔːfɪk/ *adj.* (esp. of a deity) having an animal form. [as THERIANTHROPIC + Gk *morphē* form]

therm /θɜːm/ *n.* a unit of heat, esp. as the former statutory unit of gas supplied in the UK equivalent to 100,000 British thermal units or 1.055 × 10⁸ joules. [Gk *thermē* heat]

thermae /ˈθɜːmiː/ *n.pl. Gk & Rom. Antiq.* public baths. [L f. Gk *thermai* (pl.) (as THERM)]

thermal /ˈθɜːm(ə)l/ *adj. & n.* ● *adj.* **1** of, for, or producing heat. **2** promoting the retention of heat (*thermal underwear*). ● *n.* **1** a rising current of heated air (used by gliders, balloons, and birds to gain height). **2** (in *pl.*) thermal underwear. □ **thermal capacity** the number of heat units needed to raise the temperature of a body by one degree. **thermal imaging** the technique of using the heat given off by an object or substance to produce an image of it or to locate it. **thermal neutron** *Physics* a neutron in thermal equilibrium with its surroundings. **thermal printer** a printer in which fine heated pins form characters on heat-sensitive paper. **thermal reactor** a nuclear reactor using thermal neutrons. **thermal springs** springs of naturally hot water. **thermal unit** a unit for measuring heat (see *British thermal unit*). □ **thermally** *adv.* **thermalize** *v.tr. & intr.* (also **-ise**). **thermalization** /ˌθɜːməlaɪˈzeɪʃ(ə)n/ *n.* [F (as THERM)]

thermic /ˈθɜːmɪk/ *adj.* of or relating to heat.

thermidor see *lobster thermidor*.

thermion /ˈθɜːmɪˌɒn/ *n. Physics* an ion or electron emitted by a substance at high temperature. [THERMO- + ION]

thermionic /ˌθɜːmɪˈɒnɪk/ *adj. Physics* of or relating to electrons emitted from a substance at very high temperature. □ **thermionic emission** the emission of electrons from a heated source. **thermionic valve** (US **tube**) a device consisting of a sealed evacuated tube containing two or more electrodes, one of which is heated to produce a flow of electrons in one direction. (*See note below.*)

■ The diode valve (which has two electrodes) functions as a rectifier and was so used by Sir John Ambrose Fleming in 1904. The triode valve (with two electrodes and a grid) was invented in 1906 by Lee De Forest and functioned as an amplifier — a weak signal received in the grid produced a stronger signal in the anode circuit. The thermionic valve was formerly used in all electronic equipment such as radio, radar, and computers, but it has been largely replaced by the transistor.

thermionics /ˌθɜːmɪˈɒnɪks/ *n.pl.* (treated as *sing.*) the branch of science and technology concerned with thermionic emission.

thermistor /θɜːˈmɪstə(r)/ *n. Electr.* a resistor whose resistance is greatly reduced by heating, used for measurement and control. [*thermal resistor*]

thermite /ˈθɜːmaɪt/ *n.* (also **thermit** /-mɪt/) a mixture of finely powdered aluminium and iron oxide that produces a very high temperature on combustion (used in welding and for incendiary bombs). [G *Thermit* (as THERMO-, -ITE¹)]

thermo- /ˈθɜːməʊ/ *comb. form* denoting heat. [Gk f. *thermos* hot, *thermē* heat]

thermochemistry /ˌθɜːməʊˈkemɪstrɪ/ *n.* the branch of chemistry dealing with the quantities of heat evolved or absorbed during chemical reactions. □ **thermochemical** *adj.*

thermocline /ˈθɜːməʊˌklaɪn/ *n.* **1** a temperature gradient, esp. an abrupt one in a large body of water. **2** a layer of water in the sea or a lake, marked by an abrupt temperature change. [THERMO- + Gk *klinō* to slope]

thermocouple /ˈθɜːməʊˌkʌp(ə)l/ *n.* a thermoelectric device for measuring temperature, consisting of two wires of different metals joined at two points to form a loop. A voltage is developed in the loop proportional to the amount by which the temperature of one junction differs from that of the other junction.

thermodynamics /ˌθɜːməʊdaɪˈnæmɪks/ *n.pl.* (usu. treated as *sing.*) the science of the relations between heat and other (mechanical, electrical, etc.) forms of energy, and, by extension, of the relationships and interconvertibility of all forms of energy. (*See note below.*) □ **thermodynamic** *adj.* **thermodynamical** *adj.* **thermodynamically** *adv.* **thermodynamicist** /-mɪsɪst/ *n.*

■ The laws of thermodynamics describe the general direction of physical change in the universe. The historical development of the subject is complicated; its origins were early 19th-century studies of steam engines by Nicolas Carnot and others, and its main laws were introduced (the first two in their earliest form by Rudolf Clausius in 1850) before heat was understood in terms of the random motion of atoms and molecules. The first law of thermodynamics states the equivalence of heat and work and reaffirms the principle of conservation of energy. The second law states that heat does not of itself pass from a cooler to a hotter body. Another, equivalent, formulation of the second law is that the entropy of a closed system can only increase, that is, the energy in the system will inevitably tend to become distributed in the most probable, i.e. disordered, pattern (see also ENTROPY). The third law (also called Nernst's heat theorem) states that it is impossible to reduce the temperature of a system to absolute zero in a finite number of operations. Implicit in all this is what is sometimes called the zeroth law, that thermodynamic equilibrium is manifested by equality of temperature. Among areas in which the principles and methods of thermodynamics are of great practical importance are the study and design of engines, and the rules governing the direction of chemical and biochemical reactions.

thermoelectric /ˌθɜːməʊɪˈlektrɪk/ *adj.* producing electricity by a difference of temperatures. □ **thermoelectrically** *adv.* **thermoelectricity** /-ˌɪlekˈtrɪsɪtɪ, -ˌel-/ *n.*

thermogenesis /ˌθɜːməʊˈdʒenɪsɪs/ *n.* the production of heat, esp. in a human or animal body.

thermogram /ˈθɜːməˌgræm/ *n.* a record made by a thermograph.

thermograph /ˈθɜːməˌgrɑːf/ *n.* **1** an instrument that gives a continuous record of temperature. **2** an apparatus used to obtain an image produced by infrared radiation from a human or animal body. □ **thermographic** /ˌθɜːməˈgræfɪk/ *adj.*

thermography /θɜːˈmɒgrəfɪ/ *n. Med.* the taking or use of infrared thermograms, esp. to detect tumours.

thermolabile /ˌθɜːməʊˈleɪbaɪl, -bɪl/ *adj.* (of a substance) unstable when heated.

thermoluminescence /ˌθɜːməʊˌluːmɪˈnes(ə)ns/ *n.* the property of becoming luminescent when pretreated and subjected to high temperatures, used as a means of dating ancient artefacts. □ **thermoluminescent** *adj.*

thermolysis /θɜːˈmɒlɪsɪs/ *n. Chem.* decomposition by the action of heat. □ **thermolytic** /ˌθɜːməˈlɪtɪk/ *adj.*

thermometer /θəˈmɒmɪtə(r)/ *n.* an instrument for measuring temperature by means of a substance whose expansion and contraction under different degrees of heat and cold are capable of accurate measurement. (*See note below.*) □ **thermometry** *n.* **thermometric** /ˌθɜːməˈmetrɪk/ *adj.* **thermometrical** *adj.* [F *thermomètre* or mod.L *thermometrum* (as THERMO-, -METER)]

■ The earliest form was an air-thermometer invented by Galileo before 1597; alcohol thermometers were used c.1650. The idea of a fixed zero (originally the freezing-point of water) was introduced by Robert Hooke in 1665 and developed by Gabriel Fahrenheit, who manufactured mercury thermometers from c.1720. The most familiar type of thermometer consists of a slender hermetically sealed glass tube with a fine bore and external graduations, having a bulb at the lower end filled with mercury or alcohol that rises as a column in the tube on expansion. Thermometers were first introduced into clinical medicine in the UK 1863–4 by William Aitken (1825–92).

thermonuclear /ˌθɜːməʊˈnjuːklɪə(r)/ *adj.* **1** relating to or using nuclear reactions, e.g. fusion, that occur only at very high temperatures. **2** of, relating to, or involving weapons in which explosive force is produced by thermonuclear reactions (*thermonuclear war*).

thermophile /ˈθɜːməʊˌfaɪl/ *n. & adj.* (also **thermophil** /-fɪl/) ● *n.* a bacterium etc. growing optimally at high temperatures. ● *adj.* of or being a thermophile. □ **thermophilic** /ˌθɜːməʊˈfɪlɪk/ *adj.*

thermopile /ˈθɜːməʊˌpaɪl/ *n.* a set of thermocouples esp. arranged for measuring small quantities of radiant heat.

thermoplastic /ˌθɜːməʊˈplæstɪk/ *adj. & n.* ● *adj.* (of a substance) that becomes plastic on heating and hardens on cooling, and is able to repeat these processes. ● *n.* a thermoplastic substance.

Thermopylae /θɜːˈmɒpɪˌliː/ a pass between the mountains and the sea in Greece, about 200 km (120 miles) north-west of Athens, originally narrow but now much widened by the recession of the sea. In 480 BC it was the scene of the heroic defence by 6,000 Greeks, under the command of Leonidas (died c.480 BC), against the Persian army of Xerxes I. The defenders included 300 Spartans who fought to the death to delay the Persian advance. The pass was the traditional invasion route from northern Greece and was subsequently used by the Gauls in 279 BC and by Cato the Elder in 191 BC.

Thermos /'θɜːməs/ n. (in full **Thermos flask**) propr. a vacuum flask. [Gk (as THERMO-)]

thermosetting /'θɜːməʊˌsetɪŋ/ adj. (of plastics) setting permanently when heated. □ **thermoset** adj.

thermosphere /'θɜːməˌsfɪə(r)/ n. the region of the atmosphere above the mesopause and below the height at which it ceases to have the properties of a continuous medium, characterized throughout by an increase in temperature with height.

thermostable /ˌθɜːməʊˈsteɪb(ə)l/ adj. (of a substance) stable when heated.

thermostat /'θɜːməˌstæt/ n. a device that automatically regulates temperature, or that activates a device when the temperature reaches a certain point. □ **thermostatic** /ˌθɜːməˈstætɪk/ adj. **thermostatically** adv. [THERMO- + Gk statos standing]

thermotaxis /ˌθɜːməʊˈtæksɪs/ n. Biol. movement of an organism towards or away from a source of heat. □ **thermotactic** adj. **thermotaxic** adj.

thermotropism /ˌθɜːməʊˈtrəʊpɪz(ə)m/ n. the growing or bending of a plant towards or away from a source of heat. □ **thermotropic** /-ˈtrəʊpɪk, -ˈtrɒpɪk/ adj.

theropod /'θerəˌpɒd/ n. & adj. ● n. a saurischian dinosaur of the group Theropoda, comprising mainly bipedal carnivores and including tyrannosaurs and the possible ancestors of birds. ● adj. of or relating to this group. [mod.L f. Gk thēr wild beast + pous pod- foot]

thesaurus /θɪˈsɔːrəs/ n. (pl. **thesauri** /-raɪ/ or **thesauruses**) **1** a book that lists words in groups of synonyms and related concepts. **2** a dictionary or encyclopedia. [L f. Gk thēsauros treasure]

these pl. of THIS.

Theseus /'θiːsɪəs/ Gk Mythol. the son of Poseidon (or, in another account, of Aegeus, king of Athens) and husband of Phaedra. The legendary hero of Athens, he slew the Cretan Minotaur with the help of Ariadne and was successful in numerous other exploits.

thesis /'θiːsɪs/ n. (pl. **theses** /-siːz/) **1** a proposition to be maintained or proved. **2** a dissertation, esp. by a candidate for a degree. **3** (also /'θesɪs/) an unstressed syllable or part of a metrical foot in Greek or Latin verse (opp. ARSIS). [ME f. LL f. Gk, = putting, placing, a proposition etc. f. the- root of tithēmi to place]

thesp /θesp/ n. colloq. a thespian. [abbr.]

thespian /'θespɪən/ adj. & n. ● adj. of or relating to tragedy or drama. ● n. an actor or actress. [THESPIS]

Thespis /'θespɪs/ (6th century BC), Greek dramatic poet. He is regarded as the founder of Greek tragedy, having been named by Aristotle as the originator of the role of the actor in addition to the traditional chorus.

Thess. abbr. Epistle to the Thessalonians (New Testament).

Thessalonian /ˌθesəˈləʊnɪən/ adj. & n. hist. ● adj. of ancient Thessalonica (Thessaloníki). ● n. a native of ancient Thessalonica.

Thessalonians, Epistle to the either of two books of the New Testament, the earliest letters of St Paul, written from Corinth to the new Church at Thessalonica.

Thessaloníki /ˌθesələˈniːkɪ/ (also **Salonica** /səˈlɒnɪkə/, Latin **Thessalonica** /ˌθesəˈlɒnɪkə, -ləˈnaɪkə/) a seaport in NE Greece; pop. (1991) 378,000. Founded in 315 BC, it became the capital of the Roman province of Macedonia and an important city of Byzantium. It fell to the Turks in 1430, remaining a part of the Ottoman Empire until 1912. Now a major port and the second largest city in Greece, it is the capital of the present-day Greek region of Macedonia.

Thessaly /'θesəlɪ/ (Greek **Thessalía** /ˌθesaˈlia/) a region of NE Greece. □ **Thessalian** /θeˈseɪlɪən/ adj. & n.

theta /'θiːtə/ n. the eighth letter of the Greek alphabet (Θ, θ). [Gk]

Thetis /'θetɪs/ Gk Mythol. a sea-nymph, mother of Achilles.

theurgy /'θiːɜːdʒɪ/ n. **1 a** a supernatural or divine agency esp. in human affairs. **b** the art of securing this. **2** a system of white magic practised by the early Neoplatonists. □ **theurgist** n. **theurgic** /θiːˈɜːdʒɪk/ adj. **theurgical** adj. [LL theurgia f. Gk theourgia f. theos god + -ergos working]

thew /θjuː/ n. (often in pl.) literary **1** muscular strength. **2** mental or moral vigour. [OE thēaw usage, conduct, of unkn. orig.]

they /ðeɪ/ pron. (obj. **them**; poss. **their**, **theirs**) **1** the people, animals, or things previously named or in question (pl. of HE, SHE, IT[1]). **2** people in general (they say we are wrong). **3** those in authority (they have raised the fees). **4** disp. as a third person sing. indefinite pronoun meaning 'he or she' (anyone can come if they want to). [ME thei, obj. theim, f. ON their nom. pl. masc., theim dat. pl. of sá THE that]

they'd /ðeɪd/ contr. **1** they had. **2** they would.

they'll /ðeɪl/ contr. **1** they will. **2** they shall.

they're /ðeə(r)/ contr. they are.

they've /ðeɪv/ contr. they have.

THI abbr. temperature-humidity index.

thiamine /'θaɪəmɪn, -ˌmiːn/ n. (also **thiamin** /-mɪn/) a vitamin of the B complex, found in unrefined cereals, beans, and liver, a deficiency of which causes beriberi. Also called vitamin B₁, aneurin. [THIO- + amin from VITAMIN]

thick /θɪk/ adj., n., & adv. ● adj. **1 a** of great or specified extent between opposite surfaces (a thick wall; a wall two metres thick). **b** of large diameter (a thick rope). **2 a** (of a line etc.) broad; not fine. **b** (of script or type, etc.) consisting of thick lines. **3 a** arranged closely; crowded together; dense. **b** numerous. **c** bushy, luxuriant (thick hair; thick growth). **4** (usu. foll. by with) densely covered or filled (air thick with snow). **5 a** firm in consistency; containing much solid matter; viscous (a thick paste; thick soup). **b** made of thick material (a thick coat). **6 a** muddy, cloudy; impenetrable by sight (thick darkness). **b** (of one's head) suffering from a headache, hangover, etc. **7** colloq. (of a person) stupid, dull. **8 a** (of a voice) indistinct. **b** (of an accent) pronounced, exaggerated. **9** colloq. intimate or very friendly (esp. thick as thieves). ● n. a thick part of anything. ● adv. thickly (snow was falling thick; blows rained down thick and fast). □ **a bit thick** Brit. colloq. unreasonable or intolerable. **in the thick of 1** at the busiest part of. **2** heavily occupied with. **thick ear** Brit. sl. the external ear swollen as a result of a blow (esp. give a person a thick ear). **thick-knee** = stone curlew. **thick-skinned** not sensitive to reproach or criticism. **thick-skulled** (or **-witted**) stupid, dull; slow to learn. **through thick and thin** under all conditions; in spite of all difficulties. □ **thickish** adj. **thickly** adv. [OE thicce (adj. & adv.) f. Gmc]

thicken /'θɪkən/ v. **1** tr. & intr. make or become thick or thicker. **2** intr. become more complicated (the plot thickens). □ **thickener** n.

thickening /'θɪkənɪŋ/ n. **1** the process of becoming thick or thicker. **2** a substance used to thicken liquid. **3** a thickened part.

thicket /'θɪkɪt/ n. a tangle of shrubs or trees. [OE thiccet (as THICK, -ET[1])]

thickhead /'θɪkhed/ n. **1** colloq. a stupid person; a blockhead. **2** an Asian or Australasian perching bird of the family Pachycephalidae, with a large head (cf. WHISTLER 2a). □ **thickheaded** /θɪkˈhedɪd/ adj. **thickheadedness** n.

thickness /'θɪknɪs/ n. **1** the state of being thick. **2** the extent to which a thing is thick. **3** a layer of material of a certain thickness (three thicknesses of cardboard). **4** a part that is thick or lies between opposite surfaces (steps cut in the thickness of the wall). [OE thicness (as THICK, -NESS)]

thicko /'θɪkəʊ/ n. (pl. **-os**) colloq. an unintelligent person. [THICK + -O]

thickset /'θɪkˈset/ adj. & n. ● adj. **1** heavily or solidly built. **2** set or growing close together. ● n. a thicket.

thief /θiːf/ n. (pl. **thieves** /θiːvz/) a person who steals, esp. secretly and without violence. [OE thēof f. Gmc]

thieve /θiːv/ v. **1** intr. be a thief. **2** tr. steal (a thing). [OE thēofian (as THIEF)]

thievery /'θiːvərɪ/ n. the act or practice of stealing.

thieves pl. of THIEF.

thievish /'θiːvɪʃ/ adj. given to stealing; dishonest. □ **thievishly** adv. **thievishness** n.

thigh /θaɪ/ n. **1** the part of the human leg between the hip and the knee. **2** a corresponding part in other animals. □ **thigh-bone** = FEMUR. **thigh-slapper** colloq. an exceptionally funny joke etc. □ **-thighed** adj. (in comb.). [OE thēh, thēoh, thīoh, OHG dioh, ON thjó f. Gmc]

thigmotropism /ˌθɪgməˈtrəʊpɪz(ə)m/ n. Biol. the movement of a part or the whole of an organism in response to a touch stimulus. □ **thigmotropic** /-ˈtrəʊpɪk, -ˈtrɒpɪk/ adj. [Gk thigma touch + TROPISM]

thill /θɪl/ n. a shaft of a cart or carriage, esp. one of a pair. [ME: orig. unkn.]

thill-horse /'θɪlhɔːs/ n. (also **thiller** /'θɪlə(r)/) a horse put between thills.

thimble /'θɪmb(ə)l/ n. **1** a metal or plastic cap, usu. with a closed end, worn to protect the finger and push the needle in sewing. **2** Mech. a short metal tube or ferrule etc. **3** Naut. a metal ring concave on the outside and fitting in a loop of spliced rope to prevent chafing. [OE thȳmel (as THUMB, -LE[1])]

thimbleful /ˈθɪmb(ə)lˌfʊl/ n. (pl. **-fuls**) a small quantity, esp. of liquid to drink.

thimblerig /ˈθɪmb(ə)lˌrɪg/ n. a game often involving sleight of hand, in which three inverted thimbles or cups are moved about, contestants having to spot which is the one with a pea or other object beneath. □ **thimblerigger** n. [THIMBLE + RIG² in sense 'trick, dodge']

Thimphu /ˈtɪmpuː, ˈθɪm-/ (also **Thimbu** /ˈtɪmbuː, ˈθɪm-/) the capital of Bhutan, in the Himalayas at an altitude of 2,450 m (8,000 ft); pop. (1991) 27,000.

thin /θɪn/ adj., adv., & v. ● adj. (**thinner**, **thinnest**) **1** having the opposite surfaces close together; of small thickness or diameter. **2 a** (of a line) narrow or fine. **b** (of a script or type etc.) consisting of thin lines. **3** made of thin material (a thin dress). **4** lean; not plump. **5 a** not dense or copious (thin hair; a thin haze). **b** not full or closely packed (a thin audience). **6** of slight consistency (a thin paste). **7** weak; lacking an important ingredient (thin blood; a thin voice). **8** (of an excuse, argument, disguise, etc.) flimsy or transparent. ● adv. thinly (cut the bread very thin). ● v. (**thinned**, **thinning**) **1** tr. & intr. make or become thin or thinner. **2** tr. & intr. (often foll. by out) reduce; make or become less dense or crowded or numerous. **3** tr. (often foll. by out) remove some of a crop of (seedlings, saplings, etc.) or some young fruit from (a vine or tree) to improve the growth of the rest. □ **have a thin time** colloq. have a wretched or uncomfortable time. **on thin ice** see ICE. **thin air** a state of invisibility or non-existence (vanished into thin air). **thin end of the wedge** see WEDGE¹. **thin on the ground** see GROUND¹. **thin on top** balding. **thin-skinned** sensitive to reproach or criticism; easily upset. □ **thinly** adv. **thinnish** adj. **thinness** /ˈθɪnnɪs/ n. [OE thynne f. Gmc]

thine /ðaɪn/ poss.pron. archaic or dial. **1** (predic. or absol.) of or belonging to you. **2** (attrib. before a vowel) = THY. [OE thīn f. Gmc]

thing /θɪŋ/ n. **1** a material or non-material entity, idea, action, etc., that is or may be thought about or perceived. **2** an inanimate material object (take that thing away). **3** an unspecified object or item (have a few things to buy). **4** an act, idea, or utterance (a silly thing to do). **5** an event (an unfortunate thing to happen). **6** a quality (patience is a useful thing). **7** (with ref. to a person) expressing pity, contempt, or affection (poor thing!; a dear old thing). **8** a specimen or type of something (quarks are an important thing in physics). **9** colloq. **a** one's special interest or concern (not my thing at all). **b** an obsession, fear, or prejudice (spiders are a thing of mine). **10** colloq. something remarkable (now there's a thing!). **11** (prec. by the) colloq. **a** what is conventionally proper or fashionable (the done thing). **b** what is needed or required (your suggestion was just the thing). **c** what is to be considered (the thing is, shall we go or not?). **d** what is important (the thing about them is their reliability). **12** (in pl.) personal belongings or clothing (where have I left my things?). **13** (in pl.) equipment (painting things). **14** (in pl.) affairs in general (not in the nature of things). **15** (in pl.) circumstances or conditions (things look good). **16** (in pl. with a following adjective) all that is so describable (all things Greek). **17** (in pl.) Law property. □ **do one's own thing** colloq. pursue one's own interests or inclinations. **do things to** colloq. affect remarkably. **have a thing about** colloq. be obsessed, fearful, or prejudiced about. **make a thing of** colloq. **1** regard as essential. **2** cause a fuss about. **one** (or **just one**) **of those things** colloq. something unavoidable or to be accepted. [OE f. Gmc]

thingummy /ˈθɪŋəmɪ/ n. (also **thingamy**, **thingumabob** /-məˌbɒb/, **thingumajig** /-məˌdʒɪg/) (pl. **-ies**) colloq. a person or thing whose name one has forgotten or does not know or does not wish to mention. [THING + meaningless suffix]

thingy /ˈθɪŋɪ/ n. (pl. **-ies**) = THINGUMMY.

think /θɪŋk/ v. & n. ● v. (past and past part. **thought** /θɔːt/) **1** tr. (foll. by that + clause) be of the opinion (we think that they will come). **2** tr. (foll. by that + clause or to + infin.) judge or consider (is thought to be a fraud). **3** intr. exercise the mind positively with one's ideas etc. (let me think for a moment). **4** intr. (foll. by of or about) **a** consider; be or become mentally aware of (think about you constantly). **b** form or entertain the idea of; imagine to oneself (couldn't think of such a thing). **c** choose mentally; hit upon (think of a number). **d** form or have an opinion of (what do you think of them?). **5** tr. have a half-formed intention (I think I'll stay). **6** tr. form a conception of (cannot think how you do it). **7** tr. reduce to a specified condition by thinking (cannot think away a toothache). **8** tr. recognize the presence or existence of (the child thought no harm). **9** tr. (foll. by to + infin.) intend or expect (thinks to deceive us). **10** tr. (foll. by to + infin.) remember (did not think to lock the door). ● n. colloq. an act of thinking (must have a think about that). □ **think again** revise one's plans or opinions. **think aloud** utter one's thoughts as soon as they occur. **think back to** recall (a past event or time). **think better of** change one's mind about (an intention) after reconsideration. **think big** see

BIG. **think fit** see FIT¹. **think for oneself** have an independent mind or attitude. **think little** (or **nothing**) **of** consider to be insignificant or unremarkable. **think much** (or **highly**) **of** have a high opinion of. **think on** (or **upon**) archaic think of or about. **think out 1** consider carefully. **2** produce (an idea etc.) by thinking. **think over** reflect upon in order to reach a decision. **think through** reflect fully upon (a problem etc.). **think twice** use careful consideration, avoid hasty action, etc. **think up** colloq. devise; produce by thought. □ **thinkable** adj. [OE thencan thōhte gethōht f. Gmc]

thinker /ˈθɪŋkə(r)/ n. **1** a person who thinks, esp. in a specified way (an original thinker). **2** a person with a skilled or powerful mind.

thinking /ˈθɪŋkɪŋ/ adj. & n. ● adj. using thought or rational judgement. ● n. **1** opinion or judgement. **2** thought; train of thought. □ **put on one's thinking cap** colloq. meditate on a problem.

think-tank /ˈθɪŋktæŋk/ n. a body of experts providing advice and ideas on specific national and commercial problems.

thinner /ˈθɪnə(r)/ n. a volatile liquid used to dilute paint etc.

thio- /ˈθaɪəʊ/ comb. form sulphur, esp. replacing oxygen in compounds (thio-acid). [Gk theion sulphur]

thiol /ˈθaɪɒl/ n. Chem. any organic compound containing an alcohol-like group but with sulphur in place of oxygen. [THIO- + -OL]

thiopentone /ˌθaɪəˈpentəʊn/ n. Pharm. a sulphur-containing barbiturate drug related to pentobarbitone, esp. the sodium salt used as a general anaesthetic and a hypnotic, and (reputedly) as a truth drug. Also called Pentothal. [THIO- + PENTOBARBITONE]

thiosulphate /ˌθaɪəˈsʌlfeɪt/ n. Chem. a sulphate in which one oxygen atom is replaced by sulphur.

thiourea /ˌθaɪəˈjʊərɪə/ n. a crystalline compound used in photography and the manufacture of synthetic resins. [THIO- + UREA]

Thíra see THERA.

third /θɜːd/ n. & adj. ● n. **1** the position in a sequence corresponding to that of the number 3 in the sequence 1–3. **2** something occupying this position. **3** each of three equal parts of a thing. **4** = third gear. **5** Mus. **a** an interval or chord spanning three consecutive notes in the diatonic scale (e.g. C to E). **b** a note separated from another by this interval. **6 a** a place in the third class in an examination. **b** a person having this. ● adj. that is the third. □ **third age** the period in life of active retirement. **third-best** of third quality. ● n. a thing in this category. **third class** the third-best group or category, esp. of hotel and train accommodation. **third-class** adj. **1** belonging to or travelling by the third class. **2** of lower quality; inferior. ● adv. by the third class (travels third-class). **third degree** long and severe questioning esp. by police to obtain information or a confession. **third-degree** adj. Med. denoting burns of the most severe kind, affecting lower layers of tissue. **third eye 1** Hinduism & Buddhism the 'eye of insight' in the forehead of an image of a deity, esp. the god Siva. **2** the faculty of intuitive insight or prescience. **3** the pineal gland in certain vertebrates. **third force** a political group or party acting as a check on conflict between two opposing parties. **third gear** the third lowest in a set of gears. **third man** Cricket **1** a fielder positioned near the boundary behind the slips. **2** this position. **third part** each of three equal parts into which a thing is or might be divided. **third party 1** another party besides the two principals. **2** a bystander etc. **third-party** (of insurance) covering damage or injury suffered by a person other than the insured. **third person 1** = third party. **2** Gram. see PERSON 3. **third-rate** inferior; very poor in quality. **third reading** a third presentation of a bill to a legislative assembly, in the UK to debate committee reports and in the US to consider it for the last time. □ **thirdly** adv. [OE third(d)a, thridda f. Gmc]

Third Estate n. usu. hist. the lowest of the three estates into which a society, parliament, etc., is sometimes divided, esp. the common people and bourgeoisie in France before the Revolution.

Third International see INTERNATIONAL.

Third Reich the Nazi regime in Germany 1933–45 (see also REICH¹).

Third Republic hist. **1** the republican regime in France between the fall of Napoleon III (1870) and the German occupation of 1940. **2** this period in France.

Third World the developing countries of Asia, Africa, and Latin America. The term originated with French commentators in the 1950s, who used tiers monde to distinguish the developing countries from the capitalist and Communist blocs. (See also DEVELOPING COUNTRY.)

thirst /θɜːst/ n. & v. ● n. **1** a physical need to drink liquid, or the feeling of discomfort caused by this. **2** a strong desire or craving (a thirst for *power*). ● *v.intr.* (often foll. by *for* or *after*) **1** feel thirst. **2** have a strong desire. [OE *thurst, thyrstan* f. WG]

thirsty /'θɜːstɪ/ *adj.* (**thirstier, thirstiest**) **1** feeling thirst. **2** (of land, a season, etc.) dry or parched. **3** (often foll. by *for* or *after*) eager. **4** *colloq.* causing thirst (*thirsty work*). □ **thirstily** *adv.* **thirstiness** *n.* [OE *thurstig, thyrstig* (as THIRST, -Y¹)]

thirteen /θɜːˈtiːn/ *n. & adj.* ● *n.* **1** one more than twelve, or three more than ten. **2** a symbol for this (13, xiii, XIII). **3** a size etc. denoted by thirteen. ● *adj.* that amount to thirteen. □ **thirteenth** *adj. & n.* [OE *thrēotīene* (as THREE, -TEEN)]

Thirteen Colonies *hist.* the British colonies that ratified the Declaration of Independence in 1776 and thereby became founding states of the US. The colonies were Virginia, Massachusetts, Maryland, Connecticut, Rhode Island, North Carolina, South Carolina, New York, New Jersey, Delaware, New Hampshire, Pennsylvania, and Georgia.

thirty /'θɜːtɪ/ *n. & adj.* ● *n.* (*pl.* **-ies**) **1** the product of three and ten. **2** a symbol for this (30, xxx, XXX). **3** (in *pl.*) the numbers from 30 to 39, esp. the years of a century or of a person's life. ● *adj.* that amount to thirty. □ **thirty-first, -second**, etc. the ordinal numbers between thirtieth and fortieth. **thirty-one, -two**, etc. the cardinal numbers between thirty and forty. **thirty-second note** esp. *N. Amer. Mus.* = DEMISEMIQUAVER. **thirty-two-mo** a book with thirty-two leaves to the printing-sheet. □ **thirtieth** *adj. & n.* **thirtyfold** *adj. & adv.* [OE *thrītig* (as THREE, -TY²)]

Thirty-nine Articles /'θɜːtɪˌnaɪn/ the set of points of doctrine adopted by the Church of England in 1571 as a statement of its dogmatic position. Many, perhaps intentionally, allow a wide variety of interpretation. Since 1865, clergy in the Church of England have been asked to give a general assent to them rather than the more particular subscription previously demanded; other Anglican Churches have differing attitudes to the Articles.

thirtysomething /'θɜːtɪˌsʌmθɪŋ/ *n. colloq.* **1** an undetermined age between thirty and forty. **2** a person of this age.

Thirty Years War a prolonged European war 1618–48. Beginning as a struggle between the Catholic Holy Roman emperor and some of his German Protestant states, the war gradually drew in most of the major European military powers, and developed into a fight for continental hegemony with France, Sweden, Spain, and the Holy Roman Empire as the major protagonists. The result of three decades of intermittent hostilities, which ended with the Treaty of Westphalia, was the emergence of Bourbon France as the pre-eminent European power and the devastation of much of Germany, which had throughout been the centre of military activity.

this /ðɪs/ *pron., adj., & adv.* ● *demonstr. pron.* (*pl.* **these** /ðiːz/) **1** the person or thing close at hand or indicated or already named or understood (*can you see this?; this is my cousin*). **2** (contrasted with *that*) the person or thing nearer to hand or more immediately in mind. **3** the action, behaviour, or circumstances under consideration (*this won't do at all; what do you think of this?*). **4** (on the telephone): **a** *Brit.* the person speaking. **b** *US* the person spoken to. ● *demonstr. adj.* (*pl.* **these** /ðiːz/) **1** designating the person or thing close at hand etc. (cf. senses 1, 2 of *pron.*). **2** (of time): **a** the present or current (*am busy all this week*). **b** relating to today (*this morning*). **c** just past or to come (*have been asking for it these three weeks*). **3** *colloq.* (in narrative) designating a person or thing previously unspecified (*then up came this policeman*). ● *adv.* to the degree or extent indicated (*knew him when he was this high; did not reach this far*). □ **this and that** *colloq.* various unspecified examples of things (esp. trivial). **this here** *sl.* this particular (person or thing). **this much** the amount or extent about to be stated (*I know this much, that he was not there*). **this world** mortal life. [OE, neut. of *thes*]

Thisbe /'θɪzbɪ/ *Rom. Mythol.* a Babylonian girl, lover of Pyramus.

thistle /'θɪs(ə)l/ *n.* **1** a prickly composite herbaceous plant, esp. of the genera *Cirsium, Carlina*, or *Carduus*, usu. with globular heads of purple flowers. **2** a figure of this as the heraldic emblem of Scotland, and part of the insignia of the Order of the Thistle. [OE *thistel* f. Gmc]

Thistle, Order of the a distinctively Scottish order of knighthood instituted in 1687 by James II and revived after James's exile by Queen Anne in 1703.

thistledown /'θɪs(ə)lˌdaʊn/ *n.* a light fluffy down or pappus attached to thistle-seeds and blown about in the wind.

thistly /'θɪslɪ/ *adj.* overgrown with thistles.

thither /'ðɪðə(r)/ *adv. archaic* or *formal* to or towards that place. [OE *thider*, alt. (after HITHER) of *thæder*]

Thívai see THEBES 2.

thixotropy /θɪkˈsɒtrəpɪ/ *n. Chem.* the property of becoming temporarily liquid when shaken or stirred etc., and returning to a gel on standing. □ **thixotropic** /ˌθɪksəˈtrɒpɪk, -ˈtrɒpɪk/ *adj.* [Gk *thixis* touching + *tropē* turning]

tho' var. of THOUGH.

thole¹ /θəʊl/ *n.* (in full **thole-pin**) **1** a pin in the gunwale of a boat as the fulcrum for an oar. **2** either of two such pins forming a rowlock. [OE *thol* fir-tree, peg]

thole² /θəʊl/ *v.tr. Sc.* or *archaic* **1** undergo or suffer (pain, grief, etc.). **2** permit or admit of. [OE *tholian* f. Gmc]

tholos /'θɒlɒs/ *n.* (*pl.* **tholoi** /-lɔɪ/) *Gk Antiq.* a dome-shaped tomb, esp. of the Mycenaean period. [Gk]

Thomas¹ /'tɒməs/, Dylan (Marlais) (1914–53), Welsh poet. He moved to London in 1934 and worked in journalism and broadcasting while continuing to write poetry. He won recognition with *Deaths and Entrances* (1946), and continued his success with works such as *Portrait of the Artist as a Young Dog* (1940; prose); *Adventures in the Skin Trade*, an unfinished novel, was published posthumously in 1955. Shortly before his death in New York of alcohol poisoning, he narrated on radio his best-known work, *Under Milk Wood*, a portrait of a small Welsh town, interspersing poetic alliterative prose with songs and ballads.

Thomas² /'tɒməs/, (Philip) Edward (1878–1917), English poet. He wrote mostly journalistic prose and reviews before being encouraged by Robert Frost in 1914 to concentrate on writing poetry; most of his work was written while serving in the First World War and was published posthumously after he was killed at Arras in 1917. His work offers a sympathetic but unidealized depiction of rural English life, adapting colloquial speech rhythms to poetic metre.

Thomas, St an Apostle. He said that he would not believe that Christ had risen again until he had seen and touched his wounds (John 20:24–9); the story is the origin of the nickname 'doubting Thomas'. According to tradition he preached in SW India (see MALABAR CHRISTIANS). Feast day, 21 Dec.

Thomas à Kempis /ə ˈkempɪs/ (born Thomas Hemerken) (c.1380–1471), German theologian. Born at Kempen, near Cologne (from where he gets his name), he became an Augustinian canon in Holland. He wrote a number of ascetic treatises and is the probable author of *On the Imitation of Christ* (c.1415–24), a manual of spiritual devotion.

Thomas Aquinas, St, see AQUINAS, ST THOMAS.

Thomas More, St see MORE.

Thomism /'təʊmɪz(ə)m/ *n.* the doctrine of St Thomas Aquinas or of his followers. □ **Thomist** *n.* **Thomistic** /təʊˈmɪstɪk/ *adj.* **Thomistical** *adj.*

Thompson¹ /'tɒmps(ə)n/, Daley (b.1958), English athlete. He won a number of major decathlon titles in the 1980s, including gold medals in the Olympic Games of 1980 and 1984.

Thompson² /'tɒmps(ə)n/, Francis (1859–1907), English poet. His best-known work uses powerful imagery to convey intense religious experience, and includes the poems 'The Hound of Heaven' and 'The Kingdom of God'. He published three volumes of verse (1893–7) and much literary criticism in periodicals. He died from the combined effects of opium addiction and tuberculosis.

Thomson¹ /'tɒms(ə)n/, James (1700–48), Scottish poet. His poem in four books *The Seasons* (1726–30) anticipated the romantic movement in its treatment of nature; the text was adapted for Haydn's oratorio of that name (1799–1801). He also co-wrote the masque *Alfred* (1740) with David Mallet (c.1705–65); it contains the song 'Rule, Britannia', whose words have been attributed to him.

Thomson² /'tɒms(ə)n/, James (1834–82), Scottish poet. He is chiefly remembered for the poem 'The City of Dreadful Night' (1874), a powerful evocation of a half-ruined city where the narrator encounters tormented shades wandering in a Dantesque living hell, presided over by Melancolia.

Thomson³ /'tɒms(ə)n/, Sir Joseph John (1856–1940), English physicist, discoverer of the electron. While professor of physics at Cambridge (1884–1918) he consolidated the reputation of the Cavendish Laboratory. From his experiments on the deflection of cathode rays in magnetic and electric fields, Thomson deduced that he was dealing with particles smaller than the atom. These he initially called *corpuscles* and later *electrons*, a word coined by the Irish mathematical physicist George Johnstone Stoney (1826–1911). Thomson received the 1906 Nobel Prize for physics for his researches into the electrical conductivity of gases. His son Sir George Paget Thomson (1892–1975) shared the 1937 Nobel Prize for physics for his discovery of electron diffraction by crystals.

Thomson[4] /'tɒms(ə)n/, Roy Herbert, 1st Baron Thomson of Fleet (1894–1976), Canadian-born British newspaper proprietor and media entrepreneur. In 1931 he opened his own radio station in northern Ontario. He then built up his North American press and radio holdings before acquiring his first British newspaper, the *Scotsman*, in 1952, the year he settled in Britain. He subsequently added the *Sunday Times* (1959) and *The Times* (1966) to his acquisitions, by which time the Thomson Organization had become an international corporation, with interests in publishing, printing, television, and travel.

Thomson[5] /'tɒms(ə)n/, Tom (full name Thomas John Thomson) (1877–1917), Canadian painter. A pioneering artist of Canada's wilderness, he began sketching in Algonquin Park, Ontario, in the spring of 1912, returning there in subsequent summers. Major paintings include *Northern Lake* (1913), *The West Wind* (1917), and *The Jack Pine* (1917). Thomson's premature death at Canoe Lake, Algonquin Park, remains a mystery.

Thomson[6] /'tɒms(ə)n/, Sir William, see KELVIN.

thong /θɒŋ/ *n. & v.* ● *n.* **1** a narrow strip of hide or leather used as the lash of a whip, as a halter or rein, etc. **2** *Austral., NZ, & N. Amer.* = FLIP-FLOP *n.* 1. **3** a skimpy bathing garment like a G-string. ● *v.tr.* **1** provide with a thong. **2** strike with a thong. [OE *thwang, thwong* f. Gmc]

Thor /θɔː(r)/ *Scand. Mythol.* the god of thunder, the weather, agriculture, and the home, the son of Odin and Frigga. He is represented as armed with a hammer. Thursday is named after him.

thorax /'θɔːræks/ *n.* (*pl.* **thoraces** /-rəˌsiːz/ or **thoraxes**) **1** *Anat. & Zool.* the part of the trunk between the neck and the abdomen. **2** *Gk Antiq.* a breastplate or cuirass. □ **thoracal** /-rək(ə)l/ *adj.* **thoracic** /θɔːˈræsɪk/ *adj.* [L f. Gk *thōrax -akos*]

Thoreau /'θɔːrəʊ/, Henry David (1817–62), American essayist and poet. Together with his friend and mentor Ralph Waldo Emerson he is regarded as a key figure of Transcendentalism. He is best known for his book *Walden, or Life in the Woods* (1854), an account of a two-year experiment in self-sufficiency when he built a wooden hut by Walden Pond, near his home town of Concord, Massachusetts, and sought to live according to his ideals of simplicity and closeness to nature. His essay on civil disobedience (1849), in which he argues the right of the individual to refuse to pay taxes when conscience dictates, influenced Mahatma Gandhi's policy of passive resistance.

thoria /'θɔːrɪə/ *n.* the oxide of thorium.

thorium /'θɔːrɪəm/ *n.* a radioactive white metallic chemical element (atomic number 90; symbol **Th**). A member of the actinide series, thorium was discovered by Jöns Berzelius in 1828. It became economically important after 1885 when its dioxide began to be widely used as an incandescent material in gas mantles. Thorium has a variety of industrial uses and is of increasing importance as a nuclear fuel. Thorium's radioactive properties were discovered in 1898; it is one of the most abundant radioactive elements in the earth's crust and its main source is the mineral monazite. [THOR]

Thorn see TORUŃ.

thorn /θɔːn/ *n.* **1** a stiff sharp-pointed projection on a plant. **2** a thorn-bearing shrub or tree. **3** the name of an Old English and Icelandic runic letter, þ (= th). □ **on thorns** continuously uneasy, esp. in fear of being detected. **thorn-apple 1** a poisonous solanaceous plant, *Datura stramonium*, with large funnel-shaped flowers. **2** the prickly fruit of this. **a thorn in one's flesh** (or **side**) a constant annoyance. □ **thornless** *adj.* **thornproof** *adj.* [OE f. Gmc]

thornback /'θɔːnbæk/ *n.* (in full **thornback ray**) a ray with spines on the back and tail; esp. *Raja clavata*, of European waters.

thornbill /'θɔːnbɪl/ *n.* **1** a small Australian warbler of the genus *Acanthiza*. **2** a South American hummingbird, esp. of the genus *Chalcostigma* or *Rhamphomicron*, with a short bill.

Thorndike /'θɔːndaɪk/, Dame (Agnes) Sybil (1882–1976), English actress. She gave notable performances in a wide range of Shakespearian and other roles; she was particularly memorable in the title part of the first London production of George Bernard Shaw's *St Joan* (1924). From the 1920s she appeared in a number of films, including *Nicholas Nickleby* (1947).

thorntail /'θɔːnteɪl/ *n.* a bright green South American hummingbird of the genus *Popelairia*, with projecting outer tail feathers.

thorny /'θɔːnɪ/ *adj.* (**thornier, thorniest**) **1** having many thorns. **2** (of a subject) hard to handle without offence; problematic (*thorny issue*).

□ **thorny-headed worm** = ACANTHOCEPHALAN *n.* □ **thornily** *adv.* **thorniness** *n.*

thorough /'θʌrə/ *adj.* **1** complete and unqualified; not superficial (*needs a thorough change*). **2** acting or done with great care and completeness (*the report is most thorough*). **3** absolute (*a thorough nuisance*). □ **thorough bass** *Mus.* a bass part for a keyboard player with numerals and symbols below to indicate the harmony. **thorough-paced 1** (of a horse) trained to all paces. **2** complete or unqualified. □ **thoroughly** *adv.* **thoroughness** *n.* [orig. as *adv.* and *prep.* in the senses of *through*, f. OE *thuruh* var. of *thurh* THROUGH]

thoroughbred /'θʌrəˌbred/ *adj. & n.* ● *adj.* **1** of pure breed. **2** out and out; first-class. ● *n.* **1** a thoroughbred animal, esp. a horse. **2** (**Thoroughbred**) a breed of racehorse originating from English mares and Arab stallions, with documented ancestry.

thoroughfare /'θʌrəˌfeə(r)/ *n.* a road or path open at both ends, esp. for traffic.

thoroughgoing /'θʌrəˌgəʊɪŋ/ *adj.* **1** uncompromising; not superficial. **2** (usu. *attrib.*) extreme; out and out.

thorp /θɔːp/ *n.* (also **thorpe**) *archaic* a village or hamlet. ¶ Now usually only in place-names. [OE *thorp, throp*, f. Gmc]

Thorshavn see TÓRSHAVN.

Thorvaldsen /'tɔːvæls(ə)n/, Bertel (also **Thorwaldsen**) (c.1770–1844), Danish neoclassical sculptor. From 1797 he lived and worked chiefly in Rome, where he made his name with a statue of Jason (1803); other major works include the tomb of Pius VII at St Peter's in Rome (1824–31) and a monument to Byron in Cambridge (1829).

Thos. *abbr.* Thomas.

those *pl.* of THAT.

Thoth /θəʊθ, təʊt/ *Egyptian Mythol.* a moon-god, the god of wisdom, justice, and writing, and patron of the sciences. Thoth was closely associated with Ra and was his messenger, which led the Greeks to identify him with Hermes. He is most often represented in human form with the head of an ibis surmounted by the moon's disc and crescent.

thou[1] /ðaʊ/ *pron.* (*obj.* **thee**; *poss.* **thy** or **thine**; *pl.* **ye** or **you**) second person singular pronoun. ¶ now replaced by *you* except in some formal, liturgical, dialect, and poetic uses. [OE *thu* f. Gmc]

thou[2] /θaʊ/ *n.* (*pl.* same or **thous**) *colloq.* **1** a thousand. **2** one-thousandth. [abbr.]

though /ðəʊ/ *conj. & adv.* (also **tho'**) ● *conj.* **1** despite the fact that (*though it was early we went to bed*). **2** (introducing a possibility) even if (*ask him though he may refuse; would not attend though the queen herself were there*). **3** and yet; nevertheless (*she read on, though not to the very end*). **4** in spite of being (*ready though unwilling*). ● *adv. colloq.* however; all the same (*I wish you had told me, though*). [ME *thoh* etc. f. ON *thó* etc., corresp. to OE *thēah*, f. Gmc]

thought[1] /θɔːt/ *n.* **1** the process or power of thinking; the faculty of reason. **2** a way of thinking characteristic of or associated with a particular time, people, group, etc. (*medieval European thought*). **3 a** sober reflection or consideration (*gave it much thought*). **b** care, regard, concern (*had no thought for others*). **4** an idea or piece of reasoning produced by thinking (*many good thoughts came out of the discussion*). **5** (foll. by *of* + verbal noun or *to* + infin.) a partly formed intention or hope (*gave up all thoughts of winning; had no thought to go*). **6** (usu. in *pl.*) what one is thinking; one's opinion (*have you any thoughts on this?*). **7** the subject of one's thinking (*my one thought was to get away*). **8** (prec. by a) somewhat (*seems to me a thought arrogant*). □ **give thought to** consider; think about. **in thought** thinking, meditating. **take thought** consider matters. **thought-provoking** stimulating serious thought. **thought-reader** a person supposedly able to perceive another's thoughts. **thought-reading** the supposed perception of what another is thinking. **thought transference** telepathy. **thought-wave** a wave of the supposed medium of thought transference. □ **-thoughted** *adj.* (in *comb.*). [OE *thōht* (as THINK)]

thought[2] *past* and *past part.* of THINK.

thoughtful /'θɔːtfʊl/ *adj.* **1** engaged in or given to meditation. **2** (of a book, writer, remark, etc.) giving signs of serious thought. **3** (often foll. by *of*) (of a person or conduct) considerate; not haphazard or unfeeling. □ **thoughtfully** *adv.* **thoughtfulness** *n.*

thoughtless /'θɔːtlɪs/ *adj.* **1** careless of consequences or of others' feelings. **2** due to lack of thought. □ **thoughtlessly** *adv.* **thoughtlessness** *n.*

thousand /'θaʊz(ə)nd/ *n. & adj.* ● *n.* (*pl.* **thousands** or (in sense 1) **thousand**) (in *sing.* prec. by *a* or *one*) **1** the product of a hundred and ten. **2** a symbol for this (1,000, m, M). **3** a set of a thousand things. **4** (in *sing.* or *pl.*) *colloq.* a large number. ● *adj.* that amount to a thousand. □ **thousandfold** *adj. & adv.* **thousandth** *adj. & n.* [OE *thūsend* f. Gmc]

Thousand and One Nights, The see ARABIAN NIGHTS.

Thousand Island *n.* a piquant mayonnaise salad-dressing made with tomatoes, chilli, chopped peppers, etc. [THOUSAND ISLANDS]

Thousand Islands 1 a group of about 1,500 islands in a widening of the St Lawrence River, just below Kingston. Some of the islands belong to Canada and some to the US. **2** (called in Indonesian *Pulau Seribu*) a group of about 100 small islands off the north coast of Java, forming part of Indonesia.

Thrace /θreɪs/ an ancient country lying west of the Black Sea and north of the Aegean, originally inhabited by a warlike Indo-European people. Conquered by Philip II of Macedon in 342 BC, it became a Roman province in AD 46. It is now divided between Turkey, Bulgaria, and Greece. □ **Thracian** /ˈθreɪʃ(ə)n/ *adj. & n.*

thrall /θrɔːl/ *n. literary* **1** (often foll. by *of, to*) a slave (of a person, or a power or influence). **2** a state of slavery or servitude (*in thrall*). □ **thraldom** *n.* (also **thralldom**). [OE *thrǣl* f. ON *thrǽll*, perh. f. a Gmc root = run]

thrash /θræʃ/ *v. & n.* ● *v.* **1** *tr.* beat severely, esp. with a stick or whip. **2** *tr.* defeat thoroughly in a contest. **3** *intr.* (of a paddle wheel, branch, etc.) act like a flail; deliver repeated blows. **4** *intr.* (foll. by *about, around*) move or fling the limbs about violently or in panic. **5** *intr.* (of a ship) keep striking the waves; make way against the wind or tide (*thrash to windward*). **6** *tr.* = THRESH 1. ● *n.* **1** an act of thrashing. **2** *colloq.* a party, esp. a lavish one. **3** *colloq.* a style of fast harsh-sounding rock music combining elements of punk and heavy metal. □ **thrash out** discuss to a conclusion. □ **thrashing** *n.* [OE *therscan*, later *threscan*, f. Gmc]

thrasher[1] /ˈθræʃə(r)/ *n.* **1** a person or thing that thrashes. **2** = THRESHER 1.

thrasher[2] /ˈθræʃə(r)/ *n.* a long-tailed thrushlike North American bird of the family Mimidae. [perh. f. Engl. dial. *thrusher* = THRUSH[1]]

thrawn /θrɔːn/ *adj. Sc.* **1** perverse or ill-tempered. **2** misshapen, crooked. [Sc. form of *thrown* in obs. senses]

thread /θred/ *n. & v.* ● *n.* **1 a** a spun-out filament of cotton, silk, glass, etc.; yarn. **b** a length of this. **2** a thin cord of twisted yarns used esp. in sewing and weaving. **3** anything regarded as threadlike with reference to its continuity or connectedness (*the thread of life; lost the thread of his argument*). **4** the spiral ridge of a screw. **5** (in *pl.*) *sl.* clothes. **6** a thin seam or vein of ore. ● *v.tr.* **1 a** pass a thread through the eye of (a needle). **b** (often foll. by *through*) pass (a thread, ribbon, etc.) through a hole or series of holes. **2** put (beads) on a thread. **3** arrange (material in a strip form, e.g. film or magnetic tape) in the proper position on equipment. **4** make (one's way) carefully through a crowded place, over a difficult route, etc. **5** streak (hair etc.) as with threads. **6** form a screw-thread on. □ **hang by a thread** be in a precarious state, position, etc. **thread mark** a mark in the form of a thin line made in banknote paper with highly coloured silk fibres to prevent photographic counterfeiting. □ **threader** *n.* **threadlike** *adj.* [OE *thrǣd* f. Gmc]

threadbare /ˈθredbeə(r)/ *adj.* **1** (of cloth) so worn that the nap is lost and the thread visible. **2** (of a person) wearing such clothes. **3 a** hackneyed. **b** feeble or insubstantial (*a threadbare excuse*).

threadfin /ˈθredfɪn/ *n.* a fish with long narrow fins; esp. a small tropical fish of the family Polynemidae, with long streamers from its pectoral fins.

Threadneedle Street /ˈθredˌniːd(ə)l/ a street in the City of London containing the premises of the Bank of England, which is also known as the *Old Lady of Threadneedle Street*. The name is derived from *three-needle*, possibly from a tavern with the arms of the city of London Guild of Needlemakers.

threadworm /ˈθredwɜːm/ *n.* a threadlike nematode worm, often parasitic, esp. a pinworm.

thready /ˈθredɪ/ *adj.* (**threadier, threadiest**) **1** of or like a thread. **2** (of a person's pulse) scarcely perceptible.

threat /θret/ *n.* **1 a** a declaration of an intention to punish or hurt. **b** *Law* a menace of bodily hurt or injury, such as may restrain a person's freedom of action. **2** an indication of something undesirable coming (*the threat of war*). **3** a person or thing as a likely cause of harm etc. [OE *thrēat* affliction etc. f. Gmc]

threaten /ˈθret(ə)n/ *v.tr.* **1 a** make a threat or threats against. **b** constitute a threat to; be likely to harm; put into danger. **2** be a sign or indication of (something undesirable). **3** (foll. by *to* + infin.) announce one's intention to do an undesirable or unexpected thing

(*threatened to resign*). **4** (also *absol.*) give warning of the infliction of (harm etc.) (*the clouds were threatening rain*). **5** (as **threatened** *adj.*) (of a species etc.) in danger of becoming rare or extinct. □ **threatener** *n.* **threatening** *adj.* **threateningly** *adv.* [OE *thrēatnian* (as THREAT)]

three /θriː/ *n. & adj.* ● *n.* **1 a** one more than two, or seven less than ten. **b** a symbol for this (3, iii, III). **2** a size etc. denoted by three. **3** three o'clock. **4** a set of three. **5** a card with three pips. ● *adj.* that amount to three. □ **three-card trick** a game in which bets are made on which is the queen among three cards lying face downwards. **three cheers** see CHEER. **three-colour process** a process of reproducing natural colours by combining photographic images in the three primary colours. **three-cornered 1** triangular. **2** (of a contest etc.) between three parties as individuals. **three-decker 1** *hist.* a warship with three gun-decks. **2** a novel in three volumes. **3** a sandwich with three slices of bread. **three-dimensional** having or appearing to have length, breadth, and depth. **three-handed 1** having or using three hands. **2** involving three players. **three-legged race** a running-race between pairs, one member of each pair having the left leg tied to the right leg of the other. **three-line whip** *Brit.* a written notice, underlined three times to indicate great urgency, requesting the members of Parliament of a particular party to attend a parliamentary vote. **three parts** three-quarters. **three-phase** see PHASE. **three-piece** consisting of three items (esp. of a suit of clothes or a suite of furniture). **three-ply** *adj.* of three strands, webs, or thicknesses. ● *n.* **1** three-ply wool. **2** three-ply wood made by gluing together three layers with the grain in different directions. **three-point landing** *Aeron.* the landing of an aircraft on the two main wheels and the tail wheel or skid simultaneously. **three-point turn** a method of turning a vehicle round in a narrow space by moving forwards, backwards, and forwards again in a sequence of arcs. **three-quarter** *n.* (also **three-quarter back**) *Rugby* any of three or four players just behind the half-backs. ● *adj.* **1** consisting of three-fourths of something. **2** (of a portrait) going down to the hips or showing three-fourths of the face (between full face and profile). **three-quarters** three out of four equal parts into which something may be divided. **three-ring circus** esp. *US* **1** a circus with three rings for simultaneous performances. **2** an extravagant display. **the three Rs** reading, writing, and arithmetic, regarded as the fundamentals of learning. **three-way** involving three ways or participants. **three-wheeler** a vehicle with three wheels. [OE *thrī* f. Gmc]

Three Estates the three groups constituting Parliament, now the Lords Temporal (the peers), the Lords Spiritual (the bishops), and the Commons.

threefold /ˈθriːfəʊld/ *adj. & adv.* **1** three times as much or as many. **2** consisting of three parts.

Three Mile Island an island in the Susquehanna river near Harrisburg, Pennsylvania, site of a nuclear power station. In 1979 an accident caused damage to the reactor core, an incident that provoked strong reactions against the nuclear industry in the US and precipitated a reassessment of safety standards.

threepence /ˈθrep(ə)ns, ˈθrʊp-/ *n. Brit.* the sum of three pence, esp. before decimalization.

threepenny /ˈθrep(ə)nɪ, ˈθrʊp-/ *adj. Brit.* costing three pence, esp. before decimalization. □ **threepenny bit** *hist.* a former coin worth three old pence.

threescore /ˈθriːskɔː(r)/ *n. archaic* sixty.

threesome /ˈθriːsəm/ *n.* **1** a group of three persons. **2** a game etc. for three, esp. *Golf* one against two.

thremmatology /ˌθreməˈtɒlədʒɪ/ *n.* the science of breeding animals and plants. [Gk *thremma -matos* nursling + -LOGY]

threnody /ˈθrenədɪ/ *n.* (also **threnode** /-nəʊd/) (*pl.* **-ies** or **threnodes**) **1** a lamentation, esp. on a person's death. **2** a song of lamentation. □ **threnodist** *n.* **threnodial** /θrɪˈnəʊdɪəl/ *adj.* **threnodic** /-ˈnɒdɪk/ *adj.* [Gk *thrēnōidia* f. *thrēnos* wailing + *ōidē* ODE]

threonine /ˈθriːəˌniːn, -nɪn/ *n. Biochem.* an amino acid, considered essential for growth. [*threose* (name of a tetrose sugar) ult. f. Gk *eruthros* red + -INE[4]]

thresh /θreʃ/ *v.* **1** beat out or separate grain from (corn etc.). **2** *intr.* = THRASH *v.* 4. **3** *tr.* (foll. by *over*) analyse (a problem etc.) in search of a solution. □ **threshing-floor** a hard level floor for threshing esp. with flails. **threshing-machine** a power-driven machine for separating the grain from the straw or husk. **thresh out** = thrash out. [var. of THRASH]

thresher /ˈθreʃə(r)/ *n.* **1** a person or machine that threshes. **2** a shark,

Alopias vulpinus, with a very long upper lobe to its tail with which it can lash the water to direct its prey.

threshold /ˈθreʃəʊld, ˈθreʃhəʊld/ *n.* **1** a strip of wood or stone forming the bottom of a doorway and crossed in entering a house or room etc. **2** a point of entry or beginning (*on the threshold of a new century*). **3** *Physiol. & Psychol.* a limit below which a stimulus causes no reaction (*pain threshold*). **4** *Physics* a limit below which no reaction occurs, esp. a minimum dose of radiation producing a specified effect. **5** (often *attrib.*) a step in a scale of wages or taxation, usu. operative in specified conditions. [OE *therscold*, *threscold*, etc., rel. to THRASH in the sense 'tread']

threw *past* of THROW.

thrice /θraɪs/ *adv. archaic* or *literary* **1** three times. **2** (esp. in *comb.*) highly (*thrice-blessed*). [ME *thries* f. *thrie* (adv.) f. OE *thrīwa*, *thrīga* (as THREE, -s³)]

thrift /θrɪft/ *n.* **1** frugality; economical management. **2** a plant of the genus *Armeria*, esp. the sea pink. □ **thrift shop** (or **store**) a shop selling second-hand items usu. for charity. [ME f. ON (as THRIVE)]

thriftless /ˈθrɪftlɪs/ *adj.* wasteful, improvident. □ **thriftlessly** *adv.* **thriftlessness** *n.*

thrifty /ˈθrɪftɪ/ *adj.* (**thriftier**, **thriftiest**) **1** economical, frugal. **2** thriving, prosperous. □ **thriftily** *adv.* **thriftiness** *n.*

thrill /θrɪl/ *n. & v.* ● *n.* **1 a** a wave or nervous tremor of emotion or sensation (*a thrill of joy; a thrill of recognition*). **b** a thrilling experience (*seeking new thrills*). **2** a throb or pulsation. **3** *Med.* a vibratory movement or resonance heard in auscultation. ● *v.* **1** *intr.* feel a thrill (*thrilled to the sound*). **2** *tr.* **a** cause to feel a thrill (*a voice that thrilled millions*). **b** *colloq.* (in *passive*) be pleased or delighted (*they were thrilled with their presents*). **3** *intr.* quiver or throb with or as with emotion. **4** *intr.* (foll. by *through*, *over*, *along*) (of an emotion etc.) pass with a thrill through etc. (*fear thrilled through my veins*). □ **thrills and spills** the excitement of potentially dangerous activities. **thrill to bits** *colloq.* (in *passive*) be extremely pleased or delighted. □ **thrilling** *adj.* **thrillingly** *adv.* [*thirl* (now dial.) f. OE *thyrlian* pierce f. *thyrel* hole f. *thurh* THROUGH]

thriller /ˈθrɪlə(r)/ *n.* an exciting or sensational story or play etc., esp. one involving crime or espionage.

thrips /θrɪps/ *n.* (*pl.* same) a minute blackish insect of the order Thysanoptera, some members of which are injurious to plants. Also called *thunderbug*, *thunderfly*. [L f. Gk, = woodworm]

thrive /θraɪv/ *v.intr.* (*past* **throve** /θrəʊv/ or **thrived**; *past part.* **thriven** /ˈθrɪv(ə)n/ or **thrived**) **1** prosper or flourish. **2** grow rich. **3** (of a child, animal, or plant) grow vigorously. [ME f. ON *thrífask* refl. of *thrífa* grasp]

thro' *var.* of THROUGH.

throat /θrəʊt/ *n.* **1 a** the windpipe or gullet. **b** the front part of the neck containing this. **2** *literary* **a** a voice, esp. of a songbird. **b** a thing compared to a throat, esp. a narrow passage, entrance, or exit. **3** *Naut.* the forward upper corner of a fore-and-aft sail. □ **cut one's own throat** bring about one's own downfall. **ram** (or **thrust**) **down a person's throat** force (a thing) on a person's attention. □ **-throated** *adj.* (in *comb.*). [OE *throte*, *throtu* f. Gmc]

throaty /ˈθrəʊtɪ/ *adj.* (**throatier**, **throatiest**) **1** (of a voice) deficient in clarity; hoarsely resonant. **2** guttural; uttered in the throat. **3** having a prominent or capacious throat. □ **throatily** *adv.* **throatiness** *n.*

throb /θrɒb/ *v. & n.* ● *v.intr.* (**throbbed**, **throbbing**) **1** palpitate or pulsate, esp. with more than the usual force or rapidity. **2** vibrate or quiver with a persistent rhythm or with emotion. ● *n.* **1** a throbbing. **2** a palpitation or (esp. violent) pulsation. [ME, app. imit.]

throe /θrəʊ/ *n.* (usu. in *pl.*) **1** a violent pang, esp. of childbirth or death. **2** anguish. □ **in the throes of** struggling with the task of. [ME *throwe* perh. f. OE *thrēa*, *thrawu* calamity, alt. perh. by assoc. with *woe*]

thrombi *pl.* of THROMBUS.

thrombin /ˈθrɒmbɪn/ *n.* *Biochem.* a plasma enzyme that causes the clotting of blood by converting fibrinogen to fibrin. [as THROMBUS + -IN]

thrombocyte /ˈθrɒmbəˌsaɪt/ *n.* *Physiol.* = PLATELET. [as THROMBUS + -CYTE]

thrombocytopenia /ˌθrɒmbəʊˌsaɪtəʊˈpiːnɪə/ *n.* *Med.* deficiency of platelets in the blood. [THROMBOCYTE + Gk *penia* poverty]

thrombose /θrɒmˈbəʊz/ *v.tr. & intr.* *Med.* affect with or undergo thrombosis. [back-form. f. THROMBOSIS]

thrombosis /θrɒmˈbəʊsɪs/ *n.* (*pl.* **thromboses** /-siːz/) *Med.* the coagulation or clotting of the blood in a blood vessel or the heart. □ **thrombotic** /-ˈbɒtɪk/ *adj.* [mod.L f. Gk *thrombōsis* curdling (as THROMBUS)]

thrombus /ˈθrɒmbəs/ *n.* (*pl.* **thrombi** /-baɪ/) *Med.* a blood clot formed in the vascular system and impeding blood flow. [mod.L f. Gk *thrombos* lump, blood clot]

throne /θrəʊn/ *n. & v.* ● *n.* **1** a chair of state for a sovereign or bishop etc. **2** sovereign power (*came to the throne*). **3** (in *pl.*) the third order of the ninefold celestial hierarchy (see ORDER *n.* 19). **4** *colloq.* a lavatory seat and bowl. ● *v.tr.* place on a throne. □ **throneless** *adj.* [ME f. OF *trone* f. L *thronus* f. Gk *thronos* high seat]

throng /θrɒŋ/ *n. & v.* ● *n.* **1** a crowd of people. **2** (often foll. by *of*) a multitude, esp. in a small space. ● *v.* **1** *intr.* come in great numbers (*crowds thronged to the stadium*). **2** *tr.* flock into or crowd round; fill with or as with a crowd (*crowds thronged the streets*). [ME *thrang*, *throng*, OE *gethrang*, f. verbal stem *thring- thrang-*]

throstle /ˈθrɒs(ə)l/ *n.* **1** a song thrush. **2** (in full **throstle-frame**) *hist.* a machine for continuously spinning wool or cotton etc. [OE f. Gmc: rel. to THRUSH¹]

throttle /ˈθrɒt(ə)l/ *n. & v.* ● *n.* **1 a** (in full **throttle-valve**) a valve controlling the flow of fuel or steam etc. in an engine. **b** (in full **throttle-lever**) a lever or pedal operating this valve. **2** the throat, gullet, or windpipe. ● *v.tr.* **1** choke or strangle. **2** prevent the utterance etc. of. **3** control (an engine or steam etc.) with a throttle. □ **throttle back** (or **down**) reduce the speed of (an engine or vehicle) by throttling. □ **throttler** *n.* [ME *throtel* (v.), perh. f. THROAT + -LE⁴: (n.) perh. a dimin. of THROAT]

through /θruː/ *prep., adv., & adj.* (also **thro'**, *US* **thru**) ● *prep.* **1 a** from end to end or from side to side of. **b** going in one side and out the other of. **2** between or among (*swam through the waves*). **3** from beginning to end of (*read through the letter; went through many difficulties; through the years*). **4** because of; by the agency, means, or fault of (*lost it through carelessness*). **5** *N. Amer.* up to and including (*Monday through Friday*). ● *adv.* **1** through a thing; from side to side, end to end, or beginning to end; completely, thoroughly (*went through to the garden; would not let us through*). **2** having completed (esp. successfully) (*are through their exams*). **3** so as to be connected by telephone (*will put you through*). ● *attrib.adj.* **1** (of a journey, route, etc.) done without a change of line or vehicle etc. or with one ticket. **2** (of traffic) going through a place to its destination. **3** (of a road) open at both ends. □ **be through 1** (often foll. by *with*) have finished. **2** (often foll. by *with*) cease to have dealings. **3** have no further prospects (*is through as a politician*). **no through road** no thoroughfare. **through and through 1** thoroughly, completely. **2** through again and again. [OE *thurh* f. WG]

throughout /θruːˈaʊt/ *prep. & adv.* ● *prep.* right through; from end to end of (*throughout the town; throughout the 18th century*). ● *adv.* in every part or respect (*the timber was rotten throughout*).

throughput /ˈθruːpʊt/ *n.* the amount of material put through a process, esp. in manufacturing or computing.

throughway /ˈθruːweɪ/ *n.* (also *US* **thruway**) a thoroughfare, esp. *N. Amer.* a motorway.

throve *past* of THRIVE.

throw /θrəʊ/ *v. & n.* ● *v.tr.* (*past* **threw** /θruː/; *past part.* **thrown** /θrəʊn/) **1** propel with some force through the air or in a particular direction. **2** force violently into a specified position or state (*the ship was thrown on the rocks; threw themselves down*). **3** compel suddenly to be in a specified condition (*was thrown out of work*). **4** turn or move (part of the body) quickly or suddenly (*threw an arm out*). **5** project or cast (light, a shadow, a spell, etc.). **6 a** bring to the ground in wrestling. **b** (of a horse) unseat (its rider). **7** *colloq.* disconcert (*the question threw me for a moment*). **8** (foll. by *on*, *off*, etc.) put (clothes etc.) hastily on or off etc. **9 a** cause (dice) to fall on a table. **b** obtain (a specified number) by throwing dice. **10** cause to pass or extend suddenly to another state or position (*threw in the army; threw a bridge across the river*). **11** move (a switch or lever) so as to operate it. **12 a** form (ceramic ware) on a potter's wheel. **b** turn (wood etc.) on a lathe. **13** have (a fit or tantrum etc.). **14** give (a party). **15** *colloq.* lose (a contest or race etc.) intentionally. **16** *Cricket* bowl (a ball) with an illegitimate sudden straightening of the elbow. **17** (of a snake) cast (its skin). **18** (of an animal) give birth to (young). **19** twist (silk etc.) into thread or yarn. **20** (often foll. by *into*) put into another form or language etc. ● *n.* **1** an act of throwing. **2** the distance a thing is or may be thrown (*a record throw with the hammer*). **3** the act of being thrown in wrestling. **4** *Geol. & Mining* **a** a fault in strata. **b** the amount of vertical displacement caused by this. **5** a machine or device giving rapid rotary motion. **6 a** the movement of a crank or cam etc. **b** the extent of this. **7** the distance moved by the pointer of an instrument etc. **8** esp. *N. Amer.* **a** a light fabric cover for furniture, esp. a bedspread. **b** (in full **throw rug**) a light rug. **9** (prec. by *a*) *colloq.* each; per item (*sold at £10 a throw*). □ **throw about** (or **around**) **1** throw in various directions. **2** spend (one's money) ostentatiously. **throw away** **1** discard as useless or unwanted. **2** waste or fail to make use of (an opportunity etc.). **3** discard (a card). **4** *Theatr.* speak (lines) with deliberate underemphasis. **5** (in *passive*; often foll. by *on*) be wasted (*the advice was thrown away on him*). **throw-away** *adj.* **1** meant to be thrown away after (one) use. **2** (of lines etc.) deliberately underemphasized. **3** disposed to throwing things away; wasteful (*throw-away prices*). ● *n.* a thing to be thrown away after (one) use. **throw back 1** revert to ancestral character. **2** (usu. in *passive*; foll. by *on*) compel to rely on (*was thrown back on his savings*). **throw-back** *n.* **1** reversion to ancestral character. **2** an instance of this. **throw cold water on** see COLD. **throw down** cause to fall. **throw down the gauntlet** (or **glove**) issue a challenge. **throw dust in a person's eyes** mislead a person by misrepresentation or distraction. **throw good money after bad** incur further loss in a hopeless attempt to recoup a previous loss. **throw one's hand in 1** abandon one's chances in a card-game, esp. poker. **2** give up; withdraw from a contest. **throw in 1** interpose (a word or remark). **2** include at no extra cost. **3** throw (a football) from the edge of the pitch where it has gone out of play. **4** *Cricket* return (the ball) from the outfield. **5** *Cards* give (a player) the lead, to the player's disadvantage. **throw-in** *n.* the throwing in of a football during play. **throw in one's lot with** see LOT. **throw in** (or **up**) **the sponge 1** (of a boxer or his attendant) throw the sponge used between rounds into the air as a token of defeat. **2** abandon a contest; admit defeat. **throw in the towel** admit defeat. **throw light on** see LIGHT[1]. **throw off 1** discard; contrive to get rid of. **2** write or utter in an offhand manner. **3** (of hounds or a hunt) begin hunting; make a start. **throw-off** *n.* the start in a hunt or race. **throw oneself at** seek blatantly as a spouse or sexual partner. **throw oneself into** engage vigorously in. **throw oneself on** (or **upon**) **1** rely completely on. **2** attack. **throw open** (often foll. by *to*) **1** cause to be suddenly or widely open. **2** make accessible. **3** invite general discussion of or participation in. **throw out 1** put out forcibly or suddenly. **2** discard as unwanted. **3** expel (a troublemaker etc.). **4** build (a wing of a house, a pier, or a projecting or prominent thing). **5** (of a plant) rapidly develop (a side-shoot etc.). **6** put forward tentatively. **7** reject (a proposal or bill in Parliament). **8** confuse or distract (a person speaking, thinking, or acting) from the matter in hand. **9** *Cricket & Baseball* put out (an opponent) by throwing the ball to the wicket or base. **throw over** desert or abandon. **throw stones** cast aspersions. **throw together 1** assemble hastily. **2** bring into casual contact. **throw up 1** abandon. **2** resign from. **3** *colloq.* vomit. **4** erect hastily. **5** bring to notice. **6** lift (a sash-window) quickly. **7** abandon a contest; admit defeat. **throw one's weight about** (or **around**) *colloq.* act in a domineering or over-assertive manner. □ **throwable** *adj.* **thrower** *n.* (also in *comb.*). [OE *thrāwan* twist, turn f. WG]

throwster /ˈθrəʊstə(r)/ *n.* a person who throws silk.

thru *US* var. of THROUGH.

thrum[1] /θrʌm/ *v. & n.* ● *v.* (**thrummed**, **thrumming**) **1** *tr.* play (a stringed instrument) monotonously or unskilfully. **2** *intr.* (often foll. by *on*) beat or drum idly or monotonously. ● *n.* **1** such playing. **2** the resulting sound. [imit.]

thrum[2] /θrʌm/ *n. & v.* ● *n.* **1** (in weaving) the unwoven end of a warp-thread, or the whole of such ends, left when the finished web is cut away. **2** any short loose thread. ● *v.tr.* (**thrummed**, **thrumming**) make of or cover with thrums. □ **thrummer** *n.* **thrummy** *adj.* [OE f. Gmc]

thrush[1] /θrʌʃ/ *n.* a medium-sized songbird of the family Turdidae; esp. one of the genus *Turdus*, usu. with a brown back and spotted breast (*mistle thrush, song thrush*). [OE *thrysce* f. Gmc: cf. THROSTLE]

thrush[2] /θrʌʃ/ *n.* **1** *Med.* **a** infection with the yeastlike fungus *Candida albicans*, causing white patches in the mouth and throat. **b** a similar infection of the vagina. **2** inflammation affecting the frog of a horse's foot. [17th c.: orig. unkn.]

thrust /θrʌst/ *v. & n.* ● *v.* (*past* and *past part.* **thrust**) **1** *tr.* push with a sudden impulse or with force (*thrust the letter into my pocket*). **2** *tr.* (foll. by *on*) impose (a thing) forcibly; enforce acceptance of (a thing) (*had it thrust on me*). **3** *intr.* (foll. by *at*, *through*) pierce or stab; make a sudden lunge. **4** *tr.* make (one's way) forcibly. **5** *intr.* (foll. by *through*, *past*, etc.) force oneself (*thrust past me abruptly*). **6** *intr.* (as **thrusting** *adj.*) aggressive, ambitious. ● *n.* **1** a sudden or forcible push or lunge. **2** the propulsive force developed by a jet or rocket engine. **3** a strong attempt to penetrate an enemy's line or territory. **4** a remark aimed at a person. **5** the stress between the parts of an arch etc. **6** (often foll. by *of*) the chief theme or gist of remarks etc. **7** an attack with the point of a weapon. **8** (in full **thrust fault**) *Geol.* a low-angle reverse fault, with

older strata displaced horizontally over newer. □ **thrust-block** a casting or frame carrying or containing the bearings on which the collars of a propeller shaft press. **thrust oneself** (or **one's nose**) **in** obtrude, interfere. **thrust stage** *Theatr.* a stage extending into the audience. [ME *thruste* etc. f. ON *thrýsta*]

thruster /ˈθrʌstə(r)/ *n.* **1** a person or thing that thrusts. **2** a small rocket engine used to provide extra or correcting thrust on a spacecraft.

thruway *US* var. of THROUGHWAY.

Thucydides /θjuːˈsɪdɪˌdiːz/ (*c.*455–*c.*400 BC), Greek historian. He is remembered for his *History of the Peloponnesian War*, a conflict in which he fought on the Athenian side. The work covers events up to about 411 and presents an analysis of the origins and course of the war, based on painstaking inquiry into what actually happened and including the reconstruction of political speeches of figures such as Pericles, whom he greatly admired.

thud /θʌd/ *n. & v.* ● *n.* a low dull sound as of a blow on a non-resonant surface. ● *v.intr.* (**thudded**, **thudding**) make or fall with a thud. □ **thuddingly** *adv.* [prob. f. OE *thyddan* thrust]

Thug /θʌɡ/ *n. hist.* a member of a religious organization of robbers and assassins in India. Devotees of the goddess Kali, the Thugs waylaid and strangled their victims, usually travellers, in a ritually prescribed manner. They were suppressed by the British in the 1830s. [Hindi & Marathi *ṭhag* swindler]

thug /θʌɡ/ *n.* a vicious or brutal ruffian. □ **thuggery** *n.* **thuggish** *adj.* **thuggishly** *adv.* **thuggishness** *n.* [THUG]

thuggee /θʌˈɡiː/ *n. hist.* murder practised by the Thugs. □ **thuggism** *n.* [Hindi *ṭhagī* (as THUG)]

thuja /ˈθuːjə/ *n.* (also **thuya**) an evergreen coniferous tree of the genus *Thuja*, native to North America and East Asia, usu. of pyramidal habit with flattened shoots bearing scale-leaves. Also called *arbor vitae, tree of life*. [mod.L f. Gk *thuia* as Afr. tree]

Thule 1 /ˈθjuːliː/ a country described by the ancient Greek explorer Pytheas (*c.*310 BC) as being six days' sail north of Britain, variously identified with Iceland, the Shetland Islands, and, most plausibly, Norway. It was regarded by the ancients as the northernmost part of the world (cf. ULTIMA THULE). **2** /ˈθjuːl, ˈθjuːlː/ an Eskimo culture widely distributed from Alaska to Greenland AD *c.*500–1400. Members of the culture hunted whales, seals, walrus, and caribou, lived in semi-permanent houses with whalebone frames located by the sea in winter and probably in tents made of animal skin located inland in summer, and travelled by kayak or dog-drawn sled. **3** /ˈtuːliː/ a settlement on the NW coast of Greenland, founded in 1910 by the Danish explorer Knud Rasmussen (1879–1933). A US air base is located nearby.

thulium /ˈθjuːlɪəm/ *n.* a soft silvery-white metallic chemical element (atomic number 69; symbol **Tm**). A member of the lanthanide series, thulium was discovered by the Swedish chemist Per Teodor Cleve (1840–1905) in 1879. It has few commercial uses. [f. THULE]

thumb /θʌm/ *n. & v.* ● *n.* **1 a** a short thick terminal projection on the human hand, set lower and apart from the other four and opposable to them. **b** a digit of other animals corresponding to this. **2** the part of a glove etc. intended to cover a thumb. ● *v.* **1** *tr.* wear or soil (pages etc.) with a thumb (*a well-thumbed book*). **2** *intr.* turn over pages with or as with a thumb (*thumbed through the directory*). **3** *tr.* request or obtain (a lift in a passing vehicle) by signalling with a raised thumb. **4** *tr.* gesture at (a person) with the thumb. □ **be all thumbs** be clumsy with one's hands. **thumb index** *n.* a set of lettered grooves cut down the side of a diary, dictionary, etc. for easy reference. ● *v.tr.* provide (a book etc.) with these. **thumb one's nose** = *cock a snook* (see SNOOK[1]). **thumb-nut** a nut shaped for turning with the thumb and forefinger. **thumbs down** an indication of rejection or failure. **thumbs up** an indication of satisfaction or approval. **under a person's thumb** completely dominated by a person. □ **thumbed** *adj.* (also in *comb.*). **thumbless** *adj.* [OE *thūma* f. a WG root = swell]

thumbnail /ˈθʌmneɪl/ *n.* **1** the nail of a thumb. **2** (*attrib.*) denoting conciseness (*a thumbnail sketch*).

thumbprint /ˈθʌmprɪnt/ *n.* an impression of a thumb, esp. as used for identification.

thumbscrew /ˈθʌmskruː/ *n.* **1** (usu. in *pl.*) an instrument of torture for crushing the thumbs. **2** a screw with a protruding winged or flattened head for turning with the thumb and forefinger.

thumbtack /ˈθʌmtæk/ *n. N. Amer.* a drawing-pin.

thump /θʌmp/ *v. & n.* ● *v.* **1** *tr.* beat or strike heavily, esp. with the fist (*threatened to thump me*). **2** *intr.* throb or pulsate strongly (*my heart was thumping*). **3** *intr.* (foll. by *at*, *on*, etc.) deliver blows, esp. to attract attention (*thumped on the door*). **4** *tr.* (often foll. by *out*) play (a tune on a piano etc.) with a heavy touch. **5** *intr.* tread heavily. ● *n.* **1** a heavy blow. **2** the sound of this. □ **thumper** *n.* [imit.]

thumping /ˈθʌmpɪŋ/ *adj. colloq.* big, prominent (*a thumping majority; a thumping lie*).

thunder /ˈθʌndə(r)/ *n. & v.* ● *n.* **1** a loud rumbling or crashing noise heard after a lightning flash and due to the expansion of rapidly heated air. **2** a resounding loud deep noise (*thunders of applause*). **3** strong censure or denunciation. ● *v.* **1** *intr.* (prec. by *it* as subject) thunder sounds (*it is thundering; if it thunders*). **2** *intr.* make or proceed with a noise suggestive of thunder (*the applause thundered in my ears; the traffic thundered past*). **3** *tr.* utter or communicate (approval, disapproval, etc.) loudly or impressively. **4** *intr.* (foll. by *against* etc.) **a** make violent threats etc. against. **b** criticize violently. □ **steal a person's thunder** see STEAL. **thunder-box** *colloq.* a primitive lavatory. □ **thunderer** *n.* **thunderless** *adj.* **thundery** *adj.* [OE *thunor* f. Gmc]

Thunder Bay a city on an inlet of Lake Superior in SW Ontario; pop. (1991) 109,330. Now one of Canada's major ports, Thunder Bay was created in 1970 by the amalgamation of the twin cities of Fort William and Port Arthur and two adjoining townships.

thunderbolt /ˈθʌndəˌbəʊlt/ *n.* **1 a** a flash of lightning with a simultaneous crash of thunder. **b** a stone etc. imagined to be a destructive bolt. **2** a sudden or unexpected occurrence or item of news. **3** a supposed bolt or shaft as a destructive agent, esp. as an attribute of a god.

thunderbug /ˈθʌndəˌbʌg/ *n.* = THRIPS.

thunderclap /ˈθʌndəˌklæp/ *n.* **1** a crash of thunder. **2** something startling or unexpected.

thundercloud /ˈθʌndəˌklaʊd/ *n.* a cumulus cloud with a tall diffuse top, charged with electricity and producing thunder and lightning.

thunderflash /ˈθʌndəˌflæʃ/ *n.* a noisy but harmless explosive used esp. in military exercises.

thunderfly /ˈθʌndəˌflaɪ/ *n.* (*pl.* **-flies**) = THRIPS.

thunderhead /ˈθʌndəˌhed/ *n.* esp. *US* a rounded cumulus cloud projecting upwards and heralding thunder.

thundering /ˈθʌndərɪŋ/ *adj. colloq.* very big or great (*a thundering nuisance*). □ **thunderingly** *adv.*

thunderous /ˈθʌndərəs/ *adj.* **1** like thunder. **2** very loud. □ **thunderously** *adv.* **thunderousness** *n.*

thunderstorm /ˈθʌndəˌstɔːm/ *n.* a storm with thunder and lightning and usu. heavy rain or hail.

thunderstruck /ˈθʌndəˌstrʌk/ *adj.* overwhelmingly surprised or startled; amazed.

thunk /θʌŋk/ *n. & v.intr. colloq.* = THUD. [Imit.]

Thur. *abbr.* Thursday.

Thurber /ˈθɜːbə(r)/, James (Grover) (1894–1961), American humorist and cartoonist. In 1927 he began his lifelong association with the *New Yorker* magazine, in which he published many of his essays, stories, and sketches. Among his many collections are *My Life and Hard Times* (1933) and *My World — And Welcome to It* (1942), which contains the story 'The Secret Life of Walter Mitty'.

thurible /ˈθjʊərɪb(ə)l/ *n.* a censer. [ME f. OF *thurible* or L *t(h)uribulum* f. *thus thur-* incense (as THURIFER)]

thurifer /ˈθjʊərɪfə(r)/ *n.* an acolyte carrying a censer. [LL f. *thus thuris* incense f. Gk *thuos* sacrifice + *-fer* -bearing]

Thuringia /θjʊəˈrɪndʒɪə/ (German **Thüringen** /ˈtyːrɪŋən/) a densely forested state of central Germany; capital, Erfurt.

Thurs. *abbr.* Thursday.

Thursday /ˈθɜːzdeɪ, -dɪ/ *n. & adv.* ● *n.* the fifth day of the week, following Wednesday. ● *adv. colloq.* **1** on Thursday. **2** (**Thursdays**) on Thursdays; each Thursday. [OE *thunresdæg, thur(e)sdæg*, day of thunder, named after Thor, whose name was substituted for that of Jove or Jupiter, representing LL *Jovis dies* day of Jupiter]

Thurso /ˈθɜːsəʊ/ a fishing port on the northern coast of Scotland, in Highland Region, the northernmost town on the mainland of Britain; pop. (1981) 8,900.

thus /ðʌs/ *adv. formal* **1 a** in this way. **b** as indicated. **2 a** accordingly. **b** as a result or inference. **3** to this extent; so (*thus far; thus much*). □ **thusly** *adv. colloq.* [OE (= OS *thus*), of unkn. orig.]

thuya var. of THUJA.

thwack /θwæk/ *v. & n. colloq.* ● *v.tr.* hit with a heavy blow; whack. ● *n.* a heavy blow. [imit.]

thwaite /θweɪt/ *n. Brit. dial.* a piece of wild land made arable. ¶ Now usually only in place-names. [ON *thveit(i)* paddock, rel. to OE *thwītan* to cut]

thwart /θwɔːt/ *v., n., prep., & adv.* ● *v.tr.* frustrate or foil (a person or purpose etc.). ● *n.* a rower's seat placed across a boat. ● *prep. & adv. archaic* across, athwart. [ME *thwert* (adv.) f. ON *thvert* neut. of *thverr* transverse = OE *thwe(o)rh* f. Gmc]

thy /ðaɪ/ *poss.pron.* (attrib.) (also **thine** /ðaɪn/ before a vowel) of or belonging to you. ¶ now replaced by *your* except in some formal, liturgical, dialect, and poetic uses. [ME *thī*, reduced f. *thīn* THINE]

Thyestes /θaɪˈestiːz/ *Gk Mythol.* the brother of Atreus and father of Aegisthus. □ **Thyestean** /-ˈestɪən/ *adj.*

thylacine /ˈθaɪləˌsiːn, -ˌsaɪn/ *n.* a large doglike flesh-eating marsupial, *Thylacinus cynocephalus*, with dark stripes over the rump. It has been confined to Tasmania for several centuries, but is now very rare or extinct. Also called *Tasmanian wolf* or *tiger*. [mod.L f. Gk *thulakos* pouch]

thyme /taɪm/ *n.* a dwarf labiate shrub of the genus *Thymus*, with aromatic leaves; esp. *garden thyme* (*T. vulgaris*), grown for culinary use. □ **thymy** *adj.* [ME f. OF *thym* f. L *thymum* f. Gk *thumon* f. *thuō* burn a sacrifice]

thymi *pl.* of THYMUS.

thymidine /ˈθaɪmɪˌdiːn/ *n. Biochem.* a nucleoside of thymine and deoxyribose present in DNA. [THYMINE + -IDE + -INE⁴]

thymine /ˈθaɪmiːn/ *n. Biochem.* a pyrimidine found in all living tissue as a component base of DNA. [*thymic* (as THYMUS) + -INE⁴]

thymol /ˈθaɪmɒl/ *n. Chem.* a white crystalline phenol obtained from oil of thyme and used as an antiseptic. [as THYME + -OL]

thymus /ˈθaɪməs/ *n.* (*pl.* **thymi** /-maɪ/) (in full **thymus gland**) *Anat.* a lymphoid organ situated in the neck of vertebrates (in humans becoming much smaller at the approach of puberty) producing T-lymphocytes for the immune response. [mod.L f. Gk *thumos*]

thyristor /θaɪˈrɪstə(r)/ *n. Electronics* a four-layered semiconductor rectifier in which the flow of current between two electrodes is triggered by a signal at a third electrode. [Gk *thura* gate + TRANSISTOR]

thyro- /ˈθaɪərəʊ/ *comb. form* (also **thyreo-** /-rɪəʊ/) thyroid.

thyroid /ˈθaɪrɔɪd/ *n. & adj.* ● *n.* **1** *Anat.* (in full **thyroid gland**) a large ductless gland in the neck of vertebrates secreting hormones which regulate growth and development through the rate of metabolism. **2** *Med.* an extract prepared from the thyroid gland of animals and used in treating goitre and cretinism etc. ● *adj. Anat. & Zool.* **1** connected with the thyroid cartilage (*thyroid artery*). **2** shield-shaped. □ **thyroid cartilage** *Anat.* a large cartilage of the larynx, the projection of which in humans forms the Adam's apple. [obs.F *thyroide* or mod.L *thyroides*, irreg. f. Gk *thureoeidēs* f. *thureos* oblong shield]

thyroxine /ˌθaɪəˈrɒksɪn, -siːn/ *n.* the main hormone produced by the thyroid gland, which increases the metabolic rate and regulates growth and development in animals. [THYROID + OX- + -INE⁴]

thyrsus /ˈθɜːsəs/ *n.* (*pl.* **thyrsi** /-saɪ/) **1** *Gk & Rom. Antiq.* a staff tipped with an ornament like a pine-cone, an attribute of Bacchus. **2** *Bot.* an inflorescence as in lilac, with the primary axis racemose and the secondary axis cymose. [L f. Gk *thursos*]

Thysanoptera /ˌθaɪsəˈnɒptərə/ *n.pl. Zool.* an order of insects comprising the thrips or thunderflies. □ **thysanopteran** *n. & adj.* **thysanopterous** *adj.* [mod.L f. Gk *thusanos* fringe + *pteron* wing]

Thysanura /ˌθaɪsəˈnjʊərə/ *n.pl. Zool.* an order of wingless insects comprising the bristletails and silverfish. □ **thysanuran** *n. & adj.* **thysanurous** *adj.* [mod.L f. Gk *thusanos* tassel + *oura* tail]

thyself /ðaɪˈself/ *pron. archaic & dial.* emphat. & refl. form of THOU¹, THEE.

Ti *symb. Chem.* the element titanium.

ti¹ /tiː/ *n.* = *cabbage tree* 1. [Tahitian, Maori, etc.]

ti² var. of TE.

Tiamat /ˈtɪəmɑːt/ *Babylonian Mythol.* a monstrous primeval being, a she-dragon who was the mother of the first Babylonian gods. She was slain by Marduk.

Tiananmen Square /ˈtjenənmən/ a square in the centre of Beijing adjacent to the Forbidden City, the largest public open space in the world. Long the site for festivals, rallies, and demonstrations, it attracted worldwide attention in spring 1989 when it was occupied by hundreds of thousands of student-led protesters of the emerging pro-democracy movement. The demonstration ended when government troops opened fire on the unarmed protesters, killing over 2,000. [Chin., = square of heavenly peace]

Tianjin /tjen'dʒɪn/ (also **Tientsin** /-'tʃɪn/) a port in NE China, in Hubei province; pop. (1990) 5,700,000.

Tian Shan see TIEN SHAN.

tiara /tɪ'ɑːrə/ n. **1** a jewelled ornamental band worn on the front of a woman's hair. **2** a three-crowned diadem worn by a pope. **3** hist. a turban worn by ancient Persian kings. □ **tiaraed** /-rəd/ adj. (also **tiara'd**). [L f. Gk, of unkn. orig.]

Tiber /'taɪbə(r)/ (Italian **Tevere** /'te:vere/) a river of central Italy, upon which Rome stands. It rises in the Tuscan Apennines and flows 405 km (252 miles) generally south-westwards, entering the Tyrrhenian Sea at Ostia.

Tiberias, Lake /taɪ'bɪərɪəs/ an alternative name for the Sea of Galilee (see GALILEE, SEA OF).

Tiberius /taɪ'bɪərɪəs/ (full name Tiberius Julius Caesar Augustus) (42 BC–AD 37), Roman emperor AD 14–37. He was the adopted successor of his stepfather and father-in-law Augustus, under whom he had pursued a distinguished military career. As emperor he sought to continue his stepfather's policies but he became increasingly tyrannical and his reign was marked by a growing number of treason trials and executions. In 26 he retired to Capri, never returning to Rome.

Tibesti Mountains /tɪ'bestɪ/ a mountain range in north central Africa, in the Sahara in northern Chad and southern Libya. It rises to 3,415 m (11,201 ft) at Emi Koussi, the highest point in the Sahara.

Tibet /tɪ'bet/ (Chinese **Xizang** /ʃiː'zæŋ/) a mountainous country in Asia on the northern side of the Himalayas, since 1965 forming an autonomous region in the west of China; pop. (1990) 2,196,000; official languages, Tibetan and Chinese; capital, Lhasa. Ruled by Buddhist lamas since the 7th century, Tibet was conquered by the Mongols in the 13th century and the Manchus in the 18th. China extended its authority over Tibet in 1951 but only gained full control after crushing a revolt in 1959, during which the country's spiritual leader, the Dalai Lama, escaped into India; he remains in exile and sporadic unrest has continued. Tibet is sometimes called 'the roof of the world'; most of the territory forms a high plateau with an average elevation of over 4,000 m (12,500 ft), and it is the source of some of Asia's largest rivers, including the Yangtze, Salween, and Mekong. □ **Little Tibet** see BALTISTAN.

Tibetan /tɪ'bet(ə)n/ n. & adj. ● n. **1 a** a native of Tibet. **b** a person of Tibetan descent. **2** the language of Tibet, spoken by about 4 million people in Tibet and in neighbouring areas of China, India, and Nepal. It belongs to the Sino-Tibetan language group and is most closely related to Burmese, although its alphabet is based on that of Sanskrit. ● adj. of or relating to Tibet or its language.

Tibetan Buddhism n. the religion of Tibet, a form of Mahayana Buddhism. Buddhism entered Tibet in the 7th century AD, but did not take hold until the next century, when an Indian missionary added to Buddhist beliefs elements of the indigenous Tibetan religion. The head of the religion is the Dalai Lama, who was also head of the Tibetan state until the Chinese annexation of the country. Tibetan Buddhism is notable for the degree of participation in religious activity: until the 1950s about a quarter of Tibetan men were members of a religious order.

tibia /'tɪbɪə/ n. (pl. **tibiae** /-bɪ,iː/) **1** Anat. the inner and usu. larger of two bones extending from the knee to the ankle. **2** Zool. the tibiotarsus of a bird. **3** Zool. the fourth joint of the leg in insects etc., seen as a long slender segment between the femur and tarsus. □ **tibial** adj. [L, = shin-bone]

tibiotarsus /,tɪbɪəʊ'tɑːsəs/ n. (pl. **tibiotarsi** /-saɪ/) Zool. the bone in a bird corresponding to the tibia fused at the lower end with some bones of the tarsus. [TIBIA + TARSUS]

Tibullus /tɪ'bʌləs/, Albius (c.50–19 BC), Roman poet. He is known for his elegiac love poetry and for his celebration of peaceful rural life in preference to the harsh realities of military campaigning.

tic /tɪk/ n. a habitual spasmodic contraction of the muscles esp. of the face. □ **tic douloureux** /,duːluˈruː, -ˈrɜː/ Med. trigeminal neuralgia. [F f. It. ticchio: douloureux F, = painful]

tice /taɪs/ n. **1** Cricket = YORKER. **2** Croquet a stroke tempting an opponent to aim at one's ball. [tice (now dial.), = ENTICE]

Tichborne claimant /'tɪtʃbɔːn/ see ORTON[1].

Ticino /tɪ'tʃiːnəʊ/ (French **Tessin** /tɛsɛ̃/, German **Tessin** /tɛ'siːn/) a predominantly Italian-speaking canton in southern Switzerland, on the Italian border; capital, Bellinzona. It joined the Swiss Confederation in 1803.

tick[1] /tɪk/ n. & v. ● n. **1** a slight recurring click, esp. that of a watch or clock. **2** esp. Brit. colloq. a moment; an instant. **3** a mark (✓) to denote correctness, check items in a list, etc. ● v. **1** intr. **a** (of a clock etc.) make ticks. **b** (foll. by away) (of time etc.) pass. **2** intr. **a** (of a mechanism) work, function (take it apart to see how it ticks). **3** tr. **a** mark (a written answer etc.) with a tick. **b** (often foll. by off) mark (an item in a list etc.) with a tick in checking. □ **in two ticks** Brit. colloq. in a very short time. **tick off 1** colloq. reprimand. **2** US sl. annoy, anger; dispirit. **tick over 1** (of an engine etc.) idle. **2** (of a person, project, etc.) be working or functioning at a basic or minimum level. **tick-tack** (or **tic-tac**) Brit. a kind of manual semaphore signalling used by racecourse bookmakers to exchange information. **tick-tack-toe** N. Amer. noughts and crosses. **tick-tock** the ticking of a large clock etc. **what makes a person tick** colloq. a person's motivation. □ **tickless** adj. [ME: cf. Du. tik, LG tikk touch, tick]

tick[2] /tɪk/ n. **1** a blood-sucking arachnid of the order Acarina, parasitic on the skin of dogs, cattle, etc. **2** a parasitic fly, esp. a ked. **3** colloq. an unpleasant or despicable person. □ **tick-bird** = ox-pecker. **tick fever** a bacterial or rickettsial fever transmitted by the bite of a tick. [OE ticca (recorded as ticia); ME teke, tyke: cf. MDu., MLG tēke, OHG zēcho]

tick[3] /tɪk/ n. colloq. credit (buy goods on tick). [app. an abbr. of TICKET in phr. on the ticket]

tick[4] /tɪk/ n. **1** the cover of a mattress or pillow. **2** = TICKING. [ME tikke, tēke f. WG f. L theca f. Gk thēkē case]

ticker /'tɪkə(r)/ n. colloq. **1** the heart. **2** a watch. **3** N. Amer. a tape machine. □ **ticker-tape 1** a paper strip from a tape machine. **2** (often attrib.) this or similar material thrown from windows etc. to greet a celebrity.

ticket /'tɪkɪt/ n. & v. ● n. **1** a written or printed piece of paper or card entitling the holder to enter a place, participate in an event, travel by public transport, use a public amenity, etc. **2** an official notification of a traffic offence etc. (parking ticket). **3** Brit. a certificate of discharge from the army. **4** a certificate of qualification as a ship's master, pilot, etc. **5** a label attached to a thing and giving its price or other details. **6** esp. US **a** a list of candidates put forward by one group, esp. a political party. **b** the principles of a party. **7** (prec. by the) colloq. what is correct or needed. ● v.tr. (**ticketed, ticketing**) attach a ticket to. □ **have tickets on oneself** Austral. colloq. be conceited. **just the ticket** colloq. entirely appropriate. **ticket-day** Stock Exch. (in the UK) the day before settling day, when the names of actual purchasers are handed to stockbrokers. **ticket office** an office or kiosk where tickets are sold for transport, entertainment, etc. **ticket-of-leave man** Brit. hist. a prisoner or convict who had served part of his time and was granted certain concessions, esp. leave. □ **ticketed** adj. **ticketless** adj. [obs.F étiquet f. OF estiquet(te) f. estiquier, estechier fix f. MDu. steken]

tickety-boo /,tɪkətɪ'buː/ adj. Brit. colloq. all right; in order. [20th c.: orig. uncert.]

ticking /'tɪkɪŋ/ n. a stout usu. striped material used to cover mattresses etc. [TICK[4] + -ING[1]]

tickle /'tɪk(ə)l/ v. & n. ● v. **1 a** tr. apply light touches or strokes to (a person or part of a person's body) so as to excite the nerves and usu. produce laughter and spasmodic movement. **b** intr. be subject to this sensation (my foot tickles). **c** intr. produce this sensation (this woolly jumper tickles). **2** tr. excite agreeably; amuse or divert (a person, a sense of humour, vanity, etc.) (was highly tickled at the idea; this will tickle your fancy). **3** tr. catch (a trout etc.) by rubbing it so that it moves backwards into the hand. ● n. **1** an act of tickling. **2** a tickling sensation. □ **tickled pink** (or **to death**) colloq. extremely amused or pleased. □ **tickler** n. **tickly** adj. [ME, prob. frequent. of TICK[1]]

ticklish /'tɪklɪʃ/ adj. **1** sensitive to tickling. **2** (of a matter or person to be dealt with) difficult; requiring careful handling. □ **ticklishly** adv. **ticklishness** n.

tic-tac var. of tick-tack (see TICK[1]).

tidal /'taɪd(ə)l/ adj. relating to, like, or affected by tides (tidal basin; tidal river). □ **tidal bore** see BORE[3]. **tidal energy** = tidal power. **tidal flow** the regulated movement of traffic in opposite directions on the same stretch of road at different times of the day. **tidal power** power obtained by harnessing the energy of marine tidal flows. (See note below.) **tidal wave 1** an exceptionally large ocean wave, esp. = TSUNAMI. **2** (foll. by of) a widespread manifestation of feeling etc. □ **tidally** adv.

▪ The use of tidal energy is restricted to coastlines with large differences between high and low tide levels. Though other methods have been proposed, both ancient tide-mills and modern tidal power stations operate on the same principle: the rising tide passes through sluices in a dam or barrage, and is forced out past water-wheels or turbines as the tide falls again. The only schemes currently operating are in France, Russia, and China.

tidbit *N. Amer.* var. of TITBIT.

tiddledy-wink *US* var. of TIDDLY-WINK.

tiddler /ˈtɪdlə(r)/ *n. Brit. colloq.* **1** a small fish, esp. a stickleback or minnow. **2** an unusually small thing or person. [perh. rel. to TIDDLY² and *tittlebat*, a childish form of *stickleback*]

tiddly¹ /ˈtɪdlɪ/ *adj.* (**tiddlier, tiddliest**) esp. *Brit. colloq.* slightly drunk. [19th c., earlier = a drink: orig. unkn.]

tiddly² /ˈtɪdlɪ/ *adj.* (**tiddlier, tiddliest**) *Brit. colloq.* little. [19th c.: orig. unkn.]

tiddly-wink /ˈtɪdlɪˌwɪŋk/ *n.* (*US* **tiddledy-wink** /ˈtɪd(ə)ldɪ-/) **1** a counter flicked with another into a cup etc. **2** (in *pl.*) this game. [19th c.: perh. rel. to TIDDLY¹]

tide /taɪd/ *n. & v.* ● *n.* **1 a** the periodic rise and fall of the sea due to the attraction of the moon and sun (see EBB *n.* 1, FLOOD *n.* 3). **b** the water as affected by this. **2** a time or season (usu. in *comb.*: *Whitsuntide*). **3** a marked trend of opinion, fortune, or events. ● *v.intr.* drift with or as with the tide, esp. (of a ship) work in or out of harbour with the help of the tide. □ **tide-mill** a mill with a water-wheel driven by the tide. **tide over** enable or help (a person) to deal with an awkward situation, difficult period, etc. (*the money will tide me over until Friday*). **tide-rip** (or **-rips**) rough water caused by opposing tides. **work double tides** work twice the normal time, or extra hard. □ **tideless** *adj.* [OE *tīd* f. Gmc, rel. to TIME]

tideland /ˈtaɪdlænd/ *n. N. Amer.* land that is submerged at high tide.

tideline /ˈtaɪdlaɪn/ *n.* the edge defined by the tide on the shore.

tidemark /ˈtaɪdmɑːk/ *n.* **1** a mark made by the tide at high water. **2** esp. *Brit.* **a** a mark left round a bath at the level of the water in it. **b** a line on a person's body, garment, etc. marking the extent to which it has been washed.

tidetable /ˈtaɪdˌteɪb(ə)l/ *n.* a table indicating the times of high and low tides at a place.

tidewaiter /ˈtaɪdˌweɪtə(r)/ *n. hist.* a customs officer who boarded ships on their arrival to enforce the customs regulations.

tidewater /ˈtaɪdˌwɔːtə(r)/ *n.* **1** water brought by or affected by tides. **2** (*attrib.*) *US* affected by tides (*tidewater region*).

tidewave /ˈtaɪdweɪv/ *n.* an undulation of water passing round the earth and causing high and low tides.

tideway /ˈtaɪdweɪ/ *n.* **1** a channel in which a tide runs, esp. the tidal part of a river. **2** the ebb or flow in a tidal channel.

tidings /ˈtaɪdɪŋz/ *n.* (as *sing.* or *pl.*) *literary* news, information. [OE *tīdung*, prob. f. ON *títhindi* events f. *títhr* occurring]

tidy /ˈtaɪdɪ/ *adj., n., & v.* ● *adj.* (**tidier, tidiest**) **1** neat, orderly; methodically arranged. **2** (of a person) methodically inclined. **3** *colloq.* considerable (*it cost a tidy sum*). ● *n.* (*pl.* **-ies**) **1** a receptacle for holding small objects or waste scraps, esp. in a kitchen sink. **2** an act or spell of tidying. **3** esp. *US* a detachable ornamental cover for a chair-back etc. ● *v.tr.* (**-ies, -ied**) (also *absol.*; often foll. by *up*) put in good order; make (oneself, a room, etc.) tidy. □ **tidily** *adv.* **tidiness** *n.* [ME, = timely etc., f. TIDE + -Y¹]

tie /taɪ/ *v. & n.* ● *v.* (**tying**) **1** *tr.* **a** attach or fasten with string or cord etc. (*tie the dog to the gate; tie his hands together; tied on a label*). **b** link conceptually. **2** *tr.* **a** form (a string, ribbon, shoelace, necktie, etc.) into a knot or bow. **b** form (a knot or bow) in this way. **3** *tr.* restrict or limit (a person) as to conditions, occupation, place, etc. (*is tied to his family*). **4** *intr.* (often foll. by *with*) achieve the same score or place as another competitor (*they tied at ten games each; tied with her for first place*). **5** *tr.* hold (rafters etc.) together by a crosspiece etc. **6** *tr. Mus.* **a** unite (written notes) by a tie. **b** perform (two notes) as one unbroken note. ● *n.* **1** a cord or chain etc. used for fastening. **2** a strip of material worn round the collar and tied in a knot at the front with the ends hanging down. **3** a thing that unites or restricts persons; a bond or obligation (*family ties; ties of friendship; children are a real tie*). **4** a draw, dead heat, or equality of score among competitors. **5** *Brit.* a match between any pair from a group of competing players or teams. **6** (also **tie-beam** etc.) a rod or beam holding parts of a structure together. **7** *Mus.* a curved line above or below two notes of the same pitch indicating that they are to be played for the combined duration of their time values. **8** *N. Amer.* a railway sleeper. **9** *US* a shoe tied with a lace. □ **fit to be tied** *colloq.* very angry. **tie-break** (or **-breaker**) a means of deciding a winner from competitors who have tied. **tie down** = TIE *v.* 3 above. **tie-dye** (or **tie and dye**) a method of producing dyed patterns by tying string etc. to shield parts of the fabric from the dye. **tie in** (foll. by *with*) bring

into or have a close association or agreement. **tie-in** *n.* **1** a connection or association. **2** (often *attrib.*) esp. *N. Amer.* a form of sale or advertising that offers or requires more than a single purchase. **3** the joint promotion of related commodities etc. (e.g. a book and a film). **tie the knot** *colloq.* get married. **tie-line** a transmission line connecting parts of a system, esp. a telephone line connecting two private branch exchanges. **tie-pin** (or **-clip**) an ornamental pin or clip for holding a tie in place. **tie up 1** bind or fasten securely with cord etc. **2** invest or reserve (capital etc.) so that it is not immediately available for use. **3** moor (a boat). **4** secure (an animal). **5** obstruct; prevent from acting freely. **6** secure or complete (an undertaking etc.). **7** (often foll. by *with*) = tie in. **8** (usu. in *passive*) fully occupy (a person). □ **tieless** *adj.* [OE *tīgan, tēgan* (v.), *tēah, tēg* (n.) f. Gmc]

tied /taɪd/ *adj. Brit.* **1** (of a house) occupied subject to the tenant's working for its owner. **2** (of a public house etc.) bound to supply the products of a particular brewery only.

Tien Shan /tjen ˈʃæn/ (also **Tian Shan**) a range of mountains lying to the north of the Tarim Basin in the Xinjiang autonomous region and eastern Kyrgyzstan. Extending for about 2,500 km (1,500 miles), it rises to 7,439 m (24,406 ft) at Pik Pobedy.

Tientsin see TIANJIN.

Tiepolo /ˈtjepəˌləʊ/, Giovanni Battista (1696–1770), Italian painter. One of the leading artists of the rococo style, he painted numerous frescos and altarpieces in Italy, Germany, and Spain, assisted on many commissions by his two sons. His painting is characterized by dramatic foreshortening, translucent colour, and settings of theatrical splendour. His works include the *Antony and Cleopatra* frescos in the Palazzo Labia, Venice (*c.*1750) and the decoration of the residence of the Prince-Bishop at Würzburg (1751–3). In 1762, at the request of Charles III, he moved to Madrid where he spent the rest of his life, working on the ceilings of the Royal Palace and the royal chapel at Aranjuez.

tier /tɪə(r)/ *n.* **1** a row or rank or unit of a structure, as one of several placed one above another (*tiers of seats*). **2** *Naut.* **a** a circle of coiled cable. **b** a place for a coiled cable. □ **tiered** *adj.* (also in *comb.*). [earlier *tire* f. F f. *tirer* draw, elongate f. Rmc]

tierce /tɪəs/ *n.* **1** *Eccl.* = TERCE. **2** *Mus.* an interval of two octaves and a major third. **3** a sequence of three cards. **4** *Fencing* the third of eight parrying positions. **5** *archaic* the corresponding thrust. **5** *archaic* a former wine-measure of one-third of a pipe. **b** a cask containing a certain quantity (varying with the goods) esp. of provisions. [ME f. OF *t(i)erce* f. L *tertia* fem. of *tertius* third]

tierced /tɪəst/ *adj. Heraldry* divided into three parts of different tinctures.

tiercel var. of TERCEL.

tiercet var. of TERCET.

Tierra del Fuego /tɪˌeərə del ˈfweɪgəʊ/ an island at the southern extremity of South America, separated from the mainland by the Strait of Magellan. Its name is Spanish for 'land of fire'. Discovered by Ferdinand Magellan in 1520, it is now divided between Argentina and Chile.

tiff /tɪf/ *n. & v.* ● *n.* **1** a slight or petty quarrel. **2** a fit of peevishness. ● *v.intr.* have a petty quarrel; bicker. [18th c.: orig. unkn.]

Tiffany /ˈtɪfənɪ/, Louis Comfort (1848–1933), American glass-maker and interior decorator. He was the son of Charles Louis Tiffany (1812–1902), who founded the New York jewellers Tiffany and Company. A leading exponent of American art nouveau, he established an interior decorating firm in New York in 1881 which produced stained glass and mosaic in a distinctive style, as well as iridescent glass vases and lamps.

tiffany /ˈtɪfənɪ/ *n.* (*pl.* **-ies**) thin gauze muslin. [orig. dress worn on Twelfth Night, f. OF *tifanie* f. eccl.L *theophania* f. Gk *theophaneia* Epiphany]

tiffin /ˈtɪfɪn/ *n. & v. Ind.* ● *n.* a light meal, esp. lunch. ● *v.intr.* (**tiffined, tiffining**) take lunch etc. [app. f. *tiffing* sipping]

Tiflis /ˈtɪfliːs/ the official Russian name (1845–1936) for TBILISI.

tig /tɪg/ *n.* = TAG² *n.* 1 [var. of TICK¹]

Tiger /ˈtaɪgə(r)/ *n.* a Tamil Tiger.

tiger /ˈtaɪgə(r)/ *n.* **1** a large powerful feline, *Panthera tigris*, found in India and some other parts of Asia, having a tawny yellow coat with black stripes. **2** a fierce, energetic, or formidable person. □ **tiger beetle** a carnivorous beetle of the family Cicindelidae, often flying in sunshine, with spotted or striped wing-cases. **tiger-cat 1** a moderate-sized striped feline, e.g. the ocelot, serval, or margay. **2** *Austral.* a catlike flesh-eating marsupial of the genus *Dasyurus*. **tiger's-eye** (or **tiger-**

eye) **1** a silky yellow-brown semiprecious variety of quartz, often striped. **2** *US* a pottery-glaze of similar appearance. **tiger lily** a tall garden lily, *Lilium tigrinum*, with flowers of dull orange spotted with black or purple. **tiger moth** a usu. brightly coloured moth of the family Arctiidae, with richly spotted and streaked wings suggesting a tiger's skin, esp. *Arctia caja*. **tiger shark** a voracious striped or spotted shark, esp. *Galeocerdo cuvieri* of warm seas or *Stegosoma tigrinum* of the Indian Ocean. **tiger-wood** a striped or streaked wood used for cabinet-making. □ **tigerish** *adj.* **tigerishly** *adv.* [ME f. OF *tigre* f. L *tigris* f. Gk *tigris*]

tight /taɪt/ *adj., n.,* & *adv.* ● *adj.* **1** closely held, drawn, fastened, fitting, etc. (*a tight hold; a tight skirt*). **2 a** closely and firmly put together (*a tight joint*). **b** close, evenly matched (*a tight finish*). **3** (of clothes etc.) too closely fitting (*my shoes are rather tight*). **4** impermeable, impervious, esp. (in *comb.*) to a specified thing (*watertight*). **5** tense; stretched so as to leave no slack (*a tight bowstring*). **6** *colloq.* drunk. **7** *colloq.* (of a person) mean, stingy. **8 a** (of money or materials) not easily obtainable. **b** (of a money market) in which money is tight. **9 a** (of precautions, a programme, etc.) stringent, demanding. **b** presenting difficulties (*a tight situation*). **c** (of an organization, group, or member) strict, disciplined. **10** produced by or requiring great exertion or pressure (*a tight squeeze*). **11** (of control etc.) strictly imposed. ● *adv.* tightly (*hold tight!*). □ **tight-assed** (or **-arsed**) esp. *N. Amer. sl.* inhibited, strait-laced. **tight corner** (or **place** or **spot**) a difficult situation. **tight end** *Amer. Football* an offensive end who lines up close to the tackle. **tight-fisted** stingy. **tight-fitting** (of a garment) fitting (often too) close to the body. **tight-knit** (or **tightly-**) = *close-knit* (see CLOSE¹). **tight-lipped** with or as with the lips compressed to restrain emotion or speech. □ **tightly** *adv.* **tightness** *n.* [prob. alt. of *thight* f. ON *théttr* watertight, of close texture]

tighten /ˈtaɪt(ə)n/ *v.tr.* & *intr.* (also foll. by *up*) make or become tight or tighter. □ **tighten one's belt** see BELT.

tightrope /ˈtaɪtrəʊp/ *n.* a rope stretched tightly high above the ground, on which acrobats perform.

tights /taɪts/ *n.pl.* **1** a thin close-fitting wool or nylon etc. garment covering the legs and the lower part of the torso, worn by women in place of stockings. **2** a similar garment worn by a dancer, acrobat, etc.

Tiglath-pileser /ˌtɪɡlæθpaɪˈliːzə(r)/ the name of three kings of Assyria, notably:

Tiglath-pileser I reigned *c.*1115–*c.*1077 BC. He extended Assyrian territory further into Asia Minor, taking Cappadocia and reaching Syria, as well as expanding his kingdom to the upper Euphrates and defeating the king of Babylonia.

Tiglath-pileser III (known as Pulu) reigned *c.*745–727 BC. He brought the Assyrian empire to the height of its power, subduing large parts of Syria and Palestine, and, towards the end of his reign, conquering Babylonia and ascending the Babylonian throne under the name of Pulu.

tigon /ˈtaɪɡən/ *n.* the offspring of a tiger and a lioness (cf. LIGER). [portmanteau word f. TIGER + LION]

Tigray /ˈtiːɡreɪ/ (also **Tigre**) a province of Ethiopia, in the north of the country, bordering Eritrea; capital, Mekele. An ancient kingdom, Tigray was annexed as a province of Ethiopia in 1855. The people are mainly Christian and were engaged in a bitter guerrilla war against the government of Ethiopia 1975–91, during which time the region suffered badly from drought and famine. □ **Tigrayan** *adj.* & *n.*

Tigre¹ see TIGRAY.

Tigre² /ˈtiːɡreɪ/ *n.* a Semitic language spoken in Eritrea and adjoining parts of Sudan. It is not the language of Tigray (Tigre), which is Tigrinya. [Tigre]

tigress /ˈtaɪɡrɪs/ *n.* **1** a female tiger. **2** a fierce or passionate woman.

Tigrinya /tɪˈɡriːnjə/ *n.* a Semitic language spoken in the Ethiopian province of Tigray (cf. TIGRE²). [Tigrinya]

Tigris /ˈtaɪɡrɪs/ a river in SW Asia, the more easterly of the two rivers of ancient Mesopotamia. It rises in the mountains of eastern Turkey and flows 1,850 km (1,150 miles) south-eastwards through Iraq, passing through Baghdad, to join the Euphrates, forming the Shatt al-Arab, which flows into the Persian Gulf.

Tihwa /tiːˈhwɑː/ the former name (until 1954) for URUMQI.

Tijuana /tɪˈwɑːnə/ a town in NW Mexico, situated just south of the US frontier; pop. (1990) 742,690.

Tikal /tɪˈkɑːl/ an ancient Mayan city in the tropical Petén region of northern Guatemala, with great plazas, pyramids, and palaces. It

flourished AD 300–800, reaching its peak towards the end of that period.

tike var. of TYKE.

tiki /ˈtɪki/ *n.* (*pl.* **tikis**) *NZ* a large wooden or small ornamental greenstone image representing a human figure. [Maori]

tikka /ˈtɪkə, ˈtiːkə/ *n.* an Indian dish of small pieces of meat marinaded in spices and cooked on a skewer in a tandoor (often in *comb.*: *chicken tikka*). [Punjabi *ṭikkā*]

tilapia /tɪˈleɪpɪə/ *n.* a freshwater African cichlid fish of *Tilapia* or a related genus, widely introduced as a food fish. [mod.L, orig. unkn.]

Tilburg /ˈtɪlbɜːɡ/ an industrial city in the southern Netherlands, in the province of North Brabant; pop. (1991) 158,850.

Tilbury /ˈtɪlbərɪ/ the principal container port of London and south-east England, on the north bank of the River Thames in Essex.

tilbury /ˈtɪlbərɪ/ *n.* (*pl.* **-ies**) *hist.* a light open two-wheeled carriage. [after the inventor's name]

tilde /ˈtɪldə/ *n.* a mark (˜), put over a letter, e.g. over a Spanish *n* when pronounced *ny* (as in *señor*) or a Portuguese *a* or *o* when nasalized (as in *São Paulo*). [Sp., ult. f. L *titulus* TITLE]

tile /taɪl/ *n.* & *v.* ● *n.* **1** a thin slab of concrete or baked clay etc. used in series for covering a roof or pavement etc. **2** a similar slab of glazed pottery, cork, linoleum, etc., for covering a floor, wall, etc. **3** a thin flat piece used in a game (esp. mah-jong). ● *v.tr.* cover with tiles. □ **on the tiles** *colloq.* having a spree. [OE *tigule, -ele,* f. L *tegula*]

tiler /ˈtaɪlə(r)/ *n.* **1** a person who makes or lays tiles. **2** the doorkeeper of a Freemasons' lodge.

tiling /ˈtaɪlɪŋ/ *n.* **1** the process of fixing tiles. **2** an area of tiles.

till¹ /tɪl/ *prep.* & *conj.* ● *prep.* **1** up to or as late as (*wait till o'clock; did not return till night*). **2** up to the time of (*faithful till death; waited till the end*). ● *conj.* **1** up to the time when (*wait till I return*). **2** so long that (*laughed till I cried*). ¶ *Until* is more usual when beginning a sentence. [OE & ON *til* to, rel. to TILL³]

till² /tɪl/ *n.* a drawer for money in a shop or bank etc., esp. with a device recording the amount of each purchase. [ME: orig. unkn.]

till³ /tɪl/ *v.tr.* prepare and cultivate (land) for crops. □ **tillable** *adj.* **tiller** *n.* [OE *tilian* strive for, cultivate, f. Gmc]

till⁴ /tɪl/ *n.* stiff clay containing boulders, sand, etc., deposited by melting glaciers and ice-sheets. [17th c. (Sc.): orig. unkn.]

tillage /ˈtɪlɪdʒ/ *n.* **1** the preparation of land for crop-bearing. **2** tilled land.

tiller¹ /ˈtɪlə(r)/ *n.* a horizontal bar fitted to the head of a boat's rudder to turn it in steering. [ME f. AF *telier* weaver's beam f. med.L *telarium* f. L *tela* web]

tiller² /ˈtɪlə(r)/ *n.* & *v.* ● *n.* a shoot of a plant, esp. a cereal grass, springing from the bottom of the original stalk. ● *v.intr.* develop tillers. [app. repr. OE *telgor* extended f. *telga* bough]

Tillich /ˈtɪlɪk/, Paul (Johannes) (1886–1965), German-born American theologian and philosopher. He proposed a form of Christian existentialism, outlining a reconciliation of religion and secular society, as expounded in *Systematic Theology* (1951–63).

tilt /tɪlt/ *v.* & *n.* ● *v.* **1** *intr.* & *tr.* assume or cause to assume a sloping position; heel over. **2** *intr.* (foll. by *at*) strike, thrust, or run at, with a weapon, esp. *hist.* in jousting. **3** *intr.* (foll. by *with*) engage in a contest. **4** *tr.* forge or work (steel etc.) with a tilt-hammer. ● *n.* **1** the act or an instance of tilting. **2** a sloping position. **3** *hist.* (of medieval knights etc.) the act of charging with a lance against an opponent or at a mark, done for exercise or as a sport. **4** an encounter between opponents; an attack esp. with argument or satire (*have a tilt at*). **5** = *tilt-hammer.* □ **full** (or **at full**) **tilt 1** at full speed. **2** with full force. **tilt-hammer** a heavy pivoted hammer used in forging. **tilt-yard** *hist.* a place where tilts (see sense 3 of *n.*) took place. □ **tilter** *n.* [ME *tilte* perh. f. an OE form rel. to *tealt* unsteady f. Gmc: weapon senses of unkn. orig.]

tilth /tɪlθ/ *n.* **1** tillage, cultivation. **2** the condition of tilled soil (*in good tilth*). [OE *tilth(e)* (as TILL³)]

Tim. *abbr.* Epistle to Timothy (New Testament).

Timaru /ˈtɪmə,ruː/ a port and resort on the east coast of South Island, New Zealand; pop. (1991) 27,640.

timbal /ˈtɪmb(ə)l/ *n. archaic* a kettledrum. [F *timbale,* earlier *tamballe* f. Sp. *atabal* f. Arab. *aṭ-ṭabl* the drum]

timbale /tæmˈbɑːl/ *n.* a drum-shaped dish of minced meat or fish cooked in a pastry shell or in a mould. [F: see TIMBAL]

timber /ˈtɪmbə(r)/ *n.* **1** wood prepared for building, carpentry, etc. **2** a piece of wood or beam, esp. as the rib of a ship. **3** large standing trees suitable for timber; woods or forest. **4** (esp. as *int.*) a warning cry that a tree is about to fall. □ **timber hitch** a knot used in attaching a rope to a log or spar. **timber wolf** a type of large North American grey wolf. □ **timbering** *n.* [OE, = building, f. Gmc]

timbered /ˈtɪmbəd/ *adj.* **1** (esp. of a building) made wholly or partly of timber. **2** (of country) wooded.

timberland /ˈtɪmbəˌlænd/ *n.* US land covered with forest yielding timber.

timberline /ˈtɪmbəˌlaɪn/ *n.* (on a mountain) the line or level above which no trees grow.

timberman /ˈtɪmbəmən/ *n.* (*pl.* **-men**) **1** a person who works with timber. **2** a longhorn beetle with wood-boring larvae; esp. *Acanthocinus aedilis*, which has extremely long antennae.

timbre /ˈtæmbə(r), -brə/ *n.* the distinctive character of a musical sound or voice apart from its pitch and intensity. [F f. Rmc f. med.Gk *timbanon* f. Gk *tumpanon* drum]

timbrel /ˈtɪmbrəl/ *n. archaic* a tambourine or similar instrument. [dimin. of ME *timbre* f. OF (as TIMBRE, -LE²)]

Timbuktu /ˌtɪmbʌkˈtuː/ (also **Timbuctoo**) **1** (French **Tombouctou** /tɔ̃buktu/) a town in northern Mali; pop. (1976) 20,500. Founded by the Tuareg in the 11th century, it became a Muslim centre of learning and a major trading centre for gold and salt on the trans-Saharan trade routes. It reached the height of its prosperity in the 16th century, falling into decline after its capture by the Moroccans in 1591. **2** a remote or extremely distant place (*from here to Timbuktu*).

time /taɪm/ *n. & v.* ● *n.* **1** the indefinite continued progress of existence, events, etc., in past, present, and future regarded as a whole. **2 a** the progress of this as affecting persons or things (*stood the test of time*). **b** (in full **Father Time**) the personification of time, esp. as an old man with a scythe and hourglass. **3** a more or less definite portion of time belonging to particular events or circumstances (*the time of the Plague*; *prehistoric times*; *the scientists of the time*). **4** an allotted, available, or measurable portion of time; the period of time at one's disposal (*am wasting my time*; *had no time to visit*; *how much time do you need?*). **5** a point of time esp. in hours and minutes (*the time is 7.30*; *what time is it?*). **6** (prec. by *a*) an indefinite period (*waited for a time*). **7** time or an amount of time as reckoned by a conventional standard (*the time allowed is one hour*; *ran the mile in record time*; *eight o'clock New York time*). (See note below.) **8 a** an occasion (*last time I saw you*). **b** an event or occasion qualified in some way (*had a good time*). **9** a moment or definite portion of time destined or suitable for a purpose etc. (*now is the time to act*; *shall we fix a time?*). **10** (in *pl.*) expressing multiplication (*is four times as old*; *five times six is thirty*). **11** a lifetime (*will last my time*). **12** (in *sing.* or *pl.*) **a** the conditions of life or of a period (*hard times*; *times have changed*). **b** (prec. by *the*) the present age, or that being considered. **13** *colloq.* a prison sentence (*is doing time*). **14** an apprenticeship (*served his time*). **15** a period of gestation. **16** the date or expected date of childbirth (*is near her time*) or of death (*my time is drawing near*). **17** measured time spent in work (*put them on short time*). **18** *Mus.* **a** any of several rhythmic patterns of music (*in waltz time*). **b** the duration of a note as indicated by a crotchet, minim, etc. **19** *Brit.* the moment at which the opening hours of a public house end. ● *v.tr.* **1** choose the time or occasion for (*time your remarks carefully*). **2** do at a chosen or correct time. **3** arrange the time of arrival of. **4** ascertain the time taken by (a process or activity, or a person doing it). **5** regulate the duration or interval of; set times for (*trains are timed to arrive every hour*). □ **against time** with utmost speed, so as to finish by a specified time (*working against time*). **ahead of time** earlier than expected. **ahead of one's time** having ideas too enlightened or advanced to be accepted by one's contemporaries. **all the time 1** during the whole of the time referred to (often despite some contrary expectation etc.) (*we never noticed, but she was there all the time*). **2** constantly (*nags all the time*). **3** at all times (*leaves a light on all the time*). **at one time 1** in or during a known but unspecified past period. **2** simultaneously (*ran three businesses at one time*). **at the same time 1** simultaneously; at a time that is the same for all. **2** nevertheless (*at the same time, I do not want to offend you*). **at a time** separately in the specified groups or numbers (*came three at a time*). **at times** occasionally, intermittently. **before time** (usu. prec. by *not*) before the due or expected time. **before one's time** prematurely (*old before his time*). **for the time being** for the present; until some other arrangement is made. **half the time** *colloq.* as often as not. **have no time for 1** be unable or unwilling to spend time on. **2** dislike. **have the time 1** be able to spend the time needed. **2** know from a watch etc. what time it is. **have a time of it** undergo trouble or difficulty. **in no** (or **less than no**) **time 1** very soon. **2** very quickly.

in one's own good time at a time and a rate decided by oneself. **in one's own time** outside working hours. **in time 1** not late, punctual (*was in time to catch the bus*). **2** eventually (*in time you may agree*). **3** in accordance with a given rhythm or tempo, esp. of music. **in one's time** at or during some previous period of one's life (*in her time she was a great hurdler*). **keep good** (or **bad**) **time 1** (of a clock etc.) record time accurately (or inaccurately). **2** be habitually punctual (or not punctual). **keep time** move or sing etc. in time. **know the time of day** be well-informed. **lose no time** (often foll. by *in* + verbal noun) act immediately (*lost no time in cashing the cheque*). **not before time** not too soon; timely. **on time** see ON. **no time** *colloq.* a very short interval (*it was no time before they came*). **out of time** unseasonable; unseasonably. **pass the time of day** *colloq.* exchange a greeting or casual remarks. **time after time 1** repeatedly, on many occasions. **2** in many instances. **time and** (or **time and time**) **again** on many occasions. **time and a half** a rate of payment for work at one and a half times the normal rate. **time-and-motion** (usu. *attrib.*) concerned with measuring the efficiency of industrial and other operations. **time bomb** a bomb designed to explode at a pre-set time. **time capsule** a box etc. containing objects typical of the present time, buried for discovery in the future. **time clock 1** a clock with a device for recording workers' hours of work. **2** a switch mechanism activated at pre-set times by a built-in clock. **time code** a coded signal on videotape or film giving information about frame number, time of recording, etc. **time-consuming** using much or too much time. **time exposure** the exposure of photographic film for longer than the maximum normal shutter setting. **time factor** the passage of time as a limitation on what can be achieved. **time-fuse** a fuse calculated to burn for or explode at a given time. **time-honoured** esteemed by tradition or through custom. **time immemorial** (or **out of mind**) a longer time than anyone can remember or trace. **time-lag** an interval of time between an event, a cause, etc. and its effect. **time-lapse** (of photography) using frames taken at long intervals to photograph a slow process, and shown continuously as if at normal speed. **time-limit** the limit of time within which a task must be done. **time-lock** *n.* **1** a lock that is operated by a timing device. **2** *Computing* a device built into a program to stop it operating after a certain time. ● *v.tr.* (as **time-locked** *adj.*) **1** inextricably linked to a certain period of time. **2** secured by a time lock. **the time of day** the hour by the clock. **time off** time for rest or recreation etc. **the time of one's life** a period or occasion of exceptional enjoyment. **time out** esp. *N. Amer.* **1** a brief intermission in a game etc. **2** = *time off.* **time-scale** the time allowed for or taken by a sequence of events in relation to a broader period of time. **time-served** having completed a period of apprenticeship or training. **time-server** a person who changes his or her view to suit the prevailing circumstances, fashion, etc. **time-serving** self-seeking or obsequious. **time-share** a share in a property under a time-sharing scheme. **time-sharing 1** the operation of a computer system by several users for different operations at one time. **2** the use of a holiday home at agreed different times by several joint owners. **time sheet** a sheet of paper for recording hours of work etc. **time signal** an audible (esp. broadcast) signal or announcement of the exact time of day. **time signature** *Mus.* an indication of rhythm following a clef, usu. expressed as a fraction with the denominator defining the beat as a division of a semibreve and the numerator giving the number of beats in each bar. **time-span** a period spanning a duration of time. **time switch** a switch acting automatically at a pre-set time. **time travel** (in science fiction) travel through time into the past or future. **time traveller** (in science fiction) a person who travels into the past or future. **time trial** a race in which participants are individually timed. **time warp 1** (in science fiction) an imaginary distortion of space in relation to time, whereby persons or objects of one age can be moved to another. **2** a state in which the styles, attitudes, etc. of a past period are retained (*caught in a 1950s time warp*). **time was** there was a time (*time was when I could do that*). **time-wasting** *n.* **1** (in sport) the tactic of slowing down before the end of a match to prevent further scoring by the opposition. **2** the act of wasting time. ● *adj.* that wastes time. **time-work** work paid for by the time it takes. **time-worn** impaired by age. **time zone** a range of longitudes where a common standard time is used. [OE *tīma* f. Gmc]

▪ Local solar time is four minutes later for each degree travelled westwards, and different towns originally kept their own local time. In the mid-19th century the mean time at the Royal Greenwich Observatory in London was adopted by railways throughout Great Britain, and in 1880 Greenwich Mean Time (GMT) became the legal time for the whole country. Both the Greenwich meridian and GMT were accepted as international standards in 1884, after which the earth was divided into 25 time zones each having a standard time so many hours (and half hours) in advance of or behind Greenwich. GMT is now known formally as Universal Time.

timekeeper /'taɪmˌkiːpə(r)/ n. **1** a person who records time, esp. of workers or in a game. **2 a** a watch or clock as regards accuracy (*a good timekeeper*). **b** a person as regards punctuality. □ **timekeeping** n.

timeless /'taɪmlɪs/ adj. not affected by the passage of time; eternal. □ **timelessly** adv. **timelessness** n.

timely /'taɪmlɪ/ adj. (**timelier, timeliest**) opportune; coming at the right time. □ **timeliness** n.

timepiece /'taɪmpiːs/ n. an instrument, such as a clock or watch, for measuring time.

timer /'taɪmə(r)/ n. **1** a person or device that measures or records time taken. **2** an automatic mechanism for activating a device etc. at a pre-set time.

timetable /'taɪmˌteɪb(ə)l/ n. & v. ● n. a list of times at which events are scheduled to take place, esp. the arrival and departure of buses or trains etc., or a sequence of lessons in a school or college. ● v.tr. include in or arrange to a timetable; schedule.

timid /'tɪmɪd/ adj. (**timider, timidest**) easily frightened; apprehensive, shy. □ **timidly** adv. **timidness** n. **timidity** /tɪ'mɪdɪtɪ/ n. [F timide or L timidus f. timere fear]

timing /'taɪmɪŋ/ n. **1** the way an action or process is timed, esp. in relation to others. **2** the regulation of the opening and closing of valves in an internal-combustion engine.

Timişoara /ˌtɪmɪ'ʃwɑːrə/ (Hungarian **Temesvár** /'tɛmɛʃˌvɑːr/) an industrial city in western Romania; pop. (1989) 333,365. Formerly part of Hungary, the city has substantial Hungarian and German-speaking populations. It was a focal point for demonstrations during the revolution in 1989 which brought about the collapse of the Ceauşescu regime.

timocracy /tɪ'mɒkrəsɪ, taɪ-/ n. (pl. **-ies**) **1** a form of government in which possession of property is required in order to hold office. **2** a form of government in which rulers are motivated by love of honour. □ **timocratic** /ˌtɪmə'krætɪk, ˌtaɪm-/ adj. [OF timocracie f. med.L timocratia f. Gk timokratia f. timē honour, worth + kratia -CRACY]

Timor /'tiːmɔː(r)/ the largest of the Lesser Sunda Islands, in the southern Malay Archipelago; pop. (est. 1990), East Timor 714,000, West Timor 3,383,500. The island was formerly divided into Dutch West Timor and Portuguese East Timor. In 1950 West Timor (chief town, Kupang) was absorbed into the newly formed Republic of Indonesia. In 1975 East Timor (chief town, Dili) declared itself independent but was invaded and occupied by Indonesia in 1976. (See also EAST TIMOR.) □ **Timorese** /ˌtiːmɔː'riːz/ adj. & n.

timorous /'tɪmərəs/ adj. **1** timid; easily alarmed. **2** frightened. □ **timorously** adv. **timorousness** n. [ME f. OF temoreus f. med.L timorosus f. L timor f. timere fear]

Timor Sea an arm of the Indian Ocean between Timor and NW Australia.

timothy[1] /'tɪməθɪ/ n. (in full **timothy grass**) a fodder grass, *Phleum pratense*. [Timothy Hanson, 18th-c. Amer. farmer, who is said to have introduced it in Carolina c.1720]

timothy[2] /'tɪməθɪ/ n. Austral. sl. a brothel. [20th c.: orig. unkn.]

Timothy, Epistle to /'tɪməθɪ/ either of two books of the New Testament, epistles of St Paul addressed to St Timothy.

Timothy, St (1st century AD), convert and disciple of St Paul. Traditionally he was the first bishop of Ephesus and was martyred in the reign of the Roman emperor Nerva. Feast day, Jan. 22 or 26.

timpani /'tɪmpənɪ/ n.pl. (also **tympani**) kettledrums. □ **timpanist** n. [It., pl. of timpano = TYMPANUM]

Timur /tiː'mʊə(r)/ see TAMERLANE.

tin /tɪn/ n. & v. ● n. **1 a** a silvery-white metallic chemical element (atomic number 50; symbol **Sn**). (*See note below.*) **2 a** a receptacle or container made of tin or tinned iron. **b** Brit. an airtight sealed container made of tin plate or aluminium for preserving food. (See CAN².) **3** = tin plate. **4** Brit. sl. money. **5** (prec. by the) the metal strip fitted along the bottom of the front wall of a squash court which resounds when struck by the ball, showing it to be out of play. ● v.tr. (**tinned, tinning**) **1** seal (food) in an airtight tin for preservation. **2** cover or coat with tin. □ **put the tin lid on** see LID. **tin can** a tin container (see sense 2 of n.), esp. an empty one. **tin foil** foil made of tin, aluminium, or tin alloy, used for wrapping food for cooking or storing. **tin-glaze** a glaze made white and opaque by the addition of tin oxide. **tin god 1** an object of unjustified veneration. **2** a self-important person. **tin hat** colloq. a

military steel helmet. **tin Lizzie** colloq. an old or decrepit car. **tin-opener** a tool for opening tins. **tin-pan alley** the world of composers and publishers of popular music. **tin plate** sheet iron or sheet steel coated with tin. **tin-plate** v.tr. coat with tin. **tin soldier** a toy soldier made of metal. **tin-tack** an iron tack. **tin whistle** = penny whistle. [OE f. Gmc]

▪ Tin was known in ancient times, especially in the form of bronze, an alloy of tin and copper. It is quite a rare element, occurring chiefly in the mineral cassiterite. Pure crystalline tin exists in two allotropic modifications, the metallic form (white tin), and a semi-metallic form (grey tin). Tin is used in various alloys and for electroplating iron or steel sheets to make tin plate. The symbol Sn is from the Latin stannum.

tinamou /'tɪnəˌmuː/ n. a ground-dwelling South American bird of the family Tinamidae, resembling a grouse but related to the rheas. [F f. Galibi tinamu]

Tinbergen[1] /'tɪnˌbɜːgən/, Jan (1903–94), Dutch economist. In 1969 he shared with Ragnar Frisch the first Nobel Prize for economics, awarded for his pioneering work on econometrics. He was the brother of the zoologist Nikolaas Tinbergen.

Tinbergen[2] /'tɪnˌbɜːgən/, Nikolaas (1907–88), Dutch zoologist. After the Second World War he moved to Oxford where he helped to establish ethology as a distinct discipline, with relevance also to human psychology and sociology. From his classical studies on herring gulls, sticklebacks, and digger wasps, he found that much animal behaviour was innate and stereotyped, and he introduced the concept of displacement activity. Tinbergen shared a Nobel Prize in 1973 with Karl von Frisch and Konrad Lorenz. He was the brother of the economist Jan Tinbergen.

tinctorial /tɪŋk'tɔːrɪəl/ adj. **1** of or relating to colour or dyeing. **2** producing colour. [L tinctorius f. tinctor dyer: see TINGE]

tincture /'tɪŋktʃə(r)/ n. & v. ● n. (often foll. by of) **1** a slight flavour or trace. **2** a tinge (of a colour). **3** a medicinal solution (of a drug) in alcohol (tincture of quinine). **4** Heraldry an inclusive term for the metals, colours, and furs used in coats of arms. **5** colloq. an alcoholic drink. ● v.tr. **1** colour slightly; tinge, flavour. **2** (often foll. by with) affect slightly (with a quality). [ME f. L tinctura dyeing (as TINGE)]

tinder /'tɪndə(r)/ n. a dry substance such as wood that readily catches fire from a spark. □ **tinder-box** hist. a box containing tinder, flint, and steel, formerly used for kindling fires. □ **tindery** adj. [OE tynder, tyndre f. Gmc]

tine /taɪn/ n. a prong or tooth or point of a fork, comb, antler, etc. □ **tined** adj. (also in comb.). [OE tind]

tinea /'tɪnɪə/ n. Med. ringworm or athlete's foot. [L, = moth, worm]

ting /tɪŋ/ n. & v. ● n. a tinkling sound as of a bell. ● v.intr. & tr. emit or cause to emit this sound. [imit.]

tinge /tɪndʒ/ v. & n. ● v.tr. (**tinging** or **tingeing**) (often foll. by with; often in passive) **1** colour slightly (is tinged with red). **2** affect slightly (regret tinged with satisfaction). ● n. **1** a tendency towards or trace of some colour. **2** a slight admixture of a feeling or quality. [ME f. L tingere tinct- dye, stain]

tingle /'tɪŋg(ə)l/ v. & n. ● v. **1** intr. **a** feel a slight prickling, stinging, or throbbing sensation. **b** cause this (the reply tingled in my ears). **2** tr. make (the ear etc.) tingle. ● n. a tingling sensation. [ME, perh. var. of TINKLE]

tingly /'tɪŋglɪ/ adj. (**tinglier, tingliest**) causing or characterized by tingling.

tinhorn /'tɪnhɔːn/ n. & adj. US sl. ● n. a pretentious but unimpressive person. ● adj. cheap, pretentious.

tinker /'tɪŋkə(r)/ n. & v. ● n. **1** an itinerant mender of kettles and pans etc. **2** Sc. & Ir. a gypsy. **3** colloq. a mischievous person or animal. **4** a spell of tinkering. **5** a rough-and-ready worker. ● v. **1** intr. (foll. by at, with) work in an amateurish or desultory way, esp. to adjust or mend machinery etc. **2 a** intr. work as a tinker. **b** tr. repair (pots and pans). □ **tinkerer** n. [ME: orig. unkn.]

tinkle /'tɪŋk(ə)l/ v. & n. ● v. **1** intr. & tr. make or cause to make a succession of short light ringing sounds. **2** intr. colloq. (esp. as a child's term) urinate. ● n. **1** a tinkling sound. **2** Brit. colloq. a telephone call (will give you a tinkle on Monday). **3** colloq. (a child's name for) an act of urinating. □ **tinkly** adj. [ME f. obs. tink to chink (imit.)]

tinner /'tɪnə(r)/ n. **1** a tin-miner. **2** a tinsmith.

tinnitus /'tɪnɪtəs, tɪ'naɪt-/ n. Med. a ringing in the ears. [L f. tinnire tinnit- ring, tinkle, of imit. orig.]

tinny /'tɪnɪ/ adj. & n. ● adj. (**tinnier, tinniest**) **1** of or like tin. **2** (of a metal object) flimsy, insubstantial; of poor quality. **3 a** sounding like struck tin. **b** (of reproduced sound) thin and metallic, lacking low frequencies. **4** *Austral. sl.* lucky. ● *n.* (also **tinnie**) (*pl.* **-ies**) *Austral. sl.* a can of beer. □ **tinnily** *adv.* **tinniness** *n.*

Tin Pan Alley originally the name given to a district in New York (28th Street, between 5th Avenue and Broadway) where many songwriters, arrangers, and music publishers were based. The district gave its name to the American popular music industry between the late 1880s and the mid-20th century, particularly to the work of such composers as Irving Berlin, Jerome Kern, George Gershwin, Cole Porter, and Richard Rodgers. The term was later also applied to Denmark Street in London.

tinpot /'tɪnpɒt/ *attrib.adj. Brit.* cheap, inferior.

tinsel /'tɪns(ə)l/ *n. & v.* ● *n.* **1** glittering metallic strips, threads, etc., used as decoration to give a sparkling effect. **2** a fabric adorned with tinsel. **3** superficial brilliance or splendour. **4** (*attrib.*) showy, gaudy, flashy. ● *v.tr.* (**tinselled**, **tinselling**) adorn with or as with tinsel. □ **tinselled** *adj.* **tinselly** *adj.* [OF *estincele* spark f. L *scintilla*]

tinsmith /'tɪnsmɪθ/ *n.* a worker in tin and tin plate.

tinsnips /'tɪnsnɪps/ *n.pl.* a pair of clippers for cutting sheet metal.

tinstone /'tɪnstəʊn/ *n. Geol.* = CASSITERITE.

tint /tɪnt/ *n. & v.* ● *n.* **1** a variety of a colour, esp. one made lighter by adding white. **2** a tendency towards or admixture of a different colour (*red with a blue tint*). **3** a faint colour spread over a surface, esp. as a background for printing on. **4** a set of parallel engraved lines to give uniform shading. ● *v.tr.* apply a tint to; colour. □ **tinter** *n.* [alt. of earlier *tinct* f. L *tinctus* dyeing (as TINGE), perh. infl. by It. *tinto*]

Tintagel /tɪn'tædʒəl/ a village on the coast of northern Cornwall. Nearby are the ruins of Tintagel Castle, the legendary birthplace of King Arthur and a stronghold of the Earls of Cornwall from the 12th to the 15th centuries.

tintinnabulation /ˌtɪntɪˌnæbjʊ'leɪʃ(ə)n/ *n.* a ringing or tinkling of bells. [as L *tintinnabulum* tinkling bell f. *tintinnare* redupl. form of *tinnire* ring]

Tintoretto /ˌtɪntə'retəʊ/ (born Jacopo Robusti) (1518–94), Italian painter. He acquired his name because his father was a dyer (It. *tintore*). Based in Venice, Tintoretto gained fame with the painting *St Mark Rescuing a Slave* (1548), whose bright colours were influenced by Titian. From this time, his work was typified by a mannerist style, including unusual viewpoints, striking juxtapositions in scale, and bold chiaroscuro effects. Primarily a religious painter, he is best known for the huge canvas *Paradiso* (after 1577) in the main hall of the Doges' Palace in Venice, and for his paintings of the life of the Virgin, the life of Christ, and the Passion (1576–88) in the halls of the Scuola di San Rocco, also in Venice.

tinware /'tɪnweə(r)/ *n.* articles made of tin or tin plate.

tiny /'taɪnɪ/ *adj. & n.* ● *adj.* (**tinier, tiniest**) very small or slight. ● *n.* (usu. in *pl.*) (*pl.* **-ies**) a very young child. □ **tinily** *adv.* **tininess** *n.* [obs. *tine*, *tyne* (adj. & n.) small, a little: ME, of unkn. orig.]

-tion /ʃ(ə)n/ *suffix* forming nouns of action, condition, etc. (see -ION, -ATION, -ITION, -UTION). [from or after F *-tion* or L *-tio -tionis*]

tip[1] /tɪp/ *n. & v.* ● *n.* **1** an extremity or end, esp. of a small or tapering thing (*tips of the fingers*). **2** a small piece or part attached to the end of a thing, e.g. a ferrule on a stick. **3** a leaf-bud of tea. ● *v.tr.* (**tipped**, **tipping**) **1** provide with a tip. **2** (foll. by *in*) attach (a loose sheet) to a page at the inside edge. □ **on the tip of one's tongue** about to be said, esp. after difficulty in recalling to mind. **the tip of the iceberg** a small evident part of something much larger or more significant. □ **tipless** *adj.* **tippy** *adj.* (in sense 3). [ME f. ON *typpi* (n.), *typpa* (v.), *typptr* tipped f. Gmc (rel. to TOP[1]): prob. reinforced by MDu. & MLG *tip*]

tip[2] /tɪp/ *v. & n.* ● *v.* (**tipped**, **tipping**) **1** (often foll. by *over*, *up*) **a** *intr.* lean or slant. **b** *tr.* cause to do this. **2** *tr.* (foll. by *into* etc.) **a** overturn or cause to overbalance (*was tipped into the pond*). **b** empty the contents from (a container etc.) in this way. **c** pour out (the contents of a container) in this way. ● *n.* **1 a** a slight push or tilt. **b** a light stroke, esp. in baseball. **2** *Brit.* a place where material (esp. refuse) is tipped. □ **tip the balance** make the critical difference. **tip one's hat** (or **cap**) raise or touch one's hat or cap in greeting or acknowledgement. **tip the scales** see SCALE[2]. **tip-up** able to be tipped, e.g. of a seat in a theatre to allow passage past. [17th c.: orig. uncert.]

tip[3] /tɪp/ *v. & n.* ● *v.* (**tipped**, **tipping**) **1** *tr.* make a small present of money to, esp. for a service given (*have you tipped the porter?*). **2** *tr.* name as the likely winner of a race or contest etc. **3** *tr.* strike or touch lightly. **4** *tr. sl.* give, hand, pass (esp. in *tip a person the wink* below). ● *n.* **1** a small money present, esp. for a service given. **2** a piece of private or special information, esp. regarding betting or investment. **3** a small or casual piece of advice. □ **tip off 1** give (a person) a hint or piece of special information or warning, esp. discreetly or confidentially. **2** *Basketball* start play by throwing the ball up between two opponents. **tip-off** *n.* a hint or warning etc. given discreetly or confidentially. **tip a person the wink** *sl.* give a person private information. □ **tipper** *n.* [ME: orig. uncert.]

tipcat /'tɪpkæt/ *n.* **1** a game with a short piece of wood tapering at the ends and struck with a stick. **2** this piece of wood.

tipper /'tɪpə(r)/ *n.* (often *attrib.*) a road haulage vehicle that tips at the back to discharge its load.

Tipperary /ˌtɪpə'reərɪ/ a county in the centre of the Republic of Ireland, in the province of Munster; county town, Clonmel.

tippet /'tɪpɪt/ *n.* **1** a covering of fur etc. for the shoulders formerly worn by women. **2** a similar garment worn as part of some official costumes, esp. by the clergy. **3** *hist.* a long narrow strip of cloth as part of or an attachment to a hood etc. [ME, prob. f. TIP[1]]

Tippett /'tɪpɪt/, Sir Michael (Kemp) (b.1905), English composer. He established his reputation with the oratorio *A Child of Our Time* (1941), which drew on jazz, madrigals, and spirituals besides classical sources. He is also noted for five operas, for all of which he also wrote the libretti; these include *The Midsummer Marriage* (1955). Among his other works are the oratorio *The Mask of Time* (1983), four symphonies, and several song cycles.

tipple /'tɪp(ə)l/ *v. & n.* ● *v.* **1** *intr.* drink intoxicating liquor habitually. **2** *tr.* drink (liquor) repeatedly in small amounts. ● *n. colloq.* a drink, esp. a strong one. □ **tippler** *n.* [ME, back-form. f. *tippler*, of unkn. orig.]

tipstaff /'tɪpstɑːf/ *n.* **1** a sheriff's officer. **2** a metal-tipped staff carried as a symbol of office. [contr. of *tipped staff*, i.e. tipped with metal]

tipster /'tɪpstə(r)/ *n.* a person who gives tips, esp. about betting at horse-races.

tipsy /'tɪpsɪ/ *adj.* (**tipsier, tipsiest**) **1** slightly intoxicated. **2** caused by or showing intoxication (*a tipsy leer*). □ **tipsy-cake** *Brit.* a sponge cake soaked in wine or spirits, often served with custard. □ **tipsily** *adv.* **tipsiness** *n.* [prob. f. TIP[2] = inclined to lean, unsteady: for *-sy* cf. FLIMSY, TRICKSY]

tiptoe /'tɪptəʊ/ *n., v., & adv.* ● *n.* the tips of the toes. ● *v.intr.* (**tiptoes, tiptoed, tiptoeing**) walk on tiptoe, or very stealthily. ● *adv.* (also **on tiptoe**) with the heels off the ground and the weight on the balls of the feet.

tiptop /'tɪptɒp/ *adj. & n. colloq.* ● *adj.* highest in excellence; very best. ● *n.* the highest point of excellence.

TIR *abbr.* international road transport (esp. with ref. to EC regulations). [F, = *transport international routier*]

tirade /taɪ'reɪd/ *n.* a long vehement denunciation or declamation. [F, = long speech, f. It. *tirata* volley f. *tirare* pull f. Rmc]

tirailleur /ˌtiːraɪ'jɜː(r), ˌtɪrə'lɜː(r)/ *n.* **1** a sharpshooter. **2** a skirmisher. [F f. *tirailler* shoot independently f. *tirer* shoot, draw, f. Rmc]

tiramisu /ˌtɪrəmɪ'suː/ *n.* an Italian dessert of layers of sponge soaked in coffee and brandy with mascarpone cheese. [It. *tira mi sù* pick me up]

Tirana /tɪ'rɑːnə/ (also **Tiranë**) the capital of Albania, on the Ishm river in central Albania; pop. (1990) 210,000. Founded by the Turks in the 17th century, it became capital of Albania in 1920.

tire[1] /'taɪə(r)/ *v.* **1** *tr. & intr.* make or grow weary. **2** *tr.* exhaust the patience or interest of; bore. **3** *tr.* (in *passive*; foll. by *of*) have had enough of; be fed up with (*was tired of arguing*). [OE *tēorian*, of unkn. orig.]

tire[2] *US* var. of TYRE.

tired /'taɪəd/ *adj.* **1** weary, exhausted; ready for sleep. **2** (of an idea etc.) hackneyed. □ **tiredly** *adv.* **tiredness** *n.*

Tiree /taɪ'riː/ an island to the west of Mull in the Inner Hebrides.

tireless /'taɪəlɪs/ *adj.* having inexhaustible energy. □ **tirelessly** *adv.* **tirelessness** *n.*

Tiresias /taɪ'riːsɪəs/ *Gk Mythol.* a blind Theban prophet, so wise that even his ghost had its wits and was not a mere phantom. Legends account variously for his wisdom and blindness; some stories hold also that he spent seven years as a woman.

tiresome /'taɪəsəm/ *adj.* **1** wearisome, tedious. **2** *colloq.* annoying (*how tiresome of you!*). □ **tiresomely** *adv.* **tiresomeness** *n.*

Tîrgu Mureş /ˌtɪəgu: 'mʊəreʃ/ a city in central Romania, on the River Mureş; pop. (1989) 164,780.

Tirich Mir /ˌtɪrɪtʃ ˈmɪə(r)/ the highest peak in the Hindu Kush, in NW Pakistan, rising to 7,690 m (25,230 ft).

Tir-nan-Og /ˌtɪənæˈnəʊg/ *Ir. Mythol.* a land of perpetual youth, the Irish equivalent of Elysium. [Ir., = land of the young]

tiro /ˈtaɪərəʊ/ *n.* (also **tyro**) (*pl.* **-os**) a beginner or novice. [L *tiro*, med.L *tyro*, recruit]

Tirol see TYROL.

Tiruchirapalli /ˌtɪrʊtʃɪˈrɑːpəli/ (also **Trichinopoly** /ˌtrɪtʃɪˈnɒpəli/) a city in Tamil Nadu, southern India; pop. (1991) 387,000.

'tis /tɪz/ *archaic* it is. [contr.]

Tisa see TISZA.

tisane /tɪˈzæn/ *n.* an infusion of dried herbs etc. [F: see PTISAN]

Tishri /ˈtɪʃriː/ *n.* (also **Tisri** /ˈtɪzriː/) (in the Jewish calendar) the first month of the civil and seventh of the religious year, usually coinciding with parts of September and October. [Heb. *tišrī*]

Tisiphone /tɪˈsɪfəni/ *Gk Mythol.* one of the Furies. [Gk, = the avenger of blood]

tissue /ˈtɪʃuː, ˈtɪsjuː/ *n.* **1** any of the coherent collections of specialized cells of which animals or plants are made (*muscular tissue; nervous tissue*). **2** = *tissue-paper*. **3** a disposable piece of thin soft absorbent paper for wiping, drying, etc. **4** fine woven esp. gauzy fabric. **5** (foll. by *of*) a connected series (*a tissue of lies*). □ **tissue culture** *Biol. & Med.* **1** the growth in an artificial medium of cells derived from living tissue. **2** a culture of this kind. **tissue-paper** thin soft unsized paper for wrapping or protecting fragile or delicate articles. [ME f. OF *tissu* rich material, past part. of *tistre* f. L *texere* weave]

Tisza /ˈtiːsɑ/ (Serbian **Tisa** /ˈtisɑ/) a river in SE Europe, the longest tributary of the Danube, which rises in the Carpathian Mountains of western Ukraine and flows 960 km (600 miles) westwards into Hungary, then southwards, joining the Danube in Serbia north-west of Belgrade.

Tit. *abbr.* Epistle to Titus (New Testament).

tit[1] /tɪt/ *n.* **1** a small active tree-dwelling songbird of the family Paridae (*blue tit; great tit*). Also called *titmouse.* **2** a similar small songbird (*bearded tit*). [prob. f. Scand.]

tit[2] /tɪt/ *n.* □ **tit for tat** /tæt/ blow for blow; retaliation. [= earlier *tip* (TIP[2]) *for tap*]

tit[3] /tɪt/ *n.* **1** *colloq.* a nipple. **2** *coarse sl.* a woman's breast. □ **get on a person's tits** *coarse sl.* annoy, irritate. [OE: cf. MLG *titte*]

tit[4] /tɪt/ *n. coarse sl.* a term of contempt for a person. [20th c.: perh. f. TIT[3]]

Titan /ˈtaɪt(ə)n/ *n.* **1** *Gk Mythol.* any of the older gods who preceded the Olympians and were the children of Uranus (Heaven) and Gaia (Earth). Led by Cronus, they overthrew Uranus; when Cronus' son, Zeus, rebelled against his father, most of the Titans supported Cronus, but they were eventually defeated by Zeus. **2** (often **titan**) a person of very great strength, intellect, or importance. **3** *Astron.* satellite VI of Saturn, the fifteenth closest to the planet, discovered by Christiaan Huygens in 1655. With a diameter of 5,150 km it is the largest Saturnian moon, and is unique in the solar system in having an atmosphere of nitrogen and methane under an opaque orange haze of oily hydrocarbons. It is thought that there may be lakes of liquid methane and other hydrocarbons on the surface. □ **Titaness** /-nɪs/ *n. fem.* [ME f. L f. Gk]

Titania /tɪˈtɑːnɪə, -ˈteɪnɪə/ **1** the queen of the fairies in Shakespeare's *A Midsummer Night's Dream.* **2** *Astron.* satellite III of Uranus, the fourteenth closest to the planet, discovered by William Herschel in 1787. It is the largest moon of Uranus (diameter 1,600 km), and its icy surface has many small craters and an extensive network of faults.

Titanic /taɪˈtænɪk/ a British passenger liner, supposedly unsinkable and the largest ship in the world when she was built, that struck an iceberg in the North Atlantic on her maiden voyage in April 1912 and sank with the loss of 1,490 lives. The disaster led to new regulations requiring ships to carry sufficient lifeboats for all on board, a more southerly liner track across the Atlantic, and an ice patrol which continues to this day.

titanic[1] /taɪˈtænɪk/ *adj.* **1** of or like the Titans. **2** gigantic, colossal. □ **titanically** *adv.* [Gk *titanikos* (as TITAN)]

titanic[2] /taɪˈtænɪk, tɪ-/ *adj. Chem.* of titanium, esp. with a valency of four. □ **titanate** /ˈtaɪtəˌneɪt, ˈtɪt-/ *n.*

titanium /taɪˈteɪnɪəm, tɪ-/ *n.* a hard silver-grey metallic chemical element (atomic number 22; symbol **Ti**). Titanium was identified as an element and named by Martin Klaproth in 1795, although a British amateur scientist, William Gregor (1761-1817), had found it independently in 1791. One of the transition metals, titanium is a common element in the earth's crust; the main sources are the minerals ilmenite and rutile. It is used to make strong light corrosion-resistant alloys. Very large quantities of the dioxide are manufactured for use as a white pigment in paper, paint, etc. [f. TITAN, after *uranium*]

titbit /ˈtɪtbɪt/ *n.* (*N. Amer.* **tidbit** /ˈtɪdbɪt/) **1** a dainty morsel. **2** a piquant item of news etc. [perh. f. dial. *tid* tender + BIT[1]]

titch /tɪtʃ/ *n.* (also **tich**) *colloq.* a small person. [*Tich,* stage name of Harry Relph, Engl. music-hall comedian (1868-1928)]

titchy /ˈtɪtʃi/ *adj.* (**titchier, titchiest**) *colloq.* very small.

titer *US* var. of TITRE.

titfer /ˈtɪtfə(r)/ *n. Brit. sl.* a hat. [abbr. of *tit for tat,* rhyming sl.]

tithe /taɪð/ *n. & v.* ● *n.* **1 a** one tenth of the annual produce of land or labour, formerly taken as a tax for the support of the Church and clergy. **b** (in certain religious denominations) a tenth of an individual's income, pledged to the church. **2** a tenth part. ● *v.* **1** *tr.* subject to tithes. **2** *intr.* pay tithes. □ **tithe barn** a barn built to hold tithes paid in kind. □ **tithable** *adj.* [OE *teogotha* tenth]

tithing /ˈtaɪðɪŋ/ *n.* **1** the practice of taking or paying a tithe. **2** *hist.* **a** ten householders living near together and collectively responsible for each other's behaviour. **b** the area occupied by them. [OE *tīgething* (as TITHE, -ING[1])]

Tithonus /tɪˈθəʊnəs/ *Gk Mythol.* a Trojan prince with whom the goddess Aurora fell in love. She asked Zeus to make him immortal but omitted to ask for eternal youth, and he became very old and decrepit although he talked perpetually. Tithonus prayed her to remove him from this world and she changed him into a grasshopper, which chirps ceaselessly.

titi /ˈtiːtiː/ *n.* (*pl.* **titis**) a small long-coated monkey of the genus *Callicebus,* native to South America. [Tupi]

Titian[1] /ˈtɪʃ(ə)n/ (Italian name Tiziano Vecellio) (*c.*1488-1576), Italian painter. The most important painter of the Venetian school, he was first a pupil of Bellini and Giorgione, and completed a number of the latter's unfinished paintings after his death in 1510. Titian subsequently experimented with vivid colours and often broke conventions of composition; his *Madonna with Saints and Members of the Pesaro Family* (1519-*c.*28) was innovative in not having the Madonna in the centre of the picture. He also painted many sensual mythological works, including *Bacchus and Ariadne* (*c.*1518-23). He was much in demand throughout Europe for his portraits, which were notable for the characterization of their sitters; he was appointed court painter by the Holy Roman Emperor Charles V in 1533.

Titian[2] /ˈtɪʃ(ə)n/ *adj.* (in full **Titian red**) (of hair) bright golden auburn, a colour of hair favoured by Titian in his pictures. [TITIAN[1]]

Titicaca, Lake /ˌtɪtɪˈkɑːkə/ a lake in the Andes, on the border between Peru and Bolivia. At an altitude of 3,809 m (12,497 ft), it is the highest large lake in the world.

titillate /ˈtɪtɪˌleɪt/ *v.tr.* **1** excite pleasantly. **2** tickle. □ **titillatingly** *adv.* **titillation** /ˌtɪtɪˈleɪʃ(ə)n/ *n.* [L *titillare titillat-*]

titivate /ˈtɪtɪˌveɪt/ *v.tr.* (also **tittivate**) *colloq.* **1** adorn, smarten. **2** (often *refl.*) put the finishing touches to. □ **titivation** /ˌtɪtɪˈveɪʃ(ə)n/ *n.* [earlier *tidivate,* perh. f. TIDY after *cultivate*]

titlark /ˈtɪtlɑːk/ *n.* a pipit, esp. the meadow pipit.

title /ˈtaɪt(ə)l/ *n. & v.* ● *n.* **1** the name of a book, work of art, piece of music, etc. **2** the heading of a chapter, poem, document, etc. **3 a** the contents of the title-page of a book. **b** a book regarded in terms of its title (*published 20 new titles*). **4** a caption or credit in a film, broadcast, etc. **5** a form of nomenclature indicating a person's status (e.g. *professor, queen*) or used as a form of address or reference (e.g. *Lord, Mr, Your Grace*). **6** a championship in sport. **7** *Law* **a** the right to ownership of property with or without possession. **b** the facts constituting this. **c** (foll. by *to*) a just or recognized claim. **8** *Eccl.* **a** a fixed sphere of work and source of income as a condition for ordination. **b** a parish church in Rome under a cardinal. ● *v.tr.* give a title to. □ **title-deed** a legal instrument as evidence of a right, esp. to property. **title-page** a page at the beginning of a book giving the title and particulars of authorship etc. **title role** the part in a play etc. that gives it its name (e.g. Othello). [ME f. OF f. L *titulus* placard, title]

titled /ˈtaɪt(ə)ld/ *adj.* having a title of nobility or rank.

titling[1] /ˈtaɪtlɪŋ/ *n.* the impressing of a title in gold leaf etc. on the cover of a book.

titling[2] /ˈtɪtlɪŋ/ *n.* **1** a titlark. **2** a titmouse.

titmouse /ˈtɪtmaʊs/ *n.* (*pl.* **titmice** /-maɪs/) = TIT[1] 1. [ME *titmōse* f. TIT[1] + OE *māse* titmouse, assim. to MOUSE]

Tito /ˈtiːtəʊ/ (born Josip Broz) (1892–1980), Yugoslav Marshal and statesman, Prime Minister 1945–53 and President 1953–80. Born in Croatia, he served in the Austro-Hungarian army during the First World War and was captured by the Russians in 1915. After escaping, he fought with the Bolsheviks in the Russian Revolution and became an active Communist organizer on returning to his country in 1920. Tito responded to the German invasion of Yugoslavia (1941) by organizing a Communist resistance movement using guerrilla tactics. His success in resisting the Germans earned him Allied support and he emerged as head of the new government at the end of the war. Tito defied Stalin over policy in the Balkans in 1948, proceeding to establish Yugoslavia as a non-aligned Communist state with a federal constitution. He was made President for life in 1974.

Titograd /ˈtiːtəʊˌɡræd/ a former name (1946–93) for PODGORICA.

Titoism /ˈtiːtəʊˌɪz(ə)m/ n. the Communist policies followed by Marshal Tito in the former Yugoslavia, which concentrated on the national interest independent of the Soviet Union. □ **Titoist** n. & adj.

titrate /ˈtaɪtreɪt, ˈtɪt-/ v.tr. Chem. ascertain the amount of a constituent in (a solution) by measuring the volume of a known concentration of reagent required to complete a reaction with that constituent, often marked by a change in colour. □ **titratable** adj. **titration** /taɪˈtreɪʃ(ə)n, tɪ-/ n.

titre /ˈtaɪtə(r)/ n. (US **titer**) **1** Chem. the concentration of a solution as determined by titration. **2** Med. the concentration of an antibody, determined by finding the highest dilution at which it is still active. [F, = TITLE]

titter /ˈtɪtə(r)/ v. & n. ● v.intr. laugh in a furtive or restrained way; giggle. ● n. a furtive or restrained laugh. □ **titterer** n. **titteringly** adv. [imit.]

tittivate var. of TITIVATE.

tittle /ˈtɪt(ə)l/ n. **1** a small written or printed stroke or dot. **2** a particle; a whit (esp. in not one jot or tittle). [ME f. L (as TITLE)]

tittlebat /ˈtɪt(ə)lˌbæt/ n. Brit. a stickleback. [fanciful var.]

tittle-tattle /ˈtɪt(ə)lˌtæt(ə)l/ n. & v. ● n. petty gossip. ● v.intr. gossip, chatter. [redupl. of TATTLE]

tittup /ˈtɪtəp/ v. & n. ● v.intr. (**tittuped**, **tittuping** or **tittupped**, **tittupping**) go about friskily or jerkily; bob up and down; canter. ● n. such a gait or movement. [perh. imit. of hoof-beats]

titty /ˈtɪtɪ/ n. (pl. **-ies**) sl. = TIT³ (esp. as a child's term).

titubation /ˌtɪtjʊˈbeɪʃ(ə)n/ n. Med. unsteadiness, esp. as caused by nervous disease. [L titubatio f. titubare totter]

titular /ˈtɪtjʊlə(r)/ adj. & n. ● adj. **1** of or relating to a title (the book's titular hero). **2** existing, or being what is specified, in name or title only (titular ruler; titular sovereignty). ● n. **1** the holder of an office etc., esp. a benefice, without the corresponding functions or obligations. **2** a titular saint. □ **titular bishop** a bishop, esp. in a non-Christian country, with a see named after a Christian see no longer in existence. **titular saint** the patron saint of a particular church. □ **titularly** adv. [F titulaire or mod.L titularis f. titulus TITLE]

Titus /ˈtaɪtəs/ (full name Titus Vespasianus Augustus; born Titus Flavius Vespasianus) (AD 39–81), Roman emperor 79–81, son of Vespasian. In 70 he ended a revolt in Judaea with the conquest of Jerusalem. Titus also helped to complete the Colosseum and provided relief for the survivors of the eruption of Vesuvius in 79.

Titus, Epistle to a book of the New Testament, an epistle of St Paul addressed to St Titus.

Titus, St (1st century AD), Greek churchman. A convert and helper of St Paul, he was traditionally the first bishop of Crete. Feast day (in the Eastern Church) 23 Aug.; (in the Western Church) 6 Feb.

tizzy /ˈtɪzɪ/ n. (also **tizz, tiz**) (pl. **-ies**) colloq. a state of nervous agitation (in a tizzy). [20th c.: orig. unkn.]

T-junction /ˈtiːˌdʒʌŋkʃ(ə)n/ n. a road junction at which one road joins another at right angles without crossing it.

TKO abbr. Boxing technical knockout.

Tl symb. Chem. the element thallium.

Tlaxcala /tlɑːsˈkɑːlə/ **1** a state of east central Mexico. **2** its capital city; pop. (est. 1990) 25,000.

TLC abbr. colloq. tender loving care.

Tlemcen /tlemˈsen/ a city in NW Algeria; pop. (1989) 146,000. In the 13th–15th centuries it was the capital of a Berber dynasty. It has notable mosques and medieval buildings.

Tlingit /ˈklɪŋkɪt, ˈklɪŋɡɪt, ˈtlɪŋ-/ n. & adj. ● n. (pl. **Tlingits**, same) **1** a member of an American Indian people of the coasts and islands of south-eastern Alaska. **2** the language of this people. ● adj. of or relating to this people or their language. [Tlingit]

TLS abbr. Times Literary Supplement.

T-lymphocyte /tiːˈlɪmfəˌsaɪt/ n. Physiol. a lymphocyte of a type produced or processed by the thymus gland and active in the immune system, esp. the cell-mediated immune response. Also called T-cell.

TM see TRANSCENDENTAL MEDITATION.

Tm symb. Chem. the element thulium.

tmesis /ˈtmiːsɪs/ n. (pl. **tmeses** /-siːz/) Gram. the separation of parts of a compound word by an intervening word or words (esp. in colloq. speech, e.g. can't find it any-blooming-where). [Gk tmēsis cutting f. temnō cut]

TN abbr. US Tennessee (in official postal use).

tn abbr. **1** US ton(s). **2** town.

TNT abbr. trinitrotoluene, a high explosive formed from toluene by substitution of three hydrogen atoms with nitro groups.

to /tə, before a vowel tʊ, emphat. tuː/ prep. & adv. ● prep. **1** introducing a noun: **a** expressing what is reached, approached, or touched (fell to the ground; went to Paris; put her face to the window; five minutes to six). **b** expressing what is aimed at: often introducing the indirect object of a verb (throw it to me; explained the problem to them). **c** as far as; up to (went on to the end; have to stay from Tuesday to Friday). **d** to the extent of (were all drunk to a man; was starved to death). **e** expressing what is followed (according to instructions; made to order). **f** expressing what is considered or affected (am used to that; that is nothing to me). **g** expressing what is caused or produced (turn to stone; tear to shreds). **h** expressing what is compared (nothing to what it once was; comparable to any other; equal to the occasion; won by three goals to two). **i** expressing what is increased (add it to mine). **j** expressing what is involved or composed as specified (there is nothing to it; more to him than meets the eye). **k** expressing the substance of a debit entry in accounting (to four chairs, sixty pounds). **l** archaic for; by way of (took her to wife). **2** introducing the infinitive: **a** as a verbal noun (to get there is the priority). **b** expressing purpose, consequence, or cause (we eat to live; left him to starve; am sorry to hear that). **c** as a substitute for to + infinitive (wanted to come but was unable to). ● adv. **1** in the normal or required position or condition (come to; heave to). **2** (of a door) in a nearly closed position. □ **to and fro 1** backwards and forwards. **2** repeatedly between the same points. [OE tō (adv. & prep.) f. WG]

toad /təʊd/ n. **1** a froglike amphibian of the family Bufonidae, esp. of the genus Bufo, breeding in water but living chiefly on land, and usu. having a dry warty skin. **2** any similar amphibian (Suriname toad). **3** a repulsive or detestable person. □ **toad-eater** archaic a toady. **toad-in-the-hole** Brit. sausages or other meat baked in batter. □ **toadish** adj. [OE tādige, tādde, tāda, of unkn. orig.]

toadfish /ˈtəʊdfɪʃ/ n. a marine fish of the family Batrachoididae, with a large head and wide mouth, making grunting noises by vibrating the walls of its swim-bladder.

toadflax /ˈtəʊdflæks/ n. **1** a plant of the genus Linaria or Chaenorrhinum, of the figwort family; esp. yellow toadflax (L. vulgaris), with narrow flaxlike leaves and spurred yellow flowers. **2** a related plant, Cymbalaria muralis, with lilac flowers and ivy-shaped leaves.

toadstone /ˈtəʊdstəʊn/ n. a stone, sometimes precious, supposed to resemble or to have been formed in the body of a toad, formerly used as an amulet etc.

toadstool /ˈtəʊdstuːl/ n. the spore-bearing body of various fungi, usu. with a round cap and slender stalk; esp. one that is poisonous or inedible (cf. MUSHROOM).

toady /ˈtəʊdɪ/ n. & v. ● n. (pl. **-ies**) a sycophant; an obsequious hanger-on. ● v.tr. (**-ies, -ied**) behave servilely to; fawn upon. □ **toadyish** adj. **toadyism** n. [contr. of toad-eater, a charlatan's attendant who ate toads (regarded as poisonous)]

toast /təʊst/ n. & v. ● n. **1** bread in slices browned on both sides by radiant heat. **2 a** a person (orig. esp. a woman) or thing in whose honour a company is requested to drink. **b** a call to drink or an instance of drinking in this way. ● v. **1** tr. cook or brown (bread, a teacake, cheese, etc.) by radiant heat. **2** intr. (of bread etc.) become brown in this way. **3** tr. warm (one's feet, oneself, etc.) at a fire etc. **4** tr. drink to the health or in honour of (a person or thing). □ **have a person on toast** colloq. be in a position to deal with a person as one wishes. **toasting-fork** a long-handled fork for making toast before a fire. **toast rack** a rack for holding slices of toast at table. [ME (orig. as verb) f. OF toster roast, ult. f. L torrere tost- parch: sense 2 of the noun reflects the notion that a woman's name flavours the drink as spiced toast would]

toaster /'təʊstə(r)/ n. an electrical device for making toast.

toastmaster /'təʊst,mɑːstə(r)/ n. (fem. **toastmistress** /-,mɪstrɪs/) an official responsible for announcing toasts at a public occasion.

toasty /'təʊstɪ/ adj. **1** comfortably warm. **2** of or resembling toast. **3** (of wine) characterized by the aroma of maturation in small French oak casks.

tobacco /tə'bækəʊ/ n. (pl. **-os**) **1** a solanaceous plant of the genus *Nicotiana*, of American origin, with narcotic leaves used for smoking, chewing, or snuff. **2** its leaves, esp. as prepared for smoking. (See note below.) □ **tobacco mosaic virus** a virus that causes mosaic disease in tobacco, much used in biochemical research. **tobacco-pipe** see PIPE n. 2. **tobacco plant 1** = TOBACCO 1. **2** a plant of the genus *Nicotiana* grown for its night-scented flowers. **tobacco-stopper** an instrument for pressing down the tobacco in a pipe. [Sp. *tabaco*, of Amer. Indian orig.]

■ Tobacco was originally used by American Indians as a narcotic drink, but by the time that Christopher Columbus arrived they were smoking it. They used it for ceremonial purposes and believed it to have medicinal properties, which was the main reason for taking it back to Europe. The diplomat Jean Nicot (see NICOTIANA) is said to have introduced it to France in 1556. Tobacco was originally smoked mainly in pipes and cigars; cigarettes did not become socially acceptable until the late 18th century. It is usually made from the leaves of common tobacco (*Nicotiana tabacum*) and sometimes from wild tobacco (*N. rusticum*); many varieties have been developed and a number of different additives and preparation techniques are used. The principal narcotic drug in tobacco is nicotine, which accounts for its addictive nature. Increasing concerns over the links between cigarette smoking and lung cancer, heart disease, and other health problems have caused many countries to take steps to discourage smoking. The taxing of tobacco is, however, a major source of revenue for most governments.

tobacconist /tə'bækənɪst/ n. a retail dealer in tobacco and cigarettes etc.

Tobago see TRINIDAD AND TOBAGO.

Tobit /'təʊbɪt/ **1** a pious Israelite during the Babylonian Captivity, described in the Apocrypha. **2** a book of the Apocrypha telling the story of Tobit.

toboggan /tə'bɒgən/ n. & v. ● n. a long light narrow sledge for sliding downhill esp. over compacted snow or ice. (See note below.) ● v.intr. ride on a toboggan. □ **tobogganer** n. **tobogganing** n. **tobogganist** n. [Canad. F *tabaganne* f. Algonquian]

■ The use of the toboggan for winter recreation is recorded in the 16th century, although tobogganing as a racing sport dates from the mid-19th century. Today the most famous tobogganing course is the Cresta Run in Switzerland.

Tobruk /tə'brʊk/ (Arabic **Tubruq** /tʊ'bruːk/) a port on the Mediterranean coast of NE Libya; pop. (1984) 94,000. It was the scene of fierce fighting during the North African campaign in the Second World War.

Toby /'təʊbɪ/ the name of the trained dog introduced (in the first half of the 19th century) into the Punch and Judy show, which wears a frill round its neck.

toby jug /'təʊbɪ/ n. a jug or mug for ale etc., usu. in the form of a stout old man wearing a three-cornered hat. [familiar form of the name *Tobias*]

Tocantins /,təʊkən'tiːns/ **1** a river of South America, which rises in central Brazil and flows 2,640 km (1,640 miles) northwards, joining the Pará to enter the Atlantic through a large estuary at Belém. **2** a state of central Brazil; capital, Palmas.

toccata /tə'kɑːtə/ n. a musical composition for a keyboard instrument designed to exhibit the performer's touch and technique. [It., fem. past part. of *toccare* touch]

Toc H /tɒk 'eɪtʃ/ (in the UK) a society, originally of ex-servicemen and women, founded after the First World War to promote Christian fellowship and social service. [*toc* (former telegraphy code for *T*) + *H*, for Talbot House, a soldier's club established in Belgium in 1915]

Tocharian /tə'keəriən/ n. & adj. ● n. **1** an extinct Indo-European language of a central Asian people in the first millennium AD. **2** a member of the people speaking this language. ● adj. of or in this language. [F *tocharien* f. L *Tochari* f. Gk *Tokharoi* a Scythian tribe]

tocopherol /,təʊkəʊ'fiərɒl/ n. any of a group of closely related vitamins, found in wheat-germ oil, egg yolk, and leafy vegetables,

and important in the stabilization of cell membranes etc. Also called *vitamin E*. [Gk *tokos* offspring + *pherō* bear + -OL]

tocsin /'tɒksɪn/ n. an alarm bell or signal. [F f. OF *touquesain*, *toquassen* f. Prov. *tocasenh* f. *tocar* TOUCH + *senh* signal-bell]

tod /tɒd/ n. Brit. sl. □ **on one's tod** alone; on one's own. [20th c.: perh. f. rhyming sl. *on one's Tod Sloan* (name of a jockey)]

today /tə'deɪ/ adv. & n. ● adv. **1** on or in the course of this present day (*shall we go today?*). **2** nowadays, in modern times. ● n. **1** this present day (*today is my birthday*). **2** modern times. □ **today week** (or **fortnight** etc.) a week (or fortnight etc.) from today. [OE *tō dæg* on (this) day (as TO, DAY)]

Todd[1] /tɒd/, Mark James (b.1956), New Zealand equestrian. He won individual gold medals for three-day eventing in the Olympic Games of 1984 and 1988, and was ranked number one in the world in 1984, 1988, and 1989.

Todd[2] /tɒd/, Sweeney. A barber who murdered his customers, the central character of a play by George Dibdin Pitt (1799–1855) and of later plays.

toddle /'tɒd(ə)l/ v. & n. ● v.intr. **1** walk with short unsteady steps like those of a small child. **2** colloq. **a** (often foll. by *round*, *to*, etc.) take a casual or leisurely walk. **b** (usu. foll. by *off*) depart. ● n. **1** a toddling walk. **2** colloq. a stroll or leisurely walk. [16th-c. *todle* (Sc. & N. Engl.), of unkn. orig.]

toddler /'tɒdlə(r)/ n. a child who is just beginning to walk. □ **toddlerhood** n.

toddy /'tɒdɪ/ n. (pl. **-ies**) **1** a drink of spirits with hot water and sugar or spices. **2** the sap of some kinds of palm, fermented to produce arrack. [Hind. *tāṛī* f. *tāṛ* palm f. Skr. *tāla* palmyra]

to-do /tə'duː/ n. (pl. **to-dos**) a commotion or fuss. [*to do* as in *what's to do* (= to be done)]

tody /'təʊdɪ/ n. (pl. **-ies**) a small insect-eating West Indian bird of the family Todidae, related to the kingfisher, with a green back and red throat. [F *todier* f. L *todus*, a small bird]

toe /təʊ/ n. & v. ● n. **1** any of the five terminal projections of the human foot. **2** the corresponding part of an animal. **3** the part of an item of footwear that covers the toes. **4** the lower end or tip of an implement etc. **5** Archit. a projection from the foot of a buttress etc. to give stability. **6** Austral. & NZ sl. speed, energy. ● v. (**toes**, **toed**, **toeing**) **1** tr. touch (a starting-line etc.) with the toes before starting a race. **2** tr. **a** mend the toe of (a sock etc.). **b** provide with a toe. **3** intr. (foll. by *in*, *out*) **a** walk with the toes pointed in (or out). **b** (of a pair of wheels) converge (or diverge) slightly at the front. **4** tr. Golf strike (the ball) with a part of the club too near the toe. □ **on one's toes** alert, eager. **toe-clip** a clip on a bicycle pedal to prevent the foot from slipping. **toe-hold 1** a small foothold. **2** a small beginning or advantage. **toe the line** conform to a general policy or principle, esp. unwillingly or under pressure. **turn up one's toes** colloq. die. □ **toed** adj. (also in comb.). **toeless** adj. [OE *tā* f. Gmc]

toecap /'təʊkæp/ n. the (usu. strengthened) outer covering of the toe of a boot or shoe.

toenail /'təʊneɪl/ n. **1** the nail at the tip of each toe. **2** a nail driven obliquely through the end of a board etc.

toerag /'təʊræg/ n. Brit. sl. a term of contempt for a person. [earlier = tramp, vagrant, f. the rag wrapped round the foot in place of a sock]

toey /'təʊɪ/ adj. Austral. sl. restless, nervous, touchy.

toff /tɒf/ n. & v. Brit. sl. ● n. an upper-class person; a smart or well-dressed person. ● v.tr. (foll. by *up*) dress up smartly. [perh. a perversion of *tuft* = titled undergraduate (from the gold tassel formerly worn on the cap)]

toffee /'tɒfɪ/ n. (also **toffy**) (pl. **toffees** or **toffies**) **1** a kind of firm or hard sweet which softens when sucked or chewed, made by boiling sugar, butter, etc. **2** Brit. a small piece of this. □ **for toffee** colloq. (prec. by *can't* etc.) (denoting incompetence) at all (*they couldn't sing for toffee*). **toffee-apple** an apple with a thin coating of toffee. **toffee-nosed** esp. Brit. colloq. snobbish, pretentious. [earlier TAFFY]

toft /tɒft/ n. Brit. **1** a homestead. **2** land once occupied by this. [OE f. ON *topt*]

tofu /'təʊfuː, 'tɒfuː/ n. (esp. in China and Japan) a curd made from mashed soya beans. [Jap. *tōfu* f. Chin. *dòufu*, = rotten beans]

tog[1] /tɒg/ n. & v. colloq. ● n. (usu. in pl.) **1** an item of clothing. **2** Austral. & NZ colloq. a swimming-costume. ● v.tr. & intr. (**togged**, **togging**) (foll. by *out*, *up*) dress, esp. elaborately. [app. abbr. of 16th-c. cant *togeman(s)*, *togman*, f. F *toge* or L *toga*: see TOGA]

tog[2] /tɒg/ *n.* a unit of thermal resistance used to express the insulating properties of clothes and quilts. [arbitrary, prob. f. TOG[1]]

toga /ˈtəʊgə/ *n. hist.* an ancient Roman citizen's loose flowing outer garment. □ **togaed** /-gəd/ *adj.* (also **toga'd**). [L, rel. to *tegere* cover]

together /təˈgeðə(r)/ *adv. & adj.* ● *adv.* **1** in company or conjunction (*walking together*; *built it together*; *were at school together*). **2** simultaneously; at the same time (*both shouted together*). **3** one with another (*were talking together*). **4** into conjunction; so as to unite (*tied them together*; *put two and two together*). **5** into company or companionship (*came together in friendship*). **6** uninterruptedly (*could talk for hours together*). ● *adj. colloq.* well organized or controlled. □ **together with** as well as; and also. [OE *tōgædere* f. TO + *gædre* together: cf. GATHER]

togetherness /təˈgeðənɪs/ *n.* **1** the condition of being together. **2** a feeling of comfort from being together.

toggery /ˈtɒgərɪ/ *n. colloq.* clothes, togs.

toggle /ˈtɒg(ə)l/ *n. & v.* ● *n.* **1** a device for fastening (esp. a garment), consisting of a crosspiece which can pass through a hole or loop in one position but not in another. **2** a pin or other crosspiece put through the eye of a rope, a link of a chain, etc., to keep it in place. **3** a pivoted barb on a harpoon. **4** *Computing* a switch action that is operated the same way but with opposite effect on successive occasions. ● *v.tr.* provide or fasten with a toggle. □ **toggle joint** a device for exerting pressure along two jointed rods by applying a transverse force at the joint. **toggle switch** an electric switch with a projecting lever to be moved usu. up and down. [18th-c. Naut.: orig. unkn.]

Togliatti /tɒˈljætɪ/ (Russian **Tolyatti** /taˈljattɪ/) an industrial city and river port in SW Russia, on the River Volga; pop. (1990) 642,000. It was founded in 1738 but relocated in the mid-1950s to make way for the Kuibyshev reservoir. Formerly called Stavropol, it was renamed in 1964 after Palmiro Togliatti (1893–1964), leader of the Italian Communist Party.

Togo /ˈtəʊgəʊ/ a country in West Africa with a short coastline on the Gulf of Guinea; pop. (est. 1991) 3,761,000; languages, French (official), West African languages; capital, Lomé. The region formerly known as Togoland lay between the military powers of Ashanti and Dahomey and became a centre of the slave trade. It was annexed by Germany in 1884 and divided between France and Britain after the First World War. The British western section joined Ghana on the latter's independence (1957). The remainder, administered by France under a United Nations mandate after the Second World War, became an independent republic with the name Togo in 1960. After a long period of military rule a multi-party constitution was approved amidst continuing civil unrest in 1992. The country's first multi-party elections for the presidency were held in 1993, and for the National Assembly in 1994. □ **Togolese** /ˌtəʊgəˈliːz/ *adj. & n.*

Tohoku /təʊˈhəʊkuː/ a region of Japan, on the island of Honshu; capital, Sendai.

toil /tɔɪl/ *v. & n.* ● *v.intr.* **1** work laboriously or incessantly. **2** make slow painful progress (*toiled along the path*). ● *n.* prolonged or intensive labour; drudgery. □ **toil-worn** worn or worn out by toil. □ **toiler** *n.* [ME f. AF *toiler* (v.), *toil* (n.), dispute, OF *tooilier*, *tooil*, f. L *tudiculare* stir about f. *tudicula* machine for bruising olives, rel. to *tundere* beat]

toile /twʌl/ *n.* **1** cloth esp. for garments. **2** a garment reproduced in muslin or other cheap material for fitting or for making copies. [F *toile* cloth f. L *tela* web]

toilet /ˈtɔɪlɪt/ *n.* **1** = LAVATORY. **2 a** the process of washing oneself, dressing, etc. (*make one's toilet*). **b** (usu. *attrib.*) the articles used in washing oneself, dressing, applying make-up, etc. **3** the cleansing of part of the body after an operation or at the time of childbirth. □ **toilet bag** = WASHBAG. **toilet paper** (or **tissue**) paper for cleaning oneself after excreting. **toilet roll** a roll of toilet paper. **toilet set** a set of hairbrushes, combs, etc. **toilet soap** soap for washing oneself. **toilet table** a dressing-table usu. with a mirror. **toilet-train** cause (a young child) to undergo toilet-training. **toilet-training** the training of a young child to use the lavatory. **toilet water** a dilute form of perfume. [F *toilette* cloth, wrapper, dimin. f. *toile*: see TOILE]

toiletry /ˈtɔɪlɪtrɪ/ *n.* (*pl.* **-ies**) (usu. in *pl.*) any of various articles or cosmetics used in washing, dressing, etc.

toilette /twʌˈlet/ *n.* = TOILET 2a. [F: see TOILET]

toils /tɔɪlz/ *n.pl.* a net or snare. [pl. of *toil* f. OF *toile* cloth f. L *tela* web]

toilsome /ˈtɔɪlsəm/ *adj.* involving toil; laborious. □ **toilsomely** *adv.* **toilsomeness** *n.*

toing and froing /ˈtuːɪŋ ˈfrəʊɪŋ/ *n.* constant movement to and fro; bustle; dispersed activity. [TO *adv.* + FRO + -ING[1]]

Tojo /ˈtəʊdʒəʊ/, Hideki (1884–1948), Japanese military leader and statesman, Prime Minister 1941–4. By 1937 he had risen to Chief of Staff of the Japanese army of occupation in Manchuria, and was appointed Minister of War in 1940. Shortly after becoming Premier in 1941 he initiated the Japanese attack on the US base at Pearl Harbor. By 1944 Tojo had assumed virtual control of all political and military decision-making, but was forced to resign later that year following a number of Japanese military defeats. After Japan's surrender in 1945 he was tried and hanged as a war criminal.

tokamak /ˈtəʊkəˌmæk/ *n. Physics* a toroidal apparatus for producing controlled fusion reactions in hot plasma. [Russ. acronym, f. *toroidal'naya kamera s magnitnym polem* 'toroidal chamber with magnetic field']

Tokay /təʊˈkeɪ/ *n.* **1** a sweet aromatic wine made near Tokaj in Hungary. **2** a similar wine produced elsewhere.

tokay /ˈtəʊkeɪ/ *n.* (in full **tokay gecko**) a large grey SE Asian gecko, *Gekko gecko*, with orange and blue spots, and a loud call resembling the name. [Malay dial. *toke* f. Javanese *tekèk*, imit. of this call: cf. GECKO]

Tokelau /ˌtəʊkəˈlɑːuː/ a group of three islands in the western Pacific, between Kiribati and Western Samoa, forming an overseas territory of New Zealand; pop. (1986) 1,690.

token /ˈtəʊkən/ *n.***1** a thing serving as a symbol, reminder, or distinctive mark of something (*as a token of affection*; *in token of my esteem*). **2** a thing serving as evidence of authenticity or as a guarantee. **3** a voucher exchangeable for goods (often of a specified kind), given as a gift. **4** anything used to represent something else, esp. a metal disc etc. used instead of money in coin-operated machines etc. **5** (*attrib.*) **a** nominal or perfunctory (*token effort*). **b** conducted briefly to demonstrate strength of feeling (*token resistance*; *token strike*). **c** serving to acknowledge a principle only (*token payment*). **d** chosen by way of tokenism to represent a particular group (*the token woman on the committee*). □ **by this** (or **the same**) **token 1** similarly. **2** moreover. **token money** coins having a higher face value than their worth as metal. **token vote** a parliamentary vote of money, the stipulated amount of which is not meant to be binding. [OE *tāc(e)n* f. Gmc, rel. to TEACH]

tokenism /ˈtəʊkəˌnɪz(ə)m/ *n.* **1** esp. *Polit.* the principle or practice of granting minimum concessions, esp. to appease radical demands etc. (cf. TOKEN 5d). **2** making only a token effort.

Tokugawa /ˌtɒkʊˈgɑːwə/ the last shogunate in Japan, founded by Tokugawa Ieyasu (1543–1616) and in power 1603–1867. The period was marked by internal peace, stability, and economic growth, by a rigidly feudal social order, and by almost complete international isolation: Japanese people were forbidden to travel abroad, missionaries were excluded, and a few traders in Nagasaki were the only foreigners allowed in the country. The shogunate was followed by the restoration of imperial power under Meiji Tenno.

Tokyo /ˈtəʊkɪəʊ/ the capital of Japan and capital of Kanto region; pop. (1990) 8,163,000. Formerly called Edo, it was the centre of the military government under the shoguns (1603–1867). It was renamed Tokyo in 1868, when it replaced Kyoto as the imperial capital. [Jap., = eastern capital]

tolbooth var. of TOLLBOOTH.

Tolbukhin /tɒlˈbuːkɪn/ the former name (1949–1991) for DOBRICH.

told *past* and *past part.* of TELL[1].

Toledo 1 /təˈleɪdəʊ/ a city in central Spain on the River Tagus, capital of Castilla-La Mancha region; pop. (1991) 63,560. It was a pre-eminent city and cultural centre of Castile. Toledan steel and sword-blades have been famous since the first century BC. **2** /təˈliːdəʊ/ an industrial city and port on Lake Erie, in NW Ohio; pop. (1990) 332,940. □ **Toledan** *adj. & n.*

tolerable /ˈtɒlərəb(ə)l/ *adj.* **1** able to be endured. **2** fairly good; mediocre. □ **tolerably** *adv.* **tolerableness** *n.* **tolerability** /ˌtɒlərəˈbɪlɪtɪ/ *n.* [ME f. OF f. L *tolerabilis* (as TOLERATE)]

tolerance /ˈtɒlərəns/ *n.* **1** a willingness or ability to tolerate; forbearance. **2** the capacity to tolerate something, esp. a drug, transplant, antigen, environmental condition, etc., without adverse reaction. **3** an allowable variation in any measurable property. **4** diminution in the response to a drug after continued use. [ME f. OF f. L *tolerantia* (as TOLERATE)]

tolerant /ˈtɒlərənt/ *adj.* **1** disposed or accustomed to tolerate others or their acts or opinions. **2** (foll. by *of*) enduring or patient. □ **tolerantly** *adv.* [F *tolérant* f. L *tolerare* (as TOLERATE)]

tolerate /ˈtɒləˌreɪt/ *v.tr.* **1** allow the existence or occurrence of without authoritative interference. **2** leave unmolested. **3** endure or permit, esp. with forbearance. **4** sustain or endure (suffering etc.). **5** be capable of continued subjection to (a drug, radiation, etc.) without harm. **6** find or treat as endurable. □ **tolerator** *n.* [L *tolerare tolerat-* endure]

toleration /ˌtɒləˈreɪʃ(ə)n/ *n.* the process or practice of tolerating, esp. the allowing of differences in religious opinion without discrimination. [F *tolération* f. L *toleratio* (as TOLERATE)]

Toleration Act an act of 1689 granting freedom of worship to dissenters (excluding Roman Catholics and Unitarians) on certain conditions. Its real purpose was to unite all Protestants under William III against the deposed Roman Catholic James II.

Tolkien /ˈtɒlkiːn/, J(ohn) R(onald) R(euel) (1892–1973), British novelist, born in South Africa. Professor of Anglo-Saxon and later of English Language and Literature at Oxford University, Tolkien is famous for the fantasy adventures *The Hobbit* (1937) and *The Lord of the Rings* (1954–5). These were set in Middle Earth, an imaginary land peopled by hobbits and other mythical creatures, and for which he devised languages with considerable thoroughness. *The Silmarillion*, an account of the mythology and early history of Middle Earth, was published posthumously (1977).

toll[1] /təʊl/ *n.* **1** a charge payable for permission to pass a barrier or use a bridge or road etc. **2** the cost or damage caused by a disaster, battle, etc., or incurred in an achievement (*death toll*). **3** *N. Amer.* a charge for a long distance telephone call. □ **take its toll** be accompanied by loss or injury etc. **toll-bridge** a bridge at which a toll is charged. **toll-gate** a gate preventing passage until a toll is paid. **toll-house** a house at a toll-gate or -bridge, used by a toll-collector. **toll-road** a road maintained by the tolls collected on it. [OE f. med.L *toloneum* f. LL *telonium* f. Gk *telōnion* toll-house f. *telos* tax]

toll[2] /təʊl/ *v.* & *n.* ● *v.* **1 a** *intr.* (of a bell) sound with a slow uniform succession of strokes. **b** *tr.* ring (a bell) in this way. **c** *tr.* (of a bell) announce or mark (a death etc.) in this way. **2** *tr.* strike (the hour). ● *n.* **1** the act of tolling. **2** a stroke of a bell. [ME, special use of (now dial.) *toll* entice, pull, f. an OE root *-tyllan* (recorded in *fortyllan* seduce)]

tollbooth /ˈtəʊlbuːð, -buːθ/ *n.* (also **tolbooth**) **1** a booth at the roadside from which tolls are collected. **2** *Sc. archaic* a town hall. **3** *Sc. archaic* a town jail.

Tollund Man /ˈtɒlənd/ the well-preserved corpse of an Iron Age man (*c.*500 BC–400 AD) found in 1950 in a peat bog in central Jutland, Denmark. Around the neck was a plaited leather noose, indicating that Tollund Man had met his death by hanging, a victim of murder or sacrifice.

Tolpuddle martyrs /ˈtɒlˌpʌd(ə)l/ six farm labourers from the village of Tolpuddle in Dorset who attempted to form a trade union and were sentenced in 1834 to seven years' transportation on a charge of administering unlawful oaths. Their harsh sentences caused widespread protests, and two years later they were pardoned and repatriated from Australia.

Tolstoy /ˈtɒlstɔɪ/, Leo (Russian name Count Lev Nikolaevich Tolstoi) (1828–1910), Russian writer. He is best known for the novels *War and Peace* (1863–9) and *Anna Karenina* (1873–7). The former is an epic tale of the Napoleonic invasion and the lives of three aristocratic families; the latter describes a married woman's passion for a young officer and her tragic fate. Tolstoy subsequently underwent a spiritual crisis, espousing a moral code based on self-sufficiency, non-resistance to evil, belief in God, love of humankind, renouncing property, and repudiating government and organized religion. This provoked his excommunication from the Russian Orthodox Church in 1901 and the banning of many of his works.

Toltec /ˈtɒltek/ *n.* & *adj.* ● *n.* (*pl.* same or **Toltecs**) a member of a Nahuatl-speaking people who dominated central Mexico AD *c.*900–1200. (*See note below.*) ● *adj.* of or relating to the Toltec. [Sp. f. Nahuatl]

▪ The Toltec, a Chichimec people, were a warrior aristocracy whose period of domination was violent and innovative. They founded or developed cities (their capital was Tula, founded in 968) and sacked the great city of Teotihuacán, but were unable to consolidate their hold on the conquered area, which developed into a number of states, mostly independent. In the 12th–13th centuries famine and drought (perhaps caused by climatic changes) brought catastrophe, and the disunited area fell to other invading Chichimec peoples from the north.

tolu /təˈluː, ˈtəʊluː/ *n.* a fragrant brown balsam obtained from either of two South American trees, *Myroxylon balsamum* or *M. toluifera*, and used in perfumery and medicine. [Santiago de *Tolu* in Colombia]

Toluca /təˈluːkə/ (in full **Toluca de Lerdo** /deɪ ˈleədəʊ/) a city in central Mexico, capital of the state of Mexico; pop. (1990) 488,000. It lies at the foot of the extinct volcano Nevado de Toluca, at an altitude of 2,680 m (8,793 ft).

toluene /ˈtɒljʊˌiːn/ *n. Chem.* a colourless aromatic liquid hydrocarbon derivative of benzene, orig. obtained from tolu, used in the manufacture of explosives etc. Also called *methyl benzene.* □ **toluol** /-jʊˌɒl/ *n.* **toluic** /təˈluːɪk/ *adj.* [TOLU + -ENE]

Tolyatti see TOGLIATTI.

tom /tɒm/ *n.* a male of various animals, esp. (in full **tom-cat**) a male cat. [abbr. of the name *Thomas*]

tomahawk /ˈtɒməˌhɔːk/ *n.* & *v.* ● *n.* **1** a North American Indian war-axe with a stone or iron head. **2** *Austral.* a hatchet. ● *v.tr.* strike, cut, or kill with a tomahawk. [Renape *tämähäk* f. *tämäham* he etc. cuts]

tomatillo /ˌtɒməˈtiːjəʊ/ *n.* (*pl.* **-os**) esp. *US* **1** the edible purplish fruit of a ground cherry, *Physalis philadelphica*, of Mexico. **2** this plant. [Sp., dimin. of *tomate* TOMATO]

tomato /təˈmɑːtəʊ/ *n.* (*pl.* **-oes**) **1** a glossy red or yellow pulpy edible fruit. **2** a solanaceous plant, *Lycopersicon esculentum*, bearing this. □ **tomatoey** /-təʊɪ/ *adj.* [17th-c. *tomate*, = F or Sp. & Port., f. Mex. *tomatl*]

tomb /tuːm/ *n.* **1** a large esp. underground vault for the burial of the dead. **2** an enclosure cut in the earth or in rock to receive a dead body. **3** a sepulchral monument. **4** (prec. by *the*) the state of death. [ME *t(o)umbe* f. AF *tumbe*, OF *tombe* f. LL *tumba* f. Gk *tumbos*]

tombac /ˈtɒmbæk/ *n.* an alloy of copper and zinc used esp. as material for cheap jewellery. [F f. Malay *tembaga* copper]

Tombaugh /ˈtɒmbɔː/, Clyde William (b.1906), American astronomer. His chief interest was in the search for undiscovered planets, which he carried out mainly at the Lowell Observatory in Arizona. His extensive examination of photographic plates led to his discovery of the planet Pluto on 13 March 1930. Tombaugh subsequently discovered numerous asteroids.

tombola /tɒmˈbəʊlə/ *n. Brit.* a kind of lottery with tickets usu. drawn from a turning drum-shaped container for immediate prizes, esp. at a fête or fair. [F *tombola* or It. f. *tombolare* tumble]

tombolo /ˈtɒmbəˌləʊ/ *n.* (*pl.* **-os**) a spit joining an island to the mainland. [It., = sand-dune]

Tombouctou see TIMBUKTU 1.

tomboy /ˈtɒmbɔɪ/ *n.* a girl who behaves in a boisterous or active way formerly thought more appropriate for boys. □ **tomboyish** *adj.* **tomboyishness** *n.*

tombstone /ˈtuːmstəʊn/ *n.* a stone standing or laid over a grave, usu. with an epitaph.

Tom Collins /tɒm ˈkɒlɪnz/ *n.* an iced cocktail of gin with soda, lemon or lime juice, and sugar. [20th c.: orig. unkn.]

Tom, Dick, and Harry /tɒm, dɪk, ˈhærɪ/ *n.* (usu. prec. by *any*, *every*) usu. *derog.* ordinary people taken at random.

tome /təʊm/ *n.* a large heavy book or volume. [F f. L *tomus* f. Gk *tomos* section, volume f. *temnō* cut]

-tome /təʊm/ *comb. form* forming nouns meaning: **1** an instrument for cutting (*microtome*). **2** a section or segment. [Gk *tomē* a cutting, *-tomos* -cutting, f. *temnō* cut]

tomentum /təˈmentəm/ *n.* (*pl.* **tomenta** /-tə/) **1** *Bot.* matted woolly down on stems and leaves. **2** *Anat.* the tufted inner surface of the pia mater in the brain. □ **tomentose** /-ˈmentəʊz, -təʊs/ *adj.* **tomentous** *adj.* [L, = cushion-stuffing]

tomfool /tɒmˈfuːl/ *n.* **1** a foolish person. **2** (*attrib.*) silly, foolish (*a tomfool idea*).

tomfoolery /tɒmˈfuːlərɪ/ *n.* (*pl.* **-ies**) **1** foolish behaviour; nonsense. **2** an instance of this.

Tomis /ˈtəʊmɪs/ an ancient name for CONSTANŢA.

Tommy /ˈtɒmɪ/ *n.* (*pl.* **-ies**) *colloq.* a British private soldier. [*Tommy* (*Thomas*) *Atkins*, a name used in specimens of completed official forms]

tommy-bar /ˈtɒmɪˌbɑː(r)/ *n.* a short bar for use with a box spanner.

tommy-gun /ˈtɒmɪˌɡʌn/ *n.* a type of sub-machine-gun. [John Taliaferno *Thompson*, Amer. army officer (1860–1940), its co-inventor]

tommy-rot /ˈtɒmɪˌrɒt/ *n. sl.* nonsense.

tommy ruff see RUFF[2] 2.

tomogram /ˈtɒməˌɡræm/ *n.* a record obtained by tomography.

tomography /tə'mɒɡrəfi/ *n.* esp. *Med.* a type of radiography in which the detector and radiation source are rotated around the body or other object so as to give a clear image in a selected plane. The technique was devised in the early 1930s and subsequently much improved: CAT-scanning makes use of computer processing to construct three-dimensional images from a series of sequential scans. [Gk *tomē* a cutting + -GRAPHY]

tomorrow /tə'mɒrəʊ/ *adv. & n.* ● *adv.* **1** on the day after today. **2** at some future time. ● *n.* **1** the day after today. **2** the near future. □ **tomorrow morning** (or **afternoon** etc.) in the morning (or afternoon etc.) of tomorrow. **tomorrow week** a week from tomorrow. [TO + MORROW: cf. TODAY]

Tompion /'tɒmpɪən/, Thomas (c.1639–1713), English clock and watchmaker. He made one of the first balance-spring watches to the design of Robert Hooke, and for the Royal Greenwich Observatory he made two large pendulum clocks which needed winding only once a year. Tompion also collaborated with Edward Barlow (1636–1716) in patenting the horizontal-wheel cylinder escapement needed to produce flat watches.

tompion var. of TAMPION.

Tomsk /tɒmsk/ an industrial city in southern Siberian Russia, a port on the River Tom; pop. (1990) 506,000.

Tom Thumb *n.* **1** a very short male person. Tom Thumb was the hero of an old nursery tale, the son of a ploughman in the time of King Arthur who was only as tall as his father's thumb. **2** (also **tom thumb**) **a** a dwarf variety of a cultivated flower or vegetable. **b** a small wild flower, esp. bird's-foot trefoil.

tomtit /'tɒmtɪt/ *n.* a tit, esp. a blue tit.

tom-tom /'tɒmtɒm/ *n.* **1** a primitive drum beaten with the hands. **2** a small to medium-sized drum used esp. as part of a drum-kit. [Hindi *tamtam*, imit.]

-tomy /təmi/ *comb. form* forming nouns denoting cutting, esp. in surgery (*laparotomy*). [Gk -*tomia* cutting f. *temnō* cut]

ton[1] /tʌn/ *n.* **1** (in full **long ton**) a unit of weight equal to 2,240 lb avoirdupois (1016.05 kg). **2** (in full **short ton**) a unit of weight equal to 2,000 lb avoirdupois (907.19 kg). **3** (in full **metric ton**) = TONNE. **4 a** (in full **displacement ton**) a unit of measurement of a ship's weight or volume in terms of its displacement of water with the loadline just immersed, equal to 2,240 lb or 35 cu. ft (0.99 cubic metres). **b** (in full **freight ton**) a unit of weight or volume of cargo, equal to a metric ton (1,000 kg) or 40 cu. ft. **5 a** (in full **gross ton**) a unit of gross internal capacity, equal to 100 cu. ft (2.83 cubic metres). **b** (in full **net** or **register ton**) an equivalent unit of net internal capacity. **6** a unit of refrigerating power able to freeze 2,000 lb of water at 0°C in 24 hours. **7** a measure of capacity for various materials, esp. 40 cu. ft of timber. **8** (usu. in *pl.*) *colloq.* a large number or amount (*tons of money*). **9** esp. *Brit. sl.* **a** a speed of 100 m.p.h. **b** a sum of £100. **c** a score of 100. □ **ton-mile** one ton of goods carried one mile, as a unit of traffic. **ton-up** *Brit. sl. n.* a speed of 100 m.p.h. ● *attrib.adj.* **1** (of a motorcyclist) achieving this, esp. habitually and recklessly. **2** fond or capable of travelling at high speed. **weigh a ton** *colloq.* be very heavy. [orig. the same word as TUN: differentiated in the 17th c.]

ton[2] /tɒn/ *n.* **1** a prevailing mode or fashion. **2** fashionable society. [F]

tonal /'təʊn(ə)l/ *adj.* **1** of or relating to tone or tonality. **2** *Mus.* (of a fugue etc.) having repetitions of the subject at different pitches in the same key. □ **tonally** *adv.* [med.L *tonalis* (as TONE)]

tonality /tə'nælɪti/ *n.* (*pl.* -**ies**) **1** *Mus.* **a** the relationship between the tones of a musical scale. **b** the observance of a single tonic key as the basis of a composition. **2** the tone or colour scheme of a picture. **3** *Linguistics* the differentiation of words, syllables, etc. by a change of vocal pitch.

tondo /'tɒndəʊ/ *n.* (*pl.* **tondi** /-dɪ/) a circular painting or relief. [It., = round (plate), f. *rotondo* f. L *rotundus* round]

tone /təʊn/ *n. & v.* ● *n.* **1** a musical or vocal sound, esp. with reference to its pitch, quality, and strength. **2** (often in *pl.*) modulation of the voice expressing a particular feeling or mood (*a cheerful tone*; *suspicious tones*). **3** a manner of expression in writing. **4** *Mus.* **a** a musical sound, esp. of a definite pitch and character. **b** an interval of a major second, e.g. C–D. **5 a** the general effect of colour or of light and shade in a picture. **b** the tint or shade of a colour. **6 a** the prevailing character of the morals and sentiments etc. in a group. **b** an attitude or sentiment expressed esp. in a letter etc. **7** the proper firmness or functioning of bodily organs or tissues. **8** a state of good or specified health or quality. **9** *Phonet.* **a** an accent on one syllable of a word. **b** a way of pronouncing a word or syllable with a specific intonation to distinguish it from others of a similar sound (*Mandarin Chinese has four tones*). ● *v.* **1** *tr.* give the desired tone to. **2** *tr.* modify the tone of. **3** *intr.* (often foll. by *to*) attune. **4** *intr.* (foll. by *with*) be in harmony (of colour) (*does not tone with the wallpaper*). **5** *tr. Photog.* give (a monochrome picture) an altered colour in finishing by means of a chemical solution. **6** *intr.* undergo a change in colour by toning. □ **tone-arm** the movable arm supporting the pick-up of a record-player. **tone control** a switch for varying the proportion of high and low frequencies in reproduced sound. **tone-deaf** unable to perceive differences of musical pitch accurately. **tone-deafness** the condition of being tone-deaf. **tone down 1** make or become softer in tone of sound or colour. **2** make (a statement etc.) less harsh or emphatic. **tone poem** = *symphonic poem*. **tone-row** = SERIES 8. **tone up 1** make or become stronger in tone of sound or colour. **2** make (a statement etc.) more emphatic. **3** make (muscles) firm by exercise etc.; make or become fitter. **whole-tone scale** see WHOLE. □ **toneless** *adj.* **tonelessly** *adv.* [ME f. OF *ton* or L *tonus* f. Gk *tonos* tension, tone f. *teinō* stretch]

toneburst /'təʊnbɜːst/ *n.* an audio signal used in testing the transient response of audio components.

toneme /'təʊniːm/ *n. Phonet.* a phoneme distinguished from another only by its tone. □ **tonemic** /tə'niːmɪk/ *adj.* [TONE after *phoneme*]

tonepad /'təʊnpæd/ *n. Computing* a device for communicating with a computer over a telephone line using electronically recognized sounds.

toner /'təʊnə(r)/ *n.* **1** a chemical bath for toning a photographic print. **2** powder used in xerographic copying processes. **3** a cosmetic preparation for toning the skin.

tong /tɒŋ/ *n.* a Chinese guild, association, or secret society. [Chin. *tang* meeting-place]

Tonga /'tɒŋə, 'tɒŋɡə/ a country in the South Pacific consisting of an island group south-east of Fiji; pop. (est. 1991) 100,000; official languages, Tongan and English; capital, Nuku'alofa. The kingdom of Tonga consists of about 170 volcanic and coral islands, of which thirty-six are inhabited. Visited by the Dutch in the early 17th century, the islands were later named the Friendly Islands by Captain James Cook. British influence began with Methodist missionaries in the early 19th century and the kingdom became a British protectorate in 1900. Tonga, which has been a constitutional monarchy since 1875, became an independent Commonwealth state in 1970.

tonga /'tɒŋɡə/ *n.* a light horse-drawn two-wheeled vehicle used in India. [Hindi *tāṅgā*]

Tongan /'tɒŋən, 'tɒŋɡən/ *adj. & n.* ● *adj.* of or relating to Tonga or its people or language. ● *n.* **1** a native or national of Tonga. **2** the Polynesian language spoken in Tonga.

Tongariro, Mount /ˌtɒŋə'rɪərəʊ/ a mountain in North Island, New Zealand. It rises to a height of 1,968 m (6,457 ft) and is held sacred by the Maoris.

tongs /tɒŋz/ *n.pl.* (also **pair of tongs** *sing.*) an instrument with two hinged or sprung arms for grasping and holding. [pl. of *tong* f. OE *tang(e)* f. Gmc]

Tongshan /tɒŋ'ʃæn/ a former name (1912–45) for XUZHOU.

tongue /tʌŋ/ *n. & v.* ● *n.* **1** the fleshy muscular organ in the mouth used in tasting, licking, and swallowing, and (in humans) for speech. **2** the tongue of an ox etc. as food. **3** the faculty of or a tendency in speech (*a sharp tongue*). **4** a particular language (*the German tongue*). **5** a thing like a tongue in shape or position, esp.: **a** a long low promontory. **b** a strip of leather etc., attached at one end only, under the laces in a shoe. **c** the clapper of a bell. **d** the pin of a buckle. **e** the projecting strip on a wooden etc. board fitting into the groove of another. **f** a vibrating slip in the reed of some musical instruments. **g** a jet of flame. ● *v.* (**tongues, tongued, tonguing**) *Mus.* **1** *tr.* produce staccato etc. effects with (a flute etc.) by means of tonguing. **2** *intr.* use the tongue in this way. □ **find** (or **lose**) **one's tongue** be able (or unable) to express oneself after a shock etc. **gift of tongues** the power of speaking in unknown languages, regarded as one of the gifts of the Holy Spirit (Acts 2). **keep a civil tongue in one's head** avoid rudeness. **tongue-and-groove** applied to boards in which a tongue along one edge fits into a groove along the edge of the next, each board having a tongue on one edge and a groove on the other. **tongue-in-cheek** *adj.* ironic; slyly humorous. ● *adv.* insincerely or ironically. **tongue-lashing** a severe scolding or reprimand. **tongue-tie** a speech impediment due to a malformation of the tongue. **tongue-tied 1** too shy or embarrassed to speak. **2** having a tongue-tie. **tongue-twister** a sequence of words difficult to pronounce quickly and correctly. **with**

one's **tongue hanging out** eagerly or expectantly. **with one's tongue in one's cheek** insincerely or ironically. □ **tongued** adj. (also in comb.). **tongueless** adj. [OE tunge f. Gmc, rel. to L lingua]

tonguing /ˈtʌŋɪŋ/ n. Mus. the technique of playing a wind instrument using the tongue to articulate certain notes.

tonic /ˈtɒnɪk/ n. & adj. ● n. **1** an invigorating medicine. **2** anything serving to invigorate. **3** = tonic water. **4** Mus. the first degree of a scale, forming the keynote of a piece (see KEYNOTE 3). ● adj. **1** serving as a tonic; invigorating. **2** Mus. denoting the first degree of a scale. **3** Med. **a** producing tension, esp. of the muscles. **b** restoring normal tone to organs. □ **tonic accent** Phonet. an accent marked by a change of pitch within a syllable. **tonic-clonic** Med. of or characterized by successive phases of tonic and clonic spasm. **tonic spasm** Med. continuous muscular contraction (cf. CLONUS). **tonic water** a carbonated soft drink containing quinine. □ **tonically** adv. [F tonique f. Gk tonikos (as TONE)]

tonicity /təˈnɪsɪtɪ/ n. **1** the state of being tonic. **2** Med. a healthy elasticity of muscles etc. **3** Linguistics phonetic emphasis at a certain place in an intonation pattern.

tonic sol-fa /sɒlˈfɑː/ n. a system of musical notation used esp. in teaching singing, in which the seven notes of the major scale in any key are sung to syllables written doh, ray, me, fah, soh, lah, te, with doh always denoting the tonic or keynote. Tonic sol-fa was developed in the mid-19th century from an earlier system (see GAMUT).

tonight /təˈnaɪt/ adv. & n. ● adv. on the present or approaching evening or night. ● n. the evening or night of the present day. [TO + NIGHT: cf. TODAY]

tonka bean /ˈtɒŋkə/ n. the black fragrant seed of a leguminous South American tree, Dipteryx odorata, used in perfumery etc. [Carib tonka f. Tupi tōka]

Tonkin /tɒnˈkɪn/ a mountainous region of northern Vietnam, centred on the delta of the Red River. Formerly part of China, it became a French protectorate in 1883. It was part of French Indo-China from 1887 until 1946.

Tonkin, Gulf of an arm of the South China Sea, bounded by the coasts of southern China and northern Vietnam. Its chief port is Haiphong. An incident there in 1964 led to increased US military involvement in the area prior to the Vietnam War.

Tonlé Sap /ˌtɒnleɪ ˈsæp/ a lake in central Cambodia, linked to the Mekong river by the Tonlé Sap river. The area of the lake is tripled during the wet season (June–November). On the north-west shore stand the ruins of the ancient city of Angkor, capital of the former Khmer empire.

tonnage /ˈtʌnɪdʒ/ n. **1** a ship's internal cubic capacity or freight-carrying capacity measured in tons. **2** the total carrying capacity esp. of a country's mercantile marine. **3** a charge per ton on freight or cargo. [orig. in sense 'duty on a tun of wine': OF tonnage f. tonne TUN: later f. TON¹]

tonne /tʌn/ n. a metric ton, equal to 1,000 kg. [F: see TUN]

tonneau /ˈtɒnəʊ/ n. (pl. **tonneaux** /-nəʊz/) the part of a motor car occupied by the back seats, esp. in an open car. □ **tonneau cover** a removable flexible cover for the passenger seats in an open car, boat, etc., when they are not in use. [F, lit. 'cask, tun']

tonometer /təˈnɒmɪtə(r)/ n. **1** Mus. a tuning-fork or other instrument for measuring the pitch of tones. **2** Med. an instrument for measuring the pressure in the eyeball (to test for glaucoma) or that in a blood vessel etc. [formed as TONE + -METER]

tonsil /ˈtɒns(ə)l, -sɪl/ n. either of two small masses of lymphoid tissue, one on each side of the root of the tongue. □ **tonsillar** adj. [F tonsilles or L tonsillae (pl.)]

tonsillectomy /ˌtɒnsɪˈlektəmɪ/ n. (pl. **-ies**) the surgical removal of the tonsils.

tonsillitis /ˌtɒnsɪˈlaɪtɪs/ n. inflammation of the tonsils.

tonsorial /tɒnˈsɔːrɪəl/ adj. usu. joc. of or relating to a hairdresser or hairdressing. [L tonsorius f. tonsor barber f. tondere tons- shave]

tonsure /ˈtɒnsjə(r), ˈtɒnʃə(r)/ n. & v. ● n. **1** the shaving of the crown of the head (in the Roman Catholic Church until 1972) or the whole head (in the Orthodox Church), esp. of a person entering a priesthood or monastic order. **2** a bare patch made in this way. ● v.tr. give a tonsure to. [ME f. OF tonsure or L tonsura (as TONSORIAL)]

tontine /tɒnˈtiːn/ n. an annuity shared by subscribers to a loan, the shares increasing as subscribers die until the last survivor gets all, or until a specified date when the remaining survivors share the proceeds.

[F, f. the name of Lorenzo Tonti, Neapolitan banker (1630–95), originator of tontines in France c.1653]

Tonton Macoute /ˌtɒntɒn məˈkuːt/ n. (pl. **Tontons Macoutes** pronunc. same) a member of a militia formed in 1961 by President François Duvalier of Haiti, notorious for its brutal and arbitrary actions, and disbanded in 1986 after a mass uprising forced Duvalier to flee the country. [Haitian F, app. with ref. to an ogre of folk-tales]

tony /ˈtəʊnɪ/ adj. (**tonier, toniest**) US colloq. having 'tone'; stylish, fashionable.

too /tuː/ adv. **1** to a greater extent than is desirable, permissible, or possible for a specified or understood purpose (too colourful for my taste; too large to fit). **2** colloq. extremely (you're too kind). **3** in addition (are they coming too?). **4** moreover (we must consider, too, the time of year). □ **none too 1** rather less than (feeling none too good). **2** barely. **too bad** see BAD. **too much, too much for** see MUCH. **too right** see RIGHT. **too-too** adj. & adv. colloq. extreme, excessive(ly). [stressed form of TO, f. 16th-c. spelling too]

toodle-oo /ˌtuːd(ə)lˈuː/ int. (also **toodle-pip**) colloq. goodbye. [20th c.: orig. unkn.: perh. alt. of F à tout à l'heure see you soon]

took past of TAKE.

tool /tuːl/ n. & v. ● n. **1** any device or implement used to carry out mechanical functions whether manually or by a machine. **2** a thing used in an occupation or pursuit (the tools of one's trade; literary tools). **3** a person used as a mere instrument by another. **4** coarse sl. the penis. **5 a** a distinct design in the tooling of a book. **b** a small stamp or roller used to make this. ● v.tr. **1** dress (stone) with a chisel. **2** impress a design on (a leather book-cover). **3** (foll. by along, around, etc.) sl. drive or ride, esp. in a casual or leisurely manner. **4** (often foll. by up) equip with tools. □ **tool-box** a box or container for keeping tools in. **tool kit 1** a set of tools. **2** Computing a set of software tools, esp. designed for a specific application. **tool-pusher** a worker directing the drilling on an oil rig. **tool up 1** sl. arm oneself. **2** equip oneself. □ **tooler** n. [OE tōl f. Gmc]

tooling /ˈtuːlɪŋ/ n. **1** the process of dressing stone with a chisel. **2** the ornamentation of a leather book-cover with designs impressed by heated tools.

toolmaker /ˈtuːlˌmeɪkə(r)/ n. a person who makes precision tools, esp. tools used in a press. □ **toolmaking** n.

toot¹ /tuːt/ n. & v. ● n. **1** a short sharp sound as made by a horn, trumpet, or whistle. **2** US sl. cocaine or a snort (see SNORT n. 4) of cocaine. **3** sl. a drinking session; a binge, a spree. ● v. **1** tr. sound (a horn etc.) with a short sharp sound. **2** intr. give out such a sound. □ **tooter** n. [prob. f. MLG tūten, or imit.]

toot² /tʊt/ n. Austral. sl. a lavatory. [20th c.: orig. unkn.]

tooth /tuːθ/ n. & v. ● n. (pl. **teeth** /tiːθ/) **1** each of a set of hard bony enamel-coated structures in the jaws of most vertebrates, used for biting and chewing. **2** a toothlike part or projection, e.g. the cog of a gearwheel, the point of a saw or comb, etc. **3** (often foll. by for) one's sense of taste; an appetite or liking. **4** (in pl.) force or effectiveness (the penalties give the contract teeth). ● v. **1** tr. provide with teeth. **2** intr. (of cog-wheels) engage, interlock. □ **armed to the teeth** completely and elaborately armed or equipped. **fight tooth and nail** fight very fiercely. **get one's teeth into** devote oneself seriously to. **in the teeth of 1** in spite of (opposition or difficulty etc.). **2** contrary to (instructions etc.). **3** directly against (the wind etc.). **set a person's teeth on edge** see EDGE. **tooth-billed** (of a bird) having toothlike projections on the cutting edges of the bill. **tooth-comb** = fine-tooth comb (see FINE¹). **tooth powder** powder for cleaning the teeth. **tooth shell** = tusk shell. □ **toothless** adj. **toothlike** adj. [OE tōth (pl. tēth) f. Gmc]

toothache /ˈtuːθeɪk/ n. a (usu. prolonged) pain in a tooth or teeth.

toothbrush /ˈtuːθbrʌʃ/ n. a brush for cleaning the teeth.

toothed /tuːθt/ adj. having teeth or toothlike projections. □ **toothed whale** a whale of the suborder Odontoceti, having teeth rather than baleen plates, including sperm whales, killer whales, dolphins and porpoises.

toothing /ˈtuːθɪŋ/ n. projecting bricks or stones left at the end of a wall to allow its continuation.

toothpaste /ˈtuːθpeɪst/ n. a paste for cleaning the teeth.

toothpick /ˈtuːθpɪk/ n. a small sharp instrument for removing small pieces of food lodged between the teeth.

toothsome /ˈtuːθsəm/ adj. (of food) delicious, appetizing. □ **toothsomely** adv. **toothsomeness** n.

toothwort /ˈtuːθwɜːt/ n. a leafless parasitic plant, *Lathraea squamaria*, of the broomrape family, with toothlike root scales.

toothy /ˈtuːθɪ/ adj. (**toothier, toothiest**) having or showing large, numerous, or prominent teeth (*a toothy grin*). □ **toothily** adv.

tootle /ˈtuːt(ə)l/ v.intr. **1** toot gently or repeatedly. **2** (usu. foll. by *along, around*, etc.) colloq. move casually or aimlessly. □ **tootler** n.

tootsy /ˈtʊtsɪ/ n. (also **tootsie**) (pl. **-ies**) colloq. & joc. a foot. [joc. dimin.: cf. FOOTSIE]

Toowoomba /təˈwʊmbə/ a town in Queensland, Australia, to the west of Brisbane; pop. (1991) 75,960. It was formerly known as The Swamps.

top[1] /tɒp/ n., adj., & v. ● n. **1** the highest point or part (*the top of the house*). **2 a** the highest rank or place (*at the top of the school*). **b** a person occupying this (*was top in maths*). **c** the upper end or head (*the top of the table*). **3** the upper surface of a thing, esp. of the ground, a table, etc. **4** the upper part of a thing, esp.: **a** a blouse, jumper, etc. for wearing with a skirt or trousers. **b** the upper part of a shoe or boot. **c** the stopper of a bottle. **d** the lid of a jar, saucepan, etc. **e** the creamy part of milk. **f** the folding roof of a car, pram, or carriage. **g** the upper edge or edges of a page or pages in a book (*gilt top*). **5** the utmost degree; height (*called at the top of his voice*). **6** the high-frequency component of reproduced sound. **7** (in pl.) colloq. a person or thing of the best quality (*he's tops at cricket*). **8** (esp. in pl.) the leaves etc. of a plant grown esp. for its root (*turnip-tops*). **9** (usu. in pl.) a bundle of long wool fibres prepared for spinning. **10** Naut. a platform round the head of the lower masts of a sailing-ship, serving to extend the topmast rigging or carry guns. **11** (in pl.) esp. *Bridge* the two or three highest cards of a suit. **12** = *top gear* (see GEAR). **13** = TOPSPIN. ● adj. **1** highest in position (*the top shelf*). **2** highest in degree or importance (*at top speed; the top job*). ● v.tr. (**topped, topping**) **1** provide with a top, cap, etc. (*cake topped with icing*). **2** remove the top of (a plant, fruit, etc.), esp. to improve growth, prepare for cooking, etc. **3** be higher or better than; surpass; be at the top of (*topped the list*). **4** sl. **a** execute esp. by hanging, kill. **b** (refl.) commit suicide. **5** reach the top of (a hill etc.). **6** Golf **a** hit (a ball) above the centre. **b** make (a stroke) in this way. □ **at the top** (or **at the top of the tree**) in the highest rank of a profession etc. **come to the top** colloq. win distinction. **from top to toe** from head to foot; completely. **off the top of one's head** see HEAD. **on top 1** in a superior position; above. **2** on the upper part of the head (*bald on top*). **on top of 1** fully in command of. **2** in close proximity to. **3** in addition to. **4** above, over. **on top of the world** colloq. exuberant. **over the top 1** esp. *hist.* over the parapet of a trench (and into battle). **2** into a final or decisive state. **3** to excess, beyond reasonable limits (*that joke was over the top*). **top banana 1** Theatr. sl. a comedian who tops the bill of a show. **2** US sl. a leader; the head of an organization etc. **top-boot** esp. *hist.* a boot with a high top esp. of a different material or colour. **top brass** esp. Mil. colloq. the highest-ranking officers, heads of industries, etc. **top copy** the uppermost typed copy (cf. *carbon copy*). **top dog** colloq. a victor or master. **top-down 1** proceeding from the top downwards; hierarchical. **2** (of knowledge etc.) proceeding from general theory to example. **3** Computing working from the top or root of a tree towards the branches (with or without backtracking). **top drawer 1** the uppermost drawer in a chest etc. **2** colloq. high social position or origin. **top-drawer** adj. colloq. of high social standing; of the highest level or quality. **top-dress** apply manure or fertilizer on the top of (earth) instead of ploughing it in. **top-dressing 1** this process. **2** manure so applied. **3** a superficial show. **top-flight** in the highest rank of achievement. **top fruit** Brit. fruit grown on trees, not bushes. **top gear** see GEAR. **top-hamper** an encumbrance on top, esp. the upper sails and rigging of a sailing-ship. **top hat** a man's tall silk hat. **top-hole** Brit. colloq. first-rate. **top-level** of the highest level of importance, prestige, etc. **top-line 1** of the highest quality; top of the range. **2** (of an entertainment etc.) considered worthy of top billing. **top-notch** colloq. first-rate. **top-notcher** colloq. a first-rate person or thing. **top off** (or **up**) put an end or the finishing touch to (a thing). **top out** put the highest stone on (a building). **top one's part** esp. Theatr. act or discharge one's part to perfection. **top secret** of the highest secrecy. **top-stitch** make a row of neat, esp. decorative, stitches on the right side of (a garment etc.). **top-stitching** stitching (esp. for decoration) on the right side of a garment etc. **top ten** (or **twenty** etc.) the first ten (or twenty etc.) records in the charts. **top up** esp. Brit. **1 a** complete (an amount or number). **b** fill up (a glass or other partly full container). **2** top up something for (a person) (*may I top you up with sherry?*). **top-up** n. an addition; something that serves to top up (esp. a partly full glass). □ **topmost** adj. [OE *topp*]

top[2] /tɒp/ n. a wooden or metal toy, usu. conical, spherical, or pear-

shaped, spinning on a point when set in motion by hand, string, etc. □ **top-shell 1** a marine gastropod mollusc of the family Trochidae, with a short conical shell. **2** its shell. [OE, of uncert. orig.]

topaz /ˈtəʊpæz/ n. **1** a transparent or translucent aluminium silicate mineral, usu. yellow, used as a gem. **2** a South American hummingbird of the genus *Topaza*, with a metallic yellow or green throat. [ME f. OF *topace, topaze* f. L *topazus* f. Gk *topazos*]

topazolite /təˈpæzəˌlaɪt/ n. a yellow or green kind of garnet. [TOPAZ + -LITE]

topcoat /ˈtɒpkəʊt/ n. **1** an overcoat. **2** an outer coat of paint etc.

tope[1] /təʊp/ v.intr. archaic or literary drink alcohol to excess, esp. habitually. □ **toper** n. [perh. f. obs. *top* quaff]

tope[2] /təʊp/ n. Ind. a grove, esp. of mangoes. [Telugu *tōpu*, Tamil *tōppu*]

tope[3] /təʊp/ n. = STUPA. [Punjab *tōp* f. Prakrit & Pali *thūpo* f. Skr. STUPA]

tope[4] /təʊp/ n. a small shark, *Galeorhinus galeus*. [perh. f. Cornish]

topee var. of TOPI[1].

Topeka /təˈpiːkə/ the state capital of Kansas; pop. (1990) 119,880.

topgallant /tɒpˈgælənt, təˈgæl-/ n. Naut. the mast, sail, yard, or rigging immediately above the topmast and topsail.

top-heavy /tɒpˈhevɪ/ adj. **1** disproportionately heavy at the top so as to be in danger of toppling. **2 a** (of an organization, business, etc.) having a disproportionately large number of people in senior administrative positions. **b** overcapitalized. **3** colloq. (of a woman) having a disproportionately large bust. □ **top-heavily** adv. **top-heaviness** n.

Tophet /ˈtəʊfɪt/ n. Bibl. hell. [name of a place in the Valley of Hinnom near Jerusalem used for idolatrous worship and later for burning refuse: f. Heb. *tōpet*]

tophus /ˈtəʊfəs/ n. (pl. **tophi** /-faɪ/) **1** Med. a gouty deposit of crystalline uric acid and other substances at the surface of joints. **2** Geol. = TUFA. [L, name of loose porous stones]

topi[1] /ˈtəʊpɪ/ n. (also **topee**) (pl. **topis** or **topees**) Anglo-Ind. a hat, esp. a sola topi. [Hindi *topī*]

topi[2] /ˈtəʊpɪ/ n. (pl. **topis** or same) a large African antelope, *Damaliscus lunatus*, with a sloping back and usu. reddish-brown with black patches; esp. a race of this occurring in coastal East Africa (cf. TSESSEBI). [Mande]

topiary /ˈtəʊpɪərɪ/ adj. & n. ● adj. concerned with or formed by clipping shrubs, trees, etc. into ornamental shapes. ● n. (pl. **-ies**) **1** topiary art. **2** an example of this. □ **topiarist** n. **topiarian** /ˌtəʊpɪˈeərɪən/ adj. [F *topiaire* f. L *topiarius* landscape gardener f. *topia opera* fancy gardening f. Gk *topia* pl. dimin. of *topos* place]

topic /ˈtɒpɪk/ n. **1** a theme for a book, discourse, essay, sermon, etc. **2** the subject of a conversation or argument. [L *topica* f. Gk (ta) *topika* topics, as title of a treatise by Aristotle f. *topos* a place, a commonplace]

topical /ˈtɒpɪk(ə)l/ adj. **1** dealing with the news, current affairs, etc. (*a topical song*). **2** dealing with a place; local. **3** Med. (of an ailment, medicine, etc.) affecting or applied externally to a particular part of the body. **4** of or concerning topics. □ **topically** adv. **topicality** /ˌtɒpɪˈkælɪtɪ/ n.

Topkapi Palace /tɒpˈkɑːpɪ/ the former seraglio or residence in Istanbul of the sultans of the Ottoman Empire, last occupied by Mahmut II (1808–39) and now a museum.

topknot /ˈtɒpnɒt/ n. a knot, tuft, crest, or bow of ribbon, worn or growing on the head.

topless /ˈtɒplɪs/ adj. **1** without or seeming to be without a top. **2 a** (of clothes) having no upper part. **b** (of a person) wearing such clothes; bare-breasted. **c** (of a place, esp. a beach) where women go topless. □ **toplessness** n.

toplofty /tɒpˈlɒftɪ/ adj. US colloq. haughty.

topman /ˈtɒpmən/ n. (pl. **-men**) Naut. a man doing duty in a top.

topmast /ˈtɒpmɑːst, -məst/ n. Naut. the mast next above the lower mast.

topographical /ˌtɒpəˈgræfɪk(ə)l/ adj. of or relating to topography; (in art) dealing with or depicting places (esp. towns), buildings, natural prospects, etc., in a realistic and detailed manner. □ **topographic** adj. **topographically** adv. [f. as TOPOGRAPHY]

topography /təˈpɒgrəfɪ/ n. **1 a** a detailed description, representation on a map, etc., of the natural and artificial features of a town, district, etc. **b** such features. **2** Anat. the mapping of the surface of the body with reference to the parts beneath. □ **topographer** n. [ME f. LL *topographia* f. Gk f. *topos* place]

topoi *pl.* of TOPOS.

topology /təˈpɒlədʒɪ/ *n.* Math. the study of geometrical properties and spatial relations unaffected by the continuous change of shape or size of figures. □ **topologist** *n.* **topological** /ˌtɒpəˈlɒdʒɪk(ə)l/ *adj.* **topologically** *adv.* [G *Topologie* f. Gk *topos* place]

toponym /ˈtɒpənɪm/ *n.* **1** a place-name. **2** a descriptive place-name, usu. derived from a topographical feature of the place. [TOPONYMY]

toponymy /təˈpɒnɪmɪ/ *n.* the study of the place-names of a region. □ **toponymic** /ˌtɒpəˈnɪmɪk/ *adj.* [Gk *topos* place + *onoma* name]

topos /ˈtɒpɒs/ *n.* (*pl.* **topoi** /-pɔɪ/) a stock theme in literature etc. [Gk, = commonplace]

topper /ˈtɒpə(r)/ *n.* **1** a thing that tops. **2** *colloq.* = *top hat* (see TOP¹). **3** *colloq.* an exceptionally good person or thing.

topping /ˈtɒpɪŋ/ *adj. & n.* ● *adj.* **1** pre-eminent in position, rank, etc. **2** *Brit. archaic sl.* excellent. ● *n.* anything that tops something else, esp. cream etc. on a dessert.

topple /ˈtɒp(ə)l/ *v.intr. & tr.* (often foll. by *over*) **1 a** fall or cause to fall as if top-heavy. **b** fall or cause to fall from power. **2** overthrow or be overthrown (*the government was toppled by a coup*). [TOP¹ + -LE⁴]

topsail /ˈtɒpseɪl, -s(ə)l/ *n.* Naut. **1** a square sail, or either of two, next above the lowest. **2** a fore-and-aft sail above the gaff.

topside /ˈtɒpsaɪd/ *n.* **1** *Brit.* the outer side of a round of beef. **2** the side of a ship above the water-line.

topsoil /ˈtɒpsɔɪl/ *n.* the top layer of soil (opp. SUBSOIL).

topspin /ˈtɒpspɪn/ *n.* a fast forward spinning motion imparted to a ball in tennis etc. by hitting it forward and upward.

topsy-turvy /ˌtɒpsɪˈtɜːvɪ/ *adv., adj., & n.* ● *adv. & adj.* **1** upside down. **2** in utter confusion. ● *n.* utter confusion. □ **topsy-turvily** *adv.* **topsy-turviness** *n.* [app. f. TOP¹ + obs. *terve* overturn]

toque /təʊk/ *n.* **1** a woman's small brimless hat. **2** *hist.* a small cap or bonnet for a man or woman. [F, app. = It. *tocca*, Sp. *toca*, of unkn. orig.]

toquilla /təˈkiːjə/ *n.* **1** a palmlike tree, *Carludovica palmata*, native to South America. **2** a fibre produced from the leaves of this. [Sp., = small gauze head-dress, dimin. of *toca* toque]

tor /tɔː(r)/ *n.* a hill or rocky peak, esp. in Devon or Cornwall. [OE *torr*: cf. Gael. *tòrr* bulging hill]

Torah /ˈtɔːrə/ *n.* **1** (usu. prec. by *the*) **a** the Pentateuch. **b** a scroll containing this. **2** the will of God as revealed in Mosaic law. [Heb. *tōrāh* instruction]

Torbay /tɔːˈbeɪ/ a borough in Devon, SW England; pop. (est. 1991) 121,000. A popular seaside resort, it was formed in 1968 from the amalgamation of Torquay, Paignton, and Brixham.

torc var. of TORQUE 1.

torch /tɔːtʃ/ *n. & v.* ● *n.* **1** (also **electric torch**) *Brit.* a portable battery-powered electric lamp. **2 a** a piece of wood, cloth, etc., soaked in tallow and lighted for illumination. **b** any similar lamp, e.g. an oil lamp on a pole. **3** a source of heat, illumination, or enlightenment (*bore aloft the torch of freedom*). **4** esp. N. Amer. a blowlamp. **5** *US sl.* an arsonist. ● *v.tr.* esp. N. Amer. sl. set alight with or as with a torch. □ **carry a torch for** suffer from unrequited love for. **put to the torch** destroy by burning. **torch-bearer 1** a person who leads the way in an attempt to reform, inspire, etc. **2** a person who carries a usu. ceremonial torch. **torch-fishing** catching fish by torchlight at night. **torch-race** *Gk Antiq.* a festival performance of runners handing lighted torches to others in relays. **torch singer** a person who sings torch songs. **torch song** a popular song of unrequited love. **torch-thistle** a treelike American cactus of the genus *Cereus*, with funnel-shaped flowers which open at night. [ME f. OF *torche* f. L *torqua* f. *torquere* twist]

torchère /tɔːˈʃeə(r)/ *n.* a tall stand with a small table for a candlestick etc. [F (as TORCH)]

torchlight /ˈtɔːtʃlaɪt/ *n.* the light of a torch or torches. □ **torchlit** *adj.*

torchon /ˈtɔːʃ(ə)n/ *n.* (in full **torchon lace**) coarse bobbin lace with geometrical designs. [F, = duster, dishcloth f. *torcher* wipe]

tore¹ past of TEAR¹.

tore² /tɔː(r)/ *n.* = TORUS 1, 4. [F f. L *torus*: see TORUS]

toreador /ˈtɒrɪəˌdɔː(r)/ *n.* a bullfighter, esp. on horseback. □ **toreador pants** close-fitting calf-length women's trousers. [Sp. f. *torear* fight bulls f. *toro* bull f. L *taurus*]

torero /tɒˈreərəʊ/ *n.* (*pl.* **-os**) a bullfighter. [Sp. f. *toro*: see TOREADOR]

toreutic /təˈruːtɪk/ *adj. & n.* ● *adj.* of or concerning the chasing, carving, and embossing of esp. metal. ● *n.* (in *pl.*) the art or practice of this. [Gk *toreutikos* f. *toreuō* work in relief]

torgoch /ˈtɔːɡɒx/ *n.* a kind of red-bellied char found in some Welsh lakes. [Welsh f. *tor* belly + *coch* red]

tori *pl.* of TORUS.

toric /ˈtɒrɪk/ *adj.* Geom. having the form of a torus or part of a torus.

torii /ˈtɔːrɪˌiː/ *n.* (*pl.* same) the gateway of a Shinto shrine, with two uprights and two crosspieces. [Jap.]

Torino see TURIN.

torment *n. & v.* ● *n.* /ˈtɔːment/ **1** severe physical or mental suffering (*was in torment*). **2** a cause of this. **3** *archaic* **a** torture. **b** an instrument of torture. ● *v.tr.* /tɔːˈment/ **1** subject to torment (*tormented with worry*). **2** tease or worry excessively (*enjoyed tormenting the teacher*). □ **tormentedly** /tɔːˈmentɪdlɪ/ *adv.* **tormentingly** /-ˈmentɪŋlɪ/ *adv.* **tormentor** /-ˈmentə(r)/ *n.* [ME f. OF *torment*, *tormenter* f. L *tormentum* missile-engine f. *torquere* to twist]

tormentil /ˈtɔːmənˌtɪl/ *n.* a low-growing rosaceous plant, *Potentilla erecta*, with bright yellow flowers and a highly astringent rootstock used in medicine. [ME f. OF *tormentille* f. med.L *tormentilla*, of unkn. orig.]

torn past part. of TEAR¹.

tornado /tɔːˈneɪdəʊ/ *n.* (*pl.* **-oes**) **1** a violent storm of small extent with whirling winds, esp.: **a** in West Africa at the beginning and end of the rainy season. **b** in the US etc. over a narrow path often accompanied by a funnel-shaped cloud. **2** an outburst or volley of cheers, hisses, missiles, etc. □ **tornadic** /-ˈnædɪk/ *adj.* [app. assim. of Sp. *tronada* thunderstorm (f. *tronar* to thunder) to Sp. *tornar* to turn]

Tornio /ˈtɔːnɪəʊ/ (Swedish **Torne Älv** /ˌtɔːnə ˈɛlv/) a river which rises in NE Sweden and flows 566 km (356 miles) generally southwards, forming the border between Sweden and Finland before emptying into the Gulf of Bothnia.

toroid /ˈtɔːrɔɪd/ *n.* a figure of toroidal shape.

toroidal /tɔːˈrɔɪd(ə)l/ *adj.* Geom. of or resembling a torus. □ **toroidally** *adv.*

Toronto /təˈrɒntəʊ/ the capital of Ontario and the largest city in Canada, situated on the north shore of Lake Ontario; pop. (1991) 612,300; metropolitan area pop. (1991) 3,550,000. Founded in 1793, it was originally named York but in 1834 was renamed Toronto, from a Huron word meaning 'meeting place'.

torose /ˈtɔːrəʊz, -rəʊs/ *adj.* **1** Bot. (of plants, esp. their stalks) cylindrical with bulges at intervals. **2** Zool. knobby. [L *torosus* f. *torus*: see TORUS]

torpedo /tɔːˈpiːdəʊ/ *n. & v.* ● *n.* (*pl.* **-oes**) **1 a** a self-propelled underwater missile, usu. cylindrical with a pointed or tapered nose, fired at a ship and exploding on impact. (*See note below.*) **b** (in full **aerial torpedo**) a similar device dropped from an aircraft. **2** Zool. an electric ray. **3** *US* an explosive device or firework. ● *v.tr.* (**-oes**, **-oed**) **1** destroy or attack with a torpedo. **2** make (a policy, institution, plan, etc.) ineffective or inoperative; destroy. □ **torpedo-boat** a small fast lightly armed warship for carrying or discharging torpedoes. **torpedo-net** (or **-netting**) netting of steel wire hung round a ship to intercept torpedoes. **torpedo-tube** a tube from which torpedoes are fired. □ **torpedo-like** *adj.* [L, = numbness, electric ray f. *torpere* be numb]

▪ The first self-propelled torpedo was designed in 1866 by the British engineer Robert Whitehead (1823–1905), but the first successful use of the weapon in war was by the Japanese against Russian ships in 1904. Torpedoes can be launched by ships or aircraft but have been used most successfully by submarines, and accounted for heavy shipping losses in the Second World War. Since then developments have included sophisticated acoustic devices that enable the torpedo to home in on its target.

torpefy /ˈtɔːpɪˌfaɪ/ *v.tr.* (**-ies**, **-ied**) make numb or torpid. [L *torpefacere* f. *torpere* be numb]

torpid /ˈtɔːpɪd/ *adj.* **1** sluggish, inactive, dull, apathetic. **2** numb. **3** (of a hibernating animal) dormant. □ **torpidly** *adv.* **torpidness** *n.* **torpidity** /tɔːˈpɪdɪtɪ/ *n.* [L *torpidus* (as TORPOR)]

torpor /ˈtɔːpə(r)/ *n.* torpidity. □ **torporific** /ˌtɔːpəˈrɪfɪk/ *adj.* [L f. *torpere* be sluggish]

Torquay /tɔːˈkiː/ a resort town in SW England, in Devon, administratively part of Torbay since 1968; pop. (1981) 57,500.

torque /tɔːk/ *n.* **1** (also **torc**) *hist.* a necklace of twisted metal, esp. of the ancient Gauls and Britons. **2** Mech. the moment of a system of forces tending to cause rotation. □ **torque converter** a device to transmit the correct torque from the engine to the axle in a motor vehicle. [(sense 1 F f. L *torques*) f. L *torquere* to twist]

Torquemada /ˌtɔːkɪˈmɑːdə/, Tomás de (c.1420–98), Spanish cleric and

Grand Inquisitor. A Dominican monk, he became confessor to Ferdinand and Isabella, whom he persuaded to institute the Inquisition in 1478. Torquemada was appointed Inquisitor-General of Spain in 1483, and earned a reputation for ruthlessness and ferocious suppression of heresy. He was also the prime mover behind the expulsion of the Jews from Spain in and after 1492.

torr /tɔː(r)/ *n.* (*pl.* same) a unit of pressure used in measuring partial vacuums, equal to 133.32 pascals. [TORRICELLI]

torrefy /'tɒrɪˌfaɪ/ *v.tr.* (**-ies, -ied**) **1** roast or dry (metallic ore, a drug, etc.). **2** parch or scorch with heat. □ **torrefaction** /ˌtɒrɪ'fækʃ(ə)n/ *n.* [F *torréfier* f. L *torrefacere* f. *torrere* scorch]

torrent /'tɒrənt/ *n.* **1** a rushing stream of water, lava, etc. **2** (usu. in *pl.*) a great downpour of rain (*came down in torrents*). **3** (usu. foll. by *of*) a violent or copious flow (*uttered a torrent of abuse*). □ **torrential** /tə'renʃ(ə)l/ *adj.* **torrentially** *adv.* [F f. It. *torrente* f. L *torrens -entis* scorching, boiling, roaring f. *torrere* scorch]

Torres Strait /'tɒrɪs/ a channel separating the northern tip of Queensland, Australia, from the island of New Guinea and linking the Arafura Sea and the Coral Sea. It is named after the Spanish explorer Luis Vaez de Torres, who in 1606 was the first European to sail along the south coast of New Guinea.

Torricelli /ˌtɒrɪ'tʃelɪ/, Evangelista (1608–47), Italian mathematician and physicist. He was a disciple of Galileo, whom he succeeded as mathematician to the court of Tuscany. He proposed a physical law governing the velocity of liquids flowing under the force of gravity from orifices. Torricelli's most important invention, however, was the mercury barometer, with which he demonstrated that the atmosphere exerts a pressure sufficient to support a column of mercury in an inverted closed tube. He was also the first person to produce a sustained vacuum.

Torricellian vacuum /ˌtɒrɪ'tʃelɪən, ˌtɒrɪ'sel-/ *n.* a vacuum formed when mercury in a long tube closed at one end is inverted with the open end in a reservoir of mercury (the principle on which a barometer is made). [TORRICELLI]

torrid /'tɒrɪd/ *adj.* **1 a** (of the weather) very hot and dry. **b** (of land etc.) parched by such weather. **2** (of language or actions) emotionally charged; passionate, intense. □ **torrid zone** the central belt of the earth, between the Tropics of Cancer and Capricorn. □ **torridly** *adv.* **torridness** *n.* **torridity** /tə'rɪdɪtɪ/ *n.* [F *torride* or L *torridus* f. *torrere* parch]

torse /tɔːs/ *n.* Heraldry a wreath. [obs. F *torse, torce* wreath ult. f. L *torta* fem. past part. (as TORT)]

torsel var. of TASSEL[2].

Tórshavn /'tɔːʃaʊn/ (also **Thorshavn**) the capital of the Faeroe Islands, a port on the island of Strømø; pop. (1988) 14,550.

torsion /'tɔːʃ(ə)n/ *n.* **1** twisting, esp. of one end of a body while the other is held fixed. **2** *Math.* the extent to which a curve departs from being planar. **3** *Biol.* the state of being twisted into a spiral. **4** *Med.* the twisting of the cut end of an artery after surgery etc. to impede bleeding. □ **torsion balance** an instrument for measuring very weak forces by their effect upon a system of fine twisted wire. **torsion bar** a bar forming part of a vehicle suspension, twisting in response to the motion of the wheels and absorbing their vertical movement. **torsion pendulum** a pendulum working by rotation rather than by swinging. □ **torsional** *adj.* **torsionally** *adv.* **torsionless** *adj.* [ME f. OF f. LL *torsio -onis* f. L *tortio* (as TORT)]

torsk /tɔːsk/ *n.* a fish of the cod family, *Brosmius brosme*, abundant in northern waters and often dried for food. [Norw. *to(r)sk* f. ON *tho(r)skr* prob. rel. to *thurr* dry]

torso /'tɔːsəʊ/ *n.* (*pl.* **-os**) **1** the trunk of the human body. **2** a statue of a human consisting of the trunk alone, without head or limbs. **3** an unfinished or mutilated work (esp. of art, literature, etc.). [It., = stalk, stump, torso, f. L *thyrsus*]

tort /tɔːt/ *n.* Law a breach of duty (other than under contract) leading to liability for damages. [ME f. OF f. med.L *tortum* wrong, neut. past part. of L *torquere* tort- twist]

torte /'tɔːtə/ *n.* (*pl.* **torten** /-t(ə)n/ or **tortes**) an elaborate sweet cake or tart. [G]

Tortelier /ˌtɔː'telɪˌeɪ/, Paul (1914–90), French cellist. He made his concert début as a cellist in 1931, achieving international recognition after the Second World War. He was noted for his interpretations of Bach and Elgar, and also gave recitals with his wife and children, all musicians. Tortelier was appointed professor at the Paris Conservatoire in 1957, where Jacqueline du Pré was among his pupils.

tortelli /tɔː'telɪ/ *n.pl.* small pasta parcels stuffed with a cheese or vegetable mixture etc. [It., pl. of *tortello* small cake, fritter]

tortellini /ˌtɔːtə'liːnɪ/ *n.pl.* tortelli which have been rolled and formed into small rings. [It., pl. of *tortellino* dimin. of *tortello* (see TORTELLI)]

tortfeasor /'tɔːtˌfiːzə(r)/ *n.* Law a person guilty of tort. [OF *tort-fesor, tort-faiseur*, etc. f. tort wrong, *-fesor, faiseur* doer]

torticollis /ˌtɔːtɪ'kɒlɪs/ *n.* Med. a rheumatic etc. disease of the muscles of the neck, causing twisting and stiffness. [mod.L f. L *tortus* crooked + *collum* neck]

tortilla /tɔː'tiːjə/ *n.* a thin flat orig. Mexican maize cake eaten hot or cold with or without a filling. [Sp. dimin. of *torta* cake f. LL]

tortious /'tɔːʃəs/ *adj.* Law constituting a tort; wrongful. □ **tortiously** *adv.* [AF *torcious* f. *torcion* extortion f. LL *tortio* torture: see TORSION]

tortoise /'tɔːtəs/ *n.* **1** a slow-moving plant-eating land reptile of the family Testudinidae (order Chelonia), encased in a scaly domed shell and having a retractile head and pillar-like legs. **2** Rom. Antiq. = TESTUDO. □ **tortoise-like** *adj.* & *adv.* [ME *tortuce*, OF *tortue*, f. med.L *tortuca*, of uncert. orig.]

tortoiseshell /'tɔːtəsˌʃel/ *n.* & *adj.* ● *n.* **1** the yellowish-brown mottled or clouded outer shell of some turtles, used for decorative hair-combs, jewellery, etc. **2 a** = tortoiseshell cat. **b** = tortoiseshell butterfly. ● *adj.* **1** having the colouring or appearance of tortoiseshell. **2** made of tortoiseshell or a synthetic substitute. □ **tortoiseshell butterfly** a butterfly of the genus *Aglais* or *Nymphalis*, with orange-brown wings mottled with black. **tortoiseshell cat** a domestic cat, usu. female, with a mottled black, orange, and cream or white coat.

Tortola /tɔː'təʊlə/ the principal island of the British Virgin Islands in the West Indies. Its chief town, Road Town, is the capital of the British Virgin Islands. Its name is Spanish and means 'turtle dove'.

tortrix /'tɔːtrɪks/ *n.* a small moth of the family Tortricidae, esp. of the genus *Tortrix*, the larvae of which live inside rolled leaves and are often pests. [mod.L, fem. of L *tortor* twister: see TORT]

tortuous /'tɔːtjʊəs/ *adj.* **1** full of twists and turns (*followed a tortuous route*). **2** devious, circuitous, crooked (*has a tortuous mind*). □ **tortuously** *adv.* **tortuousness** *n.* **tortuosity** /ˌtɔːtjʊ'ɒsɪtɪ/ *n.* (*pl.* **-ies**). [ME f. OF f. L *tortuosus* f. *tortus* a twist (as TORT)]

torture /'tɔːtʃə(r)/ *n.* & *v.* ● *n.* **1** the infliction of severe bodily pain esp. as a punishment or a means of persuasion. **2** severe physical or mental suffering (*the torture of defeat*). ● *v.tr.* **1** subject to torture (*torture him on the rack; tortured by guilt*). **2** force out of a natural position or state; deform; pervert. □ **torturable** *adj.* **torturer** *n.* **torturous** *adj.* **torturously** *adv.* [F f. LL *tortura* twisting (as TORT)]

torula /'tɒrʊlə/ *n.* (*pl.* **torulae** /-ˌliː/) **1** a yeast, *Candida utilis*, used medicinally and as a food additive. **2** a yeastlike fungus of the genus *Torula*, growing on dead vegetation. [mod.L, dimin. of L TORUS]

Toruń /'tɒrʊnj/ (German **Thorn** /tɔːrn/) an industrial city in northern Poland, on the River Vistula; pop. (1990) 200,820. It was founded by the Teutonic Knights in 1230, becoming part of Poland in 1454.

torus /'tɔːrəs/ *n.* (*pl.* **tori** /-raɪ/) **1** *Archit.* a large convex moulding, usu. semicircular in section, esp. as the lowest part of the base of a column. **2** *Bot.* the receptacle of a flower. **3** *Anat.* a smooth ridge of bone or muscle. **4** *Geom.* a surface or solid formed by rotating a closed curve, esp. a circle, about a line in its plane but not intersecting it. [L, = swelling, bulge, cushion]

Torvill and Dean /'tɔːvɪl, diːn/, Jayne Torvill (b.1957) and Christopher (Colin) Dean (b.1958), English ice-skaters. In partnership they won the European ice-dancing championship (1981–2), the world championships (1981–3), and the gold medal in the 1984 Winter Olympics (with a famous routine to Ravel's *Boléro*). After a spell as professionals they returned to amateur competition in 1994, winning the European championship again and then a bronze medal in the 1994 Winter Olympics.

Tory /'tɔːrɪ/ *n.* & *adj.* ● *n.* (*pl.* **-ies**) **1** a member of the Conservative Party in the UK. **2** *hist.* a member of the party that opposed the exclusion of James II and later supported the established order and gave rise to the Conservative Party (opp. WHIG). (*See note below.*) **3** US *hist.* a colonist loyal to the British during the War of American Independence. ● *adj.* of or relating to Tories or to the Conservative Party. □ **Toryism** *n.* [prob. f. Ir. f. *tóir* pursue.]

▪ The term was originally used in the 17th century of dispossessed Irish who became outlaws and lived by plundering English settlers and soldiers. It became used as an abusive nickname for those who opposed the exclusion of the Catholic Duke of York (later James II) from the succession to the Crown — reputedly because the Duke was seen to favour the Irish. As a party, however, the Tories were associated with the Church of England and with non-toleration of religious Nonconformists and Catholics. They were successful under Queen Anne but after the succession of the Hanoverian George I suffered as a result of their links with the Jacobite cause: some Tories supported the deposed Stuarts, despite their Catholicism, as the possessors of the hereditary right to rule. After 1760 the Tories accepted George III and the established order in Church and state; their fortunes rose as opponents of the French Revolution, and they held office almost continuously between 1760 and 1830. In the following decades, particularly under the influence of Sir Robert Peel, the nature of the party changed, and it became known by the name Conservative. (See also TAMWORTH MANIFESTO.)

tosa /ˈtəʊsə/ n. a large breed of dog like the mastiff, originally kept for dog-fighting. [*Tosa*, former province in Japan]

Toscana see TUSCANY.

Toscanini /ˌtɒskəˈniːnɪ/, Arturo (1867–1957), Italian conductor. Making his conducting début in 1886, he was musical director at La Scala in Milan (1898–1903; 1906–8) before becoming a conductor at the Metropolitan Opera, New York (1908–21). Toscanini later returned to La Scala (1921–9). Among the works he premièred were Puccini's *La Bohème* (1896) and *Turandot* (1926).

tosh /tɒʃ/ n. colloq. rubbish, nonsense. [19th c.: orig. unkn.]

Tosk /tɒsk/ n. & adj. ● n. (pl. same or **Tosks**) **1** a member of one of the main ethnic groups of Albania, living mainly in the south of the country. **2** the dialect of Albanian spoken by this people, forming the basis for standard Albanian. ● adj. of or relating to the Tosk or their dialect. [Albanian]

toss /tɒs/ v. & n. ● v. **1** tr. throw up (a ball etc.), esp. with the hand. **2** tr. & intr. roll about, throw, or be thrown, restlessly or from side to side (*the ship tossed on the ocean; was tossing and turning all night*). **3** tr. (usu. foll. by to, away, aside, out, etc.) throw (a thing) lightly or carelessly (*tossed the letter away*). **4** tr. **a** throw (a coin) into the air to decide a choice etc. by the side on which it lands. **b** (also absol.; often foll. by for) settle a question or dispute with (a person) in this way (*tossed him for the armchair; tossed for it*). **5** tr. **a** (of a bull etc.) throw (a person etc.) up with the horns. **b** (of a horse etc.) throw (a rider) off its back. **6** tr. coat (food) with dressing etc. by shaking. **7** tr. bandy about in debate; discuss (*tossed the question back and forth*). ● n. **1** the act or an instance of tossing (a coin, the head, etc.). **2** Brit. a fall, esp. from a horse. □ **toss one's head** throw it back esp. in anger, impatience, etc. **tossing the caber** see CABER. **toss oars** raise oars to an upright position in salute. **toss off 1** drink off at a draught. **2** dispatch (work) rapidly or without effort (*tossed off an omelette*). **3** Brit. coarse sl. masturbate. **toss a pancake** throw it up so that it flips on to the other side in the frying-pan. **toss up** toss a coin to decide a choice etc. **toss-up** n. **1** a doubtful matter; a close thing (*it's a toss-up whether he wins*). **2** the tossing of a coin. [16th c.: orig. unkn.]

tosser /ˈtɒsə(r)/ n. **1** Brit. coarse sl. an unpleasant or contemptible person. **2** a person or thing that tosses.

tot¹ /tɒt/ n. **1** a small child (*a tiny tot*). **2** a dram of liquor. [18th c., of dial. orig.]

tot² /tɒt/ v. & n. ● v. (**totted, totting**) **1** tr. (usu. foll. by up) add (figures etc.). **2** intr. (foll. by up) (of items) mount up. ● n. Brit. archaic a set of figures to be added. □ **totting-up 1** the adding of separate items. **2** Brit. the adding of convictions for driving offences to cause disqualification. **tot up to** amount to. [abbr. of TOTAL or of L *totum* the whole]

tot³ /tɒt/ v. & n. Brit. ● v.intr. (**totted, totting**) collect saleable items from refuse as an occupation. ● n. sl. an article collected from refuse. □ **totter** n. [19th c.: orig. unkn.]

total /ˈtəʊt(ə)l/ adj., n., & v. ● adj. **1** complete, comprising the whole (*the total number of people*). **2** absolute, unqualified (*in total ignorance; total abstinence*). ● n. a total number or amount. ● v. (**totalled, totalling**; US **totaled, totaling**) **1** tr. **a** amount in number to (*they totalled 131*). **b** find the total of (things, a set of figures, etc.). **2** intr. (foll. by to, up to) amount to, mount up to. **3** tr. N. Amer. sl. wreck completely. □ **total abstainer** a teetotaller. **total abstinence** abstaining completely from alcohol. **total eclipse** an eclipse in which the whole disc (of the sun, moon, etc.) is obscured. **total internal reflection** reflection without refraction of a light-ray meeting the interface between two media at more than a certain critical angle to the normal. **total recall** the ability to remember every detail of one's experience clearly. **total war** a war in which all available weapons and resources are employed. □ **totally** adv. [ME f. OF f. med.L *totalis* f. *totus* entire]

totalitarian /ˌtəʊtælɪˈteərɪən/ adj. & n. ● adj. of or relating to totalitarianism. ● n. a person advocating a totalitarian system.

totalitarianism /ˌtəʊtælɪˈteərɪəˌnɪz(ə)m/ n. an authoritarian system of government which tolerates only one political party, to which all other institutions are subordinated, and which usually demands the complete subservience of the individual to the state.

totality /təʊˈtælɪtɪ/ n. **1** the complete amount or sum. **2** Astron. the time during which an eclipse is total.

totalizator /ˈtəʊtəlaɪˌzeɪtə(r)/ n. (also **totalisator**) **1** a device showing the number and amount of bets staked on a race, to facilitate the division of the total among those backing the winner. **2** a system of betting based on this.

totalize /ˈtəʊtəˌlaɪz/ v.tr. (also **-ise**) collect into a total; find the total of. □ **totalization** /ˌtəʊtəlaɪˈzeɪʃ(ə)n/ n.

totalizer /ˈtəʊtəˌlaɪzə(r)/ n. = TOTALIZATOR.

tote¹ /təʊt/ n. colloq. **1** a totalizator. **2** a lottery. [abbr.]

tote² /təʊt/ v.tr. esp. N. Amer. colloq. carry, convey, or wield (esp. a heavy load) (*toting a gun*). □ **tote bag** a large bag for shopping etc. **tote box** N. Amer. a small container for goods. □ **toter** n. (also in comb.). [17th-c. US, prob. of dial. orig.]

totem /ˈtəʊtəm/ n. **1** a natural object, esp. an animal, adopted by North American Indians as an emblem of a clan or an individual. **2** an image of this. □ **totem-pole 1** a pole on which totems are carved or hung. **2** a hierarchy. □ **totemism** n. **totemist** n. **totemic** /təʊˈtemɪk/ adj. **totemistic** /ˌtəʊtəˈmɪstɪk/ adj. [Algonquian]

tother /ˈtʌðə(r)/ adj. & pron. (also **t'other**) dial. or joc. the other. □ **tell tother from which** joc. tell one from the other. [ME *the tother*, for earlier *thet other* = the other (thet obs. neut. of THE)]

totter /ˈtɒtə(r)/ v. & n. ● v.intr. **1** stand or walk unsteadily or feebly (*tottered out of the pub*). **2 a** (of a building etc.) shake or rock as if about to collapse. **b** (of a system of government etc.) be about to fall. ● n. an unsteady or shaky movement or gait. □ **totterer** n. **tottery** adj. [ME f. MDu. *touteren* to swing]

toucan /ˈtuːkən/ n. a tropical American fruit-eating bird of the family Ramphastidae, with an immense beak and often brightly coloured plumage. [Tupi *tucana*, Guarani *tucã*]

touch /tʌtʃ/ v. & n. ● v. **1** tr. come into or be in physical contact with (another thing) at one or more points. **2** tr. (often foll. by with) bring the hand etc. into contact with (*touched her arm*). **3 a** intr. (of two things etc.) be in or come into contact with one another (*the balls were touching*). **b** tr. bring (two things) into mutual contact (*they touched hands*). **4** tr. rouse tender or painful feelings in (*was touched by his appeal*). **5** tr. strike lightly (*just touched the wall with the back bumper*). **6** tr. (usu. with neg.) **a** disturb or interfere with (*don't touch my things*). **b** have any dealings with (*won't touch bricklaying*). **c** consume; use up; make use of (*dare not touch alcohol; has not touched her breakfast; need not touch your savings*). **d** cope with; affect; manage (*soap won't touch this dirt*). **7** tr. **a** deal with (a subject) lightly or in passing (*touched the matter of their expenses*). **b** concern (*it touches you closely*). **8** tr. **a** reach or rise as far as, esp. momentarily (*the thermometer touched 90°*). **b** (usu. with neg.) approach in excellence etc. (*can't touch him for style*). **9** tr. affect slightly; modify (*pity touched with fear*). **10** tr. (as **touched** adj.) colloq. slightly mad. **11** tr. (often foll. by in) esp. Art mark lightly, put in (features etc.) with a brush, pencil, etc. **12** tr. **a** strike (the keys, strings, etc., of a musical instrument). **b** strike the keys or strings of (a piano etc.). **13** tr. (usu. foll. by for) sl. ask for and get money etc. from (a person) as a loan or gift (*touched him for £5*). **14** tr. injure slightly (*blossom touched by frost*). **15** tr. Geom. be tangent to (a curve). ● n. **1** the act or an instance of touching, esp. with the body or hand (*felt a touch on my arm*). **2 a** the faculty of perception through physical contact, esp. with the fingers (*has no sense of touch in her right arm*). **b** the qualities of an object etc. as perceived in this way (*the soft touch of silk*). **3** a small amount; a slight trace (*a touch of salt; a touch of irony*). **4 a** a musician's manner of playing keys or strings. **b** the manner in which the keys or strings respond to touch. **c** an artist's or writer's style of workmanship, writing, etc. (*has a delicate touch*). **5 a** a distinguishing quality or trait (*a rather amateur touch*). **b** a special skill or proficiency (*have lost my touch*). **6** (esp. in pl.) **a** a light stroke with a pen, pencil, etc. **b** a slight alteration or improvement (*speech needs a few touches*). **7** = TAG² n. 1. **8** (prec. by a) slightly (*is a touch too arrogant*). **9** sl. **a** the act of asking for and getting money etc. from a person. **b** a person from whom money etc. is so obtained. **10** Football the part of the field outside the side limits. **11** archaic a test with or as if with a touchstone (*put it to the touch*). □ **at a touch** if touched, however lightly (*opened at a touch*). **easy touch** = soft touch. **finishing touch** (or **touches**) the final details completing and enhancing a piece of work etc. **get** (or **put**) **in** (or **into**) **touch with** come or cause to come into communication with; contact. **in touch** (often foll. by *with*) **1** in communication (*we're still in touch after all these years*). **2** up to date, esp. regarding news etc. (*keeps in touch with events*). **3** aware, conscious, empathetic (*not in touch with her own feelings*). **keep in touch** (often foll. by *with*) **1** remain informed (*kept in touch with the latest developments*). **2** continue correspondence, a friendship, etc. **lose touch** (often foll. by *with*) **1** cease to be informed. **2** cease to correspond with or be in contact with another person. **lose one's touch** not show one's customary skill. **the Nelson touch** a masterly or sympathetic approach to a problem (NELSON[2]). **out of touch** (often foll. by *with*) **1** not in correspondence. **2** not up to date or modern. **3** lacking in awareness or sympathy (*out of touch with his own beliefs*). **personal touch** a characteristic or individual approach to a situation. **soft touch** *colloq.* a person easily manipulated, esp. easily induced to part with money. **to the touch** when touched (*was cold to the touch*). **touch-and-go** uncertain regarding a result; risky (*it was touch-and-go whether we'd catch the train*). **touch at** (of a ship) call at (a port etc.). **touch bottom 1** reach the bottom of water with one's feet. **2** be at the lowest or worst point. **3** be in possession of the full facts. **touch down 1 a** *Rugby* touch the ground with the ball behind one's own or esp. the opponents' goal-line. **b** *Amer. Football* score by being in possession of the ball behind the opponents' goal-line. **2** (of an aircraft or spacecraft) make contact with the ground in landing. **touch football** *US* football with touching in place of tackling. **touch-hole** a small hole in a gun for igniting the charge. **touch-in-goal** *Rugby* each of the four corners enclosed by continuations of the touch-lines and goal-lines. **touch-judge** *Rugby* a linesman. **touch-line** (in various sports) either of the lines marking the side boundaries of the pitch. **touch-mark** the maker's mark on pewter. **touch-me-not** a balsam plant of the genus *Impatiens*, with ripe seed-capsules jerking open when touched. **touch-needle** a needle of gold or silver alloy of known composition used as a standard in testing other alloys on a touchstone. **touch off 1** represent exactly (in a portrait etc.). **2** explode by touching with a match etc. **3** initiate (a process) suddenly (*touched off a run on the pound*). **touch of nature 1** a natural trait. **2** *colloq.* an exhibition of human feeling with which others sympathize (from a misinterpretation of Shakespeare's *Troilus and Cressida* III. iii. 169). **touch of the sun 1** a slight attack of sunstroke. **2** a little sunlight. **touch on** (or **upon**) **1** treat (a subject) briefly, refer to or mention casually. **2** verge on (*that touches on impudence*). **touch-paper** paper impregnated with nitre, for firing gunpowder, fireworks, etc. **touch screen** a VDU screen that displays data, esp. information to customers, when it is touched. **touch the spot** *colloq.* find out or do exactly what was needed. **touch-type** type without looking at the keys. **touch-typing** this skill. **touch-typist** a person who touch-types. **touch up 1** give finishing touches to or retouch (a picture, writing, etc.). **2** *Brit. sl.* **a** caress so as to excite sexually. **b** sexually molest. **3** strike (a horse) lightly with a whip. **touch-up** *n.* a quick restoration or improvement (of paintwork, a piece of writing, etc.). **touch wood** touch something wooden with the hand to avert ill luck. **would not touch with a bargepole** see BARGEPOLE. □ **touchable** *adj.* [ME f. OF *tochier*, *tuchier* (v.), *touche* (n.): prob. imit., imitating a knock]

touchdown /ˈtʌtʃdaʊn/ *n.* **1** the act or an instance of an aircraft or spacecraft making contact with the ground during landing. **2** *Rugby* & *Amer. Football* the act or an instance of touching down.

touché /ˈtuːʃeɪ, tuːˈʃeɪ/ *int.* **1** *Fencing* the acknowledgement of a hit by an opponent. **2** the acknowledgement of a justified accusation, a witticism, or a point made in reply to one's own. [F, past part. of *toucher* TOUCH]

toucher /ˈtʌtʃə(r)/ *n.* **1** a person or thing that touches. **2** *Bowls* a wood that touches the jack.

touching /ˈtʌtʃɪŋ/ *adj. & prep.* ● *adj.* moving; pathetic (*a touching incident; touching confidence*). ● *prep. literary* concerning; about. □ **touchingly** *adv.* **touchingness** *n.* [ME f. TOUCH: (prep.) f. OF *touchant* pres. part. (as TOUCH)]

touchstone /ˈtʌtʃstəʊn/ *n.* **1** a fine-grained dark schist or jasper used for testing alloys of gold etc. by marking it with the alloy and observing the colour of the mark. **2** a standard or criterion.

touchwood /ˈtʌtʃwʊd/ *n.* readily inflammable wood, esp. when made soft by fungi, used as tinder.

touchy /ˈtʌtʃɪ/ *adj.* (**touchier**, **touchiest**) **1** apt to take offence; over-sensitive. **2** not to be touched without danger; requiring careful handling, risky, awkward. □ **touchily** *adv.* **touchiness** *n.* [perh. alt. of TETCHY after TOUCH]

tough /tʌf/ *adj. & n.* ● *adj.* **1** hard to break, cut, tear, or chew; durable;

strong. **2** (of a person) able to endure hardship; hardy. **3** unyielding, stubborn, difficult (*it was a tough job; a tough customer*). **4** *colloq.* **a** acting sternly; hard (*get tough with*). **b** (of circumstances, luck, etc.) severe, unpleasant, hard, unjust. **5** *colloq.* criminal or violent (*tough guys*). ● *n.* a tough person, esp. a ruffian or criminal. □ **tough guy** *colloq.* **1** a hard unyielding person. **2** a violent aggressive person. **tough it** (or **tough it out**) *colloq.* endure or withstand difficult conditions. **tough-minded** realistic, not sentimental. **tough-mindedness** being tough-minded. □ **toughen** *v.tr.* & *intr.* **toughener** *n.* **toughish** *adj.* **toughly** *adv.* **toughness** *n.* [OE *tōh*]

toughie /ˈtʌfɪ/ *n. colloq.* a tough person or problem.

Toulon /tuːˈlɒn/ a port and naval base on the Mediterranean coast of southern France; pop. (1990) 170,170.

Toulouse /tuːˈluːz/ a city in SW France on the Garonne river, principal city of the Midi-Pyrénées region; pop. (1990) 365,930. Capital of the Visigoths between AD 419 and 507, it was taken by Clovis in 508 and later became the chief town of Aquitaine and the Languedoc.

Toulouse-Lautrec /tuːˌluːz ləʊˈtrek/, Henri (Marie Raymond) de (1864–1901), French painter and lithographer. Toulouse-Lautrec's reputation is based on his colour lithographs from the 1890s, depicting actors, music-hall singers, prostitutes, and waitresses in Montmartre: particularly well known is the *Moulin Rouge* series (1894). His work is noted for its calligraphic line and flatness, influenced by Japanese prints.

toupee /ˈtuːpeɪ/ *n.* (also **toupet**) a wig or artificial hairpiece to cover a bald spot. [F *toupet* hair-tuft dimin. of OF *toup* tuft (as TOP[1])]

tour /tʊə(r)/ *n. & v.* ● *n.* **1 a** a journey from place to place as a holiday. **b** an excursion, ramble, or walk (*made a tour of the garden*). **2 a** a spell of duty on military or diplomatic service. **b** the time to be spent at a particular post. **3** a series of performances, matches, etc., at different places on a route through a country etc. ● *v.* **1** *intr.* (usu. foll. by *through*) make a tour (*toured through India*). **2** *tr.* make a tour of (a country etc.). □ **on tour** (esp. of a team, theatre company, etc.) touring. **touring-car** a car with room for passengers and much luggage. **tour operator** a travel agent specializing in package holidays. [ME f. OF *to(u)r* f. L *tornus* f. Gk *tornos* lathe]

touraco var. of TURACO.

tour de force /ˌtʊə də ˈfɔːs/ *n.* (*pl.* **tours de force** pronunc. same) a feat of strength or skill. [F]

Tour de France /ˌtʊə də ˈfrɑːns/ a gruelling French race for professional cyclists held annually since 1903 over approximately 4,800 km (3,000 miles) of roads, many of them in mountainous country. The race lasts for about three weeks, each day constituting a stage; the overall leader after each stage wears the famous yellow leader's jersey.

tourer /ˈtʊərə(r)/ *n.* a vehicle, esp. a car, for touring.

Tourette's syndrome /tʊəˈrets/ *n. Med.* a neurological disorder characterized by involuntary tics and vocalizations and the compulsive utterance of obscenities. [Gilles de la *Tourette*, Fr. neurologist (1857–1904)]

tourism /ˈtʊərɪz(ə)m/ *n.* the organization and operation of (esp. foreign) holidays, esp. as a commercial enterprise.

tourist /ˈtʊərɪst/ *n.* **1** a person making a visit or tour as a holiday; a traveller, esp. abroad (often *attrib.*: *tourist accommodation*). **2** a member of a touring sports team. □ **tourist class** the lowest class of passenger accommodation in a ship, aircraft, etc. □ **touristic** /tʊəˈrɪstɪk/ *adj.* **touristically** *adv.*

Tourist Trophy (abbr. **TT**) a motorcycle-racing competition held annually on roads in the Isle of Man since 1907.

touristy /ˈtʊərɪstɪ/ *adj.* usu. *derog.* appealing to or visited by many tourists.

tourmaline /ˈtʊəməlɪn, -ˌliːn/ *n. Mineral.* a boron aluminium silicate mineral of various colours, possessing unusual electrical properties, and used in electrical and optical instruments and as a gemstone. [F f. Sinh. *toramalli* porcelain]

Tournai /tʊəˈneɪ/ (Flemish **Doornik** /ˈdɔːrnɪk/) a town in Belgium, on the River Scheldt near the French frontier; pop. (1991) 67,730. It became the Merovingian capital in the 5th century and was the birthplace of the Frankish king Clovis. From the 9th century it was controlled by the counts of Flanders until taken by France in 1188, returning to the Netherlands in 1814.

tournament /ˈtʊənəmənt/ *n.* **1** a contest of skill between a number of competitors, esp. played in heats (*chess tournament; tennis tournament*). **2** a display of military exercises etc. (*Royal Tournament*). **3** *hist.* **a** a pageant in which jousting with blunted weapons took place. **b** a meeting for jousting between single knights for a prize etc. [ME f. OF *torneiement* f. *torneier* TOURNEY]

tournedos /ˈtʊənəˌdəʊ/ *n.* (*pl.* same /-ˌdəʊz/) a small round thick cut from a fillet of beef. [F]

tourney /ˈtʊənɪ/ *n. & v.* ● *n.* (*pl.* **-eys**) a tournament. ● *v.intr.* (**-eys**, **-eyed**) take part in a tournament. [ME f. OF *tornei* (n.), *torneier* (v.), ult. f. L *tornus* a turn]

tourniquet /ˈtʊənɪˌkeɪ/ *n.* a device for stopping the flow of blood through an artery by twisting a bar etc. in a ligature or bandage. [F prob. f. OF *tournicle* coat of mail, TUNICLE, infl. by *tourner* TURN]

Tours /tʊə(r)/ an industrial city in west central France, on the Loire; pop. (1990) 133,400. It became a bishopric in the mid-3rd century and was the seat of St Martin, the patron saint of France, and St Gregory of Tours, who founded the abbey.

tousle /ˈtaʊz(ə)l/ *v.tr.* **1** make (esp. the hair) untidy; rumple. **2** handle roughly or rudely. [frequent. of (now dial.) *touse*, ME f. OE rel. to OHG *-zuson*]

tous-les-mois /ˌtuːleɪˈmwʌ/ *n.* **1** food starch obtained from tubers of a canna, *Canna indica*. **2** this plant. [F, lit. 'every month', prob. corrupt. of West Indian *toloman*]

Toussaint L'Ouverture /ˌtuːsæ̃ˌluːvəˈtjʊə(r)/, Pierre Dominique (c.1743–1803), Haitian revolutionary leader. Brought up a slave in the western part of Hispaniola (now Haiti), in 1791 he became one of the leaders of a rebellion that succeeded in emancipating the island's slaves by 1793. In 1797 he was appointed Governor-General by the revolutionary government of France, and led the drive to expel the British and Spanish from western Hispaniola. In 1801 he took control of the whole island, establishing his own constitution, but the following year Napoleon (wishing to restore slavery) ordered his forces to regain the island; Toussaint was eventually taken to France, where he died in prison.

tout /taʊt/ *v. & n.* ● *v.* **1** *intr.* (usu. foll. by *for*) solicit custom persistently; pester customers (*touting for business*). **2** *tr.* solicit the custom of (a person) or for (a thing). **3** *intr.* **a** *Brit.* spy out the movements and condition of racehorses in training. **b** *US* offer racing tips for a share of the resulting profit. ● *n.* a person employed in touting. □ **touter** *n.* [ME *tūte* look out = ME (now dial.) *toot* (OE *tōtian*) f. Gmc]

tout court /tuː ˈkʊə(r)/ *adv.* without addition; simply (*called James tout court*). [F]

tout de suite /tuːt ˈswiːt/ *adv.* at once, immediately. [F]

tovarish /təˈvɑːrɪʃ/ *n.* (also **tovarich**) (in the former USSR) comrade (esp. as a form of address). [Russ. *tovarishch*]

tow[1] /təʊ/ *v. & n.* ● *v.tr.* **1** (of a motor vehicle, horse, or person controlling it) pull (a boat, another motor vehicle, a caravan, etc.) along by a rope, tow-bar, etc. **2** pull (a person or thing) along behind one. ● *n.* the act or an instance of towing; the state of being towed. □ **have in** (or **on**) **tow 1** be towing. **2** be accompanied by and often in charge of (a person). **tow-bar** a bar for towing esp. a trailer or caravan. **tow-** (or **towing-)line** (or **rope**) a line etc. used in towing. **tow-** (or **towing-)net** a net used for dragging through water to collect specimens. **tow-** (or **towing-)path** a path beside a river or canal orig. used for towing boats by horse. □ **towable** *adj.* **towage** *n.* [OE *togian* f. Gmc, rel. to TUG]

tow[2] /təʊ/ *n.* **1** the coarse and broken part of flax or hemp prepared for spinning. **2** a loose bunch of rayon etc. strands. □ **tow-coloured** (of hair) very light. **tow-head** tow-coloured or unkempt hair. **tow-headed** having very light or unkempt hair. □ **towy** *adj.* [ME f. MLG *touw* f. OS *tou*, rel. to ON *tó* wool: cf. TOOL]

toward *prep. & adj.* ● *prep.* /təˈwɔːd, twɔːd, tɔːd/ = TOWARDS. ● *adj.* /ˈtəʊəd/ *archaic* **1** about to take place; in process. **2** docile, apt. **3** promising, auspicious. □ **towardness** /ˈtəʊədnɪs/ *n.* (in sense of *adj.*).

towards /təˈwɔːdz, twɔːdz, tɔːdz/ *prep.* **1** in the direction of (*set out towards town*). **2** as regards; in relation to (*his attitude towards death*). **3** as a contribution to; for (*put this towards your expenses*). **4** near (*towards the end of our journey*). [OE *tōweard* (adj.) future (as TO, -WARD)]

towel /ˈtaʊəl/ *n. & v.* ● *n.* **1 a** a piece of rough-surfaced absorbent cloth used for drying oneself or a thing after washing. **b** absorbent paper etc. used for this. **c** a cloth used for drying plates, dishes, etc.; a tea towel. **2** *Brit.* = sanitary towel. ● *v.* (**towelled, towelling**; *US* **toweled, toweling**) **1** *tr.* (often *refl.*) wipe or dry with a towel. **2** *intr.* wipe or dry oneself with a towel. **3** *tr. sl.* thrash. □ **towel-horse** (or **-rail**) a frame for hanging towels on. [ME f. OF *toail(l)e* f. Gmc]

towelling /ˈtaʊəlɪŋ/ *n.* **1** rough-surfaced absorbent cloth, esp. cotton

or linen with uncut loops, used as material for towels. **2 a** the act of towelling something. **b** *sl.* a beating, a thrashing.

tower /ˈtaʊə(r)/ *n. & v.* ● *n.* **1 a** a tall esp. square or circular structure, often part of a church, castle, etc. **b** a fortress etc. comprising or including a tower. **c** a tall structure housing machinery, apparatus, operators, etc. (*cooling tower; control tower*). **2 a** place of defence; a protection. ● *v.intr.* **1** (usu. foll. by *above, high*) reach or be high or above; be superior. **2** (of a bird, esp. a falcon) soar or hover. **3** (as **towering** *adj.*) **a** high, lofty (*towering intellect*). **b** violent (*towering rage*). □ **the Tower** the Tower of London. **tower block** a tall building containing offices or flats. **tower of silence** a tall open-topped structure on which Parsees place their dead. **tower of strength** a person who gives strong and reliable support. □ **towered** *adj.* **towery** *adj.* [OE *torr*, & ME *tūr*, AF & OF *tur* etc., f. L *turris* f. Gk]

Tower Bridge a bridge across the Thames in London, famous for its twin towers and for the two bascules of which the roadway consists, able to be lifted to allow the passage of large ships. It was completed in 1894.

Tower of Babel (in the Bible) a tower built in an attempt to reach heaven, which God frustrated by confusing the languages of its builders so that they could not understand one another (Gen. 11:1–9). The story was probably inspired by the Babylonian ziggurat, and may be an attempt to explain the existence of different languages. The term is sometimes used in reference to a visionary or unrealistic plan.

Tower of London (also **the Tower**) a fortress by the Thames just east of the City of London. The oldest part, the White Tower, was begun in 1078. It was later used as a state prison, and is now open to the public as a repository of ancient armour, weapons, etc., and of the crown jewels (which have been kept there since the time of Henry III).

towhee /ˈtəʊiː, ˈtaʊiː/ *n.* a North American bunting of the genus *Pipilo*, found in brush and woodland; esp. the widespread *rufous-sided towhee* (*P. erythrophthalmus*), with brightly coloured plumage. [imit.]

town /taʊn/ *n.* **1 a** a large urban area with a name, defined boundaries, and local government, being larger than a village and usu. not created a city. **b** any densely populated area, esp. as opposed to the country or suburbs. **c** the people of a town (*the whole town knows of it*). **2 a** *Brit.* London or the chief city or town in one's neighbourhood (*went up to town*). **b** the central business or shopping area in a neighbourhood (*just going into town*). **3** the permanent residents of a university town as distinct from the members of the university (cf. GOWN *n.* 4). **4** *N. Amer.* = TOWNSHIP 2. □ **go to town** *colloq.* act or work with energy or enthusiasm. **on the town** *colloq.* enjoying the entertainments, esp. the night-life, of a town; celebrating. **town clerk 1** *US & hist.* the officer of the corporation of a town in charge of records etc. **2** *Brit. hist.* the secretary and legal adviser of a town corporation until 1974. **town council** the elective governing body in a municipality. **town councillor** an elected member of this. **town crier** see CRIER. **town gas** manufactured gas for domestic and commercial use. **town hall** a building for the administration of local government, having public meeting rooms etc. **town house 1** a town residence, esp. of a person with a house in the country. **2** a terrace house, esp. of a stylish modern type. **3** a house in a planned group in a town. **4** *Brit.* a town hall. **town-major** *hist.* the chief executive officer in a garrison town or fortress. **town mayor** *Brit.* the chairman of a town council. **town meeting** *US* a meeting of the voters of a town for the transaction of public business. **town planner** a person whose occupation is town planning. **town planning** the planning of the construction and growth of towns. □ **townish** *adj.* **townless** *adj.* **townlet** /ˈtaʊnlɪt/ *n.* **townward** *adj. & adv.* **townwards** *adv.* [OE *tūn* enclosure f. Gmc]

townee /taʊˈniː/ *n.* (also **townie** /ˈtaʊnɪ/) *derog.* a person living in a town, esp. as opposed to a country-dweller or (in a university town) a student etc.

Townes /taʊnz/, Charles Hard (b.1915), American physicist. Following work on radar in the Second World War he investigated the quantum electronic effects associated with microwave radiation. His development of microwave oscillators and amplifiers led to his invention of the maser in 1954. Townes later showed that an optical maser (a laser) was possible, though the first working laser was constructed by others. He shared the Nobel Prize for physics in 1964.

townscape /ˈtaʊnskeɪp/ *n.* **1** the visual appearance of a town or towns. **2** a picture of a town.

townsfolk /ˈtaʊnzfəʊk/ *n.* (treated as *pl.*) the inhabitants of a particular town or towns.

township /ˈtaʊnʃɪp/ *n.* **1** *S. Afr.* **a** esp. *hist.* an urban area set aside for black occupation. **b** a white urban area (esp. if new or about to be developed). **2** *N. Amer.* **a** a division of a county with some corporate powers. **b** a district six miles square. **3** *Brit. hist.* **a** a community inhabiting a manor, parish, etc. **b** a manor or parish as a territorial division. **c** a small town or village forming part of a large parish. **4** *Austral. & NZ* a small town; a town-site. [OE *tūnscipe* (as TOWN, -SHIP)]

townsman /ˈtaʊnzmən/ *n.* (*pl.* **-men**; *fem.* **townswoman**, *pl.* **-women**) an inhabitant of a town; a fellow citizen.

townspeople /ˈtaʊnzˌpiːp(ə)l/ *n.pl.* the people of a town.

Townsville /ˈtaʊnzvɪl/ an industrial port and resort on the coast of Queensland, NE Australia; pop. (1991) 101,400.

Townswomen's Guilds /ˈtaʊnzˌwɪmɪnz/ (in the UK) a network of women's organizations, first set up in 1929, that functions as an urban counterpart of the Women's Institute.

toxaemia /tɒkˈsiːmɪə/ *n.* (*US* **toxemia**) **1** blood-poisoning. **2** a condition in pregnancy characterized by increased blood pressure. □ **toxaemic** *adj.* [as TOXI- + -AEMIA]

toxi- /ˈtɒksɪ/ *comb. form* (also **toxico-** /ˈtɒksɪˌkəʊ/, **toxo-** /ˈtɒksəʊ/) poison; poisonous, toxic.

toxic /ˈtɒksɪk/ *adj.* **1** of or relating to poison (*toxic symptoms*). **2** poisonous (*toxic gas*). **3** caused by poison (*toxic anaemia*). □ **toxic shock syndrome** (abbr. **TSS**) *Med.* acute septicaemia in women, typically caused by bacterial infection from a retained tampon, IUD, etc. □ **toxically** *adv.* **toxicity** /tɒkˈsɪsɪtɪ/ *n.* [med.L *toxicus* poisoned f. L *toxicum* f. Gk *toxikon* (*pharmakon*) (poison for) arrows f. *toxon* bow, *toxa* arrows]

toxicology /ˌtɒksɪˈkɒlədʒɪ/ *n.* the scientific study of poisons. □ **toxicologist** *n.* **toxicological** /-kəˈlɒdʒɪk(ə)l/ *adj.*

toxin /ˈtɒksɪn/ *n.* a poison produced by a living organism, esp. one formed in the body and stimulating the production of antibodies. [TOXIC + -IN]

toxocara /ˌtɒksəʊˈkɑːrə/ *n.* a nematode worm of the genus *Toxocara*, esp. the common roundworms of dogs and cats which are transmissible to humans. □ **toxocariasis** /-kəˈraɪəsɪs/ *n.* [mod.L f. *toxo-* (see TOXI-) + Gk *kara* head]

toxophilite /tɒkˈsɒfɪˌlaɪt/ *n. & adj.* ● *n.* a student or lover of archery. ● *adj.* of or concerning archery. □ **toxophily** *n.* [f. Roger Ascham's *Toxophilus* (1545) f. Gk *toxon* bow + *-philos* -PHILE]

toy /tɔɪ/ *n. & v.* ● *n.* **1 a** a plaything, esp. for a child. **b** (often *attrib.*) a model or miniature replica of a thing, esp. as a plaything (*toy gun*). **2 a** a thing, esp. a gadget or instrument, regarded as providing amusement. **b** a task or undertaking regarded in an unserious way. **3** (usu. *attrib.*) a diminutive breed or variety of dog etc. ● *v.intr.* (usu. foll. by *with*) **1** trifle, amuse oneself, esp. with a person's affections; flirt (*toyed with the idea of going to Africa*). **2 a** move a material object idly (*toyed with her necklace*). **b** nibble at food etc. unenthusiastically (*toyed with a peach*). □ **toy-box** *n.* usu. wooden box for keeping toys in. **toyboy** *colloq.* a woman's much younger male lover. **toy soldier 1** a miniature figure of a soldier. **2** *colloq.* a soldier in a peacetime army. [16th c.: earlier = dallying, fun, jest, whim, trifle: orig. unkn.]

Toynbee /ˈtɔɪnbɪ/, Arnold (Joseph) (1889–1975), English historian. He is best known for his twelve-volume *Study of History* (1934–61), in which he surveyed the history of different civilizations, tracing in them a pattern of growth, maturity, and decay and concluding that contemporary Western civilization is in the last of these stages.

toytown /ˈtɔɪtaʊn/ *n. & adj.* ● *n.* a model of a town used as a plaything etc. ● *adj.* **1** resembling a model of a town; seemingly diminutive. **2** quaint, lightweight.

Tpr. *abbr.* Trooper.

trabeation /ˌtreɪbɪˈeɪʃ(ə)n/ *n. Archit.* the use of beams instead of arches or vaulting in construction. □ **trabeate** /ˈtreɪbɪət/ *adj.* [L *trabs trabis* beam]

trabecula /trəˈbekjʊlə/ *n.* (*pl.* **trabeculae** /-ˌliː/) **1** *Anat.* a supporting band or bar of connective or bony tissue, esp. dividing an organ into chambers. **2** *Bot.* a beamlike projection or process within a hollow structure. □ **trabecular** *adj.* **trabeculate** /-lət/ *adj.* [L, dimin. of *trabs* beam]

Trâblous see TRIPOLI 2.

Trabzon /ˈtræbz(ə)n/ (also **Trebizond** /ˈtrebɪˌzɒnd/) a port on the Black Sea in northern Turkey; pop. (1990) 143,940. Founded by Greek colonists in 756 BC, its ancient name was Trapezus. In 1204, after the sack of Constantinople by the Crusaders, an offshoot of the Byzantine Empire was founded with Trabzon as its capital, which was annexed to the Ottoman Empire in 1461.

tracasserie /trəˈkæsərɪ/ *n.* **1** a state of annoyance. **2** a fuss; a petty quarrel. [F f. *tracasser* bustle]

trace[1] /treɪs/ *v. & n.* ● *v.tr.* **1** observe, discover, or find vestiges or signs of by investigation. **b** (often foll. by *along, through, to,* etc.) follow or mark the track or position of (*traced their footprints in the mud; traced the outlines of a wall*). **c** (often foll. by *back*) follow to its origins (*can trace my family to the 12th century; the report has been traced back to you*). **2** (often foll. by *over*) copy (a drawing etc.) by drawing over its lines on a superimposed piece of translucent paper, or by using carbon paper. **3** (often foll. by *out*) mark out, delineate, sketch, or write esp. laboriously (*traced out a plan of the district; traced out his vision of the future*). **4** pursue one's way along (a path etc.). ● *n.* **1 a** a sign or mark or other indication of something having existed; a vestige (*no trace remains of the castle; has the traces of a vanished beauty*). **b** a very small quantity. **c** an amount of rainfall etc. too small to be measured. **2** a track or footprint left by a person or animal. **3** a track left by the moving pen of an instrument etc. **4** a line on the screen of a cathode-ray tube showing the path of a moving spot. **5** a curve's projection on or intersection with a plane etc. **6** a change in the brain caused by learning processes. □ **trace element 1** a chemical element occurring in minute amounts. **2** a chemical element required only in minute amounts by living organisms for normal growth. **trace fossil** a fossil that represents a burrow, footprint, etc., of an organism. □ **traceless** *adj.* **traceable** *adj.* **traceability** /ˌtreɪsəˈbɪlɪtɪ/ *n.* [ME f. OF *trace* (n.), *tracier* (v.) f. L *tractus* drawing: see TRACT[1]]

trace[2] /treɪs/ *n.* each of the two side-straps, chains, or ropes by which a horse draws a vehicle. □ **kick over the traces** become insubordinate or reckless. **trace-horse** a horse that draws in traces or by a single trace, esp. one hitched on to help uphill etc. [ME f. OF *trais*, pl. of TRAIT]

tracer /ˈtreɪsə(r)/ *n.* **1** a person or thing that traces. **2** *Mil.* **a** a bullet etc. whose course is made visible in flight because of flames etc. emitted. **b** such bullets etc. collectively, used to assist in aiming. **3** an artificially produced radioactive isotope capable of being followed through the body by the radiation it produces.

tracery /ˈtreɪsərɪ/ *n.* (*pl.* **-ies**) **1** ornamental stone openwork esp. in the upper part of a Gothic window. **2** a fine decorative pattern. **3** a natural object finely patterned. □ **traceried** *adj.*

trachea /trəˈkiːə, ˈtreɪkɪə/ *n.* (*pl.* **tracheae** /trəˈkiːiː, ˈtreɪkɪˌiː/) **1** the passage, reinforced by rings of cartilage, through which air reaches the bronchial tubes from the larynx; the windpipe. **2** each of the air passages in the body of an insect etc. **3** any duct or vessel in a plant. □ **tracheal** /trəˈkiːəl, ˈtreɪkɪəl/ *adj.* **tracheate** /trəˈkiːət, ˈtreɪkɪˌeɪt/ *adj.* [ME f. med.L, = LL *trachia* f. Gk *trakheia* (*artēria*) rough (artery), f. *trakhus* rough]

tracheo- /ˈtrækɪəʊ/ *comb. form.*

tracheotomy /ˌtrækɪˈɒtəmɪ/ *n.* (also **tracheostomy** /-ˈɒstəmɪ/) (*pl.* **-ies**) an incision made in the trachea to relieve an obstruction to breathing. □ **tracheotomy tube** a breathing-tube inserted into this incision.

trachoma /trəˈkəʊmə/ *n. Med.* a contagious disease of the eye with inflamed granulation on the inner surface of the lids, caused by chlamydiae. □ **trachomatous** /-ˈkəʊmətəs, -ˈkɒmətəs/ *adj.* [mod.L f. Gk *trakhōma* f. *trakhus* rough]

trachyte /ˈtreɪkaɪt, ˈtræk-/ *n.* a light-coloured volcanic rock rough to the touch. □ **trachytic** /trəˈkɪtɪk/ *adj.* [F f. Gk *trakhutēs* roughness (as TRACHOMA)]

tracing /ˈtreɪsɪŋ/ *n.* **1** a copy of a drawing etc. made by tracing. **2** = TRACE[1] *n.* 3. **3** the act or an instance of tracing. □ **tracing-paper** translucent paper used for making tracings.

track[1] /træk/ *n. & v.* ● *n.* **1 a** a mark or marks left by a person, animal, or thing in passing. **b** (in *pl.*) such marks, esp. footprints. **2** a rough path, esp. one beaten by use. **3** a continuous railway line (*laid three miles of track*). **4 a** a racecourse for horses, dogs, etc. **b** a prepared course for runners etc. **5 a** a groove on a gramophone record. **b** a section of a record, cassette tape, compact disc, etc., containing one song etc. (*this side has six tracks*). **c** a lengthwise strip of magnetic tape containing one sequence of signals. **6 a** a line of travel, passage, or motion (*followed the track of the hurricane; America followed in the same track*). **b** the path travelled by a ship, aircraft, etc. (cf. COURSE *n.* 2c.) **7** a continuous band round the wheels of a tank, tractor, etc. **8** the transverse distance between a vehicle's wheels. **9** = SOUNDTRACK. **10** a line of reasoning or thought (*this track proved fruitless*). ● *v.* **1** *tr.* follow the track of (an animal, person, spacecraft, etc.). **2** *tr.* make out (a course, development, etc.); trace by vestiges. **3** *intr.* (often foll. by *back, in,* etc.) (of a film or television camera) move in relation to the subject being filmed. **4** *intr.* (of wheels) run so that the back ones are exactly in the track of the front ones. **5** *intr.* (of a record-player stylus) follow a groove. **6** *tr. N. Amer.* **a** make a track with (dirt etc.) from the feet. **b** leave such a track on (a floor etc.). □ **in one's tracks** *colloq.* where one stands, there and then (*stopped him in his tracks*). **keep** (or **lose**) **track of** follow (or fail to follow) the course or development of. **make tracks** *colloq.* go or run away. **make tracks for** *colloq.* go in pursuit of or towards. **off the track** away from the subject. **on a person's track 1** in pursuit of him or her. **2** in possession of a clue to a person's conduct, plans, etc. **on the wrong side of** (or **across**) **the tracks** *colloq.* in a poor or less prestigious part of town. **on the wrong** (or **right**) **track** following the wrong (or right) line of inquiry. **track down** reach or capture by tracking. **track events** running-races as opposed to jumping etc. (cf. *field events*). **tracking station** an establishment set up to track objects in the sky. **track-laying vehicle** a vehicle having a caterpillar tread. **track record** a person's past performance or achievements. **track shoe** a spiked shoe worn by a runner. **track system** *US* streaming in education. **track with** *Austral. sl.* associate with, court. □ **trackage** *US n.* [ME f. OF *trac*, perh. f. LG or Du. *tre(c)k* draught etc.]

track² /træk/ *v.* **1** *tr.* tow (a boat) by rope etc. from a bank. **2** *intr.* travel by being towed. [app. f. Du. *trekken* to draw etc., assim. to TRACK¹]

trackbed /'trækbed/ *n.* the foundation layer on which railway tracks are laid.

tracker /'trækə(r)/ *n.* **1** a person or thing that tracks. **2** a police dog tracking by scent. **3** a wooden connecting-rod in the mechanism of an organ. **4** = *black tracker.*

tracking /'trækɪŋ/ *n. Electr.* the formation of a conducting path over the surface of an insulating material.

tracklayer /'træk,leɪə(r)/ *n.* **1** *US* = TRACKMAN. **2** a tractor or other vehicle equipped with continuous tracks (see TRACK¹ n. 7).

tracklement /'træk(ə)lmənt/ *n.* an item of food, esp. a jelly, served with meat. [20th c.: orig. unkn.]

trackless /'træklɪs/ *adj.* **1** without a track or tracks; untrodden. **2** leaving no track or trace. **3** not running on a track. □ **trackless trolley** *US* a trolleybus.

trackman /'trækmən/ *n.* (*pl.* **-men**) a platelayer.

tracksuit /'træksuːt, -sjuːt/ *n.* a loose warm suit worn by an athlete etc. for exercising or jogging. □ **tracksuited** *adj.*

trackway /'trækweɪ/ *n.* a beaten path; an ancient roadway.

tract¹ /trækt/ *n.* **1** a region or area of indefinite, esp. large, extent (*pathless desert tracts*). **2** *Anat.* an area of an organ or system (*respiratory tract*). **3** *Brit. archaic* a period of time etc. [L *tractus* drawing f. *trahere* tract-draw, pull]

tract² /trækt/ *n.* a short treatise in pamphlet form esp. on a religious subject. [app. abbr. of L *tractatus* TRACTATE]

tract³ /trækt/ *n. RC Ch. & Mus.* an anthem replacing the alleluia in some masses. [med.L *tractus* (*cantus*) drawn-out (song), past part. of L *trahere* draw]

tractable /'træktəb(ə)l/ *adj.* **1** (of a person) easily handled; manageable; docile. **2** (of material etc.) pliant, malleable. □ **tractably** *adv.* **tractableness** *n.* **tractability** /,træktə'bɪlɪtɪ/ *n.* [L *tractabilis* f. *tractare* handle, frequent. of *trahere* tract- draw]

Tractarianism /træk'teərɪə,nɪz(ə)m/ *n. hist.* = OXFORD MOVEMENT. □ **Tractarian** *adj. & n.* [after *Tracts for the Times*, pamphlets outlining the movement's principles]

tractate /'trækteɪt/ *n.* a treatise. [L *tractatus* f. *tractare*: see TRACTABLE]

traction /'trækʃ(ə)n/ *n.* **1** the act of drawing or pulling a thing over a surface, esp. a road or track (*steam traction*). **2 a** a sustained pulling on a limb, muscle, etc., by means of pulleys, weights, etc. **b** contraction, e.g. of a muscle. **3** the grip of a tyre on a road, a wheel on a rail, etc. **4** *US* public transport by rail or tram (usu. *attrib.*: *traction system*). □ **traction-engine** a steam or diesel engine for drawing heavy loads on roads, fields, etc. **traction-wheel** the driving-wheel of a locomotive etc. □ **tractional** *adj.* **tractive** /'træktɪv/ *adj.* [F *traction* or med.L *tractio* f. L *trahere* tract- draw]

tractor /'træktə(r)/ *n.* **1** a motor vehicle used for hauling esp. farm machinery, heavy loads, etc. **2** a traction-engine. □ **tractor beam** (in science fiction) a beam capable of holding or pulling a distant object. [LL *tractor* (as TRACTION)]

Tracy /'treɪsɪ/, Spencer (1900–67), American actor. After his screen début in 1930, he won his first Oscar seven years later for his performance in *Captain Courageous*. Tracy formed a successful film partnership with Katharine Hepburn, co-starring with her in films such as *Adam's Rib* (1949) and *Guess Who's Coming to Dinner?* (1967).

trad /træd/ *n. & adj. esp. Brit. colloq.* ● *n.* traditional jazz. ● *adj.* traditional. [abbr.]

trade /treɪd/ *n. & v.* ● *n.* **1 a** buying and selling. **b** buying and selling conducted between nations etc. **c** business conducted for profit (esp. as distinct from a profession) (*a butcher by trade*). **d** business of a specified nature or time (*Christmas trade; tourist trade*). **2** a skilled handicraft esp. requiring an apprenticeship (*learned a trade; his trade is plumbing*). **3** (usu. prec. by *the*) **a** the people engaged in a specific trade (*the trade will never agree to it; trade enquiries only*). **b** *Brit. colloq.* licensed victuallers. **c** *Brit. Mil. sl.* the submarine service. **4** *US* a transaction, esp. a swap. **5** (usu. in *pl.*) a trade wind. ● *v.* **1** *intr.* (often foll. by *in, with*) engage in trade; buy and sell (*trades in plastic novelties; we trade with Japan*). **2** *tr.* **a** exchange in commerce; barter (goods). **b** exchange (insults, blows, etc.). **c** *US* swap, exchange (esp. sports players). **3** *intr.* (usu. foll. by *with, for*) have a transaction with a person for a thing. **4** *intr.* (usu. foll. by *to*) carry goods to a place. **5** *intr.* (of shares, currency, etc.) be bought and sold (*the pound is trading lower this month*). □ **be in trade** esp. *derog.* be in commerce, esp. keep a shop. **foreign trade** international trade. **trade book** a book published by a commercial publisher and intended for general readership. **trade cycle** *Brit.* recurring periods of boom and recession. **trade gap** the extent by which a country's imports exceed its exports. **trade in** (often foll. by *for*) exchange (esp. a used car etc.) in esp. part payment for another. **trade-in** *n.* a thing, esp. a car, exchanged in this way. **trade journal** a periodical containing news etc. concerning a particular trade. **trade-last** *US* a compliment from a third person which is reported to the person complimented in exchange for one to the reporter. **trade name 1** a name by which a thing is called in a trade. **2** a name given to a product. **3** a name under which a business trades. **trade off** exchange, esp. as a compromise. **trade-off** *n.* such an exchange. **trade on** take advantage of (a person's credulity, one's reputation, etc.). **trade paper** = *trade journal*. **trade plates** number-plates used by a car-dealer etc. on unlicensed cars. **trade price** a wholesale price charged to the dealer before goods are retailed. **trade secret 1** a secret device or technique used esp. in a trade. **2** *joc.* any secret. **trade-weighted** (esp. of exchange rates) weighted according to the importance of the trade with the various countries involved. □ **tradable** *adj.* **tradeable** *adj.* [ME f. MLG *trade* track f. OS *trada*, OHG *trata*: cf. TREAD]

Trade Board *n. Brit. hist.* a statutory body for settling disputes etc. in certain industries.

trademark /'treɪdmɑːk/ *n. & v.* ● *n.* **1** a device, word, or combination of words secured by legal registration or established by use as representing a company etc. **2** a distinctive characteristic etc. ● *v.tr.* affix a trademark on, provide with a trademark.

trader /'treɪdə(r)/ *n.* **1** a person engaged in trade. **2** a merchant ship.

tradescantia /,trædɪ'skæntɪə/ *n.* a usu. trailing plant of the genus *Tradescantia*, with large blue, white, or pink flowers. [mod.L f. John Tradescant, Engl. naturalist (1570–1638)]

tradesman /'treɪdzmən/ *n.* (*pl.* **-men**; *fem.* **tradeswoman**, *pl.* **-women**) a person engaged in trading or a trade, esp. a shopkeeper or skilled craftsman.

tradespeople /'treɪdz,piːp(ə)l/ *n.pl.* people engaged in trade and their families.

Trades Union Congress (abbr. **TUC**) the official representative body of British trade unions, founded in 1868, which meets annually to discuss matters of common concern. It is made up of delegates of the affiliated unions.

trade union *n.* (also **trades union**) an organized association of workers in a trade, group of trades, or a profession, formed to protect and further the workers' rights and interests. Although prefigured by the medieval artisans' guilds, trade unions were a product of the Industrial Revolution in Britain. They expanded in size and importance during the 19th century, despite being subject, particularly in the earlier years, to repressive legislation such as the Combination Acts. The Trades Union Congress first met in 1868, and the position of unions in Britain was legally recognized by the Trade Union Act of 1871; the unions began assuming a more assertive and politicized outlook towards the end of the century, playing a central role in the formation of the Labour Party. The trade-union movement achieved true national importance in the early 20th century, with smaller unions tending to amalgamate into larger national organizations and white-collar workers joining in greater numbers. After confrontations between the government and unions in the 1960s and 1970s the

Thatcher government severely restricted the powers of the trade unions in Britain. □ **trade unionism** n. **trade unionist** n.

trade wind n. a wind blowing steadily towards the equator from the north-east (in the northern hemisphere) or south-east (in the southern hemisphere), especially at sea. The name was originally applied to any wind that 'blows trade', i.e. blows in a constant course or way: it had nothing to do with commercial navigation, although the term was often later understood in that sense. Two belts of trade winds encircle the earth, blowing from the tropical high-pressure belts to the low-pressure zone at the equator; the system is seasonally displaced respectively to the north and south of the equator in the northern and southern summers.

trading /ˈtreɪdɪŋ/ n. the act of engaging in trade. □ **trading estate** esp. Brit. a specially designed industrial and commercial area. **trading post** a store etc. established in a remote or unsettled region. **trading-stamp** a stamp given to customers by some stores which is exchangeable in large numbers for various articles.

tradition /trəˈdɪʃ(ə)n/ n. **1 a** a custom, opinion, or belief handed down to posterity esp. orally or by practice. **b** this process of handing down. **2** esp. joc. an established practice or custom (it's a tradition to complain about the weather). **3** artistic, literary, etc., principles based on experience and practice; any one of these (stage tradition; traditions of the Dutch School). **4** Theol. doctrine or a particular doctrine etc. claimed to have divine authority without documentary evidence, esp.: **a** the oral teaching of Christ and the Apostles. **b** the laws held by the Pharisees to have been delivered by God to Moses. **c** the words and deeds of Muhammad not in the Koran. **5** Law the formal delivery of property etc. □ **traditionary** adj. **traditionist** n. **traditionless** adj. [ME f. OF tradicion or L traditio f. tradere hand on, betray (as TRANS-, dare give)]

traditional /trəˈdɪʃən(ə)l/ adj. **1** of, based on, or obtained by tradition. **2** (of jazz) in the style of the early 20th century. □ **traditionally** adv.

traditionalism /trəˈdɪʃənəˌlɪz(ə)m/ n. **1** respect, esp. excessive, for tradition, esp. in religion. **2** a philosophical system referring all religious knowledge to divine revelation and tradition. □ **traditionalist** n. & adj. **traditionalistic** /-ˌdɪʃənəˈlɪstɪk/ adj.

traditor /ˈtrædɪtə(r)/ n. (pl. **traditors** or **traditores** /ˌtrædɪˈtɔːriːz/) hist. an early Christian who surrendered copies of Scripture or Church property to his or her persecutors to save his or her life. [L: see TRAITOR]

traduce /trəˈdjuːs/ v.tr. speak ill of; misrepresent. □ **traducement** n. **traducer** n. [L traducere disgrace (as TRANS-, ducere duct- lead)]

Trafalgar, Battle of /trəˈfælɡə(r)/ a naval battle fought on 21 Oct. 1805 off the cape of Trafalgar on the south coast of Spain, a decisive engagement of the Napoleonic Wars. The British fleet under Nelson (who was killed in the action) won a victory over the combined fleets of France and Spain, which were attempting to clear the way for Napoleon's projected invasion of Britain. Superior British seamanship and gunnery ensured the surrender of more than half the Franco-Spanish fleet after several hours of hard fighting, and Napoleon was never again able to mount a serious threat to British naval supremacy.

Trafalgar Square /trəˈfælɡə(r)/ a square in central London, planned by John Nash and built largely in the 1830s and 1840s. It is next to the National Gallery and is dominated by Nelson's Column, a memorial to Lord Nelson.

traffic /ˈtræfɪk/ n. & v. ● n. **1** (often attrib.) **a** vehicles moving in a public highway, esp. of a specified kind, density, etc. (heavy traffic on the M1; traffic accident). **b** such movement in the air or at sea. **2** (usu. foll. by in) trade, esp. illegal (the traffic in drugs). **3 a** the transportation of goods, the coming and going of people or goods by road, rail, air, sea, etc. **b** the persons or goods so transported. **4** dealings or communication between people etc. (had no traffic with them). **5** the messages, signals, etc., transmitted through a communications system; the flow or volume of such business. ● v. (**trafficked, trafficking**) **1** intr. (usu. foll. by in) deal in something, esp. illegally (trafficked in narcotics; traffics in innuendo). **2** tr. deal in; barter. □ **traffic calming** the deliberate slowing of traffic in esp. residential areas by narrowing or obstructing roads. **traffic circle** N. Amer. a roundabout. **traffic cop** colloq. a traffic policeman. **traffic island** a paved or grassed area in a road to divert traffic and provide a refuge for pedestrians. **traffic jam** a line of traffic at a standstill because of roadworks, an accident, etc. **traffic sign** a sign conveying information, a warning, etc., to vehicle-drivers. **traffic warden** Brit. a uniformed official employed to help control road traffic and esp. parking. □ **trafficker** n. **trafficless** adj. [F traf(f)ique, Sp. tráfico, It. traffico, of unkn. orig.]

trafficator /ˈtræfɪˌkeɪtə(r)/ n. Brit. hist. a signal attached to the side of a motor vehicle, raised and illuminated automatically to indicate a change of direction. [TRAFFIC + INDICATOR]

traffic light n. (also in pl.) a signal controlling road traffic by means of coloured lights, often red, amber, and green. Early road signals were manually operated, whereas modern traffic-actuated systems of control sometimes involve the computer processing of information. A traffic signal was first tried out in Westminster, London, in 1868 (it was a gas-lit modification of the railway signalling system); electric traffic lights were introduced in Cleveland, Ohio, in 1914 and in New York in 1918.

tragacanth /ˈtræɡəˌkænθ/ n. a white or reddish gum from a leguminous plant, Astragalus gummifer, used in pharmacy, dyeing, etc., as a vehicle for drugs, dye, etc. [F tragacante f. L tragacantha f. Gk tragakantha, name of a shrub, f. tragos goat + akantha thorn]

tragedian /trəˈdʒiːdɪən/ n. **1** a writer of tragedies. **2** an actor in tragedy. [ME f. OF tragediane (as TRAGEDY)]

tragedienne /trəˌdʒiːdɪˈen/ n. an actress in tragedy. [F fem. (as TRAGEDIAN)]

tragedy /ˈtrædʒɪdɪ/ n. (pl. **-ies**) **1** a serious accident, crime, or natural catastrophe. **2** a sad event; a calamity (the team's defeat is a tragedy). **3** a play, novel, etc., dealing with tragic events and with an unhappy ending, esp. concerning the downfall of the protagonist; tragic plays etc. as a genre (cf. COMEDY). (See note below.) [ME f. OF tragedie f. L tragoedia f. Gk tragōidia app. goat-song f. tragos goat + ōidē song]

▪ Tragedy in the Western tradition began in ancient Greece, where it developed in the verse plays of Aeschylus, Sophocles, and Euripides. From their works Aristotle made an influential analysis of tragedy in his Poetics (4th century BC): he stated that tragedy is the imitation of an action that is serious and complete, and that achieves a catharsis in its audience through the arousing of pity and terror; the protagonist is often brought to disaster by hamartia (variously translated as 'error' or 'fatal flaw'), often in the form of hubris (excessive pride or presumption) which is punished by the gods or by fate. Roman tragedy, as represented by Seneca, tended to be more sensational and rhetorical; it set the pattern for the bloody Elizabethan and Jacobean writers like Kyd, Marlowe, Webster, and Middleton (see also REVENGE TRAGEDY). The major tragedies of that period, however, are those by Shakespeare (Othello, Hamlet, Macbeth, King Lear), which use a larger cast of characters and wider variety of scenes than their classical predecessors. In 17th-century France Corneille and Racine wrote tragedies strictly governed by rules derived from Aristotle, in particular observance of the three unities (see UNITY 5). As a genre tragedy did not revive seriously until the late 19th century; Ibsen, Strindberg, and Chekhov, together with 20th-century dramatists such as Eugene O'Neill, wrote tragic plays in prose about ordinary rather than noble characters. Some novels, such as Hardy's The Mayor of Casterbridge, can also be described as tragedies.

tragic /ˈtrædʒɪk/ adj. **1** (also **tragical** /-k(ə)l/) sad; calamitous; greatly distressing (a tragic tale). **2** of, or in the style of, tragedy (tragic drama; a tragic actor). □ **tragic irony** a device, orig. in Greek tragedy, by which words carry a tragic, esp. prophetic, meaning to the audience, unknown to the character speaking. □ **tragically** adv. [F tragique f. L tragicus f. Gk tragikos f. tragos goat: see TRAGEDY]

tragicomedy /ˌtrædʒɪˈkɒmɪdɪ/ n. (pl. **-ies**) **1 a** a play etc. having a mixture of comedy and tragedy. **b** plays etc. of this kind as a genre. **2** an event etc. having tragic and comic elements. □ **tragicomic** adj. **tragicomically** adv. [F tragicomédie or It. tragicomedia f. LL tragicomoedia f. L tragico-comoedia (as TRAGIC, COMEDY)]

tragopan /ˈtræɡəˌpæn/ n. an Asian pheasant of the genus Tragopan, with erectile fleshy horns on its head. [L f. Gk f. tragos goat + PAN]

Traherne /trəˈhɜːn/, Thomas (1637–74), English prose writer and poet. His major prose work Centuries, originally published in 1699, faded into obscurity until it was rediscovered on a London bookstall along with some poems in manuscript form in 1896; it was republished as Centuries of Meditation (1908). Its account of his early intuitions constitutes an important contribution to the portrayal of childhood experience in English literature. His poems were republished as Poetical Works (1903).

trahison des clercs /ˌtraːiːˌzɒn deɪ ˈkleə(r)/ n. the betrayal of standards, scholarship, etc., by intellectuals. [F, title of a book by Julien Benda (1927)]

trail /treɪl/ n. & v. ● n. **1 a** a track left by a thing, person, etc., moving over a surface (left a trail of wreckage; a slug's slimy trail). **b** a track or scent followed in hunting, seeking, etc. (he's on the trail). **2** a beaten path or track, esp. through a wild region. **3** a part dragging behind a thing or person; an appendage (*a trail of smoke; a condensation trail*). **4** the rear end of a gun-carriage stock. ● *v.* **1** *tr. & intr.* draw or be drawn along behind, esp. on the ground. **2** *intr.* (often foll. by *behind*) walk wearily; lag; straggle. **3** *tr.* follow the trail of; pursue (*trailed him to his home*). **4** *intr.* be losing in a game or other contest (*trailing by three points*). **5** *intr.* (usu. foll. by *away, off*) peter out; tail off. **6** *intr.* **a** (of a plant etc.) grow or hang over a wall, along the ground etc. **b** (of a garment etc.) hang loosely. **7** *tr.* (often *refl.*) drag (oneself, one's limbs, etc.) along wearily etc. **8** *tr.* (often *refl.*) drag (oneself, one's limbs, etc.) along wearily etc. **8** *tr.* advertise (a film, a radio or television programme, etc.) in advance by showing extracts etc. **9** *tr. Pottery* apply (slip) through a nozzle or spout to decorate ceramic ware. □ **at the trail** *Mil.* with arms trailed. **trail arms** *Mil.* let a rifle etc. hang balanced in one hand and (*Brit.*) parallel to the ground. **trail bike** a light motorcycle for use in rough terrain. **trail-blazer 1** a person who marks a new track through wild country. **2** a pioneer; an innovator. **trail-blazing** *n.* the act or process of blazing a trail. ● *attrib.adj.* that blazes a trail; pioneering. **trail one's coat** deliberately provoke a quarrel, fight, etc. **trailing edge 1** the rear edge of an aircraft's wing etc. **2** *Electronics* the part of a pulse in which the amplitude diminishes (opp. *leading edge* (see LEADING[1])). **trailing wheel** a wheel not given direct motive power. **trail-net** a drag-net. [ME (earlier as verb) f. OF *traillier* to tow, or f. MLG *treilen* haul f. L *tragula* drag-net]

trailer /ˈtreɪlə(r)/ *n.* **1** a vehicle towed by another, esp.: **a** the rear section of an articulated lorry. **b** an open cart. **c** a platform for transporting a boat etc. **d** *N. Amer.* a caravan. **2** a series of brief extracts from a film etc., used to advertise it in advance. **3** a person or thing that trails. **4** a trailing plant.

train /treɪn/ *v. & n.* ● *v.* **1 a** *tr.* (often foll. by *to* + infin.) teach (a person, animal, oneself, etc.) a specified skill esp. by practice (*trained the dog to beg; was trained in midwifery*). **b** *intr.* undergo this process (*trained as a teacher*). **2** *tr. & intr.* bring or come into a state of physical efficiency by exercise, diet, etc.; undergo physical exercise, esp. for a specific purpose (*trained me for the high jump; the team trains every evening*). **3** *tr.* cause (a plant) to grow in a required shape (*trained the peach tree up the wall*). **4** (usu. as **trained** *adj.*) make (the mind, eye, etc.) sharp or discerning as a result of instruction, practice, etc. **5** *tr.* (often foll. by *on*) point or aim (a gun, camera, etc.) at an object etc. **6** *colloq.* **a** *intr.* go by train. **b** *tr.* (foll. by *it* as object) make a journey by train (*trained it to Aberdeen*). **7** *tr.* (usu. foll. by *away*) *archaic* entice, lure. ● *n.* **1** a series of railway carriages or trucks drawn by a locomotive or moved by a powered carriage. **2** something dragged along behind or forming the back part of a dress, robe, etc. (*wore a dress with a long train; the train of the peacock*). **3** a succession or series of people, things, events, etc. (*a long train of camels; interrupted my train of thought; a train of ideas*). **4** a body of followers; a retinue (*a train of admirers*). **5** a succession of military vehicles etc., including artillery, supplies, etc. (*baggage train*). **6** a line of gunpowder etc. to fire an explosive charge. **7** a series of connected wheels or parts in machinery. □ **in train** properly arranged or directed. **in a person's train** following behind a person. **in the train of** as a sequel of. **train-bearer** a person employed to hold up the train of a robe etc. **train down** treat with exercise or diet to lower one's weight. **train-ferry** (*pl.* **-ies**) a ship that conveys a railway train across water. **train-mile** one mile travelled by one train, as a unit of traffic. **train-shed** a roof supported by posts to shelter railway platforms etc. at a station. **train-spotter** a person who collects locomotive numbers as a hobby. **train-spotting** this hobby. □ **trainable** *adj.* **trainless** *adj.* **trainability** /ˌtreɪnəˈbɪlɪtɪ/ *n.* **trainee** /treɪˈniː/ *n.* [ME f. OF *trainer, trahiner,* ult. f. L *trahere* draw]

trainband /ˈtreɪnbænd/ *n. hist.* any of several divisions of London citizen soldiers, esp. in the Stuart period.

trainer /ˈtreɪnə(r)/ *n.* **1** a person who trains. **2** a person who trains horses, athletes, footballers, etc., as a profession. **3** an aircraft or device simulating it used to train pilots. **4** *Brit.* a soft sports or running shoe of leather, canvas, etc.

training /ˈtreɪnɪŋ/ *n.* the act or process of teaching or learning a skill, discipline, etc. (*physical training*). □ **go into training** begin physical training. **in training 1** undergoing physical training. **2** physically fit as a result of this. **out of training 1** no longer training. **2** physically unfit. **training-college** a college or school for training esp. prospective teachers. **training-ship** a ship on which young people are taught seamanship etc. **training shoe** = TRAINER 4.

trainload /ˈtreɪnləʊd/ *n.* a number of people, quantity of goods, etc., transported by train.

trainman /ˈtreɪnmæn/ *n.* (*pl.* **-men**) a railway employee working on trains.

train-oil /ˈtreɪnɔɪl/ *n.* oil obtained from the blubber of a whale (esp. of a right whale). [obs. *train, trane* train-oil f. MLG *trān,* MDu. *traen,* app. = TEAR[2]]

trainsick /ˈtreɪnsɪk/ *adj.* affected with nausea by the motion of a train. □ **trainsickness** *n.*

traipse /treɪps/ *v. & n.* (also **trapes**) *colloq.* ● *v.intr.* **1** tramp or trudge wearily. **2** (often foll. by *about*) go on errands. ● *n.* **1** a tedious journey on foot. **2** *archaic* a slattern. [16th-c. *trapes* (v.), of unkn. orig.]

trait /treɪt, treɪ/ *n.* a distinguishing feature or characteristic esp. of a person. [F f. L *tractus* (as TRACT[1])]

traitor /ˈtreɪtə(r)/ *n.* (*fem.* **traitress** /-trɪs/) (often foll. by *to*) a person who is treacherous or disloyal, esp. to his or her country. □ **traitorous** *adj.* **traitorously** *adv.* [ME f. OF *trait(o)ur* f. L *traditor -oris* f. *tradere*: see TRADITION]

Trajan /ˈtreɪdʒən/ (Latin name Marcus Ulpius Traianus) (AD *c.*53–117), Roman emperor 98–117. Born in Spain, he was adopted by Nerva as his successor. Trajan's reign is noted for the Dacian wars (101–6), which ended in the annexation of Dacia as a province; the campaigns are illustrated on Trajan's Column in Rome. He was also an efficient administrator and many public works were undertaken during his reign.

trajectory /trəˈdʒektərɪ, ˈtrædʒɪk-/ *n.* (*pl.* **-ies**) **1** the path described by a projectile flying or an object moving under the action of given forces. **2** *Geom.* a curve or surface cutting a system of curves or surfaces at a constant angle. [(orig. adj.) f. med.L *trajectorius* f. L *traicere traject-* (as TRANS-, *jacere* throw)]

tra-la /trɑːˈlɑː/ *int.* an expression of joy or gaiety. [imit. of song]

Tralee /trəˈliː/ the county town of Kerry, a port on the SW coast of the Republic of Ireland; pop. (1991) 17,200.

tram[1] /træm/ *n.* **1** *Brit.* an electrically powered passenger vehicle running on rails laid in a public road. Also called (*Brit.*) **tramcar,** (US) **streetcar.** (See note below.) **2** a four-wheeled vehicle used in coal mines. □ **tram-road** *hist.* a road with wooden, stone, or metal wheel-tracks. [MLG & MDu. *trame* balk, beam, barrow-shaft]

■ The tram was invented in New York in 1830 by John Stephenson, an Irish coach-builder. Trams were originally horse-drawn (see RAILWAY), and then steam power was used for either self-propulsion or hauling by cable. The great expansion came with electric traction in the 1890s, current being collected either from overhead wires or (sometimes) from a conductor rail in the road surface. The famous streetcars of San Francisco are cable-hauled. The motor bus and the electric trolleybus, with their less restricted course, virtually replaced trams in both Britain and the US. However, some transport planners advocate a return to the tram on grounds of economy and its pollution-free operation. Trams have remained in use in many European cities.

tram[2] /træm/ *n.* (in full **tram silk**) double silk thread used for the weft of some velvets and silks. [F *trame* f. L *trama* weft]

tramcar /ˈtræmkɑː(r)/ *n. Brit.* = TRAM[1] 1.

Traminer /trəˈmiːnə(r)/ *n.* **1** a white grape used in wine-making, esp. in Germany and Alsace. **2** a white wine with perfumed bouquet made from these grapes. [G f. *Tramin* (It. *Termeno*) a village in northern Italy]

tramlines /ˈtræmlaɪnz/ *n.pl.* **1** rails for a tramcar. **2** a pair of parallel lines, esp. either of two sets of long lines at the sides of a tennis-court or at the side or back of a badminton court. **3** inflexible principles or courses of action etc.

trammel /ˈtræm(ə)l/ *n. & v.* ● *n.* **1** (usu. in *pl.*) an impediment to free movement; a hindrance (*the trammels of domesticity*). **2** a triple drag-net for fish, which are trapped in a pocket formed when they attempt to swim through. **3** an instrument for drawing ellipses etc. with a bar sliding in upright grooves. **4** a beam-compass. **5** *US* a hook in a fireplace for a kettle etc. ● *v.tr.* (**trammelled, trammelling;** US **trammeled, trammeling**) confine or hamper with or as if with trammels. [in sense 'net' ME f. OF *tramail* f. med.L *tramaculum, tremaculum,* perh. formed as TRI- + *macula* (MAIL[2]): later history uncert.]

trammie /ˈtræmɪ/ *n.* esp. *Austral. & NZ colloq.* the conductor or driver of a tram.

tramontana /ˌtrɑːmɒnˈtɑːnə/ *n.* a cold north wind in the Adriatic. [It.: see TRAMONTANE]

tramontane /trəˈmɒnteɪn/ *adj. & n.* ● *adj.* **1** situated or living on the other side of mountains, esp. the Alps as seen from Italy. **2** (esp. from the Italian point of view) foreign; barbarous. ● *n.* **1** a tramontane person. **2** = TRAMONTANA. [ME f. It. *tramontano* f. L *transmontanus* beyond the mountains (as TRANS-, *mons montis* mountain)]

tramp /træmp/ *v. & n.* ● *v.* **1** *intr.* **a** walk heavily and firmly (*tramping* *about upstairs.* **b** go on foot, esp. a distance. **2** *tr.* **a** cross on foot, esp. wearily or reluctantly. **b** cover (a distance) in this way (*tramped forty miles*). **3** *tr.* (often foll. by *down*) tread on; trample; stamp on. **4** *tr. Austral. colloq.* dismiss from employment, sack. **5** *intr.* live as a tramp. ● *n.* **1** an itinerant vagrant or beggar. **2** the sound of a person, or esp. people, walking, marching, etc., or of horses' hooves. **3** a journey on foot, esp. a protracted one. **4 a** an iron plate protecting the sole of a boot used for digging. **b** the part of a spade that it strikes. **5** esp. *US sl. derog.* a promiscuous woman. **6** = *ocean tramp.* □ **tramper** *n.* **trampish** *adj.* [ME *trampe* f. Gmc]

trample /ˈtræmp(ə)l/ *v. & n.* ● *v.tr.* **1** tread underfoot. **2** press down or crush in this way. ● *n.* the sound or act of trampling. □ **trample on** (or **underfoot**) **1** tread heavily on. **2** treat roughly or with contempt; disregard (a person's feelings etc.). □ **trampler** *n.* [ME f. TRAMP + -LE⁴]

trampoline /ˈtræmpəˌliːn/ *n. & v.* ● *n.* a strong fabric sheet connected by springs to a horizontal frame, used by gymnasts etc. for somersaults, as a springboard, etc. ● *v.intr.* use a trampoline. (*See note below.*) □ **trampolinist** *n.* [It. *trampolino* f. *trampoli* stilts]

▪ Trampolining as a sport was introduced into Britain from the US, though it was known in Europe as a circus act centuries before the earliest competitions, which were held in the US in the late 1940s. Marks are given for the difficulty and style of the moves performed by the trampolinist in the air.

tramway /ˈtræmweɪ/ *n.* **1** *hist.* = *tram-road* (see TRAM¹). **2 a** rails for a tramcar. **b** a tramcar system.

trance /trɑːns/ *n. & v.* ● *n.* **1 a** a sleeplike or half-conscious state without response to stimuli. **b** a hypnotic or cataleptic state. **2** such a state as entered into by a medium. **3** a state of extreme exaltation or rapture; ecstasy. ● *v.tr. poet.* = ENTRANCE². □ **trancelike** *adj.* [ME f. OF *transe* f. *transir* depart, fall into trance f. L *transire*: see TRANSIT]

tranche /trɑːnʃ/ *n.* a portion, esp. of income or of a block of shares. [F, = slice (as TRENCH)]

tranny /ˈtræni/ *n.* (*pl.* **-ies**) esp. *Brit. colloq.* a transistor radio. [abbr.]

tranquil /ˈtræŋkwɪl/ *adj.* calm, serene, unruffled. □ **tranquilly** *adv.* **tranquillity** /træŋˈkwɪlɪti/ *n.* [F *tranquille* or L *tranquillus*]

tranquillize /ˈtræŋkwɪˌlaɪz/ *v.tr.* (also **-ise**, *US* **tranquilize**) make tranquil, esp. by a drug etc.

tranquillizer /ˈtræŋkwɪˌlaɪzə(r)/ *n.* (also **-iser**, *US* **tranquilizer**) a drug used to diminish anxiety.

trans- /træns, trɑːns, trænz, trɑːnz/ *prefix* **1** across, beyond (*transcontinental; transgress*). **2** on or to the other side of (*transatlantic*) (opp. CIS-). **3** through (*transonic*). **4** into another state or place (*transform; transcribe*). **5** surpassing, transcending (*transfinite*). **6** *Chem.* **a** (of an isomer) having the same atom or group on opposite sides of a given plane in the molecule (cf. CIS- 4). **b** having a higher atomic number than (*transuranic*). [from or after L *trans* across]

transact /trænˈzækt, trɑːn-/ *v.tr.* perform or carry through (business). □ **transactor** *n.* [L *transigere transact-* (as TRANS-, *agere* do)]

transaction /trænˈzækʃ(ə)n, trɑːn-/ *n.* **1 a** a piece of esp. commercial business done; a deal (*a profitable transaction*). **b** the management of business etc. **2** (in *pl.*) published reports of discussions, papers read, etc., at the meetings of a learned society. □ **transactional** *adj.* **transactionally** *adv.* [ME f. LL *transactio* (as TRANSACT)]

transalpine /trænzˈælpaɪn, trɑːnz-/ *adj.* beyond the Alps, esp. from the Italian point of view. [L *transalpinus* (as TRANS-, *alpinus* ALPINE)]

Transalpine Gaul see GAUL¹.

transatlantic /ˌtrænzətˈlæntɪk, ˌtrɑːnz-/ *adj.* **1** beyond the Atlantic, esp.: **a** *Brit.* American. **b** *US* European. **2** crossing the Atlantic (*a transatlantic flight*).

Transcaucasia /ˌtrænzkɔːˈkeɪʒə, ˌtrɑːnz-, -ˈkeɪzɪə/ a region lying to the south of the Caucasus mountains, between the Black Sea and the Caspian, and comprising the present-day republics of Georgia, Armenia, and Azerbaijan. During the 19th century much of the region was subjugated to the Russian empire. It was created a republic of the Soviet Union in 1922 as the Transcaucasian Soviet Federated Socialist Republic, but was broken up into its constituent republics in 1936. □ **Transcaucasian** *adj.*

transceiver /trænˈsiːvə(r), trɑːn-/ *n.* a combined radio transmitter and receiver.

transcend /trænˈsend, trɑːn-/ *v.tr.* **1** be beyond the range or grasp of (human experience, reason, belief, etc.). **2** excel; surpass. [ME f. OF *transcendre* or L *transcendere* (as TRANS-, *scandere* climb)]

transcendent /trænˈsendənt, trɑːn-/ *adj. & n.* ● *adj.* **1** excelling,

surpassing (*transcendent merit*). **2** transcending human experience. **3** *Philos.* **a** higher than or not included in any of Aristotle's ten categories in scholastic philosophy. **b** not realizable in experience in Kantian philosophy. **4** (esp. of the supreme being) existing apart from, not subject to the limitations of, the material universe (opp. IMMANENT). ● *n. Philos.* a transcendent thing. □ **transcendence** *n.* **transcendency** *n.* **transcendently** *adv.* [L *transcendent-* pres. part. stem of *transcendere* TRANSCEND]

transcendental /ˌtrænsenˈdent(ə)l, ˌtrɑːn-/ *adj. & n.* ● *adj.* **1** = TRANSCENDENT *adj.* **2** *Philos.* **a** (in Kantian philosophy) presupposed in and necessary to experience; a priori. **b** explaining matter and objective things as products of the subjective mind. **c** (esp. in Emerson's philosophy) regarding the divine as the guiding principle in humankind. **3 a** visionary, abstract. **b** vague, obscure. **4** *Math.* (of a function) not capable of being produced by the algebraical operations of addition, multiplication, and involution, or the inverse operations. ● *n.* a transcendental term, conception, etc. □ **transcendental cognition** *Philos.* a priori knowledge. **transcendental object** *Philos.* a real (unknown and unknowable) object. **transcendental unity** *Philos.* unity brought about by cognition. □ **transcendentally** *adv.* [med.L *transcendentalis* (as TRANSCENDENT)]

Transcendentalism /ˌtrænsenˈdentəˌlɪz(ə)m, ˌtrɑːn-/ *n.* a movement which developed in New England around 1836 and encompassed philosophical, religious, and social thought. Influenced by European romanticism, Platonism, and Kantian philosophy (the name comes from Kant's *Critique of Practical Reason*, 1788), the movement reacted against 18th-century rationalism and espoused an idealistic philosophy based on belief in divinity pervading the whole of nature and humankind. Members also pursued experiments in communal living and held progressive views on feminism and social issues. Central figures were Ralph Waldo Emerson and Henry David Thoreau. □ **Transcendentalist** *n. & adj.*

transcendentalism /ˌtrænsenˈdentəˌlɪz(ə)m, ˌtrɑːn-/ *n.* **1** transcendental philosophy. **2** exalted or visionary language. □ **transcendentalist** *n.* **transcendentalize** *v.tr.* (also **-ise**).

Transcendental Meditation *n.* (abbr. **TM**) a method of detaching oneself from problems, anxiety, etc., by silent meditation and repetition of a mantra. Based in part on Hindu meditation, this method became known in the West in the 1960s, especially because of the interest shown in it by members of the Beatles.

transcode /trænzˈkəʊd, trɑːnz-/ *v.tr. & intr.* convert from one form of coded representation to another.

transcontinental /ˌtrænzkɒntɪˈnent(ə)l, ˌtrɑːnz-/ *adj. & n.* ● *adj.* (of a railway etc.) extending across a continent. ● *n.* a transcontinental railway or train. □ **transcontinentally** *adv.*

transcribe /trænˈskraɪb, trɑːn-/ *v.tr.* **1** make a copy of, esp. in writing. **2** transliterate. **3** write out (shorthand, notes, etc.) in ordinary characters or continuous prose. **4 a** a record for subsequent reproduction. **b** broadcast in this form. **5** arrange (music) for a different instrument etc. □ **transcriber** *n.* **transcriptive** /-ˈskrɪptɪv/ *adj.* [L *transcribere transcript-* (as TRANS-, *scribere* write)]

transcript /ˈtrænskrɪpt, ˈtrɑːn-/ *n.* **1** a written or recorded copy. **2** any copy. [ME f. OF *transcrit* f. L *transcriptum* neut. past part.: see TRANSCRIBE]

transcription /trænˈskrɪpʃ(ə)n, trɑːn-/ *n.* **1 a** the action or process of transcribing something. **b** an instance of this. **2** a transcript or copy. **3** *Biol.* the process by which a sequence of nucleotides is copied from a DNA template during the synthesis of a molecule of RNA (cf. TRANSLATION 3). □ **transcriptional** *adj.*

transducer /trænzˈdjuːsə(r), trɑːnz-/ *n.* any device for converting a non-electrical signal into an electrical one (e.g. pressure into voltage). [L *transducere* lead across (as TRANS-, *ducere* lead)]

transect /trænˈsekt, trɑːn-/ *v.tr.* cut across or transversely. □ **transection** /-ˈsekʃ(ə)n/ *n.* [TRANS- + L *secare sect-* cut]

transept /ˈtrænsept, ˈtrɑːn-/ *n.* **1** either arm of the part of a cross-shaped church at right angles to the nave (*north transept; south transept*). **2** this part as a whole. □ **transeptal** /trænˈsept(ə)l, trɑːn-/ *adj.* [mod.L *transeptum* (as TRANS-, SEPTUM)]

transexual var. of TRANSSEXUAL.

transfer *v. & n.* ● *v.* /trænsˈfɜː(r), trɑːns-/ (**transferred, transferring**) **1** *tr.* (often foll. by *to*) convey, remove, or hand over (a thing etc.) (*transferred the bag from the car to the station*). **b** make over the possession of (property, a ticket, rights, etc.) to a person (*transferred his membership to his son*). **2** *tr. & intr.* change or move to another group, club, department, etc. **3** *intr.* change from one station, route, etc., to another on a journey.

4 *tr.* **a** convey (a drawing etc.) from one surface to another, esp. to a lithographic stone by means of transfer-paper. **b** remove (a picture) from one surface to another, esp. from wood or a wall to canvas. **5** *tr.* change (the sense of a word etc.) by extension or metaphor. ● *n.* /ˈtrænsfɜː(r), ˈtrɑːns-/ **1** the act or an instance of transferring or being transferred. **2 a** a design etc. conveyed or to be conveyed from one surface to another. **b** a small usu. coloured picture or design on paper, which is transferable to another surface. **3** a football player etc. who is or is to be transferred. **4 a** the conveyance of property, a right, etc. **b** a document effecting this. **5** *N. Amer.* a ticket allowing a journey to be continued on another route etc. □ **transfer-book** a register of transfers of property, shares, etc. **transfer company** *US* a company conveying passengers or luggage between stations. **transfer fee** a fee paid for the transfer of esp. a professional footballer. **transfer ink** ink used for making designs on a lithographic stone or transfer-paper. **transfer list** a list of footballers available for transfer. **transfer-paper** specially coated paper to receive the impression of transfer ink and transfer it to stone. **transfer price** a price charged for goods or services provided by one branch of a company to another, esp. in a different country. **transfer pricing** the manipulation of transfer prices as a means of moving profits or funds within a company. **transfer RNA** RNA conveying an amino-acid molecule from the cytoplasm to a ribosome for use in protein synthesis. □ **transferee** /ˌtrænsfɜːˈriː, ˌtrɑːns-/ *n.* **transferor** /trænsˈfɜːrə(r), trɑːns-/ *esp. Law n.* **transferrer** *n.* [ME f. F *transférer* or L *transferre* (as TRANS-, *ferre* lat- bear)]

transferable /trænsˈfɜːrəb(ə)l, trɑːns-/ *adj.* capable of being transferred. □ **transferable vote** a vote that can be transferred to another candidate if the first choice is eliminated. □ **transferability** /-ˌfɜːrəˈbɪlɪtɪ/ *n.*

transference /ˈtrænsfərəns, ˈtrɑːns-/ *n.* **1** the act or an instance of transferring; the state of being transferred. **2** *Psychol.* the redirection of childhood emotions to a new object, esp. to a psychoanalyst.

transferral /trænsˈfɜːrəl, trɑːns-/ *n.* = TRANSFER *n.* 1.

transferrin /trænsˈfɜːrɪn, trɑːns-/ *n. Biochem.* a protein transporting iron in blood serum. [TRANS- + L *ferrum* iron]

transfiguration /ˌtrænsfɪɡjʊˈreɪʃ(ə)n, ˌtrɑːns-/ *n.* **1** a change of form or appearance. **2 a** Christ's appearance in radiant glory to three of his disciples (Matt. 17:2, Mark 9:2–3). **b** (**Transfiguration**) the festival of Christ's transfiguration, 6 Aug. [ME f. OF *transfiguration* or L *transfiguratio* (as TRANSFIGURE)]

transfigure /trænsˈfɪɡə(r), trɑːns-/ *v.tr.* change in form or appearance, esp. so as to elevate or idealize. [ME f. OF *transfigurer* or L *transfigurare* (as TRANS-, FIGURE)]

transfinite /trænsˈfaɪnaɪt, trɑːns-/ *adj.* **1** beyond or surpassing the finite. **2** *Math.* (of a number) exceeding all finite numbers.

transfix /trænsˈfɪks, trɑːns-/ *v.tr.* **1** pierce with a sharp implement or weapon. **2** root (a person) to the spot with horror or astonishment; paralyse the faculties of. □ **transfixion** /-ˈfɪkʃ(ə)n/ *n.* [L *transfigere transfix-* (as TRANS-, FIX)]

transform *v. & n.* ● *v.* /trænsˈfɔːm, trɑːns-/ **1 a** *tr.* make a thorough or dramatic change in the form, outward appearance, character, etc., of. **b** *intr.* (often foll. by *into*, *to*) undergo such a change. **2** *tr. Electr.* change the voltage etc. of (a current). **3** *tr. Math.* change (a mathematical entity) by transformation. ● *n.* /ˈtrænsfɔːm, ˈtrɑːns-/ *Math. & Linguistics* the product of a transformation. □ **transformable** /trænsˈfɔːməb(ə)l, trɑːns-/ *adj.* **transformative** *adj.* [ME f. OF *transformer* or L *transformare* (as TRANS-, FORM)]

transformation /ˌtrænsfəˈmeɪʃ(ə)n, ˌtrɑːns-/ *n.* **1** the act or an instance of transforming; the state of being transformed. **2** *Zool.* a change of form at metamorphosis, esp. of insects, amphibia, etc. **3** the induced or spontaneous change of one element into another. **4** *Math. & Logic* a change of any entity in accordance with some definite rule or set of rules. **5** *Linguistics* the conversion of one syntactic structure into another by the application of specific rules. **6** *Biol.* the modification of a eukaryotic cell from its normal state to a malignant state. **7** *archaic* a woman's wig. **8** a sudden dramatic change of scene on stage. [ME f. OF *transformation* or LL *transformatio* (as TRANSFORM)]

transformational /ˌtrænsfəˈmeɪʃən(ə)l, ˌtrɑːns-/ *adj.* relating to or involving transformation. □ **transformationally** *adv.*

transformational grammar *n.* (also **transformational-generative grammar**) a grammar consisting of a set of rules which generate all and only the grammatical sentences of a language. Some of these rules are phrase-structure rules, which, when applied to the lexicon of the language, generate the logical deep structure of its sentences. Others are transformational rules, which generate the surface structure of sentences from their deep structure, and which account for logically related structures, for example active and passive sentences. Transformational grammar was first described theoretically by Noam Chomsky in *Syntactic Structures* (1957), although linguistic transformational rules had first been proposed by the American linguist Zellig S. Harris (b.1909).

transformer /trænsˈfɔːmə(r), trɑːns-/ *n.* **1** an apparatus for reducing or increasing the voltage of an alternating current. **2** a person or thing that transforms.

transfuse /trænsˈfjuːz, trɑːns-/ *v.tr.* **1 a** permeate (*purple dye transfused the water*). **b** instil (an influence, quality, etc.) into (*transfused enthusiasm into everyone*). **2 a** transfer (blood) from one person or animal to another. **b** inject (liquid) into a blood vessel to replace lost fluid. □ **transfusion** /-ˈfjuːʒ(ə)n/ *n.* [ME f. L *transfundere transfus-* (as TRANS-, *fundere* pour)]

transgenic /trænzˈdʒɛnɪk, trɑːnz-/ *adj. Biol.* (of an animal or plant) having genetic material artificially introduced from another species.

transgress /trænzˈɡres, trɑːnz-/ *v.tr.* (also *absol.*) **1** go beyond the bounds or limits set by (a commandment, law, etc.); violate; infringe. **2** *Geol.* (of the sea) spread over (the land). □ **transgressive** *adj.* **transgressor** *n.* **transgression** /-ˈɡreʃ(ə)n/ *n.* [F *transgresser* or L *transgredi transgress-* (as TRANS-, *gradi* go)]

tranship var. of TRANSSHIP.

transhumance /trænzˈhjuːməns, trɑːnz-/ *n.* the seasonal moving of livestock to a different region. [F f. *transhumer* f. L TRANS- + *humus* ground]

transient /ˈtrænzɪənt, ˈtrɑːnz-/ *adj. & n.* ● *adj.* **1** of short duration; momentary; passing; impermanent (*life is transient; of transient interest*). **2** *Mus.* serving only to connect; inessential (*a transient chord*). ● *n.* **1** a temporary visitor, worker, etc. **2** *Electr.* a brief current etc. □ **transience** *n.* **transiency** *n.* **transiently** *adv.* [L *transire* (as TRANS-, *ire* go)]

transilluminate /ˌtrænzɪˈluːmɪˌneɪt, ˌtrɑːnz-/ *v.tr.* pass a strong light through for inspection, esp. for medical diagnosis. □ **transillumination** /-ˌluːmɪˈneɪʃ(ə)n/ *n.*

transire /trænˈsaɪə(r), trɑːn-/ *n. Brit.* a customs permit for the passage of goods. [L *transire* go across (as TRANSIENT)]

transistor /trænˈzɪstə(r), trɑːn-, -ˈsɪstə(r)/ *n.* **1** a semiconductor device with usu. three connections, capable of amplification in addition to rectification. (*See note below.*) **2** (in full **transistor radio**) a portable radio with transistors. [portmanteau word, f. TRANSFER + RESISTOR]

▪ The transistor, developed by William Shockley and colleagues at the Bell Telephone Laboratories in the US in 1948, ushered in the era of solid-state electronics. In its simplest form the transistor is a sandwich of three layers of semiconducting materials, usually silicon-based, forming a system of three terminals separated by two junctions. They depend for their operation on the fact that the electrical properties of semiconductors can be varied by doping with small amounts of impurities. Transistors can be made to carry out all the functions of thermionic valves, such as amplification and rectification, with the advantages of greatly reduced size (down to the scale of miniaturized printed circuits), lower heat output and operating voltage, and much higher reliability and robustness.

transistorize /trænˈzɪstəˌraɪz, trɑːn-, -ˈsɪstəˌraɪz/ *v.tr.* (also **-ise**) design or equip with, or convert to, transistors rather than valves. □ **transistorization** /-ˌzɪstəraɪˈzeɪʃ(ə)n, -ˌsɪstə-/ *n.*

transit /ˈtrænzɪt, ˈtrɑːnz-/ *n. & v.* ● *n.* **1** the act or process of going, conveying, or being conveyed, esp. over a distance (*transit by rail; made a transit of the lake*). **2** a passage or route (*the overland transit*). **3** *Astron.* **a** the apparent passage of a celestial object across the meridian of a place. **b** the passage of an inferior planet across the face of the sun, or of a moon across the face of a planet. **4** *N. Amer.* the local conveyance of passengers on public routes. ● *v.* (**transited, transiting**) **1** *tr.* make a transit across. **2** *intr.* make a transit. □ **in transit** while going or being conveyed. **transit camp** a camp for the temporary accommodation of soldiers, refugees, etc. **transit-circle** (or **-instrument**) an instrument for observing the transit of a celestial object across the meridian. **transit-compass** (or **-theodolite**) a surveyor's instrument for measuring a horizontal angle. **transit-duty** duty paid on goods passing through a country. **transit lounge** a lounge at an airport for passengers waiting between flights. **transit visa** a visa allowing only passage through a country. [ME f. L *transitus* f. *transire* (as TRANSIENT)]

transition /trænˈzɪʃ(ə)n, trɑːn-/ *n.* **1** a passing or change from one place, state, condition, etc., to another (*an age of transition; a transition from plain to hills*). **2** *Mus.* a momentary modulation. **3** (in the arts) a change from one style to another, esp. *Archit.* from Norman to Early English. **4** *Physics* a change of an atomic nucleus or orbital electron from one quantum state to another, with emission or absorption of radiation. □ **transition point** *Physics* the point at which different phases of the same substance can be in equilibrium. □ **transitional** *adj.* **transitionally** *adv.* **transitionary** *adj.* [F *transition* or L *transitio* (as TRANSIT)]

transition metal *n.* (also **transition element**) *Chem.* any of the set of elements occupying a central block (Groups IVB–VIII, IB, and IIB, or 4–12) in the periodic table, and including iron, copper, silver, etc. They are all dense metals; chemically they show variable valency and a strong tendency to form coordination compounds, and many of their compounds are coloured.

transitive /ˈtrænzɪtɪv, ˈtrɑːnz-/ *adj.* **1** *Gram.* (of a verb or sense of a verb) that takes a direct object (whether expressed or implied), e.g. *saw* in *saw the donkey, saw that she was ill* (opp. INTRANSITIVE). **2** *Logic* (of a relation) such as to be valid for any two members of a sequence if it is valid for every pair of successive members. □ **transitively** *adv.* **transitiveness** *n.* **transitivity** /ˌtrænzɪˈtɪvɪtɪ, ˌtrɑːnz-/ *n.* [LL *transitivus* (as TRANSIT)]

transitory /ˈtrænzɪtərɪ, ˈtrɑːnz-/ *adj.* not permanent, brief, transient. □ **transitory action** *Law* an action that can be brought in any country irrespective of where the transaction etc. started. □ **transitorily** *adv.* **transitoriness** *n.* [ME f. AF *transitorie*, OF *transitoire* f. L *transitorius* (as TRANSIT)]

Transjordan /trænzˈdʒɔːd(ə)n, trɑːnz-/ the former name of an area of Palestine east of the River Jordan, now part of the Hashemite Kingdom of Jordan (see JORDAN).

Transjordanian /ˌtrænzdʒɔːˈdeɪnɪən, ˌtrɑːnz-/ *n. & adj.* ● *n.* **1** *hist.* a native or inhabitant of Transjordan. **2** a person from beyond the river Jordan. ● *adj.* of or relating to the land beyond the river Jordan or *hist.* to Transjordan or its people.

Transkei /trænˈskaɪ/ a former homeland established in South Africa for the Xhosa people, now part of the province of Eastern Cape. (See also HOMELAND.)

translate /trænzˈleɪt, trɑːnz-/ *v.* **1** *tr.* (also *absol.*) **a** (often foll. by *into*) express the sense of (a word, sentence, speech, book, etc.) in another language; produce a translation of. **b** do this as a profession etc. (*translates for the UN*). **2** *intr.* (of a literary work etc.) be translatable, bear translation (*does not translate well*). **3** *tr.* express (an idea, book, etc.) in another, esp. simpler, form. **4** *tr.* interpret the significance of; infer as (*translated his silence as dissent*). **5** *tr.* move or change, esp. from one person, place, or condition to another (*was translated by joy*). **6** *intr.* (foll. by *into*) result in; be converted into; manifest itself as. **7** *tr. Eccl.* **a** remove (a bishop) to another see. **b** remove (a saint's relics etc.) to another place. **8** *tr. Bibl.* convey to heaven without death; transform. **9** *tr. Mech.* **a** cause (a body) to move so that all its parts travel in the same direction. **b** impart motion without rotation to. □ **translatable** *adj.* **translatability** /-ˌleɪtəˈbɪlɪtɪ/ *n.* [ME f. L *translatus*, past part. of *transferre*: see TRANSFER]

translation /trænzˈleɪʃ(ə)n, trɑːnz-/ *n.* **1** the act or an instance of translating. **2** a written or spoken rendering of the meaning of a word, speech, book, etc., in another language. **3** *Biol.* the process by which a sequence of nucleotide triplets in a molecule of messenger RNA is translated into a sequence of amino acids during the synthesis of a protein or polypeptide (cf. TRANSCRIPTION 3). □ **translational** *adj.* **translationally** *adv.*

translator /trænzˈleɪtə(r), trɑːnz-/ *n.* **1** a person who translates from one language into another. **2** a television relay transmitter. **3** *Computing* a program that translates from one (esp. programming) language into another.

transliterate /trænzˈlɪtəˌreɪt, trɑːnz-/ *v.tr.* represent (a word etc.) in the closest corresponding letters or characters of a different alphabet or language. □ **transliterator** *n.* **transliteration** /-ˌlɪtəˈreɪʃ(ə)n/ *n.* [TRANS- + L *littera* letter]

translocate /ˌtrænzləʊˈkeɪt, ˌtrɑːnz-/ *v.tr.* **1** move from one place to another. **2** (usu. in *passive*) *Physiol. & Biochem.* transport (a dissolved substance) within an organism, esp. in the phloem of a plant, or actively across a cell membrane. **3** *Biol.* move (a portion of a chromosome) to a new position on the same or another chromosome. □ **translocation** /-ˈkeɪʃ(ə)n/ *n.*

translucent /trænzˈluːs(ə)nt, trɑːnz-/ *adj.* **1** allowing light to pass through diffusely; semi-transparent. **2** transparent. □ **translucence** *n.* **translucency** *n.* **translucently** *adv.* [L *translucere* (as TRANS-, *lucere* shine)]

translunar /trænzˈluːnə(r), trɑːnz-/ *adj.* **1** lying beyond the moon. **2** of or relating to space travel or a trajectory towards the moon.

transmarine /ˌtrænzməˈriːn, ˌtrɑːnz-/ *adj.* situated or going beyond the sea. [L *transmarinus* f. *marinus* MARINE]

transmigrant /trænzˈmaɪgrənt, trɑːnz-/ *adj. & n.* ● *adj.* passing through, esp. through a country on the way to another. ● *n.* a migrant or alien passing through a country etc. [L *transmigrant-*, part. stem of *transmigrare* (as TRANSMIGRATE)]

transmigrate /ˌtrænzmaɪˈgreɪt, ˌtrɑːnz-/ *v.intr.* **1** (of the soul) pass into a different body, esp. at or after death; undergo metempsychosis. **2** migrate. □ **transmigrator** *n.* **transmigration** /-ˈgreɪʃ(ə)n/ *n.* **transmigratory** /-ˈmaɪgrətərɪ/ *adj.* [ME f. L *transmigrare* (as TRANS-, MIGRATE)]

transmission /trænzˈmɪʃ(ə)n, trɑːnz-/ *n.* **1** the act or an instance of transmitting; the state of being transmitted. **2** a broadcast radio or television programme. **3** the mechanism by which power is transmitted from an engine to the axle in a motor vehicle. □ **transmission line** a conductor or conductors carrying electricity over large distances with minimum losses. [L *transmissio* (as TRANS-, MISSION)]

transmit /trænzˈmɪt, trɑːnz-/ *v.tr.* (**transmitted**, **transmitting**) **1 a** pass or hand on; transfer (*transmitted the message; how diseases are transmitted*). **b** communicate (ideas, emotions, etc.). **2 a** allow (heat, light, sound, electricity, etc.) to pass through; be a medium for. **b** be a medium for (ideas, emotions, etc.) (*his message transmits hope*). **3** broadcast (a radio or television programme). □ **transmittable** *adj.* **transmittal** *n.* **transmissible** /-ˈmɪsɪb(ə)l/ *adj.* **transmissive** *adj.* [ME f. L *transmittere* (as TRANS-, *mittere miss-* send)]

transmittance /trænzˈmɪt(ə)ns, trɑːnz-/ *n.* *Physics* a measure of the transparency of a substance, equal to the ratio of the transmitted luminous flux to the incident luminous flux (cf. ABSORBANCE). [TRANSMIT + -ANCE]

transmitter /trænzˈmɪtə(r), trɑːnz-/ *n.* **1** a person or thing that transmits. **2** a set of equipment used to generate and transmit electromagnetic waves carrying messages, signals, etc., esp. those of radio or television. **3** = NEUROTRANSMITTER.

transmogrify /trænzˈmɒgrɪˌfaɪ, trɑːnz-/ *v.tr.* (**-ies**, **-ied**) esp. *joc.* transform, esp. in a magical or surprising manner. □ **transmogrification** /-fɪˈkeɪʃ(ə)n/ *n.* [17th c.: orig. unkn.]

transmontane /trænzˈmɒnteɪn, trɑːnz-/ *adj.* = TRAMONTANE *adj.* [L *transmontanus*: see TRAMONTANE]

transmutation /ˌtrænzmjuːˈteɪʃ(ə)n, ˌtrɑːnz-/ *n.* **1** the act or an instance of transmuting or changing into another form etc. **2** *Alchemy* the supposed process of changing base metals into gold. **3** *Physics* the changing of one element into another by nuclear bombardment etc. **4** *Geom.* the changing of a figure or body into another of the same area or volume. **5** *Biol.* Lamarck's theory of the change of one species into another. □ **transmutational** *adj.* **transmutationist** *n.* [ME f. OF *transmutation* or LL *transmutatio* (as TRANSMUTE)]

transmute /trænzˈmjuːt, trɑːnz-/ *v.tr.* **1** change the form, nature, or substance of. **2** *Alchemy* subject (base metals) to transmutation. □ **transmuter** *n.* **transmutative** /-tətɪv/ *adj.* **transmutable** *adj.* **transmutability** /-ˌmjuːtəˈbɪlɪtɪ/ *n.* [ME f. L *transmutare* (as TRANS-, *mutare* change)]

transnational /trænzˈnæʃ(ə)n(ə)l, trɑːnz-/ *adj.* extending beyond national boundaries.

transoceanic /ˌtrænzəʊʃɪˈænɪk, ˌtrɑːnz-, -əʊsɪˈænɪk/ *adj.* **1** situated beyond the ocean. **2** concerned with crossing the ocean (*transoceanic flight*).

transom /ˈtrænsəm/ *n.* **1** a horizontal bar of wood or stone across a window or the top of a door (cf. MULLION). **2** each of several beams fixed across the stern-post of a ship. **3** a beam across a saw-pit to support a log. **4** a strengthening crossbar. **5** *US* = *transom window.* □ **transom window 1** a window divided by a transom. **2** a window placed above the transom of a door or larger window; a fanlight. □ **transomed** *adj.* [ME *traversayn, transyn, -ing*, f. OF *traversin* f. *traverse* TRAVERSE]

transonic /trænˈsɒnɪk, trɑːn-/ *adj.* (also **trans-sonic**) relating to speeds close to that of sound. [TRANS- + SONIC, after *supersonic* etc.]

transpacific /ˌtrænspəˈsɪfɪk, ˌtrɑːns-/ *adj.* **1** beyond the Pacific. **2** crossing the Pacific.

transparence /trænsˈpærəns, trɑːns-, -ˈpeərəns/ *n.* = TRANSPARENCY 1.

transparency /trænsˈpærənsɪ, trɑːns-, -ˈpeərənsɪ/ *n.* (*pl.* **-ies**) **1** the condition of being transparent. **2** *Photog.* a positive transparent photograph on glass or in a frame to be viewed using a slide projector etc. **3** a picture, inscription, etc., made visible by a light behind it. [med.L *transparentia* (as TRANSPARENT)]

transparent /træns'pærənt, trɑ:ns-, -'peərənt/ *adj.* **1** allowing light to pass through so that bodies can be distinctly seen (cf. TRANSLUCENT 1). **2 a** (of a disguise, pretext, etc.) easily seen through. **b** (of a motive, quality, etc.) easily discerned; evident; obvious. **3** (of a person etc.) easily understood; frank; open. **4** *Physics* transmitting heat or other electromagnetic rays without distortion. □ **transparently** *adv.* **transparentness** *n.* [ME f. OF f. med.L *transparens* f. L *transparere* shine through (as TRANS-, *parere* appear)]

transpersonal /træns'pɜːsən(ə)l, ˌtrɑ:ns-/ *adj.* **1** (in literature etc.) transcending the personal. **2** *Psychol.* (esp. in psychotherapy) of or relating to the exploration of transcendental states of consciousness beyond personal identity.

transpierce /træns'pɪəs, trɑ:ns-/ *v.tr.* pierce through.

transpire /træn'spaɪə(r), trɑ:n-/ *v.* **1** *intr.* (of a secret or something unknown) leak out; come to be known. **2** *intr.* **a** (prec. by *it* as subject) turn out; prove to be the case (*it transpired he knew nothing about it*). **b** occur; happen. **3** *tr. & intr.* emit (vapour, sweat, etc.), or be emitted, through the skin or lungs; perspire. **4** *intr.* (of a plant or leaf) release water vapour. □ **transpirable** *adj.* **transpiratory** /-'spɪrətərɪ/ *adj.* **transpiration** /ˌtrænspɪ'reɪʃ(ə)n, ˌtrɑ:n-/ *n.* [F *transpirer* or med.L *transpirare* (as TRANS-, L *spirare* breathe)]

transplant *v. & n.* ● *v.tr.* /træns'plɑ:nt, trɑ:ns-/ **1 a** plant in another place (*transplanted the daffodils*). **b** move to another place (*whole nations were transplanted*). **2** *Surgery* transfer (living tissue or an organ) and implant in another part of the body or in another body. ● *n.* /'trænsplɑ:nt, 'trɑ:ns-/ **1** *Surgery* **a** the transplanting of an organ or tissue. **b** such an organ etc. **2** a thing, esp. a plant, transplanted. □ **transplantable** /træns'plɑ:ntəb(ə)l, trɑ:ns-/ *adj.* **transplanter** /-'plɑ:ntə(r)/ *n.* **transplantation** /ˌtrænsplɑ:n'teɪʃ(ə)n, ˌtrɑ:ns-/ *n.* [ME f. LL *transplantare* (as TRANS-, PLANT)]

transponder /træn'spɒndə(r), trɑ:n-/ *n.* a device for receiving a radio signal and automatically transmitting a different signal. [TRANSMIT + RESPOND]

transpontine /træns'pɒntaɪn, trɑ:ns-/ *adj.* **1** on the other side of a bridge. **2** on or from the other side of an ocean; esp. North American. **3** *archaic* on the south side of the Thames. [TRANS- + L *pons pontis* bridge]

transport *v. & n.* ● *v.tr.* /træns'pɔ:t, trɑ:ns-/ **1** take or carry (a person, goods, troops, baggage, etc.) from one place to another. **2** *hist.* take (a criminal) to a penal colony; deport. **3** (as **transported** *adj.*) (usu. foll. by *with*) affected with strong emotion. ● *n.* /'trænspɔ:t, 'trɑ:ns-/ **1 a** a system of conveying people, goods, etc., from place to place. **b** the means of this (*our transport has arrived*). **2** a ship, aircraft, etc., used to carry soldiers, stores, etc. **3** (esp. in *pl.*) vehement emotion (*transports of joy*). **4** *hist.* a transported convict. □ **transport café** *Brit.* a roadside café catering esp. for truck-drivers. [ME f. OF *transporter* or L *transportare* (as TRANS-, *portare* carry)]

transportable /træns'pɔ:təb(ə)l, trɑ:ns-/ *adj.* **1** capable of being transported. **2** *hist.* (of an offender or an offence) punishable by transportation. □ **transportability** /-ˌpɔ:tə'bɪlɪtɪ/ *n.*

transportation /ˌtrænspɔ:'teɪʃ(ə)n, ˌtrɑ:ns-/ *n.* **1** the act of conveying or the process of being conveyed. **2 a** a system of conveying. **b** esp. *N. Amer.* the means of this. **3** *hist.* removal to a penal colony.

transporter /træns'pɔ:tə(r), trɑ:ns-/ *n.* **1** a person or device that transports. **2** a vehicle used to transport other vehicles or large pieces of machinery etc. by road. □ **transporter bridge** a bridge carrying vehicles etc. across water on a suspended moving platform.

transpose /træns'pəʊz, trɑ:ns-/ *v.tr.* **1 a** cause (two or more things) to change places. **b** change the position of (a thing) in a series. **2** change the order or position of (words or a word) in a sentence. **3** (also *absol.*) (also foll. by *up, down*) *Mus.* write or play in a different key from the original. **4** *Algebra* transfer (a term) with a changed sign to the other side of an equation. □ **transposing instrument** *Mus.* an instrument producing notes different in pitch from the written notes. **transposing piano** etc. *Mus.* a piano etc. on which a transposition may be effected mechanically. □ **transposable** *adj.* **transposal** *n.* **transposer** *n.* [ME, = transform f. OF *transposer* (as TRANS-, L *ponere* put)]

transposition /ˌtrænspə'zɪʃ(ə)n, ˌtrɑ:ns-/ *n.* the act or an instance of transposing; the state of being transposed. □ **transpositional** *adj.* **transpositive** /-'pɒzɪtɪv/ *adj.* [F *transposition* or LL *transpositio* (as TRANS-, POSITION)]

transputer /træns'pju:tə(r), trɑ:ns-/ *n.* a microprocessor with integral memory designed for parallel processing. [TRANSISTOR + COMPUTER]

transsexual /trænz'seksjʊəl, trɑ:nz-, -'sekʃʊəl/ *adj. & n.* (also **transexual**) ● *adj.* having the physical characteristics of one sex and the supposed psychological characteristics of the other. ● *n.* **1** a transsexual person. **2** a person whose sex has been changed by surgery. □ **transsexualism** *n.*

transship /trænz'ʃɪp, trɑ:nz-/ *v.tr.* (also **tranship**) *intr.* (**-shipped**, **-shipping**) transfer from one ship or form of transport to another. □ **transshipment** *n.*

Trans-Siberian Railway /ˌtrænzsaɪ'bɪərɪən, ˌtrɑ:nz-/ a railway running from Moscow east around Lake Baikal to Vladivostok on the Sea of Japan, a distance of 9,311 km (5,786 miles). Begun in 1891 and virtually completed by 1904, it opened up Siberia and advanced Russian interest in eastern Asia. A second line, taking a more northerly route across Siberia, opened in 1984.

trans-sonic var. of TRANSONIC.

transubstantiation /ˌtrænsəbˌstænʃɪ'eɪʃ(ə)n, ˌtrɑ:n-/ *n.* (in Roman Catholic and Orthodox belief) the conversion in the Eucharist, after consecration, of the whole substance of the bread and wine into the body and blood of Christ, only the appearances of bread and wine remaining. The belief was defined at the Lateran Council in 1215, and the terminology is based on Aristotelian theories on the nature of substance. The doctrine asserting transubstantiation was challenged at the Reformation by the Lutheran doctrine of consubstantiation. [med.L (as TRANS-, SUBSTANCE)]

transude /træn'sju:d, trɑ:n-/ *v.intr.* (of a fluid) pass through the pores or interstices of a membrane etc. □ **transudatory** /-dətərɪ/ *adj.* **transudation** /ˌtrænsjʊ'deɪʃ(ə)n, ˌtrɑ:n-/ *n.* [F *transsuder* f. OF *tressuer* (as TRANS-, L *sudare* sweat)]

transuranic /ˌtrænzjʊ'rænɪk, ˌtrɑ:nz-/ *adj. Chem.* (of an element) having a higher atomic number than uranium. There are at least 13 transuranic elements known, all of which are radioactive and were first obtained artificially in nuclear reactors or after nuclear explosions.

Transvaal /trænz'vɑ:l/ (also **the Transvaal**) a former province in north-eastern South Africa, lying north of the River Vaal. It was first settled by Boers *c*.1840 after the Great Trek, becoming the core of the Boer republic in 1857. Britain annexed Transvaal in 1877, resistance to British rule leading to the Boer Wars. After defeat in the Second Boer War (1899–1902), the Boer republic lost its independence and Transvaal became a British Crown Colony. Self-governing from 1906, it became a founding province of the Union of South Africa in 1910. In 1994 it was divided into the provinces of Northern Transvaal, Eastern Transvaal, Pretoria-Witwatersrand-Vereeniging, and the eastern part of North-West Province.

Transvaal daisy *n.* a South African gerbera, *Gerbera jamesonii*, grown for its large daisy-like flowers in many shades.

transversal /trænz'vɜ:s(ə)l, trɑ:nz-/ *adj. & n.* ● *adj.* (of a line) cutting a system of lines. ● *n.* a transversal line. □ **transversally** *adv.* **transversality** /ˌtrænzvɜ:'sælɪtɪ, ˌtrɑ:nz-/ *n.* [ME f. med.L *transversalis* (as TRANSVERSE)]

transverse /'trænzvɜ:s, 'trɑ:nz-/ *adj.* situated, arranged, or acting in a crosswise direction. □ **transverse magnet** a magnet with poles at the sides and not the ends. **transverse wave** *Physics* a wave in which the medium vibrates at right angles to the direction of its propagation. □ **transversely** *adv.* [L *transvertere transvers-* turn across (as TRANS-, *vertere* turn)]

transvestism /trænz'vestɪz(ə)m, trɑ:nz-/ *n.* the practice of wearing the clothes of the opposite sex, esp. as a sexual stimulus. [G *Transvestismus* f. TRANS- + L *vestire* clothe]

transvestite /trænz'vestaɪt, trɑ:nz-/ *n.* a person, esp. a man, given to transvestism.

Transylvania /ˌtrænsɪl'veɪnɪə/ a large tableland region of NW Romania, separated from the rest of the country by the Carpathian Mountains and the Transylvanian Alps. This region formed the nucleus of the Roman province of Dacia. Conquered by the Magyars at the end of the 9th century, it formed part of Hungary until it became a principality of the Ottoman Empire in the 16th century. It was returned to Hungary at the end of the 17th century and was incorporated into Romania in 1918. Its name means 'beyond the forest' in Latin. □ **Transylvanian** *adj.*

trap[1] /træp/ *n. & v.* ● *n.* **1 a** an enclosure or device, often baited, for catching animals, usu. by affording a way in but not a way out. **b** a device with bait for killing vermin, esp. = MOUSETRAP 1. **2** a trick betraying a person into speech or an act (*is this question a trap?*). **3** an arrangement to catch an unsuspecting person, e.g. a speeding motorist. **4** a device for hurling an object such as a clay pigeon into the air to be shot at. **5** a compartment from which a greyhound is released at the start of a race. **6** a shoe-shaped wooden device with a pivoted bar that sends a ball from its heel into the air on being struck at the other end with a bat. **7 a** a curve in a downpipe etc. that fills with liquid and forms a seal against the upward passage of gases. **b** a device for preventing the passage of steam etc. **8** *Golf* a bunker. **9** a device allowing pigeons to enter but not leave a loft. **10** a two-wheeled carriage (*a pony and trap*). **11** = TRAPDOOR. **12** *sl.* the mouth (esp. *shut one's trap*). **13** (esp. in *pl.*) *sl.* a percussion instrument esp. in a jazz band. ● *v.tr.* (**trapped**, **trapping**) **1** catch (an animal) in a trap. **2** catch or catch out (a person) by means of a trick, plan, etc. **3** stop and retain in or as in a trap. **4** provide (a place) with traps. □ **trap-ball** a game played with a trap (see sense 6 of *n.*). **trap-shooter** a person who practises trap-shooting. **trap-shooting** the sport of shooting at objects released from a trap. □ **traplike** *adj.* [OE *treppe, træppe*, rel. to MDu. *trappe*, med.L *trappa*, of uncert. orig.]

trap² /træp/ *v.tr.* (**trapped**, **trapping**) (often foll. by *out*) **1** provide with trappings. **2** adorn. [obs. *trap* (n.): ME f. OF *drap*: see DRAPE]

trap³ /træp/ *n.* (in full **trap-rock**) *Geol.* any dark-coloured igneous rock, fine-grained and columnar in structure, esp. basalt. [Sw. *trapp* f. *trappa* stair, f. the often stairlike appearance of its outcroppings]

trapdoor /ˈtræpdɔː(r)/ *n.* a door or hatch in a floor, ceiling, or roof, usu. made flush with the surface. □ **trapdoor spider** a spider of the family Ctenizidae, living in a burrow with a hinged trapdoor-like cover.

trapes var. of TRAIPSE.

trapeze /trəˈpiːz/ *n.* a crossbar or set of crossbars suspended by ropes and used as a swing for acrobatics etc. [F *trapèze* f. LL *trapezium*: see TRAPEZIUM]

trapezium /trəˈpiːzɪəm/ *n.* (*pl.* **trapezia** /-zɪə/ or **trapeziums**) **1** *Brit.* a quadrilateral with only one pair of sides parallel. **2** *N. Amer.* = TRAPEZOID 1. [LL f. Gk *trapezion* f. *trapeza* table]

trapezius /trəˈpiːzɪəs/ *n.* (*pl.* **trapezii** /-zɪˌaɪ/) (in full **trapezius muscle**) *Anat.* either of a pair of large flat triangular muscles extending over the back of the neck and shoulders. [mod.L f. Gk *trapezion* f. *trapeza* table]

trapezoid /ˈtræpɪˌzɔɪd/ *n.* **1** *Brit.* a quadrilateral with no two sides parallel. **2** *N. Amer.* = TRAPEZIUM 1. □ **trapezoidal** /ˌtræpɪˈzɔɪd(ə)l/ *adj.* [mod.L *trapezoides* f. Gk *trapezoeidēs* (as TRAPEZIUM)]

trapper /ˈtræpə(r)/ *n.* a person who traps wild animals, esp. to obtain furs.

trappings /ˈtræpɪŋz/ *n.pl.* **1** ornamental accessories, esp. as an indication of status (*the trappings of office*). **2** the harness of a horse esp. when ornamental. [ME (as TRAP²)]

Trappist /ˈtræpɪst/ *n.* & *adj.* ● *n.* a member of a branch of the Cistercian order of monks founded in 1664 at La Trappe in Normandy and noted for an austere rule including a vow of silence. ● *adj.* of or relating to this order. [F *trappiste* f. *La Trappe*]

traps /træps/ *n.pl. colloq.* personal belongings; baggage. [perh. contr. f. TRAPPINGS]

trash /træʃ/ *n.* & *v.* ● *n.* **1** esp. *US* **a** worthless or poor quality stuff, esp. literature; things of poor workmanship or material. **b** rubbish, refuse. **c** absurd talk or ideas; nonsense. **2** a worthless person or persons. **3** (in full **cane-trash**) *West Indies* the refuse of crushed sugar canes and dried stripped leaves and tops of sugar cane used as fuel. ● *v.tr.* **1** esp. *N. Amer. colloq.* wreck. **2** strip (sugar canes) of their outer leaves to speed up the ripening process. **3** esp. *US colloq.* expose the worthless nature of; disparage. **4** esp. *US colloq.* throw away, discard. □ **trash can** *N. Amer.* a dustbin. **trash-ice** (on a sea, lake, etc.) broken ice mixed with water. **white trash** (or **poor white trash**) esp. *US derog.* the poor white population, esp. in the southern US. [16th c.: orig. unkn.]

trashy /ˈtræʃɪ/ *adj.* (**trashier**, **trashiest**) worthless; poorly made. □ **trashily** *adv.* **trashiness** *n.*

Trás-os-Montes /ˌtræʒuːʃˈmɒntɛʃ/ a mountainous region of NE Portugal, north of the Douro river. Its name means 'beyond the mountains'.

trass /træs/ *n.* (also **tarras** /təˈræs/) a light-coloured variety of the rock tuff used in making cement. [Du. *trass*, earlier *terras, tiras* f. Rmc: cf. TERRACE]

trattoria /ˌtrætəˈrɪə/ *n.* an Italian restaurant. [It.]

trauma /ˈtrɔːmə, ˈtraʊmə/ *n.* (*pl.* **traumata** /-mətə/ or **traumas**) **1 a** *Psychol.* emotional shock following a stressful event, sometimes

leading to long-term neurosis. **b** (in general use) a distressing or emotionally disturbing experience etc. **2** any physical wound or injury. **3** physical shock following this, characterized by a drop in body temperature, mental confusion, etc. □ **traumatize** *v.tr.* (also **-ise**). **traumatization** /ˌtrɔːmətaɪˈzeɪʃ(ə)n, ˌtraʊmə-/ *n.* [Gk *trauma traumatos* wound]

traumatic /trɔːˈmætɪk, traʊˈmæt-/ *adj.* **1** of or causing trauma. **2** *colloq.* (in general use) distressing; emotionally disturbing (*a traumatic experience*). **3** of or for wounds. □ **traumatically** *adv.* [LL *traumaticus* f. Gk *traumatikos* (as TRAUMA)]

traumatism /ˈtrɔːməˌtɪz(ə)m, ˈtraʊmə-/ *n.* **1** the action of a trauma. **2** a condition produced by this.

travail /ˈtræveɪl/ *n.* & *v. literary* ● *n.* **1** painful or laborious effort. **2** the pangs of childbirth. ● *v.intr.* undergo a painful effort, esp. in childbirth. [ME f. OF *travail, travaillier* ult. f. med.L *trepalium* instrument of torture f. L *tres* three + *palus* stake]

travel /ˈtræv(ə)l/ *v.* & *n.* ● *v.intr.* & *tr.* (**travelled, travelling**; *US* **traveled, traveling**) **1** *intr.* go from one place to another; make a journey, esp. of some length or abroad. **2** *tr.* **a** journey along or through (a country). **b** cover (a distance) in travelling. **3** *intr. colloq.* withstand a long journey (*wines that do not travel*). **4** *intr.* go from place to place as a salesman. **5** *intr.* move or proceed in a specified manner or at a specified rate (*light travels faster than sound*). **6** *intr. colloq.* move quickly. **7** *intr.* pass esp. in a deliberate or systematic manner from point to point (*the photographer's eye travelled over the scene*). **8** *intr.* (of a machine or part) move or operate in a specified way. **9** *intr.* (of deer etc.) move onwards in feeding. ● *n.* **1 a** the act of travelling, esp. in foreign countries. **b** (often in *pl.*) a spell of this (*have returned from their travels*). **2** the range, rate, or mode of motion of a part in machinery. □ **travel agency** (or **bureau**) an agency that makes the necessary arrangements for travellers. **travel agent** a person or firm acting as a travel agency. **travelling crane** a crane able to move on rails, esp. along an overhead support. **travelling-rug** a rug used for warmth on a journey. **travelling salesman** = TRAVELLER 2. **travelling wave** *Physics* a wave in which the medium moves in the direction of propagation. **travel-sick** suffering from nausea caused by motion in travelling. **travel-sickness** the condition of being travel-sick. □ **travelling** *adj.* [ME, orig. = TRAVAIL]

travelled /ˈtræv(ə)ld/ *adj.* experienced in travelling (also in *comb.*: *much-travelled*).

traveller /ˈtrævələ(r)/ *n.* (*US* **traveler**) **1** a person who travels or is travelling. **2** a travelling salesman. **3** a gypsy. **4** (also **New Age traveller**) a person who embraces New Age values and leads an itinerant and unconventional lifestyle. **5** *Austral.* an itinerant workman; a swagman. **6** a moving mechanism, esp. a travelling crane. □ **traveller's cheque** (*US* **check**) a cheque for a fixed amount that may be cashed on signature, usu. internationally. **traveller's joy** a wild clematis, *Clematis vitalba*, with grey fluffy hairs round the seeds (also called *old man's beard*). **traveller's tale** an incredible and probably untrue story.

travelogue /ˈtrævəˌlɒg/ *n.* a film or illustrated lecture about travel. [TRAVEL after *monologue* etc.]

traverse /ˈtrævəs, trəˈvɜːs/ *v.* & *n.* ● *v.* **1** *tr.* travel or lie across (*traversed the country; a pit traversed by a beam*). **2** *tr.* consider or discuss the whole extent of (a subject). **3** *tr.* turn (a large gun) horizontally. **4** *tr. Law* deny (an allegation) in pleading. **5** *tr.* thwart, frustrate, or oppose (a plan or opinion). **6** *intr.* (of the needle of a compass etc.) turn on or as on a pivot. **7** *intr.* (of a horse) walk obliquely. **8** *intr.* make a traverse in climbing. ● *n.* **1** a sideways movement. **2** an act of traversing. **3** a thing, esp. part of a structure, that crosses another. **4** a gallery extending from side to side of a church or other building. **5 a** a single line of survey, usu. plotted from compass bearings and chained or paced distances between angular points. **b** a tract surveyed in this way. **6** *Naut.* a zigzag line taken by a ship because of contrary winds or currents. **7** a skier's similar movement on a slope. **8** the sideways movement of a part in a machine. **9 a** a sideways motion across a rock-face from one practicable line of ascent or descent to another. **b** a place where this is necessary. **10** *Mil.* a pair of right-angle bends in a trench to avoid enfilading fire. **11** *Law* a denial, esp. of an allegation of a matter of fact. **12** the act of turning a large gun horizontally to the required direction. □ **traversable** *adj.* **traverser** *n.* **traversal** /trəˈvɜːs(ə)l/ *n.* [OF *traverser* f. LL *traversare, transversare* (as TRANSVERSE)]

travertine /ˈtrævəˌtiːn/ *n.* a white or light-coloured calcareous rock deposited from springs. [It. *travertino, tivertino* f. L *tiburtinus* of Tibur (Tivoli) near Rome]

travesty /ˈtrævɪstɪ/ *n.* & *v.* ● *n.* (*pl.* **-ies**) a grotesque misrepresentation or imitation (*a travesty of justice*). ● *v.tr.* (**-ies, -ied**) make or be a travesty of. [(orig. adj.) f. F *travesti* past part. of *travestir* disguise, change the clothes of, f. It. *travestire* (as TRANS-, *vestire* clothe)]

travois /trə'vɔɪ/ *n.* (*pl.* same /-'vɔɪz/) a North American Indian vehicle of two joined poles pulled by a horse etc. for carrying a burden. [earlier *travail* f. F, perh. the same word as TRAVAIL]

trawl /trɔːl/ *v. & n.* ● *v.* **1** *intr.* **a** fish with a trawl or seine. **b** search thoroughly. **2** *tr.* **a** catch by trawling. **b** (often foll. by *for*) search thoroughly through (*trawled the schools for new trainees*). ● *n.* **1** an act of trawling. **2** (in full **trawl-net**) a large wide-mouthed fishing-net dragged by a boat along the bottom. **3** (in full **trawl-line**) *US* a long sea-fishing line buoyed and supporting short lines with baited hooks. [prob. f. MDu. *traghelen* to drag (cf. *traghel* drag-net), perh. f. L *tragula*]

trawler /'trɔːlə(r)/ *n.* **1** a boat used for trawling. **2** a person who trawls.

tray /treɪ/ *n.* **1** a flat shallow vessel usu. with a raised rim for carrying dishes etc. or containing small articles, papers, etc. **2** a shallow lidless box forming a compartment of a trunk. □ **trayful** *n.* (*pl.* **-fuls**). [OE *trīg* f. Gmc, rel. to TREE]

treacherous /'tretʃərəs/ *adj.* **1** guilty of or involving treachery. **2** (of the weather, ice, the memory, etc.) not to be relied on; likely to fail or give way. □ **treacherously** *adv.* **treacherousness** *n.* [ME f. OF *trecherous* f. *trecheor* a cheat f. *trechier, trichier*: see TRICK]

treachery /'tretʃərɪ/ *n.* (*pl.* **-ies**) **1** violation of faith or trust; betrayal. **2** an instance of this.

treacle /'triːk(ə)l/ *n.* **1** esp. *Brit.* **a** a syrup produced in refining sugar. **b** molasses. **2** cloying sentimentality or flattery. □ **treacly** *adj.* [ME *triacle* f. OF f. L *theriaca* f. Gk *thēriakē* antidote against venom, fem. of *thēriakos* (adj.) f. *thērion* wild beast]

tread /tred/ *v. & n.* ● *v.* (*past* **trod** /trɒd/; *past part.* **trodden** /'trɒd(ə)n/ or **trod**) **1** *intr.* (often foll. by *on*) **a** set down one's foot; walk or step (*do not tread on the grass; trod on a snail*). **b** (of the foot) be set down. **2** *tr.* **a** walk on. **b** (often foll. by *down*) press or crush with the feet. **3** *tr.* perform (steps etc.) by walking (*trod a few paces*). **4** *tr.* make (a hole etc.) by treading. **5** *intr.* (foll. by *on*) suppress; subdue mercilessly. **6** *tr.* make a track with (dirt etc.) from the feet. **7** *tr.* (often foll. by *in, into*) press down into the ground with the feet (*trod dirt into the carpet*). **8** *tr.* (also *absol.*) (of a male bird) copulate with (a hen). ● *n.* **1** a manner or sound of walking (*recognized the heavy tread*). **2** (in full **tread-board**) the top surface of a step or stair. **3** the thick moulded part of a vehicle tyre for gripping the road. **4 a** the part of a wheel that touches the ground or rail. **b** the part of a rail that the wheels touch. **5** the part of the sole of a shoe that rests on the ground. **6** (of a male bird) copulation. □ **tread the boards** (or **stage**) be or become an actor; appear on the stage. **tread on air** see AIR. **tread on a person's toes** offend a person or encroach on a person's privileges etc. **tread out 1** stamp out (a fire etc.). **2** press out (wine or grain) with the feet. **tread water** maintain an upright position in the water by moving the feet with a walking movement and the hands with a downward circular motion. □ **treader** *n.* [OE *tredan* f. WG]

treadle /'tred(ə)l/ *n. & v.* ● *n.* a lever worked by the foot and imparting motion to a machine. ● *v.intr.* work a treadle. [OE *tredel* stair (as TREAD)]

treadmill /'tredmɪl/ *n.* **1** a wide mill-wheel turned by people treading on steps fixed along its circumference, in the 19th century worked by some prisoners as a punishment. The treadmill sometimes operated machines for pumping or grinding, but was often imposed as mere labour without any other purpose. **2** a similar device used for exercise. **3** tiring monotonous routine work.

treadwheel /'tredwiːl/ *n.* a treadmill or similar appliance.

treason /'triːz(ə)n/ *n.* **1** (in full **high treason**: see note below) violation by a subject of allegiance to the sovereign or to the state, esp. by attempting to kill or overthrow the sovereign or to overthrow the government. **2** (in full **petty treason**) *hist.* murder of one's master or husband, regarded as a form of treason. ¶ The crime of *petty treason* was abolished in 1828; the term *high treason*, originally distinguished from *petty treason*, now has the same meaning as *treason*. □ **treasonous** *adj.* [ME f. AF *treisoun* etc., OF *traïson*, f. L *traditio* handing over (as TRADITION)]

treasonable /'triːzənəb(ə)l/ *adj.* involving or guilty of treason. □ **treasonably** *adv.*

treasure /'treʒə(r)/ *n. & v.* ● *n.* **1 a** precious metals or gems. **b** a hoard of these. **c** accumulated wealth. **2** a thing valued for its rarity, workmanship, associations, etc. (*art treasures*). **3** *colloq.* a much loved or highly valued person. ● *v.tr.* **1** (often foll. by *up*) store up as valuable. **2** value (esp. a long-kept possession) highly. □ **treasure hunt 1** a

search for treasure. **2** a game in which players seek a hidden object from a series of clues. **treasure trove 1** *Law* treasure of unknown ownership which is found hidden in the ground etc. and is declared the property of the Crown. **2** a collection of valuable or delightful things. [ME f. OF *tresor*, ult. f. Gk *thēsauros*: see THESAURUS]

treasurer /'treʒərə(r)/ *n.* **1** a person appointed to administer the funds of a society or municipality etc. **2** an officer authorized to receive and disburse public revenues. □ **treasurership** *n.* [ME f. AF *tresorer*, OF *tresorier* f. *tresor* (see TREASURE) after LL *thesaurarius*]

treasury /'treʒərɪ/ *n.* (*pl.* **-ies**) **1** a place or building where treasure is stored. **2** the funds or revenue of a state, institution, or society. **3 (Treasury) a** the department managing the public revenue of a country. (*See note below.*) **b** the offices and officers of this. **c** the place where the public revenues are kept. □ **Treasury bench** (in the UK) the front bench in the House of Commons occupied by the Prime Minister, Chancellor of the Exchequer, etc. **treasury bill** a bill of exchange issued by the government to raise money for temporary needs. **treasury note** *US & hist.* a note issued by the Treasury for use as currency. [ME f. OF *tresorie* (as TREASURE)]

▪ In Britain the Treasury began to take over the functions of the Exchequer from the time of Elizabeth I, under Lord Burghley, and in the 18th century the First Lord of the Treasury gradually assumed the role of Prime Minister. The office of First Lord of the Treasury is now always held by the Prime Minister, but the functions of the office are carried out by the Chancellor of the Exchequer. (See EXCHEQUER.)

treat /triːt/ *v. & n.* ● *v.* **1** *tr.* act or behave towards or deal with (a person or thing) in a certain way (*treated me kindly; treat it as a joke*). **2** *tr.* deal with or apply a process to; act upon to obtain a particular result (*treat it with acid*). **3** *tr.* apply medical care or attention to. **4** *tr.* present or deal with (a subject) in literature or art. **5** *tr.* (often foll. by *to*) provide with food or drink or entertainment, esp. at one's own expense (*treated us to dinner*). **6** *intr.* (often foll. by *with*) negotiate terms (with a person). **7** *intr.* (often foll. by *of*) give a spoken or written exposition. ● *n.* **1** an event or circumstance (esp. when unexpected or unusual) that gives great pleasure. **2** a meal, entertainment, etc., provided by one person for the enjoyment of another or others. **3** (*prec. by a*) extremely good or well (*they looked a treat; has come on a treat*). □ **treatable** *adj.* **treater** *n.* **treating** *n.* [ME f. AF *treter*, OF *traitier* f. L *tractare* handle, frequent. of *trahere tract-* draw, pull]

treatise /'triːtɪs, -tɪz/ *n.* a written work dealing formally and systematically with a subject. [ME f. AF *tretis* f. OF *traitier* TREAT]

treatment /'triːtmənt/ *n.* **1** a process or manner of behaving towards or dealing with a person or thing (*received rough treatment*). **2** the application of medical care or attention to a patient. **3** a manner of treating a subject in literature or art. **4** subjection to the action of a chemical, physical, or biological agent. **5** (*prec. by the*) *colloq.* the customary way of dealing with a person, situation, etc. (*got the full treatment*).

treaty /'triːtɪ/ *n.* (*pl.* **-ies**) **1** a formally concluded and ratified agreement between states. **2** an agreement between individuals or parties, esp. for the purchase of property. □ **treaty port** *hist.* a port that a country was bound by treaty to keep open to foreign trade. [ME f. AF *treté* f. L *tractatus* TRACTATE]

Treaty of Rome, Treaty of Versailles, etc. see ROME, TREATY OF; VERSAILLES, TREATY OF, etc.

Trebizond see TRABZON.

treble /'treb(ə)l/ *adj., n., & v.* ● *adj.* **1 a** threefold. **b** triple. **c** three times as much or many (*treble the amount*). **2** (of a voice) high-pitched. ● *n.* **1** a treble quantity or thing. **2** *Darts* a hit on the narrow ring enclosed by the two middle circles of a dartboard, scoring treble. **3 a** (often *attrib.*) = SOPRANO (esp. a boy's voice or part, or an instrument). **b** a high-pitched voice. **4** the high-frequency output of a radio, record-player, etc., corresponding to the treble in music. **5** a system of betting in which the winnings and stake from the first bet are transferred to a second and then (if successful) to a third. **6** *Sport* three victories or championships in the same season, event, etc. ● *v.* **1** *tr. & intr.* make or become three times as much or many; increase threefold; multiply by three. **2** *tr.* amount to three times as much as. □ **treble chance** a form of football pool in which different numbers of points are awarded for draws and home and away wins predicted by the competitors. **treble clef** a clef placing G above middle C on the second lowest line of the staff. **treble rhyme** a rhyme including three syllables. □ **trebly** *adv.* (in sense 1 of *adj.*). [ME f. OF f. L *triplus* TRIPLE]

Treblinka /tre'blɪŋkə/ a Nazi concentration camp in Poland in the

Second World War, where a great many of the Jews of the Warsaw ghetto were murdered.

trebuchet /ˈtrebjʊˌʃet/ n. (also **trebucket** /-ˌket/) **1** hist. a military machine used in siege warfare for throwing stones etc. **2** a tilting balance for accurately weighing light articles. [ME f. OF f. *trebucher* overthrow, ult. f. Frank.]

trecento /treɪˈtʃentəʊ/ n. the style of Italian art and literature of the 14th century. □ **trecentist** n. [It., = 300 used with ref. to the years 1300–99]

tree /triː/ n. & v. ● n. **1 a** a perennial plant with a tall woody self-supporting main stem or trunk when mature and usu. unbranched for some distance above the ground (cf. SHRUB¹). **b** any similar plant having a tall erect usu. single stem, e.g. palm tree. **2** a piece or frame of wood etc. for various purposes (*shoe-tree*). **3** archaic or poet. **a** a gibbet. **b** a cross, esp. the one used for Christ's Crucifixion. **4** (in full **tree diagram**) Math. a diagram with a structure of branching connecting lines. **5** = *family tree*. ● v.tr. (**trees, treed**) **1** force to take refuge in a tree. **2** esp. US put into a difficult position. **3** stretch on a shoe-tree. □ **grow on trees** (usu. with *neg.*) be plentiful; be easily obtainable. **tree agate** agate with treelike markings. **tree calf** a calf binding for books stained with a treelike design. **tree-fern** a tree-sized fern, esp. of the family Cyatheaceae, with an upright trunklike stem. **tree frog** a small frog of the family Hylidae, climbing by means of adhesive discs on the tips of its toes. **tree heath** = BRIER² 1. **tree hopper** a homopteran bug of the family Membracidae, living in trees. **tree house** a structure or hut in a tree, esp. one for children to play in. **tree line** = TIMBERLINE. **tree mallow** a tall woody-stemmed European mallow, *Lavatera arborea*, of coasts, cliffs, and rocks. **tree of heaven** an ornamental Asian tree, *Ailanthus altissima*, with evil-smelling flowers. **tree of knowledge 1** the tree in the Garden of Eden bearing the forbidden fruit. **2** the branches of knowledge as a whole. **tree of life** = THUJA. **tree ring** a ring in a cross-section of a tree, from one year's growth (see also DENDROCHRONOLOGY). **tree shrew** a small insect-eating arboreal mammal of the order Scandentia, native to SE Asia, having a pointed nose and bushy tail. **tree sparrow 1** a Eurasian sparrow, *Passer montanus*, inhabiting agricultural land. **2** N. Amer. a North American sparrow-like bird, *Spizella arborea*, of the bunting family, breeding on the edge of the tundra. **tree surgeon** a person who treats decayed trees in order to preserve them. **tree surgery** the art or practice of such treatment. **tree toad** = *tree frog*. **tree tomato** a solanaceous South American shrub, *Cyphomandra betacea*, with egg-shaped red fruit. **tree-trunk** the trunk of a tree. **up a tree** esp. N. Amer. cornered; nonplussed. □ **treeless** adj. **treelessness** n. **tree-like** adj. [OE *trēow* f. Gmc]

treecreeper /ˈtriːˌkriːpə(r)/ n. a small bird that creeps on the trunks and branches of trees to look for insects; esp. one of the family Certhiidae, with a down-curved bill.

treen /triːn/ n. (treated as *pl.*) small domestic wooden objects, esp. antiques. [*treen* (adj.) wooden f. OE *trēowen* (as TREE)]

treenail /ˈtriːneɪl/ n. (also **trenail**) a hard wooden pin for securing timbers etc.

treetop /ˈtriːtɒp/ n. the topmost part of a tree.

trefa /ˈtreɪfə/ adj. (also **tref** /treɪf/ and other variants) not kosher. [Heb. *ṭ'rēpāh* the flesh of an animal torn f. *ṭārap* rend]

trefoil /ˈtrefɔɪl, ˈtriːf-/ n. & adj. ● n. **1** a leguminous plant of the genus *Trifolium*, with leaves of three leaflets, esp. a clover. **2** any plant with similar leaves. **3** a three-lobed ornamentation, esp. in tracery windows. **4** a thing arranged in or with three lobes. ● adj. of or concerning a three-lobed plant, window tracery, etc. □ **trefoiled** adj. (also in comb.). [ME f. AF *trifoil* f. L *trifolium* (as TRI-, *folium* leaf)]

trek /trek/ v. & n. ● v.intr. (**trekked, trekking**) **1** travel or make one's way arduously (*trekking through the forest*). **2** esp. S. Afr. hist. migrate or journey with one's belongings by ox-wagon. **3** S. Afr. (of an ox) draw a vehicle or pull a load. ● n. **1 a** a journey or walk made by trekking (*it was a trek to the nearest launderette*). **b** each stage of such a journey. **2** an organized migration of a body of persons. □ **trekker** n. [S. Afr. Du. *trek* (n.), *trekken* (v.) draw, travel]

Trekkie /ˈtrekɪ/ n. sl. a fan of *Star Trek*, a TV science-fiction drama series created by Gene Roddenberry.

trellis /ˈtrelɪs/ n. & v. ● n. (in full **trellis-work**) a lattice or grating of light wooden or metal bars used esp. as a support for fruit-trees or creepers and often fastened against a wall. ● v.tr. (**trellised, trellising**) **1** provide with a trellis. **2** support (a vine etc.) with a trellis. [ME f. OF *trelis, trelice* ult. f. L *trilix* three-ply (as TRI-, *licium* warp-thread)]

trematode /ˈtreməˌtəʊd/ n. Zool. a parasitic flatworm of the class Trematoda, which comprises the flukes and their relatives, equipped with hooks or suckers. They are divided into two subclasses, the Monogenea, which are ectoparasites of fish etc., and the Digenea, which are endoparasites with complex life cycles. [mod.L *Trematoda* f. Gk *trēmatōdēs* perforated f. *trēma* hole]

tremble /ˈtremb(ə)l/ v. & n. ● v.intr. **1** shake involuntarily from fear, excitement, weakness, etc. **2** be in a state of extreme apprehension (*trembled at the very thought of it*). **3** move in a quivering manner (*leaves trembled in the breeze*). ● n. **1** a trembling state or movement; a quiver (*couldn't speak without a tremble*). **2** (in *pl.*) a disease (esp. of cattle) marked by trembling. □ **all of a tremble** colloq. **1** trembling all over. **2** extremely agitated. **trembling poplar** an aspen. □ **tremblingly** adv. [ME f. OF *trembler* f. med.L *tremulare* f. L *tremulus* TREMULOUS]

trembler /ˈtremblə(r)/ n. an automatic vibrator for making and breaking an electrical circuit.

trembly /ˈtremblɪ/ adj. (**tremblier, trembliest**) colloq. trembling; agitated.

tremendous /trɪˈmendəs/ adj. **1** awe-inspiring, fearful, overpowering. **2** colloq. remarkable, considerable, excellent (*a tremendous explosion; gave a tremendous performance*). □ **tremendously** adv. **tremendousness** n. [L *tremendus*, gerundive of *tremere* tremble]

tremolo /ˈtreməˌləʊ/ n. (pl. **-os**) Mus. **1** a tremulous effect produced on musical instruments or in singing: **a** esp. on bowed stringed instruments or an organ; **b** by rapid reiteration of a note by rapid alternation between two notes; **c** by rapid slight variation in the pitch of a note, e.g. on an electric guitar (cf. VIBRATO). **2 a** a device in an organ producing a tremolo. **b** (in full **tremolo arm**) a lever on an electric guitar used to produce a tremolo. [It. (as TREMULOUS)]

tremor /ˈtremə(r)/ n. & v. ● n. **1** a shaking or quivering. **2** a thrill (of fear or exultation etc.). **3** (in full **earth tremor**) a slight earthquake. ● v.intr. undergo a tremor or tremors. [ME f. OF *tremour* & L *tremor* f. *tremere* tremble]

tremulous /ˈtremjʊləs/ adj. **1** trembling or quivering (*in a tremulous voice*). **2** (of a line etc.) drawn with a tremulous hand. **3** timid or vacillating. □ **tremulously** adv. **tremulousness** n. [L *tremulus* f. *tremere* tremble]

trenail var. of TREENAIL.

trench /trentʃ/ n. & v. ● n. **1** a long narrow usu. deep depression or ditch. **2** Mil. **a** this dug by troops to stand in and be sheltered from enemy fire. **b** (in *pl.*) a defensive system of these, as used in the First World War. **3** a long narrow deep depression in the ocean bed. ● v. **1** tr. dig a trench or trenches in (the ground). **2** tr. turn over the earth of (a field, garden, etc.) by digging a succession of adjoining ditches. **3** intr. (foll. by *on, upon*) archaic **a** encroach. **b** verge or border closely. □ **trench coat 1** a soldier's lined or padded waterproof coat. **2** a loose belted raincoat. **trench fever** a highly contagious rickettsial disease transmitted by lice, that infested soldiers in the trenches in the First World War. **trench foot** a painful condition of the feet caused by long immersion in cold water or mud and marked by blackening and death of surface tissue. **trench mortar** a light simple mortar throwing a bomb from one's own into the enemy trenches. **trench warfare** hostilities carried on from more or less permanent trenches. [ME f. OF *trenche* (n.) *trenchier* (v.), ult. f. L *truncare* TRUNCATE]

trenchant /ˈtrentʃənt/ adj. **1** (of a style or language etc.) incisive, terse, vigorous. **2** archaic or poet. sharp, keen. □ **trenchancy** n. **trenchantly** adv. [ME f. OF, part. of *trenchier*: see TRENCH]

trencher /ˈtrentʃə(r)/ n. **1** hist. a wooden or earthenware platter for serving food. **2** (in full **trencher cap**) a stiff square academic cap; a mortarboard. [ME f. AF *trenchour*, OF *trencheoir* f. *trenchier*: see TRENCH]

trencherman /ˈtrentʃəmən/ n. (pl. **-men**) a person who eats well, or in a specified manner (*a good trencherman*).

trend /trend/ n. & v. ● n. a general direction and tendency (esp. of events, fashion, or opinion etc.). ● v.intr. **1** bend or turn away in a specified direction. **2** be chiefly directed; have a general and continued tendency. □ **trend-setter** a person who leads the way in fashion etc. **trend-setting** establishing trends or fashions. [ME 'revolve' etc. f. OE *trendan* f. Gmc: cf. TRUNDLE]

trendy /ˈtrendɪ/ adj. & n. colloq. ● adj. (**trendier, trendiest**) often derog. fashionable; following fashionable trends. ● n. (pl. **-ies**) a trendy person. □ **trendily** adv. **trendiness** n.

Trengganu /treŋˈɡɑːnuː/ (also **Terengganu** /ˌtereŋ-/) a state of Malaysia, on the east coast of the Malay Peninsula; capital, Kuala Trengganu.

Trent /trent/ the chief river of central England, which rises in

Staffordshire and flows 275 km (170 miles) generally north-eastwards, uniting with the River Ouse 25 km (15 miles) west of Hull to form the Humber estuary.

Trent, Council of an ecumenical council of the Roman Catholic Church, held in three sessions between 1545 and 1563 in Trento, northern Italy. Prompted by the opposition of the Reformation, the council clarified and redefined the Church's position on certain issues (rejecting, for example, Luther's doctrine of justification by faith alone), abolished many of the abuses that had been prevalent, and strengthened the authority of the papacy; it also ordered a revision of the Vulgate. The overall effect was to provide the Church with a solid foundation for the Counter-Reformation.

trente-et-quarante /ˌtrɒnteɪkæˈrɒnt/ n. = rouge-et-noir. [F, = thirty and forty]

Trentino-Alto Adige /trenˌtiːnəʊ ˌæltəʊ ˈædɪˌdʒeɪ/ a region of NE Italy; capital, Bolzano. Situated on the border with Austria, it includes the Dolomites.

Trento /ˈtrentəʊ/ a city on the Adige river in northern Italy; pop. (1990) 102,120. A medieval ecclesiastic principality, it was the scene of the Council of Trent in 1545–63.

Trenton /ˈtrentən/ the state capital of New Jersey; pop. (1990) 88,675.

trepan /trɪˈpæn/ n. & v. ● n. **1** a cylindrical saw formerly used by surgeons for removing part of the bone of the skull. **2** a borer for sinking shafts. ● v.tr. (**trepanned**, **trepanning**) perforate (the skull) with a trepan. □ **trepanning** n. **trepanation** /ˌtrepəˈneɪʃ(ə)n/ n. [ME f. med.L trepanum f. Gk trupanon f. trupaō bore f. trupē hole]

trepang /trɪˈpæŋ/ n. = BÊCHE-DE-MER 1. [Malay trīpang]

trephine /trɪˈfaɪn, -ˈfiːn/ n. & v. ● n. an improved form of trepan with a guiding centre-pin. ● v.tr. operate on with this. □ **trephination** /ˌtrefɪˈneɪʃ(ə)n/ n. [orig. trafine, f. L tres fines three ends, app. formed after TREPAN]

trepidation /ˌtrepɪˈdeɪʃ(ə)n/ n. **1** a feeling of fear or alarm; perturbation of the mind. **2** tremulous agitation. **3** the trembling of limbs, e.g. in paralysis. [L trepidatio f. trepidare be agitated, tremble, f. trepidus alarmed]

trespass /ˈtrespəs/ v. & n. ● v.intr. **1** (usu. foll. by on, upon) make an unlawful or unwarrantable intrusion (esp. on land or property). **2** (foll. by on) make unwarrantable claims (shall not trespass on your hospitality). **3** (foll. by against) literary or archaic offend. ● n. **1** Law a voluntary wrongful act against the person or property of another, esp. unlawful entry to a person's land or property. **2** archaic a sin or offence. □ **trespass on a person's preserves** meddle in another person's affairs. □ **trespasser** n. [ME f. OF trespasser pass over, trespass, trespas (n.), f. med.L transpassare (as TRANS-, PASS¹)]

tress /tres/ n. & v. ● n. **1** a long lock of human (esp. female) hair. **2** (in pl.) a woman's or girl's head of hair. ● v.tr. arrange (hair) in tresses. □ **tressed** adj. (also in comb.). **tressy** adj. [ME f. OF tresse, perh. ult. f. Gk trikha threefold]

tressure /ˈtreʃə(r)/ n. Heraldry a narrow orle. [ME, orig. = hair-ribbon, f. OF tressour etc. (as TRESS)]

trestle /ˈtres(ə)l/ n. **1** a supporting structure for a table etc., consisting of two frames fixed at an angle or hinged or of a bar supported by two divergent pairs of legs. **2** (in full **trestle-table**) a table consisting of a board or boards laid on trestles or other supports. **3** (in full **trestle-work**) an open braced framework to support a bridge etc. **4** (in full **trestle-tree**) Naut. each of a pair of horizontal pieces on a lower mast supporting the topmast etc. [ME f. OF trestel ult. f. L transtrum]

tret /tret/ n. hist. an allowance of extra weight formerly made to purchasers of some goods for waste in transportation. [ME f. AF & OF, var. of trait draught: see TRAIT]

Tretyakov Gallery /ˈtretjəˌkɒf/ an art gallery in Moscow, one of the largest in the world. It houses exhibits ranging from early Russian art to contemporary work, and has a huge collection of icons.

trevally /trɪˈvælɪ/ n. (pl. **-ies**) an Australian fish of the genus Caranx, used as food. [prob. alt. f. cavally, a kind of fish, f. Sp. caballo horse f. L (as CAVALRY)]

Trèves /trɛv/ the French name for TRIER.

Trevithick /trəˈvɪθɪk/, Richard (1771–1833), English engineer. Known as 'the Cornish Giant', he was the most notable engineer from the Cornish mining industry, where steam engines were first widely used. His chief contribution was in the use of high-pressure steam to drive a double-acting engine, which could then be both compact and portable. Trevithick built many stationary engines and a few that were self-propelled — including the world's first railway locomotive (1804),

designed for an ironworks in South Wales. He also applied the principle in other enterprises, though his attempt to introduce steam power to silver mines in Peru proved financially disastrous.

trews /truːz/ n.pl. esp. Brit. trousers, esp. close-fitting tartan trousers worn by women. [Ir. trius, Gael. triubhas (sing.): cf. TROUSERS]

trey /treɪ/ n. (pl. **treys**) the three on dice or cards. [ME f. OF trei, treis three f. L tres]

TRH abbr. Their Royal Highnesses.

tri- /traɪ/ comb. form forming nouns and adjectives meaning: **1** three or three times. **2** Chem. (forming the names of compounds) containing three atoms or groups of a specified kind (triacetate). [L & Gk f. L tres, Gk treis three]

triable /ˈtraɪəb(ə)l/ adj. **1** liable to a judicial trial. **2** that may be tried or attempted. [ME f. AF (as TRY)]

triacetate /traɪˈæsɪˌteɪt/ n. Chem. a cellulose derivative containing three acetate groups, esp. as a base for man-made fibres.

triad /ˈtraɪæd/ n. **1** a group of three (esp. notes in a chord). **2** the number three. **3** (usu. **Triad**) a Chinese secret society. (See note below.) **4** a Welsh form of literary composition with an arrangement in groups of three. □ **triadic** /traɪˈædɪk/ adj. **triadically** adv. [F triade or LL trias triad- f. Gk trias -ados f. treis three]

▪ The Triads were apparently formed in China c.1730 with the purpose of ousting the Manchu dynasty, although their origins may be far older. The name was used for various fraternal and criminal organizations sharing a similar ritual, in which a triangle was of importance in initiation ceremonies. In the mid-19th century the Triads grew in strength, acquiring a large membership in southern China and among Chinese communities in various foreign countries. Thereafter they played an erratic role in China: some societies assisted the Kuomintang leader Sun Yat-sen, while others acquired political influence in Shanghai and other cities. Today the Triads often flourish among overseas Chinese, allegedly being involved in gambling, prostitution, drugs, and other organized crime.

triadelphous /ˌtraɪəˈdelfəs/ adj. Bot. having stamens united in three bundles. [TRI- + Gk adelphos brother]

triage /ˈtraɪdʒ/ n. **1** the act of sorting according to quality. **2** the assignment of degrees of urgency to decide the order of treatment of wounds, illnesses, etc. [F f. trier: cf. TRY]

trial /ˈtraɪəl/ n. & v. ● n. **1** a judicial examination and determination of issues between parties by a judge with or without a jury (stood trial for murder). **2 a** a process or mode of testing qualities. **b** experimental treatment. **c** a test (will give you a trial). **d** an attempt. **e** (attrib.) experimental. **3** a trying thing or experience or person, esp. hardship or trouble (the trials of old age). **4** a sports match to test the ability of players eligible for selection to a team. **5** a test of individual ability on a motorcycle over rough ground or on a road. **6** any of various contests involving performance by horses, dogs, or other animals. ● v.tr. & intr. (**trialled**, **trialling**; US **trialed**, **trialing**) subject to or undergo a test to assess performance. □ **on trial 1** being tried in a court of law. **2** being tested; to be chosen or retained only if suitable. **trial and error** repeated (usu. varied and unsystematic) attempts or experiments continued until successful. **trial balance** (in double-entry bookkeeping) a comparison of the totals on either side of the ledger, the inequality of which reveals errors in posting. **trial jury** = petty jury. **trial run** a preliminary test of a vehicle, vessel, machine, etc. [AF trial, triel f. trier TRY]

trialist /ˈtraɪəlɪst/ n. **1** a person who takes part in a sports trial, motorcycle trial, etc. **2** a person involved in a judicial trial.

triandrous /traɪˈændrəs/ adj. Bot. having three stamens.

triangle /ˈtraɪˌæŋg(ə)l/ n. **1** a plane figure with three sides and angles. **2** any three things not in a straight line, with imaginary lines joining them. **3** an implement of this shape. **4** a musical instrument consisting of a steel rod bent into a triangle and sounded by striking it with a small steel rod. **5** a situation, esp. an emotional relationship, involving three people. **6** a right-angled triangle of wood etc. as a drawing-implement. **7** Naut. a device of three spars for raising weights. **8** hist. a frame of three halberds joined at the top to which a soldier was bound for flogging. □ **triangle of forces** a triangle whose sides represent in magnitude and direction three forces in equilibrium. [ME f. OF triangle or L triangulum neut. of triangulus three-cornered (as TRI-, ANGLE¹)]

triangular /traɪˈæŋgjʊlə(r)/ adj. **1** triangle-shaped, three-cornered. **2** (of a contest or treaty etc.) between three persons or parties. **3** (of a pyramid) having a three-sided base. □ **triangularly** *adv.* **triangularity** /-ˌæŋgjʊˈlærɪtɪ/ *n.* [LL *triangularis* (as TRIANGLE)]

triangulate *v. & adj.* ● *v.tr.* /traɪˈæŋgjʊˌleɪt/ **1** divide (an area) into triangles for surveying purposes. **2 a** measure and map (an area) by the use of triangles with a known base length and base angles. **b** determine (a height, distance, etc.) in this way. ● *adj.* /traɪˈæŋgjʊlət/ *Zool.* marked with triangles. □ **triangulately** /-lətlɪ/ *adv.* **triangulation** /-ˌæŋgjʊˈleɪʃ(ə)n/ *n.* [L *triangulatus* triangular (as TRIANGLE)]

Trianon /ˈtriːəˌnɒn/ either of two small palaces in the great park at Versailles in France. The larger (*Grand Trianon*) was built by Louis XIV in 1687; the smaller (*Petit Trianon*), built by Louis XV 1762–8, was used first by his mistress Madame du Barry (1743–93) and afterwards by Marie Antoinette.

Triassic /traɪˈæsɪk/ *adj. & n. Geol.* ● *adj.* of or relating to the earliest period of the Mesozoic era. (*See note below.*) ● *n.* this period or the corresponding geological system. [LL *trias* (as TRIAD), because the strata are divisible into three groups]

   ▪ The Triassic lasted from about 248 to 213 million years ago, between the Permian and Jurassic periods. Dinosaurs became numerous during this period, which also saw the first mammals.

triathlon /traɪˈæθlɒn/ *n.* an athletic contest consisting of three different events, usually swimming, cycling, and long-distance running. The triathlon was instituted in the US in 1974, and the first world championships were held in 1989. Participants in international competitions contest a 1.5 km swimming race, a cycle ride of 40 km, and a 10-km run, although there is also an even more demanding form of triathlon known as the *iron man* (see IRON). □ **triathlete** /-liːt/ *n.* [TRI- after DECATHLON]

triatomic /ˌtraɪəˈtɒmɪk/ *adj. Chem.* **1** having three atoms (of a specified kind) in the molecule. **2** having three replacement atoms or radicals.

triaxial /traɪˈæksɪəl/ *adj.* having three axes.

tribade /ˈtrɪbaːd/ *n.* a lesbian. □ **tribadism** *n.* [F *tribade* or L *tribas* f. Gk f. *tribō* rub]

tribal /ˈtraɪb(ə)l/ *adj.* of, relating to, or characteristic of a tribe or tribes. □ **tribally** *adv.*

tribalism /ˈtraɪbəˌlɪz(ə)m/ *n.* **1** tribal organization. **2** loyalty to one's own tribe or social group. □ **tribalist** *n.* **tribalistic** /ˌtraɪbəˈlɪstɪk/ *adj.*

tribasic /traɪˈbeɪsɪk/ *adj. Chem.* (of an acid) having three replaceable hydrogen atoms.

tribe /traɪb/ *n.* **1** a group of (esp. primitive) families or communities, linked by social, economic, religious, or blood ties, and usu. having a common culture and dialect and a recognized leader. **2** any similar natural or political division. **3** *Rom. Hist.* each of the political divisions of the Roman people. **4** each of the twelve divisions of the Israelites, each traditionally descended from one of the patriarchs. **5** usu. *derog.* a set or number of persons esp. of one profession etc. or family (*the whole tribe of actors*). **6** *Biol.* a group of organisms usu. ranking between genus and the subfamily. **7** (in *pl.*) large numbers. [ME, orig. in pl. form *tribuz, tribus* f. OF or L *tribus* (sing. & pl.)]

tribesman /ˈtraɪbzmən/ *n.* (*pl.* **-men**; *fem.* **tribeswoman**, *pl.* **-women**) a member of a tribe or of one's own tribe.

Tribes of Israel (also called the *Twelve Tribes of Israel*) the twelve divisions of ancient Israel, each traditionally descended from one of the twelve sons of Jacob. Ten of the tribes (Asher, Dan, Gad, Issachar, Levi, Manasseh, Naphtali, Reuben, Simeon, and Zebulun, known as the *Lost Tribes*) were deported to captivity in Assyria *c.*720 BC, leaving only the tribes of Judah and Benjamin. (See also LOST TRIBES.)

tribespeople /ˈtraɪbzˌpiːp(ə)l/ *n.pl.* the members of a tribe.

triblet /ˈtrɪblɪt/ *n.* a mandrel used in making tubes, rings, etc. [F *triboulet*, of unkn. orig.]

tribo- /ˈtrɪbəʊ, ˈtraɪbəʊ/ *comb. form* rubbing, friction. [Gk *tribos* rubbing]

triboelectricity /ˌtrɪbəʊˌɪlekˈtrɪsɪtɪ, ˌtraɪb-, -ˌelekˈtrɪsɪtɪ/ *n.* the generation of an electric charge by friction.

tribology /traɪˈbɒlədʒɪ/ *n.* the study of friction, wear, lubrication, and the design of bearings; the science of interacting surfaces in relative motion. □ **tribologist** *n.*

triboluminescence /ˌtrɪbəʊˌluːmɪˈnes(ə)ns, ˌtraɪb-/ *n.* the emission of light from a substance when rubbed, scratched, etc. □ **triboluminescent** *adj.*

tribometer /traɪˈbɒmɪtə(r)/ *n.* an instrument for measuring friction in sliding.

tribrach /ˈtraɪbræk, ˈtrɪb-/ *n. Prosody* a foot of three short or unstressed

syllables. □ **tribrachic** /traɪˈbrækɪk, trɪ-/ *adj.* [L *tribrachys* f. Gk *tribrakhus* (as TRI-, *brakhus* short)]

tribulation /ˌtrɪbjʊˈleɪʃ(ə)n/ *n.* **1** great affliction or oppression. **2** a cause of this (*was a real tribulation to me*). [ME f. OF f. eccl.L *tribulatio -onis* f. L *tribulare* press, oppress, f. *tribulum* sledge for threshing, f. *terere trit-* rub]

tribunal /traɪˈbjuːn(ə)l, trɪ-/ *n.* **1** *Brit.* a board appointed to adjudicate in some matter, esp. one appointed by the government to investigate a matter of public concern. **2** a court of justice. (See also REVOLUTIONARY TRIBUNAL.) **3** a seat or bench for a judge or judges. **4 a** a place of judgement. **b** judicial authority (*the tribunal of public opinion*). [F *tribunal* or L *tribunus* (as TRIBUNE²)]

tribune¹ /ˈtrɪbjuːn/ *n.* **1** a popular leader or demagogue. **2** *Rom. Hist.* **a** (in full **tribune of the people**) an official in ancient Rome chosen by the people to protect their interests. **b** (in full **military tribune**) a Roman legionary officer. □ **tribunate** /-nət/ *n.* **tribuneship** *n.* [ME f. L *tribunus*, prob. f. *tribus* tribe]

tribune² /ˈtrɪbjuːn/ *n.* **1 a** a bishop's throne in a basilica. **b** an apse containing this. **2** a dais or rostrum. **3** a raised area with seats. [F f. It. f. med.L *tribuna* TRIBUNAL]

Tribune Group a left-wing group within the British Labour Party consisting of supporters of the views put forward in the weekly journal *Tribune*.

tributary /ˈtrɪbjʊtərɪ/ *n. & adj.* ● *n.* (*pl.* **-ies**) **1** a river or stream flowing into a larger river or lake. **2** *hist.* a person or state paying or subject to tribute. ● *adj.* **1** (of a river etc.) that is a tributary. **2** *hist.* **a** paying tribute. **b** serving as tribute. □ **tributarily** *adv.* **tributariness** *n.* [ME f. L *tributarius* (as TRIBUTE)]

tribute /ˈtrɪbjuːt/ *n.* **1** a thing said or done or given as a mark of respect or affection etc. (*paid tribute to their achievements; floral tributes*). **2** *hist.* **a** a payment made periodically by one state or ruler to another, esp. as a sign of dependence. **b** an obligation to pay this (*was laid under tribute*). **3** (foll. by *to*) an indication of (some praiseworthy quality) (*their success is a tribute to their perseverance*). **4** a proportion of ore or its equivalent paid to a miner for his work, or to the owner of a mine. [ME f. L *tributum* neut. past part. of *tribuere tribut-* assign, orig. divide between tribes (*tribus*)]

tricar /ˈtraɪkɑː(r)/ *n. Brit.* a three-wheeled motor car.

trice /traɪs/ *n.* □ **in a trice** in a moment; instantly. [ME *trice* (v.) pull, haul f. MDu. *trīsen*, MLG *trīssen*, rel. to MDu. *trīse* windlass, pulley]

tricentenary /ˌtraɪsenˈtiːnərɪ/ *n.* (*pl.* **-ies**) = TERCENTENARY.

triceps /ˈtraɪseps/ *adj. & n.* ● *adj.* (of a muscle) having three heads or points of attachment. ● *n.* any triceps muscle, esp. the large muscle at the back of the upper arm. [L, = three-headed (as TRI-, *-ceps* f. *caput* head)]

triceratops /traɪˈserəˌtɒps/ *n.* a large quadrupedal plant-eating dinosaur of the late Cretaceous genus *Triceratops*, with two large horns, a smaller one on the snout, and a bony frill above the neck. [mod.L f. Gk *trikeratos* three-horned + *ōps* face]

trichiasis /trɪˈkaɪəsɪs/ *n. Med.* ingrowth or introversion of the eyelashes. [LL f. Gk *trikhiasis* f. *trikhiaō* be hairy]

trichina /trɪˈkaɪnə/ *n.* (*pl.* **trichinae** /-niː/) a hairlike parasitic nematode worm of the genus *Trichinella* (formerly *Trichina*); esp. *T. spiralis*, the adults of which live in the small intestine, and whose larvae become encysted in the muscle tissue of humans and flesh-eating mammals causing trichinosis. □ **trichinous** *adj.* [mod.L f. Gk *trikhinos* of hair: see TRICHO-]

Trichinopoly see TIRUCHIRAPALLI.

trichinosis /ˌtrɪkɪˈnəʊsɪs/ *n. Med.* a disease caused by trichinae, usu. ingested in meat, and characterized by digestive disturbance, fever, and muscular rigidity.

tricho- /ˈtrɪkəʊ/ *comb. form* hair. [Gk *thrix trikhos* hair]

trichogenous /trɪˈkɒdʒɪnəs/ *adj.* causing or promoting the growth of hair.

trichology /trɪˈkɒlədʒɪ, traɪ-/ *n.* the study of the structure, functions, and diseases of the hair. □ **trichologist** *n.*

trichome /ˈtraɪkəʊm/ *n. Bot.* a hair, scale, prickle, or other outgrowth from the epidermis of a plant. [Gk *trikhōma* f. *trikhoō* cover with hair (as TRICHO-)]

trichomonad /ˌtrɪkəˈmɒnæd/ *n.* a flagellate protozoan of the order Trichomonadida, parasitic in humans, cattle, and fowls. [mod.L *Trichomonas* genus name, f. as TRICHO- + Gk *monas* MONAD]

trichomoniasis /ˌtrɪkəməˈnaɪəsɪs/ *n.* an infection caused by trichomonads parasitic in the urinary tract, vagina, or digestive system.

trichopathy /trɪˈkɒpəθɪ/ n. the treatment of diseases of the hair. □ **trichopathic** /ˌtrɪkəˈpæθɪk/ adj.

Trichoptera /traɪˈkɒptərə/ n.pl. Zool. an order of insects comprising the caddis-flies. □ **trichopteran** n. & adj. **trichopterous** adj. [mod.L f. TRICHO- + Gk pteron wing]

trichotomy /traɪˈkɒtəmɪ/ n. (pl. -**ies**) a division (esp. sharply defined) into three categories, esp. of human nature into body, soul, and spirit. □ **trichotomic** /ˌtraɪkəˈtɒmɪk/ adj. [Gk trikha threefold f. treis three, after DICHOTOMY]

trichroic /traɪˈkrəʊɪk/ adj. (esp. of a crystal viewed in different directions) showing three colours. □ **trichroism** /ˈtraɪkrəʊˌɪz(ə)m/ n. [Gk trikhroos (as TRI-, khrōs colour)]

trichromatic /ˌtraɪkrəˈmætɪk/ adj. **1** having or using three colours. **2** (of vision) having the normal three colour-sensations, i.e. red, green, and purple. □ **trichromatism** /traɪˈkrəʊməˌtɪz(ə)m/ n.

trick /trɪk/ n. & v. ● n. **1** an action or scheme undertaken to fool, outwit, or deceive. **2** an optical or other illusion (a trick of the light). **3** a special technique; a knack or special way of doing something. **4 a** a feat of skill or dexterity. **b** an unusual action (e.g. begging) learned by an animal. **5** a mischievous, foolish, or discreditable act; a practical joke (a mean trick to play). **6** a peculiar or characteristic habit or mannerism (has a trick of repeating himself). **7 a** the cards played in a single round of a card-game, usu. one from each player. **b** such a round. **c** a point gained as a result of this. **8** (attrib.) done to deceive or mystify or to create an illusion (trick photography; trick question). **9** Naut. a sailor's turn at the helm, usu. two hours. ● v.tr. **1** deceive by a trick; outwit. **2** (often foll. by out of, or into + verbal noun) cheat; treat deceitfully so as to deprive (were tricked into agreeing; were tricked out of their savings). **3** (of a thing) foil or baffle; take by surprise; disappoint the calculations of. □ **do the trick** colloq. accomplish one's purpose; achieve the required result. **how's tricks?** colloq. how are you? **not miss a trick** see MISS¹. **trick cyclist 1** a cyclist who performs tricks, esp. in a circus. **2** sl. a psychiatrist. **trick of the trade** a special usu. ingenious technique or method of achieving a result in an industry or profession etc. **trick or treat** esp. N. Amer. a children's custom of calling at houses at Hallowe'en with the threat of pranks if they are not given a small gift. **trick out** (or **up**) dress, decorate, or deck out esp. showily. **up to one's tricks** colloq. misbehaving. **up to a person's tricks** colloq. aware of what a person is likely to do by way of mischief. □ **tricker** n. **trickish** adj. **trickless** adj. [ME f. OF dial. trique, OF triche f. trichier deceive, of unkn. orig.]

trickery /ˈtrɪkərɪ/ n. (pl. -**ies**) **1** the practice or an instance of deception. **2** the use of tricks.

trickle /ˈtrɪk(ə)l/ v. & n. ● v. **1** intr. & tr. flow or cause to flow in drops or a small stream (water trickled through the crack). **2** intr. come or go slowly or gradually (information trickles out). ● n. a trickling flow. □ **trickle charger** an electrical charger for batteries that works at a steady slow rate from the mains. [ME trekel, trikle, prob. imit.]

trickster /ˈtrɪkstə(r)/ n. a deceiver or rogue.

tricksy /ˈtrɪksɪ/ adj. (**tricksier**, **tricksiest**) full of tricks; playful. □ **tricksily** adv. **tricksiness** n. [TRICK: for -sy cf. FLIMSY, TIPSY]

tricky /ˈtrɪkɪ/ adj. (**trickier**, **trickiest**) **1** difficult or intricate; requiring care and adroitness (a tricky job). **2** crafty or deceitful. **3** resourceful or adroit. □ **trickily** adv. **trickiness** n.

triclinic /traɪˈklɪnɪk/ adj. **1** (of a mineral) having three unequal oblique axes. **2** denoting the system classifying triclinic crystalline substances. [Gk TRI- + klinō incline]

triclinium /traɪˈklɪnɪəm, trɪ-/ n. (pl. **triclinia** /-nɪə/) Rom. Antiq. **1** a dining-table with couches along three sides. **2** a room containing this. [L f. Gk triklinion (as TRI-, klinē couch)]

tricolour /ˈtrɪkələ(r), ˈtraɪˌkʌlə(r)/ n. & adj. (US **tricolor**) ● n. a flag of three colours, esp. the French national flag of blue, white, and red. ● adj. (also **tricoloured** /ˈtraɪˌkʌləd/) having three colours. [F tricolore f. LL tricolor (as TRI-, COLOUR)]

tricorn /ˈtraɪkɔːn/ adj. & n. (also **tricorne**) ● adj. **1** having three horns. **2** (of a hat) having a brim turned up on three sides. ● n. **1** an imaginary animal with three horns. **2** a tricorn hat. [F tricorne or L tricornis (as TRI-, cornu horn)]

tricot /ˈtrɪkəʊ, ˈtriːkəʊ/ n. **1 a** a hand-knitted woollen fabric. **b** an imitation of this. **2** a ribbed woollen cloth. [F, = knitting f. tricoter knit, of unkn. orig.]

tricrotic /traɪˈkrɒtɪk/ adj. (of the pulse) having a triple beat. [TRI- after DICROTIC]

tricuspid /traɪˈkʌspɪd/ n. & adj. ● n. **1** a tooth with three cusps or points. **2** a heart-valve formed of three triangular segments. ● adj. (of a tooth) having three cusps or points.

tricycle /ˈtraɪsɪk(ə)l/ n. & v. ● n. **1** a vehicle having three wheels, two on an axle at the back and one at the front, driven by pedals in the same way as a bicycle. **2** a three-wheeled motor vehicle for a disabled driver. ● v.intr. ride on a tricycle. □ **tricyclist** n.

tridactyl /traɪˈdæktɪl/ adj. Zool. having three digits on each foot.

trident /ˈtraɪd(ə)nt/ n. **1** a three-pronged spear, esp. as an attribute of Poseidon (Neptune) or Britannia. **2** (**Trident**) a US type of submarine-launched long-range ballistic missile, designed to carry nuclear warheads and in service with the Royal Navy from 1993. [L tridens trident- (as TRI-, dens tooth)]

tridentate /traɪˈdenteɪt/ adj. having three teeth or prongs. [TRI- + L dentatus toothed]

Tridentine /traɪˈdentaɪn, trɪ-/ adj. & n. ● adj. of or relating to the Council of Trent, esp. as the basis of Roman Catholic doctrine. ● n. a Roman Catholic adhering to this traditional doctrine. □ **Tridentine mass** the Eucharistic liturgy used by the Roman Catholic Church from 1570 to 1964. [med.L Tridentinus f. Tridentum Trent]

triduum /ˈtrɪdjʊəm/ n. RC Ch. esp. hist. a period of three days' prayer in preparation for a saint's day or other religious occasion. [L (as TRI-, dies day)]

tridymite /ˈtrɪdɪˌmaɪt/ n. Mineral. a crystallized form of silica, occurring in cavities of volcanic rocks. [G Tridymit f. Gk tridumos threefold (as TRI-, didumos twin), from its occurrence in groups of three crystals]

tried past and past part. of TRY.

triennial /traɪˈenɪəl/ adj. & n. ● adj. **1** lasting three years. **2** recurring every three years. ● n. a visitation of an Anglican diocese by its bishop every three years. □ **triennially** adv. [LL triennis (as TRI-, L annus year)]

triennium /traɪˈenɪəm/ n. (pl. **trienniums** or **triennia** /-nɪə/) a period of three years. [L (as TRIENNIAL)]

Trier /trɪə(r)/ (called in French Trèves) a city on the River Mosel in Rhineland-Palatinate, western Germany; pop. (1991) 98,750. Established by a Germanic tribe, the Treveri, c.400 BC, Trier is one of the oldest cities in Europe. The Roman town of Augusta Treverorum was founded in 15 BC by the Emperor Augustus, and was an important town of Roman Gaul. Trier was a powerful archbishopric from 815 until the 18th century, but fell into decline after the French occupation in 1797.

trier /ˈtraɪə(r)/ n. **1** a person who perseveres (is a real trier). **2** a tester, esp. of foodstuffs. **3** a person appointed to decide whether a challenge to a juror is well founded.

Trieste /trɪˈest/ a city in NE Italy, the largest port on the Adriatic and capital of Friuli-Venezia Giulia region; pop. (1990) 231,000. Formerly held by Austria (1382–1918), Trieste was annexed by Italy after the First World War. The Free Territory of Trieste was created after the Second World War but returned to Italy in 1954.

trifacial nerve /traɪˈfeɪʃ(ə)l/ n. = TRIGEMINAL NERVE.

trifecta /traɪˈfektə/ n. N. Amer., Austral., & NZ a form of betting in which the first three places in a race must be predicted in the correct order. [TRI- + PERFECTA]

trifid /ˈtraɪfɪd/ adj. esp. Biol. partly or wholly split into three divisions or lobes. [L trifidus (as TRI-, findere fid- split)]

trifle /ˈtraɪf(ə)l/ n. & v. ● n. **1** a thing of slight value or importance. **2 a** a small amount esp. of money (was sold for a trifle). **b** (prec. by a) somewhat (seems a trifle annoyed). **3** Brit. a dessert of sponge cake (often flavoured with sherry) with custard, jelly, fruit, cream, etc. ● v. **1** intr. talk or act frivolously. **2** intr. (foll. by with) **a** treat or deal with frivolously or derisively; flirt heartlessly with. **b** refuse to take seriously. **3** tr. (foll. by away) waste (time, energies, money, etc.) frivolously. □ **trifler** n. [ME f. OF truf(f)le by-form of trufe deceit, of unkn. orig.]

trifling /ˈtraɪflɪŋ/ adj. **1** unimportant, petty. **2** frivolous. □ **triflingly** adv.

trifocal /traɪˈfəʊk(ə)l/ adj. & n. ● adj. having three focuses, esp. of a lens with different focal lengths. ● n. (in pl.) trifocal spectacles.

trifoliate /traɪˈfəʊlɪət/ adj. Bot. **1** (of a compound leaf) having three leaflets. **2** (of a plant) having such leaves.

triforium /traɪˈfɔːrɪəm/ n. (pl. **triforia** /-rɪə/) a gallery or arcade above the arches of the nave, choir, and transepts of a church. [AL, of unkn. orig.]

triform /ˈtraɪfɔːm/ adj. (also **triformed**) **1** formed of three parts. **2** having three forms or bodies.

trifurcate v. & adj. ● v.tr. & intr. /ˈtraɪfəˌkeɪt/ divide into three branches. ● adj. /ˈtraɪfəkət/ divided into three branches.

trig[1] /trɪg/ n. colloq. trigonometry. [abbr.]

trig[2] /trɪg/ adj. & v. archaic or dial. ● adj. trim or spruce. ● v.tr. (**trigged**, **trigging**) make trim; smarten. [ME, = trusty, f. ON tryggr, rel. to TRUE]

trigamous /ˈtrɪgəməs/ adj. **1 a** three times married. **b** having three wives or husbands at once. **2** Bot. having male, female, and hermaphrodite flowers in the same head. □ **trigamist** n. **trigamy** n. [Gk trigamos (as TRI-, gamos marriage)]

trigeminal nerve /traɪˈdʒemɪn(ə)l/ n. Anat. the largest cranial nerve, which divides into the ophthalmic, maxillary, and mandibular nerves. □ **trigeminal neuralgia** Med. neuralgia involving one or more of these branches, and often causing severe pain. [as TRIGEMINUS]

trigeminus /traɪˈdʒemɪnəs/ n. (pl. **trigemini** /-ˌnaɪ/) Anat. the trigeminal nerve. [L, = born as a triplet (as TRI-, geminus born at the same birth)]

trigger /ˈtrɪgə(r)/ n. & v. ● n. **1** a movable device for releasing a spring or catch and so setting off a mechanism (esp. that of a gun). **2** an event, occurrence, etc., that sets off a chain reaction. ● v.tr. **1** (often foll. by off) set (an action or process) in motion; initiate, precipitate. **2** fire (a gun) by the use of a trigger. □ **quick on the trigger** quick to respond. **trigger fish** a usu. tropical marine fish of the family Balistidae, in which the first spine of the dorsal fin can only be depressed by releasing the second. **trigger-happy** apt to shoot without or with slight provocation. □ **triggered** adj. [17th-c. tricker f. Du. trekker f. trekken pull: cf. TREK]

Triglav /ˈtriːglæf/ a mountain in the Julian Alps, NW Slovenia, near the Italian border. Rising to 2,863 m (9,392 ft), it is the highest peak in the mountains east of the Adriatic.

triglyceride /traɪˈglɪsəraɪd/ n. Chem. any ester formed from glycerol and three acid radicals, including the main constituents of fats and oils.

triglyph /ˈtraɪglɪf/ n. Archit. each of a series of tablets with three vertical grooves, alternating with metopes in a Doric frieze. □ **triglyphic** /traɪˈglɪfɪk/ adj. **triglyphical** adj. [L triglyphus f. Gk trigluphos (as TRI-, gluphē carving)]

trigon /ˈtraɪgɒn/ n. **1** a triangle. **2** an ancient triangular lyre or harp. **3** the cutting region of an upper molar tooth. [L trigonum f. Gk trigōnon neuter of trigōnos three-cornered (as TRI-, -GON)]

trigonal /ˈtrɪgən(ə)l/ adj. **1** triangular; of or relating to a triangle. **2** Biol. triangular in cross-section. **3** (of a crystal etc.) having an axis with threefold symmetry. □ **trigonally** adv. [med.L trigonalis (as TRIGON)]

trigonometry /ˌtrɪgəˈnɒmɪtri/ n. the branch of mathematics dealing with the relations of the sides and angles of triangles and with the relevant functions of any angles. □ **trigonometric** /-nəˈmetrɪk/ adj. **trigonometrical** adj. [mod.L trigonometria (as TRIGON, -METRY)]

trigraph /ˈtraɪgrɑːf/ n. (also **trigram** /-græm/) **1** a group of three letters representing one sound. **2** a figure of three lines.

trigynous /ˈtrɪdʒɪnəs/ adj. Bot. having three pistils.

trihedral /traɪˈhiːdrəl/ adj. having three surfaces.

trihedron /traɪˈhiːdrən/ n. a figure of three intersecting planes.

trihydric /traɪˈhaɪdrɪk/ adj. Chem. containing three hydroxyl groups.

trike /traɪk/ n. & v.intr. colloq. a tricycle. [abbr.]

trilabiate /traɪˈleɪbɪət/ adj. Bot. & Zool. three-lipped.

trilateral /traɪˈlætərəl/ adj. & n. ● adj. **1** of, on, or with three sides. **2** shared by or involving three parties, countries, etc. (trilateral negotiations). ● n. a figure having three sides.

trilby /ˈtrɪlbɪ/ n. (pl. **-ies**) Brit. a soft felt hat with a narrow brim and indented crown. □ **trilbied** adj. [name of the heroine in George du Maurier's novel Trilby (1894), in the stage version of which such a hat was worn]

trilinear /traɪˈlɪnɪə(r)/ adj. of or having three lines.

trilingual /traɪˈlɪŋgwəl/ adj. **1** able to speak three languages, esp. fluently. **2** spoken or written in three languages. □ **trilingualism** n.

triliteral /traɪˈlɪtərəl/ adj. **1** of three letters. **2** (of a Semitic language) having (most) roots with three consonants.

trilith /ˈtraɪlɪθ/ n. (also **trilithon** /-lɪθən/) a megalithic structure consisting of three stones, esp. of two uprights and a lintel. □ **trilithic** /traɪˈlɪθɪk/ adj. [Gk trilithon (as TRI-, lithos stone)]

trill /trɪl/ n. & v. ● n. **1** a quavering or vibratory sound, esp. a rapid alternation of sung or played notes. **2** a bird's warbling sound. **3** the pronunciation of r with a vibration of the tongue. ● v. **1** intr. produce a trill. **2** tr. warble (a song) or pronounce (r etc.) with a trill. [It. trillo (n.), trillare (v.)]

trillion /ˈtrɪljən/ n. (pl. same or (in sense 3) **trillions**) **1** a million million (1,000,000,000,000 or 10¹²). **2** (formerly, esp. Brit.) a million million million (1,000,000,000,000,000,000 or 10¹⁸). **3** (in pl.) colloq. a very large number (trillions of times). ¶ Senses 1-2 correspond to the change in sense of billion. □ **trillionth** adj. & n. [F trillion or It. trilione (as TRI-, MILLION), after billion]

trillium /ˈtrɪlɪəm/ n. a liliaceous plant of the North American and Asian genus Trillium, with a solitary three-petalled flower above a whorl of three leaves. [mod.L, app. alt. of Sw. trilling triplet, f. the parts of the plant being in threes]

trilobite /ˈtraɪləbaɪt/ n. a fossil marine arthropod of the subphylum Trilobita, characterized by a three-lobed body and numerous legs, and found commonly in Palaeozoic rocks. [mod.L Trilobites former name (as TRI-, Gk lobos lobe)]

trilogy /ˈtrɪlədʒɪ/ n. (pl. **-ies**) **1** a group of three related literary or operatic works. **2** Gk Antiq. a set of three tragedies performed as a group. [Gk trilogia (as TRI-, -LOGY)]

Trim /trɪm/ a town in Meath, in the Republic of Ireland, situated to the north-west of Dublin; pop. (1991) 18,120.

trim /trɪm/ v., n., & adj. ● v. (**trimmed**, **trimming**) **1** tr. make neat or of the required size or form, esp. by cutting away irregular or unwanted parts. **2** tr. (foll. by off, away) remove by cutting off (such parts). **3** tr. **a** (often foll. by up) make (a person) neat in dress and appearance. **b** ornament or decorate (esp. clothing, a hat, etc., by adding ribbons, lace, etc.). **4** tr. adjust the balance of (a ship or aircraft) by the arrangement of its cargo etc. **5** tr. arrange (sails) to suit the wind. **6** intr. **a** associate oneself with currently prevailing views, esp. to advance oneself. **b** hold a middle course in politics or opinion. **7** tr. set in good order, prepare. **8** tr. colloq. **a** rebuke sharply. **b** thrash. **c** get the better of in a bargain etc. ● n. **1** the state or degree of readiness or fitness (found everything in perfect trim). **2** ornament or decorative material. **3** dress or equipment. **4** the act of trimming a person's hair. **5** the inclination of an aircraft to the horizontal. ● adj. (**trimmer**, **trimmest**) **1** neat, slim, or spruce. **2** in good condition or order; well arranged or equipped. □ **in trim 1** looking smart, healthy, etc. **2** Naut. in good order. □ **trimly** adv. **trimness** n. [perh. f. OE trymman, trymian make firm, arrange: but there is no connecting evidence between OE and 1500]

trimaran /ˈtraɪməˌræn/ n. a vessel like a catamaran, with three hulls side by side. [TRI- + CATAMARAN]

trimer /ˈtraɪmə(r)/ n. Chem. a polymer comprising three monomer units. □ **trimeric** /traɪˈmerɪk/ adj. [TRI- + -MER]

trimerous /ˈtraɪmərəs, ˈtrɪm-/ adj. having three parts.

trimester /traɪˈmestə(r)/ n. a period of three months, esp. of human gestation or N. Amer. as a university term. □ **trimestral** adj. **trimestrial** adj. [F trimestre f. L trimestris (as TRI-, -mestris f. mensis month)]

trimeter /ˈtrɪmɪtə(r)/ n. Prosody a verse of three measures. □ **trimetric** /traɪˈmetrɪk/ adj. **trimetrical** adj. [L trimetrus f. Gk trimetros (as TRI-, metron measure)]

trimmer /ˈtrɪmə(r)/ n. **1** a person who trims articles of dress. **2** a person who trims in politics etc.; a time-server. **3** an instrument for clipping etc. **4** Archit. a short piece of timber across an opening (e.g. for a hearth) to carry the ends of truncated joists. **5** a small capacitor etc. used to tune a radio set. **6** Austral. colloq. a striking or outstanding person or thing.

trimming /ˈtrɪmɪŋ/ n. **1** ornamentation or decoration, esp. for clothing. **2** (in pl.) colloq. the usual accompaniments, esp. of the main course of a meal. **3** (in pl.) pieces cut off in trimming.

Trimontium /traɪˈmɒntɪəm/ the Roman name for PLOVDIV.

trimorphism /traɪˈmɔːfɪz(ə)m/ n. Bot., Zool., & Crystallog. existence in three distinct forms. □ **trimorphic** adj. **trimorphous** adj.

Trimurti /trɪˈmʊətɪ/ Hinduism the triad formed by the gods Brahma, Vishnu, and Siva. [Skr. f. tri three + mūrti form]

Trincomalee /ˌtrɪŋkəməˈliː/ the principal port of Sri Lanka, on the east coast; pop. (1981) 44,300. One of the finest natural harbours in the world, Trincomalee was the chief British naval base in SE Asia during the Second World War after the fall of Singapore.

trine /traɪn/ adj. & n. ● adj. **1** threefold, triple; made up of three parts. **2** Astrol. denoting the aspect of two celestial bodies 120° (one-third of the zodiac) apart. ● *n. Astrol.* a trine aspect. □ **trinal** *adj.* [ME f. OF *trin trine* f. L *trinus* threefold f. *tres* three]

Trinidad and Tobago /ˈtrɪnɪˌdæd, təˈbeɪgəʊ/ a country in the West Indies consisting of two islands off the NE coast of Venezuela; pop. (est. 1991) 1,249,000; languages, English (official), Creoles; capital, Port-of-Spain (on Trinidad). Much the larger of the two islands is Trinidad, with Tobago to the north-east. Trinidad was visited by Columbus in 1498, the Spaniards later killing or enslaving the native Arawak population. Tobago, occupied by Caribs, was eventually colonized by the French and later the British in the 18th century. Trinidad became British during the Napoleonic Wars and was formally amalgamated with Tobago as a Crown Colony in 1888. After a short period as a member of the West Indies Federation between 1958 and 1962, Trinidad and Tobago became an independent member state of the Commonwealth in 1962 and finally a republic in 1976. Trinidad has large oil and gas reserves and is a leading producer of asphalt. □ **Trinidadian** /ˌtrɪnɪˈdeɪdɪən, -ˈdædɪən/ *adj. & n.* **Tobagan** *adj. & n.* **Tobagonian** /ˌtəʊbəˈgəʊnɪən/ *adj. & n.*

Trinitarian /ˌtrɪnɪˈteərɪən/ *n. & adj.* ● *n.* a person who believes in the Christian doctrine of the Trinity. ● *adj.* of or relating to this belief. □ **Trinitarianism** *n.*

trinitrotoluene /traɪˌnaɪtrəˈtɒljuˌiːn/ *n.* (also **trinitrotoluol** /-ˌɒl/) = TNT.

trinity /ˈtrɪnɪtɪ/ *n.* (*pl.* **-ies**) **1** the state of being three. **2** a group of three. **3** (**the Trinity** or **Holy Trinity**) (in Christian theology) the three persons of the Christian Godhead (Father, Son, and Holy Spirit). □ **Trinity term** *Brit.* a term in some Universities and the law beginning after Easter. [ME f. OF *trinité* f. L *trinitas -tatis* triad (as TRINE)]

Trinity Brethren *n.pl.* the members of Trinity House.

Trinity House a corporation founded in 1514 which has official responsibility for the licensing of ships' pilots in the UK and the erection and maintenance of lighthouses, buoys, etc., around the coasts of England and Wales. In Scotland this latter function is discharged by the Commissioners of Northern Lighthouses.

Trinity Sunday the next Sunday after Whit Sunday, celebrated in honour of the Holy Trinity.

trinket /ˈtrɪŋkɪt/ *n.* a trifling ornament, jewel, etc., esp. one worn on the person. □ **trinketry** *n.* [16th c.: orig. unkn.]

trinomial /traɪˈnəʊmɪəl/ *adj. & n.* ● *adj.* consisting of three terms. ● *n.* a scientific name or algebraic expression of three terms. [TRI- after BINOMIAL]

trio /ˈtriːəʊ/ *n.* (*pl.* **-os**) **1** a set or group of three. **2** *Mus.* **a** a composition for three performers. **b** a group of three performers. **c** the central, usu. contrastive, section of a minuet, scherzo, or march. **3** (in piquet) three aces, kings, queens, or jacks in one hand. [F & It. f. L *tres* three, after *duo*]

triode /ˈtraɪəʊd/ *n. Electr.* **1** a thermionic valve having three electrodes. **2** a semiconductor rectifier having three connections. [TRI- + ELECTRODE]

trioecious /traɪˈiːʃəs/ *adj. Bot.* having male, female, and hermaphrodite organs each on separate plants. [TRI- + Gk *oikos* house]

triolet /ˈtriːəlɪt, ˈtraɪə-/ *n.* a poem of eight (usu. eight-syllabled) lines rhyming *abaaabab*, the first line recurring as the fourth and seventh and the second as the eighth. [F (as TRIO)]

trioxide /traɪˈɒksaɪd/ *n. Chem.* an oxide containing three oxygen atoms.

trip /trɪp/ *v. & n.* ● *v.intr. & tr.* (**tripped, tripping**) **1** *intr.* **a** a walk or dance with quick light steps. **b** (of a rhythm etc.) run lightly. **2 a** *intr. & tr.* (often foll. by *up*) stumble or cause to stumble, esp. by catching or entangling the feet. **b** *intr. & tr.* (foll. by *up*) make or cause to make a slip or blunder. **3** *tr.* detect (a person) in a blunder. **4** *intr.* make an excursion to a place. **5** *tr.* release (part of a machine) suddenly by knocking aside a catch etc. **6 a** release and raise (an anchor) from the bottom by means of a cable. **b** turn (a yard etc.) from a horizontal to a vertical position for lowering. **7** *intr. colloq.* have a hallucinatory experience caused by a drug, esp. LSD. ● *n.* **1** a journey or excursion, esp. for pleasure. **2 a** a stumble or blunder. **b** the act of tripping or the state of being tripped up. **3** a nimble step. **4** *colloq.* a hallucinatory experience caused by a drug, esp. LSD. **5** a contrivance for a tripping mechanism etc. □ **trip-hammer** a large tilt-hammer operated by tripping. **trip the light fantastic** *joc.* dance. **trip-wire** a wire stretched close to the ground, operating an alarm etc. when disturbed. [ME f. OF *triper, tripper,* f. MDu. *trippen* skip, hop]

tripartite /traɪˈpɑːtaɪt/ *adj.* **1** consisting of three parts. **2** shared by or involving three parties. **3** *Bot.* (of a leaf) divided into three segments

almost to the base. □ **tripartitely** *adv.* **tripartition** /ˌtraɪpɑːˈtɪʃ(ə)n/ *n.* [ME f. L *tripartitus* (as TRI-, *partitus* past part. of *partiri* divide)]

tripe /traɪp/ *n.* **1** the first or second stomach of a ruminant, esp. an ox, as food. **2** *colloq.* nonsense, rubbish (*don't talk such tripe*). [ME f. OF, of unkn. orig.]

triphibious /traɪˈfɪbɪəs/ *adj.* (of military operations) on land, on sea, and in the air. [irreg. f. TRI- after *amphibious*]

triphthong /ˈtrɪfθɒŋ, ˈtrɪpθɒŋ/ *n.* **1** a union of three vowels (letters or sounds) pronounced in one syllable (as in *fire*). **2** three vowel characters representing the sound of a single vowel (as in *beau*). □ **triphthongal** /trɪfˈθɒŋg(ə)l, trɪpˈθɒŋ-/ *adj.* [F *triphtongue* (as TRI-, DIPHTHONG)]

Tripitaka /trɪˈpɪtəkə/ *n.* the sacred canon of Theravada Buddhism, written in the Pali language. [Skr., = the three baskets or collections]

triplane /ˈtraɪpleɪn/ *n.* an early type of aeroplane having three sets of wings, one above the other.

triple /ˈtrɪp(ə)l/ *adj., n., & v.* ● *adj.* **1** consisting of three usu. equal parts or things; threefold. **2** involving three parties. **3** three times as much or many (*triple the amount; triple thickness*). ● *n.* **1** a threefold number or amount. **2** a set of three. **3** (in *pl.*) a peal of changes on seven bells. ● *v.tr. & intr.* multiply by three; increase threefold. □ **triple crown 1** *RC Ch.* the pope's tiara. **2** the act of winning all three of a group of important events in horse-racing, rugby, etc. **triple jump** an athletic exercise or contest comprising a hop, a step, and a jump. **triple play** *Baseball* the act of putting out three runners in a row. **triple point** *Physics* the temperature and pressure at which the solid, liquid, and vapour phases of a pure substance can coexist in equilibrium. **triple rhyme** a rhyme including three syllables. **triple time** *Mus.* a rhythm with three beats to the bar; waltz time. □ **triply** /-plɪ/ *adv.* [OF *triple* or L *triplus* f. Gk *triplous*]

Triple Alliance an alliance of three states etc. Notable Triple Alliances have been made in 1668, between England, the Netherlands, and Sweden against France; in 1717, between Britain, France, and the Netherlands against Spain; in 1865, between Argentina, Brazil, and Uruguay against Paraguay; in 1882, between Germany, Austria–Hungary, and Italy against France and Russia.

Triple Entente an early 20th-century alliance between Great Britain, France, and Russia. Originally a series of loose agreements, the Triple Entente began to assume the nature of a more formal alliance as the prospect of war with the Central Powers became more likely, and formed the basis of the Allied powers in the First World War. Britain's entente with France was signed in 1904, with Russia 1907.

triplet /ˈtrɪplɪt/ *n.* **1** each of three children or animals born at one birth. **2** a set of three things, esp. of equal notes played in the time of two or of verses rhyming together. □ **triplet code** *Biol.* the standard version of the genetic code, in which a sequence of three nucleotides on a DNA or RNA molecule codes for a specific amino acid in protein synthesis. [TRIPLE + -ET[1], after *doublet*]

triplex /ˈtrɪpleks/ *adj. & n.* ● *adj.* triple or threefold. ● *n.* (**Triplex**) *Brit. propr.* toughened or laminated safety glass for car windows etc. [L *triplex -plicis* (as TRI-, *plic-* fold)]

triplicate *adj., n., & v.* ● *adj.* /ˈtrɪplɪkət/ **1** existing in three examples or copies. **2** having three corresponding parts. **3** tripled. ● *n.* /ˈtrɪplɪkət/ each of a set of three copies or corresponding parts. ● *v.tr.* /ˈtrɪplɪˌkeɪt/ **1** make in three copies. **2** multiply by three. □ **in triplicate** consisting of three exact copies. □ **triplication** /ˌtrɪplɪˈkeɪʃ(ə)n/ *n.* [ME f. L *triplicatus* past part. of *triplicare* (as TRIPLEX)]

triplicity /trɪˈplɪsɪtɪ/ *n.* (*pl.* **-ies**) **1** the state of being triple. **2** a group of three things. **3** *Astrol.* a set of three zodiacal signs. [ME f. LL *triplicitas* f. L TRIPLEX]

triploid /ˈtrɪplɔɪd/ *n. & adj. Biol.* ● *n.* an organism or cell having three times the haploid set of chromosomes. ● *adj.* of or being a triploid. [mod.L *triploides* f. Gk (as TRIPLE)]

triploidy /ˈtrɪplɔɪdɪ/ *n. Biol.* the condition of being triploid.

tripmeter /ˈtrɪpˌmiːtə(r)/ *n.* a vehicle instrument that can be set to record the distance of individual journeys.

tripod /ˈtraɪpɒd/ *n.* **1** a three-legged stand for supporting a camera etc. **2** a stool, table, or utensil resting on three feet or legs. **3** *Gk Antiq.* a bronze altar at Delphi on which a priestess sat to utter oracles. □ **tripodal** /ˈtrɪpəd(ə)l/ *adj.* [L *tripus tripodis* f. Gk *tripous* (as TRI-, *pous podos* foot)]

Tripoli /ˈtrɪpəlɪ/ **1** (Arabic **Tarabulus Al-Gharb** /təˌrɑːbələs ælˈgɑːb/, 'western Tripoli') the capital and chief port of Libya, on the Mediterranean coast in the north-west of the country; pop. (1984) 990,700. Founded by Phoenicians in the 7th century BC, its ancient name was Oea. It was held successively by Romans, Vandals, Arabs, the Spanish, the Knights of St John, and the Ottoman Turks. **2** (Arabic **Tarabulus Ash-Sham** /æʃ'ʃæm/, 'eastern Tripoli', **Trâblous** /trɑ:'blu:s/) a port in NW Lebanon; pop. (1988) 160,000. It was founded *c.*700 BC and was the capital of the Phoenician triple federation formed by the city-states Sidon, Tyre, and Arvad. Today it is a major port and commercial centre of Lebanon.

tripoli /'trɪpəlɪ/ *n.* = *rotten-stone*. [F f. TRIPOLI]

Tripolitania /ˌtrɪpəlɪ'teɪnɪə, trɪˌpɒlɪ-/ a coastal region surrounding Tripoli in North Africa, in what is now NE Libya. Its name is Latin, referring to three Phoenician cities which were established there in the 7th century BC, Oea (now Tripoli), Leptis Magna, and Sabratha. □ **Tripolitanian** *adj.* & *n.* [L *Tripolis* three cities]

tripos /'traɪpɒs/ *n.* (at Cambridge University) the honours examination for the BA degree. [as TRIPOD, with ref. to the stool on which graduates sat to deliver a satirical speech at the degree ceremony]

tripper /'trɪpə(r)/ *n.* **1** *Brit.* a person who goes on a pleasure trip or excursion. **2** *colloq.* a person experiencing hallucinatory effects of a drug.

trippy /'trɪpɪ/ *adj.* (**trippier, trippiest**) *colloq.* (of music etc.) producing an effect resembling that of a drug, esp. LSD.

triptych /'trɪptɪk/ *n.* **1 a** a picture or relief carving on three panels, usu. hinged vertically together and often used as an altarpiece. **b** a set of three associated pictures placed in this way. **2** a set of three writing-tablets hinged or tied together. **3** a set of three artistic works. [TRI-, after DIPTYCH]

triptyque /trɪp'tiːk/ *n.* a customs permit serving as a passport for a motor vehicle. [F, as TRIPTYCH (orig. having three sections)]

Tripura /'trɪpʊərə/ a small state in the far north-east of India, on the eastern border of Bangladesh; capital, Agartala. An ancient Hindu kingdom, Tripura once covered a large area including Bengal, Assam, and parts of Burma. It acceded to India after independence in 1947, was designated a Union Territory in 1957, and achieved full status as a state in 1972.

triquetra /traɪ'kwetrə/ *n.* (*pl.* **triquetrae** /-tri:/) a symmetrical ornament of three interlaced arcs. [L, fem. of *triquetrus* three-cornered]

trireme /'traɪriːm/ *n.* an ancient Greek warship, a type of galley with three banks of oars on each side. The arrangement of the oarsmen has been the subject of much controversy: the rowers, each with an oar, probably sat in threes on angled benches rather than in three superimposed banks (as was previously thought). The trireme was the foundation of Athenian sea-power. [F *trirème* or L *triremis* (as TRI-, *remus* oar)]

trisaccharide /traɪ'sækəˌraɪd/ *n.* *Chem.* a sugar consisting of three linked monosaccharides.

Trisagion /trɪ'sægɪən/ *n.* a hymn, esp. in the Orthodox Church, with a triple invocation of God as holy. [ME f. Gk, neut. of *trisagios* f. *tris* thrice + *hagios* holy]

trisect /traɪ'sekt/ *v.tr.* cut or divide into three (usu. equal) parts. □ **trisector** *n.* **trisection** /-'sekʃ(ə)n/ *n.* [TRI- + *secare sect-* cut]

trishaw /'traɪʃɔ:/ *n.* a light three-wheeled pedalled vehicle used in the Far East. [TRI- + RICKSHAW]

triskelion /trɪ'skelɪən/ *n.* a symbolic figure of three legs or lines from a common centre. [Gk TRI- + *skelos* leg]

trismus /'trɪzməs/ *n.* *Med.* a variety of tetanus with tonic spasm of the jaw muscles causing the mouth to remain tightly closed. Also called *lockjaw*. [mod.L f. Gk *trismos* = *trigmos* a scream, grinding]

trisomy /'trɪsəmɪ/ *n.* *Med.* a condition in which an extra copy of a chromosome is present in the cell nuclei, causing developmental abnormalities. □ **trisomy-21** the most common form of Down's syndrome, caused by an extra chromosome number 21. [f. TRI- + -SOME³]

Tristan var. of TRISTRAM.

Tristan da Cunha /ˌtrɪstən də 'ku:nə/ the largest of a small group of volcanic islands in the South Atlantic, 2,112 km (1,320 miles) south-west of the British colony of St Helena, of which it is a dependency; pop. (1988) 313. It was discovered in 1506 by the Portuguese admiral Tristão da Cunha and annexed to Britain in 1816. Its original inhabitants were shipwrecked sailors. In 1961 the entire population was evacuated for two years after a volcanic eruption.

triste /tri:st/ *adj. literary* sad, melancholy, dreary. [F f. L *tristis*]

Tristram /'trɪstrəm/ (also **Tristan** /'trɪstən/) (in medieval legend) a knight who was the lover of Iseult.

trisyllable /traɪ'sɪləb(ə)l, trɪ-/ *n.* a word or metrical foot of three syllables. □ **trisyllabic** /ˌtraɪsɪ'læbɪk/ *adj.*

tritagonist /traɪ'tægənɪst, trɪ-/ *n.* the third actor in an ancient Greek drama (cf. DEUTERAGONIST). [Gk *tritagōnistēs* (as TRITO-, *agōnistēs* actor)]

trite /traɪt/ *adj.* (of a phrase, opinion, etc.) hackneyed, worn out by constant repetition. □ **tritely** *adv.* **triteness** *n.* [L *tritus* past part. of *terere* rub]

tritiate /'trɪtɪˌeɪt/ *v.tr.* *Chem.* replace the ordinary hydrogen in (a substance) by tritium. □ **tritiation** /ˌtrɪtɪ'eɪʃ(ə)n/ *n.*

tritium /'trɪtɪəm/ *n.* *Chem.* a radioactive isotope of hydrogen with a mass three times that of ordinary hydrogen (symbol: **T**). Discovered in 1934, tritium has two neutrons as well as a proton in the nucleus. It occurs in minute traces in nature and can be made artificially from lithium or deuterium in nuclear reactors; it is used as a fuel in thermonuclear bombs. [f. Gk *tritos* third]

trito- /'traɪtəʊ, 'trɪtəʊ/ *comb. form* third. [Gk *tritos* third]

Triton /'traɪt(ə)n/ *n.* **1** *Gk Mythol.* the son of Poseidon and Amphitrite, usually represented as a man with a fish's (sometimes a horse's) tail carrying a trident and shell-trumpet. Sometimes Tritons are regarded as constituting a class of minor sea-gods or mermen. **2** *Astron.* the largest satellite of Neptune (diameter 2,700 km), the seventh closest to the planet, discovered in 1846. It has a retrograde orbit. Voyager II found evidence for active geyser-like vents on Triton's complex surface of ice and frozen gases; despite a surface temperature of only about 38 kelvins it has a thin nitrogen atmosphere. [L f. Gk *Tritōn*]

triton¹ /'traɪt(ə)n/ *n.* **1** a large tropical marine gastropod mollusc of the family Cymatiidae, with a pointed spiral shell. **2** a newt. [TRITON 1]

triton² /'traɪt(ə)n/ *n.* *Physics* a nucleus of a tritium atom, consisting of a proton and two neutrons. [TRITIUM]

tritone /'traɪtəʊn/ *n.* *Mus.* an interval of an augmented fourth, comprising three tones.

triturate /'trɪtjʊˌreɪt/ *v.tr.* **1** grind to a fine powder. **2** masticate thoroughly. □ **triturator** *n.* **triturable** /-rəb(ə)l/ *adj.* **trituration** /ˌtrɪtjʊ'reɪʃ(ə)n/ *n.* [L *triturare* thresh corn f. *tritura* rubbing (as TRITE)]

triumph /'traɪəmf/ *n.* & *v.* ● *n.* **1 a** the state of being victorious or successful (*returned home in triumph*). **b** a great success or achievement. **2** a supreme example (*a triumph of engineering*). **3** joy at success; exultation (*could see triumph in her face*). **4** *Rom. Hist.* the processional entry of a victorious general into ancient Rome. ● *v.intr.* **1** (often foll. by *over*) gain a victory; be successful; prevail. **2** ride in triumph. **3** (often foll. by *over*) exult. [ME f. OF *triumphe* (n.), *triumpher* (v.), f. L *triump(h)us* prob. f. Gk *thriambos* hymn to Bacchus]

triumphal /traɪ'ʌmf(ə)l/ *adj.* of or used in or celebrating a triumph. [ME f. OF *triumphal* or L *triumphalis* (as TRIUMPH)]

triumphalism /traɪ'ʌmfəˌlɪz(ə)m/ *n.* excessive exultation over the victories of one's own party, country, etc. □ **triumphalist** *adj.* & *n.*

triumphant /traɪ'ʌmf(ə)nt/ *adj.* **1** victorious or successful. **2** exultant. □ **triumphantly** *adv.* [ME f. OF *triumphant* or L *triumphare* (as TRIUMPH)]

triumvir /'traɪəmˌvɪə(r), traɪ'ʌmvə(r)/ *n.* (*pl.* **triumvirs** or **triumviri** /traɪ'ʌmvɪˌraɪ/) **1** each of three men holding a joint office. **2** a member of a triumvirate. □ **triumviral** /traɪ'ʌmvɪrəl/ *adj.* [L, orig. in pl. *triumviri*, back-form. f. *trium virorum* genitive of *tres viri* three men]

triumvirate /traɪ'ʌmvɪrət/ *n.* a board or ruling group of three men (*triumvirs*), esp. in ancient Rome; also, the office or function of a triumvir. The term is used specifically of the unofficial coalition of Julius Caesar, Pompey, and Crassus in 60 BC (the *First Triumvirate*) and of the office to which Antony, Lepidus, and Octavian were appointed in 43 BC (the *Second Triumvirate*).

triune /'traɪjuːn/ *adj.* three in one, esp. with ref. to the Trinity. □ **triunity** /traɪ'juːnɪtɪ/ *n.* (*pl.* **-ies**). [TRI- + L *unus* one]

trivalent /traɪ'veɪlənt/ *adj.* *Chem.* having a valency of three; tervalent.

Trivandrum /trɪ'vændrəm/ the capital of the state of Kerala, a port on the SW coast of India; pop. (1991) 524,000.

trivet /'trɪvɪt/ *n.* **1** an iron tripod or bracket for a cooking pot or kettle to stand on. **2** an iron bracket designed to hook on to bars of a grate for a similar purpose. □ **as right as a trivet** *colloq.* in a perfectly good state, esp. healthy. **trivet table** a table with three feet. [ME *trevet*, app. f. L *tripes* (as TRI-, *pes pedis* foot)]

trivia /'trɪvɪə/ *n.pl.* **1** trifles or trivialities. **2** miscellaneous facts, esp. as used in a game or quiz. [mod.L, pl. of TRIVIUM, infl. by TRIVIAL]

trivial /'trɪvɪəl/ *adj.* **1** of small value or importance; trifling (*raised trivial objections*). **2** (of a person etc.) concerned only with trivial things. **3** *archaic* commonplace or humdrum (*the trivial round of daily life*). **4** *Biol.* &

Chem. of a name: **a** popular; not scientific. **b** specific, as opposed to generic. **5** *Math.* giving rise to no difficulty or interest. □ **trivially** *adv.*

trivialness *n.* **triviality** /ˌtrɪvɪˈælɪtɪ/ *n.* (*pl.* **-ies**). [L *trivialis* commonplace f. *trivium*: see TRIVIUM]

trivialize /ˈtrɪvɪəˌlaɪz/ *v.tr.* (also **-ise**) make trivial or apparently trivial; minimize. □ **trivialization** /ˌtrɪvɪəlaɪˈzeɪʃ(ə)n/ *n.*

trivium /ˈtrɪvɪəm/ *n. hist.* a medieval university course of grammar, rhetoric, and logic. [L, = place where three roads meet (as TRI-, *via* road)]

tri-weekly /traɪˈwiːklɪ/ *adj.* produced or occurring three times a week or every three weeks.

-trix /trɪks/ *suffix* (*pl.* **-trices** /trɪsɪz, ˈtraɪsiːz/ or **-trixes**) forming feminine agent nouns corresponding to masculine nouns in *-tor*, esp. in Law (*executrix*). [L *-trix -tricis*]

tRNA *abbr.* transfer RNA.

Troad /ˈtrəʊæd/ an ancient region of NW Asia Minor, of which ancient Troy was the chief city.

Trobriand Islands /ˈtrəʊbrɪənd/ a small group of islands in the SW Pacific, in Papua New Guinea, situated off the south-eastern tip of the island of New Guinea.

trocar /ˈtrəʊkɑː(r)/ *n. Med.* a surgical instrument with a three-sided cutting point, used for withdrawing fluid from a body cavity, esp. in oedema etc. [F *trois-quarts*, *trocart* f. *trois* three + *carre* side, face of an instrument]

trochaic /trəˈkeɪɪk/ *adj. & n. Prosody* ● *adj.* of or using trochees. ● *n.* (usu. in *pl.*) trochaic verse. [L *trochaicus* f. Gk *trokhaïkos* (as TROCHEE)]

trochal /ˈtrəʊk(ə)l/ *adj. Zool.* wheel-shaped. □ **trochal disc** *Zool.* the retractable disc on the head of a rotifer bearing a crown of cilia, used for drawing in food or for propulsion. [Gk *trokhos* wheel]

trochanter /trəˈkæntə(r)/ *n.* **1** *Anat.* any of several bony protuberances by which muscles are attached to the upper part of the thigh-bone. **2** *Zool.* the second joint of the leg in insects etc., seen as a small segment at the base of the femur. [F f. Gk *trokhantēr* f. *trekhō* run]

troche /trəʊʃ/ *n.* a small usu. circular medicated tablet or lozenge. [obs. *trochisk* f. OF *trochisque* f. LL *trochiscus* f. Gk *trokhiskos* dimin. of *trokhos* wheel]

trochee /ˈtrəʊkiː, -kɪ/ *n. Prosody* a foot consisting of one long or stressed syllable followed by one short or unstressed syllable. [L *trochaeus* f. Gk *trokhaios* (*pous*) running (foot) f. *trekhō* run]

trochlea /ˈtrɒklɪə/ *n.* (*pl.* **trochleae** /-lɪˌiː/) *Anat.* a pulley-like structure or arrangement of parts, e.g. the groove at the lower end of the humerus. □ **trochlear** *adj.* [L, = pulley f. Gk *trokhilia*]

trochoid /ˈtrəʊkɔɪd/ *adj. & n.* ● *adj.* **1** *Anat.* rotating on its own axis. **2** *Geom.* (of a curve) traced by a point on a radius of a circle rotating along a straight line or another circle. ● *n.* a trochoid joint or curve. □ **trochoidal** /trəʊˈkɔɪd(ə)l/ *adj.* [Gk *trokhoeidēs* wheel-like f. *trokhos* wheel]

trod *past* and *past part.* of TREAD.

trodden *past part.* of TREAD.

trog /trɒg/ *n. sl.* a contemptible person; a lout or hooligan. [abbr. of TROGLODYTE]

troglodyte /ˈtrɒgləˌdaɪt/ *n.* **1** a cave-dweller, esp. of prehistoric times. **2** a hermit. **3** *derog.* a wilfully obscurantist or old-fashioned person. □ **troglodytism** *n.* **troglodytic** /ˌtrɒgləˈdɪtɪk/ *adj.* **troglodytical** *adj.* [L *troglodyta* f. Gk *trōglodutēs* f. the name of an Ethiopian people, after *trōglē* hole]

trogon /ˈtrəʊgɒn/ *n.* a tropical bird of the family Trogonidae, with brilliantly coloured plumage, a short thick bill, and often a long tail. [mod.L f. Gk *trōgōn* f. *trōgō* gnaw]

troika /ˈtrɔɪkə/ *n.* **1 a** a Russian vehicle with a team of three horses abreast. **b** this team. **2** a group of three people, esp. as an administrative council. [Russ. f. *troe* set of three]

troilism /ˈtrɔɪlɪz(ə)m/ *n.* sexual activity involving three participants. [perh. f. F *trois* three]

Troilus /ˈtrɔɪləs/ *Gk Mythol.* a Trojan prince, the son of Priam and Hecuba, killed by Achilles. In medieval legends of the Trojan war he was the forsaken lover of Cressida.

Trojan /ˈtrəʊdʒən/ *adj. & n.* ● *adj.* of or relating to ancient Troy in Asia Minor. ● *n.* **1** a native or inhabitant of Troy. **2** a person who works, fights, etc., courageously (*works like a Trojan*). [ME f. L *Troianus* f. *Troia* Troy]

Trojan Horse *n.* **1** *Gk Mythol.* a hollow wooden horse used by the Greeks

to enter Troy. **2** a person or device planted to bring about an enemy's downfall.

Trojan War *Gk Mythol.* the ten-year siege of Troy by a coalition of Greeks, described in Homer's *Iliad*. The Greeks were attempting to recover Helen, wife of Menelaus, who had been abducted by the Trojan prince Paris. The war ended with the capture of the city by a trick: the Greeks ostensibly ended the siege but left behind a group of men concealed in a hollow wooden horse so large that the city walls had to be breached for it to be drawn inside. (See also ILIAD, TROY.)

troll¹ /trəʊl/ *n.* (in Scandinavian folklore) a fabulous being, esp. a giant or dwarf dwelling in a cave. [ON & Sw. *troll*, Danish *trold*]

troll² /trəʊl/ *v. & n.* ● *v.* **1** *intr.* sing out in a carefree jovial manner. **2** *tr.* & *intr.* fish by drawing bait along in the water. **3** *intr.* esp. *Brit.* walk, stroll. ● *n.* **1** the act of trolling for fish. **2** a line or bait used in this. □ **troller** *n.* [ME 'stroll, roll': cf. OF *troller* quest, MHG *trollen* stroll]

trolley /ˈtrɒlɪ/ *n.* (*pl.* **-eys**) **1** esp. *Brit.* a table, stand, or basket on wheels or castors for serving food, transporting luggage or shopping, gathering purchases in a supermarket, etc. **2** esp. *Brit.* a low truck running on rails. **3** (in full **trolley-wheel**) a wheel attached to a pole etc. used for collecting current from an overhead electric wire to drive a vehicle. **4 a** *US* = trolley-car. **b** *Brit.* = TROLLEYBUS. □ **off one's trolley** *sl.* crazy. **trolley-car** *US* a tram powered by electricity obtained from an overhead cable by means of a trolley-wheel. [of dial. orig., perh. f. TROLL²]

trolleybus /ˈtrɒlɪˌbʌs/ *n. Brit.* a bus powered by electricity obtained from an overhead cable by means of a trolley-wheel. Also called (*Brit.*) trolley, (*US*) trackless trolley.

trollop /ˈtrɒləp/ *n.* **1** a disreputable or promiscuous girl or woman. **2** a prostitute. □ **trollopish** *adj.* **trollopy** *adj.* [17th c.: perh. rel. to TRULL]

Trollope /ˈtrɒləp/, Anthony (1815–82), English novelist. He worked for the General Post Office in London, Ireland, and other parts of the world from 1834 to 1867, during which time he introduced the pillar-box to Britain. Trollope established his reputation as a writer with his fourth novel *The Warden* (1855), the first of the six 'Barsetshire' novels which also include *Barchester Towers* (1857) and *The Last Chronicle of Barset* (1867). Set in an imaginary English West Country and with recurring characters, they portray a solid rural society of curates and landed gentry. Another novel sequence, the six political 'Palliser' novels, was published 1864–80. Trollope was a prolific writer, publishing travel books, biographies, and short stories as well as forty-seven novels.

trombone /trɒmˈbəʊn/ *n.* **1 a** a large brass wind instrument with a forward-pointing extendable slide. (*See note below.*) **b** a player of this instrument. **2** an organ stop with the quality of a trombone. □ **trombonist** *n.* [F or It. f. It. *tromba* TRUMPET]

▪The trombone is the oldest brass instrument to possess a full chromatic compass. Introduced in the mid-15th century, it became popular in both instrumental and sacred and secular vocal music, particularly in Italy. Its association with church music developed in 18th-century Germany and Austria, leading to its use by Mozart for dramatic and solemn effect; Beethoven and Wagner also favoured the instrument.

trommel /ˈtrɒm(ə)l/ *n. Mining* a revolving cylindrical sieve for cleaning ore. [G, = drum]

tromometer /trəˈmɒmɪtə(r)/ *n.* an instrument for measuring very slight earthquake shocks. [Gk *tromos* trembling + -METER]

trompe /trɒmp/ *n.* an apparatus for producing a blast in a furnace by using falling water to displace air. [F, = trumpet: see TRUMP¹]

trompe-l'oeil /trɒmpˈlɜːɪ/ *n.* (*pl.* **trompe l'oeils** *pronunc.* same) a still-life painting etc. designed to be so lifelike as to create a three-dimensional illusion of reality. [F, lit. 'deceives the eye']

Tromsø /ˈtrɒmsɜː/ the principal city of Arctic Norway, situated on an island just west of the mainland; pop. (1991) 51,330.

-tron /trɒn/ *suffix Physics* forming nouns denoting: **1** a subatomic particle (*positron*). **2** a particle accelerator. **3** a thermionic valve. [after ELECTRON]

Trondheim /ˈtrɒndhaɪm/ a fishing port in west central Norway; pop. (1991) 138,060. Founded by Olaf I Tryggvason, it was the capital of Norway during the Viking period.

Troon /truːn/ a town on the west coast of Scotland, in Strathclyde region; pop. (1981) 14,230. It is noted for its championship golf course.

troop /truːp/ *n. & v.* ● *n.* **1** an assembled company; an assemblage of people or animals. **2** (in *pl.*) soldiers or armed forces. **3** a cavalry unit commanded by a captain. **4** a unit of artillery and armoured formation. **5** a grouping of three or more Scout patrols. ● *v.intr.* (foll. by *in, out, off,* etc.) come together or move in large numbers. □ **troop carrier** a large aircraft, armoured vehicle, etc., for carrying troops. **troop the colour** (esp. in the UK) show a flag ceremonially at a public mounting of garrison guards. **troop-ship** a ship used for transporting troops. [F *troupe,* back-form. f. *troupeau* dimin. of med.L *troppus* flock, prob. of Gmc orig.]

trooper /ˈtruːpə(r)/ *n.* **1** a private soldier in a cavalry or armoured unit. **2** *Austral.* & *US* a mounted or motor-borne policeman. **3** a cavalry horse. **4** esp. *Brit.* a troop-ship. □ **swear like a trooper** swear extensively or forcefully.

tropaeolum /trəˈpiːələm/ *n.* a trailing or climbing plant of the genus *Tropaeolum.* (See also NASTURTIUM 1.) [mod.L f. L *tropaeum* trophy, with ref. to the likeness of the flower and leaf to a helmet and shield]

trope /trəʊp/ *n.* a figurative (e.g. metaphorical or ironical) use of a word. [L *tropus* f. Gk *tropos* turn, way, trope f. *trepō* turn]

trophic /ˈtrəʊfɪk, ˈtrɒf-/ *adj.* of or concerned with nutrition (*trophic nerves*). [Gk *trophikos* f. *trophē* nourishment f. *trephō* nourish]

-trophic /ˈtrəʊfɪk, ˈtrɒf-/ *comb. form* **1** relating to nutrition (*autotrophic, oligotrophic*). **2** relating to maintenance or regulation, esp. by a hormone (*corticotrophic, gonadotrophic*) (cf. -TROPIC 2). □ **-trophism, -trophy** *comb. forms.* [Gk *tropheia* nourishment]

tropho- /ˈtrɒfəʊ/ *comb. form* nourishment. [Gk *trophē:* see TROPHIC]

trophoblast /ˈtrɒfəʊˌblæst/ *n.* *Zool.* a layer of tissue on the outside of a mammalian blastula, supplying the embryo with nourishment and later forming most of the placenta.

trophy /ˈtrəʊfɪ/ *n.* (*pl.* **-ies**) **1** a cup or other decorative object awarded as a prize or memento of victory or success in a contest etc. **2** a memento or souvenir, e.g. a deer's antlers, taken in hunting. **3** *Gk* & *Rom. Antiq.* the weapons etc. of a defeated army set up as a memorial of victory. **4** an ornamental group of symbolic or typical objects arranged for display. □ **trophied** *adj.* (also in *comb.*). [F *trophée* f. L *trophaeum* f. Gk *tropaion* f. *tropē* rout f. *trepō* turn]

tropic /ˈtrɒpɪk/ *n.* & *adj.* ● *n.* **1** the parallel of latitude 23°27′ north (*tropic of Cancer*) or south (*tropic of Capricorn*) of the equator, marking the limits at which the sun is vertically overhead on at least one day in the year. **2** *Astron.* either of two corresponding circles on the celestial sphere where the sun appears to turn after reaching its greatest declination, marking the northern and southern limits of the ecliptic. **3** (**the Tropics**) the region between the tropics of Cancer and Capricorn. ● *adj.* **1** = TROPICAL 1. **2** *Biol.* of tropism. □ **tropic bird** a tropical seabird of the family Phaethontidae, with mainly white plumage and long central tail feathers. [ME f. L *tropicus* f. Gk *tropikos* f. *tropē* turning f. *trepō* turn]

-tropic /ˈtrɒpɪk, ˈtrəʊpɪk/ *comb. form* **1** turning towards (*heliotropic*). **2** affecting (*psychotropic*); (esp. in names of hormones)= -TROPHIC. [Gk *tropē* turn, turning]

tropical /ˈtrɒpɪk(ə)l/ *adj.* **1** of, peculiar to, or suggesting the Tropics (*tropical fish; tropical diseases*). **2** very hot; passionate, luxuriant. **3** of or by way of a trope. □ **tropical cyclone** (or **storm**) see CYCLONE 2. **tropical year** see YEAR 1. □ **tropically** *adv.*

tropism /ˈtrəʊpɪz(ə)m/ *n.* *Biol.* the turning of all or part of an organism in a particular direction in response to an external stimulus. [Gk *tropos* turning f. *trepō* turn]

tropology /trəˈpɒlədʒɪ/ *n.* **1** the figurative use of words. **2** figurative interpretation, esp. of the Scriptures. □ **tropological** /ˌtrɒpəˈlɒdʒɪk(ə)l/ *adj.* [LL *tropologia* f. Gk *tropologia* (as TROPE)]

tropopause /ˈtrɒpəˌpɔːz, ˈtrəʊp-/ *n.* the interface between the troposphere and the stratosphere. [TROPOSPHERE + PAUSE]

troposphere /ˈtrɒpəˌsfɪə(r), ˈtrəʊp-/ *n.* the lowest region of the atmosphere, extending to a height of between 6 and 10 km from the earth's surface, and in which the temperature falls with increasing height (cf. STRATOSPHERE, IONOSPHERE). □ **tropospheric** /ˌtrɒpəˈsferɪk, ˌtrəʊp-/ *adj.* [Gk *tropos* turning + SPHERE]

troppo[1] /ˈtrɒpəʊ/ *adv.* *Mus.* too much (qualifying a tempo indication). □ **ma non troppo** /ˌmɑː nɒn/ but not too much so. [It.]

troppo[2] /ˈtrɒpəʊ/ *adj.* *Austral. sl.* mentally disturbed from exposure to a tropical climate, crazy.

Trossachs, the /ˈtrɒsəks/ a picturesque wooded valley in central Scotland, between Loch Achray and the lower end of Loch Katrine.

Trot /trɒt/ *n.* *colloq.* usu. *derog.* a Trotskyist. [abbr.]

trot /trɒt/ *v.* & *n.* ● *v.* (**trotted, trotting**) **1** *intr.* (of a person) run at a moderate pace esp. with short strides. **2** *intr.* (of a horse) proceed at a steady pace faster than a walk lifting each diagonal pair of legs

alternately. **3** *intr. colloq.* walk or go. **4** *tr.* cause (a horse or person) to trot. **5** *tr.* traverse (a distance) at a trot. ● *n.* **1** the action or exercise of trotting (*proceed at a trot; went for a trot*). **2** (**the trots**) *sl.* an attack of diarrhoea. **3** a brisk steady movement or occupation. **4** (in *pl.*) *Austral. colloq.* **a** trotting-races. **b** a meeting for these. **5** *US sl.* a literal translation of a text used by students; a crib. □ **on the trot** *colloq.* **1** continually busy (*kept them on the trot*). **2** in succession (*five weeks on the trot*). **trot out 1** cause (a horse) to trot to show his paces. **2** *colloq.* produce or introduce (as if) for inspection and approval, esp. tediously or repeatedly. [ME f. OF *troter* f. Rmc & med.L *trottare,* of Gmc orig.]

troth /trəʊθ/ *n. archaic* **1** faith, loyalty. **2** truth. □ **pledge** (or **plight**) **one's troth** pledge one's word esp. in marriage or betrothal. [ME *trowthe,* for OE *trēowth* TRUTH]

Trotsky /ˈtrɒtskɪ/, Leon (born Lev Davidovich Bronshtein) (1879–1940), Russian revolutionary. Joining the Bolsheviks in 1917, he helped to organize the October Revolution with Lenin, and built up the Red Army that eventually defeated the White Russian forces in the Russian Civil War. After Lenin's death he alienated Stalin and others with his view that socialism within the Soviet Union could not come about until revolution had occurred in western Europe and worldwide. Trotsky was eventually defeated by Stalin in the struggle for power, being expelled from the party in 1927 and exiled in 1929. After settling in Mexico in 1937, he was murdered three years later by a Stalinist assassin.

Trotskyism /ˈtrɒtskɪˌɪz(ə)m/ *n.* the principles of Leon Trotsky, who believed in the theory of continuing revolution rather than the more pragmatic ideas of state Communism generally accepted in Russia in the post-revolutionary era. With the defeat and disgrace of their leader in the power struggle following the death of Lenin, the Trotskyists were branded by the successful Stalinists as perverters of the Revolution. Trotskyism has generally included elements of anarchism and syndicalism, but the term has come to be used indiscriminately to describe a great many forms of radical socialism. □ **Trotskyist** *n.* **Trotskyite** *n. derog.*

trotter /ˈtrɒtə(r)/ *n.* **1** a horse bred or trained for trotting. **2** (usu. in *pl.*) **a** an animal's foot as food (*pig's trotters*). **b** *joc.* a human foot.

trotting /ˈtrɒtɪŋ/ *n.* a form of horse-racing (also called *harness racing*) in which a horse pulls a two-wheeled vehicle (a *sulky*) and its driver with a trotting or pacing gait. Similar races with chariots were held in Asia Minor as early as 1350 BC, and constituted one of the favourite sports in the Roman Empire; trotting in its modern form arose in the early 19th century, and is now popular particularly in the US, Australia, and New Zealand.

troubadour /ˈtruːbəˌdɔː(r)/ *n.* **1** any of a number of French medieval lyric poets composing and singing in Provençal during the 12th and early 13th centuries (and perhaps earlier). (*See note below.*) **2** a singer or poet. [F f. Prov. *trobador* f. *trobar* find, invent, compose in verse]

▪ The medieval troubadours are famous for the complexity of their verse forms and for the conception of chivalry and courtly love which prevails in their poems. They flourished in the courts of Spain, Italy, and France, and through their influence on the northern French poets and on the German Minnesingers they had a major effect on all the subsequent development of European lyric poetry. As well as love poetry they also composed moralizing, satirical, political, and military poems.

trouble /ˈtrʌb(ə)l/ *n.* & *v.* ● *n.* **1** difficulty or distress; vexation, affliction (*am having trouble with my car*). **2 a** inconvenience; unpleasant exertion; bother (*went to a lot of trouble*). **b** a cause of this (*the child was no trouble*). **3** a cause of annoyance or concern (*the trouble with you is that you can't say no*). **4** a faulty condition or operation (*kidney trouble; engine trouble*). **5 a** fighting, disturbance (*crowd trouble; don't want any trouble*). **b** (in *pl.*) political or social unrest, public disturbances (see also TROUBLES, THE). **6** disagreement, strife (*is having trouble at home*). ● *v.* **1** *tr.* cause distress or anxiety to; disturb (*were much troubled by their debts*). **2** *intr.* be disturbed or worried (*don't trouble about it*). **3** *tr.* afflict; cause pain etc. to (*am troubled with arthritis*). **4** *tr.* & *intr.* (often *refl.*) subject or be subjected to inconvenience or unpleasant exertion (*sorry to trouble you; don't trouble yourself; don't trouble to explain*). □ **ask** (or **look**) **for trouble** *colloq.* invite trouble or difficulty by one's actions, behaviour, etc.; behave rashly or indiscreetly. **be no trouble** cause no inconvenience etc. **go to the trouble** (or **some trouble** etc.) exert oneself to do something. **in trouble 1** involved in a matter likely to bring censure or punishment. **2** *colloq.* pregnant while unmarried. **look for trouble** see *ask for trouble* above. **take trouble** (or **the trouble**) exert oneself to do something. **trouble and strife** *rhyming sl.* wife. **trouble spot** a place where difficulties regularly occur. □ **troubler** n. [ME f. OF *truble* (n.), *trubler, turbler* (v.) ult. f. L *turbidus* TURBID]

troubled /ˈtrʌb(ə)ld/ adj. showing, experiencing, or reflecting trouble, anxiety, etc. (*a troubled mind; a troubled childhood*).

troublemaker /ˈtrʌb(ə)lˌmeɪkə(r)/ n. a person who habitually causes trouble. □ **troublemaking** n.

Troubles, the any of various periods of civil war or unrest in Ireland, especially in 1919–23 and (in Northern Ireland) from 1968.

troubleshooter /ˈtrʌb(ə)lˌʃuːtə(r)/ n. **1** a mediator in industrial or diplomatic etc. disputes. **2** a person who traces and corrects faults in machinery etc. □ **troubleshooting** n.

troublesome /ˈtrʌb(ə)lsəm/ adj. **1** causing trouble. **2** vexing, annoying. □ **troublesomely** adv. **troublesomeness** n.

troublous /ˈtrʌbləs/ adj. archaic or literary full of troubles; agitated, disturbed (*troublous times*). [ME f. OF *troubleus* (as TROUBLE)]

trough /trɒf/ n. **1** a long narrow open receptacle for water, animal feed, etc. **2** a channel for conveying a liquid. **3** an elongated region of low barometric pressure. **4** a hollow between two wave crests. **5** the time of lowest economic performance etc. **6** a region around the minimum on a curve of variation of a quantity. **7** a low point or depression. [OE *trog* f. Gmc]

trounce /traʊns/ v.tr. **1** defeat heavily. **2** beat, thrash. **3** punish severely. □ **trouncer** n. **trouncing** n. [16th c., = afflict: orig. unkn.]

troupe /truːp/ n. a company of actors or acrobats etc. [F, = TROOP]

trouper /ˈtruːpə(r)/ n. **1** a member of a theatrical troupe. **2** a staunch colleague.

trousers /ˈtraʊzəz/ n.pl. **1** (also **pair of trousers**) an outer garment reaching from the waist usu. to the ankles, divided into two parts to cover the legs. **2** (**trouser**) (attrib.) designating parts of this (*trouser leg*). □ **trouser-clip** = bicycle clip. **trouser suit** a woman's suit of trousers and jacket. **wear the trousers** be the dominant partner in a marriage etc. □ **trousered** adj. **trouserless** adj. [archaic *trouse* (sing.) f. Ir. & Gael. *triubhas* TREWS: pl. form after *drawers*]

trousseau /ˈtruːsəʊ/ n. (pl. **trousseaus** or **trousseaux** /-səʊz/) the clothes collected by a bride for her marriage. [F, lit. 'bundle', dimin. of *trousse* TRUSS]

trout /traʊt/ n. (pl. same or **trouts**) **1 a** a mainly freshwater game-fish of the genus *Salmo* or *Salvelinus*, of the salmon family; esp. *Salmo trutta* of Europe, valued as food. (*See note below.*) **b** a similar but unrelated fish. **2** sl. derog. a woman, esp. an old or ill-tempered one (usu. *old trout*). □ **troutlet** /ˈtraʊtlɪt/ n. **troutling** n. **trouty** adj. [OE *truht* f. LL *tructa*]

▪ The European brown trout, *Salmo trutta*, is now known to occur in a number of races with differing appearance, habitat requirements, and migratory behaviour – the brown trout proper, the lake trout, and the sea trout. In North America there are about five species of trout in the genus *Salmo* (including the cutthroat and rainbow trouts), and about three in the related genus *Salvelinus* (including a different lake trout); the name *sea trout* there refers to marine fish of an unrelated family.

trouvaille /truːˈvaɪ/ n. a lucky find; a windfall. [F f. *trouver* find]

trouvère /truːˈvɛə(r)/ n. a medieval epic poet in Northern France in the 11th–14th centuries. [OF *trovere* f. *trover* find: cf. TROUBADOUR]

trove /trəʊv/ n. = *treasure trove*. [AF *trové* f. *trover* find]

trover /ˈtrəʊvə(r)/ n. Law **1** finding and keeping personal property. **2** common-law action to recover the value of personal property wrongfully taken etc. [OF *trover* find]

trow /traʊ, trəʊ/ v.tr. archaic think, believe. [OE *trūwian, trēowian*, rel. to TRUCE]

Trowbridge /ˈtrəʊbrɪdʒ/ a town in SW England, the county town of Wiltshire; pop. (1981) 22,980.

trowel /ˈtraʊəl/ n. & v. ● n. **1** a small hand-held tool with a flat pointed blade, used to apply and spread mortar etc. **2** a similar tool with a curved scoop for lifting plants or earth. ● v.tr. (**trowelled, trowelling**; US **troweled, troweling**) **1** apply (plaster etc.). **2** dress (a wall etc.) with a trowel. [ME f. OF *truele* f. med.L *truella* f. L *trulla* scoop, dimin. of *trua* ladle etc.]

Troy /trɔɪ/ (also called *Ilium*) in Homeric legend, the city of King Priam that was besieged for ten years by the Greeks (see TROJAN WAR). Troy was regarded as having been a purely legendary city until Heinrich Schliemann identified the mound of Hissarlik on the NE Aegean coast of Turkey as the site of Troy. Schliemann's excavations showed the mound to be composed of nine main strata, dating from the early Bronze Age to the Roman era. The stratum known as Troy VIIa is

believed to be that of the Homeric city; it was apparently sacked and destroyed by fire in the mid-13th century BC, a period coinciding with the Mycenaean civilization of Greece.

troy /trɔɪ/ n. (in full **troy weight**) a system of weights used for precious metals and gems, with a pound of 12 ounces or 5,760 grains. [ME, prob. f. TROYES[1]]

Troyes[1] /trwʌ/ a town in northern France, on the River Seine; pop. (1990) 60,755. It was capital of the former province of Champagne. It gave its name to troy weight, the system of weights and measures which was first used at the medieval fairs held in this town.

Troyes[2], Chrétien de, see CHRÉTIEN DE TROYES.

trs. abbr. transpose (letters or words etc.).

truant /ˈtruːənt/ n., adj., & v. ● n. **1** a child who stays away from school without leave or explanation. **2** a person missing from work etc. ● adj. (of a person, conduct, thoughts, etc.) shirking, idle, wandering. ● v.intr. (also **play truant**) stay away as a truant. □ **truancy** n. [ME f. OF, prob. ult. f. Celt.: cf. Welsh *truan*, Gael. *truaghan* wretched]

truce /truːs/ n. **1** a temporary agreement to cease hostilities. **2** a suspension of private feuding or bickering. □ **truceless** adj. [ME *trew(e)s* (pl.) f. OE *trēow*, rel. to TRUE]

Trucial States /ˈtruːʃ(ə)l/ hist. a group of Arab sheikhdoms on the Persian Gulf, since 1971 forming the United Arab Emirates. The Trucial States had a special treaty relationship with Britain following the signing of a maritime truce in 1836.

truck[1] /trʌk/ n. & v. ● n. **1** Brit. an open railway wagon for carrying freight. **2** a large road vehicle for carrying heavy goods, troops, supplies, etc.; a lorry. **3** a railway bogie. **4** a wheeled stand for transporting goods. **5 a** Naut. a wooden disc at the top of a mast with holes for halyards. **b** a small solid wheel. ● v. **1** tr. convey on or in a truck. **2** intr. N. Amer. drive a truck. **3** intr. N. Amer. sl. proceed; go, stroll. □ **truckage** n. [perh. short for TRUCKLE in sense 'wheel, pulley']

truck[2] /trʌk/ n. & v. ● n. **1** dealings; exchange, barter. **2** small wares. **3** N. Amer. market-garden produce (*truck farm*). **4** colloq. odds and ends. **5** hist. the payment of workers in kind or with vouchers rather than with money (see also TRUCK ACTS). ● v.tr. & intr. archaic barter, exchange. □ **have no truck with** avoid dealing with. [ME f. OF *troquer* (unrecorded) = *trocare*, of unkn. orig.]

Truck Acts (in the UK) a series of Acts directed, from 1830 onwards, against the system, common in the 19th century, whereby workers received their wages in the form of vouchers for goods redeemable only at a special shop (often run by the employer). The Acts required wages to be paid in cash.

trucker /ˈtrʌkə(r)/ n. **1** a long-distance lorry-driver. **2** a firm dealing in long-distance carriage of goods.

truckie /ˈtrʌkɪ/ n. Austral. colloq. a lorry-driver; a trucker.

trucking /ˈtrʌkɪŋ/ n. esp. US conveyance of goods by lorry.

truckle /ˈtrʌk(ə)l/ n. & v. ● n. **1** (in full **truckle-bed**) a low bed on wheels that can be stored under a larger bed. **2** orig. dial. a small barrel-shaped cheese. ● v.intr. (foll. by to) submit obsequiously. □ **truckler** n. [orig. = wheel, pulley, f. AF *trocle* f. L *trochlea* pulley]

truckload /ˈtrʌkləʊd/ n. **1** a quantity of goods etc. or a number of people that can be transported in a truck. **2** colloq. a large quantity or number. □ **by the truckload** in large quantities or numbers.

truculent /ˈtrʌkjʊlənt/ adj. **1** aggressively defiant. **2** aggressive, pugnacious. **3** fierce, savage. □ **truculence** n. **truculency** n. **truculently** adv. [L *truculentus* f. *trux trucis* fierce]

Trudeau /ˈtruːdəʊ/, Pierre (Elliott) (b.1919), Canadian Liberal statesman, Prime Minister of Canada 1968–79 and 1980–4. Trudeau was noted for his commitment to federalism; in 1980 he held a provincial referendum in Quebec, which rejected independence. He also presided over the transfer of residual constitutional powers from Britain to Canada in 1982.

trudge /trʌdʒ/ v. & n. ● v. **1** intr. go on foot esp. laboriously. **2** tr. traverse (a distance) in this way. ● n. a trudging walk. □ **trudger** n. [16th c.: orig. unkn.]

trudgen /ˈtrʌdʒən/ n. a swimming stroke like the crawl with a scissors movement of the legs. [John *Trudgen*, English swimmer (1852–1902)]

true /truː/ adj., adv., & v. ● adj. (**truer, truest**) **1** in accordance with fact or reality (*a true story*). **2** genuine; rightly or strictly so called; not spurious or counterfeit (*a true friend; the true heir to the throne*). **3** (often foll. by to) loyal or faithful (*true to one's word*). **4** (foll. by to) accurately conforming (to a standard or expectation etc.) (*true to form*). **5** correctly positioned or balanced; upright, level. **6** exact, accurate (*a true aim; a* *true copy*). **7** (*absol.*) (also **it is true**) certainly, admittedly (*true, it would cost more*). **8** (of a note) exactly in tune. **9** *archaic* honest, upright (*twelve good men and true*). ● *adv.* **1** truly (*tell me true*). **2** accurately (*aim true*). **3** without variation (*breed true*). ● *v.tr.* (**trues, trued, truing** or **trueing**) bring (a tool, wheel, frame, etc.) into the exact position or form required. □ **come true** actually happen or be the case. **out of true** (or **the true**) not in the correct or exact position. **true bill** *US & hist.* a bill of indictment endorsed by a grand jury as being sustained by evidence. **true-blue** *adj.* extremely loyal or orthodox; Conservative. ● *n.* such a person, esp. a Conservative. **true-born** genuine (*a true-born Englishman*). **true-bred** of a genuine or good breed. **true coral** see CORAL 2. **true-hearted** faithful, loyal. **true horizon** see HORIZON 1c. **true-love** a sweetheart. **true-love** (or **-lover's**) **knot** a kind of knot with interlacing bows on each side, symbolizing true love. **true north** etc. north etc. according to the earth's axis, not magnetic north. **true rib** a rib joined directly to the breastbone. **true to form** (or **type**) being or behaving etc. as expected. **true to life** accurately representing life. □ **trueish** *adj.* **trueness** *n.* [OE *trēowe, trȳwe*, f. the Gmc noun repr. by TRUCE]

Trueman /'truːmən/, Frederick Sewards ('Fred') (b.1931), English cricketer. A fast bowler, he played for England from 1952 until 1965, during which time he became the first bowler to take 300 test wickets (1964) and took 307 wickets overall. He also played for his home county, Yorkshire (1949–68), taking a total of 2,304 wickets in first-class games. After retiring from the sport he became a cricket commentator and journalist.

Truffaut /'truːfəʊ/, François (1932–84), French film director. He was an influential film critic in the 1950s and originated the idea of the director as 'auteur'. In 1959 he directed his first feature film, *Les Quatre cents coups*, a work which established him as a leading director of the *nouvelle vague*. He acted and collaborated in the scriptwriting of this and many of his other films, among which are *Jules et Jim* (1961), *La Nuit américaine* (1973; *Day for Night*), which won an Oscar for best foreign film, and *The Last Metro* (1980).

truffle /'trʌf(ə)l/ *n.* **1** a strong-smelling underground fungus of the order Tuberales, used as a culinary delicacy and found esp. in France by trained dogs or pigs. **2** a usu. round sweet made of chocolate mixture covered with cocoa etc. [prob. f. Du. *truffel* f. obs. F *truffle* ult. f. L *tubera* pl. of TUBER]

trug /trʌg/ *n. Brit.* **1** a shallow oblong garden-basket usu. of wood strips. **2** *archaic* a wooden milk-pan. [perh. a dial. var. of TROUGH]

truism /'truːɪz(ə)m/ *n.* **1** an obviously true or hackneyed statement. **2** a proposition that states nothing beyond what is implied in any of its terms. □ **truistic** /truːˈɪstɪk/ *adj.*

Trujillo¹ /truːˈhiːjəʊ/ a city on the coast of NW Peru; pop. (1990) 532,000.

Trujillo² /truːˈhiːəʊ/, Rafael (born Rafael Leónidas Trujillo Molina; known as 'Generalissimo') (1891–1961), Dominican statesman, President of the Dominican Republic 1930–8 and 1942–52. Although he was formally President for only two periods, he wielded dictatorial powers from 1930 until his death. His dictatorship was marked by some improvement in social services and material benefits for the people, but also by the deployment of a strong and ruthless police force to crush all opposition. He was assassinated in 1961.

Truk Islands /trʌk/ a group of fourteen volcanic islands and numerous atolls in the western Pacific, in the Caroline Islands group, forming part of the Federated States of Micronesia; pop. (est. 1990) 53,700. There was a Japanese naval base there during the Second World War.

trull /trʌl/ *n. archaic* a prostitute. [16th c.: cf. G *Trulle*, TROLLOP]

truly /'truːlɪ/ *adv.* **1** sincerely, genuinely (*am truly grateful*). **2** really, indeed (*truly, I do not know*). **3** faithfully, loyally (*served them truly*). **4** accurately, truthfully (*is not truly depicted; has been truly stated*). **5** rightly, properly (*well and truly*). [OE *trēowlice* (as TRUE, -LY²)]

Truman /'truːmən/, Harry S (1884–1972), American Democratic statesman, 33rd President of the US 1945–53. As Vice-President, he automatically took office on Franklin Roosevelt's death in 1945. One of his first actions was to authorize the use of the atom bomb against Hiroshima and Nagasaki in 1945 to end the war with Japan. At home Truman put forward an extensive social programme which was largely blocked by Congress, although racial segregation in the armed forces and in federally funded schools was ended. In 1948 his administration introduced the Marshall Plan of emergency aid to war-shattered European countries and helped to establish NATO the following year.

He later involved the US in the Korean War. (See also TRUMAN DOCTRINE.)

Truman Doctrine the principle that the US should give support to countries or peoples threatened by Soviet forces or Communist insurrection. First expressed in 1947 by US President Truman in a speech to Congress seeking aid for Greece and Turkey, the doctrine was seen by the Communists as an open declaration of the cold war.

trumeau /truːˈməʊ/ *n.* (*pl.* **trumeaux** /-ˈməʊz/) *Archit.* a section of wall or a pillar between two openings, e.g. a pillar dividing a large doorway. [F]

trump¹ /trʌmp/ *n. & v.* ● *n.* **1 a** a playing card of a suit ranking above the others. **b** (in *pl.*) this suit (*hearts are trumps*). **2** an advantage esp. involving surprise. **3** *colloq.* a helpful or admired person. **b** *Austral. & NZ* a person in authority. ● *v.* **1 a** *tr.* defeat (a card or its player) with a trump. **b** *intr.* play a trump card when another suit has been led. **2** *tr. colloq.* gain a surprising advantage over (a person, proposal, etc.). □ **trump card 1** a card belonging to, or turned up to determine, a trump suit. **2** *colloq.* **a** a valuable resource. **b** a surprise move to gain an advantage. **trump up** fabricate or invent (an accusation, excuse, etc.) (*on a trumped-up charge*). **turn up trumps** *Brit. colloq.* **1** turn out better than expected. **2** be greatly successful or helpful. [corrupt. of TRIUMPH in the same (now obs.) sense]

trump² /trʌmp/ *n. archaic* a trumpet-blast. □ **the last trump** the trumpet-blast to wake the dead on Judgement Day. [ME f. OF *trompe* f. Frank.: prob. imit.]

trumpery /'trʌmpərɪ/ *n. & adj.* ● *n.* (*pl.* **-ies**) **1 a** worthless finery. **b** a worthless article. **2** rubbish. ● *adj.* **1** showy but worthless (*trumpery jewels*). **2** delusive, shallow (*trumpery arguments*). [ME f. OF *tromperie* f. *tromper* deceive]

trumpet /'trʌmpɪt/ *n. & v.* ● *n.* **1 a** a tubular or conical brass instrument with a flared bell and a bright penetrating tone. (See note below.) **b** a player of this instrument. **c** an organ stop with a quality resembling a trumpet. **2 a** the tubular corona of a daffodil etc. **b** a trumpet-shaped thing (*ear-trumpet*). **3** a sound of or like a trumpet. ● *v.* (**trumpeted, trumpeting**) **1** *intr.* **a** blow a trumpet. **b** (of an elephant etc.) make a loud sound as of a trumpet. **2** *tr.* proclaim loudly (a person's or thing's merit). □ **trumpet-call** an urgent summons to action. **trumpet major** the chief trumpeter of a cavalry regiment. □ **trumpetless** *adj.* [ME f. OF *trompette* dimin. (as TRUMP²)]

▪ One of the most ancient of instruments (two are preserved from the tomb of Tutankhamen), in its early days the trumpet produced only 'natural' notes and was particularly associated with fanfares and flourishes, from pageantry to stirring military music. In the 18th century crooks were added to allow a wider selection of notes, and in the early 19th century the valved trumpet made possible a far greater and less typecast contribution to orchestral music. In the 20th century the trumpet has become one of the most widely used jazz instruments.

trumpeter /'trʌmpɪtə(r)/ *n.* **1** a person who plays or sounds a trumpet, esp. a cavalry soldier giving signals. **2** a bird making a trumpet-like sound, esp.: **a** a variety of domestic pigeon. **b** a large black South American cranelike bird of the genus *Psophia*. □ **trumpeter swan** a large North American wild swan, *Cygnus buccinator*, with a black bill.

truncal /'trʌŋk(ə)l/ *adj.* of or relating to the trunk of a body or a tree.

truncate *v. & adj.* ● *v.tr.* /trʌŋˈkeɪt, ˈtrʌŋkeɪt/ **1** cut the top or the end from (a tree, a body, a piece of writing, etc.). **2** *Crystallog.* replace (an edge or an angle) by a plane. ● *adj.* /'trʌŋkeɪt/ *Bot. & Zool.* (of a leaf or feather etc.) ending abruptly as if cut off at the base or tip. □ **truncately** /'trʌŋkeɪtlɪ/ *adv.* **truncation** /trʌŋˈkeɪʃ(ə)n/ *n.* [L *truncare truncat-* maim]

truncheon /'trʌntʃ(ə)n/ *n.* **1** esp. *Brit.* a short club or cudgel, esp. carried by a policeman. **2** a staff or baton as a symbol of authority, esp. that of the Earl Marshal. [ME f. OF *tronchon* stump ult. f. L *truncus* trunk]

trundle /'trʌnd(ə)l/ *v.tr. & intr.* roll or move heavily or noisily esp. on or as on wheels. □ **trundle-bed** = TRUCKLE *n.* 1. [var. of obs. or dial. *trendle, trindle*, f. OE *trendel* circle (as TREND)]

trunk /trʌŋk/ *n.* **1** the main stem of a tree as distinct from its branches and roots. **2** a person's or animal's body apart from the limbs and head. **3** the main part of any structure. **4** a large box with a hinged lid for transporting luggage, clothes, etc. **5** *N. Amer.* the luggage compartment of a motor car. **6** an elephant's elongated prehensile nose. **7** (in *pl.*) men's close-fitting shorts worn for swimming, boxing, etc. **8** the main body of an artery, nerve, etc. **9** an enclosed shaft or conduit for cables, ventilation, etc. □ **trunk call** esp. *Brit.* a telephone call on a trunk line with charges made according to distance. **trunk**

line a main line of a railway, telephone system, etc. **trunk road** esp. *Brit.* an important main road. □ **trunkful** *n.* (*pl.* **-fuls**). **trunkless** *adj.* [ME f. OF *tronc* f. L *truncus*]

trunking /'trʌŋkɪŋ/ *n.* **1** a system of shafts or conduits for cables, ventilation, etc. **2** the use or arrangement of trunk lines.

trunnion /'trʌnjən/ *n.* **1** a supporting cylindrical projection on each side of a cannon or mortar. **2** a hollow gudgeon supporting a cylinder in a steam engine and giving passage to the steam. [F *trognon* core, tree-trunk, of unkn. orig.]

Truro /'truərəʊ/ the county town of Cornwall; pop. (1981) 18,560.

truss /trʌs/ *n.* & *v.* ● *n.* **1** a framework, e.g. of rafters and struts, supporting a roof or bridge etc. **2** a surgical appliance worn to support a hernia. **3** *Brit.* a bundle of old hay (56 lb) or new hay (60 lb) or straw (36 lb). **4** a compact terminal cluster of flowers or fruit. **5** a large corbel supporting a monument etc. **6** *Naut.* a heavy iron ring securing the lower yards to a mast. ● *v.tr.* **1** tie up (a fowl) compactly for cooking. **2** (often foll. by *up*) tie (a person) up with the arms to the sides. **3** support (a roof or bridge etc.) with a truss or trusses. □ **trusser** *n.* [ME f. OF *trusser* (v.), *trusse* (n.), of unkn. orig.]

trust /trʌst/ *n.* & *v.* ● *n.* **1 a** a firm belief in the reliability or truth or strength etc. of a person or thing. **b** the state of being relied on. **2** a confident expectation. **3 a** a thing or person committed to one's care. **b** the resulting obligation or responsibility (*am in a position of trust; have fulfilled my trust*). **4** a person or thing confided in (*is our sole trust*). **5** reliance on the truth of a statement etc. without examination. **6** commercial credit (*obtained goods on trust*). **7** *Law* a confidence placed in a person by making that person the nominal owner of property to be used for another's benefit. **b** the right of the latter to benefit by such property. **c** the property so held. **d** the legal relation between the holder and the property so held. **8 a** a body of trustees. **b** an organization managed by trustees. **c** an organized association of several companies for the purpose of reducing or defeating competition etc., esp. one in which all or most of the stock is transferred to a central committee and shareholders lose their voting power although remaining entitled to profits. ● *v.* **1** *tr.* place trust in; believe in; rely on the character or behaviour of. **2** *tr.* (foll. by *with*) allow (a person) to have or use (a thing) from confidence in its proper use (*was reluctant to trust them with my books*). **3** *tr.* (often foll. by *that* + clause) have faith or confidence or hope that a thing will take place (*I trust you will not be late; I trust that she is recovering*). **4** *tr.* (foll. by *to*) consign (a thing) to (a person) with trust. **5** *tr.* (foll. by *for*) allow credit to (a customer) for (goods). **6** *intr.* (foll. by *in*) place reliance in (*we trust in you*). **7** *intr.* (foll. by *to*) place (esp. undue) reliance on (*shall have to trust to luck*). □ **in trust** *Law* held on the basis of trust (see sense 7 of *n.*). **on trust 1** on credit. **2** on the basis of trust or confidence. **take on trust** accept (an assertion, claim, etc.) without evidence or investigation. **trust company** a company formed to act as a trustee or to deal with trusts. **trust fund** a fund of money etc. held in trust. **trust territory** a territory under the trusteeship of the United Nations or of a state designated by them. □ **trustable** *adj.* **truster** *n.* [ME *troste, truste* (n.) f. ON *traust* f. *traustr* strong: (v.) f. ON *treysta*, assim. to the noun]

trustbuster /'trʌst,bʌstə(r)/ *n.* esp. *US* a person or agency employed to dissolve trusts.

trustee /trʌs'tiː/ *n.* **1** *Law* a person or member of a board given control or powers of administration of property in trust with a legal obligation to administer it solely for the purposes specified. **2** a state made responsible for the government of an area. □ **trusteeship** *n.*

trustful /'trʌstfʊl/ *adj.* **1** full of trust or confidence. **2** not feeling or showing suspicion. □ **trustfully** *adv.* **trustfulness** *n.*

trusting /'trʌstɪŋ/ *adj.* having trust (esp. characteristically); trustful. □ **trustingly** *adv.* **trustingness** *n.*

trustworthy /'trʌst,wɜːðɪ/ *adj.* deserving of trust; reliable. □ **trustworthily** *adv.* **trustworthiness** *n.*

trusty /'trʌstɪ/ *adj.* & *n.* ● *adj.* (**trustier, trustiest**) **1** *archaic* or *joc.* trustworthy (*a trusty steed*). **2** *archaic* loyal (to a sovereign) (*my trusty subjects*). ● *n.* (*pl.* **-ies**) a prisoner who is given special privileges for good behaviour. □ **trustily** *adv.* **trustiness** *n.*

truth /truːθ/ *n.* (*pl.* **truths** /truːðz, truːθs/) **1** the quality or a state of being true or truthful (*doubted the truth of the statement; there may be some truth in it*). **2 a** what is true (*tell us the whole truth; the truth is that I forgot*). **b** what is accepted as true (*one of the fundamental truths*). □ **in truth** *literary* truly, really. **to tell the truth** (or **truth to tell**) to be frank. **truth drug** any of various drugs supposedly able to induce a person to tell the truth. **truth table** a list indicating the truth or falsity of various propositions in logic etc. □ **truthless** *adj.* [OE *trīewth, trēowth* (as TRUE)]

truthful /'truːθfʊl/ *adj.* **1** habitually speaking the truth. **2** (of a story etc.) true. **3** (of a likeness etc.) corresponding to reality. □ **truthfully** *adv.* **truthfulness** *n.*

try /traɪ/ *v.* & *n.* ● *v.* (**-ies, -ied**) **1** *intr.* make an effort with a view to success (often foll. by *to* + infin.; *colloq.* foll. by *and* + infin.: *tried to be on time; try and be early; I shall try hard*). ¶ Use with *and* is uncommon in the past tense and in negative contexts (except in *imper.*). **2** *tr.* make an effort to achieve (*tried my best; had better try something easier*). **3** *tr.* **a** test (the quality of a thing) by use or experiment. **b** test the qualities of (a person or thing) (*try it before you buy*). **4** *tr.* make severe demands on (a person, quality, etc.) (*my patience has been sorely tried*). **5** *tr.* examine the effectiveness or usefulness of for a purpose (*try cold water; try the off-licence; have you tried kicking it?*). **6** *tr.* ascertain the state of fastening of (a door, window, etc.). **7** *tr.* **a** investigate and decide (a case or issue) judicially. **b** subject (a person) to trial (*will be tried for murder*). **8** *tr.* make an experiment in order to find out (*let us try which takes longest*). **9** *intr.* (foll. by *for*) **a** apply or compete for. **b** seek to reach or attain (*am going to try for a gold medal*). **10** *tr.* (often foll. by *out*) **a** extract (oil) from fat by heating. **b** treat (fat) in this way. **11** *tr.* (often foll. by *up*) smooth (roughly planed wood) with a plane to give an accurately flat surface. ● *n.* (*pl.* **-ies**) **1** an effort to accomplish something; an attempt (*give it a try*). **2** *Rugby* the act of touching the ball down behind the opposing goal-line, scoring points and entitling the scoring side to a kick at goal. **3** *Amer. Football* an attempt to score an extra point in various ways after a touchdown. □ **tried and tested** (or **true**) proved reliable by experience; dependable. **try conclusions with** see CONCLUSION. **try a fall with** contend with. **try for size** try out or test for suitability. **try one's hand** see how skilful one is, esp. at the first attempt. **trying-plane** a plane used in trying (see sense 11 of *v.*). **try it on** (often foll. by *with*) *colloq.* **1** test another's patience. **2** attempt to outwit, deceive, or seduce another person. **try on** put on (clothes etc.) to see if they fit or suit the wearer. **try-on** *n. Brit. colloq.* **1** an act of trying it on. **2** an attempt to fool or deceive. **try out 1** put to the test. **2** test thoroughly. **try-out** *n.* an experimental test of efficiency, popularity, etc. **try-sail** /'traɪs(ə)l/ a small strong fore-and-aft sail set on the mainmast or other mast of a sailing-vessel in heavy weather. **try-square** a carpenter's square, usu. with one wooden and one metal limb. [ME, = separate, distinguish, etc., f. OF *trier* sift, of unkn. orig.]

trying /'traɪɪŋ/ *adj.* annoying, vexatious; hard to endure. □ **tryingly** *adv.*

trypanosome /'trɪpənə,səʊm/ *n. Med.* a protozoan parasite of the genus *Trypanosoma*, having a long trailing flagellum and infesting the blood etc. where it causes trypanosomiasis. [Gk *trupanon* borer + -SOME³]

trypanosomiasis /,trɪpənəsə'maɪəsɪs/ *n. Med.* a disease caused by a trypanosome and transmitted by biting insects, e.g. sleeping sickness and Chagas' disease.

trypsin /'trɪpsɪn/ *n. Biochem.* a digestive enzyme acting on proteins and present in the pancreatic juice. □ **tryptic** *adj.* [Gk *tripsis* friction f. *tribō* rub (because it was first obtained by rubbing down the pancreas with glycerine)]

trypsinogen /trɪp'sɪnədʒən/ *n. Biochem.* a substance in the pancreas from which trypsin is formed.

tryptophan /'trɪptə,fæn/ *n. Biochem.* an amino acid essential in the diet of vertebrates. [as TRYPSIN + -*phan* f. Gk *phainō* appear]

tryst /trɪst/ *n.* & *v. archaic* ● *n.* **1** a time and place for a meeting, esp. of lovers. **2** such a meeting (*keep a tryst; break one's tryst*). ● *v.intr.* (foll. by *with*) make a tryst. □ **tryster** *n.* [ME, = obs. *trist* (= TRUST) f. OF *triste* an appointed station in hunting]

Tsao-chuang see ZAOZHUANG.

tsar /zɑː(r)/ *n.* (also **czar, tzar**) **1** *hist.* the title of the former emperor of Russia. **2** a person with great authority. □ **tsardom** *n.* **tsarism** *n.* **tsarist** *n.* & *adj.* [Russ. *tsar'*, ult. f. L *Caesar*]

tsarevich /'zɑːrɪvɪtʃ/ *n.* (also **czarevich, tzarevich**) *hist.* the eldest son of a tsar of Russia. [Russ., = son of a tsar]

tsarina /zɑː'riːnə/ *n.* (also **czarina, tzarina**) *hist.* the title of the former empress of Russia. [It. & Sp. (c)*zarina* f. G *Czarin, Zarin*, fem. of *Czar, Zar*]

Tsaritsyn /tsɑː'riːtsɪn/ a former name (until 1925) for VOLGOGRAD.

Tsavo National Park /'tsɑːvəʊ/ an extensive national park in SE Kenya, established in 1948.

tsessebi /tse'seɪbɪ/ *n.* (also **sassaby** /sə'seɪ-/) a race of the topi (antelope) occurring in eastern and southern Africa. [Setswana]

tsetse /'tsɛtsɪ, 'tɛt-/ *n.* (often **tsetse fly**) an African fly of the genus *Glossina*, that feeds on human and animal blood using a needle-like proboscis and transmits sleeping sickness. [Setswana]

TSH *abbr.* **1** thyroid-stimulating hormone. **2** Their Serene Highnesses.

T-shirt /'tiː.ʃɜːt/ *n.* (also **teeshirt**) a short-sleeved casual top, usu. of knitted cotton and having the form of a T when spread out.

Tsinan see JINAN.

Tsinghai see QINGHAI.

Tsiolkovsky /ˌtsiːɒlˈkɒfskɪ/, Konstantin (Eduardovich) (1857–1935), Russian aeronautical engineer. His early ideas for aircraft and rockets were not officially recognized until after the Russian Revolution, though his proposal for the use of liquid fuel in rockets predated Goddard's successful rocket flight by nearly forty years. During the 1920s Tsiolkovsky carried out pioneering theoretical work on multi-stage rockets, jet engines, and space flight.

Tsitsikamma Forest /ˌtsɪtsɪˈkɑːmə/ an area of dense natural forest, now a national park, on the south coast of South Africa. The name is derived from a local word meaning 'clear water'.

Tskhinvali /tskɪˈnvɑːlɪ/ the capital of South Ossetia.

Tsonga /'tsɒŋɡə/ *n.* & *adj.* ● *n.* (*pl.* same or **Tsongas**) **1** a member of the Shangaan people inhabiting parts of southern Mozambique and South Africa. **2** the Bantu language of this people. ● *adj.* of or relating to this people or their language. [Bantu]

tsp. *abbr.* (*pl.* **tsps.**) teaspoonful.

T-square /'tiːskweə(r)/ *n.* a T-shaped instrument for drawing or testing right angles.

TSS *abbr. Med.* toxic shock syndrome.

tsunami /tsuːˈnɑːmɪ/ *n.* (*pl.* **tsunamis**) a long high sea wave caused by underwater earthquakes or other disturbances. Also called *tidal wave*. [Jap. f. *tsu* harbour + *nami* wave]

Tsushima /tsuːˈʃiːmə/ a Japanese island in the Korea Strait, between South Korea and Japan. In 1905 it was the scene of a defeat for the Russian navy when their fleet was destroyed by the Japanese.

Tswana /'tswɑːnə/ *n.* & *adj.* ● *n.* **1** (*pl.* **Batswana** /bəˈtswɑːnə/, **Tswana**, or **Tswanas**) a member of a southern African people living in Botswana and neighbouring areas. **2** the Bantu language of this people, Setswana. ● *adj.* of or relating to the Batswana or their language. [Bantu]

TT *abbr.* **1** see TOURIST TROPHY. **2** tuberculin-tested. **3 a** teetotal. **b** teetotaller.

TU *abbr.* Trade Union.

Tu. *abbr.* Tuesday.

Tuamotu Archipelago /ˌtuːəˈməʊtuː/ a group of about eighty coral islands forming part of French Polynesia, in the South Pacific; pop. (1988) 12,370. It is the largest group of coral atolls in the world. The islands of Mururoa and Fangataufa have been used by the French since 1966 for nuclear testing.

Tuareg /'twɑːrɛɡ/ *n.* & *adj.* ● *n.* (*pl.* same or **Tuaregs**) **1** a member of a Berber group of nomadic pastoralists of the western and central Sahara, now concentrated mainly in Algeria, Mali, Niger, and western Libya. Tuareg men traditionally wear dark blue robes, turbans, and veils. **2** the Berber dialect of the Tuareg, the only Berber language to have an indigenous written form. Its alphabet is related to ancient Phoenician script. ● *adj.* of or relating to this people or their language. [Berber]

tuatara /ˌtuːəˈtɑːrə/ *n.* a large lizard-like reptile, *Sphenodon punctatum*, having a crest of soft spines extending along its back and found only on a few islands off New Zealand. It is regarded as the sole surviving member of the order Rhynchocephalia. [Maori f. *tua* on the back + *tara* spine]

Tuatha Dé Danaan /ˌtʊəhə deɪ ˈdænən/ *n.pl. Ir. Mythol.* the members of an ancient race who inhabited Ireland before the historical Irish. Formerly believed to have been a real people, they are credited with the possession of magical powers and great wisdom. [Ir., = people of the goddess Danaan]

tub /tʌb/ *n.* & *v.* ● *n.* **1** an open flat-bottomed usu. round container for various purposes. **2** a tub-shaped (usu. plastic) carton. **3** the amount a tub will hold. **4** *colloq.* a bath. **5 a** *colloq.* a clumsy slow boat. **b** a stout roomy boat for rowing practice. **6** *Mining* a container for conveying ore, coal, etc. ● *v.* (**tubbed**, **tubbing**) **1** *tr.* & *intr.* plant, bathe, or wash in a tub. **2** *tr.* enclose in a tub. **3** *tr. Mining* line (a shaft) with a wooden or iron casing. □ **tub chair** a chair with solid arms continuous with a usu. semicircular back. **tub-thumper** *colloq.* a ranting preacher or orator.

tub-thumping *colloq.* ranting oratory. □ **tubbable** *adj.* **tubbish** *adj.*

tubful *n.* (*pl.* **-fuls**). [ME, prob. of LG or Du. orig.: cf. MLG, MDu. *tubbe*]

tuba /'tjuːbə/ *n.* (*pl.* **tubas**) **1 a** a low-pitched brass wind instrument. (See note below.) **b** a player of this instrument. **2** an organ stop with the quality of a tuba. [It. f. L, = trumpet]

▪ The tuba, the largest valved brass instrument and normally the lowest-toned instrument in the orchestra, was developed in Germany and Austria in the 1830s. It is held upright to be played, the mouthpiece being set approximately at right angles to the upward-pointing bell.

tubal /'tjuːb(ə)l/ *adj. Anat.* of or relating to a tube, esp. the bronchial or Fallopian tubes.

tubby /'tʌbɪ/ *adj.* (**tubbier**, **tubbiest**) **1** (of a person) short and fat. **2** (of a violin) dull-sounding, lacking resonance. □ **tubbiness** *n.*

tube /tjuːb/ *n.* & *v.* ● *n.* **1** a long hollow rigid or flexible cylinder, esp. for holding or carrying air, liquids, etc. **2** a soft metal or plastic cylinder sealed at one end and having a screw cap at the other, for holding a semi-liquid substance ready for use (*a tube of toothpaste*). **3** *Anat. & Zool.* a hollow cylindrical organ in the body (*bronchial tubes; Fallopian tubes*). **4** *colloq.* (often prec. by *the*) **a** an underground railway system, esp. the one in London (*went by tube*). **b** a train on such a system. **5 a** a cathode-ray tube esp. in a television set. **b** (prec. by *the*) esp. *US colloq.* television. **6** *US* a thermionic valve. **7** = *inner tube*. **8** the cylindrical body of a wind instrument. **9** *Austral. sl.* a can of beer. ● *v.tr.* **1** equip with tubes. **2** enclose in a tube. □ **tube worm** a polychaete worm which constructs or secretes a tube in which it lives. □ **tubeless** *adj.* (esp. in sense 7 of *n.*).

tubelike *adj.* [F *tube* or L *tubus*]

tubectomy /tjuːˈbɛktəmɪ/ *n.* (*pl.* **-ies**) the surgical removal of a Fallopian tube.

tuber /'tjuːbə(r)/ *n.* **1 a** the short thick rounded part of a stem or rhizome, usu. found underground and covered with modified buds, e.g. in a potato. **b** the similar root of a dahlia etc. **2** *Anat.* a lump or swelling. [L, = hump, swelling]

tubercle /'tjuːbək(ə)l/ *n.* **1** a small rounded protuberance esp. on a bone. **2** a small rounded inflamed swelling on the body or in an organ, esp. a nodular lesion characteristic of tuberculosis in the lungs etc. **3** a small tuber; a wartlike growth. □ **tubercle bacillus** a bacterium causing tuberculosis. □ **tuberculate** /tjʊˈbɜːkjʊlət/ *adj.* **tuberculous** *adj.* [L *tuberculum*, dimin. of *tuber*: see TUBER]

tubercular /tjʊˈbɜːkjʊlə(r)/ *adj.* & *n.* ● *adj.* of or having tubercles or tuberculosis. ● *n.* a person with tuberculosis. [f. L *tuberculum* (as TUBERCLE)]

tuberculation /tjʊˌbɜːkjʊˈleɪʃ(ə)n/ *n.* **1** the formation of tubercles. **2** a growth of tubercles. [f. L *tuberculum* (as TUBERCLE)]

tuberculin /tjʊˈbɜːkjʊlɪn/ *n.* a sterile liquid protein extract from cultures of tubercle bacillus, used in the diagnosis and (formerly) the treatment of tuberculosis. □ **tuberculin test** a hypodermic injection of tuberculin to detect infection with or immunity from tuberculosis.

tuberculin-tested (of milk) from cows giving a negative response to a tuberculin test. [f. L *tuberculum* (as TUBERCLE)]

tuberculosis /tjʊˌbɜːkjʊˈləʊsɪs/ *n.* an infectious disease caused by the tubercle bacillus *Mycobacterium tuberculosis*, characterized by the growth of tubercles, especially in the lungs. The most common form, pulmonary tuberculosis (formerly popularly known as *consumption*), was common in Europe in the 19th century; it is still widespread in some developing countries, where 3 million people die from it each year. Tuberculosis was once considered incurable, but early X-ray diagnosis, new drugs, and lung operations can now arrest the disease; the use of vaccines and improved hygiene and living conditions are also important. Tuberculosis can affect other parts of the body, in particular the bones and joints and the central nervous system. Elimination of tuberculosis from dairy herds and pasteurization of milk have greatly reduced the transmission of bovine tuberculosis from cattle.

tuberose¹ /'tjuːbəˌrəʊz, -ˌrəʊs/ *adj.* **1** covered with tubers; knobby. **2** of or resembling a tuber. **3** bearing tubers. □ **tuberosity** /ˌtjuːbəˈrɒsɪtɪ/ *n.* [L *tuberosus* f. TUBER]

tuberose² /'tjuːbəˌrəʊz/ *n.* a Mexican plant, *Polianthes tuberosa*, of the agave family, cultivated for the perfume extracted from its heavily scented white funnel-shaped flowers. [L *tuberosa* fem. (as TUBEROSE¹)]

tuberous /'tjuːbərəs/ *adj.* = TUBEROSE¹. □ **tuberous root** a thick and fleshy root like a tuber but without buds. [F *tubéreux* or L *tuberosus* f. TUBER]

tubifex /'tjuːbɪˌfɛks/ *n.* a small red annelid worm of the genus *Tubifex*, found in mud at the bottom of rivers and lakes and used as food for aquarium fish. Also called *bloodworm*. [mod.L f. L *tubus* tube + *-fex* f. *facere* make]

tubiform /'tju:bɪˌfɔːm/ *adj.* tube-shaped.

tubing /'tju:bɪŋ/ *n.* **1** a length of tube. **2** a quantity of tubes.

Tubruq see TOBRUK.

Tubuai Islands /tu:b'wɑːɪ/ (also called the *Austral Islands*) a group of volcanic islands in the South Pacific, forming part of French Polynesia; chief town, Mataura (on the island of Tubuai); pop. (1988) 6,500.

tubular /'tju:bjʊlə(r)/ *adj.* **1** tube-shaped. **2** having or consisting of tubes. **3** (of furniture etc.) made of tubular pieces. □ **tubular bells** an orchestral instrument consisting of a row of vertically suspended metal tubes that are struck with a hammer.

tubule /'tju:bju:l/ *n.* a small tube in a plant or an animal body. [L *tubulus*, dimin. of *tubus* tube]

tubulous /'tju:bjʊləs/ *adj.* = TUBULAR.

TUC see TRADES UNION CONGRESS.

tuck /tʌk/ *v. & n.* ● *v.* **1** *tr.* (often foll. by *in*, *up*) **a** draw, fold, or turn the outer or end parts of (cloth or clothes etc.) close together so as to be held; thrust in the edge of (a thing) so as to confine it (*tucked his shirt into his trousers*; *tucked the sheet under the mattress*). **b** thrust in the edges of bedclothes around (a person) (*came to tuck me in*). **2** *tr.* draw together into a small space (*tucked her legs under her*; *the bird tucked its head under its wing*). **3** *tr.* stow (a thing) away in a specified place or way (*tucked it in a corner*; *tucked it out of sight*). **4** *tr.* **a** make a stitched fold in (material, a garment, etc.). **b** shorten, tighten, or ornament with stitched folds. **5** *tr.* hit (a ball) to the desired place. ● *n.* **1** a flattened usu. stitched fold in material, a garment, etc., often one of several parallel folds for shortening, tightening, or ornament. **2** *Brit. colloq.* food, esp. cakes and sweets eaten by children (also *attrib.*: *tuck box*). **3** *Naut.* the part of a ship's hull where the planks meet under the stern. **4** (in full **tuck position**) (in diving, gymnastics, etc.) a position with the knees bent upwards into the chest and the hands clasped round the shins. □ **tuck in** *colloq.* eat food heartily. **tuck-in** *n. Brit. colloq.* a large meal. **tuck into** (or **away**) *colloq.* eat (food) heartily (*tucked into their dinner*; *could really tuck it away*). **tuck-net** (or **-seine**) a small net for taking caught fish from a larger net. **tuck shop** *Brit.* a small shop, esp. near or in a school, selling food to children. [ME *tukke*, *tokke*, f. MDu., MLG *tucken*, = OHG *zucchen* pull, rel. to TUG]

tucker /'tʌkə(r)/ *n. & v.* ● *n.* **1** a person or thing that tucks. **2** *hist.* a piece of lace or linen etc. in or on a woman's bodice. **3** *Austral. colloq.* food. ● *v.tr.* (esp. in *passive*; often foll. by *out*) *N. Amer. colloq.* tire, exhaust. □ **best bib and tucker** see BIB[1]. **tucker-bag** (or **-box**) *Austral. & NZ colloq.* a container for food.

tucket /'tʌkɪt/ *n. archaic* a flourish on a trumpet. [ONF *toquer* beat (a drum)]

tucking /'tʌkɪŋ/ *n.* a series of usu. stitched tucks in material or a garment.

Tucson /'tu:sɒn/ a city in SE Arizona; pop. (1990) 405,390.

-tude /tju:d/ *suffix* forming abstract nouns (*altitude*; *attitude*; *solitude*). [from or after F *-tude* f. L *-tudo -tudinis*]

Tudor[1] /'tju:də(r)/ the English royal house which ruled England from 1485 (Henry VII) until the death of Elizabeth I (1603). The Tudors were descended from Owen Tudor, who married Catherine, widowed queen of Henry V. ● *adj.* of the architectural style prevailing in England during the reigns of the Tudors; or of resembling the domestic architecture of this period. (*See note below.*) □ **Tudor rose** a conventionalized five-lobed decorative figure of a rose, especially a combination of the red and white roses of York and Lancaster adopted as a badge by Henry VII.

▪ Tudor architecture is marked by the final development of Perpendicular or English Gothic style and the introduction of decorative elements from the Italian Renaissance. Brickwork (frequently patterned) dominated domestic architecture, which typically featured half-timbering, elaborate chimneys, many gables, rich oriel windows, and much interior panelling and moulded plasterwork. The architecture of the later Tudor period, during the reign of Elizabeth I (1558–1603), is usually distinguished as 'Elizabethan'. Stone was employed as a material for some larger buildings, and there was a richer use of Italian and Flemish ornamentation.

Tudor[2] /'tju:də(r)/, Henry, Henry VII of England (see HENRY[1]).

Tudor[3] /'tju:də(r)/, Mary, Mary I of England (see MARY[2]).

Tudorbethan /ˌtju:də'bi:θ(ə)n/ *adj.* (of a house etc.) imitating Tudor and Elizabethan styles in design. [blend of TUDOR[1] and ELIZABETHAN]

Tues. *abbr.* (also **Tue.**) Tuesday.

Tuesday /'tju:zdeɪ, -dɪ/ *n. & adv.* ● *n.* the third day of the week, following Monday. ● *adv. colloq.* **1** on Tuesday. **2** (**Tuesdays**) on Tuesdays; each Tuesday. [OE *Tiwesdæg* f. *Tiw* TYR, representing L *dies Marti* day of Mars]

tufa /'tju:fə/ *n.* **1** a porous rock composed of calcium carbonate and formed round mineral springs. **2** = TUFF. □ **tufaceous** /tju:'feɪʃəs/ *adj.* [It., var. of *tufo*: see TUFF]

tuff /tʌf/ *n.* rock formed by the consolidation of volcanic ash. □ **tuffaceous** /tʌ'feɪʃəs/ *adj.* [F *tuf*, *tuffe* f. It. *tufo* f. LL *tofus*, L TOPHUS]

tuffet /'tʌfɪt/ *n.* **1** = TUFT n. 1. **2** a low seat. [var. of TUFT]

tuft /tʌft/ *n. & v.* ● *n.* **1** a bunch or collection of threads, grass, feathers, hair, etc., held or growing together at the base. **2** *Anat.* a bunch of small blood vessels. ● *v.* **1** *tr.* provide with a tuft or tufts. **2** *tr.* make depressions at regular intervals in (a mattress etc.) by passing a thread through. **3** *intr.* grow in tufts. □ **tufty** *adj.* [ME, prob. f. OF *tofe*, *toffe*, of unkn. orig.: for *-t* cf. GRAFT[1]]

tufted /'tʌftɪd/ *adj.* **1** having or growing in a tuft or tufts. **2** (of a bird) having a tuft of feathers on the head. □ **tufted duck** a small Old World freshwater duck, *Aythya fuligula*, with black and white plumage in the male and a drooping crest.

Tu Fu /tu: 'fu:/ (also **Du Fu** /du:/) (AD 712–70), Chinese poet. He is noted for his bitter satiric poems attacking social injustice and corruption at court. The finest of these were written during the turbulent 750s, in which he suffered much personal hardship.

tug /tʌg/ *v. & n.* ● *v.* (**tugged**, **tugging**) **1** *tr.* & (foll. by *at*) *intr.* pull hard or violently; jerk (*tugged it from my grasp*; *tugged at my sleeve*). **2** *tr.* tow (a ship etc.) by means of a tugboat. ● *n.* **1** a hard, violent, or jerky pull (*gave a tug on the rope*). **2** a sudden strong emotional feeling (*felt a tug as I watched them go*). **3** a small powerful boat for towing larger boats and ships. **4** an aircraft towing a glider. **5** (of a horse's harness) a loop from a saddle supporting a shaft or trace. □ **tug of love** *colloq.* a dispute over the custody of a child. **tug of war 1** a trial of strength between two sides pulling against each other on a rope. **2** a decisive or severe contest. □ **tugger** *n.* [ME *togge*, *tugge*, intensive f. Gmc: see TOW[1]]

tugboat /'tʌgbəʊt/ *n.* = TUG n. 3.

tugrik /'tu:gri:k/ *n.* the basic monetary unit of Mongolia, equal to 100 mongos. [Mongolian]

tui /'tu:ɪ/ *n.* a large honeyeater, *Prosthemadura novaeseelandiae*, native to New Zealand, having glossy bluish-black plumage with two white tufts at the throat. [Maori]

Tuileries /'twi:lərɪ/ (in full **Tuileries Gardens**) formal gardens next to the Louvre in Paris, laid out by André Le Nôtre in the mid-17th century. The gardens are all that remain of the Tuileries Palace, a royal residence begun in 1564 and burnt down in 1871 during the Commune of Paris.

tuition /tju:'ɪʃ(ə)n/ *n.* **1** teaching or instruction, esp. if paid for (*driving tuition*; *music tuition*). **2** a fee for this. □ **tuitional** *adj.* [ME f. OF f. L *tuitio -onis* f. *tueri tuit-* watch, guard]

Tula /'tu:lə/ **1** an industrial city in European Russia, to the south of Moscow; pop. (1990) 543,000. **2** the ancient capital city of the Toltecs in Mexico, usually identified with a site near the town of Tula in Hidalgo State, central Mexico.

tularaemia /ˌtu:lə'ri:mɪə/ *n.* (US **tularemia**) a severe infectious disease of animals transmissible to humans, caused by the bacterium *Pasteurella tularense* and characterized by ulcers at the site of infection, fever, and loss of weight. □ **tularaemic** *adj.* [mod.L f. *Tulare* County in California, where it was first observed]

tulip /'tju:lɪp/ *n.* **1** a bulbous spring-flowering liliaceous plant of the genus *Tulipa*, esp. one of the many cultivated forms with showy cup-shaped flowers of various colours and markings. **2** a flower of this plant. □ **tulip-root** a disease of oats etc. causing the base of the stem to swell. **tulip tree** an ornamental North American tree, *Liriodendron tulipifera*, with tulip-like flowers and lobed leaves. **tulip-wood** a fine-grained pale timber obtained from the tulip tree. [orig. *tulipa(n)* f. mod.L *tulipa* f. Turk. *tul(i)band* f. Pers. *dulband* TURBAN (from the shape of the expanded flower)]

Tull /tʌl/, Jethro (1674–1741), English agriculturalist. He had a profound effect on agricultural practice with his invention of the seed drill (1701), which could sow seeds in accurately spaced rows at a controlled rate. This made possible the control of weeds by horse-drawn hoe, reducing the need for farm labourers.

Tullamore /ˌtʊlə'mɔː(r)/ the county town of Offaly, in the Republic of Ireland; pop. (1991) 8,620.

tulle /tjuːl/ n. a soft fine silk etc. net for veils and dresses. [*Tulle* in SW France, where it was first made]

Tulsa /ˈtʌlsə/ a port on the Arkansas river in NE Oklahoma; pop. (1990) 367,300.

Tulsidas /ˈtʊlsɪˌdɑːs/ (c.1543–1623), Indian poet. He was a leading Hindu devotional poet who is chiefly remembered for the *Ramcaritmanas* (c.1574–7), a work consisting of seven cantos and based on the Sanskrit epic the Ramayana. The poet's expression of worship or bhakti for Rama led to the cult of Rama (rather than that of Krishna) dominating the Hindu culture of northern India.

tum /tʌm/ n. colloq. stomach. [abbr. of TUMMY]

tumble /ˈtʌmb(ə)l/ v. & n. ● v. **1** intr. & tr. fall or cause to fall suddenly, clumsily, or headlong. **2** intr. fall rapidly in amount etc. (*prices tumbled*). **3** intr. (often foll. by *about, around*) roll or toss erratically or helplessly to and fro. **4** intr. move or rush in a headlong or blundering manner (*the children tumbled out of the car*). **5** intr. (often foll. by *to*) colloq. grasp the meaning or hidden implication of an idea, circumstance, etc. (*they quickly tumbled to our intentions*). **6** tr. overturn; fling or push roughly or carelessly. **7** intr. perform acrobatic feats, esp. somersaults. **8** tr. rumple or disarrange; pull about; disorder. **9** tr. dry (washing) in a tumble-drier. **10** tr. clean (castings, gemstones, etc.) in a tumbling-barrel. **11** intr. (of a pigeon) turn over backwards in flight. ● n. **1** a sudden or headlong fall. **2** a somersault or other acrobatic feat. **3** an untidy or confused state. □ **tumble-drier** (or **-dryer**) n. a machine for drying washing in a heated rotating drum. **tumble-dry** v.tr. & intr. (**-dries, -dried**) dry in a tumble-drier. **tumbling-barrel** (or **-box** etc.) a revolving device containing an abrasive substance, in which castings, gemstones, etc., are cleaned by friction. **tumbling-bay 1** the outfall of a river, reservoir, etc. **2** a pool into which this flows. [ME *tumbel* f. MLG *tummelen*, OHG *tumalōn* frequent. of *tūmōn*: cf. OE *tumbian* dance]

tumbledown /ˈtʌmb(ə)lˌdaʊn/ adj. falling or fallen into ruin; dilapidated.

tumbler /ˈtʌmblə(r)/ n. **1** a drinking-glass with no handle or foot (formerly with a rounded bottom so as not to stand upright). **2** an acrobat, esp. one performing somersaults. **3** (in full **tumbler-drier**) = tumble-drier. **4 a** a pivoted piece in a lock that holds the bolt until lifted by a key. **b** a notched pivoted plate in a gunlock. **5** a kind of pigeon that turns over backwards in flight. **6** an electrical switch worked by pushing a small sprung lever. **7** a toy figure that rocks when touched. **8** = tumbling-barrel (see TUMBLE). □ **tumblerful** n. (pl. **-fuls**).

tumbleweed /ˈtʌmb(ə)lˌwiːd/ n. N. Amer. & Austral. a plant of arid areas that forms a globular bush that breaks off in late summer and is tumbled about by the wind, esp. *Amaranthus albus*.

tumbrel /ˈtʌmbrəl/ n. (also **tumbril** /-brɪl/) hist. **1** an open cart in which condemned persons were conveyed to their execution, esp. to the guillotine during the French Revolution. **2** a two-wheeled covered cart for carrying tools, ammunition, etc. **3** a cart that tips to empty its load, esp. one carrying dung. [ME f. OF *tumberel, tomberel* f. *tomber* fall]

tumefy /ˈtjuːmɪˌfaɪ/ v. (**-ies, -ied**) **1** intr. swell, inflate; be inflated. **2** tr. cause to do this. □ **tumefacient** /ˌtjuːmɪˈfeɪʃ(ə)nt/ adj. **tumefaction** /-ˈfækʃ(ə)n/ n. [F *tuméfier* f. L *tumefacere* f. *tumere* swell]

tumescent /tjuːˈmes(ə)nt/ adj. **1** becoming tumid; swelling. **2** swelling as a response to sexual stimulation. □ **tumescence** n. **tumescently** adv. [L *tumescere* (as TUMEFY)]

tumid /ˈtjuːmɪd/ adj. **1** (of parts of the body etc.) swollen, inflated. **2** (of a style etc.) inflated, bombastic. □ **tumidly** adv. **tumidness** n. **tumidity** /tjuːˈmɪdɪtɪ/ n. [L *tumidus* f. *tumere* swell]

tummy /ˈtʌmɪ/ n. (pl. **-ies**) colloq. the stomach. □ **tummy-ache** an abdominal pain; indigestion. **tummy-button** the navel. [childish pronunc. of STOMACH]

tumour /ˈtjuːmə(r)/ n. (US **tumor**) a swelling, esp. from an abnormal growth of tissue, which may be either benign or malignant. □ **tumorous** adj. [L *tumor* f. *tumere* swell]

tump /tʌmp/ n. esp. dial. a hillock; a mound; a tumulus. [16th c.; orig. unkn.]

tumult /ˈtjuːmʌlt/ n. **1** an uproar or din, esp. of a disorderly crowd. **2** an angry demonstration by a mob; a riot; a public disturbance. **3** a conflict of emotions in the mind. [ME f. OF *tumulte* or L *tumultus*]

tumultuous /tjuːˈmʌltjʊəs/ adj. **1** noisily vehement; uproarious; making a tumult (*a tumultuous welcome*). **2** disorderly. **3** agitated. □ **tumultuously** adv. **tumultuousness** n. [OF *tumultuous* or L *tumultuosus* (as TUMULT)]

tumulus /ˈtjuːmjʊləs/ n. (pl. **tumuli** /-ˌlaɪ/) an ancient burial mound or barrow. □ **tumular** adj. [L f. *tumere* swell]

tun /tʌn/ n. & v. ● n. **1** a large beer or wine cask. **2** a brewer's fermenting-vat. **3** a measure of capacity, equal to 252 wine gallons. ● v.tr. (**tunned, tunning**) store (wine etc.) in a tun. [OE *tunne* f. med.L *tunna*, prob. of Gaulish orig.]

tuna[1] /ˈtjuːnə/ n. (pl. same or **tunas**) **1** a large marine game-fish of the family Scombridae, found in tropical and warm waters, having a rounded body and pointed snout and used for food; esp. the *common* or *bluefin tuna* (*Thunnus thynnus*). Also called *tunny*. **2** (in full **tuna-fish**) the flesh of the tuna, usu. tinned in oil or brine. [Amer. Sp., perh. f. Sp. *atún*]

tuna[2] /ˈtjuːnə/ n. **1** a prickly pear cactus, esp. *Opuntia tuna* of Central America and the West Indies. **2** the edible fruit of this. [Sp. f. Haitian]

Tunb Islands /ˈtuːnəb/ two small islands (Greater and Lesser Tunb) in the Persian Gulf, administered by the emirate of Ras al Khaimah until occupied by Iran in 1971.

Tunbridge Wells /ˌtʌnbrɪdʒ ˈwelz/ (official name **Royal Tunbridge Wells**) a spa town in Kent, SE England; pop. (1981) 58,140. Founded in the 1630s after the discovery of iron-rich springs, the town was patronized by royalty throughout the 17th and 18th centuries. It was awarded the status of a Royal borough by Edward VII in 1909.

tundish /ˈtʌndɪʃ/ n. **1** a wooden funnel used esp. in brewing. **2** an intermediate reservoir in metal-founding.

tundra /ˈtʌndrə/ n. a low-lying treeless cold region, chiefly in the Arctic, with a mainly marshy surface and underlying permafrost. □ **tundra swan** a migratory swan, *Cygnus columbianus*, breeding in the Arctic, with a black and yellow bill; esp. an American race that winters in the US (cf. BEWICK'S SWAN). [Lappish]

tune /tjuːn/ n. & v. ● n. a melody with or without harmony. ● v. **1** tr. put (a musical instrument) in tune. **2 a** tr. adjust (a radio receiver etc.) to the particular frequency of the required signals. **b** intr. (foll. by *in*) adjust a radio receiver to the required signal (*tuned in to Radio 2*). **3** tr. adjust (an engine etc.) to run smoothly and efficiently. **4** tr. (foll. by *to*) adjust or adapt to a required or different purpose, situation, etc. **5** intr. (foll. by *with*) be in harmony with. □ **in tune 1** having the correct pitch or intonation (*sings in tune*). **2** (usu. foll. by *with*) harmonizing with one's company, surroundings, etc. **out of tune 1** not having the correct pitch or intonation (*always plays out of tune*). **2** (usu. foll. by *with*) clashing with one's company etc. **to the tune of** colloq. to the considerable sum or amount of. **tuned in** (often foll. by *to*) colloq. acquainted, in rapport, or up to date (with). **tune up 1** (of a musician) bring one's instrument to the proper or uniform pitch. **2** begin to play or sing. **3** bring to the most efficient condition. □ **tunable** adj. (also **tuneable**). [ME: unexpl. var. of TONE]

tuneful /ˈtjuːnfʊl/ adj. melodious, musical. □ **tunefully** adv. **tunefulness** n.

tuneless /ˈtjuːnlɪs/ adj. **1** unmelodious, unmusical. **2** out of tune. □ **tunelessly** adv. **tunelessness** n.

tuner /ˈtjuːnə(r)/ n. **1** a person who tunes musical instruments, esp. pianos. **2** a device for tuning a radio receiver. **3** an electronic device for tuning a guitar etc.

tung /tʌŋ/ n. (in full **tung-tree**) a tree of the genus *Aleurites*, native to China and Japan, bearing poisonous fruits containing seeds that yield oil. □ **tung oil** this oil used in paints and varnishes. [Chin. *tong*]

tungstate /ˈtʌŋsteɪt/ n. Chem. a salt in which the anion contains both tungsten and oxygen, especially one of the anion WO_4^{2-}.

tungsten /ˈtʌŋstən/ n. a hard steel-grey metallic chemical element (atomic number 74; symbol **W**). (*See note below.*) □ **tungsten carbide** a very hard black solid (chem. formula: WC) used to make cutting tools and dies. [Sw. f. *tung* heavy + *sten* stone]

▪ A transition metal, tungsten was identified as an element by Scheele in 1781. The chief ores are wolframite and scheelite. Tungsten has a very high melting-point (3410°C), and is the chief material from which electric light filaments are made; it is also used in alloy steels. The symbol W is from the alternative name *wolfram*.

Tungus /ˈtʊŋʊs, tʊŋˈuːs/ n. (pl. same) **1** a member of a Mongoloid people of eastern Siberia. **2** an Altaic language or group of languages spoken in parts of Siberia, since the 1930s set down in an alphabet based on the Russian alphabet. □ **Tungusian** /tʊŋˈgjuːsɪən/ adj. & n. **Tungusic** /-ˈgjuːsɪk/ adj. & n. [Yakut]

Tunguska /tʊŋˈguːskə/ two rivers in Siberian Russia, the *Lower Tunguska* and *Stony Tunguska*, flowing westwards into the Yenisei river through the forested, sparsely populated Tunguska Basin. The area was the scene, in 1908, of a devastating explosion believed to have been due to the disintegration in the atmosphere of a meteorite or small comet.

tunic /ˈtjuːnɪk/ n. **1 a** a close-fitting short coat of police or military etc. uniform. **b** a loose often sleeveless garment usu. reaching to about the knees, as worn in ancient Greece and Rome. **c** any of various loose pleated dresses gathered at the waist with a belt or cord. **d** a tunicle. **2** Zool. the rubbery outer coat of an ascidian etc. **3** Bot. **a** any of the concentric layers of a bulb. **b** the tough covering of a part of this. **4** Anat. a membrane enclosing or lining an organ. [F tunique or L tunica]

tunica /ˈtjuːnɪkə/ n. (pl. **tunicae** /-ˌkiː/) Bot. & Anat. = TUNIC 3, 4. [L]

tunicate /ˈtjuːnɪkət, -ˌkeɪt/ n. & adj. ● n. Zool. a marine animal of the subphylum Urochordata, having a rubbery or hard outer coat and including the ascidians or sea squirts. ● adj. **1** Zool. of or relating to this subphylum. **2** (also **tunicated** /-ˌkeɪtɪd/) Bot. having concentric layers, as in an onion bulb. [L tunicatus past part. of tunicare clothe with a tunic (as TUNICA)]

tunicle /ˈtjuːnɪk(ə)l/ n. Eccl. a short vestment worn by a bishop or subdeacon at the Eucharist etc. [ME f. OF tunicle or L tunicula dimin. of TUNICA]

tuning /ˈtjuːnɪŋ/ n. the process or a system of putting a musical instrument into tune. □ **tuning-fork** a two-pronged steel fork that gives a particular note when struck, used in tuning. **tuning-peg** (or **pin** etc.) a peg or pin etc. attached to the strings of a stringed instrument and turned to alter their tension in tuning.

Tunis /ˈtjuːnɪs/ the capital of Tunisia, a port on the Mediterranean coast of North Africa; pop. (1984) 596,650.

Tunisia /tjuːˈnɪzɪə/ a country in North Africa; pop. (est. 1991) 8,223,000; official language, Arabic; capital, Tunis. Tunisia has a Mediterranean coastline and extends south into the Sahara Desert. Phoenician coastal settlements developed into the commercial empire of Carthage (near modern Tunis), which after defeat in the Punic Wars became a Roman province. The area was conquered by the Vandals in the 5th century AD and subsequently by the Arabs (7th century); in the 16th century it became part of the Ottoman Empire. Governed by beys under loose Turkish rule, Tunisia was a centre of piratical activity before the establishment of a French protectorate in 1886. The rise of nationalism led to independence and the establishment of a republic in 1956-7. □ **Tunisian** adj. & n.

tunnel /ˈtʌn(ə)l/ n. & v. ● n. **1** an artificial underground passage through a hill or under a road or river etc., esp. for a railway or road to pass through, or in a mine. **2** an underground passage dug by a burrowing animal. **3** a prolonged period of difficulty or suffering (esp. in metaphors, e.g. the end of the tunnel). **4** a tube containing a propeller shaft etc. ● v. (**tunnelled, tunnelling**; US **tunneled, tunneling**) **1** intr. (foll. by through, into, etc.) make a tunnel through (a hill etc.). **2** tr. make (one's way) by tunnelling. **3** intr. (often as **tunnelling** n.) Physics (of a particle) pass through a potential barrier, esp. by virtue of a quantum-mechanical effect whereby a particle has a finite probability of existing on the other side of the barrier even if it has less energy than the height of the barrier. □ **tunnel diode** Electronics a two-terminal semiconductor diode using tunnelling electrons to perform high-speed switching operations (see also ESAKI). **tunnel-kiln** a kiln in which ceramic ware is carried on trucks along a continuously heated passage. **tunnel-net** a fishing-net wide at the mouth and narrow at the other end. **tunnel vision 1** vision that is defective in not adequately including objects away from the centre of the field of view. **2** colloq. inability to see more than a single or limited view. □ **tunneller** n. [ME f. OF tonel dimin. of tonne TUN]

tunny /ˈtʌnɪ/ n. (pl. same or **-ies**) = TUNA[1]. [F thon f. Prov. ton, f. L thunnus f. Gk thunnos]

tup /tʌp/ n. & v. ● n. **1** esp. Brit. a male sheep; a ram. **2** the striking-head of a pile-driver etc. ● v.tr. (**tupped, tupping**) esp. Brit. (of a ram) copulate with (a ewe). [ME toje, tupe, of unkn. orig.]

Tupamaro /ˌtuːpəˈmɑːrəʊ/ n. (pl. **-os**) a member of the Movimento de Liberación Nacional, a Marxist urban guerrilla organization in Uruguay founded in 1963. [Tupac Amaru, the name of two Inca leaders]

Tupelo /ˈtjuːpəˌləʊ/ a city in NE Mississippi; pop. (1990) 30,685.

tupelo /ˈtjuːpəˌləʊ/ n. (pl. **-os**) **1** a chiefly North American deciduous tree of the genus Nyssa, with colourful foliage and growing in swampy conditions. **2** the wood of this tree. [Creek f. ito tree + opílwa swamp]

Tupi /ˈtuːpɪ/ n. & adj. ● n. (pl. same or **Tupis**) **1** a member of a native people occupying the Amazon valley. **2** the language of this people. ● adj. of or relating to this people or language. [Amer. Indian name]

Tupi-Guarani /ˌtuːpɪˈɡwɑːrənɪ/ n. & adj. ● n. (pl. same or **Tupi-**

Guaranis) **1** a member of a South American native people of Tupi, Guarani, or other related stock. **2** a South American language family whose principal members are Tupi and Guarani. ● adj. of or relating to the Tupi-Guarani or their languages.

tuppence /ˈtʌp(ə)ns/ n. Brit. = TWOPENCE. [phonet. spelling]

tuppenny /ˈtʌp(ə)nɪ/ adj. Brit. = TWOPENNY. [phonet. spelling]

Tupperware /ˈtʌpəˌweə(r)/ n. propr. a range of plastic containers for storing food. [Tupper, name of the Amer. manufacturer, + WARE[1]]

tuque /tuːk/ n. a Canadian stocking cap. [Canad. F form of TOQUE]

turaco /ˈtʊərəˌkəʊ/ n. (also **touraco**) (pl. **-os**) an African bird of the family Musophagidae, with brilliant purple, green, and crimson plumage and a prominent crest. [F f. W.Afr. name]

Turanian /tjʊˈreɪnɪən/ n. & adj. ● n. the languages of central Asia that are neither Semitic nor Indo-European, esp. the (supposed) group comprising Uralic and Altaic (formerly called Ural-Altaic). ● adj. of or relating to these languages. [Pers. Tūrān region beyond the Oxus]

turban /ˈtɜːb(ə)n/ n. **1** a man's head-dress of cotton or silk wound round a cap or the head, worn esp. by Muslims and Sikhs. **2** a woman's head-dress or hat resembling this. □ **turbaned** adj. [16th c. (also tulbant etc.), ult. f. Turk. tülbent f. Pers. dulband: cf. TULIP]

turbary /ˈtɜːbərɪ/ n. (pl. **-ies**) Brit. **1** the right of digging turf on common ground or on another's ground. **2** a place where turf or peat is dug. [ME f. AF turberie, OF tourberie f. tourbe TURF]

turbellarian /ˌtɜːbɪˈleərɪən/ n. & adj. Zool. ● n. a usu. free-living flatworm of the class Turbellaria, having a ciliated surface. ● adj. of or relating to this class. [mod.L Turbellaria f. L turbella dimin. of turba crowd: see TURBID]

turbid /ˈtɜːbɪd/ adj. **1** (of a liquid or colour) muddy, thick; not clear. **2** (of a style etc.) confused, disordered. □ **turbidly** adv. **turbidness** n. **turbidity** /tɜːˈbɪdɪtɪ/ n. [L turbidus f. turba a crowd, a disturbance]

turbinate /ˈtɜːbɪnət/ adj. **1** shaped like a spinning-top or inverted cone. **2** (of a shell) with whorls decreasing rapidly in size. **3** Anat. (esp. of some nasal bones) shaped like a scroll. □ **turbinal** adj. **turbination** /ˌtɜːbɪˈneɪʃ(ə)n/ n. [L turbinatus (as TURBINE)]

turbine /ˈtɜːbaɪn/ n. a machine for producing continuous mechanical power, in which a fluid (water, steam, air, or a gas) is accelerated to a high speed in a channel or nozzle and the resulting jet or jets directed at a rotating wheel with radial vanes or scoops. In the 19th century water turbines were developed, chiefly in France and Germany, to drive machinery in factories. The steam turbine was suggested 2,000 years ago by Hero of Alexandria, but the first practical one was devised by Sir Charles Parsons in 1884. Steam turbines have replaced the steam engine in electrical power stations, while hydroelectric power stations use water turbines. (See also PARSONS, GAS TURBINE.) [F f. L turbo -binis spinning-top, whirlwind]

turbit /ˈtɜːbɪt/ n. a breed of domestic pigeon of stout build with a neck frill and short beak. [app. f. L turbo top, from its figure]

turbo /ˈtɜːbəʊ/ n. (pl. **-os**) **1** = TURBOCHARGER. **2** a motor vehicle equipped with this.

turbo- /ˈtɜːbəʊ/ comb. form turbine.

turbocharge /ˈtɜːbəʊˌtʃɑːdʒ/ v.tr. (esp. as **turbocharged** adj.) equip with a turbocharger.

turbocharger /ˈtɜːbəʊˌtʃɑːdʒə(r)/ n. a supercharger driven by a turbine powered by the engine's exhaust gases.

turbo-diesel /ˈtɜːbəʊˌdiːz(ə)l/ n. **1** a turbocharged diesel engine. **2** a vehicle powered by this.

turbofan /ˈtɜːbəʊˌfæn/ n. Aeron. **1** a jet engine in which a turbine-driven fan provides additional thrust. (See JET PROPULSION.) **2** an aircraft powered by this.

turbojet /ˈtɜːbəʊˌdʒet/ n. Aeron. **1** a jet engine of the basic form, with a turbine-driven compressor. (See JET PROPULSION.) **2** an aircraft powered by this.

turboprop /ˈtɜːbəʊˌprɒp/ n. Aeron. **1** a jet engine in which a turbine is used as in a turbojet and also to drive a propeller. **2** an aircraft powered by this.

turboshaft /ˈtɜːbəʊˌʃɑːft/ n. a gas turbine that powers a shaft for driving heavy vehicles, generators, pumps, etc.

turbosupercharger /ˌtɜːbəʊˈsuːpəˌtʃɑːdʒə(r), -ˈsjuːpəˌtʃɑːdʒə(r)/ n. = TURBOCHARGER.

turbot /ˈtɜːbət/ n. **1** a flatfish, Scophthalmus maximus, having large bony tubercles on the body and head and prized as food. **2** a similar flatfish, e.g. the halibut. [ME f. OF f. OSw. törnbut f. törn thorn + but BUTT[3]]

turbulence /ˈtɜːbjʊləns/ n. **1** an irregularly fluctuating flow of air or fluid. **2** Meteorol. stormy conditions as a result of atmospheric disturbance. **3** a disturbance, commotion, or tumult.

turbulent /ˈtɜːbjʊlənt/ adj. **1** disturbed; in commotion. **2** (of a flow of air etc.) varying irregularly; causing disturbance. **3** tumultuous. **4** insubordinate, riotous. □ **turbulently** adv. [L turbulentus f. turba crowd]

Turco /ˈtɜːkəʊ/ n. (pl. **-os**) hist. an Algerian soldier in the French army. [Sp., Port., & It., = TURK]

Turco- /ˈtɜːkəʊ/ comb. form (also **Turko-**) Turkish; Turkish and. [med.L (as TURK)]

Turcoman var. of TURKOMAN.

turd /tɜːd/ n. coarse sl. **1** a lump of excrement. **2** a term of contempt for a person. [OE tord f. Gmc]

turdoid /ˈtɜːdɔɪd/ adj. thrushlike. [L turdus THRUSH¹]

tureen /tjʊəˈriːn, təˈriːn/ n. a deep covered dish for serving soup etc. [earlier terrine, -ene f. F terrine large circular earthenware dish, fem. of OF terrin earthen ult. f. L terra earth]

turf /tɜːf/ n. & v. ● n. (pl. **turfs** or **turves** /tɜːvz/) **1 a** a layer of grass etc. with earth and matted roots as the surface of grassland. **b** a piece of this cut from the ground. **2** a slab of peat for fuel. **3** (prec. by the) **a** horse-racing generally. **b** a general term for racecourses. **4** sl. a person's territory or sphere of influence. ● v.tr. **1** cover (ground) with turf. **2** (foll. by out) esp. Brit. colloq. expel or eject (a person or thing). □ **turf accountant** Brit. a bookmaker. [OE f. Gmc]

Turfan Depression /ˈtʊəfæn, tʊəˈfæn/ (also **Turpan**) a low-lying area in Xinjiang, western China, descending to 154 m (505 ft) below sea level, with an area of 50,000 sq. km (20,000 sq. miles). It is China's lowest point below sea level.

turfman /ˈtɜːfmən/ n. (pl. **-men**) esp. US a devotee of horse-racing.

turfy /ˈtɜːfɪ/ adj. (**turfier, turfiest**) like turf; grassy.

Turgenev /tɜːˈɡeɪnjef, tʊəˈɡenjef/, Ivan (Sergeevich) (1818–83), Russian novelist, dramatist, and short-story writer. From the 1850s he spent much of his life abroad, especially in Paris and Baden-Baden. His play A Month in the Country (1850) was followed by the prose work A Sportsman's Sketches (1852), which condemned the institution of serfdom. He subsequently wrote a series of novels examining individual lives to illuminate the social, political, and philosophical issues of the day, as in Rudin (1856) and Fathers and Sons (1862); in the latter, Turgenev depicted the rise of nihilism in Russia through his hero Bazarov.

turgescent /tɜːˈdʒes(ə)nt/ adj. becoming turgid; swelling. □ **turgescence** n.

turgid /ˈtɜːdʒɪd/ adj. **1** swollen, inflated, enlarged. **2** (of language) pompous, bombastic. □ **turgidly** adv. **turgidness** n. **turgidity** /tɜːˈdʒɪdɪtɪ/ n. [L turgidus f. turgere swell]

turgor /ˈtɜːɡə(r)/ n. Bot. the rigidity of cells due to the absorption of water. [LL (as TURGID)]

Turin /tjʊəˈrɪn/ (Italian **Torino** /toˈriːno/) a city in NW Italy on the River Po, capital of Piedmont region; pop. (1990) 991,870. Turin was the capital of the kingdom of Sardinia from 1720 and a centre of the Risorgimento in the 19th century. It was the first capital of a unified Italy (1861–4).

Turing /ˈtjʊərɪŋ/, Alan Mathison (1912–54), English mathematician. He developed the concept of a theoretical computing machine in 1937, a key step in the development of the first computer. Turing carried out important work on code-breaking during the Second World War, after which he worked on early computers at the National Physical Laboratory and the University of Manchester. He also investigated artificial intelligence and suggested criteria for a machine which would respond like a human being. Turing committed suicide after being prosecuted for homosexuality and forced to undergo hormone treatment.

Turing test n. Computing a test for intelligence in a computer, which requires that a human should be unable to distinguish it from another human by the replies to questions put to both. [TURING]

Turin Shroud a relic, preserved at Turin since 1578, venerated as the winding-sheet in which Christ's body was wrapped for burial. It bears the imprint of the front and back of a human body as well as markings that correspond to the traditional stigmata. Scientific tests carried out in 1988 dated the shroud to the 13th–14th centuries.

turion /ˈtʊərɪən/ n. Bot. **1** a young shoot or sucker arising from an

underground bud. **2** a bud formed by certain aquatic plants. [F f. L turio -onis shoot]

Turk /tɜːk/ n. **1 a** a native or national of Turkey in SE Europe and Asia Minor. **b** a person of Turkish descent. **2** a member of a central Asian people from whom the Ottomans derived, speaking Turkic languages. **3** offens. a ferocious, wild, or unmanageable person. □ **Turk's cap** a martagon lily or other plant with turban-like flowers. **Turk's head** a turban-like ornamental knot. [ME, = F Turc, It. etc. Turco, med.L Turcus, Pers. & Arab. Turk, of unkn. orig.]

Turkana /tɜːˈkɑːnə/ n. & adj. ● n. (pl. same) **1** a member of an East African people living between Lake Turkana and the Nile. **2** the Nilotic language of the Turkana. ● adj. of or relating to the Turkana or their language. [Nilotic]

Turkana, Lake a salt lake in NW Kenya, with no outlet. It was discovered in 1888 by the Hungarian explorer Count Teleki (1845–1916), who named it Lake Rudolf after the Crown Prince of Austria. It was given its present name in 1979.

Turkestan /ˌtɜːkɪˈstɑːn/ (also **Turkistan**) a region of central Asia between the Caspian Sea and the Gobi Desert, inhabited mainly by Turkic peoples. It is divided by the Pamir and Tien Shan mountains into western Turkestan, which comprises present-day Turkmenistan, Kazakhstan, Uzbekistan, Tajikistan, and Kyrgyzstan, and eastern Turkestan, which comprises the Xinjiang autonomous region of China. Western Turkestan came under the control of the Russian empire in the 1850s, later forming part of the Soviet Union. In 1924 separate republics were created for each of the main ethnic groups.

Turkey /ˈtɜːkɪ/ a country comprising the whole of the Anatolian peninsula in western Asia, with a small enclave in SE Europe to the west of Istanbul; pop. (1990) 56,473,000; official language, Turkish; capital, Ankara. Turkey was the centre of the Ottoman Empire, established in the late Middle Ages and largely maintained until its collapse at the end of the First World War, in which Turkey supported the Central Powers. The nationalist leader Kemal Atatürk drove out the Greeks who had occupied part of Anatolia and established the modern republic of Turkey in the 1920s. Turkey was neutral in the Second World War but is a member of NATO. The country has a large Kurdish minority in the east and south-east; fighting against Kurdish secessionists broke out in 1984 and has continued sporadically.

turkey /ˈtɜːkɪ/ n. (pl. **-eys**) **1** a large mainly domesticated game bird, Meleagris gallopavo, orig. of North America, with a bald head and (in the male) red wattles. (See note below.) **2** the flesh of the turkey as food. **3** esp. N. Amer. sl. **a** a theatrical failure; a flop. **b** a stupid or inept person. □ **talk turkey** N. Amer. colloq. talk frankly and straightforwardly; get down to business. **turkey vulture** a common American vulture, Cathartes aura, with a bare reddish head (also called (US) turkey buzzard). [16th c.: short for turkeycock or turkeyhen, orig. applied to the guinea-fowl which was imported through Turkey, and then erron. to the Amer. bird]

▪ The turkey had already been domesticated by the Aztecs when Mexico was first visited by Europeans in 1518. It was soon introduced into Europe and thence to Britain, where it ousted peacock and swan as a table fowl. The wild bird has dark plumage with a green or bronze sheen, though domesticated birds are often white. It is traditionally eaten on festive occasions such as Christmas and, in the US, Thanksgiving.

Turkey carpet n. = Turkish carpet.

turkeycock /ˈtɜːkɪˌkɒk/ n. **1** a male turkey. **2** a pompous or self-important person.

Turkey red n. **1** a scarlet pigment obtained from the madder or alizarin. **2** a cotton cloth dyed with this.

Turki /ˈtɜːkɪ/ n. & adj. ● n. **1** the Turkic languages, especially those of central Asia, collectively. **2** a member of a Turkic-speaking people. ● adj. of or relating to these languages or their speakers. [Pers. turkī (as TURK)]

Turkic /ˈtɜːkɪk/ adj. & n. ● adj. of, relating to, or designating a large group of Altaic languages including Turkish, Azerbaijani, Kyrgyz, Tartar, and Kazakh, or the people speaking them. ● n. the Turkic languages collectively.

Turkish /ˈtɜːkɪʃ/ adj. & n. ● adj. of or relating to Turkey or to the Turks or their language. ● n. the official language of Turkey, spoken by about 50 million people, the most important of the Turkic group. It was originally written in Arabic script but changed over to the Roman alphabet in 1928. □ **Turkish bath 1** a hot-air or steam bath followed by washing, massage, etc. **2** (in sing. or pl.) a building for this. **Turkish carpet** a wool carpet with a thick pile and traditional bold design.

Turkish coffee a strong black coffee. **Turkish delight** a sweet of flavoured gelatine coated in powdered sugar. **Turkish towel** a towel made of cotton terry.

Turkistan see TURKESTAN.

Turkmenistan /tɜːkˌmenɪˈstɑːn/ (also **Turkmenia** /-ˈmiːnɪə/) a republic in central Asia, lying between the Caspian Sea and Afghanistan; pop. (est. 1992) 3,861,000; languages, Turkoman (official), Russian; capital, Ashgabat. Turkmenistan is dominated by the Karakum Desert, which occupies about 90 per cent of the country. Previously part of Turkistan, from 1924 it formed a separate constituent republic of the USSR; Turkmenistan became an independent republic within the Commonwealth of Independent States in 1991. □ **Turkmen** /ˈtɜːkmən/ adj. & n.

Turko- var. of TURCO-.

Turkoman /ˈtɜːkəʊmən/ n. & adj. ● n. (also **Turcoman**) (pl. **-mans**) **1** a member of any of various Turkic peoples of Turkmenistan. **2** the language of these peoples. ● adj. of or relating to these peoples or their language. □ **Turkoman carpet** a traditional rich-coloured carpet with a soft long nap. [Pers. Turkumān (as TURK, mānistan resemble)]

Turks and Caicos Islands /tɜːks, ˈkeɪkɒs/ a British dependency in the West Indies, comprising two island groups between Haiti and the Bahamas; pop. (1990) 12,350; capital, Cockburn Town (on the island of Grand Turk).

Turku /ˈtʊəkuː/ (called in Swedish Åbo) an industrial port in SW Finland; pop. (1990) 159,180. Founded in the 11th century, Turku was capital of Finland until 1812.

turmeric /ˈtɜːmərɪk/ n. **1** a tropical Asian plant, Curcuma longa, of the ginger family, yielding aromatic rhizomes used as a spice and for yellow dye. **2** this powdered rhizome used as a spice esp. in curry-powder. [16th-c. forms tarmaret etc. perh. f. F terre mérite and mod.L terra merita, of unkn. orig.]

turmoil /ˈtɜːmɔɪl/ n. **1** violent confusion; agitation. **2** din and bustle. [16th c.: orig. unkn.]

turn /tɜːn/ v. & n. ● v. **1** tr. & intr. move around a point or axis so that the point or axis remains in a central position; give a rotary motion to or receive a rotary motion (turned the wheel; the wheel turns; the key turns in the lock). **2** tr. & intr. change in position so that a different side, end, or part becomes outermost or uppermost etc.; invert or reverse or cause to be inverted or reversed (turned inside out; turned it upside down). **3** a tr. give a new direction to (turn your face this way). **b** intr. take a new direction (turn left here; my thoughts have often turned to you). **4** tr. aim in a certain way (turned the hose on them). **5** intr. & tr. (foll. by into) change in nature, form, or condition to (turned into a dragon; then turned him into a frog; turned the book into a play). **6** intr. (foll. by to) **a** apply oneself to; set about (turned to doing the ironing). **b** have recourse to; begin to indulge in habitually (turned to drink; turned to me for help). **c** go on to consider next (let us now turn to your report). **7** intr. & tr. become or cause to become (turned hostile; has turned informer; your comment turned them angry). **8 a** tr. & intr. (foll. by against) make or become hostile to (has turned them against us). **b** intr. (foll. by on, upon) become hostile to; attack (suddenly turned on them). **9** intr. (of hair or leaves) change colour. **10** intr. (of milk) become sour. **11** intr. (of the stomach) be nauseated. **12** tr. twist out of position or sprain (an ankle). **13** intr. (of the head) become giddy. **14** tr. cause (milk) to become sour, (the stomach) to be nauseated, or (the head) to become giddy. **15** tr. translate (turn it into French). **16** tr. move to the other side of; go round (turned the corner). **17** tr. pass the age or time of (he has turned 40; it has now turned 4 o'clock). **18** intr. (foll. by on) depend on; be determined by; concern (it all turns on the weather tomorrow; the conversation turned on my motives). **19** tr. send or put into a specified place or condition; cause to go (was turned loose; turned the water out into a basin). **20** tr. perform (a somersault etc.) with rotary motion (turned cartwheels). **21** tr. remake (a garment or, esp., a sheet) putting the worn outer side on the inside. **22** tr. make (a profit). **23** tr. (also foll. by aside) divert, deflect (something material or immaterial). **24** tr. blunt (the edge of a knife, slot of a screw-head, etc.). **25** tr. shape (an object) on a lathe. **26** tr. give an (esp. elegant) form to (turn a compliment). **27** intr. Golf begin the second half of a round. **28** tr. (esp. as **turned** adj.) Printing invert (type) to make it appear upside down (a turned comma). **29** tr. Mil. pass round (the flank etc. of an army) so as to attack it from the side or rear. **30** intr. (of the tide) change from flood to ebb or vice versa. ● n. **1** the act or process or an instance of turning; rotary motion (a single turn of the handle). **2 a** a changed or a change of direction or tendency (took a sudden turn to the left). **b** a deflection or deflected part (full of twists and turns). **3** a point at which a turning or change occurs. **4** a turning of a road. **5** a change of the tide from ebb

to flow or from flow to ebb. **6** a change in the course of events. **7** a tendency or disposition (is of a mechanical turn of mind). **8** an opportunity or obligation etc. that comes successively to each of several persons etc. (your turn will come; my turn to read). **9** a short walk or ride (shall take a turn in the garden). **10** a short performance on stage or in a circus etc. **11** service of a specified kind (did me a good turn). **12** purpose (served my turn). **13** colloq. a momentary nervous shock or ill feeling (gave me quite a turn). **14** Mus. an ornament consisting of the principal note with those above and below it. **15** one round in a coil of rope etc. **16** Printing **a** inverted type as a temporary substitute for a missing letter. **b** a letter turned wrong side up. **17 a** Brit. the difference between the buying and selling price of stocks etc. **b** a profit made from this. □ **at every turn** continually; at each new stage etc. **by turns** in rotation of individuals or groups; alternately. **in turn** in succession; one by one. **in one's turn** when one's turn or opportunity comes. **not know which way** (or **where**) **to turn** be completely at a loss, unsure how to act, etc. **not turn a hair** see HAIR. **on the turn 1** changing. **2** (of milk) becoming sour. **3** at the turning-point. **out of turn 1** at a time when it is not one's turn. **2** inappropriately; inadvisedly or tactlessly (did I speak out of turn?). **take turns** (or **take it in turns**) act or work alternately or in succession. **to a turn** (esp. cooked) to exactly the right degree etc. **turn about** move so as to face in a new direction. **turn-about** n. **1** an act of turning about. **2** an abrupt change of policy etc. **turn and turn about** alternately. **turn around** esp. N. Amer. = turn round. **turn aside** see TURN v. 23 above. **turn away 1** turn to face in another direction. **2** refuse to accept; reject. **3** send away. **turn back 1** begin or cause to retrace one's steps. **2** fold back. **turn one's back on** see BACK. **turn-bench** a watchmaker's portable lathe. **turn-buckle** a device for tightly connecting parts of a metal rod or wire. **turn-cap** a revolving chimney-top. **turn the corner 1** pass round it into another street. **2** pass the critical point in an illness, difficulty, etc. **turn a deaf ear** see DEAF. **turn down 1** reject (a proposal, application, etc.). **2** reduce the volume or strength of (sound, heat, etc.) by turning a knob etc. **3** fold down. **4** place downwards. **turn-down** n. a rejection, a refusal. ● adj. (of a collar) turned down. **turn one's hand to** see HAND. **turn a person's head** see HEAD. **turn an honest penny** see HONEST. **turn in 1** hand in or over; deliver. **2** achieve or register (a performance, score, etc.). **3** colloq. go to bed in the evening. **4** fold inwards. **5** incline inwards (his toes turn in). **6** hand over (a suspect etc.) to the authorities. **7** colloq. abandon (a plan etc.). **turn in one's grave** see GRAVE¹. **turn inside out** see INSIDE. **turn off 1 a** stop the flow or operation of (water, electricity, etc.) by means of a tap, switch, etc. **b** operate (a tap, switch, etc.) to achieve this. **2** enter a side-road. **b** (of a side-road) lead off from another road. **3** colloq. repel; cause to lose interest (turned me right off with their complaining). **4** dismiss from employment. **turn-off** n. **1** a turning off a main road. **2** colloq. something that repels or causes a loss of interest. **turn of speed** the ability to go fast when necessary. **turn on 1 a** start the flow or operation of (water, electricity, etc.) by means of a tap, switch, etc. **b** operate (a tap, switch, etc.) to achieve this. **2** colloq. excite; stimulate the interest of, esp. sexually. **3** tr. & intr. sl. intoxicate or become intoxicated with drugs. **turn-on** n. colloq. a person or thing that causes (esp. sexual) arousal. **turn on a sixpence** see SIXPENCE. **turn on one's heel** see HEEL¹. **turn out 1** expel. **2** extinguish (an electric light etc.) **3** dress or equip (well turned out). **4** produce (manufactured goods etc.). **5** empty or clean out (a room etc.). **6** empty (a pocket) to see the contents. **7** colloq. **a** get out of bed. **b** go out of doors. **8** colloq. assemble; attend a meeting etc. **9** (often foll. by to + infin. or that + clause) prove to be the case; result (turned out to be true; we shall see how things turn out). **10** Mil. call (a guard) from the guardroom. **turn over 1** reverse or cause to reverse vertical position; bring the under or reverse side into view (turn over the page). **2** upset; fall or cause to fall over. **3 a** cause (an engine) to run. **b** (of an engine) start running. **4** consider thoroughly. **5** (foll. by to) **a** transfer the care or conduct of (a person or thing) to (a person) (shall turn it all over to my deputy). **b** = turn in 6. **6** do business to the amount of (turns over £5,000 a week). **turn over a new leaf** improve one's conduct or performance. **turn round 1** turn so as to face in a new direction. **2 a** Commerce unload and reload (a ship, vehicle, etc.). **b** receive, process, and send out again; cause to progress through a system. **3** adopt new opinions or policy. **turn-round** n. **1 a** the process of loading and unloading. **b** the process of receiving, processing, and sending out again; progress through a system. **2** the reversal of an opinion or tendency. **turn the scales** see SCALE². **turn the tables** see TABLE. **turn tail** turn one's back; run away. **turn the tide** reverse the trend of events. **turn to** set about one's work (came home and immediately turned to). **turn to account** see ACCOUNT. **turn turtle** see TURTLE. **turn up 1** increase the volume or strength of (sound, heat, etc.) by turning a knob etc. **2** place upwards. **3** discover or reveal. **4** be found, esp. by chance (*it turned up on a rubbish dump*). **5** happen or present itself; (of a person) put in an appearance (*a few people turned up late*). **6** *colloq.* cause to vomit (*the sight turned me up*). **7** shorten (a garment) by increasing the size of the hem. **turn-up** *n.* **1** *Brit.* the lower turned up end of a trouser leg. **2** (also **turn-up for the books**) *colloq.* an unexpected (esp. welcome) happening; a surprise. [OE *tyrnan, turnian* f. L *tornare* f. *tornus* lathe f. Gk *tornos* lathe, circular movement: prob. reinforced in ME f. OF *turner, torner*]

turnaround /ˈtɜːnəˌraʊnd/ *n.* **1 a** the process of receiving, processing, and sending out again; progress through a system; the time taken for this (*a seven-day turnaround*). **b** the process of unloading and reloading a ship, vehicle, etc.; the time taken for this. **2** an abrupt or unexpected change of attitude etc.

turncoat /ˈtɜːnkəʊt/ *n.* a person who changes sides in a conflict, dispute, etc.

turncock /ˈtɜːnkɒk/ *n.* an official employed to turn on water for the mains supply etc.

Turner /ˈtɜːnə(r)/, J(oseph) M(allord) W(illiam) (1775–1851), English painter. The originality of his work and the extent to which it anticipated later styles such as impressionism and abstract art made him a highly controversial figure in his day. He made his name with stormy seascapes such as *The Shipwreck* (1805) and landscapes painted in the grand Italian style of Claude Lorrain, as in *Crossing the Brook* (1815). From 1819, Turner made several visits to Italy and became increasingly concerned with depicting the power of light, using primary colours — especially yellow — often arranged in a swirling vortex. In the 1830s and 1840s he adopted watercolour techniques for his 'colour beginnings' in oil, which he later worked up to finished paintings, notably in *Norham Castle* (*c.*1845). He is perhaps best known for *Rain, Steam, Speed* (1844) and for the seascape *The Fighting Téméraire* (1838).

turner /ˈtɜːnə(r)/ *n.* **1** a person or thing that turns. **2** a person who works with a lathe. [ME f. OF *tornere -eor* f. LL *tornator* (as TURN)]

turnery /ˈtɜːnərɪ/ *n.* **1** objects made on a lathe. **2** work with a lathe.

turning /ˈtɜːnɪŋ/ *n.* **1 a** a road that branches off another. **b** a place where this occurs. **2 a** use of the lathe. **b** (in *pl.*) chips or shavings from a lathe. □ **turning-circle** the smallest circle in which a vehicle can turn without reversing. **turning-point** a point at which a decisive change occurs.

turnip /ˈtɜːnɪp/ *n.* **1 a** a cruciferous plant, *Brassica rapa*, with a large white globular root and sprouting leaves. **b** a similar or related plant, esp. a swede. **2** the root of such a plant used as a vegetable. **3** a large thick old-fashioned watch. □ **turnip-top** the leaves of the turnip eaten as a vegetable. □ **turnipy** *adj.* [earlier *turnep(e)* f. *neep* f. L *napus*: first element of uncert. orig.]

turnkey /ˈtɜːnkiː/ *n. & adj.* ● *n.* (*pl.* **-eys**) *archaic* a jailer. ● *adj.* (of a contract etc.) providing for a supply of equipment in a state ready for operation.

turnout /ˈtɜːnaʊt/ *n.* **1** the number of people attending a meeting, voting at an election, etc. (*rain reduced the turnout*). **2** the quantity of goods produced in a given time. **3** a set or display of equipment, clothes, etc.

turnover /ˈtɜːnˌəʊvə(r)/ *n.* **1** the act or an instance of turning over. **2** the amount of money taken in a business. **3** the number of people entering and leaving employment etc. **4** a small pie or tart made by folding a piece of pastry over a filling.

turnpike /ˈtɜːnpaɪk/ *n.* **1** *hist.* a defensive frame of spikes. **2** *hist.* **a** a toll-gate. **b** a road on which a toll was collected at a toll-gate. (*See note below.*) **3** *US* a motorway on which a toll is charged.

■ In Britain turnpikes were built and maintained by private companies, with parliamentary authority, in the late 17th and 18th centuries in response to the increasing need for good roads. The turnpikes fell into decline during the 19th centuries through competition from the railways and tolls were progressively abolished between the 1870s and 1890s. The road system remained in a state of neglect until the advent of the motor car in the early 20th century.

turnsick /ˈtɜːnsɪk/ *n.* = STURDY *n.*

turnside /ˈtɜːnsaɪd/ *n.* giddiness in dogs and cattle.

turnsole /ˈtɜːnsəʊl/ *n.* a plant whose flowers are said to turn with the sun; esp. the Mediterranean plant *Chrozophora tinctoria* (spurge family), grown for its violet juice. [OF *tournesole* f. Prov. *tournasol* f. L *tornare* TURN + *sol* sun]

turnspit /ˈtɜːnspɪt/ *n. hist.* a person whose job it was to turn a spit; a

small dog kept to turn a spit by running on a treadmill connected to it.

turnstile /ˈtɜːnstaɪl/ *n.* a gate for admission or exit, with revolving arms allowing people through singly.

turnstone /ˈtɜːnstəʊn/ *n.* a small wading bird of the genus *Arenaria*, which has a short bill and turns over stones to feed on small animals; esp. the widespread *A. interpres*.

turntable /ˈtɜːnˌteɪb(ə)l/ *n.* **1** a circular revolving plate supporting a gramophone record that is being played. **2** a circular revolving platform for turning a railway locomotive or other vehicle.

turpentine /ˈtɜːpənˌtaɪn/ *n. & v.* ● *n.* **1** (in full **crude** or **gum turpentine**) an oleoresin secreted by certain trees, esp. pines, distilled to make rosin and oil of turpentine. **2** (in full **oil of turpentine**) a volatile pungent oil distilled from gum turpentine or pine wood, used in mixing paints and varnishes, and in medicine. **3** a tree that yields turpentine or a similar resin; esp. various conifers, the terebinth, or *Syncarpia glomulifera*. ● *v.tr.* apply turpentine to. □ **Chian turpentine** the fragrant resin of the terebinth tree. [ME f. OF *ter(e)bentine* f. L *ter(e)binthina* (*resina* resin) (as TEREBINTH)]

turpeth /ˈtɜːpɪθ/ *n.* (in full **turpeth root**) the root of a tropical Asian plant, *Ipomoea turpethum*, of the bindweed family, used as a cathartic. [ME f. med.L *turbit(h)um* f. Arab. & Pers. *turbid*]

Turpin /ˈtɜːpɪn/, Dick (1706–39), English highwayman. He was a cattle and deer thief in Essex before entering into partnership with Tom King, a notorious highwayman. Turpin eventually fled north and was hanged at York for horse-stealing. His escapades were celebrated in the popular literature of the day.

turpitude /ˈtɜːpɪˌtjuːd/ *n. formal* baseness, depravity, wickedness. [F *turpitude* or L *turpitudo* f. *turpis* disgraceful, base]

turps /tɜːps/ *n. colloq.* oil of turpentine. [abbr.]

turquoise /ˈtɜːkwɔɪz, -kwɑːz/ *n. & adj.* ● *n.* **1** a semiprecious stone, usu. opaque and greenish or sky-blue, consisting of hydrated copper aluminium phosphate. **2** a greenish-blue colour. ● *adj.* of this colour. [ME *turkeis* etc. f. OF *turqueise* (later *-oise*) Turkish (stone)]

turret /ˈtʌrɪt/ *n.* **1** a small tower, usu. projecting from the wall of a building as a decorative addition. **2** a low flat usu. revolving armoured tower for a gun and gunners in a ship, aircraft, fort, or tank. **3** a rotating holder for tools in a lathe etc. □ **turret lathe** = *capstan lathe*. □ **turreted** *adj.* [ME f. OF *to(u)rete* dimin. of *to(u)r* TOWER]

turtle /ˈtɜːt(ə)l/ *n.* **1** an aquatic reptile of the order Chelonia, encased in a shell of bony plates; esp. a large marine reptile of the family Cheloniidae or Dermochelyidae, having a streamlined body, non-retractable head, and limbs modified as flippers. **2** the flesh of a turtle, esp. used for soup. **3** *Computing* a directional cursor in a computer graphics system which can be instructed to move around a screen. □ **turn turtle** capsize. **turtle-neck 1** a high close-fitting neck on a knitted garment. **2** *US* = *polo-neck*. [app. alt. of *tortue*: see TORTOISE]

turtle-dove /ˈtɜːt(ə)lˌdʌv/ *n.* a dove of the genus *Streptopelia*, often having chestnut and pink plumage; esp. *S. turtur*, noted for its soft purring call and its apparent affection for its mate. [archaic *turtle* (in the same sense) f. OE *turtla, turtle* f. L *turtur*, of imit. orig.]

turves *pl.* of TURF.

Tuscan /ˈtʌskən/ *n. & adj.* ● *n.* **1** an inhabitant of Tuscany in central Italy. **2** the classical Italian language of Tuscany. ● *adj.* **1** of or relating to Tuscany or the Tuscans. **2** *Archit.* denoting the least ornamented of the classical orders. □ **Tuscan straw** fine yellow wheat-straw used for hats etc. [ME f. F f. L *Tuscanus* f. *Tuscus* Etruscan]

Tuscany /ˈtʌskənɪ/ (Italian **Toscana** /tosˈkana/) a region of west central Italy, on the Ligurian Sea; capital, Florence. Consisting of several small states in medieval times, it united under Florence in the 15th and 16th centuries and became part of the kingdom of Italy in 1861.

Tuscarora /ˌtʌskəˈrɔːrə/ *n. & adj.* ● *n.* (*pl.* same or **Tuscaroras**) **1** an American Indian people forming part of the Iroquois confederacy, originally inhabiting Carolina and later New York State. **2** the Iroquoian language of this people. ● *adj.* of or relating to the Tuscarora or their language. [Iroquois]

tush[1] /tʌʃ/ *int. archaic* expressing strong disapproval or scorn. [ME: imit.]

tush[2] /tʌʃ/ *n.* **1** a long pointed tooth, esp. a canine tooth of a horse. **2** an elephant's short tusk. [OE *tusc* TUSK]

tush[3] /tʌʃ/ *n.* esp. *US sl.* the buttocks. [20th c.: abbr. or dimin. of *tokus* f. Yiddish *tokhes*]

tusk /tʌsk/ *n. & v.* ● *n.* **1** a long pointed tooth, esp. protruding from a closed mouth, as in the elephant, walrus, etc. **2** a tusklike tooth or other object. ● *v.tr.* gore, thrust at, or tear up with a tusk or tusks. □ **tusk shell 1** a mollusc of the class Scaphopoda. **2** its long tusk-shaped shell, open at both ends (cf. DENTALIUM). □ **tusked** *adj.* (also in comb.). **tusky** *adj.* [ME, alt. of OE *tux* var. of *tusc*: cf. TUSH²]

tusker /'tʌskə(r)/ *n.* an elephant or wild boar with well-developed tusks.

tussah *US* var. of TUSSORE.

Tussaud /tə'sɔːd, French tyso/, Madame (née Marie Grosholtz) (1761–1850), French founder of Madame Tussaud's waxworks, resident in Britain from 1802. After taking death masks in wax of prominent victims of the French Revolution, she toured Britain with her wax models, which came to include other famous and topical people. In 1835 she founded a permanent waxworks exhibition in Baker Street, London.

tusser var. of TUSSORE.

tussive /'tʌsɪv/ *adj.* of or relating to a cough. [L *tussis* cough]

tussle /'tʌs(ə)l/ *n.* & *v.* ● *n.* a struggle or scuffle. ● *v.intr.* engage in a tussle. [orig. Sc. & N. Engl., perh. dimin. of TOUSLE]

tussock /'tʌsək/ *n.* **1** a clump of grass etc. **2** (in full **tussock moth**) a moth of the family Lymantriidae, with hairy tufted larvae that are often pests of trees. □ **tussock grass** a grass that grows in tussocks, esp. one of the genera *Poa*, *Nassella*, or *Deschampsia*. □ **tussocky** *adj.* [16th c.: perh. alt. f. dial. *tusk* tuft]

tussore /'tʌsɔː(r), -sə(r)/ *n.* (also **tusser**, *US* **tussah** /-sə(r)/) **1** an Indian or Chinese silkworm of the genus *Antheraea*, yielding strong but coarse brown silk. **2** (in full **tussore-silk**) silk from this and some other silkworms. [Urdu f. Hindi *tasar* f. Skr. *tasara* shuttle]

tut var. of TUT-TUT.

Tutankhamen /ˌtuːt(ə)n'kɑːm(ə)n/ (also **Tutankhamun** /-kɑː'muːn/) (died *c*.1352 BC), Egyptian pharaoh of the 18th dynasty, reigned *c*.1361–*c*.1352 BC. Ascending the throne while still a boy, he abandoned the worship of the sun-god instituted by Akhenaten, reinstating the worship of Amun and making Thebes the capital city once again. Although not important in the history of Egypt, he became world famous because of the rich and varied contents of his tomb, discovered virtually intact by the English archaeologist Howard Carter in 1922.

tutelage /'tjuːtɪlɪdʒ/ *n.* **1** guardianship. **2** the state or duration of being under this. **3** instruction, tuition. [L *tutela* f. *tueri tuit-* or *tut-* watch]

tutelary /'tjuːtɪlərɪ/ *adj.* (also **tutelar** /-lə(r)/) **1 a** serving as guardian. **b** relating to a guardian (*tutelary authority*). **2** giving protection (*tutelary saint*). [LL *tutelaris*, L *-arius* f. *tutela*: see TUTELAGE]

tutenag /'tjuːtɪˌnæg/ *n.* **1** zinc imported from China and SE Asia. **2** a white alloy like German silver. [Marathi *tuttināg* perh. f. Skr. *tuttha* copper sulphate + *nāga* tin, lead]

Tuthmosis III /tʌθ'məʊsɪs/ (died *c*.1450 BC), son of Tuthmosis II, Egyptian pharaoh of the 18th dynasty *c*.1504–*c*.1450. He was initially joint ruler with his aunt Hatshepsut until her death in 1482. His reign was marked by the conquest of all of Syria and the annexation of part of Nubia, as well as by extensive building; the monuments he erected included Cleopatra's Needles (*c*.1475).

tutor /'tjuːtə(r)/ *n.* & *v.* ● *n.* **1** a private teacher, esp. one in general charge of a person's education. **2** a university teacher supervising the studies or welfare of assigned undergraduates. **3** *Brit.* a book of instruction in a subject. ● *v.* **1** *tr.* act as a tutor to. **2** *intr.* work as a tutor. **3** *tr.* restrain, discipline. **4** *intr.* *US* receive tuition. □ **tutorage** *n.* **tutorship** *n.* [ME f. AF, OF *tutour* or L *tutor* f. *tueri tut-* watch]

tutorial /tjuː'tɔːrɪəl/ *adj.* & *n.* ● *adj.* of or relating to a tutor or tuition. ● *n.* a period of tuition given by a university etc. tutor to an individual or a small group. □ **tutorially** *adv.* [L *tutorius* (as TUTOR)]

tutsan /'tʌts(ə)n/ *n.* a St John's wort, *Hypericum androsaemum*, formerly used to heal wounds etc. [ME f. AF *tutsaine* all healthy]

Tutsi /'tʊtsɪ/ *n.* (*pl.* same or **Tutsis**) a member of a Bantu-speaking people, traditionally cattle-owners, who form a minority of the population of Rwanda and Burundi but who formerly dominated the Hutu majority. [Bantu]

tutti /'tʊtɪ/ *adv.* & *n.* *Mus.* ● *adv.* with all voices or instruments together. ● *n.* (*pl.* **tuttis**) a passage to be performed in this way. [It., pl. of *tutto* all]

tutti-frutti /ˌtʊtɪ'fruːtɪ/ *n.* (*pl.* **-fruttis**) a confection, esp. ice-cream, of or flavoured with mixed fruits. [It., = all fruits]

tut-tut /tʌt'tʌt/ *int.*, *n.*, & *v.* (also **tut**) ● *int.* expressing rebuke, impatience, or contempt. ● *n.* such an exclamation. ● *v.intr.* (**-tutted**, **-tutting**) exclaim this. [imit. of a click of the tongue against the teeth]

tutty /'tʌtɪ/ *n.* impure zinc oxide or carbonate used as a polishing powder. [ME f. OF *tutie* f. med.L *tutia* f. Arab. *tūtiyā*]

Tutu /'tuːtuː/, Desmond (Mpilo) (b.1931), South African clergyman. He served as General Secretary of the South African Council of Churches (1979–84), during which time he became a leading voice in the struggle against apartheid, calling for economic sanctions against South Africa and emphasizing non-violent action. He was awarded the Nobel Peace Prize in 1984, and in the following year he became Johannesburg's first black Anglican bishop. He was made archbishop of Cape Town in 1986.

tutu¹ /'tuːtuː/ *n.* a ballet-dancer's short skirt of stiffened projecting frills. [F]

tutu² /'tuːtuː/ *n.* *NZ* a shrub, *Coriaria arborea*, native to New Zealand, bearing poisonous purplish-black berries. [Maori]

Tuva /'tuːvə/ (formerly **Tannu-Tuva** /ˌtænuː-/) an autonomous republic in south central Russia, on the border with Mongolia; pop. (1990) 314,000; capital, Kyzyl. A mountainous region that includes the upper basin of the Yenisei river, Tannu-Tuva was part of the Chinese empire between 1757 and 1911, falling to tsarist Russia in 1914. The Bolsheviks established an independent republic in Tannu-Tuva in 1921. Annexed by the USSR in 1944, it became an autonomous republic within the RSFSR in 1961, styling itself the Republic of Tuva following the breakup of the Soviet Union in 1991.

Tuvalu /tuː'vɑːluː/ a country in the SW Pacific consisting of a group of nine main islands, the former Ellice Islands; pop. (est. 1988) 8,500; official languages, English and Tuvaluan (local Malayo-Polynesian language); capital, Funafuti. The Ellice Islands (named after the owner of his ship by a visiting Dutch captain in 1819) formed part of the British colony of the Gilbert and Ellice Islands but separated from the Gilberts (later Kiribati) after a referendum in 1975. Tuvalu became independent within the Commonwealth in 1978. □ **Tuvaluan** /ˌtuːvə'luːən, tuː'vɑːluːən/ *adj.* & *n.*

tu-whit, tu-whoo /tʊˌwɪt tʊ'wuː/ *n.* a representation of the cry of an owl. [imit.]

tux /tʌks/ *n.* N. Amer. colloq. = TUXEDO.

tuxedo /tʌk'siːdəʊ/ *n.* (*pl.* **-os** or **-oes**) N. Amer. **1** a dinner-jacket. **2** a suit of clothes including this. [after a country club at *Tuxedo* Park, New York]

Tuxtla Gutiérrez /ˌtʊstlə ˌguːtɪ'eərez/ a city in SE Mexico, capital of the state of Chiapas; pop. (1990) 295,615.

tuyère /twiː'jeə(r), tuː'jeə(r)/ *n.* (also **tuyere**, **twyer** /'twaɪə(r)/) a nozzle through which air is forced into a furnace etc. [F f. *tuyau* pipe]

Tuzla /'tʊzlə/ a town in NE Bosnia; pop. (1981) 121,710. The town, a Muslim enclave which has an airport, suffered damage and heavy casualties when besieged by Bosnian Serb forces between 1992 and 1994.

TV *abbr.* television (the system or a set).

TVA see TENNESSEE VALLEY AUTHORITY.

Tver /tveə(r)/ an industrial port in European Russia, on the River Volga north-west of Moscow; pop. (1990) 454,000. It was known as Kalinin, in honour of President Kalinin, from 1931 until 1991.

TVP *abbr. propr.* textured vegetable protein (in foods made from vegetables but given a meatlike texture).

Twa /twɑː/ *n.* & *adj.* ● *n.* (*pl.* same, **Twas**, or **Batwa** /'bætwɑː/) a member of a pygmy people inhabiting parts of Burundi, Rwanda, and Zaire. ● *adj.* of or relating to the Twa. [Bantu, = foreigner, outsider]

twaddle /'twɒd(ə)l/ *n.* & *v.* ● *n.* useless, senseless, or dull writing or talk. ● *v.intr.* indulge in this. □ **twaddler** *n.* [alt. of earlier *twattle*, alt. of TATTLE]

Twain /tweɪn/, Mark (pseudonym of Samuel Langhorne Clemens) (1835–1910), American novelist and humorist. After working as a river pilot on the Mississippi he established a reputation as a humorist with early work including *The Innocents Abroad* (1869), a satirical account of an American cruise to the Mediterranean. He is best known for the novels *The Adventures of Tom Sawyer* (1876) and *The Adventures of Huckleberry Finn* (1885); both works give a vivid evocation of Mississippi frontier life, faithfully capturing Southern speech patterns and combining picaresque adventure with moral commentary.

twain /tweɪn/ *adj.* & *n.* archaic two (usu. *in twain*). [OE *twegen*, masc. form of *twā* TWO]

twang /twæŋ/ *n.* & *v.* ● *n.* **1** a strong ringing sound made by the plucked string of a musical instrument or a released bowstring. **2** the nasal quality of a voice compared to this. ● *v.* **1** *intr.* & *tr.* emit or cause to emit this sound. **2** *tr.* usu. *derog.* play (a tune or instrument) in this way. **3** *tr.* utter with a nasal twang. □ **twangy** *adj.* [imit.]

'twas /twɒz, twəz/ *archaic* it was. [contr.]

twat /twɒt/ *n. coarse sl.* **1** the female genitals. **2** *Brit.* a term of contempt for a person. [17th c.: orig. unkn.]

twayblade /'tweɪbleɪd/ *n.* an orchid with a single pair of leaves; esp. one of the genus *Listera*, with greenish flowers. [tway var. of TWAIN + BLADE]

tweak /twi:k/ *v. & n.* ● *v.tr.* **1** pinch and twist sharply; pull with a sharp jerk; twitch. **2** make fine adjustments to (a mechanism). ● *n.* an instance of tweaking. [prob. alt. of dial. *twick* & TWITCH]

twee /twi:/ *adj.* (**tweer** /'twi:ə(r)/; **tweest** /'twi:ɪst/) *Brit.* usu. *derog.* affectedly dainty or quaint. □ **tweely** *adv.* **tweeness** *n.* [childish pronunc. of SWEET]

Tweed /twi:d/ a river which rises in the Southern Uplands of Scotland and flows 155 km (97 miles) generally eastwards, crossing into NE England and entering the North Sea at Berwick-upon-Tweed. For part of its lower course it forms the border between Scotland and England.

tweed /twi:d/ *n.* **1** a rough-surfaced woollen cloth, usu. of mixed flecked colours, orig. produced in Scotland. **2** (in *pl.*) clothes made of tweed. [orig. a misreading of *tweel*, Sc. form of TWILL, infl. by assoc. with the river Tweed]

Tweedledum and Tweedledee /ˌtwi:d(ə)l'dʌm, ˌtwi:d(ə)l'di:/ two persons or things differing only or chiefly in name. *Tweedledum* and *Tweedledee* were originally names applied to the composers Handel and Bononcini (1670–1747) in a 1725 satire by John Byrom (1692–1763); they were later used for two identical characters in Lewis Carroll's *Through the Looking Glass*.

tweedy /'twi:dɪ/ *adj.* (**tweedier**, **tweediest**) **1** of or relating to tweed cloth. **2** characteristic of the country gentry, heartily informal. □ **tweedily** *adv.* **tweediness** *n.*

'tween /twi:n/ *prep. archaic* = BETWEEN. □ **'tween-decks** *Naut.* the space between decks. [contr.]

tweet /twi:t/ *n. & v.* (also **tweet tweet**) ● *n.* the chirp of a small bird. ● *v.intr.* make a chirping noise. [imit.]

tweeter /'twi:tə(r)/ *n.* a loudspeaker designed to reproduce high frequencies.

tweezers /'twi:zəz/ *n.pl.* a small pair of pincers for taking up small objects, plucking out hairs, etc. [extended form of *tweezes* (cf. *pincers* etc.) pl. of obs. *tweeze* case for small instruments, f. *etweese* = *étuis*, pl. of ÉTUI]

twelfth /twelfθ/ *n. & adj.* ● *n.* **1** the position in a sequence corresponding to the number 12 in the sequence 1–12. **2** something occupying this position. **3** each of twelve equal parts of a thing. **4** *Mus.* **a** an interval or chord spanning an octave and a fifth in the diatonic scale. **b** a note separated from another by this interval. ● *adj.* that is the twelfth. □ **twelfth man** a reserve member of a cricket team. **twelfth part** = sense 3 of *n.* □ **twelfthly** *adv.* [OE *twelfta* (as TWELVE)]

Twelfth Day 6 January, the twelfth day after Christmas, the festival of the Epiphany.

Twelfth Night *n.* **1** the evening of 5 Jan., the eve of the Epiphany, formerly the last day of Christmas festivities. **2** the evening of Twelfth Day (6 Jan.).

twelve /twelv/ *n. & adj.* ● *n.* **1** one more than eleven; the product of two units and six units. **2** a symbol for this (12, xii, XII). **3** a size etc. denoted by twelve. **4** twelve o'clock. **5** (**the Twelve**) the twelve Apostles. **6** (**12**) *Brit.* (of films) classified as suitable for persons of 12 years and over. ● *adj.* that amount to twelve. [OE *twelf(e)* f. Gmc, prob. rel. to TWO]

twelvefold /'twelvfəʊld/ *adj. & adv.* **1** twelve times as much or as many. **2** consisting of twelve parts.

twelvemo /'twelvməʊ/ *n.* = DUODECIMO.

twelvemonth /'twelvmʌnθ/ *n. archaic* a year; a period of twelve months.

twelve-note /'twelvnəʊt/ *adj.* (also **twelve-tone**) *Mus.* denoting a system of musical composition using the twelve chromatic notes of the octave on an equal basis without dependence on a key system. Developed by Schoenberg, the technique is central to serialism; indeed, serialism, especially that following the principles of Schoenberg rather than those of the later Pierre Boulez or Milton Babbitt, is sometimes known as twelve-note music.

Twelve Tables a set of laws drawn up in ancient Rome in 451 and 450 BC, embodying the most important rules of Roman law.

Twelve Tribes of Israel see TRIBES OF ISRAEL.

Twentieth Century Fox a US film production company formed in 1935 by the merger of the Fox Company with Twentieth Century. Under production head Darryl F. Zanuck (1902–79) the company made several John Ford westerns in the 1940s, later pioneering wide-screen film techniques and gaining further successes with *The Sound of Music* (1965) and *Star Wars* (1977).

twenty /'twentɪ/ *n. & adj.* ● *n.* (*pl.* **-ies**) **1** the product of two and ten. **2** a symbol for this (20, xx, XX). **3** (in *pl.*) the numbers from 20 to 29, esp. the years of a century or of a person's life. **4** *colloq.* a large indefinite number (*have told you twenty times*). ● *adj.* that amount to twenty. □ **twenty-first, -second**, etc. the ordinal numbers between twentieth and thirtieth. **twenty-one, -two**, etc. the cardinal numbers between twenty and thirty. **twenty-twenty** (or **20/20**) **1** denoting vision of normal acuity. **2** *colloq.* denoting clear perception or hindsight. □ **twentieth** *adj.* & *n.* **twentyfold** *adj.* & *adv.* [OE *twentig* (perh. as TWO, -TY²)]

'twere /twɜ:(r)/ *archaic* it were. [contr.]

twerp /twɜ:p/ *n.* (also **twirp**) *sl.* a stupid or objectionable person. [20th c.: orig. unkn.]

Twi /twi:, tʃwi:/ *n. & adj.* ● *n.* (*pl.* same or **Twis**) **1** the chief language spoken in Ghana, consisting of several mutually intelligible dialects. **2** the people speaking this language. ● *adj.* of or relating to the Twi or their language. [Kwa]

twibill /'twaɪbɪl/ *n.* a double-bladed battleaxe. [OE f. *twi-* double + BILL³]

twice /twaɪs/ *adv.* **1** two times (esp. of multiplication); on two occasions. **2** in double degree or quantity (*twice as good*). [ME *twiges* f. OE *twige* (as TWO, -S³)]

twiddle /'twɪd(ə)l/ *v. & n.* ● *v.* **1** *tr.* & (foll. by *with* etc.) *intr.* twirl, adjust, or play randomly or idly. **2** *intr.* move twirlingly. ● *n.* **1** an act of twiddling. **2** a twirled mark or sign. □ **twiddle one's thumbs 1** make them rotate round each other. **2** have nothing to do. □ **twiddler** *n.* **twiddly** *adj.* [app. imit., after *twirl*, *twist*, and *fiddle*, *piddle*]

twig¹ /twɪg/ *n.* **1** a small branch or shoot of a tree or shrub. **2** *Anat.* a small branch of an artery etc. □ **twigged** *adj.* (also in *comb.*). **twiggy** *adj.* [OE *twigge* f. a Gmc root *twi-* (unrecorded) as in TWICE, TWO]

twig² /twɪg/ *v.tr.* (**twigged**, **twigging**) *colloq.* **1** (also *absol.*) understand; grasp the meaning or nature of. **2** perceive, observe. [18th c.: orig. unkn.]

twilight /'twaɪlaɪt/ *n.* **1** the soft glowing light from the sky when the sun is below the horizon, esp. in the evening. **2** the period of this. **3** a faint light. **4** a state of imperfect knowledge or understanding. **5** a period of decline or destruction. **6** *attrib.* of, resembling, or occurring at twilight. □ **twilight sleep** *Med.* a state of partial narcosis, esp. to ease the pain of childbirth. **twilight zone 1** an urban area that is becoming dilapidated. **2** any physical or conceptual area which is undefined or intermediate. [ME f. OE *twi-* two (in uncert. sense) + LIGHT¹]

twilight of the gods *Scand. & Germanic Mythol.* the destruction of the gods and the world in a final conflict with the powers of evil. Also called *Götterdämmerung*, *Ragnarök*. [transl. Icel. *ragna rökr* (see RAGNARÖK)]

twilit /'twaɪlɪt/ *adj.* (also **twilighted** /-ˌlaɪtɪd/) dimly illuminated by or as by twilight. [past part. of *twilight* (v.) f. TWILIGHT]

twill /twɪl/ *n. & v.* ● *n.* a fabric so woven as to have a surface of diagonal parallel ridges. ● *v.tr.* (esp. as **twilled** *adj.*) weave (fabric) in this way. □ **twilled** *adj.* [N. Engl. var. of obs. *twilly*, OE *twili*, f. *twi-* double, after L *bilix* (as BI-, *licium* thread)]

'twill /twɪl/ *archaic* it will. [contr.]

twin /twɪn/ *n., adj., & v.* ● *n.* **1** each of a closely related or associated pair, esp. of children or animals born at the same birth. **2** the exact counterpart of a person or thing. **3** *Crystallog.* a compound crystal one part of which is in a reversed position with reference to the other. **4** (**the Twins**) the zodiacal sign or constellation Gemini. ● *adj.* **1** forming, or being one of, such a pair (*twin brothers*). **2** *Bot.* growing in pairs. **3** consisting of two closely connected and similar parts. ● *v.* (**twinned, twinning**) **1** *tr.* & *intr.* **a** join intimately together. **b** (foll. by *with*) pair. **2** *intr.* bear twins. **3** *intr. Crystallog.* grow as a twin crystal. **4** *intr.* & *tr. Brit.* link or cause (a town) to link with one in a different country, for the purposes of friendship and cultural exchange. □ **twin bed** each of a pair of single beds. **twin-cam** (of an engine or vehicle) having two camshafts. **twin-engined** having two engines. **twin-screw** (of a ship) having two propellers on separate shafts with opposite twists.

twin set esp. *Brit.* a woman's matching cardigan and jumper. **twin town** *Brit.* a town which is twinned with another. □ **twinning** *n.* [OE *twinn* double, f. *twi-* two: cf. ON *tvinnr*]

twine /twaɪn/ *n. & v.* ● *n.* **1** a strong thread or string of two or more strands of hemp or cotton etc. twisted together. **2** a coil or twist. **3** a tangle; an interlacing. ● *v.* **1** *tr.* form (a string or thread etc.) by twisting strands together. **2** *tr.* form (a garland etc.) of interwoven material. **3** *tr.* (often foll. by *with*) garland (a brow etc.). **4** *intr.* (often foll. by *round, about*) coil or wind. **5** *intr. & refl.* (of a plant) grow in this way. □ **twiner** *n.* [OE *twīn, twigin* linen, ult. f. the stem of *twi-* two]

twinge /twɪndʒ/ *n. & v.* ● *n.* a sharp momentary local pain or pang (*a twinge of toothache; a twinge of conscience*). ● *v.intr. & tr.* experience or cause to experience a twinge. [*twinge* (v.) pinch, wring f. OE *twengan* f. Gmc]

twinkle /ˈtwɪŋk(ə)l/ *v. & n.* ● *v.* **1** *intr.* (of a star or light etc.) shine with rapidly intermittent gleams. **2** *intr.* (of the eyes) sparkle, esp. with amusement. **3** *intr.* (of the feet in dancing) move lightly and rapidly. **4** *tr.* emit (a light or signal) in quick gleams. **5** *tr.* blink or wink (one's eyes). ● *n.* **1 a** a sparkle or gleam of the eyes. **b** a blink or wink. **2** a slight flash of light; a glimmer. **3** a short rapid movement. □ **in a twinkle** (or **a twinkling** or **the twinkling of an eye**) in an instant. □ **twinkler** *n.* **twinkly** *adj.* [OE *twinclian*]

twirl /twɜːl/ *v. & n.* ● *v.tr. & intr.* spin or swing or twist quickly and lightly round. ● *n.* **1** a twirling motion. **2** a form made by twirling, esp. a flourish made with a pen. □ **twirler** *n.* **twirly** *adj.* [16th c.: prob. alt. (after *whirl*) of obs. *tirl* TRILL]

twirp var. of TWERP.

twist /twɪst/ *v. & n.* ● *v.* **1 a** *tr.* change the form of by rotating one end and not the other or the two ends in opposite directions. **b** *intr.* undergo such a change; take a twisted position (*twisted round in his seat*). **c** *tr.* wrench or pull out of shape with a twisting action (*twisted my ankle*). **2** *tr.* **a** wind (strands etc.) about each other. **b** form (a rope etc.) by winding the strands. **c** (foll. by *with, in with*) interweave. **d** form by interweaving or twining. **3 a** *tr.* give a spiral form to (a rod, column, cord, etc.) as by rotating the ends in opposite directions. **b** *intr.* take a spiral form. **4** *tr.* (foll. by *off*) break off or separate by twisting. **5** *tr.* distort or misrepresent the meaning of (words). **6 a** *intr.* take a curved course. **b** *tr.* make (one's way) in a winding manner. **7** *tr. Brit. sl.* cheat (*twisted me out of £20*). **8** *tr. Sport* cause (the ball, esp. in billiards) to rotate while following a curved path. **9** *tr.* (as **twisted** *adj.*) (of a person or mind) emotionally unbalanced. **10** *intr.* dance the twist. ● *n.* **1** the act or an instance of twisting. **2 a** a twisted state. **b** the manner or degree in which a thing is twisted. **3** a thing formed by or as by twisting, esp. a thread or rope etc. made by winding strands together. **4** the point at which a thing twists or bends. **5** usu. *derog.* a peculiar tendency of mind or character etc. **6 a** an unexpected development of events, esp. in a story etc. **b** an unusual interpretation or variation. **c** a distortion or bias. **7** a fine strong silk thread used by tailors etc. **8** a roll of bread, tobacco etc., in the form of a twist. **9** *Brit.* a paper packet with screwed-up ends. **10** a curled piece of lemon etc. peel to flavour a drink. **11** *Sport* a spinning motion given to a ball in cricket etc. to make it take a special curve. **12 a** a twisting strain. **b** the amount of twisting of a rod etc., or the angle showing this. **c** forward motion combined with rotation about an axis. **13** *Brit.* a drink made of two ingredients mixed together. **14** *Brit. sl.* a swindle. **15** (prec. by *the*) a dance with a twisting movement of the body, popular in the 1960s. □ **round the twist** *Brit. colloq.* crazy. **twist a person's arm** *colloq.* apply coercion, esp. by moral pressure. **twist round one's finger** see FINGER. □ **twistable** *adj.* **twisty** *adj.* (**twistier, twistiest**) [ME, rel. to TWIN, TWINE]

twister /ˈtwɪstə(r)/ *n.* **1** *Brit. colloq.* a swindler; a dishonest person. **2** *Sport* a twisting ball in cricket, billiards, etc. **3** *N. Amer.* a tornado or waterspout.

twit[1] /twɪt/ *n.* esp. *Brit. colloq.* a silly or foolish person. [orig. dial.: perh. f. TWIT[2]]

twit[2] /twɪt/ *v.tr.* (**twitted, twitting**) reproach or taunt, usu. good-humouredly. [16th-c. *twite* f. *atwite* f. OE *ætwītan* reproach with f. *æt* at + *wītan* blame]

twitch /twɪtʃ/ *v. & n.* ● *v.* **1** *intr.* (of the features, muscles, limbs, etc.) move or contract spasmodically. **2** *tr.* give a short sharp pull at. ● *n.* **1 a** a sudden involuntary contraction or movement. **2** a sudden pull or jerk. **3** a state of nervousness. **4** a noose and stick for controlling a horse during a veterinary operation. □ **twitchy** *adj.* (**twitchier, twitchiest**) (in sense 3 of *n.*). [ME f. Gmc: cf. OE *twiccian*, dial. *twick*]

twitcher /ˈtwɪtʃə(r)/ *n.* **1** *sl.* a bird-watcher who tries to get sightings of rare birds. **2** a person or thing that twitches.

twitch grass *n.* = COUCH[2]. [var. of QUITCH]

twite /twaɪt/ *n.* a moorland finch, *Acanthis flavirostris*, resembling a linnet but with a pinkish rump. [imit. of its cry]

twitter /ˈtwɪtə(r)/ *v. & n.* ● *v.* **1** *intr.* **a** (of a bird) emit a succession of light tremulous sounds. **b** talk rapidly in an idle or trivial way. **2** *tr.* utter or express in this way. ● *n.* **1** the act or an instance of twittering. **2** *colloq.* a tremulously excited state. □ **twitterer** *n.* **twittery** *adj.* [ME, imit.: cf. -ER[4]]

'twixt /twɪkst/ *prep. archaic* = BETWIXT. [contr.]

twizzle /ˈtwɪz(ə)l/ *v. & n. colloq.* or *dial.* ● *v.tr. & intr.* twist, turn. ● *n.* a twist or turn. [prob. imit., infl. by TWIST]

two /tuː/ *n. & adj.* ● *n.* **1** one more than one; the sum of one unit and another unit. **2** a symbol for this (2, ii, II). **3** a size etc. denoted by two. **4** two o'clock. **5** a set of two. **6** a card with two pips. ● *adj.* that amount to two. □ **in two** in or into two pieces. **in two shakes** (or **ticks**) see SHAKE, TICK[1]. **or two** denoting several (*a thing or two* = several things). **put two and two together** make (esp. an obvious) inference from what is known or evident. **that makes two of us** *colloq.* that is true of me also. **two-bit** *N. Amer. colloq.* cheap, petty. **two-by-four** a length of timber with a rectangular cross-section 2 in. by 4 in. **two by two** (or **two and two**) in pairs. **two can play at that game** *colloq.* another person's behaviour can be copied to that person's disadvantage. **two-dimensional 1** having or appearing to have length and breadth but no depth. **2** lacking depth or substance; superficial. **two-edged** double-edged. **two-faced 1** having two faces. **2** insincere; deceitful. **2,4,5-T** see T[1]. **two-handed 1** having, using, or requiring the use of two hands. **2** (of a card-game) for two players. **two-line whip** *Brit.* a notice, underlined twice to indicate less urgency than a three-line whip, requesting the members of Parliament of a particular party to attend a parliamentary vote. **two a penny** see PENNY. **two-piece** *adj.* (of a suit etc.) consisting of two matching items. ● *n.* a two-piece suit etc. **two-ply** *adj.* of two strands, webs, or thicknesses. ● *n.* **1** two-ply wool. **2** two-ply wood made by gluing together two layers with the grain in different directions. **two-seater 1** a vehicle or aircraft with two seats. **2** a sofa etc. for two people. **two-sided 1** having two sides. **2** having two aspects; controversial. **two-step** a round dance with a sliding step in march or polka time. **two-time** *colloq.* **1** deceive or be unfaithful to (esp. a partner or lover). **2** swindle, double-cross. **two-timer** *colloq.* a person who is deceitful or unfaithful. **two-tone** having two colours or sounds. **two-up** *Austral. & NZ* a gambling game with bets placed on a showing of two heads or two tails. **two-way 1** involving two ways or participants. **2** (of a switch) permitting a current to be switched on or off from either of two points. **3** (of a radio) capable of transmitting and receiving signals. **4** (of a tap etc.) permitting fluid etc. to flow in either of two channels or directions. **5** (of traffic etc.) moving in two esp. opposite directions. **two-way mirror** a panel of glass that can be seen through from one side and is a mirror on the other. **two-wheeler** a vehicle with two wheels. [OE *twā* (fem. & neut.), *tū* (neut.), with Gmc cognates and rel. to Skr. *dwau, dwe*, Gk & L *duo*]

twofold /ˈtuːfəʊld/ *adj. & adv.* **1** twice as much or as many. **2** consisting of two parts.

twoness /ˈtuːnɪs/ *n.* the fact or state of being two; duality.

twopence /ˈtʌp(ə)ns/ *n. Brit.* **1** the sum of two pence, esp. before decimalization. **2** (esp. with *neg.*) *colloq.* a thing of little value (*don't care twopence*).

twopenn'orth /tuːˈpenəθ/ *n.* **1** as much as is worth or costs twopence. **2** a paltry or insignificant amount. □ **add** (or **put in**) **one's twopenn'orth** *colloq.* contribute one's opinion.

twopenny /ˈtʌp(ə)nɪ/ *adj. Brit.* **1** costing two pence, esp. before decimalization. **2** *colloq.* cheap, worthless. □ **twopenny-halfpenny** cheap, insignificant.

twosome /ˈtuːsəm/ *n.* **1** two persons together. **2** a game, dance, etc., for two persons.

two-stroke /ˈtuːstrəʊk/ *adj. & n.* ● *attrib.adj.* **1** (of an internal-combustion engine) having its power cycle completed in one up-and-down movement of the piston, with inlet and exhaust ports in the sides of the cylinder that are opened and closed by movements of the piston. **2** (of a vehicle) having a two-stroke engine. ● *n.* a two-stroke engine or vehicle.

'twould /twʊd/ *archaic* it would. [contr.]

twyer var. of TUYÈRE.

TX *abbr. US* Texas (in official postal use).

-ty[1] /tɪ/ *suffix* forming nouns denoting quality or condition (*cruelty; plenty*). [ME *-tie, -tee, -te* f. OF *-té, -tet* f. L *-tas -tatis*: cf. -ITY]

-ty² /tɪ/ *suffix* denoting tens (*twenty*; *thirty*; *ninety*). [OE *-tig*]

Tyburn /ˈtaɪbɜːn/ a place in London, near Marble Arch, where public hangings were held *c*.1300–1783. It is named after a tributary of the Thames, which flows in an underground culvert nearby.

tychism /ˈtaɪkɪz(ə)m/ *n. Philos.* the theory that chance controls the universe. [Gk *tukhē* chance]

Tycho Brahe see BRAHE.

tycoon /taɪˈkuːn/ *n.* **1** a business magnate. **2** *hist.* a title applied by foreigners to the shogun of Japan 1857–68. [Jap. *taikun* great lord]

tying *pres. part.* of TIE.

tyke /taɪk/ *n.* (also **tike**) **1** esp. *Brit.* an unpleasant or coarse man. **2** a mongrel. **3** a small child. **4** *Brit. sl.* a Yorkshireman. **5** *Austral. & NZ sl. offens.* a Roman Catholic. [ME f. ON *tík* bitch: sense 5 assim. from TAIG]

Tyler¹ /ˈtaɪlə(r)/, John (1790–1862), American Whig statesman, 10th President of the US 1841–5. Successor to William Henry Harrison as President, he was noted for securing the annexation of Texas (1845). Throughout his political career Tyler advocated states' rights, and his alliance with Southern Democrats on this issue helped to accentuate the divide between North and South in the years leading up to the American Civil War.

Tyler² /ˈtaɪlə(r)/, Wat (d.1381), English leader of the Peasants' Revolt of 1381. After capturing Canterbury, he led the rebels to Blackheath and took London. During a conference with the young king Richard II, he put forward the rebels' demands (including the lifting of the newly imposed poll tax), to which Richard consented. At a later conference in Smithfield he was killed by the Lord Mayor of London and several other royal supporters.

tylopod /ˈtaɪləˌpɒd/ *n. & adj. Zool.* ● *n.* any animal that bears its weight on the sole-pads of the feet rather than on the hooves, esp. the camel. ● *adj.* (of an animal) bearing its weight in this way. □ **tylopodous** /taɪˈlɒpədəs/ *adj.* [Gk *tulos* knob or *tulē* callus, cushion + *pous podos* foot]

tympan /ˈtɪmpən/ *n.* **1** *Printing* an appliance in a printing-press used to equalize pressure between the platen etc. and a printing-sheet. **2** *Archit.* = TYMPANUM 3. [F *tympan* or L *tympanum*: see TYMPANUM]

tympani *var.* of TIMPANI.

tympanic /tɪmˈpænɪk/ *adj.* **1** *Anat.* of, relating to, or having a tympanum. **2** resembling or acting like a drumhead. □ **tympanic bone** *Anat.* the bone supporting the tympanic membrane. **tympanic membrane** *Anat.* the membrane separating the outer ear and middle ear and transmitting vibrations resulting from sound waves to the inner ear.

tympanites /ˌtɪmpəˈnaɪtiːz/ *n. Med.* a swelling of the abdomen caused by gas in the intestine etc. □ **tympanitic** /-ˈnɪtɪk/ *adj.* [LL f. Gk *tumpanitēs* of a drum (as TYMPANUM)]

tympanum /ˈtɪmpənəm/ *n.* (*pl.* **tympanums** or **tympana** /-nə/) **1** *Anat.* **a** the middle ear. **b** the tympanic membrane. **2** *Zool.* the membrane covering the hearing organ on the leg of an insect. **3** *Archit.* **a** a vertical triangular space forming the centre of a pediment. **b** a similar space over a door between the lintel and the arch; a carving on this space. **4** a drum-wheel etc. for raising water from a stream. [L f. Gk *tumpanon* drum f. *tuptō* strike]

Tyndale /ˈtɪnd(ə)l/, William (*c*.1494–1536), English translator and Protestant martyr. Faced with ecclesiastical opposition to his project for translating the Bible into English, Tyndale went abroad in 1524, never to return to his own country; his translation of the New Testament (*c*.1525–6) was published in Germany. He then translated the Pentateuch (1530) and Jonah (1531), both of which were printed in Antwerp. Tyndale's translations later formed the basis of the Authorized Version. In 1535 he was arrested in Antwerp on a charge of heresy, and subsequently strangled and burnt at the stake.

Tyndall /ˈtɪnd(ə)l/, John (1820–93), Irish physicist. He is best known for his work on heat, studying such aspects as the absorbance and transmission of heat by gases and liquids, the thermal conductivity of solids, and the use of discontinuous heating as a sterilization technique. Tyndall also worked on diamagnetism, glaciers, the transmission of sound, and the scattering of light by suspended particles, becoming the first person to explain the blue colour of the sky. He was a prolific and popular lecturer and writer.

Tyne /taɪn/ **1** a river in NE England, formed by the confluence of two headstreams, the North Tyne, which rises in the Cheviot Hills, and the South Tyne, which rises in the northern Pennines. It flows 50 km (31 miles) generally eastwards, entering the North Sea at Tynemouth. **2** a shipping forecast area covering English coastal waters roughly from Flamborough Head in the south to Berwick in the north.

Tyne and Wear /wɪə(r)/ a metropolitan county of NE England; pop. (1991) 1,087,000; administrative centre, Newcastle-upon-Tyne.

Tyneside /ˈtaɪnsaɪd/ an industrial conurbation on the banks of the River Tyne, in NE England, stretching from Newcastle-upon-Tyne to the coast. □ **Tynesider** *n.*

Tynwald /ˈtɪnwɒld/ the legislative assembly of the Isle of Man, which meets annually to proclaim newly enacted laws. It consists of the governor (representing the sovereign) and council acting as the upper house, and an elected assembly called the House of Keys. [ON *thing-völlr* place of assembly f. *thing* assembly + *völlr* field]

type /taɪp/ *n. & v.* ● *n.* **1 a** a class of things or persons having common characteristics. **b** a kind or sort (*would like a different type of car*). **2** a person, thing, or event serving as an illustration, symbol, or characteristic specimen of another, or of a class. **3** (in *comb.*) made of, resembling, or functioning as (*ceramic-type material; Cheddar-type cheese*). **4** *colloq.* a person, esp. of a specified character (*is rather a quiet type; is not really my type*). **5** an object, conception, or work of art serving as a model for subsequent artists. **6** *Printing* **a** a piece of metal etc. with a raised letter or character on its upper surface for use in printing; such pieces collectively. **b** printed characters produced by type (*printed in large type*). **7** a device on either side of a medal or coin. **8** (in Christian theology) a foreshadowing in the Old Testament of a person or event of the Christian dispensation. **9** *Biol.* an organism having or chosen as having the essential characteristics of its group and giving its name to the next highest group. ● *v.* **1** *tr.* be a type or example of. **2** *tr. & intr.* write with a typewriter. **3** *tr.* esp. *Biol. & Med.* assign to a type; classify. **4** *tr.* = TYPECAST. □ **in type** *Printing* composed and ready for printing. **type-founder** a designer and maker of metal types. **type-foundry** a foundry where type is made. **type-metal** *Printing* an alloy of lead etc., used for casting printing types. **type-site** *Archaeol.* a site where objects regarded as defining the characteristics of a period etc. are found. **type specimen** *Biol.* the specimen used for naming and describing a new species. □ **typal** *adj.* [ME f. F *type* or L *typus* f. Gk *tupos* impression, figure, type, f. *tuptō* strike]

typecast /ˈtaɪpkɑːst/ *v.tr.* (*past* and *past part.* **-cast**) assign (an actor or actress) repeatedly to the same type of role, esp. one in character.

typeface /ˈtaɪpfeɪs/ *n. Printing* **1** a set of types or characters in a particular design. **2** the inked part of type, or the impression made by this.

typescript /ˈtaɪpskrɪpt/ *n.* a typewritten document.

typesetter /ˈtaɪpˌsetə(r)/ *n. Printing* **1** a person who composes type. **2** a composing-machine. □ **typesetting** *n.*

typewrite /ˈtaɪpraɪt/ *v.tr. & intr.* (*past* **-wrote** /-rəʊt/; *past part.* **-written** /-ˌrɪt(ə)n/) *formal* = TYPE *v.* 2.

typewriter /ˈtaɪpˌraɪtə(r)/ *n.* a machine with keys for producing printlike characters one at a time on paper inserted round a roller. An Englishman, Henry Mill, was granted a patent for a machine to impress or transcribe letters singly, but the first practical typewriter was produced in 1874 by the Remington Arms Company of America. It had forty-four keys arranged in the familiar 'qwerty' order, to regulate the typist's speed and prevent the bars which carried the letters from jamming — an arrangement still in use despite advances that make it unnecessary. The electric typewriter dates from 1935, and later improvements include placing the letters on a spinning 'golf ball' or 'daisy wheel', and using plastic ribbon in a cartridge instead of an inked ribbon on a spool. Typewriters have been superseded in many applications by the word processor, and modern forms usually have a limited electronic memory.

typewritten /ˈtaɪpˌrɪt(ə)n/ *adj.* produced with a typewriter.

typhlitis /tɪfˈlaɪtɪs/ *n. Med.* inflammation of the caecum. □ **typhlitic** /-ˈlɪtɪk/ *adj.* [mod.L f. Gk *tuphlon* caecum or blind gut f. *tuphlos* blind + -ITIS]

typhoid /ˈtaɪfɔɪd/ *n. & adj.* ● *n.* **1** (in full **typhoid fever**) an infectious fever caused by the bacterium *Salmonella typhi*, with a rash of red spots and severe intestinal irritation. (*See note below.*) **2** a similar disease of animals. ● *adj.* like typhus. □ **typhoid condition** (or **state**) a state of depressed vitality occurring in many acute diseases. □ **typhoidal** /taɪˈfɔɪd(ə)l/ *adj.* [TYPHUS + -OID]

▪ Typhoid fever is characterized by high fever, septicaemia, and intestinal ulceration. The bacteria are excreted in the faeces and spread through contaminated food and water. The illness used to be prolonged, incapacitating, and frequently fatal, and was a major cause of mortality in military campaigns. However, antibacterial treatment has much improved the outlook, and a vaccine has been developed which gives temporary immunity.

typhoon /taɪˈfuːn/ n. a tropical cyclone in eastern Asian seas. □ **typhonic** /-ˈfɒnɪk/ adj. [partly f. Port. tufão f. Arab. ṭūfān perh. f. Gk tuphōn whirlwind; partly f. Chin. dial. tai fung big wind]

typhus /ˈtaɪfəs/ n. Med. an infectious fever caused by rickettsiae, characterized by a purple rash, headaches, fever, and usu. delirium. Typhus refers to a group of rickettsial diseases marked by symptoms resembling those of typhoid fever, with which it was formerly confused. Unlike typhoid fever, however, it is spread by the bites of fleas, lice, ticks, and mites. It used to be associated with the crowded insanitary conditions found in jails, ships, and military camps. Typhus can now be treated with antibiotics, and a vaccine is available. Also called spotted fever. □ **typhous** adj. [mod.L f. Gk tuphos smoke, stupor f. tuphō to smoke]

typical /ˈtɪpɪk(ə)l/ adj. 1 serving as a characteristic example; representative. 2 characteristic of or serving to distinguish a type. 3 (often foll. by of) conforming to expected behaviour, attitudes, etc. (is typical of them to forget). 4 symbolic. □ **typically** adv. **typicality** /ˌtɪpɪˈkælɪtɪ/ n. [med.L typicalis f. L typicus f. Gk tupikos (as TYPE)]

typify /ˈtɪpɪˌfaɪ/ v.tr. (-ies, -ied) 1 be a representative example of; embody the characteristics of. 2 represent by a type or symbol; serve as a type, figure, or emblem of; symbolize. □ **typifier** n. **typification** /ˌtɪpɪfɪˈkeɪʃ(ə)n/ n. [L typus TYPE + -FY]

typist /ˈtaɪpɪst/ n. a person who uses a typewriter, esp. professionally.

typo /ˈtaɪpəʊ/ n. (pl. -os) colloq. 1 a typographical error. 2 a typographer. [abbr.]

typographer /taɪˈpɒɡrəfə(r)/ n. a person skilled in typography.

typography /taɪˈpɒɡrəfɪ/ n. 1 printing as an art. 2 the style and appearance of printed matter. □ **typographic** /ˌtaɪpəˈɡræfɪk/ adj. **typographical** adj. **typographically** adv. [F typographie or mod.L typographia (as TYPE, -GRAPHY)]

typology /taɪˈpɒlədʒɪ/ n. the study and interpretation of (esp. biblical) types. □ **typologist** n. **typological** /ˌtaɪpəˈlɒdʒɪk(ə)l/ adj. [Gk tupos TYPE + -LOGY]

Tyr /tɪə(r)/ Scand. Mythol. the god of battle, identified with Mars, after whom Tuesday is named.

tyramine /ˈtaɪərəˌmiːn/ n. Biochem. a derivative of tyrosine occurring in cheese and other foods, having a similar effect in the body to that of adrenaline. [TYROSINE + AMINE]

tyrannical /tɪˈrænɪk(ə)l/ adj. 1 acting like a tyrant; imperious, arbitrary. 2 given to or characteristic of tyranny. □ **tyrannically** adv. [OF tyrannique f. L tyrannicus f. Gk turannikos (as TYRANT)]

tyrannicide /tɪˈrænɪˌsaɪd/ n. 1 the act or an instance of killing a tyrant. 2 the killer of a tyrant. □ **tyrannicidal** /-ˌrænɪˈsaɪd(ə)l/ adj. [F f. L tyrannicida, -cidium (as TYRANT, -CIDE)]

tyrannize /ˈtɪrəˌnaɪz/ v.tr. & (foll. by over) intr. (also -ise) behave like a tyrant towards; rule or treat despotically or cruelly. [F tyranniser (as TYRANT)]

tyrannosaurus /tɪˌrænəˈsɔːrəs/ n. (also **tyrannosaur** /-ˈrænəˌsɔː(r)/) a very large bipedal flesh-eating dinosaur, Tyrannosaurus rex, with powerful jaws and hind legs, small clawlike front legs, and a large tail. It was the largest land carnivore in the late Cretaceous. [mod.L f. Gk turannos TYRANT + sauros lizard]

tyranny /ˈtɪrənɪ/ n. (pl. -ies) 1 the cruel and arbitrary use of authority. 2 a tyrannical act; tyrannical behaviour. 3 a a rule by a tyrant. b a period of this. c a state ruled by a tyrant. □ **tyrannous** adj. **tyrannously** adv. [ME f. OF tyrannie f. med.L tyrannia f. Gk turannia (as TYRANT)]

tyrant /ˈtaɪərənt/ n. 1 an oppressive or cruel ruler. 2 a person exercising power arbitrarily or cruelly. 3 Gk Hist. an absolute ruler who seized power without the legal right. 4 (in full **tyrant flycatcher**) a small bird of the New World family Tyrannidae, resembling the Old World flycatchers in behaviour. [ME tyran, -ant, f. OF tiran, tyrant f. L tyrannus f. Gk turannos]

Tyre /ˈtaɪə(r)/ a port on the Mediterranean in southern Lebanon; pop. (1988) 14,000. Founded in the 2nd millennium BC as a colony of Sidon, it was for centuries a Phoenician port and trading centre. Its prosperity did not decline until the 14th century.

tyre /ˈtaɪə(r)/ n. (US **tire**) 1 a rubber covering, usu. inflated, placed round a wheel to form a soft contact with the road. (See note below.) 2 a strengthening band of metal fitted round the rim of a wheel, esp. of a railway vehicle. □ **tyre-gauge** a portable device for measuring the air-pressure in a tyre. [ME, perh. = archaic tire head-dress]

▪ In the mid-19th century tyres for carriage wheels began to be made of solid rubber instead of metal, although this still gave a hard ride. The first pneumatic tyres were invented in 1845 by the Scottish engineer R. W. Thomson, but the idea lapsed until 1888 when J. B. Dunlop devised the first practical one for use on his son's tricycle. He used an air-filled tube inside a canvas cover with rubber treads, and the Michelin Tyre Company pioneered its use on motor vehicles in the 1890s. Modern car tyres have dispensed with the inner tube, though most bicycle tyres still use one. The old cross-ply tyre with its transversely arranged fabric cords has given way to the radial, which has the cords arranged radially and is usually reinforced with steel wires.

Tyrian /ˈtɪrɪən/ adj. & n. ● adj. of or relating to ancient Tyre in Phoenicia. ● n. a native or inhabitant of Tyre. □ **Tyrian purple** see PURPLE n. 2. [L Tyrius f. Tyrus Tyre]

tyro var. of TIRO.

Tyrol /tɪˈrəʊl/ (German **Tirol** /tiˈrol/) an Alpine state of western Austria; capital, Innsbruck. The southern part was ceded to Italy after the First World War. □ **Tyrolean** /ˌtɪrəˈliːən/ adj. & n. **Tyrolese** /-ˈliːz/ adj. & n.

Tyrone /tɪˈrəʊn/ one of the Six Counties of Northern Ireland, formerly an administrative area; pop. (1981) 143,900; chief town, Omagh.

tyrosine /ˈtaɪərəˌsiːn/ n. Biochem. a hydrophilic amino acid present in many proteins and important in the synthesis of some hormones etc. [irreg. f. Gk turos cheese + -INE[4]]

Tyrrhene /ˈtɪriːn/ adj. & n. archaic or poet. = ETRUSCAN. [L Tyrrhenus]

Tyrrhenian /tɪˈriːnɪən/ adj. & n. 1 of, relating to, or denoting the Tyrrhenian Sea or its region. 2 archaic or poet. = ETRUSCAN. [L Tyrrhenus Etruscan]

Tyrrhenian Sea a part of the Mediterranean Sea between mainland Italy and the islands of Sicily and Sardinia.

Tyumen /tjuːˈmen/ a city in west Siberian Russia, in the eastern foothills of the Ural Mountains; pop. (1990) 487,000. Founded in 1586 on the site of a 14th-century Tartar town, it is regarded as the oldest city in Siberia.

Tzara /ˈzɑːrə/, Tristan (born Samuel Rosenstock) (1896–1963), Romanian-born French poet. Emigrating to France at the age of 19, in 1916 he became one of the founders of the Dada movement; he edited the periodical Dada (1917–21) and wrote the movement's manifestos. His poetry, with its continuous flow of unconnected images, and his suggestion that poems should be composed from words cut from a newspaper and selected at random, helped form the basis for surrealism, a movement he became involved with from about 1930 onwards.

tzatziki /tsætˈsiːkɪ/ n. a Greek side dish of yoghurt with cucumber, garlic, and often mint. [mod.Gk]

tzigane /tsɪˈɡɑːn/ n. 1 a Hungarian gypsy. 2 (attrib.) characteristic of the tziganes or (esp.) their music. [F f. Hungarian c(z)igány]

Tzu-po see ZIBO.

Uu

U[1] /juː/ n. (also **u**) (pl. **Us** or **U's**) **1** the twenty-first letter of the alphabet. **2** a U-shaped object or curve (esp. in *comb.*: *U-bolt*).

U[2] /juː/ adj. esp. Brit. colloq. **1** upper class. **2** supposedly characteristic of the upper class. [abbr.]

U[3] /uː/ n. a Burmese title of respect before a man's name. [Burmese]

U[4] symb. **1** Chem. the element uranium. **2** Brit. universal (of films classified as suitable without restriction).

u symb. = MU 2 (μ).

U2 /juːˈtuː/ an Irish rock group formed in 1977, featuring the vocals of Bono (Paul Hewson, b.1960) and the distinctive guitar work of 'The Edge' (David Evans, b.1961). Since the recording of early songs such as 'I Will Follow' in 1980, U2 have gradually become one of the world's most popular rock groups; they were one of the first bands to play concerts in sports stadiums. Albums include *Rattle and Hum* (1988).

UAE abbr. United Arab Emirates.

UAR abbr. United Arab Republic.

Ubaid /uːˈbaɪd/ adj. & n. Archaeol. of, relating to, or denoting a mesolithic culture in Mesopotamia that flourished during the 5th millennium BC, named after Tell Al 'Ubaid near the ancient city of Ur.

Ubanghi Shari /juːˌbæŋgɪ ˈʃɑːrɪ/ a former name (until 1958) for the CENTRAL AFRICAN REPUBLIC.

Übermensch /ˈuːbəˌmenʃ/ n. (also called *superman*, *overman*) the ideal superior man of the future, originally described by Nietzsche in *Thus Spake Zarathustra* (1883–5). Nietzsche thought that such a being could arise when any man of superior potential took control over himself and shook off the conventional Christian morality of the masses to create and impose his own values. The concept was used by George Bernard Shaw in *Man and Superman* (1903). [G, = superhuman person]

ubiety /juːˈbaɪətɪ/ n. the fact or condition of being in a definite place; local relation. [med.L *ubietas* f. L *ubi* where]

-ubility /jʊˈbɪlɪtɪ/ suffix forming nouns from, or corresponding to, adjectives in *-uble* (*solubility*; *volubility*). [L *-ubilitas*: cf. -ITY]

ubiquitarian /juːˌbɪkwɪˈteərɪən/ adj. & n. (in Christian theology) ● adj. relating to or believing in the doctrine of the omnipresence of Christ's body. ● n. a believer in this. □ **ubiquitarianism** n. [mod.L *ubiquitarius* (as UBIQUITOUS)]

ubiquitous /juːˈbɪkwɪtəs/ adj. **1** present everywhere or in several places simultaneously. **2** often encountered. □ **ubiquitously** adv. **ubiquitousness** n. **ubiquity** n. [mod.L *ubiquitas* f. L *ubique* everywhere f. *ubi* where]

-uble /jʊb(ə)l/ suffix forming adjectives meaning 'that may or must (be)' (see -ABLE) (*soluble*; *voluble*). [F f. L *-ubilis*]

-ubly /jʊblɪ/ suffix forming adverbs corresponding to adjectives in *-uble*.

U-boat /ˈjuːbəʊt/ n. hist. a German submarine used in the First and Second World Wars. [G *U-boot* = *Unterseeboot* under-sea boat]

UC abbr. University College.

u.c. abbr. upper case.

UCATT abbr. (in the UK) Union of Construction, Allied Trades, and Technicians.

UCCA /ˈʌkə/ abbr. (in the UK) Universities Central Council on Admissions.

Uccello /uːˈtʃeləʊ/, Paolo (born Paolo di Dono) (c.1397–1475), Italian painter. His nickname (*uccello* = 'bird') is said to refer to his love of animals, especially birds. He was based largely in Florence and is associated with the early use of perspective in painting. His surviving works include *The Rout of San Romano* (one of a series of three panels, c.1454–7) and *The Hunt*, one of the earliest known paintings on canvas, noted for its atmosphere of fairy-tale romance.

UCW abbr. (in the UK) Union of Communication Workers.

UDA abbr. Ulster Defence Association (a Loyalist paramilitary organization).

udal /ˈjuːd(ə)l/ n. (also **odal** /ˈəʊd-/) the kind of freehold right based on uninterrupted possession prevailing in northern Europe before the feudal system and still in use in Orkney and Shetland. [ON *óthal* f. Gmc]

UDC abbr. hist. (in the UK) Urban District Council.

udder /ˈʌdə(r)/ n. the mammary gland of cattle, sheep, etc., hanging as a baglike organ with several teats. □ **uddered** adj. (also in comb.). [OE *úder* f. WG]

UDI see UNILATERAL DECLARATION OF INDEPENDENCE.

Udmurtia /ʊdˈmʊətɪə/ (also **Udmurt Republic** /ˈʊdmʊət/) an autonomous republic in central Russia; pop. (1990) 1,619,000; capital, Izhevsk.

udometer /juːˈdɒmɪtə(r)/ n. formal a rain-gauge. [F *udomètre* f. L *udus* damp]

UDR abbr. hist. Ulster Defence Regiment.

UEFA /juːˈiːfə/ abbr. Union of European Football Associations.

uey /ˈjuːɪ/ n. Austral. colloq. a U-turn.

Ufa /uːˈfɑː/ the capital of Bashkiria, in the Ural Mountains; pop. (1990) 1,094,000.

Uffizi /ʊˈfiːtsɪ/ an art gallery and museum in Florence, housing one of Europe's finest art collections. Italian Renaissance painting is particularly well represented, although the collection also contains sculptures, drawings, and Flemish, French, and Dutch paintings. The building, the Uffizi palace, was designed by Vasari c.1560 as offices for the Medici family.

UFO /ˈjuːfəʊ, ˌjuːefˈəʊ/ n. (also **ufo**) an unidentified flying object. The term is often applied to supposed vehicles ('flying saucers') piloted by beings from outer space, for which much alleged evidence has been produced but no proof obtained. Most UFO sightings are eventually identified as weather balloons, aircraft, or meteorological phenomena. [acronym]

ufology /juːˈfɒlədʒɪ/ n. the study of UFOs. □ **ufologist** n.

Uganda /juːˈgændə/ a landlocked country in East Africa; pop. (est. 1991) 16,876,000; languages, English (official), Swahili, and other languages; capital, Kampala. The greater part of Uganda is savannah, and the country has several large lakes, notably Lake Victoria. Ethnically and culturally diverse, Uganda was explored by Europeans in the mid-19th century, becoming a British protectorate in 1894. It became an independent Commonwealth state in 1962. The country was ruled (1971–9) by the brutal dictator Idi Amin, who came to power in an army coup. His overthrow, with Tanzanian military intervention, was followed by several years of conflict, partly resolved in 1986 by the formation of a government by the National Resistance Movement under President Yoweri Museveni (b.1944). □ **Ugandan** adj. & n.

Ugarit /ˈuːgəˌrɪt/ an ancient port in northern Syria, which was occupied from neolithic times until its destruction by the Sea Peoples in about the 12th century BC. Ugarit was an important commercial city during the late Bronze Age, to which period belong a palace, temples, and private residences containing legal, religious, and administrative cuneiform texts in Sumerian, Akkadian, Hurrian, Hittite, and Ugaritic languages. The last of these was written in an early form of the Phoenician alphabet. □ **Ugaritic** /ˌuːgəˈrɪtɪk/ adj. & n.

ugh /əx, ʌg, ʌx/ int. **1** expressing disgust or horror. **2** the sound of a cough or grunt. [imit.]

Ugli /ˈʌglɪ/ n. (pl. **Uglis** or **Uglies**) propr. a mottled green and yellow citrus fruit, a hybrid of a grapefruit and tangerine. [UGLY]

uglify /ˈʌglɪˌfaɪ/ v.tr. (**-ies**, **-ied**) make ugly. □ **uglification** /ˌʌglɪfɪˈkeɪʃ(ə)n/ n.

ugly /ˈʌglɪ/ adj. (**uglier**, **ugliest**) **1** unpleasing or repulsive to see or hear (an ugly scar; spoke with an ugly snarl). **2** unpleasantly suggestive; discreditable (ugly rumours are about). **3** threatening, dangerous (the sky has an ugly look; an ugly mood). **4** morally repulsive; vile (ugly vices). □ **ugly customer** an unpleasantly formidable person. **ugly duckling** a person who turns out to be beautiful or talented etc. against all expectations (with ref. to a cygnet in a brood of ducks in a tale by Hans Christian Andersen). □ **uglily** adv. **ugliness** n. [ME f. ON ugglígr to be dreaded f. ugga to dread]

Ugric /ˈuːgrɪk, ˈjuː-/ adj. & n. (also **Ugrian** /-rɪən/) ● adj. of or relating to the eastern branch of Finnic peoples, esp. the Finns and Magyars, or the group of Finno-Ugric languages that includes Hungarian. ● n. **1** a member of any of these peoples. **2** any of the languages of these peoples, esp. Hungarian. [Russ. ugry name of a people dwelling east of the Urals]

UHF abbr. ultra-high frequency.

uh-huh /ˈʌhʌ/ int. colloq. expressing assent. [imit.]

uhlan /ˈuːlɑːn, ˈjuːlən/ n. hist. a cavalryman armed with a lance in some European armies, esp. the former German army. [F & G f. Pol. (h)ulan f. Turk. oğlan youth, servant]

UHT abbr. ultra heat treated (esp. of milk, for long keeping).

Uighur /ˈwiːgə(r)/ n. & adj. (also **Uigur**) ● n. **1** a member of a Turkic people inhabiting NW China, particularly the Xinjiang region, and adjoining areas. **2** the Turkic language of this people. ● adj. of or relating to this people or their language.

Uist /ˈjuːɪst/ two small islands in the Outer Hebrides, off the west coast of Scotland, North Uist and South Uist, lying to the south of Lewis and Harris, separated from each other by the island of Benbecula.

Uitlander /ˈeɪtˌlɒndə(r)/ n. S. Afr. hist. any of the people who entered Transvaal from Britain and elsewhere after the discovery of gold (1886). These immigrants were denied citizenship and taxed heavily by the Boers, who regarded them as a cultural and economic threat. Tension over the treatment of the Uitlanders was one of the factors in the outbreak of the Second Boer War. [Afrik. f. Du. uit out + land land]

ujamaa /ˌuˈdʒɑːmɑː/ n. (usu. attrib.) a system of self-help village cooperatives established by President Nyerere in Tanzania in the 1960s. [Kiswahili, = brotherhood, f. jamaa family, f. Arab. jamā'a community]

Ujiyamada /ˌuːjɪjəˈmɑːdə/ the former name (until 1956) for ISE.

Ujjain /ˈuːdʒaɪn/ a city in west central India in Madhya Pradesh; pop. (1991) 376,000. One of the seven holy cities of Hinduism, its name means 'she who conquers'.

Ujung Pandang /ˌuːdʒʊŋ pænˈdæŋ/ the chief seaport of the island of Sulawesi in Indonesia; pop. (1980) 709,000. Known as Makassar (Macassar) until 1973, it gave its name to an oil once used as a dressing for the hair.

UK see UNITED KINGDOM.

UKAEA abbr. United Kingdom Atomic Energy Authority.

ukase /juːˈkeɪz/ n. **1** a command or edict, esp. in Russia or the former USSR. **2** an edict of the Russian government. [Russ. ukaz ordinance, edict f. ukazat' show, decree]

ukiyo-e /uːˌkiːjəʊˈeɪ/ n. a school of Japanese art using subjects from everyday life and characterized by simple treatment. It was the dominant movement in Japanese art in the 17th–19th centuries, favoured by such celebrated artists as Hokusai and Utamaro. [Jap., = genre picture]

Ukraine /juːˈkreɪn/ (also **the Ukraine**) a country in eastern Europe, to the north of the Black Sea; pop. (est. 1991) 51,999,000; languages, Ukrainian and Russian; capital, Kiev. The greater part of Ukraine consists of grain-producing steppes stretching east from the Carpathians and drained by major rivers including the Dniester, Dnieper, and Donets. It also includes the mainly Russian-populated Crimea. Often a borderland between rival powers, Ukraine was united with Russia, with the capital at Kiev, in the 9th century. After a period of division between Poland, Russia, and the Ottoman Empire it was reunited with Russia in 1785. Briefly independent following the 1917 revolution, it became one of the original constituent republics (and the third largest) of the USSR. In 1991, on the breakup of the Soviet Union, Ukraine became an independent republic within the Commonwealth of Independent States. [Russ. ukraina frontier region f. u at + KRAI]

Ukrainian /juːˈkreɪnɪən/ n. & adj. ● n. **1** a native or inhabitant of Ukraine. **2** the Slavonic language of Ukraine. ● adj. of or relating to Ukraine or its people or language.

ukulele /ˌjuːkəˈleɪlɪ/ n. a small four-stringed Hawaiian (orig. Portuguese) guitar. [Hawaiian, = jumping flea]

Ulala /ˌuːlɑːˈlɑː/ a former name (until 1932) for GORNO-ALTAISK.

Ulan Bator /ˌuːlɑːn ˈbɑːtə(r)/ (also **Ulaanbaatar**) the capital of Mongolia; pop. (1990) 575,000. It was founded in the 17th century and developed as a religious centre around a Buddhist monastery. It was known as Urga until 1924, changing to its present name, which means 'red hero', after becoming capital of the Communist state of Mongolia.

Ulanova /uːˈlɑːnəvə/, Galina (Sergeevna) (b.1910), Russian ballet-dancer. In 1928 she joined the Kirov Ballet, transferring to the Bolshoi company in 1944. She gave notable interpretations of 19th-century ballets, such as Swan Lake and Giselle, and also danced the leading roles, composed especially for her, in all three of Prokofiev's ballets. During the 1950s she became well known in the West through touring with the Bolshoi Ballet. After her retirement as a dancer in 1962 she remained with the company as a teacher.

Ulan-Ude /uːˌlɑːnuːˈdeɪ/ an industrial city in southern Siberian Russia, capital of the republic of Buryatia; pop. (1990) 359,000. Founded in 1783, it was known as Verkhneudinsk until 1934.

-ular /jʊlə(r)/ suffix forming adjectives, sometimes corresp. to nouns in -ule (pustular) but often without diminutive force (angular; granular). □ **-ularity** /jʊˈlærɪtɪ/ suffix forming nouns. [from or after L -ularis (as -ULE, -AR¹)]

ulcer /ˈʌlsə(r)/ n. **1** an open sore on an external or internal surface of the body, often forming pus. **2 a** a moral blemish. **b** a corroding or corrupting influence etc. □ **ulcered** adj. **ulcerous** adj. [ME f. L ulcus -eris, rel. to Gk helkos]

ulcerate /ˈʌlsəˌreɪt/ v.tr. & intr. form into or affect with an ulcer. □ **ulcerable** /-rəb(ə)l/ adj. **ulcerative** /-rətɪv/ adj. **ulceration** /ˌʌlsəˈreɪʃ(ə)n/ n. [ME f. L ulcerare ulcerat- (as ULCER)]

-ule /juːl/ suffix forming diminutive nouns (capsule; globule). [from or after L -ulus, -ula, -ulum]

Uleåborg /ˈuːlɪəˌbɔːrj/ the Swedish name for OULU.

ulema /ˈuːlɪmə/ n. **1** a body of Muslim doctors of sacred law and theology. **2** a member of this. [Arab. 'ulamā pl. of 'ālim learned f. 'alama know]

-ulent /jʊlənt/ suffix forming adjectives meaning 'abounding in, full of' (fraudulent; turbulent). □ **-ulence** suffix forming nouns. [L -ulentus]

Ulfilas /ˈʊlfɪˌlæs/ (also **Wulfila** /ˈwʊlfɪlə/) (c.311–c.381), bishop and translator. Believed to be of Cappadocian descent, he became bishop of the Visigoths in 341. Ulfilas is best known for his translation of the Bible from Greek into Gothic (of which fragments survive), the earliest known translation of the Bible into a Germanic language. The translation uses the Gothic alphabet, based on Latin and Greek characters, which Ulfilas is traditionally held to have invented.

Ulhasnagar /ˌuːlhəsˈnʌgə(r)/ a city in western India, in the state of Maharashtra; pop. (1991) 369,000.

uliginose /juːˈlɪdʒɪˌnəʊz, -ˌnəʊs/ adj. (also **uliginous** /-nəs/) Bot. growing in wet or swampy places. [L uliginosus f. uligo -ginis moisture]

ullage /ˈʌlɪdʒ/ n. **1** the amount by which a cask etc. falls short of being full. **2** loss by evaporation or leakage. [ME f. AF ulliage, OF ouillage f. ouiller fill up, ult. f. L oculus eye, with ref. to the bung-hole]

Ulm /ʊlm/ an industrial city on the Danube in Baden-Württemberg, southern Germany; pop. (1991) 112,170. In 1805 it was the scene of a battle, during the Napoleonic Wars, in which Napoleon defeated the Austrians.

ulna /ˈʌlnə/ n. (pl. **ulnae** /-niː/) **1** the thinner and longer bone in the forearm, on the side opposite to the thumb (cf. RADIUS 3). **2** Zool. a corresponding bone in an animal's foreleg or a bird's wing. □ **ulnar** adj. [L, rel. to Gk ōlenē and ELL]

ulotrichan /juːˈlɒtrɪkən/ adj. & n. Anthropol. ● adj. (also **ulotrichous** /-kəs/) having tightly curled hair, esp. denoting a human type. ● n. a person having such hair. [mod.L *Ulotrichi* f. Gk *oulos* woolly, crisp + *thrix trikhos* hair]

-ulous /-jʊləs/ *suffix* forming adjectives (*fabulous*; *populous*). [L *-ulosus, -ulus*]

Ulpian /'ʌlpɪən/ (Latin name Domitius Ulpianus) (died *c.*228), Roman jurist, born in Phoenicia. His numerous legal writings provided one of the chief sources for Justinian's *Digest* of 533.

Ulsan /uːlˈsɑːn/ an industrial port on the south coast of South Korea; pop. (1990) 682,980.

Ulster /'ʌlstə(r)/ **1** a former province of Ireland, in the north of the island. The nine counties of Ulster are now divided between Northern Ireland (Antrim, Down, Armagh, Londonderry, Tyrone, and Fermanagh) and the Republic of Ireland (Cavan, Donegal, and Monaghan). **2** (loosely) Northern Ireland.

ulster /'ʌlstə(r)/ *n.* a long loose overcoat of rough cloth. [ULSTER, where it was orig. sold]

Ulster Democratic Unionist Party an extreme Loyalist political party in Northern Ireland, led by Ian Paisley, who was one of its founders in 1972.

Ulsterman /'ʌlstəmən/ *n.* (*pl.* **-men**; *fem.* **Ulsterwoman**; *pl.* **-women**) a native of Ulster.

Ulster Unionist Council (abbr. **UUC**) a political party in Northern Ireland seeking to maintain the union of Northern Ireland with Britain. Founded in 1905, the UUC is regarded as being more moderate than the Ulster Democratic Unionist Party, and maintains organizational links with the Conservative Party.

ult. *abbr.* ultimo.

ulterior /ʌlˈtɪərɪə(r)/ *adj.* **1** existing in the background, or beyond what is evident or admitted; hidden, secret (esp. *ulterior motive*). **2** situated beyond. **3** more remote; not immediate; in the future. □ **ulteriorly** *adv.* [L, = further, more distant]

ultima /'ʌltɪmə/ *n. Linguistics* the last syllable of a word. [L *ultima (syllaba)*, fem. of *ultimus* last]

ultimate /'ʌltɪmət/ *adj. & n.* ● *adj.* **1** last, final. **2** beyond which no other exists or is possible (*the ultimate analysis*). **3** fundamental, primary, unanalysable (*ultimate truths*). **4** maximum (*ultimate tensile strength*). ● *n.* **1** (prec. by *the*) the best achievable or imaginable. **2** a final or fundamental fact or principle. □ **ultimately** *adj.* **ultimateness** *n.* [LL *ultimatus* past part. of *ultimare* come to an end]

ultima Thule /ˌʌltɪmə ˈθjuːliː/ *n.* **1** a far-away unknown region. **2** the limit of what is attainable. [L, = furthest Thule: see THULE]

ultimatum /ˌʌltɪˈmeɪtəm/ *n.* (*pl.* **ultimatums** or **ultimata** /-tə/) a final demand or statement of terms by one party, the rejection of which by another could cause a breakdown in relations, war, or an end of cooperation etc. [L neut. past part.: see ULTIMATE]

ultimo /'ʌltɪməʊ/ *adj. Commerce* of last month (*the 28th ultimo*). [L *ultimo mense* in the last month]

ultimogeniture /ˌʌltɪməʊˈdʒenɪtʃə(r)/ *n.* a system in which the youngest son has the right of inheritance (cf. PRIMOGENITURE 2). [L *ultimus* last, after PRIMOGENITURE]

ultra /'ʌltrə/ *adj. & n.* ● *adj.* favouring extreme views or measures, esp. in religion or politics. ● *n.* an extremist. [orig. as abbr. of F *ultra-royaliste*: see ULTRA-]

ultra- /'ʌltrə/ *comb. form* **1** beyond; on the other side of (opp. CIS-). **2** extreme(ly), excessive(ly) (*ultra-conservative*; *ultra-modern*). [L *ultra* beyond]

ultracentrifuge /ˌʌltrəˈsentrɪˌfjuːdʒ/ *n.* a very fast centrifuge used to precipitate large biological molecules from solution or separate them by their different rates of sedimentation. (See CENTRIFUGE.)

ultrafiltration /ˌʌltrəfɪlˈtreɪʃ(ə)n/ *n.* filtration using a filter fine enough to retain large molecules, colloidal particles, and viruses.

ultra-high /ˌʌltrəˈhaɪ/ *adj.* (of a frequency) in the range 300 to 3,000 megahertz.

ultraist /'ʌltrəɪst/ *n.* the holder of extreme positions in politics, religion, etc. □ **ultraism** *n.*

ultramarine /ˌʌltrəməˈriːn/ *n. & adj.* ● *n.* **1 a** a brilliant blue pigment orig. obtained from lapis lazuli. **b** an imitation of this from powdered fired clay, sodium carbonate, sulphur, and resin. **2** the colour of this. ● *adj.* **1** of this colour. **2** *archaic* situated beyond the sea. [obs. It. *oltramarino* & med.L *ultramarinus* beyond the sea (as ULTRA-, MARINE), because lapis lazuli was brought from beyond the sea]

ultramicroscope /ˌʌltrəˈmaɪkrəˌskəʊp/ *n.* an optical microscope used to reveal very small particles by means of light scattered by them.

ultramicroscopic /ˌʌltrəˌmaɪkrəˈskɒpɪk/ *adj.* **1** too small to be seen by an ordinary optical microscope. **2** of or relating to an ultramicroscope.

ultramontane /ˌʌltrəˈmɒnteɪn/ *adj. & n.* ● *adj.* **1** situated on the other side of the Alps from the point of view of the speaker. **2** of or holding a doctrine of supreme papal authority in matters of faith and discipline (cf. GALLICAN *adj.* 2). This was firmly established by the declaration of papal infallibility in 1870. ● *n.* **1** a person living on the other side of the Alps. **2** a believer in supreme papal authority. [med.L *ultramontanus* (as ULTRA-, L *mons montis* mountain)]

ultramundane /ˌʌltrəˈmʌndeɪn/ *adj.* lying beyond the world or the solar system. [L *ultramundanus* (as ULTRA-, *mundanus* f. *mundus* world)]

ultrasonic /ˌʌltrəˈsɒnɪk/ *adj.* of or involving sound waves with a frequency above the upper limit of human hearing. Many animals are able to hear sounds in the ultrasonic range, and some, such as bats and dolphins, make ultrasonic sounds for purposes of echolocation. (See also ULTRASONICS.) □ **ultrasonically** *adv.*

ultrasonics /ˌʌltrəˈsɒnɪks/ *n.pl.* (usu. treated as *sing.*) the science and application of ultrasonic waves (ultrasound). Sound of very high frequency (around 20,000 Hz and above) can be formed into a powerful beam, and ultrasound has a number of practical applications. It is used in sonar, detecting faults or cracks in metals, in cleaning processes, forming emulsions between immiscible liquids, disrupting bacterial and other cells, and as an alternative to X-rays in medical diagnosis. Cutting tools of relatively soft metals can be ultrasonically vibrated to cut shapes or holes in glassy or ceramic materials that cannot be machined by conventional techniques.

ultrasound /'ʌltrəˌsaʊnd/ *n.* **1** sound having an ultrasonic frequency. **2** ultrasonic waves. □ **ultrasound cardiography** = ECHOCARDIOGRAPHY.

ultrastructure /'ʌltrəˌstrʌktʃə(r)/ *n. Biol.* fine structure not visible with an optical microscope.

ultraviolet /ˌʌltrəˈvaɪələt/ *adj. & n.* (abbr. **UV**) ● *adj.* **1** (of electromagnetic radiation) having a wavelength less than that of the violet end of the visible light spectrum but greater than that of X-rays. (*See note below.*) **2** of or using such radiation. ● *n.* the ultraviolet part of the spectrum.

▪ Ultraviolet radiation spans wavelengths from about 10 nm to 400 nm, the longer wavelengths being visible to bees and some other insects. It is an important component of sunlight, although the ozone layer prevents much of it from reaching the earth's surface. While ultraviolet in sunlight is necessary for the production of vitamin D_2 in the skin, it is now known that excessive exposure to it can be harmful, causing skin cancer and genetic mutation. Uses of ultraviolet include sterilization, spectroscopy, and sunlamps; fluorescent substances such as optical whiteners reflect ultraviolet light at longer (and therefore visible) wavelengths. Ultraviolet radiation is divided, in order of decreasing wavelength, into UVA, UVB, and UVC, of which the last does not penetrate the ozone layer. (See also OZONE LAYER.)

ultra vires /ˌʌltrə ˈvaɪəriːz, ˌʊltrə ˈvɪəreɪz/ *adv. & predic.adj.* beyond one's legal power or authority. [L]

ululate /'juːljʊˌleɪt/ *v.intr.* howl, wail; make a hooting cry. □ **ululant** *adj.* **ululation** /ˌjuːljʊˈleɪʃ(ə)n/ *n.* [L *ululare* ululat- (imit.)]

Ulundi /ʊˈlʊndɪ/ a town in KwaZulu/Natal, South Africa. Founded in 1873 as the capital of Zululand, it was restored as the capital of the former homeland of KwaZulu in the early 1980s.

Uluru /ʊˈlʊəruː/ the Aboriginal name for AYERS ROCK.

Ulyanov /uːlˈjɑːnɒf/, Vladimir Ilich, see LENIN.

Ulyanovsk /uːlˈjɑːnəfsk/ the former name (1924–92) for SIMBIRSK.

Ulysses /'juːlɪˌsiːz/ **1** *Rom. Mythol.* the Roman name for Odysseus. **2** a space probe of the European Space Agency, launched in 1990 to investigate the polar regions of the sun.

um /ʌm, əm/ *int.* expressing hesitation or a pause in speech. [imit.]

-um var. of -IUM 1.

Umayyad /ʊˈmaɪjæd/ *adj. & n.* (also **Omayyad** /əʊˈmaɪ-/) ● *adj.* of or relating to a Muslim dynasty, which included the family of the prophet Muhammad, that ruled the Islamic world from AD 660 (or 661) to 750 and later ruled Moorish Spain 756–1031. ● *n.* a member of this dynasty.

Umayyad Mosque a mosque in Damascus, Syria, built AD 705–15 on the site of a church dedicated to St John the Baptist. Within the mosque is a shrine believed to contain the relic of the saint's head.

umbel /'ʌmb(ə)l/ *n. Bot.* a flower-cluster in which stalks nearly equal in length spring from a common centre and form a flat or curved surface, as in parsley. □ **umbellate** adj. **umbellar** /ʌm'belə(r)/ adj. **umbellule** /-'belju:l/ adj. [obs. F umbelle or L umbella sunshade, dimin. of UMBRA]

umbellifer /ʌm'belɪfə(r)/ n. Bot. a plant of the large family Umbelliferae, having a cluster of small flowers forming an umbel. The umbellifers include food plants such as carrot and parsnip and many culinary herbs like parsley, fennel, anise, and coriander. □ **umbelliferous** /ˌʌmbə'lɪfərəs/ adj. [obs. F umbellifère f. L (as UMBEL, -fer bearing)]

umber /'ʌmbə(r)/ n. & adj. ● n. **1** a natural pigment like ochre but darker and browner. **2** the colour of this. ● adj. **1** of this colour. **2** dark, dusky. [F (terre d')ombre or It. (terra di) ombra = shadow (earth), f. L UMBRA or Umbra fem. of Umber Umbrian]

umbilical /ʌm'bɪlɪk(ə)l, ˌʌmbɪ'laɪk-/ adj. **1** of, situated near, or affecting the navel. **2** centrally placed. □ **umbilical cord 1** a flexible cordlike structure containing blood vessels and attaching a foetus to the placenta. **2** a supply cable linking a missile to its launcher, or an astronaut in space to a spacecraft. [obs. F umbilical or f. UMBILICUS]

umbilicate /ʌm'bɪlɪkət/ adj. **1** shaped like a navel. **2** having an umbilicus.

umbilicus /ʌm'bɪlɪkəs, ˌʌmbɪ'laɪk-/ n. (pl. **umbilici** /ʌm'bɪlɪsaɪ, ˌʌmbɪ'laɪs-/ or **umbiliceuses**) **1** Anat. the navel. **2** Bot. & Zool. a navel-like formation. **3** Geom. a point in a surface through which all cross-sections have the same curvature. [L, rel. to Gk omphalos and to NAVEL]

umbles /'ʌmb(ə)lz/ n.pl. archaic the edible offal of deer etc. (cf. eat humble pie (see HUMBLE)). [ME var. of NUMBLES]

umbo /'ʌmbəʊ/ n. (pl. **-os** or **umbones** /ʌm'bəʊni:z/) **1** the boss of a shield, esp. in the centre. **2** Bot. & Zool. a rounded knob or protuberance. □ **umbonal** /-bən(ə)l/ adj. **umbonate** /-bənət/ adj. [L umbo -onis]

umbra /'ʌmbrə/ n. (pl. **umbras** or **umbrae** /-bri:/) **1** the fully shaded inner region of the shadow cast by an opaque object, esp. (Astron.) the area on the earth or moon experiencing the total phase of an eclipse (cf. PENUMBRA 1a). **2** Astron. the dark central part of a sunspot. □ **umbral** adj. [L, = shade]

umbrage /'ʌmbrɪdʒ/ n. **1** offence; a sense of slight or injury (esp. give or take umbrage at). **2** archaic **a** a shade. **b** what gives shade. [ME f. OF ult. f. L umbraticus f. umbra: see UMBRA]

umbrella /ʌm'brelə/ n. **1** a light portable device for protection against rain, strong sun, etc., consisting of a usu. circular canopy of fabric mounted by means of a collapsible metal frame on a central stick. **2** protection or patronage. **3** (often attrib.) a coordinating or unifying agency (umbrella organization). **4** a screen of fighter aircraft or a curtain of fire put up as a protection against enemy aircraft. **5** Zool. the gelatinous disc of a jellyfish etc., which it contracts and expands to move through the water. □ **umbrella bird** a South American cotinga of the genus Cephalopterus, with a black radiating crest and long wattles. **umbrella pine 1** = stone pine. **2** a tall Japanese evergreen conifer, Sciadopitys verticillata, with leaves in umbrella-like whorls. **umbrella plant 1** a Californian plant, Darmera peltophylla (saxifrage family), of boggy places, with large round leaves. **2** a tropical Old World sedge, Cyperus involucratus, with a whorl of bracts at the top of the stem, popular as a house plant. **umbrella stand** a stand for holding closed upright umbrellas. **umbrella tree** a small tree with leaves in a whorl like an umbrella; esp. Magnolia tripetala of North America, or Schefflera actinophylla of Australia (grown as a house plant). □ **umbrellaed** /-ləd/ adj. **umbrella-like** adj. [It. ombrella, dimin. of ombra shade f. L umbra: see UMBRA]

Umbria /'ʌmbrɪə/ a region of central Italy, in the valley of the Tiber; capital, Perugia.

Umbrian /'ʌmbrɪən/ n. & adj. ● n. **1** a native or inhabitant of Umbria, esp. a member of the Italic people inhabiting this area in pre-Roman times. **2** the extinct Italic language of ancient Umbria. ● adj. of or relating to Umbria, its inhabitants, or the Umbrian language.

Umbriel /'ʌmbrɪəl/ Astron. satellite II of Uranus, the thirteenth closest to the planet, discovered in 1851 (diameter 1,190 km). The surface is dark and heavily cratered. [a sprite in a poem by Alexander Pope]

umbriferous /ʌm'brɪfərəs/ adj. formal providing shade. [L umbrifer f. umbra shade: see -FEROUS]

Umeå /'u:mə,əʊ/ a city in NE Sweden, on an inlet of the Gulf of Bothnia; pop. (1990) 91,260.

umiak /'u:mɪ,æk/ n. (also **oomiak**) an Eskimo skin-and-wood open boat propelled by women with paddles. [Eskimo (Inuit) umiaq]

umlaut /'ʊmlaʊt/ n. & v. ● n. **1** a mark (¨) used over a vowel, esp. in Germanic languages, to indicate a vowel change. **2** such a vowel

change, e.g. German Mann, Männer, English man, men, due to i, j, etc. (now usu. lost or altered) in the following syllable. ● v.tr. modify (a form or a sound) by an umlaut. [G f. um about + Laut sound]

Umm al Qaiwain /ˌʊm æl kaɪ'waɪn/ **1** one of the seven member states of the United Arab Emirates; pop. (1985) 25,230. **2** its capital city.

umpire /'ʌmpaɪə(r)/ n. & v. ● n. **1** a person chosen to enforce the rules and settle disputes in various sports. **2** a person chosen to arbitrate between disputants, or to see fair play. ● v. **1** intr. (usu. foll. by for, in, etc.) act as umpire. **2** tr. act as umpire in (a game etc.). □ **umpirage** n. **umpireship** n. [ME, later form of noumpere f. OF nonper not equal (as NON-, PEER²): for loss of n- cf. ADDER]

umpteen /ʌmp'ti:n/ adj. & pron. colloq. ● adj. indefinitely many; a lot of. ● pron. indefinitely many. □ **umpteenth** adj. **umpty** /'ʌmptɪ/ adj. [joc. formation on -TEEN]

umpty-doo /ˌʌmptɪ'du:/ adj. Austral. colloq. drunk. [20th c.: prob. f. HUMPTY-DUMPTY]

Umtali /ʊm'tɑ:lɪ/ the former name (until 1982) for MUTARE.

UN see UNITED NATIONS.

un-¹ /ʌn/ prefix **1** added to adjectives and participles and their derivative nouns and adverbs, meaning: **a** not: denoting the absence of a quality or state (unusable; uncalled-for; uneducated; unfailing; unofficially; unhappiness). **b** the reverse of, usu. with an implication of approval or disapproval, or with some other special connotation (unselfish; unsociable; unscientific). ¶ Words formed in this way often have neutral counterparts in non- (see NON- 6) and counterparts in in- (see IN-¹), e.g. unadvisable. **2** (less often) added to nouns, meaning 'a lack of' (unrest; untruth). ¶ The number of words that can be formed with this prefix (and similarly with un-²) is potentially as large as the number of adjectives in use; consequently only a selection, being considered the most current or semantically noteworthy, can be given here. [OE f. Gmc, rel. to L in-]

un-² /ʌn/ prefix added to verbs and (less often) nouns, forming verbs denoting: **1** the reversal or cancellation of an action or state (undress; unlock; unsettle). **2** deprivation or separation (unmask). **3** release from (unburden; uncage). **4** causing to be no longer (unman). ¶ See the note at un-¹. Both un-¹ and un-² can be understood in some forms in -able, -ed (especially), and -ing: for example, undressed can mean either 'not dressed' or 'no longer dressed'. [OE un-, on- f. Gmc]

un-³ /ʌn/ prefix Chem. denoting 'one', combined with other numerical roots nil (=0), un (=1), bi (=2), etc., to form names of newly discovered elements. The roots are put together in order of the digits which make up the atomic number, and are terminated by -ium (unnilquadium = 104, ununbium = 112, etc.). The system was devised by IUPAC, to be used pending agreement on a permanent name. [L unus one]

'un /ən/ pron. colloq. one (that's a good 'un). [dial. var.]

UNA abbr. United Nations Association.

unabashed /ˌʌnə'bæʃt/ adj. not abashed. □ **unabashedly** /-ʃɪdlɪ/ adv.

unabated /ˌʌnə'beɪtɪd/ adj. not abated; undiminished. □ **unabatedly** adv.

unable /ʌn'eɪb(ə)l/ adj. (usu. foll. by to + infin.) not able; lacking ability.

unabridged /ˌʌnə'brɪdʒd/ adj. (of a text etc.) complete; not abridged.

unabsorbed /ˌʌnəb'sɔːbd, -'zɔːbd/ adj. not absorbed.

unacademic /ˌʌnækə'demɪk/ adj. **1** not academic (esp. not scholarly or theoretical). **2** (of a person) not suited to academic study.

unaccented /ˌʌnək'sentɪd/ adj. not accented; not emphasized.

unacceptable /ˌʌnək'septəb(ə)l/ adj. not acceptable, unwelcome. □ **unacceptableness** n. **unacceptably** adv.

unacclaimed /ˌʌnə'kleɪmd/ adj. not acclaimed.

unaccommodating /ˌʌnə'kɒmə,deɪtɪŋ/ adj. not accommodating; disobliging.

unaccompanied /ˌʌnə'kʌmpənɪd/ adj. **1** not accompanied. **2** Mus. without accompaniment.

unaccomplished /ˌʌnə'kʌmplɪʃt, -'kɒmplɪʃt/ adj. **1** not accomplished; uncompleted. **2** lacking accomplishments.

unaccountable /ˌʌnə'kaʊntəb(ə)l/ adj. **1** unable to be explained. **2** unpredictable or strange in behaviour. **3** not responsible. □ **unaccountably** adv. **unaccountableness** n. **unaccountability** /-ˌkaʊntə'bɪlɪtɪ/ n.

unaccounted /ˌʌnə'kaʊntɪd/ adj. of which no account is given. □ **unaccounted for** unexplained; not included in an account.

unaccustomed /ˌʌnəˈkʌstəmd/ adj. **1** (usu. foll. by to) not accustomed. **2** not usual or customary; uncharacteristic (his unaccustomed silence). □ **unaccustomedly** adv.

unachievable /ˌʌnəˈtʃiːvəb(ə)l/ adj. not achievable.

unacknowledged /ˌʌnəkˈnɒlɪdʒd/ adj. not acknowledged.

unacquainted /ˌʌnəˈkweɪntɪd/ adj. (usu. foll. by with) not acquainted.

unadaptable /ˌʌnəˈdæptəb(ə)l/ adj. not adaptable.

unadapted /ˌʌnəˈdæptɪd/ adj. not adapted.

unaddressed /ˌʌnəˈdrest/ adj. (esp. of a letter etc.) without an address.

unadjacent /ˌʌnəˈdʒeɪs(ə)nt/ adj. not adjacent.

unadjusted /ˌʌnəˈdʒʌstɪd/ adj. (esp. of figures) not adjusted; crude.

unadopted /ˌʌnəˈdɒptɪd/ adj. **1** not adopted. **2** Brit. (of a road) not taken over for maintenance by a local authority.

unadorned /ˌʌnəˈdɔːnd/ adj. not adorned; plain.

unadulterated /ˌʌnəˈdʌltəˌreɪtɪd/ adj. **1** not adulterated; pure; concentrated. **2** sheer, complete, utter (unadulterated nonsense).

unadventurous /ˌʌnədˈventʃərəs/ adj. not adventurous. □ **unadventurously** adv.

unadvertised /ʌnˈædvəˌtaɪzd/ adj. not advertised.

unadvisable /ˌʌnədˈvaɪzəb(ə)l/ adj. **1** not open to advice. **2** (of a thing) inadvisable.

unadvised /ˌʌnədˈvaɪzd/ adj. **1** indiscreet; rash. **2** not having had advice. □ **unadvisedly** /-zɪdlɪ/ adv. **unadvisedness** n.

unaffected /ˌʌnəˈfektɪd/ adj. **1** (usu. foll. by by) not affected. **2** free from affectation; genuine; sincere. □ **unaffectedly** adv. **unaffectedness** n.

unaffiliated /ˌʌnəˈfɪlɪˌeɪtɪd/ adj. not affiliated.

unaffordable /ˌʌnəˈfɔːdəb(ə)l/ adj. not affordable.

unafraid /ˌʌnəˈfreɪd/ adj. not afraid.

unaggressive /ˌʌnəˈgresɪv/ adj. not aggressive.

unaided /ʌnˈeɪdɪd/ adj. not aided; without help.

unalienable /ʌnˈeɪlɪənəb(ə)l/ adj. Law = INALIENABLE.

unaligned /ˌʌnəˈlaɪnd/ adj. **1** = NON-ALIGNED. **2** not physically aligned.

unalike /ˌʌnəˈlaɪk/ adj. not alike; different.

unalive /ˌʌnəˈlaɪv/ adj. **1** lacking in vitality. **2** (foll. by to) not fully susceptible or awake to.

unalleviated /ˌʌnəˈliːvɪˌeɪtɪd/ adj. not alleviated; relentless.

unallied /ˌʌnəˈlaɪd/ adj. not allied; having no allies.

unallowable /ˌʌnəˈlaʊəb(ə)l/ adj. not allowable.

unalloyed /ˌʌnəˈlɔɪd, ʌnˈæl-/ adj. **1** not alloyed; pure. **2** complete; utter (unalloyed joy).

unalterable /ʌnˈɔːltərəb(ə)l, ʌnˈɒl-/ adj. not alterable, unshakeable. □ **unalterableness** n. **unalterably** adv.

unaltered /ʌnˈɔːltəd, ʌnˈɒl-/ adj. not altered; remaining the same.

unamazed /ˌʌnəˈmeɪzd/ adj. not amazed.

unambiguous /ˌʌnæmˈbɪɡjʊəs/ adj. not ambiguous; clear or definite in meaning. □ **unambiguously** adv. **unambiguity** /-æmbɪˈɡjuːɪtɪ/ n.

unambitious /ˌʌnæmˈbɪʃəs/ adj. not ambitious; without ambition. □ **unambitiously** adv. **unambitiousness** n.

unambivalent /ˌʌnæmˈbɪvələnt/ adj. (of feelings etc.) not ambivalent; straightforward. □ **unambivalently** adv.

un-American /ˌʌnəˈmerɪkən/ adj. **1** not in accordance with American characteristics etc. **2** contrary to the interests of the US; (in the US) treasonable. □ **un-Americanism** n.

unamiable /ʌnˈeɪmɪəb(ə)l/ adj. not amiable.

unamplified /ʌnˈæmplɪˌfaɪd/ adj. not amplified.

unamused /ˌʌnəˈmjuːzd/ adj. not amused.

unanalysable /ʌnˈænəˌlaɪzəb(ə)l/ adj. not able to be analysed.

unanalysed /ʌnˈænəˌlaɪzd/ adj. not analysed.

unaneled /ˌʌnəˈniːld/ adj. archaic not having received extreme unction.

unanimous /juːˈnænɪməs/ adj. **1** all in agreement (the committee was unanimous). **2** (of an opinion, vote, etc.) held or given by general consent (the unanimous choice). □ **unanimously** adv. **unanimousness** n. **unanimity** /ˌjuːnəˈnɪmɪtɪ/ n. [LL unanimis, L unanimus f. unus one + animus mind]

unannounced /ˌʌnəˈnaʊnst/ adj. not announced; without warning (of arrival etc.).

unanswerable /ʌnˈɑːnsərəb(ə)l/ adj. **1** unable to be refuted (has an unanswerable case). **2** unable to be answered (an unanswerable question). □ **unanswerableness** n. **unanswerably** adv.

unanswered /ʌnˈɑːnsəd/ adj. not answered.

unanticipated /ˌʌnænˈtɪsɪˌpeɪtɪd/ adj. not anticipated.

unapologetic /ˌʌnəˌpɒləˈdʒetɪk/ adj. not apologetic or sorry.

unapparent /ˌʌnəˈpærənt/ adj. not apparent.

unappealable /ˌʌnəˈpiːləb(ə)l/ adj. esp. Law not able to be appealed against.

unappealing /ˌʌnəˈpiːlɪŋ/ adj. not appealing; unattractive. □ **unappealingly** adv.

unappeasable /ˌʌnəˈpiːzəb(ə)l/ adj. not appeasable.

unappeased /ˌʌnəˈpiːzd/ adj. not appeased.

unappetizing /ʌnˈæpɪˌtaɪzɪŋ/ adj. not appetizing. □ **unappetizingly** adv.

unapplied /ˌʌnəˈplaɪd/ adj. not applied.

unappreciated /ˌʌnəˈpriːʃɪˌeɪtɪd, ˌʌnəˈpriːsɪ-/ adj. not appreciated.

unappreciative /ˌʌnəˈpriːʃətɪv/ adj. not appreciative.

unapproachable /ˌʌnəˈprəʊtʃəb(ə)l/ adj. **1** not approachable; remote, inaccessible. **2** (of a person) unfriendly. □ **unapproachably** adv. **unapproachableness** n. **unapproachability** /-ˌprəʊtʃəˈbɪlɪtɪ/ n.

unappropriated /ˌʌnəˈprəʊprɪˌeɪtɪd/ adj. **1** not allocated or assigned. **2** not taken into possession by anyone.

unapproved /ˌʌnəˈpruːvd/ adj. not approved or sanctioned.

unapt /ʌnˈæpt/ adj. **1** (usu. foll. by for) not suitable. **2** (usu. foll. by to + infin.) not apt. □ **unaptly** adv. **unaptness** n.

unarguable /ʌnˈɑːɡjʊəb(ə)l/ adj. not arguable; certain.

unarm /ʌnˈɑːm/ v.tr. deprive or free of arms or armour.

unarmed /ʌnˈɑːmd/ adj. not armed; without weapons.

unarresting /ˌʌnəˈrestɪŋ/ adj. uninteresting, dull. □ **unarrestingly** adv.

unarticulated /ˌʌnɑːˈtɪkjʊˌleɪtɪd/ adj. not articulated or distinct.

unartistic /ˌʌnɑːˈtɪstɪk/ adj. not artistic, esp. not concerned with art. □ **unartistically** adv.

unascertainable /ˌʌnæsəˈteɪnəb(ə)l/ adj. not ascertainable.

unascertained /ˌʌnæsəˈteɪnd/ adj. not ascertained; unknown.

unashamed /ˌʌnəˈʃeɪmd/ adj. **1** feeling no guilt, shameless. **2** blatant; bold. □ **unashamedly** /-mɪdlɪ/ adv. **unashamedness** n.

unasked /ʌnˈɑːskt/ adj. (often foll. by for) not asked, requested, or invited.

unassailable /ˌʌnəˈseɪləb(ə)l/ adj. unable to be attacked or questioned; impregnable. □ **unassailably** adv. **unassailableness** n. **unassailability** /-ˌseɪləˈbɪlɪtɪ/ n.

unassertive /ˌʌnəˈsɜːtɪv/ adj. (of a person) not assertive or forthcoming; reticent. □ **unassertively** adv. **unassertiveness** n.

unassignable /ˌʌnəˈsaɪnəb(ə)l/ adj. not assignable.

unassigned /ˌʌnəˈsaɪnd/ adj. not assigned.

unassimilated /ˌʌnəˈsɪmɪˌleɪtɪd/ adj. not assimilated. □ **unassimilable** adj.

unassisted /ˌʌnəˈsɪstɪd/ adj. not assisted.

unassociated /ˌʌnəˈsəʊʃɪˌeɪtɪd, ˌʌnəˈsəʊsɪ-/ adj. (often foll. by with) having no connection or association.

unassuaged /ˌʌnəˈsweɪdʒd/ adj. not assuaged. □ **unassuageable** adj.

unassuming /ˌʌnəˈsjuːmɪŋ/ adj. not pretentious or arrogant; modest. □ **unassumingly** adv. **unassumingness** n.

unatoned /ˌʌnəˈtəʊnd/ adj. not atoned for.

unattached /ˌʌnəˈtætʃt/ adj. **1** (often foll. by to) not attached, esp. to a particular body, organization, etc. **2** not engaged or married; not having a boyfriend or girlfriend.

unattackable /ˌʌnəˈtækəb(ə)l/ adj. unable to be attacked or damaged.

unattainable /ˌʌnəˈteɪnəb(ə)l/ adj. not attainable, out of reach. □ **unattainableness** n. **unattainably** adv.

unattempted /ˌʌnəˈtemptɪd/ adj. not attempted.

unattended /ˌʌnəˈtendɪd/ adj. **1** (usu. foll. by to) not attended. **2** (of a person, vehicle, etc.) not accompanied; alone; uncared for.

unattractive /ˌʌnəˈtræktɪv/ adj. not attractive. □ **unattractively** adv. **unattractiveness** n.

unattributable /ˌʌnəˈtrɪbjʊtəb(ə)l/ adj. (esp. of information) that cannot or may not be attributed to a source etc. □ **unattributably** adv.

unauthentic /ˌʌnɔːˈθentɪk/ *adj.* not authentic. □ **unauthentically** *adv.*

unauthenticated /ˌʌnɔːˈθentɪˌkeɪtɪd/ *adj.* not authenticated.

unauthorized /ʌnˈɔːθəˌraɪzd/ *adj.* (also **unauthorised**) not authorized.

unavailable /ˌʌnəˈveɪləb(ə)l/ *adj.* not available. □ **unavailableness** *n.* **unavailability** /-ˌveɪləˈbɪlɪtɪ/ *n.*

unavailing /ˌʌnəˈveɪlɪŋ/ *adj.* not availing; achieving nothing; ineffectual. □ **unavailingly** *adv.*

unavoidable /ˌʌnəˈvɔɪdəb(ə)l/ *adj.* not avoidable; inevitable. □ **unavoidably** *adv.* **unavoidableness** *n.* **unavoidability** /-ˌvɔɪdəˈbɪlɪtɪ/ *n.*

unavowed /ˌʌnəˈvaʊd/ *adj.* not avowed.

unawakened /ˌʌnəˈweɪkənd/ *adj.* **1** (often foll. by *to*) not yet aware; dormant. **2** not awake.

unaware /ˌʌnəˈweə(r)/ *adj. & adv.* ● *adj.* **1** (usu. foll. by *of*, or *that* + clause) not aware; ignorant (*unaware of her presence*). **2** (of a person) insensitive; unperceptive. ● *adv.* = UNAWARES. □ **unawareness** *n.*

unawares /ˌʌnəˈweəz/ *adv.* **1** unexpectedly (*met them unawares*). **2** inadvertently (*dropped it unawares*). [earlier *unware(s)* f. OE *unwær(es)*: see WARE²]

unbacked /ʌnˈbækt/ *adj.* **1** not supported. **2** (of a horse etc.) having no backers. **3** (of a chair, picture, etc.) having no back or backing.

unbalance /ʌnˈbæləns/ *v. & n.* ● *v.tr.* **1** upset the physical or mental balance of (*unbalanced by the blow; the shock unbalanced him*). **2** (as **unbalanced** *adj.*) **a** not balanced. **b** (of a mind or a person) unstable or deranged. ● *n.* lack of balance; instability, esp. mental.

unban /ʌnˈbæn/ *v.tr.* (**unbanned, unbanning**) cease to ban; remove a ban from.

unbar /ʌnˈbɑː(r)/ *v.tr.* (**unbarred, unbarring**) **1** remove a bar or bars from (a gate etc.). **2** unlock, open.

unbearable /ʌnˈbeərəb(ə)l/ *adj.* not bearable. □ **unbearableness** *n.* **unbearably** *adv.*

unbeatable /ʌnˈbiːtəb(ə)l/ *adj.* not beatable; excelling.

unbeaten /ʌnˈbiːt(ə)n/ *adj.* **1** not beaten. **2** (of a record etc.) not surpassed. **3** *Cricket* (of a player) not out.

unbeautiful /ʌnˈbjuːtɪˌfʊl/ *adj.* not beautiful; ugly. □ **unbeautifully** *adv.*

unbecoming /ˌʌnbɪˈkʌmɪŋ/ *adj.* **1** (esp. of clothing) not flattering or suiting a person. **2** (usu. foll. by *to*, *for*) not fitting; indecorous or unsuitable. □ **unbecomingly** *adv.* **unbecomingness** *n.*

unbefitting /ˌʌnbɪˈfɪtɪŋ/ *adj.* not befitting; inappropriate. □ **unbefittingly** *adv.* **unbefittingness** *n.*

unbefriended /ˌʌnbɪˈfrendɪd/ *adj.* not befriended.

unbegotten /ˌʌnbɪˈɡɒt(ə)n/ *adj.* not begotten.

unbeholden /ˌʌnbɪˈhəʊldən/ *predic.adj.* (usu. foll. by *to*) under no obligation.

unbeknown /ˌʌnbɪˈnəʊn/ *adj.* (also **unbeknownst** /-ˈnəʊnst/) (foll. by *to*) without the knowledge of (*was there all the time unbeknown to us*). [UN-¹ + *beknown* (archaic) = KNOWN]

unbelief /ˌʌnbɪˈliːf/ *n.* lack of belief, esp. in religious matters. □ **unbeliever** *n.* **unbelieving** *adj.* **unbelievingly** *adv.*

unbelievable /ˌʌnbɪˈliːvəb(ə)l/ *adj.* not believable; incredible. □ **unbelievably** *adv.* **unbelievableness** *n.* **unbelievability** /-ˌliːvəˈbɪlɪtɪ/ *n.*

unbeloved /ˌʌnbɪˈlʌvd/ *adj.* not beloved.

unbelt /ʌnˈbelt/ *v.tr.* remove or undo the belt of (a garment etc.).

unbend /ʌnˈbend/ *v.* (*past* and *past part.* **unbent** /-ˈbent/) **1** *tr. & intr.* change from a bent position; straighten. **2** *intr.* relax from strain or severity; become affable. **3** *tr. Naut.* **a** unfasten (sails) from yards and stays. **b** cast (a cable) loose. **c** untie (a rope).

unbending /ʌnˈbendɪŋ/ *adj.* **1** not bending; inflexible. **2** firm; austere (*unbending rectitude*). **3** relaxing from strain, activity, or formality. □ **unbendingly** *adv.* **unbendingness** *n.*

unbiased /ʌnˈbaɪəst/ *adj.* (also **unbiassed**) not biased; impartial.

unbiblical /ʌnˈbɪblɪk(ə)l/ *adj.* **1** not in or authorized by the Bible. **2** contrary to the Bible.

unbiddable /ʌnˈbɪdəb(ə)l/ *adj. Brit.* disobedient; not docile.

unbidden /ʌnˈbɪd(ə)n/ *adj.* not commanded or invited (*arrived unbidden*).

unbind /ʌnˈbaɪnd/ *v.tr.* (*past* and *past part.* **unbound** /-ˈbaʊnd/) release from bonds or binding.

unbirthday /ʌnˈbɜːθdeɪ/ *n.* (often *attrib.*) *joc.* any day but one's birthday (*an unbirthday party*).

unbleached /ʌnˈbliːtʃt/ *adj.* not bleached.

unblemished /ʌnˈblemɪʃt/ *adj.* not blemished.

unblessed /ʌnˈblest/ *adj.* (also **unblest**) not blessed.

unblinking /ʌnˈblɪŋkɪŋ/ *adj.* **1** not blinking. **2** steadfast; not hesitating. **3** stolid; cool. □ **unblinkingly** *adv.*

unblock /ʌnˈblɒk/ *v.tr.* **1** remove an obstruction from (esp. a pipe, drain, etc.). **2** (also *absol.*) *Cards* allow the later unobstructed play of (a suit) by playing a high card.

unblown /ʌnˈbləʊn/ *adj.* **1** not blown. **2** *archaic* (of a flower) not yet in bloom.

unblushing /ʌnˈblʌʃɪŋ/ *adj.* **1** not blushing. **2** unashamed; frank. □ **unblushingly** *adv.*

unbolt /ʌnˈbəʊlt/ *v.tr.* release (a door etc.) by drawing back the bolt.

unbolted /ʌnˈbəʊltɪd/ *adj.* **1** not bolted. **2** (of flour etc.) not sifted.

unbonnet /ʌnˈbɒnɪt/ *v.* (**unbonneted, unbonneting**) **1** *tr.* remove the bonnet from. **2** *intr. archaic* remove one's hat or bonnet esp. in respect.

unbookish /ʌnˈbʊkɪʃ/ *adj.* **1** not academic; not often inclined to read. **2** free from bookishness.

unboot /ʌnˈbuːt/ *v.intr. & tr.* remove one's boots or the boots of (a person).

unborn /ʌnˈbɔːn/ *adj.* **1** not yet born (*an unborn child*). **2** never to be brought into being (*unborn hopes*).

unbosom /ʌnˈbʊz(ə)m/ *v.tr.* **1** disclose (thoughts, secrets, etc.). **2** (*refl.*) unburden (oneself) of one's thoughts, secrets, etc.

unbothered /ʌnˈbɒðəd/ *adj.* not bothered; unconcerned.

unbound¹ /ʌnˈbaʊnd/ *adj.* **1** not bound or tied up. **2** unconstrained. **3 a** (of a book) not having a binding. **b** having paper covers. **4** *Chem. & Physics* (of a substance or particle) in a loose or free state.

unbound² *past* and *past part.* of UNBIND.

unbounded /ʌnˈbaʊndɪd/ *adj.* not bounded; infinite (*unbounded optimism*). □ **unboundedly** *adv.* **unboundedness** *n.*

unbowed /ʌnˈbaʊd/ *adj.* (usu. *predic.*) undaunted.

unbrace /ʌnˈbreɪs/ *v.tr.* **1** (also *absol.*) free from tension; relax (the nerves etc.). **2** remove a brace or braces from.

unbranded /ʌnˈbrændɪd/ *adj.* **1** (of a product) not bearing a brand name. **2** (of livestock) not branded with the owner's mark.

unbreachable /ʌnˈbriːtʃəb(ə)l/ *adj.* not able to be breached.

unbreakable /ʌnˈbreɪkəb(ə)l/ *adj.* not breakable.

unbreathable /ʌnˈbriːðəb(ə)l/ *adj.* not able to be breathed.

unbribable /ʌnˈbraɪbəb(ə)l/ *adj.* not bribable.

unbridgeable /ʌnˈbrɪdʒəb(ə)l/ *adj.* not able to be bridged.

unbridle /ʌnˈbraɪd(ə)l/ *v.tr.* **1** remove a bridle from (a horse). **2** remove constraints from (one's tongue, a person, etc.). **3** (as **unbridled** *adj.*) unconstrained (*unbridled insolence*).

unbroken /ʌnˈbrəʊkən/ *adj.* **1** not broken. **2** not tamed (*an unbroken horse*). **3** not interrupted (*unbroken sleep*). **4** not surpassed (*an unbroken record*). □ **unbrokenly** *adv.* **unbrokenness** /-kənnɪs/ *n.*

unbruised /ʌnˈbruːzd/ *adj.* not bruised.

unbuckle /ʌnˈbʌk(ə)l/ *v.tr.* release the buckle of (a strap, shoe, etc.).

unbuild /ʌnˈbɪld/ *v.tr.* (*past* and *past part.* **unbuilt** /-ˈbɪlt/) **1** demolish or destroy (a building, theory, system, etc.). **2** (as **unbuilt** *adj.*) not yet built or (of land etc.) not yet built on.

unbundle /ʌnˈbʌnd(ə)l/ *v.tr.* **1** unpack; remove from a bundle. **2** market (goods or services) separately. **3** split (a company) into separate businesses. □ **unbundler** *n.* (in sense 3).

unburden /ʌnˈbɜːd(ə)n/ *v.tr.* **1** relieve of a burden. **2** (esp. *refl.*; often foll. by *to*) relieve (oneself, one's conscience, etc.) by confession etc. □ **unburdened** *adj.*

unburied /ʌnˈberɪd/ *adj.* not buried.

unburnt /ʌnˈbɜːnt/ *adj.* (also **unburned** /-ˈbɜːnd/) **1 a** not consumed by fire. **b** not affected or damaged by fire. **2** (esp. of bricks) not subjected to the action of fire.

unbury /ʌnˈberɪ/ *v.tr.* (**-ies, -ied**) **1** remove from the ground etc. after burial. **2** unearth (a secret etc.).

unbusinesslike /ʌnˈbɪznɪsˌlaɪk/ *adj.* not businesslike.

unbutton /ʌnˈbʌt(ə)n/ *v.tr.* **1 a** unfasten (a coat etc.) by taking the buttons out of the buttonholes. **b** unbutton the clothes of (a person).

2 (*absol.*) *colloq.* relax from tension or formality, become communicative.
3 (as **unbuttoned** *adj.*) **a** not buttoned. **b** *colloq.* communicative; informal.

uncage /ʌnˈkeɪdʒ/ *v.tr.* **1** release from a cage. **2** release from constraint; liberate.

uncalled /ʌnˈkɔːld/ *adj.* not summoned or invited. □ **uncalled-for** (of an opinion, action, etc.) impertinent or unnecessary (*an uncalled-for remark*).

uncandid /ʌnˈkændɪd/ *adj.* not candid; disingenuous.

uncanny /ʌnˈkænɪ/ *adj.* (**uncannier**, **uncanniest**) seemingly supernatural; mysterious. □ **uncannily** *adv.* **uncanniness** *n.* [(orig. Sc. & N. Engl.) f. UN-¹ + CANNY]

uncanonical /ˌʌnkəˈnɒnɪk(ə)l/ *adj.* not canonical. □ **uncanonically** *adv.*

uncap /ʌnˈkæp/ *v.tr.* (**uncapped**, **uncapping**) **1** remove the cap from (a jar, bottle, etc.). **2** remove a cap from (the head or another person).

uncared-for /ʌnˈkeədfɔː(r)/ *adj.* disregarded; neglected.

uncaring /ʌnˈkeərɪŋ/ *adj.* lacking compassion or concern for others.

uncarpeted /ʌnˈkɑːpɪtɪd/ *adj.* not covered with carpet.

uncase /ʌnˈkeɪs/ *v.tr.* remove from a cover or case.

uncashed /ʌnˈkæʃt/ *adj.* not cashed.

uncaught /ʌnˈkɔːt/ *adj.* not caught.

unceasing /ʌnˈsiːsɪŋ/ *adj.* not ceasing; continuous (*unceasing effort*). □ **unceasingly** *adv.*

UNCED *abbr.* United Nations Conference on Environment and Development.

uncensored /ʌnˈsensəd/ *adj.* not censored.

uncensured /ʌnˈsensjəd/ *adj.* not censured.

unceremonious /ˌʌnserɪˈməʊnɪəs/ *adj.* **1** lacking ceremony or formality. **2** abrupt; discourteous. □ **unceremoniously** *adv.* **unceremoniousness** *n.*

uncertain /ʌnˈsɜːt(ə)n, -tɪn/ *adj.* **1** not certainly knowing or known (*uncertain what it means; the result is uncertain*). **2** unreliable (*his aim is uncertain*). **3** changeable, erratic (*uncertain weather*). □ **in no uncertain terms** clearly and forcefully. □ **uncertainly** *adv.*

uncertainty /ʌnˈsɜːt(ə)ntɪ, -tɪntɪ/ *n.* (*pl.* **-ies**) **1** the fact or condition of being uncertain. **2** an uncertain matter or circumstance.

uncertainty principle *n.* **1** (also **Heisenberg uncertainty principle**) *Physics* the quantum-mechanical principle that the momentum and position of a particle cannot both be precisely determined at the same time. Also called *principle of indeterminacy*. (*See note below.*) **2** any similar restriction on the accuracy of measurement.

▪ The uncertainty principle was stated by Werner Heisenberg in 1927. In mathematical terms, the product of the uncertainties in the momentum and position of a particle cannot be less than h divided by 4π, where h is Planck's constant. The principle is a manifestation of the dual nature of matter as simultaneously particle and wave (see also QUANTUM) and of the inherent impossibility, on a quantum scale, of observing something without interacting with it.

uncertified /ʌnˈsɜːtɪˌfaɪd/ *adj.* **1** not attested as certain. **2** not guaranteed by a certificate of competence etc. **3** not certified as insane.

unchain /ʌnˈtʃeɪn/ *v.tr.* **1** remove the chains from. **2** release; liberate.

unchallengeable /ʌnˈtʃælɪndʒəb(ə)l/ *adj.* not challengeable; unassailable. □ **unchallengeably** *adv.*

unchallenged /ʌnˈtʃælɪndʒd/ *adj.* not challenged.

unchallenging /ʌnˈtʃælɪndʒɪŋ/ *adj.* not presenting a challenge or other stimulation.

unchangeable /ʌnˈtʃeɪndʒəb(ə)l/ *adj.* not changeable; immutable, invariable. □ **unchangeably** *adv.* **unchangeableness** *n.* **unchangeability** /-ˌtʃeɪndʒəˈbɪlɪtɪ/ *n.*

unchanged /ʌnˈtʃeɪndʒd/ *adj.* not changed; unaltered.

unchanging /ʌnˈtʃeɪndʒɪŋ/ *adj.* not changing; remaining the same. □ **unchangingly** *adv.* **unchangingness** *n.*

unchaperoned /ʌnˈʃæpəˌrəʊnd/ *adj.* without a chaperon.

uncharacteristic /ˌʌnkærɪktəˈrɪstɪk/ *adj.* not characteristic. □ **uncharacteristically** *adv.*

uncharged /ʌnˈtʃɑːdʒd/ *adj.* not charged (esp. in senses 3, 7, 8 of CHARGE *v.*).

uncharismatic /ˌʌnkærɪzˈmætɪk/ *adj.* lacking charisma.

uncharitable /ʌnˈtʃærɪtəb(ə)l/ *adj.* censorious, severe in judgement. □ **uncharitableness** *n.* **uncharitably** *adv.*

uncharted /ʌnˈtʃɑːtɪd/ *adj.* not charted, mapped, or surveyed.

unchartered /ʌnˈtʃɑːtəd/ *adj.* **1** not furnished with a charter; not formally privileged or constituted. **2** unauthorized; illegal.

unchaste /ʌnˈtʃeɪst/ *adj.* not chaste. □ **unchastely** *adv.* **unchasteness** *n.* **unchastity** /-ˈtʃæstɪtɪ/ *n.*

unchecked /ʌnˈtʃekt/ *adj.* **1** not checked. **2** freely allowed; unrestrained (*unchecked violence*).

unchivalrous /ʌnˈʃɪvəlrəs/ *adj.* not chivalrous; discourteous. □ **unchivalrously** *adv.*

unchosen /ʌnˈtʃəʊz(ə)n/ *adj.* not chosen.

unchristian /ʌnˈkrɪstɪən, -ˈkrɪstʃən/ *adj.* **1 a** contrary to Christian principles, esp. uncaring or selfish. **b** not Christian. **2** *colloq.* outrageous. □ **unchristianly** *adv.*

unchurch /ʌnˈtʃɜːtʃ/ *v.tr.* **1** excommunicate. **2** deprive (a building) of its status as a church.

uncial /ˈʌnsɪəl, ˈʌnʃ(ə)l/ *adj.* & *n.* ● *adj.* **1** of or written in majuscule writing with rounded unjoined letters found in manuscripts of the 4th–8th centuries, from which modern capitals are derived. **2** of or relating to an inch or an ounce. ● *n.* **1** an uncial letter. **2** an uncial style or manuscript. [L *uncialis* f. *uncia* inch: sense 1 in LL sense of *unciales litterae*, the orig. application of which is unclear]

unciform /ˈʌnsɪˌfɔːm/ *n.* esp. *Anat.* = UNCINATE.

uncinate /ˈʌnsɪnət/ *adj.* esp. *Anat.* hooked; crooked. [L *uncinatus* f. *uncinus* hook]

uncircumcised /ʌnˈsɜːkəmˌsaɪzd/ *adj.* **1** not circumcised. **2** *archaic* spiritually impure; heathen.

uncivil /ʌnˈsɪvɪl/ *adj.* **1** ill-mannered; impolite. **2** not public-spirited. □ **uncivilly** *adv.*

uncivilized /ʌnˈsɪvɪˌlaɪzd/ *adj.* (also **uncivilised**) **1** not civilized. **2** rough; uncultured.

unclad /ʌnˈklæd/ *adj.* not clad; naked.

unclaimed /ʌnˈkleɪmd/ *adj.* not claimed.

unclasp /ʌnˈklɑːsp/ *v.tr.* **1** loosen the clasp or clasps of. **2** release the grip of (a hand etc.).

unclassifiable /ʌnˈklæsɪˌfaɪəb(ə)l/ *adj.* not classifiable.

unclassified /ʌnˈklæsɪˌfaɪd/ *adj.* **1** not classified. **2** (of state information) not secret.

uncle /ˈʌŋk(ə)l/ *n.* **1 a** the brother of one's father or mother. **b** an aunt's husband. **2** *colloq.* a name given by children to a male family friend. **3** *sl.* esp. *hist.* a pawnbroker. [ME f. AF *uncle*, OF *oncle* f. LL *aunculus* f. L *avunculus* maternal uncle: see AVUNCULAR]

-uncle /ʌŋk(ə)l/ *suffix* forming nouns, usu. diminutives (*carbuncle*). [OF *-uncle*, *-oncle* or L *-unculus*, *-la*, a special form of *-ulus* -ULE]

unclean /ʌnˈkliːn/ *adj.* **1** not clean. **2** unchaste. **3** unfit to be eaten; ceremonially impure. **4** *Bibl.* (of a spirit) wicked. □ **uncleanly** *adv.* **uncleanness** *n.* **uncleanly** /-ˈklenlɪ/ *adj.* **uncleanliness** /-ˈklenlɪnɪs/ *n.* [OE *unclæne* (as UN-¹, CLEAN)]

unclear /ʌnˈklɪə(r)/ *adj.* **1** not clear or easy to understand; obscure, uncertain. **2** (of a person) doubtful, uncertain (*I'm unclear as to what you mean*). □ **unclearly** *adv.* **unclearness** *n.*

uncleared /ʌnˈklɪəd/ *adj.* **1** (of a cheque etc.) not cleared. **2** not cleared away or up. **3** (of land) not cleared of trees etc.

unclench /ʌnˈklentʃ/ *v.* **1** *tr.* release (clenched hands, features, teeth, etc.). **2** *intr.* (of clenched hands etc.) become relaxed or open.

Uncle Sam a personification of the government or people of the United States of America. The explanation that it arose as a facetious interpretation of the letters US is as old as the first recorded instance (in the early 19th century), and later statements connecting it with different government officials of the name of Samuel appear to be unfounded.

Uncle Tom *n. derog.* a black man considered to be servile, cringing, etc. (from the hero of Harriet Beecher Stowe's *Uncle Tom's Cabin*, 1852).

unclimbed /ʌnˈklaɪmd/ *adj.* (of a peak, rock-face, etc.) not previously climbed.

unclinch /ʌnˈklɪntʃ/ *v.tr. & intr.* release or become released from a clinch.

uncloak /ʌnˈkləʊk/ *v.tr.* **1** expose, reveal. **2** remove a cloak from.

unclog /ʌnˈklɒg/ *v.tr.* (**unclogged**, **unclogging**) unblock (a drain, pipe, etc.).

unclose /ʌnˈkləʊz/ *v.* **1** *tr. & intr.* open. **2** *tr.* reveal; disclose.

unclothe /ʌnˈkləʊð/ *v.tr.* **1** remove the clothes from. **2** strip of leaves or vegetation (*trees unclothed by the wind*). **3** expose, reveal. □ **unclothed** *adj.*

unclouded /ʌnˈklaʊdɪd/ *adj.* **1** not clouded; clear; bright. **2** untroubled (*unclouded serenity*).

uncluttered /ʌnˈklʌtəd/ *adj.* not cluttered; austere, simple.

unco /ˈʌŋkəʊ/ *adj., adv., & n. Sc.* ● *adj.* strange, unusual; notable. ● *adv.* remarkably; very. ● *n.* (*pl.* **-os**) **1** a stranger. **2** (in *pl.*) news. □ **the unco guid** /gɪd/ *esp. derog.* the rigidly religious. [ME, var. of UNCOUTH]

uncoil /ʌnˈkɔɪl/ *v.tr. & intr.* unwind.

uncollected /ʌnkəˈlektɪd/ *adj.* **1** left awaiting collection. **2** (of funds, money, etc.) not collected in or claimed. **3** (of literary work) not gathered into a collection for publication.

uncoloured /ʌnˈkʌləd/ *adj.* (*US* **uncolored**) **1** having no colour. **2** not influenced; impartial. **3** not exaggerated.

uncombed /ʌnˈkəʊmd/ *adj.* (of hair or a person) not combed.

uncome-at-able /ˌʌnkʌmˈætəb(ə)l/ *adj. colloq.* inaccessible; beyond reach, unattainable. [UN-¹ + *come-at-able*: see COME]

uncomely /ʌnˈkʌmlɪ/ *adj.* **1** improper; unseemly. **2** ugly.

uncomfortable /ʌnˈkʌmftəb(ə)l/ *adj.* **1** not comfortable. **2** uneasy; causing or feeling disquiet (*an uncomfortable silence*). □ **uncomfortableness** *n.* **uncomfortably** *adv.*

uncommercial /ˌʌnkəˈmɜːʃ(ə)l/ *adj.* **1** not commercial. **2** contrary to commercial principles.

uncommitted /ˌʌnkəˈmɪtɪd/ *adj.* **1** not committed. **2** unattached to any specific political cause or group.

uncommon /ʌnˈkɒmən/ *adj. & adv.* ● *adj.* **1** not common; unusual; remarkable. **2** remarkably great etc. (*an uncommon fear of spiders*). ● *adv. archaic* uncommonly (*he was uncommon fat*). □ **uncommonly** *adv.* **uncommonness** /-mənnɪs/ *n.*

uncommunicative /ˌʌnkəˈmjuːnɪkətɪv/ *adj.* not wanting to communicate; taciturn. □ **uncommunicatively** *adv.* **uncommunicativeness** *n.*

uncompanionable /ˌʌnkəmˈpænjənəb(ə)l/ *adj.* unsociable.

uncompensated /ʌnˈkɒmpenseɪtɪd/ *adj.* not compensated.

uncompetitive /ˌʌnkəmˈpetɪtɪv/ *adj.* not competitive.

uncomplaining /ˌʌnkəmˈpleɪnɪŋ/ *adj.* not complaining; resigned. □ **uncomplainingly** *adv.*

uncompleted /ˌʌnkəmˈpliːtɪd/ *adj.* not completed; incomplete.

uncomplicated /ʌnˈkɒmplɪˌkeɪtɪd/ *adj.* not complicated; simple; straightforward.

uncomplimentary /ˌʌnkɒmplɪˈmentərɪ/ *adj.* not complimentary; insulting.

uncompounded /ˌʌnkəmˈpaʊndɪd/ *adj.* not compounded; unmixed.

uncomprehending /ˌʌnkɒmprɪˈhendɪŋ/ *adj.* not comprehending. □ **uncomprehendingly** *adv.* **uncomprehension** /-ˈhenʃ(ə)n/ *n.*

uncompromising /ʌnˈkɒmprəˌmaɪzɪŋ/ *adj.* unwilling to compromise; stubborn; unyielding. □ **uncompromisingly** *adv.* **uncompromisingness** *n.*

unconcealed /ˌʌnkənˈsiːld/ *adj.* not concealed; obvious.

unconcern /ˌʌnkənˈsɜːn/ *n.* lack of concern; indifference; apathy. □ **unconcerned** *adj.* **unconcernedly** /-nɪdlɪ/ *adv.*

unconcluded /ˌʌnkənˈkluːdɪd/ *adj.* not concluded.

unconditional /ˌʌnkənˈdɪʃən(ə)l/ *adj.* not subject to conditions; complete (*unconditional surrender*). □ **unconditionally** *adv.* **unconditionality** /-ˌdɪʃəˈnælɪtɪ/ *n.*

unconditioned /ˌʌnkənˈdɪʃ(ə)nd/ *adj.* **1** not subject to conditions or to an antecedent condition. **2** (of behaviour etc.) not determined by conditioning; natural. □ **unconditioned reflex** an instinctive response to a stimulus.

unconfined /ˌʌnkənˈfaɪnd/ *adj.* not confined; boundless.

unconfirmed /ˌʌnkənˈfɜːmd/ *adj.* not confirmed.

unconformable /ˌʌnkənˈfɔːməb(ə)l/ *adj.* **1** not conformable or conforming. **2** *Geol.* (of rock strata) not having the same direction of stratification. **3** *hist.* not conforming to the provisions of the Act of Uniformity (1662). □ **unconformableness** *n.* **unconformably** *adv.*

unconformity /ˌʌnkənˈfɔːmɪtɪ/ *n. Geol.* **1** a large break in the chronological sequence of layers of rock. **2** the surface of contact between two groups of unconformable strata.

uncongenial /ˌʌnkənˈdʒiːnɪəl/ *adj.* not congenial.

unconjecturable /ˌʌnkənˈdʒektʃərəb(ə)l/ *adj.* not conjecturable.

unconnected /ˌʌnkəˈnektɪd/ *adj.* **1** not physically joined. **2** not connected or associated. **3** (of speech etc.) disconnected; not joined in order or sequence (*unconnected ideas*). **4** not related by family ties. □ **unconnectedly** *adv.* **unconnectedness** *n.*

unconquerable /ʌnˈkɒŋkərəb(ə)l/ *adj.* not conquerable; insuperable. □ **unconquerableness** *n.* **unconquerably** *adv.*

unconquered /ʌnˈkɒŋkəd/ *adj.* not conquered or defeated.

unconscionable /ʌnˈkɒnʃənəb(ə)l/ *adj.* **1 a** having no conscience. **b** contrary to conscience. **2 a** unreasonably excessive (*an unconscionable length of time*). **b** not right or reasonable. □ **unconscionableness** *n.* **unconscionably** *adv.* [UN-¹ + obs. *conscionable* f. *conscions* obs. var. of CONSCIENCE]

unconscious /ʌnˈkɒnʃəs/ *adj. & n.* ● *adj.* not conscious (*unconscious of any change; fell unconscious on the floor; an unconscious prejudice*). ● *n.* that part of the mind which is inaccessible to the conscious mind but which affects behaviour, emotions, etc. (cf. *collective unconscious*). □ **unconsciously** *adv.* **unconsciousness** *n.*

unconsecrated /ʌnˈkɒnsɪˌkreɪtɪd/ *adj.* not consecrated.

unconsenting /ˌʌnkənˈsentɪŋ/ *adj.* not consenting.

unconsidered /ˌʌnkənˈsɪdəd/ *adj.* **1** not considered; disregarded. **2** (of a response etc.) immediate; not premeditated.

unconsolable /ˌʌnkənˈsəʊləb(ə)l/ *adj.* unable to be consoled; inconsolable. □ **unconsolably** *adv.*

unconstitutional /ˌʌnkɒnstɪˈtjuːʃ(ə)n(ə)l/ *adj.* not in accordance with the political constitution or with procedural rules. □ **unconstitutionally** *adv.* **unconstitutionality** /-ˌtjuːʃəˈnælɪtɪ/ *n.*

unconstrained /ˌʌnkənˈstreɪnd/ *adj.* not constrained or compelled. □ **unconstrainedly** /-nɪdlɪ/ *adv.*

unconstraint /ˌʌnkənˈstreɪnt/ *n.* freedom from constraint.

unconstricted /ˌʌnkənˈstrɪktɪd/ *adj.* not constricted.

unconsumed /ˌʌnkənˈsjuːmd/ *adj.* not consumed.

unconsummated /ʌnˈkɒnsjʊˌmeɪtɪd/ *adj.* not consummated.

uncontainable /ˌʌnkənˈteɪnəb(ə)l/ *adj.* not containable.

uncontaminated /ˌʌnkənˈtæmɪˌneɪtɪd/ *adj.* not contaminated.

uncontentious /ˌʌnkənˈtenʃəs/ *adj.* not controversial.

uncontested /ˌʌnkənˈtestɪd/ *adj.* not contested. □ **uncontestedly** *adv.*

uncontradicted /ˌʌnkɒntrəˈdɪktɪd/ *adj.* not contradicted.

uncontrollable /ˌʌnkənˈtrəʊləb(ə)l/ *adj.* not controllable (*flew into an uncontrollable rage*). □ **uncontrollableness** *n.* **uncontrollably** *adv.*

uncontrolled /ˌʌnkənˈtrəʊld/ *adj.* not controlled; unrestrained, unchecked.

uncontroversial /ˌʌnkɒntrəˈvɜːʃ(ə)l/ *adj.* not controversial. □ **uncontroversially** *adv.*

uncontroverted /ˌʌnkɒntrəˈvɜːtɪd, ʌnˈkɒntrəˌvɜːt-/ *adj.* not controverted. □ **uncontrovertible** *adj.*

unconventional /ˌʌnkənˈvenʃən(ə)l/ *adj.* not bound by convention or custom; unusual; unorthodox. □ **unconventionalism** *n.* **unconventionally** *adv.* **unconventionality** /-ˌvenʃəˈnælɪtɪ/ *n.*

unconverted /ˌʌnkənˈvɜːtɪd/ *adj.* not converted.

unconvinced /ˌʌnkənˈvɪnst/ *adj.* not convinced.

unconvincing /ˌʌnkənˈvɪnsɪŋ/ *adj.* not convincing; unimpressive. □ **unconvincingly** *adv.*

uncooked /ʌnˈkʊkt/ *adj.* not cooked; raw.

uncool /ʌnˈkuːl/ *adj.* **1** *colloq.* not stylish or fashionable. **2** *colloq.* unrelaxed; unpleasant. **3** (of jazz) not cool.

uncooperative /ˌʌnkəʊˈɒpərətɪv/ *adj.* unwilling to cooperate, intransigent. □ **uncooperatively** *adv.*

uncoordinated /ˌʌnkəʊˈɔːdɪˌneɪtɪd/ *adj.* **1** not coordinated. **2** (of a person's movements etc.) clumsy.

uncopiable /ʌnˈkɒpɪəb(ə)l/ *adj.* not able to be copied.

uncord /ʌnˈkɔːd/ *v.tr.* remove the cord from.

uncordial /ʌnˈkɔːdɪəl/ *adj.* not congenial; unfriendly.

uncork /ʌnˈkɔːk/ *v.tr.* **1** draw the cork from (a bottle). **2** allow (feelings etc.) to be vented.

uncorroborated /ˌʌnkəˈrɒbəˌreɪtɪd/ *adj.* (esp. of evidence etc.) not corroborated.

uncorrupted /ˌʌnkəˈrʌptɪd/ *adj.* not corrupted.

uncountable /ʌnˈkaʊntəb(ə)l/ *adj.* **1** inestimable, immense (*uncountable wealth*). **2** *Gram.* (of a noun) that cannot form a plural or be used with the indefinite article (e.g. *happiness*). □ **uncountably** *adv.* **uncountability** /-ˌkaʊntəˈbɪlɪtɪ/ *n.*

uncounted /ʌnˈkaʊntɪd/ *adj.* **1** not counted. **2** very many; innumerable.

uncouple /ʌnˈkʌp(ə)l/ *v.tr.* **1** unfasten, disconnect. **2** release (wagons) from couplings. **3** release (hunting dogs etc.) from being fastened in couples. □ **uncoupled** *adj.*

uncourtly /ʌnˈkɔːtlɪ/ *adj.* not courteous; ill-mannered.

uncouth /ʌnˈkuːθ/ *adj.* **1** (of a person, manners, appearance, etc.) lacking in ease and polish; uncultured, rough (*uncouth voices; behaviour was uncouth*). **2** *archaic* not known; desolate; wild; uncivilized (*an uncouth place*). □ **uncouthly** *adv.* **uncouthness** *n.* [OE *uncūth* unknown (as UN-¹ + *cūth* past part. of *cunnan* know, CAN¹)]

uncovenanted /ʌnˈkʌvənəntɪd/ *adj.* **1** not bound by a covenant. **2** not promised by or based on a covenant, esp. God's covenant.

uncover /ʌnˈkʌvə(r)/ *v.* **1** *tr.* **a** remove a cover or covering from. **b** make known; disclose (*uncovered the truth at last*). **2** *intr. archaic* remove one's hat, cap, etc. **3** *tr.* (as **uncovered** *adj.*) **a** not covered by a roof, clothing, etc. **b** not wearing a hat.

uncreate /ˌʌnkrɪˈeɪt/ *v.tr. literary* annihilate.

uncreated /ˌʌnkrɪˈeɪtɪd/ *adj.* existing without having been created; not created. [UN-¹ + obs. *create* f. L *creatus* past part. of *creare*: see CREATE]

uncreative /ˌʌnkrɪˈeɪtɪv/ *adj.* not creative.

uncritical /ʌnˈkrɪtɪk(ə)l/ *adj.* **1** not critical; complacently accepting. **2** not in accordance with the principles of criticism. □ **uncritically** *adv.*

uncropped /ʌnˈkrɒpt/ *adj.* not cropped.

uncross /ʌnˈkrɒs/ *v.tr.* **1** remove (the limbs, knives, etc.) from a crossed position. **2** (as **uncrossed** *adj.*) **a** *Brit.* (of a cheque) not crossed. **b** not thwarted or challenged. **c** not wearing a cross.

uncrowded /ʌnˈkraʊdɪd/ *adj.* not filled or likely to fill with crowds.

uncrown /ʌnˈkraʊn/ *v.tr.* **1** deprive (a monarch etc.) of a crown. **2** deprive (a person) of a position. **3** (as **uncrowned** *adj.*) **a** not crowned. **b** having the status but not the name of (*the uncrowned king of boxing*).

uncrushable /ʌnˈkrʌʃəb(ə)l/ *adj.* not crushable.

uncrushed /ʌnˈkrʌʃt/ *adj.* not crushed.

UNCSTD *abbr.* United Nations Conference on Science and Technology for Development.

UNCTAD *abbr.* United Nations Conference on Trade and Development.

unction /ˈʌŋkʃ(ə)n/ *n.* **1 a** the act of anointing with oil etc. as a religious rite. **b** the oil etc. so used. **2 a** soothing words or thought. **b** excessive or insincere flattery. **3 a** the act of anointing for medical purposes. **b** an ointment so used. **4 a** a fervent or sympathetic quality in words or tone caused by or causing deep emotion. **b** a pretence of this. [ME f. L *unctio* f. *ung(u)ere unct-* anoint]

unctuous /ˈʌŋktjʊəs/ *adj.* **1** (of behaviour, speech, etc.) unpleasantly flattering; oily. **2** (esp. of minerals) having a greasy or soapy feel; oily. □ **unctuously** *adv.* **unctuousness** *n.* [ME f. med.L *unctuosus* f. L *unctus* anointing (as UNCTION)]

unculled /ʌnˈkʌld/ *adj.* not culled.

uncultivated /ʌnˈkʌltɪˌveɪtɪd/ *adj.* (esp. of land) not cultivated.

uncultured /ʌnˈkʌltʃəd/ *adj.* **1** not cultured, unrefined. **2** (of soil or plants) not cultivated.

uncurb /ʌnˈkɜːb/ *v.tr.* remove a curb or curbs from. □ **uncurbed** *adj.*

uncured /ʌnˈkjʊəd/ *adj.* **1** not cured. **2** (of pork etc.) not salted or smoked.

uncurl /ʌnˈkɜːl/ *v.intr. & tr.* relax from a curled position, untwist.

uncurtailed /ˌʌnkəˈteɪld/ *adj.* not curtailed.

uncurtained /ʌnˈkɜːt(ə)nd/ *adj.* not curtained.

uncut /ʌnˈkʌt/ *adj.* **1** not cut. **2** (of a book) with the pages not cut open or with untrimmed margins. **3** (of a book, film, etc.) complete; uncensored. **4** (of a stone, esp. a diamond) not shaped by cutting. **5** (of fabric) having its pile-loops intact (*uncut moquette*).

undamaged /ʌnˈdæmɪdʒd/ *adj.* not damaged; intact.

undated /ʌnˈdeɪtɪd/ *adj.* not provided or marked with a date.

undaunted /ʌnˈdɔːntɪd/ *adj.* not daunted. □ **undauntedly** *adv.* **undauntedness** *n.*

undecagon /ʌnˈdekəgən/ *n.* = HENDECAGON. [L *undecim* eleven, after *decagon*]

undeceive /ˌʌndɪˈsiːv/ *v.tr.* (often foll. by *of*) free (a person) from a misconception, deception, or error.

undecidable /ˌʌndɪˈsaɪdəb(ə)l/ *adj.* that cannot be established or refuted; uncertain.

undecided /ˌʌndɪˈsaɪdɪd/ *adj.* **1** not settled or certain (*the question is undecided*). **2** hesitating; irresolute (*undecided about their relative merits*). □ **undecidedly** *adv.*

undecipherable /ˌʌndɪˈsaɪfərəb(ə)l/ *adj.* not decipherable.

undeclared /ˌʌndɪˈkleəd/ *adj.* not declared.

undecorated /ʌnˈdekəˌreɪtɪd/ *adj.* **1** unadorned; plain. **2** not honoured with an award, esp. a military one.

undefeated /ˌʌndɪˈfiːtɪd/ *adj.* not defeated.

undefended /ˌʌndɪˈfendɪd/ *adj.* (esp. of a lawsuit) not defended.

undefiled /ˌʌndɪˈfaɪld/ *adj.* not defiled; pure.

undefined /ˌʌndɪˈfaɪnd/ *adj.* **1** not defined. **2** not clearly marked; vague, indefinite. □ **undefinable** *adj.* **undefinably** *adv.*

undelivered /ˌʌndɪˈlɪvəd/ *adj.* **1** not delivered or handed over. **2** not set free or released. **3 a** (of a pregnant woman) not yet having given birth. **b** (of a child) not yet born.

undemanding /ˌʌndɪˈmɑːndɪŋ/ *adj.* not demanding; easily satisfied. □ **undemandingness** *n.*

undemocratic /ˌʌndeməˈkrætɪk/ *adj.* not democratic, authoritarian. □ **undemocratically** *adv.*

undemonstrated /ʌnˈdemənˌstreɪtɪd/ *adj.* not demonstrated.

undemonstrative /ˌʌndɪˈmɒnstrətɪv/ *adj.* not expressing one's feelings or opinions outwardly; reserved. □ **undemonstratively** *adv.* **undemonstrativeness** *n.*

undeniable /ˌʌndɪˈnaɪəb(ə)l/ *adj.* **1** unable to be denied or disputed; certain. **2** excellent (*was of undeniable character*). □ **undeniableness** *n.* **undeniably** *adv.*

undenied /ˌʌndɪˈnaɪd/ *adj.* not denied.

undependable /ˌʌndɪˈpendəb(ə)l/ *adj.* not to be depended upon; unreliable.

under /ˈʌndə(r)/ *prep., adv., & adj.* ● *prep.* **1 a** in or to a position lower than; below; beneath (*fell under the table; under the left eye*). **b** within, on the inside of (a surface etc.) (*wore a vest under his shirt*). **2 a** inferior to; less than (*a captain is under a major; is under 18*). **b** at or for a lower cost than (*was under £20*). **3 a** subject to or liable to; controlled or bound by (*lives under oppression; under pain of death; born under Saturn; the country prospered under her*). **b** undergoing (*is under repair*). **c** classified or subsumed in (*that book goes under biology; under many names*). **4** at the foot of or sheltered by (*hid under the wall; under the cliff*). **5** planted with (a crop). **6** powered by (sail, steam, etc.). **7** following (another player in a card-game). **8** *archaic* attested by (esp. *under one's hand and seal* = signature). ● *adv.* **1** in or to a lower position or condition (*kept him under*). **2** *colloq.* in or into a state of unconsciousness (*put him under for the operation*). ● *adj.* lower (*the under jaw*). □ **under age** see AGE. **under one's arm** see ARM¹. **under arms** see ARM². **under one's belt** see BELT. **under one's breath** see BREATH. **under canvas** see CANVAS. **under a cloud** see CLOUD. **under control** see CONTROL. **under the counter** see COUNTER¹. **under cover** see COVER *n.* 4. **under fire** see FIRE. **under foot** see FOOT. **under hatches** see HATCH¹. **under a person's nose** see NOSE. **under the rose** see ROSE¹. **under separate cover** in another envelope. **under the sun** anywhere in the world. **under water** in and covered by water. **under way** in motion; in progress. **under the weather** see WEATHER. □ **undermost** *adj.* [OE f. Gmc]

under- /ˈʌndə(r)/ *prefix* in senses of UNDER: **1** below, beneath (*undercarriage; underground*). **2** lower in status; subordinate (*under-secretary*). **3** insufficiently, incompletely (*undercook; underdeveloped*). [OE (as UNDER)]

underachieve /ˌʌndərəˈtʃiːv/ *v.intr.* do less well than might be expected (esp. scholastically). □ **underachievement** *n.* **underachiever** *n.*

underact /ˌʌndərˈækt/ *v.* **1** *tr.* act (a part etc.) with insufficient force. **2** *intr.* act a part in this way.

underarm /ˈʌndərˌɑːm/ *adj., adv., & n.* ● *adj. & adv.* **1** *Sport*, esp. *Cricket* with the arm below shoulder-level. **2** under the arm. **3** in the armpit. ● *n.* the armpit.

underbelly /ˈʌndəˌbelɪ/ *n.* (*pl.* **-ies**) the under surface of an animal, vehicle, etc., esp. as an area vulnerable to attack.

underbid *v. & n.* ● *v.tr.* /ˌʌndəˈbɪd/ (**-bidding**; *past* and *past part.* **-bid**) **1** make a lower bid than (a person). **2** (also *absol.*) *Bridge* etc. bid less on (one's hand) than its strength warrants. ● *n.* /ˈʌndəˌbɪd/ **1** such a bid. **2** the act or an instance of underbidding.

underbidder /ˌʌndəˈbɪdə(r)/ n. **1** the person who makes the bid next below the highest. **2** *Bridge* etc. a player who underbids.

underbody /ˈʌndəˌbɒdɪ/ n. (pl. **-ies**) the under surface of the body of an animal, vehicle, etc.

underbred /ˌʌndəˈbred/ adj. **1** ill-bred, vulgar. **2** not of pure breeding.

underbrush /ˈʌndəˌbrʌʃ/ n. N. Amer. undergrowth in a forest.

undercapitalize /ˌʌndəˈkæpɪtəˌlaɪz/ v.tr. (also **-ise**) (esp. as **undercapitalized** adj.) provide (a business etc.) with insufficient capital to achieve a desired result.

undercarriage /ˈʌndəˌkærɪdʒ/ n. **1** a wheeled structure beneath an aircraft, usu. retracted when not in use, to receive the impact on landing and support the aircraft on the ground etc. **2** the supporting frame of a vehicle.

undercart /ˈʌndəˌkɑːt/ n. Brit. colloq. the undercarriage of an aircraft.

undercharge /ˌʌndəˈtʃɑːdʒ/ v.tr. **1** charge too little to (a person). **2** give less than the proper charge to (a gun, an electric battery, etc.).

underclass /ˈʌndəˌklɑːs/ n. a subordinate social class.

underclay /ˈʌndəˌkleɪ/ n. a clay bed under a coal seam.

undercliff /ˈʌndəˌklɪf/ n. a terrace or lower cliff formed by a landslip.

underclothes /ˈʌndəˌkləʊðz/ n.pl. clothes worn under others, esp. next to the skin.

underclothing /ˈʌndəˌkləʊðɪŋ/ n. underclothes collectively.

undercoat /ˈʌndəˌkəʊt/ n. **1 a** a preliminary layer of paint under the finishing coat. **b** the paint used for this. **2** an animal's under layer of hair or down. **3** a coat worn under another. □ **undercoating** n.

undercover /ˈʌndəˌkʌvə(r)/ adj. (usu. attrib.) **1** surreptitious. **2** engaged in spying, esp. by working with or among those to be observed (*undercover agent*).

undercroft /ˈʌndəˌkrɒft/ n. a crypt. [ME f. UNDER- + *croft* crypt f. MDu. *crofte* cave f. med.L *crupta* for L *crypta*: see CRYPT]

undercurrent /ˈʌndəˌkʌrənt/ n. **1** a current below the surface. **2** an underlying often contrary feeling, activity, or influence (*an undercurrent of protest*).

undercut v. & n. ● v.tr. /ˌʌndəˈkʌt/ (**-cutting**; past and past part. **-cut**) **1** sell or work at a lower price or lower wages than. **2** *Golf* strike (a ball) so as to make it rise high. **3 a** cut away the part below or under (a thing). **b** cut away material to show (a carved design etc.) in relief. **4** render unstable or less firm, undermine. ● n. /ˈʌndəˌkʌt/ **1** Brit. the underside of a sirloin. **2** US a notch cut in a tree-trunk to guide its fall when felled. **3** any space formed by the removal or absence of material from the lower part of something.

underdeveloped /ˌʌndədɪˈveləpt/ adj. **1** not fully developed; immature. **2** (of a country etc.) below its potential economic level. **3** *Photog.* not developed sufficiently to give a normal image. □ **underdevelopment** n.

underdog /ˈʌndəˌdɒɡ/ n. **1** a dog, or usu. a person, losing a fight. **2** a person who is in a state of inferiority or subjection.

underdone /ˌʌndəˈdʌn/ adj. **1** not thoroughly done. **2** (of food) lightly or insufficiently cooked.

underdress /ˌʌndəˈdres/ v.tr. & intr. dress with too little formality or too lightly.

underemphasis /ˌʌndərˈemfəsɪs/ n. (pl. **-emphases** /-ˌsiːz/) an insufficient degree of emphasis. □ **underemphasize** v.tr. (also **-ise**).

underemployed /ˌʌndərɪmˈplɔɪd/ adj. not fully employed. □ **underemployment** n.

underestimate v. & n. ● v.tr. /ˌʌndərˈestɪˌmeɪt/ form too low an estimate of. ● n. /ˌʌndərˈestɪmət/ an estimate that is too low. □ **underestimation** /-ˌestɪˈmeɪʃ(ə)n/ n.

underexpose /ˌʌndərɪkˈspəʊz/ v.tr. Photog. expose (film) for too short a time or with insufficient light. □ **underexposure** /-ˈspəʊʒə(r)/ n.

underfed /ˌʌndəˈfed/ adj. insufficiently fed.

underfelt /ˈʌndəˌfelt/ n. felt for laying under a carpet.

underfloor /ˈʌndəˌflɔː(r)/ attrib.adj. situated or operating beneath the floor (*underfloor heating*).

underflow /ˈʌndəˌfləʊ/ n. an undercurrent.

underfoot /ˌʌndəˈfʊt/ adv. **1** under one's feet. **2** on the ground. **3** in a state of subjection. **4** so as to obstruct or inconvenience.

underfund /ˌʌndəˈfʌnd/ v.tr. (esp. as **underfunded** adj.) provide insufficient funding for. □ **underfunding** n.

underfur /ˈʌndəˌfɜː(r)/ n. an inner layer of fine short fur or down underlying an animal's outer fur.

undergarment /ˈʌndəˌɡɑːmənt/ n. a piece of underclothing.

undergird /ˌʌndəˈɡɜːd/ v.tr. **1** make secure underneath. **2** strengthen, support.

underglaze /ˈʌndəˌɡleɪz/ adj. & n. ● adj. **1** (of painting on porcelain etc.) done before the glaze is applied. **2** (of colours) used in such painting. ● n. underglaze painting.

undergo /ˌʌndəˈɡəʊ/ v.tr. (3rd sing. present **-goes**; past **-went** /-ˈwent/; past part. **-gone** /-ˈɡɒn/) be subjected to; suffer; endure. [OE *undergān* (as UNDER-, GO[1])]

undergrad /ˌʌndəˈɡræd/ n. colloq. = UNDERGRADUATE. [abbr.]

undergraduate /ˌʌndəˈɡrædjʊət/ n. a student at a university who has not yet taken a first degree.

underground adv., adj., n., & v. ● adv. /ˌʌndəˈɡraʊnd/ **1** beneath the surface of the ground. **2** in or into secrecy or hiding. ● adj. /ˈʌndəˌɡraʊnd/ **1** situated underground. **2** secret, hidden, esp. working secretly to subvert a ruling power. **3** unconventional, experimental (*underground press*). ● n. /ˈʌndəˌɡraʊnd/ **1** an underground railway. **2** a secret group or activity, esp. aiming to subvert the established order. ● v.tr. /ˈʌndəˌɡraʊnd/ lay (cables) below ground level.

Underground Railroad (in the US) a secret network for helping slaves escape from the South to the North and Canada in the years before the American Civil War. Escaped slaves were given safe houses, transport, and other assistance, often by members of the free black community.

undergrowth /ˈʌndəˌɡrəʊθ/ n. a dense growth of shrubs etc., esp. under large trees.

underhand adj. & adv. ● adj. /ˈʌndəˌhænd/ **1** secret, clandestine, not above-board. **2** deceptive, crafty. **3** *Sport*, esp. *Cricket* underarm. ● adv. /ˌʌndəˈhænd/ in an underhand manner. [OE (as UNDER-, HAND)]

underhanded /ˌʌndəˈhændɪd/ adj. & adv. = UNDERHAND.

underhung /ˌʌndəˈhʌŋ/ adj. **1** (of the lower jaw) projecting beyond the upper jaw. **2** having an underhung jaw.

underlay[1] v. & n. ● v.tr. /ˌʌndəˈleɪ/ (past and past part. **-laid** /-ˈleɪd/) lay something under (a thing) to support or raise it. ● n. /ˈʌndəˌleɪ/ a thing laid under another, esp. material laid under a carpet or mattress as protection or support. [OE *underlecgan* (as UNDER-, LAY[1])]

underlay[2] past of UNDERLIE.

underlease n. /ˈʌndəˌliːs/ & v.tr. /ˌʌndəˈliːs/ = SUBLEASE.

underlet /ˌʌndəˈlet/ v.tr. (**-letting**; past and past part. **-let**) **1** sublet. **2** let at less than the true value.

underlie /ˌʌndəˈlaɪ/ v.tr. (**-lying**; past **-lay** /-ˈleɪ/; past part. **-lain** /-ˈleɪn/) **1** (also absol.) lie or be situated under (a stratum etc.). **2** (also absol.) (esp. as **underlying** adj.) (of a principle, reason, etc.) be the basis of (a doctrine, law, conduct, etc.). **3** exist beneath the superficial aspect of. [OE *underlicgan* (as UNDER-, LIE[1])]

underline v. & n. ● v.tr. /ˌʌndəˈlaɪn/ **1** draw a line under (a word etc.) to give emphasis or draw attention or indicate italic or other special type. **2** emphasize, stress. ● n. /ˈʌndəˌlaɪn/ **1** a line drawn under a word etc. **2** a caption below an illustration.

underlinen /ˈʌndəˌlɪnɪn/ n. underclothes esp. of linen.

underling /ˈʌndəlɪŋ/ n. usu. derog. a subordinate.

underlip /ˈʌndəˌlɪp/ n. the lower lip of a person, animal, etc.

underlying pres. part. of UNDERLIE.

undermanned /ˌʌndəˈmænd/ adj. having too few people as crew or staff.

undermentioned /ˌʌndəˈmenʃ(ə)nd/ adj. Brit. mentioned at a later place in a book etc.

undermine /ˌʌndəˈmaɪn/ v.tr. **1** injure (a person, reputation, influence, etc.) by secret or insidious means. **2** weaken, injure, or wear out (health etc.) imperceptibly or insidiously. **3** wear away the base or foundation of (*rivers undermine their banks*). **4** make a mine or excavation under. □ **underminer** n. **underminingly** adv. [ME f. UNDER- + MINE[2]]

underneath /ˌʌndəˈniːθ/ prep., adv., n., & adj. ● prep. **1** at or to a lower place than, below. **2** on the inside of, within. ● adv. **1** at or to a lower place. **2** inside. ● n. the lower surface or part. ● adj. lower. [OE *underneothan* (as UNDER + *neothan*: cf. BENEATH)]

undernourished /ˌʌndəˈnʌrɪʃt/ adj. insufficiently nourished. □ **undernourishment** n.

underpaid past and past part. of UNDERPAY.

underpants /ˈʌndəˌpænts/ n.pl. an undergarment, esp. men's, covering the lower part of the body and part of the legs.

underpart /ˈʌndəˌpɑːt/ n. **1** (usu. in pl.) a lower part, esp. a part of the underside of an animal. **2** a subordinate part in a play etc.

underpass /ˈʌndəˌpɑːs/ n. **1** a road etc. passing under another. **2** a crossing of this form.

underpay /ˌʌndəˈpeɪ/ v.tr. (past and past part. **-paid** /-ˈpeɪd/) pay too little to (a person) or for (a thing). □ **underpayment** n.

underperform /ˌʌndəpəˈfɔːm/ v. **1** intr. perform les well or be less profitable than expected. **2** tr. perform less well or be less profitable than. □ **underperformance** n.

underpin /ˌʌndəˈpɪn/ v.tr. (**-pinned**, **-pinning**) **1** support from below with masonry etc. **2** support, strengthen.

underpinning /ˈʌndəˌpɪnɪŋ/ n. **1** a physical or metaphorical foundation. **2** the action or process of supporting from below.

underplant /ˌʌndəˈplɑːnt/ v.tr. (usu. foll. by with) plant or cultivate the ground about (a tall plant) with smaller ones.

underplay /ˌʌndəˈpleɪ/ v. **1** tr. play down the importance of. **2** intr. & tr. Theatr. **a** perform with deliberate restraint. **b** underact.

underplot /ˈʌndəˌplɒt/ n. a subordinate plot in a play etc.

underpopulated /ˌʌndəˈpɒpjʊˌleɪtɪd/ adj. having an insufficient or very small population.

underpowered /ˌʌndəˈpaʊəd/ adj. **1** lacking full electrical, mechanical, etc. power; lacking sufficient amplification. **2** with insufficient authority.

underprice /ˌʌndəˈpraɪs/ v.tr. price lower than what is usual or appropriate.

underprivileged /ˌʌndəˈprɪvɪlɪdʒd/ adj. **1** less privileged than others. **2** not enjoying the normal standard of living or rights in a society.

underproduction /ˌʌndəprəˈdʌkʃ(ə)n/ n. production of less than is usual or required.

underproof /ˈʌndəˌpruːf/ adj. containing less alcohol than proof spirit does.

underprop /ˌʌndəˈprɒp/ v.tr. (**-propped**, **-propping**) **1** support with a prop. **2** support, sustain.

underquote /ˌʌndəˈkwəʊt/ v.tr. **1** quote a lower price than (a person). **2** quote a lower price than others for (goods etc.).

underrate /ˌʌndəˈreɪt/ v.tr. have too low an opinion of.

under-rehearsed /ˌʌndərɪˈhɜːst/ adj. (esp. of a performance) insufficiently rehearsed.

under-report /ˌʌndərɪˈpɔːt/ v.tr. (usu. as **under-reported** adj.) fail to report (news, data, etc.) fully.

under-represent /ˌʌndəˌreprɪˈzent/ v.tr. (usu. as **under-represented** adj.) not include (a social group, specimen, type, etc.) in sufficient numbers.

underscore v. & n. ● v.tr. /ˌʌndəˈskɔː(r)/ = UNDERLINE v. ● n. /ˈʌndəˌskɔː(r)/ = UNDERLINE n. 1.

undersea /ˈʌndəˌsiː/ adj. below the sea or the surface of the sea, submarine.

underseal /ˈʌndəˌsiːl/ v. & n. ● v.tr. seal the underpart of (esp. a motor vehicle against rust etc.). ● n. a protective coating for this.

under-secretary /ˌʌndəˈsekrɪtərɪ, -ˈsekrətrɪ/ n. (pl. **-ies**) a subordinate official, esp. a junior minister or civil servant.

undersell /ˌʌndəˈsel/ v.tr. (past and past part. **-sold** /-ˈsəʊld/) **1** sell at a lower price than (another seller). **2** sell at less than the true value.

underset v. & n. ● v.tr. /ˌʌndəˈset/ (**-setting**; past and past part. **-set**) place something under (a thing). ● n. /ˈʌndəˌset/ Naut. an undercurrent.

undersexed /ˌʌndəˈsekst/ adj. having unusually weak sexual desires.

under-sheriff /ˈʌndəˌʃerɪf/ n. a deputy sheriff.

undershirt /ˈʌndəˌʃɜːt/ n. esp. N. Amer. an undergarment worn under a shirt; a vest.

undershoot v. & n. ● v.tr. /ˌʌndəˈʃuːt/ (past and past part. **-shot** /-ˈʃɒt/) **1** (of an aircraft) land short of (a runway etc.). **2** shoot short of or below. ● n. /ˈʌndəˌʃuːt/ the act or an instance of undershooting.

undershorts /ˈʌndəˌʃɔːts/ n. US short underpants; trunks.

undershot /ˈʌndəˌʃɒt/ adj. **1** (of a water-wheel) turned by water flowing under it. **2** = UNDERHUNG.

undershrub /ˈʌndəˌʃrʌb/ n. = SUBSHRUB.

underside /ˈʌndəˌsaɪd/ n. the lower or under side or surface.

undersigned /ˈʌndəˌsaɪnd, ˌʌndəˈsaɪnd/ adj. whose signature is appended (we, the undersigned, wish to state ...).

undersized /ˈʌndəˌsaɪzd, ˌʌndəˈsaɪzd/ adj. of less than the usual size.

underskirt /ˈʌndəˌskɜːt/ n. a skirt worn under another; a petticoat.

underslung /ˌʌndəˈslʌŋ/ adj. **1** supported from above. **2** (of a vehicle chassis) hanging lower than the axles.

undersold past and past part. of UNDERSELL.

undersow /ˈʌndəˌsəʊ/ v.tr. (past part. **-sown** /-ˌsəʊn/) **1** sow (a later-growing crop) on land already seeded with another crop. **2** (foll. by with) sow land already seeded with (a crop) with a later-growing crop.

underspend /ˌʌndəˈspend/ v. (past and past part. **-spent** /-ˈspent/) **1** tr. spend less than (a specified amount). **2** intr. & refl. spend too little.

understaffed /ˌʌndəˈstɑːft/ adj. having too few staff.

understand /ˌʌndəˈstænd/ v. (past and past part. **-stood** /-ˈstʊd/) **1** tr. perceive the meaning of (words, a person, a language, etc.) (does not understand what you say; understood you perfectly; cannot understand French). **2** tr. perceive the significance or explanation or cause of (do not understand why she came; could not understand what the noise was about; do not understand the point of his remark). **3** tr. be sympathetically aware of the character or nature of, know how to deal with (quite understand your difficulty; cannot understand him at all; could never understand algebra). **4** tr. **a** (often foll. by that + clause) infer esp. from information received, take as implied, take for granted (I understand that it begins at noon; I understand her to be a distant relation; am I to understand that you refuse?). **b** (absol.) believe or assume from knowledge or inference (he is coming tomorrow, I understand). **5** tr. supply (a word) mentally (the verb may be either expressed or understood). **6** tr. accept (terms, conditions, etc.) as part of an agreement. **7** intr. have understanding (in general or in particular). □ **understand each other 1** know each other's views or feelings. **2** be in agreement or collusion. □ **understandable** adj. **understandably** adv. **understander** n. [OE understandan (as UNDER-, STAND)]

understanding /ˌʌndəˈstændɪŋ/ n. & adj. ● n. **1 a** the ability to understand or think; intelligence. **b** the power of apprehension; the power of abstract thought. **2** an individual's perception or judgement of a situation etc. **3** an agreement; a thing agreed upon, esp. informally (had an understanding with the rival company; consented only on this understanding). **4** harmony in opinion or feeling (disturbed the good understanding between them). **5** sympathetic awareness or tolerance. ● adj. **1** having understanding or insight or good judgement. **2** sympathetic to others' feelings. □ **understandingly** adv. [OE (as UNDERSTAND)]

understate /ˌʌndəˈsteɪt/ v.tr. (often as **understated** adj.) **1** express in greatly or unduly restrained terms. **2** represent as being less than it actually is. □ **understatement** n. **understater** n.

understeer n. & v. ● n. /ˈʌndəˌstɪə(r)/ a tendency of a motor vehicle to turn less sharply than was intended. ● v.intr. /ˌʌndəˈstɪə(r)/ have such a tendency.

understood past and past part. of UNDERSTAND.

understorey /ˈʌndəˌstɔːrɪ/ n. (pl. **-eys**) **1** a layer of vegetation beneath the main canopy of a forest. **2** the plants forming this.

understudy /ˈʌndəˌstʌdɪ/ n. & v. esp. Theatr. ● n. (pl. **-ies**) a person who studies another's role or duties in order to act at short notice in the absence of the other. ● v.tr. (**-ies**, **-ied**) **1** study (a role etc.) as an understudy. **2** act as an understudy to (a person).

undersubscribed /ˌʌndəsəbˈskraɪbd/ adj. without sufficient subscribers, participants, etc.

undersurface /ˈʌndəˌsɜːfɪs/ n. the lower or under surface.

undertake /ˌʌndəˈteɪk/ v.tr. (past **-took** /-ˈtʊk/; past part. **-taken** /-ˈteɪkən/) **1** bind oneself to perform, make oneself responsible for, engage in, enter upon (work, an enterprise, a responsibility). **2** (usu. foll. by to + infin.) accept an obligation, promise. **3** guarantee, affirm (I will undertake that he has not heard a word).

undertaker /ˈʌndəˌteɪkə(r)/ n. **1** a person whose business is to make arrangements for funerals. **2** (also /ˌʌndəˈteɪkə(r)/) a person who undertakes to do something. **3** hist. an influential person in 17th-century England who undertook to procure particular legislation, esp. to obtain supplies from the House of Commons if the king would grant some concession.

undertaking /ˌʌndəˈteɪkɪŋ/ n. **1** work etc. undertaken, an enterprise (a serious undertaking). **2** a pledge or promise. **3** /ˈʌndəˌteɪkɪŋ/ the management of funerals as a profession.

undertenant /ˈʌndəˌtenənt/ n. a subtenant. □ **undertenancy** n. (pl. **-ies**).

underthings /ˈʌndəˌθɪŋz/ n.pl. colloq. underclothes.

undertint /ˈʌndəˌtɪnt/ n. a subdued tint.

undertone /ˈʌndəˌtəʊn/ n. **1** a subdued tone of sound or colour. **2** an underlying quality. **3** an undercurrent of feeling.

undertook past of UNDERTAKE.

undertow /ˈʌndəˌtəʊ/ n. a current below the surface of the sea moving in the opposite direction to the surface current.

undertrick /ˈʌndəˌtrɪk/ n. Bridge a trick by which the declarer falls short of his or her contract.

underuse /ˌʌndəˈjuːz/ v.tr. (esp. as **underused** adj.) use below the optimum level.

underutilize /ˌʌndəˈjuːtɪˌlaɪz/ v.tr. (also **-ise**) (esp. as **underutilized** adj.) = UNDERUSE.

undervalue /ˌʌndəˈvæljuː/ v.tr. (**-values**, **-valued**, **-valuing**) **1** value insufficiently. **2** underestimate. □ **undervaluation** /-ˌvæljuːˈeɪʃ(ə)n/ n.

undervest /ˈʌndəˌvest/ n. Brit. an undergarment worn on the upper part of the body; a vest.

underwater /ˌʌndəˈwɔːtə(r)/ adj. & adv. ● adj. situated or done under water. ● adv. under water.

underwear /ˈʌndəˌweə(r)/ n. underclothes.

underweight adj. & n. ● adj. /ˌʌndəˈweɪt/ weighing less than is normal or desirable. ● n. /ˈʌndəˌweɪt/ insufficient weight.

underwent past of UNDERGO.

underwhelm /ˌʌndəˈwelm/ v.tr. joc. fail to impress. [after OVERWHELM]

underwing /ˈʌndəˌwɪŋ/ n. **1** the hindwing of an insect. **2** the underside of a bird's wing.

underwired /ˌʌndəˈwaɪəd/ adj. (usu. attrib.) (of a bra) having a thin semicircular support of wire inset under each cup.

underwood /ˈʌndəˌwʊd/ n. undergrowth.

underwork /ˌʌndəˈwɜːk/ v. **1** tr. impose too little work on. **2** intr. do too little work.

underworld /ˈʌndəˌwɜːld/ n. **1** the part of society comprising those who live by organized crime and immorality. **2** the mythical abode of the dead under the earth. **3** the antipodes.

underwrite /ˌʌndəˈraɪt/ v. (past **-wrote** /-ˈrəʊt/; past part. **-written** /-ˈrɪt(ə)n/) **1 a** tr. sign, and accept liability under (an insurance policy, esp. on shipping etc.). **b** tr. accept (liability) in this way. **c** intr. practise (marine) insurance. **2** tr. undertake to finance or support. **3** tr. engage to buy all the stock in (a company etc.) not bought by the public. **4** tr. write below (the underwritten names). □ **underwriter** /ˈʌndəˌraɪtə(r)/ n.

undescended /ˌʌndɪˈsendɪd/ adj. Med. (of a testicle) remaining in the abdomen instead of descending normally into the scrotum.

undeserved /ˌʌndɪˈzɜːvd/ adj. not deserved (as reward or punishment). □ **undeservedly** /-vɪdlɪ/ adv.

undeserving /ˌʌndɪˈzɜːvɪŋ/ adj. not deserving. □ **undeservingly** adv.

undesigned /ˌʌndɪˈzaɪnd/ adj. unintentional. □ **undesignedly** /-nɪdlɪ/ adv.

undesirable /ˌʌndɪˈzaɪərəb(ə)l/ adj. & n. ● adj. not desirable, objectionable, unpleasant. ● n. an undesirable person. □ **undesirably** adv. **undesirableness** n. **undesirability** /-ˌzaɪərəˈbɪlɪtɪ/ n.

undesired /ˌʌndɪˈzaɪəd/ adj. not desired.

undesirous /ˌʌndɪˈzaɪərəs/ adj. not desirous.

undetectable /ˌʌndɪˈtektəb(ə)l/ adj. not detectable. □ **undetectably** adv. **undetectability** /-ˌtektəˈbɪlɪtɪ/ n.

undetected /ˌʌndɪˈtektɪd/ adj. not detected.

undetermined /ˌʌndɪˈtɜːmɪnd/ adj. = UNDECIDED.

undeterred /ˌʌndɪˈtɜːd/ adj. not deterred.

undeveloped /ˌʌndɪˈveləpt/ adj. not developed.

undeviating /ʌnˈdiːvɪˌeɪtɪŋ/ adj. not deviating; steady, constant. □ **undeviatingly** adv.

undiagnosed /ʌnˈdaɪəɡˌnəʊzd, ˌʌndaɪəɡˈnəʊzd/ adj. not diagnosed.

undid past of UNDO.

undies /ˈʌndɪz/ n.pl. colloq. (esp. women's) underclothes. [abbr.]

undifferentiated /ˌʌndɪfəˈrenʃɪˌeɪtɪd/ adj. not differentiated; amorphous.

undigested /ˌʌndaɪˈdʒestɪd, ˌʌndɪ-/ adj. **1** not digested. **2** (esp. of information, facts, etc.) not properly arranged or considered.

undignified /ʌnˈdɪɡnɪˌfaɪd/ adj. lacking dignity.

undiluted /ˌʌndaɪˈluːtɪd, -ˈljuː-tɪd/ adj. **1** not diluted. **2** complete, utter.

undiminished /ˌʌndɪˈmɪnɪʃt/ adj. not diminished or lessened.

undine /ˈʌndiːn/ n. a female water-spirit. [mod.L undina (word invented by Paracelsus) f. L unda wave]

undiplomatic /ˌʌndɪpləˈmætɪk/ adj. tactless. □ **undiplomatically** adv.

undirected /ˌʌndaɪˈrektɪd, ˌʌndɪ-/ adj. aimless; lacking direction; unfocused.

undischarged /ˌʌndɪsˈtʃɑːdʒd/ adj. (esp. of a bankrupt or a debt) not discharged.

undiscipline /ʌnˈdɪsɪplɪn/ n. lack of discipline.

undisciplined /ʌnˈdɪsɪplɪnd/ adj. lacking discipline; not disciplined.

undisclosed /ˌʌndɪsˈkləʊzd/ adj. not revealed or made known.

undiscoverable /ˌʌndɪsˈkʌvərəb(ə)l/ adj. that cannot be discovered.

undiscovered /ˌʌndɪsˈkʌvəd/ adj. not discovered.

undiscriminating /ˌʌndɪsˈkrɪmɪˌneɪtɪŋ/ adj. not showing good judgement.

undiscussed /ˌʌndɪsˈkʌst/ adj. not discussed.

undisguised /ˌʌndɪsˈɡaɪzd/ adj. not disguised. □ **undisguisedly** /-zɪdlɪ/ adv.

undismayed /ˌʌndɪsˈmeɪd/ adj. not dismayed.

undisputed /ˌʌndɪˈspjuːtɪd/ adj. not disputed or called in question.

undissolved /ˌʌndɪˈzɒlvd/ adj. not dissolved.

undistinguishable /ˌʌndɪsˈtɪŋɡwɪʃəb(ə)l/ adj. (often foll. by from) indistinguishable.

undistinguished /ˌʌndɪsˈtɪŋɡwɪʃt/ adj. not distinguished; mediocre.

undistorted /ˌʌndɪsˈtɔːtɪd/ adj. not distorted.

undistributed /ˌʌndɪsˈtrɪbjuːtɪd/ adj. not distributed (undistributed profits). □ **undistributed middle** Logic a fallacy arising from the failure of the middle term of a syllogism to refer to all the members of a class.

undisturbed /ˌʌndɪsˈtɜːbd/ adj. not disturbed or interfered with.

undivided /ˌʌndɪˈvaɪdɪd/ adj. not divided or shared; whole, entire (gave him my undivided attention).

undo /ʌnˈduː/ v.tr. (3rd sing. present **-does** /-ˈdʌz/; past **-did** /-ˈdɪd/; past part. **-done** /-ˈdʌn/) **1 a** unfasten or untie (a coat, button, parcel, etc.). **b** unfasten the clothing of (a person). **2** annul, cancel (cannot undo the past). **3** ruin the prospects, reputation, or morals of. [OE undōn (as UN-², DO¹)]

undock /ʌnˈdɒk/ v.tr. **1** (also absol.) separate (a spacecraft) from another in space. **2** take (a ship) out of a dock.

undocumented /ʌnˈdɒkjʊˌmentɪd/ adj. **1** US not having the appropriate document. **2** not proved by or recorded in documents.

undoing /ʌnˈduːɪŋ/ n. **1** ruin or a cause of ruin. **2** the process of reversing what has been done. **3** the action of opening or unfastening.

undomesticated /ˌʌndəˈmestɪˌkeɪtɪd/ adj. not domesticated.

undone /ʌnˈdʌn/ past part. of UNDO. ● adj. **1** not done; incomplete (left the job undone). **2** not fastened (left the buttons undone). **3** archaic ruined.

undoubtable /ʌnˈdaʊtəb(ə)l/ adj. that cannot be doubted; indubitable.

undoubted /ʌnˈdaʊtɪd/ adj. certain, not questioned, not regarded as doubtful. □ **undoubtedly** adv.

UNDP abbr. United Nations Development Programme.

undrained /ʌnˈdreɪnd/ adj. not drained.

undramatic /ˌʌndrəˈmætɪk/ adj. **1** lacking a dramatic quality or effect. **2** understated; unremarkable.

undraped /ʌnˈdreɪpt/ adj. **1** not covered with drapery. **2** naked.

undreamed /ʌnˈdriːmd, -ˈdremt/ adj. (also **undreamt** /-ˈdremt/) (often foll. by of) not dreamed or thought of or imagined.

undress /ʌnˈdres/ v. & n. ● v. **1** intr. take off one's clothes. **2** tr. take the clothes off (a person). ● n. **1** ordinary dress as opposed to full dress or uniform (also attrib.: undress cap). **2** casual or informal dress. **3** the state of being naked or only partially clothed.

undressed /ʌnˈdrest/ adj. **1** not or no longer dressed; partly or wholly naked. **2** (of leather etc.) not treated. **3** (of food) not having a dressing.

undrinkable /ʌnˈdrɪŋkəb(ə)l/ adj. unfit for drinking.

UNDRO abbr. United Nations Disaster Relief Office.

undue /ʌnˈdjuː/ adj. (usu. attrib.) **1** excessive, disproportionate. **2** not suitable. **3** not owed. □ **undue influence** Law influence by which a person is induced to act otherwise than by his or her own free will, or without adequate attention to the consequences. □ **unduly** adv.

undulant /ˈʌndjʊlənt/ adj. moving like waves; fluctuating. □ **undulant fever** Med. brucellosis in humans. [L undulare (as UNDULATE)]

undulate v. & adj. ● v. /ˈʌndjʊˌleɪt/ intr. & tr. have or cause to have a wavy motion or look. ● adj. /ˈʌndjʊlət/ wavy, going alternately up and down or in and out (*leaves with undulate margins*). □ **undulately** /-lətlɪ/ *adv.* [LL *undulatus* f. L *unda* wave]

undulation /ˌʌndjʊˈleɪʃ(ə)n/ *n.* **1** a wavy motion or form, a gentle rise and fall. **2** each wave of this. **3** a set of wavy lines.

undulatory /ˈʌndjʊlətərɪ/ *adj.* **1** undulating, wavy. **2** of or due to undulation.

undutiful /ʌnˈdjuːtɪˌfʊl/ *adj.* not dutiful. □ **undutifully** *adv.* **undutifulness** *n.*

undyed /ʌnˈdaɪd/ *adj.* not dyed.

undying /ʌnˈdaɪɪŋ/ *adj.* **1** immortal. **2** never-ending (*undying love*). □ **undyingly** *adv.*

unearned /ʌnˈɜːnd/ *adj.* not earned. □ **unearned income** income from interest payments etc. as opposed to salary, wages, or fees. **unearned increment** an increase in the value of property not due to the owner's labour or outlay.

unearth /ʌnˈɜːθ/ *v.tr.* **1 a** discover by searching or in the course of digging or rummaging. **b** dig out of the earth. **2** drive (a fox etc.) from its earth.

unearthly /ʌnˈɜːθlɪ/ *adj.* **1** supernatural, mysterious. **2** *colloq.* absurdly early or inconvenient (*got up at an unearthly hour*). **3** not earthly. □ **unearthliness** *n.*

unease /ʌnˈiːz/ *n.* lack of ease, discomfort, distress.

uneasy /ʌnˈiːzɪ/ *adj.* (**uneasier, uneasiest**) **1** disturbed or uncomfortable in mind or body (*passed an uneasy night*). **2** disturbing (*had an uneasy suspicion*). □ **uneasily** *adv.* **uneasiness** *n.*

uneatable /ʌnˈiːtəb(ə)l/ *adj.* not able to be eaten, esp. because of its condition (cf. INEDIBLE).

uneaten /ʌnˈiːt(ə)n/ *adj.* not eaten; left undevoured.

uneconomic /ˌʌniːkəˈnɒmɪk, ˌʌnek-/ *adj.* not economic; incapable of being profitably operated etc. □ **uneconomically** *adv.*

uneconomical /ˌʌniːkəˈnɒmɪk(ə)l, ˌʌnek-/ *adj.* not economical; wasteful.

unedifying /ʌnˈedɪˌfaɪɪŋ/ *adj.* not edifying, esp. uninstructive or degrading. □ **unedifyingly** *adv.*

unedited /ʌnˈedɪtɪd/ *adj.* not edited.

uneducated /ʌnˈedjʊˌkeɪtɪd/ *adj.* not educated. □ **uneducable** /-kəb(ə)l/ *adj.*

unelectable /ˌʌnɪˈlektəb(ə)l/ *adj.* (of a candidate, party, etc.) associated with or holding views likely to bring defeat at an election.

unelected /ˌʌnɪˈlektɪd/ *adj.* not elected.

unembarrassed /ˌʌnɪmˈbærəst/ *adj.* not embarrassed.

unembellished /ˌʌnɪmˈbelɪʃt/ *adj.* not embellished or decorated.

unemotional /ˌʌnɪˈməʊʃən(ə)l/ *adj.* not emotional; lacking emotion. □ **unemotionally** *adv.*

unemphatic /ˌʌnɪmˈfætɪk/ *adj.* not emphatic. □ **unemphatically** *adv.*

unemployable /ˌʌnɪmˈplɔɪəb(ə)l/ *adj.* & *n.* ● *adj.* unfitted for paid employment. ● *n.* an unemployable person. □ **unemployability** /-ˌplɔɪəˈbɪlɪtɪ/ *n.*

unemployed /ˌʌnɪmˈplɔɪd/ *adj.* & *n.* ● *adj.* **1** not having paid employment; out of work. **2** not in use. ● *n.* (prec. by *the*; treated as *pl.*) unemployed people.

unemployment /ˌʌnɪmˈplɔɪmənt/ *n.* **1** the state of being unemployed. **2** the condition or extent of this in a country or region etc. (*the North has higher unemployment*). □ **unemployment benefit** a payment made by the state or (in the US) a trade union to an unemployed person.

unenclosed /ˌʌnɪnˈkləʊzd/ *adj.* not enclosed.

unencumbered /ˌʌnɪnˈkʌmbəd/ *adj.* **1** *Law* (of an estate) not having any liabilities (e.g. a mortgage) on it. **2** having no encumbrance; free.

unending /ʌnˈendɪŋ/ *adj.* having or apparently having no end. □ **unendingly** *adv.* **unendingness** *n.*

unendowed /ˌʌnɪnˈdaʊd/ *adj.* not endowed.

unendurable /ˌʌnɪnˈdjʊərəb(ə)l/ *adj.* that cannot be endured. □ **unendurably** *adv.*

unenforceable /ˌʌnɪnˈfɔːsəb(ə)l/ *adj.* (of a contract, law, etc.) impossible to enforce.

unengaged /ˌʌnɪnˈɡeɪdʒd/ *adj.* not engaged; uncommitted.

un-English /ʌnˈɪŋɡlɪʃ/ *adj.* **1** not characteristic of the English. **2** not English.

unenjoyable /ˌʌnɪnˈdʒɔɪəb(ə)l/ *adj.* not enjoyable.

unenlightened /ˌʌnɪnˈlaɪt(ə)nd/ *adj.* not enlightened, uninformed. □ **unenlightenment** *n.*

unenterprising /ʌnˈentəˌpraɪzɪŋ/ *adj.* not enterprising.

unenthusiastic /ˌʌnɪnˌθjuːzɪˈæstɪk, -ˌθuːzɪˈæstɪk/ *adj.* not enthusiastic. □ **unenthusiastically** *adv.*

unenviable /ʌnˈenvɪəb(ə)l/ *adj.* not enviable. □ **unenviably** *adv.*

unenvied /ʌnˈenvɪd/ *adj.* not envied.

UNEP *abbr.* United Nations Environment Programme.

unequal /ʌnˈiːkwəl/ *adj.* **1** (often foll. by *to*) not equal. **2** of varying quality. **3** lacking equal advantage to both sides (*an unequal bargain*). □ **unequally** *adv.*

unequalize /ʌnˈiːkwəˌlaɪz/ *v.tr.* (also **-ise**) make unequal.

unequalled /ʌnˈiːkwəld/ *adj.* superior to all others.

unequipped /ˌʌnɪˈkwɪpt/ *adj.* not equipped.

unequivocal /ˌʌnɪˈkwɪvək(ə)l/ *adj.* not ambiguous, plain, unmistakable. □ **unequivocally** *adv.* **unequivocalness** *n.*

unerring /ʌnˈɜːrɪŋ/ *adj.* not erring, failing, or missing the mark; true, certain. □ **unerringly** *adv.* **unerringness** *n.*

unescapable /ˌʌnɪˈskeɪpəb(ə)l/ *adj.* inescapable.

UNESCO /juːˈneskəʊ/ *abbr.* (also **Unesco**) United Nations Educational, Scientific, and Cultural Organization, an agency of the United Nations set up in 1945 to promote the exchange of information, ideas, and culture. In 1984 the US withdrew from the organization, followed a year later by the UK, in protest at its allegedly over-politicized nature. Its headquarters are in Paris.

unescorted /ˌʌnɪˈskɔːtɪd/ *adj.* not escorted.

unessential /ˌʌnɪˈsenʃ(ə)l/ *adj.* & *n.* ● *adj.* **1** not essential (cf. INESSENTIAL). **2** not of the first importance. ● *n.* an unessential part or thing.

unestablished /ˌʌnɪˈstæblɪʃt/ *adj.* not established.

unethical /ʌnˈeθɪk(ə)l/ *adj.* not ethical, esp. unscrupulous in business or professional conduct. □ **unethically** *adv.*

unevangelical /ˌʌniːvænˈdʒelɪk(ə)l/ *adj.* not evangelical.

uneven /ʌnˈiːv(ə)n/ *adj.* **1** not level or smooth. **2** not uniform or equable. **3** (of a contest) unequal. □ **unevenly** *adv.* **unevenness** /-v(ə)nnɪs/ *n.* [OE *unefen* (as UN-[1], EVEN[1])]

uneventful /ˌʌnɪˈventfʊl/ *adj.* not eventful. □ **uneventfully** *adv.* **uneventfulness** *n.*

unexamined /ˌʌnɪɡˈzæmɪnd/ *adj.* not examined.

unexampled /ˌʌnɪɡˈzɑːmp(ə)ld/ *adj.* having no precedent or parallel.

unexceptionable /ˌʌnɪkˈsepʃənəb(ə)l/ *adj.* with which no fault can be found; entirely satisfactory. □ **unexceptionableness** *n.* **unexceptionably** *adv.*

unexceptional /ˌʌnɪkˈsepʃən(ə)l/ *adj.* not out of the ordinary; usual, normal. □ **unexceptionally** *adv.*

unexcitable /ˌʌnɪkˈsaɪtəb(ə)l/ *adj.* not easily excited, staid. □ **unexcitability** /-ˌsaɪtəˈbɪlɪtɪ/ *n.*

unexciting /ˌʌnɪkˈsaɪtɪŋ/ *adj.* not exciting; dull.

unexecuted /ʌnˈeksɪˌkjuːtɪd/ *adj.* not carried out or put into effect.

unexhausted /ˌʌnɪɡˈzɔːstɪd/ *adj.* **1** not used up, expended, or brought to an end. **2** not emptied.

unexpected /ˌʌnɪkˈspektɪd/ *adj.* not expected; surprising. □ **unexpectedly** *adv.* **unexpectedness** *n.*

unexpired /ˌʌnɪkˈspaɪəd/ *adj.* that has not yet expired.

unexplainable /ˌʌnɪkˈspleɪnəb(ə)l/ *adj.* that cannot be explained. □ **unexplainably** *adv.*

unexplained /ˌʌnɪkˈspleɪnd/ *adj.* not explained.

unexploded /ˌʌnɪkˈspləʊdɪd/ *adj.* (usu. *attrib.*) (of a bomb etc.) that has not exploded.

unexploited /ˌʌnɪkˈsplɔɪtɪd/ *adj.* (of resources etc.) not exploited.

unexplored /ˌʌnɪkˈsplɔːd/ *adj.* not explored.

unexposed /ˌʌnɪkˈspəʊzd/ *adj.* not exposed.

unexpressed /ˌʌnɪkˈsprest/ *adj.* not expressed or made known (*unexpressed fears*).

unexpurgated /ʌnˈekspəˌɡeɪtɪd/ *adj.* (esp. of a text etc.) not expurgated; complete.

unfaceable /ʌnˈfeɪsəb(ə)l/ *adj.* that cannot be faced or confronted.

unfading /ʌnˈfeɪdɪŋ/ *adj.* that never fades. □ **unfadingly** *adv.*

unfailing /ʌnˈfeɪlɪŋ/ *adj.* **1** not failing. **2** not running short. **3** constant. **4** reliable. □ **unfailingly** *adv.* **unfailingness** *n.*

unfair /ʌnˈfeə(r)/ *adj.* **1** not equitable or honest (*obtained by unfair means*). **2** not impartial or according to the rules (*unfair play*). □ **unfairly** *adv.* **unfairness** *n.* [OE *unfæger* (as UN-¹, FAIR¹)]

unfaithful /ʌnˈfeɪθfʊl/ *adj.* **1** not faithful, esp. adulterous. **2** not loyal. **3** treacherous. □ **unfaithfully** *adv.* **unfaithfulness** *n.*

unfaltering /ʌnˈfɒltərɪŋ, ʌnˈfɔːl-/ *adj.* not faltering; steady, resolute. □ **unfalteringly** *adv.*

unfamiliar /ˌʌnfəˈmɪlɪə(r)/ *adj.* not familiar. □ **unfamiliarity** /-ˌmɪlɪˈærɪtɪ/ *n.*

unfashionable /ʌnˈfæʃənəb(ə)l/ *adj.* not fashionable or stylish. □ **unfashionableness** *n.* **unfashionably** *adv.*

unfashioned /ʌnˈfæʃ(ə)nd/ *adj.* not made into its proper shape.

unfasten /ʌnˈfɑːs(ə)n/ *v.* **1** *tr. & intr.* make or become loose. **2** *tr.* open the fastening(s) of. **3** *tr.* detach.

unfastened /ʌnˈfɑːs(ə)nd/ *adj.* **1** that has not been fastened. **2** that has been loosened, opened, or detached.

unfathered /ʌnˈfɑːðəd/ *adj.* **1** having no known or acknowledged father; illegitimate. **2** of unknown origin (*unfathered rumours*).

unfatherly /ʌnˈfɑːðəlɪ/ *adj.* not befitting a father.

unfathomable /ʌnˈfæðəməb(ə)l/ *adj.* incapable of being fathomed. □ **unfathomableness** *n.* **unfathomably** *adv.*

unfathomed /ʌnˈfæðəmd/ *adj.* **1** of unascertained depth. **2** not fully explored or known.

unfavourable /ʌnˈfeɪvərəb(ə)l/ *adj.* (*US* **unfavorable**) not favourable; adverse, hostile. □ **unfavourableness** *n.* **unfavourably** *adv.*

unfavourite /ʌnˈfeɪvərɪt/ *adj.* (*US* **unfavorite**) *colloq.* least favourite; most disliked.

unfazed /ʌnˈfeɪzd/ *adj. colloq.* untroubled; not disconcerted.

unfeasible /ʌnˈfiːzɪb(ə)l/ *adj.* not feasible; impractical. □ **unfeasibly** *adv.* **unfeasibility** /-ˌfiːzɪˈbɪlɪtɪ/ *n.*

unfed /ʌnˈfed/ *adj.* not fed.

unfeeling /ʌnˈfiːlɪŋ/ *adj.* **1** unsympathetic, harsh, not caring about others' feelings. **2** lacking sensation or sensitivity. □ **unfeelingly** *adv.* **unfeelingness** *n.* [OE *unfelende* (as UN-¹, FEELING)]

unfeigned /ʌnˈfeɪnd/ *adj.* genuine, sincere. □ **unfeignedly** /-nɪdlɪ/ *adv.*

unfelt /ʌnˈfelt/ *adj.* not felt.

unfeminine /ʌnˈfemɪnɪn/ *adj.* not in accordance with, or appropriate to, female character. □ **unfemininity** /-ˌfemɪˈnɪnɪtɪ/ *n.*

unfenced /ʌnˈfenst/ *adj.* **1** not provided with fences. **2** unprotected.

unfermented /ˌʌnfəˈmentɪd/ *adj.* not fermented.

unfertilized /ʌnˈfɜːtɪˌlaɪzd/ *adj.* (also **unfertilised**) not fertilized.

unfetter /ʌnˈfetə(r)/ *v.tr.* release from fetters.

unfettered /ʌnˈfetəd/ *adj.* unrestrained, unrestricted.

unfilial /ʌnˈfɪlɪəl/ *adj.* not befitting a son or daughter. □ **unfilially** *adv.*

unfilled /ʌnˈfɪld/ *adj.* not filled.

unfiltered /ʌnˈfɪltəd/ *adj.* **1** not filtered. **2** (of a cigarette) not provided with a filter.

unfinancial /ˌʌnfaɪˈnænʃ(ə)l, ˌʌnfɪ-/ *adj. Austral.* **1** insolvent. **2** not having paid a subscription (*some members are unfinancial*).

unfinished /ʌnˈfɪnɪʃt/ *adj.* not finished; incomplete.

unfit /ʌnˈfɪt/ *adj. & v.* ● *adj.* (often foll. by *for*, or *to* + infin.) not fit. ● *v.tr.* (**unfitted**, **unfitting**) (usu. foll. by *for*) make unsuitable. □ **unfitly** *adv.* **unfitness** *n.*

unfitted /ʌnˈfɪtɪd/ *adj.* **1** not fit. **2** not fitted or suited. **3** not provided with fittings.

unfitting /ʌnˈfɪtɪŋ/ *adj.* not fitting or suitable, unbecoming. □ **unfittingly** *adv.*

unfix /ʌnˈfɪks/ *v.tr.* **1** release or loosen from a fixed state. **2** detach.

unfixed /ʌnˈfɪkst/ *adj.* not fixed.

unflagging /ʌnˈflægɪŋ/ *adj.* tireless, persistent. □ **unflaggingly** *adv.*

unflappable /ʌnˈflæpəb(ə)l/ *adj. colloq.* imperturbable; remaining calm in a crisis. □ **unflappably** *adv.* **unflappability** /-ˌflæpəˈbɪlɪtɪ/ *n.*

unflattering /ʌnˈflætərɪŋ/ *adj.* not flattering. □ **unflatteringly** *adv.*

unflavoured /ʌnˈfleɪvəd/ *adj.* (*US* **unflavored**) not flavoured.

unfledged /ʌnˈfledʒd/ *adj.* **1** (of a person) inexperienced. **2** (of a bird) not yet fledged.

unfleshed /ʌnˈfleʃt/ *adj.* **1** not covered with flesh. **2** stripped of flesh.

unflinching /ʌnˈflɪntʃɪŋ/ *adj.* not flinching. □ **unflinchingly** *adv.*

unfocused /ʌnˈfəʊkəst/ *adj.* (also **unfocussed**) not focused.

unfold /ʌnˈfəʊld/ *v.* **1** *tr.* open the fold or folds of, spread out. **2** *tr.* reveal (thoughts etc.). **3** *intr.* become opened out. **4** *intr.* develop. □ **unfoldment** *n. US.* [OE *unfealdan* (as UN-², FOLD¹)]

unforced /ʌnˈfɔːst/ *adj.* **1** not produced by effort; easy, natural. **2** not compelled or constrained. □ **unforcedly** /-sɪdlɪ/ *adv.*

unfordable /ʌnˈfɔːdəb(ə)l/ *adj.* that cannot be forded.

unforeseeable /ˌʌnfɔːˈsiːəb(ə)l/ *adj.* not foreseeable.

unforeseen /ˌʌnfɔːˈsiːn/ *adj.* not foreseen.

unforetold /ˌʌnfɔːˈtəʊld/ *adj.* not foretold; unpredicted.

unforgettable /ˌʌnfəˈgetəb(ə)l/ *adj.* that cannot be forgotten; memorable, wonderful (*an unforgettable experience*). □ **unforgettably** *adv.*

unforgivable /ˌʌnfəˈgɪvəb(ə)l/ *adj.* that cannot be forgiven. □ **unforgivably** *adv.*

unforgiven /ˌʌnfəˈgɪv(ə)n/ *adj.* not forgiven.

unforgiving /ˌʌnfəˈgɪvɪŋ/ *adj.* not forgiving. □ **unforgivingly** *adv.* **unforgivingness** *n.*

unforgotten /ˌʌnfəˈgɒt(ə)n/ *adj.* not forgotten.

unformed /ʌnˈfɔːmd/ *adj.* **1** not formed. **2** shapeless. **3** not developed.

unformulated /ʌnˈfɔːmjʊˌleɪtɪd/ *adj.* not formulated.

unforthcoming /ˌʌnfɔːθˈkʌmɪŋ/ *adj.* not forthcoming.

unfortified /ʌnˈfɔːtɪˌfaɪd/ *adj.* not fortified.

unfortunate /ʌnˈfɔːtjʊnət, -tʃənət/ *adj. & n.* ● *adj.* **1** having bad fortune; unlucky. **2** unhappy. **3** regrettable. **4** disastrous. ● *n.* an unfortunate person.

unfortunately /ʌnˈfɔːtjʊnətlɪ, -tʃənətlɪ/ *adv.* **1** (qualifying a whole sentence) it is unfortunate that. **2** in an unfortunate manner.

unfounded /ʌnˈfaʊndɪd/ *adj.* having no foundation (*unfounded hopes*; *unfounded rumour*). □ **unfoundedly** *adv.* **unfoundedness** *n.*

UNFPA *abbr.* United Nations Fund for Population Activities.

unframed /ʌnˈfreɪmd/ *adj.* (esp. of a picture) not framed.

unfree /ʌnˈfriː/ *adj.* deprived or devoid of liberty. □ **unfreedom** *n.*

unfreeze /ʌnˈfriːz/ *v.* (*past* **unfroze** /-ˈfrəʊz/; *past part.* **unfrozen** /-ˈfrəʊz(ə)n/) **1** *tr.* cause to thaw. **2** *intr.* thaw. **3** *tr.* remove restrictions from, make (assets, credits, etc.) realizable.

unfrequented /ˌʌnfrɪˈkwentɪd/ *adj.* not frequented.

unfriended /ʌnˈfrendɪd/ *adj. literary* without friends.

unfriendly /ʌnˈfrendlɪ/ *adj.* (**unfriendlier**, **unfriendliest**) not friendly. □ **unfriendliness** *n.*

unfrock /ʌnˈfrɒk/ *v.tr.* = DEFROCK.

unfroze *past* of UNFREEZE.

unfrozen *past part.* of UNFREEZE.

unfruitful /ʌnˈfruːtfʊl/ *adj.* **1** not producing good results, unprofitable. **2** not producing fruit or crops. □ **unfruitfully** *adv.* **unfruitfulness** *n.*

unfulfilled /ˌʌnfʊlˈfɪld/ *adj.* not fulfilled. □ **unfulfillable** *adj.*

unfunded /ʌnˈfʌndɪd/ *adj.* (of a debt) not funded.

unfunny /ʌnˈfʌnɪ/ *adj.* (**unfunnier**, **unfunniest**) not amusing (though meant to be). □ **unfunnily** *adv.* **unfunniness** *n.*

unfurl /ʌnˈfɜːl/ *v.* **1** *tr.* spread out (a sail, umbrella, etc.). **2** *intr.* become spread out.

unfurnished /ʌnˈfɜːnɪʃt/ *adj.* **1** (usu. foll. by *with*) not supplied. **2** without furniture.

unfussy /ʌnˈfʌsɪ/ *adj.* not fussy.

ungainly /ʌnˈgeɪnlɪ/ *adj.* (of a person, animal, or movement) awkward, clumsy. □ **ungainliness** *n.* [UN-¹ + obs. *gainly* graceful ult. f. ON *gegn* straight]

ungallant /ʌnˈgælənt/ *adj.* not gallant. □ **ungallantly** *adv.*

ungenerous /ʌnˈdʒenərəs/ *adj.* not generous; mean. □ **ungenerously** *adv.* **ungenerousness** *n.*

ungenial /ʌnˈdʒiːnɪəl/ *adj.* not genial.

ungentle /ʌnˈdʒent(ə)l/ *adj.* not gentle. □ **ungentleness** *n.* **ungently** *adv.*

ungentlemanly /ʌnˈdʒent(ə)lmənlɪ/ *adj.* not gentlemanly, unsporting. □ **ungentlemanliness** *n.*

unget-at-able /ˌʌngetˈætəb(ə)l/ *adj. colloq.* inaccessible.

ungifted /ʌnˈgɪftɪd/ *adj.* not gifted or talented.

ungird /ʌnˈgɜːd/ v.tr. **1** release the girdle, belt, or girth of. **2** release or take off by undoing a belt or girth.

unglamorous /ʌnˈglæmərəs/ adj. **1** lacking glamour or attraction. **2** mundane.

unglazed /ʌnˈgleɪzd/ adj. not glazed.

ungloved /ʌnˈglʌvd/ adj. not wearing a glove or gloves.

ungodly /ʌnˈgɒdlɪ/ adj. **1** impious, wicked. **2** colloq. outrageous (an ungodly hour to arrive). □ **ungodliness** n.

ungovernable /ʌnˈgʌvənəb(ə)l/ adj. uncontrollable, violent. □ **ungovernably** adv. **ungovernability** /-ˌgʌvənəˈbɪlɪtɪ/ n.

ungraceful /ʌnˈgreɪsfʊl/ adj. not graceful. □ **ungracefully** adv. **ungracefulness** n.

ungracious /ʌnˈgreɪʃəs/ adj. **1** not kindly or courteous; unkind. **2** unattractive. □ **ungraciously** adv. **ungraciousness** n.

ungrammatical /ˌʌngrəˈmætɪk(ə)l/ adj. contrary to the rules of grammar. □ **ungrammatically** adv. **ungrammaticalness** n. **ungrammaticality** /-ˌmætɪˈkælɪtɪ/ n.

ungraspable /ʌnˈgrɑːspəb(ə)l/ adj. that cannot be grasped or comprehended.

ungrateful /ʌnˈgreɪtfʊl/ adj. **1** not feeling or showing gratitude. **2** not pleasant or acceptable. □ **ungratefully** adv. **ungratefulness** n.

ungrounded /ʌnˈgraʊndɪd/ adj. **1** having no basis or justification; unfounded. **2** Electr. not earthed. **3** (foll. by in a subject) not properly instructed. **4** (of an aircraft, ship, etc.) no longer grounded.

ungrudging /ʌnˈgrʌdʒɪŋ/ adj. not grudging. □ **ungrudgingly** adv.

ungual /ˈʌŋgwəl/ adj. of, like, or bearing a nail, hoof, or claw. [L UNGUIS]

unguard /ʌnˈgɑːd/ v.tr. Cards discard a low card that was protecting (a high card) from capture.

unguarded /ʌnˈgɑːdɪd/ adj. **1** incautious, thoughtless (an unguarded remark). **2** not guarded; without a guard. □ **in an unguarded moment** unawares. □ **unguardedly** adv. **unguardedness** n.

unguent /ˈʌŋgwənt/ n. a soft substance used as ointment or for lubrication. [L unguentum f. unguere anoint]

unguessable /ʌnˈgesəb(ə)l/ adj. that cannot be guessed or imagined.

unguiculate /ʌŋˈgwɪkjʊlət/ adj. **1** Zool. having one or more nails or claws. **2** Bot. (of petals) having an unguis. [mod.L unguiculatus f. unguiculus dimin. of UNGUIS]

unguided /ʌnˈgaɪdɪd/ adj. not guided in a particular path or direction; left to take its own course.

unguis /ˈʌŋgwɪs/ n. (pl. **ungues** /-wiːz/) **1** Bot. the narrow base of a petal. **2** Zool. a nail or claw. [L]

ungula /ˈʌŋgjʊlə/ n. (pl. **ungulae** /-ˌliː/) a hoof or claw. [L, dimin. of UNGUIS]

ungulate /ˈʌŋgjʊlət, -ˌleɪt/ adj. & n. Zool. ● adj. hoofed. ● n. a hoofed mammal. The ungulates are now placed in two orders, the Perissodactyla or odd-toed ungulates, which include the horses, tapirs, and rhinoceroses, and the Artiodactyla or even-toed ungulates, which include the pigs, hippos, camels, and ruminants. [LL ungulatus f. UNGULA]

unhallowed /ʌnˈhæləʊd/ adj. **1** not consecrated. **2** not sacred; unholy, wicked.

unhampered /ʌnˈhæmpəd/ adj. not hampered.

unhand /ʌnˈhænd/ v.tr. rhet. or joc. **1** take one's hands off (a person). **2** release from one's grasp.

unhandsome /ʌnˈhænsəm/ adj. not handsome.

unhandy /ʌnˈhændɪ/ adj. **1** not easy to handle or manage; awkward. **2** not skilful in using the hands. □ **unhandily** adv. **unhandiness** n.

unhang /ʌnˈhæŋ/ v.tr. (past and past part. **unhung** /-ˈhʌŋ/) take down from a hanging position.

unhappy /ʌnˈhæpɪ/ adj. (**unhappier, unhappiest**) **1** not happy, miserable. **2** unsuccessful, unfortunate. **3** causing misfortune. **4** disastrous. **5** inauspicious. □ **unhappily** adv. **unhappiness** n.

unharbour /ʌnˈhɑːbə(r)/ v.tr. Brit. dislodge (a deer) from a covert.

unharmed /ʌnˈhɑːmd/ adj. not harmed.

unharmful /ʌnˈhɑːmfʊl/ adj. not harmful.

unharmonious /ˌʌnhɑːˈməʊnɪəs/ adj. not harmonious.

unharness /ʌnˈhɑːnɪs/ v.tr. remove a harness from.

unhasp /ʌnˈhɑːsp/ v.tr. free from a hasp or catch; unfasten.

unhatched /ʌnˈhætʃt/ adj. (of an egg etc.) not hatched.

UNHCR abbr. United Nations High Commission for Refugees, an agency of the United Nations set up in 1951 to aid, protect, and monitor refugees. Its headquarters are in Geneva.

unhealed /ʌnˈhiːld/ adj. not yet healed.

unhealthful /ʌnˈhelθfʊl/ adj. harmful to health, unwholesome. □ **unhealthfulness** n.

unhealthy /ʌnˈhelθɪ/ adj. (**unhealthier, unhealthiest**) **1** not in good health. **2 a** (of a place etc.) harmful to health. **b** unwholesome. **c** sl. dangerous to life. □ **unhealthily** adv. **unhealthiness** n.

unheard /ʌnˈhɜːd/ adj. **1** not heard. **2** (usu. **unheard-of**) unprecedented, unknown.

unheated /ʌnˈhiːtɪd/ adj. not heated.

unhedged /ʌnˈhedʒd/ adj. **1** not bounded by a hedge. **2** (of a speculative investment) not hedged.

unheeded /ʌnˈhiːdɪd/ adj. not heeded; disregarded.

unheedful /ʌnˈhiːdfʊl/ adj. heedless; taking no notice.

unheeding /ʌnˈhiːdɪŋ/ adj. not giving heed; heedless. □ **unheedingly** adv.

unhelpful /ʌnˈhelpfʊl/ adj. not helpful. □ **unhelpfully** adv. **unhelpfulness** n.

unheralded /ʌnˈherəldɪd/ adj. not heralded; unannounced.

unheroic /ˌʌnhɪˈrəʊɪk/ adj. not heroic. □ **unheroically** adv.

unhesitating /ʌnˈhezɪˌteɪtɪŋ/ adj. without hesitation, immediate. □ **unhesitatingly** adv. **unhesitatingness** n.

unhindered /ʌnˈhɪndəd/ adj. not hindered.

unhinge /ʌnˈhɪndʒ/ v.tr. **1** take (a door etc.) off its hinges. **2** (esp. as **unhinged** adj.) unsettle or disorder (a person's mind etc.), make (a person) crazy.

unhip /ʌnˈhɪp/ adj. sl. unaware of current fashions.

unhistoric /ˌʌnhɪˈstɒrɪk/ adj. not historic or historical.

unhistorical /ˌʌnhɪˈstɒrɪk(ə)l/ adj. not historical. □ **unhistorically** adv.

unhitch /ʌnˈhɪtʃ/ v.tr. **1** release from a hitched state. **2** unhook, unfasten.

unholy /ʌnˈhəʊlɪ/ adj. (**unholier, unholiest**) **1** impious, profane, wicked. **2** colloq. dreadful, outrageous (made an unholy row about it). **3** not holy. □ **unholiness** n. [OE unhālig (as UN-¹, HOLY)]

unhonoured /ʌnˈɒnəd/ adj. not honoured.

unhood /ʌnˈhʊd/ v.tr. remove the hood from (a falcon, horse, captive, etc.).

unhook /ʌnˈhʊk/ v.tr. **1** remove from a hook or hooks. **2** unfasten by releasing a hook or hooks.

unhoped /ʌnˈhəʊpt/ adj. (foll. by for) not hoped for or expected.

unhorse /ʌnˈhɔːs/ v.tr. **1** throw or drag from a horse. **2** (of a horse) throw (a rider). **3** dislodge, overthrow.

unhouse /ʌnˈhaʊz/ v.tr. deprive of shelter; turn out of a house.

unhuman /ʌnˈhjuːmən/ adj. **1** not human. **2** superhuman. **3** inhuman, brutal.

unhung¹ /ʌnˈhʌŋ/ adj. **1** not (yet) executed by hanging. **2** not hung up (for exhibition).

unhung² past and past part. of UNHANG.

unhurried /ʌnˈhʌrɪd/ adj. not hurried. □ **unhurriedly** adv.

unhurt /ʌnˈhɜːt/ adj. not hurt.

unhusk /ʌnˈhʌsk/ v.tr. remove a husk or shell from.

unhygienic /ˌʌnhaɪˈdʒiːnɪk/ adj. not hygienic. □ **unhygienically** adv.

unhyphenated /ʌnˈhaɪfəˌneɪtɪd/ adj. not hyphenated.

uni /ˈjuːnɪ/ n. (pl. **unis**) esp. Brit., Austral., & NZ colloq. a university. [abbr.]

uni- /ˈjuːnɪ/ comb. form one; having or consisting of one. [L f. unus one]

Uniat /ˈjuːnɪˌæt, -nɪət/ adj. & n. (also **Uniate** /-nɪˌeɪt/) ● adj. of or relating to any of the Churches in eastern Europe and the Middle East that acknowledge papal supremacy but retain their own language, rites, and canon law. ● n. a member of such a Church. [Russ. uniyat f. uniya f. L unio UNION]

uniaxial /ˌjuːnɪˈæksɪəl/ adj. having a single axis. □ **uniaxially** adv.

unicameral /ˌjuːnɪˈkæmərəl/ adj. with a single legislative chamber.

UNICEF /ˈjuːnɪˌsef/ abbr. United Nations Children's (originally International Children's Emergency) Fund, an agency of the United Nations established in 1946 to help governments (especially in developing countries) improve the health and education of children and their mothers. Its headquarters are in New York.

unicellular /ˌjuːnɪˈseljʊlə(r)/ adj. (of an organism, organ, tissue, etc.) consisting of a single cell.

unicolour /ˈjuːnɪˌkʌlə(r)/ adj. (also **unicoloured**) of one colour.

unicorn /ˈjuːnɪˌkɔːn/ n. **1 a** a mythical animal usually represented as having the body of a horse with a single straight horn, projecting from its forehead, that was reputed to have medicinal or magical properties. **b** a heraldic representation of this, with a twisted horn, a deer's feet, a goat's beard, and a lion's tail, esp. as a supporter of the royal arms of Great Britain or Scotland. **2 a** a pair of horses and a third horse in front. **b** an equipage with these. [ME f. OF *unicorne* f. L *unicornis* f. UNI- + *cornu* horn, transl. Gk *monocerōs*]

unicuspid /juːnɪˈkʌspɪd/ adj. & n. ● adj. with one cusp. ● n. a unicuspid tooth.

unicycle /ˈjuːnɪˌsaɪk(ə)l/ n. a single-wheeled cycle, esp. as used by acrobats. □ **unicyclist** n.

unidea'd /ˌʌnaɪˈdɪəd/ adj. having no ideas.

unideal /ˌʌnaɪˈdiːl/ adj. not ideal.

unidealized /ˌʌnaɪˈdɪəlaɪzd/ adj. free from idealism; not idealized.

unidentifiable /ˌʌnaɪˈdentɪˌfaɪəb(ə)l/ adj. unable to be identified.

unidentified /ˌʌnaɪˈdentɪˌfaɪd/ adj. not identified.

unidimensional /ˌjuːnɪdaɪˈmenʃ(ə)n(ə)l, ˌjuːnɪdɪ-/ adj. having (only) one dimension.

unidirectional /ˌjuːnɪdaɪˈrekʃ(ə)n(ə)l, ˌjuːnɪdɪ-/ adj. having only one direction of motion, operation, etc. □ **unidirectionally** adv. **unidirectionality** /-ˌrekʃəˈnælɪtɪ/ n.

UNIDO /juːˈniːdəʊ/ abbr. United Nations Industrial Development Organization.

unification /ˌjuːnɪfɪˈkeɪʃ(ə)n/ n. the act or an instance of unifying; the state of being unified. □ **unificatory** /ˈjuːnɪfɪˌkeɪtərɪ/ adj.

Unification Church the Holy Spirit Association for the Unification of World Christianity, an evangelical religious and political organization founded in 1954 in Korea by Sun Myung Moon. The Church moved its headquarters to New York State in 1973, and became known as a cult religion in the US and elsewhere. Moon claims Messianic status, believing that he was chosen by God to save humankind from Satanism and regarding Communists as Satan's representatives. Much controversy has been aroused by the organization's fund-raising techniques and by the methods it uses to draw young people into the movement.

uniflow /ˈjuːnɪˌfləʊ/ adj. involving flow (esp. of steam or waste gases) in one direction only.

uniform /ˈjuːnɪˌfɔːm/ adj., n., & v. ● adj. **1** not changing in form or character; the same, unvarying (*present a uniform appearance; all of uniform size and shape*). **2** conforming to the same standard, rules, or pattern. **3** constant in the course of time (*uniform acceleration*). **4** (of a tax, law, etc.) not varying with time or place. ● n. uniform distinctive clothing worn by members of the same body, e.g. by soldiers, police, and schoolchildren. ● v.tr. **1** clothe in uniform (*a uniformed officer*). **2** make uniform. □ **uniformly** adv. [F *uniforme* or L *uniformis* (as UNI-, FORM)]

uniformitarian /ˌjuːnɪˌfɔːmɪˈteərɪən/ adj. & n. ● adj. of or relating to the theory that geological processes are always due to continuously and uniformly operating forces. ● n. a holder of this theory. □ **uniformitarianism** n.

uniformity /ˌjuːnɪˈfɔːmɪtɪ/ n. (pl. **-ies**) **1** being uniform; sameness, consistency. **2** an instance of this. [ME f. OF *uniformité* or LL *uniformitas* (as UNIFORM)]

Uniformity, Act of (in British history) any of various pieces of legislation intended to secure uniformity in public worship and use of a particular Book of Common Prayer. The first such Acts, establishing the foundations of the English Protestant Church, were passed in the reign of Edward VI but repealed under his Catholic successor Mary I; a third was passed in the reign of Elizabeth I, and a final Act in 1662 after the Restoration.

unify /ˈjuːnɪˌfaɪ/ v.tr. (also *absol.*) (**-ies, -ied**) reduce to unity or uniformity. □ **unified field theory** Physics a theory that seeks to explain all the field phenomena (e.g. gravitation and electromagnetism: see FIELD n. 9) formerly treated by separate theories. □ **unifier** n. [F *unifier* or LL *unificare* (as UNI-, -FY)]

unilateral /ˌjuːnɪˈlætərəl/ adj. **1** performed by or affecting only one person or party (*unilateral disarmament; unilateral declaration of independence*). **2** one-sided. **3** (of the parking of vehicles) restricted to one side of the street. **4** (of leaves) all on the same side of the stem. **5** (of

a line of descent) through ancestors of one sex only. □ **unilaterally** adv.

Unilateral Declaration of Independence (abbr. **UDI**) the declaration of independence from the United Kingdom made by Rhodesia under Ian Smith in 1965 (see ZIMBABWE).

unilateralism /ˌjuːnɪˈlætərəˌlɪz(ə)m/ n. **1** unilateral disarmament. **2** US the pursuit of a foreign policy without allies. □ **unilateralist** n. & adj.

unilingual /ˌjuːnɪˈlɪŋgwəl/ adj. of or in only one language. □ **unilingually** adv.

uniliteral /ˌjuːnɪˈlɪtərəl/ adj. consisting of one letter.

unilluminated /ˌʌnɪˈluːmɪˌneɪtɪd, ˌʌnɪˈljuːmɪ-/ adj. not illuminated.

unillustrated /ʌnˈɪləˌstreɪtɪd/ adj. (esp. of a book) without illustrations.

unilocular /ˌjuːnɪˈlɒkjʊlə(r)/ adj. Bot. & Zool. single-chambered.

unimaginable /ˌʌnɪˈmædʒɪnəb(ə)l/ adj. impossible to imagine. □ **unimaginably** adv.

unimaginative /ˌʌnɪˈmædʒɪnətɪv/ adj. lacking imagination; stolid, dull. □ **unimaginatively** adv. **unimaginativeness** n.

unimpaired /ˌʌnɪmˈpeəd/ adj. not impaired.

unimpassioned /ˌʌnɪmˈpæʃ(ə)nd/ adj. not impassioned.

unimpeachable /ˌʌnɪmˈpiːtʃəb(ə)l/ adj. giving no opportunity for censure; beyond reproach or question. □ **unimpeachably** adv.

unimpeded /ˌʌnɪmˈpiːdɪd/ adj. not impeded. □ **unimpededly** adv.

unimportance /ˌʌnɪmˈpɔːt(ə)ns/ n. lack of importance.

unimportant /ˌʌnɪmˈpɔːt(ə)nt/ adj. not important.

unimposing /ˌʌnɪmˈpəʊzɪŋ/ adj. unimpressive. □ **unimposingly** adv.

unimpressed /ˌʌnɪmˈprest/ adj. not impressed.

unimpressionable /ˌʌnɪmˈpreʃənəb(ə)l/ adj. not impressionable.

unimpressive /ˌʌnɪmˈpresɪv/ adj. not impressive. □ **unimpressively** adv. **unimpressiveness** n.

unimproved /ˌʌnɪmˈpruːvd/ adj. **1** not made better. **2** not made use of. **3** (of land) not used for agriculture or building; not developed.

unincorporated /ˌʌnɪnˈkɔːpəˌreɪtɪd/ adj. **1** not incorporated or united. **2** not formed into a corporation.

uninfected /ˌʌnɪnˈfektɪd/ adj. not infected.

uninflamed /ˌʌnɪnˈfleɪmd/ adj. not inflamed.

uninflammable /ˌʌnɪnˈflæməb(ə)l/ adj. not inflammable.

uninflected /ˌʌnɪnˈflektɪd/ adj. **1** Gram. (of a language) not having inflections. **2** not changing or varying. **3** not bent or deflected.

uninfluenced /ʌnˈɪnflʊənst/ adj. (often foll. by by) not influenced.

uninfluential /ˌʌnɪnflʊˈenʃ(ə)l/ adj. having little or no influence.

uninformative /ˌʌnɪnˈfɔːmətɪv/ adj. not informative; giving little information.

uninformed /ˌʌnɪnˈfɔːmd/ adj. **1** not informed or instructed. **2** ignorant, uneducated.

uninhabitable /ˌʌnɪnˈhæbɪtəb(ə)l/ adj. that cannot be inhabited. □ **uninhabitableness** n.

uninhabited /ˌʌnɪnˈhæbɪtɪd/ adj. not inhabited.

uninhibited /ˌʌnɪnˈhɪbɪtɪd/ adj. not inhibited. □ **uninhibitedly** adv. **uninhibitedness** n.

uninitiated /ˌʌnɪˈnɪʃɪˌeɪtɪd/ adj. not initiated; not admitted or instructed.

uninjured /ʌnˈɪndʒəd/ adj. not injured.

uninspired /ˌʌnɪnˈspaɪəd/ adj. **1** not inspired. **2** (of oratory etc.) commonplace.

uninspiring /ˌʌnɪnˈspaɪərɪŋ/ adj. not inspiring. □ **uninspiringly** adv.

uninstructed /ˌʌnɪnˈstrʌktɪd/ adj. not instructed or informed.

uninsulated /ʌnˈɪnsjʊˌleɪtɪd/ adj. not insulated.

uninsurable /ˌʌnɪnˈʃʊərəb(ə)l/ adj. that cannot be insured.

uninsured /ˌʌnɪnˈʃʊəd/ adj. not insured.

unintelligent /ˌʌnɪnˈtelɪdʒənt/ adj. not intelligent. □ **unintelligently** adv.

unintelligible /ˌʌnɪnˈtelɪdʒɪb(ə)l/ adj. not intelligible. □ **unintelligibly** adv. **unintelligibleness** n. **unintelligibility** /-ˌtelɪdʒɪˈbɪlɪtɪ/ n.

unintended /ˌʌnɪnˈtendɪd/ adj. not intended.

unintentional /ˌʌnɪnˈtenʃ(ə)n(ə)l/ adj. not intentional, unintended. □ **unintentionally** adv.

uninterested /ʌnˈɪntrəstɪd, -trɪstɪd/ adj. **1** not interested.

2 unconcerned, indifferent. □ **uninterestedly** *adv.* **uninterestedness** *n.*

uninteresting /ʌnˈɪntrəstɪŋ, -trɪstɪŋ/ *adj.* not interesting. □ **uninterestingly** *adv.* **uninterestingness** *n.*

uninterpretable /ˌʌnɪnˈtɜːprɪtəb(ə)l/ *adj.* that cannot be interpreted.

uninterrupted /ˌʌnɪntəˈrʌptɪd/ *adj.* not interrupted, continuous. □ **uninterruptedly** *adv.* **uninterruptedness** *n.*

uninterruptible /ˌʌnɪntəˈrʌptɪb(ə)l/ *adj.* that cannot be interrupted.

uninucleate /ˌjuːnɪˈnjuːklɪˌeɪt, -lɪət/ *adj. Biol.* having a single nucleus.

uninventive /ˌʌnɪnˈventɪv/ *adj.* not inventive. □ **uninventively** *adv.* **uninventiveness** *n.*

uninvestigated /ˌʌnɪnˈvestɪˌɡeɪtɪd/ *adj.* not investigated.

uninvited /ˌʌnɪnˈvaɪtɪd/ *adj.* not invited. □ **uninvitedly** *adv.*

uninviting /ˌʌnɪnˈvaɪtɪŋ/ *adj.* not inviting, unattractive, repellent. □ **uninvitingly** *adv.*

uninvoked /ˌʌnɪnˈvəʊkt/ *adj.* not invoked.

uninvolved /ˌʌnɪnˈvɒlvd/ *adj.* not involved.

union /ˈjuːnjən, -nɪən/ *n.* **1** the act or an instance of uniting; the state of being united. **2 a** a whole resulting from the combination of parts or members. **b** (**Union**) a political unit formed in this way, esp. the US, the UK, or South Africa. (See also UNION, THE.) **3** = TRADE UNION. **4** marriage, matrimony. **5** concord, agreement (*lived together in perfect union*). **6** (**Union**) **a** a general social club and debating society at some universities and colleges. **b** the buildings or accommodation of such a society. **7** *Math.* the totality of the members of two or more sets. **8** *Brit. hist.* **a** two or more parishes consolidated for the administration of the Poor Laws. **b** (in full **union workhouse**) a workhouse erected by this. **9** *Brit.* an association of independent (esp. Congregational or Baptist) churches for purposes of cooperation. **10** a part of a flag with a device emblematic of union, normally occupying the upper corner next to the staff. **11** a joint or coupling for pipes etc. **12** a fabric of mixed materials, e.g. cotton with linen or silk. □ **union-bashing** *Brit. colloq.* active opposition to trade unions and their rights. **union catalogue** a catalogue of the combined holdings of several libraries. **union down** (of a flag) hoisted with the union below as a signal of distress. **union shop** a shop, factory, trade, etc., in which employees must belong to a trade union or join one within an agreed time. **union suit** *N. Amer.* a single undergarment for the body and legs; combinations. [ME f. OF *union* or eccl.L *unio* unity f. L *unus* one]

Union, Act of (in British history) either of the parliamentary acts by which the countries of the United Kingdom were brought together as a political whole. By the first Act of Union (1707) Scotland was joined with England to form Great Britain, with Scotland losing its Parliament (the crowns of the two countries had been united in 1603). The second Act of Union, in 1801, established the United Kingdom of Great Britain and Ireland and abolished the free Protestant Parliament of Ireland. Wales had been incorporated with England, and given parliamentary representation, in 1536.

Union, the *hist.* **1** (also called *the Federal Union*) **a** the United States, esp. from its founding by the original thirteen states in 1787–90 to the secession of the Confederate states in 1860–1. **b** the twenty-three northern states of the US which opposed the seceding Confederate states in the American Civil War. **2** the union of the English and Scottish crowns, of the English and Scottish Parliaments, or of the British and Irish Parliaments (see UNION, ACT OF).

Union flag see UNION JACK.

unionist /ˈjuːnjənɪst, ˈjuːnɪə-/ *n.* **1 a** a member of a trade union. **b** an advocate of trade unions. **2** (usu. **Unionist**) an advocate of union, esp.: **a** (in British politics) a person (usu. Protestant) advocating or supporting maintenance of the political union between Great Britain and Ireland (later, Northern Ireland); also, a member of a political party having these aims, esp. the Ulster Unionist Council or the Ulster Democratic Unionist Party; *hist.* a Liberal Unionist. **b** *US* a supporter or advocate of the Federal Union of the United States of America, especially one who during the Civil War (1861–5) was opposed to secession. □ **unionism** *n.* **unionistic** /-ˈnɪstɪk/ *adj.*

unionize /ˈjuːnjəˌnaɪz, ˈjuːnɪə-/ *v.tr. & intr.* (also **-ise**) bring or come under trade-union organization or rules. □ **unionization** /ˌjuːnjənaɪˈzeɪʃ(ə)n, ˌjuːnɪə-/ *n.*

un-ionized /ʌnˈaɪəˌnaɪzd/ *adj.* (also **-ised**) not ionized.

Union Jack 1 (also **Union flag**) the national flag or ensign of the United Kingdom (formerly of Great Britain). (*See note below.*) **2** (**union jack**) *US* a jack consisting of the union from the national flag.

▪ This flag (originally and more properly the *Union flag*) was formed by combining the crosses of the three patron saints St George, St Andrew, and St Patrick, retaining the blue ground of the banner of St Andrew. The flags of St George and St Andrew were first combined to symbolize the union of the crowns of England and Scotland (1603), the cross of St Patrick being added on the union of the Parliaments of Great Britain and Ireland (1801). Originally the term *Union Jack* denoted a small British Union flag flown as the jack of a ship, in later and more extended use coming to denote any size or adaptation of the Union flag.

Union of Soviet Socialist Republics (abbr. **USSR**) the full name of the former SOVIET UNION.

Union Territory any of several territories of India which are administered by the central government.

uniparous /juːˈnɪpərəs/ *adj.* **1** *Zool.* producing one offspring at a birth. **2** *Bot.* having one axis or branch.

uniped /ˈjuːnɪˌped/ *n. & adj.* ● *n.* a person having only one foot or leg. ● *adj.* one-footed, one-legged. [UNI- + *pes pedis* foot]

unipersonal /ˌjuːnɪˈpɜːsən(ə)l/ *adj.* (of the Deity) existing only as one person.

uniplanar /ˌjuːnɪˈpleɪnə(r)/ *adj.* lying in one plane.

unipod /ˈjuːnɪˌpɒd/ *n.* a one-legged support for a camera etc. [UNI-, after TRIPOD]

unipolar /ˌjuːnɪˈpəʊlə(r)/ *adj.* **1** (esp. of an electric or magnetic apparatus) showing only one kind of polarity. **2** *Biol.* (of a nerve cell etc.) having only one pole or process. □ **unipolarity** /-pəˈlærɪti/ *n.*

unique /juːˈniːk/ *adj. & n.* ● *adj.* **1** of which there is only one; unequalled; having no like, equal, or parallel (*his position was unique; this vase is considered unique*). **2** *disp.* unusual, remarkable (*the most unique man I ever met*). ● *n.* a unique thing or person. □ **uniquely** *adv.* **uniqueness** *n.* [F f. L *unicus* f. *unus* one]

unironed /ʌnˈaɪənd/ *adj.* (esp. of clothing, linen, etc.) not ironed.

uniserial /ˌjuːnɪˈsɪərɪəl/ *adj. Bot. & Zool.* arranged in one row.

unisex /ˈjuːnɪˌseks/ *adj.* (of clothing, hairstyles, etc.) designed to be suitable for both sexes.

unisexual /ˌjuːnɪˈseksjʊəl, -ˈsekʃʊəl/ *adj.* **1 a** of one sex. **b** *Bot.* having stamens or pistils but not both. **2** unisex. □ **unisexually** *adv.* **unisexuality** /-ˌseksjʊˈælɪti, -ˌsekʃʊ-/ *n.*

Unison /ˈjuːnɪs(ə)n/ (in the UK) a trade union for workers in the public services, formed in July 1993 from an amalgamation of COHSE, NALGO, and NUPE.

unison /ˈjuːnɪs(ə)n/ *n. & adj.* ● *n.* **1** *Mus.* **a** a coincidence in pitch of sounds or notes. **b** this regarded as an interval. **2** *Mus.* a combination of voices or instruments at the same pitch or at pitches differing by one or more octaves (*sang in unison*). **3** agreement, concord (*acted in perfect unison*). ● *adj. Mus.* coinciding in pitch. □ **unison string** a string tuned in unison with another string and meant to be sounded with it. □ **unisonant** /juːˈnɪsənənt/ *adj.* **unisonous** *adj.* [OF *unison* or LL *unisonus* (as UNI-, *sonus* SOUND¹)]

unissued /ʌnˈɪʃuːd, -ˈɪsjuːd/ *adj.* not issued.

unit /ˈjuːnɪt/ *n.* **1 a** an individual thing, person, or group regarded as single and complete, esp. for purposes of calculation. **b** each of the separate individuals or (smallest) groups into which a complex whole may be analysed (*the family as the unit of society*). **2** a quantity chosen as a standard in terms of which other quantities may be expressed (*unit of heat; SI unit; mass per unit volume*). **3** *Brit.* the smallest share in a unit trust. **4** a device with a specified function forming part of a complex mechanism. **5** a piece of furniture for fitting with others like it or made of complementary parts. **6** a group with a special function in an organization. **7** a group of buildings, wards, etc., in a hospital. **8** the number 'one'. □ **unit cell** *Crystallog.* the smallest repeating group of atoms, ions, or molecules in a crystal. **unit cost** the cost of producing one item of manufacture. **unit-holder** *Brit.* a person with a holding in a unit trust. **unit-linked** *Brit.* (of a policy, personal equity plan, etc.) whose return is linked to the rise and fall of the price of units bought in a choice of investment funds. **unit price** the price charged for each unit of goods supplied. **unit trust** *Brit.* an investment company investing combined contributions from many persons in various securities and paying them dividends (calculated on the average return of the securities) in proportion to their holdings. [L *unus*, prob. after DIGIT]

UNITA /juːˈniːtə/ União Nacional para a Independencia Total de Angola, National Union for the Total Independence of Angola, an Angolan nationalist movement founded in 1966 by Jonas Savimbi (b.1934) to fight Portuguese rule. After independence was achieved in 1975 UNITA, still led by Savimbi, continued to fight against the ruling Marxist MPLA, with help from South Africa. [Port., acronym]

UNITAR /juːˈniːtɑː(r)/ *abbr.* United Nations Institute for Training and Research.

Unitarian /ˌjuːnɪˈteərɪən/ *n. & adj.* ● *n.* **1** a person who believes that God is not a Trinity but one being. **2** a member of a religious body maintaining this and advocating freedom from formal dogma or doctrine. (*See note below.*) ● *adj.* of or relating to the Unitarians. □ **Unitarianism** *n.* [mod.L *unitarius* f. L *unitas* UNITY]

▪ Unitarians have no formal creed; originally their teaching was based on scriptural authority, but now reason and conscience have become their criteria for belief and practice. A similar position was held in the early Church, but modern Unitarianism dates historically from the Reformation era; organized Unitarian communities became established in the 16th–17th centuries in Poland, Hungary, and England, while in the 20th century Unitarianism has been particularly significant in the US.

unitary /ˈjuːnɪtəri/ *adj.* **1** of a unit or units. **2** marked by unity or uniformity. □ **unitarily** *adv.* **unitarity** /ˌjuːnɪˈtærɪti/ *n.*

unite /juːˈnaɪt/ *v.* **1** *tr. & intr.* join together; make or become one; combine. **2** *tr. & intr.* join together for a common purpose or action (*united in their struggle against injustice*). **3** *tr. & intr.* join in marriage. **4** *tr.* possess (qualities, features, etc.) in combination (*united anger with mercy*). **5** *intr. & tr.* form or cause to form a physical or chemical whole (*oil will not unite with water*). □ **unitive** /ˈjuːnɪtɪv/ *adj.* **unitively** *adv.* [ME f. L *unire* unit- f. *unus* one]

united /juːˈnaɪtɪd/ *adj.* **1** that has united or been united. **2 a** of or produced by two more persons or things in union; joint. **b** resulting from the union of two or more parts (esp. in the names of churches, societies, and football clubs). **3** in agreement, of like mind. □ **unitedly** *adv.*

United Arab Emirates (abbr. **UAE**) an independent state on the south coast of the Persian Gulf, west of the Gulf of Oman; population (est. 1991) 1,630,000; official language, Arabic; capital, Abu Dhabi. The United Arab Emirates was formed in 1971 by the federation of the independent sheikhdoms formerly called the Trucial States. The member states are Abu Dhabi, Ajman, Dubai, Fujairah, Ras al Khaimah (which joined early in 1972), Sharjah, and Umm al Qaiwain.

United Arab Republic (abbr. **UAR**) a former political union established by Egypt and Syria in 1958. It was seen as the first step towards the creation of a pan-Arab union in the Middle East, but only Yemen (1958–66) entered into loose association with it and Syria withdrew following a coup in 1961. Egypt retained the name United Arab Republic until 1971.

United Artists a US film production company founded in 1919 by actors Charlie Chaplin, Douglas Fairbanks, and Mary Pickford and director D. W. Griffith. Formed to make films without the artistic strictures of the larger companies, United Artists owned no studios or cinemas. After silent films like Chaplin's *The Gold Rush* (1925) the company went on to produce many literary adaptations in the 1930s and 1940s; in the 1960s it produced *The Magnificent Seven* (1960), *West Side Story* (1961), and the James Bond series.

United Brethren *n.pl. Eccl.* the Moravians.

United Free Church a Presbyterian Church in Scotland formed in 1900 by the union of the Free Church of Scotland with the United Presbyterian Church. In 1929 the majority of the new Church joined with the established Church of Scotland.

United Kingdom (abbr. **UK**) (in full **United Kingdom of Great Britain and Northern Ireland**) a country of western Europe consisting of England, Wales, Scotland, and Northern Ireland; pop. (1991), 55,700,000; official language, English; capital, London. England (which had incorporated Wales in the 16th century) and Scotland have had the same monarch since 1603, when James VI of Scotland succeeded to the English crown as James I; the kingdoms were formally united by the Act of Union in 1707. An Act of Parliament joined Great Britain and Ireland in 1801, but the Irish Free State (later the Republic of Ireland) withdrew in 1921. Constitutional monarchy in Great Britain was established by the Glorious Revolution of 1688, with parliamentary supremacy guaranteed by the passing of the Bill of Rights (1689). Although the American colonies broke away in 1783, in the 18th century Great Britain became the leading naval and colonial power in the world, while the Industrial Revolution which was then beginning made it the first industrialized country. The United Kingdom fought on the victorious side through both world wars. In the 20th century it gradually lost its status as a world power, although the monarch remains titular head of the Commonwealth, an association of the former states of the British Empire. The UK became a member of the EEC in 1973. (See also BRITAIN, GREAT BRITAIN.)

United Nations (abbr. **UN**) an international organization of countries set up in 1945, in succession to the League of Nations, to promote international peace, security, and cooperation. Its headquarters are in New York. Its members, originally the countries that fought against the Axis Powers in the Second World War, now number more than 150 and include most sovereign states of the world, the chief exceptions being Switzerland and North and South Korea. Administration is by a secretariat headed by the Secretary-General. The chief deliberative body is the General Assembly, in which each member state has one vote; recommendations are passed but are not binding on members, and in general have had little effect on world politics. At first the US and its allies were usually able to command a majority in the General Assembly, but more recently the process of decolonization has led to a shift of balance in favour of the developing countries. The Security Council bears the primary responsibility for the maintenance of peace and security, and may call on members to take military or economic action to enforce its decisions; other bodies carry out the functions of the UN with regard to international economic, social, judicial, cultural, educational, health, and other matters. Forces under UN auspices engage chiefly in peacekeeping activities, as in Angola, Somalia, and the former Yugoslavia, although a US-dominated coalition forcibly expelled Iraq from Kuwait in the Gulf War of 1991. (See also SECURITY COUNCIL.)

United Presbyterian Church a Presbyterian Church in Scotland formed in 1847. In 1900 it joined with the Free Church of Scotland to form the United Free Church.

United Provinces *hist.* **1** the seven Dutch provinces of Friesland, Gelderland, Groningen, Holland, Overijssel, Utrecht, and Zeeland, which formed a union under the Treaty of Utrecht in 1579 following their successful rebellion against Spanish rule, leading to the formation of the Dutch Republic. **2** a former administrative division of British India, formed in 1902 by the union of Agra and Oudh. In 1950 it became the state of Uttar Pradesh.

United Provinces of Central America a former federal republic in Central America, formed in 1823 to unite the states of Guatemala, El Salvador, Honduras, Nicaragua, and Costa Rica, all newly independent from Spain. Political and religious disagreements between the autonomous states led to the collapse of the federation in 1838.

United Reformed Church (abbr. **URC**) a Church formed in 1972 by the union of the greater part of the Congregational Church in England and Wales with the Presbyterian Church in England.

United States (abbr. **US**) (in full **United States of America**, abbr. **USA**) a country occupying most of the southern half of North America and including also Alaska and the Hawaiian Islands; pop. (1990) 248,709,870; official language, English; capital, Washington, DC. The US is a federal republic comprising fifty states and the Federal District of Columbia. It originated in the successful rebellion of the British colonies on the east coast in 1775–83 (see AMERICAN INDEPENDENCE, WAR OF). The original thirteen states which formed the Union (see UNION, THE 1a) drew up a federal constitution in 1787, and George Washington was elected the first President in 1789. In the 19th century the territory of the US was extended across the continent through the westward spread of pioneers and settlers (at the expense of the American Indian peoples), by the purchase of the Louisiana territory from France (1803) and Florida from Spain (1819), and by the acquisition of Texas, California, and other territory from Mexico in the 1840s. The American Civil War (1861–5), between the northern states and the southern Confederacy, which wished to secede over the issues of slavery and states' rights, ended in defeat for the South and the abolition of slavery in the US. Throughout the 19th and early 20th century immigration continued, mainly from Europe, greatly increasing the population and increasing the country's cultural diversity. Arizona became the 48th state of the Union in 1912, completing the organization of the coterminous territory; Alaska, purchased from Russia in 1867, and Hawaii, annexed in 1898, achieved statehood in 1959. After a long period of isolation in foreign affairs, the US participated on the Allied side in both world wars and dominated the non-Communist world throughout the cold war as one of the two rival superpowers. In the latter part of the 20th century the United States has become the world's leading military and economic power.

unity /ˈjuːnɪti/ *n.* (*pl.* **-ies**) **1** oneness; being one, single, or individual;

being formed of parts that constitute a whole; due interconnection and coherence of parts (*disturbs the unity of the idea; the pictures lack unity; national unity*). **2** harmony or concord between persons etc. (*lived together in unity*). **3** a thing forming a complex whole (*a person regarded as a unity*). **4** *Math.* the number 'one', the factor that leaves unchanged the quantity on which it operates. **5** *Theatr.* each of the three dramatic principles requiring limitation of the supposed time of a drama to that occupied in acting it or to a single day (*unity of time*), use of one scene throughout (*unity of place*), and concentration on the development of a single plot (*unity of action*). [ME f. OF *unité* f. L *unitas -tatis* f. *unus* one]

Univ. *abbr.* University.

univalent *adj. & n.* ● *adj.* **1** /ˌjuːnɪˈveɪlənt/ *Chem.* = MONOVALENT. **2** /juːˈnɪvələnt/ *Biol.* (of a chromosome) remaining unpaired during meiosis. ● *n.* /juːˈnɪvələnt/ *Biol.* a univalent chromosome.

univalve /ˈjuːnɪˌvælv/ *adj. & n.* *Zool.* ● *adj.* having one valve or shell. ● *n.* a univalve mollusc, esp. a gastropod.

Universal /ˌjuːnɪˈvɜːs(ə)l/ a US film production company formed in 1912. One of the first studios to move from New York to the Los Angeles area, Universal produced several early horror films, many starring Boris Karloff or Bela Lugosi, in the 1930s and 1940s. A decline in fortunes was arrested by successful Steven Spielberg films such as *Jaws* (1975) and *ET* (1982).

universal /ˌjuːnɪˈvɜːs(ə)l/ *adj. & n.* ● *adj.* **1** of, belonging to, or done etc. by all persons or things in the world or in the class concerned; applicable to all cases (*the feeling was universal; met with universal approval*). **2** *Logic* (of a proposition) in which something is asserted of all of a class (opp. PARTICULAR *adj.* 5). ● *n.* **1** *Logic* a universal proposition. **2** *Philos.* **a** a term or concept of general application. **b** a nature or essence signified by a general term. (See also PLATO.) □ **universal agent** an agent empowered to do all that can be delegated. **universal compass** a pair of compasses with legs that may be extended for large circles. **universal coupling** (or **joint**) a coupling or joint which can transmit rotary power by a shaft at any selected angle. **universal donor** *Med.* a person of blood group O, who can in theory donate blood to recipients of any ABO blood group. **universal language** an artificial language intended for use by all nations. **universal recipient** *Med.* a person of blood group AB, who can in theory receive donated blood of any ABO blood group. **universal suffrage** a suffrage extending to all adults with minor exceptions. □ **universally** *adv.* **universality** /-ˈvɜːˈsælɪtɪ/ *n.* [ME f. OF *universal* or L *universalis* (as UNIVERSE)]

universalist /ˌjuːnɪˈvɜːsəlɪst/ *n.* (in Christian theology) **1** a person who holds that all humankind will eventually be saved. **2** a member of an organized body of Christians who hold this. □ **universalism** *n.* **universalistic** /-ˌvɜːsəˈlɪstɪk/ *adj.*

universalize /juːnɪˈvɜːs(ə)laɪz/ *v.tr.* (also **-ise**) **1** apply universally; give a universal character to. **2** bring into universal use; make available for all. □ **universalizability** /-zəˈbɪlɪtɪ/ *n.* **universalization** /-ˈzeɪʃ(ə)n/ *n.*

Universal Postal Union (abbr. **UPU**) an agency of the United Nations that regulates international postal affairs. Its headquarters are in Berne.

Universal Time (abbr. **UT**) (in full **Universal Time Coordinated**, abbr. **UTC**) the formal term for Greenwich Mean Time, used in worldwide communications, astronomy, etc. (See TIME.)

universe /ˈjuːnɪˌvɜːs/ *n.* **1 a** all existing matter and space considered as a whole; the cosmos. (See note below.) **b** a sphere of existence, influence, activity, etc. **2** all humankind. **3** *Statistics & Logic* all the objects under consideration. □ **universe of discourse** *Logic* = sense 3. [F *univers* f. L *universum* neut. of *universus* combined into one, whole f. UNI- + *versus* past part. of *vertere* turn]

▪ Ancient and medieval philosophers followed the Ptolemaic system, which placed the earth at the centre of the universe with the celestial bodies orbiting it on idealized crystal spheres. It later gave way to the Copernican system, placing the sun at the centre. This was in turn displaced by the realization (due to William Herschel) that the Milky Way was a great host of stars, of which our sun was but one. Only in the 20th century was it conclusively shown that there are billions of other star systems or galaxies, many bigger than our own, and the universe is now believed to be at least 10,000 million light-years in diameter. (See COSMOLOGY.)

university /ˌjuːnɪˈvɜːsɪtɪ/ *n.* (*pl.* **-ies**) **1** an educational institution designed for instruction, examination, or both, of students in many branches of advanced learning, conferring degrees in various faculties,

and often embodying colleges and similar institutions. **2** the members of this collectively. **3** a team, crew, etc., representing a university. □ **at university** studying at a university. [ME f. OF *université* f. L *universitas -tatis* the whole (world), in LL college, guild (as UNIVERSE)]

univocal /juːˈnɪvək(ə)l, ˌjuːnɪˈvəʊk-/ *adj. & n.* ● *adj.* (of a word etc.) having only one proper meaning. ● *n.* a univocal word. □ **univocally** *adv.* **univocality** /ˌjuːnɪvəʊˈkælɪtɪ/ *n.*

Unix /ˈjuːnɪks/ *n. propr. Computing* a multi-user operating system. [UNI- + -ICS, after an earlier, less compact system called *Multics*]

unjoin /ʌnˈdʒɔɪn/ *v.tr.* detach from being joined; separate.

unjoined /ʌnˈdʒɔɪnd/ *adj.* not joined.

unjoint /ʌnˈdʒɔɪnt/ *v.tr.* **1** separate the joints of. **2** disunite.

unjust /ʌnˈdʒʌst/ *adj.* not just, contrary to justice or fairness. □ **unjustly** *adv.* **unjustness** *n.*

unjustifiable /ʌnˈdʒʌstɪˌfaɪəb(ə)l/ *adj.* not justifiable. □ **unjustifiably** *adv.*

unjustified /ʌnˈdʒʌstɪˌfaɪd/ *adj.* not justified.

unkempt /ʌnˈkempt/ *adj.* **1** untidy, of neglected appearance. **2** uncombed, dishevelled. □ **unkemptly** *adv.* **unkemptness** *n.*

unkept /ʌnˈkept/ *adj.* **1** (of a promise, law, etc.) not observed; disregarded. **2** not tended; neglected.

unkillable /ʌnˈkɪləb(ə)l/ *adj.* that cannot be killed.

unkind /ʌnˈkaɪnd/ *adj.* **1** not kind. **2** harsh, cruel. **3** unpleasant. □ **unkindly** *adv.* **unkindness** *n.*

unking /ʌnˈkɪŋ/ *v.tr.* **1** deprive of the position of king; dethrone. **2** deprive (a country) of a king.

unkink /ʌnˈkɪŋk/ *v.* **1** *tr.* remove the kinks from; straighten. **2** *intr.* lose kinks; become straight.

unknit /ʌnˈnɪt/ *v.tr.* (**unknitted, unknitting**) separate (things joined, knotted, or interlocked).

unknot /ʌnˈnɒt/ *v.tr.* (**unknotted, unknotting**) release the knot or knots of, untie.

unknowable /ʌnˈnəʊəb(ə)l/ *adj. & n.* ● *adj.* that cannot be known. ● *n.* **1** an unknowable thing. **2** (**the Unknowable**) the postulated absolute or ultimate reality.

unknowing /ʌnˈnəʊɪŋ/ *adj. & n.* ● *adj.* (often foll. by *of*) not knowing; ignorant, unconscious. ● *n.* ignorance (*cloud of unknowing*). □ **unknowingly** *adv.* **unknowingness** *n.*

unknown /ʌnˈnəʊn/ *adj. & n.* ● *adj.* (often foll. by *to*) not known, unfamiliar (*his purpose was unknown to me*). ● *n.* **1** an unknown thing or person. **2** an unknown quantity (*equation in two unknowns*). □ **unknown country** see COUNTRY. **unknown quantity** a person or thing whose nature, significance, etc., cannot be determined. **unknown to** without the knowledge of (*did it unknown to me*). □ **unknownness** /-ˈnəʊnnɪs/ *n.*

Unknown Soldier *n.* an unidentified representative member of a country's armed forces killed in war, given burial with special honours in a national memorial.

Unknown Warrior *n.* = UNKNOWN SOLDIER.

unlabelled /ʌnˈleɪb(ə)ld/ *adj.* (US **unlabeled**) not labelled; without a label.

unlaboured /ʌnˈleɪbəd/ *adj.* (US **unlabored**) not laboured.

unlace /ʌnˈleɪs/ *v.tr.* **1** undo the lace or laces of. **2** unfasten or loosen in this way.

unlade /ʌnˈleɪd/ *v.tr.* **1** take the cargo out of (a ship). **2** discharge (a cargo etc.) from a ship.

unladen /ʌnˈleɪd(ə)n/ *adj.* not laden. □ **unladen weight** the weight of a vehicle etc. when not loaded with goods etc.

unladylike /ʌnˈleɪdɪˌlaɪk/ *adj.* not ladylike.

unlaid[1] /ʌnˈleɪd/ *adj.* not laid.

unlaid[2] *past and past part.* of UNLAY.

unlamented /ˌʌnləˈmentɪd/ *adj.* not lamented.

unlash /ʌnˈlæʃ/ *v.tr.* unfasten (a thing lashed down etc.).

unlatch /ʌnˈlætʃ/ *v.* **1** *tr.* release the latch of. **2** *tr. & intr.* open or be opened in this way.

unlawful /ʌnˈlɔːfʊl/ *adj.* not lawful; illegal, not permissible. □ **unlawfully** *adv.* **unlawfulness** *n.*

unlay /ʌnˈleɪ/ *v.tr.* (*past and past part.* **unlaid** /-ˈleɪd/) *Naut.* untwist (a rope). [UN-[2] + LAY[1]]

unleaded /ʌnˈledɪd/ *adj.* **1** (of petrol etc.) without added lead. **2** not covered, weighted, or framed with lead. **3** *Printing* not spaced with leads.

unlearn /ʌnˈlɜːn/ *v.tr.* (*past* and *past part.* **unlearned** or **unlearnt** /-ˈlɜːnt/) **1** discard from one's memory. **2** rid oneself of (a habit, false information, etc.).

unlearned[1] /ʌnˈlɜːnɪd/ *adj.* not well educated; untaught, ignorant. □ **unlearnedly** *adv.*

unlearned[2] /ʌnˈlɜːnd/ *adj.* (also **unlearnt** /-ˈlɜːnt/) that has not been learned.

unleash /ʌnˈliːʃ/ *v.tr.* **1** release from a leash or restraint. **2** set free to engage in pursuit or attack.

unleavened /ʌnˈlev(ə)nd/ *adj.* not leavened; made without yeast or other raising agent.

unless /ʌnˈles, ən-/ *conj.* if not; except when (*shall go unless I hear from you*; *always walked unless I had a bicycle*). [ON or IN + LESS, assim. to UN-[1]]

unlettered /ʌnˈletəd/ *adj.* **1** illiterate. **2** not well educated.

unliberated /ʌnˈlɪbəˌreɪtɪd/ *adj.* not liberated.

unlicensed /ʌnˈlaɪs(ə)nst/ *adj.* not licensed, esp. without a licence to sell alcoholic drink.

unlighted /ʌnˈlaɪtɪd/ *adj.* **1** not provided with light. **2** not set burning.

unlike /ʌnˈlaɪk/ *adj.* & *prep.* ● *adj.* **1** not like; different from (*is unlike both his parents*). **2** uncharacteristic of (*such behaviour is unlike him*). **3** dissimilar, different. ● *prep.* differently from (*acts quite unlike anyone else*). □ **unlike signs** *Math.* plus and minus. □ **unlikeness** *n.* [perh. f. ON *úlíkr*, OE *ungelic*: see LIKE[1]]

unlikeable /ʌnˈlaɪkəb(ə)l/ *adj.* (also **unlikable**) not easy to like; unpleasant.

unlikely /ʌnˈlaɪklɪ/ *adj.* (**unlikelier**, **unlikeliest**) **1** improbable (*unlikely tale*). **2** (foll. by *to* + infin.) not to be expected to do something (*he's unlikely to be available*). **3** unpromising (*an unlikely candidate*). □ **unlikelihood** *n.* **unlikeliness** *n.*

unlimited /ʌnˈlɪmɪtɪd/ *adj.* without limit; unrestricted; very great in number or quantity (*has unlimited possibilities*; *an unlimited expanse of sea*). □ **unlimitedly** *adv.* **unlimitedness** *n.*

unlined[1] /ʌnˈlaɪnd/ *adj.* **1** (of paper etc.) without lines. **2** (of a face etc.) without wrinkles.

unlined[2] /ʌnˈlaɪnd/ *adj.* (of a garment etc.) without lining.

unlink /ʌnˈlɪŋk/ *v.tr.* **1** undo the links of (a chain etc.). **2** detach or set free by undoing or unfastening a link or chain.

unliquidated /ʌnˈlɪkwɪˌdeɪtɪd/ *adj.* not liquidated.

unlisted /ʌnˈlɪstɪd/ *adj.* not included in a published list, esp. of stock exchange prices or of telephone numbers.

unlit /ʌnˈlɪt/ *adj.* not lit.

unlivable /ʌnˈlɪvəb(ə)l/ *adj.* that cannot be lived or lived in.

unlived-in /ʌnˈlɪvdɪn/ *adj.* **1** appearing to be uninhabited. **2** unused by the inhabitants.

unload /ʌnˈləʊd/ *v.tr.* **1** (also *absol.*) remove a load from (a vehicle etc.). **2** remove (a load) from a vehicle etc. **3** remove the charge from (a firearm etc.). **4** *colloq.* get rid of. **5** (often foll. by *on*) *colloq.* **a** divulge (information). **b** (also *absol.*) give vent to (feelings). □ **unloader** *n.*

unlock /ʌnˈlɒk/ *v.tr.* **1 a** release the lock of (a door, box, etc.). **b** release or disclose by unlocking. **2** release thoughts, feelings, etc., from (one's mind etc.).

unlocked /ʌnˈlɒkt/ *adj.* not locked.

unlooked-for /ʌnˈlʊktfɔː(r)/ *adj.* unexpected, unforeseen.

unloose /ʌnˈluːs/ *v.tr.* (also **unloosen** /-ˈluːs(ə)n/) loose; set free.

unlovable /ʌnˈlʌvəb(ə)l/ *adj.* not lovable.

unloved /ʌnˈlʌvd/ *adj.* not loved.

unlovely /ʌnˈlʌvlɪ/ *adj.* not attractive; unpleasant, ugly. □ **unloveliness** *n.*

unloving /ʌnˈlʌvɪŋ/ *adj.* not loving or showing love. □ **unlovingly** *adv.* **unlovingness** *n.*

unlucky /ʌnˈlʌkɪ/ *adj.* (**unluckier**, **unluckiest**) **1** not fortunate or successful. **2** wretched. **3** bringing bad luck. **4** ill-judged. □ **unluckily** *adv.* **unluckiness** *n.*

unmade /ʌnˈmeɪd/ *adj.* **1** not made. **2** destroyed, annulled.

unmake /ʌnˈmeɪk/ *v.tr.* (*past* and *past part.* **unmade** /-ˈmeɪd/) undo the making of; destroy, depose, annul.

unmalleable /ʌnˈmælɪəb(ə)l/ *adj.* not malleable.

unman /ʌnˈmæn/ *v.tr.* (**unmanned**, **unmanning**) **1** deprive of supposed manly qualities (e.g. self-control, courage); cause to weep etc., discourage. **2** deprive (a ship etc.) of men.

unmanageable /ʌnˈmænɪdʒəb(ə)l/ *adj.* not (easily) managed, manipulated, or controlled. □ **unmanageableness** *n.* **unmanageably** *adv.*

unmanaged /ʌnˈmænɪdʒd/ *adj.* **1** not handled or directed in a controlled way. **2** (of land etc.) left wild; in a natural state.

unmanly /ʌnˈmænlɪ/ *adj.* not manly. □ **unmanliness** *n.*

unmanned /ʌnˈmænd/ *adj.* **1** not manned. **2** overcome by emotion etc.

unmannered /ʌnˈmænəd/ *adj.* lacking affectation; straightforward.

unmannerly /ʌnˈmænəlɪ/ *adj.* **1** without good manners. **2** (of actions, speech, etc.) showing a lack of good manners. □ **unmannerliness** *n.*

unmapped /ʌnˈmæpt/ *adj.* **1** not represented on a usu. geographical or chromosome map. **2** unexplored.

unmarked /ʌnˈmɑːkt/ *adj.* **1** not marked. **2** not noticed.

unmarketable /ʌnˈmɑːkɪtəb(ə)l/ *adj.* not marketable.

unmarried /ʌnˈmærɪd/ *adj.* not married; single.

unmask /ʌnˈmɑːsk/ *v.* **1** *tr.* **a** remove the mask from. **b** expose the true character of. **2** *intr.* remove one's mask. □ **unmasker** *n.*

unmatchable /ʌnˈmætʃəb(ə)l/ *adj.* that cannot be matched. □ **unmatchably** *adv.*

unmatched /ʌnˈmætʃt/ *adj.* not matched or equalled.

unmatured /ˌʌnməˈtjʊəd/ *adj.* not yet matured.

unmeaning /ʌnˈmiːnɪŋ/ *adj.* having no meaning or significance; meaningless. □ **unmeaningly** *adv.* **unmeaningness** *n.*

unmeant /ʌnˈment/ *adj.* not meant or intended.

unmeasurable /ʌnˈmeʒərəb(ə)l/ *adj.* that cannot be measured. □ **unmeasurably** *adv.*

unmeasured /ʌnˈmeʒəd/ *adj.* **1** not measured. **2** limitless.

unmediated /ʌnˈmiːdɪˌeɪtɪd/ *adj.* **1** with no intervention. **2** directly perceived.

unmelodious /ˌʌnmɪˈləʊdɪəs/ *adj.* not melodious; discordant. □ **unmelodiously** *adv.*

unmelted /ʌnˈmeltɪd/ *adj.* not melted.

unmemorable /ʌnˈmemərəb(ə)l/ *adj.* not memorable, forgettable. □ **unmemorably** *adv.*

unmentionable /ʌnˈmenʃənəb(ə)l/ *adj.* & *n.* ● *adj.* that cannot (properly) be mentioned. ● *n.* **1** (in *pl.*) *joc.* **a** undergarments. **b** *archaic* trousers. **2** a person or thing not to be mentioned. □ **unmentionably** *adv.* **unmentionableness** *n.* **unmentionability** /-ˌmenʃənəˈbɪlɪtɪ/ *n.*

unmentioned /ʌnˈmenʃ(ə)nd/ *adj.* not mentioned.

unmerchantable /ʌnˈmɜːtʃəntəb(ə)l/ *adj.* not merchantable.

unmerciful /ʌnˈmɜːsɪfʊl/ *adj.* merciless. □ **unmercifully** *adv.* **unmercifulness** *n.*

unmerited /ʌnˈmerɪtɪd/ *adj.* not merited.

unmet /ʌnˈmet/ *adj.* (of a quota, demand, goal, etc.) not achieved or fulfilled.

unmetalled /ʌnˈmet(ə)ld/ *adj.* *Brit.* (of a road etc.) not surfaced with road-metal.

unmethodical /ˌʌnmɪˈθɒdɪk(ə)l/ *adj.* not methodical, unsystematic. □ **unmethodically** *adv.*

unmetrical /ʌnˈmetrɪk(ə)l/ *adj.* not metrical.

unmilitary /ʌnˈmɪlɪtərɪ/ *adj.* not military.

unmindful /ʌnˈmaɪndfʊl/ *adj.* (often foll. by *of*) not mindful. □ **unmindfully** *adv.* **unmindfulness** *n.*

unmissable /ʌnˈmɪsəb(ə)l/ *adj.* that cannot or should not be missed.

unmistakable /ˌʌnmɪˈsteɪkəb(ə)l/ *adj.* that cannot be mistaken or doubted, clear. □ **unmistakably** *adv.* **unmistakableness** *n.* **unmistakability** /-ˌsteɪkəˈbɪlɪtɪ/ *n.*

unmistaken /ˌʌnmɪˈsteɪkən/ *adj.* not mistaken; right, correct.

unmitigated /ʌnˈmɪtɪˌɡeɪtɪd/ *adj.* **1** not mitigated or modified. **2** absolute, unqualified (*an unmitigated disaster*). □ **unmitigatedly** *adv.*

unmixed /ʌnˈmɪkst/ *adj.* not mixed. □ **unmixed blessing** a thing having advantages and no disadvantages.

unmodernized /ʌnˈmɒdəˌnaɪzd/ *adj.* (also **-ised**) (of a house etc.) not modernized; retaining the original features.

unmodified /ʌnˈmɒdɪˌfaɪd/ *adj.* not modified.

unmodulated /ʌnˈmɒdjʊˌleɪtɪd/ *adj.* not modulated.

unmolested /ˌʌnməˈlestɪd/ *adj.* not molested.

unmoor /ʌnˈmʊə(r), -ˈmɔː(r)/ v.tr. **1** (also absol.) release the moorings of (a vessel). **2** weigh all but one anchor of (a vessel).

unmoral /ʌnˈmɒrəl/ adj. not concerned with morality (cf. IMMORAL). □ **unmorally** adv. **unmorality** /ˌʌnmɒˈrælɪtɪ/ n.

unmotherly /ʌnˈmʌðəlɪ/ adj. not motherly.

unmotivated /ʌnˈməʊtɪˌveɪtɪd/ adj. without motivation; without a motive.

unmounted /ʌnˈmaʊntɪd/ adj. not mounted.

unmourned /ʌnˈmɔːnd/ adj. not mourned.

unmoved /ʌnˈmuːvd/ adj. **1** not moved. **2** not changed in one's purpose. **3** not affected by emotion. □ **unmovable** adj. (also **unmoveable**).

unmoving /ʌnˈmuːvɪŋ/ adj. **1** not moving; still. **2** not emotive.

unmown /ʌnˈməʊn/ adj. not mown.

unmuffle /ʌnˈmʌf(ə)l/ v.tr. **1** remove a muffler from (a face, bell, etc.). **2** free of something that muffles or conceals.

unmurmuring /ʌnˈmɜːmərɪŋ/ adj. literary not complaining. □ **unmurmuringly** adv.

unmusical /ʌnˈmjuːzɪk(ə)l/ adj. **1** not pleasing to the ear. **2** unskilled in or indifferent to music. □ **unmusically** adv. **unmusicalness** n. **unmusicality** /-ˌmjuːzɪˈkælɪtɪ/ n.

unmutilated /ʌnˈmjuːtɪˌleɪtɪd/ adj. not mutilated.

unmuzzle /ʌnˈmʌz(ə)l/ v.tr. **1** remove a muzzle from. **2** relieve of an obligation to remain silent.

unnail /ʌnˈneɪl/ v.tr. unfasten by the removal of nails.

unnameable /ʌnˈneɪməb(ə)l/ adj. that cannot be named, esp. too bad to be named.

unnamed /ʌnˈneɪmd/ adj. not named.

unnatural /ʌnˈnætʃrəl/ adj. **1** contrary to nature or the usual course of nature; not normal. **2 a** lacking natural feelings. **b** extremely cruel or wicked. **3** artificial. **4** affected, contrived. □ **unnaturally** adv. **unnaturalness** n.

unnavigable /ʌnˈnævɪɡəb(ə)l/ adj. not navigable. □ **unnavigability** /-ˌnævɪɡəˈbɪlɪtɪ/ n.

unnecessary /ʌnˈnesəsərɪ/ adj. & n. ● adj. **1** not necessary. **2** more than is necessary (with unnecessary care). ● n. (pl. **-ies**) (usu. in pl.) an unnecessary thing. □ **unnecessarily** adv. **unnecessariness** n.

unneeded /ʌnˈniːdɪd/ adj. not needed.

unneighbourly /ʌnˈneɪbəlɪ/ adj. not neighbourly, unfriendly. □ **unneighbourliness** n.

unnerve /ʌnˈnɜːv/ v.tr. deprive of strength or resolution. □ **unnervingly** adv.

unnilpentium /ˌʌnɪlˈpentɪəm/ n. an artificial transuranic chemical element (atomic number 105; symbol **Unp**). (Cf. HAHNIUM, NIELSBOHRIUM.) [UN-³]

unnilquadium /ˌʌnɪlˈkwɒdɪəm/ n. an artificial transuranic chemical element (atomic number 104; symbol **Unq**). (Cf. KURCHATOVIUM, RUTHERFORDIUM.) [UN-³]

unnoticeable /ʌnˈnəʊtɪsəb(ə)l/ adj. not easily seen or noticed. □ **unnoticeably** adv.

unnoticed /ʌnˈnəʊtɪst/ adj. not noticed.

unnumbered /ʌnˈnʌmbəd/ adj. **1** not marked with a number. **2** not counted. **3** countless.

UNO /ˈjuːnəʊ/ abbr. United Nations Organization.

unobjectionable /ˌʌnəbˈdʒekʃənəb(ə)l/ adj. not objectionable; acceptable. □ **unobjectionableness** n. **unobjectionably** adv.

unobliging /ˌʌnəˈblaɪdʒɪŋ/ adj. not obliging; unhelpful, uncooperative.

unobscured /ˌʌnəbˈskjʊəd/ adj. not obscured.

unobservable /ˌʌnəbˈzɜːvəb(ə)l/ adj. not observable; imperceptible.

unobservant /ˌʌnəbˈzɜːv(ə)nt/ adj. not observant. □ **unobservantly** adv.

unobserved /ˌʌnəbˈzɜːvd/ adj. not observed. □ **unobservedly** /-vɪdlɪ/ adv.

unobstructed /ˌʌnəbˈstrʌktɪd/ adj. not obstructed.

unobtainable /ˌʌnəbˈteɪnəb(ə)l/ adj. that cannot be obtained.

unobtrusive /ˌʌnəbˈtruːsɪv/ adj. not making oneself or itself noticed. □ **unobtrusively** adv. **unobtrusiveness** n.

unoccupied /ʌnˈɒkjʊˌpaɪd/ adj. not occupied.

unoffending /ˌʌnəˈfendɪŋ/ adj. not offending; harmless, innocent. □ **unoffended** adj.

unofficial /ˌʌnəˈfɪʃ(ə)l/ adj. **1** not officially authorized or confirmed. **2** not characteristic of officials. □ **unofficial strike** a strike not formally approved by the strikers' trade union. □ **unofficially** adv.

unoiled /ʌnˈɔɪld/ adj. not oiled.

unopened /ʌnˈəʊp(ə)nd/ adj. not opened.

unopposed /ˌʌnəˈpəʊzd/ adj. not opposed.

unordained /ˌʌnɔːˈdeɪnd/ adj. not ordained.

unordinary /ʌnˈɔːdɪnərɪ, -d(ə)nrɪ/ adj. not ordinary.

unorganized /ʌnˈɔːɡəˌnaɪzd/ adj. (also **-ised**) not organized (cf. DISORGANIZE).

unoriginal /ˌʌnəˈrɪdʒɪn(ə)l/ adj. lacking originality; derivative. □ **unoriginally** adv. **unoriginality** /-ˌrɪdʒɪˈnælɪtɪ/ n.

unornamental /ˌʌnɔːnəˈment(ə)l/ adj. not ornamental; plain.

unornamented /ʌnˈɔːnəˌmentɪd/ adj. not ornamented.

unorthodox /ʌnˈɔːθəˌdɒks/ adj. not orthodox. □ **unorthodoxly** adv. **unorthodoxy** n.

unostentatious /ˌʌnɒstenˈteɪʃəs/ adj. not ostentatious, plain. □ **unostentatiously** adv. **unostentatiousness** n.

unowned /ʌnˈəʊnd/ adj. **1** unacknowledged. **2** having no owner.

Unp symb. Chem. the element unnilpentium.

unpack /ʌnˈpæk/ v.tr. **1** (also absol.) open and remove the contents of (a package, luggage, etc.). **2** take (a thing) out from a package etc. □ **unpacker** n.

unpaged /ʌnˈpeɪdʒd/ adj. with pages not numbered.

unpaid /ʌnˈpeɪd/ adj. (of a debt or a person) not paid.

unpainted /ʌnˈpeɪntɪd/ adj. not painted.

unpaired /ʌnˈpeəd/ adj. **1** not arranged in pairs. **2** not forming one of a pair.

unpalatable /ʌnˈpælətəb(ə)l/ adj. **1** not pleasant to taste. **2** (of an idea, suggestion, etc.) disagreeable, distasteful. □ **unpalatableness** n. **unpalatability** /-ˌpælətəˈbɪlɪtɪ/ n.

unparalleled /ʌnˈpærəˌleld/ adj. having no parallel or equal.

unpardonable /ʌnˈpɑːdənəb(ə)l/ adj. that cannot be pardoned. □ **unpardonableness** n. **unpardonably** adv.

unparliamentary /ˌʌnpɑːləˈmentərɪ/ adj. contrary to proper parliamentary usage. □ **unparliamentary language** oaths or abuse.

unpasteurized /ʌnˈpɑːstjəˌraɪzd, ʌnˈpæs-, -tʃəˌraɪzd/ adj. not pasteurized.

unpatented /ʌnˈpeɪt(ə)ntɪd, ʌnˈpæt-/ adj. not patented.

unpatriotic /ˌʌnpætrɪˈɒtɪk, ˌʌnpeɪt-/ adj. not patriotic. □ **unpatriotically** adv.

unpatronizing /ʌnˈpætrəˌnaɪzɪŋ/ adj. (also **-ising**) not showing condescension.

unpaved /ʌnˈpeɪvd/ adj. not paved.

unpeeled /ʌnˈpiːld/ adj. not peeled.

unpeg /ʌnˈpeɡ/ v.tr. (**unpegged**, **unpegging**) **1** unfasten by the removal of pegs. **2** cease to maintain or stabilize (prices etc.).

unpeople v. & n. ● v.tr. /ʌnˈpiːp(ə)l/ depopulate. ● n.pl. /ˈʌnˌpiːp(ə)l/ unpersons.

unperceived /ˌʌnpəˈsiːvd/ adj. not perceived; unobserved.

unperceptive /ˌʌnpəˈseptɪv/ adj. not perceptive. □ **unperceptively** adv. **unperceptiveness** n.

unperfected /ˌʌnpəˈfektɪd/ adj. not perfected.

unperforated /ʌnˈpɜːfəˌreɪtɪd/ adj. not perforated.

unperformed /ˌʌnpəˈfɔːmd/ adj. not performed.

unperfumed /ʌnˈpɜːfjuːmd, ˌʌnpəˈfjuːmd/ adj. not perfumed.

unperson /ˈʌnˌpɜːs(ə)n/ n. a person whose name or existence is denied or ignored.

unpersuadable /ˌʌnpəˈsweɪdəb(ə)l/ adj. not able to be persuaded; obstinate.

unpersuaded /ˌʌnpəˈsweɪdɪd/ adj. not persuaded.

unpersuasive /ˌʌnpəˈsweɪsɪv/ adj. not persuasive. □ **unpersuasively** adv.

unperturbed /ˌʌnpəˈtɜːbd/ adj. not perturbed. □ **unperturbedly** /-bɪdlɪ/ adv.

unphilosophical /ˌʌnfɪləˈsɒfɪk(ə)l/ adj. (also **unphilosophic**) **1** not according to philosophical principles. **2** lacking philosophy. □ **unphilosophically** *adv.*

unphysiological /ˌʌnfɪzɪəˈlɒdʒɪk(ə)l/ *adj.* (also **unphysiologic**) not in accordance with normal physiological functioning. □ **unphysiologically** *adv.*

unpick /ʌnˈpɪk/ *v.tr.* undo the sewing of (stitches, a garment, etc.).

unpicked /ʌnˈpɪkt/ *adj.* **1** not selected. **2** (of a flower) not plucked.

unpicturesque /ˌʌnpɪktʃəˈresk/ *adj.* not picturesque.

unpin /ʌnˈpɪn/ *v.tr.* (**unpinned**, **unpinning**) **1** unfasten or detach by removing a pin or pins. **2** *Chess* release (a piece that has been pinned).

unpitied /ʌnˈpɪtɪd/ *adj.* not pitied.

unpitying /ʌnˈpɪtɪɪŋ/ *adj.* not pitying. □ **unpityingly** *adv.*

unplaceable /ʌnˈpleɪsəb(ə)l/ *adj.* that cannot be placed or classified (*his accent was unplaceable*).

unplaced /ʌnˈpleɪst/ *adj.* not placed, esp. not placed as one of the first three finishing in a race etc.

unplanned /ʌnˈplænd/ *adj.* not planned.

unplanted /ʌnˈplɑːntɪd/ *adj.* not planted.

unplausible /ʌnˈplɔːzɪb(ə)l/ *adj.* not plausible.

unplayable /ʌnˈpleɪəb(ə)l/ *adj.* **1** *Sport* (of a ball) that cannot be struck or returned. **2** that cannot be played. □ **unplayably** *adv.*

unpleasant /ʌnˈplez(ə)nt/ *adj.* not pleasant; displeasing; disagreeable. □ **unpleasantly** *adv.* **unpleasantness** *n.*

unpleasantry /ʌnˈplez(ə)ntrɪ/ *n.* (*pl.* **-ies**) **1** unkindness. **2** (in *pl.*) **a** unpleasant comments. **b** unpleasant problems.

unpleasing /ʌnˈpliːzɪŋ/ *adj.* not pleasing. □ **unpleasingly** *adv.*

unploughed /ʌnˈplaʊd/ *adj.* not ploughed.

unplucked /ʌnˈplʌkt/ *adj.* not plucked.

unplug /ʌnˈplʌɡ/ *v.tr.* (**unplugged**, **unplugging**) **1** disconnect (an electrical device) by removing its plug from the socket. **2** unstop.

unplumbed /ʌnˈplʌmd/ *adj.* **1** not plumbed. **2** not fully explored or understood. □ **unplumbable** *adj.*

unpoetic /ˌʌnpəʊˈetɪk/ *adj.* (also **unpoetical**) not poetic.

unpointed /ʌnˈpɔɪntɪd/ *adj.* **1** having no point or points. **2 a** not punctuated. **b** (of written Hebrew etc.) without vowel points. **3** (of masonry or brickwork) not pointed.

unpolished /ʌnˈpɒlɪʃt/ *adj.* **1** not polished; rough. **2** without refinement; crude.

unpolitic /ʌnˈpɒlɪtɪk/ *adj.* impolitic, unwise.

unpolitical /ˌʌnpəˈlɪtɪk(ə)l/ *adj.* not concerned with politics. □ **unpolitically** *adv.*

unpolled /ʌnˈpəʊld/ *adj.* **1** not having voted at an election. **2** not included in an opinion poll.

unpolluted /ˌʌnpəˈluːtɪd/ *adj.* not polluted.

unpopular /ʌnˈpɒpjʊlə(r)/ *adj.* not popular; not liked by the public or by people in general. □ **unpopularly** *adv.* **unpopularity** /-ˌpɒpjʊˈlærɪtɪ/ *n.*

unpopulated /ʌnˈpɒpjʊˌleɪtɪd/ *adj.* not populated.

unposed /ʌnˈpəʊzd/ *adj.* not in a posed position, esp. for a photograph.

unpossessed /ˌʌnpəˈzest/ *adj.* **1** (foll. by *of*) not in possession of. **2** not possessed.

unpowered /ʌnˈpaʊəd/ *adj.* (of a boat, vehicle, etc.) propelled other than by fuel.

unpractical /ʌnˈpræktɪk(ə)l/ *adj.* **1** not practical. **2** (of a person) not having practical skill. □ **unpractically** *adv.* **unpracticality** /-ˌpræktɪˈkælɪtɪ/ *n.*

unpractised /ʌnˈpræktɪst/ *adj.* (*US* **unpracticed**) **1** not experienced or skilled. **2** not put into practice.

unprecedented /ʌnˈpresɪˌdentɪd/ *adj.* **1** having no precedent; unparalleled. **2** novel. □ **unprecedentedly** *adv.*

unpredictable /ˌʌnprɪˈdɪktəb(ə)l/ *adj.* that cannot be predicted. □ **unpredictably** *adv.* **unpredictableness** *n.* **unpredictability** /-ˌdɪktəˈbɪlɪtɪ/ *n.*

unpredicted /ˌʌnprɪˈdɪktɪd/ *adj.* not predicted or foretold.

unprejudiced /ʌnˈpredʒʊdɪst/ *adj.* not prejudiced.

unpremeditated /ˌʌnpriːˈmedɪˌteɪtɪd/ *adj.* not previously thought over, not deliberately planned, unintentional. □ **unpremeditatedly** *adv.*

unprepared /ˌʌnprɪˈpeəd/ *adj.* not prepared (in advance); not ready. □ **unpreparedly** /-ˈpeərɪdlɪ/ *adv.* **unpreparedness** *n.*

unprepossessing /ˌʌnpriːpəˈzesɪŋ/ *adj.* not prepossessing; unattractive.

unprescribed /ˌʌnprɪˈskraɪbd/ *adj.* (esp. of drugs) not prescribed.

unpresentable /ˌʌnprɪˈzentəb(ə)l/ *adj.* not presentable.

unpressed /ʌnˈprest/ *adj.* not pressed, esp. (of clothing) unironed.

unpressurized /ʌnˈpreʃəˌraɪzd/ *adj.* (also **-ised**) not pressurized.

unpresuming /ˌʌnprɪˈzjuːmɪŋ/ *adj.* not presuming; modest.

unpresumptuous /ˌʌnprɪˈzʌmptjʊəs/ *adj.* not presumptuous.

unpretending /ˌʌnprɪˈtendɪŋ/ *adj.* unpretentious. □ **unpretendingly** *adv.* **unpretendingness** *n.*

unpretentious /ˌʌnprɪˈtenʃəs/ *adj.* not making a great display; simple, modest. □ **unpretentiously** *adv.* **unpretentiousness** *n.*

unpriced /ʌnˈpraɪst/ *adj.* not having a price or prices fixed, marked, or stated.

unprimed /ʌnˈpraɪmd/ *adj.* not primed.

unprincipled /ʌnˈprɪnsɪp(ə)ld/ *adj.* lacking or not based on good moral principles. □ **unprincipledness** *n.*

unprintable /ʌnˈprɪntəb(ə)l/ *adj.* that cannot be printed, esp. because too indecent or libellous or blasphemous. □ **unprintably** *adv.*

unprinted /ʌnˈprɪntɪd/ *adj.* not printed.

unprivileged /ʌnˈprɪvɪlɪdʒd/ *adj.* not privileged.

unproblematic /ˌʌnprɒbləˈmætɪk/ *adj.* causing no difficulty. □ **unproblematically** *adv.*

unprocessed /ʌnˈprəʊsest/ *adj.* (esp. of food, raw materials) not processed.

unproclaimed /ˌʌnprəˈkleɪmd/ *adj.* not proclaimed.

unprocurable /ˌʌnprəˈkjʊərəb(ə)l/ *adj.* that cannot be procured.

unproductive /ˌʌnprəˈdʌktɪv/ *adj.* not productive. □ **unproductively** *adv.* **unproductiveness** *n.*

unprofessional /ˌʌnprəˈfeʃən(ə)l/ *adj.* **1** contrary to professional standards of behaviour etc. **2** not belonging to a profession; amateur. □ **unprofessionally** *adv.*

unprofitable /ʌnˈprɒfɪtəb(ə)l/ *adj.* not profitable, unremunerative. □ **unprofitableness** *n.* **unprofitably** *adv.*

Unprofor /ˈʌnprəˌfɔː(r)/ *abbr.* (also **UNPROFOR**) United Nations Protection Force.

unprogressive /ˌʌnprəˈɡresɪv/ *adj.* not progressive.

unpromising /ʌnˈprɒmɪsɪŋ/ *adj.* not likely to turn out well. □ **unpromisingly** *adv.*

unprompted /ʌnˈprɒmptɪd/ *adj.* spontaneous.

unpronounceable /ˌʌnprəˈnaʊnsəb(ə)l/ *adj.* that cannot be pronounced. □ **unpronounceably** *adv.*

unpropitious /ˌʌnprəˈpɪʃəs/ *adj.* not propitious. □ **unpropitiously** *adv.*

unprosperous /ʌnˈprɒspərəs/ *adj.* not prosperous. □ **unprosperously** *adv.*

unprotected /ˌʌnprəˈtektɪd/ *adj.* **1** not protected. **2** (of sexual intercourse) without the use of a contraceptive sheath. □ **unprotectedness** *n.*

unprotesting /ˌʌnprəˈtestɪŋ/ *adj.* not protesting. □ **unprotestingly** *adv.*

unprovable /ʌnˈpruːvəb(ə)l/ *adj.* that cannot be proved. □ **unprovableness** *n.* **unprovability** /-ˌpruːvəˈbɪlɪtɪ/ *n.*

unproved /ʌnˈpruːvd/ *adj.* (also **unproven** /-v(ə)n/) not proved.

unprovided /ˌʌnprəˈvaɪdɪd/ *adj.* (usu. foll. by *with*) not furnished, supplied, or equipped.

unprovoked /ˌʌnprəˈvəʊkt/ *adj.* (of a person or act) without provocation.

unpublicized /ʌnˈpʌblɪˌsaɪzd/ *adj.* (also **-ised**) not publicized.

unpublished /ʌnˈpʌblɪʃt/ *adj.* not published. □ **unpublishable** *adj.*

unpunctual /ʌnˈpʌŋktʃʊəl, -tjʊəl/ *adj.* not punctual. □ **unpunctuality** /-ˌpʌŋktʃʊˈælɪtɪ, -ˌpʌŋktjʊ-/ *n.*

unpunctuated /ʌnˈpʌŋktʃʊˌeɪtɪd, -tjʊˌeɪtɪd/ *adj.* not punctuated.

unpunishable /ʌnˈpʌnɪʃəb(ə)l/ *adj.* that cannot be punished.

unpunished /ʌnˈpʌnɪʃt/ *adj.* not punished.

unpurified /ʌnˈpjʊərɪˌfaɪd/ *adj.* not purified.

unputdownable /ˌʌnpʊtˈdaʊnəb(ə)l/ *adj.* *colloq.* (of a book) so engrossing that one has to go on reading it.

Unq *symb. Chem.* the element unnilquadium.

unqualified /ʌnˈkwɒlɪˌfaɪd/ *adj.* **1** not competent (*unqualified to give an answer*). **2** not legally or officially qualified (*an unqualified practitioner*). **3** not modified or restricted; complete (*unqualified assent; unqualified success*).

unquantifiable /ʌnˈkwɒntɪˌfaɪəb(ə)l/ *adj.* impossible to quantify. □ **unquantified** *adj.*

unquenchable /ʌnˈkwentʃəb(ə)l/ *adj.* that cannot be quenched. □ **unquenchably** *adv.*

unquenched /ʌnˈkwentʃt/ *adj.* not quenched.

unquestionable /ʌnˈkwestʃənəb(ə)l/ *adj.* that cannot be disputed or doubted. □ **unquestionably** *adv.* **unquestionableness** *n.* **unquestionability** /-ˌkwestʃənəˈbɪlɪtɪ/ *n.*

unquestioned /ʌnˈkwestʃənd/ *adj.* **1** not disputed or doubted; definite, certain. **2** not interrogated.

unquestioning /ʌnˈkwestʃənɪŋ/ *adj.* **1** asking no questions. **2** done etc. without asking questions. □ **unquestioningly** *adv.*

unquiet /ʌnˈkwaɪət/ *adj.* **1** restless, agitated, stirring. **2** perturbed, anxious. □ **unquietly** *adv.* **unquietness** *n.*

unquotable /ʌnˈkwəʊtəb(ə)l/ *adj.* that cannot be quoted.

unquote /ʌnˈkwəʊt/ *v.tr.* (as *int.*) (in dictation, reading aloud, etc.) indicate the presence of closing quotation marks (cf. QUOTE *v.* 5b).

unquoted /ʌnˈkwəʊtɪd/ *adj.* not quoted, esp. on the Stock Exchange.

unravel /ʌnˈræv(ə)l/ *v.* (**unravelled, unravelling**; *US* **unraveled, unraveling**) **1** *tr.* cause to be no longer ravelled, tangled, or intertwined. **2** *tr.* probe and solve (a mystery etc.). **3** *tr.* undo (a fabric, esp. a knitted one). **4** *intr.* become disentangled or unknitted.

unreachable /ʌnˈriːtʃəb(ə)l/ *adj.* that cannot be reached. □ **unreachableness** *n.* **unreachably** *adv.*

unreached /ʌnˈriːtʃt/ *adj.* **1** not reached. **2** not yet evangelized.

unread /ʌnˈred/ *adj.* **1** (of a book etc.) not read. **2** (of a person) not well-read.

unreadable /ʌnˈriːdəb(ə)l/ *adj.* **1** too dull or too difficult to be worth reading. **2** illegible. **3** inscrutable. □ **unreadably** *adv.* **unreadability** /-ˌriːdəˈbɪlɪtɪ/ *n.*

unready[1] /ʌnˈredɪ/ *adj.* **1** not ready. **2** not prompt in action. □ **unreadily** *adv.* **unreadiness** *n.*

unready[2] /ʌnˈredɪ/ *adj. archaic* lacking good advice; rash (*Ethelred the Unready*). [UN-[1] + REDE, assim. to UNREADY[1]]

unreal /ʌnˈriːl/ *adj.* **1** not real. **2** imaginary, illusory. **3** *sl.* incredible, amazing. □ **unreally** /-ˈriːllɪ/ *adv.* **unreality** /ʌnrɪˈælɪtɪ/ *n.*

unrealism /ʌnˈrɪəlɪz(ə)m/ *n.* lack of realism.

unrealistic /ʌnrɪəˈlɪstɪk/ *adj.* not realistic. □ **unrealistically** *adv.*

unrealizable /ʌnˈrɪəlaɪzəb(ə)l/ *adj.* that cannot be realized.

unrealized /ʌnˈrɪəlaɪzd/ *adj.* not realized.

unreason /ʌnˈriːz(ə)n/ *n.* lack of reasonable thought or action. [ME, = injustice, f. UN-[1] + REASON]

unreasonable /ʌnˈriːzənəb(ə)l/ *adj.* **1** going beyond the limits of what is reasonable or equitable (*unreasonable demands*). **2** not guided by or listening to reason. □ **unreasonableness** *n.* **unreasonably** *adv.*

unreasoned /ʌnˈriːz(ə)nd/ *adj.* not reasoned.

unreasoning /ʌnˈriːzənɪŋ/ *adj.* not reasoning. □ **unreasoningly** *adv.*

unreceptive /ʌnrɪˈseptɪv/ *adj.* not receptive.

unreciprocated /ʌnrɪˈsɪprəˌkeɪtɪd/ *adj.* not reciprocated.

unreckoned /ʌnˈrekənd/ *adj.* not calculated or taken into account.

unreclaimed /ʌnrɪˈkleɪmd/ *adj.* not reclaimed.

unrecognizable /ʌnˈrekəɡˌnaɪzəb(ə)l/ *adj.* (also **-isable**) that cannot be recognized. □ **unrecognizableness** *n.* **unrecognizably** *adv.*

unrecognized /ʌnˈrekəɡˌnaɪzd/ *adj.* (also **-ised**) not recognized.

unrecompensed /ʌnˈrekəmˌpenst/ *adj.* not recompensed.

unreconciled /ʌnˈrekənˌsaɪld/ *adj.* not reconciled.

unreconstructed /ʌnriːkənˈstrʌktɪd/ *adj.* **1** not reconciled or converted to the current political orthodoxy. **2** not rebuilt.

unrecorded /ʌnrɪˈkɔːdɪd/ *adj.* not recorded. □ **unrecordable** *adj.*

unrectified /ʌnˈrektɪˌfaɪd/ *adj.* not rectified.

unredeemable /ʌnrɪˈdiːməb(ə)l/ *adj.* that cannot be redeemed. □ **unredeemably** *adv.*

unredeemed /ʌnrɪˈdiːmd/ *adj.* not redeemed.

unredressed /ʌnrɪˈdrest/ *adj.* not redressed.

unreel /ʌnˈriːl/ *v.tr. & intr.* unwind from a reel.

unreeve /ʌnˈriːv/ *v.tr.* (*past* **unrove** /-ˈrəʊv/) withdraw (a rope etc.) from being reeved.

unrefined /ʌnrɪˈfaɪnd/ *adj.* not refined.

unreflecting /ʌnrɪˈflektɪŋ/ *adj.* not thoughtful. □ **unreflectingly** *adv.* **unreflectingness** *n.*

unreformed /ʌnrɪˈfɔːmd/ *adj.* not reformed.

unregarded /ʌnrɪˈɡɑːdɪd/ *adj.* not regarded.

unregenerate /ʌnrɪˈdʒenərət/ *adj.* not regenerate; obstinately wrong or bad. □ **unregeneracy** *n.* **unregenerately** *adv.*

unregistered /ʌnˈredʒɪstəd/ *adj.* not registered.

unregulated /ʌnˈreɡjʊˌleɪtɪd/ *adj.* not regulated.

unrehearsed /ʌnrɪˈhɜːst/ *adj.* not rehearsed.

unrelated /ʌnrɪˈleɪtɪd/ *adj.* not related. □ **unrelatedness** *n.*

unrelaxed /ʌnrɪˈlækst/ *adj.* not relaxed.

unreleased /ʌnrɪˈliːst/ *adj.* not released, esp. (of a recording, film, etc.) to the public.

unrelenting /ʌnrɪˈlentɪŋ/ *adj.* **1** not relenting or yielding. **2** unmerciful. **3** not abating or relaxing. □ **unrelentingly** *adv.* **unrelentingness** *n.*

unreliable /ʌnrɪˈlaɪəb(ə)l/ *adj.* not reliable; erratic. □ **unreliably** *adv.* **unreliableness** *n.* **unreliability** /-ˌlaɪəˈbɪlɪtɪ/ *n.*

unrelieved /ʌnrɪˈliːvd/ *adj.* **1** lacking the relief given by contrast or variation. **2** not aided or assisted. □ **unrelievedly** /-vɪdlɪ/ *adv.*

unreligious /ʌnrɪˈlɪdʒəs/ *adj.* **1** not concerned with religion. **2** irreligious.

unremarkable /ʌnrɪˈmɑːkəb(ə)l/ *adj.* not remarkable; uninteresting. □ **unremarkably** *adv.*

unremembered /ʌnrɪˈmembəd/ *adj.* not remembered; forgotten.

unremitting /ʌnrɪˈmɪtɪŋ/ *adj.* never relaxing or slackening, incessant. □ **unremittingly** *adv.* **unremittingness** *n.*

unremorseful /ʌnrɪˈmɔːsfʊl/ *adj.* not showing or feeling remorse. □ **unremorsefully** *adv.*

unremovable /ʌnrɪˈmuːvəb(ə)l/ *adj.* that cannot be removed.

unremunerative /ʌnrɪˈmjuːnərətɪv/ *adj.* bringing no, or not enough, profit or income; unprofitable. □ **unremuneratively** *adv.* **unremunerativeness** *n.*

unrenewable /ʌnrɪˈnjuːəb(ə)l/ *adj.* that cannot be renewed. □ **unrenewed** *adj.*

unrepealed /ʌnrɪˈpiːld/ *adj.* not repealed.

unrepeatable /ʌnrɪˈpiːtəb(ə)l/ *adj.* **1** that cannot be done, made, or said again. **2** too indecent to be said again. □ **unrepeatability** /-ˌpiːtəˈbɪlɪtɪ/ *n.*

unrepentant /ʌnrɪˈpentənt/ *adj.* not repentant, impenitent. □ **unrepentantly** *adv.*

unreported /ʌnrɪˈpɔːtɪd/ *adj.* not reported.

unrepresentative /ʌnreprɪˈzentətɪv/ *adj.* not representative. □ **unrepresentativeness** *n.*

unrepresented /ʌnreprɪˈzentɪd/ *adj.* not represented.

unreproved /ʌnrɪˈpruːvd/ *adj.* not reproved.

unrequested /ʌnrɪˈkwestɪd/ *adj.* not requested or asked for.

unrequited /ʌnrɪˈkwaɪtɪd/ *adj.* (of love etc.) not returned. □ **unrequitedly** *adv.* **unrequitedness** *n.*

unreserve /ʌnrɪˈzɜːv/ *n.* lack of reserve; frankness.

unreserved /ʌnrɪˈzɜːvd/ *adj.* **1** not reserved (*unreserved seats*). **2** without reservations; absolute (*unreserved confidence*). **3** free from reserve (*an unreserved nature*). □ **unreservedly** /-vɪdlɪ/ *adv.* **unreservedness** *n.*

unresisted /ʌnrɪˈzɪstɪd/ *adj.* not resisted. □ **unresistedly** *adv.*

unresisting /ʌnrɪˈzɪstɪŋ/ *adj.* not resisting. □ **unresistingly** *adv.* **unresistingness** *n.*

unresolvable /ʌnrɪˈzɒlvəb(ə)l/ *adj.* (of a problem, conflict, etc.) that cannot be resolved.

unresolved /ʌnrɪˈzɒlvd/ *adj.* **1 a** uncertain how to act, irresolute. **b** uncertain in opinion, undecided. **2** (of questions etc.) undetermined, undecided, unsolved. **3** not broken up or dissolved. □ **unresolvedly** /-vɪdlɪ/ *adv.* **unresolvedness** *n.*

unresponsive /ʌnrɪˈspɒnsɪv/ *adj.* not responsive. □ **unresponsively** *adv.* **unresponsiveness** *n.*

unrest /ʌnˈrest/ *n.* **1** lack of rest. **2** restlessness, disturbance, agitation.

unrested /ʌnˈrestɪd/ *adj.* not refreshed by rest.

unrestful /ʌnˈrestfʊl/ adj. not restful. □ **unrestfully** adv.

unresting /ʌnˈrestɪŋ/ adj. not resting. □ **unrestingly** adv.

unrestored /ˌʌnrɪˈstɔːd/ adj. not restored.

unrestrainable /ˌʌnrɪˈstreɪnəb(ə)l/ adj. that cannot be restrained; irrepressible, ungovernable.

unrestrained /ˌʌnrɪˈstreɪnd/ adj. not restrained. □ **unrestrainedly** /-nɪdlɪ/ adv. **unrestrainedness** n.

unrestraint /ˌʌnrɪˈstreɪnt/ n. lack of restraint.

unrestricted /ˌʌnrɪˈstrɪktɪd/ adj. not restricted. □ **unrestrictedly** adv. **unrestrictedness** n.

unreturned /ˌʌnrɪˈtɜːnd/ adj. **1** not reciprocated or responded to. **2** not having returned or been returned.

unrevealed /ˌʌnrɪˈviːld/ adj. not revealed; secret.

unreversed /ˌʌnrɪˈvɜːst/ adj. (esp. of a decision etc.) not reversed.

unrevised /ˌʌnrɪˈvaɪzd/ adj. not revised; in an original form.

unrevoked /ˌʌnrɪˈvəʊkt/ adj. not revoked or annulled; still in force.

unrewarded /ˌʌnrɪˈwɔːdɪd/ adj. not rewarded.

unrewarding /ˌʌnrɪˈwɔːdɪŋ/ adj. not rewarding or satisfying.

unrhymed /ʌnˈraɪmd/ adj. not rhymed.

unrhythmical /ʌnˈrɪðmɪk(ə)l/ adj. not rhythmical. □ **unrhythmically** adv.

unridable /ʌnˈraɪdəb(ə)l/ adj. that cannot be ridden.

unridden /ʌnˈrɪd(ə)n/ adj. not ridden.

unriddle /ʌnˈrɪd(ə)l/ v.tr. solve or explain (a mystery etc.).

unrig /ʌnˈrɪg/ v.tr. (**unrigged, unrigging**) **1** remove the rigging from (a ship). **2** dial. undress.

unrighteous /ʌnˈraɪtʃəs/ adj. not righteous; unjust, wicked, dishonest. □ **unrighteously** adv. **unrighteousness** n. [OE unrihtwīs (as UN-[1], RIGHTEOUS)]

unrip /ʌnˈrɪp/ v.tr. (**unripped, unripping**) open by ripping.

unripe /ʌnˈraɪp/ adj. not ripe. □ **unripeness** n.

unrisen /ʌnˈrɪz(ə)n/ adj. that has not risen.

unrivalled /ʌnˈraɪv(ə)ld/ adj. (US **unrivaled**) having no equal; peerless.

unrivet /ʌnˈrɪvɪt/ v.tr. (**unriveted, unriveting**) **1** undo, unfasten, or detach by the removal of rivets. **2** loosen, relax, undo, detach.

unrobe /ʌnˈrəʊb/ v.tr. & intr. **1** disrobe. **2** undress.

unroll /ʌnˈrəʊl/ v.tr. & intr. **1** open out from a rolled-up state. **2** display or be displayed in this form.

unromantic /ˌʌnrəˈmæntɪk/ adj. not romantic. □ **unromantically** adv.

unroof /ʌnˈruːf/ v.tr. remove the roof of.

unroofed /ʌnˈruːft/ adj. not provided with a roof.

unroot /ʌnˈruːt/ v.tr. **1** uproot. **2** eradicate.

unrope /ʌnˈrəʊp/ v. **1** tr. detach by undoing a rope. **2** intr. Mountaineering detach oneself from a rope.

unrounded /ʌnˈraʊndɪd/ adj. not rounded.

unrove past of UNREEVE.

unroyal /ʌnˈrɔɪəl/ adj. not royal.

unruffled /ʌnˈrʌf(ə)ld/ adj. **1** not agitated or disturbed; calm. **2** not physically ruffled.

unruled /ʌnˈruːld/ adj. **1** not ruled or governed. **2** not having ruled lines.

unruly /ʌnˈruːlɪ/ adj. (**unrulier, unruliest**) not easily controlled or disciplined, disorderly. □ **unruliness** n. [ME f. UN-[1] + ruly f. RULE]

UNRWA /ˈʌnrɑː/ abbr. United Nations Relief and Works Agency for Palestine Refugees in the Near East.

unsaddle /ʌnˈsæd(ə)l/ v.tr. **1** remove the saddle from (a horse etc.). **2** dislodge from a saddle.

unsafe /ʌnˈseɪf/ adj. **1** not safe. **2** Law (of a verdict, conviction, etc.) likely to constitute a miscarriage of justice. □ **unsafely** adv. **unsafeness** n.

unsaid[1] /ʌnˈsed/ adj. not said or uttered.

unsaid[2] past and past part. of UNSAY.

unsalaried /ʌnˈsælərɪd/ adj. not salaried.

unsaleable /ʌnˈseɪləb(ə)l/ adj. not saleable. □ **unsaleability** /-ˌseɪləˈbɪlɪtɪ/ n.

unsalted /ʌnˈsɔːltɪd, ʌnˈsɒl-/ adj. not salted.

unsanctified /ʌnˈsæŋktɪˌfaɪd/ adj. not sanctified.

unsanctioned /ʌnˈsæŋkʃ(ə)nd/ adj. not sanctioned.

unsanitary /ʌnˈsænɪtərɪ/ adj. not sanitary.

unsatisfactory /ˌʌnsætɪsˈfæktərɪ/ adj. **1** not satisfactory; poor, unacceptable. **2** Law (of a conviction etc.) likely to constitute a miscarriage of justice. □ **unsatisfactorily** adv. **unsatisfactoriness** n.

unsatisfied /ʌnˈsætɪsˌfaɪd/ adj. not satisfied. □ **unsatisfiedness** n.

unsatisfying /ʌnˈsætɪsˌfaɪɪŋ/ adj. not satisfying. □ **unsatisfyingly** adv.

unsaturated /ʌnˈsætʃəˌreɪtɪd, -ˈsætjʊˌreɪtɪd/ adj. **1** Chem. (of a compound, esp. a fat or oil) having double or triple bonds in its molecule and therefore capable of further reaction. **2** not saturated. □ **unsaturation** /-ˌsætʃəˈreɪʃ(ə)n, -ˌsætjʊˈreɪ-/ n.

unsaved /ʌnˈseɪvd/ adj. not saved.

unsavoury /ʌnˈseɪvərɪ/ adj. (US **unsavory**) **1** disagreeable to the taste, smell, or feelings; disgusting. **2** disagreeable, unpleasant (an unsavoury character). **3** morally offensive. □ **unsavourily** adv. **unsavouriness** n.

unsay /ʌnˈseɪ/ v.tr. (past and past part. **unsaid** /-ˈsed/) retract (a statement).

unsayable /ʌnˈseɪəb(ə)l/ adj. that cannot be said.

unscalable /ʌnˈskeɪləb(ə)l/ adj. that cannot be scaled.

unscarred /ʌnˈskɑːd/ adj. not scarred or damaged.

unscathed /ʌnˈskeɪðd/ adj. without suffering any injury.

unscented /ʌnˈsentɪd/ adj. not scented.

unscheduled /ʌnˈʃedjuːld, ʌnˈsked-/ adj. not scheduled.

unscholarly /ʌnˈskɒləlɪ/ adj. not scholarly. □ **unscholarliness** n.

unschooled /ʌnˈskuːld/ adj. **1** uneducated, untaught. **2** not sent to school. **3** untrained, undisciplined. **4** not made artificial by education.

unscientific /ˌʌnsaɪənˈtɪfɪk/ adj. **1** not in accordance with scientific principles. **2** not familiar with science. □ **unscientifically** adv.

unscramble /ʌnˈskræmb(ə)l/ v.tr. restore from a scrambled state, esp. interpret (a scrambled transmission etc.). □ **unscrambler** n.

unscreened /ʌnˈskriːnd/ adj. **1 a** (esp. of coal) not passed through a screen or sieve. **b** not investigated or checked, esp. for security or medical problems. **2** not provided with a screen. **3** (of a film etc.) not shown on a screen.

unscrew /ʌnˈskruː/ v. **1** tr. & intr. unfasten or be unfastened by turning or removing a screw or screws or by twisting like a screw. **2** tr. loosen (a screw).

unscripted /ʌnˈskrɪptɪd/ adj. (of a speech etc.) delivered without a prepared script.

unscriptural /ʌnˈskrɪptʃərəl/ adj. against or not in accordance with Scripture. □ **unscripturally** adv.

unscrupulous /ʌnˈskruːpjʊləs/ adj. having no scruples, unprincipled. □ **unscrupulously** adv. **unscrupulousness** n.

unseal /ʌnˈsiːl/ v.tr. break the seal of; open (a letter, receptacle, etc.).

unsealed /ʌnˈsiːld/ adj. not sealed.

unsearchable /ʌnˈsɜːtʃəb(ə)l/ adj. inscrutable, imponderable. □ **unsearchableness** n. **unsearchably** adv.

unsearched /ʌnˈsɜːtʃt/ adj. not searched.

unseasonable /ʌnˈsiːzənəb(ə)l/ adj. **1** not appropriate to the season. **2** untimely, inopportune. □ **unseasonableness** n. **unseasonably** adv.

unseasoned /ʌnˈsiːz(ə)nd/ adj. **1** not flavoured with salt, herbs, etc. **2** (esp. of timber) not matured. **3** not habituated.

unseat /ʌnˈsiːt/ v.tr. **1** remove from a seat, esp. in an election. **2** dislodge from a seat, esp. on horseback.

unseaworthy /ʌnˈsiːˌwɜːðɪ/ adj. not seaworthy.

unsecured /ˌʌnsɪˈkjʊəd/ adj. not secured.

unseeable /ʌnˈsiːəb(ə)l/ adj. that cannot be seen.

unseeded /ʌnˈsiːdɪd/ adj. Sport (of a player) not seeded.

unseeing /ʌnˈsiːɪŋ/ adj. **1** unobservant. **2** blind. □ **unseeingly** adv.

unseemly /ʌnˈsiːmlɪ/ adj. (**unseemlier, unseemliest**) **1** indecent. **2** unbecoming. □ **unseemliness** n.

unseen /ʌnˈsiːn/ adj. & n. ● adj. **1** not seen. **2** invisible. **3** (of a translation) to be done without preparation. ● n. Brit. an unseen translation.

unsegregated /ʌnˈsegrɪˌgeɪtɪd/ adj. not segregated.

unselect /ˌʌnsɪˈlekt/ adj. not select.

unselective /ˌʌnsɪˈlektɪv/ adj. not selective.

unselfconscious /ˌʌnself'kɒnʃəs/ adj. not self-conscious; uninhibited. □ **unselfconsciously** adv. **unselfconsciousness** n.

unselfish /ʌn'selfɪʃ/ adj. mindful of others' interests. □ **unselfishly** adv. **unselfishness** n.

unsensational /ˌʌnsen'seɪʃən(ə)l/ adj. not sensational, prosaic. □ **unsensationally** adv.

unsentimental /ˌʌnsentɪ'ment(ə)l/ adj. not sentimental. □ **unsentimentally** adv. **unsentimentality** /-men'tælɪtɪ/ n.

unseparated /ʌn'sepəˌreɪtɪd/ adj. not separated.

unserious /ʌn'sɪərɪəs/ adj. not serious; light-hearted.

unserviceable /ʌn'sɜːvɪsəb(ə)l/ adj. not serviceable; unfit for use. □ **unserviceability** /-ˌsɜːvɪsə'bɪlɪtɪ/ n.

unset /ʌn'set/ adj. not set.

unsettle /ʌn'set(ə)l/ v. **1** tr. disturb the settled state or arrangement of; discompose. **2** tr. derange. **3** intr. become unsettled. □ **unsettlement** n. **unsettling** adj.

unsettled /ʌn'set(ə)ld/ adj. **1** not (yet) settled. **2** liable or open to change or further discussion. **3** (of a bill etc.) unpaid. □ **unsettledness** n.

unsewn /ʌn'səʊn/ adj. not sewn. □ **unsewn binding** = perfect binding.

unsex /ʌn'seks/ v.tr. deprive (a person) of the qualities of his or her sex.

unsexed /ʌn'sekst/ adj. having no sexual characteristics.

unsexy /ʌn'seksɪ/ adj. (**unsexier**, **unsexiest**) not sexually attractive or stimulating; not appealing.

unshackle /ʌn'ʃæk(ə)l/ v.tr. **1** release from shackles. **2** set free.

unshaded /ʌn'ʃeɪdɪd/ adj. not shaded.

unshakeable /ʌn'ʃeɪkəb(ə)l/ adj. (also **unshakable**) that cannot be shaken; firm, obstinate. □ **unshakeably** adv. **unshakeability** /-ˌʃeɪkə'bɪlɪtɪ/ n.

unshaken /ʌn'ʃeɪkən/ adj. not shaken. □ **unshakenly** adv.

unshapely /ʌn'ʃeɪplɪ/ adj. not shapely. □ **unshapeliness** n.

unshared /ʌn'ʃeəd/ adj. not shared.

unsharp /ʌn'ʃɑːp/ adj. Photog. not sharp. □ **unsharpness** n.

unshaved /ʌn'ʃeɪvd/ adj. not shaved.

unshaven /ʌn'ʃeɪv(ə)n/ adj. not shaved.

unsheathe /ʌn'ʃiːð/ v.tr. remove (a knife etc.) from a sheath.

unshed /ʌn'ʃed/ adj. not shed.

unshell /ʌn'ʃel/ v.tr. (usu. as **unshelled** adj.) extract from its shell.

unsheltered /ʌn'ʃeltəd/ adj. not sheltered.

unshielded /ʌn'ʃiːldɪd/ adj. not shielded or protected.

unship /ʌn'ʃɪp/ v.tr. (**unshipped**, **unshipping**) **1** remove or discharge (a cargo or passenger) from a ship. **2** esp. Naut. remove (an object, esp. a mast or oar) from a fixed position.

unshockable /ʌn'ʃɒkəb(ə)l/ adj. that cannot be shocked. □ **unshockably** adv. **unshockability** /-ˌʃɒkə'bɪlɪtɪ/ n.

unshod /ʌn'ʃɒd/ adj. not wearing shoes.

unshorn /ʌn'ʃɔːn/ adj. not shorn.

unshrinkable /ʌn'ʃrɪŋkəb(ə)l/ adj. (of fabric etc.) not liable to shrink. □ **unshrinkability** /-ˌʃrɪŋkə'bɪlɪtɪ/ n.

unshrinking /ʌn'ʃrɪŋkɪŋ/ adj. not flinching or hesitating, fearless. □ **unshrinkingly** adv.

unsighted /ʌn'saɪtɪd/ adj. **1** not sighted or seen. **2** prevented from seeing, esp. by an obstruction.

unsightly /ʌn'saɪtlɪ/ adj. unpleasant to look at, ugly; spoiling the appearance. □ **unsightliness** n.

unsigned /ʌn'saɪnd/ adj. not signed.

unsinkable /ʌn'sɪŋkəb(ə)l/ adj. unable to be sunk. □ **unsinkability** /-ˌsɪŋkə'bɪlɪtɪ/ n.

unsized[1] /ʌn'saɪzd/ adj. **1** not made to a size. **2** not sorted by size.

unsized[2] /ʌn'saɪzd/ adj. not treated with size.

unskilful /ʌn'skɪlfʊl/ adj. (US **unskillful**) not skilful. □ **unskilfully** adv. **unskilfulness** n.

unskilled /ʌn'skɪld/ adj. lacking or not needing special skill or training.

unskimmed /ʌn'skɪmd/ adj. (of milk) not skimmed.

unslakeable /ʌn'sleɪkəb(ə)l/ adj. (also **unslakable**) that cannot be slaked or quenched.

unsleeping /ʌn'sliːpɪŋ/ adj. not or never sleeping. □ **unsleepingly** adv.

unsliced /ʌn'slaɪst/ adj. (esp. of a loaf of bread when it is bought) not having been cut into slices.

unsling /ʌn'slɪŋ/ v.tr. (past and past part. **unslung** /-'slʌŋ/) free from being slung or suspended.

unsmiling /ʌn'smaɪlɪŋ/ adj. not smiling. □ **unsmilingly** adv. **unsmilingness** n.

unsmoked /ʌn'sməʊkt/ adj. **1** not cured by smoking (*unsmoked bacon*). **2** not consumed by smoking (*an unsmoked cigar*).

unsnarl /ʌn'snɑːl/ v.tr. disentangle. [UN-[2] + SNARL[2]]

unsociable /ʌn'səʊʃəb(ə)l/ adj. not sociable, disliking the company of others. □ **unsociably** adv. **unsociableness** n. **unsociability** /-ˌsəʊʃə'bɪlɪtɪ/ n.

unsocial /ʌn'səʊʃ(ə)l/ adj. **1** not social; not suitable for, seeking, or conforming to society. **2** outside the normal working day (*unsocial hours*). **3** antisocial. □ **unsocially** adv.

unsocialist /ʌn'səʊʃəlɪst/ adj. not socialist.

unsoiled /ʌn'sɔɪld/ adj. not soiled or dirtied.

unsold /ʌn'səʊld/ adj. not sold.

unsolder /ʌn'səʊldə(r), ʌn'sɒl-/ v.tr. undo the soldering of.

unsoldierly /ʌn'səʊldʒəlɪ/ adj. not soldierly.

unsolicited /ˌʌnsə'lɪsɪtɪd/ adj. not asked for; given or done voluntarily. □ **unsolicitedly** adv.

unsolvable /ʌn'sɒlvəb(ə)l/ adj. that cannot be solved, insoluble. □ **unsolvableness** n. **unsolvability** /-ˌsɒlvə'bɪlɪtɪ/ n.

unsolved /ʌn'sɒlvd/ adj. not solved.

unsophisticated /ˌʌnsə'fɪstɪˌkeɪtɪd/ adj. **1** artless, simple, natural, ingenuous. **2** not adulterated or artificial. □ **unsophisticatedly** adv. **unsophisticatedness** n. **unsophistication** /-ˌfɪstɪ'keɪʃ(ə)n/ n.

unsorted /ʌn'sɔːtɪd/ adj. not sorted.

unsought /ʌn'sɔːt/ adj. **1** not searched out or sought for. **2** unasked; without being requested.

unsound /ʌn'saʊnd/ adj. **1** unhealthy, diseased. **2** rotten, weak. **3 a** ill-founded, fallacious. **b** unorthodox, heretical. **4** unreliable. **5** wicked. □ **of unsound mind** insane. □ **unsoundly** adv. **unsoundness** n.

unsounded[1] /ʌn'saʊndɪd/ adj. **1** not uttered or pronounced. **2** not made to sound.

unsounded[2] /ʌn'saʊndɪd/ adj. unfathomed.

unsoured /ʌn'saʊəd/ adj. not soured.

unsown /ʌn'səʊn/ adj. not sown.

unsparing /ʌn'speərɪŋ/ adj. **1** lavish, profuse. **2** merciless. □ **unsparingly** adv. **unsparingness** n.

unspeakable /ʌn'spiːkəb(ə)l/ adj. **1** that cannot be expressed in words. **2** indescribably bad or objectionable. □ **unspeakableness** n. **unspeakably** adv.

unspeaking /ʌn'spiːkɪŋ/ adj. literary not speaking; silent.

unspecialized /ʌn'speʃəˌlaɪzd/ adj. not specialized.

unspecified /ʌn'spesɪˌfaɪd/ adj. not specified.

unspectacular /ˌʌnspek'tækjʊlə(r)/ adj. not spectacular; dull. □ **unspectacularly** adv.

unspent /ʌn'spent/ adj. **1** not expended or used. **2** not exhausted or used up.

unspilled /ʌn'spɪld/ adj. not spilt.

unspilt /ʌn'spɪlt/ adj. not spilt.

unspiritual /ʌn'spɪrɪtjʊəl/ adj. not spiritual; earthly, worldly. □ **unspiritually** adv. **unspiritualness** n. **unspirituality** /-ˌspɪrɪtjʊ'ælɪtɪ/ n.

unspoiled /ʌn'spɔɪld/ adj. **1** unspoilt. **2** not plundered.

unspoilt /ʌn'spɔɪlt/ adj. not spoilt.

unspoken /ʌn'spəʊkən/ adj. **1** not expressed in speech. **2** not uttered as speech.

unsponsored /ʌn'spɒnsəd/ adj. not supported or promoted by a sponsor.

unsporting /ʌn'spɔːtɪŋ/ adj. not sportsmanlike; not fair or generous. □ **unsportingly** adv. **unsportingness** n.

unsportsmanlike /ʌn'spɔːtsmənˌlaɪk/ adj. unsporting.

unspotted /ʌn'spɒtɪd/ adj. **1 a** not marked with a spot or spots. **b** morally pure. **2** unnoticed.

unsprayed /ʌn'spreɪd/ adj. not sprayed, esp. (of crops etc.) with a pesticide.

unsprung /ʌnˈsprʌŋ/ *adj.* not provided with a spring or springs; not resilient.

unstable /ʌnˈsteɪb(ə)l/ *adj.* (**unstabler**, **unstablest**) **1** not stable. **2** changeable. **3** showing a tendency to sudden mental or emotional changes. □ **unstable equilibrium** a state in which a body when disturbed tends to move further from equilibrium. □ **unstableness** *n.* **unstably** *adv.*

unstained /ʌnˈsteɪnd/ *adj.* not stained.

unstamped /ʌnˈstæmpt/ *adj.* **1** not marked by stamping. **2** not having a stamp affixed.

unstarched /ʌnˈstɑːtʃt/ *adj.* not starched.

unstated /ʌnˈsteɪtɪd/ *adj.* not stated or declared.

unstatesmanlike /ʌnˈsteɪtsmənˌlaɪk/ *adj.* not statesmanlike.

unstatutable /ʌnˈstætjʊtəb(ə)l/ *adj.* contrary to a statute or statutes. □ **unstatutably** *adv.*

unsteadfast /ʌnˈstedfɑːst/ *adj.* not steadfast.

unsteady /ʌnˈstedɪ/ *adj.* (**unsteadier**, **unsteadiest**) **1** not steady or firm. **2** changeable, fluctuating. **3** not uniform or regular. □ **unsteadily** *adv.* **unsteadiness** *n.*

unsterile /ʌnˈsteraɪl/ *adj.* **1** (of a syringe etc.) not sterile. **2** productive.

unstick *v. & n.* ● *v.* /ʌnˈstɪk/ (*past* and *past part.* **unstuck** /-ˈstʌk/) **1** *tr.* separate (a thing stuck to another). **2** *Aeron. colloq.* **a** *intr.* take off. **b** *tr.* cause (an aircraft) to take off. ● *n.* /ˈʌnstɪk/ *Aeron. colloq.* the moment of take-off. □ **come unstuck** *colloq.* come to grief, fail.

unstinted /ʌnˈstɪntɪd/ *adj.* not stinted. □ **unstintedly** *adv.*

unstinting /ʌnˈstɪntɪŋ/ *adj.* ungrudging, lavish. □ **unstintingly** *adv.*

unstirred /ʌnˈstɜːd/ *adj.* not stirred.

unstitch /ʌnˈstɪtʃ/ *v.tr.* undo the stitches of.

unstop /ʌnˈstɒp/ *v.tr.* (**unstopped**, **unstopping**) **1** free from obstruction. **2** remove the stopper from.

unstoppable /ʌnˈstɒpəb(ə)l/ *adj.* that cannot be stopped or prevented. □ **unstoppably** *adv.* **unstoppability** /-ˌstɒpəˈbɪlɪtɪ/ *n.*

unstopper /ʌnˈstɒpə(r)/ *v.tr.* remove the stopper from.

unstrained /ʌnˈstreɪnd/ *adj.* **1** not subjected to straining or stretching. **2** not injured by overuse or excessive demands. **3** not forced or produced by effort. **4** not passed through a strainer.

unstrap /ʌnˈstræp/ *v.tr.* (**unstrapped**, **unstrapping**) undo the strap or straps of.

unstreamed /ʌnˈstriːmd/ *adj. Brit.* (of schoolchildren) not arranged in streams.

unstressed /ʌnˈstrest/ *adj.* **1** (of a word, syllable, etc.) not pronounced with stress. **2** not subjected to stress.

unstring /ʌnˈstrɪŋ/ *v.tr.* (*past* and *past part.* **unstrung** /-ˈstrʌŋ/) **1** remove or relax the string or strings of (a bow, harp, etc.). **2** remove from a string. **3** (esp. as **unstrung** *adj.*) unnerve.

unstructured /ʌnˈstrʌktʃəd/ *adj.* **1** not structured. **2** informal.

unstuck *past* and *past part.* of UNSTICK.

unstudied /ʌnˈstʌdɪd/ *adj.* easy, natural, spontaneous. □ **unstudiedly** *adv.*

unstuffed /ʌnˈstʌft/ *adj.* not stuffed.

unstuffy /ʌnˈstʌfɪ/ *adj.* **1** informal, casual. **2** not stuffy.

unstylish /ʌnˈstaɪlɪʃ/ *adj.* **1** lacking style. **2** unfashionable.

unsubdued /ˌʌnsəbˈdjuːd/ *adj.* not subdued.

unsubjugated /ʌnˈsʌbdʒʊˌɡeɪtɪd/ *adj.* not subjugated.

unsubstantial /ˌʌnsəbˈstænʃ(ə)l/ *adj.* having little or no solidity, reality, or factual basis. □ **unsubstantially** *adv.* **unsubstantiality** /-ˌstænʃɪˈælɪtɪ/ *n.*

unsubstantiated /ˌʌnsəbˈstænʃɪˌeɪtɪd/ *adj.* not substantiated.

unsubtle /ʌnˈsʌt(ə)l/ *adj.* not subtle; obvious, clumsy. □ **unsubtly** *adv.*

unsuccess /ˌʌnsəkˈses/ *n.* **1** lack of success; failure. **2** an instance of this.

unsuccessful /ˌʌnsəkˈsesfʊl/ *adj.* not successful. □ **unsuccessfully** *adv.* **unsuccessfulness** *n.*

unsugared /ʌnˈʃʊɡəd/ *adj.* not sugared.

unsuggestive /ˌʌnsəˈdʒestɪv/ *adj.* not suggestive.

unsuitable /ʌnˈsuːtəb(ə)l, ʌnˈsjuː-/ *adj.* not suitable. □ **unsuitably** *adv.* **unsuitableness** *n.* **unsuitability** /-ˌsuːtəˈbɪlɪtɪ, -ˌsjuː-/ *n.*

unsuited /ʌnˈsuːtɪd, ʌnˈsjuː-/ *adj.* **1** (usu. foll. by *for*) not fit for a purpose. **2** (usu. foll. by *to*) not adapted.

unsullied /ʌnˈsʌlɪd/ *adj.* not sullied.

unsummoned /ʌnˈsʌmənd/ *adj.* not summoned.

unsung /ʌnˈsʌŋ/ *adj.* **1** not celebrated in song; unknown. **2** not sung.

unsupervised /ʌnˈsuːpəˌvaɪzd, ʌnˈsjuː-/ *adj.* not supervised.

unsupportable /ˌʌnsəˈpɔːtəb(ə)l/ *adj.* **1** that cannot be endured. **2** indefensible. □ **unsupportably** *adv.*

unsupported /ˌʌnsəˈpɔːtɪd/ *adj.* not supported. □ **unsupportedly** *adv.*

unsupportive /ˌʌnsəˈpɔːtɪv/ *adj.* not giving support.

unsure /ʌnˈʃʊə(r), -ˈʃɔː(r)/ *adj.* not sure, uncertain (*unsure of the facts*). □ **unsurely** *adv.* **unsureness** *n.*

unsurfaced /ʌnˈsɜːfɪst/ *adj.* (of a road etc.) not provided with a surface.

unsurpassable /ˌʌnsəˈpɑːsəb(ə)l/ *adj.* that cannot be surpassed. □ **unsurpassably** *adv.*

unsurpassed /ˌʌnsəˈpɑːst/ *adj.* not surpassed.

unsurprising /ˌʌnsəˈpraɪzɪŋ/ *adj.* not surprising. □ **unsurprisingly** *adv.*

unsusceptible /ˌʌnsəˈseptɪb(ə)l/ *adj.* not susceptible. □ **unsusceptibility** /-ˌseptɪˈbɪlɪtɪ/ *n.*

unsuspected /ˌʌnsəˈspektɪd/ *adj.* not suspected. □ **unsuspectedly** *adv.*

unsuspecting /ˌʌnsəˈspektɪŋ/ *adj.* not suspecting, off guard. □ **unsuspectingly** *adv.* **unsuspectingness** *n.*

unsuspicious /ˌʌnsəˈspɪʃəs/ *adj.* not suspicious. □ **unsuspiciously** *adv.* **unsuspiciousness** *n.*

unsustainable /ˌʌnsəˈsteɪnəb(ə)l/ *adj.* not sustainable. □ **unsustainably** *adv.*

unsustained /ˌʌnsəˈsteɪnd/ *adj.* not sustained.

unswathe /ʌnˈsweɪð/ *v.tr.* free from being swathed.

unswayed /ʌnˈsweɪd/ *adj.* uninfluenced, unaffected.

unsweetened /ʌnˈswiːt(ə)nd/ *adj.* not sweetened.

unswept /ʌnˈswept/ *adj.* not swept.

unswerving /ʌnˈswɜːvɪŋ/ *adj.* **1** steady, constant. **2** not turning aside. □ **unswervingly** *adv.*

unsworn /ʌnˈswɔːn/ *adj.* **1** (of a person) not subjected to or bound by an oath. **2** not confirmed by an oath.

unsymmetrical /ˌʌnsɪˈmetrɪk(ə)l/ *adj.* not symmetrical. □ **unsymmetrically** *adv.*

unsympathetic /ˌʌnsɪmpəˈθetɪk/ *adj.* not sympathetic, unfeeling. □ **unsympathetically** *adv.*

unsystematic /ˌʌnsɪstəˈmætɪk/ *adj.* not systematic, haphazard. □ **unsystematically** *adv.*

untack /ʌnˈtæk/ *v.tr.* detach, esp. by removing tacks.

untainted /ʌnˈteɪntɪd/ *adj.* not tainted.

untalented /ʌnˈtæləntɪd/ *adj.* not talented.

untameable /ʌnˈteɪməb(ə)l/ *adj.* that cannot be tamed.

untamed /ʌnˈteɪmd/ *adj.* not tamed, wild.

untangle /ʌnˈtæŋɡ(ə)l/ *v.tr.* **1** free from a tangled state. **2** free from entanglement.

untanned /ʌnˈtænd/ *adj.* not tanned.

untapped /ʌnˈtæpt/ *adj.* not (yet) tapped or wired (*untapped resources*).

untarnished /ʌnˈtɑːnɪʃt/ *adj.* not tarnished.

untasted /ʌnˈteɪstɪd/ *adj.* not tasted.

untaught /ʌnˈtɔːt/ *adj.* **1** not instructed by teaching; ignorant. **2** not acquired by teaching; natural, spontaneous.

untaxed /ʌnˈtækst/ *adj.* not required to pay or not attracting taxes.

unteach /ʌnˈtiːtʃ/ *v.tr.* (*past* and *past part.* **untaught** /-ˈtɔːt/) **1** cause (a person) to forget or discard previous knowledge. **2** remove from the mind (something known or taught) by different teaching.

unteachable /ʌnˈtiːtʃəb(ə)l/ *adj.* **1** incapable of being instructed. **2** that cannot be imparted by teaching.

untearable /ʌnˈteərəb(ə)l/ *adj.* that cannot be torn.

untechnical /ʌnˈteknɪk(ə)l/ *adj.* not technical. □ **untechnically** *adv.*

untempered /ʌnˈtempəd/ *adj.* (of metal etc.) not brought to the proper hardness or consistency.

untenable /ʌnˈtenəb(ə)l/ *adj.* not tenable; that cannot be defended. □ **untenably** *adv.* **untenableness** *n.* **untenability** /-ˌtenəˈbɪlɪtɪ/ *n.*

untended /ʌnˈtendɪd/ *adj.* not tended; neglected.

untenured /ʌnˈtenjəd/ *adj.* not tenured.

untested /ʌn'testɪd/ adj. not tested or proved.

untether /ʌn'teðə(r)/ v.tr. release (an animal) from a tether.

untethered /ʌn'teðəd/ adj. not tethered.

unthanked /ʌn'θæŋkt/ adj. not thanked.

unthankful /ʌn'θæŋkfʊl/ adj. not thankful. □ **unthankfully** adv. **unthankfulness** n.

untheorized /ʌn'θɪəraɪzd/ adj. (also **-ised**) not elaborated from a fundamental theory.

unthinkable /ʌn'θɪŋkəb(ə)l/ adj. **1** that cannot be imagined or grasped by the mind. **2** colloq. highly unlikely or undesirable. □ **unthinkably** adv. **unthinkableness** n. **unthinkability** /-ˌθɪŋkə'bɪlɪtɪ/ n.

unthinking /ʌn'θɪŋkɪŋ/ adj. **1** thoughtless. **2** unintentional, inadvertent. □ **unthinkingly** adv. **unthinkingness** n.

unthought /ʌn'θɔːt/ adj. (often foll. by of) not thought of.

unthoughtful /ʌn'θɔːtfʊl/ adj. unthinking, unmindful; thoughtless. □ **unthoughtfulness** n.

unthread /ʌn'θred/ v.tr. **1** take the thread out of (a needle etc.). **2** find one's way out of (a maze).

unthreatening /ʌn'θret(ə)nɪŋ/ adj. not threatening or aggressive; safe.

unthrifty /ʌn'θrɪftɪ/ adj. **1** wasteful, extravagant, prodigal. **2** not thriving or flourishing. □ **unthriftily** adv. **unthriftiness** n.

unthrone /ʌn'θrəʊn/ v.tr. dethrone.

untidy /ʌn'taɪdɪ/ adj. (**untidier, untidiest**) not neat or orderly. □ **untidily** adv. **untidiness** n.

untie /ʌn'taɪ/ v.tr. (pres. part. **untying**) **1** undo (a knot etc.). **2** unfasten the cords etc. of (a package etc.). **3** release from bonds or attachment. [OE untīgan (as UN-², TIE)]

untied /ʌn'taɪd/ adj. not tied.

until /ən'tɪl, ʌn-/ prep. & conj. = TILL¹. ¶ Used esp. when beginning a sentence and in formal style, e.g. until you told me, I had no idea; he resided there until his decease. [orig. northern ME untill f. ON und as far as + TILL¹]

untilled /ʌn'tɪld/ adj. not tilled.

untimely /ʌn'taɪmlɪ/ adj. & adv. ● adj. **1** inopportune. **2** (of death) premature. ● adv. archaic **1** inopportunely. **2** prematurely. □ **untimeliness** n.

untinged /ʌn'tɪndʒd/ adj. not tinged.

untiring /ʌn'taɪərɪŋ/ adj. tireless. □ **untiringly** adv.

untitled /ʌn'taɪt(ə)ld/ adj. having no title.

unto /'ʌntuː, -tə/ prep. archaic = TO prep. (in all uses except as the sign of the infinitive); (do unto others; faithful unto death; take unto oneself). [ME f. UNTIL, with TO replacing northern TILL¹]

untold /ʌn'təʊld/ adj. **1** not told. **2** not (able to be) counted or measured (untold misery). [OE untēald (as UN-¹, TOLD)]

untouchable /ʌn'tʌtʃəb(ə)l/ adj. & n. ● adj. that may not or cannot be touched. ● n. a member of a hereditary Hindu group held to defile members of higher castes on contact. (See note below.) □ **untouchableness** n. **untouchability** /-ˌtʌtʃə'bɪlɪtɪ/ n.

  ■ Use of the term, and social restrictions accompanying it, were declared illegal under the constitution of India in 1949 and of Pakistan in 1953, although members of the group still suffer from discrimination. The term Harijan, introduced and popularized by Mahatma Gandhi, is preferred today, although some members of the group refer to themselves as Dalits ('depressed (classes)'). (See also CASTE.)

untouched /ʌn'tʌtʃt/ adj. **1** not touched. **2** not affected physically, not harmed, modified, used, or tasted. **3** not affected by emotion. **4** not discussed.

untoward /ˌʌntə'wɔːd, ʌn'təʊəd/ adj. **1** inconvenient, unlucky. **2** awkward. **3** perverse, refractory. **4** unseemly. □ **untowardly** /ˌʌntə'wɔːdlɪ/ adv. **untowardness** n.

untraceable /ʌn'treɪsəb(ə)l/ adj. that cannot be traced. □ **untraceably** adv.

untraced /ʌn'treɪst/ adj. not traced.

untracked /ʌn'trækt/ adj. **1** not marked with tracks from skis etc. **2** having no previously trodden track. **3** not traced or followed.

untraditional /ˌʌntrə'dɪʃən(ə)l/ adj. not traditional; unusual.

untrained /ʌn'treɪnd/ adj. not trained.

untrammelled /ʌn'træm(ə)ld/ adj. not trammelled, unhampered.

untransferable /ˌʌntræns'fɜːrəb(ə)l, ʌntrɑː-ns-/ adj. not transferable.

untransformed /ˌʌntræns'fɔːmd, ʌntrɑː-ns-/ adj. that has not transformed or has not been transformed.

untranslatable /ˌʌntrænz'leɪtəb(ə)l, ʌntrɑː-nz-/ adj. that cannot be translated satisfactorily. □ **untranslatably** adv. **untranslatability** /-ˌleɪtə'bɪlɪtɪ/ n.

untransportable /ˌʌntræns'pɔːtəb(ə)l, ʌntrɑː-ns-/ adj. that cannot be transported.

untravelled /ʌn'træv(ə)ld/ adj. (US **untraveled**) **1** that has not travelled. **2** that has not been travelled over or through.

untreatable /ʌn'triːtəb(ə)l/ adj. (of a disease etc.) that cannot be treated.

untreated /ʌn'triːtɪd/ adj. not treated.

untried /ʌn'traɪd/ adj. **1** not tried or tested. **2** inexperienced. **3** not yet tried by a judge.

untrimmed /ʌn'trɪmd/ adj. **1** left uncut or in an irregular shape. **2** not adorned.

untrodden /ʌn'trɒd(ə)n/ adj. not trodden, stepped on, or traversed.

untroubled /ʌn'trʌb(ə)ld/ adj. not troubled; calm, tranquil.

untrue /ʌn'truː/ adj. **1** not true, contrary to what is the fact. **2** (often foll. by to) not faithful or loyal. **3** deviating from an accepted standard. □ **untruly** adv. [OE untrēowe etc. (as UN-¹, TRUE)]

untruss /ʌn'trʌs/ v.tr. unfasten (a pack, trussed fowl, etc.).

untrusting /ʌn'trʌstɪŋ/ adj. not trusting; suspicious.

untrustworthy /ʌn'trʌstˌwɜːðɪ/ adj. not trustworthy or reliable. □ **untrustworthiness** n.

untruth /ʌn'truːθ/ n. (pl. **untruths** /-'truːðz, -'truːθs/) **1** the state of being untrue, falsehood. **2** a false statement (told me an untruth). [OE untrēowth etc. (as UN-¹, TRUTH)]

untruthful /ʌn'truːθfʊl/ adj. not truthful. □ **untruthfully** adv. **untruthfulness** n.

untuck /ʌn'tʌk/ v.tr. free (bedclothes etc.) from being tucked in or up.

untunable /ʌn'tjuːnəb(ə)l/ adj. (of a piano etc.) that cannot be tuned.

untuned /ʌn'tjuːnd/ adj. **1** not in tune, not made tuneful. **2** (of a radio receiver etc.) not tuned to any one frequency. **3** not in harmony or concord, disordered.

untuneful /ʌn'tjuːnfʊl/ adj. not tuneful. □ **untunefully** adv. **untunefulness** n.

unturned /ʌn'tɜːnd/ adj. **1** not turned over, round, away, etc. **2** not shaped by turning.

untutored /ʌn'tjuːtəd/ adj. uneducated, untaught.

untwine /ʌn'twaɪn/ v.tr. & intr. untwist, unwind.

untwist /ʌn'twɪst/ v.tr. & intr. open from a twisted or spiralled state.

untying pres. part. of UNTIE.

untypical /ʌn'tɪpɪk(ə)l/ adj. not typical, unusual. □ **untypically** adv.

unusable /ʌn'juːzəb(ə)l/ adj. not usable.

unused adj. **1** /ʌn'juːzd/ **a** not in use. **b** never having been used. **2** /ʌn'juːst/ (foll. by to) not accustomed.

unusual /ʌn'juːʒʊəl/ adj. **1** not usual. **2** exceptional, remarkable. □ **unusually** adv. **unusualness** n.

unutterable /ʌn'ʌtərəb(ə)l/ adj. inexpressible; beyond description (unutterable torment; an unutterable fool). □ **unutterableness** n. **unutterably** adv.

unuttered /ʌn'ʌtəd/ adj. not uttered or expressed.

unvaccinated /ʌn'væksɪˌneɪtɪd/ adj. not vaccinated.

unvalued /ʌn'væljuːd/ adj. **1** not regarded as valuable. **2** not having been valued.

unvanquished /ʌn'væŋkwɪʃt/ adj. not vanquished.

unvaried /ʌn'veərɪd/ adj. not varied.

unvarnished /ʌn'vɑːnɪʃt/ adj. **1** not varnished. **2** (of a statement or person) plain and straightforward (the unvarnished truth).

unvarying /ʌn'veərɪŋ/ adj. not varying. □ **unvaryingly** adv. **unvaryingness** n.

unveil /ʌn'veɪl/ v. **1** tr. remove a veil from. **2** tr. remove a covering from (a statue, plaque, etc.) as part of the ceremony of the first public display. **3** tr. disclose, reveal, make publicly known. **4** intr. remove one's veil.

unventilated /ʌn'ventɪˌleɪtɪd/ adj. **1** not provided with a means of ventilation. **2** not discussed.

unverifiable /ʌn'verɪˌfaɪəb(ə)l/ adj. that cannot be verified.

unverified /ʌn'verɪˌfaɪd/ adj. not verified.

unversed /ʌnˈvɜːst/ *adj.* (usu. foll. by *in*) not experienced or skilled.

unviable /ʌnˈvaɪəb(ə)l/ *adj.* not viable. □ **unviability** /-ˌvaɪəˈbɪlɪtɪ/ *n.*

unviolated /ʌnˈvaɪəˌleɪtɪd/ *adj.* not violated.

unvisited /ʌnˈvɪzɪtɪd/ *adj.* not visited.

unvitiated /ʌnˈvɪʃɪˌeɪtɪd/ *adj.* not vitiated.

unvoiced /ʌnˈvɔɪst/ *adj.* **1** not spoken. **2** *Phonet.* not voiced.

unwaged /ʌnˈweɪdʒd/ *adj.* not receiving a wage; out of work.

unwalled /ʌnˈwɔːld/ *adj.* without enclosing walls.

unwanted /ʌnˈwɒntɪd/ *adj.* not wanted.

unwarlike /ʌnˈwɔːlaɪk/ *adj.* not warlike.

unwarmed /ʌnˈwɔːmd/ *adj.* not warmed.

unwarned /ʌnˈwɔːnd/ *adj.* not warned or forewarned.

unwarrantable /ʌnˈwɒrəntəb(ə)l/ *adj.* indefensible, unjustifiable. □ **unwarrantableness** *n.* **unwarrantably** *adv.*

unwarranted /ʌnˈwɒrəntɪd/ *adj.* **1** unauthorized. **2** unjustified.

unwary /ʌnˈweərɪ/ *adj.* **1** not cautious. **2** (often foll. by *of*) not aware of possible danger etc. □ **unwarily** *adv.* **unwariness** *n.*

unwashed /ʌnˈwɒʃt/ *adj.* **1** not washed. **2** not usually washed or clean. □ **the great unwashed** *colloq.* the rabble.

unwatchable /ʌnˈwɒtʃəb(ə)l/ *adj.* disturbing or not interesting to watch.

unwatched /ʌnˈwɒtʃt/ *adj.* not watched.

unwatchful /ʌnˈwɒtʃfʊl/ *adj.* not watchful.

unwatered /ʌnˈwɔːtəd/ *adj.* not watered.

unwavering /ʌnˈweɪvərɪŋ/ *adj.* not wavering. □ **unwaveringly** *adv.*

unweaned /ʌnˈwiːnd/ *adj.* not weaned.

unwearable /ʌnˈweərəb(ə)l/ *adj.* that cannot be worn.

unwearied /ʌnˈwɪərɪd/ *adj.* **1** not wearied or tired. **2** never becoming weary, indefatigable. **3** unremitting. □ **unweariedly** *adv.* **unweariedness** *n.*

unweary /ʌnˈwɪərɪ/ *adj.* not weary.

unwearying /ʌnˈwɪərɪɪŋ/ *adj.* **1** persistent. **2** not causing or producing weariness. □ **unwearyingly** *adv.*

unwed /ʌnˈwed/ *adj.* unmarried.

unwedded /ʌnˈwedɪd/ *adj.* unmarried. □ **unweddedness** *n.*

unweeded /ʌnˈwiːdɪd/ *adj.* not cleared of weeds.

unweighed /ʌnˈweɪd/ *adj.* **1** not considered; hasty. **2** (of goods) not weighed.

unweight /ʌnˈweɪt/ *v.tr.* (usu. *absol.*) remove the weight from (esp. a ski, by ceasing to press). [back-form. f. *unweighted*]

unwelcome /ʌnˈwelkəm/ *adj.* not welcome or acceptable. □ **unwelcomely** *adv.* **unwelcomeness** *n.*

unwelcoming /ʌnˈwelkəmɪŋ/ *adj.* **1** having an inhospitable atmosphere. **2** hostile; unfriendly.

unwell /ʌnˈwel/ *adj.* **1** not in good health; (somewhat) ill. **2** indisposed.

unwept /ʌnˈwept/ *adj.* **1** not wept for. **2** (of tears) not wept.

unwetted /ʌnˈwetɪd/ *adj.* not wetted.

unwhipped /ʌnˈwɪpt/ *adj.* **1** not punished by or as by whipping. **2** *Brit.* not subject to a party whip.

unwholesome /ʌnˈhəʊlsəm/ *adj.* **1** not promoting, or detrimental to, physical or moral health. **2** unhealthy, insalubrious. **3** unhealthy-looking. □ **unwholesomely** *adv.* **unwholesomeness** *n.*

unwieldy /ʌnˈwiːldɪ/ *adj.* (**unwieldier, unwieldiest**) cumbersome, clumsy, or hard to manage, owing to size, shape, or weight. □ **unwieldily** *adv.* **unwieldiness** *n.* [ME f. UN-¹ + *wieldy* active (now dial.) f. WIELD]

unwilling /ʌnˈwɪlɪŋ/ *adj.* not willing or inclined; reluctant. □ **unwillingly** *adv.* **unwillingness** *n.* [OE *unwillende* (as UN-¹, WILLING)]

unwind /ʌnˈwaɪnd/ *v.* (*past* and *past part.* **unwound** /-ˈwaʊnd/) **1 a** *tr.* draw out (a thing that has been wound). **b** *intr.* become drawn out after having been wound. **2** *intr. & tr. colloq.* relax.

unwinking /ʌnˈwɪŋkɪŋ/ *adj.* **1** not winking. **2** watchful, vigilant. □ **unwinkingly** *adv.*

unwinnable /ʌnˈwɪnəb(ə)l/ *adj.* that cannot be won.

unwisdom /ʌnˈwɪzdəm/ *n.* lack of wisdom, folly, imprudence. [OE *unwīsdōm* (as UN-¹, WISDOM)]

unwise /ʌnˈwaɪz/ *adj.* **1** foolish, imprudent. **2** injudicious. □ **unwisely** *adv.* [OE *unwīs* (as UN-¹, WISE¹)]

unwished /ʌnˈwɪʃt/ *adj.* (usu. foll. by *for*) not wished for.

unwithered /ʌnˈwɪðəd/ *adj.* not withered; still vigorous or fresh.

unwitnessed /ʌnˈwɪtnɪst/ *adj.* not witnessed.

unwitting /ʌnˈwɪtɪŋ/ *adj.* **1** unaware of the state of the case (*an unwitting offender*). **2** unintentional. □ **unwittingly** *adv.* **unwittingness** *n.* [OE *unwitende* (as UN-¹, WIT²)]

unwomanly /ʌnˈwʊmənlɪ/ *adj.* not womanly; not befitting a woman. □ **unwomanliness** *n.*

unwonted /ʌnˈwəʊntɪd/ *adj.* not customary or usual. □ **unwontedly** *adv.* **unwontedness** *n.*

unwooded /ʌnˈwʊdɪd/ *adj.* not wooded, treeless.

unworkable /ʌnˈwɜːkəb(ə)l/ *adj.* not workable; impracticable. □ **unworkably** *adv.* **unworkableness** *n.* **unworkability** /-ˌwɜːkəˈbɪlɪtɪ/ *n.*

unworked /ʌnˈwɜːkt/ *adj.* **1** not wrought into shape. **2** not exploited or turned to account.

unworkmanlike /ʌnˈwɜːkmənˌlaɪk/ *adj.* badly done or made.

unworldly /ʌnˈwɜːldlɪ/ *adj.* **1** spiritually-minded. **2** spiritual. □ **unworldliness** *n.*

unworn /ʌnˈwɔːn/ *adj.* not worn or impaired by wear.

unworried /ʌnˈwʌrɪd/ *adj.* not worried; calm.

unworthy /ʌnˈwɜːðɪ/ *adj.* (**unworthier, unworthiest**) **1** (often foll. by *of*) not worthy or befitting the character of a person etc. **2** discreditable, unseemly. **3** contemptible, base. □ **unworthily** *adv.* **unworthiness** *n.*

unwound¹ /ʌnˈwaʊnd/ *adj.* not wound or wound up.

unwound² *past* and *past part.* of UNWIND.

unwounded /ʌnˈwuːndɪd/ *adj.* not wounded, unhurt.

unwoven /ʌnˈwəʊv(ə)n/ *adj.* not woven.

unwrap /ʌnˈræp/ *v.* (**unwrapped, unwrapping**) **1** *tr.* remove the wrapping from. **2** *tr.* open or unfold. **3** *intr.* become unwrapped.

unwrinkled /ʌnˈrɪŋk(ə)ld/ *adj.* free from wrinkles, smooth.

unwritable /ʌnˈraɪtəb(ə)l/ *adj.* that cannot be written.

unwritten /ʌnˈrɪt(ə)n/ *adj.* **1** not written. **2** (of a law etc.) resting originally on custom or judicial decision, not on statute.

unwrought /ʌnˈrɔːt/ *adj.* (of metals) not hammered into shape or worked into a finished condition.

unyielding /ʌnˈjiːldɪŋ/ *adj.* **1** not yielding to pressure etc. **2** firm, obstinate. □ **unyieldingly** *adv.* **unyieldingness** *n.*

unyoke /ʌnˈjəʊk/ *v.* **1** *tr.* release from a yoke. **2** *intr.* cease work.

unzip /ʌnˈzɪp/ *v.tr.* (**unzipped, unzipping**) unfasten the zip of.

up /ʌp/ *adv., prep., adj., n., & v.* ● *adv.* **1** at, in, or towards a higher place or position (*jumped up in the air; what are they doing up there?*). **2** to or in a place regarded as higher, esp.: **a** northwards (*up in Scotland*). **b** *Brit.* towards a major city or a university (*went up to London*). **3** *colloq.* ahead etc. as indicated (*went up front*). **4 a** to or in an erect position or condition (*stood it up*). **b** to or in a prepared or required position (*wound up the watch*). **c** in or into a condition of efficiency, activity, or progress (*stirred up trouble; the house is up for sale; the hunt is up*). **5** *Brit.* in a stronger or winning position or condition (*our team was three goals up; am £10 up on the transaction*). **6** (of a computer) running and available for use. **7** to the place or time in question or where the speaker etc. is (*a child came up to me; went straight up to the door; has been fine up till now*). **8** at or to a higher price or value (*our costs are up; shares are up*). **9 a** completely or effectually (*burn up; eat up; tear up; use up*). **b** more loudly or clearly (*speak up*). **10** in a state of completion; denoting the end of availability, supply, etc. (*time is up*). **11** into a compact, accumulated, or secure state (*pack up; save up; do up*). **12** out of bed (*are you up yet?*). **13** (of the sun etc.) having risen. **14** happening, esp. unusually or unexpectedly (*something is up*). **15** (usu. foll. by *on* or *in*) taught or informed (*is well up in French*). **16** (usu. foll. by *before*) appearing for trial etc. (*was up before the magistrate*). **17** (of a road etc.) being repaired. **18** (of a jockey) in the saddle. **19** towards the source of a river. **20** inland. **21** (of the points etc. in a game): **a** registered on the scoreboard. **b** forming the total score for the time being. **22** upstairs, esp. to bed (*are you going up yet?*). **23** (of a theatre-curtain) raised etc. to reveal the stage. **24** (as *int.*) get up. **25** (of a ship's helm) with rudder to leeward. **26** in rebellion. ● *prep.* **1** upwards along, through, or into (*climbed up the ladder*). **2** from the bottom to the top of. **3** along (*walked up the road*). **4 a** at or in a higher part of (*is situated up the street*). **b** towards the source of (a river). ● *adj.* **1** directed upwards (*up stroke*). **2** *Brit.* of travel towards a capital or centre (*the up train; the up platform*). **3** (of beer etc.) effervescent, frothy. ● *n.* a spell of good fortune. ● *v.* (**upped, upping**) **1** *intr. colloq.* start up; begin abruptly to say or do something (*upped and hit him*). **2** *intr.* (foll.

by *with*) raise; pick up (*upped with his stick*). **3** *tr.* increase or raise, esp. abruptly (*upped all their prices*). □ **be all up with** be disastrous or hopeless for (a person). **on the up and up** *colloq.* **1** *Brit.* steadily improving. **2** esp. *N. Amer.* honest(ly); on the level. **something is up** *colloq.* something unusual or undesirable is afoot or happening. **up against 1** close to. **2** in or into contact with. **3** *colloq.* confronted with (*up against a problem*). **up against it** *colloq.* in great difficulties. **up-anchor** *Naut.* weigh anchor. **up and about** (or **doing**) having risen from bed; active. **up-and-coming** *colloq.* (of a person) making good progress and likely to succeed. **up and down 1** to and fro (along). **2** in every direction. **3** *colloq.* in varying health or spirits. **up-and-over** (of a door) opened by being raised and pushed back into a horizontal position. **up and running** *adj. & adv.* functioning; in operation. **up-and-under** *Rugby* a high kick to allow time for fellow team members to reach the point where the ball will come down. **up draught** an upward draught, esp. in a chimney. **up for** available for or being considered for (office etc.). **up hill and down dale** up and down hills on an arduous journey. **up in arms** see ARM². **up-market** *adj. & adv.* towards or relating to the dearer or more affluent sector of the market. **up the pole** see POLE¹. **ups and downs 1** rises and falls. **2** alternate good and bad fortune. **up the spout** see SPOUT. **up stage** at or to the back of a theatre stage. **up sticks** see STICK¹. **up-stroke** a stroke made or written upwards. **up-tempo** *adj. & adv.* at a fast or an increased tempo. **up to 1** until (*up to the present*). **2** not more than (*you can have up to five*). **3** less than or equal to (*sums up to £10*). **4** incumbent on (*it is up to you to say*). **5** capable of or fit for (*am not up to a long walk*). **6** occupied or busy with (*what have you been up to?*). **up to date** see DATE¹. **up to the mark** see MARK¹. **up to the minute** see MINUTE¹. **up to snuff** see SNUFF². **up to one's tricks** see TRICK. **up to a person's tricks** see TRICK. **up with** *int.* expressing support for a stated person or thing. **up yours** *coarse sl.* expressing contemptuous defiance or rejection. **what's up?** *colloq.* **1** what is going on? **2** what is the matter? [OE *up(p), uppe*, rel. to OHG *ūf*]

up- /ʌp/ *prefix* in senses of UP, added: **1** as an adverb to verbs and verbal derivations, = 'upwards' (*upcurved*; *update*). **2** as a preposition to nouns forming adverbs and adjectives (*up-country*; *uphill*). **3** as an adjective to nouns (*upland*; *up-stroke*). [OE *up(p)-*, = UP]

Upanishad /ʊˈpænɪˌʃæd/ *n.* each of a series of Hindu sacred treatises based on the Vedas, written in Sanskrit *c.*800–200 BC. The Upanishads mark the transition from ritual sacrifice to a mystical concern with the nature of reality; polytheism is superseded by a pantheistic monism derived from the basic concepts of atman and Brahman. [Skr., = sitting near, i.e. at the feet of a master, f. *upa* near + *ni-ṣad* sit down]

upas /ˈjuːpəs/ *n.* **1** (in full **upas-tree**) **a** a Javanese tree, *Antiaris toxicaria*, yielding a milky sap used as arrow-poison. **b** *Mythol.* a Javanese tree thought to be fatal to whatever came near it. **c** a pernicious influence, practice, etc. **2** the poisonous sap of upas and other trees. [Malay *ūpas* poison]

upbeat /ˈʌpbiːt/ *n. & adj.* ● *n.* an unaccented beat in music. ● *adj. colloq.* optimistic or cheerful.

upbraid /ʌpˈbreɪd/ *v.tr.* (often foll. by *with*, *for*) chide or reproach (a person). □ **upbraiding** *n.* [OE *upbrēdan* (as UP-, *brēdan = bregdan* BRAID in obs. sense 'brandish')]

upbringing /ˈʌpˌbrɪŋɪŋ/ *n.* the bringing up of a child; education. [obs. *upbring* to rear (as UP-, BRING)]

upcast *n. & v.* ● *n.* /ˈʌpkɑːst/ **1** the act of casting up; an upward throw. **2** *Mining* a shaft through which air leaves a mine. **3** *Geol.* = UPTHROW. ● *v.tr.* /ʌpˈkɑːst/ (*past and past part.* **upcast**) cast up.

upchuck /ˈʌptʃʌk/ *v.tr. & intr.* US *sl.* vomit.

upcoming /ʌpˈkʌmɪŋ/ *adj.* esp. *N. Amer.* forthcoming; about to happen.

up-country /ʌpˈkʌntrɪ/ *adv. & adj.* inland; towards the interior of a country.

update *v. & n.* ● *v.tr.* /ʌpˈdeɪt/ bring up to date. ● *n.* /ˈʌpdeɪt/ **1** the act or an instance of updating. **2** an updated version; a set of updated information. □ **updater** /ˈʌpˌdeɪtə(r)/ *n.*

Updike /ˈʌpdaɪk/, John (Hoyer) (b.1932), American novelist, poet, and short-story writer. He is noted for his quartet of novels *Rabbit, Run* (1960), *Rabbit Redux* (1971), *Rabbit is Rich* (1981), and *Rabbit at Rest* (1990), a small-town tragicomedy tracing the career of an ex-basketball player; the latter two novels were awarded Pulitzer Prizes. Other novels include *Couples* (1968) and *The Witches of Eastwick* (1984).

up-end /ʌpˈend/ *v.tr. & intr.* set or rise up on end.

upfield /ˈʌpfiːld/ *adv.* in or to a position nearer to the opponents' end of a football etc. field.

upfold /ˈʌpfəʊld/ *n. Geol.* an anticline.

upfront /ʌpˈfrʌnt/ *adv. & adj. colloq.* ● *adv.* (usu. **up front**) **1** at the front; in front. **2** (of payments) in advance. ● *adj.* **1** honest, open, frank. **2** (of payments) made in advance. **3** at the front or most prominent.

upgrade *v. & n.* ● *v.tr.* /ʌpˈgreɪd/ **1** raise in rank etc. **2** improve (equipment, machinery, etc.) esp. by replacing components. ● *n.* /ˈʌpgreɪd/ **1** the act or an instance of upgrading. **2** an upgraded piece of equipment etc. **3** US an upward slope. □ **on the upgrade 1** improving in health etc. **2** advancing, progressing. □ **upgrader** /ʌpˈgreɪdə(r)/ *n.*

upgrowth /ˈʌpgrəʊθ/ *n.* the process or result of growing upwards.

upheaval /ʌpˈhiːv(ə)l/ *n.* **1** a violent or sudden change or disruption. **2** *Geol.* an upward displacement of part of the earth's crust. **3** the act or an instance of heaving up.

upheave /ʌpˈhiːv/ *v.* **1** *tr.* heave or lift up, esp. forcibly. **2** *intr.* rise up.

uphill *adv., adj., & n.* ● *adv.* /ʌpˈhɪl/ in an ascending direction up a hill, slope, etc. ● *adj.* /ˈʌphɪl/ **1** sloping up; ascending. **2** arduous, difficult (*an uphill task*). ● *n.* /ˈʌphɪl/ an upward slope.

uphold /ʌpˈhəʊld/ *v.tr.* (*past and past part.* **upheld** /-ˈheld/) **1** confirm or maintain (a decision etc., esp. of another). **2** give support or countenance to (a person, practice, etc.). □ **upholder** *n.*

upholster /ʌpˈhəʊlstə(r)/ *v.tr.* **1** provide (furniture) with upholstery. **2** furnish (a room etc.) with furniture, carpets, etc. □ **well-upholstered** *joc.* (of a person) fat. [back-form. f. UPHOLSTERER]

upholsterer /ʌpˈhəʊlstərə(r)/ *n.* a person who upholsters furniture, esp. professionally. [obs. *upholster* (n.) f. UPHOLD (in obs. sense 'keep in repair') + -STER]

upholstery /ʌpˈhəʊlstərɪ/ *n.* **1** textile covering, padding, springs, etc., for furniture. **2** an upholsterer's work.

upkeep /ˈʌpkiːp/ *n.* **1** maintenance in good condition. **2** the cost or means of this.

upland /ˈʌplənd/ *n. & adj.* ● *n.* high or hilly country. ● *adj.* of or relating to this.

uplift *v. & n.* ● *v.tr.* /ʌpˈlɪft/ **1** raise; lift up. **2** elevate or stimulate morally or spiritually. ● *n.* /ˈʌplɪft/ **1** the act or an instance of being raised. **2** *Geol.* the raising of part of the earth's surface. **3** *colloq.* a morally or spiritually elevating influence. **4** support for the bust etc. from a garment. □ **uplifter** /ʌpˈlɪftə(r)/ *n.* **uplifting** *adj.* (esp. in sense 2 of *v.*).

uplighter /ˈʌpˌlaɪtə(r)/ *n.* a light placed or designed to throw illumination upwards.

upload *v. & n. Computing* ● *v.tr.* /ʌpˈləʊd/ (also *absol.*) transfer (data) esp. to a larger storage device or system. ● *n.* /ˈʌpləʊd/ (usu. *attrib.*) a transfer of this type (*upload feature*).

upmost var. of UPPERMOST.

upon /əˈpɒn/ *prep.* = ON. ¶ *Upon* is sometimes more formal, and is preferred in *once upon a time* and *upon my word*, and in uses such as *row upon row of seats* and *Christmas is almost upon us*. [ME f. UP + ON *prep.*, after ON *upp á*]

upper¹ /ˈʌpə(r)/ *adj. & n.* ● *adj.* **1** higher in place; situated above another part (*the upper atmosphere*; *the upper lip*). **2** higher in rank or dignity etc. (*the upper class*). **3 a** situated on higher ground or further inland (*Upper Egypt*). **b** situated to the north (*Upper California*). **4** *Geol. & Archaeol.* (of a stratigraphic division, deposit, or period) later or younger, and usu. nearer the surface (*Upper Jurassic, upper palaeolithic*). ● *n.* the part of a boot or shoe above the sole. □ **on one's uppers** *colloq.* extremely short of money. **upper case** see CASE². **upper class** the highest class of society, esp. the aristocracy. **upper-class** *adj.* of the upper class. **the upper crust** *colloq.* the aristocracy. **upper-crust** *adj. colloq.* of the aristocracy. **upper-cut** *n.* an upward blow delivered with the arm bent. ● *v.tr.* hit with an upper-cut. **the upper hand** dominance or control. **the upper regions 1** the sky. **2** heaven. **upper works** the part of a ship that is above the water when fully laden. [ME f. UP + -ER²]

upper² /ˈʌpə(r)/ *n. sl.* a stimulant drug, esp. an amphetamine. [UP *v.* + -ER¹]

Upper Austria (called in German *Oberösterreich*) a state of NW Austria; capital, Linz.

Upper Canada the mainly English-speaking region of Canada north of the Great Lakes and west of the Ottawa river, in what is now southern Ontario. It was a British colony from 1791 to 1841, when it was united with Lower Canada.

Upper House *n.* the higher house in a legislature, esp. the House of Lords.

uppermost /ˈʌpəˌməʊst/ *adj. & adv.* ● *adj.* (also **upmost** /ˈʌpməʊst/)

1 highest in place or rank. **2** predominant. ● *adv.* at or to the highest or most prominent position.

upperpart /'ʌpə(ə)ˌpɑːt/ *n.* (usu. in *pl.*) a higher part, esp. a part of the upperside of an animal.

Upper Volta the former name (until 1984) for Burkina.

uppish /'ʌpɪʃ/ *adj.* esp. *Brit. colloq.* self-assertive or arrogant. □ **uppishly** *adv.* **uppishness** *n.*

uppity /'ʌpɪtɪ/ *adj. colloq.* uppish, snobbish. [fanciful f. UP]

Uppsala /'ʊpsɑːlə/ a city in eastern Sweden; pop. (1990) 167,500. Its university, founded in 1477, is the oldest in northern Europe.

upraise /ʌp'reɪz/ *v.tr.* raise to a higher level.

uprate /ʌp'reɪt/ *v.tr.* **1** increase the value of (a pension, benefit, etc.). **2** upgrade, esp. to improve performance.

upright /'ʌpraɪt/ *adj., adv. & n.* ● *adj.* **1** erect, vertical (*an upright posture*; *stood upright*). **2** (of a piano) with vertical strings. **3** (of a person or behaviour) righteous; strictly honourable or honest. **4** (of a picture, book, etc.) greater in height than breadth. ● *adv.* in a vertical direction, vertically upwards; into an upright position. ● *n.* **1** a post or rod fixed upright esp. as a structural support. **2** an upright piano. □ **uprightly** *adv.* **uprightness** *n.* [OE *upriht* (as UP, RIGHT)]

uprise /ʌp'raɪz/ *v.intr.* (**uprose** /-'rəʊz/, **uprisen** /-'rɪz(ə)n/) rise (to a standing position, etc.).

uprising /'ʌpˌraɪzɪŋ/ *n.* a rebellion or revolt.

upriver *adv. & adj.* ● *adv.* /ʌp'rɪvə(r)/ at or towards a point nearer to the source of a river. ● *adj.* /'ʌpˌrɪvə(r)/ situated or occurring upriver.

uproar /'ʌprɔː(r)/ *n.* a tumult; a violent disturbance. [Du. *oproer* f. *op* up + *roer* confusion, assoc. with ROAR]

uproarious /ʌp'rɔːrɪəs/ *adj.* **1** very noisy; tumultuous. **2** provoking loud laughter. □ **uproariously** *adv.* **uproariousness** *n.*

uproot /ʌp'ruːt/ *v.tr.* **1** pull (a plant etc.) up from the ground. **2** displace (a person) from an accustomed location. **3** eradicate, destroy. □ **uprooter** *n.*

uprose *past* of UPRISE.

uprush /'ʌprʌʃ/ *n.* an upward rush, esp. *Psychol.* from the subconscious.

ups-a-daisy var. of UPSY-DAISY.

upset *v., n., & adj.* ● *v.* /ʌp'set/ (**upsetting**; *past* and *past part.* **upset**) **1 a** *tr. & intr.* overturn or be overturned. **b** *tr.* overcome, defeat. **2** *tr.* disturb the composure or digestion of (*was very upset by the news*; *ate something that upset me*). **3** *tr.* disrupt. **4** *tr.* shorten and thicken (metal, esp. a tire) by hammering or pressure. ● *n.* /'ʌpset/ **1** a condition of upsetting or being upset (*a stomach upset*). **2** a surprising result in a game etc.; a giant-killing. ● *adj.* /'ʌpset/ disturbed (*an upset stomach*). □ **upset price** the lowest acceptable selling price of a property in an auction etc.; a reserve price. □ **upsetter** /ʌp'setə(r)/ *n.* **upsettingly** *adv.*

upshift /'ʌpʃɪft/ *v. & n.* ● *v.* **1** *intr.* move to a higher gear in a motor vehicle. **2** *tr.* esp. *US* increase (*upshifted the penalties*). ● *n.* a movement upwards, esp. a change to a higher gear.

upshot /'ʌpʃɒt/ *n.* the final or eventual outcome or conclusion.

upside down /ˌʌpsaɪd 'daʊn/ *adv. & adj.* ● *adv.* **1** with the upper part where the lower part should be; in an inverted position. **2** in or into total disorder (*everything was turned upside down*). ● *adj.* (also **upside-down** *attrib.*) that is positioned upside down; inverted. □ **upside-down cake** a sponge cake baked with fruit in a syrup at the bottom, and inverted for serving. [ME, orig. *up so down*, perh. = 'up as if down']

upsides /'ʌpsaɪdz/ *adv. Brit. colloq.* (foll. by *with*) equal with (a person) by revenge, retaliation, etc. [*upside* = top part]

upsilon /'ʌpsɪˌlɒn, juːp'saɪlən/ *n.* the twentieth letter of the Greek alphabet (*Y, υ*). [Gk, = slender U f. *psilos* slender, with ref. to its later coincidence in sound with Gk *oi*]

upstage /ʌp'steɪdʒ/ *adj., adv., & v.* ● *adj. & adv.* **1** nearer the back of a theatre stage. **2** snobbish(ly). ● *v.tr.* **1** (of an actor) move upstage to make (another actor) face away from the audience. **2** divert attention from (a person) to oneself; outshine.

upstairs *adv., adj., & n.* ● *adv.* /ʌp'steəz/ to or on an upper floor. ● *adj.* /'ʌpsteəz/ (also **upstair**) situated upstairs. ● *n.* /'ʌpsteəz/ an upper floor.

upstanding /ʌp'stændɪŋ/ *adj.* **1** standing up. **2** strong and healthy. **3** honest or straightforward.

upstart /'ʌpstɑːt/ *n. & adj.* ● *n.* a person who has risen suddenly to prominence, esp. one who behaves arrogantly. ● *adj.* **1** that is an upstart. **2** of or characteristic of an upstart.

upstate /'ʌpsteɪt/ *n., adj., & adv. US* ● *n.* part of a state remote from its

large cities, esp. the northern part. ● *adj.* of or relating to this part. ● *adv.* in or to this part. □ **upstater** *n.*

upstream /'ʌpstriːm/ *adv. & adj.* ● *adv.* against the flow of a stream etc. ● *adj.* moving upstream.

upsurge /'ʌpsɜːdʒ/ *n.* an upward surge; a rise (esp. in feelings etc.).

upswept /'ʌpswept/ *adj.* **1** (of the hair) combed to the top of the head. **2** curved or sloped upwards.

upswing /'ʌpswɪŋ/ *n.* an upward movement or trend.

upsy-daisy /'ʌpsɪˌdeɪzɪ/ *int.* (also **ups-a-daisy** /'ʌpsəˌdeɪzɪ/) expressing encouragement to a child who is being lifted or has fallen. [earlier *up-a-daisy*: cf. LACKADAISICAL]

uptake /'ʌpteɪk/ *n.* **1** *colloq.* understanding; comprehension (esp. *quick* or *slow on the uptake*). **2** the act or an instance of taking up.

upthrow /'ʌpθrəʊ/ *n.* **1** the act or an instance of throwing upwards. **2** *Geol.* an upward dislocation of strata.

upthrust /'ʌpθrʌst/ *n.* **1** upward thrust, e.g. of a fluid on an immersed body. **2** *Geol.* = UPHEAVAL 2.

uptight /ʌp'taɪt, 'ʌptaɪt/ *adj. colloq.* **1** nervously tense or angry. **2** rigidly conventional.

uptime /'ʌptaɪm/ *n.* time during which a machine, esp. a computer, is in operation.

uptown /'ʌptaʊn/ *adj., adv., & n. N. Amer.* ● *adj.* of or in the residential part of a town or city. ● *adv.* in or into this part. ● *n.* this part. □ **uptowner** *n.*

upturn *n. & v.* ● *n.* /'ʌptɜːn/ **1** an upward trend; an improvement. **2** an upheaval. ● *v.tr.* /ʌp'tɜːn/ turn up or upside down.

UPU see Universal Postal Union.

UPVC *abbr.* unplasticized polyvinyl chloride.

upward /'ʌpwəd/ *adv. & adj.* ● *adv.* (also **upwards** /-wədz/) towards what is higher, superior, larger in amount, more important, or earlier. ● *adj.* moving, extending, pointing, or leading upward. □ **upward mobility** social or professional advancement. **upwards of** more than (*found upwards of forty specimens*). [OE *upweard(es)* (as UP, -WARD)]

upwardly /'ʌpwədlɪ/ *adv.* in an upward direction. □ **upwardly mobile** able or aspiring to advance socially or professionally.

upwarp /'ʌpwɔːp/ *n. Geol.* a broad surface elevation; an anticline.

upwell /ʌp'wel/ *v.intr.* (esp. as **upwelling** *n.*) (of liquid, esp. sea water or magma) well upwards, rise to the surface.

upwind /'ʌpwɪnd/ *adj. & adv.* against the direction of the wind.

Ur /ɜː(r)/ an ancient Sumerian city formerly on the Euphrates, in southern Iraq. It was one of the oldest cities of Mesopotamia, dating from the 4th millennium BC, and reached its zenith in the 3rd millennium BC. It was ruled by Sargon I in the 24th century BC and according to the Bible was Abraham's place of origin. Much of the city was sacked and destroyed by the Elamites *c.*2000 BC, but it recovered and underwent a revival under Nebuchadnezzar. Ur was conquered by Cyrus the Great in the 6th century BC, after which it fell into decline and was finally abandoned in the 4th century BC. The site was excavated in 1922–34 by Sir Leonard Woolley, who discovered spectacular royal tombs and vast ziggurats dating from the period *c.*2600–2000 BC.

ur- /ʊə(r)/ *comb. form* primitive, original, earliest. [G]

uracil /'jʊərəsɪl/ *n. Biochem.* a pyrimidine found in living tissue as a component base of RNA. [UREA + ACETIC]

uraemia /jʊ'riːmɪə/ *n.* (*US* **uremia**) *Med.* a raised level in the blood of nitrogenous waste compounds (esp. urea) that are normally eliminated by the kidneys. □ **uraemic** *adj.* [Gk *ouron* urine + *haima* blood]

uraeus /jʊ'riːəs/ *n.* the sacred serpent as an emblem of power represented on the head-dress of ancient Egyptian divinities and sovereigns. [mod.L f. Gk *ouraios*, repr. the Egypt. word for 'cobra']

Ural-Altaic /ˌjʊərəlæl'teɪɪk/ *n. & adj.* ● *n.* a group of languages, formerly thought to be related, comprising the Uralic and Altaic language families. ● *adj.* **1** of or relating to this group of languages. **2** of or relating to the Ural and Altai Mountains.

Uralic /jʊ'rælɪk/ *n. & adj.* ● *n.* a family of languages comprising the Finno-Ugric group and Samoyed, spoken over a wide area in Europe, Asia, and the Scandinavian countries. ● *adj.* **1** of or relating to this family of languages. **2** of or relating to the Ural Mountains or the surrounding areas. [URAL MOUNTAINS]

Ural Mountains /'jʊərəl/ (also **Urals**) a mountain range in northern Russia, extending 1,600 km (1,000 miles) southwards from the Arctic

Ocean to the Aral Sea, in Kazakhstan, rising to 1,894 m (6,214 ft) at Mount Narodnaya. It forms the boundary between Europe and Asia.

Urania /juˈreɪnɪə/ *Gk & Rom. Mythol.* the Muse of astronomy. [Gk, = heavenly]

uraninite /juˈrænɪˌnaɪt/ *n.* a black, brown, or grey mineral consisting mainly of uranium dioxide, esp. in the form of pitchblende. It is the major ore of uranium.

uranium /juˈreɪnɪəm/ *n.* a grey dense radioactive metallic chemical element (atomic number 92; symbol **U**). Uranium was discovered by Klaproth in 1789 and first isolated in 1841. It occurs in pitchblende, uraninite, and other minerals, and is a chemically reactive metal belonging to the actinide series. Becquerel discovered radioactivity in uranium in 1896, and its capacity to undergo fission, first revealed in 1938, led to its use as a source of energy, though the fissile isotope, uranium-235, has to be separated from the more common uranium-238 by a laborious process before it can be used. The atom bomb exploded over Hiroshima in 1945 contained uranium-235. This isotope has since been used to produce power in nuclear reactors, as has uranium-238 which, though not fissile, can be transmuted into the fissile element plutonium. Uranium compounds have been used as colouring agents for ceramics and glass. □ **uranic** /-ˈrænɪk/ *adj.* **uranous** /ˈjʊərənəs/ *adj.* [f. URANUS]

urano-[1] /ˈjʊərənəʊ/ *comb. form* the heavens. [Gk *ouranos* heaven(s)]

urano-[2] /ˈjʊərənəʊ/ *comb. form* uranium.

uranography /ˌjʊərəˈnɒɡrəfɪ/ *n.* the branch of astronomy concerned with describing and mapping the stars etc. □ **uranographer** *n.* **uranographic** /-nəˈɡræfɪk/ *adj.*

Uranus /ˈjʊərənəs, jʊˈreɪn-/ **1** *Gk Mythol.* a personification of heaven or the sky, the most ancient of the Greek gods and first ruler of the universe. He was overthrown and castrated by his son Cronus. **2** *Astron.* the seventh planet from the sun in the solar system, orbiting between Jupiter and Neptune at an average distance of 2,870 million km from the sun. (*See note below.*) [L f. Gk *Ouranos* heaven]

▪ Uranus was the first planet to be discovered by telescope, by William Herschel in 1781. It is the third largest planet (but fourth in mass), with an equatorial diameter of 50,800 km, and one of the gas giants. The planet is bluish-green in colour, having an upper atmosphere consisting almost entirely of hydrogen and helium, and appears featureless. It is unusual in that its axis lies almost in the plane of the solar system. There are at least fifteen satellites, the largest of which are Oberon and Titania, and a faint ring system.

urban /ˈɜːb(ə)n/ *adj.* of, living in, or situated in a town or city (*an urban population*) (opp. RURAL 1). □ **urban district** *Brit. hist.* a group of urban communities governed by an elected council. **urban guerrilla** a terrorist operating in an urban area. **urban renewal** slum clearance and redevelopment in a city or town. **urban sprawl** the uncontrolled expansion of urban areas. [L *urbanus* f. *urbs urbis* city]

urbane /ɜːˈbeɪn/ *adj.* courteous; suave; elegant and refined in manner. □ **urbanely** *adv.* **urbaneness** *n.* [F *urbain* or L *urbanus*: see URBAN]

urbanism /ˈɜːbəˌnɪz(ə)m/ *n.* **1** urban character or way of life. **2** a study of urban life. □ **urbanist** *n.*

urbanite /ˈɜːbəˌnaɪt/ *n.* a dweller in a city or town.

urbanity /ɜːˈbænɪtɪ/ *n.* **1** an urbane quality; refinement of manner. **2** urban life. [F *urbanité* or L *urbanitas* (as URBAN)]

urbanize /ˈɜːbəˌnaɪz/ *v.tr.* (also **-ise**) **1** make urban. **2** destroy the rural quality of (a district). □ **urbanization** /ˌɜːbənaɪˈzeɪʃ(ə)n/ *n.* [F *urbaniser* (as URBAN)]

URC see UNITED REFORMED CHURCH.

urceolate /ˈɜːsɪələt/ *adj. Bot.* having the shape of a pitcher, with a large body and small mouth. [L *urceolus* dimin. of *urceus* pitcher]

urchin /ˈɜːtʃɪn/ *n.* **1** a mischievous child, esp. young and raggedly dressed. **2** = sea urchin. **3** *archaic* **a** a hedgehog. **b** a goblin. [ME *hirchon*, *urcheon* f. ONF *herichon*, OF *heriçon* ult. f. L (*h*)*ericius* hedgehog]

Urdu /ˈʊədu:, ˈɜːdu:/ *n.* an Indic language related to Hindi, which it resembles in grammar and structure, but with a large admixture of Arabic and Persian words. Urdu developed from the language adopted by the early Muslim invaders of the Indian subcontinent and is usually written in Persian script. It is an official language in Pakistan, and the major language of Muslims in India. [Hind. (*zabān i*) *urdū* (language of the) camp, f. Pers. *urdū* f. Turkic *ordū*: see HORDE]

-ure /jə(r), ə(r)/ *suffix* forming: **1** nouns of action or process (*censure*; *closure*; *seizure*). **2** nouns of result (*creature*; *scripture*). **3** collective nouns (*legislature*; *nature*). **4** nouns of function (*judicature*; *ligature*). [from or after OF *-ure* f. L *-ura*]

urea /ˈjʊərɪə, jʊˈriːə/ *n. Biochem.* a soluble colourless crystalline nitrogenous compound contained esp. in the urine of mammals. □ **ureal** /ˈjʊərɪəl/ *adj.* [mod.L f. F *urée* f. Gk *ouron* urine]

uremia *US* var. of URAEMIA.

ureter /jʊˈriːtə(r)/ *n. Anat. & Zool.* the duct by which urine passes from the kidney to the bladder or cloaca. □ **ureteral** *adj.* **ureteric** /ˌjʊərɪˈterɪk/ *adj.* **ureteritis** /jʊˌriːtəˈraɪtɪs/ *n.* [F *uretère* or mod.L *ureter* f. Gk *ourētēr* f. *oureō* urinate]

urethane /jʊˈriːθeɪn, ˈjʊərɪˌθeɪn/ *n. Chem.* a crystalline amide, ethyl carbamate, used in plastics and paints. [F *uréthane* (as UREA, ETHANE)]

urethra /jʊˈriːθrə/ *n.* (*pl.* **urethrae** /-riː/ or **urethras**) *Anat.* the duct by which urine is discharged from the bladder. □ **urethral** *adj.* **urethritis** /ˌjʊərɪˈθraɪtɪs/ *n.* [LL f. Gk *ourēthra* (as URETER)]

Urey /ˈjʊərɪ/, Harold Clayton (1893–1981), American chemist. He searched for a heavy isotope of hydrogen, discovering deuterium in 1932, and developing a technique for obtaining heavy water. Because of his work on isotope separation he was made director of the atom bomb project at Columbia University during the Second World War. Urey pioneered the use of isotope labelling; he also developed theories on the formation of the planets and of the possible synthesis of organic compounds in earth's primitive atmosphere. He was awarded the Nobel Prize for chemistry in 1934.

Urga /ˈʊəɡə/ the former name (until 1924) for ULAN BATOR.

urge /ɜːdʒ/ *v. & n.* ● *v.tr.* **1** (often foll. by *on*) drive forcibly; impel; hasten (*urged them on*; *urged the horses forward*). **2** (often foll. by *to* + infin. *or that* + clause) encourage or entreat earnestly or persistently (*urged them to go*; *urged them to action*; *urged that they should go*). **3** (often foll. by *on, upon*) advocate (an action or argument etc.) pressingly or emphatically (to a person). **4** adduce forcefully as a reason or justification (*urged the seriousness of the problem*). **5** ply (a person etc.) hard with argument or entreaty. ● *n.* **1** an urging impulse or tendency. **2** a strong desire. [L *urgere* press, drive]

urgent /ˈɜːdʒənt/ *adj.* **1** requiring immediate action or attention (*an urgent need for help*). **2** importunate; earnest and persistent in demand. □ **urgency** *n.* **urgently** *adv.* [ME f. F (as URGE)]

urger /ˈɜːdʒə(r)/ *n.* **1** a person who urges or incites. **2** *Austral. sl.* a person who obtains money dishonestly, esp. as a racing tipster.

URI *abbr.* upper respiratory infection.

-uria /ˈjʊərɪə/ *comb. form* forming nouns denoting that a substance is (esp. excessively) present in the urine. [mod.L f. Gk *-ouria* (as URINE)]

Uriah /jʊˈraɪə/ (in the Bible) a Hittite officer in David's army, husband of Bathsheba, whom David caused to be killed in battle (2 Sam. 11).

uric /ˈjʊərɪk/ *adj.* of or relating to urine. □ **uric acid** an almost insoluble acid forming the main nitrogenous waste product in birds, reptiles, and insects. [F *urique* (as URINE)]

urinal /jʊˈraɪn(ə)l, ˈjʊərɪn-/ *n.* **1** a sanitary fitting, usu. against a wall, for men to urinate into. **2** a place or receptacle for urination. [ME f. OF f. LL *urinal* neut. of *urinalis* (as URINE)]

urinalysis /ˌjʊərɪˈnælɪsɪs/ *n.* (*pl.* **urinalyses** /-ˌsiːz/) the chemical analysis of urine esp. for diagnostic purposes.

urinary /ˈjʊərɪnərɪ/ *adj.* **1** of or relating to urine. **2** affecting or occurring in the urinary system (*urinary diseases*).

urinate /ˈjʊərɪˌneɪt/ *v.intr.* discharge urine. □ **urination** /ˌjʊərɪˈneɪʃ(ə)n/ *n.* [med.L *urinare* (as URINE)]

urine /ˈjʊərɪn/ *n.* a pale-yellow fluid secreted as waste from the blood by the kidneys, stored in the bladder, and discharged through the urethra. □ **urinous** *adj.* [ME f. OF f. L *urina*]

urn /ɜːn/ *n. & v.* ● *n.* **1** a vase with a foot and usu. a rounded body, esp. for storing the ashes of the cremated dead or as a vessel or measure. **2** a large vessel with a tap, in which tea or coffee etc. is made or kept hot. **3** *poet.* anything in which a dead body or its remains are preserved, e.g. a grave. ● *v.tr.* enclose in an urn. □ **urnful** *n.* (*pl.* **-fuls**). [ME f. L *urna*, rel. to *urceus* pitcher]

urnfield /ˈɜːnfiːld/ *n. & adj. Archaeol.* ● *n.* a cemetery in which cremated remains were placed in pottery vessels (cinerary urns) and buried, especially one belonging to a group of Bronze Age cultures of central Europe and associated with Celtic peoples. ● *adj.* designating a people or culture characterized by burial in this form of cemetery.

uro-[1] /ˈjʊərəʊ/ *comb. form* urine. [Gk *ouron* urine]

uro-[2] /ˈjʊərəʊ/ *comb. form* tail. [Gk *oura* tail]

urochord /ˈjʊərəʊˌkɔːd/ *n. Zool.* the notochord of a tunicate. [URO-[2] + CHORD[2]]

urodele /ˈjʊərəʊˌdiːl/ *n. Zool.* an amphibian of the order Urodela, having a tail when in the adult form, including newts and salamanders. [URO-² + Gk *dēlos* evident]

urogenital /ˌjʊərəʊˈdʒɛnɪt(ə)l/ *adj.* of or relating to urinary and genital products or organs.

urology /jʊˈrɒlədʒɪ/ *n.* the scientific study of the urinary system. □ **urologist** *n.* **urologic** /ˌjʊərəˈlɒdʒɪk/ *adj.*

uropygium /ˌjʊərəʊˈpɪdʒɪəm/ *n.* Zool. the rump of a bird. [med.L f. Gk *ouropugion*]

uroscopy /jʊˈrɒskəpɪ/ *n.* Med. hist. the examination of urine by simple inspection, esp. in diagnosis.

Ursa Major /ˌɜːsə ˈmeɪdʒə(r)/ Astron. one of the largest and most prominent northern constellations (the Great Bear). The seven brightest stars form a familiar formation variously called the Plough, Big Dipper, or Charles's Wain, and include the Pointers. [L, = greater (she-)bear, from the story in Greek mythology that the nymph Callisto was turned into a bear and placed by Zeus in the heavens in the form of the constellation]

Ursa Minor /ˌɜːsə ˈmaɪnə(r)/ Astron. a northern constellation (the Little Bear, also known as the Little Dipper). It contains the north celestial pole and Polaris. [L, = lesser (she-)bear]

ursine /ˈɜːsaɪn/ *adj.* of or like a bear. [L *ursinus* f. *ursus* bear]

Ursula, St /ˈɜːsjʊlə/ a legendary British saint and martyr, said to have been put to death with 11,000 virgins after being captured by Huns near Cologne while on a pilgrimage. The legend probably developed from an incident of the 4th century or earlier.

Ursuline /ˈɜːsjʊˌlaɪn, -lɪn/ *n. & adj.* ● *n.* a nun of an order founded by St Angela Merici (1470–1540) at Brescia in northern Italy in 1535 for nursing the sick and teaching girls. It is the oldest teaching order of women in the Roman Catholic Church. ● *adj.* of or relating to this order. [St *Ursula*, the founder's patron saint (see URSULA, ST)]

urticaria /ˌɜːtɪˈkɛərɪə/ *n.* Med. nettle-rash. [mod.L f. L *urtica* nettle f. *urere* burn]

urticate /ˈɜːtɪˌkeɪt/ *v.tr.* sting like a nettle. □ **urtication** /ˌɜːtɪˈkeɪʃ(ə)n/ *n.* [med.L *urticare* f. L *urtica*: see URTICARIA]

Uruguay /ˈjʊərəˌɡwaɪ/ a country on the Atlantic coast of South America south of Brazil; pop. (est. 1991) 3,110,000; official language, Spanish; capital, Montevideo. Much of the country consists of low hills and grassy plains, suited to stock raising. Not permanently settled by Europeans until the 17th century, Uruguay was liberated from Spanish colonial rule in 1825. In the early 20th century it was moulded into South America's first welfare state. Civil unrest beginning in the 1960s, and particularly fighting against the Marxist Tupamaro guerrillas, led to a period of military rule, but civilian government was restored in 1985. □ **Uruguayan** /ˌjʊərəˈɡwaɪən/ adj. & n.

Uruk /ˈʊrʊk/ (Arabic **Warka** /wəˈkɑː/; biblical name **Erech** /ˈerek/) an ancient city in southern Mesopotamia, to the north-west of Ur. One of the greatest cities of Sumeria, it was built in the 5th millennium BC and was the seat of the legendary hero Gilgamesh. Excavations began in 1928 and revealed ziggurats and temples dedicated to the sky-god Anu.

Urumqi /ʊˈrʊmtʃɪ/ (also **Urumchi**) the capital of Xinjiang autonomous region in NW China; pop. (1990) 1,110,000. It was a major trading centre on the ancient caravan routes of central Asia, and developed during the 20th century into the main industrial centre of the region. In 1954 it was given the Mongolian name Urumqi, which means 'fine pasture', replacing its Chinese name, Tihwa.

urus /ˈjʊərəs/ *n.* = AUROCHS. [L f. Gmc]

US *abbr.* **1** United States (of America). **2** Under-Secretary. **3** unserviceable.

us /ʌs, əs/ *pron.* **1** objective case of WE (they saw us). **2** = WE (are older than us). ¶ Strictly, the pronoun following the verb *to be* should be in the subjective rather than the objective case, i.e. *are older than we* rather than *are older than us*. **3** colloq. = ME¹ (give us a kiss). [OE *ūs* f. Gmc]

USA *abbr.* **1** United States of America. **2** US United States Army.

usable /ˈjuːzəb(ə)l/ *adj.* that can be used. □ **usableness** *n.* **usability** /ˌjuːzəˈbɪlɪtɪ/ *n.*

USAF *abbr.* United States Air Force.

usage /ˈjuːsɪdʒ/ *n.* **1** a manner of using or treating; treatment (damaged by rough usage). **2** habitual or customary practice, esp. as creating a right, obligation, or standard. [ME f. OF f. *us* USE n.]

usance /ˈjuːz(ə)ns/ *n.* the time allowed by commercial usage for the payment of foreign bills of exchange. [ME f. OF (as USE)]

USDAW /ˈʌzdɔː/ *abbr.* (in the UK) Union of Shop, Distributive, and Allied Workers.

use *v. & n.* ● *v.tr.* /juːz/ **1 a** cause to act or serve for a purpose; bring into service; avail oneself of (rarely uses the car; use your discretion). **b** consume by eating or drinking; take (alcohol, a drug, etc.), esp. habitually. **2** treat (a person) in a specified manner (they used him shamefully). **3** exploit for one's own ends (they are just using you; used his position). **4** (in past /juːst/; foll. by to + infin.) did or had in the past (but no longer) as a customary practice or state (I used to be an archaeologist; it used not (or did not use) to rain so often). **5** (as **used** adj.) second-hand. **6** (as **used** /juːst/ predic.adj.) (foll. by to) familiar by habit; accustomed (not used to hard work). **7** apply (a name or title etc.) to oneself. ● *n.* /juːs/ **1** the act of using or the state of being used; application to a purpose (put it to good use; is in daily use; worn and polished with use). **2** the right or power of using (lost the use of my right arm). **3 a** the ability to be used (a torch would be of use). **b** the purpose for which a thing can be used (it's no use talking). **4** custom or usage (long use has reconciled me to it). **5** the characteristic ritual and liturgy of a church or diocese etc. **6** Law hist. the benefit or profit of lands, esp. in the possession of another who holds them solely for the beneficiary. □ **could use** colloq. would be glad to have; would be improved by having. **have no use for 1** be unable to find a use for. **2** dislike or be impatient with. **make use of 1** employ, apply. **2** benefit from. **use and wont** established custom. **use-by date** the latest recommended date of use marked on the packaging of perishable food etc. **use a person's name** quote a person as an authority or reference etc. **use up 1** consume completely, use the whole of. **2** find a use for (something remaining). **3** exhaust or wear out e.g. with overwork. [ME f. OF *us, user*, ult. f. L *uti us-* use]

useful /ˈjuːsfʊl/ *adj.* **1 a** of use; serviceable. **b** producing or able to produce good results (gave me some useful hints). **2** colloq. highly creditable or efficient (a useful performance). □ **make oneself useful** perform useful services. **useful load** the load carried by an aircraft etc. in addition to its own weight. □ **usefully** adv. **usefulness** n.

useless /ˈjuːslɪs/ *adj.* **1** serving no purpose; unavailing (the contents were made useless by damp; protest is useless). **2** colloq. feeble or ineffectual (am useless at swimming; a useless gadget). □ **uselessly** adv. **uselessness** n.

user /ˈjuːzə(r)/ *n.* **1** a person who uses (esp. a particular commodity or service, or a computer). **2** colloq. a drug addict. **3** Law the continued use or enjoyment of a right etc. □ **right of user** Law **1** a right to use. **2** a presumptive right arising from the user. **user-friendly** esp. Computing (of a machine or program) designed to be easy to use.

ushabti /ʊˈʃæbtɪ/ *n.* each of a set of wooden, stone, or faience figurines, in the form of mummies, placed in an ancient Egyptian tomb to do any work that the dead person might be called upon to do in the afterlife. They were often 365 in number, one for each day of the year. [Egypt., = answerer]

usher /ˈʌʃə(r)/ *n. & v.* ● *n.* **1** a person who shows people to their seats in a hall or theatre etc. **2** a doorkeeper at a court etc. **3** Brit. an officer walking before a person of rank. **4** archaic or joc. an assistant teacher. ● *v.tr.* **1** act as usher to. **2** (usu. foll. by in) announce or show in etc. (ushered us into the room; ushered in a new era). □ **ushership** n. [ME f. AF *usser*, OF *uissier*, var. of *huissier* f. med.L *ustiarius* for L *ostiarius* f. *ostium* door]

usherette /ˌʌʃəˈret/ *n.* a female usher esp. in a cinema.

Ushuaia /uːˈswaɪə/ a port in Argentina, in Tierra del Fuego; pop. (1980) 11,000. Founded by English missionaries in the 1870s, it is the southernmost town in the world. In 1884 it was taken over by Argentinian naval forces and used as a penal colony.

Üsküdar /ˌuːskʊˈdɑː(r)/ a suburb of Istanbul, on the eastern side of the Bosporus where it joins the Sea of Marmara; pop. (1990) 395,620. It served as a British army base during the Crimean War, when it was known as Scutari. It is noted as the site of a hospital created in the barracks by Florence Nightingale.

USM *abbr.* Stock Exch. Unlisted Securities Market.

US Marines a US armed service (part of the US navy) trained to operate at land and at sea. Founded in 1775, the US Marines are now constituted of infantry and air support units specializing in amphibious landings and commando raids.

USN *abbr.* United States Navy.

Uspallata Pass /ˌuːspəˈjɑːtə/ a pass over the Andes near Santiago, in southern South America. The principal route across the Andes, it links Argentina with Chile. At its highest point stands a statue, 'Christ of the Andes', erected in 1904.

usquebaugh /'ʌskwɪˌbɔː/ n. esp. Ir. & Sc. whisky. [Ir. & Sc. Gael. *uisge beatha* water of life: cf. WHISKY]

USS abbr. United States Ship.

USSR abbr. hist. Union of Soviet Socialist Republics.

Ustabakanskoe /ˌuːstæbəˈkɑːnskəʊje/ the former name (until 1931) for ABAKAN.

Ustashe /uːˈstɑːʃɪ/ n.pl. (also treated as sing.) (also **Ustasha(s)**, **Ustashi**) the members of a Croatian extreme nationalist movement that engaged in terrorist activity before the Second World War and ruled Croatia with Nazi support after Yugoslavia was invaded and divided by the Germans in 1941. During the war the Ustashe massacred hundreds of thousands of Serbs, Jews, and members of the resistance movement before being forced to flee at the end of the war. [Serbo-Croat *Ustaše* rebels]

Ustinov[1] /uːˈstiːnɒf/ the former name (1984–7) for IZHEVSK.

Ustinov[2] /'juːstɪˌnɒf/, Sir Peter (Alexander) (b.1921), British actor, director, and dramatist, of Russian descent. He has written and acted in a number of plays including *Romanoff and Juliet* (1956), and his many films include *Spartacus* (1960) and *Death on the Nile* (1978). Ustinov is also well known as a mimic, raconteur, broadcaster, and novelist.

usual /'juːʒʊəl/ adj. **1** such as commonly occurs, or is observed or done; customary, habitual (*the usual formalities; it is usual to tip them; forgot my keys as usual*). **2** (prec. by the, my, etc.) colloq. a person's usual drink etc. □ **usually** adv. **usualness** n. [ME f. OF *usual, usuel* or LL *usualis* (as USE)]

usucaption /ˌjuːzjuːˈkæpʃ(ə)n/ n. (also **usucapion** /-ˈkeɪpɪən/) (in Roman and Scots law) the acquisition of a title or right to property by uninterrupted and undisputed possession for a prescribed term. [OF *usucap(t)ion* or L *usucap(t)io* f. *usucapere* acquire by prescription f. *usu* by use + *capere capt-* take]

usufruct /'juːzjuːˌfrʌkt/ n. & v. ● n. (in Roman and Scots law) the right of enjoying the use and advantages of another's property short of the destruction or waste of its substance. ● v.tr. hold in usufruct. □ **usufructuary** /ˌjuːzjuːˈfrʌktʃʊərɪ, -tjʊərɪ/ adj. & n. [med.L *usufructus* f. L *usus (et) fructus* f. *usus* USE + *fructus* FRUIT]

Usumbura /ˌuːzəmˈbʊərə/ the former name (until 1962) for BUJUMBURA.

usurer /'juːʒərə(r)/ n. a person who practises usury. [ME f. AF *usurer*, OF *usureor* f. *usure* f. L *usura*: see USURY]

usurious /jʊˈʒʊərɪəs/ adj. of, involving, or practising usury. □ **usuriously** adv.

usurp /jʊˈzɜːp/ v. **1** tr. seize or assume (a throne or power etc.) wrongfully. **2** intr. (foll. by on, upon) encroach. □ **usurper** n. **usurpation** /ˌjuːzɜːˈpeɪʃ(ə)n/ n. [ME f. OF *usurper* f. L *usurpare* seize for use]

usury /'juːʒərɪ/ n. **1** the act or practice of lending money at interest, esp. Law at an exorbitant rate. **2** interest at this rate. □ **with usury** rhet. or poet. with increased force etc. [ME f. med.L *usuria* f. L *usura* (as USE)]

UT abbr. **1** Universal Time. **2** US Utah (in official postal use).

Utah /'juːtɔː, -tɑː/ a state in the western US; pop. (1990) 1,722,850; capital, Salt Lake City. The region became part of Mexico in 1821 and was ceded to the US in 1848. The first permanent settlers, who arrived in 1847, were Mormons fleeing persecution. Statehood was refused until these renounced polygamy — a dispute which led to the Utah War (1857–58). Utah became the 45th state of the US in 1896. □ **Utahan** /juːˈtɔːən, -ˈtɑːən/ adj. & n.

Utamaro /ˌuːtəˈmɑːrəʊ/, Kitagawa (born Kitagawa Nebsuyoshi) (1753–1806), Japanese painter and printmaker. A leading exponent of the ukiyo-e school, he created many books of woodblock prints and was noted for his sensual depictions of women. His technique of portraying his subjects seemingly cut off by the margin of a print was admired by the impressionists.

UTC abbr. Universal Time Coordinated.

Utd abbr. United (esp. in the name of a football team).

Ute /juːt/ n. & adj. ● n. (pl. same or **Utes**) **1** a member of an American Indian people living chiefly in Colorado, Utah, and New Mexico. **2** the Shoshonean language of this people. ● adj. of or relating to this people or their language. [Sp. *Yuta*: cf. PAIUTE]

ute /juːt/ n. Austral. & NZ sl. a utility truck. [abbr.]

utensil /juːˈtens(ə)l/ n. an implement or vessel, esp. for domestic use (*cooking utensils*). [ME f. OF *utensile* f. med.L, neut. of L *utensilis* usable (as USE)]

uterine /'juːtəˌraɪm, -rɪn/ adj. **1** of or relating to the uterus. **2** born of the same mother but not the same father (*sister uterine*). [ME f. LL *uterinus* (as UTERUS)]

uterus /'juːtərəs/ n. (pl. **uteri** /-ˌraɪ/) the womb. □ **uteritis** /ˌjuːtəˈraɪtɪs/ n. [L]

Uther Pendragon /ˌjuːθə penˈdrægən/ (in Arthurian legend) king of the Britons and father of Arthur.

utile /'juːtaɪl/ adj. useful; having utility. [ME f. OF f. L *utilis* f. *uti* use]

utilitarian /juːˌtɪlɪˈteərɪən/ adj. & n. ● adj. **1** designed to be useful for a purpose rather than attractive; severely practical. **2** of utilitarianism. ● n. an adherent of utilitarianism.

utilitarianism /juːˌtɪlɪˈteərɪəˌnɪz(ə)m/ n. Philos. the theory that an action is right in so far as it promotes happiness, and that the guiding principle of conduct should be to achieve the greatest happiness or benefit of the greatest number. Utilitarianism was advanced most famously by Jeremy Bentham and J. S. Mill (although an important antecedent was David Hume), and has developed to become one of the most influential of philosophical theories. Its concern with the consequences of an action, rather than with the action's intrinsic nature or with the motives of the agent, has aroused opposition; other problems facing the theory include the difficulty of comparing quantitatively the happiness of one person with that of another, and of accounting for the value placed on concepts such as justice and equality.

utility /juːˈtɪlɪtɪ/ n. (pl. **-ies**) **1** the condition of being useful or profitable. **2** a useful thing. **3** = public utility. **4** (attrib.) **a** severely practical and standardized (*utility furniture*). **b** made or serving for utility. □ **utility program** Computing a program for carrying out a routine function. **utility room** a room equipped with appliances for washing, ironing, and other domestic work. **utility vehicle** (or **truck** etc.) a vehicle capable of serving various functions. [ME f. OF *utilité* f. L *utilitas -tatis* (as UTILE)]

utilize /'juːtɪˌlaɪz/ v.tr. (also **-ise**) make practical use of; turn to account; use effectively. □ **utilizable** adj. **utilizer** n. **utilization** /ˌjuːtɪlaɪˈzeɪʃ(ə)n/ n. [F *utiliser* f. It. *utilizzare* (as UTILE)]

-ution /'juːʃ(ə)n, 'uːʃ-/ suffix forming nouns, = -ATION (*solution*). [F f. L *-utio*]

utmost /'ʌtməʊst/ adj. & n. ● adj. furthest, extreme, or greatest (*the utmost limits; showed the utmost reluctance*). ● n. (prec. by the) the utmost point or degree etc. □ **do one's utmost** do all that one can. [OE *ūt(e)mest* (as OUT, -MOST)]

utopia /juːˈtəʊpɪə/ n. (also **Utopia**) an imagined perfect place or state of things (opp. DYSTOPIA). The word was first used as the name of an imaginary island, governed on a perfect political and social system, in the book *Utopia* (1516) by Sir Thomas More. [mod.L f. Gk *ou* not + *topos* place]

utopian /juːˈtəʊpɪən/ adj. & n. (also **Utopian**) ● adj. characteristic of a utopia; idealistic. ● n. an idealistic reformer. □ **utopianism** n.

Utrecht /juːˈtrext/ a city in the central Netherlands, capital of a province of the same name; pop. (1991) 231,230.

Utrecht, Peace of a series of treaties (1713–14) ending the War of the Spanish Succession. By their terms the disputed throne of Spain was given to the French aspirant Philip V, but the union of the French and Spanish thrones was forbidden, while the succession of the House of Hanover to the British throne was secured and the former Spanish territories in Italy were ceded to the Habsburgs.

utricle /'juːtrɪk(ə)l/ n. Zool. & Bot. a small cell or sac in an animal or plant, esp. one in the inner ear. □ **utricular** /juːˈtrɪkjʊlə(r)/ adj. [F *utricule* or L *utriculus* dimin. of *uter* leather bag]

Utrillo /juːˈtrɪləʊ, French ytrijo/, Maurice (1883–1955), French painter. The son of the French painter Susan Valadon (1867–1938), he was adopted by the Spanish architect and writer Miguel Utrillo in 1891. Utrillo is chiefly known for his depictions of Paris street scenes, especially the Montmartre district; the works of his 'white period' (1909–14), when he made extensive use of white pigment, are considered particularly notable.

Utsire /ʊtˈsɪərə/ a small island off the coast of southern Norway to the north-west of Stavanger. The shipping forecast area *North Utsire* covers Norwegian coastal waters immediately to the north of the island, while *South Utsire* covers the area to the south, as far as the mouth of the Skagerrak.

Uttar Pradesh /ˌʊtə prəˈdeʃ/ a large state in northern India, bordering on Tibet and Nepal; capital, Lucknow. The most populous state, it was formed in 1950 from the United Provinces of Agra and Oudh.

utter[1] /'ʌtə(r)/ attrib.adj. complete, total, absolute (*utter misery; saw the* *utter absurdity of it).* □ **utterly** *adv.* **utterness** *n.* [OE *ūtera, ūttra,* compar. adj. f. *ūt* OUT: cf. OUTER]

utter[2] /ˈʌtə(r)/ *v.tr.* **1** emit audibly (*uttered a startled cry*). **2** express in spoken or written words. **3** *Law* put (esp. forged money) into circulation. □ **utterable** *adj.* **utterer** *n.* [ME f. MDu. *ūteren* make known, assim. to UTTER[1]]

utterance /ˈʌtərəns/ *n.* **1** the act or an instance of uttering. **2** a thing spoken. **3 a** the power of speaking. **b** a manner of speaking. **4** *Linguistics* an uninterrupted chain of spoken or written words not necessarily corresponding to a single or complete grammatical unit.

uttermost /ˈʌtəˌməʊst/ *adj.* furthest, extreme.

Uttley /ˈʌtlɪ/, Alison (1884–1976), English author. She is remembered for her children's books, particularly the 'Little Grey Rabbit' series (1929 onwards) and the 'Sam Pig' stories (1940 onwards).

U-turn /ˈjuːtɜːn/ *n.* **1** the turning of a vehicle in a U-shaped course so as to face in the opposite direction. **2** a reversal of policy.

UUC see ULSTER UNIONIST COUNCIL.

UV *abbr.* ultraviolet.

UVA *abbr.* ultraviolet radiation of relatively long wavelengths.

UVB *abbr.* ultraviolet radiation of relatively short wavelengths.

UVC *abbr.* ultraviolet radiation of very short wavelengths, which does not penetrate the earth's ozone layer.

uvea /ˈjuːvɪə/ *n.* the pigmented layer of the eye, lying beneath the outer layer. [med.L f. L *uva* grape]

uvula /ˈjuːvjʊlə/ *n.* (pl. **uvulae** /-ˌliː/) *Anat.* **1** a fleshy extension of the soft palate hanging above the throat. **2** a similar process in the bladder or cerebellum. [ME f. LL, dimin. of L *uva* grape]

uvular /ˈjuːvjʊlə(r)/ *adj. & n.* ● *adj.* **1** of or relating to the uvula. **2** articulated with the back of the tongue and the uvula, as in *r* in French. ● *n.* a uvular consonant.

uxorial /ʌkˈsɔːrɪəl/ *adj.* of or relating to a wife.

uxoricide /ʌkˈsɒrɪˌsaɪd/ *n.* **1** the killing of one's wife. **2** a person who does this. □ **uxoricidal** /-ˌsɒrɪˈsaɪd(ə)l/ *adj.* [L *uxor* wife + -CIDE]

uxorious /ʌkˈsɔːrɪəs/ *adj.* **1** greatly or excessively fond of one's wife. **2** (of behaviour etc.) showing such fondness. □ **uxoriously** *adv.* **uxoriousness** *n.* [L *uxoriosus* f. *uxor* wife]

Uzbek /ˈʊzbek, ˈʌz-/ *n. & adj.* (also **Uzbeg** /-beg/) ● *n.* **1 a** a member of a Turkic people living in Uzbekistan and also in Turkmenistan, Tajikistan, Kazakhstan, and Afghanistan. **b** a native or national of Uzbekistan. **2** the Turkic language of Uzbekistan, having some 16 million speakers. ● *adj.* of or relating to Uzbekistan, the Uzbeks, or their language. [Uzbek]

Uzbekistan /ʊzˌbekɪˈstaːn, ʌz-/ an independent republic in central Asia, lying south and south-east of the Aral Sea; pop. (est. 1991) 20,955,000; official language, Uzbek; capital, Tashkent. Largely flat irrigated land to the west and north and mountainous in the east and south, Uzbekistan was formerly a constituent republic of the USSR (established in 1924). It became independent within the Commonwealth of Independent States on the breakup of the Soviet Union in 1991.

Vv

V¹ /viː/ n. (also **v**) (pl. **Vs** or **V's**) **1** the twenty-second letter of the alphabet. **2** a V-shaped thing. **3** (as a Roman numeral) five.

V² symb. **1** Chem. the element vanadium. **2 a** volt(s). **b** voltage, potential difference. **3** volume.

V³ abbr. Vergeltungswaffe, used esp. in V1 and V2, German missile weapons used in the latter stages of the Second World War. The V1 was a small flying bomb powered by a simple jet engine (also called (colloq.) doodlebug). The V2, a rocket designed under the direction of Wernher von Braun, was the first ballistic missile and formed the basis of postwar American and Russian rocket technology. Several hundred V-weapons of both kinds were launched by the Germans against London, Antwerp, and other cities, and caused many casualties. [G Vergeltungswaffe, = reprisal weapon]

v. abbr. **1** verse. **2** verso. **3** versus. **4** very. **5** vide. **6** verb. **7** velocity.

V1 /viːˈwʌn/ n. see V³.

V2 /viːˈtuː/ n. see V³.

VA abbr. **1** US (Department of) Veterans' Affairs (formerly Veterans' Administration). **2** Vicar Apostolic. **3** Vice Admiral. **4** US Virginia (in official postal use). **5** (in the UK) Order of Victoria and Albert.

Va. abbr. Virginia.

Vaal /vɑːl/ a river of South Africa, the chief tributary of the Orange River, rising in the Drakensberg Mountains, Eastern Transvaal, and flowing 1,200 km (750 miles) south-westwards to the Orange River near Douglas, in Northern Cape. For much of its length it forms the border between North-West Province and Orange Free State.

Vaasa /ˈvɑːsə/ (Swedish **Vasa** /ˈvɑːsa/) a port in western Finland, on the Gulf of Bothnia; pop. (1990) 53,430.

vac /væk/ n. colloq. **1** Brit. vacation (esp. of universities). **2** vacuum cleaner. [abbr.]

vacancy /ˈveɪkənsɪ/ n. (pl. **-ies**) **1 a** the state of being vacant or empty. **b** an instance of this; empty space. **2** an unoccupied post or job (there are three vacancies for typists). **3** an available room in a hotel etc. **4** emptiness of mind; idleness, listlessness.

vacant /ˈveɪkənt/ adj. **1** not filled or occupied; empty. **2** not mentally active; showing no interest (had a vacant stare). □ **vacant possession** Brit. ownership of a house etc. with any previous occupant having moved out. □ **vacantly** adv. [ME f. OF vacant or L vacare (as VACATE)]

vacate /vəˈkeɪt/ v.tr. **1** leave vacant or cease to occupy (a house, room, etc.). **2** give up tenure of (a post etc.). **3** Law annul (a judgement or contract etc.). □ **vacatable** adj. [L vacare vacat- be empty]

vacation /vəˈkeɪʃ(ə)n/ n. & v. ● n. **1** a fixed period of cessation from work, esp. in universities and lawcourts. **2** N. Amer. a holiday. **3** the act of vacating (a house or post etc.). ● v.intr. US take a holiday. □ **vacation land** US an area providing attractions for holiday-makers. □ **vacationer** n. **vacationist** n. [ME f. OF vacation or L vacatio (as VACATE)]

vaccinate /ˈvæksɪˌneɪt/ v.tr. inoculate with a vaccine to procure immunity from a disease; immunize. (See note below.) □ **vaccinator** n. **vaccination** /ˌvæksɪˈneɪʃ(ə)n/ n.

▪ The practice of vaccination was established by Edward Jenner in the late 18th century (see SMALLPOX). The body is stimulated to produce antibodies by being inoculated with a vaccine which contains antigens from the disease-causing organism. The vaccine is generally a preparation of dead or weakened organisms. Vaccines have been prepared for many diseases and for non-infectious agents such as snake venom.

vaccine /ˈvæksiːn/ n. & adj. ● n. **1** an antigenic preparation used to stimulate the production of antibodies and procure immunity from one or several diseases. **2** hist. the cowpox virus used in vaccination against smallpox. ● adj. of or relating to cowpox or vaccination. □ **vaccinal** /-sɪn(ə)l/ adj. [L vaccinus f. vacca cow]

vaccinia /vækˈsɪnɪə/ n. Med. = COWPOX. [mod.L (as VACCINE)]

vacillate /ˈvæsɪˌleɪt/ v.intr. **1** fluctuate in opinion or resolution. **2** move from side to side; oscillate, waver. □ **vacillator** n. **vacillation** /ˌvæsɪˈleɪʃ(ə)n/ n. [L vacillare vacillat- sway]

vacua pl. of VACUUM.

vacuole /ˈvækjʊˌəʊl/ n. Biol. a small membrane-bound space or vesicle within the cytoplasm of a cell, usu. filled with fluid. □ **vacuolar** /ˌvækjʊˈəʊlə(r)/ adj. **vacuolation** /ˌvækjʊəˈleɪʃ(ə)n/ n. [F, dimin. of L vacuus empty]

vacuous /ˈvækjʊəs/ adj. **1** lacking expression (a vacuous stare). **2** unintelligent (a vacuous remark). **3** empty. □ **vacuously** adv. **vacuousness** n. **vacuity** /vəˈkjuːɪtɪ/ n. [L vacuus empty (as VACATE)]

vacuum /ˈvækjʊəm, -juːm/ n. & v. ● n. (pl. **vacuums** or **vacua** /-jʊə/) **1** a space entirely devoid of matter. **2** a space or vessel from which the air has been completely or partly removed by a pump etc. **3 a** the absence of the normal or previous content of a place, environment, etc. **b** the absence of former circumstances, activities, etc. **4** (pl. **vacuums**) colloq. a vacuum cleaner. **5** a decrease of pressure below the normal atmospheric value. ● v. colloq. **1** tr. clean with a vacuum cleaner. **2** intr. use a vacuum cleaner. □ **vacuum brake** a brake in which pressure is caused by the exhaustion of air. **vacuum-clean** clean with a vacuum cleaner. **vacuum gauge** a gauge for testing the pressure after the production of a vacuum. **vacuum-packed** sealed after the partial removal of air. **vacuum pump** a pump for producing a vacuum. **vacuum tube** a tube with a near-vacuum for the free passage of electric current. [mod.L, neut. of L vacuus empty]

vacuum cleaner n. an electric appliance for taking up dust, dirt, etc., by suction. The first vacuum cleaner was invented and named by the English engineer Hubert Cecil Booth (1871–1955) in 1901. In contrast with today's compact domestic models, early machines were large and cumbersome and required a horse-drawn cart to convey the electric motor which powered them.

vacuum flask n. Brit. a vessel with a double wall enclosing a vacuum so that the liquid in the inner receptacle retains its temperature. The basic design of the modern flask was devised by the chemist Sir James Dewar in the 1890s.

VAD abbr. **1** Voluntary Aid Detachment. **2** a member of this.

vade-mecum /ˌvɑːdɪˈmeɪkəm, ˌveɪdɪˈmiːkəm/ n. a handbook etc. carried constantly for use. [F f. mod.L, = go with me]

Vadodara /vəˈdəʊdərə/ a city in the state of Gujarat, western India; pop. (1991) 1,021,000. The capital of the former state of Baroda, the city was known as Baroda until 1976.

Vaduz /væˈdʊts/ the capital of Liechtenstein; pop. (est. 1990) 4,870.

vag /væg/ n. & v. Austral. & US sl. ● n. a vagrant. ● v.tr. charge with vagrancy. [abbr.]

vagabond /ˈvægəˌbɒnd/ n., adj., & v. ● n. **1** a wanderer or vagrant, esp. an idle one. **2** colloq. a scamp or rascal. ● adj. having no fixed habitation;

wandering. ● *v.intr.* wander about as a vagabond. □ **vagabondage** *n.* [ME f. OF *vagabond* or L *vagabundus* f. *vagari* wander]

vagal see VAGUS.

vagary /'veɪgərɪ/ *n.* (*pl.* **-ies**) a caprice; an eccentric idea or act (*the vagaries of Fortune*). □ **vagarious** /və'geərɪəs/ *adj.* [L *vagari* wander]

vagi *pl.* of VAGUS.

vagina /və'dʒaɪnə/ *n.* (*pl.* **vaginas** or **vaginae** /-niː/) **1** the canal between the uterus and vulva of a woman or other female mammal. **2** a sheath formed round a stem by the base of a leaf. □ **vaginal** /və'dʒaɪn(ə)l, 'vædʒɪn-/ *adj.* **vaginitis** /ˌvædʒɪ'naɪtɪs/ *n.* [L, = sheath, scabbard]

vaginismus /ˌvædʒɪ'nɪzməs/ *n.* a painful spasmodic contraction of the vagina in response to physical contact or pressure (esp. in sexual intercourse). [mod.L (as VAGINA)]

vagrant /'veɪgrənt/ *n. & adj.* ● *n.* **1** a person without a settled home or regular work; a wanderer or tramp. **2** a bird that has wandered away from its normal range or migratory route. Also called *accidental.* ● *adj.* **1** wandering or roving (*a vagrant musician*). **2** being a vagrant. □ **vagrancy** *n.* **vagrantly** *adv.* [ME f. AF *vag(a)raunt*, perh. alt. f. AF *wakerant* etc. by assoc. with L *vagari* wander]

vague /veɪg/ *adj.* **1** of uncertain or ill-defined meaning or character (*gave a vague answer; has some vague idea of emigrating*). **2** (of a person or mind) imprecise; inexact in thought, expression, or understanding. □ **vaguely** *adv.* **vagueness** *n.* **vaguish** *adj.* [F *vague* or L *vagus* wandering, uncertain]

vagus /'veɪgəs/ *n.* (*pl.* **vagi** /-gaɪ/) *Anat.* either of the tenth pair of cranial nerves with branches to the heart, lungs, and viscera. □ **vagal** *adj.* [L: see VAGUE]

vail /veɪl/ *v. archaic* **1** *tr.* lower or doff (one's plumes, pride, crown, etc.) esp. in token of submission. **2** *intr.* yield; give place; remove one's hat as a sign of respect etc. [ME f. obs. *avale* f. OF *avaler* to lower f. *a val* down, f. *val* VALE[1]]

vain /veɪn/ *adj.* **1** excessively proud or conceited, esp. about one's own attributes. **2** empty, trivial, unsubstantial (*vain boasts; vain triumphs*). **3** useless; followed by no good result (*in the vain hope of dissuading them*). □ **in vain** without result or success (*it was in vain that we protested*). **take a person's name in vain** use it lightly or profanely. □ **vainly** *adv.* **vainness** /'veɪnnɪs/ *n.* [ME f. OF f. L *vanus* empty, without substance]

vainglory /veɪn'glɔːrɪ/ *n.* *literary* boastfulness; extreme vanity. □ **vainglorious** *adj.* **vaingloriously** *adv.* **vaingloriousness** *n.* [ME, after OF *vaine gloire*, L *vana gloria*]

vair /veə(r)/ *n.* **1** *archaic* or *hist.* a squirrel-fur widely used for medieval linings and trimmings. **2** *Heraldry* fur represented by small shield-shaped or bell-shaped figures usu. alternately azure and argent. [ME f. OF f. L (as VARIOUS)]

Vaishnava /'vaɪʃnəvə/ *n.* *Hinduism* a devotee of Vishnu. [Skr. *vaiṣṇava*]

Vaisya /'vaɪʃjə/ *n.* a member of the third of the four great Hindu classes or varnas, comprising the merchants and farmers. [Skr. *vaiśya* peasant, labourer]

valance /'væləns/ *n.* (also **valence**) a short curtain round the frame or canopy of a bedstead, above a window, or under a shelf. □ **valanced** *adj.* [ME ult. f. OF *avaler* descend: see VAIL]

vale[1] /veɪl/ *n.* *archaic* or *poet.* (except in place-names) a valley (*Vale of Glamorgan*). □ **vale of tears** *literary* the world as a scene of life, trouble, etc. [ME f. OF *val* f. L *vallis, valles*]

vale[2] /'vɑːleɪ/ *int. & n.* ● *int.* farewell. ● *n.* a farewell. [L, imper. of *valere* be well or strong]

valediction /ˌvælɪ'dɪkʃ(ə)n/ *n.* **1** the act or an instance of bidding farewell. **2** the words used in this. [L *valedicere valedict-* (as VALE[2], *dicere* say), after *benediction*]

valedictorian /ˌvælɪdɪk'tɔːrɪən/ *n.* *N. Amer.* a person who gives a valedictory, esp. the highest-ranking member of a graduating class.

valedictory /ˌvælɪ'dɪktərɪ/ *adj. & n.* ● *adj.* serving as a farewell. ● *n.* (*pl.* **-ies**) a farewell address.

valence[1] /'veɪləns/ *n.* *Chem.* = VALENCY. □ **valence electron** an electron in the outermost shell of an atom involved in forming a chemical bond.

valence[2] var. of VALANCE.

Valencia /və'lensɪə/ **1** an autonomous region of eastern Spain, on the Mediterranean coast. It was formerly a Moorish kingdom (1021–1238). **2** its capital, a port on the Mediterranean coast; pop. (1991) 777,430. **3** a city in northern Venezuela; pop. (1991) 903,080.

Valenciennes /vəˌlɒnsɪ'en/ *n.* a rich kind of lace. [*Valenciennes* in NE France, where it was made in the 17th and 18th c.]

valency /'veɪlənsɪ/ *n.* (*pl.* **-ies**) *Brit. Chem.* the combining power of an atom measured by the number of hydrogen atoms it can displace or combine with. [LL *valentia* power, competence f. *valere* be well or strong]

-valent /'veɪlənt/ *suffix Chem.* forming adjectives meaning 'having a valency of the specified number' (*monovalent, divalent*).

valentine /'vælən͵taɪn/ *n.* a card or gift sent, often anonymously, as a mark of love or affection on St Valentine's Day (14 Feb.); also, a sweetheart chosen on this day. Valentine cards, with drawings and verses, first appeared in the 18th century and were made by the sender. In the 19th century shop-made valentines were first produced and became increasingly elaborate, adorned with lace, real flowers, feathers, and moss; their dispatch was aided by the introduction of the penny post. [ME f. OF *Valentin* f. L *Valentinus* Valentine (see VALENTINE, ST)]

Valentine, St /'vælən͵taɪn/ either of two early Italian saints (who may have been the same person) traditionally commemorated on 14 Feb. — a Roman priest martyred *c.*269, and a bishop of Terni martyred at Rome. St Valentine was regarded as the patron of lovers, a tradition which may be connected with the old belief that birds pair on 14 Feb. or with the pagan fertility festival of Lupercalia (15 Feb.).

Valentino /ˌvælən'tiːnəʊ/, Rudolph (born Rodolfo Guglielmi di Valentina d'Antonguolla) (1895–1926), Italian-born American actor. He became a leading star of silent films in the 1920s, playing the romantic hero in films such as *The Sheikh* (1921) and *Blood and Sand* (1922). After his death from a perforated ulcer thousands of women attended his funeral, at which there were scenes of mass hysteria and reports of several suicides.

Valera, Eamon de, see DE VALERA.

Valerian /və'lɪərɪən/ (Latin name Publius Licinius Valerianus) (d.260), Roman emperor 253–60. He became emperor following the murder of Gallus (reigned 251–3), and appointed his son Gallienus as joint ruler. During his reign Valerian renewed the persecution of the Christians initiated by Decius. He was captured while campaigning against the Persians of the Sassanian dynasty and died in captivity. Gallienus continued to rule as sole emperor until 268.

valerian /və'lɪərɪən/ *n.* **1** a flowering plant of the family Valerianaceae, with small tubular flowers. **2** the strong-smelling root of this, esp. of common valerian, used as a medicinal sedative. □ **common valerian** a valerian, *Valeriana officinalis*, with pink or white flowers and a strong smell liked by cats. [ME f. OF *valeriane* f. med.L *valeriana (herba)*, app. fem. of *Valerianus* of Valerius]

valeric acid /və'lerɪk, -'lɪərɪk/ *n.* *Chem.* = PENTANOIC ACID. [VALERIAN + -IC]

Valéry /'vælerɪ/, Paul (Ambroise) (1871–1945), French poet, essayist, and critic. His poetry, influenced by symbolist poets such as Mallarmé and blending lyricism, rich imagery, and intellectual eloquence, includes *La Jeune parque* (1917) and *Le Cimetière marin* (1922). He later concentrated on prose, publishing essays on a variety of literary, philosophical, and aesthetic subjects. He is also known for his notebooks, published posthumously as *Cahiers* (1958–62).

valet /'vælɪt, -leɪ/ *n. & v.* ● *n.* **1** a man's personal (usu. male) attendant who looks after his clothes etc. **2** a hotel etc. employee with similar duties. ● *v.* (**valeted, valeting**) **1** *intr.* work as a valet. **2** *tr.* act as a valet to. **3** *tr.* clean or clean out (a car). [F, = OF *valet, vaslet*, VARLET: rel. to VASSAL]

valeta var. of VELETA.

valetudinarian /ˌvælɪˌtjuːdɪ'neərɪən/ *n. & adj.* ● *n.* a person of poor health or unduly anxious about health. ● *adj.* **1** of or being a valetudinarian. **2** of poor health. **3** seeking to recover one's health. □ **valetudinarianism** *n.* [L *valetudinarius* in ill health f. *valetudo -dinis* health f. *valere* be well]

valetudinary /ˌvælɪ'tjuːdɪnərɪ/ *adj. & n.* (*pl.* **-ies**) = VALETUDINARIAN.

valgus /'vælgəs/ *n.* a deformity involving the outward displacement of the foot or hand from the midline. [L, = knock-kneed]

Valhalla /væl'hælə/ *n.* **1** *Scand. Mythol.* the hall in which heroes who have died in battle feast with Odin. **2** a building used for honouring the illustrious. [mod.L f. ON *Valhǫll* f. *valr* the slain + *hǫll* HALL]

valiant /'vælɪənt/ *adj.* (of a person or conduct) brave, courageous. □ **valiantly** *adv.* [ME f. AF *valiaunt*, OF *vaillant* ult. f. L *valere* be strong]

valid /'vælɪd/ *adj.* **1** (of a reason, objection, etc.) sound or defensible; well-grounded. **2 a** executed with the proper formalities (*a valid contract*). **b** legally acceptable (*a valid passport*). **c** not having reached its expiry date. □ **validly** adv. **validity** /vəˈlɪdɪtɪ/ n. [F valide or L validus strong (as VALIANT)]

validate /ˈvælɪˌdeɪt/ v.tr. make valid; ratify, confirm. □ **validation** /ˌvælɪˈdeɪʃ(ə)n/ n. [med.L validare f. L (as VALID)]

valine /ˈveɪliːn/ n. Biochem. an amino acid that is an essential nutrient for vertebrates and a general constituent of proteins. [VALERIC ACID + -INE⁴]

valise /vəˈliːz/ n. **1** a kitbag. **2** US a small portmanteau. [F f. It. valigia corresp. to med.L valisia, of unkn. orig.]

Valium /ˈvælɪəm/ n. propr. the drug diazepam used as a tranquillizer and relaxant. [20th c.: orig. uncert.]

Valkyrie /vælˈkɪərɪ, ˈvælkɪrɪ/ n. Scand. Mythol. each of Odin's twelve handmaidens who hovered over battlefields and carried slain warriors designated by the gods to Valhalla. [ON Valkyrja, lit. 'chooser of the slain' f. valr the slain + (unrecorded) kur-, kuz- rel. to CHOOSE]

Valladolid /ˌvælədəˈliːd/ **1** a city in northern Spain, capital of Castilla-León region; pop. (1991) 345,260. It was the principal residence of the kings of Castile in the 15th century and capital of Spain until 1561. **2** the former name (until 1828) for MORELIA.

vallecula /vəˈlekjʊlə/ n. (pl. **valleculae** /-ˌliː/) Anat. & Bot. a groove or furrow. □ **vallecular** adj. **valleculate** /-lət/ adj. [LL, dimin. of L vallis valley]

Valle d'Aosta /ˌvæleɪ dɑːˈɒstə/ a mountainous region in the north-western corner of Italy; capital, Aosta.

Valletta /vəˈletə/ the capital and chief port of Malta; pop. (1987) 9,240; urban harbour area pop. (1992) 102,000. It is named after Jean de Valette, Grand Master of the Knights of St John, who built the town after the victory over the Turks in 1565.

valley /ˈvælɪ/ n. (pl. **-eys**) **1** a low area more or less enclosed by hills and usu. with a stream flowing through it. **2** any depression compared to this. **3** Archit. an internal angle formed by the intersecting planes of a roof. [ME f. AF valey, OF valee ult. f. L vallis, valles: cf. VALE¹]

Valley Forge the site on the Schuylkill river in Pennsylvania, about 32 km (20 miles) to the north-west of Philadelphia, where George Washington's Continental Army spent the bitterly cold winter of 1777–8, during the War of American Independence, in conditions of extreme hardship.

Valley of the Kings a valley near ancient Thebes in Egypt where the pharaohs of the New Kingdom (c.1550–1070 BC) were buried. Although the tombs, typically consisting of a richly decorated chamber at the end of a long series of descending corridors, were hidden in the lonely valley at least partly to preserve them from thieves, most were robbed in antiquity; the exception was the tomb of Tutankhamen, almost untouched until discovered in 1922.

vallum /ˈvæləm/ n. Rom. Antiq. a rampart and stockade as a defence. [L, collect. f. vallus stake]

Valois¹ /ˈvælwɑː/ **1** a medieval duchy of northern France, home of the Valois dynasty. **2** the French royal house from the accession of Philip VI, successor to the last Capetian king, in 1328 to the death of Henry III (1589), when the throne passed to the Bourbons.

Valois², Dame Ninette de, see DE VALOIS.

Valona see VLORË.

valonia /vəˈləʊnɪə/ n. the acorn-cups and acorns of an evergreen oak, Quercus macrolepis, used in tanning, dyeing, and making ink. [It. vallonia ult. f. Gk balanos acorn]

valor US var. of VALOUR.

valorize /ˈvælərˌaɪz/ v.tr. (also **-ise**) raise or fix the price of (a commodity etc.) by artificial means, esp. by government action. □ **valorization** /ˌvælraɪˈzeɪʃ(ə)n/ n. [back-form. f. valorization f. F valorisation (as VALOUR)]

valour /ˈvælə(r)/ n. (US **valor**) personal courage, esp. in battle. □ **valorous** adj. [ME f. OF f. LL valor -oris f. valere be strong]

Valparaíso /ˌvælpəˈraɪzəʊ/ the principal port of Chile, in the centre of the country, near the capital Santiago; pop. (est. 1987) 278,760.

valse /vɑːls, vɔːls/ n. a waltz. [F f. G (as WALTZ)]

valuable /ˈvæljʊəb(ə)l/ adj. & n. ● adj. of great value, price, or worth (a valuable property; valuable information). ● n. (usu. in pl.) a valuable thing, esp. a small article of personal property. □ **valuably** adv.

valuation /ˌvæljʊˈeɪʃ(ə)n/ n. **1 a** an estimation (esp. by a professional valuer) of a thing's worth. **b** the worth estimated. **2** the price set on a thing. □ **valuate** /ˈvæljʊˌeɪt/ v.tr. esp. US.

valuator /ˈvæljʊˌeɪtə(r)/ n. a person who makes valuations; a valuer.

value /ˈvæljuː/ n. & v. ● n. **1** the worth, desirability, or utility of a thing, or the qualities on which these depend (the value of regular exercise). **2** worth as estimated; valuation (set a high value on my time). **3** the amount of money or goods for which a thing can be exchanged in the open market; purchasing power. **4** the equivalent of a thing; what represents or is represented by or may be substituted for a thing (paid them the value of their lost property). **5** (in full **value for money**) something well worth the money spent (usu. hyphenated when attrib.; value-for-money purchases). **6** the ability of a thing to serve a purpose or cause an effect (news value; nuisance value). **7** (in pl.) one's principles or standards; one's judgement of what is valuable or important in life. **8** Mus. the duration of the sound signified by a note. **9** Math. the amount denoted by an algebraic term or expression. **10** (foll. by of) **a** the meaning (of a word etc.). **b** the quality (of a spoken sound). **11** the relative rank or importance of a playing card, chess piece, etc., according to the rules of the game. **12** the relation of one part of a picture to others in respect of light and shade; the part being characterized by a particular tone. **13** Physics & Chem. the numerical measure of a quantity or a number denoting magnitude on some conventional scale (the value of gravity at the equator). ● v.tr. (**values**, **valued**, **valuing**) **1** estimate the value of; appraise (esp. professionally) (valued the property at £200,000). **2** have a high or specified opinion of; attach importance to (a valued friend). □ **value judgement** a subjective estimate of quality etc. **value received** money or its equivalent given for a bill of exchange. [ME f. OF, fem. past part. of valoir be worth f. L valere]

value-added /ˌvæljuːˈædɪd/ attrib.adj. **1** Econ. designating an increase in value of an article at each stage of its production, exclusive of initial costs. **2 a** (of goods etc.) having enhancements etc. added to a basic line esp. to increase profit margins. **b** (of a company) offering specialized or extended services in a commercial area.

value-added tax n. (abbr. **VAT**) a tax nominally on the amount by which the value of an article has been increased at each stage of its production. In the UK VAT was introduced on certain goods and services in 1973, in conformity with other EC countries, in place of purchase tax.

valueless /ˈvæljʊlɪs/ adj. having no value. □ **valuelessness** n.

valuer /ˈvæljʊə(r)/ n. a person who estimates or assesses values, esp. professionally.

valuta /vəˈljuːtə/ n. **1** the value of one currency with respect to another. **2** a currency considered in this way. [It., = VALUE]

valve /vælv/ n. **1** a device for controlling the passage of fluid through a pipe etc., esp. an automatic device allowing movement in one direction only. **2** Anat. & Zool. a membranous part of an organ etc. allowing a flow of blood etc. in one direction only. **3** Brit. = thermionic valve. **4** a device to vary the effective length of the tube in a brass musical instrument. **5** each of the two shells of an oyster, mussel, etc. **6** Bot. each of the segments into which a capsule or dry fruit dehisces. **7** archaic a leaf of a folding door. □ **valvate** /-veɪt/ adj. **valved** adj. (also in comb.). **valveless** adj. **valvule** /-vjuːl/ n. [ME f. L valva leaf of a folding door]

valvular /ˈvælvjʊlə(r)/ adj. **1** having a valve or valves. **2** having the form or function of a valve. [mod.L valvula, dimin. of L valva]

valvulitis /ˌvælvjʊˈlaɪtɪs/ n. Med. inflammation of the valves of the heart.

vambrace /ˈvæmbreɪs/ n. hist. defensive armour for the forearm. [ME f. AF vaunt-bras, OF avant-bras f. avant before (see AVAUNT) + bras arm]

vamoose /vəˈmuːs/ v.intr. US (esp. as int.) sl. depart hurriedly. [Sp. vamos let us go]

vamp¹ /væmp/ n. & v. ● n. **1** the upper front part of a boot or shoe. **2** a patched-up article. **3** an improvised musical accompaniment. ● v. **1** tr. (often foll. by up) repair or furbish. **2** tr. (foll. by up) make by patching or from odds and ends. **3 a** tr. & intr. improvise a musical accompaniment (to). **b** tr. improvise (a musical accompaniment). **4** tr. put a new vamp to (a boot or shoe). [ME f. OF avantpié f. avant before (see AVAUNT) + pied foot]

vamp² /væmp/ n. & v. colloq. ● n. a woman who uses sexual attraction to exploit men; an unscrupulous flirt. ● v. **1** tr. allure or exploit (a man). **2** intr. act as a vamp. [abbr. of VAMPIRE]

vampire /ˈvæmpaɪə(r)/ n. **1** a ghost or reanimated corpse supposed to leave its grave at night to suck the blood of persons sleeping, often represented as a human figure with long pointed canine teeth. (See note below.) **2** a person who preys ruthlessly on others. **3** (in full **vampire bat**) a Central and South American bat of the family Desmodontidae; esp. Desmodus rotundus, which has sharp incisors for piercing flesh and laps the blood of large animals. **4** Theatr. a small spring trapdoor used for sudden disappearances. □ **vampiric** /væmˈpɪrɪk/ *adj.* [F *vampire* or G *Vampir* f. Hungarian *vampir* perh. f. Turk. *uber* witch]

▪ The vampire legend is widespread in Europe and Asia but is particularly associated with the folklore of eastern Europe, especially Transylvania. Its popularity in the 20th century is largely due to the success of Bram Stoker's novel *Dracula* (1897), which has given rise to what almost amounts to a genre of fiction in literature and film. A number of superstitions are associated with vampires: that vampires cast no shadow, are not reflected in mirrors, attack only at night, and can be warded off by garlic or by a crucifix; the body of a vampire can be destroyed by being beheaded or burnt or by a stake being driven through its heart.

vampirism /ˈvæmpaɪəˌrɪz(ə)m/ *n.* **1** belief in the existence of vampires. **2** the practices of a vampire.

vamplate /ˈvæmpleɪt/ *n. hist.* an iron plate on a lance protecting the hand when the lance was couched. [ME f. AF *vauntplate* (as VAMBRACE, PLATE)]

van[1] /væn/ *n.* **1** a covered vehicle for conveying goods etc. **2** *Brit.* a railway carriage for luggage or for the use of the guard. **3** *Brit.* a gypsy caravan. [abbr. of CARAVAN]

van[2] /væn/ *n.* **1** a vanguard. **2** the forefront (*in the van of progress*). [abbr. of VANGUARD]

van[3] /væn/ *n.* **1** the testing of ore quality by washing on a shovel or by machine. **2** *archaic* a winnowing fan. **3** *archaic* or *poet.* a wing. [ME, southern & western var. of FAN[1], perh. partly f. OF *van* or L *vannus*]

van[4] /væn/ *n. Brit. Tennis colloq.* = ADVANTAGE. [abbr.]

Van, Lake /væn/ a large salt lake in the mountains of eastern Turkey.

vanadium /vəˈneɪdɪəm/ *n.* a hard grey metallic chemical element (atomic number 23; symbol **V**). A transition metal, vanadium was discovered by the Swedish chemist N. G. Sefström in 1830. The most important source is the uranium mineral carnotite. The metal is used to make alloy steels, where it improves strength and hardness; vanadium compounds are used as catalysts and in the ceramic and glass industries. □ **vanadic** /-ˈnædɪk/ *adj.* **vanadous** /ˈvænədəs/ *adj.* [f. ON *Vanadís* name of FREYA]

Van Allen belt /væn ˈælən/ *n.* (also **Van Allen layer**) either of two regions of intense radiation partly surrounding the earth at heights of several thousand kilometres. [James A. *Van Allen*, Amer. physicist (b.1914)]

Vanbrugh /ˈvænbrə/, Sir John (1664–1726), English architect and dramatist. In his early life he gained success as a dramatist with his comedies, including *The Relapse* (1696) and *The Provok'd Wife* (1697). After 1699 he became known as an architect and as one of the chief exponents of the English baroque; major works include Castle Howard in Yorkshire (1702) and Blenheim Palace in Oxfordshire (1705), both produced in collaboration with Nicholas Hawksmoor, and Seaton Delaval Hall in Northumberland (1720).

Van Buren /væn ˈbjʊərən/, Martin (1782–1862), American Democratic statesman, 8th President of the US 1837–41. He was appointed Andrew Jackson's Vice-President in 1832 and became President five years later. His measure of placing government funds, previously held in private banks, in an independent treasury caused many Democrats to join the Whig party.

Vancouver[1] /vænˈkuːvə(r)/ a city and port in British Columbia, SW Canada, situated on the mainland opposite Vancouver Island; pop. (1991) 471,840; metropolitan area pop. 1,602,500. Named after the explorer George Vancouver, it is the largest city in western Canada and its chief Pacific port.

Vancouver[2] /vænˈkuːvə(r)/, George (1757–98), English navigator. After accompanying Captain James Cook on his second and third voyages, he took command of a naval expedition exploring the coasts of Australia, New Zealand, and Hawaii (1791–2). He later charted much of the west coast of North America between southern Alaska and California. Vancouver Island and the city of Vancouver are named after him.

Vancouver Island a large island off the Pacific coast of Canada, in SW British Columbia, named after the explorer George Vancouver. Its capital, Victoria, is the capital of British Columbia. It became a British Crown Colony in 1849, later uniting with British Columbia to join the Dominion of Canada.

V. & A. see VICTORIA AND ALBERT MUSEUM.

Vanda see VANTAA.

Vandal /ˈvænd(ə)l/ *n.* a member of a Germanic people that overran

part of Roman Europe in the 4th and 5th centuries AD, establishing kingdoms in Gaul and Spain, and finally (428–9) migrated to North Africa. They sacked Rome in 455 in a marauding expedition, but were eventually defeated by the Byzantine general Belisarius (*c.*505–65), after which their North African kingdom fell prey to Muslim invaders. □ **Vandalic** /vænˈdælɪk/ *adj.* [L *Vandalus* f. Gmc]

vandal /ˈvænd(ə)l/ *n.* a person who wilfully or maliciously destroys or damages property or things of beauty. [VANDAL]

vandalism /ˈvændəˌlɪz(ə)m/ *n.* wilful or malicious destruction or damage to property etc. □ **vandalistic** /ˌvændəˈlɪstɪk/ *adj.* **vandalistically** *adv.*

vandalize /ˈvændəˌlaɪz/ *v.tr.* (also **-ise**) destroy or damage wilfully or maliciously.

van de Graaff generator /ˌvæn də ˈɡrɑːf/ *n. Electr.* a machine devised to generate electrostatic charge by means of a vertical endless belt collecting charge from a voltage source and transferring it to a large insulated metal dome, where a high voltage is produced. [R. J. *van de Graaff*, Amer. physicist (1901–67)]

Vanderbijlpark /ˈvændəbaɪlˌpɑːk/ a steel-manufacturing city in South Africa, in the province of Pretoria-Witwatersrand-Vereeniging, south of Johannesburg; pop. (1985) 540,140 (with Vereeniging).

Vanderbilt /ˈvændəˌbɪlt/, Cornelius (1794–1877), American businessman and philanthropist. Vanderbilt amassed a fortune from shipping and railroads, and from this made an endowment to found Vanderbilt University in Nashville, Tennessee (1873). Subsequent generations of his family, including his son William Henry Vanderbilt (1821–85), increased the family wealth and continued his philanthropy.

Van der Post /ˌvæn də ˈpɒst/, Sir Laurens (Jan) (b.1906), South African explorer and writer. In 1949 Van der Post was sent to explore Nyasaland (now Malawi) for the British; his book of the journey, *Venture to the Interior* (1952), was the first of several books combining travel writing and descriptions of fauna with philosophical speculation based on his Jungian ideas of the necessary balance between 'unconscious, feminine' Africa, and 'conscious, masculine' Europe. He developed this theme after travels among the Bushmen of the Kalahari Desert, described in books such as *The Lost World of the Kalahari* (1958).

van der Waals forces /ˌvæn də ˈwɑːlz/ *n.pl. Chem.* short-range attractive forces between uncharged molecules arising from the interaction of dipole moments. [Johannes *van der Waals*, Dutch physicist (1837–1923)]

van de Velde[1] /ˌvæn də ˈvelt, ˈveldə/, a family of Dutch painters. Willem (known as Willem van de Velde the Elder) (1611–93) painted marine subjects and was for a time official artist to the Dutch fleet. He also worked for Charles II. His sons were Willem (known as Willem van de Velde the Younger) (1633–1707) and Adriaen (1636–72). Like his father, Willem the Younger was a notable marine artist who painted for Charles II, while Adriaen's works included landscapes, portraits, and biblical and genre scenes.

van de Velde[2] /ˌvæn də ˈvelt, ˈveldə/, Henri (Clemens) (1863–1957), Belgian architect, designer, and teacher. He was influenced by the Arts and Crafts Movement and pioneered the development of art nouveau design and architecture in Europe. In 1906 he became head of the Weimar School of Arts and Crafts (which developed into the Bauhaus), Walter Gropius being among his pupils. Van de Velde's buildings include the Werkbund Theatre in Cologne (1914) and his own house near Brussels (1895). He also designed furniture, ceramics, and graphics.

Van Diemen's Land /væn ˈdiːmənz/ the former name (until 1855) for TASMANIA.

Van Dyck /væn ˈdaɪk/, Sir Anthony (also **Vandyke**) (1599–1641), Flemish painter. Having worked as an assistant in Rubens's workshop in Antwerp (1618–20), he visited Italy, where he based himself in Genoa and studied Titian and other Venetian painters. While in Genoa he painted a number of portraits which marked the onset of his artistic maturity. Thereafter he received commissions from several royal clients, including Charles I, who invited Van Dyck to England and knighted him in 1632. His subsequent portraits of members of the English court, noted for their refinement of style and elegant composition, determined the course of portraiture in England for more than 200 years.

vandyke /vænˈdaɪk/ *n. & adj.* ● *n.* **1** each of a series of large points forming a border to lace or cloth etc. **2** a cape or collar etc. with these. **3** (**Vandyke**) a Vandyke beard. ● *adj.* (**Vandyke**) in the style of dress, esp. with pointed borders, common in portraits by Van Dyck.

□ **Vandyke beard** a neat pointed beard. **Vandyke brown** a deep rich brown. [VAN DYCK]

vane /veɪn/ n. **1** (in full **weather-vane**) **a** a revolving pointer mounted on a church spire or other high place to show the direction of the wind (cf. WEATHERCOCK). **b** an inconstant person. **2** a blade of a screw propeller or a windmill etc. **3** the sight of surveying instruments, a quadrant, etc. **4** the flat part of a bird's feather formed by the barbs. □ **vaned** adj. **vaneless** adj. [ME, southern & western var. of obs. *fane* f. OE *fana* banner f. Gmc]

Vänern /ˈveɪnən/ a lake in SW Sweden, the largest lake in Sweden and the third largest in Europe.

Van Eyck /væn ˈaɪk/, Jan (c.1370–1441), Flemish painter. He is said by Vasari to have invented oil-painting, and though it is known that oils were used before his time there is no doubt that he made an innovative contribution to the technique of their use, bringing greater flexibility, richer and denser colour, and a wider range from light to dark. His best-known works include the altarpiece *The Adoration of the Lamb* (known as the Ghent Altarpiece, 1432) in the church of St Bavon in Ghent and the portrait *The Arnolfini Marriage* (1434).

vang /væŋ/ n. Naut. either of two guy-ropes running from the end of a gaff to the deck. [earlier *fang* = gripping-device: OE f. ON *fang* grasp f. Gmc]

Van Gogh /væn ˈɡɒx, ˈɡɒf, ˈɡəʊ/, Vincent (Willem) (1853–90), Dutch painter. Although he is most closely identified with post-impressionism, his early paintings, such as *The Potato Eaters* (1885), were often of peasants and are characterized by the use of dark colours. A move to Paris in 1886 brought him into contact with the works of the impressionists and with Japanese woodcuts, instigating a change in his technique. He began to use brighter colours and adopted broad, vigorous, swirling brushstrokes. In 1888 he settled at Arles, where he was briefly joined by Gauguin, but they quarrelled, and Van Gogh, suffering from depression, cut off part of his own ear. After a year spent in an asylum he entered into a period of intense creative activity but continued to suffer from severe depression, and eventually committed suicide. Among his best-known works are several studies of sunflowers and *A Starry Night* (1889).

vanguard /ˈvænɡɑːd/ n. **1** the foremost part of an army or fleet advancing or ready to advance. **2** the leaders of a movement or of opinion etc. [earlier *vandgard*, (*a*)*vantgard*, f. OF *avan*(*t*)*garde* f. *avant* before (see AVAUNT) + *garde* GUARD]

vanilla /vəˈnɪlə/ n. **1 a** a tropical climbing orchid of the genus *Vanilla*; esp. *V. planifolia*, with fragrant flowers. **b** (in full **vanilla-pod**) the fruit of this. **2** a substance obtained from the vanilla-pod or synthesized, used to flavour ice-cream, chocolate, etc. [Sp. *vainilla* pod, dimin. of *vaina* sheath, pod, f. L VAGINA]

vanillin /vəˈnɪlɪn/ n. Chem. the fragrant principle of vanilla.

vanish /ˈvænɪʃ/ v. **1** intr. **a** disappear suddenly. **b** disappear gradually; fade away. **2** intr. cease to exist. **3** intr. Math. become zero. **4** tr. cause to disappear. □ **vanishing cream** an ointment that leaves no visible trace when rubbed into the skin. **vanishing-point 1** the point at which receding parallel lines viewed in perspective appear to meet. **2** the state of complete disappearance of something. [ME f. OF *e*(*s*)*vaniss-* stem of *e*(*s*)*vanir* ult. f. L *evanescere* (as EX-¹, *vanus* empty)]

Vanitory /ˈvænɪtərɪ/ n. (pl. **-ies**) propr. = vanity unit. [f. VANITY + -ORY¹, after *lavatory*]

vanity /ˈvænɪtɪ/ n. (pl. **-ies**) **1** conceit and desire for admiration of one's personal attainments or attractions. **2 a** a futility or unsubstantiality (*the vanity of human achievement*). **b** an unreal thing. **3** ostentatious display. **4** N. Amer. a dressing-table. □ **vanity bag** (or **case**) a bag or case carried by a woman and containing a small mirror, make-up, etc. **vanity unit** a unit consisting of a wash-basin set into a flat top with cupboards beneath. [ME f. OF *vanité* f. L *vanitas -tatis* (as VAIN)]

Vanity Fair the world (allegorized in John Bunyan's *The Pilgrim's Progress*) as a scene of vanity.

van Leyden, Lucas, see LUCAS VAN LEYDEN.

vanquish /ˈvæŋkwɪʃ/ v.tr. literary conquer, defeat, overcome. □ **vanquishable** adj. **vanquisher** n. [ME *venkus*, *-quis*, etc., f. OF *vencus* past part. and *venquis* past tenses of *veintre* f. L *vincere*: assim. to -ISH²]

Vantaa /ˈvæntɑː/ (Swedish **Vanda** /ˈvandɑ/) a city in southern Finland; pop. (1990) 154,930. It was a suburb of Helsinki until 1972, when it was established as a separate city.

vantage /ˈvɑːntɪdʒ/ n. **1** (also **vantage point** or **ground**) a place affording a good view or prospect. **2** Tennis = ADVANTAGE. **3** archaic an advantage or gain. [ME f. AF f. OF *avantage* ADVANTAGE]

Vanuatu /ˌvænuːˈɑːtuː/ a country consisting of a group of islands in the SW Pacific; pop. (est. 1991) 156,000; official languages, Bislama, English, and French; capital, Vila. Encountered by the Portuguese in the early 17th century, the islands were administered jointly by Britain and France as the condominium of the New Hebrides. The islands were used by the British as a source of forced labour for Australian sugar plantations in the 19th century, severely reducing the indigenous Melanesian population. Vanuatu became an independent republic within the Commonwealth in 1980. □ **Vanuatuan** adj. & n.

vapid /ˈvæpɪd/ adj. insipid; lacking interest; flat, dull (*vapid moralizing*). □ **vapidly** adv. **vapidness** n. **vapidity** /vəˈpɪdɪtɪ/ n. [L *vapidus*]

vapor US var. of VAPOUR.

vaporetto /ˌvæpəˈretəʊ/ n. (pl. **vaporetti** /-tɪ/ or **-os**) (in Venice) a canal boat (originally a steamboat, now a motor boat) used for public transport. [It., = small steamboat, dimin. of *vapore* f. L *vapor* steam]

vaporific /ˌveɪpəˈrɪfɪk/ adj. concerned with or causing vapour or vaporization.

vaporimeter /ˌveɪpəˈrɪmɪtə(r)/ n. an instrument for measuring the amount of vapour.

vaporize /ˈveɪpəˌraɪz/ v.tr. & intr. (also **-ise**) convert or be converted into vapour. □ **vaporizable** adj. (also **vaporable**). **vaporization** /ˌveɪpəraɪˈzeɪʃ(ə)n/ n.

vaporizer /ˈveɪpəˌraɪzə(r)/ n. a device that vaporizes substances, esp. for medicinal inhalation.

vapour /ˈveɪpə(r)/ n. & v. (US **vapor**) ● n. **1** moisture or another substance diffused or suspended in air, e.g. mist or smoke. **2** Physics a gaseous form of a normally liquid or solid substance (cf. GAS). **3** a medicinal agent for inhaling. **4** (in pl.) archaic a state of depression or melancholy thought to be caused by exhalations of vapour from the stomach. ● v.intr. **1** rise as vapour. **2** make idle boasts or empty talk. □ **vapour density** the density of a gas or vapour relative to hydrogen etc. **vapour pressure** the pressure of a vapour in contact with its liquid or solid form. **vapour trail** a trail of condensed water from an aircraft or rocket at high altitude, seen as a white streak against the sky. □ **vaporous** adj. **vaporously** adv. **vaporousness** n. **vapourer** n. **vapouring** n. **vapourish** adj. **vapoury** adj. [ME f. OF *vapour* or L *vapor* steam, heat]

vapourer /ˈveɪpərə(r)/ n. (in full **vapourer moth**) a European tussock moth of the genus *Orgyia*, with hairy larvae and wingless females, esp. the reddish-brown *O. antiqua*.

var. abbr. **1** variant. **2** variety.

varactor /vəˈræktə(r)/ n. Electr. a semiconductor diode with a capacitance dependent on the applied voltage. [*varying* re*actor*]

Varah /ˈvɑːrə/, (Edward) Chad (1911–93), English clergyman, founder of the Samaritans. He founded the Samaritans (see SAMARITAN 2) in 1953 after recognizing a widespread need for an anonymous counselling service. He was president of Befrienders International (Samaritans Worldwide) from 1983 to 1986 and travelled widely abroad to spread the organization's principles.

Varanasi /vəˈrɑːnəsɪ/ (formerly **Benares** /bɪˈnɑːrɪz/) a city on the Ganges, in Uttar Pradesh, northern India; pop. (1991) 926,000. It is a holy city and a place of pilgrimage for Hindus, who undergo ritual purification in the Ganges.

Varangian /vəˈrændʒɪən/ n. & adj. ● n. any of the Scandinavian rovers who penetrated into Russia in the 9th and 10th centuries AD, establishing the Rurik dynasty and reaching Constantinople. ● adj. of or relating to the Varangians. □ **Varangian guard** the bodyguard of the later Byzantine emperors, comprising Varangians and later also Anglo-Saxons. [med. L *Varangus*, ult. f. ON, = confederate]

varec /ˈværek/ n. **1** seaweed. **2** = KELP. [F *varec*(h) f. ON: rel. to WRECK]

Varese /vəˈreɪzɪ/ a town in Lombardy, northern Italy; pop. (1990) 87,970.

Varèse /væˈrez/, Edgar(d) (1883–1965), French-born American composer. Varèse emigrated to the US in 1915, and most of his works from before this date have been lost. His compositions were predominantly orchestral in nature until the 1950s, when he began to experiment with tape-recordings and electronic instruments (see MUSIQUE CONCRÈTE). His works are known for their use of dissonance and for their experimentation with unusual sounds and instrument combinations; they include *Ionisation* (1931), for percussion, piano, and two sirens, and *Poème électronique* (1958).

Vargas /ˈvɑːɡəs/, Getúlio Dornelles (1883–1954), Brazilian statesman, President 1930–45 and 1951–4. Although defeated in the presidential elections of 1930, Vargas seized power in the ensuing revolution, overthrowing the republic and ruling as a virtual dictator for the next fifteen years. He furthered Brazil's modernization by the introduction of fiscal, educational, electoral, and land reforms, but his regime was totalitarian and repressive. He was overthrown in a coup in 1945. He was returned to power after elections in 1951, but his government was unpopular and he committed suicide after widespread calls for his resignation.

Vargas Llosa /ˈvɑːɡəs ˈjəʊsə/, (Jorge) Mario (Pedro) (b.1936), Peruvian novelist, dramatist, and essayist. His fiction often contains elements of myth and fantasy and has been associated with magic realism; it is frequently critical of the political situation in Peru. Novels include *The Time of the Hero* (1963), satirizing Peruvian society via the microcosm of a corrupt military academy, *Aunt Julia and the Scriptwriter* (1977), and *The War of the End of the World* (1982). He returned to Peru in 1974, after living abroad for many years, and stood unsuccessfully for the presidency in 1990.

variable /ˈveərɪəb(ə)l/ adj. & n. ● adj. **1 a** that can be varied or adapted (*a rod of variable length; the pressure is variable*). **b** (of a gear) designed to give varying speeds. **2** apt to vary; not constant; unsteady (*a variable mood; variable fortunes*). **3** *Math.* (of a quantity) indeterminate; able to assume different numerical values. **4** (of wind or currents) tending to change direction. **5** *Bot. & Zool.* (of a species) including individuals or groups that depart from the type. **6** *Biol.* (of an organism or part of it) tending to change in structure or function. ● n. **1** a variable thing or quantity. **2** *Math.* a variable quantity. **3** *Naut.* **a** a shifting wind. **b** (in *pl.*) the region between the north-east and south-east trade winds. **4** *Astron.* = VARIABLE STAR. □ **variably** adv. **variableness** n. **variability** /ˌveərɪəˈbɪlɪtɪ/ n. [ME f. OF f. L *variabilis* (as VARY)]

variable star n. *Astron.* a star whose brightness changes, either irregularly or regularly. Irregular variables are generally stars undergoing drastic physical changes and include those which undergo a single outburst, such as novae and supernovae. Some regular or periodic variables, such as Algol, are eclipsing binaries; others fluctuate because of regular changes within the stellar interior, as in the case of Mira and the cepheids. The period of variability may range from a few hours to several years.

variance /ˈveərɪəns/ n. **1** difference of opinion; dispute, disagreement; lack of harmony (*at variance among ourselves; a theory at variance with all known facts*). **2** *Law* a discrepancy between statements or documents. **3** *Statistics* a quantity equal to the square of the standard deviation. [ME f. OF f. L *variantia* difference (as VARY)]

variant /ˈveərɪənt/ adj. & n. ● adj. **1** differing in form or details from the main one (*a variant spelling*). **2** having different forms (*forty variant types of pigeon*). **3** variable or changing. ● n. a variant form, spelling, type, reading, etc. [ME f. OF (as VARY)]

variate /ˈveərɪət/ n. *Statistics* **1** a quantity having a numerical value for each member of a group. **2** a variable quantity, esp. one whose values occur according to a frequency distribution. [past part. of L *variare* (as VARY)]

variation /ˌveərɪˈeɪʃ(ə)n/ n. **1** the act or an instance of varying. **2** departure from a former or normal condition, action, or amount, or from a standard or type (*prices are subject to variation*). **3** the extent of this. **4** a thing that varies from a type. **5** *Mus.* a repetition (usu. one of several) of a theme in a changed or elaborated form. **6** *Astron.* a deviation of a celestial body from its mean orbit or motion. **7** *Math.* a change in a function etc. due to small changes in the values of constants etc. **8** *Ballet* a solo dance. □ **variational** adj. [ME f. OF *variation* or L *variatio* (as VARY)]

varicella /ˌværɪˈselə/ n. *Med.* **1** = CHICKENPOX. **2** (in full **varicella-zoster** /-ˈzɒstə(r)/) a herpesvirus that causes both chickenpox and shingles. [mod.L, irreg. dimin. of VARIOLA]

varices pl. of VARIX.

varicocele /ˈværɪkəˌsiːl/ n. *Med.* a mass of varicose veins in the spermatic cord. [formed as VARIX + -CELE]

varicoloured /ˈveərɪˌkʌləd/ adj. (US **varicolored**) **1** variegated in colour. **2** of various or different colours. [L *varius* VARIOUS + COLOURED]

varicose /ˈværɪˌkəʊs/ adj. (esp. of the veins of the legs) affected by a condition causing them to become dilated and swollen. □ **varicosity** /ˌværɪˈkɒsɪtɪ/ n. [L *varicosus* f. VARIX]

varied /ˈveərɪd/ adj. showing variety; diverse. □ **variedly** adv.

variegate /ˈveərɪˌɡeɪt, -rɪəˌɡeɪt/ v.tr. **1** (often as **variegated** adj.) mark with irregular patches of different colours. **2** diversify in appearance, esp. in colour. **3** (as **variegated** adj.) *Bot.* (of plants) having leaves

containing two or more colours. □ **variegation** /ˌveərɪˈɡeɪʃ(ə)n, ˌveərɪəˈɡeɪ-/ n. [L *variegare variegat-* f. *varius* various]

varietal /vəˈraɪət(ə)l/ adj. **1** esp. *Bot. & Zool.* of, forming, or designating a variety. **2** (of wine) made from a single designated variety of grape. □ **varietally** adv.

varietist /vəˈraɪətɪst/ n. a person whose habits etc. differ from what is normal.

variety /vəˈraɪətɪ/ n. (pl. **-ies**) **1** diversity; absence of uniformity; many-sidedness; the condition of being various (*not enough variety in our lives*). **2** a quantity or collection of different things (*for a variety of reasons*). **3 a** a class of things different in some common qualities from the rest of a larger class to which they belong. **b** a specimen or member of such a class. **4** (foll. by *of*) a different form of a thing, quality, etc. **5** *Biol.* **a** a subspecies. **b** a cultivar. **c** an individual or group usually fertile within the species to which it belongs but differing from the species type in some qualities capable of perpetuation. **6** a mixed sequence of dances, songs, comedy acts, etc. (usu. *attrib.: a variety show*). □ **variety store** N. *Amer.* a shop selling many kinds of small items. [F *variété* or L *varietas* (as VARIOUS)]

varifocal /ˌveərɪˈfəʊk(ə)l/ adj. & n. ● adj. having a focal length that can be varied, esp. of a lens that allows an infinite number of focusing distances for near, intermediate, and far vision. ● n. (in *pl.*) varifocal spectacles.

variform /ˈveərɪˌfɔːm/ adj. having various forms. [L *varius* + -FORM]

variola /vəˈraɪələ/ n. *Med.* smallpox. □ **variolar** adj. **variolous** adj.

varioloid /ˈveərɪəˌlɔɪd/ adj. [med.L, = VARIOUS]

variole /ˈveərɪˌəʊl/ n. **1** a shallow pit like a smallpox mark. **2** *Geol.* a small spherical mass in variolite. [med.L *variola*: see VARIOLA]

variolite /ˈveərɪəˌlaɪt/ n. *Geol.* a rock with embedded small spherical masses causing on its surface an appearance like smallpox pustules. □ **variolitic** /ˌveərɪəˈlɪtɪk/ adj. [as VARIOLE + -ITE[1]]

variometer /ˌveərɪˈɒmɪtə(r)/ n. **1** a device for varying the inductance in an electric circuit. **2** a device for indicating an aircraft's rate of change of altitude. [as VARIOUS + -METER]

variorum /ˌveərɪˈɔːrəm/ adj. & n. ● adj. **1** (of an edition of a text) having notes by various editors or commentators. **2** (of an edition of an author's works) including variant readings. ● n. a variorum edition. [L f. *editio cum notis variorum* edition with notes by various (commentators): genitive pl. of *varius* VARIOUS]

various /ˈveərɪəs/ adj. **1** different, diverse (*too various to form a group*). **2** more than one, several (*for various reasons*). □ **variously** adv. **variousness** n. [L *varius* changing, diverse]

varistor /vəˈrɪstə(r)/ n. *Electr.* a semiconductor diode with resistance dependent on the applied voltage. [*varying resistor*]

varix /ˈveərɪks/ n. (pl. **varices** /ˈværɪˌsiːz/) **1** *Med.* **a** a permanent abnormal dilation of a vein or artery. **b** a vein etc. dilated in this way. **2** each of the ridges across the whorls of a univalve shell. [ME f. L *varix -icis*]

varlet /ˈvɑːlɪt/ n. **1** *archaic* or *joc.* a menial or rascal. **2** *hist.* a knight's attendant. □ **varletry** n. [ME f. OF, var. of *vaslet*: see VALET]

varmint /ˈvɑːmɪnt/ n. N. *Amer.* or *dial.* a mischievous or discreditable person or animal, esp. a fox. [var. of *varmin*, VERMIN]

Varna /ˈvɑːnə/ a port and resort in eastern Bulgaria, on the western shores of the Black Sea; pop. (1990) 320,640. It was established by the Greeks in 585 BC as Odessos, and developed by the Romans as a spa.

varna /ˈvɑːnə/ n. any of four great Hindu classes that form the basis for the caste system in the Indian subcontinent. In the Vedic religion of the ancient Aryans, the first three classes of society were Brahman, Kshatriya, and Vaisya (priests, warriors, merchants and farmers). The fourth class, Sudra (labourers), was probably added later after contact with the indigenous people of the subcontinent. Each class was considered equally necessary to the social order, the separate functions being complementary. [Skr. = colour, class]

varnish /ˈvɑːnɪʃ/ n. & v. ● n. **1** a resinous solution used to give a hard shiny transparent coating to wood, metal, paintings, etc. **2** any other preparation for a similar purpose (*nail varnish*). **3** external appearance or display without an underlying reality. **4** artificial or natural glossiness. **5** a superficial polish of manner. ● v.tr. **1** apply varnish to. **2** gloss over (a fact). □ **varnisher** n. [ME f. OF *vernis* f. med.L *veronix* fragrant resin, sandarac or med.Gk *berenikē* prob. f. *Berenice* in Cyrenaica]

Varro /ˈværəʊ/, Marcus Terentius (116–27 BC), Roman scholar and satirist. He was a prolific author and although most of his writings are now lost, his prose works are known to have covered many subjects, including philosophy, agriculture, the Latin language, and education. His satires (*Saturae Menippeae*) presented critical sketches of Roman life in a mixture of verse and prose.

varsity /'vɑːsɪtɪ/ *n.* (*pl.* **-ies**) **1** *Brit. colloq.* (esp. with ref. to sports) university. **2** *N. Amer.* a university etc. first team in a sport. [abbr.]

Varuna /'vʌrʊnə/ *Hinduism* one of the gods in the Rig-veda. Originally the sovereign lord of the universe and guardian of cosmic law, he is known in later Hinduism as god of the waters.

varus /'veərəs/ *n. Med.* a deformity involving the inward displacement of the foot or hand from the midline. [L, = bent, crooked]

varve /vɑːv/ *n. Geol.* a pair of layers of silt deposited annually in lakes where a glacier melts, one being of fine silt (deposited in winter, when there is little melting) and one of coarser silt (deposited in summer, when the ice melts more freely). Varves can be used to determine the chronology of glacial sediments. □ **varved** *adj.* [Sw. *varv* layer]

vary /'veərɪ/ *v.* (**-ies**, **-ied**) **1** *tr.* make different; modify, diversify (*seldom varies the routine; the style is not sufficiently varied*). **2** *intr.* **a** undergo change; become or be different (*the temperature varies from 30° to 70°*). **b** be of different kinds (*his mood varies*). **3** *intr.* (foll. by *as*) be in proportion to. □ **varyingly** *adv.* [ME f. OF *varier* or L *variare* (as VARIOUS)]

vas /væs/ *n.* (*pl.* **vasa** /'veɪsə/) *Anat.* a vessel or duct. □ **vas deferens** /'defə,renz/ (*pl.* **vasa deferentia** /,defə'renʃɪə/) *Anat.* the spermatic duct from the testicle to the urethra. □ **vasal** /'veɪs(ə)l/ *adj.* [L, = vessel]

Vasa see VAASA.

Vasarely /,væsə'relɪ/, Viktor (b.1908), Hungarian-born French painter. He settled in Paris in 1930 after studying art in Budapest. A pioneer of op art, he began experimenting with the use of optical illusion during the 1930s, although the style of geometric abstraction for which he is best known dates from the late 1940s. His paintings are characterized by their repeated geometric forms and interacting vibrant colours which create a visually disorientating effect of movement.

Vasari /və'sɑːrɪ/, Giorgio (1511–74), Italian painter, architect, and biographer. He wrote *Lives of the Most Excellent Painters, Sculptors and Architects* (1550, enlarged 1568), a work which laid the basis for later study of art history in the West; the book traces the development of Renaissance painting from the work of painters such as Giotto to that of Leonardo da Vinci, Raphael, and particularly Michelangelo, whose work he regarded as the pinnacle of achievement. His own work was mannerist in style and includes the vast frescos depicting the history of Florence and the Medici family in the Palazzo Vecchio in Florence, as well as the design of the Uffizi palace.

Vasco da Gama /'væskəʊ/ see DA GAMA.

vascular /'væskjʊlə(r)/ *adj.* of, made up of, or containing vessels for conveying blood or sap etc. (*vascular functions; vascular tissue*). □ **vascular bundle** *Bot.* a strand of conducting vessels in the stem or leaves of a plant, usu. with phloem on the outside and xylem on the inside. **vascular plant** *Bot.* a plant with conducting tissue. □ **vascularly** *adv.* **vascularity** /,væskjʊ'lærɪtɪ/ *n.* [mod.L *vascularis* f. L VASCULUM]

vascularize /'væskjʊlə,raɪz/ *v.tr.* (also **-ise**) (usu. in *passive*) *Med.* & *Anat.* make vascular, develop (esp. blood) vessels in. □ **vascularization** /,væskjʊlərər'zeɪʃ(ə)n/ *n.*

vasculum /'væskjʊləm/ *n.* (*pl.* **vascula** /-lə/) a botanist's (usu. metal) collecting-case with a lengthwise opening. [L, dimin. of VAS]

vase /vɑːz/ *n.* a vessel, usu. tall and of circular cross-section, used as an ornament or container, esp. for flowers. □ **vaseful** *n.* (*pl.* **-fuls**). [F f. L VAS]

vasectomy /və'sektəmɪ/ *n.* (*pl.* **-ies**) the surgical removal of part of each vas deferens, esp. as a means of sterilization. □ **vasectomize** *v.tr.* (also **-ise**).

Vaseline /'væsɪ,liːn/ *n.* & *v.* ● *n. propr.* a type of petroleum jelly used as an ointment, lubricant, etc. ● *v.tr.* (**vaseline**) treat with Vaseline. [irreg. f. G *Wasser* + Gk *elaion* oil]

vasiform /'veɪzɪ,fɔːm/ *adj.* **1** duct-shaped. **2** vase-shaped. [L *vasi-* f. VAS + -FORM]

vaso- /'veɪzəʊ/ *comb. form* a vessel, esp. a blood vessel (*vasoconstrictive*). [L *vas*: see VAS]

vasoactive /,veɪzəʊ'æktɪv/ *adj.* = VASOMOTOR.

vasoconstrictive /,veɪzəʊkən'strɪktɪv/ *adj.* causing constriction of blood vessels.

vasodilating /,veɪzəʊdaɪ'leɪtɪŋ/ *adj.* causing dilatation of blood vessels. □ **vasodilation** /-'leɪʃ(ə)n/ *n.*

vasomotor /'veɪzəʊ,məʊtə(r)/ *adj.* causing constriction or dilatation of blood vessels.

vasopressin /,veɪzəʊ'presɪn/ *n.* a pituitary hormone acting to reduce diuresis and increase blood pressure. Also called *antidiuretic hormone*.

vassal /'væs(ə)l/ *n.* **1** *hist.* a holder of land by feudal tenure on conditions of homage and allegiance. **2** *rhet.* a humble dependant. □ **vassalage** *n.* [ME f. OF f. med.L *vassallus* retainer, of Celt. orig.: the root *vassus* corresp. to Breton *gwaz*, Welsh *gwas*, Ir. *foss*: cf. VAVASOUR]

vast /vɑːst/ *adj.* & *n.* ● *adj.* **1** immense, huge; very great (*a vast expanse of water; a vast crowd*). **2** *colloq.* great, considerable (*makes a vast difference*). ● *n. poet.* or *rhet.* a vast space (*the vast of heaven*). □ **vastly** *adv.* **vastness** *n.* [L *vastus* void, immense]

Västerås /,vestə'rɔːs/ a port on Lake Mälaren in eastern Sweden; pop. (1990) 119,760.

VAT /,viːeɪ'tiː, væt/ see VALUE-ADDED TAX.

vat /væt/ *n.* & *v.* ● *n.* **1** a large tank or other vessel, esp. for holding liquids or something in liquid in the process of brewing, tanning, dyeing, etc. **2** a dyeing liquor in which a textile is soaked to take up a colourless soluble dye afterwards coloured by oxidation in air. ● *v.tr.* (**vatted**, **vatting**) place or treat in a vat. □ **vatful** *n.* (*pl.* **-fuls**). [ME, southern & western var. of *fat*, OE *fæt* f. Gmc]

vatic /'vætɪk/ *adj. formal* prophetic or inspired. [L *vates* prophet]

Vatican /'vætɪkən/ *n.* **1** the palace and official residence of the pope in the Vatican City, Rome. **2** papal government. □ **Vaticanism** *n.* **Vaticanist** *n.* [F *Vatican* or L *Vaticanus* name of a hill in Rome]

Vatican City an independent papal state in the city of Rome, the seat of government of the Roman Catholic Church; pop. (est. 1991) 1,000. It covers an area of 44 hectares (109 acres) around St Peter's Basilica and the palace of the Vatican, and is extended to some outlying buildings such as the pope's summer residence, Castel Gandolfo. After the incorporation of the former Papal States into a unified Italy in 1870, the temporal power of the pope was suspended until the Lateran Treaty of 1929, signed by Pope Pius XI and Mussolini. This recognized the full and independent sovereignty of the Holy See in the Vatican City.

Vatican Council either of two ecumenical councils of the Roman Catholic Church which met in the Vatican. The first, held in 1869–70, proclaimed the infallibility of the pope when speaking *ex cathedra*; the second, which took place in 1962–5, sought to bring about liturgical and organizational reforms and to promote unity between different Christian Churches.

vaticinate /væ'tɪsɪ,neɪt/ *v.tr.* & *intr. formal* prophesy. □ **vaticinator** *n.* **vaticinal** /-n(ə)l/ *adj.* **vaticination** /-,tɪsɪ'neɪʃ(ə)n/ *n.* [L *vaticinari* f. *vates* prophet]

VATman /'vætmæn/ *n.* (*pl.* **-men**) *colloq.* a customs and excise officer who administers VAT.

Vättern /'vet(ə)n/ a large lake in southern Sweden.

Vaud /vəʊ/ (German **Waadt** /vaːt/) a canton on the shores of Lake Geneva in western Switzerland; capital, Lausanne. Noted for its wines, it joined the Swiss Confederation in 1803.

vaudeville /'vɔːdə,vɪl, 'vəʊd-/ *n.* **1** a form of variety entertainment popular particularly in the US from about 1880 until the early 1930s. Among stars to have emerged from vaudeville were W. C. Fields and Al Jolson. **2** a stage play on a trivial theme with interspersed songs. **3** a satirical or topical song with a refrain. □ **vaudevillian** /,vɔːdə'vɪlɪən, ,vəʊd-/ *adj.* & *n.* [F, orig. of convivial song, esp. any of those composed by O. Basselin, 15th-c. poet born at *Vau de Vire* in Normandy]

Vaudois[1] /'vəʊdwʌ/ *n.* & *adj.* ● *n.* (*pl.* same) **1** a native of Vaud in Switzerland. **2** the French dialect spoken in Vaud. ● *adj.* of or relating to Vaud or its dialect. [F]

Vaudois[2] /'vəʊdwʌ/ *n.* & *adj.* ● *n.* (*pl.* same) a member of the Waldenses. ● *adj.* of or relating to the Waldenses. [F, repr. med.L *Valdensis*: see WALDENSES]

Vaughan[1] /vɔːn/, Henry (1621–95), Welsh poet. One of the group of metaphysical poets, his poems have a distinctive ethereal quality, which has led him to be described as a mystic. His volumes of religious poetry include *Silex Scintillans* (1650, 1655), in which he acknowledges his debt to George Herbert. Among his prose works is *The Mount of Olives, or Solitary Devotions* (1652).

Vaughan[2] /vɔːn/, Sarah (Lois) (1924–90), American jazz singer and pianist. She began singing with jazz bands in the early 1940s and was chiefly associated with bebop. She performed as a soloist from 1945, and became internationally famous in the early 1950s. Her style of singing was notable for her vocal range, her use of vibrato, and her improvisational skills.

Vaughan Williams /vɔːn 'wɪljəmz/, Ralph (1872–1958), English composer. He composed strongly melodic music in almost every genre, including nine symphonies, operas, choral works, and many songs. His compositions frequently reflect his interest in Tudor composers and English folk-songs, which he collected and arranged from about 1903 onwards. Among his most notable works are the *Fantasia on a Theme by Thomas Tallis* (1910), *A London Symphony* (1914), and the Mass in G minor (1922).

vault /vɔːlt, vɒlt/ *n. & v.* ● *n.* **1 a** an arched roof. **b** a continuous arch. **c** a set or series of arches whose joints radiate from a central point or line. **2** a vaultlike covering (*the vault of heaven*). **3** an underground chamber: **a** as a place of storage (*bank vaults*). **b** as a place of interment beneath a church or in a cemetery etc. (*family vault*). **4** an act of vaulting. **5** *Anat.* the arched roof of a cavity. ● *v.* **1** *intr.* leap or spring, esp. while resting on one or both hands or with the help of a pole. **2** *tr.* spring over (a gate etc.) in this way. **3** *tr.* (esp. as **vaulted**) **a** make in the form of a vault. **b** provide with a vault or vaults. □ **vaulter** *n.* [OF *voute*, *vaute*, ult. f. L *volvere* roll]

vaulting /'vɔːltɪŋ, 'vɒlt-/ *n.* **1** arched work in a vaulted roof or ceiling. **2** a gymnastic or athletic exercise in which participants vault over obstacles. □ **vaulting-horse** a wooden block to be vaulted over by gymnasts.

vaunt /vɔːnt/ *v. & n. literary* ● *v.* **1** *intr.* boast, brag. **2** *tr.* boast of; extol boastfully. ● *n.* a boast. □ **vaunter** *n.* **vauntingly** *adv.* [ME f. AF *vaunter*, OF *vanter* f. LL *vantare* f. L *vanus* VAIN: partly obs. *avaunt* (v.) f. *avanter* f. *a-* intensive + *vanter*]

vavasory /'vævəsərɪ/ *n.* (*pl.* **-ies**) *hist.* the estate of a vavasour. [OF *vavasorie* or med.L *vavasoria* (as VAVASOUR)]

vavasour /'vævəˌsuːə(r)/ *n. hist.* a vassal owing allegiance to a great lord and having other vassals under him. [ME f. OF *vavas(s)our* f. med.L *vavassor*, perh. f. *vassus vassorum* VASSAL of vassals]

Vavilov /'vævɪˌlɒf/, Nikolai (Ivanovich) (1887–c.1943), Soviet plant geneticist. He travelled extensively on botanical expeditions and amassed a considerable collection of new plants, with the aim of utilizing their genetic resources for crop improvement. He did much to improve the yields of Soviet agriculture, and located the centres of origin of many cultivated plants. However, Vavilov's views conflicted with official Soviet ideology (dominated by the theories of T. D. Lysenko) and he was arrested in 1940, dying later in a labour camp. His reputation was subsequently restored, a research institute being named after him.

VC *abbr.* **1** see VICTORIA CROSS. **2** Vice-Chairman. **3** Vice-Chancellor. **4** Vice-Consul.

VCR *abbr.* video cassette recorder.

VD *abbr.* venereal disease.

VDU *abbr.* visual display unit.

VE *abbr.* Victory in Europe (in 1945, the victory of the Allies over the Germans in the Second World War). □ **VE day** 8 May, the day marking this.

've /v/ *abbr.* (chiefly after pronouns) = HAVE (*I've*; *they've*).

veal /viːl/ *n.* calf's flesh as food. □ **vealy** *adj.* [ME f. AF *ve(e)l*, OF *veiaus* *veel* f. L *vitellus* dimin. of *vitulus* calf]

Veblen /'veblən/, Thorstein (Bunde) (1857–1929), American economist and social scientist. He is best known as the author of *The Theory of the Leisure Class* (1899), a critique of capitalism in which he coined the phrase 'conspicuous consumption'. This and subsequent works, such as *The Theory of Business Enterprise* (1904), had a significant influence on later economists such as J. K. Galbraith.

vector /'vektə(r)/ *n. & v.* ● *n.* **1** *Math. & Physics* a quantity having direction as well as magnitude, esp. as determining the position of one point in space relative to another (*radius vector*) (cf. SCALAR). **2** an organism, esp. an insect, that transmits a disease or parasite from one animal or plant to another. **3** a course to be taken by an aircraft. ● *v.tr.* direct (an aircraft in flight) to a desired point. □ **vectorize** *v.tr.* (also **-ise**) (in sense 1 of *n.*). **vectorization** /ˌvektəraɪ'zeɪʃ(ə)n/ *n.* **vectorial** /vek'tɔːrɪəl/ *adj.* [L, = carrier, f. *vehere vect-* convey]

Veda /'veɪdə, 'viːdə/ *n.* (in *sing.* or *pl.*) the most ancient and sacred literature of the Hindus, handed down in four collections and believed to have been directly revealed to seers among the early Aryans in India. It contains hymns, ritual, and philosophy composed in early Sanskrit and preserved by oral tradition as a guide for the priests of Vedic religion. The oldest and most important collection is the Rig-veda,

followed by the Sama-veda and Yajur-veda; the Atharva-veda was added to the canon later. In its wider sense, the term also includes the Brahmanas and the mystical Aranyakas and Upanishads. [Skr. *veda*, lit. '(sacred) knowledge']

Vedanta /veɪ'dæntə, -'dɑːntə/ *n.* **1** the Upanishads. **2** the Hindu philosophy based on these, esp. in its monistic form. □ **Vedantic** *adj.* **Vedantist** *n.* [Skr. *vedānta* (as VEDA, *anta* end)]

Vedda /'vedə/ *n.* a member of an aboriginal people of Sri Lanka. [Sinh. *veddā* hunter]

vedette /vɪ'det/ *n.* a mounted sentry positioned beyond an army's outposts to observe the movements of the enemy. [F, = scout, f. It. *vedetta*, *veletta* f. Sp. *vela(r)* watch f. L *vigilare*]

Vedic /'veɪdɪk, 'viːd-/ *adj. & n.* ● *adj.* of, relating to, or contemporary with the Veda or Vedas; of or relating to the period when Aryans invaded India from Persia (Iran), *c.*2000–1200 BC. ● *n.* an early form of Sanskrit used in the Vedas. Also called *Vedic Sanskrit* (see SANSKRIT). [F *védique* or G *vedisch* (as VEDA)]

Vedic religion *n.* the ancient religion of the Aryan peoples who entered NW India *c.*2000–1200 BC from Persia (Iran), the precursor of Hinduism; the religious beliefs and practices contained in the Veda. It was a religion of ritual sacrifice to many gods, especially Indra, Varuna, and Agni; animal, vegetable, and possibly human sacrifice are described. Vedic society was divided into distinct classes (varnas) that formed the basis of the caste system and are still in evidence in Hinduism today. The increasing complexity of Vedic ritual led to dominance by a specialist priesthood, the composition of the Brahmanas, and the rigidity of orthodox Brahmanism (*c.*900 BC onwards). The transition to classical Hinduism began in about the 5th century BC.

vee /viː/ *n.* **1** the letter V. **2** a thing shaped like a V. [name of the letter]

veep /viːp/ *n.* US *colloq.* a vice-president. [f. the initials *VP*]

veer[1] /vɪə(r)/ *v. & n.* ● *v.intr.* **1** change direction, esp. (of the wind) clockwise (cf. BACK *v.* 5). **2** change in course, opinion, conduct, emotions, etc. **3** *Naut.* = WEAR[2]. ● *n.* a change of course or direction. [F *virer* f. Rmc, perh. alt. f. L *gyrare* GYRATE]

veer[2] /vɪə(r)/ *v.tr. Naut.* slacken or let out (a rope, cable, etc.). [ME f. MDu. *vieren*]

veery /'vɪərɪ/ *n.* a North American woodland thrush, *Catharus fuscescens*. [perh. imit.]

veg[1] /vedʒ/ *n. colloq.* a vegetable or vegetables. [abbr.]

veg[2] /vedʒ/ *v. intr.* (**vegged, vegging**) (usu. foll. by *out*) *sl.* spend time idly or mindlessly; relax, esp. by watching television. [abbr. of VEGETATE]

Vega[1] /ˌveɪɡə/, Lope de (full name Lope Felix de Vega Carpio) (1562–1635), Spanish dramatist and poet. He is regarded as the founder of Spanish drama and is said to have written 1,500 plays, of which several hundred survive. His dramas cover a wide range of genres from the historical and sacred to contemporary plays of intrigue and chivalry. His other works include epic poems and pastoral romances.

Vega[2] /'veɪɡə/ *Astron.* the fifth brightest star in the sky, and the brightest in the constellation of Lyra, overhead in summer to observers in the northern hemisphere. It is a hot blue star, and is surrounded by a disc of material in which planets may be forming. [Sp. or med.L *Vega* f. Arab., = the falling vulture (i.e. the constellation Lyra)]

vegan /'viːɡən/ *n. & adj.* ● *n.* a person who does not eat or use animal products. ● *adj.* using or containing no animal products. [VEGETABLE + -AN]

Vegeburger /'vedʒɪˌbɜːɡə(r)/ *n. propr.* a flat savoury cake made with vegetable protein, soya, etc. [VEGETABLE + BURGER]

Vegemite /'vedʒɪˌmaɪt/ *n. Austral. propr.* concentrated yeast extract, used as a spread in sandwiches and for flavouring. [blend of VEGETABLE and MARMITE]

vegetable /'vedʒɪtəb(ə)l, 'vedʒtə-/ *n. & adj.* ● *n.* **1** any plant, esp. a herbaceous plant used wholly or partly for food, e.g. a cabbage, potato, turnip, or bean. **2** *colloq.* **a** often *offens.* a person who is incapable of normal intellectual activity, esp. through brain injury etc. **b** *derog.* a person lacking in animation or living a monotonous life. ● *adj.* **1** of, derived from, relating to, or comprising plants or plant life, esp. as distinct from animal life or mineral substances. **2** of or relating to vegetables as food. **3 a** unresponsive to stimulus (*vegetable behaviour*). **b** uneventful, monotonous (*a vegetable existence*). □ **vegetable butter** a vegetable fat with the consistency of butter. **vegetable ivory** see IVORY. **vegetable marrow** see MARROW 1. **vegetable oil** an oil derived from plants, e.g. rapeseed oil, olive oil, sunflower oil. **vegetable oyster** = SALSIFY. **vegetable parchment** see PARCHMENT 2. **vegetable spaghetti 1** a variety of marrow with flesh resembling spaghetti. **2** its flesh. **vegetable sponge** = LOOFAH 2. **vegetable tallow** see TALLOW. **vegetable wax** an exudation of certain plants such as sumac. [ME f. OF *vegetable* or LL *vegetabilis* animating (as VEGETATE)]

vegetal /'vedʒɪt(ə)l/ *adj.* **1** of or having the nature of plants (*vegetal growth*). **2** vegetative. [med.L *vegetalis* f. L *vegetare* animate]

vegetarian /ˌvedʒɪ'teərɪən/ *n. & adj.* ● *n.* a person who abstains from animal food, esp. one who avoids meat but will consume eggs and dairy products and sometimes also fish. ● *adj.* excluding animal food, esp. meat (*a vegetarian diet*). □ **vegetarianism** *n.* [irreg. f. VEGETABLE + -ARIAN]

vegetate /'vedʒɪˌteɪt/ *v.intr.* **1** live an uneventful or monotonous life. **2** grow as plants do; fulfil vegetal functions. [L *vegetare* animate f. *vegetus* f. *vegere* be active]

vegetation /ˌvedʒɪ'teɪʃ(ə)n/ *n.* **1** plants collectively; plant life (*luxuriant vegetation; no sign of vegetation*). **2** the process of vegetating. □ **vegetational** *adj.* [med.L *vegetatio* growth (as VEGETATE)]

vegetative /'vedʒɪtətɪv/ *adj.* **1** concerned with growth and development as distinct from sexual reproduction. **2** of or relating to vegetation or plant life. **3** *Med.* alive but not showing evidence of brain activity or responsiveness to stimuli. □ **vegetatively** *adv.* **vegetativeness** *n.* [ME f. OF *vegetatif -ive* or med.L *vegetativus* (as VEGETATE)]

veggie /'vedʒɪ/ *n. & adj.* (also **vegie**) *colloq.* (a) vegetarian. [abbr.]

vehement /'viːəmənt/ *adj.* showing or caused by strong feeling; forceful, ardent (*a vehement protest; vehement desire*). □ **vehemence** *n.* **vehemently** *adv.* [ME f. F *véhément* or L *vehemens -entis*, perh. f. *vemens* (unrecorded) deprived of mind, assoc. with *vehere* carry]

vehicle /'viːɪk(ə)l/ *n.* **1** any conveyance for transporting people, goods, etc., esp. on land. **2** a medium for thought, feeling, or action (*the stage is the best vehicle for their talents*). **3** a liquid etc. as a medium for suspending pigments, drugs, etc. **4** the literal meaning of a word or words used metaphorically (opp. TENOR 6). □ **vehicular** /vɪ'hɪkjʊlə(r)/ *adj.* [F *véhicule* or L *vehiculum* f. *vehere* carry]

veil /veɪl/ *n. & v.* ● *n.* **1** a piece of usu. more or less transparent fabric attached to a woman's hat etc., esp. to conceal the face or protect against the sun, dust, etc. **2** a piece of linen etc. as part of a nun's head-dress, resting on the head and shoulders. **3** a curtain, esp. that separating the sanctuary in the Jewish Temple. **4** a disguise; a pretext; a thing that conceals (*under the veil of friendship; a veil of mist*). **5** *Photog.* slight fogging. **6** huskiness of the voice. **7** = VELUM. ● *v.tr.* **1** cover with a veil. **2** (esp. as **veiled** *adj.*) partly conceal (*veiled threats*). □ **beyond the veil** in the unknown state of life after death. **draw a veil over** avoid discussing or calling attention to. **take the veil** become a nun. □ **veilless** /'veɪllɪs/ *adj.* [ME f. AF *veil(e)*, OF *voil(e)* f. L *vela* pl. of VELUM]

veiling /'veɪlɪŋ/ *n.* light fabric used for veils etc.

vein /veɪn/ *n. & v.* ● *n.* **1 a** any of the tubes by which blood is conveyed to the heart (cf. ARTERY 1). **b** (in general use) any blood vessel (*has royal blood in his veins*). **2** *Zool.* a nervure of an insect's wing. **3** *Bot.* a slender bundle of tissue forming a rib in the framework of a leaf. **4** a streak or stripe of a different colour in wood, marble, cheese, etc. **5** a fissure in rock filled with ore or other deposited material. **6** a source of a particular characteristic (*a rich vein of humour*). **7** a distinctive character or tendency; a cast of mind or disposition; a mood (*spoke in a sarcastic vein*). ● *v.tr.* fill or cover with or as with veins. □ **veinless** *n.* **veinlet** /'veɪnlɪt/ *n.* **veinlike** *adj.* **veiny** *adj.* (**veinier, veiniest**). [ME f. OF *veine* f. L *vena*]

veining /'veɪnɪŋ/ *n.* a pattern of streaks or veins.

veinstone /'veɪnstəʊn/ *n.* = GANGUE.

vela *pl.* of VELUM.

velamen /vɪ'leɪmən/ *n.* (*pl.* **velamina** /-'læmɪnə/) *Bot.* an enveloping membrane, esp. of an aerial root of an orchid. [L f. *velare* cover]

velar /'viːlə(r)/ *adj.* **1** of a veil or velum. **2** *Phonet.* pronounced with the back of the tongue near the soft palate. [L *velaris* f. *velum*: see VELUM]

Velázquez /vɪ'læskwɪz/, Diego Rodríguez de Silva y (1599–1660), Spanish painter. His early paintings consisted chiefly of naturalistic religious works and domestic genre scenes. After his appointment as court painter to Philip IV in 1623, he painted many notable portraits, paintings which humanized the stiff and formal Spanish tradition of idealized figures and tended towards naturalness and simplicity. Among the best known are *Pope Innocent X* (1650) and *Las Meninas*

(*c.*1656). Other notable works include *The Toilet of Venus* (known as The Rokeby Venus, *c.*1651).

Velázquez de Cuéllar /deɪ 'kweɪjɑː(r)/, Diego (*c.*1465–1524), Spanish conquistador. After sailing with Columbus to the New World in 1493, he began the conquest of Cuba in 1511; he founded a number of settlements including Havana (1515), and later initiated expeditions to conquer Mexico.

Velcro /'velkrəʊ/ *n. propr.* a fastener for clothes etc. consisting of two strips of nylon fabric, one looped and one burred, which adhere when pressed together. □ **Velcroed** *adj.* [F *velours croché* hooked velvet]

veld /felt, velt/ *n.* (also **veldt**) *S. Afr.* open country; grassland. [Afrik. f. Du., = FIELD]

Velde, van de[1], Henri, see VAN DE VELDE[2].

Velde, van de[2], Willem and sons, see VAN DE VELDE[1].

veldskoen /'feltskuːn, 'velt-/ *n.* a strong suede or leather shoe or boot. [Afrik., = field-shoe]

veleta /vəˈliːtə/ *n.* (also **valeta**) a ballroom dance in triple time. [Sp., = weather-vane]

veliger /'viːlɪdʒə(r)/ *n. Zool.* the free-swimming larva of a mollusc, with a ciliated velum. [VELUM + L *-ger* bearing]

velitation /ˌvelɪ'teɪʃ(ə)n/ *n. archaic* a slight skirmish or controversy. [L *velitatio* f. *velitari* skirmish f. *veles velitis* light-armed skirmisher]

velleity /ve'liːɪtɪ/ *n. literary* **1** a low degree of volition not conducive to action. **2** a slight wish or inclination. [med.L *velleitas* f. L *velle* to wish]

Velleius Paterculus /veˌleɪəs pə'tɜːkjʊləs/ (*c.*19 BC–AD *c.*30), Roman historian and soldier. His *Roman History* in two volumes, covering the period from the early history of Rome to AD 30, is notable for its rhetorical manner and for its eulogistic depiction of Tiberius, with whom Velleius had served in Germany before Tiberius became emperor.

vellum /'veləm/ *n.* **1 a** a fine parchment orig. from the skin of a calf. **b** a manuscript written on this. **2** smooth writing-paper imitating vellum. [ME f. OF *velin* (as VEAL)]

velocimeter /ˌveləˈsɪmɪtə(r)/ *n.* an instrument for measuring velocity.

velocipede /vɪ'lɒsɪˌpiːd/ *n.* **1** *hist.* an early form of bicycle propelled by pressure from the rider's feet on the ground. **2** *US* a child's tricycle. □ **velocipedist** *n.* [F *vélocipède* f. L *velox -ocis* swift + *pes pedis* foot]

velociraptor /vɪˈlɒsɪˌræptə(r)/ *n.* a usu. small bipedal carnivorous dinosaur of the genus *Velociraptor*, of the Cretaceous period, with an enlarged curved claw on each hind foot. [mod.L f. L *velox -ocis* swift + RAPTOR]

velocity /vɪ'lɒsɪtɪ/ *n.* (*pl.* **-ies**) **1** the measure of the rate of movement of a usu. inanimate object in a given direction. **2** speed in a given direction. **3** (in general use) speed. □ **velocity of escape** = *escape velocity*. [F *vélocité* or L *velocitas* f. *velox -ocis* swift]

velodrome /'veləˌdrəʊm/ *n.* a special place or building with a track for cycle-racing. [F *vélodrome* f. *vélo* bicycle (as VELOCITY, -DROME)]

velour /və'lʊə(r)/ *n.* (also **velours**) **1** a plush woven fabric or felt resembling velvet. **2** *archaic* a hat of this felt. [F *velours* velvet f. OF *velour, velous* f. L *villosus* hairy f. *villus*: see VELVET]

velouté /və'luːteɪ/ *n.* a sauce made from a roux of butter and flour with white stock. [F, = velvety]

velum /'viːləm/ *n.* (*pl.* **vela** /-lə/) a membrane, membranous covering, or flap. [L, = sail, curtain, covering, veil]

velutinous /vɪ'luːtɪnəs/ *adj.* esp. *Bot.* covered with soft fine hairs. [perh. f. It. *vellutino* f. *velluto* VELVET]

velvet /'velvɪt/ *n. & adj.* ● *n.* **1** a closely woven fabric of silk, cotton, etc., with a thick short pile on one side. **2** the furry skin on a deer's growing antler. **3** anything smooth and soft like velvet. ● *adj.* of, like, or soft as velvet. □ **on** (or **in**) **velvet** in an advantageous or prosperous position. **velvet ant** a parasitic wasp of the family Mutillidae, with wingless females and a velvety body. **an iron hand in a velvet glove** see IRON. □ **velveted** *adj.* **velvety** *adj.* [ME f. OF *veluotte* f. *velu* velvety f. med.L *villutus* f. *villus* tuft, down]

velveteen /ˌvelvɪ'tiːn/ *n.* **1** a cotton fabric with a pile like velvet. **2** (in *pl.*) trousers etc. made of this.

velvet revolution *n.* a non-violent political revolution, especially the relatively smooth change from Communism to a Western-style democracy in Czechoslovakia at the end of 1989. [transl. Czech *sametová revoluce*]

Velvet Underground, the a US rock group formed in 1965, unsuccessful in the 1960s but extremely influential since their dissolution. They are probably best known for their first album, *The Velvet Underground and Nico* (1967), produced in association with Andy Warhol and featuring the German model and singer Nico. The songs of vocalist Lou Reed (b.1944) dealt with hitherto taboo subjects such as heroin addiction and sado-masochism, to a backing that displayed the avant-garde influences of classically trained viola player John Cale (b.1942).

Ven. *abbr.* Venerable (as the title of an archdeacon).

vena cava /ˌviːnə ˈkeɪvə/ n. (pl. **venae cavae** /ˌviːniː ˈkeɪviː/) each of usu. two veins carrying blood into the heart. [L, = hollow vein]

venal /ˈviːn(ə)l/ adj. **1** (of a person) able to be bribed or corrupted. **2** (of conduct etc.) characteristic of a venal person. □ **venally** adv. **venality** /vɪˈnælɪtɪ/ n. [L venalis f. venum thing for sale]

venation /vɪˈneɪʃ(ə)n/ n. the arrangement of veins in a leaf or an insect's wing etc., or the system of venous blood vessels in an organism. □ **venational** adj. [L vena vein]

vend /vend/ v.tr. **1** offer (small wares) for sale. **2** Law sell. □ **vending-machine** a machine that dispenses small articles for sale when a coin or token is inserted. □ **vendible** adj. [F vendre or L vendere sell (as VENAL, dare give)]

Venda[1] /ˈvendə/ a former homeland established in South Africa for the Venda people, now part of the province of Northern Transvaal. (See also HOMELAND.)

Venda[2] /ˈvendə/ n. & adj. ● n. **1** a member of a people inhabiting NE Transvaal and southern Zimbabwe. **2** the Bantu language of this people. ● adj. of or relating to this people or their language. [Bantu]

vendace /ˈvendeɪs/ n. a form of the freshwater whitefish *Coregonus albula*, found in some British lakes (cf. POLLAN). [OF vendese, -oise f. Gaulish]

vendee /venˈdiː/ n. Law the buying party in a sale, esp. of property.

vendetta /venˈdetə/ n. **1 a** a blood feud in which the family of a murdered person seeks vengeance on the murderer or the murderer's family. **b** this practice as prevalent in Corsica and Sicily. **2** a prolonged bitter quarrel. [It. f. L vindicta: see VINDICTIVE]

vendeuse /vɒnˈdɜːz/ n. a saleswoman, esp. in a fashionable dress-shop. [F]

vendor /ˈvendə(r), -dɔː(r)/ n. **1** Law the seller in a sale, esp. of property. **2** = vending-machine (see VEND). [AF vendour (as VEND)]

vendue /venˈdjuː/ n. US a public auction. [Du. vendu(e) f. F vendue sale f. vendre VEND]

veneer /vɪˈnɪə(r)/ n. & v. ● n. **1 a** a thin covering of fine wood or other surface material applied to a coarser wood. **b** a layer in plywood. **2** (often foll. by of) a deceptive outward appearance of a good quality etc. ● v.tr. **1** apply a veneer to (wood, furniture, etc.). **2** disguise (an unattractive character etc.) with a more attractive manner etc. [earlier fineer f. G furni(e)ren f. OF fournir FURNISH]

veneering /vɪˈnɪərɪŋ/ n. material used as veneer.

venepuncture /ˈviːnɪˌpʌŋktʃə(r), ˈven-/ n. (also **venipuncture**) Med. the puncture of a vein with a hypodermic needle, to withdraw blood or for an intravenous injection. [L vena vein + PUNCTURE]

venerable /ˈvenərəb(ə)l/ adj. **1** entitled to veneration on account of character, age, associations, etc. (a venerable priest; venerable relics). **2** as the title of an archdeacon in the Church of England. **3** RC Ch. as the title of a deceased person who has attained a certain degree of sanctity but has not been fully beatified or canonized. □ **venerably** adv. **venerableness** n. **venerability** /ˌvenərəˈbɪlɪtɪ/ n. [ME f. OF venerable or L venerabilis (as VENERATE)]

venerate /ˈvenəˌreɪt/ v.tr. **1** regard with deep respect. **2** revere on account of sanctity etc. □ **venerator** n. **veneration** /ˌvenəˈreɪʃ(ə)n/ n. [L venerari adore, revere]

venereal /vɪˈnɪərɪəl/ adj. **1** of or relating to sexual desire or intercourse. **2** relating to venereal disease. □ **venereal disease** any disease contracted chiefly through sexual intercourse with an infected person, a sexually transmitted disease. □ **venereally** adv. [ME f. L venereus f. venus veneris sexual love]

venereology /vɪˌnɪərɪˈɒlədʒɪ/ n. the scientific study of venereal diseases. □ **venereologist** n. **venereological** /-rɪəˈlɒdʒɪk(ə)l/ adj.

venery[1] /ˈvenərɪ/ n. archaic sexual indulgence. [med.L veneria (as VENEREAL)]

venery[2] /ˈvenərɪ/ n. archaic hunting. [ME f. OF venerie f. vener to hunt ult. f. L venari]

venesection /ˈviːnɪˌsekʃ(ə)n, ˈven-/ n. (also **venisection**) Med. phlebotomy. [med.L venae sectio cutting of a vein (as VEIN, SECTION)]

Venetia /vɪˈniːʃə/ (Italian **Veneto** /ˈvɛnɛto/) a region of NE Italy; capital, Venice. The region takes its name from the pre-Roman inhabitants, the Veneti.

Venetian /vɪˈniːʃ(ə)n/ n. & adj. ● n. **1** a native or inhabitant of Venice in NE Italy. **2** the Italian dialect of Venice. **3** (**venetian**) = venetian blind. ● adj. of or relating to Venice. □ **venetian blind** a window-blind of adjustable horizontal slats to control the amount of light excluded. **Venetian glass** a kind of delicate glassware originally made at Murano near Venice. **Venetian red** a reddish pigment of ferric oxides. **Venetian window** a window with three separate openings, the central one being arched and highest. □ **venetianed** adj. (in sense 3 of n.). [ME f. OF Venicien, assim. to med.L Venetianus f. Venetia Venice]

Venezia see VENICE.

Venezuela /ˌvenɪˈzweɪlə/ a republic on the north coast of South America, with a coastline on the Caribbean Sea; pop. (est. 1991) 20,191,000; official language, Spanish; capital, Caracas. Columbus discovered the mouth of the Orinoco river in 1498, and in the following year Vespucci explored the coast. The early Spanish explorers gave the country its name (= little Venice), when they saw native houses built on stilts over water. Colonized by the Spanish in the 16th century, Venezuela won its independence in 1821 after a ten-year struggle, but did not finally emerge as a separate nation until its secession from federation with Colombia (1830). It is a major oil-exporting country, with the industry based on the area around Lake Maracaibo in the north-west. □ **Venezuelan** adj. & n.

vengeance /ˈvendʒəns/ n. punishment inflicted or retribution exacted for wrong to oneself or to a person etc. whose cause one supports. □ **with a vengeance** in a higher degree than was expected or desired; in the fullest sense (punctuality with a vengeance). [ME f. OF f. venger avenge f. L (as VINDICATE)]

vengeful /ˈvendʒfʊl/ adj. vindictive; seeking vengeance. □ **vengefully** adv. **vengefulness** n. [obs. venge avenge (as VENGEANCE)]

venial /ˈviːnɪəl/ adj. (of a sin or fault) pardonable, excusable; (in Catholic theology) not mortal. □ **venially** adv. **venialness** n. **veniality** /ˌviːnɪˈælɪtɪ/ n. [ME f. OF f. LL venialis f. venia forgiveness]

Venice /ˈvenɪs/ (Italian **Venezia** /veˈnetsja/) a city in NE Italy, capital of Venetia region; pop. (1990) 317,840. Situated on a lagoon of the Adriatic, it is built on numerous islands that are separated by canals and linked by bridges. It was a powerful republic in the Middle Ages and from the 13th to the 16th centuries a leading sea-power, controlling trade to the Levant and ruling parts of the eastern Mediterranean. Its commercial importance declined after the Cape route to India was discovered at the end of the 16th century, but it remained an important centre of art and music. After the Napoleonic Wars Venice was placed under Austrian rule and was incorporated into a unified Italy in 1866.

Vening Meinesz /ˌvenɪŋ ˈmaɪnɛʃ/, Felix Andries (1887–1966), Dutch geophysicist. He devised a technique for making accurate gravity measurements with the aid of a pendulum, using it first for a gravity survey of the Netherlands. He then pioneered the use of submarines for marine gravity surveys, locating negative gravity anomalies in the deep trenches near island arcs in the Pacific and interpreting them as being due to the downward buckling of the oceanic crust. This was eventually confirmed; Vening Meinesz, however, never supported the idea of continental drift.

venipuncture var. of VENEPUNCTURE.

venisection var. of VENESECTION.

venison /ˈvenɪs(ə)n, -ɪz(ə)n/ n. a deer's flesh as food. [ME f. OF veneso(u)n f. L venatio -onis hunting f. venari to hunt]

Venite /vɪˈnaɪtɪ/ n. **1** a canticle consisting of Psalm 95. **2** a musical setting of this. [ME f. L, = 'come ye', its first word]

Venn diagram /ven/ n. a diagram of usu. circular intersecting areas used to represent mathematical sets and show the relationships between them. [John Venn, Engl. logician (1834–1923)]

venom /ˈvenəm/ n. **1** a poisonous fluid secreted by snakes, scorpions, etc., usu. transmitted by a bite or sting. **2** malignity; virulence of feeling, language, or conduct. □ **venomed** adj. [ME f. OF venim, var. of venin ult. f. L venenum poison]

venomous /ˈvenəməs/ adj. **1 a** containing, secreting, or injecting venom. **b** (of a snake etc.) inflicting poisonous wounds by this means. **2** (of a person etc.) virulent, spiteful, malignant. □ **venomously** adv. **venomousness** n. [ME f. OF venimeux f. venim: see VENOM]

venose /ˈviːnəʊz, -nəʊs/ adj. having many or very marked veins. [L venosus f. vena vein]

venous /'vi:nəs/ adj. of, full of, or contained in veins. □ **venously** adv. **venosity** /vɪ'nɒsɪtɪ/ n. [L venosus VENOSE or L vena vein + -OUS]

vent[1] /vent/ n. & v. ● n. **1** (also **vent-hole**) a hole or opening allowing motion of air etc. out of or into a confined space. **2** an outlet; free passage or play (gave vent to their indignation). **3** the anus esp. of a lower animal, serving for both excretion and reproduction. **4** the venting of an otter, beaver, etc. **5** an aperture or outlet through which volcanic products are discharged at the earth's surface. **6** a touch-hole of a gun. **7** a finger-hole in a musical instrument. **8** a flue of a chimney. ● v. **1** tr. **a** make a vent in (a cask etc.). **b** provide (a machine) with a vent. **2** tr. give vent or free expression to (vented my anger on the cat). **3** intr. (of an otter, beaver, etc.) come to the surface for breath. **4** tr. & intr. discharge. □ **vent one's spleen on** scold or ill-treat without cause. □ **ventless** adj. [partly F vent f. L ventus wind, partly F évent f. éventer expose to air f. OF esventer ult. f. L ventus wind]

vent[2] /vent/ n. a slit in a garment, esp. in the lower edge of the back of a coat. [ME, var. of fent f. OF fente slip ult. f. L findere cleave]

ventiduct /'ventɪˌdʌkt/ n. Archit. an air-passage, esp. for ventilation. [L ventus wind + ductus DUCT]

ventifact /'ventɪˌfækt/ n. a stone shaped by wind-blown sand. [L ventus wind + factum neut. past part. of facere make]

ventil /'ventɪl/ n. Mus. **1** a valve in a wind instrument. **2** a shutter for regulating the airflow in an organ. [G f. It. ventile f. med.L ventile sluice f. L ventus wind]

ventilate /'ventɪˌleɪt/ v.tr. **1** cause air to circulate freely in (a room etc.). **2** submit (a question, grievance, etc.) to public consideration and discussion. **3** Med. **a** oxygenate (the blood). **b** admit or force air into (the lungs). □ **ventilative** /-ˌleɪtɪv/ adj. **ventilation** /ˌventɪ'leɪʃ(ə)n/ n. [L ventilare ventilat- blow, winnow, f. ventus wind]

ventilator /'ventɪˌleɪtə(r)/ n. **1** an appliance or aperture for ventilating a room etc. **2** Med. = RESPIRATOR 2.

ventral /'ventrəl/ adj. **1** Anat. & Zool. of or on the abdomen (cf. DORSAL). **2** Bot. of the front or lower surface. □ **ventral fin** either of the ventrally placed fins on a fish. □ **ventrally** adv. [obs. venter abdomen f. L venter ventr-]

ventre à terre /ˌvɒntrə æ 'teə(r)/ adv. at full speed. [F, lit. 'with belly to the ground']

ventricle /'ventrɪk(ə)l/ n. Anat. **1** a cavity in the body. **2** a hollow part of an organ, esp. in the brain or heart. □ **ventricular** /ven'trɪkjʊlə(r)/ adj. [ME f. L ventriculus dimin. of venter belly]

ventricose /'ventrɪˌkəʊz, -ˌkəʊs/ adj. **1** having a protruding belly. **2** Bot. distended, inflated. [irreg. f. VENTRICLE + -OSE[1]]

ventriloquism /ven'trɪləˌkwɪz(ə)m/ n. the skill of speaking or uttering sounds so that they seem to come from the speaker's dummy or a source other than the speaker. □ **ventriloquist** n. **ventriloquize** v.intr. (also **-ise**). **ventriloquial** /ˌventrɪ'ləʊkwɪəl/ adj. [ult. f. L ventriloquus ventriloquist f. venter belly + loqui speak]

ventriloquy /ven'trɪləkwɪ/ n. = VENTRILOQUISM.

venture /'ventʃə(r)/ n. & v. ● n. **1** an undertaking of a risk; a risky enterprise. **2** a commercial speculation. ● v. **1** intr. dare; not be afraid (did not venture to stop them). **2** intr. (usu. foll. by out etc.) dare to go (out), esp. outdoors. **3** tr. dare to put forward (an opinion, suggestion, etc.). **4 a** tr. expose to risk; stake (a bet etc.). **b** intr. take risks. **5** intr. (foll. by on, upon) dare to engage in etc. (ventured on a longer journey). □ **at a venture** at random; without previous consideration. **venture capital** = risk capital. [aventure = ADVENTURE]

venturer /'ventʃərə(r)/ n. hist. a person who undertakes or shares in a trading venture.

Venture Scout n. a member of the Scout Association aged between 16 and 20.

venturesome /'ventʃəsəm/ adj. **1** disposed to take risks. **2** risky. □ **venturesomely** adv. **venturesomeness** n.

Venturi /ven'tjʊərɪ/, Robert (Charles) (b.1925), American architect. He reacted against the prevailing international style of the 1960s and pioneered the development of post-modernist architecture. His work is eclectic, making use of historical references and sometimes of kitsch. Among his buildings are the Humanities Classroom Building of the State University of New York (1973) and the Sainsbury Wing of the National Gallery in London (1991). His writings include Complexity and Contradiction in Architecture (1966).

venturi /ven'tjʊərɪ/ n. (pl. **venturis**) (also **venturi tube**) a device consisting of a short section of tube which is narrower in the middle than at the ends, so that gas or liquid flowing through it experiences a drop in pressure. It is used to produce an effect of suction or in measuring the rate of flow. [Giovanni Battista Venturi, Italian physicist (1746–1822)]

venue /'venju:/ n. **1 a** an appointed site or meeting-place, esp. for a sports event, meeting, concert, etc. **b** a rendezvous. **2** Law hist. the county or other place within which a jury must be gathered and a cause tried (orig. the neighbourhood of the crime etc.). [F, = a coming, fem. past part. of venir come f. L venire]

venule /'venju:l/ n. Anat. a small vein, esp. one that collects blood from the capillaries. [L venula dimin. of vena vein]

Venus /'vi:nəs/ n. **1** Rom. Mythol. a goddess, identified with Aphrodite (the Greek goddess of love) in classical Rome though apparently a spirit of kitchen gardens in earlier times. **2** poet. **a** a beautiful woman. **b** sexual love; amorous influences or desires. **3** Astron. the second planet from the sun in the solar system, orbiting between Mercury and the earth at an average distance of 108 million km from the sun. (See note below.) □ **Venus** (or **Venus's**) **fly-trap** an American carnivorous plant, Dionaea muscipula, with hinged leaves that spring shut on insects etc. **Venus's comb** = shepherd's needle (see SHEPHERD). **Venus's looking-glass** a plant of the genus Legousia, of the bellflower family, with small blue flowers.

▪ The planet Venus is almost equal in size to the earth, with a diameter of 12,104 km. It is the brightest celestial object after the sun and moon, frequently appearing in the twilight sky as the evening or morning star (Hesperus and Lucifer to the ancient Greeks and Romans), and it shows phases similar to the moon. The planet's disc is completely covered by clouds consisting chiefly of sulphuric acid droplets, and no surface detail can be seen by telescope. There is a dense atmosphere of carbon dioxide, which traps the heat of the sun by the greenhouse effect to produce a surface temperature of 460°C. Space probes have shown the surface to be a vast rocky plain with two raised 'continental' plateaux and a very high mountain massif, baking under a dull orange sky. The planet has no natural satellite.

Venus de Milo /də 'maɪləʊ, 'mi:l-/ a classical sculpture of Aphrodite dated to c.100 BC, perhaps the most famous sculpture of antiquity. It was discovered on the Greek island of Melos in 1820 and is now in the Louvre in Paris, having formed part of the war loot acquired by Napoleon on his campaigns. [F, = Venus of Melos]

Venusian /vɪ'nju:sɪən/ adj. & n. ● adj. of the planet Venus. ● n. a hypothetical inhabitant of the planet Venus. [VENUS + -IAN]

veracious /və'reɪʃəs/ adj. formal **1** speaking or disposed to speak the truth. **2** (of a statement etc.) true or meant to be true. □ **veraciously** adv. **veraciousness** n. [L verax veracis f. verus true]

veracity /və'ræsɪtɪ/ n. **1** truthfulness, honesty. **2** accuracy (of a statement etc.). [F veracité or med.L veracitas (as VERACIOUS)]

Veracruz /ˌveərə'kru:z/ **1** a state of east central Mexico, with a long coastline on the Gulf of Mexico; capital, Jalapa Enriquez. **2** a city and port of Mexico, in Veracruz state, on the Gulf of Mexico; pop. (1990) 327,520.

veranda /və'rændə/ n. (also **verandah**) **1** a portico or external gallery, usu. with a roof, along the side of a house. **2** Austral. & NZ a roof over a pavement in front of a shop. [Port. varanda prob. f. Hindi varaṇḍā]

veratrine /'verəˌtri:n, -trɪn/ n. a poisonous compound obtained from sabadilla etc., and used esp. as a local irritant in the treatment of neuralgia and rheumatism. [F vératrine f. L veratrum hellebore]

verb /vɜ:b/ n. Gram. a word used to indicate an action, state, or occurrence, and forming the main part of the predicate of a sentence (e.g. hear, become, happen). [ME f. OF verbe or L verbum word, verb]

verbal /'vɜ:b(ə)l/ adj., n., & v. ● adj. **1** of or concerned with words (made a verbal distinction; verbal reasoning). **2** oral, not written (gave a verbal statement). **3** Gram. of or in the nature of a verb (verbal inflections). **4** literal (a verbal translation). **5** talkative, articulate. ● n. **1** Gram. **a** a verbal noun. **b** a word or words functioning as a verb. **2** sl. a verbal statement, esp. one made to the police. **3** (usu. in pl.) sl. an insult; abuse (gave them a lot of verbals). ● v.tr. (**verballed**, **verballing**) Brit. sl. attribute a damaging statement to (a suspect). □ **verbal noun** Gram. a noun formed as an inflection of a verb and partly sharing its constructions (e.g. smoking in smoking is forbidden: see -ING[1]). □ **verbally** adv. [ME f. F verbal or LL verbalis (as VERB)]

verbalism /'vɜ:bəˌlɪz(ə)m/ n. **1** minute attention to words: verbal criticism. **2** merely verbal expression. □ **verbalist** n. **verbalistic** /ˌvɜ:bə'lɪstɪk/ adj.

verbalize /'vɜ:bəˌlaɪz/ v. (also **-ise**) **1** tr. express in words. **2** intr. be verbose. **3** *tr.* make (a noun etc.) into a verb. □ **verbalizable** *adj.* **verbalizer** *n.* **verbalization** /ˌvɜːbəlaɪˈzeɪʃ(ə)n/ *n.*

verbascum /vəˈbæskəm/ *n.* a plant of the genus *Verbascum*, a mullein. [L]

verbatim /vɜːˈbeɪtɪm/ *adv. & adj.* in exactly the same words; word for word (*copied it verbatim; a verbatim report*). [ME f. med.L (adv.), f. L *verbum* word: cf. LITERATIM]

verbena /vɜːˈbiːnə/ *n.* a plant of the genus *Verbena*, bearing clusters of fragrant flowers (cf. VERVAIN). [L, = sacred bough of olive etc., in med.L vervain]

verbiage /ˈvɜːbiɪdʒ/ *n.* needless accumulation of words; verbosity. [F f. obs. *verbeier* chatter f. *verbe* word: see VERB]

verbose /vɜːˈbəʊs/ *adj.* using or expressed in more words than are needed; wordy, long-winded. □ **verbosely** *adv.* **verboseness** *n.* **verbosity** /-ˈbɒsɪtɪ/ *n.* [L *verbosus* f. *verbum* word]

verboten /feəˈbəʊt(ə)n/ *adj.* forbidden, esp. by an authority. [G]

verb. sap. /vɜːb ˈsæp/ *int.* expressing the absence of the need for a further explicit statement. [abbr. of L *verbum sapienti sat est* a word is enough for the wise person]

verdant /ˈvɜːd(ə)nt/ *adj.* **1** (of grass etc.) green, fresh-coloured. **2** (of a field etc.) green with grass and vegetation, lush. **3** (of a person) unsophisticated, raw, green. □ **verdancy** *n.* **verdantly** *adv.* [perh. f. OF *verdeant* part. of *verdoier* be green ult. f. L *viridis* green]

verd-antique /ˌvɜːdænˈtiːk/ *n.* **1** ornamental usu. green serpentine. **2** a green incrustation on ancient bronze. **3** green porphyry. [obs. F, = antique green]

verderer /ˈvɜːdərə(r)/ *n. Brit.* a judicial officer of royal forests. [AF (earlier *verder*), OF *verdier* ult. f. L *viridis* green]

Verdi /ˈveədɪ/, Giuseppe (Fortunino Francesco) (1813–1901), Italian composer. He is chiefly remembered for his many operas, which are notable for strong characterization, original orchestration, and memorable tunes. Verdi was a supporter of the movement for Italian unity, and several of his early works were identified with the nationalist cause, including *Nabucco* (1842), with which he established his reputation. Other operas include *Rigoletto* (1851), *La Traviata* (1853), and *Otello* (1887); *Aida*, commissioned to be staged in Cairo to celebrate the opening of the Suez Canal, was delayed due to the Franco-Prussian War and eventually first performed there in 1871. Among his other compositions are sacred choral works such as the Requiem Mass of 1874.

verdict /ˈvɜːdɪkt/ *n.* **1** a decision on an issue of fact in a civil or criminal cause or an inquest. **2** a decision; a judgement. [ME f. AF *verdit*, OF *voirdit* f. *voir, veir* true f. L *verus* + *dit* L DICTUM saying]

verdigris /ˈvɜːdɪˌɡriː, -ˌɡriːs/ *n.* **1 a** a green crystallized substance formed on copper by the action of acetic acid. **b** this used as a medicine or pigment. **2** green rust on copper or brass. [ME f. OF *verte-gres, vert de Grece* green of Greece]

Verdon-Roe /ˌvɜːd(ə)nˈrəʊ/, Sir (Edwin) Alliott Verdon (1877–1958), English engineer and aircraft designer. He built the first British seaplane to take off from the water and (in 1912) the first aircraft with an enclosed cabin. With his brother H. V. Roe he founded the Avro Company and built a number of planes, of which the Avro 504 trainer biplane was the most successful; in 1928 he formed the Saunders-Roe Company to design and manufacture flying boats. Verdon-Roe also invented anti-dazzle car headlights.

Verdun, Battle of /vɜːˈdʌn/ a long and severe battle in 1916, during the First World War, at the fortified town of Verdun in NE France. The French, initially unprepared, eventually repelled a prolonged German offensive but suffered heavy losses.

verdure /ˈvɜːdjə(r)/ *n.* **1** green vegetation. **2** the greenness of this. **3** *poet.* freshness. □ **verdured** *adj.* **verdurous** *adj.* [ME f. OF f. *verd* green f. L *viridis*]

Vereeniging /fəˈriːnɪkɪŋ, -ˈreɪnɪxɪŋ/ a city in South Africa, in the province of Pretoria-Witwatersrand-Vereeniging; pop. (1985) 540,140 (with Vanderbijlpark). It was the scene in 1902 of the signing of the treaty terminating the Second Boer War.

verge¹ /vɜːdʒ/ *n.* **1** an edge or border. **2** an extreme limit beyond which something happens (*on the verge of tears*). **3** *Brit.* a grass edging of a road, flower-bed, etc. **4** *Archit.* an edge of tiles projecting over a gable. **5** a wand or rod carried before a bishop, dean, etc., as an emblem of office. [ME f. OF f. L *virga* rod]

verge² /vɜːdʒ/ *v.intr.* **1** incline downwards or in a specified direction (*the now verging sun; verge to a close*). **2** (foll. by *on*) border on; approach closely (*verging on the ridiculous*). [L *vergere* bend, incline]

verger /ˈvɜːdʒə(r)/ *n.* (also **virger**) **1** an official in a church who acts as caretaker and attendant. **2** an officer who bears the staff before a bishop etc. □ **vergership** *n.* [ME f. AF (as VERGE¹)]

Vergil see VIRGIL.

verglas /ˈveəɡlɑː/ *n.* a thin coating of ice or frozen rain. [F]

veridical /vɪˈrɪdɪk(ə)l/ *adj.* **1** *formal* truthful. **2** *Psychol.* (of visions etc.) coinciding with reality. □ **veridically** *adv.* **veridicality** /-ˌrɪdɪˈkælɪtɪ/ *n.* [L *veridicus* f. *verus* true + *dicere* say]

veriest /ˈverɪɪst/ *adj. archaic* most complete or extreme; absolute, real (*the veriest fool knows that*). [superl. of VERY]

verification /ˌverɪfɪˈkeɪʃ(ə)n/ *n.* **1** the process or an instance of establishing the truth or validity of something. **2** *Philos.* the establishment of the validity of a proposition empirically. **3** the process of verifying procedures laid down in weapons agreements.

verify /ˈverɪˌfaɪ/ *v.tr.* (**-ies, -ied**) **1** establish the truth or correctness of by examination or demonstration (*must verify the statement; verified my figures*). **2** (of an event etc.) bear out or fulfil (a prediction or promise). **3** *Law* append an affidavit to (pleadings); support (a statement) by testimony or proofs. □ **verifiable** *adj.* **verifiably** *adv.* **verifier** *n.* [ME f. OF *verifier* f. med.L *verificare* f. *verus* true]

verily /ˈverɪlɪ/ *adv. archaic* really, truly. [ME f. VERY + -LY², after OF & AF]

verisimilitude /ˌverɪsɪˈmɪlɪˌtjuːd/ *n.* **1** the appearance or semblance of being true or real. **2** a statement etc. that seems true. □ **verisimilar** /-ˈsɪmɪlə(r)/ *adj.* [L *verisimilitudo* f. *verisimilis* probable f. *veri* genitive of *verus* true + *similis* like]

verism /ˈvɪərɪz(ə)m/ *n.* realism in literature or art. □ **verist** *n.* **veristic** /vɪəˈrɪstɪk/ *adj.* [L *verus* or It. *vero* true + -ISM]

verismo /veˈrɪzməʊ/ *n.* (esp. of opera) realism. [It. (as VERISM)]

veritable /ˈverɪtəb(ə)l/ *adj.* real; rightly so called (*a veritable feast*). □ **veritably** *adv.* [OF (as VERITY)]

verity /ˈverɪtɪ/ *n.* (*pl.* **-ies**) **1** a true statement, esp. one of fundamental import. **2** truth. **3** a really existent thing. [ME f. OF *verité, verté* f. L *veritas -tatis* f. *verus* true]

verjuice /ˈvɜːdʒuːs/ *n.* **1** an acid liquor obtained from crab-apples, sour grapes, etc., and formerly used in cooking and medicine. **2** bitter feelings, thoughts, etc. [ME f. OF *vertjus* f. VERT green + *jus* JUICE]

Verkhneudinsk /ˌveəxnjeˈuːdɪnsk/ the former name (until 1934) for ULAN-UDE.

verkrampte /fəˈkræmptə/ *adj. & n. S. Afr.* ● *adj.* politically or socially conservative or reactionary, esp. as regards apartheid. ● *n.* a person holding such views. [Afrik., lit. 'narrow, cramped']

Verlaine /veəˈleɪn/, Paul (1844–96), French poet. Initially a member of the Parnassian group of poets, he later became prominent among the symbolists, especially with the publication of his influential essay 'Art poétique' (1882). Notable collections of poetry include *Poèmes saturniens* (1867), *Fêtes galantes* (1869), and *Romances san paroles* (1874), a work characterized by an intense musicality and metrical inventiveness. The last was written in prison, where Verlaine was serving a two-year sentence for wounding his lover, the poet Arthur Rimbaud, during a quarrel.

verligte /fəˈlɪxtə/ *adj. & n. S. Afr.* ● *adj.* progressive or enlightened, esp. as regards racial issues or (formerly) apartheid. ● *n.* a person holding such views. [Afrik., = enlightened]

Vermeer /vɜːˈmɪə(r)/, Jan (1632–75), Dutch painter. He spent his life in his native town of Delft, where he generally painted domestic genre scenes, often depicting a single figure engaged in an ordinary task (e.g. *The Kitchen-Maid*, *c.*1658). His work is distinguished by its clear design and simple form, and by its harmonious balance of predominant yellows, blues, and greys. His other works include two views of Delft and a self-portrait, the *Allegory of Painting* (*c.*1665). His paintings were long neglected and only began to receive full recognition in the later 19th century.

vermeil /ˈvɜːmeɪl, -mɪl/ *n.* **1** silver gilt. **2** an orange-red garnet. **3** *poet.* vermilion. [ME f. OF: see VERMILION]

vermi- /ˈvɜːmɪ/ *comb. form* worm. [L *vermis* worm]

vermian /ˈvɜːmɪən/ *adj.* of worms; wormlike. [L *vermis* worm]

vermicelli /ˌvɜːmɪˈselɪ, -ˈtʃelɪ/ *n.pl.* **1** pasta made in long slender threads. **2** shreds of chocolate used as cake decoration etc. [It., pl. of *vermicello* dimin. of *verme* f. L *vermis* worm]

vermicide /ˈvɜːmɪˌsaɪd/ *n.* a substance that kills worms.

vermicular /vəˈmɪkjʊlə(r)/ adj. **1** like a worm in form or movement; vermiform. **2** Med. of or caused by intestinal worms. **3** marked with close wavy lines. [med.L vermicularis f. L vermiculus dimin. of vermis worm]

vermiculate /vəˈmɪkjʊlət/ adj. **1** = VERMICULAR. **2** wormeaten. [L vermiculatus past part. of vermiculari be full of worms (as VERMICULAR)]

vermiculation /vəˌmɪkjʊˈleɪʃ(ə)n/ n. **1** the state or process of being eaten or infested by or converted into worms. **2** a vermicular marking. **3** a wormeaten state. [L vermiculatio (as VERMICULATE)]

vermiculite /vəˈmɪkjʊˌlaɪt/ n. **1** Mineral. a hydrated silicate resulting from the alteration of mica etc., esp. an aluminosilicate of magnesium. **2** flakes of this used as a medium for growing plants, as an insulation material, etc. [as VERMICULATE + -ITE¹]

vermiform /ˈvɜːmɪˌfɔːm/ adj. worm-shaped. □ **vermiform appendix** see APPENDIX 1.

vermifuge /ˈvɜːmɪˌfjuːdʒ/ adj. & n. Med. ● adj. that expels intestinal worms. ● n. a drug that does this.

vermilion /vəˈmɪljən/ n. & adj. ● n. **1** cinnabar. **2 a** a brilliant red pigment made by grinding this or artificially. **b** the colour of this. ● adj. of this colour. [ME f. OF vermeillon f. vermeil f. L vermiculus dimin. of vermis worm]

vermin /ˈvɜːmɪn/ n. (usu. treated as pl.) **1** mammals and birds injurious to game, crops, etc., e.g. foxes, rodents, and noxious insects. **2** parasitic worms or insects. **3** vile or contemptible persons. □ **verminous** adj. [ME f. OF vermin, -ine ult. f. L vermis worm]

verminate /ˈvɜːmɪˌneɪt/ v.intr. **1** breed vermin. **2** become infested with parasites. □ **vermination** /ˌvɜːmɪˈneɪʃ(ə)n/ n. [L verminare verminat- f. vermis worm]

vermivorous /vəˈmɪvərəs/ adj. feeding on worms.

Vermont /vəˈmɒnt/ a state in the north-eastern US, on the border with Canada; pop. (1990) 562,760; capital, Montpelier. Explored and settled by the French during the 17th and 18th centuries, it became an independent republic in 1777 and the 14th state of the US in 1791. □ **Vermonter** n.

vermouth /ˈvɜːməθ, vəˈmuːθ/ n. a wine flavoured with aromatic herbs. [F vermout f. G Wermut WORMWOOD]

vernacular /vəˈnækjʊlə(r)/ n. & adj. ● n. **1** the language or dialect of a particular country (Latin gave place to the vernacular). **2** the language of a particular clan or group. **3** homely speech. ● adj. **1** (of language) of one's native country; not of foreign origin or of learned formation. **2** (of architecture) concerned with ordinary rather than monumental buildings. □ **vernacularism** n. **vernacularize** v.tr. (also **-ise**). **vernacularly** adv. **vernacularity** /-ˌnækjʊˈlærɪtɪ/ n. [L vernaculus domestic, native f. verna home-born slave]

vernal /ˈvɜːn(ə)l/ adj. of, in, or appropriate to spring (vernal equinox; vernal breezes). □ **vernal equinox** see EQUINOX. **vernal grass** a sweet-scented European grass, Anthoxanthum odoratum, grown for hay. □ **vernally** adv. [L vernalis f. vernus f. ver spring]

vernalization /ˌvɜːnəlaɪˈzeɪʃ(ə)n/ n. (also **-isation**) the cooling of seed before planting, in order to accelerate flowering. □ **vernalize** /ˈvɜːnəˌlaɪz/ v.tr. (also **-ise**). [VERNAL]

vernation /vɜːˈneɪʃ(ə)n/ n. Bot. the arrangement of leaves in a leaf-bud (cf. AESTIVATION 1). [mod.L vernatio f. L vernare bloom (as VERNAL)]

Verne /vɜːn/, Jules (1828–1905), French novelist. Regarded as one of the first writers of science fiction, in his adventure stories he often anticipated later scientific and technological developments. He explored the possibilities of space travel in From the Earth to the Moon (1865) and the use of submarines in Twenty Thousand Leagues under the Sea (1870). Other novels include Journey to the Centre of the Earth (1864) and Around the World in Eighty Days (1873).

vernicle /ˈvɜːnɪk(ə)l/ n. = VERONICA 2. [ME f. OF (earlier ver(o)nique), f. med.L VERONICA]

vernier /ˈvɜːnɪə(r)/ n. a small movable graduated scale for obtaining fractional parts of subdivisions on a fixed main scale of a barometer, sextant, etc. □ **vernier engine** an auxiliary engine for slight changes in the motion of a space rocket etc. [Pierre Vernier, Fr. mathematician (1580–1637)]

Verny /ˈveənɪ/ the former name (until 1921) for ALMATY.

Verona /vəˈrəʊnə/ a city on the River Adige, in NE Italy; pop. (1990) 258,950.

veronal /ˈverən(ə)l/ n. propr. a sedative drug, a derivative of barbituric acid. [G, f. Verona in Italy + -AL²]

Veronese /ˌverəˈneɪzɪ/, Paolo (born Paolo Caliari) (c.1528–88), Italian painter. Born in Verona, by about 1553 he had established himself in Venice, where he gained many commissions, including the painting of frescos in a number of churches and in the Doges' Palace. Assisted by the staff of a large workshop, he produced numerous paintings, mainly dealing with religious, allegorical, and historical subjects; he is particularly known for his richly coloured feast-scenes (e.g. The Marriage at Cana, 1562). Other notable works include his series of frescos in Palladio's villa at Maser near Treviso (1561).

veronica /vəˈrɒnɪkə/ n. **1** a plant now or formerly of the genus Veronica, esp. speedwell. **2 a** a cloth supposedly impressed with an image of Christ's face. **b** any similar picture of Christ's face. (See VERONICA, ST.) **3** Bullfighting the movement of a matador's cape away from a charging bull. [med.L f. the name Veronica]

Veronica, St /vəˈrɒnɪkə/ (in the Bible) a woman of Jerusalem who offered her head-cloth to Christ on the way to Calvary, to wipe the blood and sweat from his face. The cloth is said to have retained the image of his features.

Verrazano-Narrows Bridge /ˌverəˌzɑːnəʊˈnærəʊz/ a suspension bridge across New York harbour between Brooklyn and Staten Island, the longest in the world when it was completed in 1964. It was named after the Italian explorer Giovanni da Verrazano (1485–1528).

verruca /vəˈruːkə/ n. (pl. **verrucae** /-ˈruːsiː/ or **verrucas**) a wart or similar growth, esp. a contagious wart on the sole of the foot. □ **verrucose** /ˈverʊˌkəʊz, -ˌkəʊs/ adj. **verrucous** adj. [L]

Versailles /veəˈsaɪ/ a palace built for Louis XIV near the town of Versailles, south-west of Paris. The palace was built around a château belonging to Louis XIII, which was transformed by the addition of wings and other edifices in the grand French classical style. Much splendid interior work was carried out by Charles Lebrun and André-Charles Boulle (1642–1732), while the elaborate gardens, full of fountains and statuary, were laid out by André Le Nôtre. Later extensions, such as the Trianon palaces in the grounds, were added during the reigns of Louis XIV and Louis XV. The palace's active life terminated when Louis XVI was forced out by the Paris revolutionaries in October 1789, but Versailles was restored and turned into a museum by King Louis Philippe.

Versailles, Treaty of **1** a treaty which terminated the War of American Independence in 1783. **2** a treaty signed in 1919 which, along with a series of associated agreements, brought a formal end to the First World War, redivided the territory of the defeated Central Powers, restricted Germany's armed forces, and established the League of Nations. The treaty in fact represented an unhappy compromise between conciliation and punishment, leaving Germany smarting under what it considered a vindictive settlement while not sufficiently restricting its ability eventually to rearm and seek forcible redress.

versant /ˈvɜːs(ə)nt/ n. **1** the extent of land sloping in one direction. **2** the general slope of land. [F f. verser f. L versare frequent. of vertere vers- turn]

versatile /ˈvɜːsəˌtaɪl/ adj. **1** turning easily or readily from one subject or occupation to another; capable of dealing with many subjects (a versatile mind). **2** (of a device etc.) having many uses. **3** Bot. & Zool. moving freely about or up and down on a support (versatile antenna). **4** archaic changeable, inconstant. □ **versatilely** adv. **versatility** /ˌvɜːsəˈtɪlɪtɪ/ n. [F versatile or L versatilis (as VERSANT)]

verse /vɜːs/ n. & v. ● n. **1 a** a metrical composition in general (opp. PROSE) (wrote pages of verse). **b** a particular type of this (English verse). **2 a** a metrical line in accordance with the rules of prosody. **b** a group of a definite number of such lines. **c** a stanza of a poem with or without refrain. **3** a part of a popular song which leads into the chorus or separates one chorus from another. **4** each of the short numbered divisions of a chapter in the Bible or other scripture. **5 a** a versicle. **b** Mus. a passage (of an anthem etc.) for solo voice. ● v.tr. **1** express in verse. **2** (usu. refl.; foll. by in) instruct; make knowledgeable. □ **verselet** /ˈvɜːslɪt/ n. [OE fers f. L versus a turn of the plough, a furrow, a line of writing f. vertere vers- turn: in ME reinforced by OF vers f. L versus]

versed¹ /vɜːst/ predic.adj. (foll. by in) experienced or skilled in; knowledgeable about. [F versé or L versatus past part. of versari be engaged in (as VERSANT)]

versed² /vɜːst/ adj. Math. reversed. □ **versed sine** unity minus cosine. [mod.L (sinus) versus turned (sine), formed as VERSE]

verset /ˈvɜːsɪt/ n. Mus. a short prelude or interlude for organ. [F: dimin. of vers VERSE]

versicle /ˈvɜːsɪk(ə)l/ n. each of the short sentences in a liturgy said or sung by a priest etc. and alternating with responses. □ **versicular**

/vəˈsɪkjʊlə(r)/ adj. [ME f. OF versicule or L versiculus dimin. of versus: see VERSE]

versicoloured /ˈvɜːsɪˌkʌləd/ adj. **1** changing from one colour to another in different lights. **2** variegated. [L versicolor f. versus past part. of vertere turn + color colour]

versify /ˈvɜːsɪˌfaɪ/ v. (-ies, -ied) **1** tr. turn into or express in verse. **2** intr. compose verses. □ **versifier** n. **versification** /ˌvɜːsɪfɪˈkeɪʃ(ə)n/ n. [ME f. OF versifier f. L versificare (as VERSE)]

versin /ˈvɜːsaɪn/ n. (also **versine**) Math. = versed sine (see VERSED²).

version /ˈvɜːʃ(ə)n/ n. **1** an account of a matter from a particular person's point of view (told them my version of the incident). **2** a book or work etc. in a particular edition or translation (Authorized Version). **3** a form or variant of a thing as performed, adapted, etc. **4** a piece of translation, esp. as a school exercise. **5** Med. the manual turning of a foetus in the womb to improve presentation. □ **versional** adj. [F version or med.L versio f. L vertere vers- turn]

vers libre /veə ˈliːbrə/ n. irregular or unrhymed verse in which the traditional rules of prosody are disregarded. [F, = free verse]

verso /ˈvɜːsəʊ/ n. (pl. **-os**) **1 a** the left-hand page of an open book. **b** the back of a printed leaf of paper or manuscript (opp. RECTO 2). **2** the reverse of a coin. [L verso (folio) on the turned (leaf)]

verst /vɜːst/ n. a Russian measure of length, about 1.1 km (0.66 mile). [Russ. versta]

versus /ˈvɜːsəs/ prep. (abbr. **v.**, **vs.**) against (esp. in legal and sports use). [L, = towards, in med.L against]

vert /vɜːt/ n. & (usu. placed after noun) adj. Heraldry green. [ME f. OF f. L viridis green]

vertebra /ˈvɜːtɪbrə/ n. (pl. **vertebrae** /-ˌbriː/) **1** each segment of the backbone. **2** (in pl.) the backbone. □ **vertebral** adj. [L f. vertere turn]

vertebrate /ˈvɜːtɪbrət, -ˌbreɪt/ n. & adj. ● n. an animal of the subphylum Vertebrata, having a spinal column. (See note below.) ● adj. of or relating to the vertebrates. [L vertebratus jointed (as VERTEBRA)]

▪ Though the vertebrates often dominate the human view of the animal world, they comprise only 5 per cent of all animal species. Most classification schemes include about seven classes in this subphylum of the phylum Chordata: the jawless fishes, cartilaginous fishes, bony fishes, amphibians, reptiles, birds, and mammals. Birds and mammals are warm-blooded, and some Mesozoic reptiles may also have been so.

vertebration /ˌvɜːtɪˈbreɪʃ(ə)n/ n. division into vertebrae or similar segments.

vertex /ˈvɜːteks/ n. (pl. **vertices** /-tɪˌsiːz/ or **vertexes**) **1** the highest point; the top or apex. **2** Geom. **a** each angular point of a polygon, polyhedron, etc. **b** a meeting-point of two lines that form an angle. **c** the point at which an axis meets a curve or surface. **d** the point opposite the base of a figure. **3** Anat. the crown of the head. [L vertex -ticis whirlpool, crown of a head, vertex, f. vertere turn]

vertical /ˈvɜːtɪk(ə)l/ adj. & n. ● adj. **1** at right angles to a horizontal plane, perpendicular. **2** in a direction from top to bottom of a picture etc. **3** of or at the vertex or highest point. **4** at, or passing through, the zenith. **5** Anat. of or relating to the crown of the head. **6** involving all the levels in an organizational hierarchy or stages in the production of a class of goods (vertical integration). ● n. a vertical line or plane. □ **out of the vertical** not vertical. **vertical angles** Math. each pair of opposite angles made by two intersecting lines. **vertical fin** Zool. a dorsal, anal, or caudal fin. **vertical plane** a plane at right angles to the horizontal. **vertical take-off** the take-off of an aircraft directly upwards. □ **verticalize** v.tr. (also **-ise**). **vertically** adv. **verticality** /ˌvɜːtɪˈkælɪtɪ/ n. [F vertical or LL verticalis (as VERTEX)]

verticil /ˈvɜːtɪsɪl/ n. Bot. & Zool. a whorl; a set of parts arranged in a circle round an axis. □ **verticillate** /vəˈtɪsɪlət/ adj. [L verticillus whorl of a spindle, dimin. of VERTEX]

vertiginous /vɜːˈtɪdʒɪnəs/ adj. of or causing vertigo. □ **vertiginously** adv. [L vertiginosus (as VERTIGO)]

vertigo /ˈvɜːtɪˌɡəʊ/ n. a condition with a sensation of whirling and a tendency to lose balance; dizziness, giddiness. [L vertigo -ginis whirling f. vertere turn]

vertu var. of VIRTU.

Verulamium /ˌverʊˈleɪmɪəm/ see ST ALBANS.

vervain /ˈvɜːveɪn/ n. a herbaceous plant of the genus Verbena, esp. V. officinalis, with small blue, white, or purple flowers (cf. VERBENA). [ME f. OF verveine f. L VERBENA]

verve /vɜːv/ n. enthusiasm, vigour, spirit, esp. in artistic or literary work. [F, earlier = a form of expression, f. L verba words]

vervet /ˈvɜːvɪt/ n. a common small African monkey, Cercopithecus aethiops, esp. one of a black-faced race of southern and eastern Africa. [F]

Verviers /ˈveəvɪˌeɪ/ a manufacturing town in eastern Belgium; pop. (1991) 53,480.

Verwoerd /fəˈvʊət/, Hendrik (Frensch) (1901–66), South African statesman, Prime Minister 1958–66. Born in the Netherlands, he was active in Afrikaner groups during the Second World War. He joined the National Party and served as Minister of Bantu Affairs from 1950 to 1958, developing the segregation policy of apartheid. As Premier, Verwoerd banned the ANC and the Pan-Africanist Congress in 1960, following the Sharpeville massacre. He withdrew South Africa from the Commonwealth and declared it a republic in 1961. Verwoerd was assassinated by a parliamentary messenger.

very /ˈverɪ/ adv. & adj. ● adv. **1** in a high degree (did it very easily; had a very bad cough; am very much better). **2** in the fullest sense (foll. by own or superl. adj.: at the very latest; do your very best; my very own room). ● adj. **1** (usu. prec. by the, this, his, etc.) **a** real, true, actual; truly such (emphasizing identity, significance, or extreme degree: the very thing we need; those were his very words). **b** mere, sheer (the very idea of it was horrible). **2** archaic real, genuine (very God). □ **not very 1** in a low degree. **2** far from being. **very good** (or **well**) a formula of consent or approval. **very high frequency** (of radio frequency) in the range 30–300 megahertz. **the very same** see SAME. [ME f. OF verai ult. f. L verus true]

Very light /ˈverɪ, ˈvɪərɪ/ n. a flare projected from a pistol for signalling or temporarily illuminating the surroundings. [Edward W. Very, Amer. inventor (1847–1910)]

Very pistol /ˈverɪ, ˈvɪərɪ/ n. a gun for firing a Very light.

Very Reverend n. the title of a dean.

Vesalius /vɪˈseɪlɪəs/, Andreas (1514–64), Flemish anatomist, the founder of modern anatomy. He challenged traditional theories of anatomy, holding them to be seriously flawed in being based on the bodies of apes — a view borne out by later studies. His major work, De Humani Corporis Fabrica (1543), contained accurate descriptions of human anatomy; it owed much of its great historical impact, however, to the woodcuts of his dissections, which were drawn and engraved by others. He became unofficial physician to the Emperor Charles V, followed by a period with Philip II of Spain. The medical profession there still supported Galen, whose works he had strongly criticized, and Vesalius died while returning to his old post at Padua.

vesica /ˈvesɪkə/ n. **1** Anat. & Zool. a bladder, esp. the urinary bladder. **2** (in full **vesica piscis** /ˈpɪsɪs/ or **piscium** /ˈpɪsɪəm/) Art a pointed oval used as an aureole in medieval sculpture and painting. □ **vesical** adj. [L]

vesicate /ˈvesɪˌkeɪt/ v.tr. esp. Med. raise blisters on. □ **vesicant** adj. & n. **vesicatory** /-ˌkeɪtərɪ/ adj. & n. **vesication** /ˌvesɪˈkeɪʃ(ə)n/ n. [LL vesicare vesicat- (as VESICA)]

vesicle /ˈvesɪk(ə)l/ n. **1 a** Anat. & Zool. a small fluid-filled bladder, sac, or vacuole. **b** Bot. an air-filled swelling in a seaweed etc. **2** Geol. a small cavity in volcanic rock produced by gas bubbles. **3** Med. a blister. □ **vesicular** /vɪˈsɪkjʊlə(r)/ adj. **vesiculate** /-lət/ adj. **vesiculation** /-ˌsɪkjʊˈleɪʃ(ə)n/ n. [F vésicule or L vesicula dimin. of VESICA]

Vespasian /veˈspeɪʒ(ə)n/ (Latin name Titus Flavius Vespasianus) (AD 9–79), Roman emperor 69–79 and founder of the Flavian dynasty. A distinguished general, he was acclaimed emperor by the legions in Egypt during the civil wars that followed the death of Nero and gained control of Italy after the defeat of Vitellius. His reign saw the restoration of financial and military order and the initiation of a public building programme which included the rebuilding of the Capitol and the beginning of the construction of the Colosseum (75).

vesper /ˈvespə(r)/ n. **1** poet. Venus as the evening star. **2** poet. evening. **3** (in pl.) **a** the sixth of the canonical hours of prayer. **b** evensong. [L vesper evening (star): sense 3 partly f. OF vespres f. eccl.L vesperas f. L vespera evening]

vespertine /ˈvespəˌtaɪn, -tɪn/ adj. **1** Bot. (of a flower) opening in the evening. **2** Zool. active in the evening. **3** Astron. setting near the time of sunset. **4** of or occurring in the evening. [L vespertinus f. vesper evening]

vespiary /ˈvespɪərɪ/ n. (pl. **-ies**) a nest of wasps. [irreg. f. L vespa wasp, after apiary]

vespine /ˈvespaɪn/ adj. of or relating to wasps. [L vespa wasp]

Vespucci /veˈspuːtʃɪ/, Amerigo (1451–1512), Italian merchant and explorer. He is said to have made a number of voyages to the New World, although only two have been authenticated. His first such voyage, for the Spanish, was probably made between 1499 and 1500 and saw the explorer reach the coast of Venezuela. Having entered Portuguese service, he embarked on another voyage (1501–2), during which he explored the Brazilian coastline and sailed at least as far south as the River Plate. The Latin form of his first name is believed to have given rise to the name of America.

vessel /'vesəl/ *n.* **1** a hollow receptacle esp. for liquid, e.g. a cask, cup, pot, bottle, or dish. **2** a ship or boat, esp. a large one. **3 a** *Anat.* a duct or canal etc. holding or conveying blood or other fluid, esp. = *blood vessel.* **b** *Bot.* a woody duct carrying or containing sap etc. **4** *Bibl.* or *joc.* a person regarded as the recipient or exponent of a quality (*a weak vessel*). [ME f. AF *vessel(e)*, OF *vaissel(le)* f. LL *vascellum* dimin. of *vas* vessel]

vest /vest/ *n. & v.* ● *n.* **1** an undergarment worn on the upper part of the body. **2** *N. Amer. & Austral.* a waistcoat. **3** a usu. V-shaped piece of material to fill the opening at the neck of a woman's dress. ● *v.* **1** *tr.* (esp. in *passive*; foll. by *with*) bestow or confer (powers, authority, etc.) on (a person). **2** *tr.* (foll. by *in*) confer (property or power) on (a person) with an immediate fixed right of immediate or future possession. **3** *intr.* (foll. by *in*) (of property, a right, etc.) come into the possession of (a person). **4 a** *tr. poet.* clothe. **b** *intr. Eccl.* put on vestments. □ **vested interest 1** *Law* an interest (usu. in land or money held in trust) recognized as belonging to a person. **2** a personal interest in a state of affairs, usu. with an expectation of gain. [(n.) F *veste* f. It. *veste* f. L *vestis* garment: (v.) ME, orig. past part. f. OF *vestu* f. *vestir* f. L *vestire vestit-* clothe]

Vesta /'vestə/ **1** *Rom. Mythol.* the goddess of the hearth and household. She was worshipped in a round building in the Forum at Rome, probably an imitation in stone of an ancient round hut, which contained no image but a fire which was kept constantly burning and was tended by the Vestal Virgins. **2** *Astron.* asteroid 4, discovered in 1807. It is the brightest asteroid and the third largest (diameter 501 km), and appears to consist of basaltic rock. [L f. Gk *hestia* hearth]

vesta /'vestə/ *n.* esp. *hist.* a short wooden or wax match. [VESTA 1]

Vestal /'vest(ə)l/ *adj. & n.* ● *adj.* **1** *Rom. Hist.* of or relating to the goddess Vesta. **2** (**vestal**) chaste, pure. ● *n.* **1** (in full **Vestal Virgin**) *Rom. Hist.* a virgin consecrated to the goddess Vesta and vowed to chastity. (*See note below.*) **2** (**vestal**) a chaste woman, esp. a nun. [ME f. L *vestalis* (adj. & n.) (as VESTA)]

■ There were normally six Vestals, each of whom served for thirty years. Vestals wore an old style of sacral dress otherwise reserved for brides. Their duty was to tend the undying fire burning in the shrine of Vesta in Rome, and to remain virgin: an unchaste Vestal was punished by being entombed alive.

vestee /ve'stiː/ *n.* = VEST *n.* 3.

Vesterålen /'vestə,rɔːlən/ a group of islands of Norway, north of the Arctic Circle.

vestiary /'vestɪərɪ/ *n. & adj.* ● *n.* (*pl.* **-ies**) **1** a vestry. **2** a robing-room; a cloakroom. ● *adj.* of or relating to clothes or dress. [ME f. OF *vestiarie, vestiaire*: see VESTRY]

vestibule /'vestɪ,bjuːl/ *n.* **1 a** an antechamber, hall, or lobby next to the outer door of a building. **b** a porch of a church etc. **2** *US* an enclosed entrance to a railway carriage. **3** *Anat.* **a** a chamber or channel communicating with others. **b** part of the mouth outside the teeth. **c** the central cavity of the labyrinth of the inner ear. □ **vestibular** /ve'stɪbjʊlə(r)/ *adj.* [F *vestibule* or L *vestibulum* entrance-court]

vestige /'vestɪdʒ/ *n.* **1** a trace or piece of evidence; a sign (*vestiges of an earlier civilization; found no vestige of their presence*). **2** a slight amount; a particle (*without a vestige of clothing; showed not a vestige of decency*). **3** *Biol.* a part or organ of an organism that is reduced or functionless but was well developed in its ancestors. [F f. L *vestigium* footprint]

vestigial /ve'stɪdʒɪəl, -dʒəl/ *adj.* **1** being a vestige or trace. **2** *Biol.* (of an organ or appendage) degenerate or atrophied, having become functionless in the course of evolution (*a vestigial wing*). □ **vestigially** *adv.*

vestiture /'vestɪtʃə(r)/ *n.* **1** *Zool.* hair, scales, etc., covering a surface. **2** *archaic* **a** clothing. **b** investiture. [ME f. med.L *vestitura* f. L *vestire*: see VEST]

Vestmannaeyjar /,vestmanə'eɪjaːr/ the Icelandic name for the WESTMAN ISLANDS.

vestment /'vestmənt/ *n.* **1** any of the official robes of clergy, choristers, etc., worn during divine service, esp. a chasuble. **2** a garment, esp. an official or state robe. [ME f. OF *vestiment, vestement* f. L *vestimentum* (as VEST)]

vestry /'vestrɪ/ *n.* (*pl.* **-ies**) **1** a room or building attached to a church for keeping vestments in. **2** *hist.* **a** a meeting of parishioners, usu. in a vestry, for parochial business. **b** a body of parishioners meeting in this way. □ **vestral** *adj.* [ME f. OF *vestiaire, vestiarie*, f. L *vestiarium* (as VEST)]

vestryman /'vestrɪmən/ *n.* (*pl.* **-men**) a member of a vestry.

vesture /'vestʃə(r)/ *n. & v.* ● *n. poet.* **1** garments, dress. **2** a covering. ● *v.tr.* clothe. [ME f. OF f. med.L *vestitura* (as VEST)]

Vesuvius /vɪ'suːvɪəs/ an active volcano near Naples, in southern Italy, 1,277 m (4,190 ft) high. A violent eruption in AD 79 buried the towns of Pompeii and Herculaneum.

vet¹ /vet/ *n. & v.* ● *n. colloq.* a veterinary surgeon. ● *v.tr.* (**vetted, vetting**) **1** make a careful and critical examination of (a scheme, work, candidate, etc.). **2** examine or treat (an animal). [abbr.]

vet² /vet/ *n. N. Amer. colloq.* a veteran. [abbr.]

vetch /vetʃ/ *n.* a leguminous plant of the genus *Vicia*, with numerous leaflets; esp. *V. sativa*, largely used for silage or fodder. Also called *tare*. □ **vetchy** *adj.* [ME f. AF & ONF *veche* f. L *vicia*]

vetchling /'vetʃlɪŋ/ *n.* a leguminous plant of the genus *Lathyrus*, related to vetches but with fewer leaflets.

veteran /'vetərən/ *n.* **1** a person who has grown old in or had long experience of esp. military service or an occupation (*a war veteran; a veteran of the theatre; a veteran marksman*). **2** *N. Amer.* an ex-serviceman or servicewoman. **3** (*attrib.*) of or for veterans. □ **veteran car** *Brit.* a car made before 1916, or (strictly) before 1905. [F *vétéran* or L *veteranus* (adj. & n.) f. *vetus -eris* old]

Veterans Day (in the US) 11 November, a public holiday held on the anniversary of the end of the First World War to honour US veterans and victims of all wars. It replaced Armistice Day in 1954.

veterinarian /,vetərɪ'neərɪən/ *n. N. Amer.* a veterinary surgeon. [L *veterinarius* (as VETERINARY)]

veterinary /'vetərɪnərɪ/ *adj. & n.* ● *adj.* of or for diseases and injuries of farm and domestic animals, or their treatment. ● *n.* (*pl.* **-ies**) a veterinary surgeon. □ **veterinary surgeon** *Brit.* a person qualified to treat diseased or injured animals. [L *veterinarius* f. *veterinae* cattle]

vetiver /'vetɪvə(r)/ *n.* = CUSCUS¹. [F *vétiver* f. Tamil *veṭṭivēru* f. *vēr* root]

veto /'viːtəʊ/ *n. & v.* ● *n.* (*pl.* **-oes**) **1 a** a constitutional right to reject a legislative enactment. **b** the right of a permanent member of the UN Security Council to reject a resolution. **c** such a rejection. **d** an official message conveying this. **2** a prohibition (*put one's veto on a proposal*). ● *v.tr.* (**-oes, -oed**) **1** exercise a veto against (a measure etc.). **2** forbid authoritatively. □ **vetoer** *n.* [L, = I forbid, with ref. to its use by Roman tribunes of the people in opposing measures of the Senate]

vex /veks/ *v.tr.* **1** anger by a slight or a petty annoyance; irritate. **2** *archaic* grieve, afflict. □ **vexer** *n.* **vexing** *adj.* **vexingly** *adv.* [ME f. OF *vexer* f. L *vexare* shake, disturb]

vexation /vek'seɪʃ(ə)n/ *n.* **1** the act or an instance of vexing; the state of being vexed. **2** an annoying or distressing thing. [ME f. OF *vexation* or L *vexatio -onis* (as VEX)]

vexatious /vek'seɪʃəs/ *adj.* **1** such as to cause vexation. **2** *Law* not having sufficient grounds for action and seeking only to annoy the defendant. □ **vexatiously** *adv.* **vexatiousness** *n.*

vexed /vekst/ *adj.* **1** irritated, angered. **2** (of a problem, issue, etc.) difficult and much discussed; problematic. □ **vexedly** /-sɪdlɪ/ *adv.*

vexillology /,veksɪ'lɒlədʒɪ/ *n.* the study of flags. □ **vexillologist** *n.* **vexillological** /-lə'lɒdʒɪk(ə)l/ *adj.* [VEXILLUM + -LOGY]

vexillum /vek'sɪləm/ *n.* (*pl.* **vexilla** /-lə/) **1** *Rom. Antiq.* **a** a military standard, esp. of a maniple. **b** a body of troops under this. **2** *Bot.* the large upper petal of a papilionaceous flower. **3** *Zool.* the vane of a feather. **4** *Eccl.* **a** a flag attached to a bishop's staff. **b** a processional banner or cross. [L = flag f. *vehere vect-* carry]

VG *abbr.* **1** very good. **2** Vicar-General.

VHF *abbr.* very high frequency.

VHS *abbr. propr.* Video Home System (one of the standard formats for video cassettes).

VI *abbr.* Virgin Islands.

via /'vaɪə/ *prep.* by way of; through (*London to Rome via Paris; send it via your secretary*). [L, ablat. of *via* way, road]

Via Appia see APPIAN WAY.

viable /'vaɪəb(ə)l/ *adj.* **1** (of a plan etc.) feasible; practicable, esp. from an economic standpoint. **2 a** (of a seed or spore) able to germinate. **b** (of a plant, animal, etc.) capable of living or developing normally under particular environmental conditions. **3** *Med.* (of a foetus or unborn child) able to live after birth. □ **viably** *adv.* **viability** /ˌvaɪəˈbɪlɪtɪ/ *n.* [F f. *vie* life f. L *vita*]

viaduct /ˈvaɪədʌkt/ *n.* **1** a long bridgelike structure, esp. a series of arches, carrying a road or railway across a valley or dip in the ground. **2** such a road or railway. [L *via* way, after AQUEDUCT]

vial /ˈvaɪəl/ *n.* a small (usu. cylindrical glass) vessel, esp. for holding liquid medicines. □ **vialful** *n.* (*pl.* **-fuls**). [ME, var. of *fiole* etc.: see PHIAL]

via media /ˌvaɪə ˈmiːdɪə, ˌviːə ˈmedɪə/ *n. literary* a middle way or compromise between extremes. [L]

viand /ˈvaɪənd/ *n. formal* **1** an article of food. **2** (in *pl.*) provisions, victuals. [ME f. OF *viande* food, ult. f. L *vivenda*, neut. pl. gerundive of *vivere* to live]

viaticum /vaɪˈætɪkəm/ *n.* (*pl.* **viatica** /-kə/) **1** the Eucharist as given to a person near or in danger of death. **2** provisions or an official allowance of money for a journey. [L, neut. of *viaticus* f. *via* road]

vibes /vaɪbz/ *n.pl. colloq.* **1** vibrations, esp. in the sense of feelings or atmosphere communicated (*the house had bad vibes*). **2** = VIBRAPHONE. [abbr.]

vibraculum /vaɪˈbrækjʊləm/ *n.* (*pl.* **vibracula** /-lə/) *Zool.* a whiplike structure of bryozoans used to bring food within reach by lashing movements. □ **vibracular** *adj.* [mod.L (as VIBRATE)]

vibrant /ˈvaɪbrənt/ *adj.* **1** vibrating. **2** (often foll. by *with*) (of a person or thing) thrilling, quivering (*vibrant with emotion*). **3** (of sound) resonant. **4** (of colour) bright and striking. □ **vibrancy** *n.* **vibrantly** *adv.* [L *vibrare*: see VIBRATE]

vibraphone /ˈvaɪbrəˌfəʊn/ *n.* a percussion instrument of tuned metal bars with motor-driven resonators and metal tubes giving a vibrato effect. □ **vibraphonist** *n.* [VIBRATO + -PHONE]

vibrate /vaɪˈbreɪt/ *v.* **1** *intr. & tr.* move or cause to move continuously and rapidly to and fro; oscillate. **2** *intr. Physics* move unceasingly to and fro, esp. rapidly. **3** *intr.* (of a sound) throb; continue to be heard. **4** *intr.* (foll. by *with*) quiver, thrill (*vibrating with passion*). **5** *intr.* (of a pendulum) swing to and fro. □ **vibrative** /ˈvaɪbrətɪv, vaɪˈbreɪt-/ *adj.* [L *vibrare vibrat-* shake, swing]

vibratile /ˈvaɪbrəˌtaɪl/ *adj.* **1** capable of vibrating. **2** *Biol.* (of cilia etc.) used in vibratory motion. [VIBRATORY, after *pulsatile* etc.]

vibration /vaɪˈbreɪʃ(ə)n/ *n.* **1** the act or an instance of vibrating; oscillation. **2** *Physics* (esp. rapid) motion to and fro, esp. of the parts of a fluid or an elastic solid whose equilibrium has been disturbed or of an electromagnetic wave. **3** (in *pl.*) **a** a mental (esp. occult) influence. **b** a characteristic atmosphere or feeling in a place, regarded as communicable to people present in it. □ **vibrational** *adj.* [L *vibratio* (as VIBRATE)]

vibrato /vɪˈbrɑːtəʊ/ *n. Mus.* a rapid slight variation in pitch in singing or playing a stringed or wind instrument, producing a tremulous effect (cf. TREMOLO). [It., past part. of *vibrare* VIBRATE]

vibrator /vaɪˈbreɪtə(r)/ *n.* **1** a device that vibrates or causes vibration, esp. an electric or other instrument used in massage or for sexual stimulation. **2** *Mus.* a reed in a reed-organ.

vibratory /ˈvaɪbrətərɪ, vaɪˈbreɪt-/ *adj.* causing vibration.

vibrio /ˈvɪbrɪˌəʊ/ *n.* (*pl.* **vibrios** or **vibriones** /ˌvɪbrɪˈəʊniːz/) *Biol. & Med.* a water-borne bacterium of the genus *Vibrio* etc., typically shaped like a curved rod with a flagellum, and including the cholera bacterium. [mod.L f. L *vibrare* VIBRATE]

vibrissae /vaɪˈbrɪsiː/ *n.pl.* **1** stiff coarse hairs near the mouth of most mammals (e.g. a cat's whiskers) and in the human nostrils. **2** bristle-like feathers near the mouth of insect-eating birds. [L (as VIBRATE)]

viburnum /vaɪˈbɜːnəm/ *n.* a shrub of the genus *Viburnum* (honeysuckle family), usu. with white flowers, esp. the guelder rose and wayfaring-tree. [L, = wayfaring-tree]

Vic. *abbr.* Victoria.

vicar /ˈvɪkə(r)/ *n.* **1 a** (in the Church of England) an incumbent of a parish where tithes formerly passed to a chapter or religious house or layman (cf. RECTOR 1). **b** (in an Episcopal Church) a member of the clergy deputizing for another. **2** *RC Ch.* a representative or deputy of a bishop. **3** (in full **lay vicar** or **vicar choral**) a cleric or choir member appointed to sing certain parts of a cathedral service. □ **vicar apostolic** *RC Ch.* a Roman Catholic missionary or titular bishop. **vicar-general** (*pl.* **vicars-general**) **1** an Anglican official assisting or representing a bishop esp. in administrative matters. **2** *RC Ch.* a bishop's assistant in matters of jurisdiction etc. □ **vicarship** *n.* **vicariate** /vɪˈkeərɪət/ *n.* [ME f. AF *viker(e)*, OF *vicaire* f. L *vicarius* substitute f. *vicis*: see VICE³]

vicarage /ˈvɪkərɪdʒ/ *n.* the residence or benefice of a vicar.

vicarial /vɪˈkeərɪəl/ *adj.* of or serving as a vicar.

vicarious /vɪˈkeərɪəs/ *adj.* **1** experienced in the imagination through another person (*vicarious pleasure*). **2** acting or done for another (*vicarious suffering*). **3** deputed, delegated (*vicarious authority*). □ **vicariously** *adv.* **vicariousness** *n.* [L *vicarius*: see VICAR]

Vicar of Christ a name for the pope.

vice¹ /vaɪs/ *n.* **1 a** evil or grossly immoral conduct, depravity. **b** a particular form of this, esp. involving prostitution, drugs, etc. **c** an immoral or dissolute habit or practice (*has the vice of gluttony*). **2** a defect or weakness of character or behaviour, a flaw (*drunkenness was not among his vices*). **3** a fault or bad habit in a horse etc. □ **vice ring** a group of criminals involved in organizing illegal prostitution. **vice squad** a police department enforcing laws against prostitution, drug abuse, etc. □ **viceless** *adj.* [ME f. OF f. L *vitium*]

vice² /vaɪs/ *n. & v.* ● *n.* (US **vise**) an instrument, esp. attached to a workbench, with two movable jaws between which an object may be clamped so as to leave the hands free to work on it. ● *v.tr.* secure in a vice. □ **vicelike** *adj.* [ME, = winding stair, screw, f. OF *vis* f. L *vitis* vine]

vice³ /ˈvaɪsɪ/ *prep.* in the place of; in succession to. [L, ablat. of *vix* (recorded in oblique forms in *vic-*) change]

vice⁴ /vaɪs/ *n. colloq.* a vice-president, vice admiral, etc. [abbr.]

vice- /vaɪs/ *comb. form* forming nouns meaning: **1** acting as a substitute or deputy for (*vice-president*). **2** next in rank to (*vice admiral*). [as VICE⁴]

vice admiral *n.* a naval officer ranking below admiral and above rear admiral. □ **vice-admiralty** *n.* (*pl.* **-ies**).

vice-chamberlain /vaɪsˈtʃeɪmbəlɪn/ *n.* a deputy chamberlain, esp. the deputy of the Lord Chamberlain.

vice-chancellor /vaɪsˈtʃɑːnsələ(r)/ *n.* a deputy chancellor (esp. of a British university, discharging most of the administrative duties).

vicegerent /vaɪsˈdʒerənt/ *adj. & n.* ● *adj.* exercising delegated power. ● *n.* a vicegerent person; a deputy. □ **vicegerency** *n.* (*pl.* **-ies**). [med.L *vicegerens* (as VICE³, L *gerere* carry on)]

vicennial /vaɪˈsenɪəl/ *adj.* lasting for or occurring every twenty years. [LL *vicennium* period of 20 years f. *vicies* 20 times f. *viginti* 20 + *annus* year]

Vicente /vɪˈsentɪ/, Gil (c.1465–c.1536), Portuguese dramatist and poet. Vicente is regarded as Portugal's most important dramatist. He enjoyed royal patronage for much of his life, and many of his poems and plays were written to commemorate national or court events. His works (some written in Portuguese, some in Spanish) include dramas on religious themes, farces, pastoral plays, and comedies satirizing the nobility and clergy.

Vicenza /vɪˈtʃentsə/ a city in NE Italy; pop. (1990) 109,330.

vice-president /vaɪsˈprezɪd(ə)nt/ *n.* an official ranking below and deputizing for a president. □ **vice-presidency** *n.* (*pl.* **-ies**). **vice-presidential** /-ˌprezɪˈdenʃ(ə)l/ *adj.*

viceregal /vaɪsˈriːg(ə)l/ *adj.* of or relating to a viceroy. □ **viceregally** *adv.*

vicereine /ˈvaɪsreɪn/ *n.* **1** the wife of a viceroy. **2** a woman viceroy. [F (as VICE-, *reine* queen)]

viceroy /ˈvaɪsrɔɪ/ *n.* a ruler exercising authority on behalf of a sovereign in a colony, province, etc. □ **viceroyship** *n.* **viceroyal** /vaɪsˈrɔɪəl/ *adj.* **viceroyalty** *n.* [F (as VICE-, *roy* king)]

vicesimal /vaɪˈsezɪm(ə)l/ *adj.* = VIGESIMAL. [L *vicesimus* twentieth]

vice versa /ˌvaɪsɪ ˈvɜːsə, vaɪs/ *adv.* with the order of the terms or conditions changed; the other way round, conversely (*could go from left to right or vice versa*). [L, = the position being reversed (as VICE³, *versa* ablat. fem. past part. of *vertere* turn)]

Vichy /ˈviːʃɪ/ a town in south central France; pop. (1990) 28,050. A noted spa town, it is the source of an effervescent mineral water. The town was the headquarters during the Second World War of the regime (1940–4) that was set up under Marshal Pétain after the Franco-German armistice and the occupation of northern France, to administer unoccupied France and the colonies. Never recognized by the Allies, the regime was an authoritarian one that functioned as a puppet government for the Nazis and continued to collaborate with the Germans after they had moved into the unoccupied parts of France in 1942.

vichyssoise /ˌviːʃɪˈswɑːz/ *n.* a creamy soup of leeks and potatoes, usu. served chilled. [F *vichyssois -oise* of VICHY]

Vichy water *n.* an effervescent mineral water from Vichy.

vicinage /'vɪsɪnɪdʒ/ n. **1** a neighbourhood; a surrounding district. **2** relation in terms of nearness etc. to neighbours. [ME f. OF vis(e)nage ult. f. L vicinus neighbour]

vicinal /'vɪsɪn(ə)l, vɪ'saɪn-/ adj. **1** neighbouring, adjacent. **2** of a neighbourhood; local. [F vicinal or L vicinalis f. vicinus neighbour]

vicinity /vɪ'sɪnɪtɪ/ n. (pl. **-ies**) **1** a surrounding district. **2** (foll. by *to*) nearness or closeness of place or relationship. □ **in the vicinity** (often foll. by *of*) near (to). [L vicinitas (as VICINAL)]

vicious /'vɪʃəs/ adj. **1** bad-tempered, spiteful (*vicious remarks*). **2** violent, severe, savage (*a vicious attack*). **3** of the nature of or addicted to vice. **4** (of language or reasoning etc.) faulty or unsound. □ **vicious circle** see CIRCLE n. 11. **vicious spiral** continual harmful interaction of causes and effects, esp. as causing repeated rises in both prices and wages. □ **viciously** adv. **viciousness** n. [ME f. OF vicious or L vitiosus f. vitium VICE[1]]

vicissitude /vɪ'sɪsɪˌtjuːd, vaɪ-/ n. **1** a change of circumstances, esp. variation of fortune. **2** archaic or poet. regular change; alternation. □ **vicissitudinous** /-ˌsɪsɪ'tjuːdɪnəs/ adj. [F vicissitude or L vicissitudo -dinis f. vicissim by turns (as VICE[3])]

Vicksburg /'vɪksbɜːg/ a city on the Mississippi river, in western Mississippi; pop. (1990) 20,910. In 1863, during the American Civil War, it was successfully besieged by Union forces under General Grant. It was the last Confederate-held outpost on the river and its loss effectively split the secessionist states in half, bringing the end of the war much nearer.

Vico /'viːkəʊ/, Giambattista (1668–1744), Italian philosopher. His work championed the philosophy of history rather than the mathematical and scientific philosophy favoured by his contemporaries. In *Scienza Nuova* (1725) he proposed that civilizations are subject to recurring cycles of barbarism, heroism, and reason. Vico argued that these cycles are accompanied by corresponding cultural, linguistic, and political modes. He claimed that in literature, for example, poetry flourishes in the heroic age, while prose enters in the age of reason. His historicist philosophy made an impression on later philosophers such as Marx and semioticians such as Umberto Eco.

victim /'vɪktɪm/ n. **1** a person injured or killed as a result of an event or circumstance (*a road victim; the victims of war*). **2** a person or thing harmed or destroyed in pursuit of an object or in gratification of a passion etc. (*the victim of their ruthless ambition*). **3** a prey; a dupe (*fell victim to a confidence trick*). **4** a living creature sacrificed to a deity or in a religious rite. [L victima]

victimize /'vɪktɪˌmaɪz/ v.tr. (also **-ise**) **1** single out (a person) for punishment or unfair treatment, esp. dismissal from employment. **2** make (a person etc.) a victim. □ **victimizer** n. **victimization** /ˌvɪktɪmaɪ'zeɪʃ(ə)n/ n.

victimless /'vɪktɪmlɪs/ adj. (of a crime) in which there is no injured party.

victor /'vɪktə(r)/ n. a winner in battle or in a contest. [ME f. AF victo(u)r or L victor f. vincere vict- conquer]

Victor Emmanuel II /ˌvɪktər ɪ'mænjʊəl/ (1820–78), ruler of the kingdom of Sardinia 1849–61 and king of Italy 1861–78. His appointment of Cavour as Premier in 1852 hastened the drive towards Italian unification. In 1859 Victor Emmanuel led his Piedmontese army to victory against the Austrians at the battles of Magenta and Solferino, and in 1860 entered the papal territories around French-held Rome to join his forces with those of Garibaldi. After being crowned first king of a united Italy in Turin in 1861, Victor Emmanuel continued to add to his kingdom, acquiring Venetia in 1866 and Rome in 1870.

Victor Emmanuel III /ˌvɪktər ɪ'mænjʊəl/ (1869–1947), king of Italy 1900–46. He succeeded to the throne after his father's assassination. Under Mussolini, whom he had invited to form a government in 1922 in order to forestall civil war, Victor Emmanuel lost all political power. However, during the Second World War, after the loss of Sicily to the Allies (1943), he acted to dismiss Mussolini and conclude an armistice. Victor Emmanuel abdicated in favour of his son in 1946, but a republic was established the same year by popular vote and both he and his son went into exile.

Victoria[1] /vɪk'tɔːrɪə/ **1** a state of SE Australia; pop. (est. 1990) 4,394,000; capital, Melbourne. Originally a district of New South Wales, it became a separate colony in 1851 and was federated with the other states of Australia in 1901. **2** a port at the southern tip of Vancouver Island, capital of British Columbia; pop. (1991) 262,220. **3** the capital of the Seychelles, a port on the island of Mahé; pop. (est. 1987) 24,000. **4** the capital of Hong Kong; pop. (1981) 590,771.

Victoria[2] /vɪk'tɔːrɪə/ (1819–1901), queen of Great Britain and Ireland 1837–1901 and empress of India 1876–1901. She succeeded to the throne on the death of her uncle, William IV, and married her cousin Prince Albert in 1840; they had nine children. As queen she took an active interest in the policies of her ministers, although she did not align the Crown with any one political party. She largely retired from public life and went into seclusion after Albert's death in 1861, but lived to achieve the longest reign in British history, a time during which Britain became a powerful and prosperous imperial nation. Her golden jubilee (1887) and diamond jubilee (1897) were marked with popular celebration.

Victoria[3] /vɪk'tɔːrɪə/, Tomás Luis de (1548–1611), Spanish composer. In 1565 he went to Rome, where he may have studied with Palestrina; he eventually returned to Spain, settling in Madrid in 1594. His music, all of it religious, is characterized by its dramatic vigour and colour and resembles that of Palestrina in its contrapuntal nature; it includes motets, masses, and hymns.

Victoria[4] /vɪk'tɔːrɪə/ n. (also **victoria**) **1** a low light four-wheeled carriage with a collapsible top, seats for two passengers, and a raised driver's seat. **2** (in full **Victoria water lily**) a South American water lily, *Victoria amazonica* or *V. cruziana*, with gigantic floating leaves. **3** (in full **Victoria plum**) *Brit.* a large red luscious variety of plum. □ **Victoria crowned pigeon** a large blue crested pigeon, *Goura victoria*, of New Guinea. **Victoria sandwich** (or **sponge**) a sponge cake consisting of two layers of sponge with a jam filling. [VICTORIA[2]]

Victoria, Lake (also **Victoria Nyanza** /nɪ'ænzə/) the largest lake in Africa, with shores in Uganda, Tanzania, and Kenya, and drained by the Nile. It was named in 1858 by John Hanning Speke.

Victoria and Albert Museum (abbr. **V. & A.**) a national museum of fine and applied art in South Kensington, London, having collections principally of pictures, textiles, ceramics, and furniture. Created in 1852 out of the surplus funds of the Great Exhibition, the museum moved to its present site in 1857.

Victoria Cross n. (abbr. **VC**) a decoration awarded to members of the Commonwealth armed services for a conspicuous act of bravery, instituted by Queen Victoria in 1856. The medals were originally struck from the metal of guns captured at Sebastopol during the Crimean War.

Victoria Falls a spectacular waterfall 109 m (355 ft) high, on the River Zambezi, on the Zimbabwe–Zambia border. It was discovered in 1855 by David Livingstone. Its native name is Mosi-oa-tunya, 'the smoke that thunders'.

Victoria Island an island in the Canadian Arctic, in the Northwest Territories.

Victorian /vɪk'tɔːrɪən/ adj. & n. ● adj. **1** of or characteristic of the time of Queen Victoria (reigned 1837–1901). (See note below.) **2** associated with attitudes attributed to this time, esp. of prudery and moral strictness. ● n. a person, esp. a writer, of this time. □ **Victorianism** n.
- The Victorian age was a time of increasing prosperity for the nation as a whole, of immense industrial and scientific development, and of the furthest extent of imperial expansion. In the arts, in reaction to the growth of industrialism and mass production, there was a vogue for Renaissance and medieval styles, as exemplified by the work of the Pre-Raphaelite painters and the Arts and Crafts Movement. Victorian style was in general lavish and highly ornate, particularly in architecture, where the prevailing forms were those of neoclassicism and of the Gothic revival. Architects such as Pugin advocated the Gothic style as the only one suitable for churches, but it was also used for civic buildings and even for small terraced houses.

Victoriana /vɪkˌtɔːrɪ'ɑːnə/ n.pl. **1** articles, esp. collectors' items, of the Victorian period. **2** attitudes characteristic of this period.

Victoria Nile the upper part of the White Nile, between Lake Victoria and Lake Albert. (See NILE[1].)

Victoria Peak a mountain on Hong Kong Island, rising to 554 m (1,818 ft).

victorious /vɪk'tɔːrɪəs/ adj. **1** having won a victory; conquering, triumphant. **2** marked by victory (*victorious day*). □ **victoriously** adv. **victoriousness** n. [ME f. AF victorious, OF victorieux, f. L victoriosus (as VICTORY)]

victor ludorum /ˌvɪktə luː'dɔːrəm/ n. the overall champion in a sports competition. [L, = victor of the games]

Victory /'vɪktərɪ/ the flagship of Lord Nelson at the Battle of Trafalgar, launched in 1765. It has been restored, and is now on display in dry dock at Portsmouth.

victory /'vɪktərɪ/ n. (pl. **-ies**) **1** the process of defeating an enemy in battle or war or an opponent in a contest. **2** an instance of this; a triumph. □ **victory roll** a roll performed by an aircraft as a sign of triumph, esp. after a successful mission. [ME f. AF victorie, OF victoire, f. L victoria (as VICTOR)]

victual /'vɪt(ə)l/ n. & v. ● n. (usu. in pl.) food, provisions, esp. as prepared for use. ● v. (**victualled, victualling**; US **victualed, victualing**) **1** tr. supply with victuals. **2** intr. obtain stores. **3** intr. eat victuals. [ME f. OF vitaille f. LL victualia, neut. pl. of L victualis f. victus food, rel. to vivere live]

victualler /'vɪtlə(r)/ n. (US **victualer**) **1 a** a person etc. who supplies victuals. **b** (in full **licensed victualler**) Brit. a publican etc. licensed to sell alcoholic liquor. **2** a ship carrying stores for other ships. [ME f. OF vitaill(i)er, vitaillour (as VICTUAL)]

vicuña /vɪ'kju:nə/ n. **1** a South American hoofed mammal, Vicugna vicugna, found in the high Andes, related to the llama but with fine silky wool. **2 a** a cloth made from its wool. **b** an imitation of this. [Sp. f. Quechua]

Vic-Wells Ballet /vɪk'welz/ (in the UK) a ballet company set up by Ninette de Valois, based first at the Old Vic and from 1931 established at Sadler's Wells Theatre. The company later became the Sadler's Wells Ballet and then became part of the newly formed Royal Ballet.

vid /vɪd/ n. colloq. = VIDEO n. 4. [abbr.]

Vidal /vɪ'dɑ:l/, Gore (born Eugene Luther Vidal) (b.1925), American novelist, dramatist, and essayist. His first novel Williwaw (1946) was based on his wartime experiences. His other novels, usually satirical comedies, include Myra Breckenridge (1968) and Creation (1981). Among his plays are Suddenly Last Summer (1958). His essays, published in a number of collections, form a satirical commentary on American political and cultural life.

vide /'vɪdeɪ, 'vi:deɪ, 'vaɪdɪ/ v.tr. (as an instruction in a reference to a passage in a book etc.) see, consult. [L, imper. of videre see]

videlicet /vɪ'delɪˌset/ adv. = VIZ. [ME f. L f. videre see + licet it is permissible]

video /'vɪdɪəʊ/ n., adj., & v. ● n. (pl. **-os**) **1** the process of recording, reproducing, or broadcasting visual images on magnetic tape. (See note below.) **2** the visual element of television broadcasts. **3** colloq. = video recorder. **4** (in full **video film**) a film etc. recorded on a videotape. ● adj. **1** relating to the recording, reproducing, or broadcasting of visual images on magnetic tape. **2** relating to the broadcasting of television pictures. ● v.tr. (**-oes, -oed**) make a video recording of. □ **video cassette** a cassette of videotape. **video frequency** a frequency in the range used for video signals in television. **video game** a game played by electronically manipulating images produced by a computer program on a television screen. **video nasty** colloq. an explicitly horrific or pornographic video film. **video** (or **video cassette**) **recorder** an apparatus for recording and playing videotapes. **video signal** a signal containing information for producing a television image. [L videre see, after AUDIO]

▪ Video recorders operate in a similar way to audio tape recorders, except that the recording head rotates at high speed. The first video tape recorder was developed by the Ampex Corporation of America in about 1956, but the first machine for domestic use (by the Sony Corporation) was not introduced until 1969. Video tape recorders are used mainly in the broadcasting industry, which uses tape 2 inches (50 mm) wide. The video cassette recorder was pioneered by Peter Goldmark, and the first, using Sony's Betamax system, was put on the market in 1972. This has since given way to the popular VHS system, designed by the Matsushita Corporation. Most video cassette machines use half-inch (13 mm) tape, but in 1985 small cassettes were introduced with 0.3 inch (8 mm) tape suitable for use in camcorders.

videodisc /'vɪdɪəʊˌdɪsk/ n. a metal-coated disc on which visual material is recorded for reproduction on a television screen.

videofit /'vɪdɪəʊˌfɪt/ n. a reconstructed picture of a person (esp. one sought by the police) built up on a computer screen by selecting and combining facial features according to witnesses' descriptions (cf. PHOTOFIT).

videophone /'vɪdɪəʊˌfəʊn/ n. a telephone device transmitting a visual image as well as sound.

videotape /'vɪdɪəʊˌteɪp/ n. & v. ● n. magnetic tape for recording television pictures and sound. ● v.tr. make a recording of (broadcast material etc.) with this. □ **videotape recorder** = video recorder.

videotex /'vɪdɪəʊˌteks/ n. (also **videotext** /-ˌtekst/) any electronic information system, esp. teletext or viewdata.

vidimus /'vaɪdɪməs/ n. an inspection or certified copy of accounts etc. [L, = we have seen f. videre see]

vie /vaɪ/ v.intr. (**vying**) (often foll. by with) compete; strive for superiority (vied with each other for recognition). [prob. f. ME (as ENVY)]

vielle /vɪ'el/ n. a hurdy-gurdy. [F f. OF viel(l)e: see VIOL]

Vienna /vɪ'enə/ (German **Wien** /vi:n/) the capital of Austria, situated in the north-east of the country on the River Danube; pop. (1991) 1,533,180. As Vindobona, it was an important military centre under the Romans. From 1278 to 1918 it was the seat of the Habsburgs. It has long been a centre of the arts and especially music, Mozart, Beethoven, and the Strauss family being among the great composers who lived and worked there. It has many fine buildings, including St Stephen's cathedral, the Opera House, and the Hofburg (now the residence of the President of Austria). □ **Viennese** /vɪə'ni:z/ adj. & n.

Vienna, Congress of an international conference held 1814–15 to agree the settlement of Europe after the Napoleonic Wars. Attended by all the major European powers, it was dominated by Prussia, Russia, Britain, Austria, and France. The guiding principle of the settlement was the restoration and strengthening of hereditary and sometimes despotic rulers; the result was a political stability that lasted for three or four decades but often at the expense of liberal and nationalistic feeling.

Vienna Circle a group of empiricist philosophers, scientists, and mathematicians active in Vienna from the 1920s until 1938. The work of the circle, members of which included Rudolf Carnap and Kurt Gödel, laid the foundation of the school of logical positivism.

Vienna schnitzel see SCHNITZEL.

Vienna Secession a group of painters, architects, and designers in Vienna, headed by Gustav Klimt, who broke away from the official Vienna Academy of Fine Arts in 1897 to organize their own more avant-garde exhibitions. Aside from Klimt, probably the best-known figure associated with the group is Egon Schiele.

Vientiane /ˌvjentɪ'ɑ:n/ the capital and chief port of Laos, on the Mekong river; pop. (1985) 377,400.

Vierwaldstättersee /ˌfi:ər'valtˌʃtɛtərˌze:/ the German name for Lake Lucerne (see LUCERNE, LAKE).

Vietcong /vjet'kɒŋ/ n. (pl. same) a member of the Communist guerrilla movement in Vietnam which fought the South Vietnamese government forces 1954–75 with the support of the North Vietnamese army and opposed the South Vietnam and US forces in the Vietnam War. [Vietnamese, lit. 'Vietnamese Communist']

Vietminh /vjet'mɪn/ n. (pl. same) a member of a Communist-dominated nationalist movement, formed in 1941, that fought for Vietnamese independence from French rule. Members of the Vietminh later joined with the Vietcong in the Vietnam War. [Vietnamese Viet-Nam Dôc-Lâp Dong-Minh Vietnamese Independence League]

Vietnam /vjet'næm/ a country in SE Asia, with a coastline on the South China Sea; pop. (est. 1991) 67,843,000; official language, Vietnamese; capital, Hanoi. Traditionally dominated by China, Vietnam came under increasing French influence and formed part of French Indo-China between 1862 and 1954. The country was occupied by the Japanese during the Second World War. After the war, hostilities between the French and the Communist Vietminh ended with French defeat and withdrawal. Vietnam was partitioned along the 17th parallel between Communist North Vietnam (capital, Hanoi) and non-Communist South Vietnam (capital, Saigon). A prolonged conflict between the North and the US-backed South ended in the victory of the North in 1975 (see VIETNAM WAR) and the reunification of the country under a Communist regime in the following year. Since then Vietnam has been involved in border disputes with China and military intervention to oust the Khmer Rouge regime in Cambodia, while its predominantly agricultural economy has been slowly recovering from wartime destruction and dislocation. [Vietnamese Viet name of the inhabitants, nam south]

Vietnamese /ˌvjetnə'mi:z/ adj. & n. ● adj. of or relating to Vietnam or its people or language. ● n. (pl. same) **1** a native or national of Vietnam. **2** the language of Vietnam, spoken by about 60 million people. Its origin is uncertain but it may be distantly related to Chinese, from which it derives about half its vocabulary.

Vietnam War a war between Communist North Vietnam and US-backed South Vietnam. Since the partition of Vietnam in 1954 the Communist North had attempted, largely through a guerrilla campaign in the South, to unite the country as a Communist state, fuelling US concern over the possible spread of Communism in SE Asia. After two US destroyers were reportedly fired on in the Gulf of Tonkin in 1964, President Johnson was given Congressional approval for military intervention. A US army (numbering 400,000, including many conscripts, by 1967) was sent to Vietnam, supported by contingents from South Korea, Australia, New Zealand, and Thailand, while American aircraft bombed North Vietnamese forces and (from 1969) areas of Cambodia. The Tet Offensive of 1968 damaged American confidence in its ability to win the war, which was arousing immense controversy and resentment at home, and US forces began to be withdrawn, finally leaving in 1973 during Richard Nixon's presidency; the North Vietnamese captured the southern capital Saigon to end the war in 1975. As well as causing millions of Vietnamese and Cambodian casualties (as against some 55,000 American dead) the war contributed to the destabilization of Cambodia and the rise of the Khmer Rouge movement.

vieux jeu /vjɜː ˈʒɜː/ *adj.* old-fashioned, hackneyed. [F, lit. 'old game']

view /vjuː/ *n. & v.* ● *n.* **1** range of vision; extent of visibility (*came into view; in full view of the crowd*). **2 a** what is seen from a particular point; a scene or prospect (*a fine view of the downs; a room with a view*). **b** a picture etc. representing this. **3** an inspection by the eye or mind; a visual or mental survey. **4** an opportunity for visual inspection; a viewing (*a private view of the exhibition*). **5 a** an opinion (*holds strong views on morality*). **b** a mental attitude (*took a favourable view of the matter*). **c** a manner of considering a thing (*took a long-term view of it*). ● *v.* **1** *tr.* look at; survey visually; inspect (*we are going to view the house*). **2** *tr.* examine; survey mentally (*different ways of viewing a subject*). **3** *tr.* form a mental impression or opinion of; consider (*does not view the matter in the same light*). **4** *intr.* watch television. **5** *tr. Hunting* see (a fox) break cover. □ **have in view 1** have as one's object. **2** bear (a circumstance) in mind in forming a judgement etc. **in view of** having regard to; considering. **on view** being shown (for observation or inspection); being exhibited. **view halloo** *Hunting* a shout on seeing a fox break cover. **with a view to 1** with the hope or intention of. **2** with the aim of attaining (*with a view to marriage*). □ **viewable** *adj.* [ME f. AF v(i)ewe, OF vēue fem. past part. f. vēoir see f. L videre]

viewdata /ˈvjuːˌdeɪtə/ *n.* a news and information service from a computer source to which a television screen is connected by telephone link.

viewer /ˈvjuːə(r)/ *n.* **1** a person who views. **2** a person watching television. **3** a device for looking at film transparencies etc.

viewership /ˈvjuːəˌʃɪp/ *n.* **1** the audience for a television programme, channel, etc. **2** the number of such viewers.

viewfinder /ˈvjuːˌfaɪndə(r)/ *n.* a device on a camera showing the area covered by the lens in taking a photograph.

viewgraph /ˈvjuːgrɑːf/ *n.* a graph produced as a transparency, for projection on to a screen, or for transmission during a teleconference.

viewing /ˈvjuːɪŋ/ *n.* **1** an opportunity or occasion to view; an exhibition. **2** the act or practice of watching television.

viewless /ˈvjuːlɪs/ *adj.* **1** not having or affording a view. **2** lacking opinions.

viewpoint /ˈvjuːpɔɪnt/ *n.* a point of view, a standpoint.

Vigée-Lebrun /ˌviːʒeɪləˈbrɜːn/, (Marie Louise) Élisabeth (1755–1842), French painter. In 1779 she was commissioned to paint Marie Antoinette, whom she painted about twenty-five times over the next ten years. Vigée-Lebrun became a member of the Royal Academy of Painting and Sculpture in Paris in 1783. On the outbreak of the French Revolution in 1789 she fled to Italy, where she was acclaimed for her portraits of Lady Hamilton. Vigée-Lebrun worked in many countries throughout Europe, chiefly as a portraitist of women and children, before returning to France in 1810.

vigesimal /vɪˈdʒesɪm(ə)l, vaɪ-/ *adj.* **1** of twentieths or twenty. **2** reckoning or reckoned by twenties. □ **vigesimally** *adv.* [L vigesimus f. viginti twenty]

vigil /ˈvɪdʒɪl/ *n.* **1 a** keeping awake during the time usually given to sleep, esp. to keep watch or pray (*keep vigil*). **b** a period of this. **2** a stationary peaceful demonstration in support of a particular cause, usu. without speeches. **3** *Eccl.* **a** the eve of a festival or holy day. **b** (in *pl.*) nocturnal devotions. [ME f. OF vigile f. L vigilia f. vigil awake]

vigilance /ˈvɪdʒɪləns/ *n.* watchfulness, caution, circumspection. □ **vigilance committee** *US* a self-appointed body for the maintenance of order etc. [F *vigilance* or L *vigilantia* f. *vigilare* keep awake (as VIGIL)]

vigilant /ˈvɪdʒɪlənt/ *adj.* watchful against danger, difficulty, etc. □ **vigilantly** *adv.* [L *vigilans -antis* (as VIGILANCE)]

vigilante /ˌvɪdʒɪˈlæntɪ/ *n.* a person or member of a group undertaking law enforcement without legal authority. [Sp., = vigilant]

vigneron /ˈviːnjəˌrɒn/ *n.* a vine-grower. [F f. *vigne* VINE]

vignette /viːˈnjet/ *n. & v.* ● *n.* **1** a short descriptive essay or character sketch. **2** an illustration or decorative design, esp. on the title-page of a book, not enclosed in a definite border. **3** a photograph or portrait showing only the head and shoulders with the background gradually shaded off. ● *v.tr.* **1** make a portrait of (a person) in vignette style. **2** shade off (a photograph or portrait). □ **vignettist** *n.* [F, dimin. of *vigne* VINE]

Vignola /viːˈnjəʊlə/, Giacomo Barozzi da (1507–73), Italian architect. His designs were mannerist in style and include a number of churches in Rome as well as private residences such as the Palazzo Farnese near Viterbo (1559–73). One of his most influential designs was that of the church of Il Gesù in Rome (begun 1568), the headquarters of the Jesuit order; based on Alberti's church of San Andrea, Mantua, it has a Latin cross plan, with the nave broadened and the dome area increased, in accordance with Counter-Reformation ideas and the new importance attached to preaching. He also wrote a significant treatise on the five orders of architecture.

Vigny /ˈviːnjɪ/, Alfred Victor, Comte de (1797–1863), French poet, novelist, and dramatist. From 1822 he published several volumes of verse which reveal his philosophy of stoic resignation; later poems, published posthumously in 1863, assert his faith in 'man's unconquerable mind'. Other works include his historical novel *Cinq-Mars* (1826) and the play *Chatterton* (1835), whose hero epitomizes the romantic notion of the poet as an isolated genius.

Vigo[1] /ˈviːgəʊ/ a port on the Atlantic in Galicia, NW Spain; pop. (1991) 276,570.

Vigo[2] /ˈviːgəʊ/, Jean (1905–34), French film director. He is noted for his experimental films, which combine lyrical, surrealist, and realist elements. These include two short films and the two feature films *Zéro de conduite* (1933) and *L'Atalante* (1934). The former's indictment of repressive authority in a French boarding-school caused it to be banned in France until 1945.

vigor *US* var. of VIGOUR.

vigorish /ˈvɪgərɪʃ/ *n. US sl.* **1** the percentage deducted by the organizers of a game from a gambler's winnings. **2** an excessive rate of interest on a loan. [prob. f. Yiddish f. Russ. *vyigrysh* gain, winnings]

vigoro /ˈvɪgərəʊ/ *n. Austral.* a team ball game combining elements of cricket and baseball. [app. f. VIGOROUS]

vigorous /ˈvɪgərəs/ *adj.* **1** strong and active; robust. **2** (of a plant) growing strongly. **3** forceful; acting or done with physical or mental vigour; energetic. **4** full of vigour; showing or requiring physical strength or activity. □ **vigorously** *adv.* **vigorousness** *n.* [ME f. OF f. med.L vigorosus f. L vigor (as VIGOUR)]

vigour /ˈvɪgə(r)/ *n.* (*US* **vigor**) **1** active physical strength or energy. **2** a flourishing physical condition. **3** healthy growth; vitality; vital force. **4 a** mental strength or activity shown in thought or speech or in literary style. **b** forcefulness; trenchancy, animation. □ **vigourless** *adj.* [ME f. OF vigour f. L vigor -oris f. vigere be lively]

vihara /vɪˈhɑːrə/ *n.* a Buddhist temple or monastery. [Skr.]

Vijayawada /ˌvɪdʒaɪəˈwɑːdə/ a city on the Krishna River in Andhra Pradesh, SE India; pop. (1991) 701,000.

Viking[1] /ˈvaɪkɪŋ/ *n. & adj.* ● *n.* any of the Scandinavian traders and pirates who ravaged much of northern Europe in the 8th–11th centuries AD. (*See note below.*) ● *adj.* of or relating to the Vikings or their time. [ON *víkingr*, perh. f. OE *wīcing* f. *wīc* camp]

■ The Vikings set sail from Norway, Sweden, and Denmark in their characteristic longships, reaching as far west as Newfoundland, south to the Mediterranean, and eastwards to Russia and the Byzantine Empire, where they threatened Constantinople. In Europe they were able to strike far inland, sailing up the Rhine, Loire, and other rivers and often being bought off rather than fought by local rulers. While the early Viking expeditions were generally little more than raids in search of plunder, in later years they tended to end in conquest and colonization. Areas in which Vikings settled included Normandy, part of Ireland, and much of northern and eastern England (which they administered as the Danelaw): eventually a king of Denmark, Canute, succeeded to the English throne.

Viking[2] /ˈvaɪkɪŋ/ a shipping forecast area covering the central North Sea between southern Norway and the Shetland Islands.

Viking[3] /'vaɪkɪŋ/ either of two American space probes sent to Mars in 1975, each of which consisted of a soft lander that conducted experiments on the surface and an orbiter.

Vila /'viːlə/ (also **Port Vila**) the capital of Vanuatu, on the SW coast of the island of Efate; pop. (est. 1989) 19,400.

vile /vaɪl/ adj. **1** disgusting. **2** morally base; depraved, shameful. **3** colloq. abominably bad (vile weather). **4** archaic worthless. □ **vilely** adv. **vileness** n. [ME f. OF vil vile f. L vilis cheap, base]

vilify /'vɪlɪ.faɪ/ v.tr. (**-ies, -ied**) defame; speak evil of. □ **vilifier** n. **vilification** /ˌvɪlɪfɪ'keɪʃ(ə)n/ n. [ME in sense 'lower in value', f. LL vilificare (as VILE)]

vill /vɪl/ n. hist. a feudal township. [AF f. OF vile, ville farm f. L (as VILLA)]

Villa /'viːjə/, Pancho (born Doroteo Arango) (1878–1923), Mexican revolutionary. He played a prominent role in the revolution of 1910–11 led by Francisco Madero (1873–1913), and together with Venustiano Carranza (1859–1920) overthrew the dictatorial regime of General Victoriano Huerta (1854–1916) in 1914. Later that year, however, he and Emiliano Zapata rebelled against Carranza and fled to the north of the country after suffering a series of defeats. Villa invaded the US in 1916 but was forced back into Mexico by the American army. He continued to oppose Carranza's regime until the latter's overthrow in 1920. Villa was eventually assassinated.

villa /'vɪlə/ n. **1** Rom. Antiq. a large country house with an estate. **2** a country residence. **3** Brit. a detached or semi-detached house in a residential district. **4** a rented holiday home, esp. abroad. [It. f. L]

village /'vɪlɪdʒ/ n. **1 a** a group of houses and associated buildings, larger than a hamlet and smaller than a town, esp. in a rural area. **b** the inhabitants of a village regarded as a community. **2** Brit. a self-contained district or community within a town or city, regarded as having features characteristic of village life. **3** US a small municipality with limited corporate powers. **4** Austral. a select suburban shopping centre. □ **villager** n. **villagey** adj. [ME f. OF f. L villa]

villagization /ˌvɪlɪdʒaɪ'zeɪʃ(ə)n/ n. (also **-isation**) **1 a** (in Africa and Asia) the concentration of the population in villages. **b** the transfer of land to the communal control of villagers. **2** = UJAMAA.

Villahermosa /ˌviːjəə'məʊsə/ (in full **Villahermosa de San Juan Bautista** /deɪ sæn 'hwɑːn baʊ'tiːstə/) a city in SE Mexico, capital of the state of Tabasco; pop. (1990) 390,160.

villain /'vɪlən/ n. **1** a person guilty or capable of great wickedness. **2** colloq. usu. joc. a rascal or rogue. **3** (also **villain of the piece**) (in a play etc.) a character whose evil actions or motives are important in the plot. **4** Brit. colloq. a professional criminal. **5** archaic a rustic; a boor. [ME f. OF vilein, vilain ult. f. L villa: see VILLA]

villainous /'vɪlənəs/ adj. **1** characteristic of a villain; wicked. **2** colloq. abominably bad; atrocious (villainous weather). □ **villainously** adv. **villainousness** n.

villainy /'vɪlənɪ/ n. (pl. **-ies**) **1** villainous behaviour. **2** a wicked act. [OF vilenie (as VILLAIN)]

Villa-Lobos /ˌvɪlə'ləʊbɒs/, Heitor (1887–1959), Brazilian composer. At the age of 18 he journeyed into the Brazilian interior collecting folk music. He later wove this music into many of his instrumental compositions, notably the series of fourteen Chôros (1920–9), scored in the style of Puccini, and the nine Bachianas brasileiras (1930–45), arranged in counterpoint after the manner of Bach. Villa-Lobos was a prolific composer, writing at least two thousand works; he was also a major force in music education in Brazil, founding the Academy of Music in Rio de Janeiro in 1945 and serving as its president until his death.

villanelle /ˌvɪlə'nel/ n. a usu. pastoral or lyrical poem of nineteen lines, with only two rhymes throughout and some lines repeated. [F f. It. villanella fem. of villanello rural, dimin. of villano (as VILLAIN)]

-ville /vɪl/ comb. form colloq. forming the names of fictitious places with ref. to a particular quality etc. (dragsville; squaresville). [F ville town, as in many US town-names]

villein /'vɪlɪn/ n. (under the feudal system) a tenant entirely subject to a lord or attached to a manor; a serf. [ME, var. of VILLAIN]

villeinage /'vɪlɪnɪdʒ/ n. hist. the tenure or status of a villein.

villus /'vɪləs/ n. (usu. pl. **villi** /-laɪ/) **1** Anat. a small vascular hairlike or finger-like process, esp. as lining the small intestine in large numbers to form the absorptive surface. **2** Bot. a long slender fine hair. □ **villiform** adj. **villose** /-ləʊz, -ləʊs/ adj. **villous** adj. **villosity** /vɪ'lɒsɪtɪ/ n. [L, = shaggy hair]

Vilnius /'vɪlnɪəs/ the capital of Lithuania; pop. (1991) 593,000.

vim /vɪm/ n. colloq. vigour. [perh. f. L, accus. of vis energy]

vimineous /vɪ'mɪnɪəs/ adj. Bot. of or producing twigs or shoots. [L vimineus f. vimen viminis osier]

Vimy Ridge, Battle of /'viːmɪ/ an Allied attack on the German position of Vimy Ridge, near Arras in France, during the First World War. One of the key points on the Western Front, it had long resisted assaults, but on 9 April 1917 it was taken by Canadian troops in some fifteen minutes, at the cost of heavy casualties.

vina /'viːnə/ n. a plucked Indian musical instrument of the zither type, fretted and with four steel and brass strings. It has a half-gourd, acting as a resonator, at each end. [Skr. & Hindi viṇā]

vinaceous /vaɪ'neɪʃəs/ adj. wine-red. [L vinaceus f. vinum wine]

vinaigrette /ˌvɪnɪ'gret/ n. **1** (in full **vinaigrette sauce**) a salad dressing of oil, wine vinegar, and seasoning. **2** a small ornamental bottle for holding smelling-salts. [F, dimin. of vinaigre VINEGAR]

vinca /'vɪŋkə/ n. a plant of (or formerly of) the genus Vinca, including Catharanthus roseus, the source of several alkaloids used to treat cancer; a periwinkle. [mod.L f. LL pervinca: see PERIWINKLE[1]]

Vincent de Paul, St /ˌvɪns(ə)nt də 'pɔːl/ (1581–1660), French priest. He devoted his life to work among the poor and the sick and established a number of institutions to continue his work. In 1625 he established the Congregation of the Mission (see LAZARIST), and in 1633 was one of the founders of the Daughters of Charity (Sisters of Charity of St Vincent de Paul), an unenclosed women's order devoted to the care of the poor and the sick. Feast day, 19 July.

Vincentian /vɪn'senʃ(ə)n/ see LAZARIST.

Vinci, Leonardo da, see LEONARDO DA VINCI.

vincible /'vɪnsɪb(ə)l/ adj. literary that can be overcome or conquered. □ **vincibility** /ˌvɪnsɪ'bɪlɪtɪ/ n. [L vincibilis f. vincere overcome]

vinculum /'vɪŋkjʊləm/ n. (pl. **vincula** /-lə/) **1** Algebra a horizontal line drawn over a group of terms to show they have a common relation to what follows or precedes (e.g. $\overline{a+b} \times c = ac + bc$, but $a + b \times c = a + bc$). **2** Anat. a ligament; a fraenum. [L, = bond, f. vincire bind]

vindaloo /ˌvɪndə'luː/ n. a highly spiced hot Indian curry dish. [prob. f. Port. vin d'alho 'wine and garlic (sauce)']

vindicate /'vɪndɪ.keɪt/ v.tr. **1** clear of blame or suspicion. **2** establish the existence, merits, or justice of (one's courage, conduct, assertion, etc.). **3** justify (a person, oneself, etc.) by evidence or argument. □ **vindicator** n. **vindicable** /-kəb(ə)l/ adj. **vindicative** /-kətɪv/ adj. **vindication** /ˌvɪndɪ'keɪʃ(ə)n/ n. [L vindicare claim, avenge f. vindex -dicis claimant, avenger]

vindicatory /'vɪndɪ.keɪtərɪ/ adj. **1** tending to vindicate. **2** (of laws) punitive.

vindictive /vɪn'dɪktɪv/ adj. **1** tending to seek revenge. **2** spiteful. □ **vindictive damages** Law damages exceeding simple compensation and awarded to punish the defendant. □ **vindictively** adv. **vindictiveness** n. [L vindicta vengeance (as VINDICATE)]

Vine /vaɪn/, Barbara, the pseudonym used by Ruth Rendell.

vine /vaɪn/ n. **1** a climbing or trailing woody-stemmed plant; esp. one of the genus Vitis, bearing grapes. **2** a slender trailing or climbing stem. □ **vine-dresser** a person who prunes, trains, and cultivates vines. □ **viny** adj. [ME f. OF vi(g)ne f. L vinea vineyard f. vinum wine]

vinegar /'vɪnɪgə(r)/ n. **1** a sour liquid obtained from wine, cider, etc., by fermentation and used as a condiment or for pickling. **2** sour behaviour or character. □ **vinegarish** adj. **vinegary** adj. [ME f. OF vyn egre ult. f. L vinum wine + acer, acre sour]

Vinegar Bible a 1717 edition of the English Bible with parable of the vinegar (for vineyard) in the heading above Luke 20.

vinery /'vaɪnərɪ/ n. (pl. **-ies**) **1** a greenhouse for grapevines. **2** a vineyard.

vineyard /'vɪnjɑːd, -jəd/ n. **1** a plantation of grapevines, esp. for wine-making. **2** Bibl. a sphere of action or labour (see Matt. 20:1). [ME f. VINE + YARD[2]]

vingt-et-un /ˌvænteɪ'ɜːn/ n. = PONTOON[1]. [F, = twenty-one]

vinho verde /ˌviːnuː 'veədɪ/ n. a young Portuguese wine, not allowed to mature. [Port., = green wine]

vini- /'vɪnɪ/ comb. form wine. [L vinum]

viniculture /'vɪnɪ.kʌltʃə(r)/ n. the cultivation of grapevines. □ **vinicultural** /ˌvɪnɪ'kʌltʃərəl/ adj. **viniculturist** n.

vinification /ˌvɪnɪfɪ'keɪʃ(ə)n/ n. the conversion of grape juice etc. into wine.

vining /'vaɪnɪŋ/ n. the separation of leguminous crops from their vines and pods.

Vinland /'vɪnlənd/ the region of the NE coast of North America which was visited in the 11th century by Norsemen led by Leif Ericsson. It was so named from the report that grapevines were found growing there. The exact location is uncertain: sites from the northernmost tip of Newfoundland, where Viking remains have been found, to Cape Cod and even Virginia have been proposed.

Vinnytsya /'viːnɪtsjə/ (Russian **Vinnitsa** /'viːnɪtsə/) a city in central Ukraine; pop. (1990) 379,000.

vino /'viːnəʊ/ n. sl. wine, esp. of an inferior kind. [Sp. & It., = wine]

vin ordinaire /ˌvæn ɔːdɪ'neə(r)/ n. cheap (usu. red) wine for everyday use. [F, = ordinary wine]

vinous /'vaɪnəs/ adj. **1** of, like, or associated with wine. **2** addicted to wine. □ **vinosity** /vɪ'nɒsɪti/ n. [L vinum wine]

vin rosé /væn 'rəʊzeɪ/ n. = ROSÉ. [F]

Vinson Massif /'vɪns(ə)n/ the highest mountain range in Antarctica, in Ellsworth Land, rising to 5,140 m (16,863 ft).

vint[1] /vɪnt/ v.tr. make (wine). [back-form. f. VINTAGE]

vint[2] /vɪnt/ n. a Russian card-game like auction bridge. [Russ., = screw]

vintage /'vɪntɪdʒ/ n. & adj. ● n. **1 a** a season's produce of grapes. **b** the wine made from this. **2 a** the gathering of grapes for wine-making. **b** the season of this. **3** a wine of high quality from a single identified year and district. **4 a** the year etc. when a thing was made etc. **b** a thing made etc. in a particular year etc. **5** poet. or rhet. wine. ● adj. **1** of high quality, esp. from the past or characteristic of the best period of a person's work. **2** of a past season. □ **vintage car** Brit. a car made between 1917 and 1930. **vintage festival** a carnival to celebrate the beginning of the vintage. [alt. (after VINTNER) of ME vendage, vindage f. OF vendange f. L vindemia f. vinum wine + demere remove]

vintager /'vɪntɪdʒə(r)/ n. a grape-gatherer.

vintner /'vɪntnə(r)/ n. a wine-merchant. [ME f. AL vintenarius, vinetarius f. AF vineter, OF vinetier f. med.L vinetarius f. L vinetum vineyard f. vinum wine]

viny see VINE.

vinyl /'vaɪnɪl, -naɪl/ n. **1** (attrib.) Chem. the radical –CHCH₂, derived from ethylene by removal of a hydrogen atom. **2** a plastic made by polymerizing a compound containing the vinyl group, esp. polyvinyl chloride. **3 a** a gramophone record. **b** records collectively. [L vinum wine + -YL]

viol /'vaɪəl/ n. a stringed musical instrument of the Renaissance and baroque periods, with six strings and (unlike the instruments of the violin family) gut frets tied round the neck. The music for viol consort by 16th and 17th-century English composers such as William Byrd and Henry Purcell represents the heyday of the instrument, but already in other countries the violin was assuming greater popularity, and the viol fell into disuse until its revival in the early 20th century. [ME viel etc. f. OF viel(l)e, alt. of viole f. Prov. viola, viula, prob. ult. f. L vitulare be joyful: cf. FIDDLE]

viola[1] /vɪ'əʊlə/ n. **1** an instrument of the violin family, slightly larger than the violin and of lower pitch. Sometimes known as the 'alto' or 'tenor' violin, it is a standard member of the orchestra and the string quartet. **2** a viol. □ **viola da braccio** /də 'brɑːtʃəʊ/ a viol corresponding to the modern viola. **viola da gamba** /də 'gæmbə/ a viol held between the player's legs, esp. one corresponding to the modern cello. **viola d'amore** /dæ'mɔːreɪ/ a sweet-toned tenor viol. [It. & Sp., prob. f. Prov.: see VIOL]

viola[2] /'vaɪələ/ n. a plant of the genus Viola, which includes the pansy and violet; esp. a cultivated hybrid of this genus. [L, = violet]

violaceous /ˌvaɪə'leɪʃəs/ adj. **1** of a violet colour. **2** Bot. of the violet family Violaceae. [L violaceus (as VIOLA²)]

violate /'vaɪəleɪt/ v.tr. **1** disregard; fail to comply with (an oath, treaty, law, etc.). **2** treat (a sanctuary etc.) profanely or with disrespect. **3** break in upon, disturb (a person's privacy etc.). **4** assault sexually; rape. □ **violator** n. **violable** /-ləb(ə)l/ adj. **violation** /ˌvaɪə'leɪʃ(ə)n/ n. [ME f. L violare treat violently]

violence /'vaɪələns/ n. **1** the quality of being violent. **2** violent conduct or treatment, outrage, injury. **3** Law **a** the unlawful exercise of physical force. **b** intimidation by the exhibition of this. □ **do violence to 1** act contrary to; outrage. **2** distort. [ME f. OF f. L violentia (as VIOLENT)]

violent /'vaɪələnt/ adj. **1** involving or using great physical force, esp. aggressively (a violent person; a violent storm; came into violent collision). **2 a** intense, vehement, passionate, furious (a violent contrast; violent

dislike). **b** vivid (violent colours). **3** (of death) resulting from external force or from poison (cf. NATURAL adj. 2). **4** involving an unlawful exercise of force (laid violent hands on him). □ **violently** adv. [ME f. OF f. L violentus]

violet /'vaɪələt/ n. & adj. ● n. **1 a** a plant of the genus Viola, esp. the sweet violet, with usu. purple, blue, or white flowers. **b** any plant resembling the sweet violet. **2** the bluish-purple colour seen at the end of the spectrum opposite red. **3 a** a pigment of this colour. **b** clothes or material of this colour. ● adj. of this colour. [ME f. OF violet(te) dimin. of viole f. L VIOLA²]

violin /ˌvaɪə'lɪn/ n. **1** a musical instrument with four strings of treble pitch played with a bow. (See note below.) **2** a player of this instrument. □ **violinist** n. [It. violino dimin. of VIOLA¹]

▪ The violin family comprises the double bass, cello, viola, and the violin itself. All the instruments are bowed and have four strings, smooth unfretted necks (unlike the viol), and f-shaped sound-holes. They developed in Italy from the early 16th century. The violin is the treble of the family, with a compass of over three-and-a-half octaves allowing a wide range of expression. It became widely popular with the advent of baroque music, and soon replaced the viol, first in Italy and then in France and elsewhere. It is an indispensable member of the orchestra and the leading instrument in a great deal of chamber music, although its true solo repertoire is small; the violin is also an important folk instrument, and is occasionally used in jazz. Many of the instruments of the great Italian makers such as the Amati family and Antonio Stradivari are still in use today and are taken as models for new instruments.

violist /'vaɪəlɪst/ n. a viol- or viola-player.

violoncello /ˌvaɪələn'tʃeləʊ, ˌviː-ə-/ n. (pl. **-os**) formal = CELLO. □ **violoncellist** n. [It., dimin. of VIOLONE]

violone /vɪə'ləʊni, ˌvaɪəˌləʊn/ n. a double bass viol. [It., augment. of VIOLA¹]

VIP abbr. very important person.

viper /'vaɪpə(r)/ n. **1** a venomous snake of the family Viperidae, esp. the common viper (see ADDER). **2** a malignant or treacherous person. □ **viper in one's bosom** a person who betrays those who have helped him or her. **viper's bugloss** a stiff bristly blue-flowered plant, Echium vulgare, of the borage family. **viper's grass** = SCORZONERA. □ **viperine** /-ˌraɪn/ adj. **viperish** adj. **viper-like** adj. **viperous** adj. [F vipère or L vipera f. vivus alive + parere bring forth]

virago /vɪ'rɑːgəʊ, -'reɪgəʊ/ n. (pl. **-os**) **1** a fierce or abusive woman. **2** archaic a woman of masculine strength or spirit. [OE f. L, = female warrior, f. vir man]

viral /'vaɪər(ə)l/ adj. of or caused by a virus. □ **virally** adv.

Virchow /'vɜːkəʊ, German 'fɪrço/, Rudolf Karl (1821–1902), German physician and pathologist, founder of cellular pathology. He saw the cell as the basis of life, and believed that diseases were reflected in specific cellular abnormalities. He set these views out in Die Cellularpathologie (1858), thus giving a scientific basis to pathology. Virchow also worked on improving sanitary conditions in Berlin, and believed that environmental factors such as poor living conditions could be as much a cause of disease as germs. He helped to make Berlin a European centre of medicine.

virelay /'vɪrɪleɪ/ n. a short (esp. old French) lyric poem with two rhymes to a stanza variously arranged. [ME f. OF virelai]

virement /'vaɪəmənt, 'vɪəmɒn/ n. the transfer of items from one financial account to another. [F f. virer turn: see VEER¹]

vireo /'vɪrɪəʊ/ n. (pl. **-os**) a small American songbird of the family Vireonidae, often with greenish upperparts. [L, perh. = greenfinch]

virescence /vɪ'res(ə)ns/ n. **1** greenness. **2** Bot. abnormal greenness in petals etc. normally of some bright colour. □ **virescent** adj. [L virescere, inceptive of virere be green]

virga /'vɜːgə/ n. (treated as sing. or pl.) Meteorol. streaks of rain etc. appearing to hang under a cloud and evaporating before reaching the ground. [L, = rod, stripe]

virgate[1] /'vɜːgət/ adj. Bot. & Zool. slim, straight, and erect. [L virgatus f. virga rod]

virgate[2] /'vɜːgət/ n. Brit. hist. a varying measure of land, esp. 30 acres. [med.L virgata (rendering OE gierd-land yard-land) f. L virga rod]

virger var. of VERGER.

Virgil /'vɜːdʒɪl/ (also **Vergil**) (Latin name Publius Vergilius Maro) (70–19 BC), Roman poet. Virgil is one of the most important poets of the Augustan period. His first major work was the Eclogues, ten pastoral poems, modelled on those of Theocritus, in which the traditional themes of Greek bucolic poetry are blended with contemporary political and literary themes. His next work, the *Georgics*, is a didactic poem on farming, which also treats the wider themes of the relationship of human beings to nature and outlines an ideal of national revival after civil war. His last work was the *Aeneid*, an epic poem in twelve books (see AENEID). His works quickly established themselves as classics of Latin poetry and exerted great influence on later classical and post-classical literature.

virgin /ˈvɜːdʒɪn/ n. & adj. ● n. **1** a person (esp. a woman) who has never had sexual intercourse. **2 a (the Virgin)** Christ's mother the Blessed Virgin Mary. **b** a picture or statue of the Virgin. **3 (the Virgin)** the zodiacal sign or constellation Virgo. **4** *colloq.* a naive, innocent, or inexperienced person (*a political virgin*). **5** a member of any order of women under a vow to remain virgins. **6** a female insect producing eggs without impregnation. ● adj. **1** that is a virgin. **2** of or befitting a virgin (*virgin modesty*). **3** not yet used, penetrated, or tried (*virgin soil*). **4** undefiled, spotless. **5** (of olive oil etc.) obtained from the first pressing of the olives. **6** (of clay) not fired. **7** (of metal) made from ore by smelting. **8** (of wool) not yet, or only once, spun or woven. **9** (of an insect) producing eggs without impregnation. □ **virgin birth 1** the doctrine that Christ had no human father but was conceived by the Virgin Mary by the power of the Holy Spirit. **2** parthenogenesis. **virgin comb** a honeycomb that has been used only once for honey and never for brood. **virgin forest** a forest in an untouched natural state. **virgin honey** honey taken from a virgin comb, or drained from the comb without heat or pressure. **virgin queen** an unfertilized queen bee. **virgin's bower** a clematis, esp. the American *Clematis virginiana*. □ **virginhood** n. [ME f. AF & OF *virgine* f. L *virgo -ginis*]

virginal /ˈvɜːdʒɪn(ə)l/ adj. & n. ● adj. that is or befits or belongs to a virgin. ● n. (usu. in *pl.*) (in full **pair of virginals**) an early form of keyboard instrument resembling a spinet and set in a box, used in the 16th and 17th centuries. □ **virginalist** n. **virginally** adv. [ME f. OF *virginal* or L *virginalis* (as VIRGIN): name of the instrument perh. from its use by young women]

Virginia[1] /vəˈdʒɪnɪə/ a state of the eastern US, on the Atlantic coast; pop. (1990) 6,187,360; capital, Richmond. It was the site of the first permanent European settlement in North America in 1607, and was named in honour of Elizabeth I, the 'Virgin Queen'. It was one of the original thirteen states of the Union (1788). □ **Virginian** n. & adj.

Virginia[2] /vəˈdʒɪnɪə/ n. **1** tobacco from Virginia. **2** a cigarette made of this. [VIRGINIA[1]]

Virginia Beach a city and resort on the Atlantic coast of SE Virginia; pop. (1990) 393,000.

Virginia creeper n. a North American vine, *Parthenocissus quinquefolia*, cultivated esp. for its red autumn foliage.

Virginia reel n. N. Amer. a country dance.

Virginia stock n. (also **Virginian stock**) a cruciferous plant, *Malcolmia maritima*, with white or pink flowers.

Virgin Islands a group of Caribbean islands at the eastern extremity of the Greater Antilles, divided between British and US administration. The islands were encountered by Columbus in 1493 and settled, mainly in the 17th century, by British and Danish sugar planters, who introduced African slaves. The British Virgin Islands consists of about forty islands in the north-east of the group; pop. (est. 1991) 16,750; capital, Road Town (on Tortola). They have constituted a separate British Crown Colony since 1956, having previously been part of the Leeward Islands. The remaining islands (about fifty) make up the US unincorporated territory of the Virgin Islands; pop. (1990) 101,800; capital, Charlotte Amalie (on St Thomas). They were purchased from Denmark in 1917 because of their strategic position.

virginity /vəˈdʒɪnɪtɪ/ n. the state of being a virgin. [OF *virginité* f. L *virginitas* (as VIRGIN)]

Virgin Mary, the mother of Jesus (see MARY[1]).

Virgin Queen Queen Elizabeth I of England.

Virgo /ˈvɜːgəʊ/ n. **1** *Astron.* a large constellation (the Virgin), said to represent a maiden or goddess associated with the harvest. It contains several bright stars, the brightest of which is Spica, and a dense cluster of galaxies. **2** *Astrol.* **a** the sixth sign of the zodiac, which the sun enters about 23 Aug. **b** a person born when the sun is in this sign. □ **Virgoan** /ˈvɜːgəʊən, vɜːˈgəʊ-/ n. & adj. [OE f. L, = virgin]

virgule /ˈvɜːgjuːl/ n. **1** a slanting line used to mark division of words or lines. **2** = SOLIDUS 1. [F, = comma, f. L *virgula* dimin. of *virga* rod]

viridescent /ˌvɪrɪˈdes(ə)nt/ adj. greenish, tending to become green. □ **viridescence** n. [LL *viridescere* f. L *viridis*: see VIRIDIAN]

viridian /vɪˈrɪdɪən/ n. & adj. ● n. **1** a bluish-green chromium oxide pigment. **2** the colour of this. ● adj. bluish-green. [L *viridis* green f. *virere* be green]

viridity /vɪˈrɪdɪtɪ/ n. *literary* greenness, verdancy. [ME f. OF *viridité* or L *viriditas* f. *viridis*: see VIRIDIAN]

virile /ˈvɪraɪl/ adj. **1** of or characteristic of a man; having masculine (esp. sexual) vigour or strength. **2** of or having procreative power. **3** of a man as distinct from a woman or child. □ **virility** /vɪˈrɪlɪtɪ/ n. [ME f. F *viril* or L *virilis* f. *vir* man]

virilism /ˈvɪrɪˌlɪz(ə)m/ n. *Med.* the development of secondary male characteristics in a female or precociously in a male.

virino /vɪˈriːnəʊ/ n. (pl. **-os**) *Med.* a hypothetical infectious particle consisting of nucleic acid in a protective coat of host cell proteins, postulated as the cause of scrapie, BSE, CJD, and kuru (cf. PRION[2]). [VIRUS + *-ino* dimin. suffix]

virion /ˈvɪrɪən/ n. *Biol.* a single infective virus particle as it exists outside a host cell, consisting of a nucleic acid core within a protein coat.

viroid /ˈvaɪərɔɪd/ n. an infectious entity affecting plants, similar to a virus but smaller and consisting only of nucleic acid without a protein coat.

virology /ˌvaɪəˈrɒlədʒɪ/ n. the scientific study of viruses. □ **virologist** n. **virological** /-rəˈlɒdʒɪk(ə)l/ adj. **virologically** adv.

virtu /vɜːˈtuː/ n. (also **vertu**) **1 a** a knowledge of or expertise in the fine arts. **b** *collect.* objects of art; curios. **2** inherent moral worth or virtue. □ **article** (or **object**) **of virtu** an article interesting because of its workmanship, antiquity, rarity, etc. [It. *virtù* VIRTUE, virtu]

virtual /ˈvɜːtjʊəl/ adj. **1** that is such for practical purposes though not in name or according to strict definition (*is the virtual manager of the business; take this as a virtual promise*). **2** *Optics* relating to the points at which rays would meet if produced backwards (*virtual focus; virtual image*). **3** *Mech.* relating to an infinitesimal displacement of a point in a system. **4** *Computing* not physically existing as such but made by software to appear to do so. □ **virtual memory** *Computing* a memory in which required data is automatically transferred from a secondary data store so that it appears to be in the main memory. **virtual reality** an image or environment generated by computer software with which a user can interact realistically using a helmet containing a screen, gloves fitted with sensors, etc. □ **virtually** adv. **virtuality** /ˌvɜːtjʊˈælɪtɪ/ n. [ME f. med.L *virtualis* f. L *virtus* after LL *virtuosus*]

virtue /ˈvɜːtjuː/ n. **1** moral excellence; uprightness, goodness. **2** a particular form of this (*patience is a virtue*). **3** chastity, esp. of a woman. **4** a good quality (*has the virtue of being adjustable*). **5** efficacy; inherent power (*no virtue in such drugs*). **6** an angelic being of the seventh order of the celestial hierarchy (see ORDER n. 19). □ **by** (or **in**) **virtue of** on the strength or ground of (*got the job by virtue of his experience*). **make a virtue of necessity** derive some credit or benefit from an unwelcome obligation. □ **virtueless** adj. [ME f. OF *vertu* f. L *virtus -tutis* f. *vir* man]

virtuoso /ˌvɜːtjʊˈəʊsəʊ, -ˈəʊzəʊ/ n. (pl. **virtuosi** /-sɪ, -zɪ/ or **-os**) **1 a** a person highly skilled in the technique of a fine art, esp. music. **b** (*attrib.*) displaying the skills of a virtuoso. **2** a person with a special knowledge of or taste for works of art or virtu. □ **virtuosoship** n. **virtuosic** /-ˈɒsɪk/ adj. **virtuosity** /-ˈɒsɪtɪ/ n. [It., = learned, skilful, f. LL (as VIRTUOUS)]

virtuous /ˈvɜːtjʊəs/ adj. **1** possessing or showing moral rectitude. **2** (esp. of a woman) chaste. □ **virtuous circle** a beneficial recurring cycle of cause and effect (cf. *vicious circle* (see CIRCLE n. 11)). □ **virtuously** adv. **virtuousness** n. [ME f. OF *vertuous* f. LL *virtuosus* f. *virtus* VIRTUE]

virulent /ˈvɪrʊlənt, ˈvɪrjʊ-/ adj. **1** strongly poisonous. **2** (of a disease) violent or malignant. **3** bitterly hostile (*virulent animosity; virulent abuse*). □ **virulence** n. **virulently** adv. [ME, orig. of a poisoned wound, f. L *virulentus* (as VIRUS)]

virus /ˈvaɪərəs/ n. **1** a submicroscopic infective agent that usually consists of a nucleic acid molecule inside a protein coat, only able to multiply within the living cells of a host. (*See note below.*) **2** *Computing* = *computer virus*. **3** *archaic* a poison, a source of disease. **4** a harmful or corrupting influence. [L, = slimy liquid, poison]

▪ It was discovered in the 1890s that some diseases are caused by agents that are not retained by filters and are too small to be seen under a light microscope, but it was not until 1915–17 that the existence of viruses was confirmed. Examples of the viral diseases affecting humans are the common cold, influenza, measles, mumps, polio, rabies, and smallpox; many others affect animals, plants, and bacteria. Viruses utilize the host cell's own biochemical mechanisms to assemble replica viruses, which are then released into the host. Outside the cell they are metabolically inert – plant and insect viruses can even be crystallized – so they are not strictly organisms in the usual sense. The various cylindrical, polyhedral, and complex shapes that viruses exhibit under the electron microscope are used in their classification, as is the type of nucleic acid and the nature of the protein coat.

Vis. *abbr.* Viscount.

visa /ˈviːzə/ *n. & v.* ● *n.* an endorsement on a passport etc. showing that it has been found correct, esp. as allowing the holder to enter or leave a country. ● *v.tr.* (**visas, visaed** /-zəd/ or **visa'd, visaing** /-zeɪŋ/) mark with a visa. [F f. L *visa* neut. pl. past part. of *videre* see]

visage /ˈvɪzɪdʒ/ *n. literary* a face, a countenance. □ **visaged** *adj.* (also in *comb.*). [ME f. OF f. L *visus* sight (as VISA)]

Visakhapatnam /vɪˌsɑːkəˈpʌtnəm/ a port on the coast of Andhra Pradesh, in SE India; pop. (1991) 750,000.

vis-à-vis /ˌviːzæˈviː/ *prep., adv., & n.* ● *prep.* **1** in relation to. **2** opposite to. ● *adv.* facing one another. ● *n.* (*pl.* same) **1** a person or thing facing another, esp. in some dances. **2** a person occupying a corresponding position in another group. **3** *US* a social partner. [F, = face to face, f. *vis* face f. L (as VISAGE)]

Visby /ˈvɪzbɪ, Swedish ˈviːsbʏ/ a port on the west coast of the Swedish island of Gotland, of which it is the capital; pop. (1990) 57,110. Closely linked with the cities of the Hanseatic League, it was a flourishing trading city in the 13th and 14th centuries.

Visc. *abbr.* Viscount.

viscacha /vɪsˈkætʃə/ *n.* (also **vizcacha** /vɪzˈkætʃə/) a large South American burrowing rodent of the genus *Lagidium* or *Lagostomus*, related to the chinchilla. [Sp. f. Quechua (h)*uiscacha*]

viscera /ˈvɪsərə/ *n.pl. Anat.* the interior organs in the great cavities of the body (e.g. brain, heart, liver), esp. in the abdomen (e.g. the intestines). [L, pl. of *viscus*: see VISCUS]

visceral /ˈvɪsərəl/ *adj.* **1** *Anat.* of the viscera. **2** relating to inward feelings rather than conscious reasoning. □ **visceral muscle** = *smooth muscle.* **visceral nerve** a sympathetic nerve (see SYMPATHETIC *adj.* 9). □ **viscerally** *adv.*

viscid /ˈvɪsɪd/ *adj.* **1** glutinous, sticky. **2** semifluid. □ **viscidity** /vɪˈsɪdɪtɪ/ *n.* [LL *viscidus* f. L *viscum* birdlime]

viscometer /vɪsˈkɒmɪtə(r)/ *n.* an instrument for measuring the viscosity of liquids. □ **viscometry** *n.* **viscometric** /ˌvɪskəˈmetrɪk/ *adj.* **viscometrically** *adv.* [var. of *viscosimeter* (as VISCOSITY)]

Visconti /vɪsˈkɒntɪ/, Luchino (full name Don Luchino Visconti, Conte di Modrone) (1906–76), Italian film and theatre director. Born into an aristocratic family, he became a Marxist and his films reflect his commitment to social issues. He worked for a time with Jean Renoir, the influence of whose naturalistic technique is seen in his first film *Obsession* (1942), which was later regarded as the forerunner of neo-realism. Other notable films include *The Leopard* (1963) and *Death in Venice* (1971). In the theatre Visconti directed many successful dramatic and operatic productions, including works by Jean-Paul Sartre, Jean Cocteau, and Verdi.

viscose /ˈvɪskəʊz, -kəʊs/ *n.* **1** a form of cellulose in a highly viscous state suitable for drawing into yarn. **2** rayon made from this. [LL *viscosus* (as VISCOUS)]

viscosity /vɪsˈkɒsɪtɪ/ *n.* (*pl.* **-ies**) **1** the quality or degree of being viscous. **2** *Physics* **a** (of a fluid) internal friction, the resistance to flow. **b** a quantity expressing this. □ **dynamic viscosity** *Physics* a quantity measuring the force needed to overcome internal friction. **kinematic viscosity** *Physics* a quantity measuring the dynamic viscosity per unit density. □ **viscosimeter** /ˌvɪskəˈsɪmɪtə(r)/ *n.* [ME f. OF *viscosité* or med.L *viscositas* (as VISCOUS)]

viscount /ˈvaɪkaʊnt/ *n.* a British nobleman ranking between an earl and a baron. This use of the title dates from 1440, during the reign of Henry VI. □ **viscountcy** *n.* (*pl.* **-ies**). **viscountship** *n.* **viscounty** *n.* (*pl.* **-ies**). [ME f. AF *viscounte*, OF *vi(s)conte* f. med.L *vicecomes -mitis* (as VICE-, COUNT[2])]

viscountess /ˈvaɪkaʊntɪs/ *n.* **1** a viscount's wife or widow. **2** a woman holding the rank of viscount in her own right.

viscous /ˈvɪskəs/ *adj.* **1** glutinous, sticky. **2** semifluid. **3** *Physics* having a high viscosity; not flowing freely. □ **viscously** *adv.* **viscousness** *n.* [ME f. AF *viscous* or LL *viscosus* (as VISCID)]

viscus /ˈvɪskəs/ *n.* (*pl.* **viscera** /ˈvɪsərə/) (usu. in *pl.*) *Anat.* any of the soft internal organs of the body. [L]

vise *US* var. of VICE[2].

Vishnu /ˈvɪʃnuː/ *Hinduism* one of the major gods, originally a minor Vedic god but now the preserver of the cosmos in the Hindu triad with Siva and Brahma. His consort is Lakshmi, his mount the eagle Garuda. Vishnu is considered to have descended to earth nine times to save the world in various incarnations or avatars such as Rama, Krishna, and the historical Buddha; the tenth incarnation will herald the end of the world. □ **Vishnuism** *n.* **Vishnuite** *n. & adj.* [Skr. *Viṣṇu*]

visibility /ˌvɪzɪˈbɪlɪtɪ/ *n.* **1** the state of being visible. **2** the range or possibility of vision as determined by the conditions of light and atmosphere (*visibility was down to 50 yards*). [F *visibilité* or LL *visibilitas* f. L *visibilis*: see VISIBLE]

visible /ˈvɪzɪb(ə)l/ *adj.* **1 a** that can be seen by the eye. **b** (of light) within the range of wavelengths to which the eye is sensitive. **2** that can be perceived or ascertained; apparent, open (*has no visible means of support; spoke with visible impatience*). **3** (of exports etc.) consisting of actual goods (cf. *invisible exports*). □ **the Church visible** the whole body of professed Christian believers. **visible horizon** see HORIZON 1b. □ **visibleness** *n.* **visibly** *adv.* [ME f. OF *visible* or L *visibilis* f. *videre* vis- see]

Visigoth /ˈvɪzɪˌɡɒθ/ *n.* a member of the western branch of the Goths, who invaded the Roman Empire between the 3rd and 5th centuries AD and eventually established in Spain a kingdom that was overthrown by the Moors in 711–12. [LL *Visigothus*]

vision /ˈvɪʒ(ə)n/ *n. & v.* ● *n.* **1** the act or faculty of seeing, sight (*has impaired his vision*). **2 a** a thing or person seen in a dream or trance. **b** a supernatural or prophetic apparition. **3** a thing or idea perceived vividly in the imagination (*the romantic visions of youth; had visions of warm sandy beaches*). **4** imaginative insight. **5** statesmanlike foresight; sagacity in planning. **6** a person etc. of unusual beauty. **7** what is seen on a television screen; television images collectively. ● *v.tr.* see or present in or as in a vision. □ **field of vision** all that comes into view when the eyes are turned in some direction. **vision-mixer** a person whose job is to switch from one image to another in television broadcasting or recording. □ **visional** *adj.* **visionless** *adj.* [ME f. OF f. L *visio -onis* (as VISIBLE)]

visionary /ˈvɪʒ(ə)nərɪ/ *adj. & n.* ● *adj.* **1** informed or inspired by visions; indulging in fanciful theories. **2** having vision or foresight. **3** existing in or characteristic of a vision or the imagination. **4** not practicable. ● *n.* (*pl.* **-ies**) a visionary person. □ **visionariness** *n.*

visit /ˈvɪzɪt/ *v. & n.* ● *v.* (**visited, visiting**) **1 a** *tr.* (also *absol.*) go or come to see (a person, place, etc.) as an act of friendship or ceremony, on business or for a purpose, or from interest. **b** *tr.* go or come to see for the purpose of official inspection, supervision, consultation, or correction. **2** *tr.* reside temporarily with (a person) or at (a place). **3** *intr.* be a visitor. **4** *tr.* (of a disease, calamity, etc.) come upon, attack. **5** *tr. Bibl.* **a** (foll. by *with*) punish (a person). **b** (often foll. by *upon*) inflict punishment for (a sin). **6** *intr.* **a** (foll. by *with*) N. Amer. go to see (a person) esp. socially. **b** (usu. foll. by *with*) US converse, chat. **7** *tr.* (often foll. by *with*) *archaic* comfort, bless (with salvation etc.). ● *n.* **1 a** an act of visiting, a call on a person or at a place (*was on a visit to some friends; paid her a long visit*). **b** temporary residence with a person or at a place. **2** (foll. by *to*) an occasion of going to a doctor, dentist, etc. **3** a formal or official call for the purpose of inspection etc. **4** *US* a chat. □ **right of visit** = *right of visitation* (see VISITATION). □ **visitable** *adj.* [ME f. OF *visiter* or L *visitare* go to see, frequent. of *visare* view f. *videre* vis- see: (n.) perh. f. F *visite*]

visitant /ˈvɪzɪt(ə)nt/ *n. & adj.* ● *n.* **1** a visitor, esp. a supposedly supernatural one. **2** = VISITOR 2. ● *adj. archaic* or *poet.* visiting. [F *visitant* or L *visitare* (as VISIT)]

visitation /ˌvɪzɪˈteɪʃ(ə)n/ *n.* **1** an official visit of inspection, esp. a bishop's examination of a church in his diocese. **2** trouble or difficulty regarded as a divine punishment. **3** (**Visitation**) **a** the visit of the Virgin Mary to Elizabeth related in Luke 1:39–56. **b** the festival commemorating this on 2 July. **4** *colloq.* an unduly protracted visit or social call. **5** the boarding of a vessel belonging to another state to learn its character and purpose. □ **right of visitation** the right to conduct a visitation of a vessel, not including the right of search. [ME f. OF *visitation* or LL *visitatio* (as VISIT)]

visitatorial /ˌvɪzɪtəˈtɔːrɪəl/ *adj.* of an official visitor or visitation. [ult. f. L *visitare* (see VISIT)]

visiting /ˈvɪzɪtɪŋ/ *n. & adj.* ● *n.* paying a visit or visits. ● *attrib.adj.* (of an academic) spending some time at another institution (*a visiting professor*). □ **visiting-card** a card with a person's name etc., sent or left in lieu of a formal visit. **visiting fireman** (*pl.* **-men**) *US sl.* a visitor given especially cordial treatment. **visiting hours** a time designated when visitors may call, esp. to see a person in a hospital or other institution.

visitor /ˈvɪzɪtə(r)/ n. **1** a person who visits a person or place. **2** a migratory bird present in a locality for part of the year (*winter visitor*). **3** *Brit.* (in a college etc.) an official with the right or duty of occasionally inspecting and reporting. □ **visitors' book** a book in which visitors to a hotel, church, embassy, etc., write their names and addresses and sometimes remarks. [ME f. AF *visitour*, OF *visiteur* (as VISIT)]

visitorial /ˌvɪzɪˈtɔːrɪəl/ adj. of an official visitor or visitation.

visor /ˈvaɪzə(r)/ n. (also **vizor**) **1 a** a movable part of a helmet covering the face. **b** *hist.* a mask. **c** the projecting front part of a cap. **2** a shield (fixed or movable) to protect the eyes from unwanted light, esp. one at the top of a vehicle windscreen. □ **visored** adj. **visorless** adj. [ME f. AF *viser*, OF *visiere* f. *vis* face f. L *visus*: see VISAGE]

vista /ˈvɪstə/ n. **1** a long narrow view as between rows of trees. **2** a mental view of a long succession of remembered or anticipated events (*opened up new vistas to his ambition*). □ **vistaed** /-təd/ adj. [It., = view, f. *visto* seen, past part. of *vedere* see f. L *videre*]

Vistula /ˈvɪstjʊlə/ (Polish **Wisła** /ˈviswa/) a river in Poland which rises in the Carpathian Mountains and flows 940 km (592 miles) generally northwards, through Cracow and Warsaw, to the Baltic near Gdańsk.

visual /ˈvɪzjʊəl, ˈvɪʒʊ-/ adj. & n. ● adj. of, concerned with, or used in seeing. ● n. (usu. in *pl.*) a visual image or display, a picture. □ **visual aid** a film, model, etc., as an aid to learning. **visual angle** *Optics* the angle formed at the eye by rays from the extremities of an object viewed. **visual display unit** *Computing* a device displaying data as characters on a screen and usu. incorporating a keyboard. **visual field** field of vision. **visual purple** a light-sensitive pigment in the retina, rhodopsin. **visual ray** *Optics* a line extended from an object to the eye. □ **visually** adv. **visuality** /ˌvɪzjʊˈælɪtɪ, ˌvɪʒʊ-/ n. [ME f. LL *visualis* f. L *visus* sight f. *videre* see]

visualize /ˈvɪzjʊəˌlaɪz, ˈvɪʒʊ-/ v.tr. (also **-ise**) **1** make visible, esp. to one's mind (a thing not visible to the eye). **2** make visible to the eye. □ **visualizable** adj. **visualization** /ˌvɪzjʊəlaɪˈzeɪʃ(ə)n, ˌvɪʒʊ-/ n.

vital /ˈvaɪt(ə)l/ adj. & n. ● adj. **1** of, concerned with, or essential to organic life (*vital functions*). **2** essential to the existence of a thing or to the matter in hand (*a vital question*; *secrecy is vital*). **3** full of life or activity. **4** affecting life. **5** fatal to life or to success etc. (*a vital error*). **6** *disp.* important. ● n. (in *pl.*) the body's vital organs, e.g. the heart and brain. □ **vital capacity** the volume of air that can be expelled from the lungs after taking the deepest possible breath. **vital force 1** (in Bergson's philosophy) life-force. **2** any mysterious vital principle. **vital power** the power to sustain life. **vital statistics 1** the number of births, marriages, deaths, etc. **2** *colloq. offens.* the measurements of a woman's bust, waist, and hips. □ **vitally** adv. [ME f. OF f. L *vitalis* f. *vita* life]

vitalism /ˈvaɪtəˌlɪz(ə)m/ n. *Biol.* the doctrine that life originates in a vital principle distinct from chemical and other physical forces. □ **vitalist** n. **vitalistic** /ˌvaɪtəˈlɪstɪk/ adj. [F *vitalisme* or f. VITAL]

vitality /vaɪˈtælɪtɪ/ n. **1** liveliness, animation. **2** the ability to sustain life, vital power. **3** (of an institution, language, etc.) the ability to endure and to perform its functions. [L *vitalitas* (as VITAL)]

vitalize /ˈvaɪtəˌlaɪz/ v.tr. (also **-ise**) **1** endow with life. **2** infuse with vigour. □ **vitalization** /ˌvaɪtəlaɪˈzeɪʃ(ə)n/ n.

vitally /ˈvaɪtəlɪ/ adv. essentially, indispensably.

vitamin /ˈvɪtəmɪn, ˈvaɪt-/ n. any of a group of organic compounds, essential for normal growth and nutrition, which are required in small quantities in the diet because they cannot be synthesized by the body. (*See note below.*) [orig. *vitamine* f. L *vita* life + AMINE, because orig. thought to contain an amino acid]

▪ Several vitamins (the first of which were identified by Casimir Funk, who coined the term in 1912) are now recognized to be needed in the human diet: *vitamin A* (retinol) contributes to the pigments of the eye. Deficiency causes night-blindness and other symptoms. It can be obtained from various animal products, especially liver, and from vegetables such as carrots. *vitamin B* is a complex of substances, including *vitamin B₁* (thiamine), *vitamin B₂* (riboflavin), *vitamin B₆* (pyridoxine), and *vitamin B₁₂* (cyanocobalamin), that are essential for the working of certain enzymes. They are found in a wide variety of animal and plant foods. *vitamin C* (ascorbic acid) serves to aid cells in adhering together and to maintain connective tissue. Deficiency results in scurvy. It is found mainly in fresh fruits and vegetables. *vitamin D* is a group of vitamins, including *vitamin D₂* (calciferol) and *vitamin D₃* (cholecalciferol), involved in the absorption and

deposition in the bone of calcium and phosphorus. They are obtained especially from fish oils and liver, and cholecalciferol is produced in the skin in the presence of sunlight. Deficiency results in rickets in children and osteomalacia in adults. *vitamin E* (tocopherol) protects some types of molecule in the body from oxidation. It is found in a wide variety of foods. *vitamin H* is a name used especially in the US for biotin, otherwise regarded as part of the vitamin B complex, and involved in the metabolism of carbohydrates, fats, and proteins. *vitamin K* is a group of compounds, including *vitamin K₁* (phylloquinone) and *vitamin K₂* (menaquinone), essential for blood clotting. *vitamin M* is a name used especially in the US for folic acid, otherwise regarded as part of the vitamin B complex, and essential for red blood cell formation.

vitaminize /ˈvɪtəmɪˌnaɪz, ˈvaɪt-/ v.tr. (also **-ise**) add vitamins to.

Vitebsk see VITSEBSK.

vitelli *pl.* of VITELLUS.

vitellin /vɪˈtelɪn, vaɪ-/ n. *Chem.* the chief protein constituent of the yolk of egg. [VITELLUS + -IN]

vitelline /vɪˈtelaɪn, vaɪ-, -lɪn/ adj. *Zool.* of the vitellus. □ **vitelline membrane** the yolk-sac. [med.L *vitellinus* (as VITELLUS)]

Vitellius /vɪˈtelɪəs/, Aulus (15–69), Roman emperor. He was acclaimed emperor in January 69 by the legions in Germany during the civil wars that followed the death of Nero. Vitellius defeated his main rival Otho and briefly reigned as emperor but was in turn defeated and killed by the supporters of Vespasian in December of the same year.

vitellus /vɪˈteləs, vaɪ-/ n. (pl. **vitelli** /-laɪ/) *Zool.* **1** the yolk of an egg. **2** the contents of the ovum. [L, = yolk]

vitiate /ˈvɪʃɪˌeɪt/ v.tr. **1** impair the quality or efficiency of; corrupt, debase, contaminate. **2** make invalid or ineffectual. □ **vitiator** n. **vitiation** /ˌvɪʃɪˈeɪʃ(ə)n/ n. [L *vitiare* f. *vitium* VICE¹]

viticulture /ˈvɪtɪˌkʌltʃə(r)/ n. the cultivation of grapevines; the science or study of this. □ **viticultural** /ˌvɪtɪˈkʌltʃərəl/ adj. **viticulturist** n. [L *vitis* vine + CULTURE]

Viti Levu /ˌviːtɪ ˈlevuː/ the largest of the Fiji islands. Its chief settlement is Suva.

vitiligo /ˌvɪtɪˈlaɪgəʊ/ n. *Med.* a condition in which the pigment is lost from areas of the skin, causing whitish patches. [L, = tetter]

Vitoria /vɪˈtɔːrɪə/ a city in NE Spain, capital of the Basque Provinces; pop. (1991) 208,570. In 1813 a British army under Wellington defeated a French force there under Napoleon's brother, Joseph Bonaparte (1768–1844), and thus freed Spain from French domination.

Vitória /vɪˈtɔːrɪə/ a port in eastern Brazil, capital of the state of Espírito Santo; pop. (1990) 276,170.

Vitosha /ˈviːtɒʃə/ the largest ski resort in Bulgaria, situated in the mountains to the east of Sofia.

vitreous /ˈvɪtrɪəs/ adj. **1** of, or of the nature of, glass. **2** like glass in hardness, brittleness, transparency, structure, etc. (*vitreous enamel*). □ **vitreous humour** (or **body**) see HUMOUR. □ **vitreousness** n. [L *vitreus* f. *vitrum* glass]

vitrescent /vɪˈtres(ə)nt/ adj. tending to become glass; turned into glass, glassy. □ **vitrescence** n.

vitriform /ˈvɪtrɪˌfɔːm/ adj. having the form or appearance of glass.

vitrify /ˈvɪtrɪˌfaɪ/ v.tr. & intr. (**-ies, -ied**) convert or be converted into glass or a glasslike substance, esp. by heat. □ **vitrifiable** adj. **vitrifaction** /ˌvɪtrɪˈfækʃ(ə)n/ n. **vitrification** /-fɪˈkeɪʃ(ə)n/ n. [F *vitrifier* or med.L *vitrificare* (as VITREOUS)]

vitriol /ˈvɪtrɪəl/ n. **1** sulphuric acid or a sulphate, orig. one of glassy appearance. **2** caustic or hostile speech, criticism, or feeling. □ **copper vitriol** concentrated sulphuric acid. **oil of vitriol** concentrated sulphuric acid. [ME f. OF *vitriol* or med.L *vitriolum* f. L *vitrum* glass]

vitriolic /ˌvɪtrɪˈɒlɪk/ adj. (of speech or criticism) caustic or hostile.

Vitruvius /vɪˈtruːvɪəs/ (full name Marcus Vitruvius Pollio) (fl. 1st century BC), Roman architect and military engineer. He wrote a comprehensive ten-volume treatise on architecture, largely based on Greek sources. This deals with all aspects of building, including matters such as acoustics and water supply as well as the more obvious aspects of architectural design, decoration, and building. His influence was considerable, both during his own time and in the Renaissance.

Vitsebsk /ˈviːtsjebsk/ (Russian **Vitebsk** /ˈvitjɪpsk/) a city in NE Belarus; pop. (1990) 356,000.

vitta /ˈvɪtə/ n. (pl. **vittae** /-tiː/) **1** *Bot.* an oil-tube in the fruit of some plants. **2** *Zool.* a stripe of colour. □ **vittate** /-teɪt/ adj. [L, = band, chaplet]

vituperate /vɪˈtjuːpəˌreɪt, vaɪ-/ v.tr. & intr. revile, abuse. □ **vituperator** n. **vituperative** /-rətɪv/ adj. **vituperation** /-ˌtjuːpəˈreɪʃ(ə)n/ n. [L vituperare f. vitium VICE[1]]

Vitus, St /ˈvaɪtəs/ (died c.300), Christian martyr. Little is known of his life, but he is said to have been martyred during the reign of Diocletian. He was invoked against rabies and as the patron of those who suffered from epilepsy and certain nervous disorders, including St Vitus's dance (Sydenham's chorea). Feast day, 15 June.

viva[1] /ˈvaɪvə/ n. & v. Brit. colloq. ● n. = VIVA VOCE. ● v.tr. (**vivas, vivaed** /-vəd/ or **viva'd, vivaing**) give (a person) a viva voce. [abbr.]

viva[2] /ˈviːvə/ int. & n. ● int. long live. ● n. a cry of this as a salute etc. [It., 3rd sing. pres. subjunctive of vivere live f. L]

vivace /vɪˈvɑːtʃɪ/ adv. Mus. in a lively brisk manner. [It. f. L (as VIVACIOUS)]

vivacious /vɪˈveɪʃəs/ adj. lively, sprightly, animated. □ **vivaciously** adv. **vivaciousness** n. **vivacity** /-ˈvæsɪtɪ/ n. [L vivax -acis f. vivere live]

Vivaldi /vɪˈvældɪ/, Antonio (Lucio) (1678–1741), Italian composer and violinist. Throughout his life he worked in an orphanage in Venice as violin teacher and composer, and he composed many of his works for performance there. He emerged as one of the most important baroque composers; his feeling for texture and melody is evident not only in his best-known works, such as *The Four Seasons* (1725), but also in his surviving operas and solo motets. His other numerous compositions include hundreds of concertos (several of which were later arranged by J. S. Bach) and seventy-three sonatas. His work has enjoyed a re-evaluation during the 20th century, especially since the revival of interest in authentic methods of performing baroque music.

vivarium /vaɪˈvɛərɪəm, vɪ-/ n. (pl. **vivaria** /-rɪə/) a place artificially prepared for keeping animals in (nearly) their natural state; an aquarium or terrarium. [L, = warren, fishpond, f. vivus living f. vivere live]

vivat /ˈvaɪvæt, ˈviːv-/ int. & n. = VIVA[2]. [L, 3rd sing. pres. subjunctive of vivere live]

viva voce /ˌvaɪvə ˈvəʊtʃɪ, ˈvəʊsɪ/ adj., adv., & n. ● adj. oral. ● adv. orally. ● n. an oral examination for an academic qualification. [med.L, = with the living voice]

Vivekananda /ˌvɪveɪkɑːˈnʌndə/, Swami (born Narendranath Datta) (1863–1902), Indian spiritual leader and reformer. He was a disciple of the Indian mystic Ramakrishna and did much to spread his teachings; during his extensive travels he was also responsible for introducing Vedantic philosophy to the US and Europe. On his return to India in 1897 he founded the Ramakrishna Mission near Calcutta, devoted to charitable work among the poor.

viverrid /vɪˈverɪd, vaɪ-/ n. & adj. Zool. ● n. a flesh-eating mammal of the family Viverridae, which includes civets, genets, and (in some classifications) mongooses. ● adj. of or relating to this family. [L viverra ferret + -ID[3]]

vivers /ˈvaɪvəz/ n.pl. Sc. food, victuals. [F vivres f. vivre live f. L vivere]

vivid /ˈvɪvɪd/ adj. **1** (of light or colour) strong, intense, glaring (a vivid flash of lightning; of a vivid green). **2** (of a mental faculty, impression, or description) clear, lively, graphic (has a vivid imagination; have a vivid recollection of the scene). **3** (of a person) lively, vigorous. □ **vividly** adv. **vividness** n. [L vividus f. vivere live]

vivify /ˈvɪvɪˌfaɪ/ v.tr. (**-ies, -ied**) enliven, animate, make lively or living. □ **vivification** /ˌvɪvɪfɪˈkeɪʃ(ə)n/ n. [F vivifier f. LL vivificare f. L vivus living f. vivere live]

viviparous /vɪˈvɪpərəs, vaɪ-/ adj. **1** Zool. bringing forth young alive, not hatching them by means of eggs (cf. OVIPAROUS, OVOVIVIPAROUS). **2** Bot. producing bulbs or seeds that germinate while still attached to the parent plant. □ **viviparously** adv. **viviparousness** n. **viviparity** /ˌvɪvɪˈpærɪtɪ/ n. [L viviparus f. vivus: see VIVIFY]

vivisect /ˈvɪvɪˌsekt/ v.tr. perform vivisection on. [back-form. f. VIVISECTION]

vivisection /ˌvɪvɪˈsekʃ(ə)n/ n. **1** dissection or other painful treatment of living animals for purposes of scientific research. **2** unduly detailed or ruthless criticism. □ **vivisectional** adj. **vivisectionist** n. **vivisector** /ˈvɪvɪˌsektə(r)/ n. [L vivus living (see VIVIFY), after DISSECTION (as DISSECT)]

vixen /ˈvɪks(ə)n/ n. **1** a female fox. **2** a spiteful or quarrelsome woman. □ **vixenish** adj. **vixenly** adj. [ME fixen f. OE, fem. of FOX]

Viyella /vaɪˈelə/ n. propr. a fabric made from a twilled mixture of cotton and wool. [f. Via Gellia a valley in Derbyshire where first made]

viz. /vɪz, or by substitution ˈneɪmlɪ/ adv. (usu. introducing a gloss or explanation) namely; that is to say; in other words (came to a firm conclusion, viz. that we were right). [abbr. of VIDELICET, z being med.L symbol for abbr. of -et]

vizard /ˈvɪzəd/ n. archaic a mask or disguise. [VISOR + -ARD]

vizcacha var. of VISCACHA.

vizier /vɪˈzɪə(r), ˈvɪzɪə(r)/ n. hist. a high official in some Muslim countries, esp. in Turkey under Ottoman rule. □ **viziership** n. **vizierate** /-rət/ n. **vizierial** /vɪˈzɪərɪəl/ adj. [ult. f. Arab. wazīr caliph's chief counsellor]

vizor var. of VISOR.

VJ abbr. **1** Victory over Japan (in 1945, ending the Second World War). **2** video jockey. □ **VJ day** 15 August, the day when Japan ceased fighting in 1945, or 2 September, when the country formally surrendered.

Vlach /vlæk/ n. & adj. ● n. a member of a people inhabiting Romania and parts of the former Soviet Union. ● adj. of or relating to this people. [Bulg. f. Old Ch. Slav. Vlachŭ Romanian etc. f. Gmc, = foreigner]

Vladikavkaz /ˌvlædɪkæfˈkɑːs/ a city in SW Russia, capital of the autonomous republic of North Ossetia; pop. (1990) 306,000. It was formerly known as Ordzhonikidze (1931–44 and 1954–93) and Dzaudzhikau (1944–54).

Vladimir /ˈvlædɪˌmɪə(r), vləˈdiːmɪə(r)/ a city in European Russia, east of Moscow; pop. (1990) 353,000.

Vladimir I /ˈvlædɪˌmɪə(r)/ (known as Vladimir the Great; canonized as St Vladimir) (956–1015), grand prince of Kiev 980–1015. By 980 he had seized Kiev from his elder brother and had extended his rule from Ukraine to the Baltic Sea. After marrying a sister of the Byzantine emperor Basil II (c.987) he converted to Christianity, a move which resulted in Christianity in Russia developing in close association with the Orthodox rather than the Western Church. Feast day, 15 July.

Vladivostok /ˌvlædɪˈvɒstɒk/ a city in the extreme south-east of Russia, on the coast of the Sea of Japan, capital of Primorsky krai; pop. (1990) 643,000. It is the chief port of Russia's Pacific coast and terminus of the Trans-Siberian Railway.

Vlaminck /vlæˈmæŋk/, Maurice de (1876–1958), French painter and writer. Largely self-taught, he met Derain and Matisse in the early 1900s and with them became a leading exponent of fauvism, painting mainly landscapes. He was later influenced by Cézanne and from about 1908 his colour and brushwork became more subdued. He also wrote novels and memoirs and was a pioneer collector of African art.

vlei /vleɪ, fleɪ/ n. S. Afr. a hollow in which water collects during the rainy season. [Du. dial. f. Du. vallei valley]

VLF abbr. very low frequency.

Vlissingen see FLUSHING.

Vlorë /ˈvlɔːrə/ (also **Vlona** /ˈvləʊnə/, Italian **Valona** /vaˈloːna/) a port in SW Albania, on the Adriatic coast; pop. (1988) 61,000.

Vltava /ˈvəltəvə/ (called in German Moldau) a river of the Czech Republic, which rises in the Bohemian Forest on the German–Czech border and flows 435 km (270 miles) generally northwards, passing through Prague before joining the Elbe north of the city.

V-neck /ˈviːnek/ n. (often attrib.) **1** a neck of a pullover etc. with straight sides meeting at an angle in the front to form a V. **2** a garment with this.

VO abbr. (in the UK) (Royal) Victorian Order.

vocable /ˈvəʊkəb(ə)l/ n. a word, esp. with reference to form rather than meaning. [F vocable or L vocabulum f. vocare call]

vocabulary /vəˈkæbjʊlərɪ/ n. (pl. **-ies**) **1** the (principal) words used in a language or a particular book or branch of science etc. or by a particular author (scientific vocabulary; the vocabulary of Shakespeare). **2** a list of these, arranged alphabetically with definitions or translations. **3** the range of words known to an individual (his vocabulary is limited). **4** a set of artistic or stylistic forms or techniques, esp. a range of set movements in ballet etc. [med.L vocabularius, -um (as VOCABLE)]

vocal /ˈvəʊk(ə)l/ adj. & n. ● adj. **1** of or concerned with or uttered by the voice (a vocal communication). **2** expressing one's feelings freely in speech (was very vocal about her rights). **3** Phonet. voiced. **4** poet. (of trees, water, etc.) endowed with a voice or a similar faculty. **5** (of music) written for or produced by the voice with or without accompaniment (cf. INSTRUMENTAL adj. 2). ● n. **1** (in sing. or pl.) the sung part of a musical composition. **2** a musical performance with singing. □ **vocal cords** folds of the lining membrane of the larynx near the opening of the glottis, with edges vibrating in the air-stream to produce the voice. **vocal score** a musical score showing the voice parts in full. □ **vocally** adv. **vocality** /vəˈkælɪtɪ/ n. [ME f. L vocalis (as VOICE)]

vocalic /vəˈkælɪk/ adj. of or consisting of a vowel or vowels.

vocalism /ˈvəʊkəˌlɪz(ə)m/ n. **1** the use of the voice in speaking or singing. **2** a vowel sound or system.

vocalist /ˈvəʊkəlɪst/ n. a singer, esp. of jazz or popular songs.

vocalize /ˈvəʊkəˌlaɪz/ v. (also **-ise**) **1** tr. **a** form (a sound) or utter (a word) with the voice. **b** Phonet. = VOICE v. 2. **2** intr. utter a vocal sound. **3** tr. articulate, express. **4** tr. write (Hebrew etc.) with vowel points. **5** intr. Mus. sing with several notes to one vowel. □ **vocalizer** n. **vocalization** /ˌvəʊkəlaɪˈzeɪʃ(ə)n/ n.

vocation /vəˈkeɪʃ(ə)n/ n. **1** a strong feeling of fitness for a particular career or occupation (in religious contexts regarded as a divine call). **2 a** a person's employment, esp. regarded as requiring dedication. **b** a trade or profession. [ME f. OF vocation or L vocatio f. vocare call]

vocational /vəˈkeɪʃən(ə)l/ adj. **1** of or relating to an occupation or employment. **2** (of education or training) directed at a particular occupation and its skills. □ **vocationalism** n. **vocationalize** v.tr. (also **-ise**). **vocationally** adv.

vocative /ˈvɒkətɪv/ n. & adj. Gram. ● n. the case of nouns, pronouns, and adjectives used in addressing or invoking a person or thing. ● adj. of or in this case. [ME f. OF vocatif -ive or L vocativus f. vocare call]

vociferate /vəˈsɪfəˌreɪt/ v. **1** tr. utter (words etc.) noisily. **2** intr. shout, bawl. □ **vociferance** n. **vociferant** adj. & n. **vociferator** n. **vociferation** /-ˌsɪfəˈreɪʃ(ə)n/ n. [L vociferari f. vox voice + ferre bear]

vociferous /vəˈsɪfərəs/ adj. **1** (of a person, speech, etc.) noisy, clamorous. **2** insistently and forcibly expressing one's views. □ **vociferously** adv. **vociferousness** n.

vocoder /vəʊˈkəʊdə(r)/ n. a synthesizer that produces sounds from an analysis of speech input. [VOICE + CODE]

vodka /ˈvɒdkə/ n. an alcoholic spirit made esp. in Russia by distillation of rye etc. [Russ., dimin. of voda water]

voe /vəʊ/ n. a small bay or creek in Orkney or Shetland. [Norw. våg, ON vágr]

vogue /vəʊg/ n. **1** (prec. by the) the prevailing fashion. **2** popular use or currency (has had a great vogue). □ **in vogue** in fashion, generally current. **vogue-word** a word currently fashionable. □ **voguish** adj. [F f. It. voga rowing, fashion f. vogare row, go well]

voice /vɔɪs/ n. & v. ● n. **1 a** sound formed in the larynx etc. and uttered by the mouth, esp. human utterance in speaking, shouting, singing, etc. (heard a voice; spoke in a low voice). **b** the ability to produce this (has lost her voice). **2 a** the use of the voice; utterance, esp. in spoken or written words (esp. give voice). **b** an opinion so expressed. **c** the right to express an opinion (I have no voice in the matter). **d** an agency by which an opinion is expressed. **3** Gram. a form or set of forms of a verb showing the relation of the subject to the action (active voice; passive voice). **4** Mus. **a** a vocal part in a composition. **b** a constituent part in a fugue. **5** Phonet. sound uttered with resonance of the vocal cords, not with mere breath. **6** (usu. in pl.) the supposed utterance of an invisible guiding or directing spirit. ● v.tr. **1** give utterance to; express (the letter voices our opinion). **2** (esp. as **voiced** adj.) Phonet. utter with vibration of the vocal cords (e.g. b, d, g, v, z). **3** Mus. regulate the tone-quality of (organ-pipes). □ **in voice** (or **good voice**) in proper vocal condition for singing or speaking. **voice-box** the larynx. **the voice of God** the expression of God's will, wrath, etc. **voice-over** n. narration in a film etc. not accompanied by a picture of the speaker. ● v.tr. & intr. speak a voice-over (for). **voice-print** a visual record of speech, analysed with respect to frequency, duration, and amplitude. **voice vote** a vote taken by noting the relative strength of calls of aye and no. **with one voice** unanimously. □ **-voiced** adj. **voicer** n. (in sense 3 of v.). [ME f. AF voiz, OF vois f. L vox vocis]

voiceful /ˈvɔɪsfʊl/ adj. poet. or rhet. **1** vocal. **2** sonorous.

voiceless /ˈvɔɪslɪs/ adj. **1** dumb, mute, speechless. **2** Phonet. uttered without vibration of the vocal cords (e.g. f, k, p, s, t). □ **voicelessly** adv. **voicelessness** n.

Voice of America an official US radio station, operated by the Board for International Broadcasting, that broadcasts around the world in English and many other languages. It was founded in 1942.

void /vɔɪd/ adj., n., & v. ● adj. **1 a** empty, vacant. **b** (foll. by of) lacking; free from (a style void of affectation). **2** esp. Law (of a contract, deed, promise, etc.) invalid, not binding (null and void). **3** useless, ineffectual. **4** (often foll. by in) Cards (of a hand) having no cards in a given suit. **5** (of an office) vacant (esp. fall void). ● n. **1** an empty space, a vacuum (vanished into the void; cannot fill the void made by death). **2** an unfilled space in a wall or building. **3** (often foll. by in) Cards the absence of cards in a particular suit. ● v.tr. **1** render invalid. **2** (also absol.) excrete.

□ **voidable** adj. **voidness** n. [ME f. OF dial. voide, OF vuide, vuit, rel. to L vacare VACATE: v. partly f. AVOID, partly f. OF voider]

voidance /ˈvɔɪd(ə)ns/ n. **1** Eccl. a vacancy in a benefice. **2** the act or an instance of voiding; the state of being voided. [ME f. OF (as VOID)]

voided /ˈvɔɪdɪd/ adj. Heraldry (of a bearing) having the central area cut away so as to show the field.

voile /vɔɪl, vwɑːl/ n. a thin semi-transparent dress-material of cotton, wool, or silk. [F, = VEIL]

Vojvodina /vɔɪˈvɒdɪnə/ a mainly Hungarian-speaking province of northern Serbia, on the Hungarian border; pop. (1981) 2,034,700; capital, Novi Sad.

vol. abbr. volume.

volant /ˈvəʊlənt/ adj. **1** Zool. flying, able to fly. **2** Heraldry represented as flying. **3** literary nimble, rapid. [F f. voler f. L volare fly]

volar /ˈvəʊlə(r)/ adj. Anat. of the palm or sole. [L vola hollow of hand or foot]

volatile /ˈvɒləˌtaɪl/ adj. & n. ● adj. **1** evaporating rapidly (volatile salts). **2** changeable, fickle. **3** lively, light-hearted. **4** apt to break out into violence. **5** transient. ● n. a volatile substance. □ **volatile oil** = essential oil. □ **volatileness** n. **volatility** /ˌvɒləˈtɪlɪtɪ/ n. [OF volatil or L volatilis f. volare volat- fly]

volatilize /vəˈlætɪˌlaɪz/ v. (also **-ise**) **1** tr. cause to evaporate. **2** intr. evaporate. □ **volatilizable** adj. **volatilization** /-ˌlætɪlaɪˈzeɪʃ(ə)n/ n.

vol-au-vent /ˈvɒləʊˌvɒn/ n. a (usu. small) round case of puff pastry filled with meat, fish, etc., and sauce. [F, lit. 'flight in the wind']

volcanic /vɒlˈkænɪk/ adj. (also **vulcanic** /vʌl-/) of, like, or produced by a volcano. □ **volcanic bomb** a mass of ejected lava usu. rounded and sometimes hollow. **volcanic glass** obsidian. □ **volcanically** adv. **volcanicity** /ˌvɒlkəˈnɪsɪtɪ/ n. [F volcanique f. volcan VOLCANO]

volcanism /ˈvɒlkəˌnɪz(ə)m/ n. (also **vulcanism** /ˈvʌl-/) volcanic activity or phenomena.

volcano /vɒlˈkeɪnəʊ/ n. (pl. **-oes**) **1** a mountain or hill having a crater or vent from which molten lava, rock fragments, steam, gases, etc., are or have been expelled continuously or at intervals. (See note below.) **2 a** a state of things likely to cause a violent outburst. **b** a violent esp. suppressed feeling. [It. f. L Volcanus VULCAN]

▪ Volcanoes occur where molten magma is formed as a result of stresses in the earth's crust, and is able to rise to the earth's surface to be erupted as lava. They are found mainly at the margins of the crustal plates such as the Mid-Atlantic Ridge and around the rim of the Pacific Ocean. Much of the driving force for eruptions comes from the release of dissolved gas and vapour from the magma as it rises. The kind of activity shown by a volcano depends on the chemistry of its magma. At one extreme, basic (silica-poor) lavas are fluid and are erupted smoothly and effusively, as in the Hawaiian volcanoes. Acidic (silica-rich) lavas are viscous, leading to violently explosive and destructive activity, as shown by the eruptions of Mount St Helens (1980) and Pinatubo (1991). Attempts to predict a volcano's eruptions so as to reduce hazards to people have focused on studying past eruptions and monitoring seismic activity and gas output. Spacecraft have discovered volcanoes on Mars, Venus, and Jupiter's satellite Io.

volcanology var. of VULCANOLOGY.

vole[1] /vəʊl/ n. a small ratlike or mouselike plant-eating rodent, esp. of the genus Microtus or Clethrionomys, having a blunt snout and inconspicuous ears. [orig. vole-mouse f. Norw. f. voll field + mus mouse]

vole[2] /vəʊl/ n. archaic the winning of all tricks at cards. [F f. voler fly f. L volare]

volet /ˈvɒleɪ/ n. a panel or wing of a triptych. [F f. voler fly f. L volare]

Volga /ˈvɒlgə/ the longest river in Europe, which rises in NW Russia and flows 3,688 km (2,292 miles) generally eastwards to Kazan, where it turns south-eastwards to the Caspian Sea. It flows through many lakes and reservoirs and has been dammed at several points to provide hydroelectric power. Navigable for most of its length, it forms the central part of the Russian inland waterway system.

Volgograd /ˈvɒlgəˌgræd/ an industrial city in SW Russia, situated at the junction of the Don and Volga rivers; pop. (1990) 1,005,000. The city was called Tsaritsyn until 1925, and Stalingrad from then until 1961. (See also STALINGRAD, BATTLE OF.)

volitant /ˈvɒlɪt(ə)nt/ adj. Zool. volant. [L volitare frequent. of volare fly]

volition /vəˈlɪʃ(ə)n/ n. **1** the exercise of the will. **2** the power of willing. □ **of** (or **by**) **one's own volition** voluntarily. □ **volitional** adj.

volitionally *adv.* **volitive** /ˈvɒlɪtɪv/ *adj.* [F *volition* or med.L *volitio* f. *volo* I wish]

Völkerwanderung /ˈfɜːlkəˌvɑːndəˌrʊŋ/ *n.* a migration of peoples, esp. that of Germanic and Slavic peoples in Europe from the 2nd to the 11th centuries. [G]

volley /ˈvɒlɪ/ *n. & v.* ● *n.* (*pl.* **-eys**) **1 a** the simultaneous discharge of a number of weapons. **b** the bullets etc. discharged in a volley. **2** (usu. foll. by *of*) a noisy emission of oaths etc. in quick succession. **3** *Tennis* the return of a ball in play before it touches the ground. **4** *Football* the kicking of a ball in play before it touches the ground. **5** *Cricket* **a** a ball pitched right up to the batsman or the stumps without bouncing. **b** the pitching of the ball in this way. ● *v.* (**-eys, -eyed**) **1** *tr.* (also *absol.*) *Tennis & Football* return or send (a ball) by a volley. **2** *tr. & absol.* discharge (bullets, abuse, etc.) in a volley. **3** *intr.* (of bullets etc.) fly in a volley. **4** *intr.* (of guns etc.) sound together. **5** *intr.* make a sound like a volley of artillery. □ **volleyer** *n.* [F *volée* ult. f. L *volare* fly]

volleyball /ˈvɒlɪˌbɔːl/ *n.* a game played by two teams (now usually of six players) hitting a large ball by hand over a high net. The aim is to score points by making the ball reach the ground on the opponent's side of the court. Volleyball was developed in the US from 1895.

Vologda /ˈvɒləɡdə/ a city in northern Russia; pop. (1990) 286,000.

Volos /ˈvɒlɒs/ (Greek **Vólos**) /ˈvɒlɒs/) a port on an inlet of the Aegean Sea, in Thessaly, eastern Greece; pop. (1981) 107,400.

volplane /ˈvɒlpleɪn/ *n. & v. Aeron.* ● *n.* a glide. ● *v.intr.* glide. [F *vol plané* f. *vol* flight + *plané* past part. of *planer* hover, rel. to PLANE[1]]

vols. *abbr.* volumes.

Volscian /ˈvɒlʃ(ə)n/ *n. & adj.* ● *n.* **1** a member of an ancient people formerly inhabiting eastern Latium, in conflict with Rome from the 5th century BC until finally defeated in 304 BC. **2** the Italic language of the Volscians. ● *adj.* of or relating to the Volscians or their language. [L *Volscus*, pl. *Volsci*]

volt[1] /vəʊlt/ *n. Electr.* the SI unit of electromotive force (symbol: **V**), equal to the potential difference that would send one ampere of current through a resistance of one ohm. [VOLTA[2]]

volt[2] /vɒlt, vəʊlt/ *v. & n.* ● *v.intr. Fencing* make a volte. ● *n.* var. of VOLTE. [F *volter* (as VOLTE)]

Volta[1] /ˈvɒltə/ a river of West Africa, which is formed in central Ghana by the junction of its headwaters, the Black Volta, the White Volta, and the Red Volta, which rise in Burkina. At Akosombo in SE Ghana the river has been dammed, creating Lake Volta, which is one of the world's largest man-made lakes. The lake serves for navigation, fishing, irrigation, and hydroelectric power.

Volta[2] /ˈvɒltə/, Alessandro Giuseppe Antonio Anastasio, Count (1745–1827), Italian physicist. Volta was the inventor of a number of important electrical instruments, including the electrophorus and the condensing electroscope, but is best known for the voltaic pile or electrochemical battery (1800) – the first device to produce a continuous electric current. The impetus for this was Luigi Galvani's claim to have discovered a new kind of electricity produced in animal tissue, which Volta ascribed to normal electricity produced by the contact of two dissimilar metals.

voltage /ˈvəʊltɪdʒ/ *n.* electromotive force or potential difference expressed in volts.

Voltaic /vɒlˈteɪɪk/ *adj. & n.* ● *adj.* of or relating to a group of Niger-Congo languages of West Africa. ● *n.* the Voltaic languages. [VOLTA[1]]

voltaic /vɒlˈteɪɪk/ *adj. archaic* of electricity from a primary battery; galvanic (*voltaic battery*).

Voltaire /vɒlˈteə(r)/ (pseudonym of François-Marie Arouet) (1694–1778), French writer, dramatist, and poet. He was a leading figure of the Enlightenment, and frequently came into conflict with the Establishment as a result of his radical political and religious views and satirical writings. He spent a period in exile in England (1726–9) and was introduced there to the theories of Isaac Newton and the empiricist philosophy of John Locke. He also became acquainted with British political institutions, and extolled them as against the royal autocracy of France. Voltaire lived in Switzerland from 1754, only returning to Paris just before his death. Major works include *Lettres philosophiques* (1734) and *Candide* (1758), a satirical tale attacking Leibniz's optimism; he also wrote plays, poetry, and historical works, and was a contributor to the *Encyclopédie* (see ENCYCLOPEDIA).

voltameter /vɒlˈtæmɪtə(r)/ *n.* an instrument for measuring an electric charge.

volte /vɒlt, vəʊlt/ *n.* (also **volt**) **1** *Fencing* a quick movement to escape a

thrust. **2** a sideways circular movement of a horse. [F f. It. *volta* turn, fem. past part. of *volgere* turn f. L *volvere* roll]

volte-face /vɒltˈfɑːs, vəʊlt-/ *n.* **1** a complete reversal of position in argument or opinion. **2** the act or an instance of turning round. [F f. It. *voltafaccia*, ult. f. L *volvere* roll + *facies* appearance, face]

voltmeter /ˈvəʊltˌmiːtə(r)/ *n.* an instrument for measuring electric potential in volts.

voluble /ˈvɒljʊb(ə)l/ *adj.* **1** speaking or spoken vehemently, incessantly, or fluently (*voluble spokesman; voluble excuses*). **2** *Bot.* twisting round a support, twining. □ **volubly** *adv.* **volubleness** *n.* **volubility** /ˌvɒljʊˈbɪlɪtɪ/ *n.* [F *voluble* or L *volubilis* f. *volvere* roll]

volume /ˈvɒljuːm/ *n.* **1 a** a set of sheets of paper, usu. printed, bound together and forming part or the whole of a work or comprising several works (*issued in three volumes; a library of 12,000 volumes*). **b** *hist.* a scroll of papyrus etc., an ancient form of book. **2 a** solid content, bulk. **b** the space occupied by a gas or liquid. **c** (foll. by *of*) an amount or quantity (*large volume of business*). **3 a** quantity or power of sound. **b** fullness of tone. **4** (foll. by *of*) **a** a moving mass of water etc. **b** (usu. in *pl.*) a wreath or coil or rounded mass of smoke etc. □ **volumed** *adj.* (also in *comb.*). [ME f. OF *volum(e)* f. L *volumen -minis* roll f. *volvere* to roll]

volumetric /ˌvɒljʊˈmetrɪk/ *adj.* of or relating to measurement by volume. □ **volumetrically** *adv.* [VOLUME + METRIC]

voluminous /vəˈluːmɪnəs, -ˈljuːmɪnəs/ *adj.* **1** large in volume; bulky. **2** (of drapery etc.) loose and ample. **3** consisting of many volumes. **4** (of a writer) producing many books, writing at great length. □ **voluminously** *adv.* **voluminousness** *n.* **voluminosity** /-ˌluːmɪˈnɒsɪtɪ, -ˌljuː-/ *n.* [LL *voluminosus* (as VOLUME)]

voluntarism /ˈvɒləntəˌrɪz(ə)m/ *n.* **1** the principle of relying on voluntary action rather than compulsion. **2** *Philos.* the doctrine that the will is a fundamental or dominant factor in the individual or the universe. **3** *hist.* the doctrine that the Church or schools should be independent of the state and supported by voluntary contributions. □ **voluntarist** *n.* [irreg. f. VOLUNTARY]

voluntary /ˈvɒləntərɪ/ *adj. & n.* ● *adj.* **1** done, acting, or able to act of one's own free will; not constrained or compulsory, intentional (*a voluntary gift*). **2** unpaid (*voluntary work*). **3** (of an institution) supported by voluntary contributions. **4** *Brit.* (of a school) built by a voluntary institution but maintained by a local education authority. **5** brought about, produced, etc., by voluntary action. **6** (of a movement, muscle, or limb) controlled by the will. **7** (of a confession by a criminal) not prompted by a promise or threat. **8** *Law* (of a conveyance or disposition) made without return in money or other consideration. ● *n.* (*pl.* **-ies**) **1 a** an organ solo played before, during, or after a church service. **b** the music for this. **c** *archaic* an extempore performance, esp. as a prelude to other music. **2** (in competitions) a special performance left to the performer's choice. **3** *hist.* a person who holds that the Church or schools should be independent of the state and supported by voluntary contributions. □ **voluntary aided** (usu. *attrib.*) (in the UK) designating a voluntary school funded mainly by the local authority. **voluntary-controlled** (usu. *attrib.*) (in the UK) designating a voluntary school fully funded by the local authority. □ **voluntarily** *adv.* **voluntariness** *n.* [ME f. OF *volontaire* or L *voluntarius* f. *voluntas* will]

Voluntary Aid Detachment (abbr. **VAD**) (in the UK) a group of organized voluntary first-aid and nursing workers, founded in 1909.

voluntaryism /ˈvɒləntərɪˌɪz(ə)m/ *n.* = VOLUNTARISM 1, 3. □ **voluntaryist** *n.*

Voluntary Service Overseas (abbr. **VSO**) (in the UK) a charitable organization founded in 1958 to promote voluntary work in developing countries. Its headquarters are in London.

volunteer /ˌvɒlənˈtɪə(r)/ *n. & v.* ● *n.* **1** a person who voluntarily takes part in an enterprise or who offers his or her services in any capacity. **2** a person who voluntarily enrols for military service, esp. *Mil. hist.* a member of any of the corps of voluntary soldiers formerly organized in the UK and provided with instructors, arms, etc., by the state. **3** (usu. *attrib.*) a self-sown plant. ● *v.* **1** *tr.* (often foll. by *to* + infin.) undertake or offer (one's services, a remark or explanation, etc.) voluntarily. **2** *intr.* (often foll. by *for*) make a voluntary offer of one's services; be a volunteer. **3** *tr.* (usu. in *passive*; often foll. by *for*, or *to* + infin.) assign or commit (a person) to a particular undertaking, esp. without consultation. [F *volontaire* (as VOLUNTARY), assim. to -EER]

volunteerism /ˈvɒlənˌtɪərɪz(ə)m/ *n.* esp. *N. Amer.* the involvement of volunteers, esp. in community service.

voluptuary /vəˈlʌptjʊərɪ/ *n. & adj.* ● *n.* (*pl.* **-ies**) a person given up to luxury and sensual pleasure. ● *adj.* concerned with luxury and sensual pleasure. [L *volupt(u)arius* (as VOLUPTUOUS)]

voluptuous /vəˈlʌptjʊəs/ *adj.* **1** of, tending to, occupied with, or derived from, sensuous or sensual pleasure. **2** (of a woman) curvaceous and sexually desirable. □ **voluptuously** *adv.* **voluptuousness** *n.* [ME f. OF *voluptueux* or L *voluptuosus* f. *voluptas* pleasure]

volute /vəˈluːt, -ˈljuːt/ *n. & adj.* ● *n.* **1** *Archit.* a spiral scroll characteristic of Ionic capitals and also used in Corinthian and composite capitals. **2 a** a marine gastropod mollusc of the genus *Voluta*. **b** the spiral shell of this. ● *adj.* esp. *Bot.* rolled up. □ **voluted** *adj.* [F *volute* or L *voluta* fem. past part. of *volvere* roll]

volution /vəˈluːʃ(ə)n, -ˈljuːʃ(ə)n/ *n.* **1** a rolling motion. **2** a spiral turn. **3** a whorl of a spiral shell. **4** *Anat.* a convolution. [as VOLUTE, after REVOLUTION etc.]

volvox /ˈvɒlvɒks/ *n. Biol.* an aquatic flagellate unicellular plant which forms colonies resembling minute green globes. [mod.L, as if f. L *volvere* to roll, though orig. a misreading of Pliny]

Volzhsky /ˈvɒlʒskɪ/ an industrial city in SW Russia, on the Volga; pop. (1990) 275,000.

vomer /ˈvəʊmə(r)/ *n. Anat.* the small thin bone separating the nostrils in humans and most vertebrates. [L, = ploughshare]

vomit /ˈvɒmɪt/ *v. & n.* ● *v.tr.* (**vomited, vomiting**) **1** (also *absol.*) eject (matter) from the stomach through the mouth. **2** (of a volcano, chimney, etc.) eject violently, belch (forth). ● *n.* **1** matter vomited from the stomach. **2** *archaic* an emetic. □ **vomiter** *n.* [ME ult. f. L *vomere* vomit- or frequent. L *vomitare*]

vomitorium /ˌvɒmɪˈtɔːrɪəm/ *n.* (*pl.* **vomitoria** /-rɪə/) *Rom. Antiq.* a vomitory. [L: see VOMITORY]

vomitory /ˈvɒmɪtərɪ/ *adj. & n.* ● *adj.* emetic. ● *n.* (*pl.* **-ies**) *Rom. Antiq.* each of a series of passages for entrance and exit in an amphitheatre or theatre. [L *vomitorius* (adj.), *-um* (n.) (as VOMIT)]

von Braun see BRAUN².

Vonnegut /ˈvɒnɪgət/, Kurt (b.1922), American novelist and short-story writer. His works are experimental in nature and blend elements of realism, science fiction, fantasy, and satire. They include *Cat's Cradle* (1963) and *Slaughterhouse-Five, or The Children's Crusade* (1969), based on the fire-bombing of Dresden in 1945, which Vonnegut himself experienced as a prisoner of war. Among his notable recent writings is the novel *Galapagos* (1985).

von Neumann see NEUMANN.

von Sternberg /fɒn ˈstɜːnbɜːg/, Josef (1894–1969), Austrian-born American film director. After emigrating to the US in his childhood, he made a series of silent films about the criminal underworld in the 1920s, before directing his best-known film *Der Blaue Engel* (1930; *The Blue Angel*), starring Marlene Dietrich, in Germany. Capturing the decadence of Berlin in the 1920s, the film made Dietrich an international star; von Sternberg proceeded to collaborate with her on a series of Hollywood films, including *Dishonored* (1931) and *Shanghai Express* (1932).

voodoo /ˈvuːduː/ *n. & v.* ● *n.* **1** a religious cult practised in parts of the West Indies (esp. Haiti), the southern states of the US, and Brazil, and characterized by sorcery and spirit possession. (*See note below.*) **2** a person who practises voodoo. **3** a voodoo spell; voodoo ritual. ● *v.tr.* (**voodoos, voodooed**) affect by voodoo; bewitch. □ **voodooism** *n.* **voodooist** *n.* [Louisiana French from Kwa *vodu* tutelary deity, fetish]

■ In voodoo some elements of Roman Catholic ritual, dating from the French colonial period before 1804, are blended with African religious and magical elements derived from the former slave population. Belief in a supreme God is combined with service to the *loa*, who are local (or African) gods, deified ancestors, or deities corresponding to certain Catholic saints. The *loa* are invoked at ceremonies by a priest or priestess and possess worshippers, who go into a trance.

voracious /vəˈreɪʃəs/ *adj.* **1** greedy in eating, ravenous. **2** very eager in some activity (*a voracious reader*). □ **voraciously** *adv.* **voraciousness** *n.* **voracity** /-ˈræsɪtɪ/ *n.* [L *vorax* f. *vorare* devour]

Vorarlberg /ˈfɔːrɑːlˌbɜːg/ an Alpine state of western Austria; capital, Bregenz.

Voronezh /vəˈrɒneʒ/ a city in Russia, south of Moscow; pop. (1990) 895,000.

Voroshilovgrad /ˌvɒrəˈʃiːləfˌgræd/ a former name (1935–91) for LUHANSK.

-vorous /vərəs/ *comb. form* forming adjectives meaning 'feeding on'

(*carnivorous*). □ **-vora** /vərə/ *comb. form* forming names of groups. **-vore** /vɔː(r)/ *comb. form* forming names of individuals. [L *-vorus* f. *vorare* devour]

vortex /ˈvɔːteks/ *n.* (*pl.* **vortexes** or **vortices** /-tɪˌsiːz/) **1** a mass of whirling fluid, esp. a whirlpool or whirlwind. **2** any whirling motion or mass. **3** a system, occupation, pursuit, etc., viewed as swallowing up or engrossing those who approach it (*the vortex of society*). **4** *Physics* a portion of fluid whose particles have rotatory motion. □ **vortex-ring** a vortex whose axis is a closed curve, e.g. a smoke-ring. □ **vortical** /-tɪk(ə)l/ *adj.* **vortically** *adv.* **vorticose** /-tɪˌkəʊs/ *adj.* **vorticity** /vɔːˈtɪsɪtɪ/ *n.* **vorticular** /-ˈtɪkjʊlə(r)/ *adj.* [L *vortex -icis* eddy, var. of VERTEX]

vorticella /ˌvɔːtɪˈselə/ *n.* a sedentary protozoan of the genus *Vorticella*, having a long contractile stalk with a ciliated body shaped like an inverted bell. [mod.L, dimin. of VORTEX]

vorticism /ˈvɔːtɪˌsɪz(ə)m/ *n.* **1** *Metaphysics* the theory, as advanced by Descartes, that the universe is a plenum in which motion propagates itself in circles. **2** an avant-garde literary and artistic movement that flourished from 1912 to 1915, attacking the sentimentality of 19th-century art and characterized by a harsh, mechanistic style. (*See note below.*) □ **vorticist** *n.*

■ At the centre of the movement was Wyndham Lewis, but vorticism also encompassed Ezra Pound (who coined the term, and edited the magazine *Blast* with Lewis) and the sculptors Henri Gaudier-Brzeska and Jacob Epstein. In the visual arts it produced bold abstract compositions and was indebted to cubism and futurism. The movement was short-lived, having only one exhibition, in 1915, but it nevertheless had a significant influence on the growth of modernism.

Vosges /vəʊʒ/ a mountain system of eastern France, in Alsace near the border with Germany.

Vostok /ˈvɒstɒk/ a series of six manned Soviet orbiting spacecraft, the first of which, launched in April 1961, carried the first man in space (Yuri Gagarin).

votary /ˈvəʊtərɪ/ *n.* (*pl.* **-ies**; *fem.* **votaress** /-rɪs/) (usu. foll. by *of*) **1** a person vowed to the service of God or a god or cult. **2** a devoted follower, adherent, or advocate of a person, system, occupation, etc. □ **votarist** *n.* [L *vot-*: see VOTE]

vote /vəʊt/ *n. & v.* ● *n.* **1** a formal expression of choice or opinion by means of a ballot, show of hands, etc., concerning a choice of candidate, approval of a motion or resolution, etc. (*let us take a vote on it; gave my vote to the independent candidate*). **2** (usu. prec. by *the*) the right to vote, esp. in a state election. **3 a** an opinion expressed by a majority of votes. **b** *Brit.* money granted by a majority of votes. **4** the collective votes that are or may be given by or for a particular group (*will lose the Welsh vote; the Conservative vote increased*). **5** a ticket etc. used for recording a vote. ● *v.* **1** *intr.* (often foll. by *for, against,* or *to* + infin.) give a vote. **2** *tr.* **a** (often foll. by *that* + clause) enact or resolve by a majority of votes. **b** grant (a sum of money) by a majority of votes. **c** cause to be in a specified position by a majority of votes (*was voted off the committee*). **3** *tr.* pronounce or declare by general consent (*the event was voted a success*). **4** *tr.* (often foll. by *that* + clause) *colloq.* announce one's proposal (*I vote that we all go home*). □ **put to** (or **the**) **vote** submit to a decision by voting. **vote down** defeat (a proposal etc.) in a vote. **vote in** elect by voting. **vote of censure** = *vote of no confidence.* **vote of confidence** (or **no confidence**) a vote showing that the majority support (or do not support) the policy of the governing body etc. **vote off** dismiss from (a committee etc.) by voting. **vote out** dismiss from office etc. by voting. **vote with one's feet** *colloq.* indicate an opinion by one's presence or absence. **voting-machine** (esp. in the US) a machine for the automatic registering of votes. **voting-paper** a paper used in voting by ballot. **voting stock** *Stock Exch.* stock entitling the holder to a vote. □ **votable** *adj.* **voteless** *adj.* [ME f. past part. stem *vot-* of L *vovēre* vow]

voter /ˈvəʊtə(r)/ *n.* **1** a person with the right to vote at an election. **2** a person who votes.

votive /ˈvəʊtɪv/ *adj.* offered or consecrated in fulfilment of a vow (*votive offering; votive picture*). □ **votive mass** *Eccl.* a mass celebrated for a special purpose or occasion. [L *votivus* (as VOTE)]

vouch /vaʊtʃ/ *v.* **1** *intr.* (foll. by *for*) answer for, be surety for (*will vouch for the truth of this; can vouch for him; could not vouch for his honesty*). **2** *tr. archaic* cite as an authority. **3** *tr. archaic* confirm or uphold (a statement) by evidence or assertion. [ME f. OF *vo(u)cher* summon etc., ult. f. L *vocare* call]

voucher /ˈvaʊtʃə(r)/ *n.* **1** a document which can be exchanged for goods or services as a token of payment made or promised by the holder or another. **2** a document establishing the payment of money or the truth of accounts. **3** a person who vouches for a person, statement, etc. [AF *voucher* (as VOUCH) or f. VOUCH]

vouchsafe /vaʊtʃˈseɪf/ *v.tr. formal* **1** condescend to give or grant (*vouchsafed me no answer*). **2** (foll. by *to* + infin.) condescend. [ME f. VOUCH in sense 'warrant' + SAFE]

voussoir /ˈvuːswɑː(r)/ *n. Archit.* each of the wedge-shaped or tapered stones forming an arch. [OF *vossoir* etc. f. pop.L *volsorium* ult. f. L *volvere* roll]

vow /vaʊ/ *n. & v.* ● *n.* **1** *Relig.* a solemn promise, esp. in the form of an oath to God or another deity or to a saint. **2** (in *pl.*) the promises by which a monk or nun is bound to poverty, chastity, and obedience. **3** a promise of fidelity (*lovers' vows; marriage vows*). **4** (usu. as **baptismal vows**) the promises given at baptism by the baptized person or by sponsors. ● *v.tr.* **1** promise solemnly (*vowed obedience*). **2** dedicate to a deity. **3** (also *absol.*) *archaic* declare solemnly. □ **under a vow** having made a vow. [ME f. AF *v(o)u*, OF *vo(u)*, f. L (as VOTE): (v.) f. OF *vouer*, in sense 2 partly f. AVOW]

vowel /ˈvaʊəl/ *n.* **1** a speech sound made with vibration of the vocal cords but without audible friction, more open than a consonant and capable of forming a syllable. **2** a letter or letters representing this, as *a, e, i, o, u, aw, ah*. □ **vowel gradation** = ABLAUT. **vowel harmony** a feature of a language whereby successive syllables in a word are limited to a particular class of vowel. **vowel mutation** = UMLAUT 2. **vowel-point** each of a set of marks indicating vowels in Hebrew etc. □ **vowelled** *adj.* (also in *comb.*). **vowelly** *adj.* **vowelless** /ˈvaʊəllɪs/ *adj.* [ME f. OF *vouel, voiel* f. L *vocalis* (*littera*) VOCAL (letter)]

vowelize /ˈvaʊəlaɪz/ *v.tr.* (also **-ise**) insert the vowels in (shorthand, Hebrew, etc.).

vox angelica /ˌvɒks ænˈdʒelɪkə/ *n. Mus.* an organ-stop with a soft tremulous tone. [LL, = angelic voice]

vox humana /ˌvɒks hjuːˈmɑːnə/ *n. Mus.* an organ-stop with a tone supposed to resemble a human voice. [L, = human voice]

vox pop /vɒks ˈpɒp/ *n. Broadcasting colloq.* popular opinion as represented by informal comments from members of the public; statements or interviews of this kind. [abbr. of vox POPULI]

vox populi /vɒks ˈpɒpjʊˌliː, -ˌlaɪ/ *n.* public opinion, the general verdict, popular belief or rumour. [L, = the people's voice]

voyage /ˈvɔɪdʒ/ *n. & v.* ● *n.* **1** a journey, esp. a long one by water, air, or in space. **2** an account of this. ● *v.* **1** *intr.* make a voyage. **2** *tr.* traverse, esp. by water or air. □ **voyageable** *adj.* **voyager** *n.* [ME f. AF & OF *veiage, voiage* f. L *viaticum*]

Voyager /ˈvɔɪdʒə(r)/ either of two American space probes launched in 1977 to investigate the outer planets. Voyager 1 encountered Jupiter (1979) and Saturn (1980), while Voyager 2 reached Jupiter (1979), Saturn (1981), Uranus (1986), and finally Neptune (1989). Both craft sent back spectacular photographs of these planets and their moons.

voyageur /ˌvwʌjʌˈʒɜː(r)/ *n.* a Canadian boatman, esp. *hist.* one employed in transporting goods and passengers between trading posts. [F, = voyager (as VOYAGE)]

voyeur /vwʌˈjɜː(r)/ *n.* **1** a person who obtains sexual gratification from observing others' sexual actions or organs. **2** a powerless or passive spectator. □ **voyeurism** *n.* **voyeuristic** /ˌvwʌjɜːˈrɪstɪk/ *adj.* **voyeuristically** *adj.* [F f. *voir* see]

VP *abbr.* Vice-President.

VR *abbr.* **1** Queen Victoria. **2** variant reading. [sense 1 f. L *Victoria Regina*]

vroom /vruːm, vrʊm/ *v., n., & int.* ● *v.* **1** *intr.* (esp. of an engine) make a roaring noise. **2** *intr.* (of a motor vehicle) travel at speed. **3** *tr.* rev (an engine). ● *n.* the roaring sound of an engine. ● *int.* an imitation of such a sound. [imit.]

VS *abbr.* Veterinary Surgeon.

vs. *abbr.* versus.

V-sign /ˈviːsaɪn/ *n.* **1** *Brit.* a sign of the letter V made with the first two fingers pointing up and the back of the hand facing outwards, as a gesture of abuse, contempt, etc. **2** a similar sign made with the palm of the hand facing outwards, as a symbol of victory.

VSO see VOLUNTARY SERVICE OVERSEAS.

VSOP *abbr.* Very Special (or Superior) Old Pale (brandy).

VT *abbr. US* Vermont (in official postal use).

Vt. *abbr.* Vermont.

VTO *abbr.* vertical take-off.

VTOL /ˈviːtɒl/ *abbr.* vertical take-off and landing.

vug /vʌɡ/ *n.* a rock-cavity lined with crystals. □ **vuggy** *adj.* **vugular** /ˈvʌɡjʊlə(r)/ *adj.* [Cornish *vooga*]

Vuillard /ˈviːɑː(r)/, (Jean) Edouard (1868–1940), French painter and graphic artist. A member of the Nabi Group, he produced decorative panels, murals, paintings, and lithographs; domestic interiors and portraits were his most typical subjects. His early work in particular, with its flat areas of colour, reflects the influence of Japanese prints.

Vulcan /ˈvʌlkən/ *Rom. Mythol.* the god of fire, identified with the Greek Hephaestus.

vulcanic var. of VOLCANIC.

vulcanism var. of VOLCANISM.

Vulcanist /ˈvʌlkənɪst/ *n. Geol. hist.* a holder of the Plutonic theory in geology (see PLUTONIST).

vulcanite /ˈvʌlkəˌnaɪt/ *n.* a hard black vulcanized rubber, ebonite. [as VULCANIZE]

vulcanize /ˈvʌlkəˌnaɪz/ *v.tr.* (also **-ise**) treat (rubber or rubberlike material) with sulphur etc. at a high temperature to increase its strength. □ **vulcanizable** *adj.* **vulcanizer** *n.* **vulcanization** /ˌvʌlkənaɪˈzeɪʃ(ə)n/ *n.* [*Vulcan*, Roman god of fire and metal-working]

vulcanology /ˌvʌlkəˈnɒlədʒɪ/ *n.* (also **volcanology** /ˌvɒl-/) the scientific study of volcanoes. □ **vulcanologist** *n.* **vulcanological** /-nəˈlɒdʒɪk(ə)l/ *adj.*

vulgar /ˈvʌlɡə(r)/ *adj.* **1 a** of or characteristic of the common people, plebeian. **b** coarse in manners; low (*vulgar expressions; vulgar tastes*). **2** in common use; generally prevalent (*vulgar errors*). □ **vulgar fraction** *Math.* a fraction expressed by numerator and denominator, not decimally. **vulgar Latin** see LATIN. **the vulgar tongue** the national or vernacular language, esp. formerly as opposed to Latin. □ **vulgarly** *adv.* [ME f. L *vulgaris* f. *vulgus* common people]

vulgarian /vʌlˈɡeərɪən/ *n.* a vulgar (esp. rich) person.

vulgarism /ˈvʌlɡəˌrɪz(ə)m/ *n.* **1** a word or expression in coarse or uneducated use. **2** an instance of coarse or uneducated behaviour.

vulgarity /vʌlˈɡærɪtɪ/ *n.* (*pl.* **-ies**) **1** the quality of being vulgar. **2** an instance of this.

vulgarize /ˈvʌlɡəˌraɪz/ *v.tr.* (also **-ise**) **1** make (a person, manners, etc.) vulgar, infect with vulgarity. **2** spoil (a scene, sentiment, etc.) by making it too common, frequented, or well known. **3** popularize. □ **vulgarization** /ˌvʌlɡəraɪˈzeɪʃ(ə)n/ *n.*

Vulgate /ˈvʌlɡeɪt, -ɡət/ *n.* **1** the Latin version of the Bible prepared mainly by St Jerome in the late 4th century (it was finished *c.*405), containing a version of the Old Testament translated directly from the Hebrew text. The Council of Trent decreed in 1546 that the Vulgate was to be the sole Latin authority for the Bible, and it was used in the revision of 1592 until the recent preparation of a new Latin translation. **2** (**vulgate**) the traditionally accepted text of any author. **3** (**vulgate**) common or colloquial speech. [L *vulgata* (*editio* edition), fem. past part. of *vulgare* make public f. *vulgus*: see VULGAR]

vulnerable /ˈvʌlnərəb(ə)l, *disp.* ˈvʌnə-/ *adj.* **1** that may be wounded or harmed. **2** (foll. by *to*) exposed to damage by a weapon, criticism, etc. **3** *Bridge* having won one game towards rubber and therefore liable to higher penalties. □ **vulnerably** *adv.* **vulnerableness** *n.* **vulnerability** /ˌvʌlnərəˈbɪlɪtɪ/ *n.* [LL *vulnerabilis* f. L *vulnerare* to wound f. *vulnus -eris* wound]

vulnerary /ˈvʌlnərərɪ/ *adj. & n.* ● *adj.* useful or used for the healing of wounds. ● *n.* (*pl.* **-ies**) a vulnerary drug, plant, etc. [L *vulnerarius* f. *vulnus*: see VULNERABLE]

vulpine /ˈvʌlpaɪn/ *adj.* **1** of or like a fox. **2** crafty, cunning. [L *vulpinus* f. *vulpes* fox]

vulture /ˈvʌltʃə(r)/ *n.* **1 a** a large Old World bird of prey of the family Accipitridae, with the head and neck more or less bare of feathers, feeding chiefly on carrion and reputed to gather with others in anticipation of a death. **b** a similar bird of prey of the family Cathartidae, of the New World. **2** a rapacious person. □ **vulturine** /-ˌraɪn/ *adj.* **vulturish** *adj.* **vulturous** *adj.* [ME f. AF *vultur*, OF *voltour* etc., f. L *vulturius*]

vulva /ˈvʌlvə/ *n.* (*pl.* **vulvas**) *Anat.* the external female genitals, esp. the external opening of the vagina. □ **vulvar** *adj.* **vulvitis** /vʌlˈvaɪtɪs/ *n.* [L, = womb]

vv. *abbr.* **1** verses. **2** volumes.

Vyatka /ˈvjɑːtkə/ an industrial town in north central European Russia, on the Vyatka river; pop. (est. 1989) 441,000. It was formerly known as Kirov (1934–92).

vying *pres. part.* of VIE.

Ww

W[1] /ˈdʌb(ə)lˌjuː/ n. (also **w**) (pl. **Ws** or **W's**) the twenty-third letter of the alphabet.

W[2] abbr. (also **W.**) **1** West; Western. **2** women's (size).

W[3] symb. **1** Chem. the element tungsten. **2** watt(s). [sense 1 from WOLFRAM]

w. abbr. **1** Cricket **a** wicket(s). **b** wide(s). **2** with. **3** wife. **4** weight. **5** width.

WA abbr. **1** Western Australia. **2** US Washington (State) (in official postal use).

Waac /wæk/ n. hist. a member of the Women's Army Auxiliary Corps (Brit. 1917–19 or US 1942–8). [initials WAAC]

Waadt see VAUD.

Waaf /wæf/ n. Brit. hist. a member of the Women's Auxiliary Air Force (1939–48). [initials WAAF]

Waal /wɑːl/ a river of the south central Netherlands. The most southerly of two major distributaries of the Rhine, it flows for 84 km (52 miles) from the point where the Rhine forks, just west of the border with Germany, to the estuary of the Meuse (Maas) on the North Sea.

WAC abbr. (in the US) Women's Army Corps.

wack[1] /wæk/ n. esp. US sl. a crazy person. [prob. back-form. f. WACKY]

wack[2] /wæk/ n. dial. a familiar term of address. [perh. f. wacker Liverpudlian]

wacke /ˈwækə/ n. Geol. any sandstone containing between 15 and 75 per cent mud matrix (cf. GREYWACKE). [G f. MHG wacke large stone, OHG wacko pebble]

wacko /ˈwækəʊ/ adj. & n. esp. N. Amer. sl. ● adj. crazy. ● n. (pl. **-os** or **-oes**) a crazy person. [WACKY + -o]

wacky /ˈwæki/ adj. & n. (also **whacky**) sl. ● adj. (**-ier, -iest**) crazy. ● n. (pl. **-ies**) a crazy person. □ **wackily** adv. **wackiness** n. [orig. dial., = left-handed, f. WHACK]

wad /wɒd/ n. & v. ● n. **1** a lump or bundle of soft material used esp. to keep things apart or in place or to stuff up an opening. **2** a disc of felt etc. keeping powder or shot in place in a gun. **3** a number of banknotes or documents placed together. **4** Brit. sl. a bun, sandwich, etc. **5** (in sing. or pl.) a large quantity esp. of money. ● v.tr. (**wadded, wadding**) **1** stop up (an aperture or a gun-barrel) with a wad. **2** keep (powder etc.) in place with a wad. **3** line or stuff (a garment or coverlet) with wadding. **4** protect (a person, walls, etc.) with wadding. **5** press (cotton etc.) into a wad or wadding. [perh. rel. to Du. watten, F ouate padding, cotton wool]

wadding /ˈwɒdɪŋ/ n. **1** soft pliable material of cotton or wool etc. used to line or stuff garments, quilts, etc., or to pack fragile articles. **2** any material from which gun-wads are made.

waddle /ˈwɒd(ə)l/ v. & n. ● v.intr. walk with short steps and a swaying motion, like a stout short-legged person or a bird with short legs set far apart (e.g. a duck or goose). ● n. a waddling gait. □ **waddler** n. [perh. frequent. of WADE]

waddy /ˈwɒdi/ n. (pl. **-ies**) **1** an Australian Aboriginal's war-club. **2** Austral. & NZ any club or stick. [Dharuk wadi tree, stick, club]

Wade[1] /weɪd/, George (1673–1748), English soldier. After serving with distinction in the British army in Spain 1704–10, he was posted to the Scottish Highlands in 1724. There he was responsible for the construction of a network of roads and bridges to facilitate government control of the Jacobite clans after the 1715 uprising.

Wade[2] /weɪd/, (Sarah) Virginia (b.1945), English tennis player. She won many singles titles, including the US Open (1968), the Italian championship (1971), the Australian Open (1972), and Wimbledon (1977).

wade /weɪd/ v. & n. ● v. **1** intr. walk through water or some impeding medium, e.g. snow, mud, or sand. **2** intr. make one's way with difficulty or by force. **3** intr. (foll. by through) read (a book etc.) in spite of its dullness etc. **4** intr. (foll. by into) colloq. attack (a person or task) vigorously. **5** tr. ford (a stream etc.) on foot. ● n. a spell of wading. □ **wade in** colloq. make a vigorous attack or intervention. **wading bird** any long-legged waterbird that wades. □ **wadable** adj. (also **wadeable**). [OE wadan f. Gmc, = go (through)]

Wade–Giles /weɪdˈdʒaɪlz/ n. a system of romanized spelling for transliterating Chinese, devised by the British diplomat and academic Sir Thomas Francis Wade (1818–95) and subsequently modified by Herbert Allen Giles (1845–1935), Wade's successor as professor of Chinese at Cambridge University. The system, which produces the spelling Mao Tse-tung rather than Mao Zedong, has been widely superseded by Pinyin.

wader /ˈweɪdə(r)/ n. **1 a** a person who wades. **b** a wading bird, esp. a bird of the order Charadriiformes such as a sandpiper or plover (cf. SHOREBIRD). **2** (in pl.) high waterproof boots, or a waterproof garment for the legs and body, worn in fishing etc.

wadgula /ˈwɑːdjʊˌlɑː/ n. Austral. (in Aboriginal English) a white person. [alt. of white fellow]

wadi /ˈwɒdi/ n. (also **wady**) (pl. **wadis** or **wadies**) a rocky watercourse in North Africa etc., dry except in the rainy season. [Arab. wādī]

Wadi Halfa /ˌwɒdi ˈhælfə/ a town in northern Sudan, on the border with Egypt. It is situated on the Nile at the southern end of Lake Nasser and is the terminus of the railway from Khartoum.

WAF abbr. (in the US) Women in the Air Force.

w.a.f. abbr. with all faults.

wafer /ˈweɪfə(r)/ n. & v. ● n. **1** a very thin light crisp sweet biscuit, esp. of a kind eaten with ice-cream. **2** Eccl. a thin disc of unleavened bread used in the Eucharist. **3** a disc of red paper stuck on a legal document instead of a seal. **4** Electronics a very thin slice of a semiconductor crystal used as the substrate for solid-state circuitry. **5** hist. a small disc of dried paste formerly used for fastening letters, holding papers together, etc. ● v.tr. fasten or seal with a wafer. □ **wafer-thin** very thin. □ **wafery** adj. [ME f. AF wafre, ONF waufre, OF gaufre (cf. GOFFER) f. MLG wāfel waffle: cf. WAFFLE[2]]

waffle[1] /ˈwɒf(ə)l/ n. & v. esp. Brit. colloq. ● n. verbose but aimless, misleading, or ignorant talk or writing. ● v.intr. (often foll. by on) indulge in waffle. □ **waffler** n. **waffly** adj. [orig. dial., frequent. of waff = yelp, yap (imit.)]

waffle[2] /ˈwɒf(ə)l/ n. a small crisp batter cake. □ **waffle-iron** a utensil, usu. of two shallow metal pans hinged together, for baking waffles. [Du. wafel, waefel f. MLG wāfel: cf. WAFER]

waft /wɒft, wɑːft/ v. & n. ● v.tr. & intr. convey or travel easily as through air or over water; sweep smoothly and lightly along. ● n. **1** (usu. foll. by of) a whiff or scent. **2** a transient sensation of peace, joy, etc. **3** (also **weft** /weft/) Naut. a distress signal, e.g. an ensign rolled or knotted or a garment flown in the rigging. [orig. 'convoy (ship etc.)', back-form. f. obs. waughter, wafter armed convoy-ship, f. Du. or LG wachter f. wachten to guard]

wag[1] /wæg/ v. & n. ● v. (**wagged, wagging**) **1** tr. & intr. shake or wave rapidly or energetically to and fro. **2** *intr. archaic* (of the world, times, etc.) go along with varied fortune or characteristics. ● *n.* a single wagging motion (*with a wag of his tail*). □ **the tail wags the dog** the less or least important member of a society, section of a party, or part of a structure has control. **tongues** (or **beards** or **chins** or **jaws**) **wag** there is talk. [ME *waggen* f. root of OE *wagian* sway]

wag² /wæg/ *n.* **1** a facetious person, a joker. **2** *Brit. sl.* a truant (*play the wag*). [prob. f. obs. *waghalter* one likely to be hanged (as WAG¹, HALTER)]

wage /weɪdʒ/ *n. & v.* ● *n.* **1** (in *sing.* or *pl.*) a fixed regular payment, usu. daily or weekly, made by an employer to an employee, esp. to a manual or unskilled worker (cf. SALARY). **2** (in *sing.* or *pl.*) requital (*the wages of sin is death*). **3** (in *pl.*) *Econ.* the part of total production that rewards labour rather than remunerating capital. ● *v.tr.* carry on (a war, conflict, or contest). □ **living wage** a wage that affords the means of normal subsistence. **wage bill** the amount paid in wages to employees. **wage-claim** = *pay-claim* (see PAY¹). **wage-earner** a person who works for wages. **wages council** a board of workers' and employers' representatives determining wages where there is no collective bargaining. **wage slave** a person dependent on income from labour in conditions likened to slavery. [ME f. AF & ONF *wage*, OF *g(u)age*, f. Gmc, rel. to GAGE¹, WED]

wager /ˈweɪdʒə(r)/ *n. & v.tr. & intr.* = BET. □ **wager of battle** *hist.* an ancient form of trial by personal combat between the parties or their champions. **wager of law** *hist.* a form of trial in which the defendant was required to produce witnesses who would swear to his or her innocence. [ME f. AF *wageure* f. *wager* (as WAGE)]

Wagga Wagga /ˌwɒɡə ˈwɒɡə/ a town on the Murrumbidgee river, in New South Wales, SE Australia; pop. (1991) 40,875.

waggery /ˈwæɡərɪ/ *n.* (*pl.* **-ies**) **1** waggish behaviour, joking. **2** a waggish action or remark, a joke.

waggish /ˈwæɡɪʃ/ *adj.* playful, facetious. □ **waggishly** *adv.* **waggishness** *n.*

waggle /ˈwæɡ(ə)l/ *v. & n. colloq.* ● *v.* **1** *intr. & tr.* wag. **2** *intr. Golf* swing the club-head to and fro over the ball before playing a shot. ● *n.* a waggling motion. [WAG¹ + -LE⁴]

waggly /ˈwæɡlɪ/ *adj.* unsteady.

Wagner /ˈvɑːɡnə(r)/, (Wilhelm) Richard (1813–83), German composer. He developed an operatic genre which he called music drama, synthesizing music, drama, verse, legend, and spectacle. In the late 1840s and early 1850s, Wagner propounded his theories in a series of essays which polarized European musical opinion for decades. *The Flying Dutchman* (1841) was innovative in using scenes rather than numbers to form the structure. Wagner was forced into exile after supporting the German nationalist uprising in Dresden in 1848; in the same year he began writing the text of *Der Ring des Nibelungen* (*The Ring of the Nibelungs*), a cycle of four operas (*Das Rheingold*, *Die Walküre*, *Siegfried*, and *Götterdämmerung*) based loosely on ancient Germanic sagas. He wrote the accompanying music from 1854 to 1874, during which time he returned to Germany (1860). The *Ring* cycle is notable for its use of leitmotifs and orchestral colour to unify the music, dramatic narrative, and characterization. It was first staged in 1876 at Wagner's new Bayreuth theatre. Other works include the music drama *Tristan and Isolde* (1859) and the *Siegfried Idyll* (1870) for orchestra. □ **Wagnerian** /vɑːɡˈnɪərɪən/ *adj. & n.*

wagon /ˈwæɡən/ *n.* (also *Brit.* **waggon**) **1 a** a four-wheeled vehicle for heavy loads, often with a removable tilt or cover. **b** a lorry or truck. **2** *Brit.* a railway vehicle for goods, esp. an open truck. **3** a trolley for conveying tea etc. **4** (in full **water-wagon**) a vehicle for carrying water. **5** *N. Amer.* a light horse-drawn vehicle. **6** *colloq.* a motor car, esp. an estate car. □ **on the wagon** (or **water-wagon**) *colloq.* teetotal. **wagon-roof** (or **-vault**) = *barrel vault*. **wagon-train** a succession of wagons, esp. *hist.* as used by pioneers or settlers in North America. [earlier *wagon*, *wag(h)en*, f. Du. *wag(h)en*, rel. to OE *wægn* WAIN]

wagoner /ˈwæɡənə(r)/ *n.* (also *Brit.* **waggoner**) the driver of a wagon. [Du. *wagenaar* (as WAGON)]

wagonette /ˌwæɡəˈnet/ *n.* (also *Brit.* **waggonette**) a four-wheeled horse-drawn pleasure vehicle, usu. open, with facing side-seats.

wagon-lit /ˌvæɡɒ̃ˈliː/ *n.* (*pl.* **wagons-lits** *pronunc.* same) a sleeping-car on a Continental railway. [F]

wagtail /ˈwæɡteɪl/ *n.* a songbird that habitually wags its tail; esp. one of the genus *Motacilla*, with a long tail that is continually wagged up and down.

Wahabi /wəˈhɑːbɪ/ *n.* (also **Wahhabi**) (*pl.* **-is**) a member of a strictly orthodox Sunni Muslim sect founded by Muhammad ibn Abd al-Wahab

(1703–92). Abd al-Wahab called for a return to the earliest doctrines and practices of Islam as embodied in the Koran and Sunna, and opposed mystical groups such as the Sufis. Followers and allies of Abd al-Wahab forged a state which came to encompass most of the Arabian peninsula; this was eventually conquered by the Ottomans, but formed the basis for the state of Saudi Arabia (1932), which is still dominated by Wahabi doctrines.

wahine /wɑːˈhiːnɪ/ *n. NZ* a woman or wife. [Maori]

wahoo¹ /wɑːˈhuː/ *n. N. Amer.* **1 a** (in full **wahoo elm**) a North American elm, *Ulmus alata*. **b** a North American spindle tree, *Euonymus atropurpureus*. **2** a large fast-swimming tropical marine fish, *Acanthocybium solanderi*, of the tuna family. **3** = YAHOO. [sense 1b f. Dakota *waⁿhu* 'arrow-wood'; sense 3 prob. f. WAHOO²; senses 1a (18th c.) and 3 (20th c.) of unkn. orig.]

wahoo² /wɑːˈhuː/ *int. N. Amer.* expressing exuberance or triumph. [natural exclamation]

wah-wah /ˈwɑːwɑː/ *n.* (also **wa-wa**) *Mus.* an effect achieved on brass instruments by alternately applying and removing a mute, and on an electric guitar by controlling the output from the amplifier with a pedal. [imit.]

waif /weɪf/ *n.* **1** a homeless and helpless person, esp. an abandoned child. **2** an ownerless object or animal, a thing cast up by or drifting in the sea or brought by an unknown agency. □ **waifs and strays** **1** homeless or neglected children. **2** odds and ends. □ **waifish** *adj.* [ME f. AF *waif*, *weif*, ONF *gaif*, prob. of Scand. orig.]

Waikato /waɪˈkɑːtəʊ/ the longest river of New Zealand, which flows 434 km (270 miles) generally north-westwards from Lake Taupo, at the centre of North Island, to the Tasman Sea.

Waikiki /ˈwaɪkɪˌkiː/ a beach resort, a suburb of Honolulu, on the island of Oahu, in Hawaii.

wail /weɪl/ *n. & v.* ● *n.* **1** a prolonged and plaintive loud high-pitched cry of pain, grief, etc. **2** a sound like or suggestive of this. ● *v.* **1** *intr.* utter a wail. **2** *intr.* lament or complain persistently or bitterly. **3** *intr.* (of the wind etc.) make a sound like a person wailing. **4** *tr. poet.* or *rhet.* bewail; wail over. □ **wailer** *n.* **wailful** *adj. poet.* **wailingly** *adv.* [ME f. ON, rel. to WOE]

Wailing Wall (also called *Western Wall*) an ancient wall in Jerusalem sacred to both Jews and Muslims. The wall is the sole remaining part of the Temple, the rest of which was destroyed in AD 70: Jews traditionally lament there the destruction of the Temple and pray for its restoration. Since the 7th century the wall has formed the western wall of the sanctuary enclosing the Dome of the Rock and other buildings, the third most holy place to Muslims after Mecca and Medina.

wain /weɪn/ *n. archaic* a wagon. [OE *wæg(e)n*, *wæn*, f. Gmc, rel. to WAY, WEIGH¹]

wainscot /ˈweɪnskət/ *n. & v.* ● *n.* **1** boarding or wooden panelling on the lower part of a room-wall. **2** *Brit. hist.* imported oak of fine quality. ● *v.tr.* (**wainscoted**, **wainscoting**) line with wainscot. [ME f. MLG *wagenschot*, app. f. *wagen* WAGON + *schot* of uncert. meaning]

wainscoting /ˈweɪnskətɪŋ/ *n.* **1** a wainscot. **2** material for this.

wainwright /ˈweɪnraɪt/ *n. archaic* a wagon-builder.

waist /weɪst/ *n.* **1 a** the part of the human body below the ribs and above the hips, usu. of smaller circumference than these; the narrower middle part of the normal human figure. **b** the circumference of this. **2** a similar narrow part in the middle of a violin, hourglass, wasp, etc. **3 a** the part of a garment encircling or covering the waist. **b** the narrow middle part of a woman's dress etc. **c** *US* a blouse or bodice. **4** the middle part of a ship, between the forecastle and the quarterdeck. □ **waist-cloth** a loincloth. **waist-deep** (or **-high**) up to the waist (*waist-deep in water*). □ **waisted** *adj.* (also in *comb.*). **waistless** *adj.* [ME *wast*, perh. f. OE f. the root of WAX²]

waistband /ˈweɪstbænd/ *n.* a strip of cloth forming the waist of a garment.

waistcoat /ˈweɪskəʊt, ˈweɪstkəʊt, ˈweskət/ *n. Brit.* a close-fitting waist-length garment, without sleeves or collar but usu. buttoned, worn usu. by men over a shirt and under a jacket.

waistline /ˈweɪstlaɪn/ *n.* the outline or the size of a person's body at the waist.

wait /weɪt/ *v. & n.* ● *v.* **1** *intr.* **a** defer action or departure for a specified time or until some expected event occurs (*wait a minute*; *wait till I come*; *wait for a fine day*). **b** be expectant or on the watch (*waited to see what would happen*). **c** (foll. by *for*) refrain from going so fast that (a person) is left behind (*wait for me!*). **2** *tr.* await (an opportunity, one's turn, etc.).

3 *tr.* defer (an activity) until a person's arrival or until some expected event occurs. **4** *intr.* (usu. as **waiting** *n.*) park a vehicle for a short time at the side of a road etc. (*no waiting*). **5** *intr.* **a** (in full **wait at** (or *N. Amer.* **on**) **table**) act as a waiter or as a servant with similar functions. **b** act as an attendant. **6** *intr.* (foll. by *on, upon*) **a** await the convenience of. **b** serve as an attendant to. **c** pay a respectful visit to. ● *n.* **1** a period of waiting (*had a long wait for the train*). **2** (usu. foll. by *for*) watching for an enemy; ambush (*lie in wait; lay wait*). **3** (in *pl.*) *Brit.* **a** *archaic* street singers of Christmas carols. **b** *hist.* official bands of musicians maintained by a city or town. □ **cannot wait 1** is impatient. **2** needs to be dealt with immediately. **can wait** need not be dealt with immediately. **wait-a-bit** a plant with hooked thorns etc. that catch the clothing. **wait and see** await the progress of events. **wait for it!** *colloq.* **1** do not begin before the proper moment. **2** used to create an interval of suspense before saying something unexpected or amusing. **wait on** *Austral., NZ, & N. Engl.* be patient, wait. **wait up** (often foll. by *for*) not go to bed until a person arrives or an event happens. **you wait!** used to imply a threat, warning, or promise. [ME f. ONF *waitier* f. Gmc, rel. to WAKE¹]

Waitangi, Treaty of /waɪˈtæŋɪ/ a treaty signed in 1840 at the settlement of Waitangi in New Zealand, which formed the basis of the British annexation of New Zealand. The Maori chiefs of North Island accepted British sovereignty in exchange for protection, and direct purchase of land from the Maoris was forbidden. Subsequent contraventions of the treaty by the British led eventually to the Maori Wars (1845–8 and 1860–72) and the final destruction of Maori independence.

Waitangi Day /waɪˈtæŋɪ/ the anniversary of the signing of the Treaty of Waitangi, celebrated as a public holiday in New Zealand on 6 February since 1960.

waiter /ˈweɪtə(r)/ *n.* **1** a man who serves at table in a hotel or restaurant etc. **2** a person who waits for a time, event, or opportunity. **3** a tray or salver.

waiting /ˈweɪtɪŋ/ *n.* **1** in senses of WAIT *v.* **2 a** official attendance at court. **b** one's period of this. □ **waiting game** abstention from early action in a contest etc. so as to act more effectively later. **waiting-list** a list of people waiting for a thing not immediately available. **waiting-room** a room provided for people to wait in, esp. at a doctor's, dentist's, etc., or at a railway or bus station.

waitress /ˈweɪtrɪs/ *n.* a woman who serves at table in a hotel or restaurant etc.

waive /weɪv/ *v.tr.* refrain from insisting on or using (a right, claim, opportunity, legitimate plea, rule, etc.). [ME f. AF *weyver*, OF *gaiver* allow to become a WAIF, abandon]

waiver /ˈweɪvə(r)/ *n. Law* **1** the act or an instance of waiving. **2** a document recording this.

Wajda /ˈvaɪdə/, Andrzej (b.1929), Polish film director. He came to prominence in the 1950s with a trilogy (including *Ashes and Diamonds*, 1958) about the disaffected younger generation in Poland during and after the Second World War. A recurrent theme in his work is the conflict between individual choice and the march of political events, as in *Man of Iron* (1981), which draws on the early history of the trade-union movement Solidarity, and *Danton* (1983), which traces developments in the French Revolution leading up to the Reign of Terror.

wake¹ /weɪk/ *v. & n.* ● *v.* (*past* **woke** /wəʊk/ or **waked**; *past part.* **woken** /ˈwəʊkən/ or **waked**) **1** *intr. & tr.* (often foll. by *up*) cease or cause to cease to sleep. **2** *intr. & tr.* (often foll. by *up*) become or cause to become alert, attentive, or active (*needs something to wake him up*). **3** *intr.* (archaic except as **waking** *adj. & n.*) be awake (*in her waking hours; waking or sleeping*). **4** *tr.* disturb (silence or a place) with noise; make re-echo. **5** *tr.* evoke (an echo). **6** *intr. & tr.* rise or raise from the dead. ● *n.* **1** a watch or vigil beside a corpse before burial; lamentation and (less often) merrymaking in connection with this. **2** (usu. in *pl.*) an annual holiday in (industrial) northern England. **3** *hist.* **a** a vigil commemorating the dedication of a church. **b** a fair or merrymaking on this occasion. □ **be a wake-up** (often foll. by *to*) *Austral. sl.* be alert or aware. **wake-robin 1** *Brit.* the cuckoo-pint. **2** *N. Amer.* a liliaceous plant of the genus *Trillium.* □ **waker** *n.* [OE *wacan* (recorded only in past *woc*), *wacian* (weak form), rel. to WATCH: sense 'vigil' perh. f. ON]

wake² /weɪk/ *n.* **1** the track left on the water's surface by a moving ship. **2** turbulent air left behind a moving aircraft etc. □ **in the wake of** behind, following, as a result of, in imitation of. [prob. f. MLG f. ON *vök* hole or opening in ice]

Wakefield /ˈweɪkfiːld/ the county town of West Yorkshire; pop. (1981) 75,840.

wakeful /ˈweɪkfʊl/ *adj.* **1** unable to sleep. **2** (of a night etc.) passed with little or no sleep. **3** vigilant. □ **wakefully** *adv.* **wakefulness** *n.*

waken /ˈweɪkən/ *v.tr. & intr.* make or become awake. [ON *vakna* f. Gmc, rel. to WAKE¹]

Waksman /ˈwæksmən/, Selman Abraham (1888–1973), Russian-born American microbiologist. He searched for potential antibiotics in soil micro-organisms, discovering the bacterium *Streptomyces griseus* in 1915 and isolating streptomycin from it in 1943. This was developed into the first effective drug against tuberculosis, which became a much less serious public health problem through its use. Waksman was awarded a Nobel Prize in 1952.

Walachia see WALLACHIA.

Waldenses /wɒlˈdensiːz/ *n.pl.* the adherents of a puritan religious sect founded in southern France *c.*1173 by Peter Valdes (d.1205), a merchant of Lyons. The sect suffered persecution throughout the Middle Ages; after the Reformation it survived as a Protestant sect and now exists chiefly in northern Italy (Piedmont) and North America. □ **Waldensian** *adj. & n.*

Waldheim /ˈvælthaɪm/, Kurt (b.1918), Austrian diplomat and statesman, President 1986–92. He was Secretary-General of the United Nations (1972–81), and, five years later, he stood as the right-wing People's Party candidate for the presidency of Austria; this he secured after a run-off election. During the campaign, he denied allegations that as an army intelligence officer he had direct knowledge of Nazi atrocities during the Second World War; he was subsequently cleared in court of charges relating to his war record.

wale /weɪl/ *n. & v.* ● *n.* **1** = WEAL¹ *n.* **2** a ridge on a woven fabric, e.g. corduroy. **3** *Naut.* a broad thick timber along a ship's side. **4** a specially woven strong band round a woven basket. ● *v.tr.* provide or mark with wales; thrash, whip. □ **wale-knot** a knot made at the end of a rope by intertwining strands to prevent unravelling or act as a stopper. [OE *walu* stripe, ridge]

Wales /weɪlz/ (called in Welsh *Cymru*) a principality of Great Britain and the United Kingdom, to the west of central England; pop. (1991) 2,798,200; capital, Cardiff. Wales is mainly mountainous and forms a large peninsula extending into the Irish Sea. The earliest inhabitants appear to have been overrun by Celtic peoples in the Bronze and Iron Ages. The Romans established forts and outposts in the north and south, linked by a system of roads. After the Roman withdrawal in the 5th century, the Celtic inhabitants of Wales successfully maintained independence against the Anglo-Saxons who settled in England, and in the 8th century Offa, king of Mercia, built an earthwork (see OFFA'S DYKE) marking the frontier. In the Dark Ages Christianity was spread through Wales by missionaries, but the country remained somewhat isolated and divided between rival rulers and chieftains. Norman colonization from England began in the 11th century, and their control over the country was assured by Edward I's conquest (1277–84). Edward began the custom of making the English sovereign's eldest son Prince of Wales. Sporadic resistance to English rule continued, but Wales was formally brought into the English legal and parliamentary system by Henry VIII (1536). Wales has retained a distinct cultural identity, and the Welsh language, spoken by about half a million people, is widely used.

Wales, Prince of see PRINCE OF WALES; CHARLES, PRINCE.

Wałęsa /vəˈwensə, -ˈlensə/, Lech (b.1943), Polish trade unionist and statesman, President 1990–5. A former electrician at the Lenin Shipyards in Gdańsk, he emerged as a strike leader at the port in 1980, founding the trade-union movement Solidarity later the same year. Following the banning of Solidarity in 1981, Wałęsa was imprisoned from 1981 to 1982. He was awarded the Nobel Peace Prize in 1983. Wałęsa led Solidarity to a landslide victory in the 1989 free elections and successfully contested the 1990 presidential elections. He was defeated in the 1995 election by Aleksander Kwaśniewski, a former Communist Party official.

Walian /ˈweɪlɪən/ *n. & adj.* (with defining word, as **North Walian**, **South Walian**, etc.) ● *n.* a native or inhabitant of (a specified region of) Wales. ● *adj.* of or relating to (a specified region of) Wales.

walk /wɔːk/ *v. & n.* ● *v.* **1** *intr.* **a** (of a person or other biped) progress by lifting and setting down each foot in turn, never having both feet off the ground at once. **b** progress with similar movements (*walked on his hands*). **c** go with the gait usual except when speed is desired. **d** (of a quadruped) go with the slowest gait, always having at least two feet on the ground at once. **2** *intr.* **a** travel or go on foot. **b** take exercise in this way (*walks for two hours each day*). **3** *tr.* **a** perambulate, traverse on foot at walking speed, tread the floor or surface of. **b** traverse or cover (a specified distance) on foot (*walks five miles a day*). **4** *tr.* **a** cause to walk with one. **b** accompany in walking. **c** ride or lead (a horse, dog, etc.) at walking pace. **d** take charge of (a puppy) at walk (see sense 4 of *n.*). **5** *intr.* (of a ghost) appear. **6** *intr. Cricket* leave the wicket on being out. **7** *Baseball* **a** *intr.* reach first base on being entitled to do so after not hitting at four balls pitched outside specified limits. **b** *tr.* allow to do this. **8** *intr. archaic* live in a specified manner, conduct oneself (*walk humbly; walk with God*). **9** *intr. N. Amer. sl.* be released from suspicion or from a charge. ● *n.* **1 a** an act of walking, the ordinary human gait (*go at a walk*). **b** the slowest gait of an animal. **c** a person's manner of walking (*know him by his walk*). **2 a** (an act or instance of) travelling a specified distance on foot (*is only ten minutes' walk from here; it's quite a walk to the bus-stop*). **b** an excursion on foot; a stroll (*go for a walk*). **c** a journey on foot completed to earn money promised for a charity etc. **3 a** a place, track, or route intended or suitable for walking; a promenade, colonnade, or footpath. **b** a person's favourite place or route for walking. **c** the round of a postman, hawker, etc. **4** a farm etc. where a hound-puppy is sent to accustom it to various surroundings. **5** the place where a gamecock is kept. **6** a part of a forest under one keeper. □ **in a walk** without effort (*won in a walk*). **walk about** stroll. **walk all over** *colloq.* **1** defeat easily. **2** take advantage of. **walk away from 1** easily outdistance. **2** refuse to become involved with; fail to deal with. **walk away with** *colloq.* = *walk off with*. **walk the boards** be an actor. **walk the hospitals** = *walk the wards*. **walk in** (often foll. by *on*) enter or arrive, esp. unexpectedly or easily. **walk into** *colloq.* **1** encounter through unwariness (*walked into the trap*). **2** get (a job) easily. **walk it 1** make a journey on foot, not ride. **2** *colloq.* achieve something (esp. a victory) easily. **walk Matilda** see MATILDA². **walk off 1** depart (esp. abruptly). **2** get rid of the effects of (a meal, ailment, etc.) by walking. **walk a person off his** (or **her**) **feet** (or **legs**) exhaust a person with walking. **walk off with** *colloq.* **1** steal. **2** win easily. **walk of life** an occupation, profession, or calling. **walk-on 1** (in full **walk-on part**) a non-speaking dramatic role. **2** an actor playing this. **walk on air** see AIR. **walk out 1** depart suddenly or angrily. **2** (usu. foll. by *with*) *Brit. archaic* go for walks in courtship. **3** cease work, esp. to go on strike. **walk-out** *n.* a sudden angry departure, esp. as a protest or strike. **walk out on** desert, abandon. **walk over** *colloq.* = *walk all over*. **2** (often *absol.*) traverse (a racecourse) without needing to hurry, because one has no opponents or only inferior ones. **walk-over** *n.* an easy victory or achievement. **walk the plank** see PLANK. **walk the streets 1** be a prostitute. **2** traverse the streets esp. in search of work etc. **walk tall** *colloq.* feel justifiable pride. **walk up!** a showman's invitation to a circus etc. **walk-up** *N. Amer. adj.* (of a building) allowing access to the upper floors only by stairs. ● *n.* a walk-up building. **walk up to** approach (a person) for a talk etc. **walk the wards** be a medical student. □ **walkable** *adj.* [OE *wealcan* roll, toss, wander, f. Gmc]

walkabout /ˈwɔːkəˌbaʊt/ *n.* **1** an informal stroll among a crowd by a visiting dignitary. **2** a period of wandering in the bush by an Australian Aboriginal. □ **go walkabout** go on a walkabout.

walkathon /ˈwɔːkəˌθɒn/ *n.* an organized fund-raising walk. [WALK, after MARATHON]

Walker¹ /ˈwɔːkə(r)/, Alice (Malsenior) (b.1944), American writer and critic. She won international acclaim for *The Color Purple* (1982), an epistolary novel about a young black woman faced with recreating her life after being raped by her supposed father. It won her the Pulitzer Prize and was made into a successful film by Steven Spielberg (1985). Other works include *In Search of Our Mothers' Gardens: Womanist Prose* (1983), a collection of her critical essays, and the novel *Possessing the Secret of Joy* (1992), an indictment of female circumcision.

Walker² /ˈwɔːkə(r)/, John (b.1952), New Zealand athlete. He was the first athlete to run a mile in less than 3 minutes 50 seconds (1975) and was also the first to run one hundred sub-four-minute miles.

walker /ˈwɔːkə(r)/ *n.* **1** a person or animal that walks. **2 a** a wheeled or footed framework in which a baby can learn to walk. **b** = *walking frame*.

Walker Cup a golf tournament played between teams of male amateurs representing the US and Great Britain and Ireland, held every second year in May, alternately in the US and Great Britain. The trophy for the tournament was donated by George Herbert Walker, a former President of the US Golf Association, who instituted the competition in 1922.

walkies /ˈwɔːkɪz/ *int. & n.* ● *int.* a command to a dog to prepare for a walk. ● *n.pl. colloq.* or *joc.* a walk or a spell of walking, esp. with a dog. □ **go walkies** *colloq.* or *joc.* **1** go for a walk. **2** go missing (*my address book's gone walkies*).

walkie-talkie /ˌwɔːkɪˈtɔːkɪ/ *n.* a two-way radio carried on the person, esp. by policemen etc.

walking /ˈwɔːkɪŋ/ *n. & adj.* in senses of WALK *v.* □ **walking delegate** a trade-union official who visits members and their employers for discussions. **walking dictionary** (or **encyclopedia**) *colloq.* a person having a wide general knowledge. **walking fern** an American evergreen fern of the genus *Camptosorus*, with fronds that root at the ends. **walking frame** a usu. tubular metal frame with rubberized ferrules, used by disabled or old people to help them walk. **walking gentleman** (or **lady**) *Theatr.* a non-speaking extra; a supernumerary. **walking leaf** = *walking fern*. **walking-on part** a non-speaking dramatic role. **walking papers** *colloq.* dismissal (*gave him his walking papers*). **walking-tour** a holiday journey on foot, esp. of several days. **walking wounded 1** (of soldiers etc.) able to walk despite injuries; not bedridden. **2** *colloq.* a person or people having esp. mental or emotional difficulties.

walking-stick /ˈwɔːkɪŋˌstɪk/ *n.* **1** a stick held or used as a support when walking. **2** *N. Amer.* = *stick insect* (see STICK¹).

Walkman /ˈwɔːkmən/ *n.* (*pl.* **-mans** or **-men**) *propr.* a type of personal stereo.

walkway /ˈwɔːkweɪ/ *n.* a passage or path for walking along, esp.: **1** a raised passageway connecting different sections of a building. **2** a wide path in a garden etc.

wall /wɔːl/ *n. & v.* ● *n.* **1 a** a continuous and usu. vertical structure usu. of brick or stone, having little width in proportion to its length and height and esp. enclosing, protecting, or dividing a space or supporting a roof. **b** the surface of a wall, esp. inside a room (*hung the picture on the wall*). **2** anything like a wall in appearance or effect, esp.: **a** the steep side of a mountain. **b** a protection or obstacle (*a wall of steel bayonets; a wall of indifference*). **c** *Anat.* the outermost layer or enclosing membrane etc. of an organ, structure, etc. **d** the outermost part of a hollow structure (*stomach wall*). **e** *Mining* rock enclosing a lode or seam. **3** (in full **wall brown**) an orange-brown satyrid butterfly of the genus *Lasiommata*. ● *v.tr.* **1** (esp. as **walled** *adj.*) surround or protect with a wall (*walled garden*). **2 a** (usu. foll. by *up, off*) block or seal (a space etc.) with a wall. **b** (foll. by *up*) enclose (a person) within a sealed space (*walled them up in the dungeon*). □ **drive a person up the wall** *colloq.* **1** make a person angry; infuriate. **2** drive a person mad. **go to the wall** be defeated or ruined. **go up the wall** *colloq.* become crazy or furious (*went up the wall when he heard*). **off the wall** *esp. N. Amer. sl.* unorthodox, unconventional; crazy, outlandish. **wall bar** any of a set of parallel bars, attached to the wall of a gymnasium, on which exercises are performed. **wall-barley** wild barley as a weed. **wall-board** a type of wall-covering made from wood pulp etc. **wall cress** = ARABIS. **wall fern** an evergreen polypody, *Polypodium vulgare*, with very large leaves. **wall-fruit** fruit grown on trees trained against a wall for protection and warmth. **wall game** see ETON WALL GAME. **wall-knot** = *wale-knot*. **wall-painting** a mural or fresco. **wall pepper** a succulent stonecrop, *Sedum acre*, with a pungent taste. **wall-plate** timber laid in or on a wall to distribute the pressure of a girder etc. **wall rocket** see ROCKET². **wall rue** a small fern, *Asplenium ruta-muraria*, with leaves like rue, growing on walls and rocks. **walls have ears** it is unsafe to speak openly, as there may be eavesdroppers. **wall space** space on the surface of a wall, available for use. **wall-to-wall 1** (of a carpet) fitted to cover a whole room etc. **2** *colloq.* profuse, ubiquitous (*wall-to-wall pop music*). □ **walling** *n.* **wall-less** *adj.* [OE f. L *vallum* rampart f. *vallus* stake]

wallaby /ˈwɒləbɪ/ *n.* (*pl.* **-ies**) **1** a marsupial of the family Macropodidae, smaller than a kangaroo, with large hind legs and a long tail. **2** (**Wallabies**) the Australian international Rugby Union team. □ **on the wallaby** (or **wallaby track**) *Austral.* vagrant; unemployed. [Dharuk *walabi* or *waliba*]

Wallace¹ /ˈwɒlɪs/, Alfred Russel (1823–1913), English naturalist and a founder of zoogeography. He independently formulated a theory of the origin of species that was very similar to that of Charles Darwin, to whom he communicated his conclusions. He travelled extensively in South America and the East Indies, collecting specimens and studying the geographical distribution of animals. In 1858 a summary of the joint views of Wallace and Darwin concerning natural selection was read to the Linnaean Society in London, but credit for the theory has been attached somewhat arbitrarily to Darwin.

Wallace[2] /'wɒlɪs/, (Richard Horatio) Edgar (1875–1932), English novelist, screenwriter, and dramatist. He is noted for his crime novels, including *The Four Just Men* (1905) and *The Crimson Circle* (1922). Based in Hollywood from 1931, he wrote the screenplay for the film of *King Kong*, which was made shortly after his death.

Wallace[3] /'wɒlɪs/, Sir William (c.1270–1305), Scottish national hero. He was a leader of Scottish resistance to Edward I, defeating the English army at Stirling in 1297. In the same year he mounted military campaigns against the north of England and was appointed Guardian of the Realm of Scotland. After Edward's second invasion of Scotland in 1298, Wallace was defeated at the battle of Falkirk; he was subsequently captured and executed by the English.

Wallace Collection a museum in Manchester Square, London, containing French 18th-century paintings and furniture, English 18th-century portraits, and medieval armour. The collection takes its name from Lord Richard Wallace (1819–90), who built up the collection with his father and whose widow gave it to the nation in 1897.

Wallace's line *n. Zool.* a hypothetical line, proposed by A. R. Wallace, marking the boundary between the oriental and Australasian zoogeographical regions. It is now placed along the continental shelf of SE Asia, east of the islands of Borneo, Bali, and the Philippines. To the west of the line Asian animals such as monkeys predominate, while to the east of it the fauna is dominated by marsupials; there is some overlap between the two faunas in Sulawesi and neighbouring islands.

Wallachia /wɒˈleɪkɪə/ (also **Walachia**) a former principality of SE Europe, between the Danube and the Transylvanian Alps. In 1861 it was united with Moldavia to form Romania. □ **Wallachian** *adj. & n.* [as VLACH]

wallah /'wɒlə/ *n.* orig. *Anglo-Ind.*, now *sl.* **1** a person concerned with or in charge of a usu. specified thing, business, etc. (*asked the ticket wallah*). **2** a person doing a routine administrative job; a bureaucrat. [Hindi -*wālā* suffix = -ER[1]]

wallaroo /ˌwɒləˈruː/ *n.* a large brownish-black kangaroo, *Macropus robustus*. [Dharuk *walaru*]

Wallasey /'wɒləsɪ/ a town in Merseyside, NW England, on the Wirral Peninsula; pop. (1981) 62,530.

wallchart /'wɔːltʃɑːt/ *n.* a chart or poster designed for display on a wall as a teaching aid, source of information, etc.

Wallenberg /'vɑːlənˌbɜːɡ/, Raoul (1912–?), Swedish diplomat. While working as a businessman in Budapest in 1944, he was entrusted by the Swedish government with the protection of Hungarian Jews from the Nazis. Wallenberg helped many thousands of Jews (according to estimates, as many as 95,000) to escape death by issuing them with Swedish passports, but when Soviet forces took control of Budapest in 1945, he was arrested, taken to Moscow, and imprisoned. Although the Soviet authorities stated that Wallenberg had died in prison in 1947, his fate remains uncertain and there were claims that he was still alive in the 1970s.

Waller /'wɒlə(r)/, Fats (born Thomas Wright Waller) (1904–43), American jazz pianist, songwriter, band-leader, and singer. He composed the songs 'Ain't Misbehavin'' (1928) and 'Honeysuckle Rose' (1929), and was the foremost exponent of the New York 'stride school' of piano playing, which used tenths in the left hand to give a strong bass line. From 1934 onwards Waller achieved popular success as the leader of the group Fats Waller and His Rhythm.

wallet /'wɒlɪt/ *n.* **1** a small flat esp. leather case for holding banknotes etc. **2** *archaic* a bag for carrying food etc. on a journey, esp. as used by a pilgrim or beggar. [ME *walet*, prob. f. AF *walet* (unrecorded), perh. f. Gmc]

wall-eye /'wɔːlaɪ/ *n.* **1 a** an eye with a streaked or opaque white iris. **b** an eye squinting outwards. **2** a pikelike American perch, *Stizostedion vitreum*, with large prominent eyes. □ **wall-eyed** *adj.* [back-form. f. *wall-eyed*: ME f. ON *vagleygr* f. *vagl* (unrecorded: cf. Icel. *vagl* film over the eye) + *auga* EYE]

wallflower /'wɔːlˌflaʊə(r)/ *n.* **1** a spring-flowering garden plant, *Erysimum cheiri*, with fragrant yellow, orange-red or dark red flowers, sometimes naturalized on old walls. **2** *colloq.* a neglected or socially awkward person, esp. a woman sitting out at a dance for lack of partners.

Wallis /'wɒlɪs/, Sir Barnes Neville (1887–1979), English inventor. Working for the Vickers company, he pioneered geodetic construction in his designs for the R100 airship (1930) and the Wellington bomber used in the Second World War. During the war he designed more effective bombs, including the bouncing bomb used against the Ruhr dams in Germany in 1943. His main postwar work was on guided missiles and supersonic aircraft: he pioneered variable geometry (swing-wing) designs (although these were not fully developed at the time). Altogether he patented more than 140 designs.

Wallis and Futuna Islands /'wɒlɪs, fəˈtjuːnə/ an overseas territory of France comprising two groups of islands to the west of Samoa in the central Pacific; pop. (est. 1988) 15,400; capital, Mata-Utu.

Walloon /wɒˈluːn/ *n. & adj.* ● *n.* **1** a member of a people speaking a French dialect, living in southern Belgium (where they form the majority of the population) and neighbouring parts of France. (See also FLEMING[4].) **2** the French dialect spoken by this people. ● *adj.* of or concerning the Walloons or their language. [F *Wallon* f. med.L *Wallo -onis* f. Gmc: cf. WELSH]

wallop /'wɒləp/ *v. & n. sl.* ● *v.tr.* (**walloped**, **walloping**) **1 a** thrash; beat. **b** hit hard. **2** (as **walloping** *adj.*) big; strapping; thumping (*a walloping profit*). ● *n.* **1** a heavy blow; a thump. **2** *Brit.* beer or any alcoholic drink. □ **walloping** *n.* [earlier senses 'gallop', 'boil', f. ONF (*walop* *n.* f.) *waloper*, OF *galoper*: cf. GALLOP]

walloper /'wɒləpə(r)/ *n.* **1** a person or thing that wallops. **2** *Austral. sl.* a policeman.

wallow /'wɒləʊ/ *v. & n.* ● *v.intr.* **1** (esp. of an animal) roll about in mud, sand, water, etc. **2** (usu. foll. by *in*) indulge in unrestrained sensuality, pleasure, misery, etc. (*wallowing in nostalgia*). ● *n.* **1** the act or an instance of wallowing. **2 a** a place used by buffalo etc. for wallowing. **b** the depression in the ground caused by this. □ **wallower** *n.* [OE *walwian* roll f. Gmc]

wallpaper /'wɔːlˌpeɪpə(r)/ *n. & v.* ● *n.* **1** paper sold in rolls for pasting on to interior walls as decoration. **2** an unobtrusive background, esp. (usu. *derog.*) with ref. to sound, music, etc. ● *v.tr.* (often *absol.*) decorate with wallpaper.

Wall Street a street at the south end of Manhattan, where the New York Stock Exchange and other leading American financial institutions are located. The name is used allusively to refer to the American money market or financial interests. It is named after a wooden stockade which was built in 1653 around the original Dutch settlement of New Amsterdam.

Wall Street Crash the collapse of prices on the New York Stock Exchange in Oct. 1929. It was a major factor giving impetus to the early stages of the Depression.

wally /'wɒlɪ/ *n.* (*pl.* **-ies**) *Brit. sl.* a foolish or inept person. [orig. uncert., perh. shortened form of *Walter*]

walnut /'wɔːlnʌt/ *n.* **1** a tree of the genus *Juglans*, esp. *J. regia*, having aromatic leaves and drooping catkins. **2** the nut of this tree, containing an edible wrinkled kernel in a roughly spherical shell. **3** the timber of the walnut tree used in cabinet-making. [OE *walh-hnutu* f. Gmc NUT]

Walpole[1] /'wɔːlpəʊl/, Horace, 4th Earl of Orford (1717–97), English writer and Whig politician, son of Sir Robert Walpole. He wrote *The Castle of Otranto* (1764), which is regarded as one of the first Gothic novels. Walpole is also noted for his contribution to the Gothic revival in architecture, converting his Strawberry Hill home at Twickenham, near London, into a Gothic castle (c.1753–76). He served as an MP from 1741 to 1767.

Walpole[2] /'wɔːlpəʊl/, Sir Hugh (Seymour) (1884–1941), British novelist, born in New Zealand. His third novel *Mr Perrin and Mr Traill* (1911) reflects his experiences as a schoolmaster. He is best known for *The Herries Chronicle* (1930–3), a historical sequence set in the Lake District.

Walpole[3] /'wɔːlpəʊl/, Sir Robert, 1st Earl of Orford (1676–1745), British Whig statesman, First Lord of the Treasury and Chancellor of the Exchequer 1715–17 and 1721–42. Walpole is generally regarded as the first British Prime Minister in the modern sense, having presided over the Cabinet for George I and George II during his second term as First Lord of the Treasury and Chancellor. His period of office was marked by considerable peace and prosperity, although Walpole failed to prevent war with Spain in 1739. He was the father of Horace Walpole.

Walpurgis night /vælˈpʊəɡɪs/ the eve of 1 May, on which, according to German legend, a witches' sabbath took place on the Brocken in the Harz Mountains. It is named after St Walburga, an English nun who in the 8th century helped to convert the Germans to Christianity; one of her feast days coincided with an ancient pagan feast with rites intended to give protection from witchcraft.

walrus /'wɔːlrəs, 'wɒl-/ *n.* a large amphibious Arctic mammal, *Odobenus rosmarus*, related to the sea lions and having two long tusks. □ **walrus moustache** a long thick drooping moustache. [prob. f. Du. *walrus*,

-ros, perh. by metathesis after *walvisch* 'whale-fish' f. word repr. by OE *horschwæl* 'horse-whale']

Walsall /ˈwɔːlsɔːl, ˈwɒl-/ an industrial town in the West Midlands, England; pop. (1991) 255,600.

Walsingham /ˈwɔːlsɪŋəm, ˈwɒl-/, Sir Francis (c.1530–90), English politician. From 1573 to 1590 he served as Secretary of State to Queen Elizabeth I. He developed a domestic and foreign spy network that led to the detection of numerous Catholic plots against Elizabeth I and the gathering of intelligence about the Spanish Armada. In 1586 Walsingham uncovered a plot against Elizabeth involving Mary, Queen of Scots; he subsequently exerted his judicial power to have Mary executed.

Walton[1] /ˈwɔːlt(ə)n, ˈwɒl-/, Ernest Thomas Sinton (1903–95), Irish physicist. In 1932 he succeeded, with Sir John Cockcroft, in splitting the atom (see COCKCROFT).

Walton[2] /ˈwɔːlt(ə)n, ˈwɒl-/, Izaak (1593–1683), English writer. He is chiefly known for *The Compleat Angler* (1653; largely rewritten, 1655), which combines practical information on fishing with folklore, interspersed with pastoral songs and ballads. He also wrote biographies of John Donne (1640) and George Herbert (1670).

Walton[3] /ˈwɔːlt(ə)n, ˈwɒl-/, Sir William (Turner) (1902–83), English composer. He lived for a time with the Sitwells, and gained fame with *Façade* (1921–3), a setting of poems by Edith Sitwell for recitation. The work of Stravinsky and Hindemith strongly influenced his Viola Concerto (1928–9). Other works include two symphonies, two operas, the oratorio *Belshazzar's Feast* (1930–1), and film scores for three Shakespeare plays and *The Battle of Britain* (1969).

waltz /wɔːls, wɔːlts, wɒls, wɒlts/ n. & v. ● n. **1** a ballroom dance in triple time performed by couples who rotate and progress round the floor. It evolved in the 18th century, probably from German and Austrian folk-dances. **2** the usu. flowing and melodious music for this. The most famous examples are those by 19th-century Viennese composers, particularly Johann Strauss the Elder and Younger. ● v. **1** intr. dance a waltz. **2** intr. (often foll. by in, out, round, etc.) colloq. move lightly, casually, with deceptive ease, etc. (*waltzed in and took first prize*). **3** tr. move (a person) in or as if in a waltz, with ease (*was waltzed off to Paris*). □ **waltz Matilda** see WALTZING MATILDA. **waltz off with** colloq. **1** steal. **2** win (a prize etc.) easily. [G *Walzer* f. *walzen* revolve]

waltzer /ˈwɔːlsə(r), ˈwɔːlts-, ˈwɒls-, ˈwɒlts-/ n. **1** a person who dances the waltz. **2** a fairground ride in which cars spin round as they are carried on an undulating track.

Waltzing Matilda an Australian song with words by Andrew Barton Paterson (1864–1941). A 'Matilda' is a bushman's bundle of belongings; to 'waltz Matilda' is to travel with this. The tune has been identified as that of a march arranged from an old Scottish ballad, which in turn was an adaptation of 'The Bold Fusilier', a song that was popular with British soldiers in the wars of the early 18th century.

Walvis Bay /ˈwɔːlvɪs/ a port in Namibia; pop. (1980) 25,000. For many years it was administratively an exclave of the former Cape Province, South Africa. In 1994 it was transferred to Namibia by the South African government.

wampum /ˈwɒmpəm/ n. beads made from shells and strung together for use as money, decoration, or as aids to memory by North American Indians. [Algonquian *wampumpeag* f. *wap* white + *umpe* string + *-ag* pl. suffix]

wan /wɒn/ adj. (**wanner**, **wannest**) **1** (of a person's complexion or appearance) pale; exhausted; weak; worn. **2** (of a star etc. or its light) partly obscured; faint. **3** archaic (of night, water, etc.) dark, black. □ **wanly** adv. **wanness** /ˈwɒnnɪs/ n. [OE *wann* dark, black, of unkn. orig.]

wand /wɒnd/ n. **1 a** a supposedly magic stick used in casting spells by a fairy, magician, etc. **b** a stick used by a conjuror for effect. **2** a slender rod carried or used as a marker in the ground. **3** a staff symbolizing some officials' authority. **4** colloq. a conductor's baton. **5** a hand-held electronic device which can be passed over a bar-code to read the data this represents. [ME prob. f. Gmc: cf. WEND, WIND[2]]

wander /ˈwɒndə(r)/ v. & n. ● v. **1** intr. (often foll. by in, off, etc.) go about from place to place aimlessly. **2** intr. **a** (of a person, river, road, etc.) wind about; diverge; meander. **b** (of esp. a person) get lost; leave home; stray from a path etc. **3** intr. talk or think incoherently; be inattentive or delirious. **4** tr. cover while wandering (*wanders the world*). ● n. the act or an instance of wandering (*went for a wander round the garden*). □ **wandering albatross** a very large white albatross, *Diomedea exulans*, of southern oceans, having black-tipped wings. **wandering sailor** the moneywort. **wander-plug** a plug that can be fitted into any

of various sockets in an electrical device. □ **wanderer** n. **wandering** n. (esp. in pl.). [OE *wandrian* (as WEND)]

wandering Jew n. **1 a** (in medieval legend) a man condemned to wander the earth until the Day of Judgement, as a punishment for having insulted Christ on the way to the Crucifixion. **b** a person who never settles down. **2 a** a climbing plant, *Tradescantia albiflora*, with stemless variegated leaves. **b** a trailing plant, *Zebrina pendula*, with pink flowers.

wanderlust /ˈwɒndəˌlʌst, ˈvændəˌlʊst/ n. an eagerness for travelling or wandering. [G]

wanderoo /ˌwɒndəˈruː/ n. the hanuman monkey. [Sinh. *wanderu* monkey]

wane /weɪn/ v. & n. ● v.intr. **1** (of the moon) decrease in apparent size after the full moon (cf. WAX[2] 1). **2** decrease in power, vigour, importance, brilliance, size, etc.; decline. ● n. **1** the process of waning. **2** a defect of a plank etc. that lacks square corners. □ **on the wane** waning; declining. □ **waney** adj. (in sense 2 of n.). [OE *wanian* lessen f. Gmc]

Wanganui /ˌwɒŋɡəˈnuːɪ/ a port in New Zealand, on the west coast of North Island; pop. (1991) 41,210.

wangle /ˈwæŋɡ(ə)l/ v. & n. colloq. ● v.tr. **1** (often refl.) to obtain (a favour etc.) by scheming etc. (*wangled himself a free trip*). **2** alter or fake (a report etc.) to appear more favourable. ● n. the act or an instance of wangling. □ **wangler** n. [19th-c. printers' sl.: orig. unkn.]

wank /wæŋk/ v. & n. coarse sl. ● v.intr. & tr. masturbate. ● n. an act of masturbating. [20th c.: orig. unkn.]

Wankel engine /ˈwæŋk(ə)l, ˈvæŋ-/ n. a rotary internal-combustion engine in which a curved, triangular, eccentrically pivoted piston rotates in an elliptical chamber, forming three combustion spaces that vary in volume as it turns. Its main advantage over the conventional reciprocating-piston design is its freedom from vibration, which enables it to be used at a higher rotational speed, and it is more compact and lighter for a given capacity. It has several problems, however, including sealing at the tips of the rotor lobes, higher fuel consumption, and increased pollutants in the exhaust, and few cars have been fitted with it. [Felix *Wankel*, Ger. engineer (1902–88)]

wanker /ˈwæŋkə(r)/ n. coarse sl. **1** a contemptible or ineffectual person. **2** a person who masturbates.

Wankie /ˈwɑːŋkɪ/ the former name (until 1982) for HWANGE.

wanna /ˈwɒnə/ contr. colloq. **1** want to (*I wanna win*). **2** want a (*I wanna biscuit*). ¶ *Wanna* is non-standard and should generally be avoided in both speech and writing.

wannabe /ˈwɒnəbɪ/ n. sl. an avid fan who tries to emulate the person he or she admires; a person who aspires to be someone else.

want /wɒnt/ v. & n. ● v. **1 a** (often foll. by to + infin.) desire; wish for possession of; need (*wants a toy train; wants it done immediately; wanted to leave; wanted him to leave*). **b** need or desire (a person, esp. sexually). **c** esp. Brit. require to be attended to in esp. a specified way (*the garden wants weeding*). **d** (foll. by to + infin.) colloq. ought; should; need (*you want to pull yourself together; you don't want to overdo it*). **2** intr. (usu. foll. by for) lack; be deficient (*wants for nothing*). **3** tr. be without or fall short by (esp. a specified amount or thing) (*the drawer wants a handle*). **4** intr. (foll. by in, out) esp. US colloq. desire to be in, out, etc. (*wants in on the deal*). **5** tr. (as **wanted** adj.) (of a suspected criminal etc.) sought by the police. ● n. **1** (often foll. by of) **a** a lack, absence, or deficiency (*could not go for want of time; shows great want of judgement*). **b** poverty; need (*living in great want; in want of necessities*). **2 a** a desire for a thing etc. (*meets a long-felt want*). **b** a thing so desired (*can supply your wants*). □ **do not want to** am unwilling to. **want ad** US a classified newspaper advertisement for something wanted. □ **wanter** n. [ME f. ON *vant* neut. of *vanr* lacking = OE *wana*, formed as WANE]

wanting /ˈwɒntɪŋ/ adj. **1** lacking (in quality or quantity); deficient, not equal to requirements (*wanting in judgement; the standard is sadly wanting*). **2** absent, not supplied or provided. □ **be found wanting** fail to meet requirements.

wanton /ˈwɒntən/ adj., n., & v. ● adj. **1** licentious; lewd; sexually promiscuous. **2** capricious; random; arbitrary; motiveless (*wanton destruction; wanton wind*). **3** luxuriant; unrestrained; unruly (*wanton extravagance; wanton behaviour*). **4** archaic playful; sportive (*a wanton child*). ● n. literary an immoral or licentious person, esp. a woman. ● v.intr. literary **1** gambol; sport; move capriciously. **2** (foll. by with) behave licentiously. □ **wantonly** adv. **wantonness** /-tənnɪs/ n. [ME *wantowen* (wan- UN-[1] + *towen* f. OE *togen* past part. of *tēon* discipline, rel. to TEAM)]

wapentake /ˈwɒp(ə)nˌteɪk, ˈwæp-/ n. Brit. hist. a subdivision of certain northern and midland counties of England; a hundred. [OE *wǽpen(ge)tæc* f. ON *vápnatak* f. *vápn* weapon + *tak* taking f. *taka* TAKE: perh. with ref. to voting in assembly by show of weapons]

wapiti /'wɒpɪtɪ/ *n.* (*pl.* **wapitis**) a large North American deer, *Cervus canadensis*, now often regarded as a large race of the red deer. [Cree *wapitik* white deer]

War. *abbr.* Warwickshire.

war /wɔː(r)/ *n. & v.* ● *n.* **1 a** armed hostilities between esp. nations; conflict (*war broke out*). **b** a specific conflict or the period of time during which such conflict exists (*was before the war*). **c** the suspension of international law etc. during such a conflict. **2** (as **the War**) a war in progress or recently ended; the most recent major war. **3 a** a hostility or contention between people, groups, etc. (*war of words*). **b** (often foll. by *on*, *against*) a sustained campaign against crime, disease, poverty, etc. ● *v.intr.* (**warred, warring**) **1** (as **warring** *adj.*) a rival; fighting (*warring factions*). **b** conflicting (*warring principles*). **2** make war. □ **art of war** strategy and tactics. **at war** (often foll. by *with*) engaged in a war. **go to war 1** declare or begin a war. **2** (of a soldier etc.) see active service. **go to the wars** *archaic* serve as a soldier. **have been in the wars** *colloq.* appear injured, bruised, unkempt, etc. **war baby** a child, esp. illegitimate, born in wartime. **war bride** a woman who marries a serviceman met during a war. **war chest** funds for a war or any other campaign. **war-cloud** a threatening international situation. **war correspondent** a correspondent reporting from a scene of war. **war crime** a crime violating the international laws of war. **war criminal** a person committing or sentenced for such crimes. **war cry 1** a phrase or name shouted to rally one's troops. **2** a party slogan etc. **war damage** damage to property etc. caused by bombing, shelling, etc. **war dance** a dance performed before a battle or to celebrate victory. **war department** the state office in charge of the army etc. **war-game 1** a military exercise testing or improving tactical knowledge etc. **2** a battle etc. conducted with toy soldiers. **war-gaming** the playing of war-games. **war grave** the grave of a serviceman who died on active service, esp. one in a special cemetery etc. **war loan** stock issued by the British government to raise funds in wartime. **war memorial** a monument etc. commemorating those killed in a war. **war of attrition** a war in which each side seeks to wear out the other over a long period. **war of the elements** *poet.* storms or natural catastrophes. **war of nerves** an attempt to wear down an opponent by psychological means. **war pension** a pension paid to someone disabled or widowed by war. **war-plane** a military aircraft. **war poet** a poet writing on war themes, esp. of the two world wars. **war-weary** (esp. of a population) exhausted and dispirited by war. **war widow** a woman whose husband has been killed in war. **war-worn** = *war-weary*. **war zone** an area in which a war takes place. [ME *werre* f. AF, ONF var. of OF *guerre*: cf. WORSE]

waratah /'wɒrətə/ *n.* an Australian crimson-flowered shrub, *Telopea speciosissima*. [Dharuk *warrada*]

warb /wɔːb/ *n. Austral. sl.* an idle, unkempt, or disreputable person. [20th c.: orig. unkn.]

Warbeck /'wɔːbek/, Perkin (1474–99), Flemish claimant to the English throne. Encouraged by Yorkists in England and on the Continent, he claimed to be Richard, Duke of York (see PRINCES IN THE TOWER) in an attempt to overthrow Henry VII. After a series of attempts to enter the country and begin a revolt he was captured and imprisoned in the Tower of London in 1497; he was later executed.

warble[1] /'wɔːb(ə)l/ *v. & n.* ● *v.* **1** *intr. & tr.* sing in a gentle trilling manner. **2** *tr.* **a** speak or utter in a warbling manner. **b** express in a song or verse (*warbled his love*). ● *n.* a warbled song or utterance. [ME f. ONF *werble(r)* f. Frank. *hwirbilōn* whirl, trill]

warble[2] /'wɔːb(ə)l/ *n.* **1** a hard lump on a horse's back caused by the galling of a saddle. **2 a** a swelling or abscess caused by the larva of a warble fly beneath the skin of cattle etc. **b** the larva that causes this. □ **warble fly** a large fly of the genus *Hypoderma*, with parasitic larvae that burrow under the skin of cattle and horses. [16th c.: orig. uncert.]

warbler /'wɔːblə(r)/ *n.* **1** a person, bird, etc. that warbles. **2 a** a small insect-eating Old World bird of the family Sylviidae, which includes the blackcap, whitethroat, and chiffchaff, usu. having drab plumage and a warbling song. **b** any other small bird with a warbling song, esp. (*US*) of the family Parulidae, or (*Austral.*) of the family Acanthizidae or Maluridae.

Warburg[1] /'wɔːbɜːg/, Aby (Moritz) (1866–1929), German art historian. From 1905 he built up a library in Hamburg, dedicated to preserving the classical heritage of Western culture. It became part of the new University of Hamburg in 1919; four years after his death, it was

transferred to England and housed in the Warburg Institute (part of the University of London).

Warburg[2] /'wɔːbɜːg/, Otto Heinrich (1883–1970), German biochemist. He pioneered the use of the techniques of chemistry for biochemical investigations, especially for his chief work on intracellular respiration. He devised a manometer for this research, enabling him to study the action of respiratory enzymes and poisons in detail. Warburg was awarded a Nobel Prize in 1931, but the Hitler regime prevented him from accepting a second one in 1944 because of his Jewish ancestry.

warby /'wɔːbɪ/ *adj. Austral. sl.* shabby, decrepit. [f. WARB]

Ward /wɔːd/, Mrs Humphry (née Mary Augusta Arnold) (1851–1920), English writer and anti-suffrage campaigner. The niece of Matthew Arnold, she is best known for several novels dealing with social and religious themes, especially *Robert Elsmere* (1888). Although she supported higher education for women, she was an active opponent of the women's suffrage movement, becoming the first president of the Anti-Suffrage League in 1908.

ward /wɔːd/ *n. & v.* ● *n.* **1** a separate room or division of a hospital, prison, etc. (*maternity ward*). **2 a** *Brit.* an administrative division of a constituency, usu. electing a councillor or councillors etc. **b** esp. *US* a similar administrative division. **3 a** a minor under the care of a guardian appointed by the parents or a court. **b** (in full **ward of court**) a minor or mentally deficient person placed under the protection of a court. **4** (in *pl.*) the corresponding notches and projections in a key and a lock. **5** *archaic* **a** the act of guarding or defending a place etc. **b** the bailey of a castle. **c** a guardian's control; confinement; custody. ● *v.tr. archaic* guard; protect. □ **ward-heeler** *US* a party worker in elections etc. **ward off 1** parry (a blow). **2** avert (danger, poverty, etc.). [OE *weard*, *weardian* f. Gmc: cf. GUARD]

-ward /wəd/ *suffix* (also **-wards** /wədz/) added to nouns of place or destination and to adverbs of direction and forming: **1** adverbs (usu. **-wards**) meaning 'towards the place etc.' (*moving backwards; set off homewards*). **2** adjectives (usu. **-ward**) meaning 'turned or tending towards' (*a downward look; an onward rush*). **3** (less commonly) nouns meaning 'the region towards or about' (*look to the eastward*). [from or after OE *-weard* f. a Gmc root meaning 'turn']

warden /'wɔːd(ə)n/ *n.* **1** (usu. in *comb.*) a supervising official (*churchwarden; traffic warden*). **2 a** *Brit.* a president or governor of a college, school, hospital, youth hostel, etc. **b** esp. *US* a prison governor. □ **wardenship** *n.* [ME f. AF & ONF *wardein* var. of OF *g(u)arden* GUARDIAN]

warder /'wɔːdə(r)/ *n.* **1** *Brit.* (*fem.* **wardress** /-drɪs/) a prison officer. **2** a guard. [ME f. AF *wardere*, *-our* f. ONF *warder*, OF *garder* to GUARD]

wardrobe /'wɔːdrəʊb/ *n.* **1** a large movable or built-in cupboard with rails, shelves, hooks, etc., for storing clothes. **2** a person's entire stock of clothes. **3** the costume department or costumes of a theatre, a film company, etc. **4** a department of a royal household in charge of clothing. □ **wardrobe mistress** (or **master**) a person in charge of a theatrical or film wardrobe. **wardrobe trunk** a trunk fitted with rails, shelves, etc. for use as a travelling wardrobe. [ME f. ONF *warderobe*, OF *garderobe* (as GUARD, ROBE)]

wardroom /'wɔːdruːm, -rʊm/ *n.* a room in a warship for the use of commissioned officers.

-wards var. of -WARD.

wardship /'wɔːdʃɪp/ *n.* **1** a guardian's care or tutelage (*under his wardship*). **2** the condition of being a ward.

ware[1] /weə(r)/ *n.* **1** (esp. in *comb.*) things of the same kind, esp. ceramics, made usu. for sale (*chinaware; hardware*). **2** (usu. in *pl.*) **a** articles for sale (*displayed his wares*). **b** a person's skills, talents, etc. **3** ceramics etc. of a specified material, factory, or kind (*Wedgwood ware; Delft ware*). [OE *waru* f. Gmc, perh. orig. = 'object of care', rel. to WARE[3]]

ware[2] /weə(r)/ *v.tr.* (also **'ware**) (esp. in hunting) look out for; avoid (usu. in *imper.*: *ware hounds!*). [OE *warian* f. Gmc (as WARE[3]), & f. ONF *warer*]

ware[3] /weə(r)/ *predic.adj. poet.* aware. [OE *wær* f. Gmc: cf. WARD]

warehouse /'weəhaʊs/ *n. & v.* ● *n.* **1** a building in which esp. retail goods are stored; a repository. **2** esp. *Brit.* a wholesale or large retail store. ● *v.tr.* /'weəhaʊz/ store (esp. furniture or bonded goods) temporarily in a repository. □ **warehouse party** a large, usu. illegal, organized public party with dancing, held in a warehouse or similar large building. □ **warehouseman** *n.* (*pl.* **-men**).

warfare /'wɔːfeə(r)/ *n.* a state of war; campaigning, engaging in war (*chemical warfare*).

warfarin /ˈwɔːfərɪn/ n. a water-soluble anticoagulant used as a rat poison and in the treatment of thrombosis. [Wisconsin Alumni Research Foundation + -arin, after COUMARIN]

warhead /ˈwɔːhed/ n. the explosive head of a missile, torpedo, or similar weapon.

Warhol /ˈwɔːhəʊl/ Andy (born Andrew Warhola) (c.1928–87), American painter, graphic artist, and film-maker. His work played a definitive role in New York pop art of the 1960s. Warhol's background in commercial art and advertising illustration was central to pop art's concern with the imagery of the mass media; his famous statement 'I like boring things' was expressed in the standardized, consciously banal, nature of his work. In the early 1960s he achieved fame for a series of silk-screen prints and acrylic paintings of familiar objects (such as Campbell's soup tins), car accidents, and Marilyn Monroe, treated with objectivity and precision. From 1965, Warhol increasingly devoted himself to film, playing an important part in the emergence of the new American underground cinema. He also managed the rock group Velvet Underground.

warhorse /ˈwɔːhɔːs/ n. **1** hist. a knight's or trooper's powerful horse. **2** colloq. a veteran soldier, politician, etc.; a reliable hack.

Warka see URUK.

warlike /ˈwɔːlaɪk/ adj. **1** threatening war; hostile. **2** martial; soldierly. **3** of or for war; military (warlike preparations).

warlock /ˈwɔːlɒk/ n. archaic a sorcerer or wizard. [OE wǣr-loga traitor f. wǣr covenant: loga rel. to LIE²]

warlord /ˈwɔːlɔːd/ n. a military commander or commander-in-chief; esp. a regional military ruler in China during the early 20th century holding power independently of central government. After the revolution of 1911 and the deposition of the last emperor in 1912 and then the death of the republic's first President (1916), China was divided between rival warlords until Chiang Kai-shek's Kuomintang re-established central control in 1928; after this, warlords held on in some remote western areas until the 1940s.

warm /wɔːm/ adj., v., & n. ● adj. **1** of or at a fairly or comfortably high temperature. **2** (of clothes etc.) affording warmth (needs warm gloves). **3 a** (of a person, action, feelings, etc.) sympathetic; cordial; friendly; loving (a warm welcome; has a warm heart). **b** enthusiastic; hearty (was warm in her praise). **4** animated, heated, excited. **5** dangerous, difficult, or hostile. **6** colloq. **a** (of a participant in esp. a children's game of seeking) close to the object etc. sought. **b** near to guessing or finding out a secret. **7** (of a colour, light, etc.) reddish, pink, or yellowish, etc., suggestive of warmth. **8** Hunting (of a scent) fresh and strong. **9 a** (of a person's temperament) amorous; sexually demanding. **b** erotic; arousing. ● v. **1** tr. **a** make warm (warm your feet by the fire). **b** excite; make cheerful (warms the heart). **2** intr. **a** (often foll. by up) warm oneself at a fire etc. (warmed himself up). **b** (often foll. by to) become animated, enthusiastic, or sympathetic (warmed to his subject). ● n. **1** the act of warming; the state of being warmed (gave it a warm; had a nice warm by the fire). **2** the warmth of the atmosphere etc. **3** Brit. archaic a warm garment, esp. an army greatcoat. □ **warmed-up** (N. Amer. **-over**) **1** (of food etc.) reheated. **2** stale; second-hand. **warm front** an advancing mass of warm air. **warming-pan** hist. a usu. brass container for live coals with a flat body and a long handle, used for warming a bed. **warm up 1** (of an athlete, performer, etc.) prepare for a contest, performance, etc. by practising. **2** become or cause to become warmer (the room warmed up gradually; warm your feet up). **3** (of a person) become enthusiastic etc. **4** (of a radio, engine, etc.) reach a temperature for efficient working. **5** reheat (food). **warm-up** n. a period of preparatory exercise for a contest or performance. **warm work 1** work etc. that makes one warm through exertion. **2** dangerous conflict etc. □ **warmer** n. (also in comb.). **warmish** adj. **warmly** adv. **warmness** n. **warmth** n. [OE wearm f. Gmc]

warm-blooded /wɔːmˈblʌdɪd/ adj. **1** (of an organism) having warm blood; mammalian (see HOMEOTHERM). **2** ardent, passionate. □ **warm-bloodedness** n.

warm-hearted /wɔːmˈhɑːtɪd/ adj. having a warm heart; kind, friendly. □ **warm-heartedly** adv. **warm-heartedness** n.

warmonger /ˈwɔːˌmʌŋgə(r)/ n. a person who seeks to bring about or promote war. □ **warmongering** n. & adj.

warn /wɔːn/ v.tr. **1** (also absol.) **a** (often foll. by of, or that + clause, or to + infin.) inform of danger, unknown circumstances, etc. (warned them of the danger; warned her that she was being watched; warned him to expect a visit). **b** (often foll. by against) inform (a person etc.) about a specific danger, hostile person, etc. (warned her against trusting him). **2** (usu. with neg.) admonish; tell forcefully (has been warned not to go). **3** give (a person) cautionary notice regarding conduct etc. (shall not warn you again). □ **warn off 1** tell (a person) to keep away (from). **2** prohibit from attending races, esp. at a specified course. □ **warner** n. [OE war(e)nian, wearnian ult. f. Gmc: cf. WARE³]

Warner Brothers /ˈwɔːnə(r)/ a US film production company founded in 1923 by the brothers Harry, Jack, Sam, and Albert Warner. The company produced the first full-length sound film, The Jazz Singer, in 1927 and went on to release successful gangster films with Humphrey Bogart, and Busby Berkeley musicals in the 1930s and 1940s.

warning /ˈwɔːnɪŋ/ n. & adj. ● n. **1** in senses of WARN v. **2** anything that serves to warn; a hint or indication of difficulty, danger, etc. **3** archaic = NOTICE n. 3b. ● adj. serving to warn. □ **warning coloration** Biol. conspicuous colouring that warns a predator etc. against attacking. □ **warningly** adv. [OE war(e)nung etc. (as WARN, -ING¹)]

War of 1812 a conflict between the US and the UK 1812–14. The war was prompted by restrictions on US trade because of the British blockade of French and allied ports during the Napoleonic Wars, and by British and Canadian support for American Indians trying to resist westward expansion. The main area of operations was the Canadian border, which US forces tried repeatedly but unsuccessfully to breach. The Americans captured Detroit and won a series of engagements between single ships, while the British took Washington and burned public buildings (including the White House), but the war was eventually terminated by a treaty which restored all conquered territories to their owners before outbreak of war.

War of American Independence see AMERICAN INDEPENDENCE, WAR OF.

War Office n. hist. the British state department in charge of the army.

War of Jenkins's Ear see JENKINS's EAR, WAR OF.

warp /wɔːp/ v. & n. ● v. **1** tr. & intr. **a** make or become bent or twisted out of shape, esp. by the action of heat, damp, etc. **b** make or become perverted, bitter, or strange (a warped sense of humour). **2 a** tr. haul (a ship) by a rope attached to a fixed point. **b** intr. progress in this way. **3** tr. silt over (land) by flooding. **4** tr. (foll. by up) choke (a channel) with an alluvial deposit etc. **5** tr. arrange (threads) as a warp. ● n. **1 a** a state of being warped, esp. of shrunken or expanded timber. **b** perversion, bitterness, etc. of the mind or character. **2** the threads stretched lengthwise in a loom to be crossed by the weft. **3** a rope used in towing or warping, or attached to a trawl-net. **4** sediment etc. left esp. on poor land by standing water. □ **time warp** see TIME. □ **warpage** n. (esp. in sense 1a of v.). **warper** n. (in sense 5 of v.). [OE weorpan throw, wearp f. Gmc]

warpaint /ˈwɔːpeɪnt/ n. **1** paint used to adorn the body before battle, esp. by North American Indians. **2** colloq. elaborate make-up.

warpath /ˈwɔːpɑːθ/ n. **1** a warlike expedition of North American Indians. **2** colloq. any hostile course or attitude (is on the warpath again).

warragal var. of WARRIGAL.

warrant /ˈwɒrənt/ n. & v. ● n. **1 a** anything that authorizes a person or an action (have no warrant for this). **b** a person so authorizing (I will be your warrant). **2 a** a written authorization, money voucher, travel document, etc. (a dividend warrant). **b** a written authorization allowing police to search premises, arrest a suspect, etc. **3** a document authorizing counsel to represent the principal in a lawsuit (warrant of attorney). **4** a certificate of service rank held by a warrant-officer. ● v.tr. **1** serve as a warrant for; justify (nothing can warrant his behaviour). **2** guarantee or attest to esp. the genuineness of an article, the worth of a person, etc. □ **I** (or **I'll**) **warrant** I am certain; no doubt (he'll be sorry, I'll warrant). **warrant-officer** an officer ranking between commissioned officers and NCOs. □ **warranter** n. **warrantor** /ˌwɒrənˈtɔː(r)/ n. [ME f. ONF warant, var. of OF guarant, -and f. Frank. werēnd (unrecorded) f. giwerēn be surety for]

warrantable /ˈwɒrəntəb(ə)l/ adj. **1** able to be warranted. **2** (of a stag) old enough to be hunted (5 or 6 years). □ **warrantableness** n. **warrantably** adv.

warrantee /ˌwɒrənˈtiː/ n. a person to whom a warranty is given.

warranty /ˈwɒrəntɪ/ n. (pl. **-ies**) **1** an undertaking as to the ownership or quality of a thing sold, hired, etc., often accepting responsibility for defects or liability for repairs needed over a specified period. **2** (usu. foll. by for + verbal noun) an authority or justification. **3** an undertaking by an insured person of the truth of a statement or fulfilment of a condition. [ME f. AF warantie, var. of garantie (as WARRANT)]

Warren¹ /ˈwɒrən/, Earl (1891–1974), American judge. During his time as Chief Justice of the US Supreme Court (1953–69) he did much to improve civil liberties, including the prohibition of segregation in US schools, effected by his decision of 1954. He is also remembered for heading the commission of inquiry (known as the Warren Commission) held in 1964 into the assassination of President Kennedy; the commission found that Lee Harvey Oswald was the sole gunman, a decision that has since been much disputed.

Warren[2] /'wɒrən/, Robert Penn (1905–89), American poet, novelist, and critic. An advocate of New Criticism, he collaborated with the American critic Cleanth Brooks (1906–94) in writing such critical works as *Understanding Poetry* (1938) and *Understanding Fiction* (1943). Warren also wrote several novels, including *All the King's Men* (1946), and many volumes of poetry; he became the first to win Pulitzer Prizes in both fiction and poetry categories and in 1986 he was made the first American Poet Laureate.

warren /'wɒrən/ n. **1 a** a network of interconnecting rabbit burrows. **b** a piece of ground occupied by this. **2** a densely populated or labyrinthine building or district. **3** *hist.* a piece of ground on which game is preserved. [ME f. AF & ONF *warenne*, OF *garenne* game-park f. Gmc]

warrigal /'wɒrɪg(ə)l/ n. & adj. (also **warragal** /'wɒrəg-/) *Austral.* ● n. **1** a dingo dog. **2** an untamed horse. **3** a wild Aboriginal. ● adj. wild, untamed. [Dharuk]

warring /'wɔːrɪŋ/ adj. rival, antagonistic.

Warrington /'wɒrɪŋtən/ an industrial town on the River Mersey in Cheshire, central England; pop. (1981) 82,520.

warrior /'wɒrɪə(r)/ n. **1** a person experienced or distinguished in fighting. **2** a courageous or aggressive person. **3** (attrib.) of or characteristic of a warrior; martial (*a warrior nation*). [ME f. ONF *werreior* etc., OF *guerreior* etc. f. *werreier*, *guerreier* make WAR]

Warsaw /'wɔːsɔː/ (Polish **Warszawa** /var'ʃava/) the capital of Poland, on the River Vistula; pop. (1990) 1,655,660. The city suffered severe damage and the loss of 700,000 lives during the Second World War and was almost completely rebuilt.

Warsaw Pact a treaty of mutual defence and military aid signed at Warsaw on 14 May 1955 by Communist states of Europe under Soviet influence, in response to the admission of West Germany to NATO; collectively, the group of states which signed the treaty. Following changes in eastern Europe and the collapse of the Communist system the Pact was dissolved in 1991.

warship /'wɔːʃɪp/ n. an armoured ship used in war.

Wars of Religion see FRENCH WARS OF RELIGION.

Wars of the Roses the civil wars in England during the 15th century arising from the dynastic struggle between the followers of the House of York (with the white rose as its emblem) and the House of Lancaster (with the red rose) during the reigns of Henry VI, Edward IV, and Richard III. The struggle was largely ended in 1485 by the defeat and death of the Yorkist king Richard III at the Battle of Bosworth and the accession of the Lancastrian Henry Tudor (Henry VII), who united the two houses by marrying Elizabeth, daughter of Edward IV. The name *Wars of the Roses* was popularized by Sir Walter Scott in the 19th century.

wart /wɔːt/ n. **1** a small benign growth on the skin, usu. hard and rounded, caused by a virus-induced abnormal growth of skin cells and thickening of the epidermis. **2** a protuberance on the skin of an animal, surface of a plant, etc. **3** *colloq.* an objectionable person. □ **warts and all** *colloq.* with no attempt to conceal blemishes or inadequacies. □ **warty** adj. [OE *wearte* f. Gmc]

warthog /'wɔːthɒg/ n. an African wild pig, *Phacochoerus aethiopicus*, with a large head and warty lumps on its face, and large curved tusks.

wartime /'wɔːtaɪm/ n. the period during which a war is waged.

Warwick[1] /'wɒrɪk/ the county town of Warwickshire, in central England, on the River Avon; pop. (1981) 21,990.

Warwick[2] /'wɒrɪk/, Earl of (title of Richard Neville; known as 'the Kingmaker') (1428–71), English statesman. During the Wars of the Roses he fought first on the Yorkist side, helping Edward IV to gain the throne in 1461. Having lost influence at court he then fought on the Lancastrian side, briefly restoring Henry VI to the throne in 1470. Warwick was killed at the battle of Barnet.

Warwickshire /'wɒrɪk,ʃɪə(r)/ a county of central England; county town, Warwick.

wary /'weərɪ/ adj. (**warier, wariest**) **1** on one's guard; given to caution; circumspect. **2** (foll. by *of*) cautious, suspicious (*am wary of using lifts*). **3** showing or done with caution or suspicion (*a wary expression*). □ **warily** adv. **wariness** n. [WARE[2] + -Y[1]]

was *1st* & *3rd sing.* past of BE.

Wash. abbr. Washington.

wash /wɒʃ/ v. & n. ● v. **1** tr. cleanse (oneself or a part of oneself, clothes, etc.) with liquid, esp. water. **2** tr. (foll. by *out, off, away*, etc.) remove (a stain or dirt, a surface, or some physical feature of the surface) in this way; eradicate all traces of. **3** *intr.* wash oneself or esp. one's hands and face. **4** *intr.* wash clothes etc. **5** *intr.* (of fabric or dye) bear washing without damage. **6** *intr.* (foll. by *off, out*) (of a stain etc.) be removed by washing. **7** tr. *poet.* moisten, water (*tear-washed eyes; a rose washed with dew*). **8** tr. (of a river, sea, etc.) touch (a country, coast, etc.) with its waters. **9** a tr. (of moving liquid) carry along in a specified direction (*a wave washed him overboard; was washed up on the shore*). **b** tr. be carried thus (*shells wash up on the beaches*). **10** tr. (also foll. by *away, out*) **a** scoop out (*the water had washed a channel*). **b** erode, denude (*sea-washed cliffs*). **11** *intr.* (foll. by *over, along*, etc.) sweep, move, or splash. **12** *intr.* (foll. by *over*) occur all around without greatly affecting (a person). **13** tr. sift (ore) by the action of water. **14** tr. **a** brush a thin coat of watery paint or ink over (paper in watercolour painting etc., or a wall). **b** (foll. by *with*) coat (inferior metal) with gold etc. ● n. **1 a** the act or an instance of washing; the process of being washed (*give them a good wash; only needed one wash*). **b** (prec. by *the*) treatment at a laundry etc. (*sent them to the wash*). **2** a quantity of clothes for washing or just washed. **3** the visible or audible motion of agitated water or air, esp. due to the passage of a ship etc. or aircraft. **4 a** soil swept off by water; alluvium. **b** a sandbank exposed only at low tide. **5** kitchen slops and scraps given to pigs. **6 a** thin, weak, or inferior liquid food. **b** liquid food for animals. **7** a liquid to spread over a surface to cleanse, heal, or colour. **8** a thin coating of watercolour, wall-colouring, or metal. **9** malt etc. fermenting before distillation. **10** a lotion or cosmetic. □ **come out in the wash** *colloq.* be clarified, or (of contingent difficulties) be resolved or removed, in the course of time. **wash-and-wear** adj. (of a fabric or garment) easily and quickly laundered. **wash-basin** a basin for washing one's hands, face, etc., esp. fixed to a wall and connected to a water supply and a drain. **wash one's dirty linen in public** see LINEN. **wash down 1** wash completely (esp. a large surface or object). **2** (usu. foll. by *with*) accompany or follow (food) with a drink. **washed out 1** faded by washing. **2** pale. **3** *colloq.* limp, enfeebled. **washed up** esp. *N. Amer. sl.* defeated, having failed. **wash one's hands** *euphem.* go to the lavatory. **wash one's hands of** renounce responsibility for (orig. with allusion to Pontius Pilate in Matt. 27:24). **wash-hand stand** = WASHSTAND. **wash-house** a building where clothes are washed. **wash-leather** chamois or similar leather for washing windows etc. **wash one's mouth out with soap** suffer punishment or rebuke (orig. literally by washing out the mouth), esp. for swearing or bad language. **wash out 1** clean the inside of (a thing) by washing. **2** clean (a garment etc.) by brief washing. **3 a** call off (an event), esp. owing to rain. **b** *colloq.* cancel. **4** (of a flood, downpour, etc.) make a breach in (a road etc.). **5** = sense 2 of v. **wash-out** n. **1** *colloq.* a fiasco; a complete failure. **2** a breach in a road, railway track, etc., caused by flooding (see also WASHOUT). **wash up 1** tr. (also *absol.*) esp. *Brit.* wash (crockery and cutlery) after use. **2** *intr.* *N. Amer.* wash one's face and hands. **won't wash** esp. *Brit. colloq.* (of an argument etc.) will not be believed or accepted. [OE *wæscan* etc. f. Gmc, rel. to WATER]

Wash, the an inlet of the North Sea on the east coast of England between Norfolk and Lincolnshire.

washable /'wɒʃəb(ə)l/ adj. that can be washed, esp. without damage. □ **washability** /,wɒʃə'bɪlɪtɪ/ n.

washbag /'wɒʃbæg/ n. a waterproof bag for toilet articles.

washboard /'wɒʃbɔːd/ n. **1** a board of ribbed wood or a sheet of corrugated zinc on which clothes are scrubbed in washing. **2** this used as a percussion instrument, played with the fingers.

washday /'wɒʃdeɪ/ n. a day on which clothes etc. are washed.

washer /'wɒʃə(r)/ n. **1 a** a person or thing that washes. **b** a washing-machine. **2** a flat ring of rubber, metal, leather, etc., inserted at a joint to tighten it and prevent leakage. **3** a similar ring placed under the head of a screw, bolt, etc., or under a nut, to disperse its pressure. **4** *Austral.* a cloth for washing the face. □ **washer-drier** (or **-dryer**) a washing machine with an inbuilt tumble-drier. **washer-up** (pl. **washers-up**) a person who washes up dishes etc.

washerwoman /'wɒʃə,wʊmən/ n. (pl. **-women**) a woman whose occupation is washing clothes; a laundress.

washeteria /,wɒʃə'tɪərɪə/ n. = LAUNDERETTE.

washing /'wɒʃɪŋ/ n. **1** a quantity of clothes for washing or just washed. **2** the act of washing clothes etc. □ **washing machine** a machine for washing clothes and linen etc. **washing-powder** powder of soap or detergent for washing clothes. **washing soda** sodium carbonate, used dissolved in water for washing and cleaning. **washing-up** *Brit.* **1** the process of washing dishes etc. after use. **2** used dishes etc. for washing.

Washington[1] /'wɒʃɪŋtən/ **1** a state of the north-western US, on the Pacific coast; pop. (1990) 4,866,690; capital, Olympia. It became the 42nd state in 1889. **2** (in full **Washington, DC**) the capital of the US; pop. (1990) 606,900. It is coextensive with the District of Columbia, a Federal district on the Potomac river with boundaries on the states of Virginia and Maryland. It was founded in 1790, during the presidency of George Washington, after whom it is named. Washington is one of the few capital cities that was planned and built as such: many of its most distinctive features, including the complex of parks and circular roads and the impressive siting of the Capitol, date from the earliest designs by the French engineer Pierre-Charles L'Enfant (1754–1825). **3** an industrial town in Tyne and Wear, NE England; pop. (1981) 48,830. It was designated as a new town in 1964. The original village of Washington was the ancestral home of George Washington, first President of the United States. □ **Washingtonian** /ˌwɒʃɪŋ'təʊnɪən/ n. & adj.

Washington[2] /'wɒʃɪŋtən/, Booker T(aliaferro) (1856–1915), American educationist. An emancipated slave, he pursued a career in teaching and was appointed head in 1881 of the newly founded Tuskegee Institute in Alabama for the training of black teachers. Washington emerged as a leading commentator for black Americans at the turn of the century and published his influential autobiography, *Up from Slavery*, in 1901. His emphasis on vocational skills and financial independence for blacks rather than on intellectual development or political rights, combined with his support for segregation, brought harsh criticism from other black leaders such as W. E. B. Du Bois, and from civil-rights activists of the 20th century.

Washington[3] /'wɒʃɪŋtən/, George (1732–99), American soldier and statesman, 1st President of the US 1789–97. After serving as a soldier 1754–9 in the war against the French, Washington took part in the Continental Congresses of 1774 and 1775, and in 1775 was chosen as commander of the Continental Army. He served in that capacity throughout the War of Independence, bringing about the eventual American victory by keeping the army together through the bitter winter of 1777–8 at Valley Forge and winning a decisive battle at Yorktown (1781). Washington chaired the convention at Philadelphia (1787) that drew up the American Constitution and, two years later, he was unanimously elected President, initially remaining unaligned to any of the newly emerging political parties but later joining the Federalist Party. He served two terms, following a policy of neutrality in international affairs, before declining a third term and retiring to private life.

washland /'wɒʃlænd/ n. land periodically flooded by a stream.

washout /'wɒʃaʊt/ n. *Geol.* a narrow river-channel that cuts into pre-existing sediments (see also *wash-out*).

washroom /'wɒʃruːm, -rʊm/ n. *N. Amer.* a room with washing and toilet facilities.

washstand /'wɒʃstænd/ n. *hist.* a piece of furniture to hold a jug of water, a basin, soap, etc. for washing oneself with.

washtub /'wɒʃtʌb/ n. a tub or vessel for washing clothes etc.

washy /'wɒʃɪ/ adj. (**washier, washiest**) **1** (of liquid food) too watery or weak; insipid. **2** (of colour) faded-looking, thin, faint. **3** (of a style, sentiment, etc.) lacking vigour or intensity. □ **washily** adv. **washiness** n.

wasn't /'wɒz(ə)nt/ contr. was not.

Wasp /wɒsp/ n. (also **WASP**) *N. Amer.* usu. *derog.* a middle-class American white Protestant descended from early European settlers. □ **Waspy** adj. [White Anglo-Saxon Protestant]

wasp /wɒsp/ n. **1** a stinging insect of the order Hymenoptera; esp. a social species of the genus *Vespula*, having black and yellow stripes and feeding on other insects and sweet material. **2** (in comb.) any insect resembling a wasp in some way (*wasp beetle*). □ **wasp-waist** a very slender waist. **wasp-waisted** having a very slender waist. □ **wasplike** adj. [OE wæfs, wæps, wæsp, f. WG: perh. rel. to WEAVE[1] (from the weblike form of its nest)]

waspish /'wɒspɪʃ/ adj. irritable, petulant; sharp in retort. □ **waspishly** adv. **waspishness** n.

wassail /'wɒseɪl, -s(ə)l/ n. & v. archaic ● n. **1** a festive occasion; a drinking-bout. **2** a kind of liquor drunk on such an occasion. ● v.intr. make merry; celebrate with drinking etc. □ **wassail-bowl** (or **-cup**) a bowl or cup

from which healths were drunk, esp. on Christmas Eve and Twelfth Night. □ **wassailer** n. [ME wæs hæil etc. f. ON ves heill, corresp. to OE wes hāl 'be in health', a form of salutation: cf. HALE[1]]

Wassermann test /'væsəmən/ n. *Med.* a diagnostic test for syphilis which uses a specific antibody reaction in the patient's blood serum. [A. von *Wassermann*, Ger. pathologist (1866–1925)]

wast /wɒst, wəst/ archaic or dial. 2nd sing. past of BE.

wastage /'weɪstɪdʒ/ n. **1** an amount wasted. **2** loss by use, wear, or leakage. **3** (also **natural wastage**) loss of employees other than by redundancy (esp. by retirement or resignation).

waste /weɪst/ v., adj., & n. ● v. **1** tr. use to no purpose or for inadequate result or extravagantly (*waste time*). **2** tr. fail to use (esp. an opportunity). **3** tr. (often foll. by on) **a** give (advice etc.), utter (words etc.), without effect. **b** (often in passive) fail to be appreciated or used properly (*she was wasted on him*). **4** tr. & intr. (usu. foll. by away) wear gradually away; make or become weak; wither. **5** tr. **a** literary ravage, devastate. **b** US sl. murder, kill. **6** tr. treat as wasted or valueless. **7** intr. be expended without useful effect. **8** (as **wasted** adj.) sl. **a** tired, exhausted. **b** intoxicated by alcohol or drugs. ● adj. **1** superfluous; no longer serving a purpose. **2** (of a district etc.) not inhabited or cultivated; desolate (*waste ground*). **3** presenting no features of interest. ● n. **1** the act or an instance of wasting; extravagant or ineffectual use of an asset, of time, etc. **2** waste material or food; refuse; useless remains or by-products. **3** a waste region; a desert etc. **4** the state of being used up; diminution by wear and tear. **5** *Law* damage to an estate caused by an act or by neglect, esp. by a life-tenant. **6** = *waste pipe*. □ **go** (or **run**) **to waste** be wasted. **lay waste** (or **lay to waste**) ravage, devastate. **waste-basket** esp. *N. Amer.* = *waste-paper basket*. **waste bin** a bin for waste paper, refuse, etc., used indoors or outdoors. **waste one's breath** see BREATH. **waste disposal** (often attrib.) the disposing of waste products, rubbish, etc., esp. as a public or corporate process. **waste not, want not** extravagance leads to poverty. **waste paper** used or valueless paper. **waste-paper basket** esp. *Brit.* a receptacle for waste paper. **waste pipe** a pipe to carry off waste water, e.g. from a sink. **waste products** useless by-products of manufacture or of an organism or organisms. **waste words** see WORD. □ **wastable** adj. **wasteless** adj. [ME f. ONF wast(e), var. of OF g(u)ast(e), f. L vastus]

wasteful /'weɪstfʊl/ adj. **1** extravagant. **2** causing or showing waste. □ **wastefully** adv. **wastefulness** n.

wasteland /'weɪstlænd/ n. **1** an unproductive or useless area of land. **2** a place or time considered spiritually or intellectually barren.

waster /'weɪstə(r)/ n. **1** a wasteful person. **2** colloq. a worthless person; an idler.

wastrel /'weɪstrəl/ n. **1** a wasteful or good-for-nothing person. **2** a waif; a neglected child.

watch /wɒtʃ/ v. & n. ● v. **1** tr. keep the eyes fixed on; observe by looking at; look at attentively. **2** tr. **a** keep under observation; follow observantly. **b** monitor or consider carefully; pay attention to (*have to watch my weight; watched their progress with interest*). **3** intr. (often foll. by for) be in an alert state; be vigilant; take heed (*watch for the holes in the road; watch for an opportunity*). **4** intr. (foll. by over) look after; take care of. ● n. **1** a small portable timepiece for carrying on one's person. (See note below.) **2** a state of alert or constant observation or attention. **3** *Naut.* **a** usu. four-hour spell of duty. **b** (in full **starboard** or **port watch**) each of the halves, divided according to the position of the bunks, into which a ship's crew is divided to take alternate watches. **4** *hist.* a watchman or group of watchmen, esp. patrolling the streets at night. **5** *hist.* a band of irregular Highland troops in the 18th century. □ **on the watch** (or **on watch**) waiting for an expected or feared occurrence. **set the watch** *Naut.* station sentinels etc. **watch-case** the outer metal case enclosing the works of a watch. **watch-chain** a metal chain for securing a pocket-watch. **watch-glass 1** a glass disc covering the dial of a watch. **2** a similar disc used in a laboratory etc. to hold material for use in experiments. **watching brief** see BRIEF. **watch it** (or **oneself**) colloq. be careful. **watch-night 1** the last night of the year. **2** a religious service held on this night. **watch out 1** (often foll. by for) be on one's guard. **2** as a warning of immediate danger. **watch-spring** the mainspring of a watch. **watch one's step** proceed cautiously. **watch-strap** esp. *Brit.* a strap for fastening a watch on the wrist. **watch-tower** a tower from which observation can be kept. □ **watchable** adj. **watcher** n. (also in comb.). [OE wæcce (n.), rel. to WAKE[1]]

 ▪ Portable clocks first appeared in the early 16th century, but pocket or fob watches did not become practicable until the late 17th century with the advent of the coiled mainspring and the balance spring.

Most mechanical watches now use the lever escapement invented by Thomas Mudge (1717–94) in 1765. Wrist-watches first appeared in the 1880s and self-winding watches in the 1920s. The second half of the 20th century has seen the introduction of both electric and electronic watches, which now achieve great accuracy from the vibrations of a quartz crystal. Electronics has enabled the use of digital readouts and a variety of timing functions. (See also CLOCK[1].)

watchband /'wɒtʃbænd/ n. N. Amer. = watch-strap.

Watch Committee n. hist. (in the UK) the committee of a county borough council dealing with policing etc.

watchdog /'wɒtʃdɒg/ n. & v. ● n. **1** a dog kept to guard property etc. **2** a person or body monitoring others' rights, behaviour, etc. ● v.tr. (-**dogged**, -**dogging**) maintain surveillance over.

watchful /'wɒtʃfʊl/ adj. **1** accustomed to watching. **2** on the watch. **3** showing vigilance. **4** archaic wakeful. □ **watchfully** adv. **watchfulness** n.

watchmaker /'wɒtʃˌmeɪkə(r)/ n. a person who makes and repairs watches and clocks. □ **watchmaking** n.

watchman /'wɒtʃmən/ n. (pl. -**men**) **1** a man employed to look after an empty building etc. at night. **2** archaic or hist. a member of a night-watch.

watchword /'wɒtʃwɜːd/ n. **1** a phrase summarizing a guiding principle; a slogan. **2** hist. a military password.

water /'wɔːtə(r)/ n. & v. ● n. **1** a colourless transparent odourless tasteless liquid compound of oxygen and hydrogen (chem. formula: H_2O). (See note below.) **2** a liquid consisting chiefly of this and found in seas, lakes, and rivers, in rain, and in the fluids of living organisms. **3** an expanse of water; a sea, lake, river, etc. **4** (in pl.) part of a sea or river (in Icelandic waters). **5** (often as **the waters**) mineral water at a spa etc. **6** the state of a tide (high water). **7** a solution of a specified substance in water (lavender-water). **8** the quality of the transparency and brilliance of a gem, esp. a diamond. **9** Finance an amount of nominal capital added by watering (see sense 10 of v.). **10** (attrib.) **a** found in, on, or near water. **b** of, for, or worked by water. **c** involving, using, or yielding water. **11 a** urine. **b** (usu. in pl.) the amniotic fluid discharged from the womb before childbirth. ● v. **1** tr. sprinkle or soak with water. **2** tr. supply (a plant) with water. **3** tr. give water to (an animal) to drink. **4** intr. (of the mouth or eyes) secrete water as saliva or tears. **5** tr. (as **watered** adj.) (of silk etc.) having irregular wavy glossy markings. **6** tr. adulterate (milk, beer, etc.) with water. **7** tr. (of a river etc.) supply (a place) with water. **8** intr. (of an animal) go to a pool etc. to drink. **9** intr. (of a ship, engine, etc., or the person in charge of it) take in a supply of water. **10** tr. Finance increase (a company's debt, or nominal capital) by the issue of new shares without a corresponding addition to assets. □ **by water** using a ship etc. for travel or transport. **cast one's bread upon the waters** see BREAD. **like water** lavishly, profusely. **like water off a duck's back** see DUCK[1]. **make one's mouth water** cause one's saliva to flow, stimulate one's appetite or anticipation. **of the first water 1** (of a diamond) of the greatest brilliance and transparency. **2** of the finest quality or extreme degree. **on the water** on a ship etc. **on the water-wagon** see WAGON. **water-bag** a bag of leather, canvas, etc., for holding water. **water bailiff 1** an official enforcing fishing laws. **2** hist. a custom-house officer at a port. **water-based** (usu. attrib.) **1** (of a solution etc.) having water as the main ingredient. **2** (of a sport) practised on water. **water bear** = TARDIGRADE n. **water-bed** a bed with a mattress of rubber or plastic etc. filled with water. **water-biscuit** a thin crisp unsweetened biscuit made from flour and water. **water blister** a blister containing a colourless fluid, not blood or pus. **water-boatman** an aquatic bug of the family Notonectidae or Corixidae, swimming with oarlike hind legs. **water-borne 1** (of goods etc.) conveyed by or travelling on water. **2** (of a disease) communicated or propagated by contaminated water. **water-buck** a large African antelope, Kobus ellipsiprymnus, frequenting watered districts. **water buffalo** the common domestic Indian buffalo, Bubalus arnee, now rare in the wild. **water bus** a boat carrying passengers on a regular run on a river, lake, etc. **water-butt** a barrel used to catch rainwater. **water-cannon** a device giving a powerful jet of water to disperse a crowd etc. **water chestnut 1** an aquatic plant, Trapa natans, bearing an edible seed. **2 a** (in full **Chinese water chestnut**) a sedge, Eleocharis tuberosa, with rushlike leaves arising from a starchy tuber. **b** this tuber used as food. **water-closet 1** a lavatory with a means for flushing the pan by water. **2** a room containing this. **water-cooled** cooled by the circulation of water. **water-cooler** a tank of cooled drinking-water. **water cure** = HYDROPATHY. **water-diviner** Brit. a person who dowses (see DOWSE[1]) for water. **water down 1** dilute with water. **2** (esp. as **watered-down**

adj.) make less vivid, forceful, or horrifying. **water flea** = DAPHNIA. **water gauge 1** a glass tube etc. indicating the height of water in a reservoir, boiler, etc. **2** pressure expressed in terms of a head of water. **water-glass 1** a solution of sodium or potassium silicate used for preserving eggs, as a vehicle for fresco-painting, and for hardening artificial stone. **2** a tube with a glass bottom enabling objects under water to be observed. **water-hammer** a knocking noise in a water-pipe when a tap is suddenly turned off. **water-heater** a device for heating (esp. domestic) water. **water hemlock** a poisonous umbelliferous plant, Cicuta maculata, found in marshes (also called cowbane). **water-hole** a shallow depression in which water collects (esp. in the bed of a river otherwise dry). **water hyacinth** a tropical American aquatic plant, Eichhornia crassipes, which is a serious weed of waterways in warmer parts of the world. **water-ice** a confection of flavoured and frozen water and sugar etc.; a sorbet. **water jump** a place where a horse in a steeplechase etc. must jump over water. **water-level 1 a** the surface of the water in a reservoir etc. **b** the height of this. **2** a level below which the ground is saturated with water. **3** a level using water to determine the horizontal. **water lily** an aquatic plant of the family Nymphaeaceae, with broad flat floating leaves and large usu. cup-shaped floating flowers. **water-line 1** the line along which the surface of water touches a ship's side (marked on a ship for use in loading). **2** a linear watermark. **water main** the main pipe in a water-supply system. **water-meadow** a meadow periodically flooded by a stream or river. **water measurer** a long thin aquatic bug of the family Hydrometridae which walks slowly on the surface film of water. **water melon** a large smooth green melon, Citrullus lanatus, with red pulp and watery juice. **water meter** a device for measuring and recording the amount of water supplied to a house etc. **water milfoil** see MILFOIL 2. **water-mill** a mill worked by a water-wheel. **water moccasin** see MOCCASIN 2. **water-nymph** Mythol. a nymph living in water; a naiad. **water of crystallization** water forming an essential part of the structure of some crystals. **water of life** rhet. spiritual enlightenment. **water opossum** = YAPOK. **water ouzel** = DIPPER 1. **water-pepper** a knotweed of damp ground, Persicaria hydropiper, the leaves of which have a peppery taste. **water-pipe 1** a pipe for conveying water. **2** a hookah. **water-pistol** a toy pistol shooting a jet of water. **water plantain** a ditch plant of the genus Alisma, with plantain-like leaves. **water-power 1** mechanical force derived from the weight or motion of water. **2** a fall in the level of a river, as a source of this force. **water purslane** a creeping plant, Lythrum portula, growing in damp places. **water rail** a Eurasian rail, Rallus aquaticus, frequenting marshes. **water-rat** a water vole or other large semiaquatic rodent. **water-rate** a charge made for the use of the public water-supply. **water-repellent** not easily penetrated by water. **water-resistant** able to resist, but not entirely prevent, the penetration of water. **water-scorpion** a flattened predatory aquatic bug of the family Nepidae, breathing through a bristle-like tubular tail. **water slide** a usu. high or long slide, down which water cascades, esp. into a swimming-pool. **water-softener** an apparatus or substance for softening hard water. **water soldier** a European aquatic plant, Stratiotes aloides, of the frogbit family, with rigid spiny sword-shaped leaves. **water-soluble** soluble in water. **water-splash** part of a road submerged by a stream or pool. **water sport** a sport practised on water, such as windsurfing, water-skiing, etc. **water starwort** a small aquatic plant of the genus Callitriche, with star-shaped leaf rosettes. **water-supply** the provision and storage of water, or the amount of water stored, for the use of a town, house, etc. **water-table** = water-level 2. **water torture** a form of torture in which the victim is exposed to the incessant dripping of water on the head, or the sound of dripping. **water-tower** a tower with an elevated tank to give pressure for distributing water. **water under the bridge** past events accepted as past and irrevocable. **water vole** a large semiaquatic vole, esp. Arvicola terrestris, living in holes in river-banks. **water-weed** any aquatic plant. **water-wheel** a wheel driven by water to work machinery, or to raise water. **water-wings** inflated floats fixed on the arms of a person learning to swim. □ **waterer** n. **waterless** adj. [OE wæter f. Gmc, rel. to WET]

■ Water is one of the most abundant substances on earth, with oceans covering a large proportion of the planet's surface. It is also vital to life — which is believed to have first evolved in the sea — and almost all the metabolic processes of living organisms take place in watery solutions. Weather systems are powered by the evaporation of water by solar energy, and its condensation as liquid in clouds and, ultimately, rain, snow, etc. The importance of water is due to its physical and chemical properties, some of which are highly distinctive: it is able to dissolve many other substances; its solid form (ice) is less dense than the liquid form; its boiling-point is unusually high for its molecular weight; it has high viscosity and surface tension; and it is partially dissociated into hydrogen and hydroxyl ions. Heavy water contains the hydrogen isotope deuterium.

waterbird /ˈwɔːtəˌbɜːd/ n. a bird frequenting esp. fresh water.

waterbrash /ˈwɔːtəˌbræʃ/ n. pyrosis. [WATER + BRASH[3]]

Water-carrier /ˈwɔːtəˌkærɪə(r)/ n. (or **-bearer**) the zodiacal sign or constellation Aquarius.

water-clock /ˈwɔːtəˌklɒk/ n. a device for measuring time by the flow of water. In ancient Egypt (c.1400 BC) such clocks consisted of round vessels from which water flowed through a hole in the base, with a time-scale marked on the inside of the vessel. Similar vessels were used in ancient Greece. A more sophisticated device, using a float linked to a cogwheel to which was fixed a pointer moving over a dial, was developed by the Romans and used in Europe until the 16th century.

watercolour /ˈwɔːtəˌkʌlə(r)/ n. (US **watercolor**) **1** (usu. in pl.) artists' paint made of pigment to be diluted with water and not oil. **2** a picture painted with this. **3** the art of painting with watercolours. □ **watercolourist** n.

watercourse /ˈwɔːtəˌkɔːs/ n. **1** a brook, stream, or artificial water-channel. **2** the bed along which this flows.

watercress /ˈwɔːtəˌkres/ n. a hardy perennial cress, *Nasturtium officinale*, growing in running water, with pungent leaves used in salad.

waterfall /ˈwɔːtəˌfɔːl/ n. a vertically descending part of a stream or river where it falls from a height over a precipice or down a steep hillside.

Waterford /ˈwɔːtəfəd/ **1** a county of the Republic of Ireland, in the south-east, in the province of Munster; county town, Waterford; main administrative centre, Dungarvan. **2** its county town, a port on an inlet of St George's Channel; pop. (1991) 40,345. It is noted for its clear, colourless flint glass, known as Waterford crystal.

waterfowl /ˈwɔːtəˌfaʊl/ n. (usu. collect. as pl.) birds frequenting water, esp. ducks and geese, regarded as game.

waterfront /ˈwɔːtəˌfrʌnt/ n. the part of a town adjoining a river, lake, harbour, etc.

Watergate /ˈwɔːtəˌgeɪt/ a US political scandal deriving from a bungled attempt to bug the national headquarters of the Democratic Party (in the Watergate building in Washington, DC) during the US election campaign of 1972. Five men hired by the Republican organization campaigning to re-elect Richard Nixon President were caught with electronic bugging equipment at the offices. The attempted cover-up and subsequent inquiry gravely weakened the prestige of the government and finally led to the resignation of President Nixon in Aug. 1974, before impeachment proceedings against him could begin.

watergate /ˈwɔːtəˌgeɪt/ n. **1** a floodgate. **2** a gate giving access to a river etc.

watering /ˈwɔːtərɪŋ/ n. the act or an instance of supplying water or (of an animal) obtaining water. □ **watering-can** a portable container with a long spout usu. ending in a perforated sprinkler, for watering plants. **watering-hole 1** a pool of water from which animals regularly drink. **2** sl. a bar. **watering-place 1** = watering-hole. **2** a spa or seaside resort. **3** a place where water is obtained. [OE wæterung (as WATER, -ING[1])]

waterlogged /ˈwɔːtəˌlɒgd/ adj. **1** saturated with water. **2** (of a boat etc.) hardly able to float from being saturated or filled with water. **3** (of ground) made useless by being saturated with water. [waterlog (v.), f. WATER + LOG[1], prob. orig. = 'reduce (a ship) to the condition of a log']

Waterloo, Battle of /ˌwɔːtəˈluː/ a battle fought on 18 June 1815 near the village of Waterloo (in what is now Belgium), in which Napoleon's army was defeated by the British (under the Duke of Wellington) and Prussians. Attempting to exploit a temporary separation of the British and Prussian armies, Napoleon attacked the outnumbered British force, but Wellington's forces were able to hold off the French until the arrival of the Prussians on his left flank forced the French to retreat. Under the pressure of the allied pursuit Napoleon's army disintegrated completely, effectively ending his bid to return to power. □ **meet one's Waterloo** come up against a decisive defeat or contest.

waterman /ˈwɔːtəmən/ n. (pl. **-men**) **1** a boatman plying for hire. **2** an oarsman as regards skill in keeping the boat balanced.

watermark /ˈwɔːtəˌmɑːk/ n. & v. ● n. a faint design made in some paper during manufacture, visible when held against the light, identifying the maker etc. ● v.tr. mark with this.

water polo n. a seven-a-side game played by teams of swimmers, who swim around a pool passing a ball the size of a football to each other and attempting to throw it into their opponents' net. The game developed in Britain from about 1870 and the first rules were drafted in 1876. It was first included in the Olympic Games in 1900.

waterproof /ˈwɔːtəˌpruːf/ adj., n., & v. ● adj. impervious to water. ● n. a waterproof garment or material. ● v.tr. make waterproof.

Waters /ˈwɔːtəz/, Muddy (born McKinley Morganfield) (1915–83), American blues singer and guitarist. Based in Chicago from 1943, he became famous with his song 'Rollin' Stone' (1950). In the same year he formed a blues band, with which he recorded such hits as 'Got My Mojo Working' (1957). Waters impressed new rhythm-and-blues bands such as the Rolling Stones, who took their name from his 1950 song.

watershed /ˈwɔːtəˌʃed/ n. **1** a line of separation between waters flowing to different rivers, basins, or seas. **2** a turning-point in affairs; a crucial time or occurrence. [WATER + shed ridge of high ground (rel. to SHED[2]), after G Wasserscheide]

waterside /ˈwɔːtəˌsaɪd/ n. the margin of a sea, lake, or river.

water-ski /ˈwɔːtəˌskiː/ n. & v. ● n. (pl. **-skis**) each of a pair of skis for skimming the surface of the water when towed by a motor boat. ● v.intr. (**-skis**, **-ski'd** or **-skied** /-ˌskiːd/; **-skiing**) travel on water-skis. □ **water-skier** n.

waterspout /ˈwɔːtəˌspaʊt/ n. a gyrating column of water and spray formed by a whirlwind occurring over a body of water.

waterthrush /ˈwɔːtəˌθrʌʃ/ n. a small thrushlike North American warbler of the genus *Seiurus* (family Parulidae), found near woodland streams and swamps.

watertight /ˈwɔːtəˌtaɪt/ adj. **1** (of a joint, container, vessel, etc.) closely fastened or fitted or made so as to prevent the passage of water. **2** (of an argument etc.) unassailable.

waterway /ˈwɔːtəˌweɪ/ n. **1** a navigable channel. **2** a route for travel by water. **3** a thick plank at the outer edge of a deck along which a channel is hollowed for water to run off by.

waterworks /ˈwɔːtəˌwɜːks/ n. **1** an establishment for managing a water-supply. **2** colloq. the shedding of tears (esp. as turn on the waterworks). **3** Brit. the urinary system.

watery /ˈwɔːtərɪ/ adj. **1** containing too much water. **2** too thin in consistency. **3** of or consisting of water. **4** (of the eyes) suffused or running with water. **5** (of conversation, style, etc.) vapid, uninteresting. **6** (of colour) pale. **7** (of the sun, moon, or sky) rainy-looking. □ **watery grave** the bottom of the sea as a place where a person lies drowned. □ **wateriness** n. [OE wæterig (as WATER, -Y[1])]

Watford /ˈwɒtfəd/ a town in Hertfordshire, SE England; pop. (1981) 74,460.

watjin /wɑːˈdʒɪn/ n. Austral. (in Aboriginal English) a white woman. [alt. of white gin: see GIN[3]]

Watling Street /ˈwɒtlɪŋ/ a Roman road (now largely underlying modern roads) running north-westwards across England, from Richborough in Kent through London and St Albans to Wroxeter in Shropshire.

Watson[1] /ˈwɒts(ə)n/, James Dewey (b.1928), American biologist. Together with Francis Crick he proposed a model for the structure of the DNA molecule (see CRICK), later recounting the discovery in The Double Helix (1968). He shared a Nobel Prize with Crick and Maurice Wilkins in 1962. He became director of the molecular biology laboratory at Cold Spring Harbor on Long Island in 1968, concentrating efforts on cancer research; he also served as director of the National Center for Human Genome Research (1989–92).

Watson[2] /ˈwɒts(ə)n/, John Broadus (1878–1958), American psychologist, founder of the school of behaviourism. He viewed behaviour as determined by an interplay between genetic endowment and environmental influences, and held that the role of the psychologist was to discern, through observation and experimentation, which behaviour was innate and which was acquired. In seeking an objective study of psychology, he set the stage for the empirical study of animal and human behaviour which was to dominate 20th-century psychology, particularly in the US.

Watson, Dr /ˈwɒts(ə)n/ a doctor who is the companion and assistant of Sherlock Holmes in detective stories by Sir Arthur Conan Doyle.

Watson-Watt /ˌwɒts(ə)nˈwɒt/, Sir Robert Alexander (1892–1973), Scottish physicist. He produced a system for locating thunderstorms by means of their radio emissions, and went on to lead a team that developed radar into a practical system for locating aircraft. This was improved and rapidly deployed in Britain for use in the Second World War, in which it was to play a vital role. (See RADAR.)

Watt /wɒt/, James (1736–1819), Scottish engineer. He greatly improved the efficiency of the Newcomen beam engine by condensing the spent steam in a separate chamber, allowing the cylinder to remain hot. The improved engines were adopted for a variety of purposes, especially after Watt entered into a business partnership with the engineer Matthew Boulton. Watt continued inventing until the end of his life, introducing rotatory engines, controlled by a centrifugal governor, and devising a chemical method of copying documents. He also introduced the term *horsepower*.

watt /wɒt/ *n.* the SI unit of power (symbol **W**), equal to one joule per second and corresponding to the power in an electric circuit where the potential difference is one volt and the current one ampere. □ **watt-hour** the energy used when one watt is applied for one hour. [Watt]

wattage /'wɒtɪdʒ/ *n.* an amount of electrical power expressed in watts.

Watteau /'wɒtəʊ/, Jean Antoine (1684–1721), French painter, of Flemish descent. An initiator of the rococo style in painting, he is also known for his invention of the pastoral genre known as the *fête galante*. Watteau deliberately created an imaginary and rather theatrical world; the light-hearted imagery of his painting contrasted with the serious religious and classical subject-matter approved by the Royal Academy of Painting and Sculpture. His best-known painting is *L'Embarquement pour l'Île de Cythère* (1717).

wattle[1] /'wɒt(ə)l/ *n.* & *v.* ● *n.* **1 a** interlaced rods and split rods as a material for making fences, walls, etc. **b** (in *sing.* or *pl.*) rods and twigs for this use. **2** an Australian acacia with long pliant branches, with bark used in tanning and golden flowers used as the national emblem. **3** *dial.* a wicker hurdle. ● *v.tr.* **1** make of wattle. **2** enclose or fill up with wattles. □ **wattle and daub** a network of rods and twigs plastered with mud or clay as a building material. [OE *watul*, of unkn. orig.]

wattle[2] /'wɒt(ə)l/ *n.* **1** a loose fleshy appendage on the head or throat of a turkey or other birds. **2** = BARB *n.* 3. □ **wattled** *adj.* [16th c.: orig. unkn.]

wattlebird /'wɒt(ə)l,bɜːd/ *n.* **1** an Australian honeyeater of the genus *Anthochaera* or *Melidectes*, with a wattle hanging from each cheek. **2** a New Zealand bird of the family Callaeidae, with wattles hanging from the base of the bill, e.g. the saddleback.

wattmeter /'wɒt,miːtə(r)/ *n.* a meter for measuring the amount of electricity in watts.

Watts[1] /wɒts/, George Frederick (1817–1904), English painter and sculptor. His work reflects his view of art as a vehicle for moral purpose; this is most evident in his allegorical paintings such as *Hope* (1886). He is best known for his portraits of public figures, including Gladstone, Tennyson, and J. S. Mill. He married the actress Ellen Terry in 1864, but the marriage lasted less than a year and the couple were finally divorced in 1877.

Watts[2] /wɒts/, Isaac (1674–1748), English hymn-writer and poet. His songs for children, which included 'How Doth the Little Busy Bee' (1715), anticipated those of William Blake. Watts is also remembered for hymns such as 'O God, Our Help in Ages Past' (1719).

Watusi /wə'tuːsɪ/ *n.*, *adj.*, & *v.* (also **Watutsi** /-'tʊtsɪ/) ● *n.* (*pl.* same or **Watusis**) **1** = TUTSI *n.* **2** an energetic popular dance of the 1960s. ● *adj.* = TUTSI *adj.* ● *v.intr.* dance the Watusi. [Bantu, f. *wa-* pl. prefix + TUTSI]

Waugh /wɔː/, Evelyn (Arthur St John) (1903–66), English novelist. His first novel *Decline and Fall*, a satire on life in a boys' preparatory school, was published in 1928 and was immensely successful. Waugh's other early novels, all social satires, include *Vile Bodies* (1930), *A Handful of Dust* (1934), and *Scoop* (1938). His work was profoundly influenced by his conversion to Roman Catholicism in 1930. *Brideshead Revisited* (1945), a complex story of an aristocratic Roman Catholic family, was more sombre in tone. Later works include *The Loved One* (1948) and his trilogy (based on his wartime experiences) *Men at Arms* (1952), *Officers and Gentlemen* (1955), and *Unconditional Surrender* (1961).

waul /wɔːl/ *v.intr.* (also **wawl**) give a loud plaintive cry like a cat. [imit.]

wave /weɪv/ *v.* & *n.* ● *v.* **1 a** *intr.* (often foll. by *to*) move a hand etc. to and fro in greeting or as a signal (*waved to me across the street*). **b** *tr.* move (a hand etc.) in this way. **2 a** *intr.* show a sinuous or sweeping motion as of a flag, tree, or a cornfield in the wind; flutter, undulate. **b** *tr.* impart a waving motion to. **3** *tr.* brandish (a sword etc.) as an encouragement to followers etc. **4** *tr.* tell or direct (a person) by waving (*waved them away; waved them to follow*). **5** *tr.* express (a greeting etc.) by waving (*waved goodbye to them*). **6** *tr.* give an undulating form to (hair, drawn lines, etc.); make wavy. **7** *intr.* (of hair etc.) have such a form; be wavy. ● *n.* **1** a ridge of water between two depressions. **2** a long body of water curling into an arched form and breaking on the shore. **3** a thing compared

to this, e.g. a body of persons in one of successive advancing groups. **4** a gesture of waving. **5 a** the process of waving the hair. **b** an undulating form produced in the hair by waving. **6 a** a temporary occurrence or increase of a condition, emotion, or influence (*a wave of enthusiasm*). **b** (in phrases **heatwave**, **cold wave**) a spell of abnormal heat or cold. **7** *Physics* **a** a periodic disturbance of the particles of a substance which may be propagated without net movement of the particles, as in the passage of undulating motion, heat, sound, etc. (*See note below.*) **b** a single curve in the course of this motion (see also *travelling wave* (see TRAVEL), *standing wave*). **c** a similar variation of an electromagnetic field in the propagation of light or other radiation through a medium or vacuum. **8** (in *pl.*; prec. by *the*) *poet.* the sea; water. □ **make waves** *colloq.* **1** cause trouble. **2** create a significant impression. **wave aside** dismiss as intrusive or irrelevant. **wave down** wave to (a vehicle or its driver) as a signal to stop. **wave energy** = *wave power*. **wave equation** a differential equation expressing the properties of motion in waves. **wave-form** *Physics* a curve showing the shape of a wave at a given time. **wave-front** *Physics* a surface containing points affected in the same way by a wave at a given time. **wave function** a function satisfying a wave equation and describing the properties of a wave. **wave mechanics** a method of analysis of the behaviour esp. of atomic phenomena with particles represented by wave equations; quantum mechanics. **wave number** *Physics* the number of waves in a unit distance. **wave power** power obtained by harnessing the energy of waves. (*See note below.*) **wave theory** *hist.* the theory that light is propagated through the ether by a wave-motion imparted to the ether by the molecular vibrations of the radiant body. **wave train** a group of waves of equal or similar wavelengths travelling in the same direction. □ **waveless** *adj.* **wavelike** *adj.* & *adv.* [OE *wafian* (v.) f. Gmc: (n.) also alt. of ME *wawe*, *wage*]

- Waves can be longitudinal, transverse, or torsional vibrations. In longitudinal waves (e.g. sound waves) the to-and-fro vibration is parallel to the direction of propagation; in transverse waves (e.g. water waves or light waves) the vibration is at right angles to the direction of propagation; in torsional waves a rotatory motion is propagated. Waves display characteristic properties of reflection, refraction, diffraction, and interference, under appropriate conditions. When two waves travelling in opposite directions meet, a standing wave can result, which vibrates up and down but is not propagated. Wave phenomena are of great importance throughout physics. In quantum theory particles such as photons and electrons are considered to have a dual nature, sharing characteristics of both waves and particles.

- Though the first practical device for generating electricity from ocean waves was built a century ago, such generators are still not economically and technically proven on a large scale. Most have been based on the oscillation of water in vertical columns, driving turbines either directly or via air compressed by the water. Recently other techniques have been investigated, mainly in the UK, using hinged rafts, compressible air-bags, or chains of rocking devices.

waveband /'weɪvbænd/ *n.* a range of (esp. radio) wavelengths between certain limits.

waveguide /'weɪvgaɪd/ *n.* *Electr.* a metal tube etc. confining and conveying microwaves.

wavelength /'weɪvleŋθ, -leŋkθ/ *n.* **1** the distance between successive crests of a wave, esp. points in a sound wave or electromagnetic wave (symbol: λ). **2** this as a distinctive feature of radio waves from a transmitter. **3** *colloq.* a particular mode or range of thinking and communicating (*we don't seem to be on the same wavelength*).

wavelet /'weɪvlɪt/ *n.* a small wave on water.

waver /'weɪvə(r)/ *v.intr.* **1** be or become unsteady; falter; begin to give way. **2** be irresolute or undecided between different courses or opinions; be shaken in resolution or belief. **3** (of a light) flicker. □ **waverer** *n.* **waveringly** *adv.* [ME f. ON *vafra* flicker f. Gmc, rel. to WAVE]

wavy /'weɪvɪ/ *adj.* (**wavier**, **waviest**) (of a line or surface) having waves or alternate contrary curves (*wavy hair*). □ **wavily** *adv.* **waviness** *n.*

wa-wa var. of WAH-WAH.

wawl var. of WAUL.

wax[1] /wæks/ *n.* & *v.* ● *n.* **1** a sticky mouldable yellowish substance secreted by bees as the material of honeycomb cells; beeswax. **2** a white translucent material obtained from this by bleaching and purifying and used for candles, in modelling, as a basis of polishes, and for other purposes. **3** any similar substance, typically a lipid or hydrocarbon (*earwax*, *paraffin wax*). **4** *colloq.* **a** a gramophone record.

b material for the manufacture of this. **5** (*attrib.*) made of wax. ● *v.tr.* **1** cover or treat with wax. **2** remove unwanted hair from (legs etc.) by applying wax and peeling off the wax and hairs together. **3** *colloq.* record for the gramophone. □ **be wax in a person's hands** be overcompliant or easily influenced by a person. **lost wax** = CIRE PERDUE. **wax-light** a taper or candle of wax. **wax moth** a pyralid moth whose larvae feed on beeswax, esp. *Galleria mellonella*, a pest of beehives. **wax myrtle** = BAYBERRY. **wax-painting** = ENCAUSTIC. **wax palm 1** a South American palm, *Ceroxylon alpinum*, with its stem coated in a mixture of resin and wax. **2** the carnauba. **wax paper** paper waterproofed with a layer of wax. **wax-pod** a yellow-podded bean. **wax-tree** a tree that yields wax, esp. *Rhus succedanea* of eastern Asia, having white berries used as a substitute for beeswax. □ **waxer** *n.* [OE *wæx, weax* f. Gmc]

wax² /wæks/ *v.intr.* **1** (of the moon between new and full) have a progressively larger part of its visible surface illuminated, increasing in apparent size (cf. WANE *v.* 1). **2** *literary* become larger or stronger. **3** (foll. by compl.) pass into a specified state or mood (*wax lyrical*). □ **wax and wane** undergo alternate increases and decreases. [OE *weaxan* f. Gmc]

wax³ /wæks/ *n. sl.* (esp. **be in a wax**) a fit of anger. [19th c.: orig. uncert.: perh. f. WAX² *wroth* etc.]

waxberry /ˈwæksbəri/ *n.* (*pl.* **-ies**) **1** a wax-myrtle. **2** the fruit of this.

waxbill /ˈwæksbɪl/ *n.* a small finchlike bird of the family Estrildidae, esp. *Estrilda astrild* of Africa, with a red bill resembling the colour of sealing wax.

waxcloth /ˈwæksklɒθ/ *n.* oilcloth.

waxen /ˈwæks(ə)n/ *adj.* **1** having a smooth pale translucent surface as of wax. **2** able to receive impressions like wax; plastic. **3** *archaic* made of wax.

waxwing /ˈwækswɪŋ/ *n.* a songbird of the genus *Bombycilla*, with red waxlike tips to some wing-feathers.

waxwork /ˈwækswɜːk/ *n.* **1 a** an object, esp. a lifelike dummy, modelled in wax. **b** the making of waxworks. **2** (in *pl.*) an exhibition of wax dummies.

waxy¹ /ˈwæksɪ/ *adj.* (**waxier, waxiest**) resembling wax in consistency or in its surface. □ **waxily** *adv.* **waxiness** *n.* [WAX¹ + -Y¹]

waxy² /ˈwæksɪ/ *adj.* (**waxier, waxiest**) *Brit. sl.* angry, quick-tempered. [WAX³ + -Y¹]

way /weɪ/ *n. & adv.* ● *n.* **1** a road, track, path, etc., for passing along. **2** a course or route for reaching a place, esp. the best one (*asked the way to London*). **3** a place of passage into a building, through a door, etc. (*could not find the way out*). **4 a** a method or plan for attaining an object (*that is not the way to do it*). **b** the ability to obtain one's object (*has a way with him*). **5 a** a person's desired or chosen course of action. **b** a custom or manner of behaving; a personal peculiarity (*has a way of forgetting things; things had a way of going badly*). **6** a specific manner of life or procedure (*soon got into the way of it*). **7** the normal course of events (*that is always the way*). **8** a travelling distance; a length traversed or to be traversed (*is a long way away*). **9 a** an unimpeded opportunity of advance. **b** a space free of obstacles. **10** a region or ground over which advance is desired or natural. **11** advance in some direction; impetus, progress (*pushed my way through*). **12** movement of a ship etc. (*gather way; lose way*). **13** the state of being engaged in movement from place to place; time spent in this (*met them on the way home; with songs to cheer the way*). **14** a specified direction (*step this way; which way are you going?*). **15** (in *pl.*) parts into which a thing is divided (*split it three ways*). **16** *colloq.* the scope or range of something (*want a few things in the stationery way*). **17** a person's line of occupation or business. **18** a specified condition or state (*things are in a bad way*). **19** a respect (*is useful in some ways*). **20 a** (in *pl.*) a structure of timber etc. down which a new ship is launched. **b** parallel rails etc. as a track for the movement of a machine. ● *adv. colloq.* to a considerable extent; far (*you're way off the mark*). □ **across** (or **over**) **the way** opposite. **any way** = ANYWAY. **be on one's way** set off; depart. **by the way 1** incidentally; as a more or less irrelevant comment. **2** during a journey. **by way of 1** through; by means of. **2** as a substitute for or as a form of (*did it by way of apology*). **3** with the intention of (*asked by way of discovering the truth*). **come one's way** become available to one; become one's lot. **find a way** discover a means of obtaining one's object. **get** (or **have**) **one's way** (or **have it one's own way** etc.) get what one wants; ensure one's wishes are met. **give way 1 a** make concessions. **b** fail to resist; yield. **2** (often foll. by *to*) concede precedence (to); (at a road junction) give right of way to other traffic. **3** (of a structure etc.) be dislodged or broken under a load; collapse. **4** (foll. by *to*) be superseded by. **5** (foll.

by *to*) be overcome by (an emotion etc.). **6** (of rowers) row hard. **go out of one's way** (often foll. by *to* + infin.) make a special effort; act gratuitously or without compulsion (*went out of their way to help*). **go one's own way** act independently, esp. against contrary advice. **go one's way 1** leave, depart. **2** (of events, circumstances, etc.) be favourable to one. **go a person's way** travel in the same direction as a person (*are you going my way?*). **have it both ways** see BOTH. **in its way** if regarded from a particular standpoint appropriate to it. **in no way** not at all; by no means. **in a way** in a certain respect but not altogether or completely. **in the** (or **one's**) **way** forming an obstacle or hindrance. **lead the way 1** act as guide or leader. **2** show how to do something. **look the other way 1** ignore what one should notice. **2** disregard an acquaintance etc. whom one sees. **one way and another** taking various considerations into account. **one way or another** by some means. **on the** (or **one's**) **way 1** in the course of a journey etc. **2** having progressed (*is well on the way to completion*); in the pipeline. **3** *colloq.* (of a child) conceived but not yet born. **on the way out** *colloq.* going down in status, estimation, or favour; going out of fashion. **the other way about** (or **round**) in an inverted or reversed position or direction. **out of the way 1** no longer an obstacle or hindrance. **2** disposed of; settled. **3** (of a person) imprisoned or killed. **4** unusual or remarkable (*nothing out of the way*). **5** (of a place) remote, inaccessible. **out of one's way** not on one's intended route. **put a person in the way of** give a person the opportunity of. **way back** *colloq.* long ago. **way-leave** a right of way rented to another. **the way of the Cross** a series of images representing the Stations of the Cross, esp. in a church or on the road to a church or shrine. **way of life** the principles or habits governing all one's actions etc. **way of thinking** one's customary opinion of matters. **way of the world** the normal state of things. **way-out** *colloq.* **1** unusual, eccentric. **2** avant-garde, progressive. **3** excellent, exciting. **ways and means 1** methods of achieving something. **2** methods of raising government revenue. **way station** *N. Amer.* **1** a minor station on a railway. **2** a point marking progress in a certain course of action etc. **way-worn** tired with travel. [OE *weg* f. Gmc: (adv.) f. AWAY]

-way /weɪ/ *suffix* = -WAYS.

wayback /ˈweɪbæk/ *n. esp. Austral.* = OUTBACK.

waybill /ˈweɪbɪl/ *n.* a list of passengers or parcels on a vehicle.

waybread /ˈweɪbred/ *n. Brit. archaic* a broad-leaved plantain (see PLANTAIN¹). [OE *wegbræde* (as WAY, BROAD)]

wayfarer /ˈweɪˌfeərə(r)/ *n.* a traveller, esp. on foot.

wayfaring /ˈweɪˌfeərɪŋ/ *n.* travelling, esp. on foot. □ **wayfaring-tree** a white-flowered Eurasian shrub, *Viburnum lantana*, common along roadsides, with berries turning from green through red to black.

Wayland the Smith /ˈweɪlənd/ (also **Weland** /ˈwel-/) *Scand. & Anglo-Saxon Mythol.* a smith with supernatural powers, in English legend supposed to have his forge in a neolithic barrow (*Wayland's Smithy*) on the downs in SW Oxfordshire.

waylay /weɪˈleɪ/ *v.tr.* (*past* and *past part.* **waylaid** /-ˈleɪd/) **1** lie in wait for. **2** stop to rob or interview. □ **waylayer** *n.*

waymark /ˈweɪmɑːk/ *n.* a natural or artificial object as a guide to travellers, esp. walkers.

Wayne /weɪn/, John (born Marion Michael Morrison; known as 'the Duke') (1907–79), American actor. Associated with the film director John Ford from 1930, Wayne became a Hollywood star with his performance in Ford's western *Stagecoach* (1939). Remembered as the archetypal cowboy hero, he appeared in many other classic westerns, notably *Red River* (1948) and *True Grit* (1969), for which he won an Oscar. Wayne's fierce patriotism was reflected in his other main roles, e.g. as a US marine in Ford's *Sands of Iwo Jima* (1949) and as Davy Crockett in *The Alamo* (1960), which he also produced and directed.

waypoint /ˈweɪpɔɪnt/ *n.* **1** a stopping-place, esp. on a journey. **2** the computer-checked coordinates of each stage of a flight, sea journey, etc. (also *attrib.*: *waypoint navigation facility*).

-ways /weɪz/ *suffix* forming adjectives and adverbs of direction or manner (*sideways*) (cf. -WISE). [WAY + -'s¹]

wayside /ˈweɪsaɪd/ *n.* **1** the side or margin of a road. **2** the land at the side of a road. □ **fall by the wayside** fail to continue in an endeavour or undertaking (after Luke 8:5).

wayward /ˈweɪwəd/ *adj.* **1** childishly self-willed or perverse; capricious. **2** unaccountable or freakish. □ **waywardly** *adv.* **waywardness** *n.* [ME f. obs. *awayward* turned away f. AWAY + -WARD: cf. FROWARD]

wayzgoose /ˈweɪzguːs/ *n.* (*pl.* **-gooses**) an annual summer dinner or outing held by a printing-house for its employees. [17th c. (earlier *waygoose*): orig. unkn.]

Wb *abbr.* weber(s).

WC *abbr.* **1** water-closet. **2** West Central.

WCC see World Council of Churches.

W/Cdr. *abbr.* wing commander.

WD *abbr.* **1** War Department. **2** Works Department.

we /wiː, wɪ/ *pron. (obj.* **us**; *poss.* **our, ours**) **1** (*pl.* of I²) used by and with reference to more than one person speaking or writing, or one such person and one or more associated persons. **2** used for or by a royal person in a proclamation etc. and by a writer or editor in a formal context. **3** people in general (cf. ONE *pron.* 2). **4** *colloq.* = I² (*give us a chance*). **5** *colloq.* (often implying condescension) you (*how are we feeling today?*). [OE f. Gmc]

WEA *abbr.* (in the UK) Workers' Educational Association.

weak /wiːk/ *adj.* **1** deficient in strength, power, or number; fragile; easily broken or bent or defeated. **2** deficient in vigour; sickly, feeble (*weak health*; *a weak imagination*). **3 a** deficient in resolution; easily led (*a weak character*). **b** indicating a lack of resolution (*a weak surrender*). **4** unconvincing or logically deficient (*weak evidence*; *a weak argument*). **5** (of a mixed liquid or solution) watery, thin, dilute (*weak tea*). **6** (of a style etc.) not vigorous or well-knit; diffuse, slipshod. **7** (of a crew) short-handed. **8** (of a syllable etc.) unstressed. **9** *Gram.* in Germanic languages: **a** (of a verb) forming inflections by the addition of a suffix to the stem. **b** (of a noun or adjective) belonging to a declension in which the stem originally ended in -*n* (opp. STRONG *adj.* 22). □ **weak chin** a chin that is not well-defined or prominent, esp. as supposedly indicative of lack of firmness of character. **weak ending** an unstressed syllable in a normally stressed place at the end of a verse-line. **the weaker sex** *derog.* women. **weak grade** *Gram.* an unstressed ablaut-form. **weak interaction** *Physics* the weakest form of interaction between subatomic particles. **weak-kneed** *colloq.* lacking resolution. **weak-minded 1** mentally deficient. **2** lacking in resolution. **weak-mindedness** the state of being weak-minded. **weak moment** a time when one is unusually compliant or temptable. **weak point** (or **spot**) **1** a place where defences are assailable. **2** a flaw in an argument or character or in resistance to temptation. □ **weakish** *adj.* [ME f. ON *veikr* f. Gmc]

weaken /ˈwiːkən/ *v.* **1** *tr.* & *intr.* make or become weak or weaker. **2** *intr.* relent, give way; succumb to temptation etc. □ **weakener** *n.*

weakfish /ˈwiːkfɪʃ/ *n.* US a marine fish of the genus *Cynoscion* (family Sciaenidae), used as food. [obs. Du. *weekvisch* f. *week* soft (formed as WEAK) + *visch* FISH¹]

weakling /ˈwiːklɪŋ/ *n.* a feeble person or animal.

weakly /ˈwiːklɪ/ *adv.* & *adj.* ● *adv.* in a weak manner. ● *adj.* (**weaklier, weakliest**) sickly, not robust. □ **weakliness** *n.*

weakness /ˈwiːknɪs/ *n.* **1** the state or condition of being weak. **2** a weak point; a defect. **3** the inability to resist a particular temptation. **4** (foll. by *for*) a self-indulgent liking (*have a weakness for chocolate*).

weal¹ /wiːl/ *n.* & *v.* ● *n.* **1** a ridge raised on the flesh by a stroke of a rod or whip. **2** *Med.* a raised and reddened area of the skin, usu. accompanied by itching. ● *v.tr.* mark with a weal. [var. of WALE, infl. by obs. *wheal* suppurate]

weal² /wiːl/ *n.* literary welfare, prosperity; good fortune. [OE *wela* f. WG (as WELL¹)]

Weald /wiːld/ *n.* (also **weald**) (prec. by *the*) a formerly heavily wooded district of SE England, including parts of Kent, Surrey, and East Sussex. □ **weald-clay** beds of clay, sandstone, limestone, and ironstone, forming the top of Wealden strata, with abundant fossil remains. [OE, = *wald* WOLD]

Wealden /ˈwiːld(ə)n/ *adj.* & *n.* ● *adj.* **1** of the Weald. **2** resembling the Weald geologically. ● *n.* a series of Lower Cretaceous freshwater deposits above Jurassic strata and below chalk, best exemplified in the Weald.

wealth /welθ/ *n.* **1** riches; abundant possessions; opulence. **2** the state of being rich. **3** (foll. by *of*) an abundance or profusion (*a wealth of new material*). **4** *archaic* welfare or prosperity. □ **wealth tax** a tax on personal capital. [ME *welthe*, f. WELL¹ or WEAL² + -TH², after *health*]

wealthy /ˈwelθɪ/ *adj.* (**wealthier, wealthiest**) having an abundance esp. of money. □ **wealthily** *adv.* **wealthiness** *n.*

wean¹ /wiːn/ *v.tr.* **1** accustom (an infant or other young mammal) to food other than (esp. its mother's) milk. **2** (often foll. by *from, away from*) disengage (from a habit etc.) by enforced discontinuance. [OE *wenian* accustom f. Gmc: cf. WONT]

wean² /wiːn/ *n. Sc.* a young child. [contr. of *wee ane* little one]

weaner /ˈwiːnə(r)/ *n.* a young animal recently weaned.

weanling /ˈwiːnlɪŋ/ *n.* a newly weaned child etc.

weapon /ˈwep(ə)n/ *n.* **1** a thing designed or used or usable for inflicting bodily harm (e.g. a gun or cosh). **2** a means employed for trying to gain the advantage in a conflict (*irony is a double-edged weapon*). □ **weaponed** *adj.* (also in *comb.*). **weaponless** *adj.* [OE *wǣp(e)n* f. Gmc]

weaponry /ˈwep(ə)nrɪ/ *n.* weapons collectively.

wear¹ /weə(r)/ *v.* & *n.* ● *v.* (*past* **wore** /wɔː(r)/; *past part.* **worn**) **1** *tr.* have on one's person as clothing or an ornament etc. (*is wearing shorts*; *wears earrings*). **2** *tr.* be dressed habitually in (*wears green*). **3** *tr.* exhibit or present (a facial expression or appearance) (*wore a frown*; *the day wore a different aspect*). **4** *tr. Brit.* (usu. with *neg.*) *colloq.* tolerate, accept (*they won't wear that excuse*). **5** (often foll. by *away*) **a** *tr.* injure the surface of, or partly obliterate or alter, by rubbing, stress, or use. **b** *intr.* undergo such injury or change. **6** *tr.* & *intr.* (foll. by *off, away*) rub or be rubbed off. **7** *tr.* make (a hole etc.) by constant rubbing or dripping etc. **8** *tr.* & *intr.* (often foll. by *out*) exhaust, tire or be tired. **9** *tr.* (foll. by *down*) overcome by persistence. **10** *intr.* **a** remain for a specified time in working order or a presentable state; last long. **b** (foll. by *well, badly,* etc.) endure continued use or life. **11 a** *intr.* (usu. foll. by *on*) (of time) pass, esp. tediously (*the day wore on*). **b** *tr.* pass (time) gradually away. **12** *tr.* (of a ship) fly (a flag). ● *n.* **1** the act of wearing or the state of being worn (*suitable for informal wear*). **2** things worn; fashionable or suitable clothing (*sportswear*; *footwear*). **3** (in full **wear and tear**) damage sustained from continuous use. **4** the capacity for resisting wear and tear (*still a great deal of wear left in it*). □ **in wear** being regularly worn. **wear one's heart on one's sleeve** see HEART. **wear off** lose effectiveness or intensity. **wear out 1** use or be used until no longer usable. **2** tire or be tired out. **wear thin** (of patience, excuses, etc.) begin to fail. **wear the trousers** see TROUSERS. **wear** (or **wear one's years**) **well** *colloq.* remain young-looking. □ **wearer** *n.* **wearable** *adj.* **wearability** /ˌweərəˈbɪlɪtɪ/ *n.* [OE *werian* f. Gmc]

wear² /weə(r)/ *v.* (*past* and *past part.* **wore** /wɔː(r)/) **1** *tr.* bring (a ship) about by turning its head away from the wind. **2** *intr.* (of a ship) come about in this way (cf. TACK¹ *v.* 4a). [17th c.: orig. unkn.]

wearing /ˈweərɪŋ/ *adj.* & *n.* ● *predic.adj.* tiresome, tedious. ● *n.* in senses of WEAR¹ *v.* □ **wearingly** *adv.*

wearisome /ˈwɪərɪsəm/ *adj.* tedious; tiring by monotony or length. □ **wearisomely** *adv.* **wearisomeness** *n.*

weary /ˈwɪərɪ/ *adj.* & *v.* ● *adj.* (**wearier, weariest**) **1** unequal to or disinclined for further exertion or endurance; tired. **2** (foll. by *of*) dismayed at the continuing of; impatient of. **3** tiring or tedious. ● *v.* (**-ies, -ied**) **1** *tr.* & *intr.* make or grow weary. **2** *intr.* esp. *Sc.* long. □ **weariless** *adj.* **wearily** *adv.* **weariness** *n.* **wearyingly** *adv.* [OE *wērig, wǣrig* f. WG]

weasel /ˈwiːz(ə)l/ *n.* & *v.* ● *n.* **1** a small brown and white flesh-eating mammal, *Mustela nivalis*, with a slender body, related to the stoat. **2** *colloq.* a deceitful or treacherous person. ● *v.intr.* (**weaselled, weaselling;** *US* **weaseled, weaseling**) **1** esp. *US* equivocate or quibble. **2** (foll. by *on, out*) default on an obligation. □ **weasel-faced** having thin sharp features. **weasel word** (usu. in *pl.*) a word that is intentionally ambiguous or misleading. □ **weaselly** *adj.* [OE *wesle, wesule* f. WG]

weather /ˈweðə(r)/ *n.* & *v.* ● *n.* **1** the state of the atmosphere at a place and time as regards heat, cloudiness, dryness, sunshine, wind, and rain etc. **2** (*attrib.*) *Naut.* windward (*on the weather side*). ● *v.* **1** *tr.* expose to or affect by atmospheric changes, esp. deliberately to dry, season, etc. (*weathered timber*). **2 a** *tr.* (usu. in *passive*) discolour or partly disintegrate (rock or stones) by exposure to air. **b** *intr.* be discoloured or worn in this way. **3** *tr.* make (boards or tiles) overlap downwards to keep out rain etc. **4** *tr.* **a** come safely through (a storm). **b** survive (a difficult period etc.). **5** *tr.* (of a ship or its crew) get to the windward of (a cape etc.). □ **keep a** (or **one's**) **weather eye open** be watchful. **make good** (or **bad**) **weather of it** *Naut.* (of a ship) behave well (or badly) in a storm. **make heavy weather of** *colloq.* exaggerate the difficulty or burden presented by (a problem, course of action, etc.). **under the weather** *colloq.* indisposed or slightly unwell. **weather-beaten** affected by exposure to the weather. **weather-bound** unable to proceed owing to bad weather. **weather-chart** (or **-map**) a diagram showing the state of the weather over a large area. **weather forecast** an analysis of the state of the weather with an assessment of likely developments over a certain time. **weather-glass** a barometer. **weather side** the side from which the wind is blowing (opp. *lee side*, see LEE 2). **weather station** an observation post for recording meteorological data. **weather-strip** a piece of material used to make a door or window proof against rain or wind. **weather-tiles** tiles arranged to overlap like weatherboards. **weather-vane** see VANE 1. **weather-worn** damaged by storms etc. [OE *weder* f. Gmc]

weatherboard /ˈweðəˌbɔːd/ *n. & v.* ● *n.* **1** a sloping board attached to the bottom of an outside door to keep out the rain etc. **2** each of a series of horizontal boards with edges overlapping to keep out the rain etc. ● *v.tr.* fit or supply with weatherboards. ☐ **weatherboarding** *n.* (in sense 2 of *n.*).

weathercock /ˈweðəˌkɒk/ *n.* **1** a weather-vane (see VANE 1) in the form of a cock. **2** an inconstant person.

weathering /ˈweðərɪŋ/ *n.* **1** the action of the weather on materials, rocks, etc. exposed to it. **2** exposure to adverse weather conditions (see WEATHER *v.* 1).

weatherly /ˈweðəlɪ/ *adj. Naut.* **1** (of a ship) making little leeway. **2** capable of keeping close to the wind. ☐ **weatherliness** *n.*

weatherman /ˈweðəˌmæn/ *n.* (*pl.* **-men**) a meteorologist, esp. one who broadcasts a weather forecast.

weatherproof /ˈweðəˌpruːf/ *adj. & v.* ● *adj.* resistant to the effects of bad weather, esp. rain. ● *v.tr.* make weatherproof. ☐ **weatherproofed** *adj.*

weave[1] /wiːv/ *v. & n.* ● *v.* (*past* **wove** /wəʊv/; *past part.* **woven** /ˈwəʊv(ə)n/ or **wove**) **1** *tr.* **a** form (fabric) by interlacing long threads in two directions. **b** form (thread) into fabric in this way. **2** *intr.* **a** make fabric in this way. **b** work at a loom. **3** *tr.* make (a basket or wreath etc.) by interlacing rods or flowers etc. **4** *tr.* **a** (foll. by *into*) make (facts etc.) into a story or connected whole. **b** make (a story) in this way. ● *n.* a style of weaving. [OE *wefan* f. Gmc]

weave[2] /wiːv/ *v.intr.* **1** move repeatedly from side to side; take an intricate course to avoid obstructions. **2** *colloq.* manoeuvre an aircraft in this way; take evasive action. ☐ **get weaving** *sl.* begin action; hurry. [prob. f. ME *weve*, var. of *waive* f. ON *veifa* WAVE]

weaver /ˈwiːvə(r)/ *n.* **1** a person whose occupation is weaving. **2** (in full **weaver-bird**) a tropical sparrow-like bird of the family Ploceidae, building elaborately woven nests. ☐ **weaver's knot** a sheet bend (see SHEET[2]) used in weaving.

web /web/ *n. & v.* ● *n.* **1** a network of fine threads constructed by a spider to catch its prey and secreted from its spinnerets; a cobweb or similar product of a spinning creature. **2 a** a complete structure or connected series (*a web of lies*). **b** a snare or trap (*trapped in his own web of deceit*). **3 a** a membrane between the toes of a swimming animal or bird. **b** the vane of a bird's feather. **4** a woven fabric; an amount woven in one piece. **5 a** a large roll of paper used in a continuous printing process. **b** an endless wire mesh on rollers, on which this is made. **6** a thin flat part connecting thicker or more solid parts in machinery etc. ● *v.* (**webbed**, **webbing**) **1** *tr.* weave a web on. **2** *intr.* weave a web. ☐ **web-foot** a foot with webbed toes. **web-footed** having web-feet. **web offset** offset printing on a web of paper. **web-wheel** a wheel having a plate or web instead of spokes, or with rim, spokes, and centre in one piece as in watch-wheels. **web-worm** *US* a gregarious caterpillar spinning a large web in which to sleep or to feed on enclosed foliage. ☐ **webbed** *adj.* [OE *web*, *webb* f. Gmc]

Webb[1] /web/, (Gladys) Mary (1881–1927), English novelist. Her novels, representative of much regional English fiction popular at the beginning of the century, include *Gone to Earth* (1917) and *Precious Bane* (1924). The earthy subject-matter and purple prose typical of her writing were satirized in *Cold Comfort Farm* by the English novelist Stella Gibbons (1902–89).

Webb[2] /web/, (Martha) Beatrice (née Potter) (1858–1943) and Sidney (James), Baron Passfield (1859–1947), English socialists, economists, and historians. After their marriage in 1892, they were prominent members of the Fabian Society, and helped to establish the London School of Economics (1895). Together they wrote several important books on socio-political theory and history, including *The History of Trade Unionism* (1894) and *Industrial Democracy* (1897), as well as founding the weekly magazine the *New Statesman* (1913). Sidney Webb became a Labour MP in 1922 and served in the first two Labour governments.

webbing /ˈwebɪŋ/ *n.* strong narrow closely woven fabric used for supporting upholstery, for belts, etc.

Weber[1] /ˈveɪbə(r)/, Carl Maria (Friedrich Ernst) von (1786–1826), German composer. He is regarded as the founder of the German romantic school of opera. In contrast to the prevailing classical school,

his opera *Der Freischütz* (1817–21) turned to German folklore for its subject-matter, and was immediately successful. His German grand opera *Euryanthe* (1822–3) gave prominence to the orchestration and made use of leitmotifs; both operas were to influence Wagner.

Weber[2] /ˈveɪbə(r)/, Max (1864–1920), German economist and sociologist. His writings on the relationship between economy and society established him as one of the founders of modern sociology. In his celebrated book *The Protestant Ethic and the Spirit of Capitalism* (1904), Weber argued that there was a direct relationship between the Protestant work ethic and the rise of Western capitalism. Throughout his work he stressed that the bureaucratization of political and economic society was the most significant development in the modernization of Western civilization. Other works include *Economy and Society* (1922).

Weber[3] /ˈveɪbə(r)/, Wilhelm Eduard (1804–91), German physicist. His early researches were in acoustics and animal locomotion, but he is chiefly remembered for his contributions in the fields of electricity and magnetism. Weber proposed a unified system for electrical units, determined the ratio between the units of electrostatic and electromagnetic charge, and devised a law of electrical force (later replaced by Maxwell's field theory). He went on to investigate electrodynamics and the nature and role of the electric charge.

weber /ˈveɪbə(r)/ *n. Physics* the SI unit of magnetic flux (symbol: **Wb**), equal to 100 million maxwells. One weber of magnetic flux would induce an EMF of one volt in a circuit of one turn when generated or removed in one second. [WEBER[3]]

Webern /ˈveɪb(ə)n/, Anton (Friedrich Ernst) von (1883–1945), Austrian composer. He followed the development of the work of his teacher Schoenberg from tonality to atonality; this departure is evident in his 1908–9 setting of songs by the German poet Stefan George (1868–1933). His music is marked by its brevity; the atonal *Five Pieces for Orchestra* (1911–13) lasts under a minute. Together with Berg, Webern became the leading exponent of the serialism developed by Schoenberg. Important serial works include the *Symphony* (1928) and the *Variations for Orchestra* (1940). During the Second World War his work was denounced by the Nazis. He was shot, accidentally, by an American soldier during the postwar occupation of Austria.

Webster[1] /ˈwebstə(r)/, John (*c.*1580–*c.*1625), English dramatist. He wrote several plays in collaboration with other dramatists but his reputation rests chiefly on two revenge tragedies, *The White Devil* (1612) and *The Duchess of Malfi* (1623). His plays were not popular in his own day; Charles Lamb, in particular, was responsible for his revival in the 19th century.

Webster[2] /ˈwebstə(r)/, Noah (1758–1843), American lexicographer. His *American Dictionary of the English Language* (1828) in two volumes was the first dictionary to give comprehensive coverage of American usage and his name survives in the many dictionaries produced by the American publishing house Merriam–Webster. These include the collegiate dictionaries, published since 1898, and *Webster's Third New International Dictionary of the English Language* (1961).

Wed. *abbr.* Wednesday.

wed /wed/ *v.* (**wedding**; *past* and *past part.* **wedded** or **wed**) **1** usu. *formal* or *literary* **a** *tr. & intr.* marry. **b** *tr.* join in marriage. **2** *tr.* unite (*wed efficiency to economy*). **3** *tr.* (as **wedded** *adj.*) of or in marriage (*wedded bliss*). **4** *tr.* (as **wedded** *adj.*) (foll. by *to*) obstinately attached or devoted (to a pursuit etc.). [OE *weddian* to pledge f. Gmc]

we'd /wiːd, wɪd/ *contr.* **1** we had. **2** we should; we would.

Weddell Sea /ˈwed(ə)l/ an arm of the Atlantic Ocean, off the coast of Antarctica. It is named after the British explorer James Weddell (1787–1834), who visited it in 1823.

wedding /ˈwedɪŋ/ *n.* a marriage ceremony (considered by itself or with the associated celebrations). ☐ **wedding breakfast** a meal etc. usually served between a wedding and the departure for the honeymoon. **wedding cake** a rich iced cake served at a wedding reception. **wedding day** the day or anniversary of a wedding. **wedding march** a march played at the entrance of the bride or the exit of the couple at a wedding. **wedding night** the night after a wedding (esp. with ref. to its consummation). **wedding ring** a ring worn by a married person. [OE *weddung* (as WED, -ING[1])]

Wedekind /ˈveɪdəˌkɪnt/, Frank (1864–1918), German dramatist. He was a key figure in the emergence of expressionist drama. His play *The Awakening of Spring* (1891) scandalized contemporary German society with its explicit and sardonic portrayal of sexual awakening. Wedekind later attacked the bourgeois sexual code at the turn of the century in his two tragedies *Earth Spirit* (1895) and *Pandora's Box* (1904), both featuring the femme fatale Lulu. The plays form the basis of Berg's opera *Lulu*.

wedge[1] /wedʒ/ *n. & v.* ● *n.* **1 a** a piece of wood or metal etc. tapering to a thin edge, that is driven between two objects or parts of an object to secure or separate them. **b** a thing separating two people or groups of people. **2** anything resembling a wedge in shape etc. (*a wedge of cheese*; *troops formed a wedge*). **3** a golf club with a wedge-shaped head. **4 a** a wedge-shaped heel. **b** a shoe with this. ● *v.tr.* **1** tighten, secure, or fasten by means of a wedge (*wedged the door open*). **2** force open or apart with a wedge. **3** (foll. by *in*, *into*) pack or thrust (a thing or oneself) tightly in or into. □ **thin end of the wedge** *colloq.* an action or procedure of little importance in itself, but likely to lead to more serious developments. **wedge-shaped 1** shaped like a solid wedge. **2** V-shaped. □ **wedgelike** *adj.* **wedgewise** *adv.* [OE *wecg* f. Gmc]

wedge[2] /wedʒ/ *v.tr. Pottery* prepare (clay) for use by cutting, kneading, and throwing down. [17th c.: orig. uncert.]

wedgie /wedʒɪ/ *n. colloq.* a shoe with a wedge-shaped heel or platform sole.

Wedgwood[1] /ˈwedʒwʊd/ Josiah (1730–95), English potter. He earned an international reputation with the pottery factories that he established in Staffordshire in the 1760s. These produced both practical and ornamental ware, maintaining high standards of quality despite large-scale production. Wedgwood's designs, and those of his chief designer John Flaxman, were often based on antique relief sculptures and contributed to the rise of neoclassical taste in England. His name is perhaps most associated with the powder-blue stoneware pieces with white embossed cameos that first appeared in 1775.

Wedgwood[2] /ˈwedʒwʊd/ *n. propr.* **1** ceramic ware made by Josiah Wedgwood and his successors, esp. a kind of fine stoneware usu. with a white cameo design. Cf. JASPER 2. **2** the characteristic pale blue colour of some Wedgwood stoneware.

wedlock /ˈwedlɒk/ *n.* the married state. □ **born in** (or **out of**) **wedlock** born of married (or unmarried) parents. [OE *wedlāc* marriage vow f. *wed* pledge (rel. to WED) + *-lāc* suffix denoting action]

Wednesday /ˈwenzdeɪ, -dɪ/ *n. & adv.* ● *n.* the fourth day of the week, following Tuesday. ● *adv. colloq.* **1** on Wednesday. **2** (**Wednesdays**) on Wednesdays; each Wednesday. [ME *wednesdei*, OE *wōdnesdæg* day of Odin (or Woden), transl. LL *Mercurii dies* day of Mercury]

Weds. *abbr.* Wednesday.

wee[1] /wiː/ *adj.* (**weer** /ˈwiːə(r)/; **weest** /ˈwiːɪst/) **1** esp. *Sc.* little; very small. **2** *colloq.* tiny; extremely small (*a wee bit*). [orig. Sc. noun, f. north. ME *wei* (small) quantity f. Anglian *wēg*: cf. WEY]

wee[2] /wiː/ *n.* esp. *Brit. sl.* = WEE-WEE.

weed /wiːd/ *n. & v.* ● *n.* **1** a wild plant growing where it is not wanted. **2** *colloq.* a thin weak-looking person; a feeble or contemptible person. **3** *sl.* an inferior horse. **4** (prec. by *the*) *sl.* **a** marijuana. **b** tobacco. ● *v.* **1** *tr.* a clear (an area) of weeds. **b** remove unwanted parts from. **2** *tr.* (foll. by *out*) a sort out (inferior or unwanted parts etc.) for removal. **b** rid (a quantity or company) of inferior or unwanted members etc. **3** *intr.* cut off or uproot weeds. □ **weed-grown** overgrown with weeds. **weed-killer** a substance used to destroy weeds. □ **weeder** *n.* **weedless** *adj.* [OE *wēod*, of unkn. orig.]

weeds /wiːdz/ *n.pl.* (in full **widow's weeds**) *archaic* deep mourning worn by a widow. [OE *wǣd(e)* garment f. Gmc]

weedy /ˈwiːdɪ/ *adj.* (**weedier**, **weediest**) **1** having many weeds. **2** (esp. of a person) weak, of poor physique; feeble, lacking strength of character. □ **weediness** *n.*

Wee Free *n.* a member of the minority group nicknamed the *Wee Free Kirk* which stood apart from the Free Church of Scotland when the majority amalgamated with the United Presbyterian Church to form the United Free Church in 1900. The group continued to call itself the Free Church of Scotland after this date.

week /wiːk/ *n.* **1** a period of seven days reckoned usu. from and to midnight on Saturday–Sunday. (*See note below.*) **2** a period of seven days reckoned from any point (*would like to stay for a week*). **3** the six days between Sundays. **4 a** the five days Monday to Friday. **b** the time spent working in this period (*a 35-hour week*). **5** (in *pl.*) a long time; several weeks (*have not seen you for weeks*; *did it weeks ago*). **6** (prec. by a specified day) a week after (that day) (*Tuesday week*; *tomorrow week*). [OE *wice* f. Gmc, prob. orig. = sequence]

▪ Days were grouped in sevens by the ancient Babylonians, who named them after the five planets known in ancient astronomy and the sun and the moon; each day bore the name of the planet supposed to govern its first hour. The Romans later adopted a similar system, introducing planetary names shortly before the Christian era; the English names for Saturday, Sunday, and Monday are based on these planetary names (Saturn, Sun, and Moon), while names for the other days of the week are derived from Anglo-Saxon words for Norse gods (see individual entries). The concept of a day of rest specially dedicated to God was retained from Judaism by the first Christians, and soon transferred to the first day of the week in honour of the Resurrection, which was said to have taken place on that day. The five-day working week was introduced in Britain and the US from the 1930s.

weekday /ˈwiːkdeɪ/ *n.* a day other than Sunday or other than at a weekend.

weekend /wiːkˈend/ *n. & v.* ● *n.* **1** Sunday and Saturday or part of Saturday. **2** this period extended slightly esp. for a holiday or visit etc. (*going away for the weekend*; *a weekend cottage*). ● *v.intr.* spend a weekend (*decided to weekend in the country*).

weekender /wiːkˈendə(r)/ *n.* **1** a person who spends weekends away from home. **2** *colloq.* a holiday cottage. **3** a small pleasure boat, esp. one designed for private use at weekends. **4** a travel bag for weekend use.

weeklong /ˈwiːklɒŋ/ *adj.* lasting for a week.

weekly /ˈwiːklɪ/ *adj., adv., & n.* ● *adj.* done, produced, or occurring once a week. ● *adv.* once a week; from week to week. ● *n.* (*pl.* **-ies**) a weekly newspaper or periodical.

ween /wiːn/ *v.tr. archaic* be of the opinion; think, suppose. [OE *wēnan* f. Gmc]

weenie var. of WIENER.

weeny /ˈwiːnɪ/ *adj.* (**weenier**, **weeniest**) *colloq.* tiny. □ **weeny-bopper** a girl (or boy) like a teeny-bopper but younger. [WEE[1] after *tiny*, *teeny*]

weep /wiːp/ *v. & n.* ● *v.* (*past* and *past part.* **wept** /wept/) **1** *intr.* shed tears. **2 a** *tr. &* (foll. by *for*) *intr.* shed tears for; bewail, lament over. **b** *tr.* utter or express with tears (*'Don't go,' he wept*; *wept her thanks*). **3 a** *intr.* be covered with or send forth drops. **b** *intr. & tr.* come or send forth in drops; exude liquid (*weeping sore*). **4** *intr.* (as **weeping** *adj.*) (of a tree or plant) having drooping branches (*weeping willow*, *weeping fig*). ● *n.* a fit or spell of weeping. □ **weep out** utter with tears. □ **weepingly** *adv.* [OE *wēpan* f. Gmc (prob. imit.)]

weeper /ˈwiːpə(r)/ *n.* **1 a** a person who weeps. **b** *hist.* a hired mourner at a funeral. **2** a small image of a mourner on a monument. **3** (in *pl.*) *hist.* **a** a man's crape hatband for funerals. **b** a widow's black crape veil or white cuffs.

weepie /ˈwiːpɪ/ *n.* (also **weepy**) (*pl.* **-ies**) *colloq.* a sentimental or emotional film, play, etc.

Weeping Cross *n. hist.* a wayside cross for penitents to pray at.

weepy /ˈwiːpɪ/ *adj.* (**weepier**, **weepiest**) *colloq.* inclined to weep; tearful. □ **weepily** *adv.* **weepiness** *n.*

weever /ˈwiːvə(r)/ *n.* an Atlantic fish of the genus *Trachinus*, which lies half buried in sand and has sharp venomous dorsal spines. [perh. f. OF *wivre*, *guivre*, serpent, dragon, f. L *vipera* VIPER]

weevil /ˈwiːvɪl/ *n.* **1** a beetle of the large family Curculionidae or a related family, with its head extended into a beak or rostrum. Weevils feed on seeds etc. and many are pests of stored grain and other plant products. **2** any insect damaging stored grain. □ **weevily** *adj.* [ME f. MLG *wevel* f. Gmc]

wee-wee /ˈwiːwiː/ *n. & v.* esp. *Brit. sl.* ● *n.* **1** the act or an instance of urinating. **2** urine. ● *v.intr.* (**-wees**, **-weed**) urinate. [20th c.: orig. unkn.]

w.e.f. *abbr.* with effect from.

weft[1] /weft/ *n.* **1 a** the threads woven across a warp to make fabric. **b** yarn for these. **c** a thing woven. **2** filling-strips in basket-weaving. [OE *weft(a)* f. Gmc: rel. to WEAVE[1]]

weft[2] var. of WAFT n. 3.

Wegener /ˈveɪɡənə(r)/, Alfred Lothar (1880–1930), German meteorologist and geologist. He was the first serious proponent of the theory of continental drift, but this was not accepted by most geologists during his lifetime, partly because he could not suggest a convincing motive force to account for continental movements. It is, however, now accepted as correct in principle. Wegener also wrote a standard textbook of meteorology. He died on the Greenland ice-cap in 1930 during an expedition.

Wehrmacht /ˈveəmɑːxt/ *n. hist.* the German armed forces, esp. the army, from 1921 to 1945. [G, = defensive force]

Wei /weɪ/ the name of several dynasties which ruled in China, especially that of AD 386–535.

Weifang /weɪˈfæŋ/ a city in Shandong province, eastern China; pop. (est. 1986) 1,310,000. It was formerly called Weihsien.

weigela /waɪˈdʒiːlə/ n. an oriental shrub of the genus *Weigela* (honeysuckle family), with red, pink, or white flowers; esp. *W. florida*, grown for ornament. [mod.L f. C. E. *Weigel*, Ger. physician (1748–1831)]

weigh[1] /weɪ/ v. **1** tr. find the weight of. **2** tr. balance in the hands to guess or as if to guess the weight of. **3** tr. (often foll. by *out*) **a** take a definite weight of; take a specified weight from a larger quantity. **b** distribute in exact amounts by weight. **4** tr. **a** estimate the relative value, importance, or desirability of; consider with a view to choice, rejection, or preference (*weighed the consequences*; *weighed the merits of the candidates*). **b** (foll. by *with*, *against*) compare (one consideration with another). **5** tr. be equal to (a specified weight) (*weighs three kilos*; *weighs very little*). **6** intr. **a** have (esp. a specified) importance; exert an influence. **b** (foll. by *with*) be regarded as important by (*the point that weighs with me*). **7** intr. (often foll. by *on*) be heavy or burdensome (to); be depressing (to). □ **weigh anchor** see ANCHOR. **weigh down 1** bring or keep down by exerting weight. **2** be oppressive or burdensome to (*weighed down with worries*). **weigh in** (of a boxer before a contest, or a jockey after a race) be weighed. **weigh-in** n. the weighing of a boxer before a fight. **weighing-machine** a machine for weighing persons or large weights. **weigh into** *colloq.* attack (physically or verbally). **weigh in with** *colloq.* advance (an argument etc.) assertively or boldly. **weigh out** (of a jockey) be weighed before a race. **weigh up** form an estimate of; consider carefully. **weigh one's words** carefully choose the way one expresses something. □ **weighable** adj. **weigher** n. [OE *wegan* f. Gmc, rel. to WAY]

weigh[2] /weɪ/ n. □ **under weigh** (of a vessel etc.) in motion, under way. [18th c.: f. WAY, from an erron. assoc. with *weigh anchor*]

weighbridge /ˈweɪbrɪdʒ/ n. a weighing-machine for vehicles, usu. having a plate set into the road for vehicles to drive on to.

weight /weɪt/ n. & v. ● n. **1** *Physics* **a** the force experienced by a body as a result of the earth's gravitation (cf. MASS[1] n. 8). **b** any similar force with which a body tends to a centre of attraction. **2** the heaviness of a body regarded as a property of it; its relative mass or the quantity of matter contained in it giving rise to a downward force (*is twice your weight*; *kept in position by its weight*). **3 a** the quantitative expression of a body's weight (*has a weight of three pounds*). **b** a scale of such weights (*troy weight*). **4** a body of a known weight for use in weighing. **5** a heavy body esp. used in a mechanism etc. (*a clock worked by weights*). **6** a load or burden (*a weight off my mind*). **7 a** influence, importance (*carried weight with the public*). **b** preponderance (*the weight of evidence was against them*). **8** (in *pl.*) blocks or discs of metal etc. used in weightlifting or weight-training. **9** the surface density of cloth etc. as a measure of its suitability. ● v.tr. **1** attach a weight to. **b** hold down with a weight or weights. **2** (foll. by *with*) impede or burden. **3** *Statistics* multiply the components of (an average) by factors to take account of their importance. **4** assign a handicap weight to (a horse). **5** treat (a fabric) with a mineral etc. to make it seem stouter. □ **put on weight 1** increase one's weight. **2** become fat or fatter. **throw one's weight about** (or **around**) *colloq.* be unpleasantly self-assertive. **weight training** a system of physical training using weights in the form of barbells or dumb-bells. **worth one's weight in gold** (of a person) exceedingly useful or helpful. [OE (ge)*wiht* f. Gmc: cf. WEIGH[1]]

weighting /ˈweɪtɪŋ/ n. an extra allowance paid in special cases, esp. to allow for a higher cost of living (*London weighting*).

weightless /ˈweɪtlɪs/ adj. (of a body, esp. in an orbiting spacecraft etc.) not apparently acted on by gravity. □ **weightlessly** adv. **weightlessness** n.

weightlifting /ˈweɪtˌlɪftɪŋ/ n. the sport of lifting heavy objects. The ancient Greeks practised weightlifting with heavy stones and the sport was included in the ancient Olympic Games; the activity persisted in many parts of Europe throughout the Middle Ages. During the 19th century circuses and music-halls included strongman acts, but the modern sport using barbells and dumb-bells developed at the end of the 19th century and was included in the first modern Olympic Games (1896). There are two standard lifts in modern weightlifting: the single-movement lift from floor to extended position, called the *snatch*, and the two-movement lift from floor to shoulder position, and from shoulders to extended position, called the *jerk*. □ **weightlifter** n.

weighty /ˈweɪtɪ/ adj. (**weightier**, **weightiest**) **1** weighing much; heavy. **2** momentous, important. **3** (of utterances etc.) deserving consideration; careful and serious. **4** influential, authoritative. □ **weightily** adv. **weightiness** n.

Weihsien /weɪˈʃjen/ a former name for WEIFANG.

Weil /vaɪl, French vɛj/, Simone (1909–43), French essayist, philosopher, and mystic. Weil deliberately chose to work in a Renault car factory (1934–5) and to serve in the Spanish Civil War on the Republican side (1936). Two years later she had the first in a series of mystical experiences which had a profound influence on her writing; she did not, however, affiliate herself to any established religion. During the Second World War, Weil joined the resistance movement in England, where she died from tuberculosis while weakened by voluntary privations. Her reputation is based on the posthumous publication of such autobiographical works as *Waiting for God* (1949) and *Notebooks* (1951–6).

Weill /vaɪl/, Kurt (1900–50), German composer, resident in the US from 1935. In 1926 Weill married the Austrian singer Lotte Lenya (1900–81), for whom many of his songs were written. He was based in Berlin until 1933 and is best known for the operas he wrote in collaboration with Bertolt Brecht, political satires which evoke the harsh decadence of the pre-war period in Germany and which are marked by his direct and harmonically simple style of composition. These include *The Rise and Fall of the City of Mahagonny* (1927) and *The Threepenny Opera* (1928).

Weil's disease /vaɪlz/ n. *Med.* a severe, sometimes fatal, form of leptospirosis with fever, jaundice, and muscle pain, transmitted by rats via contaminated water. [H. Adolf *Weil*, Ger. physician (1848–1916)]

Weimar /ˈvaɪmɑː(r)/ a city in Thuringia, central Germany; pop. (1991) 59,100. It was famous in the late 18th and early 19th century for its intellectual and cultural life; both Goethe and Schiller lived and worked there. (See also WEIMAR REPUBLIC.)

Weimaraner /ˈvaɪməˌrɑːnə, ˈwaɪ-/ n. a usu. grey dog of a variety of pointer used as a gun dog. [G, f. WEIMAR]

Weimar Republic the German republic of 1919–33, so called because its constitution was drawn up at Weimar. The republic was faced with huge reparation costs deriving from the Treaty of Versailles as well as soaring inflation and high unemployment. The 1920s saw a growth in support for right-wing groups and the Republic was eventually overthrown by the Nazi Party of Adolf Hitler, after his appointment as Chancellor by President Hindenburg.

weir /wɪə(r)/ n. **1** a dam built across a river to raise the level of water upstream or regulate its flow. **2** an enclosure of stakes etc. set in a stream as a trap for fish. [OE *wer* f. *werian* dam up]

weird /wɪəd/ adj. & n. ● adj. **1** uncanny, supernatural. **2** *colloq.* strange, queer, incomprehensible. **3** *archaic* connected with fate. ● n. esp. *Sc. archaic* fate, destiny. □ **the weird sisters 1** the Fates. **2** witches. □ **weirdly** adv. **weirdness** n. [(earlier as noun) f. OE *wyrd* destiny f. Gmc]

weirdie /ˈwɪədɪ/ n. (also **weirdy**) (pl. **-ies**) *colloq.* = WEIRDO.

weirdo /ˈwɪədəʊ/ n. (pl. **-os**) *colloq.* an odd or eccentric person.

Weismann /ˈvaɪsmən/, August Friedrich Leopold (1834–1914), German biologist, one of the founders of modern genetics. He expounded the theory of germ plasm, a substance which he postulated bore the factors that determine the transmission of characters from parent to offspring. The theory ruled out the transmission of acquired characteristics. Weismann realized that reproduction involved the halving and later reuniting of chromosomes, and suggested that variability came from their recombination.

Weissmuller /ˈvaɪsˌmʊlə(r)/, John Peter ('Johnny') (1904–84), American swimmer and actor. He won three Olympic gold medals in 1924 and two in 1928. He was the first man to swim 100 metres in under a minute and set twenty-eight world records in freestyle events. He later achieved wider recognition as the star of the Tarzan films of the 1930s and 1940s.

Weizmann /ˈvaɪtsmən, ˈwaɪzmən/, Chaim (Azriel) (1874–1952), Israeli statesman, President 1949–52. Born in Russia, he became a British citizen in 1910. A supporter of Zionism from the early 1900s, Weizmann participated in the negotiations that led to the Balfour Declaration (1917), which outlined British support for a Jewish homeland in Palestine. He later served as president of the World Zionist Organization (1920–31; 1935–46), facilitating Jewish immigration into Palestine in the 1930s. Weizmann also played an important role in persuading the US government to recognize the new state of Israel (1948) and became its first President in 1949.

weka /ˈweɪkə/ n. a large flightless New Zealand rail, *Gallirallus australis*, with brown and black plumage. [Maori: imit. of its cry]

Weland see WAYLAND THE SMITH.

Welch /welʃ/ var. of WELSH (now only in *Royal Welch Fusiliers*).

welch var. of WELSH.

welcome /'welkəm/ *n., int., v., & adj.* ● *n.* the act or an instance of greeting or receiving (a person, idea, etc.) gladly; a kind or glad reception (*gave them a warm welcome*). ● *int.* expressing such a greeting (*welcome!; welcome home!*). ● *v.tr.* receive with a welcome (*welcomed them home; would welcome the opportunity*). ● *adj.* **1** that one receives with pleasure (*a welcome guest; welcome news*). **2** (foll. by *to*, or *to* + infin.) **a** cordially allowed or invited; released of obligation (*you are welcome to use my car*). **b** *iron.* gladly given (*an unwelcome task, thing, etc.*) (*here's my work and you are welcome to it*). □ **make welcome** receive hospitably. **outstay one's welcome** stay too long as a visitor etc. **you are welcome** there is no need for thanks. □ **welcomely** *adv.* **welcomeness** *n.* **welcomer** *n.* **welcomingly** *adv.* [orig. OE *wilcuma* a person whose coming is pleasing f. *wil-* desire, pleasure + *cuma* comer, with later change to *wel-* WELL[1] after OF *bien venu* or ON *velkominn*]

weld[1] /weld/ *v. & n.* ● *v.tr.* **1 a** hammer or press (pieces of iron or other metal usu. heated but not melted) into one piece. **b** join by fusion with an electric arc etc. **c** form by welding into some article. **2** fashion (arguments, members of a group, etc.) into an effectual or homogeneous whole. ● *n.* a welded joint. □ **welder** *n.* **weldable** *adj.* **weldability** /ˌweldə'bɪlɪtɪ/ *n.* [alt. of WELL[2] *v.* in obs. sense 'melt or weld (heated metal)', prob. infl. by past part.]

weld[2] /weld/ *n.* **1** a plant, *Reseda luteola*, related to mignonette and yielding a yellow dye. **2** *hist.* this dye. [ME f. OE *w(e)alde* (unrecorded): cf. MDu. *woude*, MLG *walde*]

welfare /'welfeə(r)/ *n.* **1** well-being, happiness; health and prosperity (of a person or a community etc.). **2** (**Welfare**) **a** the maintenance of persons in such a condition esp. by statutory procedure or social effort. **b** financial support given for this purpose. □ **welfare work** organized effort for the welfare of the poor, disabled, etc. [ME f. WELL[1] + FARE]

welfare state *n.* a comprehensive system whereby the state undertakes to protect the health and well-being of its citizens, by providing pensions, hospitals, sickness and unemployment benefit, etc.; a country practising this system. Although there were social insurance schemes in Germany and other European states in the late 19th century, and in Britain before the First World War, the foundations for the modern welfare state were laid by the Beveridge Report of 1942. Most of William Beveridge's proposals, including the establishment of a National Health Service and the National Insurance scheme, were implemented by Attlee's Labour administration in 1948 and have been widely imitated, especially in Europe. In recent years high unemployment and poor social conditions creating demand have led to concern in Britain and elsewhere over the direct cost to the state of welfare provision.

welfarism /'welfeəˌrɪz(ə)m/ *n.* principles characteristic of a welfare state. □ **welfarist** *n.*

welkin /'welkɪn/ *n. poet.* the sky; the upper air. [OE *wolcen* cloud, sky]

Welkom /'welkəm, 'vel-/ a town in central South Africa, in Orange Free State; pop. (1985) 185,500.

well[1] /wel/ *adv., adj., & int.* ● *adv.* (**better, best**) **1** in a satisfactory way (*you have worked well*). **2** in the right way (*well said; you did well to tell me*). **3** with some talent or distinction (*plays the piano well*). **4** in a kind way (*treated me well*). **5 a** thoroughly, extensively, soundly (*polish it well; team was well beaten*). **b** intimately, closely (*knew them well*). **6** with heartiness or approval; favourably (*speak well of; the book was well reviewed*). **7** probably, reasonably, advisably (*you may well be right; you may well ask; we might well take the risk*). **8** to a considerable extent (*is well over forty*). **9** successfully, fortunately (*it turned out well*). **10** luckily, opportunely (*well met!*). **11** with a fortunate outcome; without disaster (*were well rid of them*). **12** profitably (*did well for themselves*). **13** comfortably, abundantly, liberally (*we live well here; the job pays well*). **14** *sl.* (as an intensifier) very, extremely (*was kept well busy*). ● *adj.* (**better, best**) **1** (usu. *predic.*) in good health (*are you well?; was not a well person*). **2** (*predic.*) **a** in a satisfactory state or position (*all is well*). **b** advisable (*it would be well to enquire*). ● *int.* expressing surprise, resignation, insistence, etc., or used merely to introduce a remark, though sometimes also to express acceptance of or to qualify a situation or foregoing remark (*well I never!; well, I suppose so; well, who was it?*). □ **as well 1** in addition; to an equal extent. **2** (also **just as well**) with equal reason; with no loss of advantage or need for regret (*may as well give up; it would be just as well to stop now*). **as well as** in addition to. **leave** (or **let**) **well alone** avoid needless change or disturbance. **take well** react calmly to (a thing, esp. bad news). **well-acquainted** (usu. foll. by *with*) familiar. **well-adjusted 1** mentally and emotionally stable. **2** in a good state

of adjustment. **well-advised** (usu. foll. by *to* + infin.) (of a person) prudent (*would be well-advised to wait*). **well-affected** (often foll. by *to, towards*) favourably disposed. **well and good** expressing dispassionate acceptance of a decision etc. **well and truly** decisively, completely. **well-appointed** having all the necessary equipment. **well aware** certainly aware (*well aware of the danger*). **well away 1** having made considerable progress. **2** *colloq.* fast asleep or drunk. **well-balanced 1** sane, sensible. **2** equally matched. **3** having a symmetrical or orderly arrangement of parts. **well-behaved** see BEHAVE. **well-being** a state of being well, healthy, contented, etc. **well-beloved** *adj.* dearly loved. ● *n.* (*pl.* same) a dearly loved person. **well-born** of noble family. **well-bred** having or showing good breeding or manners. **well-built 1** of good construction. **2** (of a person) big and strong and well-proportioned. **well-chosen** (of words etc.) carefully selected for effect. **well-conditioned** in good physical or moral condition. **well-conducted** (of a meeting etc.) properly organized and controlled. **well-connected** see CONNECTED. **well-covered** *colloq.* plump, corpulent. **well-cut 1** (of a garment) well-tailored. **2** (esp. of hair) cut skilfully. **well-defined** clearly indicated or determined. **well-deserved** rightfully merited or earned. **well-disposed** (often foll. by *towards*) having a good disposition or friendly feeling (for). **well done 1** (of meat etc.) thoroughly cooked. **2** (of a task etc.) performed well (also as *int.*). **well-dressed** fashionably smart. **well-earned** fully deserved. **well-endowed 1** well provided with talent etc. **2** *colloq.* **a** (of a man) having large genitals. **b** (of a woman) large-breasted. **well-established** long-standing, familiar, traditional. **well-favoured** good-looking. **well-fed** having or having had plenty to eat. **well-formed** (*attrib.*) **1** correctly or attractively formed. **2** conforming to the formation rules of a logical system. **well-found** = *well-appointed*. **well-founded** (of suspicions etc.) based on good evidence; having a foundation in fact or reason. **well-groomed** (of a person) with carefully tended hair, clothes, etc. **well-grounded 1** = *well-founded*. **2** having a good training in or knowledge of the groundwork of a subject. **well-heeled** *colloq.* wealthy. **well-hung** *colloq.* (of a man) having large genitals. **well-informed** having much knowledge or information about a subject. **well-intentioned** having or showing good intentions. **well-judged** opportunely, skilfully, or discreetly done. **well-kept** kept in good order or condition. **well-knit** (esp. of a person) compact; not loose-jointed or sprawling. **well-known 1** known to many. **2** known thoroughly. **well-liked** liked; popular. **well-made 1** strongly or skilfully manufactured. **2** (of a person or animal) having a good build. **well-maintained 1** kept in good repair. **2** kept up to date. **well-mannered** having good manners. **well-marked** distinct; easy to detect. **well-matched** see MATCH[1]. **well-meaning** (or **-meant**) well-intentioned (but ineffective or unwise). **well off 1** having plenty of money. **2** in a fortunate situation or circumstances. **well-oiled** *colloq.* **1** drunk. **2** (of a compliment etc.) easily expressed through habitual use. **well-ordered** arranged in an orderly manner. **well-organized 1** skilfully or carefully organized. **2** (of a person) orderly; able to organize. **well-paid 1** (of a job) that pays well. **2** (of a person) amply rewarded for a job. **well-pleased** highly gratified or satisfied. **well-prepared 1** prepared with care. **2** having prepared thoroughly (for an interview etc.). **well-preserved** see PRESERVE. **well-read** knowledgeable through much reading. **well-received** welcomed; favourably received. **well-rounded 1** complete and symmetrical. **2** (of a phrase etc.) complete and well expressed. **3** (of a person) having or showing a fully developed personality, ability, etc. **4** fleshy, plump. **well-spent** (esp. of money or time) used profitably. **well-spoken** articulate or refined in speech. **well-thought-of** having a good reputation; esteemed, respected. **well-thought-out** carefully devised. **well-thumbed** bearing marks of frequent handling. **well-timed** opportune, timely. **well-to-do** prosperous. **well-trodden** much frequented. **well-tried** often tested with good results. **well-turned 1** (of a compliment, phrase, or verse) elegantly expressed. **2** (of a leg, ankle, etc.) elegantly shaped or displayed. **well-upholstered** see UPHOLSTER. **well-used 1** much frequented. **2** much handled or worn. **well-wisher** a person who wishes one well. **well-woman** of or relating to health advice and check-ups for problems specific to women (*well-woman clinic*). **well-worn 1** much worn by use. **2** (of a phrase etc.) trite, hackneyed. **well worth** certainly worth (*well worth a visit; well worth visiting*). ¶ A hyphen is normally used in combinations of *well-* when used attributively, but not when used predicatively, e.g. *a well-made coat* but *the coat is well made*. □ **wellness** *n.* [OE *wel, well* prob. f. the same stem as WILL[1]]

well[2] /wel/ *n. & v.* ● *n.* **1** a shaft sunk into the ground to obtain water, oil, etc. **2** an enclosed space like a well-shaft, e.g. in the middle of a building for stairs or a lift, or for light or ventilation. **3** (foll. by *of*) a source, esp. a copious one (*a well of information*). **4 a** a mineral spring. **b** (in *pl.*) a spa. **5** = *ink-well*. **6** *archaic* a water-spring or fountain. **7** *Brit.* a railed space for solicitors etc. in a lawcourt. **8** a depression for gravy etc. in a dish or tray, or for a mat in the floor. **9** *Physics* a region of minimum potential etc. ● *v.intr.* (foll. by *out*, *up*) spring as from a fountain; flow copiously. □ **well-head** (or **-spring**) a source. [OE *wella* (= OHG *wella* wave, ON *vella* boiling heat), *wellan* boil, melt f. Gmc]

we'll /wiːl, wɪl/ *contr.* we shall; we will.

Welland Canal /'welənd/ (also **Welland Ship Canal**) a canal in southern Canada, 42 km (26 miles) long, linking Lake Erie with Lake Ontario, bypassing Niagara Falls and forming part of the St Lawrence Seaway. It has eight locks, providing navigation for ocean-going vessels.

Welles /welz/, (George) Orson (1915–85), American film director and actor. Welles formed his own Mercury Theatre company and caused a public sensation in 1938 with a radio dramatization of H. G. Wells's *The War of the Worlds*, whose realism and contemporary American setting persuaded many listeners that a Martian invasion was really happening. Turning to films, he produced, directed, wrote, and acted in the critically acclaimed *Citizen Kane* (1941), based on the life of the newspaper tycoon William Randolph Hearst. Welles was an important figure in the *film noir* genre as shown by such films as *The Lady from Shanghai* (1948), in which he co-starred with his second wife Rita Hayworth. His best-known film performance was as Harry Lime in *The Third Man* (1949).

Wellington[1] /'welɪŋtən/ the capital of New Zealand, situated at the southern tip of North Island; pop. (1991) 150,300. The city was established in 1840 and named after the Duke of Wellington. It became capital in 1865, when the seat of government was moved from Auckland.

Wellington[2] /'welɪŋtən/, 1st Duke of (title of Arthur Wellesley; also known as the 'Iron Duke') (1769–1852), British soldier and Tory statesman, Prime Minister 1828–30 and 1834. Born in Ireland, he served as commander of British forces in the Peninsular War, winning a series of victories against the French and finally driving them across the Pyrenees into southern France (1814). The following year Wellington defeated Napoleon at the Battle of Waterloo, so ending the Napoleonic Wars. During his first term as Prime Minister he granted Catholic Emancipation under pressure from Daniel O'Connell.

wellington /'welɪŋtən/ *n.* (in full **wellington boot**) *Brit.* a waterproof rubber or plastic boot usu. reaching the knee. [WELLINGTON[2]]

well-nigh /'welnaɪ/ *adv. archaic* or *rhet.* almost (*well-nigh impossible*).

Wells /welz/, H(erbert) G(eorge) (1866–1946), English novelist. After studying biology with T. H. Huxley, he wrote some of the earliest science-fiction novels, such as *The Time Machine* (1895) and *The War of the Worlds* (1898). These combined political satire, warnings about the dangerous new powers of science, and a hope for the future. In 1903 Wells joined the Fabian Society; his socialism was reflected in several comic novels about lower-middle-class life, including *Kipps* (1905) and *The History of Mr Polly* (1910). He is also noted for much speculative writing about the future of society, particularly in *The Shape of Things to Come* (1933).

Wells, Fargo, & Co. /welz, 'fɑːɡəʊ/ a US transportation company founded in 1852 by the businessmen Henry Wells (1805–78) and William Fargo (1818–81) and others. It carried mail to and from the newly developed West, founded a San Francisco bank, and later ran a stagecoach service (having bought the Pony Express system) until the development of a transcontinental railway service. In 1918 the express carrier business was taken over by the American Railway Express Company.

welly /'welɪ/ *n.* (also **wellie**) (*pl.* **-ies**) *colloq.* = WELLINGTON. □ **give it some welly** an expression of encouragement to use more effort, strength, etc. [abbr.]

wels /welz, vels/ *n.* a very large freshwater catfish, *Silurus glanis*, found from central Europe to central Asia. [G]

Welsbach /'welzbæk/, Carl Auer von, see AUER.

Welsh /welʃ/ *adj. & n.* ● *adj.* of or relating to Wales or its people or language. ● *n.* **1** the Celtic language of Wales. (*See note below.*) **2** (prec. by *the*; treated as *pl.*) the people of Wales. □ **Welsh corgi** see CORGI. **Welsh dresser** a type of dresser with open shelves above a cupboard. **Welsh harp** a harp with three rows of strings. **Welsh onion** a kind of onion, *Allium fistulosum*, forming clusters of bulbs. **Welsh rabbit** (or **rarebit** by folk etymology) a dish of melted cheese on toast. [OE *Welisc*, *Wælisc*, etc., f. Gmc f. L *Volcae*, the name of a Celtic people]

■ Welsh belongs to the Brythonic group of the Celtic languages. It is

spoken by about 500,000 people in Wales, mostly in the north and west, and has a substantial literature dating from the medieval period. After the Act of Union with England in 1536 the number of Welsh speakers began to decline; its survival was assisted by the publication of a Bible in Welsh in 1588. Revivals in the 18th, 19th, and particularly the 20th century led to a halt in the decline; Welsh now has official status (with English) in Wales, and is taught in schools. In addition, a Welsh television channel was established in the 1970s.

welsh /welʃ/ *v.intr.* (also **welch** /weltʃ/) **1** (of a loser of a bet, esp. a bookmaker) decamp without paying. **2** evade an obligation. **3** (foll. by *on*) **a** fail to carry out a promise to (a person). **b** fail to honour (an obligation). □ **welsher** *n.* [19th c.: orig. unkn.]

Welshman /'welʃmən/ *n.* (*pl.* **-men**) a man who is Welsh by birth or descent.

Welshwoman /'welʃˌwʊmən/ *n.* (*pl.* **-women**) a woman who is Welsh by birth or descent.

welt /welt/ *n. & v.* ● *n.* **1** a leather rim sewn round the edge of a shoe-upper for the sole to be attached to. **2** = WEAL[1] *n.* **3** a ribbed or reinforced border of a garment; a trimming. **4** a heavy blow. ● *v.tr.* **1** provide with a welt. **2** rain welts on; thrash. [ME *welte*, *walt*, of unkn. orig.]

Weltanschauung /'veltanˌʃaʊʊŋ/ *n.* a particular philosophy or view of life; a conception of the world. [G f. *Welt* world + *Anschauung* perception]

welter[1] /'weltə(r)/ *v. & n.* ● *v.intr.* **1** roll, wallow; be washed about. **2** (foll. by *in*) lie prostrate or be soaked or steeped in blood etc. ● *n.* **1** a state of general confusion. **2** (foll. by *of*) a disorderly mixture or contrast of beliefs, policies, etc. [ME f. MDu., MLG *welteren*]

welter[2] /'weltə(r)/ *n.* **1** a heavy rider or boxer. **2** *colloq.* a heavy blow. **3** *colloq.* a big person or thing. [19th c.: orig. unkn.]

welterweight /'weltəˌweɪt/ *n.* **1** a weight in certain sports intermediate between lightweight and middleweight, in the amateur boxing scale 63.5–67 kg but differing for professionals, wrestlers, and weightlifters. **2** a sportsman of this weight. □ **junior welterweight 1** a weight in professional boxing of 61.2–63.5 kg. **2** a professional boxer of this weight. **light welterweight 1** a weight in amateur boxing of 60–63.5 kg. **2** an amateur boxer of this weight.

Weltschmerz /'veltʃmeəts/ *n.* a feeling of pessimism; an apathetic or vaguely yearning outlook on life. [G f. *Welt* world + *Schmerz* pain]

Welty /'weltɪ/, Eudora (b.1909), American novelist, short-story writer, and critic. During the Depression of the mid-1930s she worked in her native Mississippi as an official photographer for the US government. She published her first short-story collection in 1941; several collections followed including *The Golden Apples* (1949), about three generations of Mississippi families. Welty's novels chiefly focus on life in the South and contain Gothic elements; they include *Delta Wedding* (1946) and *The Optimist's Daughter* (1972), which won the Pulitzer Prize.

Wembley Stadium /'wemblɪ/ a sports stadium in Wembley, NW London, now used also for pop concerts, religious events, etc. The FA Cup Final and the England football team's home matches are played there, and it was the venue for the 1966 World Cup Final.

wen[1] /wen/ *n.* **1** a benign tumour on the skin esp. of the scalp. **2** an outstandingly large or congested city. □ **the great wen** London. [OE *wen*, *wenn*, of unkn. orig.: cf. Du. *wen*, MLG *wene*, LG *wehne* tumour, wart]

wen[2] /wen/ *n.* (also **wyn** /wɪn/) a runic letter in Old and Middle English, later replaced by *w*. [OE, var. of *wyn* joy (see WINSOME), used because it begins with this letter: cf. THORN 3]

Wenceslas /'wensɪsləs/ (also **Wenceslaus**) (1361–1419), king of Bohemia (as Wenceslas IV) 1378–1419. He became king of Germany and Holy Roman emperor in the same year as he succeeded to the throne of Bohemia, but was deposed by the German Electors in 1400. As king of Bohemia, Wenceslas supported the growth of the Hussite movement, but could not prevent the execution of John Huss in 1415.

Wenceslas, St (also **Wenceslaus**; known as 'Good King Wenceslas') (c.907–29), Duke of Bohemia and patron saint of the Czech Republic. He worked to Christianize the people of Bohemia but was murdered by his brother Boleslaus; he later became venerated as a martyr and hero of Bohemia. The story told in the Christmas carol 'Good King Wenceslas', by J. M. Neale (1818–66), appears to have no basis in fact. Feast day, 28 Sept.

wench /wentʃ/ *n. & v.* ● *n.* **1** *joc.* a girl or young woman. **2** *archaic* a prostitute. ● *v.intr. archaic* (of a man) consort with prostitutes.

□ **wencher** *n.* [ME *wenche, wenchel* f. OE *wencel* child: cf. OE *wancol* weak, tottering]

Wen-Chou see WENZHOU.

Wend /wend/ *n.* a member of a Slavic people of northern Germany, now inhabiting eastern Saxony. □ **Wendic** *adj.* **Wendish** *adj.* [G *Wende* f. OHG *Winida*, of unkn. orig.]

wend /wend/ *v.tr.* & *intr. literary* or *archaic* go. □ **wend one's way** make one's way. [OE *wendan* turn f. Gmc, rel. to WIND²]

Wendy house /'wendɪ/ *n.* a children's small houselike tent or structure for playing in. [after the house built around *Wendy* in J. M. Barrie's *Peter Pan*]

Wensleydale /'wenzlɪˌdeɪl/ *n.* **1** a variety of white or blue cheese. **2** a breed of sheep with long wool. [*Wensleydale* in North Yorkshire]

went *past* of GO¹.

wentletrap /'went(ə)lˌtræp/ *n.* a marine gastropod mollusc of the family Epitoniidae, having a spiral conical shell with ridged whorls. [Du. *wenteltrap* winding stair, spiral shell]

Wenzhou /wɛnˈdʒəʊ/ (also **Wen-Chou** /-'tʃəʊ/) an industrial city in Zhejiang province, eastern China; pop. (est. 1984) 519,100.

wept *past* of WEEP.

were *2nd sing. past, pl. past,* and *past subjunctive* of BE.

we're /wɪə(r)/ *contr.* we are.

weren't /wɜːnt/ *contr.* were not.

werewolf /'wɪəwʊlf, 'weə-/ *n.* (also **werwolf** /'wɜːwʊlf/) (*pl.* **-wolves**) a mythical being who at times changes from a person to a wolf. [OE *werewulf*: first element perh. f. OE *wer* man = L *vir*]

Werner¹ /'vɜːnə(r)/, Abraham Gottlob (1749–1817), German geologist. He was the chief exponent of the Neptunian theory, which included the belief that rocks such as granites (now known to be of igneous origin) were formed as crystalline precipitates from a primeval ocean. Although this theory was invalid, the controversy that it stimulated prompted a rapid increase in geological research, and Werner's was probably the first attempt to establish a universal stratigraphic sequence.

Werner² /'vɜːnə(r)/, Alfred (1866–1919), French-born Swiss chemist, founder of coordination chemistry. He demonstrated that stereochemistry was not just the property of carbon compounds, but was general to the whole of chemistry. In 1893 he announced his theory of chemical coordination, proposing a secondary or residual form of valency to explain the structures of coordination compounds. He was awarded the Nobel Prize for chemistry in 1913.

wert /wɜːt, wət/ *archaic 2nd sing. past* of BE.

Weser /'veɪzə(r)/ a river of NW Germany, which is formed at the junction of the Werra and Fulda rivers in Lower Saxony and flows 292 km (182 miles) northwards to the North Sea near Bremerhaven.

Wesker /'weskə(r)/, Arnold (b.1932), English dramatist. His writing is associated with British kitchen-sink drama of the 1950s and his plays, reflecting his commitment to socialism, often deal with the working-class search for cultural identity, as in *Roots* (1959). He is also noted for *Chips with Everything* (1962), a study of class attitudes in the RAF during national service, and *The Merchant* (1977), which reworks the story of Shylock in an indictment of anti-Semitism.

Wesley /'wezlɪ/, John (1703–91), English preacher and co-founder of Methodism. He became the leader of a small group in Oxford, formed in 1729 by his brother Charles (1707–88); its members were nicknamed the 'Methodists'. In 1738 John Wesley experienced a spiritual conversion as a result of a reading in London of Luther's preface to the Epistle to the Romans. He resolved to devote his life to evangelistic work; however, when Anglican opposition caused the churches to be closed to him, he and his followers began preaching out of doors. Wesley subsequently travelled throughout Britain winning many working-class converts and widespread support of the Anglican clergy. Despite his wish for Methodism to remain within the Church of England, his practice of ordaining his missionaries himself (since the Church refused to do so) brought him increasing opposition from the Anglican establishment and eventual exclusion; the Methodists formally separated from the Church of England in 1791.

Wesleyan /'wezlɪən/ *adj.* & *n.* ● *adj.* of or relating to John Wesley, his doctrines, or Christian denominations based on them. ● *n.* a member of any of these denominations. □ **Wesleyanism** *n.*

Wessex /'wesɪks/ the kingdom of the West Saxons, established in Hampshire in the early 6th century and gradually extended by conquest to include much of southern England. Under Alfred the Great and his successors it formed the nucleus of the Anglo-Saxon kingdom of England. Athelstan, Alfred's grandson, became king of England. The name was revived in the 19th century by Thomas Hardy to designate the south-western counties of England (especially Dorset) in which his novels are set.

West¹ /west/, Benjamin (1738–1820), American painter, resident in Britain from 1763. He was appointed historical painter to George III in 1769 and became the second president of the Royal Academy on Joshua Reynolds's death in 1792. His portrait *The Death of General Wolfe* (1771) depicted its subject in contemporary rather than classical dress, signifying a new departure in English historical painting.

West² /west/, Mae (1892–1980), American actress and dramatist. She made her name on Broadway appearing in her own comedies *Sex* (1926) and *Diamond Lil* (1928), which were memorable for their frank and spirited approach to sexual matters; the former resulted in a short period of imprisonment for alleged obscenity. In the early 1930s West began her long and successful Hollywood career; major films included *She Done Him Wrong* (1933) and *Klondike Annie* (1936). She is also noted for the autobiography *Goodness Had Nothing To Do With It* (1959). (Cf. MAE WEST.)

West³ /west/, Dame Rebecca (born Cicily Isabel Fairfield) (1892–1983), British writer and feminist, born in Ireland. From 1911 she wrote journalistic articles in support of women's suffrage, adopting 'Rebecca West' as her pseudonym after one of Ibsen's heroines. She was sent to report on Yugoslavia in 1937, publishing her observations in the *Black Lamb and Grey Falcon* (1942) in two volumes. Her other works include *The Meaning of Treason* (1949), a study of the psychology of traitors, and *A Train of Power* (1955), a critique of the Nuremberg war trials. West's many novels include *The Fountain Overflows* (1957).

west /west/ *n., adj.,* & *adv.* ● *n.* **1 a** the point of the horizon where the sun sets at the equinoxes (cardinal point 90° to the left of north). **b** the compass point corresponding to this. **c** the direction in which this lies. **2** (usu. **the West**) **a** European in contrast to oriental civilization. **b** *hist.* the non-Communist States of Europe and North America. **c** non-Communist States or regions in general. **d** the western part of the late Roman Empire. **e** the western part of a country, town, etc. **3** *Bridge* a player occupying the position designated 'west'. ● *adj.* **1** towards, at, near, or facing west. **2** coming from the west (*west wind*). ● *adv.* **1** towards, at, or near the west. **2** (foll. by *of*) further west than. □ **go west** *sl.* be killed or destroyed etc. **west-north-** (or **south-**) **west** the direction or compass-point midway between west and north-west (or south-west). [OE f. Gmc]

West Africa the western part of the African continent, especially the countries bounded by and including Mauritania, Mali, and Niger in the north and Gabon in the south.

West Bank a region west of the River Jordan and north-west of the Dead Sea. It contains Jericho, Hebron, Nablus, Bethlehem, and other settlements; 97 per cent of its inhabitants are Palestinian Arabs. It became part of Jordan in 1948 and was occupied by Israel following the Six Day War of 1967. In 1993 an agreement was signed which granted limited autonomy to the Palestinians, including the withdrawal of Israeli troops and the establishment of a Palestinian police force.

West Bengal a state in eastern India; capital, Calcutta. It was formed in 1947 from the predominantly Hindu area of former Bengal.

westbound /'westbaʊnd/ *adj.* travelling or leading westwards.

West Bromwich /'brɒmɪtʃ/ an industrial town in the West Midlands of England; pop. (1981) 154,530.

West Country the south-western counties of England.

West End the entertainment and shopping area of London to the west of the City.

westering /'westərɪŋ/ *adj.* (of the sun) nearing the west. [*wester* (v.) ME f. WEST]

westerly /'westəlɪ/ *adj., adv.,* & *n.* ● *adj.* & *adv.* **1** in a western position or direction. **2** (of a wind) blowing from the west. ● *n.* (*pl.* **-ies**) a wind blowing from the west. [*wester* (adj.) f. OE *westra* f. WEST]

western /'westən/ *adj.* & *n.* ● *adj.* **1** of or in the west; inhabiting the west. **2** lying or directed towards the west. **3** (**Western**) of or relating to the West (see WEST *n.* 2). ● *n.* a film, television drama, or novel belonging to a genre depicting life in the western US during the 19th and early 20th centuries, usu. featuring cowboys in heroic roles, gunfights, etc. (See note below.) □ **westernmost** *adj.* [OE *westerne* (as WEST, -ERN)]

▪The western developed in the US pulp fiction of the 19th century, glamorizing the lawless world of the Wild West, and became an established form with novels such as *The Virginian* (1902) by Owen Wister (1860–1938) and *Riders of the Purple Sage* (1912) by Zane Grey (1872–1939). Westerns constitute the oldest and most enduring genre in cinema: in 1908–14 D. W. Griffith made a number of one-reel westerns, and by the 1920s the western was internationally accepted. In the 1930s many of the conventions of the genre became established, such as the opposition between heroic cowboy and villainous Indian, and popular characters such as the Lone Ranger and Hopalong Cassidy emerged. One of the best-known westerns of this period is *Stagecoach* (1939), directed by John Ford and starring John Wayne. Other famous films from the western's heyday include *High Noon* (1952) and *Shane* (1953), and there were many television drama series. From the 1960s the western became less popular, and films such as the spaghetti westerns of the Italian director Sergio Leone (1928–89), mainly starring Clint Eastwood, and films by Sam Peckinpah (1925–84), and Eastwood himself, have often portrayed violence and brutality much more explicitly and in other ways departed from earlier conventions of the genre.

Western Australia a state comprising the western part of Australia; pop. (1990) 1,642,700; capital, Perth. It was colonized by the British in 1826, and was federated with the other states of Australia in 1901.

Western Cape a province of south-western South Africa, formerly part of Cape Province; capital, Cape Town.

Western Church the Churches of Western Christendom as distinct from the Eastern or Orthodox Church.

Western Empire the western part of the Roman Empire, after its division in AD 395 (see ROMAN EMPIRE).

westerner /ˈwestənə(r)/ n. a native or inhabitant of the west.

Western Front the zone of fighting in western Europe in the First World War, in which the German army engaged the armies to its west, i.e. France, the UK (and its dominions), and, from 1917, the US. For most of the war the front line stretched from the Vosges mountains in eastern France through Amiens to Ostend in Belgium. Fighting began in Aug. 1914: German forces attacking through Belgium were checked in the first battle of the Marne and then at Ypres, after which both sides engaged in trench warfare, the distinctive feature of warfare on this front. Battles were inconclusive with heavy casualties on both sides, notably at Verdun, the Somme, and Passchendaele. Early in 1917 the Germans withdrew to the Hindenburg Line, a fortified line of trenches that was not breached for more than a year; after the entry of the US into the war in 1917, US forces in the region helped to tip the balance in the Allies' favour.

Western Ghats see GHATS, THE.

western hemisphere the half of the earth containing the Americas.

Western Isles 1 an alternative name for the Hebrides (see HEBRIDES, THE). **2** an administrative region of Scotland, consisting of the Outer Hebrides; administrative centre, Stornoway.

westernize /ˈwestənaɪz/ v.tr. (also **Westernize, -ise**) influence with or convert to the ideas and customs etc. of the West. □ **westernizer** n. **westernization** /ˌwestənaɪˈzeɪʃ(ə)n/ n.

Western roll n. a technique of turning the body over the bar in the high jump.

Western Roman Empire see ROMAN EMPIRE.

Western Sahara a region of NW Africa, on the Atlantic coast between Morocco and Mauritania; pop. (est. 1989) 186,500; capital, La'youn. The region was formerly an overseas Spanish province, called Spanish Sahara. After the Spanish withdrew in 1976 it was renamed and annexed by Morocco and Mauritania. Mauritania withdrew in 1979 and Morocco extended its control over the entire region. A liberation movement, the Polisario Front (*Frente Polisario*), which launched a guerrilla war against the Spanish in 1973, has continued its struggle against Morocco in an attempt to establish an independent Saharawi Arab Democratic Republic.

Western Samoa a country consisting of a group of nine islands in the SW Pacific; pop. (1991) 159,860; official languages, Samoan and English; capital, Apia. Visited by the Dutch in the early 18th century, the islands were administered by Germany from 1899. After the First World War they were mandated to New Zealand, and became an independent republic within the Commonwealth in 1962. The country is a constitutional monarchy with a legislative assembly of elected clan leaders.

Western Wall see WAILING WALL.

Western Zhou see ZHOU.

Westfalen see WESTPHALIA.

West Flanders a province of NW Belgium; capital, Bruges. (See also FLANDERS.)

West Frisian Islands see FRISIAN ISLANDS.

West Germany *hist.* the Federal Republic of Germany (see GERMANY).

West Glamorgan a county of South Wales; administrative centre, Swansea. It was formed in 1974 from part of the former county of Glamorgan.

West Indies 1 a chain of islands extending from the Florida peninsula to the coast of Venezuela, lying between the Caribbean and the Atlantic. They consist of three main island groups, the Greater and Lesser Antilles and the Bahamas, with Bermuda lying further to the north. Originally inhabited by Arawak and Carib Indians, the islands were visited by Columbus in 1492 and named by him in the belief that he had discovered the coast of India. The Spanish settlements of the 16th and 17th centuries were fiercely contested by other European powers, principally the British, the French, and the Dutch; the region became notorious for buccaneering and piracy in the 17th century. Many of the islands became British colonies in the 17th and 18th centuries, and were centres for the cultivation of sugar on large plantations worked by slaves imported from West Africa. The islands now consist of a number of independent states and British, French, Dutch, and US dependencies. **2** an international cricket team drawn from Trinidad, Jamaica, Barbados, Guyana, and other former British dependencies in the region. □ **West Indian** *adj. & n.*

westing /ˈwestɪŋ/ n. *Naut.* the distance travelled or the angle of longitude measured westward from either a defined north–south grid line or a meridian.

Westinghouse /ˈwestɪŋˌhaʊs/, George (1846–1914), American engineer. His achievements covered several fields and he held over 400 patents, but he is best known for developing vacuum-operated safety brakes and electrically controlled signals for railways. He was concerned with the generation and transmission of electric power; he championed the use of alternating current (through Nikola Tesla), and built up a huge company to manufacture his products. Westinghouse also pioneered the use of natural gas and compressed air, and installed water turbines to generate electric power at Niagara Falls.

West Irian see IRIAN JAYA.

Westmann Islands /ˈvestmən, ˈwest-/ (called in Icelandic *Vestmannaeyjar*) a group of fifteen volcanic islands off the south coast of Iceland.

Westmeath /westˈmiːð/ a county of the Republic of Ireland, in the province of Leinster; county town, Mullingar.

West Midlands a metropolitan county of central England; administrative centre, Birmingham.

Westminster /ˈwestˌmɪnstə(r)/ (also **City of Westminster**) an inner London borough, which contains the Houses of Parliament and many government offices. The name is also used allusively to refer to the British Parliament.

Westminster, Palace of the building in Westminster in which the British Parliament meets, the Houses of Parliament. The original palace, supposed to date from the time of Edward the Confessor, was damaged by fire in 1512 and ceased to be a royal residence, but a great part of it remained. The Houses of Lords and Commons for a long time sat in these buildings, until they were destroyed by fire in 1834. The present building, designed by Sir Charles Barry, was formally opened in 1852.

Westminster, Statute of a statute of 1931 recognizing the equality of status of the dominions as autonomous communities within the British Empire, and giving their legislatures independence from British control.

Westminster Abbey the collegiate church of St Peter in Westminster, London, originally the abbey church of a Benedictine monastery. The present building, begun by Henry III in 1245 and altered and added to by successive rulers, replaced an earlier church built by Edward the Confessor. Nearly all the kings and queens of England have been crowned in Westminster Abbey; it is also the burial place of many of England's monarchs (up to George II), and of some of the nation's leading statesmen, poets (in the section called Poets' Corner), and other celebrities.

Westmorland /ˈwestmələnd/ a former county of NW England. In 1974 it was united with Cumberland and northern parts of Lancashire to form the county of Cumbria.

Weston-super-Mare /ˌwestən.suːpəˈmeə(r), -ˌsjuːpəˈmeə(r)/ a resort on the Bristol Channel, in Avon, SW England; pop. (1981) 62,260.

Westphalia /westˈfeɪlɪə/ (German **Westfalen** /vɛstˈfaːlən/) a former province of NW Germany. Settled by the Saxons from about 700 AD, the region became a duchy of the archbishop of Cologne in the 12th century. Held briefly by Napoleon in the early 19th century, it became a province of Prussia in 1815. In 1946 the major part was incorporated in the state of North Rhine–Westphalia, the northern portion becoming part of Lower Saxony. □ **Westphalian** *adj. & n.*

Westphalia, Treaty of the peace accord (1648) which ended the Thirty Years War, signed simultaneously in Osnabrück and Münster.

West Point (in full **West Point Academy**) the US Military Academy, founded in 1802, located on the site of a former strategic fort on the west bank of the Hudson River in New York State.

West Side the western part of any of several North American cities or boroughs, especially the island borough of Manhattan, New York.

West Sussex a county of SE England; county town, Chichester. It was formed in 1974 from part of the former county of Sussex.

West Virginia a state of the eastern US; pop. (1990) 1,793,480; capital, Charleston. It separated from Virginia during the American Civil War (1861) and became the 35th state of the US in 1863.

westward /ˈwestwəd/ *adj., adv., & n.* ● *adj. & adv.* (also **westwards** /-wədz/) towards the west. ● *n.* a westward direction or region.

West Yorkshire a metropolitan county of northern England; administrative centre, Wakefield.

wet /wet/ *adj., v., & n.* ● *adj.* (**wetter, wettest**) **1** soaked, covered, or dampened with water or other liquid (*a wet sponge; a wet surface; got my feet wet*). **2** (of the weather etc.) rainy (*a wet day*). **3** (of paint, ink, etc.) not yet dried. **4** used with water (*wet shampoo*). **5** *Brit. colloq.* feeble, inept. **6** *Brit. Polit. colloq.* Conservative with liberal tendencies, esp. as regarded by right-wing Conservatives. **7** *colloq.* (of a country, of legislation, etc.) allowing the free sale of alcoholic drink. **8** (of a baby or young child) incontinent (*is still wet at night*). ● *v.tr.* (**wetting**; *past* and *past part.* **wet** or **wetted**) **1** make wet. **2 a** urinate in or on (*wet the bed*). **b** *refl.* urinate involuntarily. ● *n.* **1** moisture; liquid that wets something. **2** rainy weather; a time of rain. **3** *Brit. colloq.* a feeble or inept person. **4** *Brit. Polit. colloq.* a Conservative with liberal tendencies (see sense 6 of *adj.*). **5** *colloq.* a drink. □ **wet the baby's head** *colloq.* celebrate its birth with a (usu. alcoholic) drink. **wet behind the ears** immature, inexperienced. **wet blanket** see BLANKET. **wet dock** a dock in which a ship can float. **wet dream** an erotic dream with involuntary ejaculation of semen. **wet fly** an artificial fly used under water by an angler. **wet look** a shiny surface given to clothing materials. **wet-nurse** *n.* a woman employed to suckle another's child. ● *v.tr.* **1** act as a wet-nurse to. **2** *colloq.* treat as if helpless. **wet pack** the therapeutic wrapping of the body in wet cloths etc. **wet rot 1** a brown rot affecting moist timber. **2** the fungus causing this, often *Coniophora puteana*. **wet suit** a close-fitting rubber garment worn by skin-divers etc. to keep warm. **wet through** (or **to the skin**) with one's clothes soaked. **wetting agent** a substance that helps water etc. to spread or penetrate. **wet-weather 1** for use in wet weather. **2** occurring in wet weather. **wet one's whistle** see WHISTLE. □ **wetly** *adv.* **wetness** *n.* **wettable** *adj.* **wetting** *n.* **wettish** *adj.* [OE *wǣt* (adj. & n.), *wǣtan* (v.), rel. to WATER: in ME replaced by past part. of the verb]

weta /ˈwetə/ *n.* a large wingless grasshopper-like insect of the family Stenopelmatidae, with long spiny legs, confined to New Zealand. There are some giant ground-living kinds, such as *Deinacrita*. [Maori]

wetback /ˈwetbæk/ *n. US colloq.* an illegal immigrant from Mexico to the US. [WET + BACK: from the practice of swimming the Rio Grande to reach the US]

wether /ˈweðə(r)/ *n.* a castrated ram. [OE f. Gmc]

wetland /ˈwetlənd/ *n.* (usu. in *pl.*) a marsh, swamp, or other usually saturated area of land.

we've /wiːv, wɪv/ *contr.* we have.

Wexford /ˈweksfəd/ **1** a county of the Republic of Ireland, in the south-east in the province of Leinster. **2** its county town, a port on the Irish Sea; pop. (1991) 9,540.

wey /weɪ/ *n.* a former unit of weight or volume varying with different kinds of goods, e.g. 3 cwt. of cheese. [OE *wǣg(e)* balance, weight f. Gmc, rel. to WEIGH¹]

Weyden /ˈvaɪd(ə)n/, Rogier van der (French name Rogier de la Pasture) (c.1400–64), Flemish painter. He was based in Brussels from about 1435, when he was appointed official painter to the city. His work, mostly portraits and religious paintings, became widely known in Europe during his lifetime; he was particularly influential in the development of Dutch portrait painting. Major works include *The Last Judgement* and *The Deposition in the Tomb* (both c.1450).

Weymouth /ˈweɪməθ/ a resort and port on the coast of Dorset, southern England; pop. (1981) 38,400.

w.f. *abbr. Printing* wrong fount.

WFC *abbr.* World Food Council.

WFP *abbr.* World Food Programme.

WFTU see WORLD FEDERATION OF TRADE UNIONS.

Wg. Cdr. *abbr.* wing commander.

whack /wæk/ *v. & n. colloq.* ● *v.tr.* **1** strike or beat forcefully with a sharp blow. **2** (as **whacked** *adj.*) esp. *Brit.* tired out; exhausted. ● *n.* **1** a sharp or resounding blow. **2** *sl.* a turn; an attempt. **3** *sl.* a share or portion. □ **have a whack at** *sl.* attempt. **out of whack** esp. *N. Amer. sl.* out of order; malfunctioning. **top** (or **full** or **the full**) **whack** the maximum price or rate. □ **whacker** *n.* **whacking** *n.* [imit., or alt. of THWACK]

whacking /ˈwækɪŋ/ *adj. & adv. colloq.* ● *adj.* very large. ● *adv.* very (*a whacking great skyscraper*).

whacko /ˈwækəʊ/ *int. sl.* expressing delight or enjoyment.

whacky var. of WACKY.

whale¹ /weɪl/ *n.* (*pl.* same or **whales**) a large marine mammal of the order Cetacea, having a streamlined body and horizontal tail, and breathing air through a blowhole on the head. (*See note below.*) □ **whale-headed stork** an African stork, *Balaeniceps rex*, with a large clog-shaped bill and grey plumage (also called *shoebill*). **a whale of a** *colloq.* an exceedingly good or fine etc. **whale-oil** oil from the blubber of whales. **whale shark** a very large tropical shark, *Rhincodon typus*, that feeds on plankton and is the largest living fish. [OE *hwæl*]

■ Whales are adapted for an exclusively aquatic life, having a somewhat fishlike body with forelimbs modified as flippers and no hindlimbs. There are two main groups: the toothed whales (suborder Odontoceti), which include the sperm whales, killer whales, dolphins, and porpoises, and the baleen whales (suborder Mysticeti), which feed on plankton and include the blue whale, the largest of all animals. Whales are intelligent and sociable creatures and communicate by sound. Toothed whales use a form of echolocation, especially for detecting prey, and some can dive to great depths and stay under water for an hour or more. In recent decades commercial whaling has endangered many species, especially the blue, right, and humpback whales. (See also WHALING.)

whale² /weɪl/ *v.tr.* esp. *US colloq.* beat, thrash. [var. of WALE]

whaleback /ˈweɪlbæk/ *n.* something shaped like a whale's back.

whaleboat /ˈweɪlbəʊt/ *n.* a double-bowed boat of a kind orig. used in whaling.

whalebone /ˈweɪlbəʊn/ *n.* an elastic horny substance growing in thin parallel plates in the upper jaw of some whales, used by them for filter-feeding and by humans for stiffening etc. Also called *baleen*. □ **whalebone whale** a baleen whale.

whaler /ˈweɪlə(r)/ *n.* **1** a whaling ship or a seaman engaged in whaling. **2** (also **whaler shark**) a large shark, *Carcharhinus brachyurus*, of Australian waters. **3** *Austral. sl.* a tramp.

whaling /ˈweɪlɪŋ/ *n.* the practice or industry of hunting and killing whales, especially for their oil, meat, ambergris, or whalebone. Native American and Eskimo peoples hunted whales in early times, while in Europe Basque whalers operated in the Bay of Biscay during the Middle Ages. However, it was the development in the 18th century of facilities to process whale products at sea, and of the explosive harpoon in the mid-19th century, that led to whales being killed in far greater numbers. Alarm over the rapidly declining whale population led the International Whaling Commission (which had been established in 1947) to introduce a moratorium on commercial whaling in 1985, a decision ratified in subsequent conventions. Several countries opposed the commercial ban: Japan, South Korea, and Iceland continued to kill whales, under a provision for small numbers to be killed for scientific purposes. In 1992 Iceland withdrew from the IWC, while Norway decided to resume commercial whaling in defiance of the moratorium. In 1994 the IWC voted to establish a whale sanctuary in the Antarctic Ocean south of the 40th parallel, a decision opposed only by Japan, although Norway and others abstained from the vote.

wham /wæm/ *int., n., & v. colloq.* ● *int.* expressing the sound of a forcible impact. ● *n.* such a sound. ● *v.* (**whammed, whamming**) **1** *intr.* make such a sound or impact. **2** *tr.* strike forcibly. [imit.]

whammy /ˈwæmɪ/ *n.* (*pl.* **-ies**) esp. *US colloq.* an evil or unlucky influence. □ **double whammy** *colloq.* see DOUBLE. [20th c.: f. WHAM + -Y¹]

whang /wæŋ/ *v.* & *n. colloq.* ● *v.* **1** *tr.* strike heavily and loudly; whack. **2** *intr.* (of a drum etc.) sound under or as under a blow. ● *n.* a whanging sound or blow. [imit.]

Whangarei /ˌwæŋgəˈreɪ/ a port on the NE coast of North Island, New Zealand; pop. (1991) 44,180.

whangee /wæŋˈgiː/ *n.* **1** a Chinese bamboo of the genus *Phyllostachys*, esp. *P. nigra.* **2** a yellow knobbed cane made from this. [Chin. *huang* old bamboo-sprouts]

whap esp. *US* var. of WHOP.

whare /ˈwɒrɪ/ *n.* a Maori hut or house. [Maori]

wharf /wɔːf/ *n.* & *v.* ● *n.* (pl. **wharves** /wɔːvz/ or **wharfs**) a level quayside area to which a ship may be moored to load and unload. ● *v.tr.* **1** moor (a ship) at a wharf. **2** store (goods) on a wharf. [OE *hwearf*]

wharfage /ˈwɔːfɪdʒ/ *n.* **1** accommodation at a wharf. **2** a fee for this.

wharfie /ˈwɔːfɪ/ *n. Austral.* & *NZ colloq.* a waterside worker; a wharf-labourer.

wharfinger /ˈwɔːfɪndʒə(r)/ *n.* an owner or keeper of a wharf. [prob. ult. f. WHARFAGE]

Wharton /ˈwɔːt(ə)n/, Edith (Newbold) (1862–1937), American novelist and short-story writer, resident in France from 1907. She established her reputation with the novel *The House of Mirth* (1905). Her novels, many of them set in New York high society, show the influence of Henry James and are chiefly preoccupied with the often tragic conflict between social and individual fulfilment. They include *Ethan Frome* (1911) and *The Age of Innocence* (1920), which won a Pulitzer Prize.

wharves *pl.* of WHARF.

what /wɒt/ *adj., pron.,* & *adv.* ● *interrog.adj.* **1** asking for a choice from an indefinite number or for a statement of amount, number, or kind (*what books have you read?; what news have you?*). **2** *colloq.* = WHICH *interrog.adj.* (*what book have you chosen?*). ● *adj.* (usu. in exclamations) how great or remarkable (*what luck!*). ● *rel.adj.* the or any . . . that (*will give you what help I can*). ● *pron.* (corresp. to the functions of the *adj.*) **1** what thing or things? (*what is your name?; I don't know what you mean*). **2** (asking for a remark to be repeated) = what did you say? **3** asking for confirmation or agreement of something not completely understood (*you did what?; what, you really mean it?*). **4** how much (*what you must have suffered!*). **5** (as *rel.pron.*) that or those which; a or the or any thing which (*what followed was worse; tell me what you think*). ● *adv.* to what extent (*what does it matter?*). □ **what about** what is the news or position or your opinion of (*what about me?; what about a game of tennis?*). **what-d'you-call-it** (or **whatchamacallit** or **what's-its-name**) *colloq.* a substitute for a name not recalled. **what ever** what at all or in any way (*what ever do you mean?*) (see also WHATEVER). **what for** *colloq.* **1** for what reason? **2** a severe reprimand (esp. *give a person what for*). **what have you** *colloq.* (prec. by *or*) anything else similar. **what if? 1** what would result etc. if. **2** what would it matter if. **what is more** and as an additional point; moreover. **what next?** *colloq.* what more absurd, shocking, or surprising thing is possible? **what not** (prec. by *and*) other similar things. **what of?** what is the news concerning? **what of it?** why should that be considered significant? **what's-his** (or **-its**) **-name** see *what-d'you-call-it*. **what's what** *colloq.* what is useful or important etc. **what with** *colloq.* because of (usu. several things). [OE *hwæt* f. Gmc]

whate'er /wɒtˈeə(r)/ *poet.* var. of WHATEVER.

whatever /wɒtˈevə(r)/ *adj.* & *pron.* **1** = WHAT (in relative uses) with the emphasis on indefiniteness (*lend me whatever you can; whatever money you have*). **2** though anything (*we are safe whatever happens*). **3** (with *neg.* or *interrog.*) at all; of any kind (*there is no doubt whatever*). **4** *colloq.* = *what ever.* □ **or whatever** *colloq.* or anything similar.

whatnot /ˈwɒtnɒt/ *n.* **1** an indefinite or trivial thing. **2** a stand with shelves for small objects.

whatsit /ˈwɒtsɪt/ *n. colloq.* a person or thing whose name one cannot recall or does not know.

whatso /ˈwɒtsəʊ/ *adj.* & *pron. archaic* = WHATEVER 1, 2. [ME, = WHAT + SO¹, f. OE *swā hwæt swā*]

whatsoe'er /ˌwɒtsəʊˈeə(r)/ *poet.* var. of WHATSOEVER.

whatsoever /ˌwɒtsəʊˈevə(r)/ *adj.* & *pron.* = WHATEVER 1–3.

whaup /wɔːp/ *n.* esp. *Sc.* a curlew. [imit. of its cry]

wheal var. of WEAL¹.

wheat /wiːt/ *n.* **1** a cereal plant of the genus *Triticum*, bearing dense four-sided seed-spikes. (*See note below.*) **2** its grain, used in making flour

etc. □ **separate the wheat from the chaff** see CHAFF. **wheat-belt** a region where wheat is the chief agricultural product. **wheat germ** the embryo of the wheat grain, extracted as a source of vitamins. **wheat-grass** couch grass (see COUCH²). [OE *hwæte* f. Gmc, rel. to WHITE]

▪ Wheat was probably one of the first plants to be cultivated by humans, remains of the primitive einkorn wheat being found at a site in Iraq dated to 6750 BC. Einkorn, together with other early forms of wheat such as emmer and spelt, are now generally used only for animal fodder. The main species grown for flour, *Triticum aestivum*, has been greatly improved in recent decades, and has become the world's most important food plant in economic terms. Pasta is derived from durum wheat, *T. durum*.

wheatear /ˈwiːtɪə(r)/ *n.* a songbird of the genus *Oenanthe*, with a black tail and white rump, esp. the migratory *O. oenanthe* of Europe. [app. f. *wheatears* (as WHITE, ARSE)]

wheaten /ˈwiːt(ə)n/ *adj.* made of wheat.

wheatmeal /ˈwiːtmiːl/ *n.* flour made from wheat with some of the bran and germ removed.

Wheatstone /ˈwiːtstən/, Sir Charles (1802–75), English physicist and inventor. He was a member of a family of musical instrument-makers and his interests included acoustics, optics, electricity, and telegraphy. He invented the stereoscope and concertina, but is best known for his electrical inventions which included an electric clock. In the 1830s Wheatstone collaborated with Sir W. F. Cooke to develop the electric telegraph, and he later devised the Wheatstone bridge (based on an idea of the mathematician Samuel Christie) and the rheostat.

Wheatstone bridge *n.* an apparatus for measuring electrical resistances by equalizing the potential at two points of a circuit. [WHEATSTONE]

whee /wiː/ *int.* expressing delight or excitement, esp. on travelling downhill or at speed. [imit.]

wheedle /ˈwiːd(ə)l/ *v.tr.* **1** coax by flattery or endearments. **2** (foll. by *out*) **a** get (a thing) out of a person by wheedling. **b** cheat (a person) out of a thing by wheedling. □ **wheedler** *n.* **wheedling** *adj.* **wheedlingly** *adv.* [perh. f. G *wedeln* fawn, cringe f. *Wedel* tail]

wheel /wiːl/ *n.* & *v.* ● *n.* **1** a circular frame or disc arranged to revolve on an axle and used to facilitate the motion of a vehicle or for various mechanical purposes. (*See note below.*) **2** a wheel-like thing (*Catherine wheel; potter's wheel; steering wheel*). **3** motion as of a wheel, esp. the movement of a line of people with one end as a pivot. **4** a machine etc. of which a wheel is an essential part. **5** (in *pl.*) *sl.* a car. **6** *US sl.* = big wheel 2. **7** a set of short lines concluding a stanza. ● *v.* **1** *intr.* & *tr.* **a** turn on an axis or pivot. **b** swing round in line with one end as a pivot. **2 a** *intr.* (often foll. by *about, round*) change direction or face another way. **b** *tr.* cause to do this. **3** *tr.* push or pull (a wheeled thing esp. a barrow, bicycle, or pram, or its load or occupant). **4** *intr.* go in circles or curves (*seagulls wheeling overhead*). □ **at the wheel 1** driving a vehicle. **2** directing a ship. **3** in control of affairs. **on wheels** (or **oiled wheels**) smoothly. **wheel and deal** engage in political or commercial scheming. **wheel-back** *adj.* (of a chair) with a back shaped like or containing the design of a wheel. **wheel clamp** *n.* = CLAMP¹ *n.* 2. ● *v.tr.* = CLAMP¹ *v.* 3. **wheel-lock** *hist.* **1** a kind of gunlock having a steel wheel to rub against flint etc. **2** a gun with this. **wheel of Fortune** the wheel which Fortune is held to turn, as an emblem of mutability and luck. **wheel-spin** rotation of a vehicle's wheels without traction. **wheels within wheels 1** intricate machinery. **2** *colloq.* indirect or secret agencies. □ **wheeled** *adj.* (also in *comb.*).

wheelless /ˈwiːllɪs/ *adj.* [OE *hwēol, hwēogol* f. Gmc]

▪ The wheel is regarded as one of the most important inventions, first recorded in Mesopotamia about 3500 BC, made of solid wood; it was unknown in the pre-Columbian civilizations of Central America other than Mexico. Spoked wheels appeared soon after 2000 BC; iron tyres were used to protect and strengthen the rim. A wheel using wire spokes in tension was evolved for the bicycle: by arranging alternate spokes to lead to opposite flanges of the hub and be tangential to the hub in the forward and backward directions the result was a strong lightweight wheel able to transmit a torque from hub to rim. One of the more recent developments is that of pneumatic tyres, making it easier for vehicles to traverse uneven surfaces.

wheelbarrow /ˈwiːlˌbærəʊ/ *n.* a small cart with one wheel and two shafts for carrying garden loads etc.

wheelbase /ˈwiːlbeɪs/ *n.* the distance between the front and rear axles of a vehicle.

wheelchair /ˈwiːltʃeə(r)/ n. a chair on wheels for an invalid or disabled person.

wheeler /ˈwiːlə(r)/ n. **1** (in comb.) a vehicle having a specified number of wheels. **2** a wheelwright. **3** a horse harnessed next to the wheels and behind another. □ **wheeler-dealer** colloq. a person who wheels and deals.

wheel-house /ˈwiːlhaʊs/ n. **1** a steersman's shelter. **2** Archaeol. a stone-built circular house with inner partition walls radiating like the spokes of a wheel, found in western and northern Scotland and dating chiefly from c.100 BC–AD c.100.

wheelie /ˈwiːlɪ/ n. sl. the stunt of riding a bicycle or motorcycle for a short distance with the front wheel off the ground.

wheelie bin n. (also **wheely bin**) Brit. colloq. a large refuse bin on wheels. ¶ The official name is wheeled bin.

wheelman /ˈwiːlmən/ n. (pl. **-men**) esp. US **1** a driver of a wheeled vehicle. **2** a helmsman.

wheelsman /ˈwiːlzmən/ n. (pl. **-men**) US a steersman.

wheelwright /ˈwiːlraɪt/ n. a person who makes or repairs esp. wooden wheels.

wheely bin var. of WHEELIE BIN.

wheeze /wiːz/ v. & n. ● v. **1** intr. breathe with an audible chesty whistling sound. **2** tr. (often foll. by out) utter in this way. ● n. **1** a sound of wheezing. **2** colloq. **a** Brit. a clever scheme. **b** a catch-phrase. □ **wheezer** n. **wheezingly** adv. **wheezy** adj. **wheezily** adv. **wheeziness** n. [prob. f. ON hvæsa to hiss]

whelk¹ /welk/ n. a predatory marine gastropod mollusc of the family Buccinidae, with a spiral shell, esp. Buccinum undatum, used as food. [OE wioloc, weoloc, of unkn. orig.: perh. infl. by WHELK²]

whelk² /welk/ n. a pimple. [OE hwylca f. hwelian suppurate]

whelm /welm/ v.tr. poet. **1** engulf, submerge. **2** crush with weight, overwhelm. [OE hwelman (unrecorded) = hwylfan overturn]

whelp /welp/ n. & v. ● n. **1** a young dog; a puppy. **2** archaic a cub. **3** an ill-mannered child or youth. **4** (esp. in pl.) a projection on the barrel of a capstan or windlass. ● v.tr. (also absol.) **1** bring forth (a whelp or whelps). **2** derog. (of a human mother) give birth to. **3** originate (an evil scheme etc.). [OE hwelp]

when /wen/ adv., conj., pron., & n. ● interrog.adv. **1** at what time? **2** on what occasion? **3** how soon? **4** how long ago? ● rel.adv. (prec. by time etc.) at or on which (there are times when I could cry). ● conj. **1** at the or any time that; as soon as (come when you like; come when ready; when I was your age). **2** although; considering that (why stand up when you could sit down?). **3** after which; and then; but just then (was nearly asleep when the bell rang). ● pron. what time? which time (till when can you stay?; since when it has been better). ● n. time, occasion, date (fixed the where and when). [OE hwanne, hwenne]

whence /wens/ adv. & conj. formal ● adv. from what place? (whence did they come?). ● conj. **1** to the place from which (return whence you came). **2** (often prec. by place etc.) from which (the source whence these errors arise). **3** and thence (whence it follows that). ¶ Use of from whence as in the place from whence they came, though common, is generally considered incorrect. [ME whannes, whennes f. whanne, whenne f. OE hwanon(e) whence, formed as WHEN + -s³: cf. THENCE]

whencesoever /ˌwensəʊˈevə(r)/ adv. & conj. formal from whatever place or source.

whene'er /wenˈeə(r)/ poet. var. of WHENEVER.

whenever /wenˈevə(r)/ conj. & adv. **1** at whatever time; on whatever occasion. **2** every time that. □ **or whenever** colloq. or at any similar time.

whensoe'er /ˌwensəʊˈeə(r)/ poet. var. of WHENSOEVER.

whensoever /ˌwensəʊˈevə(r)/ conj. & adv. formal = WHENEVER.

where /weə(r)/ adv., conj., pron., & n. ● interrog.adv. **1** in or to what place or position? (where is the milk?; where are you going?). **2** in what direction or respect? (where does the argument lead?; where does it concern us?). **3** in what book etc.?; from whom? (where did you read that?; where did you hear that?). **4** in what situation or condition? (where does that leave us?). ● rel.adv. (prec. by place etc.) in or to which (places where they meet). ● conj. **1** in or to the or any place, direction, or respect in which (go where you like; that is where you are wrong; delete where applicable). **2** and there (reached Crewe, where the car broke down). ● pron. what place? (where do you come from?; where are you going to?). ● n. a place; a or the scene of something (see WHEN n.). [OE hwær, hwār]

whereabouts adv. & n. ● adv. /ˌweərəˈbaʊts/ where or approximately where (whereabouts are they?; show me whereabouts to look). ● n.

/ˈweərəˌbaʊts/ (as sing. or pl.) a person's or thing's approximate location.

whereafter /weərˈɑːftə(r)/ conj. formal after which.

whereas /weərˈæz/ conj. **1** in contrast or comparison with the fact that. **2** (esp. in legal preambles) taking into consideration the fact that.

whereat /weərˈæt/ conj. archaic **1** at which place or point. **2** for which reason.

whereby /weəˈbaɪ/ conj. by what or which means.

where'er /weərˈeə(r)/ poet. var. of WHEREVER.

wherefore /ˈweəfɔː(r)/ adv. & n. ● adv. archaic **1** for what reason? **2** for which reason. ● n. a reason (the whys and wherefores).

wherefrom /weəˈfrɒm/ conj. archaic from which, from where.

wherein /weərˈɪn/ conj. & adv. formal ● conj. in what or which place or respect. ● adv. in what place or respect?

whereof /weərˈɒv/ conj. & adv. formal ● conj. of what or which (the means whereof). ● adv. of what?

whereon /weərˈɒn/ conj. & adv. archaic ● conj. on what or which. ● adv. on what?

wheresoe'er /ˌweəsəʊˈeə(r)/ poet. var. of WHERESOEVER.

wheresoever /ˌweəsəʊˈevə(r)/ conj. & adv. formal or literary = WHEREVER.

whereto /weəˈtuː/ conj. & adv. formal ● conj. to what or which. ● adv. to what?

whereupon /ˌweərəˈpɒn/ conj. immediately after which.

wherever /weərˈevə(r)/ adv. & conj. ● adv. in or to whatever place. ● conj. in every place that. □ **or wherever** colloq. or in any similar place.

wherewith /weəˈwɪð/ conj. formal or archaic with or by which.

wherewithal /ˈweəwɪˌðɔːl/ n. colloq. money etc. needed for a purpose (has not the wherewithal to do it).

wherry /ˈwerɪ/ n. (pl. **-ies**) **1** a light rowing-boat usu. for carrying passengers. **2** a large light barge. [ME: orig. unkn.]

wherryman /ˈwerɪmən/ n. (pl. **-men**) a man employed on a wherry.

whet /wet/ v. & n. ● v.tr. (**whetted**, **whetting**) **1** sharpen (a tool or weapon) by grinding. **2** stimulate (the appetite or a desire, interest, etc.). ● n. **1** the act or an instance of whetting. **2** a small quantity stimulating one's appetite for more. □ **whet one's whistle** see WHISTLE. □ **whetter** n. (also in comb.). [OE hwettan f. Gmc]

whether /ˈweðə(r)/ conj. introducing the first or both of alternative possibilities (I doubt whether it matters; I do not know whether they have arrived or not). □ **whether or no** see NO². [OE hwæther, hwether f. Gmc]

whetstone /ˈwetstəʊn/ n. **1** a fine-grained stone used esp. with water to sharpen cutting tools (cf. OILSTONE). **2** a thing that sharpens the senses etc.

whew /hwjuː/ int. expressing surprise, consternation, or relief. [imit.: cf. PHEW]

whey /weɪ/ n. the watery liquid left when milk forms curds. □ **whey-faced** pale esp. with fear. [OE hwæg, hweg f. LG]

which /wɪtʃ/ adj. & pron. ● interrog.adj. asking for choice from a definite set of alternatives (which John do you mean?; say which book you prefer; which way shall we go?). ● rel.adj. being the one just referred to; and this or these (ten years, during which time they admitted nothing). ● interrog.pron. **1** which person or persons (which of you is responsible?). **2** which thing or things (say which you prefer). ● rel.pron. (poss. **of which**, **whose** /huːz/) **1** which thing or things, usu. introducing a clause not essential for identification (cf. THAT pron. 7) (the house, which is empty, has been damaged). **2** used in place of that after in or that (there is the house in which I was born; that which you have just seen). □ **which is which** a phrase used when two or more persons or things are difficult to distinguish from each other. [OE hwilc f. Gmc]

whichever /wɪtʃˈevə(r)/ adj. & pron. **1** any which (take whichever you like; whichever one you like). **2** no matter which (whichever one wins, they both get a prize).

whichsoever /ˌwɪtʃsəʊˈevə(r)/ adj. & pron. archaic = WHICHEVER.

whidah var. of WHYDAH.

whiff /wɪf/ n. & v. ● n. **1** a puff or breath of air, smoke, etc. (went outside for a whiff of fresh air). **2** a smell (caught the whiff of a cigar). **3** (foll. by of) a trace or suggestion of scandal etc. **4** a small cigar. **5** a minor discharge (of grapeshot etc.). **6** a light narrow outrigged sculling-boat. ● v. **1** tr. & intr. blow or puff lightly. **2** intr. Brit. smell (esp. unpleasant). **3** tr. get a slight smell of. [imit.]

whiffle /ˈwɪf(ə)l/ v. & n. ● v. **1** intr. & tr. (of the wind) blow lightly, shift about. **2** intr. be variable or evasive. **3** intr. (of a flame, leaves, etc.) flicker, flutter. **4** *intr.* make the sound of a light wind in breathing etc. ● *n.* a slight movement of air. □ **whiffler** *n.* [WHIFF + -LE⁴]

whiffletree /ˈwɪf(ə)lˌtriː/ *n. US* = SWINGLETREE. [var. of WHIPPLETREE]

whiffy /ˈwɪfɪ/ *adj. colloq.* having an unpleasant smell.

Whig /wɪg/ *n. hist.* **1** a 17th-century Scottish Presbyterian. **2** a member of a former British political group. (*See note below.*) **3** *US* a supporter of the War of American Independence. **4** *US* a member of a political party of 1834–56, succeeded by the Republicans. □ **Whiggery** *n.* **Whiggish** *adj.* **Whiggism** *n.* [prob. a shortening of Sc. *whiggamer, -more*, nickname of 17th-c. Sc. rebels (see below), f. *whig* to drive + MARE¹]

▪ The Whigs acquired their name as a derogatory reference to a group of Scottish Covenanters by their opponents, the Tories, during the exclusion crisis (1679) of Charles II's reign, in which they attempted to prevent James, Duke of York (later James II) from succeeding to the English throne. Until the second half of the 18th century, the Whigs constituted a political interest group which was composed of a loose alliance of the country aristocracy and various trading interests, functioning largely through patronage. Opponents of Jacobitism and advocates of the supremacy of Parliament and the Hanoverian succession, the Whigs dominated the English political scene in the late 17th and first half of the 18th century. After 1784, in the aftermath of the War of American Independence, there was considerable realignment of party interests and a new Whig party emerged, championing the interests of religious dissenters and those who sought electoral and social reform. There was further regrouping after the French Revolution and in the early 19th century the Whigs continued to advocate electoral reform, passing the Reform Act of 1832. In the middle of the century the term *Liberal* gradually replaced *Whig*.

while /waɪl/ *n., conj., v., & adv.* ● *n.* **1** a space of time, time spent in some action (*a long while ago; waited a while; all this while*). **2** (*prec. by the*) **a** during some other process. **b** *poet.* during the time that. **3** (*prec. by a*) for some time (*have not seen you a while*). ● *conj.* **1** during the time that; for as long as; at the same time as (*while I was away, the house was burgled; fell asleep while reading*). **2** in spite of the fact that; although, whereas (*while I want to believe it, I cannot*). **3** *N. Engl.* until (*wait while Monday*). ● *v.tr.* (foll. by *away*) pass (time etc.) in a leisurely or interesting manner. ● *rel.adv.* (prec. by *time* etc.) during which (*the summer while I was abroad*). □ **all the while** during the whole time (that). **for a long while** for a long time past. **for a while** for some time. **a good** (or **great**) **while** a considerable time. **in a while** (or **little while**) soon, shortly. **worth while** (or **one's while**) worth the time or effort spent. [OE *hwīl* f. Gmc: (conj.) abbr. of OE *thā hwīle the*, ME *the while that*]

whiles /waɪlz/ *conj. archaic* = WHILE. [orig. in the adverbs *somewhiles, otherwhiles*]

whilom /ˈwaɪləm/ *adv. & adj. archaic* ● *adv.* formerly, once. ● *adj.* former, erstwhile (*my whilom friend*). [OE *hwīlum* dative pl. of *hwīl* WHILE]

whilst /waɪlst/ *adv. & conj. esp. Brit.* while. [ME f. WHILES: cf. AGAINST]

whim /wɪm/ *n.* **1 a** a sudden fancy; a caprice. **b** capriciousness. **2** *archaic* a kind of windlass for raising ore or water from a mine. [17th c.: orig. unkn.]

whimbrel /ˈwɪmbrɪl/ *n.* a small migratory curlew, *Numenius phaeopus*, with a trilling call, breeding in northern Eurasia and northern Canada. [WHIMPER (imit.): cf. *dotterel*]

whimper /ˈwɪmpə(r)/ *v. & n.* ● *v.* **1** *intr.* make feeble, querulous, or frightened sounds; cry and whine softly. **2** *tr.* utter whimperingly. ● *n.* **1** a whimpering sound. **2** a feeble note or tone (*the conference ended on a whimper*). □ **whimperer** *n.* **whimperingly** *adv.* [imit., f. dial. *whimp*]

whimsical /ˈwɪmzɪk(ə)l/ *adj.* **1** odd or quaint; fanciful; humorous. **2** capricious. □ **whimsically** *adv.* **whimsicalness** *n.* **whimsicality** /ˌwɪmzɪˈkælɪtɪ/ *n.*

whimsy /ˈwɪmzɪ/ *n.* (also **whimsey**) (*pl.* **-ies** or **-eys**) **1** a whim; a capricious notion or fancy. **2** capricious or quaint humour. [rel. to WHIM-WHAM: cf. *flimsy*]

whim-wham /ˈwɪmwæm/ *n. archaic* **1** a toy or plaything. **2** = WHIM 1. [redupl.: orig. uncert.]

whin¹ /wɪn/ *n.* (in *sing.* or *pl.*) furze, gorse. [prob. Scand.: cf. Norw. *hvine*, Sw. *hven*]

whin² /wɪn/ *n.* **1** hard dark esp. basaltic rock or stone. **2** a piece of this. [ME: orig. unkn.]

whinchat /ˈwɪntʃæt/ *n.* a small migratory thrush, *Saxicola rubetra*, resembling the stonechat. [WHIN¹ + CHAT²]

whine /waɪn/ *n. & v.* ● *n.* **1** a complaining long-drawn wail as of a dog. **2** a similar shrill prolonged sound. **3 a** a querulous tone. **b** an instance of feeble or undignified complaining. ● *v.* **1** *intr.* emit or utter a whine. **2** *intr.* complain in a querulous tone or in a feeble or undignified way. **3** *tr.* utter in a whining tone. □ **whiner** *n.* **whiningly** *adv.* **whiny** *adj.* (**whinier, whiniest**). [OE *hwīnan*]

whinge /wɪndʒ/ *v. & n. colloq.* ● *v.intr.* (**whingeing**) whine; grumble peevishly. ● *n.* a whining complaint; a peevish grumbling. □ **whinger** *n.* **whingingly** *adv.* **whingy** *adj.* [OE *hwinsian* f. Gmc]

whinny /ˈwɪnɪ/ *n. & v.* ● *n.* (*pl.* **-ies**) a gentle or joyful neigh. ● *v.intr.* (**-ies, -ied**) give a whinny. [imit.: cf. WHINE]

whinstone /ˈwɪnstəʊn/ *n.* = WHIN².

whip /wɪp/ *n. & v.* ● *n.* **1** a lash attached to a stick for urging on animals or punishing etc. **2 a** a member of a political party in Parliament appointed to control its parliamentary discipline and tactics, esp. ensuring attendance and voting in debates. **b** *Brit.* the whips' written notice requesting or requiring attendance for voting at a division etc., variously underlined according to the degree of urgency (see *three-line whip, two-line whip*). **c** (prec. by *the*) party discipline and instructions (*asked for the Labour whip*). **3** a dessert made with whipped cream etc. **4** the action of beating cream, eggs, etc., into a froth. **5** = WHIPPER-IN. **6** a rope-and-pulley hoisting apparatus. ● *v.* (**whipped, whipping**) **1** *tr.* beat or urge on with a whip. **2** *tr.* beat (cream or eggs etc.) into a froth. **3** *tr. & intr.* take or move suddenly, unexpectedly, or rapidly (*whipped away the tablecloth; whipped out a knife*). **4** *tr. Brit. sl.* steal (*who's whipped my pen?*). **5** *tr. sl.* **a** excel. **b** defeat. **6** *tr.* bind with spirally wound twine. **7** *tr.* sew with overcast stitches. □ **whip-crane** a light derrick with tackle for hoisting. **whip-graft** (in horticulture) a graft with the tongue of the scion in a slot in the stock and vice versa. **whip hand 1** a hand that holds the whip (in riding etc.). **2** (usu. prec. by *the*) the advantage or control in any situation. **whip in** bring (hounds) together. **whip on** urge into action. **whip-round** *esp. Brit. colloq.* an informal collection of money from a group of people. **whip scorpion** an arachnid of the order Uropygi, resembling a scorpion but with a long slender tail-like appendage. **whip snake** a long slender snake, esp of the family Colubridae. **whip-stitch** an overcast stitch. **whip up 1** excite or stir up (feeling etc.). **2** gather, summon up. **3** prepare (a meal etc.) hurriedly. □ **whipless** *adj.* **whiplike** *adj.* **whipper** *n.* [ME (h)*wippen* (v.), prob. f. MLG & MDu. *wippen* swing, leap, dance]

whipbird /ˈwɪpbɜːd/ *n.* an Australian songbird of the genus *Psophodes*, with a call like the crack of a whip.

whipcord /ˈwɪpkɔːd/ *n.* **1** a tightly twisted cord such as is used for making whiplashes. **2** a closely woven worsted fabric.

whiplash /ˈwɪplæʃ/ *n.* **1** the flexible end of a whip. **2** a blow with a whip. **3** (in full **whiplash injury**) an injury to the neck caused by a severe jerk of the head, esp. as in a motor accident.

whipper-in /ˌwɪpərˈɪn/ *n.* (*pl.* **whippers-in**) a huntsman's assistant who manages the hounds.

whippersnapper /ˈwɪpəˌsnæpə(r)/ *n.* **1** a small child. **2** an insignificant but presumptuous or intrusive (esp. young) person. [perh. for *whipsnapper*, implying noise and unimportance]

whippet /ˈwɪpɪt/ *n.* a cross-bred dog of the greyhound type used for racing. [prob. f. obs. *whippet* move briskly, f. *whip* it]

whipping /ˈwɪpɪŋ/ *n.* **1** a beating, esp. with a whip. **2** cord wound round in binding. □ **whipping-boy 1** a scapegoat. **2** *hist.* a boy educated with a young prince and punished instead of him. **whipping-cream** medium-thick cream suitable for whipping. **whipping-post** *hist.* a post used for public whippings. **whipping-top** a top kept spinning by blows of a lash.

whippletree /ˈwɪp(ə)lˌtriː/ *n.* = SWINGLETREE. [app. f. WHIP + TREE]

whippoorwill /ˈwɪpʊəˌwɪl/ *n.* a nightjar, *Caprimulgus vociferus*, of North and Central America. [imit. of its call]

whippy /ˈwɪpɪ/ *adj.* flexible, springy. □ **whippiness** *n.*

whipsaw /ˈwɪpsɔː/ *n. & v.* ● *n.* a saw with a narrow blade held at each end by a frame. ● *v.* (*past part.* **-sawn** /-sɔːn/ or **-sawed**) **1** *tr.* cut with a whipsaw. **2** *US sl.* **a** *tr.* cheat by joint action on two others. **b** *intr.* be cheated in this way.

whipstock /ˈwɪpstɒk/ *n.* the handle of a whip.

whir var. of WHIRR.

whirl /wɜːl/ *v. & n.* ● *v.* **1** *tr. & intr.* swing round and round; revolve rapidly. **2** *tr. & intr.* (foll. by *away*) convey or go rapidly in a vehicle etc. **3** *tr. & intr.* send or travel swiftly in an orbit or a curve. **4** *intr.* **a** (of the brain, senses, etc.) seem to spin round. **b** (of thoughts etc.) be confused; follow each other in bewildering succession. ● *n.* **1** a whirling movement (*vanished in a whirl of dust*). **2** a state of intense activity (*the social whirl*). **3** a state of confusion (*my mind is in a whirl*). **4** *colloq.* an attempt (*give it a whirl*).

□ **whirling dervish** see DERVISH. □ **whirler** n. **whirlingly** adv. [ME: (v.) ON hvirfla: (n.) f. MLG & MDu. wervel spindle & ON hvirfill circle f. Gmc]

whirligig /'wɜːlɪˌgɪg/ n. **1** a spinning or whirling toy. **2** a merry-go-round. **3** a revolving motion. **4** anything regarded as hectic or constantly changing (the whirligig of time). **5** (in full **whirligig beetle**) a small water beetle of the family Gyrinidae that swims rapidly in circles over the water surface. [ME f. WHIRL + obs. gig whipping-top]

whirlpool /'wɜːlpuːl/ n. a powerful circular eddy in the sea etc. often causing suction to its centre.

whirlwind /'wɜːlwɪnd/ n. **1** a mass or column of air whirling rapidly round and round in a cylindrical or funnel shape over land or water. **2** a confused tumultuous process. **3** (attrib.) very rapid (a whirlwind romance). □ **reap the whirlwind** suffer worse results of a bad action.

whirlybird /'wɜːlɪˌbɜːd/ n. colloq. a helicopter.

whirr /wɜː(r)/ n. & v. (also **whir**) ● n. a continuous rapid buzzing or softly clicking sound as of a bird's wings or of cog-wheels in constant motion. ● v.intr. (**whirred, whirring**) make this sound. [ME, prob. Scand.: cf. Danish hvirre, Norw. kvirra, perh. rel. to WHIRL]

whisht /hwɪʃt/ v. (also **whist** /hwɪst/) esp. Sc. & Ir. dial. **1** intr. (esp. as int.) be quiet; hush. **2** tr. quieten. [imit.]

whisk /wɪsk/ v. & n. ● v. **1** tr. (foll. by away, off) **a** brush with a sweeping movement. **b** take with a sudden motion (whisked the plate away). **2** tr. whip (cream, eggs, etc.). **3** tr. & intr. convey or go (esp. out of sight) lightly or quickly (whisked me off to the doctor). **4** tr. wave or lightly brandish. ● n. **1** a whisking action or motion. **2** a utensil for whisking eggs or cream etc. **3** a bunch of grass, twigs, bristles, etc., for removing dust or flies. [ME wisk, prob. Scand.: cf. ON visk wisp]

whisker /'wɪskə(r)/ n. **1** (usu. in pl.) the hair growing on a man's face, esp. on the cheek. **2** each of the bristles on the face of a cat etc. **3** colloq. a small distance (within a whisker of; won by a whisker). **4** a strong hairlike crystal of metal etc. □ **have** (or **have grown**) **whiskers** colloq. (esp. of a story etc.) be very old. □ **whiskered** adj. **whiskery** adj. [WHISK + -ER¹]

whisky /'wɪskɪ/ n. (US & Ir. **whiskey**) (pl. -**ies** or -**eys**) **1** a spirit distilled esp. from malted barley, other grains, or potatoes, etc. **2** a drink of this. [abbr. of obs. whiskybae, var. of USQUEBAUGH]

whisper /'wɪspə(r)/ v. & n. ● v. **1** intr. speak very softly without vibration of the vocal cords. **b** intr. & tr. talk or say in a barely audible tone or in a secret or confidential way. **2** intr. speak privately or conspiratorially. **3** intr. (of leaves, wind, or water) rustle or murmur. ● n. **1** an act of whispering or speaking softly (talking in whispers). **2** a whispering sound. **3** a thing whispered. **4** a rumour or piece of gossip. **5** a brief mention; a hint or suggestion. □ **it is whispered** there is a rumour. **whispering-gallery** a gallery esp. under a dome with acoustic properties such that a whisper may be heard round its entire circumference. □ **whisperer** n. **whispering** n. [OE hwisprian f. Gmc]

whist¹ /wɪst/ n. a card-game, usually played between two pairs of players, in which points are scored according to the number of tricks won and (in some forms) by the highest trumps or honours held by each pair. □ **whist drive** a social occasion with the playing of whist. [earlier whisk, perh. f. WHISK (with ref. to whisking away the tricks): perh. assoc. with WHIST²]

whist² var. of WHISHT.

whistle /'wɪs(ə)l/ n. & v. ● n. **1** a clear shrill sound made by forcing breath through a small hole between nearly closed lips. **2** a similar sound made by a bird, the wind, a missile, etc. **3 a** an instrument producing a shrill note by means of breath, air, steam, etc., used esp. for signalling. **b** a simple musical instrument resembling a recorder. **4** rhyming sl. a suit (short for whistle and flute). **5** colloq. the mouth or throat. ● v. **1** intr. emit a whistle. **2 a** intr. give a signal or express surprise or derision by whistling. **b** tr. (often foll. by up) summon or give a signal to (a dog etc.) by whistling. **3** tr. (also absol.) produce (a tune) by whistling. **4** intr. (foll. by for) vainly seek or desire. □ **as clean** (or **clear** or **dry**) **as a whistle** very clean or clear or dry. **blow the whistle on** bring (an activity) to an end; inform on (those responsible). **wet** (or **whet**) **one's whistle** colloq. or joc. have a drink esp. an alcoholic one. **whistle-blower** a person who blows the whistle on someone or something. **whistle down the wind 1** let go, abandon. **2** turn (a hawk) loose. **whistle in the dark** pretend to be unafraid. **whistle-stop 1** US a small unimportant town on a railway. **2** a politician's brief pause for an electioneering speech on tour. **3** (attrib.) with brief pauses; very fast (a whistle-stop tour). **whistling kettle** a kettle fitted with a whistle sounded by steam when the kettle is boiling. [OE

(h)wistlian (v.), (h)wistle (n.) of imit. orig.: cf. ON hvísla whisper, MSw. hvisla whistle]

Whistler /'wɪslə(r)/, James (Abbott) McNeill (1834–1903), American painter and etcher. In 1855 he went to Paris, where he was influenced by the realist painter Gustave Courbet. Based mainly in London from 1859, Whistler retained his contact with Paris and shared the admiration of contemporary French painters for Japanese prints; his work reflected this in the attention he gave to the composition of subtle patterns of light. Whistler mainly painted in one or two colours, and sought to achieve harmony of colour and tone, as in the portrait Arrangement in Grey and Black: The Artist's Mother (1872) and the landscape Old Battersea Bridge: Nocturne – Blue and Gold (c.1872–5).

whistler /'wɪslə(r)/ n. **1 a** a person who whistles. **b** a thing which makes a whistling sound. **2 a** a bird of the thickhead family, esp. of the genus Pachycephala, with a loud melodious call. **b** a North American marmot, Marmota caligata, with a whistling call.

Whit /wɪt/ n. & adj. ● n. = WHITSUNTIDE. ● adj. connected with, belonging to, or following Whit Sunday (Whit Monday; Whit weekend). [See WHIT SUNDAY]

whit /wɪt/ n. a particle; a least possible amount (not a whit better). □ **every whit** the whole; wholly. **no** (or **never a** or **not a**) **whit** not at all. [earlier w(h)yt app. alt. f. WIGHT in phr. no wight etc.]

Whitby /'wɪtbɪ/ a town on the coast of North Yorkshire, NE England; pop. (1981) 13,380.

Whitby, Synod of a conference held in Whitby in 664 that resolved the differences between the Celtic and Roman forms of Christian worship in England, in particular the method of calculating the date of Easter. The Northumbrian Christians had followed the Celtic method of fixing the date while those of the south had adopted the Roman system. King Oswy (612–70) of Northumbria decided in favour of Rome, and England as a result effectively severed the connection with the Celtic Church. An account of the proceedings is given by Bede.

White¹ /waɪt/, Gilbert (1720–93), English clergyman and naturalist. He spent most of his life in his native village of Selborne, Hampshire, becoming curate there in 1784. White is best known for the many letters he wrote to friends, sharing his acute observations on all aspects of natural history, especially ornithology; these were published in 1789 as The Natural History and Antiquities of Selborne, which has remained in print ever since. He was the first to identify the harvest mouse and the noctule bat, and the first to recognize the chiffchaff, willow warbler, and wood warbler as different species.

White² /waɪt/, Patrick (Victor Martindale) (1912–90), Australian novelist, born in Britain. White's international reputation is based on his two novels The Tree of Man (1955) and Voss (1957); the latter relates the doomed attempt made in 1845 by a German explorer to cross the Australian continent. He was awarded the Nobel Prize for literature in 1973.

white /waɪt/ adj., n., & v. ● adj. **1** resembling a surface reflecting sunlight without absorbing any of the visible rays; of the colour of milk or fresh snow. **2** approaching such a colour; pale esp. in the face (turned as white as a sheet). **3** less dark than other things of the same kind. **4 a** of the human group having light-coloured skin. **b** of or relating to white people. **5** albino (white mouse). **6 a** (of hair) having lost its colour esp. in old age. **b** (of a person) white-haired. **7** colloq. innocent, untainted. **8 a** (of a plant) having white flowers or pale-coloured fruit etc. (white hyacinth; white cauliflower). **b** (of a tree) having light-coloured bark etc. (white ash; white poplar). **9** (of wine) made from white grapes or dark grapes with the skins removed. **10** (of coffee) with milk or cream added. **11** transparent, colourless (white glass). **12** hist. counter-revolutionary or reactionary (white guard; white army). ● n. **1** a white colour or pigment. **2 a** white clothes or material (dressed in white). **b** (in pl.) white garments as worn in cricket, tennis, etc. **c** (in pl.) white clothing for washing. **3 a** (in a game or sport) a white piece, ball, etc. **b** the player using such pieces. **4** the white part or albumen round the yolk of an egg. **5** the visible part of the eyeball round the iris. **6** a member of a light-skinned race. **7** a white butterfly. **8** a blank space in printing. ● v.tr. archaic make white. □ **bleed white** drain (a person, country, etc.) of wealth etc. **white admiral** a butterfly of the genus Limenitis, having blackish wings with white markings; esp. L. camilla of Europe. **white ant** a termite. **white blood cell** (or **corpuscle**) = LEUCOCYTE. **white-bread** N. Amer. colloq. having or reflecting white middle-class values (as conservatism etc.). **white cell** (or **corpuscle**) = LEUCOCYTE. **white Christmas** Christmas with snow on the ground. **white coal** water as a source of power. **white-collar** (of a worker) engaged in clerical or administrative rather than manual work. **white currant** a cultivar of the redcurrant with pale edible berries. **whited sepulchre** see SEPULCHRE. **white elephant** a possession, building, etc. that is no longer useful or wanted, esp. one that is expensive or difficult to maintain. **white ensign** see ENSIGN. **white-eye** a bird with a white iris or white plumage around the eyes, esp. a small Old World songbird of the mainly tropical family Zosteropidae. **white feather** a symbol of cowardice (a white feather in the tail of a game bird being a mark of bad breeding). **white flag** a symbol of surrender or a period of truce. **white frost** see FROST n. 1a. **white gold** any of various silver-coloured alloys of gold, used as substitutes for platinum in jewellery. **white goods 1** domestic linen. **2** large domestic electrical equipment. **white heat 1** the temperature at which metal emits white light. **2** a state of intense passion or activity. **white hope** a person expected to achieve much for a group, organization, etc. **white horses** white-crested waves at sea. **white-hot** at white heat. **white-knuckle** *colloq.* (esp. of a fairground ride) causing fear. **white lead** a mixture of lead carbonate and hydrated lead oxide used as pigment. **white lie** a harmless or trivial untruth. **white light** apparently colourless light, e.g. ordinary daylight. **white lime** lime mixed with water as a coating for walls; whitewash. **white magic** magic used only for beneficent purposes. **white matter** the part of the brain and spinal cord consisting mainly of nerve fibres (see also *grey matter*). **white meat** poultry, veal, rabbit, and pork. **white metal** a white or silvery alloy. **white monk** a Cistercian monk. **white night** a sleepless night. **white noise** noise containing many frequencies with equal intensities. **white out** obliterate, esp. with white correction fluid, paint, etc. **white-out** n. **1** a dense blizzard esp. in polar regions. **2** a weather condition in which the shadows, horizon, and features of snow-covered country are indistinguishable. **3** a white correction fluid. **white ox-eye** = ox-eye daisy. **white pepper** see PEPPER. **white poplar** = ABELE. **white rose** the emblem of Yorkshire or the House of York. **white sale** a sale of household linen. **white sauce** a sauce of flour, melted butter, and milk or cream. **white slave** a woman tricked or forced into prostitution, usu. abroad. **white slavery** traffic in white slaves. **white sock** = STOCKING 3. **white spirit** light petroleum as a solvent. **white sugar** purified sugar. **white tie** a man's white bow-tie as part of full evening dress. **white vitriol** *Chem.* zinc sulphate. **white wagtail** a grey, black, and white Eurasian wagtail, *Motacilla alba*, esp. a widespread grey-backed race (cf. *pied wagtail*). **white water** a shallow or foamy stretch of water. **white wedding** a wedding at which the bride wears a formal white dress. **white whale** a small northern whale, *Delphinapterus leucas*, which is white when adult (also called *beluga*). □ **whitely** *adv.* **whiteness** *n.* **whitish** *adj.* [OE *hwīt* f. Gmc]

whitebait /ˈwaɪtbeɪt/ n. (pl. same) **1** (usu. pl.) the small silvery-white young of herrings and sprats esp. as food. **2** NZ a young inanga.

whitebeam /ˈwaɪtbiːm/ n. a rosaceous tree, *Sorbus aria*, having red berries and leaves with a white downy underside.

whiteboard /ˈwaɪtbɔːd/ n. a board with a white surface for writing or drawing on with felt-tipped pens.

white dwarf n. *Astron.* a star of high density, formed when a low-mass star has exhausted all its central nuclear fuel and lost its outer layers as a planetary nebula. Although they have a mass as great as that of many normal stars, white dwarfs typically have a radius no greater than that of the earth, implying densities as high as several tons per cubic centimetre.

whiteface /ˈwaɪtfeɪs/ n. the white make-up of an actor etc.

whitefish /ˈwaɪtfɪʃ/ n. **1** food fish with pale flesh, e.g plaice, cod, etc. **2** a lake fish of the genus *Coregonus* or *Prosopium*, of the trout family, used (esp. in North America) for food.

whitefly /ˈwaɪtflaɪ/ n. (pl. **-flies**) a small homopterous insect of the family Aleyrodidae, having waxy white wings and often a pest of garden and greenhouse plants.

White Friar n. a Carmelite monk.

Whitehall /ˈwaɪthɔːl/ a street in Westminster, London, in which many government offices are located. The name is taken from the former royal palace of White Hall, originally a residence of Cardinal Wolsey, and is used allusively to refer to the British Civil Service.

Whitehead /ˈwaɪthed/, Alfred North (1861–1947), English philosopher and mathematician. He is remembered chiefly for *Principia Mathematica* (1910–13; see MATHEMATICAL LOGIC), on which he collaborated with his pupil Bertrand Russell, but he was concerned to explain more generally the connections between mathematics, theoretical science, and ordinary experience. Whitehead's work on geometry led to an interest in the philosophy of science; he proposed an alternative to Einstein's theories of relativity, and later developed a general and systematic metaphysical view.

whitehead /ˈwaɪthed/ n. *colloq.* a white or white-topped skin-pustule.

Whitehorse /ˈwaɪthɔːs/ the capital of Yukon Territory in NW Canada; pop. (1991) 21,650. Situated on the Alaska Highway, Whitehorse is the centre of a copper-mining and fur-trapping region. It became the capital, replacing Dawson, in 1953.

White House the official residence of the US President in Washington, DC; by extension, the President or presidency of the US. The White House was built 1792–9 of greyish-white limestone from designs of the Irish-born architect James Hoban (c.1762–1831), on a site chosen by George Washington; President John Adams took up residence there in 1800. The building was restored in 1814 after being burnt by British troops during the War of 1812, the smoke-stained walls being painted white. Although known informally as the White House from the early 19th century, it was not formally so designated until the time of Theodore Roosevelt (1902).

whiten /ˈwaɪt(ə)n/ v.tr. & intr. make or become white. □ **whitener** n. **whitening** n.

White Nile the name for the main, western branch of the Nile between the Ugandan–Sudanese border and its confluence with the Blue Nile at Khartoum (see NILE¹).

White Paper n. (in the UK) a government report giving information or proposals on an issue.

White Russia an alternative name for BELARUS.

White Russian n. & adj. ● n. **1** a Belorussian. **2** *hist.* an opponent of the Bolsheviks during the Russian Civil War (1918–21). ● adj. **1** Belorussian. **2** *hist.* of or relating to the White Russians.

White Sands an area of white gypsum salt flats in central New Mexico, designated a national monument in 1933. It is surrounded by a large missile testing range, which, in 1945, as part of the Manhattan Project, was the site of the detonation of the first nuclear weapon.

White Sea an inlet of the Barents Sea on the coast of NW Russia.

whitesmith /ˈwaɪtsmɪθ/ n. **1** a worker in tin. **2** a polisher or finisher of metal goods.

whitethorn /ˈwaɪtθɔːn/ n. the hawthorn.

whitethroat /ˈwaɪtθrəʊt/ n. a white-throated Eurasian warbler of the genus *Sylvia*, esp. the *common whitethroat* (*S. communis*).

whitewash /ˈwaɪtwɒʃ/ n. & v. ● n. **1** a solution of quicklime or of whiting and size for whitening walls etc. **2** a means employed to conceal mistakes or faults in order to clear a person or institution of imputations. ● v.tr. **1** cover with whitewash. **2** attempt by concealment to clear the reputation of. **3** (in *passive*) (of an insolvent) get a fresh start by passage through a bankruptcy court. **4** defeat (an opponent) without allowing any opposing score. □ **whitewasher** n.

whitewood /ˈwaɪtwʊd/ n. a light-coloured wood esp. prepared for staining etc.

Whitey /ˈwaɪtɪ/ n. (pl. **-eys**) *sl. offens.* **1** a white person. **2** white people collectively.

whither /ˈwɪðə(r)/ adv. & conj. archaic ● adv. **1** to what place, position, or state? **2** (prec. by *place* etc.) to which (*the house whither we were walking*). ● conj. **1** to the or any place to which (*go whither you will*). **2** and thither (*we saw a house, whither we walked*). [OE *hwider* f. Gmc: cf. WHICH, HITHER, THITHER]

whithersoever /ˌwɪðəsəʊˈevə(r)/ adj. & conj. archaic to any place to which.

whiting¹ /ˈwaɪtɪŋ/ n. (pl. same) a small white-fleshed marine fish, *Merlangus merlangus*, used as food. [ME f. MDu. *wijting*, app. formed as WHITE + -ING³]

whiting² /ˈwaɪtɪŋ/ n. ground chalk used in whitewashing, plate-cleaning, etc.

Whitlam /ˈwɪtləm/, (Edward) Gough (b.1916), Australian Labor statesman, Prime Minister 1972–5. While in office he ended compulsory military service in Australia and relaxed the laws for Asian and African immigrants. When, in 1975, the opposition blocked finance bills in the Senate, he refused to call a general election and was dismissed by the Governor-General Sir John Kerr, the first occasion in 200 years that the British Crown had removed an elected Prime Minister. Whitlam remained leader of the Labor Party until 1977.

whitleather /ˈwɪtˌleðə(r)/ n. tawed leather. [ME f. WHITE + LEATHER]

whitlow /ˈwɪtləʊ/ n. an inflammation near a fingernail or toenail. □ **whitlow-grass** a dwarf cruciferous plant of the genus *Erophila* or

Draba, of rocks, walls, etc., esp. *E. verna*, formerly thought to cure whitlows. [ME *whitflaw*, *-flow*, app. = WHITE + FLAW[1] in the sense 'crack', but perh. of LG orig.: cf. Du. *fijt*, LG *fit* whitlow]

Whitman /'wɪtmən/, Walt (1819–92), American poet. His philosophical outlook was profoundly influenced by his experience of frontier life in his travels to Chicago and St Louis in 1848, as well as by transcendentalist ideas, in particular the work of Emerson. He published the first edition of *Leaves of Grass*, incorporating 'I Sing the Body Electric' and 'Song of Myself', in 1855; eight further editions followed in Whitman's lifetime, the work enlarging as the poet developed. Written in free verse, the collection celebrates democracy, sexuality, the self, and the liberated American spirit in union with nature. It was criticized as immoral during his lifetime and later formed the basis of the claims that he was homosexual. In *Drum-Taps* (1865), Whitman recorded his impressions as a hospital visitor during the American Civil War. The *Sequel* (1865) to these poems included his elegy on Lincoln, 'When Lilacs Last in the Dooryard Bloom'd'.

Whitney /'wɪtnɪ/, Eli (1765–1825), American inventor. He devised the mechanical cotton-gin (patented 1794), as well as conceiving the idea of mass-producing interchangeable parts. This he applied in his fulfilment of a US government contract (1797) to supply muskets; Whitney manufactured these in standardized parts for reassembly, meaning that for the first time worn parts could be replaced by spares rather than requiring special replacements to be made.

Whitney, Mount /'wɪtnɪ/ a mountain in the Sierra Nevada in California. Rising to 4,418 m (14,495 ft), it is the highest peak in the continental US outside Alaska.

Whitsun /'wɪts(ə)n/ *n. & adj.* ● *n.* = WHITSUNTIDE. ● *adj.* = WHIT. [ME, f. *Whitsun Day* = Whit Sunday]

Whit Sunday in the Christian Church, the seventh Sunday after Easter, commemorating the descent of the Holy Spirit on the Apostles at Pentecost (Acts 2). 'Whit' derives from an earlier form of 'White', and probably refers to the white robes of the newly baptized at Pentecost; in the Western Church the festival became a date for baptisms.

Whitsuntide /'wɪts(ə)n,taɪd/ *n.* the weekend or week including Whit Sunday.

Whittier /'wɪtɪə(r)/, John Greenleaf (1807–92), American poet and abolitionist. From the early 1840s he edited various periodicals and wrote poetry for the abolitionist cause. He is best known for his poems on rural themes, especially 'Snow-Bound' (1866).

Whittington /'wɪtɪŋtən/, Sir Richard ('Dick') (d.1423), English merchant and Lord Mayor of London. Whittington was a London mercer who became Lord Mayor three times (1397–8; 1406–7; 1419–20). He left legacies for rebuilding Newgate Prison and for establishing a city library. The popular legend of Dick Whittington's early life as an orphan from a lowly background, his only possession a cat, is first recorded in 1605.

Whittle /'wɪt(ə)l/, Sir Frank (b.1907), English aeronautical engineer, test pilot, and inventor of the jet aircraft engine. He took out the first patent for a turbojet engine in 1930 while still a student. A Gloster aircraft made the first British flight using Whittle's jet engine in May 1941 (two years after the first German jet aircraft), and similar machines later entered service with the RAF. He took up a post at the US Naval Academy in Maryland in 1977.

whittle /'wɪt(ə)l/ *v.* **1** *tr.* & (foll. by *at*) *intr.* pare (wood etc.) with repeated slicing with a knife. **2** *tr.* (often foll. by *away*, *down*) reduce by repeated subtractions. [var. of ME *thwitel* long knife f. OE *thwitan* to cut off]

Whitworth /'wɪtwəθ/ *adj.* (of a screw etc.) having a standard screw-thread. [Sir Joseph *Whitworth*, Engl. engineer (1803–87)]

whity /'waɪtɪ/ *adj.* whitish; rather white (usu. in *comb.*: *whity-brown*) (cf. WHITEY).

whiz /wɪz/ *n. & v.* (also **whizz**) *colloq.* ● *n.* **1** the sound made by the friction of a body moving through the air at great speed. **2** (also **wiz**) *colloq.* a person who is remarkable or skilful in some respect (*is a whiz at chess*). ● *v.intr.* (**whizzed**, **whizzing**) move with or make a whiz. □ **whiz-bang** *colloq.* **1** a high-velocity shell from a small-calibre gun, whose passage is heard before the gun's report. **2** a jumping kind of firework. **whiz-kid** *colloq.* a brilliant or highly successful young person. [imit.: in sense 2 infl. by WIZARD]

WHO see WORLD HEALTH ORGANIZATION.

who /hu:/ *pron.* (*obj.* **whom** /hu:m/ or *colloq.* **who**; *poss.* **whose**) **1 a** what or which person or persons? (*who called?*; *you know who it was*; *whom* or *who did you see?*). ¶ In the last example *whom* is correct but *who* is common in less formal contexts. **b** what sort of person or persons?

(*who am I to object?*). **2** (a person) that (*anyone who wishes can come*; *the woman whom you met*; *the man who you saw*). ¶ In the last two examples *whom* is correct but *who* is common in less formal contexts. **3** and (or but) he, she, they, etc. (*gave it to Tom, who sold it to Jim*). **4** *archaic* the or any person or persons that (*whom the gods love die young*). □ **as who should say** like a person who said; as though one said. **who-does-what** (of a dispute etc.) about which group of workers should do a particular job. **who goes there?** see GO[1]. **who's who 1** who or what each person is (*know who's who*). **2** a list or directory with facts about notable persons. [OE *hwā* f. Gmc: *whom* f. OE dative *hwām*, *hwǣm*: *whose* f. genitive *hwæs*]

Who, the an English rock group formed in 1964, including Pete Townshend (b.1945), Roger Daltrey (b.1944), and Keith Moon (1947–78). The Who made their name with a swaggering 'mod' image, being known as much for their wild performances and for destroying their instruments on stage as for the songs of guitarist Pete Townshend; among the best known of these songs are 'My Generation' (1965) and 'Substitute' (1966). Later the group recorded Townshend's rock opera *Tommy* (1969), one of the first of its kind.

whoa /wəʊ/ *int.* used as a command to stop or slow a horse etc. [var. of HO]

who'd /hu:d/ *contr.* **1** who had. **2** who would.

whodunit /hu:'dʌnɪt/ *n.* (also **whodunnit**) *colloq.* a story or play about the detection of a crime etc., esp. murder. [f. *who done it?* non-standard form of *who did it?*]

whoe'er /hu:'eə(r)/ *poet.* var. of WHOEVER.

whoever /hu:'evə(r)/ *pron.* (*obj.* **whomever** /hu:m'evə(r)/ or *colloq.* **whoever**; *poss.* **whosever** /hu:z'evə(r)/) **1** the or any person or persons who (*whoever comes is welcome*). **2** though anyone (*whoever else objects, I do not*; *whosever it is, I want it*). **3** *colloq.* (as an intensive) who ever; who at all (*whoever heard of such a thing?*).

whole /həʊl/ *adj. & n.* ● *adj.* **1** in an uninjured, unbroken, intact, or undiminished state (*swallowed it whole*). **2** not less than; all there is of; entire, complete (*waited a whole year*; *tell the whole truth*; *the whole school knows*). **3** (of blood or milk etc.) with no part removed. **4** (of a person) healthy, recovered from illness or injury. ● *n.* **1** a thing complete in itself. **2** all there is of a thing (*spent the whole of the summer by the sea*). **3** (foll. by *of*) all members, inhabitants, etc., of (*the whole of London knows it*). □ **as a whole** as a unity; not as separate parts. **go the whole hog** see HOG. **on the whole** taking everything relevant into account; in general (*it was, on the whole, a good report*; *they behaved well on the whole*). **whole cloth** cloth of full size as manufactured. **whole holiday** a whole day taken as a holiday (cf. *half holiday*). **whole-life insurance** life insurance for which premiums are payable throughout the remaining life of the person insured. **a** (or **the**) **whole lot** see LOT. **whole note** esp. *N. Amer. Mus.* = SEMIBREVE. **whole number** a number without fractions; an integer. **whole-tone scale** *Mus.* a scale consisting entirely of tones, with no semitones. □ **wholeness** *n.* [OE *hāl* f. Gmc]

wholefood /'həʊlfu:d/ *n.* food which has not been processed or refined more than is minimally necessary (often *attrib.*: *wholefood restaurant*; *wholefood diet*).

wholegrain /'həʊlgreɪn/ *adj.* made with or containing whole grains (*wholegrain bread*).

wholehearted /həʊl'hɑ:tɪd/ *adj.* **1** (of a person) completely devoted or committed. **2** (of an action etc.) done with all possible effort, attention, or sincerity; thorough. □ **wholeheartedly** *adv.* **wholeheartedness** *n.*

wholemeal /'həʊlmi:l/ *n.* (usu. *attrib.*) *Brit.* meal or flour of wheat or other cereals with none of the bran or germ removed.

wholesale /'həʊlseɪl/ *n., adj., adv., & v.* ● *n.* the selling of things in large quantities to be retailed by others (cf. RETAIL). ● *adj. & adv.* **1** by wholesale; at a wholesale price (*can get it for you wholesale*). **2** on a large scale (*wholesale destruction occurred*; *was handing out samples wholesale*). ● *v.tr.* sell wholesale. □ **wholesaler** *n.* [ME: orig. *by whole sale*]

wholesome /'həʊlsəm/ *adj.* **1** promoting or indicating physical, mental, or moral health (*wholesome meal*; *a wholesome appearance*). **2** prudent (*wholesome respect*). □ **wholesomely** *adv.* **wholesomeness** *n.* [ME, prob. f. OE (unrecorded) *hālsum* (as WHOLE, -SOME[1])]

wholewheat /'həʊlwi:t/ *n.* (usu. *attrib.*) wheat with none of the bran or germ removed; wholemeal.

wholism var. of HOLISM.

wholly /'həʊllɪ/ *adv.* **1** entirely; without limitation or diminution (*I am* *wholly at a loss*). **2** purely, exclusively (*a wholly bad example*). [ME, f. OE (unrecorded) *hāllīce* (as WHOLE, -LY²)]

whom *objective case of* WHO.

whomever *objective case of* WHOEVER.

whomso *archaic objective case of* WHOSO.

whomsoever *objective case of* WHOSOEVER.

whoop /wuːp, huːp/ *n. & v.* (also **hoop** /huːp/) ● *n.* **1** a loud cry of or as of excitement etc. **2** a long rasping indrawn breath in whooping cough. ● *v.intr.* utter a whoop. □ **whooping cough** /ˈhuːpɪŋ/ an infectious bacterial disease, esp. of children, with a series of short violent coughs followed by a whoop (also called *pertussis*). **whooping crane** a large mainly white North American crane, *Grus americana*, now an endangered species. **whoop it up** *colloq.* **1** engage in revelry. **2** *US* make a stir. [ME: imit.]

whoopee *int. & n. colloq.* ● *int.* /wʊˈpiː/ expressing exuberant joy. ● *n.* /ˈwʊpɪ/ exuberant enjoyment or revelry. □ **make whoopee** *colloq.* **1** rejoice noisily or hilariously. **2** make love. **whoopee cushion** a rubber cushion that when sat on makes a sound like the breaking of wind.

whooper /ˈhuːpə(r)/ *n.* (in full **whooper swan**) a large migratory swan, *Cygnus cygnus*, with a black and yellow bill and a loud whooping call.

whoops /wʊps/ *int. colloq.* expressing surprise or apology, esp. on making an obvious mistake. [var. of OOPS]

whoosh /wʊʃ/ *v., n., & int.* (also **woosh**) ● *v.intr. & tr.* move or cause to move with a rushing sound. ● *n.* a sudden movement accompanied by a rushing sound. ● *int.* an exclamation imitating this. [imit.]

whop /wɒp/ *v. & n.* (also esp. *US* **whap**) ● *v.tr.* (**whopped, whopping**) *sl.* **1** thrash. **2** defeat, overcome. ● *n.* **1** the sound of a blow or sudden thud. **2** the regular pulsing sound of a helicopter rotor. [ME: var. of dial. *wap*, of unkn. orig.]

whopper /ˈwɒpə(r)/ *n. sl.* **1** something big of its kind. **2** a blatant or gross lie.

whopping /ˈwɒpɪŋ/ *adj. sl.* very big (*a whopping lie; a whopping great fish*).

whore /hɔː(r)/ *n. & v.* ● *n.* **1** a prostitute. **2** *derog.* a promiscuous woman. ● *v.intr.* **1** (of a man) seek or chase after whores. **2** (foll. by *after*) *archaic* commit idolatry or iniquity. □ **whore-house** a brothel. □ **whoredom** *n.* **whorer** *n.* [OE *hōre* f. Gmc]

whoremaster /ˈhɔːˌmɑːstə(r)/ *n. archaic* = WHOREMONGER.

whoremonger /ˈhɔːˌmʌŋgə(r)/ *n. archaic* a sexually promiscuous man; a lecher.

whoreson /ˈhɔːs(ə)n/ *n. archaic* **1** a disliked person. **2** (*attrib.*) (of a person or thing) vile.

whorish /ˈhɔːrɪʃ/ *adj.* of or like a whore. □ **whorishly** *adv.* **whorishness** *n.*

whorl /wɜːl/ *n.* **1** a ring of leaves or other organs round a stem of a plant. **2** one turn of a spiral, esp. on a shell. **3** a complete circle in a fingerprint. **4** *archaic* a small wheel on a spindle steadying its motion. □ **whorled** *adj.* [ME *wharwyl, whorwil*, app. var. of WHIRL: infl. by *wharve* (n.) = whorl of a spindle]

whortleberry /ˈwɜːt(ə)lˌberɪ/ *n.* (*pl.* **-ies**) a bilberry. [16th c.: dial. form of *hurtleberry*, ME, of unkn. orig.]

whose /huːz/ *pron. & adj.* ● *pron.* of or belonging to which person (*whose is this book?*). ● *adj.* of whom or which (*whose book is this?; the man, whose name was Tim; the house whose roof was damaged*).

whoseso *archaic poss. of* WHOSO.

whosesoever *poss. of* WHOSOEVER.

whosever *poss. of* WHOEVER.

whoso /ˈhuːsəʊ/ *pron.* (*obj.* **whomso** /ˈhuːmsəʊ/; *poss.* **whoseso** /ˈhuːzsəʊ/) *archaic* = WHOEVER. [ME = WHO + SO¹, f. OE *swā hwā swā*]

whosoever /ˌhuːsəʊˈevə(r)/ *pron.* (*obj.* **whomsoever** /ˌhuːmsəʊ-/; *poss.* **whosesoever** /ˌhuːzsəʊ-/) *archaic* = WHOEVER.

whump /wʌmp/ *n. & v.* ● *n.* a dull thud. ● *v.* **1** *intr.* make such a sound. **2** *tr.* strike heavily with a whump. [imit.]

why /waɪ/ *adv., int., & n.* ● *adv.* **1 a** for what reason or purpose (*why did you do it?; I do not know why you came*). **b** on what grounds (*why do you say that?*). **2** (*prec. by reason* etc.) for which (*the reasons why I did it*). ● *int.* expressing: **1** surprised discovery or recognition (*why, it's you!*). **2** impatience (*why, of course I do!*). **3** reflection (*why, yes, I think so*). **4** objection (*why, what is wrong with it?*). ● *n.* (*pl.* **whys**) a reason or explanation (esp. *whys and wherefores*). □ **why so?** on what grounds?;

for what reason or purpose? [OE *hwī, hwȳ* instrumental of *hwæt* WHAT f. Gmc]

Whyalla /waɪˈælə/ a steel-manufacturing town on the coast of South Australia, on the Spencer Gulf; pop. (1991) 25,525.

whydah /ˈwɪdə/ *n.* (also **whidah**) **1** an African weaver-bird of the genus *Vidua*, the male having mainly black plumage and tail feathers of great length. Also called *widow-bird*. **2** = *widow-bird* 1. [*widow-bird*, altered f. assoc. with *Whidah* (now Ouidah), a town in Benin]

Whymper /ˈwɪmpə(r)/, Edward (1840–1911), English mountaineer. In 1860 he was commissioned to make drawings of the Alps, and in the following year he returned to attempt to climb the Matterhorn. After seven attempts, he finally succeeded in 1865, at the age of 25. On the way down, four of his fellow climbers fell to their deaths, raising contemporary public doubts about the sport.

WI *abbr.* **1** West Indies. **2** see WOMEN'S INSTITUTE. **3** *US* Wisconsin (in official postal use).

Wicca /ˈwɪkə/ *n.* the religious cult of modern witchcraft. □ **Wiccan** *adj. & n.* [OE *wicca* WITCH]

Wichita /ˈwɪtʃɪˌtɔː/ a city in southern Kansas, on the River Arkansas; pop. (1990) 304,000.

wick¹ /wɪk/ *n.* **1** a strip or thread of fibrous or spongy material feeding a flame with fuel in a candle, lamp, etc. **2** *Surgery* a gauze strip inserted in a wound to drain it. □ **dip one's wick** *coarse sl.* (of a man) have sexual intercourse. **get on a person's wick** *Brit. colloq.* annoy a person. [OE *wēoce, -wēoc* (cf. MDu. *wiecke*, MLG *wēke*), of unkn. orig.]

wick² /wɪk/ *n. dial.* except in compounds e.g. *bailiwick*, and in place-names e.g. *Hampton Wick, Warwick* **1** a town, hamlet, or district. **2** a dairy farm. [OE *wīc*, prob. f. Gmc f. L *vicus* street, village]

wicked /ˈwɪkɪd/ *adj.* (**wickeder, wickedest**) **1** sinful, iniquitous, given to or involving immorality. **2** spiteful, ill-tempered; intending or intended to give pain. **3** playfully malicious. **4** *colloq.* foul; very bad; formidable (*wicked weather; a wicked cough*). **5** *sl.* excellent, remarkable. □ **wickedly** *adv.* **wickedness** *n.* [ME f. obs. *wick* (perh. adj. use of OE *wicca* wizard) + -ED² as in *wretched*]

Wicked Bible an edition of the Bible of 1631, with the misprinted commandment 'thou shalt commit adultery'.

wicker /ˈwɪkə(r)/ *n.* plaited twigs or osiers etc. as material for chairs, baskets, mats, etc. [ME, f. E.Scand.: cf. Sw. *viker* willow, rel. to *vika* bend]

wickerwork /ˈwɪkəˌwɜːk/ *n.* **1** wicker. **2** things made of wicker.

wicket /ˈwɪkɪt/ *n.* **1** *Cricket* **a** a set of three stumps with the bails in position defended by a batsman. **b** the ground between two wickets. **c** the state of this (*a slow wicket*). **d** an instance of a batsman being got out (*bowler has taken four wickets*). **e** a pair of batsmen batting at the same time (*a third-wicket partnership*). **2** (in full **wicket-door** or **-gate**) a small door or gate esp. beside or in a larger one or closing the lower part only of a doorway. **3** *US* an aperture in a door or wall usu. closed with a sliding panel. **4** *US* a croquet hoop. □ **at the wicket** *Cricket* **1** batting. **2** by the wicket-keeper (*caught at the wicket*). **keep wicket** *Cricket* be a wicket-keeper. **on a good** (or **sticky**) **wicket** *colloq.* in a favourable (or unfavourable) position. **wicket-keeper** *Cricket* the fielder stationed close behind a batsman's wicket, usu. equipped with gloves and pads. [ME f. AF & ONF *wiket*, OF *guichet*, of uncert. orig.]

wickiup /ˈwɪkɪˌʌp/ *n.* a North American Indian hut of a frame covered with grass etc. [Fox *wikiyap*]

Wicklow /ˈwɪkləʊ/ **1** a county of the Republic of Ireland, in the east, in the province of Leinster. **2** its county town, on the Irish Sea; pop. (1991) 5,850.

widdershins /ˈwɪdəˌʃɪnz/ *adv.* (also **withershins** /ˈwɪðə-/) esp. *Sc.* in a direction contrary to the sun's course (considered as unlucky). [MLG *weddersins* f. MHG *widersinnes* f. *wider* against + *sin* direction]

wide /waɪd/ *adj., adv., & n.* ● *adj.* **1 a** measuring much or more than other things of the same kind across or from side to side. **b** considerable; more than is needed (*a wide margin*). **2** (following a measurement) in width (*a metre wide*). **3** extending far; embracing much; of great extent (*has a wide range; has wide experience; reached a wide public*). **4** not tight or close or restricted; loose. **5 a** free, liberal; unprejudiced (*takes wide views*). **b** not especialized; general. **6** open to the full extent (*staring with wide eyes*). **7 a** (foll. by *of*) not within a reasonable distance of. **b** at a considerable distance from a point or mark. **8** *Brit. sl.* shrewd; skilled in sharp practice. **9** (in *comb.*) extending over the whole of (*nationwide*). ● *adv.* **1** widely. **2** to the full extent (*wide awake*). **3** far from the target etc. (*is shooting wide*). ● *n.* **1** *Cricket* a ball judged to pass the wicket beyond the batsman's reach and so scoring a run. **2** (*prec. by the*) the wide world. □ **give a wide berth to** see

BERTH. **wide-angle 1** (of a lens) having a shorter focal length than standard and so giving a wide field of view. **2** (of a photograph etc.) showing a wide field of view. **wide awake 1** fully awake. **2** *colloq.* wary, knowing. **wide ball** *Cricket* = WIDE *n.* 1. **wide boy** *Brit. sl.* a man skilled in sharp practice; a spiv. **wide-eyed** surprised or naive. **wide of the mark** see MARK[1]. **wide open** (often foll. by *to*) exposed or vulnerable (to attack etc.). **wide-ranging** covering an extensive range. **wide receiver** *Amer. Football* a receiver who lines up several yards to the side of an offensive formation. **wide screen** for or designed with a screen presenting a wide field of vision in relation to its height. **the wide world** all the world great as it is. □ **wideness** *n.* **widish** *adj.* [OE *wīd* (adj.), *wīde* (adv.) f. Gmc]

wideawake /ˈwaɪdəˌweɪk/ *n.* a soft felt hat with a low crown and wide brim. [punningly so named as not having a nap]

widely /ˈwaɪdlɪ/ *adv.* **1** to a wide extent; far apart. **2** extensively (*widely read*; *widely distributed*). **3** by many people (*it is widely thought that*). **4** considerably; to a large degree (*holds a widely different view*).

widen /ˈwaɪd(ə)n/ *v.tr. & intr.* make or become wider. □ **widener** *n.*

widespread /ˈwaɪdspred/ *adj.* widely distributed or disseminated.

widgeon var. of WIGEON.

widget /ˈwɪdʒɪt/ *n. colloq.* any gadget or device. [perh. alt. of GADGET]

widgie /ˈwɪdʒɪ/ *n. Austral. colloq.* the female counterpart of a bodgie (see BODGIE 1). [20th c.: orig. unkn.]

Widnes /ˈwɪdnɪs/ a town on the River Mersey in Cheshire, NW England; pop. (1981) 55,930.

widow /ˈwɪdəʊ/ *n. & v.* ● *n.* **1** a woman who has lost her husband by death and has not married again. **2** a woman whose husband is often away on a specified activity (*golf widow*). **3** extra cards dealt separately and taken by the highest bidder. **4** *Printing* the short last line of a paragraph at the top of a page or column. ● *v.tr.* **1** make into a widow or widower. **2** (as **widowed** *adj.*) bereft by the death of a spouse (*my widowed mother*). **3** (foll. by *of*) deprive of. □ **widow-bird 1** a weaver-bird of the genus *Euplectes*, the male of which has mainly black plumage (also called *whydah*). **2** = WHYDAH 1. **widow's cruse** an apparently small supply that proves or seems inexhaustible (see 1 Kings 17:10–16). **widow's mite** a small money contribution (see Mark 12:42). **widow's peak** a V-shaped growth of hair towards the centre of the forehead. **widow's weeds** see WEEDS. [OE *widewe*, rel. to OHG *wituwa*, Skr. *vidhávā*, L *viduus* bereft, widowed, Gk *ēitheos* unmarried man]

widower /ˈwɪdəʊə(r)/ *n.* a man who has lost his wife by death and has not married again.

widowhood /ˈwɪdəʊˌhʊd/ *n.* the state or period of being a widow.

width /wɪtθ, wɪdθ/ *n.* **1** measurement or distance from side to side. **2** a large extent. **3** breadth or liberality of thought, views, etc. **4** a strip of material of full width as woven. □ **widthways** *adv.* **widthwise** *adv.* [17th c. (as WIDE, -TH[2]) replacing *wideness*]

wield /wiːld/ *v.tr.* **1** hold and use (a weapon or tool). **2** exert or command (power or authority etc.). □ **wielder** *n.* [OE *wealdan*, *wieldan* f. Gmc]

wieldy /ˈwiːldɪ/ *adj.* (**wieldier**, **wieldiest**) easily wielded, controlled, or handled.

Wien see VIENNA.

Wiener /ˈwiːnə(r)/, Norbert (1894–1964), American mathematician. He is best known for establishing the science of cybernetics in the late 1940s. Wiener spent most of his working life at the Massachusetts Institute of Technology, making major contributions to the study of stochastic processes, integral equations, harmonic analysis, and related fields.

wiener /ˈwiːnə(r)/ *n. N. Amer.* (also *colloq.* **wienie**, **weenie** /-nɪ/) **1** a frankfurter. **2** *coarse sl.* the penis. □ **Wiener schnitzel** see SCHNITZEL. [G, abbr. of *Wienerwurst* Vienna sausage]

Wiesbaden /ˈviːsˌbɑːd(ə)n/ a city in western Germany, the capital of the state of Hesse, situated on the Rhine opposite Mainz; pop. (1991) 264,020. A capital of the duchy of Nassau in the early 19th century, it passed to Prussia in 1866. It has been a popular spa town since Roman times.

Wiesel /ˈwiːz(ə)l/, Elie (full name Eliezer Wiesel) (b.1928), Romanian-born American human-rights campaigner, novelist, and academic. A survivor of Auschwitz and Buchenwald concentration camps, he emigrated to the US in 1956 and subsequently pursued a career as a humanities lecturer. Wiesel emerged as a leading authority on the Holocaust, documenting and publicizing Nazi war crimes perpetrated against Jews and others during the Second World War. Genocide,

violence, and race hate were also the subjects of several acclaimed novels and short-story collections. Wiesel was awarded the Nobel Peace Prize in 1986.

Wiesenthal /ˈviːz(ə)nˌtɑːl/, Simon (b.1908), Austrian Jewish investigator of Nazi war crimes. After spending 1942 to 1945 in Nazi labour and concentration camps, he began his long campaign to bring Nazi war criminals to justice. Enlisting the help of West German, Israeli, and other government agents, he traced some 1,000 unprosecuted criminals, including Adolf Eichmann. In 1961 he opened the Jewish Documentation Centre, also known as the Wiesenthal Centre, in Vienna and continued to track down Nazi criminals when other countries had ceased to pursue their cases.

wife /waɪf/ *n.* (pl. **wives** /waɪvz/) **1** a married woman esp. in relation to her husband. **2** *archaic* a woman, esp. an old or uneducated one. **3** (in *comb.*) a woman engaged in a specified activity (*fishwife*; *housewife*; *midwife*). □ **have** (or **take**) **to wife** *archaic* marry (a woman). **wife-swapping** *colloq.* exchanging wives for sexual relations. □ **wifehood** *n.* **wifeless** *adj.* **wifelike** *adj.* **wifely** *adj.* **wifeliness** *n.* **wifish** *adj.* [OE *wīf* woman: ult. orig. unkn.]

wig[1] /wɪg/ *n.* an artificial head of hair esp. to conceal baldness or as a disguise, or worn by a judge or barrister or as period dress. □ **wigged** *adj.* (also in *comb.*). **wigless** *adj.* [abbr. of PERIWIG: cf. WINKLE]

wig[2] /wɪg/ *v.tr.* (**wigged**, **wigging**) *colloq.* rebuke sharply; rate. [app. f. WIG[1] in sl. or colloq. sense 'rebuke' (19th c.)]

Wigan /ˈwɪgən/ a town in Greater Manchester, NW England; pop. (1981) 88,900.

wigeon /ˈwɪdʒən/ *n.* (also **widgeon**) a dabbling duck of the genus *Anas*, the male of which has a whitish forehead and crown and a whistling call; esp. *Anas penelope* of Eurasia, or *A. americana* of North America. [16th c.: orig. uncert.]

wigging /ˈwɪgɪŋ/ *n. colloq.* a reprimand.

wiggle /ˈwɪg(ə)l/ *v. & n. colloq.* ● *v.intr. & tr.* move or cause to move quickly from side to side etc. ● *n.* an act of wiggling; a kink in a line etc. □ **get a wiggle on** esp. *US sl.* hurry up. □ **wiggler** *n.* [ME f. MLG & MDu. *wiggelen*: cf. WAG[1], WAGGLE]

wiggly /ˈwɪglɪ/ *adj.* (**wigglier**, **wiggliest**) *colloq.* **1** characteristic or suggestive of wiggling movement. **2** having small irregular undulations.

Wight /waɪt/ a shipping forecast area covering the English Channel roughly between the Strait of Dover and the meridian of Poole. (See ISLE OF WIGHT.)

wight /waɪt/ *n. archaic* a person (*wretched wight*). [OE *wiht* = thing, creature, of unkn. orig.]

Wight, Isle of see ISLE OF WIGHT.

Wightman Cup /ˈwaɪtmən/ an annual lawn tennis contest between women players of the US and Britain, inaugurated in 1923. The trophy was donated by the US player Mrs H. H. Wightman (1886–1974). The competition was suspended in 1990 because the standard of British tennis was felt to be insufficiently high.

Wigtownshire /ˈwɪgtənˌʃɪə(r)/ a former county of SW Scotland. It became a part of the region of Dumfries and Galloway in 1975.

wigwag /ˈwɪgwæg/ *v.intr.* (**wigwagged**, **wigwagging**) *colloq.* **1** move lightly to and fro. **2** wave flags in this way in signalling. [redupl. f. WAG[1]]

wigwam /ˈwɪgwæm/ *n.* **1 a** a domed hut or tent of skins, mats, or bark on poles, used by North American Indians. **b** any conical tent or tepee. **2** a similar structure for children etc. [Ojibwa *wigwaum*, Algonquian *wikiwam* their house]

Wilberforce /ˈwɪlbəˌfɔːs/, William (1759–1833), English politician and social reformer. An MP and close associate of Pitt the Younger, he was a prominent campaigner for the abolition of the slave trade, successfully promoting a bill outlawing its practice in the British West Indies (1807). Later he pushed for the abolition of slavery throughout the British Empire, his efforts resulting in the 1833 Slavery Abolition Act.

wilco /ˈwɪlkəʊ/ *int. colloq.* expressing compliance or agreement, esp. acceptance of instructions received by radio. [abbr. of *will comply*]

Wilcox /ˈwɪlkɒks/, Ella Wheeler (1850–1919), American poet, novelist, and short-story writer. She wrote many volumes of romantic verse, the most successful one being *Poems of Passion* (1883).

wild /waɪld/ *adj., adv., & n.* ● *adj.* **1** (of an animal or plant) in its original natural state; not domesticated or cultivated (esp. of species or varieties allied to others that are not wild). **2** not civilized; barbarous. **3** (of scenery etc.) having a conspicuously desolate appearance.

4 unrestrained, disorderly, uncontrolled (*a wild youth*; *wild hair*). **5** tempestuous, violent (*a wild night*). **6 a** intensely eager; excited, frantic (*wild with excitement*; *wild delight*). **b** (of looks, appearance, etc.) indicating distraction. **c** (foll. by *about*) *colloq.* enthusiastically devoted to (a person or subject). **7** *colloq.* infuriated, angry (*makes me wild*). **8** haphazard, ill-aimed, rash (*a wild guess*; *a wild shot*; *a wild venture*). **9** (of a horse, game bird, etc.) shy; easily startled. **10** *colloq.* exciting, delightful. **11** (of a card) having any rank chosen by the player holding it (*the joker is wild*). ● *adv.* in a wild manner (*shooting wild*). ● *n.* **1** a wild tract. **2** a desert. □ **in the wild** in an uncultivated etc. state. **in** (or **out in**) **the wilds** *colloq.* far from normal habitation. **run wild** grow or stray unchecked or undisciplined. **sow one's wild oats** see OAT. **wild and woolly** uncouth; lacking refinement. **wild arum** = *cuckoo-pint*. **wild boar** see BOAR 1. **wild card 1** see sense 11 of *adj.* **2** *Computing* a character that will match any character or sequence of characters in a file name etc. **3** *Sport* an extra player or team chosen to enter a competition at the selectors' discretion. **wild duck** an undomesticated duck, esp. the mallard. **wild-goose chase** a foolish or hopeless and unproductive quest. **wild horse 1** a horse not domesticated or broken in. **2** (in *pl.* with *neg.*) *colloq.* not even the most powerful force (*wild horses would not drag the secret from me*). **wild hyacinth** = BLUEBELL 1. **wild man of the woods** *colloq.* an orang-utan. **wild oat** a grass, *Avena fatua*, related to the cultivated oat and found as a weed in cornfields. **wild oats** see OAT. **wild rice** a tall grass of the genus *Zizania*, yielding edible grains. **wild silk 1** silk from wild silkworms. **2** an imitation of this from short silk fibres. **wild type** *Biol.* a strain or characteristic which prevails in natural conditions, as distinct from an atypical mutant. □ **wildish** *adj.* **wildly** *adv.* **wildness** *n.* [OE *wilde* f. Gmc]

wildcat /ˈwaɪldkæt/ *n. & adj.* ● *n.* (usu. **wild cat**) **1** a smallish cat of a non-domesticated kind; esp. *Felis silvestris* of Eurasia and Africa, with a grey and black coat and bushy tail, or (*US*) a bobcat. **2** a hot-tempered or violent person. **3** an exploratory oil well. ● *adj.* (*attrib.*) **1** esp. *US* reckless; financially unsound. **2** (of a strike) sudden and unofficial.

Wilde /waɪld/, Oscar (Fingal O'Flahertie Wills) (1854–1900), Irish dramatist, novelist, poet, and wit. He was born in Dublin but from the early 1880s spent most of his time in London. He became known for his flamboyant aestheticism and his advocacy of 'art for art's sake' regardless of moral stance; this is particularly evident in his only novel *The Picture of Dorian Gray*, published in 1890. As a dramatist he achieved considerable success with a series of comedies noted for their epigrammatic wit and sharp social observation, including *Lady Windermere's Fan* (1892) and *The Importance of Being Earnest* (1895). However, Wilde's drama *Salomé* (1893), written in French, was refused a performing licence; it was published in English in 1894 and first performed in Paris in 1896. Wilde's relationship with Lord Alfred Douglas (1870–1945) resulted in his imprisonment for homosexual offences (1895–7); he died in exile in Paris.

wildebeest /ˈwɪldəˌbiːst, ˈvɪl-/ *n.* (*pl.* same or **wildebeests**) = GNU. [Afrik. (as WILD, BEAST)]

Wilder[1] /ˈwaɪldə(r)/, Billy (born Samuel Wilder) (b.1906), American film director and screenwriter, born in Austria. After emigrating to the US in 1934, he wrote screenplays for a number of Hollywood films before earning recognition as a writer-director with *Double Indemnity* (1944); written with Raymond Chandler, it is regarded as a *film noir* classic. He subsequently co-wrote and directed *Sunset Boulevard* (1950), *Some Like It Hot* (1959), and *The Apartment* (1960); the latter won Oscars for best script, director, and picture.

Wilder[2] /ˈwaɪldə(r)/, Thornton (Niven) (1897–1975), American novelist and dramatist. His work is especially concerned with the universality of human experience, irrespective of time or place. He established his reputation as a novelist with *The Bridge of San Luis Rey* (1927), for which he won a Pulitzer Prize. His plays, often experimental in form, include *Our Town* (1938) and *The Skin of Our Teeth* (1942), both of which received Pulitzer Prizes. His comic drama *The Matchmaker* (1954) provided the basis for the musical *Hello, Dolly!* (1964).

wilder /ˈwɪldə(r)/ *v.tr. archaic* **1** lead astray. **2** bewilder. [perh. based on WILDERNESS]

wilderness /ˈwɪldənɪs/ *n.* **1** a desert; an uncultivated and uninhabited region. **2** part of a garden left with an uncultivated appearance. **3** (foll. by *of*) a confused assemblage of things. □ **in the wilderness** out of political office. **voice in the wilderness** an unheeded advocate of reform (see Matt. 3:3 etc.). [OE *wildēornes* f. *wild dēor* wild deer]

wildfire /ˈwaɪldˌfaɪə(r)/ *n. hist.* **1** a combustible liquid, esp. Greek fire, formerly used in warfare. **2** = WILL-O'-THE-WISP 1. □ **spread like wildfire** spread with great speed.

wildfowl /ˈwaɪldfaʊl/ *n.* (*pl.* same) (usu. in *pl.*) a game bird, esp. an aquatic one; a duck, goose, or swan.

wilding /ˈwaɪldɪŋ/ *n.* (also **wildling** /ˈwaɪldlɪŋ/) **1** a plant sown by natural agency, esp. a wild crab-apple. **2** the fruit of such a plant. [WILD + -ING[3]]

wildlife /ˈwaɪldlaɪf/ *n.* wild animals collectively.

Wild West the western regions of the US in the 19th century, when they were lawless frontier districts. As settlers moved gradually further west North America had a succession of frontiers, of which that known as the Wild West was the last. The frontier was officially declared closed in 1890, and the Wild West disappeared with the ending of hostilities with the American Indians, the building of the railways, and the establishment of settled communities.

wildwood /ˈwaɪldwʊd/ *n. poet.* uncultivated or unfrequented woodland.

wile /waɪl/ *n. & v.* ● *n.* (usu. in *pl.*) a stratagem; a trick or cunning procedure. ● *v.tr.* (foll. by *away*, *into*, etc.) lure or entice. [ME *wil*, perh. f. Scand. (ON *vél* craft)]

wilful /ˈwɪlfʊl/ *adj.* (*US* **willful**) **1** (of an action or state) intentional, deliberate (*wilful murder*; *wilful neglect*; *wilful disobedience*). **2** (of a person) obstinate, headstrong. □ **wilfully** *adv.* **wilfulness** *n.* [ME f. WILL[2] + -FUL]

wilga /ˈwɪlgə/ *n. Austral.* a small drought-resistant tree of the genus *Geijera*, with white flowers. [Wiradhuri *wilgar*]

Wilhelm I /ˈvɪlhelm/ (1797–1888), king of Prussia 1861–88 and emperor of Germany 1871–88. His reign saw the unification of Germany, the driving force behind which was Bismarck, his chief minister. He became the first emperor of Germany after Prussia's victory against France in 1871. The latter part of his reign was marked by the rise of German socialism, to which he responded with harsh repressive measures.

Wilhelm II /ˈvɪlhelm/ (known as Kaiser Wilhelm) (1859–1941), emperor of Germany 1888–1918, grandson of Queen Victoria. After forcing his chief minister, Bismarck, to resign in 1890 he proved unable to exercise a strong or consistent influence over German policies, which became increasingly militaristic in foreign affairs. He was unable to prevent the outbreak of the First World War (1914), and was vilified by Allied propaganda as the author of the conflict. In 1918 he went into exile in Holland and abdicated his throne.

Wilhelmshaven /ˈvɪlhelmzˌhɑːv(ə)n/ a port and resort in NW Germany, on the North Sea; pop. (1991) 91,150. It was a major naval base until 1945.

wiliness see WILY.

Wilkes Land /wɪlks/ a region of Antarctica with a coast on the Indian Ocean. It is named after the American naval officer Charles Wilkes (1798–1877), who sighted and surveyed it between 1838 and 1842. It is claimed by Australia.

Wilkie /ˈwɪlki/, Sir David (1785–1841), Scottish painter. He made his name with the painting *Village Politicians* (1806); influenced by 17th-century Dutch and Flemish genre painters, it defined Wilkie's style for the next twenty years. Wilkie's paintings were popular and contributed to the growing prestige of genre painting in Britain.

Wilkins /ˈwɪlkɪnz/, Maurice Hugh Frederick (b.1916), New Zealand-born British biochemist and molecular biologist. Studying the structure of the DNA molecule by means of X-ray diffraction analysis, he and his colleague Rosalind Franklin provided the evidence for and confirmed the double helix structure proposed by Francis Crick and James Watson in 1953. Wilkins, Crick, and Watson shared a Nobel Prize for their work on DNA in 1962.

will[1] /wɪl/ *v.aux. & tr.* (3rd sing. present **will**; archaic 2nd sing. present **wilt** /wɪlt/; past **would**) (foll. by infin. without *to*, or *absol.*; present and past only in use) **1** (in the 2nd and 3rd persons, and often in the 1st: see SHALL) expressing the future tense in statements, commands, or questions (*you will regret this*; *they will leave at once*; *will you go to the party?*). **2** (in the 1st person) expressing a wish or intention (*I will return soon*). ¶ For the other persons in senses 1, 2, see SHALL. **3** expressing desire, consent, or inclination (*will you have a sandwich?*; *come when you will*; *the door will not open*). **4** expressing a request as a question (*will you please open the window?*). **5** expressing ability or capacity (*the jar will hold a kilo*). **6** expressing habitual or inevitable tendency (*accidents will happen*; *will sit there for hours*). **7** expressing probability or expectation (*that will be my wife*). □ **will do** *colloq.* expressing willingness to carry out a request. [OE *wyllan*, (unrecorded) *willan* f. Gmc: rel. to L *volo*]

will[2] /wɪl/ *n. & v.* ● *n.* **1** the faculty by which a person decides or is regarded as deciding on and initiating action (*the mind consists of the* *understanding and the will*). **2** (also **will-power**) control exercised by deliberate purpose over impulse; self-control (*has a strong will; overcame his shyness by will-power*). **3** a deliberate or fixed desire or intention (*a will to live*). **4** energy of intention; the power of effecting one's intentions or dominating others. **5** directions (usu. written) in legal form for the disposition of one's property after death (*make one's will*). **6** disposition towards others (*good will*). **7** *archaic* what one desires or ordains (*thy will be done*). ● *v.tr.* **1** have as the object of one's will; intend unconditionally (*what God wills; willed that we should succeed*). **2** (*absol.*) exercise will-power. **3** instigate or impel or compel by the exercise of will-power (*you can will yourself into contentment*). **4** bequeath by the terms of a will (*shall will my money to charity*). □ **at will 1** whenever one pleases. **2** *Law* able to be evicted without notice (*tenant at will*). **have one's will** obtain what one wants. **what is your will?** what do you wish done? **where there's a will there's a way** determination will overcome any obstacle. **a will of one's own** obstinacy; wilfulness of character. **will-power** see sense 2 of *n.* **with the best will in the world** however good one's intentions. **with a will** energetically or resolutely. □ **willed** *adj.* (also in *comb.*). **willer** *n.* **will-less** *adj.* [OE *willa* f. Gmc]

Willard /'wɪlɑːd/, Emma (1787–1870), American educational reformer. A pioneer of women's education, she founded a boarding-school in Vermont (1814) to teach subjects not then available to women (e.g. mathematics and philosophy). Willard moved the school to Troy, New York (1821), where it became known as the Troy Female Seminary; the college education that it offered served as a model for subsequent women's colleges in the US and Europe.

Willemstad /'wɪləm,stɑːt, 'vɪl-/ the capital of the Netherlands Antilles, situated on the SW coast of the island of Curaçao; pop. (est. 1986) 50,000.

willet /'wɪlɪt/ *n.* (*pl.* same) a large North American sandpiper, *Catoptophorus semipalmatus*. [*pill-will-willet*, imit. of its call]

willful *US* var. of WILFUL.

William /'wɪljəm/ the name of two kings of England, and two of Great Britain and Ireland:

William I (known as William the Conqueror) (*c.*1027–87), reigned 1066–87, the first Norman king of England. He was the illegitimate son of Robert, Duke of Normandy and claimed the English throne on the death of Edward the Confessor, stating that Edward had promised it to him. He landed in England at the head of an invasion force, defeated Harold II at the Battle of Hastings (1066), and was crowned king. Having repressed a series of uprisings, he imposed his rule on England, introducing Norman institutions and customs (including feudalism and administrative and legal practices). He also instigated the property survey of England known as the Domesday Book.

William II (known as William Rufus = 'red-faced') (*c.*1060–1100), son of William I, reigned 1087–1100. His succession was challenged by a group of Norman barons in England who wanted William's elder brother Robert Curthose (*c.*1054–1134), Duke of Normandy, to rule England instead. However, William's forces crushed their rebellions in 1088 and 1095. William also campaigned against his brother Robert in Normandy (1089–96), ultimately acquiring the duchy in 1096 when Robert mortgaged it to William before leaving to go on the First Crusade. In the north of England William secured the frontier against the Scots along a line from the Solway Firth to the Tweed. He was killed by an arrow while out hunting; whether he was assassinated or whether his death was an accident remains unclear.

William III (known as William of Orange) (1650–1702), grandson of Charles I, reigned 1689–1702. Son of the Prince of Orange and Mary, daughter of Charles I, William was stadtholder (chief magistrate) of the Netherlands from 1672 and married Mary, daughter of the future James II, in 1677. In 1688 he landed in England at the invitation of disaffected politicians, deposed James II, and, having accepted the Declaration of Rights, was crowned along with his wife the following year. He defeated James's supporters in Scotland and Ireland (1689–90), and thereafter devoted most of his energies towards opposing the territorial ambitions of Louis XIV of France.

William IV (known as 'the Sailor King') (1765–1837), son of George III, reigned 1830–7. He served in the Royal Navy from 1779, rising to Lord High Admiral in 1827, and came to the throne after the death of his brother George III. Although an opponent of the first Reform Bill (1832), William reluctantly agreed to create fifty new peers to overcome opposition to it in the House of Lords. In 1834 he intervened in political affairs by imposing his own choice of Prime Minister (the Conservative Robert Peel), despite a Whig majority in Parliament.

William I (known as William the Lion) (1143–1214), grandson of David I, king of Scotland 1165–1214. He attempted to reassert Scottish independence but was forced to pay homage to Henry II of England after being captured by him in 1174.

William of Occam /'ɒkəm/ (also **Ockham**) (*c.*1285–1349), English philosopher and Franciscan friar. He studied and taught philosophy at Oxford until 1324; in the later part of his life (1333–47) he lived in Munich and was active in writing anti-papal pamphlets. He was the last of the major scholastic philosophers; his form of nominalist philosophy saw God as beyond human powers of reasoning, and things as provable only by experience or by scriptural authority — hence his maxim, known as *Occam's razor*, that explanations should be examined and unnecessary assumptions stripped away. Occam distinguished between faith and reason, advocated a radical separation of the Church from the world, denied the pope all temporal authority, and conceded large powers to the laity and their representatives. He was also a firm supporter of the Franciscan doctrine of poverty; this brought strong papal opposition and he was excommunicated in 1328. His ideas had a significant influence on Luther.

William of Orange, William III of Great Britain and Ireland (see WILLIAM).

William Rufus, William II of England (see WILLIAM).

Williams[1] /'wɪljəmz/, John (Christopher) (b.1941), Australian guitarist and composer. He studied with Andrés Segovia before making his London début in 1958. Based in Britain, he became much in demand as a recitalist, noted for an eclectic repertoire that includes both classical and popular music. In 1979 he founded the pop group Sky, playing with them until 1984.

Williams[2] /'wɪljəmz/, J(ohn) P(eter) R(hys) (b.1949), Welsh Rugby Union player. A former junior Wimbledon tennis champion (1966), he made his rugby début for Wales in 1969 and became one of the leading full-backs of the 1970s. Williams played sixty-three times for his country (1969–81), as well as for the British Lions (1971–7). In 1981 he retired from rugby to pursue a career in orthopaedic medicine.

Williams[3] /'wɪljəmz/, Tennessee (born Thomas Lanier Williams) (1911–83), American dramatist. He was brought up in the South, the setting for many of his plays. He achieved success with the semi-autobiographical *The Glass Menagerie* (1944) and *A Streetcar Named Desire* (1947), plays which deal with the tragedy of vulnerable heroines living in fragile fantasy worlds shattered by brutal reality. Williams's later plays, while still dealing with strong passions and family tensions, increasingly feature Gothic and macabre elements; they include *Cat on a Hot Tin Roof* (1955), *Suddenly Last Summer* (1958), and *The Night of the Iguana* (1962).

Williams[4] /'wɪljəmz/, William Carlos (1883–1963), American poet, essayist, novelist, and short-story writer. He worked throughout his life as a paediatrician in his native New Jersey. His poetry illuminates the ordinary by vivid, direct observation; it is characterized by avoidance of emotional content and the use of the American vernacular. Collections of his poetry include *Spring and All* (1923) and *Pictures from Brueghel* (1963). His long poem *Paterson* (1946–58), written in free verse, draws on and explores the cultural, historical, and mythic resonances associated with the industrial New Jersey city of Paterson. Non-fiction works include *In the American Grain* (1925), essays exploring the nature of American literature and the influence of Puritanism in American culture.

Williamsburg /'wɪljəmz,bɜːg/ a city in SE Virginia, between the James and York rivers; pop. (1990) 11,530. First settled as Middle Plantation in 1633, it was the state capital of Virginia from 1699, when it was renamed in honour of William III, until 1799, when Richmond became the capital. A large part of the town has been restored and reconstructed so that it appears as it was during the colonial era.

William the Conqueror, William I of England (see WILLIAM).

willie var. of WILLY.

willies /'wɪlɪz/ *n.pl. colloq.* nervous discomfort (esp. *give* or *get the willies*). [19th c.: orig. unkn.]

willing /'wɪlɪŋ/ *adj. & n.* ● *adj.* **1** ready to consent or undertake (*a willing ally; am willing to do it*). **2** given or done etc. by a willing person (*willing hands; willing help*). ● *n.* cheerful intention (*show willing*). □ **willingly** *adv.* **willingness** *n.*

will-o'-the-wisp /,wɪlədə'wɪsp/ *n.* **1** a phosphorescent light seen on marshy ground, perhaps resulting from the combustion of gases. **2** an elusive person. **3** a delusive hope or plan. [orig. *Will with the wisp: wisp* = handful of (lighted) hay etc.]

willow /'wɪləʊ/ *n.* **1** a tree or shrub of the genus *Salix*, growing usu. near water in temperate climates, with small flowers borne on catkins, and pliant branches yielding osiers and timber for cricket-bats, baskets, etc. **2** a cricket-bat. □ **willow grouse** (or *N. Amer.* **ptarmigan**) a common European and North American grouse, *Lagopus lagopus*, with brown breeding plumage and white winter plumage (cf. *red grouse*). **willow tit** a Eurasian black-capped tit, *Parus montanus*. **willow warbler** (or **willow-wren**) a small Eurasian woodland warbler, *Phylloscopus trochilus*, with a tuneful song. [OE *welig*]

willowherb /ˈwɪləʊˌhɜːb/ *n.* a plant of the genus *Epilobium* or a related genus, usu. with willow-like leaves and pinkish flowers.

willow pattern *n.* a conventional English pottery design of blue on white china etc., representing a Chinese scene in which a willow tree is a prominent feature. Originating *c.*1780, the pattern became standardized and was widely copied in the 19th century. It shows a pagoda, two birds in the sky, a fence in the foreground, a boat, a three-arched bridge across which walk three Chinese figures, and a willow tree overhanging the bridge.

willowy /ˈwɪləʊɪ/ *adj.* **1** having or bordered by willows. **2** lithe and slender.

Wills /wɪlz/, William John (1834–61), English explorer. Having emigrated to Australia, in 1860 he was a member, with two others, of Robert Burke's expedition to cross the continent from south to north. They became the first white people to make this journey, but Wills, Burke, and one of their companions died of starvation on the return journey.

willy /ˈwɪlɪ/ *n.* (also **willie**) (*pl.* **-ies**) *Brit. colloq.* the penis.

willy-nilly /ˌwɪlɪˈnɪlɪ/ *adv.* & *adj.* ● *adv.* whether one likes it or not. ● *adj.* existing or occurring willy-nilly. [later spelling of *will I, nill I* I am willing, I am unwilling]

willy wagtail *n.* **1** an Australian fantail flycatcher, *Rhapidura leucophrys*, with striking black and white plumage. **2** *dial.* the pied wagtail.

willy-willy /ˈwɪlɪˌwɪlɪ/ *n.* (*pl.* **-ies**) *Austral.* a cyclone or dust-storm. [Aboriginal]

Wilson[1] /ˈwɪls(ə)n/, Sir Angus (Frank Johnstone) (1913–91), English novelist and short-story writer. His works include the novels *Anglo-Saxon Attitudes* (1956), *The Old Men at the Zoo* (1961), and *Setting the World on Fire* (1980). These, together with several volumes of short stories, display his satiric wit, acute social observation, and a love of the macabre and the farcical. He also wrote studies of Zola, Dickens, and Kipling.

Wilson[2] /ˈwɪls(ə)n/, Charles Thomson Rees (1869–1959), Scottish physicist. He is chiefly remembered for inventing the cloud chamber, building his first one in 1895 in an attempt to reproduce the conditions in which clouds are formed in nature. He later improved the design and by 1911 had a chamber in which the track of an ion could be made visible. This became a major tool of particle physicists in subsequent years, and Wilson shared a Nobel Prize for physics in 1927. He also investigated atmospheric electricity and thundercloud formation, and predicted the discovery of cosmic rays.

Wilson[3] /ˈwɪls(ə)n/, Edmund (1895–1972), American critic, essayist, and short-story writer. He is remembered chiefly for works of literary and social criticism, which include *Axel's Castle* (1931), a study of symbolist literature, *To the Finland Station* (1940), tracing socialist and revolutionary theory, and *Patriotic Gore: Studies in the Literature of the American Civil War* (1962). He was a friend of F. Scott Fitzgerald, and edited the latter's unfinished novel *The Last Tycoon* (published posthumously in 1941). Wilson's third wife was the novelist Mary McCarthy.

Wilson[4] /ˈwɪls(ə)n/, Edward Osborne (b.1929), American social biologist. He has worked principally on social insects, notably ants and termites, extrapolating his findings to the social behaviour of other animals including humans. His book *Sociobiology: the New Synthesis* (1975) integrated these ideas and effectively launched the science of sociobiology. Wilson has also studied the colonization of islands by insects, showing that the number of species reaches a dynamic equilibrium for any given island.

Wilson[5] /ˈwɪls(ə)n/, (James) Harold, Baron Wilson of Rievaulx (1916–95), British Labour statesman, Prime Minister 1964–70 and 1974–6. His administrations were pragmatic in outlook rather than rigidly socialist. In both terms of office he faced severe economic problems; repeated sterling crises led to devaluation in 1967, while he attempted unsuccessfully to deal with high inflation in 1974–6 by seeking an agreement with trade unions over limiting pay increases. His government introduced a number of social reforms, including reducing the voting age to 18, liberalizing the laws on divorce,

homosexuality, and abortion, and introducing comprehensive schooling. Overseas, he was unable to persuade the regime of Ian Smith in Rhodesia (Zimbabwe) to back down over its declaration of independence (1965), and therefore introduced economic sanctions against Rhodesia. In 1974 Wilson renegotiated Britain's terms of entry into the European Economic Community, confirming British membership after a referendum in 1975. He resigned as leader of the Labour Party the following year and was replaced as Prime Minister by James Callaghan.

Wilson[6] /ˈwɪls(ə)n/, (Thomas) Woodrow (1856–1924), American Democratic statesman, 28th President of the US 1913–21. He was a prominent academic in the field of law and political economy prior to his election victory. As President he carried out a series of successful administrative and fiscal reforms. He initially kept America out of the First World War, but, following the German reintroduction of unrestricted submarine warfare, entered the war on the Allied side in April 1917. Wilson's conditions for a peace treaty, as set out in his 'Fourteen Points' speech (1918), and his plan for the formation of the League of Nations were crucial in the international negotiations surrounding the end of the war, and he was awarded the Nobel Peace Prize in 1919. However, he was unable to obtain the Senate's ratification of the Treaty of Versailles, his health collapsed, and he lost the 1920 presidential election.

wilt[1] /wɪlt/ *v.* & *n.* ● *v.* **1** *intr.* (of a plant, leaf, or flower) wither, droop. **2** *intr.* (of a person) lose one's energy, flag, tire, droop. **3** *tr.* cause to wilt. ● *n.* a plant-disease causing wilting. [orig. dial.: perh. alt. f. *wilk, welk*, of LG or Du. orig.]

wilt[2] /wɪlt/ *archaic 2nd person sing.* of WILL[1].

Wilton /ˈwɪlt(ə)n/ *n.* a kind of woven carpet with a thick pile. [*Wilton* in southern England]

Wilts. *abbr.* Wiltshire.

Wiltshire /ˈwɪltʃɪə(r)/ a county of southern England; county town, Trowbridge.

wily /ˈwaɪlɪ/ *adj.* (**wilier, wiliest**) full of wiles; crafty, cunning. □ **wilily** *adv.* **wiliness** *n.*

Wimbledon /ˈwɪmb(ə)ldən/ an annual international tennis championship on grass for individual players and pairs, held at the headquarters of the All England Lawn Tennis and Croquet Club in the London suburb of Wimbledon. Now one of the world's major tennis championships, it has been played since 1877; women were first admitted in 1884, and professionals in 1968.

wimp /wɪmp/ *n.* & *v. colloq.* ● *n.* a feeble or ineffectual person. ● *v.intr.* (foll. by *out*) feebly withdraw from an undertaking. □ **wimpish** *adj.* **wimpishly** *adv.* **wimpishness** *n.* **wimpy** *adj.* [20th c.: orig. uncert.]

wimple /ˈwɪmp(ə)l/ *n.* & *v.* ● *n.* a linen or silk head-dress covering the neck and the sides of the face, formerly worn by women and still worn by some nuns. ● *v.tr.* & *intr.* arrange or fall in folds. [OE *wimpel*]

Wimpy /ˈwɪmpɪ/ *n.* (*pl.* **-ies**) *Brit. propr.* a hamburger served in a plain bun.

Wimshurst machine /ˈwɪmzhɜːst/ *n.* a device for generating an electric charge by turning glass discs in opposite directions. [James Wimshurst, Engl. engineer (1832–1903)]

win /wɪn/ *v.* & *n.* ● *v.* (**winning**; *past* and *past part.* **won** /wʌn/) **1** *tr.* acquire or secure as a result of a fight, contest, bet, litigation, or some other effort (*won some money; won my admiration*). **2** *tr.* be victorious in (a fight, game, race, etc.). **3** *intr.* **a** be the victor; win a race or contest etc. (*who won?; persevere, and you will win*). **b** (foll. by *through, free*, etc.) make one's way or become by successful effort. **4** *tr.* reach by effort (*win the summit; win the shore*). **5** *tr.* obtain (ore) from a mine. **6** *tr.* dry (hay etc.) by exposure to the air. ● *n.* a victory in a game, contest, bet, etc. □ **win the day** be victorious in battle, argument, etc. **win over** persuade, gain the support of. **win one's spurs 1** *colloq.* gain distinction or fame. **2** *hist.* gain a knighthood. **win through** (or **out**) overcome obstacles. **you can't win** *colloq.* there is no way to succeed. **you can't win them all** *colloq.* a resigned expression of consolation on failure. □ **winnable** *adj.* [OE *winnan* toil, endure: cf. OHG *winnan*, ON *vinna*]

wince[1] /wɪns/ *n.* & *v.* ● *n.* a start or involuntary shrinking movement showing pain or distress. ● *v.intr.* give a wince. □ **wincer** *n.* **wincingly** *adv.* [ME f. OF *guenchir* turn aside: cf. WINCH, WINK]

wince[2] /wɪns/ *n.* a roller for moving textile fabric through a dyeing-vat. [var. of WINCH]

wincey /ˈwɪnsɪ/ *n.* (*pl.* **winceys**) a strong lightweight fabric of wool and cotton or linen. [orig. Sc.: app. f. *woolsey* in LINSEY-WOOLSEY]

winceyette /ˌwɪnsɪˈet/ *n. Brit.* a lightweight napped flannelette used esp. for nightclothes.

winch /wɪntʃ/ *n. & v.* ● *n.* **1** the crank of a wheel or axle. **2** a windlass. **3** the reel of a fishing-rod. **4** = WINCE². ● *v.tr.* lift with a winch. □ **wincher** *n.* [OE *wince* f. Gmc (base also of WINK): cf. WINCE¹]

Winchester¹ /ˈwɪntʃɪstə(r)/ a city in southern England, the county town of Hampshire; pop. (1981) 35,660. Known to the Romans as Venta Belgarum, it became capital of the West Saxon kingdom of Wessex in 519. It is the site of Winchester College, the oldest public school in England, founded by the bishop of Winchester William of Wykeham (1324–1404) in 1382.

Winchester² /ˈwɪntʃɪstə(r)/ *n. propr.* a breech-loading repeating rifle. □ **Winchester drive** (or **disk**) *Computing* a (usu. fixed) disk drive with a sealed unit that contains a high-capacity hard disk and the read-write heads (so called because its original numerical designation corresponded to that of the rifle's calibre). [Oliver Fisher *Winchester*, Amer. manufacturer of the rifle (1810–80)]

Winckelmann /ˈvɪŋk(ə)lmən/, Johann (Joachim) (1717–68), German archaeologist and art historian, born in Prussia. In 1755 he was appointed librarian to a cardinal in Rome; there he took part in the excavations at Pompeii and Herculaneum and was instrumental in enforcing professional archaeological standards. Winckelmann became superintendent of Roman antiquities in 1763. His best-known work, *History of the Art of Antiquity* (1764), was a seminal text in the neoclassical movement; it was particularly influential in popularizing the art and culture of ancient Greece.

wind¹ /wɪnd/ *n. & v.* ● *n.* **1 a** air in more or less rapid natural motion, esp. from an area of high pressure to one of low pressure. **b** a current of wind blowing from a specified direction or otherwise defined (*north wind*; *contrary wind*). **2 a** breath as needed in physical exertion or in speech. **b** the power of breathing without difficulty while running or making a similar continuous effort (*let me recover my wind*). **c** a spot below the centre of the chest where a blow temporarily paralyses breathing. **3** mere empty words; meaningless rhetoric. **4** gas generated in the bowels etc. by indigestion; flatulence. **5 a** an artificially produced current of air, esp. for sounding an organ or other wind instrument. **b** air stored for use or used as a current. **c** the wind instruments of an orchestra collectively (*poor balance between wind and strings*). **6** a scent carried by the wind, indicating the presence or proximity of an animal etc. ● *v.tr.* **1** exhaust the wind of by exertion or a blow. **2** renew the wind of by rest (*stopped to wind the horses*). **3** make breathe quickly and deeply by exercise. **4** make (a baby) bring up wind after feeding. **5** detect the presence of by a scent. **6** /waɪnd/ (*past and past part.* **winded** or **wound** /waʊnd/) *poet.* sound (a bugle or call) by blowing. □ **before the wind** helped by the wind's force. **between wind and water** at a vulnerable point. **close to** (or **near**) **the wind 1** sailing as nearly against the wind as is consistent with using its force. **2** *colloq.* verging on indecency or dishonesty. **get wind of 1** smell out. **2** begin to suspect; hear a rumour of. **get** (or **have**) **the wind up** *colloq.* be alarmed or frightened. **how** (or **which way**) **the wind blows** (or **lies**) **1** what is the state of opinion. **2** what developments are likely. **in the wind** happening or about to happen. **in the wind's eye** directly against the wind. **like the wind** swiftly. **off the wind** *Naut.* with the wind on the quarter. **on a wind** *Naut.* against a wind on either bow. **on the wind** (of a sound or scent) carried by the wind. **put the wind up** *colloq.* alarm or frighten. **take wind** be rumoured; become known. **take the wind out of a person's sails** frustrate a person by anticipating an action or remark etc. **throw caution to the winds** not worry about taking risks; be reckless. **to the winds** (or **four winds**) **1** in all directions. **2** into a state of abandonment or neglect. **wind and weather** exposure to the effects of the elements. **wind band** a group of wind instruments as a band or section of an orchestra. **wind-blown** exposed to or blown about by the wind. **wind-break** a row of trees or a fence or wall etc. serving to break the force of the wind. **wind-chill** the cooling effect of wind blowing on a surface. **wind-cone** = *wind-sock*. **wind energy** = *wind power*. **wind farm** a group of energy-producing windmills or wind turbines. **wind-force** the force of the wind, esp. as measured on the Beaufort scale. **wind-gap** a dried-up former river valley through ridges or hills. **wind-gauge 1** an anemometer. **2** an apparatus attached to the sights of a gun enabling allowance to be made for the wind in shooting. **3** *Mus.* a device for indicating the wind-pressure in an organ. **wind instrument** a musical instrument in which sound is produced by a current of air, esp. the breath. **wind-jammer** a merchant sailing-ship. **wind machine** a device for producing a blast of air or the sound of wind. **wind** (or **winds**) **of change** a force or influence for reform.

wind power power obtained by harnessing the wind. (*See note below.*) **wind pump** a water pump driven by the wind. **wind-rose** a diagram of the relative frequency of wind directions at a place. **wind-row** a line of raked hay, corn-sheaves, peats, etc., for drying by the wind. **wind-sail** a canvas funnel conveying air to the lower parts of a ship. **wind shear** a variation in wind velocity at right angles to the wind's direction. **wind-sleeve** = *wind-sock*. **wind-sock** a canvas cylinder or cone on a mast to show the direction of the wind at an airfield etc. **wind-tunnel** a tunnel-like device to produce an air-stream past models of aircraft etc. for the study of wind effects on them. **wind turbine** a turbine driven by wind, esp. one generating electricity. □ **windless** *adj.* [OE f. Gmc]

● Wind power has been used for centuries to drive windmills for grinding corn (see WINDMILL); the wind pump, first developed in the 15th century in the Netherlands for drainage purposes, was adapted for raising water from deep wells in the US and Australia. The first wind-powered electric generator was built in Denmark in 1890, and in the next fifty years many low-power machines were erected on farms etc. After the Second World War more powerful generators were developed, using two or three aerofoil blades, sometimes with a vertical axis. Wind power is a relatively expensive method of generating electricity, though some modern wind turbines are capable of producing up to three megawatts and commercial wind farms have been developed in the US (particularly in California and the Midwest) and elsewhere.

wind² /waɪnd/ *v. & n.* ● *v.* (*past and past part.* **wound** /waʊnd/) **1** *intr.* go in a circular, spiral, curved, or crooked course (*a winding staircase*; *the path winds up the hill*). **2** *tr.* make (one's way) by such a course (*wind your way up to bed*; *wound their way into our affections*). **3** *tr.* wrap closely; surround with or as with a coil (*wound the blanket round me*; *wound my arms round the child*; *wound the child in my arms*). **4 a** *tr.* coil; provide with a coiled thread etc. (*wind the ribbon on to the card*; *wound cotton on a reel*; *winding wool into a ball*). **b** *intr.* coil; (of wool etc.) coil into a ball (*the creeper winds round the pole*; *the wool wound into a ball*). **5** *tr.* wind up (a clock etc.). **6** *tr.* hoist or draw with a windlass etc. (*wound the cable-car up the mountain*). ● *n.* **1** a bend or turn in a course. **2** a single turn when winding. □ **wind down 1** lower by winding. **2** (of a mechanism) unwind. **3** (of a person) relax. **4** draw gradually to a close. **wind-down** *n. colloq.* a gradual lessening of excitement or reduction of activity. **wind off** unwind (string, wool, etc.). **wind round one's finger** see FINGER. **wind up 1** coil the whole of (a piece of string etc.). **2** tighten the coiling or coiled spring of (esp. a clock etc.). **3 a** *colloq.* increase the tension or intensity of (*wound myself up to fever pitch*). **b** irritate or provoke (a person) to the point of anger. **4** bring to a conclusion; end (*wound up his speech*). **5** *Commerce* **a** arrange the affairs of and dissolve (a company). **b** (of a company) cease business and go into liquidation. **6** *colloq.* arrive finally; end in a specified state or circumstance (*you'll wind up in prison*; *wound up owing £100*). **wind-up** *n.* **1** a conclusion; a finish. **2** *colloq.* an attempt to provoke someone. **wound up** *adj.* (of a person) excited or tense or angry. [OE *windan* f. Gmc, rel. to WANDER, WEND]

windage /ˈwɪndɪdʒ/ *n.* **1** the friction of air against the moving part of a machine. **2 a** the effect of the wind in deflecting a missile. **b** an allowance for this. **3** the difference between the diameter of a gun's bore and its projectile, allowing the escape of gas.

Windaus /ˈvɪndaʊs/, Adolf (1876–1959), German organic chemist. He did pioneering work on the chemistry and structure of steroids and their derivatives, notably cholesterol. He also investigated the D vitamins and vitamin B₁, and discovered the important substance histamine. Windaus was awarded the Nobel Prize for chemistry in 1928.

windbag /ˈwɪndbæg/ *n. colloq.* a person who talks a lot but says little of any value.

windbound /ˈwɪndbaʊnd/ *adj.* unable to sail because of contrary winds.

windbreaker /ˈwɪndˌbreɪkə(r)/ *n. US* = WINDCHEATER.

windburn /ˈwɪndbɜːn/ *n.* inflammation of the skin caused by exposure to the wind.

windcheater /ˈwɪndˌtʃiːtə(r)/ *n.* a kind of wind-resistant outer jacket with close-fitting neck, cuffs, and lower edge.

winder /ˈwaɪndə(r)/ *n.* a winding mechanism, esp. of a clock or watch.

Windermere /ˈwɪndəˌmɪə(r)/ a lake in Cumbria, in the south-eastern part of the Lake District. At about 17 km (10 miles) in length, it is the largest lake in England. The town of Windermere lies on its eastern shores.

windfall /'wɪndfɔːl/ *n.* **1** an apple or other fruit blown to the ground by the wind. **2** a piece of unexpected good fortune, esp. a legacy.

windflower /'wɪnd,flaʊə(r)/ *n.* an anemone.

Windhoek /'wɪnthʊk, 'vɪnt-/ the capital of Namibia, situated in the centre of the country; pop. (1992) 58,600. It was the capital of the former German protectorate of South West Africa from 1892 until 1919, emerging as capital of independent Namibia in 1990.

windhover /'wɪnd,hɒvə(r)/ *n. Brit.* a kestrel.

winding /'waɪndɪŋ/ *n.* **1** in senses of WIND[2] *v.* **2** curved or sinuous motion or movement. **3 a** a thing that is wound round or coiled. **b** *Electr.* coils of wire as a conductor round an armature etc. □ **winding-engine** a machine for hoisting. **winding-sheet** a sheet in which a corpse is wrapped for burial.

windlass /'wɪndləs/ *n. & v.* ● *n.* a machine with a horizontal axle for hauling or hoisting. ● *v.tr.* hoist or haul with a windlass. [alt. (perh. by assoc. with dial. *windle* to wind) of obs. *windas* f. OF *guindas* f. ON *vindáss* f. *vinda* WIND[2] + *áss* pole]

windlestraw /'wɪnd(ə)l,strɔː/ *n. archaic* an old dry stalk of grass. [OE *windelstrēaw* grass for plaiting f. *windel* basket (as WIND[2], -LE[1]) + *strēaw* STRAW]

windmill /'wɪndmɪl/ *n.* **1** a mill, pump, or generator driven by the action of the wind on its rotating sails or blades. (*See note below.*) **2** esp. *Brit.* a toy consisting of a stick with curved vanes attached that revolve in a wind. □ **throw one's cap** (or **bonnet**) **over the windmill** act recklessly or unconventionally. **tilt at** (or **fight**) **windmills** attack an imaginary enemy or source of grievance (with ref. to Don Quixote, who attacked windmills, thinking that they were giants).

▪ Windmills for grinding corn originated in Persia in the 7th century, those with horizontal geared shafts appearing in Europe in the 12th century. In early post-mills the entire body of the mill was rotated manually to face the wind; the development of the fantail in the 18th century enabled the cap and sails to face the wind automatically. The sails were originally of canvas, English windmills adopting wooden sails with pivoted slats for greater control. Few windmills are used for grinding corn today, though many small multi-bladed windmills for pumping water are still in use (see WIND[1]).

window /'wɪndəʊ/ *n.* **1 a** an opening in a wall, roof, vehicle, etc., usu. with glass in fixed, sliding, or hinged frames, to admit light or air etc. and allow the occupants to see out. **b** the glass filling this opening (*have broken the window*). **2** a space for display behind the front window of a shop. **3** an aperture in a wall etc. through which customers are served in a bank, ticket office, etc. **4** an opportunity to observe or learn. **5** an opening or transparent part in an envelope to show an address. **6** *Computing* a part of a VDU display selected to show a particular category or part of the data. **7 a** an interval during which atmospheric and astronomical circumstances are suitable for the launch of a spacecraft. **b** any interval or opportunity for action. **8** strips of metal foil dispersed in the air to obstruct radar detection. **9** *Physics* a range of electromagnetic wavelengths for which a medium, esp. the atmosphere, is transparent. □ **out of the window** *colloq.* no longer taken into account. **window-box** a box placed on an outside window-sill for growing flowers. **window-cleaner** a person who is employed to clean windows. **window-dressing 1** the art of arranging a display in a shop-window etc. **2** an adroit presentation of facts etc. to give a deceptively favourable impression. **window-ledge** = *window-sill.* **window-pane** a pane of glass in a window. **window-seat 1** a seat below a window, esp. in a bay or alcove. **2** a seat next to a window in an aircraft, train, etc. **window-shop** (**-shopped, -shopping**) look at goods displayed in shop-windows, usu. without buying anything. **window-shopper** a person who window-shops. **window-sill** a sill below a window. **window tax** *Brit. hist.* a tax on windows or similar openings (abolished in 1851). □ **windowed** *adj.* (also in *comb.*). **windowless** *adj.* [ME f. ON *vindauga* (as WIND[1], EYE)]

windowing /'wɪndəʊɪŋ/ *n. Computing* the selection of part of a stored image for display or enlargement.

windpipe /'wɪndpaɪp/ *n. Anat.* the air-passage from the throat to the lungs; the trachea.

Windscale /'wɪndskeɪl/ a former name for SELLAFIELD.

windscreen /'wɪndskriːn/ *n. Brit.* a screen of glass at the front of a motor vehicle. □ **windscreen wiper** a device consisting of a rubber blade on an arm, moving in an arc, for keeping a windscreen clear of rain etc.

windshield /'wɪndʃiːld/ *n. N. Amer.* = WINDSCREEN.

Windsor[1] /'wɪnzə(r)/ an industrial city and port in Ontario, southern Canada, opposite the US city of Detroit; pop. (1991) 223,240.

Windsor[2] /'wɪnzə(r)/ the name in current use by the British royal house. The name, previously Saxe-Coburg-Gotha, was changed in 1917 in response to wartime anti-German feeling.

Windsor, Duke of the title conferred on Edward VIII on his abdication in 1936.

Windsor Castle a royal residence at Windsor, Berkshire, founded by William the Conqueror on the site of an earlier fortress and extended by his successors, particularly Edward III. It includes St George's Chapel, built in the Perpendicular style, and lies on the edge of Windsor Great Park. The castle was severely damaged by fire in 1992.

Windsor chair *n.* a wooden dining chair with a semicircular back supported by upright rods.

windsurfing /'wɪnd,sɜːfɪŋ/ *n.* the sport of riding on water on a sailboard. □ **windsurf** *v.intr.* **windsurfer** *n.*

windswept /'wɪndswept/ *adj.* exposed to or swept back by the wind.

windward /'wɪndwəd/ *adj., adv., & n.* ● *adj. & adv.* on the side from which the wind is blowing (opp. LEEWARD). ● *n.* the windward region, side, or direction (*to windward; on the windward of*). □ **get to windward of 1** place oneself there to avoid the smell of. **2** gain an advantage over.

Windward Islands a group of islands in the eastern Caribbean. Constituting the southern part of the Lesser Antilles, they include Martinique, Dominica, St Lucia, Barbados, St Vincent and the Grenadines, and Grenada. Their name refers to their position further upwind, in terms of the prevailing south-easterly winds, than the Leeward Islands.

windy /'wɪndɪ/ *adj.* (**windier, windiest**) **1** stormy with wind (*a windy night*). **2** exposed to wind; windswept (*a windy plain*). **3** generating or characterized by flatulence. **4** *colloq.* wordy, verbose, empty (*a windy speech*). **5** *colloq.* nervous, frightened. □ **windily** *adv.* **windiness** *n.* [OE *windig* (as WIND[1], -Y[1])]

wine /waɪn/ *n. & v.* ● *n.* **1** fermented grape juice as an alcoholic drink. **2** a fermented drink resembling this made from other fruits etc. as specified (*elderberry wine; ginger wine*). **3** = wine red *n.* ● *v.* **1** *intr.* drink wine. **2** *tr.* entertain to wine. □ **wine and dine** entertain to or have a meal with wine. **wine bar** a bar or small restaurant where wine is the main drink available. **wine bottle** a glass bottle for wine, the standard size holding 75 cl or 26²⁄₃ fl. oz. **wine box** a square carton of wine with a dispensing tap. **wine cellar 1** a cellar for storing wine. **2** the contents of this. **wine-grower** a cultivator of grapes for wine. **wine list** a list of wines available in a restaurant etc. **wine red** *n.* the dark red colour of red wine. ● *adj.* (hyphenated when *attrib.*) of this colour. **wine-tasting 1** judging the quality of wine by tasting it. **2** an occasion for this. **wine vinegar** vinegar made from wine as distinct from malt. **wine waiter** a waiter responsible for serving wine. □ **wineless** *adj.* [OE *wīn* f. Gmc f. L *vinum*]

wineberry /'waɪnbərɪ/ *n.* (*pl.* **-ies**) **1 a** a deciduous bristly shrub, *Rubus phoenicolasius*, from China and Japan, producing scarlet berries used in cookery. **b** this berry. **2** = MAKO[2].

winebibber /'waɪn,bɪbə(r)/ *n. archaic* or *literary* a tippler or drunkard. □ **winebibbing** *n. & adj.* [WINE + *bib* to tipple]

wineglass /'waɪnglɑːs/ *n.* **1** a glass for wine, usu. with a stem and foot. **2** the contents of this, a wineglassful.

wineglassful /'waɪnglɑːs,fʊl/ *n.* (*pl.* **-fuls**) **1** the capacity of a wineglass, esp. of the size used for sherry, as a measure of liquid, about four tablespoons. **2** the contents of a wineglass.

winepress /'waɪnpres/ *n.* a press in which grapes are squeezed in making wine.

winery /'waɪnərɪ/ *n.* (*pl.* **-ies**) esp. *US* an establishment where wine is made.

wineskin /'waɪnskɪn/ *n.* a whole skin of a goat etc. sewn up and used to hold wine.

wing /wɪŋ/ *n. & v.* ● *n.* **1** each of the limbs or organs by which a bird, bat, or insect is able to fly. **2** a rigid horizontal winglike structure forming a supporting part of an aircraft. **3** part of a building etc. which projects or is extended in a certain direction (*lived in the north wing*). **4 a** a forward player at either end of a line in football, hockey, etc. **b** the side part of a playing-area. **5** (in *pl.*) the sides of a theatre stage out of view of the audience. **6** a section of a political party in terms of the extremity of its views. **7** a flank of a battle array (*the cavalry were massed on the left wing*). **8** *Brit.* the part of a motor vehicle covering a wheel. **9 a** an air-force unit of several squadrons or groups. **b** (in *pl.*) a pilot's badge in the RAF etc. (*get one's wings*). **10** *Anat. & Bot.* a lateral part or projection of an organ or structure. ● v. **1** intr. & tr. travel or traverse on wings or in an aircraft (*winging through the air; am winging my way home*). **2** tr. wound in a wing or an arm. **3** tr. equip with wings. **4** tr. enable to fly; send in flight (*fear winged my steps; winged an arrow towards them*). □ **give** (or **lend**) **wings to** speed up (a person or a thing). **on the wing** flying or in flight. **on a wing and a prayer** with only the slightest chance of success. **spread** (or **stretch**) **one's wings** develop one's powers fully. **take under one's wing** treat as a protégé. **take wing** fly away; soar. **waiting in the wings** holding oneself in readiness. **wing-beat** one complete set of motions with a wing in flying. **wing-case** the horny cover of an insect's wing. **wing-chair** a chair with side-pieces projecting forwards at the top of a high back. **wing-collar** a high stiff shirt collar with turned-down corners. **wing commander** an RAF officer next below group captain. **winged words** highly apposite or significant words. **wing forward** Rugby a forward who plays on the wing. **wing-game** game birds. **wing-half** Football a right or left half-back. **wing-nut** a nut with projections for the fingers to turn it on a screw. **wing-span** (or **-spread**) measurement right across the wings of a bird or aircraft. **wing-stroke** = wing-beat. **wing-tip** the outer end of an aircraft's or a bird's wing. □ **winged** adj. (also in comb.). **wingless** adj. **winglet** /ˈwɪŋlɪt/ n. **winglike** adj. [ME pl. wenge, -en, -es f. ON vængir, pl. of vængr]

wingding /ˈwɪŋdɪŋ/ n. sl. **1** esp. N. Amer. a wild party. **2** US a drug addict's real or feigned seizure. [20th c.: orig. unkn.]

Winged Victory n. a winged statue of Nike, the Greek goddess of victory, especially the Nike of Samothrace (c.200 BC) preserved in the Louvre in Paris.

winger /ˈwɪŋə(r)/ n. **1** a player on a wing in football, hockey, etc. **2** (in comb.) a member of a specified political wing (*left-winger*).

wink /wɪŋk/ v. & n. ● v. **1 a** tr. close and open (one eye or both eyes) quickly. **b** intr. close and open an eye. **2** intr. (often foll. by at) wink one eye as a signal of friendship or greeting or to convey a message to a person. **3** intr. (of a light etc.) twinkle; shine or flash intermittently. ● n. **1** the act or an instance of winking, esp. as a signal etc. **2** a brief moment of sleep (*didn't sleep a wink*). □ **as easy as winking** colloq. very easy. **in a wink** very quickly. **wink at 1** purposely avoid seeing; pretend not to notice. **2** connive at (a wrongdoing etc.). [OE wincian f. Gmc: cf. WINCE[1], WINCH]

winker /ˈwɪŋkə(r)/ n. **1** colloq. a flashing indicator light on a motor vehicle. **2** (usu. in pl.) a horse's blinker.

winkle /ˈwɪŋk(ə)l/ n. & v. ● n. a small marine gastropod mollusc of the genus Littorina, used for food. Also called periwinkle. ● v.tr. (foll. by out) esp. Brit. extract or eject (*winkled the information out of them*). □ **winkle-picker** sl. a shoe with a long pointed toe. □ **winkler** n. [abbr. of PERIWINKLE[2]: cf. WIG[1]]

winless /ˈwɪnlɪs/ adj. esp. N. Amer. without having won usu. a match in a series of contests.

winner /ˈwɪnə(r)/ n. **1** a person, racehorse, etc. that wins. **2** colloq. a successful or highly promising idea, enterprise, etc. (*the new scheme seemed a winner*).

winning /ˈwɪnɪŋ/ adj. & n. ● adj. **1** having or bringing victory or an advantage (*the winning entry; a winning stroke*). **2** attractive, persuasive (*a winning smile; winning ways*). ● n. (in pl.) money won, esp. in betting etc. □ **winning-post** a post marking the end of a race. □ **winningly** adv. **winningness** n.

Winnipeg /ˈwɪnɪˌpeg/ the capital of Manitoba, situated in the south of the province at the confluence of the Assiniboine and Red rivers, to the south of Lake Winnipeg; pop. (1991) 612,770. First settled as a French trading post in 1738, it was the site of a Scottish colony founded in 1811 and of a trading post established by the Hudson's Bay Company in 1821. It grew rapidly after the arrival of the railway in 1881.

Winnipeg, Lake a large lake in the province of Manitoba in Canada, to the north of the city of Winnipeg. Fed by the Saskatchewan, Winnipeg, and Red rivers from the east and south, the lake is drained by the Nelson river, which flows north-eastwards to Hudson Bay.

winnow /ˈwɪnəʊ/ v.tr. **1** blow (grain) free of chaff etc. by an air-current. **2** (foll. by out, away, from, etc.) get rid of (chaff etc.) from grain. **3 a** sift, separate; clear of refuse or inferior specimens. **b** sift or examine (evidence for falsehood etc.). **c** clear, sort, or weed out (rubbish etc.). **4** poet. **a** fan (the air with wings). **b** flap (wings). **c** stir (the hair etc.). □ **winnower** n. (in senses 1, 2). [OE windwian (as WIND[1])]

wino /ˈwaɪnəʊ/ n. (pl. **-os**) sl. a habitual excessive drinker of cheap wine; an alcoholic.

winsome /ˈwɪnsəm/ adj. (of a person, looks, or manner) winning,

attractive, engaging. □ **winsomely** adv. **winsomeness** n. [OE wynsum f. wyn JOY + -SOME[1]]

winter /ˈwɪntə(r)/ n. & v. ● n. **1** the coldest season of the year, in the northern hemisphere from December to February and in the southern hemisphere from June to August. **2** Astron. the period from the winter solstice to the vernal equinox. **3** a bleak or lifeless period or region etc. (*nuclear winter*). **4** poet. a year (esp. of a person's age) (*a man of fifty winters*). **5** (attrib.) **a** characteristic of or suitable for winter (*winter light; winter clothes*). **b** (of fruit) ripening late or keeping until or during winter. **c** (of wheat or other crops) sown in autumn for harvesting the following year. ● v. **1** intr. (usu. foll. by at, in) pass the winter (*likes to winter in the Canaries*). **2** tr. keep or feed (plants, cattle) during winter. □ **winter aconite** see ACONITE 2. **winter cress** a bitter-tasting cress of the genus Barbarea, esp. B. vulgaris. **winter garden** a garden or conservatory of plants flourishing in winter. **winter heliotrope** see HELIOTROPE 2. **winter jasmine** a jasmine, Jasminum nudiflorum, with yellow flowers in winter. **winter quarters** a place where soldiers spend the winter. **winter sleep** hibernation. **winter solstice** see SOLSTICE. **winter sports** sports performed on snow or ice esp. in winter (e.g. skiing and ice-skating). **winter-tide** poet. = WINTERTIME. **winter wren** N. Amer. see WREN 1. □ **winterer** n. **winterless** adj. **winterly** adj. [OE f. Gmc, prob. rel. to WET]

wintergreen /ˈwɪntəˌgriːn/ n. **1** a low-growing woodland plant of the genus Pyrola, with drooping spikes of white bell-shaped flowers. **2 a** N. Amer. the checkerberry, Gaultheria procumbens. **b** (in full **oil of wintergreen**) a pungent oil obtained from the leaves of this, now usu. made synthetically and used medicinally and as a flavouring.

Winterhalter /ˈvɪntəˌhæltə(r)/, Franz Xavier (1806–73), German painter. He painted many portraits of European royalty and aristocracy. His subjects included Napoleon III, the emperor Franz Josef, and Queen Victoria and her family.

winterize /ˈwɪntəˌraɪz/ v.tr. (also **-ise**) esp. N. Amer. adapt for operation or use in cold weather. □ **winterization** /ˌwɪntəraɪˈzeɪʃ(ə)n/ n.

Winter Olympics an international contest of winter sports held every four years. The first Winter Olympics took place in 1924, at Chamonix, France, after which they were held in the same year as the main Olympics until 1992: starting with the Lillehammer (Norway) competition of 1994 they now take place at a two-year interval from the summer games. Events in the Winter Olympics include skiing, ski-jumping, ice hockey, ice-skating, bob-sleigh, and biathlon.

Winter Palace the former Russian imperial residence in St Petersburg, stormed in the Revolution of 1917 and later used as a museum and art gallery.

Winterthur /ˈvɪntəˌtʊə(r)/ an industrial town in northern Switzerland; pop. (1990) 85,680.

wintertime /ˈwɪntəˌtaɪm/ n. the season of winter.

wintry /ˈwɪntrɪ/ adj. (also **wintery** /-tərɪ/; **wintrier**, **wintriest**) **1** characteristic of winter (*wintry weather; a wintry sun; a wintry landscape*). **2** (of a smile, greeting, etc.) lacking warmth or enthusiasm. □ **wintrily** adv. **wintriness** n. [OE wintrig, or f. WINTER]

winy /ˈwaɪnɪ/ adj. (**winier**, **winiest**) resembling wine in taste or appearance. □ **wininess** n.

wipe /waɪp/ v. & n. ● v.tr. **1** clean or dry the surface of by rubbing with the hands or a cloth etc. **2** rub (a cloth) over a surface. **3** spread (a liquid etc.) over a surface by rubbing. **4** (often foll. by away, off, etc.) **a** clear or remove by wiping (*wiped the mess off the table; wipe away your tears*). **b** remove or eliminate completely (*the village was wiped off the map*). **5 a** erase (data, a recording, etc., from a magnetic medium). **b** erase data from (the medium). **6** Austral. & NZ sl. reject or dismiss (a person or idea). ● n. **1** an act of wiping (*give the floor a wipe*). **2** a piece of disposable absorbent cloth, usu. treated with a cleansing agent, for wiping something clean (*antiseptic wipes*). □ **wipe down** clean (esp. a vertical surface) by wiping. **wipe a person's eye** colloq. get the better of a person. **wipe the floor with** colloq. inflict a humiliating defeat on. **wipe off** annul (a debt etc.). **wipe out 1 a** destroy, annihilate (*the whole population was wiped out*). **b** obliterate (*wiped it out of my memory*). **2** sl. murder. **3** clean the inside of. **4** avenge (an insult etc.). **wipe-out** n. **1** the obliteration of one radio signal by another. **2** an instance of destruction or annihilation. **3** sl. a fall from a surfboard. **wiped out** adj. sl. tired out, exhausted. **wipe the slate clean** see SLATE. **wipe up 1** Brit. dry (dishes etc.). **2** take up (a liquid etc.) by wiping. □ **wipeable** adj. [OE wipian: cf. OHG wifan wind round, Goth. weipan crown: rel. to WHIP]

wiper /ˈwaɪpə(r)/ n. **1** = windscreen wiper. **2** Electr. a moving contact. **3** a cam or tappet.

WIPO see WORLD INTELLECTUAL PROPERTY ORGANIZATION.

Wiradhuri /wɪˈrædʒərɪ/ *n.* an Aboriginal language of SE Australia, now extinct.

wire /waɪə(r)/ *n. & v.* ● *n.* **1 a** a metal drawn out into the form of a thread or thin flexible rod. **b** a piece of this. **c** (*attrib.*) made of wire. **2** a length or quantity of wire used for fencing or to carry an electric current etc. **3** *colloq.* a telegram or cablegram. ● *v.tr.* **1** provide, fasten, strengthen, etc., with wire. **2** (often foll. by *up*) *Electr.* install electrical circuits in (a building, piece of equipment, etc.). **3** *colloq.* telegraph (*wired me that they were coming*). **4** snare (an animal etc.) with wire. **5** (usu. in *passive*) *Croquet* obstruct (a ball, shot, or player) by a hoop. □ **by wire** by telegraph. **get one's wires crossed** become confused and misunderstood. **wire brush 1** a brush with tough wire bristles for cleaning hard surfaces, esp. metal. **2** a brush with wire strands brushed against cymbals to produce a soft metallic sound. **wire cloth** cloth woven from wire. **wire-cutter** a tool for cutting wire. **wire gauge 1** a gauge for measuring the diameter of wire etc. **2** a standard series of sizes in which wire etc. is made. **wire gauze** a stiff gauze woven from wire. **wire grass** a grass with a tough wiry stem. **wire-haired** (esp. of a dog) having stiff or wiry hair. **wire mattress** a mattress supported by wires stretched in a frame. **wire netting** netting of wire twisted into meshes. **wire rope** rope made by twisting wires together as strands. **wire-tapper** a person who indulges in wire-tapping. **wire-tapping** the practice of tapping (see TAP¹ *v.* 4) a telephone or telegraph line to eavesdrop. **wire-walker** an acrobat performing feats on a wire rope. **wire wheel** a vehicle-wheel with spokes of wire. **wire wool** a mass of fine wire for cleaning. □ **wirer** *n.* [OE *wīr*]

wiredraw /ˈwaɪəˌdrɔː/ *v.tr.* (*past* **-drew** /-ˌdruː/; *past part.* **-drawn** /-ˌdrɔːn/) **1** draw (metal) out into wire. **2** elongate; protract unduly. **3** (esp. as **wiredrawn** *adj.*) refine or apply or press (an argument etc.) with idle or excessive subtlety.

wireless /ˈwaɪəlɪs/ *n. & adj.* ● *n. archaic or hist.* **1** esp. *Brit.* **a** (in full **wireless set**) a radio receiving set. **b** the transmission and reception of radio signals. **2** (in full **wireless telegraphy**) = RADIO-TELEGRAPHY. ● *adj.* lacking or not requiring wires.

wireman /ˈwaɪəmən/ *n.* (*pl.* **-men**) **1** esp. *US* an installer or repairer of electric wires. **2** a journalist working for a telegraphic news agency.

wirepuller /ˈwaɪəˌpʊlə(r)/ *n.* esp. *US* a politician etc. who exerts a hidden influence. □ **wirepulling** *n.*

wireworm /ˈwaɪəˌwɜːm/ *n.* the larva of the click beetle causing damage to crop plants.

wiring /ˈwaɪərɪŋ/ *n.* **1** a system of wires providing electrical circuits. **2** the installation of this (*came to do the wiring*).

Wirral /ˈwɪrəl/ (in full **the Wirral Peninsula**) a peninsula on the coast of NW England, between the estuaries of the rivers Dee and Mersey.

wiry /ˈwaɪərɪ/ *adj.* (**wirier, wiriest**) **1** tough and flexible as wire. **2** (of a person) thin and sinewy; untiring. **3** made of wire. □ **wirily** *adv.* **wiriness** *n.*

Wis. *abbr.* Wisconsin.

wis /wɪs/ *v.intr. archaic* know well. [orig. *I wis* = obs. *iwis* 'certainly' f. OE *gewis*, erron. taken as 'I know' and as pres. tense of *wist* (WIT²)]

Wisconsin /wɪsˈkɒnsɪn/ a state in the northern US, bordering on Lakes Superior (in the north-west) and Michigan (in the east); pop. (1990) 4,891,770; capital, Madison. Ceded to Britain by the French in 1763 and acquired by the US in 1783 as part of the former Northwest Territory, it became the 30th state of the US in 1848.

Wisd. *abbr.* Wisdom of Solomon (Apocrypha).

Wisden /ˈwɪzdən/, John (1826–84), English cricketer. He is remembered as the publisher of *Wisden Cricketers' Almanack*, an annual publication which first appeared in 1864.

wisdom /ˈwɪzdəm/ *n.* **1** the state of being wise. **2** experience and knowledge together with the power of applying them critically or practically. **3** sagacity, prudence; common sense. **4** wise sayings, thoughts, etc., regarded collectively. □ **in his** (or **her** etc.) **wisdom** usu. *iron.* thinking it would be best (*the committee in its wisdom decided to abandon the project*). **wisdom tooth** each of four hindmost molars not usu. cut before 20 years of age. [OE *wīsdōm* (as WISE¹, -DOM)]

Wisdom of Solomon a book of the Apocrypha ascribed to Solomon and containing a meditation on wisdom. The book is thought actually to date from about 1st century BC–1st century AD.

wise¹ /waɪz/ *adj. & v.* ● *adj.* **1 a** having experience and knowledge and judiciously applying them. **b** (of an action, behaviour, etc.) determined by or showing or in harmony with such experience and knowledge.

2 sagacious, prudent, sensible, discreet. **3** having knowledge. **4** suggestive of wisdom (*with a wise nod of the head*). **5** *US colloq.* **a** alert, crafty. **b** (often foll. by *to*) having (usu. confidential) information (about). ● *v.tr. & intr.* (foll. by *up*) esp. *US colloq.* put or get wise. □ **be** (or **get**) **wise to** *colloq.* be (or become) aware of. **no** (or **none the** or **not much**) **wiser** knowing no more than before. **put a person wise** (often foll. by *to*) *colloq.* inform a person (about). **wise after the event** able to understand and assess an event or circumstance after its implications have become obvious. **wise guy** *colloq.* a know-all. **wise man** a man versed in magic, astrology, etc., esp. one of the Magi. **wise saw** a proverbial saying. **without anyone's being the wiser** undetected. □ **wisely** *adv.* [OE *wīs* f. Gmc: see WIT²]

wise² /waɪz/ *n. archaic* way, manner, or degree (*in solemn wise*; *on this wise*). □ **in no wise** not at all. [OE *wīse* f. Gmc f. WIT²]

-wise /waɪz/ *suffix* forming adjectives and adverbs of manner (*crosswise*; *clockwise*; *lengthwise*) or respect (*moneywise*) (cf. -WAYS). ¶ More fanciful phrase-based combinations, such as *employment-wise* (= as regards employment) are *colloq.*, and restricted to informal contexts. [as WISE²]

wiseacre /ˈwaɪzˌeɪkə(r)/ *n.* a person who affects a wise manner. [MDu. *wijsseggher* soothsayer, prob. f. OHG *wīssago*, *wīzago*, assim. to WISE¹, ACRE]

wisecrack /ˈwaɪzkræk/ *n. & v. colloq.* ● *n.* a smart pithy remark. ● *v.intr.* make a wisecrack. □ **wisecracker** *n.*

wisent /ˈwiːz(ə)nt/ *n.* the European bison, *Bison bonasus*. [G: cf. BISON]

wish /wɪʃ/ *v. & n.* ● *v.* **1** *intr.* (often foll. by *for*) have or express a desire or aspiration for (*wish for happiness*). **2** *tr.* (often foll. by *that* + clause, usu. with *that* omitted) have as a desire or aspiration (*I wish I could sing*; *I wished that I was dead*). **3** *tr.* want or demand, usu. so as to bring about what is wanted (*I wish to go*; *I wish you to do it*; *I wish it done*). **4** *tr.* express one's hopes for (*we wish you well*; *wish them no harm*; *wished us a pleasant journey*). **5** *tr.* (foll. by *on*, *upon*) *colloq.* foist on a person. ● *n.* **1 a** a desire, request, or aspiration. **b** an expression of this. **2** a thing desired (*got my wish*). □ **best** (or **good**) **wishes** hopes felt or expressed for another's happiness etc. **wish-fulfilment** a tendency for subconscious desire to be satisfied in fantasy. **wishing-well** a well into which coins are dropped and a wish is made. **the wish is father to the thought** we believe a thing because we wish it true. **wish-list** a mental or hypothetical list of wishes or desires. □ **wisher** *n.* (in sense 4 of *v.*); (also in *comb.*). [OE *wȳscan*, OHG *wunsken* f. Gmc, ult. rel. to WEEN, WONT]

wishbone /ˈwɪʃbəʊn/ *n.* **1** a forked bone between the neck and breast of a cooked bird: when broken between two people the longer portion entitles the holder to make a wish. **2** an object of similar shape.

wishful /ˈwɪʃfʊl/ *adj.* **1** (often foll. by *to* + infin.) desiring, wishing. **2** having or expressing a wish. □ **wishful thinking** belief founded on wishes rather than facts. □ **wishfully** *adv.* **wishfulness** *n.*

wish-wash /ˈwɪʃwɒʃ/ *n.* **1** a weak or watery drink. **2** insipid talk or writing. [redupl. of WASH]

wishy-washy /ˈwɪʃɪˌwɒʃɪ/ *adj.* **1** feeble, insipid, or indecisive in quality or character. **2** (of tea, soup, etc.) weak, watery, sloppy. [redupl. of WASHY]

Wisła see VISTULA.

wisp /wɪsp/ *n.* **1** a small bundle or twist of straw etc. **2** a small separate quantity of smoke, hair, etc. **3** a small thin person etc. **4** a flock (of snipe). □ **wispy** *adj.* (**wispier, wispiest**). **wispily** *adv.* **wispiness** *n.* [ME: orig. uncert.: cf. West Frisian *wisp*, and WHISK]

wist *past* and *past part.* of WIT².

wisteria /wɪˈstɪərɪə/ *n.* (also **wistaria** /-ˈsteərɪə/) a climbing leguminous shrub of the genus *Wisteria*, with hanging clusters of pale bluish-lilac flowers; esp. *W. floribunda* and *W. sinensis*, of Japan and China respectively, much planted for ornament. [Caspar *Wistar* (or *Wister*), Amer. anatomist (1761–1818)]

wistful /ˈwɪstfʊl/ *adj.* (of a person, looks, etc.) yearningly or mournfully expectant, thoughtful, or wishful. □ **wistfully** *adv.* **wistfulness** *n.* [app. assim. of obs. *wistly* (adv.) intently (cf. WHISHT) to *wishful*, with corresp. change of sense]

wit¹ /wɪt/ *n.* **1** (in *sing.* or *pl.*) intelligence; quick understanding (*has quick wits*; *a nimble wit*). **2 a** the unexpected, quick, and humorous combining or contrasting of ideas or expressions (*conversation sparkling with wit*). **b** the power of giving intellectual pleasure by this. **3** a person possessing such a power, esp. a cleverly humorous person. □ **at one's wit's** (or **wits'**) **end** utterly at a loss or in despair. **have** (or **keep**) **one's wits about one** be alert or vigilant or of lively intelligence. **live by one's wits** live by ingenious or crafty expedients, without a settled occupation. **out of one's wits** mad, distracted. **set one's**

wits to argue with. □ **witted** *adj.* (in sense 1); (also in *comb.*). [OE *wit*(*t*), *gewit*(*t*) f. Gmc]

wit[2] /wɪt/ *v.tr. & intr.* (*1st & 3rd sing. present* **wot** /wɒt/; *past and past part.* **wist** /wɪst/) (often foll. by *of*) *archaic* know. □ **to wit** that is to say; namely. [OE *witan* f. Gmc]

witan /ˈwɪt(ə)n/ *n.* = WITENAGEMOT. [OE]

witch /wɪtʃ/ *n. & v.* ● *n.* **1 a** a sorceress, esp. a woman supposed to have dealings with the Devil or evil spirits. **b** a follower or practitioner of the religious cult of modern witchcraft. **2** an ugly old woman; a hag. **3** a fascinating girl or woman. **4** a flatfish, *Glyptocephalus cynoglossus*, resembling the lemon sole. ● *v.tr. archaic* **1** bewitch. **2** fascinate, charm, lure. □ **witch ball** a coloured glass ball of the kind formerly hung up to keep witches away. **witch-doctor** a tribal magician of primitive people. **witches' broom** a dense twiggy growth in a tree, esp. birch, caused by infection with fungus, mites, or viruses. **witches' sabbath** see SABBATH 3. **witch-hunt 1** *hist.* a search for and persecution of supposed witches. **2** a campaign directed against a particular group of those holding unpopular or unorthodox views, esp. communists. **the witching hour** midnight, when witches are supposedly active (after Shakespeare's *Hamlet* III. ii. 377 *the witching time of night*). □ **witching** *adj.* **witchlike** *adj.* **witchy** *adj.* [OE *wicca* (masc.), *wicce* (fem.), rel. to *wiccian* (v.) practise magic arts]

witch- var. of WYCH-.

witch-alder /ˈwɪtʃˌɔːldə(r)/ *n.* (also **wych-alder**) a North American shrub, *Fothergilla gardenii*, with alder-like leaves.

witchcraft /ˈwɪtʃkrɑːft/ *n.* **1** the use of magic or sorcery. (*See note below.*) **2** power or influence like that of a witch; bewitching charm.

▪ Witchcraft, which has a long history around the world, became associated in Christianity with demonic possession and rejection of God in Europe in the early Middle Ages; by the late 12th century cases were being dealt with by the Inquisition. Mass persecutions began to take place in the 15th century in Europe; many people (usually women) were tried as witches, often after accusations from neighbours claiming to have suffered harm; witches were defined by Catholics and later also by Protestants as heretics who obtained their power through a pact with the Devil. After a papal bull of 1484, roughly a thousand people in England were hanged or burned for witchcraft, mostly under Elizabeth I and James I; the last execution was in 1685. On the Continent, and in Scotland, the use of torture produced far more victims and bizarre confessions of sabbaths and night-flying. Belief in witchcraft was also widespread in Puritan communities in North America; more than thirty people were convicted of witchcraft in the notorious trials at Salem, Massachusetts, in 1692. Although belief in witchcraft declined in Europe and America after the 17th century, it continues to be a powerful force in many preliterate and tribal societies. In the 20th century a new kind of witchcraft (known as *Wicca*) has emerged in parts of Europe and North America, claiming its origins in pre-Christian pagan religions.

witchery /ˈwɪtʃərɪ/ *n.* **1** witchcraft. **2** power exercised by beauty or eloquence or the like.

witchetty /ˈwɪtʃətɪ/ *n.* (*pl.* **-ies**) *Austral.* (in full **witchetty grub**) a large white larva of a beetle or moth, eaten as food by Aboriginals. [Aboriginal]

witch-hazel /ˈwɪtʃˌheɪz(ə)l/ *n.* (also **wych-hazel**) **1** an ornamental East Asian or American shrub of the genus *Hamamelis*, with yellow flowers in winter or spring. **2** an astringent lotion obtained from the leaves and bark of *H. virginiana* of North America.

witenagemot /ˌwɪtənəɡəˈməʊt/ *n. hist.* the council that advised the Anglo-Saxon kings, consisting of nobles and churchmen chosen by the king himself. [OE f. *witena* genitive pl. of *wita* wise man (as WIT[2]) + *gemōt* meeting: cf. MOOT]

with /wɪð/ *prep.* expressing: **1** an instrument or means used (*cut with a knife*; *can walk with assistance*). **2** association or company (*lives with his mother*; *works with Shell*; *lamb with mint sauce*). **3** cause or origin (*shiver with fear*; *in bed with measles*). **4** possession, attribution (*the man with dark hair*; *a vase with handles*). **5** circumstances; accompanying conditions (*sleep with the window open*; *a holiday with all expenses paid*). **6** manner adopted or displayed (*behaved with dignity*; *spoke with vehemence*; *handle with care*; *won with ease*). **7** agreement or harmony (*sympathize with*; *I believe with you that it can be done*). **8** disagreement, antagonism, competition (*incompatible with*; *stop arguing with me*). **9** responsibility or care for (*the decision rests with you*; *leave the child with me*). **10** material (*made with gold*). **11** addition or supply; possession of as a material, attribute, circumstance, etc. (*fill it with water*; *threaten with dismissal*; *decorate with holly*). **12** reference or regard (*be patient with them*; *how are things with you?*; *what do you want with me?*; *there's nothing wrong with expressing one's opinion*). **13** relation or causative association (*changes with the weather*; *keeps pace with the cost of living*). **14** an accepted circumstance or consideration (*with all your faults, we like you*). □ **away** (or **in** or **out** etc.) **with** (as *int.*) take, send, or put (a person or thing) away, in, out, etc. **be with a person 1** agree with and support a person. **2** *colloq.* follow a person's meaning (*are you with me?*). **one with** part of the same whole as. **with child** (or **young**) *literary* pregnant. **with it** *colloq. adj.* (hyphenated when *attrib.*) **1** up to date; conversant with modern ideas etc. **2** alert and comprehending. **3** (of clothes etc.) fashionable. ● *adv.* besides, in addition. **with-profits** (usu. *attrib.*) *Brit.* (of an insurance policy) allowing the holder a share of profits made by the company, usu. in the form of a bonus. **with that** thereupon. [OE, prob. shortened f. a Gmc prep. corresp. to OE *wither*, OHG *widar* against]

withal /wɪˈðɔːl/ *adv. & prep. archaic* ● *adv.* moreover; as well; at the same time. ● *prep.* (placed after its expressed or omitted object) with (*what shall he fill his belly withal?*). [ME f. WITH + ALL]

withdraw /wɪðˈdrɔː/ *v.* (*past* **withdrew** /-ˈdruː/; *past part.* **withdrawn** /-ˈdrɔːn/) **1** *tr.* pull or take aside or back (*withdrew my hand*). **2** *tr.* discontinue, cancel, retract (*withdrew my support*; *the promise was later withdrawn*). **3** *tr.* remove; take away (*withdrew the child from school*; *withdrew their troops*). **4** *tr.* take (money) out of an account. **5** *intr.* retire or go away; move away or back. **6** *intr.* (as **withdrawn** *adj.*) abnormally shy and unsociable; mentally detached. □ **withdrawing-room** *archaic* = DRAWING-ROOM 1. □ **withdrawer** *n.* [ME f. *with-* away (as WITH) + DRAW]

withdrawal /wɪðˈdrɔːəl/ *n.* **1** the act or an instance of withdrawing or being withdrawn. **2** a process of ceasing to take addictive drugs, often with an unpleasant physical reaction (*withdrawal symptoms*). **3** = *coitus interruptus*.

withe /wɪθ, wɪð, waɪð/ (also **withy** /ˈwɪðɪ/) (*pl.* **withes** or **-ies**) *n.* a tough flexible shoot, esp. of willow or osier, used for tying a bundle of wood etc. [OE *withthe*, *withig* f. Gmc, rel. to WIRE]

wither /ˈwɪðə(r)/ *v.* **1** *tr. & intr.* (often foll. by *up*) make or become dry and shrivelled (*withered flowers*). **2** *tr. & intr.* (often foll. by *away*) deprive of or lose vigour, vitality, freshness, or importance. **3** *intr.* decay, decline. **4** *tr.* **a** blight with scorn etc. **b** (as **withering** *adj.*) scornful (*a withering look*). □ **witheringly** *adv.* [ME, app. var. of WEATHER differentiated for certain senses]

withers /ˈwɪðəz/ *n.pl.* the ridge between a horse's shoulder-blades. [shortening of (16th-c.) *widersome* (or *-sone*) f. *wider-*, *wither-* against (cf. WITH), as the part that resists the strain of the collar: second element obscure]

withershins var. of WIDDERSHINS.

withhold /wɪðˈhəʊld/ *v.tr.* (*past and past part.* **-held** /-ˈheld/) **1** (often foll. by *from*) hold back; restrain. **2** refuse to give, grant, or allow (*withhold one's consent*; *withhold the truth*). □ **withholder** *n.* [ME f. *with-* away (as WITH) + HOLD[1]]

within /wɪˈðɪn/ *adv. & prep.* ● *adv. archaic* or *literary* **1** inside; to, at, or on the inside; internally. **2** indoors (*is anyone within?*). **3** in spirit (*make me pure within*). **4** inside the city walls (*Bishopsgate within*). ● *prep.* **1** inside; enclosed or contained by. **2 a** not beyond or exceeding (*within one's means*). **b** not transgressing (*within the law*; *within reason*). **3** not further off than (*within three miles of a station*; *within shouting distance*; *within ten days*). □ **within doors** in or into a house. **within one's grasp** see GRASP. **within reach** (or **sight**) **of** near enough to be reached or seen. [OE *withinnan* on the inside (as WITH, *innan* (adv. & prep.) within, formed as IN)]

without /wɪˈðaʊt/ *prep. & adv.* ● *prep.* **1** not having, feeling, or showing (*came without any money*; *without hesitation*; *without any emotion*). **2** with freedom from (*without fear*; *without embarrassment*). **3** in the absence of (*cannot live without you*; *the train left without us*). **4** with neglect or avoidance of (*do not leave without telling me*). ¶ Use as a *conj.*, as in *do not leave without you tell me*, is non-standard. **5** *archaic* outside (*without the city wall*). ● *adv. archaic* or *literary* **1** outside (*seen from without*). **2** out of doors (*remained shivering without*). **3** in outward appearance (*rough without but kind within*). **4** outside the city walls (*Bishopsgate without*). □ **without end** infinite, eternal. [OE *withūtan* (as WITH, *ūtan* from outside, formed as OUT)]

withstand /wɪðˈstænd/ *v.* (*past and past part.* **-stood** /-ˈstʊd/) **1** *tr.* oppose, resist, hold out against (a person, force, etc.). **2** *intr.* make opposition; offer resistance. □ **withstander** *n.* [OE *withstandan* f. *with-* against (as WITH) + STAND]

withy /ˈwɪðɪ/ *n.* (*pl.* **-ies**) **1** a willow of any species. **2** var. of WITHE.

witless /ˈwɪtlɪs/ adj. **1** lacking wits; foolish, stupid. **2** crazy. □ **witlessly** adv. **witlessness** n. [OE witlēas (as WIT[1], -LESS)]

witling /ˈwɪtlɪŋ/ n. archaic usu. derog. a person who fancies himself or herself as a wit.

witness /ˈwɪtnɪs/ n. & v. ● n. **1** a person present at some event and able to give information about it (cf. EYEWITNESS). **2 a** a person giving sworn testimony. **b** a person attesting another's signature to a document. **3** (foll. by to, of) a person or thing whose existence, condition, etc., attests or proves something (is a living witness to their generosity). **4** testimony, evidence, confirmation. ● v. **1** tr. be a witness of (an event etc.) (did you witness the accident?). **2** tr. be witness to the authenticity of (a document or signature). **3** tr. serve as evidence or an indication of. **4** intr. (foll. by against, for, to) give or serve as evidence. □ **bear witness to** (or **of**) **1** attest the truth of. **2** state one's belief in. **call to witness** appeal to for confirmation etc. **witness-box** (US **-stand**) an enclosure in a lawcourt from which witnesses give evidence. [OE witnes (as WIT[1], -NESS)]

Wittenberg /ˈvɪt(ə)nˌbɜːg/ a town in eastern Germany, on the River Elbe north-east of Leipzig; pop. (1991) 87,000. It was the scene in 1517 of Martin Luther's campaign against the Roman Catholic Church, which was a major factor in the rise of the Reformation.

witter /ˈwɪtə(r)/ v.intr. (often foll. by on) colloq. speak tediously on trivial matters. [20th c.: prob. imit.]

Wittgenstein /ˈvɪtɡənˌstaɪn/, Ludwig (Josef Johann) (1889–1951), Austrian-born philosopher. He came to England in 1911 and studied mathematical logic at Cambridge under Bertrand Russell (1912–13). He then turned to the study of language and its relationship to the world, and in the Tractatus Logico-philosophicus (1921) contended that language achieves meaning by 'picturing' things by established conventions. He also pointed out that logical truths are tautologous because they are necessarily true within their own system and argued that metaphysical speculation is meaningless, theories which influenced the development of logical positivism. He returned to Cambridge in 1929, where he was professor of philosophy (1939–47); he became a British citizen in 1938. Principal among his later works (all published posthumously) is Philosophical Investigations (1953). In this he argues that words take on different roles according to the different human activities in which they are used, and that they do not have definite intrinsic meanings. He showed that some philosophical problems are simply a result of a misunderstanding of the nature of language, as for example the assumption by some earlier philosophers that individual human beings have a private language in which their thoughts as well as their utterances are composed.

witticism /ˈwɪtɪˌsɪz(ə)m/ n. a witty remark. [coined by John Dryden (1677) f. WITTY, after criticism]

witting /ˈwɪtɪŋ/ adj. **1** aware. **2** intentional. □ **wittingly** adv. [ME f. WIT[2] + -ING[2]]

witty /ˈwɪtɪ/ adj. (**wittier, wittiest**) **1** showing verbal wit. **2** characterized by wit or humour. □ **wittily** adv. **wittiness** n. [OE witig, wittig (as WIT[1], -Y[1])]

Witwatersrand, the /wɪtˈwɔːtəzˌrænd/ (also called the Rand) a region of South Africa, around the city of Johannesburg. Consisting of a series of parallel rocky ridges, it forms a watershed between the Vaal and Olifant rivers. The region contains rich gold deposits, first discovered in 1886. Its Afrikaans name means 'ridge of white waters'.

wivern var. of WYVERN.

wives pl. of WIFE.

wiz var. of WHIZ n. 2.

wizard /ˈwɪzəd/ n. & adj. ● n. **1** a sorcerer; a magician. **2** a person of remarkable powers, a genius. **3** a conjuror. ● adj. archaic sl. esp. Brit. wonderful, excellent. □ **wizardly** adj. **wizardry** n. [ME f. WISE[1] + -ARD]

wizened /ˈwɪz(ə)nd/ adj. (also **wizen**) (of a person or face etc.) shrivelled-looking. [past part. of wizen shrivel f. OE wisnian f. Gmc]

wk. abbr. **1** week. **2** work. **3** weak.

wks. abbr. weeks.

Władysław II see LADISLAUS II.

Wm. abbr. William.

WMO see WORLD METEOROLOGICAL ORGANIZATION.

WNW abbr. west-north-west.

WO abbr. Warrant Officer.

wo /wəʊ/ int. = WHOA. [var. of who (int.), HO]

w.o. abbr. walk-over.

woad /wəʊd/ n. **1** a cruciferous plant, Isatis tinctoria, yielding a blue dye now superseded by indigo. **2** the dye obtained from this. [OE wād f. Gmc]

wobbegong /ˈwɒbɪˌɡɒŋ/ n. a brown shark, Orectolobus maculatus, of Australian waters, patterned with buff markings. [Aboriginal]

wobble /ˈwɒb(ə)l/ v. & n. ● v. **1 a** intr. sway or vibrate unsteadily from side to side. **b** tr. cause to do this. **2** intr. stand or go unsteadily; stagger. **3** intr. waver, vacillate; act inconsistently. **4** intr. (of the voice or sound) quaver, pulsate. ● n. **1** a wobbling movement. **2** an instance of vacillation or pulsation. □ **wobble-board** Austral. a piece of fibreboard used as a musical instrument with a low booming sound. □ **wobbler** n. [earlier wabble, corresp. to LG wabbeln, ON vafla waver f. Gmc: cf. WAVE, WAVER, -LE[4]]

Wobbly /ˈwɒblɪ/ see INDUSTRIAL WORKERS OF THE WORLD.

wobbly /ˈwɒblɪ/ adj. & n. ● adj. (**wobblier, wobbliest**) **1** wobbling or tending to wobble. **2** wavy, undulating (a wobbly line). **3** unsteady; weak after illness (feeling wobbly). **4** wavering, vacillating, insecure (the economy was wobbly). ● n. colloq. a fit of panic or temper. □ **throw a wobbly** colloq. have a fit of panic or temper. □ **wobbliness** n.

Wodehouse /ˈwʊdhaʊs/, Sir P(elham) G(renville) (1881–1975), British-born writer. He was a prolific writer of humorous novels and short stories and his writing career spanned more than seventy years. His best-known works are those set in the leisured English upper-class world of Bertie Wooster and his valet, Jeeves, the first of which appeared in 1917. During the Second World War Wodehouse was interned by the Germans; when released in 1941 he made five radio broadcasts from Berlin, which, although they were comic and non-political in nature, led to accusations in Britain that he was a traitor. He eventually settled in the US, becoming an American citizen in 1955; he was knighted a few weeks before his death.

Woden /ˈwəʊd(ə)n/ see ODIN.

wodge /wɒdʒ/ n. Brit. colloq. a chunk or lump. [alt. of WEDGE[1]]

woe /wəʊ/ n. archaic or literary **1** affliction; bitter grief; distress. **2** (in pl.) calamities, troubles. **3** joc. problems (told me a tale of woe). □ **woe betide** there will be unfortunate consequences for (woe betide you if you are late). **woe is me** an exclamation of distress. [OE wā, wǣ f. Gmc, a natural exclam. of lament]

woebegone /ˈwəʊbɪˌɡɒn/ adj. dismal-looking. [WOE + begone = surrounded f. OE begān (as BE-, GO[1])]

woeful /ˈwəʊfʊl/ adj. **1** sorrowful; afflicted with distress (a woeful expression). **2** causing sorrow or affliction. **3** very bad; wretched (woeful ignorance). □ **woefully** adv. **woefulness** n.

wog[1] /wɒɡ/ n. sl. offens. a foreigner, esp. a non-white one. [20th c.: orig. unkn.]

wog[2] /wɒɡ/ n. Austral. sl. an illness or infection. [20th c.: orig. unkn.]

woggle /ˈwɒɡ(ə)l/ n. a leather etc. ring through which the ends of a Scout's neckerchief are passed at the neck. [20th c.: orig. unkn.]

Wöhler /ˈvɜːlə(r)/, Friedrich (1800–82), German chemist. With his synthesis of urea from ammonium cyanate in 1828, he demonstrated that organic compounds could be made from inorganic compounds, ending for many scientists their belief in the doctrine of vitalism. He went on to make a number of new inorganic and organic compounds, and was the first to isolate the elements aluminium and beryllium. Wöhler and Justus von Liebig discovered the existence of chemical radicals (see LIEBIG).

wok /wɒk/ n. a bowl-shaped frying-pan used in esp. Chinese cookery. [Cantonese]

woke past of WAKE[1].

woken past part. of WAKE[1].

Woking /ˈwəʊkɪŋ/ a town in Surrey, SE England; pop. (1981) 81,770.

wold /wəʊld/ n. a piece of high open uncultivated land or moor. [OE wald f. Gmc, perh. rel. to WILD: cf. WEALD]

Wolf /vɒlf/, Hugo (Philipp Jakob) (1860–1903), Austrian composer. He is chiefly known as a composer of lieder and from about 1883 onwards produced some 300 songs. His early songs are settings of German poets, especially Goethe and Heinrich Heine. He turned to translations of Spanish and Italian verse for the three volumes of his Spanish Songbook (1891) and the two volumes of his Italian Songbook (1892–6). He also wrote an opera, Der Corregidor (1895). His career was cut short by mental illness resulting from syphilis, and his last years were spent in an asylum.

wolf /wʊlf/ n. & v. ● n. (pl. **wolves** /wʊlvz/) **1** a large doglike flesh-eating mammal; esp. Canis lupus of Eurasia and North America, which usu. has a greyish coat and hunts in packs, and is the main ancestor of the domestic dog. **2** *colloq.* a man given to seducing women. **3** a rapacious or greedy person. **4** *Mus.* **a** a jarring sound from some notes in a bowed instrument. **b** an out-of-tune effect when playing certain chords on old organs (before the present 'equal temperament' was in use). ● *v.tr.* (often foll. by *down*) devour (food) greedily. □ **cry wolf** raise repeated false alarms (so that a genuine one is disregarded). **have** (or **hold) a wolf by the ears** be in a precarious position. **keep the wolf from the door** avert hunger or starvation. **lone wolf** a person who prefers to act alone. **throw to the wolves** sacrifice without compunction. **wolf-cub** a young wolf. **wolf-fish** a large deep-water marine fish of the genus *Anarrhichas*, with large doglike teeth. **wolf in sheep's clothing** a hostile person who pretends friendship. **wolf-pack** an attacking group of submarines or aircraft. **wolf's-milk** spurge. **wolf-spider** a ground-dwelling spider of the family Lycosidae, which runs after and springs on its prey. **wolf-whistle** *n.* a sexually admiring whistle. ● *v.intr.* make a wolf-whistle. □ **wolfish** *adj.* **wolfishly** *adv.* **wolflike** *adj. & adv.* [OE *wulf* f. Gmc]

Wolf Cub *n.* the former name for a Cub Scout. (See SCOUT ASSOCIATION.)

Wolfe /wʊlf/, James (1727–59), British general. As one of the leaders of the expedition sent to seize French Canada, he played a vital role in the capture of Louisbourg on Cape Breton Island in 1758. The following year he commanded the attack on the French capital, the city of Quebec. He was fatally wounded while leading his troops to victory on the Plains of Abraham, the scene of the battle which led to British control of Canada.

wolfhound /ˈwʊlfhaʊnd/ *n.* any of several breeds of large dog originally bred to hunt wolves.

wolfram /ˈwʊlfrəm/ *n.* **1** tungsten. **2** tungsten ore; a native tungstate of iron and manganese. [G: perh. f. *Wolf* WOLF + *Rahm* cream, or MHG *rām* dirt, soot]

wolframite /ˈwʊlfrəˌmaɪt/ *n.* = WOLFRAM 2.

wolfsbane /ˈwʊlfsbeɪn/ *n.* an aconite, esp. *Aconitum lycoctonum*.

Wolfsburg /ˈvɒlfsbɜːg/ an industrial city on the Mittelland Canal in Lower Saxony, NW Germany; pop. (1991) 128,995.

wolfskin /ˈwʊlfskɪn/ *n.* **1** the skin of a wolf. **2** a mat, cloak, etc., made from this.

Wolfson /ˈwʊlfs(ə)n/, Sir Isaac (1897–1991), Scottish businessman and philanthropist. He was appointed managing director of Great Universal Stores in 1934, later becoming its chairman (1946) and honorary life president (1987). In 1955 he established the Wolfson Foundation for promoting and funding medical research and education. In 1966 Wolfson endowed the Oxford college that now bears his name; University College, Cambridge, changed its name to Wolfson College in 1973 in recognition of grants received from the foundation.

Wollaston /ˈwʊləstən/, William Hyde (1766–1828), English chemist and physicist. He pioneered powder metallurgy, developed while he was attempting to produce malleable platinum; in the course of this work he discovered palladium and rhodium. The income he derived from his platinum process allowed him to devote himself to scientific research. Wollaston demonstrated that static and current electricity were the same, and was the first to observe the dark lines in the solar spectrum. He also invented a kind of slide-rule for use in chemistry, and several optical instruments. He supported John Dalton's atomic theory and the wave theory of light.

Wollongong /ˈwʊlənˌgɒŋ/ a city on the coast of New South Wales, SE Australia; pop. (1991) 211,420.

Wollstonecraft /ˈwʊlstənˌkrɑːft/, Mary (1759–97), English writer and feminist, of Irish descent. She was associated with a radical circle known as the 'English Jacobins', whose members included Thomas Paine and William Godwin. In 1790 she published *A Vindication of the Rights of Man* in reply to Edmund Burke's *Reflections on the Revolution in France*. Her best-known work, *A Vindication of the Rights of Woman* (1792), defied Jean-Jacques Rousseau's assumptions about male supremacy and championed educational equality for women. In 1797 she married Godwin and died shortly after giving birth to their daughter Mary Shelley.

Wolof /ˈwəʊlɒf/ *n. & adj.* ● *n.* (*pl.* same or **Wolofs**) **1** a member of a people of Senegal and the Gambia. **2** the Niger-Congo language of this people. ● *adj.* of or relating to the Wolof or their language. [Wolof]

Wolsey /ˈwʊlzɪ/, Thomas (known as Cardinal Wolsey) (c.1474–1530), English prelate and statesman. Favoured by Henry VIII, he dominated foreign and domestic policy in the early part of Henry's reign and served to foster the development of royal absolutism in politics and Church affairs; he held positions as Archbishop of York (1514–30), cardinal (1515–30), and Lord Chancellor (1515–29). His main interest was foreign politics, in which he sought to increase England's influence in European affairs by holding the balance of power between the Holy Roman Empire and France. Wolsey incurred royal displeasure through his failure to secure the papal dispensation necessary for Henry's divorce from Catherine of Aragon; he was arrested on a charge of treason and died on his way to trial in London.

Wolverhampton /ˌwʊlvəˈhæmptən/ an industrial city in the West Midlands, north-west of Birmingham; pop. (1991) 239,800.

wolverine /ˈwʊlvəˌriːn/ *n.* (also **wolverene**) a heavily built flesh-eating animal, *Gulo gulo*, the largest member of the weasel family, found in the cold coniferous forests of northern Eurasia and North America. Its thick fur sheds moisture and is prized for trimming Eskimo clothing. Also called *glutton*. [16th-c. *wolvering*, somehow derived f. *wolv-*, stem of WOLF]

wolves *pl.* of WOLF.

woman /ˈwʊmən/ *n.* (*pl.* **women** /ˈwɪmɪn/) **1** an adult human female. **2** the female sex; any or an average woman (*how does woman differ from man?*). **3** *colloq.* a wife or female sexual partner. **4** (*prec. by the*) emotions or characteristics traditionally associated with women (*brought out the woman in him*). **5** a man with characteristics traditionally associated with women. **6** (*attrib.*) female (*woman driver; women friends*). **7** (as second element in *comb.*) a woman of a specified nationality, profession, skill, etc. (*Englishwoman; horsewoman*). **8** *colloq.* a female domestic help. **9** *archaic* or *hist.* a queen's etc. female attendant ranking below lady (*woman of the bedchamber*). □ **woman of the streets** a prostitute. **women's lib** = *women's liberation*. **women's libber** *colloq.* a supporter of women's liberation. **women's liberation 1** the liberation of women from inequalities and subservient status in relation to men, and from attitudes causing these. **2** (also **women's liberation movement**) = *women's movement*. **women's movement** a broad movement campaigning for women's liberation and rights. **women's rights** rights that promote a position of legal and social equality of women with men. □ **womanless** *adj.* **womanlike** *adj.* [OE *wīfmon*, -*men* (as WIFE, MAN), a formation peculiar to English, the ancient word being WIFE]

womanhood /ˈwʊmənˌhʊd/ *n.* **1** female maturity. **2** womanly instinct. **3** womankind.

womanish /ˈwʊmənɪʃ/ *adj.* usu. *derog.* **1** (of a man) effeminate, unmanly. **2** suitable to or characteristic of a woman. □ **womanishly** *adv.* **womanishness** *n.*

womanize /ˈwʊməˌnaɪz/ *v.* (also **-ise**) **1** *intr.* chase after women; philander. **2** *tr.* make womanish. □ **womanizer** *n.*

womankind /ˈwʊmənˌkaɪnd/ *n.* (also **womenkind** /ˈwɪmɪn-/) women in general.

womanly /ˈwʊmənlɪ/ *adj.* (of a woman) having or showing qualities traditionally associated with women; not masculine or girlish. □ **womanliness** *n.*

womb /wuːm/ *n.* **1** the organ of conception and gestation in a woman and other female mammals; the uterus. **2** a place of origination and development. □ **womblike** *adj.* [OE *wamb, womb*]

wombat /ˈwɒmbæt/ *n.* a burrowing plant-eating Australian marsupial of the family Vombatidae, with a thick heavy bearlike body and short legs. [Dharuk]

women *pl.* of WOMAN.

womenfolk /ˈwɪmɪnˌfəʊk/ *n.* **1** women in general. **2** the women in a family.

womenkind var. of WOMANKIND.

Women's Institute /ˈwɪmɪnz/ (abbr. **WI**) an organization established to enable women in rural areas to meet regularly and engage in crafts, cultural activities, social work, etc. Now worldwide, it was first set up in Ontario, Canada, in 1895, and in Britain in 1915.

won[1] /wɒn/ *n.* (*pl.* same) the basic monetary unit of North and South Korea, equal to 100 jun in North Korea and 100 jeon in South Korea. [Korean]

won[2] past and past part. of WIN.

Wonder /ˈwʌndə(r)/, Stevie (born Steveland Judkins Morris) (b.1950), American singer, songwriter, and musician. He was blind from birth, but his musical gifts were recognized at an early age and he became a recording artist with Motown in 1961. Although at first a soul singer, from the 1970s his repertoire has broadened to include rock, funk, and romantic ballads. Among his albums are *Innervisions* (1973) and *Songs in the Key of Life* (1976).

wonder /ˈwʌndə(r)/ n. & v. ● n. **1** an emotion excited by what is unexpected, unfamiliar, or inexplicable, esp. surprise mingled with admiration or curiosity etc. **2** a strange or remarkable person or thing, specimen, event, etc. **3** (attrib.) having marvellous or amazing properties etc. (a wonder drug). **4** a surprising thing (it is a wonder you were not hurt). ● v. **1** intr. (often foll. by at, or to + infin.) be filled with wonder or great surprise. **2** tr. (foll. by that + clause) be surprised to find. **3** tr. desire or be curious to know (I wonder what the time is). **4** tr. expressing a tentative enquiry (I wonder whether you would mind?). **5** intr. (foll. by about) ask oneself with puzzlement or doubt about; question (wondered about the sense of the decision). □ **I shouldn't wonder** colloq. I think it likely. **I wonder** I very much doubt it. **no** (or **small**) **wonder** (often foll. by that + clause) one cannot be surprised; one might have guessed; it is natural. **wonder-struck** (or **-stricken**) reduced to silence by wonder. **wonders will never cease** an exclamation of extreme (usu. agreeable) surprise. **wonder-worker** a person who performs wonders. **work** (or **do**) **wonders 1** do miracles. **2** succeed remarkably. □ **wonderer** n. [OE wundor, or to + infin.]

wonderful /ˈwʌndəˌfʊl/ adj. **1** very remarkable or admirable. **2** arousing wonder. □ **wonderfully** adv. **wonderfulness** n. [OE wunderfull (as WONDER, -FUL)]

wondering /ˈwʌndərɪŋ/ adj. filled with wonder; marvelling (their wondering gaze). □ **wonderingly** adv.

wonderland /ˈwʌndəˌlænd/ n. **1** a fairyland. **2** a land of surprises or marvels.

wonderment /ˈwʌndəmənt/ n. surprise, awe.

wondrous /ˈwʌndrəs/ adj. & adv. poet. ● adj. wonderful. ● adv. wonderfully (wondrous kind). □ **wondrously** adv. **wondrousness** n. [alt. of obs. wonders (adj. & adv.), = genitive of WONDER (cf. -s³) after marvellous]

wonky /ˈwɒŋkɪ/ adj. (**wonkier**, **wonkiest**) Brit. colloq. **1** crooked, off-centre, askew. **2** loose, unsteady. **3** unreliable. □ **wonkily** adv. **wonkiness** n. [fanciful formation]

wont /wəʊnt/ adj., n., & v. ● predic.adj. (foll. by to + infin.) archaic or literary accustomed (as we were wont to say). ● n. formal or joc. what is customary, one's habit (as is my wont). ● v.tr. & intr. (3rd sing. present **wonts** or **wont**; past **wont** or **wonted**) archaic make or become accustomed. [OE gewunod past part. of gewunian f. wunian dwell]

won't /wəʊnt/ contr. will not.

wonted /ˈwəʊntɪd/ attrib.adj. habitual, accustomed, usual.

wonton /wɒnˈtɒn/ n. a Chinese snack of thin dough wrapped around a savoury filling and deep-fried or boiled. [Cantonese wān t'ān]

woo /wuː/ v.tr. (**woos**, **wooed**) **1** court; seek the hand or love of (a woman). **2** try to win (fame, fortune, etc.). **3** seek the favour or support of. **4** coax or importune. □ **wooable** adj. **wooer** n. [OE wōgian (intr.), āwōgian (tr.), of unkn. orig.]

Wood /wʊd/, Sir Henry (Joseph) (1869–1944), English conductor. In 1895 he instituted the first of the Promenade Concerts at the Queen's Hall in London, and conducted these every year until he died (see also PROMENADE CONCERT). During this time he introduced music by composers such as Schoenberg, Janáček, and Scriabin to British audiences. He made many orchestral transcriptions and arranged the Fantasia on British Sea Songs (including 'Rule, Britannia'), which became a regular feature of the last night of each year's promenade concert season.

wood /wʊd/ n. **1 a** a hard fibrous material that forms the main substance of the trunk or branches of a tree or shrub. **b** this cut for timber or for fuel, or for use in crafts, manufacture, etc. **2** (in sing. or pl.) growing trees densely occupying a tract of land. **3** (prec. by the) wooden storage, esp. a cask, for wine etc. (poured straight from the wood). **4** a wooden-headed golf club. **5** = BOWL² n. 1. □ **not see the wood for the trees** fail to grasp the main issue from over-attention to details. **out of the wood** (or **woods**) out of danger or difficulty. **wood alcohol** methanol. **wood anemone** a wild spring-flowering anemone, Anemone nemorosa. **wood-engraver** a maker of wood-engravings. **wood-engraving 1** a relief cut on a block of wood sawn across the grain. **2** a print made from this. **3** the technique of making such reliefs and prints. **wood-fibre** fibre obtained from wood esp. as material for paper. **wood hyacinth** = BLUEBELL 1. **wood mouse** a Eurasian mouse, Apodemus sylvaticus, with large eyes and ears, and a long tail (also called field mouse). **wood nymph** Mythol. a dryad or hamadryad. **wood pulp** wood-fibre reduced chemically or mechanically to pulp as raw material for paper. **wood rat** a rat of the North American genus Neotoma, esp. the pack rat. **wood-screw** a metal male screw with a slotted head and sharp point. **wood sorrel**

a small woodland plant, Oxalis acetosella, with trifoliate leaves and white flowers streaked with purple. **wood spirit** crude methanol obtained from wood. **wood warbler 1** a small yellow-breasted Eurasian woodland warbler, Phylloscopus sibilatrix, with a trilling song. **2** an American warbler of the family Parulidae. **wood wool** fine pine etc. shavings used as a surgical dressing or for packing. □ **woodless** adj. [OE wudu, wi(o)du f. Gmc]

woodbind /ˈwʊdbaɪnd/ n. = WOODBINE.

woodbine /ˈwʊdbaɪn/ n. **1** wild honeysuckle. **2** US Virginia creeper.

woodblock /ˈwʊdblɒk/ n. a block from which woodcuts are made.

woodcarver /ˈwʊdˌkɑːvə(r)/ n. **1** a person who carves designs in relief on wood. **2** a tool for carving wood.

woodcarving /ˈwʊdˌkɑːvɪŋ/ n. **1** (also attrib.) the act or process of carving wood. **2** the art or skill of a woodcarver. **3** a design in wood produced by this art.

woodchat /ˈwʊdtʃæt/ n. (in full **woodchat shrike**) a shrike, Lanius senator, of southern Europe, North Africa, and the Middle East, having black and white plumage with a chestnut head. [WOOD + CHAT²]

woodchuck /ˈwʊdtʃʌk/ n. a North American marmot, Marmota monax, with reddish-brown and grey fur. Also called groundhog. [prob. Algonquian: cf. Cree wuchak, otchock]

woodcock /ˈwʊdkɒk/ n. (pl. same) a migratory long-billed woodland bird of the genus Scolopax, with cryptic brown plumage, related to the snipe.

woodcraft /ˈwʊdkrɑːft/ n. esp. N. Amer. **1** skill in woodwork. **2** knowledge of woodland esp. in camping, scouting, etc.

woodcut /ˈwʊdkʌt/ n. **1** a relief cut on a block of wood sawn along the grain. **2** a print made from this, esp. as an illustration in a book. **3** the technique of making such reliefs and prints.

woodcutter /ˈwʊdˌkʌtə(r)/ n. **1** a person who cuts wood. **2** a maker of woodcuts.

wooded /ˈwʊdɪd/ adj. having woods or many trees.

wooden /ˈwʊd(ə)n/ adj. **1** made of wood. **2** like wood. **3 a** stiff, clumsy, or stilted; without animation or flexibility (wooden movements; a wooden performance). **b** expressionless (a wooden stare). □ **wooden-head** colloq. a stupid person. **wooden-headed** colloq. stupid. **wooden-headedness** colloq. stupidity. **wooden horse** = TROJAN HORSE. **wooden spoon** a booby prize (orig. a spoon given to the candidate coming last in the Cambridge mathematical tripos). □ **woodenly** adv. **woodenness** /-d(ə)nnɪs/ n.

woodgrouse /ˈwʊdgraʊs/ n. = CAPERCAILLIE.

woodland /ˈwʊdlənd/ n. wooded country, woods (often attrib.: woodland scenery). □ **woodlander** n.

woodlark /ˈwʊdlɑːk/ n. a small lark, Lullula arborea, with a melodious song.

woodlouse /ˈwʊdlaʊs/ n. (pl. **-lice** /-laɪs/) a small terrestrial isopod crustacean of the suborder Oniscoidea, living in damp shady places.

woodman /ˈwʊdmən/ n. (pl. **-men**) **1** a forester. **2** a woodcutter.

woodmouse /ˈwʊdmaʊs/ n. (pl. **-mice** /-maɪs/) a fieldmouse.

woodnote /ˈwʊdnəʊt/ n. (often in pl.) a natural or spontaneous note of a bird etc.

woodpecker /ˈwʊdˌpekə(r)/ n. a bird of the family Picidae that taps or pecks holes in tree-trunks in search of insects.

woodpigeon /ˈwʊdˌpɪdʒɪn/ n. a large grey pigeon, Columba palumbus, having white patches that form a partial ring round its neck. Also called ring-dove.

woodpile /ˈwʊdpaɪl/ n. a pile of wood, esp. for fuel.

woodruff /ˈwʊdrʌf/ n. a low-growing woodland plant of the genus Galium; esp. G. odoratum, with whorled leaves and small white flowers, smelling of hay when dried.

woodrush /ˈwʊdrʌʃ/ n. a grasslike herbaceous plant of the genus Luzula, with long flat leaves.

woodshed /ˈwʊdʃed/ n. a shed where wood for fuel is stored. □ **something nasty in the woodshed** colloq. a shocking or distasteful thing kept secret.

woodsman /ˈwʊdzmən/ n. (pl. **-men**) **1** a person who lives in or is familiar with woodland. **2** a person skilled in woodcraft.

Woodstock /ˈwʊdstɒk/ a small town in New York State, situated in the south-east near Albany. Noted as an artists' colony, it gave its name in the summer of 1969 to a huge rock festival held some 96 km (60 miles) to the south-west, which came to symbolize the youth culture of the period.

woodsy /ˈwʊdzɪ/ adj. N. Amer. like or characteristic of woods. [irreg. f. WOOD + -Y¹]

Woodward /ˈwʊdwəd/, Robert Burns (1917–79), American organic chemist. He was the first to synthesize a wide range of complex organic compounds, including quinine, cholesterol, cortisone, strychnine, chlorophyll, and vitamin B₁₂. In 1965, with the Polish-born American chemist Roald Hoffmann (b.1937), he devised the symmetry-based rules which govern the course of concerted rearrangement reactions involving cyclic intermediates. He was awarded the Nobel Prize for chemistry in 1965.

woodwasp /ˈwʊdwɒsp/ n. a large sawfly of the family Siricidae, with a long ovipositor and wood-boring larvae; esp. the very large *Urocerus gigas*, that resembles a hornet and is a pest of conifers.

woodwind /ˈwʊdwɪnd/ n. (often attrib.) **1** (collect.) the wind instruments of the orchestra that were (mostly) originally made of wood, including the flute, clarinet, oboe, and bassoon. **2** (usu. in pl.) an individual instrument of this kind or its player (*the woodwinds are out of tune*).

woodwork /ˈwʊdwɜːk/ n. **1** the making of things in wood. **2** things made of wood, esp. the wooden parts of a building. □ **crawl** (or **come**) **out of the woodwork** colloq. (of something unwelcome) appear; become known. □ **woodworker** n. **woodworking** n.

woodworm /ˈwʊdwɜːm/ n. **1** the wood-boring larva of the furniture beetle. **2** the damaged condition of wood affected by this.

woody /ˈwʊdɪ/ adj. (**woodier**, **woodiest**) **1** (of a region) wooded; abounding in woods. **2** like or of wood (*woody tissue*). □ **woody nightshade** see NIGHTSHADE. □ **woodiness** n.

woodyard /ˈwʊdjɑːd/ n. a yard where wood is used or stored.

woof¹ /wʊf/ n. & v. ● n. the gruff bark of a dog. ● v.intr. give a woof. [imit.]

woof² /wuːf/ n. = WEFT¹. [OE ōwef, alt. of ōwebb (after wefan WEAVE¹), formed as A-², WEB: infl. by warp]

woofer /ˈwuːfə(r)/ n. a loudspeaker designed to reproduce low frequencies (cf. TWEETER). [WOOF¹ + -ER¹]

wool /wʊl/ n. **1** fine soft wavy hair from the fleece of sheep, goats, etc. **2 a** yarn produced from this hair. **b** cloth or clothing made from it. **3** any of various wool-like substances (*steel wool*). **4** soft short under-fur or down. **5** colloq. a person's hair, esp. when short and curly. □ **pull the wool over a person's eyes** deceive a person. **wool-fat** lanolin. **wool-fell** Brit. the skin of a sheep etc. with the fleece still on. **wool-gathering** absent-mindedness; dreamy inattention. **wool-grower** a breeder of sheep for wool. **wool-oil** suint. **wool-pack 1** a fleecy cumulus cloud. **2** hist. a bale of wool. **wool-skin** = wool-fell. **wool-sorters' disease** anthrax. **wool-stapler** a person who grades wool. □ **wool-like** adj. [OE wull f. Gmc]

Woolf /wʊlf/, (Adeline) Virginia (née Stephen) (1882–1941), English novelist, essayist, and critic. From 1904 her family's London house became the centre of the Bloomsbury Group, among whose members was Leonard Woolf (1880–1969), whom she married in 1912. She and her husband founded the Hogarth Press in 1917. She gained recognition with her third novel, *Jacob's Room* (1922); subsequent novels, such as *Mrs Dalloway* (1925), *To the Lighthouse* (1927), and *The Waves* (1931), characterized by their stream-of-consciousness technique and poetic impressionism, established her as a principal exponent of modernism. Her non-fiction includes *A Room of One's Own* (1929), a major work of the women's movement, and collections of essays and letters. She suffered from acute depression throughout her life, and drowned herself shortly after completing her final and most experimental novel, *Between the Acts* (published posthumously in 1941).

woollen /ˈwʊlən/ adj. & n. (US **woolen**) ● adj. made wholly or partly of wool, esp. from short fibres. ● n. **1** a fabric produced from wool. **2** (in pl.) woollen garments. [OE wullen (as WOOL, -EN²)]

Woolley /ˈwʊlɪ/, Sir (Charles) Leonard (1880–1960), English archaeologist. Between 1922 and 1934 he was director of a joint British–American archaeological expedition to excavate the Sumerian city of Ur (in what is now southern Iraq). His discoveries included rich royal tombs and thousands of clay tablets providing valuable information on everyday life of the period.

woolly /ˈwʊlɪ/ adj. & n. ● adj. (**woollier**, **woolliest**) **1** bearing or naturally covered with wool or wool-like hair; downy. **2** resembling or suggesting wool (*woolly clouds*). **3** made of wool, woollen. **4** (of a sound) indistinct. **5** (of thought) vague or confused. **6** lacking in definition, luminosity, or incisiveness. ● n. (pl. **-ies**) colloq. a woollen garment, esp. a knitted pullover. □ **woolly-bear 1** a large hairy caterpillar, esp. of the tiger moth. **2** the small hairy larva of a carpet beetle, which is

destructive to carpets, textiles, insect collections, etc. **woolly mammoth** a mammoth, *Mammuthus primigenius*, with a coat of long woolly hair, sometimes found preserved in the Siberian permafrost. □ **woolliness** n.

Woolsack /ˈwʊlsæk/ (in the UK) the usual seat of the Lord Chancellor in the House of Lords, made of a large square bag of wool and covered with cloth. It is said to have been adopted in Edward III's reign as a reminder to the Lords of the importance to England of the wool trade. The term is also used to denote the office or function of Lord Chancellor.

woolshed /ˈwʊlʃed/ n. Austral. & NZ a large shed for shearing and baling wool.

Woolworth /ˈwʊlwəθ/, Frank Winfield (1852–1919), American businessman. He opened his first shop selling low-priced goods in 1879. He gradually built up a large chain of US stores selling a wide variety of items, and the business later became an international retail organization.

Woomera /ˈwuːmərə/ a town in central South Australia, the site of a vast military testing ground used in the 1950s for nuclear tests and since the 1960s for tracking space satellites.

woomera /ˈwuːmərə/ n. Austral. **1** an Aboriginal stick for throwing a dart or spear more forcibly. **2** a club used as a missile. [Dharuk wamara]

woop woop /ˈwuːp wuːp/ n. Austral. & NZ sl. **1** a jocular name for a remote outback town or district. **2** (**Woop Woop**) an imaginary remote place. [mock Aboriginal]

woosh var. of WHOOSH.

woozy /ˈwuːzɪ/ adj. (**woozier**, **wooziest**) colloq. **1** dizzy or unsteady. **2** dazed or slightly drunk. **3** vague. □ **woozily** adv. **wooziness** n. [19th c.: orig. unkn.]

wop /wɒp/ n. sl. offens. an Italian or other southern European. [20th c.: orig. uncert.: perh. f. It. guappo bold, showy, f. Sp. guapo dandy]

Worcester¹ /ˈwʊstə(r)/ a cathedral city in western England, on the River Severn, the administrative centre of Hereford and Worcester; pop. (1991) 81,000. It was the scene in 1651, during the English Civil War, of a battle in which Oliver Cromwell defeated a Scottish army under Charles II. It has been a centre of porcelain manufacture since 1751.

Worcester² /ˈwʊstə(r)/ n. (in full **Royal Worcester**) porcelain made at Worcester in a factory founded in 1751. The porcelain (largely tableware) at first showed strong influence from Chinese and Dresden designs before being produced in a wider variety of designs.

Worcester sauce n. (also **Worcestershire sauce**) a pungent sauce containing soy, vinegar, and seasoning, first made at Worcester.

Worcestershire /ˈwʊstəˌʃɪə(r)/ a former county of west central England. It became part of Hereford and Worcester in 1974.

Worcs. abbr. Worcestershire.

word /wɜːd/ n. & v. ● n. **1** a sound or combination of sounds forming a meaningful element of speech, usu. shown with a space on either side of it when written or printed, used as part (or sometimes as the whole) of a sentence. **2** speech, esp. as distinct from action (*bold in word only*). **3** one's promise or assurance (*gave us their word*). **4** (in sing. or pl.) a thing said, a remark or conversation. **5** (in pl.) the text of a song or an actor's part. **6** (in pl.) angry talk (*they had words*). **7** news, intelligence; a message. **8** a command, password, or motto (*gave the word to begin*). **9** Computing a basic unit of the expression of data in a computer. ● v.tr. put into words; select words to express (*how shall we word that?*). □ **at a word** as soon as requested. **be as good as** (or **better than**) **one's word** fulfil (or exceed) what one has promised. **break one's word** fail to do what one has promised. **have no words for** be unable to express. **have a word** (often foll. by with) speak briefly (to). **in other words** expressing the same thing differently. **in so many words** explicitly or bluntly. **in a** (or **one**) **word** briefly. **keep one's word** do what one has promised. **my** (or **upon my**) **word** an exclamation of surprise or consternation. **not the word for it** not an adequate or appropriate description. **of few words** taciturn. **of one's word** reliable in keeping promises (*a woman of her word*). **on** (or **upon**) **my word** a form of asseveration. **put into words** express in speech or writing. **take a person at his** or **her word** interpret a person's words literally or exactly. **take a person's word for it** believe a person's statement without investigation etc. **too ... for words** too ... to be adequately described (*was too funny for words*). **waste words** talk in vain. **the Word** (or **Word of God**) **1** the Bible or a part of it. **2** = LOGOS. **word-blind** affected by word-blindness. **word-blindness** the inability to read, esp. alexia, in which the ability to write and speak is unimpaired.

word-class Linguistics a category of words of similar form or function, esp. a part of speech. **word-deaf** incapable of identifying spoken words owing to a brain defect. **word-deafness** this condition. **word for word** in exactly the same or (of translation) corresponding words. **word-game** a game involving the making or selection etc. of words. **word of honour** an assurance given upon one's honour. **word of mouth** speech (only). **word-of-mouth** verbal, unwritten. **word order** the sequence of words in a sentence, esp. affecting meaning etc. **word-painting** a vivid description in writing. **word-perfect** knowing one's part etc. by heart. **word-picture** a piece of word-painting. **word processor** a purpose-built computer system for electronically storing text entered from a keyboard, incorporating corrections, and providing a printout. **words fail me** an expression of disbelief, dismay, etc. **word-square** a set of words of equal length written one under another to read the same down as across (e.g. *too old ode*). **a word to the wise** = VERB. SAP. **word wrap** Computing in word-processing, the automatic shifting of a word too long to fit on a line to the beginning of the next line. □ **wordage** n. **wordless** adj. **wordlessly** adv. **wordlessness** n. [OE f. Gmc]

wordbook /'wɜːdbʊk/ n. a book with lists of words; a vocabulary or dictionary.

wording /'wɜːdɪŋ/ n. **1** a form of words used. **2** the way in which something is expressed.

wordplay /'wɜːdpleɪ/ n. use of words to witty effect, esp. by punning.

wordsearch /'wɜːdsɜːtʃ/ n. a grid-shaped puzzle of letters in columns, containing several hidden words written in any direction.

wordsmith /'wɜːdsmɪθ/ n. a skilled user or maker of words.

Wordsworth /'wɜːdzwəθ/, William (1770–1850), English poet. He was born in the Lake District and much of his work was inspired by the landscape of this region. He spent some time in France (1790–1) and became an enthusiastic supporter of the French Revolution, although he later became disillusioned by the excesses of the Terror and in later life assumed a more conservative stance. From 1795 to 1799 he lived in Somerset, where, with Coleridge, he composed the *Lyrical Ballads* (1798), a landmark in the history of English romantic poetry, containing in particular his poem 'Tintern Abbey'. In 1799 he returned to the Lake District, settling in Grasmere with his sister Dorothy (1771–1855); his wife joined them after their marriage in 1802. Among his many poems are the ode 'Intimations of Immortality' (1807), sonnets, such as 'Surprised by Joy' and 'I Wandered Lonely as a Cloud' (both 1815), and the posthumously published autobiographical *The Prelude* (1850). He was appointed Poet Laureate in 1843.

wordy /'wɜːdɪ/ adj. (**wordier, wordiest**) **1** using or expressed in many or too many words; verbose. **2** consisting of words. □ **wordily** adv. **wordiness** n. [OE *wordig* (as WORD, -Y¹)]

wore¹ past of WEAR¹.

wore² past and past part. of WEAR².

work /wɜːk/ n. & v. ● n. **1** the application of mental or physical effort to a purpose; the use of energy. **2 a** a task to be undertaken. **b** the materials for this. **c** (prec. by *the*; foll. by *of*) a task occupying (no more than) a specified time (*the work of a moment*). **3** a thing done or made by work; the result of an action; an achievement; a thing made. **4** a person's employment or occupation etc., esp. as a means of earning income (*looked for work*; *is out of work*). **5 a** a literary or musical composition. **b** (in pl.) all such by an author or composer etc. **6** actions or experiences of a specified kind (*good work!*; *this is thirsty work*). **7 a** (in comb.) things or parts made of a specified material or with specified tools etc. (*ironwork*; *needlework*). **b** archaic needlework. **8** (in pl.) the operative part of a clock or machine. **9** Physics the exertion of force overcoming resistance or producing molecular change (*convert heat into work*). **10** (in pl.) colloq. all that is available; everything needed. **11** (in pl.) operations of building or repair (*roadworks*). **12** (in pl.; often treated as sing.) a place where manufacture is carried on. **13** (usu. in pl.) (in Christian theology) a meritorious act. **14** (usu. in pl. or in comb.) a defensive structure (*earthworks*). **15** (in comb.) **a** ornamentation of a specified kind (*poker-work*). **b** articles having this. ● v. (past and past part. **worked** or (esp. as adj.) **wrought**) **1** intr. (often foll. by *at, on*) do work; be engaged in bodily or mental activity. **2** intr. **a** be employed in certain work (*works in industry*; *works as a secretary*). **b** (foll. by *with*) be the workmate of (a person). **3** intr. (often foll. by *for*) make efforts; conduct a campaign (*works for peace*). **4** intr. (foll. by *in*) be a craftsman (in a material). **5** intr. operate or function, esp. effectively (*how does this machine work?*; *your idea will not work*). **6** intr. (of a part of a machine) run, revolve; go through regular motions. **7** tr. carry on, manage, or control (*cannot work the machine*). **8** tr. **a** put or keep in operation or at work; cause to toil

(*this mine is no longer worked*; *works the staff very hard*). **b** cultivate (land). **9** tr. **a** bring about; produce as a result (*worked miracles*). **b** colloq. arrange (matters) (*worked it so that we could go*; *can you work things for us?*). **10** tr. knead, hammer; bring to a desired shape or consistency. **11** tr. do, or make by, needlework etc. **12** tr. & intr. (cause to) progress or penetrate, or make (one's way), gradually or with difficulty in a specified way (*worked our way through the crowd*; *worked the peg into the hole*). **13** intr. (foll. by *loose* etc.) gradually become (loose etc.) by constant movement. **14** tr. artificially excite (*worked themselves into a rage*). **15** tr. solve (a sum) by mathematics. **16** tr. **a** purchase with one's labour instead of money (*work one's passage*). **b** obtain by labour the money for (one's way through university etc.). **17** intr. (foll. by *on, upon*) have influence. **18** intr. be in motion or agitated; cause agitation, ferment (*his features worked violently*; *the yeast began to work*). **19** intr. Naut. sail against the wind. □ **at work** in action or engaged in work. **get worked up** become angry, excited, or tense. **give a person the works** colloq. give or tell a person everything. **2** colloq. treat a person harshly. **3** sl. kill a person. **have one's work cut out** be faced with a hard task. **in the works** colloq. in progress, in the pipeline. **in work 1** having a job. **2** US = **in the works**. **out of work** unemployed. **set to work** begin or cause to begin operations. **work away** (or **on**) continue to work. **work-basket** (or **-bag** etc.) a basket or bag etc. containing sewing materials. **work camp** a camp at which community work is done, esp. by young volunteers. **work ethic** see *Protestant ethic*. **work one's fingers to the bone** see BONE. **work group 1** a group within a workforce. **2** a small group of usu. trainees, formed to practise or develop a skill etc. **work in** find a place for. **work it** colloq. bring it about; achieve a desired result. **work of art** a fine picture, poem, or building etc. **work off** get rid of by work or activity. **work out 1 a** solve (a sum) or find out (an amount) by calculation. **b** solve or understand (a problem, person, etc.). **2** (foll. by *at*) be calculated (*the total works out at 230*). **3** give a definite result (*this sum will not work out*). **4** have a specified or satisfactory result (*the plan worked out well*; *glad the arrangement worked out*). **5** provide for the details of (*has worked out a scheme*). **6** accomplish or attain with difficulty (*work out one's salvation*). **7** exhaust with work (*the mine is worked out*). **8** engage in physical exercise or training. **work over 1** examine thoroughly. **2** colloq. treat with violence. **work rate 1** the rate at which work is done or produced. **2** Sport the rate at which energy is expended. **works council** esp. Brit. a group of employees representing those employed in a works etc. in discussions with their employers. **work-shy** disinclined to work. **works of supererogation** see SUPEREROGATION. **work study** a system of assessing methods of working so as to achieve the maximum output and efficiency. **work table** a table for working at, esp. with a sewing machine. **work to rule** (esp. as a form of industrial action) follow official working rules exactly in order to reduce output and efficiency. **work-to-rule** n. the act or an instance of working to rule. **work up 1** bring gradually to an efficient state. **2** (foll. by *to*) advance gradually to a climax. **3** elaborate or excite by degrees; bring to a state of agitation. **4** mingle (ingredients) into a whole. **5** learn (a subject) by study. **work one's will** (foll. by *on, upon*) archaic accomplish one's purpose on (a person or thing). **work wonders** see WONDER. □ **workless** adj. [OE *weorc* etc. f. Gmc]

workable /'wɜːkəb(ə)l/ adj. **1** that can be worked or will work. **2** that is worth working; practicable, feasible (*a workable quarry*; *a workable scheme*). □ **workably** adv. **workableness** n. **workability** /ˌwɜːkə'bɪlɪtɪ/ n.

workaday /'wɜːkədeɪ/ adj. **1** ordinary, everyday, practical. **2** fit for, used, or seen on workdays.

workaholic /ˌwɜːkə'hɒlɪk/ n. & adj. colloq. ● n. a person addicted to working. ● adj. addicted to working.

workbench /'wɜːkbentʃ/ n. a bench for doing mechanical or practical work, esp. carpentry.

workbox /'wɜːkbɒks/ n. a box for holding tools, materials for sewing, etc.

workday /'wɜːkdeɪ/ n. esp. US a day on which work is usually done.

worker /'wɜːkə(r)/ n. **1** a person who works, esp. a manual or industrial employee. **2** a neuter or undeveloped female of various social insects, esp. a bee or ant, that does the basic work of its colony. **3** a person who works hard. □ **worker priest** a French Roman Catholic or an Anglican priest who engages part-time in secular work.

workforce /'wɜːkfɔːs/ n. **1** the workers engaged or available in an industry etc. **2** the number of such workers.

workhorse /'wɜːkhɔːs/ n. a horse, person, or machine that performs hard work.

workhouse /'wɜːkhaʊs/ n. **1** Brit. hist. a public institution where people of a parish unable to support themselves were housed and (if able-bodied) made to work. (See note below.) **2** US a house of correction for petty offenders.

▪ Workhouses were made the responsibility of the parish from 1601. Groups of parishes combined to build workhouses for the poor, the aged, the disabled, and for orphaned children from about 1815. Under the Poor Law Amendment Act of 1834 the able-bodied poor could obtain assistance only by staying at workhouses, where conditions were made as uncomfortable as possible and families were split up with a view to discouraging all but the really needy. Conditions gradually improved in the face of public protest; following legislation passed in the early 20th century (particularly by the Liberal government 1906–14) they finally disappeared c.1930.

working /'wɜːkɪŋ/ adj. & n. ● adj. **1 a** engaged in work, esp. in manual or industrial labour. **b** spent in work or employment (all his working life). **2** functioning or able to function. ● n. **1** the activity of work. **2** the act or manner of functioning of a thing. **3 a** a mine or quarry. **b** the part of this in which work is being or has been done (a disused working). □ **working capital** the capital actually used in a business. **working class** the class of people who are employed for wages, esp. in manual or industrial work. **working-class** adj. of the working class. **working day** Brit. **1** a workday. **2** the part of the day devoted to work. **working drawing** a drawing to scale, serving as a guide for construction or manufacture. **working girl** US sl. a prostitute. **working hours** hours normally devoted to work. **working hypothesis** a hypothesis used as a basis for action. **working knowledge** knowledge adequate to work with. **working lunch** etc. a meal at which business is conducted. **working man** a man employed esp. in a manual or industrial job. **working order** the condition in which a machine works (satisfactorily or as specified). **working-out 1** the calculation of results. **2** the elaboration of details. **working party** a group of people appointed to study a particular problem or advise on some question. **working woman** a woman employed esp. in a manual or industrial job.

Workington /'wɜːkɪŋtən/ a port on the coast of Cumbria, NW England; pop. (1981) 26,120.

workload /'wɜːkləʊd/ n. the amount of work to be done by an individual etc.

workman /'wɜːkmən/ n. (pl. **-men**) **1** a man employed to do manual labour. **2** a person considered with regard to skill in a job (a good workman).

workmanlike /'wɜːkmən,laɪk/ adj. characteristic of a good workman; showing practised skill.

workmanship /'wɜːkmən,ʃɪp/ n. **1** the degree of skill in doing a task or of finish in the product made. **2** a thing made or created by a specified person etc.

workmate /'wɜːkmeɪt/ n. a person who works with another; a colleague.

workout /'wɜːkaʊt/ n. a session of physical exercise or training.

workpeople /'wɜːk,piːp(ə)l/ n.pl. people in paid employment.

workpiece /'wɜːkpiːs/ n. a thing worked on with a tool or machine.

workplace /'wɜːkpleɪs/ n. a place at which a person works; an office, factory, etc.

workroom /'wɜːkruːm, -rʊm/ n. a room for working in, esp. one equipped for a certain kind of work.

worksheet /'wɜːkʃiːt/ n. **1** a paper for recording work done or in progress. **2** a paper listing questions or activities for students etc. to work through.

workshop /'wɜːkʃɒp/ n. **1** a room or building in which goods are manufactured. **2 a** a meeting for concerted discussion or activity (a dance workshop). **b** the members of such a meeting.

workspace /'wɜːkspeɪs/ n. **1** space in which to work. **2** an area rented or sold for commercial purposes. **3** Computing a memory storage facility for temporary use.

workstation /'wɜːk,steɪʃ(ə)n/ n. **1** the location of a stage in a manufacturing process. **2** a computer terminal or the desk etc. where this is located.

worktop /'wɜːktɒp/ n. a flat surface for working on, esp. in a kitchen.

workwoman /'wɜːk,wʊmən/ n. (pl. **-women**) a female worker or operative.

world /wɜːld/ n. **1 a** the earth, or a planetary body like it. **b** its countries and their inhabitants. **c** all people; the earth as known or in some particular respect. **2 a** the universe or all that exists; everything. **b** everything that exists outside oneself (dead to the world). **3 a** the time, state, or scene of human existence. **b** (prec. by the, this) mortal life. **4** secular interests and affairs. **5** human affairs; their course and conditions; active life (how goes the world with you?). **6** average, respectable, or fashionable people or their customs or opinions. **7** all that concerns or all who belong to a specified class, time, domain, or sphere of activity (the medieval world; the world of sport). **8** (foll. by of) a vast amount (that makes a world of difference). **9** (attrib.) affecting many nations, of all nations (world politics; a world champion). □ **all the world and his wife 1** any large mixed gathering of people. **2** all with pretensions to fashion. **be worlds apart** differ greatly, esp. in nature or opinion. **bring into the world** give birth to or attend at the birth of. **carry the world before one** have rapid and complete success. **come into the world** be born. **for all the world** (foll. by like, as if) precisely (looked for all the world as if they were real). **get the best of both worlds** benefit from two incompatible sets of ideas, circumstances, etc. **in the world** of all; at all (used as an intensifier in questions) (what in the world is it?). **man** (or **woman**) **of the world** a person experienced and practical in human affairs. **the next** (or **other**) **world** a supposed life after death. **not for the world** not whatever the inducement. **out of this world** colloq. extremely good etc. (the food was out of this world). **see the world** travel widely; gain wide experience. **think the world of** have a very high regard for. **world-beater** a person or thing surpassing all others. **world-class** of a quality or standard regarded as high throughout the world. **world famous** known throughout the world. **the world, the flesh, and the devil** the various kinds of temptation. **world language 1** an artificial language for international use. **2** a language spoken in many countries. **world-line** Physics a curve in space-time joining the positions of a particle throughout its existence. **world music 1** traditional local or ethnic music, especially from the developing world, that is promoted or played in the West. **2** popular music influenced by or incorporating elements of such music. **the** (or **all the**) **world over** throughout the world. **world power** a nation having power and influence in world affairs. **the world's end** the farthest attainable point of travel. **world-shaking** of supreme importance. **the world to come** supposed life after death. **world-view** = WELTANSCHAUUNG. **world-weariness** being world-weary. **world-weary** weary of the world and life on it. **world without end** for ever. [OE w(e)orold, world f. a Gmc root meaning 'age': rel. to OLD]

World Bank an alternative name for the INTERNATIONAL BANK FOR RECONSTRUCTION AND DEVELOPMENT.

World Council of Churches (abbr. **WCC**) an association established in 1948 to promote unity among the many different Christian Churches. Its member Churches number over 300, and include virtually all Christian traditions except Roman Catholicism and Unitarianism. Its headquarters are in Geneva.

World Cup n. any of various international sports competitions (or the trophies awarded for them), especially a competition in Association football instituted in 1930 and held every fourth year between national teams who qualify from regional preliminary rounds. Among recent winners are Germany, Italy, Argentina, and Brazil. The original trophy was named after Jules Rimet (1873–1956), the French football administrator who proposed the competition; this was awarded permanently to Brazil after their third World Cup win in 1970, and a new trophy substituted.

world fair n. an international exhibition of the industrial, scientific, technological, and artistic achievements of the participating nations. Early examples were the Great Exhibition of 1851 (held at the Crystal Palace in London), the New York Fair in 1853–4, and the Paris Exhibition of 1861. Important world fairs of the 20th century include those in New York in 1939–40 and 1964–5, the Brussels World Fair in 1958, the Seattle Century 21 Exposition of 1962, and Expo 67 in Montreal.

World Federation of Trade Unions (abbr. **WFTU**) an association of trade unions founded in 1945, with headquarters in Prague. In 1949 most of the representatives from non-Communist countries withdrew and founded the International Confederation of Free Trade Unions; the WFTU is now composed largely of groups from developing countries.

World Health Organization (abbr. **WHO**) an agency of the United Nations, established in 1948 to promote health and control communicable diseases. It assists in the efforts of member governments, and pursues biomedical research through some 500 collaborating research centres throughout the world. Its headquarters are in Geneva.

World Heritage Site *n.* a natural or man-made site, area, or structure recognized as being of outstanding international importance and therefore as deserving special protection. Sites are nominated by countries to the World Heritage Convention (an organization of UNESCO), which decides whether or not to declare them World Heritage Sites.

World Intellectual Property Organization (abbr. **WIPO**) an organization, established in 1967 and an agency of the United Nations from 1974, for cooperation between governments in matters concerning patents, trademarks, copyright, etc., and the transfer of technology between countries. Its headquarters are in Geneva.

worldly /ˈwɜːldlɪ/ *adj.* (**worldlier, worldliest**) **1** temporal or earthly (*worldly goods*). **2** engrossed in temporal affairs, esp. the pursuit of wealth and pleasure. **3** experienced in life, sophisticated, practical. □ **worldly-minded** intent on worldly things. **worldly wisdom** prudence as regards one's own interests. **worldly-wise** having worldly wisdom. □ **worldliness** *n.* [OE *woruldlic* (as WORLD, -LY¹)]

World Meteorological Organization (abbr. **WMO**) an agency of the United Nations, established in 1950 with the aim of facilitating worldwide cooperation in meteorological observations, research, and services. Its headquarters are in Geneva.

world order *n.* (esp. in phr. **new world order**) a system controlling events in the world, esp. an international set of arrangements for preserving global political security.

World Series *n.* a series of baseball championship matches played after the end of the season between the champions of the two major North American professional baseball leagues, the American League and the National League. It was first played in 1903.

World Service a service of the British Broadcasting Corporation that transmits radio programmes in English and over thirty other languages around the world twenty-four hours a day. A worldwide television station was established in 1991 on a similar basis.

World Trade Center a complex of buildings in New York featuring twin skyscrapers 110 storeys high, designed by Minoru Yamasaki and completed in 1972.

world war *n.* a war involving many important nations. The name is commonly given to the wars of 1914–18 and 1939–45, although only the second of these was truly global. (See FIRST WORLD WAR, SECOND WORLD WAR.)

World War I see FIRST WORLD WAR.

World War II see SECOND WORLD WAR.

worldwide *adj.* & *adv.* ● *adj.* /ˈwɜːldwaɪd/ affecting, occurring in, or known in all parts of the world. ● *adv.* /wɜːldˈwaɪd/ throughout the world.

World Wide Fund for Nature (abbr. **WWF**) an international organization established (as the World Wildlife Fund) in 1961 to raise funds for projects including the conservation of endangered species or of valuable habitats. Its headquarters are in Gland, Switzerland.

World Wide Web an international computer network incorporating multimedia techniques and using hypertext links to access and retrieve information.

worm /wɜːm/ *n.* & *v.* ● *n.* **1** a slender elongated invertebrate animal, usu. with a soft moist body, inhabiting marine or fresh water, burrowing in the soil, or parasitic in the bodies of vertebrates; esp. an earthworm. (*See note below.*) **2** the elongated larva of some insects, esp. in fruit or wood. **3** (in *pl.*) intestinal or other internal parasites. **4** a small limbless reptile, esp. the blindworm or slow-worm. **5** a maggot supposed to eat dead bodies in the grave. **6** *colloq.* an insignificant or contemptible person. **7 a** the spiral part of a screw. **b** a short screw working in a worm-gear. **8** the spiral pipe of a still in which the vapour is cooled and condensed. **9** the ligament under a dog's tongue. ● *v.* **1** *intr.* & *tr.* (often *refl.*) move with a crawling motion (*wormed through the bushes*; *wormed our way through the bushes*). **2** *intr.* & *refl.* (foll. by *into*) insinuate oneself into a person's favour, confidence, etc. **3** *tr.* (foll. by *out*) obtain (a secret etc.) by cunning persistence (*managed to worm the truth out of them*). **4** *tr.* cut the worm of (a dog's tongue). **5** *tr.* rid (a plant or dog etc.) of worms. **6** *tr. Naut.* make (a rope etc.) smooth by winding thread between the strands. □ **food for worms** *colloq.* a dead person. **worm-cast** a convoluted mass of earth left on the surface by a burrowing earthworm. **worm-fishing** fishing with worms for bait. **worm-gear** an arrangement of a toothed wheel worked by a revolving spiral. **worm-seed 1** seed used to expel intestinal worms. **2** a plant e.g. santonica bearing this seed. **worm's-eye view** a view as seen from below or from a humble position. **worm-wheel** the wheel of a worm-

gear. **a** (or **even a**) **worm will turn** the meekest will resist or retaliate if pushed too far. □ **wormer** *n.* **wormlike** *adj.* [OE *wyrm* f. Gmc]

■ The term *worm* was formerly used for anything that creeps or crawls, and even zoologists applied it to a wide range of soft-bodied invertebrates. Three important animal phyla include worms — the annelids, which comprise segmented worms such as earthworms, leeches, and ragworms; the nematodes or roundworms; and the platyhelminths, which include the flatworms, tapeworms, and flukes. There are also several minor phyla of wormlike animals, such as the arrow worms and acorn worms.

wormeaten /ˈwɜːmˌiːt(ə)n/ *adj.* **1 a** eaten into by worms. **b** rotten, decayed. **2** old and dilapidated.

wormhole /ˈwɜːmhəʊl/ *n.* **1** a hole made by a burrowing worm or insect in wood, fruit, books, etc. **2** *Physics* a hypothetical connection between widely separated regions of space–time.

Worms /vɜːmz, German vɔrms/ an industrial town in western Germany, on the Rhine north-west of Mannheim; pop. (1991) 77,430. It was the scene in 1521 of the condemnation of Martin Luther's teaching, at the Diet of Worms.

Worms, Diet of see DIET OF WORMS.

wormwood /ˈwɜːmwʊd/ *n.* **1** a woody aromatic composite plant of the genus *Artemisia*, with a very bitter taste, used in the preparation of vermouth and absinthe and in medicine. **2** bitter mortification or a source of this. [ME, alt. f. obs. *wormod* f. OE *wormōd*, *wermōd*, after *worm*, *wood*: cf. VERMOUTH]

wormy /ˈwɜːmɪ/ *adj.* (**wormier, wormiest**) **1** full of worms. **2** wormeaten. □ **worminess** *n.*

worn /wɔːn/ *past part.* of WEAR¹. ● *adj.* **1** damaged by use or wear. **2** looking tired and exhausted. **3** (in full **well-worn**) (of a joke etc.) stale; often heard. □ **worn-out 1** exhausted. **2** worn, esp. to the point of being no longer usable (hyphenated when *attrib.*: *worn-out engine*).

worriment /ˈwʌrɪmənt/ *n.* esp. *US* **1** the act of worrying or state of being worried. **2** a cause of worry.

worrisome /ˈwʌrɪsəm/ *adj.* causing or apt to cause worry or distress. □ **worrisomely** *adv.*

worrit /ˈwʌrɪt/ *v.intr. colloq.* worry, fret. [alt. of WORRY]

worry /ˈwʌrɪ/ *v.* & *n.* ● *v.* (**-ies, -ied**) **1** *intr.* give way to anxiety or unease; allow one's mind to dwell on difficulty or troubles. **2** *tr.* harass, importune; be a trouble or anxiety to. **3** *tr.* **a** (of a dog etc.) shake or pull about with the teeth. **b** attack repeatedly. **4** (as **worried** *adj.*) **a** uneasy, troubled in the mind. **b** suggesting worry (*a worried look*). ● *n.* (*pl.* **-ies**) **1** a thing that causes anxiety or disturbs a person's tranquillity. **2** a disturbed state of mind; anxiety; a worried state. **3** a dog's worrying of its quarry. □ **not to worry** *colloq.* there is no need to worry. **worry along** (or **through**) manage to advance by persistence in spite of obstacles. **worry beads** a string of beads manipulated with the fingers to occupy or calm oneself. **worry-guts** (or **-wart**) *colloq.* a person who habitually worries unduly. **worry oneself** (usu. in *neg.*) take needless trouble. **worry out** obtain (the solution to a problem etc.) by dogged effort. □ **worriedly** *adv.* **worrier** *n.* **worryingly** *adv.* [OE *wyrgan* strangle f. WG]

worse /wɜːs/ *adj., adv.,* & *n.* ● *adj.* **1** more bad. **2** (*predic.*) in or into worse health or a worse condition (*is getting worse*; *is none the worse for it*). ● *adv.* more badly or more ill. ● *n.* **1** a worse thing or things (*you might do worse than accept*). **2** (*prec. by the*) a worse condition (*a change for the worse*). □ **none the worse** (often foll. by *for*) not adversely affected (by). **or worse** or as an even worse alternative. **the worse for drink** fairly drunk. **the worse for wear 1** damaged by use. **2** injured. **3** *joc.* drunk. **worse luck** see LUCK. **worse off** in a worse (esp. financial) position. [OE *wyrsa, wiersa* f. Gmc]

worsen /ˈwɜːs(ə)n/ *v.tr.* & *intr.* make or become worse.

worship /ˈwɜːʃɪp/ *n.* & *v.* ● *n.* **1 a** homage or reverence paid to a deity, esp. in a formal service. **b** the acts, rites, or ceremonies of worship. **2** adoration or devotion comparable to religious homage shown towards a person or principle (*the worship of wealth; regarded them with worship in their eyes*). **3** *archaic* worthiness, merit; recognition given or due to these; honour and respect. ● *v.* (**worshipped, worshipping**; *US* **worshiped, worshiping**) **1** *tr.* adore as divine; honour with religious rites. **2** *tr.* idolize or regard with adoration (*he worships the ground she walks on*). **3** *intr.* attend public worship. **4** *intr.* be full of adoration. □ **Your** (or **His** or **Her**) **Worship** esp. *Brit.* a title of respect used to or of a mayor, certain magistrates, etc. □ **worshipper** *n.* (*US* **worshiper**). [OE *weorthscipe* (as WORTH, -SHIP)]

worshipful /ˈwɜːʃɪpˌfʊl/ *adj.* **1** (usu. **Worshipful**) *Brit.* a title given to justices of the peace and to certain old companies or their officers etc. **2** *archaic* entitled to honour or respect. **3** *archaic* imbued with a spirit of veneration. □ **worshipfully** *adv.* **worshipfulness** *n.*

worst /wɜːst/ *adj., adv., n., & v.* ● *adj.* most bad. ● *adv.* most badly. ● *n.* the worst part, event, circumstance, or possibility (*the worst of the storm is over*; *prepare for the worst*). ● *v.tr.* get the better of; defeat, outdo. □ **at its** etc. **worst** in the worst state. **at worst** (or **the worst**) in the worst possible case. **do your worst** an expression of defiance. **get** (or **have**) **the worst of** it be defeated. **if the worst comes to the worst** if the worst happens. [OE *wierresta*, *wyrresta* (adj.), *wyrst*, *wyrrest* (adv.), f. Gmc]

worsted /ˈwʊstɪd/ *n.* **1** a fine smooth yarn spun from combed long staple wool. **2** fabric made from this. [*Worstead* in Norfolk, England]

wort /wɜːt/ *n.* **1** *archaic* (except in names) a plant or herb (*liverwort*; *St John's wort*). **2** the infusion of malt which after fermentation becomes beer. [OE *wyrt*: rel. to ROOT[1]]

Worth /wɜːθ/, Charles Frederick (1825–95), English couturier, resident in France from 1845. He opened his own establishment in Paris in 1858, and soon gained the patronage of the empress Eugénie, wife of Napoleon III. Regarded as the founder of Parisian *haute couture*, he is noted for designing gowns with crinolines, making extensive use of rich fabrics, and for introducing the bustle.

worth /wɜːθ/ *adj. & n.* ● *predic.adj.* (governing a noun like a preposition) **1** of a value equivalent to (*is worth £50*; *is worth very little*). **2** such as to justify or repay; deserving; bringing compensation for (*worth doing*; *not worth the trouble*). **3** possessing or having property amounting to (*is worth a million pounds*). ● *n.* **1** what a person or thing is worth; the (usu. specified) merit of (*of great worth*; *persons of worth*). **2** the equivalent of money in a commodity (*ten pounds' worth of petrol*). □ **for all one is worth** *colloq.* with one's utmost efforts; without reserve. **for what it is worth** without a guarantee of its truth or value. **worth it** *colloq.* worth the time or effort spent. **worth one's salt** see SALT. **worth while** (or **one's while**) see WHILE. [OE *w(e)orth*]

Worthing /ˈwɜːðɪŋ/ a town on the south coast of England, in West Sussex; pop. (1980) 92,050.

worthless /ˈwɜːθlɪs/ *adj.* without value or merit. □ **worthlessly** *adv.* **worthlessness** *n.*

worthwhile /wɜːθˈwaɪl/ *adj.* that is worth the time or effort spent; of value or importance. □ **worthwhileness** *n.*

worthy /ˈwɜːðɪ/ *adj. & n.* ● *adj.* (**worthier**, **worthiest**) **1** estimable; having some moral worth; deserving respect (*lived a worthy life*). **2** (of a person) entitled to (esp. condescending) recognition (*a worthy old couple*). **3 a** (foll. by *of* or to +infin.) deserving (*worthy of a mention*; *worthy to be remembered*). **b** (foll. by *of*) adequate or suitable to the dignity etc. of (*in words worthy of the occasion*). ● *n.* (pl. **-ies**) **1** a worthy person. **2** a person of some distinction. **3** *joc.* a person. □ **worthily** *adv.* **worthiness** *n.* [ME *wurthi* etc. f. WORTH]

-worthy /ˌwɜːðɪ/ *comb. form* forming adjectives meaning: **1** deserving of (*blameworthy*; *noteworthy*). **2** suitable or fit for (*newsworthy*; *roadworthy*).

wot see WIT[2].

Wotan /ˈwəʊtɑːn/ see ODIN.

wotcher /ˈwɒtʃə(r)/ *int. Brit. sl.* a form of casual greeting. [corrupt. of *what cheer*]

would /wʊd, wəd/ *v.aux.* (3rd sing. **would**) past of WILL[1], used esp.: **1** (in the 2nd and 3rd persons, and often in the 1st: see SHOULD): **a** in reported speech (*he said he would be home by evening*). **b** to express the conditional mood (*they would have been killed if they had gone*). **2** to express habitual action (*would wait for her every evening*). **3** to express a question or polite request (*would they like it?*; *would you come in, please?*). **4** to express probability (*I guess she would be over fifty by now*). **5** (foll. by *that* + clause) *literary* to express a wish (*would that you were here*). **6** to express consent (*they would not help*). □ **would-be** often *derog.* desiring or aspiring to be (*a would-be politician*). [OE *wolde*, past of *wyllan*: see WILL[1]]

wouldn't /ˈwʊd(ə)nt/ *contr.* would not. □ **I wouldn't know** *colloq.* (as is to be expected) I do not know. **wouldn't it!** esp. *Austral. & NZ sl.* an interjection expressing dismay or disgust.

wouldst /wʊdst/ *archaic* 2nd sing. past of WOULD.

Woulfe bottle /wʊlf/ *n. Chem.* a glass bottle with more than one neck, used for passing a gas through a liquid etc. [Peter *Woulfe*, Engl. chemist (c.1727–1803)]

wound[1] /wuːnd/ *n. & v.* ● *n.* **1** an injury done to living tissue by a cut or blow etc., esp. beyond the cutting or piercing of the skin. **2** an injury to a person's reputation or a pain inflicted on a person's feelings. **3** *poet.*

the pangs of love. ● *v.tr.* inflict a wound on (*wounded soldiers*; *wounded feelings*). □ **woundingly** *adv.* **woundless** *adj.* [OE *wund* (n.), *wundian* (v.)]

wound[2] past and past part. of WIND[2] (cf. WIND[1] *v.* 6).

Wounded Knee, Battle of the last major confrontation (1890) between the US army and American Indians, in which more than 150 largely unarmed Sioux men, women, and children were massacred at the village of Wounded Knee on a reservation in South Dakota. The event was recalled in 1973 when members of the American Indian Movement (an organization founded in 1968 to campaign for civil rights) occupied the site; after a skirmish with federal marshals in which two protesters were killed they agreed to evacuate the area in exchange for negotiations on their grievances.

woundwort /ˈwuːndwɜːt/ *n.* a labiate plant of the genus *Stachys*, formerly supposed to have healing properties.

wove[1] past of WEAVE[1].

wove[2] /wəʊv/ *adj.* (of paper) made on a wire-gauze mesh and so having a uniform unlined surface. [var. of *woven*, past part. of WEAVE[1]]

woven past part. of WEAVE[1].

wow[1] /waʊ/ *int., n., & v.* ● *int.* expressing astonishment or admiration. ● *n. sl.* a sensational success. ● *v.tr. sl.* impress or excite greatly. [orig. Sc.: imit.]

wow[2] /waʊ/ *n.* a slow pitch-fluctuation in sound-reproduction, perceptible in long notes. [imit.]

wowser /ˈwaʊzə(r)/ *n. Austral. sl.* **1** a puritanical fanatic. **2** a spoilsport. **3** a teetotaller. [20th c.: orig. uncert.]

WP *abbr.* word processor or processing.

w.p. *abbr.* weather permitting.

w.p.b. *abbr.* waste-paper basket.

WPC *abbr.* (in the UK) woman police constable.

w.p.m. *abbr.* words per minute.

WRAC *abbr.* (in the UK) Women's Royal Army Corps.

wrack[1] /ræk/ *n.* **1** seaweed cast up or growing on the shore. **2** destruction. **3** a wreck or wreckage. **4** = RACK[5] *n.* [ME f. MDu. *wrak* or MLG *wra(c)k*, a parallel formation to OE *wræc*, rel. to *wrecan* WREAK: cf. WRECK, RACK[5]]

wrack[2] var. of RACK[1] *v.*

WRAF *abbr.* (in the UK) Women's Royal Air Force.

wraggle-taggle var. of RAGGLE-TAGGLE.

wraith /reɪθ/ *n.* **1** a ghost or apparition. **2** the spectral appearance of a living person supposed to portend that person's death. □ **wraithlike** *adj.* [16th-c. Sc.: orig. unkn.]

Wrangel Island /ˈræŋg(ə)l/ an island in the East Siberian Sea, off the coast of NE Russia. It was named after the Russian admiral and explorer Baron Ferdinand Wrangel (1794–1870).

wrangle /ˈræŋg(ə)l/ *n. & v.* ● *n.* a noisy argument, altercation, or dispute. ● *v.* **1** *intr.* engage in a wrangle. **2** *tr. US* herd (cattle). [ME, prob. f. LG or Du.: cf. LG *wrangelen*, frequent. of *wrangen* to struggle, rel. to WRING]

wrangler /ˈræŋglə(r)/ *n.* **1** a person who wrangles. **2** *US* a cowboy. **3** (at Cambridge University) a person placed in the first class of the mathematical tripos (*senior wrangler*).

wrap /ræp/ *v. & n.* ● *v.tr.* (**wrapped**, **wrapping**) **1** (often foll. by *up*) envelop in folded or soft encircling material (*wrap it up in paper*; *wrap up a parcel*). **2** (foll. by *round*, *about*) arrange or draw (a pliant covering) round (a person) (*wrapped the scarf closer around me*). **3** (foll. by *round*) *sl.* crash (a vehicle) into a stationary object. **4** (foll. by *on to*) *Computing* cause (text) to be carried over to a new line automatically as the right margin is reached. ● *n.* **1** a shawl or scarf or other such addition to clothing; a wrapper. **2** esp. *US* material used for wrapping. □ **take the wraps off** disclose. **under wraps** in secrecy. **wrap-over** *adj.* (attrib.) (of a garment) having no seam at one side but wrapped around the body and fastened. ● *n.* such a garment. **wrapped up in** engrossed or absorbed in. **wrap up 1** finish off, bring to completion (*wrapped up the deal in two days*). **2** put on warm clothes (*mind you wrap up well*). **3** (in *imper.*) *sl.* be quiet. [ME: orig. unkn.]

wraparound /ˈræpəˌraʊnd/ *adj. & n.* (also **wrapround** /ˈræpraʊnd/) ● *adj.* **1** (esp. of clothing) designed to wrap round. **2** curving or extending round at the edges or sides. ● *n.* **1** anything that wraps round. **2** *Computing* a facility by which a linear sequence of memory locations or screen positions is treated as a continuous circular series.

wrappage /ˈræpɪdʒ/ *n.* a wrapping or wrappings.

wrapper /ˈræpə(r)/ n. **1** a cover for a sweet, chocolate, etc. **2** a cover enclosing a newspaper or similar packet for posting. **3** a paper cover of a book, usu. detachable. **4** a loose enveloping robe or gown. **5** a tobacco-leaf of superior quality enclosing a cigar.

wrapping /ˈræpɪŋ/ n. (esp. in pl.) material used to wrap; wraps, wrappers. □ **wrapping paper** strong or decorative paper for wrapping parcels.

wraparound var. of WRAPAROUND.

wrasse /ræs/ n. a brightly coloured marine fish of the family Labridae, with thick lips and strong teeth. [Cornish wrah, var. of gwrah, = Welsh gwrach, lit. 'old woman']

wrath /rɒθ, rɔːθ/ n. literary extreme anger. [OE wrǣththu f. wrāth WROTH]

Wrath, Cape /rɒθ, rɔːθ/ a headland at the north-western tip of the mainland of Scotland.

wrathful /ˈrɒθfʊl, ˈrɔːθ-/ adj. literary extremely angry. □ **wrathfully** adv. **wrathfulness** n.

wrathy /ˈræθɪ/ adj. N. Amer. = WRATHFUL.

wreak /riːk/ v.tr. **1** (usu. foll. by upon) give play or satisfaction to; put in operation (vengeance or one's anger etc.). **2** cause (damage etc.) (the hurricane wreaked havoc on the crops). **3** archaic avenge (a wrong or wronged person). □ **wreaker** n. [OE wrecan drive, avenge, etc., f. Gmc: cf. WRACK¹, WRECK, WRETCH]

wreath /riːθ/ n. (pl. **wreaths** /riːðz, riːθs/) **1** flowers or leaves fastened in a ring esp. as an ornament for a person's head or a building or for laying on a grave etc. as a mark of honour or respect. **2 a** a similar ring of soft twisted material such as silk. **b** Heraldry a representation of this below a crest. **3** a carved representation of a wreath. **4** (foll. by of) a curl or ring of smoke or cloud. **5** a light drifting mass of snow etc. [OE writha f. weak grade of wrīthan WRITHE]

wreathe /riːð/ v. **1** tr. encircle as, with, or like a wreath. **2** tr. (foll. by round) put (one's arms etc.) round (a person etc.). **3** intr. (of smoke etc.) move in the shape of wreaths. **4** tr. form (flowers, silk, etc.) into a wreath. **5** tr. make (a garland). [partly back-form. f. archaic wrethen past part. of WRITHE; partly f. WREATH]

wreck /rek/ n. & v. ● n. **1** the destruction or disablement esp. of a ship. **2** a ship that has suffered a wreck (the shores are strewn with wrecks). **3** a greatly damaged or disabled building, thing, or person (had become a physical and mental wreck). **4** (foll. by of) a wretched remnant or disorganized set of remains. **5** Law goods etc. cast up by the sea. ● v. **1** tr. cause the wreck of (a ship etc.). **2** tr. completely ruin (hopes, chances, etc.). **3** intr. suffer a wreck. **4** tr. (as **wrecked** adj.) involved in a shipwreck (wrecked sailors). **5** intr. US deal with wrecked vehicles etc. □ **wreckmaster** an officer appointed to take charge of goods etc. cast up from a wrecked ship. [ME f. AF wrec etc. (cf. VAREC) f. a Gmc root meaning 'to drive': cf. WREAK]

wreckage /ˈrekɪdʒ/ n. **1** wrecked material. **2** the remnants of a wreck. **3** the action or process of wrecking.

wrecker /ˈrekə(r)/ n. **1** a person or thing that wrecks or destroys. **2** esp. hist. a person on the shore who tries to bring about a shipwreck in order to plunder or profit by the wreckage. **3** esp. N. Amer. a person employed in demolition, or in recovering a wrecked ship or its contents. **4** US a person who breaks up damaged vehicles for spares and scrap. **5** N. Amer. a vehicle or train used in recovering a damaged one.

Wren¹ /ren/, Sir Christopher (1632–1723), English architect. He turned to architecture in the 1660s after an academic career as a scientist. Following the Great Fire of London (1666) he submitted plans for the rebuilding of the city; although these were never realized, Wren was appointed Surveyor-General of the King's Works in 1669 and was responsible for the design of the new St Paul's Cathedral (1675–1711) and many of the city's churches. The influence of the baroque, seen in elements of St Paul's, is particularly apparent in his Greenwich Hospital (begun 1696). Among Wren's other works are Greenwich Observatory (1675) and a partial rebuilding of Hampton Court (1689–94). He was a founder member and later president of the Royal Society (1680–2).

Wren² /ren/, P(ercival) C(hristopher) (1885–1941), English novelist. He is best known for his romantic adventure stories dealing with life in the French Foreign Legion, the first of which was Beau Geste (1924).

Wren³ /ren/ n. (in the UK) a member of the Women's Royal Naval Service. [orig. in pl., f. abbr. WRNS]

wren /ren/ n. **1** a short-winged songbird of the family Troglodytidae; esp. the very small Troglodytes troglodytes of Eurasia and North America, having brown barred plumage and a short cocked tail (also called (N.

Amer.) winter wren). **2** Brit. colloq. a small warbler or kinglet (willow-wren). **3** a small Australasian or South American bird resembling a wren (fairy-wren). [OE wrenna, rel. to OHG wrendo, wrendilo, Icel. rindill]

wrench /rentʃ/ n. & v. ● n. **1** a violent twist or oblique pull or act of tearing off. **2** an adjustable tool like a spanner for gripping and turning nuts etc. **3** an instance of painful uprooting or parting (leaving home was a great wrench). **4** Physics a combination of a couple with a force along its axis. ● v.tr. **1 a** twist or pull violently round or sideways. **b** injure (a limb etc.) by undue straining; sprain. **2** (often foll. by off, away, etc.) pull off with a wrench. **3** seize or take forcibly. **4** distort (facts) to suit a theory etc. [(earlier as verb:) OE wrencan twist]

wrest /rest/ v. & n. ● v.tr. **1** force or wrench away from a person's grasp. **2** (foll. by from) obtain by effort or with difficulty. **3** distort into accordance with one's interests or views (wrest the law to suit themselves). ● n. archaic a key for tuning a harp or piano etc. □ **wrest-block** (or **-plank**) the part of a piano or harpsichord holding the wrest-pins. **wrest-pin** each of the pins to which the strings of a piano or harpsichord are attached. [OE wrǣstan f. Gmc, rel. to WRIST]

wrestle /ˈres(ə)l/ n. & v. ● n. **1** a contest in which two opponents grapple and try to throw each other to the ground or hold the opponent down on the ground, esp. as an athletic sport under a code of rules. (See WRESTLING.) **2** a hard struggle. ● v. **1** intr. (often foll. by with) take part in a wrestle. **2** tr. fight (a person) in a wrestle (wrestled his opponent to the ground). **3** intr. **a** (foll. by with, against) struggle, contend. **b** (foll. by with) do one's utmost to deal with (a task, difficulty, etc.). **4** tr. move with efforts as if wrestling. □ **arm-wrestling** see ARM¹. □ **wrestler** n. [OE (unrecorded) wrǣstlian: cf. MLG wrosteln, OE wraxlian]

wrestling /ˈreslɪŋ/ n. the sport or activity of taking part in a wrestle. Wrestling is one of the oldest and most basic of all sports. Popular in ancient Egypt, China, and Greece, it was introduced to the Olympic Games in 704 BC; many of the holds and throws used now are the same as those of antiquity. The two main competition styles are Graeco-Roman (in which holds below the waist are prohibited) and freestyle. Wrestling was included in the first modern Olympic Games in 1896, and world championships have been held since 1921; freestyle wrestling has also, with a substantial show-business element, become a popular televised sport. (See also SUMO.)

wretch /retʃ/ n. **1** an unfortunate or pitiable person. **2** (often as a playful term of depreciation) a reprehensible or contemptible person. [OE wrecca f. Gmc]

wretched /ˈretʃɪd/ adj. (**wretcheder**, **wretchedest**) **1** unhappy or miserable. **2** of bad quality or no merit; contemptible. **3** unsatisfactory or displeasing. □ **feel wretched 1** be unwell. **2** be much embarrassed. □ **wretchedly** adv. **wretchedness** n. [ME, irreg. f. WRETCH + -ED¹: cf. WICKED]

Wrexham /ˈreksəm/ a mining and industrial town in Clwyd, NE Wales; pop. (1981) 40,930.

wrick Brit. var. of RICK².

wriggle /ˈrɪg(ə)l/ v. & n. ● v. **1** intr. (of a worm etc.) twist or turn its body with short writhing movements. **2** intr. (of a person or animal) make similar motions. **3** tr. & intr. (foll. by along) move or go in this way (wriggled into the corner; wriggled his hand into the hole). **4** tr. make (one's way) by wriggling. **5** intr. practise evasion. ● n. an act of wriggling. □ **wriggle out of** colloq. avoid on a contrived pretext. □ **wriggler** n. **wriggly** adj. [ME f. MLG wriggelen frequent. of wriggen]

Wright¹ /raɪt/, Frank Lloyd (1869–1959), American architect. His early work, with its use of new building materials and cubic forms, was particularly significant for the development of modernist architecture, in particular the international style. His 'prairie-style' houses in Chicago revolutionized American domestic architecture in the first decade of the 20th century with their long low horizontal lines and intercommunicating interior spaces. He advocated an 'organic' architecture, characterized by a close relationship between building and landscape and the nature of the materials used, as can be seen in the Kaufmann House (known as 'Falling Water') in Pennsylvania (1935–9), which incorporates natural features such as a waterfall into its design. Other notable buildings include the Johnson Wax office block in Racine, Wisconsin (1936) and the Guggenheim Museum of Art in New York (1956–9).

Wright² /raɪt/, Orville (1871–1948) and Wilbur (1867–1912), American aviation pioneers. In 1903 at Kitty Hawk, North Carolina, the Wright brothers were the first to make brief powered sustained and controlled flights in an aeroplane which they had designed and built themselves, having first experimented with gliders. They were also the first to make and fly a fully practical powered aeroplane (1905) and passenger-carrying aeroplane (1908).

wright /raɪt/ n. a maker or builder (usu. in comb.: playwright; shipwright). [OE wryhta, wyrhta f. WG: cf. WORK]

wring /rɪŋ/ v. & n. ● v.tr. (past and past part. **wrung** /rʌŋ/) **1 a** squeeze tightly. **b** (often foll. by out) squeeze and turn esp. to remove liquid. **2** twist forcibly; break by twisting. **3** distress or torture. **4** extract by squeezing. **5** (foll. by out, from) obtain by pressure or importunity; extort. ● n. an act of wringing; a squeeze. □ **wring a person's hand** clasp it forcibly or press it with emotion. **wring one's hands** clasp them as a gesture of great distress. **wring the neck of** kill (a chicken etc.) by twisting its neck. [OE wringan, rel. to WRONG]

wringer /ˈrɪŋə(r)/ n. a device for wringing water from washed clothes etc. □ **put through the wringer** colloq. subject to a very stressful experience.

wringing /ˈrɪŋɪŋ/ adj. (in full **wringing wet**) so wet that water can be wrung out.

wrinkle /ˈrɪŋk(ə)l/ n. & v. ● n. **1** a slight crease or depression in the skin such as is produced by age. **2** a similar mark in another flexible surface. **3** colloq. a detail, esp. a useful tip or technical innovation. ● v. **1** tr. make wrinkles in. **2** intr. form wrinkles; become marked with wrinkles. [orig. repr. OE gewrinclod sinuous]

wrinkly /ˈrɪŋklɪ/ adj. & n. ● adj. (**wrinklier**, **wrinkliest**) having many wrinkles. ● n. (also **wrinklie**) (pl. -**ies**) sl. offens. an old or middle-aged person.

wrist /rɪst/ n. **1** the part connecting the hand with the forearm. **2** the corresponding part in an animal. **3** the part of a garment covering the wrist. **4 a** (in full **wrist-work**) the act or practice of working the hand without moving the arm. **b** the effect produced in fencing, ball games, sleight of hand, etc., by this. **5** (in full **wrist-pin**) Mech. a stud projecting from a crank etc. as an attachment for a connecting-rod. □ **wrist-drop** the inability to extend the hand through paralysis of the forearm muscles. **wrist-watch** a small watch worn on a strap round the wrist. [OE f. Gmc, prob. f. a root rel. to WRITHE]

wristband /ˈrɪstbænd/ n. a band forming or concealing the end of a shirt-sleeve; a cuff.

wristlet /ˈrɪstlɪt/ n. a band or ring worn on the wrist to strengthen or guard it or as an ornament, bracelet, handcuff, etc.

wristy /ˈrɪstɪ/ adj. (esp. of a shot in cricket, tennis, etc.) involving or characterized by movement of the wrist.

writ¹ /rɪt/ n. **1** a form of written command in the name of a court, state, sovereign, etc., to act or abstain from acting in some way. **2** a Crown document summoning a peer to Parliament or ordering the election of a member or members of Parliament. □ **serve a writ on** deliver a writ to (a person). **one's writ runs** one has authority (as specified). [OE (as WRITE)]

writ² /rɪt/ archaic past part. of WRITE. □ **writ large** in magnified or emphasized form.

write /raɪt/ v. (past **wrote** /rəʊt/; past part. **written** /ˈrɪt(ə)n/) **1** intr. mark paper or some other surface by means of a pen, pencil, etc., with symbols, letters, or words. **2** tr. form or mark (such symbols etc.). **3** tr. form or mark the symbols that represent or constitute (a word or sentence, or a document etc.) with writing. **5** tr. put (data) into a computer store. **6** tr. (esp. in passive) indicate (a quality or condition) by one's or its appearance (guilt was written on his face). **7** tr. compose (a text, article, novel, etc.) for written or printed reproduction or publication; put into literary etc. form and set down in writing. **8** intr. be engaged in composing a text, article, etc. (writes for the local newspaper). **9** intr. (foll. by to) write and send a letter (to a recipient). **10** tr. US or colloq. write and send a letter to (a person) (wrote him last week). **11** tr. convey (news, information, etc.) by letter (wrote that they would arrive next Friday). **12** tr. state in written or printed form (it is written that). **13** tr. cause to be recorded. **14** tr. underwrite (an insurance policy). **15** tr. (foll. by into, out of) include or exclude (a character or episode) in a story by suitable changes of the text. **16** tr. archaic describe in writing. □ **nothing to write home about** colloq. of little interest or value. **write down 1** record or take note of in writing. **2** write as if for those considered inferior. **3** disparage in writing. **4** reduce the nominal value of (stock, goods, etc.). **write-down** n. a reduction in the estimated or nominal value of stock, assets, etc. **write in 1** send a suggestion, query, etc., in writing to an organization, esp. a broadcasting station. **2** US add (an extra name) on a list of candidates when voting. **write-in** n. US an instance of writing in (see write in 2). **write off 1** write and send a letter. **2** cancel the record of (a bad debt etc.); acknowledge the loss of or failure to

recover (an asset). **3** damage (a vehicle etc.) so badly that it cannot be repaired. **4** compose with facility. **5** dismiss as insignificant. **write-off** n. a thing written off, esp. a vehicle too badly damaged to be repaired. **write out 1** write in full or in finished form. **2** exhaust (oneself) by writing (have written myself out). **write up 1** write a full account of. **2** praise in writing. **3** make entries to bring (a diary etc.) up to date. **write-up** n. colloq. a written or published account, a review. □ **writable** adj. [OE wrītan scratch, score, write, f. Gmc: orig. used of symbols inscribed with sharp tools on stone or wood]

writer /ˈraɪtə(r)/ n. **1** a person who writes or has written something. **2** a person who writes books; an author. **3** a clerk, esp. in the navy or in government offices. **4** a scribe. □ **writer's block** a periodic lack of inspiration afflicting creative writers etc. **writer's cramp** a muscular spasm due to excessive writing. [OE wrītere (as WRITE)]

writerly adj. **1** characteristic of a professional author. **2** consciously literary.

Writer to the Signet n. a Scottish solicitor conducting cases in the Court of Session.

writhe /raɪð/ v. & n. ● v. **1** intr. twist or roll oneself about in or as if in acute pain. **2** intr. suffer severe mental discomfort or embarrassment (writhed with shame; writhed at the thought of it). **3** tr. twist (one's body etc.) about. ● n. an act of writhing. [OE wrīthan, rel. to WREATHE]

writing /ˈraɪtɪŋ/ n. **1** a group or sequence of letters or symbols. (See note below.) **2** = HANDWRITING. **3** the art or profession of literary composition. **4** (usu. in pl.) a piece of literary work done; a book, article, etc. **5** (**Writings**) the Hagiographa. □ **in writing** in written form. **writing-desk** a desk for writing at, esp. with compartments for papers etc. **the writing on the wall** an ominously significant event or sign that something is doomed (with allusion to the biblical story of the writing that appeared on the wall at Belshazzar's feast (Dan. 5:5, 25–8), foretelling his doom). **writing-pad** a pad of paper for writing on. **writing-paper** paper for writing (esp. letters) on.

▪ Writing is thought to have begun in Mesopotamia with pictographic signs and to have developed into a formalized system of linear and then cuneiform signs representing first words or ideas, then syllables, and finally individual sounds. The cuneiform system was adopted from the Sumerians by the Akkadians and passed on to Syria in the 3rd millennium BC, to the Assyrians, Hittites, Canaanites, and Egyptians in the 2nd millennium BC, and to the Achaemenid Persians in the 1st millennium BC, after which alphabetic systems of writing began to develop (see ALPHABET). In the Chinese script, pictographic signs developed into ideographs which then proliferated and became standardized into a system of characters similar to that used today.

written past part. of WRITE.

WRNS abbr. (in the UK) Women's Royal Naval Service.

Wrocław /ˈvrɒtswɑːf/ (German **Breslau** /ˈbrɛslaʊ/) an industrial city on the Oder river in western Poland; pop. (1990) 643,220. Held by the Habsburgs from the 16th century, it was taken by Prussia in 1741. It passed to Poland in 1945.

wrong /rɒŋ/ adj., adv., n., & v. ● adj. **1** mistaken; not true; incorrect (gave a wrong answer; we were wrong to think that). **2** unsuitable; less or least desirable (the wrong road; a wrong decision). **3** contrary to law or morality (it is wrong to steal). **4** amiss; out of order, in or into a bad or abnormal condition (something wrong with my heart; my watch has gone wrong). ● adv. (usually placed last) in a wrong manner or direction; with an incorrect result (guessed wrong; told them wrong). ● n. **1** what is morally wrong; a wrong action. **2** injustice; unjust action or treatment (suffer wrong). ● v.tr. **1** treat unjustly; do wrong to. **2** mistakenly attribute bad motives to; discredit. □ **do wrong** commit sin; transgress, offend. **do wrong to** malign or mistreat (a person). **get in wrong with** colloq. incur the dislike or disapproval of (a person). **get on the wrong side of** fall into disfavour with. **get wrong 1** misunderstand (a person, statement, etc.). **2** obtain an incorrect answer to. **get** (or **get hold of**) **the wrong end of the stick** misunderstand completely. **go down the wrong way** (of food) enter the windpipe instead of the gullet. **go wrong 1** take the wrong path. **2** stop functioning properly. **3** depart from virtuous or suitable behaviour. **in the wrong** responsible for a quarrel, mistake, or offence. **on the wrong side of 1** out of favour with (a person). **2** somewhat more than (a stated age). **wrong-foot** colloq. **1** (in tennis, football, etc.) play so as to catch (an opponent) off balance. **2** disconcert; catch unprepared. **wrong-headed** perverse and obstinate. **wrong-headedly** in a wrong-headed manner. **wrong-headedness** the state of being wrong-headed. **wrong side** the worse or undesired or unusable side of something, esp. fabric. **wrong side out** inside out. **wrong way round** in the opposite or reverse of the normal or desirable orientation or sequence etc. □ **wronger** /'rɒŋə(r)/ n. **wrongly** adv. **wrongness** n. [OE wrang f. ON rangr awry, unjust, rel. to WRING]

wrongdoer /'rɒŋˌduːə(r)/ n. a person who behaves immorally or illegally. □ **wrongdoing** n.

wrongful /'rɒŋfʊl/ adj. **1** characterized by unfairness or injustice. **2** contrary to law. **3** (of a person) not entitled to the position etc. occupied. □ **wrongfully** adv. **wrongfulness** n.

wrong'un /'rɒŋən/ n. colloq. a person of bad character. [contr. of wrong one]

wrote past of WRITE.

wroth /rəʊθ, rɒθ/ predic.adj. archaic angry. [OE wrāth f. Gmc]

wrought /rɔːt/ adj. archaic past and past part. of WORK. **1** (of metals) beaten out or shaped by hammering. **2** (of timber) planed on one or more sides.

wrought iron n. a tough malleable form of iron suitable for forging or rolling rather than casting, obtained by puddling pig-iron while molten. It is nearly pure but contains some slag in the form of filaments and therefore shows a fibrous structure.

wrung past and past part. of WRING.

WRVS abbr. (in the UK) Women's Royal Voluntary Service.

wry /raɪ/ adj. (**wryer**, **wryest** or **wrier**, **wriest**) **1** distorted or turned to one side. **2** (of a face or smile etc.) contorted in disgust, disappointment, or mockery. **3** (of humour) dry and mocking. □ **wryly** adv. **wryness** n. [wry (v.) f. OE wrīgian tend, incline, in ME deviate, swerve, contort]

wryneck /'raɪnek/ n. **1** a bird of the genus Jynx, of the woodpecker family, with cryptic coloration and able to turn its head over its shoulder. **2** = TORTICOLLIS.

WSW abbr. west-south-west.

wt. abbr. weight.

WTO abbr. World Trade Organization.

Wu /wuː/ n. a dialect of Chinese spoken in the Kiangsu and Zhejiang provinces. [Chin.]

Wuhan /wuː'hæn/ a port in eastern China, the capital of Hubei province; pop. (1990) 3,710,000. Situated at the confluence of the Han and the Yangtze rivers, it is a conurbation consisting of three adjacent towns (Hankow, Hanyang, and Wuchang), administered jointly since 1950.

Wu-hsi see WUXI.

Wulfila see ULFILAS.

wunderkind /'vʊndəˌkɪnt/ n. colloq. a person who achieves great success while relatively young. [G f. Wunder wonder + Kind child]

Wundt /vʊnt/, Wilhelm (1832–1920), German psychologist. Working in Leipzig, he was the founder of psychology as a separate discipline, establishing a laboratory devoted to its study. He felt that the major task of the psychologist was to analyse human consciousness, which could be broken down into simpler fundamental units. Wundt required subjects to report their sensory impressions under controlled conditions, and although this method of inquiry was later rejected, his legacy includes the rigorous methodology upon which he insisted.

Wuppertal /'vʊpəˌtɑːl/ an industrial city in western Germany, in North Rhine-Westphalia north-east of Düsseldorf; pop. (1991) 385,460.

Wurlitzer /'wɜːlɪtsə(r)/ n. propr. a musical instrument or jukebox made in the US by the company founded by the German-born instrument-maker Rudolf Wurlitzer (1831–1914), esp. a type of large pipe organ used esp. in the cinemas of the 1930s.

wurst /vʊəst, vɜːst/ n. German or Austrian sausage. [G]

Würzburg /'vʊətsbɜːg/ an industrial city on the River Main in Bavaria, southern Germany; pop. (1991) 128,500.

wu-wei /wuː'weɪ/ n. Taoism the doctrine of letting things follow their natural course. [Chin., f. wú no, without + wei doing, action]

Wuxi /wuː'ʃiː/ (also **Wu-hsi**) a city on the Grand Canal in Jiangsu province, eastern China; pop. (est. 1990) 930,000.

WV abbr. US West Virginia (in official postal use).

W.Va. abbr. West Virginia.

WW abbr. US World War (I, II).

WWF see WORLD WIDE FUND FOR NATURE.

WX abbr. women's extra-large (size).

WY abbr. US Wyoming (in official postal use).

Wyandot /'waɪənˌdɒt/ n. & adj. (also **Wyandotte**, esp. in sense 2 of noun) ● n. **1 a** a member of a Huron people originally of Ontario. **b** the language of this people. **2** a medium-sized American breed of domestic fowl. ● adj. of or relating to the Wyandots or their language. [F Ouendat f. Huron Wendat]

Wyatt[1] /'waɪət/, James (1746–1813), English architect. Following a six-year stay in Italy, he returned to England, where he built the neoclassical Pantheon in London (1772), later destroyed by fire. Although he continued to build in a neoclassical style, Wyatt became a leading figure in the Gothic revival, most notably with his design for Fonthill Abbey in Wiltshire (1796–1807). He was also involved in the restoration of several English medieval cathedrals, work which was later strongly criticized by Pugin.

Wyatt[2] /'waɪət/, Sir Thomas (1503–42), English poet. He held various diplomatic posts in the service of Henry VIII, one of which took him to Italy (1527), a visit which probably stimulated him to translate and imitate the poems of Petrarch. His work includes sonnets, rondeaux, songs for the lute, and satires. His son, also named Sir Thomas Wyatt (c.1521–54), was executed after leading an unsuccessful rebellion against the proposed marriage of Mary I to the future Philip II of Spain.

wych- /wɪtʃ/ comb. form (also **witch-**) in names of trees with pliant branches. [OE wic(e) app. f. a Gmc root meaning 'bend': rel. to WEAK]

wych-elm /'wɪtʃelm/ n. (also **witch-elm**) a Eurasian elm, Ulmus glabra, with large rough leaves.

Wycherley /'wɪtʃəlɪ/, William (c.1640–1716), English dramatist. His Restoration comedies are characterized by their acute social criticism, particularly of sexual morality and marriage conventions. They include The Gentleman Dancing-Master (1672), The Country Wife (1675), and The Plain-Dealer (1676).

Wyclif /'wɪklɪf/, John (also **Wycliffe**) (c.1330–84), English religious reformer. He was a lecturer at Oxford (1361–82) and a prolific writer, whose attacks on medieval theocracy are regarded as precursors of the Reformation. He criticized the wealth and power of the Church, upheld the Bible as the sole guide for doctrine, and questioned the scriptural basis of the papacy; his teachings were disseminated by itinerant preachers. In accordance with his belief that such texts should be accessible to ordinary people, he instituted the first English translation of the complete Bible. He was compelled to retire from Oxford after his attack on the doctrine of transubstantiation and after the Peasants' Revolt (1381), which was blamed on his teaching. The followers of Wyclif were known as Lollards.

Wye /waɪ/ a river which rises in the mountains of western Wales and flows 208 km (132 miles) generally south-eastwards, entering the Severn estuary at Chepstow. In its lower reaches it forms part of the border between Wales and England.

Wykehamist /'wɪkəmɪst/ n. & adj. ● n. a past or present member of Winchester College. ● adj. of or concerning Winchester College. [mod.L Wykehamista f. William of Wykeham, bishop of Winchester and founder of the college (1324–1404)]

wyn var. of WEN[2].

wynd /waɪnd/ n. Sc. & N. Engl. a narrow street or alley. [ME, app. f. the stem of WIND[2]]

Wyndham /'wɪndəm/, John (pseudonym of John Wyndham Parkes Lucas Beynon Harris) (1903–69), English writer of science fiction. He is noted for several novels, including The Day of the Triffids (1951), The Chrysalids (1955), and The Midwich Cuckoos (1957). His fiction often deals with a sudden invasion of catastrophe, usually fantastic rather than technological in nature, and analyses its psychological impact.

Wyo. abbr. Wyoming.

Wyoming /waɪ'əʊmɪŋ/ a state in the west central US; pop. (1990) 453,590; capital, Cheyenne. Acquired as part of the Louisiana Purchase in 1803, it became the 44th state of the US in 1890.

WYSIWYG /'wɪzɪˌwɪg/ adj. (also **wysiwyg**) Computing denoting the representation of text onscreen in a form exactly corresponding to its appearance on a printout. [acronym of what you see is what you get]

wyvern /'waɪv(ə)n/ n. (also **wivern**) Heraldry a winged two-legged dragon with a barbed tail. [ME wyver f. OF wivre, guivre f. L vipera: for -n cf. BITTERN[1]]

Xx

X[1] /eks/ *n.* (also **x**) (*pl.* **Xs** or **X's**) **1** the twenty-fourth letter of the alphabet. **2** (as a Roman numeral) ten. **3** (usu. **x**) *Algebra* the first unknown quantity. **4** *Geom.* the first coordinate. **5** an unknown or unspecified number or person etc. **6** a cross-shaped symbol esp. used to indicate position (*X marks the spot*) or incorrectness, or to symbolize a kiss or a vote, or as the signature of a person who cannot write. □ **x-axis** the horizontal axis in a set of rectangular coordinates, a graph, etc.

X[2] *symb.* (of films) classified as suitable for adults only. ¶ Formerly used to indicate that persons under 18 would not be admitted; it was replaced in the UK in 1983 by *18*, and in the US in 1990 by *NC-17* (no children under 17).

-x *suffix* forming the plural of many nouns in *-u* taken from French (*beaux; tableaux*). [F]

Xankändi /ˌxɑːnkənˈdiː/ (called in Russian *Stepanakert*) the capital of Nagorno-Karabakh in southern Azerbaijan; pop. (1990) 58,000. The city was founded after the Russian Revolution; the Russian form was named after Stepan Shaumyan, a Communist leader from Baku.

Xanthian Marbles /ˈzænθɪən/ sculptures found in 1838 at the site of the ancient Lycian town of Xanthus (now in Turkey), which are now in the British Museum. The figures are Assyrian in character and are believed to have been executed before 500 BC; the subjects include processions, athletic activity, sieges, and tomb scenes.

xanthic /ˈzænθɪk/ *adj.* yellowish. □ **xanthic acid** *Chem.* an organic acid containing the group –OCS₂H. □ **xanthate** /-θeɪt/ *n.* [Gk *xanthos* yellow]

Xanthippe /zænˈθɪpɪ/ (also **Xantippe** /-ˈtɪpɪ/) (5th century BC), wife of the philosopher Socrates. Her bad-tempered behaviour towards her husband has made her proverbial as a shrew.

xanthoma /zænˈθəʊmə/ *n.* (*pl.* **xanthomas** or **xanthomata** /-mətə/) *Med.* **1** a skin disease characterized by irregular yellow patches. **2** such a patch. [as XANTHIC + -OMA]

xanthophyll /ˈzænθəfɪl/ *n. Biochem.* an oxygen-containing carotenoid found in plants, one of the pigments involved in photosynthesis. Xanthophylls are responsible for the yellow colour of leaves in the autumn. [as XANTHIC + Gk *phullon* leaf]

Xavier, St Francis /ˈzævɪə(r), ˈzeɪv-/ (known as the 'Apostle of the Indies') (1506–52), Spanish missionary. While studying in Paris in 1529 he met St Ignatius Loyola and five years later became with him one of the original seven Jesuits. He was ordained in 1537, and from 1540 onwards made a series of missionary journeys to southern India, Malacca, the Moluccas, Sri Lanka, and Japan, during which he made many thousands of converts. He died while on his way to China. Feast day, 3 Dec.

X chromosome *n.* a mammalian sex chromosome of which the number in female cells is twice that in male cells. [X as an arbitrary label + CHROMOSOME]

x.d. *abbr.* ex dividend.

Xe *symb. Chem.* the element xenon.

xebec /ˈziːbek/ *n.* (also **zebec, zebeck**) a small three-masted Mediterranean vessel with lateen and usu. some square sails. [alt. (after Sp. *xabeque*) of F *chebec* f. It. *sciabecco* f. Arab. *šabāk*]

Xenakis /zeˈnɑːkɪs/, Iannis (b.1922), French composer and architect, of Greek descent. Born in Romania, he later moved to Greece, where he studied engineering, and settled in Paris in 1947. He became a

French citizen and worked for twelve years for the architect Le Corbusier, during which time he began to compose music. He evolved a stochastic style of composition, in which a random sequence of notes is produced according to mathematical probabilities, as in *Pithoprakta* (1955–6). His music also makes use of computer-aided calculations and electronic instruments.

xeno- /ˈzenəʊ/ *comb. form* **1 a** foreign. **b** a foreigner. **2** other. [Gk *xenos* strange, foreign, stranger]

xenogamy /zeˈnɒɡəmɪ/ *n. Bot.* cross-fertilization. □ **xenogamous** *adj.*

xenograft /ˈzenəˌɡrɑːft/ *n. Med. & Biol.* a tissue graft from a donor of a different species from the recipient (cf. ALLOGRAFT). Also called *heterograft, heterotransplant.* [XENO- + GRAFT[1]]

xenolith /ˈzenəlɪθ/ *n. Geol.* an inclusion within an igneous rock mass, usu. derived from the immediately surrounding rock.

xenon /ˈzenɒn/ *n.* an unreactive gaseous chemical element (atomic number 54; symbol **Xe**). One of the noble gases, xenon was discovered by William Ramsay and M. W. Travers (1872–1961) in 1898. It is obtained by distilling liquefied air and is used in certain specialized electric lamps. Since 1962 several compounds of xenon have been prepared. [Gk, neut. of *xenos* strange]

Xenophanes /zeˈnɒfəˌniːz/ (*c.*570–*c.*480 BC), Greek philosopher. He was a member of the Eleatic school of philosophers and a critic of the belief that the gods resembled human beings, whether in conduct, physical appearance, or understanding. He was a proponent of a form of monotheism, arguing that there is a single eternal self-sufficient Consciousness which influences the universe (with which it is identical) through thought.

xenophobe /ˈzenəˌfəʊb/ *n. & adj.* ● *n.* a person given to xenophobia. ● *adj.* being or characteristic of a xenophobe.

xenophobia /ˌzenəˈfəʊbɪə/ *n.* a deep dislike of foreigners. □ **xenophobic** *adj.*

Xenophon /ˈzenəf(ə)n/ (*c.*435–*c.*354 BC), Greek historian, writer, and military leader. He was born in Athens and became a disciple and friend of Socrates. In 401 he joined the campaign of the Persian prince Cyrus the Younger (see CYRUS[2]) against Artaxerxes II; when Cyrus was killed north of Babylon, Xenophon led an army of 10,000 Greek mercenaries in their retreat to the Black Sea, a journey of about 1,500 km (900 miles). His historical works include the *Anabasis*, an account of the campaign with Cyrus and its aftermath, and the *Hellenica*, a history of Greece. Among his other writings are three works concerning the life and teachings of Socrates, and the *Cyropaedia*, a historical romance on the education of Cyrus the Younger.

xeranthemum /zɪəˈrænθɪməm/ *n.* a composite plant of the genus *Xeranthemum*, with dry everlasting daisy-like flowers. [mod.L f. Gk *xēros* dry + *anthemon* flower]

xeric /ˈzɪərɪk/ *adj. Ecol.* (of habitat) very dry. [as XERO- + -IC]

xero- /ˈzɪərəʊ, ˈzerəʊ/ *comb. form* dry. [Gk *xēros* dry]

xeroderma /ˌzɪərəˈdɜːmə, ˌzer-/ *n. Med.* a disease characterized by extreme dryness of the skin, esp. a mild form of ichthyosis. [mod.L (as XERO-, Gk *derma* skin)]

xerograph /ˈzɪərəˌɡrɑːf, ˈzer-/ *n.* a copy produced by xerography.

xerography /zɪəˈrɒɡrəfɪ, zeˈrɒɡ-/ *n.* a dry copying process in which black or coloured powder adheres to parts of a surface remaining electrically charged after exposure of the surface to light from an image of the document to be copied. □ **xerographic** /ˌzɪərə'græfɪk, ˌzer-/ adj. **xerographically** adv.

xerophilous /zɪə'rɒfɪləs, ze'rɒf-/ adj. (of a plant) adapted to extremely dry conditions.

xerophyte /'zɪərəˌfaɪt, 'zer-/ n. (also **xerophile** /-ˌfaɪl/) a plant able to grow in very dry conditions, e.g. in a desert.

Xerox /'zɪərɒks, 'zer-/ n. & v. ● n. propr. **1** a machine for copying by xerography. **2** a copy made using this machine. ● v.tr. (**xerox**) reproduce by this process. [invented f. XEROGRAPHY]

Xerxes I /'zɜːksiːz/ (c.519–465 BC), son of Darius I, king of Persia 486–465. When his father died he inherited the task of taking revenge on the Greeks for their support of the Ionian cities that had revolted against Persian rule. His invasion force crossed the Hellespont with a bridge of boats, and in 480 won victories by sea at Artemisium and by land at Thermopylae. However, he was subsequently forced to withdraw his forces from Greece after suffering defeat by sea at Salamis (480) and by land at Plataea (479). Xerxes was later murdered, together with his eldest son.

Xhosa /'kəʊsə, 'kɔːsə/ n. & adj. ● n. **1** (pl. same or **Xhosas**) a member of a Bantu-speaking people forming the second largest ethnic group in South Africa after the Zulus. **2** the language of this people, forming part of the Nguni language group (which also includes Zulu). ● adj. of or relating to this people or language. [Nguni]

xi /saɪ, gzaɪ, zaɪ/ n. the fourteenth letter of the Greek alphabet (Ξ, ξ). [Gk]

Xiamen /ʃɑː'men/ (also **Hsia-men**; also called *Amoy*) a port in Fujian province, SE China; pop. (est. 1986) 962,000.

Xian /ʃiː'æn/ (also **Hsian, Sian**) an industrial city in central China, capital of Shaanxi province; pop. (1990) 2,710,000. The city has been inhabited since the 11th century BC and was the capital of several ruling dynasties of China. The present city was founded in 202 BC and, named Changan, was the capital of the Han dynasty (206 BC–AD 220) and the Sui dynasty (581–618). Named Siking it was the capital of the Tang dynasty (618–907). The capital moved away after the fall of the Tang dynasty, and the city declined. The name Xian 'western peace', first adopted during the Ming dynasty (1368–1644), replaced the official name Siking in 1943. In 1974 the discovery was made nearby of the burial place of the Qin emperor Shi Huangdi (c.259–210 BC), who is buried in an elaborate tomb complex guarded by a 'terracotta army', 10,000 life-size pottery soldiers and horses.

Ximenes de Cisneros see JIMÉNEZ DE CISNEROS.

Xingtai /ʃɪŋ'taɪ/ a city in NE China, situated to the south of Shijiazhuang in the province of Hebei; pop. (1986) 1,167,000.

Xingú /ʃɪŋ'guː/ a South American river, which rises in the Mato Grosso of western Brazil and flows 1,979 km (1,230 miles) generally northwards to join the Amazon delta.

Xining /ʃiː'nɪŋ/ (also **Hsining, Sining**) a city in north central China, capital of Qinghai province; pop. (est. 1986) 927,000.

Xinjiang /ʃɪn'dʒjæŋ/ (also **Sinkiang** /-'kjæŋ/) an autonomous region of NW China, on the border with Mongolia and Kazakhstan; pop. (1990) 15,156,000; capital, Urumqi. A remote mountainous region, it includes the Tien Shan and Kunlun Shan mountains, the Taklimakan Desert, and the arid Tarim Basin. Controlled for centuries by the Uygur people, the region was traversed by the Silk Road, China's trade route with the west.

-xion /kʃ(ə)n/ suffix forming nouns (see -ION) from Latin participial stems in -x- (*fluxion*).

Xiphias /'zɪfɪˌæs/ Astron. an old name for the constellation Dorado. [Gk *xiphos* sword]

xiphisternum /ˌzɪfɪ'stɜːnəm/ n. Anat. = xiphoid process. [as XIPHOID + STERNUM]

xiphoid /'zɪfɔɪd/ adj. Biol. sword-shaped. □ **xiphoid process** Anat. the cartilaginous process at the lower end of the sternum. [Gk *xiphoeidēs* f. *xiphos* sword]

Xizang see TIBET.

Xmas /'krɪsməs, 'eksməs/ n. colloq. = CHRISTMAS. [16th c., with X for the initial chi of Gk *Khristos* Christ]

xoanon /'zəʊəˌnɒn/ n. (pl. **xoana** /-nə/) Gk Antiq. a primitive usu.

wooden image of a deity, often said to have fallen from heaven. [Gk f. *xeō* carve]

X-rated /'eksreɪtɪd/ attrib.adj. **1** indecent, pornographic (*X-rated humour*). **2** hist. relating to films given an X classification.

X-ray /'eksreɪ/ n. & v. (also **x-ray**) ● n. **1** (in pl.) electromagnetic radiation of short wavelength, able to pass through opaque bodies. (*See note below.*) **2** an image made by the effect of X-rays on a photographic plate, esp. showing the position of bones etc. by their greater absorption of the rays. ● v.tr. photograph, examine, or treat with X-rays. □ **X-ray crystallography** the study of crystals and their structure by means of the diffraction of X-rays by the regularly spaced atoms of a crystalline material. **X-ray tube** a device for generating X-rays by accelerating electrons to high energies and causing them to strike a metal target from which the X-rays are emitted. [transl. G *x-Strahlen* (pl.) f. *Strahl* ray, so called because when discovered in 1895 the nature of the rays was unknown]

▪ X-rays were discovered in 1895 by Wilhelm Röntgen, who noticed the fluorescence of certain crystals near to an operating cathode-ray tube. He attributed this to a new kind of radiation from the tube which he called X-rays. He discovered that the rays travel in straight lines, that they are not electrical, that they affect photographic plates, and that they have considerable penetrating power. He produced the first X-ray photograph or radiograph of the human hand. By 1912 it was recognized that X-rays are electromagnetic radiation of a very high frequency with wavelengths more than a thousand times smaller than those of visible light. X-rays are produced when very high-velocity electrons strike a target, as in an X-ray tube; the rapid deceleration of the electrons produces a pulse of X-radiation. High doses of X-rays are damaging to living tissue but may be used to treat cancer.

X-ray astronomy n. the observation of celestial objects with instruments capable of detecting and measuring high-energy electromagnetic radiation. Because the atmosphere absorbs practically all cosmic X-rays, it is necessary to place X-ray telescopes and spectrometers in orbiting satellites. Objects known to emit X-rays include the sun, certain cataclysmic variable stars, and galaxies containing extremely hot gas. The detection of X-rays from certain close binary stars offers a potential method of detecting black holes.

Xuzhou /ʃuː'dʒəʊ/ (also **Hsu-chou** /-'tʃəʊ/, **Suchow** /suː'tʃəʊ/) a city in Jiangsu province, eastern China; pop. (est. 1990) 910,000. It was formerly known as Tongshan (1912–45).

xylem /'zaɪlem/ n. Bot. woody tissue (cf. PHLOEM). [Gk *xulon* wood]

xylene /'zaɪliːn/ n. Chem. each of three isomeric liquid hydrocarbons (chem. formula: $C_6H_4(CH_3)_2$), obtained by distilling wood, coal tar, etc. [f. as XYLEM + -ENE]

xylo- /'zaɪləʊ/ comb. form wood. [Gk *xulon* wood]

xylocarp /'zaɪləˌkɑːp/ n. Bot. a hard woody fruit. □ **xylocarpous** /ˌzaɪlə'kɑːpəs/ adj.

xylograph /'zaɪləˌɡrɑːf/ n. a woodcut or wood-engraving (esp. an early one).

xylography /zaɪ'lɒɡrəfɪ/ n. **1** the (esp. early or primitive) practice of making woodcuts or wood-engravings. **2** the use of wood blocks in printing.

Xylonite /'zaɪləˌnaɪt/ n. propr. a kind of celluloid. [irreg. f. *xyloidin* (as XYLO-) + -ITE¹]

xylophagous /zaɪ'lɒfəɡəs/ adj. (of an insect or mollusc) eating, or boring into, wood.

xylophone /'zaɪləˌfəʊn/ n. a musical instrument of graduated wooden bars with tubular resonators suspended vertically beneath them, struck with small hammers. The instrument was known before the 9th century in Africa and in Java; in central Europe a different type was a popular folk instrument from the 15th century. It was first used in orchestral music by Saint-Saëns in *Danse macabre* (1874), its sound representing rattling skeletons. □ **xylophonic** /ˌzaɪlə'fɒnɪk/ adj. **xylophonist** /zaɪ'lɒfənɪst/ n. [Gk *xulon* wood + -PHONE]

xystus /'zɪstəs/ n. (pl. **xysti** /-taɪ/) **1** a covered portico used by athletes in ancient Greece for exercise. **2** Rom. Antiq. a garden walk or terrace. [L f. Gk *xustos* smooth f. *xuō* scrape]

Yy

Y¹ /waɪ/ n. (also **y**) (pl. **Ys** or **Y's**) **1** the twenty-fifth letter of the alphabet. **2** (usu. **y**) *Algebra* the second unknown quantity. **3** *Geom.* the second coordinate. **4 a** a Y-shaped thing, esp. an arrangement of lines, piping, roads, etc. **b** a forked clamp or support. □ **y-axis** the vertical axis in a set of rectangular coordinates, a graph, etc.

Y² *abbr.* (also **Y.**) **1** yen. **2** *N. Amer.* = YMCA, YWCA.

Y³ *symb. Chem.* the element yttrium.

y. *abbr.* year(s).

y- /ɪ/ *prefix archaic* forming past participles, collective nouns, etc. (*yclept*). [OE *ge-* f. Gmc]

-y¹ /ɪ/ *suffix* forming adjectives: **1** from nouns and adjectives, meaning: **a** full of; having the quality of (*messy; icy; horsy*). **b** addicted to (*boozy*). **2** from verbs, meaning 'inclined to', 'apt to' (*runny; sticky*). [from or after OE *-ig* f. Gmc]

-y² /ɪ/ *suffix* (also **-ey, -ie**) forming diminutive nouns, pet names, etc. (*granny; Sally; nightie; Mickey*). [ME (orig. Sc.)]

-y³ /ɪ/ *suffix* forming nouns denoting: **1** state, condition, or quality (*courtesy; orthodoxy; modesty*). **2** an action or its result (*colloquy; remedy; subsidy*). [from or after F *-ie* f. L *-ia, -ium*, Gk *-eia, -ia*: cf. -ACY, -ERY, -GRAPHY, and others]

yabber /ˈjæbə(r)/ v. & n. *Austral. & NZ colloq.* ● *v.intr. & tr.* talk, conversation, language. □ **yabber stick** = *message stick.* [perh. f. an Aboriginal language]

yabby /ˈjæbɪ/ n. (also **yabbie**) (pl. **-ies**) *Austral.* **1** a small edible freshwater crayfish, esp. of the genus *Cherax.* **2** = NIPPER 5. [Aboriginal]

yacht /jɒt/ n. & v. ● n. **1** a light sailing-vessel, esp. equipped for racing. **2** a larger usu. power-driven vessel equipped for cruising. **3** a light vessel for travel on sand or ice. ● *v.intr.* race or cruise in a yacht. □ **yacht-club** a club esp. for yacht-racing. [early mod.Du. *jaghte* = *jaghtschip* fast pirate-ship f. *jag(h)t* chase f. *jagen* to hunt + *schip* SHIP]

yachtie /ˈjɒtɪ/ n. a yachtsman or yachtswoman.

yachting /ˈjɒtɪŋ/ n. the sport or pastime of racing or cruising in yachts. Yachting was pioneered in the 17th century in the Netherlands, where the sheltered waterways provided the main communication network. Charles II acquired a liking for the sport during his Dutch exile, but it remained an eccentric pursuit in England until yacht racing began, supported by royal patronage, in the mid-18th century. The first English club was formed at Cowes on the Isle of Wight in 1815. Yachting became an Olympic sport in 1900, with events in various classes for yachts having one, two, or three crew members. The most important yachting trophy is the America's Cup; in recent years regular single-handed transatlantic races have also been established.

yachtsman /ˈjɒtsmən/ n. (pl. **-men**; *fem.* **yachtswoman**, pl. **-women**) a person who sails yachts.

yack /jæk/ n. & v. (also **yackety-yack**, /ˌjækətɪˈjæk/) *sl. derog.* ● n. trivial or unduly persistent conversation. ● *v.intr.* engage in this. [imit.]

yacka (also **yacker**) var. of YAKKA.

yaffle /ˈjæf(ə)l/ n. *dial.* the green woodpecker. [imit. of its laughing cry]

Yafo see JAFFA¹.

Yagi aerial /ˈjɑːgɪ/ n. a highly directional radio aerial made of several short rods mounted across an insulating support and transmitting or receiving a narrow band of frequencies. [Hidetsugu *Yagi*, Jap. engineer (1886–1976)]

yah /jɑː/ int. expressing derision or defiance. [imit.]

yahoo /jəˈhuː/ n. a coarse person; a lout, hooligan. [name of an imaginary race of brutish creatures in Swift's *Gulliver's Travels* (1726)]

Yahweh /ˈjɑːweɪ/ n. (also **Yahveh** /ˈjɑːveɪ/) a form of the Hebrew name of God (*YHVH*) used in the Bible. The name came to be regarded by Jews (*c.*300 BC) as too sacred to be spoken, and the vowel sounds are uncertain. (See also JEHOVAH.)

Yahwist /ˈjɑːwɪst/ n. (also **Yahvist** /ˈjɑːvɪst/) the postulated author or authors of parts of the first six books of the Bible in which God is regularly named *Yahweh* (cf. ELOHIST). [YAHWEH]

Yajur-veda /ˌjʌdʒʊəˈveɪdə, -ˈviːdə/ *Hinduism* one of the four Vedas, based on a collection of sacrificial formulae in early Sanskrit used in the Vedic religion by the priest in charge of sacrificial ritual. [Skr., f. *yajus* sacrificial formula + VEDA]

yak /jæk/ n. a long-haired humped domesticated ox, *Bos grunniens*. It is kept in the uplands of central Asia for its milk and soft under-fur, and as a pack-animal. [Tibetan *gyag*]

yakka /ˈjækə/ n. (also **yacka, yacker** /-kə(r)/) *Austral. sl.* work. [Aboriginal]

Yakut /jæˈkʊt/ n. & adj. ● n. (pl. same or **Yakuts**) **1** a member of a Mongoloid people of Yakutia in Siberia. **2** the Turkic language of the Yakuts. ● adj. of or relating to the Yakuts or their language. [Russ. f. Yakut]

Yakutia /jæˈkʊtɪə/ (official name *Republic of Sakha*) an autonomous republic in eastern Russia; pop. (1989) 1,081,000; capital, Yakutsk. It is the largest of the sovereign republics and the coldest inhabited region of the world, with 40 per cent of its territory lying to the north of the Arctic Circle.

Yakutsk /jæˈkʊtsk/ a city in eastern Russia, on the Lena river, capital of the republic of Yakutia; pop. (1990) 187,000.

yakuza /jəˈkuːzə/ n. (pl. same) **1** a Japanese gangster or racketeer. **2** such gangsters collectively. [Jap., f. *ya* eight + *ku* nine + *za* three, with ref. to the worst kind of hand in a gambling game]

Yale lock /jeɪl/ n. propr. a type of lock for doors etc. with a cylindrical barrel turned by a flat key with a serrated edge. [Linus *Yale*, Jr., Amer. locksmith (1821–68)]

Yale University /jeɪl/ an American university, founded in 1701 at Killingworth and Saybrook, Connecticut, by a group of Congregational ministers. In 1716 it moved to its present site at New Haven and soon afterwards was renamed Yale College after Elihu Yale (1649–1721), a notable benefactor. In 1887 it became Yale University.

Yalta Conference /ˈjæltə/ a meeting which took place in Feb. 1945 at Yalta, a Crimean port on the Black Sea. The Allied leaders Churchill, Roosevelt, and Stalin met to plan the final stages of the Second World War and to agree the subsequent territorial division of Europe.

Yalu /ˈjɑːluː/ a river of eastern Asia, which rises in the mountains of Jilin province in NE China and flows about 800 km (500 miles) generally south-westwards to the Yellow Sea, forming most of the border between China and North Korea. In Nov. 1950 the advance of UN troops towards the Yalu river precipitated the Chinese invasion of North Korea.

yam /jæm/ n. **1 a** a tropical or subtropical climbing plant of the genus *Dioscorea.* **b** the edible starchy tuber of this, a staple food in many tropical countries. **2** *N. Amer.* a variety of the sweet potato. [Port. *inhame* or Sp. *iñame*, prob. of W. Afr. origin]

Yama /ˈjæmə/ *Hindu Mythol.* the first man to die. He became the guardian, judge, and ruler of the dead, and is represented as carrying a noose and riding a buffalo. [Skr. *yama* restraint (*yam* restrain)]

Yamamoto /ˌjæməˈməʊtəʊ/, Isoroku (1884–1943), Japanese admiral. Although he initially opposed his country's involvement in the Second World War, as Commander-in-Chief of the Combined Fleet (air and naval forces) from 1939, he was responsible for planning the successful Japanese attack on the US naval base at Pearl Harbor (1941). He then directed his forces in Japanese operations to gain control of the Pacific, but was thwarted by the defeat of his fleet at the Battle of Midway (1942). He was killed when the Allies shot down his plane over the Solomon Islands.

Yamasaki /ˌjæməˈsɑːkɪ/, Minoru (1912–86), American architect. He designed the St Louis Municipal Airport Terminal (1956), a barrel-vaulted building which influenced much subsequent American air-terminal design. Other notable designs include the World Trade Center in New York (1972), a skyscraper consisting of twin towers 110 storeys high.

Yamato-e /jæˈmætəʊ͵eɪ/ *n.* a style of decorative painting in Japan during the 12th and early 13th centuries, characterized by strong colour and flowing lines. [Jap., = Japanese]

yammer /ˈjæmə(r)/ *n. & v. colloq. or dial.* ● *n.* **1** a lament, wail, or grumble. **2** voluble talk. ● *v.intr.* **1** utter a yammer. **2** talk volubly. □ **yammerer** *n.* [OE *geōmrian* f. *geōmor* sorrowful]

Yamoussoukro /ˌjæmuːˈsuːkrəʊ/ the capital of the Ivory Coast; pop. (est. 1986) 120,000. It was originally a small village, the birthplace and headquarters of President Félix Houphouët-Boigny (1905–93), and was designated as capital in 1983, replacing Abidjan.

Yamuna see JUMNA.

Yancheng /jæŋˈtʃɛŋ/ (also **Yen-cheng** /jen-/) a city in Jiangsu province, eastern China; pop. (1986) 1,265,000.

yandy /ˈjændɪ/ *v. & n. Austral.* ● *v.tr.* (**-ies, -ied**) separate (grass seed) from refuse by special shaking. ● *n.* (*pl.* **-ies**) a shallow dish used for this. [Aboriginal]

yang /jæŋ/ *n.* (in Chinese philosophy) the active principle of the universe, characterized as male and creative and associated with heaven, heat, and light (complemented by YIN). [Chin.]

Yangon see RANGOON.

Yangshao /jæŋˈʃaʊ/ an ancient civilization of northern China during the 3rd millennium BC, characterized by painted pottery with naturalistic designs of fish and human faces and abstract patterns of triangles, spirals, arcs, and dots.

Yangtze /ˈjæŋtsɪ/ (also called *Chang Jiang*) the principal river of China, which rises as the Jinsha in the Tibetan highlands and flows 6,380 km (3,964 miles) southwards then generally eastwards through central China, entering the East China Sea at Shanghai.

Yank /jæŋk/ *n. colloq.* often *derog.* an inhabitant of the US; an American. [abbr.]

yank /jæŋk/ *v. & n. colloq.* ● *v.tr.* pull with a jerk. ● *n.* a sudden hard pull. [19th c.: orig. unkn.]

Yankee /ˈjæŋkɪ/ *n. colloq.* **1** often *derog.* = YANK. **2** *US* an inhabitant of New England or one of the northern states. **3** *hist.* a Federal soldier in the Civil War. **4** a type of bet on four or more horses to win (or be placed) in different races. **5** (*attrib.*) of or as of the Yankees. [18th c.: orig. uncert.: perh. f. Du. *Janke* dimin. of *Jan* John, attested (17th c.) as a nickname]

Yankee Doodle *n.* **1** a popular American tune, first used by British troops in the War of American Independence to deride the American colonial revolutionaries and subsequently adopted by the Americans. **2** = YANKEE.

Yantai /jænˈtaɪ/ (also **Yen-tai** /jen-/) a port on the Yellow Sea in Shandong province, eastern China; pop. (est. 1986) 734,000. Founded in the 3rd century BC, it was traditionally known as Chefoo, being named Yantai in the 15th century.

Yaoundé /jæˈʊndeɪ/ the capital of Cameroon; pop. (est. 1984) 552,000.

yap /jæp/ *v. & n.* ● *v.intr.* (**yapped, yapping**) **1** bark shrilly or fussily. **2** *colloq.* talk noisily, foolishly, or complainingly. ● *n.* **1** a sound of yapping. **2** *US sl.* the mouth. □ **yapper** *n.* [imit.]

yapok /ˈjæpɒk/ *n.* a semiaquatic tropical American opossum, *Chironectes minimus*, having grey fur with dark bands. Also called *water opossum*. [*Oyapok, Oiapoque*, N. Brazilian river]

yapp /jæp/ *n. Brit.* a form of bookbinding with a limp leather cover projecting to fold over the edges of the leaves. [William *Yapp*, London bookseller (1854–75), for whom first made]

yarborough /ˈjɑːbərə/ *n.* a whist or bridge hand with no card above a 9. [Earl of *Yarborough* (1809–97), said to have betted against its occurrence]

yard[1] /jɑːd/ *n.* **1** a unit of linear measure equal to 3 feet (0.9144 metre). **2** this length of material (*a yard and a half of cloth*). **3** a square or cubic yard esp. (in building) of sand etc. **4** a cylindrical spar tapering to each end slung across a mast for a sail to hang from. **5** (in *pl.*; foll. by *of*) *colloq.* a great length (*yards of spare wallpaper*). □ **by the yard** at great length. **yard-arm** the outer extremity of a ship's yard. **yard of ale** *Brit.* **1** a deep slender beer glass, about a yard long and holding two to three pints. **2** the contents of this. [OE *gerd* f. WG]

yard[2] /jɑːd/ *n.* ● *n.* **1** a piece of enclosed ground esp. attached to a building or used for a particular purpose. **2** *N. Amer.* the garden of a house. ● *v.tr.* put (cattle) into a stockyard. □ **the Yard** *Brit. colloq.* = SCOTLAND YARD. **yard-man 1** a person working in a railway-yard or timber-yard. **2** *US* a gardener or a person who does various outdoor jobs. [OE *geard* enclosure, region, f. Gmc: cf. GARDEN]

yardage /ˈjɑːdɪdʒ/ *n.* **1** a number of yards of material etc. **2 a** the use of a stockyard etc. **b** payment for this.

yardbird /ˈjɑːdbɜːd/ *n. US sl.* **1** a new military recruit. **2** a convict.

Yardie /ˈjɑːdɪ/ *n.* a member of any of various West Indian (esp. Jamaican) gangs engaging in organized crime, esp. illegal drug-trafficking. [Jamaican Engl. *yard* home, Jamaica]

yardstick /ˈjɑːdstɪk/ *n.* **1** a standard used for comparison. **2** a measuring rod a yard long, usu. divided into inches etc.

yarmulke /ˈjɑːmʊlkə/ *n.* (also **yarmulka**) a skullcap worn by Jewish men. [Yiddish]

yarn /jɑːn/ *n. & v.* ● *n.* **1** spun thread, esp. for knitting, weaving, rope-making, etc. **2** *colloq.* a long or rambling story or discourse. ● *v.intr. colloq.* tell yarns. [OE *gearn*]

Yaroslavl /ˌjɑːrəsˈlɑːv(ə)l/ a port in European Russia, on the River Volga north-east of Nizhni Novgorod; pop. (1990) 636,000.

yarran /ˈjærən/ *n.* an Australian acacia; esp. *Acacia omalophylla*, a small tree with scented wood used for fencing, fuel, etc. [Aboriginal]

yarrow /ˈjærəʊ/ *n.* a perennial composite plant of the genus *Achillea*; esp. *A. millefolium*, with finely divided leaves and clusters of small white or pink flowers (also called *milfoil*). [OE *gearwe*, of unkn. orig.]

yashmak /ˈjæʃmæk/ *n.* a veil concealing the face except the eyes, worn by some Muslim women when in public. [Arab. *yašmaḳ*, Turk. *yaşmak*]

yataghan /ˈjætə͵gæn/ *n.* a sword without a guard and often with a double-curved blade, used in Muslim countries. [Turk. *yātāğan*]

yatter /ˈjætə(r)/ *v. & n. colloq. or dial.* ● *v.intr.* (often foll. by *on*) talk idly or incessantly; chatter. ● *n.* idle talk; incessant chatter. □ **yattering** *n. & adj.*

yaw /jɔː/ *v. & n.* ● *v.intr.* (of a ship or aircraft etc.) fail to hold a straight course; fall off; go unsteadily (esp. turning from side to side). ● *n.* the yawing of a ship etc. from its course. [16th c.: orig. unkn.]

yawl /jɔːl/ *n.* **1** a two-masted fore-and-aft sailing-boat with the mizen-mast stepped far aft. **2** a small kind of fishing-boat. **3** *hist.* a ship's jolly boat with four or six oars. [MLG *jolle* or Du. *jol*, of unkn. orig.: cf. JOLLY[2]]

yawn /jɔːn/ *v. & n.* ● *v.* **1** *intr.* (as a reflex) open the mouth wide and inhale esp. when sleepy or bored. **2** *intr.* (of a chasm etc.) gape, be wide open. **3** *tr.* utter or say with a yawn. ● *n.* **1** an act of yawning. **2** *colloq.* a boring or tedious idea, activity, etc. □ **yawner** *n.* **yawningly** *adv.* [OE *ginian, geonian*]

yawp /jɔːp/ *v. & n.* ● *n.* **1** a harsh or hoarse cry. **2** *US* foolish talk. ● *v.intr.* utter these. □ **yawper** *n.* [ME (imit.)]

yaws /jɔːz/ *n.pl.* (usu. treated as *sing.*) a contagious tropical skin disease with large red swellings, caused by a spirochaete bacterium. [17th c.: orig. unkn.]

Yayoi /ˈjɑːjɔɪ/ a neolithic culture of Japan dating from the 3rd century BC. It was marked by the introduction of rice cultivation to Japan, and the appearance of large burial mounds has suggested the emergence of an increasingly powerful ruling class. [Jap., name of a street in Tokyo where its characteristic chiefly wheel-made pottery was first discovered]

Yb *symb. Chem.* the element ytterbium.

Y chromosome *n.* a mammalian sex chromosome occurring only in male cells. [*Y* as an arbitrary label + CHROMOSOME]

yclept /ɪˈklept/ *adj. archaic or joc.* called (by the name of). [OE *gecleopod* past part. of *cleopian* call f. Gmc]

yd. *abbr.* yard (measure).

yds *abbr.* yards (measure).

ye[1] /jiː/ *pron. archaic* pl. of THOU[1]. □ **ye gods!** *joc.* an exclamation of astonishment. [OE *ge* f. Gmc]

ye[2] /jiː/ *adj. pseudo-archaic* = THE (*ye olde tea-shoppe*). [var. spelling f. the *y*-shaped letter THORN (representing *th*) in the 14th c.]

yea /jeɪ/ *adv. & n. archaic* ● *adv.* **1** yes. **2** indeed, nay (*ready, yea eager*). ● *n.* the word *yea*. □ **yea and nay** shilly-shally. **yeas and nays** affirmative and negative votes. [OE *gea, ge* f. Gmc]

yeah /jeə/ *adv. colloq.* yes. □ **oh yeah?** expressing incredulity. [casual pronunc. of YES]

yean /jiːn/ *v.tr. & intr. archaic* bring forth (a lamb or kid). [perh. f. OE *geēanian* (unrecorded, as Y-, *ēanian* to lamb)]

yeanling /ˈjiːnlɪŋ/ *n. archaic* a young lamb or kid.

year /jɪə(r)/ *n.* **1** (also **astronomical year, equinoctial year, natural year, solar year, tropical year**) the time occupied by the earth in one revolution round the sun, 365 days, 5 hours, 48 minutes, and 46 seconds in length (cf. *sidereal year*). **2** (also **calendar year, civil year**) the period of 365 days (**common year**) or 366 days (see *leap year*) from 1 Jan. to 31 Dec., used for reckoning time in ordinary affairs. **3 a** a period of the same length as this starting at any point (*four years ago*). **b** such a period in terms of a particular activity etc. occupying its duration (*school year; tax year*). **4** (in *pl.*) age or time of life (*young for his years*). **5** (usu. in *pl.*) *colloq.* a very long time (*it took years to get served*). **6** a group of students entering college etc. in the same academic year. □ **in the year of Our Lord** (foll. by the year) in a specified year AD. **of the year** chosen as outstanding in a particular year (*sportsman of the year*). **a year and a day** the period specified in some legal matters to ensure the completion of a full year. **the year dot** see DOT[1]. **year-end** *n.* the end of esp. the financial year (often *attrib.*: *year-end profits*). **year in, year out** continually over a period of years. **year-long** lasting a year or the whole year. **year of grace** the year AD. **year-round** existing etc. throughout the year. [OE *gē(a)r* f. Gmc]

yearbook /ˈjɪəbʊk/ *n.* an annual publication dealing with events or aspects of the (usu. preceding) year.

yearling /ˈjɪəlɪŋ/ *n. & adj.* ● *n.* **1** an animal between one and two years old. **2** a racehorse in the calendar year after the year of foaling. ● *adj.* **1** a year old; having existed or been such for a year (*a yearling heifer*). **2** intended to terminate after one year (*yearling bonds*).

yearly /ˈjɪəlɪ/ *adj. & adv.* ● *adj.* **1** done, produced, or occurring once a year. **2** of or lasting a year. ● *adv.* once a year; from year to year. [OE *gēarlic, -lice* (as YEAR)]

yearn /jɜːn/ *v.intr.* **1** (usu. foll. by *for, after,* or *to* + infin.) have a strong emotional longing. **2** (usu. foll. by *to, towards*) be filled with compassion or tenderness. □ **yearner** *n.* **yearning** *n. & adj.* **yearningly** *adv.* [OE *giernan* f. a Gmc root meaning 'eager']

yeast /jiːst/ *n.* **1 a** a microscopic fungus of the genus *Saccharomyces*, consisting of single oval cells that reproduce by budding, and capable of converting sugar into alcohol and carbon dioxide. **b** a greyish-yellow preparation of this obtained esp. from fermented malt liquors, used as a fermenting agent and to raise bread etc. **2** any unicellular fungus in which vegetative reproduction takes place by budding or fission. □ **yeastless** *adj.* **yeastlike** *adj.* [OE *gist, giest* (unrecorded): cf. MDu. *ghist*, MHG *jist*, ON *jöstr*]

yeasty /ˈjiːstɪ/ *adj.* (**yeastier, yeastiest**) **1** frothy or tasting like yeast. **2** in a ferment. **3** working like yeast. **4** (of talk etc.) light and superficial. □ **yeastily** *adv.* **yeastiness** *n.*

Yeats /jeɪts/, W(illiam) B(utler) (1865–1939), Irish poet and dramatist. He spent a large part of his life in London, although his interest in Irish cultural and political life remained constant. He was a co-founder of the Irish National Theatre Company (later based at the Abbey Theatre) and his play *The Countess Cathleen* (1892) began the Irish theatrical revival. He was also prominent in Ireland's cultural and literary revival, which takes its name from his collection of stories *The Celtic Twilight* (1893). In his poetry, the elaborate style of his earlier work was influenced by the Pre-Raphaelites, while his later work used a sparser, more lyrical, style and was influenced by symbolism, as well as by his interest in mysticism and the occult. His best-known collections from this later period are *The Tower* (1928), including the poems 'Sailing to Byzantium' and 'Leda and the Swan', and *The Winding Stair* (1929). Yeats served as a senator of the Irish Free State (1922–8) and was awarded the Nobel Prize for literature in 1923.

yegg /jeg/ *n. US sl.* a travelling burglar or safe-breaker. [20th c.: perh. a surname]

Yekaterinburg /jeˌkætəˈriːnbɜːg/ (also **Ekaterinburg**) an industrial city in central Russia, in the eastern foothills of the Urals; pop. (1990) 1,372,000. It was founded by Peter the Great in 1721 and named after his wife Catherine (Russian *Ekaterina*). Between 1924 and 1991 it was known as Sverdlovsk.

Yekaterinodar /jəˌkætəˈriːnəˌdɑː(r)/ (also **Ekaterinodar**) a former name (until 1922) for KRASNODAR.

Yekaterinoslav /jəˌkætəˈriːnəˌslɑːf/ (also **Ekaterinoslav**) a former name (1787–1926) for DNIPROPETROVSK.

Yelizavetpol /jəˌliːzəˈvjetpɒl/ (also **Elizavetpol**) the former Russian name (1804–1918) for GÄNCÄ.

yell /jel/ *n. & v.* ● *n.* **1** a loud sharp cry of pain, anger, fright, encouragement, delight, etc. **2** a shout. **3** *US* an organized shout, used esp. to support a sports team. **4** *sl.* an amusing person or thing. ● *v.tr. & intr.* utter with or make a yell. [OE *g(i)ellan* f. Gmc]

yellow /ˈjeləʊ/ *adj., n., & v.* ● *adj.* **1** of the colour between green and orange in the spectrum, of buttercups, lemons, egg-yolks, or gold. **2** of the duller colour of faded leaves, ripe wheat, old paper, etc. **3** having a yellow skin or complexion. **4** *colloq.* cowardly. **5** (of looks, feelings, etc.) jealous, envious, or suspicious. **6** (of newspapers etc.) unscrupulously sensational. ● *n.* **1** a yellow colour or pigment. **2** yellow clothes or material (*dressed in yellow*). **3 a** a yellow ball, piece, etc., in a game or sport. **b** the player using such pieces. **4** (usu. in *comb.*) a yellow moth or butterfly. **5** (in *pl.*) jaundice of horses etc. **6** *US* a peach-disease with yellowed leaves. ● *v.tr. & intr.* make or become yellow. □ **yellow archangel** a yellow-flowered Eurasian dead-nettle, *Lamiastrum galeobdolon*. **yellow arsenic** = ORPIMENT 1. **yellow-bellied 1** *colloq.* cowardly. **2** (of an animal) having yellow underparts. **yellow-belly 1** *colloq.* a coward. **2** a fish or turtle with yellow underparts. **yellow card** *Football* a card shown by the referee to a player being cautioned. **yellow flag 1** a flag displayed by a ship in quarantine. **2** a yellow-flowered European iris, *Iris pseudacorus*, common in streams and pools. **yellow jack 1** = YELLOW FEVER. **2** = *yellow flag* 1. **yellow line** (in the UK) a line painted along the side of the road in yellow either singly or in pairs to denote parking restrictions. **yellow metal** brass of 60 parts copper and 40 parts zinc. **yellow peril** *derog. offens.* the political or military threat regarded as emanating from Asian peoples, esp. the Chinese. **yellow rattle** a yellow-flowered partly parasitic plant, *Rhinanthus minor*, of the figwort family. **yellow rocket** see ROCKET[2]. **yellow spot** the point of acutest vision in the retina. **yellow streak** *colloq.* a trait of cowardice. □ **yellowish** *adj.* **yellowly** *adv.* **yellowness** *n.* **yellowy** *adj.* [OE *geolu, geolo* f. WG, rel. to GOLD]

yellowback /ˈjeləʊˌbæk/ *n.* a cheap novel etc. in a yellow cover.

Yellow Book an illustrated literary periodical published quarterly in the UK between 1894 and 1897, so called because of its distinctive yellow binding. Associated with the Aesthetic Movement and often controversial, the *Yellow Book* contained contributions from writers including Max Beerbohm, Henry James, Edmund Gosse, Arnold Bennett, and H. G. Wells. The art editor was Aubrey Beardsley.

yellowcake /ˈjeləʊˌkeɪk/ *n.* impure uranium oxide obtained during the processing of uranium ore.

yellow fever *n. Med.* a tropical viral disease with fever and jaundice, transmitted by the bite of the mosquito *Aedes aegypti*. Primarily a disease of monkeys, it became established in humans many centuries ago and was probably carried from Africa to the West Indies when African slaves were first taken there. Yellow fever is the notorious 'yellow jack' of the old sea-stories. It is often fatal if untreated, and is best controlled by prevention and inoculation.

yellowhammer /ˈjeləʊˌhæmə(r)/ *n.* a common Eurasian bunting, *Emberiza citrinella*, the male of which has a bright yellow head and underparts. [16th c.: orig. of *hammer* uncert.]

Yellowknife /ˈjeləʊˌnaɪf/ the capital of the Northwest Territories in Canada, on the north shore of the Great Slave Lake; pop. (1991) 11,860. It was founded in 1935 as a gold-mining town and became capital in 1967.

yellowlegs /ˈjeləʊˌlegz/ *n.* a migratory American sandpiper with yellow legs. There are two species, the *lesser yellowlegs* (*Tringa flavipes*) and the *greater yellowlegs* (*T. melanoleuca*).

Yellow Pages *n.pl. propr.* a telephone directory, or a section of a directory, printed on yellow paper and listing business subscribers according to the goods or services they offer.

Yellow River (called in Chinese *Huang Ho*) the second largest river in China, which rises in the mountains of west central China and flows over 4,830 km (3,000 miles) in a huge semicircle before entering the gulf of Bo Hai.

Yellow Sea (called in Chinese *Huang Hai*) an arm of the East China Sea, separating the Korean peninsula from the east coast of China.

Yellowstone National Park /ˈjeləʊˌstəʊn/ a national park in NW Wyoming and Montana. Named after the Yellowstone river, a tributary of the Missouri which runs through it, the park was established in 1872 and was the first national park in the US. It is noted for its many geysers, hot springs, and mud volcanoes, especially Old Faithful, a geyser which erupts every 45 to 80 minutes to a height of about one hundred feet.

yelp /jelp/ *n. & v.* ● *n.* a sharp shrill cry of or as of a dog in pain or excitement. ● *v.intr.* utter a yelp. □ **yelper** *n.* [OE *gielp(an)* boast (imit.): cf. YAWP]

Yeltsin /ˈjeltsɪn/, Boris (Nikolaevich) (b.1931), Russian statesman, President of the Russian Federation since 1991. At first a supporter of Mikhail Gorbachev's reform programme, he soon became his leading radical opponent, impatient with the slow pace of change. In 1990 Yeltsin was elected President of the Russian Soviet Federative Socialist Republic; shortly afterwards he and his supporters resigned from the Communist Party, creating a powerful opposition movement. He emerged with new stature after the attempted coup of 1991 during which he rallied support for Gorbachev; on the breakup of the USSR at the end of that year he became President of the independent Russian Federation. He continued to face opposition to his reforms and in 1993 survived an attempted coup.

Yemen /ˈjemən/ a country in the south and south-west of the Arabian peninsula; pop. (est. 1991) 12,533,000; official language, Arabic; capital, Sana'a. An Islamic country since the mid-7th century, Yemen was part of the Ottoman Empire from the 16th century. It came under increasing British influence in the 19th century as a result of its strategic importance at the mouth of the Red Sea; the port of Aden was developed as a British military base. After the Second World War civil war between royalist and republican forces ended with British withdrawal and the partition of the country (1967). South Yemen declared itself independent as the People's Democratic Republic of Yemen, the North becoming the Yemen Arab Republic. The countries reunited to form the Republic of Yemen in 1990, the former capital of South Yemen, Aden, functioning as the country's commercial centre. In 1994, however, a civil war erupted and the South proclaimed its secession. □ **Yemeni** *adj. & n.* **Yemenite** *adj. & n.*

yen[1] /jen/ *n.* (*pl.* same) the chief monetary unit of Japan. [Jap. *en* round]

yen[2] /jen/ *n. & v. colloq.* ● *n.* a longing or yearning. ● *v.intr.* (**yenned**, **yenning**) feel a longing. [Chin. *yăn* craving for opium]

Yen-cheng see YANCHENG.

Yenisei /ˌjenɪˈseɪ/ a river in Siberia, which rises in the mountains on the Mongolian border and flows 4,106 km (2,566 miles) generally northwards to the Arctic coast, emptying into the Kara Sea.

Yen-tai see YANTAI.

yeoman /ˈjəʊmən/ *n.* (*pl.* **-men**) **1** esp. *hist.* a man holding and cultivating a small landed estate. **2** *hist.* a person qualified by possessing free land of an annual value of 40 shillings to serve on juries, vote for the knight of the shire, etc. **3** *Brit.* a member of the yeomanry force. **4** *hist.* a servant in a royal or noble household. **5** (in full **yeoman of signals**) a petty officer in the navy, concerned with visual signalling. **6** *US* a petty officer performing clerical duties on board ship. □ **yeoman** (or **yeoman's**) **service** efficient or useful help in need. [ME *yoman*, *yeman*, etc., prob. f. YOUNG + MAN]

yeomanly /ˈjəʊmənlɪ/ *adj.* **1** of the rank of yeoman. **2** characteristic of or befitting a yeoman; sturdy, reliable.

Yeoman of the Guard *n.* a member of the British sovereign's bodyguard, first established by Henry VII, now having only ceremonial duties. Along with the warders of the Tower of London (who, like them, wear Tudor dress as uniform) they are commonly known as beefeaters.

yeomanry /ˈjəʊmənrɪ/ *n.* (*pl.* **-ies**) **1** a body of yeomen. **2** *Brit. hist.* a volunteer cavalry force raised from the yeoman class (1794–1908).

Yeoman Usher *n. Brit.* the deputy of Black Rod.

yep /jep/ *adv. & n.* (also **yup** /jʌp/) *US colloq.* = YES. [corrupt.]

-yer /jə(r)/ *suffix* var. of -IER esp. after *w* (*bowyer*; *lawyer*).

yerba maté /ˈjɜːbə/ *n.* = MATÉ. [Sp., = herb maté]

Yerevan /ˌjerɪˈvæn/ (also **Erevan**) the capital of Armenia; pop. (1990) 1,202,000.

yes /jes/ *adv. & n.* ● *adv.* **1** equivalent to an affirmative sentence: the answer to your question is affirmative, it is as you say or as I have said, the statement etc. made is correct, the request or command will be complied with, the negative statement etc. made is not correct. **2** (in answer to a summons or address) an acknowledgement of one's presence. ● *n.* **1** an utterance of the word *yes*. **2** an affirmation or assent. **3** a vote in favour of a proposition. □ **say yes** grant a request or confirm a statement. **yes?** **1** indeed? is that so? **2** what do you want? **yes, and** a form for introducing a stronger phrase (*he came home drunk — yes, and was sick*). **yes and no** that is partly true and partly untrue. **yes-man** (*pl.* **-men**) *colloq.* a weakly acquiescent person, an obsequious subordinate. [OE *gēse*, *gīse*, prob. f. *gīa sīe* may it be (*gīa* is unrecorded)]

yester- /ˈjestə(r)/ *comb. form poet.* or *archaic* of yesterday; that is the last past (*yester-eve*). [OE *geostran*]

yesterday /ˈjestəˌdeɪ, -dɪ/ *adv. & n.* ● *adv.* **1** on the day before today. **2** in the recent past. **3** *colloq.* immediately, very urgently (*need that report yesterday*). ● *n.* **1** the day before today. **2** the recent past. □ **not born yesterday** not naive or a fool. **yesterday morning** (or **afternoon** etc.) in the morning (or afternoon etc.) of yesterday. [OE *giestran dæg* (as YESTER-, DAY)]

yesteryear /ˈjestəˌjɪə(r)/ *n. literary* **1** last year. **2** the recent past.

yet /jet/ *adv. & conj.* ● *adv.* **1** as late as now (or then), until now (or then) (*there is yet time; your best work yet*). **2** (with *neg.* or *interrog.*) so soon as now (or then), by now (or then) (*it is not time yet; have you finished yet?*). **3** again; in addition (*more and yet more*). **4** in the remaining time available; before all is over (*I will do it yet*). **5** (foll. by *compar.*) even (*a yet more difficult task*). **6** nevertheless; and in spite of that; but for all that (*it is strange, and yet it is true*). ● *conj.* but at the same time; but nevertheless (*I won, yet what good has it done?*). □ **nor yet** and also not (*won't listen to me nor yet to you*). [OE *gīet(a)*, = OFris. *iēta*, of unkn. orig.]

yeti /ˈjetɪ/ *n.* = abominable snowman. [Tibetan]

Yevtushenko /ˌjeftʊˈʃeŋkəʊ/, Yevgeni (Aleksandrovich) (b.1933), Russian poet. He gained recognition in the 1950s with works such as *Third Snow* (1955) and *Zima Junction* (1956), which were regarded as encapsulating the feelings and aspirations of the post-Stalin generation. He also wrote love poetry and personal lyrics, which had been out of favour during the Stalin era. A champion of greater artistic freedom, he incurred official hostility because of the outspokenness of some of his poetry, notably *Babi Yar* (1961), which strongly criticized Russian anti-Semitism.

yew /juː/ *n.* **1** an evergreen coniferous tree of the genus *Taxus*, often planted in churchyards, having dark green needle-like leaves and poisonous seeds enclosed in a fleshy red aril. **2** its dense red wood, used in cabinet-making and formerly for making bows. [OE *īw*, *ēow* f. Gmc]

Y-fronts /ˈwaɪfrʌnts/ *n. propr.* men's or boys' briefs with a Y-shaped seam at the front.

Yggdrasil /ˈɪgdrəsɪl/ *Scand. Mythol.* a huge ash tree located at the centre of the earth, with three roots, one extending to Niflheim (the underworld), one to Jotunheim (land of the giants), and one to Asgard (land of the gods). Although threatened by a malevolent serpent that gnaws at its roots and by deer eating its foliage, the tree survives because it is watered by the Norns from the well of fate. [ON *yg(g)drasill* f. *Yggr* Odin + *drasill* horse]

YHA *abbr.* (in the UK) Youth Hostels Association.

Yichun /jiːˈtʃʊn/ (also **I-chun** /iː-/) a city in Heilongjiang province, NE China; pop. (est. 1986) 1,167,000.

Yid /jɪd/ *n. sl. offens.* a Jew. [back-form. f. YIDDISH]

Yiddish /ˈjɪdɪʃ/ *n. & adj.* ● *n.* a vernacular language related to German, used by Jews, originally in central and eastern Europe but now also in the US, Israel, and elsewhere. (*See note below.*) ● *adj.* of or relating to this language. [G *jüdisch* Jewish]

▪ Yiddish is written in Hebrew characters, with the difference that vowels are written with separate signs. It originated in the 9th century among the Jews (Ashkenazim) who settled in cities along the Rhine in Germany and adopted the local German dialect, albeit in a form heavily influenced by Hebrew, which remained their literary language. In the 14th century the dialect was carried eastwards, where it took on influences from the Slavonic tongues and developed into the distinct language known as Yiddish. In 1908 it was declared a national Jewish language; despite severe persecution by Germany in the 1930s and 1940s and by the Soviet Union in the 1950s the language survived. It is spoken today by about 200,000 people as a first language and by a further 2 million (largely in the US and Israel) as a second language. It has also had a significant effect on English vocabulary and idiom in the US, providing words such as *bagel*, *schmaltz*, and *schmuck*.

Yiddisher /ˈjɪdɪʃə(r)/ *n. & adj.* ● *n.* a person speaking Yiddish. ● *adj.* Yiddish-speaking.

Yiddishism /ˈjɪdɪˌʃɪz(ə)m/ *n.* **1** a Yiddish word, idiom, etc., esp. one adopted into another language. **2** advocacy of Yiddish culture.

yield /jiːld/ *v. & n.* ● *v.* **1** *tr.* (also *absol.*) produce or return as a fruit, profit, or result (*the land yields crops*; *the land yields poorly*; *the investment yields 15%*). **2** *tr.* give up; surrender, concede; comply with a demand for (*yielded the fortress*; *yielded themselves prisoners*). **3** *intr.* (often foll. by *to*) **a** surrender; make submission. **b** give consent or change one's course of action in deference to; respond as required to (*yielded to persuasion*). **4** *intr.* (foll. by *to*) be inferior or confess inferiority to (*I yield to none in understanding the problem*). **5** *intr.* (foll. by *to*) give right of way to other traffic. **6** *intr.* US allow another the right to speak in a debate etc. ● *n.* an amount yielded or produced; an output or return. □ **yield point** *Physics* the stress beyond which a material becomes plastic. □ **yielder** *n.* [OE *g(i)eldan* pay f. Gmc]

yielding /ˈjiːldɪŋ/ *adj.* **1** compliant, submissive. **2** (of a substance) able to bend; soft and pliable, not stiff or rigid. □ **yieldingly** *adv.* **yieldingness** *n.*

yikes /jaɪks/ *int. sl.* expressing surprise and sudden apprehension. [orig. unkn.]

yin /jɪn/ *n.* (in Chinese philosophy) the passive principle of the universe, characterized as female and sustaining and associated with earth, dark, and cold (complemented by YANG). [Chin.]

Yinchuan /jɪnˈtʃwɑːn/ the capital of Ningxia, an autonomous region of China, on the Yellow River; pop. (est. 1986) 658,000.

yip /jɪp/ *v. & n.* US ● *v.intr.* (**yipped**, **yipping**) = YELP *v.* ● *n.* = YELP *n.* [imit.]

yippee /ˈjɪpiː, jɪˈpiː/ *int.* expressing delight or excitement. [natural excl.]

yips /jɪps/ *n.pl.* (usu. prec. by *the*) *colloq.* extreme nervousness causing a golfer to miss an easy putt. [orig. unkn.]

-yl /aɪl, ɪl/ *suffix Chem.* forming nouns denoting a radical (*ethyl*; *hydroxyl*; *phenyl*).

ylang-ylang /ˈiːlæŋˌiːlæŋ/ *n.* (also **ilang-ilang**) **1** a tropical Asian tree, *Cananga odorata*, with fragrant greenish-yellow flowers from which a perfume is distilled. **2** this perfume. [Tagalog *álang-ílang*]

YMCA 1 Young Men's Christian Association, a welfare movement that began in London in 1844 and now has branches all over the world. The World Alliance of YMCAs was established in 1855; its headquarters are in Geneva. **2** a hostel run by this association. [abbr.]

Ymir /ˈiːmə(r)/ *Scand. Mythol.* the primeval giant from whose body the gods created the world.

-yne /aɪn/ *suffix Chem.* forming names of unsaturated compounds containing a triple bond (*ethyne* = acetylene).

yo /jəʊ/ *int. sl.* as a greeting or calling attention. [natural exclam.]

yob /jɒb/ *n. Brit. sl.* a lout or hooligan. □ **yobbish** *adj.* **yobbishly** *adv.* **yobbishness** *n.* [back sl. for BOY]

yobbo /ˈjɒbəʊ/ *n.* (pl. **-os**) *Brit. sl.* = YOB.

yocto- /ˈjɒktəʊ/ *comb. form* denoting a factor of 10^{-24} (*yoctometre*). [adapted f. OCTO-, after PETA-, EXA-, etc.]

yod /jɒd/ *n.* **1** the tenth and smallest letter of the Hebrew alphabet. **2** its semivowel sound /j/. [Heb. *yōd* f. *yad* hand]

yodel /ˈjəʊd(ə)l/ *v. & n.* ● *v.tr. & intr.* (**yodelled**, **yodelling**; US **yodeled**, **yodeling**) sing with melodious inarticulate sounds and frequent changes between falsetto and the normal voice in the manner of the Swiss mountain-dwellers. ● *n.* a yodelling cry. □ **yodeller** *n.* [G *jodeln*]

yoga /ˈjəʊgə/ *n.* **1** a Hindu system of philosophic meditation and asceticism designed to effect reunion with the universal spirit. (*See note below.*) **2** = HATHA YOGA. □ **yogic** /ˈjəʊgɪk/ *adj.* [Hind. f. Skr., = union]

■ Yoga dates back to the 2nd century BC or earlier. The highest form, *raja yoga*, aims at the spiritual purification of the practitioner through a series of eight stages, requiring expert teaching and guidance. The first four stages teach restraint and religious observance, followed by physical preparations and breathing exercises; the next stages involve the withdrawal of the senses, concentration of the mind, and meditation, until the final stage, *samadhi* or union with the divine, is reached. In the West the first four stages, which form the basis of *hatha yoga*, have become popular as a form of exercise and relaxation.

yogh /jɒg/ *n.* a Middle English letter (ʒ) used for certain values of *g* and *y*. [ME]

yoghurt /ˈjɒgət/ *n.* (also **yogurt**) a semi-solid sourish food prepared from milk fermented by added bacteria and often flavoured with fruit etc. [Turk. *yoğurt*]

yogi /ˈjəʊgɪ/ *n.* (pl. **yogis**) a person proficient in yoga. □ **yogism** *n.* [Hind. f. YOGA]

Yogyakarta /ˌjɒgjəˈkɑːtə/ (also **Jogjakarta** /ˌdʒɒgdʒə-/) a city in south central Java, Indonesia; pop. (1980) 398,700. It was formerly the capital of Indonesia (1945–9).

yo-heave-ho /ˌjəʊhiːvˈhəʊ/ *int. & n.* = heave-ho.

yohi /ˈjəʊaɪ/ *n.* (also **youi**) *Austral.* (in Aboriginal English) an affirmative reply, 'yes'. [Aboriginal *yaway*]

yo-ho /jəʊˈhəʊ/ *int.* (also **yo-ho-ho** /ˌjəʊhəʊˈhəʊ/) **1** used to attract attention. **2** = YO-HEAVE-HO. [cf. YO-HEAVE-HO & YOO-HOO]

yoicks /jɔɪks/ *int.* (also **hoicks** /hɔɪks/) a cry used by fox-hunters to urge on the hounds. [orig. unkn.: cf. *hyke* call to hounds, HEY[1]]

yoke /jəʊk/ *n. & v.* ● *n.* **1** a wooden crosspiece fastened over the necks of two oxen etc. and attached to the plough or wagon to be drawn. **2** (pl. same or **yokes**) a pair (of oxen etc.). **3** an object like a yoke in form or function, e.g. a wooden shoulder-piece for carrying a pair of pails, the top section of a dress or skirt etc. from which the rest hangs. **4** sway, dominion, or servitude, esp. when oppressive. **5** a bond of union, esp. that of marriage. **6** *Rom. Hist.* an uplifted yoke, or an arch of three spears symbolizing it, under which a defeated army was made to march. **7** *archaic* the amount of land that one yoke of oxen could plough in a day. **8** a crossbar on which a bell swings. **9** the crossbar of a rudder to whose ends ropes are fastened. **10** a bar of soft iron between the poles of an electromagnet. ● *v.* **1** *tr.* put a yoke on. **2** *tr.* couple or unite (a pair). **3** *tr.* (foll. by *to*) link (one thing) to (another). **4** *intr.* match or work together. [OE *geoc* f. Gmc]

yokel /ˈjəʊk(ə)l/ *n.* a rustic; a country bumpkin. [perh. f. dial. *yokel* green woodpecker]

Yokohama /ˌjəʊkəʊˈhɑːmə/ a seaport on the island of Honshu, Japan; pop. (1990) 3,220,000. Originally a small fishing village, Yokohama developed as a major port and is now the second largest city in Japan.

yolk[1] /jəʊk/ *n.* **1** the yellow inner part of an egg, rich in fat and protein, that nourishes the young before it hatches. **2** *Biol.* the corresponding part of any animal ovum. □ **yolk-bag** (or **-sac**) a membrane enclosing the yolk of an egg. □ **yolked** *adj.* (also in *comb.*). **yolkless** *adj.* **yolky** *adj.* [OE *geol(o)ca* f. *geolu* YELLOW]

yolk[2] /jəʊk/ *n.* = SUINT. [OE *eowoca* (unrecorded)]

Yom Kippur /ˌjɒm ˈkɪpə(r), kɪˈpʊə(r)/ the most solemn religious fast of the Jewish year, the last of the ten days of penitence that begin with Rosh Hashana (the Jewish New Year). Also called *Day of Atonement*. [Heb.]

Yom Kippur War the Israeli name for the Arab–Israeli conflict in 1973, called by the Arabs the *October War*. The war lasted for less than three weeks; it started on the festival of Yom Kippur (in that year, 6 Oct.) when Egyptian forces crossed the Suez Canal and attacked Israel from the south, while Syrian forces attacked Israeli forces in the Golan Heights (occupied by Israel since the Six Day War) from the north. The Syrians were repulsed and the Egyptians were surrounded. A cease-fire followed and disengagement agreements over the Suez area were signed in 1974 and 1975.

yomp /jɒmp/ *v.intr. Brit. sl.* march with heavy equipment over difficult terrain. [20th c.: orig. unkn.]

yon /jɒn/ *adj., adv., & pron. literary & dial.* ● *adj. & adv.* yonder. ● *pron.* yonder person or thing. [OE *geon*]

yonder /ˈjɒndə(r)/ *adv. & adj.* ● *adv.* over there; at some distance in that direction; in the place indicated by pointing etc. ● *adj.* situated yonder. [ME: cf. OS *gendra*, Goth. *jaindrē*]

yoni /ˈjəʊnɪ/ *n.* a symbol of the female genitals venerated by Hindus etc. [Skr., = source, womb, female genitals]

yonks /jɒŋks/ *n.pl. sl.* a long time (*haven't seen them for yonks*). [20th c.: orig. unkn.]

yoo-hoo /ˈjuːhuː/ *int.* used to attract a person's attention. [natural excl.]

yore /jɔː(r)/ *n. literary* □ **of yore** formerly; in or of old days. [OE *geāra*, *geāre* etc., adv. forms of uncert. orig.]

York /jɔːk/ a city in North Yorkshire, on the River Ouse; pop. (1991) 100,600. The Romans occupied the site, known as Eboracum, in AD 71 and it served as their northern military headquarters in Britain until about AD 400. In AD 867 it was taken by the Vikings; the city's present name is derived from the Danish Jorvik (Yorvik). It is the seat of the Archbishop of York and is noted for its magnificent cathedral, York Minster. An important railway junction, York is now the home of the National Railway Museum.

york /jɔːk/ v.tr. Cricket bowl (a batsman) with a yorker. [back-form. f. YORKER]

York, Archbishop of the archbishop of the northern province of England, Primate of England in the Anglican Church. The office of Archbishop of York dates back to the year 735.

York, Cape a cape extending into the Torres Strait at the north-east tip of Australia, in Queensland. It is the northernmost point of the continent of Australia.

York, House of the English royal house descended from Edmund of Langley (1341–1402), 1st Duke of York and 5th son of Edward III, which ruled England from 1461 (Edward IV) until the defeat and death of Richard III in 1485, with a short break in 1470–1 (the restoration of Henry VI). The House of York fought the Wars of the Roses with the House of Lancaster, both houses being branches of the Plantagenet line. Lancaster eventually prevailed, through their descendants, the Tudors, but the houses were united when the victorious Henry VII married Elizabeth, the eldest daughter of Edward IV (1486).

yorker /ˈjɔːkə(r)/ n. Cricket a ball bowled so that it pitches immediately under the bat. [prob. f. YORK, as having been introduced by Yorkshire players]

Yorkist /ˈjɔːkɪst/ n. & adj. ● n. hist. a supporter of the House of York, esp. during the Wars of the Roses (cf. LANCASTRIAN). ● adj. of or concerning the House of York.

Yorks. abbr. Yorkshire.

Yorkshire /ˈjɔːkʃɪə(r)/ a former county of northern England, divided administratively into East, West, and North Ridings. In 1974 it was divided into the new counties of North, West, and South Yorkshire, while part of the East Riding went to Humberside and part of the North Riding became Cleveland.

Yorkshire fog n. a pasture grass, Holcus lanatus, with soft hairy leaves. [FOG²]

Yorkshireman /ˈjɔːkʃəmən/ n. (pl. **-men**; fem. **Yorkshirewoman**, pl. **-women**) a native of Yorkshire.

Yorkshire pudding n. a baked batter pudding usu. eaten with roast beef.

Yorkshire terrier n. a small long-haired blue-grey and tan kind of terrier.

Yoruba /ˈjɒrʊbə/ n. & adj. ● n.**1** (pl. same or **Yorubas**) a member of a people living mainly on the coast of West Africa, especially in Nigeria where they are numerically the largest ethnic group. **2** the Niger-Congo language of the Yoruba, which is one of the major languages of Nigeria and is spoken by about 20 million people in the south-west of the country. ● adj. of or relating to the Yoruba or their language. [Yoruba]

Yorvik /ˈjɔːvɪk/ (also **Jorvik**) the Viking name for YORK.

Yosemite National Park /jəʊˈsemɪtɪ/ a national park in the Sierra Nevada in central California. Yosemite Valley, in the centre of the park, is noted for its sheer granite cliffs such as the mile-high rock face of El Capitan and several spectacular waterfalls, including Yosemite Falls, the highest waterfall in the US, with a drop of 739 m (2,425 ft).

Yoshkar-Ola /jæʃˌkɑːrəˈlɑː/ the capital of the republic of Mari El, in Russia; pop. (1990) 246,000.

yotta- /ˈjɒtə/ comb. form denoting a factor of 10^{24} (yottametre). [formed like YOCTO-, app. adapted after It. otto eight]

you /juː, jʊ, jə/ pron. (obj. **you**; poss. **your, yours**) **1** used with reference to the person or persons addressed or one such person and one or more associated persons. **2** (as int. with a noun) in an exclamatory statement (you fools!). **3** (in general statements) one, a person, anyone, or everyone (it's bad at first, but you get used to it). □ **you-all** US colloq. you (usu. more than one person). **you and yours** you together with your family, property, etc. **you-know-what** (or **-who**) something or someone unspecified but understood. [OE ēow accus. & dative of gē YE¹ f. WG: supplanting ye because of the more frequent use of the obj. case, and thou and thee as the more courteous form]

you'd /juːd, jʊd, jəd/ contr. **1** you had. **2** you would.

youi var. of YOHI.

you'll /juːl, jɔːl, jʊl, jəl/ contr. you will; you shall.

Young¹ /jʌŋ/, Brigham (1801–77), American Mormon leader. He

became a Mormon in 1832 and succeeded Joseph Smith as the movement's leader in 1844, establishing its headquarters at Salt Lake City, Utah, three years later. He served as governor of the territory of Utah from 1850 until 1857, and retained his position as leader of the Mormons until his death.

Young² /jʌŋ/, Thomas (1773–1829), English physicist, physician, and Egyptologist. Young learned thirteen languages while still a child, and pursued diverse interests in his career. His major work in physics concerned the wave theory of light, which he supported with the help of advanced experiments in optical interference. Young also devised a modulus of elasticity derived from Hooke's Law, investigated the optics of the human eye, and played a major part in the deciphering of the Rosetta Stone.

young /jʌŋ/ adj. & n. ● adj. (**younger** /ˈjʌŋgə(r)/; **youngest** /ˈjʌŋgɪst/) **1** not far advanced in life, development, or existence; not yet old. **2 a** immature or inexperienced. **b** youthful. **3** felt in or characteristic of youth (young love; young ambition). **4** representing young people (Young Conservatives; Young England). **5** distinguishing a son from his father (young Jones). **6** (**younger**) **a** distinguishing one person from another of the same name (the younger Pitt). **b** (**the Younger**) as a title denoting the younger holder of the same name (Pitt the Younger). **c** Sc. the heir of a landed commoner. ● n. (collect.) offspring, esp. of animals before or soon after birth. □ **with young** (of an animal) pregnant. **young blood** see BLOOD. **younger hand** Cards the second player of two. **young fogey** see FOGEY. **young fustic** see FUSTIC. **young hopeful** see HOPEFUL n. **young idea** the child's mind. **young lady 1** a girlfriend or sweetheart. **2** (as a patronizing form of address) a girl. **young man 1** a boyfriend or sweetheart. **2** (as a patronizing form of address) a boy. **young offender** a young criminal esp.: **a** Brit. Law a young criminal between 14 and 17 years of age. **b** Can. Law a young criminal between 12 and 18 years of age. **young person** Law (in the UK) a person generally between 14 and 17 years of age. **young thing** archaic or colloq. an indulgent term for a young person. **young 'un** colloq. a youngster. **young woman** colloq. a girlfriend or sweetheart. □ **youngish** /ˈjʌŋɪʃ/ adj. **youngling** n. [OE g(e)ong f. Gmc]

Young Italy a movement founded by Giuseppe Mazzini in 1831 to work towards a united Italian republic. In the 1830s and 1840s the movement plotted against the Italian governments but failed to gain popular support, appealing mainly to the middle classes. It was nevertheless a significant stimulus to the Risorgimento.

Young Pretender, the see STUART².

Young's modulus n. Physics a measure of elasticity, equal to the ratio of the stress acting on a substance to the strain produced. [YOUNG²]

youngster /ˈjʌŋstə(r)/ n. a child or young person.

Young Turk n. **1** a member of a group of reformers in the Ottoman Empire in the late 19th and early 20th centuries, who carried out the revolution of 1908 and deposed the sultan Abdul Hamid II. **2** a young person eager for radical change to the established order. **3** (**young turk**) offens. a violent child or youth.

younker /ˈjʌŋkə(r)/ n. archaic = YOUNGSTER. [MDu. jonckher f. jonc YOUNG + hēre lord: cf. JUNKER]

your /jɔː(r), jʊə(r), jə(r)/ poss.pron. (attrib.) **1** of or belonging to you or yourself or yourselves (your house; your own business). **2** colloq. usu. derog. much talked of; well known (none so fallible as your self-styled expert). [OE ēower genitive of gē YE¹]

you're /jɔː(r), jʊə(r), jə(r)/ contr. you are.

yours /jɔːz, jʊəz/ poss.pron. **1** the one or ones belonging to or associated with you (it is yours; yours are over there). **2** your letter (yours of the 10th). **3** introducing a formula ending a letter (yours ever; yours truly). □ **of yours** of or belonging to you (a friend of yours). **up yours** see UP.

yourself /jɔːˈself, jʊəˈself, jə-/ pron. (pl. **yourselves** /-ˈselvz/) **1 a** emphat. form of YOU. **b** refl. form of YOU. **2** in your normal state of body or mind (are quite yourself again). □ **be yourself** act in your normal, unconstrained manner. **how's yourself?** sl. how are you? (esp. after answering a similar enquiry).

youth /juːθ/ n. (pl. **youths** /juːðz/) **1** the state of being young; the period between childhood and adult age. **2** the vigour or enthusiasm, inexperience, or other characteristic of this period. **3** an early stage of development etc. **4** a young person (esp. male). **5** (pl.) young people collectively (the youth of the country). □ **youth club** (or **centre**) a place or organization provided for young people's leisure activities. **youth hostel** a place where (esp. young) holiday-makers can put up cheaply for the night. **youth hosteller** a user of a youth hostel. [OE geoguth f. Gmc, rel. to YOUNG]

youthful /ˈjuːθfʊl/ adj. **1** young, esp. in appearance or manner. **2** having the characteristics of youth (*youthful impatience*). **3** having the freshness or vigour of youth (*a youthful complexion*). □ **youthfully** adv. **youthfulness** n.

you've /juːv, jʊv, jəv/ contr. you have.

yowl /jaʊl/ n. & v. ● n. a loud wailing cry of or as of a cat or dog in pain or distress. ● v.intr. utter a yowl. [imit.]

yo-yo /ˈjəʊjəʊ/ n. & v. ● n. (also **Yo-Yo**) (pl. **yo-yos**) propr. **1** a toy consisting of a pair of discs with a deep groove between them in which string is attached and wound, and which can be spun alternately downward and upward by its weight and momentum as the string unwinds and rewinds. **2** a thing that repeatedly falls and rises again. ● v.intr. (**yo-yoes**, **yo-yoed**) **1** play with a yo-yo. **2** move up and down; fluctuate. [20th c.: orig. unkn.]

Ypres /ˈiːprə/ (Flemish **Ieper** /ˈiːpər/) a town in NW Belgium, near the border with France, in the province of West Flanders; pop. (1990) 35,235. Ypres was the scene of some of the bitterest fighting of the First World War (see YPRES, BATTLE OF).

Ypres, Battle of each of three battles on the Western Front near Ypres during the First World War. In the first battle (Oct.–Nov. 1914) Allied forces prevented the Germans breaking through to the Channel ports; the second battle (Apr.–May 1915) was an inconclusive trench conflict in which poison gas was used for the first time, while the third battle (1917) was the slaughter of Passchendaele (see PASSCHENDAELE, BATTLE OF).

yr. abbr. **1** year(s). **2** younger. **3** your.

yrs. abbr. **1** years. **2** yours.

YTS abbr. (in the UK) Youth Training Scheme.

ytterbium /ɪˈtɜːbɪəm/ n. a silvery-white metallic chemical element (atomic number 70; symbol **Yb**). A member of the lanthanide series, ytterbium was discovered by the Swiss chemist Jean Charles Galissard de Marignac (1817–94) in 1886. It has few commercial uses. [f. *Ytterby* in Sweden, site of deposits of rare-earth minerals]

yttrium /ˈɪtrɪəm/ n. a greyish-white metallic chemical element (atomic number 39; symbol **Y**). Its oxide was identified as a component of rare-earth minerals by the Finnish mineralogist Johan Gadolin (1760–1852) in 1794, and the element is usually included among the rare-earth metals. Yttrium compounds are used to make red phosphors for colour-television tubes. The metal itself is used in some alloys. Some mixed oxides of yttrium, copper, and barium are superconducting at relatively high temperatures (see SUPERCONDUCTIVITY). [f. as YTTERBIUM]

Yuan /juːˈɑːn/ a dynasty that ruled China AD 1259–1368, established by the Mongols under Kublai Khan. It was succeeded by the Ming dynasty.

yuan /juːˈɑːn/ n. (pl. same) the chief monetary unit of China. [Chin.: see YEN¹]

Yuan Jiang /juːˌɑːn ˈdʒjæŋ/ the Chinese name for the Red River in SE Asia (see RED RIVER 1).

Yucatán /ˌjuːkəˈtɑːn/ a state of SE Mexico, at the northern tip of the Yucatán Peninsula; capital, Mérida.

Yucatán Peninsula a peninsula in southern Mexico, lying between the Gulf of Mexico and the Caribbean Sea.

yucca /ˈjʌkə/ n. a woody-stemmed American plant of the genus *Yucca*, of the agave family, with rigid sword-shaped leaves. [Carib]

yuck /jʌk/ int. & n. (also **yuk**) sl. ● int. an expression of strong distaste or disgust. ● n. something messy or repellent. [imit.]

yucky /ˈjʌkɪ/ adj. (also **yukky**) (-ier, -iest) sl. **1** messy, repellent. **2** sickly, sentimental.

Yugoslav /ˈjuːɡəˌslɑːv/ n. & adj. (also **Jugoslav**) hist. ● n. **1** a native or national of Yugoslavia. **2** a person of Yugoslav descent. ● adj. of or relating to Yugoslavia or its people. [Austrian G *Jugoslav* f. Serb. *jugo-* f. *jug* south + SLAV]

Yugoslavia /ˌjuːɡəˈslɑːvɪə/ (also **Jugoslavia**) a former federation of Slavic states in SE Europe, in the Balkans. The country was formed as the Kingdom of Serbs, Croats, and Slovenes in the peace settlements at the end of the First World War. It comprised Serbia, Montenegro, and the former South Slavic provinces of the Austro-Hungarian empire, and assumed the name of Yugoslavia in 1929; its capital was Belgrade. Invaded by the Axis Powers during the Second World War, Yugoslavia emerged from a long guerrilla war as a Communist federal republic. Under the postwar leadership of Marshal Tito, Yugoslavia refused to accept Soviet domination and pursued a policy of non-alignment. Communist authority weakened after Tito's death in 1980, and in 1990 single-party rule was formally ended. Four of the six constituent republics (Slovenia, Croatia, Bosnia–Herzegovina, and Macedonia) then rapidly seceded, amid worsening civil and ethnic conflict. The two remaining republics, Serbia and Montenegro, declared a new federal republic of Yugoslavia in 1992, but this has not received widespread international recognition. □ **Yugoslavian** adj. & n.

Yuit /ˈjuːɪt/ n. (pl. same or **Yuits**) an Eskimo of Siberia, the Aleutian Islands, and parts of Alaska. [Eskimo (Yupik), = people]

yuk var. of YUCK.

yukky var. of YUCKY.

Yukon /ˈjuːkɒn/ a river of NW North America, which rises in Yukon Territory, NW Canada, and flows 3,020 km (1,870 miles) westwards through central Alaska to the Bering Sea.

Yukon Territory a territory of NW Canada, on the border with Alaska; pop. (1991) 26,900; capital, Whitehorse. The capital was at Dawson until 1953. It is a sparsely inhabited, largely undeveloped Arctic region, isolated by mountains. The population increased briefly during the Klondike gold rush (1897–9).

yule /juːl/ n. (in full **yule-tide**) archaic the Christmas festival. □ **yule-log 1** a large log burnt in the hearth on Christmas Eve. **2** a log-shaped chocolate cake eaten at Christmas. [OE *ġēol(a)*: cf. ON *jól*]

yummy /ˈjʌmɪ/ adj. (**yummier, yummiest**) colloq. tasty, delicious. [YUM-YUM + -Y¹]

yum-yum /jʌmˈjʌm/ int. expressing pleasure from eating or the prospect of eating. [natural excl.]

Yunnan /juːˈnæn/ a province of SW China, on the border with Vietnam, Laos, and Burma; capital, Kunming.

yup var. of YEP.

Yupik /ˈjuːpɪk/ n. the language of the Yuit, a major division of the Eskimo-Aleut family, comprising dialects spoken in Siberia and parts of Alaska (cf. INUPIAQ). [Eskimo (Yupik), = real person]

yuppie /ˈjʌpɪ/ n. & adj. (also **yuppy**; pl. **-ies**) colloq., usu. derog. ● n. a young middle-class professional person working in a city. ● adj. characteristic of a yuppie or yuppies. [young urban professional]

yuppify /ˈjʌpɪfaɪ/ v.tr. (**-ies, -ied**) (esp. as **yuppified** adj.) colloq. make typical of or suitable for yuppies. □ **yuppification** /ˌjʌpɪfɪˈkeɪʃ(ə)n/ n.

yurt /jʊət/ n. **1** a circular tent of felt, skins, etc., on a collapsible framework, used by nomads in Mongolia and Siberia. **2** a semi-subterranean hut, usu. of timber covered with earth or turf. [Russ. *yurta* via F *yourte* or G *Jurte* f. Turk. *jurt*]

Yuzovka /ˈjuːzəfkə/ a former name (1872–1924) for DONETSK.

YWCA 1 Young Women's Christian Association, a welfare movement that began in Britain in 1855 and now has branches in many countries. The world organization, with headquarters in Geneva, was established in 1894. **2** a hostel run by this association. [abbr.]

Zz

Z /zed/ *n.* (also **z**) (*pl.* **Zs** or **Z's**) **1** the twenty-sixth letter of the alphabet. **2** (usu. **z**) *Algebra* the third unknown quantity. **3** *Geom.* the third coordinate. **4** *Chem.* atomic number.

zabaglione /ˌzæbə'ljəʊnɪ/ *n.* an Italian sweet of whipped and heated egg yolks, sugar, and (esp. Marsala) wine. [It.]

Zabrze /'zɑːbʒə/ an industrial and mining city in Upper Silesia, southern Poland; pop. (1990) 205,000. From 1915 to 1945, when it passed back to Poland, it was a German city bearing the name Hindenburg, after the Field Marshal.

Zacatecas /ˌzækə'teɪkəs/ **1** a state of north central Mexico. **2** its capital, a silver-mining city situated at an altitude of 2,500 m (8,200 ft); pop. (1980) 165,000.

zaffre /'zæfə(r)/ *n.* (*US* **zaffer**) an impure cobalt oxide used as a blue pigment. [It. *zaffera* or F *safre*]

zag /zæg/ *n.* a sharp change of direction in a zigzag course. [ZIGZAG]

Zagazig /'zægə,zɪg/ (also **Zaqaziq**) a city in the Nile delta, northern Egypt; pop. (1991) 279,000.

Zagreb /'zɑːgreb/ the capital of Croatia; pop. (1991) 706,700.

Zagros Mountains /'zægrɒs/ a range of mountains in western Iran, rising to 4,548 m (14,921 ft) at Zard Kuh. Most of Iran's oilfields lie along the western foothills.

Zaire /zɑː'ɪə(r)/ (also **Zaïre**) a large equatorial country in central Africa with a short coastline on the Atlantic Ocean; pop. (est. 1991) 38,473,000; languages, French (official), Kongo, Lingala, Swahili, and other languages; capital, Kinshasa. Zaire is largely a low-lying forested region, encompassing the greater part of the Congo river basin. Only gradually opened up by European exploration, the area became a Belgian colony, and was known as the Congo Free State (1885–1908) and the Belgian Congo (1908–60). Independence in 1960 was followed by civil war and UN intervention. General Mobutu seized control in a coup in 1965 and has remained in power; he changed the name of the country from the Republic of the Congo to Zaire in 1971. The economy is largely based on mineral exports, particularly of copper, of which Zaire is one of the world's major producers. In 1994 the country experienced a huge influx of refugees from the violence in Rwanda. □ **Zairean** /-'ɪərɪən/ *adj. & n.*

zaire /zɑː'ɪə(r)/ *n.* the basic monetary unit of Zaire, equal to 100 makuta. [ZAIRE]

Zaire River (also called the *Congo*) a major river of central Africa, which rises as the Lualaba, to the south of Kisangani, and flows 4,630 km (2,880 miles) in a great curve westwards, turning south-westwards to form the border between the Congo and Zaire, before emptying into the Atlantic.

Zákinthos /'zækɪn,θɒs/ (also **Zakynthos**, **Zánte** /'zæntɪ/) a Greek island off the SW coast of mainland Greece, in the Ionian Sea, one of the Ionian Islands; pop. (1991) 32,750.

Zakopane /ˌzækə'pɑːneɪ/ a winter-sports resort in the Tatra Mountains of southern Poland; pop. (1990) 28,630.

Zambezi /zæm'biːzɪ/ a river of East Africa, which rises in NW Zambia and flows for 2,560 km (1,600 miles) southwards through Angola and Zaire to the Victoria Falls near Maramba, turning eastwards along the border between Zambia and Zimbabwe, before crossing central Mozambique and entering the Indian Ocean. It is dammed in two places to form the Kariba and Cabora Bassa lakes.

Zambia /'zæmbɪə/ a landlocked country in central Africa, divided from Zimbabwe by the Zambezi river; pop. (est. 1991) 8,373,000; languages, English (official), various Bantu languages; capital, Lusaka. Formerly the British protectorate of Northern Rhodesia (see RHODESIA), Zambia became an independent republic within the Commonwealth in 1964, under Kenneth Kaunda (President 1964–91). Zambia's economy was adversely affected by its involvement in the Zimbabwe independence struggle (1965–79), and a railway was constructed (with Chinese help) into Tanzania to provide an alternative route for Zambia's important copper exports. □ **Zambian** *adj. & n.*

Zamboanga /ˌzæmbəʊ'æŋgə/ a port on the west coast of Mindanao, in the southern Philippines; pop. (1990) 442,000.

zander /'zændə(r)/ *n.* a large pike-perch, *Stizostedion lucioperca*, native to central and northern Europe and introduced into western Europe. [Ger.]

Zánte see ZÁKINTHOS.

ZANU /'zɑːnuː/ see ZIMBABWE AFRICAN NATIONAL UNION.

zany /'zeɪnɪ/ *adj. & n.* ● *adj.* (**zanier**, **zaniest**) comically idiotic; crazily ridiculous. ● *n.* **1** a buffoon or jester. **2** *hist.* an attendant clown awkwardly mimicking a chief clown in shows; a merry andrew. □ **zanily** *adv.* **zaniness** *n.* [F *zani* or It. *zan(n)i*, Venetian form of *Gianni*, *Giovanni* John]

Zanzibar /ˌzænzɪ'bɑː(r)/ an island off the coast of East Africa, part of Tanzania; pop. (1988) 640,580. Coming under Omani Arab rule in the 17th century, Zanzibar was developed as a trading port. It became a sultanate in 1856 and a British protectorate in 1890. Zanzibar became an independent Commonwealth state in 1963, but in the following year the sultan was overthrown and the country became a republic, uniting with Tanganyika to form Tanzania. The island has its own administration and retains considerable autonomy. □ **Zanzibari** *adj. & n.*

Zaozhuang /zaʊ'dʒwæŋ/ (also **Tsao-chuang** /tʃaʊ'tʃwæŋ/) a city in Shandong province, eastern China; pop. (est. 1986) 1,612,000.

zap /zæp/ *v., n., & int. sl.* ● *v.* (**zapped**, **zapping**) **1** *tr.* **a** kill or destroy; deal a sudden blow to. **b** hit forcibly (*zapped the ball over the net*). **2** *intr.* move quickly and vigorously. **3** *tr.* overwhelm emotionally. **4** *tr. Computing* erase or change (an item in a program). **5** *intr.* (foll. by *through*) fast-wind a videotape to skip a section. ● *n.* **1** energy, vigour. **2** a strong emotional effect. ● *int.* expressing the sound or impact of a bullet, ray gun, etc., or any sudden event. [imit.]

Zapata /zə'pɑːtə/, Emiliano (1879–1919), Mexican revolutionary. In 1911 he participated in the revolution led by Francisco Madero (1873–1913); when Madero failed to redistribute land to the peasants, Zapata intitiated his own programme of agrarian reform and attempted to implement this by means of guerrilla warfare. He later joined forces with Pancho Villa and others, overthrowing General Huerta (1854–1916) in 1914; from 1914 to 1919 he and Villa fought against the regime of Venustiano Carranza (1859–1920). Zapata was ambushed and killed by Carranza's soldiers in 1919.

zapateado /ˌzæpətɪ'ɑːdəʊ/ *n.* (*pl.* **-os**) **1** a flamenco dance with rhythmic stamping of the feet. **2** this technique or action. [Sp. f. *zapato* shoe]

Zaporizhzhya /ˌzæpɒ'rɪʒjə/ (Russian **Zaporozhe** /zəpɑ'rɔʒjе/) an industrial city of Ukraine, on the Dnieper river; pop. (1990) 891,000. It was known until 1921 as Aleksandrovsk and developed as a major industrial centre after the construction of a hydroelectric dam in 1932.

Zapotec /'zæpə,tek/ *n. & adj.* ● *n.* (*pl.* same or **Zapotecs**) **1** a member of a native people inhabiting the region around Oaxaca in SW Mexico. **2** the language of this people. ● *adj.* of or relating to the Zapotec or their language. [Sp. f. Nahuatl]

Zappa /'zæpə/, Frank (1940–93), American rock singer, musician, and songwriter. In 1965 he formed the Mothers of Invention, who released their first album, *Freak-Out!*, in 1966. They played psychedelic rock with elements of jazz, satire, and parodies of 1950s pop, while their stage performances set out to shock. Zappa later pursued a solo career, in which he frequently combined flowing guitar improvisations with scatological humour; he also became a respected composer of avant-garde orchestral and electronic music.

zappy /'zæpɪ/ *adj.* (**zappier, zappiest**) *colloq.* **1** lively, energetic. **2** striking.

ZAPU /'zɑːpuː/ see ZIMBABWE AFRICAN PEOPLE'S UNION.

Zaqaziq see ZAGAZIG.

Zaragoza see SARAGOSSA.

zarape var. of SERAPE.

Zarathustra /ˌzærəˈθuːstrə/ the Avestan name for the Persian prophet Zoroaster. ☐ **Zarathustrian** *adj. & n.*

Zaria /'zɑːrɪə/ a city in northern Nigeria; pop. (1983) 274,000.

zariba /zəˈriːbə/ *n.* (also **zareba**) **1** a hedged or palisaded enclosure for the protection of a camp or village in Sudan etc. **2** a restricting or confining influence. [Arab. *zarība* cattle-pen]

Zarqa /'zɑːkə/ (also **Az Zarqa** /æz/) a city in northern Jordan, north-east of Amman; pop. (est. 1983) 215,000.

zarzuela /zɑːˈzweɪlə/ *n.* a Spanish traditional form of musical comedy. [Sp.: app. f. a place-name]

Zatopek /'zætəˌpek/, Emil (b.1922), Czech long-distance runner. During his career he set world records for nine different distances and in the 1952 Olympic Games won gold medals in the 5,000 metres, 10,000 metres, and the marathon.

zax var. of SAX².

zeal /ziːl/ *n.* **1** earnestness or fervour in advancing a cause or rendering service. **2** hearty and persistent endeavour. [ME *zele* f. eccl.L *zelus* f. Gk *zēlos*]

Zealand /'ziːlənd/ (Danish **Sjælland** /'sjɛlɑn/) the principal island of Denmark, situated between the Jutland peninsula and the southern tip of Sweden. Its chief city is Copenhagen.

zealot /'zelət/ *n.* **1** an uncompromising or extreme partisan; a fanatic. **2** (**Zealot**) *hist.* a member of an ancient Jewish sect aiming at a world Jewish theocracy and resisting the Romans until AD 70. ☐ **zealotry** *n.* [eccl.L *zelotes* f. Gk *zēlōtēs* (as ZEAL)]

zealous /'zeləs/ *adj.* full of zeal; enthusiastic. ☐ **zealously** *adv.* **zealousness** *n.*

zebec (also **zebeck**) var. of XEBEC.

zebra /'zebrə, 'ziːb-/ *n.* **1** a wild African horse with black and white stripes and an erect mane, esp. the common *Equus burchelli.* **2** (*attrib.*) with alternate dark and pale stripes. ☐ **zebra crossing** *Brit.* a pedestrian crossing marked by a black and white strip on the road. **zebra finch** a small Australian waxbill, *Poephila guttata*, with black and white stripes on the face, popular as a cage-bird. ☐ **zebrine** /-raɪn/ *adj.* [It. or Port. f. Congolese]

zebu /'ziːbuː/ *n.* a humped domesticated ox, *Bos indicus*, tolerant of heat and drought, orig. from India and widely kept in eastern Asia and Africa. (Cf. BRAHMAN 3; see also CATTLE.) [F *zébu*, of unkn. orig.]

Zebulun /'zebjʊlən/ (also **Zebulon**) **1** a Hebrew patriarch, son of Jacob and Leah (Gen. 30:20). **2** the tribe of Israel traditionally descended from him.

Zech. *abbr.* (in the Bible) Zechariah.

Zechariah /ˌzekəˈraɪə/ **1** a Hebrew minor prophet of the 6th century BC. **2** a book of the Bible containing his prophecies, urging the restoration of the Temple, and some later material.

zed /zed/ *n. Brit.* the letter Z. [F *zède* f. LL *zeta* f. Gk ZETA]

Zedekiah /ˌzedɪˈkaɪə/ the last king of Judaea, who rebelled against Nebuchadnezzar and was carried off to Babylon into captivity (2 Kings 24–5, 2 Chron. 36).

zedoary /'zedəʊərɪ/ *n.* **1** an Indian plant, *Curcuma zedoaria*, allied to turmeric and with an aromatic rhizome. **2** a ginger-like substance made from this rhizome, used in medicine, perfumery, and dyeing. [ME f. med.L *zedoarium* f. Pers. *zidwār*]

zee /ziː/ *n. US* the letter Z. [17th c.: var. of ZED]

Zeebrugge /'ziːˌbrʊgə/ a seaport on the coast of Belgium, linked by canal to Bruges and by ferry to Hull and Dover in England.

Zeeland /'ziːlənd/ an agricultural province of the south-western Netherlands, at the estuary of the Maas and Scheldt rivers; capital, Middelburg.

Zeeman effect /'ziːmən/ *n. Physics* the splitting of a spectrum line into several components by a magnetic field. The effect has been used in detecting and measuring the magnetic fields of stars, and in studying subatomic particles and nuclei. [Pieter *Zeeman*, Dutch physicist (1865–1943)]

Zeffirelli /ˌzefəˈrelɪ/, Franco (born Gianfranco Corsi) (b.1923), Italian film and theatre director. He began to direct his own operatic productions in the early 1950s, becoming known for the opulence of his sets and costumes, and working in many of the world's leading opera houses. He began his film career working as Visconti's assistant, and made his directorial début in the late 1960s. Among his films are *Romeo and Juliet* (1968), *Brother Sun, Sister Moon* (1973), and a film version of the opera *La Traviata* (1983).

zein /'ziːɪn/ *n. Biochem.* the principal protein of maize. [mod.L *Zea*, genus name of maize + -IN]

Zeiss /zaɪs/, Carl (1816–88), German optical instrument-maker. He established a workshop in Jena for the production of precision optical instruments in 1846, quickly establishing a reputation for products of the highest quality. Twenty years later he went into partnership with Ernst Abbe, who further enhanced the reputation of the company and eventually became its sole owner. After the Second World War a separate Zeiss company was formed in West Germany, producing microscopes, lenses, binoculars, and cameras; following German reunification, the two companies merged but the original site at Jena was closed down.

Zeitgeist /'tsaɪtgaɪst/ *n.* **1** the spirit of the times. **2** the trend of thought and feeling in a period. [G f. *Zeit* time + *Geist* spirit]

Zen /zen/ *n.* (also **Zen Buddhism**) a Japanese school of Mahayana Buddhism that teaches the attainment of enlightenment through meditation and intuition rather than through ritual worship or study of scriptures. Originating in India, Zen Buddhism was introduced to Japan via China in the 12th century, where it has had a profound cultural influence. The aim of Zen is to achieve sudden enlightenment (*satori*), usually under the guidance of a teacher; different branches of Zen teach various methods of attaining this, such as meditation on paradoxical statements (*koans*) or maintenance of a seated posture. ☐ **Zenist** *n.* (also **Zennist**). [Jap., = meditation]

zenana /zɪˈnɑːnə/ *n.* the part of a house for the seclusion of women of high-caste families in India and Iran. [Hind. *zenāna* f. Pers. *zanāna* f. *zan* woman]

Zend /zend/ *n.* an interpretation of the Avesta, each Zend being part of the Zend-Avesta. ☐ **Zend-Avesta** the Zoroastrian sacred writings of the Avesta or text and Zend or commentary. [Pers. *zand* interpretation]

Zener cards /'ziːnə(r)/ *n.* a set of 25 cards each with one of five different symbols, used in ESP research. [Karl Edward *Zener*, Amer. psychologist (1903–61)]

Zener diode /'ziːnə(r)/ *n. Electronics* a form of diode in which a certain reverse voltage produces a sudden increase in reverse current. [Clarence Melvin *Zener*, Amer. physicist (1905–93)]

zenith /'zenɪθ, 'ziːn-/ *n.* **1** *Astron.* the part of the celestial sphere directly above an observer (opp. NADIR). **2** the highest point in one's fortunes; a time of great prosperity etc. ☐ **zenith distance** an arc intercepted between a celestial object and its zenith; the complement of an object's altitude. [ME f. OF *cenit* or med.L *cenit* ult. f. Arab. *samt (ar-ra's)* path (over the head)]

zenithal /'zenɪθəl, 'ziːn-/ *adj.* of or relating to a zenith. ☐ **zenithal projection** a projection of part of a globe on to a plane tangential to the centre of the part, showing the correct directions of all points from the centre.

Zeno¹ /'ziːnəʊ/ (fl. 5th century BC), Greek philosopher. Born in Elea in SW Italy, he was a member of the Eleatic school of philosophers and a pupil of Parmenides. He defended Parmenides' theories by formulating paradoxes which set out to demonstrate that motion and plurality are illusions. One of the best known of these is the paradox of Achilles and the tortoise, apparently refuting the existence of motion; it shows that once Achilles has given the tortoise a start he can never overtake it, since by the time he arrives where it was it has already moved on.

Zeno² /'ziːnəʊ/ (known as Zeno of Citium) (c.335–c.263 BC), Greek philosopher, founder of Stoicism. Based in Athens from about 312, he was a pupil of Cynic philosophers before founding the school of Stoic philosophy in about 300 (see STOICISM). Zeno's influence on the development of Stoicism was considerable, particularly in the field of ethics, although all that remains of his treatises are fragments of quotations.

Zenobia /zeˈnəʊbɪə/ (3rd century AD), queen of Palmyra *c*.267–272. She succeeded her murdered husband as ruler and then conquered Egypt and much of Asia Minor. When she proclaimed her son emperor, the Roman emperor Aurelian marched against her and eventually defeated and captured her. She was later given a pension and a villa in Italy.

zeolite /ˈziːəˌlaɪt/ *n*. each of a number of minerals consisting mainly of hydrous silicates of calcium, sodium, and aluminium. Their crystal structures contain extensive cavities and they are able to act as molecular sieves and ion exchangers. □ **zeolitic** /ˌziːəˈlɪtɪk/ *adj*. [Sw. & G *zeolit* f. Gk *zeō* boil + -LITE (from their characteristic swelling and fusing under the blowpipe)]

Zeph. *abbr.* (in the Bible) Zephaniah.

Zephaniah /ˌzefəˈnaɪə/ **1** a Hebrew minor prophet of the 7th century BC. **2** a book of the Bible containing his prophecies.

zephyr /ˈzefə(r)/ *n*. **1** *literary* a mild gentle wind or breeze. **2** a fine cotton fabric. **3** an athlete's thin gauzy jersey. [F *zéphyr* or L *zephyrus* f. Gk *zephuros* (god of the) west wind]

Zeppelin[1] /ˈzepəlɪn/, Ferdinand (Adolf August Heinrich), Count von (1838–1917), German aviation pioneer. An army officer until his retirement in 1890, he devoted the rest of his life to the development of the dirigible airship for which he is known (see ZEPPELIN[2]). After his airship's maiden flight in 1900, Zeppelin continued to develop and produce airships at his factory at Friedrichshafen; one of his craft achieved the first 24-hour flight in 1906.

Zeppelin[2] /ˈzepəlɪn/ *n. hist.* a German dirigible airship, long and cylindrical in shape and with a rigid framework. Zeppelins were named after Count Ferdinand von Zeppelin, who developed them from the 1890s. During the First World War Zeppelins were used for reconnaissance and bombing, and after the war functioned as passenger-carrying transports until such airships were largely abandoned in the 1930s. (See also AIRSHIP.)

zepto- /ˈzeptəʊ/ *comb. form* denoting a factor of 10^{-21} (*zeptometre*). [adapted f. SEPTI-, after PETA-, EXA-, etc.]

Zermatt /ˈzɜːmæt/ an Alpine ski resort and mountaineering centre near the Matterhorn, in southern Switzerland.

zero /ˈzɪərəʊ/ *n. & v.* ● *n.* (*pl.* **-os**) **1 a** the figure 0; nought. **b** no quantity or number; nil. **2** a point on the scale of an instrument from which a positive or negative quantity is reckoned. **3** (*attrib.*) having a value of zero; no, not any (*zero population growth*). **4** (in full **zero-hour**) **a** the hour at which a planned, esp. military, operation is timed to begin. **b** a crucial moment. **5** the lowest point; a nullity or nonentity. ● *v.tr.* (**-oes**, **-oed**) **1** adjust (an instrument etc.) to zero point. **2** set the sights of (a gun) for firing. □ **zero in on 1** take aim at. **2** focus one's attention on. **zero option** a disarmament proposal for the total removal of certain types of weapons on both sides. **zero-rated** on which no value-added tax is charged. **zero-sum** (of a game, political situation, etc.) in which whatever is gained by one side is lost by the other so that the net change is always zero. [F *zéro* or It. *zero* f. OSp. f. Arab. *ṣifr* CIPHER]

zeroth /ˈzɪərəʊθ/ *adj.* immediately preceding what is regarded as 'first' in a series (*zeroth law of thermodynamics*).

zest /zest/ *n*. **1** piquancy; a stimulating flavour or quality. **2 a** keen enjoyment or interest. **b** (often foll. by *for*) relish. **c** gusto (*entered into it with zest*). **3** a scraping of orange or lemon peel as flavouring. □ **zestful** *adj.* **zestfully** *adv.* **zestfulness** *n*. **zestiness** *n*. **zesty** *adj.* (**zestier**, **zestiest**). [F *zeste* orange or lemon peel, of unkn. orig.]

zeta /ˈziːtə/ *n*. the sixth letter of the Greek alphabet (Ζ, ζ). [Gk *zēta*]

zetetic /ziːˈtetɪk/ *adj.* proceeding by inquiry. [Gk *zētētikos* f. *zēteō* seek]

zetta- /ˈzetə/ *comb. form* denoting a factor of 10^{21} (*zettametre*). [formed like ZEPTO-, app. adapted after It. *sette* seven]

zeugma /ˈzjuːɡmə/ *n*. a figure of speech using a verb or adjective with two nouns, to one of which it is strictly applicable while the word appropriate to the other is not used (e.g. *with weeping eyes and* [sc. *grieving*] *hearts*) (cf. SYLLEPSIS). □ **zeugmatic** /zjuːˈɡmætɪk/ *adj.* [L f. Gk *zeugma -atos* f. *zeugnumi* to yoke, *zugon* yoke]

Zeus /zjuːs/ *Gk Mythol.* the supreme god, the son of Cronus (whom he dethroned) and Rhea, and husband of Hera. Zeus was the protector and ruler of humankind, the dispenser of good and evil, and the god of weather and atmospheric phenomena (rain, thunder, etc.). He was identified by the Romans with Jupiter. [Gk, cognate with Skr. *dyauḥ*, = sky]

Zeuxis /ˈzjuːksɪs/ (fl. late 5th century BC), Greek painter, born at Heraclea in southern Italy. His works (none of which survive) are only known through the records of ancient writers, who make reference to monochrome techniques and his use of shading to create an illusion of depth. His verisimilitude is the subject of many anecdotes; his paintings of grapes are said to have deceived the birds.

Zhangjiakou /ˌdʒæŋdʒjaːˈkəʊ/ (also **Chang-chiakow** /ˌtʃæŋtʃjaːˈkaʊ/; called in Mongolian *Kalgan*) a city situated on the Great Wall in Hebei province, NE central China; pop. (est. 1990) 1,670,000.

Zhanjiang /dʒænˈdʒjæŋ/ (also **Chan-chiang** /tʃænˈtʃjæŋ/) a port in Guangdong province, southern China; pop. (est. 1986) 947,000.

Zhdanov /ˈʒdɑːnɒf/ the former name (1948–89) for MARIUPOL.

Zhejiang /dʒeˈdʒjæŋ/ (also **Chekiang** /tʃeˈkjæŋ/) a province of eastern China; capital, Hangzhou.

Zhengzhou /dʒeŋˈdʒəʊ/ (also **Chengchow** /tʃeŋˈtʃaʊ/) the capital of Henan province, in NE central China; pop. (1990) 1,660,000.

Zhenjiang /dʒenˈdʒjæŋ/ (also **Chen-chiang** /tʃenˈtʃjæŋ/, **Chinkiang** /tʃɪnˈkjæŋ/) a port in Jiangsu province, on the Yangtze river, eastern China; pop. (est. 1986) 422,300.

Zhitomir see ZHYTOMYR.

zho var. of DZO.

Zhongshan /dʒɒŋˈʃæn/ (also **Chung-shan** /tʃʊŋ-/) a city in Guangdong province, southern China; pop. (est. 1986) 1,073,000.

Zhou /dʒəʊ/ (also **Chou** /tʃəʊ/) a dynasty which ruled in China from the 11th century BC to 256 BC. The dynasty's rule is commonly divided into *Western Zhou* (which ruled from a capital in the west of the region near Xian until 771 BC) and *Eastern Zhou* (which ruled after 771 BC from a capital based in the east). The rule of the Eastern Zhou, although weak and characterized by strife, saw the Chinese classical age of Confucius and Lao-tzu.

Zhou Enlai /ˌdʒəʊ enˈlaɪ/ (also **Chou En-lai** /tʃəʊ/) (1898–1976), Chinese Communist statesman, Prime Minister of China 1949–76. One of the founders of the Chinese Communist Party, he joined Sun Yat-sen in 1924. In 1927 he organized a Communist workers' revolt in Shanghai in support of the Kuomintang forces surrounding the city. From the early 1930s he formed a partnership with Mao Zedong, supporting his rise to power within the Communist Party in 1935. On the formation of the People's Republic of China in 1949 Zhou became Premier and also served as Foreign Minister (1949–58). During the 1960s he continued to keep open communication channels with the US, and he presided over the moves towards détente in 1972–3. He was also a moderating influence during the Cultural Revolution.

Zhukov /ˈʒuːkɒf/, Georgi (Konstantinovich) (1896–1974), Soviet military leader, born in Russia. He was responsible for much of the planning of the Soviet Union's campaigns in the Second World War. He defeated the Germans at Stalingrad (1943), lifted the siege of Leningrad (1944), led the final assault on Germany and the capture of Berlin (1945), and became commander of the Soviet zone in occupied Germany after the war.

Zhytomyr /ʒɪˈtɒmɪə(r)/ (Russian **Zhitomir** /ʒɪˈtɔmir/) an industrial city in central Ukraine; pop. (1990) 296,000.

Zia ul-Haq /ˌzɪə ʊlˈhæk/, Muhammad (1924–88), Pakistani general and statesman, President 1978–88. As Chief of Staff he led the bloodless coup which deposed President Zulfikar Bhutto in 1977. After being sworn in as President in 1978, he banned all political parties and began to introduce strict Islamic laws. Re-appointed President in 1984, Zia ul-Haq lifted martial law but continued to maintain strict political control. He died in an air crash, possibly as the result of sabotage.

Zibo /ziːˈbəʊ, dzəˈbəʊ/ (also **Tzu-po** /tsuːˈpəʊ/) a city in Shandong province, eastern China; pop. (est. 1990) 2,460,000.

zidovudine /ˌzaɪdəʊˈvjuːdiːn/ *n*. a derivative of thymidine used to treat HIV and other viral infections. Also called *azidothymidine*, AZT. [arbitrary alt. of *azidothymidine*, chem. name]

Ziegfeld /ˈziːɡfeld/, Florenz (1869–1932), American theatre manager. In 1907 he produced the first of a series of revues in New York, based on those of the Folies-Bergère, entitled the *Ziegfeld Follies*. These continued annually until his death, being seen intermittently thereafter until 1957. Among the many famous performers promoted by Ziegfeld were W. C. Fields and Fred Astaire.

ziff /zɪf/ *n. Austral. sl.* a beard. [20th c.: orig. unkn.]

ziggurat /ˈzɪɡəˌræt/ n. (in ancient Mesopotamia) a pyramidal stepped tower built in several stages which diminished in size towards the summit, on which there may have been a shrine. Possibly derived from earlier platform temples, ziggurats are first attested in the late 3rd millennium BC; the one at Babylon may have been the biblical Tower of Babel (Gen. 11:1–9). [Assyr. *ziqquratu* pinnacle]

zigzag /ˈzɪɡzæɡ/ n., adj., adv. & v. ● n. **1** a line or course having abrupt alternate right and left turns. **2** (often in pl.) each of these turns. ● adj. having the form of a zigzag; alternating right and left. ● adv. with a zigzag course. ● v.intr. (**zigzagged**, **zigzagging**) move in a zigzag course. □ **zigzaggedly** /-ˌzæɡɪdlɪ/ adv. [F f. G zickzack]

zilch /zɪltʃ/ n. esp. N. Amer. sl. nothing. [20th c.: orig. uncert.]

zillah /ˈzɪlə/ n. an administrative district in India, containing several parganas. [Hind. *ḍilah* division]

zillion /ˈzɪljən/ n. colloq. an indefinite large number. □ **zillionth** adj. & n. [Z (perh. = unknown quantity) + MILLION]

Zimbabwe /zɪmˈbɑːbwɪ, -weɪ/ a landlocked country in SE Africa, divided from Zambia by the Zambezi river; pop. (est. 1991) 10,080,000; languages, English (official), Shona, Ndebele, and other languages; capital, Harare. Formerly known as Southern Rhodesia (see RHODESIA), the country was a self-governing British colony from 1923. In 1965 the white minority government of the colony (then called Rhodesia) issued a unilateral declaration of independence (UDI) under its Prime Minister, Ian Smith. Despite UN sanctions, illegal independence lasted until 1979, when the Lancaster House Agreement led to all-party elections and black majority rule under Robert Mugabe. The country then became an independent republic and a member of the Commonwealth, taking its name from the centre of a black civilization of the 14th–15th centuries (see GREAT ZIMBABWE). □ **Zimbabwean** /-wɪən, -weɪən/ adj. & n.

Zimbabwe African National Union (abbr. **ZANU, ZANU–PF**) a Zimbabwean political party formed in 1963 as a guerrilla organization and led from 1975 by Robert Mugabe. Having formed an alliance (called the Patriotic Front) with ZAPU in 1976 to coordinate opposition to white rule, ZANU won a large majority in the first post-independence elections (1980). ZANU and ZAPU ruled Zimbabwe as a coalition until a rift developed between them in 1982; in 1987, however, the parties agreed formally to merge, adopting the name ZANU–PF on merger in 1989.

Zimbabwe African People's Union (abbr. **ZAPU**) a Zimbabwean political party formed in 1961 as a guerrilla organization and led by Joshua Nkomo. It merged with ZANU in 1989 (see also ZIMBABWE AFRICAN NATIONAL UNION).

Zimmer /ˈzɪmə(r)/ n. propr. (also **zimmer**) (in full **Zimmer frame**) a kind of walking frame. [*Zimmer* Orthopaedic Ltd., manufacturer]

zinc /zɪŋk/ n. a silvery-white metallic chemical element (atomic number 30; symbol **Zn**). (See note below.) ● v.t. (usu. as **zinced** /zɪŋkt/ adj.) coat (iron etc.) with zinc or a zinc compound to prevent rust. □ **flowers of zinc** = zinc oxide. **zinc blende** see BLENDE. **zinc cream** esp. Austral. a heavy cream used as a preventative against sunburn. **zinc oxide** a powder (chem. formula: ZnO) used as a white pigment and in medicinal ointments. [G *Zink*, of unkn. orig.]

■ Zinc only began to be well known as a metal in the early 18th century, although brass, an alloy of zinc and copper, had been known since ancient times. The chief zinc ore is sphalerite or zinc blende. The metal is used as a component in various alloys, as well as for coating (galvanizing) iron and steel to protect against corrosion.

zinco /ˈzɪŋkəʊ/ n. & v. ● n. (pl. **-os**) = ZINCOGRAPH. ● v.tr. & intr. (**-oes, -oed**) = ZINCOGRAPH. [abbr.]

zincograph /ˈzɪŋkəˌɡrɑːf/ n. & v. ● n. **1** a zinc plate with a design etched in relief on it for printing from. **2** a print taken from this. ● v. **1** tr. & intr. etch on zinc. **2** tr. reproduce (a design) in this way. □ **zincography** /zɪŋˈkɒɡrəfɪ/ n.

zincotype /ˈzɪŋkəˌtaɪp/ n. = ZINCOGRAPH.

zing /zɪŋ/ n. & v. colloq. ● n. vigour, energy. ● v.intr. move swiftly or with a shrill sound. □ **zingy** adj. (**zingier, zingiest**). [imit.]

Zingaro /ˈzɪŋɡəˌrəʊ/ n. (pl. **Zingari** /-rɪ/) a gypsy. [It.]

zinger /ˈzɪŋə(r)/ n. US sl. an outstanding person or thing.

Zinjanthropus /zɪnˈdʒænθrəpəs/ n. a genus name sometimes applied to the fossil hominid *Australopithecus boisei*, named from remains found in Olduvai Gorge in 1959. (See AUSTRALOPITHECUS, NUTCRACKER MAN.) [mod.L f. Arab. *Zinj* East Africa + Gk *anthrōpos* man]

zinnia /ˈzɪnɪə/ n. a Mexican composite plant of the genus *Zinnia*, with showy daisy-like flowers of deep red and other colours. [Johann Gottfried *Zinn*, Ger. physician and botanist (1727–59)]

Zion /ˈzaɪən/ n. (also **Sion** /ˈsaɪ-/) **1** one of the two hills and also the citadel of ancient Jerusalem. The name came to signify Jerusalem itself and, allegorically, the heavenly city or kingdom of heaven. **2** the Jewish religion or people. **3** the Christian Church. **4** a Nonconformist chapel. [OE f. eccl.L *Sion* f. Heb. *ṣiyôn*]

Zionism /ˈzaɪəˌnɪz(ə)m/ n. a movement for the establishment of a Jewish nation in Palestine, an area claimed as the biblical territory of the Jews. The Zionist movement was established as a political organization in 1897 under the leadership of the writer Theodor Herzl, and later led by Chaim Weizmann. Under Weizmann the movement's aims were achieved with the foundation of the state of Israel in 1948, but Zionism remains an important factor in the politics of Israel and the Middle East. The movement is now concerned with the development and protection of Israel as a Jewish state and the encouragement of Jewish immigration. □ **Zionist** n. & adj.

zip /zɪp/ n. & v. ● n. **1** a light fast sound, as of a bullet passing through air. **2** energy, vigour. **3** esp. Brit. **a** a zip-fastener. **b** (attrib.) having a zip-fastener (zip bag). ● v. (**zipped, zipping**) **1** tr. & intr. (often foll. by up) fasten with a zip-fastener. **2** intr. move with zip or at high speed. □ **zip-fastener** a fastening device of two flexible strips with interlocking projections closed or opened by a sliding clip pulled along them. **zip-up** attrib.adj. having a zip-fastener. [imit.]

Zip code /zɪp/ n. US a system of postal codes consisting of five-digit numbers. [zone improvement plan]

zipper /ˈzɪpə(r)/ n. & v. esp. US ● n. a zip-fastener. ● v.tr. (often foll. by up) fasten with a zip-fastener. □ **zippered** adj.

zippy /ˈzɪpɪ/ adj. (**zippier, zippiest**) colloq. **1** bright, fresh, lively. **2** fast, speedy. □ **zippily** adv. **zippiness** n.

zircon /ˈzɜːkən/ n. a zirconium silicate of which some translucent varieties are cut into gems (see HYACINTH 4, JARGON²). [G *Zirkon*: cf. JARGON²]

zirconia /zɜːˈkəʊnɪə/ n. zirconium dioxide (chem. formula: ZrO_2), used in ceramics, refractory coatings, etc., and in fused form as a synthetic substitute for diamonds in jewellery.

zirconium /zɜːˈkəʊnɪəm/ n. a hard silver-grey metallic chemical element (atomic number 40; symbol **Zr**). A transition metal, zirconium was discovered by Martin Klaproth in 1789. It is a component of zircon and other minerals. Because of its resistance to heat and corrosion and its transparency to neutrons the metal is used as a structural material in nuclear reactors. [ZIRCON]

zit /zɪt/ n. esp. US sl. a pimple. [20th c.: orig. unkn.]

zither /ˈzɪðə(r)/ n. **1** a plucked stringed folk instrument of Austria and Bavaria, in its usual modern form consisting of a shallow wooden soundbox over which are stretched five melody strings and two sets of accompaniment strings. **2** any of various instruments in which the strings are stretched between the two ends of a flat board, stick, etc., e.g. the psaltery, dulcimer, and vina. □ **zitherist** n. [G (as CITTERN)]

zizz /zɪz/ n. & v. colloq. ● n. **1** a whizzing or buzzing sound. **2** a short sleep. ● v.intr. **1** make a whizzing sound. **2** doze or sleep. [imit.]

zloty /ˈzlɒtɪ/ n. (pl. same or **zlotys**) the chief monetary unit of Poland. [Pol., lit. 'golden']

Zn symb. Chem. the element zinc.

zodiac /ˈzəʊdɪˌæk/ n. **1 a** a belt of the heavens within about 8° of the ecliptic, including all the apparent positions of the sun, moon, and most familiar planets, and divided into twelve signs. (See note below.) **b** a representation of the signs of the zodiac or of a similar astrological system. **2** a complete cycle, circuit, or compass. □ **sign of the zodiac** each of twelve equal divisions of the zodiac (Aries, Taurus, Gemini, Cancer, Leo, Virgo, Libra, Scorpio, Sagittarius, Capricorn, Aquarius, and Pisces), originally containing the similarly named constellations. [ME f. OF *zodiaque* f. L *zodiacus* f. Gk *zōidiakos* f. *zōidion* sculptured animal-figure, dimin. of *zōion* animal]

■ The seven moving celestial bodies known before modern times — the Sun, Moon, Mercury, Venus, Mars, Jupiter, and Saturn — all move within the zodiacal band, constantly changing their positions in relation to Earth and to each other, and the supposed interpretation of these movements forms the basis of astrology. However, the modern constellations no longer represent equal divisions of the zodiac, and the ecliptic now passes through a thirteenth (Ophiuchus). Also, owing to movement of the earth's axis (see PRECESSION), the signs of the zodiac now roughly correspond to the constellations that bear the names of the *preceding* signs (though astronomers use the names Scorpius and Capricornus).

zodiacal /zəˈdaɪək(ə)l/ *adj.* of or in the zodiac. [F (as ZODIAC)]

zodiacal light *n.* a very faint elongated cone of light sometimes seen in the night sky, extending from the horizon along the ecliptic. It is thought to be due to reflection of sunlight from minute particles of ice and dust within the plane of the solar system, and is most often seen in the west after evening twilight and in the east before morning twilight, esp. in the darkness of tropical skies.

zoetrope /ˈzəʊɪˌtrəʊp/ *n. hist.* an optical toy in the form of a cylinder with a series of pictures on the inner surface which give an impression of continuous motion when viewed through slits with the cylinder rotating. [irreg. f. Gk *zōē* life + *-tropos* turning]

Zoffany /ˈzɒfənɪ/, Johann (c.1733–1810), German-born painter, resident in England from 1758. Many of his earlier paintings depict scenes from the contemporary theatre, and feature the actor David Garrick (e.g. *The Farmer's Return*, 1762). Zoffany received the patronage of George III and painted several portraits of the royal family. The king also paid for Zoffany to visit Italy (1772–9), where he painted one of his best-known works, *The Tribuna of the Uffizi* (1772–80).

Zog I /zɒg/ (full name Ahmed Bey Zogu) (1895–1961), Albanian statesman and ruler, Prime Minister 1922–4, President 1925–8, and king 1928–39. A leader of the reformist Popular Party, he headed a republican government as Premier and later President, ultimately proclaiming himself king in 1928. Zog's autocratic rule resulted in a period of relative political stability, but the close links which he had cultivated with Italy from 1925 onwards led to increasing Italian domination of Albania, and when the country was invaded by Italy in 1939, Zog went into exile. He abdicated in 1946 after Albania became a Communist state, and died in France.

zoic /ˈzəʊɪk/ *adj.* **1** of or relating to animals. **2** *Geol.* (of rock etc.) containing fossils; with traces of animal or plant life. [prob. back-form. f. AZOIC]

Zola /ˈzəʊlə/, Émile (Édouard Charles Antoine) (1840–1902), French novelist and critic. Between 1871 and 1893 he published a series of twenty novels collectively entitled *Les Rougon-Macquart*; it includes *Nana* (1880), *Germinal* (1885), and *La Terre* (1887). The series chronicles in great detail the lives of the Rougon and Macquart families over several generations, and sets out to show how human behaviour is determined by environment and heredity. His collection of essays *Le Roman expérimental* (1880), which establishes an analogy between the novelist's aims and practices and those of the scientist, is regarded as the manifesto of naturalism. Zola is also remembered for his outspoken support of Alfred Dreyfus, most notably for his pamphlet *J'accuse* (1898).

zollverein /ˈtsɒlfəˌraɪn/ *n. hist.* a customs union, esp. of German states in the 19th century. [G]

zombie /ˈzɒmbɪ/ *n.* **1** *colloq.* a dull or apathetic person. **2** a corpse said to be revived by witchcraft. [W. Afr. *zumbi* fetish]

zonation /zəʊˈneɪʃ(ə)n/ *n.* distribution in zones, esp. (*Ecol.*) of plants into zones characterized by the dominant species.

zonda /ˈzɒndə/ *n.* a hot dusty north wind in Argentina. [Amer. Sp.]

zone /zəʊn/ *n. & v.* ● *n.* **1** an area having particular features, properties, purpose, or use (*danger zone; erogenous zone; smokeless zone*). **2** any well-defined region of more or less beltlike form. **3 a** an area between two exact or approximate concentric circles. **b** a part of the surface of a sphere enclosed between two parallel planes, or of a cone or cylinder etc., between such planes cutting it perpendicularly to the axis. **4** (in full **time zone**) a range of longitudes where a common standard time is used. **5** *Geol.* etc. a range between specified limits of depth, height, etc., esp. a section of strata distinguished by characteristic fossils. **6** *Geog.* any of five divisions of the earth bounded by circles parallel to the equator (see FRIGID, TEMPERATE, TORRID). **7** an encircling band or stripe distinguishable in colour, texture, or character from the rest of the object encircled. **8** *archaic* a belt or girdle worn round the body. ● *v.tr.* **1** encircle as or with a zone. **2** arrange or distribute by zones. **3** assign as or to a particular area. □ **zonal** *adj.* **zoning** *n.* (in sense 3 of *v.*). [F *zone* or L *zona* girdle f. Gk *zōnē*]

zonk /zɒŋk/ *v. & n. sl.* ● *v.* **1** *tr.* hit or strike. **2** (often foll. by *out*) **a** *tr.* overcome with sleep; intoxicate. **b** *intr.* fall heavily asleep. ● *n.* (often as *int.*) the sound of a blow or heavy impact. [imit.]

zonked /zɒŋkt/ *adj. sl.* (often foll by *out*) exhausted; intoxicated.

zoo /zuː/ *n.* **1** an establishment which maintains a collection of animals in a park, gardens, etc. for display to the public and often also for research and conservation. **2** the premises where these animals are kept. [short for *zoological garden*]

zoo- /ˈzəʊə/ *comb. form* of animals or animal life. [Gk *zōio-* f. *zōion* animal]

zoogeography /ˌzəʊədʒɪˈɒɡrəfɪ/ *n.* the branch of zoology dealing with the geographical distribution of animals. □ **zoogeographic** /-ˌdʒiːəˈɡræfɪk/ *adj.* **zoogeographical** *adj.* **zoogeographically** *adv.*

zoography /zəʊˈɒɡrəfɪ/ *n.* descriptive zoology.

zooid /ˈzəʊɔɪd/ *n. Zool.* **1** a more or less independent invertebrate organism arising by budding or fission. **2** a distinct member of an invertebrate colony. □ **zooidal** /zəʊˈɔɪd(ə)l/ *adj.* [formed as ZOO- + -OID]

zookeeper /ˈzuːˌkiːpə(r)/ *n.* **1** a person in charge of animals in a zoo. **2** a zoo owner or director.

zoolatry /zəʊˈɒlətrɪ/ *n.* the worship of animals.

zoological /ˌzuːəˈlɒdʒɪk(ə)l, ˌzəʊə-/ *adj.* of or relating to zoology. □ **zoological garden** (or **gardens**) a zoo. □ **zoologically** *adv.*

zoology /zuːˈɒlədʒɪ, zəʊˈɒl-/ *n.* the scientific study of animals, esp. with reference to their structure, physiology, classification, and distribution. □ **zoologist** *n.* [mod.L *zoologia* (as ZOO-, -LOGY)]

▪ The systematic study of zoology began with the Greeks, stimulated by the works of Aristotle, and five of the eleven volumes of Pliny's *Historia naturalis* were on zoology. From the 17th to the early 19th centuries it became more scientifically based, due largely to the work of Harvey, Linnaeus, Buffon, and Cuvier. In the 19th century cell theory, histology, and biochemistry emerged, with major contributions also from Darwin, Wallace, and Mendel in the fields of evolution, zoogeography, and genetics. Today many zoologists are involved in laboratory-based studies in physiology, cytology, and embryology, while others study animal ecology or behaviour in the field. Classification and nomenclature remain a major field, particularly in entomology, where millions of species have yet to be discovered and there are important applications in pest research.

zoom /zuːm/ *v. & n.* ● *v.* **1** *intr.* move quickly, esp. with a buzzing sound. **2 a** *intr.* cause an aeroplane to mount at high speed and a steep angle. **b** *tr.* cause (an aeroplane) to do this. **3 a** *intr.* alter the field of view of a camera by varying the focal length of a zoom lens, esp. (foll. by *in* (on)) so as to close up on a subject without losing focus. **b** *tr.* cause (a lens or camera) to do this. **4** *intr.* (of prices) rise sharply. ● *n.* **1** an aeroplane's steep climb. **2** a zooming camera shot. □ **zoom lens** a lens with a variable focal length, esp. one able to change focal length without losing focus. [imit.]

zoomancy /ˈzəʊəˌmænsɪ/ *n.* divination from the appearances or behaviour of animals.

zoomorphic /ˌzəʊəˈmɔːfɪk/ *adj.* **1** dealing with or represented in animal forms. **2** having gods of animal form. □ **zoomorphism** *n.*

zoonosis /ˌzəʊəˈnəʊsɪs/ *n.* (pl. **-noses** /-siːz/) *Med.* a disease which can be transmitted to humans from animals. [ZOO- + Gk *nosos* disease]

zoophyte /ˈzəʊəˌfaɪt/ *n.* a plantlike animal, esp. a coral, sea anemone, or sponge. □ **zoophytic** /ˌzəʊəˈfɪtɪk/ *adj.* [Gk *zōophuton* (as ZOO-, -PHYTE)]

zooplankton /ˌzəʊəˈplæŋktən/ *n.* plankton consisting of animals.

zoospore /ˈzəʊəˌspɔː(r)/ *n.* a spore of fungi, algae, etc. capable of motion. □ **zoosporic** /ˌzəʊəˈspɔːrɪk/ *adj.*

zootomy /zəʊˈɒtəmɪ/ *n.* the dissection or anatomy of animals.

zoot suit /zuːt/ *n. colloq.* a man's suit with a long loose jacket and high-waisted tapering trousers. [rhyming on SUIT]

zori /ˈzɔːrɪ/ *n.* (pl. **zoris**) a Japanese straw or rubber etc. sandal. [Jap.]

zorilla /zɒˈrɪlə/ *n.* (also **zorille** /ˈzɒrɪl/) a striped flesh-eating African mammal, *Ictonyx striatus*, of the weasel family, resembling a skunk. [F *zorille* f. Sp. *zorrilla* dimin. of *zorro* fox]

Zoroaster /ˌzɒrəʊˈæstə(r)/ (Avestan name Zarathustra) (c.628–c.551 BC), Persian prophet and founder of Zoroastrianism. Little is known of his life, but according to tradition he was born in Persia, and began to preach the tenets of what was later called Zoroastrianism after receiving a vision from Ahura Mazda. After his death he became the subject of many legends and was variously believed to have been a magician, an astrologer, a mathematician, and a philosopher.

Zoroastrianism /ˌzɒrəʊˈæstrɪəˌnɪz(ə)m/ *n.* a monotheistic pre-Islamic religion of ancient Persia (Iran) founded by Zoroaster (or Zarathustra) in the 6th century BC. According to the teachings of Zoroaster the supreme god, Ahura Mazda, created twin spirits, one of which chose truth and light, the other untruth and darkness. Later writings present a more dualistic cosmology in which the struggle is between Ahura Mazda (now called Ormazd) and the evil spirit Ahriman. The scriptures of Zoroastrianism are the Zend-Avesta. Zoroastrianism survives today in isolated areas of Iran and in India, where followers are known as Parsees. □ **Zoroastrian** *adj. & n.*

Zouave /zuːˈɑːv, zwɑːˈv/ *n.* **1** a member of a French light-infantry corps originally formed of Algerians and retaining their oriental uniform. **2** (in *pl.*) women's trousers with wide tops, tapering to a narrow ankle. [F f. *Zouaoua*, name of a tribe]

Zoug see ZUG.

zouk /zuːk/ *n.* a style of popular music combining Caribbean and Western elements and having a fast heavy beat. [French Creole (Guadeloupe), lit. 'party']

zounds /zaʊndz/ *int. archaic* expressing surprise or indignation. [(*God*)'s *wounds* (i.e. those of Christ on the Cross)]

ZPG *abbr.* zero population growth.

Zr *symb. Chem.* the element zirconium.

Zsigmondy /ˈʃɪgmɒndɪ/, Richard Adolph (1865–1929), Austrian-born German chemist. His research began with a study of the colours of glass, which developed into his main work on colloids. He investigated the properties of various colloidal solutions, especially of gold in glass or water, and invented the ultramicroscope for counting colloidal particles. Zsigmondy was awarded the Nobel Prize for chemistry in 1925.

zucchetto /tsʊˈketəʊ/ *n.* (*pl.* **-os**) a Roman Catholic ecclesiastic's skullcap, black for a priest, purple for a bishop, red for a cardinal, and white for a pope. [It. *zucchetta* dimin. of *zucca* gourd, head]

zucchini /zuːˈkiːnɪ/ *n.* (*pl.* same or **zucchinis**) esp. *N. Amer. & Austral.* a courgette. [It., pl. of *zucchino* dimin. of *zucca* gourd]

Zug /tsuːg/ (French **Zoug** /zug/) **1** a mainly German-speaking canton in central Switzerland. The smallest canton, it joined the confederation in 1352. **2** its capital; pop. (1990) 21,500.

zugzwang /ˈtsuːktsvɑːŋ/ *n. Chess* an obligation to move in one's turn even when this must be disadvantageous. [G f. *Zug* move + *Zwang* compulsion]

Zuider Zee /ˌzaɪdə ˈziː, ˈzeɪ/ a former large shallow inlet of the North Sea, in the Netherlands. In 1932 a dam across its entrance was completed, and since then large parts of it have been drained and reclaimed as agricultural land. The remainder forms the freshwater lake of the IJsselmeer. [Du., = southern sea]

Zulu /ˈzuːluː/ *n. & adj.* ● *n.* **1** a member of a Bantu-speaking people forming the largest ethnic group in South Africa. (*See note below.*) **2** the language of this people, one of the major Bantu languages of southern Africa, spoken by about 6 million people. ● *adj.* of or relating to the Zulus or their language. [Zulu *umzulu*, pl. *amazulu*]

▪ The Zulus formed a powerful military empire in southern Africa during the 19th century before being defeated in a series of engagements with white Afrikaner and British settlers. Some Zulus still live under the traditional clan system in the province of KwaZulu/Natal, but many now work in the cities. In recent years the Zulu Inkatha movement has been drawn into violent clashes with other black groups in South Africa, particularly the Xhosa, and also into conflict with the outgoing government and the African National Congress in the period of transfer to majority rule, over future constitutional arrangements and demands for Zulu autonomy.

Zululand see KWAZULU.

Zurbarán /ˌzʊəbəˈrɑːn/, Francisco de (1598–1664), Spanish painter. In 1628 he became official painter to the town of Seville, where he spent much of his life; he also carried out commissions for many churches and for Philip IV, for whom he painted a series of mythological pictures *The Labours of Hercules* (1634) and a historical scene *The Defence of Cadiz* (1634). His work reflects the influence of Caravaggio and much of his subject-matter is religious; his other works include narrative series of scenes from the lives of the saints, painted with simple colour and form in a realistic style.

Zurich /ˈzjʊərɪk/ a city in north central Switzerland, situated on Lake Zurich; pop. (1990) 342,860. The largest city in Switzerland, it is a major international financial centre.

Zwickau /ˈtsvɪkaʊ/ a mining and industrial city in SE Germany, in Saxony; pop. (1991) 112,565.

zwieback /ˈzwiːbæk, ˈtsviːbɑːk/ *n.* a kind of biscuit rusk or sweet cake toasted in slices. [G, = twice baked]

Zwingli /ˈtsvɪŋlɪ/, Ulrich (1484–1531), Swiss Protestant reformer, the principal figure of the Swiss Reformation. He was minister of Zurich from 1518, where he sought to carry through his political and religious reforms and met with strong local support. From 1522 he published articles advocating the liberation of believers from the control of the papacy and bishops, and upholding the Gospel as the sole basis of truth. He attacked the idea of purgatory, the invocation of saints, monasticism, and other orthodox doctrines. His beliefs differed most markedly from Luther's in his rejection of the latter's doctrine of consubstantiation: at a conference in Marburg in 1529, he upheld his belief that the significance of the Eucharist was purely symbolic. The spread of Zwingli's ideas in Switzerland met with fierce resistance in some regions, and Zwingli was killed in the resulting civil war.

zwitterion /ˈzwɪtəˌraɪən, ˈtsvɪt-/ *n. Chem.* a molecule or ion having separate positively and negatively charged groups. [G f. *Zwitter* a hybrid]

Zwolle /ˈzwɒlə/ a town in the eastern Netherlands, capital of Overijssel province; pop. (1991) 95,570.

Zworykin /ˈzwɔːrɪkɪn, zvəˈriːkɪn/, Vladimir (Kuzmich) (1889–1982), Russian-born American physicist and television pioneer. He invented an electronic television input device, which incorporated a screen scanned by an electron beam and sent an electric signal to a cathode-ray tube adapted to reproduce the image. This had been developed into the first practical television camera by about 1929. Zworykin continued to be involved in television development, introducing photomultipliers to make cameras more sensitive.

zydeco /ˈzaɪdekəʊ/ *n.* a style of dance music combining elements of Cajun music with blues and rock and roll. [Louisiana Creole]

zygo- /ˈzaɪgəʊ, ˈzɪgəʊ/ *comb. form* joining, pairing. [Gk *zugo-* f. *zugon* yoke]

zygodactyl /ˌzaɪgəʊˈdæktɪl, ˌzɪgə-/ *adj. Zool.* (of a bird) having two toes pointing forward and two backward. □ **zygodactyly** *n.*

zygoma /zaɪˈgəʊmə, zɪˈgəʊ-/ *n.* (*pl.* **zygomas** or **zygomata** /-mətə/) *Anat.* the bony arch of the cheek formed by connection of the zygomatic and temporal bones. [Gk *zugōma -atos* f. *zugon* yoke]

zygomatic /ˌzaɪgəˈmætɪk, ˌzɪg-/ *adj. Anat.* of or relating to the zygoma. □ **zygomatic arch** = ZYGOMA. **zygomatic bone** the bone that forms the prominent part of the cheek.

zygomorphic /ˌzaɪgəˈmɔːfɪk, ˌzɪg-/ *adj.* (also **zygomorphous** /-ˈmɔːfəs/) *Bot.* (of a flower) divisible into similar halves by only one plane of symmetry.

zygospore /ˈzaɪgəˌspɔː(r), ˈzɪg-/ *n.* a thick-walled spore formed by certain fungi.

zygote /ˈzaɪgəʊt, ˈzɪg-/ *n. Biol.* a cell formed by the union of two gametes. □ **zygotic** /zaɪˈgɒtɪk, zɪˈgɒt-/ *adj.* **zygotically** *adv.* [Gk *zugōtos* yoked f. *zugoō* to yoke]

zygotene /ˈzaɪgəˌtiːn, ˈzɪg-/ *n. Biol.* a stage during the prophase of meiosis when homologous chromosomes begin to pair. [ZYGO- + Gk *tainia* band]

zymase /ˈzaɪmeɪs/ *n.* the enzyme fraction in yeast which catalyses the alcoholic fermentation of glucose. [F f. Gk *zumē* leaven]

zymology /zaɪˈmɒlədʒɪ/ *n.* the scientific study of fermentation. □ **zymologist** *n.* **zymological** /ˌzaɪməˈlɒdʒɪk(ə)l/ *adj.* [as ZYMASE + -LOGY]

zymosis /zaɪˈməʊsɪs, zɪˈməʊ-/ *n. archaic* fermentation. [mod.L f. Gk *zumōsis* (as ZYMASE)]

zymotic /zaɪˈmɒtɪk, zɪˈmɒt-/ *adj. archaic* of or relating to fermentation. □ **zymotic disease** *archaic* a disease regarded as caused by the multiplication of germs introduced from outside. [Gk *zumōtikos* (as ZYMOSIS)]

zymurgy /ˈzaɪmɜːdʒɪ, ˈzɪm-/ *n.* the branch of applied chemistry dealing with the use of fermentation in brewing etc. [Gk *zumē* leaven, after *metallurgy*]

Appendices

Appendices

Appendix 1 **Chronology of World Events**

All dates BP (before present)

c.4,000,000   Early hominids (*Australopithecus*) evolving in East and southern Africa.

LOWER PALAEOLITHIC

c.2,000,000   *Homo habilis* evolving in Africa; shapes and uses stone as tools; omnivore, killing small game; also a scavenger.

c.1,500,000   *Homo erectus* (formerly *Pithecanthropus*) appears in Africa; uses fabricated stone tools; omnivore, killing small animals, scavenging remains of larger ones; camps by lakes and in river valleys. Acheulian hand-axes and cleavers in East and southern Africa.

c.1,000,000   *Homo erectus* in East and SE Asia; *Australopithecus* dies out.

c.700,000   *Homo erectus* in Europe.

c.400,000   Archaic *Homo sapiens* (the earliest form of modern human) appears in Europe.

MIDDLE PALAEOLITHIC

c.120,000   Neanderthal man emerging in Europe; cave-dweller, using flint scrapers for preparing furs; burial of dead.

UPPER PALAEOLITHIC

c.35,000   Cro-Magnon man, the first anatomically modern human (*Homo sapiens sapiens*) spreading from Africa through Asia, China, Australia, America (?c.25,000). Variety of tools (knives, axes, harpoons, needles, etc.) and materials (wood, bone, stone, reed, leather, fur, etc.).

c.27,000   First painting on stone tablets in southern Africa (Namibia).

c.20,000   Finger-drawings on clay walls, e.g. Koonalda Cave, Australia. Paintings on rock surfaces in Australia.

c.17,000   First cave-painting and carving SW France (Lascaux). Female figurines across Europe and Russia.

c.15,000   Tools made on elongate blades of stone in Europe, western Asia, East and southern Africa. High level of art. Personal adornment (beads, pendants).

MESOLITHIC

c.12,000   Siberia first peopled as ice age ends. Earliest potters in Japan. Microlithic tools widely made throughout the world. Bows, spears, knives in use.

NEOLITHIC

c.10,000   Climatic changes stimulate new economies and techniques. Beginning of farming in several parts of world.

All dates BC	**NEAR EAST, MEDITERRANEAN, AND EUROPE**	**REST OF THE WORLD**
9000	Sedentary societies emerging; Natufian culture in Syria and Palestine; collection of wild cereals, first domestication of dog, pig, and goat.	
8300	Post-glacial warming and spread of forests.	
8000	Pre-pottery neolithic societies in Syria and Palestine, with cultivated cereals and mud-brick villages (Jericho). Rapid retreat of ice in northern Europe and spread of light forest; mesolithic societies, e.g. Maglemosian culture; hunting elk and wild cattle; first evidence of dugout canoe.	Cattle-keeping groups, making pottery, spread into Sahara, then wetter than today.
7000	Domestication of sheep and cattle in Near East; animals used mainly for meat, not milk or wool. Linen textiles. First use of copper for small ornaments, made by hammering and heating pure copper. Obsidian imported to mainland Greece from island of Melos.	Cultivation of root crops in New Guinea and South America.
6000	First pottery in Near East; use of smelting (lead). Major site of Çatal Hüyük in central Turkey. Domestication of cattle. Farming spreads to SE Europe; first neolithic cultures there. Oak forests spread to northern Europe; deer and pig hunted. Cattle first used for traction in Near East (plough, sledge). First evidence of irrigation, in Iraq.	Tropical millet first cultivated in the southern Sahara; temperate millet in China. Wheat and barley introduced to Pakistan.
5000	Copper-smelting in Turkey and Iran. Woollen textiles; use of animals for milk; domestication of horse and donkey. Tree crops (olive, fig, vine) cultivated in the eastern Mediterranean. Growth of population in lowland Mesopotamia; date-palm cultivated. Farming spreads into central Europe. Ertebølle culture in Baltic.	Beginning of maize cultivation in Mexico; rice cultivation in China and India.
4000	Copper-casting and alloying in Near East; development of simple copper metallurgy in SE Europe. Gold-working in Near East and Europe. First urban civilization develops in Sumer, with extensive irrigation. Trading colonies established in Syria; temple-building, craft workshops with extensive importation of raw materials such as metals and precious stones. First wheeled vehicles and sailing-boats on Euphrates and Nile; use of writing (cuneiform script); spread of advanced farming (plough, tree-crops, wool) and technology (wheel, alloy metallurgy) to SE Europe. Farming spreads to western and northern Europe;	Llama domesticated in highland Peru as pack animal; cotton cultivated in lowland Peru. Pottery comes into use in South America. Jade traded in China.

All dates BC   **NEAR EAST, MEDITERRANEAN, AND EUROPE**

	construction of monumental tombs in Portugal, Brittany, British Isles, Scandinavia. Use of horse leads to expansion of first pastoralist communities on steppes north of Black Sea; burial mounds covering pit-graves. Plough and cart widely adopted in Europe.	
3100	Unification of Upper and Lower Egypt; trading expeditions up into the countries of the eastern Mediterranean, where urban societies now exist. Troy an important trading centre in north Aegean; Cycladic culture in the Greek islands. Copper-working in Iberia (at Los Millares).	
3000	Egyptian hieroglyphic script develops; pyramid building begins. Royal tombs in Mesopotamia (e.g. Ur) demonstrate high level of craftsmanship in secular city-states. Spread of burial mounds in northern Europe replaces megalithic tradition, though ceremonial monuments (Avebury, first phase of Stonehenge) continue in British Isles. Stone-built temples in Malta.	Introduction of dog to Australia. Copper and bronze metallurgy in China and SE Asia; silk production.
2500	Extensive Egyptian maritime trade with Phoenician city of Byblos (Lebanon); Ebla a major centre in Syria, in contact with both Byblos and Mesopotamia. Exploration of eastern Mediterranean maritime routes along southern coast of Turkey to Crete, using boats with sails. Beaker cultures bring innovations (copper-working, horse, drinking-cups, woollen textiles) to Atlantic coast.	Permanent villages with temple mounds and ceremonial centres in Peru. First towns in China (Longshan culture) with trade and specialized production.
2300	Empire of Akkad unites Mesopotamian city-states. Trade with Indus valley civilization in Pakistan. Akkadian becomes diplomatic language of the Near East.	Spread of pottery-making and maize cultivation in Middle and South America.
2134	Collapse of Old Kingdom in Egypt, and of empire of Akkad.	
2040	Middle Kingdom established in Egypt; beginnings of middle Minoan (palatial) period in Crete. Revival of northern Mesopotamian centres (Assur and Mari); Assyrian merchant colonies established in Anatolia. Development of Hittite culture. Babylonian empire expands.	
1700	Egypt dominated by Asiatic rulers (Hyksos). Cretan palaces reconstructed after damage caused by earthquake or warfare; expanded trade-links with mainland Greece; growth of Mycenaean civilization. Hittite empire expands; Assyria dominated by Mitanni. Use of bronze now standard in Europe. Appearance of chariot.	Emergence of Shang civilization in China.
1500	Kassite dynasty in Babylonia; New Kingdom in Egypt following expulsion of the Hyksos; expansion of Egyptian empire in the Middle East. Rulers buried in Valley of the Kings. Akhenaten introduces monotheistic cult c.1360 and founds new capital. Canaanite cities flourish in the Middle East; use of alphabet evolving.	Metal-working (copper, gold) in Peru. Expansion of Lapita culture into western Polynesia.
1200	General recession and political collapse in many parts of eastern Mediterranean; end of Mycenaean and Hittite palace centres, decline of Egyptian power; invasion of Sea Peoples. Spread of iron metallurgy. Expansion of nomadic Aramaean tribes in Middle East. Temporary expansion of Assyria, and capture of Babylon. Expansion of agriculture and bronze-working in temperate Europe, associated with expansion of urnfield cultures.	Olmec civilization, with temple mounds and massive stone sculptures, in Mexico.

	EUROPE AND THE MEDITERRANEAN	**REST OF THE WORLD**	**CULTURE/TECHNOLOGY**
1000	Development of spice route to Arabia; growth of coastal trade in eastern Mediterranean under Phoenicians; colonization of Cyprus and exploration of central and western Mediterranean. David king of Israel (c.1000–c.962); makes Jerusalem his capital. Solomon king of Israel (c.970–930); extends his kingdom to Egypt and Euphrates.	Zhou dynasty in China. Adena culture with rich burials under large mounds in Ohio and Mississippi valleys. Chavín civilization in Andes. Early cities in Ganges valley.	Hebrew and Greek alphabets developing from Phoenician. Worship of Dionysus enters Greece from Thrace. 957 The Temple built in Jerusalem.
930	Israel divides into kingdom of Israel in the north (c.930–721) and kingdom of Judah in the south. Phoenician contacts with Crete and Euboea.		Early Hebrew texts (Psalms, Ecclesiastes). Early version of great Hindu epic the Mahabharata.
900	Celts move west to Austria and Germany.	Farming villages on Amazon floodplain.	Geometric-style pottery in Greece.
858	Assyrian empire reaches Mediterranean.		
814	Legendary date at which Phoenicians found Carthage.		
753	Legendary date for foundation of Rome.		First Olympic Games held (776).

All dates BC	**EUROPE AND THE MEDITERRANEAN**	**REST OF THE WORLD**	**CULTURE/TECHNOLOGY**
750	Greek colonies in southern Italy (Magna Graecia, Cumae) and Sicily. Greek city-state culture through Aegean; Lydia pioneers coinage. Greek colonies spread through Mediterranean.		
721	Assyrians under Sargon II (721–705) conquer Israel; under Sennacherib (705–681) empire expands. Nineveh becomes Assyrian capital.		
712	Cushite king Shabaka conquers Egypt.		
700	Hallstatt culture (Celtic iron-age warriors) in Austria moves west and down Rhône to Spain. Carthage expands through western Mediterranean, occupying Sardinia and Ibiza.	Zhou dynasty in China establishes legal system.	The *Iliad* and *Odyssey* emerge from oral tradition.
660	First Celtic hill-forts.	660 Traditional date for Jimmu, first Japanese emperor.	Dionysiac festivals in Greece leading to drama.
640	Assyrians under Ashurbanipal conquer Elamites.		Library established in Nineveh under Ashurbanipal (*c.*668–627).
625	Babylon under Chaldean dynasty (625–539).		
612	Assyria defeated by Medes and Babylonians; sack of Nineveh.		Zoroaster in Persia.
*c.*600	Ionian Greeks found Massilia (Marseilles).	Magadha kingdom on Ganges in India.	Doric order appears in Greek architecture. Trireme (Greek warship) evolves. Hanging Gardens of Babylon built.
594	Solon begins reforms of Athenian law.	New Nubian kingdom of Cush established at Meroe on upper Nile.	Greek lyric poetry (Sappho, Alcaeus). Jeremiah writing.
586	Chaldean Nebuchadnezzar conquers Jerusalem. Israelites taken into Babylonian Captivity. Greeks colonize Spain; import Cornish tin.	The king of Persia (Vishtaspa) converted to Zoroastrianism.	Thales of Miletus developing physical science and geometry.
561	Pisistratus controls Athens 561–*c.*527.	550 Cyrus the Great defeats Medes; establishes Persian empire from Susa; captures Babylonia (539).	Aesop: fables.
546	Persian empire extends to Aegean. Ionian Greek cities captured.		Taoism founded in China, traditionally by Lao-tzu (6th–5th century). Pythagoras teaching in southern Italy.
539	Israelites return from Babylonian Captivity and rebuild Temple in Jerusalem.		Ionic order appears in Greek architecture.
509	Roman Republic proclaimed.	Persian empire reaches India.	Siddhartha Gautama (Buddha) teaching.
508	Cleisthenes establishes democratic constitution in Athens.	Chinese bronze coinage.	Athenian pottery at its zenith. Iron-working in China.
500	Etruscans at the height of their power.		Emergence of Greek drama; theatres built.
499	Revolt of Ionian Greek cities against the Persians.	First inscriptions of Monte Albán, Mexico.	Greek philosophical thought emerging (Heraclitus).
490	Persian emperor Darius invades Greece; his army defeated at Marathon.		Confucius teaching in China.
480	Xerxes I, son of Darius, invades Greece with army and navy. Allied with Thebes, he wins land battle at Thermopylae, devastates Attica, but is defeated by Greeks in sea battle at Salamis. Xerxes retreats.		
472	Athens controls Aegean through Delian League.		458 Aeschylus: *Oresteia*.

All dates BC	**EUROPE AND THE MEDITERRANEAN**	**REST OF THE WORLD**	**CULTURE/TECHNOLOGY**
450	Rome extending power in Latium and against Etruscans. Twelve Tables (set of laws) drawn up.	Persian empire in decline.	
443	Pericles dominates Athenian democracy until 429. Athens rebuilt. Parthenon built. Celtic La Tène culture flourishes in Switzerland.	The Carthaginian explorer Hanno sails to Senegal. Coinage reaches India. Extensive trade links between Mediterranean and Asia.	Sophocles: *Antigone.* Herodotus: *History.* Phidias leading Greek sculptor. Solar calendar in China.
431	Peloponnesian War begins; Athens against Sparta and her allies (431–404, with brief interlude 421–415).		Democritus: atomic theory. Euripides: *Medea.*
415	Alcibiades leads Athenian expedition to Sicily. Disastrous siege of Syracuse; many Athenians put to death.	Nok culture in West Africa (northern Nigeria), lasting until AD *c.*200. Iron metallurgy; clay figurines.	Aristophanes: comedies.
405	Spartan naval victory at Aegospotami on the Hellespont; Athens sues for peace (404).		Thucydides: *History of the Peloponnesian War.*
390	Celts sack Rome.	In China Zhou dynasty in decline.	399 Socrates condemned to death in Athens. 387 Plato founds Academy in Athens.
371	Thebes defeats Sparta at battle of Leuctra and briefly dominates Greece. Rome dominates Latium, building roads and aqueducts.		Hippocrates developing medicine.
338	Macedonia defeats Thebes at battle of Charonea; controls all Greece under Philip II (359–336).		Iron metallurgy in central Africa.
336	Alexander the Great king of Macedonia (336–323).		
334	Alexander crosses Hellespont, defeats Persians at battle of Granicus, liberates Ionian cities; captures Tyre and Egypt; marches east.	Alexander master of Persian empire; invades India.	335 Aristotle founds Lyceum at Athens.
323	Death of Alexander at Babylon. Empire disintegrates. Hellenistic kingdom of Attalids established at Pergamum.	Mauryan dynasty established in India.	Epicurean and atomistic theory fashionable. Hellenistic art spreads throughout Asia. Menander: comedies.
311	Seleucid power established in Babylon.		
304	Ptolemy founds dynasty in Egypt.		
280	Pyrrhus of Epirus campaigns in Italy; defeats Romans in several battles but is unable to exploit victory and suffers heavy losses. Rome continues to advance into southern Italy.	Early Maya culture developing in Guatemala.	292–280 Colossus of Rhodes built. 284 Library founded (100,000 volumes) at Alexandria. Euclid teaching there. Catapult and quinquereme warship invented at Syracuse. Elephants first used in battle. Theocritus writes idylls idealizing bucolic life.
264	First Punic War (264–241), Rome against Carthage. Rome expands navy and wins control of Sicily (but not Syracuse), Sardinia, and Corsica.	269 Asoka (*c.*269–*c.*232), Mauryan emperor of India. Enthusiastic convert to Buddhism. 256 Zhou dynasty ends in China.	Zoroastrianism spreading in Persia. Greek and Oriental cultures fusing in Hellenistic period.
237	Hamilcar of Carthage conquers SE Iberia.	Extensive trade between China and Hellenistic world.	Latin literature beginning to emerge.
218	Second Punic War (218–201). Hannibal crosses Alps from Spain and defeats Romans but fails to take Rome itself.	221 Qin dynasty established in China. Zhou provinces conquered and country united politically.	214 Archimedes' inventions resist Romans in siege of Syracuse.
214	Romans fight first Macedonian War (214–205).	*c.*210 Construction of Great Wall of China begun.	
202	Scipio Africanus defeats Carthage at Zama and second Punic War ends (201).	206 Han dynasty established in China.	Horse collar and harness in China.

All dates BC	EUROPE AND THE MEDITERRANEAN	REST OF THE WORLD	CULTURE/TECHNOLOGY
200	Second Macedonian War (200–196). Rome defeats Philip V (197).		
192	Seleucid Antiochus III occupies Athens and Greece.	c.184 Mauryan empire ends in India.	Plautus (later Terence) writing comedies in Rome.
171	Third Macedonian War (171–168).		
167	Revolt of Maccabees in Judaea against Seleucids; Judas Maccabaeus establishes Jewish dynasty in Jerusalem (165).	Parthian empire (c.250–AD c.230) at its height, from Caspian Sea and Euphrates to the Indus.	
149	Third Punic War begins (149–146).		
146	Carthage destroyed. Rome dominates western Mediterranean. Macedonia becomes Roman province.		Buddhism spreading throughout SE Asia.
133	Rome master of Iberia, occupies Balearic Islands (123). The Gracchus brothers attempt social and legal reforms in Rome 133–121.	127–101 Han armies from China conquer central Asia. Drift west of Asiatic tribes.	Polybius: *Histories* (40 volumes of Roman history, 220–146).
112	Outbreak of war between Rome and Jugurtha, king of Numidia.	Roman envoys to Han China.	Parchment invented in Pergamum to replace papyrus.
107	Gaius Marius elected to the first of his seven consulships at Rome.	Teotihuacán and Monte Albán developing in Mexico.	
105	Jugurtha captured by the Romans.		
c.100	Celtic Belgae first settle in SE Britain.		Water-mill first described in Greek writings; came from China.
91	Confederacy of Italian tribes on Adriatic and in Apennines rebels against Rome; civil war.		
90	Roman citizenship extended to the Latin and some Italian cities.		
88	Mithridates VI, king of Pontus, invades Greece; defeated by the Roman general Sulla (85).		Buddhism spreading in China.
73	Spartacus leads revolt of slaves against Rome; suppressed by Crassus (71).		
66	Mithridates finally defeated by Pompey.	Civilization in Peru emerging (pyramids, palaces, etc.).	Cicero pleading and writing in Rome.
63	Catiline's conspiracy at Rome.	Romans under Pompey conquer Syria and Palestine; end of Seleucid empire.	
60	First Triumvirate in Rome; Crassus, Pompey, and Julius Caesar coordinate their political activities.		
58	Caesar fights Gallic Wars (58–51).		
55	Caesar invades Britain (55, 54).		
53	Crassus killed in battle against Parthians.		
51	Transalpine Gauls defeated by Caesar.		
49	Caesar crosses Rubicon and begins civil war.		
48	Caesar defeats Pompey in battle of Pharsalus. Pompey murdered in Egypt.		
45	Caesar dictator of Rome.		46 Caesar introduces Julian calendar.
44	Caesar assassinated.		
43	Second Triumvirate (Octavian, Antony, Lepidus).		42 Virgil begins to write *Eclogues*.
31	Battle of Actium. Antony and Cleopatra commit suicide (30). Octavian sole ruler of Rome.		

EUROPE AND THE MEDITERRANEAN	REST OF THE WORLD	CULTURE/TECHNOLOGY
27 BC Octavian accepts title of Augustus. Date traditionally marks beginning of Roman Empire (to AD 476).		Pantheon built in Rome. 19 Virgil dies, leaving *Aeneid* unfinished. 17 Herod the Great rebuilds Temple at Jerusalem. Vitruvius: treatise on architecture.
*c.*6 BC Jesus of Nazareth born.		Strabo: *Geographica.*
All dates AD		
9 Germans annihilate three Roman legions. Rome withdraws to Rhine.		
14 Death of Augustus. Tiberius emperor (14–37).		Ovid: *Metamorphoses.* *c.*30 Crucifixion of Jesus.
37 Caligula emperor; assassinated 41.		
41 Claudius emperor (41–54).		
43 Britain occupied under Claudius.		
54 Nero emperor (54–68).		
61 Boudicca's revolt in Britain crushed by Suetonius Paulinus.	Kingdom of Aksum (Ethiopia) flourishes.	
64 Great fire of Rome: Christians blamed and martyred in Rome (64–7), including St Paul and St Peter.	Dead Sea Scrolls hidden at Qumran near the Dead Sea. 66 Revolt of Jews in Judaea.	Hero invents various machines in Alexandria.
69 Vespasian emperor (69–79); first of Flavian emperors; restores imperial economy.	70 Destruction of the Temple at Jerusalem under Titus, son of Vespasian. Jewish Diaspora.	First Gospel (St Matthew).
79 Eruption of Vesuvius. Pompeii and Herculaneum buried.		*c.*75 Colosseum in Rome begun. Paper, magnetic compass, and fireworks invented in China. 90 Plutarch: *Parallel Lives.*
98 Trajan emperor (98–117); extends empire, defeating Dacians, Armenians, and Parthians. Empire at its fullest extent.	Christianity spreading.	Reform of Buddhism in India. Iron-working in Zambia.
100 Europe, North Africa, and Middle East under Roman control.		Juvenal: *Satires.* Canon of Hebrew Bible (Old Testament) fixed for Judaism.
117 Hadrian emperor (117–38).		Tacitus: *Annals.*
122 Hadrian's Wall built in Britain against northern tribes.		Suetonius: *Lives of the Caesars.*
132 Bar-Cochba leads revolt of Jews against the Romans.	Teotihuacán civilization in Mexico flourishing.	
138 Antoninus Pius emperor (138–61). Founds Antonine dynasty; streamlines imperial government.		
*c.*140 Antonine Wall built in Britain.		
161 Marcus Aurelius emperor (161–80); a Stoic philosopher, he campaigns on eastern and northern frontiers of empire.	Parthian empire weakening.	Astronomy developing in school of Alexandria. Marcus Aurelius: *Meditations.*
180 Commodus emperor (180–92); murdered for his wild extravagance and cruelty.	184 Revolt in China as Han dynasty declines.	Galen practising medicine in Rome.
193 Septimius Severus emperor (193–211); resumes persecution of Christians; long campaign in Britain against Picts.		Tertullian: Christian apologetics.
211 Caracalla emperor (211–17); his reign one of cruelty and extortion.		
212 Roman citizenship extended to all freemen of the Roman Empire.		
218 Heliogabalus emperor (218–22); wild and decadent. Murdered in Rome.	220 Han dynasty ends in China.	Neoplatonism in Alexandria.

EUROPE AND THE MEDITERRANEAN	REST OF THE WORLD	CULTURE/TECHNOLOGY
222 Alexander Severus emperor (222–35); rule remembered as just; re-established authority of Rome.	**224** Persians defeat Parthians, whose empire collapses. The Persian Sassanian empire established.	Indian art (sculpture and painting) flourishes. Chinese literature developing.
235 Political tensions in Rome as empire begins to decline.		Gnostic Manichaeism spreads from Persia.
249 Decius emperor (249–51); intense persecution of Christians; Danube and the Balkans overrun by Goths.	*c.*250 Syrian kingdom of Palmyra rises in power.	Roman architecture covers Europe and Mediterranean.
253 Valerian (253–60) and Gallienus emperors (253–68); Franks invade empire and Sassanians take Syria.	260 Sassanian emperor captures and imprisons Valerian. 265 Foundation of Western Jin dynasty in China (265–317). Bantu-speaking peoples move into southern Africa.	Christian theology emerging in Asia Minor and Egypt.
270 Emperor Aurelian (270–5) abandons Dacia to Goths but regains Rhine and Danube.	272 Kingdom of Palmyra conquered by Aurelian.	
284 Diocletian emperor (284–305); re-establishes frontiers and reorganizes government; divides empire with Maximian (286).	Classic Maya civilization emerging (Tikal, Palenque).	Monastic ideal (hermits) becoming popular (St Anthony).
293 Diocletian establishes tetrarchy; he rules with Galerius in east, Maximian and Constantius in the west.	292 First Mayan stela from Tikal.	Arius of Alexandria founds Arianism, denying divinity of Christ.
303 Diocletian persecutes Christians.		
306 Constantine proclaimed emperor in York (306–37); engages in long and complex civil war.		
313 Edict of Milan allows freedom of worship to Christians in Roman Empire.	317 Foundation of Eastern Jin dynasty in China (317–420).	Pappus last great mathematician of Alexandria.
324 Byzantium rebuilt by Constantine; becomes capital of empire (330); named Constantinople.	320 Foundation of Gupta empire (to *c.*550) in India (golden age of religion, philosophy, literature, and architecture).	325 Council of Nicaea denounces Arianism and agrees Nicene Creed. Chinese mathematics reducing fractions and solving linear equations.
353 Constantius II emperor (353–60). Saxons invading coasts of Britain.		Chinese bucolic literature flourishing.
360 Julian (the Apostate) emperor (360–3); restores paganism briefly.		
374 St Ambrose elected bishop of Milan; influences emperor and dominates Western Church.		371–97 St Martin bishop of Tours.
378 Visigoths defeat Roman army.		
379 Theodosius I (the Great) emperor (379–95); pious Christian; defeats usurpers and makes treaty with Visigoths (382).	Huns from Asia concentrate on River Volga; moving west.	
391 All pagan worship banned in Roman Empire by Theodosius.		
395 Roman Empire divided into East and West on death of Theodosius; Honorius (395–423) rules from Milan, Arcadius (395–408) from Constantinople.		
396 Roman victories in Britain against Picts, Scots, and Saxons (396–8).		*c.*400 Text of Palestinian Talmud finalized.
402 Western capital of Roman Empire moves from Milan to Ravenna.		*c.*405 St Jerome completes Vulgate (Latin Bible).
408 Visigoths invade Italy.		
410 Visigoths, led by Alaric, sack Rome. Romans evacuate Britain. Franks occupy northern Gaul and Celts move into Breton peninsula.	420 End of Eastern Jin dynasty in China.	Christian theology continues to be defined; Athanasian Creed agreed.

	EUROPE AND THE MEDITERRANEAN	REST OF THE WORLD	CULTURE/TECHNOLOGY
476	Romulus Augustulus, last Western emperor (475–6), is deposed; date traditionally marks the end of the Roman Empire. Saxon settlement in Sussex.		Shinto religion in Japan.
481	Clovis king of Salian Franks (481–511).		Buddhism dominant in China.
488	Theodoric and Ostrogoths invade Italy.		
496	Clovis baptized; establishes Merovingian Frankish kingdom. Wessex occupied by Saxons.	Ecuadorean pottery dated c.500 found in Galapagos Islands. Evidence of Pacific trade?	
516	British victory (King Arthur) over Saxons at Badon Hill. Visigoths established in Spain; Vandals in Africa (429–534).		
527	Justinian Byzantine emperor (527–65); attempts to recover western part of Roman Empire; drives Ostrogoths from Italy (552).	c.550 End of Gupta empire in India following attacks by Huns.	c.550 St David founds monasteries.
560	Ethelbert (d.616) king of Kent.		Buddhism in Japan along with Shinto.
c.563	St Columba founds monastery at Iona.		
568	Lombards invade northern Italy. Anglo-Saxon kingdoms in Britain emerging.	c.570 Birth of Muhammad.	Byzantine architecture spreads throughout Eastern Empire and southern Italy.
577	West Saxons take Bath and Gloucester.	Chinese Sui dynasty 581–618; reunites country and rebuilds Great Wall.	
590	Election of Gregory the Great as pope (590–604).	607 Unification of Tibet, which becomes centre of Buddhism. 618 Tang dynasty established in China (618–907); strong centralizing power restores order.	Gregorian chant and Roman ritual imposed by Gregory the Great.
627	Angle and Saxon kingdoms in Britain (Mercia, Wessex, and Northumbria) struggle for power.	622 Hegira of Muhammad and friends; Mecca to Medina. In Mexico Mayan civilization at height; temples and palaces in stone (complex astronomical and mathematical knowledge).	Christian missionaries to Germany and England (St Augustine, 597). Sutton Hoo burial. c.625 Isidore of Seville: *Etymologies*. Parchment displacing papyrus.
629	Dagobert, king of Franks (629–39), reunites all Franks.	632 Death of Muhammad. 634 Rapid spread of Islam in Arabia, Syria, Iran, North Africa under Caliph Umar (634–44). 637–51 Collapse of Sassanian empire.	c.632 The 114 chapters or suras of the Koran collected. c.641 The great library of Alexandria destroyed by Arabs. 644 Windmills in Persia.
664	Synod of Whitby.	660 Damascus capital of Umayyad empire of Islam.	
678	Constantinople successfully resists Arabs.		
679	Mercia becomes major British power.	Teotihuacán civilization of Mexico declining.	680 Divisions within Islam produce Sunnis and Shiites.
685	Battle of Nechtansmere; Picts defeat Northumbrians. Wessex expanding; Kent, Surrey, Sussex taken by Saxons.	Zapotec civilization, Monte Albán, flourishing; influenced by Teotihuacán. Afghanistan conquered by Arabs who cross Khyber Pass and conquer the Punjab.	Dome of the Rock in Jerusalem (692) and Umayyad Mosque in Damascus (705–15) built.
711	Muslim Arabs enter Spain, conquer Seville (712).	Tang emperor Xuan Zong (712–56) suffers drastic incursions by Arabs and revolts.	
718	Bulgars pressing south towards Constantinople.		726 Icons banned in Byzantine Church.

EUROPE AND THE MEDITERRANEAN	REST OF THE WORLD	CULTURE/TECHNOLOGY
732 Battle of Poitiers; decisive victory of Charles Martel, Frankish king, over Constantinople.	Kingdom of Ghana established; to last till 1240.	Block printing in China for Buddhist texts. Bede: *The Ecclesiastical History of the English People.*
751 Pepin III (the Short), son of Charles Martel, ousts last Merovingian, Childeric III, and founds Carolingian dynasty.	Establishment of Abbasid caliphate in Baghdad, 750–1258.	*Beowulf:* Anglo-Saxon poem. Golden age of Chinese poetry (Li Po, 701–61) and art.
754 Pepin the Short crowned by Pope Stephen II; recognizes Papal States (756).		Cordoba centre of Muslim culture in Spain. Irish Book of Kells.
757 Offa king of Mercia (757–96).		
768 Charlemagne king of Franks (768–814); campaigns against Avars and Saxons in east.		Offa's Dyke built. Dravidian temples in India.
774 Charlemagne annexes Lombardy, but is checked in Spain (death of Roland at Roncesvalles 778).	Caliph Harun al-Raschid establishes close links with Constantinople and with Charlemagne. Patronage of learning and the arts (*Arabian Nights*).	Cotton grown in Spain.
794 Viking raids on England and Ireland; Jarrow and Iona (795) sacked.	Kyoto capital of Japan.	
800 Charlemagne crowned in Rome by Pope Leo III.		Alcuin at court of Charlemagne; Carolingian Renaissance. 805 Aachen cathedral inspired by Byzantine models.
812 Charlemagne recognized emperor of the West by Byzantine emperor Michael I.		
813 Byzantine army defeats Bulgars. Constantinople besieged by Bulgars and Arab army.		
825 Wessex annexes Essex.		
827 Byzantine loss of Sicily and Crete to Arab Saracens.		833 Observatory in Baghdad; Arabs develop astronomy, mathematics (algebra from India), optics, medicine.
843 Carolingian empire divided; East Franks, West Franks, and Lotharingia Lorraine). Vikings trading to Volga and Baghdad.		843 Restoration of cult of images in Byzantine Church. Icon art to influence West through Venice 845 Buddhism outlawed in China. *c.*850 Windmills in Europe.
*c.*859 Novgorod founded as trading centre; Viking and Byzantine merchants. St Cyril's mission to Moravia; Bulgars accept Christianity.		
866 Danes land on east coast of Britain; Northumbria conquered; Danelaw established.		
867 Danes conquer York.		
870 St Edmund the Martyr murdered by Danes.		Romanesque architecture developing in West.
871 Alfred the Great king of Wessex (871–99).		
*c.*880 Kingdom of Kiev established in Russia.		
885 Paris besieged by Vikings.		
896 Magyars settle in Hungary.	Classic Mayan civilization in Mexico ending. *c.*900 Teotihuacán civilization ends; rise of Toltecs in Mexico based on Tula.	Benedictine Order spreads through Europe.

EUROPE AND THE MEDITERRANEAN	REST OF THE WORLD	CULTURE/TECHNOLOGY
912 Rollo the Norseman established as first Duke of Normandy.	907 Tang dynasty in China ends; China fragments. 909 Shiite dynasty, the Fatimids, conquers North Africa.	910 Monastery of Cluny established.
925 Athelstan first king of all England (925–39).	Arabs trading along East African coast.	Stone replacing wood as building material in western Europe.
936 Otto I king of Germans (936–73).		Expansion of European agriculture. Buddhism flourishes in Korea.
937 Athelstan defeats Vikings and Scots.		
939 Edmund I king of England (939–46).		
945 Norsemen in Constantinople and Kiev.	947 Liao dynasty from Manchuria (947–1125) extends control over northern China.	943 Dunstan abbot of Glastonbury; under his leadership monasticism re-established in England.
955 Otto I defeats Magyars.		
959 Edgar king of England (959–75).		
960 Dunstan archbishop of Canterbury.	Sung dynasty in China (960–1279); gradually reunites country; high levels of art and literature.	
962 Otto I crowned Holy Roman Emperor in Rome by Pope John XII. Seeks to establish power in Italy.		
971 Bulgaria and Phoenicia conquered by Byzantine armies.	969 Fatimids conquer western Arabia, Syria, and Egypt; Cairo capital.	
975 Edward the Martyr king (975–8); murdered by half-brother Ethelred.		
978 Ethelred II (the Unready) king of England (978–1016).		
987 Hugh Capet king of France (987–96).	986 Viking settlements in Greenland. New Mayan empire emerging under Toltec influence.	988 Baptism of Vladimir, prince of Kiev (956–1015).
991 English treaty with Normans.		
996 Otto III crowned Holy Roman Emperor (996–1002); establishes capital in Rome (999); with Pope Sylvester II aims to create universal Christian empire.		Avicenna has lasting influence on West: philosophy and medicine.
1001 Christian kingdom of Hungary established.		c.1000 Arabic description of magnifying properties of glass lens.
1003 Sweyn Forkbeard, king of Denmark, attacks Britain.		
1013 Sweyn invades England and takes London.		
1014 Death of Sweyn. His son Canute elected king by Danes in England.		
1016 Death of Ethelred and then of his son Edmund Ironside; Canute accepted as king of England (1017–35). Byzantine Empire at height of power and influence under Basil II (976–1025).	Ghaznavid empire extends into Persia and Punjab from Afghanistan.	
1035 Harold I king of England (1035–40).		
1040 Hardecanute king of England and Denmark (1040–2).		
1042 Edward the Confessor king of England (1042–66).		c.1045 Printing by movable type in China.
1050 Bohemia, Poland, and Hungary become fiefs of Holy Roman Empire.	Toltecs flourish in Mexico; conspicuous Maya influence.	Salerno medical school emerging; Arabic expertise. Cult of Quetzalcóatl in Toltec Mexico.

EUROPE AND THE MEDITERRANEAN	REST OF THE WORLD	CULTURE/TECHNOLOGY
1054 Great Schism within Christian Church; Orthodox Eastern Churches split from Catholic Rome.		
1066 Harold II (1066) king of England. Normans conquer England; William I king (1066–87).	**1064** Seljuk Turks menace Byzantine Empire.	**1063** St Mark's Cathedral, Venice, rebuilt. Pisa Cathedral built.
1069 North of England ravaged by William I's troops.	**1068** Almoravid Berber dynasty in North Africa; fanatical Muslims; build Marrakesh.	
1071 Normans established in southern Italy.	Seljuk Turks rout Byzantine army at battle of Manzikert and threaten Asia Minor.	Bayeux Tapestry.
1081 Normans invade Balkans; Venice aids Constantinople against Normans; Venetian trade expands.	**1080** Turks control Asia Minor and interrupt Christian pilgrim routes to Jerusalem.	
1084 Foundation of Carthusian order of monks.		
1086 Domesday Book in England.		
1087 William II king of England (1087–1100).		Omar Khayyám: algebra, astronomy, poetry (*The Rubáiyát of Omar Khayyám*). **1090** Water-powered mechanical clock in China.
1095 Pope Urban II urges crusade to rescue holy places from Turks.	First Crusade (1096–9).	**1094** St Anselm: *Cur Deus Homo?*
1098 Cistercian order of monks founded.	**1099** Jerusalem captured by Crusaders.	Feudal system well established throughout Europe. Urban society developing in Flanders, Germany, and Italy.
1100 Henry I king of England (1100–35).	Christian states in Palestine.	*Chanson de Roland.*
1108 Louis the Fat king of France (1108–37); Capetian power expanding.		
1120 Henry I's son William drowned.	**1115** Jin dynasty in Manchuria (1115–1234). **1118** Knights Templars founded in Jerusalem. **1121** Almohads dynasty founded in North Africa. **1126** Sung capital Kaifeng in China sacked by Jin horsemen.	**1115** St Bernard abbot of Clairvaux; stresses spiritualism of monasticism. Rediscovery of Aristotle. Abelard teaching in Paris.
1137 Catalonia linked by marriage with Aragon.		
1138 Beginning of long medieval struggle; Guelphs (for pope) against Ghibellines (for Holy Roman Emperor).		Gothic architecture beginning.
1139 Civil war in England: Stephen against Matilda.		*c.*1139 Geoffrey of Monmouth: *Historia Regum Britanniae.*
1147 The Almohads rule in southern Spain.	Second Crusade (1147–9).	
1152 Henry of Anjou marries Eleanor of Aquitaine. Frederick I ('Barbarossa') crowned Holy Roman Emperor (1152–90); seeks to extend power in Italy.	*c.*1150 Khmer temple Angkor Wat built in Cambodia.	*c.*1150 Toledo school of translators transmits Arab learning to West.
1154 Henry II king of England (1154–89). Adrian IV pope (1154–9); only Englishman to have been pope.	Toltec capital Tula overrun by Chichimecs from northern Mexico. Toltec power declines.	**1158** Bologna University granted charter by Frederick I.
1165 William I ('the Lion') king of Scotland (1165–1214).		
1166 Jury system established in England.	Aztecs moving into Mexico; destroy Toltec empire.	Catharist Manichaean heresy spreading. *c.*1167 Oxford University founded.

EUROPE AND THE MEDITERRANEAN	REST OF THE WORLD	CULTURE/TECHNOLOGY
1170 Murder of St Thomas à Becket.	1169 Saladin conquers Egypt. Drives Christians from Acre and Jerusalem.	c.1170 Paris University founded. Troubadour songs in France. Early polyphonic music. 1171 Averroës teaching in Cordoba.
1180 Philip II ('Augustus') king of France (1180–1223); greatly expands kingdom.		
1182 Massacre of Latin merchants in Constantinople.		
1189 Richard I king of England (1189–99).	Third Crusade (1189–92); Acre retaken (1191). Saladin grants pilgrims access to holy places.	1193 Zen Buddhism in Japan.
1198 Innocent III pope (1198–1216); papacy has maximum authority during these years.		1194 Chartres Cathedral rebuilding begun in Gothic style.
1199 John king of England (1199–1216).	1200 Incas developing civilization based on Cuzco. Chimu civilization in Peru (1200–1465); large urban centres; elaborate irrigation. Fourth Crusade (1202–4); Constantinople sacked.	1202 Arabic mathematics in Pisa.
1204 Philip Augustus victorious in Normandy.	1206 Genghis Khan proclaims Mongol empire.	
1212 Children's Crusade; thousands enslaved.		1209 Cambridge University established. Franciscan (1209) and Dominican (1215–16) orders founded.
1214 Philip Augustus defeats King John and German emperor Otto IV; gains Normandy, Anjou, and Poitou for France.		
1215 John accepts Magna Carta from barons. Fourth Lateran Council condemns Albigensian heresy; transubstantiation doctrine.	Beijing sacked and Jin empire overrun by Mongols.	Islam spreading into SE Asia and Africa.
1216 Henry III king of England aged 9 (1216–72); Earl of Pembroke regent until 1227.		
1220 Frederick II Holy Roman Emperor (1220–50); inherits southern Italy and Sicily, which he makes power base.		
1223 Louis VIII of France (1223–6) conquers Languedoc in crusade against Cathars (1224–6). Mongols invade Russia.		1224 Frederick II founds Naples University: Jews, Christians, and Arabs.
1226 Louis XI king of France (1226–70).	1229 Frederick II negotiates access for pilgrims to Jerusalem, Bethlehem, and Nazareth.	
1236 Cordoba falls to Castile.	c.1235 Mali empire established in West Africa.	Roman de la rose.
1241 Formation of Hanseatic League.	Mongol Golden Horde emerges. 1244 Christians are driven from Jerusalem.	
1250 Italian cities gain power on collapse of Frederick II's empire.	Mamelukes establish dynasty in Egypt.	1248 Alhambra begun. Cologne Cathedral begun.
1258 Catalans expel Moors from Balearics.	Mongols take Baghdad.	Thomas Aquinas: Summa Contra Gentiles.
1261 Byzantines regain Constantinople.	1259 Kublai Khan elected khan. Establishes capital in Beijing and founds Yuan dynasty.	c.1260 Nicola Pisano: pulpit in Pisa Baptistery; renaissance of classical style.
1264 Battle of Lewes: Simon de Montfort effective ruler of England until 1265.		

EUROPE AND THE MEDITERRANEAN	REST OF THE WORLD	CULTURE/TECHNOLOGY
1270 Louis IX dies on crusade outside Tunis. Philip III succeeds (1270–85).	Marco Polo travelling 1271–95.	Gothic architecture throughout Europe.
1272 Edward I of England (1272–1307) begins conquest of Wales.		Mechanical clock developed.
1282 Sicilian Vespers; revolt against Angevins (ruled since 1266). Sicily goes to Aragon. Welsh resistance collapses.		Duccio painting in Siena.
1284 English rule of Wales confirmed.		Giotto active in Florence; significant in development of modern painting.
1285 Philip IV king of France (1285–1314).	Inca empire expanding in Peru.	
1290 Jews expelled from England. Scottish throne vacant.		
1297 William Wallace defeats English army at Stirling.	New Mayan empire flourishing in Yucatán (Chichén Itzá).	Roger Bacon teaching philosophy, science, technology; eventually imprisoned for heresy.
1305 Edward I executes William Wallace. Clement V pope (1305–14); papacy moves to Avignon (1309).		Spectacles invented.
1306 Jews expelled from France.		Duns Scotus and nominalists oppose Aquinas's theology.
1307 Knights Templars suppressed in France. Edward II king of England (1307–27). Italian cities flourish as German empire abandons control.	Kingdom of Benin emerging in southern Nigeria.	Dante: *The Divine Comedy* begun *c.*1309.
1314 Robert the Bruce defeats English at Bannockburn.		
1327 Edward II imprisoned and murdered. Edward III king of England (1327–77); his mother Isabella and her lover Mortimer rule till 1330.	Aztecs adopt cult of Quetzalcóatl from Toltecs.	Spinning-wheel from India in Europe.
1328 Scottish independence recognized. Capetian line of kings ends. Philip VI first Valois king of France (1328–50).	Disease (plague) and famine weaken Yuan dynasty in China.	
1337 Edward III claims French throne and Hundred Years War begins.		
1340 English gain control of Channel.		Paper-mill in Italy. Bruges centre of wool trade; Flemish art emerging.
1346 Battle of Crécy; English victory; cannon used; Calais occupied.	1347 Black Death reaches Europe from China.	*c.*1344 Order of Garter in England.
1348 Black Death arrives in England; one-third of population dies.	Aztec empire thriving.	Boccaccio: *Decameron.*
1353 Ottoman Turks enter Europe.		Petrarch: *Canzoniere.*
1356 Edward the Black Prince wins Poitiers; French king John II captured.		
1360 England makes peace with France, keeps western France.		
1369 Hundred Years War resumed.	Mongol Tamerlane conquers Persia, Syria, and Egypt (1364–1405). Ming dynasty in China founded (1368–1644); great period for pottery and bronze.	William Langland: *Piers Plowman.* Jean Froissart: *Chronicles.* Siena artists flourish.
1371 Robert II king of Scotland (1371–90); first Stuart.		
1377 Richard II king of England (1377–99). Papacy returns to Rome from Avignon.		
1378 Great Schism in Church; two popes.		Flamboyant architecture in Europe: Beauvais Cathedral in France.
1380 Charles VI king of France (1380–1422).		

EUROPE AND THE MEDITERRANEAN	REST OF THE WORLD	CULTURE/TECHNOLOGY
1381 Peasants' Revolt in England; Wat Tyler defeated; poll tax withdrawn.		1382 Lollards (John Wyclif) condemned.
1386 Poland and Lithuania unite.		1387 Chaucer: *Canterbury Tales.*
1396 Truce in Hundred Years War.		Ghiberti in Florence: baptistery doors.
1397 Union of Kalmar: crowns of Denmark, Norway, and Sweden unite (1397–1523).		
1399 Richard II deposed; Bolingbroke becomes Henry IV (1399–1413).	*c.*1400 Foundation of Malacca sultanate.	
1402 Owen Glendower defeats English.	Tamerlane defeats Ottomans at Ankara.	
1403 Prince Henry (later Henry V) defeats Sir Henry Percy ('Hotspur') and Glendower rebellions (1408).		
1413 Henry V king of England (1413–22).	Portuguese voyages of exploration begin under Henry the Navigator.	1410 St Andrews University founded.
1415 Henry V wins Battle of Agincourt; occupies Normandy. John Huss burnt at the stake; Hussites seek revenge for martyr Huss. Great Schism ends (1417).		Painting in oil begins.
1422 Henry VI king of England (1422–61, 1470–1). Charles VII king of France (1422–61).		Thomas à Kempis: *On the Imitation of Christ.* Masaccio: frescoes. 1420 Dome of Florence Cathedral begun by Brunelleschi.
1429 Joan of Arc relieves Orleans.		
1431 Joan of Arc burnt at stake in Rouen.	Thais of Siam take Angkor. Phnom Penh new Khmer city.	
1434 Cosimo de' Medici rules in Florence; patron of learning. Burgundy emerging under strong dukes.	Inca ascendancy in Peru; high level of astronomical and surgical knowledge; cotton and potato grown.	Donatello: *David.* Van Eyck: *The Arnolfini Marriage.*
1452 Frederick III (1452–93) first Habsburg Holy Roman Emperor.		1452–66 Piero della Francesca: *Arezzo* frescoes.
1453 Constantinople falls to Ottoman Turks. Hundred Years War ends. English Wars of Roses begin.		Alberti: façade of Santa Maria Novella; Florence centre of artistic activity. *c.*1455 Gutenberg Bible. *c.*1454–7 Uccello: *The Rout of San Romano.*
1461 Edward of York seizes English throne; Edward IV (1461–83). Louis XI king of France (1461–83).		
1469 Ferdinand and Isabella I marry; unite Aragon and Castile (1479).		
1470 Henry VI of England restored; Edward IV exiled.		
1471 Lancastrians defeated at Tewkesbury; Henry VI killed; Edward IV accepted.		
1478 Spanish Inquisition established.	1476 Incas conquer Chimu.	Caxton printing at Westminster. Topkapi Palace built in Constantinople. Botticelli: *Primavera.*
1480 Ivan III overthrows Mongol Golden Horde.		
1483 Edward V; Richard III king of England (1483–5).		
1485 Henry Tudor victorious at Bosworth Field; Henry VII (1485–1509).		Leonardo da Vinci: anatomy, mechanics, painting, etc.
1492 Reconquest of Spain from Moors complete.	Columbus reaches the New World.	

EUROPE AND THE MEDITERRANEAN	REST OF THE WORLD	CULTURE/TECHNOLOGY
1498 Louis XII king of France (1498–1515).	1497 John Cabot reaches mainland North America from Bristol. 1498 Vasco da Gama rounds Cape of Good Hope and reaches Calicut; beginning of Portuguese empire.	Nanak founds Sikh religion.
1503 Julius II pope (1503–13).		1501–4 Michelangelo: *David*. *c*.1503–6 Leonardo da Vinci: *Mona Lisa*. 1505–7 Dürer in Italy. 1506 Bramante designs St Peter's, Rome.
1509 Henry VIII king of England (1509–47).		1508–12 Michelangelo: Sistine chapel ceiling. 1509 Watch invented in Nuremberg.
1513 Scots defeated at Flodden; death of James IV.	1511 Portuguese conquer Malacca (Melaka).	
1515 Francis I king of France (1515–47).		1516 Grünewald: Isenheim Altar. King's College Chapel, Cambridge, completed.
1517 Start of Reformation in Germany.	Ottoman Turks conquer Egypt.	
1519 Charles V elected Holy Roman Emperor.	Cortés conquers Aztecs. Magellan crosses Pacific.	
1521 Diet of Worms condemns Luther's teaching.	Suleiman the Magnificent sultan of the Ottoman Empire.	
1525 Reformation moves to Switzerland.	*c*.1525 Babur invades India and founds Mogul dynasty.	Titian: *Bacchus and Ariadne*.
1526 Battle of Mohács. Ottoman Turks occupy Hungary.		
1527 Charles V's troops sack Rome.		Paracelsus in Basle (new concept of disease).
1529 Ottomans besiege Vienna. Fall of English chancellor Wolsey.	Franciscan mission to Mexico. European spice trade with Asia; trade in sugar/slaves with America.	Early Italian madrigal.
1530 Augsburg Confession states Lutheran position.		
1533 Henry VIII marries Anne Boleyn.	1531–3 Pizarro conquers Inca empire.	
1534 English Act of Supremacy; break with papacy. Anabaptists revolt in Münster, Germany.	Persia conquered by Ottoman Turks.	Luther's translation of the Bible. Rabelais: *Gargantua*. Jesuits founded.
1535 John Calvin in Geneva.		
1536 Dissolution of English and Welsh monasteries.		Calvin: *Institutes of the Christian Religion*.
1538 Pope Paul III excommunicates Henry VIII.		
1540 Henry VIII tries to impose political and religious settlement on Ireland.	1542 St Francis Xavier in India, Sri Lanka, Japan (1549).	1543 Copernicus: *De Revolutionibus Orbium Coelestium*. 1543 Vesalius: *De Humani Corporis Fabrica*.
1545 Council of Trent begins (1545–63).		
1547 Edward VI king of England (1547–53). Ivan the Terrible tsar of Russia (1547–84).	Portuguese settling coast of Brazil.	1548 Ignatius Loyola: *Spiritual Exercises*.
1553 Mary Tudor queen of England (1553–8); persecution of Protestants.		1550 Vasari: *Lives of the Most Excellent Painters, Sculptors, and Architects*.
1556 Charles V retires; Philip II king of Spain (1556–98).	Akbar Mogul emperor of India (1556–1605). Expands empire and unites its peoples.	
1558 France recaptures Calais from English. Elizabeth I queen of England (1558–1603).	1557 Portuguese found Macao.	

EUROPE AND THE MEDITERRANEAN	REST OF THE WORLD	CULTURE/TECHNOLOGY
1559 John Knox active in Scotland.		Tobacco enters Europe.
1562 Start of French Wars of Religion.		
1567 Dutch revolt against Spanish begins. Mary Queen of Scots flees to England.		1566 Palladio starts to build church of San Giorgio Maggiore in Venice.
1571 Battle of Lepanto; Turkish domination of eastern Mediterranean ends. Religious settlement in England; Thirty-nine Articles.		1569 Mercator invents map projection.
1572 Massacre of St Bartholomew (French Huguenots).		
1577 Drake's voyage round the world begins.	Spanish expanding in Mexico and Colombia.	El Greco to Toledo. Tycho Brahe: *De Nova Stella.*
1579 Protestant Dutch unite, form United Provinces. Irish rebels massacred; plantation of English settlers.		
1580 Spain occupies Portugal.	1582 Warrior Hideyoshi unites Japan and campaigns in Korea (1592, 1598).	1580–95 Montaigne: *Essays.*
1585 England and Spain at war.	1584 Walter Raleigh attempts to found colony of Virginia.	
1587 Mary Queen of Scots executed.		
1588 Spanish Armada defeated.		
1589 Henry IV king of France (1589–1610).		Early ballet in France. c.1590 Marlowe: *Doctor Faustus.* 1590, 1596 Spenser: *Faerie Queene.*
1593 Henry IV accepts Catholicism in France.		Early microscope; thermometer; water-closet. 1594 Death of composers Palestrina and Lassus.
1598 Edict of Nantes ends Wars of Religion in France. Boris Godunov Russian tsar (1598–1605). 1601 Irish revolt suppressed.	1600 East India Company formed.	Shakespeare: *Romeo and Juliet.* Globe theatre built (1599). Giordano Bruno burnt at stake for heretical theory of universe.
1603 James VI of Scotland (1567–1625) becomes James I of England (1603–25).		Shakespeare: *Hamlet.*
1604 Anglo-Spanish peace treaty.		
1605 Gunpowder Plot in English Parliament.	1607 Virginia settled by British. 1608 Quebec settled by Champlain for France.	1606 Ben Jonson: *Volpone.* 1605–15 Cervantes: *Don Quixote.* 1607 Monteverdi: *Orfeo.* 1609 Shakespeare: sonnets. Galileo: telescope. Kepler: laws of planetary motion.
1610 Ulster planted with English and Scottish settlers. Louis XIII king of France (1610–43).		1610 Caravaggio dies. 1611 Authorized Version of the Bible. Shakespeare: *The Tempest.*
1613 Russian Romanov dynasty established.		1614 John Napier: work on logarithms.
1618 Thirty Years War begins in Europe.	1616 Japan ejects Christian missionaries. Tobacco plantations in Virginia expanding.	1619–21 Inigo Jones designs Banqueting Hall at Whitehall. Francis Bacon: *Novum Organum.*
1621 Philip IV king of Spain (1621–65).	1620 Pilgrim Fathers arrive at Cape Cod on the *Mayflower.*	
1624 Richelieu in power in France. Britain and Spain renew war (1624–30).		Frans Hals: *The Laughing Cavalier.*
1625 Charles I king of England (1625–49).	1626 Dutch purchase Manhattan (New Amsterdam).	
1627 Britain and France at war (1627–9). Richelieu defeats Huguenots.		
1629 Charles I governs without Parliament.	1630–42 Large-scale British emigration to Massachusetts.	1628 William Harvey: *De Motu Cordis.*

EUROPE AND THE MEDITERRANEAN	REST OF THE WORLD	CULTURE/TECHNOLOGY
1630 Gustavus Adolphus of Sweden joins Thirty Years War.		1632 Van Dyck to England.
1633 William Laud elected Archbishop of Canterbury; opposes Puritans in Britain.		Galileo before Inquisition; recants.
1635 France joins Thirty Years War.		
1637 Charles I faces crisis in Scotland over new liturgy.	Japan closed to Europeans.	Corneille: *Le Cid.* Descartes active in Holland.
1640 Long Parliament begins. Braganza dynasty rules in Portugal.		
1642 English Civil War begins (1642–9). Death of Richelieu.	1642–3 Tasman explores Antipodes.	Rembrandt: *Night Watch.* Pascal's calculating machine.
1643 Louis XIV king of France (1643–1715). Mazarin in power.		Torricelli invents barometer.
1644 Charles I defeated at Marston Moor.	Qing dynasty established in China.	
1645 New Model Army formed. Charles I defeated at Naseby. Main phase of war ends (1646).		
1647 Leveller influence in New Model Army.		
1648 Charles I's attempt to regain power with Scottish help defeated. Treaty of Westphalia ends Thirty Years War. Independence of Dutch from Spain recognized.	Atlantic slave trade expanding.	
1649 Charles I tried and executed. Cromwell defeats Irish at Wexford, massacres garrison at Drogheda.		*c.*1649 Taj Mahal completed. 1650 Guericke invents air pump.
1651 Cromwell defeats Scots at Worcester.		1651 Hobbes: *Leviathan.*
1652 English and Dutch at war (1652–7).	Dutch found Cape Colony.	
1653 Oliver Cromwell 'Protector' (1653–8).		*c.*1655 Poussin: *Et in Arcadia Ego.* 1656 Bernini completes piazza of St Peter's, Rome. 1656 Huygens invents pendulum clock. *c.*1656 Velázquez: *Las Meninas.*
1660 Restoration of Charles II as king (1660–85).	Aurangzeb Mogul emperor (1658–1707); expansion followed by decline after his death.	Vermeer at work. 1662 Royal Society founded in London. Robert Boyle: Boyle's Law.
1664 English and Dutch at war (1664–7).		Molière: *Tartuffe.*
1665 Great Plague of London. Colbert chief minister in France (1665–83).		
1666 Fire of London.	Hindu Maratha kingdom rises in western India; challenges Moguls.	1666–7 Newton invents differential calculus. Spirit-level invented. 1667 Milton: *Paradise Lost.*
1670 French troops occupy Lorraine.	Rise of Ashanti in West Africa.	Versailles palace built. 1670–1720 Wren rebuilds London churches and St Paul's.
1672 English and Dutch at war again (1672–4).	1675 In India Sikhism becomes military theocracy to resist Mogul power.	
1678 Franche-Compté annexed to France.		1676 Van Leeuwenhoek identifies microbes with aid of microscope. 1677 Racine: *Phèdre.* 1678–84 Bunyan: *The Pilgrim's Progress.*
1682 Peter the Great tsar of Russia (1682–1725).	Pennsylvania founded.	1681 Pressure-cooker invented.
1683 Turks besiege Vienna.		

EUROPE AND THE MEDITERRANEAN	REST OF THE WORLD	CULTURE/TECHNOLOGY
1685 Edict of Nantes revoked. James II king of England, Scotland, and Ireland (1685–8).		
1688 English 'Glorious Revolution'; William (1689–1702) and Mary (1689–94) reign. Spain recognizes Portugal's independence.		1687 Newton: *Principia Mathematica*.
1689 England at war with France.		Purcell: *Dido and Aeneas*.
1690 Battle of Boyne; forces of James II defeated.	1692 Witch trials in Salem, Massachusetts.	Locke: *An Essay concerning Human Understanding*. 1693 François Couperin to Versailles as court organist. 1698 Savery: steam engine.
1700 Charles II (Habsburg) of Spain dies childless. Philip V (Bourbon) king of Spain (1700–46). Great Northern War (1700–21); Russia and allies oppose Sweden.		Congreve: *The Way of the World*. Stradivarius: violins.
1701 War of the Spanish Succession (1701–14): Britain, Holland, and Holy Roman Empire against France.		
1702 Anne queen of England, Scotland, and Ireland (1702–14).		
1704 Marlborough wins Battle of Blenheim. British take Gibraltar.		Newton: *Opticks*. 1705 Halley predicts return of his comet.
1706 Battle of Ramillies: Marlborough routs French.		
1707 Act of Union between England and Scotland.		
1708 Battle of Oudenarde: Marlborough's third victory.		
1709 Battle of Poltava in Great Northern War ends Swedish hegemony in the Baltic. Last of Marlborough's victories, at Malplaquet.		First piano in Italy. Newcomen: piston-operated steam engine.
1713 Peace of Utrecht ends War of the Spanish Succession. Frederick William I king of Prussia (1713–40).	Newfoundland, Nova Scotia, St Kitts, and Hudson Bay awarded to Britain.	1712 Handel to London.
1714 George I king of Great Britain and Ireland (1714–27).		Fahrenheit devises thermometer scale.
1715 Louis XV king of France (1715–74). Jacobite rebellion suppressed in Scotland and England.	1717 Shenandoah Valley settled. American Indians evicted.	1716 Couperin: treatise on harpsichord-playing. 1717 Watteau: *L'Embarquement pour l'île de Cythère*.
1720 South Sea Bubble; major financial collapse in London.	Chinese invade Tibet.	1719 Defoe: *Robinson Crusoe*.
1721 Robert Walpole first Prime Minister in Britain (1721–42).	French and English rivals in India.	1724 Bourse opens in Paris. 1726 Swift: *Gulliver's Travels*. 1726 Voltaire goes to England.
1727 George II king of Great Britain and Ireland (1727–60).	1728 Danish explorer Bering discovers Straits. 1729 North and South Carolina become Crown Colonies.	1728 Gay: *Beggar's Opera*. Pope: *Dunciad*. 1729 J. S. Bach: *The Passion according to St Matthew*. 1734 Voltaire: *Lettres philosophiques*. 1735 Linnaeus: *Systema Naturae*.
1739 War of Jenkins's Ear between England and Spain.		

EUROPE AND THE MEDITERRANEAN	REST OF THE WORLD	CULTURE/TECHNOLOGY
1740 Frederick II king of Prussia (1740–86); claims Silesia, causing War of the Austrian Succession (1740–8). Archduchess Maria Theresa, queen of Hungary and Bohemia (1740–80).	1741 Bering discovers Alaska.	Richardson: *Pamela*. 1742 Celsius devises centigrade scale. 1742 Handel: *Messiah*.
1745 Jacobite Rebellion under 'Bonnie Prince Charlie'.		
1746 Battle of Culloden in Scotland: defeat of Jacobites; rebellion ruthlessly suppressed.	English, French, Dutch trading extensively in Asia.	1748 Pompeii excavated. 1749 Bow Street Runners formed in London. 1749 Fielding: *Tom Jones*. 1750 Death of J. S. Bach. Symphonic form emerging in music. Ideas of the Enlightenment influential in Europe.
1752 England and Wales adopt Gregorian calendar.		1751 Diderot publishes first volume of *Encyclopédie*. Franklin devises lightning conductor. Jewish naturalization in Britain. Buffon: *Histoire naturelle* (36 vols., 1749–88).
1755 Lisbon earthquake.		Samuel Johnson: *Dictionary of the English Language*. Neoclassical art fashionable.
1756 Outbreak of Seven Years War (1756–63). Pitt the Elder Prime Minister (1756–61).		
1757 Frederick II wins victories for Prussia.	Robert Clive commands East India Company army. Battle of Plassey: Clive defeats Nawab of Bengal and controls state.	Sextant designed in England. French physiocrats active (early economists).
1758 Frederick II defeats Russians.		Voltaire: *Candide*.
1759 French defeated by Prussians at Minden and by British at Quiberon Bay.	Wolfe captures city of Quebec; British then take Montreal and whole colony of Quebec.	1759–67 Sterne: *Tristram Shandy*.
1760 George III king of Great Britain and Ireland (1760–1820).		1761 Haydn to court of Prince Esterházy in Hungary.
1762 Catherine the Great empress of Russia (1762–96).	African slave trade begins to attract criticism.	Jean-Jacques Rousseau: *Émile* and *Social Contract*. Gluck: *Orfeo ed Euridice*.
1763 Seven Years War ends.	American colonists move west into Ohio basin.	Compulsory education in Prussia.
1765 American Stamp Act to finance defence of colonies. Joseph II Habsburg emperor, supported by mother Maria Theresa.		
1766 Stamp Act repealed. Pitt the Elder Prime Minister (1766–8).	1766–9 Bougainville's voyage round world, exploring many Pacific islands. Increasing opposition in British American colonies to control from London through royal governors. West African kingdom of Benin declining in power. 1768–71 Cook's voyage in *Endeavour*; charts New Zealand and eastern Australia.	Cavendish isolates hydrogen. Robert and James Adam influential architects.
1769 Russia advancing against Turks and Tartars. Birth of Napoleon.		Improved steam engine using condenser patented by James Watt.
1770 Lord North Prime Minister in Britain (1770–82).	Policies of North government cause growing resentment in Virginia and Massachusetts.	Spinning-jenny patented by Hargreaves. Gainsborough: *Blue Boy*.
1772 First partition of Poland between Russia, Prussia, and Austria.		Horace Walpole rebuilds Strawberry Hill; helps to inspire Gothic revival.
1774 Financial and administrative chaos in France at the end of Louis XV's reign. Louis XVI king (1774–93).	Warren Hastings Governor-General in India; consolidates Clive's conquests.	Goethe: *The Sorrows of Young Werther*. Priestley discovers oxygen.

EUROPE AND THE MEDITERRANEAN	REST OF THE WORLD	CULTURE/TECHNOLOGY
1776 French support American colonies.	1775 War of American Independence begins (1775–83); battle of Bunker Hill. 1776 Declaration of Independence signed 4 July by thirteen rebel colonies.	1775 Beaumarchais: *The Barber of Seville.* Jenner discovers principle of vaccination. 1776–88 Gibbon: *The History of the Decline and Fall of the Roman Empire.* 1776 Adam Smith: *Inquiry into the Nature and Causes of the Wealth of Nations.*
1777 Rapid growth of British textile industry. Necker French finance minister (1777–81)	British take New York and Philadelphia but are defeated at Saratoga.	Sheridan: *The School for Scandal.* Lavoisier: work on combustion. *Sturm und Drang* literary movement.
1778 France joins American war.		La Scala in Milan built.
1779 Spain joins American war. Riots against machines in England.		
1780 Gordon Riots in London.	1781 British take Charleston but surrender at Yorktown.	1781 Kant: *Critique of Pure Reason.* Planet Uranus discovered by Herschel.
1782 Political crisis in Britain. Fall of Lord North.		Laclos: *Les Liaisons dangereuses.* Watt: improvements to the steam engine.
1783 Pitt the Younger Prime Minister (1783–1801). Peace of Paris ends War of American Independence; Britain accepts independence of US colonies but retains West Indies and Canada.	1785 Warren Hastings returns to Britain; faces seven-year trial for corruption in India.	First manned flight in a hot-air balloon (France). Cavendish determines composition of water. David: *The Oath of the Horatii.* 1786 Mozart: *The Marriage of Figaro.*
1788 George III's first period of mental illness. France bankrupt; Necker restored.	1787 US constitution written. First British convicts sent to Botany Bay; settle in Sydney Bay and create New South Wales. US constitution agreed.	1787 Schiller: *Don Carlos.* Charles's Law formulated. Improved horse-drawn threshing machine patented.
1789 French Revolution: States General summoned; Bastille stormed (14 July); National Assembly formed; French aristocrats flee to England and Germany.	George Washington first US President (1789–97). Alexander Hamilton to US Treasury.	Blake: *Songs of Innocence.* Lavoisier establishes modern chemistry.
1790 France: Church lands nationalized; country organized into departments.		Burke: *Reflections on the Revolution in France.* Ambulances in France.
1791 Louis XVI and Marie Antoinette under restraint in Paris. Attempted escape foiled; arrested at Varennes.	US Congress meets in Philadelphia. Slave revolt in Haiti under Toussaint L'Ouverture.	Mozart dies in poverty. Thomas Paine: *The Rights of Man.* Methodists separate from Church of England.
1792 Russian empire extends beyond Black Sea. French Republic proclaimed (Sept.).		Mary Wollstonecraft: *A Vindication of the Rights of Woman.*
1793 Louis XVI executed (Jan.). The Terror begins. Britain declares war on France. Second partition of Poland. Committee of Public Safety under Robespierre.	George Washington's second term as US President.	
1794 Robespierre executed (July).		Eli Whitney patents cotton-gin in US.
1795 Third partition of Poland. Rural depression and high inflation in Britain. France under Directory.		Hydraulic press in England.
1796 French campaign in Italy; Napoleon victor.		Jenner succeeds with smallpox vaccine.
1797 Talleyrand French foreign minister.	John Adams US President (1797–1801).	

EUROPE AND THE MEDITERRANEAN	REST OF THE WORLD	CULTURE/TECHNOLOGY
1798 Napoleon to Egypt. Irish rebellion suppressed.	French invade Egypt. Nelson destroys French fleet at Aboukir Bay.	Malthus: *Essay on Population*. Lithography invented. Wordsworth: *Lyrical Ballads*. Coal-gas lighting patented in England.
1799 Napoleon returns to Paris and seizes power as First Consul. European coalition against France. Income tax introduced in Britain.		Gas fire patented (France).
1800 Napoleon defeats Austrians at Marengo. France dominates Italy except for Sicily and Sardinia.		Volta makes first battery. Fichte: *The Destiny of Man*.
1801 Irish Act of Union. First census in Britain. Alexander I tsar of Russia (1801–25).	Thomas Jefferson US President (1801–9).	Chateaubriand: *Atala*. Gauss: theory of numbers. 1802 *Charlotte Dundas* first steamship.
1803 Britain declares war on France and forms new coalition. Rebellion in Ireland suppressed.	Jefferson purchases Louisiana from Napoleon.	Dalton: atomic theory.
1804 Pitt the Younger again Prime Minister (1804–6). Pope crowns Napoleon emperor.	Haiti independent. 1804–6 Lewis and Clark expedition across US.	Beethoven: *Eroica* symphony. Trevithick: first steam rail locomotive (Wales).
1805 Nelson wins Battle of Trafalgar. French defeat Austrians and Russians at Austerlitz.		
1806 Death of Pitt the Younger. Holy Roman Empire ends. Prussians defeated at Jena.	Britain seizes Cape Province.	Beaufort scale of wind velocity. Fulton: paddle-steamer.
1808 Spanish rising against French occupation; Peninsular War begins (1808–14).	South American independence movement from Spain begins in Venezuela under Bolívar.	Ingres: *Bather*. Gay-Lussac: law of gas expansion. 1808–32 Goethe: *Faust*.
1809 British victory at Corunna. France occupies Papal States.	James Madison US President (1809–17). Macquarie governor of New South Wales (1809–21).	
1810 Wellington in command of British in Peninsular War.		1810–14 Goya: *The Disasters of War*.
1811 George III mentally ill; Prince Regent installed (1811–20). Economic depression in Britain; Luddite riots against machines.	Muhammad Ali overthrows Mamelukes in Egypt. Paraguay independent.	Krupp factory at Essen. Avogadro's law.
1812 Napoleon marches on Russia; forced to retreat from Moscow. British Prime Minister Spencer Perceval assassinated. Succeeded by Lord Liverpool (1812–27).	War of 1812: US against Britain (1812–14).	Brothers Grimm: first volume of fairy tales. 1812–18 Byron: *Childe Harold's Pilgrimage*. 1812–16 Hegel: *Science of Logic*. 1813 Jane Austen: *Pride and Prejudice*.
1814 Napoleon abdicates; sent to Elba. Louis XVIII (1814–24) and Ferdinand VII (1808, 1814–33) restored as kings of France and Spain. Congress of Vienna convened.		Beethoven: *Fidelio*.
1815 Napoleon returns; raises army; defeated at Waterloo; banished to St Helena.	US westward expansion begins. 1816 Argentina declares independence. Shaka forms Zulu kingdom. 1817 James Monroe US President (1817–25).	Davy lamp invented. 1816 Rossini: *The Barber of Seville*. Stethoscope invented (France). 1817–21 Weber: *Der Freischütz*. 1818 Keats: poems. Mary Shelley: *Frankenstein*.
1819 Peterloo massacre in Manchester.	Raffles founds Singapore. US purchases Florida.	Schubert: 'Trout' quintet. Géricault: *The Raft of the Medusa*. Scott: *Ivanhoe*. Macadamized roads stimulate coach travel.
1820 Abortive risings in Portugal, Sicily, Germany, and Spain. George IV king of Great Britain and Ireland (1820–30).	US settlers beyond the Mississippi.	Shelley: *Prometheus Unbound*. Street lighting on Pall Mall, London.

EUROPE AND THE MEDITERRANEAN	REST OF THE WORLD	CULTURE/TECHNOLOGY
1821 Famine in Ireland. Greek war of independence from Turkey starts (1821–30).	Mexico independent.	Constable: *The Hay Wain.*
1822 Castlereagh dies; Canning foreign secretary.	Liberia founded for US freed slaves. Brazil independent. 1823 Monroe Doctrine extends US protection to Spanish-American republics. Costa Rica, Ecuador independent.	De Quincey: *Confessions of an English Opium Eater.* Champollion deciphers Egyptian hieroglyphics (Rosetta Stone). Rapid industrialization of NW England and Lowland Scotland (textiles).
1824 British Combination Acts repealed to allow trade unions. Charles X king of France (1824–30).	Peru independent.	Beethoven: Ninth Symphony. National Gallery founded.
1825 Nicholas I tsar of Russia (1825–55). Decembrist revolt suppressed.		Stockton and Darlington railway opened. Saint-Simon: *Nouveau Christianisme.*
1826 Muhammad Ali reconquers Peloponnese in Greek war.	Britain establishes Straits Settlements: Penang, Malacca, and Singapore.	James Fenimore Cooper: *The Last of the Mohicans.*
1827 Britain, Russia, and France send navies and destroy Turkish fleet at Navarino. Canning British Prime Minister; dies and succeeded by Wellington (1828–30).	France intervenes in Algeria.	Beethoven dies in Vienna.
1828 Daniel O'Connell (Catholic) elected to Parliament.		Webster: *American Dictionary of the English Language.*
1829 Full Catholic emancipation granted in Britain. Metropolitan Police formed.	Andrew Jackson US President (1829–37). Western Australia founded.	Braille invented (France). Sewing machine (France). Stephenson: *Rocket.*
1830 First cholera epidemics in Europe. Charles X deposed; Louis Philippe king of France (1830–48). William IV king of Great Britain and Ireland (1830–7). Lord Grey Prime Minister (1830–4). Belgium fights for independence.	Muhammad Ali encourages revival of Arabic culture. France takes Algeria. Colombia, Venezuela independent.	Stendhal: *Le Rouge et le noir.* Berlioz: *Symphonie fantastique.* Joseph Smith founds Mormon Church.
1831 Major cholera epidemic in Britain. Mazzini founds Young Italy movement. Greece gains independence.		Pushkin: *Boris Godunov.* Darwin begins voyage on *Beagle* (1831–6). Bellini: *Norma.*
1832 First Reform Act in Britain.	Forcible settlement of American Indians in Oklahoma.	Industrialization in Belgium and NW France; railways.
1833 British Factory Act: child labour regulated.	Abolition of slavery in British Empire.	Oxford Movement to restore Anglicanism.
1834 British Poor Law Amendment Act. Peel's Tamworth Manifesto. Tolpuddle martyrs transported to Australia.	1835–7 Great Trek by Afrikaners in Africa. 1836 South Australia becomes British colony. Texas independent from Mexico.	1835 Donizetti: *Lucia di Lammermoor.* Revived Gothic architecture widespread. Fox Talbot: first photographic negative. Chopin: *Preludes.* Morse code.
1837 Victoria queen of Great Britain and Ireland (1837–1901).		Railway boom in Britain. Electric telegraph. Dickens: *Oliver Twist.*
1838 *The People's Charter* drawn up in London by Chartists. Anti-Corn-Law League founded in Manchester.		Turner: *The Fighting Téméraire.* Isambard Kingdom Brunel: *Great Western.*
1839 Treaty granting Belgian independence.	First Opium War in China (1839–42).	Daguerreotype photograph. Faraday: theory of electromagnetism.
1840 Penny post in Britain.	Canadian Provinces Act of Union. Treaty of Waitangi, NZ: Maoris and settlers.	Edgar Allan Poe: *Tales of the Grotesque and Arabesque.* First bicycle in Scotland. Proudhon: *What is Property?* Schumann: songs.

	EUROPE AND THE MEDITERRANEAN	REST OF THE WORLD	CULTURE/TECHNOLOGY
1841	Robert Peel British Prime Minister (1841–6).	Britain takes Hong Kong. France occupies Tahiti, Guinea, and Gabon.	1842 Verdi's *Nabucco* encourages Italian nationalism. Gogol: *Dead Souls*. Balzac: *La Comédie humaine;* (1842–8).
1843	Free Church of Scotland formed in protest against established Church of Scotland.	Britain annexes Sind.	Joule: theory of thermodynamics.
1844	Cooperative Society formed in Rochdale.		Dumas: *The Count of Monte Cristo.* Kierkegaard: *The Concept of Dread.*
1845	Potato famine in Ireland.	US annexes Texas. Sikh Wars (1845 and 1848–9).	Engels: *The Condition of the Working Classes in England in 1844.* Disraeli: *Sybil.* Galvanized corrugated iron patented (England).
1846	Corn Laws repealed; Peel resigns; Lord Russell Prime Minister (1846–52).	Mexican–US War (1846–8).	Planet Neptune discovered.
1847	Risorgimento in Italy. Factory Act in Britain: ten-hour day.		Charlotte Brontë: *Jane Eyre.* Emily Brontë: *Wuthering Heights.*
1848	Last Chartist petition. Revolutions in Europe. Second Republic in France (1848–52). Cholera in Europe.	Gold discovered in California. Irish emigrants to US. Abolition of slavery in French West Indies.	Marx/Engels: *Communist Manifesto.* Pre-Raphaelite Brotherhood formed.
1849	Revolutions suppressed around Europe.	California gold rush. Britain annexes Punjab and subdues Sikhs. 1850–64 Taiping Rebellion in China. Livingstone crosses Africa.	Safety pin patented. 1850 Courbet: *Burial at Ornans.* Tennyson: *In Memoriam.* Nathaniel Hawthorne: *The Scarlet Letter.* Kelvin: second law of thermodynamics.
1851	Coup by Louis-Napoleon in France.	Gold in Australia; settlers moving into Victoria.	Great Exhibition in Crystal Palace.
1852	Aberdeen British Prime Minister (1852–5). Napoleon III founds French Second Empire (1852–70). Cavour premier of Piedmont (1852–9; 1860–1).		Harriet Beecher Stowe: *Uncle Tom's Cabin.*
1853	Crimean War begins (1853–6); Russia against Turkey.	First railways and telegraph in India. Gadsden Purchase in US.	Hypodermic syringe (France). Verdi: *La Traviata.*
1854	Crimean War develops; France and Britain join Turkey; siege of Sebastopol; Florence Nightingale to hospital at Scutari.	US Republican Party founded.	William Holman Hunt: *The Light of the World.* Catholic dogma of Immaculate Conception.
1855	Palmerston British Prime Minister (1855–8).	Railways in South America: Chile (1851), Brazil (1854), Argentina (1857).	Telegraph news stories of Crimean War. Walt Whitman: *Leaves of Grass.* Mendel outlines laws of heredity.
1856	Crimean War ends.	Britain annexes Oudh. Second Opium War (1856–60).	Synthetic dyes invented (England).
1857	Italian nationalism growing.	1857–8 Indian Mutiny.	Baudelaire: *Les Fleurs du mal.* Flaubert: *Madame Bovary.* Trollope: *Barchester Towers.*
1858	Lionel Rothschild first Jewish MP.	Fenians founded in US. Government of India transferred from East India Company to British Crown.	Lourdes miracles reported.
1859	French support Italians in struggle for independence from Austria.	John Brown at Harpers Ferry.	J. S. Mill: *On Liberty.* Darwin: *On the Origin of Species.* Wagner: *Tristan und Isolde.* Oil pumped in Pennsylvania.
1860	Garibaldi victorious in southern Italy.	1860–72 War in New Zealand; Maoris against settlers. South Carolina secedes from US Union.	Bessemer: mass production of steel. Huxley defends theory of evolution. George Eliot: *The Mill on the Floss.* Dickens: *Great Expectations.*

EUROPE AND THE MEDITERRANEAN	REST OF THE WORLD	CULTURE/TECHNOLOGY
1861 Victor Emmanuel II king of Italy (1861–78). Death of Prince Albert. Russia abolishes serfdom.	Lincoln US President (1861–5). American Civil War (1861–5). First battle at Bull Run.	Siemens developing open-hearth steel production.
1862 Bismarck Minister-President of Prussia.	French Indo-China established. 1863 Battle of Gettysburg; Union victory. Emancipation of US slaves. French protectorate of Cambodia.	Victor Hugo: *Les Misérables*. 1863 Manet: *Déjeuner sur l'herbe*. 1863–9 Tolstoy: *War and Peace*. *c.*1863 Rossetti: *Beata Beatrix*. Maxwell: theory of electromagnetism.
1864 First International formed in London (Karl Marx organizes). Red Cross founded.	French install Maximilian emperor of Mexico. Sherman marches through Georgia.	Jules Verne: *Journey to the Centre of the Earth*.
1865 Lord Russell British Prime Minister again (1865–6).	1865–70 Brazil, Argentina, and Uruguay at war with Paraguay; ends in disaster for Paraguay. Confederate commander Lee surrenders; American Civil War ends. Lincoln assassinated; Andrew Johnson US President (1865–9).	Lewis Carroll: *Alice's Adventures in Wonderland*. Pasteur publishes theory of germs causing disease. Whymper ascends the Matterhorn.
1866 Lord Derby British Prime Minister (1866–8).	Livingstone begins third journey in Africa. US Reconstruction under way.	Dostoevsky: *Crime and Punishment*.
1867 Austro-Hungarian empire formed. Fenian rising in Ireland. Second British Reform Act.	US purchases Alaska from Russia. Canada becomes a British dominion. Restoration of imperial power in Japan; end of shogunates.	Karl Marx: first volume of *Das Kapital*. Ibsen: *Peer Gynt*. Japanese art arrives in West.
1868 Gladstone British Prime Minister (1868–74). British TUC formed.		Wilkie Collins: *The Moonstone*. Helium discovered. 1869 Suez Canal opens. Liquefaction of gases (Andrews).
1870 Irish Land Act. Franco-Prussian War (1870–1): Napoleon III defeated at Sedan; dethroned and exiled.	Rockefeller establishes Standard Oil Company in US.	Doctrine of papal infallibility. British elementary education.
1871 French Third Republic suppresses Paris Commune and loses Alsace-Lorraine to German Empire. Wilhelm I first German emperor (1871–88). British trade unions gain legality. Rome made capital of Italy.	Stanley finds Livingstone. Ku Klux Klan suppressed in US. Feudalism suppressed in Japan; modernization begins.	1871–2 George Eliot: *Middlemarch*.
1872 Voting by secret ballot in Britain. Bismarck opposes Catholic church in *Kulturkampf*.		Air brakes patented (Westinghouse). 1873 Rimbaud: *Une Saison en enfer*. 1873–7 Tolstoy: *Anna Karenina*.
1874 Disraeli British Prime Minister (1874–80).	Stanley charts Lake Victoria and traces the course of the Congo (1874–7).	First Impressionist exhibition: Monet, Sisley, Renoir, Pissarro, Degas, Cézanne. Remington typewriter.
1875 Britain buys control of Suez Canal.		Bizet: *Carmen*.
1876 Queen Victoria Empress of India.	Battle of Little Bighorn.	Plimsoll line for ships. Bell patents telephone. Brahms: First Symphony. Wagner's *Ring* cycle first performed.
1877 Russo-Turkish War (1877–8).	US Reconstruction collapses; southern states impose racist legislation. Britain annexes Transvaal.	Phonograph (Edison). Tchaikovsky: *Swan Lake*.
1878 Salvation Army created in Britain. Serbian independence. Britain gains Cyprus. Romania independent.		Gilbert and Sullivan: *HMS Pinafore*.
1879 Land League formed in Ireland.		Mary Baker Eddy founds the Church of Christ, Scientist.
1880 Gladstone British Prime Minister (1880–5).	First Boer War (1880–1).	Swan perfects carbon-filament lamp. Burne-Jones: *The Golden Stairs*. Development of seismograph. Rodin: *The Gate of Hell*.

EUROPE AND THE MEDITERRANEAN	REST OF THE WORLD	CULTURE/TECHNOLOGY
1881 Second Irish Land Act. Assassination of Tsar Alexander II. Jewish pogroms in eastern Europe. Alexander III tsar (1881–94).	Mahdi leads revolt against Egyptian rulers of Sudan.	Ibsen: *Ghosts.* First public electricity supply (Godalming, Surrey).
1882 Triple Alliance: Germany, Austria–Hungary, Italy.	British occupy Egypt.	Manet: *A Bar at the Folies-Bergère.*
1883 Social insurance in Germany.	Jewish immigration to Palestine.	Robert Louis Stevenson: *Treasure Island.* 1883–5 Nietzsche: *Thus Spake Zarathustra.*
1884 Third British Reform Act.	Germany acquires South West Africa.	First part of *Oxford English Dictionary.* Rayon artificial fibres (France).
1885 Lord Salisbury British Prime Minister (1885–6).	Mahdi takes Khartoum; death of General Gordon. Canadian Pacific Railway complete. Indian National Congress formed.	Motor car (Benz). Motorcycle (Daimler). Pasteur: anti-rabies vaccine. Zola: *Germinal.*
1886 Gladstone British Prime Minister but resigns over Irish Home Rule. Liberal Unionists formed in protest against Home Rule. Lord Salisbury Prime Minister (1886–92).	Slavery ends in Cuba. Tunisia under French protectorate. All Burma occupied by British.	Hardy: *The Mayor of Casterbridge.*
1887 Queen Victoria's Golden Jubilee celebrated. Industrial unrest in Britain; 'Bloody Sunday' (Trafalgar Square riots).	Gold discovered in Kalgoorlie, Western Australia.	Conan Doyle: *A Study in Scarlet* (first Sherlock Holmes stories). Radio waves demonstrated (Hertz).
1888 Wilhelm II emperor of Germany (1888–1918).	Slavery ends in Brazil.	Pneumatic tyre (Dunlop). Strindberg: *Miss Julie.*
1889 Second International formed in Paris.		Van Gogh: *A Starry Night.* Eiffel Tower, Paris.
1890 Parnell resigns from Parliament. Chancellor Bismarck forced to resign.	Battle of Wounded Knee.	Death of Van Gogh. Mascagni: *Cavalleria Rusticana.* Dewar: vacuum flask.
1891 Trans-Siberian Railway begun.	Young Turk movement founded.	Gauguin to Tahiti.
1892 Tsar Alexander III forms alliance with France. Gladstone's fourth ministry (1892–4).		1892–5 Dvořák: Ninth Symphony ('From the New World'). Diesel engine patented.
1893 Independent Labour Party formed in Britain.		Tchaikovsky: Sixth Symphony. Edvard Munch: *The Scream.*
1894 Nicholas II tsar (1894–1917). Lord Rosebery British Prime Minister (1894–5). Death duties in Britain. Dreyfus imprisoned in France.	First Sino-Japanese War (1894–5).	Kipling: *The Jungle Book.* Toulouse-Lautrec: *Moulin Rouge* lithographs.
1895 *Lord Salisbury British Prime Minister (1895–1902); strongly imperialist government.	Japan takes Taiwan (Formosa). Jameson Raid in South Africa (1895–6).	Lumière brothers: cinema. X-rays discovered (Röntgen). Oscar Wilde: *The Importance of Being Earnest.* Safety razor.
1896 First modern Olympic Games in Athens.	Laurier elected Canada's first French-Canadian and Roman Catholic Prime Minister.	Hardy: *Jude the Obscure.* A. E. Housman: *A Shropshire Lad.* Radioactivity of uranium (Becquerel).
1897 Queen Victoria's Diamond Jubilee celebrated. Zionist movement founded.	Klondike gold rush (1897–9).	Thomson: the electron. Bram Stoker: *Dracula.* Henri Rousseau: *Sleeping Gypsy.* Aspirin marketed.
1898 German naval expansion rapid.	Spanish–American War. British reconquer Sudan.	Zola: *J'accuse.* Henry James: *The Turn of the Screw.* H. G. Wells: *The War of the Worlds.* Curies: radium.
1899 Britain fights Second Boer War (1899–1902).	Siege of Mafeking. Boxer Rising in China.	Freud: *The Interpretation of Dreams.* Monet: first in *Water-lilies* sequence. Scott Joplin: 'Maple Leaf Rag'.

	EUROPE AND THE MEDITERRANEAN	**REST OF THE WORLD**	**CULTURE/TECHNOLOGY**
1900	Boer War concentration camps arouse European criticism.	Mafeking relieved.	First Zeppelin. Planck: quantum theory. Conrad: *Lord Jim*. Puccini: *Tosca*. First agricultural tractor.
1901	Death of Queen Victoria. Edward VII king of Great Britain and Ireland (1901–10).	Commonwealth of Australia formed. US President McKinley assassinated. Theodore Roosevelt President (1901–9).	1901–4 Picasso: Blue Period. Gorky: *The Lower Depths*.
1902	Balfour British Prime Minister (1902–5).	Treaty of Vereeniging ends Boer War.	Gide: *The Immoralist*. John Masefield: *Salt-Water Ballads*.
1903	Bolsheviks form majority of Social Democratic party in Russia. Pogroms in Russia. Suffragette movement in Britain; Pankhursts.	Mass European emigration to US. Panama independent; Canal Zone to US.	Electrocardiograph. Wright brothers: powered flight.
1904	Franco-British *entente cordiale*.	Russo-Japanese War (1904–5).	J. M. Barrie: *Peter Pan*. Chekhov: *The Cherry Orchard*. Photoelectric cell.
1905	First Russian Revolution. Liberal government in Britain; Campbell-Bannerman Prime Minister (1905–8).	Japanese destroy Russian fleet at Tsushima.	Richard Strauss: *Salome*. Einstein: special theory of relativity. Early brain surgery (Cushing in US).
1906	Labour Party formed in Britain. Russian Duma. Clemenceau French Prime Minister (1906–9).	Muslim League founded in India. Earthquake in San Francisco.	Vitamins discovered (Hopkins).
1907	Triple Entente (Great Britain, France, and Russia).	New Zealand gains dominion status.	George Bernard Shaw: *Major Barbara*. J. M. Synge: *The Playboy of the Western World*. Electric washing machine.
1908	Bulgaria independent. Bosnia–Herzegovina occupied by Austria–Hungary. Old-age pensions introduced in Britain. Asquith British Prime Minister (1908–16).	Young Turk revolution in Turkey.	Klimt: *The Kiss*. Cubist movement. E. M. Forster: *A Room with a View*. Mahler: *Das Lied von der Erde*. Geiger counter invented. Borstal system in Britain.
1909	Lloyd George budget rejected by Lords: constitutional crisis.	Oil-drilling in Persia by British.	Model T car (Ford). Matisse: *The Dance*. Ballets Russes formed (Diaghilev). Schoenberg: *Three Piano Pieces*.
1910	George V king of Great Britain and Ireland (1910–36).	Union of South Africa. Japan annexes Korea.	1910–13 A. N. Whitehead and Bertrand Russell: *Principia Mathematica*. Post-impressionist exhibition, London.
1911	Industrial unrest in Britain.	Amundsen reaches South Pole. Chinese revolution; Sun Yat-sen establishes republic (1912).	Rutherford: nuclear model of atom.
1912	Balkan Wars (1912–13).	ANC formed in South Africa. *Titanic* sinks. French protectorate in Morocco.	Jung: *The Psychology of the Unconscious*. First parachute descent from an aircraft. Stainless steel.
1913	Ottoman Turks lose European lands except Constantinople. Crisis in Ireland. British suffragettes.	Woodrow Wilson US President (1913–21).	D. H. Lawrence: *Sons and Lovers*. Stravinsky: *The Rite of Spring*. Alain-Fournier: *Le Grand Meaulnes*. 1913–27 Proust: *À la recherche du temps perdu*.
1914	Assassination at Sarajevo of Archduke Franz Ferdinand. Outbreak of First World War (1914–18). Trench warfare on the Western Front.	Panama Canal opens. Egypt British protectorate.	1914–21 Berg: *Wozzeck*. Joyce: *Dubliners*. Vaughan Williams: *A London Symphony*.
1915	Heavy fighting at Gallipoli. Asquith forms coalition in Britain. Zeppelin raids.	*Lusitania* sunk by German submarine.	Einstein: general theory of relativity. W. G. Griffith: *The Birth of a Nation*.

EUROPE AND THE MEDITERRANEAN	REST OF THE WORLD	CULTURE/TECHNOLOGY
1916 Battles of Verdun, Somme, and Jutland. Irish Easter Rising, Dublin. Lloyd George British Prime Minister (1916–22).		Dada movement. Parry: musical setting of 'Jerusalem'.
1917 Russian Revolution: Tsar Nicholas II abdicates. Battle of Passchendaele.	US enters First World War. Balfour Declaration on Palestine.	1917–70 Pound: *Cantos.* Kafka: *Metamorphosis.* First jazz recordings.
1918 Second battle of the Marne; tanks help final Allied advance. Armistice (Nov.). Women's suffrage in UK.	British take Palestine and Syria; battle of Megiddo. Ottomans make peace.	Gerard Manley Hopkins: *Poems.* 1918–22 Spengler: *The Decline of the West.*
1919 Spartacist revolt in Germany. Sinn Fein sets up independent parliament in Ireland. Third International formed, Moscow. Treaty of Versailles. League of Nations established. Weimar Republic formed.	Arab rebellion in Egypt against British. Massacre at Amritsar in India.	Elgar: *Cello Concerto.* Alcock and Brown fly Atlantic.
1920 Separate parliaments established in northern and southern Ireland.	Prohibition comes into effect in US. Gandhi dominates Indian Congress. US Senate rejects Treaty of Versailles. Palestine mandated to Britain.	Edith Wharton: *The Age of Innocence.* Duchamp: *Mona Lisa.*
1921 Southern Ireland becomes Irish Free State; civil war (1921–3).	Chinese Communist Party founded by Mao Zedong and others. King Faisal I in Iraq (1921–33).	Charlie Chaplin: *The Kid.* Pirandello: *Six Characters in Search of an Author.* 1921–3 William Walton: *Façade.*
1922 USSR formed. Italian Fascists march on Rome; Mussolini forms government. Bonar Law British Prime Minister (1922–3).	Egypt independent.	Joyce: *Ulysses.* T. S. Eliot: *The Waste Land.* Max Weber: *Economy and Society.*
1923 Rampant inflation in Germany. French occupy Ruhr. Putsch organized by Hitler fails. Baldwin British Prime Minister.	Ottoman Empire ends.	Le Corbusier: *Towards a New Architecture.*
1924 MacDonald first Labour Prime Minister in Britain. Death of Lenin. Baldwin Prime Minister again (1924–9).	Northern Rhodesia British protectorate. US economy booming.	Honegger: *Pacific 231.* André Breton: surrealist manifesto.
1925 Locarno Pact.	Reza Pahlavi shah in Iran (1925–41). Chiang Kai-shek launches campaign to unify China.	Hitler: *Mein Kampf.* F. Scott Fitzgerald: *The Great Gatsby.* Eisenstein: *The Battleship Potemkin.* Discovery of ionosphere.
1926 British general strike. Gramsci imprisoned in Italy.	Emperor Hirohito in Japan (1926–89).	First television. Stanley Spencer: *Resurrection: Cookham.* Buster Keaton: *The General.*
1927 Stalin comes to power; Trotsky expelled from Communist Party.	Lindbergh flies Atlantic.	BBC founded. Al Jolson: *The Jazz Singer.* Walt Disney: Mickey Mouse.
1928 Stalin launches Soviet collectivization. Nazis and Communists compete in Germany. Women over 21 enfranchised in UK.	Kellogg Pact for peace.	W. B. Yeats: *The Tower.* Ravel: *Boléro.* Penicillin discovered.
1929 Depression begins. Trotsky exiled. Yugoslavia formed. Second Labour government in Britain under MacDonald (1929–31). Lateran Treaty in Italy.	Wall Street Crash. St Valentine's Day Massacre.	Hemingway: *A Farewell to Arms.* Robert Graves: *Good-bye to All That.*
1930 Airship disaster (*R101*) in France.	Revolution in Brazil; Vargas President. Haile Selassie emperor in Ethiopia (1930–74).	Evelyn Waugh: *Vile Bodies.* Noel Coward: *Private Lives.* Turbojet engine patented by Whittle. Planet Pluto discovered. 1930–1 Empire State Building in New York. First World Cup in football.

EUROPE AND THE MEDITERRANEAN	REST OF THE WORLD	CULTURE/TECHNOLOGY
1931 Depression worsens. King Alfonso XIII of Spain flees; republic formed. British national government under MacDonald (1931–5).	New Zealand independent. Japan occupies Manchuria.	Virginia Woolf: *The Waves.* Boris Karloff: *Frankenstein.* Salvador Dali: *The Persistence of Memory.*
1932 Oswald Mosley's British Union of Fascists formed. Dictator Salazar in Portugal.	14 million unemployed in US. Kingdoms of Saudi Arabia and Iraq independent.	Cockcroft and Walton split the atom. James Chadwick: discovery of the neutron. First autobahn, Cologne-Bonn, opened.
1933 Nazi Party wins German elections. Hitler appointed Chancellor. Third Reich formed.	Franklin D. Roosevelt US President (1933–45); New Deal.	Lorca: *Blood Wedding.* Dorothy L. Sayers: *Murder Must Advertise.*
1934 Night of the long knives in Germany. Stalin purges begin.	1934–5 Long March in China.	Henry Miller: *Tropic of Cancer.* Agatha Christie: *Murder on the Orient Express.*
1935 German Jews lose citizenship. Baldwin British Prime Minister (1935–7).	Italy invades Ethiopia.	Marx Brothers: *A Night at the Opera.* Paul Nash: *Equivalents for the Megaliths.*
1936 Edward VIII abdicates; George VI king of United Kingdom (1936–52). Léon Blum Popular Front government in France. Spanish Civil War (1936–9). Rhineland reoccupied by Germany.	King Farouk in Egypt (1936–52). Haile Selassie in exile (1936–41).	A. J. Ayer: *Language, Truth, and Logic.* Keynes: *The General Theory of Employment, Interest, and Money.* Prokofiev: *Peter and the Wolf.* BBC begins television broadcasts.
1937 Chamberlain British Prime Minister (1937–40).	Sino-Japanese War (1937–45).	Picasso: *Guernica.* George Orwell: *The Road to Wigan Pier.* Walt Disney: *Snow White and the Seven Dwarfs.*
1938 Austrian *Anschluss* with Germany. Munich Agreement. Czechoslovakia cedes Sudetenland.		Nuclear fission discovered. Graham Greene: *Brighton Rock.* Fluorescent lighting US. Nylon patented US.
1939 Molotov–Ribbentrop pact between Germany and USSR. Poland invaded. Franco leader of Spain. Britain and France declare war on Germany; Second World War (1939–45).	King Faisal II in Iraq (1939–58).	Pauling: *The Nature of the Chemical Bond.* Judy Garland: *The Wizard of Oz.* John Steinbeck: *The Grapes of Wrath.*
1940 Occupation by Germany of France, Belgium, the Netherlands, Norway, Denmark. British retreat from Dunkirk. Vichy government in France. Churchill British Prime Minister (1940–5). Battle of Britain.	Trotsky assassinated in Mexico.	Charlie Chaplin: *The Great Dictator.*
1941 Germany occupies Balkans and invades Soviet Union.	Lend-Lease by US. Atlantic Charter. Pearl Harbor. US enters war. Malaya, Singapore, and Burma to Japan.	Orson Welles: *Citizen Kane.*
1942 Beveridge Report. Battle of Stalingrad begins.	Midway and El Alamein key Allied victories.	Fermi: nuclear chain reaction. Camus: *The Outsider.* Bogart/Bergman: *Casablanca.*
1943 Allied bombing of Germany (1943–5). Allies invade Italy; Mussolini deposed. Germans surrender at Stalingrad.	Lebanon independent.	Sartre: *Being and Nothingness.* T. S. Eliot: *Four Quartets.*
1944 D-Day invasion; Paris liberated. Civil war in Greece (1944–9).	1944–5 Burma recaptured.	Holmes: *Principles of Physical Geology.* 1944–6 Eisenstein: *Ivan the Terrible.* Olivier: *Henry V.*
1945 Yalta Conference. War ends in Europe (May). United Nations formed. Labour wins British election; Attlee Prime Minister (1945–51). Potsdam Conference.	Death of Roosevelt; Truman US President (1945–53). Atom bombs on Hiroshima and Nagasaki. War ends (Sept.).	Benjamin Britten: *Peter Grimes.*
1946 Cold war begins. Italian Republic formed.	Perón President in Argentina. Jordan independent. Civil war in China (1946–9).	Cocteau: *La Belle et la bête.*

EUROPE AND THE MEDITERRANEAN	REST OF THE WORLD	CULTURE/TECHNOLOGY
1947 Attlee government nationalizes fuel, power, transport in Britain. Puppet Communist states in eastern Europe.	India independent; Pakistan becomes separate state.	Tennessee Williams: *A Streetcar Named Desire.* Jackson Pollock: action painting.
1948 Berlin airlift. British welfare state. Marshall Plan of aid for western Europe approved by US Congress.	State of Israel established. Ceylon in Commonwealth. 1948–60 Malayan Emergency. Apartheid legislation in South Africa. Gandhi assassinated.	Brecht: *The Caucasian Chalk Circle.* De Sica: *Bicycle Thieves.* 1948–51 Stravinsky: *The Rake's Progress.* Transistor developed in the US.
1949 Comecon and NATO formed. Republic of Ireland established. Communist regime in Hungary.	Indonesia independent. People's Republic of China declared.	George Orwell: *1984.* Simone de Beauvoir: *The Second Sex.* Genet: *Journal du voleur.*
1950 Labour Party wins election in Britain; retains power.	1950–3 Korean War.	Ionesco: *The Bald Prima Donna.* Kurosawa: *Rashomon.* Stereophonic sound developed. First successful kidney transplant.
1951 Conservatives win election in Britain; Churchill Prime Minister (1951–5).	China extends rule over Tibet. Anzus pact in Pacific.	Ray Bradbury: *Fahrenheit 451.* J. D. Salinger: *The Catcher in the Rye.*
1952 European Coal and Steel Community formed. Elizabeth II queen of United Kingdom (1952–). Conservative government in UK denationalizes iron and steel and road transport (1953).	US tests hydrogen bomb.	Hemingway: *The Old Man and the Sea.* Beckett: *Waiting for Godot.*
1953 Death of Stalin.	Egyptian Republic formed. 1953–7 Mau Mau in Kenya. McCarthy era in US. Eisenhower US President (1953–61). Korean War ends.	Crick and Watson: double helix structure of DNA. Mount Everest climbed for first time (Hillary, Tenzing Norgay). L. P. Hartley: *The Go-Between.* Matisse: *The Snail.*
1954 First commercial TV station in Britain.	Vietnam independent, partitioned. Algerian War of Independence (1954–62). Nasser in power in Egypt.	William Golding: *Lord of the Flies.* Kingsley Amis: *Lucky Jim.* 1954–5 J. R. R. Tolkien: *The Lord of the Rings.* Bill Haley: 'Rock Around the Clock'. Fortran in US. Roger Bannister runs first four-minute mile.
1955 West Germany joins NATO. Warsaw Pact formed. Anthony Eden British Prime Minister (1955–7).		Marcuse: *Eros and Civilization.* James Dean: *Rebel Without a Cause.* Hovercraft patented.
1956 Khrushchev denounces Stalin. Suez crisis. Polish and Hungarian revolts.	Hundred Flowers in China (1956–7). Morocco independent. Guerrilla conflict in Vietnam.	John Osborne: *Look Back in Anger.* Elvis Presley: 'Heartbreak Hotel'. First commercial nuclear power stations (Britain 1956, US 1957).
1957 Macmillan British Prime Minister (1957–63). Treaty of Rome; EEC formed. Soviet Sputnik flight.	Ghana independent.	Robbe-Grillet: *Jealousy.* Jack Kerouac: *On the Road.*
1958 Fifth Republic in France. Life peerages introduced in Britain.	Great Leap Forward in China. NASA set up.	Silicon chip invented in US. Harold Pinter: *The Birthday Party.* Lévi-Strauss: *Structural Anthropology.* Achebe: *Things Fall Apart.*
1959 De Gaulle French President (1959–69). Conservatives win British election. North Sea natural gas discovered.	Cuban revolution (Castro).	Günter Grass: *The Tin Drum.* Marilyn Monroe: *Some Like it Hot.*
1960 EFTA formed. Cyprus independent.	End of Malayan Emergency. Sharpeville massacre in South Africa. Belgian Congo (Zaire) independent. OPEC formed. Nigeria independent.	Hitchcock: *Psycho.* Fellini: *La Dolce vita.* Godard: *Breathless.* Lasers built (US). Oral contraceptive marketed.
1961 Berlin Wall erected.	John F. Kennedy US President (1961–3). Bay of Pigs invasion. South Africa becomes a republic, leaves Commonwealth.	Joseph Heller: *Catch-22.* Manned space flight (Yuri Gagarin).

EUROPE AND THE MEDITERRANEAN	REST OF THE WORLD	CULTURE/TECHNOLOGY
1962 Vatican Council (1962–5).	Cuban Missile Crisis. Jamaica, Trinidad and Tobago, and Uganda independent.	The Beatles: 'Love Me Do'. Andy Warhol: Marilyn Monroe prints.
1963 De Gaulle vetoes Britain's bid to enter EEC. Alec Douglas-Home British Prime Minister (1963–4).	Test-Ban Treaty signed. Kenya independent. President Kennedy assassinated. Lyndon Johnson US President (1963–9). OAU formed.	Betty Friedan: *The Feminine Mystique*. Sylvia Plath: *The Bell Jar*.
1964 Labour Party wins British election; Harold Wilson Prime Minister (1964–70). Khrushchev ousted by Brezhnev (1964–82).	US enters Vietnam War. Nelson Mandela sentenced to life imprisonment in South Africa. PLO formed.	Saul Bellow: *Herzog*. Word processor.
1965 De Gaulle re-elected French President.	UDI in Rhodesia. Conflict between India and Pakistan over Kashmir. Military takeover in Indonesia.	Joe Orton: *Loot*. The Rolling Stones: 'Satisfaction'. Bob Dylan: *Highway 61 Revisited*.
1966 Labour Party gains bigger majority in Britain.	Cultural Revolution in China (1966–8). Indira Gandhi Indian Prime Minister (1966–77).	England win World Cup. Tom Stoppard: *Rosencrantz and Guildenstern are Dead*.
1967 De Gaulle vetoes Britain's second bid to enter EEC. Abortion legalized in Britain.	Biafran war (1967–70). Six Day War between Israel and Arabs.	The Beatles: *Sergeant Pepper's Lonely Hearts Club Band*. First heart transplant (Barnard). First pulsar discovered.
1968 Soviet forces invade Czechoslovakia. Student protest in Paris and throughout Europe. Violent protest erupts in Northern Ireland.	Tet Offensive in Vietnam. Martin Luther King assassinated.	Kubrick: *2001: A Space Odyssey*. Plate tectonics.
1969 British army to Northern Ireland. De Gaulle resigns; Pompidou French President (1969–74). Brandt West German Chancellor (1969–74).	Richard Nixon US President (1969–74). Sino-Soviet frontier war. Gaddafi in power in Libya.	First man on moon (Neil Armstrong). First Concorde flight. Woodstock pop festival.
1970 Conservatives win British election; Edward Heath Prime Minister (1970–4). West German policy of Ostpolitik.	Allende President in Chile (1970–3). Biafran war ends.	Germaine Greer: *The Female Eunuch*. Beatles split up.
1971 Internment without trial in Northern Ireland. Decimal currency in UK.	Amin seizes power in Uganda. Bangladesh becomes a state.	Visconti: *Death in Venice*. Open University in UK.
1972 Bloody Sunday in Northern Ireland; direct rule from Westminster imposed. Many Ugandan Asian refugees to UK.	Nixon visits China.	Bertolucci: *Last Tango in Paris*. World Trade Center completed. Video cassette recorder marketed.
1973 Denmark, Ireland, and UK enter EEC. Widespread industrial unrest in UK.	Allende killed in Chilean coup. US withdraws from Vietnam War. OPEC raises oil prices. Yom Kippur War.	US Skylab missions. Schumacher: *Small is Beautiful*. Pink Floyd: *Dark Side of the Moon*.
1974 Harold Wilson British Prime Minister (1974–6). IRA bombings of mainland Britain. Dictatorship ended in Portugal. Northern Ireland Assembly fails.	Watergate scandal; Nixon resigns. Gerald Ford US President (1974–7). Cyprus invaded by Turkey. Haile Selassie deposed in Ethiopia.	Philip Larkin: *High Windows*.
1975 Franco dies; Juan Carlos becomes king of Spain.	Angola and Mozambique independent. End of Vietnam War. Khmer Rouge in Cambodia. Civil War in Lebanon.	Apollo and Soyuz dock in space.
1976 James Callaghan British Prime Minister (1976–9).	Death of Mao Zedong; Gang of Four. Soweto massacre.	Richard Dawkins: *The Selfish Gene*.
1977 Democratic elections in Spain. Terrorist activities in Germany and Italy.	Deng Xiaoping gains power in China. Steve Biko dies in police custody in South Africa. Jimmy Carter US President (1977–81).	Pompidou Centre opened. Woody Allen: *Annie Hall*.
1978 Pope John Paul II elected.	Camp David agreement. Boat people leaving Vietnam.	Iris Murdoch: *The Sea, The Sea*. First test-tube baby born.

	EUROPE AND THE MEDITERRANEAN	**REST OF THE WORLD**	**CULTURE/TECHNOLOGY**
1979	European Parliament direct elections. Strikes in Britain. Conservatives win British election; Margaret Thatcher Prime Minister (1979–90).	Shah of Iran deposed by Khomeini's Islamic revolution. USSR invades Afghanistan. Pol Pot deposed in Cambodia. SALT II signed. Sandinistas take power in Nicaragua.	Coppola: *Apocalypse Now.* Smallpox eradicated.
1980	Solidarity in Poland.	Zimbabwe independent. Iran–Iraq War (1980–8). US funds Contras in Nicaragua. Indira Gandhi Indian Prime Minister (1980–4).	Anthony Burgess: *Earthly Powers.* Anglican Alternative Service Book.
1981	Privatization of public corporations in Britain. High unemployment. SDP formed. Mitterrand French President.	Ronald Reagan US President (1981–9). President Sadat assassinated in Egypt.	US space shuttle. 1981–6 Pierre Boulez: *Répons.* John Updike: *Rabbit is Rich.* Microprocessors in a variety of domestic appliances and gadgets.
1982	British victory over Argentina in Falklands War.	Israel invades Lebanon. Famine in Ethiopia.	Alice Walker: *The Color Purple.* Richard Attenborough: *Gandhi.* Steven Spielberg: *ET.* Compact discs introduced. Aids identified.
1983	Thatcher re-elected Prime Minister. Cruise missiles installed in UK and Germany; peace movements active.	US troops invade Grenada. Civilian government restored in Argentina.	Wajda: *Danton.*
1984	Miners' strike in UK (1984–5). IRA bomb attack during Conservative Party conference at Brighton.	Hong Kong agreement between UK and China. Assassination of Indira Gandhi.	Milan Kundera: *The Unbearable Lightness of Being.* Madonna: *Like a Virgin.*
1985	Gorbachev General Secretary of Soviet Communist Party (1985–91); begins policy of liberalization. Anglo-Irish Agreement.	Greenpeace ship *Rainbow Warrior* sunk by French.	Primo Levi: *The Periodic Table.*
1986	Spain and Portugal join EEC. British unemployment peaks at 3.5 million. 'Big bang' in UK Stock Exchange.	US bombs Libya.	Chernobyl disaster. Wole Soyinka gains Nobel Prize. Paul Simon: *Graceland.*
1987	Single European Act. UK stock-market crisis (Black Monday). Third Thatcher government. EEC becomes EC.	Palestinian intifada.	Genetic fingerprinting in forensic science.
1988	SDP and Liberals merge to form Liberal Democrats in Britain. Mitterrand re-elected in France. Lockerbie disaster.	Iran–Iraq War ends. PLO recognizes Israel. USSR begins withdrawal from Afghanistan.	Salman Rushdie: *The Satanic Verses.* Stephen Hawking: *A Brief History of Time.* U2: *Rattle and Hum.*
1989	Berlin Wall broken. Communist regimes deposed in Hungary, Poland, East Germany, Czechoslovakia, Bulgaria, Romania.	George Bush US President (1989–93). Tiananmen Square massacre. De Klerk South African President (1989–94).	France celebrates 200th anniversary of the Revolution.
1990	East and West Germany reunited. Thatcher resigns; John Major becomes British Prime Minister. Britain joins ERM.	Iraq invades and annexes Kuwait. Nelson Mandela released; ANC talks with South African President de Klerk. Cold war formally ended. Namibia independent.	A. S. Byatt: *Possession.*
1991	USSR breaks up; attempted coup fails. Boris Yeltsin President of the Russian Federation. Croatia and Slovenia break away from Yugoslavia; civil war. Latvia, Lithuania, Estonia independent. Warsaw Pact dissolved. Maastricht Treaty agreed.	Gulf War. START agreement signed. Rajiv Gandhi assassinated.	Martin Amis: *Time's Arrow.* Nirvana: *Nevermind.*
1992	Conservatives win fourth term in Britain. Britain withdraws from ERM. Bosnia declares independence; Sarajevo besieged by Serbs.	UN intervenes in Somalia. Rioting in Los Angeles.	Merchant/Ivory: *Howard's End.* Steven Spielberg: *Jurassic Park.* David Mamet: *Oleanna.*

	EUROPE AND THE MEDITERRANEAN	**REST OF THE WORLD**	**CULTURE/TECHNOLOGY**
1993	Single market in Europe; EC becomes EU. Attempted coup against Yeltsin in Russia fails. Czechoslovakia splits into the Czech Republic and Slovakia.	Israeli and PLO leaders sign peace accord. Bill Clinton US President. Eritrea independent.	Toni Morrison wins Nobel Prize for literature.
1994	Channel Tunnel opens to rail traffic. IRA announces cease-fire. Austria, Finland, and Sweden vote to join EU; Norway rejects membership. Russian forces invade Chechen Republic.	First true democratic elections in South Africa; ANC wins, Mandela becomes President. Ethnic massacres in Rwanda. GATT agreement signed.	Steven Spielberg: *Schindler's List.* wins Oscars. Quentin Tarantino: *Pulp Fiction.* R.E.M.: *Monster.*
1995	Chirac wins French presidential election. Bosnia peace accord signed in Dayton, Ohio; NATO Implementation Force deployed. Republic of Ireland votes to lift constitutional ban on divorce. Clinton visits Northern Ireland on peace mission.	World Trade Organization established. Bomb attack on Federal Building in Oklahoma City kills 168. Yitzhak Rabin assassinated. France continues nuclear tests in Pacific despite international opposition. Ken Saro-Wiwa, writer and political campaigner, hanged in Nigeria; Nigeria suspended from Commonwealth.	Estimated 30–40 million Internet users worldwide. UN-sponsored Intergovernmental Panel on Climate Change confirms global warming is occurring. Galileo probe arrives at Jupiter.

MEDICAL SCIENCE

c.400 BC	Hippocrates brings medicine from the realm of magic and the supernatural into that of natural phenomena (Greece)
c.160	Galen shows that arteries contain blood, not air, and founds experimental physiology (Greece & Italy)
c.1050	Salerno medical school: Arabic expertise (Italy)
1363	Guy de Chauliac *Chirurgia Magna*: influential treatise on surgery (France)
1527	Paracelsus introduces chemical treatment of disease (Switzerland)
1628	circulation of the blood described by W. Harvey (Britain)
c.1676	presence of microbes first detected by A. van Leeuwenhoek (Holland)
1796	first effective vaccine (against smallpox) developed by E. Jenner (Britain)
1816	monaural stethoscope designed by R. Laënnec (France)
1842	ether first used as an anaesthetic by C. Long (US)
1860	pasteurization technique developed by L. Pasteur (France)
1863–4	clinical thermometer introduced by W. Aitken (UK)
1865	germ theory of disease published by L. Pasteur (France)
1867	first antiseptic operation performed by J. Lister (UK)
1885	cholera bacillus identified by R. Koch (Germany)
c.1891	P. Ehrlich pioneers chemotherapy (Germany)
1895	X-rays discovered by W. Röntgen (Germany)
1897	first synthetic aspirin produced (Germany)
1898	medical effects of radiation discovered by P. and M. Curie (France)
1901	existence of blood groups discovered by K. Landsteiner (Austria)
1921–2	insulin isolated by F. Banting and C. Best (Canada)
1922	tuberculosis (BCG) vaccine developed by L. Calmette and G. Guérin (France)
1928	penicillin discovered by A. Fleming (UK)
1929	first iron lung designed by P. Drinker and C. McKhann (US)
1932	first sulphonamide antibiotic developed by G. Domagk (Germany)
1938–40	penicillin isolated by H. Florey (Australia) and E. Chain (Germany)
1945	artificial kidney first used on a human (Holland)
1950	first successful kidney transplant performed by R. Lawler (US)
1955	ultrasound successfully used in body scanning by I. Donald (UK)
1958	first internal cardiac pacemaker implanted (Sweden)
1960	contraceptive pill first available (US)
1967	first heart transplant performed by C. Barnard (S. Africa)
1971	CAT (computerized axial tomography) scanner developed (UK)
1978	first test-tube baby born (UK)
1980	World Health Organization declares the world free of smallpox from 1 January 1980
1983	HIV or human immunodeficiency virus identified as responsible for causing Aids (US & France)
1991	implantation of artificial lung (UK)

TELECOMMUNICATIONS REVOLUTION

1835	Morse code invented by S. Morse (US)
1837	electric telegraph patented by W. Cooke and C. Wheatstone (UK)
1839	first commercial telegraph line installed in London (UK)
1848	Associated Press news wire service begins in New York (US)
1851	first underwater cable laid under the English Channel
1851	Reuters news wire service begins in London (UK)
1858	first transatlantic cable laid between Ireland and Newfoundland
1858	automatic telegraph system patented by C. Wheatstone (UK)
1872	duplex telegraphy patented by J. Stearns (US)
1876	telephone patented by A. G. Bell (US)
1877	phonograph developed by T. Edison (US)
1878	first telephone exchange opened in Connecticut (US)
1887	existence of radio waves demonstrated by H. Hertz (Germany)
1889	first automatic telephone exchange introduced (US)
1891	discs for recording sound pioneered by E. Berliner (Germany)
1895	G. Marconi (Italy) demonstrates radio transmission
1897	cathode-ray tube invented by K. F. Braun (Germany)
1898	machine for magnetic recording of sound patented by V. Poulsen (Denmark)
1901	G. Marconi (Italy) sends first radio signals across the Atlantic
1907	triode valve patented by L. De Forest (US)
1921–3	regular public radio broadcasting established in Britain and US
1926	television first demonstrated by J. L. Baird (UK)
1927	first transatlantic telephone links opened between London and New York
1928	magnetic tape introduced by F. Pfleumer (Germany)
1936	black and white television service introduced by the BBC (UK)
1950	stereophonic sound reproduction from magnetic tape
1953	colour television successfully transmitted in US
1954	first transistor radio developed by Regency Company (US)
1956	video tape recorder developed by Ampex (US)
1960	*Echo* and *Courier* satellites (US) launched to relay first satellite telephone calls between US and Europe
1962	*Telstar* satellite (US) relays first transatlantic television broadcast
1963	cassette tapes introduced
1966	fibre-optic cables for telephone links pioneered by K. Kao and G. Hockham (UK)
1968	pulse code modulation system installed in London (UK)
1970	videodisc introduced by Decca (UK) and AEG (Germany)
1972	first domestic video-cassette system (Betamax) introduced by Sony (Japan)
1980s	teletext system developed and introduced in Europe
1980s	direct broadcasting by satellite
1982	compact disc produced by Philips (Holland) and Sony (Japan)

1985 cellular telephone system launched in UK

1988 Switch EFTPOS system introduced (UK)

c.1988 worldwide interconnection of computer networks through Internet, enabling rapid electronic mail communication

SPACE EXPLORATION

1903 theory of rocket propulsion published by K. Tsiolkovsky (Russia)

1926 first liquid-fuel rocket launched by R. Goddard (US)

1944 V-2 rocket, forerunner of space rockets, first used (Germany)

1949 first multi-stage rocket launched in US

1957 a satellite, *Sputnik 1* (USSR), first put into earth orbit

1957 satellite *Sputnik 2* (USSR) carries the first space traveller, the dog Laika

1958 US puts a satellite, *Explorer 1*, into orbit

1959 space probe *Luna 2* (USSR) reaches the moon

1961 Y. Gagarin becomes the first man in space in *Vostok 1* (USSR)

1962–77 *Mariner* space probes (US) to Venus, Mars, and Mercury

1963 V. Tereshkova becomes the first woman in space in *Vostok 6* (USSR)

1964 space probe *Ranger 7* (US) takes close-up pictures of the moon

1965 A. Leonov makes the first walk in space from *Voskhod 2* (USSR)

1966 first moon landing, by space probe *Luna 9* (USSR)

1966 N. Armstrong and D. Scott in *Gemini 8* (US) make the first docking in space

1968 F. Borman, J. Lovell, and W. Anders orbit the moon in *Apollo 8* (US)

1969 N. Armstrong and E. Aldrin make the first manned lunar landing in the lunar module *Eagle*, with M. Collins in the command module of *Apollo 11* (US)

1971 *Soyuz 10* and *Salyut 1* (USSR) dock in orbit to form the first space station

1972 *Apollo 17*, last of six US moon landings

1973 *Skylab* (US) launched for a series of zero-gravity experiments

1975 *Soyuz 19* (USSR) and *Apollo 18* (US) dock in orbit for the Apollo-Soyuz Test Project

1976 *Viking 1* (US) lands on Mars and sends back information, but finds no conclusive evidence of life

1977 Space Shuttle *Enterprise* (US) makes the first shuttle test flight

1977 *Voyagers 1* and *2* (US) leave to fly by Jupiter (1979) and Saturn (1981). *Voyager 2* continues to Uranus (1986) and Neptune (1989)

1981 Space Shuttle *Columbia* (US) makes the first shuttle mission

1983 European *Spacelab* launched into orbit

1986 European space probe *Giotto* investigates Halley's Comet

1986 Space Shuttle *Challenger* (US) explodes killing seven astronauts

1990 Hubble Space Telescope launched into orbit

1990–4 *Magellan* (US) sends back radar images of Venus

1991–3 *Galileo* (US), on way to Jupiter, takes close-up pictures of two asteroids

COMPUTER TECHNOLOGY

1642 machine for adding and subtracting designed by B. Pascal (France)

1674 machine for multiplying and dividing designed by G. Leibniz (Germany)

1834 'analytical engine' designed by C. Babbage (UK)

1849 representation of logical events by symbols enumerated by G. Boole (UK)

1855 calculating engine constructed by G. Schentz (Sweden) and exhibited at the Paris Exhibition

1889 punched-card machine patented by H. Hollerith (US) and used to tabulate the results of the US census

1924 Tabulating Machine Company of the US becomes International Business Machines (IBM)

1943 code-breaking machine *Colossus* conceived by A. Turing (UK)

1945 Electronic Numerical Integrator and Calculator (ENIAC) designed by P. Eckert and J. Maunchly (US)

1947 transistor invented by W. Shockley and others at Bell Laboratories (US)

1948 first computer, *Manchester Mark I*, installed at Manchester University (UK)

1949–50 printed electronic circuits developed

1954 first high-level programming language (Fortran) published by IBM (US)

1958 first integrated circuit (or silicon chip) produced (US)

1964 first word processor introduced by IBM (US)

1965 first minicomputer produced (US)

1971 microprocessor introduced by Intel (US)

1972 first pocket calculator introduced (UK)

1975 first portable computer produced by Altair (US)

1978 magnetic discs first introduced by Oyz (US)

1979 first videotext information system, Prestel, launched by British Telecom (UK)

1980s RISC (Reduced Instruction Set Computer) chip developed by Intel (US)

1981 desktop microcomputer (IBM-PC) introduced by IBM (US)

1983 Esprit (European Strategic Programme for Research and Information Technology) established by the EEC to coordinate the European Information Technology industry

1984 Apple Macintosh microcomputer with mouse and window user interface (US)

1984 CD-ROM (compact disc read-only memory) introduced as data storage device

1985 transputer manufactured by Inmos (UK)

1985 optical fibres first used to link mainframe computers (US)

1986 high-temperature superconductors discovered at IBM laboratories (Zurich), with potential for superconducting computers in the future

1991 Apple Macintosh PowerBook laptop microcomputer, able to function as a desktop (US)

1993 Microsoft (US) launches Windows NT, self-contained portable operating system with networking and multitasking

Appendix 3 **Countries of the World**

Country	Capital *• see entry*	Area sq. km	Population *for date etc. see entry*	Currency unit ¤ *not settled at present (1994)*
Afghanistan	Kabul	648,000	16,600,000	afghani = 100 puls
Albania	Tirana	28,700	3,300,000	lek = 100 qindarka
Algeria	Algiers	2,319,000	25,800,000	dinar = 100 centimes
Andorra	Andorra la Vella	468	50,000	French franc, Spanish peseta
Angola	Luanda	1,246,000	10,301,000	kwanza = 100 lwei
Antigua and Barbuda	St John's	442	80,000	dollar = 100 cents
Argentina	Buenos Aires	2,780,000	32,646,000	peso = 10,000 australes
Armenia	Yerevan	29,800	3,360,000	dram = 100 luma
Australia	Canberra	7,692,000	17,500,000	dollar = 100 cents
Austria	Vienna	83,900	7,700,000	schilling = 100 groschen
Azerbaijan	Baku	86,600	7,219,000	manat = 100 gopik
Bahamas	Nassau	13,900	254,680	dollar = 100 cents
Bahrain	Manama	620	518,000	dinar = 1,000 fils
Bangladesh	Dhaka	144,000	107,992,140	taka = 100 poisha
Barbados	Bridgetown	431	258,000	dollar = 100 cents
Belarus	Minsk	208,000	10,328,000	Belorussian rouble
Belgium	Brussels	30,500	9,978,700	franc = 100 centimes
Belize	Belmopan	23,000	190,800	dollar = 100 cents
Benin	Porto Novo	113,000	4,883,000	African franc
Bhutan	Thimphu	46,600	1,467,000	ngultrum = 100 chetrum, Indian rupee
Bolivia	La Paz•	1,099,000	7,356,000	boliviano = 100 centavos
Bosnia–Herzegovina	Sarajevo	51,100	4,365,000	dinar = 100 paras
Botswana	Gaborone	600,000	1,300,000	pula = 100 thebe
Brazil	Brasilia	8,512,000	150,400,000	real = 100 centavos
Brunei	Bandar Seri Begawan	5,770	264,000	dollar = 100 sen
Bulgaria	Sofia	111,000	8,798,000	lev = 100 stotinki
Burkina	Ouagadougou	274,000	9,271,000	African franc
Burma (Myanmar)	Rangoon	677,000	42,528,000	kyat = 100 pyas
Burundi	Bujumbura	27,800	5,800,000	franc = 100 centimes
Cambodia	Phnom Penh	181,000	8,660,000	riel = 100 sen
Cameroon	Yaoundé	475,000	12,081,000	African franc
Canada	Ottawa	9,976,000	26,832,400	dollar = 100 cents
Cape Verde Islands	Praia	4,030	383,000	escudo = 100 centavos
Central African Republic	Bangui	625,000	3,113,000	African franc
Chad	N'Djamena	1,284,000	5,828,000	African franc
Chile	Santiago	757,000	13,360,000	peso = 100 centavos
China	Beijing	9,561,000	1,151,200,000	yuan = 10 jiao or 100 fen
Colombia	Bogotá	1,140,000	32,873,000	peso = 100 centavos
Comoros	Moroni	1,790	492,000	African franc
Congo	Brazzaville	342,000	2,351,000	African franc
Costa Rica	San José	51,000	2,875,000	colón = 100 centimos
Croatia	Zagreb	56,500	4,760,000	kuna = 100 lipas
Cuba	Havana	111,000	10,800,000	peso = 100 centavos
Cyprus	Nicosia	9,250	708,000	pound = 100 cents
Czech Republic	Prague	78,900	10,298,700	koruna = 100 haleru
Denmark	Copenhagen	43,100	5,100,000	krone = 100 øre
Djibouti	Djibouti	23,300	441,000	franc = 100 centimes
Dominica	Roseau	751	72,000	dollar = 100 cents
Dominican Republic	Santo Domingo	48,400	7,197,000	peso = 100 centavos
Ecuador	Quito	271,000	10,503,000	sucre = 100 centavos
Egypt	Cairo	1,002,000	53,087,000	pound = 100 piastres or 1,000 milliemes
El Salvador	San Salvador	21,400	5,308,000	colón = 100 centavos
Equatorial Guinea	Malabo	28,100	426,000	African franc
Eritrea	Asmara	118,000	3,500,000	Ethiopian birr
Estonia	Tallinn	45,100	1,591,000	kroon = 100 sents
Ethiopia	Addis Ababa	1,224,000	45,892,000	birr = 100 cents
Fiji	Suva	18,300	800,000	dollar = 100 cents
Finland	Helsinki	338,000	4,998,500	markka = 100 penniä
France	Paris	547,000	56,556,000	franc = 100 centimes
Gabon	Libreville	268,000	1,200,000	African franc
Gambia, the	Banjul	11,300	900,000	dalasi = 100 butut
Georgia	Tbilisi	69,700	5,478,000	¤
Germany	Berlin•	357,000	78,700,000	Deutschmark = 100 pfennig
Ghana	Accra	239,000	16,500,000	cedi = 100 pesewas
Greece	Athens	131,000	10,269,000	drachma = 100 leptae
Grenada	St George's	345	94,000	dollar = 100 cents
Guatemala	Guatemala City	109,000	9,200,000	quetzal = 100 centavos
Guinea	Conakry	246,000	6,909,300	franc = 100 centimes
Guinea-Bissau	Bissau	36,000	1,000,000	peso = 100 centavos
Guyana	Georgetown	215,000	800,000	dollar = 100 cents
Haiti	Port-au-Prince	27,800	6,500,000	gourde = 100 centimes
Honduras	Tegucigalpa	112,000	5,100,000	lempira = 100 centavos
Hungary	Budapest	93,000	10,600,000	forint = 100 filler

Country	Capital * see entry	Area sq. km	Population for date etc. see entry	Currency unit ¤ not settled at present (1994)
Iceland	Reykjavik	103,000	300,000	krona = 100 aurar
India	New Delhi	3,185,000	859,200,000	rupee = 100 paisa
Indonesia	Djakarta	1,905,000	184,300,000	rupiah = 100 sen
Iran	Tehran	1,648,000	54,600,000	rial = 100 dinars
Iraq	Baghdad	438,000	17,583,450	dinar = 1,000 fils
Ireland, Republic of	Dublin	70,300	3,523,400	pound (punt) = 100 pence
Israel	Jerusalem	20,800	4,600,000	shekel = 100 agora
Italy	Rome	301,000	57,746,160	lira = 100 centesemi
Ivory Coast	Yamoussoukro	322,000	12,000,000	African franc
Jamaica	Kingston	11,000	2,500,000	dollar = 100 cents
Japan	Tokyo	378,000	122,626,000	yen = 100 sen
Jordan	Amman	97,700	4,000,000	dinar = 1,000 fils
Kazakhstan	Almaty	2,717,000	16,899,000	tenge = 100 teins
Kenya	Nairobi	583,000	25,016,000	shilling = 100 cents
Kiribati	Bairiki	717	71,000	dollar = 100 cents
Kuwait	Kuwait City	17,800	1,200,000	dinar = 1,000 fils
Kyrgyzstan	Bishkek	199,000	4,448,000	som = 100 tiyin
Laos	Vientiane	237,000	4,279,000	kip = 100 at
Latvia	Riga	64,600	2,693,000	lat = 100 santims
Lebanon	Beirut	10,500	2,700,000	pound = 100 piastres
Lesotho	Maseru	30,300	1,816,000	loti = 100 lisente
Liberia	Monrovia	111,000	2,639,000	dollar = 100 cents
Libya	Tripoli	1,776,000	4,714,000	dinar = 1,000 dirhams
Liechtenstein	Vaduz	160	28,880	franc = 100 centimes or rappen
Lithuania	Vilnius	65,200	3,765,000	litas = 100 centas
Luxembourg	Luxembourg	2,590	378,000	franc = 100 centimes
Macedonia	Skopje	25,700	2,038,000	dinar = 100 paras
Madagascar	Antananarivo	587,000	12,016,000	franc malgache = 100 centimes
Malawi	Lilongwe	118,000	8,796,000	kwacha = 100 tambala
Malaysia	Kuala Lumpur	330,000	18,294,000	dollar (ringgit) = 100 sen
Maldives	Male	298	221,000	rufiyaa = 100 laris
Mali	Bamako	1,240,000	8,706,000	African franc
Malta	Valletta	316	356,000	lira = 100 cents
Marshall Islands	Majuro	181	43,420	US dollar
Mauritania	Nouakchott	1,031,000	2,023,000	ouguiya = 5 khoums
Mauritius	Port Louis	2,040	1,083,000	rupee = 100 cents
Mexico	Mexico City	1,958,000	81,140,920	peso = 100 centavos
Micronesia, Federated States of	Kolonia	701	107,900	US dollar
Moldova	Chişinău	33,700	4,384,000	leu = 100 bani
Monaco	–	1.5	29,880	franc = 100 centimes
Mongolia	Ulan Bator	1,565,000	2,184,000	tugrik = 100 mongo
Montenegro	Podgorica	13,800	632,000	dinar = 100 paras
Morocco	Rabat	459,000	25,731,000	dirham = 100 centimes
Mozambique	Maputo	799,000	16,142,000	metical = 100 centavos
Myanmar (see Burma)				
Namibia	Windhoek	824,000	1,834,000	dollar = 100 cents
Nauru	–	21	9,000	Australian dollar
Nepal	Kathmandu	147,000	19,406,000	rupee = 100 paisa
Netherlands, the	Amsterdam*	37,000	15,010,445	guilder = 100 cents
New Zealand	Wellington	268,000	3,434,950	dollar = 100 cents
Nicaragua	Managua	120,000	3,975,000	cordoba = 100 centavos
Niger	Niamey	1,267,000	7,909,000	African franc
Nigeria	Abuja	924,000	88,514,500	naira = 100 kobo
North Korea	Pyongyang	121,000	22,227,000	won = 100 jun
Norway	Oslo	324,000	4,249,830	krone = 100 øre
Oman	Muscat	212,000	1,600,000	rial = 1,000 baiza
Pakistan	Islamabad	804,000	115,588,000	rupee = 100 paisa
Panama	Panama City	77,100	2,329,330	balboa = 100 centésimos
Papua New Guinea	Port Moresby	463,000	3,529,540	kina = 100 toea
Paraguay	Asunción	407,000	4,441,000	guarani = 100 centimos
Peru	Lima	1,285,000	22,135,000	nuevo sol = 100 cents
Philippines	Manila	300,000	60,684,890	peso = 100 centavos
Poland	Warsaw	304,000	38,183,160	zloty = 100 groszy
Portugal	Lisbon	92,000	10,393,000	escudo = 100 centavos
Qatar	Doha	11,400	402,000	riyal = 100 dirhams
Romania	Bucharest	229,000	23,276,000	leu = 100 bani
Russia	Moscow	17,075,000	148,930,000	rouble = 100 copecks
Rwanda	Kigali	26,300	7,403,000	franc = 100 centimes
St Kitts and Nevis	Basseterre	261	39,000	dollar = 100 cents
St Lucia	Castries	616	152,000	dollar = 100 cents

Country	Capital * see entry	Area sq. km	Population for date etc. see entry	Currency unit ¤ not settled at present (1994)
St Vincent and the Grenadines	Kingstown	389	108,000	dollar = 100 cents
San Marino	San Marino	61	22,680	Italian lira
São Tomé and Príncipe	São Tomé	964	120,000	dobra = 100 centavos
Saudi Arabia	Riyadh	2,150,000	15,431,000	riyal = 20 qursh or 100 halala
Senegal	Dakar	197,000	7,632,000	African franc
Serbia	Belgrade	88,400	9,660,000	dinar = 100 paras
Seychelles, the	Victoria	453	69,000	rupee = 100 cents
Sierra Leone	Freetown	71,700	4,239,000	leone = 100 cents
Singapore	Singapore	618	3,045,000	dollar = 100 cents
Slovakia	Bratislava	49,000	5,268,935	koruna = 100 haleru
Slovenia	Ljubljana	20,300	1,962,000	tolar = 100 stotins
Solomon Islands	Honiara	276,000	326,000	dollar = 100 cents
Somalia	Mogadishu	638,000	8,041,000	shilling = 100 cents
South Africa	Pretoria*	1,221,000	36,762,000	rand = 100 cents
South Korea	Seoul	99,300	42,793,000	won = 100 jeon
Spain	Madrid	505,000	39,045,000	peseta = 100 centimos
Sri Lanka	Colombo	64,000	17,194,000	rupee = 100 cents
Sudan	Khartoum	2,506,000	25,855,000	dinar = 10 pounds
Suriname	Paramaribo	163,000	457,000	guilder = 100 cents
Swaziland	Mbabane	17,000	825,000	lilangeni = 100 cents
Sweden	Stockholm	450,000	8,590,630	krona = 100 öre
Switzerland	Berne	41,000	6,673,850	franc = 100 centimes or rappen
Syria	Damascus	184,000	12,824,000	pound = 100 piastres
Taiwan	Taipei	36,000	20,400,000	New Taiwan dollar = 100 cents
Tajikistan	Dushanbe	143,000	5,412,000	rouble = 100 copecks
Tanzania	Dodoma	940,000	27,270,000	shilling = 100 cents
Thailand	Bangkok	513,000	56,303,270	baht = 100 satang
Togo	Lomé	57,000	3,761,000	African franc
Tonga	Nuku'alofa	668	100,000	pa'anga = 100 seniti
Trinidad and Tobago	Port-of-Spain	5,130	1,249,000	dollar = 100 cents
Tunisia	Tunis	164,000	8,223,000	dinar = 1,000 milliemes
Turkey	Ankara	779,000	56,473,000	lira = 100 kurus
Turkmenistan	Ashgabat	488,000	3,861,000	manat = 100 tenge
Tuvalu	Funafuti	26	8,500	dollar = 100 cents
Uganda	Kampala	241,000	16,876,000	shilling = 100 cents
Ukraine	Kiev	604,000	51,999,000	¤
United Arab Emirates	Abu Dhabi	77,700	1,630,000	dirham = 100 fils
United Kingdom	London	244,000	55,700,000	pound = 100 pence
United States	Washington, DC	9,373,000	248,709,870	dollar = 100 cents
Uruguay	Montevideo	176,000	3,110,000	peso = 100 centésimos
Uzbekistan	Tashkent	447,000	20,955,000	som
Vanuatu	Vila	14,800	156,000	vatu = 100 centimes
Vatican City	—	0.44	1,000	Italian lira
Venezuela	Caracas	912,000	20,191,000	bolivar = 100 centimos
Vietnam	Hanoi	330,000	67,843,000	dong = 10 hao or 100 xu
Western Samoa	Apia	2,840	159,860	tala = 100 sene
Yemen	Sana'a	540,000	12,533,000	riyal = 100 fils
Yugoslavia (see Serbia, Montenegro)				
Zaire	Kinshasa	2,344,000	38,473,000	zaire = 100 makuta
Zambia	Lusaka	753,000	8,373,000	kwacha = 100 ngwee
Zimbabwe	Harare	391,000	10,080,000	dollar = 100 cents

Principal dependencies

Dependency	Capital	Area sq. km	Population for date etc. see entry	Currency unit
American Samoa (US)	Fagatogo	197	46,770	US dollar
Anguilla (UK)	The Valley	155	7,020	East Caribbean dollar
Aruba (Netherlands)	Oranjestad	193	60,000	florin
Bermuda (UK)	Hamilton	53	58,000	dollar
Cayman Islands (UK)	George Town	259	27,000	dollar
Christmas Island (Australia)	—	135	1,275	Australian dollar
Cocos Islands (Australia)	—	14	603	Australian dollar
Cook Islands (NZ)	Avarua	238	17,185	NZ dollar
Faeroe Islands (Denmark)	Tórshavn	1,400	47,660	Danish krone
Falkland Islands (UK)	Stanley	12,200	2,121	pound

Dependency	Capital	Area sq. km	Population for date etc. see entry	Currency unit
French Guiana (France)	Cayenne	91,000	96,000	French franc
French Polynesia (France)	Papeete	3,940	200,000	Pacific franc
Gibraltar (UK)	Gibraltar	5.9	29,140	pound
Greenland (Denmark)	Nuuk	c. 2,180,000	54,800	Danish krone
Guadeloupe (France)	Basse-Terre	1,780	386,990	French franc
Guam (US)	Agaña	541	132,000	US dollar
Hong Kong (UK)	Victoria	1,070	5,900,000	dollar
Macao (Portugal)	Macao City	17	467,000	pataca = 100 avos
Martinique (France)	Fort-de-France	1,080	359,570	French franc
Mayotte (France)	Mamoutzu	362	95,000	French franc
Montserrat (UK)	Plymouth	102	12,000	East Caribbean dollar
Netherlands Antilles (Netherlands)	Willemstad	800	200,000	guilder
New Caledonia (France)	Nouméa	18,600	164,170	Pacific franc
Niue (NZ)	Alofi	263	2,530	NZ dollar
Norfolk Island (Australia)	—	35	1,977	Australian dollar
Northern Marianas (US)	Chalan Kanoa	477	43,345	US dollar
Palau (US)	Koror	497	14,100	US dollar
Pitcairn Islands (UK)	—	4.6	61	NZ dollar
Puerto Rico (US)	San Juan	8,960	3,522,040	US dollar
Réunion (France)	Saint-Denis	2,510	596,700	French franc
St Helena and dependencies (UK)	Jamestown	420	6,880	pound
St Pierre and Miquelon (France)	St Pierre	242	6,390	French franc
Svalbard (Norway)	Longyearbyen	62,000	3,540	Norwegian krone
Turks and Caicos Islands (UK)	Cockburn Town	430	12,350	US dollar
Virgin Islands (US)	Charlotte Amalie	342	101,800	US dollar
Virgin Islands, British (UK)	Road Town	153	16,750	US dollar
Wallis and Futuna Islands (France)	Mata-Utu	274	15,400	Pacific franc
Western Sahara (Morocco)	La'youn	252,000	186,500	Moroccan dirham

THE COMMONWEALTH

The Commonwealth of Nations is a free association of the fifty sovereign independent states listed below, together with their associated states and dependencies

Antigua and Barbuda	Dominica	Malawi
Australia	Gambia, the	Malaysia
Bahamas	Ghana	Maldives
Bangladesh	Grenada	Malta
Barbados	Guyana	Mauritius
Belize	India	Nauru
Botswana	Jamaica	New Zealand
Brunei	Kenya	Nigeria (suspended 1995)
Cameroon	Kiribati	Pakistan
Canada	Lesotho	Papua New Guinea
Cyprus		St Kitts and Nevis

St Lucia
St Vincent and the Grenadines
Seychelles, the
Sierra Leone
Singapore
Solomon Islands
South Africa
Sri Lanka
Swaziland
Tanzania
Tonga
Trinidad and Tobago
Tuvalu
Uganda
United Kingdom
Vanuatu
Western Samoa
Zambia
Zimbabwe

THE COMMONWEALTH OF AUSTRALIA

States and territories

State	Capital
New South Wales	Sydney
Northern Territory	Darwin
Queensland	Brisbane
South Australia	Adelaide
Tasmania	Hobart
Victoria	Melbourne
Western Australia	Perth
Australian Capital Territory	Canberra (*federal capital*)

CANADA

Federal capital: Ottawa

Provinces and territories

Province	Postal abbr.	Capital
Alberta	AB	Edmonton
British Columbia	BC	Victoria
Manitoba	MB	Winnipeg
New Brunswick	NB	Fredericton
Newfoundland and Labrador	NF	St John's
Nova Scotia	NS	Halifax
Ontario	ON	Toronto
Prince Edward Island	PE	Charlottetown
Quebec	QC	Quebec
Saskatchewan	SK	Regina
Northwest Territories	NT	Yellowknife (*seat of government*)
Yukon Territory	YT	Whitehorse (*seat of government*)

INDIA

States and Union Territories

State	Capital
Andhra Pradesh	Hyderabad
Arunachal Pradesh	Itanagar
Assam	Dispur
Bihar	Patna
Goa	Panaji
Gujarat	Gandhinagar
Haryana	Chandigarh
Himachal Pradesh	Simla
Jammu and Kashmir	Srinagar (summer), Jammu (winter)
Karnataka	Bangalore
Kerala	Trivandrum
Madhya Pradesh	Bhopal
Maharashtra	Bombay
Manipur	Imphal
Meghalaya	Shillong
Mizoram	Aizawl
Nagaland	Kohima
Orissa	Bhubaneswar
Punjab	Chandigarh
Rajasthan	Jaipur
Sikkim	Gangtok
Tamil Nadu	Madras
Tripura	Agartala
Uttar Pradesh	Lucknow
West Bengal	Calcutta

Union Territory	
Andaman and Nicobar Islands	Port Blair
Chandigarh	Chandigarh
Dadra and Nagar Haveli	Silvassa
Daman and Diu	Daman
Delhi	Delhi
Lakshadweep	Kavaratti
Pondicherry	Pondicherry

State	Postal abbr.	Capital	Popular name
Alabama	AL	Montgomery	Yellowhammer State, Heart of Dixie, Cotton State
Alaska	AK	Juneau	Great Land
Arizona	AZ	Phoenix	Grand Canyon State
Arkansas	AR	Little Rock	Land of Opportunity
California	CA	Sacramento	Golden State
Colorado	CO	Denver	Centennial State
Connecticut	CT	Hartford	Constitution State, Nutmeg State
Delaware	DE	Dover	First State, Diamond State
Florida	FL	Tallahassee	Sunshine State
Georgia	GA	Atlanta	Empire State of the South, Peach State
Hawaii	HI	Honolulu	Aloha State
Idaho	ID	Boise	Gem State
Illinois	IL	Springfield	Prairie State
Indiana	IN	Indianapolis	Hoosier State
Iowa	IA	Des Moines	Hawkeye State
Kansas	KS	Topeka	Sunflower State
Kentucky	KY	Frankfort	Bluegrass State
Louisiana	LA	Baton Rouge	Pelican State
Maine	ME	Augusta	Pine Tree State
Maryland	MD	Annapolis	Old Line State, Free State
Massachusetts	MA	Boston	Bay State, Old Colony
Michigan	MI	Lansing	Great Lake State, Wolverine State
Minnesota	MN	St Paul	North Star State, Gopher State
Mississippi	MS	Jackson	Magnolia State
Missouri	MO	Jefferson City	Show Me State
Montana	MT	Helena	Treasure State
Nebraska	NE	Lincoln	Cornhusker State
Nevada	NV	Carson City	Sagebrush State, Battleborn State, Silver State
New Hampshire	NH	Concord	Granite State
New Jersey	NJ	Trenton	Garden State
New Mexico	NM	Sante Fe	Land of Enchantment
New York	NY	Albany	Empire State
North Carolina	NC	Raleigh	Tar Heel State, Old North State
North Dakota	ND	Bismarck	Peace Garden State
Ohio	OH	Columbus	Buckeye State
Oklahoma	OK	Oklahoma City	Sooner State
Oregon	OR	Salem	Beaver State
Pennsylvania	PA	Harrisburg	Keystone State
Rhode Island	RI	Providence	Little Rhody, Ocean State
South Carolina	SC	Columbia	Palmetto State
South Dakota	SD	Pierre	Coyote State, Sunshine State
Tennessee	TN	Nashville	Volunteer State
Texas	TX	Austin	Lone Star State
Utah	UT	Salt Lake City	Beehive State
Vermont	VT	Montpelier	Green Mountain State
Virginia	VA	Richmond	Old Dominion
Washington	WA	Olympia	Evergreen State
West Virginia	WV	Charleston	Mountain State
Wisconsin	WI	Madison	Badger State
Wyoming	WY	Cheyenne	Equality State

Appendix 6 The British Constitution

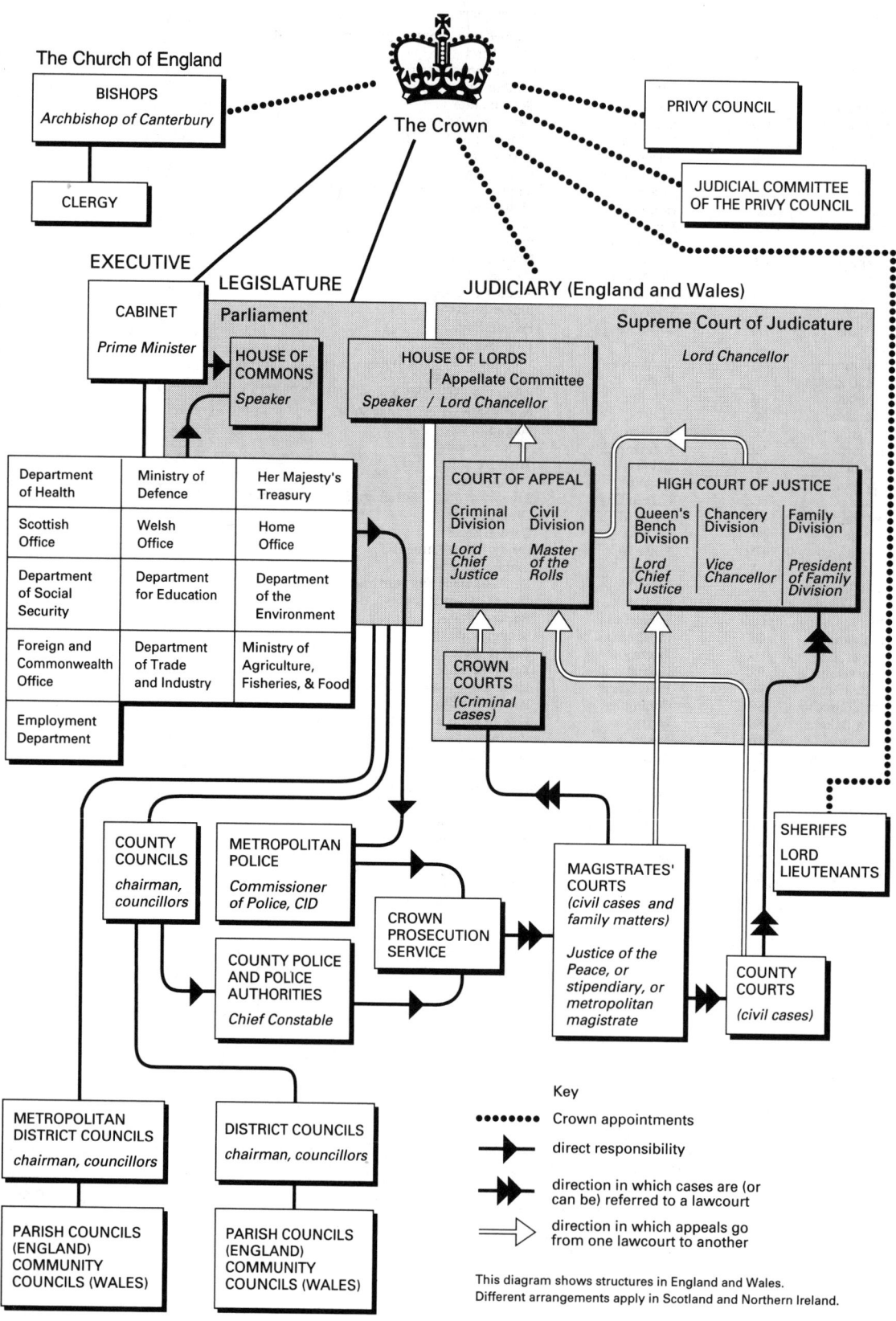

The Church of England

BISHOPS
Archbishop of Canterbury

CLERGY

The Crown

PRIVY COUNCIL

JUDICIAL COMMITTEE
OF THE PRIVY COUNCIL

EXECUTIVE

CABINET
Prime Minister

LEGISLATURE

Parliament

HOUSE OF
COMMONS
Speaker

HOUSE OF LORDS
Appellate Committee
Speaker / Lord Chancellor

JUDICIARY (England and Wales)

Supreme Court of Judicature
Lord Chancellor

Department of Health	Ministry of Defence	Her Majesty's Treasury
Scottish Office	Welsh Office	Home Office
Department of Social Security	Department for Education	Department of the Environment
Foreign and Commonwealth Office	Department of Trade and Industry	Ministry of Agriculture, Fisheries, & Food
Employment Department		

COURT OF APPEAL

Criminal Division — *Lord Chief Justice*
Civil Division — *Master of the Rolls*

HIGH COURT OF JUSTICE

Queen's Bench Division — *Lord Chief Justice*
Chancery Division — *Vice Chancellor*
Family Division — *President of Family Division*

CROWN COURTS
(Criminal cases)

COUNTY COUNCILS
chairman, councillors

METROPOLITAN POLICE
Commissioner of Police, CID

COUNTY POLICE AND POLICE AUTHORITIES
Chief Constable

CROWN PROSECUTION SERVICE

MAGISTRATES' COURTS
(civil cases and family matters)
Justice of the Peace, or stipendiary, or metropolitan magistrate

COUNTY COURTS
(civil cases)

SHERIFFS
LORD LIEUTENANTS

METROPOLITAN DISTRICT COUNCILS
chairman, councillors

DISTRICT COUNCILS
chairman, councillors

PARISH COUNCILS (ENGLAND) COMMUNITY COUNCILS (WALES)

PARISH COUNCILS (ENGLAND) COMMUNITY COUNCILS (WALES)

Key

••••••• Crown appointments

——▶ direct responsibility

——▶▶ direction in which cases are (or can be) referred to a lawcourt

——▷ direction in which appeals go from one lawcourt to another

This diagram shows structures in England and Wales.
Different arrangements apply in Scotland and Northern Ireland.

Federal Government

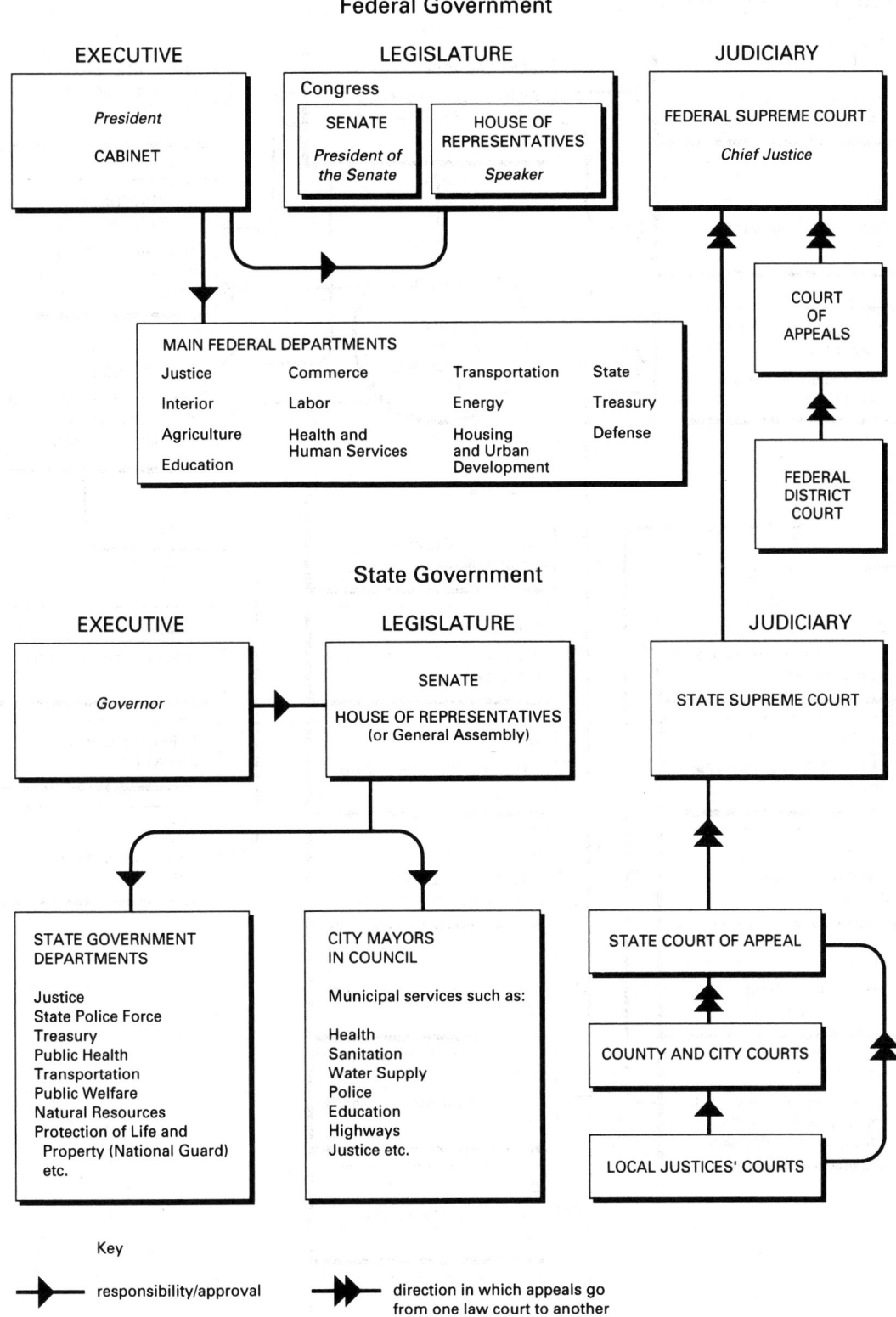

EXECUTIVE

President

CABINET

LEGISLATURE

Congress

SENATE

President of
the Senate

HOUSE OF
REPRESENTATIVES

Speaker

JUDICIARY

FEDERAL SUPREME COURT

Chief Justice

COURT
OF
APPEALS

FEDERAL
DISTRICT
COURT

MAIN FEDERAL DEPARTMENTS

Justice	Commerce	Transportation	State
Interior	Labor	Energy	Treasury
Agriculture	Health and Human Services	Housing and Urban Development	Defense
Education			

State Government

EXECUTIVE

Governor

LEGISLATURE

SENATE

HOUSE OF REPRESENTATIVES
(or General Assembly)

JUDICIARY

STATE SUPREME COURT

STATE GOVERNMENT
DEPARTMENTS

Justice
State Police Force
Treasury
Public Health
Transportation
Public Welfare
Natural Resources
Protection of Life and
  Property (National Guard)
etc.

CITY MAYORS
IN COUNCIL

Municipal services such as:

Health
Sanitation
Water Supply
Police
Education
Highways
Justice etc.

STATE COURT OF APPEAL

COUNTY AND CITY COURTS

LOCAL JUSTICES' COURTS

Key

→ responsibility/approval

⇒ direction in which appeals go
from one law court to another

Structure of organization

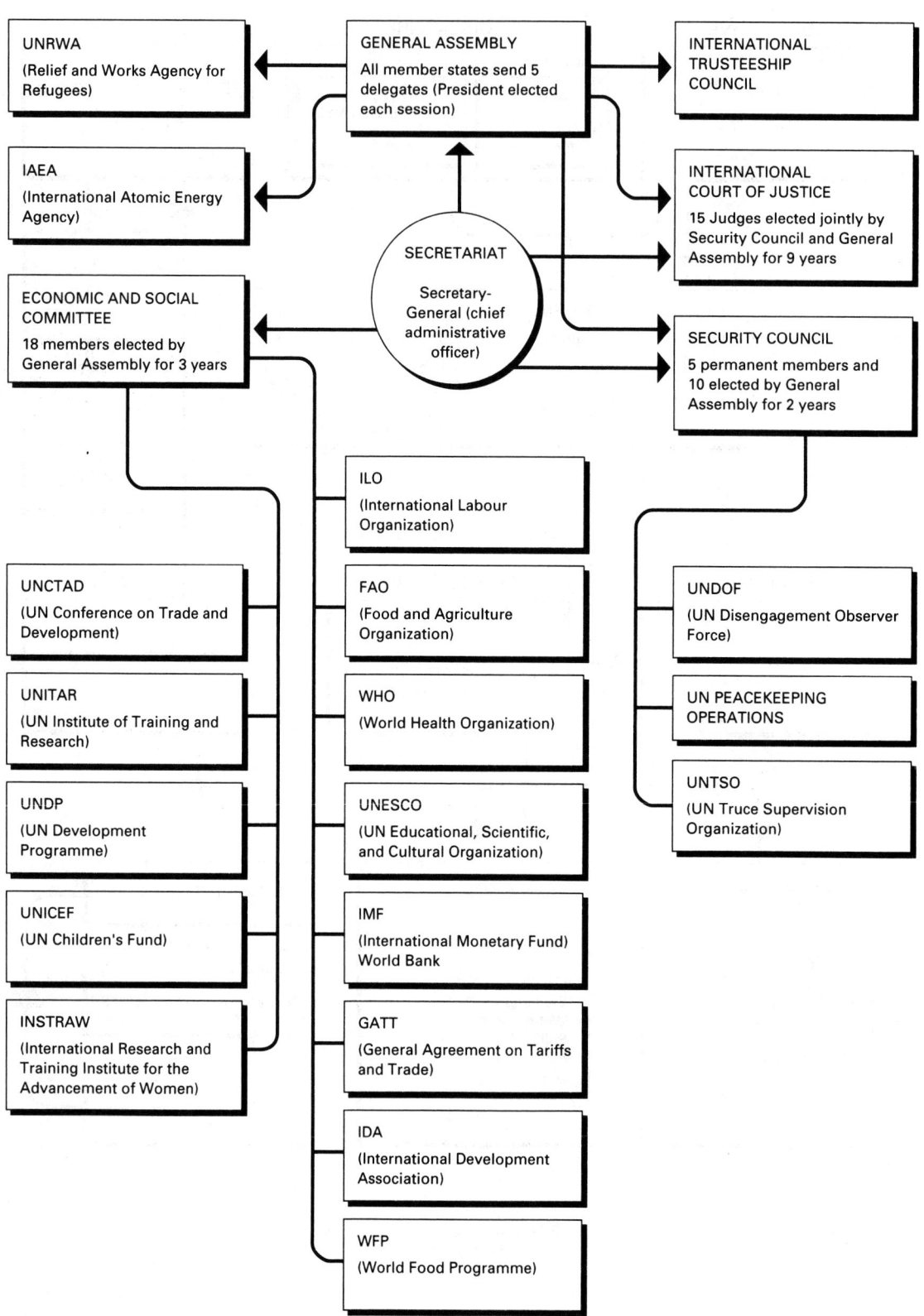

Procedure for legislation

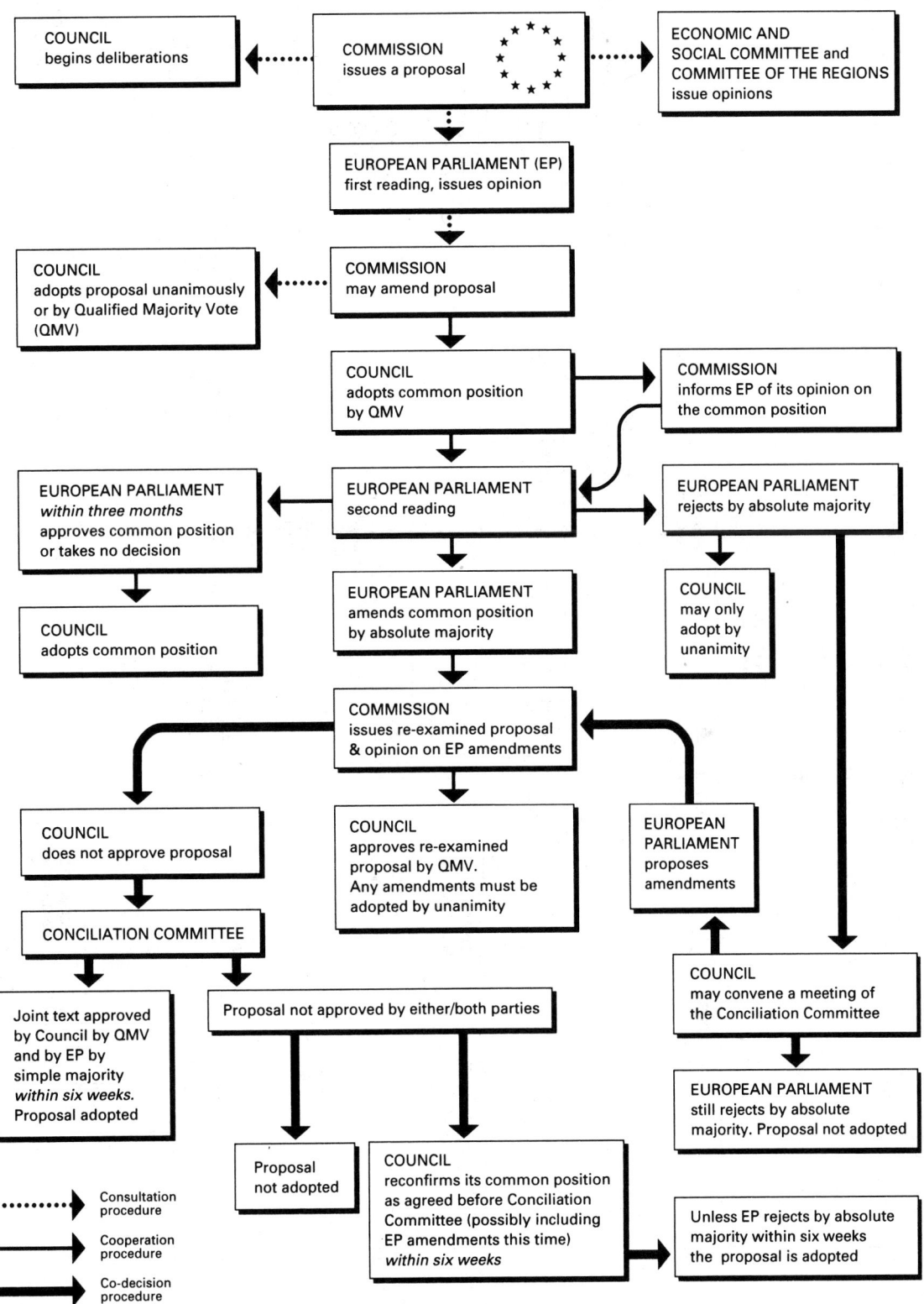

The House of Wessex 802–1066

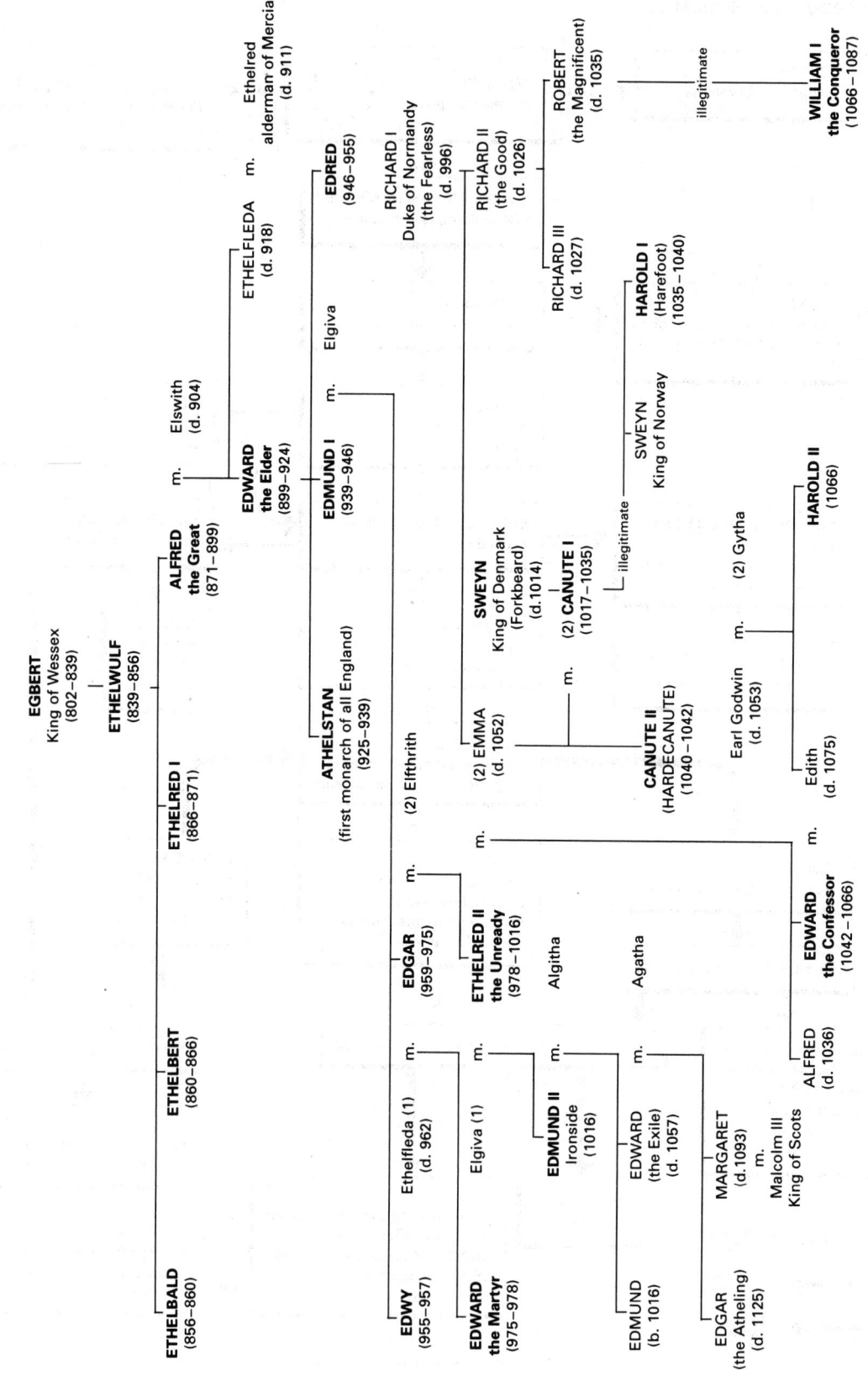

Norman and Plantagenet 1066–1399

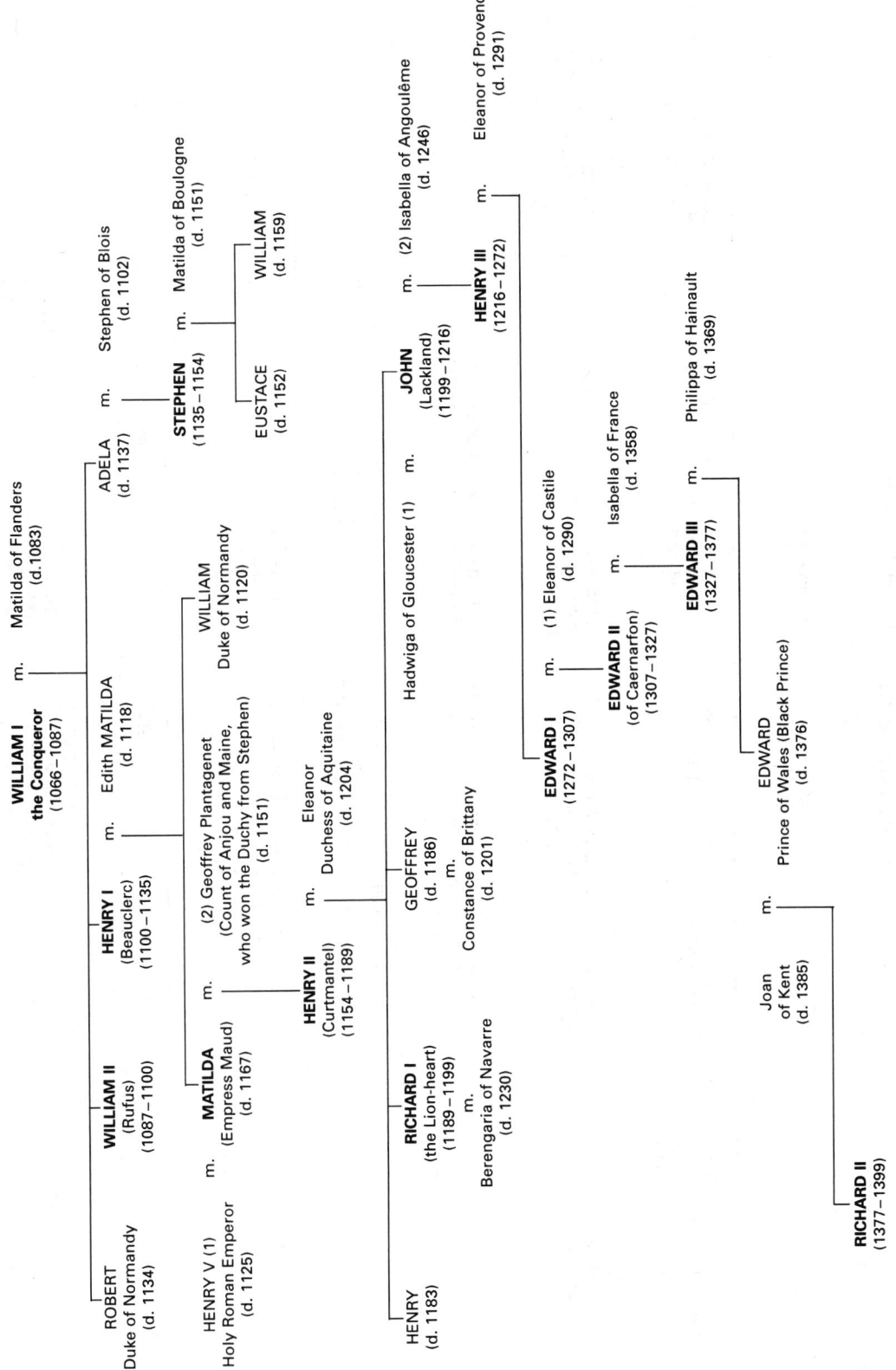

Lancaster and York 1399–1485

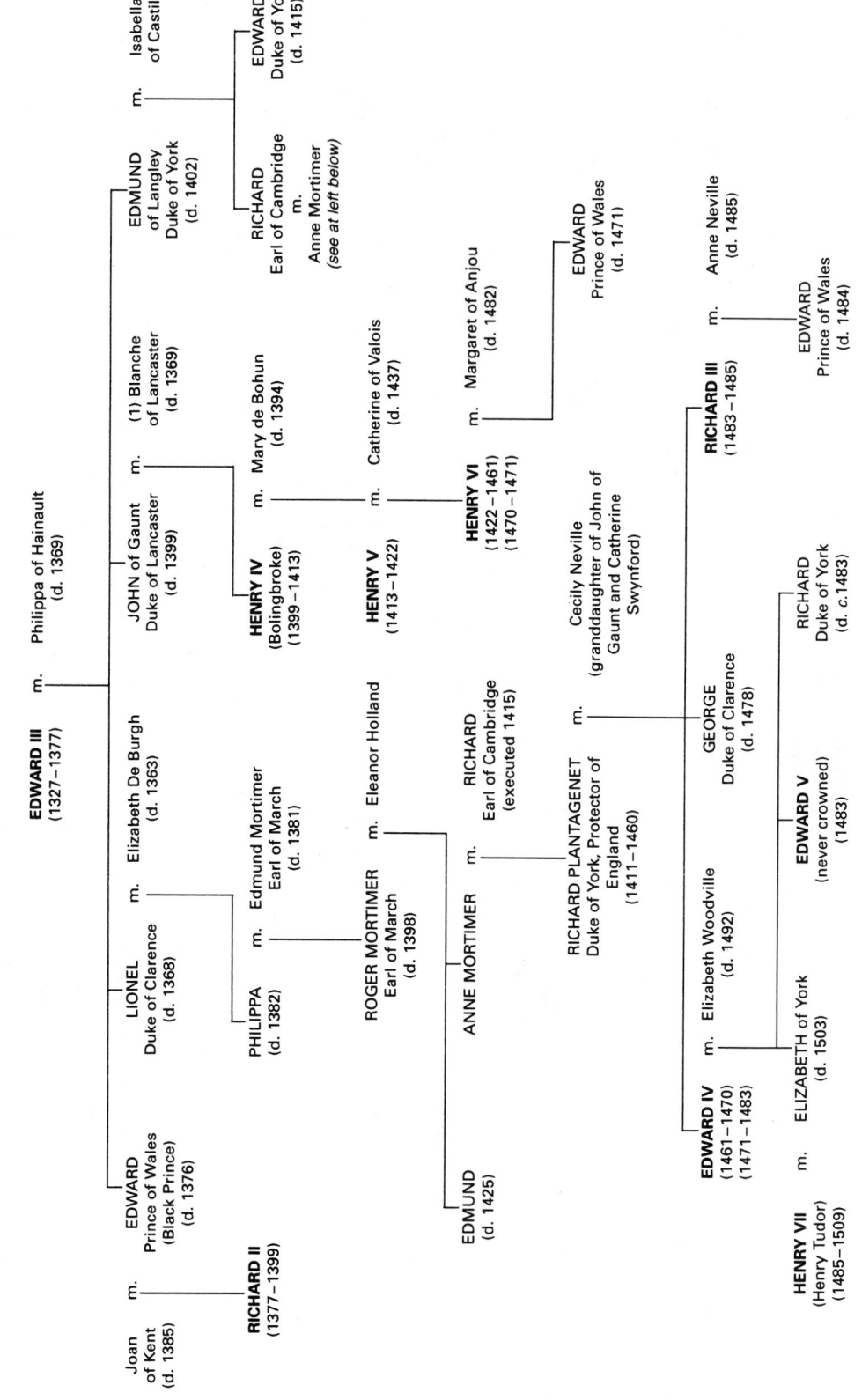

Tudors 1485–1603

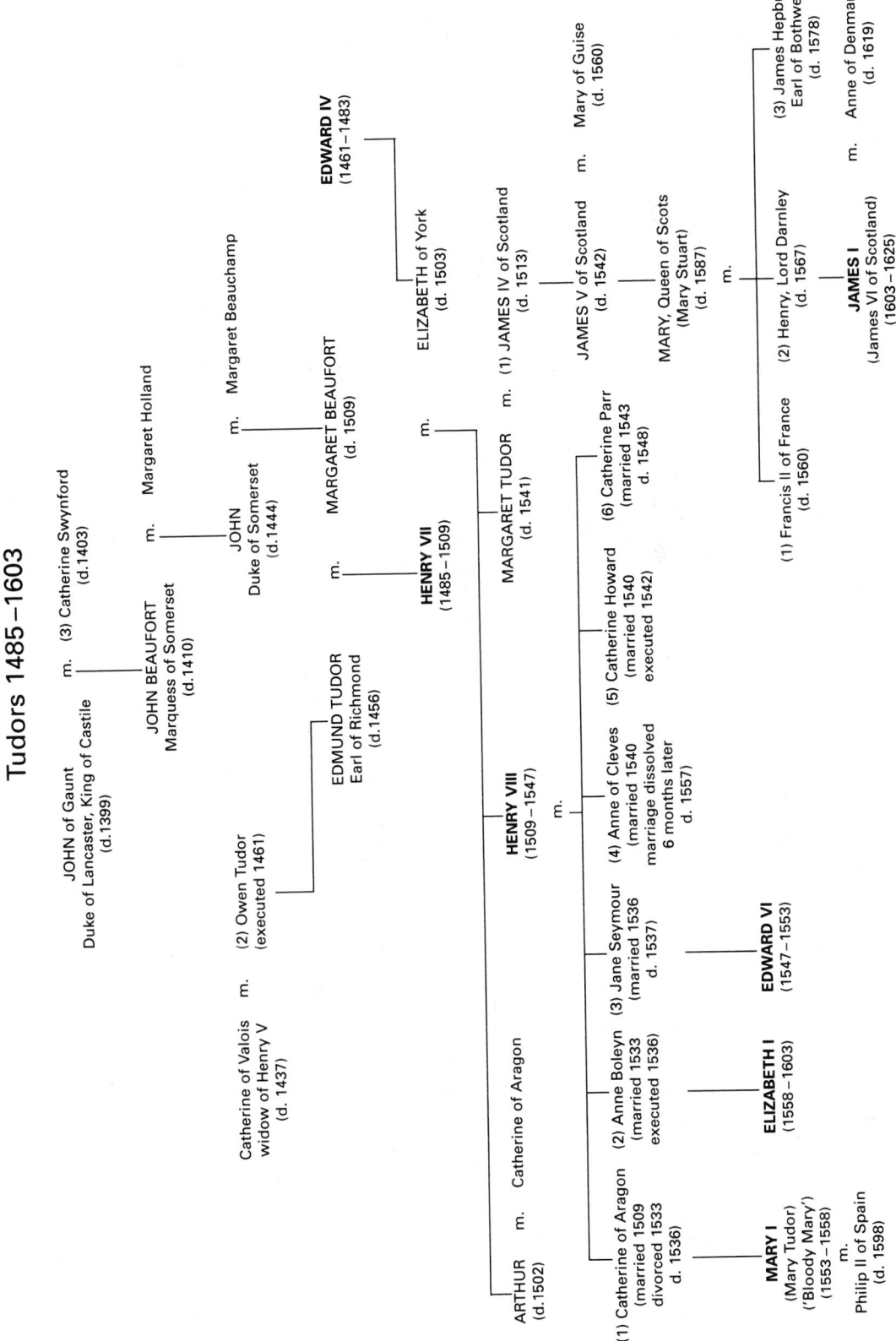

Stuarts and Hanoverians 1603–1837

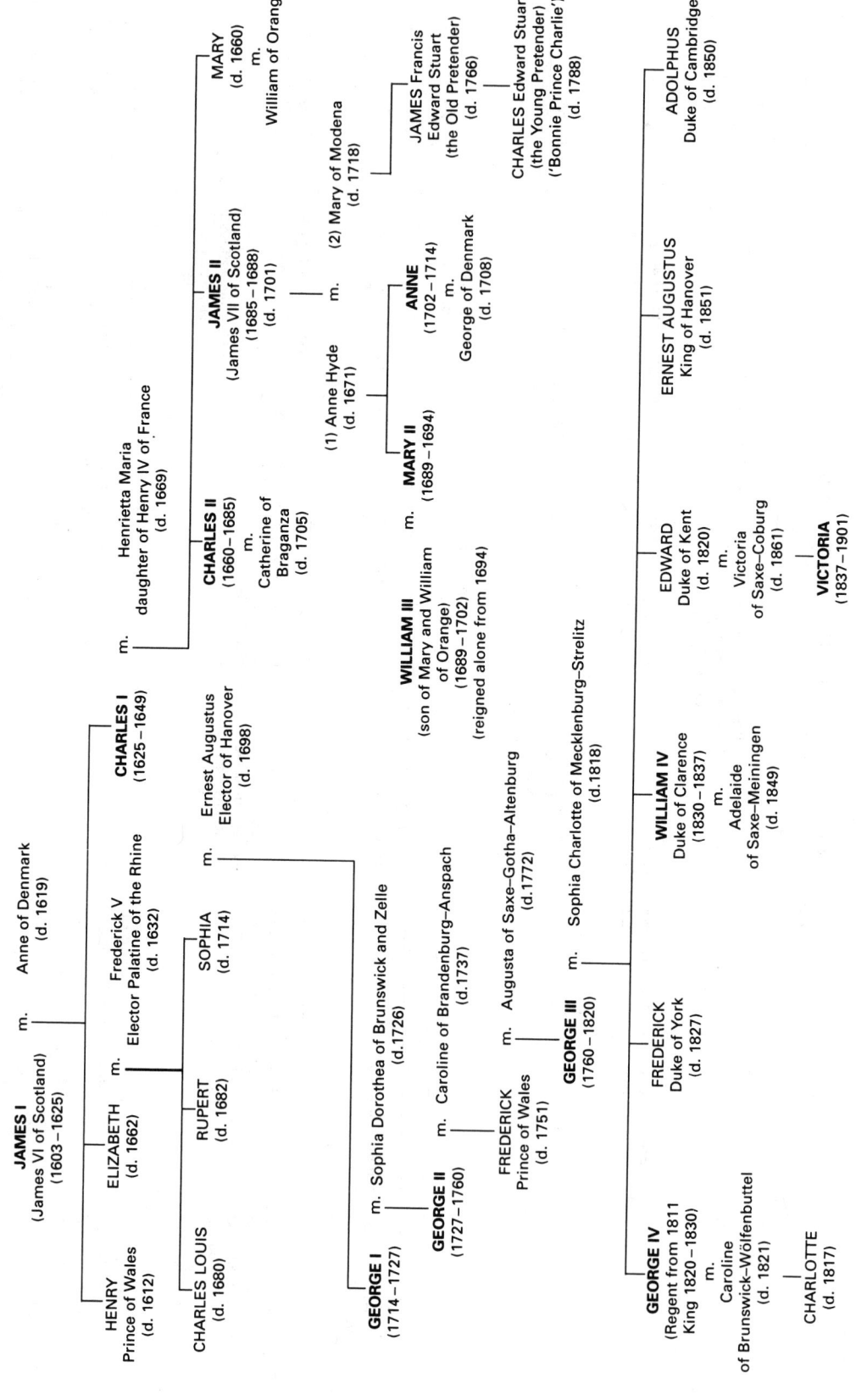

Victoria and her descendants (1837–)

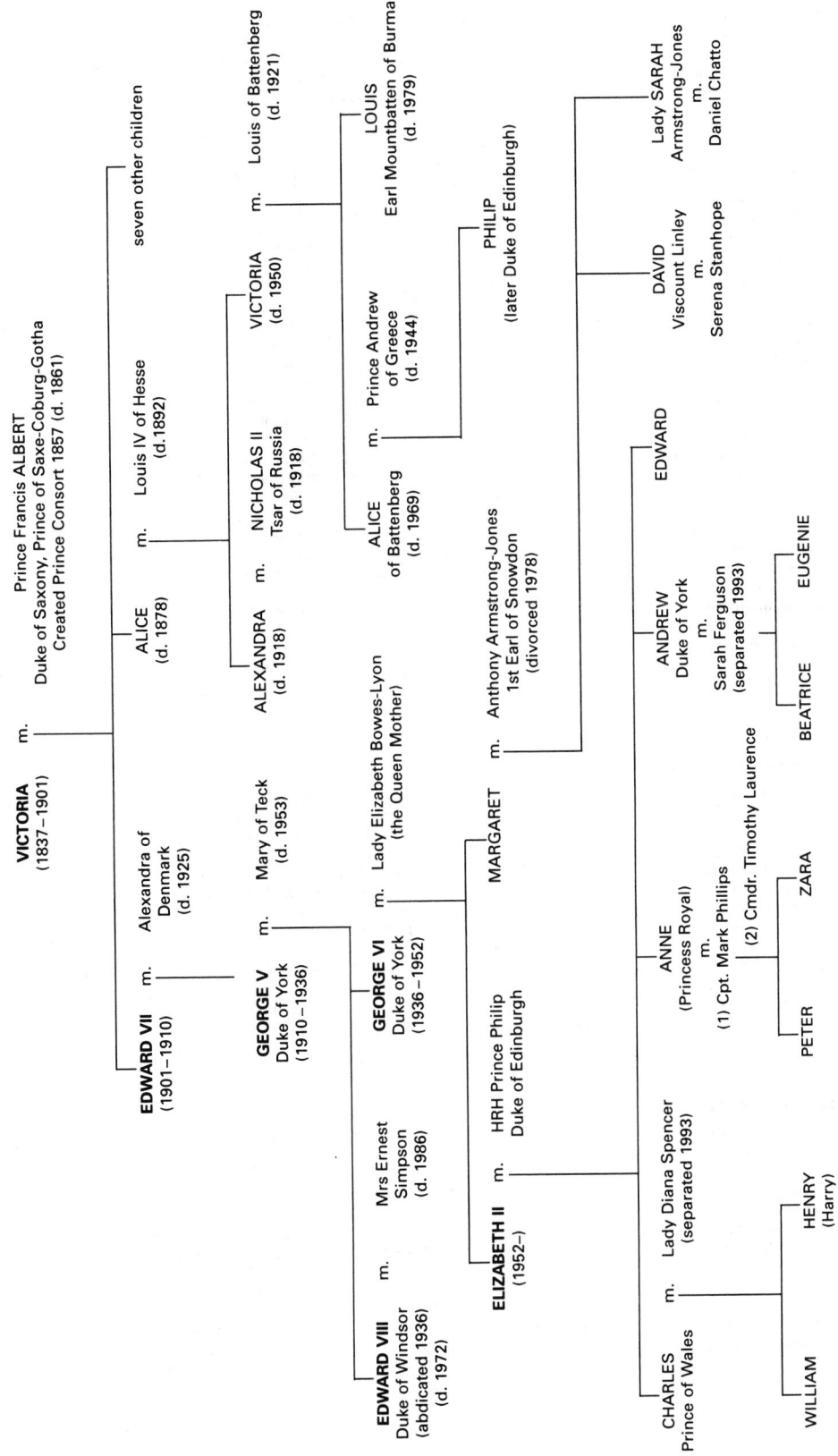

Appendix 11 Prime Ministers and Presidents

PRIME MINISTERS OF GREAT BRITAIN AND OF THE UNITED KINGDOM

[1721]–1742	Sir Robert Walpole	Whig	1859–1865	Viscount Palmerston	Liberal
1742–1743	Earl of Wilmington	"	1865–1866	Earl Russell	"
1743–1754	Henry Pelham	"	1866–1868	Earl of Derby	Conservative
1754–1756	Duke of Newcastle	"	1868	Benjamin Disraeli	
1756–1757	Duke of Devonshire	"	1868–1874	William Ewart Gladstone	Liberal
1757–1762	Duke of Newcastle	"	1874–1880	Benjamin Disraeli	Conservative
1762–1763	Earl of Bute	Tory	1880–1885	William Ewart Gladstone	Liberal
1763–1765	George Grenville	Whig	1885–1886	Marquis of Salisbury	Conservative
1765–1766	Marquis of Rockingham	"	1886	William Ewart Gladstone	Liberal
1766–1768	Earl of Chatham	"	1886–1892	Marquis of Salisbury	Conservative
1768–1770	Duke of Grafton	"	1892–1894	William Ewart Gladstone	Liberal
1770–1782	Lord North	Tory	1894–1895	Earl of Rosebery	"
1782	Marquis of Rockingham	Whig	1895–1902	Marquis of Salisbury	Conservative
1782–1783	Earl of Shelburne	"	1902–1905	Arthur James Balfour	"
1783	Duke of Portland	coalition	1905–1908	Sir Henry Campbell-Bannerman	Liberal
1783–1801	William Pitt	Tory	1908–1916	Herbert Henry Asquith	"
1801–1804	Henry Addington	"	1916–1922	David Lloyd George	coalition
1804–1806	William Pitt	"	1922–1923	Andrew Bonar Law	Conservative
1806–1807	Lord William Grenville	Whig	1923–1924	Stanley Baldwin	"
1807–1809	Duke of Portland	Tory	1924	James Ramsay MacDonald	Labour
1809–1812	Spencer Perceval	"	1924–1929	Stanley Baldwin	Conservative
1812–1827	Earl of Liverpool	"	1929–1935	James Ramsay MacDonald	coalition
1827	George Canning	"	1935–1937	Stanley Baldwin	"
1827–1828	Viscount Goderich	"	1937–1940	Neville Chamberlain	"
1828–1830	Duke of Wellington	"	1940–1945	Winston Spencer Churchill	"
1830–1834	Earl Grey	Whig	1945–1951	Clement Attlee	Labour
1834	Viscount Melbourne	"	1951–1955	Sir Winston Spencer Churchill	Conservative
1834	Duke of Wellington	Tory	1955–1957	Sir Anthony Eden	"
1834–1835	Sir Robert Peel	Conservative	1957–1963	Harold Macmillan	"
1835–1841	Viscount Melbourne	Whig	1963–1964	Sir Alec Douglas-Home	"
1841–1846	Sir Robert Peel	Conservative	1964–1970	Harold Wilson	Labour
1846–1852	Lord John Russell	Whig	1970–1974	Edward Heath	Conservative
1852	Earl of Derby	Conservative	1974–1976	Harold Wilson	Labour
1852–1855	Earl of Aberdeen	coalition	1976–1979	James Callaghan	"
1855–1858	Viscount Palmerston	Liberal	1979–1990	Margaret Thatcher	Conservative
1858–1859	Earl of Derby	Conservative	1990–	John Major	"

PRIME MINISTERS OF CANADA

1867–1873	John A. Macdonald	Conservative	1930–1935	Richard B. Bennett	Conservative
1873–1878	Alexander Mackenzie	Liberal/Reform	1935–1948	W. L. Mackenzie King	Liberal
1878–1891	John A. Macdonald	Conservative	1948–1957	Louis Stephen St Laurent	"
1891–1892	John J. C. Abbott	Liberal-Conservative	1957–1963	John George Diefenbaker	Progressive Conservative
1892–1894	John S. D. Thompson	Conservative	1963–1968	Lester B. Pearson	Liberal
1894–1896	Mackenzie Bowell	"	1968–1979	Pierre Elliott Trudeau	"
1896	Charles Tupper	"	1979–1980	Joseph Clark	Progressive Conservative
1896–1911	Wilfrid Laurier	Liberal	1980–1984	Pierre Elliott Trudeau	Liberal
1911–1920	Robert L. Borden	Conservative	1984	John Turner	Conservative
1920–1921	Arthur Meighen	Liberal	1984–1993	Brian Mulroney	Progressive Conservative
1921–1926	W. L. Mackenzie King	"	1993	Kim Campbell	"
1926	Arthur Meighen	Conservative	1993–	Jean Chrétien	Liberal
1926–1930	W. L. Mackenzie King	Liberal			

PRIME MINISTERS OF AUSTRALIA

1901–1903	Edmund Barton	–	1932–1939	Joseph A. Lyons	United Australia Party
1903–1904	Alfred Deakin	–	1939–1941	Robert Gordon Menzies	"
1904	John C. Watson	Labor	1941	Arthur William Fadden	Country Party
1904–1905	George Houstoun Reid	Free Trade	1941–1945	John Curtin	Labor
1905–1908	Alfred Deakin	Liberal	1945–1949	Joseph Benedict Chifley	"
1908–1909	Andrew Fisher	Labor	1949–1966	Robert Gordon Menzies	Liberal
1909–1910	Alfred Deakin	Liberal	1966–1967	Harold Edward Holt	"
1910–1913	Andrew Fisher	Labor	1967–1968 (Dec.–Jan.)	John McEwen	"
1913–1914	Joseph Cook	Liberal	1968–1971	John Grey Gorton	"
1914–1915	Andrew Fisher	Labor	1971–1972	William McMahon	"
1915–1923	William M. Hughes	Nationalist	1972–1975	Gough Whitlam	Labor
1923–1929	Stanley M. Bruce	"	1975–1983	J. Malcolm Fraser	Liberal
1929–1931	James H. Scullin	Labor	1983–1991	Robert J. L. Hawke	Labor
			1991–	Paul Keating	"

PRIME MINISTERS OF NEW ZEALAND

(since the emergence of party government in 1891)

1891–1893	John Ballance	Liberal	1957	Keith J. Holyoake	National Party
1893–1906	Richard John Seddon	"	(Aug.–Nov.)		
1906	William Hall-Jones	"	1957–1960	Walter Nash	Labour
1906–1912	Joseph George Ward	"	1960–1972	Keith J. Holyoake	National Party
1912	Thomas Mackenzie	"	1972	John R. Marshall	"
1912–1925	William Ferguson Massey	Reform	1972–1974	Norman Kirk	Labour
1925	Francis Henry Dillon Bell	"	1974–1975	Wallace Rowling	"
1925–1928	Joseph Gordon Coates	"	1975–1984	Robert D. Muldoon	National Party
1928–1930	Joseph George Ward	Liberal	1984–1989	David Lange	Labour
1930–1935	George William Forbes	"	1989–1990	Geoffrey Palmer	"
1935–1940	Michael J. Savage	Labour	1990	Mike Moore	"
1940–1949	Peter Fraser	"	(Sept.–Oct.)		
1949–1957	Sidney G. Holland	National Party	1990–	James B. Bolger	National Party

PRESIDENTS OF THE UNITED STATES OF AMERICA

1789–1797	1. George Washington	Federalist	1885–1889	22. Grover Cleveland	Democrat
1797–1801	2. John Adams	"	1889–1893	23. Benjamin Harrison	Republican
1801–1809	3. Thomas Jefferson	Democratic Republican	1893–1897	24. Grover Cleveland	Democrat
1809–1817	4. James Madison	"	1897–1901	25. William McKinley	Republican
1817–1825	5. James Monroe	"	1901–1909	26. Theodore Roosevelt	"
1825–1829	6. John Quincy Adams	Independent	1909–1913	27. William H. Taft	"
1829–1837	7. Andrew Jackson	Democrat	1913–1921	28. Woodrow Wilson	Democrat
1837–1841	8. Martin Van Buren	"	1921–1923	29. Warren G. Harding	Republican
1841	9. William H. Harrison	Whig	1923–1929	30. Calvin Coolidge	"
1841–1845	10. John Tyler	Whig, then Democrat	1929–1933	31. Herbert Hoover	"
1845–1849	11. James K. Polk	Democrat	1933–1945	32. Franklin D. Roosevelt	Democrat
1849–1850	12. Zachary Taylor	Whig	1945–1953	33. Harry S Truman	"
1850–1853	13. Millard Fillmore	"	1953–1961	34. Dwight D. Eisenhower	Republican
1853–1857	14. Franklin Pierce	Democrat	1961–1963	35. John F. Kennedy	Democrat
1857–1861	15. James Buchanan	"	1963–1969	36. Lyndon B. Johnson	"
1861–1865	16. Abraham Lincoln	Republican	1969–1974	37. Richard M. Nixon	Republican
1865–1869	17. Andrew Johnson	Democrat	1974–1977	38. Gerald R. Ford	"
1869–1877	18. Ulysses S. Grant	Republican	1977–1981	39. James Earl Carter	Democrat
1877–1881	19. Rutherford B. Hayes	"	1981–1989	40. Ronald W. Reagan	Republican
1881	20. James A. Garfield	"	1989–1993	41. George H. W. Bush	"
1881–1885	21. Chester A. Arthur	"	1993–	42. William J. Clinton	Democrat

Appendix 12 **The Beaufort Scale of Wind Force**

Beaufort number	Equivalent speed at 10m above ground		Description of wind	Specifications for use at sea	Specifications for use on land
	Knots	Kilometres per hour			
0	<1	<1	Calm	Sea like a mirror	Calm, smoke rises vertically
1	1–3	1–6	Light air	Ripples with the appearance of scales formed but without foam crests	Direction of wind shown by smoke drift, but not by wind vanes
2	4–6	7–12	Light breeze	Small wavelets, still short but more pronounced; crests have a glassy appearance and do not break	Wind felt on face; leaves rustle; ordinary vane moved by wind
3	7–10	13–19	Gentle breeze	Large wavelets; crests begin to break; foam of glassy appearance; perhaps scattered white horses	Leaves and small twigs in constant motion; wind extends light flag
4	11–16	20–30	Moderate breeze	Small waves, becoming longer; fairly frequent white horses	Raises dust and loose paper; small branches moved
5	17–21	31–39	Fresh breeze	Moderate waves, taking a more pronounced long form; many white horses are formed; chance of some spray	Small trees in leaf begin to sway; crested wavelets form on inland waters
6	22–27	40–50	Strong breeze	Large waves begin to form; the white foam crests are more extensive everywhere; probably some spray	Large branches in motion; whistling heard in telegraph wires; umbrellas used with difficulty
7	28–33	51–62	Near gale	Sea heaps up and white foam from breaking waves begins to be blown in streaks along the direction of the wind	Whole trees in motion; inconvenience felt when walking against wind
8	34–40	63–74	Gale	Moderately high waves of greater length; edges of crests begin to break into spindrift; the foam is blown in well-marked streaks along the direction of the wind	Breaks twigs off trees; generally impedes progress
9	41–47	75–87	Strong gale	High waves; dense streaks of foam along the direction of the wind; crests of waves begin to topple, tumble, and roll over; spray may affect visibility	Slight structural damage occurs (chimney-pots and slates removed)
10	48–55	88–102	Storm	Very high waves with long overhanging crests; the resulting foam, in great patches, is blown in dense white streaks along the direction of the wind; on the whole, the surface of the sea takes a white appearance; the tumbling of the sea becomes heavy and shocklike; visibility affected	Seldom experienced inland; trees uprooted; considerable structural damage occurs
11	56–63	103–117	Violent storm	Exceptionally high waves (small and medium-sized ships might be for a time lost to view behind the waves); the sea is completely covered with long white patches of foam lying along the direction of the wind; everywhere the edges of the wave crests are blown into froth; visibility affected	Very rarely experienced; accompanied by widespread damage
12	≥64	≥118	Hurricane	The air is filled with foam and spray; sea completely white with driving spray; visibility very seriously affected	—

The conversion factors are not exact unless so marked. They are given only to the accuracy likely to be needed in everyday calculations.

1. BRITISH AND AMERICAN, WITH METRIC EQUIVALENTS

Linear measure

1 inch	= 25.4 millimetres exactly
1 foot = 12 inches	= 0.3048 metre exactly
1 yard = 3 feet	= 0.9144 metre exactly
1 (statute) mile = 1,760 yards	= 1.609 kilometres
1 int. nautical mile = 1.150779 miles	= 1.852 km exactly

Square measure

1 square inch	= 6.45 sq. centimetres
1 square foot = 144 sq. in.	= 9.29 sq. decimetres
1 square yard = 9 sq. ft.	= 0.836 sq. metre
1 acre = 4,840 sq. yd.	= 0.405 hectare
1 square mile = 640 acres	= 259 hectares

Cubic measure

1 cubic inch	= 16.4 cu. centimetres
1 cubic foot = 1,728 cu. in.	= 0.0283 cu. metre
1 cubic yard = 27 cu. ft.	= 0.765 cu. metre

Capacity measure

British

1 fluid oz. = 1.8047 cu. in.	= 0.0284 litre
1 gill = 5 fluid oz.	= 0.1421 litre
1 pint = 20 fluid oz.	= 0.568 litre
= 34.68 cu. in.	
1 quart = 2 pints	= 1.136 litres
1 gallon = 4 quarts	= 4.546 litres
1 peck = 2 gallons	= 9.092 litres
1 bushel = 4 pecks	= 36.4 litres
1 quarter = 8 bushels	= 2.91 hectolitres

American dry

1 pint = 33.60 cu. in.	= 0.550 litre
1 quart = 2 pints	= 1.101 litres
1 peck = 8 quarts	= 8.81 litres
1 bushel = 4 pecks	= 35.3 litres

American liquid

1 pint = 16 fluid oz.	= 0.473 litre
= 28.88 cu. in.	
1 quart = 2 pints	= 0.946 litre
1 gallon = 4 quarts	= 3.785 litres

Avoirdupois weight

1 grain	= 0.065 gram
1 dram	= 1.772 grams
1 ounce = 16 drams	= 28.35 grams
1 pound = 16 ounces	= 0.4536 kilogram
= 7,000 grains	(0.45359237 exactly)
1 stone = 14 pounds	= 6.35 kilograms
1 quarter = 2 stones	= 12.70 kilograms
1 hundredweight = 4 quarters	= 50.80 kilograms
1 (long) ton = 20 hundredweight	= 1.016 tonnes
1 short ton = 2,000 pounds	= 0.907 tonne

2. METRIC, WITH BRITISH EQUIVALENTS

Linear measure

1 millimetre	= 0.039 inch
1 centimetre = 10 mm	= 0.394 inch
1 decimetre = 10 cm	= 3.94 inches
1 metre = 100 cm	= 1.094 yards
1 decametre = 10 m	= 10.94 yards
1 hectometre = 100 m	= 109.4 yards
1 kilometre = 1,000 m	= 0.6214 mile

Square measure

1 square centimetre	= 0.155 sq. inch
1 square metre = 10,000 sq. cm	= 1.196 sq. yards
1 are = 100 sq. metres	= 119.6 sq. yards
1 hectare = 100 ares	= 2.471 acres
1 square kilometre = 100 hectares	= 0.386 sq. mile

Cubic measure

1 cubic centimetre	= 0.061 cu. inch
1 cubic metre = 1,000,000 cu. cm	= 1.308 cu. yards

Capacity measure

1 millilitre	= 0.002 pint (British)
1 centilitre = 10 ml	= 0.018 pint
1 decilitre = 100 ml	= 0.176 pint
1 litre = 1000 ml	= 1.76 pints
1 decalitre = 10 l	= 2.20 gallons
1 hectolitre = 100 l	= 2.75 bushels
1 kilolitre = 1,000 l	= 3.44 quarters

Weight

1 milligram	= 0.015 grain
1 centigram = 10 mg	= 0.154 grain
1 decigram = 100 mg	= 1.543 grain
1 gram = 1,000 mg	= 15.43 grain
1 decagram = 10 g	= 5.64 drams
1 hectogram = 100 g	= 3.527 ounces
1 kilogram = 1,000 g	= 2.205 pounds
1 tonne (metric ton) = 1,000 kg	= 0.984 (long) ton

3. SI UNITS

Base units

Physical quantity	Name	Abbreviation or symbol
length	metre	m
mass	kilogram	kg
time	second	s
electric current	ampere	A
temperature	kelvin	K
amount of substance	mole	mol
luminous intensity	candela	cd

Supplementary units

Physical quantity	Name	Abbreviation or symbol
plane angle	radian	rad
solid angle	steradian	sr

Derived units with special names

Physical quantity	Name	Abbreviation or symbol
frequency	hertz	Hz
energy	joule	J
force	newton	N
power	watt	W
pressure	pascal	Pa
electric charge	coulomb	C
electromotive force	volt	V
electric resistance	ohm	Ω
electric conductance	siemens	S
electric capacitance	farad	F
magnetic flux	weber	Wb
inductance	henry	H
magnetic flux density	tesla	T
luminous flux	lumen	lm
illumination	lux	lx

4. TEMPERATURE

Fahrenheit: Water boils (under standard conditions) at 212° and freezes at 32°.
Celsius or Centigrade: Water boils at 100° and freezes at 0°.
Kelvin: Water boils at 373.15 kelvins and freezes at 273.15 kelvins.

Celsius	Fahrenheit
−17.8°	0°
−10°	14°
0°	32°
10°	50°
20°	68°
30°	86°
40°	104°
50°	122°
60°	140°
70°	158°
80°	176°
90°	194°
100°	212°

To convert Celsius into Fahrenheit: multiply by 9, divide by 5, and add 32.
To convert Fahrenheit into Celsius: subtract 32, multiply by 5, and divide by 9.

5. METRIC PREFIXES

	Abbreviation or symbol	Factor
deca-	da	10
hecto-	h	10^2
kilo-	k	10^3
mega-	M	10^6
giga-	G	10^9
tera-	T	10^{12}
peta-	P	10^{15}
exa-	E	10^{18}
deci-	d	10^{-1}
centi-	c	10^{-2}
milli-	m	10^{-3}
micro-	μ	10^{-6}
nano-	n	10^{-9}
pico-	p	10^{-12}
femto-	f	10^{-15}
atto-	a	10^{-18}

Pronunciations and derivations of these are given at their alphabetical places in the dictionary. They may be applied to any units of the metric system: hectogram (abbr. hg) = 100 grams; kilowatt (abbr. kW) = 1,000 watts; megahertz (MHz) = 1 million hertz; centimetre (cm) = $^1/_{100}$ metre; microvolt (μV) = one millionth of a volt; picofarad (pF) = 10^{-12} farad, and are sometimes applied to other units (megabit, microinch).

6. POWER NOTATION

This expresses concisely any power of ten (any number that is composed of factors 10), and is sometimes used in the dictionary. 10^2 or ten squared = $10 \times 10 = 100$; 10^3 or ten cubed = $10 \times 10 \times 10 = 1,000$. Similarly, $10^4 = 10,000$ and $10^{10} = 1$ followed by ten noughts = 10,000,000,000. Proceeding in the opposite direction, dividing by ten and subtracting one from the index, we have $10^2 = 100$, $10^1 = 10$, $10^0 = 1$, $10^{-1} = {}^1/_{10}$, $10^{-2} = {}^1/_{100}$, and so on; $10^{-10} = 1/10^{10} = 1/10,000,000,000$.

7. BINARY SYSTEM

Only two units (0 and 1) are used, and the position of each unit indicates a power of two.

One to ten written in binary form:

	eights (2^3)	fours (2^2)	twos (2^1)	one
1				1
2			1	0
3			1	1
4		1	0	0
5		1	0	1
6		1	1	0
7		1	1	1
8	1	0	0	0
9	1	0	0	1
10	1	0	1	0

i.e. ten is written as 1010 ($2^3 + 0 + 2^1 + 0$); one hundred is written as 1100100 ($2^6 + 2^5 + 0 + 0 + 2^2 + 0 + 0$)

Element	Symbol	Atomic no.	Element	Symbol	Atomic no.	Element	Symbol	Atomic no.
actinium	Ac	89	hafnium	Hf	72	promethium	Pm	61
aluminium	Al	13	helium	He	2	protactinium	Pa	91
americium	Am	95	holmium	Ho	67	radium	Ra	88
antimony	Sb	51	hydrogen	H	1	radon	Rn	86
argon	Ar	18	indium	In	49	rhenium	Re	75
arsenic	As	33	iodine	I	53	rhodium	Rh	45
astatine	At	85	iridium	Ir	77	rubidium	Rb	37
barium	Ba	56	iron	Fe	26	ruthenium	Ru	44
berkelium	Bk	97	krypton	Kr	36	samarium	Sm	62
beryllium	Be	4	lanthanum	La	57	scandium	Sc	21
bismuth	Bi	83	lawrencium	Lr	103	selenium	Se	34
boron	B	5	lead	Pb	82	silicon	Si	14
bromine	Br	35	lithium	Li	3	silver	Ag	47
cadmium	Cd	48	lutetium	Lu	71	sodium	Na	11
caesium	Cs	55	magnesium	Mg	12	strontium	Sr	38
calcium	Ca	20	manganese	Mn	25	sulphur	S	16
californium	Cf	98	mendelevium	Md	101	tantalum	Ta	73
carbon	C	6	mercury	Hg	80	technetium	Tc	43
cerium	Ce	58	molybdenum	Mo	42	tellurium	Te	52
chlorine	Cl	17	neodymium	Nd	60	terbium	Tb	65
chromium	Cr	24	neon	Ne	10	thallium	Tl	81
cobalt	Co	27	neptunium	Np	93	thorium	Th	90
copper	Cu	29	nickel	Ni	28	thulium	Tm	69
curium	Cm	96	niobium	Nb	41	tin	Sn	50
dysprosium	Dy	66	nitrogen	N	7	titanium	Ti	22
einsteinium	Es	99	nobelium	No	102	tungsten	W	74
erbium	Er	68	osmium	Os	76	unnilpentium[1]	Unp	105
europium	Eu	63	oxygen	O	8	unnilquadium[1]	Unq	104
fermium	Fm	100	palladium	Pd	46	uranium	U	92
fluorine	F	9	phosphorus	P	15	vanadium	V	23
francium	Fr	87	platinum	Pt	78	xenon	Xe	54
gadolinium	Gd	64	plutonium	Pu	94	ytterbium	Yb	70
gallium	Ga	31	polonium	Po	84	yttrium	Y	39
germanium	Ge	32	potassium	K	19	zinc	Zn	30
gold	Au	79	praseodymium	Pr	59	zirconium	Zr	40

[1] Provisional name (IUPAC) used pending agreement on a permanent name. Such names are based on the atomic number and are formed systematically from the numerical roots *nil* (= 0), *un* (= 1), *bi* (= 2), etc. (e.g. *unnilquadium* = 104, *ununbium* = 112, etc.).

The periodic table

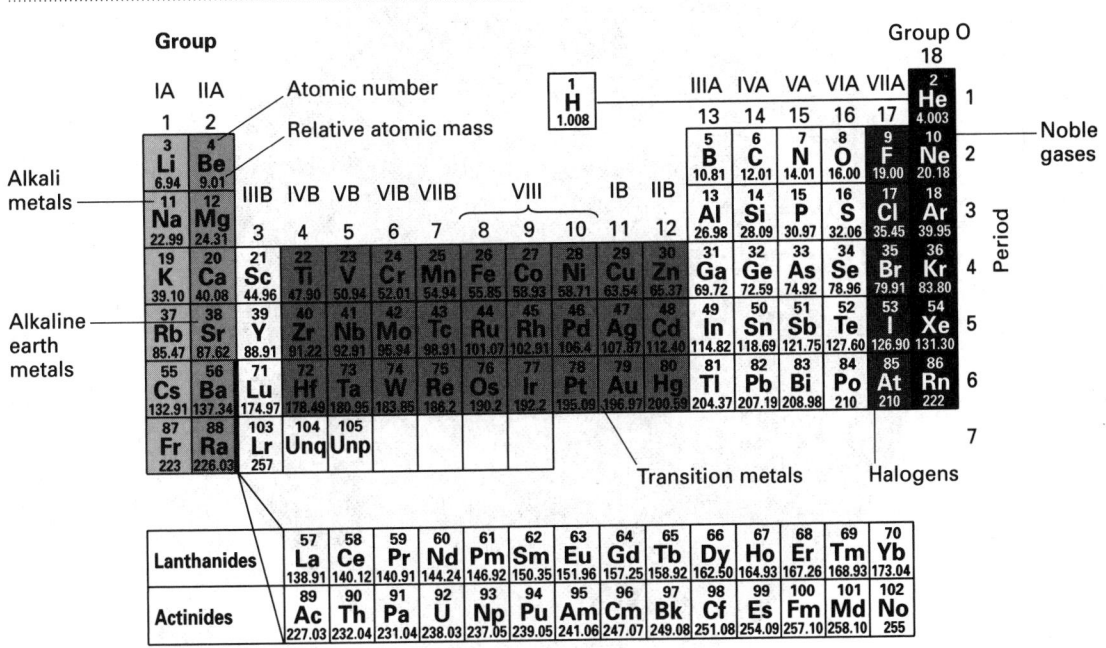

NORTHERN HEMISPHERE

SOUTHERN HEMISPHERE

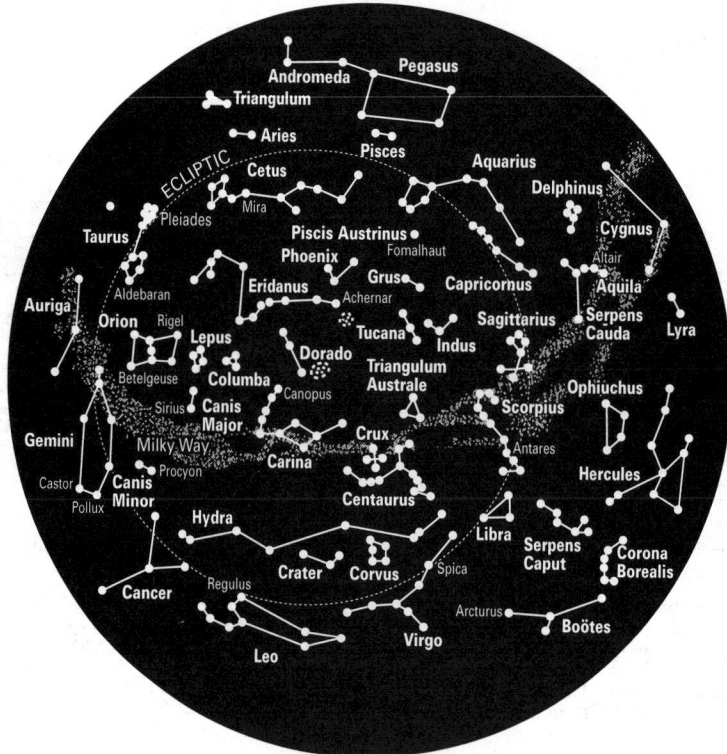

THE SUN AND PLANETS

Planet	Mean distance from sun (10^6 km)	Equatorial diameter (km)	Mass (earth = 1)	Volume (earth = 1)	Orbital period or 'year'	Rotation period or 'day'
Sun	—	1,400,000	330,000	1,300,000	—	25d*
Mercury	57.91	4,878	0.06	0.06	87.97d	58.65d
Venus	108.2	12,102	0.81	0.86	224.7d	243.0d(R)
Earth	149.6	12,756	1.00	1.00	365.3d	23.93h
Mars	227.9	6,786	0.11	0.15	687.0d	24.62h
Jupiter	778.3	142,980	318	1,323	11.86y	9.93h*
Saturn	1,427	120,540	95.2	752	29.46y	10.66h*
Uranus	2,871	51,120	14.5	64	84.01y	17.24h*(R)
Neptune	4,497	49,530	17.1	54	164.8y	16.11h*
Pluto	5,914	2,280	0.002	0.01	248.5y	6.39d(R)

* at equator   (R) retrograde

PRINCIPAL PLANETARY SATELLITES

Planet (total number of known satellites)	Satellite	Year of discovery	Diameter (km)	Mean distance from centre of planet (10^3 km)	Orbital period (d)
Earth (1)	Moon	—	3,476	384.4	27.32
Mars (2)	Phobos	1877	27*	9.4	0.319
	Deimos	1877	15*	23.5	1.262
Jupiter (16)	Amalthea	1892	262*	181	0.498
	Io	1610	3,630	422	1.769
	Europa	1610	3,138	671	3.551
	Ganymede	1610	5,262	1,070	7.155
	Callisto	1610	4,800	1,883	16.69
Saturn (18)	Mimas	1789	390	199	0.942
	Enceladus	1789	500	238	1.370
	Tethys	1684	1,050	295	1.888
	Dione	1684	1,120	377	2.737
	Rhea	1672	1,530	527	4.518
	Titan	1655	5,150	1,222	15.95
	Hyperion	1848	340*	1,481	21.28
	Iapetus	1671	1,440	3,561	79.33
	Phoebe	1898	220	12,952	550.5(R)
Uranus (15)	Miranda	1948	480	130	1.414
	Ariel	1851	1,160	191	2.520
	Umbriel	1851	1,190	266	4.144
	Titania	1787	1,600	436	8.706
	Oberon	1787	1,550	583	13.46
Neptune (8)	Proteus	1989	400	118	1.12
	Triton	1846	2,700	354	5.877(R)
	Nereid	1949	340	551	360.2
Pluto (1)	Charon	1978	1,190	20	6.387

* irregular: maximum dimension   (R) retrograde

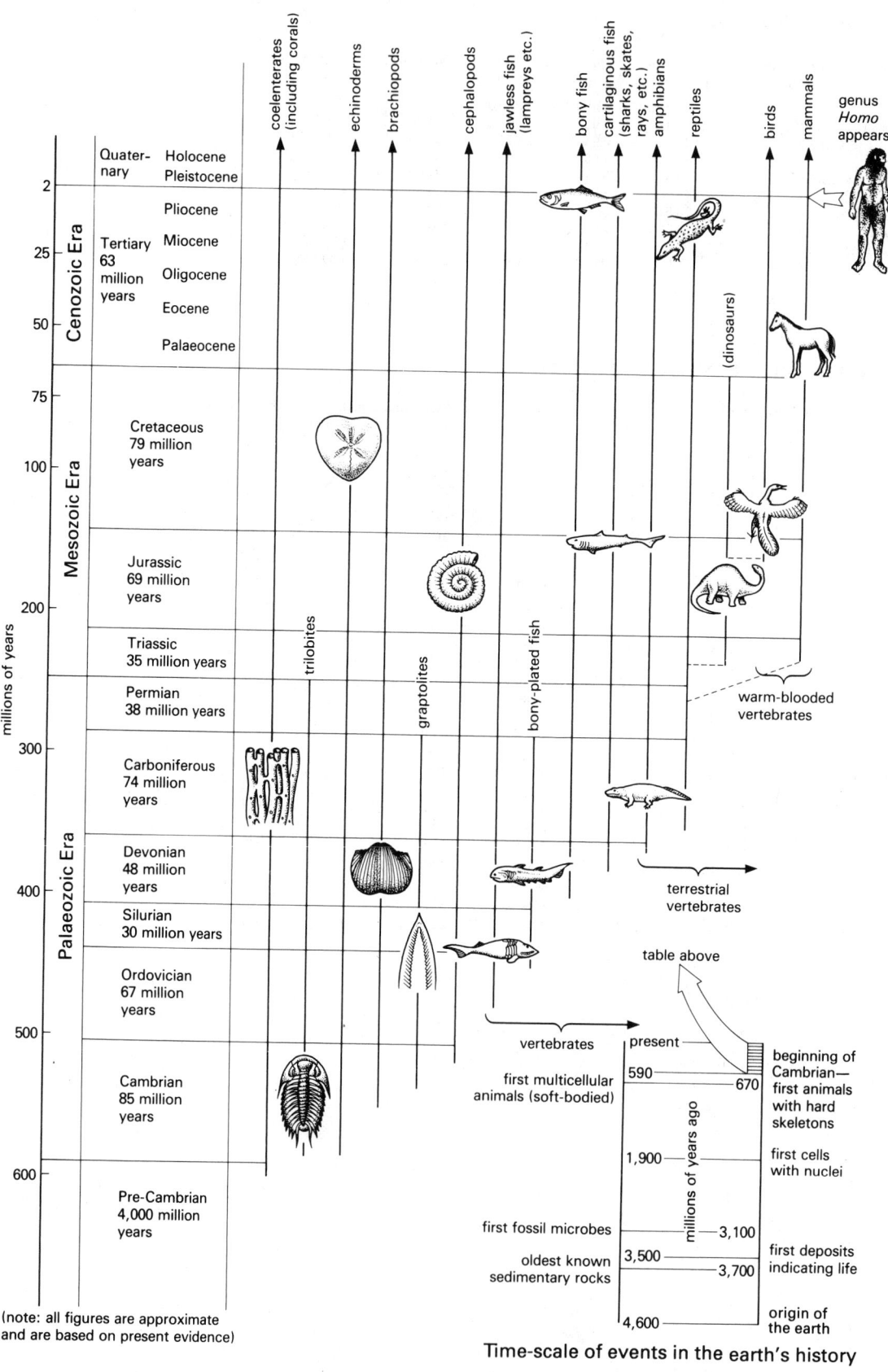

Time-scale of events in the earth's history

(note: all figures are approximate and are based on present evidence)

Terms marked † belong to 15th-century lists of 'proper terms', notably that in the *Book of St Albans* attributed to Dame Juliana Barnes (1486). Many of these are fanciful or humorous terms which probably never had any real currency, but have been taken up by Joseph Strutt in *Sports and Pastimes of England* (1801) and by other antiquarian writers.

a †shrewdness of apes
a herd or †pace of asses
a †cete of badgers
a †sloth or †sleuth of bears
a hive of bees; a swarm, drift, or bike of bees
a flock, flight, (*dial.*) parcel, pod, †fleet, or †dissimulation of (small) birds; a volary of birds in an aviary
a sounder of wild boar
a †blush of boys
a herd or gang of buffalo
a †clowder or †glaring of cats; a †dowt (= ?do-out) or †destruction of wild cats
a herd, drove, (*dial.*) drift, or (*US & Austral.*) mob of cattle
a brood, (*dial.*) cletch or clutch, or †peep of chickens
a †chattering or †clattering of choughs
a †drunkship of cobblers
a †rag or †rake of colts
a †hastiness of cooks
a †covert of coots
a herd of cranes
a litter of cubs
a herd of curlew
a †cowardice of curs
a herd or mob of deer
a pack or kennel of dogs
a trip of dotterel
a flight, †dole, or †piteousness of doves
a raft, bunch, or †paddling of ducks on water; a team of wild ducks in flight
a fling of dunlins
a herd of elephants
a herd or (*US*) gang of elk
a †business of ferrets
a charm or †chirm of finches
a shoal of fish; a run of fish in motion
a cloud of flies
a †stalk of foresters
a †skulk of foxes
a gaggle or (in the air) a skein, team, or wedge of geese
a herd of giraffes
a flock, herd, or (*dial.*) trip of goats
a pack or covey of grouse
a †husk or †down of hares
a cast of hawks let fly
an †observance of hermits
a †siege of herons
a stud or †haras of (breeding) horses; (*dial.*) a team of horses
a kennel, pack, cry or †mute of hounds
a flight or swarm of insects
a mob or troop of kangaroos
a kindle of kittens
a bevy of ladies
a †desert of lapwing

an †exaltation or bevy of larks
a †leap of leopards
a pride of lions
a †tiding of magpies
a †sord or †sute (= suit) of mallard
a †richesse of martens
a †faith of merchants
a †labour of moles
a troop of monkeys
a †barren of mules
a †watch of nightingales
a †superfluity of nuns
a covey of partridges
a †muster of peacocks
a †malapertness (= impertinence) of pedlars
a rookery of penguins
a head or (*dial.*) nye of pheasants
a kit of pigeons flying together
a herd of pigs
a stand, wing, or †congregation of plovers
a rush or flight of pochards
a herd, pod, or school of porpoises
a †pity of prisoners
a covey of ptarmigan
a litter of pups
a bevy or drift of quail
a string of racehorses
an †unkindness of ravens
a bevy of roes
a parliament or †building of rooks
a hill of ruffs
a herd or rookery of seals; a pod of seals
a flock, herd, (*dial.*) drift or trip, or (*Austral.*) mob of sheep
a †dopping of sheldrake
a wisp or †walk of snipe
a †host of sparrows
a †murmuration of starlings
a flight of swallows
a game or herd of swans; a wedge of swans in the air
a herd of swine; a †sounder of tame swine, a †drift of wild swine
a †glozing (= fawning) of taverners
a †spring of teal
a bunch or knob of waterfowl
a school, herd, or gam of whales; a pod of whales; a grind of bottle-nosed whales
a company or trip of wigeon
a bunch, trip, or plump of wildfowl; a knob (less than 30) of wildfowl
a pack or †rout of wolves
a gaggle of women (*derog.*)
a †fall of woodcock
a herd of wrens

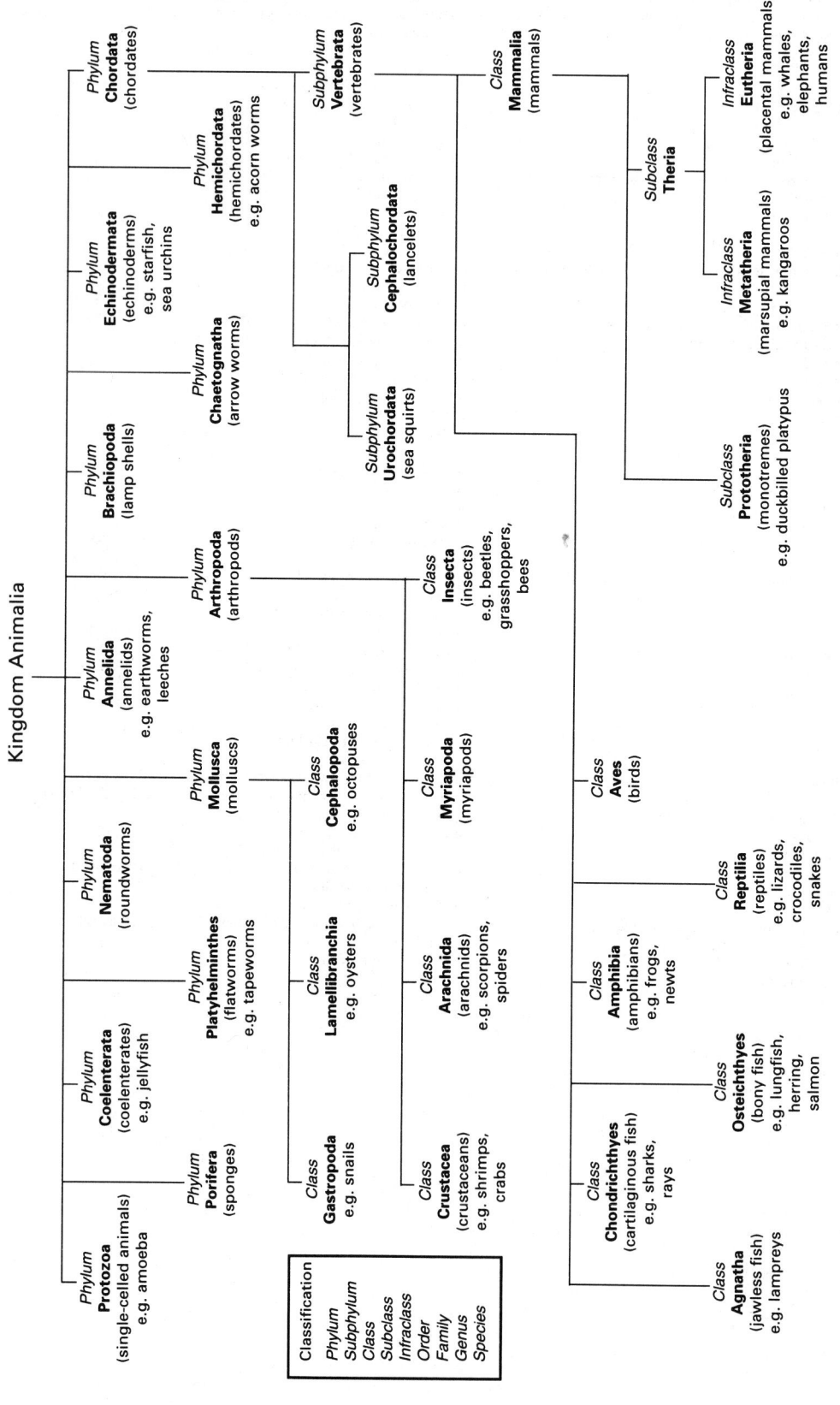

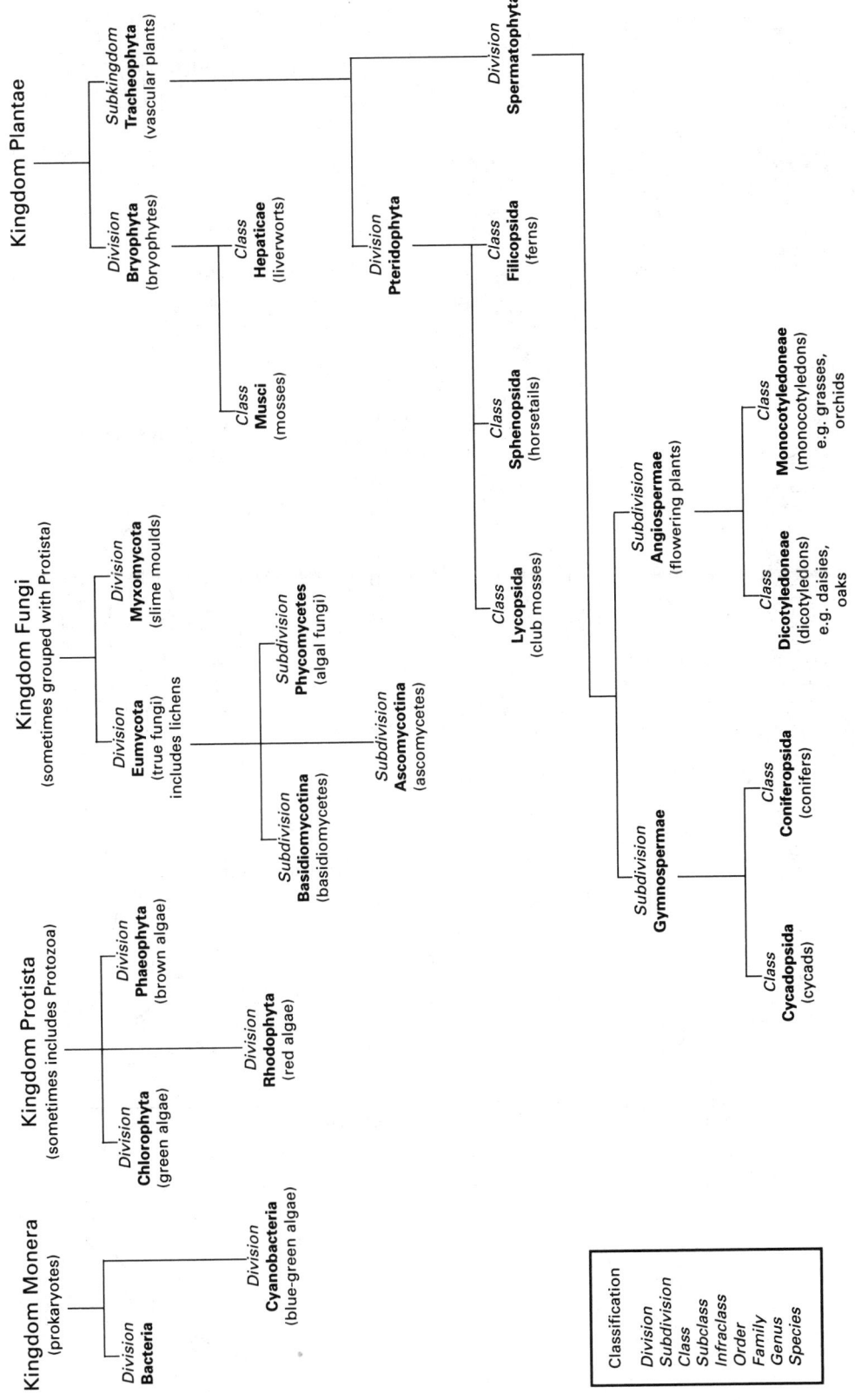

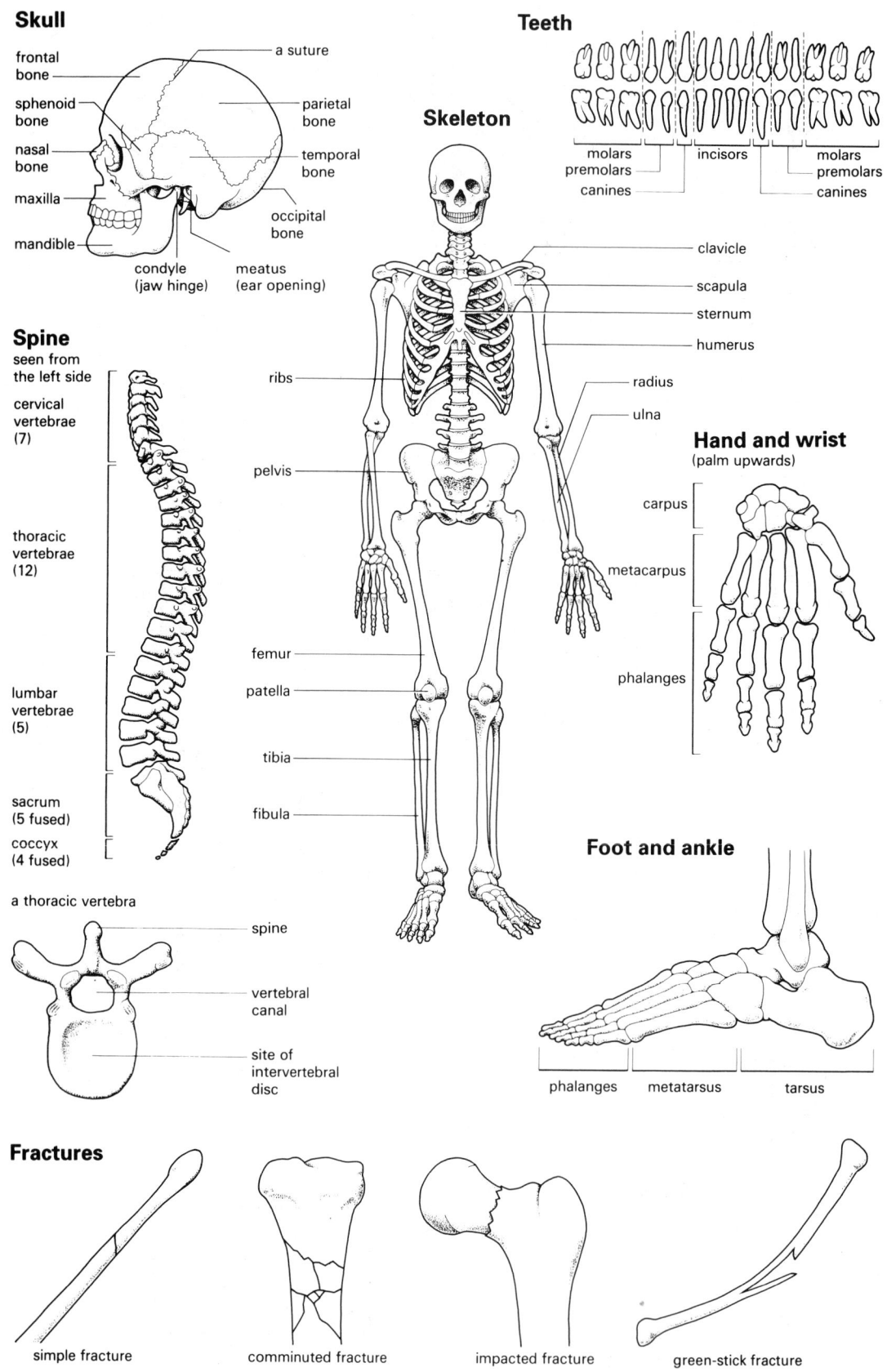

Skull

frontal bone

sphenoid bone

nasal bone

maxilla

mandible

a suture

parietal bone

temporal bone

occipital bone

condyle (jaw hinge)

meatus (ear opening)

Teeth

molars
premolars
canines

incisors

molars
premolars
canines

Skeleton

clavicle

scapula

sternum

humerus

radius

ulna

ribs

pelvis

femur

patella

tibia

fibula

Spine

seen from the left side

cervical vertebrae (7)

thoracic vertebrae (12)

lumbar vertebrae (5)

sacrum (5 fused)

coccyx (4 fused)

a thoracic vertebra

spine

vertebral canal

site of intervertebral disc

Hand and wrist
(palm upwards)

carpus

metacarpus

phalanges

Foot and ankle

phalanges

metatarsus

tarsus

Fractures

simple fracture

comminuted fracture

impacted fracture

green-stick fracture

Ball-and-socket joint

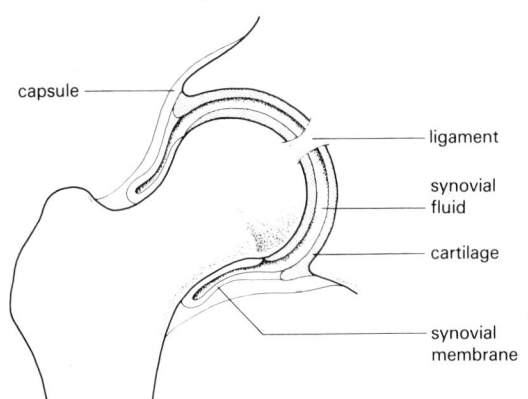

- capsule
- ligament
- synovial fluid
- cartilage
- synovial membrane

Lower abdomen

male

spine   ureter

- vas deferens
- bladder
- prostate gland
- penis
- scrotum enclosing testicles

rectum   urethra

Parts of a muscle

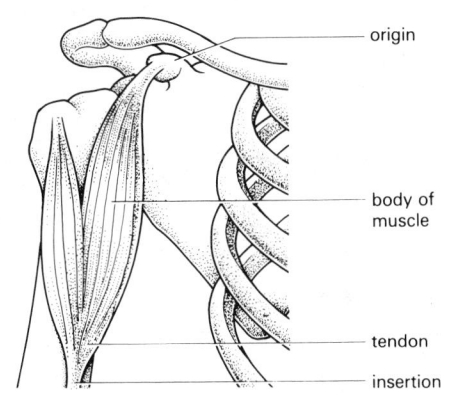

- origin
- body of muscle
- tendon
- insertion

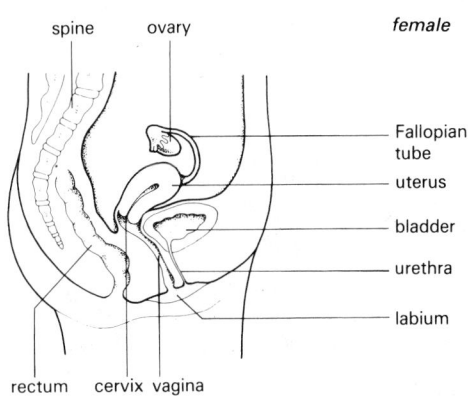

female

spine   ovary

- Fallopian tube
- uterus
- bladder
- urethra
- labium

rectum   cervix vagina

The alimentary canal

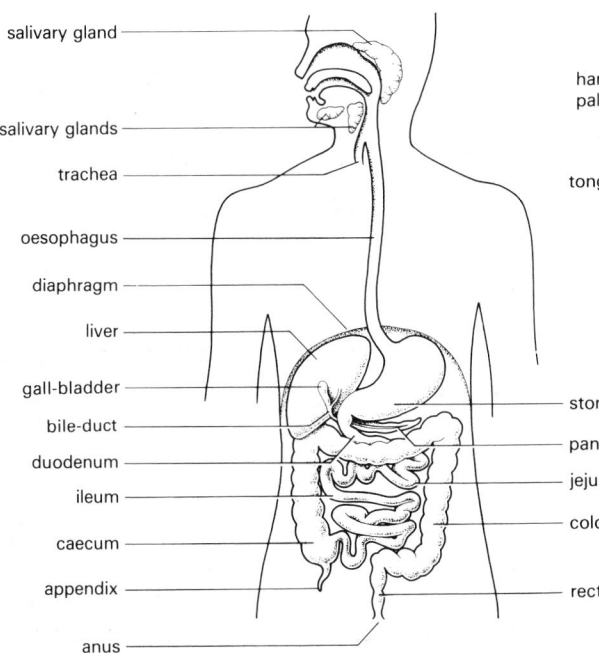

- salivary gland
- salivary glands
- trachea
- oesophagus
- diaphragm
- liver
- gall-bladder
- bile-duct
- duodenum
- ileum
- caecum
- appendix
- anus
- stomach
- pancreas
- jejunum
- colon
- rectum

Nose, mouth, and throat

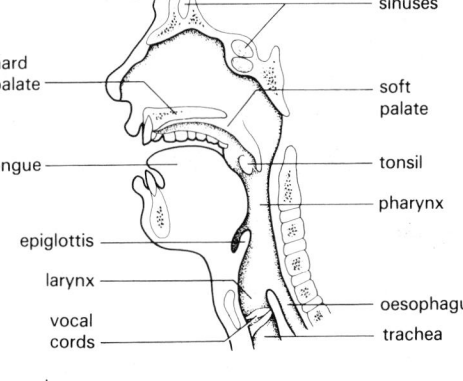

- sinuses
- hard palate
- soft palate
- tongue
- tonsil
- pharynx
- epiglottis
- larynx
- vocal cords
- oesophagus
- trachea

Respiration

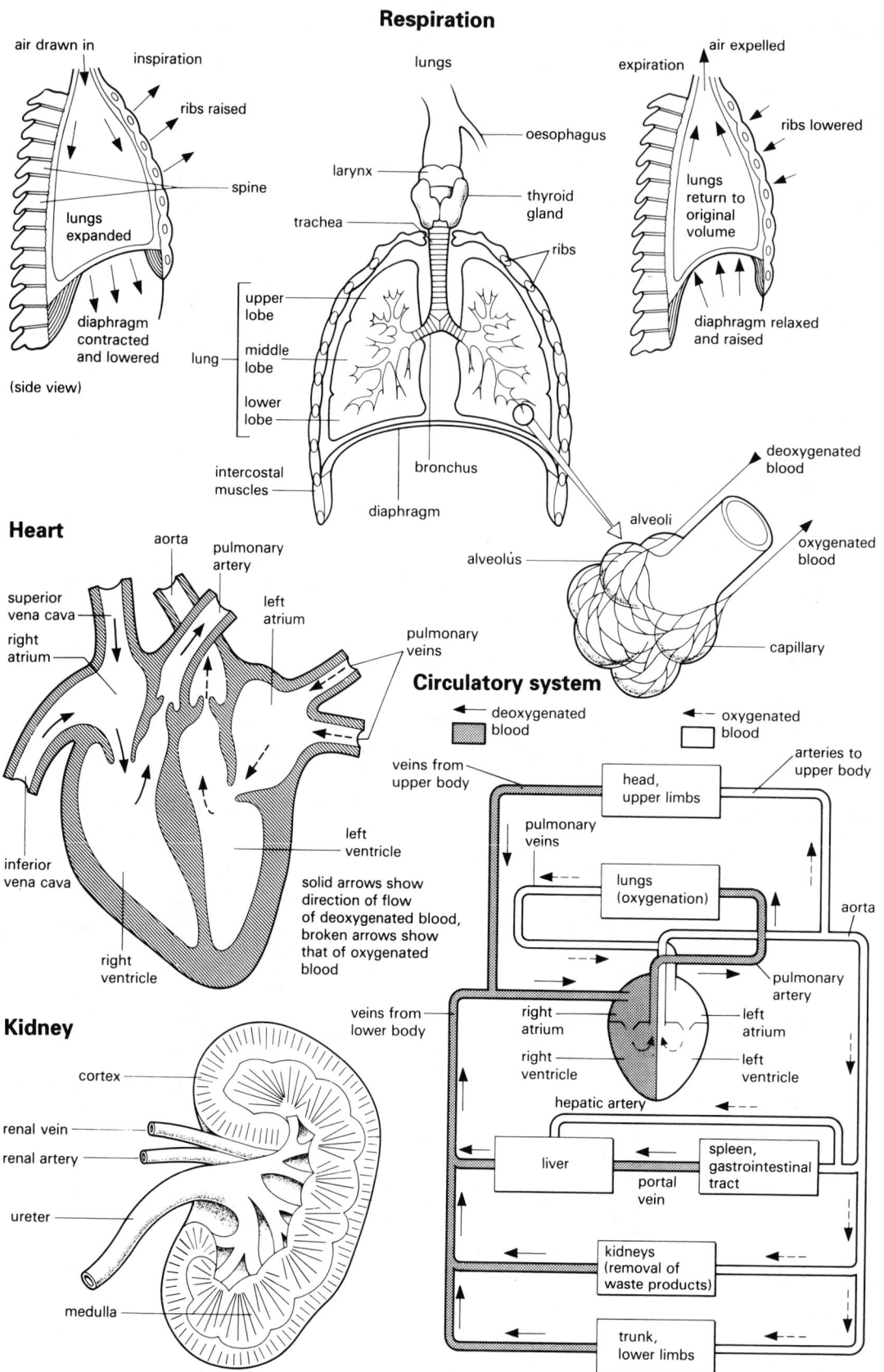

air drawn in
inspiration
ribs raised
spine
lungs expanded
diaphragm contracted and lowered
(side view)

lungs
oesophagus
larynx
thyroid gland
trachea
ribs
upper lobe
middle lobe
lung
lower lobe
intercostal muscles
bronchus
diaphragm

expiration
air expelled
ribs lowered
lungs return to original volume
diaphragm relaxed and raised

deoxygenated blood
oxygenated blood
alveoli
alveolus
capillary

Heart

aorta
pulmonary artery
superior vena cava
right atrium
left atrium
pulmonary veins
inferior vena cava
left ventricle
right ventricle

solid arrows show direction of flow of deoxygenated blood, broken arrows show that of oxygenated blood

Circulatory system

deoxygenated blood
oxygenated blood
veins from upper body
pulmonary veins
arteries to upper body
head, upper limbs
lungs (oxygenation)
aorta
pulmonary artery
veins from lower body
right atrium
right ventricle
left atrium
left ventricle
hepatic artery
liver
spleen, gastrointestinal tract
portal vein
kidneys (removal of waste products)
trunk, lower limbs

Kidney

cortex
renal vein
renal artery
ureter
medulla

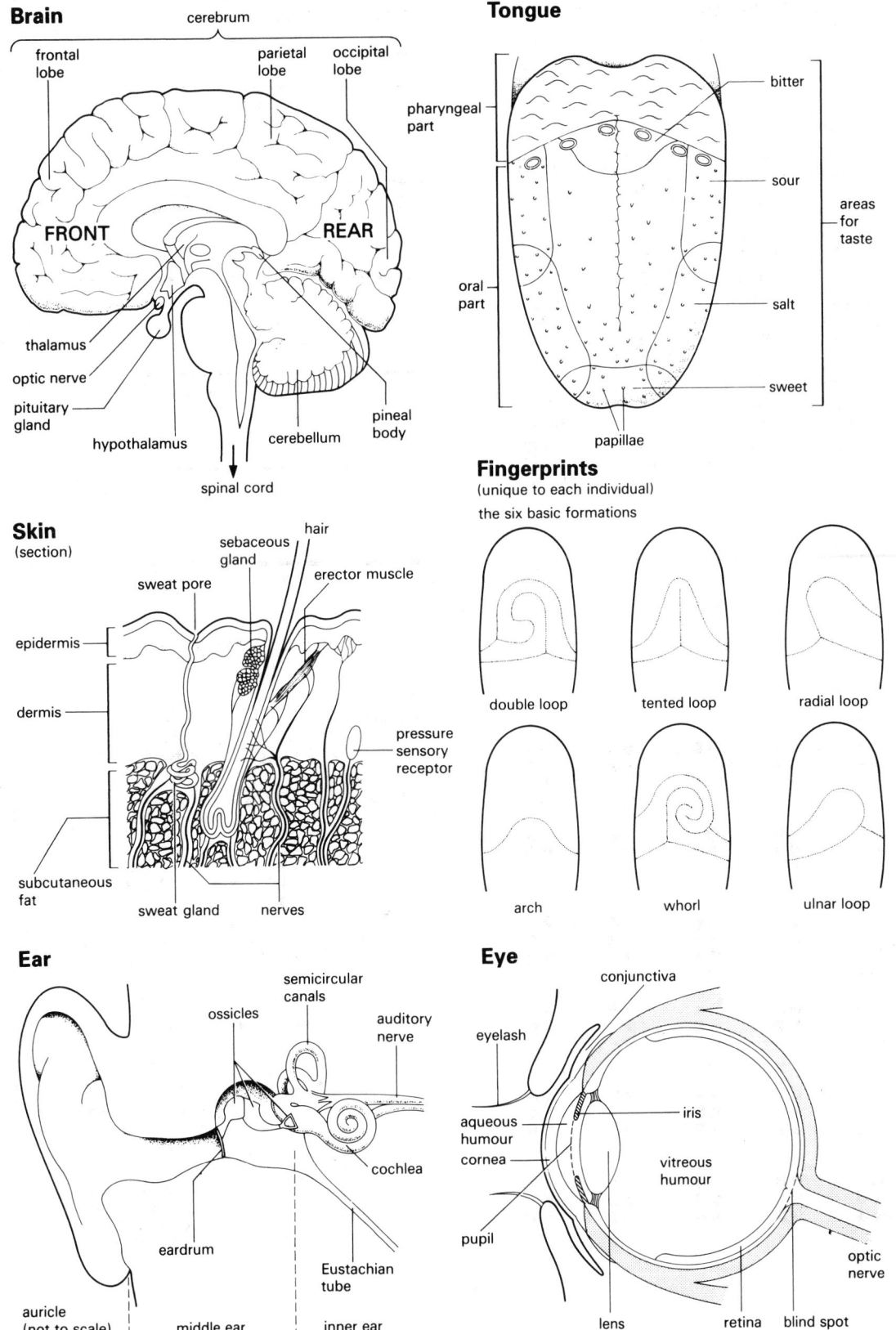

Brain

cerebrum

frontal lobe · parietal lobe · occipital lobe

FRONT REAR

thalamus
optic nerve
pituitary gland
hypothalamus
cerebellum
pineal body
spinal cord

Tongue

pharyngeal part
oral part

bitter
sour
salt
sweet

areas for taste

papillae

Skin
(section)

sebaceous gland
hair
sweat pore
erector muscle
epidermis
dermis
pressure sensory receptor
subcutaneous fat
sweat gland
nerves

Fingerprints
(unique to each individual)
the six basic formations

double loop   tented loop   radial loop

arch   whorl   ulnar loop

Ear

ossicles
semicircular canals
auditory nerve
cochlea
eardrum
Eustachian tube
auricle (not to scale)
middle ear
inner ear

Eye

conjunctiva
eyelash
aqueous humour
cornea
iris
vitreous humour
pupil
lens
retina
blind spot
optic nerve

1753

Alphabets for the deaf and for the blind

finger spelling

Alphabets for signalling

Morse code

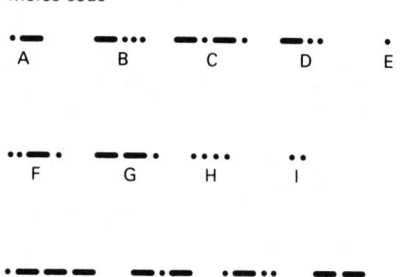

Braille

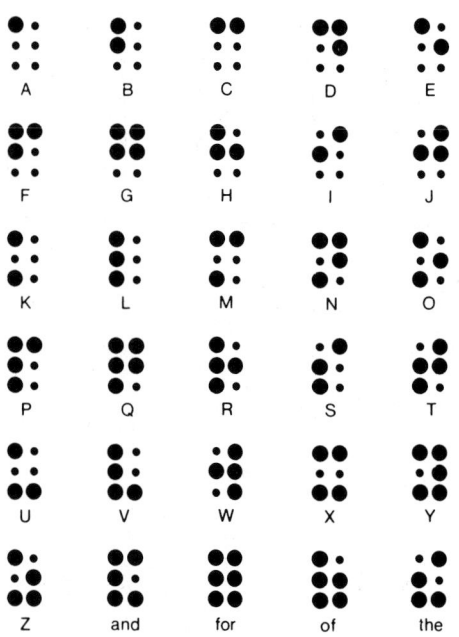

semaphore

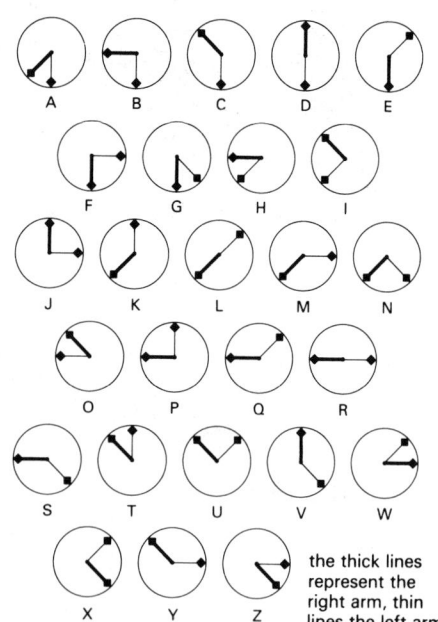

the thick lines represent the right arm, thin lines the left arm

Arabic

Alone	Final	Medial	Initial		
ا	ل			'alif	'
ب	ب	ـبـ	بـ	bā'	b
ت	ت	ـتـ	تـ	tā'	t
ث	ث	ـثـ	ثـ	thā'	th
ج	ج	ـجـ	جـ	jīm	j
ح	ح	ـحـ	حـ	ḥā'	ḥ
خ	خ	ـخـ	خـ	khā'	kh
د	ـد			dāl	d
ذ	ـذ			dhāl	dh
ر	ـر			rā'	r
ز	ـز			zay	z
س	ـس	ـسـ	سـ	sīn	s
ش	ـش	ـشـ	شـ	shīn	sh
ص	ـص	ـصـ	صـ	ṣād	ṣ
ض	ـض	ـضـ	ضـ	ḍād	ḍ
ط	ـط	ـطـ	طـ	ṭā'	ṭ
ظ	ـظ	ـظـ	ظـ	ẓā'	ẓ
ع	ـع	ـعـ	عـ	'ayn	'
غ	ـغ	ـغـ	غـ	ghayn	gh
ف	ـف	ـفـ	فـ	fā'	f
ق	ـق	ـقـ	قـ	qāf	q
ك	ـك	ـكـ	كـ	kāf	k
ل	ـل	ـلـ	لـ	lām	l
م	ـم	ـمـ	مـ	mīm	m
ن	ـن	ـنـ	نـ	nūn	n
ه	ـه	ـهـ	هـ	hā'	h
و	ـو			wāw	w
ى	ـى	ـيـ	يـ	yā'	y

Hebrew

א	aleph	'
ב	beth	b, bh
ג	gimel	g, gh
ד	daleth	d, dh
ה	he	h
ו	waw	w
ז	zayin	z
ח	ḥeth	ḥ
ט	ṭeth	ṭ
י	yodh	y
כ ך	kaph	k, kh
ל	lamedh	l
מ ם	mem	m
נ ן	nun	n
ס	samekh	s
ע	'ayin	'
פ ף	pe	p, ph
צ ץ	ṣadhe	ṣ
ק	qoph	q
ר	resh	r
שׂ	śin	ś
שׁ	shin	sh
ת	taw	t, th

Greek

Α α	alpha	a
Β β	beta	b
Γ γ	gamma	g
Δ δ	delta	d
Ε ε	epsilon	e
Ζ ζ	zeta	z
Η η	eta	ē
Θ θ	theta	th
Ι ι	iota	i
Κ κ	kappa	k
Λ λ	lambda	l
Μ μ	mu	m
Ν ν	nu	n
Ξ ξ	xi	x
Ο ο	omicron	o
Π π	pi	p
Ρ ρ	rho	r, rh
Σ σ ς	sigma	s
Τ τ	tau	t
Υ υ	upsilon	u
Φ φ	phi	ph
Χ χ	chi	kh
Ψ ψ	psi	ps
Ω ω	omega	ō

Russian

А а	a
Б б	b
В в	v
Г г	g
Д д	d
Е е	e, ye
Ё ё	yo
Ж ж	zh
З з	z
И и	i
Й й	i
К к	k
Л л	l
М м	m
Н н	n
О о	o
П п	p
Р р	r
С с	s
Т т	t
У у	u
Ф ф	f
Х х	kh
Ц ц	ts
Ч ч	ch
Ш ш	sh
Щ щ	shch
Ъ ъ	″ ('hard sign')
Ы ы	y
Ь ь	′ ('soft sign')
Э э	e
Ю ю	yu
Я я	ya

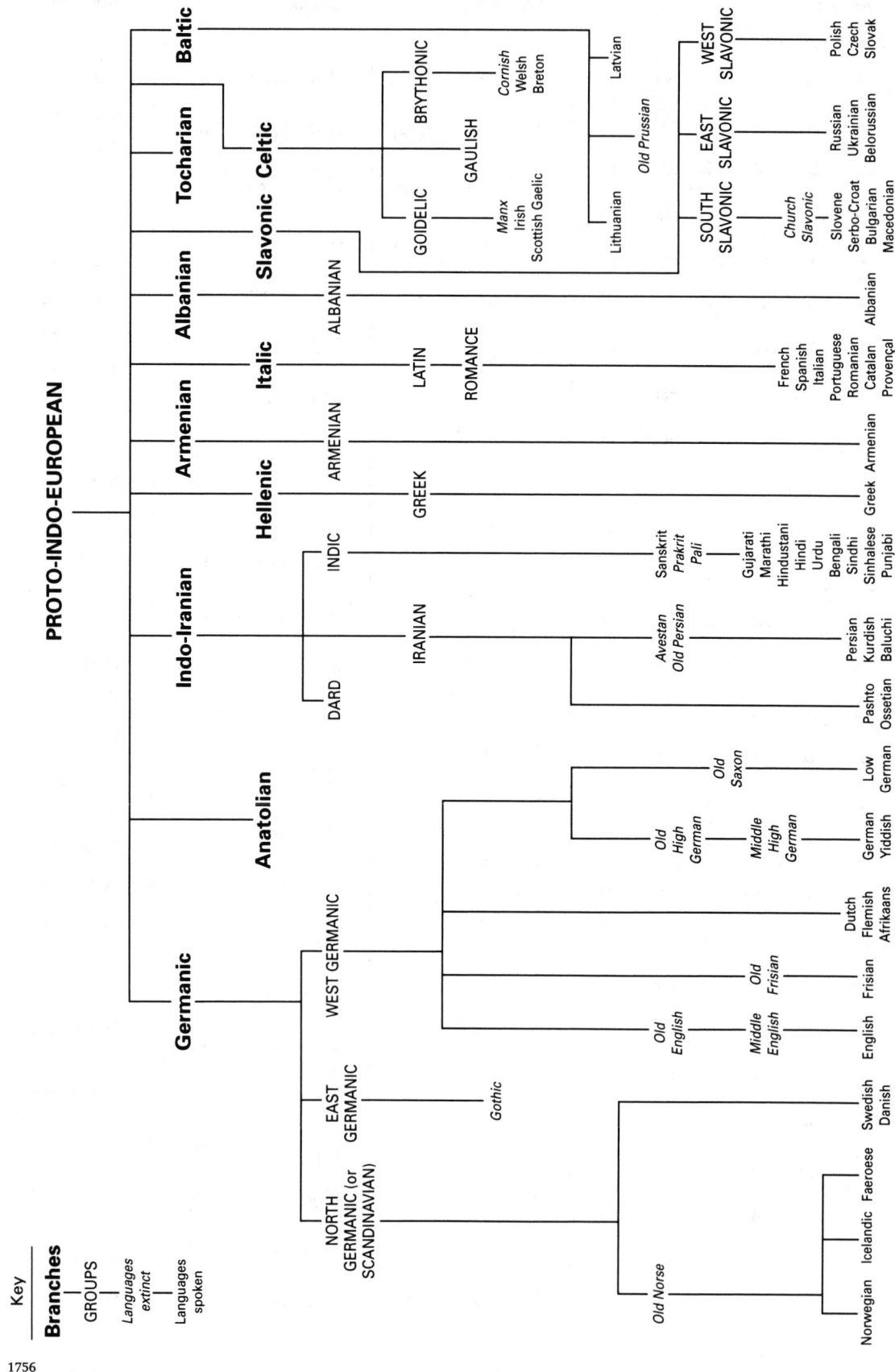

PROTO-INDO-EUROPEAN

Branches

Key

GROUPS

*Languages
extinct*

Languages
spoken

All marks shown relate to silver except where otherwise indicated

A hallmark

maker's mark   standard mark   Assay Office mark   date letter

Maker's mark (from 1363)
originally symbols, now initials

 symbol

  symbol and initials

  initials

Some earlier Assay Office marks (with dates of closure)

Norwich (1702)   York (1856)   Exeter (1883)   Newcastle (1884)   Chester (1962)   Glasgow (1964)

    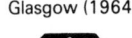 

Assay Office mark (from 1300)
now only London, Birmingham, Sheffield, and Edinburgh

London

gold and silver (leopard's head uncrowned from 1821; mark includes platinum from 1975)

Britannia silver (prior to 1975)

Edinburgh

gold and silver (also platinum from 1975)

Birmingham

gold (also platinum from 1975)

silver

Sheffield

silver (prior to 1975)

gold (also silver and platinum from 1975)

Standard mark (from 1544)
Marks guaranteeing pure metal content of the percentage shown

sterling silver 92.5%

 marked in England

 marked in Scotland (from 1975)

 marked in Scotland (prior to 1975)

Britannia standard silver (1697–1720, also occasional use since) 95.8%

gold (crown followed by millesimal figure of the standard)

 i.e. 18 carat 75%

 22 carat 91.6%

 14 carat 58.5%

 9 carat 37.5%

(prior to 1975 marks incorporated the carat figure, and Scottish 18 and 22 carat gold bore a thistle mark instead of the crown)

Date letter (from 1478)
one letter per year before changing to next style of letter and/or shield

cycles vary between Assay Offices

London date letters (A-U used, excluding J) showing style of first letter and years of cycle

	1498–1518[1]		1598–1618		1697[3]–1716		1796–1816		1896–1916
	1518–1538		1618–1638		1716–1736		1816–1836		1916–1936
	1538–1558		1638–1658		1736–1756		1836–1856		1936–1956
	1558–1578		1658–1678		1756–1776		1856–1876		1956–1974[2]
	1578–1598		1678–1697[2]		1776–1796		1876–1896		1975[4] –

Notes   1. Letter changed on 19 May until 1697
   2. No U used in these cycles
   3. A from 27 March–28 May 1697; year letters then changed on 29 May until 1975
   4. Year letter changed with each calendar year; from 1975 all UK Offices use the same date letters and shield shape

Values of notes and rests

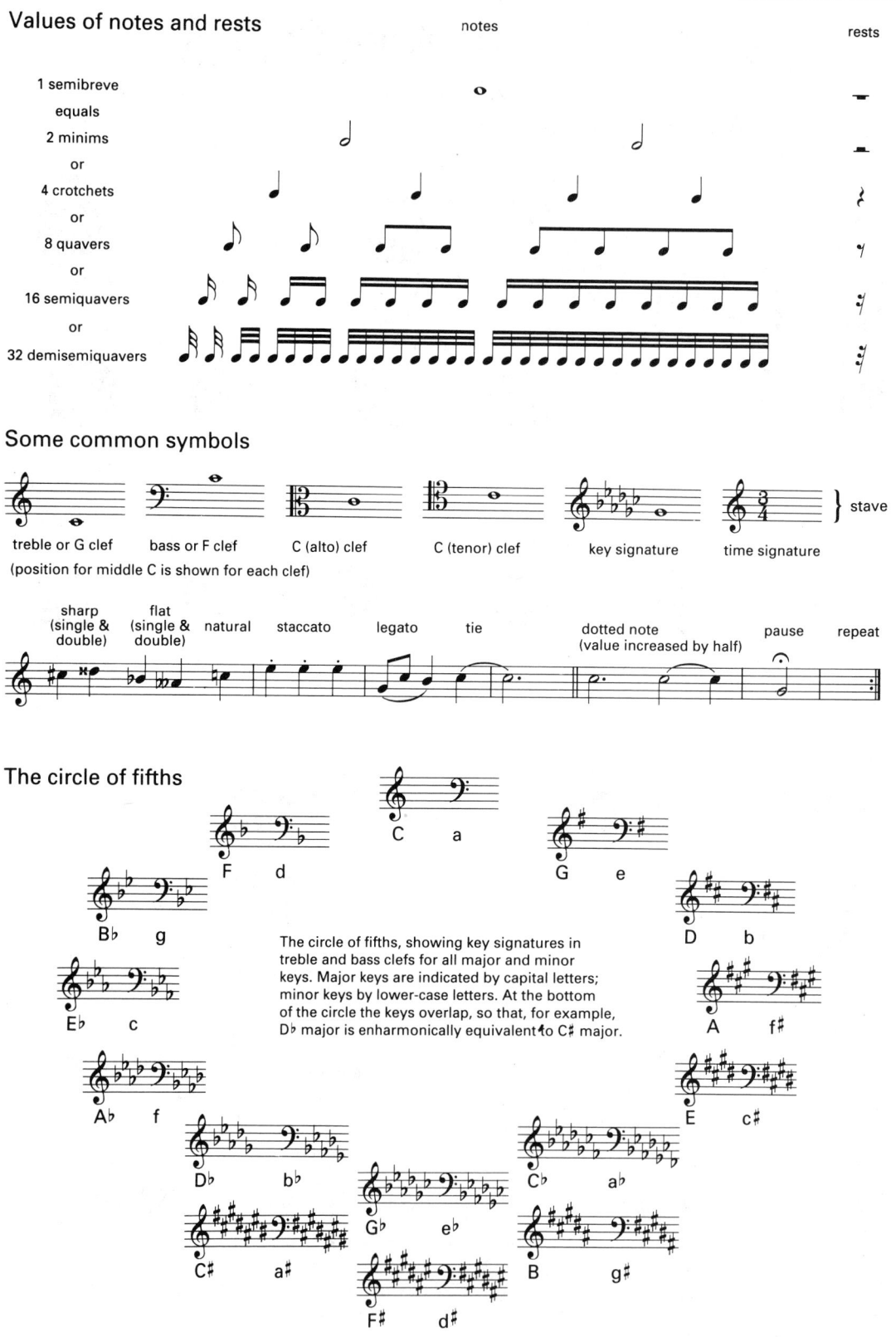

notes

rests

1 semibreve
equals
2 minims
or
4 crotchets
or
8 quavers
or
16 semiquavers
or
32 demisemiquavers

Some common symbols

treble or G clef   bass or F clef   C (alto) clef   C (tenor) clef   key signature   time signature   } stave

(position for middle C is shown for each clef)

sharp (single & double)   flat (single & double)   natural   staccato   legato   tie   dotted note (value increased by half)   pause   repeat

The circle of fifths

The circle of fifths, showing key signatures in treble and bass clefs for all major and minor keys. Major keys are indicated by capital letters; minor keys by lower-case letters. At the bottom of the circle the keys overlap, so that, for example, D♭ major is enharmonically equivalent to C♯ major.

C   a
F   d
G   e
B♭   g
D   b
E♭   c
A   f♯
A♭   f
E   c♯
D♭   b♭
C♭   a♭
G♭   e♭
B   g♯
C♯   a♯
F♯   d♯

An orchestral layout

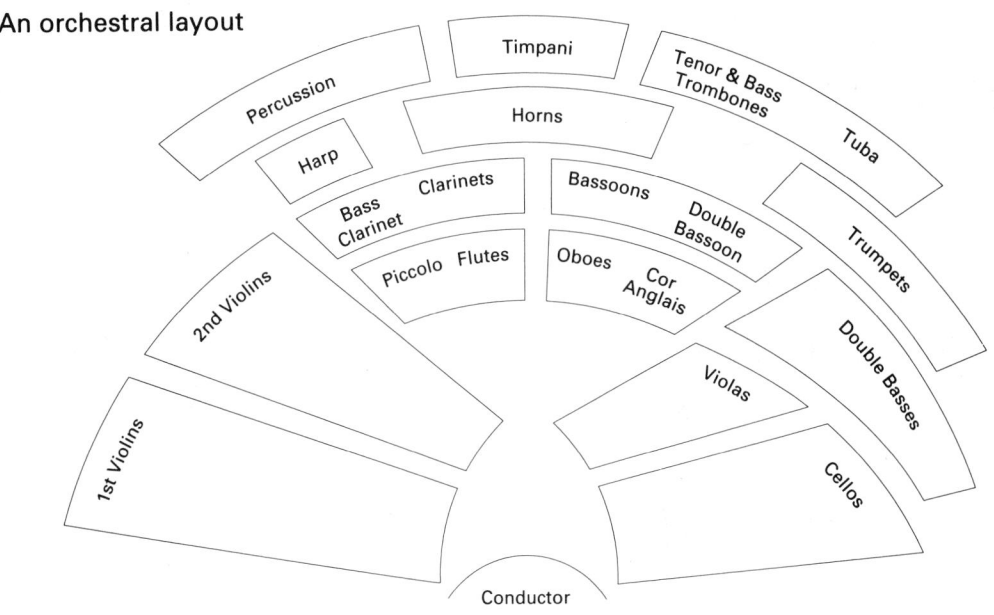

Dynamics

<	*crescendo*	get louder
>	*diminuendo*	get quieter
ppp		very, very quiet
pp	*pianissimo*	very quiet
p	*piano*	quiet
mp	*mezzopiano*	quite quiet
mf	*mezzoforte*	quite loud
f	*forte*	loud
ff	*fortissimo*	very loud
fff		very, very loud
sf	*sforzando*	suddenly very loud

Tempo indicators

adagio	slow
largo	slow and dignified
andante	flowing, at a walking pace
allegro	quick and bright
allegretto	not as quick as allegro
vivace	fast and lively
presto	very quick
accelerando	getting faster
ritenuto (rit.)	holding back
rallentando (rall.)	getting slower
rubato	flexible tempo

Interpretive indicators

cantabile	singing style
dolce	soft and sweet
espressivo	expressively

legato	smooth
staccato	detached

Appendix 26 **Architecture**

Classical

A Greek Doric temple

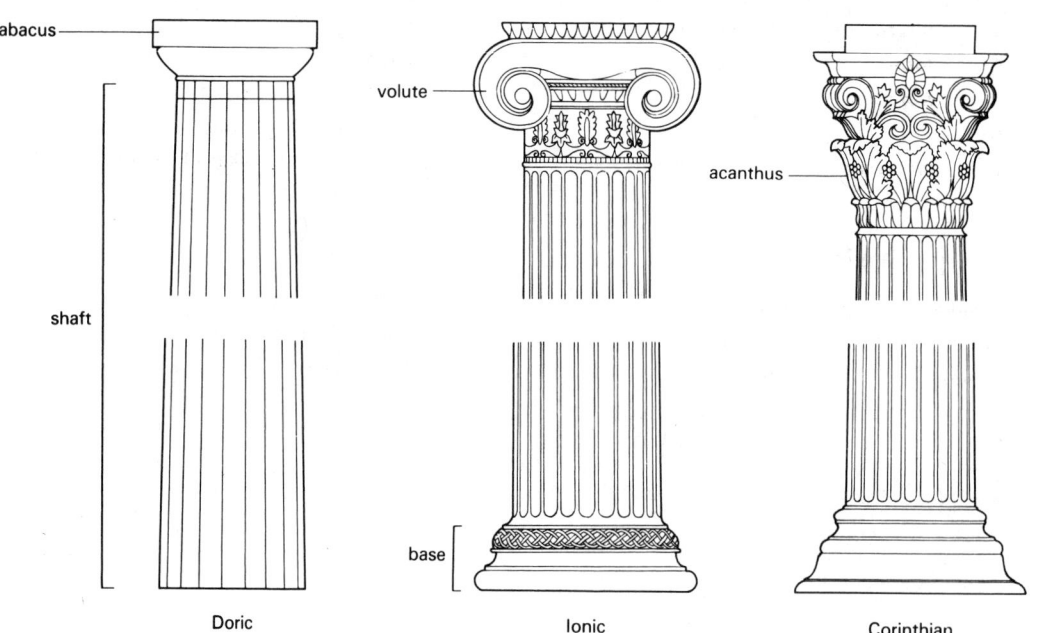

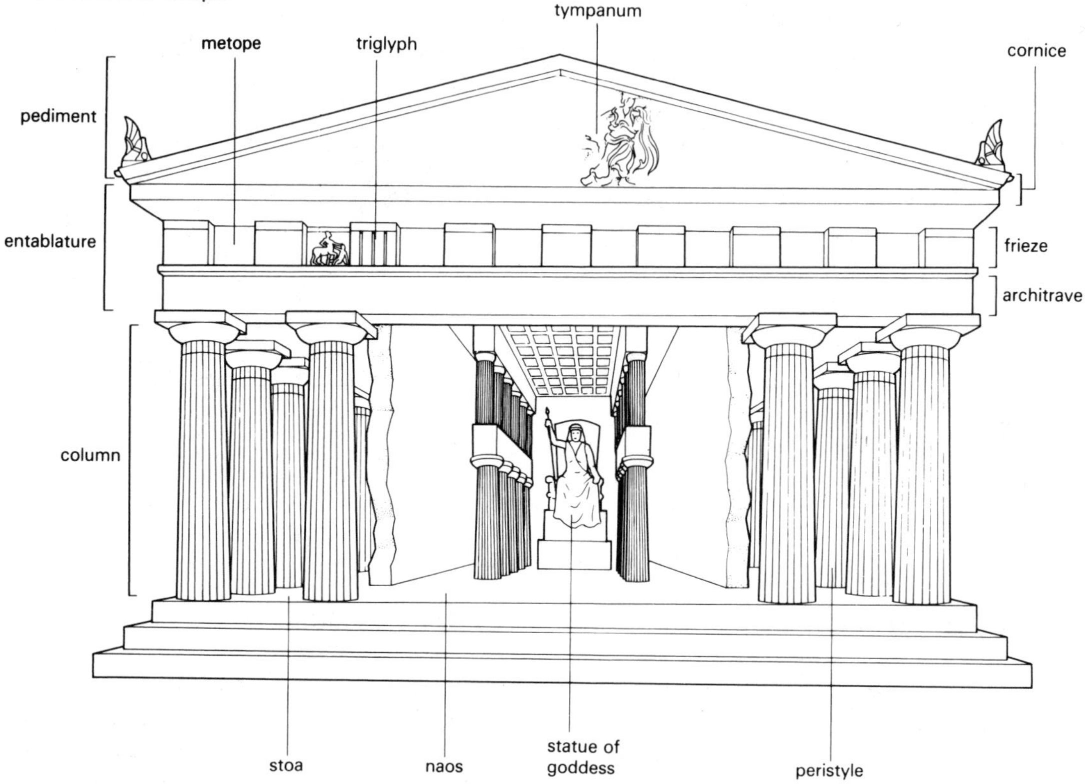
tympanum
metope
triglyph
cornice
pediment
entablature
frieze
architrave
column
stoa
naos
statue of goddess
peristyle

Orders of architecture: Greek origin

abacus
volute
acanthus
shaft
base

Doric

Ionic

Corinthian

Structure

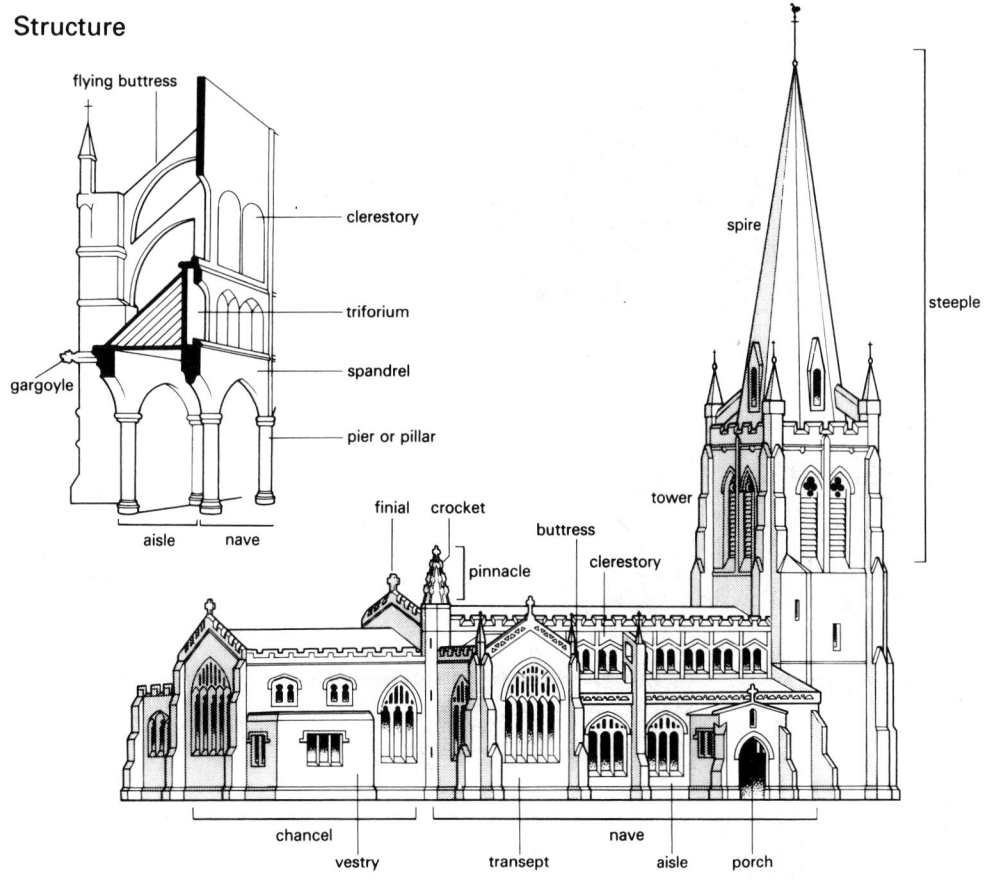

flying buttress · *clerestory* · *triforium* · *spandrel* · *gargoyle* · *pier or pillar* · *aisle* · *nave*

finial · *crocket* · *buttress* · *pinnacle* · *clerestory* · *tower* · *spire* · *steeple* · *chancel* · *vestry* · *transept* · *nave* · *aisle* · *porch*

Windows

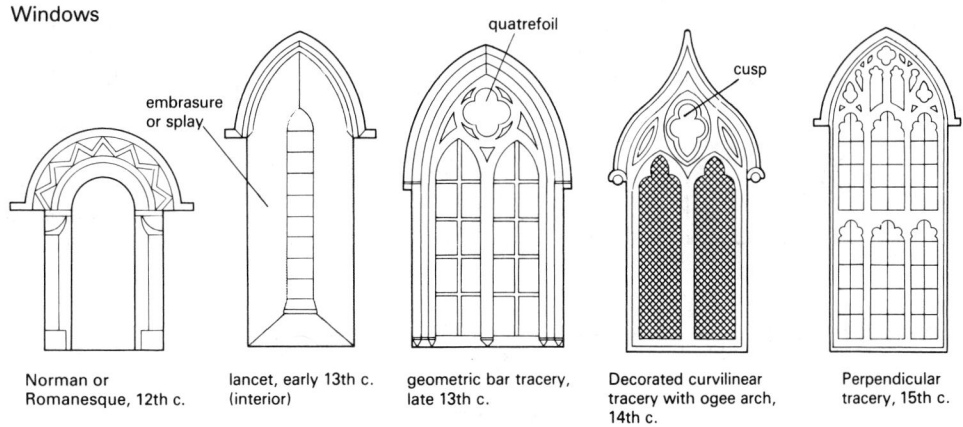

Norman or Romanesque, 12th c.

lancet, early 13th c. (interior) — *embrasure or splay*

geometric bar tracery, late 13th c. — *quatrefoil*

Decorated curvilinear tracery with ogee arch, 14th c. — *cusp*

Perpendicular tracery, 15th c.

Vaults

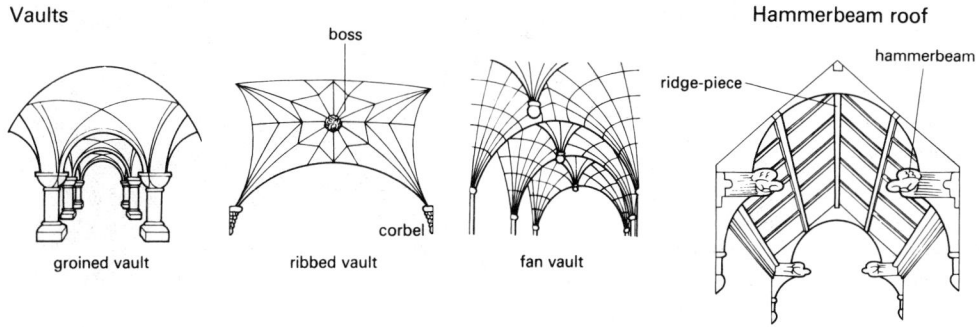

groined vault

ribbed vault — *boss* · *corbel*

fan vault

Hammerbeam roof

ridge-piece · *hammerbeam*

Appendix 27 **Sports and Games**

Association football

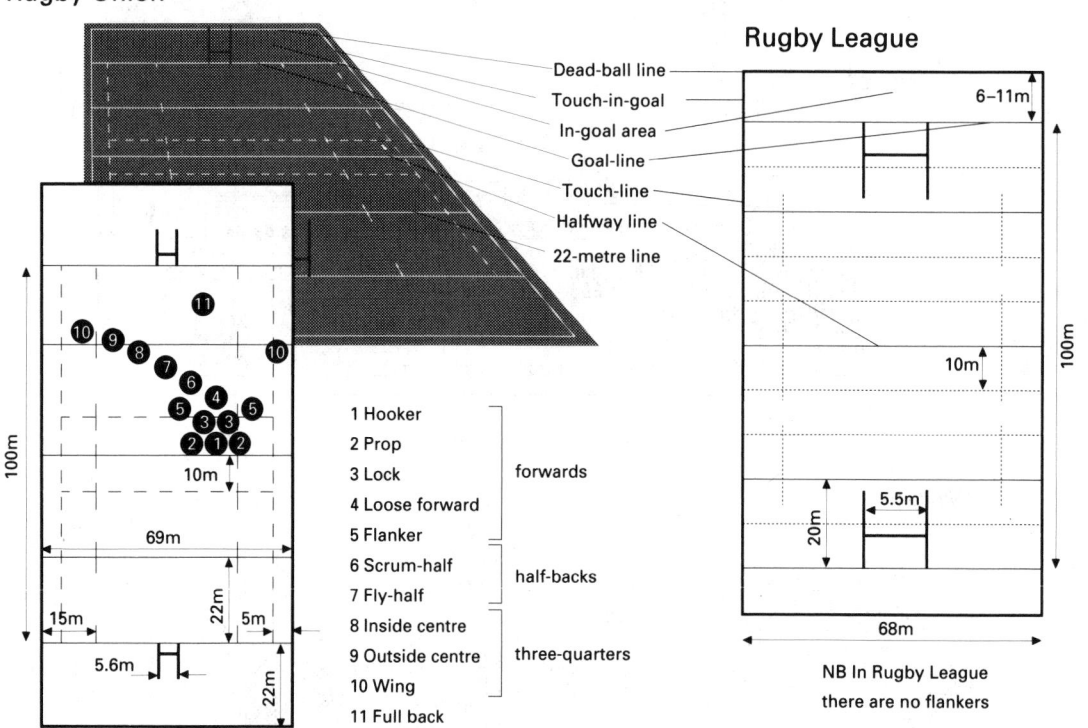

Goal
Penalty spot
Centre circle
Touch-line
Halfway line
Penalty area
Goal area
Goal-line

Corner-kick quadrant

90-120m

16.5m
5.5m
40.32m
18.32m
45-90m
9.15m

A possible team formation:

1 Forward
2 Midfield player
3 Defender
4 Sweeper
5 Goalkeeper

Rugby Union

Rugby League

Dead-ball line
Touch-in-goal
In-goal area
Goal-line
Touch-line
Halfway line
22-metre line

6–11m
10m
100m
20m
5.5m
68m

NB In Rugby League there are no flankers

100m
10m
69m
15m
22m
5m
5.6m
22m

1 Hooker
2 Prop
3 Lock } forwards
4 Loose forward
5 Flanker
6 Scrum-half } half-backs
7 Fly-half
8 Inside centre
9 Outside centre } three-quarters
10 Wing
11 Full back

American football

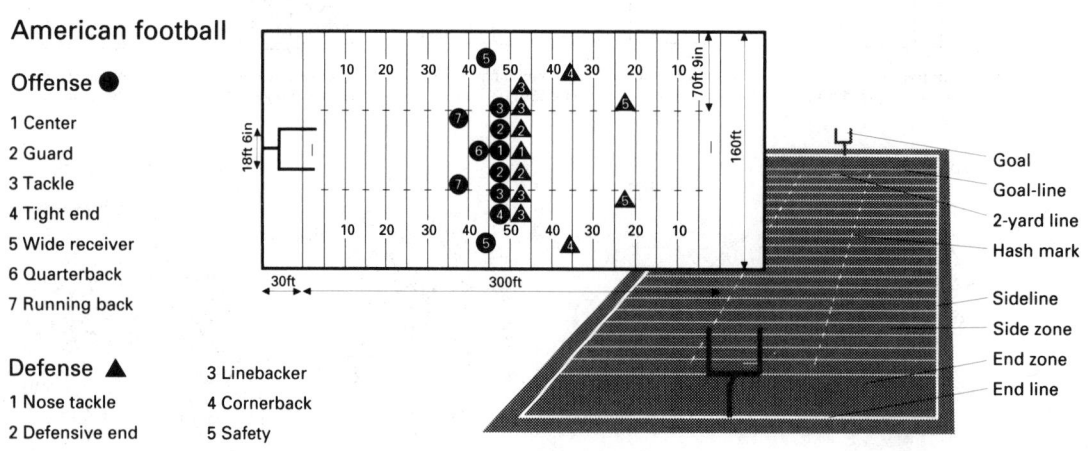

Offense ●

1 Center
2 Guard
3 Tackle
4 Tight end
5 Wide receiver
6 Quarterback
7 Running back

Defense ▲

1 Nose tackle
2 Defensive end
3 Linebacker
4 Cornerback
5 Safety

10 20 30 40 50 40 30 20 10
70ft 9in
18ft 6in
160ft
30ft
300ft

Goal
Goal-line
2-yard line
Hash mark
Sideline
Side zone
End zone
End line

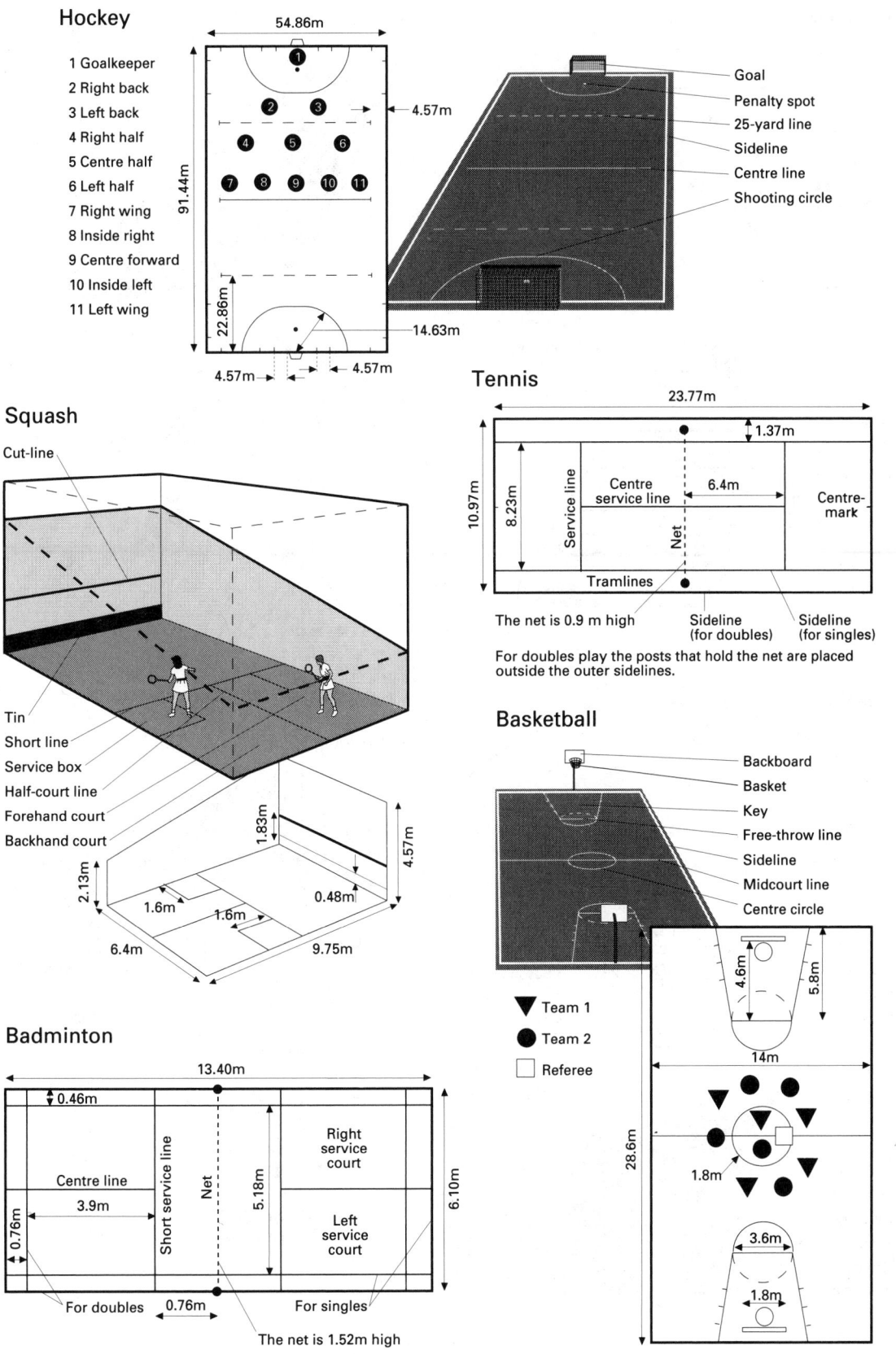

Hockey

54.86m

1 Goalkeeper
2 Right back
3 Left back
4 Right half
5 Centre half
6 Left half
7 Right wing
8 Inside right
9 Centre forward
10 Inside left
11 Left wing

91.44m

4.57m

22.86m

14.63m

4.57m   4.57m

Goal
Penalty spot
25-yard line
Sideline
Centre line
Shooting circle

Squash

Cut-line

Tin
Short line
Service box
Half-court line
Forehand court
Backhand court

1.83m
2.13m
0.48m
4.57m
1.6m
1.6m
6.4m
9.75m

Badminton

13.40m

0.46m

Centre line
3.9m

Short service line

Net

Right service court

Left service court

5.18m

6.10m

0.76m

For doubles   0.76m   For singles

The net is 1.52m high

Tennis

23.77m

1.37m

10.97m

8.23m

Service line

Centre service line

6.4m

Net

Centre-mark

Tramlines

The net is 0.9 m high   Sideline (for doubles)   Sideline (for singles)

For doubles play the posts that hold the net are placed outside the outer sidelines.

Basketball

Backboard
Basket
Key
Free-throw line
Sideline
Midcourt line
Centre circle

▼ Team 1
● Team 2
□ Referee

4.6m
5.8m
14m
28.6m
1.8m
3.6m
1.8m

Ice hockey

56–61m

26–30m

1.83m

2.4m

1.2m

4.6m

9.2m

1 Goalkeeper
2 Defenceman
3 Winger
4 Centre forward

Boards
Goal-line
Goal
Goal crease
Referee's crease
Centre (red) line
Centre face-off circle
Blue line
Face-off circle

Cricket

Wicket-keeper

Wicket

Batsman

Bowler

Crease

Typical field for right-handed batsman (B1)

1 Bowler
2 Wicket-keeper
3 First slip
4 Second slip
5 Gully
6 Third man
7 Extra cover
8 Midwicket
9 Deep backward square leg
10 Mid-on
11 Mid-off

B2

B1

22yds

Baseball

8

7

9

approx 350ft

5   4

6   90ft   3

2

60ft

1

Positions for right-handed batter

1 Catcher
2 Pitcher
3 First baseman

4 Second baseman
5 Shortstop
6 Third baseman
7 Left-fielder
8 Centre-fielder
9 Right-fielder

Foul line
Second base
Third base
Pitcher's mound
First base
Home plate

Fence

Australian Rules football

6.4m

Behind line
Goal
Boundary line

45.7m   45.7m

135–185m

6.4m

110–155m

1 Full forward
2 Full back
3 Half-back
4 Half-back
5 Centre
6 Follower
7 Rover

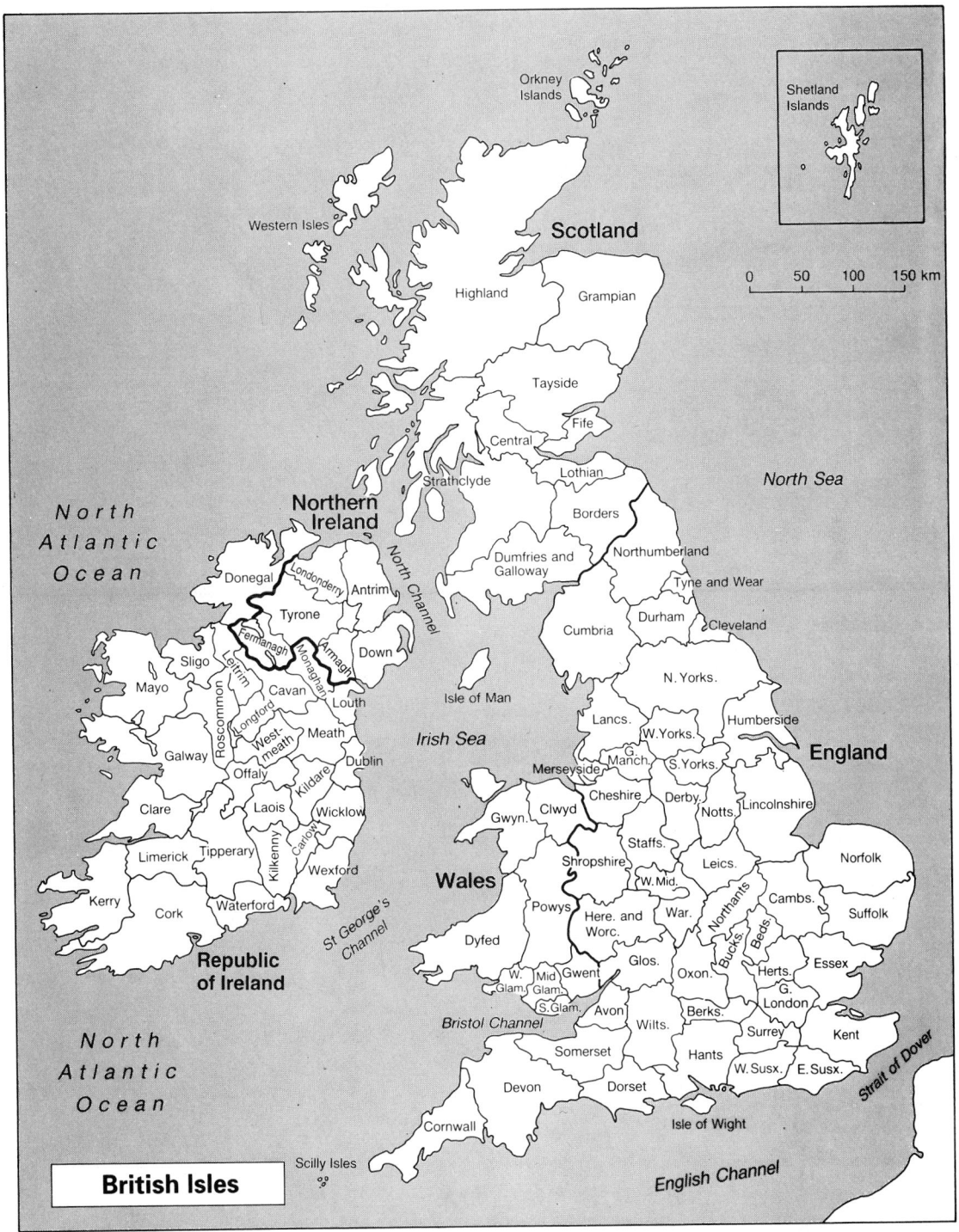

Orkney Islands

Shetland Islands

Western Isles

Scotland

Highland

Grampian

0 50 100 150 km

Tayside

Fife

Central

Lothian

Strathclyde

Northern Ireland

Borders

North Sea

North Atlantic Ocean

Donegal

Londonderry

Antrim

Dumfries and Galloway

Northumberland

North Channel

Tyrone

Tyne and Wear

Fermanagh

Armagh

Down

Durham

Cleveland

Sligo

Leitrim

Monaghan

Cumbria

Mayo

Roscommon

Cavan

Louth

N. Yorks.

Longford

West-meath

Meath

Isle of Man

Lancs.

W. Yorks.

Humberside

Galway

Offaly

Dublin

Irish Sea

G. Manch.

S. Yorks.

England

Merseyside

Clare

Laois

Kildare

Wicklow

Cheshire

Derby.

Notts.

Lincolnshire

Gwyn.

Clwyd

Limerick

Tipperary

Kilkenny

Carlow

Wexford

Staffs.

Leics.

Norfolk

Wales

Shropshire

W. Mid.

Kerry

Cork

Waterford

St George's Channel

Powys

Here. and Worc.

War.

Northants

Cambs.

Suffolk

Republic of Ireland

Dyfed

Glos.

Oxon.

Bucks.

Beds

Herts.

Essex

W. Glam.

Mid Glam.

Gwent

Berks.

G. London

North Atlantic Ocean

S. Glam.

Avon

Wilts.

Surrey

Kent

Bristol Channel

Somerset

Hants

W. Susx.

E. Susx.

Strait of Dover

Devon

Dorset

Cornwall

Isle of Wight

English Channel

Scilly Isles

British Isles

Boundaries
international
internal

Communications
motorway
other major road
railway

Cities and towns
major built-up areas
■ over 1 million inhabitants
● more than 100,000 inhabitants
• smaller towns

Land height
metres
1000
500
200
100
sea level
land below sea level
▲ spot height in metres

Scale 1: 5 160 000
0   50   100 km

Shetland Islands
Yell Unst
Fetlar
Mainland
Lerwick
Foula
Sumburgh Head
Fair Isle

Westray Sanday
Mainland Stronsay
Kirkwall Orkney Islands
Hoy
Pentland Firth John o'Groats
Duncansby Head
Thurso Wick
Cape Wrath

ATLANTIC OCEAN

Rona
Butt of Lewis
North Minch
Lewis Stornoway
St Kilda Harris
North Uist
Benbecula
South Uist
Barra
Outer Hebrides
Little Minch
Skye Kyle of Lochalsh
▲1009 ▲1183
Rhum Eigg ▲1310
Mallaig
Coll
Tiree Loch Linnhe
Iona Fort William ▲1344
Mull Ben Nevis
Firth of Lorn Oban
Colonsay
Islay Jura Sound of Jura

Inner Hebrides

Northwest Highlands
Ullapool
Loch Shin
Dornoch Firth
Ben Wyvis ▲1046
Dingwall Moray Firth
Inverness Elgin
Monadhliath Mtns.
Cairngorms
Ben Macdhui
Spey Don
Dee Aberdeen
Fraserburgh
Peterhead

SCOTLAND
Grampian Mountains

North Sea

Tay Sidlaw Hills Arbroath
Loch Tay Perth Dundee
Loch Lomond Stirling St Andrews
Greenock Alloa Kirkcaldy Firth of Forth
Clydebank Falkirk Dunfermline
Paisley Glasgow Cumbernauld Edinburgh
Bute East Kilbride Motherwell Lammermuir Hills St Abb's Head
Arran Kilmarnock Lindisfarne
Campbeltown Ayr Southern Uplands Berwick-upon-Tweed
Mull of Kintyre ▲840 Hawick Galashiels
Rathlin I. Cheviot Hills
North Channel Dumfries Coquet

Malin Head
Bloody Foreland
Lough Foyle
Coleraine
Londonderry Antrim Mtns.
Ballymena Larne
NORTHERN IRELAND Newtownabbey
Omagh Lough Neagh Bangor
Enniskillen Lisburn Belfast
Lower Lough Erne Portadown Lurgan
Armagh
Upper Lough Erne Newry ▲852
Mourne Mtns. Slieve Donard
Dundalk

Stranraer
Kirkcudbright Solway Firth Carlisle
Mull of Galloway Workington Penrith
Whitehaven St Bees Head ▲893
Cumbrian Mtns. ▲978
Scafell Pike Durham
Barrow-in-Furness Kendal
Isle of Man Douglas
Barrow-in-Furness

Newcastle-upon-Tyne
Blyth
Gateshead Sunderland
Hartlepool
Stockton-on-Tees Middlesbrough
Darlington North York Moors
Scarborough

Aran I.
Donegal Mtns.
Donegal
Donegal Bay
Erris Head
Achill I.
Blacksod Bay
Westport Castlebar
Lough Conn Lough Mask
Lough Corrib
Galway Galway Bay
Aran Is.
Clew

REPUBLIC OF IRELAND

Sligo
Lough Allen
Longford Boyne
Athlone Mullingar
Lough Ree
Lough Derg
Tullamore
Portlaoise Naas
Nore Barrow
Clonmel Suir

Dublin
Dun Laoghaire
Bray
Wicklow Mtns.
▲926
Wexford
Rosslare
Carnsore Point

Irish Sea

Anglesey
Holyhead
Holy I.
Caernarfon Bangor Bay Colwyn
Snowdon ▲1085
Wrexham
Conwy

Blackpool
Southport
Liverpool Birkenhead St Helens
Chester Crewe
Stoke-on-Trent

Preston Blackburn Bradford Leeds York
Bolton Wigan Huddersfield
Oldham Manchester Yorkshire Wolds
Stockport Sheffield Doncaster
Chesterfield
Lincoln Wolds

Pennines
Lancaster
Ribble
Wharfe
Aire
Ouse
Hull
Grimsby
Humber Spurn Head
Scunthorpe
Harrogate
Flamborough Head

ENGLAND

UNITED KINGDOM

Shannon
Limerick
Galty Mtns.
Tralee Killarney
▲1041
Carrauntoohill
Caha Mtns.
Lee Cork
Blackwater Kilkenny Waterford
Youghal
Cape Clear
Old Head of Kinsale

Lundy
Hartland Point
Barnstaple
Exmoor Hills
Taw
Quantock Hills
Bridgwater
Bodmin Moor
Dartmoor ▲619
Torbay
Penzance Truro
Land's End
Lizard Point
Plymouth Start Point
Scilly Isles
Lyme Bay Weymouth

Celtic Sea

St George's Channel
Fishguard
St David's Head
Milford Haven
Carmarthen Teifi
Llanelli
Swansea Neath Merthyr Tydfil
Rhondda
Cardiff Newport Cwmbran
Barry Bristol Channel
Brecon Beacons Black Mtns.
Cardigan Bay
Aberystwyth
Cambrian Mtns. ▲892

WALES

Shrewsbury Telford
Wolverhampton Dudley Walsall
Solihull Birmingham Coventry
Worcester Rugby
Cheltenham Banbury
Gloucester Cotswold Hills
Hereford Wye
Teme Severn

Derby Nottingham
Stafford
Leicester
Peterborough The Fens
Northampton Bedford
Milton Keynes Cambridge
Luton St Albans
Aylesbury
Chiltern Hills
Oxford Watford
Swindon Slough London
Reading Basingstoke
Guildford
Salisbury Plain Andover
Salisbury Winchester Crawley Maidstone
Taunton The Weald
Southampton Portsmouth Brighton
Poole Bournemouth Worthing Eastbourne
Isle of Wight South Downs Beachy Head

Boston
The Wash
King's Lynn
Great Yarmouth
Norwich
Lowestoft
Ipswich Felixstowe Harwich
Colchester
Chelmsford Basildon
Southend-on-Sea
Gillingham Margate
Canterbury
North Downs Dover
Folkestone
Hastings Calais
Boulogne

Skegness
Lincoln

Nene
Great Ouse
Thames
Mendip Hills
Bath Bristol

le Touquet
Paris–Plage

English Channel

Strait of Dover

Guernsey Alderney
Channel Islands
St Peter Port Sark
Jersey St Helier
Cherbourg
le Havre
FRANCE
Rouen
le Tréport Dieppe
Caen Seine

Transverse Mercator Projection
Oxford University Press

Boundaries

international

disputed
~~~~~~~~~~

internal
- - - - - - -

Communications

motorway

other major road

railway

canal

Cities and towns

■ over 1 million inhabitants

● more than 100 000 inhabitants

• smaller towns

Physical features

seasonal river/lake

marsh

salt pan

ice cap

sand dunes

Land height

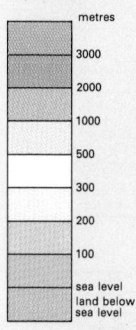

metres

3000
2000
1000
500
300
200
100
sea level
land below sea level

▲ spot height in metres

Scale 1: 14 000 000

0   125   250 km

ICELAND

Arctic Circle
20°W   15°W
Grimsey
Siglufjördhur
Akureyri
Stykkishólmur
ófn
Pjórsá ▲2000
Hekla
Vatnajökull
Reykjavik
Keflavik Hafnarfjördur 1491
20°W   15°W   65°N

NORWAY
Bergen Voss ▲1862
Hardangerfjord ▲1660
Haugesund Stavanger Boknafjord
Skager
Fred
Aal

Shetland Is.
Lerwick
Orkney Is. Kirkwall
Thurso Wick
Moray Firth
Inverness SCOTLAND Aberdeen
Outer Hebrides The Minch
Dundee
Glasgow Edinburgh Firth of Forth
Clyde UNITED Newcastle-upon-Tyne
Londonderry NORTHERN Carlisle Middlesbrough
Sligo IRELAND Belfast Isle of Man KINGDOM
Galway Leeds Hull
REPUBLIC Dublin Liverpool Sheffield
OF Manchester Nottingham
IRELAND Stoke-on-Trent Derby Leicester Norwich
Limerick Holyhead WALES ENGLAND
Tralee Birmingham
Waterford Fishguard Thames NETHERLANDS Bremen
Cork Swansea Cardiff Bristol London Amsterdam
Cherbourg Southampton Dover The Hague Utrecht
Land's End Brighton Rotterdam Duisburg Essen
Plymouth Calais Ostend Antwerp BELGIUM Aachen
English Channel Dunkirk Brussels Liège Cologne Bonn
le Havre Lille Luxembourg LUX GE
Brest St Malo Rouen Amiens Mainz
Caen Seine Lens Reims Metz Saarbrücken
Channel Is. Somme Oise Nancy Strasbourg
St Malo Chartres Paris Marne Vosges
Rennes le Mans Troyes Jura
Lorient Angers Orleans FRANCE Basle Zu
Nantes Tours Nevers Dijon Lucerne Berne
Loire SWITZERLAND
Bay of la Rochelle Poitiers Vichy Lyons 4807 Geneva
Limoges Clermont-Ferrand Saône Mont Blanc
Biscay 1886 Massif Rhône Grenoble
Bordeaux Central Valence Turin
Corunna Gironde Lot Dordogne Nîmes Avignon
C. Finisterre Gijón Santander Dordogne Montpellier Aix-en-Provence MONACO
Santiago de Oviedo Bilbao Biarritz Bayonne Toulouse Nice
Compostela 2321 Cordillera Cantabrica San Sebastián Lourdes Perpignan Marseilles Côte d'Azur
Vigo León Vitoria Pyrénées Cannes Toulon Bast
Minho Pamplona 3404 ANDORRA
Braga Duero Burgos Ebro Costa Brava Corsica (France) Ajaccio
Oporto Valladolid Saragossa Lérida Sabadell
PORTUGAL Douro Barcelona Sassari
Coimbra 2468 Hospitalet
Salamanca Tarragona
Tagus SPAIN Madrid C. de Tortosa Minorca
Cascais Toledo Castellón de la Plana
Lisbon Guadiana Majorca Sardinia (Italy)
Badajoz Albacete Júcar Valencia Palma
Sierra Morena Ibiza Cabrera Balearic Is.
Huelva Guadalquivir Segura Elche Alicante Formentera
Faro Córdoba Murcia Costa Blanca
Gulf of Cadiz Seville Granada Cartagena
Jerez 3482 Almería Mediter
Cadiz Malaga C. de Gata
Algeciras GIBRALTAR (U.K.) M Algiers Tizi Ouzou Skikda Annaba
ATLANTIC Strait of Tangier Ceuta (Sp.) Blida Bejaïa Constantine
Tétouan Ech Cheliff Médéa Sétif
OCEAN Melilla (Sp.) Mostaganem Ksar El Boukhari TUN
Kenitra Nador Oran Sidi bel Abbès Bou Saâda Batna
Casablanca Rabat 2455 Tlemcen 2326
El Jadida Fez Taza Oujda Djelfa Biskra
MOROCCO Meknès Khemisset Mecheria Laghouat Chott Melrhir -31
Settat Khouribga Sahara Atlas Chott Ech Chergui Nefta Gafsa
Safi Beni Mellal ALGERIA Touggourt
Marrakesh 4165 High Atlas Bouârfa Chott El Jerid
Aïn Sefra

Conical Orthomorphic Pro

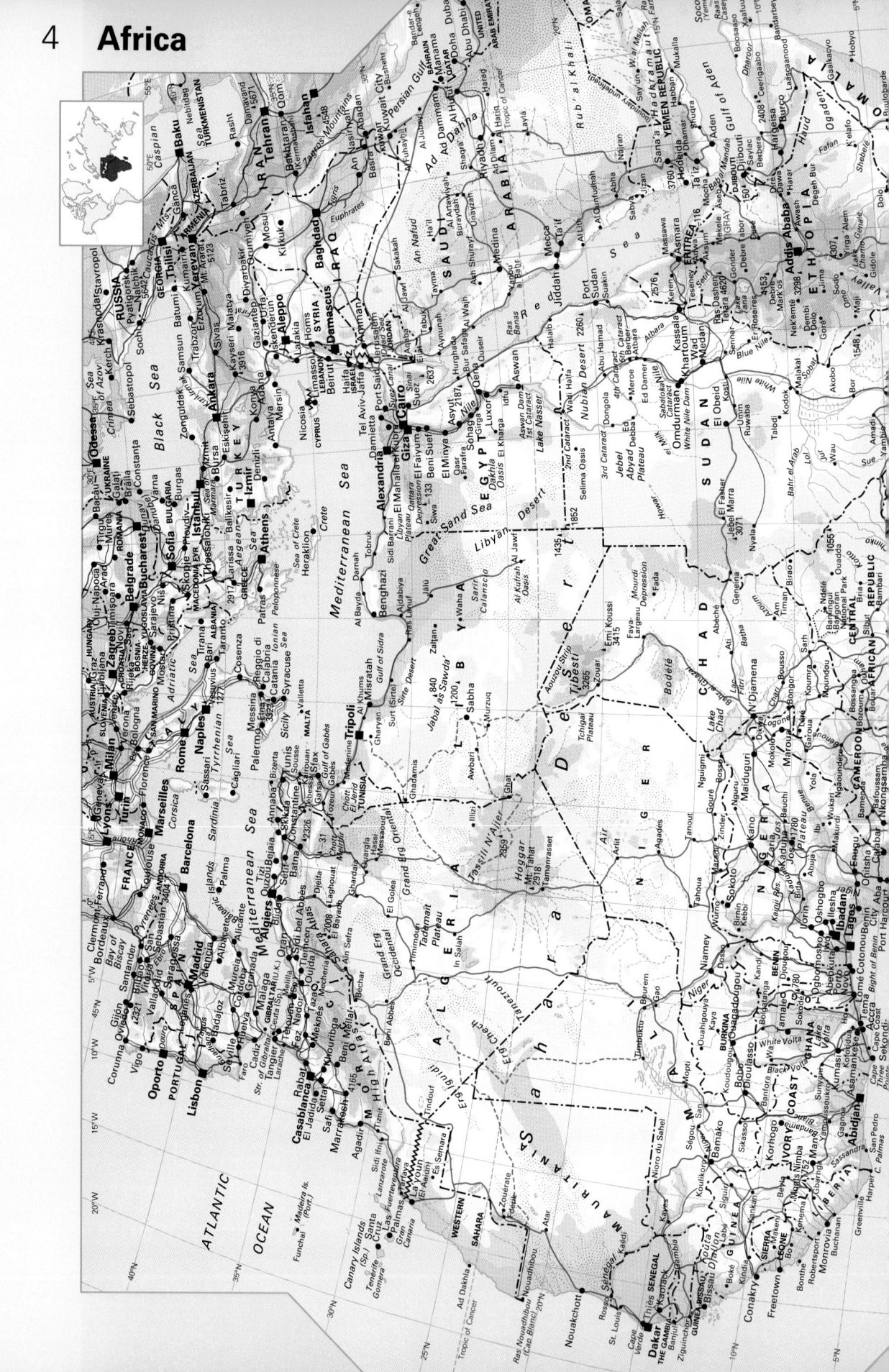

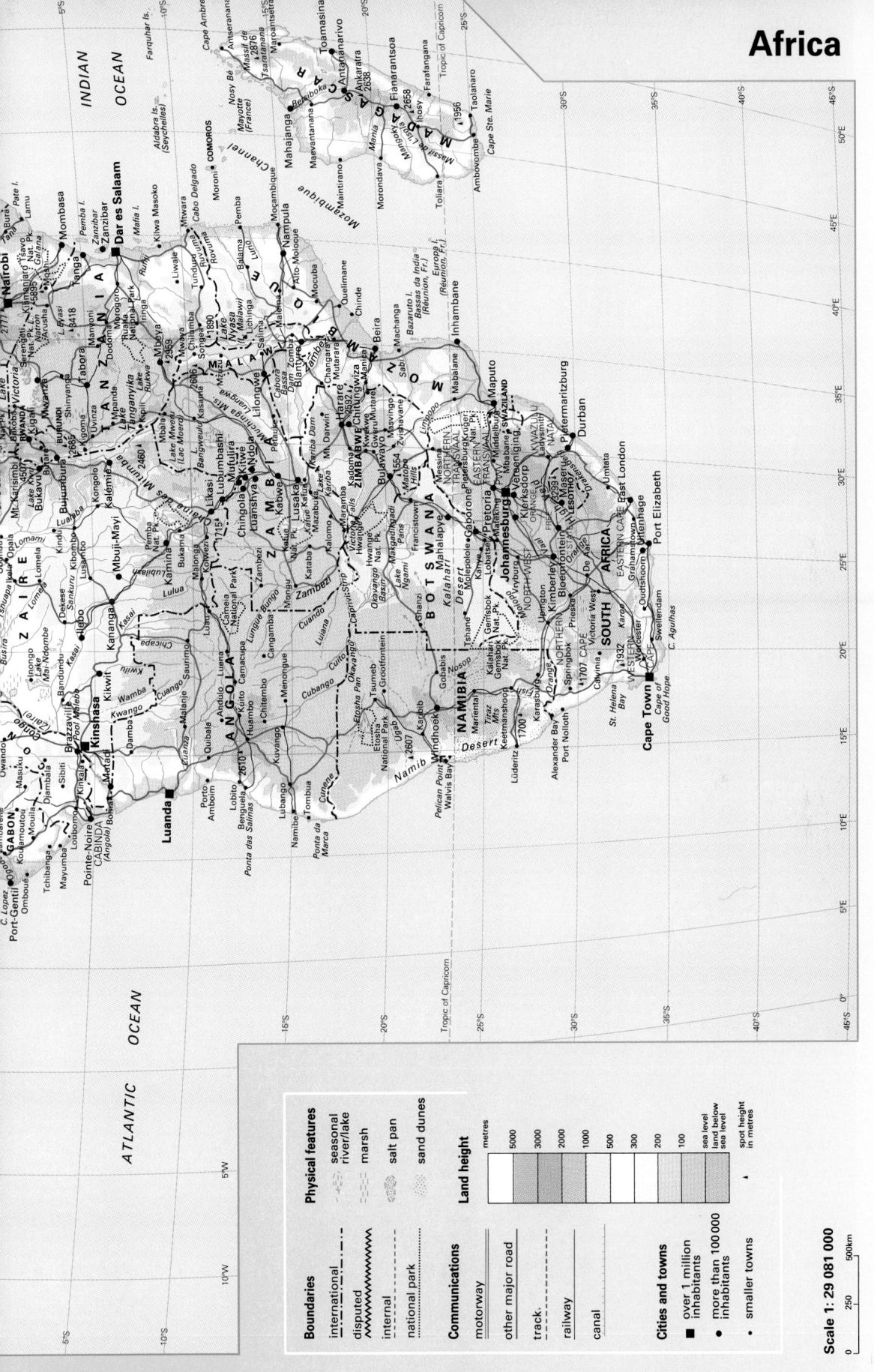

Boundaries

international
disputed
internal
national park

Communications

motorway
other major road
track
railway
canal

Cities and towns

■ over 1 million
  inhabitants
● more than 100 000
  inhabitants
· smaller towns

Physical features

seasonal
river/lake
marsh
salt pan
sand dunes

Land height

metres
5000
3000
2000
1000
500
300
200
100
sea level
land below
sea level
spot height
in metres

Scale 1: 29 081 000

0   250   500km

Northern Asia

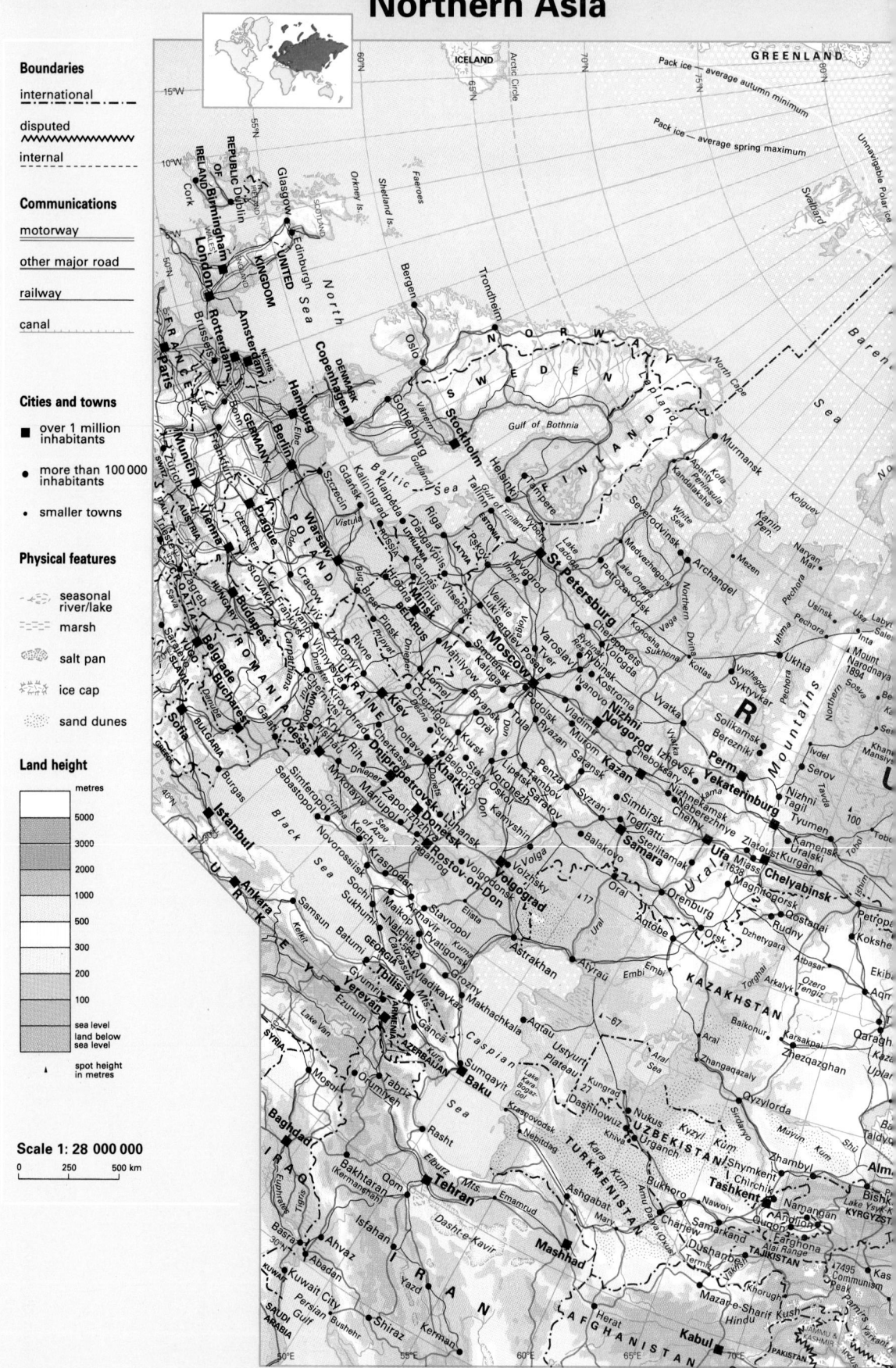

Boundaries

international — ·· — ··

disputed ∿∿∿∿∿∿

internal — — — —

Communications

motorway ══════

other major road ──────

railway ┼┼┼┼┼┼

canal ──────

Cities and towns

■ over 1 million inhabitants

● more than 100 000 inhabitants

• smaller towns

Physical features

seasonal river/lake

marsh

salt pan

ice cap

sand dunes

Land height

metres
5000
3000
2000
1000
500
300
200
100
sea level
land below sea level

▲ spot height in metres

Scale 1: 28 000 000

0   250   500 km

Conical Orthomorphic P

Middle East, Pakistan and India

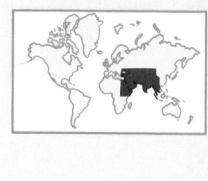

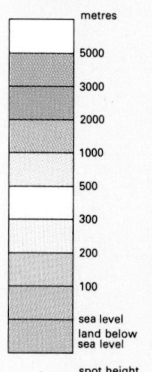

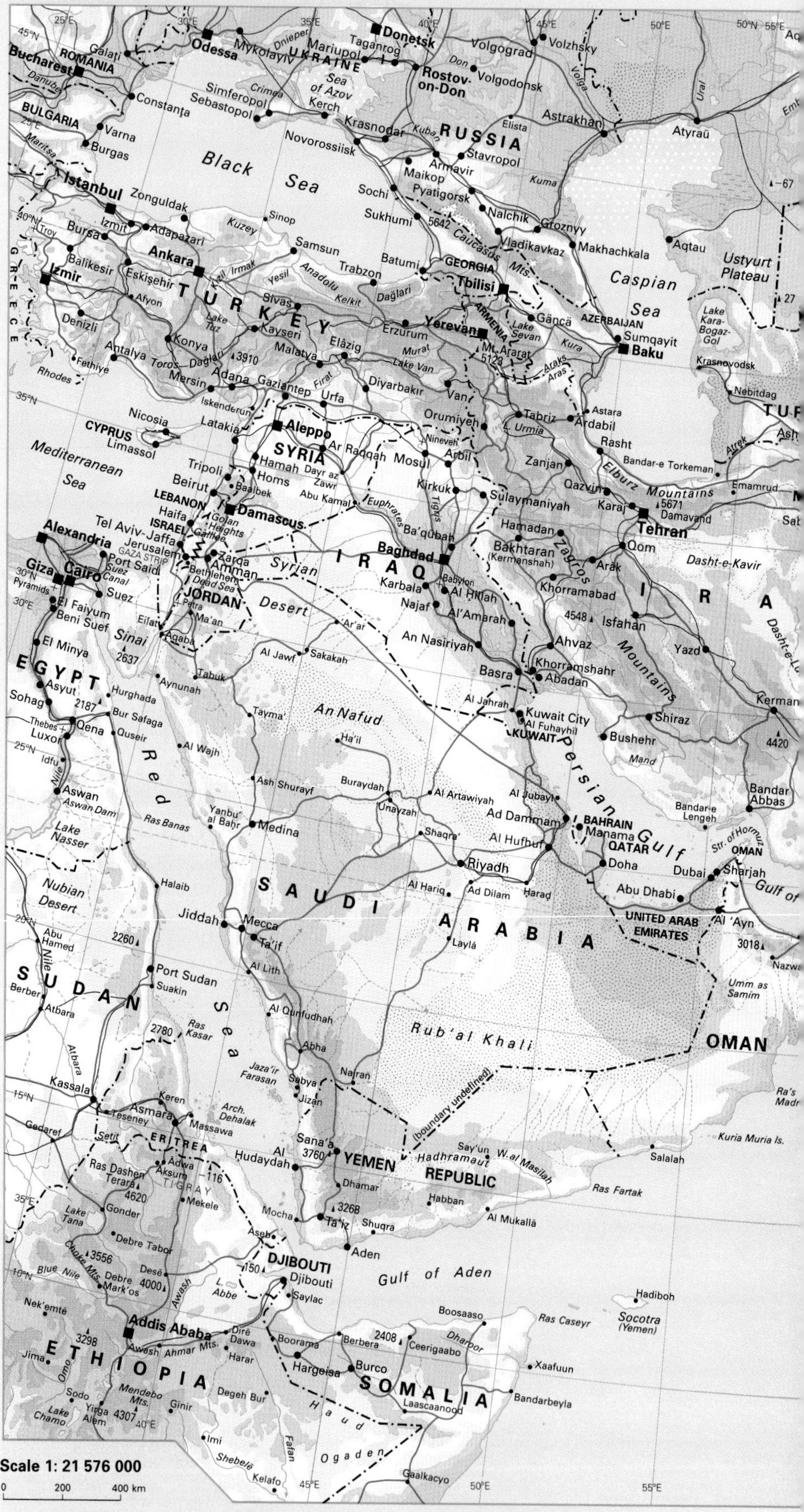

Boundaries

international

disputed

internal

Communications

motorway

other major road

railway

canal

Cities and towns

■ over 1 million inhabitants

● more than 100 000 inhabitants

• smaller towns

+ historic sites

Physical features

seasonal river/lake

marsh

salt pan

ice cap

sand dunes

Land height

metres

5000

3000

2000

1000

500

300

200

100

sea level

land below sea level

▲ spot height in metres

Scale 1: 21 576 000

0 200 400 km

Conical Orthomorphic P

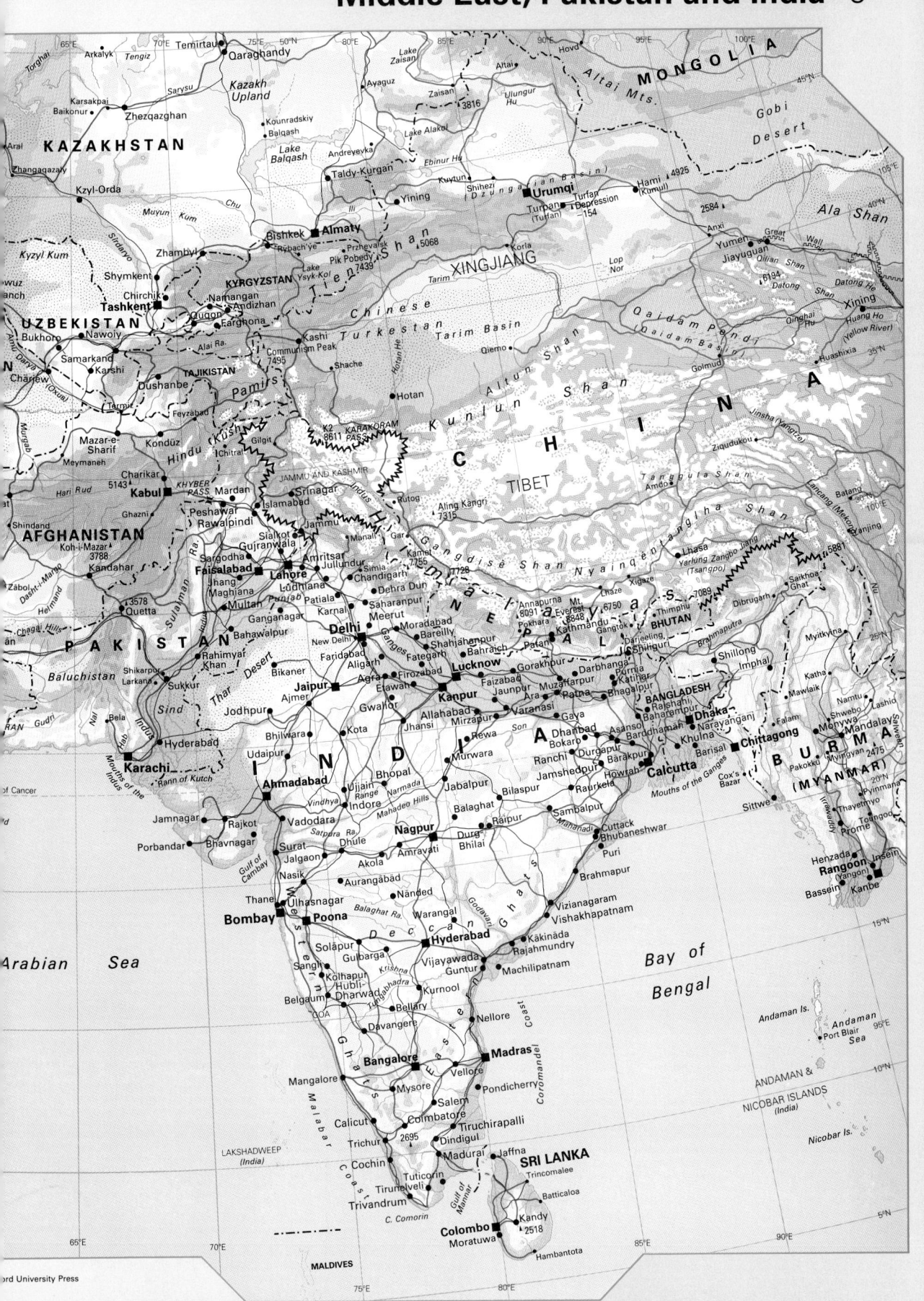

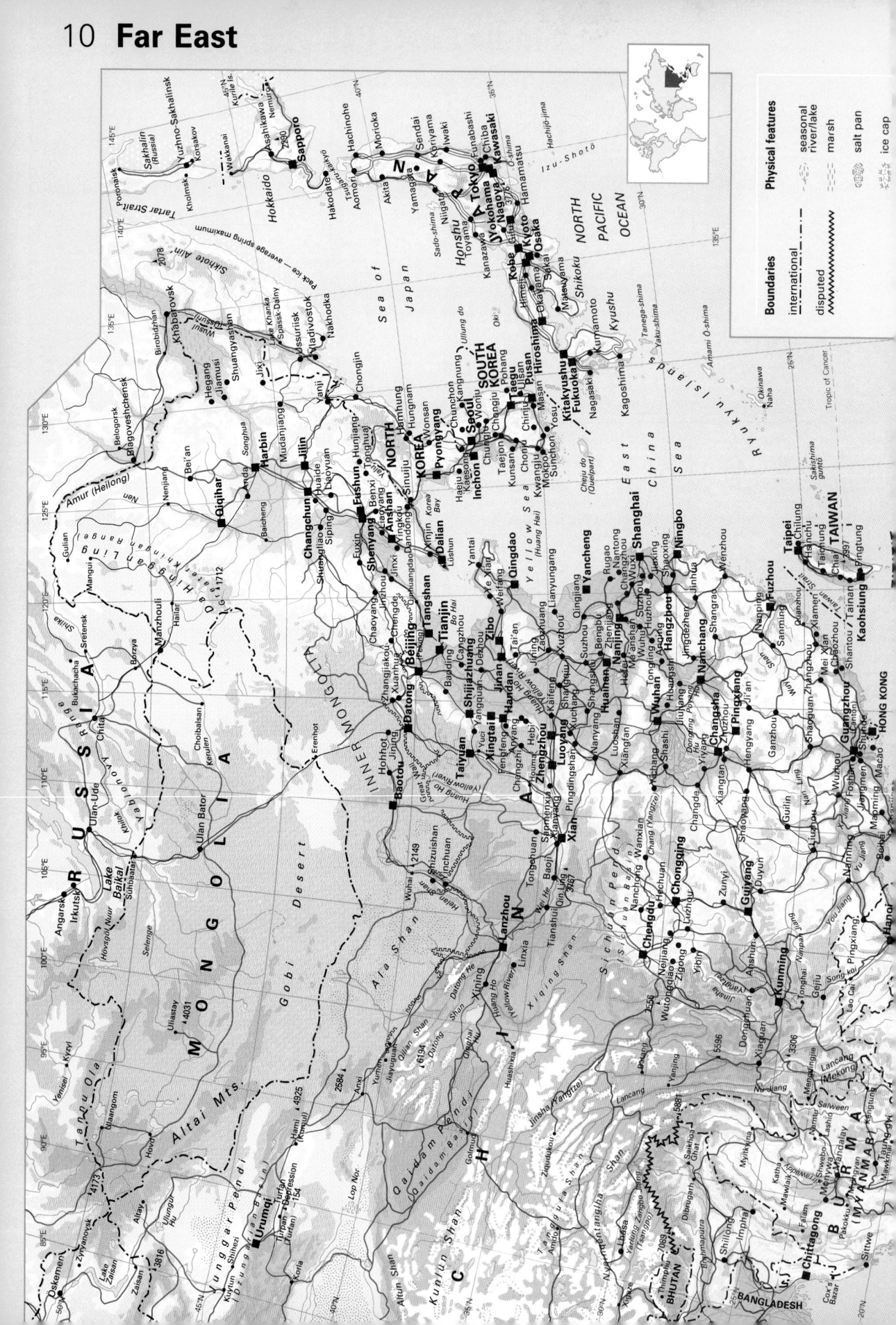

Physical features

seasonal
river/lake
marsh
salt pan
ice cap

Boundaries

international
disputed

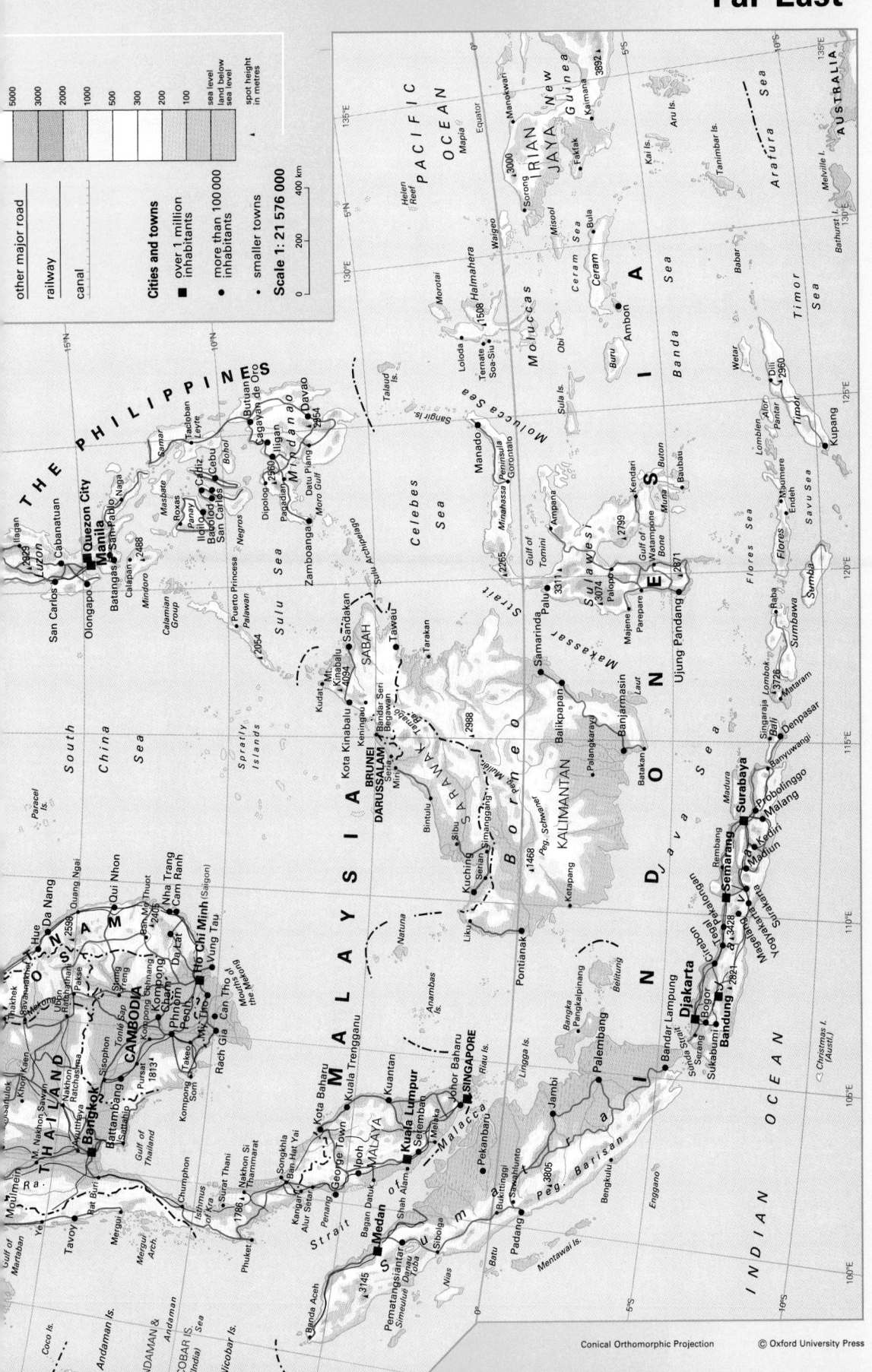

Celebes Sea
Morotai
Helen Reef
150°E
120°E
125°E
130°E
135°E
140°E
145°E

Manado
Loloda
Halmahera
Mapia
Equator
Kaniet Is.
Minahassa Penin.
2565
Ternate 1508
Soa-Siu
Waigeo
Ninigo Group
Hermit Is.
Saint Matthias Group
Lyr
Samarinda
Gulf of Tomini
Gorontalo
Molucca Sea
Biak
Wuvulu
Admiralty Is.
Bismarck Archipelago
New Ireland
Palu
Ampana
Sula Is.
Moluccas
Obi
Sorong
3000
Manokwari
Jayapura
Wewak
Bismarck Sea
Rabaul
Balikpapan 3311
Sulawesi 3074
Ceram Sea
Misool
Gulf of Cenderawasih
Pegunungan
IRIAN
Sepik
Madang
New Britain
Borneo
1892
Palopo 2799
Kendari
Buru
Ceram
Bula
Fakfak
Kaimana
3892
Pegunungan Maoke
Pk. Jaya 5030 JAYA
New 3993
Mount Hagen
Goroka
Lae
NEW
GUINEA
Banjarmasin
Majene 3726
Parepare
Gulf of Bone
Ambon
Kai Is.
P. Dolak
Mendi
Wabag
Kikori
Kerema
PAPUA
Solomon Sea
Batakan
Watampone 2871
Muna
Buton
Baubau
Banda Sea
Tanimbar Is.
Aru Is.
Tanahmerah
Merauke
Daru
Gulf of Papua
Popondetta
Woodla
Ujung Pandang
INDONESIA
Wetar
Babar
Arafura Sea
Port Moresby
Owen Stanley Ra.
D'Entrecasteaux Islands
Louisiade Archipelago
Java Sea
Flores Sea
Lomblen
Alor
Dili 2960
Timor
Savu Sea
Kupang
Timor Sea
Torres Strait
C. York
Singaraja Lombok
Flores
Pantar
Endeh
Maumere
Sumba
Bathurst I.
Melville I.
Nhulunbuy
Weipa
Cape York Peninsula
Great
C. Melville
Coral S
Bali Denpasar Sumbawa
Raba

INDIAN
OCEAN

Darwin
Arnhem Land
Gulf of Carpentaria
Cooktown
Dividing
Barrier
CORAL SEA ISLANDS TERR.
Bonaparte Archipelago
C. Talbot
Joseph Bonaparte Gulf
Daly
Katherine
Groote Eylandt
Wellesley Is.
Mitchell
Cairns
Reef
Wyndham
Kununurra
Argyle
Daly Waters
Normanton
Gilbert
Innisfail
786
Kimberley Plateau
NORTHERN
Barkly Tableland
Croydon
Ingham
Townsville
King Sound
Derby
Halls Creek
Stuart Highway
Richmond
Flinders
Bowen
Charters Towers
Broome
Fitzroy
Tennant Creek
Cloncurry
Hughenden
Mackay
Eighty Mile Beach
Sturt Creek
TERRITORY
Mount Isa
Winton
Capricorn Channel
Goldsworthy
Port Hedland
Lake Mackay
Georgina
QUEENSLAND
Longreach
Emerald
Yeppoon
Rockhampton
Dampier
Marble Bar
Alice Springs
Diamantina
Barcaldine
Springsure
Mount Morgan
Gladstone
Barrow I.
Onslow
Newman
Amadeus
MacDonnell Ranges
Simpson Desert
Thomson
Blackall
Monto
Bundab
N.W. Cape
Fortescue
Hamersley Ra.
Mt. Tom Price 1235
Great Sandy Desert
Ayers Rock 867
Musgrave Ranges
Birdsville
Barcoo
Charleville
Roma
Taroom
Maryb
Exmouth
Ashburton
Paraburdoo
WESTERN
Gibson Desert
AUSTRALIA
Eyre Creek
Grey Ra.
Quilpie
Mitchell
Chinchilla
Dalby
Gymp
Macleod
Gascoyne
722
SOUTH
Cooper Creek
Cunnamula
Warego
Toowoomba
Bris
Carnarvon
Murchison
Meekatharra
Wiluna
Carnegie
Lake Eyre
AUSTRALIA
Wompah
Paroo
Bourke
Goondiwindi
Warwick
Go
Lisr
Mt. Magnet
Leonora
Laverton
Great Victoria Desert
Torrens
Flinders Ra.
Broken Hill
Cobar
Moree
NEW
Tamworth
Armidale
Port M
L. Barlee
L. Moore
L. Raeside
Forrest
L. Gairdner
Iron Knob
Peterborough
Nyngan
Graf
Taree
Northampton
Geraldton
Zanthus
Eucla
Ceduna
Port Augusta
Dubbo
SOUTH
Orange
Newcastl
Maitland
Moora
Kalgoorlie
Nullarbor Plain
Eyre Penin.
Whyalla
Wallaroo
Port Pirie
Mildura
Murrumbidgee
WALES
Bathurst
Lithgow
Sydney
Perth
Fremantle
Coolgardie
Norseman
Great Australian Bight
Port Lincoln
Spencer Gulf
Elizabeth
Murray
Lachlan
Goulburn
A.C.T.
Canberra
Wollongong
Northam
Narrogin
Esperance
Kangaroo I.
Adelaide
Murray Bridge
Wagga Wagga
Queanbeyan
Collie
Bunbury
Wagin
Ravensthorpe
Horsham
Bendigo
Albury
Wangaratta
Mt. Kosciusko 2230
Cape Howe
C. Naturaliste
Augusta
1042
Albany
VICTORIA
Ballarat
Melbourne
Gippsland
C. Leeuwin
Mount Gambier
Geelong
Moe-Yallourn
Portland
Warrnambool
King I.
Bass Strait
Furneaux Group
Burnie
Devonport
Launceston
Queenstown
Mt. Ossa 1617
TASMANIA
Hobart
S.E. Cape

INDIAN
OCEAN

110°E
45°S
115°E
120°E
125°E
130°E
135°E
140°E
145°E
150°E
155°E

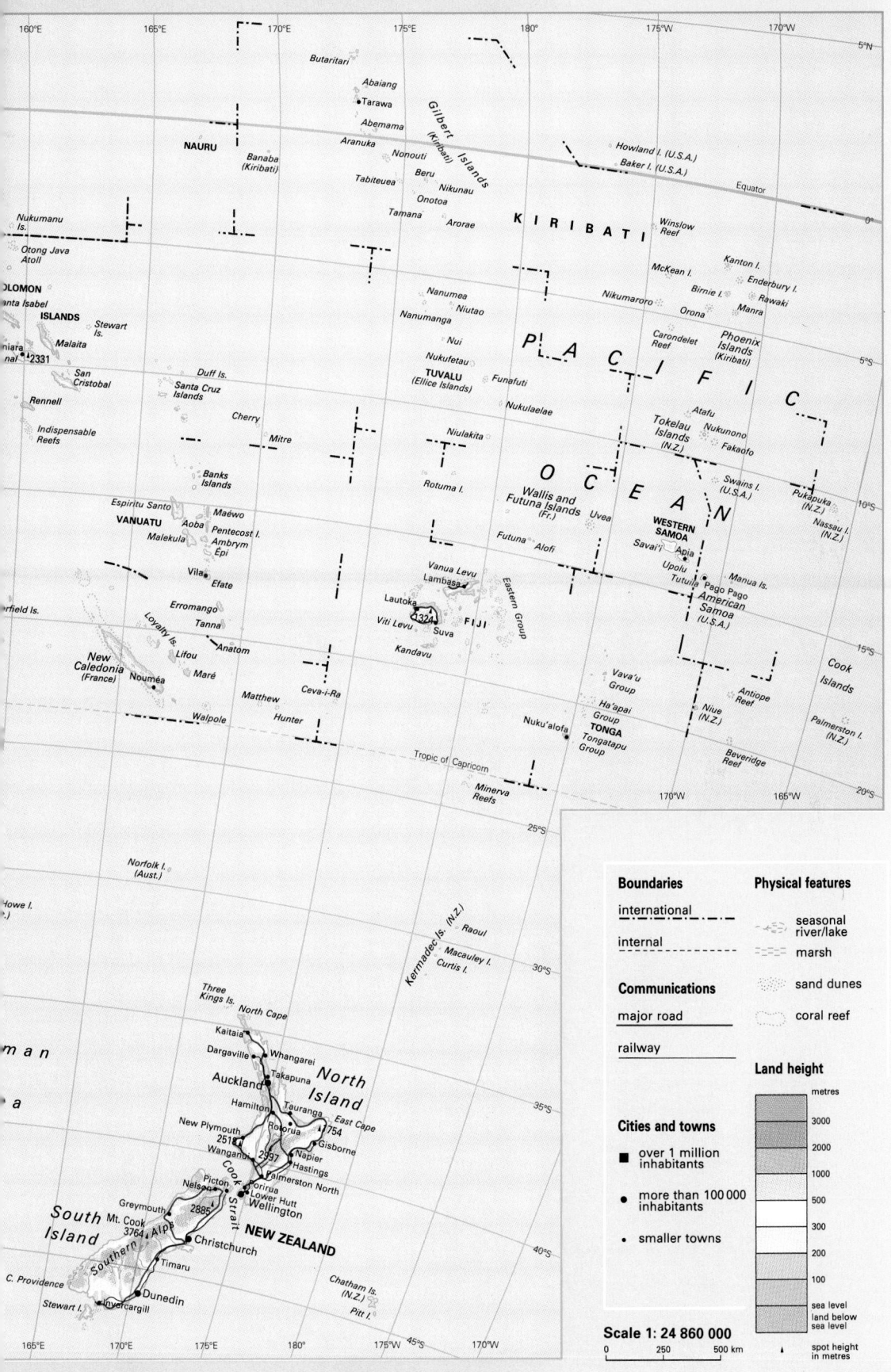

Boundaries

international

internal

Communications

major road

railway

Cities and towns

■ over 1 million inhabitants

● more than 100 000 inhabitants

· smaller towns

Physical features

seasonal river/lake

marsh

sand dunes

coral reef

Land height

	metres
	3000
	2000
	1000
	500
	300
	200
	100
	sea level
	land below sea level

▲ spot height in metres

Scale 1: 24 860 000

0   250   500 km